For Reference

Not to be taken from this room

SCIENTIFIC ENCYCLOPEDIA

Tenth Edition

VOLUME 2

VAN NOSTRAND'S

SCIENTIFIC ENCYCLOPEDIA

Tenth Edition

VOLUME 2

Glenn D. Considine
Editor-in-Chief

Peter H. Kulik
Associate Editor

WILEY-INTERSCIENCE

A John Wiley & Sons, Inc., Publication

Library of Congress Cataloging-in-Publication Data:

Van Nostrand's scientific encyclopedia. — 10th ed. / edited by Glenn D. Considine.
     p. cm.
  Includes index.
  "Three volume set."
  ISBN 978-0-471-74338-5 (cloth)
 1. Science — Encyclopedias. 2. Engineering — Encyclopedias. I. Considene, Glenn D. II. Title: Scientific encyclopedia.
  Q121.V3 2008
  503 − dc22

2007046658

PREFACE

The editors are pleased to introduce *Van Nostrand's Scientific Encyclopedia*, Tenth Edition, thus building upon the long tradition of excellence that dates back some seventy years to the First Edition, published in 1938 as a single volume over 1,200 pages and updated over the decades in one volume (1947, 1958, 1968, 1976, 1983), and then in two volumes (1989, 1995, 2002). This Tenth Edition, in response to exponential growth, both in scientific knowledge and in the electronic availability of that knowledge, has grown perforce to three volumes over 6,100 pages.

The essence of *VNSE* is enduring, and it remains a fine, concise, comprehensive, and accessible general science work. Its intellectual scope ranges from the introductory to the highly technical in a vast and ever-expanding array of topical coverage in the sciences, engineering, mathematics, medicine, and more. As has long been the case, the editors have designed the book to be approachable by students of many ages. An important feature continued in this work, therefore, is the progressive development of the discussion of each topic, beginning with a simple definition expressed in plain terms, developing into a more detailed treatment, and augmented by often-extensive Additional Reading suggestions.

Contemporary readers can continue to turn to *VNSE* for information about how their daily lives are increasingly affected by the sophistication of today's science and the complexity of modern technology. They will be reminded that knowledge and discovery exist in a continuum, and that often, but not always, what is new depends entirely on what came before. As our esteemed, late editor of more than 30 years, Douglas M. Considine, was wont to say, "Science is history". With that mantra in mind, and noting that it has already been six years since the Ninth Edition in 2002, it is time to examine this new edition.

The five major changes noted in the Preface to the Ninth Edition are still very much a part of the book and are likely to remain so. First, the way that the editors wrote, gathered and assembled articles for this book is almost entirely electronic — from query letters to individuals or to academic, industrial or technical societies or entities; to primary research for new entries and updates for mature ones, now conducted almost entirely on the Internet; to communication between editor and publisher and the transmission of text between them in the partnership that results in the book itself. Second, the mushrooming Age of Discovery is still very much with us and is reflected in updates to articles and families of articles, with special emphasis on the many life sciences, space science, and computer technology, and much, much more, necessitating the expansion to a third volume. Third, the use of Internet references as Additional Reading suggestions, a groundbreaking feature of the Ninth edition, has been retained and expanded; even print articles are often archived electronically by magazines and journals and thus the quest for additional material is increasingly electronic in nature, with readers having the advantage of much of the culling of thousands of "hits" on a given topic having been done for them by the editors. Fourth, the editors have inculcated the previously new feature of Time Lines and Glossaries, to add at-a-glance information and historical perspective, into a staple of the book; new coverage includes virology, physiology, pharmaceuticals, the Internet, weather technology, and more. And fifth, the sense of history is further developed by expansion of another once-new feature, brief biographies of many influential scientists including many Nobel Prize winners in Medicine, Chemistry, and Physics; a history of their times is not complete without mention of their works. Science *is* history.

A statistical summary of the Tenth Edition would include more than 10,000 entries and 9,500 cross-references; 4,800 diagrams, graphs, and photographs; more than 600 tables; and an alphabetical Index of more than 100 printed pages that finds few rivals in the realm of technical literature. These are large increases, staggering particularly in the number of entries, and they are the result of an exhaustive, ongoing selection process for content perhaps best described so presciently by Douglas Considine in the Preface to the Sixth Edition in 1983: "Obviously, an encyclopedia of this type cannot serve the same purpose as a news medium. Science, too, has its own noise level. This is particularly evident from the hundreds of prematurely announced and exaggerated claims one frequently finds in the general communications media. Consequently, the authors and editors of this encyclopedia must carefully sift through the vast scientific data bank and sort out the trivia from real progress... there is no room for rumor and the untried and very little room indeed for the controversial in a permanent scientific reference such as this." To which the present editors say, Amen. But even the strictest selection criteria, applied to the gargantuan growth of knowledge, lead inevitably to the paradox that in print, what is new today will be, if not obsolete, at least in need of immediate updating tomorrow. To that end the editors have included thousands of Internet references to foster contemporaneity in research and Additional Reading. And further to that end, the editors have a wonderful announcement.

Van Nostrand's Scientific Encyclopedia, Tenth Edition, is now available on the Internet through the Wiley Interscience website. The online version uses embedded Internet references as hyperlinks both within the book itself and without — to the greater online community and also to other titles owned by John Wiley and Sons. At the click of a mouse the student or reader can now navigate VNSE instantly; this reflects how modern students actually pursue knowledge, and their parents are not far behind. The possibilities for ready acquisition of topical knowledge in, simply put, what is known, thus grow logarithmically through the interconnectedness of the VNSE and its links to the cyberworld. The editors feel strongly that, while there will always be an important place for printed literature of all kinds, as evidenced by the beautiful volumes at hand, the realm of scientific knowledge, compelled by the ongoing effects of that knowledge itself as expressed through information technology, will be accessed increasingly via electronic means.

In closing, one trusts that the reader will indulge a personal observation. In 1976 at the age of 61, Douglas M. Considine wrote and edited the Fifth Edition of *VNSE*. In 2008 at a similar age, his son Glenn D. Considine, also a writer and editor, presents the Tenth Edition of *VNSE*. Science is history. Family is history, too. Thus, this book is dedicated with abiding affection and gratitude to Douglas M. Considine.

GLENN D. CONSIDINE, Editor
PETER H. KULIK, Associate Editor

REPRESENTATIVE TOPICAL COVERAGE

ANIMAL LIFE

Amphibians	Coelenterates	Mammals	Protozoa
Annelida	Echinoderms	Mesozoa	Reptiles
Arthropods	Fishes	Mollusks	Rotifers
Birds	Insects	Paleontology	Zoology

BIOSCIENCES

Amino Acids	Biophysics	Genetics	Proteins
Bacteriology	Cytology	Hormones	Recombinant DNA
Biochemistry	Enzymes	Microbiology	Viruses
Biology	Fermentation	Molecular Biology	Vitamins

CHEMISTRY

Acids and Bases	Corrosion	Inorganic Chemistry	Oxidation-Reduction
Catalysts	Crystals	Ions	Photochemistry
Chemical Elements	Electrochemistry	Macromolecular Science	Physical Chemistry
Colloid Systems	Free Radicals	Organic Chemistry	Solutions and Sales

EARTH AND ATMOSPHERIC SCIENCES

Climatology	Geodynamics	Hydrology	Tectonics
Ecology	Geology	Meteorology	Seismology
Geochemistry	Geophysics	Oceanography	Volcanology

ENERGY SOURCES AND POWER TECHNOLOGY

Batteries	Electric Power	Nuclear Energy	Steam Generation
Biomass and Wastes	Geothermal Energy	Ocean Energy Resources	Tidal Energy
Coal	Hydroelectric Power	Petroleum	Turbines
Combustion	Natural Gas	Solar Energy	Wind Power

MATHEMATICS AND INFORMATION SCIENCES

Automatic Control	Computing	Measurements	Statistics
Communications	Data Processing	Navigation and Guidance	Units and Standards

MATERIALS AND ENGINEERING SCIENCES

Chemical Engineering	Laser Technology	Mining	Process Engineering
Civil Engineering	Mechanical Engineering	Microelectronics	Structural Engineering
Glass and Ceramics	Metallurgy	Plastics and Fibers	Transportation

MEDICINE, ANATOMY, AND PHYSIOLOGY

Brain and Nervous System	Genetic Disorders	Ophthalmology
Cancer and Oncology	Gerontology	Otorhinolaryngology/Dental
Cardiovascular System	Hematology	Parasitology
Chemotherapy	Immunology	Pharmacology
Dermatology	Infectious Diseases	Reproductive System
Diagnostics	Kidney and Urinary Tract	Respiratory System
Digestive System	Mental Illness	Rheumatology
Endocrine System	Muscular System	Skeletal System

PHYSICS

Atoms and Molecules	Gravitation	Optics	Subatomic Particles
Electricity	Magnetism	Radiation	Surfaces
Electronics	Mechanics	Solid State	Theoretical Physics
Fluid State	Motion	Sound	Waves

PLANT SCIENCES

Agriculture	Diseases and Pests	Growth Modifiers	Seeds and Germ Plasm
Algae	Fruits	Nutritional Values	Trees
Botany	Fungi	Plant Breeding	Yeasts and Molds

SPACE AND PLANETARY SCIENCES

Astrochemistry	Astronautics	Astrophysics	Probes and Satellites
Astrodynamics	Astronomy	Cosmology	Solar Systems

ACKNOWLEDGMENTS

Specialists in all disciplines of the scientific community have contributed in many ways to the preparation of this Tenth Edition of the *Van Nostrand's Scientific Encyclopedia*. Their inputs ranged from the preparation of manuscripts on complex topics, the submittal of new data for the first time, advice and counsel to the Editorial Board, the seeking out of obscure and discrete information, and the reporting of research findings. Inputs represent those of scientists, engineers, technologists, located worldwide. In addition to numerous academic institutions and private industries, the editors are much indebted to numerous governmental departments, agencies, and field organizations for their cooperation. It has always been in the best tradition of the history of science to share knowledge. It is therefore no mere coincidence that so many contributors are teachers at the university level, for they not only have deep knowledge in their respective fields, but they also can communicate that knowledge effectively. The great improvements to the substance of this book would not have been possible without them, and the editors have preserved the individual styles of the authors in keeping with the tradition of *VNSE* as an eminently personal, and, one hopes, more accessible work of general science. The editors and staff of this encyclopedia gratefully acknowledge their excellent cooperation and stress that the following abridged list of over 650 individuals and groups could be much longer.

Special appreciation must be extended for the efforts of Michael Ladisch of Purdue University, for his home article on Bioprocess Engineering (Biotechnology). — David Leake of Indiana University, both for his home article on Artificial Intelligence, and for quarterbacking the entire family of twelve AI "sidebar" articles. — The American Meteorological Society for their gracious permission to use numerous articles from the glossary of meteorology. — Ramon A. Mata-Toledo, *James Madison University*, who reviewed the Computer Sciences and authored several articles. — Joseph Castellano, President and CEO of *Stanford Resources*, who prepared numerous entries on Flat Panel Display Technology. — Dr. Thomas J. Harrison, who prepared numerous articles on computers and digital technology. — Dr. Steven N. Shore, who authored and arranged several entries dealing with astronomy and related sciences. — Dr. Ann C. DeBaldo, University of South Florida, who prepared numerous entries in the areas of immunology, oncology, and infectious diseases. — Drs. M. L. and W. L. Dilling, who skillfully summarized the complex world of organic chemistry, its nomenclature and equations. — Richard Q. Hofacker Jr., who authored articles on microelectronics and telephony and who rendered invaluable assistance toward creating comprehensive, yet concise, inputs concerning the broad field of telecommunications. — Peter E. Kraght, who not only authored several articles, but who also prepared the foundation for other descriptions in the spheres of meteorology and climatology. — Elmer Rowley, who made the coverage of mineralogy and crystallography in this encyclopedia truly outstanding. — *VisionRx*, Elmsford, NY., for the numerous entries on Vision and eye related disorders. — Jeanne Maree Iacono, who authored and rendered invaluable assistance toward creating brief biographies on scores of scientists. Without exaggeration, the list of such very special efforts could be extended by several additional paragraphs.

NOTE: In the cases of relatively short articles, the authors' initials may be used instead of their full names. In the following list, an asterisk indicates such authors. For example: *American Meteorological Society (AMS).

R. C. Aalberse, *Central Laboratory of the Blood Transfusion Service of the Netherlands Red Cross, Amsterdam, The Netherlands.* http://stinet. dtic.mil/oai/oai?verb=getRecord&metadataPrefix=html&identifier=AD0 727999

Bernard W. Agranoff, *University of Michigan, Ann Arbor, MI.* http://www.biochem.med.umich.edu/biochem/research/profiles/agranoff.html

Mark Adams, *Fisher Controls International, Inc., Marshalltown, IA.* http://www.emersonprocess.com/fisher/

O. J. Adlhart, *Engelhard Corporation, Iselin, NJ.* www.engelhard.com

H. J. Albert, *Parr Instrument Company, Moline, IL.* http://www.parrinst. com/

P. S. Albright, *Wichita, KS.*

W. Randall Albury, *University of New England, Armidale, New South Wales, Australia.* http://www.une.edu.au/

R. A. Alfano, *City University of New York (CUNY), New York City, NY.* http://portal.cuny.edu/portal/site/cuny/index.jsp?front_ door=true

Ulf Alkner, *AstraZeneca R&D, Lund, Sweden.* http://www.astrazeneca. com/

K. G. Alle, *Imperial College, London, UK.* http://www3.imperial.ac.uk/

D. Allen, *NCR Corporation, Fort Collins, CO.*

David E. Allen, *Wellcome Institute for the History of Medicine, London, UK.* http://www.wellcome.ac.uk/

American Forests, *Washington, DC .* http://www.americanforests.org/

American Gas Association (The), *Washington, DC.* http://www.aga.org

***American Meteorological Society, (AMS),** *Boston, MA.* http://www. ametsoc.org/; and http://amsglossary.allenpress.com/glossary/browse? s=A&p=1

Ames Research Center, *National Aeronautics and Space Administration Moffett Field, CA.* http://www.arc.nasa.gov/

M. J. Aminoff, *University of California, San Francisco, CA.*

Edward S. Amis, *University of Arkansas, Fayetteville, AR.*

R. C. Anderson, *Jet Propulsion Laboratory, Pasadena, CA.* http://www.jpl. nasa.gov/index.cfm

Lorella Angelini, *NASA/Goddard Space Flight Center, Greenbelt, MD,* BeppoSAX (Satellite). http://www.gsfc.nasa.gov/

F. Arnold, *Kollmorgen Corporation, Northampton, MA.* http://www.eo. kollmorgen.com/

H. R. Arum, *Designatronics, Inc., New Hyde Park, NY.* http://www.design-atronics.com/

P. Auvray, *Levallois-Perret-Cedex, France.*

J. Bakos, *J. H. Fletcher & Company, Huntington, WV.* http://www.jhfletcher.com/

M. S. Baldwin, *Westinghouse Electric Corporation, East Pittsburgh, PA.*

D. Bane, *Jet Propulsion Laboratory/California Institute of Technology, Pasadena, CA.* http://www.jpl.nasa.gov/index.cfm

Gary A. Bannon, *Monsanto Company, St Louis, MO.*

Gregor Barclay, *University of the West Indies, St Augustine, Trinidad and Tobago.* http://sta.uwi.edu/

R. Q. Barr, *Climax Molybdenum Company, (A subsidiary of the Phelps Dodge Corporation), Phoenix, AZ.* http://www.climaxmolybdenum.com/

Alan D. T. Barrett, *University of Texas, Galveston, TX.*

Alan J. Barrett, *The Babraham Institute, Babraham, Cambridge, UK.* http://www.babraham.ac.uk/

W. T. Barrett, *Foote Mineral Company, Exton, PA.*

Derrick Baxby, *University of Liverpool, UK.*

Trudy E. Bell, *Science@NASA.* http://science.nasa.gov/

James Bellows, *Westinghouse Electric Corporation, Cranberry Township, Butler County, PA.* http://www.westinghouse.com/home.html

E. Bendel, *McDonnell Douglas Corporation, Long Beach, CA.*

Richard E. Benedick, *Pacific Northwest National Laboratory (PNNL), Richland, WA.* http://www.pnl.gov/

R. J. Benke, *Westinghouse Electric Corporation, Pittsburgh, PA.* http://www.westinghouse.com/

W. O. Bennett, *American Time Products, Woodside, NY.*

Michael J. Benton, *University of Bristol, Bristol, UK.*

Jeremy Berg, *National Institute of General Medical Sciences, Bethesda, MD.* http://www.nigms.nih.gov/

Kathleen L. Berkner, *The Cleveland Clinic Foundation, Cleveland, OH.* http://www.clevelandclinic.org/

M. S. Bernath, *Gould, Inc., Andover, MA.*

Ravi Bhagavathula, *Wichita State University, Wichita, KS.*

Deepak Bhatnagar, *United States Department of Agriculture (USDA), New Orleans, LA.*

Neil W. Blackstone, *Northern Illinois University, De Kalb, IL.*

J. Blackwell, *Department of Macromolecular Science, Case Western Reserve University, Cleveland, OH.*

J. A. Blaeser, *Gould, Inc., Andover, MA.*

Suzanne Board, *Toronto, Ontario, Canada.*

Robert E. Bodenheimer, Jr., *Georgia Institute of Technology, Atlanta, GA.*

Giles Boland, *Harvard Medical School, Boston, MA.* http://hms.harvard. edu/hms/home.asp

Katherine R. Bonson, *National Institute of Mental Health, Bethesda, MD.* http://www.nimh.nih.gov/

Fred L. Bookstein, *University of Michigan, Ann Arbor, MI.*

BorgWarner Chemicals, *Engineering Staff, Washington, WV.*

G. Bouissières, *University of Paris, Orsay, France.*

R. S. Boulton, *Ministry of Works, Wellington, New Zealand.*

C. O. Bounds, *St. Joe Minerals Corporation, Monaca, PA.*

R. G. Bowen, *Consulting Geologist, Portland, OR.*

Peter J. Bowler, *Queen's University of Belfast, Belfast, Ireland, UK.*

Patricia T. Boyd, Ph.D., *U. Maryland Baltimore County, and NASA's Goddard Space Flight Center, Greenbelt, MD.*

J. Boyle, *Giddings & Lewis Electronics Company, Fond Du Lac, WI.*

J. M. Breen, *Adaptive Intelligence Corporation, Milpitas, CA.*

Emera Bridger, *SUNNY-ESF, Syracuse, NY.*

E. H. Bristol, *The Foxboro Company, Foxboro, MA.* http://www.foxboro. com/us/eng/Homepage

W. H. Brock, *University of Leicester, Leicester, UK.*

Aaron L. Brody, *Packaging/Brody, Inc., Duluth, GA.*

Cecelia M. Brown, *University of Oklahoma, Norman, OK.*

Joh H. Brown, *Fort Worth, TX.*

P. M. Brown, *Foote Mineral Company, Exton, PA.*

Janet Browne, *Wellcome Institute for the History of Medicine, London, UK.* http://www.wellcome.ac.uk/

N. W. Browne, *Davy McKee (Oil & Chemicals) Ltd., London, UK.*

Donald E. Brownlee, *University of Washington, Seattle, WA.*

Nils Brünner, *Copenhagen University Hospital, Copenhagen, Denmark.* http://www.ku.dk/english/

R. Brunner, *Semiconductor Products Sector, Motorola Inc., Phoenix, AZ, Bureau International de l'Heure, Paris, France*

Bruce G. Buchanan, *University of Pittsburgh, Pittsburgh, PA.*

Robert Bud, *Science Museum, London, UK.* http://www.sciencemuseum. org.uk/

Alan Buis, *Jet Propulsion Laboratory (JPL), Pasadena, CA.* http://www. jpl.nasa.gov/index.cfm

B. M. Burns, *Coal Technology Association, Gaithersburg, MA.* http://www. coaltechnologies.com/

L. H. Busker, *Beloit Corporation, Beloit, WI.*

W. F. Bynum, *Wellcome Trust Centre for the History of Medicine at UCL, London, UK.* http://www.wellcome.ac.uk/

Ross N. P. Cahill, *University of Melbourne, Melbourne, Australia.*

E. R. Caianiello, *Instituto di Fisica Teorica, Università di Napoli, Naples, Italy*

Joe Cain, *University College London, London, UK.*

Charles H. Calisher, *Colorado State University, Fort Collins, CO.*

Canadian Association of Petroleum Producers, *Calgary, Alberta, Canada.* http://www.capp.ca/

Stefano Canali, *University of Rome, Rome, Italy.*

Donald Canton, *University of Florida, Gainesville, FL.*

J. Caraceni, *International Fuel Cells, Inc., South Windsor, CT.*

S. C. Carapella, Jr., *ASARCO LLC, Tucson, AZ.* http://www.asarco.com/

J. J. Carpenter, *American Time Products, Woodside, NY.*

Kenneth Carpenter, *Denver Museum of Natural History, Denver, CO.* http://www.dmns.org/main/en/

K. J. Carpenter, *University of California, Berkeley, CA.*

Ann Koh Carolyn, *Colorado School of Mines, Golden, CO.* http://www. mines.edu/index_js.shtml

M. S. Carrigy, *Alberta Oil Sands Technology and Research Authority, Edmonton, Alberta, Canada.* www.asra.gov.ab.ca/strategic/energy.htm

R. T. Carson, *Eaton Corporation, Milwaukee, WI.*

Arturo Casadevall, *Albert-Einstein College of Medicine, New York, NY.* http://www.aecom.yu.edu/home/

Joseph Castellano, *Stanford Resources, Inc., San Jose, CA.*

Centers for Disease Control and Prevention (CDC), *Atlanta, GA.* http://www.cdc.gov/health/diseases.htm

Centre National de la Recherche Scientifique, *Solar Energy Laboratory, Font Romeau, France.* http://www.cnrs.fr/index.html

C. G. Chaggaris, *ORS Automation, Inc., Princeton, NJ.* http://namc-hitech.com/portfolio/ORS/

Perng-Kuang Chang, *United States Department of Agriculture (USDA), New Orleans, LA.*

Francis F. Chen, *University of California, Los Angeles, CA.*

Vinton G. Cherf, *Internet Architecture and Technology, at MCI World-Com.*

R. H. Cherry, *Consultant, Huntington Valley, PA.*

Boris Chertok, *ENERGIA Space Association, Russian Federation.* http://www.energia.ru/english/

Thomas M. Chiang, *University of Tennessee, Memphis, TN.*

A. Chiavello, *Satellite Communications, Denver, CO.* http://www.space-colorado.org/index.html

W. Chow, *Electric Power Research Institute, Palo Alto, CA.* http://www. epri.com/

Neil D. Christensen, *Pennsylvania State University, Hershey, PA.*

Henrik I. Christensen, *Royal Institute of Technology, Stockholm, Sweden.* http://www.kth.se/?l=en

Dennon Clardy, *National Aeronautics and Space Administration (NASA).* http://discovery.nasa.gov/

David D. Clark, *MIT Laboratory for Computer Science, Cambridge, MA.* http://www.csail.mit.edu/index.php

David L. Clark, *Department of Geology and Geophysics, University of Wisconsin, Madison, WI.* http://www.geology.wisc.edu/home.html

Euan N. K. Clarkson, *University of Edinburgh, Edinburgh, Scotland, UK.*

J. Cobb, *Cognex Corporation, Needham, MA.* http://www.cognex.com/

Noel G. Coley, *The Open University, Milton Keynes, UK.*

Desmond H. Collins, *Royal Ontario Museum, Toronto, Ontario, Canada.* http://www.rom.on.ca/index.php

R. L. Colona, *General Scanning Inc., Watertown, MA.*

David P. Commander, *Water and Rivers Commission, Perth, Australia.* http://www.wrc.wa.gov.au/waterinf/index.html

R. K. Conolly, *American Petroleum Institute, Washington, DC.* http://api-ec.api.org/frontpage.cfm

P. J. Constantino, *Jervis B. Webb Company, Farmington Hills, MI.* http://www.jervisbwebb.com/jbw/jerviswebbhomepage_def.htm

Aldo Conti, *Frascati(RM), Italy.*

Jimmy G. Converse, *Sterling Chemicals Inc., Texas City, TX.*

C. Sharp Cook, *University of Texas, El Paso, TX.*

P. H. Cook, *The Dow Chemical Company, Freeport, TX.*

T. E. Cook, *The Procter & Gamble Company, Cincinnati, OH.* http://www. pg.com/main.jhtml

A. B. Coon, *University of Illinois, Urbana, IL.*

George R. Cooper, *School of Electrical Engineering, Purdue University, West Lafayette, IN.* https://engineering.purdue.edu/ECE/

Giberto Corbellni, *University 'La Sapienza', Rome, Italy.*

Katrina Cornish, *United States Department of Agriculture (USDA), Washington, D.C.* http://www.usda.gov/wps/portal/usdahome

D. A. Corrigan, *Handy & Harman, Fairfield, CT.*

A. T. Coscia, *American Cyanamid Company, Stamford, CT.*

Keith A. Crandall, *Brigham Young University, Provo, UT.*

David L. Crawford, Ph.D., *International Dark-Sky Association, (Emeritus Astronomer at National Optical Astronomy Observatories/Kitt Peak National Observatory), Tuscon, AZ, Light Pollution.* http://www.darksky. org/

J. H. Cronin, *Westinghouse Electric Corporation, East Pittsburgh, PA*

A. B. Crossman, *Brown & Root, Inc., Houston, TX.*

J. M. Cruse, *University of Mississippi Medical Center, Jackson, MS.*

F. A. Cucinotta, *NASA-Johnson Space Center, Houston, TX.* http://www. nasa.gov/centers/johnson/home/index.html

W. J. Culhane, *Mead Corporation, Chillicothe, OH.* http://www. meadwestvaco.com/

V. Cullen, *Woods Hole Oceanographic Institution, Woods Hole, MA.* http://www.whoi.edu/

Emma J. A. Cunningham, *University of Cambridge, Cambridge, UK.*

Robert A. Daene, *Beloit Corporation, Beloit, WI.*

Eli Dahi, *Environmental Development Corporation, Søborg, Denmark.*

R. M. Dahlgren, *The Procter & Gamble Company, Cincinnati, OH.* http://www.pg.com/en_US/index.jhtml

E. E. David, Jr., *Exxon Research and Engineering Company, Annandale, NJ.*

R. Davis, *NCR Corporation, Fort Collins, CO.*

R. Dean, *GA Technologies, Inc. San Diego, CA.*

Ann. C. DeBaldo, Ph.D., *College of Public Health, University of South Florida, Tampa, FL.* http://health.usf.edu/publichealth/homepage.html

D. F. DeCraene, *Chemetals Corporation, Baltimore, MD.*

Alexander F. Dedus, *Russian Aviation and Space Agency, Russian Federation.*

W. E. Degenhard, *Carl Zeiss, Inc., New York, NY.* http://www.zeiss.de/us/micro/home.nsf

Steven R. Deitcher, *Cleveland Clinic Foundation, Cleveland, OH.* http://www.clevelandclinic.org/

Francesca Dellacasa, *Università di Pisa, Pisa, Italy.*

Ramon López de Mántaras, *Artificial Intelligence Research Institute, Spanish Council for Scientific Research.* http://www.iiia.csic.es/lang-en/

W. F. Dennen, *University of Kentucky, Lexington, KY.*

S. E. Desai, *Davy McKee Iron & Steel, Stockton-on-Tees, UK.*

Marie desJardins, *Department of Computer Science and Electrical Engineering Department, University of Maryland, Baltimore, MD.* http://www.umbc.edu/engineering/csee/faculty/desjardins.html

D. L. Dexter, *University of Rochester, Rochester, NY.*

Ivan Diamond, *University of California, San Francisco, CA.*

J. F. Dicello, *Johns Hopkins University School of Medicine, Baltimore, MD.*

B. Dickie, *Ministry of Mines and Minerals, Edmonton, Alberta, Canada.*

J. Dietl, *Wacker Chemie, GMBH, Munich, Germany.* http://www.chemie.de/firmen/e/2279/wacker_chemie_ag.html

E. D. Dietz, *Consultant, Toledo, OH.*

W. Dietz, *Wacker Chemie, GMBH, Munich, Germany.*

M. L., and W. L. Dilling, *The Dow Chemical Company Midland, MI.*

Adrian K. Dixon, *University of Cambridge, Cambridge, UK.*

Z. C. Dobrowolski, *Kinney Vacuum Company, Canton, MA.* http://vacuum.tuthill.com/; and http://vacuum.tuthill.com/About/about_history_kinney.asp

V. J. Dobson, *Dynapath System Inc., Livonia, MI.* http://www.dynapath.com/

Stephen K. Donovan, *The Natural History Museum, London, UK.* http://www.nhm.ac.uk/

F. Dostal, *American Time Products, Woodside, NY.*

Jim Douglas, *Dammeron Valley, UT.*

R. G. Douglas, *State University of New York at Stony Brook, Stony Brook, NY.*

E. A. Draeger, *McNally Pittsburg Mfg. Corp., Pittsburg, PA.*

H. Dressler, *Koppers Company, Inc., Monroeville, PA.* http://www.koppers.com/about.htm

Keith Dreyer, *Harvard Medical School, Boston, MA.*

R. M. Durham, *Infrared Industries, Inc., Santa Barbara, CA.* www.electro-optical.com

Gareth J. Dyke, *University College Dublin, Belfield, Ireland.*

C. J. Easton, *Sensotec, Inc., Columbus, OH.* http://www.sensotec.com/index.html

Kenneth C. Ehrlich, *United States Department of Agriculture (USDA), New Orleans, LA.*

Jan-Olof Eklundh, *Royal Institute of Technology, Stockholm, Sweden.* http://www.kth.se/?l=en

Gabriel Elkaim, *Stanford University, Stanford, CA.*

R. A. Elliott, *Qualiplus USA, Inc., Stamford, CT.*

Stanley B. Elliott, *Bedford, OH.*

Mohgah Elsheikh, *Radcliffe Infirmary, Oxford, UK.* http://www.oxford-radcliffe.nhs.uk/home.aspx

Theodore A. Endrenny, *SUNNY-ESF, Syracuse, NY.*

Eurotunnel Exhibition Centre, *Victoria Plaza, 111 Buckingham Palace Road, London SW1W OST, UK.*

Eurotunnel Information Centre, *St. Martin's Plain, Cheriton High Street, Folkstone, Kent CT19 4QD, UK.* http://ww1.eurotunnel.com/rcs/etun/pb_english/en_wp_corp/index.jsp

B. Evans, *Rare-earth Information Center, Institute for Physical Research and Technology. Iowa State University, Ames, IA.* http://www.external.ameslab.gov/RIC/index.html.

Maxime A. Faget, *NASA-Johnson Space Center, Houston, TX.* http://www.nasa.gov/centers/johnson/home/index.html

Christopher G. Fairburn, *University of Oxford, Oxford, UK.*

J. J. Faran, Jr., (retired), *Lincoln, MA.*

Daniel F. Farkas (retired), *Oregon State University, Eugene, OR.*

Gene Carl Feldman, *Goddard Space Flight Center, Greenbelt, MD.* http://www.gsfc.nasa.gov/

Dale Fenn, *Orbital Sciences Corporation, Dulles, VA.* http://www.orbital.com/

H. Fenninger, *Wacher Wachere Chemie, GMBH Munich, Germany.* http://www.chemie.de/firmen/e/2279/wacker_chemie_ag.html

L. Fieser, *Harvard University, Cambridge, MA.*

M. Fieser, *Harvard University, Cambridge, MA.*

J. File, *Plasma Physics Laboratory, Princeton University, Princeton, NJ.* http://www.pppl.gov/

T. Flack, *Westinghouse Electric Corporation, Madison Heights, MI.*

R. Fletcher, *J. H Fletcher & Company. Huntington, WV.* http://www.jhfletcher.com/

P. A. Flinn, *GMF Robotics Corporation, Troy, MI.*

Kevin Flurkey, Ph.D., *The Jackson Laboratory, Bar Harbor, ME.* http://www.jax.org

Charles T. Force, *Tracy's Landing, MD.*

Donald R. Forsdyke, *Queen's University, Kingston, Ontario, Canada.*

Jennifer M. Fostel, *Pharmacia Corporation, Kalamazoo, MI.*

Susan Eileen Fox, *Macalester College, St. Paul, MN.*

Thomas Leth Frandsen, *Copenhagen University Hospital, Copenhagen, Denmark.* http://www.ku.dk/english/

Christian D. Frazar, *Silver Spring, MD.*

John C. Freeman, *Certified Consulting Meteorologist, Weather Research Center, TX.* http://www.wxresearch.com/

Bettina C. Fries, *Albert-Einstein College of Medicine, New York, NY.* http://www.aecom.yu.edu/home/

Watson Fuller, *Keele University, Keele, UK.*

K. Galle, *Imperial College, London, UK.*

U. L. Gantenbein, *Institute for History of Medicine, Zurich, Switzerland.* http://www.dur.ac.uk/chmd/sauerteig/sexedu/

Jonathan P. Gardner, *National Aeronautics and Space Administration (NASA).*

J. A. Garman, *Great Lakes Chemical Corporation, West Lafayette, IN.*

Gas Research Institute, *DesPlaines, IL.* http://www.gri.org/

Stephen M. Gatesy, *Brown University, Providence, RI.*

R. E. Gebelein, *Moore Products Company, Spring House, PA.*

Walter Geller, *UFZ-Centre for Environmental Research, Magdeburg, Germany.* http://www.ufz.de/index.php?en=11385

F. B. Gerhard, Jr., *GTE Laboratories Incorporated, Waltham, MA.*

Sigmar German, *Physikalisch-Technische Bundesanstalt, Braunschweig, Germany.* http://www.ptb.de/index_en.html

H. P. Gerrish, *National Hurricane Center, Coral Gables, FL.* http://www.nhc.noaa.gov/

I. Gilmour, *Polaroid Corporation, Cambridge, MA.*

K. F. Glasser, *Consolidated Edison Company of New York, Inc., New York, NY.* http://www.coned.com/

Beverley J. Glover, *University of Cambridge, Cambridge, UK.*

Goddard Institute for Space Studies, *Columbia University, New York, NY.* http://www.giss.nasa.gov/

J. Golden, *National Oceanic and Atmospheric Administration, Boulder, CO.* http://www.noaa.gov/

D. T. Goldman, *National Bureau of Standards, Washington, DC.* http://www.100.nist.gov/

James E. Goldman, *Purdue University, West Lafayette, Indiana*

David S. Goldstein, *National Institutes of Health, Bethesda, MD.* www.nih.gov/

Teresa Gomez, *NASA Johnson Space Center, Houston, TX.* http://www.nasa.gov/centers/johnson/home/index.html

Avelino, J. Gonzalez, *University of Central Florida, Department of Electrical Engineering and Computer Science, Orlando, FL.* http://www.eecs.ucf.edu/

Louis H. Goodson, *Midwest Research Institute, Kansas City, MO.* http://www.mriresearch.org/

Michael D. Gottfried, *Michigan State University Museum, East Lansing, MI.*

Cristoph Gradmann, *University of Heidelberg, Heidelberg, Germany.*

Gregory Gregoriadis, *University of London, London, UK.*

Rita Groβ-Hardt, *Center of Plant Molecular Biology (ZMBP), Tübingen, Germany.* http://www.zmbp.uni-tuebingen.de/

Mario A. Di Gregori, *University of L'Aguila, Italy.*

D. L. Gregory, *Boeing Aerospace Company, Seattle, WA.* http://www.boeing.com/flash.html

E. A. Groh, *Geologist, Portland, OR.*

L. Groszek, *Technical Center, Ford Motor Company Dearborn, MI.*

K. A. Gschneidner, Jr., *Rare-earth Information Center, Institute for Physical Research and Technology. Iowa State University, Ames, IA.* http://www.external.ameslab.gov/RIC/index.html

David A. Gustafson, *James Madison University, Harrisonburg, VA.*

Colleen Hadigan, *Children's Hospital, Boston, MA.*

G. A. Hall, Jr., *Westinghouse Electric Corporation, Pittsburgh, PA.*

R. C. Hamilton, (retired), *Cornell University, Ithaca, NY.*

R. J. Hamilton, *Liverpool John Moores University, Liverpool, UK.*

William J. Hankley, *Department of Computing and Information Science, Kansas State University, Manhattan, KS.* http://www.cis.ksu.edu/

P. S. Hansen, *The Foxboro Company, Invensys Process Systems, Foxboro, MA.* http://www.foxboro.com/

P. S. Hansen, *Iowa State University, Ames, Iowa*

A. O. Hanson, *University of Illinois, Urbana, IL.*

Navraj S. Hanspal, *Loughborough University, Loughborough, UK.*

Ulla-Britt Hansson, *Lund University, Lund, Sweden.*

V. A. Harden, *National Institutes of Health, Bethesda, MD.* http://www.nih.gov/

Anne Hardy, *Wellcome Institute for the History of Medicine, London, UK.* http://www.wellcome.ac.uk/

P. W. Harland, *Ametek, Inc., Paoli, PA.* http://www.ametek.com/

***Thomas J. Harrison, (T.J.H) (retired),** *IBM Corporation, Boca Raton, FL.*

Thomas Harter, *University of California, Davis, CA.*

Martin Harwitt, *Cornell University, Ithaca, NY.*

W. Havemann, *Carl Zeiss, Inc., New York, NY.* http://www.zeiss.de/us/micro/home.nsf

John B. Hay, *University of Toronto, Toronto, Canada.*

Michael J. Hayes, *National Drought Mitigation Center, University of Nebraska, Lincoln, NE.* http://www.drought.unl.edu/

B. W. Heinemeyer, *The Dow Chemical Company, Freeport, TX.*

L. L. Hench, *University of Florida, Gainesville, FL.*

William R. Hendee, *Medical College of Wisconsin, Milwaukee, WI.* http://www.mcw.edu/display/router.asp?DocID=1

Meenhard Herlyn, *Wistar Institute, Philadelphia, PA.* http://www.wistar.org/

Claire E. J. Herrick, *Rockefeller University, London, UK.*

E. W. Hewson, *Oregon State University, Corvallis, OR.*

S. P. Higgins, Jr., *Honeywell, Inc., Phoenix, AZ.* www.honeywell.com/

Katherine A. High, *The Children's Hospital of Philadelphia, Philadelphia, PA.* http://www.chop.edu/consumer/index.jsp

D. Hines, *New Mexico Institute of Mining and Technology, Socorro, NM.* http://www.nmt.edu/

Joel H. Hildebrand, *University of California Berkeley, CA.*

Geoffrey Hinton, *Department of Computer Science, University of Toronto, Toronto, Canada.* http://web.cs.toronto.edu/dcs/

Rebecca Hitchin, *University of Bristol, Bristol, UK.*

S. E. Hluchan, *Pfizer, Inc., Wallingford, CT.* http://www.pfizer.com/main.html

Donald R. Hodge, *The BDM Corporation, Vienna, VA.*

Jessica K. Hodgins, *Georgia Institute of Technology, Atlanta, GA.*

D. M. Hoelzl, *GTE Laboratories, Incorporated, Waltham, MA.*

Richard Q. Hofacker, Jr., (retired), *Bell Laboratories, Short Hills, NJ,* Satellites (Communications and Navigation); Telephony (Telecommunications).

Josephine Hoh, *Rockefeller University, New York, NY.* http://www.rockefeller.edu/

Michael T. Holbrook, *Dow Chemical, U.S.A.*

Joseph Holden, *University of Leeds, Leeds, UK.*

Norman E. Holden, *National Nuclear Data Center, Brookhaven National Laboratory, Upton, NY.* http://www.nndc.bnl.gov/

Stephen T. Holgate, *University of Southampton, Southampton, UK.*

Arthur Hollman, *Pett, East Sussex, UK.*

Gordon Holman, *Laboratory for Astronomy and Solar Physics, NASA/Goddard Space Center, Greenbelt, MD.* http://astrophysics.gsfc.nasa.gov/astroparticles/

Arthur M. Holst, *Philadelphia Water Department, Philadelphia, PA.* http://www.phila.gov/water/

Claus Holst-Hansen, *Copenhagen University Hospital, Copenhagen, Denmark.* http://www.ku.dk/english/

K. Honchell, *Cincinnati Milacron, Lebanon, OH.* http://www.milacron.com/

J. C. Hoogendorn, *South African Coal, Oil and Gas Corp., Ltd., Sasolburg, Republic of South Africa.*

L. Hoover, *American Geological Institute (AGI), Washington, DC.* http://www.agiweb.org/

H. S. Hopkins, (retired), *Olin Corporation, Norwalk, CT.* http://www.olin.com/

Stephen Horan, *New Mexico State University, NM.*

Tim Horder, *University of Oxford, Oxford, UK.*

David W. Howard, *Brookfield Engineering Laboratories, Inc. Stoughton, MA.* http://www.brookfieldengineering.com/index.asp

Samuel C. Hsieh, *Department of Computer Science, Ball State University, Muncie, IN.* http://www.bsu.edu/cs/

Patrick Hughes, *Earth Observatory, NASA, Washington, DC.* http://earthobservatory.nasa.gov/

Martin Hülskamp, *University of Tübingen, Tübingen, Baden-Württemberg, Germany.*

G. C. Humphreys, *Davy McKee (Oil & Chemicals) Ltd., London, UK.*

Christopher J. Humphries, *The Natural History Museum, London, UK.* http://www.nhm.ac.uk/index.html

Michael Hunter, *Birkbeck University of London, London, UK.*

Charles D. Hurd, *Northwestern University, Evanston, IL.*

T. N. Hurst, *Hewlett-Packard Company, Boise, ID.*

John R. Hutchinson, *University of California, Berkeley, CA.*

***Jeanne Maree Iacono, (J. M. I.),** *Dammeron Valley, UT.*

R. P. Iacono, M.D., F.A.C.S., *Redlands, CA.*

J. Ingle, *Caterpillar, Inc., Peoria, IL .* http://www.caterpillar.com/

Martin Ingrouille, *Birkbeck College, University of London, London, UK.*

Institute of Gas Technology, *Chicago, IL.*

John Issitt, *University of York, York, UK.*

Jyrki Jaakkola, *Valmet Corporation, Charlotte, NC.*

R. B. Jacques, *Black Mesa Pipeline, Inc., Flagstaff, AZ.* http://www.black-mesapipeline.com/

Fred Jansen, *Space Science Department, ESA Directorate of Scientific Programmes, ESTEC, Noordwijk, The Netherlands.* http://eu.spaceref.com/

Michael C. Jarvis, *Glasgow University, Glasgow, UK.*

A. Jayaraman, *AT&T Bell Laboratories, Murray Hill, NJ.*

W. D. Jensen, *GTE Laboratories Incorporated, Waltham, MA.*

Jolyon Jesty, *State University of New York, Stony Brook, NY.*

Graham B. Jones, *Northeastern University, Boston, MA.*

Ross M. Jones, *Jet Propulsion Laboratory (JPL), Pasadena, CA.*

Andrew Juhl, *Lamont-Doherty Earth Observatory of Columbia University, Palisades, NY.* http://www.ldeo.columbia.edu/

Pierre Y. Julien, *Colorado State University, Fort Collins, CO.*

Deborah O. Jung, *Southern Illinois University, Carbondale, IL.*

Robert E. Kahn, *Corporation for National Research Initiatives, Reston VA.* http://www.cnri.reston.va.us/

D. Kaiser, *Parker Hannifin Corporation, Richmond, CA.* http://www.parker.com/

G. J. Kaminsky, *The Procter & Gamble Company, Cincinnati, OH.* http://www.pg.com/main.jhtml

M. L. Kapsenberg, *Academic Medical Center, Amsterdam, The Netherlands.* http://www.onderzoekinformatie.nl/en/oi/nod/organisatie/ORG12-38482/

J. N. Karlberg, *The Procter & Gamble Company, Cincinnati, OH.* http://www.pg.com/main.jhtml

Gholam A. Kazemi, *Shahrood University of Technology, Shahrood, Iran.*

David W. Kelley, *University of St. Thomas, Saint Paul, MN.*

Sir Maurice Kendall, *International Statistical Institute, London, UK.* http://isi.cbs.nl/index.htm

E. W. Kent, *National Bureau of Standards, Washington, DC.* http://www.nist.gov/

Gerhard Kerstiens, *Lancaster University, Lancaster, UK.*

Daniel J. Kevles, *California Institute of Technology, Pasadena, CA.* http://www.hss.caltech.edu/people/faculty/daniel.kevles@yale.edu

R. W. Keyes, *IBM Corporation, Yorktown Heights, NY.*

B. P. Kibble, *National Physical Laboratory, Middlesex, England.* http://www.npl.co.uk/server.php?show=nav.1

K. E. Kimball, *Siemens Capital Corporation, Iselin, NJ.*

Wayne G. Kimpton, *University of Toronto, Toronto, Canada.*

J. P. King, *The Foxboro Company, Rahway, NJ.*

Thereza L. Kipnis, *State University of Fluminense, Rio de Janeiro, Brazil.*

Gerry G. B. Klaus, *National Institute for Medical Research, London, UK.* http://www.nimr.mrc.ac.uk/

Daniel L. Klayman, *Walter Reed Army Institute of Research, Silver Spring, MD.* http://wrair-www.army.mil/

Leonard Kleinrock, *Professor of Computer Science, University of California, Los Angeles, CA.* http://www.lk.cs.ucla.edu/

Aaron Klug, *Medical Research Council, Cambridge, UK.* http://www.mrc.ac.uk/index.htm

Timothy W. Kneeland, *Nazareth College, Rochester, NY.*

George S. Kobayashi, *Washington University School of Medicine, St. Louis, MO.* http://medschool.wustl.edu/

D. M. Koffman, *GTE Laboratories Incorporated, Waltham, MA.*

Michael Kohlhase, *Department of Computer Science, Carnegie Mellon University, Pittsburgh, PA.* http://www.cs.cmu.edu/

George Kontaxakis, *Universidad Politécnica de Madrid, Madrid, Spain.*

Jean Kovalevsky, *Cerga-Observatoire de la Côte d' Azur, Grasse, France.*

Peter E. Kraght, (retired), *Consulting Meteorologist, Mabank, TX.*

P. A. Kraska, *Pattern Processing Technologies, Inc., Minneapolis, MN.*

T. W. Krauss, *Intec Controls Corporation, Foxboro, MA.*

G. Kuebler, *GLI International, Inc., (formerly Great Lakes Instruments), Milwaukee, WI.* http://www.gliint.com/

I. A. Kunasz, *Foote Mineral Company, Exton, PA.*

W. Kupper, *Mettler Instrument Corporation, Hightstown, NJ.*

Alexander N. Kuznetsov, *Russian Aviation and Space Agency, Russian Federation.*

Hyuck Kwon, *Wichita State University, Wichita, KS.*

Michael R. Ladisch, *Director, Laboratory of Renewable Resources Engineering;* http://fairway.ecn.purdue.edu/IIES/LORRE/index *and Department of Agricultural and Biological Engineering;* http://abe.www.ecn.purdue.edu/ABE/Fac_Staff/ladisch, *Purdue University, West Lafayette, IN.*

Jennifer Lagier, *Hartnell College, Salinas, CA.*

Oliver Lagueux, *Yale University, New Haven, CT.*

A. H. Lalas, *Chrysler Corporation, Detroit, MI.* http://www.chrysler.com/

Joseph B. Lambert, *Northwestern University, Evanston, IL.*

Thomas K. Landers, *AT&T Bell Laboratories, Short Hills, NJ.*

M. D. Laubichler, *Princeton University, Princeton, NJ.*

G. G. Lauer, (retired), *Koppers Company, Inc., Monroeville, PA.*

Thomas Laux, *Center of Plant Molecular Biology (ZMBP), Tübingen, Germany.* http://www.zmbp.uni-tuebingen.de/

R. F. Lawrence, (retired), *Westinghouse Electric Corporation, East Pittsburgh, PA.*

W. W. Lawrence, Jr., *Ethyl Corporation, Baton Rouge, LA.*

David B. Leake, *Computer Science Department, Indiana University, Bloomington, IN.* http://www.cs.indiana.edu/~leake/

C. Lebarbier, *Electricité de France, Paris, France.* http://www.edf.com/20403i/Home-com.html

J. M. Lee, *The M. W. Kellogg Company, Houston, TX.*

Bernard Le Guenno, *Institut Pasteur, Paris, France.*

Barry M. Leiner, *Research Institute for Advanced Computer Science, Moffett Field, CA.*

Nancy J. Leon, *Jet Propulsion Laboratory, Pasadena, CA.* http://www.jpl.nasa.gov/index.cfm

Leona M. Leonard, *University College Dublin, Belfast, Ireland.*

Leo S. Leonhart, *Hargis + Associates, Inc., Tucson, AZ.* http://www.hargis.com/index2.cfm

John S. Lewis, *University of Arizona, Tucson, AZ.*

R. E. Lewis, *University of Mississippi Medical Center, Jackson, MS.* http://www.umc.edu/

Jiayin Li, *National Institute of General Medical Sciences, Bethesda, MD.* http://www.nigms.nih.gov/

L. Libby, *Simmons Refining Company, Chicago, IL.*

Daniel V. Lim, *University of South Florida, Tampa, FL.*

Zhi-Qing Lin, *Southern Illinois University at Edwardsville, Edwardsville, IL.*

B. Lindal, *Virkir Consulting Group Ltd., Reykjavik, Iceland.*

Rebecca Lindsey, *NASA's Goddard Space Flight Center, Greenbelt, MD.*

N. C. Liston, *U. S. Department of Army Cold Regions Research and Engineering Laboratory, Hanover, NH.*

B. Lohff, *Medizinischen Hochschule, Hannover, Germany.*

Jamie Love, *Science Explained, Cloning (Mammals); and Cloning (The Story of Dolly the Sheep).* www.synapses.co.uk/science/index.html

S. Lovejoy, *McGill University, Montreal, Quebec.*

B. A. Loyer, *Motorola, Inc., Phoenix, AZ.*

Lucent Technologies, *Optical Fiber Solutions, Norcross, GA.* http://www.ofsoptics.com/

David C. Lynch, *CyberCash Inc., New York, NY.*

Steven L. Lytinen, *School of Computer Science, Telecommunications, and Information Systems, DePaul University, Chicago, IL.*

John B. Macauley, Ph.D., *The Jackson Laboratory, Bar Harbor, ME.* http://www.jax.org

Fred T. Mackenzie, *Northwestern University, Evanston, IL.*

Ralph E. Mackiewicz, *Sisco, Inc., Sterling Heights, MI.* http://www.sisconet.com/

Michael T. Madigan, *Southern Illinois University, Carbondale, IL.*

E. C. Magison, *Consulting Engineer, Ambler, PA.*

M. E. Magnello, *Wellcome Institute for the History of Medicine, London, UK.* http://www.wellcome.ac.uk/

Lois N. Magner, *Purdue University, West Lafayette, West Lafayette, IN.*

C. L. Mamzic, *Siemens Energy & Automation Inc., (formally Moore Products Company, Spring House, PA.* http://www.mooreproducts.com/

Jack Maniloff, *University of Rochester, Rochester, NY.* http://www.urmc.rochester.edu/gebs/faculty/jack_maniloff.htm

Diana E. Manuel, *Wellcome Institute for the History of Medicine, London, UK.* http://www.wellcome.ac.uk/

Jerry W. Manweiler, *Lawrence, KS.*

John Marafino, *Department of Mathematics, James Madison University, Harrisonburg, VA.* http://www.math.jmu.edu/

Julie R. Mariga, *Purdue University, West Lafayette, IN.*

Hans Mark, *Austin, TX.*

D. L. Marrin, *Hanalei, HI.*

Cathie Martin, *John Innes Centre, Norwich, UK.* http://www.jic.ac.uk/science/cdb/Index.htm

J. R. Masson, *Davy McKee (Oil and Chemicals) Ltd., London, UK.*

Ramon A. Mata-Toledo, *James Madison University, Harrisonburg, VA.*

Prabhaker Mateti, *Wright State University, Dayton, OH.*

Brian A. Maurer, *Michigan State University, East Lansing, MI.*

H. L. Mayer, *Hydro-Quebec, Montreal, Quebec, Canada.* http://www.hydroquebec.com/en/index.html

J. Mazurkiewicz, *Pacific Scientific, Rockford, IL.* http://www.pacsci.com/

Dennis J. McCance, *University of Rochester, Rochester, NY.*

Robert W. McCarley, *Harvard University, Boston, MA.*

Maclyn McCarty, *Rockefeller University, New York, NY.*

Sheila McCormic, *University of California, Berkeley, CA.*

W. R. McCown, *Westinghouse Electric Corporation, Pittsburgh, PA.*

W. F. McIlhenny, *The Dow Chemical Company, Midland, MI.*

Ian R. McNab, *The University of Texas at Austin, Austin, TX.*

Terence Meaden, *Oxford University, Oxford, UK.*

Lisa Meeden, *Associate Professor and Director, Computer Science Program, Swarthmore College, Swarthmore, PA.* http://www.cs.swarthmore.edu/

Roger W. Melvold, *University of North Dakota, Grand Forks, ND.*

Robert O. Messing, *University of California, San Francisco, CA.*

Amit Metha, *Harvard Medical School, Boston, MA.*

R. W. Miller, *Consultant, Foxboro, MA.*

Andrew R. Milner, *Birkbeck College, London, UK.*

E. D. Mohr, *Unimation (Westinghouse Electric Corporation), Danbury, CT.*

John E. Moore (retired), *USGS, Denver, CO.*

S. M. Moore, *Lawrence Berkeley Laboratory, Berkeley, CA.* http://www.lbl.gov/

Michel Morange, *Ecole Normale Superieure, Paris, France.* http://www.ens.fr/index_ en.php

Gregory J. Morgan, *Johns Hopkins University, Baltimore, MD.*

J. A. Morgan, *North American Electric Reliability Council, Princeton, NJ.*

V. I. Moroz, *Russian Academy of Sciences, Moscow, Russian Federation.*

Kevin Mulrooney, *Newark, DE. Index*

T. Murphy, *IBM Corporation, Yorktown Heights, NY.*

J. Nagy, *Beckman Industrial Corporation, Cedar Grove, NJ.*

NASA Astrobiology Institute (NAI), *Washington DC.* http://nai.arc.nasa.gov/

NASA/Goddard Space Flight Center, *Greenbelt, MD.* http://www.gsfc.nasa.gov/

NASA's Jet Propulsion Laboratory/California Institute of Technology, *Pasadena, CA.* http://www.jpl.nasa.gov/index.cfm

National Indoor Environmental Institute, *Plymouth Meeting, PA.*

National Institute of Neurological Disorders and Stroke, *Bethesda, MD.* http://www.ninds.nih.gov/index.htm

National Institutes of Health (NIH), *Bethesda, MD.* http://www.nih.gov/

Robert J. Naumann, *University of Alabama in Huntsville, Huntsville, AL.*

William T. Nearn, *Weyerhaeuser Company, Seattle, WA.* http://www.weyerhaeuser.com/

M. M. Nelson, *Honeywell Inc., Billerica, MA.*

Amiko Nevills, *National Aeronautics and Space Administration (NASA).*

L. R. Newitt, *Geological Survey of Canada, Ottawa, Ontario.* http://gsc.nrcan.gc.ca/contact_ e.php

E. R. Niblett, *Geological Survey of Canada, Ottawa, Ontario.*

Claus Nielsen, *Zoologisk Museum, Copenhagen, Denmark.* http://zoologi.snm.ku.dk/english/

S. Nojiima, *Japan Gasoline Company, Ltd., Tokyo, Japan.* http://www.tokyo-gas.co.jp/index_ e.html

Northeastern Forest Experiment Station, *U.S. Department of Agriculture (USDA), Darby, PA.*

John Norvell, *National Institute of General Medical Sciences, Bethesda, MD.* http://www.nigms.nih.gov/

Gustav J. V. Nossal, *University of Melbourne, Melbourne, Australia.*

V. Nutton, *Wellcome Institute for the History of Medicine, London, UK.* http://www.wellcome.ac.uk/

Oak Ridge National Laboratory, *Oak Ridge, TN.* http://www.ornl.gov/ornlhome/index.htm

James F. O'Brien, *Georgia Institute of Technology, Atlanta, GA.*

H. Oeda, *Ojinomoto Co., Inc., Kawaski, Japan.*

E. A. Ogryzlo, *University of British Columbia, Vancouver, British Columbia, Canada.*

Ronald J. Oldfield, *Macquarie University, Sydney, Australia.*

Robert S. Oldham, *De Montfort University, Leicester, UK.*

David Oldroyd, *The University of New South Wales, Sydney, New South Wales, Australia.*

Régis Olry, *University of Quebec at Trois-Rivières, Quebec, Canada.*

R. L. Osborne, *Honeywell Inc., Billerica, MA.*

R. H. Osman, *Robicon Corporation, (A Subsidiary of High Voltage Engineering Corporation), New Kensington, PA.* http://www.robicon.com/

Jurg Ott, *Rockefeller University, New York, NY.*

John S. Oxford, *St Bartholomew's and the Royal London School of Medicine and Dentistry, London, UK.* http://www.smd.qmul.ac.uk/

V. C. Oxley, *GTE Laboratories Incorporated, Waltham, MA.*

S. T. Oyama, *Lawrence Berkeley Laboratory, Berkeley, CA.* http://www.lbl.gov/

Pacific Gas and Electric Company, *(a subsidiary of PG&E Corporation), San Francisco, CA.* http://www.pge.com/

Stefano Pagliara, *Università di Pisa, Pisa, Italy.* http://www.unipi.it/english/index.htm

Panel on Mathematical Sciences, *Commission on Physical Sciences, Mathematics, and Resources, National Research Council, Washington, DC.* http://sites.nationalacademies.org/nrc/index.htm

John Parascandola, *U.S. Department of Health and Human Services, Rockville, MD.* http://www.hhs.gov/

B. S. Park, *National Institutes of Health, Bethesda, MD.* http://www.nih.gov/

Bradford Parkinson, *Stanford University, Stanford, CA.*

Ohad Parnes, *Wellcome Trust Centre for the History of Medicine at UCL, London, UK.* http://www.wellcome.ac.uk/

Ohad S. Parnes, *Max Planck Institute for the History of Science, Berlin, Germany.* http://www.mpiwg-berlin.mpg.de/en/index.html

Judith Totman Parrish, *University of Arizona, Tucson, AZ.*

J. M. Pasachoff, *Hopkins Observatory, Williams College, Williamstown, MA.* http://www.williams.edu/Astronomy/

Jose O. Payero, *University of Nebraska-Lincoln, North Platte, NE.*

R. Peacock, *LTV Steel Company, Inc. Independence, OH.* http://www.ltvsteel.com/htmfiles/glance.htm

Amanda R. Perry, *Institute of Cancer Research, Sutton, Surrey, UK.* http://www.icr.ac.uk/

Max Perutz, *Cambridge University, Cambridge, UK.*

Peter Pesch, *Astronomy Department, Case Western Reserve University, Cleveland, OH.* http://burro.astr.cwru.edu/dept/

Alan Petersen, *University of Plymouth, Plymouth, UK.*

L. V. Pfaender, *Owens-Illinois, Toledo, OH.*

Sir David Phillips, *University of Oxford, Oxford, UK.*

A. K. Pierce, *Kitt Peak National Observatory (a division of the National Optical Astronomy Observatories which is operated by the Association of Universities for Research in Astronomy (AURA), Inc. under cooperative agreement with the National Science Foundation, Tucson, AZ.* http://www.noao.edu/kpno/

W. T. Plass, *U.S. Department of Agriculture Forest Service, Northeastern Forest Experimentation, Princeton, WV.* http://www.na.fs.fed.us/

Benjamin R. Pobanz, *Purdue University, West Lafayette, IN.*

Howard W. Post, *Williamsville, NY.*

D. Postma, *General Motors Corporation, Detroit, MI.*

H. J. Power, *Wellcome Trust Centre for the History of Medicine at UCL, London, UK.* http://www.wellcome.ac.uk/

D. B. Priddy, *The Dow Chemical Company, Midland, MI.*

Nitish Priyadarshi, *Ranchi University, Ranchi, Jharkhand, India.*

Donald R. Prothero, *Occidental College, Los Angeles, CA.*

Michele L. Pruyn, *Oregon State University, Corvallis, OR.*

J. H. Purnell, *Department of Chemistry, University of Swansea, Swansea, UK.* http://www.swan.ac.uk/

Viviane M. Quirke, *The Royal Institution of Great Britain, London, UK.* http://www.rigb.org/registrationControl?action=home

Kanury V. S. Rao, *International Center for Genetic Engineering and Biotechnology, New Delhi, India.* http://www.icgeb.trieste.it/RESEARCH/ND/ndrsprg.htm

P. Krishna Rao, *National Oceanic and Atmospheric Administration, Silver Springs, MD.* http://www.noaa.gov/

Jeremy Rasmussen, *Sypris Electronics, LLC., Tampa, FL.*

Todd Rasmussen, *The University of Georgia, Athens, GA.*

M. J. Ratcliff, *Wellcome Institute for the History of Medicine, London, UK.* http://www.wellcome.ac.uk/

N. Razo, *National Center for Atmospheric Research, Boulder, CO.* http://www.ncar.ucar.edu/ncar/

Philip F. Rehbock, *University of Hawaii, Honolulu, HI.*

R. D. Reincke, *Caterpillar Inc., Peoria, IL.* http://www.caterpillar.com/

R. G. Reip, *Consulting Engineer, Sawyer, MI.*

Victor I. Reus, *University of California, San Francisco, CA.*

Vladimir V. Riabov, *River College, Nashua, NH.* http://www.rivier.edu

R. P. Rich, *Eastman Chemical Company, Kingsport, TN.* http://www.eastman.com/Markets/Textiles/Textiles_intro.asp

E. H. Richardson, *Herzberg Institute of Astrophysics Dominion Astrophysical Observatory, Victoria, British Columbia, Canada.* http://www.hia.nrc.ca/

J. A. Riddick, *Baton Rouge, LA.*

J. C. Riley, *Consulting Engineer, Portland, OR.*

G. G. Robert, *University of Oxford, Oxford, UK.*

Gareth Roberts, *FRS, Thorn EMI plc and University of Oxford, Oxford, UK.*

Lawrence G. Roberts, *Caspian Networks, San Jose, CA.* http://www.caspian.com/home.asp

Robert L. Roberts, *University of California at Los Angeles, Los Angeles, CA.*

T. H. Rogers, (retired), *Elastomers Consultant, Clearwater, FL.*

Nils Roll-Hansen, *University of Oslo, Oslo, Norway.*

G. R. Romovacek, *Koppers Company, Inc., Monroeville, PA.*

B. A. Ross, *General Motors Corporation, Indianapolis, IN.*

Duane L. Ross, *NASA Johnson Space Center, Houston, TX.* http://www.nasa.gov/centers/johnson/home/index.html

D. M. Ross, *Propellants Consultant, Lancaster, CA.*

Alex T. Rowland, *Gettysburg College, Gettysburg, PA.*

Elmer B. Rowley, (retired), *Union College, Schenectady, NY.*

P. F. H. Rudolph, *Lurgi Mineralotechnik, GMBH, Frankfurt (Main), West Germany.*

Edward G. Ruestow, *University of Colorado, Boulder, CO.*

Nicolaas A. Rupke, *Göttingen University, Göttingen, Germany*

G. A. Russell, *Texas A&M University System Health Science Center, College Station, TX.* http://medicine.tamhsc.edu/

L. Russell, *MTS Systems Corporation, Eden Prairie, MN.* http://www.mts.com/

Jack J. Rutledge, *Professor and Chair Department of Meat and Animal Science, University of Wisconsin-Madison, Madison, WI.* http://www.ansci.wisc.edu/

Kirstie Saltsman, *National Institute of General Medical Sciences, Bethesda, MD.* http://www.nigms.nih.gov/

Paul M. Salvaterra, *Beckman Research Institute, Duarte, CA.* http://www.cityofhope.org/bricoh

Sundeep S. Salvi, *University of Southampton, Southampton, UK.*

Anthony P. Sampson, *University of Southampton, Southampton, UK.*

Neeraja Sankaran, *Yale University, New Haven, CT.*

S. J. Sansonetti, *Consultant, Reynolds Metals Company (ALCOA), Richmond. VA.* http://www.alcoa.com/

R. P. Santandrea, *Los Alamos National Laboratory, Los Alamos, NM.* http://www.lanl.gov/worldview/

E. J. Sare, *PPG Industries Inc., Barberton, OH.*

W. L. W. Sargent, *Royal Greenwich Laboratory, Sussex, UK.* http://www.the-observatory.org/

Kapaettu Satyamoorthy, *Wistar Institute, Philadelphia, PA.* http://www.wistar.org/

Jonathan Schaeffer, Ph.D., *Department of Computer Science, University of Alberta, Edmonton, Alberta, Canada.* http://www.cs.ualberta.ca/

D. Schertzer, *Météorologie Nationale, Paris, France.*

William T. Schiano, *Bentley College, Waltham, MA.*

C. E. Schildknecht, *Gettysburg College, Gettysburg, PA.*

W. R. Schiller, *Wacher Chemie, GMBH, Munich, Germany.*

Lukas Schreiber, *University of Bonn, Bonn, Germany.*

M. Schussler, *Fansteel, North Chicago, IL.*

Birgit Schwab, *University of Tübingen, Tübingen, Baden-Württemberg, Germany.*

James H. Schwartz, *Columbia University College of Physicians and Surgeons, New York, NY.* http://www.cumc.columbia.edu/dept/ps/

M. Sekino, *Toyobo Co., Ltd., Iwakuni, Yamaguch-Pref., Japan.*

Raj Sharma, *University of KwaZulu-Natal, Durban, South Africa.*

W. G. Shequen, (retired), *Bausch & Lomb, Sunland, CA.* http://www.bausch.com/

*Steven N. Shore, (S.N.S),** *University of Indiana South Bend, South Bend, IN.*

E. C. Shuman, *Consulting Engineering, State College, PA.*

Siemens Aktiengesellschaft Engineering Staff, *Erlangen, Germany.*

W. Dias Da Silva, *Universidade Estadual do Norte Fluminense, Rio De Janeiro, Brazil.*

Milton A. Silveira, *NASA-Johnson Space Center, Houston, TX.* http://www.nasa.gov/centers/johnson/home/index.html

Arthur M. Silverstein, *John Hopkins University School of Medicine, Baltimore, MD.* http://www.jhu.edu/

L. E. Simmons, *Simmons Refining Company, Chicago, IL.*

S. Fred Singer, *The Science & Environmental Policy Project (SEEP), Arlington, VA.* http://www.sepp.org/

Pratap Singh, *National Institute of Hydrology, Roorkee, India.* http://www.nih.ernet.in/

Christopher M. Sinton, *Harvard University, Boston, MA.*

D. C. Sleeman, *Davy McKee (Oil & Chemicals) Ltd., London, UK.*

L. F. Small, *Oregon State University, Corvallis, OR.*

Mark D. Smith, *Allied Signal Aerospace Company, Phoenix, AZ.* http://ludb.clui.org/ex/i/AZ3132/

David R. Smyth, *Monash University, Melbourne, Australia.*

Walter E. Sneader, *University of Strathclyde, Glasgow, Scotland, UK.* http://www.strath.ac.uk/

James S. Sochacki, *James Madison University, Harrisonburg, VA.*

G. A. Somorjai, *Lawrence Berkeley Laboratory, Berkeley, CA.* http://www.lbl.gov/

P. E. Spargo, *University of Cape Town, Rondebosch, South Africa.*

E. Sperry, *Beckman Industrial Corporation, Cedar Grove, NJ.*

James Spiker, *Stanford University, Stanford, CA.*

M. A. Stadtherr, *Department of Chemical Engineering, University of Illinois, Urbana, IL.* http://www.engr.uiuc.edu/

S. Stamas, *Exxon Corporation, New York, NY.* http://www.exxon.com/index_flash.html

Susan-Marie Stedman, *NMFS F/HC, Silver Spring, MD.*

C. Bruce Stephenson, *Department of Astronomy, Case Western University, Cleveland, OH.* http://burro.astr.cwru.edu/dept/

Peter F. Stevens, *Missouri Botanical Gardens, St. Louis, MO.* http://www.mobot.org

J. Stevenson, *West Instruments, East Greenwich, RI.*

Richard E. Stiehm, *University of California at Los Angeles, Los Angeles, CA.*

S. Stoddard, *Waugh Controls Corp., Chatsworth, CA.*

T. S. Storer, *Hewlett-Packard Company, Palo Alto, CA.* www.hp.com/

E. Sulzer, *Siemens Energy & Automation, Inc., Peabody, MA.*

J. C. Summers, *Automotive Catalyst Company, Tulsa, OK.*

Kenneth S. Suslick, *University of Illinois at Urbana-Champaign, Urbana, IL.*

Michael A. Sutton, *University of Northumbria, Newcastle upon Tyne, UK.*

H. F. Szepan, (retired), *Ingersoll-Rand Co., Impco Division, Nashua, NH.* http://www.ingersoll-rand.com/

Michael Szyscher, *PolyMedica Industries, Inc.*

E. M. Tansey, *Wellcome Institute for the History of Medicine, London, UK.* http://www.wellcome.ac.uk/

Barry L. Tarmy, *TBD Technology.*

D. G. Terry, (retired), *Ingersoll-Rand Co., Impco Division, Nashua, NH.*

James Thrall, *Harvard Medical School, Boston, MA.*

Tokyo Electric Power Company, *Tokyo, Japan.*

Wesley F. Tree, *The College of Wooster, Wooster, OH.*

W. A. Troeger, *Weston (Sangamo-Weston, Inc.), Newark, NJ.*

Joachim Truemper, Ph.D., *Professor, Max Planck Institute (MPE), Germany.* http://www.mpe-garching.mpg.de/

Karen Tucker, *Chandra X-ray Observatory Center, Harvard-Smithsonian Center for Astrophysics, Cambridge, MA, X-Ray Astronomy.* http://cfa-www.harvard.edu/

Wallace Tucker, *Chandra X-ray Observatory Center, Harvard-Smithsonian Center for Astrophysics, Cambridge, MA, X-Ray Astronomy.* http://cfa-www.harvard.edu/

S. Turner, *National Bureau of Standards, Gaithersburg, MD.* http://www.nist.gov/

Izrail S. Turovsky, *Jacksonville, FL.*

David Twell, *University of Leicester, Leicester, UK.*

Mohsen G. Uizani, *Western Michigan University, Kalamazoo, MI.*

David M. Unwin, *Museum für Naturkunde, Berlin, Germany.*

L. F. Urry, *Eveready Battery Company, Ltd., Westlake, OH.* http://www.eveready.com/

U. S. Department of Energy, *Office of Health and Environmental Research, Oak Ridge, TN.* http://www.atsdr.cdc.gov/hac/oakridge/index.html

U. S. Environmental Protection Agency (EPA), *Washington, D.C.* http://www.epa.gov/

G. V. Van denBerg, *Shell Internationale Petroleum Maatschappij B. V., The Hague, Netherlands.* http://www.shell.com/

O. Vandermarcq, *Ambassade de France aux Etats-Unis Services de la Mission Scientifique, Houston, TX.*

E. Van Haaften, *American Time Products, Woodside, NY.*

J. A. Vegeasis, *Shell Development Company, Houston, TX.* http://www.shell.com/home/Framework?siteId=us-en

Manual G. Venegas, *The Procter & Gamble Company, Cincinnati, OH.* http://www.pg.com/en_ US/index.jhtml

Paul Verrell, *Washington State University, Pullman, WA.* http://www.wsu.edu/~verrelab/

*R. C. Vickery, (R.C.V), *Blanton/Dade City, FL.*

Video Logic Corporation, *Sunnyvale, CA.*

Roger, Vignelles, *Corbeil-Essonnes, France.*

R. Villalobos, *The Foxboro Company (A Siebe Company), Foxboro, MA.*

Ray Villard, *Space Telescope Science Institute, Baltimore, MD, Hubble Space Telescope.* http://www.stsci.edu/resources/

VisionRx, Inc., *Elmsford, NY.* http://visionrx.com/

G. T. Volpe, *University of Bridgeport, Bridgeport, CT.*

Robert Volpé, *University of Toronto, Toronto Ontario, Canada*

Kyle Wagner, Ph.D., NIH Fellow, *University Maryland at Baltimore, Baltimore, MD., and University of Maryland Institute for Advanced Computer Studies, College Park, MD.* http://www.umiacs.umd.edu/

J. Walker, *Ontario Hydro, Toronto, Ontario, Canada.*

W. Allan Walker, *Children's Hospital Boston, Boston, MA.* http://www.childrenshospital.org/

John Waller, *University of London, London, UK.*

K. A. Walsh, *Brush Wellman Inc., Elmore, OH.* http://www.brushwellman.com/index.asp

Johannes Walter, *Kaiser Franz Josef Spital, Vienna, Austria.*

J. D. Warnock, *Siemens Energy & Automation Inc., (formally Moore Products Company), Spring House, PA.* http://www.mooreproducts.com/

Albin H. Warth, *Cape May, NJ.*

John A. H. Wass, *Radcliffe Infirmary, Oxford, UK.* http://www.oxford-radcliffe.nhs.uk/home.aspx

C. Kenneth Waters, *Minnesota Center for Philosophy of Science, University of Minnesota, MN.* http://www.mcps.umn.edu/

Katherine D. Watson, *University of Oxford, Oxford, UK.*

Byron H. Webb, *U.S. Department of Agriculture (USDA), Washington, DC.* http://www.usda.gov/wps/portal/usdahome

Martin C. Weisskopf, *Marshall Space Flight Center, Huntsville, AL.* http://www.msfc.nasa.gov/

J. Wells, *Edison International, parent company of (Southern California Edison Company), Rosemead, CA.* http://www.edisonx.com/

J. Y. Welsh, *Chemetals Corporation, Baltimore, MD.*

Michael Werner, *Jet Propulsion Laboratory (JPL), Pasadena, CA.* http://www.jpl.nasa.gov/index.cfm

L. Werth, *Pattern Processing Technologies, Inc., Minneapolis, MN.*

J. R. Whiteway, *Ontario Hydro, Toronto, Ontario.*

Darrell Whitley, *Department of Computer Science, Colorado State University, Fort Collins, CO.* http://www.cs.colostate.edu/

Richard J. Whitley, *University of Alabama at Birmingham, Birmingham, AL.*

Robert M. Whittier, *Endevco Corporation, San Juan Capistrano, CA.* http://www.endevco.com/

P. R. Wiederhold, *General Eastern Instruments Corporation, Watertown, MA.*

Lise Wilkinson, *Wellcome Institute for the History of Medicine, London, UK.* http://www.wellcome.ac.uk/

R. N. Wilkinson, *The Procter & Gamble Company, Cincinnati, OH.* http://www.pg.com/en_ US/index.jhtml

Adrian E. Williams, *APEM Ltd., Manchester, UK.* http://www.apemltd.co.uk/aquatics/

David R. Williams, *NASA Goddard Space Flight Center, Greenbelt, MD.* http://www.gsfc.nasa.gov/

E. Williams, *Cobalt Information Centre, London, UK.*

R. L. Wilson, *Honeywell, Inc., Fort Washington, PA.*

E. G. Winchester, *Wellcome Trust Centre for the History of Medicine, London, UK.* http://www.wellcome.ac.uk/

A. T. Winfree, *Professor Ecology and Evolutionary Biology, University of Arizona, Tucson, AZ.* http://eebweb.arizona.edu/

Christer Wingren, *Lund University, Lund, Sweden.* http://www.createhealth.lth.se/research/carl_ borrebaeck/research_ groups/christer_ wingren/

J. A. Witkowski, *Cold Spring Harbor Laboratory, Cold Spring Harbor, NY.* http://www.cshl.edu/

Wolfgang Wöger, *Physikalisch-Technische Bundesanstalt, Braunschweig, Germany.*

Stephen Wolff, *Cisco Systems, Inc., San Jose, CA.* http://www.cisco.com/

A. S. Wood, *Jet Propulsion Laboratory/California Institute of Technology, Pasadena, CA.* http://www.jpl.nasa.gov/index.cfm

G. R. Woodcock, *Boeing Aerospace Company, Seattle, WA.* http://www.boeing.com/flash.html

Michael Worboys, *Sheffield Hallam University, Sheffield, UK.*

Brian S. Worthington, *University of Nottingham, Nottingham, UK.*

Edward L. (Ned) Wright, *Professor of Physics and Astronomy, UCLA, Westwood, CA.* http://www.astro.ucla.edu/%7Ewright/intro.html Cosmology.

Mike Wright, *Marshall Space Flight Center, Huntsville, AL.* http://www.nasa.gov/centers/marshall/home/index.html

Simcha Lev-Yadun, *University of Haifa-Oranim, Tivon, Israel.*

Chih Ted Yang, *Colorado State University, Fort Collins, CO.*

Michael I. Yarymovych, (retired), *Boeing Space and Communications, Seal Beach, CA.*

G. Yazbak, *MetriCor, Inc., Monument Beach, MA.*

Timothy J. Yeatman, *University of South Florida, Tampa, FL.*

Alexander V, Zakharow, *Russian Academy of Sciences, Moscow, Russian Federation.*

C. K. Zimmerman, *E. I. DuPont de Nemours & Company, Inc., Wilmington, DE.* http://www.dupont.com/

Heddy Zola, *Child Health Research Institute, Adelaide, Australia.* http://www.cafhri.org.au/

Arie J. Zuckerman, *University of London, London, UK.*

VAN NOSTRAND'S
SCIENTIFIC ENCYCLOPEDIA
Tenth Edition

VOLUME 2

G

GABBRO. Gabbro is a deep-seated and often very coarse-grained igneous rock composed of plagioclase feldspar, usually labradorite or bytownite and monoclinic pyroxene, with occasionally as accessories olivine (when it is then called olivine gabbro), biotite, magnetite, ilmenite, and hornblende. Norite is a variety of gabbro, carrying orthorhombic pyroxene, usually hypersthene instead of the monoclinic sort. Troctolite is essentially olivine and plagioclase. Quartz gabbros are known and have probably been derived from magmas somewhat oversaturated with silica. On the other hand, essexites represent gabbros whose parent magma doubtless had an insufficiency of silica resulting in the formation of nephelite. Gabbros are frequently rich in sulfides that may be of commercial value, a notable occurrence of which is at Sudbury, Canada. Here a norite carrying chalcopyrite and nickeliferous pyrrhotite forms the most important deposits of nickel known. Gold, silver and platinum are also recovered from this ore.

GABOON VIPER. See **Snakes**.

GADOLINIUM. [CAS: 7440-54-2] Chemical element symbol Gd, at. no. 64, at. wt. 157.25, seventh in the Lanthanide series in the periodic table, mp. 1,313 °C, bp 3,273 °C, density 7.901 g/cm³ (20 °C). Elemental gadolinium has a close-packed hexagonal crystal structure at 25 °C. The pure metallic gadolinium is silver-gray in color, slow to tarnish in normal atmospheres. The metal is soft, malleable, and easy to fabricate with normal tools provided that processing temperatures are maintained below 150 °C. The turnings and chips of gadolinium are mildly pyrophoric and care must be exercised in their handling. There are seven natural isotopes of gadolinium: ^{152}Gd, ^{154}Gd through ^{158}Gd, and ^{160}Gd. Eleven artificial isotopes have been prepared. The natural isotopes are not radioactive. In terms of abundance, gadolinium is present on the average of 5.4 ppm in the earth's crust, making it potentially more available than tantalum, tin, or tungsten. The element was first identified by J.C.G. Marignac in 1880. The natural isotopic mixture of gadolinium has the greatest thermal-neutron-absorption cross section of all elements, 40,000 barns. This is approximately 10 times greater than the next two elements, samarium (5,800 barns) and europium (4,300 barns). However, gadolinium is limited to nuclear applications mainly as a start-up and shutdown material because only two of the natural isotopes ^{155}Gd and ^{157}Gd behave in this manner. ^{155}Gd and ^{157}Gd make up 31% of the total weight of elemental gadolinium. The metal has a low acute-toxicity rating. Electronic configuration

$$1s^2 2s^2 2p^6 3s^2 3d^{10} 4s^2 4p^6 4d^{10} 4f^7 5s^2 5p^6 5d^1 6s^2.$$

Ionic radius Gd^{3+} 0.938 Å. Metallic radius 1.801 Å. First ionization potential 6.1 eV; second 12.09 eV.

Other important physical properties of gadolinium are given under **Rare-Earth Elements and Metals**.

Gadolinium reacts vigorously with dilute mineral acids, but is practically inert to strong bases and boiling H_2O. Gadolinium is an active reducing agent for metals, including iron, chromium, manganese, tin, lead, and zinc. The major sources of gadolinium are xenotime, monazite, gadolinite, residues from uranium mining, and ion-exchange clays found in Southern China.

Although the nuclear properties of the element are attractive, gadolinium has enjoyed rather limited applications in reactor technology. A major use of gadolinium is in amorphous Gd-Co(Fe) alloys for magnetic recording and information storage. An important discovery in the 1960s showed that gadolinium iron garnets (called GIGs) $Gd_6Fe_5O_{12}$ possess a crystalline structure which finds useful application in microwave frequency control, circulators, isolators, and bandpass filters in electronic circuitry.

Gadolinium oxide also is used as the host matrix in the red phosphor for color television picture tubes, where it is activated by europium. Gadolinium oxysulfide Gd_2O_2S is used as an x-ray image intensifier making it possible to reduce the exposure of patients to x-rays. Along with yttrium and lanthanum activated by cerium, gadolinium is used in a phosphor for single-gun beam-indexing flying-spot scanning cathode ray tubes. Gadolinium complexes are used as MRI (Magnetic Resonance Imaging) contrasting agents to improve the images obtained in MRI scans of various organs of the human body. Gadolinium also provides magnetic properties when alloyed with cobalt, cerium, iron, and copper ($Co_{3.5}CuFe_{0.5}Ce$) in permanent magnets, imparting a desirable negative temperature coefficient of magnetic saturation. Gadolinium metal and compounds are under consideration for use in a variety of magnetic refrigeration and cooling applications ranging from the liquifaction of hydrogen and natural gases, to refrigerator/freezers, supermarket chillers, and air conditioners. See also **Refrigeration**.

See references listed at ends of entries on **Chemical Elements**; and **Rare-Earth Elements and Metals**.

K.A. GSCHNEIDNER, Jr.
B. EVANS, Iowa State University, Ames, IA

GAGE (Device). An instrument or device for measuring or comparing some physical characteristics, such as size, pressure, temperature, force, water level, and surface quality. As contrasted with sophisticated recording and controlling instruments, gages are frequently manually read and often hand-applied, as in the case of the gages used in the machining and metalworking field. There are instances, however, where the term *gage* is applied to costly, complex instruments, as in the vacuum-measurement field. Gaging also can be fully automated as in the application of pneumatic and electrical gages for the continuous "go/no-go" inspection of parts. No fixed rules have been established for guidance in use of the term.

GAGE LINE. A gage line marks the limits of any standard distance used repeatedly. Structural shapes are punched or drilled on lines called gage lines. The gage lines may be varied to suit the details so long as the minimum required edge distance and clearance for punching and drilling are maintained. In some fabricating shops, holes are made with multiple punches or drills.

GAHNITE. The mineral gahnite, also known as *zinc spinel*, is isometric with an octahedral habit but may appear as dodecahedrons or modified cubes. Chemically it is zinc aluminate corresponding to the formula $ZnAl_2O_4$. There is a tendency for cleavage parallel to the octahedron; fracture varies from conchoidal to uneven; brittle; hardness 7.5–8; specific gravity 4.6; luster, vitreous; color ranges from dark green through various shades of greenish- or bluish-black, yellowish-black or grayish, subtransparent to almost opaque. Gahnite is found in association with other zinc minerals at several European localities, notably in Bavaria and Sweden. In the United States it is found at Franklin and Sterling Hill, New Jersey; at Rowe, Massachusetts and in Maryland, North Carolina, Georgia and Colorado. Gahnite was named in honor of the Swedish chemist, J.G. Gahn.

GAIN (Antenna). See **Antenna**.

GAIN BANDWIDTH PRODUCT. The gain bandwidth product is equal to the product of amplification of an amplifier stage at midband, multiplied by the bandwidth of the amplifier. The *bandwidth* is defined as the difference Δf between the two frequencies at which the power output is a specified fraction, usually one-half, of the midband (resonance) value.

GAIN (Magnitude Ratio). With reference to industrial and scientific instruments, the Instrument Society of America defines gain for a linear system or element as: the ratio of the magnitude (amplitude) of a steady-state sinusoidal output relative to the causal input; the length of a phasor from the origin to a point of the transfer locus in a complex plane.

The quantity may be separated into two factors: (1) a proportional amplification often denoted as K which is frequency-independent, and associated with a dimensioned scale factor relating to the units of input and output; and (2) a dimensionless factor often denoted as $G(j\omega)$ which is frequency-dependent. Frequency, conditions of operation, and conditions of measurement must be specified. A loop gain characteristic is a plot of log gain versus log frequency. In nonlinear systems, gains are often amplitude-dependent.

Closed Loop Gain. The gain of a closed loop system, expressed as the ratio of the output change to the input change at a specified frequency.

Derivative Action Gain (Rate Gain). The ratio of maximum gain resulting from proportional plus derivative control action to the gain due to proportional control action alone.

Dynamic Gain. The magnitude ratio of the steady-state amplitude of the output signal from an element or system to the amplitude of the input signal to that element or system, for a sinusoidal signal. It may be expressed as a ratio, or in decibels as 20 times the $\log_{10}$ of that ratio for a specified frequency.

Loop Gain. The ratio of the change in the return signal to the change in its corresponding error signal at a specified frequency. The gain of the loop elements is frequently measured by opening of the loop, with appropriate terminations. The gain so measured is often called the open loop gain.

Proportional Gain. The ratio of the change in output due to proportional control action to the change in input. Illustration: $Y = \pm PX$, where P = proportional gain; X = input transform; Y = output transform.

Static Gain. The value of the gain approached as a limit as frequency approaches zero.

GAIN (Transmission).   1. A general term used to denote an increase in signal power in transmission from one point to another; usually expressed in decibels and used to denote transducer gain.

2. The ratio of the output of a transducer to the input, even when these quantities are not measured in terms of power. Thus reference is made to the voltage gain or current gain of an amplifier.

GAJDUSEK, DANIEL CARLETON (1923–Present).   Daniel Gajdusek is an American physician and medical researcher, whose pioneering research led the way to understanding the mechanisms underlying neurodegenerative diseases such as kuru, Creutzfeldt–Jakob disease and scrapie.

Born and raised in Yonkers, New York, Gajdusek graduated from the University of Rochester with a Bachelor's degree in physics in 1943 and with an MD from Harvard Medical School in 1946. After completing his internship and residency, he held positions at Caltech, the Institute of Research of the Walter Reed Army Medical Center in Washington, DC, and the Institute Pasteur in both Paris and Teheran, Iran. In 1955 he received funding from the National Institutes of Health to go to the Walter and Eliza Hall Institute of Medical Research in Melbourne, Australia, as a visiting scholar. In 1958 he became a full-time researcher at NIH's National Institute of Neurological Diseases and Stroke in Bethesda, MD.

It was in Melbourne that Gajdusek began to investigate a unique, neurodegenerative disease seen to occur with a high frequency among certain native populations of New Guinea. The native name for the disease was 'kuru', which reflected the characteristic symptoms of trembling of the limbs brought on by progressive damage to the brain tissue. The disease always culminated in death within 6–12 months from the appearance of the first symptoms. However, nothing was known about what caused kuru or how it spread. In an attempt to get at these problems, Gajdusek began a careful analysis of the disease symptoms and its epidemiology. He was

successful in obtaining tissue from the deceased victims, and thus was able to perform microscopic examinations of the diseased brains. This led to the realization that the brain lesions in kuru patients were very similar to those found in sheep suffering from a disease called scrapie, and that; in fact, the two diseases appeared to share many common features. Gajdusek also made the inadvertent discovery that a disease identical to kuru could be induced in chimpanzees by injecting extracts of diseased brain tissue into the brains of these animals. This gave scientists the first animal model for kuru. These findings together provided very important clues for understanding the nature of not only kuru but also other neurodegenerative diseases of uncertain origins, such as Creutzfeldt–Jakob disease (CJD) and the now infamous mad cow disease (bovine spongiform encephalomyelitis, BSE). For instance, the finding that the animal models developed disease symptoms only 1–3 years after inoculation with brain extracts implied a new type of extremely slow-acting infectious agent that could be transmitted either horizontally—as was kuru—or vertically (CJD). Gajdusek's hypothesis that the infectious agent for these diseases (collectively called the spongiform encephalopathies) was a virus, which in addition to acting slowly was also capable of latency, has been replaced for the most part by the prion theory proposed in the 1990s by Stanley Prusiner. However, there is no doubt that it was Gajdusek's work that laid the foundations for the latter's discoveries. The Nobel Prize committee, who named Gajdusek a co-winner of the 1976 Prize in Physiology or Medicine along with Baruch Blumberg, recognized the two scientists for their research and discoveries regarding 'new mechanisms for the origin and dissemination of infectious diseases'. See also **Blumberg, Baruch Samuel (1925–Present)**; **Creutzfeldt-Jakob Disease and Related Diseases**; and **Virology (The History)**.

NEERAJA SANKARAN, Yale University, New Haven, CT

GAL.   See **Units and Standards**.

GALACTIC COSMIC RAYS.   Galactic cosmic rays (GCRs) are the high-energy particles that come from outside the solar system but generally from within our Milky Way galaxy. GCRs are atomic nuclei from which all of the surrounding electrons have been stripped away during their high-speed passage through the galaxy. They have probably been accelerated within the last few million years, and have traveled many times across the galaxy, trapped by the galactic magnetic field. GCRs have been accelerated to nearly the speed of light, probably by supernova remnants. As they travel through the very thin gas of interstellar space, some of the GCRs interact and emit gamma rays, which is how we know that they pass through the Milky Way and other galaxies. See also **Cosmic Rays**; and **Galaxy**.

The *elemental* makeup of GCRs has been studied in detail, and is very similar to the composition of the Earth and solar system, but studies of the composition of the *isotopes* in GCRs may indicate that the seed population for GCRs is neither the interstellar gas nor the shards of giant stars that went supernova. This is an area of current study.

Included in the cosmic rays are a number of radioactive nuclei whose numbers decrease over time. As in the carbon-14 dating technique, measurements of these nuclei can be used to determine how long it has been since cosmic ray material was synthesized in the galactic magnetic field before leaking out into the vast void between the galaxies. These nuclei are called "cosmic ray clocks".

One of the indirect observations we can make, the "composition" of GCRs, can tell us a lot about the sources and the cosmic rays' trip through the Galaxy. The "composition" of cosmic rays is the way in which the cosmic rays are divided up into each of the different types, what fraction is protons, what fraction is helium nuclei, etc. All of the natural elements in the periodic table are present in cosmic rays, in roughly the same proportion as they occur in the solar system. But detailed differences provide a "fingerprint" of the cosmic ray's source. Measuring the quantity of each different element is relatively easy, since the different charges of each nucleus give very different signatures. Harder to measure, but a better fingerprint, is the isotopic composition (nuclei of the same element but with different numbers of neutrons). To tell the isotopes apart involves, in effect, weighing each atomic nucleus that enters the cosmic ray detector.

About 90% of the cosmic ray nuclei are hydrogen (protons), about 9% are helium (alpha particles), and all of the rest of the elements make up only 1%. Even in this one percent there are very rare elements and isotopes. These require large detectors to collect enough particles to say something meaningful about the "fingerprint" of their source. The HEAO Heavy Nuclei Experiment, launched in 1979, collected only about 100 cosmic rays between element 75 and element 87 (the group of elements that includes platinum, mercury, and lead), in almost a year and a half of flight, and it was much bigger than most scientific instruments flown by NASA today. To make better measurements requires an even larger instrument, and the bigger the instrument, the greater the cost.

Oddly, the isotope ^{7}Be is found in galactic cosmic rays. In common environments this nuclide decays to 7Li by electron capture, meaning that one of the orbital electrons combines with the nucleus, reducing its charge from 4 to 3 units (a neutrino is also emitted in this process.) Cosmic rays are fully *stripped*, meaning that they have lost all their electrons; thus ^{7}Be persists and can be observed. Other radioactive elements, called *clock isotopes* are present: Be-10 (1.6 million year halflife), Al-26 (0.87 Myr), Cl-36 (0.30 Myr), and Mn-54 (0.8 Myr estimated). The abundances of all of these radioactive species are small compared to those of neighboring isotopes because significant fractions have been lost by decay during the ~10 Myr the cosmic rays have spent in the Galaxy before arriving at Earth. Even so, small quantities of these nuclides are clearly observed and can be used to investigate the confinement time of cosmic rays in the Galaxy and the distribution of matter with which cosmic ray particles interact to produce secondary nuclei. Thus the name *clock isotopes*; they tell us how long since the cosmic rays were produced.

GALACTOSEMIA. A disease caused by an inborn error of carbohydrate metabolism. Mental retardation is a clinical feature of the disease. The normal conversion of galactose, a sugar found in milk, is prevented by the absence of the enzyme galactose-1-phosphate uridyl transferase. Removal of milk, the only food source of galactose, from the diet in infancy prevents development of the condition. In the treatment of older patients by diet changes, all symptoms of the disease disappear except intellectual impairment. Success in treating galactosemia by dietary means has spurred the search for other inborn errors of metabolism among the mentally retarded. See **Genetics and Gene Science (Classical)**; and **Protein**.

GALAGO. See **Lemur**.

GALAPAGOS RISE. See **Ocean**; **Ocean Resources (Energy)**; **Ocean Resources (Living)**; and **Ocean Resources (Mineral)**.

GALAXIIDS (*Osteichthyes*). Of the order *Isospondyli*, family *Galaxiidae*, the galaxiids are scaleless, elongated fishes, quite small, seldom exceeding 6 inches (15 centimeters) in length. The *Galaxias alepidotus* (New Zealand species) was first identified in the late 1700s. It is an exception among the galaxiids in that it can attain a length up to 12 inches (30 centimeters) with some specimens recorded up to about 23 inches (58 centimeters). Even this long variety weighs but about 3 pounds (1.4 kilograms). The *Neochanna apoda* (New Zealand brown mudfish) is well known for its absence of ventral fins and is reminiscent of various lungfishes which can survive for many weeks in dried mud. Adult mudfish attain a length of about 6 inches (15 centimeters). The *Galaxias attenuatus* (whitebait) also occurs in New Zealand as well as Australia. Of interest is the fact that this species is the only galaxiid that is found in two or more locations, probably explained by its ability to tolerate fresh and brackish water, with a preference for salt water when an adult.

The fact that galaxiids are not found in the Northern Hemisphere has puzzled naturalists for many years. The matter is even more puzzling because habitats of practically the exact nature preferred by the galaxiids in the Southern Hemisphere are also found in many locations in the Northern Hemisphere.

GALAXY. Over the few centuries since the first telescopes were developed, philosophers and scientists have proposed numerous hypotheses as to what galaxies really are, how they are formed, how they differ, how they evolve, and how they may perish. Even today with what is essentially the beginnings of 21st century instrumentation, there is no general consensus pertaining to most of the foregoing factors among the experts. Hypotheses of the past have been altered or abandoned with the finding of new critical data. Seeking knowledge of the galaxies and

the cosmos is the epitome of man's "thirst to know." Probing the secrets of the galaxies is the bailiwick of the cosmologist and the astrophysicist. See also **Cosmology**.

Galaxies, aptly defined by M.J. Rees (University of Cambridge), are "the basic building blocks of the universe." As we see a galaxy from Earth, we witness the combined output of light from "their tens of hundreds of billions of constituent stars."

Traditionally, observations of the cosmos were limited to what astronomers could learn from energy emitted within the visual light portion of the electromagnetic spectrum. During the World War II era, much was learned pertaining to infrared, ultraviolet, and microwave radiation, and shortly thereafter astronomers adapted these techniques to their pursuits. Astronomy with new names began to appear — infrared astronomy, ultraviolet astronomy, radio astronomy. Gamma-ray and X-ray astronomy soon followed.

This "new" astronomy made it possible to determine previously unknown characteristics of celestial objects, including the galaxies. As an example, the Russian-French gamma-ray satellite (GRANAT), launched in the spring of 1990, fortuitously discovered on the nights of October 13 and 14 a flare of gamma-ray energy emanating at a point (offset from the center of the Milky Way galaxy by about 100 light years) estimated to exceed by a factor of 10,000 the total luminosity of the sun. The event tentatively was considered to be a sign of annihilation and emanated from a black hole. The term, great annihilator has been used.

As further innovations in observing (measuring) the galaxies are developed, perhaps by the mid-21st century, new and more accurate information will become the basis for vastly improved theories of cosmos cause and effect.

Even with its severely impaired vision, the initial Hubble Space Telescope in 1992 yielded images of what astrophysicists termed a "zoo" of ancient galaxies, estimated by some experts as being a large fraction of the way back to the big-bang, the latter being the basis of one current concept of the origination of the universe. Much was expected of the space telescope in the wake of its repair in late 1993. See also **Hubble Space Telescope (HST)**.

Earth's Favored Position. The earth and the solar system represent but miniscule components of a great spiral galaxy that is familiarly known as the Milky Way. This galaxy is some 100,000 light-years (~30,000 parsecs) across (linear diameter)[1] and is estimated to contain over 100 billion stars. The earth and solar system are located in a relatively unpopulated part of the galaxy in a position well away from the center. This is an excellent location for observing much of the galaxy. The optical telescope reveals only a small portion of the Milky Way because regions of the galaxy are obscured by great dust clouds and other interfering phenomena. With the initiation of radio astronomy, it became possible to study radio-emitting stars, radio galaxies, and hydrogen in space.

In an interesting study of the professional literature on galaxies, P.W. Hodge (University of Washington) observed that only one article on galaxies (by E.P. Hubble) appeared in the Astrophysical J. during the entire year, 1930, as compared with an increase to 276 articles in the same publication in 1980. Hodge estimated that the space assigned to galaxies exceeded that of astronomical research in the overall by a factor of ten. And, since that time, the literature on galaxies has continued to expand.

A Universe of Galaxies

A fascinating view of the universe at magnitude 27 is given in Fig. 1. Objects of this magnitude are an estimated billion times more faint than can be seen with the naked eye. They are estimated to be 10 billion light years away. When making such probes of the universe, one not only looks outward in space great distances, but also *backward in time* — toward the time of creation of the universe. Beyond a certain point in the astronomer's outreach, according to one theory, the galaxies must begin to "thin out" because during time frames this far back, one should encounter galaxies in

---

[1] Many authorities acknowledge that the diameter is probably even greater than this figure, but available optical data have not been sufficient to ascertain the details of structure and composition, let alone exact dimensions. It still has not been established whether the Milky Way is a Type Sb or Sc spiral galaxy. See also **Milky Way**.

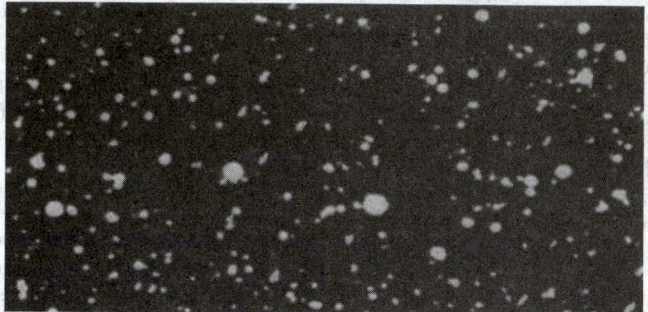

Fig. 1.   View of the universe (27th magnitude) which is at a distance of over 10 billion light-years away from Earth. Objects at the 27th magnitude are approximately one billion times fainter than those which can be seen with the naked eye. This is an approximate black-and-white facsimile of a color-enhanced image made by Tyson (AT & T Bell Laboratories) and Seitzer (National Optical Astronomy Observatories), using the 4-meter telescope at the Cerro Tololo Inter-American Observatory in Chile. The telescope was pointed at the South Galactic Pole (in Southern Hemisphere constellation *Sculptor*), making the line of sight perpendicular to the plane of the Milky Way. The image was exposed for 6 hours. CCD detector was used. The color information (not shown here) was obtained by using filters for three different wavelengths.

the process of formation. The significance of the image is that magnitude 27 may be getting close to that point.

An interesting point pertaining to the image relates to the *background light of the sky*. The light from all galaxies in the universe, it is assumed, blends into a diffuse background light in much the same way that the sound of individual raindrops blends into the diffuse sound of a rainstorm. Scientists in the past, by way of various observations and calculations, have established upper limits on this background light — approximately equivalent to the light from a single magnitude 10 star spread over a square degree of the sky. In the magnitude 27 image, it is estimated that the integrated light of all galaxies at that level represents between 70 and 80 percent of the established limit. Thus, as instruments probe further, the number of new galaxies encountered should diminish markedly.

It also has been pointed out that in looking at the very faintest galaxies in the image, on the average, one may expect them to be quite a bit smaller than those galaxies that are not so faint, in as much as they are presumably farther away. But, in fact, they are not much smaller. The galaxy images are several times larger than might be caused by blurring, due to atmospheric turbulence — so this factor can be discounted. What this may confirm is that we really are in a non-Euclidian universe — because, beyond a certain point, it turns out that Einsteinian curvature actually causes images to get larger with distance. The magnitude 27 galaxies appear to be near that point.

The color of the faint galaxies in the magnitude 27 image (obtained through the use of color filters) is extremely blue when compared with the brighter galaxies, even though they are presumed to be further away and thus have a larger red shift. It has been postulated that these galaxies must have an enormous ultraviolet enhancement, thus suggesting that the galaxies are producing hot, massive young stars at a very high rate, thus indicative of what would happen in galaxies that are themselves quite young.

Using the image in connection with a computer-based model of galactic evolution suggests that the faintest galaxies in the image are from 1 to 2 billion years old, or practically newborn in cosmic terms. Researchers admit that the conclusions are tentative and that, with improved instrumentation, such deep-space images may lead to the answer to the central question of cosmology — when did the galaxies form?

The image shown in Fig. 1 is one of a series of similar images produced during the mid-1980s. Analyzers of these views suggest that if you look at the faintest images, you can see that the sky is filling in. In the subject image, the coverage approaches 30 percent. Current aims are to increase the sensitivity of such images by a factor of 10, ultimately making it possible to view magnitude 30 galaxies. As sky coverage approaches 100 percent, new problems with no known answers will evolve.

Role of Quasars. Tentatively, a quasar is considered a "quasi-stellar radio source" and one of the most distant objects visible in the universe. See also **Quasars**. One hypothesis holds that quasars may be rejuvenated by a fresh supply of fuel from interaction with a galaxy. The remnants

of a galaxy sometimes are difficult to determine from a quasar image. Such quasars usually are comparatively close. More distant quasars are associated with an earlier stage of universe formation, at a time when they had their own energy supply. Some investigators have considered a quasar as a galaxy with a compact core that may be a black hole. See also **Black Hole**. In 1986, researchers Hazard (University of Pittsburgh), McMahon (University of Cambridge), and Sargent (California Institute of Technology) reported a quasar that may be one of the most distant objects in the universe to be found thus far. The quasar (QS01208 + 1011) was estimated to be some 12.4 billion light-years from Earth. The object was found to have a red shift of 3.8, which is 0.02 unit greater than the second most distant quasar (PKS2000 − 330). The researchers at that time used a special photographic emulsion sensitive to infrared light. The instrument was a 5-meter Hale Telescope located at Palomar Observatory. The quasar-galaxy relationship is explored in scholarly detail by M.J. Rees (University of Cambridge). (See reference listed.) Rees observes, "The first quasars appeared surprisingly soon after the big bang. Astronomers have found at least a few quasars whose light has been stretched by nearly a factor of six, revealing that they existed when the universe was younger than one billion years. These old, distant quasars place tight constraints on theories of galaxy formation. . . . Observational surveys alone cannot reveal whether quasar activity was a brief feature of all young galaxies or just a highly visible aberration in a few unusual ones. To settle this question, one needs to know how long a typical quasar lives."

Early-1990 estimates indicate that less than one quasar exists for every 100,000 galaxies. Present study indicates that, even during the hypothetical quasar era some 11 billion years ago, quasars were about 100 times less common than normal galaxies. Rees further observes, "Cygnus A galaxy, the most intense radio source in the sky, radiates primarily from two lobes of plasma (ionized gas) hundreds of thousands of light-years across. The lobes probably are powered by hot jets that squirt out when gaseous matter falls toward a large black hole at the galaxy's center. The energy in the lobes is equivalent to millions of solar masses; the size and structure of the lobes imply that Cygnus A has been active for a few tens of millions of years." The GRANAT gamma-ray satellite, previously mentioned, detected what may be a black hole in our galaxy, the Milky Way.

Formation of Galaxies. When scientists depended entirely upon optical telescopes, it is estimated that no more than 10 percent of the matter making up a galaxy could be detected. As one investigator has put it, the luminous matter that appears so impressive to the human eye may be little more than a trace element by comparison with the dark matter of a galaxy. Then, what is the dark matter?

Some researchers have proposed that the dark matter could be made up of baryonic matter that resulted from the so-called *big bang* event. However, the microwave background radiation (2.7 K) emitted from cosmic plasma is uniform to a few parts in 10^4. Pure baryonic dark matter essentially has been ruled out because of the lack of explanation as to how such matter would have been distributed non-uniformly as well as relatively promptly to galaxies. Possibly, this is a logical conclusion. However, if dark matter in galaxies is nonbaryonic, what is it?

Perhaps this nonbaryonic matter, as suggested by Zeldovich (Russia), is made up of massive neutrinos (invisible) generated in large numbers during the *big bang*. Zeldovich further has suggested that gravitational clumping of massive neutrinos may have occurred, thus creating "traps" for baryonic matter. Some studies have indicated that this concept is consistent with many of the observed characteristics of galaxies, including their streaming and large-scale structure. The massive neutrino hypothesis, however, has been faulted in several respects: (1) The age of some galaxies is estimated to extend back to 16 to 17 billion years (the present estimate of the age of the universe — hence the *big bang* — is 18 billion years). The massive neutrino concept required that galaxies would not have been formed out of superclusters until some 14 to 15 billion years ago. (2) The manner in which dark halos have formed around individual galaxies, instead of collecting in large clusters, also is inconsistent with the massive neutrino concept. Further, it has been found that the ratio of dark mass to luminous mass is impressively constant at a value of about 10 (including dwarf spheroidal galaxies and large superclusters). Thus, the massive neutrino concept has fallen out of favor with many researchers. More recent theories involve "warm" and "cold" dark matter. Models of such systems have been developed and explained in some detail.

Proposals and arguments such as the foregoing currently are awaiting the gathering of much additional information, as well as future intellectual breakthroughs.

Motion of Galaxies. In 1986, a team of seven astronomers[2] associated with observatories in both hemispheres was established to improve the way scientists estimate the distance to elliptical galaxies. A fascinating side discovery was made: that is, the apparent, large-scale bulk motions among the galaxies. These motions are leading to new hypotheses concerning the origin and development of large-scale structure in the universe. In all, some 390 elliptical galaxies were surveyed during the study—in all directions and encompassing a volume of space estimated to be about 100 million parsecs (3.26 light-years) in diameter. The investigators announced that a new distance calibration for ellipticals was derived and accurate to about 23%, considered quite acceptable by most contemporary cosmologists.

As reported by the investigators, once having determined the distance to each galaxy, Hubble's law is applied to ascertain how fast the galaxy should be receding from Earth as the result of cosmic expansion. Then, subtracted from this figure is the observed recession velocity as determined from the galaxy's red shift. The remainder, then, represents a purely local motion, one that presumably indicates how a given galaxy is interacting with its neighbors. To determine how the galaxies in their sample are moving relative to the universe as a whole, other calculations are included. When all the foregoing factors were taken into consideration, the researchers found that for approximately 50 million parsecs in all directions, clusters and superclusters of galaxies are streaming through the cosmos, as a group, at an estimated 700 km per second. It was also found that the superclusters that are a part of the overall stream behavior appear to lie in a reasonably well-defined plane (supergalactic plane). The bulk motion of the galaxies is essentially parallel to this plane. A third finding: superimposed on the bulk motion is a patchwork pattern of motions on a scale of 10 to 30 million parsecs (about the size of a single supercluster).

Burstein observes that the views of the investigative team are still unfolding, but that qualitatively the smaller scale patchwork motions are not surprising, in as much as the galaxies themselves are distributed in a patchwork pattern and that one would expect the lighter clumps to be falling toward the more massive clumps.

The most surprising of the aforementioned findings is that concerning the large-scale streaming motion. In subsequent studies, it appears that this motion is in the general direction of the Hydra-Centaurus supercluster (near the Southern Cross constellation in the Earth's sky). Hydra-Centaurus is also moving. Burstein queries—is the motion of a relic of whatever processes formed the galaxies in the first place? or is there some huge, undiscovered concentration of mass on the other side? Most likely, prospective answers to these questions will envelop a number of scenarios.

A number of sub- or ancillary hypotheses have appeared in recent years. One concept is based upon the possibility that a mass so large as to dwarf the superclusters is causing the aforementioned coherent, large-scale motions of the galaxies. This unproved mass was designated by some researchers as the *great attractor*. Support for this concept was provided by a team of British, American, and Australian scientists who worked with the Parkes radio telescope in Australia. The group observed the same particular velocity for Hydra-Centaurus as discovered by the team of seven researchers previously mentioned. Different targets and different methodologies were used by the two groups. See Fig. 2.

Although much interested in, and becoming less skeptical of, the concept, one well-known authority has posed two possible constraints. (1) Perhaps there is some peculiarity of galactic evolution that is in some way interfering with the standard distance indicators in the Hydra-Centaurus region—in some way that may have skewed the surveys into a distribution that only looks like a large scale flow. (2) If the *great attractor* is not some agglomeration of invisible "dark matter," why has not a "supercluster of galaxies" previously shown up on the sky maps? Doubtless, the concept will be debated well into the future.

[2] The team consisted of: D. Burstein (Arizona State University); R.L. Davies (Kitt Peak National Observatories); Alan Dressler (Mount Wilson and Las Campanas Observatory; Sandra M. Faber (Lick Observatory); Donald Lynden-Bell (Cambridge University); Roberto Terlevich (Royal Greenwich Observatory); and Gary Wegner (Dartmouth College). The report was presented at a workshop on the *Extra-Galactic Distance Scale and Deviations from Hubble Expansion*, given at Kona, Hawaii in January 1986.

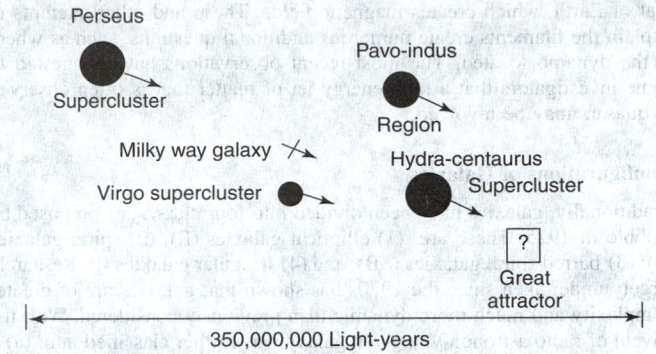

Fig. 2.   Large-scale streaming motion of galaxies as related to a possible great attractor. Diagram assumes observer is at rest with respect to the 2.7 K microwave background radiation. Motions depicted are estimated at rate of 700 km per second in direction as shown by arrows. (*After David Burstein, Arizona State University.*)

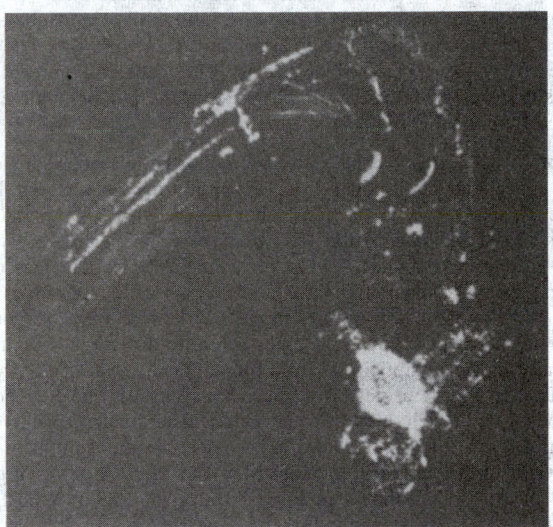

Fig. 3.   As of the late 1980s, the mysterious Continuum Arc as noted by the Very Large Array (VLA) at 20 cm wavelength and extending from Sagittarius A West, the radio source that marks the center of the Milky Way galaxy. Current knowledge indicates that the arc (upper left) and other associated filamentous patterns are due to magnetic causes. (*Facsimile of image made by the VLA, Socorro, New Mexico.*)

Magnetic Fields and Galaxies. As early as 1959, in a radio survey of the Milky Way, an arc lying perpendicular to the plane of the galaxy about 40 parsecs out from the center and extending some 20 to 30 parsecs above and below the plane, was detected. It was then reported as a narrow strip of radio emission. Later, in 1984, Morris (University of California), Yusef-Zadeh (now at Columbia University), and Chance (Columbia University), using the Very Large Array (VLA) at Socorro, New Mexico, observed the Continuum Arc and found that it is comprised of thin, parallel filaments. They also found that, at the northern end of the arc, the filaments merge with a second, rather irregular set of filaments that curve back down into Sagittarius A West (radio source marking the center of the galaxy). Sieradakis more recently has suggested that the arc may be part of a still larger structure. See Fig. 3.

The filaments suggest to the scientists that they are shaped by a magnetic field. Formation as part of an interstellar shock wave has been ruled out because the filaments are quite uniform over long distances. Independent polarization measurements of the arc (Japanese and German radio astronomers) indicate that the phenomenon is consistent with synchrotron radiation, which is produced by electrons spiraling around magnetic field lines. The magnetic field strength has been estimated at 10^{-4} gauss, which seems quite large with respect to the overall magnetic field of the galaxy. Thus, a major question is posed. What is producing the magnetic field? Candidate explanations include: (1) a black hole, but the size of the arcs tends to negate this cause; (2) a dynamo process similar to that of Earth, which creates magnetic fields. These and other attempts to explain the filaments create numerous additional questions, such as where is the dynamo located. The most recent observations have suggested to some investigators that a high-energy jet of matter that is often observed in quasars may be involved.

Configurations of Galaxies

Traditionally, galaxies have been divided into four classes, as proposed by Hubble in 1925. These are: (1) elliptical galaxies (E); (2) spiral galaxies (S); (3) barred spiral galaxies (SB); and (4) irregular galaxies (I). Research, largely undertaken since the 1970s has shown that galaxies are of greater complexity and much more dynamic than previously considered. With the advent of radio astronomy, a given galaxy was further classified into: (a) a weak emitter, or ordinary galaxy, such as the Milky Way, which typically radiates 10^{38} erg/second in radio waves, compared with 10^{44} erg/second in the optical region; and (b) a strong emitter, or radio galaxy, which may produce up to 10^{45} erg/second in the radio range alone. In both types of galaxy, the radio continuum is accounted for on the synchrotron theory. Further refinements in classification are described a bit later.

Diagrams of various classes of galaxies closely following Hubble's early proposal are given in Fig. 4. Radio astronomy observations to date suggest that the Milky Way is of the Sb or Sc class. The shape-predominant method of classifying galaxies derived logically from the photographic evidence provided by optical equipment.

Elliptical Galaxies. When viewed through the telescope, this class of galaxy appears as an elliptical disk. No spiral arms are apparent. Per unit volume of space, this is the most abundant type of galaxy and ranges from the most massive to the least massive of the galaxies. Typically, a disk galaxy incorporates few or no young stars and no gas or dust. Elliptical galaxies apparently have a smooth structure—with a smooth center, extending out to a diffuse, irregularly defined edge. Although not fully understood, the elliptical galaxies appear to differ one from the other principally in their ellipticity—from round (Type E0) to a 3:1 axis ratio (Type E7). In Fig. 5, a cluster of galaxies in Virgo is shown. Thousands of galaxies form a rich, loose, and irregular cluster, which appears to have no central concentration. The two large objects in the right-hand portion of the view are the elliptical galaxies M84 and M86. A group of galaxies located in Pegasus is shown in Fig. 6. The giant M49 elliptical galaxy NGC 4472 (Type E1), shown in Fig. 7, is a nearly circular elliptical galaxy. Typical of the elliptical galaxy, this object shows no evidence of recent star formation and little or no gas between the stars. Some 60 million light-years (18.4 million parsecs) away from earth, the mass of M49 is approximately 10^{12} times that of the sun and from 5 to 10 times more massive than the Milky Way.

Galaxy M87 (NGC 4486), a type E0 elliptical galaxy, was the first extragalactic x-ray source to be found by rocket astronomy. Optically, M87

Fig. 5.   Group of clusters in Virgo, including M84 (NGC 4374) and M86. Thousands of galaxies form a rich, loose irregular cluster which appears to have no central concentration. Many subcondensations of galaxies are seen in it. This type of cluster comprises most types of galaxies. (*National Optical Astronomy Observatories.*)

is a large elliptical galaxy characterized by an extended jet, which emerges from the nucleus to a distance of about 5000 light-years (1500 parsecs). The jet is clearly visible in the lower, right-hand inset of Fig. 8. The bluish light of the jet is highly polarized, indicating synchrotron radiation. Radio astronomers discovered an intense compact core only 4 light-months in diameter, which may be the origin of the x-radiation (inverse Compton interactions between relativistic electrons and the high density of radio photons in the nuclear region). Other features of the radio image include an extended halo and a fan jet. The fan jet may suggest that gas clouds are expelled from a compact rotating body. One observer estimates that repeated releases from the central body may furnish a few million solar masses to the various jet forms. It is postulated that to sustain a reservoir of material and energy, the gas may be constantly accreting onto a large rotating mass in the nuclear region. It is further postulated that the origin of the gas could be planetary nebulas separated from old-population red giants. If one compares this situation with our own galaxy, the rate of evolution of planetary nebulas in M87 should be about 30 per year.

More recently, the jet has been explained as having been ejected from the nucleus of the galaxy in one or a series of explosions about a million

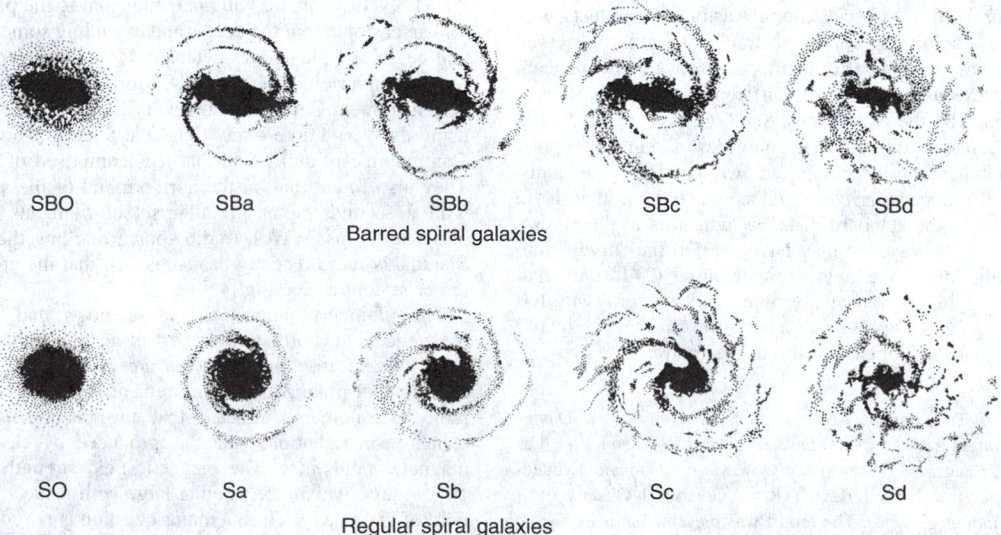

SBO   SBa   SBb   SBc   SBd

Barred spiral galaxies

SO   Sa   Sb   Sc   Sd

Regular spiral galaxies

Fig. 4.   Generalized and schematic configurations of principal types of galaxies as proposed by Hubble (1925).

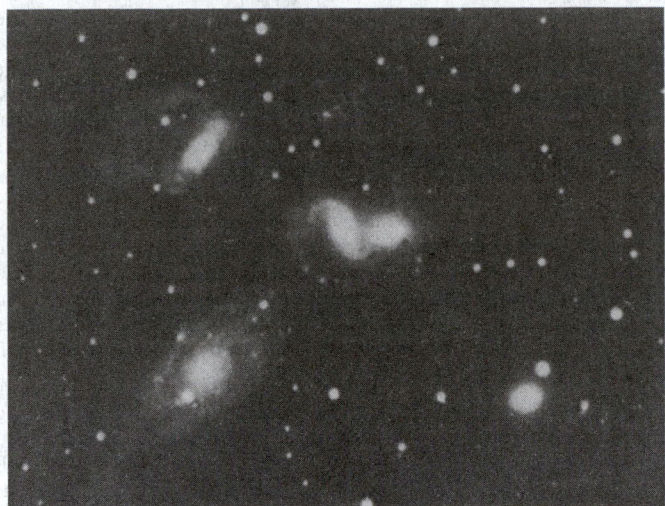

Fig. 6.   Known as Stephan's Quartet, this group of galaxies (NGC 7317; 7318A; 7318B; 7319; 7320) is located in Pegasus. Four of these galaxies have a red shift of about 6000 kilometers/second, while one (the largest) has a redshift of only 800 kilometers/second. (*National Optical Astronomy Observatories*.)

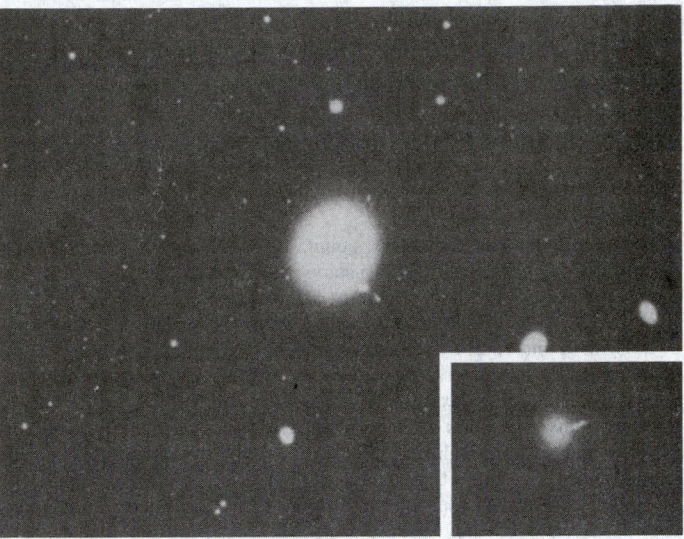

Fig. 8.   Galaxy M87 (NGC 4486), Type E0, located in Virgo. The poorly understood jet extending from the galaxy (see photo inset) is a strong radio source. (*National Optical Astronomy Observatories*.)

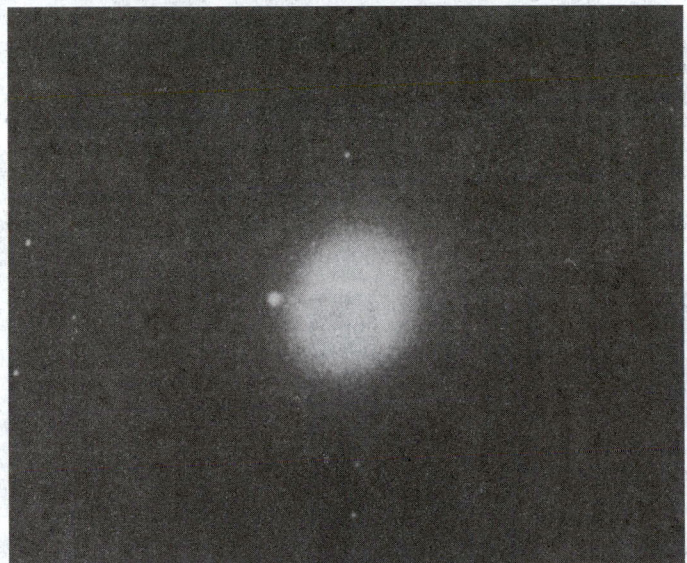

Fig. 7.   The nearly-circular elliptical galaxy M49 (NGC 4472), Type E1, located in Virgo. Such galaxies have nearly no dust or gas between their stars, and show no evidence of recent star formation. (*National Optical Astronomy Observatories*.)

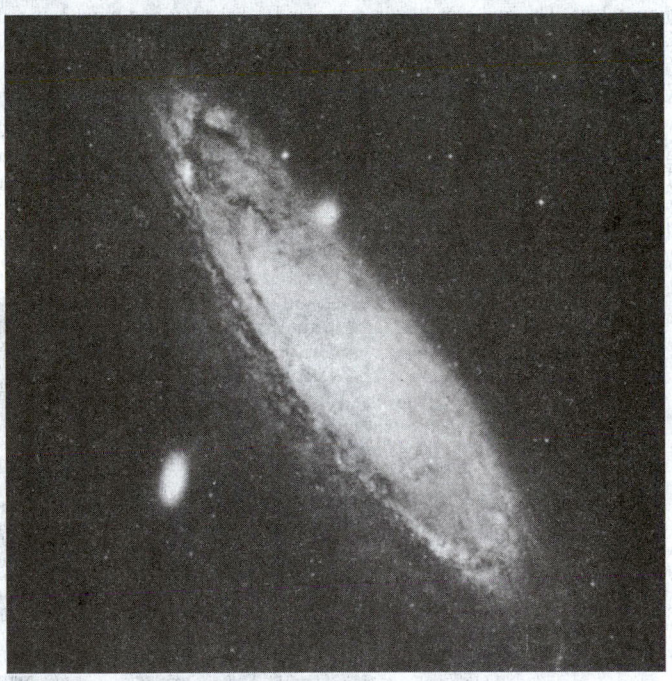

Fig. 9.   Spiral galaxy (M31; NGC 224) in Andromeda. Elliptical companion galaxy (M32; NGC 221) appears just above the central region; another elliptical companion (NGC 205) appears below to the left. (*Palomar Obs/Caltech*.)

years prior to the state that is presently being observed from earth. X-ray emission provides evidence of explosive activity, commonly encountered in elliptical galaxies as well as in other poorly understood astronomical objects, such as quasars. Energy release in such an explosion may be at a rate that is a trillion times that of the sun, a phenomenon that puzzles modern astronomers.

Early *Uhuru* satellite observations revealed that an x-ray emitting cloud about 1 million light-years (over 300,000 parsecs) across envelops the galaxy. At a later date, analysis of the x-ray spectrum was made by the British *Ariel* 5 satellite and also by the NASA OSO-8 satellite. These data, which show the presence of highly ionized iron, indicated that the x-ray radiation emits from a diffuse gas at a temperature of 30 million degrees Kelvin. These findings indicated that the x-ray radiation arises from a thermal and not a nonthermal source.

Investigators, in attempting to explain the presence of this cloud, currently offer three postulates. In one, the gas is considered as forming continuously, so that the cloud is constantly replenished. In another, the galaxy as it is currently observed is at a point in its developmental history such that there has not been sufficient time for the gas to dissipate; in the third, some force is confining the gas to the galaxy. A number of

researchers tentatively accept the last of these postulates, assuming the holding force to be gravity. Phenomena of this kind are further described in entries on **Black Hole; Cosmology**; and **Quasars**.

In an early study of bright galaxies, observers concluded that all normal spiral and irregular galaxies are probably weak radio sources. In contrast, the strong emitters tend to be elliptical galaxies, often distinguished by peculiar optical phenomena, and some observers believe that the elliptical galaxies constitute the bulk of the 1000+ discrete sources cataloged. Early studies indicated that these sources have power-law spectra, but more recent and reliable data indicate the presence of curvature in many spectra (as displayed in a log frequency-log flux diagram, where a power law is a straight line). The value of observations below a wavelength of 4 centimeters in defining the spectral shape was demonstrated. Later, interferometric studies enabled the dividing of the resolved sources into three groups on the basis of brightness distribution: simple, double, and core-halo sources.

Spiral Galaxies. Probably the most familiar and some of the most optically observed of the galaxies are the spiral galaxies, of which the Milky Way is one (Type Sb or Sc). The Milky Way is among the larger of the regular (not barred) spiral galaxies, as is also the Andromeda Galaxy (M31; NGC 224), which is one of the closest spiral systems and one of the most easily visible from earth. Distance from earth approximates 2.2 million light-years (0.7 million parsecs). See Fig. 9. Companion elliptical galaxies are also shown in the view. See also **X-Ray Astronomy**.

The spiral arms of these galaxies contain abundant dust, gas, and newly formed, bright, massive, hot, bluish stars, which often occur in clusters. The central regions have little gas and dust and are dominated by old, red giant stars. Generally, spiral galaxies have the outlines of flattish, lens-shaped disks with a maximum thickness at the center equal to approximately 10–15% of the diameter. Mass calculations and brightness observations indicate that spiral galaxies may contain from 1 billion to 100 billion or more individual stars. With reference to the diagram of Fig. 4, the Type Sb and Sc spiral galaxies are among the most frequently studied and photographed. See Fig. 10(a)–(f).

Among the observable spiral galaxies, there are relatively few that provide a view of the edge of the disk. Some of the galaxies that can be observed in a reasonably edge-on position are shown in Fig. 11(a)–(d).

Barred Spiral Galaxies. From observational data to date, it appears that a small minority of spiral galaxies have a bright bar that slices across the nucleus. The two arms begin at the ends of the bar and wind outward. Contrast this with the normal spiral galaxies, which have a central region (nucleus) to which a number of spiral arms appear to be attached. See Fig. 4 and Fig. 10(f).

Type SO Galaxies. It will be noted from Fig. 4 that Types SO and SBO galaxies differ considerably from the other spiral galaxies depicted. The Two morphologically distinct parts of a disk galaxy are: (1) the *central bulge*, which in many cases is roughly spheroidal; and (2) a comparatively thin *disk* that extends outward. Observations show a great variation in

the characteristics of these two parts or subsystems. The relative size and extent of the bulge vary from one galaxy to the next. In some galaxies, the bulge predominates; in others, the disk. Research has shown that the bulge in most disk galaxies is completely or essentially devoid of young stars, with star-formation occurring in the disk. The SO galaxies differ in that the disks are smooth and lack young stars and star-forming complexes. As pointed out by Strom and Strom, the disks of SO galaxies lack evidence of the gas needed for future star formation. SO galaxies are common in large galactic clusters, whereas the other types of spiral galaxies, such as the Milky Way, tend to be located in regions that are relatively unpopulated by galaxies. Within the proper environment, some authorities propose that a spiral galaxy can become a smooth disk without spiral arms, and that this is most likely to occur in regions with large clusters of galaxies, rather than in isolated, widely separated galaxies which are not part of a rich cluster. See Figs. 11 and 12.

Ring Galaxies. A galaxy of this type has a prominent, bright ring surrounding the center. In some cases, this center is faint; in others, bright. See Fig. 13. It has been postulated that the ring galaxies may be the result of collisions between pairs of galaxies. Once formed, the configurations appear to be stable.

Irregular and Peculiar Galaxies. The observed galaxies that do not fit well into established criteria (principally shape) are usually termed *irregular galaxies*. Some authorities have broken the irregular class into two categories—the Magellanic Cloud type, and all others, Irregular galaxies (with Q and B stars and emission nebulae) are designated Irr I; those which cannot be resolved into stars are designated Irr II. The closest of the irregular galaxies to earth are the two large, cloudlike objects in the souther sky (the Magellanic Clouds), which are actually companion galaxies to the Milky Way galaxy. See Fig. 14. The larger of these clouds has sometimes been called a barred spiral (Type SBm) with one arm. The largest gaseous nebula in the Large Magellanic Cloud is called 30 Dordus. See also **Nebula**.

The two Magellanic Cloud galaxies are rather irregular in shape and are considerably smaller than the Milky Way. Their distance is on the order

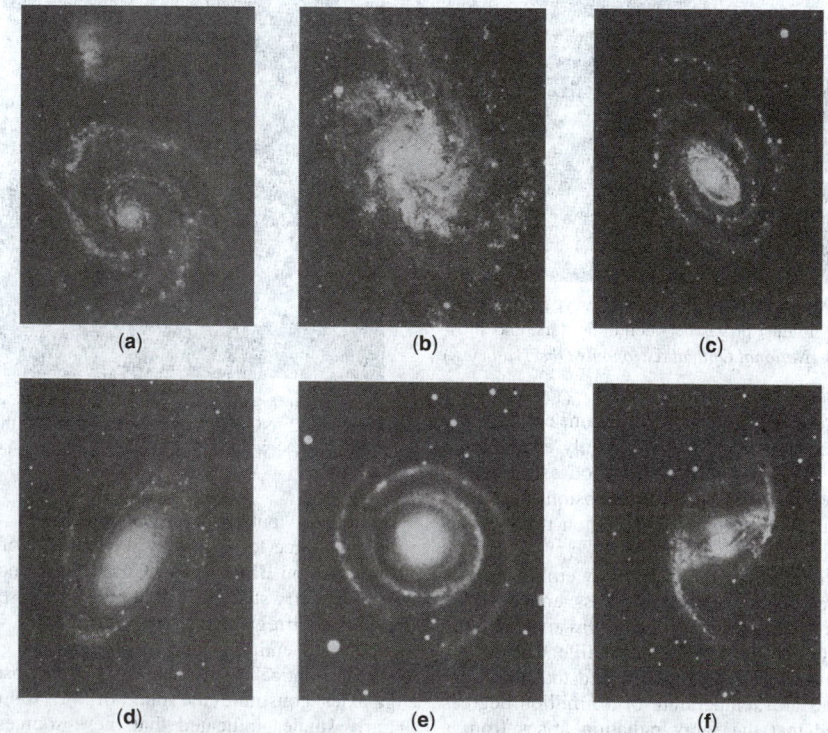

Fig. 10.   Representative spiral galaxies; (**a**) Whirlpool galaxy (M51; NGC 5194) and its satellite galaxy (NGC 5195) in Canes Venatici. Note sharp, bright nucleus. The companion is classified as an irregular galaxy. (**b**) Galaxy M33 (NGC 598) in Triangulum. This is one of the nearest of the spiral galaxies to the Milky Way, some 2.3 million light-years (~0.7 million parsecs) distant. It is a Type Sc galaxy. (**c**) Galaxy NGC 5364, a Type Sc galaxy located in Canes Venatici. (**d**) Galaxy M81 (NGC 3031), a Type Sb galaxy located in Ursa Major. (**e**) Galaxy NGC 4622, a Type Sb galaxy located in Centaurus. This is a member of the Centaurus cluster of galaxies. Within its remarkably smooth and thin spiral arms there are million of bright young stars. Distance from earth is about 2000 million light-years (~61 million parsecs). (**f**) Galaxy NGC 1530, a Type SBb galaxy, located in Camelopardalis. Note that this is a barred-type spiral—with a barlike, elongated center as compared with the more circular or elliptical centers. (*Sources of illustrations: a, Lick Observatory*; b, c, d, e, and f. *National Optical Astronomy Observatories.*)

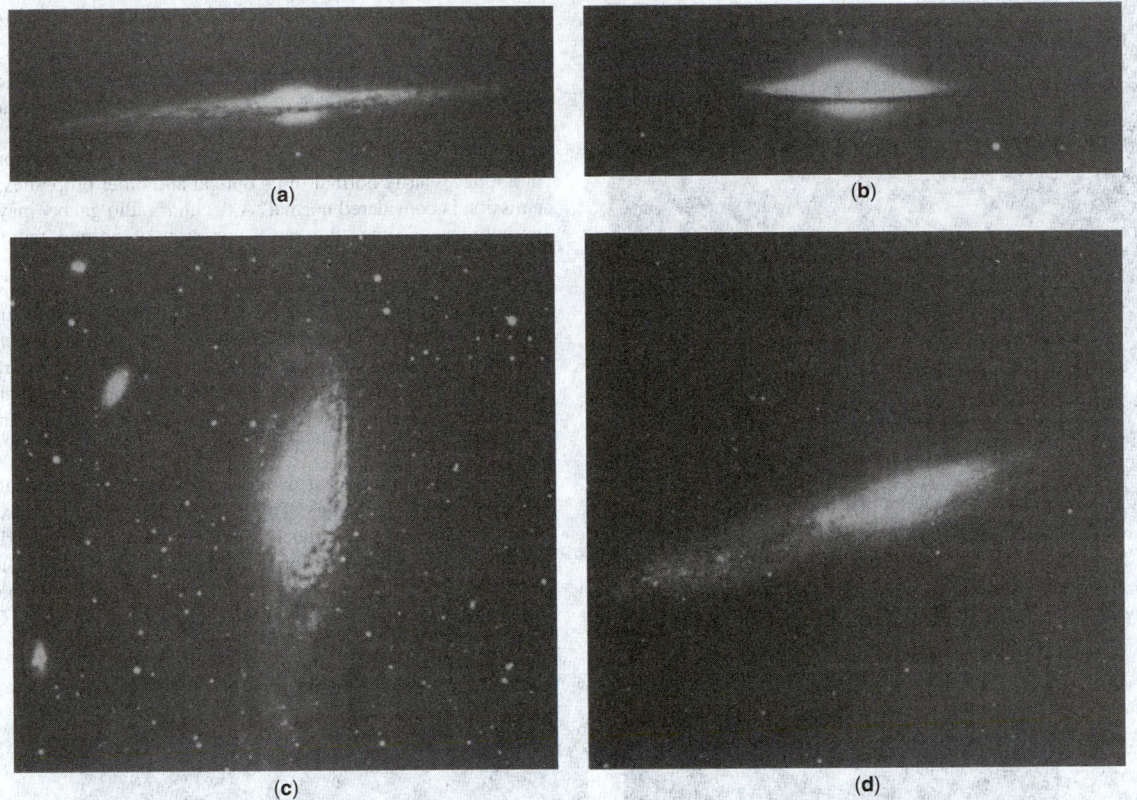

Fig. 11. Views of a few galaxies as seen edge-on or nearly so: (**a**) Spiral galaxy NGC 4565, a Type SB galaxy in Coma Berenices. Photographed on an unfiltered red-sensitive plate. (**b**) Galaxy M104 (NGC 4594), a spiral galaxy of Type Sa/Sb, located in Virgo. This object is known as the "Sombrero" for its edge-on appearance. This galaxy is inclined only about 6° to line of sight. The dark band across the galaxy's center is composed of dust and gas. (**c**) Galaxy NGC 7331, a spiral galaxy of Type Sb, located in Pegasus. (**d**) Galaxy NGC 55, a spiral galaxy of the SBm (barred) type and located in Sculptor. (*Sources of Illustrations*: a, *HaleObservatories*; b, c, and d, National Optical Astronomy Observatories.)

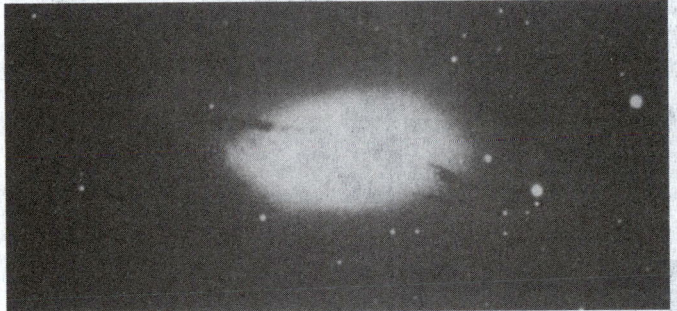

Fig. 12. Galaxy NGC 4753, a Type S0 galaxy, located in Virgo. The underlying galaxy is nearly elliptical, but the dust lanes are peculiar in that they do not appear to occur in spiral arms. (*National Optical Astronomy Observatories.*)

of 800,000 light-years (~245,000 parsecs) from earth. It is estimated that these galaxies contain on the order of 10 billion stars each. The Magellanic Clouds primarily contain Population I type stars, with lots of gas and dust, although they also exhibit such Population II type objects as globular clusters and cluster-type variables stars.

Examples of other irregular and peculiar galaxies are shown in Fig. 15.

Seyfert Galaxies. In general terms, a Seyfert galaxy is any galaxy that has a very bright nucleus showing a high excitation spectrum with broad emission lines. More specifically, a Seyfert galaxy has a small nucleus, often bluish in color and emitting radio energy. Some authorities believe that they are related to quasars and may have explosive activity proceeding in their centers. Two Seyfert galaxies are shown in Fig. 16(a) and (b).

A small percentage of all galaxies are Seyfert galaxies. These objects may contain from 10^9 to 10^{10} stars within a diameter of about 1000 light-years (~300 parsecs). High-velocity gas clouds, hot gas, and

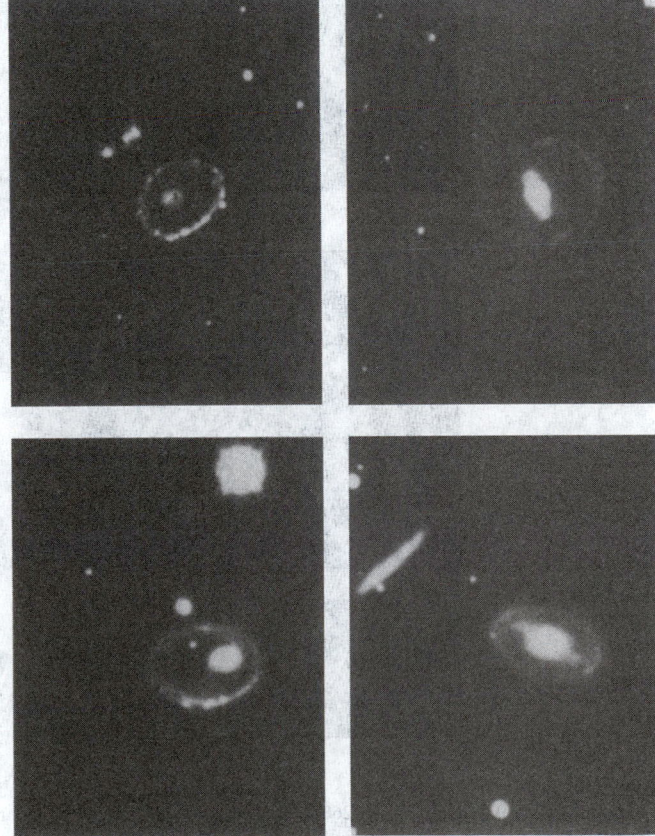

Fig. 13. Four ring galaxies. (*National Optical Astronomy Observatories.*)

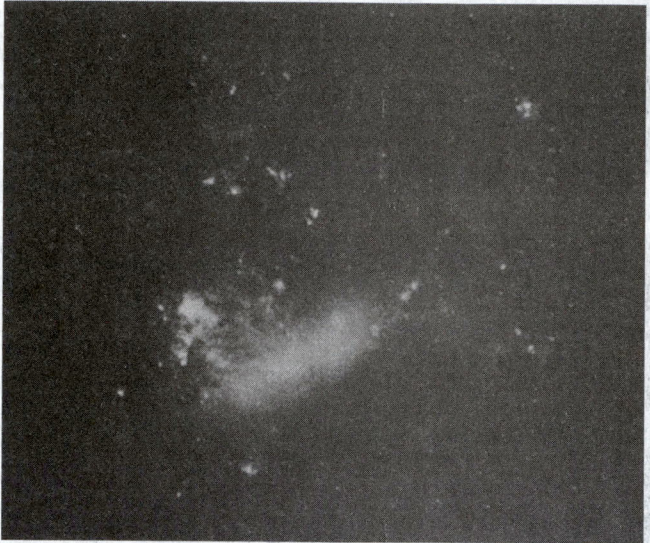

Fig. 14.   Large Magellanic Cloud photographed in Hα light. (*Photographed by Karl G. Heinze with the Mt. Wilson 10-inch reflector at the Lamont-Hussey Observatory, Bloemfontain, South Africa*).

non-thermal processes are indicated by strong, widened optical emission lines and a polarized continuum.

Radio Galaxies. Any galaxy, including Seyfert galaxies, that emits measurable amounts of radio radiation may be called a *radio galaxy*. Numbers of these have been identified in recent years. They fall into both the normal and peculiar classes of galaxy. A normal radio galaxy in not necessarily normal in its optical and other properties; rather, it radio emission is considered normal. A peculiar radio galaxy may emit hundreds to millions of times the radio emission of normal radio galaxy. Some galaxies are peculiar in terms of both radio and optical characteristics. Frequently, these are single galaxies that show evidence of explosive activity in their centers. A jet extending from the nucleus, as shown in Fig. 8, is indicative of instability. Radio galaxies that appear to involve two or more interacting of colliding galaxies also have strong radio emissions. The term "violently active galaxy" has been introduced into the literature in recent years for those objects with strong emissions in the radio and sometimes the X-ray spectrum.

Clusters of Galaxies. It is not unusual for many physical phenomena in nature to occur in clusters. This is indeed the case of galaxies. Even prior to the accumulation of much knowledge of galaxies, early investigators recognized the predisposition of nebulae to collect in bunches, so to speak. In the late 1800s, over 11,000 "nebular objects" were mapped for the "New General Catalogue," published by J.L.E. Dreyer. In 1921, C.V.L. Charlier

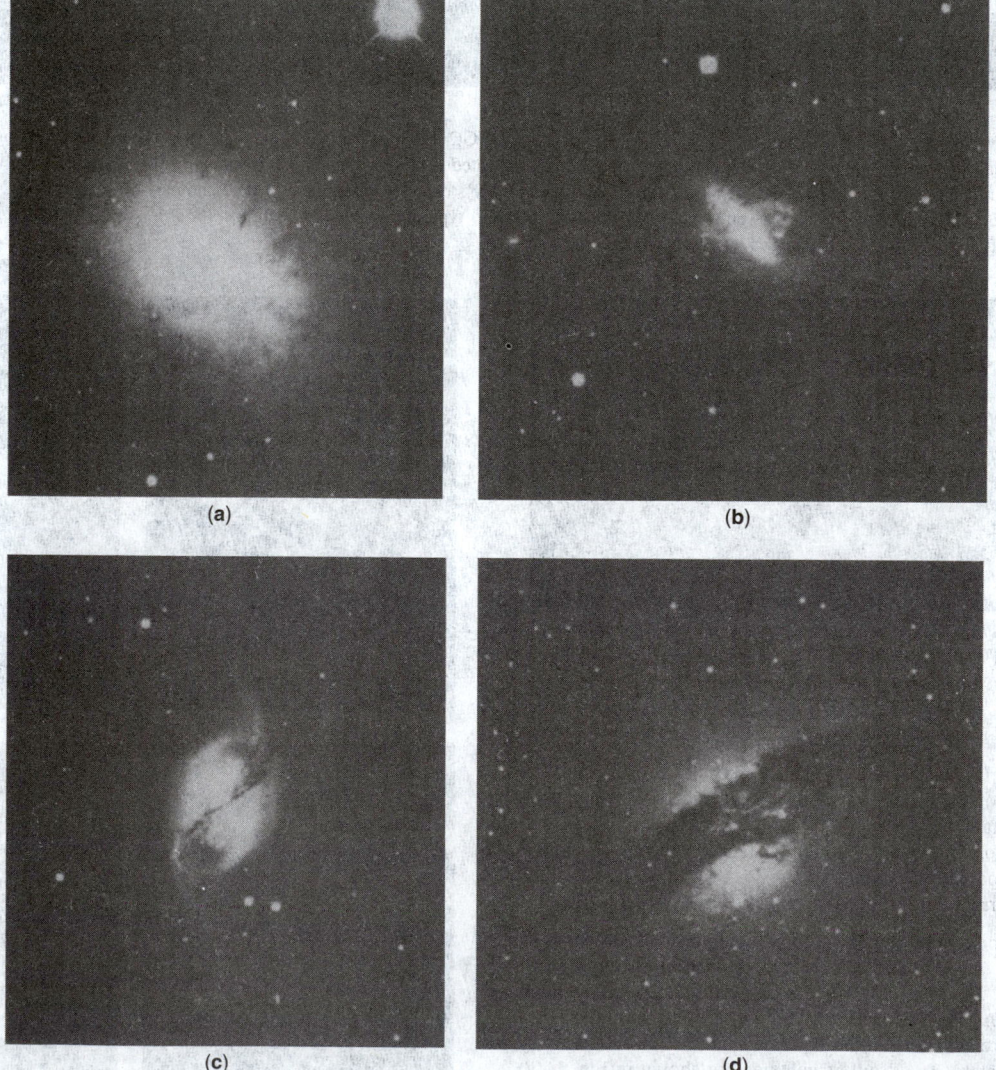

Fig. 15.   Irregular and peculiar galaxies: (**a**) Irregular (Type II) galaxy (NGC 3077), located in Ursa Major. Note that the dust lanes do not follow the usual pattern. (**b**) A peculiar Type S0 galaxy (NGC 2685), located in Ursa Major. Note that there are two axes of symmetry. (**c**) Another peculiar Type S0 galaxy, located in Ursa Major. Note the unusual absorption features. (**d**) A Type E0 elliptical galaxy located in Centaurus. This is a strong radio source and is the nearest known violent galaxy. It is also an x-ray source. (*National Optical Astronomy Observatories.*)

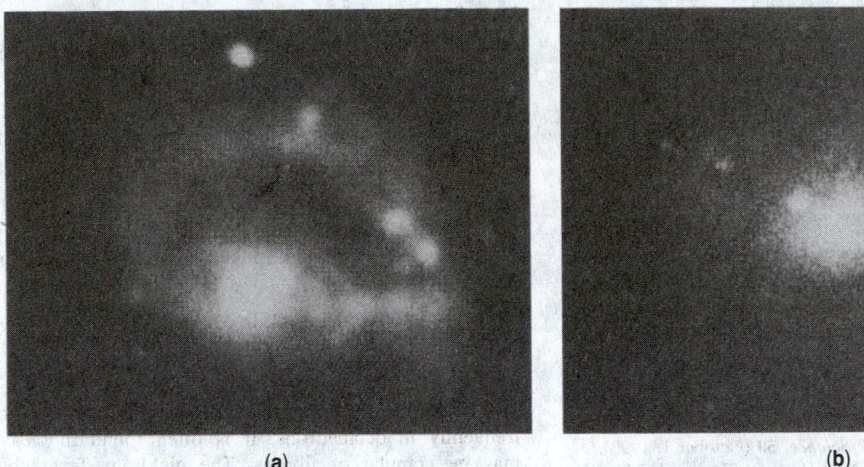

Fig. 16.   Seyfert galaxies: (**a**) Distorted ring galaxy that has a violently active Seyfert nucleus. (**b**) A peculiar galaxy (NGC 1275), located in Perseus and known to astronomers as Perseus A. This galaxy is called a Seyfert galaxy because it has large amounts of hot plasma in it. It is a strong x-ray source. (*National Optical Astronomy Observatories.*)

published the sky map shown in Fig. 17 from these cataloged objects. Most of the previously listed nebulae were found to be galaxies. The equator of the map corresponds with the central plane of the Milky Way. Because this plane is obscured by dust, few galaxies are shown in the central plane. Later knowledge indicated that clusters of galaxies are much more evenly distributed, as contrasted with the polar concentrations of the map. However, the map portrayal is significant in the manner in which it emphasizes the general clustering of galaxies, rather than uniform spacing.

As early as 1935, Shapley cataloged 25 clusters of galaxies, suggesting that clustering was related to the evolutionary processes of the Universe. Clusters of galaxies have already been shown in Figs. 5 and 6. See also Fig. 18(a and b). By definition a cluster of galaxies is a group of associated galaxies, usually within 10–100 galaxy diameters of each other.

Globular clusters of stars do not exhibit the usual features of galaxies and are believed to be much older objects. These clusters are described in the entry on **Star**.

Local Group of Galaxies. Approximately 20 of the nearest galaxies, which appear to form a cluster, are sometimes referred to as the Local Group. However, most of the mass is contained in the Milky Way and the Andromeda galaxies. In terms of increasing distance from earth, the local Group galaxies include: Milky Way (Earth is part of this); Large and Small Magellanic Clouds; Ursa Minor System; Draco System; Sculptor

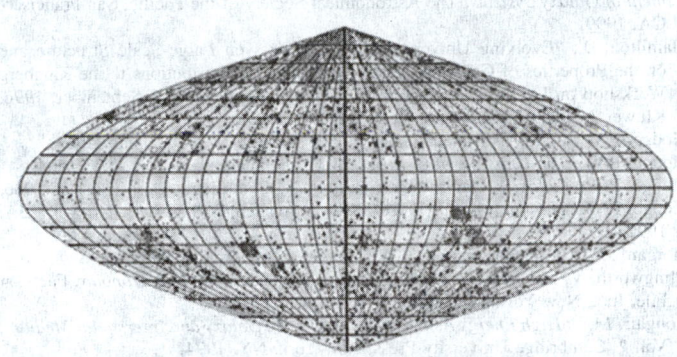

Fig. 17.   Many years prior to much detailed knowledge of galaxies, this map portraying clustering of galaxies was prepared by Charlier (1921).

System; Formax System; Leo I System; Leo II System; NGC 6822; NGC 185; NGC 147; IC 1613; M31; M32; NGC 205; and M33.

Additional Reading

Abramowicz, M.A.: "Relativity of Inwards and Outwards: An Example," *Monthly Notices of the Royal Astronomical Society,* Vol. 256, No. 4, **710** (June 15, 1992).
Abramowicz, M.A.: "Black Holes and the Centrifugal Force Paradox," *Sci. Amer.,* **74** (March 1993).

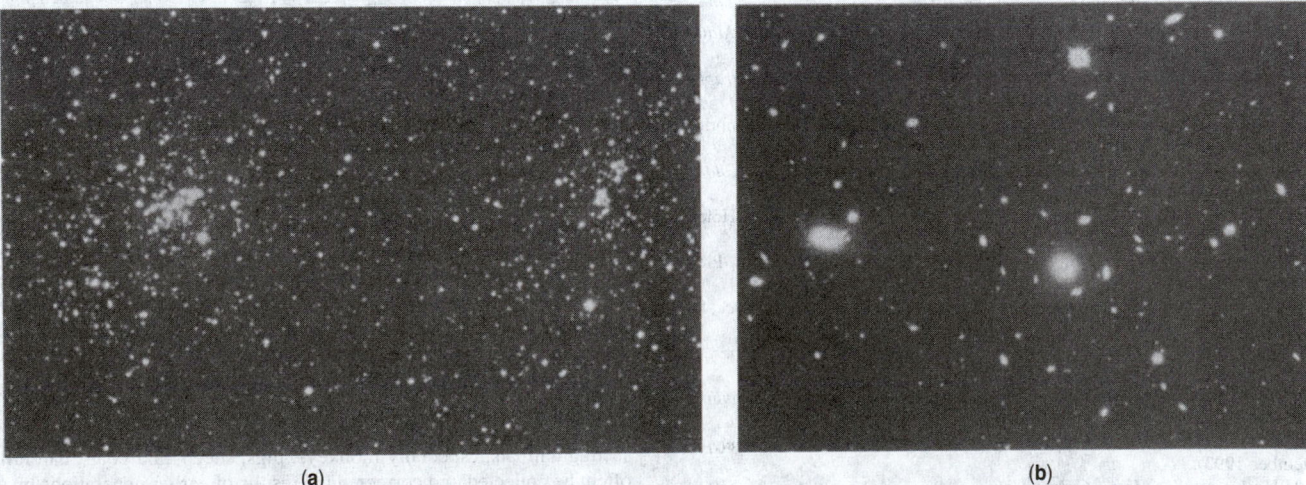

Fig. 18.   Examples of clusters of galaxies: (**a**) Two clusters of Perseus: *h* and *X* Persei. (**b**) Large clusters of galaxies in Coma Berenices. This huge cluster contains more than 100 galaxies, each a large system of stars in itself. Such regular-type clusters generally include a large number of SO and E galaxies, and are often sources of x-ray radiation. (*National Optical Astronomy Observatories.*)

Allen, B.: "Reversing Centrifugal Forces," *Nature,* Vol. 347, No. 6294, **615** (October 18, 1990).

Belfort, M.: "An Expanding Universe of Introns," *Science,* **1009** (November 12, 1993).

Cowen, R.: "Astro Eyes New Signs of Black Holes," *Science News,* **372** (December 15, 1990).

Cowen, R.: "Researchers Probe the 'Great Annihilator'," *Sci. News,* **294** (May 11, 1991).

Danks, A.C., W.H. Waller, M.N. Fanelli, and J.E. Holllis: *Ultraviolet Universe at Law and High Redshift: Probing the Progress of Galaxy Evolution,* Springer-Verlag, Inc., New York, NY, 1997.

Finkbeiner, A.: "The Life History of Galaxy Clusters," *Science,* **28** (July 2, 1993).

Flam, F.: "What Kind of a Galaxy is this, Anyway?" *Science,* **81** (November 6, 1992).

Flam, F.: "Hubble Sees a Zoo of Ancient Galaxies," *Science,* **173** (December 11, 1992).

Flam, F.: "Galaxies Keep Going with the Flow," *Science,* **31** (January 1, 1993).

Flam, F.: "Closing in On X-Ray Background Origins," *Science,* **1520** (September 17, 1993).

Flam, F.: "Are Dark Stars the Silent Majority?" *Science,* **30** (October 1, 1993).

Flam, F.: "A New Form of Strange Matter and New Hope for Finding It," *Science,* **177** (October 8, 1993).

Gehrels, N., et al.: "The Compton Gamma Ray Observatory," *Sci. Amer.,* **68** (December 1993).

Giuriein, G., P. Salucci, and M. Mezzetti: *Observational Cosmology: The Development of Galaxy Systems,* The Astronomical Society of the Pacific, San Francisco, CA, 1999.

Hamilton, D.: "Evolving Universe: Selected Topics on Large-Scale Structure and on the Properties of Galaxies Based upon, in Part, Contributions to the Ringberg Workshop on Large-Scale Structure," *Ringberg Castle,* 23–28 September, 1996, Kluwer Academic Publishers, New York, NY, 1998.

Hodge, P.W.: "The Andromeda Galaxy," *Sci. Amer.,* 92–101 (January 1981).

Hodge, P.W.: *The Universe of Galaxies,* W.H. Freeman, New York, NY, 1984.

Hodge, P.W.: *Galaxies,* Harvard University Press, Cambridge, Massachusetts, 1986.

Hodge, P.W.: *The Andromeda Galaxy,* Kluwer Academic Publishers, Norwell, MA, 1992.

Horgan, J.: "COBE Corroborated," *Sci. Amer.,* **22** (February 1993).

Illingworth, V. and J.O. Clark: *The Facts on File Dictionary of Astronomy,* Facts on File, Inc., New York, NY, 2000.

Longair, M.S.: *High Energy Astrophysics: Stars, the Galaxy and Interstellar Medium,* Vol. 2, Cambridge University Press, New York, NY, 1994.

Longair, M.S., I. Appenzeller, and R. Kippenhahn: *Galaxy Formation,* Springer-Verlag, Inc., New York, NY, 1998.

Malin, D.F.: "A Universe of Color," *Sci. Amer.,* **72** (August 1993).

Miley, G.K. and K.C. Chambers: "The Most Distant Radio Galaxies," *Sci. Amer.,* **54** (June 1993).

Moran, S.: "I've Got the Right Name for the Big Bang—and That Spells Big Bucks," *Smithsonian,* **138** (March 1994).

Murdin, P.: *Encyclopedia of Astronomy and Astrophysics,* Groves Dictionaries, Inc., New York, NY, 2000.

Osterbrock, D.E., Gwinn, J.A., and R.S. Brashear: "Edwin Hubble and the Expanding Universe," *Sci. Amer.,* **84** (July 1993).

Powell, C.S.: "Inconstant Cosmos," *Sci. Amer.,* **110** (May 1993).

Powell, C.S.: "Cosmic SNUs," *Sci. Amer.,* **50** (December 1993).

Rees, M.J.: "Black Holes in Galactic Centers," *Sci. Amer.,* **56** (November 1990).

Rfoser, H., K. Meisenheimer: *Galaxies in the Young Universe,* Springer-Verlag, New York, Inc., New York, NY, 1995.

Scheffler, H. and H. Elsasser: *Physics of the Galaxy and Interstellar Matter,* Springer-Verlag, Inc., New York, NY, 1988.

Smith, B.A. and R.H. Ressmeyer: "New Eyes on the Universe," *Nat'l. Geographic,* **2** (January 1994).

Sparke, L. and J. Gallagher: *Galaxies in the Universe: An Introduction,* Cambridge University Press, New York, NY, 2000.

Stern, S.A.: *Our Universe: The Thrill of Extragalactic Exploration as Told by Leading Experts,* Cambridge University Press, New York, NY, 2001.

Taubes, G.: "How Collapsing Stars Might Hide Their Tracks in Black Holes," *Science,* **831** (August 13, 1993).

Tucker, W. and R. Giacconi: *The X-Ray Universe,* Harvard University Press, Cambridge, MA, 1986.

Turner, M.S.: "Why is the Temperature of the Universe 2.726 Kelvin?" *Science,* **861** (November 5, 1993).

van den Bergh, S. and J.E. Hesser: "How the Milky Way Formed," *Sci. Amer.,* **72** (January 1993).

van den Bergh, S.: *Galaxy Morphology and Classification,* Cambridge University Press, New York, NY, 1998.

van den Heuvel, E.P.J. and J. van Paradijs: "X-Ray Binaries," *Sci. Amer.,* **67** (November 1993).

Voight, H.H.: *Interstellar Matter Galaxy Universe,* Springer-Verlag, Inc., New York, NY, 1999.

Waldrop, M.M.: "Black Holes Swarming at the Galactic Center?" *Science,* **166** (January 11, 1991).

Web References

Department of Astronomy, University of Texas at Austin: http://galileo.as.utexas.edu/research.html
Galaxies: http://galacticsurf.free.fr/galaxieGB.htm
Galaxies.com: www.galaxies.com
Space Telescope and Science Institute: http://www.arval.org.ve/galaxfrm.htm
The Galaxy Catalog: http://astro.princeton.edu/~frei/galaxy_catalog.html

GALAXY EVOLUTION EXPLORER (GALEX) MISSION. See **Explorers Program**; and **Space Science Missions: Universe**.

GALE. See **Winds and Air Movement**.

GALENA. The mineral galena, lead sulfide, PbS, crystallizes in the isometric system, usually in cubes or cube-octahedron combinations, less frequently in octahedrons. It is often found in cleavable masses, but may be granular or fibrous. The highly perfect cubic cleavage is an important characteristic of this mineral: it may, however, sometimes show an octahedral parting. Its hardness is 2.5; specific gravity, 7.58; luster, metallic; color, lead gray; streak, grayish-black; opaque. Galena is the most important ore of lead and in addition often carries values of silver; it is then known as argentiferous galena. It occasionally is actually mined as a silver ore. Sometimes galena contains small amounts of zinc, cadmium, antimony, bismuth, and copper as sulfides.

Galena is a very common and widely spread mineral, it occurs in veins and beds in various rocks, both crystalline and sedimentary. Some of these deposits are doubtless replacements, others seem to show a close connection with intrusive igneous rocks. Of the many European localities, the classics are Freiberg, Saxony, and the silver mines of the Harz Mountains. This mineral has been found in the lavas of Vesuvius, in Italy, and fine specimens came from Cornwall and Cumberland, England. Australia, South America, Chile, and Peru produce galena. In the United States, Missouri, Illinois, Iowa, and Wisconsin contain large and important galena deposits. In Colorado and Idaho it has been mined for its silver content. Galena is usually associated with sphalerite, smithsonite, and at Phoenixville, Pennsylvania, with beautiful pyromorphite crystals. The name is derived from the Latin galena, a term which was applied both to the lead ore and slag from refining.

ELMER B. ROWLEY, Union College, Schenectady, NY

GALEN OF PERGAMUM (AD129–216). Galen was a Greek physician, anatomist and physiologist. Galen, the son of a wealthy architect, enjoyed an excellent education in rhetoric and philosophy in his native Pergamum (Bergama, West Turkey) before turning to medicine. After studying medicine further in Smyrna and Alexandria, he began practising in Pergamum in 157, and went to Rome in 162. Driven out in 166 by hostile competitors, or fearing the plague, he returned in 169, remaining in imperial service until his death. A prodigious polymath, he wrote more than 350 treatises, of which one third survive today in Greek, or in translations in Latin or Arabic. See also **Physiology (The History)**.

His assertive personality is all-pervasive. Although Plato and Hippocrates were his gods, and Aristotle ranked only slightly below them, he was anxious to form his own independent judgments. He made ambitious efforts to encompass the entirety of medicine, deriding those who were mere specialists, and demanding excellence in both theory and practice. Galen reports some spectacular surgical successes, especially treating wounded gladiators. Although contemporaries credited him with almost miraculous skills in prognosis (which incorporated diagnosis), especially in stress-related diseases, he argued that a sound prognosis depended on detailed, close observation, and accurate reasoning. His authoritative bedside manner also contributed to his success with patients. See also **Aristotle of Stagira (384 BC–322 BC)**; and **Hippocrates of Cos (460 BC–370 BC)**.

Galen was a particularly productive anatomist and physiologist. Dissecting animals, especially monkeys, pigs, sheep, and goats, carefully and often, he collected and corrected the results of earlier generations by experiment, superior factual information, and logic. His physiological research was at times masterly, particularly in his series of experiments on the spinal cord. But his reliance largely on nonhuman anatomy, coupled with his belief that the basic structures of the human body had been described by Hippocrates, led him to 'see' things that were not there.

His pathology, founded on the doctrines of the four humors and of three organic systems, heart, brain and liver, explained disease mainly as an imbalance, detectable through qualitative changes in the body. In pharmacology and dietetics he largely codified earlier learning, adding personal observations and occasional novel ideas, e.g. that drugs could be classified according to 12 grades of activity. See also **Drug Discovery (The History)**.

His philosophy was eclectic. His major enterprise to create a logic of scientific demonstration survives only in fragments, although authors credited him with innovations in logic. In his psychology, he favored a tripartite soul, situated in the three major bodily organs, bringing the evidence of anatomy to support his case.

Galen's monotheistic views, his ardent belief that everything had been created for a purpose (teleology), and his personal piety — even anatomy was a veneration of God — foreshadow the Middle Ages. His dominant influence on later generations, comparable only to that of Aristotle, was based on his achievements as scientist, logician and universal scholar, and on his own self-proclaimed insistence to have established a universally valid medicine.

His authority, supreme in Islam and, from 1100 on, in Western Europe, was challenged in the mid-sixteenth century, first by the anatomical discoveries of Andreas Vesalius, and then by the new chemical philosophy of the Paracelsians. The discovery of the circulation of the blood by William Harvey (1628) overthrew his physiology, but his ideas on therapy lasted until the nineteenth century in European medicine, and are still used in the Yunani medicine of the Islamic world. See also **Harvey, William (1578–1657)**; and **Vesalius, Andreas (1514–1564)**.

Additional Reading

Conrad, L.I., M. Neve, V. Nutton, R. Porter, and A. Wear: *The Western Medical Tradition, 800BC–1800AD*, Cambridge University Press, New York, NY, 1995.
Nutton, V.: *Ancient Medicine (Series of Antiquity)*, Taylor & Francis, Inc., Philadelphia, PA, 2004.
Singer, P.N.: *Galen: Selected Works*, Oxford University Press, New York, NY, 1997.

V. NUTTON, Wellcome Institute for the History of Medicine, London, UK

GALILEAN TELESCOPE. A form of telescope that has a divergent lens for ocular and in which no real image is formed. The field of view is small, but the whole telescope is shorter than conventional telescopes of comparable power. Commonly used in opera glasses. See also **Telescope (Astronomical-Optical)**.

GALILEAN TRANSFORMATION. The transformation to a system moving with constant relative velocity according to nonrelativistic kinematics:

$$dx' = dx - v_x\, dt$$
$$dy' = dy - v_y\, dt$$
$$dz' = dz - v_z\, dt$$
$$dt' = dt$$

GALILEO, GALILEI (1564–1642). Galileo was an Italian astronomer and mathematician often referred to as the first modern scientist. He is known for improving the refracting telescope and using it for astronomy. Galileo turned his telescope toward the moon. He did not see Aristotle's perfect, smooth moon, but instead a moon like our earth. Other significant telescopic discoveries by Galileo include four of Jupiter's moons and the rings of Saturn.

He brought about ecclesiastical censure when he supported the Copernican theory that the Earth moves around the Sun, an idea that was opposite of what most people then believed.

Early in his life, Galileo was interested in motion especially in the way it was explained by Aristotle. He believed Aristotle was wrong. Galileo achieved great success for his experiment at the Leaning Tower of Pisa.

During this famous experiment, he proved that objects fall at a constant rate regardless of their weight.

See also **Gravitation**; **Jupiter**; **Telescope (Astronomical-Optical)**; and **Saturn**.

J. M. I.

GALILEO MISSION TO JUPITER. NASA's *Galileo* spacecraft was designed to study the large, gaseous planet Jupiter, its moons and its surrounding magnetosphere, a magnetic bubble surrounding the planet's magnetic field. It was named for the Italian Renaissance scientist Galileo Galilei, who discovered Jupiter's major moons in 1610 with the first astronomical telescope.

Mission Overview

Galileo's primary mission at Jupiter began when the spacecraft entered into orbit around Jupiter in December 1995, and its descent probe, which had been released five months earlier, dove into the giant planet's atmosphere. The primary mission included a 23-month, 11-orbit tour of the Jovian system, including 10 close encounters of Jupiter's major natural satellites, or moons.

Although the primary mission was completed in December 1997, the mission was extended for an additional two years. The extended mission, known as the Galileo Europa Mission, included up to 14 additional encounters of Jupiter's major moons — eight with Europa, four with Callisto and one or two with Io. The two-year extended mission concluded in December 1999. *Galileo* is now continuing its studies under yet another extension called the Galileo Millennium Mission. JPL, a division of the California Institute of Technology in Pasadena manages the Galileo mission for NASA's Office of Space Science, Washington, DC. On December 28, 2000 *Galileo* completed its 29th orbit as it flew by the moon Ganymede.

During its primary mission, Galileo was the first spacecraft ever to measure Jupiter's atmosphere directly with a descent probe, and the first to conduct long term observations of the Jovian system from orbit around Jupiter and during its flight to Jupiter. During the interplanetary cruise, *Galileo* became the first spacecraft to fly by an asteroid and also the first to discover the moon of another asteroid.

Launch

The *Galileo* spacecraft and its two-stage Inertial Upper Stage (IUS) were carried into Earth orbit on October 18, 1989 by space shuttle Atlantis on mission STS-34. The solid-fuel upper stage then accelerated the spacecraft out of Earth orbit toward the planet Venus for the first of three planetary flybys, or "gravity assists," designed to boost *Galileo* toward Jupiter. In a gravity assist, the spacecraft flies close enough to a planet to be propelled by its gravity, creating a "slingshot" effect for the spacecraft. The Galileo mission had originally been designed for a direct flight of about $3\frac{1}{2}$ years to Jupiter, using a planetary three-stage IUS. When this vehicle was canceled, plans were changed to use a liquid-fuel Centaur upper stage. Due to safety concerns after the *Challenger* accident, NASA cancelled use of the Centaur on the space shuttle, and *Galileo* was moved to the two-stage IUS. This, however, made it impossible for the spacecraft to fly directly to Jupiter. To save the project, Galileo engineers designed a new and remarkable six-year interplanetary flight path using planetary gravity assists.

Venus and Earth Flybys. After flying past Venus at an altitude of 16,000 kilometers (nearly 10,000 miles) on February 10, 1990, the spacecraft swung past Earth at an altitude of 960 kilometers (597 miles) on December 8, 1990. The spacecraft returned for a second Earth swingby on December 8, 1992, at an altitude of 303 kilometers (188 miles). With this, Galileo left Earth for the third and final time and headed toward Jupiter. See Fig. 1.

The flight path provided opportunities for scientific observations. At Venus, scientists obtained the first views of mid-level clouds and confirmed the presence of lightning on that planet. They also made many Earth observations, mapped the surface of Earth's Moon, and observed its north polar regions.

Because of the modification in Galileo's trajectory, the spacecraft was exposed to a hotter environment than originally planned. To protect it from

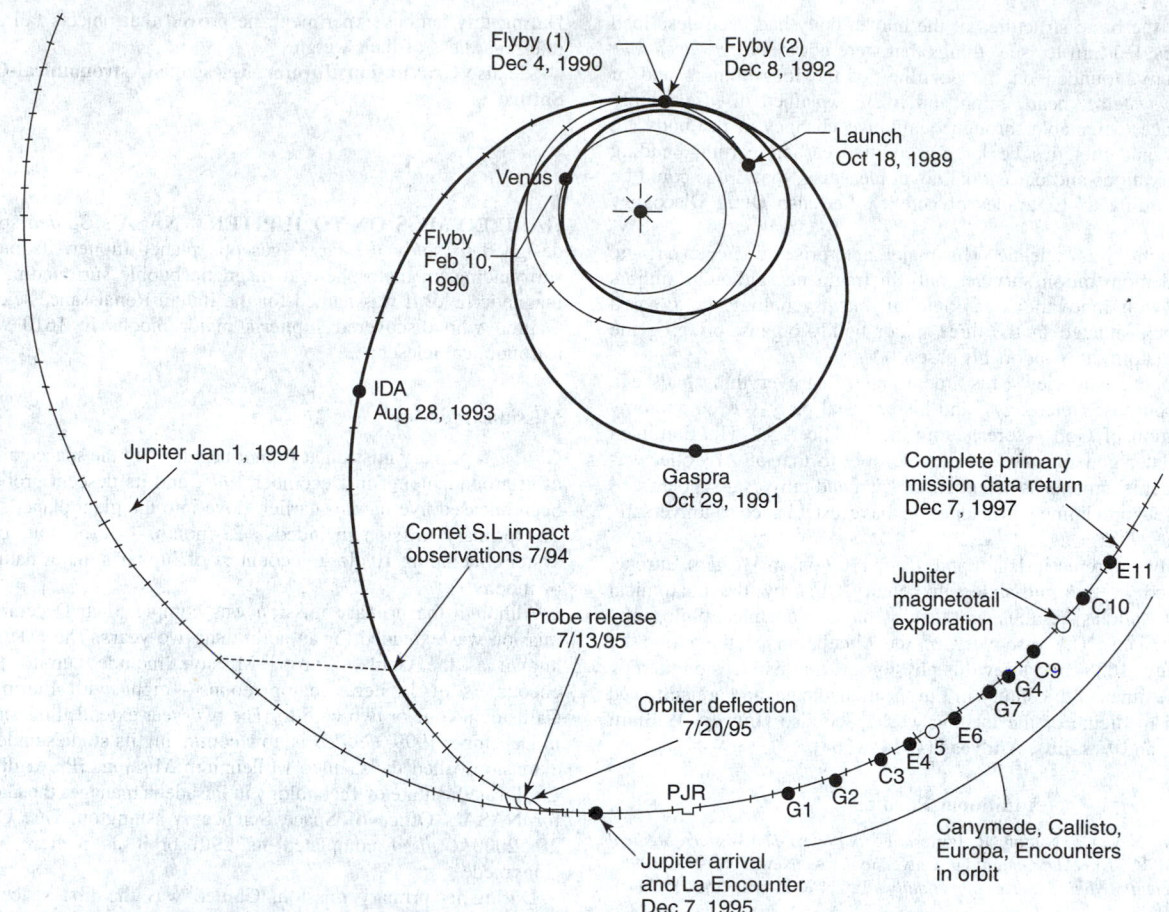

Fig. 1. Complex trajectory of *Galileo* spacecraft, which involves two flybys of Earth, one flyby of Venus, and one relatively close pass of Venus.

the sun, project engineers devised a set of sun shades and pointed the top of the spacecraft toward the Sun, with the umbrella-like high-gain antenna stowed until well after the first Earth flyby in December 1990. Flight controllers stayed in touch with the spacecraft through a pair of low-gain antennas, which send and receive data at a much slower rate.

High-gain Antenna Problem. On April 11, 1991, after Galileo had traveled far enough from the heat of the Sun, the spacecraft executed stored computer commands designed to unfurl the large high-gain antenna. But telemetry received minutes later showed that something went wrong. The motors had stalled and the antenna had only partially opened.

In a crash effort over the next several weeks, a team of more than 100 technical experts from JPL and industry analyzed *Galileo's* telemetry and conducted ground testing with an identical spare antenna. They deduced that the problem was most likely due to the sticking of a few antenna ribs, caused by friction between their standoff pins and sockets.

The excessive friction between the pins and sockets has been attributed to etching of the surfaces that occurred after the loss of a dry lubricant that had been bonded to the standoff pins during the antenna's manufacture in Florida. The antenna was originally shipped to JPL by truck in its own special shipping container. In December 1985, the antenna, again in its own shipping container, was sent by truck to NASA's Kennedy Space Center (KSC) in Florida to await launch. After the Challenger accident, *Galileo* and its antenna had to be shipped back to JPL in late 1986. Finally, they were reshipped to KSC for integration and launch in 1989. The loss of lubricant is believed to have occurred due to vibration the antenna experienced during those cross-country truck trips.

Extensive analysis has shown that, in any case, the problem existed at launch and went undetected; it is not related to sending the spacecraft on the VEEGA trajectory or the resulting delay in antenna deployment.

While diagnosis of the problem continued, the Galileo team sent a variety of commands intended to free the antenna. Most involved turning the spacecraft toward and away from the Sun, in the hope that warming and cooling the apparatus would free the stuck hardware through thermal

expansion and contraction. None of these attempts succeeded in releasing the ribs.

Further engineering analysis and testing suggested that "hammering" the antenna deployment motors (turning them on and off repeatedly) might deliver the force needed to free the stuck pins and open the antenna. After more than 13,000 hammerings between December 1992 and January 1993, engineering telemetry from the spacecraft showed that additional deployment force had been generated, but it had not freed the ribs. Other approaches were tried, such as spinning the spacecraft up to its fastest rotation rate of 10 rpm and hammering the motors again, but these efforts also failed to free the antenna.

Project engineers believe the state of the antenna has been as well defined as long-distance telemetry and laboratory tests will allow. After the years-long campaign to try to free the stuck hardware, the project has determined, there is no longer any significant prospect of the antenna being deployed.

Nevertheless, one last attempt was made in March 1996, after the orbiter's main engine is fired to raise *Galileo's* orbit around Jupiter. This "perijove raise maneuver" delivered the largest acceleration the spacecraft had experienced since launch, and it follows three other mildly jarring events: the release of the atmospheric probe, the orbiter deflection maneuver that follows probe release, and the Jupiter orbit insertion engine firing. This was the last attempt to open the antenna before radioing the new software to the spacecraft to inaugurate the advanced data compression techniques designed specifically for use with the low-gain antenna.

The Low-Gain Antenna. The difference between Galileo sending its data to Earth using the high-gain antenna and the low-gain is like the difference between the concentrated light from a spotlight versus the light emitted diffusely from a bare bulb. If unfurled, the high-gain would transmit data back to ground-based Deep Space Network (DSN) collecting antennas in a narrowly focused beam. The low-gain antenna transmits in a comparatively unfocused broadcast, and only a tiny fraction of the signal actually reaches DSN receivers. Because the received signal is

10,000 times fainter, data must be sent at a lower rate to ensure that the contents are clearly understood.

New Software on the Spacecraft. Key to the success of the mission was the first of two sets of new flight software sent to the spacecraft. The first set, called Phase 1, began operating in March 1995 and was designed expressly to partially back-up and ensure receipt of the most important data collected from the atmospheric probe. Once the critical scientific data from the probe was safely returned to Earth, the second set of new software was radioed and loaded onto the spacecraft in March 1996.

This Phase 2 software provided programs to shrink the voluminous science data the *Galileo* orbiter collected and stored on its tape recorder during its two-year mission, while retaining the scientifically important information, and return that data at the lower data rate.

Without any new enhancements, the low-gain antenna's data transmission rate at Jupiter would be limited to only 8–16 bits per second (bps), compared to the high-gain's 134,400 bps. However, the innovative Phase 2 software changes, when coupled with hardware and software adaptations at Earth-based receiving stations, increased the data rate from Jupiter by as much as 10 times, to 160 bps. The data compression methods allowed retention of the most interesting and scientifically valuable information, while minimizing or eliminating less valuable data (such as the dark background of space) before transmission. Two different methods of data compression were used. In both methods, the data are compressed onboard the spacecraft before being transmitted to Earth.

The first method, called "lossless" compression, allows the data to be reformatted back to its original state once on the ground. This technique is routinely used in personal computer modems to increase their effective transmission rates. The second compression method is called "lossy," a term used to describe the dissipation of electrical energy, but which in this case refers to the loss of some original data through mathematical approximations used to abbreviate the total amount of data to be sent to the ground. Lossy compression was used to shrink imaging and plasma wave data down to as little as $\frac{1}{80}$th of its original volume.

The Tape Recorder Problem. Galileo's tape recorder is a key link in techniques developed to compensate for the loss of use of Galileo's high-gain antenna. The tape recorder is to be used to store information, particularly imaging data, until it can be compressed and edited by spacecraft computers and radioed via *Galileo's* low-gain antenna back to Earth.

On Oct. 11, 1995, with just weeks to go before Jupiter arrival, the tape recorder malfunctioned. Data from the spacecraft showed the recorder failed to cease rewinding after recording an image of Jupiter.

A week later, following extensive analysis, the spacecraft tape recorder was tested and proved still operational, but detailed study of engineering data indicates that the tape recorder can be unreliable under some operating conditions. The problem appeared to be manageable, however, and would not jeopardize return of the full complement of images of Jupiter and its moons that are to be stored on the recorder for playback over the course of the mission.

On Oct. 24, the spacecraft executed commands for the tape recorder to wind an extra 25 times around a section of tape possibly weakened when the recorder had been stuck in rewind mode with the tape immobilized for about 15 hours. Due to uncertainty about its condition, spacecraft engineers have declared that portion near the end of the tape reel is "off-limits" for future data recording. The extra tape wound over it secures that area of tape, eliminating any stresses that could tear the tape at this potential weak spot. Unfortunately, the approach image of Jupiter that *Galileo* took October 11 was stored on the portion of tape that is now off-limits, and will not be played back. More significantly, project officials also decided not to take pictures of Io and Europa on Dec. 7, including what would have been the closest encounter of Io (from a distance of 600 miles or 1,000 kilometers). Instead, the tape recorder was completely devoted that day to gathering data from *Galileo's* Jupiter atmospheric probe.

Asteroid Flybys. Galileo became the first spacecraft ever to encounter an asteroid when it passed Gaspra on October 29, 1991. It flew within just 1,601 kilometers (1,000 miles) of the stony asteroid's center, at a relative speed of about 8 kilometers per second (18,000 miles per hour). Pictures and other data revealed a cratered, complex, irregular body about 20 by 12 by 11 kilometers (12.4 by 7.4 by 6.8 miles), with a thin covering of dirt-like "regolith" and a possible magnetic field.

On August 28, 1993, *Galileo* flew by a second asteroid, this time a larger, more distant asteroid named Ida. Ida is about 55 kilometers (34 miles) long, by 20 and 24 kilometers (12 by 15 miles). Observations indicated that both Ida and Gaspra have magnetic fields and cratered surfaces. Scientists made a dramatic discovery when they found that Ida boasts its own moon, making it the first asteroid known to have a natural satellite. The tiny moon, named Dactyl, has a diameter of only about 1.5 kilometers in diameter (less than a mile). Scientists studied Dactyl's orbit in order to estimate Ida's density.

The discovery of Comet Shoemaker-Levy 9 in March 1993 provided an exciting opportunity for Galileo's science teams and other astronomers. The comet was breaking up as it orbited Jupiter, and was headed to dive into the giant planet's atmosphere in July 1994.

The *Galileo* spacecraft, approaching Jupiter, was the only observation platform with a direct view of the impact area on Jupiter's far side. Despite the uncertainty of the predicted impact times, Galileo team members preprogrammed the spacecraft's science instruments to collect data and were able to obtain spectacular images of the comet impacts.

Jupiter Arrival. On July 13, 1995, Galileo's descent probe, which had been carried aboard the parent spacecraft, was released and began a five-month freefall toward Jupiter. The probe had no engine or thrusters, so its flight path was established by pointing of the Galileo orbiter before the probe was released. Two weeks later, Galileo used its 400-newton main rocket engine for the first time as it readjusted its flight path to arrive at the proper point at Jupiter.

Arrival day on December 7, 1995, turned out to be an extremely busy 24-hour period. When Galileo first reached Jupiter and while the probe was still approaching the planet, the orbiter flew by two of Jupiter's major moons: Europa and Io. *Galileo* passed Europa at an altitude of about 33,000 kilometers (20,000 miles), while the Io approach was at an altitude of about 900 kilometers (600 miles). About four hours after leaving Io, the orbiter made its closest approach to Jupiter, encountering 25 times more radiation than the level considered deadly for humans.

Descent Probe. Eight minutes later, the orbiter started receiving data from the descent probe, which slammed into the top of the Jovian atmosphere at a cometlike speed of 170,000 kilometers per hour (106,000 miles per hour). In the process the probe withstood temperatures twice as hot as the Sun's surface. The probe slowed by aerodynamic braking for about two minutes, then deployed its parachute and dropped its heat shield.

The wok-shaped probe floated down about 200 kilometers (125 miles) through the clouds, transmitting data to the orbiter on sunlight and heat flux, pressure, temperature, winds, lightning and atmospheric composition.

Like the other gas giant planets, Jupiter has no solid surface like Earth or Mars; heat and pressure built up to sizzling and crushing levels as the probe descended. Fifty-eight minutes into its descent, high temperatures silenced the probe's transmitters. The probe sent data from a depth with a pressure 23 times that of the average on Earth's surface, more than twice the mission requirement.

An hour after receiving the last transmission from the probe, at a point about 200,000 kilometers (130,000 miles) above the planet, the *Galileo* spacecraft fired its main engine to brake into orbit around Jupiter.

This first loop around Jupiter lasted about seven months. *Galileo* fired its thrusters at its farthest point in the orbit to keep it from coming so close to the giant planet on later orbits. This adjustment prevented possible damage to spacecraft sensors and computer chips from Jupiter's intense radiation environment.

During this first orbit, new software was installed which gave the spacecraft extensive new onboard data processing capabilities. It enabled data compression, permitting the spacecraft to transmit up to 10 times the number of pictures and other measurements that would have been possible otherwise.

In addition, hardware changes on the ground and adjustments to the spacecraft-to-Earth communication system increased the average telemetry rate tenfold. Although the problem with the high-gain antenna prevented all of the mission's original objectives from being met, the great majority of them were. So many objectives were achieved that scientists feel *Galileo* has produced considerably more science than ever envisioned at the project's start 20 years ago.

Orbital Tour. During its primary mission orbital tour, *Galileo's* itinerary included four flybys of Jupiter's moon Ganymede, three of Callisto and three of Europa. These encounters were about 100 to 1,000 times closer than those performed by NASA's *Voyager 1* and 2 spacecraft during their Jupiter flybys in 1979. During each flyby, *Galileo's* instruments scanned and scrutinized the surface and features of each moon. After about a week of intensive observation, with its tape recorder full of data, the spacecraft spent the next one to two months in orbital "cruise," sending to Earth data stored on the onboard tape recorder.

Extended Mission. A two-year extension of the *Galileo* mission began in December 1997 and ended December 31, 1999. This extension, known as the *Galileo Europa* Mission, includes intensive study of Europa through eight consecutive close encounters. Scientists anticipate that these observations will enhance their knowledge of Europa's frozen surface and offer evidence on whether liquid oceans lie underneath. In addition, there will be four encounters of Jupiter's moon Callisto and one or two flybys of Io. Galileo is now continuing its studies under yet another extension, called the Galileo Millennium Mission.

Spacecraft

The Galileo mission and systems were designed to investigate three broad aspects of the Jovian system: the planet's atmosphere, the satellites, and the magnetosphere. The spacecraft was constructed in three segments, which help focus on these areas: (1) the atmospheric probe; (2) a non-spinning section of the orbiter carrying cameras and other remote sensors; (3) the spinning main section of the orbiter spacecraft which includes the fields and particles instruments, designed to sense and measure the environment directly as the spacecraft flies through it. The spinning section also carries the communications antennas, the propulsion module, flight computers and most support systems.

This innovative "dual spin" design allows part of the orbiter to rotate constantly at three rpm, and part of the spacecraft to remain fixed. This means that the orbiter can easily accommodate magnetospheric experiments (which need to take measurements while rapidly sweeping about) while also providing stability and a fixed orientation for cameras

and other sensors. The spin rate can be increased to 10 revolutions per minute for additional stability during major propulsive maneuvers.

Galileo's atmospheric probe weighs 339 kilograms (746 pounds), and includes a deceleration module to slow and protect the descent module, which carries out the scientific mission.

The deceleration module consists of an aeroshell and an aft cover, designed to block the heat generated by friction during the sharp deceleration of atmospheric entry. Inside the shells are the descent module and its 2.5-meter (8-foot) parachute. The descent module carries a radio-relay transmitter and six scientific instruments. Operating at 128 bits per second, each of the dual L-band transmitters send nearly identical streams of scientific data to the orbiter. Probe electronics, are powered by batteries with an estimated capacity of about 18 amp-hours on arrival at Jupiter.

Probe instruments include, an atmospheric structure group of sensors measuring temperature, pressure, and deceleration; a neutral mass spectrometer and a helium abundance detector supporting atmospheric composition studies; a nephelometer for cloud location and cloud-particle observations; a net-flux radiometer measuring the difference, upward versus downward, in radiant energy flux at each altitude; and a lightning/radio-emission instrument with an energetic-particle detector, measuring light and radio emissions associated with lightning and energetic particles in Jupiter's radiation belts.

The *Galileo* orbiter spacecraft, in addition to supporting the probe activities, will support all the scientific investigations of Jupiter's satellites and magnetosphere, and remote observation of the giant planet itself. See Fig. 2.

At launch, the orbiter weighed about 2,223 kilograms (4,900 pounds), not counting the upper-stage-rocket adapter but including about 925 kilograms of usable rocket propellant. This propellant is used in almost 30 relatively small maneuvers during the long gravity-assisted flight to Jupiter, three large thrust maneuvers including the one that puts the craft into its Jupiter orbit, and the 30 or so trim maneuvers planned for the satellite tour phase. It is also consumed in the small pulses that turn and orient the spacecraft.

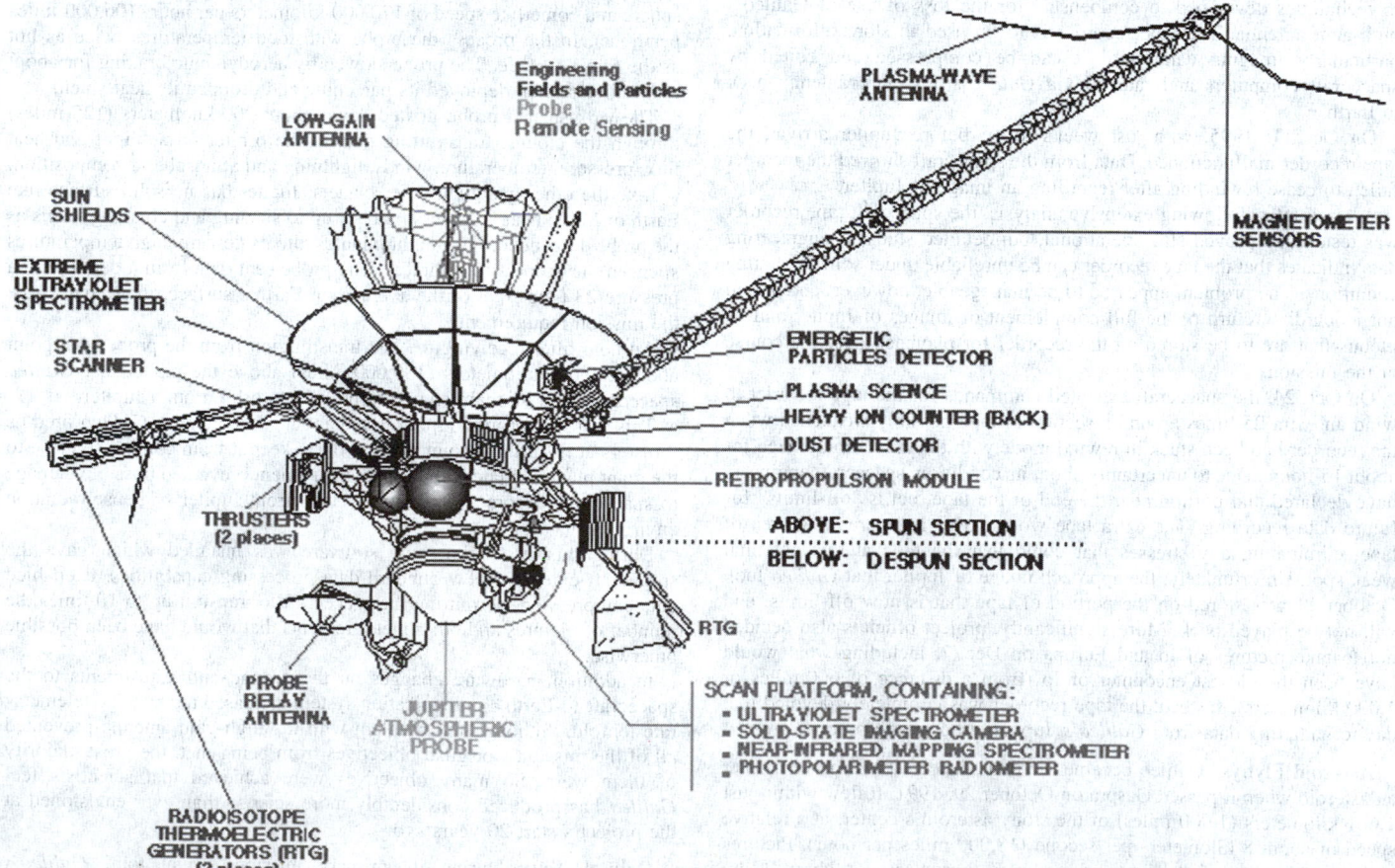

Fig. 2.   Schematic diagram of *Galileo* spacecraft indicating location of principal instrumentation and communication systems. (*Jet Propulsion Laboratory.*)

The propulsion module consists of twelve 10-newton thrusters, a single 400-newton engine, the monomethyl hydrazine fuel, nitrogen tetroxide oxidizer, and pressurizing-gas tanks, tubing, valves and control equipment. (A thrust of 10 newtons would support a weight of about one kilogram or 2.2 pounds at Earth's surface.) The propulsion system was developed and built by Messerschmitt-Bolkow-Blohm (MBB) and provided by Germany as a partner in Project Galileo.

In addition to the scientific data acquired by its 10 instruments, the *Galileo* orbiter acquires and can transmit a total of 1,418 engineering measurements of internal operating conditions including temperatures, voltages, computer states and counts. The spacecraft transmitters will operate at S-band frequency (2,295 megahertz).

Two low-gain antennas (one pointed upward or toward the Sun, and one on a deployable arm to point down, both mounted on the spinning section) supported communications during the Earth–Venus–Earth leg of the flight. The top-mounted antenna is currently carrying the communications load, including science data and playbacks, in place of the high-gain antenna, and is the basis of the redesigned Jupiter sequences. The other low-gain antenna has been re-stowed after supporting operations during the early VEEGA phase.

Because radio signals take more than one hour to travel from Earth to Jupiter and back, the *Galileo* spacecraft was designed to operate from computer instructions sent to it in advance and stored in spacecraft memory. A single master sequence of commands can cover from weeks to months of quiet operations between planetary and satellite encounters. During busy encounter operations, one sequence of commands covers only about a week.

These sequences operate through flight software installed in the principal spacecraft computers. In the command and data subsystem software, there are about 35,000 lines of code, including 7,000 lines of automatic fault protection software, which operates to put the spacecraft in a safe state if an untoward event such as an onboard computer glitch were to occur. The articulation and attitude control software has about 37,000 lines of code, including 5,500 lines devoted to fault protection.

Electrical power is provided to *Galileo's* equipment by two radioisotope thermoelectric generators. Heat produced by natural radioactive decay of plutonium 238 dioxide is converted to electricity (570 watts at launch, 485 at the end of the mission) to operate the orbiter equipment for its eight + years baseline mission. This is the same type of power source used by the two *Voyager* spacecraft missions to the outer planets, the *Pioneer* Jupiter spacecraft, and the twin *Viking* Mars landers.

Scientific instruments to measure fields and particles, together with the main antenna, the power supply, the propulsion module, most of the computers and control electronics, are mounted on the spinning section. The instruments include magnetometer sensors, mounted on an 11-meter (36-foot) boom to minimize interference from the spacecraft; a plasma instrument detecting low-energy charged particles and a plasma-wave detector to study waves generated by the particles; a high-energy particle detector; and a detector of cosmic and Jovian dust. It also carries the Heavy Ion Counter, an engineering experiment added to assess the potentially hazardous charged-particle environments the spacecraft flies through, and an added extreme ultraviolet detector associated with the UV spectrometer on the scan platform.

The despun section carries instruments and other equipment whose operation depends on a steady pointing capability. The instruments include the camera system; the near- infrared mapping spectrometer to make multispectral images for atmospheric and moon surface chemical analysis; the ultraviolet spectrometer to study gases; and the photopolarimeter–radiometer to measure radiant and reflected energy. The camera system obtained images of Jupiter's satellites at resolutions from 20 to 1,000 times better than *Voyager's* best, largely because it will be closer. The CCD sensor in *Galileo's* camera is more sensitive and has a broader color detection band than the vidicons of *Voyager*. This section also carries an articulated dish antenna to track the atmospheric probe and pick up its signals for recording and relay to Earth.

Science Results

Among the Galileo mission's key science findings so far are the following:

- Jupiter has numerous, large thunderstorms concentrated in specific zones above and below the equator, where winds are highly turbulent. Although individual lightning strokes appear less frequently than on Earth, they are up to 1,000 times more powerful than terrestrial lightning.
- The descent probe that entered Jupiter's atmosphere encountered lower humidity, or atmospheric water levels, than anticipated. In fact, it entered a "dry spot" that was relatively free of clouds. Like Earth, Jupiter has a wide range of cloudiness, and the amount of water vapor in its atmosphere can vary greatly from place to place. The presence of water, composed of hydrogen and oxygen, indicates that oxygen is typically found in greater relative abundance on Jupiter than in the Sun.
- Evidence supports a theory that liquid oceans exist under Europa's icy surface. There are places where recognizable features that once were whole have been separated from each other by new, smooth ice. These areas indicate that when the older features were separated, they floated for a time in liquid water, much as icebergs float in Earth's polar regions. Scientists believe the water later froze solid, creating "rafts" of ice, similar to some seen on a smaller scale at Earth's polar regions. There are also indications of volcanic ice flows, with liquid water flowing from Europa's volcanoes. These discoveries are particularly intriguing, since liquid water is a key ingredient in the process that may lead to the formation of life. Europa is criss-crossed by faults and ridges, with regions where large pieces of crust have separated and shifted. Faults are breaks in the moon's outer crust where the crust on either side has shifted. Ridges indicate that sections of the crust have also moved. See Fig. 3.
- Europa, Io, and Ganymede all have metallic cores. Processes on these three inner moons have permitted denser elements to separate out and sink to the moons' respective centers. On the other hand, the composition of the more distant moon Callisto is fairly uniform throughout, indicating it did not follow the same evolutionary path as the other three moons.
- Europa has an ionosphere, which is a cloud of electrically charged gases surrounding the moon. Europa's ionosphere is generated by the effect of ultraviolet radiation from the Sun and collisions with charged particles in Jupiter's magnetosphere. Oxygen atoms in Europa's thin atmosphere lose electrons, leaving them positively charged particles in the ionosphere.
- Ganymede has a very thin hydrogen atmosphere. Since lightweight hydrogen escapes easily from Ganymede's low gravity, it must be replenished continuously.
- Ganymede has a magnetic field and magnetosphere, just as Earth does. In fact, Ganymede is the first moon of any planet known to possess a magnetic field. Its magnetosphere—the small magnetic "bubble" created by its field within the surrounding, more powerful, magnetic field of Jupiter—is actually some-what larger than Mercury's, a planet of similar size. There even appears to be trapped radiation in a miniature radiation belt similar to Earth's Van Allen radiation belts on magnetic field lines close to Ganymede. The discovery of a magnetic field within Ganymede challenges theoretical models of how planetary magnetic fields arose on Earth and other celestial bodies. Its presence may imply that tidal forces—the push and pull of gravity as the other moons pass Ganymede—have heated and stirred up Ganymede's interior in recent geological times. See Fig. 4.
- Ganymede's surface shows high tectonic activity, with faulting and fracturing observed on its surface. There is some evidence of volcanic ice flows, but much of the satellite's resurfacing has been accomplished by faulting and fracturing.
- Callisto's surface shows evidence for extensive, though still mysterious, erosional processes which smooth out features on the surface. Small features are blanketed by powderlike debris.
- Io's extensive volcanic activity may be 100 times greater than that found on Earth's. It is continually modifying its surface. Many changes have been recorded since *Voyager's* visit in 1979, and other major changes have been seen during the course of the *Galileo* mission. During one four-month interval, an area the size of the state of Arizona, was blanketed by volcanic debris thrown out of the volcano Pillan. Imaging and spectral analyses have shown that most of the eruptions on Io must be composed of liquid silicate rock, which contain silicon–oxygen compounds. The temperatures of these lavas are too high for other materials, such as sulfur, and in fact are even significantly hotter than most eruptions on Earth today. The composition of these hot lavas may be more similar to a type of volcanism which occurred on the Earth more than three billion years ago. See Fig. 5.
- Some evidence exists to support the premise of a liquid ocean under Callisto's surface. This is based on data from *Galileo's* magnetometer

Fig. 3.   Possessing more water than the total amount found on earth, Europa appears to have had a salty ocean beneath its icy cracked and frozen surface. *Galileo* images show ice "rafts" the size of cities that appear to have broken off and drifted apart, a frozen "puddle" smooths over older cracks, warmer material bubbles up from below to blister the surface, evaporative-type salts are exposed. A remarkable lack of craters show the surface to be relatively young. Europa has a thin oxygen atmosphere and an ionosphere. (*PIRL: University of Arizona.*)

Fig. 4.   Internal tidal friction again causes surprising effects on the solar system's largest moon. *Galileo* revealed that Ganymede has its own magnetic field. Perhaps from a slightly different orbit in its past, enough heat from tidal friction caused the separation of material inside Ganymede and this stirring of a molten core or iron sulfide is believed to generate Ganymede's magnetic field. (*DLR: German Aerospace Center.*)

indicating that electrical currents flowing near Callisto's surface might cause changes in Callisto's magnetic field. These changes suggest that something, possibly an ocean, may be conducting the electricity.

• Jupiter's ring system is formed by dust kicked up as interplanetary meteoroids smash into the planet's four small inner moons. The outermost ring is actually composed of two rings, one embedded within the other. See Fig. 6 on p. 1560.

• *Galileo* has found a strike-slip fault on Europa as long as the California segment of the San Andreas Fault. Galileo images show evidence of movement along the fault.

Fig. 5. These images were targeted to provide the first close-up view of a chain of huge calderas (large volcanic collapse pits). These calderas are some of the largest on Io and they dwarf other calderas across the solar system. At 290 by 100 kilometers (180 by 60 miles), this chain of calderas covers an area seven times larger than the largest caldera on the Earth. The new images show the complex nature of this giant caldera on Io, with smaller collapses occurring within the elongated caldera.

Also of great interest is the flat-topped mesa on the right. The scalloped margins are typical of a process geologists call "sapping," which occurs when erosion is caused by a fluid escaping from the base of a cliff. On Earth, such sapping features are caused by springs of groundwater. Similar features on Mars are one of the key pieces of evidence for past water on the Martian surface. However, on Io, the liquid is presumed to be pressurized sulfur dioxide. The liquid sulfur dioxide should change to a gas almost instantaneously upon reaching the near-vacuum of Io's surface, blasting away material at the base of the cliff. The sulfur dioxide gas eventually freezes out on the surface of Io in the form of a frost. As the frost is buried by later deposits, it can be heated and pressurized until it becomes a liquid. This liquid then flows out of the ground, completing Io's version of the water cycle. (*PIRL: University of Arizona.*)

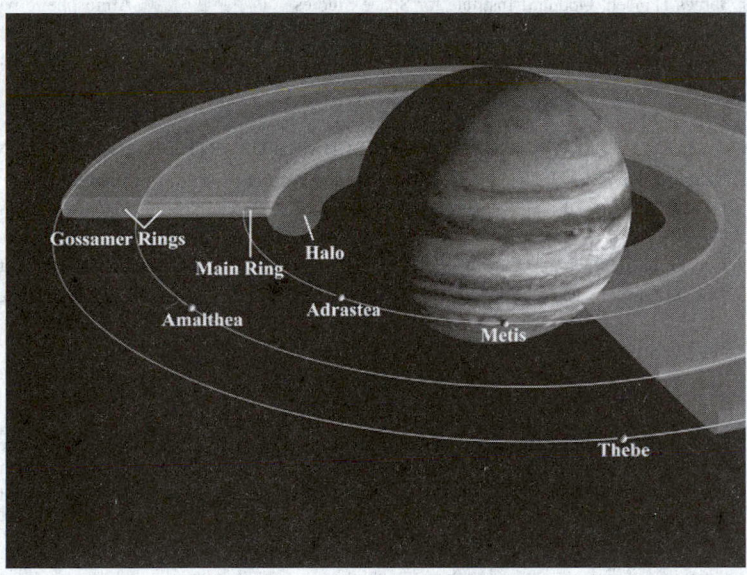

Fig. 6. Jupiter's inner satellites and ring components. (*Cornell University.*)

Scientific Experiments

Table 1 is a list of the science instrument on each part of the spacecraft, along with the name of the principal investigator responsible for the instrument and notes on the experiment's main object of study.

Ground System

Galileo communicates with Earth via NASA's Deep Space Network, a worldwide system of large antenna complexes with receivers and transmitters located in Australia, Spain and California's Mojave Desert, linked to a network control center at the Jet Propulsion Laboratory in Pasadena, CA. Through this network, the spacecraft receives commands, sends science and engineering data, and is tracked by Doppler and ranging measurements.

Doppler measurements detect changes in the frequency of the spacecraft's radio signal that reveal how fast the spacecraft is moving toward or away from Earth. In ranging measurements, radio signals transmitted from Earth are marked with a code which is returned by the spacecraft, enabling ground controllers to keep track of the distance to the spacecraft. See also **Deep Space Network**; and **Jupiter**.

Management

The *Galileo* project is managed for NASA's Office of Space Science, Washington, DC, by the Jet Propulsion Laboratory, Pasadena, CA, a division of the California Institute of Technology. JPL designed and built the *Galileo orbiter*, and operates the mission.

Galileo's scientific experiments are being carried out by more than 100 scientists from the United States, Great Britain, Germany, France, Canada and Sweden.

Technology Benefits Derived from *Galileo*. The research and development necessary to build and fly *Galileo* has produced several technological innovations.

Charge-coupled devices like those in *Galileo's* television systems are used in some home video cameras, yielding sharper images than ever conceived of in the days before the project began. In addition, radiation-resistant components developed for Galileo are now used in research, businesses, and military applications where radiation environment is a concern. Another advance, integrated circuits resistant to cosmic rays, has helped to handle disturbances to computer memory that are caused by high-energy particles; these disturbances plague extremely high-speed computers on Earth and all spacecraft.

TABLE 1. INSTRUMENTS, INVESTIGATORS, AND OBJECTIVES

Experiment/Instrument	Principal Investigator	Objectives
DESCENT PROBE		
Atmospheric structure	Dr. Alvin Seiff, San Jose State University Foundation	Temperature, pressure, density, molecular weight profiles
Neutral mass spectrometer	Dr. Hasso Niemann, NASA Goddard Space Flight Center	Chemical composition
Helium abundance	Dr. Ulf von Zahn, Institut fur Atmospharenphysik, Universitat Rostock	Helium/hydrogen ratio
Nephelometer	Dr. Boris Ragent, San Jose State University Foundation	Colds, solid/liquid particles
Net flux radiometer	Dr. Larry Sromovsky, University of Wisconsin	Thermal/solar energy profiles
Lightning and radio emissions/energetic particles	Dr. Louis Lanzerotti, Bell Laboratories, and Dr. Klaus Rinnert, Max Planck-Institut fur Aeronomie; Harald Fischer, Institut fur Reine und Angewandte Kernphysik, Universitat Kiel	Lightning detection, energetic particles
Doppler wind experiment	Dr. David Atkinson, University of Idaho	Measure winds, learn their energy source
ORBITER (REMOTE SENSING INSTRUMENTS ON NON-SPINNING SECTION):		
Solid-state imaging camera	Dr. Michael Belton, National Optical Astronomy Observatories	Galilean satellites, high resolution, atmospheric small-scale dynamics
Near-infrared mapping spectrometer	Dr. Robert Carlson, Jet Propulsion Laboratory	Surface/atmospheric composition thermal mapping
Photopolarimeter-radiometer	Dr. James Hansen, Goddard Institute for Space Studies	Atmospheric particles, thermal/reflected radiation.
Ultraviolet spectrometer/extreme ultraviolet explorer	Dr. Ian Stewart, University of Colorado	Atmospheric gases, aerosols.
ORBITER (INSTRUMENTS STUDYING MAGNETIC FIELDS AND CHARGED PARTICLES, LOCATED ON SPINNING SECTION):		
Magnetometer	Dr. Margaret Kivelson, University of California, Los Angeles	Strength and fluctuations of magnetic fields
Energetic particle detector	Dr. Donald Williams, Johns Hopkins Applied Physics Laboratory	Electrons, protons, heavy ions in atmosphere
Plasma investigation	Dr. Lou Frank, University of Iowa	Composition, energy, distribution of ions
Plasma wave subsystem	Dr. Donald Gurnett, University of Iowa	Electromagnetic waves and wave—particle interactions
Dust-detection subsystem	Dr. Harald Krueger, Max Planck Institut fur Kernphysik	Mass, velocity, charge of particles smaller than a micrometer in size
RADIO SCIENCE		
Celestial mechanics	Dr. John Anderson, Jet Propulsion Laboratory	Masses and internal structures of bodies from spacecraft tracking
Propagation	Dr. H. Taylor Howard, Stanford University	Size and atmospheric structure of Jupiter's moons from radio propagation
ENGINEERING EXPERIMENT		
Heavy ion counter	Dr. Edward Stone, California Institute of Technology	Spacecraft's charged-particle environment

Web Reference

Galileo Website: http://www.jpl.nasa.gov:80/galileo/map.html

GALILEO NUMBER.   This is defined as

$$N_{Ga} = d_p^3 \rho_f (\rho_s - \rho_f) g / \mu^2,$$

where d_p is the mean particle diameter; ρ_f is the fluid density; ρ_s is the solid density; g is the gravitational constant; and μ is the fluid viscosity.

GALLBLADDER AND BILIARY TRACT DISEASES.   The gallbladder is a pear-shaped organ (Fig. 1) situated on the underside of the liver on the right side of the body just below the ribs. This organ serves as a reservoir for the bile and by means of the cystic duct it communicates with the common duct through which the bile secreted by the liver passes to the duodenum. The gallbladder is about 3 inches (7.5 centimeters) in length and $1-1\frac{1}{4}$ inches (2.5–3 centimeters) in diameter. It holds about $1\frac{1}{2}$ ounces (29.5 milliliters) of bile. When fatty substances are ingested, the normal gallbladder empties the stored, concentrated bile into the common duct. Upon passing into the duodenum, the bile participates in a very important way in the digestion of food, notably fats. The characteristics

and role of bile in the process of digestion are described in considerable detail in the entry on **Bile**.

Principal diseases of the gallbladder and biliary tract are: (1) *cholelithiasis* (presence of stones in the gallbladder); (2) *cholecystitis* (inflammation of the gallbladder due to obstruction of the cystic duct); (3) *choledocholithiasis* (stone lodged in the common bile duct after passing from the gallbladder and through the cystic duct); any of these three diseases may be acute or chronic; (4) *chronic cholangitis* (chronic inflammation in the hepatic biliary tree); and (5) *idiopathic hyperbilirubinemia* (defect in bilirubin transport). Gilbert's syndrome (most common), the Crigler-Najjar syndrome, and the Dubin-Johnson syndrome are examples of idiopathic hyperbilirubinemia and are described in the entry on **Bile**.

The formation of gallstones derives from physicochemical changes that occur in or produce a change in the composition of the bile. Although the root causes of these changes remain poorly understood, a pathway has been described to demonstrate the alteration in bile required to produce cholesterol gallstones, but to date this theoretical approach has not led to effective ways to prevent gallstone formation and subsequent diseases. Gallstones are rather commonly found in otherwise healthy persons, particularly between ages of 55 and 65 years, where their occurrence is found in 10% of the males and about 20% of the females. It is estimated

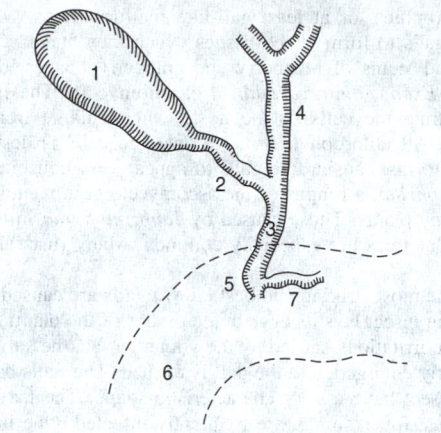

Fig. 1. Biliary tract: (1) gallbladder; (2) cystic duct; (3) common bile duct; (4) hepatic duct; (5) opening of bile duct into the duodenum; (6) duodenum; (7) duct from the pancreas. The biliary system sometimes is referred to as the biliary tree.

that 15 million persons in the United States alone have gallstones. Of these people, about 300,000 undergo surgery for gallstone removal each year. The occurrence is higher among persons with Crohn's disease (exceeds 20%). See also **Colitis and Other Inflammatory Bowel Diseases**. There is no hard evidence to the effect that gallbladder and biliary tract diseases are related to heredity. There has been an interesting finding, however, to the effect that Pima and Chippewa Indians have a much higher occurrence of gallstones. For example, it is estimated that 70% of Pima women over 25 years old have cholelithiasis, although it may be asymptomatic.

Although cholesterol is present in normal bile only to the extent of about 5%, it is the major cause of gallstones because of its insolubility in water. Precipitation of cholesterol occurs unless it is maintained in solution by the action of bile salts. The somewhat complex physical chemistry of bile is described under **Bile**. From 85 to 90% of gallbladder stones seen in patients in the United States and Europe are predominantly composed of cholesterol, which has formed on a nidus of cholesterol. Stones may range in size from a few millimeters to one or more centimeters in diameter. In contrast, the gallstones found in patients in the Orient are bilirubinate stones, which are formed by an entirely different process. Bilirubinate stones, believed to be caused by deconjugation of bilirubin diglucuronide by the action of β-glucuronidase from the microorganism *Escherichia coli*, are uniformly associated with *E. coli* infections.

Gallstones may be present in a person for many years without symptoms and may not be discovered except by a routine abdominal x-ray made to explore some other complaint. The presence of gallstones, even when asymptomatic, poses a threat, due to the risk of gallstones causing acute cholecystitis. However, knowledge of the presence of stones by no means suggests surgical removal, particularly in persons beyond middle age. Elective surgery (*cholecystectomy*) is frequently suggested for asymptomatic patients under 60 years of age. The overall mortality for persons under 60 years of age is about 0.4%, where it ranges from 1 to 4% in persons over that age. The long-term risks of not operating, in terms of ultimate development of acute or chronic disease, is unknown and apparently differs much from one individual to the next. These mortality statistics reflect the practice in large medical centers; percentages may be greater in smaller, less well equipped hospitals.

In recent years, an alternative operative procedure (laparoscopic) may be elected for certain patients. Also, drug therapy (chenodeoxycholic acid) may be effective without requiring surgery. Such decisions must made on a patient-by-patient basis because of the numerous variables involved. The physician and surgeon is guided by numerous laboratory tests, including ultrasound imaging, which can be determine the kinds and locations of gallstones that may be present as well as the general health of the patient.

Traditional surgical procedures that include a subcostal (under rib) incision continue to be favored in a number of cases. Full removal of the organ is effective because it removes all kinds of stones, not just the cholesterol stones, and prevents possible cancer of the organ if not removed, as well as the recurrence of gallstones. On the other hand, traditional surgery entails considerable postoperative discomfort, somewhat

delays the resumption of regular patient activity, and, infrequently, may cause ileus (bowel obstruction).

Laparoscopic Cholecystectomy. In some countries, this procedure is now considerably more popular than traditional surgery. In an excellent paper by L.W. Way, who describes the changing therapy for gallstone disease, he says, "The laparoscopic procedure, performed under general anesthesia, involves the creation of a pneumoperitoneum and the insertion of a laparoscope[1] and operating instruments through four small (0.5 to 1.0 centimeter) incisions in the abdomen. The cystic duct and artery are secured with clips and divided, and the gallbladder is dissected from the undersurface of the liver and removed through one of the laparoscopic ports. The procedure requires approximately $1\frac{1}{2}$ hours." Compared with traditional surgery, the patient is able to eat on the evening after surgery and may feel sufficiently well to exit the hospital on the day following the procedure. Complications that deter this surgical approach include cases that present severe adhesions (approximately 5% of cases), and the operation must be converted to an open laparotomy.

The procedure first was used in France and became popular in the United States shortly thereafter. As reported by Way in November 1990, "Although its efficacy is not in doubt, the safety of laparoscopic cholecystectomy has not yet been fully established." A survey of 1518 laparoscopic cholecystectomies was made by the Southern Surgeons Club and published in April 1991, with the conclusion: "The results of laparoscopic cholecystectomy compare favorably with those of conventional cholecystectomy with respect to mortality, complications, and length of hospital stay. A slightly higher incidence of biliary injury with the laparoscopic procedure is probably offset by the low incidence of other Complications."

Drug Therapy for Gallstones. In a review of changing therapy for gallstone disease, Way observes that the concept of using the primary bile acid (chenodeoxycholic acid) could be used to dissolve cholesterol gallstones in humans when administered orally over a period of 6 months. The original announcement was made by Danziger and coworkers (Mayo Clinic). To evaluate the efficacy of the therapy, a randomized, large, multicenter trial (the National Cooperative Gallstone Study) was conducted in the United Sates and published in 1981. It was confirmed that chenodeoxycholic acid (later known as chenodiol) could eliminate gallstones in some selected patients (persons who might be expected to respond favorably), but only in 13% of the cases over a period of treatment of 2 years. As the result of these disappointing findings, the profession continued to rely on cholecystectomy as the principal therapy for gallstone disease.

Shock-wave Lithotripsy. Researchers in Munich in 1986 reported that, in conjunction with oral therapy with bile acid, gallstone could be eliminated by lithotripsy. Preparatory to the procedure, patients were given orally administered dissolution agents to increase the effectiveness of later lithotripsy. It was found that the administration of ursodil prior to lithotripsy doubled the rate of gallstone elimination by lithotripsy. In this latter procedure, extracorporeal shock waves are administered to generate sudden bursts of high pressure that is focused on the gallstones. Through a process of compression, internal pressure-wave reflection, and cavitational forces, the stones are broken into fragments, the objective being that of producing fragments less than 5 millimeters in diameter. Initially the shock-wave procedure was performed in vitro and later in animals. The use of lithotripsy also has been performed without the aid of oral intake of bile acid ursodiol. A study on the effects of lithotripsy of gallstones was undertaken by a large group of physicians (The Dornier National Biliary Lithotripsy Study) and published in late 1990. The conclusions of that study: "Extracorporeal shock-wave lithotripsy with ursodiol was more effective than lithotripsy alone for the treatment of symptomatic gallstones, and equally safe. Treatment was more effective for solitary than multiple stones, radiolucent than slightly calcified stones, and smaller than larger stones."

Symptoms and Diagnosis. At one time, it was believed that various dyspeptic symptoms (flatulence, heartburn, fat food intolerance) were early symptoms of cholecystitis, but it has been found that these symptoms

---

[1] A laparoscope is a long, slender optical instrument that is inserted through the abdominal wall to visualize the interior of the peritoneal cavity. Modern laparoscopes include a tiny television camera so that procedures can be viewed in the operating room on a screen.

tend to occur in the normal population to about the same degree. *Acute cholecystitis* is manifested by very severe acute abdominal pain. The pain is frequently characterized by undulations, i.e., by rising to a very marked intensity for a few seconds, followed by a relatively few minutes of subsidence before the intensity returns. However, a general level of pain may persist between the peak intensities. Some authorities consider the waxing and waning nature of the pain as an essential characteristic of biliary tract disease. Description of the pain varies from one patient to the next — from excruciating to a deep ache or cramp. The pain and associated sensations motivate most persons to immediately seek medical attention. In addition to pain, there is loss of appetite, mild to rather severe nausea and vomiting, and fever in the range of 100 to 102 °F (38 to 39 °C).

Diagnosis — to differentiate acute cholecystitis (cystic duct obstruction) from *emphysematous cholecystitis* (gas forming in gallbladder from bacteria, such as *Clostridium perfringens*, other clostridia, *Escherichia coli*, and anaerobic streptococci), and other acute intra-abdominal processes, such as acute appendicitis, pancreatitis, and severe acute viral hepatitis, includes abdominal X-rays, intravenous cholangiography, and abdominal ultrasonography. Some physicians regard the latter noninvasive technique very highly and consider them to be about as reliable as oral cholecystography. Although there are some disadvantages in the use of narcotics, a drug such as meperidine (Demerol®) may be given for extreme pain. In the absence of evidence of sepsis or localized infection, antibiotics usually are not administered. However, if subsidence of the attack has not occurred within a period of several days, antibiotics may be given as a precautionary measure.

Additional Reading

Adam, A. and P. Rossi: *Biliary Tract Radiology*, Springer-Verlag, Inc., New York, NY, 1996.

Afdhal, N.H.: *Gallbadder and Biliary Tract Diseases*, Marcel Dekker, Inc., New York, NY, 2000.

Bateson, M.C.: *Gallstone Disease and Its Management*, Kluwer Academic Publishers, Norwell, MA, 1986.

Berci, G. and A. Cuschieri: *Bile Ducts and Bile Stones*, W.B. Saunders Company, Philadelphia, PA, 1996.

Buchler, M.W., E. Frei, and L. Krahenbuhl: *Five Years of Laparoscopic Cholecystectomy*, S. Karger Publishers, Inc., Farmington, CT, 1996.

Cohen, M.M., T.K. Young, and K.M. Hammarstrand: "Ethnic Variations in Cholecystectomy Rates and Outcomes, Manitoba, Canada," *Am. J. Public Health*, **124**, 79 (1989).

Cotton, P.B., J. Baillie, T. Pappas, and W.C. Meyers: "Laparoscopic Cholecystectomy and the Biliary Endoscopist," *Gastrointest Endose*, **37**, 94 (1991).

Danzinger, R.G. et al.: "Dissolution of Cholesterol Gallstones by Chenodeoxycholic Acid," *N. Eng. J. Med.*, **286**, 1 (1972).

Fromm, H. and U. Leuschner: *Bile Acids, Cholestasis, Gallstones: Advances in Basic and Clinical Bile Acid Research*, Kluwer Academic Publishers, Norwell, MA, 1996.

Kelly, J.: *Pathology of the Gallbladder, Biliary Tract and Pancreas*, W.B. Saunders Company, Philadelphia, PA, 2000.

Kremer, K., W. Lierse, S. Weller et al.: *Atlas of Operative Surgery: Gallbladder, Bile Ducts, Pancreas*, Thieme Medical Publishers, Inc., New York, NY, 1997.

Larusso, N.F.: *Gallbladder and Bile Ducts*, Vol. 6, Churchill Livingstone, Inc., Philadelphia, PA, 1997.

Meilstrup, J.W.: *Imaging Atlas of the Normal Gallbladder and Its Variants*, CRC Press, LLC., Boca Raton, FL, 1994.

Meyers, W.C. and R.S. Jones, Editors: *Textbook of Liver and Biliary Surgery*, Lippincott Williams & Wilkens, Philadelphia, PA, 1990.

Schoenfield, L.J. et al.: "The Effect of Ursodiol on the Efficacy and Safety of Extracorporeal Shock-Wave Lithotripsy of Gallstones (The Dornier National Biliary Lithotripsy Study)," *New. Eng. J. Med.*, **1239** (November 1, 1990).

Southern Surgeons Club: "A Prospective Analysis of 1518 Laparoscopic Cholecystectomies," *N. Eng. J. Med.*, **1073** (April 18, 1991).

Way, L.W.: "Changing Therapy for Gallstone Disease," *New Eng. J. Med.*, **1273** (November 1, 1990).

GALL (Botany). Abnormal outgrowths in plants caused by plant or animal parasites or induced by certain chemicals, which attack various parts of the plant. While no part of the plant is immune, galls most frequently occur in those regions composed of actively growing cells, such as leaves, or the cortical tissue of the stem, or young roots. The irritation caused by the parasite may cause numerous cell divisions, which result in a tremendous increase in the affected tissues.

The organisms that cause gall formation are many. Nematode worms often enter the roots of plants and cause the formation of irregular tumorous growth. These same organisms often infect the larger brown algae and

cause hypertrophies, or at least gall-like malformations. Many parasitic fungi cause galls to form in the tissues which they attack. Galls occur in the leaves and stems of blueberry and cranberry bushes, due to fungus infection by *Exobasidium vaccinii*, a basidiomycete. The hyphae of the fungus penetrate the cells of the host, which enlarge tremendously in consequence. All chlorophyll in these enlarged cells is destroyed, and a red pigment forms, causing the galls to appear very conspicuous. Several species of *Taphrina*, a fungus of the ascomycete group, cause galls in the leaves of many plants. Those caused by *Taphrina aurea* in the leaves and fruits of poplar trees are especially common. Many rusts also cause gall formation.

Possibly the most striking and best known galls are caused by insects. A gall-producing insect lays its eggs in the tissues of the plant. Apparently as a result of the irritations caused by the young larvae, the surrounding cells become greatly enlarged, and the gall is formed. The galls caused by each species of insect have a very characteristic shape. The leaves and stems of rose bushes, for example, are frequently infected. One insect causes a smoothly spherical gall to form; the gall produced by another is similarly shaped but studded with stiff spines; while a third causes the formation of a dense growth of matted, branched hairs, forming a structure an inch or more in diameter. Within, there may be a single insect larvae, or many, feeding on the loose parenchymatous inner tissues of the gall and protected from enemies by the firm outer layers. Often the young buds of willow twigs are parasitized, causing bud galls to form. As the bud grows older, the internodes enlarge tremendously in diameter but elongate very little, so that a gigantic bud is formed.

The leaves of oak trees are very commonly parasitized by gall-forming organisms, both fungus and insect. Considerable value attaches to these galls, because of the large accumulation of tannin occurring in the developing gall.

GALLERY FOREST. See **Biome**.

GALL GNAT (*Insecta, Diptera*). Small 2-winged flies of many species constituting the family *Cecidomyiidae*. Most are plant feeders as larvae and produce galls on the plants that they attack. Others are predacious or scavengers.

GALLIFORMES (*Aves*). This order of gallinaceous birds (ancestors of modern poultry) are mostly medium to large in size, with only a few small species. The length is 12–235 centimeters (5–92$\frac{1}{2}$ inches), and the weight is 45–11,000 grams (1$\frac{1}{2}$ ounces to 24 pounds); in domesticated forms the weight reaches 22,500 grams (49$\frac{1}{2}$ pounds). There are 10 primaries; the outer secondaries are generally very short. The feathers often have a well-developed aftershaft. Generally downy feathers are found only on the pterylae. There are no powder downs, but the preen gland is present. The males of many species are often very colorful, with widespread iridescent colors. The females generally have a protective coloration. They have very strong breast muscles which enable them to fly up quickly (except for the hoatzins). They are predominantly ground birds with strong feet. They have a strong beak, and almost always, a roomy, distensible crop, which acts as a food reservoir. There is a very strong gizzard between whose grinding surfaces, with the help of small stones swallowed for this purpose, grains and green food are ground up. They generally have a long caecum for cellulose digestion. There is a gall bladder in all species.

There are two suborders: (1) *Galli*, including the families Mound Builders, Curassows, and Pheasants and pheasant-like birds; and (2) *Opisthocomi*, with the crested fowl (the hoatzin) as the sole species. There are 94 genera and 263 species in total. They are distributed over most of the world, in semideserts, steppes, savannahs, forests, and cultivated country, and mountains up to far above the tree line (6,000 meters; 19,686 feet). All gallinaceous birds like to bathe in dust or sand, but not in water.

The *Galli* are of importance to humans, for they include four widely distributed domestic birds, including the domestic chicken. See also **Poultry**.

The great majority of *Galli* can reproduce when one year old. Most species lay many eggs. In the European partridge, up to 26 eggs have been found in one clutch. Incubation is performed almost without exception by the hen alone. Mound-building birds do not incubate at all. Newly hatched chicks have a dense, protectively colored down plumage and are soon able to feed themselves. They can fly in the first few weeks, sometimes even in the first few days. The wings of young of the true *Galli* are, however, still incomplete, having only seven short primaries. They lack secondaries. This "first wing" is much smaller than that of adults but suffices for the chicks' flight. With the increase of the bird's weight, the primaries and secondaries that were lacking grow. The inner primaries, which are too short, are replaced by longer ones. The replacements fit in with the outer primaries of later growth, which from the start are about the final length, and so are not necessarily replaced.

All true *Galli* not only have a "first wing" of short duration, but they also have a smaller, still incomplete "first tail" in many species. Its surface area is in accordance with the needs of the chick during the first weeks. As adults, many species moult the tail from the inside towards the outside (centrifugally). Others moult from the outside towards the inside (centripetally). Still others begin the moult in each half of the tail with a feather that lies between the central one and the outermost one.

All other *Galliformes* (excepting the hoatzin) are united in the large family Phasianidae. The size and weight are quite variable, ranging from only 45 grams ($1\frac{1}{2}$ ounces; Chinese painted quail) to 22.5 kilograms ($49\frac{1}{2}$ pounds; domestic turkey). There are primitive species and highly specialized ones, as well as many intermediate ones. In species that have remained primitive, males and females both have a uniform, camouflaging plumage. In highly specialized species, males have bright plumage colors, decorative ornaments, excessively large decorative feathers, and colorful distensible structures on the head and neck. These decorative feathers are important in courtship display.

They are ground dwellers. Their food consists mainly of vegetation, grains, berries, roots, conifer needles, etc., but many insects and other small animals are also eaten. The union of the sexes is extraordinarily variable, including monogamy, polygamy, or virtually no bond at all. As with birds in general, the more complicated the male's decoration and courtship behavior, the less is its participation in the rearing of offspring. Their nests are built on the ground and rarely, with the exception of the tragopans, in trees. In most cases only females incubate. The young are precocial. They are distributed over most of the world, but are absent on many islands. In America they are represented by grouse, one tribe or group of *Perdicinae*, the toothed quails, and the turkeys.

There are nine subfamilies (grouse, tragopans, pheasants, turkeys, argus pheasants, and peafowl). Altogether there are 75 genera with 204 species. See also **Poultry**.

GALLIUM. [CAS: 7440-55-3] Chemical element symbol Ga, at. no. 31, at. wt. 69.72, periodic table group 13, mp 29.78°C, bp 2403 ± 0.5°C, density 5.90 (solid at 20°C), 6.095 (liquid at 29.8°C), 5.445 (liquid at 1100°C). Elemental gallium has a one-face-centered orthorhombic crystal structure. Among the elements, gallium (like mercury) is liquid at ordinary temperatures. Gallium is a white, tough metal, but so soft that it can be cut with a knife. A freshly exposed surface soon oxidizes superficially to a bluish-gray color. When heated about 500°C, the metal burns in air. Gallium is only slightly affected by H_2O at room temperature, but reacts vigorously in boiling H_2O. The metal is only slowly attacked by concentrated acids, but does dissolve readily in aqua regia. The two stable isotopes of gallium are ^{69}Ga and ^{71}Ga. The eight radioactive isotopes include ^{64}Ga through ^{68}Ga, ^{70}Ga, ^{72}Ga, and ^{73}Ga. All have a relatively short half-life, the longest, ^{67}Ga with a half-life of 78 hours. See also **Radioactivity**. Gallium was one of the elements predicted by Mendeleev from his early periodic arrangement of the chemical elements. The element first was identified by Francois Lecoz de Boisbaudran in 1875 from observations in a spectroscopic study of zinc blende. In terms of abundance, gallium ranks 31st among the elements, with about 15 ppm in the earth's crust.

First ionization potential 6.00 eV; second, 20.43 eV; third, 30.6 eV. Oxidation potentials $Ga \rightarrow Ga^{3+} + 3e^-$, -0.52 V; $Ga + 4OH^- \rightarrow H_2GaO_3^- + H_2O + 3e^-$, 1.22 V.

Other important physical characteristics of gallium are given under **Chemical Elements**.

Gallium's renown as a valuable chemical element stems from its increasing use over the past decade in electronic devices. See also **Semiconductors**; and **Solid-State Physics**.

Gallium occurs in very small amount in zinc blende, magnetite, pyrite, bauxite, and kaolin of certain localities. A few parts per million is present in Oklahoma zinc ores. The recovery of gallium from zinc flue dust is effected by solution of the dust in excess of HCl, addition of potassium chlorate,

and distillation to remove germanium. When the residue is converted into sulfate, fractional electrolysis of the slightly acid solution removes zinc, and the gallium is obtained almost free from indium. The only known deposit of gallite, $CuGaS_2$, is in southwest Africa. The mineral contains about 1% gallium. The most important commercial source of gallium is bauxite, which contains up to 0.01% gallium. The metal is recovered from the sodium aluminate used in the extraction of aluminum from bauxite. In one process, calcium hydroxide is mixed with the sodium aluminate solution. At this juncture the ratio of gallium to aluminum is about 1 to 3,000. By precipitating and filtering out calcium aluminate, a gallium-rich solution remains. The filtrate then is agitated with CO_2 which precipitates more aluminum out as aluminum hydroxide. At this point, the enriched gallate-in-caustic solution contains approximately 0.2 grams of gallium per liter. This solution is used as an electrolyte in a mercury cathode cell. The gallium amalgamates with the mercury. It is dissolved out of the mercury with boiling NaOH in the presence of iron, which serves as a catalyst. At this point, the concentration is approximately 80 g of gallium per liter. The process is repeated several times, after which the gallium concentrate is electrolyzed, using a stainless steel cathode on which the gallium plates out. The gallium is easily removed from the cathode by raising the temperature above the melting point. For highly-pure metal, subsequent purification processes are required, including (1) crystallization as monocrystals, (2) chemical treatment with acids or oxygen at high temperatures, or (3) repeat resolution in pure boiling NaOH and reelectrolyzing. A metal of 99.99999% purity thus can be obtained.

Uses. The availability of gallium in very high purity is important to its use as a semiconductor in various electronic devices, such as diodes, laser diodes, and electroluminescent diodes. The compound usually used in these applications is gallium arsenide GaAs which is prepared by reacting hydrogen and arsenic vapor with gallium oxide Ga_2O_3 (prepared from very pure metal) at a temperature of about 600°C. Properties of the GaAs so produced include: intrinsic electron concentration, 10^7; energy gap, 1.38 eV at 20°C; electron mobility, 8,800 cm^2/V-s.

Gallium arsenide also is used in solar batteries. See also **Solar Energy**. Gallium metal is used as an activator in luminous paints and phosphors, as well as in arc rectifiers, dental amalgams, as a sealant in vacuum systems, in transistors, and in some organic syntheses. Because the metal expands upon solidifying (3.1%), it should not be stored in fragile containers. Although potentially useful in high-temperature thermometers because of its liquidity over a wide temperature range, these applications have been limited, partially because of the high cost of the element.

Chemistry and Compounds. Gallium metal is quite corrosive to most other metals because of the rapidity with which it diffuses into the crystal lattices of metals. For example, only a very small amount of gallium in contact with an aluminum plate or sheet will result in immediate embrittlement as the result of the diffusion of gallium through the grain boundaries separating them. Gallium readily forms alloys with most metals over 600°C, including barium, copper, gold, iron, lead, lithium, magnesium, manganese, nickel, platinum, silver, sodium, titanium, vanadium, zirconium, and zinc. The few metals that tend to resist attack by gallium are molybdenum, niobium, tantalum, and tungsten.

Gallium trihalides include the trifluoride, tribromide, triiodide, and the trichloride. The trichloride is readily formed by heating the metal with chlorine or HCl, is soluble in ether, and, like aluminum chloride, is effective as a catalyst in various organic reactions. Both the trichloride and the tribromide are dimeric in the vapor state. Other known trivalent gallium compounds are the sesquisulfide, sesquisulfate (which forms double salts analogous to the alums), trinitrate, nitride, sesquioxide (which is polymorphic like alumina), and trihydroxide, which is, however, of variable composition, and which forms salts, the gallates, in alkaline solution.

Known gallium(II) compounds include the sulfide, selenide, telluride, dichloride, and dibromide. The last two are unstable, reacting vigorously with water to give hydrogen, and also undergoing oxidation, or dispro-portionation to the metal and the gallium(III) compound. They also are diamagnetic and their structure is $Ga^+[GaX_4]^-$.

Simple gallium(I) compounds are also unstable, but Ga^+ may be stabilized in the presence of large anions, e.g., in $Ga[AlCl_4]$. The sulfur and selenium compounds Ga_2S and Ga_2Se have been shown to exist, but the oxide is uncertain.

Triethylgallium and trimethylgallium have been prepared, but are extremely reactive, even with air and H_2O. Like aluminum and indium, gallium forms a number of chelated oxy compounds, almost all of which are of 6-coordinate type. They include the stable crystalline inner complexes of which the β-diketones coordinate in the proportion of 3 molecules of diketone per atom of gallium. Trioxalato as well as dioxalato salts are known, and compounds such as 8-quinolinol and substituted 8-quinolinols form trimolecular chelate rings involving nitrogen and donor oxygen.

Gallium, like boron, forms a dimeric hydride, Ga_2H_6, from which a series of tetrahydrogallates, containing the GaH_4^- ion, is derived.

Gallium and most of its compounds are not highly toxic. For rats and rabbits, the LD_{100} has been established at approximately 100 mg of gallium per kilogram.

See list of references at end of entry on **Chemical Elements**.

Gallium is also described in several of the electronic component entries throughout this encyclopedia.

Additional Reading

Davis, J.R.: *Metals Handbook*, 2nd Edition, ASM International, Materials Park, OH, 1998.

Greenwood, N.N. and A. Earnshaw: *Chemistry of the Elements*, 2nd Edition, Butterworth-Heinemann, Inc., Woburn, MA, 1997.

Krebs, R.E.: *The History and Use of Our Earth's Chemical Elements, A Reference Guide*, Greenwood Publishing Group, Inc., Westport, CT, 1998.

Lide, D.R.: *CRC Handbook of Chemistry and Physics*, 88th Edition, CRC Press, LLC., Boca Raton, FL, 2007.

Mahajan, S. and L.C. Kimerling: *Concise Encyclopedia of Semiconducting Materials and Related Technologies*, Elsevier Science, New York, NY, 1992.

Parker, P.: *McGraw-Hill Encyclopedia of Chemistry*, 2nd Edition, The McGraw-Hill Companies, Inc., New York, NY, 1993.

Sandroff, C.J. et al.: "Gas Clusters in the Quantum Size Regime: Growth on High Surface Area Silica by Molecular Beam Epitaxy," *Science*, 391 (July 28, 1989).

Vander Veen, M.R.: "Gallium Arsenide Sandwich Lasers," *Advanced Materials 7 Processes*, 39 (May 1988).

Westbrook, J.H.: "Electrical Materials," in "*Encyclopedia of Materials Science and Engineering*," MIT Press, Cambridge, MA, 1986.

Willardson, R.K.: "Advances in Gallium Arsenide Crystal Growth," *Advanced Materials & Processes*, 24 (June 1986).

Wolsky, A.M., R.F. Giese, and E.J. Daniels: "The New Superconductors: Prospects for Applications," *Sci. Amer.*, 61 (February 1989).

Yablonovitch, E.: "The Chemistry of Solid-State Electronics," *Science*, 347 (October 20, 1989).

GALLIUM ARSENIDE SOLAR CELL. See **Solar Energy**.

GALLIUM LASER. See **Telephony**.

GALLO, ROBERT CHARLES (1937-Present). Robert C. Gallo best known as co-discoverer of the acquired immunoefficiency syndrome (AIDS) virus was born on 23 March 1937 in Waterbury, CT. The death of his younger sister from leukemia steered him towards a career in medicine. Educated at Providence College and Thomas Jefferson Medical College, Gallo spent his internship and residency at the University of Chicago. In 1965, he joined the National Institutes of Health, where he learned to do rigorous biochemical research. In the 1970s, as Chief of the Laboratory of Tumor Cell Biology in the National Cancer Institute, he and his colleagues discovered interleukin-2, a growth factor for T cells, and at the end of the decade identified the first human retrovirus, which caused a type of cancer. In late 1982, as epidemiological information about AIDS indicated that it was caused by a transmissible agent, Gallo turned his laboratory's efforts towards discovering the etiology of AIDS. In 1984, he published fourpapers that provided persuasive evidence for a single retrovirus as the cause of AIDS. Later that retrovirus was determined to be the one isolated by Luc Montagnier of the Pasteur Institute in 1983, but whose relation to AIDS was not proven. Gallo's laboratory continued to research the properties of the AIDS virus, named in 1986 as 'human immunodeficiency virus' (HIV). The laboratory discovered one of HIV's nine genes and produced an AIDS blood test widely used to protect the blood supply around the world. In 1996, Gallo moved his laboratory to Baltimore, where the University of Maryland created an Institute of Human Virology for him. See also **Acquired Immune Deficiency Syndrome (AIDS)**; and **Human Immunodeficiency Viruses (HIV)**.

VICTORIA A. HARDEN, National Institute of Health, Bethesda, MD

GALLSTONES. See **Gallbladder and Biliary Tract Diseases**.

GALL WASP (*Insecta, Hymenoptera*). A minute insect, whose attack on plants produces galls. They are of many species, making up the subfamily *Cynipinae*.

GALTON BOARD. See **Probability**.

GALTON, SIR FRANCIS (1822–1911). Sir Francis Galton was a scientific Renaissance man responsible for major breakthroughs in statistics and the science of heredity.

Francis Galton was born in Birmingham, the third son of Samuel Tertius Galton, a wealthy banker whose fortune mainly derived from his father's musket manufactories. Francis was also the grandson of Erasmus Darwin, in whose scientific and evolutionist footstepshe later assiduously followed. However, his entry into these fields was largely fortuitous. Aged sixteen, at his father's behest, he had enrolled as a medical student at the Birmingham Free Hospital, from where he moved to King's College London to study physiology and anatomy in 1839, and to Trinity College Cambridge in 1840 to read mathematics. Having achieved only a pass degree at Cambridge, due to the first of the mental breakdowns that punctuated his career, he persisted with his medical instruction despite a growing aversion to it. Coming into a considerable fortune on the death of his father in 1844, he immediately abandoned medicine and spent several years "hunting, fishing and shooting," the typical bucolic pursuits of a member of his privileged social class. Then, in 1850, he led and financed an exploratory expedition to Southwest Africa (modern Namibia), returning in 1852. Although the results of his travels were by no means spectacular, he was lionized by the scientific establishment and subsequently married the daughter of Dr GeorgeButler, the Dean of Peterborough. Having achieved this entrée into the scientificcommunity, by increments Galton established himself as one of Victorian Britain's most respected scientific luminaries.

Although now chiefly remembered for his innovations in the field of statistics and for advocating the practice of eugenics, Galton's interest in both of these areas initially arose from his desire to provide scientific evidence for the inheritance of human mental attributes. Indeed, Galton devoted most of his professional energies to investigatingvaried aspects of heredity, a predilection largely inspired by reading Charles Darwin's (his half-cousin) *On the Origin of Species* in 1859, a book that transformed his worldview. Shifting his focus away from the highly innovative meteorological research to which he had been devoting himself, in 1869 Galton published *Hereditary Genius*, a bookthat immediately established his Darwinian credentials. His views of mind and morality were unequivocally materialist and evolutionary, his central contention being that as the frequency of high social distinction could be shown to be considerably greater among the close family of illustrious men than among those of mediocre parentage, the personal attributes comprising eminence ("intelligence, zeal, and a capacity for work") must, *ipso facto*, be strongly determined by heredity.

On this basis Galton proposed that measures be implemented to ensure that the more able and accomplished members of society achieve higher rates of fertility than the irresponsible and intellectually subnormal. This was considered impracticable during the 1860s, and it was only during Galton's last decade that he began championing eugenics. Foreign challenges to Britain's industrial and mercantile hegemony, and fears of an increasingly volatile urban underclass rendered eugenics an unprecedentedly appealing idea to many *fin-de-siècle* Britons. In 1908, after initial reluctance occasioned by his advanced age, Galton became President of the newly formed Eugenics Education Society, nowthe Galton Institute. The term itself was coined by Galton in 1883, in his third book, *Inquiries into Human Faculty*, from the Greek *eugenes* meaning "good in stock." Today this word has the ugliest of connotations, but it is important to remember that neither Galton nor his contemporaries could have had any notion of where the idea of selective human breeding might one day lead. Moreover, eugenical thinking was far from new: Galton's originality lay only in buttressing the idea with apparently rigorous scientific research.

Galton was the first scientist systematically to apply mathematical tools to the study of the inheritance of human mental and bodily traits. Beginning with rather rudimentary methods, during the 1870s and 1880s he went on to develop the fundamental statisticaltechniques of both correlation and regression; procedures later systematized and more fully explicated by Galton's admirer and first biographer, Karl Pearson. Galton was also the first statistician to concentrate attention upon variation around the mean, an aspect virtually ignored by his predecessors as representing "error" rather than a natural phenomenon worthy of investigation. From 1904 he privately financed research fellowships in statistics and eugenics at University College London, a commitment he honored and extended upon his demise. See also **Pearson, Karl (1857–1936)**.

Galton's fascination with heredity was as much practical and experimental as it wasmathematical. Collaborating with Darwin, he subjected his kinsman's physiological theoryof heredity, *pangenesis*, to (ultimately inconclusive) experiments involving the transfusion of blood among different breeds of rabbit. He also bred several generations of sweet-pea plants, in an attempt to establish the mechanics and mathematics of heredity, from which he formulated the "ancestral law of heredity." According to this law an offspring's traits are statistically more likely to be closer to the population mean than those of their parents because the germline matter contains contributions from previous generations as well as the parents themselves. This theory would be eclipsed during the twentieth century by the advent of Mendelian genetics. Galton's interest in heredity also stimulated his work on finger-prints (and their use in criminal investigations); his novel researches into unconscious mental phenomena through "psychometric experiments," his founding of an anthropometric laboratory in 1884 to quantify and record human physiological variation; and his attempts to use composite photography to analyze group similarities and differences. Yet, although most of Galton's work did involve the study of heredity, his scientific curiosity was nonetheless polymathic in its breadth. He was able to make significant contributions to the fields of geographical exploration; the design of navigational instruments; and meteorology, including the discovery of the "anti-cyclone" in 1862, and the introduction of weather maps. See also **Genetics and Gene Science (Classical)**; and **Mendel, Gregor Johann (1822–1884)**.

Galton was one of Victorian Britain's most prominent scientists: honored with nearly all of its prestigious scientific awards, he acted as president of most of the scientific bodies of which he was a member, and was knighted in 1909. As a scientist he was idiosyncratic, innovative, utterly dedicated, and though frequently naive he was often strikingly effective. His maxim was "Whenever you can, count," and his researches suggest a man with an almost obsessive compulsion to do so. At the personal level, Galton sufferedfrom an unusually low self-esteem, social awkwardness and recurrent mental health problems. Yet, he also displayed a willingness to scandalize popular opinion, particularly with his views on religion, and push the materialism of his fellow Darwinian protagonists further than most would have dared. In both theoretical and methodological terms, Galton's contribution to the biological and social sciences can hardly be overestimated.

Additional Reading

Bulmer, M.G.: *Francis Galton: Pioneer of Heredity and Biometry*, John Hopkins University Press, Baltimore, MD, 2003.

Gillham, N.W.: *A Life of Sir Francis Galton: From African Exploration to the Birth of Eugenics*, Oxford University Press, New York, NY, 2001.

Kevles, D.: *In the Name of Eugenics: Genetics and the Uses of Human Heredity*, Penguin, Harmondsworth, UK, 1986.

Keynes, M.: *Sir Francis Galton, FRS: The Legacy of His Ideas*, Macmillan Press, Basingstoke, UK, 1993.

JOHN WALLER, University of London, London, UK

GALVANIC ACTION (Corrosion).   See **Corrosion**.

GALVANIC CELL.

Also known as a voltaic cell, an electrolytic cell that produces electric energy by electrochemical action. Although a battery may comprise only one cell, there may be several cells making up a battery and thus cell and battery are not fully synonymous. See also **Battery**.

There are many ways in which a voltage difference can be produced in an electrochemical cell. The simplest cell, thermodynamically, is the "concentration cell" in which electrolyte or electrode materials are incorporated into half-cells in differing concentrations; a half-cell is a system involving an electrolyte and a single electrode. When the half-cells are connected, the free energy change accompanying the transfer of one substance from high to low concentration results in the liberation of electrical energy. The gravity cell is a type of two-electrolyte cell in which the separation between the two ionic solutions is maintained by means of gravity. An example is the Daniel cell in which a cupric sulfate solution in contact with a copper electrode is below a zinc sulfate solution in contact with a zinc electrode. The difference in specific gravity of the solutions prevents, or at least retards, mixing. The Daniel cell also belongs to the classification of displacement cells in which the essential chemical reaction is the ionization and entry into solution of atoms of one element, and the discharge and deposition from solution of the ions of another. Concentration cells, although interesting theoretically, are not important commercially.

The majority of economically important cells consists of two dissimilar electrodes of metal or metal compounds, immersed in an aqueous solution of an acid, base, or in some cases a salt. The negative of a fresh cell is typically in the metallic state, while the positive is usually an oxide, or occasionally, a salt of the metal. During discharge, the negative electrode is oxidized as electrons leave it via the external circuit, and the positive is reduced. Since by definition an anode is an oxidation electrode, in the literature the negative is generally called the "anode" and the positive the "cathode." This conforms to accepted electrochemical terminology, although it is the cause of some confusion.

Although galvanic cells theoretically might look more attractive than heat engines as sources of electric power, since the energy changes are not subject to the limitations of the Carnot cycle, the cost comparisons of delivered power to date do not work out that way. In fact, on a kilowatt hour basis, the cost spread is 70 to 100:1 for a rechargeable battery, and many hundreds to 1 for primary cells, such as flashlight cells. This is because of inefficiencies in electrochemical operation, high material costs, high cost of the tightly controlled production operations necessary, etc. Galvanic cells have grown in importance because of the strength of other needs, such as that for a portable supply of power, for power at a place far distant from the prime power source, for a reserve or emergency source, etc. There are also needs for a source of pure direct current or for a stable reference voltage that can be provided by galvanic cells. Since the early 1970s, much interest has been regenerated in the use of galvanic cells (battery power) for small electric automobiles in an effort to reduce emissions from internal-combustion engines.

Fuel cells, although operating as galvanic cells at the electrodes, are in a separate class in that they provide direct, single-site conversion of original raw materials into electrical power, obviating the boiler-turbine-transmission-rectifier chain that precedes the production and use of ordinary batteries. See also **Fuel Cells**.

Corrosion also results from the action of oftentimes numerous galvanic cells where dissimilar metals and electrolyte (as from excessive moisture, humidity, acidic atmospheric ingredients, etc.) provide all of the electrochemical necessities for a transfer of material that causes metals to gradually "waste away" and weaken various structures. See also **Corrosion**.

GALVANIZING.

A process for rustproofing, and otherwise protecting, iron and steel, by applying a metallic zinc coating. The process can be used with nearly any size or shape of product, including large structural assemblies and steel sheet in coils and cut lengths. Millions of tons of new steel are galvanized each year, much of which is used prior to the application of other coatings, such as paint. Metallic zinc is applied to iron and steel by three processes: (1) hot dip galvanizing, (2) electrogalvanizing, and (3) zinc spraying. Most galvanized sheet steel is coated by the hot dip process. See also **Zinc**.

Hot-Dip Galvanizing. In the hot dip process, the sheets or other articles to be coated must be free from scale, dirt, grease, etc., and are usually prepared by pickling and washing before immersion in molten zinc commercially known as spelter. Articles fabricated from iron and steel sheets and wire are hand-dipped. Sheets and wire are handled mechanically.

An increasing proportion of sheet-metal products is being coated as sheet or strip before fabrication. This requires a tightly adhering coating to prevent peeling during stamping or forming operations. In order to obtain good adherence in hot-dipped coatings special processing is necessary, especially with the heavier weights of coating which give longer protection. For lighter weight coatings a duplex bath consisting of a layer of molten lead under the molten zinc is often used. The steel sheet passes through the lead, which does not adhere, and up through the zinc. The time in which the steel is in contact with the spelter is greatly reduced and consequently less zinc is deposited.

Some galvanized sheets are annealed after dipping in order to form a coating consisting entirely of iron-zinc compounds, a process which tends to increase resistance to peeling.

Electrogalvanizing. Electrodeposited zinc coatings are simpler in structure than hot dip coatings. They are composed of pure zinc, have a homogeneous structure, and are highly adherent. These coatings generally are not as thick as those produced by hot dipping. Coatings range in thickness up to 13.7 micrometer (0.065 mil). This process is particularly suitable for very thin, formable products. The electrogalvanized surface is smooth and fine and can readily be prepared for painting by phosphatizing. The coating is free of the characteristic spangled pattern of hot-dipped surfaces.

Electrogalvanizing can be done essentially at room temperature, thus the process does not alter the mechanical properties that could result from the higher temperatures encountered in dipping.

Zinc Spraying. This process involves the projection of atomized particles of molten zinc onto a prepared surface. Three types of spraying pistols are currently in use: (1) the molten metal pistol, (2) the powder pistol, and (3) the wire pistol. Sprayed coatings are slightly rough and porous. The slight porosity, however, does not adversely affect the protective value of the coating because zinc is anodic to steel. The zinc corrosion products that form during service fill the pores of the coating, giving a solid appearance. The slight roughness of the surface makes it an ideal basis for paint when properly pretreated. Spraying can be applied to nearly any shape or size of product—at the factory or at the site of final use. Spraying is the only satisfactory method of deposition available for applying very heavy zinc coatings up to 0.25 mm (0.01 in.) and greater in thickness.

GALVANOMETER.

An instrument for measuring electric currents, usually by means of their magnetic effect. Observations are made by noting the deflection produced by the reactive torque exerted between an electric current and a magnet. Galvanometers may be divided broadly into two classes, according to whether the coil is stationary and the magnet turns, or vice versa.

Perhaps the most highly developed of the first type is the Kelvin astatic galvanometer. This has two magnets equally magnetized but antiparallel mounted on the same suspension, one above the other, and each magnet is surrounded by a coil. The two coils are joined in series and are oppositely wound, so that a current through them will turn their respective magnets in the same direction. The earth's uniform field has no effect upon such an astatic pair of magnets; but there is a large control magnet, placed above the pair, against whose field the current turns the suspended system. The movement is observed by the usual mirror-and-scale or optical lever device. Galvanometers of this type are now used essentially for teaching and demonstration.

Among galvanometers of the second type, that of d'Arsonval is best known. See Fig. 1. The magnet in this instrument is a fixed, permanent magnet of the horseshoe or double-horseshoe form, with a light, rectangular coil suspended in the strong field between its poles, the suspension carrying the feeble current. The current causes the coil to turn in the field. Often a fixed iron core is supported inside the movable coil to concentrate the field.

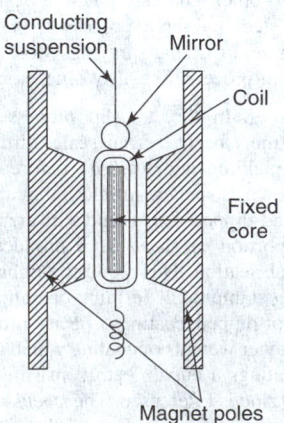

Fig. 1. Essential parts of a d'Arsonval galvanometer. This was a classic instrument that led to a better understanding of electric currents.

If these galvanometers are undamped, they will give a "throw" when a charge of electricity is sent through them, and the charge can be thereby measured. Such an instrument, with a heavy coil, called a ballistic galvanometer, is useful in capacitance measurements. The oscillations may be damped by shunting.

There are also string galvanometers, in which a straight, slender wire carrying the current is thrust to one side by a magnetic field; and vibration galvanometers, in which the string vibrates in synchronism with the alternating current traversing it.

Many of these instruments are classics and principally of historical interest today. See also **Electrical Instruments**.

GAMBLING ODDS. See **Game Theory**.

GAME CLOCK. See **Clocks**.

GAMETE.

A sexual reproductive cell or germ cell, which normally unites with another to produce a new individual. In humans and other mammals, gametes are produced in the gonads — the ovaries and testes — where meiosis takes place. By this process, the number of chromosomes is reduced from the diploid of somatic cells to the haploid number. The diploid number of chromosomes is restored at fertilization when the egg and sperm fuse to form the zygote.

The gametes of some primitive organisms are of one form; these are single cells that swim about in the water. Such organisms are said to be isogamous and the germ cells are called isogametes. In most species, however, only the male gametes retain the power of locomotion. The female gametes are larger inert cells and the organisms are called heterogamous. The male cells of these species are known as sperms or spermatozoa and the female cells as ova or eggs. Because of its smaller size the male gamete is also known as a microgamete, and the female gamete as a megagamete.

The union of two unlike gametes (heterogametes) is called heterogamy. The cell formed by the union of two gametes is called the zygote; from it a new plant or animal develops. The cells or organs in which gametes are formed in plants are called gametangia. In heterogamous plants, the gametangia containing sperms are called antheridia; those containing eggs, either oögonia or archegonia. In animals, the gamete-producing bodies are known as gonads. The male gonads are the testes and the female gonads are the ovaries.

The development of two forms of gametes permits both the freedom of movement necessary to bring the two cells together for fertilization and the storage of the protoplasm and food necessary for the development of any body of reasonable size and complexity to a stage in which it can secure more materials for itself. By the delegation of one function to each kind of cell neither is subject to harmful restriction.

In most species of animals, the sperm is a minute cell with a slender flagellum or tail whose undulating movements propel it through the water or the seminal fluid. The main part of the sperm is the head, which contains an apical body and the nucleus. Behind the head is a neck, or a middle piece of more complex structure, from which the sperm aster involved in the fertilization process sometimes develops. The sperms of some worms and arthropods lack the flagellum although many bear processes of other kinds. They are much less motile than flagellate sperms but they are said to move slowly by amoeboid action or by means of their processes. See also **Amphibia: Reproduction**.

Ova are more compact cells, often spherical in form. They contain abundant cytoplasm and in many species an enormous amount of food material (yolk, deutoplasm), as in the egg of a bird. Here the yolk is the egg cell or ovum proper, but the living protoplasm is a tiny mass at some point on its periphery. Ova may also have special envelopes such as the albumen or white of the bird's egg, the shell membrane, and the shell. In nearly all mammals the ovum has little yolk and becomes separated except at one point from the surrounding layers by a large space filled with fluid. The whole structure is known as a Graafian follicle. Insect eggs are enclosed by a shell called the chorion, which is often beautifully sculptured and strangely shaped. Where such coverings occur, a minute opening, the micropyle, sometimes provides an entrance for the sperm.

ANN C. DEBALDO Ph.D., University of South Florida, Tampa FL

GAMETOPHYTE AND SPOROPHYTE. Plants have two distinct phases in their life cycle. Unlike in animals, where the gametes (sperm and eggs) are produced in reproductive organs (testis and ovary) within anorganism, in plantsthe gametes are produced by a distinct organism, the gametophyte (gamet = gamete, phyte = plant). Gametophytes produce gametes that fuse to form the zygote. The zygote develops into the sporophyte. The sporophyte has no germline, but a few specializedcells on the sporophyte undergo meiosis and the meiotic products (spores) form the gametophyte. Female spores are produced by structures termed megasporangia and male spores are produced by structures called microsporangia. If the megasporangia and microsporangia are morphologically distinct, the plant is termed heterosporous, while if the spores are morphologically identical, the plant is termed homosporous. See also **Gamete**; and **Zygote**.

All angiosperms and gymnosperms are heterosporous. In many homosporous species thehaploid gametophytes that develop from the spores are sexually dimorphic. For example, in the fern *Ceratopteris*, there are two types of gametophyte, hermaphrodites and males, which are morphologically distinguishable. The sex of the *Ceratopteris* gametophyte is determined by a pheromone called antheridiogen. In the presence of this pheromone gametophytes are almost all males, and in its absence the gametophytes are hermaphrodites [Banks]. See also **Angiosperms**; **Ferns**; and **Gymnosperms**.

Alteration of Generations

In order to reproduce sexually, eukaryotic organisms necessarily undergo an alteration between haploid and diploid states. The diploid state, termed the sporophyte, and the haploid state, termed the gametophyte, can be multicellular and free-living, or one phase of the life cycle can be essentially wholly dependent upon the other phase. Perhaps because humans are diploid, there is a tendency to consider a predominantly diploid phase as more advanced. The life cycles of angiosperms and gymnosperms, sometimes termed higher plants, are predominantly sporophytic. However, there are many successful plants with predominantly gametophytic phases (some green algae, mosses), and others in which both phases of the life cycle are substantial (red algae, ferns). See also **Algae**; and **Plant Reproduction**.

Evolutionary Origins

The evolution of alteration of generations in land plants is generally considered to have progressed from a predominantly haplontic life cycle towards a predominantly diplontic life cycle [Valero, et al]. Why is a predominantly diploid life cycle considered more advanced? One advantage of diploidy is that deleterious mutations will be masked, as will any mutations that arise somatically. One advantage of haploidy is that deleteriousmutations will be removed from the population. Mable and Otto review recent theoretical work and state that current understanding is not sufficient to explain the persistence of haploid–diploid life cycles throughout evolution. Comparisons of global gene expression patterns in organisms that maintain isomorphic haploid and diploid phases might prove useful.

Structure: Algae, Mosses and Ferns

Representative life cycles of a green alga, a brown alga, a moss, a fern and an angiosperm are shown in Figures 1, 2, 3, 4 and 5. The gametes in algae are flagellated, and the sperm in mosses and ferns are flagellated and swim to the nonmotile egg. In gymnosperms and angiosperms, the sperm are nonmotile and are carried to the egg by the growth of the pollen tube.

Reduction of Gametophyte

The gametophyte and sporophyte can be isomorphic and equivalently sized and able to exist as free-living organisms (some algae, mosses). In manymosses, the gametophyte phase of the life cycle is dominant, while in ferns, the sporophyte is the dominant phase of the life cycle although early in its development it is dependent on the gametophyte. In angiosperms, the gametophyte generation is highly reduced and develops within tissues of the sporophyte.

Pollen Grains and Embryo Sacs

The male gametophyte in angiosperms is the pollen grain. Male gametophyte development starts with the anther, when sporogenous initial cells divide to give rise to a tapetal initial cell and a microspore mother cell.

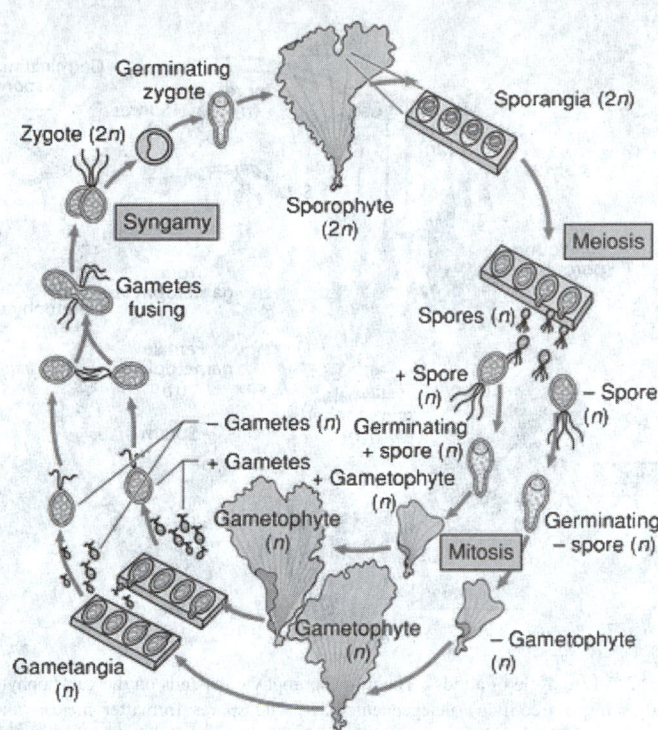

Fig. 1. Life cycle of a green alga, *Ulva*. The sporophyte and gametophyte are multicellular and look identical (isomorphic). The gametophyte and sporophyte are both two cell-layers thick. Cells on the sporophyte undergo meiosis and produce two types of spores, + and −. Each spore divides mitotically to give rise to a gametophyte (either + or −). The gametophyte produce (+) and (−) gametes. A (+) gamete fuses with a (−) gamete and forms the zygote. The zygote divides mitotically to give rise to the sporophyte.

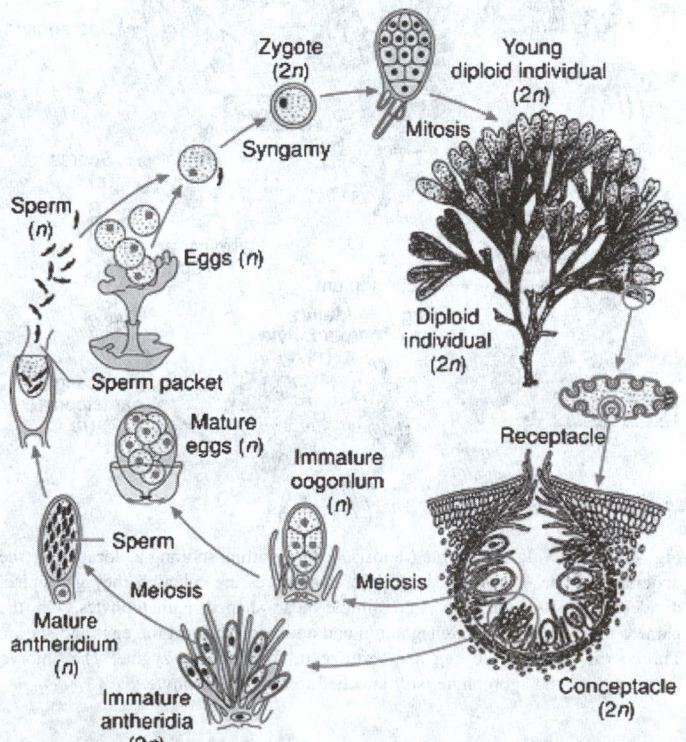

Fig. 2. Life cycle of a brown alga, *Fucus*. The diploid individuals are large and branched. The tips of the branches are called receptacles. A cross-section of the receptacle reveals its structure — meiosis occurs within the semi-circular sacscalled conceptacles. The oogonia form large eggs and the antheridia form small sperm. These gametes are released and fertilization occurs. The fertilized eggs (zygotes) attach and grow into diploid individuals.

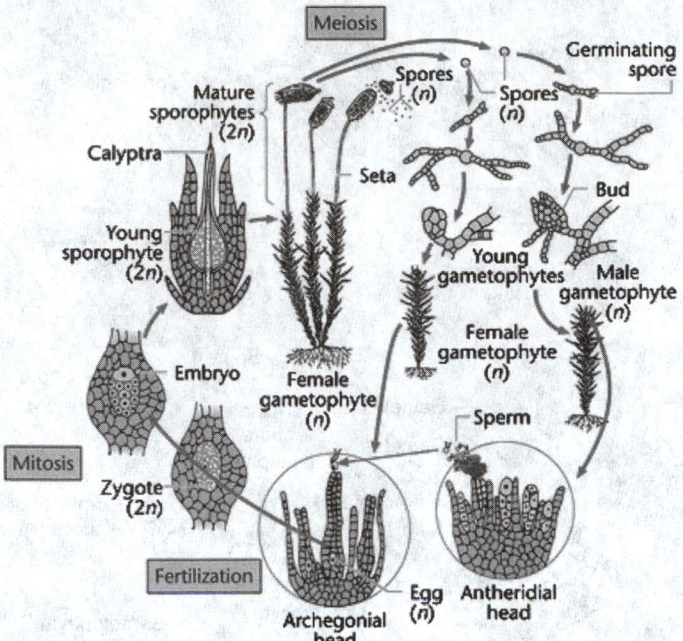

Fig. 3.   Life cycle of a moss. The moss sporophyte depends on the gametophyte and is not a free-living independent plant. The spores form after meiosis and germinate in order to develop into male or female gametophytes. Moss sperm are produced from cells in the antheridia of the male gametophyte. The moss eggs are located at the base of the archegonia, on the female gametophyte. Each sperm has two flagella. Sperm swim to the eggs and after fertilization, the zygote develops into the sporophyte.

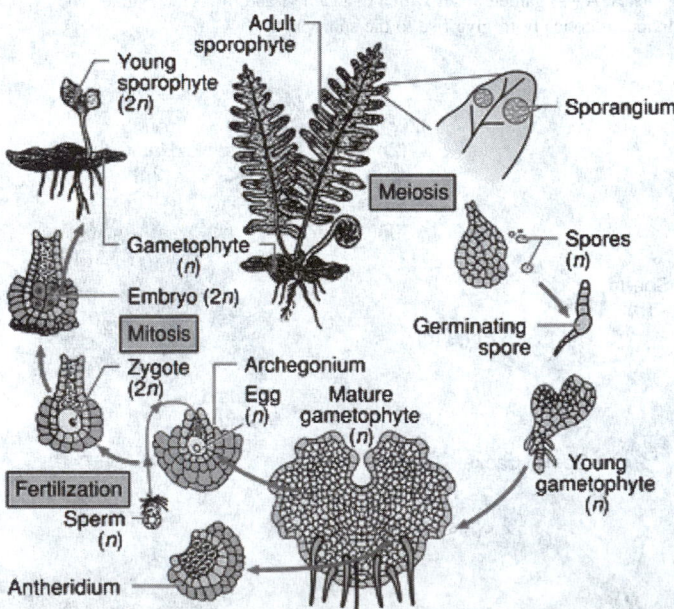

Fig. 4.   Life cycle of a fern. Meiosis occurs within sporangia, located on the underside of the sporophyte leaf. After the spores are released they germinate, divide by mitosis and grow into simple heart-shaped gametophytes. On the gametophyte, cells in the archegonium and antheridium form the eggs and sperm. The sperm swim to the egg and fertilize it, forming the zygote. The embryo develops into the sporophyte, still attached to the gametophyte.

Each tapetal initial cell will divide mitotically to eventually give rise to the tapetum, a nurse tissue that lines the locule containing the developing microspores. The diploid microspore mother cell undergoes meiosis. Meiosis yields a tetrad of four haploid cells enclosed within a callose wall. After the callose wall is dissolved, each haploid cell is termed a uninucleate microspore. Each microspore undergoes two subsequent mitoses to form the three-celled gametophyte, the pollen grain, composed of two sperm

cells enclosed within the cytoplasm of the vegetative cell. Pollen grains are released from the anther, usually in a partially dehydrated state, and disperse by wind, insect vectors or other means. See also **Pollen: Structure, Development and Function**.

When the pollen grain contacts the female, the pollen grain hydrates and the vegetative cell extends a tube through one of the apertures and grows, via tip growth, through the extracellular matrix of the pistil, in order to deliver the two sperm cells to the embryo sac. In many plants the pollen grain is released from the microsporangia (anther) in the two-celled state, containing the vegetative cell and the enclosed generative cell, and the generative cell undergoes the second mitosis while the pollen tube is growing through the female tissue.

The female gametophyte in angiosperms is the embryo sac. Female gametophyte development initiates within the ovule. The megaspore mother cell undergoes meiosis, but, in contrast to male meiosis, in most plants only one of the four resulting megaspores continues to develop, while the other three degenerate.

The final cellular composition of the embryo sac is quite diverse among plants, but about 70% of plant species follow the pattern described for *Polygonum*. The megaspore undergoes three mitoses, followed by nuclear migration and cellularization, to give rise to a seven-celled embryo sac. The embryo sac is polar, with the antipodal cells at one end, and the egg cell and two synergid cells at the other end, and a 2 N (bi-nucleate) centralcell. In some plant species the nuclei in the central cell partially fuse before the sperm arrives. Other variations within this basic *Polygonum* division pattern can occur — in some plant species the antipodal cells can undergo extra mitoses (maize) or can degenerate (*Arabidopsis*). The pollen tube enters the embryo sac through one of the two synergid cells, releasing its two passenger sperm cells.

It is interesting that the basic plant life cycle can be modified and some steps bypassed. For example, in certain angiosperms, removing the developing male gametophyte from the anther and culturing it in a medium depleted for nutrients can induce the formation of a haploid plant with the morphological appearance of a sporophyte [Reynolds]. The chromosome number of such haploid plants can be doubled, either by chemical treatment or spontaneously, and the resulting doubled haploid plants are homozygous and therefore useful in plant breeding. Thus, in angiosperms, ploidy changes are clearly not necessary for the drastic changes in morphology seen between the sporophyte and gametophyte.

Embryo and endosperm development can occur without fertilization, a process termed apomixis [Koltunow, et al.]. In sporophytic apomixis, unreduced (i.e. 2N) cells of the embryo can form an embryo, a process termed adventitious embryony; this is common in citrus species. In gametophytic apomixis, an unreduced cell in the embryo sac forms an embryo. In some cases, pollination is required to trigger apomictic development, and frequently one sperm must fuse with the central cell, although the embryo is able to develop without fertilization.

Hieracium is an apomictic plant. It is interesting to note that Mendel tried to use *Hieracium* to repeat the genetic segregations he had observed with pea, but was completely confused by the results because *Hieracium* reproduces by apomixis. Apomixis produces plants that are genetically identical to the parent plant and thus is of great interest to plant breeders — desirable genotypes, even hybrids, can be maintained without crossing. Apomixis is characteristic of polyploid plants, but in at least some species appears to be under the control of one or a few genes [Kindiger, et al.]. Genetic screens for apomictic mutants in *Arabidopsis* have thus far yielded mutants in which the endosperm develops without fertilization [Chaudhury, et al.]. See also **Mendel, Gregor Johann (1822–1884)**; and **Plant Breeding** and **Crop Improvement**.

Angiosperms undergo double fertilization. The sperm are carried to the embryo sac by the pollen tube. The sperm cells are frequently closely associated with the vegetative (tube) nucleus as they move down the tube, and this associated group is termed the male germ unit. The pollen tube enters through one of the two synergid cells adjacent to the egg cell. It is believed that there are both sporophytic and gametophytic factors that determine the selection of which synergid will be entered by the pollen tube [Russell]. In some plants one of the two synergids begins to degenerate before the pollen tube arrives. One sperm fertilizes the egg cell to form the zygote, the beginning of the sporophyte generation. The other sperm cell fuses with the diploid central cell to give rise to a triploid endosperm cell. This initial endosperm cell divides mitotically to give rise to the endosperm, a nutritive tissue for the developing zygote.

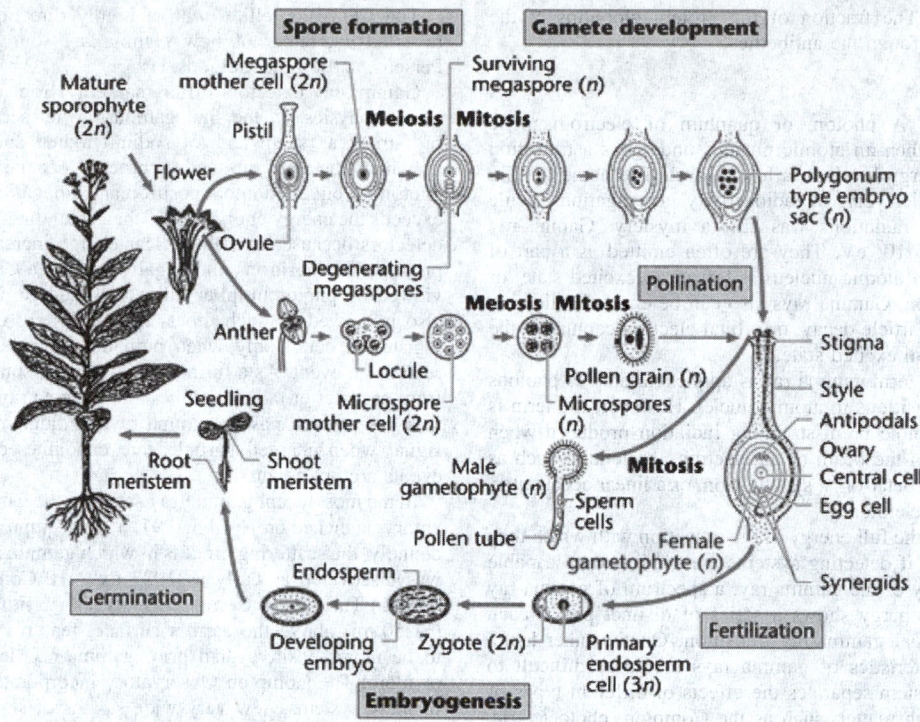

Fig. 5. Life cycle of an angiosperm. The mature sporophyte produces flowers. Flowers contain male (stamens) and female (pistil) reproductive organs. In the pistil, the megaspore mother-cells in the ovule undergo meiosis. Only one of the four products of meiosis survives and eventually develops into the eight-celled female gametophyte, containing the egg cell, the central cell, and a few other cells (antipodals and synergids). Within the anther, microspore mother-cells undergo meiosis and form four haploid microspores. Unlike in the female, each microspore develops into a male gametophyte, or pollen grain, composed of only three cells. The pollen grain contains two sperm cells. Pollen grains land on the stigma surface of the pistil, grow a tube through the extracellular matrix of the style and arrive at the ovary by bursting through one of the synergid cells. The sperm cells cannot swim, but are delivered to the embryo sac by growth of the pollen tube. One sperm cell fuses with the central cell to form the primary endosperm cell and the other sperm cell fuses with the egg to form the zygote. The zygote develops into the sporophyte. Growth of the sporophyte is the result of mitotic divisions of cells in the root and shoot meristems.

Additional Reading

Banks, J.A.: "Gametophyte Development in Ferns," *Annual Review of Plant Physiology and Plant Molecular Biology*, **50**, 163–186 (1999).

Chaudhury, A.M., S. Craig, E.S. Dennis, and W.J. Peacock: "Ovule and Embryo Development, Apomixis and Fertilization," *Current Opinion in Plant Biology*, **1**, 26–31 (1998).

Kasha, K.J., M. Maluszynski, and B.P. Forster: *Doubled Haploid Production in Crop Plants*, Springer-Verlag New York LLC, New York, NY, 2004.

Kindiger, B., D. Bai, and V. Sokolov: "Assignment of a Gene Conferring Apomixis in *Tripsacum* to a Chromosome Arm: Cytological and Molecular Evidence," *Genome*, **39**, 1133–1141 (1996).

Koltunow, A.M., R.A. Bicknell, and A.M. Chaudhury: "Apomixis: Molecular Strategies for the Generation of Genetically Identical Seeds without Fertilization," *Plant Physiology*, **108**, 1345–1352 (1995).

Lesyer, O., and S. Day: *Mechanisms in Plant Development*, Blackwell Publishers, Malden, MA, 2003.

Mable, B.K., and S.P. Otto: "The Evolution of Life Cycles with Haploid and Diploid Phases," *BioEssays*, **20**, 453–462 (1998).

Raghavan, V., and J.B. Bard: *Developmental Biology of Fern Gametophytes*, Cambridge University Press, New York, NY, 2005.

Reynolds, T.L.: "Pollen Embryogenesis," *Plant Molecular Biology*, **33**, 1–10 (1997).

Roberts, J.A., and Z. Gonzalez-Carranza: *Plant Cell Separation and Adhesion*, Blackwell Publishing, Malden, MA, 2007.

Russell, S.D.: "Double Fertilization," *International Review of Cytology*, **140**, 357–388 (1992).

Valero, M., S. Richard, V. Perrot, and C. Destombe: "Evolution of Alternation of Haploid and Diploid Phases in Life Cycles," *Trends in Ecology and Evolution*, **7**, 25–29 (1992).

SHEILA MCCORMICK, University of California, CA

GAMMA DISTRIBUTION. See **Pearson Distributions**.

GAMMA FUNCTION. The infinite integral, sometimes called Euler's second integral:

$$\Gamma(z) = \int_0^\infty e^{-t} t^{z-1} \, dt$$

It converges for all positive, real values of z. Its properties include:

$$\Gamma(z+1) = z\Gamma(z)$$

$$\Gamma(z)\Gamma(1-z) = \pi \csc \pi z$$

$$\Gamma(\tfrac{1}{2}) = \sqrt{\pi}$$

when $z = n$, a positive integer, $\Gamma(n) = (n-1)!$, hence this is often called the factorial function.

The Weierstrass definition of the function is

$$1/\Gamma(z) = ze^{Cz} \prod_{n=1}^\infty (1 + z/n) e^{-z/n}$$

where C is the Euler-Mascheroni constant:

$$C = \lim_{n\to\infty} (1 + \tfrac{1}{2} + \cdots + 1/n - \ln n)$$

$$= 0.577215$$

Another definition is that of Euler:

$$\Gamma(z) = \lim_{n\to\infty} \frac{(n-1)!}{z(z+1)(z+2)\cdots(z+n-1)} n^z$$

See also **Beta Function**, which is Euler's first integral.

SIR MAURICE KENDALL

GAMMA GLOBULIN. The fraction of the protein globulins of the blood plasma in which are found the antibodies.

See also **Blood**.

GAMMA RADIATION. A photon, or quantum of electromagnetic radiation, that is emitted when an atomic nucleus undergoes a transition from one of its excited energy levels to a lower level. The name *gamma ray* was applied in the earlier years of radioactivity investigations, while the exact nature of these radiations was still a mystery. Gamma-ray energies range from 10^4 to 10^7 eV. They are often emitted as a part of a nuclear reaction, when an atomic nucleus is left in an excited state, or during an isomeric transition. Gamma rays also can be emitted following alpha-particle decay, beta-particle decay, or orbital electron capture, if the daughter nuclide is left in an excited state.

In the strictest sense, the term gamma ray is applicable only to photons produced as a result of transitions in atomic nuclei. However, the term is also sometimes used to denote bremsstrahlung radiation produced when the high energy electrons in the beam of an electron accelerator, such as an electrostatic generator, a betatron, a synchrotron, or a linear accelerator, strike the target of that accelerator.

Gamma rays carry away the full energy of the transition with which they are associated. As a result, if detecting systems are used that are capable of absorbing the full energy of the gamma ray, a spectrum of gamma-ray numbers as a function of energy shows a series of distinct peaks, each associated with an individual gamma-ray transition. On the other hand, the discrete energy characteristics of gamma rays are more difficult to observe if the detecting system separates the effects of different types of gamma-ray interactions with matter, such as the Compton, photoelectric, and pair-production interactions. Under certain circumstances, a transition that would normally be expected to emit a gamma ray may sometimes release its energy through an internal conversion process.

See also **Particles (Subatomic)**; and **Radioactivity**.

GAMMA-RAY ASTRONOMY. The study of cosmic objects and systems based upon the detection and measurement of gamma-ray emissions received from such objects. Gamma rays are in the high-energy portion of the electromagnetic spectrum, in the frequency range of 10^{20}–10^{21} and wavelength range of 10^{-9}–10^{-11} and thus a position in the spectrum between cosmic rays and x-rays. Whereas the photon energy of visible light rays is 23 eV or 10 eV for ultraviolet rays, gammas rays carry energy ranging from 10,000 to trillions of eV. See also **Electromagnetic Phenomena**.

Gamma radiation from celestial objects represents a vast wealth of information that cannot be revealed by instruments that operate in other portions of the electromagnetic spectrum. Further, gamma rays from interstellar space are difficult to measure reliably from locations on Earth because the emissions are lost amid the confusion of gamma rays created in the atmosphere by cosmic-ray bombardment. Thus, in the beginning of gamma-ray astronomy, useful information had to be collected from instruments sent aloft in high-altitude balloons. Later, a number of satellites were launched. These included the Second Small Astronomy Satellite (SAS-2), the COS-B satellite launched by the European Space Agency, and the High Energy Astronomy Observatory (HEAO-1). Researchers in the 1970s and 1980s became fully aware of the value of gamma-ray exploration in finding and learning of new fundamental concepts of the universe. As explained in the article on **Galaxy**, gamma-ray information is indispensable to formulating concepts on how the universe was formed and how it has operated during intervening eons of time. Among the most interesting of early observations from satellites were the very bright emissions received from the plane of our own galaxy, the Milky Way.

The Crab Nebula, of which the fastest radio pulsar known is also a part, was one of the first point sources of gamma rays observed. Data from SAS-2 indicated that there is diffuse gamma radiation coming from throughout the sky. Much remains to be understood about this phenomenon. With the combination of radio and gamma ray observations, some of the most active regions of star formation and cosmic ray production have been identified. These regions appear to be located about midway between the Sun and the center of our galaxy. This ring is located some 15–20,000 light years from the galactic center. This distribution of cosmic rays correlates well with the region of highest concentration of supernova remnants and pulsars — the latter believed also to have resulted from supernovae. Thus, cosmic rays are no longer considered to be mainly extragalactic in origin, but rather they are generated within the galaxy as well.

The COS-B satellite, with a highly directional gamma ray detector, revealed a number of new gamma ray sources in the Gemini-Taurus, Perseus, and Cygnus regions.

Gamma-ray detectors differ markedly from other energy sensors used in astrophysics. In the first gamma-ray telescopes, an incoming gamma ray struck a "sandwich" of sodium iodide and cesium iodide material, in which the pair production process occurs, whereby the gamma-ray photon is converted into a positron and a negatron, provided that its energy exceeds the energy equivalent of their total mass. Cesium was used in early detectors because its high nuclear charge increases the probability of the process. The positrons and negatrons enter a Cerenkov detector, which is viewed by photomultiplier tubes. The sodium-iodide-cesium-iodide layer also is viewed by such tubes. Both units are connected to a circuit that registers a count only when both the layers and the Cerenkov detector records an event. As a further guard against spurious counts (those arising from particles or radiation other than gamma rays), the system is contained in a case, which passes gamma rays without reaction, but which gives a signal when charged particles are encountered. Thus, such coincidence events are not counted.

In the most recent gamma-ray satellite, the Compton Gamma Ray Observatory, launched on April 5,1991, a much-improved detector is used. (Incidentally, the scattering process in which gamma rays richochet off electrons was discovered as early as 1923 by A.H. Compton. See also **Compton Effect**.) The current Compton Observatory, situated in orbit some 400 km ($\sim$250 mi) above the earth's surface, features instrumentation estimated to be more sensitive than prior gamma-ray detectors by a factor of ten or more. The Compton Observatory incorporates four instruments. Three of these systems view very wide swaths of sky, and they are pointed by turning the entire spacecraft. As pointed out by researchers N. Gehrels (National Aeronautics and Space Administration, Goddard Space Flight Center) and a team of other experts (See reference listed), the COMPTEL (Imaging Compton Telescope) views a 64-degree wide circular patch of sky. The EGRET (Energetic Gamma Ray Experiment Telescope), which gathers the highest-energy gamma rays, has a slightly smaller (45°) field. The OSSE (Oriented Scintillation Spectrometer Experiment) surveys a relatively small, $4 \times 11°$ field of view. As explained by the researchers, "The OSSE can quickly point toward and away from a particular gamma-ray source, thereby enabling researchers to subtract the background noise in OSSE's detectors from the source signal." See also **Compton Gamma-Ray Observatory (CGRO)**.

The BATSE (Burst and Transient Source Experiment) is comprised of eight units, one on each corner of the satellite. These view half of the sky that is not blocked by the earth. The BATSE is designed to explore those mysterious *gamma bursts* that have been detected several times since the beginnings of gamma-ray astronomy, but to date have defied explanation. See also **Cosmic Rays**. Initial research has shown that the satellite may be viewing the edge of the population of such bursts. As pointed out by the Compton Observatory team, "Theorists have proposed many exotic explanations for the BATSE results. A few workers have suggested that the bursts result from collisions between comets or from other events lying just outside the planets in our solar system, but the mechanisms by which cometary collisions would generate gamma rays seems rather implausible. Another, more widely held possibility is that bursts occur on neutron stars that lie not in the disk of the galaxy but in a huge, outlying halo. Such models require elaborate ad hoc assumptions about the size and shape of the halo, however. They also raise the question of why neutron stars in the galactic disk do not produce significant numbers of bursts."

Much more detail is given in the Gehrels et al. reference listed.

Additional Reading

Dermer, C.D. and R. Schlickeiser: "Quasars, Blazars, and Gamma Rays," *Science*, 1642 (September 18, 1992).

Flam, F.: "Gamma-Ray Observatory: Bursting with New Results," *Science*, 34 (October 4, 1991).

Gehrels, N., et al.: "The Compton Gamma Ray Observatory," *Sci. Amer.*, 68 (December 1993).

Hurley, K.: "Probing the Gamma-Ray Sky," *Sky and Telescope*, 631 (December 1992).

Kniffen, D.A.: "The Gamma-Ray Universe," *Amer. Scientist*, 342 (July-August 1993).

Kurfess, J.D. et al.: "Oriented Scintillation Spectrometer Experiment Observations of 57CO," *Astrophysical J.*, **399**(2), L137 (November 10, 1992).

Meegan, C.A., et al.: "Spatial Distribution of Gamma-Ray Bursts Observed by BATSE," *Nature*, **355, 6356**, 143 (January 9, 1992).

Powell, C.S.: "Live from Off-Center: Astronomers Follow the Energetic Trail of the Great Annihilator," *Sci. Amer.*, 29 (July 1991).

Powell, C.S.: "Star Bursts: The Deepening Mystery of the Gamma-Ray Sky," *Sci. Amer.*, 32 (December 1991).

Primack, J.R.: "Gamma-Ray Observations of Orbiting Nuclear Reactors," *Science*, 407 (April 28, 1989).

Rieger, E. et al.: "Man-Made Transients Observed by the Gamma-Ray Spectrometer on the Solar Maximum Mission Satellite," *Science*, 441 (April 28, 1989).

Share, G.H. et al.: "Geomagnetic Origin for Transient Particle Events from Nuclear Reactor-Powered Satellites," *Science*, 444 (April 28, 1989).

Waldrop, M.W.: "Space Reactors Hinder Gamma-Ray Astronomy," *Science*, 1119 (November 25, 1988).

GAMMA-RAY BURSTS. Gamma ray bursts (GRBs for short) are intense and short (approximately 0.1–100 seconds long) bursts of gamma-ray radiation that occur all over the sky approximately once per day at very large distances from Earth. Gamma rays are very energetic photons (E > 10^5 eV), which represent the most extreme portion of the electromagnetic spectrum (ranging from radio waves at the lowest energies through visible optical light at higher energies, to gamma rays at the highest energies). They are detected at the rate of about once a day, and while they are on, they outshine every other gamma-ray source in the sky, including the sun. The naming system for gamma ray bursts is very simple: "GRB yymmdd". For example, a gamma ray burst which occurred on July 4, 1999 is called GRB 990704. If there is more than one gamma ray burst on the same day, the letter a, b, c, etc. are added to the name (for example, the second gamma ray burst on July 4, 1999 is called GRB 990704b).

GAMMA-RAY LARGE AREA SPACE TELESCOPE (GLAST) MISSION. See **Space Science Missions: Universe**.

GAMMA-RAY SPECTROSCOPY. Gamma rays of concern here originate in the nucleus of radioactive isotopes, i.e., chemical elements whose nuclei are unstable and emit radiation as they decay to stable states. Such radioactive isotope disintegration follows rules that are always the same for the nucleus. These rules can be set down in a so-called decay scheme. An example is shown in Fig. 1 for the case of the radioisotope ^{137}Cs (cesium-137). The basic decay scheme shown indicates that cesium-137 decays into ^{137}Ba (barium-137) by emitting beta particles (electrons). Eight percent of the cesium nuclei decay directly into barium-137 nuclei; then about 2.5 minutes later, the excited nuclei decay to the lowest energy or ground state by emitting gamma rays having an energy level of 662 keV. Some heavy nuclei emit alpha particles. An alpha particle is a ^{4}He (helium-4) nucleus (two protons and two neutrons). The cesium-137 isotope, with a nucleus containing a total of 137 neutrons and protons, disintegrates with a half-life of 30 years. Since the number of nuclei is halved, the amount of radiation (intensity) is halved. With existing electronic systems, half-lives between 10^{-10} second and 10^{10} years can be measured.

calculate the probable accuracy of his result, assuming no instrumentation inaccuracy.

Gamma Ray Detection. Gamma rays are high energy electromagnetic radiation with very short wavelengths (10^{-18} to 10^{-11} cm). They penetrate matter deeply—on the average much more deeply than do alpha and beta rays, which are charged particles. It is their deep penetration that makes gamma rays useful in the laboratory and industry, in much the same way as X-rays. X-rays originate from shell transitions by orbital electrons, whereas gamma rays originate in the nucleus. Gamma rays usually are detected by observing effects that they produce in matter and when they encounter an atom. Important among these effects are: (1) the photoelectric effect; and (2) the Compton effect. The photoelectric effect occurs when the gamma ray strikes one of the orbital electrons of the atom, transferring its energy to the electron. This process produces a free electron and an ionized atom. The Compton effect arises in the case where the gamma ray strikes an orbital electron without imparting all of its energy to the electron. The electron is detached from the atom but receives only part of the gamma energy. The remaining energy persists as a scattered gamma ray with lower energy than the initial ray. This scattered ray may further collide with one or more other atoms, freeing other electrons. These types of interactions occur variously in nuclear radiation detectors. In each detector type, some observable reaction results, and in one manner or another produces an electrical output charge suitable as input for an electronic measuring system.

Gamma Ray Spectra. Measurements of gamma radiation are chiefly made in two ways: (1) a record is made of the number of counts as a function of energy, in which case a gamma ray spectrum is obtained; and (2) time relations are observed, in which case several types of information may be desired. A gamma spectrum, as measured by an ideal system, might appear as in Fig. 2. This is the ideal spectrum of the cesium-137 gamma radiation phenomena discussed earlier. In this spectrum, a large peak appears at 662 keV—caused by the gamma energy radiated when the metastable barium-137 nucleus returns to its ground state. There is also a continuum representing the energies imparted to Compton-scattered electrons. In practice, the spectra measured are not so well defined. See Fig. 3. Most noticeable is that the peaks of the spectrum are broadened to a greater or lesser extent by the characteristics of the devices used to detect gamma rays. Relating to this broadening as a measure of system quality, is its "resolution." This is a function both of the detector and of the associated circuitry. Resolution commonly is defined as: the ratio of the full width at half the maximum height of the peak (FWHM) to the energy of the center of the peak. Thus, resolution indicates how well the detector can separate or resolve two different energy peaks. Typical resolutions for common gamma ray detectors range from about 10% to a few tenths of a percent. Also evident in Fig. 3 is a backscatter peak, which results because

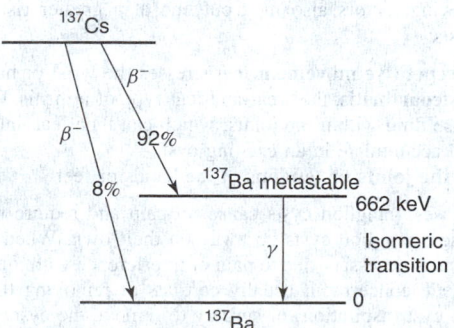

Fig. 1.   Decay scheme for ^{137}Cs.

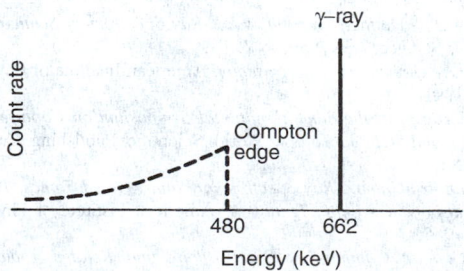

Fig. 2.   Ideal gamma spectrum of ^{137}Cs.

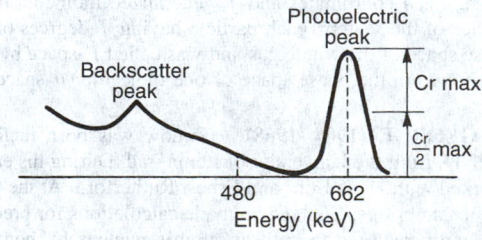

Fig. 3.   Typical gamma spectrum of ^{137}Cs.

Like most natural events, radioactive decay is not a uniform function. Consequently, the term *half-life* is meant to describe the value that would result if an infinite number of half-life measurements were made and the average calculated. Individual decays, however, follow a Poisson distribution, i.e., the standard deviation is equal to the square root of the number of observed decay events. This fact enables the experimenter to a large number of gamma rays squarely strike matter between the source and the detector, losing much of their energy before detection.

Energy Measurements. The measurements usually made in gamma ray work fall into two broad groups: (1) those made of the energy of the radiation; and (2) those made of its timing relative to another event. In addition, counting without regard to energy (often called gross counting) is also done to measure the intensity of the radiation. See Fig. 4. Intensity is measured in terms of counts/minute (or second).

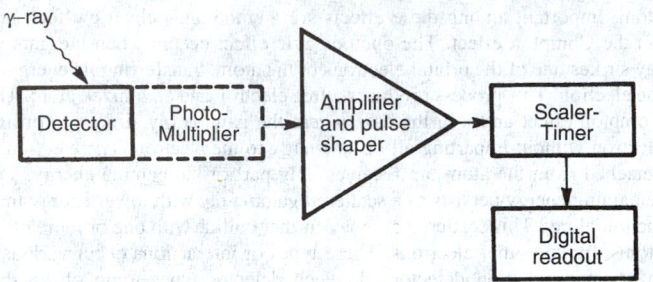

Fig. 4.   Gross counting measures radiation intensity of gamma ray regardless of energy.

Time Measurements. The second general class of measurements is one in which the time of occurrence of the gamma ray relative to a reference event is of interest to the experimenter. Such situations occur when gamma radiation is known to occur a specific interval of time after a trigger event.

Detectors. Commonly used detectors include scintillation, semiconductor, and gas proportional detectors. The scintillation detector often is preferred where high efficiency is more important than resolution — efficiency defined as a measure of the probability that an incident gamma ray will interact with the material in the detector. Semiconductor types are used increasingly, particularly where high resolution is required.

Signal Processing. The signal from the detector is a relatively short current pulse; the time integral of this current impulse is a charge proportional to the energy of the absorbed radiation. The preamplifiers and amplifiers which follow these detectors convert this impulse of a charge into a voltage pulse whose height (peak amplitude) is proportional to energy. Thus, signal processing prepares the charge from the detector for the final step, pulse height analysis. In the case of a timing measurement, signal processing prepares the charge signal for use with a timing pick-off (time discriminator). See also **Radioactivity**.

Additional Reading

Debertin, K. and R. Helmer: *Gamma and X-Ray Spectromety Semiconductor Dectors*, Elsevier Science, New York, NY, 1999.

Hoff, R.: *Capture Gamma-ray Spectroscopy*, American Institute of Physics, College Park, MD, 1991.

Kern, J.: *Proceedings of the 8th International Symposium on Capture Gamma-Ray Spectroscopy and Related Topics*, World Scientific Publishing Company, Inc., Riveredge, NJ, 1994.

Raman, S.: *Capture Gamma-Ray Spectroscopy and Related Topics, 1984: International Symposium, Knoxville, Tennessee*, American Institute of Physics, College Park, MD, 1985.

Wender, S.: *Capture Gamma-Ray Spectroscopy and Related Topics*, American Institute of Physics, College Park, MD, 2000.

GAMMA SPACE. Phase space of $2fN$ dimensions, the coordinates being f generalized coordinates and f generalized momenta for each of the N particles of the system, each particle having f degrees of freedom. It is the phase space of the whole gas and was called Γ-space by Ehrenfest to distinguish it from the phase space of one molecule (μ-space).

GAMOW, GEORGE (1904–1968). Gamow was born in Russia and earned his Ph.D. from the University of Leningrad. During his early career years he worked with Niels Bohr and Ernest Rutherford. At the Cavendish Laboratory in Cambridge, England he made calculations for predicting the amount of energy required to split an atom's nucleus by bombarding it with protons.

In 1933 he defected from the Soviet Union and came to America and began teaching physics at George Washington University and began researching in astrophysics. He began to explain the dynamics of supernovas and behaviors of red stars. In 1952 published *The Creation of the Universe*.

In his latter years he worked on molecules and hereditary information proposing a coding theory for protein synthesis.

See also **Gamow-Teller Selection Rules**.

J. M. I.

GAMOW-TELLER SELECTION RULES. A set of selection rules for beta decay which state that an allowed transition between parent and daughter states must have no change of parity but can have a spin-quantum-number change of either 0 or 1, except that no $0 \rightarrow 0$ transitions are allowed. See also **Beta Decay**.

GANGLIA. See **Central and Peripheral Nervous Systems**.

GANGLION. In zoology, a ganglion is a small mass of nervous tissue isolated from the central system but containing cell bodies as well as fibers. Many ganglia bear special names. The brain of many invertebrates, for example, is also called the cerebral ganglion, and the more numerous centers of the molluscan nervous system bear names, such as the visceral ganglia and the pedal ganglia. The dorsal root of each nerve arising from the vertebrate spinal cord bears a spinal ganglion and the sympathetic system contains numerous ganglia.

GANGLION CYSTS. Ganglia are cystic swellings under the skin, generally in the wrist or the upper surface of the foot. They may be soft or quite hard, and they are usually either painless or only somewhat bothersome. A ganglion develops when a jelly-like substance accumulates in one of two places — a joint capsule or a tendon sheath — and causes it to balloon out. It results from an accumulation of the jelly-like substance that has leaked from a joint or tendon sheath.

In most cases, ganglion cysts appear as raised lumps near the wrist or finger joints. Though they are fixed in one place, they may "give" a little when pushed.

On average, ganglion cysts are about 3 centimeters (1 inch) in diameter. They may change somewhat in size depending on one's activity level, becoming larger when the affected joint is used and growing smaller when it is at rest.

Most ganglion cysts are painless. In some cases, however, the cysts may put pressure on the nerves near the joint, which can cause pain, weakness or numbness in the hand.

Occasionally, the telltale lump that indicates a ganglion cyst is not visible. Often the only indication of these smaller, "hidden" ganglion cysts (occult ganglions) is pain.

Ganglion cysts are more common in women than they are in men.

The following factors also may put one at a greater risk of forming ganglion cysts:

- **Regular, repetitive movement** that stresses the wrist or hand.
- **Having osteoarthritis**, the wear-and-tear type of arthritis. Osteoarthritis can increase fluid within the joints, which can then leak into the tendon sheath and accumulate in an existing cyst.
- **Injury to the joints or tendons** of the hands or feet.

In most cases, ganglion cysts cause no pain and require no treatment. In many cases, ganglion cysts go away on their own. When treatment of ganglion cysts is necessary due to pain or interference with joint movement or for cosmetic concerns, it usually consists of removing the fluid from the ganglion cyst (aspiration) or surgery to remove the cyst.

GANGLIOSIDES. Identified by Kleng in 1935, the gangliosides are a family of acidic glycolipids that are characterized by the presence of sialic acid. The compounds bear a strong negative charge and are unusual in that they contain both hydrophobic and hydrophilic regions. These compounds are membrane components. Plasma cell membranes are rich with gangliosides. It has been suggested that gangliosides participate in the transmission of membrane-mediated information in living systems. As described by Fishman and Brady, "the carbohydrate portion of gangliosides is made up of molecules of sialic acid, hexoses, and *N*-acetylated

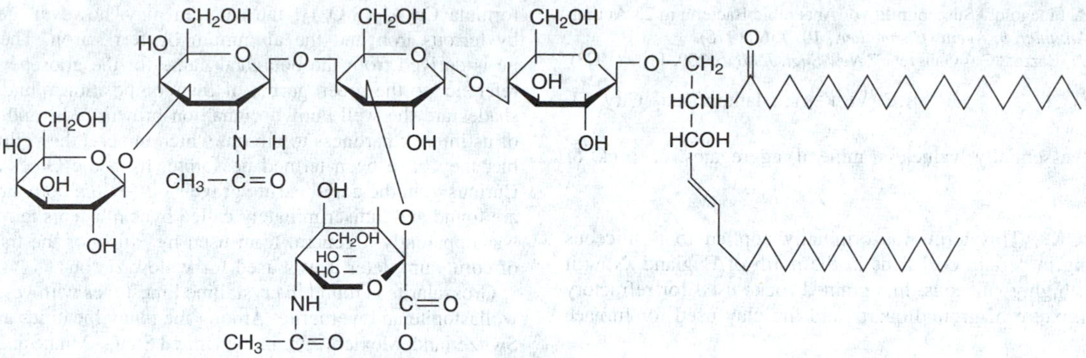

Fig. 1.   Configuration of monosialoganglioside G_{M1} as suggested by Svennerholm.

hexosamines. The hydrophobic moiety is called ceramide, and it consists of a long-chain fatty acid linked through an amide bond to the nitrogen atom on carbon 2 (C-2) of the amino alcohol, sphingosine. Oligosaccharides are linked through a glycosidic bond to C-1 of the sphingosine portion of ceramide." Svennerholm (1963) suggested the configuration given in Fig. 1. The role of gangliosides is still rather discrete, but Fishman and Brady (1976) studied in some detail the interaction of cholera toxin with ganglioside-deficient cells. They also studied, the interaction of cholera toxin with ganglioside-deficient cells, as well as the interaction with glycoprotein hormones and their effect on the action of these hormones.

Additional Reading

Fishman, P.H. and R.O. Brady: "Biosynthesis and Function of Gangliosides," *Science*, **194**, 906–915 (1976).

Kleng, E.: *Z. Physiol. Chem.*, **235**, 24 (1935).

GANGRENE.   The death of localized tissue, frequently involving the extremities—fingers, arms, toes, feet, and, in some instances, the ears, nose, and cheeks. Depending upon the cause, however, gangrene may occur in several parts of the body, including lungs, colon, among others. Gangrene may result from physical causes, where in some manner the circulation of blood is stopped or greatly impaired to certain organs. Thus, in cases of injury, severe crushing of tissues may destroy their viability by interfering with the circulation. Inflammation of an area may be so intense as to shut off circulation by strangulation of blood vessels. Circulation may be impaired by thrombosis or clotting. Gangrene may result from arrest of circulation, however produced, as is seen in various diseases causing obstruction of arteries or veins. Examples are severe hardening of the arteries (arteriosclerosis). Chemical and physical agents, including corrosives such as phenol, or prolonged exposure to heat or cold (frostbite) may cause local death of tissue. Nearly all forms of gangrene are accompanied by some kind of infection, which spreads the condition. Diabetics with tissue damage in the extremities are at special risk because gangrene can spread through local endarteritis obliterans, causing vascular damage and leading to "wet" gangrene (as distinct from "dry" gangrene resulting from simple ischemia of uninfected tissues). In serious cases, amputation of extremities or parts thereof may be indicated.

In cases of gas *gangrene*, there is infection of tissues around a wound by certain anaerobic bacteria, commonly *Clostridium perfringens* (formerly known as *C. welchii*). The infection is necrotic and rapidly spreading, usually accompanied by massive edema, gaseous infiltration, and discoloration of tissues. The organisms liberate a toxin, a phospholipase, which destroys tissue, particularly muscle, and they produce gas by fermenting muscle sugars. Much of the early information on gas gangrene, also referred to as clostridial myositis, was obtained during World War I. Many of the war wounds were infected with gas-producing organisms, often from contamination either directly or indirectly with fecal matter contained in the soil. The incidence of gas gangrene after trauma largely reflects the speed with which wounded people can be evacuated and receive surgical debridement. During the Vietnam war, there were eight cases among the 139,000 American casualties. However, when a jet airliner crashed into the Florida Everglades, 8 of the 77 injured survivors developed the disease. Wounds involving large muscle masses, wounds from high-velocity

projectiles, contamination with dirt or clothing, or wounds near fecal-contaminated skin are all attended by increased risk. In peacetime, gas gangrene may be precipitated by extensive industrial and transportation injuries, surgery on biliary tract or colon, infarcted bowel (incarcerated hernia), and arterial disease, among other causes. In rare instances, it may result from intramuscular injection, as of epinephrine.

Abscess and gangrene of the lung may be secondary complications of more severe cases of pneumonia. The signs of lung abscess include fever, sweating, and the production of a thin, brown, puslike, foul-smelling sputum.

Progressive bacterial synergistic gangrene may occur around a colostomy or ileostomy opening, in proximity to a chronic ulcer, and sometimes in association with the use of wire stay sutures in surgery. Lesions produced are painful and are surrounded by a rim of gangrenous skin. Causation is usually mixed bacteria, with microaerophilic streptococci, such as Streptococcus aureus, or Gram-negative bacilli, such as Proteus, being implicated. The condition will spread without treatment. Antibiotics, particularly penicillin G, are effective in arresting the process. However, the condition is most frequently controlled by combining wide surgical excision with parenteral antibiotics.

Anaerobic (Clostridial) cellulitis, in itself relatively benign gas-forming infection of the skin and subcutaneous tissues without involvement of muscle or toxemia, may predispose the patient to streptococcal gangrene, also called necrotizing fasciitis. In such situations, gangrene reaches the subcutaneous tissue with necrosis of the overlying skin. Again, it is usually the extremities that are involved. The region of involvement may take on a dusky blue coloration. Bullae (blisters) which exude a reddish-black fluid are present. Bursting of the bullae is followed by extensive cutaneous gangrene. Treatment is by removal of necrotic tissue, combined with antibiotic therapy for the streptococcal infection. There is a short incubation of cases of gas gangrene (clostridial myositis), ranging from 8 to 72 hours.

For many years, polyvalent (*C. perfringens*, *C. septicum*, and *C. novyi*) equine antitoxin has been used in the treatment of clostridial myonecrosis. However, because the efficacy of the antitoxin was not proved conclusively, there is no longer production of the antitoxin for clinical use. Surgical debridement of affected tissue is commonly practiced. Hyperbaric oxygen therapy (100% oxygen at 3 atmospheres of pressure over periods of about 2 hours) is frequently used in conjunction with debridement procedures and antibiotic therapy. Inasmuch as the infective agents are anaerobic, the presence of concentrated oxygen slows or even stops spread of the infection. The mortality rate in cases of gas gangrene ranges from 15 to 30%, but is as high as 50% when the abdominal wall is involved. Supportive measures include blood transfusions, plasma infusions, and electrolyte replacement to counteract any anemia, hypovolemia, and shock involved. The action of various antimicrobial agents against anaerobic bacteria is described in the Sutter (1976) reference listed.

Additional Reading

Altemeier, W.A. and W.D. Fullen: "Prevention and Treatment of Gas Gangrene," *J. Amer. Med. Assn.*, **217**, 806 (1971).

Holdeman, L.V., E.P. Cato, and W.E.C. Moore: "Current Classification of Clinically Important Anaerobes," in *Anaerobic Bacteria: Role in Disease*, (R.M. DeHaan and V.R.Dowell, Jr., editors), Charles C. Thomas, Springfield Illinois, 1974.

Sutter, V.L. and S.M. Finegold: "Susceptibility of Anaerobic Bacteria to 23 Antimicrobial Agents," *Antimicrob. Agents Chemother.*, **10**, 736 (1976).

Winstein, L. and M.A. Barza: "Gas Gangrene," *New Engl. J. Med.*, **289**, 1129 (1973).

R.C. VICKERY, Blanton/Dade City, FL

GANGUE. The essentially valueless mineral aggregates or rock of an ore.

GANISTER ROCK. This term was originally applied to a siliceous underclay occurring in certain coal beds in the north of England. Now it is often applied to highly siliceous, fine-grained rocks used for refractory purposes or to a mixture of ground quartz and fire-clay used for furnace linings.

GANNET. See **Pelecaniformes**; and **Pelicans and Cormorants**.

GANOID SCALES. See **Fishes**.

GANTRY. A frame structure that spans over something, as an elevated platform that runs astride a work area, supported by wheels on each side; short for gantry crane or scaffold.

Gantry Crane. A large crane mounted on a platform that usually runs back and forth on parallel tracks astride the work area. Often shortened to gantry.

Gantry Scaffold. A massive scaffolding structure mounted on a bridge or platform supported by a pair of towers or trestles that normally run back and forth on parallel tracks, often used to assemble and service a large rocket as the rocket rests on its launching pad. Often shortened to gantry.

GANYMEDE. See **Jupiter**.

GAP. In geology, a gap is an opening through a ridge connecting the valleys or lowlands on either side. Gaps may be formed by a river that earlier in the cycle of erosion was able to cut its way through the hard rocks now making up the ridge. If the stream is still flowing through this opening, it is spoken of as a water gap; if the stream has disappeared because of its diversion or for other reasons, it is then spoken of as a wind gap.

An electric gap is the distance separating two electrodes between which a spark or arc is caused to pass.

A magnetic gap is the distance across an air gap separating two parts of a magnetic circuit. The clearance between pole pieces and rotor of dynamo machinery is such a gap.

GAREFOWL. See **Waders, Shorebirds, and Gulls**.

GARGANEY. See **Waterfowl**.

GARLIC POWDER. See **Spices**.

GARNET. The name garnet is now applied to a group of very important minerals crystallizing in the isometric system and showing the same habitat of dodecahedrons and trapezohedrons. Garnets belong to the nesosilicate group of silicate minerals and conform to the general formula $A_3B_2(SiO_4)_3$. The elements represented by A and B, respectively, may include calcium, magnesium, manganese, and ferrous iron; aluminum, ferric iron, chromium or titanium. While garnets show no cleavage, a dodecahedral parting is rarely noted; fracture conchoidal to uneven; some varieties very tough and valuable for abrasive purposes and for polishing eyeglass lenses. The hardness of garnet varies between the different varieties from 6.5 to 7.5, and the specific gravity from 3.4 to 4.3. Luster, vitreous to resinous; colors, red, yellow, brown, black, green, or colorless; transparent to opaque. The word garnet is derived from the Latin granatus, a grain.

In general, six varieties of garnet are recognized, based on their chemical composition: grossularite (which is also called hessonite and cinnamonstone); pyrope; almandine or carbuncle; spessartine; uvarovite; and andradite. Grossularite is a calcium-aluminium garnet which corresponds to the

formula $Ca_3Al_2(SiO_4)_3$; the calcium may, however, be in part replaced by ferrous iron and the aluminum by ferric iron. The name grossularite is derived from the botanical name for the gooseberry, grossularia, in reference to the green garnet of this composition found in Siberia. Other shades are the well-known cinnamon brown, reds, and yellows. Because of its inferior hardness to zircon, which mineral the yellow crystals resemble, they have been termed hessonite, from the Greek meaning inferior. Curiously, in the gem-bearing gravels of Ceylon, both zircon and hessonite are found and indiscriminately called hyacinth. This term, from the Greek, was apparently a general term used by Pliny for the transparent varieties of corundum; later it was used for yellow zircons.

Grossularite is found in crystalline limestones with vesuvianite, diopside, wollastonite and wernerite. Among the many localities are the Urals, Italy, Switzerland, Mexico, and, in the United States, Maine and New Hampshire. Fine specimens are obtained from the Jeffrey Mine, Asbestos, Quebec, Canada.

Pyrope, sometimes called Cape ruby, is ruby-red in color and chemically a magnesium aluminum silicate with the formula $(Mg,Fe)_3Al_2(SiO)_3$; the magnesium may be replaced in part by calcium and ferrous iron. The color of pyrope varies from deep red to almost black. The transparent pyropes are used as gems, but some have a slight tinge of yellow. The name pyrope is derived from the Greek word meaning *fire-like*. A sub-variety of pyrope from Macon County, North Carolina, is of a violet-red shade and has been called rhodolite, from the Greek meaning *a rose*. In chemical composition it may be considered as essentially an isomorphous mixture of pyrope and almandine, in the proportion of two molecules of pyrope to one molecule of almandine. Pyrope is found at Teplitz and Aussig, Bohemia; in the Kimberley diamond mines in the Republic of South Africa; in Australia and elsewhere. In the United States, important localities are in Arizona, New Mexico, and Utah.

Almandine is the modern gem the carbuncle, although in Pliny's time this term was used for almost any red stone. The term carbuncle is derived from the Latin *carbunculus*, meaning a little spark. The name almandine is a corruption of Alabanda, a locality in the Middle East where, in ancient times, these red stones were cut. Chemically almandine is an iron-aluminum garnet corresponding to the formula $Fe_3Al_2(SiO_4)_3$. The deep red transparent stones are often called precious garnet and used for gems. Almandine occurs in metamorphic rocks like mica schists usually associated with typically metamorphic minerals such as staurolite, kyanite, and andalusite. Good gem material comes from India and Brazil. Almandine is also found in Australia, Alaska, Africa, Norway, Sweden, Madagascar, and Japan. In the United States almandine with 11.48% MgO pyrope content is found in the gneisses of the Adirondack region of New York, sometimes of very large size, in New England, and elsewhere.

Spessartine is manganese aluminum garnet, $Mn_3Al_2(SiO_4)_3$. The name of this mineral is derived from Spessart in Bavaria, a well-known European locality. Spessartine of a beautiful orange-yellow comes from Madagascar. Violet-red spessartine has occurred in rhyolites in Colorado and Maine. Uvarovite is a calcium chromium silicate the formula being $Ca_3Cr_2(SiO_4)_3$. It is a rather rare garnet, bright green in color, usually in small crystals associated with chromite in serpentines, sometimes in crystalline limestones or schists. It is found in the Urals, the Republic of South Africa, Canada, and, in the United States, in California and Pennsylvania. Andradite, calcium-iron garnet, $Ca_3Fe_2(SiO_4)_3$, is of variable composition and may be red, yellow, brown, green, or black, or of intermediate shades. The subvarieties topazolite, yellow or green, demantoid, green, and melanite, a black sort, are recognized. Andradite is found both in deep-seated igneous rocks like syenite as well as in serpentines, schists, and crystalline limestones. Demantoid has been called the "emerald of the Urals" from its occurrence there. Varieties of andradite are found in many localities in Europe: Italy, Switzerland, Norway, and Saxony. In the United States it is found at Franklin, New Jersey; Magnet Cove, Arkansas; and elsewhere.

ELMER B. ROWLEY, Union College, Schenectady, NY

GARNET (Synthetic). See **YAG and YIG**.

GARNIERITE. This mineral occurs as amorphous masses, presumably as a product of secondary alteration of nickel-bearing peridotites. It is a hydrous silicate of nickel and magnesium, $(Ni, Mg)_3Si_2O_5(OH)_4$. Hardness is 2–3; specific gravity 2.2–2.8, and characterized by its apple green color with dull-to-earthy luster. An important nickel-ore mineral is found with chromite and serpentine in New Caledonia. Additional localities include the Republic of South Africa, the former U.S.S.R., Madagascar, and Oregon and North Carolina in the United States.

GARROD, ARCHIBALD EDWARD (1857–1936).
Archibald Garrod was British physician-scientist whose work on human biochemical individuality, expressed as rare inherited diseases, linked the emerging disciplines of biochemistry and genetics in the 1910s.

He began his career with a series of junior appointments at St Bartholomew's and at the Hospital for Sick Children, Great Ormond Street, London. His experience with children with alkaptonuria (darkening of excreted urine after the urine has been exposed to air) led him to examine the pigments present in normal urine and that from alkaptonurics. In the late 1890s, his biochemical analysis suggested that rather than this darkening being a symptom of disease, it was an indication that an enzyme was missing from the metabolic pathway, which in normal people allows the conversion of the reddening agent, alkapton, to another substance. Garrod regarded this congenital condition as a familial one and linked its occurrence among the children of first cousin marriages with Mendel's laws of genetic inheritance, William Bateson suggesting in 1901 that this was an example of a recessive hereditary trait. Garrod also argued hypothetically that one gene is responsible for one enzyme, although he did not use this language (1902). See also **Biochemistry (The History)**.

Garrod published his Croonian Lectures of 1908 as *Inborn Errors of Metabolism*. He extended his ideas in *Inborn Factors in Disease* (1931), suggesting that chemical individuality could result in predisposition to certain diseases. Although he had recently retired from the Regius Chair of Medicine at Oxford (1920–1928) his ideas were not well received by his contemporaries; in the era of modern genetics, they have found resonance.

Additional Reading

Bearn, A.G.: *Archibald Garrod and the Individuality of Man*, Clarendon, Oxford, UK, 1993.

Garrod, A.E.: *Inborn Errors of Metabolism*, Hodder and Stoughton, London, UK, 1909.

Hopkins, F.G.: "Archibald Edward Garrod, 1857–1936," *Obituary Notices of Fellows of the Royal Society*, **2**, 225–228 (1936–1938).

Scriver, C.R., and B. Childs: *Garrod's Inborn Factors in Disease*, Oxford University Press, New York, NY, [including an annotated facsimile reprint of *The Inborn Factors in Disease* by Archibald Garrod].

HELEN J.POWER, Wellcome Trust Centre, London, UK

GARS (*Osteichthyes*).
Of the order *Ginglymodi*, there are approximately eight species, all of which have what might be termed a "crocodilian" appearance. They are heavily armored with ganoin scales, usually in the form of diamonds or rhomboids. These are flat plates with no interlocking as found in conventional fish scales. They move slowly under normal circumstances, but are capable of very fast movements when striking for food. Much as a crocodile, the gar is a slasher, with rapid sidewise movements in its efforts to tear away at its food. Gars are well known for stealing bait from the fisherman's hook. They prefer shallow areas with lots of underwater vegetation and thus it is not surprising to find that one of their natural habitats is in the Florida Everglades. Seminole Indians eat smaller gars.

Gars, like crocodiles, have ball-and-socket joints, unlike most other fishes, which have concave vertebrae. The long-nosed gar (*Lepisosteus osseus*) is found in waters eastward from the Mississippi basin. This species is easily identified by its very long jaws and by length of head and location of eyes which are large. As with other gars, this species prefers salt or brackish water, although it will survive for several years in fresh water. Alligator gars, of which there are a couple of species, definitely prefer fresh water and cannot survive for long periods in salt water. The largest of the gars, the tropical gar (*Lepisosteus tristoechus*) can attain a length of from 10 to 12 feet (3 to 3.6 meters) and is eaten in parts of Mexico. The scales also can be used in ornamental jewelry.

Gars have not been found west of the Rocky Mountains, but are found mostly in the eastern United States, up into southeastern Canada and as far south as Costa Rica. Fig. 1 shows the long-nosed gar.

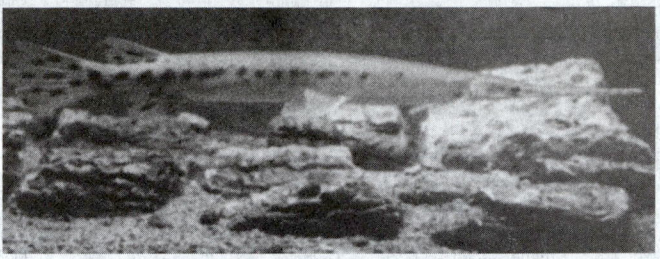

Fig. 1.   Long-nosed gar. (*A.M. Winchester.*)

GAS.
1. A state of matter, in which the molecules move freely and consequently the entire mass tends to expand indefinitely, occupying the total volume of any vessel into which it is introduced. Gases follow, within considerable degree of fidelity, certain laws relating their conditions of pressure, volume, and temperature. Gases mix freely with each other, and they can be liquefied.

2. The term is sometimes used as distinct from vapor, particularly to indicate a substance having a critical temperature below room temperature.

The fundamental gas laws are described elsewhere in this volume. In particular, see also **Equation of State**.

An inert gas is a gas that does not react chemically. The rare gases of the atmosphere were long considered to be completely inert. Also known as noble gases, these included argon, helium, krypton, neon, radon, and xenon. Definite compounds of radon and xenon, for example, have been identified in recent years, but generally their identification as being inert is well justified. Some gases are termed permanent gases, including oxygen, nitrogen, and hydrogen, which require low temperatures and, in practice, high pressures for their liquefaction. The term arises from the fact that in the early years of scientific investigation of these materials, long before the conditions of liquefaction were obtainable, it was believed that these gases could not be liquefied under any circumstances, and hence termed permanent gases.

The laws pertaining to the forces of gas pressure and to the flow of gases are based ultimately upon the kinetic theory, but certain principles can be stated without analyzing their origin to that extent. To a first approximation, the ideal gas law, or the Boyle-Charles law, represents the dynamics of gases at rest. At a given temperature, the pressure of a body of gas varies inversely as its volume, and hence directly as its density (Boyle law). And at a fixed volume, the pressure is a linear function of the temperature, varying at the same rate ($\frac{1}{273}$ per centigrade degree) for all gases (Charles law). But dynamic processes in a gas are complicated by the fact that change in volume is, in general, accompanied by change in temperature, so that simple dynamics is overshadowed by thermodynamics. It was for this reason, for example, that the correct formula for the speed of sound in air proved, for a time, elusive. A gas is highly compressible, and this property affords ready opportunity for the energy of mechanical impulses, which would be merely transmitted by a noncompressible fluid, to be transformed into heat, or for the gas to use its thermal energy to create impulses of its own. The same circumstance complicates the effect of gravity. The atmosphere is not an ocean of uniform density and definite depth; its pressure and density are logarithmic functions of the altitude. The forces associated with moving gases form the subject matter of aerodynamics. See also **Atmosphere (Earth)**.

GASAHOL.   See **Ethyl Alcohol**.

GAS ANALYZERS (Combustion-Type).
The concentration of combustible gases must be determined and controlled in manufacturing operations and other industrial situations for several reasons, including: (1) safety — to avoid explosions by maintaining concentrations well below the lower explosive limit; also to avoid the toxic effects of most combustible gases on operating personnel; (2) efficiency — to maintain optimum concentrations for combustion and other chemical reactions where such gases may be used; and (3) detection of faulty operating equipment and procedures. In combustion-type analyzers, the very quality one is seeking (combustibility) is used as the basis of instrumentation.

The most commonly used method employs a self-heated "hot wire" detector, usually platinum. The wire also serves as a combustion catalyst.

TABLE 1. HEATS OF COMBUSTION OF TYPICAL COMBUSTIBLE GASES[a]

| | | Heat of Combustion Hc at 25 °C and Constant Pressure to Form | | | | | |
| | | H_2O (gas) and CO_2 (gas) | | | H_2O (liq) and CO_2 (gas) | | |
Gas	Formula	kcal/mole	cal/g	Btu/lb	kcal/mole	cal/g	Btu/lb
Hydrogen	H_2	57.7979	28,669.6	51,571.4	68.3174	33,887.6	60,957.7
Carbon monoxide	CO				67.6361	2,414.7	4,343.6
Methane	CH_4	191.759	11,953.6	21,502	212.798	13,265.1	23,861
Ethane	C_2H_6	341.261	11,349.6	20,416	372.820	12,399.2	22,304
Propane	C_3H_8	488.527	11,079.2	19,929	530.605	12,033.5	21,646
n-Butane	C_4H_{10}	635.384	10,932.3	19,665	687.982	11,837.3	21,293
Isobutane	C_4H_{10}	633.744	10,904.1	19,614	686.342	11,809.1	21,242
n-Pentane	C_5H_{12}	782.04	10,839.7	19,499	845.16	11,714.6	21,072
Isopentane	C_5H_{12}	780.12	10,813.1	19,451	843.24	11,688.0	21,025
Neopentane	C_5H_{12}	777.37	10,775.0	19,382	840.49	11,649.8	20,956
n-Hexane	C_6H_{14}	928.93	10,780.0	19,391	1,002.57	11,634.5	20,928
n-Heptane	C_7H_{16}	1,075.85	10,737.2	19,314	1,160.01	11,577.2	20,825
n-Octane	C_8H_{18}	1,222.77	10,705.0	19,256	1,317.45	11,533.9	20,747
n-Nonane	C_9H_{20}	1,369.70	10,680.0	19,211	1,474.90	11,500.2	20,687
n-Decane	$C_{10}H_{22}$	1,516.63	10,659.7	19,175	1,632.34	11,473.0	20,638
Benzene	C_6H_6	757.52	9,698.4	17,446	789.08	10,102.4	18,172
Toluene	C_7H_8	901.50	9,784.7	17,601	943.58	10,241.4	18,422
Ethylene	C_2H_4	316.195	11,271.7	20,276	337.234	12,021.7	21,625
Acetylene	C_2H_2	300.096	11,526.2	20,734	310.615	11,930.2	21,460

[a] Values for additional gases and vapors may be obtained from the National Institute of Standards and Technology, Gaithersburg, Maryland, and the American Petroleum Institute, New York.

Where the combustible gas to be measured also contains air, the mixture simply is fed to a "hot wire" detector whereupon combustion occurs. A temperature sensor, such as a thermocouple, may detect the temperature rise and this, in turn, is a measure of the concentration of the gas. More frequently, the electrical resistance of the "hot wire" itself is measured as the means for detecting temperature rise, much as occurs in a typical electrical resistance thermometer. Where the sample does not contain an excess of oxygen, then air or oxygen must be added to the sample line in carefully controlled quantities, but added well in excess of combustion requirements, so that the reaction occurring within the detector will be limited only by the amount of combustible gases or vapors present. Wheatstone bridge circuitry usually is used in these instruments.

The quantity of heat released is related to the concentration of combustibles by reference to a of heats of combustion. See Table 1. It is important to note that an analyzer of this type is nonspecific, that is, the instrument is not capable of differentiating between different compositions of combustibles. Inasmuch as the output of the instrument is a function of the rate of combustion and heat of reaction, such analyzers frequently are calibrated in terms of *percent combustibles expressed as percent hydrogen*. However, where it is known in advance that a specific combustible will predominate in the gas stream, the instrument may be calibrated specifically in terms of that component.

Where a bridge circuit is used, a reference detector is required. The reference gas may be air, or the sample gas also may be used if the catalytic characteristics of the "hot wire" are poisoned or destroyed purposely. The latter method has the advantage of compensating for thermal conductivity changes that may occur in the sample as the result of changing sample compositions.

In another type of combustible gas analyzer, the sample gas is burned in a small pilot flame, the temperature of which is detected by a thermocouple. The presence of combustibles in the supply of gas to the pilot causes the flame temperature to increase proportionally with concentration. This method is preferred where substances may be present in the gas stream that may poison the catalytic properties of the other form of detector.

Combustible-type gas analyzers are obtainable in combination with oxygen analyzers. In portable form, this combination of instruments is used for testing various types of combustion processes.

See also **Pollution (Air)**.

GAS ANALYZERS (Thermal-Conductivity Type). Different gases vary considerably in their ability to conduct heat. These variations make it

possible to determine the concentrations of a number of gases commonly encountered in laboratory research and industrial processes. Although the relationship between thermal conductivity and gas composition has been investigated widely and so reported in the literature, in general it is not practical or profitable to make detailed calculations of thermal conductivity in designing and applying this type of instrument in gas analysis work. Data available on the thermal conductivity of gases normally is reliable only to within ±5% and, therefore, such calculations usually are confined to obtaining a broad estimate of likely sensitivity of an instrument over a limited range of composition. Further, thermal-conductivity gas analyzers normally are confined to determinations of binary gas mixtures. The method is nonspecific and nonabsolute and thus depends upon empirical calibration. Because the method is so simple, reliable, relatively fast, and convenient to adapt to continuous recording and control, however, this is one of the most widely used gas analysis methods.

The hot-wire gas analysis cell was introduced by Koepsal in 1908 and the principle of the hot wire (in various forms) remains the key approach to thermal-conductivity gas analysis. A typical cell is comprised of an electrically conductive, elongated sensing element that is mounted coaxially inside a cylindrical chamber which contains the gas. By passage of an electric current through the element, the cell is maintained at a temperature considerably higher than the cell walls. The equilibrium temperature is reached when all thermal losses from the wire are equalized by electric power input to the element. If the element is made of a material with a suitable temperature coefficient of resistance, it may serve the dual role of heat source and sensor of the equilibrium temperature. The difference of temperature between the element and the cell walls, reflected by the temperature rise of the element at equilibrium, is a function of electric power input and combined rate of heat loss from the wire by gaseous conduction, convection, radiation, and conduction through the solid supports of the element. Proper cell design and geometry makes it possible to maximize the heat loss due to gaseous conduction. Thus, a rise in the temperature of the element at constant electric power input is inversely related to the thermal conductivity of the gas within the cell.

Normally, a Wheatstone bridge is used to measure the resistance change of the sensing element. The electric current required to energize the bridge also is used to heat the wire. A single hot-wire cell is impractical because of the delicate sensitivity of such an arrangement to changes in ambient temperature and bridge-supply voltage. Commonly, two cells are used in adjacent arms of the bridge. A reference gas is contained in one of these cells. Thus, the bridge responds to the difference in temperature rise of the two cells and consequently depends only upon the difference in thermal conductivities of the sample gas and the reference gas.

While thermal-conductivity gas analyzers are widely used directly for on-line process measurements, they also find wide application in gas chromatographs for determining gas concentration after chromatographic separations. For the quantitative analysis of a binary gas mixture, a useful sensitivity of 1% of full-scale or better is obtainable. The full-scale range varies with the gas mixture and is indicated for several binary mixtures in Table 1. The practical limits of the method are given in Table 2.

TABLE 1. PRACTICAL RANGE OF THERMAL-CONDUCTIVITY METHOD TO BINARY GAS MIXTURES

Mixture	Practical Full-scale Range
Air-carbon dioxide	0–5.3% air in CO_2
	0–7.3% CO_2 in air
Air-sulfur dioxide	0–1% air in SO_2
	0–3% SO_2 in air
Air-oxygen	0–40% air in O_2
	0–38% O_2 in air
Air-helium	0–2.4% air in He
	0–0.4% He in air
Nitrogen-carbon dioxide	0–5% N_2 in CO_2
	0–7% CO_2 in N_2
Nitrogen-hydrogen	0–2.3% N_2 in H_2
	0–0.3% H_2 in N_2
Nitrogen-oxygen	0–55% N_2 in O_2
	0–52% O_2 in N_2
Nitrogen-argon	0–5% N_2 in Ar
	0–7% Ar in N_2
Hydrogen-helium	0–10% H_2 in He
	0–12% He in H_2
Carbon dioxide-oxygen	0–6.4% CO_2 in O_2
	0–4.4% O_2 in CO_2

TABLE 2. REPRESENTATIVE APPLICATIONS OF THERMAL-CONDUCTIVITY METHOD

Mixture	Appropriate Comparison Gas
H_2 in CO_2	H_2, CO_2, or $H_2 + CO_2$
H_2 in O_2	O_2, air, or H_2
H_2 in N_2	H_2, N_2, or air
H_2 in Cl_2	H_2 or Cl_2
H_2 in air	H_2 or air
H_2 in CH_4	H_2, CH_4, or $H_2 + CH_4$
H_2 in water gas ($H_2 + CO$)	H_2, or $H_2 + N_2$
Ne in air	Air
He in air, N_2, or O_2	He, air, H_2, or O_2
Cl_2 in air	Air
HCl in air	Air
Acetone in air	Air
O_2 in enriched air	Air
NH_3 in air	Air
SO_2 in air or N_2	Air or N_2
Water vapor in air, N_2, or O_2	Air, N_2, or O_2
Ar in N_2, air, or O_2	N_2, air, or O_2
CO_2 in air, N_2, or flue gas	Air
Benzol in air[a]	Air or N_2

[a] Requires pretreatment by combustion, converting benzol to CO_2 and H_2O.

The variation of thermal conductivity of binary mixtures does not always follow a simple linear law. Water vapor in air and ammonia in air are examples of nonlinear cases.

GAS AND EXPANSION TURBINES. Fundamentally, the gas turbine operates on the concept of the Brayton or Joule cycle (constant-pressure cycle) which was originally used to describe the operation of an air engine, a compressor and a combustion chamber. In the air engine, air entered the compressor wherein the pressure was increased. Fuel burning in the combustion chamber raised the temperature of the compressed air under

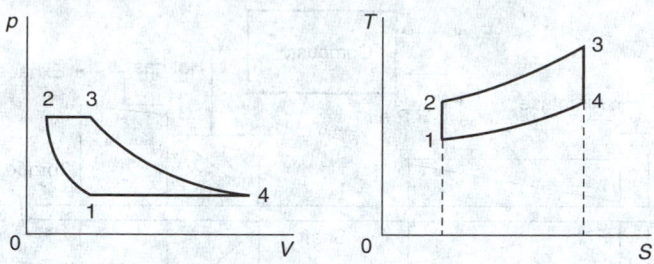

Fig. 1.   Brayton or Joule cycle.

constant-pressure conditions. The resulting high-temperature gases were then introduced to the engine where they expanded and performed work. The excess work of the engine over that required to compress the air was available for operating other devices, such as a generator. The cycle is illustrated in Fig. 1, with the following equations applying:

$$V_3/V_2 = V_4/V_1 = T_3/T_2 = T_4/T_1$$

where V = total   volume; $T = t + 459.69$ = absolute temperature = deg R

$$\frac{T_2}{T_1} = \frac{T_3}{T_4} = \left(\frac{V_1}{V_2}\right)^{k-1} = \left(\frac{V_4}{V_3}\right)^{k-1} = \left(\frac{p_2}{p_2}\right)^{k-1/k}$$

where $k = c_p/c_v$; c_p = specific heat at constant pressure; c_v = specific heat at constant volume; p = absolute pressure, pounds per square foot (1 pound/square foot = 47.88 Pascals = 4.88 kilograms/square meter);

$$(W) = Jmc_p(T_3 - T_2 - T_4 + T_1)$$

where W = external work performed on surroundings during change of state, foot-pounds; J = mechanical equivalent of heat = 778.26 foot-pounds per Btu = 4.1861 joules per cal; m = mass of substance under consideration, lb_m;

$$\text{Efficiency} = (W)/JQ_{23} = 1 - (T_1/T_2)$$

where Q = quantity of heat absorbed by the system from the surroundings, Btu.

In the gas turbine, the air compressor and engine of the foregoing scheme are replaced by an axial flow compressor and gas turbine. Although the turbine is only part of the whole assembly, in modern terminology, the complete assembly is commonly referred to simply as a gas turbine. Air is compressed in the compressor after which it enters a combustion chamber where the temperature is increased while the pressure remains constant. The resulting high-temperature air then enters the turbine, thereby performing work.

Gas turbines usually are rated according to power output (sea level and 80 °F; 26.7 °C). Some European designs are rated at 60 °F (15.6 °C). The power output and efficiency are larger for those fuels that produce larger volumes of products of combustion, inasmuch as the compressor does not do any work on additional volume. Gas turbines are classified by the physical arrangements of the component parts, and categories include: (1) single-shaft; (2) two-shaft; (3) regenerative (heat exchanger is used to recover exhaust losses and heat air to the combustor(s)); (4) intercooled (heat removed between compressors); and (5) reheat (heat added between turbines). Various configurations of gas-turbine systems are shown in Figs. 2, through 8.

Efficiency. The overall efficiency of a gas turbine is a function of the compressor and turbine efficiencies, ambient air temperature, nozzle inlet temperature, and the type of cycle used. The compressor and turbine are designed for high efficiency. The first-stage gas temperature establishes material and stress conditions for the first set of rotating blades. To the gas temperature at these blades is added the temperature drop across the first-stage nozzles to determine the inlet temperature of the turbine. This may vary from 704 to 816 °C for industrial turbines and usually will be higher for aviation gas turbines. The higher values are usually used in impulse turbines.

In a simple-cycle turbine, there is (for each turbine inlet temperature) an optimum pressure ratio producing the highest possible efficiency. The efficiency and optimum pressure ratio increases with increasing turbine

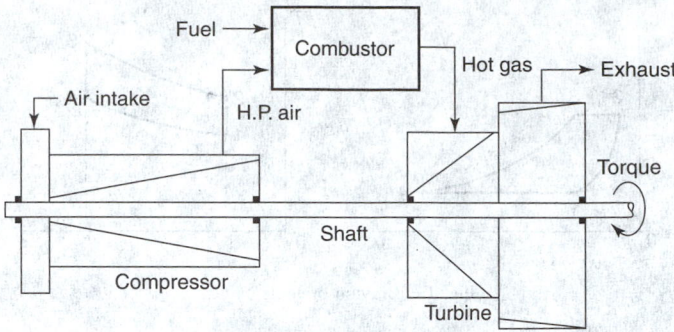

Fig. 2.   Gas-turbine configuration exhibiting basic Brayton or Joule cycle.

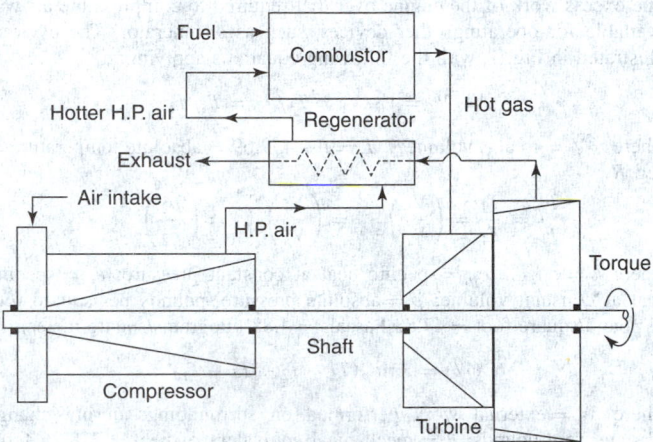

Fig. 3.   Gas-turbine configuration with regeneration.

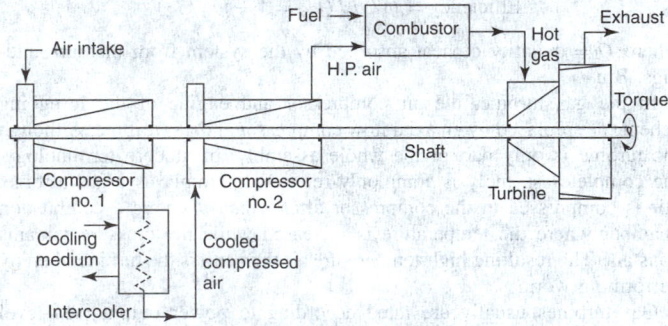

Fig. 4.   Gas-turbine configuration with intercooling.

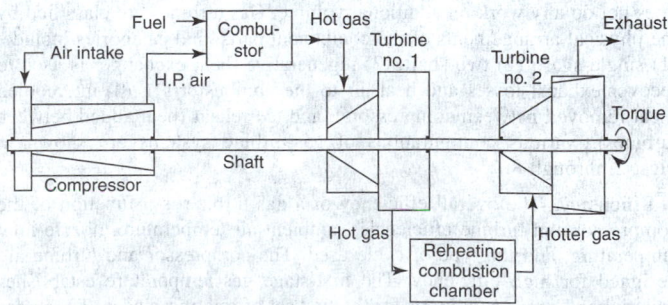

Fig. 5.   Gas-turbine configuration with reheating.

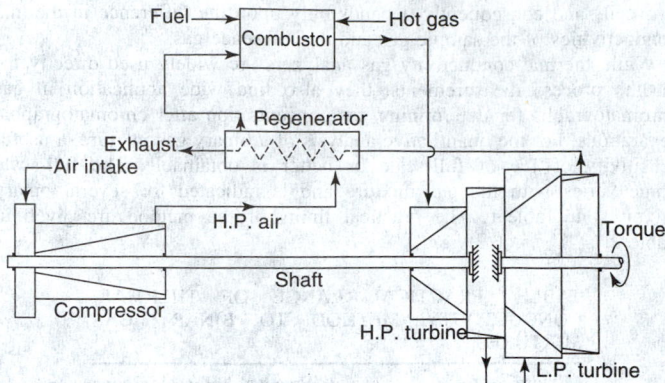

Fig. 6.   Regenerative-cycle gas turbine. Two-shaft arrangement, with separate power turbines in series.

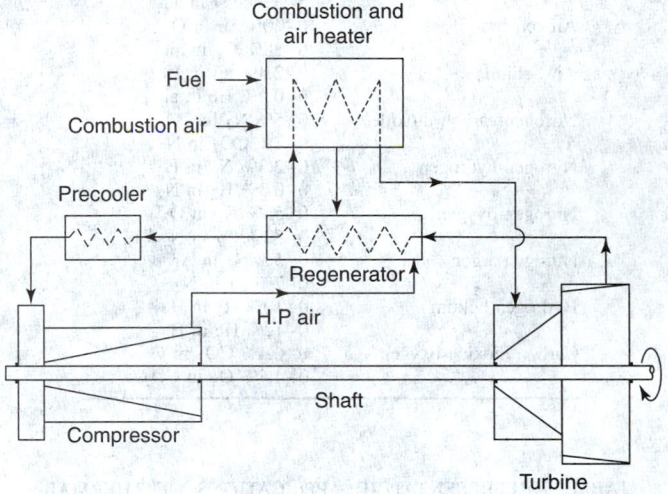

Fig. 7.   Closed-cycle gas turbine.

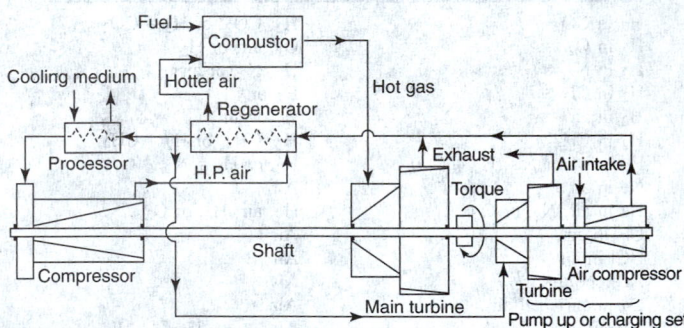

Fig. 8.   Semiclosed, internally fired gas-turbine cycle.

inlet temperatures. These pressure ratios vary from 4 (at 704 °C) up to 6 (at 816 °C).

Regenerative cycles favor lower pressure ratios which result in low compressor discharge temperatures, thus allowing greater recovery of heat from the turbine exhaust gases. High-ratio regenerative plants use intercoolers in the compressor circuit to lower the compressor discharge air temperature.

Although any type of efficient compressor can be used, such as positive displacement (Lysholm), centrifugal, and axial flow, most industrial gas turbines use axial-flow compressors. The turbine may have impulse or reaction blading. To minimize losses, air from the compressor discharge flows through the combustor directly into the turbine nozzle. Throttle valves are not used because the resulting pressure drop decreases overall efficiency.

A gas turbine has a large amount of excess air. The combustor is designed with an inner portion burning only part of the air to achieve high combustion temperatures and efficiency. Products of combustion are effectively mixed with the remainder of the air to minimize temperature stratification. Each turbine may have one large combustor or several smaller combustors operating in parallel.

Open- and Closed-Cycle Types. Most gas-turbine installations are of the open-cycle type, using atmospheric air as the working medium and burning relatively clean fuels. Where dirty fuels are used, it is possible to locate the burner in the gas-turbine discharge, using a heat exchanger to heat the air discharged by the compressor. In closed-cycle installations (Figs. 7 and 8), it may be desirable to use other gases, inasmuch as efficiency increases as the specific heat ratio (c_p/c_v) decreases. Optimum plant efficiency occurs at increasingly higher pressure ratios with decreasing values of (c_p/c_v). However, for convenience, most closed systems use air.

Closed systems can provide a high plant efficiency over a power range from 25 to 100% by varying the turbine exhaust and compressor inlet pressure from atmospheric to about 60 psig. These installations require costly heaters, located between compressor discharge and turbine inlet, and large coolers, located between the turbine exhaust and the compressor suction. Usually, combustion of a fuel provides the heat source, and cooling water the cooling medium.

Overloads. Even if only temporary in nature, a large overload can cause a single-shaft gas turbine to shut down, inasmuch as its fuel input is limited by the inlet overtemperature protective system. If the torque requirements of the driven machine do not decrease sufficiently with speed reduction, then the gas turbine will continue to slow down. This results in higher exhaust temperatures. The exhaust temperature control system will either shut off the fuel valve, or further reduce fuel input, causing the turbine to decrease its speed and finally shut down. Carefully matching the load characteristics of the driven equipment with those of the driver can prevent such occurrences.

Single-Shaft Gas Turbines. The wide acceptance of the single-shaft turbine arises from its low cost and compactness in terms of power output per cubic foot of machinery space. Disadvantages include a relatively low operating speed range and sensitivity to atmospheric temperature. The low operating speed range arises from: (1) the quantity of air flow induced by the compressor is proportional to its speed; and (2) the back pressure produced by the turbine nozzles is proportional to air flow. At low speeds, the turbine power is decreased by low air flows and secondarily by the effect of low pressures on allowable inlet air temperatures. At low flows, the decreased pressure at the turbine inlet may require a reduction of turbine inlet temperature to maintain the exhaust temperature within design limitations. This results in a further reduction in power. In most applications, it is necessary to unload the turbine during startup.

Two-Shaft Gas Turbines. A wider operating speed range is provided by the more costly two-shaft machine, which consists of a high-pressure turbine driving the air compressor and a low-pressure turbine on a separate shaft to provide output power. See Fig. 9. A variable-area nozzle can be used in the low-pressure turbine to increase the operating speed range. Change in the fuel input to the high-pressure turbine causes the speed and quality of air flow to change. The low-pressure turbine power output is changed by varying the quantity of air flow and the nozzle area of the power turbine.

Air/Temperature Relationships. The air flow to a gas turbine is inversely proportional to the absolute air temperature at the compressor inlet. Inasmuch as the compressor discharge pressure is set by the turbine nozzles (proportional to flow), this results in decreased turbine power output during hot weather, and increased power during cold weather. In hot, dry areas, hot incoming air can be cooled by evaporation using water injection. In locations where the summer season is short, it may be possible to obtain rated power by increasing the turbine temperature for a short period without appreciably shortening the life of the equipment. In extremely cold temperatures, high air pressures will exist at the turbine inlet and should be considered in the design of the gas turbine.

Since the power required by the compressor is approximately twice as great as the shaft output, a 1% change in compressor efficiency will result in a 2% change in shaft power. A 1% change in turbine efficiency will produce a 3% change in shaft power. Therefore, it is important that all losses be minimized and that sufficiently large inlet and exhaust piping or ducts be used.

Startup of Gas Turbines. A gas turbine is started by bringing it up to starting speed by application of external power (electric, air, gas) and maintaining this speed for several minutes in order to purge the casing. Some machines require that the casing or rotor be heated slowly by burning

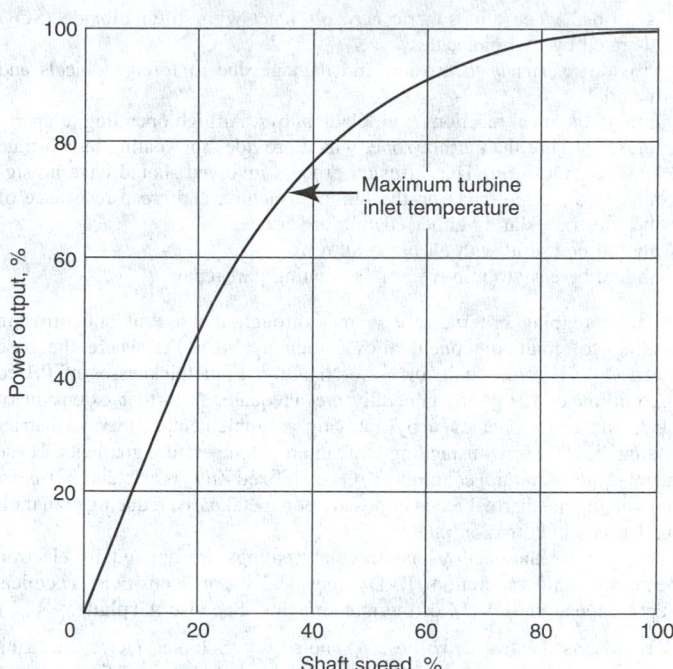

Fig. 9.   Operating range for various speeds and loads for two-shaft gas turbine.

a nominal amount of fuel in the combustors for several minutes. The turbine inlet temperature is then increased rapidly to a value above the design temperature, thus producing sufficient power in the turbine to bring the set up to full speed. Some installations will require a blowoff valve to prevent surging during startup. The starting power requirements of an unloaded gas turbine will range between 5 and 10% of the full-load speed. Two-shaft turbines will require slightly more starting power than single-shaft machines. By opening the nozzles of the low-pressure turbine, the load is not driven during startup.

Fuels. A wide variety of fuels can be used in gas turbines. The major fuel requirements are that: (1) the fuel does not form ashes that will deposit on the blades and interfere with operation; (2) the fuel does not contain dust which will erode the bladed; and (3) the fuel does not contain uninhibited vanadium. Commonly used fuels include natural and refinery gas, blast-furnace gas, fuel oils (including heavy residuals), and the growing application of gas turbines in cycles involving gases derived from coal and other previously nontraditional sources.

The simple cycle-gas turbine is relatively inefficient with almost all of its losses in the hot exhaust gases. When exhaust gases can be used in a boiler or for process heating, the combination of turbine and heat-recovery apparatus results in a high-efficiency plant. Integration of the gas turbine with process requirements also can result in high efficiency.

Improved Turbine Materials

Gas turbines find their principal uses in aircraft and for land-based applications. There is a large demand in both of these areas for improved turbine performance, including optimal energy utilization. These needs cannot be met without improved materials of construction.

Aircraft Gas-Turbine Engines. Significant reduction in fuel consumption is obtainable only by increasing gas temperature at turbine-section inlet throats. Even the super alloys used for turbine blades only provide reliable service up to about 1000 °C (1830 °F). In advanced propulsion systems, turbine-inlet temperatures may reach 1700 °C (3090 °F), and metallic structures are expected to withstand temperatures as high as 1200 °C (2190 °F). In uncoated, traditional rotor or stator blades, high-temperature, corrosive environments cannot be tolerated for long periods. Special coating methodologies have been developed during the past decade to resist high temperature and corrosion at the surface of critical parts.

Criteria established for improved turbine-blade coatings, as developed by Lämmermann and Kienel, include:

- Provide resistance to high-temperature oxidation and hot-gas corrosion. Hot-gas corrosion usually is caused by sodium sulfate (Na_2SO_4), which is formed when atmospheric aerosols react with sulfur dioxide (SO_2) liberated by fuel molecules.

- Provide resistance to erosion and damage due to foreign objects and materials.
- Inhibit chemical reactions with blade alloys. At high operating temperatures, no more than a thin zone, which provides for coating-to-substrate bond, should form. The diffusion process involved should have no significant adverse effects on the fatigue, fracture, and creep resistance of the substrate, during either coating or service.
- Should be useful with all blade alloys.
- Should be easy to remove, for facilitating part repair.

The foregoing criteria can be met through the use of anticorrosion coatings of multicomponent alloys, such as MCrAlY, where the base metal (M) is iron, cobalt, or a Co-Ni alloy. Film thicknesses of 0.1 to 0.2 millimeter (0.004 in) generally are adequate. A further extension of blade life can be achieved by applying a supplementary thermal barrier coating capable of withstanding large internal temperature gradients. These metal-oxide (ceramic) coatings (Y_2O_3-stabilized ZrO_2 is an example) assist in reducing the thermal loads imposed upon metal parts, reducing both their oxidation and corrosion rates.

Frequently, these alloy and thermal coatings are applied by electron beam vacuum evaporation (PVD = physical vapor deposition), chemical vapor deposition (CVD), or thermal spraying. See also **Airplane**.

Land-based Gas Turbines. As noted by Schilke, increased firing temperatures and pressures have helped to boost the fuel efficiency of gas turbine–based power generation systems past the 50% mark. An increase of 55 °C (100 °F) can provide corresponding increases of 10 to 13% in output and 2 to 4% in simple cycle efficiency. The cost benefit is obvious. For example, in a combined-cycle power plant, new technology that meets these rigorous demands can generate savings of hundreds of thousands of dollars per year, as compared with earlier technology.

Manufacturing processes that have contributed to stronger and high-performance turbine parts include directional solidification (DS). This process eliminates transverse grain boundaries, with a resulting increase in creep-rupture strength. Secondary operations performed on investment castings also contribute. These include electrical discharge machining (ECM and EDM), laser beam surfacing on some parts, and creep-feed grinding.

In addition to building higher performance into the turbine hardware, much greater attention is now being given to maintenance. With their modularity, ease of installation, low installed cost, and increasing efficiency and reliability, gas turbines are becoming a major source of new generating capacity for utilities in the United States. One of the challenges facing operators is that management and maintenance programs to optimize unit reliability and performance have not kept pace with rapid growth in the installed base. To extend and enhance utility maintenance capabilities, the Electric Power Research Institute has developed a variety of resources, including products and services for unit efficiency analysis, outage management, troubleshooting, technician training, turbine blade refurbishment, and information exchange. This program is well described in the Frischmuth reference. An encapsulated outline of the EPRI program is given in Table 1.

Expansion Turbines

An expansion turbine converts the energy of a gas or vapor stream into mechanical work as the gas or vapor expands through the turbine. The expansion process occurs rapidly and the heat transferred to or from the gas is usually very small. Consequently, in accordance with the first law of thermodynamics, the internal energy of the gas decreases as work is done and the resultant temperature of the gas may be quite low, thus giving the expander the ability to act as a refrigerator as well as a work-producing device. As a result, turbo-expanders have been widely used in the cryogenic field to produce the refrigeration needed for the separation and liquefaction of gases. By common usage, the terms *turboexpanders* and *expansion turbines* specifically exclude steam turbines and combustion gas turbines.

Turboexpanders may be classed into two broad categories: (1) axial-flow; and (2) radial-flow. Axial-flow turbines are those in which the gas flow is essentially parallel to the axis or shaft of the turbine. Turbines of this type resemble a conventional steam turbine and may be single-stage or multistage with impulse or reaction blading, or combination of impulse and reaction blading. Turbines of this type are not usually used for producing low temperatures, but are basically power-recovery devices and find application where flow rates, inlet temperatures, or total energy drops are quite high. Radial-flow turbines are those in which the gas flow is essentially at right angles to the turbine shaft. Flow may be radially inward or outward, but commercially available turbines are usually the radial-inward-flow type. Radial-flow turbines are usually single-stage and have combination impulse-reaction blades and a rotor that resembles a centrifugal-pump impeller. The gas is jetted tangentially into the outer periphery of the rotor and flows radially inward to the "eye," from which the gas is jetted backward by the angle of the blades so that it leaves the rotor without spin and flows axially away.

TABLE 1. GAS TURBINE MANAGEMENT AND MAINTENANCE RESOURCES
(A Development of the Electric Power Research Institute)

Service/Product	Description
Outage Management	
Gas Turbine Overhaul Plan (GTOP)	Outage planning database (available for GE MS7001 and Westinghouse 501 turbines; under development for GE MS5001 and Asea Brown Boveri 11N turbines)
Efficiency Maintenance Analysis Program	Thermal performance analysis program
Training and Expert Systems	
SA•VANT	Expert system for troubleshooting operational problems
Plant Improvement Course	Two-day seminar on improving operations and maintenance
Compressor Blade Walk Inspection	Videotape and checklist (available mid-1992)
Hot Gas Path Maintenance	
REMLIFE	Computerized algorithm to estimate remaining life of first-stage blading
Advisor for Blade Coating (ABC)	Computerized selection of blade coatings
SPECS	Specifications for repair of nozzles and turbine blades
BLADE-CT	Finite-element analysis program to assess stress, heat transfer, and vibration of blading
Blade Life Assessment and Repair Guidebook	Manual of methods for determining condition of blading
Technology Transfer	
Combustion Turbine Center (Charlotte, North Carolina)	Technology transfer and advisory center; electronic bulletin board
Data Applications Center	Service that provides easy access to databases for customized reliability information
Inventory of Gas Turbines (INTURB)	Database of gas turbine engines, sites, and personnel
Standard Equipment Code	Standardized equipment breakdown for combustion turbine and combined-cycle plants

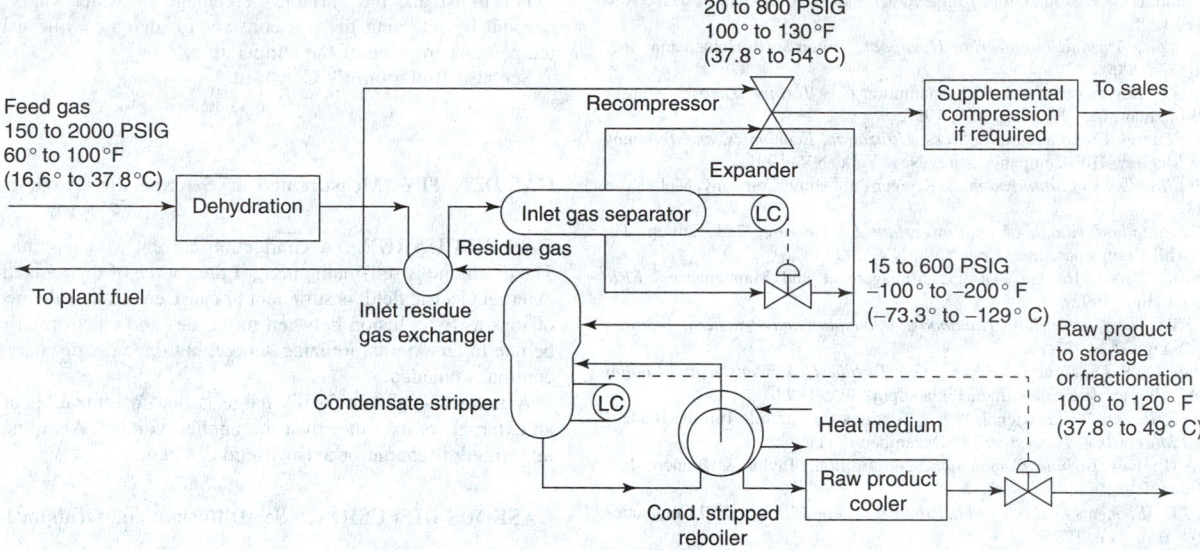

Fig. 10.   Expander system for gas processing. LC = level controller. (*Fluor Corp.*)

These latter machines usually have an efficiency of from 75 to 88%, usually operate at very low temperature, operate often on small or moderate streams, dictating a comparatively high rotating speed, and incorporate effective shaft seals to conserve the process stream. Commonly established operating limitations for turboexpanders are an enthalpy drop of 40 to 50 Btu (10–12.6 Calories)/pound/stage of expansion, and a rotor-tip speed of 1,000 feet (300 meters) per second. Commercial turboexpanders are available up to 2,500 psig inlet pressure and inlet temperatures of over 538°C. The permissible liquid production in the expanding stream varies with discharge pressure; it may be as high as 20% (weight) in the discharge, provided the turboexpander has been specifically designed to handle liquids.

Power Recovery. A potential application for the turboexpander exists whenever a large flow of gas is reduced from a high pressure to some lower pressure, or when high-temperature process streams (waste heat) are available at moderate pressures. When such conditions exist, they should be examined to determine if the use of a turboexpander is justified. In such cases, a turbine can be used to drive a pump, compressor, or electric generator, thus recovering a large portion of the otherwise wasted energy. In applications of this type, careful consideration should be given to the temperature drop that will occur in the expander. Sometimes it may be necessary to heat or dry the inlet gas to avoid low exhaust temperatures, or the formation of liquids.

Refrigeration. Turboexpanders used as components of refrigeration systems offer many possibilities to the designer of refrigeration cycles. They may be used in closed cycles with a pure gas, such as nitrogen, which is alternately compressed and expanded to provide the required refrigeration through a heat exchanger. Various types of open cycles also can be devised so that the process stream to be cooled passes through the expander, thus eliminating the need for the low-temperature heat exchanger. Liquid products can be produced directly from the turboexpander in this manner provided that the expander is specifically designed for this type of service.

The first turboexpander designs took advantage of a "free" pressure drop that was available at particular locations. Since then, the technique has been refined and enlarged to embrace practically every situation encountered in extracting hydrocarbons from a mixed gas stream; even where "free" pressure drop is not available. Designs were improved through better utilization of construction materials, design of the control system by simulation of operations in a computer, and the development of interlocking instrumentation.

The foregoing refinements enabled the utilization of the turboexpander economically in plants where full pressure restoration is required. The turboexpander system also has been used for recovery of propane alone and, in some cases, has been found to be more economical than the oil absorption process. Other turboexpander processes include dehydration and

dew point control of wellhead gas streams, utilizing pressure reduction of 2,000 to 10,000 psig, (136 to 680 atmospheres) and nitrogen rejection together with heavier hydrocarbon recovery.

Some inherent advantages of turboexpander systems include: (1) the final cryogenic operating temperature is obtained from the turboexpander and not achieved by costly low-level external refrigeration; (2) product separation pressure is set to give the most desirable equilibrium conditions; (3) plants are compact and inherently simple; (4) capital investment and operating costs are usually low, as much as 40% savings over conventional ethane recovery method; and (5) maintenance requirements are low.

With reference to Fig. 10, the turboexpander system operates as follows: feed gas is first dehydrated and sometimes alcohol is injected at strategic points to protect further against the formation of ice or hydrates. Next, the feed is chilled by heat exchange with the residue gas. Condensed liquids are then separated and the vapors delivered to the expander. A direct-connected compressor recovers the expander energy, boosting the pressure of the condensate stripper overhead gas. Residue gas can be further compressed to any delivery pressure. Alternatively, the compressor can be used to boost the pressure of the feed gas to obtain additional refrigeration. By using a turboexpander to remove energy from the gas, refrigeration is materially increased and the temperature is lowered below those conditions obtainable from simple adiabatic expansion.

Liquids condensed at the expander outlet and in the feed chilling step are fed to a fractionation system. The bottoms from fractionation represent the desired product mixture, which can have an ethane content equivalent to 90% or more of the ethane in the feed gas. Virtually 100% of propane and heavier hydrocarbons in the feed gas can be recovered. Flexibility in product composition is obtained either by adjusting the expander outlet pressure, or by stripping undesirable components overhead in the condensate stripper. Pressure difference across the expander is the principal energy source. Thus, minimal amounts of electric power and fuel are required for pumps, dehydrator regeneration heat, and stripper reboiler heat.

Because of the small size of turboexpander rotating elements, the forces and stresses at high speed (7,500 to 45,000 revolutions per minute) are equal to or less than those encountered in lower-speed rotating machinery. Any part of a turboexpander usually can be replaced in about 3 hours downtime.

Additional Reading

Anderson, J.D., Jr.: *Modern Compressible Flow*, 2nd Edition, The McGraw-Hill Companies, Inc., New York, NY, 1990.

Avallone, E.A. and T. Baumeister, III, "Marks," *Standard Handbook for Mechanical Engineers*," 10th Edition, The McGraw-Hill Companies, Inc., New York, NY, 1996.

Bathie, W.W.: *Fundamentals of Gas Turbines*, 2nd Edition, John Wiley & Sons, Inc., New York, NY, 1995.

Beck, D.S. and D.G. Wilson: *Gas-Turbine Regenerators*, Chapman & Hall, New York, NY, 1996.

Boyce, M.P.: *Gas Turbine Engineering Handbook*, Butterworth-Heinemann, Inc., Woburn, MA, 2000.

Cohen, H., G.F. Rogers, and H.I. Saravanamutto: *Gas Turbine Theory*, Addison Wesley Longman, Inc., Reading, MA, 1996.

Culp, A.W.: *Energy Classification, Sources, Utilization, Economics, and Terminology*, The McGraw-Hill Companies, Inc., New York, NY, 1991.

Ehrich, F.F.: *Handbook of Rotordynamics*, Krieger Publishing Company, Melbourne, FL, 1998.

Elliott, T.C.: *Standard Handbook of Powerplant Engineering*, 2nd Edition, The McGraw-Hill Companies, Inc., New York, NY, 1994.

Frischmuth, R.: "Tools for Gas Turbine Management and Maintenance," *EPRI Journal*, 38 (June 1992).

Giampaolo, T.: *The Gas Turbine Handbook: Principles and Practices*, Fairmont Press, Lilburn, GA, 1997.

Han, Je-Chin, C., S. Dutta, and S. Ekkad: *Gas Turbine Heat Transfer and Cooling Technology*, Taylor & Francis, Inc., Philadelphia, PA, 1999.

Lämmermann, H. and G.KG. Kienel: "PVD Coatings for Aircraft Turbine Blades," *Advanced Materials & Processes*, 18 (December 1991).

Lefebvre, A.H.: *Gas Turbine Combustion*, 2nd Edition, Taylor & Francis, Inc., Philadelphia, PA, 1998.

Mattingly, J.D.: *Elements of Gas Turbine Propulsion*, The McGraw-Hill Companies, Inc., New York, NY, 1995.

Polonyl, M.J.: *Power and Process Control Systems*, McGraw-Hill, New York, NY, 1991.

Schilke, P.W. et al.: "Advanced Materials Propel Progress in Land-Based Gas Turbines," *Advanced Materials & Processes*, 22 (April 1992).

Treager, I.E.: *Aircraft Gas Turbine Engine Technology*, 3rd Edition, The McGraw-Hill Companies, Inc., New York, NY, 1996.

Wilson, D.G.: *The Design of High-Efficiency Turbomachinery and Gas Turbines*, MIT Press, Cambridge, MA, 1990.

GAS BURNER. See **Burner**

GAS CHROMATOGRAPHY. Analytical separation technique where the minor components in a mixture of gases are separated and resolved into individual components. The technique requires the transmission of the gas sample through a column in the chromatograph using a mobile phase or carrier gas. The column is either packed or coated with a material for which the gases to be separated have an affinity and the strength of this affinity largely determines the time any individual component is retained in the column. Various detectors are employed in gas chromatography, from very specific compound-responsive detectors (flame photometric detector, electron capture detector, photoionization detector, etc.) to some very generally sensitive detectors (flame ionization detector, thermal conductivity detector, atomic emission detector, etc.) Gas chromatography is commonly used for the quantification of halocarbons and hydrocarbons in the atmosphere. See also **Chromatography**.

GAS CONSTANT. The constant factor in the equation of state for ideal gases.

The universal gas constant is

$$R^* = 8.316963 \text{ J mol}^{-1} \text{ K}^{-1}$$

The gas constant for a particular gas is

$$R = R^*/m,$$

where m is the molecular weight of the gas. For a mixture, the "molecular weight" is a weighted mean of the molecular weights of the components:

$$m = \left(\frac{f_1}{m_1} + \cdots + \frac{f_n}{m_n} \right)^{-1},$$

where $m_1, \cdots, m_n$ are the molecular weights of the n gases, and $f_1, \cdots, f_n$ are their masses relative to the total mass of the mixture. The gas constant for dry air is

$$R_d = 2.870 \times 10^2 \text{ W Kg}^{-1} \text{ K}^{-1}.$$

The gas constant for water vapor is

$$R_v = 4.615 \times 10^2 \text{ W Kg}^{-1} \text{ K}^{-1}.$$

For moist air, the variable percentage of water vapor is taken into account by retaining the gas constant for dry air while using the virtual temperature in place of the temperature.

See also **Boltzmann's Constant**.

AMS

GAS DENSITY (Measurement). See **Specific Gravity**.

GAS DISCHARGE. A conduction current in a gas due to ionization. The discharge is self-maintaining if the source of the ionization, such as an external electric field, is sufficient to cause creation of the necessary supply of ions as by collision between molecules and electrons. Breakdown may be due to an external ionizing source, but once the discharge is initiated it continues unaided.

A non-self-maintaining discharge is due to ionization of the gas from an external source other than the applied voltage. Also known as a field-intensified discharge or a Townsend discharge.

GASEOUS DIFFUSION. See **Diffusion**; and **Graham Law**.

GASES (FOODS). See **Food Additives**.

GAS HYDRATE. A clathrate compound formed by gas (either noble or reactive) and water. The compounds are cyrstalline solids and are insoluble in water. They usually form (only at relatively low temperatures and high pressures) directly by contact of gas and liquid water. From 6 to 18 molecules of water may combine with each molecule of gas, depending on the nature of the gas.

The best-known gas hydrates are those of ethane, [CAS: 74-84-0] C_2H_6, ethylene, [CAS: 74-85-1] C_2H_4, propane, [CAS: 74-98-6] C_3H_8, and isobutane, [CAS: 75-28-5] C_4H_{10}. Others include methane, [CAS: 74-82-8] CH_4, and 1-butene, [CAS: 106-98-9] C_4H_8, most of the fluro-carbon refrigerant gases, nitrous oxide [CAS: 10024-97-2] N_2O, acety-lene, [CAS: 74-86-2] C_2H_2, vinyl chloride, [CAS: 75-01-4] C_2H_3Cl, carbon dioxide, [CAS: 124-38-9] CO_2, methyl, [CAS: 74-87-3] CH_3Cl, and ethyl, [CAS: 75-00-3] C_2H_5Cl, chloride, methyl, [CAS: 74-83-9] CH_3Br, and ethyl, [CAS: 74-96-4] C_2H_5Br, bromide, cyclopropane, [CAS: 75-19-4] C_3H_6, hydrogen sulfide, [CAS: 7783-06-4] H_2S, methyl mercaptan, [CAS: 74-93-1] CH_4S, and sulfur dioxide, [CAS: 7446-09-5] O_2S.

Interest in the gas hydrates originated mainly because of the nuisance of such compound formation in gas pipelines. In recent years, propane has been used successfully to precipitate water from salt solution (or seawater), thus yielding potable water.

GAS (Ideal). See **Ideal Gas Law**.

GASIFICATION PROCESSES. See **Coal Conversion (Clean Coal) Processes**.

GAS LAWS. The thermodynamic laws applying to perfect gases: Boyle-Mariotte law, Charles-Gay-Lussac law, Dalton law, equation of state. Also called *perfect-gas laws*, or *ideal-gas laws*.

GAS METER. See **Flow Measurement**.

GASOHOL. Gasoline blended with alcohol. Typically 10% methanol or ethanol is blended with the gasoline.

See also **Petroleum**; and **Wastes as Energy Sources**.

GASOLINE. See **Petroleum**.

GAS (Perfect). See **Perfect Gas**.

GAS RESEARCH INSTITUTE (GRI). Headquartered in Chicago, Illinois, the mission of GRI is to plan, manage, and develop financing for a cooperative research and development program addressing improvements in production, transport, storage, and end use of gaseous fuels for the mutual benefit of the gas industry (producers, pipelines, and distributors) and its present and future customers. Pursuing benefits by applying new gas technology is the major element of GRI's mission. Developing new and improved technologies that maximize the value of gas energy services, while minimizing the cost of supplying and delivering gaseous fuels, is the most effective way to serve the mutual interests of both the gas industry and its customers. These mutual benefits can only be realized if the results of R&D are used. Consequently, GRI gives substantial attention to *technology transfer* and commercialization of its R&D results starting from the inception of each concept.

GRI implements its mission through a contractor-performed R&D program. The objectives and goals for these programs are reviewed annually. Integral to this review process, the proposed R&D program elements are subjected each year to a rigorous benefit-cost analysis to ensure that current and future gas consumers and the companies that serve them will realize, in a timely fashion, the expected benefits of GRI R&D.

Specifically, GRI programs are designed to:

1. Decrease the cost of producing and transporting gas.
2. Assist in assuring the adequate deliverability of natural gas.
3. Enhance the role of gas in providing *least-cost*, environmentally benign energy services.
4. Facilitate the transfer of new technology and technical and scientific information to the gas industry, its customers, gas equipment manufacturers, and the interested public.
5. Stimulate innovation in gas-related technologies through a mission-oriented basic research program.
6. Provide important scientific information on new technology performance and potential applications.
7. In the overall, provide net benefits for gas rate-payers.

GRI programs are subject to review and approval by the (U.S.) Federal Energy Regulatory Commission (FERC), with state regulatory commissions and that of the District of Columbia. GRI member companies and other interested parties are afforded an opportunity to participate in the reviewing procedure.

GAS SCRUBBING. The contacting of a gaseous mixture with a liquid for the purpose of removing gaseous contaminants or entrained liquids or solids.

GASSER, HERBERT SPENCER (1888-1963). Herbert Spencer Gasser was born in Platteville, Wisconsin in 1888. His father, a country doctor, was fascinated by the controversies surrounding the concept of evolution, and named his son after the British philosopher Herbert Spencer. Gasser inherited this enthusiasm for biology and studied zoology at the University of Wisconsin. He easily passed the BA degree, and then proceeded to study medicine there. He was taught physiology by Joseph Erlanger, with whom he would later collaborate. Whilst still a student he was appointed instructor in physiology. In 1913 he completed his clinical training at the Johns Hopkins University, graduating in 1915. He then returned to the University of Wisconsin as instructor in pharmacology. See also **Spencer, Herbert (1820-1903)**.

In 1916 he was invited by Erlanger to take up a post at the Department of Physiology at Washington University, St Louis, where Erlanger was Professor. During the First World War they studied a number of war-related problems, including shock and blood loss. In 1918 Gasser became a member of the Armed Forces Chemical Warfare Service, established in Washington, DC.

As a researcher Gasser was interested in the physiology of nerves. At Johns Hopkins he and H. Sidney Newcomer built a vacuum tube amplifier to record the progress of electrical impulses along a nerve. This was to prove an important research tool. The fruits of this early research were published in 1921, the same year that Gasser became Professorof Pharmacology at Washington University. In collaboration with Erlanger, Gasser broughtadvances in electronics and physics into the physiological laboratory. Together they used a cathode ray tube to display the amplified activity of nerve trunks. Using this technology they were able to show that the recorded wave forms were made up of individual waves representing the activity of the individual nerve fibers within the trunk. With the cathode ray oscilloscope they began to analyze the compound nature of nerve fibers,

and to categorize the role and size of the different fibers. Such studies were fraught with technical problems, but were to lead to the establishment of a theory of differentiated function.

In 1923 he traveled to Europe with the support of a grant from the Rockefeller Foundation. In his two years abroad he worked with A.V. Hill and Sir Henry Dale in London, and Walter Straub in Munich. In 1931 he was appointed Professor of Physiology at Cornell University Medical College in New York City. Then in 1935 he succeeded Simon Flexner as Director of the Rockefeller Institute for Medical Research. This role as an administrator suited his talents, but left him little time for his own researches. In 1944 Gasser and Erlanger were awarded the Nobel Prize for Medicine or Physiology for their discoveries relating to the nature of nerve fibers. On retirement in 1953 the Rockefeller Institute provided him with a small laboratory where he continued his earlier researches using another technological innovation, the electron microscope, to study nerve fibers. See also **Dale, Henry Hallett (1875-1968)**; **Electron Microscope**; **Flexner, Simon (1863-1946)**; and **Hill, Archibald Vivian (1886-1977)**.

Additional Reading

Gillispie, C.C.: *Dictionary of Scientific Biography (1970–1980)*, Vol. 5, Charles Scribner's Sons, New York, NY, 1981.

CLAIRE E. J. HERRICK, London, UK

GASTEROPODA (or *Gastropoda*). The snails, slugs, and allied forms, constituting a class of the phylum *Mollusca*. This group includes a large number of marine and freshwater species and many that are terrestrial.

The chief structural characteristics of the gasteropods are these: (1) most species have a dorsal visceral hump, which is often spirally twisted. (2) A head is present, bearing eyes and tentacles. (3) The mouth is provided with a toothed organ called the radula. (4) The foot is usually a broad creeping organ. (5) Respiratory ctenidia lie in the mantle cavity of some species, and in others, the walls of the cavity are the respiratory organ. (6) In many species, a shell, conical or spirally coiled, encloses the visceral hump.

Gasteropods are of relatively little economic importance. Snails are eaten in Europe and the abalones of the Pacific Coast are also used as food. The shell of the abalones furnishes beautifully iridescent mother-of-pearl for costume jewelry and there is an extensive traffic in the shells of many species among collectors.

The group is classified as follows:

Subclass *Streptoneura*. Usually with a shell closed by a horny shield, the operculum, when the animal is retracted.
   Order *Diotocardia* (*Aspidobranchiata*). Abalones, limpets, and other marine species. A few freshwater forms.
   Order *Monotocardia* (*Pectinibranchiata*). Whelk, periwinkle, and many other marine forms and a few freshwater species.
Subclass *Opisthobranchiata*. Shell small and internal, sometimes lacking.
   Order *Tectibranchiata*. Sea hare, sea butterflies or pteropods with the foot expanded into wing-like lobes. All marine.
   Order *Nudibranchiata*. Marine species without shells. Often with complex dorsal processes. Sea lemon; nudibranchs.
Subclass *Pulmonata*. Shell usually present but without an operculum. Mantle cavity sometimes the only respiratory organ. Mostly freshwater and terrestrial species, a few marine.
   Order *Basommatophora*. Eyes at the bases of the posterior tentacles. Many common snails.
   Order *Stylommatophora*. Eyes at the tips of the posterior tentacles. Common snails and slugs. (See also **Invertebrate Paleontology**.)

GASTRECTOMY. See **Ulcer**.

GASTRIC JUICE. See **Bile**; and **Digestive System (Human)**.

GASTRIC ULCER. See **Ulcer**.

GASTRITIS. An inflammation of the lining membrane of the stomach, occurring in an acute or chronic form. The various gastric juices, such as enzymes, pepsin, hydrochloric acid, rennin, and lipase, as well as a heavy, protective mucus, are secreted by the stomach lining. In gastritis, these functions are disturbed and the digestive process is impeded. Gastritis may result from the ingestion of poisonous corrosives. Toxic substances associated with certain infections also may initiate or aggravate gastritis. Therapy depends upon the type and source of the gastritis. Differential diagnosis is frequently required to determine the exact causative factor. Disturbance of the stomach lining in gastritis may range from tiny hemorrhagic areas to ulceration. Duration of therapy may range from several days to several weeks.

GASTROENTERITIS. An inflammation of the gastrointestinal tract which may be caused by a number of factors. One of the most frequent causes is foodborne disease, notably salmonellosis. See also **Foodborne Diseases**. A number of viruses, collectively called enteroviruses, such as the echoviruses, the rotoviruses, and specific related agents, such as the Norwalk agent, can cause gastroenteritis of varying severity and time span. See also **Coxsackie Virus**; and **Norwalk Virus**. *Bacillus cereus* can cause outbreaks of self-limited gastroenteritis that lasts 12 to 15 hours. The disease has been associated with the ingestion of a number of foods, particularly fried rice, sauces, and meat, which contain enterotoxins as the result of inadequate refrigeration. One of the major symptoms of gastroenteritis is diarrhea. See also **Diarrhea**.

GASTROINTESTINAL CANCER.   See **Cancer and Oncology**.

GASTROINTESTINAL TRACT.   See **Digestive System (Human)**.

GASTROTRICHA. A group of minute animals found in fresh and salt water on the bottom and among the debris accumulated there. They move chiefly by means of cilia and have cement glands whose secretion attaches them temporarily to supports. They have a tubular alimentary tract and an excretory system consisting of two tubules with flame cells. The group is ranked by some writers as a class in the same phylum as the rotifers and by others as of uncertain relationship.

Two orders are recognized: *Macrodasyoidea*, made up of marine species with numerous cement glands and *Chaetonotoidea*, made up of marine and freshwater species with a single pair of cement glands at the caudal end of the body, or none.

GASTRULA. The stage in embryonic development in which the initial differentiation of tissues is evident. The gastrula is typically a sac whose wall is composed of the two germ layers, an outer ectoderm and an inner endoderm. The cavity lined by the endoderm is the archenteron and the opening to the exterior is the blastophore.

The gastrula is formed from the blastula by the process of gastrulation. Typically the wall of the spherical blastula caves in on one side and the invagination progresses until this side is in contact with the opposite wall. In some coelenterates, however, the two germ layers appear as a solid mass of endoderm surrounded by a layer of ectoderm, and the archenteron forms by the splitting of the inner mass. In animals with abundant yolk, modifications also appear. In birds, for example, the stage approximating the blastula is a disk of cells on the surface of the yolk and the endoderm may be formed by the folding under of this layer at one point on the margin or by a more diffuse process of polyinvagination, recently discovered. Later the folded edge undergoes a concrescent growth until it doubles on itself and fuses to form the primitive streak, equivalent to a closed blastophore.

The mesodermal layer also appears in the gastrula of triploblastic animals. Its formation is extremely variable but it usually grows out from the indeterminate zone about the blastophore where ectoderm and endoderm join.

GAS WELDING.   See **Welding**.

GATE CIRCUIT. A circuit that amplifies or passes a signal only in the presence of an appropriate synchronizing or "gating" pulse that "opens" the gate. Also used to refer to the various logic functions and circuits used to realize computer designs, such as the AND, OR, NOT, NOR, and NAND. See Fig. 1. See also **AND (Circuit)**; **Gate (Computer System)**; **NAND (Circuit)**; **NOR (Circuit)**; **NOT (Circuit)**; and **OR (Circuit)**.

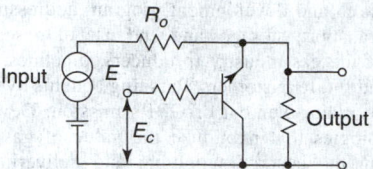

Fig. 1.   Transistor transmission gate.

GATE (Computer System). A circuit having a binary output which is fully determined by the binary state of its input signals, such as in the AND and OR gate circuits. Also, a signal which permits an AND circuit to pass a signal. Usually the gate signal is of longer duration than the signal to make certain that coincidence occurs. In conditioning the set pulse of a flip-flop, for example, the gate must precede the set signal in order that the negative shift will be recognized by the transistor. See also **Flip-Flop**.

GATE-TURNOFF SWITCH. An electronic device (GTO) that operates like a silicon-controlled rectifier with exception that the high current conduction state can be interrupted by a negative pulse applied to the gate electrode. Used in D.C. switching applications.

GATING. 1. The process of selecting those portions of a wave which exist during one or more selected time-intervals, or which have magnitudes between selected limits.

2. The function or operation of a saturable reactor or magnetic amplifier that causes it, during the first portion of the conducting alteration of the A.C. supply voltage, to block substantially all of the supply voltage from the load, and during latter portion allows substantially all of the supply voltage to appear across the load, is called gating or gating action. The "gate" is said to be virtually closed before firing and to be substantially open after firing.

GATTERMANN ALDEHYDE SYNTHESIS. Preparation of aldehydes of phenols, phenol ethers, or heterocyclic compounds by treatment of the aromatic substrate with hydrogen cyanide and hydrochloric acid in the presence of Lewis acid catalysts.

GATTERMANN-KOCH REACTION. Formulatin of benzene, alkylbenzenes, or polycyclic aromatic hydrocarbons with carbon monoxide and hydrochloric acid in the presence of aluminum chloride at high pressure. Addition of cuprous chloride allows the reaction to proceed at atmospheric pressure.

GAUCHER'S DISEASE. Gaucher disease is a rare inherited enzyme deficiency, which researchers estimate may be present in 10,000–20,000 Americans. It is a panethnic disorder, with highest prevalence in the Ashkenazi Jewish population.

During the past decade, much progress has been made in understanding the molecular biology of the disease and in the ability to treat patients with the disorder. However, many issues regarding diagnosis, population screening, and therapy for patients with Gaucher disease are controversial.

Gaucher disease is characterized by a remarkable degree of variability in its clinical signs and symptoms, ranging from severely affected infants to asymptomatic adults. Many patients suffer from anemia, bone damage, and enlarged livers and spleens; a few develop severe central nervous system damage. Gaucher disease is a potentially lethal disorder. All patients with Gaucher disease have a genetic defect in the enzyme glucocerebrosidase, which results in the accumulation of the lipid glucocerebroside within intracellular structures known as lysosomes.

Patients with Gaucher disease have been classified into three major types on the basis of clinical signs and symptoms: type 1, non-neuronopathic (adult); type 2, acute neuronopathic (infantile); and type 3, subacute neuronopathic (juvenile). All types of Gaucher disease can be diagnosed by demonstrating a deficiency of glucocerebrosidase activity.

The most striking differences among the three types are the presence or absence of neurologic manifestations and the rate of their progression. However, people with the same type of the disorder may differ in their clinical presentation. For example, certain patients with type 1 Gaucher disease, which is by far the most common type, may display some combination of anemia, low blood platelet levels, massively enlarged livers and spleens, and extensive skeletal disease. In contrast, other type 1 patients may have no symptoms and can be identified only by screening or during evaluation for other diseases.

The gene for glucocerebrosidase, which is located on chromosome 1q21, has been characterized and sequenced. Many mutations in the glucocerebrosidase gene have been identified in DNA from different patients; several of these mutations are frequent. Although some patients with the same DNA mutations have similar clinical courses, other patients with the same mutations have very different clinical manifestations. It is still not clear to what extent a person's clinical features (phenotype) or prognosis can be accurately predicted through current mutation analysis. Furthermore, although the molecular techniques can be used for early prenatal diagnosis, detection of individuals carrying the disease gene, and population screening, the appropriate clinical application of these molecular techniques remains unresolved.

Gaucher disease has been traditionally managed by supportive therapy including total or partial removal of the spleen, blood transfusions, orthopedic procedures, and occasionally bone marrow transplantation. More recently, enzyme replacement therapy has become available and has proven effective in many patients. Enzyme replacement therapy has successfully reversed many of the manifestations of the disorder, including abnormal blood counts, increased liver and spleen size, and some skeletal abnormalities. The therapy is very costly, however, ranging from $100,000 to $400,000 annually for each patient.

Natural History

The natural history of Gaucher disease is incompletely documented. The progression and outcome are well understood only in type 2 disease (infantile form). Type 3 disease (juvenile form) has a more variable course. The type 1 (adult) form is most common, especially variable, and least well characterized. Furthermore splenectomy, orthopedic intervention, and enzyme replacement therapy for type 1 Gaucher disease alter its course, natural progression, and outcome. Thus, it is important to standardize the reporting of the effects of these interventions.

Type 1 disease typically presents after infancy and often not until adult life. Indeed, some genotypically affected individuals may never come to medical attention, and their number is unknown. With DNA analysis of family members, many such individuals will be diagnosed. Current technologies may unmask and identify organ-specific manifestations in these asymptomatic individuals. Simple hematologic and biochemical assays and imaging techniques can be used to assess disease progression. Skeletal disease is especially difficult to assess. Mutation analysis provides precise diagnosis but may not give information concerning the severity or progression of the disease. In addition, there are considerable differences in the degree to which organ systems are affected. Furthermore, there are reports of intrafamilial variation. Differences in disease severity have been demonstrated even in identical twins. Thus, other genetic and nongenetic factors appear to be involved in the expression of the disease.

Prenatal diagnosis now affords an opportunity to assess the natural progression of the disease from before birth. Such information may be critical in choosing appropriate technology for prognosis and therapy. Appropriate systematic and quantitative description of the disease is essential to understand its natural course. Patient characterization requires clarification of the terminology used to describe patients, which at this time is confused (e.g., "asymptomatic" versus "asymptomatic but with physical signs and laboratory evidence of disease").

Roles of Current Molecular and Enzymatic Assays

Enzyme analysis of leukocyte or fibroblast extracts is appropriate to confirm or exclude the diagnosis of Gaucher disease. Several methodologies for enzymatic diagnosis are currently available and are reliable in experienced hands. No consensus has yet been reached on a single most appropriate method, which makes it essential that each laboratory have rigid internal quality assurance and quality control of the method it uses.

The prognosis for patients with type 1 disease cannot be predicted from the residual enzyme activity measured in tissues. Enzymatic analysis cannot be used to detect carriers reliably.

Analysis of DNA for mutations by molecular methods (genotyping) is appropriate in all individuals with glucocerebrosidase deficiency. Genotyping of siblings and parents of affected individuals is important to ascertain other potentially affected individuals who may be asymptomatic and to identify carriers for genetic counseling. Enzyme analysis of parents of affected individuals is also valuable to exclude the possibility of asymptomatic glucocerebrosidase deficiency in a parent with two mutant alleles, only one of which was identified by genotyping. Although current genotype/phenotype correlations are imperfect, genotyping may indicate that neurologic complications are unlikely. It has less value in predicting the likelihood of other complications.

Molecular methods can provide accurate carrier detection, particularly in defined populations. For example, in the Ashkenazi Jewish population, screening for five mutations allows detection of approximately 95% of heterozygous individuals. The greater variety of mutations in non-Jewish populations makes carrier detection in these populations more challenging with currently available technology. Analysis of some mutations by DNA amplification can be complicated by the presence of a highly homologous pseudogene that is located nearby. Quality control of the molecular techniques is important, as is awareness of the complexities in interpreting data produced by these amplification methods.

Widespread application of genetic screening to detect either presymptomatic patients with Gaucher disease or heterozygous carriers is not appropriate at this time. The medical value of presymptomatic diagnosis of patients with Gaucher disease and carrier testing has not been established. For this reason, pilot studies examining the potential benefits and/or harms of such screening programs should be encouraged. Ideally, the target community should be involved in the implementation and evaluation of such pilot studies.

Indications for Treatment and Appropriate Modes of Therapy

The clinical features of type 1 glucocerebrosidase deficiency are highly variable, ranging from serious multisystem involvement to the absence of signs or symptoms. In addition, the age of onset of clinical features in those who develop symptoms is variable. This degree of variability raises several important issues that must be considered before initiating treatment. First, the characteristic signs of the disorder, which include anemia, thrombocytopenia without bleeding, hepatosplenomegaly without pain or discomfort, and radiologic changes without evidence of fractures or bone pain, must be differentiated from the symptoms of the disorder, such as bleeding, somatic pain, bone crises, and fractures. Second, knowledge is inadequate on the effect of treatment for patients who display signs but no symptoms of the disease. There is a reasonable consensus to treat those who exhibit symptoms; however, no agreement exists on the clinical criteria for initiating treatment. No consistent guidelines are available at this time because of the lack of sufficient information about the natural history of the disease.

In addition, a group of individuals, of unknown number, have the enzyme deficiency but have not developed signs or symptoms. Because we cannot predict whether these individuals will ever become symptomatic, the appropriateness of prophylactic therapy has not been determined.

A systematic evaluation of enzyme-deficient individuals to define the natural history of the disease is lacking. For symptomatic patients, there should be sufficient extant data given the number of patients who have already been identified, treated, and extensively followed. For asymptomatic individuals, it is necessary to develop protocols for longitudinal evaluation.

Conservative therapy has a role in Gaucher disease, such as hydration, analgesics, and narcotics for pain in bone crises and orthopedic surgical intervention for fractures. The use of vitamin D, calcium, and bisphosphonate in bone crises and for bone growth requires further study.

Although bone marrow transplantation is an effective form of therapy, the risk of mortality and morbidity makes this mode of treatment less desirable.

In type 1 disease, there is good evidence that enzyme replacement therapy with mannose-terminated placental or recombinant glucocerebrosidase is beneficial in reducing hepatosplenomegaly, improving hematologic parameters, and, to a lesser extent, in alleviating bone disease. Enzyme therapy appears to obviate the need for splenectomy in most cases.

Several patients with type 2 disease are reported to have been treated with enzyme replacement therapy, and there was no substantial improvement in their neurologic problems. With current technology, enzyme replacement therapy is unlikely to prove efficacious for patients with type 2 disease. The efficacy of enzyme replacement for neurologic abnormalities in type 3 disease remains to be established.

For individuals with type 1 disease, controversies continue over aspects of enzyme replacement therapy, such as dosage, methods and frequency of administration of the enzyme, and cost. The most contentious issue, and potentially the most difficult for patients and their physicians, is enzyme dosage. Clinical successes have been observed with both the "high" and "low" dosage regimens (described as amount of enzyme administered during a 4-week interval for purposes of comparison, independent of dosage schedule): the 120 U/kg/4 weeks and the 30 U/kg/4 weeks, respectively. Inadequate clinical responses were also reported for all dosage regimens tested. The debate about dosage is complicated by the failure to compare data adequately and by the diversity of protocols. Review of the data indicates two salient points. First, patients vary considerably and unpredictably in their responses. Second, many patients do well on lower dosage regimens. The use of low-dosage regimens for such patients would markedly reduce costs. Debates focusing on minimal differences in degrees and rate of improvement have detracted from the appreciation of the treatment's value.

Current studies are evaluating regimens with dosages even lower than 30 U/kg/4 weeks. The patients in these studies may respond well, but some respond more slowly. Initial and maintenance therapy should be directed at achieving sustained benefit with the lowest possible dosage. The choice of dosage and frequency of enzyme administration will have to be adjusted individually while each patient's progress is monitored. Response may be slow regardless of dosage.

Given the limited number of patients, the treatment strategies, including criteria for intake, dosage, and periodic reevaluation, should be standardized to ensure that data from multiple centers can be pooled to evaluate the proposed treatment regimens. The resolution of these treatment issues can be addressed best through carefully designed, cooperative, clinical trials. The questions to be answered by such trials will be refined if existing data sets are pooled and analyzed without preconceived constraints. In addition, further studies should include the development of more efficient cellular targeting and uptake of the enzyme. The clinical and ethical ramifications of enzyme therapy and the funding of clinical trials must be considered.

Studies to evaluate alternative forms of enzyme replacement therapy and alternative approaches, such as the use of inhibitors of sphingolipid biosynthesis, should be encouraged. Moreover, Gaucher disease is an excellent candidate for gene therapy, and continued research on this modality, including the use of animals models, is therefore indicated.

Goals

The goals of treatment are the amelioration of the manifestations of Gaucher disease and the overall improvement of the health and quality of life of patients.

Although alglucerase has been shown to ameliorate many of the manifestations of type 1 Gaucher disease, the major current concerns are the proper indications to begin treatment, the most appropriate treatment regimens, and cost. Answering the many questions concerning the management of Gaucher disease will require a cooperative effort of considerable scale. For this cooperative effort to have its maximum impact, the organizer of the cooperative effort must be free of real or perceived bias. It is suggested that the National Institutes of Health should take the initiative and foster the establishment of a cooperative group of investigators involved in the diagnosis and treatment of patients with Gaucher disease. Three phases in the operation of the proposed group are (1) establishment of a patient registry, (2) analysis of the existing data on natural history and response to therapy, and (3) design and conduct of clinical trials to address unanswered questions.

It would be advantageous to enter all patients with Gaucher disease into a registry. Such a registry would provide a valuable resource for increasing knowledge of the natural history of the disease, help to identify predictors of response, and facilitate clinical trials to answer specific questions about therapy.

Clinical trials will be most informative if recruitment numbers are adequate to answer the questions addressed, if individuals are stratified for the most relevant variables (e.g., genotypes or baseline levels) to ensure comparability of the various subject groups, and if patients are randomized to treatment arms where appropriate.

A high priority for an early clinical trial is comparison of the dosage and frequency of enzyme administration to symptomatic patients. Outcomes to be assessed should include not only hemoglobin concentration, platelet count, spleen and liver size, and bone integrity but also the patients' functional state, convenience, satisfaction, quality of life, impact on the family, and cost.

A second priority for a clinical trial is to assess the need for enzymatic treatment of asymptomatic patients. Such a trial might initially be confined to high-risk, asymptomatic patients to increase the likelihood of observing a preventive effect of treatment.

Additional Reading

Barton, N.W. et al.: "Replacement Therapy for Inherited Enzyme Deficiency—Macrophage, Targeted Glucocerebrosidase for Gaucher's Disease," *N. Eng. J. Med.*, 1464 (May 23, 1991).

Bellenir, K.: *Genetic Disorders Sourcebook: Such as Down Syndrome, Pku, Hemophilia, VonWillebrand Disease, Gaucher Disease, and Tay-Sachs D*, Omnigraphics, Inc., Detroit, MI, 1996.

Beutler, E. et al.: "Enzyme Replacement Therapy for Gaucher's Disease," *N. Eng. J. Med.*, 1809 (December 19, 1991).

Beutler, E.: "Gaucher's Disease: New Molecular Approaches to Diagnosis and Treatment," *Science*, 794 (May 5, 1992).

Dulbecco, R.: *Encyclopedia of Human Biology*, Academic Press, San Diego, CA, 1977.

Web References

Gaucher Association of Canada: http://www.gaucher.org

Gauchers Association (UK): http://www.gaucher.org.uk/first.htm

National Gaucher Foundation (USA). http://www.gaucherdisease.org

National Center for Biotechnology Information: http://www.ncbi.nlm.nih.gov/disease/Gaucher.html

National Institutes of Health, Bethesda, MD

GAUGE THEORIES. An excellent and brief description is given by Hung and Quigg (1980): "At the base of the unification of interactions (particles) is the idea of gauge invariance, which draws its name from some early investigation by Weyl (1951) into a possible connection between scale changes and the laws of electromagnetism. Weyl's specific attempt to deduce electromagnetism from a symmetry principle—invariance under a change of length scale at every position of space-time independently—ran afoul of quantum mechanics, but the general strategy and the name have survived. Indeed, gauge theories constructed to embody various symmetry principles are now believed to provide the correct quantum descriptions of the strong, weak, and electromagnetic interactions. The simplest example of a gauge theory is electromagnetism itself." More details can be found in *Science*, **210**, 1208–1209 (1980). The Weyl reference cited is "Space-Time Matter," Chap. 4, sect. 35, page 282, translation published by Dover, New York (1951). Reference to "A Unified Theory of Elementary Particles and Forces," by H. Georgi, *Sci. Amer.*, **244**, 4 (1981). See also **Conservation Laws and Symmetry**; and **Particles (Subatomic)**.

Additional Reading

Adair, R.K.: *The Great Design—Particles, Fields, and Creation*, Oxford University Press, New York, NY, 1989.

Davies, P.: *The New Physics*, Cambridge University Press, New York, NY, 1992.

Ellis, P.J. and Y.C. Tang: *Trends in Theoretical Physics*, Addison-Wesley, Redwood City, CA, 1990.

Falomir, H., R.E. Gomboa, and F.A. Schaposnki: *Trends in Theoretical Physics*, American Institute of Physics, College Park, MD, 1998.

Frampton, P.H.: *Gauge Field Theories*, John Wiley & Sons, Inc., New York, NY, 2000.

Guidry, M.: *Gauge Field Theories: An Introduction with Applications*, John Wiley & Sons, Inc., New York, NY, 2000.

Leader, E. and E. Predazzi: *An Introduction to Gauge Theories and Modern Particle Physics*, Vol. 2, Cambridge University Press, New York, NY, 1995.

Pokorski, S.: *Gauge Field Theories*, Cambridge University Press, New York, NY, 2000.

GAUSS. A cgs unit of magnetic induction (or magnetic flux per unit area across an area at right angles to the magnetic field), equal to one maxwell per square centimeter.

Cgs and mks units of magnetic induction are related by 10^4 gauss = 1 weber m^{-2} = 1 Tesla (T), where 1 weber = 1 (newton meter)/amp = 1 volt s. The induction of the earth's magnetic field in the United States is of order 0.5 gauss, with the magnetic field oriented about 20° from zenith. Magnetic induction inside superconducting magnets can be as high as 20 T (2×10^5 gauss), while magnetic induction produced by the human spine is of order 15×10^{-15} T (1.5×10^{-10} gauss).

See also **Units and Standards**.

GAUSS, CARL FRIEDRICH (1777–1855). Gauss was a German mathematician, astronomer, and physicist. He contributed to most branches of mathematics. He was the founder of the modern number theory and worked in geometry without use of Euclid's fifth postulate. Gauss made advances in the mathematics of curved surfaces and in surveying techniques. He invented the heliotrope. In astronomy, he was able to calculate the minor planet Ceres' orbit path. Later in life, he became interested in electromagnetism. He is known for Gauss' laws of electricity and magnetism and Principle of Least Constraint.

See also **Gauss Theorem**; **Geometry**; and **Units and Standards**.

J. M. I.

GAUSS CONFORMAL. See **Lambert Projection**.

GAUSSIAN NOISE. See **Noise**.

GAUSS-MARKOFF THEOREM. A theorem in statistics to the general effect that the best estimator of a parameter from a population, among the class of estimators, which are linear in the sample values, is obtained by the method of least squares. "Best" in this sense means that the estimator is unbiased and has minimal variance.

GAUSS THEOREM. A relation between multiple integrals which in Cartesian coordinates is

$$\int_\tau \left(\frac{\partial u}{\partial x} + \frac{\partial v}{\partial y} + \frac{\partial w}{\partial z} \right) dx\,dy\,dz = \int_S (\lambda u + \mu v + vw)\,dS$$

The quantities u, v, w are functions of x, y, z having continuous first derivatives within a volume τ and they approach their values on the bounding surface continuously. The outward normal to the surface has direction cosines λ, μ, v.

A vector form of the theorem, often known as the divergence theorem, is

$$\int_\tau \nabla \cdot \mathbf{V}\,dt = \int_S \mathbf{V} \cdot d\mathbf{S}$$

where the vector $\mathbf{V}$ has components (u, v, w). However, the first form of the theorem holds even when u, v, w are not components of a vector.

A physical interpretation of the vector equation may be made, for if $\mathbf{V}$ represents the flux density of an incompressible fluid, $\nabla \cdot \mathbf{V}$ is the amount of fluid which flows from a volume dt per second. The volume integral is thus the total loss of fluid, which must equal the rate of flow across all boundaries of the volume, and that equals the surface integral.

This theorem is also called the Green lemma or theorem. See also **Green Function**.

GAVIIFORMES *(Aves).* Large birds with submarine-like swimming habits, much larger than most ducks and with shorter necks than geese. They are powerful swimmers with short legs, webbed feet and characteristic strong, sharp beaks. They have a thick neck, heavy body, black head, and are fast in flight, at times achieving a speed up to 60 miles (97 kilometers) per hour. In flight, the outline is hunch-backed and gangly, with a slight downward sweep to the neck and the big feet projecting beyond the tail. They build large and bulky nests close to water, mounding together grass and weeds. There are usually two olive green eggs with black spots. The incubation period is 28 days. The young take to the water almost immediately after hatching. See also **Loon**.

GAY-LUSSAC, JOSEPH LOUIS (1778–1850). A French chemist and physicist noted for the brilliance and accuracy of his reasoning and experimental work. He contributed greatly to the knowledge of gases in his discovery (1808) of the law of combining volumes and his independent discovery (1802) of the law of Charles, the relationship of temperature to the volume of gases. He graduated from and taught at the Ecole Polytechnique, becoming a full professor in 1810. His work in chemistry

was extensive, resulting in the discovery of boron [CAS: 7440-42-8], which he named, with Louis-Jacque Thenard, and a variety of compounds such as boron trifluoride [CAS: 7637-07-2] BF_3, chloric acid [CAS: 7790-93-4] $ClHO_3$, and dithionic acid ($H_2S_2O_6$). He identified iodine as an element, named it, and studied its properties. He investigated the relationship of acids and bases and introduced many analytical techniques (such as the use of litmus as an indicator). Among his many contributions to industrial chemistry were improvements in the production of sulfuric acid. Much of the progress of chemistry in the early 19th century is associated with his career.

GAY-LUSSAC'S LAW. A modification of Charles' law to state the following: At constant pressure the volume of a confined gas is proportional to its absolute temperature. The volumes of gases involved in a chemical change can always be represented by the ratio of small whole numbers.

G-DISPLAY. In radar, a rectangular display in which a target appears as a laterally centralized blip when the radar antenna is aimed at it in azimuth and wings appear to grow on the blip as the distance to the target is diminished. Horizontal and vertical aiming errors are respectively indicated by horizontal and vertical displacement of the blip. Also called *G-scan*, *G-scope*, or *G-indicator*.

GEAR TRAIN. Two or more gears, transmitting motion from one shaft to another, constitute a gear train. See also **Machine (Simple)**. If spur, bevel, or worm gears are used, the velocity ratio is inversely proportional to the numbers of teeth in the gears. A pair of spur gears, directly connected, results in a reversal of direction. If the driving gear drives an intermediate idler, which in turn drives the driven gear, the only effect of the idler is to cause the driven gear to rotate in the same direction as the driver. (The same effect can also be obtained by using an internal gear and a pinion.) If a two-gear idler, or compound gear, in which both idlers are fastened either to the idler shaft or to each other, is used, and, where the driver engages one of the compound gears, and the driven gear engages the other compound gear, the velocity ratio is equal to the product of the two trains. The back gearing of a lathe or a milling machine is a familiar example of a compound gear train; a gear attached to the driving pulley drives a large gear mounted on the back gear shaft; a small gear on this shaft in turn drives the spindle gear. See also **Epicyclic Gear Train**.

GECKOS. Of the class *Reptilia* (reptiles), subclass *Lepidosauria*, order *Squamata* (scaly reptiles), suborder *Sauria* (lizards), infraorder *Gekkota*, according to classification of Grzimek (1972). The infraorder *Gekkota* comprises the families *Gekkonidae* (geckos), *Pytopodidae* (snake lizards), and *Dibamidae*. Even though the geckos do not resemble these other families externally, certain anatomical characteristics indicate a close relationship to the snake-like pytopodids and the almost worm-shaped dibamids.

The geckos are lizards of extraordinary diverse form; furthermore, the group is quite old on the evolutionary scale. This can be inferred from the structure of the vertebrae, the presence throughout their lifetime of vestiges of the notochord, the shape of the hyoid bone, the fleshy tongue, and peculiarities of the scales. In the course of their development, the geckos in the subtropics and tropics conquered a variety of habitats—so that today they are found from the desert to the rain forest—in each case suitably modified in form. The geckos are small animals—at most about 40 centimeters (15.7 inches) long. The body is flattened. The large eyes are covered by a transparent scale, and nocturnal species have a slit pupil. The feet are specially constructed, the fingers and toes often bearing broad clinging lamellae on the underside. In contrast with other lizards, the geckos are quite vocal; the sounds they make range from quiet chirping and squeaking to loud barking. The geckos are the only living reptiles that really make extensive use of their voices, and in this respect they stand comparison with amphibians, birds, and mammals. There are 83 genera and about 670 species.

It is not uncommon to find geckos hanging head down on walls or ceilings when they are hunting insects. In managing these acrobatic feats, the geckos employ their toes, which are broadened on the underside, forming lamellate cushions. Here there are countless microscopic hook cells which, like the bristles of a brush, engage in the tiniest irregularities of a surface. This enables the geckos to run even on vertical surfaces. It was thought at one time that the lamellae exerted a sort of suction

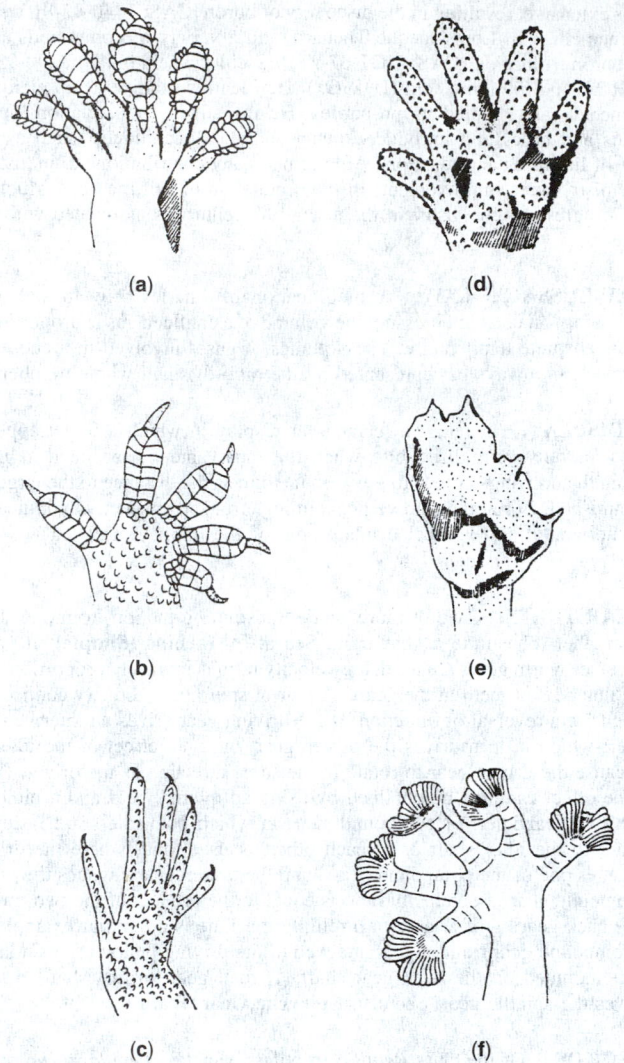

(a)

(d)

(b)

(e)

(c)

(f)

Fig. 1.   Various forms of gecko feet: (**a**) common gecko (*Tarentola maurisatannica*) the lamellae of which are equipped with tiny hooks to facilitate vertical climbing on very smooth vertical surfaces; (**b**) foot structure of the tropical gecko (*Hemidactylus mabouia*); (**c**) foot of the *Gymnodactylus kotschvi*; (**d**) sand grecko (*Chondrodactylus*); (**e**) web-footed gecko (*Palmatogecko rangei*); (**f**) bottom of right forefoot of the house gecko (*Ptyodactyluse hasselguistii*).

or even secreted a sticky substance, but investigation has proved these concepts to be incorrect. On a surface polished to a very high degree, even a gecko cannot adhere and slips like a person walking on ice. The normal mechanism of release and reattachment of the hook cells proceeds so rapidly that one cannot follow it by eye. See Fig. 1.

Many geckos have a striking ability to change color. Usually, they are lighter by day and darker by night. When one picks up a common gecko (not a simple feat, because of its agility) the skin feels soft and velvety. Like that of all reptiles, it is covered with scales, but their edges lie flush with one another, and do not overlap as in the other lizards and snakes. The gecko molts at intervals; as a rule, the skin first breaks open at the head, and it is stripped off toward the back. Most geckos eat the shed skin, at least in part. If one tries to catch a common gecko, its tail may suddenly break off (autotomy). The breaks tend to occur at specialized places in the bodies of the tail vertebrae, and they are brought about by muscular contraction. The cast-off parts regenerates, but cannot be automized again, since it now is supported by an unsegmented rod of cartilage without preformed break points. Discarding the tail affords the gecko a certain protection, for the violently jerking cast-off piece can distract a predator and give the gecko a chance to flee. The loss of the tail occurs so frequently among geckos that one often has trouble finding animals with the original tail.

Like most geckos, the common gecko (*Tarentola mauritanica*) is active in the twilight and night. Geckos prefer spiders, beetles, butterflies,

Fig. 2.   Madagascar gecko (*Phelsuma madagascariensis*). Average length, 15 centimeters. (*After Grobimunn.*)

millipedes, crickets, and cockroaches, although larger species, such as the Caledonian gecko (*Rhacodactylus leachianus*) also takes young lizards, mice, and small birds. The Madagascar gecko (*Phelsuma* spp.) is active during daylight and prefers plants, particularly fruit. See Fig. 2. The Japanese gecko (*Gekko japonicus*), which becomes very tame in a terrarium, readily accepts fruit and candy. The *Gehyra mutilata* has been called the "sugar lizard" because of its preference for sweet, fermenting substances.

Many geckos have become established in the vicinity of human dwellings and often are specifically adapted to coexistence with humans. The common gecko is frequently encountered on the rough stone walls of Mediterranean houses, and even within the house. Having become accustomed to the presence of humans, the nocturnal geckos have, to some extent, lost their fear of people. As a result, geckos living on the coasts or in ports sometimes "stow away" on ships. Thus transported, several species of geckos have greatly expanded their ranges to quite remote parts of the earth. The common gecko has moved from northern Africa to the port cities of southern France and to the Canary Islands, and in some cases will be found in the South Pacific.

The common gecko has large eyes with vertical pupils that are closed to a narrow slit in abundant bright daytime light. At night, particularly when hunting insects at twilight and early evening, the pupils open wide. The eyelids are not movable, a characteristic of most geckos. The lids are transparent and form what could be termed a "contact lens" over the eye. Only a few geckos, such as the banded gecko, have functional eyelids. The common gecko uses its tongue to clean the eye covering. See Fig. 3. The vision of the common gecko and other geckos is excellent and is specialized to see moving objects. Thus, insects that remain in place are not attacked, but they become prey when they move.

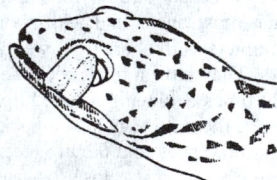

Fig. 3.   A gecko cleaning its eye with its tongue.

Behavioral Characteristics. Although not necessarily typical of all geckos, the behavioral cycle of the genus *Tetratoscincus* has been described in some detail. This species inhabits the dry regions of southwestern and central Asia and is primarily nocturnal. When threatened, it displays an interesting warning behavior. First, it raises itself high on its legs and stands rigid for a few moments, while blowing up its throat sac and rattling its tail with increasing intensity. Sound is created by rubbing together the scales on the upper surface of the tail. The animal fixes its eye on the enemy and suddenly leaps forward, simultaneously hissing, squeaking, and snapping. The tail whips the ground, throwing up fine sand, while the hissing continues. Then, suddenly, the attack is converted into flight. This sudden action bewilders even large enemies so effectively that the gecko can escape before the enemy recovers from this startling series of events.

As is typical of most reptiles, geckos lay eggs. The eggs often adhere to one another and are soft shelled initially, but soon harden after being laid. Geckos do not lay their eggs in the ground, but stick them to walls of cracks and holes. Sometimes, several families of geckos, even those of different species, will use the same location.

Many geckos have the striking ability to change color. Usually, they are lighter by day and darker by night. The reverse effect can also occur. For example, the banded leaf-toed gecko (*Hemidactylus fasciatus*) is dark brown in daytime and appears yellow to pale brown at night.

Like some other geckos, the common gecko has a "regenerative" tail. When its tail is grasped, as by a human predator, part of the tail may snap off, allowing the gecko to escape. The breaks tend to occur at specialized places in the bodies of the tail vertebrae, and they are brought about by muscular contraction. The cast-off part will regenerate, but the autotonomic action cannot occur again, because the replaced parts are now supported by an unsegmented rod of cartilage, without preformed break points.

Snake Lizards (Pygopodidae). There are relatively few species of snake lizards, and these are restricted to the region of Australia and Papua. The biological development of this family apparently has been limited to these regions. Although they have a snake-like appearance, they are closely related to the geckos. The forelimbs apparently have been lost over eons, during which the hind limbs were greatly reduced. Two-thirds of the slender body is taken by the tail. In their native habitat, humans often mistake geckos for snakes and thus they are often killed. Even experienced herpetologists can be deceived by these animals. Some snake lizards mimic the behavior of an elipid snake to ward off predators. Comparatively little research has been conducted pertaining to the reproduction, growth, and general behavior in their natural habitat. Some research has been pursued in terrariums. Length of these creatures ranges from about 20 to 60 centimeters. It has been established that the female lays two markedly elongated, cylindrical eggs at a time. The egg has a parchment-like, partially calcified shell. The snake lizard's motion is like that of the serpentines, with no motion derived from the remaining stumps of the hind legs.

Diabimids (Dibamidae). These animals constitute the third family of geckos, with only one genus and three species. Zoologists have not doubted that this small group of saurians represents a divergent group, which is fairly classified in its own right. However, it is difficult to interpret the proper systematic position of the diabimids with respect to other reptiles. The diabimids are burrowers. The bony elements of their massive skulls have lost all ability to move with respect to one another. The teeth are small; the eyes are reduced, with no external openings for either eyes or ears. Body shape is that of a long worm. Diabimids can discard the tail when danger threatens. Each tail vertebra, from the fifth on, has a break point at which autotomy can occur.

See also tabular summary, Classification of Lizards, in entry on **Lizards**.

Extensive information and excellent illustrations of the vast variety of geckos can be found in "Grzimek's Animal Life Encyclopedia," Vol. 6 (Reptiles), Van Nostrand Reinhold, New York.

GEGENSCHEIN. A slight increase in intensity of the zodiacal light at a point on the ecliptic 3° west of the antisolar point. The gegenschein appears as a soft glow against the sky, oval in shape, a few degrees wide and 10–15° in length. It is so faint that it cannot be observed on a night when there is any moon or when the patch falls in the vicinity of the Milky Way. A dust tail of the earth under radiation pressure would explain the 3° lag, and an ordinary photometric function would explain the photometric properties.

GEIGER COUNTER. Also called a Geiger-Müller or G-M counter, the name Geiger counter is now rather commonly applied to a gas-filled detector of ionizing radiations of the general design indicated in Fig. 1. When operating in the Geiger region the tube produces an output voltage pulse of approximately constant magnitude for each ionizing event that takes place within the cylindrical electrode. The development of this output

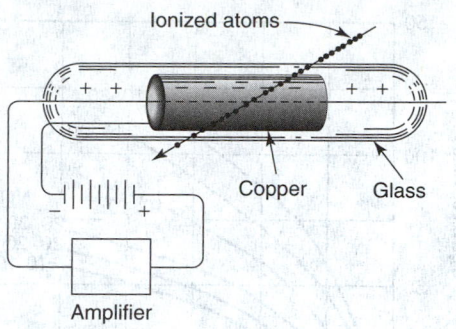

Fig. 1. Geiger counter.

pulse depends on the production of an avalanche of ionization along the central wire electrode, possible only if the central wire is of such a sufficiently small diameter (typically less than 0.010 inch), that a very high field gradient exists in the immediate vicinity of the wire. This high field gradient causes electrons attracted toward the central electrode to attain sufficiently large kinetic energies that they can ionize additional atoms of the gas inside the tube. Additional electrons, probably produced by photons emitted when some of the ion-electron pairs recombine, are attracted toward the central electrode, and they produce complete ionization of the entire region immediately surrounding the central wire in about 10 microseconds. Because of their low mobility, the positive ions produced near the central wire build up as a sheath to destroy the high voltage gradient and render the tube inactive. This action results in a pulse of approximately constant magnitude for each ionizing event.

After the Geiger counter discharges and produces a pulse, it remains inoperative for a period of time called the *dead time*. This is the time required for the positive-ion sheath to move out from the wire to a position where the electric field can recover so that another avalanche can form. The *resolving time* of the counter is larger than the dead time and is determined by the point at which the pulse size becomes large enough to again trigger the electronic equipment. See Fig. 2. The *recovery time*, larger still than the resolving time, is that point where the pulse again gains its original amplitude. All these factors determine the speed at which a counter can operate without losing a large number of counts. The dead time and recovery time are of the order of 100 to 200 microseconds for the typical Geiger counter.

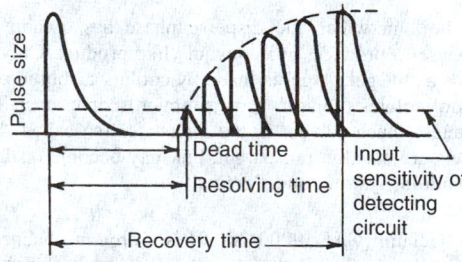

Fig. 2. Dead time and subsequent recovery of pulse size in a Geiger counter.

A large number of different quenching vapors may be used for filling Geiger counters. Amyl acetate, ether, and alcohol have had wide use. Halogen gas has been used as the quench vapor. The halogen molecule does not dissociate as does the polyatomic molecule. In actual practice, the useful life of an organic quenched counter is of the order of 10^8 counts whereas that of a halogen quenched counter may be 10^{10} counts or more. Halogen counters, unlike organic counters, are not damaged when subjected to voltages above the plateau region.

Geiger counters are made in a variety of shapes other than the most common shape of the cylindrical shell and axial wire. Cleanliness cannot be overstressed in counter construction. The counter should be washed with a detergent or alcohol, followed with several rinses of distilled water and then dried by heating. The central wire should have no sharp projections and be free of dust and lint. A typical counter of 1-inch (2.54 centimeter) diameter shell and 0.001-inch (0.025 millimeter) wire, filled with ethyl

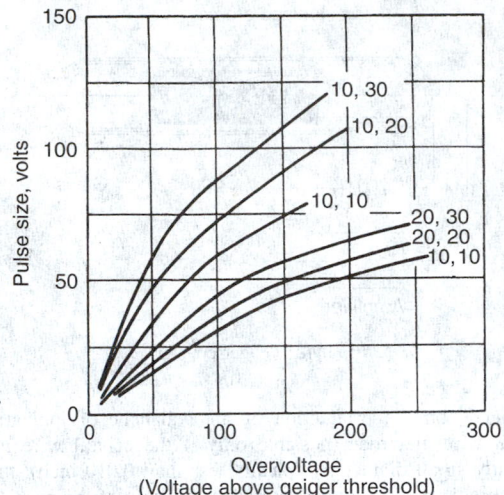

Fig. 3. Pulse size as a function of counter dimension. Tube dimensions are indicated as cathode radius (millimeters) and wire radius (thousandths of a millimeter).

alcohol to a pressure of 1 cm of Hg and argon to a pressure of 9 cm of Hg, will operate at approximately 1,000 V. The pulse size varies, with the counter dimensions and with the voltage above the Geiger threshold. See Fig. 3.

GEIGER, JOHANNES WILHELM HANS (1882–1945). Geiger was a German physicist who studied physics at the Universities of Munich and Erlangen. His thesis was on the ionization of gases and later work in this field in collaboration with Walther Muller would result in a radiation detector now known as the *Geiger Counter*, a gas-filled detector of ionizing radiations.

In 1929, Geiger become a physics professor at the University of Tubingen. During his research here he first detected cosmic-ray showers. Later at Technische Hochschule in Berline he was used a circular array of geiger counters to determine the angular distribution of particles in a cosmic-ray shower.

See also **Geiger Counter**.

J. M. I.

GEL. A colloid in which the disperse phase has combined with the continuous phase to produce a viscous jellylike product. Only 2% gelatin in water forms a stiff gel. A gel is made by cooling a solution, whereupon certain kinds of solutes (gelatin) form submicroscopic crystalline particle groups that retain much solvent in the interstices (so-called "brush-heap" structure). Gels are usually transparent, but may become opalescent.

See also **Colloid System**; and **Pectins**.

GELATIN. Gelatin [CAS: 9000-70-80] is a protein obtained by partial hydrolysis of collagen, the chief protein component in skin, bones, hides, and white connective tissues of the animal body. See also **Collagen**. Type A gelatin is produced by acid processing of collagenous raw material; type B is produced by alkaline or lime processing. Because it is obtained from collagen by a controlled partial hydrolysis and does not exist in nature, gelatin is classified as a derived protein. Animal glue and gelatin hydrolysate, sometimes referred to as liquid protein, are products obtained by a more complete hydrolysis of collagen and can thus be considered as containing lower molecular-weight fractions of gelatin.

Use of animal glues was first recorded ca 4000 BC in ancient Egypt [Koepff]. Throughout subsequent centuries, glue and crude gelatin extracts with poor organoleptic properties were prepared by boiling bone and hide pieces and allowing the solution to cool and gel. Late in the seventeenth century, the first commercial gelatin manufacturing began. At the beginning of the nineteenth century, commercial production methods gradually were improved to achieve the manufacture of high molecular weight collagen extracts with good quality that form characteristic gelatin gels.

Uses of gelatin are based on its combination of properties; reversible gel-to-sol transition of aqueous solution; viscosity of warm aqueous solutions;

ability to act as a protective colloid; water permeability; and insolubility in cold water, but complete solubility in hot water. It is also nutritious. These properties are utilized in the food, pharmaceutical, and photographic industries. In addition, gelatin forms strong, uniform, clear, moderately flexible coatings which readily swell and absorb water and are ideal for the manufacture of photographic films and pharmaceutical capsules.

Chemical Composition and Structure

Gelatin is not a single chemical substance. The main constituents of gelatin are large and complex polypeptide molecules of the same amino acid composition as the parent collagen, covering a broad molecular weight distribution range. In the parent collagen, the 18 different amino acids are arranged in ordered, long chains, each having ~95,000 mol wt. These chains are arranged in a rod-like, triple-helix structure consisting of two identical chains, called α_1, and one slightly different chain called α_2. These chains are partially separated and broken, i.e., hydrolyzed, in the gelatin manufacturing process. Different grades of gelatin have average molecular weight ranging from ~20,000 to 250,000.

Analysis shows the presence of amino acids from 0.2% tyrosine to 30.5% glycine. The five most common amino acids are glycine [CAS: 56-40-6], 26.4–30.5%; proline [CAS: 147-85-3], 14.8–18%; hydroxyproline [CAS: 51-35-4], 13.3–14.5%; glutamic acid [CAS: 56-86-0], 11.1–11.7%; and alanine [CAS: 56-41-7], 8.6–11.3%. The remaining amino acids in decreasing order are arginine [CAS: 74-79-3], aspartic acid [56-84-1], lysine [CAS: 56-87-1], serine [CAS: 56-45-1], leucine [CAS: 61-90-5], valine [CAS: 72-18-4], phenylalanine [CAS: 63-91-2], threonine [CAS: 72-19-5], isoleucine [CAS: 73-32-5], hydroxylysine [CAS: 13204-98-3], histidine [CAS: 71-00-1], methionine [CAS: 63-68-3], and tyrosine [CAS: 60-18-4].

Stability. Dry gelatin stored in airtight containers at room temperature has a shelf life of many years. However, it decomposes above 100 °C (212 °F). For complete combustion, temperatures above 500 °C (932 °F) are required. When dry gelatin is heated in air at relatively high humidity, <60% rh, and at moderate temperatures, ie, above 45 °C (113 °F), it gradually loses its ability to swell and dissolve. Aqueous solutions or gels of gelatin are highly susceptible to microbial growth and breakdown by proteolytic enzymes. Stability is a function of pH and electrolytes and decreases with increasing temperature because of hydrolysis.

Properties

Commercial gelatin is produced in mesh sizes ranging from coarse granules to fine powder. In Europe, gelatin is also produced in thin sheets for use in cooking. It is a vitreous, brittle solid, faintly yellow in color. Dry commercial gelatin contains about 9–13% moisture and is essentially tasteless and odorless with specific gravity between 1.3 and 1.4. Most physical and chemical properties of gelatin are measured on aqueous solutions and are functions of the source of collagen, method of manufacture, conditions during extraction and concentration, thermal history, pH, and chemical nature of impurities or additives.

Perhaps the most useful property of gelatin solution is its capability to form heat reversible gel–sols. When an aqueous solution of gelatin with a concentration greater than about 0.5% is cooled to about 35–40 °C (95–104 °F), it first increases in viscosity, and then forms a gel. The gelation process is thought to proceed through three stages: (1) rearrangement of individual molecular chains into ordered, helical arrangement, or collagen fold; (2) association of two or three ordered segments to create crystallites; and (3) stabilization of the structure by lateral interchain hydrogen bonding within the helical regions. The rigidity or jelly strength of the gel depends on the concentration, the intrinsic strength of the gelatin sample, pH, temperature, and additives.

In most commercial applications, gelatin is used as a solution. Gelatin is soluble in water and in aqueous solutions of polyhydric alcohols such as glycerol [CAS: 56-81-5] and propylene glycol [CAS: 57-55-6]. Examples of highly polar, hydrogen-bonding organic solvents in which gelatin dissolves are acetic acid [CAS: 64-19-7], trifluoroethanol [CAS: 75-89-8], and formamide [CAS: 75-12-7]. Gelatin is practically insoluble in less polar organic solvents such as acetone [CAS: 67-64-1], carbon tetrachloride [CAS: 56-23-5], ethanol [CAS: 64-17-5], ether [CAS: 60-29-7], benzene [CAS: 71-43-2], dimethylformamide [CAS: 68-12-2], and most other nonpolar organic solvents. Many water-soluble organic solvents are compatible with gelatin, but interfere with gelling properties.

The viscosity of gelatin solutions is affected by gelatin concentration, temperature, molecular weight of the gelatin sample, pH, additives, and impurities. In aqueous solution above $40\,^{\circ}C$ ($104\,^{\circ}F$), gelatin exhibits Newtonian behavior. Standard testing methods employ the use of a capillary viscometer.

Gelatin is an effective protective colloid that can prevent crystal, or particle, aggregation, thereby stabilizing a heterogeneous suspension. It acts as an emulsifying agent in cosmetics and pharmaceuticals involving oil-in-water dispersions. The anionic or cationic behavior of gelatin is important when used in conjunction with other ionic materials. The protective colloid property is important in photographic applications where it stabilizes and protects silver halide crystals while still allowing for their normal growth and sensitization during physical and chemical ripening processes.

A phenomenon associated with colloids wherein dispersed particles separate from solution to form a second liquid phase is termed coacervation. Gelatin solutions form coacervates with the addition of salt such as sodium sulfate [CAS: 7757-82-6], especially at pH below the isoionic point. In addition, gelatin solutions coacervate with solutions of oppositely charged polymers or macromolecules such as acacia. This property is useful for microencapsulation and photographic applications.

Swelling. The swelling property of gelatin is not only important in its solvation but also in photographic film processing and the dissolution of pharmaceutical capsules.

Manufacture and Processing

Although new methods for processing gelatin, including ion exchange and crossflow membrane filtration, have been introduced since 1960, the basic technology for modern gelatin manufacture was developed in the early 1920s. Acid and lime processes have separate facilities and are not interchangeable. In the past, bones and ossein, ie, decalcified bone, have been supplied by India and South America. In the 1990s, slaughterhouses and meat-packing houses are an important source of bones. The supply of bones has been greatly increased since the meat-packing industry introduced packaged and fabricated meats, assisted by the growth of fast-food restaurants. Dried and rendered bones yield about 14–18% gelatin, whereas pork skins yield about 18–22%.

Most type A gelatin is made from pork skins, yielding grease as a marketable by-product.

Type B gelatin is made mostly from bones, but also from bovine hides and pork skins.

Uses

Gelatin formulations in the food industry use almost exclusively water or aqueous polyhydric alcohols as solvents for candy, marshmallow, or dessert preparations. In dairy products and frozen foods, gelatin's protective colloid property prevents crystallization of ice and sugar. Gelatin products having a wide range of Bloom and viscosity values are utilized in the manufacture of food products, specific properties being selected depending on the needs of the application. For example, a 250-Bloom gelatin may be utilized at concentrations ranging from 0.25% in frozen pies to 0.5% in ice cream; the use of gelatin in ice cream has greatly diminished. In cottage cheese and sour cream, gelatin inhibits water separation, ie, syneresis. Marshmallows contain as much as 1.5% gelatin to restrain the crystallization of sugar, thereby keeping the marshmallows soft and plastic; gelatin also increases viscosity and stabilizes the foam in the manufacturing process. Many lozenges, wafers, and candy coatings contain up to 1% gelatin. In these instances, gelatin decreases the dissolution rate. In meat products, such as canned hams, various luncheon meats, corned beef, chicken rolls, jellied beef, and other similar products, gelatin in 1–5% concentration helps to retain the natural juices and enhance texture and flavor. Use of gelatin to form soft, chewy candies, so-called gummi candies, has increased worldwide gelatin demand significantly (ca 1992). Gelatin has also found new uses as an emulsifier and extender in the production of reduced-fat margarine products. The largest use of edible gelatin in the United States, however, is in the preparation of gelatin desserts in 1.5–2.5% concentrations. For this use, gelatin is sold either premixed with sugar and flavorings or as unflavored gelatin packets. Most edible gelatin is type A, but type B is also used.

Gelatin is used in the pharmaceutical industry for the manufacture of soft and hard capsules. The formulations are made with water or aqueous polyhydric alcohols. Capsules are usually preferred over tablets in administering medicine. Elastic or soft capsules are made with a rotary die from two plasticized gelatin sheets which form a sealed capsule around the material being encapsulated. Methods have been developed to encapsulate dry powders and water-soluble materials which may first be mixed with oil. The gelatin for soft capsules is low bloom type A, 170–180 g; type B, 150–175 g; or a mixture of type A and B. Hard capsules consisting of two parts are first formed and then filled. The manufacturing process is highly mechanized and sophisticated in order to produce capsules of uniform capacity and thickness. Medium-to-high bloom type A, 250–280 g; type B, 225–250 g; or the combination of type A and B gelatin are used for hard capsules. Usage of gelatin as a coating for tablets has increased dramatically. In a process similar to formation of gelatin capsules, tablets are coated by dipping in colored gelatin solutions, thereby giving the appearance and appeal of a capsule, but with some protection from adulteration of the medication. The use of glycerinated gelatin as a base for suppositories offers advantages over carbowax or cocoa butter base. Coated or cross-linked gelatin is used for enteric capsules. Gelatin is used as a carrier or binder in tablets, pastilles, and troches.

For arresting hemorrhage during surgery, a special sterile gelatin sponge known as absorbable gelatin sponge or Gelfoam is used. The gelatin is partially insolubilized by a cross-linking process. When moistened with a thrombin or sterile physiological salt solution, the gelatin sponge, left in place after bleeding stops, is slowly dissolved by tissue enzymes. Special fractionated and prepared type B gelatin can be used as a plasma expander.

Gelatin can be a source of essential amino acids when used as a diet supplement and therapeutic agent. As such, it has been widely used in muscular disorders, peptic ulcers, and infant feeding, and to spur nail growth. Gelatin is not a complete protein for mammalian nutrition, however, since it is lacking in the essential amino acid tryptophan [CAS: 73-22-3] and is deficient in sulfur-containing amino acids.

Gelatin has been used for over 100 years as a binder in light-sensitive products. The useful functions of gelatin in photographic film manufacture are a result of its protective colloidal properties during the precipitation and chemical ripening of silver halide crystals, setting and film-forming properties during coating, and swelling properties during processing of exposed film or paper. Quality requirements of photographic gelatin may be very elaborate and can include over 40 chemical and physical tests, in addition to photographic evaluation.

Gelatin is also used in so-called subbing formulations to prepare film bases such as polyester, cellulose acetate [CAS: 9004-35-7], cellulose butyrate, and polyethylene-coated paper base for coating by aqueous formulations. Solvents such as methanol [CAS: 67-56-1], acetone [CAS: 67-64-1], or chlorinated solvents are used with small amounts of water. Gelatin containing low ash, low grease, and having good solubility in mixed solvents is required for these applications. In certain lithographic printing, light-sensitive dichromated gelatin is used. Photographic technology offers a rapidly changing, highly sophisticated, very competitive market for photographic gelatin manufacturers.

See also **Colloid Systems**.

Additional Reading

Band, S.J.: "Photographic Gelatin," *Proceedings of the Fifth RPS Symposium*, Oxford, UK, 1985, The Imaging Science and Technology Group of the Royal Photographic Society, 1987.

Bogue, R.H.: *The Chemistry and Technology of Gelatin and Glue*, McGraw-Hill Book Co., Inc., New York, NY, 1922.

Johns, P., in A.G. Ward and A. Courts: *The Science and Technology of Gelatin*, Academic Press, Inc., New York, NY, 1977, pp. 475–506.

R.T. Jones, in K. Ridgway: *Hard Capsules Development and Technology*, The Pharmaceutical Press, London, 1987, pp. 41–42.

J. Photogr. Sci., **40**(5,6), 122–251 (1992).

Koepff, P., in H. Ammann-Brass and J. Pouradier, eds., "Photographic Gelatin," in *Proceedings of the Fourth IAG Conference, 1983*, Internationale Arbeitsgemeinschaft für Photogelatine, Fribourg, Switzerland, 1985, pp. 3–35.

Libicky, A., and D.I. Bermane, in R.J. Cox: *Photographic Gelatin*, Academic Press, Inc., New York, NY, 1972, pp. 29–48.

Ohno, T.Y. et al., in M. DeClercq, ed., "Photographic Gelatin," in *Proceedings of the Seventh IAG Conference*, Louvain-la-Neuve, 1999, The International Working Group for Photographic Gelatin, Fribourg, Switzerland, 1999, pp. 64–71.

Ridgway, K.: *Hard Capsules Development and Technology*, The Pharmaceutical Press, London, UK, 1987.

Standard Methods for Sampling and Testing Gelatins, Gelatin Manufacturers' Institute of America, Inc., New York, NY, 1986. http://www.gelatin-gmia.com/

Ward, A.G. and A. Courts: *The Science and Technology of Gelatin*, Academic Press, Inc., New York, NY, 1977.

GELLAN GUM.   See **Food Additives**.

GEMINI (the twins).   A constellation, marking the third sign of the zodiac, which has been recognized as a pair of twins from remote antiquity. The twins have not always been human, however; the Egyptians considered them as a pair of kids, and the Arabians as a pair of peacocks. By far the most familiar names for the two bright stars of this constellation are the names of the warrior brothers, Castor and Pollux, sons of Jupiter and Leda. Both these stars are interesting objects as seen through a 3-inch telescope, Castor being a fine binary and Pollux being a multiple star having at least six components. There is also a fine star cluster in this constellation, which can easily be seen with a field glass and can be detected with the unaided eye on a clear moonless night. (See map accompanying entry on **Constellations**.)

GEMSTONES.   A gemstone is a mineral substance which because of its beauty or rarity is in demand for ornamental purposes, chiefly personal adornment. The origin of such use for what we now call gem minerals is lost in the dim vistas of early human history. Ancient records describe the various gemstones, and archaeologists find them in their investigations of bygone peoples. When we look at a collection of minerals with their bright colors and varying degrees of transparency or light-reflecting power, we cannot doubt that primitive man was much attracted by them and valued them greatly. We may imagine, too, that the occasionally found crystals with their regular geometric forms were more highly prized than broken fragments of the same minerals. Later they learned to polish them. Apparently, the oldest form into which stones were shaped is that known as *en cabochon*, a French term derived from the Latin word for head and referring to its rounded shape. The forms were either hemispherical or hemi-ellipsoidal. The Emperor Nero is supposed to have had a large emerald cut en cabochon, and, indeed, for several centuries after his time this seems to have been the only sort of cutting employed. The supposedly accidental discovery in 1475 that diamonds would mutually scratch each other began the era of modern gem cutting. Previously it had been believed that diamonds were so hard that they could not be artificially shaped. At first, however, little progress was made in fashioning gems other than polishing a number of facets without any definite arrangement.

We owe to Vicenzio Peruzzi, a Venetian, the credit for devising the so-called "brilliant cut," the style of the modern diamond cutting. Diamond cutting except for certain refinements due to a more thorough understanding of the behavior of minerals toward light, remains the same as in Peruzzi's day. At the present time, transparent stones of all sorts are usually "brilliant cut," whereas translucent or opaque are cut en cabochon.

Since time immemorial, dealers in gems have used as the unit of weight the carat, undoubtedly introduced from the east. The word is derived from the Greek meaning a small horn. This refers to the pods of the locust tree, *Ceratonia siliqua*, a common Mediterranean tree whose seeds were said to have been taken as the unit of weight in buying and selling gems. In the nineteenth century, the actual weight of the carat differed slightly in different countries of Europe, from a little under to somewhat over $\frac{1}{5}$ of a gram. The metric carat is exactly $\frac{1}{5}$ of a gram.

There are three types of gemstone materials as defined by the U.S. Federal Trade Commission: *(1)* natural gemstones are found in nature and at most are enhanced; *(2)* imitation or simulated, fake, faux, etc., material resembles the natural material in appearance only and is frequently only colored glass or even plastic; and *(3)* synthetic material is the exact duplicate of the natural material, having the same chemical composition, optical properties, etc., as the natural, but made in the laboratory. Moreover, the word gem cannot be used for synthetic gemstone material. The synthetic equivalent of a natural material may be used as an imitation of another, e.g., synthetic cubic zirconia is widely used as a diamond imitation.

Synthetic gemstone materials often have multiple uses. Synthetic ruby and colorless sapphire are used for watch bearings, unscratchable watch crystals, and bar-code reader windows. Synthetic quartz oscillators are used for precision timekeeping, citizen's band radio (CB) crystals, and filters. Synthetic ruby, emerald, and garnets are used for masers and lasers.

In the gemstone jewelry market, synthetics provide a less expensive alternative to natural gemstones, but of a better quality than that available in costume jewelry. In general, a synthetic should be available for no more than 10% of the cost of equivalent-quality natural gemstone to be commercially viable. Synthetics are frequently divided into three groups: *(1)* luxury synthetics, involving slow and difficult growth processes, produced in small quantities for a price-restricted market; *(2)* intermediates; and *(3)* low cost synthetics, produced on a large scale.

Properties

The important properties are those of importance in natural gemstones. First is hardness, H. A value of 7 or greater on Mohs' scale is desirable to avoid scratches from the quartz (H = 7) sand present in dust. Next is color or a total lack of color, as in diamond and its simulants. A high refractive index (RI) permits the return by total internal reflection of most of the light falling onto a well-cut gemstone, giving brilliance, and a high dispersion (DISP) spreads the internally reflected light into spectral colors, resulting in fire.

Several gemstone species occur in various colors, depending on the presence of impurities or irradiation-induced color centers. Any material can have its color modified by the addition of various impurities: synthetic ruby, sapphires, and spinel are produced commercially in over 100 colors.

Manufacture

The most frequently used techniques for the commercial manufacture of synthetic gemstone materials are crystal growth from the melt, crystal growth from solution, and complex chemical procedure. Only rarely used for synthetics are such alternative growth techniques as the Bridgman technique of solidification in a crucible and the float zone technique, both involving growth from the melt.

Materials

Alexandrite [CAS: 12252-02-7], which is a colorless chrysoberyl, $BeAl_2O_4$, when pure, has a color change derived from Cr. See also **Alexandrite**.

Beryl [CAS: 1302-52-9], $Be_3Al_2Si_6O_{18}$, is called aquamarine when pale green or blue from the presence of Fe, emerald when dark green from Cr or at times V, and morganite or red beryl when pink or red, respectively, from Mn. Only synthetic emerald is in commercial production. See also **Beryl**.

Crystalline Al_2O_3, corundum [CAS: 1302-74-5], is called ruby [CAS: 12174-49-1] when colored red by about 1% Cr, and sapphire [CAS: 1317-82-4] for colorless and other colors particularly when blue from charge transfer between about 0.01% each of Fe^{2+} and Ti^{4+}. See also **Corundum**.

As of this writing, cubic zirconia [CAS: 1314-23-4], ZrO_2, is the best diamond imitation available. It is marketed under such names as CZ, Cerene, Cubic zirconium, Diamonair II, Diamonique III, Fianite, etc, and grown by the skull melting technique. This material can also be made in almost any color.

The synthesis by a high pressure process of single-crystal diamond [CAS: 7782-40-3] large enough for gemstone use was revealed by the General Electric Company in 1971. The yellow color (containing N) is grown much more easily than colorless (pure) and blue (B). None of these is likely to be viable for use in jewelry in the near future. See also **Diamond**.

Both YAG, yttrium aluminum garnet [CAS: 12005-21-9], $Y_3Al_5O_{12}$, and GGG, gadolinium gallium garnet [CAS: 12024-36-1], $Gd_3Ga_5O_{12}$, have the garnet structure and were used at one time as diamond imitations under trade names such as Diamonaire, Diamonique, Diamonite, Kimberly, Triamond, and YAlG for YAG and Diamonique II, Galliant, and Triple G for GGG.. These have been supplanted by cubic zirconia. See also **Garnet**.

Opal [CAS: 14639-88-4] is the only commercial synthetic gemstone material that is not a single crystal. It consists of a three-dimensional diffraction grating of geometrically aligned spheres of $SiO_2 \cdot xH_2O$, where x is usually <10%. See also **Opal**.

When colorless, quartz [CAS: 14808-60-7] is also known as rock crystal; irradiation of this produces smoky quartz. The name citrine [CAS: 14832-92-9] is used when quartz is colored by Fe, and irradiation of this can produce purple-colored amethyst [CAS: 14832-91-8] under certain circumstances. See also **Quartz**.

Rutile [CAS: 1317-80-2], a form of TiO_2, was at one time used as a rather poor diamond imitation. Small amounts of Al_2O_3 or Ga_2O_3 lighten the yellow color. Related is strontium titanate [CAS: 12060-59-2], $SrTiO_3$, now more properly called synthetic tausonite. See also **Rutile**.

Although the composition of natural spinel [CAS: 1302-67-6] is $MgAl_2O_4$, crystal growth is much eased by growing Al_2O_3-rich material in the solid solution region. Colorless (pure), blue (Co), and other colored synthetic spinels made by the Verneuil process are widely seen in class rings and in other jewelry uses, where the blue is often mislabeled as synthetic sapphire. See also **Spinel**.

There are several materials that have in the past been considered to be synthetics, but were found on closer examination not to deserve such a designation, being merely imitations. Examples include imitation coral, lapis lazuli, and turquoise, all made by ceramic processes.

Gemstone Treatment

Color and clarity are two of the attributes that give gemstones used in jewelry value. Gemstones deficient in either color or clarity can be enhanced. Almost worthless material can at times be converted into valuable-appearing gemstones. An estimated two-thirds of all colored gemstones used in jewelry have been treated. Accordingly, the identification of the use of treatments and the disclosure of enhancements to the purchaser are important.

Some treatments are practiced so widely that untreated material is essentially unknown in the jewelry trade. The heating of pale Fe-containing chalcedony to produce red-brown carnelian is one of these.

The stability of a particular treatment is also important. The enhancement should survive during normal wear or display conditions.

The Gemstones can be enhanced by heat treatments. Parameters for specifying the conditions for heat treatment of a gemstone material include the maximum temperature reached and the time for which the maximum temperature is sustained; the rate of heating to temperature, the rate of cooling down from temperature, and any holding stages while heating and cooling; the chemistry and pressure of the atmosphere; and any material in contact with the gemstone. Exact conditions for heat treatments vary widely according to the natural materials used.

The process of irradiation involves the exposure of a specimen to one of a variety of radiations.

When radiation interacts with matter, a displacement of the outermost electrons in atoms occurs. This displacement can lead to the formation of color centers or to valence state changes. When properly performed, there is no significant residual radioactivity.

Other treatments fall into three groups: impregnations, surface modifications and composite gemstones.

Identification of Treated Gems

A trained gemologist, taught by the Gemological Institute of America of Carlsbad, California, http://www.gia.edu/, The Gemological Association of Great Britain of London, http://www.gagtl.ac.uk/ or elsewhere, is needed for identification of treated gems. This topic is also discussed in textbooks. In some materials the induced change is the exact equivalent of a process that also occurs naturally, so that such treatments cannot be identified.

Additional Reading

Anderson, B.W. and E.A. Jobbins: *Gem Testing*, 10th Edition, Butterworths, London, UK, 1990.

Guides for the Jewelry Industry, U.S. Federal Trade Commission, Washington, DC, Feb. 27, 1979 (under revision in 1993).

Hurlbut, C.S. Jr. and R.C. Kammerling: *Gemology*, 2nd Edition, John Wiley & Sons, Inc., New York, NY, 1991.

Liddicoat, R.T. Jr.: *Handbook of Gem Identification*, 12th Edition, Gemological Institute of America, Santa Monica, CA, 1989.

Nassau, K. *Gemstone Enhancement*, 2nd Edition, Butterworths, Boston, MA, 1994.

Nassau, K.: *The Physics and Chemistry of Color*, John Wiley & Sons, Inc., New York, NY, 1983.

Nassau, K.: *Gems Made by Man*, Gemological Institute of America, Santa Monica, CA, 1980.

GENERAL CIRCULATION.   See **Meteorology**.

GENERAL CIRCULATION MODEL.   See **Meteorology**.

GENERALIZED COORDINATES.   Any set of coordinates specifying the state of the system under consideration. Usually employed in problems involving a finite number of degrees of freedom, the generalized coordinates are chosen so as to take advantage of the constraints of the system in reducing the total number of coordinates. Also called *Lagrangian coordinates*. See also **Lagrangian Coordinates**.

GENERALIZED TRANSMISSION FUNCTION.   In atmospheric-radiation theory, a set of values, variable with wavelength, each one of which represents an average transmission coefficient for a small wavelength interval and for a specified optical path through the absorbing gas in question.

GENERAL PACKET RADIO SERVICE (GPRS).   See **Wireless Communications Applications**.

GENERAL PRECESSION.   The resultant motion of the components causing precession of the equinoxes. The general precession is westward along the ecliptic at the rate of about 50.3 seconds of arc per year. The effect of the sun and moon, called lunisolar precession, is to produce a westward motion of the equinoxes along the ecliptic. The effect of other planets, called planetary precession, tends to produce a much smaller motion eastward along the ecliptic. The component of general precession along the celestial equator, called precession in right ascension, is about 46.1 seconds of arc per year; and the component along a celestial meridian, called precession in declination, is about 20.0 seconds of arc per year.

GENERATING FUNCTION.   A method of representing a function in terms of another function containing one or more variables. These generating functions give the function to be generated as coefficients involving the new variable parametrically. Each of the polynomials, cited as examples, is the solution of a differential equation generally known by the name of the mathematician indicated.

1. Legendre polynomials

$$(1 - 2xy + y^2)^{-1/2} = \sum_{n=0}^{\infty} P_n(x) y^n$$

2. Associated Legendre polynomials

$$\frac{(2m)!(1 - x^2)^{m/2} y^m}{2^m m!(1 - 2xy + y^2)^{m+1/2}} = \sum_{n=m}^{\infty} P_n^m(x) y^n$$

3. Bessel function of integral order

$$\exp\left[\frac{z}{2}(u - 1/u)\right] = \sum_{n=0}^{\infty} J_n(x) u^n$$

4. Hermite polynomials

$$\exp[x^2 - (z - x)^2] = \sum_{n=0}^{\infty} \frac{H_n(x) z^n}{n!}$$

The Hermite polynomials find applications in statistics in the Gram-Charlier Type A series and are also useful in certain applications in physics to problems of heat and quantum mechanics.

5. Laguerre polynomials

$$(1 - z)^{-1} \exp\left(\frac{-xz}{1 - z}\right) = \sum_{n=0}^{\infty} \frac{L_n(x) z^n}{n!}$$

6. Associated Laguerre polynomials

$$(-1)^k (1 - z)^{-1} \left(\frac{z}{1 - z}\right)^k \exp\left(\frac{-xz}{1 - z}\right) = \sum_{n=k}^{\infty} \frac{L_n^k(x) z^n}{n!}$$

7. Chebyshev polynomials

$$\frac{1 - xy}{1 - 2xy + y^2} = \sum_{n=0}^{\infty} T_n(x)y^n$$

SIR MAURICE KENDALL, International Statistics Institute, London

GENERIC OBJECT EXCHANGE PROFILE (GOEP). See **Bluetooth**^{TM} **Wireless Technology**.

GENESIS MISSION. See **Discovery Mission**; and **Space Science Missions: Solar System**.

GENETIC ENGINEERING. Biotechnological methods of genetic engineering are relatively new techniques that plant breeders have to make direct modifications of DNA, a living thing's genetic materials. Scientists make copies to genes for desired traits and introduce the gene copy into an organism such as a food crop. The new gene is usually a single gene whose function is well understood. These new techniques avoid one of the major problems encountered by plant breeders who use cross hybridization, no unwanted or undesirable genes are introduced with the desired gene. In addition, scientists can make copies of genes from any organism, plant, animal, or microbe, that may yield a desired trait and introduce that gene into a food crop. This greatly expands the pool of potentially useful traits available to plant breeders to improve food crops.

Once a desired gene has been introduced into a crop via genetic engineering, the gene is usually crossed into other crop lines that have desired commercial traits. Such crossing also permits the breeder to evaluate the genetic stability of the new gene. Overall, genetic engineering allows breeders to develop new varieties more rapidly, but at this stage of the technology, the new methods are used in conjunction with other methods of plant breeding, such as cross-hybridization.

The time required to evaluate new varieties and the number of field trials will vary depending on the need to confirm performance, to evaluate characteristics of the food, to evaluate environmental effects, and to produce the required amount of seed before the new plant variety can be grown commercially by farmers.

Genetic engineering is used to achieve the same goals of agronomic and quality characteristics as traditional techniques and allows the breeder to make some modifications that would not be possible through other methods of plant breeding.

Genetically engineered food crops have been developed to: resist pests and disease and to tolerate chemical herbicides; exhibit improved food processing traits; exhibit improved nutritional content; resist adverse soil and weather conditions; and to exhibit improved fruit ripening or softening, texture, or flavor.

In 1992, the FDA published a policy statement that explains how foods, fruits, vegetables, grains, and their by products such as vegetable oils, are regulated under the FD&C Act. This policy applies to foods and food ingredients, including animal feeds, derived from plants modified through all methods of plant breeding, including genetic engineering.

FDA's guidance to industry is the following:

Genetic Modification

The introduced genetic materials should be well-characterized to ensure that any introduced genes do not encode harmful substances and should be inserted stably in the plant genome to minimize the chance for subsequent undesired genetic rearrangements.

Toxicants

Plants are known to produce toxicants and antinutritional factors, such as protease inhibitors, hemolytic agents, and alkaloids, which often protect the plant against pests and disease. Many of these toxicants are present in today's crops at levels that do not cause acute toxicity or do not affect humans or animals when the food is properly prepared. New plant varieties should not contain levels of such toxicants that are above the range that exists in today's crops.

Nutrients

Another unintended consequence of genetic modification of the plant may be an alteration (relative to the total diet) in levels of important nutrients and bioavailability of a nutrient due to changes in the form of the nutrient or of other constituents that effect absorption or metabolism of nutrients.

New Substance

In some cases using genetic engineering, plant breeders may introduce genes into food crops that encode substances that differ substantially in structure and function from substances currently found in food. Based on current developments, such substances would be expected to be proteins or protein enzymes that modify carbohydrates and fatty acids in the food. In some cases, such substances will require premarket approval as food additives; in other cases, the food may require new labeling to properly inform consumers of the new attributes of the food. However, in most cases to date, the substances that occur in food as a result of gene transfer have been safely consumed as food previously or are substantially similar to food substances and would not require premarket review by FDA.

Allergenicity

There are thousands of different proteins in our food supply, and only a few cause food allergy reactions. However, because genetic engineering can result in the introduction of genetic material from essentially any source (plant, animal, or microbe) into food, there is a possibility that a protein encoded by the newly introduced genetic material will be an allergen and produce an allergic response in some members of the population. FDA has raised this issue in its guidance to industry, especially in cases where the transferred genetic material is derived from a source that is known to be commonly allergenic. Examples of such foods that affect the U.S. population include milk, eggs, fish crustacea, mollusks, tree nuts, wheat, and legumes (particularly peanuts and soybeans). FDA believes that proteins derived from commonly allergenic sources should be presumed to be allergens and special labeling would be required, unless scientific evidence demonstrates otherwise.

In cases where a protein is derived from a source that is not known to be allergenic, it is not possible to predict definitively allergenic potential. While it is unlikely that a new protein that occurs in very low concentrations in food will be an allergen (as is the case for most proteins introduced via genetic engineering at this time), developers have taken steps to minimize the likelihood that a new protein will be an allergen by evaluating whether the new proteins exhibit characteristics typical of allergenic proteins (such as stability to heat, acid, and enzyme degradation).

The FDA encourages developers to discuss questions regarding allergenicity with agency scientists.

Antibiotic Resistance Markers

In experiments involving genetic engineering, only a few plants cells take up the desired new gene. Developers use selectable marker genes during gene transfer experiments to improve their chances of selecting plants that have successfully incorporated the desired gene. The most widely used marker is kanamycin resistance gene that produces the enzyme, aminoglycoside $3'$-phosphotransferase II (also referred to as APH(3')II and neomycin phosphotransferase II). Plant cells are normally killed by antibiotics. APH3'II inactivates the antibiotics kanamycin and neomycin and permits plant cells to grow in culture that have incorporated gene and express the APH(3')II enzyme.

Once the desired plant variety has been selected, the marker gene serves no useful purpose in the new plant, but it does continue to produce the gene product, APH(3')II in the case of kanamycin resistance. This enzyme is present at very low concentrations in food.

The use of marker genes that encode resistance to clinically important antibiotics raises questions regarding whether the enzyme in the food could inactivate oral doses of the antibiotic or whether the gene present in the plant DNA could be transferred to pathogenic microbes in the GI tract or in soil rendering them resistant to treatment with the antibiotic. FDA evaluated these questions for the use of kanamycin resistance in tomato, cotton, canola.

The FDA found that kanamycin and neomycin are very toxic antibiotics and as such have very limited oral clinical use and are used only in situations where patients are not consuming food. There is also too little of the essential cofactor, ATP, present in food for the enzyme to degrade a significant amount of antibiotic.

There is no known mechanism by which a gene can be transferred from a plant chromosome to a microbe. Thus, the possibility of that such transfer would generate new resistant organisms is very small, especially when compared to the high rate of spread of resistance through known mechanisms of microbe to microbe transfer to antibiotic resistance. FDA believes that the use of marker genes that encode resistance to other clinically useful antibiotics can be evaluated by similar criteria that were used for kanamycin resistance.

Animal Feeds

Feeds developed for animals raised as food sources must be meet the same safety standards as human food under the FD&C Act. In contrast to the human diet, an animal feed derived from a single plant may make up over half of the animal's diet. Further, animals consume plants and plant parts that are not part of the human diet. Nutrient composition and availability of nutrients are important considerations for animal health.

Labeling

The FD&C Act defines the information that must be disclosed in labeling (including the food label). The Act requires that all labeling be truthful and not misleading. The Act does not require disclosure in labeling of information solely on the basis of consumer desire to know. The Act does require that a food be given a common or usual name, and that the label disclose information that is material to representations made or suggested about the product and consequences that may arise from the use of the product.

The FDA will require special labeling if the composition of a food developed through genetic engineering differs significantly from its conventional counterpart. For example, if a food contained a major new sweetener as a result of genetic modification, new common or usual name or other labeling may be required. Similarly, if a new food contains an allergen that consumers would not expect in that food, labeling would be necessary to alert sensitive consumer. However, if a protein commonly produces very serious allergic reactions (e.g., peanut protein) and is transferred to another food, FDA would need to evaluate whether labeling would provide sufficient consumer protection.

To date, the FDA is not aware of information that would distinguish genetically engineered foods as a class from foods developed through other methods of plant breeding and, thus, require such foods to be specially labeled to disclose the method of development. The agency has not required labeling for other methods of plant breeding such as chemical, or radiation-induced, not required to be labeled "hybrid sweet corn" because it was developed through cross-hybridization.

The FDA is reviewing public comments on labeling issues. One issue that is particularly difficult is the question of whether special labeling should be required for a food derived from a plant that has been modified to express a gene derived from an animal and whether the presence of such a gene or its product affects certain ethical or religious beliefs. Currently, no foods are approaching the market that raise this issue. However, the issue is very complex, and FDA believes that further discussion is warranted.

Summary

The FDA has provided guidance for developers, which establishes a standard to care to ensure that foods derived from new plant varieties are safe and wholesome. Irrespective of the method by which a food is produced, all foods must meet the same stringent safety standards and be properly labeled in accordance with the FD&C Act.

See also **Bioprocess Engineering (Biotechnology)**; **Genetics and Gene Science (Classical)**; and **Industrial Biotechnology**.

Additional Reading

Shannon, T.A.: *Genetic Engineering: A Documentary History*, Greenwood Publishing Group, Inc., Westport, CT, 1999.

Steinberg, M.L., and S.D. Cosloy: *The Facts on File Dictionary of Biotechnology and Genetic Engineering*, Revised Edition, Facts on File, Inc., New York, NY, 2000.

Yount, L.: *Biotechnology and Genetic Engineering*, Facts on File, Inc., New York, NY, 2000.

Web References

Food and Agriculture Organization of the United Nations: http://www.fao.org/

Food and Drug Administration: http://www.fda.gov/

Glossary of Biotechnology and Genetic Engineering: http://www.fao.org/DOCREP/003/X3910E/X3910E00.HTM

One World Guide to Biotechnology and Genetic Engineering: http://www.oneworld.org/guides/biotech/index.html

The Institute of Science, Technology and Public Policy: http://www.istpp.org/genetic_engineering.htm

GENETIC LINKAGE MAPPING.

Random occurrences of crossovers along a chromosome permit definition of a genetic distance. Such distances are used to create genetic maps and to localize disease genes on these maps.

Basic Principles

Human chromosomes occur in pairs in each cell, where one member of each pair is inherited from the mother and the other member from the father. This was recognized early in the twentieth century [Sturtevant, 1913]. It was also recognized at about the same time that chromosomes carry the material responsible for heredity. A connection was made between the parental origin of chromosomes and the laws of inheritance that Mendel had established in his experiments on peas some 40 years earlier. As we now know, chromosomes harbor deoxyribonucleic acid (DNA) strands, and pieces of DNA that code for proteins are called *genes*. For our purposes, we may think of genes as beads on a string (chromosome) with pairs of a given gene transmitted from parents to offspring according to the Mendelian laws. (In general, any entity that is inherited in a Mendelian fashion is called a *locus*). Several modes of inheritance are now known that do not follow these Mendelian laws. For example, trinucleotide repeat polymorphisms underlie traits such as Huntington disease, where disease occurs when the number of repeats exceeds a certain threshold. Also, traits are known in plants that result from the joint action of two genes [Strickberger, 1985].

Meiosis is the cell division that leads to the formation of gametes, that is, egg and sperm cells. At a particular stage in meiosis, the two members of each chromosome pair are aligned next to each other and form a total of four strands due to each single chromosome splitting length-wise into two chromatids. At one or several places along the chromosomes, crossing over occurs, in that chromatids form reciprocal exchanges between the two parental chromosomes. Eventually, meiosis leads to four gametic products as shown in Figure 1. Two of these will show the effects of a crossover, while the remaining two will not. Assume two loci on a chromosome, each with two alleles, A and a for locus 1, B and b for locus 2, where an allele is one of possibly many variants (e.g., slightly different DNA sequences) of a gene (Fig. 1). If a single crossover occurs between the two loci in a doubly heterozygous parent, two offspring types may be distinguished by the grandparental origin of the alleles at the two loci straddling the crossover point. Recombinant offspring are those for which the two alleles originated in different grandparents. For example, the a and B alleles (last offspring) originated in the grandfather and grandmother respectively.

The salient feature is now that crossovers occur more or less randomly along each chromosome. Therefore, when two loci are located in close vicinity to each other, it is rare that a crossover occurs between them so that most of the offspring in the family shown in Figure 1 will be nonrecombinants. The *recombination fraction* (proportion of recombint) θ will then be small. Multiple crossovers may occur on the same chromosome. If two crossovers involve the same two chromosome strands, their effects cancel in the sense that they will result in a nonrecombinant offspring. Consequently, with increasing distance between two loci on a chromosome, the recombination fraction will increase from zero to a maximum of 0.50 for each chromosome, irrespective of its length. Thus, the recombination fraction is not a suitable measure for genetic distance. Instead, genetic distance between two loci is defined as the average number of crossover points on a gamete between the two loci. The unit of measurement for genetic distance is the morgan (M) or centimorgan (cM). Small genetic distances are equivalent to recombination fractions, because more than one crossover is unlikely to occur. For example, a recombination fraction of $\theta = 0.02$ corresponds to 0.02 M or 2 cM. Because crossover intensity tends to be higher in females than in males, we distinguish different genetic distances for the two sexes. The longest human chromosome, number 1, has a genetic length of 221 cM in males and 376 cM in females [Collins, et al.: 1996]. That is, on average each gamete produced by a woman carries close to four crossover points on chromosome number 1.

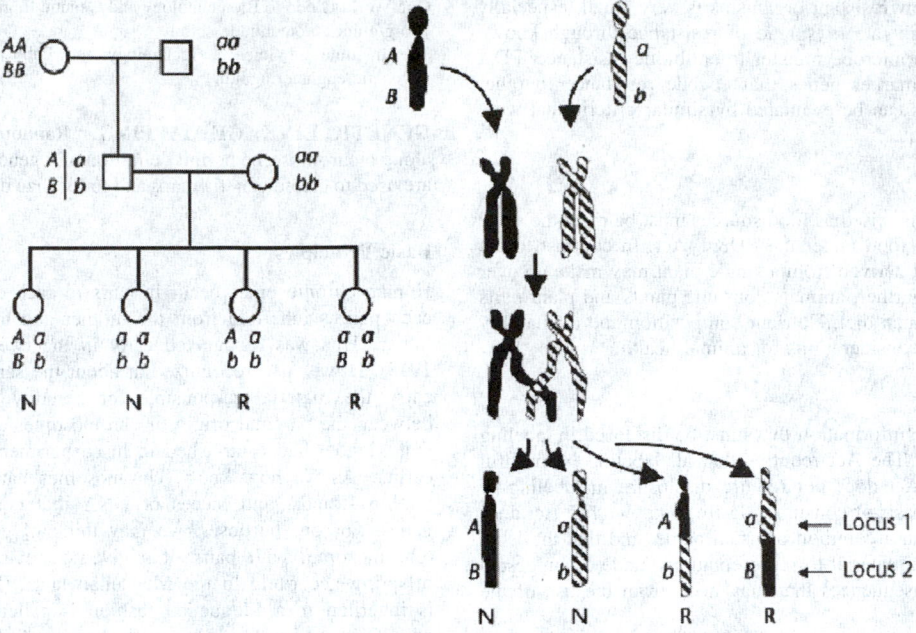

Fig. 1. Effects of a crossover between two loci on resulting gametes (right-hand side) and corresponding genetic phenotypes (left side), where R = recombinant and N = nonrecombinant gamete.

Creating Genetic Maps

A genetic map is an ordered set of genetic loci with a known Mendelian mode of inheritance and known (or estimated) distances between adjacent loci. Loci with stable Mendelian inheritance are called *marker loci* or just *markers*. Thus, a genetic map is characterized by two main aspects: (1) the order of the markers and (2) intermarker distances. Each of these two aspects requires special estimation methods. Once the DNA sequence of an organism (such as the human) is known, marker order is given and need no longer be estimated. However, estimating genetic distances between markers is still required even though physical distances in base pairs may be known. While there exists a rough proportionality between physical and genetic distance, at least on a per chromosome basis [Ott, 1999], there can be strong deviations between the two measures in certain genomic regions. Particularly for questions of genetic counseling and risk predictions, it is the genetic distance between loci that is relevant.

Identification of Suitable Markers. The earliest genetic markers were blood polymorphisms and enzyme variants. Up to about 60 of these markers were available to researchers before DNA sequence variations became available. The latter were introduced to human gene mapping by Botstein, *et al.* [1980] and have the great advantage over classical markers that they generally do not have any specific function. Also, many of these marker types can be created in large numbers.

The first modern marker type was the restriction fragment length polymorphism (RFLP), which is based on single nucleotide changes that are detected by restriction enzymes and made visible by a technique introduced by Southern [1975]. RFLPs have two alleles each and are, thus, not very polymorphic. They are not often used today and have largely been superseded by repeat polymorphisms. These so-called *microsatellites* are polymerase chain reaction (PCR) based and consist of multiple occurrences of short sequences, for example CA, where a given number of repeats represents an allele. Microsatellites are currently the markers used most often and tend to be very polymorphic with up to several dozen alleles. For genetic mapping, as for genetic linkage analysis in general, higher marker polymorphism means that a parent is more likely to be heterozygous and, thus, potentially informative for linkage.

The newest type of marker in human genetics is the single nucleotide polymorphism (SNP). Its properties are similar to the earlier RFLPs, but SNPs are PCR based and have been developed in large numbers. Currently, many 100000s of SNPs are known. The main purpose of these markers is that very dense marker maps allow the localization of disease genes by association (linkage disequilibrium) studies. SNP markers are developed that have minor allele frequencies exceeding, for example, 0.20. SNPs with appreciable allele frequencies are relatively old and tend to occur in multiple populations. Those with rare allele frequencies tend to get lost through genetic drift in some populations.

Researchers often do not choose the type of markers they want to work with. This is pretty much a matter of what is available and what genotyping laboratories are offering in terms of services.

Choice of Pedigrees and Genotyping. Estimating marker order and marker distances must be carried out with genetic linkage analysis. That is, one needs family data to estimate recombination fractions between adjacent markers. To create marker maps, a standard set of families with DNA of family members available to researchers worldwide has turned out to be extremely fruitful. This approach was pioneered in human genetics by the Centre d'Etude du Polymorphisme Humain (CEPH) in Paris [Dausset, et al.: 1990]. The standard set of 40 so-called CEPH families consists of a pair of parents each, with relatively large numbers of children and up to four grandparents. With the highly polymorphic markers used today, most researchers use only a small subset (e.g. eight) of the most informative CEPH families for genetic mapping.

DNA was made available to scientists collaborating with the CEPH and working on specific areas of the genome. These scientists would then genotype the DNA of CEPH family members for their specific markers and feed results back to CEPH. With the advent of microsatellites, a central genotyping facility, Généthon, was created in Paris. These efforts resulted in a large collection of marker genotypes for the members of the CEPH families.

Only after the human genotyping and mapping effort became very successful did genetic mapping begin in other organisms, for example, the dog. There is now a dog map, created mostly on the basis of dog reference families.

Map-Building Algorithms. The genotypes by themselves, of course, do not yet constitute a map — that had to be created (estimated) on the basis of the genotypes. Ideally, when marker order is unknown, one would want to consider all possible marker orders and, for each given marker order, estimate map distances between adjacent markers. Based on such estimates, the likelihood for a given map is then calculated. That marker order leading to the highest likelihood is the best map. Because the sheer number of markers precludes looking at all possible orders, map-building algorithms must be used. The most well known of such algorithms are the one built into the Mapmaker program [Lander, et al.: 1987] and the one used for creating the Généthon map. They typically start with two markers that are well separated and provide good information on the distance between them. Then, one marker after another is added to the existing map with re-estimation of map distances at each step. These algorithms are not guaranteed to find the maximum-likelihood map, but they have done an excellent job in creating marker maps. As mentioned above, the task of estimating map orders is no longer required once the DNA sequence is known. Genetic mapping is then reduced to the simpler task of estimating map distances between adjacent markers.

SNPs are much less polymorphic than microsatellites but the hope is that streamlined (microarray based) genotyping of very large numbers of SNPs will become feasible and economical. An increased number of SNP markers with higher marker density can compensate for the higher polymorphism of microsatellites [Terwilliger, et al.: 1992; and Kruglyak, 1997].

Map Resolution and Possible Artifacts

To establish the order of markers on a chromosome, one must be able to observe crossovers between them. Specifically, to confirm the order of three marker loci, at least one crossover in each of the two intervals is required. The number of phase-known meioses needed for this observation is approximately given by

$$n = \sqrt{6P}/h^2 r$$

where P is the power, h is the heterozygosity of the markers and r is the recombination fraction between the flanking markers [Ott, 1999]. Clearly, ordering SNPs genetically requires large numbers of observations. However, as mentioned above, the known DNA sequence also establishes marker order so that this is no longer a major issue.

In practice, the useful map resolution is determined by the rate of marker errors, which includes genotyping and other errors such as sample swaps. Errors tend to be detected through the occurrence of Mendelian inconsistencies. However, in certain small family types, more than half of all errors will go undetected [Gordon, et al.: 1999]. Errors have a deleterious effect on gene mapping and tend to bias estimates of map locations for disease loci. The recommended approach for disease gene mapping is to work under a suitable error model, that is, to analyze all data (including errors) and estimate error rates along with other parameters, rather than trying to eliminate errors and analyze the "cleaned" data, as if it contained no errors [Gordon and Ott, 2001].

Disease Gene Mapping

The main *raison d'être* of a genetic marker map is that it serves as a tool for localizing susceptibility genes on the human gene map. Of the two predominant statistical mapping tools, linkage analysis and disequilibrium analysis, the latter is gaining increased importance, particularly for case–control studies of association between disease genetic marker alleles or genotypes. Specialized statistical analysis tools are being developed to handle the large number of SNP genotypes per individual and the comparatively small number of observations. For example, in order to be able to analyze sets of SNPs that are each in the vicinity of genes jointly causing disease, a two-step procedure has been proposed: (1) to select a small subset of SNPs and (2) to analyze the selected subset by statistically sophisticated approaches [Hoh, et al.: 2001].

See also **Genome Map**.

Additional Reading

Botstein, D., R.L. White, M.H. Skolnick, and R.W. Davis: "Construction of a Genetic Linkage Map in Man Using Restriction Fragment Length Polymorphisms," *American Journal of Human Genetics*, **32**, 314–331 (1980).

Collins, A., J. Frézal, J. Teague, and N.E. Morton: "A Metric Map of Humans: 23500 loci in 850 Bands," *Proceedings of the National Academy of Sciences of the United States of America*, **93**, 14771–14775 (1996).

Dausset, J., H. Cann, D. Cohen, et al.: "Centre d'Etude du Polymorphisme Humain (CEPH): Collaborative Genetic Mapping of the Human Genome," *Genomics*, **6**, 575–577 (1990).

Gordon, D.S. C. Heath, and J. Ott: "True Pedigree Errors more Frequent than Apparent Errors for Single Nucleotide Polymorphisms," *Human Heredity*, **49**, 65–70 (1999).

Gordon, D., and J. Ott: "Assessment and Management of Single Nucleotide Polymorphism Genotype Errors in Genetic Association Analysis," *Pacific Symposia in Biocomputing 2001*, 18–29 (2001).

Hoh, J. A. Wille, and J. Ott: "Trimming, Weighting, and Grouping SNPs in Human Case–control Association Studies," *Genome Research*, **11**, 2115–2119 (2001).

Kruglyak, L.: "The Use of a Genetic Map of Biallelic Markers in Linkage Studies," *Nature Genetics*, **17**, 21–24 (1997).

Lander, E.S., P. Green, J. Abrahamson, et al.: "MAPMAKER: An Interactive Computer Package for Constructing Primary Genetic Linkage Maps of Experimental and Natural Populations," *Genomics*, **1**, 174–181 (1987).

Ott, J.: *Analysis of Human Genetic Linkage*, Johns Hopkins University Press, Baltimore, MD, 1999.

Southern, E.M.: "Detection of Specific Sequences Among DNA Fragments Separated by Gel Electrophoresis," *Journal of Molecular Biology*, **98**, 503–517 (1975).

Strachan, T., and A.P. Read: *Human Molecular Genetics*, Wiley-Liss, New York, NY, 1996.

Strickberger, M.W.: *Genetics*, 3rd Edition, Macmillan, New York, NY, 1985.

Sturtevant, A.H.: "The Linear Arrangement of Six sex-linked Factors in *Drosophila*, as Shown by their Mode of Association," *Journal of Experimental Zoology*, **14**, 43–59 (1913).

Sturtevant, A.H.: *A History of Genetics*, Cold Spring Harbor Laboratory Press, Cold Spring Harbor, NY, 2001.

Terwilliger, J.D., Y. Ding, and J. Ott: "On the Relative Importance of Marker Heterozygosity and Intermarker Distance in Gene Mapping," *Genomics*, **13**, 951–956 (1992).

JOSEPHINE HOH, JURG OTT, Rockefeller University, New York, NY

GENETICS AND GENE SCIENCE (CLASSICAL). Early biologists essentially targeted the hereditary aspects of life (animals and plants) and depended on empiricism and statistics for their knowledge. Toward the late 1880s, with the availability of improved microscopy and an increased interest in biochemistry, researchers turned much of their attention to the structure and performance of the individual cell. Soon cytology became an important biological discipline and from this, over a further time span, cells and their components were reduced to the molecular level. Although even in modern times hereditary processes remain very important research objectives, particularly as they relate to diseases, the new gene-based sciences also encompass such diverse fields as crop improvement and criminology. (See Table 1 for Chronology of Genetic Science.)

Fundamentals of Genetics

Gregor Johann Mendel (1866) sometimes is referred to as the father of genetics. By studying the crosses of garden peas in his garden, Mendel worked out the basic principles of inheritance. Over the years, genetics and the gene sciences have proceeded along six major pathways:

1. *Experimental Breeding*, a procedure dating back several centuries, requires considerable time and patience because the animals or plants studied must experience a number of lifetimes (generations). Statistical methods typify this kind of genetic research.
2. *Pedigree Analysis*, an approach widely used where experimental breeding is not practical. Pedigrees show the inheritance of specific traits, which can be traced, in all of the members of a family line. Human pedigrees have been very useful in terms of tracing the familial aspects of certain diseases. One of the first diseases so traced was hemophilia. Stock breeders keep careful pedigree records as breeding guides. Horses and other high-performing animals are bought and sold based upon their pedigrees.
3. *Cytogenetics* (cytology) is a study of the chromosomes and cellular infrastructure that are keys to heredity. This field now embraces the study of individual genes.
4. *Biochemistry* and, in particular, molecular biology is a study of the genes — what they are, how they perform, and how they reproduce. Through an analysis of gene action, biochemical geneticists — working with such diverse organisms as molds, bacteria, viruses, fruit-flies, mice, and human cells — have been able to trace the course of the breakdown of particular amino acids in the cells and to learn of abnormalities that arise when a gene fails to produce a particular enzyme. See also **Biochemistry (The History)**.
5. Population genetics deals with the distribution of genes in various populations. Human population geneticists have traced population migration and the intermixing of races through an analysis of the frequency of the various blood antigens. Within recent years, some geneticists have turned to analyzing fossil genetic material to trace the process of heredity over many thousands of years.

TABLE 1. AN ABRIDGED CHRONOLOGY OF PROGRESS IN GENETIC SCIENCE. (Early Years to Commencement of the Human Genome Project)

1543	Andreas Vesalius, Belgian anatomist, produced the first anatomical map of humans in the paper, "De Humani Corporis Fabrica." This publication is recognized as a first step toward what may be termed, *intellectual medicine*. For many decades thereafter, chromosomes and genes (prior to their identification and naming) were simply considered the *minutia* of life and beyond research — prior to the introduction of improved microscopes. See also **Andreas Vesalius (1514–1546)**.
1665	Cytology (science of cells) had its beginnings when Robert Hooke, English physicist, described the nature of cork cells. See also **Hooke, Robert (1635–1702)**.
1820	Robert Brown, Scottish botanist, postulated the "nucleus" of individual cells. See also **Brown, Robert (1773–1858)**.
1838	Mathias Jacob Schleiden, German botanist, adopted Brown's views of the nucleus and proposed the general concept that living organisms are made up of cells and that the nucleus is essential to the formation of new cells. See also **Schleiden, Matthias Jacob (1804–1881)**.
1839	Theodor Schwann, German physiologist, published a paper, "Microscopic Investigations on the Accordance in the Structure and Growth of Plants and Animals," this leading to the first acceptance of the cellular origin and structure of animals and plants. Schwann observed, "The entire animal or plant is composed either of cells or of substances thrown off by cells; cells have a life that is somewhat independent, and this individual life of all the cells is subject to that of the organism as a whole." See also **Schwann, Theodor Ambrose Hubert (1810–1882)**.
1866	Gregor Johann Mendel, Austrian naturalist and botanist, published a paper entitled, "Experiments in Plant Hybridization," which was based upon his personal experimentation with garden plants (mostly peas) for tracing the dominance of traits from one generation to the next — and, in this sense, was the father of traditional genetic science. His work, however, was not received with acclaim, but rather was considered of little importance by Karl Näageli, a revered botanist during that period. Ironically, Mendel's principles were "rediscovered" independently many years later (1900) by Hugo De Vries, a Dutch botanist, by Karl Correns, a German biologist, and by Von S. Tschermak, an Austrian naturalist. This group put to final rest the prior concept that "heredity is transmitted by fusable parental *bloods*." See also **Mendel, Gregor Johann (1822–1884)**.
1876	Johann Friedrich Horner, Swiss ophthalmologist, was the first researcher to establish a connection between a cellular deformity (gene abnormality) and the familial aspects of color blindness. See also **Horner, Johann Friedrich (1831–1886)**.
1900	Hugo De Vries observed that heredity is a conservative force and that, if heredity were perfect, all organisms would carry the same genotype and evolution would not occur. De Vries pointed out, however, that this conservatism is opposed by a factor of change, that is, *mutation*. He suggested that mutational changes must be drastic and sudden, whereas it was soon to be learned that mutations range widely in their cause and effect. See also **De Vries, Hugo (1848–1935)**.
1903	Camillio Golgi, Italian pathologist, while researching malarial parasites, demonstrated the nervous system as being interlaced rather than connected in a complete network. Golgi developed a method for staining nerve cells. Previously, Golgi had described the Golgi complex (apparatus) of the cell and considered that to be a cytoplasmic organelle occurring in almost every type of vertebrate cell. Golgi also described the importance of membranes in cells. See also **Golgi, Camillo (1843–1926)**.
1903	Wilhelm Ludwig Johannsen, Dutch geneticist, introduced the words *gene, genotype*, and *phenotype* to the literature of genetics. See also **Johannsen, Wilhelm Ludwig (1857–1927)**.
1909	A. E. Garrod, British physician, pioneered the field of developmental genetics and visualized development as a network of chemical reactions, many of which are facilitated by specific catalysts or enzymes. By extrapolation, he surmised that each enzyme is produced by just one gene and that each gene produces just one enzyme (Garrod-Beadle concept). See also **Garrod, Archibald Edward (1857–1936)**.
1910	Thomas Hunt Morgan and E. B. Wilson (Johns Hopkins University) became interested in using the fruit fly as an experimental model for heredity studies after having discovered a fly with white eyes, as contrasted with the normal red coloration. Subsequently, because of its very short reproductive span allowing many generations to be studied over a brief time period, the fruit fly (*Drosophila melanogaster*)

	became the focus of thousands of genetic studies continuing to the present. The genome of the fruit fly is approaching completion as of 1993. As a somewhat later date, mice became a model for geneticists and its genome is nearing completion as of 1994. See also **Morgan, Thomas Hunt (1866–1945)** and **Wilson, Edmund Beecher (1856–1939)**.
1911	E.B. Wilson (Columbia University) confirmed link of color blindness with the X-chromosome.
1946	Frederick Sanger, British biochemist, determined the complete amino acid sequence in the protein insulin. In prior years, Sanger had developed the use of 2,4-dinitrofluorobenzene (Sanger's reagent) which became an important tool for protein analysis. Sanger was first researcher to show that proteins are polypeptides in which alpha amino acids and imino acids are bound together by peptide bonds between their alpha-amino and alpha-carboxyl groups. Sanger was awarded the Nobel Prize (chemistry) in 1948 & 1980. See also **Sanger, Frederick (1918–)**.
1950s	Linus Carl Pauling, American physical chemist and 1954 Nobelist (chemistry), contributed new knowledge to the understanding of proteins, enzymes, and nucleic acids. Pauling also proposed the gene structure of hemoglobin, particularly as it relates to sickle cell anemia. Pauling and others also pioneered procedures for sequencing amino acids. See also **Pauling, Linus Carl (1901–1994)**.
1952	Alexander Robertus Todd (Lord), British biochemist first researcher to synthesize adenosine diphosphate (ADP and adenosine triphosphate (ATP). Todd was awarded the Nobel Prize (chemistry) in 1957. See also **Todd, Alexander Robertus (1907–1997)**.
1953	J. D. Watson, American chemist and Nobelist (1962), and Francis Harry Compton Crick, American scientist and Nobelist (1962), proposed that the molecular structure of DNA is composed of deoxyribonucleic acid and proteins (histones and high-molecular-weight proteins). These researchers proposed that the molecular structure of DNA is a double spiral helical chain. James H. White, American mathematician, shared the 1962 Nobel Prize. See also **Crick, Francis Harry Compton (1916–1004)** and **Watson, James Dewey (1928–Present)**.
1968	It was reported that 68 human genes had been mapped to the X-chromosome.
1970	Restriction enzymes, which cut DNA in specific places, were discovered and when coupled with recombinant DNA technology, made it possible to identify a specific stretch of genetic material.
1970	The concept of Recombinant DNA was proposed by several geneticists. Thus, new DNA structures could be created. Both positive results and negative concerns were expressed. For example, the addition of new genes to bacteria and viruses could confer qualities that could be harmful to other forms of life, including humans, with possibly epidemic, even catastrophic proportions. Researchers attending 1973 Gordon Research Council proposed that the National Academy of Sciences address these concerns. Guidelines and regulatory actions were initiated, some of which continue to the present. Regulations vary somewhat between one country and the next.
1970s	Torbjorn Caspersson and Lore Zech (Karolinska Institute, Stockholm) developed a staining technique (using quinacrine mustard) that fluoresces under ultraviolet light, revealing that each chromosome has a unique banding pattern.
1976	A.M. McKusick (then at University of Washington) published a catalog of 1,487 genetic disorders. This was revised in 1990 to include nearly 5,000 inherited characteristics. About a decade later, McKusick became the first head of the International Genome Organization.
1988	The National Academy of Sciences (U.S.) endorsed a massive national effort to map and sequence the human genome. The project target — to produce genetic and physical maps of increasing resolution, with a fully detailed map of the chromosomes — the project to be completed within a decade and at a cost estimated to be $3 billion.

Further details are given within text of article.

6. *Genetic Recombination*, made possible by the discovery of the recombinant DNA procedure in the 1970s, makes it possible to develop extensive and detailed maps of the nucleotide sequences of gene molecules — to the point where, in 1990, plans were outlined for mapping the complete human genome, a program that was well underway as of 1994.

Defining the Gene

Genes are the physical units of heredity. The precise definition for gene has changed over the years as more has been learned about the chemical nature of genetic material and function. In modern terms, a gene may be defined as a segment of genetic material that determines the sequence of amino acids in specific polypeptides. In lieu of additional findings, geneticists have noted a one-to-one relation between gene and polypeptide. It appears that this definition applies at least to those genes called *structural* genes because they determine the primary structure of proteins.

Structural Genes. So far as known, structural genes in all organisms are composed of nucleic acids. In the RNA viruses, the genes are RNA (ribonucleic acid) only, but in all other organisms, the DNA viruses and the cellular forms, which all possess both DNA (deoxyribonucleic acid) and RNA, the gene material is either known to be DNA, or assumed to be for good reason.

The genes of viruses and bacteria appear to consist of nucleic acid unaccompanied by closely bound protein. Ordinarily this naked nucleic acid is in the two-stranded condition; exceptions are known among both the RNA and DNA viruses, some of which possess single-stranded genetic material. In those organisms with true nuclei, the genetic material is always double-stranded DNA associated with protein ordinarily of the histone type. The function of the protein is not considered to be genetic. It probably controls DNA in its role of determining protein structure. Also it may serve to hold genes together and attached to the chromosomes of which they are a part.

Structural genes carry out their role of dictating protein structure by producing a messenger RNA (mRNA) which is a single strand of RNA containing nucleotide bases complementary to one of the strands of the double-stranded DNA of the gene from which it is copied or "transcribed." The evidence is that the same DNA strand of a gene is always transcribed into mRNA. In this way, only one kind of mRNA is made for each gene. In the transcription process the C, T, A and G bases of the DNA determine G, A, U and C, respectively, in the mRNA strand. Transcription effectively constitutes *gene action*. By definition, if a gene is not actively forming mRNA, it is inactive or "turned off."

Each kind of gene is different from every other gene in its DNA sequence. Hence, as many different kinds of mRNA are formed as there are different genes in the organism.

Genes in eukaryotic cells are often not collinear with their products. Instead genes contain intervening sequences of DNA (*introns*) which result in a gene that is much longer than required for the simple coding of amino acid sequence. An enzymatic reaction, gene splicing, is required for the expression of the genes. That is, the entire gene, including introns, is transcribed as a long mRNA precursor. The intervening sequences are clipped out and the ends rejoined to yield the mRNA with the correct coding sequence for the gene product. After their formation, the mRNA strands attach to ribosomes in the cytoplasm, and the process of protein biosynthesis commences. The significant point to be emphasized here is that the sequence of nucleotide bases, of the "genetic code," in a particular gene is reflected in a specific sequence of amino acids in the polypeptide produced through the protein synthetic mechanism.

The one-to-one relation between gene and polypeptide is a more accurate statement of the situation than the earlier one gene-one enzyme hypothesis. It is now known that a number of proteins are constituted in their functional state of subunits, which are polypeptides. When subunits are all identical, the one gene-one protein statement holds with certain exceptions. However, proteins such as vertebrate lactic acid dehydrogenase (LDH) and hemoglobin are made up of different subunits. For example, the dominant adult hemoglobin in man contains both α and β polypeptides as subunits. These have somewhat different amino acid sequences, and each has been shown to be under the control of a different gene. The genes are not even on the same chromosome. A similar situation has been found for LDH which may be made up of at least two different subunits, each one again under the control of a separate gene.

A term currently used by many synonymously with structural gene is *cistron*. Its original definition was based on complementation tests. If two chromosomes bearing the same kinds of genes (homologous chromosomes) are introduced into the same cell, "product interactions" may be observed between the genes of the same type, *i.e.*, genes which control the same kind of polypeptide. If two genes of the same type are mutant, but mutant at different sites, they may *complement* and produce a protein that has an activity comparable to the nonmutant, even though each mutation alone, or together on the same chromosome, can produce only a mutant, inactive protein. Those mutants that do not complement with the production of an active protein are said to have mutational sites within the same cistron.

Controlling Genes. Genes which do not carry codes for the synthesis of proteins which constitute the enzymes, structural components, etc., of the cell almost certainly exist. These genes may produce proteins, but the proteins presumably act by the regulation of the activity of the structural genes, turning them on and off according to circumstances within the cell.

Examples of such genes are found in *Escherichia coli*. These, termed *regulator genes*, presumably produce substances, possibly proteins, which prevent or repress structural genes from synthesizing mRNA unless other substances, the inducers, are present to inhibit the repressor substances. Alternatively, repressor substances from other types of regulator genes are active in repression only when certain substances activate the repressor substances. The reason for the existence of these genes would seem to be for the regulation of metabolism by preventing the overproduction of enzymes when their substrates are not present, or of end products such as amino acids. In the latter case, the end product is usually considered to be the substance which activates the repressor produced by the regulator.

Genetic Code. Genetic information stored in the genes, as a linear sequence of the bases (A, C, G, and T) in deoxyribonucleic acid molecules, is transcribed into a complementary base sequence (U, G, C, and A, respectively) in the messenger RNA molecules. This "coded message" contained in the mRNA, as a linear sequence or 4-letter "language," is "translated" in the process of protein biosynthesis into a linear sequence of the 20 amino acids within the protein polypeptide chain synthesized. Each nucleotide triplet or "code word," consisting of one of the 64 possible triplet combinations of U, G, C, and A nucleotides in a messenger RNA molecule, may specify one particular amino acid for incorporation into the polypeptide chain. It appears that certain amino acids may be specified by more than one of the 64 nucleotide triplets; in this respect, the genetic code is said to be "degenerate." A few particular triplet "words" may have special functions, such as to signal polypeptide-chain initiation, or chain termination. The first identification of a particular triplet as the code word for a particular amino acid was the discovery that the sequence UUU (in the form of polyuridylate) appears to be the "code word" specifying incorporation of phenylalanine into a polypeptide, in a cell-free, *in vitro* system containing ribosomes and other required components.

Evidence that a nucleotide *triplet* (and not some smaller or larger run of nucleotides) is the "code word" for incorporation of a specific amino acid has come from studies of the fine structure of genes or DNA of a bacteriophage (virus). Many tentative formulations of a "code dictionary" of messenger RNA triplets, with the corresponding amino acid specified by each triplet, have been proposed, on the basis of both experimental results (primarily those of the Nirenberg group and of the Ochoa group) and theoretical considerations. The exact determination of the genetic code, or pattern of correspondence between each possible nucleotide triplet of mRNA and the amino acid specified by that triplet for incorporation into proteins, has been an active field.

Deoxyribonucleic Acid (DNA)

DNA is a complex sugar-protein polymer of nucleoprotein, which contains the genetic code for enzymes in the cell. It occurs as a major component of the genes, which are located on the chromosomes in the cell nucleus. The DNA molecule is a unique and vastly intricate structure. It is comprised of from 3000 to several million nucleotide units arranged in a double helix containing phosphoric acid, 2-deoxyribose, and the nitrogenous bases adenine, guanine, cytosine, and thymine. The spiral (see Fig. 1) consists of two chains of alternating phosphate and deoxyribose units in continuous linkages. See Fig. 2. The nitrogenous bases project toward the axis of the spiral; they are joined to the chains by hydrogen bonds. Adenine units pair with thymine, and cytosine units with guanine. The complementarity of the bases on the joined chains allows each chain to act as a template for replication of the other when the chains are separated, thus producing two new strands of DNA. See Fig. 3. The sequence of the bases on the chains varies with the individual, and it is this sequence that governs the genetic code. DNA works in conjunction with ribonucleic acid (RNA). Genes are found in pieces that are spread out along DNA. Between gene fragments, there are long stretches of DNA, the functions of which are only recently being clarified. See also **Deoxyribonucleic Acid (DNA)** and **Recombinant DNA**.

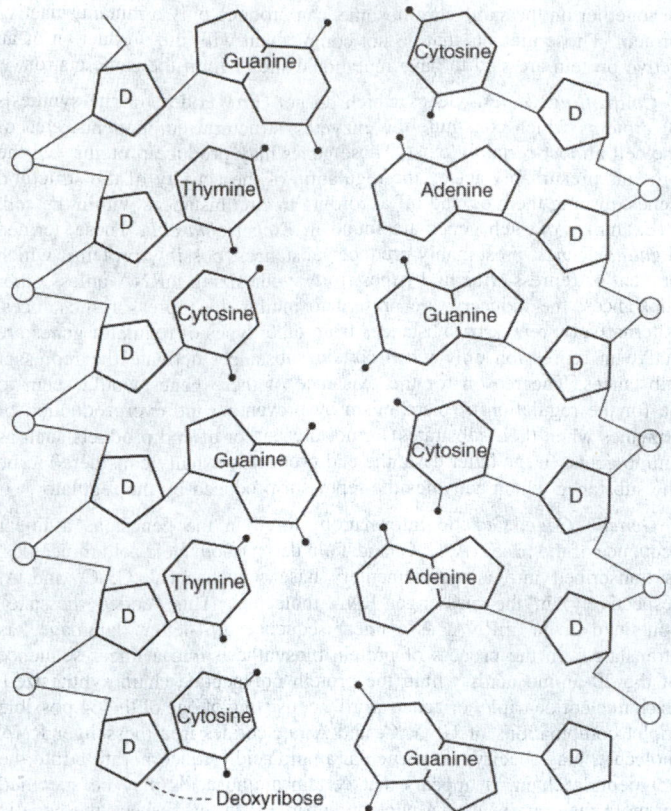

Fig. 1. Consisting of two helically intertwined strands, the DNA molecule is composed of deoxyribose and phosphate. As shown here, at periodic intervals the sugar-phosphate backbones are joined together by the complementary purine and pyrimidine bases. A single base linked to a deoxyribose-phosphate moiety constitutes a deoxyribonucleotide. Legend: Solid black circles = Thymine; Vertical bars = Adenine; Horizontal bars = Guanine; Dotted circles = Cytosine.

DNA in Perspective. As early as 1838, Schleiden and Schwann proposed that large organisms, as represented by the complete animal, are constructed from large numbers of very small cells, all of which are derived from a single original cell by the repeated process of cell division. The nature of the molecular processes underlying cell division did not emerge until the early 1950s. The chemical nature of DNA and RNA was not established until 1952 by Brown and Todd. In that same year, through a detailed analysis of insulin, Sanger showed that proteins are polypeptides in which alpha-amino and imino acids are bound together by peptide bonds between their alpha-amino and alpha-carboxyl groups. These molecules were shown to be polymers in which limited numbers of monomers linked together to form molecules having complex properties.

For several years, the biological roles of these substances were controversial topics in the scientific community. In 1944, investigators Avery, McLeod, and McCarty suggested an essential distinction between DNA and RNA; they were joined in 1952 by Hershey and Chase in this opinion. It was concluded at that time that DNA is the fundamental storehouse of genetic information.

Phillips suggests that during the early 1950s, molecular biologists were seeking the answers to three fundamental questions: (1) How is the information-embedded in the DNA of the genes copied for transmission to successive generations of cells? (2) How does this information direct the synthesis of proteins? and (3) How do proteins, essentially having simple structures, acquire their diverse and subtle chemical properties?

Very shortly, the first question was answered in principle by Watson and Crick who proposed the three-dimensional structure of DNA in 1953. Their proposal that DNA is composed of two polynucleotide chains forming a double helix was based upon studies of x-ray diffraction patterns of DNA fibers.

Ribonucleic Acid (RNA)

Ribonucleic acids comprise a group of natural polymers consisting of long chains of alternating phosphate and D-ribose units, with the bases

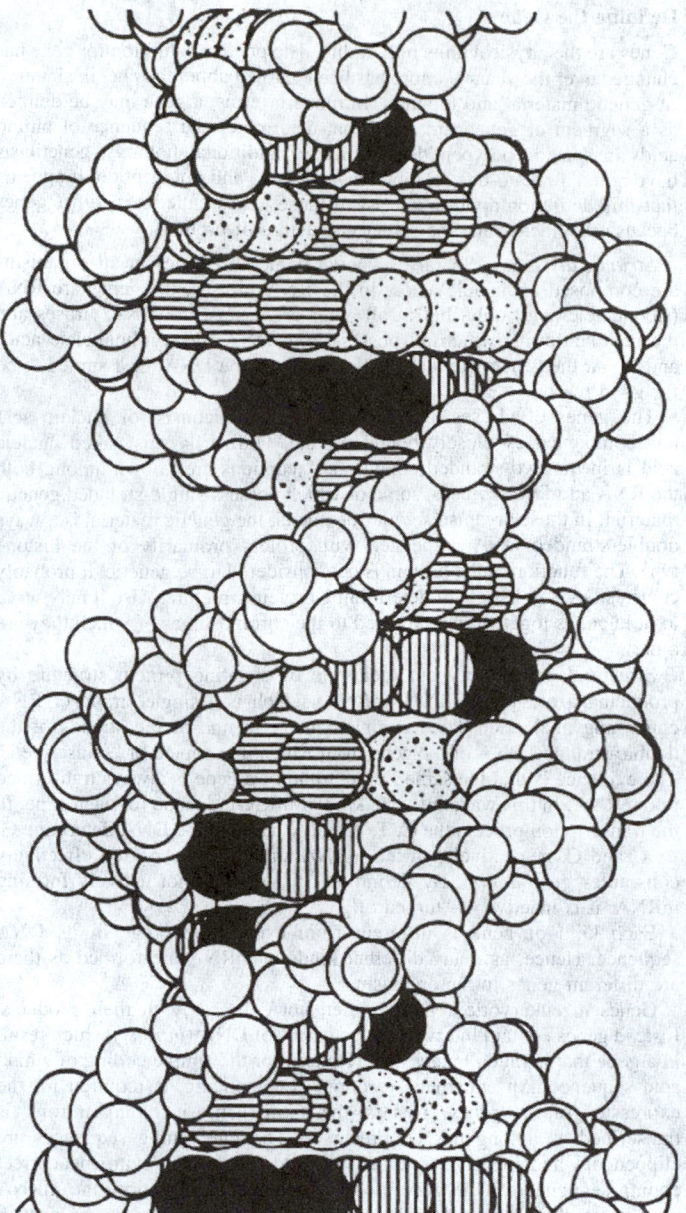

Fig. 2. Schematic of DNA molecule showing repeating sequences of deoxyribose (white pentagons) and phosphodiester units that provide structural support. The varying sequences of pyrimidine and purine bases encode genetic information. The purines are guanine and adenine; the pyrimidines are thymine and cysine. Note that guanine pairs with cytosine; adenine pairs with thymine.

adenine, guanine, cytosine, and uracil bonded to the 1-position of the ribose. Ribonucleic acid is universally present in living cells and has a functional genetic specificity due to the sequence of bases along the polyribonucleotide chain.

Types of RNA include the following. (1) *Messenger RNA*, synthesized in the living cell by the action of an enzyme that carries out the polymerization of ribonucleotides on a DNA template region which carries the information for the primary sequence of amino acids in a structural protein. It is a ribonucleotide copy of the deoxynucleotide sequences in the primary genetic material. (2) *Ribosomal RNA*, which exists as a part of a functional unit within living cells called the ribosome, a particle containing protein and ribosomal RNA in roughly 1:2 parts by weight, having a particle weight of about 3 million. Messenger RNA combines with ribosomes to form polysomes containing several ribosome units, usually five (e.g., during hemoglobin synthesis), complexed to the messenger RNA molecule. This aggregate structure is the active template for protein biosynthesis. (3) *Transfer RNA*, the smallest and best-characterized RNA class. Its molecules contain only about 80 nucleotides per chain. Within the class of transfer

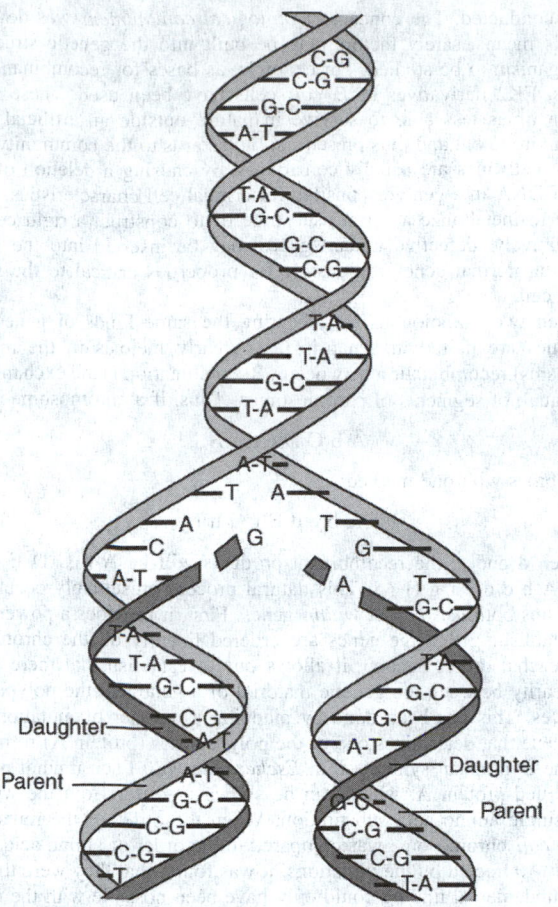

Fig. 3. For replication, the two strands of the parent DNA molecule (light gray) separate as the base pairs detach. The replicated (daughter) strands (dark gray) form as guanine (G) pairs with cytosine (C) and adenine (A) pairs with thymine (T).

RNA molecules, there must be at least 20 separate kinds, correspondingly related to each of the 20 amino acids naturally occurring in proteins. Transfer RNA must have at least two kinds of specificity. (1) It must recognize (or be recognized by) the proper amino acid activating enzyme so that the proper amino acid will be transferred to its free 2′ or 3′ OH group. (2) it must recognize the proper triplet on the messenger RNA-ribosome aggregate. Having these properties, the transfer RNA accepts or forms an intermediate transfer RNA-amino acid that finds its way to the polysome, complexes at a triplet coding for the activated amino acid, and allows transfer of the amino acid into peptide linkage.

Mutations of Genes

New organisms in nature normally are formed by very slow processes. A change in the base sequence of the DNA constituting a gene results in an inherited alteration in the code and is called a *gene mutation*. Mutations are genetic changes that occur suddenly and are thereafter heritable.

Mutations arise through three general mechanisms: (1) chemical modification of preformed DNA, such as breakage and aberrant reunion of molecules or the changes elicited by ultraviolet light, for example; (2) errors in incorporation of the purine and pyrimidine bases, or additions and subtractions of bases, during DNA replication; and (3) unequal exchange between two identical or similar DNA molecules ("unequal crossing over") during recombination. These chemical changes normally occur with low frequency (spontaneous mutations), but the frequency can be increased by means of various chemical and physical treatments (induced mutations). Even when so induced, the frequency of bacterial mutants for a particular trait, for example, is low, e.g., one mutant in 10^4 to 10^{10} bacteria. Thus, any biological evolutionary alterations brought about by the mechanism of mutation represent a very slow pathway. Such procedures do not comprise effective tools for what has been referred to as *genetic engineering*

(genetic manipulation) wherein gene structures can be willfully directed under laboratory conditions.

A change in the base sequence of the DNA constituting a gene results in an inherited alteration in the code and is called a gene mutation. Changes in base sequence may conceivably result from: (1) the deletion or addition of one or more nucleotide pairs in the DNA chain; (2) changes in one or more bases along the chain; or (3) inversion of a segment of the chain.

Good evidence for the occurrence of the first type of mutation exists at least in bacteriophage of the T series, which infect *Escherichia coli*. The deletion or addition of a single base pair into a DNA chain of a gene should be expected to cause considerable difficulties in the translation of the code in the derived mRNA, into an amino acid sequence. For example, if the mRNA of the nonmutant strain has the sequence:

$$\overline{GCU}\ \overline{AAU}\ \overline{GAA}\ \overline{UUU}\ \overline{AAA}\ \overline{CAU}\ L$$

which is read in triplets from left to right to give a particular sequence of amino acids, say ala · asp NH^2 · glu · phe · lys · his, a deletion of a base in the mutant would change the "reading frame" starting at the point of deletion. Thus, the sequence

$$\overline{GCU}\ \overline{AAU}\ \overline{G\!\downarrow\!AU}\ \overline{UUA}\ \overline{AAC}\ \overline{AUL}$$
$$A$$

would produce the sequence ala · asp NH$_2$ · asp · leu · asp NH$_2$ as one possibility. A similar result would be expected from a duplication, by shifting the reading frame. Such mutations as these should be expected to produce "nonsense" sequences of amino acids after the point of change, and they are referred to as frame shift mutations.

Mutations which are the result of the simple changing of bases, say G ⇌ A, C ⇌ T, or G ⇌ C, or A ⇌ T should obviously cause changes in a single triplet rather than a whole sequence. As a result only a single amino acid change should occur in the polypeptide, if but a single base is changed. Many mutant proteins from a variety of organisms are now known which have but a single amino acid change from the nonmutant, and are therefore presumably the result of a single, or adjacent changes within a single triplet. The nonoccurrence of single mutations causing the substitution of two adjacent amino acids within a chain is evidence that the genetic code is not overlapping. As might be expected, a number of different amino acids may be substituted for the nonmutant acid, but the number of substitutions has been found to be limited for any particular amino acid. This also has connotations for the nature of the genetic code.

Gene mutations of other types such as inversions probably occur in addition to the two discussed above, but techniques have yet to be devised to analyze them.

For the present it is enough to say that a mutation may occur at any point within a gene. Theoretically there should be as many "mutational sites" within a gene as there are nucleotide pairs.

Gene instability, the sudden occurrence of high mutability of a normally stable gene, has been described at many different loci in maize, the galactose region of *Escherichia coli*, and in the white locus of *Drosophila*. Studies of the molecular basis for this instability in bacteria have identified transposable elements (*transposons*) as the agents responsible. A transposable element is a segment of DNA capable of transposition intact from one position in the genome to another. In addition to promoting their own transposition, transposable elements can also promote inversion, deletion, and transposition of adjacent chromosomal DNA sequences, resulting in increased occurrence of mutations. Indeed transposable elements have now been found adjacent to many mutant genes in bacteria. Such transposable elements are also thought to occur in eukaryotic cells. See also **Mutation.**

Point Mutation. A classical definition of a genetic disease is one that results from the mutation of a single gene, either by inheritance or by some environmental factor, such as ionizing radiation. This situation is sometimes called *point mutation*. A mutant gene will generally cause one of two happenings: (1) it will synthesize an abnormal protein that has an altered primary amino acid sequence, or it will alter the level of production of a normal protein. Natural substances known to be affected by inherited point mutation include collagen, insulin, myoglobin, a large number of enzymes, clotting factors, albumin, and others. For example, hemolytic anemia results from underproduction of a normal form of the enzyme, glucose-6-phosphate dehydrogenase (G6PD). In this case, there are decreased levels of a normal stable enzyme. It is interesting to note that many genetic defects are not observed in utero because the mother may generate sufficient required enzymes. After birth, several weeks may elapse before the newborn indicates a lack of a given enzyme. There are other situations where years may be required for the abnormality to be detected. This may be true, for example, in the case of a degrading enzyme that very slowly causes the accumulation of metabolic waste products. In the case of Gaucher's disease, undegraded macromolecules in the liver and spleen will cause these organs to enlarge over a period of time. The time span of detection may range from the development of gross mental deficiency, blindness, and death in a child's first year of life; or, much later in life, to the detection of an enlarged spleen during surgery for some other condition.

Thus, it has been found that genetic diseases run the gamut of time and of severity. In Pompe's disease, a deficient enzyme (alpha-1,4-glucosidase) may range from death (total deficiency) to the progressive manifestation of cardiac or peripheral myopathy in later life (mild deficiency).

Recombinant DNA Technology

In the early 1970s, there was an interesting observation of great significance, that is, the discovery of certain enzymes that have the ability to cut and splice hereditary material. The cut pieces are about the order of a gene in length. Also, some of these enzymes have the further ability to cut a few bases further down than the others, so that what sometimes are known as "sticky ends" are produced. Thus, any species of DNA, if cut by the same enzyme, will possess the same type of sticky ends, and fragments of differing DNAs, through a form of biological "scissors and paste" process, can cause the lower part of one DNA molecule to stick well onto the upper part of another molecule. The result is a hybrid molecule. Theoretically, the technique can cross the boundaries of species by selecting DNA material from fully different sources. The ability to cut and recombine is the basis for the term *recombinant DNA*. See also **Recombinant DNA**.

A useful modification of the basic clip-and-paste process involves inserting the DNA fragments into a DNA molecule, which has the power of self-replication. Many bacteria contain small circular cytoplasmic DNA molecules called *plasmids*, which are capable of self-replication inside the bacterial cell. The characteristics of rapid bacterial growth and multiplication allow quantity replication of the recombinant plasmids in short periods of time. This technique thus offers an obvious advantage over the slow and laborious chemical methods.

However, obtaining sufficient quantities of a specific gene in purified form, for insertion into a plasmid, is difficult when one considers the genetic complexity of living organisms. An approach to the problem has been the use of an enzyme known as *reverse transcriptase*. This enzyme synthesizes DNA from RNA. The primary product of genes is mRNA, which possesses base sequences complementary to the genes. The large quantities of specific mRNA available, coded for by the single gene, allow biochemical purification of the mRNA. Thus, if one can isolate the mRNA coded from a particular gene, the corresponding DNA sequence, identical to the gene, can be reconstructed using reverse transcriptase. This synthesized DNA then can be inserted into a plasmid by standard recombinant DNA methods and amplified by growing the plasmid in bacteria.

The advantages of recombinant techniques for increasing knowledge of the genetic construction of any organism are immediately recognized. A number of practical findings from such investigations can be envisaged. These include the incorporation of nitrogen-fixing genes in agricultural plants, to eliminate the need for nitrogen fertilizers; and the bacterial manufacture of large quantities of polypeptide hormones, such as insulin; and the bacterial production of vaccines and enzymes, as well as the treatment of genetic diseases. Possible production of fermentation products (alcohol, methane, etc.) as fossil fuel substitutes may be aided by this technique.

It should be stressed that recombinant DNA methodology is *not* a way of constructing new forms of life in vitro. Even the simplest organisms are extremely complex and the maximum alteration of the simplest genome would be of the order of 1%. Also, the genomes of the simplest organisms are highly ordered and the random insertion of a few genes from an unrelated organism is unlikely to create a whole new organism.

In the initial stages of recombinant DNA research, there was considerable concern regarding possible serious consequences of producing biologically hazardous DNA molecules. Both self-policing and governmental guidelines, which are under continuous review, were established, and they continue in most countries where recombinant DNA research is being conducted. The concept of *biological containment* was developed. By this means, safety factors may be built into the genetic structure of the organism to be studied. For example, as bases for recombinant experiments, EK2 derivatives of *E. coli* cells have been used. These are 100 million times less able to survive in nature outside an artificial laboratory environment and thus present no biohazards to the community. These mutant cell lines are usually constructed by causing a deletion of a portion of DNA in a gene responsible for critical cell characteristics, such as ability to metabolize a certain substrate or to construct a rigid cell wall. Alternatively, defective mutant genes may be inserted into the genome replacing normal genes responsible for properties critical to the survival of the cell.

When two homologous (i.e., bearing the same kinds of genes) chromosomes are paired in synapsis (as in early meiosis in the nucleated organisms) recombination may occur. Recombination is the exchange, usually equal, of segments of chromosomes. Thus, if a chromosome marked:

$$A\,b\,C\,D\,e\,f\,g\,H\,i\,J$$

recombines with one marked:

$$A\,b\,c\,d\,E\,f\,G\,h\,I\,j$$

between d and e, the recombinant products will be $A\,b\,C\,D\,E\,f\,G\,h\,I\,j$ and $A\,b\,c\,d\,e\,f\,g\,H\,i\,J$. This natural process presumably occurs in all organisms both *between* or *within* genes. First, it provides a powerful tool for establishing that the genes are ordered linearly on the chromosome, and in what order. Second, it allows one to establish that there exists a collinearity between the genetic material of a gene and the polypeptide it produces. This has been done by mapping a number of mutational sites for a gene that determines one of the polypeptides (protein A) forming the enzyme tryptophan synthetase in *Escherichia coli*. Each mutant produces a modified protein A, which can be shown to differ from the wild type by a single amino acid substitution. When the order of the mutant sites on the *coli* chromosome was compared to the order of amino acids within protein A affected by the mutations, it was found that they were the same. This fundamental finding could only have been possible with the use of a recombination analysis.

Laboratory equipment and reagents for accelerating the manipulation of genetic material have improved markedly in recent years, but the details are beyond the scope of this encyclopedia.

Genes and Diseases

A considerable burden of human disease is attributable to an individual's genetic inheritance. Advances have enabled the detection of an increasing variety of diseases in fetal development and, in some cases, provide a basis for successful treatment.

Studies on human genetics have long been confined to observations of pedigrees and populations with respect to phenotypic traits. Most recently, however, advances in cell biology, biochemistry, cytogenetics and immunology have enabled geneticists to study the human genome more directly and techniques utilizing recombinant DNA have revolutionized these studies. Additionally, technologies employing monoclonal antibodies, hybrid cells, sophisticated protein chemistry, and prophase chromosome banding are all being brought to bear on a variety of problems in human genetics.

At the cytological level, the power and resolution of a variety of chromosome staining and banding techniques has been increased by their application to prophase chromosomes and the genetic map now locates over one thousand bands. At the nucleosomal level, the association of DNA with histone proteins is reasonably well understood. However, knowledge of the higher order structure and the nature of the association between DNA and the acidic structural scaffold, or core proteins, of the chromosome, remains unresolved.

A number of recent surprises have been: the discovery of the split nature of the gene, with its intervening introns; and the later findings of nonfunctioning gene copies or *pseudogenes* and, particularly, the finding of scattered pseudogenes representing DNA copies of processed mRNAs that had become incorporated into the genome. Large and clinically important gene clusters, such as those of the major histocompatibility complex, beta-globulins and the immunoglobulins, have been the subjects of much recent study.

The mechanisms involved in gene activation and inactivation are major problems in biology, so that transient, or permanent structures associated with such phenomena will continue to attract much attention. There is now evidence for changes in chromatin structure at chromosome sites prior to their becoming transcriptionally active; nuclease sensitive sites, enhancers, and promoters have also been identified at various loci.

Defining the location and association of genes and gene clusters in the genome is essential for the understanding of genome organization and in order that genetic techniques may identify and enlighten inherited diseases. Both family (meiotic) and somatic (mitotic) approaches have been dramatically extended, not only by introduction of recombinant DNA technology, but also through use of restriction fragment length polymorphisms and the isolation and cloning of DNA sequences of known and unknown function.

Many genes coding for proteins involved in the disease process have been isolated and cloned. A direct comparison between genomic DNAs of individuals with and without a specific inherited disease is, however, not at present practical because of the large size of the genome and the multitude of nonrandom base changes in on-coding DNA. If the disease is a consequence of lack of expression of a given gene in a specific tissue, then tissue-specific cDNA libraries can be made from mRNAs from the tissues of normal and affected individuals and the libraries compared by crosshybridization to identify a missing sequence.

Specific DNA probes exist for a number of chromosomes so that diagnosis of fetal sex, sex chromosome anomalies, trisomes, and other aneuploidies will shortly be available. Diagnosis of hemoglobulinopathies by fetal blood sampling has already been superceded by DNA analysis. Recombinant DNA technology is obviously going to play a major role in antenatal diagnoses.

Although most human cancers are acquired diseases, all types may occur in heritable or nonheritable forms, and heritability may be associated with a dominant or recessive expression at a single locus, or with a constitutional chromosome anomaly. The changes associated with inherited predisposition to cancer must involve genetic alterations or mutational events at the sites of chromosome anomalies. There is now evidence for this in retinoblastomas.

In acquired malignancies, oncogene activity appears to occur in association with chromosomal rearrangement. There is some evidence that the cooperation of two or more oncogenes, acting in concert, or in sequence, may effect transformation of a normal state to a malignant one. However, further studies are needed to clarify this situation.

Diseases arising from genetic causes may be metabolic, endocrinologic, neurologic, or may develop as the result of mutation, organ implantation, and other factors.

Metabolic Disorders. These fall into four general categories:

- Lipid — hyperlipoproteinemias.
- Purine — gout and Lesch-Nyhan syndrome.
- Metal — Wilson's disease (hepatolenticular degeneration), and hemo chromatosis.
- Porphyrin — porphyrias and idiopathic hyperbilirubinemia.

Generally, metabolic disorders result from:

- *Carbohydrate abnormalities*, such as renal glycosuria (a transport defect), pentosuria (enzyme deficiency, xylitol dehydrogenase), lactase deficiencies, fructose intolerance, galactosemia, galactokinase deficiency, oxalosis, and several glycogenoses (von Gierke's, Forbes', Andersen's, Hers's, and Tarui's diseases).
- *Lysosomal storage abnormalities*, such as glycogenosis (Pompe's disease), Tay-Sachs, Krabbe's, Gaucher's, and Fabry's diseases, as well as metachromatic leukodystrophy, aspartylglycosaminuria, and Niemann-Pick disease. Also included in this category are mucopolysaccharidoses, Hunter's, Schele's, and Hurler's syndromes.
- *Amino acid abnormalities*, such as phenylketonuria, tyrosinemia, alkaptonuria, albinism, histidinemia, hyperprolinemia, homocystinuria, cystinuria, and ketoaciduria. Note that these names, in general, imply the germane amino acid.
- *Urea cycle abnormalities* including hyperammonemia, cirtullinemia, argininosuccinicaciduria, and argininemia.
- *Collagen abnormalities*, such as Ehlers-Damlos syndrome, Marfan's syndrome, pseudoxanthoma elasticum, and osteogenesis imperfecta.

Endocrinologic disorders. These fall into two general categories:

- *Polypeptide hormonal dysfunctions*, such as diabetes mellitus, familial goiter, pseudohypoparathyroidism, and congenital adrenal hyperplasia.

- *Steroid hormonal dysfunctions*, including male pseudobermaphro ditism and testicular feminization.

Neurological Disorders. Although there are other disorders that are suspect, but fully connected to genetic causes, te principal connections already positively made are the muscular dystrophies.

Hematological Disorders. Blood related diseases include hereditary spherocytosis, pyruvate kinase deficiency, glucose-6-phosphate dehydrogenase deficiency, and hemoglobinopathies, such as thalassemias.

Renal Disorders. Kidney and urinary tract diseases include hypophosphatemic and vitamin D-resistant rickets, renal tubular acidosis, and Fanconi's syndrome.

Immunological Disorders. There are several kinds, for example, amyloidosis.

Genetic diagnosis and therapy are discussed in several articles on specific diseases throughout this encyclopedia.

See also **Human Gene Therapy**.

New Advances in Understanding DNA Replication

Many important details have emerged concerning the mechanisms of DNA replication in both bacteria (Prokaryotes) and higher cells (Eukaryotes). These mechanisms are vital in understanding how a cell duplicates its genetic material (DNA), and how this duplication is related to cell division. For these reasons, cells have evolved elaborate mechanisms to ensure that the process of duplication (DNA replication) is error free. This level of control is so important that cells will actually cease cell division if errors become too frequent and wait until the DNA is repaired.

Two relatively recent developments have added to our knowledge significantly concerning how DNA replication occurs with fidelity or in what molecular biologists and biochemists call a processive polymerase activity. DNA polymerase is the enzyme which actually polymerizes (adds DNA precursors or building blocks) DNA. There are many such DNA polymerases in pro- and eukaryotic cells that have different functions but the main enzyme in prokaryotes is DNA polymerase III and in Eukaryotes, DNA polymerases alpha, delta, and epsilon. All four of these DNA polymerases are made of subunits.

The most important of these subunits for processive activities are termed the "brace" and the "clamp." They are both remarkable structures, themselves composed of protein subunits. The brace is a complex of proteins which binds to DNA at a site where DNA replication will begin (a double stranded primed DNA template), and in the presence of the high energy compound ATP (adenosine triphosphate) will help load the clamp. The clamp is a ring-shaped DNA polymerase accessory protein complex that looks like a "doughnut" and slips over the double stranded primed template like a curtain rod sing. Once this occurs, the rest of the DNA polymerase will associate to form the processive polymerase. The clamp-DNA polymerase complex will then slide down the template DNA strand (with the clamp behind the polymerase pushing it) easily, picking up DNA precursors to be polymerized into newly synthesized DNA, until the entire DNA molecule has been duplicated faithfully.

This remarkable mechanism relates structure to function in DNA replication in ways not easily noted previously.

New Advances in Understanding DNA Replication written by William Firshein, Professor, Department of Molecular Biology and Biochemistry, Wesleyan University

Human Genome Project (HGP)

After considerable initial persuasion by the biochemical and genetic sciences community, the National Academy of Sciences (U.S.), in 1988, endorsed an effort to map and sequence the human genome.[1] Genetic maps had been constructed from many different types of data, using different metrics, ranging back to the first genetic linkage map made as early as 1913. See **Human Genome Project (The)**.

As pointed out by J.C. Stephens (National Cancer Institute) and a team of researchers (See reference listed), "Genetic linkage maps are based on the coinheritance of allele combinations across multiple polymorphic loci.

---

[1] The genetic constitution of an organism. One full set of the 24 distinct human chromosomes is estimated to contain $\sim 3 \times 10^9$ base pairs of DNA, throughout which are distributed $\sim 1 \times 10^5$ genes.

The primary source of linkage data is the observation of gametic allele combinations."

The allelic constitution of gametes for *human linkage* studies traditionally has been determined indirectly by family studies and statistical inference. Improvements in analytical methods in recent years have made possible the direct molecular analysis of gametes and single chromosomes. The highest level of resolution for a molecularly-based physical map is the DNA sequence. This yields the linear order of nucleotides for each of the 24 distinct human chromosomes. Thus, a complete reference sequence will contain $\sim 3 \times 10^9$ bp of DNA.

As of early 1994, most scientists interested in the HGP are satisfied with the progress made to date, and some forecast that the project can be completed ahead of the original target date of about the year 2010. Much of the progress is attributed to the use of advanced, automated sequencing equipment.

A major thrust of HGP is the ultimate development of *gene therapy* for diseases that derive from faults in the human gene system.

Additional Reading

Adler, R.G.: "Genome Research: Fulfilling the Public's Expectations for Knowledge and Commercialization," *Science*, **908** (August 14, 1992).

Adolph, K.W.: *Genome Research in Molecular Medicine and Virology*, Academic Press, Orlando, FL, 1993.

Aldhous, P.: "Managing the Genome Data Deluge," *Science*, **502** (October 22, 1993).

Anderson W.F.: "Human Gene Therapy," *Science*, **808** (May 8, 1992).

Bauer, W.R., F.H.C. Crick, and J.H. White: "Supercoiled DNA (*A Classic Nobel Laureate Paper*)," *Sci. Amer.*, (July 1980).

Beardsley, T.: "From Mice to Men," *Sci. Amer.*, **18** (December 1993).

Bodmer, Sir Walter: "Genome Research in Europe," *Science*, **480** (April 24, 1992).

Bull, J.J., I.J. Molineux, and J.H. Werren: "Selfish Genes," *Science*, **65** (April 3, 1992).

Calladine, C. and H. Drew: *Understanding DNA*, Academic Press, Orlando, FL, 1992.

Collins, F. and D. Galas: "A New Five-Year Plan for the U.S. Human Genome Project," *Science*, **43** (October 19, 1993).

Copeland, N.G. et al.: "A Genetic Linkage Map of the Mouse: Current Applications and Future Prospects," *Science*, **57** (October 1, 1993).

Copper, S.B.: *Human Gene Evolution*, Elsevier Science & Technology Books, New York, NY, 2000.

Culliton, B.J.: "Mapping Terra Incognita (Humani Corporis)," *Science*, **210** (October 12, 1990).

Cuticchia, A.J. et al.: "Managing All Those Bytes: The Human Genome Project," *Science*, **47** (October 1, 1993).

Dale, J.W., and M. von Schantz: *From Genes to Genomes: Concepts and Applications of DNA Technology*, John Wiley & Sons, Inc., New York, NY, 2002.

Desalle, R., and M. Yudell: *Welcome to the Genome: A User's Guide to Your Genetic Past, Present, and Future*, John Wiley & Sons, Inc., New York, NY, 2004.

Eigen, M., W. Gardiner, Schuster, P., and R. Winkler-Oswatitsch: "The Origin of Genetic Information (*A Classic Nobel Laureate Paper*)," *Sci. Amer.*, (April 1981).

Eisenberg, R.S.: "Genes, Patents, and Product Development," *Science*, **903** (August 14, 1992).

Erickson, D.: "Hacking the Genome," *Sci. Amer.*, **128** (April 1992).

Erickson, D.: "Diagnosis by DNA," *Sci. Amer.*, **116** (October 1992).

Everson, T.: *The Gene: A Historical Perspective*, Greenwood Publishing Group, Inc., Westport, CT, 2007.

Farkas, D.H.: *Molecular Biology and Pathology*, Academic Press, Orlando, FL, 1993.

Farr, C.J. and P.N. Goodfellow: "Hidden Messages in Genetic Maps," *Science*, **49** (October 2, 1992).

Farrell, R.E., Jr.: *RNA Methodologies: A Laboratory Guide for Isolation and Characterization*, 3rd Edition, Elsevier Science & Technology Books, New York, NY, 2005.

Fischman, J.: "Going for the Old: Ancient DNA Draws a Crowd," *Science*, **655** (October 29, 1993).

Fox, S.: "Applications for Synthesizing and Sequencing DNA Beyond the Genome Project," *Genetic Eng. News*, **6** (June 1991).

Friedberg, E.C., G.C. Walker, and W. Siede: *DNA Repair and Mutagenesis*, 2nd Edition, ASM Press, Washington, DC, 2006.

Friedmann, T.: *Molecular Genetic Medicine*, Academic Press, Orlando, FL, 1992.

Grunstein, M.: "Histones as Regulators of Genes," *Sci. Amer.*, **68** (October 1992).

Hall, J.C.: *Advances in Genetics*, Elsevier Science & Technology Books, New York, NY, 2007.

Hartwell, L., L. Hood, M.L. Goldberg, and A. Reynolds: *Genetics: From Genes to Genomes*, 3rd Edition, The McGraw-Hill Companies, New York, NY, 2000.

Jasny, B.R.: "Genome Delight," *Science*, **11** (October 2, 1992).

Jürgens, G.: "Genes to Greens: Embryonic Pattern Formation in Plants," *Science*, **487** (April 24, 1992).

Karlin, S. and V. Brendel: "Chance and Statistical Significance in Protein and DNA Sequence Analysis," *Science*, **39** (July 3, 1992).

Kessler, D.A. et al.: "The Safety of Foods Developed by Biotechnology," *Science*, (June 26, 1992).

Kevles, D.J. and L. Hood: *The Code of Codes: Scientific and Social Issues in the Human Genome Project*, Harvard University Press, Cambridge, MA, 1992.

Kiley, T.D.: "Patents on Random Complementary DNA Fragments?" *Science*, **915** (August 14, 1992).

Klug, A. and R.D. Kornberg: "The Nucleosome (*A Classic Nobel Laureate Paper*)," *Sci. Amer.*, (February 1981).

Klug, W.S., M. Cummings, and C. Spencer: *Essentials of Genetics*, 6th Edition, Prentice Hall, Upper Saddle River, NJ, 2006.

Marx, J.: "Genome Project Plans Described," *Science*, **152** (April 9, 1993).

Morell, V.: "30-Million-Year-Old DNA Boosts an Emerging Field," *Science*, **1860** (September 25, 1992).

Pääbo, S.: "Ancient DNA," *Sci. Amer.*, **86** (November 1993).

Palladino, M.A.: *Gene Therapy*, Pearson Education, Upper Saddle River, NJ, 2006.

Pearson, P.L. et al.: "The Human Genome Initiative — Do Databases Reflect Current Progress?" *Science*, **214** (October 11, 1991).

Primose, S.B., G. Bertola, B. Old, and R. Twyman: *Principles of Gene Manipulation*, 7th Edition, Blackwell Publishers, Malden, MA, 2005.

Rajewsky, K.: "A Phenotype or Not: Targeting Genes in the Immune System," *Science*, **483** (April 24, 1992).

Reece, R.J.: *Analysis of Genes and Genomes*, John Wiley & Sons, Inc., New York, NY, 2004.

Rhodes, D. and A. Klug: "Zinc Fingers," *Sci. Amer.*, **56** (February 1993).

Risch, N.L.: "Genetic Linkage: Interpreting Lod Scores," *Science*, **803** (February 14, 1992).

Roberts, L.: "Academy Backs Genome Project," *Science*, **725** (February 12, 1988).

Roberts, L.: "Taking Stock of the Genome Project," *Science*, **20** (October 1, 1993).

Roberts, L.: "NIH, DOE Battle for Custody of DNA Sequence Data," *Science*, **504** (October 22, 1993).

Selvin, P.R. et al.: "Torsional Rigidity of Positively and Negatively Supercoiled DNA," *Science*, **82** (January 3, 1992).

Singer, M. and P. Berg: *Genes and Genomes*, University Science Books, Mill Valley, CA, 1990.

Stephens, J.C. et al.: "Mapping the Human Genome: Current Status," *Science*, **237** (October 12, 1990).

Suzuki, D.T. et al.: *An Introduction to Genetic Analysis*, 8th Edition, W. H. Freeman, Company, New York, NY. 2004.

Thompson, L.: "At Age 2, Gene Therapy Enters a Growth Phase," *Science*, **744** (October 30, 1992).

Varmus, H.: "Reverse Transcription (*A Classic Nobel Laureate Paper*)," *Sci. Amer.*, (November 1978).

von Hippel, P.H. and T.D. Yager: "The Elongation-Termination Decision in Transcription," *Science*, **809** (February 14, 1992).

Watson, J.D. D., J. Witkowski, A. Caudy, and R. Myers: *Recombinant DNA: Genes and Genomes: A Short Course*, W. H. Freeman Company, New York, NY. 2006.

Withka, J.M.: "Toward a Dynamical Structure of DNA," *Science*, **597** (January 31, 1992).

Wolffe, A.: *Chromatin*, Academic Press, Orlando, FL, 1992.

Zyskind, J. and S.I. Bernstein: *Recombinant DNA Laboratory Manual*, Academic Press, Orlando, FL, 1992.

A.C. DEBALDO Ph.D., University of South Florida, Tampa

GENITAL WARTS. See **Sexually Transmitted Diseases**; and **Wart (*Verrucae*)**.

GENOME MAP. A genome map is a diagram, chart, table or database that shows the location in a genome of a number of identifiable features. Many methods are available for making genome maps, differing in their accuracy and in the types of feature that can be mapped.

Genetic linkage mapping relies on following the pattern of inheritance of polymorphic (variable) parts of the genome through large, multigeneration families. The polymorphisms may be known at the deoxyribonucleic acid (DNA) level (for example, microsatellites — regions of sequence containing a simple, tandemly repeated motif such as CACACACA in which the number of repetitions varies from allele to allele); or they may be phenotypic traits (such as eye color), which are assumed to reflect the presence of an underlying but unknown variable gene. In either case, the polymorphism can be regarded as a *marker*, whose inheritance can be observed. As the chromosomes are broken and shuffled in meiosis, only those markers that are close together in the genome will tend to remain linked and therefore be inherited together. This provides a means of estimating the distances between markers, allowing a map to be made.

HAPPY mapping and radiation-hybrid (RH) mapping rely on a similar principle to genetic mapping. In these cases, the genome is artificially broken by radiation or other means (rather than by meiosis), and the fragments are distributed among a number of samples. Again, DNA sequences (markers, also known as sequence-tagged sites, STSs) that are close together in the genome will tend to remain together on the same fragment after breakage and hence tend to be found together (cosegregate) in the same samples. As in linkage mapping, the frequency of cosegregation reflects the proximity of one STS to another. See also **Genetic Linkage Mapping**.

In physical mapping, the starting point is a library of cloned fragments of genomic DNA. Each clone is characterized, for example, by testing it for the presence of a number of different STSs. If two clones are found to contain the same STS, it can be assumed that they represent overlapping regions of the genome. In this way, the aim is to build up a contiguous series (*contig*) of overlapping clones, each one connected to the next by a shared STS. A physical map therefore gives the relative locations of both the cloned fragments and the STS markers that they contain.

Fluorescence in situ hybridization (FISH) is the most direct method of genome mapping. Cloned fragments of DNA are labeled and allowed to hybridize to their complementary sequence in metaphase chromosomes fixed on the surface of a microscope slide. Fluorescence microscopy is then used to directly observe the location of the hybridized probes within the chromosome.

Genome maps have many applications. Genetic linkage mapping is uniquely capable of finding the genomic location of a gene that is known only through its phenotypic effect, and it is therefore often the first step in hunting for genes responsible for genetic diseases and variable traits. Physical mapping is often a prelude to genome sequencing, because each of the mapped clones may be sequenced in turn to reconstruct the sequence of the complete genome. However, physical mapping alone is tedious and error-prone; RH or HAPPY maps are therefore often used to provide a *scaffold* of independently mapped markers on which the physical map can be assembled. The lower-resolution techniques such as genetic mapping and FISH are particularly useful in make large-scale comparisons between the genomes of different species

PAUL H. DEAR, Medical Research Council Laboratory of Molecular Biology, Cambridge, UK

GENOTYPE. The actual gene constitution of an organism as opposed to the phenotype or visible expression arising from those genes. The gene that causes albinism in humans is recessive and is represented by a. The dominant allele of this gene is A. Since each person is diploid with respect to his or her genes, there are three possible genotypes of this particular gene locus: *AA, Aa*, and *aa*. The resultant albinism or normal pigment production would be the phenotype produced. See also **Cell (Biology)**.

GENUS. See **Taxonomy**.

GEOCENTRIC COORDINATES (Astronomy). Any system of coordinates on the celestial sphere that uses for its origin, or reference point, the center of the earth. Practically all coordinates published in an ephemeris or almanac are geocentric in character. See also **Celestial Sphere and Astronomical Triangle**.

GEOCENTRIC PARALLAX. The origin of the apparent systems of spherical coordinates is a point on the surface of the earth, whereas the origin of the geocentric systems is at the center of the earth. For obvious reasons, all observations must be taken in the apparent system. For the solution of most problems, geocentric coordinates are desired. The transfer from one system to the other is made by applying a correction for geocentric parallax.

In Fig. 1, C is the center of the earth, of radius R, and O is the position of an observer on the surface. OC is the direction of gravity at O; OH the direction of the astronomical horizon; and CH a parallel direction drawn through the center of the earth. S and S' represent two positions of an object at distance ρ from the center of the earth, S being the position when the object is on the horizon. At S'', the object has an apparent altitude h' and a geocentric altitude h. P is defined as the horizontal parallax of the object and is the angle subtended at the object by the radius of the earth. For rigor, the quantity usually defined is the mean equatorial horizontal parallax. This is the angle subtended by an equatorial radius of the earth at the object, when the object is on the horizon, and at its mean or average distance from

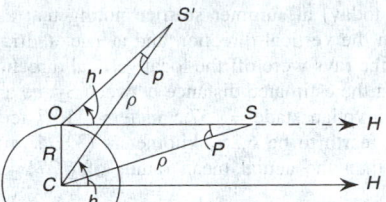

Fig. 1.   Geocentric parallel in altitude.

the earth. The equatorial horizontal parallax is tabulated in Ephemerides for all members of the solar system for selected dates. Inspection of the figure indicates that $\sin P = R/\rho$.

The geocentric altitude h is greater than the apparent altitude h' by the angle p, which is defined as the geocentric parallax in altitude. In the oblique plane triangle COS', we have

$$\frac{R}{\rho} = \frac{\sin p}{\cos h'}$$

but we have already seen that $R/\rho = \sin P$, whence

$$\sin P \cos h' = \sin p$$

Now both P and p are such small angles that, without sensible errors for most problems, except those dealing with the moon, we have $p = P \cos h'$, giving the geocentric parallax in altitude in terms of the equatorial horizontal parallax and the apparent altitude of the object. For objects outside the solar system, the value of P is far too small to be appreciable in even the most refined observations.

If other spherical coordinates than altitude are to be used, the geocentric parallax in altitude may be transformed to the desired quantities by solution of the astronomical triangle or other triangles on the celestial sphere.

GEOCHEMISTRY. The study of the chemical composition of the earth in terms of the physiochemical and geological processes and principles that produce and modify minerals and rocks. Of practical importance in discovering and establishing the limits of ore deposits, petroleum, tar sands, salt, sulfur, and other valuable resources.

GEOCHRONOLOGY. See **Geologic Time Scale**.

GEOCRONITE. A mineral sulfide of lead, antimony and arsenic, Pb_5SbAsS_8. Crystallizes in the monoclinic system. Hardness, 2.5; specific gravity, 6.4±; color, gray to blue with metallic luster; opaque.

GEODE. A hollow concretion or nodule whose inside walls are lined with crystals, commonly of quartz or calcite.

GEODESIC. That curve on a surface connecting two fixed points which has an extreme length (maximal or minimal). In three-dimensional Euclidean geometry, the geodesic is clearly a straight line; if the path is constrained to the two-dimensional surface of a sphere, it is a segment of a great circle. In the non-Euclidean geometries appropriate to the general relativity theory, the geodesic is the path followed by a particle upon which no electromagnetic forces act. Such a path is the straightest path in a four-dimensional space-time continuum.

Geodesic Parallels on a Surface. Consider a singly infinite family of geodesics. The singly infinite family of curves on the surface, which cuts these orthogonally at each point, is called geodesic parallels. The distance between two geodesic parallels measured on the surface along any geodesic of the family is the same and is called the *geodesic distance* between the two geodesic parallels.

GEODESY. Geodesy is traditionally defined as the science of studying the figure of the earth and its gravity field. Now that geodetic measurements have become very accurate and easily repeatable, the traditional definition must be extended to include temporal variations as well.

Although inquiry into the figure of the earth can be traced to antiquity, the recognized founder of scientific geodesy is Eratosthenes (276–195 B.C.) of Alexandria. Eratosthenes studied the rays of the sun at Alexandria and

Syene (Assuan, today) at summer solstice noting that at Syene the rays entered a well in the vertical direction and at Alexandria (roughly on the same meridian) the rays were off the local vertical direction by about 1/50 of a circle. Using the estimated distance between Syene and Alexandria of 5000 stadia (1 Egyptian stade = 157.5 meters (516.7 feet), he calculated the radius of the earth to be 6,267 kilometers (3,894 miles). This is less than 2% lower than the actual mean radius of 6,371 kilometers (3,959 miles).

Historically, other estimates differ by 3%–10% from the value known today. In 1669–1670, at the urging of the newly formed Academy of Sciences in Paris (1666), J. Picard carried out arc measurements on the meridian through Paris between Malvoisine and Amiens with the aid of a triangulation network. He was the first to use a telescope with cross hairs. Picard's value for the Earth radius was good to +0.01%, which aided Newton in the verification of the law of gravitation which he formulated in 1665–1666.

From observations of pendulum motion (the period), one can infer an increase of gravity from the equator to the poles. Using such observations and their own works, Isaac Newton (1643–1727) and Christian Huygens (1629–1695) suggested Earth models flattened at the poles.

If we let the letter a stand for the equatorial radius and b the polar radius, the normalized deviation or flattening is defined as $f = (a - b)/a$. Today we know that $1/f = 298.257$. Historically this value also was attempted to be found through meridian arc measurements. Depending on where the arc measurements were obtained, the determination of the flattening could have even the wrong sign (polar radius greater than equatorial). In 1743, A.C. Clairaut (1713–1765) synthesized the physical and geodetic concerns in a theorem named for him that permits the computation of flattening from two gravity measurements at different latitudes. However, the practical implementation of this theorem suffered until the twentieth century from the lack of precise and well-distributed gravity measurements and the inability to reduce these data to the Earth ellipsoid.

Historically, the task of determining the figure of the Earth was, for simplicity, subdivided into two almost independent tasks: horizontal surveys for determination of latitudes and longitudes and vertical surveys for determination of heights. The reference surface for horizontal surveys is a suitable but otherwise arbitrary rotational ellipsoid. The heights, however, are referred to an equipotential surface or mean sea level (which later will be called the geoid). The use of different reference surfaces for horizontal and for vertical coordinates makes the accurate computations of three-dimensional Cartesian coordinates impossible, unless the separation between the mean sea level and the reference ellipsoid is known accurately. The ability to compute Cartesian coordinates is of importance because modern survey methods, such as those using artificial satellites, provide such coordinates and their comparison and compatibility with older surveys is essential.

Geodesy can be subdivided into the following groups: geometric geodesy, gravimetric geodesy, and space geodesy. Geometric geodesy deals with the measuring and computational methods that are applicable in the surveys of large areas, thus beyond the scope of plane surveying, and with the geometric determination of the size and shape of the reference ellipsoid. The geometric effort results in horizontal coordinates (geodetic latitude and longitude) referred to a suitable reference ellipsoid, and in heights above sea level. These results are mainly based on distance, direction measurements, and height differences. Gravimetric geodesy deals with the determination of the gravity equipotential surface most closely approximating mean sea level (known as the geoid) as a graphical depiction of the gravity field of the Earth, and with related problems. See also **Geoid**. Geodetic astronomy provides the link between the "arbitrary" geometric reference system and the real world (nature) and thus serves to control the distorting effects of arbitrariness. This is achieved by attempting to keep the directions that are referred to the arbitrarily chosen reference ellipsoid close to "natural" directions determined from astronomical observations. Space geodesy provides local, regional, and global geodetic control through three-dimensional coordinates and vectors determined mostly from observations on artificial satellites and other space techniques. Since the orbits of satellites are affected by, among other things, the gravity field of the Earth, these observations also provide a means to improve our knowledge of this field.

With the introduction of artificial satellites in the late 1950s, the geodetic sciences have been under continuous and rapid change. Today the impacts of very high-quality measurement systems such as lasers,

very long-baseline interferometry, precise absolute gravimeters, satellite-based altimetry, and the Global Positioning System (GPS) are bringing allied sciences such as geophysics, oceanography, remote sensing and photogrammetry, and navigation into much closer contact. For example, one would not consider using traditional measurements such as angles and distances from theodolites and electronic distance-measuring devices to establish a modern national geodetic network as was done as recently as 1987. Today the increased precision with greatly reduced costs of GPS, strengthened by satellite laser ranging and/or VLBI, would be almost the exclusive techniques employed. Real-time millimeter measurements are being collected in search of earthquake and volcanic precursors. Similar measures of ice-sheet thicknesses and sea-level monitoring along with precise satellite altimetry are opening up new vistas for studying energy transfers between oceans and the atmosphere, ocean circulation, climate, and internal Earth structure.

See also **Earth**.

Additional Reading

Hwang, C., C. Shum, and J. Li: *Satellite Altimetry for Geodesy, Geophysics and Oceanography*, Springer-Verlag New York, LLC, New York, NY, 2004.
Hofmann-Wellenhof, B., and H. Moritz: *Physical Geodesy*, Springer-Verlag New York, LLC, New York, NY, 2005.
Kaula, W.M.: *Theory of Satellite Geodesy: Applications of Satellites to Geodesy*, Dover Publications, Mineola, NY, 2000.
Smith, J.R.: *Introduction to Geodesy: The History and Concepts of Modern Geodesy*, John Wiley & Sons, Inc., New York, NY, 1997.
Torge, W.: *Geodesy*, 3rd Edition, Walter de Gruyter, Inc., Ossining, NY, 2001.

GEODETIC COORDINATES. Quantities that define the position of a point on the spheroid of reference with respect to the planes of the geodetic equator and of a reference meridian.

GEODETIC DATUM. A datum consisting of five quantities, the latitude, longitude and elevation above the reference spheroid of an initial point, a line from this point, and two constants which define the reference spheroid. Azimuth or orientation of the line, given the longitude, is determined by astronomic observations. Alternatively, the datum may be considered as three rectangular coordinates fixing the origin of a coordinate system whose orientation is determined by the fixed stars, and the reference spheroid is an arbitrary coordinate surface of an orbiting ellipsoidal coordinate system.

GEODIMETER. An electronic-optical device that measures ground distances precisely by electronic timing and phase comparison of modulated light waves that travel from a master unit to a reflector and return to a light-sensitive detector where an electric current is established. It is frequently used at night and is effective with first-order accuracy up to a distance of 5–40 kilometers (3–25 miles). The ultimate precision of the geodimeter over that of the tellurometer is roughly by a factor of 3. A *tellurometer* is a rugged, lightweight portable electronic device that measures ground distances by determining the velocity of a phase-modulated, continuous microwave radio signal transmitted between two instruments operating alternately as a master station and a remote station. The instrument has a range up to 65 kilometers (35–40 miles).

GEODUCK *(Mollusca, Lamellibranchiata).* A giant clam found on the Pacific coast of North America. It attains a weight of more than 6 pounds (2.7 kilograms) and is edible.

GEOGRAPHIC COORDINATES. Geographic coordinates provide a method for determining the position of a point on the surface of the earth by means of a system of spherical coordinates. Because of the fact that the earth is not a sphere, but is, in reality, an oblate spheroid, technically, the system of coordinates cannot be strictly spherical. The geographic method of representation of the position of points on a spherical earth by means of latitude and longitude was first applied by Ptolemy in the construction of his atlas of the world during the second century of the Christian era. Also called *geographical coordinates* or *terrestrial coordinates*.

GEOGRAPHY. Literally, the study, description, and mapping of the surface phenomena of the earth or other planets without, necessarily, a consideration of the origin of the phenomena. See also **Earth**.

GEOID. The particular geopotential surface that most nearly coincides with the mean level of the oceans of the earth. For mapping purposes, it is customary to use an ellipsoid of revolution as an adequate and convenient approximation to the geoid. The dimensions and orientation of the assumed ellipsoid may represent an attempt to find the ellipsoid that most nearly fits the geoid as a whole, or they may represent an attempt to fit only a particular part of the geoid without regard to the remainder of it. When mention is made of the dimensions of the earth, the reference is usually to the dimensions of the ellipsoid most nearly representing the geoid as a whole. See also **Earth**.

GEOLOGICAL TIME: DATING TECHNIQUES. Numerical dating ('absolute dating') uses the process of radioactive decay as a clock. When a radioactive atom, such as uranium or rubidium, spontaneously decays by nuclear reactions, it gives off heat, radioactive particles, and leaves a nonradioactive daughter atom. The rate of this nuclear reaction is well known, and half of the original parent atoms decay to their daughter atoms in a fixed interval of time, known as the half-life. If we can measure the ratio of parent to daughter atoms, we can determine when this atomic decay reaction began, and date the material that contains these atoms. Because the parent atoms must be locked in at the time of their crystallization, this system only works for minerals which have cooled down from a very hot state (igneous or high-grade metamorphic minerals). See also **Geological Time: History of Ideas**.

There are many different radioactive elements in the Earth, but only a few are sufficiently abundant in crustal rocks, and have a long enough half-life, that they can be used for measuring geological events. The elements used are potassium, rubidium, uranium and carbon.

Potassium-Argon Dating

This is the mostly widely used technique, since potassium is one of the more abundant elements in the Earth's crust, found in potassium feldspars, micas, and many other common minerals. Even though it has a long half-life of 1250 million years, many minerals contain enough potassium that there is measurable daughter product even at this slow rate of decay. Thus, it can be used for measuring ages as old as the oldest rocks, although rocks younger than a million years old rarely have measurable daughter products. The major limitation of this method is that the parent atom (potassium-40, a solid) decays to argon-40 (a gas), which has a tendency to leak out of the crystal and give ages that are too old. To circumvent this problem, laboratories have now begun to use a variant called argon-40/argon-39 dating, which uses argon-39 as a proxy for potassium-40. This method measures two similar gases, so there is less problem with leakage or contamination, and has become the method of choice for obtaining ages with very small error estimates, and also for screening out problems with leakage or contamination that might give a bad date. See also **Rutherford, Ernest (1871–1937)**; and **Soddy, Frederick (1877–1956)**.

Rubidium-Strontium Dating

This method uses the decay of rubidium-87 to strontium-87 as the basis for its analysis. The main limitation of this method is that rubidium is a relatively rare element in the Earth's crust, so most minerals have no measurable rubidium-87. Rubidium ions have a +1 charge and a large ionic radius, so they can substitute for potassium or sodium in some minerals, but only in trace amounts. Another problem is that the daughter product, strontium-87, has a +2 charge, so it tends to migrate out of the crystal lattice to sites where it is electrically stable. Rubidium–strontium also has a very long half-life (48 800 million years), so very little material has decayed except in the very oldest rocks. For this reason, the method is used primarily in dating potassium- or sodium-rich granitic rocks that are at least 500 million years old (Precambrian), or occasionally early Paleozoic rocks. See also **Terrestrialization (Precambrian–Devonian)**.

Uranium-Lead Dating

There are two different radioactive isotopes of uranium, uranium-235 and uranium-238, which decay to lead-207 and lead-206, respectively. These elements are also relatively rare in crustal rocks, and have long half-lives (703 million years and 4468 million years, respectively), so they are primarily used for Precambrian rocks, and for meteorites and moon rocks. Uranium is most often found in minerals such as zircon, sphene and apatite, which form in the final stages of crystallization of a granitic magma, and have large enough lattice spaces that they can accommodate huge cations like those of uranium.

A related method is known as fission-track dating. Crystals of zircon may have damaged crystal lattices as the uranium decays and emits radiation, which leaves a pathway of destruction in the crystal. These fission tracks can be seen under the microscope if the crystal has been etched in hydrofluoric acid. The geochronologist will count the fission tracks in the etched crystal, then irradiate it in a nuclear reactor to stimulate new fission tracks and determine how much parent material was originally present. The main limitation of the method is that the old tracks close up and disappear, giving a false track count in crystals that have been reheated. In addition, fission-track dating tends to give relatively large error estimates (typically ±1–2 million years, even for relatively recent dates), and zircons are very stable as sedimentary minerals, so there is a good chance of contamination from sand-sized grains of zircon.

Carbon-14 Dating

All of the methods mentioned so far only work on crystals that have cooled down from a molten state, such as in igneous or metamorphic rocks, and thus cannot be used directly on fossils or sedimentary rocks. Although living things are made largely of the stable isotope of carbon, carbon-12, they also contain carbon-14. This isotope is produced in the atmosphere when nitrogen-14 is bombarded by cosmic radiation, and is taken up by plants during photosynthesis, so that animals and plants all have trace amounts. When an organism dies, it no longer takes up carbon-14, and that material begins its radioactive decay. Thus, a geologist or archaeologist can take any carbon-bearing material — wood and other vegetable matter, coal, baskets, bones, pottery, shells — and determine the amount of carbon-14 present and its age since death. The main limitation of the system is that the half-life of carbon-14 is only 5730 years. Its decay is very rapid, and there is no measurable parent material after about 80 000 years. Thus, this method is of practical use for archaeologists and geologists who study the last Ice Age, but of no use to geologists who study anything older than about 60 000–80 000 years old. See also **Carbon**.

Precision and Confidence

There are several constraints on radioisotopic dating. The method only works on crystals, which have locked in all their parent and daughter material, and preserved their ratio faithfully. Anything that disturbs the crystal and changes the ratio will lead to erroneous dates. If the crystal has been weathered, so that either parent or daughter atoms can leak out, then the ratio has been altered. In other cases, atoms of parent or daughter material can percolate through the groundwater and seep into the crystal, contaminating it and altering the ratio. Thus, geochronologists only use the freshest samples, and often examine each crystal under the microscope before analysis to detect any sign of alteration (although it may be undetectable, and still give erroneous dates). In some cases, an igneous rock (such as an ash flow) may have picked up ancient sedimentary mineral grains as it settled and cooled, and these detrital contaminants will give an artificially old age for the volcanic event. Geochronologists control for these problems by running the analysis several times, often using different decay systems, so that if each independent system gives a consistent age, and they are also consistent with the biostratigraphic and other age estimates, then they are considered acceptable.

Each radiometric date is actually an estimate of age, not an 'absolute' measure of time. Consequently, it is important to distinguish between the precision (reproducibility) of the age estimate, and its accuracy (closeness to truth). A radiometric age estimate can be highly precise (have very small error estimates), but be nowhere near the actual age if all the samples have been contaminated. Likewise, a number of ages could be relatively imprecise (have a large error estimate) but still be close to the true age.

Another constraint is the analytical error inherent in the system. The ratios of different atoms are measured on a mass spectrometer (which sorts isotopes by their masses), which has a certain minimum level of sensitivity. The date can only be estimated to a certain level of precision, given by the 'plus or minus' error bars that are attached to every date. For example, if the date is 100 ± 2 million years, then there is a 95% chance that the true age lies between 98 and 102 million years. Most laboratories have errors of ±1–2%; this analytical precision gets progressively worse with increasing age, since it is a percentage of that age. For example, the error on typical dates ranging from 1 to 100 million years ago is 500 000 years, but for billion-year-old Precambrian rocks, it is typically on the order of ±10–20 million years. In recent years, the development of SHRIMP (super-high-resolution ion microprobe) dating has reduced this error tremendously, so that dates as old as 4000 million years have errors as low as ±4 million years, or about 0.001%.

Effects of Precision of Dating on Macroevolutionary Conclusions

Prior to 1905, there were many estimates of the age of the Earth, but most assumed that the Earth was no more than 100 million years old. When the first Uranium–Lead dates were published between 1905 and 1915, the Earth was immediately shown to be at least 2000 million years old. Within a few decades, older and older rocks were found, and currently the oldest rocks on Earth are dated at over 4000 million years old, and some mineral grains have been dated as old as 4400 million years. As these dates emerged, the earliest fossils of bacteria were dated as old as 3500 million years, and geochemical evidence from organic carbon in grains 3800 million years old suggest life may have already been present at this early date. Recently, grains dated as old as 4400 million years suggest that the Earth had liquid water on its surface not long after its formation. These surprising dates show that the cooling of the planet, and the origin of life, was a relatively rapid process, possibly taking less than 100 million years. Even more surprising is the conclusion that once bacteria had evolved at least 3500 million years ago, they showed no further visible evolutionary change for almost 2000 million years, when the oldest evidence of eukaryotic cells appears in the fossil record. Thus, life originated much more rapidly than previously supposed, but then spent most of its early history in evolutionary stasis, as judged from its morphology.

Likewise, the timing of the 'Cambrian explosion' of life, when most multicellular, skeletonized animals originated, has been considerably revised and shortened. At one time, it was thought that evolution of complex multicellular animals from single-celled eukaryotic life was a long, protracted process, taking as much as 100 million years. Recent dates have shown that the entire 'Cambrian explosion' took place in less than 15 million years, and some phases were much more rapid than that. See also **Terrestrialization (Precambrian–Devonian)**.

The timing of mass extinctions is critical to hypotheses about their causes. For example, at one time it was suggested that the major mass extinctions operated on a 26-million-year cycle, and might have been caused by cyclic extraterrestrial impacts. However, recent dating of these mass extinctions shows that they do not have a true 26-million-year periodicity. In some cases, the potential causes of mass extinction can be critically tested by precise dating. For example, at one time the greatest mass extinction in Earth history, the Permo–Triassic extinction about 250 million year ago, was thought to be very long and protracted. Recent dating has now shown that it was a much more rapid, abrupt event, allowing for hypotheses of rapid, catastrophic extinction to be considered. See also **Mass Extinctions**.

On a more refined scale, the presence of numerous new radiometric dates allows us to estimate the duration of speciation processes, and the duration of individual species before their extinction. For example, recent dates have shown that most species have durations as long as millions of years, much longer than previously appreciated. More importantly, studies of these fossils show no visible morphological changes through these long durations of time and across major climatic changes, suggesting that species are much more static and insensitive to climatic change than previously supposed.

Conclusion

The numerical age of geological and biological events is determined by the process of radiometric dating, which uses the clock-like decay of unstable isotopes of uranium, rubidium, potassium, carbon, and other elements to determine the age of a decay process. Numerical dates are age estimates, which have varying degrees of precision (reproducibility) and accuracy (closeness to truth). Some systems, like uranium–lead and rubidium–strontium, have such long half-lives and occur in such small amounts, that they are only practical for measuring very old events (older than 1000 million years). Other systems, such as potassium–argon, occur in large enough amounts in most rocks that they can be used to date nearly any geological event older than a million years. However, the dates only work on crystals taken from igneous rocks, so the events and fossils preserved in sediments must be dated indirectly by igneous intrusions, ash

falls and lava flows. The radiocarbon dating method does allow for the direct dating of fossils and other organic material, but it has such a short half-life that it is impractical for events older than 80 000 years.

Radiometric dates have been critical not only for dating geological events, but also for dating events in the evolution of life. In some cases, new radiometric dates have shown that biological processes are much slower than previously supposed (e.g. the 2000 million years of visibly unchanged bacteria before eukaryotic cells appeared, or the long durations of species despite climatic changes). In other cases, they have shown processes to act much faster than previously supposed (e.g. the origin of life, the Cambrian explosion, the major mass extinction events). See also **Earth**.

Additional Reading

Berggren, W.A., D.V. Kent, and J. Hardenbo: *Geochronology, Time Scales, and Stratigraphic Correlation*, SEPM Special Publication 54, 1995.

Dalrymple, G.B.: *The Age of the Earth*, Stanford University Press, Palo Alto, CA, 1991.

Dalrymple, G.B., and M.A. Lanphere: *Potassium–Argon Dating*, WH Freeman, New York, NY, 1969.

Easterbrook, D.J.: "Dating Quaternary Sediments," *Geological Society of America Special Paper 227*, 1988.

Eicher, D.L.: *Geologic Time*, 2nd Edition, Prentice-Hall, Inc., Englewood Cliffs, NJ, 1976.

MacDougall, I., and T.M. Harrison: *Geochronology and Thermochronology by the $^{40}Ar/^{39}Ar$ Method*, Oxford University Press, New York, NY, 1988.

Parrish, R., and J.C. Roddick: "Geochronology and Isotope Geology for the Geologist and Explorationist," *Geological Society of Canada Short Course 4*, 1985.

Prothero, D.R.: *Interpreting the Stratigraphic Record*, WH Freeman, New York, NY, 1990.

Prothero, D.R., and R.H. Dott Jr.: *Evolution of the Earth*, 6th Edition, The McGraw-Hill Companies, Inc., New York, NY, 2001.

York, D., and R.M. Farquhar: *The Earth's Age and Geochronology*, Pergamon Press, Oxford, UK, 1972.

DONALD R. PROTHERO, Occidental College, Los Angeles, CA

GEOLOGICAL TIME: HISTORY OF IDEAS. Early estimates of the Earth's age were restricted by literal interpretations of Genesis, but the geological time scale had been greatly extended by the early nineteenth century. Calculations by physicists gave rise to very limited estimates of the Earth's age in the late nineteenth century, and it was only after the discovery of radioactivity that it became possible both to extend the scale of geological time and develop techniques for providing exact dates. See also **Geologic Time Scale**.

Some Eastern civilizations have long assumed that the universe exhibits cycles of growth and decay spread over vast periods of time, and have thus been prepared to accept that the Earth itself has a great antiquity. Hindu cosmology has certainly encouraged this extended view of natural time. Among the ancient Greeks, Aristotle taught that the Earth was indefinitely old, and was aware of lengthy cycles of change in the environment. In contrast, materialist philosophers such as Empedocles and the Roman Lucretius thought that the Earth had been formed by physical processes in the distant past, although they agreed that the planet was subject to change. The situation following the rise of Christianity was complicated by the fact that the Bible contains a creation story that links the appearance of the universe and of the Earth very closely to the appearance of the first humans (Adam and Eve). It was thus difficult for Christians to accept the idea that there was an extended period of prehistory before Adam, while the genealogies of the patriarchs seemed to imply a fairly limited period for human history itself. See also **Aristotle of Stagira (384 BC–322 BC)**.

Not all of the Church fathers adopted a literal reading of the Genesis creation story and some were prepared to admit that the creation might have been a complex process. But most estimates of the period since creation were extremely limited. Theophilus of Antioch in the second century put the date of creation at 5529 BC and in the next century Julius Africanus placed it around 5500 BC. The medieval Church was not bound by such limited estimates because it retained the right to interpret the scriptures in a non-literal manner, but this situation changed with the Protestant Reformation, which rekindled an enthusiasm for the Word of God and led many to assume that the Bible must be read literally. Martin Luther himself was led to assume that the date of creation must have been around

4000 BC. It was a Protestant cleric, James Ussher, Archbishop of Armagh, who has gone down in history as the author of the most widely quoted (and nowadays most widely ridiculed) estimate. Writing in the 1650s, Ussher placed the date of creation at 4004 BC and even identified the day of the month and the hour on which creation first began. In fact, Ussher was a respected scholar of the time, and his effort to date creation can only be understood in terms of the religious and political debates unleashed by the Reformation. Those Fundamentalist religious writers associated with what is called 'young earth creationism' still follow this intepretation of the text.

Buffon and the "Dark Abyss of Time"

By the mid-seventeenth century acceptance of the Copernican cosmology made it possible to believe that the Earth was a material structure formed by physical processes. The philosopher René Descartes proposed an account in which the Earth was the burnt-out remnants of a star. This view was elaborated in one of the most celebrated "theories of the earth," Thomas Burnet's Sacred Theory of the Earth of 1681, which also included an explanation of the deluge based on natural events. At the same time, natural philosophers such as John Ray and John Woodward were becoming aware of the evidence suggesting that the present surface of the Earth was not original—the fossil-bearing rocks having evidently been laid down under water. Woodward at least was convinced that these rocks were formed at the time of the deluge, although Robert Hooke's Discourse on Earthquakes (delivered 1668, published posthumously in 1705) proposed that areas of land could be elevated above the ocean by earth movements. Ray and Hooke were aware of evidence that geological time was more extensive than the few thousand years allowed by Ussher, but were unwilling to provide a more extended time scale. Some natural philosophers thought that the 'chaos' from which the Earth had been created could have existed for a considerable period of time. See also **Ray, John (1627–1705)**; and **Hooke, Robert (1635–1702)**.

GEOLOGIC TIME SCALE.   The geologic time scale (Table 1) combines the traditional classical time classification of rocks long used by geologists with numerical time boundaries based on radiometric age measurements.

This time scale is sometimes referred to as the "absolute" time scale, but more appropriate are the terms "geochronologic" or "radiometric" time scale to distinguish it from the older relative time classification. In the latter, the principles of stratigraphic and faunal successions have played predominant roles, and three major time divisions, the Cenozoic, Mesozoic, and Paleozoic eras, were named for recent, middle, and ancient life, respectively. This classification with subdivisions into periods and epochs was well worked out prior to 1850. See also **Mass Extinctions**.

Early speculations about the time rates of geologic processes led to attempts to place limits in years on subdivisions of geologic time as well as on the age of the earth. Various methods were tried to calculate intervals of geologic time based on the rates of deposition, erosion, development of life, accumulation of salt in the ocean, and so forth. In all these calculations, various assumptions had to be made, commonly on the most meager information. It is not surprising, therefore, that time intervals calculated by different individuals varied greatly.

Many prominent geologists, physicists, and chemists have concerned themselves with the problem of measuring geologic time with the objective of attaining some quantitative values for the intervals represented in the geological classification. One of the foremost is Arthur Holmes whose papers on the subject span half a century (1911–1960). His outstanding work in this field has won him wide recognition. In 1956 Holmes was awarded the Penrose Medal of the Geological Society of America, and in 1964 he shared the prize of the G. Unger Vetlesen Foundation at Columbia University for his contributions to the geologic time scale. A special volume, "The Phanerozoic Time Scale," was dedicated to Holmes by the Geological Society of London, and Table 1 is adapted from a time scale that resulted from the Holmes Symposium.

Phanerozoic Time Scale. In 1913, in a small volume entitled, "The Age of the Earth," Holmes outlined how age determinations based on the principles of radioactive decay, in conjunction with geological data on the maximum known thicknesses of rocks assigned to the various geological periods, might be used to construct a quantitative time scale. The ratios of the daughter products, helium and lead, to the parent uranium, were used to calculate these early radioactivity ages. This approach was used

by Joseph Barrell in a monumental paper "Rhythms and the Measurements of Geologic Time" in 1917, in which he presented a time scale (Table 2). Holmes' first extended time scale for the Phanerozoic was published in 1933 (Table 2). European scientists who made important contributions up to this time include, among many others, such illustrious names as Charles Lyell, Charles Darwin, Archibald Geikie, Lord Kelvin, Lord Rayleigh, Lord Rutherford, W.J. Sollas, John Joly, and A. de Lapparent. In addition to Barrell, Americans who made significant contributions were chemists such as B.B. Boltwood and F.W. Clarke, and geologists including G.F. Becker, J.D. Dana, G.K. Gilbert, Charles Schuchert, and C.D. Walcott.

TABLE 1. PHANEROZOIC TIME SCALE

ERA	PERIOD	EPOCH	BEGINNING OF INTERVAL (MILLION YEARS)	
CENOZOIC	QUATERNARY	PLEISTOCEME		0
		PLIOCEME — 1.5,2		
	TERTIARY	MIOCEME — 24		
		OLISOCENE — 37-36		50
		EOCENE — 53-54		
		PALEOCENE — 65		
MESOZOIC	CRETACEOUS	Upper — 100		100
		Lower		
		— 136		150
	JURASSIC	Upper — 162		
		MIDDLE — 172		
		Lower — 190-195		200
	TRIASSIC	Upper — 205		
		Middle — 215		
		Lower — 225		
PALEOZOIC	PERMIAN	Upper — 240		250
		Lower		
		— 200		
	CARBONIFEROUS PENNSYLVANIAN			300
	CARBONIFEROUS MISSISSIPPIAN	— 325		
		— 345		350
	DEVONIAN	Upper — 359		
		Middle — 370		
		Lower — 395		400
	SILURIAN	Upper — 430-440		
		— 445		450
	ORDOVICIAN	Lower		
		— 500		500
	CAMBRIAN	Upper — 515		
		MIDDLE — 540		550
		Lower		
		— 570		
	PRECAMBRIAN			600

The Barrell and Holmes time scales were based on U-Pb age calculations based on chemical determinations. In 1933, F.W. Aston showed by mass spectrographic analyses that lead is composed of a number of isotopes whose abundance ratios he determined in several samples of common lead. This work was followed in 1939–1941 by papers by A.O. Nier. His U-Pb age calculations based on mass spectrometric measurements heralded modern geochronology. The precise isotopic measurements of Aston and Nier provided Holmes with the information he needed for his 1947 geologic time scale (Table 2).

TABLE 2. VERSIONS OF THE POST-PRECAMBRIAN TIME SCALE (Millions of Years)

Geologic Division	Barrell (1917)	Holmes (1933)	Holmes (1947)	Russia (1960)	Kulp (1961)
Pleistocene	1–1.5	1	1	—	1
Pliocene	7–9	15	12–15	10	13
Miocene	19–23	32	26–32	25	25
Oligocene	35–39	42	38–47	—	36
Eocene	55–65	60	58–68	70[a]	63[a]
Cretaceous	120–150	128	127–140	140	135
Jurassic	155–195	158	152–167	185	181
Triassic	190–240	192	182–196	225	230
Permian	215–280	220	203–220	270	280
Carboniferous	300–370	285	255–275	320	345
Devonian	350–420	350	313–318	400	405
Silurian	390–460	375	350	420	425
Ordovician	480–590	440	430	480	500
Cambrian	550–700	510	510	570	600

[a] Paleocene.

Since World War II, great progress has been made in geochronology with the introduction of the K-Ar and Rb-Sr techniques for age determinations and with the application of U-Th-Pb isotopic age determinations to minerals such as zircon.

New data from a number of geochronology laboratories were used in Kulp's (1961) time scale. Table 2 also includes the geochronologic scale compiled by the Commission on Absolute-Age Determination of Geologic Formations of the Russian Academy of Sciences.

Precambrian Time Scale. Although geologists surmised that a great deal of time is represented in the Precambrian rocks, the immensity of this interval was not fully comprehended or accepted until isotopic age determinations became available. The age of the earth is now commonly taken as 4.55 billion (10^9) years, the age obtained in 1956 by C.C. Patterson by comparing the abundance ratios of lead isotopes for meteorites with terrestrial lead. Many isotopic ages in the range from 2,500 to 3,600 million years have been determined on mineral and rock samples from the Precambrian shield areas of the Americas, Africa, Australia, and Eurasia.

The lack of fossils in the Precambrian rocks and the metamorphic changes they have undergone have made extremely difficult the task of deciphering the stratigraphic succession. Locally, as for example, in the Lake Superior region, the succession and a classification of Precambrian rocks have been worked out, but there is no universally accepted classification. Similarly, the radiometric time scale for the Precambrian succession is still in an elementary form compared to that for the Phanerozoic. The metamorphic processes that have affected in varying degree the minerals of the Precambrian rocks also affected the parent-daughter nuclide ratios. Isotopic ages, therefore, are difficult to interpret in areas of complex metamorphic history and reflect metamorphic events rather than the time of first emplacement or first crystallization. Most of the progress that has been made in Precambrian geochronology has come through the dating of major periods of orogeny. See Table 3.

The development of an ordered stratigraphic succession with the use of fossils to correlate rocks in widely separated areas is one of the remarkable achievements of the geological profession. Paleontological and stratigraphic methods remain the most applicable and reliable for correlation in Phanerozoic rocks. Isotopic age measurements now provide a long-needed method for deciphering the succession of Precambrian rocks. In addition, radioactivity age measurements make

TABLE 3. PRECAMBRIAN TIME SCALES

possible a quantitative approach to the study of geologic history and processes.

Terms. Some terms used in the field of *geochronology* (the study of the earth and other components of the cosmos with relation to the passage of time) include time periods which are designated by various terms (units) that have a relationship with each other, but which are not of precise or consistent span, as say the units of time, temperature, pressure, etc. used in other measurements. This relationship is shown in a relative way by the following:

Eon — A very large part or grand division of geologic time; the longest of the geologic time units. Sometimes defined as one billion (10^9) years.

Era — An era includes two or more *periods*, during which rocks of the corresponding erathem were formed.

Period — A subdivision of an era, during which the rocks of the corresponding system were formed.

Epoch — A subdivision of a period, during which the rocks of the corresponding series were formed.

Age — A geologic time-unit shorter than an epoch and longer than a subage, during which the rocks of the corresponding stage were formed.

Subage — A rarely used term, shorter than age, during which the rocks of the corresponding substage were formed.

Additional Reading

Badash, L.: "The Age-of-the-Earth Debate," *Sci. Amer.*, 90 (August 1989).
Barrell, J.: "Rhythms and the Measurements of Geologic Time," *Geol. Soc. Amer. Bull*, **28**, 745–904 (1977).
Bowen, D.Q.: "The Last 130,000 Years," *Review (Univ. of Wales)*, **39** (Spring 1989).
Harland, W.B., D.G. Smith, R.L. Armstrong, et al.: *Geologic Timescale, 1989*, Cambridge University Press, New York, NY, 1990.
Kulp, J.L.: "Geologic Time Scale," *Science*, **133**, 1105–1114 (1961).
McElhinny, M.W.: *The Earth*, Academic Press, Inc., San Diego, CA, 1979.
Newsom, H.E. et al.: *Origin of the Earth*, Oxford University Press, Inc., New York, NY, 1990.
Press, F. and R. Siever: *Earth*, 2nd edition, W.H. Freeman and Company, New York, NY, 1979.

Schindewolf, O.H.: *Basic Questions in Paleontology: Geologic Time, Organic Evolution, and Biological Systematics*, University of Chicago Press, Chicago, IL, 1993.

Staff: "The Phanerozoic Time-Scale," Symposium of Geological Society of London, *Quart. J. Geol. Soc. London*, **120s** (1964).

Staff: "The Earth: Its Mass, Dimensions, and Other Related Quantities," in *Handbook of Chemistry and Physics*, F-193, CRC Press, LLC, Boca Raton, FL, 73rd Edition, 1992–1993.

Stockwell, C.H.: "Geochronology of Stratified Rocks of the Canadian Shield," *Can. J. Earth Sci.*, **5**, 693–698 (1968).

Wetherill, G.W. and C.L. Drake: "The Earth and Planetary Sciences," *Science*, **209**, 96–104 (1980).

Web References

Geologic Time: http://www.ucmp.berkeley.edu/help/timeformold.html
The Geological Society of America: http://www.geosociety.org/

GEOLOGY.   As defined by the American Geological Institute in its excellent "Glossary of Geology," geology is ... "The study of the planet Earth. It is concerned with the origin of the planet, the material and morphology of the Earth, and its history and the processes that acted (and act) upon it to affect its historic and present forms. In the pursuit of that knowledge, the science considers the physical forces that influenced, and continue to influence, and change. Also considered are the chemistry of its constituent materials; the record and age of its past as revealed by the organic remains that are preserved in the layers of its crust or by interpretation of relic morphology and environment. Clues to the origin of the Earth are sought through the study of extraterrestrial bodies and their atmospheres that may reflect an earlier stage of this planet, or whose history may share the events and forces that created the Earth. All of the knowledge obtained through the study of the planet is placed at the service of man, to discover useful materials within the Earth; to identify stable environments for the support of his constructed arts and utilities; and to provide him with a foreknowledge of dangers associated with the mobile being."

There are several hundred entries relating directly or indirectly to geology in this encyclopedia. In particular, see also **Earth**; and **Earth Tectonics and Earthquakes**.

GEOLOGY (Lunar).   See **Moon (Earth's)**.

GEOLOGY (Petroleum).   See **Petroleum**.

GEOMAGNETIC EQUATOR.   The terrestrial great circle everywhere 90° from the geomagnetic poles. Geomagnetic equator should not be confused with magnetic equator, the line connecting all points of zero magnetic dip. See also **Aclinic Line**.

GEOMAGNETIC LATITUDE.   Angular distance from the geomagnetic equator, measured northward or southward through 90° and labeled N or S to indicate the direction of measurement. Geomagnetic latitude should not be confused with magnetic latitude, the magnetic dip. Phenomena closely related to the earth's magnetic field are often plotted according to geomagnetic latitude rather than geographic latitude.

GEOMAGNETIC POLE.   Either of two antipodal points marking the intersection of the earth's surface with the extended axis of a dipole assumed to be located at the center of the earth and approximating the source of the actual magnetic field of the earth. That pole in the Northern Hemisphere (latitude, $78\frac{1}{2}$°N; longitude, 69°W) is designated north geomagnetic pole, and that pole in the Southern Hemisphere (latitude, $78\frac{1}{2}$°S, longitude, 111°E) is designated south geomagnetic pole. The great circle midway between these poles is called geomagnetic equator. The expression geomagnetic pole should not be confused with magnetic pole, which relates to the actual magnetic field of the earth. See also **Earth**.

GEOMAGNETISM.   See **Meteorology**.

GEOMETRICAL OPTICS.   This branch of physics treats light as if it were actually composed of "rays" diverging in various directions from the source and abruptly bent by refraction or turned back by reflection into paths determined by well-known laws. The idea that light travels in straight lines is here uppermost, while its wave character and other physical aspects are disregarded. Thus the image of a point A, if "real," is simply another point B through which the rays diverging from A ultimately pass after the several reflections or refractions produced by the mirrors, lenses, etc., of the optical system. If the image B is "virtual," the rays appear to be diverging from it, but only because their direction has been so changed that, if produced backward, the lines along which they now travel would intersect at B. A real image of a lamp may easily be formed by a reading glass; a virtual image, by a plane mirror.

The chief advantage of this mode of visualizing the behavior of light is the simplicity with which problems may be solved by geometrical constructions. The same formulae deduced by the methods of geometrical optics may be arrived at, but often with much more labor, by treating light as composed of waves and studying the changes of wave front.

GEOMETRIC DISTORTION.   Any aberration which causes the reproduced image to be geometrically dissimilar to the perspective plane-projection of the object.

GEOMETRIC PROGRESSION.   See **Progression**.

GEOMETRY.   A comprehensive branch of mathematics which is concerned with the properties, measurement, and relations between lines, angles, surfaces, and solids. Classically, the methods of Euclid, who probably lived from 330–275 B.C., were used. These were based on a number of definitions, five postulates, and nine general axioms. The definitions, which were accepted without proof, included statements on point, line, solid, proposition, hypothesis, theorem, etc. The axioms were also accepted without proof, the following being typical. If equals are added to or subtracted from equals, the sums or remainders are equal; the whole is equal to the sum of all its parts and greater than any of its parts. Among the postulates, a construction admitted to be possible, two typical examples are as follows: (1) a straight line can be drawn from one point to another and can be produced indefinitely; and (2) a circumference can be described from any point as a center and with any given radius.

Plane geometry is mostly the study of angles, triangles, polygons, circles, and other Figures which can be drawn with ruler and compass; solid geometry involves Figures in three dimensions, such as planes, spheres, cubes, polyhedra. Trigonometry is a specialized geometry of the triangle.

Until the nineteenth century, the geometry of Euclid was unquestioned, even though mathematicians had always been unable to prove his fifth postulate. In its classical statement, this postulate takes the form: "If a straight line falling on two straight lines makes the interior angles on the same side less than two right angles, the two straight lines, if produced indefinitely, meet on that side on which are the angles less than two right angles."

An equivalent, and shorter, statement of this postulate is: "Through a point outside a line only one line can be drawn parallel to the given line."

In the nineteenth century, the conclusion was reached that a logical system of geometry could be constructed without use of the fifth postulate and consistent systems were constructed denying it. In one of these, it was assumed that, through any point, there are two or more parallel lines that do not intersect a given line in the plane. This system, which was developed by C.F. Gauss (1777–1855), Wolfgang and John Balyai, and N.I. Lobachevsky (1793–1856) was named *Lobachevskian* or (later) *hyperbolic geometry*. It leads to the conclusion that the sum of the three angles in a triangle is less than two right angles. This system uses, directly or indirectly, all of Euclid's axioms and postulates except the fifth postulate, and all of his theorems that do not depend upon it.

Another non-Euclidean geometry was developed by G. Riemann (1826–1866) and was named *Riemannian* or (later) *elliptic geometry*. It replaced Euclid's fifth postulate by one denying the existence of *any* parallel lines. Therefore, unlike hyperbolic geometry, it rejected a number of Euclid's first 28 propositions, such as the sixteenth and its consequences. (The sixteenth proposition of Book I of Euclid is stated as: "In any triangle, if one of the sides is produced, the exterior angle is greater than either of the interior and opposite angles.") Riemann rejected the infinitude of the line.

In 1872 at the University of Erlangen, Felix Klein presented the so-called Erlangen Programme embodying a definition of geometry which would embrace, as subgeometries of projective geometry, the various non-Euclidean geometries as well as Euclidean geometry itself. This definition of Klein's was as follows:

"A geometry is the study of those properties of a set S which remain invariant when the elements of S are subjected to the transformations of some transformation group." A transformation T of a set A onto a set B is defined as a one-to-one correspondence between the elements of A and those of B.

To formulate a geometry by this definition, we need only choose a fundamental element (e.g., a point, line or circle), a set (or space) S of these elements (e.g., a plane or a spherical surface of points, a plane of lines, or a pencil of circles), and a group of transformations to which the fundamental elements are to be subjected. Then the definitions and theorems of the geometry consist of the properties invariant under the group of transformations.

A further development that followed the discovery of non-Euclidean geometry was the formulation of precise sets of axioms for Euclidean and non-Euclidean geometries. Axiom sets for the former include those of Pasch (1882), Peano (1889), Hilbert (1899), Veblen (1904), Forder (1927), Birkhoff (1932), Robinson (1940) and Levi (1960). Their common purpose was to place the entire structure of Euclidean geometry upon the simplest possible foundation, that is, to choose a minimum number of undefined elements and relations, and a set of axioms concerning them, with the property that all of Euclidean geometry can be deduced logically from them without any further appeal to intuition.

There are several specialized branches of classical geometry or mathematical techniques related to it. *Analytical geometry*, developed by the French mathematician and philosopher, Rene Descartes (1596–1650) and Pierre Fermat (1601–1665), another Frenchman, is an application of algebraic results in geometry. In two dimensions, considerable attention is given to conic sections; in three dimensions, to the quadric surfaces. It is also called *coordinate geometry*.

Descriptive and *projective geometries* developed from the interest of both painters and mathematicians in the problem of describing three-dimensional Figures on a plane. Prior to the time of the Renaissance, artists in general had been satisfied with symbolic representations of persons and objects. Subsequently, they became increasingly desirous of greater realism in their work. Albrecht Dürer, the German painter and engraver (1471–1528), is thought by some mathematical historians to be the inventor of descriptive geometry. Somewhat later, the French mathematician Gaspard Monge (1746–1818) placed the subject on a firm mathematical basis. As indicated before, it is the graphical description of objects in three dimensions and the mathematical technique of mechanical drawing.

A more generalized development of geometric Figures is *projective geometry*. Its founders include Gaspard Desargues, a French engineer (1593–1662); Blaise Pascal, French geometrician and philosopher (1623–1662); Jean Victor Poncelet, French mathematician and general in the armies of Napoleon (1788–1867). Although projective geometry, like descriptive geometry, uses projection and section, it is different from the latter. One of its important objects is the study of properties invariant under projection and section.

Differential geometry is essentially the application of differential and integral calculus to the study of curves and surfaces. Methods of tensor calculus are frequently used and it is the chief mathematical apparatus of relativity theory. Its developers include Gauss and Riemann.

Still other geometries follow readily from Klein's definition given earlier in this entry. Let us take a more specific statement of that definition. "Let S be a set of points P such that a unique point P corresponds to an ordered pair of real numbers (x, y) and let S' be a set of points $P'(x', y')$. Let a one-to-one correspondence between points P and P' be given by the transformations $x' = f(x, y); y' = g(x, y)$. If a set of these transformations forms a group, then the properties of the associated geometry are determined by the functions f and g. For example, if f and g are linear functions, we have *affine geometry similarity geometry*, etc., depending upon the particular group of linear functions formed. If the group of transformations consists of the rational functions f and g such that the inverse transformations are also rational and if those functions are continuous, the associated geometry is called *algebraic geometry*.

Topology is a study of one-to-one bicontinuous transformations. The usual description of topology as rubber-steel geometry is sufficiently suggestive for two-dimensional space only, but the topological spaces of most importance in applied (or pure) mathematics have an infinite number of dimensions. See also **Linear Topological Space**; and **Topology**. Numerous geometrical objects, shapes, and bodies are described in detail in this volume. See also **Fractal Geometry**; and **Mathematics**.

GEOMORPHOLOGY.   This term has replaced the earlier term physiography to denote the full scientific interpretation of the origin of topographic features, or the purely physical attributes of scenery. This relatively distinct department of the earth sciences includes the study of the origin of all topographic features in terms of process or processes of erosion and in their relation to geologic structure.

GEOPHYSICS.   Geophysics is an interdisciplinary science that integrates the observations, hypotheses, and laws of geology with the techniques and principles of physics to understand the composition, nature, structure, and processes of the Earth. The world of geophysics extends from the magnetosphere — the outermost limits of the universe surrounding the Earth that are affected by the magnetic field of the Earth — to the center of the Earth some 6,372 kilometers (3,959 miles) from the Earth's surface. The term *solid-Earth geophysics* is applied to the branch of the science of geophysics that relates to the solid Earth including the liquid outer core of the Earth.

Geophysics is the "remote sensing" science that permits us to investigate the Earth's interior indirectly. Just as a physician uses our pulse, heart rate, temperature, x rays, and other physical evidence to investigate the composition, structure, and nature of the interior of our bodies, the solid-earth geophysicist uses a range of measurements on a variety of force fields that traverse the Earth to obtain information on the "internal condition" of the Earth. This approach is necessary because we have neither the resources nor technology to sample adequately the Earth's third dimension in a more direct fashion.

The geophysical methods are based on measuring force fields, which in turn respond to the physical properties of Earth materials that they encounter. For example, seismic methods utilize elastic properties and the density of the Earth and thus the velocity of propagation of elastic waves, while electric methods are based on a variety of electrical properties of the Earth that affect its carrying of electrical current. In a similar manner, gravity and magnetic methods are dependent on the mass (density) and magnetization, respectively, of the Earth. Thus the measurements of force fields such as seismic (mechanical), gravitational, magnetic, and electric fields provide significant information about the properties of the Earth. This is especially true when the observations are taken over the surface of the Earth. Spatial variations in the measurements are ample proof that the Earth is not homogeneous, although to a first degree it is spherically symmetric in structure and composition.

The variety of force fields that are available to geophysicists for studying the Earth is advantageous in decreasing the ambiguity of geologic interpretation. In addition, the methods contrast significantly in depth of investigation, resolution of discrete anomalous Earth materials, efficiency, and ease of observation. The common characteristics of each of the methods is the strong base on physical principles, the need for geologic information in all phases of conducting the methods, and the intensity of computational aspects necessary to make force-field observations geologically significant.

Classically, geophysics is concerned with the nature of the physical occurrences at and below the surface of the earth including, therefore, geology, oceanography, geodesy, seismology, hydrology, etc. The trend is to extend the scope of geophysics to include meteorology, geomagnetism, astrophysics, and other sciences concerned with the physical nature of the universe. Geophysics uses analytical and mathematical, rather than purely descriptive, techniques.

See also **Astrophysics**; **Earth**; **Earth Tectonics and Earthquakes**; **Geodesy**; **Hydrology**; and **Ocean Resources (Energy)**.

GEOPOTENTIAL.   The potential energy of a unit mass relative to sea level, numerically equal to the work that would be done in lifting the unit mass from sea level to the height at which the mass is located; commonly expressed in terms of dynamic height or geopotential height. Unit geopotential is equal to the potential of unit mass lifted a unit distance in a force field of unit strength. Distance upward can be measured in terms of differences in geopotential.

The geopotential Φ at height z is given mathematically by the expression

$$\Phi = \int_0^z g\,dz$$

where g is the acceleration of gravity. Distance upward can be measured in terms of differences in geopotential.

GEOSCIENCE. The collective disciplines of the geological sciences; thus, the term is synonymous with geology.

GEOSPACE MISSIONS. See **Living With a Star Program (LWS)**.

GEOSPHERE. The solid and liquid portions of the earth; the lithosphere plus the hydrosphere. Above the geosphere lies the atmosphere and at the interface between these two regions is found almost all of the biosphere, or zone of life.

GEOSTATIONARY OPERATIONAL ENVIRONMENTAL SATEL-LITE (GOES). The geostationary satellite experiment began in 1966 with the launch of the first satellite of the Applications Technology Satellite (ATS) series. ATS-1, launched on December 7, 1966, carried an instrument capable of providing continuous images of the earth, and an instrument that enabled the transmission of data to and from ground stations.

Six ATS satellites were launched between 1966 and 1974. In 1967, ATS-3 was launched and provided the first color image of the entire Earth. After the success of the meteorological experiments performed aboard these satellites, the investigation of geostationary satellites became an official, operational program.

The ATS was followed by the Synchronous Meteorological Satellite (SMS), the first series of geosynchronous weather satellites. SMS-1 was launched from Cape Canaveral, FL on May 17, 1974. It was the first operational satellite capable of detecting meteorological conditions from a fixed location. SMS-1 carried a Visible Infrared Spin Scan Radiometer (VISSR), a Space Environment Monitor (SEM), and a Data Collection System (DCS). The satellite continuously monitored broad areas of the Earth, obtained both day and night data, and collected and relayed data from over 10,000 central ground stations.

After the successful launch of two experimental SMS satellites, SMS-1 and SMS-2, the Geostationary Operational Environmental Satellite (GOES) program formally began in 1975 as a joint effort of NOAA and NASA.

On October 16, 1975, the first satellite under the GOES program was launched from Cape Canaveral, FL. GOES-A was renamed GOES-I once it reached orbit. GOES-2 and GOES-3 followed in 1977 and 1978. GOES-1 through GOES-3 were almost identical to the design of the SMS satellites, spin stabilized and carrying the VISSR, SEM, and DCS.

GOES 4-7 were designed similar to the first three GOES satellites. GOES-4 was launched on September 9, 1980. It was the first satellite to carry the Visible Infrared Spin Scan Radiometer (VISSR) Atmospheric Sounder (VAS), enabling the measurement of temperature and moisture. The data derived from this instrument enabled scientists to determine the altitudes and temperatures of clouds and draw a three-dimensional picture of their distribution in the atmosphere, leading to more accurate weather predictions.

On May 3, 1986, GOES-G, was lost when its launch vehicle was struck by lightning shortly after lift-off. GOES-G, which would have been GOES-7, was designed to replace GOES-4.

GOES-7 was launched on April 28, 1987. This satellite was the first GOES satellite capable of detecting 406 MHz distress signals from emergency beacons carried aboard aircraft and vessels and sending them to ground stations. This was the last of the spin stabilized geosynchronous satellites.

The launch of GOES-8 on April 13, 1994 introduced a new generation of spacecraft, the GOES I-M series. The three-axis, body-stabilized design provided significant improvements. The Imager and the Sounder were separate instruments and operated independently, which enabled the satellites to continuously obtain both imaging and sounding data instead of alternating between the two operating modes. The satellites were also capable of capturing higher resolution images. Starting with GOES-8, the GOES search and rescue system became operational.

On July 23, 2001, GOES-12 (GOES-M), the last satellite of the GOES I-M series, was launched from Cape Canaveral, FL. It was the first satellite to carry a Solar X-Ray Imager (SXI) type instrument.

Purpose

Unlike the polar orbiting satellites, the GOES satellites can provide continuous monitoring of the Earth's atmosphere and surface over a large region of the Western Hemisphere. They circle above the Earth in a geosynchronous orbit on Earth's equatorial plane, matching exactly the Earth's rotation about its axis. This configuration allows each satellite to view the same areas of the Earth at all times from 35,800 km (22,300 miles) above the Earth's surface. These satellites monitor potential severe weather conditions, such as tornadoes, flash floods, hail storms, and hurricanes. When these conditions develop, the GOES satellites track their movements as frequently as possible.

GOES satellite imagery is also used to estimate rainfall during thunderstorms and hurricanes for flash flood warnings, as well as estimate snowfall accumulations and overall extent of snow cover. Such data help meteorologists issue winter storm warnings and spring snow melt advisories. Satellite sensors can also detect ice fields and map the movements of sea and lake ice and slower moving icebergs. With the improved resolution on the infrared channels of today's GOES satellites, detection of forest fires, fog formation, volcano plumes, and the ability to distinguish between water and ice clouds are now possible.

The Imager instrument consists of five channels ranging from the visible to the longwave infrared channel. The visible channel has a resolution of 1 km while most of the infrared channels has a resolution of 4 km at nadir. The sounder, carrying 18 thermal infrared channels, is capable of making over 50,000 soundings per hour, which is particularly useful over data sparse regions of the Western Hemisphere. Each of the GOES satellites scans predetermined areas of the earth from the mid Pacific region to the eastern Atlantic region. During routine mode, observations are taken over the United States four times every hour, but when severe weather threatens the GOES Imager is capable of one minute interval scanning over a smaller area. A variety of products from the Sounder and Imager are created operationally to improve near real-time and long range forecasts. These products are archived at the National Climatic Data Center.

GOES I-M Spacecraft

The GOES I-M satellites are the primary element of U.S. weather monitoring and forecast operations and is a key component of NOAA's National Weather Service operations and modernization program. Spacecraft and ground-based systems work together to accomplish the mission of providing weather imagery and quantitative sounding data that form a continuous and reliable stream of environmental information used for weather forecasting and related services.

The GOES I-M series of satellites has provided significant improvements over previous GOES systems in weather imagery and atmospheric sounding information, particularly in the forecasting of life- and property-threatening severe storms.

The GOES I-M satellites introduced two new features. The first feature, flexible scan, offers small-scale area imaging that lets meteorologists take pictures of local weather trouble spots. This allows them to improve short-term forecasts over local areas. The second feature, simultaneous and independent imaging and sounding, is designed to allow weather forecasters to use multiple measurements of weather phenomena to increase the accuracy of their forecasts. See Table 1 for design data.

TABLE 1. MAIN SPACECRAFT DESIGN ELEMENTS

Mission Life	5 years, minimum
Dimensions	Main Body
	2 meters (7 foot) cube
	Deployed Length
	27 meters (88 feet)
Weight	2,100 kg (4,600 lb)
Orbit	Altitude
Geosynchronous	36,000 km (22,000 mi)
	Longitude
	75 W and 135 W
	Latitude
	Equatorial, within 0.5 degrees
Power	1050 watts @ 42 volts, solar array; battery backup
Launch Vehicle	Atlas-I/Centaur (GOES I/ K)
	Atlas-IIa/Centaur (GOES-L/M)

Current Status. The Geostationary Operational Environmental Satellite Program (GOES) is a joint effort of NASA and the National Oceanic and Atmospheric Administration (NOAA).

Currently, the GOES system consists of GOES-12 operating as GOES-East in the eastern part of the constellation at 75° west longitude, and GOES-10 operating as GOES-West at 135° west longitude. These spacecraft help meteorologists observe and predict local weather events, including thunderstorms, tornadoes, fog, flash floods, and other severe weather. In addition, GOES observations have proven helpful in monitoring dust storms, volcanic eruptions, and forest fires.

The benefits that directly enhance the quality of human life and protection of Earth's environment include:

- Supports the search and rescue satellite aided system (SARSAT)
- Contributes to the development of worldwide environmental warning services and enhancements of basic environmental services
- Improves the capability for forecasting and providing real-time warning of solar disturbances
- Provides data that may be used to extend knowledge and understanding of the atmosphere and its processes

GOES — NO/P is the next series of GOES satellites. GOES-N is scheduled to launch first.

GOES — N, O, and P Spacecraft. GOES — N-P is the next series of GOES satellites (Fig. 1 and 2). The multimission GOES series N-P will be a vital contributor to weather, solar, and space operations and science. The National Aeronautics and Space Administration (NASA) and the National Oceanic and Atmospheric Administration (NOAA) are actively engaged in a cooperative program to expand the existing GOES system with the launch of the GOES N-P satellites. (see Table 2 for main design elements).

Fig. 1.   Artist conception of GOES-N and the Earth. (*NASA*).

The Goddard Space Flight Center (GSFC) is responsible for procuring, developing, and testing the spacecraft, instruments and unique ground equipment. NOAA is responsible for overall program, funding, system in-orbit operation, and determining satellite replacement needs.

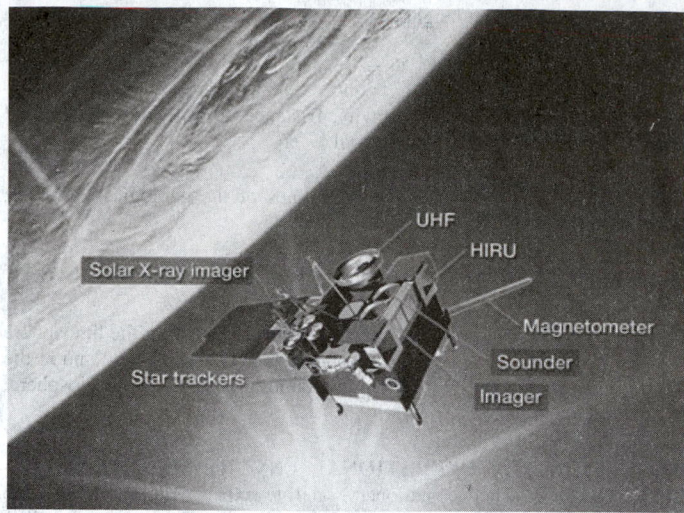

Fig. 2.   GOES-N's instruments provide high resolution visible and infrared data, as well as temperature and moisture profiles of the atmosphere. (*Image courtesy of The Boeing Company*).

TABLE 2. MAIN SPACECRAFT DESIGN ELEMENTS

Mission Life	10 years
Dimensions	Length
	13.75 ft. (4.2 m)
	Width
	6.2 ft. (1.88 m)
Weight	Launch
	6,908 lbs. (3,133 kg.)
	In orbit (beginning of life)
	3,969 lbs. (1,800 kg.)
Propulsion	Liquid Apogee Motor
	110 lbf (490 N)
	Stationkeeping Thrusters
	75 W and 135 W
Power — Solar	Beginning of Life
	2.3 kw
	End of Life
	2 kw
	Panels
	1 wing, w/1 panel of dual-junction gallium arsenide solar cells
Payload	S-Band
	1 downlink, 5 uplinks (include telemetry and command data)
	L-Band
	8 downlinks
	UHF
	1 downlink, 2 uplinks
Antennas	3 S/L Band, cup-shaped with dipole
	2 T&C antennas and 1 Omni antenna (aft)
	1 UHF, cup-shaped with dipole
Launch Vehicle	Delta IV

Future Status. GOES R Series satellites with hyperspectral imaging capability are planned for the next decade. The major instruments planned for the next generation of GOES satellites are: the Advanced Baseline Imager (ABI); the Hyperspectral Environmental Suite (HES); the Space Environment In-Situ Suite (SEISS), which includes a Magnetospheric Particle Sensor (MPS), an Energetic Heavy Ion Sensor (EHIS), and a Solar and Galactic Proton Sensor (SGPS); the Solar Imaging Suite (SIS), which includes the Solar X-Ray Imager (SXI), the Solar X-Ray Sensor (SXS), and the Extreme Ultraviolet Sensor (EUVS); the GEO *Lightning Mapper (GLM); and the Magnetometer*. For a description of these instruments, please go to the Goddard Space Flight Center GOES Program web site. http://goespoes.gsfc.nasa.gov/goes/index.html. The first launch is scheduled for the 2012 timeframe.

Advanced Baseline Imager (ABI): The ABI is a multispectral channel, two-axis scanning radiometer designed to provide variable area imagery and radiometric information of the Earth's surface, atmosphere and cloud cover. The ABI collects data on a three-axis body stabilized satellite in geosynchronous orbit, and has the capability for star sensing. The ABI is designed to measure emitted and solar reflected radiance simultaneously in all spectral channels. Data availability, radiometric quality, simultaneous data collection, coverage rates, scan flexibility, and minimizing data loss due to the sun, are prime requirements of the ABI system.

The ABI requires primary power and command input data from the spacecraft. ABI output data contains ABI information and other data. The sensor module contains the optical system, scanner, detectors and their cooling systems and directly related electronics. The electronics module contains the power supply module, command, control, and data processing circuitry. If required, an auxiliary electronics module may be used for active detector cooling. The ABI sensor module shall be mounted external to the spacecraft body. An electronics module and auxiliary electronics module will be mounted either external or internal to the spacecraft body.

The HES instruments are designed to sense emitted thermal energy and reflected solar energy from sampled areas of the Earth's surface and atmosphere. These data are used to compute vertical profiles of temperature and moisture, surface and cloud-top temperatures, and winds, and provide information about the Earth surface and oceans. The HES is part of a 3-axis stabilized, geostationary satellite system that collects weather and environmental data, in conjunction with data from an imaging instrument, to aid in the prediction of weather and climate monitoring. The HES data, depending on the task, provide moderate to high spatial resolution, high spectral, temporal and radiometric resolution to accurately monitor rapidly changing environmental conditions, including coastal waters and rapidly changing weather. The HES will provide two sounding tasks, which include disk sounding (DS) and severe weather/mesoscale sounding (SW/M), and a shelf and coastal waters imaging task (CW). These tasks will:

1. Provide vertical moisture and temperature information, and other environmental data that will be used by NOAA and other public and private agencies to produce routine meteorological analyses and forecasts (DS, SW/M).

2. Provide environmental data that can be used to expand knowledge of mesoscale and synoptic scale storm development and provide data that may be used to help in forecasting severe weather events (SW/M).

3. Provide data that may be used to extend knowledge and understanding of the atmosphere and its processes (e.g., by viewing the evolution and motion of storms and other atmospheric phenomena) in order to improve short/long-term weather forecasts (DS, SW/M).

4. Provide information about ocean color and optical properties, and as a goal, sea surface temperature (CW).

5. Provide information about ocean current, offshore ocean color, offshore optical properties, and offshore sea surface winds (CW).

The SIS is comprised of the Solar X-Ray Imager (SXI), the Solar X-Ray Sensor (XRS), and the Extreme Ultraviolet Sensor (EUVS). Following is a brief description of each of these instruments.

The SXI is a coronal imager capable of operating in the soft X-ray to EUV wavelength range. It provides full-disk solar images at high cadence around the clock except for brief periods during an eclipse. Available combinations of exposures and filters allows the coverage of the entire dynamic range of solar X-ray features, from coronal holes to X-class flares, as well as the estimate of temperature and emission measure. The operational goals are to: locate coronal holes for geomagnetic storm forecasts, detect and locate flares for forecasts of solar energetic particle (SEP) events related to flares, monitor changes in the corona that indicate coronal mass ejections (CMEs), detect active regions beyond east limb for F10.7 forecasts, and analyze active region complexity for flare forecasts.

The XRS is the primary measure of, and standard for, solar flare magnitude. Its primary function is to provide a means of detecting the beginning, duration, and magnitude of solar X-ray flares. Many space weather phenomena are preceded by a solar event such as a solar flare. In addition, the XRS is used as an input for the empirical model of Solar Energetic Proton events that can have severe impacts on satellites and astronauts. Two X-ray channels are required to monitor the disk-integrated solar fluxes in the 0.05 to 0.8 nm wavelength range at 3-second intervals. The sensor shall be sensitive enough to permit quiet sun background measurements at low levels of solar activity as well as very large solar flares. For calculations of threshold sensitivity, a 2×10^6 K solar spectrum such as the one presented by "Mewe and Groenshild, (1981)" shall be assumed.

Solar EUV radiation is a dominant energy source for the upper atmosphere and the ionizing radiation produces the ionosphere. Solar variability at these wavelengths is one of the primary drivers of thermospheric/ionospheric variability. Uncertainties in the solar EUV flux are a major source of errors in specification and modeling of the thermosphere and ionosphere. To provide adequate knowledge of this ionizing radiation, knowledge of the full EUV spectrum from 5 to 127 nm is required. The EUVS is being designed to provide this knowledge.

Space Environment In-Situ Suite (SEISS): The Space Environment In-Situ Suite (SEISS) consists of energetic particle sensors. The particle sensors will monitor the proton, electron and alpha particle fluxes. These particle fluxes roughly consist of three components: 1) a geomagnetically trapped and highly variable population of electrons and protons; 2) sporadic fluxes of electrons, protons, and alpha particles of direct solar origin; and 3) a background of galactic cosmic rays ranging from several MeV to highly relativistic energies. Knowledge of the near-Earth energetic particle environment is important in establishing the natural radiation hazard to humans at high altitudes and in space, as well as risk assessment and warning of episodes of surface charging, deep dielectric charging, and single event upset of satellite systems. Energetic particle precipitation into the Earth's ionosphere also causes disturbance and disruption of radio communications and navigation systems, and the resulting damages may be mitigated by early warnings of high flux episodes. The particle sensors include a Magnetospheric Particle Sensor (MPS), an Energetic Heavy Ion Sensor (EHIS) and a Solar and Galactic Proton Sensor (SGPS).

GEO Lightning Mapper (GLM): The GEO Lightning Mapper is a sensor, capable of continuously mapping lightning discharges during both day and night, in a geostationary orbit. From this orbit, the sensor will be capable of detecting all forms of lightning with a high spatial resolution and detection efficiency. This sensor will continuously monitor lightning activity over the United States and provide a more complete dataset than previously possible. Specific mission objectives of the GLM are:

1. Measure total lightning activity over large areas of the Americas and nearby oceans on a continuous basis.

2. Develop an extensive lightning climatology to be used for global change research.

3. Demonstrate ability to deliver, on a real-time basis, lightning measurements that are of sufficient quality and quantity for operational storm monitoring and severe weather warnings.

This instrument will allow scientists to study the electrosphere over dimensions ranging from the Earth's radius all the way down to individual thunderstorms. The sensor will detect all types of lightning phenomena, and will provide nearly uniform spatial coverage.

Disseminating this information in near real time, these measurements could be related on a continuous basis to other observable data, such as radar returns, cloud images, and other meteorological variables. Since this data would be distributed in real time, it will be an invaluable tool to aid weather forecasters in detecting severe storms in time to give advance warning to the public.

Magnetometer: The magnetometer will measure the magnitude and direction of the Earth's ambient magnetic field in three orthogonal directions in an Earth referenced coordinate system. The magnetometer will provide a map of the space environment that controls charged particle dynamics in the outer region of the magnetosphere. Magnetic field measurements provide information on the general level of geomagnetic activity, monitor current systems in space, and permit detection of magnetopause crossings, sudden storm commencements, and substorms.

GOES Data Online. The dream of NOAA to provide GOES data to online users become a reality on December 1, 2003, when the Comprehensive Large Array-data Stewardship System (CLASS) began ingesting GOES data http://www.class.noaa.gov/nsaa/products/welcome.Users can now preview inventories, browse images, and select GOES data using the web interface. Efforts are underway to migrate the entire historical database, which is well over 200 Tb into CLASS. The Remote Sensing and Applications Division (RSAD) at NCDC is developing GOES data statistic algorithms as part of the Scientific Data Stewardship Initiative to assure users that the highest quality GOES data are provided for accurate long-term climate studies. Also, the RSAD Satellite Team is developing GOES and POES data translators to enable users to import the raw scientific data into popular GIS software.

Other resources on GOES satellites and products include:

Geostationary Satellite Server—Current GOES images with an archive of up to 21 days; http://www.goes.noaa.gov; Historical GOES Browser–Daily GOES images over North America from December 1996 to present; http://cdo.ncdc.noaa.gov/GOESBrowser/goesbrowser; Operational Significant Event Imagery–NOAA's premier website for event imagery; http://www.osei.noaa.gov/; NOAA Satellite and Information Services-a wealth of information on all NOAA satellites including status, schedules, calibration, navigation, and much more!; http://noaasis.noaa.gov/NOAASIS/; Cooperative Institute for Meteorological Satellite Studies (CIMASS); http://cimss.ssec.wisc.edu/; Cooperative Institute for Research in the Atmosphere(CIRA) http://www.cira.colostate.edu/ramm/overview.htm; and the GOES Comprehensive Technical Data; http://rsd.gsfc.nasa.gov/goes/text/goestechnotes.

Applications Technology Satellite

The Applications Technology Satellite (ATS) program was established by the National Aeronautics and Space Administration (NASA) to flight test experimental payloads and investigate the space environment with the aim of developing technology of practical future benefit.

Five flight spacecraft of three configurations were built by Hughes from 1966 to 1969. In October 2000, Hughes became Boeing Satellite Systems, Inc. The satellites were designed as basic buses capable of carrying a variety of scientific payloads. A score of experiments were flown to conduct investigations in the fields of space and communications, satellite stabilization, meteorology, and the orbital environment.

ATS-1. Launch Date/Time: December 7, 1966 at 02:09:00 UTC, from Cape Canaveral, FL, aboard an Atlas Agena D launch vehicle. The (Applications Technology Satellite) was designed and launched for the purpose of (1) testing new concepts in spacecraft design, propulsion, and stabilization, (2) collecting high-quality cloud cover pictures and relaying processed meteorological data via an earth-synchronous satellite, (3) providing in situ measurements of the aerospace environment, and (4) testing improved communication systems. The spin-stabilized spacecraft was cylindrically shaped and measured 135 cm (53 inches) long and 142 cm (56 inches) in diameter. See Fig. 3. The primary structural members were a honeycombed equipment shelf and thrust tube. Support rods extended radially outward from the thrust tube. Solar panels were affixed to the support rods and formed the outer walls of the spacecraft. Equipment components and payload were mounted in the annular space between the thrust tube and solar panels. In addition to solar panels, the spacecraft was equipped with two rechargeable nickel-cadmium batteries to provide electrical power. Eight 150-cm-long (59-in-long) VHF experiment whip antennas were mounted around the aft end of the spacecraft, while eight telemetry and command antennas were placed on the forward end. Spacecraft guidance and orbital corrections were accomplished by 2.3-kg hydrogen peroxide and hydrazine thrusters, which were activated by ground command. The satellite was initially placed at 151.16 deg W longitude over the Pacific Ocean in a geosynchronous orbit. In general, most of the experiments were successful. Data coverage was nominal until about 1970, after which limited real-time data acquisition was carried out by NOAA until the May 1974 launch of SMS-1. Limited ATS-1 data acquisition was begun by NASA at about that time for ATS-1—ATS-6 correlative studies. The spacecraft has served as a communications satellite for a number of state, federal, and public organizations up to the present. It is planned to continue operations at its final longitude of 164 deg E until September 1983 and then move the spacecraft out of the geostationary orbit.

ATS-2. Launch Date/ Time: April 6, 1967 at 03:21:00 UTC, from Cape Canaveral, FL, aboard an Atlas Agena D launch vehicle. The second stage of the ATS-2 launch vehicle failed to ignite, resulting in an unplanned elliptical orbit. Stresses induced by this orbit eventually induced spacecraft tumbling. In spite of these conditions, useful data were obtained from some of the experiments, most notably the cosmic-ray and particle experiments and the field detection experiments. The satellite reentered the atmosphere on September 2, 1969, after 880 days in orbit.

ATS-3. Launch Date/Time: November 5, 1967 at 23:31:00 UTC, from Cape Canaveral, FL, aboard an Atlas Agnea D launch vehicle. The spacecraft was slightly larger than ATS-1. ATS-3 was a cylinder 60 inches (152 cm) in diameter, 72 inches (183 cm) high and weighed 805 pounds

Fig. 3.   First Geosynchronous Meteorological Satellite; Diameter 1.46 m (4 ft 9 in), Panel height 1.34 m (4 ft 5 in), and Weight in orbit 304 kg (670 lb). (*Image courtesy of The Boeing Company*).

(1,775 lbs). A phased array of eight whip antennas extended from the top, and a phased array of eight VHF antennas extended from the base. The sides of the cylinder were covered by 23,870 solar cells which, along with nicad batteries, provided the power for the craft.

Three meteorological experiments were on board. One was a Multicolor Spin-Scan Cloud-cover Camera (MSSCC) which provided continuous, full-disk hemispheric images of the sun-lit Earth every half hour. The spinning motion of the satellite would generate line scans with a spatial resolution of 3.2 km (2 miles). This process took approximately twenty minutes for the full image, and then ten minutes to reset the camera for a new image. The second experiment was an Image Dissector Camera (IDC) which scanned the full-disk electronically rather than mechanically. The third experiment called "Meteorological Data Relay System" (WEather FAcsimile (WEFAX), was a data relay and re-transmission instrument. This instrument relayed data from the central ESSA data processing facility to APT ground stations located around the western hemisphere. In addition, images from the spin scan camera were also transmitted over WEFAX to APT stations.

ATS-3 was placed in a transfer orbit directly over the equator over 45 degrees west (ATS was over the eastern Pacific at this time). The transfer orbit meant that the satellite would drift slowly westward with time. The satellite eventually reached 95 degrees west where it was deactivated with ATS-1 on December 1, 1978. Of the satellite's eleven year life span, useful data was received for the first eight (1967-1975).

Multicolor Spin-Scan Cloud-cover Camera (MSSCC). The ATS-3 Multicolor Spin-Scan Cloud-cover Camera (MSSCC) represented a significant advance over a similar but monochromatic spin-scan camera on ATS-1. The MSSCC was mounted with its optical axis perpendicular to the spacecraft's spin axis and viewed the earth through a special aperture in the spacecraft's side. The camera consisted of a high-resolution telescope, three photomultiplier light detectors (red, blue, and green), and a precision latitude step mechanism. Light entering the system was focused alternately on a set of three 0.038-mm aperture plates and then passed through various filters to impinge on the appropriate photodetector. The telescope multiplier assembly could be tilted in discrete steps to provide pole-to-pole coverage in 2400 scan lines. East-to-west scan was provided by the spin of the satellite itself. A total time of 24 min was required to scan one frame and 2.4 min to retrace with a nominal satellite rotation of 100 rpm. From its geosynchronous orbit, the camera had a ground resolution of better than 4 km at nadir. The experiment was successful, with ATS 3 being the first spacecraft to transmit operational multicolor earth-cloud photographs. Approximately 3 months after launch, however, the red and blue channels failed, and the system subsequently was limited to producing black-and-white pictures. Good quality black-and-white pictures were received daily until December 11, 1974, when operations were curtailed to three pictures a week. Experiment operation was completely discontinued on October 30, 1975. For a listing and description of the different forms of photographic data, see the "Meteorological Data Catalog for The Applications Technology Satellites" (TRF B09264), available from NSSDC. Data can be obtained through SDSD.

Image Dissector Camera (IDC). The ATS 3 Image Dissector Camera (IDC) was a camera system designed to (1) test the feasibility of using electrical scanning techniques in an earth-cloud camera and (2) provide daylight cloudcover data on a real-time basis with full earth coverage. The camera was mounted with its optical axis perpendicular to the spacecraft spin axis in such a manner that the camera produced a scan line with each revolution of the spacecraft. The direction of the scan, north to south or east to west, was determined by ground command. The image dissector tube consisted of a visible wavelength electrically scanning photocathode, a 0.018-mm scanning aperture, and a 12-stage electron multiplier. Light entering the camera was focused on the face of the photocathode, causing photoelectrons to be emitted from the surface in proportion to the number of impinging light photons. The emitted photoelectrons were propelled past the aperture by means of an external magnetic deflection coil. After passing through the aperture, the signal current was amplified by the 12-stage multiplier. The signal was further amplified and then transmitted at 28 KHz to a ground acquisition station. The 2.54-cm image dissector tube had a resolution capability of 1300 TV lines, which, at nominal spacecraft altitude, corresponded to a ground resolution of about 7 km at nadir. Successfully flown for the first time, the IDC system on ATS 3 served as a prototype for similar experiments on Nimbus 3 and 4. The camera performed normally until May 1969, when the IDC system was beset by erratic spacecraft antenna performance. Routine data acquisition ceased after May 30, 1969. The IDC system, although still capable of operation, was left in an operationally off mode since that time except for periodic engineering tests. For a listing and description of the different forms of photographic data available from this experiment, see the "Meteorological Data Catalog for the Applications Technology Satellites" (TRF B09264), available from NSSDC. Data can be obtained through SDSD.

Meteorological Data Relay System. The primary objective of the ATS-3 meteorological data relay system (WEather FAcsimile (WEFAX) experiment) was to test satellite retransmission of facsimile products prepared by NOAA to participating ground stations. Secondary objectives included (1) transmitting selected spin-scan camera pictures via satellite to APT ground stations and (2) exploring the feasibility of increasing the amount of data available to APT ground stations from ESSA and Nimbus satellites. The experiment had no unique hardware on board. It was part of the ATS-3 VHF experiment and used the VHF transponder for data relay. The transponder transmitted at 135.60 MHz and received at 149.22 MHz. Weather facsimile charts and satellite cloud cover pictures were sent via land line from NOAA, Suitland, MD, to the ATS WEFAX field station at Mojave, California. The charts and data were then transmitted to the spacecraft for relay to participating APT stations. Cloud cover photographs from the ATS-3 spin-scan camera were retransmitted through the spacecraft directly from the Mojave ATS ground station. The experiment was a success and continued to operate as of May 1972. A similar experiment possessing poorer data reproduction capabilities was flown on ATS-1.

ATS-4. Launch Date/Time: August 10, 1968 at 22:34:00 UTC, from Cape Canaveral, FL, aboard an Atlas Centaur launch vehicle. The primary objective was to evaluate gravity-gradient stabilization and new imaging techniques for meteorological data retrieval. The spacecraft was deactivated after 61 minutes when the Centaur rocket failed to ignite. The satellite's orbit was too low and the resulting atmospheric drag caused it to re-enter the earth's atmosphere and break-up on October 17, 1968.

ATS-5. Launch Date/Time: August 12, 1969 at 11:01:00 UTC, from Cape Canaveral, FL, aboard an Atlas Centaur launch vehicle. ATS-5, the

last spacecraft in the Hughes/NASA ATS program, was launched in a near-perfect trajectory for insertion into synchronous orbit. Although injected successfully into orbit, the spacecraft's reverse spin (counterclockwise) prevented successful deployment of the 124 foot (38 m) gravity gradient booms for the stabilization experiment. However nine of the other 13 experiments aboard the spacecraft returned useful data. ATS-5 was retired in March 1984.

ATS-6. The ATS-6 spacecraft was designed to accomplish a wide spectrum of experiments dealing with communications, technology, meteorology, and science using a three-axes stabilized platform in geostationary orbit. Launch occurred from Cape Canaveral, FL, on May 30, 1974, at 13:00:01UTC, aboard a Titan 3 C launch vehicle and the science experiments were activated by June 15. The mission was divided into three major phases: (1) Operations during the first mission "year" from 94 degrees west longitude over the USA; (2) Operations during the second mission "year" from 35 degrees east longitude with an emphasis on the subcontinent of India; and (3) Operations from 140 degrees west longitude after return to the U.S. An end-of-mission activity occurred at the end of July, 1979 in which the satellite was moved 450 km out of the geostationary orbit and began an eastward drift of ~6.1 per day. At the same time the satellite was spun-up around one of its equatorial-plane axes and scientifically entered a fourth phase.

The primary objectives of ATS 6 were to erect in orbit a large high-gain steerable antenna structure capable of providing a good-quality TV signal to a ground-based receiver and to measure and evaluate the performance of such an antenna. A secondary objective was to demonstrate new concepts on space technology in the areas of aircraft control, laser communications, and visual and infrared mapping of the earth/atmosphere system. The spacecraft was also capable of (1) measuring radio frequency interference in shared frequency bands and propagation characteristics of millimeter waves, (2) performing spacecraft-to-spacecraft communication and tracking experiments, and (3) making particle and radiation measurements of the geosynchronous environment. Configured somewhat like an open parasol, the ATS 6 spacecraft consisted of four major assemblies: (1) a 9.15 m (30 ft) diameter dish antenna, (2) two solar cell paddles mounted at right angles to each other on opposite sides of an upper equipment module, (3) an earth-viewing equipment module (EVM) connected by a tubular mast to the upper equipment module, and (4) an attitude control and stabilization system. See Fig. 4. The EVM, in addition to housing the earth-viewing experiments, provided support for the propulsion system and tanks, batteries, a multifrequency transponder, and the telemetry, command, and thermal control systems. The upper equipment module provided a platform for the space-viewing experiments.

Fig. 4. ATS-6 was the first 3-axis stabilized geosynchronous communications satellite. (*NASA*).

Inertia wheels were the prime means for torquing the spacecraft, with both hydrazine and ammonia multijet thruster systems included to provide the necessary torques for unloading the wheels. Also included was a small environment measurement package containing a MAG and several particle experiments. The satellite was turned off on June 30, 1979 and boosted into a higher orbit.

In addition to its technology experiments, ATS-6 became the world's first educational satellite. During its 5 year life, ATS-6 transmitted educational programming to India, the US and other countries. The vehicle also conducted air traffic control tests, was used to practice satellite-assisted search and rescue techniques, carried an experimental radiometer subsequently carried as a standard instrument aboard weather satellites, and pioneered direct broadcast TV. The satellite also played a major role in the Apollo/Soyuz docking in 1975 when it relayed signals to the Houston Control centre. It was boosted above GEO when thruster failures threatened to prevent further control of the spacecraft.

For a detailed description of the individual experiments see list below.

Measurement of Low-Energy Protons
http://nssdc.gsfc.nasa.gov/database/MasterCatalog?sc=1974-039A&ex=1

Magnetometer Experiment
http://nssdc.gsfc.nasa.gov/database/MasterCatalog?sc=1974-039A&ex=2

Low-Energy Proton/Electron Experiment
http://nssdc.gsfc.nasa.gov/database/MasterCatalog?sc=1974-039A&ex=3

Particle Acceleration Mechanisms and Dynamics of the Outer Trapping Region
http://nssdc.gsfc.nasa.gov/database/MasterCatalog?sc=1974-039A&ex=4

Auroral Particles Experiment
http://nssdc.gsfc.nasa.gov/database/MasterCatalog?sc=1974-039A&ex=5

Solar Cosmic Rays and Geomagnetically Trapped Radiation
http://nssdc.gsfc.nasa.gov/database/MasterCatalog?sc=1974-039A&ex=6

Omnidirectional Spectrometer
http://nssdc.gsfc.nasa.gov/database/MasterCatalog?sc=1974-039A&ex=7

Geosynchronous Very High Resolution Radiometer (GVHRR)
http://nssdc.gsfc.nasa.gov/database/MasterCatalog?sc=1974-039A&ex=8

Radio Beacon
http://nssdc.gsfc.nasa.gov/database/MasterCatalog?sc=1974-039A&ex=9

Radio Frequency Interference
http://nssdc.gsfc.nasa.gov/database/MasterCatalog?sc=1974-039A&ex=11

Millimeter Wave Propagation
http://nssdc.gsfc.nasa.gov/database/MasterCatalog?sc=1974-039A&ex=13

Cesium Bombardment Ion Engine Experiment
http://nssdc.gsfc.nasa.gov/database/MasterCatalog?sc=1974-039A&ex=14

Solar Cell Radiation Damage
http://nssdc.gsfc.nasa.gov/database/MasterCatalog?sc=1974-039A&ex=16

Satellite Instructional TV
http://nssdc.gsfc.nasa.gov/database/MasterCatalog?sc=1974-039A&ex=17

Tracking and Data Relay
http://nssdc.gsfc.nasa.gov/database/MasterCatalog?sc=1974-039A&ex=18

Spacecraft Attitude Control
http://nssdc.gsfc.nasa.gov/database/MasterCatalog?sc=1974-039A&ex=20

COMSAT Propagation 913-and 18-GHz
http://nssdc.gsfc.nasa.gov/database/MasterCatalog?sc=1974-039A&ex=21

Advanced Thermal Control Flight
http://nssdc.gsfc.nasa.gov/database/MasterCatalog?sc=1974-039A&ex=22

Quartz Crystal Microbalance
http://nssdc.gsfc.nasa.gov/database/MasterCatalog?sc=1974-039A&ex=23

Health and Education Telecommunications
http://nssdc.gsfc.nasa.gov/database/MasterCatalog?sc=1974-039A&ex=24

R. F. Interferometer Subsystem
http://nssdc.gsfc.nasa.gov/database/MasterCatalog?sc=1974-039A&ex=29

Television Camera
http://nssdc.gsfc.nasa.gov/database/MasterCatalog?sc=1974-039A&ex=31

Television Relay Using Small Terminals
http://nssdc.gsfc.nasa.gov/database/MasterCatalog?sc=1974-039A&ex=28

Spacecraft Vibration Accelerometer
http://nssdc.gsfc.nasa.gov/database/MasterCatalog?sc=1974-039A&ex=30

Position, Location and Aircraft Communication
http://nssdc.gsfc.nasa.gov/database/MasterCatalog?sc=1974-039A&ex=19

See also IEEE Trans. on Aerosp. "Electron Syst.," v. AES-11, n. 6, November, 1975, and the "ATS-6 Final Engineering Performance Report," NASA, RP-1080, Wash., D.C., November, 1981 (TRF B33477).

Synchronous Meteorological Satellite (SMS)

The success of the meteorological experiments carried aboard the ATS-1 and -3 satellites led to NASA's development of a satellite specifically designed to make atmospheric observations. SMS-1 and SMS-2, operational prototypes, were launched in 1974 and 1975. SMS-1 and -2, and GOES-1, -2, and -3 were essentially identical. They carried instrumentation for visible and international remote imaging, collection of data from automated remote platforms, relay of weather products (WEFAX), and measurement of a number of characteristics of the near space environment. The objective of the SMS-series was to provide improved meteorological data on weather phenomena to improve forecasting. After the successful launch of these satellites, NASA turned over the geostationary weather satellite program to NOAA for operation. NOAA bought additional spacecraft identical to SMS with the new name Geostationary Operational Environmental Satellite (GOES). The SMS series included the first operational satellite designed to sense meteorological conditions from a fixed location above the Earth, and to provide this data to operational forecasters and private interests on the ground.

SMS-1. Launch Date/Time: May 17, 1974 at 09:31:00 UTC, from Cape Canaveral, FL, aboard a Delta 2914 launch vehicle. The spin-stabilized, earth-synchronous spacecraft carried (1) a visible infrared spin-scan radiometer (VISSR) which provided high-quality day/night cloud cover data and made radiance temperatures of the earth/atmosphere system, (2) a meteorological data collection and transmission system which relayed processed data from central weather facilities to small APT-equipped regional stations and collected and retransmitted data from remotely located earth-based platforms, and (3) a space environmental monitor (SEM) which measured proton, electron, and solar X-ray fluxes and magnetic fields. The cylindrically shaped spacecraft measured 190.5 cm (75 inches) in diameter and 230 cm (91 inches) in length, exclusive of a magnetometer that extended an additional 83 cm (33 inches) beyond the cylinder shell. See Fig. 5. The primary structural members were a honeycombed equipment shelf and a thrust tube. The VISSR telescope was mounted on the equipment shelf and viewed the earth through a special aperture in the side of the spacecraft. See Fig. 6. A support structure extended radially out from the thrust tube and was affixed to the solar panels, which formed the outer walls of the spacecraft and provided the primary source of electrical power. Located in the annulus-shaped space between the thrust tube and the solar panels were station-keeping and dynamics control equipment, batteries, and most of the SEM equipment. Proper spacecraft attitude and spin rate (approximately 100 rpm) were maintained by two separate sets of jet thrusters mounted around the spacecraft equator and activated by ground command. The spacecraft used both UHF and S-band frequencies in its telemetry and command subsystem. A low-power VHF transponder provided telemetry and command during launch and then served as a backup for the primary subsystem once the spacecraft had attained synchronous orbit.

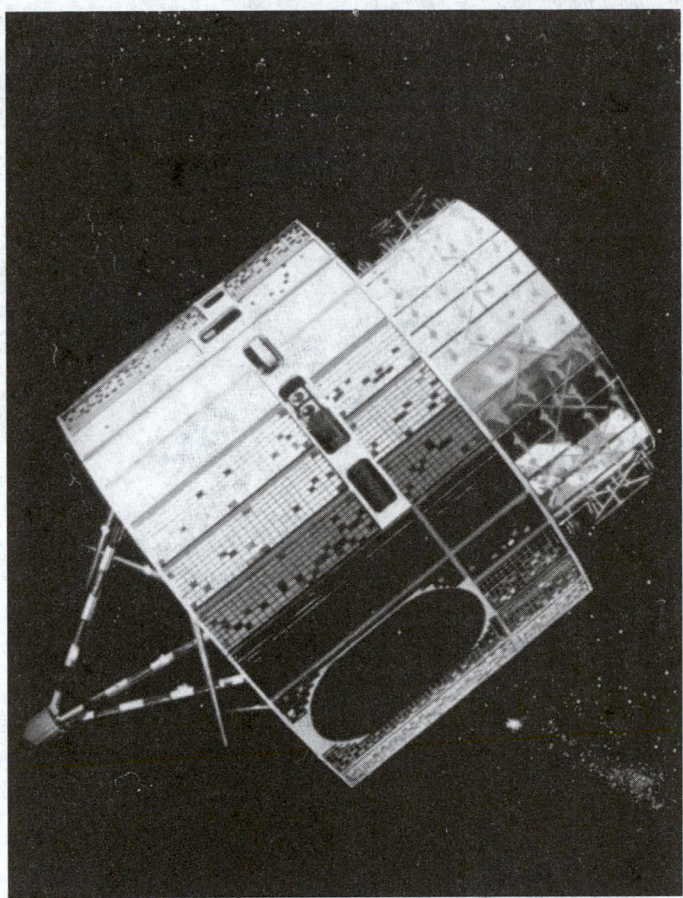

Fig. 5. Graphic of the Synchronous Meteorological Satellite, the forerunner of the GOES satellites. SMS-1 viewed the Eastern US while its sister spacecraft SMS-2 viewed the Western US. (*courtesy of NOAA*).

SMS I was placed in a geostationary orbit directly over the equator at 45 W (over the central Atlantic). This location provided continuous coverage of the Central and Eastern U.S. and out over the Atlantic Ocean. SMS I remained operational until deactivated by NASA on January 21, 1981.

Visible Infrared Spin-Scan Radiometer. The Visible Infrared Spin-Scan Radiometer (VISSR) flown on SMS-1 provided day/night observations of cloud cover and earth/cloud radiance temperature measurements from a synchronous, spin-stabilized, geostationary satellite for use in operational weather analysis and forecasting. The two-channel instrument was able to take both full and partial pictures of the earth's disk. The infrared channel (10.5 to 12.6 micrometers) and the visible channel (0.55 to 0.70 micrometer) used a common optics system. Incoming radiation was received by an elliptically shaped scan mirror and collected by a Ritchey-Chretien optical system. The scan mirror was set at a nominal angle of 45 deg to the VISSR optical axis, which was aligned parallel to the spin axis of the spacecraft. The spinning motion of the spacecraft (approximately 100 rpm) provided a west-to-east scan motion when the spin axis of the spacecraft was oriented parallel to the earth's axis. The latitudinal scan was accomplished by sequentially tilting the scanning mirror north to south at the completion of each spin. A full picture took 18.2 min to complete and about 2 min to retrace. During each scan, the field of view on the earth was swept by a linear array of eight visible-spectrum detectors, each with a ground resolution of 0.9 km at zero nadir angle. A mercury-cadmium-telluride detector sensed the infrared portion of the spectrum with a horizontal resolution of approximately 8 km at zero nadir angle. The infrared portion of the detector measured radiance temperatures between 180 and 315 deg K with a proposed sensitivity between 0.4 and 1.4 deg K. The VISSR output was digitized and transmitted to the National Oceanographic and Atmospheric Administration (NOAA) Command Data Acquisition Station (CDA), Wallops Island, Va. There the signal was fed into a "line stretcher," where it was stored and time-stretched for transmission back to the satellite at reduced bandwidth for re-broadcast to data utilization stations (DUS). The VISSR data were handled by NOAA and the majority of data were archived by the Satellite Data Service Division, National Climatic Center, NOAA, Washington, D.C. Limited amounts of research-oriented data were collected by NASA and are maintained at NSSDC.

Solar X-Ray Monitor. The X-ray counter was composed of a collimator, two ionization chambers, and two electrometers. A small angular aperture was chosen for the telescope collimator, which was mounted so that the declination of its axis could be controlled by ground command to ensure that the sun was viewed by the telescope once during every vehicle rotation. One ion chamber was filled with argon at 1 atm for detection of 1- to 8-A X-rays, and had a 1.27.E-4 m beryllium window to exclude X-rays of longer wavelengths. The other chamber was filled with xenon at 1.5

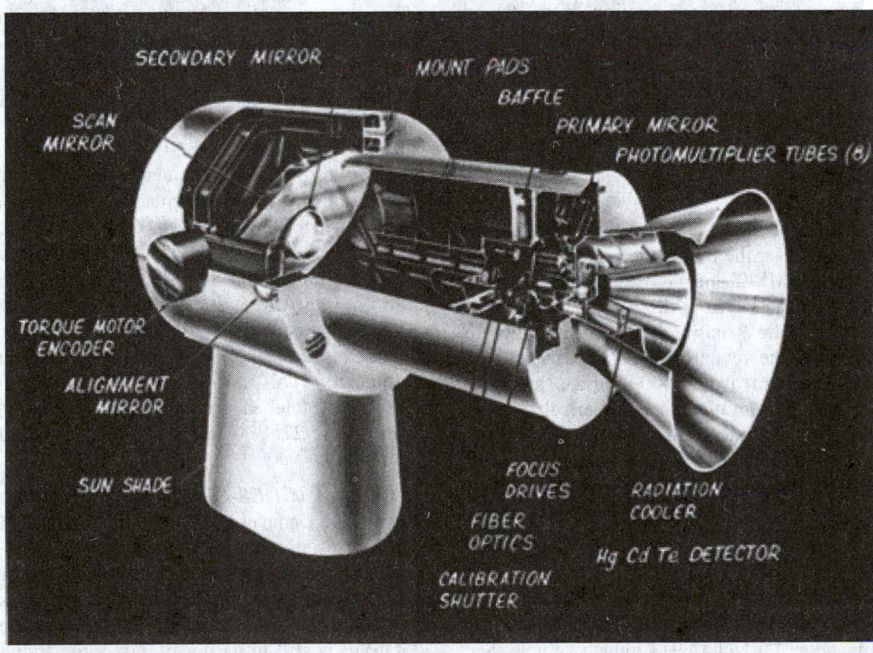

Fig. 6. Telescope onboard SMS-1 and 2, Visible Infrared Spin-Scan Radiometer, provides infrared and high-resolution visible photography. (*NASA*).

to 2 atm and had a 1.27 E-3 m beryllium window for measurements of X-rays in the wavelength range 0.5 to 3 A.

Energetic Particle Monitor. A number of separate silicon solid-state detectors, each with a tailored moderator thickness and a separate electronics unit for pulse amplification and pulse-height discrimination, were used to obtain the following particle type/energy measurements: seven channels measured protons in the range 1 to 500 MeV, six channels measured alpha particles in the range 4 to 400 MeV, and one channel measured electrons greater than 0.5 MeV.

Magnetic Field Monitor. A biaxial, short-boom-mounted (2-ft) closed-loop, fluxgate magnetometer was oriented with one axis along the S/C spin axis, and one in the spin plane. Each sensor had a selectable range (+50, 100, 200, or 400 nT), an offset field capability (plus or minus 1200 nT in 40-nT steps), and an inflight calibration capability.

Data Collection System (DCS). The meteorological data collection and transmission system was an experimental communications and data handling system designed to receive and process meteorological data collected from remotely located, earth-based, data collection (observation) platforms (DCP). The collected data were retransmitted from the satellite to small, ground-based, regional data utilization centers. Data from up to 10,000 DCP stations could be handled by the system. The system also allowed for the retransmission of narrow-band (WEFAX type) data to existing small ground-based APT receiving stations from a larger weather central facility. This communications system operated on S-band frequencies. The minimum data collection system for one SMS consisted of approximately 3500 DCP stations to be contacted in a 6-h period. The total amount of data collected during the 6-h period was between 350k and 600k bits, depending on the coding techniques. Data received from individual stations varied from 50 to 3000 bits, depending on the type and variety of sensors used at an individual DCP station.

SMS-2

Launch Date: February 6, 1975, from Cape Canaveral, FL, aboard a Delta 2914 launch vehicle. SMS-2 was the second operational satellite designed to sense meteorological conditions from a fixed location above the Earth, and to provide this data to operational forecasters and private interests on the ground. The satellite was designed to compliment SMS-1 and cover the Western U.S. and Pacific basin.

The principal instrument on board was the Visible Infrared Spin Scan Radiometer (VISSR) which provided day and night imagery of cloud conditions over the full-disk. The satellite had the capability to continuously monitor cataclysmic weather events such as hurricanes and typhoons, relay meteorological data from over 10,000 surface locations into a central processing center for incorporation into numerical weather prediction models, and to perform facsimile transmission of processed images and weather maps to WEFAX field stations. In addition, a Space Environment Monitor (SEM) and Data Collection System (DCS) similar to those on the NOAA polar orbiters were installed.

SMS-2 was placed in a geostationary orbit directly over the equator at 135 W (over the east-central Pacific) where it remained operational until deactivated by NASA on August 5, 1982.

GOES-1 (GOES-A and SMS-C). Launch Date/Time: October 16, 1975 at 22:40:00 UTC, from Cape Canaveral, FL, aboard a Delta-2914 launch vehicle. The satellite was to be placed over the Indian Ocean (west of SMS-2) so that the coverage of SMS-1, SMS-2 and GOES-1 would include nearly 60 percent of the Earth's surface. See SMS-1 for a description of instruments carried on GOES-1. The satellite had the capability to continuously monitor cataclysmic weather events such as hurricanes and typhoons, relay meteorological data from over 10,000 surface locations into a central processing center for incorporation into numerical weather prediction models, and to perform facsimile transmission of processed images and weather maps to WEFAX field stations.

GOES-1 was placed in a geostationary orbit directly over the equator over the Indian Ocean to gather data for the Global Atmospheric Research Program (GARP). The satellite was moved to replace SMS-2 (Pacific) when GOES-3 was launched. It remained operational over the equatorial Pacific until deactivated by NASA on March 7, 1985.

GOES-2 (GOES-B). Launch Date/Time: June 16, 1977 at 10:51:00 UTC, from Cape Canaveral, FL, aboard a Delta 2914 launch vehicle. See SMS-1 for a description of instruments carried on GOES-2. See Fig. 7.

Fig. 7.   GOES-2 Satellite before launch. (*NOAA*).

GOES-2 was placed in a geostationary orbit directly over the equator over 60 W in order to replace SMS-1. The WEFAX system on this satellite is still operational, although cloud images are no longer being received from this system.

After a long and distinguished career spanning almost 24 years, one of the nation's workhorse satellites was boosted into higher orbit and removed from service, announced the Commerce Department's National Oceanic and Atmospheric Administration. GOES-2 was operational as an imaging satellite until 1993, when it stopped giving imagery of cloud conditions across the United States. At that time GOES-2 was deactivated, but left in its old orbit position.

In 1995, it was re-activated to broadcast the Pan-Pacific Education and Communication Experiments by Satellite program administered by the University of Hawaii. PEACESAT is a public service satellite telecommunications network that links educational institutions, regional organizations, and governments in the Pacific Islands region.

From May 1 to 5, 2001, NOAA performed de-orbit maneuvers designed to boost the satellite into super-synchronous orbit, about 186 miles (300 kilometers) above the geosynchronous altitude of 22,300 miles (35,680 kilometers). This maneuver made room for another geosynchronous satellite to be launched. Because there are many satellites in geosynchronous orbit at the 22, 300 mile altitude, it is important to make room for new satellites.

GOES-3 (GOES-C). Launch Date/Time: June 16, 1978 at 10:49:00, from Cape Canaveral, FL, aboard a Delta 2914 launch vehicle. See SMS-1 for a description of instruments carried on GOES-3.

GOES 3 was placed in a geostationary orbit directly over the equator over the Indian Ocean to replace GOES-1 as part of Global Atmospheric Research Program GARP (GOES-1 was then moved to replace SMS-2). GOES-3 is still in its position over the Indian Ocean but is of limited use due to decaying parts.

GOES (Geostationary Operational Environmental Satellites)

Provision of timely global weather information, including advance warning of developing storms, is the primary function of the U.S. GOES meteorological program. To provide more complete data, a trio of Geostationary Operational Environmental Satellites known as GOES D, E, F (GOES-4, 5, and 6) was launched and operated by the National Oceanic and Atmospheric Administration (NOAA) as part of the Global Weather Watch. Two new satellites of a similar design, GOES G and H (GOES-7), were also built by Hughes Space and Communications Company at the Integrated Satellite Factory in El Segundo, California.

NOAA's earlier GOES spacecraft and their predecessors, the Synchronous Meteorological Satellites, employed a visible and infrared spin scan radiometer (VISSR) to produce two-dimensional imagery from the visible and infrared spectral regions. The VISSR senses radiation from Earth and its atmosphere. From the satellite's orbital altitude approximately 22,300 miles (36,000 km) above the equator, the VISSR transmits black and white, television-like images of one-third of Earth every 20 minutes.

Using GOES imagery, meteorologists were able to measure the frame-to-frame movement of selected clouds at different altitudes and to obtain their wind direction and speed in order to better understand atmospheric circulation patterns.

The GOES satellites also carried a space environment monitor (SEM) which investigated solar particle emissions and helped study the effect of solar activity on Earth's telecommunications systems. The SEM detected solar protons, alpha particles, solar electrons, solar X-rays, and magnetic fields.

A data collection system on GOES received and relayed environmental data sensed by widely dispersed surface platforms such as river and rain gauges, seismometers, tide gauges, buoys, ships and automatic weather stations. Platforms transmit sensor data to the satellite at regular intervals, upon interrogation by the satellite, or in an emergency alarm mode whenever a sensor receives information exceeding a preset level.

The GOES spacecraft transmitted meteorological imagery and data to the Wallops Island, Virginia, ground station, which relayed the data to the World Weather Building near Suitland, Maryland. The information was then processed and distributed to users throughout the United States. The GOES program was managed by the Office of Applications, NASA Headquarters, Washington, D.C. Project management of the GOES system was assigned to NASA's Goddard Space Flight Center, Greenbelt, Md.

GOES 4, 5, 6, G, and 7 [1980-1987]

Provision of timely global weather information, including advance warning of developing storms, is the primary function of the U.S. GOES meteorological program. To provide more complete data, a trio of Geostationary Operational Environmental Satellites known as GOES D, E, F was launched and operated by the National Oceanic and Atmospheric Administration (NOAA) as part of the Global Weather Watch. Two new satellites of a similar design, GOES G and H, were also built by Hughes Space and Communications Company at the Integrated Satellite Factory in El Segundo, California.

NOAA's earlier GOES spacecraft and their predecessors, the Synchronous Meteorological Satellites, employed a visible and infrared spin scan radiometer (VISSR) to produce two-dimensional imagery from the visible and infrared spectral regions. The VISSR senses radiation from Earth and its atmosphere. From the satellite's orbital altitude approximately 22,300 miles (36,000 km) above the equator, the VISSR transmits black and white, television-like images of one-third of Earth every 20 minutes.

GOES H, like other Hughes-built GOES, was equipped with an improved VISSR incorporating a visible and infrared atmospheric sounder (VAS). The VAS adds a vital third dimension to the imagery. Aboard the GOES, the VAS measured vertical temperature versus altitude cross sections of the atmosphere. From these cross sections the altitudes and temperatures of clouds were determined and a three-dimensional picture of their distribution was drawn for more accurate weather prediction.

GOES-4 (GOES-D). Launch Date/Time: September 9, 1980 at 22:27:00 UTC, from Cape Canaveral, FL, aboard a Delta 3914 launch vehicle. GOES-4 was the first geostationary satellite to provide continuous vertical profiles of atmospheric temperature and moisture.

The spacecraft was a cylinder 85 inches (216 cm) in diameter, 138 inches (351 cm) high and weighed 874 pounds (3296 kg). The sides

of the cylinder were covered by 15,000 solar cells which, along with nicad batteries, provided the power for the craft. Contained within, but protruding from the base was the primary instrument — the VAS (Visible Infrared Spin Scan Radiometer (VISSR) Atmospheric Sounder) and its sunshade. This instrument provided both day and night imagery of cloud conditions as well as temperature and moisture profiles over the full-disk. Unfortunately, the dwell times of sounder versus imager do not permit these two operations to occur simultaneously; however, soundings are still available on an hourly basis. See SMS-1 for a description of instruments carried on GOES-4.

The satellite also used new despun S-band and UHF antennas to improve the relay of meteorological data from over 10,000 surface locations into a central processing center for incorporation into numerical weather prediction models, and to perform facsimile transmission of processed images and weather maps to WEFAX field stations. In addition, a Space Environment Monitor (SEM) and Data Collection System (DCS) similar to those on the previous GOES were installed. GOES 4 was placed in a geostationary orbit directly over the equator over the Pacific (135 W). GOES 4 was deactivated November 22, 1988.

GOES-5 (GOES-E). Launch Date/Time: May 22, 1981 at 22:29:00 UTC, from Cape Canaveral, FL, aboard a Delta 3914 launch vehicle. GOES-5 was the seventh in a series of NASA-developed, NOAA-operated, geosynchronous, and operational spacecraft. GOES-5 was placed in a geostationary orbit directly over the equator over the western Atlantic (75 W). GOES-5 was deactivated July 18, 1990. The satellite was designed to sense meteorological conditions from a fixed location above the Earth, and to provide this data to operational forecasters and private interests on the ground. It was the second (designed to work in coordination with GOES-4) geostationary satellite to provide continuous vertical profiles of atmospheric temperature and moisture. See SMS-1 for a description of instruments carried on GOES-5.

GOES-6 (GOES-F). Launch Date: April 28, 1983, from Cape Canaveral, FL, aboard a Delta 3914 launch vehicle. GOES-6 was designed to replace GOES-4 and provide continuous vertical profiles of atmospheric temperature and moisture. See SMS-1 for a description of instruments carried on GOES-6.

GOES-6 was placed in a geostationary orbit directly over the equator over the Pacific (136 W) and was referred to as GOES-WEST. The VAS imager failed on January 21, 1989, so direct readout images and soundings are no longer available. It is, however, still acting as the west WEFAX relay satellite, although it's orbit is unstable.

GOES-G. Launch Date: May 3, 1986, from Cape Canaveral, FL, aboard a Delta 3914 launch vehicle. The spacecraft (to be named GOES-7, or GOES EAST) was lost in a Delta 3914 launch vehicle failure. The rocket was struck by lightning shortly after liftoff. GOES-G was designed to replace GOES-5 and provide continuous vertical profiles of atmospheric temperature and moisture.

GOES-7 (GOES-H). Launch Date/Time: February 26, 1987 at 23:05:00 UTC, from Cape Canaveral, FL, aboard a Delta 3914 launch vehicle. GOES-7 was the tenth in a series of NASA-developed, NOAA-operated, geosynchronous and operational spacecraft and was designed to replace GOES-5 in order to provide continuous vertical profiles of atmospheric temperature and moisture. See SMS-1 for a description of instruments carried on GOES-7.

GOES 7 was placed in a geostationary orbit directly over the equator over the Atlantic (75 W) and was referred to as GOES-EAST. The satellite is still operational; however, it has been moved several times to cover both the west and east coasts of the U.S. due to the failure of the imager on GOES-WEST. The current position (112 W) allows coverage of the US west coast, while the METEOSAT 3 geostationary satellite is currently being leased from the European Space Agency for coverage of the US east coast. Unfortunately, the spin of the GOES-7 satellite is no longer stable resulting in a 'figure-8' orbit which grows by 0.9 deg. latitude each year. After GOES-I was deployed in the Spring of 1994, as GOES-8 (GOES-East), GOES-7 remained in geostationary orbit, at 105 W, and is still used for satellite communications.

GOES (GOES-NEXT) 8, 9, 10, 11, and 12 [1994-2001]

Space Systems/Loral (SS/L) designed and built a series of five meteorological and environmental satellites to provide scientists with vital data to forecast weather, identify and track severe local storm conditions, gather data for meteorological research, monitor solar activity, and relay signals from ships and aircraft in distress.

Named GOES (Geostationary Operational Environmental Satellite) I through M, the spacecraft are being built for NASA, as manager of the program for NOAA. The first GOES in this series was launched in 1994 aboard an Atlas-1, the last was launched in 2001 on an Atlas-2A. The launches are staggered to keep two operational satellites on orbit at all times.

All five satellites are on station, delivering data of such high quality that it has surprised and gratified the scientists that depend on its information. They provide 24-hour monitoring and measurement of dynamic weather events in real-time, and are the first to deliver simultaneous independent imaging and sounding from geostationary orbit. GOES provides the extremely accurate image navigation capability (within 2.1 miles) necessary to successfully track local area severe storms. Over the course of the contract, signed in late 1985, many design improvements have been made to improve spacecraft performance and to allow an increase in size. Such efforts continue. New instrumentation has been developed and added to the payload.

On-board instruments include a new-generation Imager that operates in both visible and infrared spectra, and a Sounder which continuously measures vertical temperature profiles. The Space Environment Monitor includes a magnetometer, energetic-particle sensor, high energy proton and alpha-particle detector, and a solar X-ray sensor. The X-ray sensor regularly monitors solar flare activity that could cause high-level electron bombardment of people in space and satellites, affect the weather, and cause static.

This series of GOES spacecraft is based on SS/L's three-axis, body stabilized FS-1300 bus.

FS-1300 Bus. The 1300 was Space Systems/Loral's space-proven platform for a wide range of services. It underwent constant evolutionary development to deliver greater power, communications capability, and longer useful life. Total satellite power was from 5 to 12 kW, continuously throughout the life of the spacecraft. On-board transmitter power—exceeding 5,000 RF watts—could accommodate as many as 70 active transponders. Launch masses ranged up to 5,500 kg (12,125 lbs) and the spacecraft was equipped with a bipropellant propulsion system. The 1300S version provided up to 40 percent greater capacity than the 1300. Total satellite power ranged from 12 to 18 kW throughout the life

of the spacecraft. On-board transmitter power—approaching 10,000 RF watts—could accommodate as many as 90 active transponders. Launch masses up to 6,700 kilograms (14,771 lbs) were possible and the spacecraft was equipped with an ion station-keeping engine to supplement the bipropellant propulsion system. Some FS 1300 series satellites were retired after 17 years of successful service, more than twice their design life. With all of their features and power, satellites in the 1300 line will fit into a 5-meter (16 ft) launch-vehicle fairing.

GOES-8 (GOES-I, GOES-NEXT). Launch Date/Time: April 13, 1994 at 06:04:00 UTC, from Cape Canaveral, FL, aboard an Atlas-1/Centaur launch vehicle. GOES-8 is the 11th in a series of NASA-developed, NOAA-operated, geosynchronous and operational spacecraft. The triaxis-stabilized spacecraft carries (1) Imager and Sounder system to provide visible and infrared images of cloud cover, and to determine atmospheric temperature and water vapor content at various levels, (2) a meteorological data collection system to relay processed data from central weather facilities to regional stations equipped with APT and to collect and retransmit data from remotely located earth-based platforms, (3) a space environment monitor (SEM) system to measure proton, electron, and solar X-ray fluxes and magnetic fields, (4) a Search and Rescue (SAR) system to detect and relay distress calls from land and ocean, and (5) a WEFAX system to disseminate weather information to the user community via FAX. The cylindrically shaped spacecraft measures 190.5 cm (75 inches) in diameter and 230 cm (90.5 inches) in length, exclusive of a magnetometer that extends an additional 300 cm (118 inches) beyond the cylindrical shell. The imaging telescope is mounted on the equipment shelf and views the earth through a special aperture in the side of the spacecraft. The solar array of 1,057 W supplies two nickel-cadmium batteries of 12 Ampere-hour each. The CCSDS-compliant telemetry is in real-time at 2.0 kbs through S-bands. See Fig. 8. The eventual parking longitude of the spacecraft will be over 75 deg W. The SEC package (X-rays, H+, e−, monitors, and the Magnetometer) became in operational on or before April 2003.

GOES I/M Imager. The GOES Imager is an advanced Earth imaging system on the GOES-NEXT (GOES I to M) series of geostationary meteorological satellites. The Imager will monitor changing weather patterns over the United States and provide imagery throughout the 24-hour day to assist weather forecasters. The GOES Imager

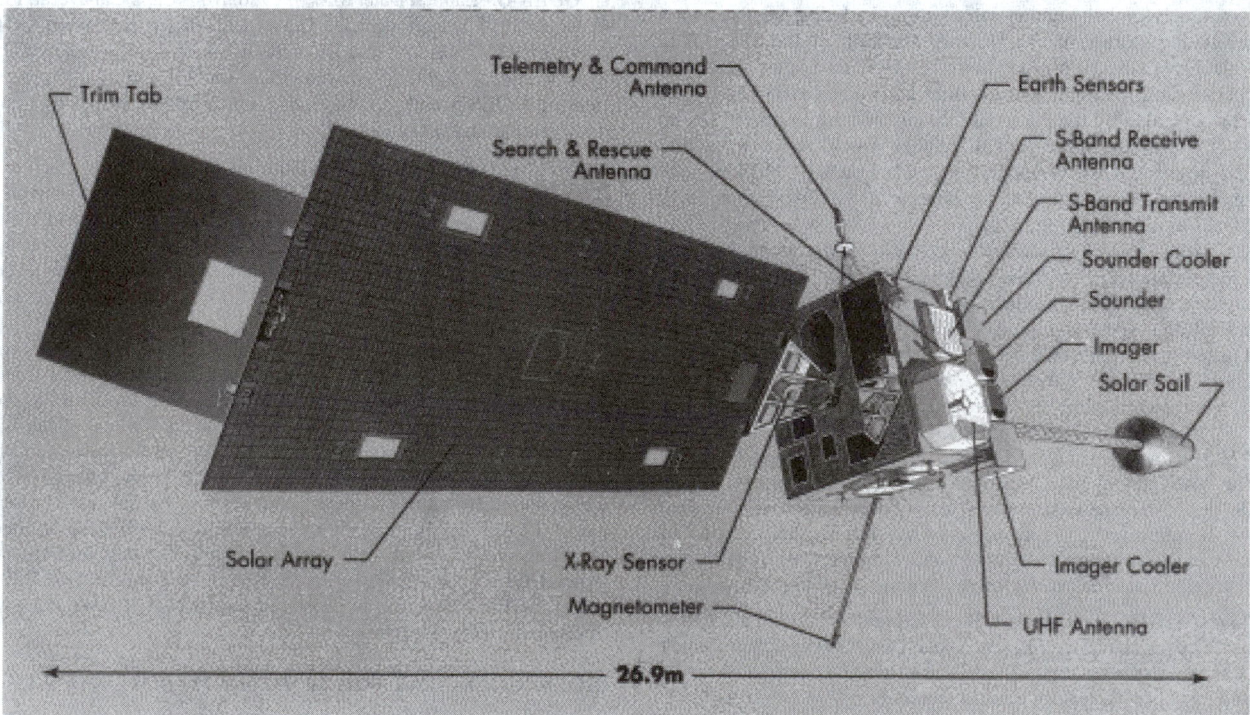

Fig. 8.   Artist rendering of GOES-8-12 satellite. (*courtesy of NASA*).

consists of 5 spectral channels: 1 visible and four infrared (including a water vapor channel). The visible channel (0.55–0.75 micron) has a resolution of 1 square kilometer at nadir. The infrared channels (3.9, 10.7, and 12.0 microns) have ground resolutions of 4 square kilometers. The water vapor channel (6.7 microns) has a ground resolution of 8 square kilometers. The Imager optical system splits the scene radiation into 5 spectral bands by dichroic beamsplitters and registers the channels to permit simultaneous viewing of the same scene. The Imager telescope is an f/12.25 Cassegrain with a focal length of 381 cm (150 inches). The four infrared channel detectors are located within a passive radiative cooler (which is thermally isolated from the rest of the sensor) and the visible detector is located on the aft optics subassembly. The visible detector is an 8 element array. Each IR channel has two active detectors for redundancy except the 6.7 micron channel which has one active and one redundant detector. The detectors convert the radiation into electrical signals which are directed through a preamplifier, a postamplifier, filters, and a 10-bit Analog-to-Digital Converter. All channels are sampled simultaneously by a sample and hold circuit and then stored in registers and read serially to the GOES data transmission system. All of the channels sweep a swath of 8 km (5 miles) as the mirror scans East to West or West to East. The GOES Imager uses an inductosyn as the servo error generator to compensate for spacecraft and orbit induced errors. The instrument channel-to-channel registration is within +/− 1 km (+/− 0.6 mi). The Earth rectangle swath from 60 North to 60 South is imaged in 25 minutes or less with three images possible in 1.25 hours. Any $3,000 \times 3,000$ km ($1,864 \times 1,864$ mile) area can be imaged within 5 minutes or less. Areas as small as $1,000 \times 1,000$ km (621×621 miles) can be obtained in two minutes or less. The GOES Imager will take no more then 3 minutes between the time a pixel is sensed and transmitted to users. The GOES Imager provides information on cloud imaging, snow and ice cover, low clouds and fog at night, cloud/snow discrimination, convection and synoptic analysis, cloud top temperatures, sea surface temperatures, and volcanic activity and forest fire detection. The GOES Imager produces images in either full disk, partial full disk or sectorized format in two basic formats: GOES projection for transmission on the GOES-TAP telephone line system and remapped imagery available (mid-1990's) to the National Weather Service (NWS) Advanced Weather Interactive Processing System (AWIPS-90). Most users will be able to obtain GOES-I images through GOES-TAP: http://goes.gsfc.nasa.gov/ams/goesinitial.html, Weather Facsimile (WEFAX) transmission: http://www.hffax.de/html/hauptteil_beginners_guide2.htmlGlobal, Telecommunications System (GTS): http://www.wmo.ch/web/www/TEM/gts.html, or through the NOAA/NESDIS Satellite Data Services Division: http://www.ssd.noaa.gov/

Energetic Particle Monitor. The energetic particle sensor (now called EP8 instead of EPS) consisted of three independent detectors: (1) EP8 Telescope, (2) Dome Assembly, and (3) High Energy Proton and Alpha Detector (HEPAD). EP8 telescope operated on the dE/dX — E mode, each of the detectors being a surface barrier semiconductor; pulse height analyzers could identify a particle either as a proton or as an alpha, besides binning them into narrower energy ranges. The Dome detector carried three separate windows of differing thicknesses, behind which lay a pair of 1500 micron thick surface barrier silicon detectors. Outputs from this three pairs of detectors passed through pulse height analyzers to provide counts in narrower bands. HEPAD is a Cerenkov counter, backed by pulse height analyzers. Over all, there were 11 energy channels for protons, eight for protons, and one for electrons of energy >2 MeV. However each such channel carried nontrivial contamination by other species. The counts from each of the 20 channels were accumulated for a few seconds (3 to 12 seconds, depending on the channel) before sampling the accumulated total for telemetry. There was also saturation limits to the level of accumulated counts, varying from 1,200 to 25,000 counts, depending upon the channel. The proton and Alpha channels covered the energy range of several hundred keV to several hundred MeV.

GOES I/M Sounder. The GOES Sounder is a 19-channel discrete-filter radiometer on the GOES-NEXT series of operational geostationary satellites (GOES-I through GOES-M). The GOES Sounder is designed to provide measurements that will allow the determination of surface skin temperature and the temperature and moisture profile of the atmosphere. The GOES Sounder will provide the National Weather Service (NWS)

with frequent updating of atmospheric stability and moisture content for severe storm forecasting applications and will also provide temperature and moisture soundings to be used as input to numerical weather prediction models. The key features of the GOES Sounder design are: (1) high sensitivity in each channel for high quality soundings; (2) a small instantaneous geometric field-of-view (IGFOV) for increased capability for clear column sounding; (3) multiple simultaneous samples and a high sampling rate; and (4) a full aperature blackbody calibration, space reference, and 13-bit quantization. The GOES Sounder measures radiation in four spectral bands: longwave, midwave, shortwave, and visible. Four detectors are simultaneously irradiated in each band providing output from four separate IGFOVs. The system samples four 8 km IGFOVs each 0.1 second through a rotating filter wheel providing data sampling at 10 steps per second. The scan system is totally digitally controlled permitting selective areas and time, providing high location accuracy, and increased dwell for sensitivity improvement. The scanner scans West-East in 10 km steps, dropping 40 km at the end of a line and returning East-West. The system allows stopping at one location for 0.1, 0.2, or 0.4 seconds in normal scan. A skip-line mode provides faster large area coverage. The scan control unit incorporates Inductosyn (as with the GOES Imager), which compensates for spacecraft and orbital motion. The Sounder consists of 7 longwave channels for temperature soundings (12.02-14.71 micrometers), 5 midwave channels for surface temperature, total ozone, and water vapor soundings (6.51-11.03 micrometers), 6 shortwave channels for temperature soundings and surface temperature (3.74-4.57 micrometers), and 1 visible channel for cloud detection at 0.969 micrometers. The Sounder location calibration system is aided by a star sensing capability at 0.8 micrometers for detecting 4th magnitude stars. The normal sounding area on the Earth is from 60 degrees N/S and 60 degrees E/W. A $3,000 \times 3,000$ km ($1,864 \times 1,864$ mile) area is sounded in 42 minutes. The maximum time between energy reaching the sensor and data transmitted to the ground is 30 seconds. The output products are similar to those provided by the HIRS/2 and VISSR Atmospheric Sounder (VAS) instruments on previous GOES satellites.

Solar X-ray Monitor. The X-ray monitor consisted of two ion chambers, mounted behind a slim rectangular field-of-view (48 deg × 3 deg) collimator made of lead-lined aluminum. The chamber for the lower wavelength band of 0.05 to 0.40 nanometer was filled with Xe-He mixture with an entry aperture made of 20 mil Be sheet. For the other band, 0.1-0.8 nm, the gas was Ar-He mixture and the aperture was a 2 mil Be. The threshold sensitivities were 1.0E-12 J per sq cm per s for the lower wavelength band, and 1.0E-11 J per sq cm per s for the higher band; each had a dynamic range of four decades. Entries of charged particles were prevented by the strong magnetic field located at the chamber windows.

GOES-8, Triaxial Fluxgate Magnetometer. The magnitude and direction of the magnetic field are measured by two redundant Schonstedt triaxial magnetometers located on a boom 3 m, and 2.7 m away from the spacecraft body. The electronics are located inside the body. The X, Y, Z component signals from the three axes are digitized by a 16-bit converter, at a sampling rate of 0.512 s. The sensitivity is 0.1 nT, and the range +/− 1000 nT. After temperature correction, and before (undescribed) stray-field correction, the accuracy is at about 1 nT level.

Space Environment Monitor. The Space Environment Monitor (SEM) System on the GOES-NEXT series of geostationary meteorological satellites (GOES-I through GOES-M) is designed to provide direct real-time measurement of solar activity. The SEM consists of a Magnetic Field Sensor, a Solar X-ray Sensor, and an Energetic Particle Sensor (formerly called EPS, now called EP8) / High Energy Proton and Alpha Detector (HEPAD). The Magnetic Field Sensor (MFS) allows for the real-time determination of the magnitude and orientation of the magnetic field. Data will be telemetered twice a second for magnetic fields having a magnitude of +/− 1000 nanotesla (nT). The Solar X-Ray Sensor permits real-time determination of the solar x-ray emission in two spectral bands: 0.5-5 angstroms and 1-8 angstroms. The EPS makes flux measurements of protons in the 0.8 to 500 MeV range. The HEPAD monitors protons in four energy ranges above 350 MeV and alpha particles in two energy ranges above 640 MeV/nucleon.

Search and Rescue Satellite Aided Tracking System (SAESAT). The Search and Rescue Satellite System (SARSAT) on the GOES-NEXT series of geostationary meteorological spacecraft receives transmissions from 406 MHz distress beacons and relays the signals to the ground terminal where the signals are processed to produce distress alerts. The SARSAT system is similar to the SARSAT system flown on the NOAA polar orbiting ATN series of spacecraft. Distress emergency beacon signals at 406 MHz are transmitted to the 406 MHz SARSAT receiver on GOES. The signal is then relayed from a 1544.5 MHz transmitter to a geosync ground processor, where the signals are decoded and sent to the USA/SARSAT mission control center for processing, identification, and subsequent action.

Data Collection System (DCS). The Data Collection System (DCS) on the GOES-NEXT series of operational geostationary meteorological spacecraft (GOES-I through GOES-M) is designed to collect and relay meteorological data from approximately 2000 Data Collection Platforms (DCPs). The DCPs are buoys, free-floating balloons, and remote weather stations which provide near real-time environmental data. The DCP information is used to provide warnings and forecasts of weather events such as tornadoes, tsunamis, and topical cyclones. The GOES DCS receives the signal from a DCP and then retransmits the data to the ground for processing. The DCS can also relay an interrogated signal from the ground station to the DCPs to activate the platform at specific times.

See also NOAA Satellite and Information Service for GOES-8 Images: http://hurricane.ncdc.noaa.gov/cgi-bin/hsei/hsei.pl and NASA http://goes. gsfc.nasa.gov/text/goes8results.html#eclipse.images.

GOES-9 (GOES-J, GOES-NEXT). Launch Date/Time: May 23, 1995 at 05:52:02 UTC, from Cape Canaveral, FL, aboard an Atlas-I/Centaur launch vehicle. The GOES 9 system performs the following basic functions: Acquisition, processing, and dissemination of imaging and sounding data. Acquisition and dissemination of Space Environment Monitor (SEM) data. Reception and relay of data from ground-based Data Collection Platforms (DCPs) that are situated in carefully selected urban and remote areas to the NOAA Command and Data Acquisition (CDA) station. Continuous relay of Weather Facsimile (WEFAX) and other data to users, independent of all other functions. Relay of distress signals from people, aircraft, or marine vessels to the search and rescue ground stations of the Search and Rescue Satellite Aided Tracking (SARSAT) system. Many small improvements were made to GOES-9 to correct problems discovered during the construction, launch and operation of GOES-8 during the previous years. The earth sensors on GOES-9 are less sensitive to sunglint. The GOES-9 spacecraft is operational as GOES-WEST at 135 W. See GOES-8 for satellite specifications and instruments flow on GOES-9. The first official GOES-9 image of Earth is shown in Fig. 9.

See also NOAA Satellite and Information Service for GOES-9 Images: http://hurricane.ncdc.noaa.gov/cgi-bin/hsei/hsei.pl and NASA http:// goes.gsfc.nasa.gov/text/goes9results.html.

GOES-10 (GOES-K, GOES-NEXT). Launch Date/Time: April 25, 1997 at 05:49:00 UTC, from Cape Canaveral, FL, aboard an Atlas-I/Centaur launch vehicle. The Geostationary Operational Environmental

Fig. 9. First official image of Earth taken by GOES-9. (*courtesy of NASA-GSFC*).

Satellite GOES-10 is the third satellite in a series of next generation geosynchronous spacecraft. See **GOES-8** for satellite specifications and instruments flow on GOES-10.

GOES-10 is located at 105 W longitude, and in post-launch testing of the instruments. After testing, NOAA-NESDIS shut GOES-10 down in June 1998, planning to keep it as an "on orbit spare" until GOES-8 or GOES-9 failed, which they expected around the year 2000 AD. However, right after NESDIS shut down GOES-10, (GOES-9's) momentum wheels threatened to fail, so GOES-10 was called right back out of storage in on 9 July 1998, for a week of spin-up and a month of orbital drift from 105 W to 135 W during August 1998 to take over as GOES-WEST.

As of February 2006, GOES-10 is still operating as GOES-WEST, but is running low on inclination-keeping fuel. See also http://goes.gsfc.nasa.gov/text/goeskstatus.html

After the launch of GOES-N (to be GOES-13) in 2006, NOAA plans to move GOES-10 to 60 W and devote more scheduled scan time to covering South America, as part of the Global Earth Observation System of Systems (GEOSS). See also **Global Earth Observation System Of Systems (GEOSS).**

See also NOAA Satellite and Information Service for GOES-10 Images: http://hurricane.ncdc.noaa.gov/cgi-bin/hsei/hsei.pl, and NASA http:// goes.gsfc.nasa.gov/text/goes10results.html.

GOES-11 (GOES-L, GOES-NEXT). Launch Date/Time: May 3, 2000 at 07:07:00 UTC, from Cape Canaveral, FL, aboard an Atlas 2A launch vehicle. The GOES-11 spacecraft is the primary replacement in the event of a failed operational spacecraft. In the event that GOES-10 or GOES-12 should fail or run out of fuel, GOES-11 could be activated and be made operational within 48 hours. At the present time GOES-11 is in Operational Storage mode at 114 W. See **GOES-8** for satellite specifications and instruments flow on GOES-11. The first GOES-11 infrared test images on June 10, 2000 are shown in Fig. 10.

GOES-12 (GOES-M, GOES-NEXT). Launch Date: July 23, 2001 at 3:23 EDT, from Cape Canaveral, FL, aboard an Atlas 2A launch vehicle. The satellite, GOES-12, will monitor hurricanes, severe thunderstorms, flash floods and other severe weather. It is the first of the GOES satellites equipped with a Solar X-ray Imager which will be used to forecast earth space weather due to solar activity. After undergoing a four-month checkout, GOES-12 was placed into on-orbit storage. Controllers at the NOAA Satellite Operations Control Center in Suitland, Md., (Fig. 11) on April 1, 2003 activated GOES-12 from an on-orbit storage mode replacing the older GOES-8 as GOES-EAST at 75 W. GOES-8 was launched April 13, 1994, to overlook the eastern part of the United States and well out into the Atlantic Ocean for nearly 10 years. GOES-10 is currently overlooking the West Coast, the Pacific Ocean and Hawaii — 22,300 miles over the equator.

Among the sophisticated instruments onboard GOES-12 is the world's most advanced solar storm detector, the Solar X-ray Imager (SXI). The Solar X-ray Imager will take a full-disk image of the Sun's atmosphere once every minute. The images will be used by NOAA and the U.S. Air Force to monitor and forecast solar flares, coronal mass ejections, coronal holes and active regions in the X-ray region of the electromagnetic spectrum from 6 to 60 Å (Angstroms). These features are the dominant sources of disturbances in space weather that lead to geomagnetic storms. The Imager will also examine flare properties, newly emerging active regions and X-ray bright points on the Sun. "The SXI will provide the kind of improvements in space weather forecasting that satellite imagery did for tracking hurricanes," said Steven Hill, SXI Program Manager at NOAA's Space Environment Center in Boulder, CO. The information will also help pinpoint when solar activity might harm billions of dollars worth of commercial and government assets in space and on land. On September 7, 2001 the GOES-12 Solar X-ray Imager took it first official image. See Fig. 12. The SXI will fly on future GOES satellites.

Solar X-ray Imager. The Solar X-ray Imager was added as part of the Space Environment Monitor suite of instruments on the GOES weather satellites. See GOES-8 for suite of instruments flow on GOES-12 in addition to the Solar X-ray Imager.

The imager was developed, tested, and calibrated by NASA's Marshall Space Flight Center in Huntsville, Ala., in conjunction with the NASA Goddard Space Flight Center in Greenbelt, Md., NOAA, and the Air Force.

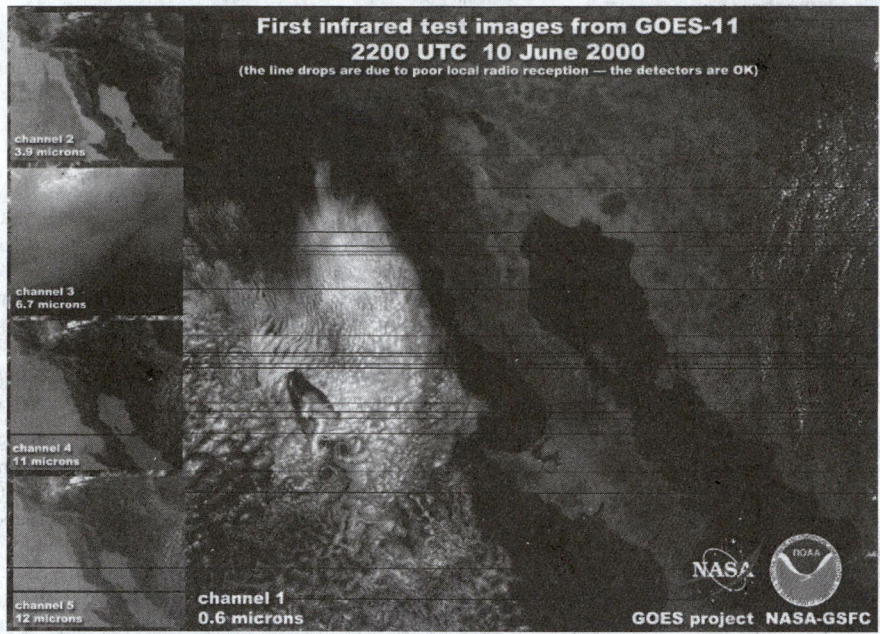

Fig. 10. First GOES-11 images. A weak broadcast signal caused static and line-drops in the reception at GSFC. (*courtesy of NASA-Goddard Space Flight Center, data from NOAA GOES*).

Fig. 11. NOAA Satellite Operations Control Center in Suitland, MD. (*photo courtesy of NOAA*).

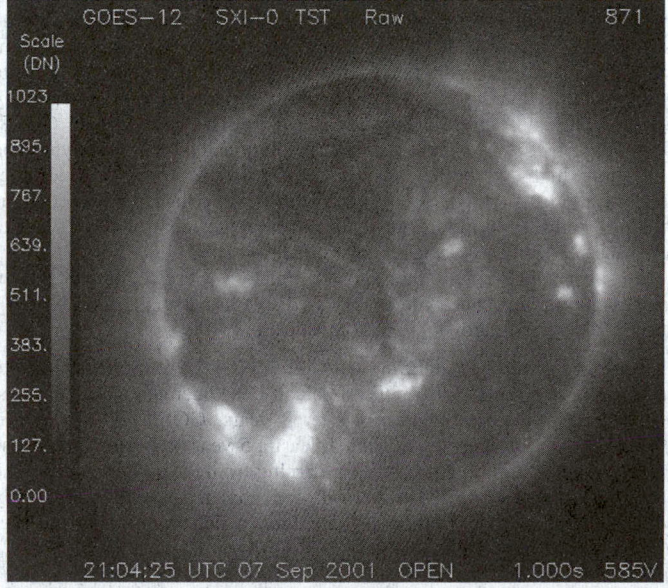

Fig. 12. First released image from the GOES-12 SXI on September 7, 2001 (unprocessed, false color). (*image courtesy of NOAA*).

The imager instrument consists of a telescope assembly with a 6.3-inch (16-centimeter) diameter gazing incidence mirror and a detector system. See Fig. 13. Incoming X-rays graze the mirror's surface at very shallow angles and are brought to a focus on the detector system.

As long as the grazing angles are very shallow, about one degree, the X-rays do not penetrate the surface but are reflected, just like visible light. The detector system contains a micro-channel plate which converts the X-rays to visible light which is then recorded using a CCD camera. Resulting data are electronically packaged for transfer to NOAA ground stations in Suitland, Md., and Boulder, CO. The images are processed and distributed to space weather forecast centers by the NOAA Space Environment Center in Boulder. The images are made immediately available to the public via the World Wide Web by the NOAA National Geophysical Data Center, also in Boulder.

The imager will provide continuous, near real-time observation of the Sun's corona. The images cover a 42 arc-minute field of view with five are-second pixels. The Sun, as viewed from Earth, is approximately 32 arc-minutes in diameter.

By recording solar images every minute, NOAA observers will be able to detect and locate the occurrence of solar flares. See also **Solar Flares**; and

Sun (The). This is the name given to explosive releases of vast amounts of magnetic energy in the solar atmosphere. Since scientists are not yet able to predict the occurrence, magnitude or location of solar flares, it is necessary to continually observe the Sun to know when they are happening.

When a flare erupts, it throws out large clouds of ionized, or electrically charged, gas. A small fraction of the cloud is very energetic and can reach the Earth within a few minutes to hours of the flare being observed. These energetic particles pose a hazard to both astronauts and spacecraft.

Coronal mass ejections, which are often associated with flares, take several days to reach the Earth. Fast, powerful ejections give rise to geomagnetic storms, which can disrupt radio transmissions and induce large currents in power transmission lines and oil pipelines. They have resulted in large-scale failures of the North American power grid and greatly increased pipeline erosion. SXI also will monitor coronal holes — persistent sources of high-speed solar wind. See also **Solar Wind**. As the Sun rotates every 27 days, these sources spray across the Earth like a lawn sprinkler and cause recurring geomagnetic storms.

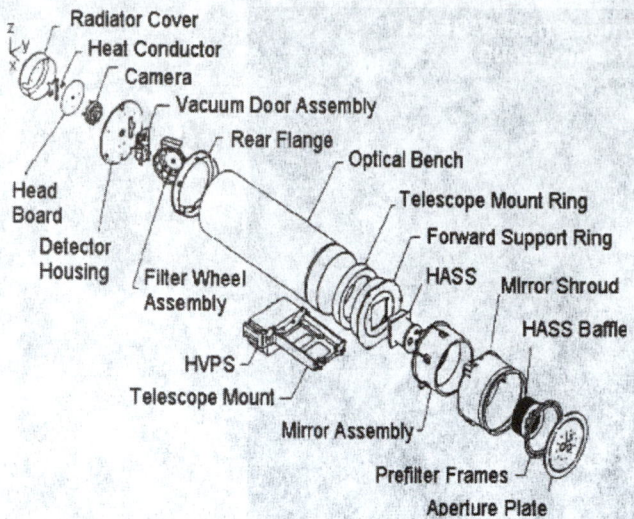

Fig. 13. Solar X-ray Imager telescope assembly. (*courtesy of NOAA*).

For more information about NOAA's Space Environment Center and the SXI visit: Space Environment Center: http://sec.noaa.gov/ SXI Home Page: http://sec.noaa.gov/sxi/ Today's Space Weather: http://sec.noaa.gov/today.html Solar-Terrestrial Physics: http://www.ngdc.noaa.gov/.

The new SXI instrument, designed and built at the Lockheed Martin [NYSE: LMT] Space Systems Advanced Technology Center (ATC) for the NASA Goddard Space Flight Center (GSFC) in Greenbelt, Md., is awaiting launch on the NOAA GOES-N (GOES-13) spacecraft from the Cape Canaveral Air Force Station, Fla. SXI is one of a suite of instruments that resides on the current generation of Geostationary Operational Environmental Satellites (GOES). The new SXI on GOES-N has a factor of two greater spatial resolution than the prototype, and like some high-end home video cameras, it has active internal jitter compensation that provides a stable picture even when the spacecraft is moving. Additionally, more sophisticated computer control allows SXI to react automatically to changing solar conditions.

See also NOAA Satellite and Information Service for GOES-10 Images: http://hurricane.ncdc.noaa.gov/cgi-bin/hsei/hsei.pl, and NASA http://goes.gsfc.nasa.gov/text/goes12results.html.

Summary. GOES-I (GOES-8) and GOES-J (GOES-9) were expected to have a 3 year on-orbit life. GOES-8 actually has operated over 8.5 years, but GOES-9 had to be shutdown after 3 years. GOES-K/L/M (GOES-10,11, and 12) are expected to have at least a 5 year on-orbit life, because the later models are more robust after learning from the earlier models. The heavy-launcher used on GOES-11/-12 delivered the satellites efficiently with about 10-years of station-keeping fuel, which is normally the life-limiting factor for GOES satellites.

Construction and launch dates for GOES-I/M have been arranged to assure two-satellite operations while minimizing storage costs by using "free" on-orbit storage. GOES-12 was activated as GOES-EAST in the spring of 2003, before GOES-11, in order to use the unique Solar X-ray Imager (SXI) on GOES-12.

Next Generation Weather Satellites (GOES-N/-13, GOES-O/-14 and GOES-P/-15)

In January 1998, The Boeing Company of El Segundo, California was awarded a contract from NASA's Goddard Space Flight Center in Greenbelt, Md. The contract currently includes the design, manufacture, integration and launch of three Geostationary Operational Environmental Satellites, GOES N, GOES O and GOES P. Upon completion of N through P, the company will have built a total of eight spacecraft in the GOES series.

GOES N-P represents the next generation of GOES satellites. Based on the highly successful Boeing 601 spacecraft, the new satellites will provide more accurate location of severe storms and other weather phenomena, resulting in more precise warnings to the public. The three-axis modified Boeing 601 body-stabilized spacecraft design enables the primary sensors to "stare" at Earth and thus frequently image clouds,

monitor Earth's surface temperature, and sound Earth's atmosphere for its vertical temperature and water vapor distribution. Atmospheric phenomena can be tracked, ensuring real-time coverage of short-lived dynamic events, such as severe local storms and tropical hurricanes and cyclones, two types of meteorological events that directly affect public safety, property, and ultimately, economic health and development. NASA and NOAA have set a high standard of accuracy for GOES N-P, including data pixel location to two kilometers from geosynchronous orbit. GOES-N will be renamed GOES-13 after achieving orbit.

Boeing will furnish the communications subsystem with a search-and-rescue capability to detect distress signals from ships and airplanes, and will also furnish space environmental monitoring instruments and operator training. Ground station upgrades will be provided by Boeing's teammate Integral Systems Inc. Boeing will also integrate three government-furnished instruments: an Imager and Sounder built by ITT Industries, Inc., and a Solar X-Ray Imager built by Lockheed Martin.

Instruments. The Imager, developed by ITT SSD (ITT Space System Division), in Fort Wayne, Indiana, is an imaging radiometer that uses data obtained from its five channels to continuously produce images of the Earth's surface, oceans, severe storm development, cloud cover, cloud temperature and height, surface temperature, and water vapor.

The Sounder, also built by ITT Space Systems Division (SSD), provides meteorologists with a detailed description of conditions in the atmosphere at any time. It gathers data over an approximately circular area extending from 60 degrees north to 60 degrees south latitude, allowing meteorologists to deduce atmospheric temperature and moisture profiles, surface and cloud-top temperatures, and ozone distributions by mathematical analysis and by adding to data from the Imager.

The Space Environment Monitor (SEM) consists of three instrument groups: 1) an energetic particle sensor (EPS) package, 2) two magnetometer sensors, and 3) a solar x-ray sensor (XRS). The EPS, developed by Assurance Technology Corporation (ATC), (Formerly GE Panametrics Corp.) in Carlisle, Massachusetts measures the energetic particles at geosynchronous orbit, including protons, electrons, and alpha particles. The magnetometers, provided by Science Applications International Corporation (SAIC), Inc., can operate independently and simultaneously. They measure the magnitude and direction of the Earth's geomagnetic field, detect variations in the magnetic field near the spacecraft, provide alerts of solar wind shocks or sudden impulses that impact the magnetosphere, and assess the level of geomagnetic activity. The XRS is an X-ray telescope that observes and measures solar x-ray emissions in two ranges. In real-time, it measures the intensity and duration of solar flares in order to provide alerts and warnings of potential geophysical responses, such as changes in ionospheric conditions, which can disrupt radio communications and Global Positioning System (GPS) signals.

A new Solar X-Ray imager will monitor the sun's X-rays for the early detection of solar flares. This early warning is important because these solar flares affect not only the safety of humans in high-altitude missions, such as the Space Shuttle, but also military and commercial satellite communications. The GOES satellites also carry space environment monitoring instruments, built by Assurance Technology Corp., which monitor X-rays, extreme ultraviolet and particle emissions including solar protons, alpha particles, and electrons. These space environment monitoring instruments also include a magnetometer, built by Science Applications International Corporation (SAIC), which samples the Earth's magnetosphere.

A data collection system on GOES receives and relays environmental data sensed by widely dispersed surface platforms such as river and rain gauges, seismometers, tide gauges, buoys, ships, and automatic weather stations. Platforms transmit sensor data to the satellite at regular or self-timed intervals, upon interrogation by the satellite, or in an emergency alarm mode whenever a sensor receives information exceeding a preset level.

GOES-N (GOES-13) Status. GOES-N will be launched from Cape Canaveral Air Force Station Space Launch Complex (SLC) 37B on a Boeing Delta IV (4,2) using a 4 meter fairing common booster core configuration, with two solid strap on motors.

In the spring of 2004, the GOES-N spacecraft completed construction, underwent thermal-vac testing, and was ready to ship to the Cape for launch. Delays with Boeing's tests of heavy-Delta IV rockets delayed shipment until March 2005.

Launch Schedule. GOES-N was originally scheduled for launch in 2001, but construction and launch were extended to 2003, since the on-orbit satellites were working well. In 2003, the GOES-N was scheduled for launch in December 2004, and then January 2005.

In mid-February 2005, the GOES-N launch date was reset to May 2005, to avoid the risk of launching during the spring eclipse season.

In April 2005, there was a concern about the Delta IV rocket, and so the GOES-N launch was postponed to June while some tanks on the rocket were replaced.

In mid-June, launch was slipped to late June to allow technicians time to check for possible damage to the Delta IV rocket's electrical systems from nearby lightning strikes.

In late June, there were uncertainties about a battery in the rocket, so the launch was postponed to July.

In late July, concerns about the satellite caused Boeing to slip-delay the launch several times to mid-August. On August 16th, launch was aborted with 4 minutes and 22 seconds to go.

Because launch slipped past mid-August, it was rescheduled for the first weekend in November, to avoid the risk of deployment during the autumnal eclipse season in geosynchronous orbit.

At the end of October 2005, Boeing's union voted to strike, putting the launch of GOES-N on hold until the union voted to accept a new contract at the beginning of February 2006.

As of mid-February, the first launch opportunity is 6 April 2006. The early date is a departure from GOES tradition of never launching during eclipse season, which is a month on either side of the equinoxes. To prepare for that, Boeing will take the satellite off the fueled rocket, rehab it and the rocket, and put them back together. See Fig. 14.

Fig. 14.   On Launch Complex 37 at Cape Canaveral Air Force Station in Florida, the GOES-N satellite is prepared for demating from its Boeing Delta IV rocket. Launch of the satellite was postponed in August 2005 due to technical issues and postponed to a later date. Due to the extended length of time the spacecraft has been atop the rocket without launching, the weather satellite is being returned to the Astrotech Space Operations payload processing facility for some precautionary retesting and state-of-health checks.

GOES-N will be launched on a Delta IV with two solids to nearly direct injection to geo-orbit. The use of the main rocket to get to orbit will save fuel on the spacecraft, and achieve at least 10 years of fuel lifetime (nominally 5 years of on orbit storage, and 5 years of operations).

The GOES-N spacecraft needs to have its folded solar panels facing outward towards the Sun during apogee-raising maneuvers around 1200 UTC on the other side of the Earth, which results in a Delta IV launch window around local sunset (2300 UTC) in Florida.

If GOES-N were stored on the Earth, it would have to be to be called out of storage to replace an on-orbit failure. There would be 9 to 12 months of preparation between call-up and launch, followed by 3 months of post-launch deployment and testing before it could become operational. On-orbit storage reduces this delay from one year to less than one week, and avoids the chance of a launch failure when you can least afford it.

GOES-O (GOES-14) Status. GOES-O is in ground storage at Boeing, in El Segundo, California and can be launched as early as July 2007, if needed by NOAA. GOES-O is nominally being planned for launch in April 2008.

GOES-P (GOES-15) Status. As of March 2006, GOES-P has completed vibration and acoustics testing and is being prepared for thermal vacuum testing. GOES-P will go into ground storage following the completion of environmental testing and should be prepared for an April 2008 launch readiness with an October 2009 planned launch date.

See also **Weather Satellites**; and **Weather Technology**.

Web References

GOES-News: http://goes.gsfc.nasa.gov/text/goesnew.html
GOES Office of Satellite Operations: http://www.oso.noaa.gov/goes/index.htm
GOES Program: http://www.osd.noaa.gov/GOES/index.htm
GOES Technical Notes: http://goes.gsfc.nasa.gov/text/goestechnotes.html

GEOSTATIONARY ORBIT (GEO). A geostationary orbit, often referred to as a GEO orbit, circles the Earth above the equator from west to east at a height of 36,000 km (22,369 miles). As it follows the Earth's rotation, which takes 23 hours 56 minutes and 4 seconds, satellites in a GEO orbit appear to be "stationary" over a fixed position. Their speed is about 3 km (1.9 miles) per second. As a result, an antenna can point in a fixed direction and maintain a link with the satellite.

As satellites in geostationary orbit continuously cover a large portion of the Earth, this makes it an ideal orbit for telecommunications or for monitoring continent-wide weather patterns and environmental conditions. It also decreases costs as ground stations do not need to track the satellite. A constellation of three equally spaced satellites can provide full coverage of the Earth, except for the polar regions.

Geostationary Transfer Orbit

This is an elliptical Earth orbit used to transfer a spacecraft from a low altitude orbit or flight trajectory to geostationary orbit. The apogee is at 36,000 km (22,369 miles). When a spacecraft reaches this point, its apogee kick motor is fired to inject it into geostationary orbit.

A worldwide network of operational geostationary satellites are used by meteorological satellites to provide visible, as well as infrared images of Earth's surface and atmosphere. These satellite systems include:

- the US GOES
- Meteosat, launched by the European Space Agency and operated by the European Weather Satellite Organization, EUMETSAT
- the Japanese GMS
- India's INSAT series

Most commercial communications satellites and television satellites operate in geostationary orbits. (Russian television satellites have used elliptical Molniya and Tundra orbits due to the high latitudes of the receiving audience.)

A statite, a hypothetical satellite that uses a solar sail to modify its orbit, could theoretically hold itself in a "geostationary" orbit with different altitude and/or inclination from the "traditional" equatorial geostationary orbit. However, this would rely on using the solar wind at high altitude outside the Earth's magnetosphere.

See also **Earth-Orbiting Satellites (Data Receiving And Handling Facilities)**; **Orbit (Astronomy)**; **Satellites (Communications and Navigation)**; and **Satellites (Scientific and Reconnaissance)**.

GEOSTROPHIC CURRENT.   A current in which the balance in the horizontal components of the equations of motion is between the horizontal pressure gradient and the Coriolis force. See also **Coriolis Force**. The vertical component is in hydrostatic balance and the pressure increases with depth in proportion to the mass of water above. If pressure is mapped on a level surface (geopotential), then geostrophic flow is parallel to the isobars, with high pressure to the right (left) of the flow in the Northern (Southern) Hemisphere. For the geostrophic balance to hold, the flow must be steady, very weak, large-scale, and friction-free.

GEOSTROPHIC WIND.   See **Winds and Air Movement**.

GEOSTROPHIC WIND SCALE.   A graphical device used for the determination of the speed of the geostrophic wind from the isobar or contour-line spacing on a synoptic chart.

It is a nomogram representing solutions of the geostrophic wind equation:

$$V_g = \frac{1}{f_\rho} \frac{\partial p}{\partial n},$$

where geopotential height is the vertical coordinate; or

$$V_g = \frac{g}{f} \frac{\partial z}{\partial n},$$

where atmospheric pressure is the vertical coordinate. In the above equations, V_g is the speed of the geostrophic wind, the density of the air, f the Coriolis parameter, p the pressure at a fixed geopotential height, z the height of a constant-pressure surface, and n horizontal distance measured normal to the flow. The n axis is directed to the right of the flow in the Northern Hemisphere and to the left of the flow in the Southern Hemisphere. In the nomogram, standard values of or g are usually adopted. The gradient of pressure or height is approximated by the finite difference ratio, $\Delta p / \Delta n$ or $\Delta z / \Delta n$, in which a standard difference in pressure or height is adopted; Δn then represents the normal distance between isobars or contour lines drawn. The nomogram often utilizes Δn as the abscissa and the latitude as ordinate, so that the speed of the geostrophic wind may be read from a family of lines of the graph.

See also **Coriolis Parameter**.

AMS

GEOSYNCHRONOUS ORBIT.   A satellite in geosynchronous orbit circles the earth once each day. The time it takes for a satellite to orbit the earth is called its period. For a satellite's orbit period to be one sidereal day, it must be approximately 35,786 kilometers (19,323 nautical miles or 22,241 statute miles) above the earth's surface. That is a lot higher than the Shuttle ever goes (usually about 300 kilometers (186 miles). We calculate this height using, what are today, common geometric formulas.

To stay over the same spot on earth, a geostationary satellite also has to be directly above the equator. Otherwise, from the earth the satellite would appear to move in a north-south line every day. We call that "orbiting in the equatorial plane."

The Shuttle's orbit is always inclined to the equator by at least 28.5 degrees. Given this and the Shuttle's relatively low orbit, getting a satellite from its deployment orbit to its final geosynchronous orbit takes an Inertial Upper Stage (IUS) for a boost, and something called a Hohmann transfer.

History

Author Arthur C. Clarke is credited with popularizing the notion of using a geostationary orbit for communications satellites. The orbit is also known as the *Clarke Orbit*. Together, the collection of artificial satellites in these orbits is known as the *Clarke Belt*.

The first communications satellite placed in a geosynchronous orbit was Syncom 2, launched in 1963. Geosynchronous orbits have been in common use ever since, including satellite television.

Initially, geostationary satellites also carried telephone calls but are no longer used so predominantly for voice communication, partly due to the inherent disconcerting delay in getting information to the satellite and back (it takes light or radio about a quarter of a second to make the round trip). Similarly, international Internet connectivity has shifted away from satellite links.

Nearly all land locations on the planet now have terrestrial communications facilities (microwave, fiber-optics), even undersea, with more than sufficient capacity. Satellite telephony is now mainly limited to small, isolated locations that have no terrestrial facilities, such as Canada's arctic islands, Antarctica, the far reaches of Alaska and Greenland, and ships at sea (via Inmarsat).

See also **Earth-Orbiting Satellites (Data Receiving And Handling Facilities)**; **Geostationary Orbit (GEO)**; **Orbit (Astronomy)**; **Satellites (Communications and Navigation)**; and **Satellites (Scientific and Reconnaissance)**.

GEOSYNCLINE.   Dana's definition of a geosyncline (1873) is a depression which has been produced by lateral compression and which is filled with sediments. Although Dana, in his original definition, suggested that subordinate ridges might be formed in the bottom of the geosyncline during its formation, it remained for Emile Haug, in his "*Traité de Géologie*," to emphasize these ridges (geanticlines) in relation to the tectonics of the Alps. According to L.W. Collet, "A geosyncline is situated between two continental masses and is destined to be filled with sediments, some of which are derived from the geanticlines which develop in it." According to R.M. Field, "A geosyncline originates in a continental block as a great trough, the locus for the accumulation of marine and terrestrial sediments, which are derived from concomitant geanticlines formed in or on the margins of the geosyncline." The geophysical and geological study of the great island arcs, such as the East Indies and West Indies, strongly intimates that the foredeeps in front of the arcs represent geosynclines that have not been filled with sediments while they were being formed. The pronounced deficiency of gravity associated with these foredeeps suggests great down buckle of the crustal or continental type of rocks called Sial into the more basic subcrustal couch called Sima. See also **Earthquakes**; **Seismology**; and **Plate Tectonics**.

GEOTECHNICAL ENGINEERING.   Simply defined Geotechnical Engineering is a branch of Civil Engineering concerned with the behavior of soils and the design of foundations.

Geotechnical engineering studies include subsurface investigations and laboratory testing, enabling the design of slopes, foundations for structures, embankments, shoring and bracing for deep excavations, earth retaining structures and other foundation systems.

Geotechnical engineering services include feasibility studies, preliminary and detailed geotechnical reports, compaction testing, foundation inspection during construction, final inspection reports, investigation of causes and repair of earth failure mechanisms, critique of proposed grading and foundation plans, assistance and presentation of findings to controlling agencies.

Earthquake Geotechnical Engineering studies include the evaluation of earthquake related distress, seismicity, site response analysis, study of the dynamic behavior of soils, and the evaluation of dynamic loads on slopes, retaining walls, earth structures and foundations.

In adverse soil conditions, Geotechnical Engineers utilize many soil improvement methods, including: shear pins, dewatering, underpinning, grouting and stabilization fill designs.

Engineering Geology is the science concerned with the application of geologic knowledge, in the investigation and evaluation of naturally occurring soils and rock, for use in the design of civil works.

Engineering Geology studies include review of geologic maps, reports and aerial photographs, geologic mapping, subsurface exploration, and monitoring of ground conditions, such as movement, by instrumentation.

Engineering Geology services include evaluation of parameters critical to slope stability, seismicity, fault investigations, ground water studies, professional geotechnical opinions and input for Environmental Impact Reports.

In adverse soil conditions, Engineering Geologists provide Geotechnical Engineers with pertinent geologic information for the design of soil improvement methods, including; shear pins, dewatering, underpinning, grouting and stabilization fill designs.

GEOTECTOCLINE. Term proposed by H.H. Hess and R.M. Field (1938) for the deformed prism of sediments in (of) the geosyncline. See also **Earth Tectonics and Earthquakes**.

GEOTHERMAL ENERGY. In the usual sense, geothermal energy is regarded as useful energy that can be extracted from naturally occurring steam and hot water found in the volcanic and young orogenic zones of the earth. Surface manifestations include hot springs, fumaroles, steam vents, and geysers. Such regions may exist without surface manifestation and astute geologists can forecast with some reliability where test bores may be made. Frequently these areas will be found close or relatively close to those areas where natural manifestations are present.

Until the last decade, geothermal energy sources were considered almost exclusively in terms of the kinds of natural phenomena just mentioned. These are the geothermal resources, such as Larderello (Italy), Wairakei (New Zealand), Geysers (California), and Reykjavik (Iceland), which have been successfully exploited for a number of years and whose outputs generally have been expanded in recent years. These regions are characterized by a unique combination of geologic and hydrologic features which brings a supply of water close to rock magma and which is capable of generating very large quantities of steam, hot water, or both. The specific characteristics of any given source range rather widely and thus generalizations are difficult to make.

Natural geothermally active zones are found in regions of frequent plate tectonic activity. Reference to maps in articles on **Earth Tectonics and Earthquakes** and on **Volcano** is suggested.

These are the familiar areas of geothermal activity and lie in those belts along the west coasts of North and South America, as far north as Alaska; then around the western Pacific to locations such as Kamchatka, Japan, the Philippines, Indonesia, and along into southern Asia and southern Europe. These are regions where seismic and volcanic activity are relatively common and where major and minor seismic events have usually occurred within the past few decades.

In these belt-situated areas, magma works close to the surface of the earth. The areas are characterized by crustal weakness. In these areas of crustal weakness, the normal geothermal gradient may be exceeded by a factor of ten or more.[1] At present drilling depths, even along these plate boundaries, there are only a relatively few locations where geothermal energy is sufficiently close to the surface have been located and/or exploited. It is also true, of course, that seismic activity is far from uniform along the plate boundaries, another factor which makes generalization difficult. It is suspected, however, that with much greater drilling depths, numerous additional geothermal energy sources could be located and exploited.

Within the past decade mainly, another category of geothermal energy source has been seriously considered by a number of researchers and long-range energy planners. This source is described as *hot dry rock* geothermal technology. It considers nearly all of the rocks that underlie the earth's surface, inasmuch as there is a geothermal gradient essentially universally present. This technology is described later in this entry. It is a recent technology and in its very early stages of investigation and development, compared with the conventional geothermal energy sources. In the long term, it could offer an extremely large and valuable source of energy.

Expanding upon the prior definition of thermal gradient, the rate of heat conduction outward from the interior of the earth to the surface is estimated to average about 1.5 calories per centimeter per second. One estimate indicates that, over a one-year period, this flux to the total surface of the earth amounts to over 10^{20} calories. Heat stored in rocks beneath the United States alone (to a depth of 10 kilometers) has been estimated to be on the order of 6×10^{24} calories. Other estimates are given later in this entry. In terms of current technology, however, these large numbers are not as exciting as they appear. It has been observed that, over the next few decades, to be practically retrievable, heat from the earth must

[1] According to Smithsonian tables, the rate of variation of temperature in soil and rock from the surface of the earth down to depths of the order of kilometers is, on the average, about $+10\,°C$ per kilometer. The thermal gradient varies greatly from place to place, depending on the geological history of the region, the thickness and strength of the crustal rocks, the conductivity of the upper rocks, and, in some regions, the radioactivity of underlying rocks. In terms of exploitation, the thermal gradient is of the utmost importance because it is directly related to drilling depth.

be concentrated in geothermal reservoirs, as previously mentioned, where the energy has accumulated and been in storage over long periods of time through geological processes.

Geothermal Energy in Italy

The use of geothermal energy for purposes other than the heating of bathing pools began in Italy in the late eighteenth and early nineteenth centuries near the present site of the Larderello field. Larderello, and the more recently exploited area, Mt. Amiata, are located on the west side of Italy, not far from Pisa. Steam from fumaroles and shallow bore holes was first used to aid the extraction of boric acid from the hot pools. That industry persisted for many years. In 1904, as a result of a dispute with the local electric utility, Prince Piero Conti, owner of the fields, decided to connect a generator to a steam engine driven by the natural steam. The success of that operation led to the installation of the first geothermal power plant, with a capacity of 250 kilowatts, installed in 1913. Increasing exploitation led to an installed capacity of approximately 385 megawatts.

Development of the Larderello field has been characterized by a lot of innovation, along with multipurpose utilization. Early developments were directed toward combining electric power production with the extraction of boron and other chemicals in the geothermal fluids. By using heat exchangers, a clean fluid could be used in the turbines. But, as the value of the chemicals declined, and as turbines were improved in construction to resist corrosion and abrasion, plants using the intermediate heat exchangers were replaced by direct-intake turbines. The direct intake turbines could be constructed at lower costs and, because there were no losses at the heat exchangers, more power could be produced per unit of steam.

Another innovation practiced in Italy has been the installation of relatively small (1.5 to 5 megawatt) back-pressure turbines that exhaust directly to the atmosphere. These are frequently used on individual wells very early in the development of new fields. Some of the advantages of using back-pressure turbines include: (1) they will handle steam containing large quantities of noncondensable gases, such as carbon dioxide, which sometimes exceed 30% (weight) of gases in a newly opened field. Thus, gas that has become concentrated over a long period of time in the upper part of the reservoir is released, and the ratio of noncondensable gases to steam is improved, to the point where it can be used in conventional condensing turbines. (2) Another advantage is that reservoir temperature-pressure-volume relationships can be determined by production testing, and reservoir life predictions can be made prior to commitment of funding for more extensive developments. Revenues obtained from the sale of electricity during the testing period, sometimes extending over 2 to 3 years, can make a significant return of exploration costs. In recent years, other geothermal reservoirs have been discovered south of Larderello.

Geothermal Energy in New Zealand

As early as 1932, scientists in New Zealand commenced investigation of thermal manifestations, such as hot springs and geysers, on North Island. It was not until 1948, however, that serious study began to appraise the geothermal resources, with the target of building a geothermal power station. By 1953, it was shown that the Wairakei area showed sufficient steam for construction of a power plant. The first Wairakei station was completed in 1958, and a second in 1963. New Zealand is well endowed with geothermal resources. A thermal area extends over a belt about 250 kilometers long and up to about 50 kilometers wide across the North Island between the central group of volcanic mountains (Mount Ruapehu, Mount Ngauruhoe, and Mount Tongariro) and the White Island volcano in the Bay of Plenty. See Fig. 1. Within this area is to be found a diversity of thermal activity—geysers, fumaroles, hot springs, and pools of boiling mud. Wairakei is one of several active areas where it is known that aquifers containing water up to and exceeding a temperature of $300\,°C$ exist. A view of the Wairakei Valley from the air is shown in Fig. 2.

Some 60 bores supply steam to the power station. Half of them are high-pressure bores producing steam at about 180 psi (12.2 atmospheres); others have intermediate pressure at about 80 psi (5.4 atmospheres). Well over 100 bores have been drilled, including those required for exploration.

Any extensive exploitation of underground water usually results in a slow decline in output from all bores, and those at Wairakei are no exception. Output is still tending to fall off, but power generation is not lessened. When pressure at the well head is lowered, a greater mass of

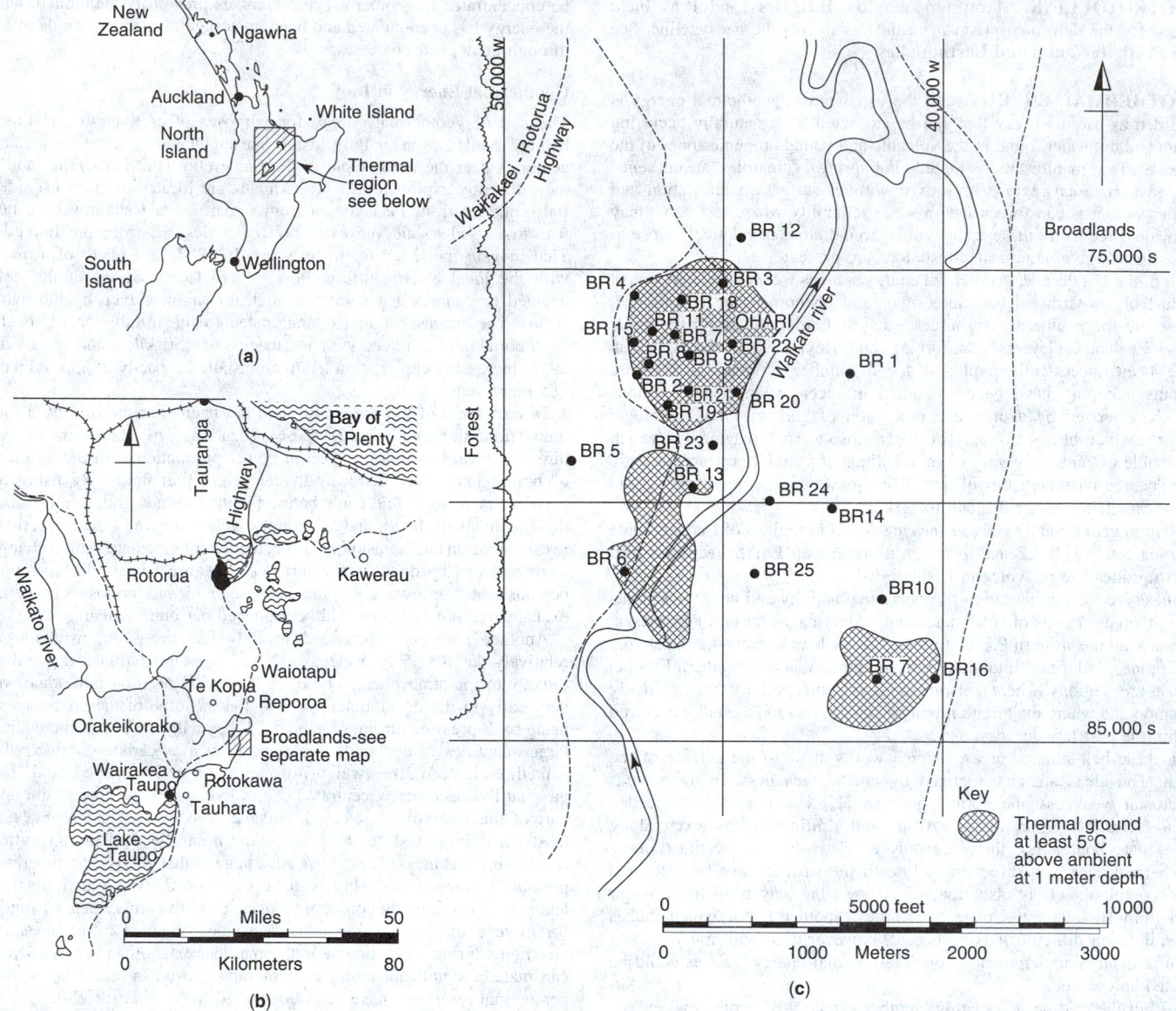

Fig. 1.   Geothermal energy in New Zealand: (**a**) North and South Island, showing thermal region; (**b**) thermal area; (**c**) Broadlands area. (*Ministry of Works and Development, Wellington North, New Zealand*.)

steam can be obtained. This, together with the use of hot water, has enabled power station output to be maintained. Some high-pressure bores with a very low level of output have been reduced as far as intermediate pressure and connected to the intermediate pressure system.

Scientists have built up a picture of underground conditions at Wairakei. It shows a vast area of hot fissured rock, thousands of feet deep and several miles wide, filled with water and the products of a million years or so of intense volcanic activity. Over a large part of this area, water slowly seeps down to the bottom, where it is heated. The source of this water is thought to be rain.

Thus, there is a large hot water reservoir, which is likely to be long lasting, but continued scientific vigilance and further intensive investigation will be needed. There is no evidence of direct interaction between widely spaced bores. Bores fed directly by large fissures do not interact even if fairly close together, but those depending mostly one permeable ground may do so. Tests of two bores 90 feet (27 meters) apart in porous ground have shown that they react to each other slightly, the output of one being about 10% higher with the other closed. But another two bores that are 60 feet (18 meters) apart and penetrate fissured formation do not affect each other. There is direct communication between the bottoms of yet other adjacent bores, but no reduction in output has been observed. General effects observed so far, over the whole steam field, are a fall in pressure and temperature at depth and subsidence of the ground surface over an

area of about $1\frac{1}{2}$ square miles (3.9 square kilometers). General indications, in spite of substantial decline in both pressure and temperature, are that full-scale production can continue for many years.

Hot water makes up about 80% (weight) of output from bores and is an important source of generating capacity although most of it is run to waste. Water at high temperature and high pressure boils and produces steam when the pressure is reduced. This process takes place in the steam bores and also in well separators. As hot water leaves the separator under pressure, it discharges to the silencer through a controlling orifice in the bypass pipe. See Figs. 3 and 4. As the pressure falls to near-atmospheric, large volumes of steam are generated. The water is led to waste and all the steam billows out from the top of the towers. This accounts for the waste steam, which puzzles many visitors and is the most noticeable feature of the whole steam field. What appears to be a great waste of energy, however, is in reality steam that cannot be used. When the pressure of hot water at the high-pressure wells is lowered suddenly, steam for the intermediate pressure system can be evolved. This presents a method of using some of the hot water.

The total length of the main steam lines at Wairakei is more than 12 miles (19 kilometers) mostly from 20 to 30 inches (51 to 76 centimeters) in diameter. There are many additional miles of branch lines.

The work at and near Wairakei has touched only a small part of New Zealand's geothermal resource, which is estimated to approach

Fig. 2. Wairakei Valley from the air.

2000 megawatts, of which only about 10% of that capacity is exploited for electricity generation. In addition to the generation of electricity at Wairakei, geothermal energy is used industrially at a pulp and paper mill at Kawerau. It has been used for many years for domestic and small-scale commercial and industrial use in the City of Rotorua, and other parts of the thermal region.

During recent years, exploration drilling has been carried out in several other geothermal fields. One of these, Broadlands, has been drilled up to a proven 150 megawatts. The other areas are Orakeikorako, Reporoa, Rotokawa, Tauhara, Te Kopia, Waiotapu, and Ngawha. With the exception of Ngawha in the extreme north, these areas all lie within the thermal region of the North Island.

All New Zealand geothermal fields so far drilled are classified as hot water fields. Formation temperatures and consequently the enthalpy of discharge vary from field to field. That at Wairakei is about 260°C; the highest temperature so far measured is 307°C at Broadlands. See Fig. 5. Generally, the chemistry of the fields is similar, although there are differences in detail. A common feature is a low total dissolved solid content of about 4,000 parts per million. This eases a number of problems in utilization found in other fields throughout the world.

Utilization of future fields probably will be for electric power generation, although other possibilities cannot be overlooked. Future developments will be quite different as compared with Wairakei. Current work, both in New Zealand and in other countries, on techniques, such as reinjection, chemical recovery, two-phase transmission (steam and water), and the binary cycle, will have marked effects on the appearance and efficiency of future schemes.

Geothermal Energy in the United States

The *total geothermal energy* resource base of the United States has been estimated by the U.S. Geological Survey (USGS) to be approximately 1.2×10^{21} Btu to a depth of 10 kilometers. This is sufficient to provide 1500 years of energy at present U.S. energy needs. Magma and hot rock resources, the most difficult to use, comprise about 85%; geopressured resources, about 14%; hydrothermal convection resources (natural steam and hot water), the only resource now in commercial use, account for the remaining 1%. Figure 6 shows the various types of geothermal resources. Approximately 24 GW of electricity-grade (>150°C) hydrothermal resources, with an expected 30-year life, have been specifically identified, located principally in the western United States. In addition to the 24 GW of identified resources, the USGS estimates 96 GW for 30 years of inferred hydrothermal resources.

Dry steam is the easiest resource to use. The United States has only three known dry steam reservoirs—located in Yellowstone and Lassen National Parks and The Geysers in northern California. Only the latter is commercially developed with some 1.4 GW of existing capacity and potential capacity estimated to be 2 to 3 GW for 30 years.

Geopressured resources are solutions of natural gas in hot water (lower than 210°C) trapped at high pressure under a sediment overburden. Magma resources result from molten igneous material that has intruded relatively close to the Earth's surface by geologically recent volcanic activity. Hot rock resources result from crystalline rock that is no longer molten. Geopressured, magma, and hot rock resources were in the early stages of research and development as of late 1980. The locations of these resources are generally known. Their potential has not been fully assessed and the

Fig. 3.   A typical wellhead set-up using a bottom outlet cyclone. The twin tower silencer to the left provides complete control over the water, as well as reducing noise to an acceptable level. The well is located in the foreground and is connected to the separator by the large sweep bend. The steam and water are separated by simple centrifugal action, the steam flowing from the cyclone, to the ball check vessel located at the left of the cyclone and thence to the steam mains through the branch line running out to the right of the photo. The tangential water outlet is connected to the water drum, and thence to the silencer.

Fig. 4.   A double flash unit. A single intermediate pressure separator is used, but because of the higher specific volume of steam at lower pressures, two intermediate low-pressure separators are required. These are the two taller vessels on the left. The two squat vessels are the water drums. Also, two silencers are necessary to cope with the amount of water finally discharged to waste.

technology required for using them reliably and economically has not been fully developed.

The Geysers, California. In the United States, the major geothermal development is at The Geysers, about 90 miles (145 kilometers) north of San Francisco. This development commenced in 1960 with a 12,500 kilowatt generating plant. Installed capacity is 1.4 GW, which makes it the largest geothermal development in the world as of the late 1980s. In Mexico, just south of the California border, a 150 megawatt geothermal plant is powered by hot water rather than steam. The hot water is derived from a zone that extends northward under the Imperial Valley and the Salton Sea. Until the technology for exploitation of hot dry rock

Fig. 5.   Well 20 Broadlands discharging 400 metric tons per hour at an enthalpy of 305 kcal per kilogram. Two silencers are necessary to provide adequate control of the water.

regions, as mentioned later, is developed, the Salton Sea region appears to be the only other geothermal region in the United States that is attractive from an electricity-generating standpoint. There are numerous other zones, however, where sufficient energy may be derived for local heating purposes (such as exploited in Iceland).

The geological situation at The Geysers is envisioned about as shown in Fig. 7. About 20 miles (32 kilometers) below the crust of the earth, a molten mass or magma is still in the process of cooling. In some places, earth tremors of the early Cenozoic era have caused fissures to open and the magma to come quite close to the surface. This process can cause active volcanoes and, where there is surface water, hot springs and geysers. The hot magma is also responsible for steam vents, like those found at The Geysers. The steam thrown off by cooling magma is called magmatic steam. Where surface water seeps down into porous rock heated by magma, the steam formed is called meteoritic steam, probably the biggest source of geothermal steam. Scientific investigators are still not entirely certain how the steam is formed at The Geysers. See map (Fig. 8).

At The Geysers, the early steam wells were drilled adjacent to the original nature steam vents (on 200- to 500-foot (61–152-meter) centers) to depths of 400 feet to 1,000 feet (122–305 meters). These wells produced steam flows in the range of 40,000 to 80,000 pounds (18,144–36,288 kilograms) per hour. Employing improved drilling techniques, wells are now deeper and tap into higher-pressure steam zones at depths between 2,000 and 7,000 feet (610–2,134 meters). Many of these deep wells are far removed from the natural steam outcroppings, and they produce considerably greater flows. One was tested at 380,000 pounds (172,368 kilograms) per hour.

The steam supplying a typical 53,000-kilowatt capacity unit is 36 inches outside diameter, $\frac{3}{8}$-inch (9.5-millimeter) wall carbon steel pipe. This would typically be connected to about seven producing wells. Centrifugal steam separators are installed in the steam pipes to remove any particulate matter and moisture. The steam contains about 1% noncondensable gases in the following approximate amounts: carbon dioxide, 0.79%; ammonia, 0.07%; methane, 0.05%; hydrogen sulfide, 0.05%; nitrogen and argon, 0.03%; and hydrogen, 0.01%.

The steam also contains powder-like dust which deposits out in protected areas of the turbines. This dust builds up on the inside of the turbine blade shrouds in the first two stages. In lower stages, the buildup appears to be washed away by water in the steam. This shroud buildup has caused blade and shroud failures. Earlier units have had heavier-duty replacement blades and shrouds installed to mitigate the problem. A turbine water-wash program also may improve the situation.

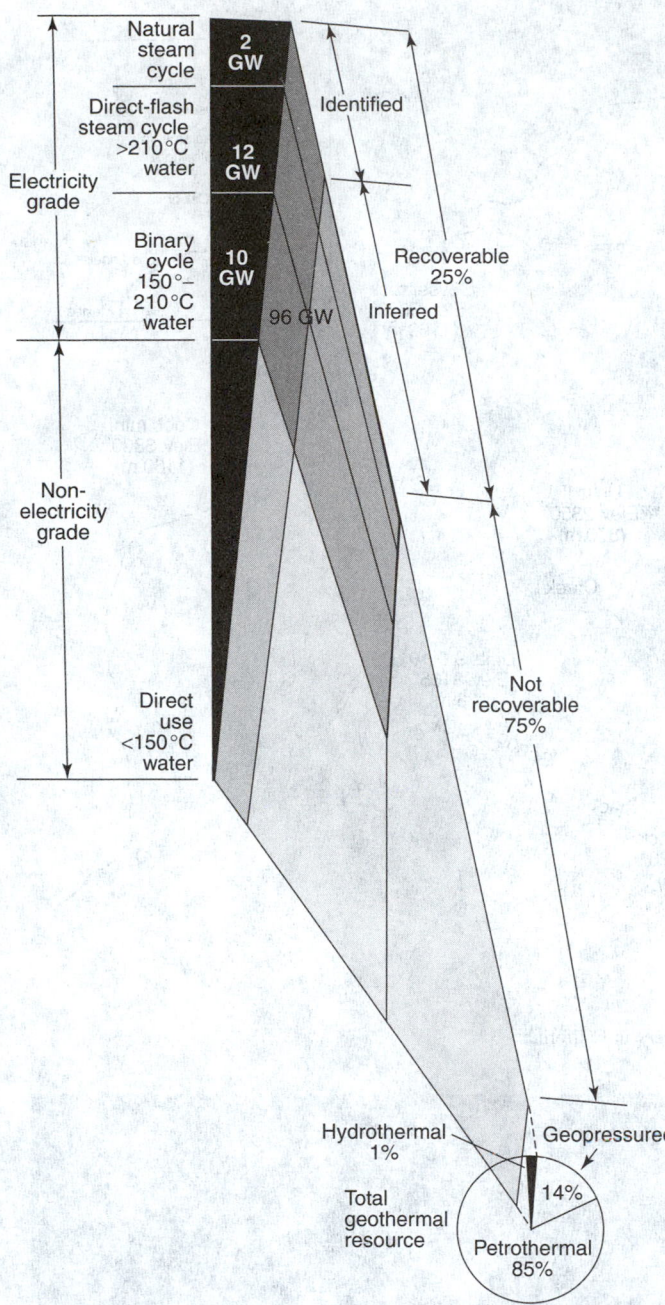

Fig. 6. Portrayal of geothermal energy resources of the United States. There are some 1.4 million quadrillion $t(1.2 \times 10^{21})$ Btu of geothermal resources to a depth of 10 km according to a U.S. Geological Survey estimate. (*Electric Power Research Institute.*)

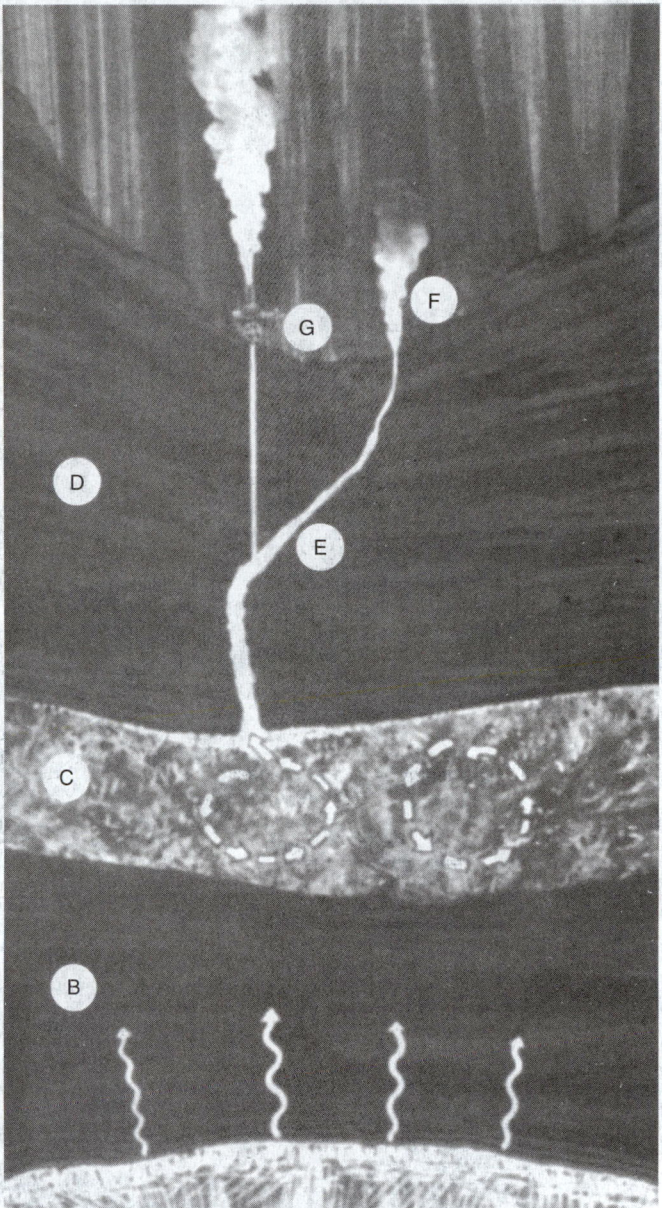

Fig. 7. Cross section of geothermal field envisioned at *The Geysers*. (A) Magma (molten mass, still in process of cooling); (B) solid rock, conducts heat upward; (C) porous rock, contains water that is boiled by heat from below; (D) solid rock, prevents steam from escaping; (E) fissure, allows steam to escape; (F) geyser, fumarole, or hot spring; (G) well, taps steam in fissure. (*Pacific Gas and Electric Co.*)

Hydrogen sulfide in the air causes serious problems in the electrical equipment because it is corrosive to copper, copper alloys, and silver. Tin alloy coatings have been found to resist corrosion effectively although they have not been satisfactory on current-carrying contact surfaces. Aluminum seems to be particularly impervious to attack, as are stainless steel and some of the precious metals. Platinum inserts or plating appear to be a good solution to the problem with contacts. Protective relays are particularly vulnerable to attack and special relays constructed with noncorrosive materials must be used. Also in the newer units, the relays, communication equipment switchgear, and generator excitation cubicle are placed in a clean-room environment. The multilevel room is maintained at slightly positive pressure with clean air from activated carbon filters.

Where two units are housed in one building, they share the same high-voltage transmission line, and have a common 480-volt station service bus. Any electrical faults that occur beyond either of the two generator

breakers requires that both units be tripped and, in addition, that the oil circuit breakers be opened.

The power cycle for all units is essentially similar. Steam from the wells is introduced into the turbines; it exhausts to direct-contact condensers located directly below the turbine. The combined condensed steam and cooling water are pumped by two condensate pumps to the cooling water tower. The turbine back pressure on all units is about 4 inches (100 millimeters) of mercury absolute. Cooled water from the tower basin is returned to the condenser by gravity and the vacuum head developed by the condenser. Since the cooling tower evaporation rate is less than the turbine steam flow, an excess of water is developed in the cycle. This flow is dependent upon the dry-bulb temperature and relative humidity, but there is a surplus under all operating conditions. For several years, this excess water from the units has been returned to the wells for reinjection into the steam reservoir. The reinjection method was tried initially with some concern over the effect that it might have in quenching the producing steam

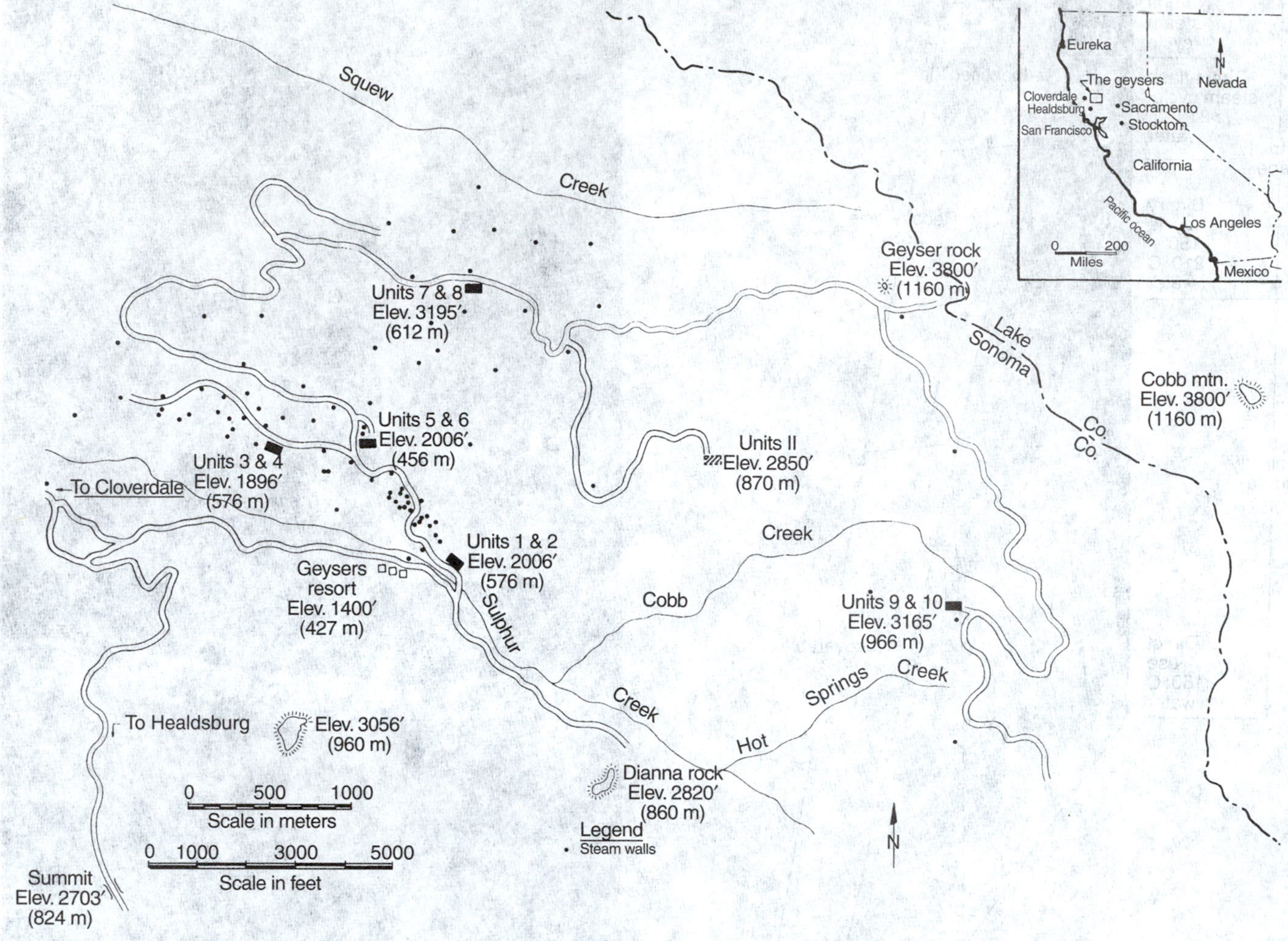

Units 7 & 8
Elev. 3195'
(612 m)

Squew

Creek

Geyser rock
Elev. 3800'
☼ (1160 m)

Units 5 & 6
Elev. 2006'.
(456 m)

Units 3 & 4
Elev. 1896'
(576 m)

To Cloverdale

Units II
Elev. 2850'
(870 m)

Cobb mtn.
Elev. 3800'
(1160 m)

Lake
Sonoma
Co.
Co.

Units 1 & 2
Elev. 2006'
(576 m)

Creek

Geysers
resort
Elev. 1400'
(427 m)

Cobb

Units 9 & 10
Elev. 3165'
(966 m)

Sulphur

Creek

Hot
Springs

Creek

To Healdsburg   Elev. 3056'
(960 m)

Creek

Dianna rock
Elev. 2820'
(860 m)

0   500   1000
Scale in meters

Legend
• Steam walls

N

0 1000   3000   5000
Scale in feet

Summit
Elev. 2703'
(824 m)

Eureka
The geysers
Cloverdale
Healdsburg
San Francisco
Sacramento
Stockton
Nevada
California
Pacific ocean
Los Angeles
Mexico
N
0   200
Miles

Fig. 8.   Area of *The Geysers* in California.

Fig. 9.   Hot water reservoirs serving the Reykjavik geothermal heating system. (*Photo by Mats Wibe Lund*.)

wells. However, it has proven successful. It is believed that reinjection can extend the productive life of the steam reservoirs since it is felt that there may be more heat in the reservoir than there is vapor to extract it. Two-stage steam-jet ejectors are used to purge the noncondensable gases from the turbine condenser. The condensers for these ejectors are also of the direct-contact design.

The steam turbines are fabricated largely of the manufacturers' standard materials for low-pressure, low-temperature service. Blades and nozzles are typically of 11–13% chrome steel. Carbon steel is used for the turbine casings. Austenitic stainless steel inserts are provided in the casings opposite the rotating blades to prevent moisture erosion of the casing.

Are the Geysers Winding Down? Once considered an excellent investment contemplating years of continuing electrical energy production, questions have been raised relatively recently concerning the long-range outlook for the installation. Perhaps, during the earlier planning stage, there was a gross miscalculation of the planet's ability to extract geothermal energy for a long period of years. It was reported in mid-1991 that the world's largest geothermal field may be rapidly running out of steam!

Analysis of the situation to date indicates that the field simply was developed at an overly accelerated rate and that its ultimate potential already may have been reached. Should this be the case, it is an exception among geothermal installations because most major installations worldwide have exceeded their original expectations.

Geothermal Energy in Iceland[2]

Geothermal energy for space heating is of great importance in some countries, notably Iceland, where about 85% of the population enjoys such heating for its homes. The geothermal fluids for such applications usually come from geothermal reservoirs at temperatures ranging from 60° to as high as 150°C (140–302°F). Thermal fluids within this temperature range occur at economically acceptable depths in Iceland and some other parts of the world.

Although geothermal space heating may serve a single house in the rural area, the most usual approach in Iceland is one of district heating services that serve whole population centers. As a rule, space heating by geothermal energy causes minimal pollution problems, inasmuch as there is no smoke and the warm effluents are distributed widely to the sewage system. In many areas where such systems have been installed, the cost of energy provided is very low when compared with fossil fuels. A depreciation time for equipment of from 20 to 30 years is usually used for economic evaluations.

Most distribution systems for hot water for space heating are single-pipe systems, which involve the discharge of the water to the sewage system after use. The distribution temperature of the water is preferably in the 80 to 90°C (176 to 194°F) range and will cool down to around 40°C (104°F) upon use. The supply mains to the distribution system will ordinarily discharge into storage tanks, which help in taking care of daily fluctuations in hot water load. See Fig. 9. Booster pumping is usually necessary in order to maintain sufficient pressure in the distribution system. The distribution network in towns is installed underground in the streets. Street mains larger than 3 inches in diameter may be placed in concrete channels and are insulated by rock wool or aerated concrete. The channels are embedded in a hard core, together with concrete drainpipes. Minimum inclination of these channels is kept at 5%. At street junctions, the channels may meet in concrete chambers, where valves, fastening bolts, and expansion joints are placed. These chambers are ventilated and they are drained from the bottom, or if that is not possible, they will have a pump pit. Smaller street mains and house connections from street mains may be insulated with polyurethane foam insulation.

A district heating system must be tailored to the local climate. The most important characteristics in this regard is the variation in daily outside temperature over the year. Since every heating arrangement for houses must ultimately have a capacity to provide comfort on the coldest day, obviously there must exist some overcapacity most of the time.

The ultimate cost of geothermal energy for space heating in such systems is usually nearly proportional to the maximum capacity required. Therefore, various approaches are used in order to increase the annual load factor.

[2] This section prepared by Baldur Lindal, **VBL Consulting Engineers**, Reykjavik, Iceland.

The latter term is defined as the ratio of total energy used to the basic design capacity. Some of the methods used include:

1. The system is designed for an outside temperature somewhat higher than that of the coldest day of the year, assuming the need for boosting from other sources for a few days each year.
2. The system may include a fossil-fuel booster intended for raising the temperature of the water during the coldest spells.
3. The system may include a local geothermal underground reservoir, where deep well pumps are installed in the drill holes. This arrangement may yield increased production for a limited time by pumping at a draw-down of the water level.

Generally, central heating systems are used for houses. The hot water is usually admitted directly to these systems and discharged to sewage after use. Hot domestic water for faucets is also supplied directly. Inferential water meters with a magnetic coupling between the flow sensor and register mechanism are frequently used. The maximum flow of hot water is also controlled by sealed maximum-flow regulators. Sometimes only maximum-flow regulators are used. When direct supply is not advisable, as in the case of water with high mineral content that would cause much scaling, heat exchangers may be used between the hot water and the water circulating in the central heating system.

In addition to use for house heating, most public buildings in Iceland are geothermally heated. A geothermally heated swimming pool in Reykjavik is shown in Fig. 10.

Fig. 10.   Geothermally heated swimming pool in Reykjavik, Iceland. (*Photo by Mats Wibe Lund*.)

Agricultural and Related Applications. Geothermal energy for heating greenhouses is important in Iceland and some other countries. Since the temperature of the heat source will vary greatly from one location to the next, as well as variations in heating requirements, the surface area of the radiator system (often consisting of bare pipes) must be carefully tailored to local conditions. Heating fluid temperature somewhat exceeding 100°C is used where steam is available. Small greenhouses may take advantage of heat in the effluent from ordinary space-heating systems. The most important crops of heated greenhouses of this type include cut flowers, tomatoes, cucumbers, and seedlings of many varieties. Animal husbandry, fish farming, and hatching stations also frequently take advantage of available geothermal hot water.

Process Heating. Since geothermal energy resources exist in a number of countries, it is of interest to point out some common factors that affect the viability of exploitations. Since the applications for heating in cold climates and the generation of electrical power are obvious uses and receiving considerable attention, it may be well to concentrate on the possibilities for process heating uses in any climate. Perhaps the three most important questions are: (1) What products may utilize the heat in geothermal fluids? (2) What are the potential savings or advantages as compared with competitive energy approaches? and (3) If there is a logistic disadvantage involved in site location, can this be offset by the lower-cost energy?

Because present technology is largely tailored to the use of fossil fuels, no conclusive answers can be sought directly from present engineering and economic practice. It may be helpful, however, to begin a search by studying the conventional processes which use fossil fuel-generated steam. And there also will be found cases where geothermal fluids may be used with an advantage even for cases where no steam is used in the conventional processing of today. Some examples include: (1) the use of indirect heating in a process instead of direct contact heating—for example, a steam-tube dryer instead of a direct-fired dryer; (2) there may exist a choice of several processes for any one specific objective. One process may permit the use of geothermal energy with a great advantage, while another may not require any heat, but entail other high-cost categories; and (3) the availability of geothermal energy may call for a completely new process. General application areas are given in Fig. 11 and in Table 1. The steam requirements and steam per unit product value for several products of the chemical and process industries are given in Table 2.

The economic importance of geothermal energy in a specific process may be judged by the share it has in the value of a product. This often can be roughly evaluated in terms of the steam or the amount of fossil fuel that would otherwise be required. The effect of a different design, and hence different investment, also enters into the calculation. Numerous cases are known where the equivalent share of thermal energy may be from 5 to 20% of the value of a product. Examples of existing and planned application of geothermal energy for process uses are given in Table 3. An industrial plant for processing diatomite, located in northern Iceland, is shown in Fig. 12.

The major industrial plants currently in operation have amply demonstrated that geothermal energy is a versatile source of energy. There are examples where process heating, space heating, and electric power production have been integrated into the same overall system. Because there is a large variation in geothermal energy sources, optimal utilization of that energy can be achieved only through individual analysis of each source. There are, however, a few generalizations:

1. When electric power is the main objective, there generally are ample opportunities for use of waste heat, at least in those plants using wet steam. In such instances, geothermal water may be rejected at elevated temperatures, which subsequently can be used for space heating, fresh water production, and some industrial applications. See Fig. 13 on p. 1608.

2. When space heating is the main objective, secondary electric power generation is possible in some cases. There are numerous secondary applications (greenhouses, soil warming, heating of swimming pools, etc.).

3. When process heating is the main objective, depending upon the geothermal source, some generation of needed electrical power may be possible and, as in the other cases, there usually is ample opportunity for secondary heating applications. See Fig. 14 on p. 1608.

Geological Aspects of Geothermal Systems

Within the last few years, a new concept of the outer few hundred kilometers of the earth has developed. This concept is embodied in the plate tectonic model of the earth. See also **Earth Tectonics and Earthquakes**. It is conceptualized that the surface of the earth, including the sea floor, is divided into several rigid plates that are moving relative to each other. The plates are composed of lithosphere which includes oceanic or continental crust or both, veneering and combined with the uppermost part of the mantle. Oceanic lithosphere is from 75 to 100 kilometers thick, while continental lithosphere is about 150 kilometers thick. Beneath the lithosphere lies the athenosphere.

The composition of the athenosphere is not known, but seismic data indicate that it is a zone of partial melting (upper portion) with several probable density transitions in the lower portion. Along the belt of oceanic ridges, the plates are moving apart at a rate of a few centimeters per year, causing gaps. New mantle material (*magma*) fills these gaps. In the direction of plate motion away from the ridges, plates must converge, one plate sinking or subducting beneath the other. Deep oceanic trenches

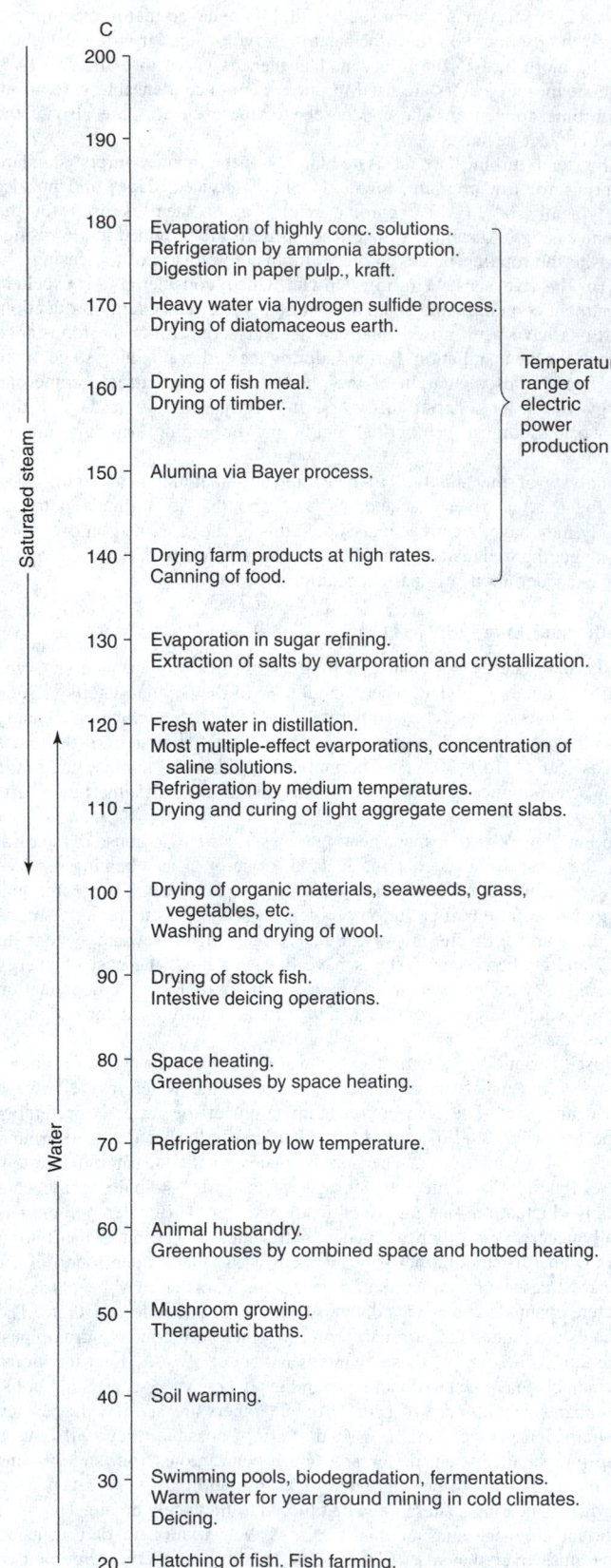

Fig. 11.   Applications versus temperature range of geothermal water and steam. (*Baldur Lindal.*)

form at these boundaries. Beyond the trenches, volcanic arcs are produced. These are accompanied by shallow to deep seismicity. Such boundaries are typified by Japan, Indonesia, Kamchatcka, the Aleutian peninsulas, and the Andes of South America. Where plates are converging, both

TABLE 1. EXAMPLES OF PROCESS DESIGN FEATURES FOR GEOTHERMAL STEAM AND WATER

Operation	Geothermal Steam		Geothermal Water	
	Type	Examples	Type	Examples
Drying	Indirect heating	Steam tube dryers Drum dryers	Indirect heating	Multideck conveyor dryer Preheaters
Evaporation	Primary heat exchangers accessible	Forced circulation evaporators	Counter-current heaters	
Distillation	Steam	General distillation	—	—
Refrigeration	Freezing	Ammonia absorption	Comfort cooling	Lithium bromide absorption
Deicing and Snow Melting	—	—	Direct application Indirect heating	Dredging and pavement deicing

TABLE 2. CONSUMPTION OF STEAM AND STEAM USED PER PRODUCT VALUE IN SOME ESTABLISHED FUEL-BASED PROCESSES

Product and Process	Steam Requirements Kilograms Steam/Kilogram Product
Heavy water by hydrogen sulfide process	10,000
Ascorbic acid	250
Viscose rayon	(70)
Lactose	40
Acetic acid from wood via Suida process	35
Ethyl alcohol from sulfite liquor	22
Ethyl alcohol from wood waste	19
Ethylene glycol via chlorohydrin	13
Casein	13
Ethylene oxide	11
Basic magnesium carbonate	9
35% hydrogen peroxide	9
85% hydrogen peroxide from 35% H_2O_2	$4\frac{3}{4}$
Solid caustic soda via diaphragm cells	8
Acetic acid from wood via solvent extraction	$7\frac{1}{2}$
Alumina via Bayer process	(7)
Ethyl alcohol from molasses	7
Beet sugar	$5\frac{3}{4}$
Sodium chlorate	$5\frac{1}{4}$
Kraft pulp	$4\frac{1}{5}$
Dissolving pulp	$4\frac{1}{5}$
Sulfite pulp	$3\frac{1}{5}$
Aluminum sulfate	$3\frac{1}{2}$
Synthetic ethyl alcohol from ethylene	3
Calcium hypochloride high–strength	$3\frac{1}{3}$
Acetic acid from wood via Othmer process	$2\frac{3}{4}$
Ammonium chloride	$2\frac{3}{4}$
Boric acid	$2\frac{1}{4}$
Soda ash via Solvay process	2
Cotton seed oil	2
Natural sodium sulfate	$1\frac{4}{5}$
Cane sugar refining	$1\frac{2}{5}$
Ammonium nitrate from ammonia	$1\frac{1}{2}$
Ammonium sulfate	$1\frac{1}{6}$

TABLE 3. EXAMPLES OF SOME EXISTING APPLICATIONS OF GEOTHERMAL ENERGY FOR PROCESS USE

Product	Country	Applications	Form of Geothermal Energy
Pulp and paper	New Zealand	Evaporating, digesting, drying	Primary and secondary steam
Timber drying and seasoning	New Zealand, Iceland	Drying, seasoning	Steam, hot water
Diatomite processing	Iceland	Drying, heating, diecing	Steam
Hay drying	Iceland	Drying	Hot water
Seaweed drying	Iceland	Drying	Hot water
Washing of wool	Iceland Russia	Heating and drying	Steam
Curing and drying of building material	Iceland	Heating and drying	Steam, hot water
Salt fish drying	Iceland	Drying	Hot water
Salt from geothermal brine	Iceland	Evaporation	Steam
Boric acid recovery	Italy	Evaporation	Steam
Brewing and distillation	Japan	Heating and evaporation	Steam

boundary between the *American Plate* and the *Pacific Plate*. Spreading of the plates is largely confined to the ocean ridges, which lie in the deep ocean. The Red Sea and the Gulf of California rifts probably developed only within the last few million years and deep ocean is yet to be attained. The great East African Rift is unusual in that separation is occurring within the continental lithosphere. Continued spreading of this rift may eventually split the African continent and produce new ocean floor. The driving mechanism for plate motion is not understood, but it appears to be associated with convective movement of the mantle. *The energy is supplied*, whatever the mechanism, *by the internal heat of the earth*.

It is along the spreading and converging plate boundaries that abnormal terrestrial heat flow occurs. Mass transfer of heat by magmas generated from the mantle brings heat to shallower levels of the crust. From these heat sources, geothermal systems are developed. All of the prospective high-enthalpy geothermal areas of the world are found within the belts of geologically young volcanism and crustal deformation produced by moving lithospheric plates.

Fundamental Geothermal Systems. These systems develop in the upper few kilometers of the earth's crust from a source of heat at some greater depth. The geothermal fluid, which contains dissolved minerals and salts, is heated and it becomes less dense. Where the overlying rock is permeable, a convection cell or system is created. For containment, a cover of impervious rock must overlie the system, thus preventing escape of the fluid to the surface. The thermal gradient is high in the covering rock and decreases rapidly within the upper part of the geothermal system where convection becomes pronounced. The temperature then varies little with depth; it is called the base temperature. This portion of the system constitutes the reservoir. Leaks from the reservoir to the surface are manifested by steam vents, hot springs, geysers, and fumaroles.

Vapor-Dominated Systems. In this type of system, saturated to slightly superheated steam (temperature about 250°C; pressures of 30 to 35 bars) is produced. The reservoir generally consists of highly fractured or porous rocks. Well flows may range from a few thousand kilograms per hour to over 250,000 kilograms per hour from depths ranging from 1,000 to 2,500 meters. Noncondensable gases in the steam may range from considerably less than 1% of the steam to 5% or more. Noncondensable gas content may be much higher initially, but diminishes with production and indicates past accumulation in the reservoir.

The hydrostatic pressure, abnormally low, in these reservoirs indicates they are sealed from groundwater infiltration. It is believed that they developed from high-temperature, liquid-dominated systems that seal their cooler margins through time by precipitation of dissolved material, mainly silica. Further, slow escape of water forms a steam space and a deep liquid phase, probably a very hot brine. Heat is received from a source beneath the system, probably a magmatic intrusion.

having a veneer of continental crust, the crust is less dense and cannot sink. Thrust faulting, folding, and thickening of the crust marks these boundaries. Examples are the Himalayan and Alpine mountain regions. At plate boundaries where neither spreading nor subduction is occurring, the plates slide past each other along great fractures which are called transform faults. The San Andreas Fault system is a prime example, as it connects the East Pacific Ridge, which enters the Gulf of California, to the Gorda Ridge, lying off the Oregon-California coast. This marks the

Fig. 12.   Diatomite processing plant operating on geothermal energy in northern Iceland. (*Photo by Mats Wibe Lund*.)

The steam fields at The Geysers, California, Larderello, Italy, and Matsukawa, Japan are typical examples of the vapor-dominated system. Reservoir characteristics are similar for all. The Geysers reservoir rocks are indurated, highly fractured graywacke sandstone and volcanic rocks. Porous limestone and dolomite are the reservoir rocks of the Larderello region, and fractured volcanic rocks serve as the reservoir at Matsukawa.

Liquid-Dominated Systems. These may be conveniently divided into two types: one having high enthalpy fluids above 200 calories per gram and one having low enthalpy fluids below this point. This division tends to separate fluids useful for generating electric power from those most useful for other purposes.

An important physical difference between the liquid and the vapor-dominated systems is the fact that the reservoir pressures in the liquid systems are near hydrostatic pressures, or around 0.1 bar per meter of depth. So at depths of 1,000 to 2,500 meters pressures are 100 to 250 bars, contrasting to the 30 to 35 bars in the vapor-dominated system.

High-enthalpy systems contain waters with dissolved solids ranging from around 2,000 ppm to as much as 260,000 ppm and temperatures of 200 °C to as high as 388 °C. The predominant anion of the dissolved solids is chloride along with lesser amounts of sulfate and carbonate. Sodium and potassium are the main cations with a smaller amount of calcium and sometimes magnesium. Up to 800 ppm of silica may be present which, along with several ppm of fluoride and several tens of ppm of boron, are troublesome in the disposal of these high enthalpy fluids.

Wells drilled into this type of reservoir produce a mixture of water and steam; the steam may be separated at a suitable pressure to operate a turbine. Noncondensable gas in the separated steam is usually below 1%.

The best-developed high-enthalpy liquid-dominated reservoir is located at Wairakei, New Zealand, where wells are drilled into a permeable pumiceous volcanic rock capped by an impermeable sedimentary formation. Temperature of the fluid is about 260 °C and about 20% is flashed to steam for power production. Another such system still being developed is the Cerro Prieto reservoir in the Mexicali Valley of Mexico north of the Gulf of California. Electric power is now being produced at Cerro Prieto from a fluid having a temperature of 300 °C or more and salinities of 15,000 to 25,000 ppm, in a reservoir of permeable sedimentary rock. To the north in the Imperial Valley of California, wells have been drilled to 2,500 meters into reservoirs with similar characteristics except

the Salton Sea reservoir which contains a concentrated brine having as much as 260,000 ppm total solids.

Low-enthalpy liquid-dominated systems have properties more variable than those known for the high enthalpy systems. In some the sulfate anion may be dominant and in others carbonate-bicarbonate. The salinities tend to be lower and some could be considered potable. Dissolved silica content, which is a function of temperature, is less and the toxic elements fluorine and boron also are generally diminished. Temperatures in low-enthalpy systems range from about 10 °C above average annual temperature to the previously mentioned arbitrary division at 200 °C.

Included in this category are low-enthalpy waters found in some deep sedimentary basins where the overlying rocks have a low conductivity. Temperatures may range from 50 to 60 °C to 120 °C, but the reservoirs are very large. The Hungarian basin and several in Russia are examples of this type. Along the Gulf Coast of the United States similar reservoirs exist within sands in undercompacted sediments. Temperatures above 200 °C, and pressures much above hydrostatic, have been reported. Deep wells, 2,000 meters or more in depth, are required to tap the thermal waters of these basins. Since no connection exists with young volcanism in these basins, heat is thought to be supplied by a slightly above normal terrestrial heat flow coupled with the insulating effect of the overlying sediments.

Iceland is perhaps best known for the many low-enthalpy reservoirs, which have been discovered and are being utilized. Numerous other reservoirs are known throughout the world, among which several are being utilized in the United States, notably in Oregon, Idaho, and California. In general, the close association of these reservoirs with young volcanism suggests magmatic heat as the source.

Exploration for Geothermal Energy Sources. The known higher-enthalpy geothermal systems or resources of the world are located where faulting has created uplift and subsidence of the crust with attendant mass transfer of heat from depth by magmas and geothermal convection systems. These activities are closely associated with geologically recent movement of the lithospheric plates.

The United States has a broad region covering the western conterminous states that has been distributed by recent interaction of the plates and changes in direction of their motions. Investigations over the last few years show large areas of this region to have above normal heat flow with numerous hot springs and wells.

Fig. 13.   Krafla geothermal power plant in northern Iceland. (*Photo by Mats Wibe Lund.*)

from a reservoir and drilling a deep test well at a spring may prove nonproductive. Also, some reservoirs may have little or no surface display. Therefore, geologic and geophysical methods must be used to enhance the chances of discovery.

Geologic studies can help to show the structure and stratigraphy that may outline domed areas, grabens, and calderas prospective for geothermal resources. Aerial and satellite photography and imagery are very important in the geologic investigations of such things as fault patterns and recent volcanism.

Of the many geophysical methods, measurement of the geothermal gradient and determination of heat flow from shallow drill holes is most valuable. Care must be used in extrapolating the data for greater depths, and groundwater migration can introduce serious discrepancies, but the method is direct in outlining thermal anomalies.

Gravimetric studies can indicate the presence of intrusive rock, which may be a heat source, or contrasting densities that may define a caldera or graben. Small gravity anomalies are associated with several known thermal anomalies in the Imperial Valley, California.

Rocks that are both hot and saturated with saline waters have very low electrical resistivities. Low resistivities are characteristic of high-temperature, liquid-dominated systems and electrical and electromagnetic methods are useful in the search for and delineation of their size. Practical results have been obtained on newly discovered reservoirs in the Imperial Valley and in New Zealand.

Passive seismic surveys, including ground noise, have been performed over a number of known geothermal reservoirs. One method involves the recording of microearthquakes, which many geothermal systems seem to generate. The activity probably arises from the highly faulted nature of the reservoirs and their association with regions of young tectonism. On the other hand, ground noise or geothermal noise surveys, record the acoustic signals within a narrow range of amplitude and frequency. Results so far suggest that individual geothermal systems produce characteristic signals and are related to reservoir depth and temperature gradients. If the reliability of this method can be demonstrated, it could become very important in geothermal exploration because of its simplicity and economy.

Active seismic surveys, generating and recording seismic waves produced by explosions or shock, are useful in determining subsurface structure and faults. Either the reflection or refraction method can be employed depending upon which is most suitable for a particular location and problem. Some recent work indicates that attenuation of seismic waves may

Surface displays of heat offer the simplest and easiest means of exploring for geothermal resources. Yet hot springs or geysers may be some distance

Fig. 14.   Svartsengi geothermal power and heating plant, Iceland. (*Photo by Mats Wibe Lund.*)

occur in geothermal systems. Further development of this procedure may increase the usefulness of active seismic investigations for geothermal resources.

Magnetic surveys involve measuring the magnetic properties of the underlying rocks. Positive magnetic anomalies often are associated with intrusive rocks and negative anomalies occur over rocks in which the magnetic minerals have been altered by geothermal fluids. A magnetic survey would thus seem to be useful for seeking geothermal reservoirs but so many complicating factors arise that results are generally very difficult to interpret.

Many geochemical and isotopic investigations have been performed on samples of spring waters and geothermal fluids throughout the world. As a result, certain constituents or ratios of these constituents may be used to indicate probable reservoir temperatures of liquid-dominated systems. Silica content and the sodium-potassium-calcium rates are the best indicators. High chloride content (above 50 ppm) in springs suggests that the system is liquid-dominated. Springs associated with vapor-dominated systems are said to contain less than 20 ppm chloride.

Isotopic analyses of hydrogen and oxygen in geothermal waters provide a means of determining the origin of the waters. It is now known that geothermal fluids are meteoric in origin and any volcanic or magmatic addition is minor. By this method, the hydrology of an area can be appraised concerning the recharge of water to a geothermal reservoir.

None of these exploration methods can prove the existence and size of a geothermal reservoir. Only the drilling of deep wells and testing of the product found will determine if successful development and utilization can be obtained.

Hot Dry Rock Geothermal Technology

Of a much longer-range potential is the possible exploitation of heat energy contained in hot dry rocks (HDR) of the earth's crust. These HDR regions are not directly related to the belts of geothermal energy activity previously described, but are reasonably well distributed under the land areas of continents, rather than concentrated in earthquake and volcano belts. While much effort remains in the development of exploitive technology, the gross estimates of HDR energy resources tend to be almost unbelievably high. Exploitation of such resources, of course, is essentially a matter of the geothermal gradient. This factor determines the depth of drilling required to reach a specified temperature. The HDR bases generally has been defined to include crustal rock that is hotter than $150\,^\circ$C and at depths of less than 10 kilometers, which is essentially at the edge of present commercially feasible drilling and recovery technology. Although a temperature of only $100\,^\circ$C would be attractive for space heating needs, a minimum temperature of $200-250\,^\circ$C is desirable for using such energy in the generation of electricity. Scientists at the Los Alamos Scientific Laboratory (L.A.S.L.) have developed a chart which indicates the best sector of useful heat, with depth plotted against temperature.

From temperature data obtained from deep gas wells, the geothermal gradient has been determined or estimated for several regions of the United States. Such regions have been found or are postulated for central and eastern Oregon, southern California, southwestern and central Arizona, western South Dakota and eastern Montana, Nebraska, much of Colorado, as well as some pockets in Indiana and Illinois, additional pockets in southeastern Texas along the Gulf Coast, various pockets in eastern Pennsylvania and New England, among others. High geothermal gradients occur on the Atlantic Coast from the Delmarva Peninsula southward to Georgia. These are regions where, it is believed, the geothermal temperature gradient is $36.5\,^\circ$C or higher per kilometer of depth. Immediately adjacent to these regions, and frequently of much larger area, the gradient lies between 29.2 and $36.5\,^\circ$C. Some scientists at the U.S. Geological Survey (U.S.G.S.) have observed that rock underlying about 5% of the total United States land area may have geothermal gradients of $40\,^\circ$C or greater. They also believe that it can be assumed conservatively that over a third of the land area in the United States has above-average heat flow with thermal gradients ranging from 30 to $36\,^\circ$C per kilometer. It has been pointed out that igneous rock systems to depths of 10 kilometers under the United States (excluding Alaska and Hawaii) contain some 105×11^{21} joules (J), which is equivalent to 105,000 quads. (A quad equals 1 quadrillion Btus.)

If one uses an average geothermal temperature gradient of $22\,^\circ$C for the entire United States, it is further estimated that the energy, if available, would amount to 13×10^{24} J. This is equivalent to 13,000,000 quads. By comparison, the current annual energy consumption of the United States approximates 80 quads. Scientists postulate that if only 0.2% of such energy were made available, it would be comparable to all of the coal remaining in the United States.

Basically, two techniques have been suggested for mining HDR geothermal heat: one method for rock formations with low permeability, and one for highly permeable rocks. As pointed out by Cummings et al. (1979), "If the permeability of the formation is low, an artificial circulation system can be created by fracturing the rock in the reservoir to provide many flow passages with a large heat-transfer surface area. A fluid—for example, water—is then circulated through the fractured reservoir to recover the energy. Most of the injected fluid is recovered in a second production wellbore simply because of the low natural permeability of the formation. Large fracture surface areas are required because rock conducts heat rather poorly, and it quickly controls the rate of heat transfer to the fluid contained in the fracture zone. Such and HDR reservoir will most likely be formed by injecting fluid through a wellbore at pressures sufficient to fracture the rock. Under ideal conditions, the fracture would be vertically oriented, circular in shape, with a maximum radius of typically 100 meters or more, and a width or opening of only a few millimeters."

Some of the facets related to the technical feasibility of HDR systems have been demonstrated in experiments conducted by L.A.S.L. at the Fenton Hill site in the Jemez Mountains of New Mexico. A hydraulically fractured reservoir in low-permeability crystalline basement rock at about $185\,^\circ$C was created and flow tested for 75 days at an energy extraction rate of about 5 thermal megawatts (MWt). Additional tests are underway on an expanded scale. For greater detail on this program, see the Cummings et al. reference.

Obviously, considerable further research, particularly of an engineering nature, is required to fully demonstrate the potential of HDR as a future major energy source. Environmental impact studies also will be required.

Acknowledgments. This technical summary on geothermal energy was made possible by information and portions of the text furnished by: R.G. Bowen, consulting geologist, Portland, Oregon and E.A. Groh, private geologist, Portland, Oregon (geological aspects and exploration); R.S. Bolton, chief geothermal engineer, Ministry of Works New Zealand, Wellington North, New Zealand; Pacific Gas and Electric Company, San Francisco, California (The Geysers); Baldur Lindal, VBL Consulting Engineers, Reykjavik, Iceland (geothermal energy in Iceland—and space and process heating); and W. Chow, Electric Power Research Institute, Palo Alto, California.

Additional Reading

Braun, J.E.: "Reducing Energy Costs and Peak Electrical Demand Through Optimal Control of Building Thermal Storage," *American Society of Heating, Refrigerating, and Air-Conditioning Engineers Transactions*, 1990.

Bresee, J.C.: *Geothermal Energy in Europe: The Soultz Hot Dry Rock Project*, Gordon & Breach Publishing Group, Newark, NJ, 1992.

Butler, E.W. and J.B. Pick: *Geothermal Energy Development*, Plenum Publishing Corporation, New York, NY, 1982.

Chandler, W.U., A.A. Makarov, and Z. Dadi: "Energy for Russia, Eastern Europe and China," *Sci. Amer.*, 121 (September 1990).

Coury, A.: "Upstream H_2S Removal from Geothermal Steam," *Rept. 1197–2*, Electric Power Research Institute, Palo Alto, CA (November 1981).

Cummings, R.G. et al.: "Mining Earth's Heat: Hot Dry Rock Geothermal Energy," *Technology Review (MIT)*, 58–78 (March 1979).

Dickson, M.H. and M. Fanelli: *Geothermal Energy*, John Wiley & Sons, Inc., New York, NY, 1995.

Graham, I.S.: *Geothermal and Bio-Energy*, Raintree Steck-Vaughn Publishers, Austin, TX, 1999.

Johansson, T.B., B. Bodlund, and R.H. Williams, Eds.: *Electricity*, American Council for an Energy Efficient Economy, Washington, DC, 1989.

Kerr, R.A.: "Geothermal Tragedy of the Commons," *Science*, 134 (July 12, 1991).

Kerr, R.A.: "The Back Burner of Geothermal Energy," *Science*, 135 (July 12, 1991).

Staff: *Geothermal Energy: Bet on It!*, Geothermal Resources Council, Davis, CA, 1984.

Staff: *Efficient Electricity Use: Estimates of Maximum Energy Savings*, Electric Power Research Institute, Palo Alto, CA, March 1990.

Staff: *Annual Book of ASTM Standards 1999: Section 12: Nuclear (ll), Solar, and Geothermal Energy*, American Society for Testing & Materials, West Conshohocken, PA, 1999.

Veziroglu, T.N.: *Energy and Environmental Progress 1: Indirect Solar, Geothermal and Nuclear Energy*, Nova Science Publishers, Inc., Huntington, NY, 1991.

Wohletz, K. and G. Heiken: *Volcanology and Geothermal Energy*, University of California Press, Los Angeles, 1992.

Web References

ANSA News Agency. http://www.mi.cnr.it

California Department of Conservation. http://www.consrv.ca.gov

California Energy Commission, Geothermal Energy. http://www.energy.ca.gov

Energy & Geoscience Institute, University of Utah. http://www.egi.utah.edu

Environmental Protection Agency. http://www.epa.gov/globalwarming/

Geothermal Energy Association. http://www.geotherm.org

Geothermal Heat Pump Consortium. http://www.ghpc.org

Geothermal Resources Council. http://www.geothermal.org/index.html

International District Energy Association. http://www.energy.rochester.edu/idea/

International Geothermal Association (IGA). http://www.demon.co.uk:80/geosci/igahome.html

International Ground Source Heat Pump Association (IGSHPA). http://www.igshpa.okstate.edu

Lawrence Berkeley National Laboratory. http://www-esd.lbl.gov/ER/geothermal.shtml

New Zealand Geothermal Association. http://www.voyager.co.nz/~tking/nzgahome.html

Oregon Institute of Technology Geo-Heat Center OIT-GHC. http://www.oit.osshe.edu/~geoheat/

Renewable Energy Policy Project (REPP). http://solstice.crest.org/renewables/repp

Sandia National Laboratories, Geothermal Research Department. http://www.sandia.gov/geothermal/

Southern Methodist University Geothermal Program. http://www.smu.edu/~geothermal

Stanford University Geothermal Program. http://ekofisk.stanford.edu/geotherm.html

University of Auckland New Zealand. http://www.auckland.ac.nz/gei

U.S. Department of Energy, Energy Information Administration. http://www.eia.doe.gov

U.S. Department of Energy Technical Site. http://geothermal.id.doe.gov/

U.S. DOE, Energy Efficiency and Renewable Energy Network (EREN). http://www.eren.doe.gov

U.S. Geological Survey. http://www.usgs.gov

Virginia Tech Geothermal. http://rglsun1.geol.vt.edu

GEOTROPISM. The response of living things to the effects of gravity. The term is used especially with reference to the response of roots and stems of plants. Most roots are said to be positively geotropic, that is, they grow in the direction of the pull of gravity. See accompanying diagram. Most stems are negatively geotropic, that is, they grow away from the pull of gravity. Such a combination of responses causes the roots to grow down and the stems to grow in an upward direction.

These responses are explained on the basis of auxins. These are produced in the tips of the stems, in young leaves, and, in lesser quantities, in the tips of the roots. When a plant is turned on its side, gravity causes auxin to accumulate on the lower side of the stem. This extra auxin acts as a stimulant to the cells in the region of elongation and the stem turns upward. The roots are much more sensitive to auxin than stems, and the same concentration that stimulates stems acts to inhibit the growth of the cells of the root in the region of elongation. Thus the growth is greater on the upper surface and the root turns down. The possible relationship between geotropism and ethylene is described by Wheeler and Salisbury (*Science*, **209**, 1126–1127, 1980).

See also **Plant Growth Modification and Regulation**.

GEPHYREA. Large marine worms. As adults they are not segmented but since the young show evidence of metameric segmentation they have been placed with the segmented worms in the phylum *Annelida*, but the three groups are now classified as separate phyla. They have a large body cavity, nephridia, and in some species a few setae. The internal organs are not metamerically arranged.

The group is divided into:

Phylum *Echinoidea*. With a pair of setae near the anterior end. Body cylindrical with a slender anterior protuberance, the prostomium.

Phylum *Sipunculoidea*. No setae. Body slender, with a protrusible proboscis and a group of tentacles near the mouth.

Phylum *Priapuloidea*. No setae or tentacles.

GERIATRICS. Geriatrics is the branch of medical science concerned with the prevention and treatment of diseases of older people. Over the past 30 years, life expectancy in the United States increased due to a greater survival of older people; this appears to result from two factors: 1) improvements in diagnosis and treatment of diseases of the elderly; and 2) a reduced age-specific incidence of some diseases, which indicates that healthful changes in life-style have contributed (See **Gerontology**).

The Relationship of Normative Aging to Geriatrics

Senescence results from inevitable physiological changes that are separable from disease. As Rowe points out, there is no plateau of the middle years during which time physiological functions are stabilized, but rather the reduction in function of many organs is progressive, even if it is not manifested dramatically. Losses in renal, pulmonary, and immune functions may occur over a long period of time; such linear reductions in homeostatic capacity (the ability to maintain steady-state conditions) in several organs result in a geometric reduction of the total homeostatic capacity, thus markedly increasing the vulnerability of the elderly to morbidity under stressful conditions. Stresses that can speed up the aging process include acute illness, trauma as precipitated by serious burns or falls, major surgery, and side effects from drug treatments.

Menopause can be used as an example that further illustrates the relationship of normative aging to geriatrics. Menopause is defined as the first one-year period in which a woman has no menstrual cycle as a result of age-related ovarian failure. It is considered normative aging, typically begins when women are in their early to mid-50s, and is complete for almost all women by the time they are 60. An important endocrine consequence of menopause is a dramatic reduction in circulating estrogen. Relatively common and usually transitory clinical manifestations include hot flashes and sleep disturbances. Geriatric consequences of menopause that affect *some* women over the remainder of their life span (and indeed affect the life span itself) include increased risks of osteoporosis and atherosclerosis, diseases with primary etiologies that are unrelated to menopause. Estrogen-treatment has been used successfully over a number of years to delay the onset of osteoporosis.

Age-Related Disorders

The major diseases of the elderly are described under specific headings elsewhere in this encyclopedia. Check the alphabetical index. Other important disorders include: incontinence, general dementia, sleep disorders, digestive system disorders, and balance and gait disorders.

Incontinence

Incontinence is classified as reversible or fixed. Reversible incontinence is frequently found among hospitalized elderly patients, with the causes usually related to acute confusional states (particularly after surgery), immobility that interferes with normal urination habits, fecal impaction, acute symptomatic bladder infection, metabolic abnormalities related to diuresis (e.g., hypercalcemia and hyperglycemia), and medications, notably sedatives or anticholinergic agents that decrease the strength of the bladder detrusor contraction. With careful attention to the patient, some of the formerly classified fixed or chronic forms of incontinence can be reversed. One of these is *urge incontinence*, where the patient senses the need to void and cannot prevent voiding. This condition can be markedly improved by the administration of smooth muscle relaxants, such as calcium channel blockers, or by anti-cholinergic medications, such as oxybutynin. These substances reduce bladder contractions. There are, however, a number of serious side effects in some patients. *Stress incontinence* also affects the elderly and may be described as an involuntary loss of urine only when the intraabdominal pressure is transiently increased. The underlying cause is usually found to be overstretching of pelvic musculature during childbirth or damage from prior surgery. Local (vaginal cream) or systemic estrogens may improve this form of incontinence. Pelvic floor exercises (voluntarily discontinuing urination several times during each void cycle) have been effective in some cases. When the exercise regimen is ceased, the incontinence usually returns.

Dementia

Despite the common stereotype, dementia is not a characteristic of normative aging. Severe dementia is present in only 2.5 to 5% of people over 65, with mild to moderate forms in an additional 10%. Dementia is the most common reason for institutionalization, and over 50 percent of nursing home residents has some form of dementia. In at least 15 percent of the elderly with chronic losses in mental function, the dementia is reversible by treating the precipitating cause, which can include a change of medication, renal problems, anemia, congestive heart failure, thyroid disease, vitamin B_{12} deficiency, and depression. Another 15–25 percent of elderly patients with dementia have suffered some cerebrovascular accident. Of the elderly presenting with dementia, 50–70% suffer from

Alzheimer's disease, which increases in prevalence with advancing age. Currently, there are no specific non-histological diagnostic tests or effective treatments for Alzheimer's disease. See also **Alzheimer's Disease and Other Dementias**.

Sleep Disorders

Although some elderly individuals may spend more time in bed than younger adults, they may sleep less and are aroused from sleep more readily. Prinz et al. describe causes of sleep disorders in the elderly.

Nocturnal Respiratory Dysfunction. Sometimes referred to as *sleep apnea syndrome*, it occurs more frequently in males than in females and is more common among the elderly than in the young of either sex. Sleep apnea is characterized by a repeated cessation of breathing during sleep for a period of several seconds or more. This produces hypoxemia (blood oxygen saturation frequently is lowered below 80%) and accompanying sleep interruption. Treatment includes the avoidance of sleeping on one's back, a reduction of body weight, the use of respiratory stimulants (acetazolamide), and the avoidance of respiratory depressant drugs such as hypnotics and alcohol. Treatments may include administration of continuous positive pressure (breathing machine), or surgical procedures to modify the upper airway. Prinz et al. (1990) report that, "There is little evidence to support the treatment of mild obstructive sleep apnea in the elderly in the absence of excessive sleepiness, cognitive impairment, or associated cardiorespiratory abnormalities."

Restless Leg Syndrome. There is a tendency among some elderly people to move the legs repeatedly, making it difficult to fall asleep. This is a poorly understood condition, but is believed to be associated with metabolic, vascular, or neurological factors, and, in some cases, with sleep apnea. Treatment usually addresses the causative factor rather than the syndrome.

Secondary Manifestations of Other Illness. Other conditions that may contribute to disorderly sleep include arthritic and other major pain, as well as respiratory, cardiac and neurologic disease, and drugs taken for other complaints. Psychiatric illnesses, including depressive reactions to severe or chronic illness, also contribute to disturbed sleep patterns in the elderly. Prescription drugs are available to relieve such sleep disorders.

Other causes for sleep disorders among the elderly include persistent psychophysiologic insomnia, secondary aspects of dementia and delirium, alcoholism, self-administered drug habits, changes in circadian rhythms, and REM (rapid eye movement) sleep behavior disorders. See also **Biological Timing and Rhythmicity**; and **Sleep**.

Sedative Hypnotic Agents. Special note must be made pertaining to the disproportionate prescription of these drugs for the elderly. For example, in 1985, over 20 million prescriptions were written for sedative-hypnotic benzodiazapines, primarily flurazapam, temazepam, and triazolam, representing an increase of 38 percent over 1980. Of these medications, 66 percent were prescribed for patients who were 60 years old or older. Older women were 1.7 times more likely to receive a prescription for such drugs than older men (Baum et al., 1985). Long-term use often results in habituation, tolerance to the drug, and drug-induced insomnia. Various health organizations, including the National Institutes of Health (U.S.) have urged that greater restraint should be used in prescribing such drugs (Freedman et al., 1984). Although a physician may encounter a distraught patient with insomnia and find it tempting to prescribe a hypnotic drug, many experts now believe that such drugs should be used for chronic sleep disturbances only very cautiously.

Digestive System Disorders

Age-related diseases are known for most organs of the digestive system (Shamburek and Farrar):

Esophagus. Disorders of swallowing are quite common in the elderly and can influence morbidity and mortality from malnutrition and aspiration pneumonia. Diseases that may affect the oropharynx, and result in dysphagia, include Parkinson's disease, stroke, diabetic neuropathy, and polymyositis (see listings elsewhere in the encyclopedia).

Stomach. During the past 20 years, the incidence of peptic ulcers requiring hospitalization has decreased markedly in all age groups except the elderly. The rate of duodenal ulcer disease, however, does not increase with age.

Gallbladder. The prevalence of gallstones is greater in women of all ages than in men, and it increases in both sexes with age. This may be attributed to the formation of cholesterol stones because of increased secretion of cholesterol by the liver. The incidence of pigmented stones also increases with age, particularly after age 70. Biliary disease in the elderly is associated with a higher mortality and a higher rate of complications than in younger patients. For symptomatic cholelithiasis, early surgery is indicated; the mortality rate of elective surgery is 1.7 percent, whereas emergency or urgent surgery has a mortality rate of 11 percent. See also **Gallbladder and Biliary Tract Diseases**.

Large Intestine. From 3 to 10 percent of idiopathic inflammatory bowel disease cases occur after age 65. Symptoms resemble those of younger patients, although, in the elderly, symptoms sometimes are mistaken for those of diverticular disease, infectious diarrhea, and ischemic colitis.

Diverticulosis increases progressively with age. The condition increases from approximately 5% of persons in their 50s to nearly 50% of persons in their 90s. The formation of diverticula has been attributed to a diet low in fiber. The effectiveness of dietary fiber supplementation has largely been confirmed, but the mechanism involved now requires re-explanation.

Constipation is one of the most common disorders in the elderly, particularly for women. Constipation is usually defined as less than three bowel movements per week, although other definitions are used. Constipation may result from a low fiber diet, sedentary habits, medications, and diseases that impair neural and motor control. A principal result, fecal impaction, occurs most frequently in the elderly who are hospitalized or confined to nursing homes. If unattended, fecal impaction can precipitate numerous complications.

Medication. The indiscriminate use of medications to treat gastrointestinal disorders in the elderly should be avoided, lest adverse reactions should occur. These can include delirium from cimetidine, constipation from iron supplements and aluminum-containing antacids, and diarrhea from magnesium-containing antacids.

Gait Disorders of the Elderly

In the general population, about 15% of those over 60 have some abnormality of gait, and about 50–60% of the patients in nursing homes have some difficulty walking. These problems tend to worsen with age. A principal concern with gait disorders is their contribution to falling. In 1990 in the United States, a majority of the 200,000 hip fractures were caused by falls of older people. Accidental injury is the sixth leading cause of death among the elderly, and the majority of injuries result from falls.

The gait can reflect musculoskeletal as well as neurological abnormalities. The most common cause of gait disorders is degenerative arthritis of the cervical spine (cervical spondylosis). The second most frequent cause is myelopathy. In Parkinson's disease, which affects about 1.5% of the population over 65, patients develop axial rigidity and gait disorders at some stage of the disease, along with an impaired sense of balance. Some drugs can alleviate the gait problem partially, but not necessarily restore the sense of balance. See also **Parkinson's Disease**.

Stroke is also a frequent cause of gait and balance disorders. These usually result from damage to the basal ganglia or the periventricular white matter of the brain, as observable by computed tomography or magnetic resonance imaging. Patients with toxic or metabolic encephalopathy also may suffer from a disturbance of motor function.

Various techniques of physical therapy usually are prescribed for patients with serious gait and balance problems.

Special Problems with Diagnosis in the Elderly

Common clinical measures that are *not* affected directly by aging include fasting glucose level, serum electrolyte concentrations, blood gas values, and hematocrit. Some authorities have observed that too often clinicians may ascribe a disability or an abnormal laboratory finding simply to "old age," when the actual cause may be a specific disease essentially unrelated to the age of the patient. For example, a person found to have a low hematocrit may be carelessly categorized as having "anemia of old age." It is possible in such instances for the physician to fail to investigate the basis of the anemia and conclude that no treatment is warranted.

There may also be a tendency on the part of the older patient to ignore abnormal conditions because "it is expected in old age." Other considerations of particular concern with elderly patients that may result in underreporting of symptoms include: cognitive impairment, fear of the severity of the illness, concern over costs, and other negative images of hospitalization. These concerns can be so great for some that they do not seek medical assistance even when needed.

Rowe summarizes, in his advice to professionals, the special considerations that should be made when working with the elderly patient, "One must obtain a thorough medication history, and be aware of the special vulnerability of the elderly to the development of adverse effects from medication. Special consideration should be given to the detection of thyroid, breast and cervical cancer, occult bleeding, hypertension, postural hypotension, disease in the oral cavity that may affect nutritional status, wax impaction in the ears that may limit hearing, and serious auditory or ophthalmic disorders. Attention should be paid to bowel function and the possible presence of varying degrees of urinary incontinence and sleep disturbance. Specific questions regarding postural stability are mandatory in view of the high prevalence and serious consequences of falls in the elderly."

Life Style Considerations

Certainly, exercise may benefit the patient when it is prescribed for specific age-related disorders such as cardiovascular disease. However, exercise may have relatively little impact on underlying aging processes. Schneider and Reed observe that," Although there are few if any studies of lifelong exercise, there have been numerous retrospective studies of the longevity of athletes, ranging from Oxford oarsmen to New Zealand rugby players. The majority of these studies have indicated that there is no relation between a history of athletic competition and longevity. However, athletic competition lasting a few decades may not be sufficient to influence longevity." Effects of lifelong aerobic exercise in laboratory animals; when husbandry conditions are controlled to eliminate respiratory infections, surprisingly, no relationship is found between exercise and life span.

Special foods and diets for the elderly in the interest of creating healthy longevity are mentioned in the entry on **Diet**. See also **Gerontology**; and **Biochemical Theories of Aging**.

Additional Reading

Baum, C. et al.: *Drug Utilization in the U.S.—1985, Seventh Annual Review*, Food and Drug Administration, Center for Drugs and Biologics, Rockville, MD, 1986.

Beers, M.H. and R. Berkov: *Merck Manual of Geriatrics*, 3rd Edition, Merck & Company, Inc., Whitehouse Station, NJ, 2000.

Freedman, D.X. et al.: "Drugs and Insomnia," *National Institutes of Health (U.S.)*, Consensus Development Report, Washington, DC, 1984.

Hazzard, W.R., et al.: *Principles of Geriatric Medicine and Gerontology*, 4th Edition, The McGraw-Hill Companies, Inc., New York, NY, 1998.

Holloszy, J.O.: *Exercise*, in *Handbook of Physiology Section 11: Aging*, Masoro, E.J., Editor, Oxford University Press, Inc., New York, NY, 1995.

Osterweil, D., J.C. Beck, and K. Brummel-Smith: *Comprehensive Geriatric Assessment*, The McGraw-Hill Companies, Inc., New York, NY, 2000.

Pathy, M.S.: *Principles and Practice of Geriatric Medicine: 2 Volumes*, 3rd Edition, John Wiley & Sons, Inc., New York, NY, 1998.

Prinz, P.N. et al.: "Geriatrics: Sleep Disorders and Aging," *N. Eng. J. Med.*, **520** (August 23, 1990).

Reuben, D.B., et al.: *Geriatrics at Your Fingertips*, 3rd Edition, Elsevier Science, New York, NY, 2001.

Rowe, J.W. and R.L. Kahn: *Successful Aging*, Pantheon Books, New York, NY, 1998.

Schamburek, R.D. and J.T. Farrar: "Disorders of the Digestive System in the Elderly," *N. Eng. J. Med.*, **438** (February 15, 1990).

Kevin Flurkey, Ph.D., The Jackson Laboratory, Bar Harbor, ME

GERM. 1. A microorganism (microbe), commonly used to refer to bacteria and their relations.

2. The reproductive material; for instance, the germ plasm is the plasm of reproductive material that links the generations. 3. The embryo of seeds; for instance, wheat germ is made from the region of the wheat kernel that contains the embryo.

GERMANIUM. [CAS: 7440-56-4] Chemical element symbol Ge, at. no. 32, at. wt. 72.59, periodic table group 14, mp 937 °C, bp 2830 °C, density 5.36 g/cm^3 (20 °C). Elemental germanium has a diamond cubic crystal structure. Germanium is a silver-white, lustrous, hard, brittle metal. When heated in oxygen to 730 °C, the metal is partially oxidized to dioxide. The element is unaffected by solutions of acids and bases, but is soluble in fused NaOH. In the form of powder (dull gray), combines readily

with chlorine to form the volatile tetrachloride. Although predicted by Mendeleev as early as 1871, the element was not fully identified until 1886 by Winkler. Mendeleev had previously termed the missing element *eka-silicon*. There are five natural isotopes ^{70}Ge, ^{72}Ge through ^{74}Ge, and ^{76}Ge. Seven radioactive isotopes include ^{67}Ge through ^{69}Ge, ^{71}Ge, ^{75}Ge, ^{77}Ge, and ^{78}Ge. All have a relatively short half-life, the longest, ^{68}Ge with a half-life of 275 days. In terms of abundance, germanium ranks 32nd among the element and thus is about as abundant as gallium, selenium, arsenic, and bromine. First ionization potential 8.13 eV; second, 15.86 eV; third, 31.97 eV; fourth, 45.5 eV. Other important physical characteristics of germanium are given under **Chemical Elements**.

Germanium occurs in very small amounts in many sulfide ores, such as American zinc ores (0.25% GeO_2), and the rare mineral argyrodite (silver germanium sulfide) of Saxony and Bolivia. The primary source is flue dust from the zinc industry. Also, it may be obtained from the reduction of oxide and sulfide ores. A major ore is germanite, a copper ore found in southwest Africa. The ore is quite complex, containing some 20 different elements. The copper content ranges as high as 45%, sulfur up to 30%, whereas the germanium content is from 6 to 9%. The ore also contains up to 1% gallium. A major sulfide ore is renierite, which contains up to about 8% germanium. Small quantities of germanium are found in lepidolite, sphalerite, and spodumene. Some English coals contain as much as 1.6% germanium oxide. The germanium metal of 99.99+% purity is obtained by zone melting. In this system, electric heating coils are moved slowly along the length of an ingot. Impurities in the metal tend to raise or lower the freezing point of the molten alloy. By progressively melting the metal along the length of the ingot, the impurities that tend to lower the melting point will be swept to the last portion of the ingot to freeze, whereas the impurities that tend to raise the melting point will concentrate in the first region to freeze.

Uses. The principal uses of germanium have been in solid-state electronic devices, notably transistors, which can be used as amplifiers and oscillators. The electrical properties of germanium metal, which have brought about its wide use in semiconductors, are its high specific resistance at ordinary temperatures and the narrow gap between its filled energy band and its conduction band. Thus, germanium is an intrinsic semiconductor, wherein an increase of temperature or the addition of very small amounts of group 3 or group 5 elements can cause electrons to move readily to the conduction band to form "holes," thus making the material conductive. A key to the manufacture of semiconductor devices is making materials of high purity, great uniformity, and in sufficient quantity. See also **Semiconductors**.

The addition of as little as 0.35% germanium to tin doubles the hardness of tin. Similarly, germanium improves the strength and hardness of aluminum and magnesium alloys. These applications are limited, however, because of the current high costs of germanium. Germanium-silicon alloys are under intensive study for use in thermoelectric generators. Advantages claimed for these metals include better thermoelectric qualities above 600 °C, an improved efficiency per unit weight factor, and virtually no corrosion or decomposition.

Chemistry and Compounds. Germanium forms compounds in which the oxidation states are (II) and (IV). The divalent ones are unstable. Thus, the monoxide is readily oxidized by air when hydrated. However, when completely dehydrated it resists the action of H_2SO_4 and potassium hydroxide, and reacts only slowly with fuming HNO_3. On heating in an inert atmosphere, it disproportionates to the elements and germanium dioxide, GeO_2. The latter resembles silicon dioxide in existing in more than one form, with a difference in chemical properties. The stable form at room temperature has the rutile structure, but just below the melting point the stable form has the cristobalite structure. Germanium(IV) oxide, [CAS: 1310-53-8] GeO_2, prepared by hydrolysis of germanium(IV) chloride, [CAS: 10038-98-9] $GeCl_4$, is somewhat soluble in water, acids, and alkalis, but GeO_2 from heating of germanic acid is insoluble. Like silicon dioxide, GeO_2 forms gels readily.

Germanium(II) hydroxide, $Ge(OH)_2$, [CAS: 10060-11-4] is obtained by action of alkali hydroxides upon germanium(II) chloride, $GeCl_2$, solutions; it is amphiprotic, dissolving in excess of the alkali. Moreover, the acid form, sometimes called germanous acid, is obtained upon heating the hydroxide: $Ge(OH)_2 \rightarrow HGe(O)H$. GeO_2 is slightly acid in solution and when freshly precipitated ($pK_A = 9.4$). There is no experimental evidence for the existence of a definite hydrate, although melting point diagrams of germanate salts have indicated the existence of ortho($\equiv GeO_4$), meta($= GeO_3$), and tetra($\equiv Ge_4O_9$) compounds.

Germanium forms dihalides and tetrahalides with all four of the common halogens. In general, the dihalides readily react with halogens or other oxidizing agents to form tetravalent germanium compounds, and some, e.g., the iodide, disproportionate to the metal and tetravalent compound.

Suggestive of carbon and silicon is the existence of hydrides of germanium, though they are much fewer in number. The compound GeH_4 [CAS: 7782-65-2] is called germane (mp $-165\,°C$, bp $-90\,°C$). Compounds having the general formula Ge_nH_{2n+2} ($n = 2$, 3, etc.) are called digermane, trigermane, etc., according to the number of germanium atoms present. The first three compounds in this series have been obtained by treatment of magnesium germanide with ammonium bromide in liquid ammonia. Compounds such as $GeHCl_3$ and alkylgermanes are also known. Germane and the alkyl- and aryl-substituted germanes retaining at least one hydrogen atom are somewhat more acidic than the corresponding silanes in nonaqueous media, easily forming alkali salts, R_3GeM and even dialkali salts R_2GeM_2 under some circumstances. Germane, GeH_4, appears to be thermodynamically stable, although no quantitative data are available on its heat of formation. It decomposes at about $285\,°C$.

Germanium also forms organometallic compounds. Over two hundred have been reported, from chloromethyl trichlorogermane, $ClCH_2GeCl_3$ to cyclotetrakis (diphenyl germanoxane), $[(C_6H_5)_2GeO]_4$.

See list of references at end of entry on **Chemical Elements**.

Germanium is also described in some of the entries on electronic components in this encyclopedia.

Additional Reading

Anderson, D.L.: "Composition of the Earth," *Science*, **367** (January 20, 1989).

Avallone, E.A. and T. Baumeister: *Mark's Standard Handbook for Mechanical Engineers*, 10th Edition, The McGraw-Hill Companies, Inc., New York, NY, 1996.

Belz, L.H.: "Special Metals in Electronics," *Advanced Materials & Processes*, 65 (November 1987).

Dahmen, U. and K.H. Westmacott: "Observations of Pentagonally Twinned Precipitate Needles of Germanium in Aluminum," *Science*, **233**, 875–876 (1986).

Davis, J.R.: *Metals Handbook*, 2nd Edition, ASM International, Materials Park, OH, 1998.

DiSalvo, F.J.: "Solid-State Chemistry: A Rediscovered Chemical Frontier," *Science*, 649 (February 9, 1990).

Greenwood, N.N. and A. Earnshaw: *Chemistry of the Elements*, 2nd Edition, Butterworth-Heinemann, Inc., Woburn, MA, 1997.

Hull, R. and J.C. Bean: *Germanium Silicon: Physics and Materials*, Vol. 5, Academic Press, Inc., San Diego, CA, 1999.

Krebs, R.E.: *The History and Use of Our Earth's Chemical Elements: A Reference Guide*, Greenwood Publishing Group, Inc., Westport, CT, 1998.

Lide, D.R.: *CRC Handbook of Chemistry and Physics*, 88th Edition, CRC Press, LLC, Boca Raton, FL, 2007.

Parker, P.: *McGraw-Hill Encyclopedia of Chemistry*, 2nd Edition, The McGraw-Hill Companies, Inc., New York, NY, 1993.

Patai, S.E.: *The Chemistry of Organic Germanium, Tin and Lead Compounds*, John Wiley & Sons, Inc., New York, NY, 1995.

Staff: *ASM Handbook: Properties and Selection of Nonferrous Alloys and Pure Metals*, ASM International, Materials Park, OH, 1990.

Westbrook, J.H.: "Electrical Properties," in *Encyclopedia of Materials Science and Engineering*, MIT Press, Cambridge, MA, 1986.

GERMICIDE.   Any substance or agent, physical or chemical, which is destructive to germs (bacteria).

GERMINATION (Seed).   See **Seed**.

GERM LAYER.   The three tissues resulting from the first differentiation in the embryonic development of multicellular animals. They are formed from the presumptive areas of the blastula during gastrulation by a redistribution, which results in three layers. The three are an outer ectoderm, an inner endoderm, and between the two the mesoderm.

Animals of the phyla *Porifera, Coelenterata*, and according to one interpretation the *Ctenophora*, develop only the first two germ layers and are said to be diploblastic. Multicellular forms of all other phyla have all three and are therefore triploblastic.The chief parts of the body formed from the various germ layers are as follows (in these lists the terms are chosen to embrace both vertebrates and invertebrates and so do not all apply to the same animal):

Ectoderm: Outer cellular layers of the integument, their glandular derivatives, and the cuticula. Exoskeleton and exoskeletal structures, such as setae, scales, feathers, hair, claws, hoofs and nails. Parts of sensory organs, including the cornea and lenses of all types of eyes, and the external and internal ears of vertebrates. Lining of oral cavities and salivary glands, and in vertebrates the enamel of the teeth. Lining of the posterior part of the alimentary tract. The entire nervous system of most animals, including the nervous structures in the sense organs. A limited amount of muscular tissue. Organs of reproduction of some animals. Lining or covering of organs of respiration of many invertebrates.

Mesoderm: Lining of body cavity, circulatory system, water vascular system of echinoderms, and parts of excretory system. Muscular tissue. Bone. Teeth, except the enamel. The mesenchymal tissues such as cartilage, connective tissue, adipose tissue, and tendon. Blood. A limited part of the nervous system of starfishes. Reproductive organs.

Endoderm: Lining of the enteric cavity, including most of the alimentary system of vertebrates and the limited midintestine of arthropods. Respiratory epithelium of vertebrates. Lining of parts of vertebrate excretory system. Reproductive organs and cells.

GERM PLASM.   The essential reproductive tissue and the germ cells that it produces.

The concept of the germ plasm has been emphasized chiefly in the field of organic development. Since the germ cells of one generation produce both the body (soma, somatoplasm) and the germ plasm of the next, the continuity of this material is evident. It has been interpreted as the perpetual living substance, whereas the material of the body appears as an offshoot in each generation. In the one-celled animals, however, there is no differentiation.

Obtaining and retaining germ plasm from very old plants is essential to the development of new species and strains of crop plants. See also **Genes and Genetics**; and **Plant Breeding**.

GERONTOLOGY.   During the 20th century, advancements in public health and medicine have had a major impact on health during old age and on life expectancy. However, these advancements did not significantly affect maximum human life span see Fig. 1. Until recently, it seemed that maximum potential life span was limited by an immutable underlying aging rate: No matter how fortunate or how healthy a person was, he or she could not escape the gradual physiological deterioration of senescence, nor postpone its inevitable consequence, death, beyond the limit of about 120 years. Now, even traditionally conservative scientists accept the possibility of an intervention to retard aging rates in humans. This new expectation results from recent discoveries in the genetics of aging, generated from both traditional approaches and from the application of new advancements, demonstrating that maximum life span potential in animals, including mammals, is alterable.

With "discoveries" of the causes of senescence and anti-aging cures enthusiastically reported at increasing frequency, it is difficult to evaluate the objectivity of the reports and the validity of the discoveries. In this article reviews theories of senescence, interventions, and the impact the new genetics is having on our understanding of the biological causes of senescence. At also provides some guidelines for evaluating emerging gerontologic data and their implications.

It is important to define several terms used throughout this article. Technically, *aging* refers to any time-dependent change, including development; however, it is often used synonymously with senescence. *Senescence*, sometimes called "normative aging," is aging that begins after maturation and that affects all individuals in a species. Senescence can be neutral or detrimental; it is rarely beneficial. It may continue a developmental process, such as the progressive, orderly loss of irreplaceable oocytes in mammals (a process called atresia), which begins before birth and continues throughout the female's reproductive life. Or it may be unique to adults, such as the graying of hair in aging humans. *Disease* is impairment in one system that affects another system. To illustrate: The age related loss of calcium in humans, osteopenia, is characteristic of all older humans, and is considered senescence. When the calcium loss from osteopenia becomes so severe that bones fracture and movement is impaired, the geriatric disease osteoporosis results. *Gerontology* is the study of senescence. *Geriatrics* is the study of disease for which age is a risk factor. Gerontology is the study of normative aging (senescence), some of which predisposes individuals to disease, whereas geriatrics is the study of the age-related diseases themselves, their etiology and consequences. See also **Geriatrics**.

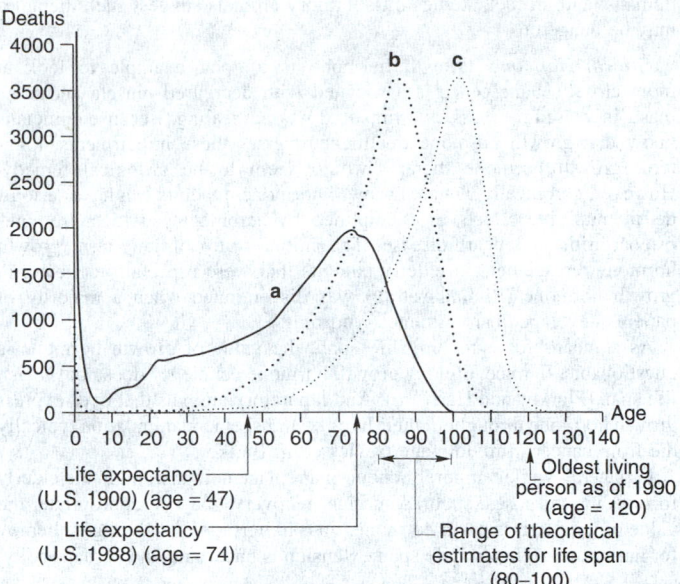

Fig. 1.   Life expectancy curves for females in the United States. (**a**) This curve represents the best estimate of life expectancy that prevailed in 1900. It reflects a high initial mortality rate (IMR), which includes a first year mortality of 12 percent and the marked mortality among women during their reproductive years. These factors result in a life expectancy at birth for all women of 47 years. (**b**) This curve represents the life expectancy in the late 1980s, when life expectancy at birth was 74 years. Improvements resulted from public health measures and medical advancements such as antibiotics and immunization. (**c**) This curve represents predicted life expectancy in 2010. Medical progress continues and most natural causes of death are being treated to further prolong life, notably after middle age. Experts predict that within another quarter-century life expectancy will range between 80–100 years, with a few people reaching 120 years. It was once thought that this was the biological limit of human life span; however, recent genetic research indicates that it may be possible to increase maximum life span, even in humans.

Theories of Senescence

Theories of senescence can be categorized according to the four biological levels of organization: population, physiological/organismal, cellular, and molecular. Although the theories across these levels are not necessarily related, they are not mutually exclusive.

Population Senescence and Evolutionary Biology

The hallmark of population senescence is the post-maturational age-associated increase in the risk of mortality, which is determined from survival statistics. In populations that do not senesce (children, for example), mortality risk remains constant or declines over time. In populations that do senesce, mortality risk increases logarithmically, as described by the Gompertz equation: $m(t) = Ae^{G(t)}$, where $m(t)$ is the age-specific mortality rate at age t, A is the age-independent mortality rate (related to the minimum risk of mortality that results from accidents, predation, and disease), and G is the exponential mortality rate coefficient. The applicability of an exponential model that relates age with mortality demonstrates that mortality rate not only increases with age, it accelerates.

Finch and colleagues (1990) redefined the two Gompertz parameters A and G as the conceptually more useful parameters initial mortality rate (IMR) and mortality rate doubling time (MRDT). Although differences in IMR can have profound effects on the population's survival characteristics, only the MRDT is considered to reflect the underlying aging rate of a population. For a given species, different environmental conditions can result in many-fold differences in IMR, whereas the MRDT is relatively stable. For example, almost the entire difference in survival between men and women is due to a difference in IMR. At the onset of maturity, women start at a lower mortality rate than men; but, as women age, their MRDT is the same as for men (Masoro). Specifically, human mortality rate doubles about every 8–9 years after age 30 (Finch et al.).

Reasons for Senescence. A major objective of population biology is to "explain" how the great diversity of species came to be. This endeavor is the realm of evolutionary biology. The primary challenge that gerontology presents to evolutionary biologists is to explain why senescence, a genetically regulated trait that clearly diminishes reproductive fitness, is so pervasive across species: If senescence is a negative trait, how does it evade selection?

The argument that senescence exists to remove feeble old individuals from a population is a tautology: If senescence did not exist—and it doesn't for most asexual species—old individuals would not be feeble. The more sophisticated idea that senescence somehow benefits the species overall at the cost of the individual (for example, by removing adults to permit the continual emergence of new genetic recombinations) generally is rejected by biologists, although it is encountered occasionally in popular literature. The difficulty with this "group selection" explanation is that only the individual carrying the trait can pass it on; and, if expression of the trait diminishes the individual's reproductive fitness, by definition, the likelihood that it will be passed on shrinks with each succeeding generation. Eventually the trait—senescence—will be lost, even if it would have benefited the species. Yet senescence persists as a characteristic of most metazoan species. This quandary remained a major paradox for biologists until the 1950s.

Mutation Accumulation Theory. In 1952, Medawar published a theory building on Haldane's earlier observation that late-onset hereditary diseases, such as Huntington's disease, are examples of genetic escape from natural selection. These deleterious mutations may be propagated if they are first expressed at an age few individuals reach, precluding the opportunity for selection to act. Medawar expanded Haldane's concept to explain how senescence might have evolved. Even if senescence didn't exist initially, over time, the initial mortality rate (IMR) would take its cumulative toll; the force of selection would weaken; and, the late-expressed, deleterious trait would escape selection as carriers died (for other reasons) before the trait was expressed. This declining force of selection during aging would permit the accumulation, over generations, of late-acting deleterious mutations in what some have termed "the genetic dustbin of senescence." Eventually, this collection of deleterious mutations would become genetically fixed, producing senescence in even the previously "ageless" population.

Antagonistic Pleiotropy Theory. Pleiotropy is the concept that a single gene can control multiple effects. Williams refined Medewar's theory of mutation accumulation by pointing out that some pleiotropic genes that *promote* reproductive fitness, when expressed before and during adulthood, may eventually *impair* reproductive fitness at a later age. These pleiotropic genes with *beneficial* effects during initial expression, but *antagonistic* effects later, when the force of selection is weak, will be selected for if the overall average effect benefits reproductive fitness. A dramatic illustration is provided by the Pacific salmon, which experience extreme elevations of corticosteroid stress hormones that facilitate their upstream struggle to spawning sites, but which die from hypercorticosteronism shortly after spawning. Salmon that do not lethally elevate their corticosterone can survive for multiple seasons, but they are not found in the wild, presumably because they are less effective competitors for spawning sites.

Antagonistic pleiotropy links genes that regulate senescence directly to their primary functions in growth, maturation, maintenance, and repair. If pleiotropic genes regulate senescence, tradeoffs should exist between vigor in youth and relative vigor during aging. For example, earlier maturation or larger litter size should be associated with more rapid reproductive senescence, and the genetic capacity to achieve extreme old age should *not* be associated with exceptional vitality in youth. Because his theory predicts that genes which regulate senescence can be found among those that regulate growth and maturation, Williams proposed a way to prevent senescence: Arrest development.

Tests of the Theories. The most direct tests of antagonistic pleiotropy involved the predicted genetic association of greater reproductive life span with greater overall life span. Initial attempts to increase life span in animals [fruit flies (*drosophila*)] by selection failed. Some scientists speculated that within-species senescence rates are so tightly constrained that selection for increased life span could not succeed. However, by the mid 1980s two independent research groups solved a number of technical problems and demonstrated that selection in fruit flies for late female fertility resulted in a genetically correlated increase in life span both for males (15%) and for females (30%), as predicted by evolutionary theories (Luckinbill et al., Rose). This was the first successful application of a genetic theory of aging, at any level of organization, to extend life span in animals. Of course, direct application to humans was impractical, because the treatment involved artificial selection and multiple generations.

Today, in an effort to identify physiological and genetic mechanisms that mediate increased life span in flies, and which may be directly applicable to humans, scientists are studying the physiological differences between control lines and lines selected for extended life. One result has shown that long-lived flies have a 70 percent greater activity of superoxide dismutase, an enzyme that could protect against free radical damage (Arking and Dudas). The laborious procedure of identifying the specific genes that control the life span difference between the lines is in progress.

Based on this success with lower organisms, scientists have started similar experiments with mice during the past five years; but, because the studies involve multiple generations, and because the trait, longevity, requires at least 2.5 years for expression, progress is very slow. Recent studies using mice are attempting to identify traits that correlate with life span but are expressed earlier, such as developmental rate, body size, or immune aging, to provide selection criteria that can be evaluated much sooner than life span. Such selection experiments may "capture" longevity alleles at genes that regulate mammalian senescence, thus tagging the genes so they can be identified.

How These Theories Relate to Others. Even if the primary function of genes that regulate senescence is the regulation of growth, development, or repair, the association of age with increased mortality risk must result from the accumulation of age-associated impairments. The study of these impairments at the system/organ, cellular, and molecular levels of organization has generated independent, but non-exclusive theories of aging. The principal debate today among gerontologists concerns which impairments are primary, and to what degree the impairment can account for the age-associated increased risk of mortality.

Organismal Senescence

At the level of the organism, senescence is expressed as a progressive loss in the ability to maintain homeostasis. Often, senescent changes are not detected in the basal undisturbed state, but become evident during response to a stress. Theories of senescence at this level postulate that primary deficiencies that occur in an integrative organ system, usually the neuroendocrine system, produce secondary dysfunction in other systems.

"Death Hormone" Theories. The idea that a single hormone causes senescence, leading eventually to death, has precedence in annual plants, the Pacific salmon, and, among mammals, the male Australian marsupial mouse (*Antichinus*), which, like the Pacific salmon, dies of hypercorticosteronism associated with reproductive stress. The relevant common characteristic linking these species is that they are semelparous, i.e., they have one cycle of reproduction followed by imminent death. Presumably, for these species, the cost of a physiological compromise necessary for survival beyond one reproductive cycle would be diminished reproductive fitness, and eventual extinction of the line. Because no eutherian mammals are known to be exclusively semelparous (Finch), it is unlikely that any advantage can be gained if the "improvement" limits reproduction to one cycle. Thus, it is doubtful that any eutherian species senesces because of a precipitous lethal elevation of some hormone.

Deficiency Theories. Various characteristic endocrine *deficiencies* contribute to senescence in mammalian species as diverse as mice and humans. Perhaps the best known example is the loss of estrogen following the age-related depletion of ovarian follicles that results in menopause in women. Besides the withdrawal of trophic support for reproductive tissues such as the uterus, this deficiency, promotes osteopenia and fat deposition, and increases the risk for heart disease. Other less precipitous endocrine deficiencies include decreased circulating growth hormone, and in males, decreased testosterone, which contribute significantly to the increased fat deposition, declining muscle mass, and diminished protein synthesis typically associated with mammalian senescence.

The success of estrogen replacement therapy opened the door to the promotion of replacement therapies using other hormones that are deficient in the elderly, particularly growth hormone, thyroid hormone, and androgens. However, with the exception of estrogen replacement, effects of long-term endocrine treatment have not been thoroughly evaluated in humans; and, in fact, studies in laboratory animals suggest such therapies may be dangerous.

Growth Hormone. Growth hormone is a good example to look at more closely. Senescence is associated with decreased muscle and bone mass, increased fat mass, and impaired wound healing. Because clinicians know that growth hormone treatment reverses these impairments, long-term growth hormone therapy would seem to be a logical remedy. However, chronically elevated growth hormone in adults can have serious, detrimental side effects as exemplified by acromegaly, which promotes osteoarthritis and heart disease. The initial study of long-term growth hormone replacement in elderly patients that were especially deficient in growth hormone (Rudman et al.) was discontinued when a majority of patients developed carpal tunnel syndrome.

As a means for extending life span, the value of growth hormone is questionable. In mice, lifelong growth hormone deficiency does not *shorten* life span (Flurkey and Harrison). And laboratory animals that over-express growth hormone throughout their lives by transgenic manipulation typically die from cancer in middle-age (reviewed in Bartke).

In general, while temporary endocrine replacement therapy in the elderly for specific purposes, such as surgical recovery, may be appropriate, the value of long-term growth hormone, thyroid hormone, or androgen therapy for normative aging and life span extension is uncertain, and perhaps risky.

Dehydroepiandrosterone (DHEA). The most remarkable age-related endocrine deficiency in humans occurs with adrenal dehydroepiandrosterone (DHEA), a weak androgen with no known unique function. In young adult humans its circulating concentration (in its sulfated form) is greater than that of any other hormone. Its concentration declines progressively after age 25 until, by age 65, it is at 20–30% of its peak level (Orentreich et al.). Circulating levels of DHEA at middle age are positively associated with protection from breast cancer and from heart disease (reviewed in Finch). While such studies were encouraging, they did not unequivocally demonstrate that maintenance of DHEA levels prevented these diseases. To demonstrate this connection studies employing direct treatment with DHEA are necessary. Such DHEA-intervention studies have been performed in rodents.

In rodents, DHEA is not produced by the adrenal gland, and normal circulating levels are low. Treatment with extremely high doses (0.5 percent of the dry weight of the food), inhibits obesity and maturity onset diabetes, and diminishes both spontaneous and induced tumorigenesis. While rodents prone to these diseases may live longer when treated with DHEA, it is unclear whether their increased life span results from retardation of senescence or solely from the prevention of these specific diseases. Because research has not shown a definitive effect of DHEA-treatment on senescence in rodents, enthusiasm in the scientific community for its potential as an anti-senescence treatment for humans has waned. Among geriatricians, DHEA currently is viewed as a model for potential anti-diabetic and anti-carcinogenic agents; however, concern exists regarding its potential to promote virilization in women and prostate cancer in men.

Melatonin. It is important to note that not all age-associated endocrine deficiencies result from normal senescence. A timely example is the age-related "deficiency" of melatonin, a hormone, produced by the pineal gland, that controls some neuroendocrine functions, including the timing of sleep cycles and seasonal reproductive cycles. Original reports of age-related declines of melatonin in humans led to a hypothesis that such declines were normative aging, and in fact might regulate senescence. In an unreplicated experiment, transplantation of pineal glands from young to middle-aged mice greatly extended the recipients' life span (Pierpaoli and Regelson). However, numerous strains of mice that have permanently lost the capacity to synthesize melatonin (due to domestication) live as long as lines that have retained that capacity. Indeed, it is likely that the transplanted pineal glands in Pierpaoli's study were from such a melatonin-deficient strain, leaving open the question of what exactly *did* produce the life span extension in their mice. Recently a study reevaluating the effect of aging on melatonin levels in humans demonstrated that, while levels may be diminished in sick people (which can account for earlier reports of age-related deficits), healthy older people (65–81 years) maintain circulating melatonin levels indistinguishable from levels in 18- to 30-year-olds (Zeitzer et al.).

Although studies of chronic treatment of melatonin in lab animals had not been done, the public eagerly embraced melatonin as an antisenescence intervention. In response to public demand, the National Institute of Aging currently is supporting studies of chronic melatonin-treatment in aging rodents to identify its potential benefits and dangers.

The Cumulative Exposure Hypothesis. In contrast with the endocrine deficiency theories, another hypothesis is that senescence may result from long-term *exposure* to normal hormone levels. This phenomenon was most thoroughly demonstrated by Finch and colleagues (reviewed in Finch, 1990) using the murine female reproductive system as a model. Ovariectomy in young adults delayed or prevented numerous age-related impairments including: the dysregulation of gonadotrophins, the elevation of glial hyperactivity in the hypothalamus, and the formation of pituitary adenomas. Superphysiological elevation of a principal ovarian hormone, estradiol, in young females accelerated the appearance of these impairments. Finch proposed that the cumulative effect on target tissue of exposure to normal levels of ovarian estradiol throughout adulthood produced many of the impairments that characterize female reproductive senescence.

Extension of this cumulative exposure model to other endocrine axes led to the recent discovery that the expression of many aspects of senescence is delayed in dwarf mice with a congenital deficiency of growth hormone, thyroid hormone, and prolactin, and that their life span is increased by 25–45%, depending on genotype and husbandry (Flurkey et al., 2000). This is the first genetic treatment that both retards senescence *and* increases life span in mammals. Along with dietary restriction (see below), it exposes the surprising plasticity of mammalian life span. Unfortunately, these panhypopituitary dwarf mice also are quite frail, demonstrating the tradeoff between youthful vigor and extreme long life that the theory of antagonistic pleiotropy predicts.

The apparent general conflict between endocrine deficiency models and cumulative exposure models results from the dual nature of hormones such as estradiol, growth hormone and thyroid hormone that are important for both homeostasis and for maturation and growth. Hormones that produce the structural and biochemical remodeling necessary for maturation often continue to promote aspects of this remodeling throughout adulthood, eventually pushing the "maturational" change past its optimal setting at advanced ages, when the force of selection against this process is weak. We might say that senescence results from hypermorphosis, the exaggeration of a trait—in this case, the exaggeration of maturational characteristics. Senescence may be slower in individuals with low levels or relative insensitivities to maturational hormones, but the homeostatic functions performed by these same hormones also will be diminished. Thus, the dual nature of many maturational hormones results in a tradeoff during aging between endocrine-support vigor and endocrine-directed senescence.

Other Integrative Systems. Immunologic aging has been proposed as a "pacemaker" of senescence in vertebrates (Walford, 1969), primarily through age-associated elevation of autoimmune reactions, through diminished protection from pathogens, and through impaired immune surveillance against tumor progression. According to Miller (1995), "Nearly all of the evidence on this attractive hypothesis is lamentably indirect.... A more powerful approach, involving longitudinal analyses of morbidity and mortality in individuals shown to differ in immune indices thought to influence disease susceptibility, has seldom been attempted and even less frequently been successful."

One careful study, using the subject pool of the Baltimore Longitudinal Study on Aging (Bender et al.), found that healthy middle-aged men with lower blood lymphocyte counts were at greater risk for subsequent mortality. Whether the lymphopenia was of the type that could mediate the proposed mechanism of immune-directed senescence has not been investigated further in humans. More recently, Miller et al. (1997) have shown that a specific marker of T lymphocyte aging (the elevation of the ratio of memory to naive T cells), when measured in young adult mice, predicts survival when they are old. To date, it has not been demonstrated that restoration of youthful immune function in old mice retards senescence in other systems or increases life span.

Evidence accumulated since about 1985 indicates that the acute phase components of the inflammatory reaction, a part of the non-specific immune response, are hyperactivated in older mammals (Papaconstantinou). These acute phase reactants include inflammatory cytokines and a wide range of stress response proteins; both basal and activated states are elevated. While this may be a response to chronic stress, for example, oxidative stress from uncoupled mitochondria, it also could result from

dysregulation of the inflammatory response. Elevated acute phase reactants could promote senescence in other systems, including osteopenia and hyperproliferative disorders such as atherosclerosis; however, it is not yet known if chronic treatment with anti-inflammatory agents affects senescence and life span.

Cellular Senescence

Until the 1960s, it was generally thought that senescence was an epiphenomenon of metazoan organization: Bacteria and most protozoan species do not exhibit senescence; transformed cell lines are immortal; and, even non-transformed cell lines (specifically, chick fibroblasts) were thought to be immortal. In 1961, Hayflick and Moorehead published their classic paper on the limited proliferation potential of cells in vitro. Their demonstration that fibroblasts from human fetal lung tissue, when cultured, could divide on the average only 50 times established cellular senescence as an important concept in gerontology.

***In Vitro* Cellular Senescence.** It is now known that, except for stem cells, virtually all mammalian cell types that can be cultured have a proliferative limit unless they transform to neoplastic cells. Although 50 population doublings is far more than necessary to supply a sufficient number of fibroblasts for even the longest human lifetime, late passage fibroblasts, i.e., cells that have already undergone a majority of their doublings, are distinctive morphologically (they are larger and more granular), functionally (they migrate and replicate more slowly), and biochemically (they increase production of extracellular matrix proteins and decrease production of growth-related proteins). Such "senescent" fibroblasts are present in explants from both young and old adults, although they are in greater concentration in old tissues (reviewed in Wolf and Pendergrass). The theory that senescence results from limited cellular replicative potential maintains that, as cells approach their proliferative limit during aging, the resulting accumulation of the senescent cells eventually impairs normal function.

Hayflick and Moorehead demonstrated that culture conditions or toxic metabolic side products could not account for the proliferative limitation. Later studies employing microtransfer of the cell nucleus demonstrated that the proliferative limit was not due to some cytoplasmic factor, but was directed by the nucleus, suggesting that cellular senescence was either programmed or a result of accumulated genetic damage. It now seems likely that the proliferation limit results from cell division-associated shortening of the chromosomal terminus structures—telomeres—which are docking sites for DNA polymerase. As a cell progresses through multiple divisions, its telomeres shrink to a size that is unrecognizable by the polymerase and replication ceases (Bodner et al.). The question of whether this process contributes to senescence in vivo, or if it should even be considered senescence *in vitro*, is still unresolved.

A recent observation demonstrates that the accumulation of "senescent" fibroblasts in vivo during adult aging may not result from the approach of a proliferative limit (Cristofalo et al.) Human fetal skin fibroblasts have a replication potential of about 60, and explants from young adults have about half that potential. However, the proliferation potential of skin fibroblasts *does not* decrease with age in healthy human adults, despite the predominance of morphologically and functionally senescent cells in explants from old people. Thus, senescent-like fibroblasts accumulate in vivo during aging in the absence of a change in proliferative potential.

One emerging view is that the maximum proliferation potential reflects the function of a relatively small number of undifferentiated and long-lived fibroblast stem cells that produce much greater numbers of transiently proliferative, but committed and shorter-lived, daughter cells. In this context, tissues from old donors would contain a greater proportion of more highly differentiated daughter cells, and the expression of their distinct, but normal, functions would be considered senescence. It is noteworthy that all fibroblasts that reach their proliferative limit do not die; they can be maintained in culture indefinitely and continue to produce the extracellular matrix proteins that are characteristic of mature fibroblasts. Thus, the proliferative limit in late passage fibroblasts may be a type of cell cycle arrest, which is a common end point of cellular developmental programs; and, the process of cellular senescence may be indistinct from normal terminal differentiation.

***In Vivo* Cellular Senescence.** Senescence of T lymphocytes can be studied *in vivo*. Most "senescent" characteristics, such as the diminished responses to non-specific mitogens, actually reflect the function of differentiated, antigen-activated T cells and memory T cells (Lerner et al. 1989), which selectively respond only to their cognate antigen and not to non-specific mitogens (Flurkey et al., 1992). These activated and memory T cells progressively replace the less mature "naive" T cells during aging (reviewed in Miller, 1995). Thus, cellular aging *in vivo* may reflect the gradual accumulation of more mature, committed cells that have a proliferative limit and that express more differentiated, not less functional, characteristics.

Because "senescent" cells represent one end of a spectrum of differentiated cells that exists in young individuals, as well as old, the most relevant, unaddressed question for cellular senescence may be, "Why do the more highly differentiated cells accumulate during senescence *in vivo*?" Cellular senescence could be simply an extension throughout adulthood of a maturational process designed to produce an optimal mix of cells, which includes less differentiated cells that retain the capacity to proliferate when needed, and more differentiated cells that best perform the functions of the cell lineage. As the animal ages, the more differentiated cells seem to accumulate eventually shifting the ratio past the optimum — another example of senescence resulting from hypermorphosis.

Molecular Senescence

The focus of gerontology at the molecular level is on age-related changes in gene expression and on the mechanisms that can damage biomolecules. Because random factors in the environment are considered the major sources of damage, theories of senescence at this level are often called stochastic theories. These stochastic theories are discussed in greater detail in **Biochemical Theories of Aging**. Definitive testing of these theories requires that a specific intervention, which is shown to diminish damage from the stochastic agent, also retards senescence and increases life span, as do the dwarf mutations (above) or diet restriction (below). To date, no such intervention has been demonstrated to retard senescence in mammals, possibly because of technical problems, for example, an inability to deliver the putative anti-senescence agent to appropriate sites in effective doses. Recently, transgenic overexpression of p66[shc], a gene that mediates response to free radical damage, was reported to increase life span in mice (Migliaccio et al.), a finding consistent with the possibility that free radicals promote senescence in mammals. If this finding is replicated, it will demonstrate the potential that the "new genetics" provides for tests of stochastic theories of senescence.

Hundreds of studies have been published on the effect of age on protein and RNA expression (reviewed in Van Remmen et al.). Whereas aging consistently may increase expression of some specific genes and decrease expression of others, the majority of research indicates that both total RNA synthesis and total protein synthesis decline with age in most mammalian tissues. Nonetheless, there does not appear to be a significant decrease in the protein or DNA content of cells, perhaps because of the demonstrated decline in the degradation of protein and RNA with age, resulting in slower turnover. Despite a general decline of macromolecular synthesis rates, the specific elevation of some gene expression during aging indicates that there is no limiting impairment in the machinery of transcription or translation. Although the decreased cycling of macromolecules could contribute to the decreased ability of senescent organisms to respond to stresses, Van Remmen et al. state, "...it is not yet known if the changes in gene expression and/or protein degradation play a causal role in senescence or if the changes are simply secondary to the aging process."

Caloric Restriction

In 1935, McCay reported that lifelong reduced food intake increased life span in rats. This finding, one of the most robust in gerontologic literature, applies throughout the animal kingdom, where restriction of food intake of 25 percent or more typically increases life span (reviewed in Yu). The critical component is restriction of calories from fats or carbohydrates; protein-or micronutrient-restriction impairs health. In supportive laboratory environments, caloric restriction does not affect IMR, but it increases MRDT by about 1.7-fold (Finch), indicating that restriction acts by retarding senescence in general. In fact, caloric restriction delays the onset of virtually all age-related diseases, and it appears to retard senescence in almost all physiological systems studied.

Diminished metabolic rate does not seem to account for the effect of caloric restriction. Oxygen consumption per gram of lean body mass, a measure of fuel usage, does not appear to be affected in caloric-restricted rodents following the first two months of adjustment. Because caloric restriction retards so many aspects of senescence, including cellular senescence and free radical and glycation damage to protein, RNA, or DNA, it is difficult to identify a specific mechanism that mediates its effects.

The effectiveness of caloric restriction is not linked to a critical developmental period analogous to the critical period for imprinting. Although caloric restriction for rats is usually begun at weaning, it is almost as effective when begun in young adults (6 months), and it still can retard senescence and increase life span when started in early middle-age.

The effectiveness of caloric restriction in rodents illustrates the remarkable degree to which senescence rates and life span can be manipulated in at least some mammalian species, and it promotes optimism concerning the potential for discovery of an intervention that is comparably effective in humans. These results also bear on the current debate over whether there is one, few, or many causes of aging in mammals. Whereas the majority of evidence indicates that there is not a single pacemaker of aging in mammals (Costa and McCrae), caloric deficiency, a single biological parameter, nonetheless has profound effects on senescence, demonstrating that a single mechanism *is* possible. However, until we discover exactly how caloric restriction retards senescence, it will be impossible to resolve this issue.

Because studies of caloric restriction in mammals have been completed only for laboratory rodents, it is possible that restriction will not be effective in mammalian species with longer life spans. Currently, caloric restriction is being tested on primates at three sites in the United States, including the National Institute on Aging. Preliminary results demonstrate that caloric restriction retards the expression of age-related disease in primates, as it does in rodents. Within 10 years enough data should be available to determine if caloric restriction is a valid anti-senescence treatment for primates.

Genetic Strategies

Classical Genetics. Until the advent of DNA sequencing and genetic engineering in the 1980s, studies in the genetics of senescence were limited to classical genetics. Overall, these studies demonstrated that: life span is heritable, i.e., genetically regulated, although within a species, non-genetic factors were generally more important; the genetic regulation of life span could be influenced by artificial selection (for invertebrates); genes could be altered (by random mutagenesis) to retard senescence and increase life span (again in invertebrates); genetic regulation of senescence in mammals involved many genes; genetic expression was confounded by environmental interactions; and, specific genes would be very difficult to identify. Table 1 summarizes several of these studies.

The New Genetics. The invention in the mid-1970s of a method to sequence DNA rapidly led to the techniques of molecular cloning and genetic engineering that have revolutionized biology and medicine. Scientists are using these techniques to identify the specific genes that regulate senescence by extending the older genetic designs and by performing experiments they never imagined possible. Even the initial results have excited the most conservative gerontologists.

Genetic Regulation of Senescence in Roundworms. Earlier mutagenesis studies in the simple soil nematode *C. elegans* created genetic variants that were not found in nature, and demonstrated that alterations of specific genes could retard senescence and more than double life span in these worms. (Such studies are impractical in mammals because it would be necessary to maintain thousands of lines for many years to evaluate the effects of each randomly induced mutation on life span.) The genes that increased life span in the worms were found to belong to a single genetic pathway, known as the dauer pathway, that regulates the response to food scarcity and overcrowding during development. Over the past five years, these genes have been sequenced (for a detailed discussion, see G. Ruvkun's webpage at http://xanadu.mgh.harvard.edu/ruvkunweb/ruvkun.html). They have surprising homology to mammalian genes for the intracellular signal transduction pathway of insulin and insulin-like growth factor (IGF-1).

Because the mutations in the worm are reduction-of-function mutations, the implication for mammals is that reduced signaling through the insulin and IGF-1 signal transduction pathways may retard senescence and increase life span. Unfortunately, unlike their beneficial effect in worms,

TABLE 1. STUDIES OF AGING USING CLASSICAL GENETICS

Strategy	Result	Reference
Analyses of heritability	For a wide range of animal species, 10–35% of the variance in life span could be accounted for by inheritance.	Finch et al., 1990
Selection studies	These studies, restricted to short lived invertebrates, were unsuccessful until the 1980s, when a 15–30% increase in life span was achieved (see "Tests of Theories" in the "Population Senescence and Evolutionary Biology" section).	Rose 1991
Analyses of genetic variants	Using special strains of mice, the major histocompatibility locus was identified as a chromosomal region that interacted with unknown environmental conditions to influence life span and senescence.	Walford (1987); Lerner and Finch 1991
Analyses of induced mutants	Induced mutations increased life span by 50–100%; studies are practical only in short lived species such as round worms (see "Genetic Strategies" section below).	Johnson et al., 1990
Analyses of spontaneous mutants	Dwarfing mutations appeared to retard aspects of senescence in mice. The first report of increased life span in mammals due to a mutation (Ames dwarf mouse).	Flurkey and Harrison, 1990; Brown-Borg et al., 1996
Analyses of gene expression	Induction of specific and general protein synthesis was slower in older animals, but the mechanisms were not known.	Van Remmen et al., 1995

the direct application of these mutations to the insulin signal transduction pathway in mammals produces severe insulin resistance and maturity onset diabetes. However, the potential benefit of indirect applications is suggested by the other treatments known to increase life span in mammals: both caloric restriction and the panhypopituitary dwarfing mutations (discussed earlier) reduce circulating levels of insulin and IGF-1, which would be predicted to diminish signaling through their receptors.

Current initiatives involve further application of advanced genetic tools to: improve our understanding of the genetic and physiological mechanisms by which the dauer pathway genes regulate senescence; evaluate the potential role of the homologous mammalian pathways in mediating the effects of caloric restriction and the dwarfing mutations on life span; and, identify the means to selectively reduce aspects of insulin signaling without inducing insulin resistance.

Genetic Regulation of Senescence in Flies and Mice. Advanced genetic tools also have greatly facilitated mapping studies, i.e., studies to identify the chromosomal location of a gene so that it may be sequenced. Such mapping studies are now being applied to the lines of long-lived flies that were produced by the selection studies, and to identify the genes that determine natural variation in life span among mice. In flies and in mice (Klebanov et al.), chromosomal regions that have major influences on maximum life span have been identified, and the process of "fine mapping," which will lead to gene-identification, has been initiated. Because it is more difficult to identify specific genes in these systems than in *C. elegans*, progress will be considerably slower. However, over the next decade, these lines of research are expected to identify the specific genes in flies that respond to selection by increasing life span, and the genes in mice that regulate normal variation in maximum life span.

Besides the mapping and sequencing studies described above, direct genetic intervention using transgenic and knock out strategies to introduce new genes or eliminate known genes is now being used to critically test theories of senescence. This strategy is most applicable to the stochastic theories, for which genes that can diminish the molecular damage or enhance the repair are known. The first report of a transgenic treatment that increases life span in mice, the overexpression of p66[shc] (Migliaccio), a gene that mediates the response to free radicals, illustrates the potential for this approach, if it can be replicated.

Evaluating Aging Theories and Anti-Aging "Cures"

The pace of gerontological discoveries has greatly accelerated since about 1990. It seems that a new mechanism of aging, a new "geronto-gene," or even a new treatment for senescence is reported from respected sources at least every other month. Despite the promise, implied and often stated, for an intervention that will retard senescence in humans, the vast majority of the research will not contribute to this end. Although such dead ends are inherent to the scientific process, especially for a problem as complex as the causes of senescence, the contrast between the fanfare and the follow-through can engender public cynicism.

One of the purposes of this article is to help the public fairly evaluate scientific discoveries related to senescence. The examples of gerontological research given in this article illustrate the following guidelines for evaluation of future reports:

1. *Results of a single study that are not replicated by an independent laboratory are unreliable, no matter how eminent the scientist.* (This is true for all biomedicine, not just gerontology.) Although unreplicated findings are not necessarily wrong, scientists generally do not accept the validity of a finding without independent replication. Even when a discovery has "the potential to lead to a treatment for senescence," the public should be as skeptical as the scientists.

2. *A valid finding in a model system does not necessarily mean it will apply to humans.* Model systems encompass lower organisms (from yeast to rodents) and in vitro cultures (such as the human fibroblast cultures). Scientists use model systems to simplify the research designs and to permit experiments that are impossible to perform on humans. Two model systems discussed in this article illustrate the point:

   a. The roundworm, *C. elegans*, is ideal for mutagenesis studies, which permit the discovery of genes that would be impossible with traditional genetic methods that depend on naturally occurring variation. Mutagenesis studies in *C. elegans* have identified three sets of genes that regulate life span.

      i. The "clock" genes probably regulate mitochondrial function and metabolic rate. While such genes are of general interest to biologists, they have no known homology to mammalian genes. It certainly is possible that an understanding of the mechanism by which clock genes regulate life span can lead to an intervention for humans, but it is by no means likely.

      ii. The dauer pathway genes (see Genetic Regulation of Senescence in Roundworms, above) do have structural and some functional homology to genes in mammals. This has led to new gerontological research strategies that have considerable potential to enhance our understanding of senescence in humans.

      iii. The "unc" genes probably increase life span by reducing food intake. Scientists do not anticipate that knowledge of their sequence will provide any new insight into mechanisms of senescence.

   b. Senescence in human lung fibroblasts in culture was prevented by transferring a gene (telomerase) into the fibroblasts that maintain telomere length, demonstrating that telomere shortening was a critical regulator of cellular senescence in vitro. Nonetheless, the idea that the proliferative limit, the defining characteristic of in vitro senescence, is relevant to senescence in vivo is severely challenged by another experiment that utilized genetic engineering. Completely knocking out the telomerase gene in mice, which results in unusually short telomeres, has no effect on life span, demonstrating that a treatment that accelerates "senescence" in vitro may have no discernible impact on survival *in vivo* (Rudolph et al.).

3. *The use of life span as a marker of senescence, although common among gerontologists, involves special precautions:*

a. A treatment that *shortens* life span probably does not accelerate aging, and therefore is not useful in identifying mechanisms of senescence. Many things can kill an animal (including unintended consequences of a treatment) without accelerating aging. Thus, tests of theories of aging that use experimental treatments that are predicted to shorten life span are meaningful only if they do not work, in which case the theory may be rejected, as in the case of the telomerase knock out mice mentioned above.

b. A treatment that *increases* life span must retard the expression of *all* the age-associated diseases and impairments that normally kill the animal. This seems unlikely in populations that die from many different causes, unless some fundamental timing mechanism that regulates disease-expression is affected; senescence is one such mechanism. Nonetheless, studies using treatments that increase life span still must be interpreted cautiously. Many strains of inbred mice die in middle age from some strain-specific inherited disease, before senescence can have an impact. If a treatment increases life span in short-lived, inbred strains of mice, it could do so through a mechanism that is independent of senescence. DHEA-treatment is an example: although it increases life span in mice that get type II diabetes, or that die from specific cancers, it has never been reported to increase life span in long-lived strains of mice. The potential for this type of misinterpretation can be minimized by the use of hybrid or genetically heterogeneous mice, a trend that is increasing among gerontologists.

c. Even treatments that increase life span in long-lived strains may do so indirectly by reducing food intake, which contributes nothing more to our understanding of senescence. Thus, data on food intake or body weight are crucial to the interpretation of any treatment purported to retard senescence.

4. *A published effect of age on some hormone, enzyme, or function does not necessarily mean the effect can be considered normative aging.* Because of the relationship of senescence with disease, the group of old subjects may include sick subjects; therefore, differences between young and old subjects may reflect an effect of diseases rather than of normal aging. When care is not taken to exclude unhealthy subjects, the results can be misleading, as illustrated by studies that led to the widespread misconception that melatonin declines with age (discussed above). For an age effect to be considered normative, the researchers must have shown that their subjects, whether human or laboratory animal, were free of disease.

5. *Although many dietary supplements, hormones, and drugs are touted in popular literature as "anti-aging" treatments, very few have been tested in lab animals or humans for long-term effects.* Even transitory long-term treatment could cause permanent damage. For example, it is not known if the chronic joint pain caused by months of growth hormone treatment in adults is reversible after the treatment is withdrawn (Papadakis et al.), or if years, or even months, of DHEA-treatment elevates forever a man's risk for prostate cancer.

Conclusion

Unlocking the mystery of senescence has never been a more exciting prospect. At no other time in history have so many theories provided so much promise—for the causes of aging *and* for credible intervention. But scientists must be responsible, tempering their excitement with the reality of replication and harmful trade-offs. And the public must be prudent consumers, accepting no less than the entire story and refusing to be misled, even if doing so lengthens the time between initial discovery and valid treatment. Some theories will collapse; some will fade slowly. But some theories will persist...and perhaps lead to successful interventions. Although we never will achieve immortality, we very well might live more vital and longer lives.

Additional Reading

Arking, R. and S.P. Dudas: "Review of genetic investigations into the aging processes of *Drosophila*," *J. Am. Geriatr. Soc.*, **37**, 757–773 (1989).
Bartke, A.: "Delayed aging in Ames Dwarf Mice. Relationships to Endocrine Function and body Size," in Hekimi, ed., *The Molecular Genetics of Aging*, Springer-Verlag, Berlin, 2000, pp. 181–202.
Bender, B.S., J.E. Nagel, W.H. Alder, and R. Andres: "Absolute Blood Peripheral Lymphocyte Count and Subsequent Mortality of Elderly men. The Baltimore Longitudinal Study of Aging," *J. Am. Geriatr. Soc.*, **34**, 649–654 (1986).

Bodner, A.G. and M. Ouellette, et al.: "Extension of Life Span by Introduction of telomerase into Normal Human Cells," *Science*, **279**, 349–352 (1998).
Brown-Borg, H.M., K.E. Borg, C.J. Meliska, and A. Bartke: "Dwarf Mice and the Aging Process," *Nature*, **384**, 33 (1996).
Costa, P.T. and R.R. McCrae: "Design and Analysis of Aging Studies," in Masoro, E.J.; ed., *Handbook of Physiology Section 11: Aging*, Oxford University Press, Oxford, 1995.
Cristofalo, V.J., R.G. Allen, R.J. Pignolo, et al.: "Relationship Between Donor Age and the Replicative Lifespan of Human Cells in Culture: A Reevaluation," *Proc. Natl. Acad. Sci. USA.*, **95**, 10614–10619 (1998).
Finch, C.E.: *Longevity, Senescence and the Genome*, The University of Chicago Press, Chicago, IL, 1990.
Finch, C.E., M.C. Pike, and M. Witten: "Slow Increases of the Gompertz Mortality Rate during Aging in Certain Animals Approximate that of Humans," *Science*, **249**, 902–905 (1990).
Flurkey, K. and D.E. Harrison: "The Use of Genetic Models to Investigate the Hypophyseal Regulation of Senescence," in D.E. Harrison, ed.: *Genetic Effects on Aging II*, Telford Press, Inc., Caldwell, NJ, 1990.
Flurkey, K., M. Stadecker, and R.A. Miller: "Memory T lymphocyte Hyporesponsiveness to Noncognate Stimuli: A Key Factor in Age-related Immunodeficiency," *Eur. J. Immunol.*, **22**, 931–1035 (1992).
Flurkey, K., R.A. Miller, J. Papaconstantinou, and D.E. Harrison: *Retarded Aging and Increased Life Span in the Snell Dwarf*, 2000.
Hayflick, L. and P. Moorhead: "The Serial Cultivation of Human Diploid Cell Strains," *Exp. Cell Res.*, **25**, 585–621 (1961).
Johnson, T.E., D.B. Friedman, N. Foltz, et al.: "Genetic Variants and Mutations of *Caenorhabditis elegans* Provide Tools for Dissecting the Aging Process," in D.E. Harrison, ed.: *Genetic Effects on Aging II*, Telford Press, Inc., Caldwell, NJ, 1990.
Klebanov, S.E., C.M. Astle, T.H. Roderick, et al.: "Evidence for Mammalian Genes that Retard Aging," *Genetics* (2000), (in press).
Lerner, A., T. Yamada, and R.A. Miller: "PGP-1[hi] Lymphocytes Accumulate with Age and Respond Poorly to Concanavalin A," *Eur. J. Immunol.*, **19**, 977–982 (1989).
Lerner, S.P. and C.E. Finch: "The Major Histocompatibility Complex and Reproductive Functions," *Endocrine Reviews*, **12**, 78–90 (1991).
Luckinbill, L.S., R. Arking, et al.: "Selection for Delayed Senescence in *Drosophila melanogaster*," *Evolution*, **38**, 996–1003 (1984).
Masoro, E.J.: "Aging, Current Concepts," in E.J. Masoro, ed.: *Handbook of Physiology Section 11: Aging*, Oxford University Press, Oxford, 1995.
Medawar, P.B. *An Unsolved Problem of Biology*, H.K. Lewis, London, 1952.
Migliaccio, E., M. Giorgio, S. Mele, et al.: "The p66shc Adaptor Protein Controls Oxidative Stress Response and Life Span in Mammals," *Nature*, **402**, 309–313 (1999).
Miller, R.A.: "Immune System," in E.J. Masoro, ed.: *Handbook of Physiology Section 11: Aging*, Oxford University Press, Oxford, 1995.
Miller, R.A., C. Chrisp, and A. Galecki: "CD4 Memory T Cell Levels Predict Life Span in genetically heterogeneous mice," *FASEB J.*, **11**, 775–783 1997.
Orentreich, N., J.L. Brind, R.L. Rizer, and J.H. Vogelman: "Age Changes and Sex Differences in Serum Dehydroepiandrosterone Sulfate Concentrations Throughout Adulthood," *J. Clin. Endocrinol. Metab.*, **59**, 551–555 (1984).
Papaconstantinou, J.: "Unifying Model of the Programmed (Intrinsic) and Stochastic (Extrinsic) Theories of Aging," *Ann. New York Acad. Sci.*, **719**, 195–211 (1994).
Papadakis, M.A., D. Grady, D. Black, et al.: Growth Hormone Replacement in Healthy Older Men Improves Body Composition but not Functional Ability, *Ann. Int. Med.*, **124**, 708–716 (1996).
Pierpaoli, W. and W. Regelson: "Pineal Control of Aging: Effect of Melatonin and Pineal Grafting on Aging Mice," *Proc. Soc. Nat. Acad. Sci. USA.*, **91**, 787–791 (1994).
Rose, M.R.: "Laboratory Evolution of Postponed Senescence in *Drosophila melanogaster*," *Evolution*, **38**, 1004–1010 (1984).
Rose, M.R.: *Evolutionary Biology of Aging*, Oxford University Press, New York, NY, 1991.
Rudman, D., A.G. Feller et al.: "Effects of Human Growth Hormone in Men Over 60 Years Old," *New Eng. J. Med.*, **323**, 1–6 (1990).
Rudolph, K.L., S. Chang, H.W. Lee et al.: "Longevity, Stress Response, and Cancer in Aging Telomerase-deficient Mice," *Cell*, **96**, 701–712 (1999).
Van Remmen, H., W. Ward, R.V. Sabia, and A. Richardson: "Gene Expression and Protein Degradation," in E.J. Masoro, ed.: *Handbook of Physiology Section 11: Aging*, Oxford University Press, Oxford, 1995.
Walford, R.L.: *The Immunological Theory of Aging*, Williams and Wilkins, 1969.
Walford, R.L.: "MHC-regulation of Aging: An Extension of the Immunologic Theory of Aging," in H.R. Warner et al., eds.: *Modern Biological Theories of Aging*, Raven Press, New York, NY, 1987.
Williams, G.C.: "Pleiotropy, Natural Selection, and the Evolution of Senescence," *Evolution*, **11**, 398–411 (1957).
Wolf N.S. and W.R. Pendergrass: "The Relationships of Animal Age and Caloric Intake to Cellular replication *in vivo* and *in vitro*: A Review," *J. Gerontol. A Biol. Sci. Med. Sci.*, **54**, B502–B517 (1999).
Yu, B.P.: "Putative Interventions," in E.J. Masoro, ed.: *Handbook of Physiology Section 11: Aging*, Oxford University Press, Oxford, 1995.

Zeitzer, J.M., J.E. Daniels, J.F. Duffy, et al.: "Do Plasma Melatonin Levels Decline with Age?" *Am. J. Med.*, **107**, 432–436 (1999).

Web References

British Society for Research on Aging, Links: http://www.bsra.org.uk/Links/links.html

The Aging Research Center: http://www.arclab.org/

The Institute of Gerontology, University of Michigan: http://www.iog.umich.edu/research/index.html

The Ellison Medical Foundation: http://www.ellison-med-fn.org/

KEVIN FLURKEY, Ph.D., The Jackson Laboratory, Bar Harbor, ME

GERSDORFFITE. A mineral related to cobaltite and ullmannite in the cobaltite group. A sulfide-arsenide of nickel, NiAsS. Crystallizes in the isometric system. Hardness, 5.5; specific gravity, 5.9; color, white to gray with metallic luster; opaque.

GESNER, CONRAD (KONRAD)1516–1565. Conrad Gesner was a Swiss physician, naturalist and philologist whose publications were among the first to emphasize plant and animal structures as keys to classification and provide extensive bibliographies.

Gesner, one of many children of Ursus Gesner, a Zurich furrier, and his wife Agathe Frick, was raised by an extended family that in addition to his parents included a great-uncle who was a Reformed minister, the Latinist Johann Amman who was a friend of Erasmus, and his godfather Ulrich Zwingli, the controversial Swiss Protestant theologian. Gesner initially intended to be a theologian, too, and studied at the Carolinum in Zurich and the Fraumünster seminary. After his father and Zwingli died in battle in 1531, Gesner left Zurich to study Hebrew in Strasbourg before taking up medical studies at Bourges, Paris, and eventually Basel, where he earned his doctorate in 1541. During this period Gesner continued to study ancient languages and gained such a reputation as a philologist that he was appointed at age 21 to the first chair of Greek at the new Lausanne Academy, a post he held until 1540. Gesner settled permanently in his native Zurich, where he combined an active medical career as Zurich's chief physician and then *canonicus* with classical studies and extensive travels to the Alps and Adriatic coast to study plants and animals.

Throughout his career Gesner combined an interest in ancient languages and authors—Greek and Pliny were his favorites—with extensive study of the natural world. The immediate success of his four-volume *Bibliotheca universalis* (1545), which was an index of Greek, Latin and Hebrew writers and texts, secured his scholarly reputation throughout Europe and is the reason modern bibliographers claim Gesner as the founder of their field. Gesner extended his bibliographic approach to his medical scholarship. For example, his *Chirurgia scriptores* (1555) combines an index of surgical authors and texts with a bibliographic essay on great surgeons and an historical sketch of the development of surgery.

Gesner's *Historia animalium (1551–1587)*, an encyclopedia of animal descriptions, was largely derivative, but its breadth and meticulous presentation nonetheless mark him as an early figure in what became veterinary science subsequently. Interested as well in paleontology, Gesner sketched fossils and studied crystal structures. However, his major project, a natural history of plants he intended to be the most extensive yet produced, remained unfinished at his death despite containing almost 1,500 of his plant drawings and descriptions. Some contemporary naturalists criticized Gesner's liberal inclusion of others' work in his texts even though Gesner usually credited his sources. For example, Leonhart Fuchs, one of the other great sixteenth-century herbalists and physicians, wrote to a colleague that he was not "worried" by possible publication of Gesner's herbal because he "has a mind that flits around through almost every type of author and rejoices to weave together new wreaths fashioned from the unraveled garlands of others." (Fuchs to J. Camerarius the Elder, 24 November 1565). In 1759 C. C. Schmiedel published two folio volumes, or about one-third of Gesner's manuscripts, with hand-colored illustrations of Gesner's drawings as *Conradi gesneri . . . Historiae plantarum . . .* Between 1972 and 1980, a selection of Gesner's descriptions and drawings was published in eight folio volumes, *Conradi Gesneri historia plantarum*, with commentary and translations by H. Zoller, M. Steinmann, and K. Schmid. Compared to other herbalists of the time such as Fuchs, Gesner emphasized differences in plant structures as a method that provided for systematic classification. Also, he drew attention to structures in seeds, which led him to establish kinship between plants that otherwise

seemed dissimilar. Linnaeus later acknowledged Gesner's influence on his own approach. See also **Linnaeus, Carl (Linné) (1707–1778)**; **Fuchs, Leonhart (1501–1566)**; and **Taxonomy (The History)**.

ROBERT L.MARTENSEN, University of Kansas, Kansas City, KS

GESTATION. The period of intrauterine fetal development. See also **Pregnancy**. Pregnancy in humans is usually about 280 days. The period varies considerably over the spectrum of mammals—from a few weeks to well over a year. See Table 1.

TABLE 1. GESTATION PERIOD OF VARIOUS SPECIES (Days)

Species	Days	Species	Days
Elephant, African	640	Goat	150
Elephant, Indian	630	Sheep	150
Rhinoceros	530	Armadillo	150
Giraffe	430	Chinchilla	115
Tapir	390	Pig	115
Ass (domestic)	365	Porcupine	112
Whale (sperm)	365	Lion	108
Zebra	365	Tiger	106
Sea lion	342	Jaguar	100
Whale (blue)	335	Leopard	95
Horse	335	Cheetah	92
Walrus	330	Hyena	91
Cow (domestic)	283	Ermine	65
Dolphin	276	Coyote	65
Bison (American)	270	Raccoon	65
Sable	250	Dog	63
Chimpanzee	245	Guinea pig	63
Seal	245	Wolf	63
Alpaca	240	Bat (brown)	55
Elk	240	Fox	55
Hippopotamus	240	Mink	42
Deer	225	Ferret	42
Reindeer	220	Kangaroo	39
Orangutan	218	Weasel	35
Gibbon	210	Rabbit	31
Bear	208	Hamster	21
Baboon	186	Rat	21
Badger	183	Mouse	20
Porpoise	183	Shrew	18
Monkey	165		

GETTERING. The absorption of gas by a getter film. When this process occurs during the dispersal of the getter through an evacuated system (such as an electron tube), it is called dispersal gettering; when by action of the already dispersed film, it is called contact gettering. In electric-discharge gettering, the process is accelerated by passing an ionizing electron discharge through the gas. The gas is ionized, and the ions are neutralized when they impinge on an electrode, so that the final product is neutral gas atoms. These are then easily absorbed by the getter.

A getter film is a metallic deposit in a vacuum system with the function of absorbing residual gas. Electropositive metals, such as sodium, potassium, magnesium, calcium, strontium, and barium have been used as getters. The process of depositing a getter film upon a surface may be done in various ways. In the distillation method, the metal to be deposited is volatilized into the vacuum system from a side tube provided with constructions for sealing-off when the process is completed. The electrolytic method is applicable where the metal to be deposited is sodium, and where the system is made of soda-lime glass. It is well known that sodium may be electrolyzed through soda-lime glass. If, therefore, a thermionic source of electrons is provided inside an evacuated sealed-off vessel, part of which is dipped into a suitable liquid kept at a high potential relative to the source of electrons, a current will pass, carried by electrons between the thermionic cathode and the inner surface of the glass, and by ions within the glass. The only ions in the glass that are mobile are sodium ions, and thus pure sodium is released at the inner surface of the envelope.

Other modern getter materials include cesium-rubidium alloys, tantalum, titanium, zirconium, and several of the rare-earth elements, such as hafnium.

GEYSER. Derived from the Icelandic word geysa, meaning gush and descriptive of hot springs which at regular, or irregular, intervals throw a column of steam and hot water into the air. Geyser waters usually build up tubes or conduits of siliceous sinter. Geyser waters have been proved to be mainly vadose with approximately 10% of juvenile or magmatic water. Geyser action is the result of vadose water coming in contact with steam arising from the solidifying magma, and periodically returning to the surface through the geyser tube. This is the same reason that water is suddenly expelled from a test tube when heated too rapidly. The mechanics of geyser action are shown in Fig. 1.

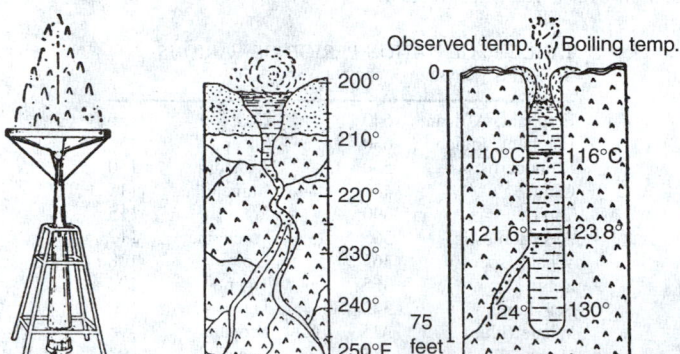

Fig. 1. The mechanics of geyser action, as illustrated by laboratory experiment, and the hypothetical cross sections of natural geysers. (*Field, "Outline of Geology," Barnes & Noble.*)

The principal geyser fields are in the western United States, notably Wyoming (Yellowstone National Park) and California, and in New Zealand and Iceland. Geysers and other sources of geothermal energy are receiving increasing attention as alternative energy supplies. Such exploitation of geothermal energy, of course, is not recent, but extends back for many years in Iceland, New Zealand, and Italy. See also **Geothermal Energy**.

Yellowstone Park claims the world's largest geyser area with approximately 3,000 geysers and hot springs.

GEYSERUTE. A loose or compact, sometimes concretionary, siliceous deposit, formed by geysers and hot springs from the material held in solution by the thermal waters.

GHATTI GUM.   See Gums and Mucilages

GHIBLI.   See Winds and Air Movement.

GHOST IMAGE. Two of the uses in science of this term are: 1. In spectroscopy, false images of a spectral line produced by irregularities in the ruling of diffraction gratings. Rowland ghosts are false images grouped symmetrically on both sides of the true line. Lyman ghosts are false orders of spectra for which the order is not an integer.

2. In television, a second image appearing on the receiver screen, superimposed on the desired signal. These images are caused by reflected rays arriving at the receiving antenna some small interval after the desired wave. A single, reflected ray from a stationary object will produce a single, clear ghost, while a number of reflected rays arriving at assorted times creates an effect known as "smearing" or "smear ghost." Ghosts may also be produced with intensity reversal (white becomes black and vice versa) due to a suitable phase of the secondary signal with respect to the primary signal, occurring on a suitable amplitude range of the received primary signal. This ghost is customarily called a negative ghost.

GIANT AND DWARF STARS. During the first two decades of this century, it was found, on the basis of parallax and photometric studies, that stars of similar spectral characteristics and temperatures diversified into two essentially distinct classes. This separation being on the basis of absolute magnitude, it was surmised by E. Hertzprung and independently by H.N. Russell that the difference must be due to a larger radius for the brighter stars at the same color, or effective temperature. The terms *giant* and *dwarf* were applied to the two groups. Intermediate groupings

are now also recognized, which are *supergiants* and *subgiants* for the most luminous stars, and *white dwarfs* and *subdwarfs* for those of lower luminosity.

Largely due to the work of Adams at Mount Wilson and Morgan at Yerkes, it was recognized by the 1930s that the spectral characteristics of the giants also differ from dwarfs, in that the giants always show narrower lines and often, at the same effective temperature, appear to have an earlier spectral type. In addition, there is a steady progression in the strength of certain lines on the basis of increasing or decreasing strength with increasing luminosity.

This is the basis of the second dimension of spectral classification in the MK (Morgan-Keenan) system, which adds a "luminosity class" to the temperature class of the Harvard (HD) system. In the MK system, luminosity classes run from I (supergiants) through V (dwarfs), and have temperature classes (in order of decreasing temperature) of O, B, A, F, G, K, M with additional classes R and S being reserved for the carbon stars. The sun, with an absolute magnitude of +4.6 and a surface temperature of about 5800 K is a G2V star, that is, a G2 dwarf, while δ Cygni is of a similar temperature, but has an absolute magnitude of −4.7 and is an F8Ib supergiant. The standard star for photometry, Vega (α Lyrae) is defined to be an AOV star having an absolute magnitude of +0.5. The MK system of classification proceeds by comparison of a given unknown stellar spectrum with agreed-upon standards and so is internally consistent. This behavior can be explained by a difference in surface gravity, and consequently pressure in the atmospheres of these stars. The lower pressure of the giant envelope produces less line broadening due to fewer perturbing collisions between radiating atoms, while the lower electron density causes an increase in the ionization at the same temperature.

The dwarf stars correspond to members of the main sequence, which is the hydrogen core burning stage of stellar evolution. It should be noted that the number of stars in any region of the Hertzprung-Russell (H-R) diagrams is approximately proportional to the period of a star's life during which it resides at that temperature and luminosity. The main sequence can thus be shown to be the longest lived stage of a star's life. The subdwarf population corresponds to the older, more metal poor, main sequence of the halo and old disk and is similar in characteristics to that observed in the globular clusters like 47 Tuc and ω Centauri. The subgiants, which are the first "post-main sequence" phase, are hydrogen core exhaustion and shell burning stars, and represent a transition between the main sequence and the giants. The brightness of giants is not as regularly correlated with mass as in the main sequence, for which a mass-luminosity relation exists (the more massive main sequence stars are brighter). The giants and supergiants are helium core and shell burning stars (and possibly double-shell sources), having ignited the spent helium core relic from the main sequence stage. These stars will eventually (depending upon mass) evolve into planetary nebulae and white dwarfs, supernovae, or if massive enough perhaps into black holes.

The giant and dwarf stars differ also in other important characteristics. While only the most massive main sequence stars show any evidence of stellar winds of any appreciable strength (greater than 10^{-9} solar masses/year), many red and blue supergiants show evidence of substantial mass loss. Blue supergiants like P Cygni, and red supergiants like α Ori and α Her show considerable envelopes, with characteristic velocities of hundreds of kilometers per second. Some also display radio continuum emission, another indication of mass loss. Only the δ Sct and β Cep stars are on or near the main sequence, while the giants and supergiants show most of the other classes of variable stars. See also **Variable Star**.

While the majority of dwarf stars in the galactic disk show abundances similar to the Sun (to within a factor of 2), giants show a wide range, indicative of considerable mixing of interior material which has undergone nuclear processing. At least one giant, FG Sge, has shown atmospheric abundance changes with time, an increase of heavy metals (rare earths) which are produced by neutron irradiation with subsequent mixing. The giant stars in globular clusters also show evidence for some time-dependent mixing processes.

The giants are best studied in globular clusters, where they form the horizontal branch population. Brighter stars, observed in several of the oldest of the clusters, lie on the asymptotic branch, parallel to the giant branch, but slightly bluer and brighter. The main sequence stars in these clusters are often too faint for careful study. The main sequence is best

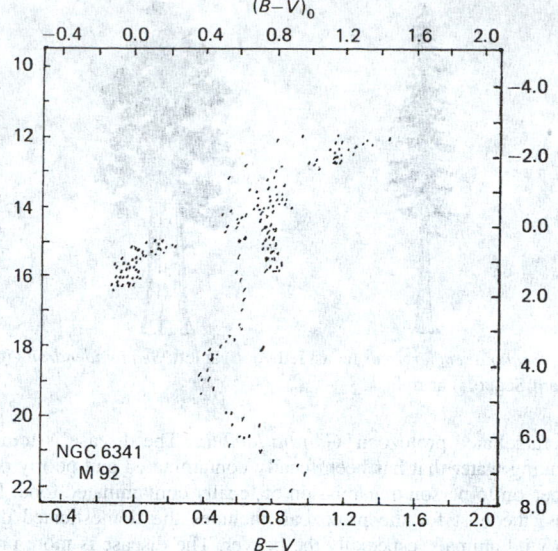

Fig. 1.   Typical, metal-poor, globular cluster M92 (=NGC 6341). Note the well-populated giant and horizontal branches. (*After Alcaino.*)

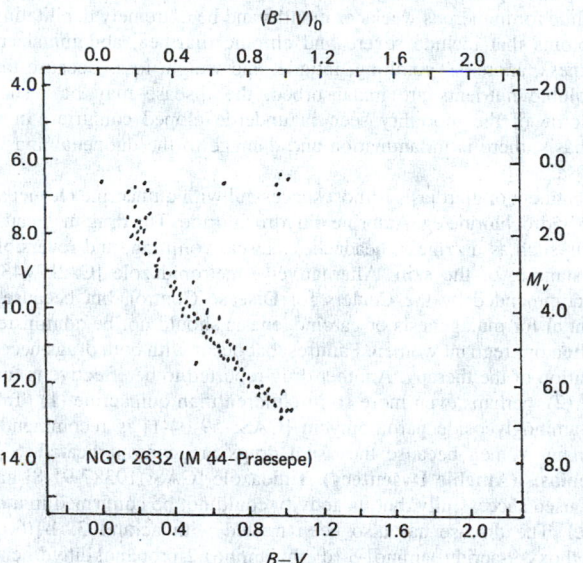

Fig. 2.   Typical young open cluster, showing a well-populated main sequence and a few late-type giants. Slight curvature at upper end of main sequence indicates these stars have begun to exhaust their hydrogen cores and evolve away from the main sequence stage. (*After Hagen.*)

observed in galactic or open clusters, like η and χ Per, Coma, the Pleiades, and the Hyades. The H-R diagrams of a typical globular cluster is shown in Fig. 1, while a diagram for a typical galactic cluster is shown in Fig. 2.

Additional Reading

De Yolung, D.S.: "Astrophysical Jets," *Science*, 389 (April 19, 1991).
Dupree, A.K. and M.T.V.T. Lago, Editors: *Formation and Evolution of Low Mass Stars*, Kluwer, Norwell, Massachusetts, 1988.
Gibbons, A.: "Astronomers Get a Whiff of Methanol," *Science*, 1094 (September 6, 1991).
Harding, A.K.: "Physics in Strong Magnetic Fields Near Neutron Stars," *Science*, 1033 (March 1, 1991).
Hunter, D.A. and J.S. Gallather, III; "Star Formation in Irregular Galaxies," *Science*, 1557 (March 24, 1989).
Lada, C.J. and F.H. Shu: "The Formation of Sunlike Stars," *Science*, 564 (May 4, 1990).
McKee, C.F. and B.T. Draine: "Interstellar Shock Waves," *Science*, 397 (April 19, 1991).

Merritt, D., Editor: *Dynamics of Dense Stellar Systems*, Cambridge University Press, New York, NY, 1989.
Stahler, S.W.: "The Early Life of Stars," *Sci. Amer.*, 48 (July 1991).
Van Horn, H.M.: "Dense Astrophysical Plasmas," *Science*, 384 (April 19, 1991).

S. N. S.

GIANTISM.   See **Hormones**; and **Pituitary Gland**.

GIANT PANDA.   See **Raccoons**; and **Pandas**.

GIANT SEQUOIA.   Of the family *Taxodiaceae* (swamp cypress family), genus *Sequoiadendron*, the Giant Sequoia (*S. giganteum*) is the only species of this genus. The champion tree, as selected by The American Forestry Association, is the "General Sherman," located in Sequoia National Park, California. See Fig. 1. This specimen has a circumference of 83 feet (25.6 meters) 11 inches at $4\frac{1}{2}$ feet (1.4 meters) above ground level, a height of 272 feet (82.9 meters) and a spread of 90 feet (27.4 meters)—as measured in 1972. In dimensions, the Giant Sequoia is rivaled only by the coast redwoods. See also **Redwood (Coast)**.

The Giant Sequoias are found on the western slopes of the Sierra Nevada Mountains of California at an altitude of from 4,500 to 8,000 feet (1,372 to 2,438 meters). There are over 25 isolated groves in which the trees occur, the taller and more dense trees being found on the northwestern slopes. The first grove was found in 1852 by a miner, A.T. Dowd. Now known as the Calaveras North Grove, it consists of about 50 acres (20 hectares) of these trees.

The bark is from 1 to 2 feet (0.3 to 0.6 meters) thick with furrows 4 to 5 inches (10 to 12.5 centimeters) wide. The bark is a red-brown color. The outer scales are fibrous and grayish-purple in color; the inner scales are a cinnamon red. The bark provides outstanding protection against the hazards of fire. The cones are deeply pitted and are of a red-brown color. The twig also is a cinnamon color and is scaly. The flower is green-gold, with pollen raining down profusely when in bloom. However, most new trees rise from shoots from stumps or roots. The leaf is from $\frac{1}{8}$ to $\frac{1}{4}$ inch (3 to 6 millimeters) long, overlapping the twig. The leaf is sharply pointed, dark green, and glossy. The wood is light in weight and not considered prime timber because it is soft, brittle but spongy, weak, and coarse-grained. At one time, the trees were cut for timber, but are now protected. Timbering operations are now concentrated on the coastal redwoods where extensive reforestation programs have been in effect for a number of years.

The Giant Sequoias also are referred to as the "Big Trees." It is important that a distinction be drawn between the coastal redwoods and the Giant Sequoias because the nomenclature can be quite confusing. Collectively, both genera are frequently referred to as redwoods. The physical differences between the two genera, however, are clearly obvious from Fig. 2.

GIANT SEQUOIA. Genus: *Sequoiadendron*; Species: *giganteum*
—also called "Big Tree," or Sierra Redwood
—Grows inland on the slopes of the Sierra Nevada Mountains
COAST REDWOOD. Genus: *Sequoia*; Species: *sempervirens*
—Sometimes also called California Redwood. The timber usually is simply referred to as redwood.
—Grows along a comparatively narrow coastal fog strip

Although the coast redwoods are taller, the extremely large girth of the Giant Sequoia qualifies it as the largest, most massive of living things. It is estimated that the tree lives for 3,000 to 4,000 years or more and thus is second only to the bristlecone pines as among the oldest living species. See also **Pine Trees**.

The "Big Tree" is highly regarded in Europe, where it was introduced shortly after its discovery in California. It is known in Europe as the "Wellingtonia." Weather and soil conditions in Britain in particular appear to be well suited to the growth of the tree. As of the mid-1970s, the tallest of these introduced trees had attained a height of over 165 feet (50.3 meters). It is located in Devonshire.

In 1864, President Abraham Lincoln authorized a federal grant transferring the area known as the Yosemite Valley and the Mariposa Grove of Redwoods to California. This act marked the beginning of the state

Fig. 1.   The "General Sherman" tree, revered specimen of Giant Sequoia, located in Sequoia National Park, California. (*National Park Service photo.*)

park concept, not just for California, but the entire nation. These properties were subsequently returned to the federal government to become part of Yosemite National Park. The first of California's present-day parks, the California Redwood Park at Big Basin, Santa Cruz County, was created in 1902, following public pressure to preserve the redwoods. This was followed by state acquisition of other notable redwood groves.

See also **Conifers**; and **Redwood (Coast)**. For references, see also **Tree**.

GIARDIASIS.   Giardiasis is a water-borne enteric disease of protozoan origin that occurs throughout the world. It is the most prevalent protozoal disease found in humans in the United States; the Rocky Mountain region is a particularly highly endemic area. It is also common in Central America, the former USSR, and India. The infection of the small intestine is caused

Fig. 2.   *Sequoia sempervirens* (coast redwood) at left; *Sequoiadendron giganteum* (the Giant Sequoia) at right.

by the flagellated protozoan, *Giardia lamblia*. The disease is transmitted by drinking water that has been fecally contaminated and poorly purified, or by person-to-person transmission of fecally contaminated food. Besides humans, the host for the protozoan includes the domesticated dog and certain wild animals, especially the beaver. The disease is more prevalent in children than in adults, and is especially common in those attending daycare centers. It is increasing in prevalence among male homosexuals.

Untreated giardiasis may be self-limiting and may produce no apparent symptoms, regardless of the duration of the infection. Alternatively, it may continue for numerous weeks or months and be extremely debilitating with symptoms that include severe and chronic diarrhea, abdominal cramps, weakness, anorexia, vomiting, fatigue, and weight loss. Because fats and fat-soluble nutrients are malabsorbed, the disease may be responsible for some of the mortality seen in underdeveloped countries. In severe giardiasis, there is inflammation and damage to the duodenal and jejunal mucosa.

Treatment of giardiasis is most successful with quinacrine (**1**, mepacrine) or its hydrochloride eg, Atabrine dihydrochloride. The drug may cause side effects such as dizziness, headache, nausea, vomiting, and reversible yellow staining of the skin. Alternatively, metronidazole [CAS: 443-48-1] is recommended by the Centers for Disease Control, but because of its potential for mutagenesis or carcinogenesis should not be administered to children or pregnant women. Failures that occur with both drugs necessitate repetition of the therapy. Another drug reported to be effective is furazolidone (**2**), perhaps even more so in children than quinacrine (**1**) (Table 1). The aminoglycoside paromomycin [CAS: 59-04-1] is recommended for pregnant women because the usual drugs are contraindicated. See also **Amebiasis (Amebic Dysentery)**. Tinidazole [CAS: 19387-91-8] has also been used successfully, but its activity could not be confirmed in a murine model. The disease has also been treated with acranil 3, 1-[(6-chloro-2-methoxy-9-acridyl)amino]-3-(diethylamino)-2-propanol dihydrochloride. Further details pertaining to giardiasis are available [Erlandsen, and Meyer 2001].

See also **Antiparasitic Agents, Amtiprotozoals**.

Additional Reading

Erlandsen, S.L., and E.A. Meyer: *Giardia and Giardiasis: Biology, Pathogenesis and Epidemiology*, Springer-Verlag New York, LLC, New York, NY, 2001.

Meyer, E.A., E.J. Ruitenberg, and A.J. Macinnis: *Giardiasis*, Elsevier Science, New York, NY, 1990.

Staff: Giardiasis: *A Medical Dictionary, Bibliography, and Annotated Research Guide to Internet References*, Icon Group International, Inc., San Diego, CA, 2004.

Thompson, R.C. A., J.A. Reynoldson, and A.J. Lmbery: *Giardia from Molecules to Disease*, CAB International, New York, NY, 2001.

Wallis, P.M., M. Olson, and S. Olson: Giardia: *The Cosmopolitan Parasite*, CAB International, New York, NY, 2002.

Web Reference

Centers for Disease Control and Prevention (CDC): http://www.cdc.gov/Ncidod/dpd/parasites/giardiasis/factsht_giardia.htm

DANIEL L.KLAYMAN, Walter Reed Army Institute of Research

GIAUQUE, WILLIAM F. (1895–1982).   An American chemist who achieved distinction for his studies of the properties of matter at temperatures approaching absolute zero (−273 °C). This research established the

TABLE 1. GIARDIASIS ANTIPROTOZOAL AGENTS[a]

Structure number	Compound name	CAS RegistryNumber	Molecular formula	Structure
(1)	Quinacrine[b]	[83-89-6]	$C_{23}H_{30}ClN_3O$	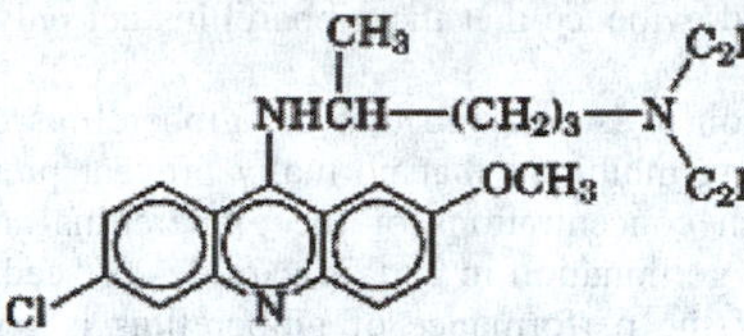
(2)	Furazolidine[c]	[67-45-8]	$C_8H_7N_3O_5$	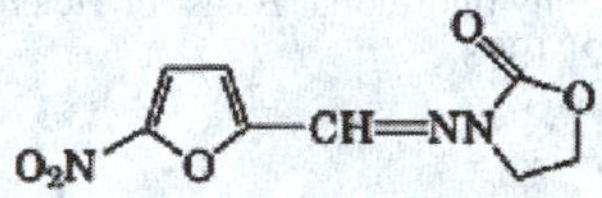
(3)	Acranil[d]	[1684-42-0]	$C_{21}H_{28}Cl_3N_3O_2$	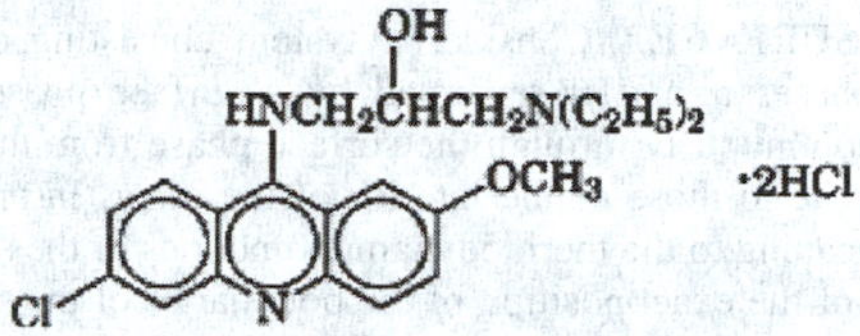

[a] Other applications are indicated in footnotes.
[b] Malaria.
[c] Histomoniasis.
[d] Hexamitosis.

science of cryogenics. Giauque received the Nobel Prize in chemistry in 1949. He was professor and research director at the University of California at Berkeley. One of his most significant contributions was the invention of a magnetic cooling device that made it possible to attain cryogenic temperatures. An important property of matter discovered as a result of his work in superconductivity.

See also **Cryogenics**; and **Superconductivity**.

GIBBERELLIC ACID AND GIBBERELLIN PLANT GROWTH HORMONES.

These organic chemical compounds, first isolated from the parasitic fungus *Gibberella fujikuori* in Japan in the late 1930s, produce unusual results when applied to plants, including various food crops. The results can be advantageous or disadvantageous. The phenomena of the gibberellins were uncovered as the result of studying the excessive leaf elongation in rice plants. This fungus disease of rice is sometimes referred to as the "foolish seedling" disease in rice. When infected with this fungus, the rice plants grow ridiculously tall and the stems break before the plants can flower and produce seed. When experimentally applied to higher plants, the gibberellins have varied effects. The most common reaction is the rapid lengthening of the stems. The stems of citrus trees, for example, have been stimulated to grow at a rate six times greater than normal. When applied to the young fruit of seedless grapes, the gibberellins cause the fruit to grow much larger and to stay on the vine longer. Although some results can be predicted from experience with other species, generally results must be observed through long trial-and-error experimentation with many plants and many different concentrations and forms of the chemical growth hormones. The gibberellins are but one category of several kinds of plant hormones that affect food crop production. See also **Plant Growth Modification and Regulation**.

Since the 1960s, commercial gibberellin formulations have been available. These take several forms, ranging from liquid concentrates through tablets and powders. In some countries, registration is required of these compounds. The following practical results, among others, have been achieved when gibberellins are used properly on certain food plants:

Artichoke: prolongs picking period
Barley: enzyme content increased
Bean: more rapid emergence of plant
Blueberry: better fruit set
Celery: extends winter crop
Cherry (sour): combats cherry yellow virus
Cucumber: produces staminate flowers
Grape: loosens and elongates clusters; increases grape size
Hops: increases yields; aids harvesting
Lemon: delays yellow color development
Lettuce: increases seed production; effects uniform bolting
Oats: promotes more rapid emergence of plant
Orange (navel): retards aging of rind
Potato: stimulates sprouting
Prune (Italian): increases yield; reduces internal browning
Rhubarb: for forced crops, increases yield
Rye: promotes more rapid emergence of plant
Soyabean: promotes more rapid emergence of plant
Sugarcane: increases sucrose yield
Tangerine: increases yield and fruit set
Wheat: promotes more rapid emergence of plant

The gibberellins are actually a family of closely related substances. To date, structures have been determined for well over a dozen of these and a number have been isolated from higher plants. See Structures 1 and 2. The structure of three fused saturated or nearly saturated rings, with two additional rings perpendicular to them, suggests relationship to the diterpens for which there is strong isotopic evidence. For example, C^{14}-kaurene is readily converted to gibberellic acid (GA$_3$) by *Gibberella* cultures. The biosynthesis is apparently inhibited by chlorocholine, which is suspected as the basis for the dwarfing action of this compound. GA$_7$ to date has had the highest activity in most tests.

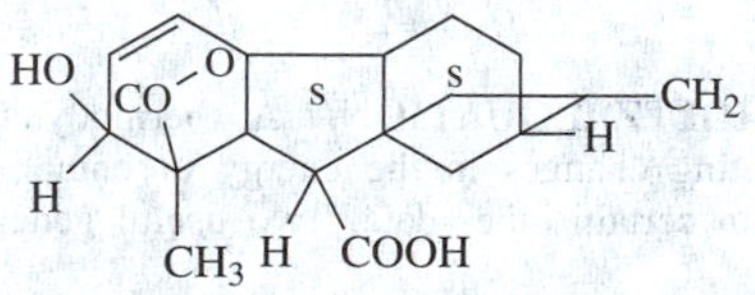

Gibberellic acid (GA$_3$)

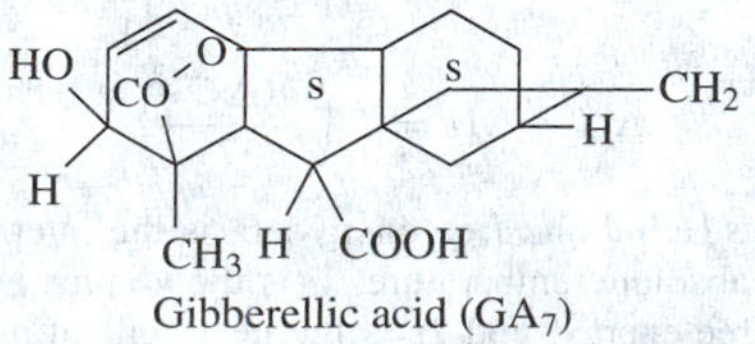

Gibberellic acid (GA$_7$)

Gibberellins cause rapid elongation of shoots; many of the dwarf forms of maize (corn), bean, pea, and morning glory (closely allied to sweet potato) are caused to grow into tall forms indistinguishable from their tall genetic relatives. Many long-day plants are brought into flower in short days by gibberellin, and some biennials, including *Hyoscyamus* (henbane), are made to flower in one year. This process depends on the activation of cell divisions in the shoot apex. Like auxins (other plant hormones), gibberellins produce parthenocarpic fruits, especially on tomato, but unlike auxins, they do not inhibit lateral bud development, but they inhibit rooting of cuttings and promote the germination of many seeds. Their transport shows no polarity. They are active at concentrations comparable to those of the auxins. There is good evidence that the gibberellins act only when auxin is present.

In their biological function, it is believed that the gibberellins destroy or bypass naturally occurring inhibitors that normally prevent premature germination. However, high concentrations of the gibberellins and like substances actually prevent germination in certain varieties of seed.

An excellent example of the performance of gibberellins is given by S.B. Ross, et al. in "Gibberellins: A Phytohormonal Basis for Heterosis in Maize," *Science*, **1216** (September 2, 1988).

GIBBON.   See **Anthropoids.**

GIBBS DIVISION SURFACE.   Consider a system consisting of two homogeneous bulk phases α and β separated by a surface phase. The concentrations vary continuously through the surface phase from those of the interior of one phase to those of the interior of the other. In order to give a well-defined meaning to the thermodynamic functions of the surface phase, independently of the exact position of the boundaries of the surface layer, it is useful, following Gibbs, to replace the real surface phase by a geometrical surface. The bulk phases are considered to be homogeneous up to this geometrical surface, which is called the Gibbs division surface. See Fig. 1.

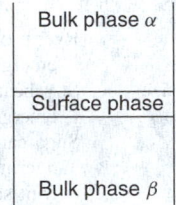

Fig. 1.   Gibbs division surface.

GIBBS-DUHEM EQUATION.   In a system of two or more components at constant temperature and pressure, the sum of the changes for the various components, of any partial molar quantity, each multiplied by the number of moles of the component present, is zero. The special case of two components is the basis of the Gibbs-Duhem equation of the form:

$$n_1\, d\overline{X}_1 = -n_2\, d\overline{X}_2$$

in which n_1 and n_2 are the number of moles of the respective components and $\overline{X}_1$ and $\overline{X}_2$ are the partial molar values of any extensive property of the components.

GIBBS-HELMHOLTZ EQUATION.   A thermodynamic relationship useful in calculating changes in the energy or enthalpy (heat content) of a system, from certain other data. Two useful general forms of this equation are:

$$\Delta A - \Delta U = T\left(\frac{\partial(\Delta A)}{\partial T}\right)_V$$

$$\Delta G - \Delta H = T\left(\frac{\partial(\Delta G)}{\partial T}\right)_P$$

in which A is the Helmholtz free energy, U is the internal energy of the system, T is the absolute temperature, V is the volume, P is the pressure, G is the Gibbs free energy, and H is the heat content of the system. See also **Free Energy.**

For a reversible cell, if the heat of the chemical reaction taking place in the cell is ΔH, F is the Faraday constant and the reaction takes place by the migration of an ion bearing a charge j, then

$$\Delta H = jF\left(\in - T\frac{d\in}{dT}\right)$$

where $\in$ is the emf of the cell.

GIBBS, JOSIAH WILLARD (1839–1903).   The father of modern thermodynamics. During his lifelong post as professor of mathematical physics at Yale, he stated the fundamental concepts embraced by the three laws of thermodynamics, especially the nature of entropy. A theorist rather than an experimenter, Gibbs was the first to expound with mathematical rigor the "relation between chemical, electrical, and thermal energy and capacity for work." It has been said that throughout his adult life Gibbs did nothing but think. The results established him as a great creative scientist.

See also **Thermodynamics.**

GIBBS-KONOVALOV THEOREMS.   Consider a binary system containing two phases (e.g., liquid and vapor). Both components can pass from one phase to another. The Gibbs-Konovalov theorems refer to the properties of the phase diagrams of such systems (see also **Azeotropic System**). The first theorem is: *At constant pressure, the temperature of coexistence passes through an extreme value (maximum, minimum or inflexion with a horizontal value), if the composition of the two phases is the same. Conversely, at a point at which the temperature passes through an extreme value, the phases have the same composition.* The second theorem is similar. It refers to the coexistence pressure at constant temperature.

GIBBS PARADOX.   When two samples of the same gas at a given temperature and pressure are allowed to mingle by the removal of a separating partition, the entropy of the resulting system is equal to the sum of the entropies of the two original parts. And, there is no extra term which arises when the two original systems are composed of different gases. This paradoxical absence is called the Gibbs paradox; it can be explained by using the theory of grand canonical ensembles.

GIBBS PHASE RULE.   See **Phase Rule.**

GIF (GRAPHICS INTERCHANGE FORMAT).   See **Data Compression.**

GILBERT.   See **Units and Standards.**

GILBERT, WALTER (1932–).   An American molecular biochemist who won the Nobel Prize for chemistry in 1980 along with Paul Berg and Frederick Sanger for their studies of the chemical structure of nucleic acid [CAS: 9008-72-4]. Author of many papers on theoretical physics and molecular biology. He has been at Harvard since 1972.

GILBERT, WILLIAM (1544–1651).   Gilbert was a British scientist most known for his experiments associated with magnets, which lead to his realization that the Earth acts as a huge magnet. The motivation for his experiments was his desire to understand the force governing the motion of the planets.

Gilbert was educated at St. John's College at Cambridge University. For awhile he was court physician for Queen Elizabeth I. In this position he had the time and money to pursue his scientific interests and he tried to analyze electric and magnetic phenomena. He was the first to use the term "electricity". He invented an electroscope to detect the presence of electric charge. His work showing the Earth acts as an enormous magnet was praised by Galileo and incorporated into the theories of Kepler. Gilbert's idea prepared the way for the theory of gravitation.

See also **Gravitation.**

J. M. I.

GILL.   A respiratory organ for the extraction of oxygen from the water and for the liberation of carbon dioxide.

Many small aquatic animals absorb oxygen through the surface of the body generally but the more complex forms have localized respiratory organs formed to present an adequate surface. They are usually thin plates of tissue or slender tufted processes and, with the exception of some aquatic insects, they contain blood or coelomic fluid which absorbs oxygen through their thin walls. In the insects a unique type of respiratory organ is the tracheal gill which contains air tubes. The oxygen of these tubes is renewed in the gills.

Gills are developed in starfishes and sea urchins (see also **Echinoidea**) as thin protuberances on the surface of the body containing diverticula of the water vascular system. In the crustaceans, mollusks, and some insects they are tufted or plate-like structures at the surface of the body in which blood circulates. The gills of other insects are of the tracheal type and also include both thin plates and tufted structures. In the larval dragonfly the wall of the caudal end of the alimentary tract (rectum) is richly supplied with tracheae as a rectal gill. Water pumped into and out of the rectum supplies oxygen to the closed tracheae.

Gills of vertebrates are developed in the walls of the pharynx along a series of gill slits opening to the exterior. Water taken into the mouth passes out of the slits, bathing the gills as it passes. Some fishes utilize the gills for the excretion of electrolytes. In some of the amphibians the gills occupy a similar position on the body but protrude as external tufts.

Gill Chamber. A partially enclosed space containing gills. In many invertebrates external gills project from the surface of the body. Such structures are very delicate and in many species are protected by folds of the body wall. The crayfish offers a good example, with the carapace extended down on each side of the body to form the outer wall of a chamber in which the gills lie.

Gill Filament. A thread-like component of a gill. Also the ciliated ridges of the gills of bivalve mollusks.

Gill Plate. The respiratory organ of some bivalve mollusks. It is formed of two thin plates or lamellae, each made up of united ctenidial filaments (ctenidium), and contains passages communicating with the mantle cavity and with the chamber above the gills. Water passes into these passages from the mantle cavity.

Gill Raker. A comb-like structure along the inner margin of the gill arches of fishes. These combs prevent the passage of food into the gill slits and direct it toward the esophagus.

Gill Slit. A perforation of the body wall of vertebrates opening into the pharynx. In the fishes and amphibians the slits are associated with the gills, but in terrestrial vertebrates they occur only in the embryo, and in mammals they usually fail to open. The gill slits are paired, opening as a series on each side of the body. In the lampreys and most cartilaginous fishes (sharks, etc.), the openings are externally separate. In the bony fishes, those of each side are covered by an operculum.

See also **Fishes**.

GILL ANEMOMETER.   See **Meteorology**.

GILSONITE.
The mineral Gilsonite, named for S.H. Gilson of Salt Lake City, is a variety of asphaltum that occurs in Uinta County, Utah. It is found in black lustrous masses that ignite easily. A less frequently used name for it is uintaite.

GIMBAL.
1. A device with two mutually perpendicular and intersecting axes of rotation, thus giving free angular movement in two directions, on which an engine or other object may be mounted.

2. In a gyroscope, a support that provides the spin axis with a degree of freedom. The outer and inner gimbals of a pendulous two-axis gyro are shown in Fig. 1. See also **Gyroscope**.

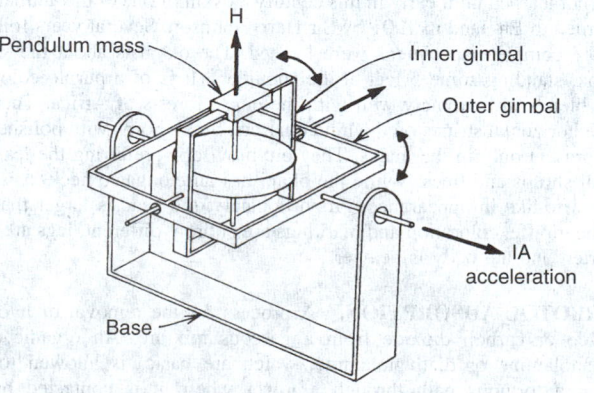

Fig. 1.   Gimbal arrangement in a pendulous two-axis gyro.

GIN.
A mixture of ethyl alcohol, water, and a flavoring agent. Although gin is probably most frequently identified with the English as producers and consumers, gin originated in Holland in the mid-1600s, and it was developed by a professor of medicine at Leyden University. The first gin was flavored with essence from the juniper berry and was promptly given the French name (*genievre*) for juniper berry. A bit later it was called Geneva and then abbreviated still further by the English to *gin*.

Although the juniper berry has traditionally been the most popular flavoring agent for gin, other substances have been used to a limited extent. These include coriander, angelica root, anise, caraway seeds, lime, lemon, and orange peel, and licorice, among others. Quite popular for many years and still produced is sloe *gin*, flavored with sloes (small blue-black, plum-like fruits from the blackthorn), which impart a reddish color to the *gin*.

Possibly of all alcoholic beverages, gin enjoys the most stained reputation. Part of this stems from the fact that the word *gin* has been and is still sometimes used incorrectly to designate any inferior liquor, flavored or not–for example, "Gin Lane" in London made famous by Hogarth; "gin mill" for a tavern or bar, and "bath tub gin," a product of the Prohibition era in the United States. In connection with an Act introduced in Parliament in 1871, which would have required the reduction of pubs in Britain, Gladstone mentioned in a speech that he had "been borne down in a torrent of Gin."

Because of the unavailability of quality distilled spirits in England during the 1600s, the flavored spirits from Holland soon became popular, thus encouraging expansion of local production in England. Growth in consumption is reflected by the figure for 1690, when about $\frac{1}{2}$ million gallons of gin were consumed in London and environs, against a figure of over 5 million gallons by 1729. Because of a rising social problem resulting from widespread drunkenness, a tax on gin and gin-selling establishments was imposed in 1729. However, a loophole in the regulation allowed widespread production of unflavored, usually poor-quality gin, called Parliamentary Brandy. The condition existing then is exemplified by a sign which appeared on a Shoreditch grog-shop: "Drunk for a penny, dead drunk for tuppence, clean straw for nothing." Thus, it required many years (until about World War I) for gin to gain respectability in Britain.

Despite the apparent relative simplicity of gin as a product, it is interesting to observe that quality gin, like any other alcoholic beverage of quality, is not easy to manufacture. There are several variations in its production. In the present-day manufacture of Dutch Geneva or Hollands Geneva by distilleries located in Schiedam (Rotterdam), the product is distilled from barley grain. Malt produced from barley is added to a mixture of grains in a large vessel in which fermentation takes place. Within several hours, after a carefully controlled temperature cycle has been completed, Dutch yeast is formed on top of the fermented mass. As a byproduct, this yeast is marketed to bakers (after further processing). The liquid is distilled a minimum of 3 times in a pot still to produce a distillate known as Malt Wine. The final Geneva is prepared by rectifying the malt wine to which juniper and other ingredients are added. The entire process is proprietary.

Gin also can be made by introducing the flavoring ingredients directly into the mash prior to distillation. More commonly, the flavoring agents are either added directly to the base of the still, allowing for liquid extraction, the vapors thus carrying flavorants with them as they rise in the still. Or the flavoring ingredients can be suspended in a basket near the head of the still or placed on trays near the top of the still where the extraction proceeds by the vapors of alcohol. To produce a smoother product, some distillers add a few plates near the top of the still to effect a degree of rectification. Still other distillers, to prevent any thermal degradation of flavoring agents, will operate the still under a vacuum, where the temperature is maintained in the region of 130° to 140°F (54° to 60°C). So-called *compounded gin* involves the simple procedure of adding essential oils directly to grain spirits.

GINGER.
The dried rhizome (rootlike stem) of a perennial monocotyledonous plant, probably native to tropical Asia. Ginger is used mainly as a condiment and as an aromatic stimulant. Several volatile oils are responsible for the characteristic odor. Ginger is used in the preparation of ginger ale and a variety of food products. Ginger also appears on the market as preserved ginger, a Chinese product made from uncured rhizomes.

The species of ginger plant used is *Zingiber officinale*, a member of the family *Zingiberaceae* (ginger family). The plant has a fleshy, irregularly branched rhizome from which arise erect leafy stems from 2 to 3 feet (0.6 to 0.9 meter) in height. The leaves are grass-like. The flowers, borne on a separate stem, are yellow and of a distinctive shape resembling those of orchids. The inside tissues of the rhizome are white and richly spotted with resin dots. The plant is propagated by means of rhizome-cuttings, each cutting having an eye or bud which produces an erect stem.

When the leaves begin to turn yellow the plant is ready to harvest. The rhizomes are dug up and cleaned, then immersed in boiling water to kill the buds or eyes, and also to loosen the periderm or outer portion. The rhizomes are then peeled and dried. See also **Spices**.

GINGIVITIS. See **Periodontitis**.

GINI MEAN DIFFERENCE. A measure of dispersion, defined as the average absolute difference between all possible pairs of observations in a sample. As an estimate of the standard deviation of a normal distribution, it is slightly more efficient than the mean deviation but much more difficult to compute.

GINKGO TREE. See **Maidenhair Tree**.

GINSENG. Of the family *Araliaceae* (ginseng family), this is a relatively small group of herbs and a few small shrubs (or trees). It is probably best known because of the curative powers that over many centuries the Chinese have attributed to the roots of notably two species, *Panax schinseng* and *P. quinquefolium*. Scientifically, these powers have not been dramatically proved or disproved. The plants are low-growing perennial herbs having compound leaves and compound umbels of small white flowers. They grow best in rich shady woods of hardwood trees. When mature, the thick fleshy roots are removed from the ground, very carefully to avoid any damage. They are then dried and marketed for use in making various brews.

Other members of the ginseng family growing in North America include the *Aralia spinosa*, the Angelica tree or Hercules' club. This plant can grow to a height of nearly 40 feet (12 meters), and it has very large doubly compounded leaves, ranging from 2 to 4 feet (0.6 to 1.2 meters) in length. The flowers also are large, white, in clusters approaching 20 inches (51 centimeters) in length. The tree bears a very small black berry, which occurs in clusters. The tree is sometimes planted in gardens and for landscaping effects. It occurs naturally from New York, south to Florida and Texas, and it is found in the midwestern states. The devil's club, *Fatsia horrida*, is a rather high shrub, ranging up to about 15 feet (4.5 meters) in height. The leaves are large, the flowers occur in terminal clusters and are of a greenish-white coloration. The shrub prefers rocky soils and ranges widely from the Great Lakes region westward into California, Oregon, and southern Alaska. The devil's club is also found in Japan.

GIRAFFE AND OKAPI (*Mammalia, Artiodactyla*). The group of *Giraffines* is one of the smaller in the order of *Artiodactyla* (even-toed hoofed animals). There are two types of giraffines remaining today: (1) Giraffes (*Giraffinae*) and (2) Okapis (*Palaeotraginae*). Because of their extremely long necks, they represent unusual natural solutions to anatomical and physiological problems.

The giraffe is the tallest of all mammals, the head rising about 18 1/2 feet (5.5 meters) above the ground. The head is long with a wide range of movements. The tail is long, slender, and tufted. There are seven cervical vertebrae, each of extra length, giving the animal its greatly elongated neck. The tongue is up to 18 inches (46 centimeters) in length and quite elastic; it can be shaped to a point to reach tiny branches. The animal prefers the leaves of the mimosa and acacia trees. Because of its great height and long legs, the giraffe must stand with its legs far apart when grazing or drinking. The animal has an ambling walk, with the legs on the same side moving together. When galloping the giraffe can attain a speed of some 30 miles (48 kilometers) per hour. For protection against certain types of predators, the giraffe can kick hard and fast with its front legs. Most giraffes are of a white-to-sandy color with darker map-like patterning. In both sexes, there are two protuberances between the ears; they appear much like horns, but they are more like raised lumps with skin and tufts of hair on them. However, in some species, a third protuberance or horn is present, making a total of three "bumps" in all. These horns are used only

Fig. 1.   Giraffes. (*A.M. Winchester.*)

in sparring when rival males engage in what might be termed "necking" combat. See Fig. 1.

Giraffes are found over most of tropical Africa. They do not frequent the closed-canopy forest, but prefer to remain on the drier savannas. They live in communities. Considered to be of a mild disposition, giraffes rely essentially on their keen vision and speed to avoid and escape danger.

There is a misconception that giraffes are voiceless. They can make whimpering and whistling sounds used when calling their young. It is interesting to note that giraffes can go for extended periods without water, essentially rivaling the camel in this respect. These animals cannot swim and are not known to wade even the shallowest of streams or ponds.

The gestation period of the giraffe is 15 months. Well-developed before birth, the young giraffe can stand within a few minutes and run within two days. A baby giraffe weighs about 85 pounds (38.5 kilograms). Multiple births are rare.

The Okapi is quite different from the giraffe. Existence of this animal was not learned until early in this century. A skin of one of the animals was returned to England in 1901 by Sir Harry Johnson. Several years followed before complete specimens were located. The okapi is about the size of an ox, standing some 5 feet at the shoulders. It is of a purple coloration that blends in extremely well with the dense forests of Africa. There are wide horizontal stripes on the hind quarters. Small horns with polished tips are present only in the males. They are browsers, preferring the leaves of small shrubs and trees. While the okapi has an elongated neck, it is quite ungiraffe-like in appearance. Proportionately, the head is larger than that of the giraffe, coloration and markings are entirely different, legs are much shorter, and the body is heavier.

GIRBOTOL ADSORPTION. A process for the removal of hydrogen sulfide or carbon dioxide from a gaseous mixture. An organic amine (ethanolamine or diethanolamine, which are basic) is allowed to flow down a tortuous path through a tower where it is contacted by and absorbs (acidic) hydrogen sulfide or carbon dioxide from the gas to be purified as it moves up the tower. The amine, contaminated with these products, is then sent from the bottom of the tower to a steam stripper where it flows countercurrent to steam, which strips the hydrogen sulfide or carbon dioxide from it. The amine is then returned to the top of the tower. The process is widely used in the petroleum industry for purifying refinery and natural gases and for recovery of hydrogen sulfide for sulfur manufacture. Removal of carbon dioxide from gases is usually done with nonoethanolamine.

See also **Petroleum Refining**.

GIRDER. A girder is a large heavy beam capable of carrying both concentrated and uniformly distributed loads. Large rolled steel beams are frequently called girders although the name is generally applied to large beams made up of rolled steel sections connected by rivets or welding. In concrete construction, the large beams used to support smaller beams are called girders. A girder, like a beam, resists transverse bending, and is loaded, ordinarily, by gravity load that is transferred by the girder to its supports. The common plate girder is a compound steel structure composed of plates and angles, bound together in one structure by the use of rivets or welding. Plate girders are used where strength requirements cannot be met by the largest available rolled steel sections. Due to their adaptability, plate girders are to be found in almost every form of construction embodying steel. Bridges, cranes, and buildings, show many examples of the plate girder.

The built-up plate girder roughly resembles an I-beam in shape. Its area may be thought of as subdivided into area of flanges and area of web. The flange sections are most useful in withstanding the bending, and the web resists most of the shear to which a girder is subjected. The arrangement of plates and angles in a plate girder is shown in Fig. 1. The girder is built up of a web plate, whose depth is nearly equal to the full depth of the girder, flange angles, which are riveted near the top and bottom of the web plate, and cover plates riveted to the flange angles. Since the flange chiefly resists bending, and bending moment is greatest at the center of a girder (for ordinary load conditions), the cover plate could be of a thickness increasing from minimum at the abutment to maximum at midspan. It is not practicable to specify a tapered plate, but the same effect is achieved by subdividing the total maximum required cover plate area into a number of plates in laminar arrangement, and achieving the taper effect by cutting off the plates where reduction of bending stress permits. Localized buckling of the web must be resisted in order to permit the girder to develop its full strength. For this purpose, stiffeners, consisting of angles arranged vertically, and riveted to the web and to the flange angles, are spaced periodically along the length of the girder. These are called stiffener angles, and may be smaller than the flange angles.

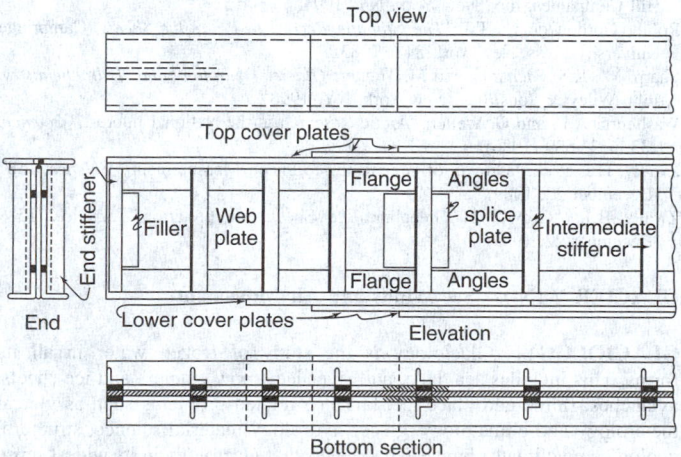

Top view

Top cover plates

Flange | Angles

End stiffener | Filler | Web plate | splice plate | Intermediate stiffener

Flange | Angles

End | Lower cover plates | Elevation

Bottom section

Fig. 1. Principal parts of a plate girder.

As the girder carries load by beam action, the flexure theory applies. The problem of design of plate girders begins with the computation of bending moment and shear. Generally, bending moment governs the design. A cross-section of the girder is then assumed and the moment of inertia of the same computed. The value of the moment of inertia must be such that the unit stress on the extreme fiber, as computed by the common flexure formula, is not greater than the allowable.

Most authorities require that the design of an important girder be carried through with an exact computation of the moment of inertia of some assumed section. If a determination of an economic section is made by trial and error, this moment of inertia method of design may become quite tedious. The number of trials can be greatly shortened if some approximation, which would guide the designer toward a correct selection of the proper structural shapes, could be employed. Such a method is outlined below. It is based on the assumption that a girder is made up of a simple rectangular web connecting rectangular flanges. Let the area of the web be A_w and the area of each flange A_F, while the distance between the centers of gravity of the area of the flanges is h. The moment of inertia of this assumed area about the neutral axis which is taken to be on the axis of symmetry is

$$I = \frac{h^2}{2}(A_F + A_w/6)$$

If this expression is substituted in the flexure formula the flange area is found to be given by the following equation:

$$A_F = \frac{M}{fh} - \frac{A_w}{6}$$

in which f represents the allowable stress.

As ordinarily given in structural texts, this formula represents the net flange area (area with rivet holes deducted). Consequently, it has A_w divided by 8 instead of 6, the difference being accounted for by deduction of a certain amount of web area to account for rivet holes. If the approximate flange area is obtained by some rapid estimating system, such as this flange area method, an arrangement of commercially procurable steel shapes can be set up, and the exact moment of inertia can be accurately established by the principles of mechanics.

The complete design of a steel plate girder includes also such problems as determining the riveting pitch in the flanges, the design of splices in the web plate, the spacing and riveting of stiffeners, and the strengthening of the ends by end stiffeners where the girder bears on its supports.

GIZZARD. In some animals, a region of the alimentary tract with thick muscular walls and some adaptation for grinding food. The gizzards of birds are the best known examples. They have a tough lining and their grinding action depends on the movements of hard particles such as gravel contained in them. One of the fishes, the gizzard shad, has a stomach of similar nature. Many insects also have a gizzard but in this organ the supposed grinding structures are chitinous folds and teeth projecting into the cavity. The grinding action of the organ has been questioned by some observers.

GIZZARD SHAD (*Osteichthyes*). A widely distributed North American fish whose stomach is developed like the gizzard of a bird. It occurs in both fresh and salt water. This fish is of the order *Isospondyli*, family *Dorosomidae*. Maximum length is usually about 20 inches (51 centimeters). They are deep-bodied and appear something like a herring. The Atlantic gizzard shad (*Dorosoma cepedianum*) has been successfully introduced as a forage fish in several areas of the United States, notably in the central and eastern regions. The small gizzard shad (*Dorosoma nasus*) is mainly a saltwater species. Found in Australian waters, they are from 6 to 15 inches (15 to 38 centimeters) in length.

GLACIAL DEPOSITS (or Drift). The general term for glacial deposits, or sands, gravels, boulders, etc., which are the result of mountain or continental glaciation. Drift is classified as either stratified drift, the result of deposition by waters from the melting glacier, or, till (unstratified drift) which is apt to be coarsely graded sediments composed of clay, sand, gravel and boulders. Till may grade, in places, into stratified drift, but is principally transported and deposited by the ice. Both stratified drift and till also form distinctive topographic features, to such an extent that both mountain ranges and even broad continental areas that have been subjected to glaciation cannot be described as having been subjected to the normal cycle of erosion. When a glacier advances over old drift, it may form cigar-shaped hills, called drumlins, whose longer axes are relatively parallel with the movement of the ice. Till, which is built up into long mounds and ridges at the frontal margin of the ice sheet, forms significant topographic features called moraines. The waters coming off from the front of a melting ice-sheet deposit great sheets of stratified gravels, sands and clays. If ice-blocks have been covered by the outwash, when these ice-blocks finally melt they leave depressions in the outwash plain, which fill with ground water to form ponds and lakes. These depressions are called kettle holes.

GLACIAL LAKES.   See Lakes.

GLACIAL PERIOD.   1.   Any of the geologic periods that embraced an ice age. For example, the Quaternary period may be called a "glacial period."

2.   Generally, an interval of geologic time that was marked by a major equatorward advance of ice. This may be applied to an entire ice age or (rarely) to the individual glacier "stages" that make up an ice age. The term "period" here is not used in the most technical sense of a geologic period.

See also **Geologic Time Scale**.

GLACIER.   A large mass of ice. Formed, at least in part, on land by the compaction and recrystallization of snow, moving slowly by creep downslope or outward in all directions due to the stress of its own weight, and surviving from year to year. Included are small mountain glaciers as well as ice sheets continental in size, and ice shelves which float on the ocean, but are fed in part by ice formed on land. The word is derived from the French *glace* (ice). (*American Geological Institute*.)

Wherever upon the earth's surface the temperature is sufficiently low and there is sufficient precipitation to produce a permanent snow field, glaciers may be found. Other things being equal, perpetual snow is more likely to be found in high latitudes and high altitudes. As examples, we have the extensive snow and ice field on Greenland and the Antarctic continent as well as valley glaciers of the Alps, of Alaska, the Rocky Mountains, the Andes, the Himalayas and elsewhere. Repeated thawing and freezing of the snow in perpetual snow fields permit the formation of coarse granular ice called nevé which passes into ice of the usual sort. On slopes, the accumulated ice will eventually begin to move, and as it fills a mountain valley, becoming literally a river of ice, it may be called a valley glacier. Even in the absence of great slopes, ice will only accumulate to a limited thickness before it commences to spread out in all directions from its place of accumulation. Such a mass of ice is called a continental ice sheet or continental glacier; Greenland is an example of such a sheet of continental ice.

Among well-known glaciers are the Zermatt, Stechelberg, Grindelwald, Trient, Les Diablerets, and Rhone in Switzerland; the Nigards, Gaupne, Fanarak, Lom, and Bøver in Norway; the Lambert, Wright, Taylor, and Wilson Piedmont glaciers in Antarctica; the Bossons Glacier in France; and, in the United States, the Emmons and Nisqually glaciers on Mt. Rainier, Washington; Grinnell glacier in Glacier National Park, Montana, the Dinwoody glacier in the Wind River Mountains, Wyoming, the Teton glacier in Teton National Park, Wyoming. And, of course, there are numerous glaciers in the Canadian Rockies.

In 1980, Meier (U.S. Geological Survey) predicted that the Columbia glacier, which enters Prince William Sound near Valdez, Alaska (southern terminus of the Trans-Alaska oil pipeline) would begin a drastic retreat. The glacier had been stable throughout the 20th Century, but in 1978, its tongue had retreated slightly. In 1985, the prediction was confirmed. The forward flow of the glacier is increasing rapidly. Iceberg production (termed "calving" of icebergs) from the terminus of the glacier is also increasing. In the summer of 1984, it is estimated that the glacier discharged about 14 million cubic meters of ice. The annual rate of decline of the glacier has been well over a million cubic meters per year. It is not likely that the resultant icebergs will affect shipping because a submerged ridge in Prince William sound bars the icebergs from floating out to sea. The retreat is expected to continue into the year 2000, during which time a fjord will be exposed. Repopulation of the fjord is a target for study by ecologists.

The Lambert glacier, a feature of the East Antarctic ice sheet, is the largest known glacier in the world. The glacier flows into the Amery Ice Shelf. More detail concerning this glacier will be found in the entry on **Polar Research**. Also see Radok reference listed.

Scientists at the U.S. Army Cold Regions Research and Engineering Laboratory and other colleagues have been studying the rheology of glacier ice. Glaciers flow under gravitationally induced stresses. The weight of the ice causes the glacier to spread and thin in a manner dictated by surface conditions, basal conditions, and the ice constitutive relation between strain rate and applied stress. As pointed out by Jezek et al., because of the complex interaction of these three elements within the glacier and because of the difficulty of simulating intraglacial conditions in the laboratory, the constitutive relation is still an issue in glaciology. The researchers have developed a new method for calculating the stress field in bounded ice shelves and this has been compared with the strain rate and deviatoric stress on the Ross Ice Shelf, Antarctica. The analysis shows that strain rate (per second) increases as the third power of deviatoric stress (in newtons per square meter), with a constant of proportionality equal to 2.3×10^{-25}.

Additional Reading

Ahnert, F.: *Introduction to Geomorphology*, Oxford University Press, Inc., New York, NY, 1998.

Benn, D.I.: *Glaciers and Glaciation*, John Wiley & Sons, Inc., New York, NY, 1998.

Bloom, A.L.: *Geomorphology: A Systematic Analysis of Late Cenozoic Landforms*, Prentice-Hall, Inc., Upper Saddle River, NJ, 1997.

Bogorodsky, V.V. et al.: *Radioglaciology*, Kluwer Academic Publishers, Norwell, MA, 1985.

Broecker, W.S. and G.H. Denton: "What Drives Glacial Cycles?" *Sci. Amer.*, 48 (January 1990).

Engelhardt, H. et al.: "Physical Conditions at the Base of a Fast-moving Ice Stream," *Science*, **57** (April 6, 1990).

Finkl, C.W., Jr.: *The Encyclopedia of Applied Geology*, Van Nostrand Reinhold, New York, NY, 1984.

Guyton, B.: *Glaciers of California: Modern Glaciers, Ice Age Glaciers*, University of California Press, Los Angeles, CA, 2001.

Kendall, D.L.: *Glaciers & Granite*, North Country Press, Unity, ME, 1996.

Kerr, R.A.: "Marking the Ice Ages in Coral Instead of Mud," *Science*, 31 (April 6, 1990).

Koerner, R.M.: "Ice Core Evidence for Extensive Melting of the Greenland Ice Sheet in the Last Interglacial," *Science*, 964 (May 26, 1989).

Lindstrom, D.R. and D.R. MacAueal: "Scandinavian, Siberian, and Arctic Ocean Glaciation: Effect of Holocene Atmospheric CO_2 Variations," *Science*, 628 (August 11, 1989).

Llewellyn, C.: *Glaciers (Geography Starts*, Heinemann Library, Portsmouth, NH, 2000.

Paterson, W.S.: *The Physics of Glaciers*, 3rd Edition, Butterworth-Heinemann, Inc., Woburn, MA, 1999.

Post, A., E.R. Lachapelle: *Glacier Ice*, University of Washington Press, Seattle, WA, 2000.

Radok, U.: "The Antarctic Ice," *Sci. Amer.*, 98–105 (August 1985).

Ritter, D.F.F., R.C. Kochel, and J.R. Miller: *Process Geomorphology*, The McGraw-Hill Companies, Inc., New York, NY, 1994.

Robin, Gordon de Q., Ed.: *The Climatic Record in Polar Ice Sheets*, Cambridge University Press, New York, NY, 1983.

Sharp, M., K.S. Richards, and M. Tranter: *Glacier Hydrology and Hydrochemistry*, John Wiley & Sons, Inc., New York, NY, 1998.

Washburn, A.L. and G. Weller. "Arctic Research in the National Interest," *Science*, **233**, 633–639 (1986).

Zwally, H.J. et al.: "Growth of Greenland Ice Sheet: Measurement," *Science*, 1587 (December 22, 1989).

Zwally, H.J.: "Growth of Greenland Ice Sheet: Interpretation," *Science*, 1589 (December 22, 1989).

GLACIER WIND.   See Winds and Air Movement.

GLACIOLOGY.   Glaciology is the study of frozen water in all its forms. This includes sea, lake, and river ice, snow glaciers and ice sheets, avlanches, rhime and glace ice, but also related problems such as glacial meterology and climatology, glacer hydrology material science, structural geology, and climate change. This means that glaciology in its widest sense encompasses much of the natural sciences and that many aspects of earth science, chemistry and physics are relevant to glaciological studies.

See also **Avalanche (Electronics)**; **Climate**; **Glacier**; **Hydrology**; and **Polar Research**.

GLANCING ANGLE.   Two common uses of the term glancing angle are: 1. The angle between a ray and the tangent plane to a surface. The complement of the angle of incidence. 2. The term is often used as a modifier, to indicate the incidence of a beam at a very small angle with the surface.

GLAND. In valve and piping terminology, a gland is a movable part that compresses the packing on a stuffing box.

In biology and medicine, a gland is an organ of epithelial structure which produces secretions necessary to the body, or which excretes waste materials from the system. Glands vary greatly in form and complexity and in the nature of their products.

The simplest glands are unicellular. In the glandular lining of the intestine, for example, are isolated cells that secrete mucus. They are known as goblet cells because the mucus accumulates in a clear ovoid mass above the constricted base of the cell, approximating the form of a goblet.

Multicellular glands develop from the epithelial layers by local increase of cells and consequent expansion of the layer into adjacent spaces or tissues. They include tubular, acinous, and alveolar structures. Tubular glands are slender tubes lined with glandular epithelium; acini are rounded groups of cells with a small central cavity; and alveoli are larger rounded chambers lined with glandular cells. Many of the larger glands of the body, including the pancreas and salivary glands, are made of great numbers of acini borne by complex branching ducts. These glands are said to be compound. In the most complex forms, the secretion may leave the cells by minute canals, or similar canals between the cells may conduct it to the cavity of the acinus. This cavity empties into a short secretory duct lined with gland cells, and this in turn into the excretory duct. These smaller ducts join to form larger and larger passages, ultimately reaching the main duct which delivers the secretion of the entire gland to its destination.

Classification of Glands. Glands may be divided into three major types: (1) glands of external secretion, whose products are discharged through ducts—also identified as *exocrine glands*; these include such glands as sweat, stomach, and salivary glands; (2) glands of internal secretion, the *endocrine* or *ductless glands* (see also **Endocrine System**; and **Hormones**); and (3) glands that have both external and internal secretion

Glands that produce cells are known as cytogenic glands. They include the reproductive glands, which produce germ cells, and the spleen, lymph glands, and red bone marrow, in which blood cells develop. See also **Blood**; and **Gonads**.

Special glands are derived from all germ layers and are associated with all organic systems. They serve for hormone production, for lubrication, to prevent drying, for defense, in reproduction, and in numerous other biochemical ways in practically all forms of life. The general term for any disease of a gland is *adenosis*.

GLASS. Traditional glass is an inorganic product of fusion that has cooled to a rigid solid without undergoing crystallization. Within the last few years, sol-gel glass has been introduced to the commercial market. Sol-gel processing is a chemically based method for producing glass at temperatures much lower than the traditional melting methods. Sol-gel glasses are described later in this article.

Glass may be transparent, translucent, or opaque, and it may be colored. The chemical composition and corresponding properties may vary over a wide range. Glass will support a load, and it may be shaped, broken, or cut. It is much like other solid materials, and yet it is unique.

Its uniqueness becomes obvious when it is examined on a submicroscopic level. Most solids have regular, orderly patterns for the arrangement of atoms, molecules, and ions, but glassy materials are highly disordered. There is some short-range order in glass, but beyond one or two atoms or ions the ordering may be described as random. Thus, on a submicroscopic level, glassy solids look more like liquids than solids.

Since glasses do not have ordered structures with correspondingly specific bonding energies between rows, stacks, planes, or discrete ions, they do not have definite melting points. When a glassy material is heated, it softens slowly and transforms to the liquid state. Crystalline solids generally transform from a solid to a liquid at a single specific temperature, the melting point. On cooling, a material that has a tendency to crystallize to solid will do so at the same temperature at which it transformed to a liquid. When a glass is cooled from a high temperature, it becomes increasingly viscous in a manner related to the inverse of the temperature until it becomes a rigid solid again. Thus, a specific temperature where

melting or freezing takes place cannot be found for glass; i.e., glass does not have a melting point.

Most glasses can be made to crystallize if they are subjected to the right conditions of temperature and rate of cooling, which suggests that the glassy state is like a supercooled liquid. This is not borne out by measurements of density and other volume properties, which do not decrease in a linear manner as glass is cooled below its crystallization temperature.

Why is it that some melts, when cooled through a crystallization temperature, form glasses while others do not? It is simply a question of whether the melt can be cooled through the temperature range of maximum crystal growth rate faster than the crystals can grow. Thus table salt cannot be formed as a glass, but sand, or SiO_2 can be. The maximum crystal growth rate is normally just below the melting point of the material, but materials that tend to form glasses easily are much more viscous at these temperatures. For example, in the extreme cases of salt and sand, the differences in viscosities at their respective melting points is about eight orders!

The two-dimensional drawing in Fig. 1 shows SiO_2 in the ordered, or crystalline, and in the random, or glassy, state to illustrate the difference on a submicroscopic scale. Figure 2 shows how the volume properties of a material would respond to temperature if they could be prepared as a glass, a supercooled liquid, or crystalline material.[1]

Most glasses are composed of inorganic oxides, and most commercial glasses contain SiO_2 as their major constituent, but there are organic glasses

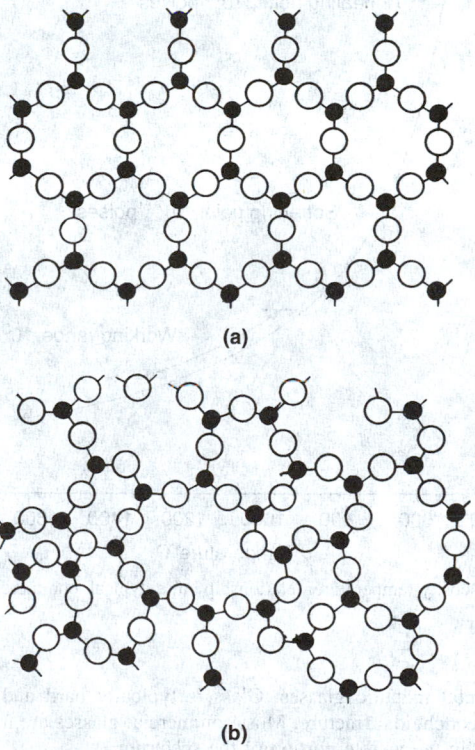

(a)

(b)

Fig. 1. Silicon dioxide (SiO_2): (**a**) crystalline, and (**b**) glassy state. (Course structure is shown. Some authorities have recently suggested that, when studied at a much finer structure (such as by neutron scattering techniques), glass shows a much more orderly structure.)

[1] *Note*: Traditionally, the structure of glass has been determined by means of x-ray crystallography. Which reveals a random network of disorderly structure. Neutron scattering of glass, however, makes it possible to examine the much finer structure of the material. It has been found that in glasses the angles between bonds that link atomic or molecular building blocks vary, whereas in crystals, of course, the links are orderly—that is, an endless repetition of a regular atomic or molecular geometry. In recent experiments at Grenoble, neutrons were beamed at samples of silicate glass. From these studies at this much finer scale, researchers now believe that the molecular structure of glass is far from random. As pointed out in 1991 by Nicholas Borrelli (Corning), "Normally glass is considered a random network, but that really is a misnomer." See Amato reference listed.

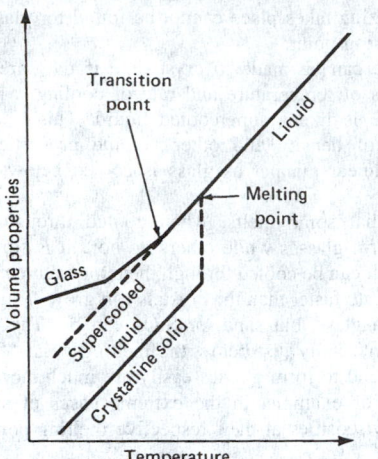

Fig. 2.   Volume properties of glass in contrast with crystalline solids as a function of temperature.

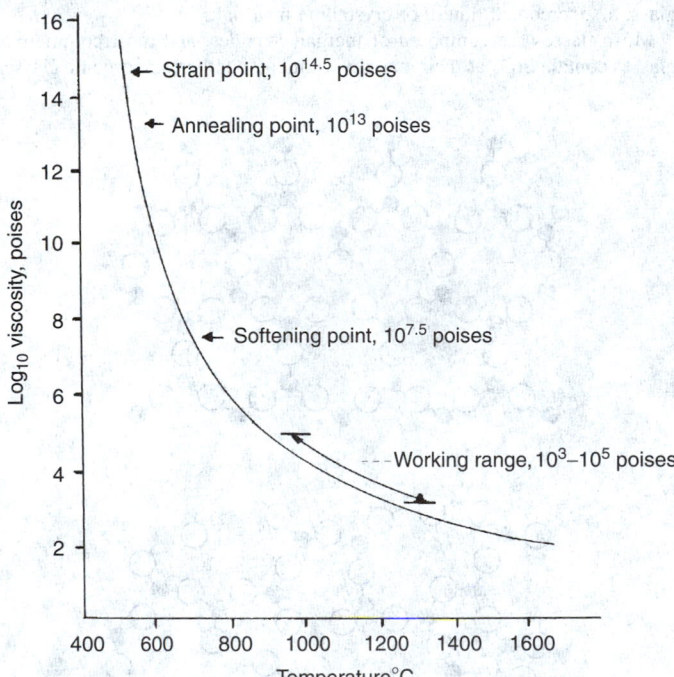

Fig. 3.   Viscosity-temperature relationship of a typical commercial soda-lime glass.

and elemental metallic glasses. Glass is typically hard and brittle, and exhibits a conchoidal fracture. Most commercial glasses are transparent or translucent in the visible portion of the spectrum.

The continuous and smooth relationship of the viscosity of glass with its temperature is an important property. Figure 3 shows a typical viscosity versus temperature curve for a commercial glass. The working range is the viscosity in which most commercial glasses are formed. Glassware formed by automatic forming equipment would be made from glass at a temperature such that the glass will have a viscosity in the lower portion of this range (10^3 to 10^5). Some other operations, such as hand working, might be done at higher viscosities.

Generally, freshly formed glasses are in danger of deforming under their own weight when they are at viscosity below the softening point. At the annealing point, the glass is rigid and at this viscosity (temperature) the internal strains caused by the forming and nonuniform cooling would be decreased to an acceptable commercial level in 15 minutes. At the strain point, the glass is substantially rigid, and at the temperature equivalent to this viscosity, the internal stresses would be reduced to very low values if the temperature were maintained for four hours.

Types of Traditional Glass

A wide range of glass products exists, each type having special properties. The properties of glass are determined primarily by chemical composition, and since the composition may be varied almost infinitely, there are many thousands of different glasses. However, they may be generally classified into soda-lime-silica glasses; lead glasses; borosilicate glasses; and a number of special glasses, including solder glasses, laser glasses, silica glass, glass-ceramics, and colored glasses. These types essentially bracket the commercial glasses.

Soda-Lime-Silica Glasses. This is the most important group in terms of tonnage melted and variety of use. The combination of silica sand, soda ash, and limestone produces a glass that is easily melted and shaped and has good chemical durability. The raw materials are indigenous to most areas of the world and inexpensive. Soda-lime glasses are particularly suited to automatic-machine-forming methods and are the basis for most of the bottle-, sheet-, and window-glass industry. Very small amounts (often less than 3% of the total batch) of alumina, magnesia, boric oxide, and other chemicals are added to act as stabilizers and to increase durability.

Lead Glasses

The glasses of this group, composed basically of silica sand and lead oxide, have a high refractive index and high electrical resistivity. Potash is present as a significant constituent in most of these glasses. The slow rate of increase in viscosity with decrease in temperature makes lead glass particularly suitable to hand fabrication. The amount of lead may vary considerably, even up to 92% lead oxide; it is a more expensive glass, as the raw materials are relatively expensive and special care is needed in melting to avoid bubbles and seeds. Glasses of this type are used in high-quality art and tableware and for special electrical applications.

Borosilicate Glasses

This group of glasses is basically a combination of silica sand with boric oxide and soda ash. The glasses have excellent chemical durability and electrical properties, and their low thermal expansion yields a glass with a high resistance to thermal shock. High durability makes them ideal for demanding industrial and domestic use, such as chemical laboratory ware, cook ware, and pharmaceutical ware. These glasses were developed in the early part of this century to cope with the problem of cold rain on hot railway-signal lights.

Special-Purpose Traditional Glasses

Solder Glasses. These glasses have low softening and annealing temperatures together with expansion characteristics which permit them to be used as intermediate glasses in making seals between two glass surfaces, between a glass and a metal, or between two ceramic surfaces. In fact, solder glass might be described as a high-grade glass glue. Normally, sealing temperatures are well below the annealing temperature of the glass being sealed, and there is little permanent effect on the glass parts being joined. The major constituents of these glasses include lead oxide, boric oxide, and zinc oxide.

Laser Glasses. Glass has various characteristics that make it an ideal laser host material. Its random structure permits broad emission and absorption bands, which provide higher efficiency, more energy storage, and greater energy per pulse than any other material. In addition, most lasing ions are easily soluble in the glass, and rods, fibers, or disks of any size and of high optical quality are easily fabricated. Of the several rare-earth ions that have been made to lase in a glass host, only neodymium has received commercial application. When a neodymium glass lases, it emits light at a rather fixed wavelength of 1.06 nm. Neodymium-doped silicate, phosphate, and fluoride have been used to provide the energy source for laser fusion research throughout the world.

Silica Glass. A glass composed of silicon dioxide as the only constituent has a very high softening temperature and a very low thermal expansion. It is costly to make and fabricate because temperature in excess of 1800 °C is required to manufacture it. However, its refractory character coupled with its very high resistance to thermal shock makes it ideal for special laboratory equipment, windows in high-temperature environments, and instruments.

Glass-Ceramics. These materials are formed in the same manner as conventional glasses and then subjected to heat treatments which caused controlled nucleation and crystallization. Although nearly completely crystalline, their properties can range from transparent to opaque; electrically insulating to weakly conducting; hard to machineable; and with positive, zero, or negative thermal expansions depending upon the composition and heat treatment. This family of materials is based on glasses whose major constituents are magnesium oxide, lithium oxide, aluminum oxide, and silicon dioxide. The crystalline phase or phases and their morphology control the properties of the materials, but the starting chemical composition and the heat treatment determine which crystalline phases will result. Glass-ceramics, which are the result of recent research efforts, have found applications as household cooking ware, reflective optics and laser gyro substrates, chemical processing components, and cooking-stove tops.

Colored Glasses. Nearly all glasses can be colored by adding one or more colorants to the batch in correct amounts. Production of some colors requires, or is enhanced by, the state of oxidation of the coloring agents and by the atmospheres in which the glasses are melted. Table 1 indicates the colors obtainable, colorants used, and chemical states required or utilized.

TABLE 1. COMMONLY USED INGREDIENTS FOR COLORING GLASS

Glass Color	Coloring Agent	State
Red	Cadmium sulfide, cadmium selenide	Reduced
	Cuprous oxide	Reduced
	Gold (metal)	
Yellow	Cerium oxide with titanium oxide	
Yellow-green	Chromic oxide	Oxidized
Blue-green	Iron chromite	Reduced
Blue	Cobalt oxide	
Purple	Neodymium oxide	
Gray	Nickel oxide with titanium oxide	
Black	Copper, cobalt, nickel, and iron oxides in combinations of two or more	
Amber	Iron sulfide	Reduced
Flint (or colorless)	Selenium and cobalt oxide*	Oxidized

*Selenium and cobalt are used in flint glass to add red and blue hues in amounts only sufficient to balance the green hue resulting from iron oxide present as impurity in most naturally occurring raw materials. The intended result is an even light transmission over the whole visible spectrum.

While the preceding paragraphs describe several classes of glass, within each class there can be infinite composition variations to fit the exact requirements of the user. Table 2 shows typical composition ranges for commercial glasses.

Traditional Manufacturing Process

Glass products are many and varied, and glass compositions range rather widely, depending on the desired products. Figure 4 shows a typical cross section of a glass manufacturing facility. Raw materials weighing, mixing, charging, and melting are common requirements regardless of the forming operation that is to follow. Most melting furnaces have a primary melting area, followed by a refining or homogenizing section, which is connected to the forming operation by channels called feeders. Although fiber glass is not passed through an annealing furnace after it is formed, most other glass products are annealed to relieve stresses caused by uneven cooling during and immediately after forming.

It is apparent that although there are several similar steps in all glass-manufacturing processes, the forming operations are the most diverse.

Batch Preparation. This begins with the selection, procurement, and storage of an adequate quantity of the raw materials. Selection is made on the basis of the oxides that each material contains and will provide to a glass and on the basis of purity and grain size. Naturally occurring raw materials are used wherever possible for economy, e.g., silica sand, limestones, feldspars, borates, soda ash, boric acid, potash, and barium carbonate. The prescribed quantities of these raw materials, depending on their chemical composition, are measured carefully and mixed together to provide a homogeneous batch. Such mixing is done on an intermittent or a continuous basis, depending on the volume of batch needed to charge the furnaces. The batch is conveyed by a variety of means to the furnaces but always in such a way that segregation is avoided. The importance of grain

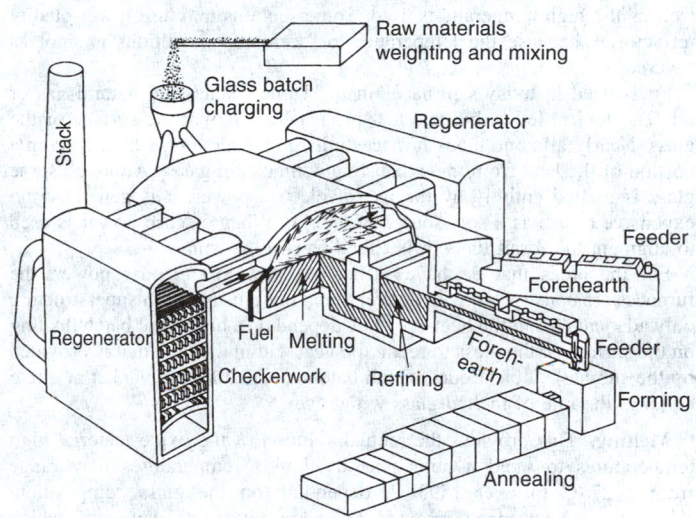

Fig. 4.   Representative glass-producing facility.

TABLE 2. COMPOSITION OF COMMERCIAL GLASSES (Weight Percent)

	Soda-Lime Silica Glass			Fiber Glass Fabrics and Insulation	Borosilicate Glass	Laser Glass	Solder Glass	Lead Glass	Glass-Ceramics
	Containers	Plate and Window Glass	Tableware						
SiO_2	70–74	71–74	71–74	65–74	70–82	61–69	0.5–16	35–70	62–70
Al_2O_3	1.5–2.5	1–2	0.5–2	2–4.5	2–7.5	0–5	0.1–4	0.5–2.0	17–22
B_2O_3				3–5.5	9–14		7–20		
Li_2O									3–5
Na_2O	13–16	12–15	13–15	8–16	3–8	12–24		4–8	
K_2O				0–1				5–10	
CaO	10–14	8–12	5.5–7.5	5–16	0.1–1.2	3–10			0–5
MgO			4.0–6.5	3–5.5					0–7
BaO					0–2.5		0–4		
ZnO							7–62		
PbO							4–77	12–60	
CuO							0–10		
Nd_2O_3						1–6			
CeO_2						0.1–1			
F_2							0–2		
$ZrO_2\ TiO_2$									3–10

size of the various raw materials becomes evident in preventing dusting and/or segregation.

Furnaces. A variety of furnaces is used in the industry to melt the batch to produce glass. They must all accomplish the two purposes of confining the heat to the necessary area and containing the melted glass within the furnace. Crucibles or pots are sometimes used to contain the batch and the melted glass, in which cases the furnace merely retains heat; however, tank furnaces (Fig. 3) are far more common. They are so constructed that the lower portion contains the glass and the superstructure retains the heat and provides combustion space for the fuels used. "Day" tanks are used in some instances where the operation is intermittent and the quantity of glass is small. The great majority of glass produced is melted in continuous furnaces, which are charged initially with batch and cullet (broken-up pieces of previously melted glass) which are melted, filling the tank to a specified depth. Thereafter, batch and cullet are charged continuously at a rate equal to that at which the molten glass is withdrawn from the working end.

Continuous tank furnaces are designed to provide for a separate melter section and a refiner or conditioning section. The melting end is maintained at the necessary high temperatures to accomplish the melting and chemical reactions of the batch materials. The refining, or conditioning, section retains the glass long enough for it to cool to the necessary lower working temperatures.

Glass-melting furnaces are built of refractory materials of various types, which will withstand the severe conditions to which they are exposed. The lower portion of the melter section, for instance, must be of the highest quality to withstand the corrosive action of the glass as well as the high temperatures used. Some sections may use lower-quality refractories because the temperature or corrosion conditions are not as severe.

Fuels used in today's furnaces in the United States are natural gas or oil. The fuel is fed to burners that project flames over the surface of the glass. Nearly all continuous furnaces utilize regenerators, which reclaim a portion of the heat from the exhausting combustion gases. Although some glass is melted entirely by the use of electric power, it is generally too expensive to use as a sole source of energy. When electric power is used to augment the fossil fuels, it is called electric boosting.

For the areas that do have sufficiently low-cost electric power, the furnaces are constructed with conventional bottoms but with superstructure only adequate for initial heat-up. They depend on a blanket of batch floating on the surface of the glass to retain the heat within the tank that is provided by the submerged electrodes. Fresh batch is added to the blanket at a rate equal to the rate of melted glass withdrawn.

Melting. This provides the mutual solution of the oxide material high temperatures to yield a homogeneous liquid. Temperatures may range from $1427\,°C$ to over $1593\,°C$, depending on the glass composition. Water vapor, entrapped air, and CO_2 are given off, some of which become entrapped in the glass, resulting, initially, in a foamy mass. As the melt moves to the higher-temperature regions, the viscosity is lowered and the gases escape. Deliberate hot spots enhance the natural convection currents, promoting homogeneity. More modern furnaces utilize bubblers, which introduce controlled pulses of air through furnace bottom, further enhancing convection. This is particularly valuable for increasing temperatures near the tank bottom in melting those glasses that are more opaque to infrared radiation.

The glass is essentially free from bubbles (or seeds) when it reaches the end of the melting chamber. It then passes under floaters in some furnaces, or through submerged throats in most, to the so-called refining section (more properly, the conditioning section). Here the refining conditioning consists of allowing the glass to increase to a more useable viscosity level by uniformly lowering the temperature, which also allows the remaining tiny seeds or gaseous inclusions to dissolve.

Furnaces supply glass to up to eight forming machines. Forehearths or alcoves serve to channel the glass to the individual machines or machine locations and to further change the temperature and viscosity.

Forming Operations. These are many and varied, involving two three, or four major steps. The first is a further temperature conditioning to place the glass in the exact viscosity range, sometimes wide but often quite narrow, suitable for the selected primary forming operation. The second step is the primary forming itself, followed usually, but not always, by an annealing step. Single or multiple secondary operations may ensue. Only the major forming processes of drawing, pressing, blowing, and casting will be discussed.

Drawing is one of the simpler forming methods by which thousands of tons of window glass and millions of feet of rod and tubing are produced annually. Drawing window glass frequently utilizes a rectangular refractory frame, called a debiteuse, placed on the surface of the conditioned glass. It has a slot roughly 4–8 in. (10–20 cm) wide and 8 ft (2.4 m) or more long through which the glass is pulled vertically. The width and length of the slot in the debiteuse, together with the drawing speed, aid materially in controlling the width and thickness of the sheet. The upward draw may continue until the sheet is nearly cold, when it can be stored and cracked off in suitable lengths. Or, it may be bent over a large roller at nearly the last moment it will withstand bending and conveyed horizontally into the annealing lehr. This method of making window glass has been largely replaced by the float glass process described below.

Glass tubing may be drawn vertically in a manner similar to that for window glass. Another common method is the Danner process, in which a suitable stream of glass is flowed onto a conical rotating mandrel supported with its small end downward and its axis at a suitable angle to the horizontal. The tubing is drawn from the small end, through which sufficient air is blown to retain the desired cross section of the tubing. Drawing continues horizontally over rollers until the tubing can be cracked off in lengths at the cold end. Glass tubing is also made by the downdraw process, where air is blown into the tube as it is drawn from the bottom of a refractory bowl of molten glass.

Plate glass may be formed by flowing the molten glass over the lip of the discharge end of the furnace between a set of large water-cooled rollers and then pulling it away by means of driven rollers. The resulting sheet is up to 1 in. (2.5 cm) or more thick and 10–12 ft. (3–3.7 m) wide. However, most flat glass made throughout the world today is made by the recently developed float-glass process. In this process the molten glass is formed into a sheet by floating it on a bath of molten metal such as tin. The glass flowing onto the bath of tin is pulled across the surface and cooled to the temperature at which it is rigid while still on the molten metal. The outstanding advantage of this process is that it produces a plate of glass both surfaces of which require no further polishing.

Modern methods of pressing, blowing, and casting usually involve an intermediate step, the formation of a suitable charge of glass, or gob, for the ensuing operation. The most common method involves a gob feeder located at the end of the forehearth. This consists of a bowl, or spout, kept full of glass by flow from the forehearth and having an orifice in its bottom and a refractory tube suspended in the bowl over the spout. The tube may be lowered to shut off the flow of glass or raised to permit flow at a selected rate. A refractory plunger operates vertically inside the tube. It provides a pumping action on its upstroke, momentarily restraining the flow of the glass. Its downstroke forces the accumulated glass out of the orifice, where it is sheared off. The result is a charge of glass, called a gob, of controlled size, which is delivered to the forming machine by gravity.

Pressing, or press-forming, operations normally are used for relatively shallow, heavy-walled products. Pressing is accomplished by means of a metal mold (usually iron or steel), a ring, which is centered on top of the mold, and a plunger, which is forced into the mold through the ring. The mold shapes the exterior of the product, the ring the sides, and the plunger the interior. A pressing machine may have many molds mounted on its circular, rotating table, a ring for each mold or, more commonly, a single ring mounted on the same mechanism as the plunger, and a single plunger. After a gob is charged into the mold, the machine indexes one station under the plunger and the plunger moves down into the mold, dwells momentarily, then retracts. It is noteworthy that the plunger action flows the glass into the mold cavity rather than stamping out the product by a quick movement. Since considerable heat is removed from the glass by the plunger, it is cooled with water internally. The product remains in the mold for about half the revolution of the press table before removal to allow it to cool below its deformation temperature. The molds may be cooled by forced air.

Blowing methods work best for deep products and it frequently must be used for thin-walled items. A common procedure, called the blow and blow, involves two steps, of which the first is shaping the glass charge into a form called a blank or parison. Gob-fed machines receive the gob in the parison mold, where it is shaped into a cylinder about two-thirds the height of the bottle. The finish, or top, of the bottle is formed in the same operation at the bottom of the mold by action of a small plunger entering the mold from below and delivering a puff of air. A transfer mechanism holding the parison by the completed finish then swings and inverts it into a second mold for the second step, blowing the glass into its final shape. A cross section of the molds shows this process in Fig. 5. The most modern machinery for rapidly forming containers and bottles commercially are individual section (IS) machines. Each section is capable of forming up to four gobs at the same time and there are as many as ten sections per machine. The individual sections can be sequenced electronically to produce more than 400 bottles per minute on a 10-section machine. See also Fig. 6.

The Owens process employs vacuum to charge the glass into the blank or parison mold. Here, a blank mold dips into a shallow pot of molten glass, a vacuum is applied, and a charge of viscous glass is pulled into the blank mold. The finish is formed simultaneously at the top of the blank. This blank or parison is subsequently transferred into the blow mold, where the bottle is blown into its final form. See Fig. 7.

In another modern machine, the glass flows downward from an orifice in a continuous stream that passes between rollers that flatten it into a ribbon with alternate thick and thin spots. The ribbon is picked up by a horizontally moving support in which voids coincide with the thick portions

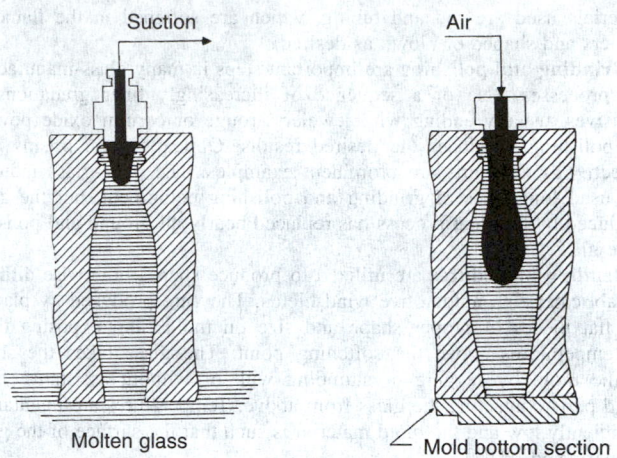

Fig. 7. Owens process. *Left*: Blank mold is dipped into the surface of molten glass, where it is filled by a vacuum suction. As the mold is lifted from the glass, a knife cuts off the glass and closes the mold. *Right*: The blank mold opens, and a puff of air is introduced to shape the parison before transferring it to the blow mold, where it is blown to its final shape.

Fig. 5.   A high-productivity IS machine manufacturing three bottles on each section at the same time. (*Owens-Illinois, Inc.*)

Fig. 6.   Three white-hot bottles immediately after being formed on a section of an IS machine. The bottles will be transferred immediately to an annealing lehr for cooling and annealing. (*Owens-Illinois, Inc.*)

of the ribbon. Blow heads on an endless belt operating from above the ribbon provide puffs of air to aid in producing a bulbous sagging in the thick portion of the ribbon. After sufficient sagging, molds on an endless belt close around the sagging glass from below, and air from the blow heads blows the glass into the shape of the mold. After the molds open, the product, frequently light bulbs or Christmas ornaments, can be cracked off the ribbon.

Casting is usually restricted to two types of operations. The first involves the simple pouring of molten glass into molds. Examples include such massive shapes as the borosilicate mirror blank for the Mt. Palomar telescope and the large glass-ceramic mirror blanks for observatories in Australia and South America. The molds are specially constructed for refractory materials.

The second type of casting is spin casting, in which a gob from a gob feeder is fed into the bottom of a metal mold supported so that it can be rotated rapidly or spun on its vertical axis. The centrifugal force thus generated causes the glass to flow up the inclined sides of the mold, producing a conical shape. The initial movement of the glass is aided by insertion of a conical plunger into the glass at the bottom of the mold when spinning is begun. Mold speeds of up to 1,600 rpm are attained within one second. The funnel portion of television tubes is sometimes produced by this method.

Annealing. As with most substances on cooling, the temperature differential between the surface and interior layers of a piece of glass establishes temporary stresses, and the higher this differential the greater the stresses. Fracturing can occur when the stresses exceed the tensile strength of the glass. Permanent stresses can be avoided by carefully controlled cooling from a little below the annealing point to the strain point. This is the annealing range. Thereafter, the rate of cooling need only be such that the temporary stresses do not exceed the tensile strength of the glass. Glass manufacturers have learned to take advantage of these phenomena.

Annealing immediately follows glass-forming operations. In continuous processes, the ware is placed on an endless belt, which carries it through the lehr, a tunnel in which the temperature is carefully controlled. Temperature of the ware is raised initially to near the softening point, then lowered slowly through the annealing range and thereafter at a more rapid rate to the point where it can be packed or stored. The process is designed to result in the degree of permanent stresses desired. Optical glass must be annealed very thoroughly to produce an essentially distortion- and strain-free lens; however, some stresses can be tolerated or become beneficial to most other products. Small rods and tubing, for instance, are strong enough because of their regular cross section to require no annealing, while tempered glass has uniformly controlled stresses to increase its mechanical performance.

Secondary Operations. Lampworking is one of the many and varied operations utilized to produce glassware following the initial forming. The materials used are rod and tubing, which are softened in the flame of burners and shaped or blown as desired.

Grinding and polishing are important steps in many glass-manufacturing processes. Use of a sequence of increasingly finer gradations of abrasives, usually ending with jewelers' rouge or cerium oxide powder for polishing, produces the desired results. Optical lenses, prisms, and reflective optics parts are prominent examples. The plate-glass industry has used long lines of grinding and polishing equipment, but the glass produced by the float process has replaced nearly all ground and polished plate glass.

Bending procedures are utilized to produce shapes otherwise difficult to fabricate, e.g., automotive windshields. They are produced by placing the flat pieces of proper shape and size on molds and exposing them to temperatures above the softening point. The glass takes the shape of the mold by sagging or slumping with or without assistance from mold parts contacting the glass from above. Temperatures are maintained sufficiently low and the mold material is such that the surface of the glass is unaffected.

Laminating to produce safety-glass parts, as for automotive windows, is a common practice. A sheet of resin, such as polyvinyl butyral, is placed between properly sized sheets of glass; the whole is exposed to slightly elevated temperatures and pressures, to bond the glass tightly to the resin.

Coating of glass products such as containers is quite common, the objective being to protect the container from abuse to which it is subjected in handling during filling and shipping. A coating which is not visible, can be labeled, protects the surface, and provides lubricity is required and usually calls for a two-layer coating such as tin or titanium oxide, followed by a lubricious coating such as polyethylene. The oxide coatings are obtained by subjecting the hot container to a vapor of chloride, which oxidizes to the oxide. Thick opaque or translucent oxide and metallic coatings are sometimes used to provide attractive color effects or light protection. Many precision optical lenses are coated with thin, vapor-deposited layers, which reduce the light lost by reflection from the surface, and some architectural glass is coated to provide attractive colors and reflect undesirable infrared radiation.

Decorating glass or glassware is an old art that takes many and varied forms. Cutting, grinding, and mechanical or chemical polishing or etching, are well known. Opaque, translucent, and transparent enamels can be applied by silk screens or other means in multiple colors and in almost any pattern. Low-melting vitreous enamels have been used for many years, and when properly fired, they provide good durability. More recently, organic polymers have been substituted for the vitreous enamel. They are not quite as durable as vitreous enamels, but they do not require high curing temperatures.

Tempering is the direct reverse of annealing; i.e., high permanent stress is induced in the glass. Rapid cooling or quenching is applied to the glass surfaces at a temperature slightly below the softening point, placing the surfaces in a high degree of compression while the balancing tensile forces are confined to the interior. Since glass always breaks in tension, very considerable strength is incorporated. Typical products are glass doors, automotive glass, windows, goggles, spectacles, and table ware. Tempering must be the final step in the production line. Other products can be strengthened by judicious control of the degree of annealing if their shapes permit it.

Sealing glasses to each other or to other materials must take into account the thermal expansion-and-contraction characteristics. Many glasses have thermal-expansion properties that allow them to be sealed to metals, but each metal usually requires a different glass composition. Solder glasses are used to seal two pieces of glass to each other, two pieces of metal, or a piece of metal and a piece of glass. The glass seals on light bulbs and vacuum tubes are examples of commercial glass-metal seals, while color TV tubes are sealed together with solder glass at a temperature at which the phosphors are not degraded.

Sealing glasses used for color television tubes are devitrifying or crystallizing sealing glasses. They crystallize during the sealing process to produce a seal that will not soften during the processing of the bulb — because the crystallized glass has a higher melting temperature than the starting sealing glass.

See also **Ceramics**

EARL D. DIETZ, Toledo, OH

Glass Blocks

Introduced during the art deco period (1920s–1930s), glass blocks for structural and decorative purposes were quite popular. Interest faded, but has returned within the last few years.

In addition to their decorative appeal, glass blocks are claimed to provide better energy conservation (solar reflective blocks are available), lower sound transmission, aesthetic flexibility, minimal maintenance, and enhanced security. In addition to plain blocks, they are available with various decorative designs. See Fig. 8. An effective use of glass blocks for an external wall is shown in Fig. 9. Design schemes for obtaining architectural effects are shown in Fig. 10.

Fig. 8. Decorative glass block with pattern Decora®. (*Pittsburgh Corning Corporation.*)

Fig. 9. Use of glass blocks for entrance to a high-rise building. (*Pittsburgh Corning Corporation.*)

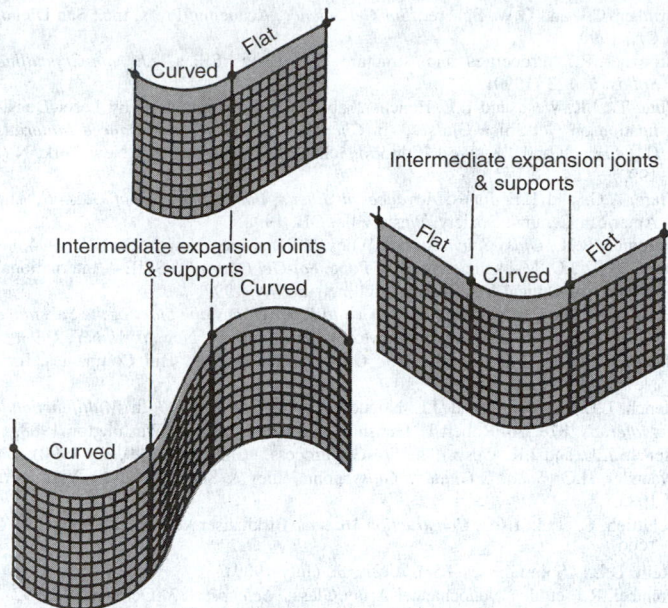

Fig. 10. Various ways to arrange glass block walls for interior or exterior. (*Pittsburgh Corning Corporation.*)

In making glass blocks, molten glass is extruded in "gobs" that are poured into an open half-block mold. A plunger, which creates the pattern on the inner surface of the block, presses the glass into the mold to produce a half-block. Using direct heat, the two halves are fused together to form a complete hollow block unit. This process thus creates an insulating air space, which makes the blocks energy efficient.

Sol-Gel Glass

Sol-gel processing is a new, chemically based method for producing glass at much lower temperatures than traditional melting methods (see above). Due to the low temperatures there are many advantages of sol-gel glass processing, such as casting of net shapes and net surfaces, improved physical properties, and the production of a new type of material, transparent porous glass matrices. See Table 3.

Three methods can be used to make sol-gel glasses:

1. Gelation of colloidal powders.
2. Hypercritical drying.

TABLE 3. ADVANTAGES OF SOL-GEL GLASS

Net-Shape/Surface Casting
Complex geometries
Lightweight optics
Aspheric optics
Surface replication (e.g., fresnel lenses)
Binary/diffractive optics
Internal structures
Reduced grinding
Reduced polishing

Improved Physical Properties (Type V Silica)*
Lower coefficient of thermal expansion
Lower vacuum ultraviolet cutoff wavelength
Higher optical transmission
No absorption due to H_2O or OH bands
Lower solarization
Higher homogeneity
Fewer defects

Transparent Porous Structures (Type VI Silica)*
Impregnated with optically active organics, such as laser dyes, NLO molecules
Graded refractive index (GRIN) lenses
Laser-enhanced densification
Laser-written microoptical arrays and wavelengths
Controlled chemical doping
Control of variable oxidation states of dopants

*Types I–IV silicas are discussed in Bruckner reference listed.

3. Controlled hydrolysis and condensation of metal alkoxide precursors, followed by drying at ambient pressure and temperature.

Definitions. Colloids are solid particles with diameters of $1 \ll 100$ nanometers. A sol is a dispersion of colloidal particles in a liquid. A gel is an interconnected rigid network of sub-micrometer dimensions. A gel can be formed from an array of discrete colloidal particles (Method 1) or the 3-D network can be formed from the hydrolysis and condensation of liquid metal alkoxide precursors (Methods 2 and 3), shown in Fig. 11. The metal alkoxide precursors used in Methods 2 and 3 are usually $Si(OR)_4$ where R is CH_3, C_2H_5, or C_3H_7. The metal ions can be Si, Ti, Sn, Al, and so on.

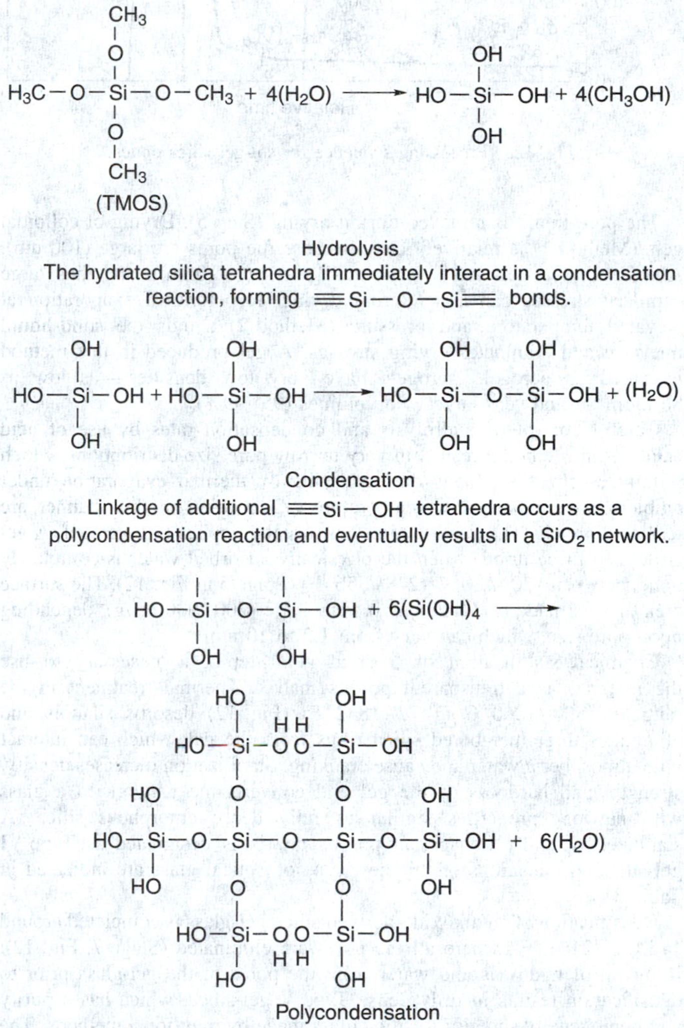

Hydrolysis
The hydrated silica tetrahedra immediately interact in a condensation reaction, forming $\equiv$Si—O—Si$\equiv$ bonds.

Condensation
Linkage of additional $\equiv$Si—OH tetrahedra occurs as a polycondensation reaction and eventually results in a SiO_2 network.

Polycondensation

Fig. 11. Chemical reactions involved in sol-gel alkoxide processing of silica gel-glass.

Processing Steps. Seven steps are involved in making glass by the sol-gel method. See Fig. 12. A low-viscosity sol is formed by mixing (Step 1). The viscosity of the sol increases greatly as a gel begins to form. Prior to gelation the sol is applied as a coating, pulled into a fiber, or cast into a mold with a precise shape and surface features (Step 2). Gelation (Step 3) occurs in the mold, forming a solid object with the desired shape and surface. Low-cost polymer molds can be used, but interfacial bubbles must be prevented and contamination must be avoided, since it can nucleate cracks in the weak gel.

After gelation, the interconnected 3-D gel network is completely filled with pore liquid. Holding the gel in its pore liquid for several hours at 25–80 °C leads to localized solution and precipitation of the solid network, called aging (Step 4). The thickness of the interparticle necks increases during aging, as does density and strength of the gel. Aging must continue until the gel is strong enough to resist cracking during drying.

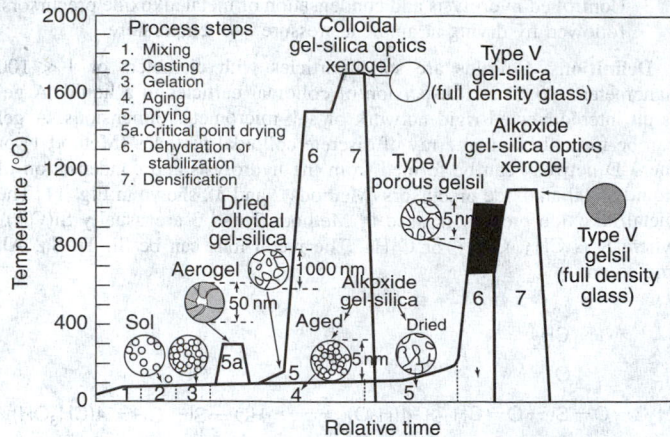

Fig. 12.   Processing sequence for sol-gel silica optics.

The pore liquid is removed during drying (Step 5). Drying of colloidal gels (Method 1) is relatively easy because the pores are large (100 nm). Alkoxide-based gels have very small pores (1–10 nm), and thus large capillary stresses can arise during drying. Hypercritical evaporation at elevated temperature and pressure (Method 2) avoids the solid-liquid interface and eliminates drying stresses. A gel produced in this method is called an aerogel. Aerogels have very low densities—as low as 80 kg/m^3—and very large void volumes (95–99%).

Careful control of hydrolysis and condensation rates by use of acid catalysts in Method 3 results in very narrow pore size distributions, which minimizes stress gradients during drying by thermal evaporation under ambient pressure and low temperatures. Gels dried in this manner are termed xerogels. The generic term gel usually applies to a xerogel. A gel is defined to be dried when the physically adsorbed water is completely gone, between 120–180 °C (248–356 °F) (Stage 5 in Fig. 12). The surface area of gels made by Method 3 is very large (200–900 m^2/g), depending upon pore size, which can vary from 1.2 to 10 nm.

Chemical stabilization of a dried gel, Step 6, is necessary to use the material as a transparent porous matrix. Thermal treatment in the range of 800–1000 °C (1472–1832 °F) (Fig. 12) desorbs silanols and eliminates three-membered silica rings from the gel, which can interact with atmospheric water and cause cracking. Stabilization increases density, strength, and hardness of the gel and converts the network to a glass with network properties similar to fully dense amorphous silica. A stabilized optically transparent porous matrix is designated as Type VI gel-silica. Applications of this new type of optical glass are indicated in Table 3.

Densification of an alkoxide-derived silica gel-glass is completed around 1150 °C (2101 °F), where all the pores are eliminated (Stage 7, Fig. 12). Removal of hydoxyls and water from the pores of the gel-glass prior to densification results in fully dense Type V gel-silica, which has a purity and homogeneity superior to silica glass made by traditional methods. The density becomes equivalent to that of (Types I and II) fused quartz or (Types VI and IV) fused silica (i.e., 2.2 g/cc).

The ability to make optics without grinding or polishing and to replicate surface features from a master mold with high accuracy (1 part in 10^4) is an important advance in optical glass technology offered by sol-gel processing of Type V gel-silica. The new Type VI porous optical matrices made by sol-gel processing make it possible to achieve multifunctional optical components, also an important advance in the field.

Additional Reading

Amato, I.: "A New Order in Glass," *Science*, **1377** (June 7, 1991).

Amstock, J.S.: *Handbook of Glass in Construction*, The McGraw-Hill Companies, Inc., New York, NY, 1997.

Bach, H. and D. Krause: *Analysis of the Composition and Structure of Glass and Glass Ceramics*, Springer-Verlag, Inc., New York, NY, 1999.

Bach, H. and D. Krause: *Thin Films on Glass*, Springer-Verlag, Inc., New York, NY, 2000.

Behling, S. and S. Behling: *Glass: Structure and Technology in Architecture*, Prestel Publishing, New York, NY, 2000.

Brinker, C.J. and G.W. Scherer: *Sol-Gel Science*, Academic Press, Inc., San Diego, CA, 1990.

Bruckner, R.: "Properties and Structure of Vitreous Silica," *J. Non-Crystalline Solids*, **5**, 123 (1990).

Chia, T., J.K. West, and L.L. Hench: "Fabrication of Microlenses by Laser Densification on Gel-Silica Glasses," in *Chemical Processing of Advanced Materials* (L.L. Hench and J.K. West, Editors), John Wiley & Sons, Inc., New York, NY, 1992.

Clare, A.G. and L.E. Jones: *Advances in Fusion and Processing of Glass II*, The American Ceramic Society, Westerville, OH, 1998.

Doremus, R.H.: *Glass Science*, John Wiley & Sons, Inc., New York, NY, 1994.

Dunn, B.S., J.D. Mackenzie, and J.E. Pope: *Sol-Gel Optics IV*, SPIE — International Society for Optical Engineering, Bellingham WA, 1997.

Ellis, W.S.: *Glass: From the First Mirror to Fiber Optics, the Story of the Substance that Changed the World*, William Morrow & Company, New York, NY, 1998.

Harper, C.A.: *Handbook of Ceramic Glasses*, The McGraw-Hill Companies, Inc., New York, NY, 2000.

Hench, L.L., S. Wang, and J.L. Noguès: "Gel-Silica Optics," in *Multifunctional Materials*, **878**, 76, Robert L. Gunshor, Editor, Bellingham, Washington 1988.

Hench, L.L. and J.K. West: "The Sol-Gel Process," *Chem. Rev.*, **90**, 33 (1990).

Pfaender, H.C.: *Schott's Guide to Glass*, John Wiley & Sons, Inc., New York, NY, 1983.

Schittich, C. et al.: *Glass Construction Manual*, Birkhauser Verlag, Cambridge, MA, 1999.

Stein, D.L.: "Spin Glasses," *Sci. Amer.*, 52 (July 1989).

Tonucci, R.J. et al.: "Nanochannel Array Glass," *Science*, 783 (October 30, 1992).

Tse, J.S. and D.D. Klug: "Structural Memory in Pressure-Amorphized AlPO4," *Science*, 1559 (March 20, 1992).

Varshneya, A.K.: *Fundamentals of Inorganic Glasses*, Academic Press, Inc., San Diego, CA, 1997.

L.L. HENCH, University of Florida, Gainesville, FL

GLASS FIBERS.   See **Fiber Glass**.

GLASS FIBERS (Optical).   See **Optical Fiber Systems**.

GLASS-IONOMER CEMENT.   See **Dental Materials**.

GLASS SNAKE (*Reptilia, Sauria*).   A legless lizard, *Ophisaurus ventralis*, whose tail is exceptionally brittle. Although snake-like, it may be recognized as a lizard by its small ventral scales and its eyelids. Its habitat is chiefly the central and southern part of the United States.

GLAUBERITE.   This anhydrous sulfate of sodium and calcium mineral, $Na_2Ca(SO_4)_2$, crystallizes in the monoclinic system. Hardness of 2.5–3, specific gravity of 2.8, with vitreous luster and pale yellow to gray in color. Grades from transparent to translucent. Perfect basal pinacoidal cleavage with conchoidal fracture. Glauberite is a product of salt lake evaporation. World occurrences include the Stassfurt, Germany saline deposits, and Borax Lake in San Bernardino County, California.

GLAUCOMA.   A disease of the eye characterized by atrophy in varying degrees of the optic-nerve head through the enlargement of the optic cup. The disease takes several forms, which are described later in this article.

In the United States, it is estimated that 2% of persons over 35 years of age have chronic glaucoma. Because glaucoma is not always discovered and treated promptly, some 3000 to 4000 persons per year become fully or partially blind. However, where glaucoma is discovered and treated early, the prognosis for useful vision over the life span is excellent. The disease is rare among young people, but the incidence increases with age. Diabetics run a twofold greater risk of having glaucoma than nondiabetics. Infrequently, the appearance of glaucoma is related to glucocorticoid therapy.

Intraocular Pressure

In the eye, there is a constant flow of fluid (*aqueous humor*) into and out of the eye. This fluid keeps the eye firm and clear so that the eyeball functions well visually. There is also a constant flow of blood into and out of the eye. The relative state of inflow and outflow of blood and of aqueous humor largely determines how firm the eye is. If the outflow of aqueous humor is blocked, the pressure inside the eye increases. The constant flow of aqueous humor is indicated in Fig. 1. This flow may be blocked at any point. Nerve damage occurs first at the optic disc. Elevated intraocular pressure can

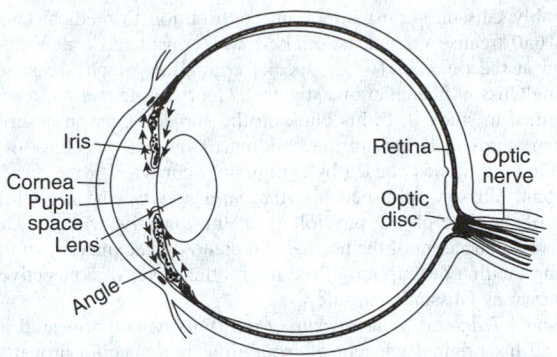

Fig. 1. Constant flow of aqueous fluid is indicated by arrows. (*Wills Eye Hospital, Philadelphia, Pennsylvania.*)

directly damage the nerves that transmit the electrical impulses from the light-sensitive element of the eye (*retina*) to the brain, where the electrical impulses are processed into images. This pressure also can squeeze out of the eye the blood required to keep the nerves healthy and can, in this fashion, damage the nerves.

The outflow of aqueous humor can be impeded in several ways. (1) The hole (*pupil*), through which the aqueous humor flows as it passes from the back to the front of the iris (colored part of eye) can be blocked by adhesions, or by a cataract. (2) The sieve, out of which the aqueous humor exists, can become blocked by debris caused by inflammation, or by deposits due to aging, or by abnormal material which is the result of certain drugs, or by the iris itself. (3) The veins into which the aqueous humor flows when it leaves the eye can be partially blocked, by heart disease, or by pressure on the large veins in the orbit.

Forms of Glaucoma

Some authorities place glaucoma into four categories: (1) *chronic-open-angle glaucoma*; (2) *acute angle-closure glaucoma*; (3) *congenital glaucoma*; and (4) *secondary glaucoma*. Obviously, the treatment of these various types of glaucoma will be different. Some types require surgery; some need medication; some require attention to other organs of the body; some require that certain medications be halted.

Chronic open-angle glaucoma represents nearly 90% of all cases of the disease and is slow and insidious in its onset. The condition can destroy vision without causing any symptoms of blurring or discomfort. The eye pressure usually rises gradually over a long time span (months or years) due to increased resistance to the outflow of aqueous humor. Loss of vision commences in the periphery or edge of the field of vision, and it is often not noticed until it has nearly reached the center. The optic nerve may permanently lose most of its function without any discomfort, blurring, or other symptoms. There are, however, subtle symptoms, such as a vague aching around the eyes, haloes, watery eyes, and frequent need to change glasses. When presented with such complaints, the physician will inquire about past incidences of glaucoma among family members. Firm diagnosis will require notation of changes in the optic nerve inside the eye as seen with an ophthalmoscope, changes in the field of vision, usually shrinkage of side vision, and elevated intra-ocular pressure. The latter measurement is made with a *tonometer*, a simple device with a footplate that rests gently on the cornea (after administration of a local anesthetic). This instrument accurately gages the pressure within the eyeball. Medication is directed toward (1) improving the drainage of fluid from the eye and (2) slowing down the production of fluid. When a patient does not respond to medication, a surgical procedure (*iridectomy*) usually will be performed to relieve the pressure.

Occurring less frequently, *acute angle-closure glaucoma* constitutes a true medical emergency. In this form of the disease, the pressure rises abruptly in one or both eyes from a normal to a very high level. Ocular pain, blurring of vision, haloes around lights, and vomiting are usually, but not always, present in varying degrees. Unless the pressure is relieved in a matter of hours, irretrievable visual loss may occur. Even one day of delay may have a disastrous effect upon a given eye. Treatment is directed first toward reducing the pressure by medical means. As soon as it has been brought to a safe level, a simple but delicate surgical operation is usually performed. If the patient receives expert treatment within a few hours after onset of attack, this operation is likely to result in a permanent cure and the eye may remain glaucoma-free from that time forward. In some cases, however, chronic glaucoma may persist for months or years.

Very rarely, *congenital glaucoma* is present in newborn infants, or appears shortly after birth. Infants thus afflicted are frequently, but not always, born with enlarged eyes. Tearing and unusual sensitivity to light are important signs of infantile glaucoma.

Secondary glaucoma occurs in connection with certain ocular inflammations, tumors, injuries, and hypermature cataracts, among others.

Timely detection of certain forms of glaucoma is difficult. Fluid pressure generated within the eye may not be present. Within the last few years, digital imaging techniques have been developed to map the topography of the optic-nerve head. Called the laser tomographic scanner, the technique provides a sensitive and precise tool for measuring and tracking nerve-head deformations. Although the technique is in the last stages of development, experimental results show that this type of examination may alter the current management of glaucoma in a dramatic fashion.

In common forms of glaucoma, where pressure of the aqueous humor is abnormally high, the physician may prescribe a drug that represses production of the aqueous humor. These substances include ophthalmic beta-blockers, such as timolol maleate, levobuolol, and beta xololol. However, a degree of systemic absorption is likely to occur—in the opposite eye, in the lungs (bronchospasms), in the central nervous system (depression), and in the heart (bradycardia). Consequently, the physician will prescribe the *lowest* effective concentration. Physicians are particularly cautious in treating pregnant patients. Although more powerful drugs are excellent for controlling elevated intraocular pressure, they do produce serious side effects, such as fatigue, weight loss, sensory neuropathy, and calcium phosphate nephrolithiasis.

See also **Vision and the Eye**.

Additional Reading

Appleton, L. and T.L. Lewis: *Primary Care of the Glaucomas*, 2nd Edition, Appleton & Lange, Old Tappan, NJ, 2000.
Bienfang, D.C. et al.: "Ophthalmology," *N. Eng. J. Med.*, 956 (October 4, 1990).
Choplin, N.T.: *Glaucoma Atlas*, Mosby-Year Book, Inc., St. Louis, MO, 1998.
Gross, R.L.: *Clinical Glaucoma Management: Critical Signs in Diagnosis and Therapy*, W.B. Saunders Company, Philadelphia, PA, 2000.
Hubel, D.H.: *Eye, Brain, and Vision*, Scientific American Library, W.H. Freeman and Company, New York, NY, 1995.
Kanski, J.J., J. Salmon, and J.A. McAllister: *Glaucoma: A Color Manual of Diagnosis and Treatment*, Butterworth-Heinemann, Inc., Woburn, MA, 1996.
Krieglstein, G.K.: *Glaucoma Update VI*, Springer-Verlag, Inc., New York, NY, 1999.
Litwak, A.B.: *Glaucoma Handbook*, Butterworth-Heinemann, Inc., Woburn, MA, 2000.
McKee, S.P. and K. Nakayama: *Optics, Physiology, and Vision*, Elsevier Science, New York, NY, 1990.
Shields, M.B.: *Color Atlas of Glaucoma*, Lippincott Williams & Wilkins, Philadelphia, PA, 1997.
Shields, M.B.: *Testbook of Glaucoma*, 4th Edition, Lippincott Williams & Wilkins, Philadelphia, PA, 1997.
Shields, M.B. and E.M. Van Buskirk: *100 Years of Progress in Glaucoma*, Lippincott Williams & Wilkins, Philadelphia, PA, 1998.
Spaeth, G.L. and T.M. Eid: *The Glaucomas: Concepts and Fundamentals*, Lippincott Williams & Wilkins, Philadelphia, PA, 1999.
Staff: "Inhibiting Capillaries," *Technology Review (MIT)*, 11 (August/September 1990).
Staff: "Glaucoma," in *Special Bulletin*, American Academy of Ophthalmology, San Francisco, California, 1990.
Weiss, R.: "Eyeing the Optic Nerve: Laser Finds Early Signs of Blinding Glaucoma," *Science News*, 330 (November 23, 1990).
Zimmerman, T.J. and K. Kooner: *Clinical Pathways in Glaucoma*, Thieme Medical Publishers, Inc., New York, NY, 2000.

Web References

All About Vision: http://www.allaboutvision.com/conditions/glaucoma.htm
American Academy of Ophthalmology EyeNet: http://www.eyenet.org/
Chicago Eye Institute: http://www.chicagoeyeinstitute.com/
International Glaucoma Association: http://www.iga.org.uk/home.htm

GLAUCONITE. Glauconite is a hydrous silicate of potassium, iron, and aluminum with considerable ionic substitution, crystallizing in the monoclinic system. A general formula is $(K, Na)(Al, Fe^{3+}, Mg)_2(Al, Si)_4O_{10}(OH)_2$. It possesses perfect basal cleavage; hardness, 2; specific gravity, 2.4–2.95; color dull green to blue-green; and is often a constituent of marine deposits, forming "green sands." It is believed to have been produced through the alteration of iron-bearing silicates, chiefly biotite and possibly augite and hornblende. It occurs along the Atlantic Coastal Plain of the United States. Frequently found filling the interiors of the shells of *Globigerina*, a common genus of the foraminifera (*Protozoa*). Since *Globigerina* occurs as a deep-sea deposit, some European geologists have claimed that glauconite is only found in deep water. On the other hand, typical green sands occur associated with sand and clays which are certainly of shallow marine origin. Glauconite derives its name from the Greek word meaning *bluish-gray*.

Glauconite, being of sedimentary origin, can be used to determine the age of those sediments by evaluating its $^{40}K/^{40}Ar$ ratio (potassium-argon isotope ratio). See also **Ocean Resources (Mineral)**.

GLAUCOPHANE. Glaucophane, essentially a complex silicate of sodium, and iron or aluminum, $Na_2(Mg, Fe)_3Al_2Si_8O_{22}(OH)_2$, is a rather rare mineral although it has been noted from widely separated occurrences. It is monoclinic and ordinarily is fibrous or granular. It is brittle; hardness, 6; specific gravity, 3–3.1; color, azure blue, blackish-blue or gray; luster, vitreous to pearly; translucent to opaque. Glaucophane is found only in the metamorphic rocks sometimes forming glaucophane schists. It is found in Switzerland, Italy, Siberia, Japan and in the United States chiefly in the rocks of the Coast Ranges in California and Oregon. The name glaucophane is derived from the Greek words meaning *bluish-gray*, and *appear*.

GLAZE. (Also called *glaze ice, glazed frost, verglas*). A coating of ice, generally clear and smooth, formed on exposed objects by the freezing of a film of supercooled water deposited by rain, drizzle, fog, or possibly condensed from supercooled water vapor. Glaze is denser, harder, and more transparent than either rime or hoarfrost. Its density may be as high as 0.8 or 0.9 g cm^{-3}. Factors that favor glaze formation are large drop size, rapid accretion, slight supercooling, and slow dissipation of heat of fusion. The opposite effects favor rime formation. The accretion of glaze on terrestrial objects constitutes an ice storm; as a type of aircraft icing it is called *clear ice*. Glaze, as well as rime, may form on ice particles in the atmosphere. Ordinary hail is composed entirely (or nearly so) of glaze; the alternating clear and opaque layers of some hailstones represent glaze and rime, deposited under varying conditions around the growing hailstone. See also **Glaze Icing**.

GLAZE ICING. Glaze aircraft icing (also known as *clear icing*) is hard and transparent. It is formed by the relatively slow freezing of supercooled water droplets as they spread over the surface. It tends to accumulate more rapidly than rime icing, and is often very hard and therefore more difficult to remove. Since it does not freeze instantly, the ice can form into shapes that cause significant aerodynamic penalties, and it therefore the most hazardous form of icing. However, it is the least frequent type of ice encountered, responsible for about 10% of icing reports.

GLAZING AND POLISHING AGENTS (FOODS). See **Food Additives**.

GLIDE PLANE. In solid state physics, this term denotes: 1. a symmetry element of a space lattice, such that the lattice remains unchanged after a reflection in the plane, followed by a translation parallel to the same plane; 2. a slip plane as defined in the theory of dislocations.

GLISSON, FRANCIS (1597–1677). Francis Glisson was a British physician who made the first accurate clinical description of rickets as well as important contributions to the anatomy and physiology of the digestive system.

Glisson was born in Dorset and studied in Cambridge where he received his MD in 1634. The following year, he was elected a fellow of the College of Physicians of London and in 1636 was appointed Regis Professor of Physic at Cambridge, a position he held for over forty years. Glisson was a founder as well as one of the first fellows of the Royal Society and was elected president of the College of Physicians in 1667, 1668 and 1669.

Probably Glisson's most important contribution to medical knowledge is his 1650 treatise on rickets. Rickets was considered a new disease in England at the time. In 1645 a special committee of physician-scholars, including Glisson (the chief investigator), George Bate and A. Regemorter, was formed to study it. In his book on the subject, Glisson described the main symptoms of rickets in the children of his native county of Dorset and denied that it was caused by syphilis or scurvy.

In 1654, Glisson published his *Anatomia hepatis*, the most important treatise of the time on the physiology of the digestive system. The work contains a full account of the normal and pathological anatomy of the liver, including the first description of the liver's thin layer of connective tissue, now known as Glisson's capsule.

Glisson's *Tractatus de ventriculo et intestinis*, which appeared in 1677, contained his original concept of irritability as a specific property of all human tissue. Experiments conducted by Albrecht von Haller in 1753 proved this hypothesis to be correct. See also **Haller, Albrecht von (1708–1777)**.

RÉGIS OLRY, University of Quebec at Trois-Rivières, Quebec, Canada

GLOBAL CHANGE. The cosmos, our galaxy, our solar system, and our planet, Earth, all are part of a dynamic system and consequently subject to change. Global change, of course, relates to Earth and much has been written about this changing planet over the last several decades. Awareness of the consequences of some of these changes has increased in recent years. These concerns relate to how humans may be influencing Earth as contrasted with those factors that are beyond the reach of human intervention. Of the basic components of Earth, its atmosphere, including the hydrosphere, is most vulnerable to changes resulting from anthropogenic activities, including pollution. Misuse and denigration of Earth's great land areas—the plains, mountains, forests, wetlands—also occurs sometimes at an alarming rate. Destruction by humans of other life forms (endangered species) is another important element of global change. Several articles in this encyclopedia deal with numerous aspects of the aforementioned topics.

GLOBAL EARTH OBSERVATION SYSTEM OF SYSTEMS (GEOSS). Over the next decade, a global Earth Observation System will revolutionize our understanding of the Earth and how it works. With benefits as broad as the planet itself, this U.S.-led initiative promises to make peoples and economies around the globe healthier, safer and better equipped to manage basic daily needs. The aim is to make 21st century technology as interrelated as the planet it observes, predicts and protects, providing the science on which sound policy and decision-making must be built.

Building an integrated, comprehensive and sustained global Earth Observation System opens a world of possibilities. Imagine a world in which we could:

- Forecast next winter's weather months in advance
- Predict where and when malaria, West Nile virus, SARS and other diseases are likely to strike
- Reduce U.S. energy costs by about $1 billion yearly
- More effectively monitor forest fires and predict the effect of air quality on sensitive populations in near real-time
- Provide farmers with immediate forecasts essential to maximizing crops yields
- Predict the pattern of the North American monsoon—Arizona derives two-thirds of its water from the monsoon weather pattern

Global architecture that reflects how our world actually works is key to making such visions operational.

Challenges

Right now many thousands of individual pieces of technology are gathering earth observations around the globe. They are demonstrating their value in estimating crop yields, monitoring water and air quality, and improving airline safety. U.S. farmers gain about $15 of value for each $1 spent on weather forecasting. Benefits to U.S. agriculture from altering planting decisions are estimated at over $250 million. The annual economic return to the U.S. of NOAA's El Nino ocean observing and forecast system is between 13% and 26%.

But while there are thousands of moored and free floating data buoys in the world's oceans, thousands of land-based environmental stations, and over 50 environmental satellites orbiting the globe, all providing millions of data sets, most of these technologies do not yet talk to each other. Until they do—and all of the individual technology is connected as one comprehensive system of systems—there will always be blind spots and scientific uncertainty. Just as a doctor can't diagnose health by taking just one measurement, neither can scientists really know what's happening on our planet without taking earth's pulse everywhere it beats—which is all over the globe.

The challenge is to connect the scientific dots—to build a system of systems that will yield the science on which sound policy must be built.

U.S. and Global Plans

The U.S. is spearheading such a system, in our country and around the world. As a collaborative effort, 15 federal agencies and three White House offices have drafted a *Strategic Plan for the U.S. Integrated Earth Observation System.* http://highstrangeness.tv/library/ieos.pdf. The draft plan was released on September 8 and, after public review, it will be finalized by late 2004.

On a parallel track, the U.S. is also spearheading the development of a Global Earth Observation System of Systems, of which the U.S. plan will be a key component. Launched at the first-ever global Earth Observation Summit, held on July 31, 2003 in Washington, D.C., the pioneering Group on Earth Observations is now supported by 61 countries, the European Commission, and over 43 international organizations. The 10-year implementation plan was adopted at the Third Global Earth Observation Summit, held on February 16, 2005 in Brussels. http://earthobservations.org/docs/GEOSS%2010-Year%20Implementation%20Plan%20(GEO%201000).pdf.

Current members of the GEO include: Algeria, Argentina, Australia, Belgium, Belize, Brazil, Cameroon, Canada, Central African Republic, Chile, China, Croatia, Cyprus, Denmark, Egypt, Finland, France, Germany, Greece, Guinea-Bissau, Honduras, Iceland, India, Indonesia, Iran, Ireland, Israel, Italy, Japan, Kazakhstan, Luxembourg, Malaysia, Mali, Mauritius, Mexico, Morocco, Nepal, Netherlands, New Zealand, Niger, Nigeria, Norway, Philippines, Portugal, Republic of Korea, Republic of the Congo, Russian Federation, Slovak Republic, Slovenia, South Africa, Spain, Sudan, Sweden, Switzerland, Thailand, Tunisia, Ukraine, United Kingdom, United States and Uzbekistan.

The above list of Members is current as of the Second Plenary Session of the GEO (GEO-II) held 14-15 December 2005 in Geneva, Switzerland. The 43 international organizations include:

- AARSE: African Association of Remote Sensing of the Environment: http://www.geolsoc.org.uk/template.cfm?name=ARSE
- ADIE: Association for the Development of Environmental Information: earthobservations.org/docs%5CGEO%20Brochure%202004.pdf –
- APN: Asia-Pacific Network for Global Change Research: http://www.apn-gcr.org/en/indexe.html
- CEOS: Committee on Earth Observation Satellites: http://www.ceos.org/
- ECMWF: European Centre for Medium-Range Weather Forecasts: http://www.ecmwf.int/
- EEA: European Environmental Agency: http://www.eea.eu.int/main_html
- ESA: European Space Agency: http://www.esa.int/esaCP/index.html
- ESEAS: European Sea Level Service: http://www.eseas.org/
- EUMETNET: Network of European Meteorological Services/Composite Observing System: http://europa.eu.int/comm/research/environment/geo/breakarticle_2451_en.htm
- EUMETSAT: European Organization for the Exploitation of Meteorological Satellites: http://www.eumetsat.int/idcplg?IdcService=SS_GET_PAGE&nodeId=31&l=en
- EuroGeoSurveys: The Association of the Geological Surveys of the European Union:
- http://www.welcomeurope.com/default.asp?id=1520&idreseau=98
- FAO: Food and Agriculture Organization of the United Nations: http://www.fao.org/
- FDSN: Federation of Digital Broad-Band Seismograph Networks: http://www.fdsn.org/
- GCOS: Global Climate Observing System: http://www.wmo.ch/web/gcos/gcoshome.html
- GSDI: Global Spatial Data Infrastructure: http://www.gsdi.org/

- GOOS: Global Ocean Observing System: http://ioc.unesco.org/goos/
- GTOS: Global Terrestrial Observing System: http://www.fao.org/gtos/
- IAG: International Association of Geodesy: http://www.gfy.ku.dk/~iag/
- ICSU: International Council for Science: http://www.icsu.org/index.php
- IEEE: Institute of Electrical and Electronics Engineers: http://ieee.org/portal/site/iportals/
- IGBP: International Geosphere-Biosphere Program:
- http://www.igbp.kva.se/cgi-bin/php/frameset.php
- IGFA: International Group of Funding Agencies for Global Change Research: http://www.igfagcr.org/
- IGOS-P: Integrated Global Observing Strategy Partnership: http://www.igospartners.org/
- IISL: International Institute for Space Law: http://www.iafastro-iisl.com/
- INCOSE: International Council on Systems Engineering: http://www.incose.org/
- IO3C: International Ozone Commission: http://ioc.atmos.uiuc.edu/about.html
- IOC: Intergovernmental Oceanographic Commission: http://ioc.unesco.org/iocweb/index.php
- ISCGM: International Steering Committee for Global Mapping: http://www.iscgm.org/cgi-bin/fswiki/wiki.cgi
- ISDR: International Strategy for Disaster Reduction: http://www.unisdr.org/
- ISPRS: International Society for Photogrammetry and Remote Sensing: http://www.isprs.org/
- OGC: Open Geospatial Consortium: http://www.opengeospatial.org/
- POGO: Partnership for Observation of the Global Ocean:
- http://www.ocean-partners.org/
- SICA/CCAD: Central American Commission for the Environment and Development: http://wbln0018.worldbank.org/MesoAm/UmbpubHP.nsf/HomePage/1?OpenDocument
- SOPAC: South Pacific Applied Geoscience Commission:www.sopac.org/
- UNCBD: United Nations Convention on Biodiversity: http://www.biodiv.org/default.shtml
- UNEP: United Nations Environment Programme: http://www.unep.org/
- UNESCO: United Nations Educational, Scientific and Cultural Organization: http://portal.unesco.org/en/ev.php-URL_ID=29008&URL_DO=DO_TOPIC&URL_SECTION=201.html
- UNFCCC: United Nations Framework Convention on Climate Change: http://unfccc.int/2860.php
- UNITAR: United Nations Institute for Training and Research: http://www.unitar.org/
- UNOOSA: United Nations Office for Outer Space Affairs: http://www.unoosa.org/oosa/index.html
- UNU-EHS: United Nations University Institute for Environment and Human Security: http://www.ehs.unu.edu/
- WCRP: World Climate Research Programme:
- http://www.wmo.ch/web/wcrp/wcrp-home.html
- WMO: World Meteorological Organization: http://www.wmo.ch/index-en.html

By adopting an implementation plan for the GEOSS, the nations have accomplished the first phase of realizing the goal of a comprehensive, integrated and sustained Earth Observation System that will improve our global ability to predict weather and climate, prepare for natural hazards, and strengthen the health and well-being of people and economies around the globe.

GEOSS will link existing technology in space, in the ocean, and on land that is already demonstrating value around the globe. GEOSS also supports the building of new observation capacities where required. It will provide a planning framework for systems, data and vital information so scientists and policy makers in many different countries can design, implement and operate systems in a compatible way.

Commerce Secretary Carlos M. Gutierrez led the U.S. delegation at the summit in Brussels. With four co-chairs, the Group on Earth Observations is helping to advance the development of the emerging global system. The co-chairs are: Vice Admiral (Ret.) Conrad C. Lautenbacher, Under Secretary of Commerce for Oceans and Atmosphere and Administrator of the National Oceanic and Atmospheric Administration (NOAA); Mr. Tetuhisa Shirakawa, of Japan, Deputy Minister of Education, Culture, Sports, Science and Technology; Mr. Achilleas Mitsos, of the European Commission, Director General for Research; and Dr. Rob Adam, of South

Africa, Director-General of Science and Technology. South Africa co-chairs on behalf of the developing world, reflecting the vital importance of developing nations to the success of the global system.

Nine Societal Benefits

In the U.S. and globally, the emerging system will focus on nine societal benefit areas:

- Improve Weather Forecasting
- Reduce Loss of Life and Property from Disasters
- Protect and Monitor Our Ocean Resource
- Understand, Assess, Predict, Mitigate and Adapt to Climate Variability and Change
- Support Sustainable Agriculture and Forestry and Combat Land Degradation
- Understand the Effect of Environmental Factors on Human Health and Well-Being
- Develop the Capacity to Make Ecological Forecasts
- Protect and Monitor Water Resources
- Monitor and Manage Energy Resources

The benefits of building global observing architecture are enormous.

Substantial Socio-Economic Payoffs

- We could more accurately know how severe next winter's weather, with strong implications for emergency managers, transportation, energy and medically personnel, farmers, families, manufacturers, storeowners, etc., etc. Weather and climate sensitive industries account for one-third of the nation's GDP, or $3 trillion.
- We could forecast weather with just one degree F more accuracy, saving at least $1 billion annually in U.S. electricity costs.
- With coastal storms reflecting 71 percent, or $7 billion, of U.S. disaster losses every year, improved forecasting will have a major favorable impact on preparedness.
- In the U.S., at a cost of $4 billion annually, weather is responsible for about two-thirds of aviation delays—$1.7 billion of which would be avoidable with better observations and forecasts.
- Benefits from more effective air quality monitoring could provide real-time information as well as accurate forecasts that, days in advance, could enable us to mitigate the effects of poor quality through proper transportation and energy use.
- Benefits from ocean instrumentation that, combined with improved satellite Earth observing coverage, could provide revolutionary decadal worldwide and regional climate forecasts, enabling us, for example, to predict years of drought.
- Benefits from real-time monitoring and forecasting of the water quality in every watershed and accompanying coastal areas could provide agricultural interests with immediate feedback and forecasts of the correct amount of fertilizers and pesticides to apply to maximize crop generation at minimum cost, helping to support both healthy eco-systems and greatly increased U.S. fishery output and value from coastal tourism.
- Globally, an estimated 300-500 million people worldwide are infected with malaria each year and about one million die from this largely preventable disease—with a linked international system, we could pinpoint where the next outbreak of SARS or West Nile virus, or malaria is likely to hit.

Web References

Canadian Group on Earth Observations (CGEO): http://www.cgeo-gcot.gc.ca/

Global Climate Observing System (GCOS) Implementation Plan–October 2004: http://www.wmo.ch/web/gcos/gcoshome.html

State-by-State Benefits and Value to Tribal Nations: http://www.epa.gov/geoss/whereyoulive.html

The Group on Earth Observations (GEO): http://earthobservations.org/

U.S. Integrated Earth Observation System: http://iwgeo.ssc.nasa.gov

1st Earth Observation Summit-July 2003: http://earthobservations.org/summit/index.html

Integrated Global Observing Strategy (IGOS): http://www.igospartners.org/

Earth Observation Partnership Meeting–October 2004: http://www.europa.eu.int/comm/space/events/article_1598_en.html

Earth Observation Summit II in Tokyo–April 2004: http://www.mext.go.jp/a_menu/kaihatu/earth/index.htm

Earth Observation Summit III in Brussels–February 2005: http://europa.eu.int/comm/research/environment/geo/article_2449_en.htm

GLOBAL POSITIONING SYSTEM (GPS). The Global Positioning System, commonly referred to as GPS, is a worldwide, satellite-based positioning and timing system that allows suitably equipped radio receivers to locate themselves in four dimensions, latitude, longitude, altitude, and time, anywhere there is a reasonably clear view of the sky. The system is also known as NAVSTAR, a convenient nickname that is not an acronym. The GPS system was developed, deployed, and is currently operated by the U.S. Air Force. GPS enables precision weapon delivery for all branches of the U.S. Department of Defense, as well as allied nations. Additionally, GPS supports civilian positioning and was always intended to support civil operations. The complete satellite constellation and ground support equipment that make up GPS was declared "operational" in December 1994, although civil use of the developmental signals started in the early 1980s. Initially, the civil signal was deliberately perturbed to prevent hostile use; this greatly degraded the civilian signal accuracy. This perturbation was called selective availability (SA), but the widespread advent of differential GPS, which calibrated these errors in real time, rendered this totally ineffective. In 1996, the President ordered this perturbation stopped, pending justification from the Department of Defense. In 2000, the selective availability perturbations in the signal were completely removed.

The fundamental operation is as follows: the 24 GPS satellites (see Fig. 1) are uploaded from the ground with their current and predicted positions (called ephemeris or orbital parameters). Small corrections of their space-borne atomic clocks are also uploaded. This information is broadcast to the user as a data modulation on an L-band signal (1575 MHz for most civilian users) that doubles as a precise, one-way ranging signal. Ranging is achieved by synchronizing the start time of a pseudorandom sequence of bits transmitted from the GPS satellites at an accuracy of about one nanosecond (10^{-9} s). Three very important results are achieved by this implementation. First, this makes GPS ranging a one-way signal that allows an infinite number of users to receive the signal and compute their position without saturating the GPS system. Additionally, this makes the GPS receiver passive, so that it does not radiate radio-frequency (RF) energy. Last, by receiving four or more satellite signals, users can synchronize their local clocks to GPS time, obviating the need for a very high quality and very expensive atomic clock in the receivers.

An important feature of the system is that all satellites broadcast on the same nominal frequency but use different modulation codes that are nearly orthogonal to each other. This technique is referred to as code division multiple access (CDMA). These codes are called pseudorandom noise or PRN codes. The user separates the signal from each satellite by correlating the incoming signal with an internally generated replica of the code for each of the satellites in view. The actual range measurement is

Fig. 1.   Early GPS satellite. A phase one GPS satellite built by Rockwell (now Boeing).

the corrected difference between the phase of these codes and the local user's clock (called *pseudorange* because the true range is offset due to the local clock bias). Algebraically, four (or more) measurements allow the user to solve simultaneously for the four dimensions of location x, y, z, and t. More than four measurements allow improved accuracy and can also be used to monitor the integrity of the computed solution. A more detailed description of the Concept of Operations can be found in a later section.

GPS system accuracy, it can be shown, is a function of both the ranging accuracy from the satellites and the geometry of the satellite constellation being received. Typically 6 to 11 satellites are in view for users anywhere in the world who have clear views of the sky. Errors will be discussed later; typical civilian GPS positioning accuracies for nominal satellite geometries are summarized in Table 1. A full capacity GPS receiver can measure position, *velocity*, and *attitude* using multiple antennae. Thus, thirteen quantities can be measured by GPS: time (t) and the three dimensional position (x, y, z) and velocity (u, v, w), as well as the three attitude rotations (ψ, ϕ, θ) and the associated attitude rotational rates (p, q, r).

TABLE 1. NOMINAL GPS MEDIAN ACCURACIES FOR CIVILIAN USERS[a]

		Operation				
Dimension	Nominal	Local differential	Wide-area differential	Carrier differential	Survey	Time transfer
Horizontal	10 m	0.5 m	1.0 m	0.01 m	0.001 m	NA
Vertical	20 m	1.0 m	2.0 m	0.02 m	0.002 m	NA
Time	50 ns	NA	NA	NA	NA	3 ns

[a] This table displays the GPS accuracies for civilian users with four or more satellites in view and reasonable geometries as a function of type of receiver and aiding. Nominal accuracies are indicative of a stand-alone, single-frequency code receiver. Differential aiding improves the accuracy of the position but does not affect the time, and the improvement in position is a strong function of the distance from the differential station. Carrier phase techniques provide enormous gains in positioning accuracy but require additional time and computation to solve for the unknown carrier cycles between the differential station and the user.

GPS is made up of three logically different systems that are commonly referred to as segments. The three fundamental GPS segments are the space segment, the ground control segment, and the user segment. The space segment (see Fig. 2) consists of approximately 24 satellites in six inclined orbital planes that have periods of 12 sidereal hours (11 hours, 56 minutes, and 4 seconds). Except for small perturbations, each satellite has a ground trace that is repeated twice per (sidereal) day. The corresponding altitude above the mean equatorial radius of the earth is 20,163 km (12,529 miles). The orbits are nearly circular to keep the received power of the signal constant, and the orbital planes are nominally inclined 55° to the equator. All operational satellites have been launched from Kennedy Space Flight Center on Delta rockets, but the advent of the evolved expendable launch vehicle (EELV) will cause a switch to those vehicles. The space segment receives uploaded predictions of location and time corrections from the control segment and stores them for transmission to users. Three navigation signals are currently being broadcast: a civilian signal (called L1 C) at 1575.42 MHz that has a modulation bit rate of 1.023 MHz; a military signal (called L1P/Y) also at 1575.42 MHz that has a modulation bit rate of 10.23 MHz; and a military signal (called L2P/Y) at 1227.6 MHz that has a modulation bit rate of 10.23 MHz. Details of the signal structure are given in Operational Concepts.

The control segment consists of five or more monitor stations, four ground antenna upload stations, and the Operational Control Center (OCS—Located at Schriever AFB outside Colorado Springs, Colorado). A backup control center is planned for Vandenberg AFB, California. Each monitor station measures the ranges to all satellites in view, smooths these measurements, and transmits these data to the OCS for further processing. The OCS predicts future satellite locations and satellite clock corrections. These data are then appropriately formatted and sent to the upload stations for relay to the satellites. The information is retained in satellite memory and sent to users as part of the data modulation scheme, at 50 bits per second. GPS is designed to retain its functionality, albeit at a degraded level, in the unlikely event that the ground stations cannot upload the data to the satellites. Modernization plans for the GPS constellation include

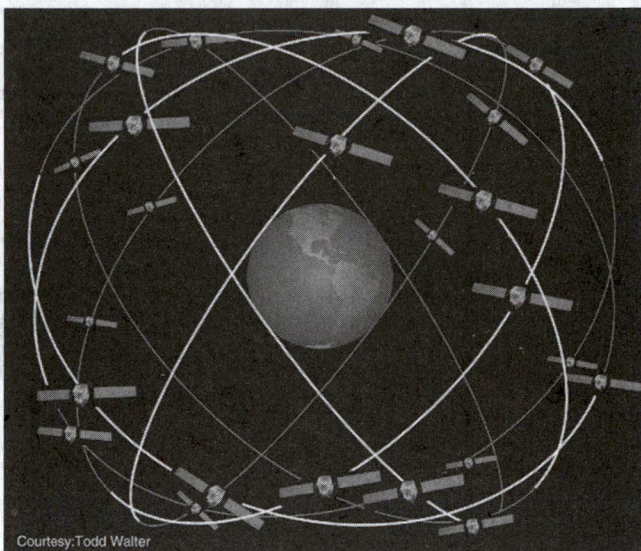

Courtesy: Todd Walter

Fig. 2. The GPS constellation consists of 24 satellites in six orbital planes. The orbital planes are nominally inclined at 55° and contain four satellites each. The satellites are not placed symmetrically around the orbital plane, but instead are placed in such a way that any single satellite failure has minimal impact on GPS. The orbital altitudes are 20,163 km (12,529 miles) above the equator. Some of the orbital planes may have extra satellites as on-orbit spares.

satellites that can communicate directly with each other at higher data rates. This will provide greater capability in the event of loss of ground contact.

The user segment consists of the receivers, which lock on the signal, demodulate the data, calculate the corrected ranges, and transform this into position, velocity, and time. Differential transmitting stations (see Fig. 3) are considered part of the user segment, even though some of these may be other satellites (e.g., Wide Area Augmentation System—WAAS, Fig. 4) or transmission towers operated by the government (e.g., National Differential GPS—NDGPS, Fig. 5).

History

For 6,000 years, humans have been developing ways to navigate to remote destinations. Driven mostly by the desire to transport goods by ship, early navigators remained within sight of land using a technique known as "piloting" that relied on navigators' recognition of coastal features. The magnetic compass appeared in China around the 1100 C.E. and in Europe approximately a century later. When forced to traverse a stretch of water outside the view of land or in inclement weather, navigators kept track of their position by "dead reckoning." Navigators would record their heading and distance traveled by hourglass timing the passage of wooden logs thrown off the bow. Needless to say, the technique was notoriously inaccurate. The development of a sextant by 1731 (early versions existed in the thirteenth century) made determining latitude fairly routine. Early efforts to navigate precisely at sea led to so many deaths that in 1714 a King's ransom was offered to anyone who could solve the problem of computing longitude. During the eighteenth and nineteenth centuries, the navies of the world refined optical instruments and timekeeping. This allowed reliance on the stars and planets to locate their ships precisely. These celestial navigation techniques fundamentally required angular measurements between the local horizon and the Sun, stars, or planets to find lines of position. Due to the motion of Earth, each angular measurement had to be carefully timed to attain the required accuracy. Earth's "rim speed" at the equator is about 1500 km/h, or 24 km min; thus a 1-second error translates to about one-half mile. (On the other hand, because GPS uses the time of flight of a radio signal, a 1-second error for GPS translates into 300,000 km error in position.)

At the turn of the twentieth century, Marconi successfully transmitted radio waves across the Atlantic. By the 1930s, aircraft navigation was becoming a concern, and radio-navigation techniques were in their infancy. Early aircraft navigation aids consisted of direction-finding equipment, which gave a bearing to the transmitting station. Radio techniques such as radio beacons and LORAN were invented to overcome the limitations of

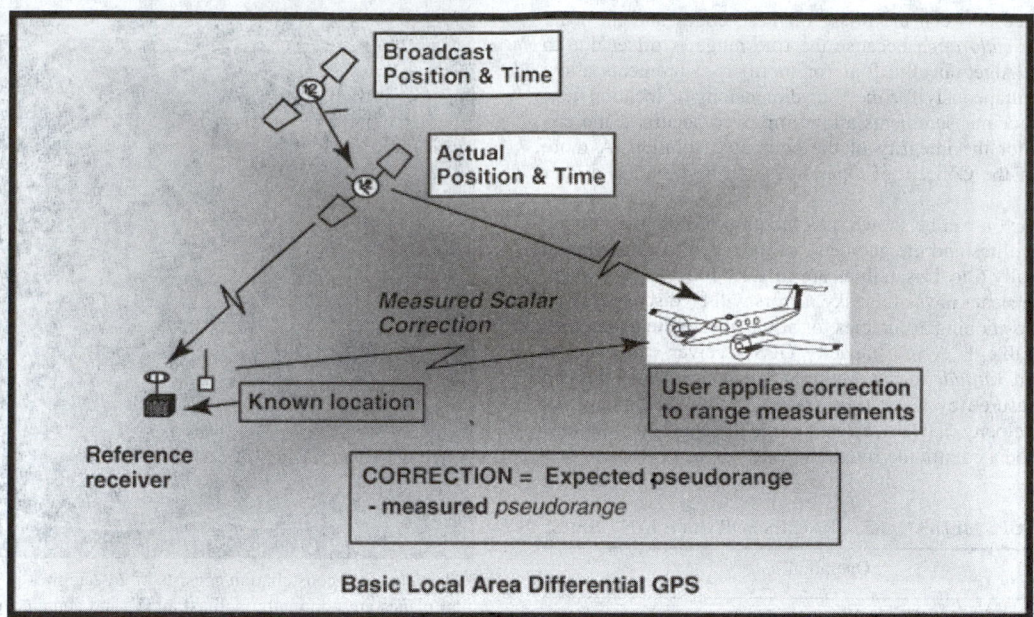

Fig. 3. Block diagram of Basic Differential GPS. Differential corrections are broadcast to the user from a receiver in a known location that computes the correction from the difference between its known location and the GPS-measured position; hence the term "differential" corrections. The error at the reference receiver and the user are correlated across distance and time so that great improvement can be achieved across smart distances and small time lags (typically 5–10 km and several minutes of latency). Note that the primary reason for using differential had been to reduce the effects of selective availability; when SA is off, much better accuracies and integrities are achieved.

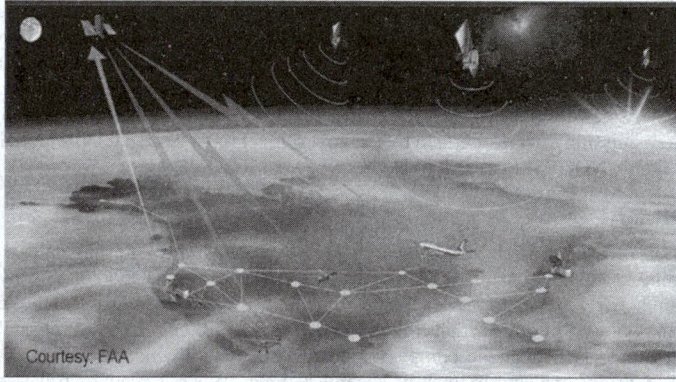

Fig. 4. The Wide Area Augmentation System (WAAS) architecture. The basic functionality of the WAAS system is to use a widely spaced set of reference stations to produce a set of vector corrections for all users within the coverage space. Data is aggregated at each station and processed into a global set of corrections at redundant WAAS Master Stations (WMS) and in turn uplinked to satellites in GEO orbits. These satellites broadcast a message that allows users in the coverage area to compute their own corrections based on the WAAS data and rough knowledge of their own positions. Courtesy: FAA.

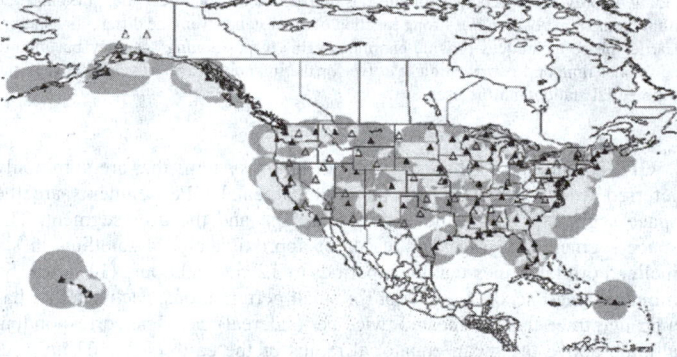

Fig. 5. U.S. National Differential GPS System. This map shows the coverage of the NDGPS system as of 2004. Existing stations broadcast corrections in the 300 kHz band and generally have a range of 100–250 km (62–155 miles). The current system covers the entire coastline and navigable rivers. Future upgrades are being deployed that will remove the gaps in coverage of the entire continental United States.

celestial navigation and were largely deployed by the end of World War II. These techniques provided all-weather, 24-hour navigation service, but only within range of the signals.

GPS Predecessor: Transit. In October 1957, the Soviet Union launched "Sputnik," the world's first space satellite. This triggered a flurry of activity within the United States to discover the exact details, especially the nature of its orbit. Two researchers, Dr. Guier and Dr. Wieffenbach, at the Johns Hopkins Applied Physics Laboratory (APL) had carefully studied Sputnik's radio signal and noted certain regular features. The most interesting of these features was the Doppler shift as the satellite passed overhead. This was caused by the change in the length of the line of sight and was enhanced by the satellite's high speed and low altitude. These scientists developed a computer program to determine Sputnik's orbit. Dr. McClure of APL, a colleague, realized that the problem could be turned on its head; the process could be reversed. By measuring the

Doppler shift to a satellite of known orbit, listeners could calculate their own positions. This solved an important problem for the U.S. Navy that yielded precise all-weather positions for submarines and other ships. After speedy approval, a program was initiated under APL's management. The first two developmental TRANSIT satellites were launched by 1960, and the system became operational by 1964.

TRANSIT eventually deployed an operational constellation that included about five polar orbiting satellites. They produced fixes every 35 to 100 minutes and provided horizontal accuracies of 100 meters or better for a stationary user. A moving receiver could compensate for velocity with some degradation in accuracy. TRANSIT was not generally used by aircraft due to the incompatibility of TRANSIT with the rapid platform motion of an aircraft. Additionally, aircraft require the third dimension (altitude) that the TRANSIT system did not provide. TRANSIT was, however, an important predecessor to GPS and pioneered a number of key technologies and concepts. TRANSIT led to a great refinement of Earth's gravity field model, successfully tested dual-frequency correction techniques for ionospheric induced delays, and was crucial in developing stable and reliable frequency sources. TRANSIT provided only periodic updates and the degradation for a moving user made it unsuitable for aircraft. By the late 1960s, better systems were being explored by the Navy.

Additional GPS Predecessors: Timation and 621B. Timation was a program under the Naval Research Laboratory (NRL) http://ncst-www.nrl.navy.mil/NCSTOrigin/Timation.html, whose goal was orbiting very accurate clocks. These clocks were to be used to transfer precise time among various laboratories around Earth. Under certain circumstances, users could also determine their positions by using the Timation signal. The approach was somewhat different from TRANSIT in that the radio signal allowed direct ranging by using a technique known as side tone ranging. Two satellites had been launched prior to the approval of GPS phase one in 1973. After that date, the Timation research effort was folded into the GPS development program. The NRL expertise played a key role in developing the atomic clocks used on GPS.

The third predecessor to GPS was a U.S. Air Force program called 621B. This effort was directed by an office in the Advanced Plans group at the Air Force's Space and Missile Systems Organization (SAMSO) in El Segundo, California. This concept was strongly supported and advocated by Dr. Getting of the Aerospace Corporation. This program evolved directly into GPS, although not before significant modifications were made to the original U.S. Air Force-only concept. By 1972, 621B had already demonstrated operation of a new type of satellite ranging signal based on pseudorandom noise (PRN). Successful aircraft tests had demonstrated the PRN technique using ground-based "simulated" satellites located on the floor of the New Mexican desert.

The PRN modulation used for ranging was essentially a repeated digital sequence of fairly random bits (ones or zeros) that possessed certain useful properties. The sequence could be generated by using a shift register or for shorter sequences, could be stored in very little memory. Given the limited capabilities of computers then, this was a crucial feature. A navigation user could detect the "phase" or start of the signal sequence and use this for determining the range to the satellite. The PRN signal also has powerful noise rejection features and can be detected even when its power density is less than one-hundredth that of ambient radio noise. Furthermore, all satellites could broadcast on the same nominal frequency because properly selected PRN codes were nearly orthogonal.

When "tuned in" to a particular PRN sequence, all other PRN sequences appear to the user as simple noise. The PRN sequence can be tracked even in the presence of large amounts of noise, so other signals on the same frequency do not generally jam the signal of interest. The ability to reject noise also implied a powerful ability to reject most forms of jamming or unintentional interference. In addition, a communication channel could be included by inverting groups of the repeated sequences at a slow rate (50 bits per second is used in GPS). This communications channel allowed the user to receive the ephemeris, clock, and health information directly as part of the single navigation signal. The original Air Force concept visualized several constellations of satellites in highly elliptical orbits with 24-hour periods. This constellation design allowed deploying the satellites gradually (for example, to cover North America first) but complicated signal tracking due to the very high line-of-sight accelerations. Initially, the concept relied on continuous signal generation on the ground with continuous monitoring and compensation for ionospheric delays.

Program 621B was the immediate predecessor of the GPS effort, but the program came perilously close to cancellation several times. In the early 1970s, Dr. David Packard, the Deputy Secretary of Defense, instituted important changes at the Department of Defense. One of these changes was to encourage joint programs that had multiple service participation. It turned out that GPS was the first "Joint Service" program. The first program director was Col. (Dr.) Bradford Parkinson, one of the authors of this article (one of the other authors is Dr. Jim Spilker who played a lead role in the design of the GPS signal structure). Dr. Parkinson was assigned to 621B in November 1972 and was directed to gain approval for the concept validation phase of the Defense Navigation Satellite System (DNSS), as the new DOD satellite navigation system was originally known. After many briefings of senior personnel in the Pentagon, a Defense Systems Acquisition Review Council (DSARC) meeting was held in August 1973, at which Dr. Parkinson presented a brief on the Air Force 621B program approval was denied.

Meanwhile, Dr. Parkinson had presented the concept to the Director of Defense Research and Engineering, Dr. Malcomb Currie, who quickly

appreciated the value of a three-dimensional, continuous, 10-meter positioning system. After the failure to gain approval, Dr. Currie invited Dr. Parkinson into his office and asked him to rethink the system and to ensure that it truly was a Joint Program, one that incorporated the best technology and concepts across DOD. He wanted a synthesis, a new all-encompassing concept. Dr. Parkinson assembled about 10 of his key program members in the halls of the Pentagon during Labor Day weekend 1973. The result was a new system concept that was later named GPS, or NAVSTAR. By mid-December 1973, the senior DOD officials had been briefed and the reconvened DSARC gave approval. By June of 1974, the satellites, ground control system, and user equipment was on contract.

The first GPS satellite was launched in February 1978 and led to successful validation of the concept. The subsequent operational satellites incorporated certain additional nonnavigation payloads, which enhanced their value, but also undoubtedly delayed full operation. At the same time, the U.S. Air Force was not comfortable with having to shoulder the whole financial burden for the program and attempted to cancel GPS at least three times. In each case, civilian leadership and the former Secretary of the Air Force, Hans Mark, overruled the suggestion. GPS was declared operational in December 1995, although both civilian and military users had been using the available developmental system for more than 10 years.

Due to the possibility that potential enemies might use GPS positioning against the United States or her allies, the civil signal was intentionally degraded through a process known as selective availability (SA). SA reduced accuracy for civilian users and remained part of GPS as a holdover from its original military history. SA was generally active, although ironically it was turned off during several national emergencies and international military campaigns due to the widespread military use of civilian receivers. It was slowly realized that the proliferation of differential corrections in the form of augmentations rendered these perturbations totally ineffective. As a result, a Presidential Decision Memorandum (PDM) was signed in 1996, which ordered the military to discontinue its use, pending justification from the DOD. In early 2000, SA was removed from the signals of all orbiting satellites.

During the first 25 years of GPS, several generations of satellite designs have been developed or are under development. These include I, II, IIA, IIR, IIRM, and IIF. In addition, there are plans for an upgraded version of GPS, known as GPS III, which is currently being defined. The Block IIRM and Block IIF satellites add additional civil GPS signals at other microwave band frequencies (see Fig. 6), which should materially improve the accuracy and robustness of the service.

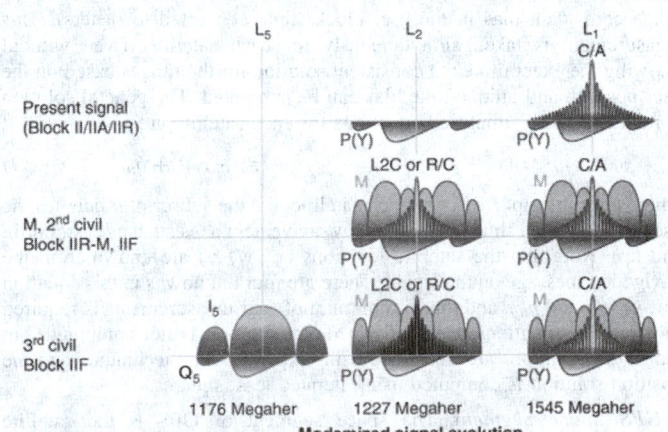

Courtesy: Aerospace Corporation

Fig. 6. The GPS signal is undergoing modernization in preparation for GPS III. The current system is shown in the topmost frame with P/Y code on both the L1 and L2 frequencies, and the C/A (civil) code only on L1. The signal modernization calls for broadcasting a second copy of the C/A code on L2, and the military will get a new spread-spectrum code, called M-code, on both L1 and L2. The M-code is structured to broadcast most of its power into the nulls of the C/A code, maximizing spectral separation. A third civil frequency on L5 is set to be implemented on the late Block II-F satellites. Courtesy: Aerospace Corporation

GPS Concept of Operation

The design objectives of the GPS system were to provide a continuously available, worldwide, all-weather, three-dimensional precision navigation system for both military and civilian users on land, at sea, or in the air (or even in space). The GPS system had to operate, even on an accelerating platform such as a maneuvering aircraft or missile. Additionally, the system had to be passive, or one-way, so that it could service an unlimited number of users. As a military system, the signal is required to be both jam-resistant and antispoof.

Each of these requirements drives a certain set of constraints. To be worldwide and continuously available, only a satellite system can provide global coverage, especially over the oceans and polar regions. As a satellite system, frequencies less than 1 MHz skip off the ionosphere, and frequencies higher than 10 GHz are very heavily attenuated by atmospheric moisture. Satellite signal frequency was a compromise among accuracy (ionospheric delay), attenuation, and the power to be received by an omnidirectional user antenna. Thus, the selected signal was placed within the L band for best performance. Two additional constraints were established by the military: that the satellites could be totally serviced from the continental United States (CONUS) and that the constellation could be tested by using a small number of satellites to minimize project risk. These constraints led to satellites in MEO orbit, which costs significantly less in energy than a GEO orbit.

Multilateration Positioning System. GPS functions as a multilateration, or rho-rho ($\rho - \rho$), system, that is, the range from at least three known locations is determined and the resulting intersection of the three spheres defines a single point that is the user location. In GPS, the system is complicated by the fact that the transmitters are moving and that the range cannot be measured directly. As a simplification, assume that the GPS satellites are stationary and that the user is upon a flat nonrotating Earth. All of the satellites are synchronized and transmit a signal at the exact same time.

The user will receive the signal from each satellite at a different time due to the time of flight of the signal from the satellite to the user across the various ranges to each satellite. If the user possessed a very accurate clock that was time synchronized with the satellites, then the product of the time of flight and the speed of light would be the true range to the satellites. However, because the user is unlikely to have an atomic clock (a requirement which would make the receivers far too expensive), the user is not synchronized to GPS time. Thus, the measured range is offset by a consistent bias and is thus referred to as pseudorange (ρ):

$$\rho i = c \times t_i + b \tag{1}$$

where c is the speed of light, t_i is the true arrival time, and b is the range equivalent bias in the user clock (time converted to meters). This measurement is taken simultaneously for each satellite. Even without knowing the exact time, the consistent solution for the ranges based on the user position and unique time bias can be computed. The general solution is nonlinear. The simplified equations for each satellite are

$$\rho = (x_u - x_i)^2 + (y_u - y_i)^2 + (z_u - z_i)^2 + b_u \tag{2}$$

where the subscript i denotes each satellite and the subscript u denotes the users location and time (x, y, z are any convenient axes such as east, north, and up). Note that the satellite locations (x_i, y_i, z_i) are known from the navigation message on the signal. There are four unknowns in this equation (x_u, y_u, z_u, and b_u), and thus a minimum of four measurements is required to solve the equations. Generally, a direct solution is not computed, but rather the equations are linearized using a perturbation technique, and the position solution is computed using iterated least squares.

GPS Space Segment. The space segment of GPS is the satellite constellation (see Fig. 2) that consists of 24 or more vehicles in six orbital planes. The planes are inclined at 55° and are spaced 60° apart. There are four satellites in each of the orbital planes, but they are not evenly spaced. This was done to minimize the impact of any single satellite failure. Additionally, there are typically on-orbit spares in some of the six planes. The satellites are in a MEO orbit at a radius of 26,561.75 km (16,505 miles) (a mean equatorial altitude of 20,163 km (12, 529 miles). The orbits are almost perfectly circular and have an eccentricity of less than 0.01. The orbital period of these orbits is 12 hours of mean sidereal time (a mean sidereal day is the rotation of Earth to the same position with respect to inertial space, as opposed to a solar day, and is approximately

four minutes shorter than a solar day). Thus, each GPS satellite repeats the same ground track, but passes the same location four minutes earlier each (solar) day.

The GPS payload consists of redundant atomic clocks, telemetry and control sections, and the signal generation subsystem. The atomic clocks are rubidium and/or cesium standards that typically have long-term stability of 1 part in 10^{13} per day (or roughly a drift of 9 nanoseconds per day). The master control station monitors the atomic clock drift rates and models them as a quadratic,

$$\delta t = a_{f0} + a_{f1}(t - t_{0c}) + a_{f2}(t - t_{0c})2 + \Delta t_r \tag{3}$$

where t_{0c} refers to the master clock, t is the satellite clock, and the various parameters a_{f0} through a_{f2} are parameters for the polynomial fit to the satellite clock drift. The last term, Δt_r, compensates for relativistic effects caused by the motion of the satellites and their position within the gravity well, which has the effect of making the satellites gain 38 microseconds per day. This is compensated for by setting the main satellite frequency standard (10.23 MHz) slower by 0.00455 Hz. GPS is the first operational system known to require a correction for relativistic effects. All of these parameters are sent in the navigation message.

The satellites' telemetry subsections are responsible for receiving the uploaded navigation data from the Master Control Station (MCS). The data is encrypted before upload to ensure than no spoofing can occur. Internal status and health is also monitored and relayed back to the MCS. The signal generation subsection is detailed later in the discussion of signal structure. Currently, the nominal signal power is set at a minimum of −160 dBw for the Coarse Acquisition (C/A) code, −163 dBw for the L1 P/Y code, and −166 dBw for the L2 P/Y code, as shown in Table 2. Note that these power levels are well below the ambient noise level. From the satellites' location, Earth subtends an angle of approximately 14°. A user at the limb of Earth is significantly farther away than one directly under the satellites. To compensate for this greater "space loss," the antenna gain pattern on the GPS satellites is such that approximately 2.1 dB more gain is at the edges than at the boresight of the beam. The beam is also slightly wider than the 14° of Earth to allow non-GPS satellites on the other side of Earth to use GPS for positioning.

TABLE 2. MINIMUM GPS BROADCAST POWER[a]

Frequency	L1 (1575.42 MHz)	L2 (1227.60 MHz)
C/A	−160 dBw	NA
P-code	−163 dBw	−166 dBw

[a]The specification for both the C/A code and the P/Y code (military) is such that the minimum broadcast power is well below the noise floor of the in-band radiation. Using the correlation properties of the PRN codes, a GPS receiver can reconstruct the phase of the signal and use this for position and temporal information.

GPS Signal Structure. The PRN spread-spectrum coding that was originally pioneered for the Air Force 621B program contributes a great deal to the functionality of GPS. GPS uses a technique called *code division multiple access* (CDMA) such that each satellite broadcasts its message simultaneously on the same frequency, and yet the receiver can select each signal separately. The L1 signal is centered at 1575.42 MHz. This frequency is modulated with the satellite's civilian PRN code using a biphase shift key (BPSK) modulation, that is, the phase of the carrier is reversed to indicate a "chip" transition (the military signal is in quadrature and the composite signal is called QPSK). A "chip" is the BPSK analog of a bit, and in the L1 C signal, is exactly 1540 carrier cycles long (or exactly 0.9775 µs). The C/A (e.g., L1 C) PRN codes are 1023 chips long, which means that the code repeats every millisecond. Last, the code itself is inverted every 20 ms to indicate a bit transition on the navigation message that is broadcast at 50 bits per second. An illustration of the signal structure is shown in Fig. 7.

The GPS C/A PRN codes are very carefully chosen for specific properties. The first property is that they can be easily generated by using a simple shift register. This was an important consideration during the development stages of GPS but is no longer relevant to modern CDMA design. The two main advantages of PRN codes are the signal spreading and the correlation properties. The unique properties of the set of PRN codes are that they have very good code-to-code and cross-correlation (multiple access) properties, even in the presence of large Doppler offsets.

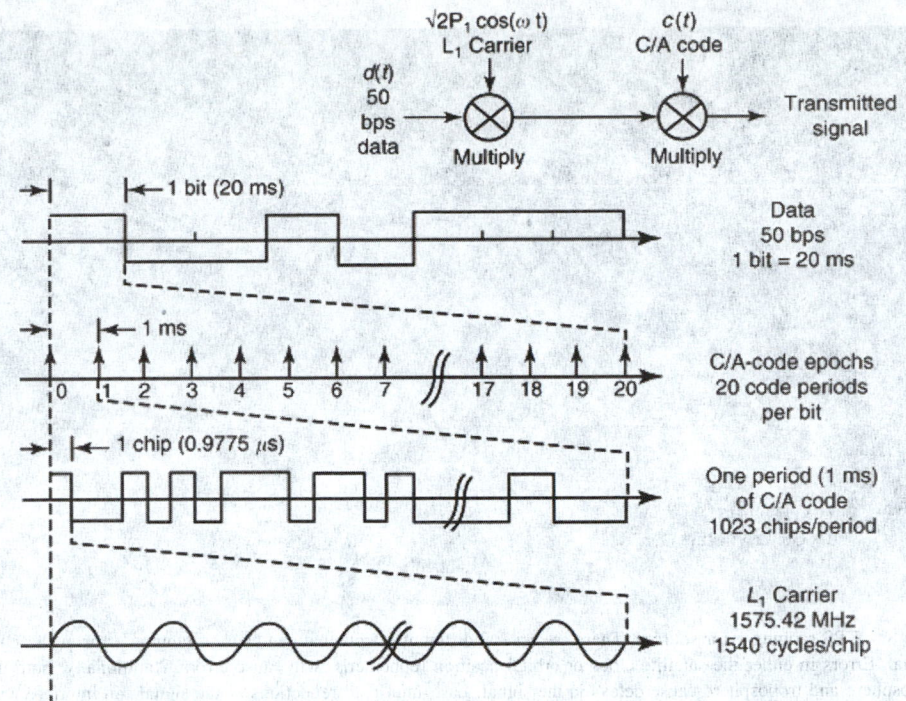

Fig. 7.   The GPS L1 signal structure is based on a carrier frequency at 1575.42 MHz. This frequency is modulated by biphase shift keying (BPSK) that uses phase reversal to indicate a changed "chip." Each chip is 1540 cycles wide (or 0.9775 µs). The code length is 1023 chips and repeats each ms. Last, every 20 ms, the entire code may invert to indicate a data bit reversal, modulated at 50 bits per second.

If a PRN code is multiplied and integrated (i.e., correlated) against a local copy of itself, it produces a large correlation coefficient when the start (or phase) of the two codes line up. If the codes are out of phase, it produces a very small value. Likewise, correlating one of the PRN codes with a different PRN code produces a very small value for all relative phases. The implication of this is that the phase of any given PRN signal within a receiver can be found by correlating a local copy at different phase offsets until a large signal is discovered. In this operation, all other PRN codes appear as noise. The worst case cross-correlation is −21.6 dB and is even lower at −23.8 if there is no Doppler offset. Pseudorange is the local clock reading (divided by the speed of light) at the start of the local code sequence when it is maximally correlated with the incoming signal.

Ground Control Segment. The GPS control segment consists of six or more monitoring stations around Earth, a Master Control Station (MCS), and upload ground-antenna stations. Each of the monitoring stations has a set of accurate atomic clocks and tracks both the code and carrier of each GPS satellite as it traverses overhead from horizon to horizon. The monitoring stations operate at both L1 and L2 frequencies to permit removing excess ionospheric delay. They also monitor atmospheric parameters such as temperature, atmospheric pressure, and humidity to permit estimating the tropospheric delay. By tracking the L-band carriers from horizon to horizon to a small fraction of a cycle (1% of an L2 carrier cycle is only 0.19 cm), a series of 15-minute averages is created and sent to the Master Control Station.

The Master Control Station receives the monitoring station tracking and ground antenna telemetry information and computes the current and predicted satellite clock offsets and satellite positions. It then converts this data to the navigation data formats described later. These rather complex satellite orbit/time filter estimating algorithms must also model the satellite solar radiation pressure, atmospheric drag on the satellite, Sun/Moon gravitational effects, including solid Earth and ocean tides, and Earth's geopotential model. Improved GPS satellite-to-satellite cross-link ranging data may also be used in the future. The navigation data are uploaded from several 10-m S-band ground antenna upload stations.

Navigation Data. The navigation data are encoded on the L1 C/A signal. This data message is transmitted at the rate of 50 bits per second and consists of a set of 6-second subframes (ten 30-bit words) and 30-second

frames. The data encoded include the full ephemeris required to calculate the current satellite position, the satellite clock quadratic polynomial model and corrections to GPS time, almanac data used to position all the other satellites, and a hand-over-word for P/Y-code users. The almanac data allow a user to compute the rough positions of the satellite and thus narrow the search space both in terms of PRN codes and Doppler bins.

User Segment. The user segment or the GPS receiver is a very sophisticated digital signal tracking device that allows converting the faint signals from the GPS satellites into an accurate position solution. The GPS receiver must process the almanac (either stored or newly acquired) to generate a search space in terms of PRN codes and Doppler frequency bins. The incoming RF signal must be amplified, downconverted through an intermediate frequency (using a mixing process), and sampled into the digital domain. The PRN codes are correlated against the incoming digitized stream, and usually a delay lock loop (DLL) is implemented to keep the signal locked.

Once the signals are tracked, the corrections are applied to the raw pseudoranges, and the position and time bias are computed through an iterated least squares calculation. The positions are now reconverted to a useful coordinate frame such as latitude, longitude, and altitude. The original GPS "manpack" receivers were backpack-sized devices that cost more than $50,000. GPS has benefited greatly from the semiconductor revolution, as has the typical consumer. A modern GPS receiver costs as little as $100 and is small enough to be embedded into at least one wristwatch. Additionally, the computer that calculates the position solution can support many additional features such as map displays and waypoint guidance at minimal additional cost.

GPS Ranging Errors. There are several error sources that can corrupt the pseudorange and carrier phase measurements, as shown in Fig. 8. Thermal noise and interference effects degrade the performance of a typical receiver. Over the years, receivers have improved in noise performance. The free electrons in the ionosphere cause a code delay but a carrier advance (the so called code–carrier divergence). The ionosphere also varies in total electron count (TEC) depending on the state of solar activity and time of day. Delays are also associated with the troposphere that are a function of the slant range and moisture content below an altitude of 40 km. Errors in satellite position and clock directly cause errors in user ranging. Foliage can attenuate the signal, and more massive

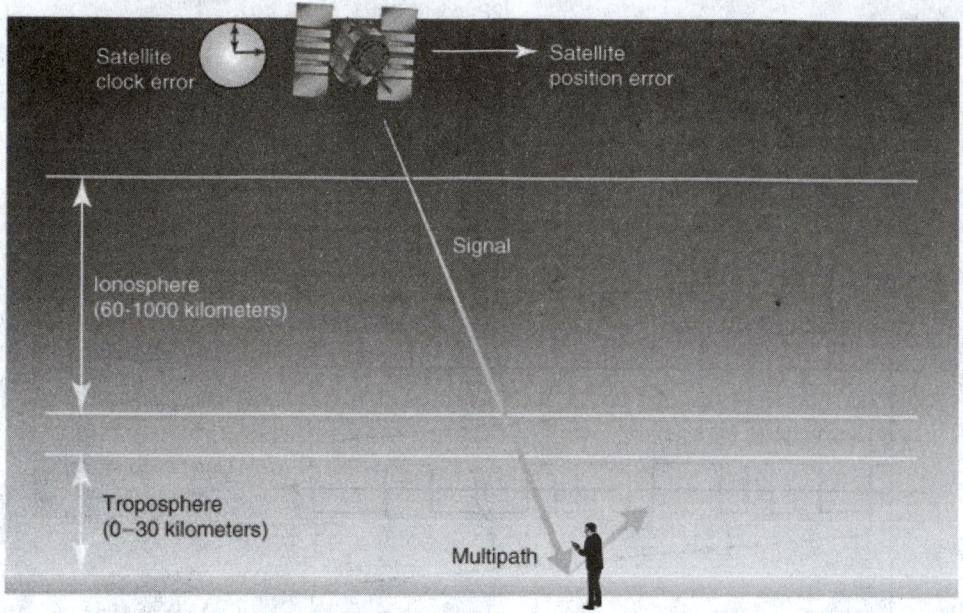

Fig. 8. GPS ranging error sources. There are several different effects that can cause a ranging error in the GPS signal. Errors in either the satellite clock or orbital position (ephemeris) will cause errors. Additionally, both the ionosphere and troposphere cause delays in the signal. Last, multipath reflections of the signal can interfere with the original signal and distort the range information.

obstructions such as buildings or hills will block the signal completely. The latter is the origin of the urban canyon problem whereby GPS position is significantly degraded in cities that have tall buildings. User motion can cause the delay lock loops to be thrown off due to rapid changes in Doppler, though most terrestrial users will not experience such high rates of acceleration.

GPS Error Analysis. To understand the potential of GPS, it is worthwhile to analyze the effect of the errors that occur when using it. In general, the errors are associated with measuring the range to the satellite. The ranges to four satellites must be processed to find the user's position, taking into account the locations of these satellites. Depending on geometry, the positioning error may be much higher than the typical ranging error. The ratio of the *positioning* error to the *ranging* error is called the geometric dilution of precision (GDOP). If all ranging errors are zero mean, uncorrelated, and have the same variance, the general relationship is

$$\sigma_P = \sigma_R \bullet \text{DOP} \tag{4}$$

where

$$\sigma_P = \text{positioning error}$$
$$\sigma_P = \text{ranging error}$$
$$\text{DOP} = \text{a multiplier due to geometry}$$

The DOPs can be calculated by forming an array of unit vectors pointed at each satellite from the user's position, e_j, using three convenient coordinate directions such as east, north, and up.

$$G = \begin{array}{|c|c|} \hline e_1^{-T} & 1 \\ \hline e_2^{-T} & 1 \\ \hline e_3^{-T} & 1 \\ \hline e_4^{-T} & 1 \\ \hline \end{array} \tag{5}$$

The DOPs are then the square roots of the diagonal terms of the resulting 4×4 matrix:

$$\text{GDOP} = (G^T G)^{-1} \tag{6}$$

and

$$\text{Covariance (position)} = (G^T G)^{-1} \bullet \sigma_R^2 \tag{7}$$

The first three diagonal terms of GDOP refer to the coordinate directions selected above (e.g., east error factor, north error factor, and up error factor). The fourth diagonal term is the dilution for the range equivalent of the timing error. By dividing by the speed of light, one can change the value to the equivalent dilution in seconds.

The major sources of ranging error were discussed previously. Typical values are provided in Table 3. The typical dilution values (VDOP and HDOP) shown in Table 3 must be used with caution. If the satellite geometry is poor, it is not uncommon to find DOP multipliers of 10 or more. This is usually caused by a reduced number of satellites due to obstructions in the satellite line of sight. Typical causes are buildings, trees, and/or terrain. Modern receivers usually state the estimated error as part of the location message. Of course, the range of errors can be much greater than shown in Table 3, depending on age of update, atmospheric conditions, magnitude of multipath reflections, etc.

TABLE 3. TYPICAL GPS RANGING ERRORS FOR VARIOUS SOURCES[a]

Error source	Typical root-mean-square ranging errors single-frequency code-tracking user		
	High	Low	Typical
Ephemeris data	3.0	0.7	1
Satellite clock	0.5	3.0	0.9
Ionosphere (after modeling)	6.0	2.0	4
Troposphere	2.0	0.3	0.5
Multipath	15.0	0.2	1.2
Receiver measurement and noise	1.0	0.2	0.5
User equivalent range error (UERE)			4.4
Vertical rms error with VDOP of 3.0 =			13.2 meters
Horizontal rms error with HDOP of 2.0 =			8.8 meters

[a]The typical dilution values (VDOP and HDOP) shown above must be used with caution. If the satellite geometry is poor, it is not uncommon to find DOP multipliers of 10 or more. This is usually caused by obstructions in the satellite line of sight due to buildings, trees, or terrain. Modern receivers usually state the estimated error as part of the location message.

As can be seen in Table 3, the largest typical error is for ionosphere transmission delays even after modeling for a single frequency receiver. Ionospheric delays are caused by the interaction of free electrons in the ionosphere with the radio signal. One of the key observations is that most of the delay through the ionosphere is proportional to the inverse square of the carrier frequency. Thus, a dual-frequency user can directly estimate the ionospheric delay and substantially reduce or eliminate this error. Currently, only military receivers are truly dual frequency. These receivers have a current user equivalent ranging error (UERE) less than 2 meters (6.5 ft), when multipath errors are small. Note that currently scheduled improvements in the GPS signal include two new civil signals at L2 (1227.6 MHz) and L5 (1176.45 MHz). By using the new second and third civil signals, all users will be able to calibrate the ionospheric delay directly. This is the largest error for most users, so accuracy will improve substantially as this error category is reduced to near zero.

For a number of years, the DOD deliberately perturbed the timing signal on GPS (a technique called selective availability or SA). This increased the UERE to about four times the typical values shown in Table 3. Of course, this also resulted in positioning errors that were about four times larger. The extensive use of real-time differential calibration of these errors made this technique ineffective, and it was discontinued by Presidential order. Additionally, over the years, the ground station has become much more skilled at calibrating the errors in the signal-in-space (i.e., ephemeris and satellite clock errors). Improvements in predicting orbits and clock drifts, plus increased uplink frequency, have reduced signal-in-space errors from 6 meters (19.7 ft) to less than 2 meters (6.5 ft).

The next largest category of error in Table 3 is multipath error. Multipath error is the misleading interference of the delayed reflection of the GPS signal. In fact, this error will sometimes exceed the ionospheric error. Several techniques have been developed to mitigate the multipath problem. These range from better antenna designs whose gain patterns strongly attenuate signals coming from below the horizon to very narrow correlators that are immune to a large class of reflectors. Additionally, code measurements can be combined with carrier phase measurements that have a very different multipath response. As technology advances in receiver electronics and signal tracking, the receiver measurement noise will improve to the point of diminishing returns. Using dual-frequency receivers, multipath will be the dominant error source for GPS. Much of the future development in GPS receivers will be directed at eliminating the distortion from reflected signals.

Differential GPS. One technique used to augment GPS is known as "differential." The basic idea is to locate one or more reference GPS receivers at known locations in users' vicinities and calibrate ranging errors as they occur (see Fig. 3). These errors are transmitted to users in near real time. The errors (or their negative, which are corrections) are highly correlated across tens of kilometers and across many minutes. Use of such corrections can greatly improve accuracy and integrity. Several large-scale differential networks have been deployed in the Unites States and elsewhere.

Overview of DGPS System

The U.S. Coast Guard (USCG) within the United States and the International Association of Lighthouse Authorities (IALA) have deployed a marine beacon differential system internationally, known in the United States as National Differential GPS (NDGPS). The Army Corps of Engineers is currently deploying additional beacons that are compatible with the U.S. Coast Guard differential system and cover the entire continental United States (see Fig. 5). The Federal Aviation Administration (FAA) is currently deploying the Wide Area Augmentation System (WAAS). WAAS is intended to provide enroute navigation and nonprecision approaches for aviation users (see Fig. 4).

The FAA is also developing a Local Area Augmentation System (LAAS) for Category I, II, and III precision landing capability at airports (see Fig. 9). This will require local ground monitoring stations to ensure the integrity of the system in addition to the nominal reference receivers. The U.S. Department of Defense is currently developing a new Military Landing System (MLS) to operate like the LAAS system but will be used on aircraft carriers and at forward bases. The system, called the Joint Precision Approach and Landing System (JPALS), has already demonstrated fully autonomous carrier landing using a specially equipped Navy F/A-18.

There are many additional private and international systems under development or deployed. Various private companies sell their own

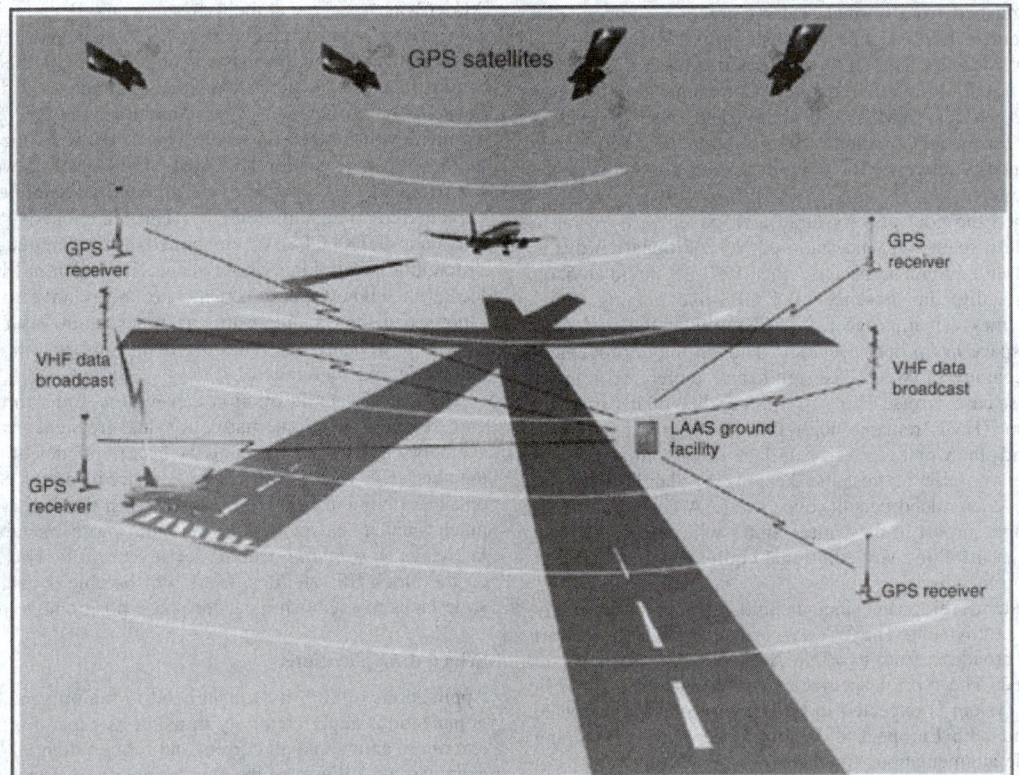

Fig. 9. The LAAS system under current development by the FAA will provide precision approach capability using GPS. Due to the exacting requirements of Category II and III landings, the LAAS requires many cross-checks of the GPS system to ensure integrity. If one of these cross-checks fails, the time to alarm of the LAAS is specified at less than 6 seconds.

proprietary carrier phase differential GPS systems for use in such diverse areas as construction, surveying, and archeology. Commercial wide area corrections are carried by at least one commercial C-band satellite broadcast, and several oil companies have put their own differential stations on oil drilling platforms to ensure accurate positions for the helicopters and ships that service these platforms. The next subsections will further explain these examples.

National Differential GPS (NDGPS). NDGPS is a system that has been developed primarily for marine use. Both the U.S. and European equivalent systems use marine radio beacons transmitting in the 300 KHz band as communication links for GPS corrections; ranges are of the order of 100 to 200 kilometers (62 to 124 miles). The applications are mostly for ships operating in coastal waters or upon navigable rivers. Typical accuracies are of the order of 1 to 2 meters (3.3 to 6.5 ft) horizontally; many commercial GPS sets now offer small additional radios to receive these corrections. Although the initial deployment has focused on U.S. Coast Guard applications, both the Army Corps of Engineers and the Department of Transportation are extending the NDGPS system to cover the entire continental United States. To expedite this full rollout, the NDGPS system will take over Ground Wave Emergency Network (GWEN) transmission stations from the U.S. Air Force because these stations are no longer necessary and were to be decommissioned. Figure 5 shows the current nominal coverage of the NDGPS network.

Wide Area Augmentation System (WAAS). The WAAS, developed by the U.S. FAA, is specifically designed to ensure integrity and improve accuracy for civil aviation users. GPS, augmented by WAAS, offers its capability for both enroute navigation and nonprecision approaches (NPA). Fig. 4 shows the general architecture of the WAAS system.

GPS satellite data is received and processed at widely dispersed Wide-Area Reference Stations (WRS) that are strategically located to provide redundant coverage across the required WAAS area. Data is forwarded to redundant Wide-Area Master Stations (WMS) that process the data from multiple WRSs to determine the integrity, differential corrections, and residual errors for each monitored satellite and each predetermined ionospheric grid point. The multiple WMSs are provided to eliminate single point failures within the WAAS network. The differential corrections are allocated to satellite, clock, and ionosphere, so they are called "vector" corrections as distinguished from normal scalar corrections. Information from all WMSs is sent to GEO Uplink Subsystems (GUS), where it is uplinked to the GEO satellites. The GEO satellites downlink this data to the users via a GPS signal at the L1 frequency. Communication between ground-based stations (WRSs, WMSs, and GUSs) and other systems is accomplished via the Terrestrial Communications Subsystem (TCS), which provides two independent networks for redundant data communications among WAAS components.

WAAS accomplished the goal of very large area coverage by using a small number of widely spaced ground stations. No additional hardware is required on the user equipment because the data are modulated on an L1 signal. Additionally, the presence of a GPS-like ranging signal on GEO WAAS satellites can improve the availability of the system if the WAAS signal-in-space has proper accuracy. Though the improvement in the positioning accuracy of WAAS is significant, more important is the bounding of worst case errors. Thus, the probability of hazardously misleading information (HMI) remains negligible ($<10^{-9}$). At the same time that WAAS bounds the worst case error, it does so without generating an unacceptable level of false alarms and keeping availability high at the nation's airports. As an added benefit, once the FAA declares WAAS operational, every single airport in the Unites States will potentially have NPA capability without installing any additional equipment at the airport (this will require new procedures).

WAAS has been under development since the mid-1990s and is currently in the final phase of deployment. The WAAS corrections, which are part of an additional GPS broadcast from two INMARSAT GEO satellites, are being extensively used. The typical accuracies for WAAS are shown in Table 4. The WAAS system is expected to be a boon to civil aviation in the United States, and both Europe and Japan are currently developing compatible nationwide augmentations for their own airspace.

Local Area Augmentation System (LAAS). Also being developed by the U.S. FAA is LAAS. It is designed to allow commercial aircraft landings down to Category II and possibly Category III. It is a highly redundant and reliable differential system that has several reference and monitoring

TABLE 4. WAAS PROJECTED SYSTEM ACCURACIES[a]

WAAS accuracies	50th percentile	95th percentile	99th percentile
Horizontal	1 meter (3.3 ft)	2 meters (6.5 ft)	5 meters (16.4 ft)
Vertical	2 meters (6.5 ft)	5 meters (16.4 ft)	10 meters (32.8 ft)

[a]The WAAS system was developed by the FAA to augment the GPS signal for civil aviation. The system is in its final stage of development, and many users are already using the corrections coming from the GEO satellites. WAAS excels in its ability to bound the worst case error and ensure that the probability of hazardously misleading information (HMI) remains very low while at the same time reducing the number of false alarms below the nuisance threshold.

stations and a very high standard of integrity. The LAAS system is meant to replace the current Instrument Landing System (ILS) at most large commercial airports (see Fig. 9). A Category III landing consists of a "zero-zero" landing (e.g., the visibility ceiling is at ground level and horizontal visibility is also zero). In practice, this means that the aircraft is landed by the autopilot. This is referred to as an "autocoupled" landing. Due to the automated nature of the landing, any landing system failure can be hazardous. This places an extremely high burden on LAAS to ensure that aircraft location is always within a well-defined error bound.

The LAAS system is designed with a very extensive set of cross-checks and verifications to ensure that no portion the system is operating outside its nominal parameters. These checks include validating the orbit of a given GPS satellite against a prediction based on the previous pass (from 12 hours before), checking the clock drift and both the range and range-rate of a rising satellite, plus many other checks. Time to alarm is vital for protecting any landing aircraft from misleading information and is specified at less than 6 seconds.

Carrier Tracking Differential (CDGPS). Differential carrier tracking is another GPS technique that has been used by surveyors since the mid-1980s. By reconstructing the L-band radio-frequency (RF) carrier signal, a GPS receiver can attain tracking precisions of 1 to 10 millimeters (0.04 to 0.39 inches). Specifically, a reference receiver (at a known location) measures the phase of the incoming carrier wave and transmits this information to a user. The user then compares this to the phase of the carrier wave received at the user's antenna. Because the wavelength of the L1 carrier is approximately 19 cm (7.5 inches), a reasonable receiver can resolve this to 1% of the phase, or about 2 mm (0.08 in). Unfortunately, this is not accuracy. To attain equivalent accuracy, it is necessary to resolve the number of integer wavelengths along the RF path, that is, there are an unknown number of whole waves between the wave front arriving at the reference station and that at the user. Several techniques exist for resolving this integer cycle ambiguity. Satellite motion that can be exploited to do this differentially. This technique is referred to as real-time kinematic (RTK) GPS. When applied, this technique provides survey-level differential positioning whose accuracies are in millimeters. Thus one can locate an unknown point on the ground relative to a survey mark very rapidly and then maintain this accuracy as the user's receiver is moved. This is now being exploited for both construction survey and real-time, automatic, machine control.

The use of satellite motion can require some time to converge on the correct solution. An alternative for dual-frequency receivers is to set up a synthetic carrier wave by using the beat frequency of the L1 and L2 carriers together. The wavelength of the beat frequency is 86 cm, so the number of integer combinations to be searched in the position volume is typically much smaller and makes the problem more tractable. This technique is known as *wide laning*. Due to the advent of the two new civil frequencies on the block IIF satellites, users will be able to walk through a series of wide lanes to establish a carrier phase positioning solution very quickly.

Selected Applications

Applications of GPS have continued to multiply, as commercial and civil organizations apply creativity in using its capability. This section will not attempt to enumerate all current and future potential uses. Instead, selected examples will illustrate the revolutionary advances that have been made possible by this remarkable system. Many of the topics presented are at the cutting edge of current research and may yield profound improvement in our understanding of our world, as well as improved productivity and safety.

Survey and Crustal Motion. Until the advent of carrier phase differential GPS, measuring the relative distance or motion of large objects accurately over time required painstaking surveys using laser interferometry and tended to be one-dimensional. However, carrier phase differential GPS that can track 3-D relative positions down to millimeter levels across very long distances is revolutionizing the field of geomatics. Currently, experiments are underway that monitor the relative positions of the mountainsides of several volcanoes in the states of Hawaii and Washington. Previous attempts at these kinds of experiments proved difficult due to the requirement for consistent line-of-sight measurements using optical sensors. Data recorded by using survey-quality GPS receivers have detected bulging of the mountains and are providing insights that may one day enable scientists to predict volcanic eruptions.

Similarly, hundreds of GPS receivers have been placed along fault lines throughout California and other parts of the world to validate theories about plate motion and gain valuable information on preconditions to earthquakes. Again, research in this area is still in its infancy, but it has never before been so economical or in some cases even possible, to measure the distance across large geographic features down to the millimeter level. At this time, data are being gathered to validate crustal motion models that will certainly lead to refinements in these models.

Aviation. The aviation industry has been an early adopter of GPS technologies and remains at the forefront of developing and implementing new GPS advances. In the early 1990s, a prototype GPS landing system for Category III (zero ft ceiling, zero miles visibility) was developed and demonstrated by Stanford University under an FAA grant. This system used carrier phase differential GPS to ensure a correct position. To resolve the integer cycle ambiguities quickly and robustly, two ground transmitters that broadcast GPS-like signals were used to augment the system. These "pseudolites" exhibited a large change in Doppler shift due to the rapid geometric change. The resulting system demonstrated more than 100 autocoupled landings at Crows Landing Airport in California; data were independently validated by using the Crows Landing laser tracker. The data showed an accuracy of better than 0.5 meter (1.6 ft) (3-D) in the final phase of landing.

During one of the autocoupled landings, a satellite upload from the Master Control Station caused the satellite to interrupt its transmission for approximately 1 millisecond. The Stanford system detected this glitch in the space segment and called off the landing in real time.

Though the FAA has not yet declared GPS operational as a *precision* navigation aid, most General Aviation and Commercial pilots use GPS as a backup system for navigation. Additionally, modern aviation GPS units are programmed with a full aviation database and can notify the user of airspace violations. In an emergency, these units can guide the pilot to the closest airport at the touch of a button.

GPS, as a full 13-state sensor for an aircraft, provides a powerful suite of information at a relatively low cost. Combined with inexpensive computer graphics, a synthetic "out-the-window" perspective display can be used to improve vastly the presentation of critical data to the pilot. The futuristic vision of tunnels-in-the-sky for improved navigation is being tested today in various laboratories around the world. Pilots who have experimented with these systems report a much reduced workload and greater situational awareness. The potential to reduce controlled flight into terrain (CFIT) could save many lives currently lost due to such accidents. Likewise, if all other aircraft are prominently displayed, it can reduce midair collisions. These displays have also shown great promise in enabling closely spaced parallel approaches (CSPA) in inclement weather. This alone can save the United States billions of dollars in runway expansions and avoiding environmental impact that such construction would have on surrounding areas.

Vehicle Tracking. The so-called "urban canyon" can adversely affect GPS, but vehicle tracking remains a very important application. During urban canyon outages, most vehicle tracking systems use inertial augmentation to provide a position solution. Commercial companies have great interest in knowing where their equipment is currently located, and GPS provides an ideal answer. Many cities now have buses equipped with GPS receivers and radio transmitters. Each bus stop has a display of the current location of the next bus, and an estimate of the time to arrival. Likewise, many cities have GPS equipment on their emergency service vehicles to manage the response better. This has been shown very effective in reducing response time and managing these scarce resources during a large-scale disaster.

Vehicle tracking yields a great competitive advantage to a corporation. In one case, a cement company in Guadalajara, Mexico, would send fully loaded cement trucks into the city every morning, even though orders had not yet been placed. Using simple radio communication, this company responded to orders in less than half the time of any of its competitors. Though several trucks of cement would go to waste at the end of each day, within a short time, this company dominated the cement delivery market.

Last, law enforcement officials have been able to use GPS to increase their effective manpower by remotely monitoring suspects. After obtaining a court order allowing them to install a GPS receiver surreptitiously on a suspect's car, Seattle police were able later to reconstruct the time and path of the location during a 2-week period, without alerting the suspect to the surveillance. This information led directly to evidence that convicted the suspect.

Precision Munitions. No discussion of GPS would be complete without a brief discussion of military applications. In spite of its explosive use for many civil applications, GPS was designed primarily as a military system, and to continue development, GPS must fulfill its primary mission. Several military applications for GPS were developed in recent years. An example is the JDAM. This precision-guided munition has demonstrated a battlefield accuracy of less than 10 meters. The trend in the future is to reduce the explosive warhead size of these kinds of munitions, which can be done only if the guidance system is capable of pinpoint accuracy.

On purely defensive military applications, the DOD recently deployed a Combat Survivor/Evader Locator (CSEL) radio for servicemen/women. This radio allows downed pilots to relay their positions to rescuers directly to enable rapid rescue and minimal exposure to hostile forces. The CSEL replaces four different individual devices with a single integrated package.

Space Applications. Some of the most innovative and unusual applications of GPS occur in the area of Earth sensing and space applications. Low Earth orbiting satellites can use GPS to measure both position and attitude. Precise satellite data can be used to refine gravitational models of Earth, and can be used as a sensor for attitude control. A soon-to-fly satellite experiment, the Gravity Probe B (GPB), uses very precise spherical gyroscopes to yield a quantitative measurement of Einstein's theory of relativity. For the experiment to be valid, GPB needs to fly a "drag-free" polar orbit to within 100 meters (328 ft). GPS is used to provide guidance information to position the orbit of the satellite initially. Last, one of the most unusual applications of GPS is using the reflection of GPS signals from waves at sea to detect wave height in the open ocean.

Relationships to Galileo

Galileo is the European version of GPS. The European Union is committed to building a 30-satellite civil space-based navigation system at an estimated cost of 3.4 billion euros. The initial funding of 547 million euros is intended to fund the study and development phase, which is expected to take approximately 3 years. Galileo will provide a highly accurate, guaranteed global positioning service under civilian control. It will be inter-operable with GPS and GLONASS, the two other global satellite navigation systems.

A user will be able to take a position with the same receiver from any of the satellites in any combination. By offering dual frequencies as standard, however, Galileo will deliver real-time positioning accuracy down to the metre range, which is unprecedented for a publicly available system.

It will guarantee availability of the service under all but the most extreme circumstances and will inform users within seconds of a failure of any satellite. This will make it suitable for applications where safety is crucial, such as running trains, guiding cars and landing aircraft.

The first experimental satellite, part of the so-called Galileo System Test Bed (GSTB) was launched December 28, 2005. The objective of this satellite is to characterize the critical technologies, which are already under development under ESA contracts. Thereafter up to four operational satellites will be launched in the timeframe 2005-2006 to validate the basic Galileo space and related ground segment. Once this In-Orbit Validation (IOV) phase has been completed, the remaining satellites will be installed to reach the Full Operational Capability (FOC).

The fully deployed Galileo system consists of 30 satellites (27 operational + 3 active spares), positioned in three circular Medium Earth Orbit (MEO) planes at 23,222 km (14,429 miles) altitude above the Earth, and at an inclination of the orbital planes of 56 degrees with reference to the equatorial plane. Once this is achieved, the Galileo navigation signals will provide good coverage even at latitudes up to 75 degrees north, which corresponds to the North Cape, and beyond. The large number of satellites together with the optimisation of the constellation, and the availability of the three active spare satellites, will ensure that the loss of one satellite has no discernible effect on the user.

Two Galileo Control Centres (GCC) will be implemented on European ground to provide for the control of the satellites and to perform the navigation mission management. The data provided by a global network of twenty Galileo Sensor Stations (GSS) will be sent to the Galileo Control Centres through a redundant communications network. The GCC's will use the data of the Sensor Stations to compute the integrity information and to synchronize the time signal of all satellites and of the ground station clocks. The exchange of the data between the Control Centres and the satellites will be performed through so-called up-link stations. Five S-band up-link stations and 10 C-band up-link stations will be installed around the globe for this purpose.

As a further feature, Galileo will provide a global Search and Rescue (SAR) function, based on the operational Cospas-Sarsat system. To do so, each satellite will be equipped with a transponder, which is able to transfer the distress signals from the user transmitters to the Rescue Co-ordination Centre, which will then initiate the rescue operation. At the same time, the system will provide a signal to the user, informing him that his situation has been detected and that help is under way. This latter feature is new and is considered a major upgrade compared to the existing system, which does not provide a feedback to the user.

See also http://www.esa.int/esaNA/galileo.html.

Future Improvements

The first block II-R GPS satellite was launched in 1997. Though the later versions of block IIs will be a bridge to a future GPS system, known as GPS III, the next generation of GPS is still being defined. Future improvements in the GPS system are driven by competing civil and military requirements. All users desire more signal power to ensure resistance to interference and/or jamming. In the last decade, GPS has become essential to virtually all DOD operations. International constraints on RF spectrum availability dictate that improvements remain within the radio navigation bands. On the civil side, the expectation has become that GPS will remain continuously available across the globe for the foreseeable future. Civilian users are urgently requesting the second and third frequencies to calibrate ionospheric delays and provide a backup if the L1 signal is jammed.

Several key advances are planned for the end of the block II series of satellites. The most important are two additional signals on the II-RMs and three on the II-Fs. The first additional signal is a replica of the C/A code but at the L2 frequency. This will allow direct measurement of ionospheric errors for civilian users. Military users will have a new split spectrum code (called M-code) on both L1 and L2. This code has the advantage of transmitting most of its power in the nulls of the C/A code, maximizing spectral separation. The signal modernization is shown in Fig. 6.

The II-Fs will include yet another civil signal at L5 (1176 MHz). This signal is intended to be a higher accuracy signal, which implies a higher chipping rate and a longer code sequence. Likely, it will include an unmodulated channel to enable much longer integration time for superior noise rejection. Other technical advances for the late II-Fs include intersatellite communication, as well as improvements in the rubidium/cesium clocks on board. Likewise, upgrades in the ground station facilities will reduce the errors in ephemeris predictions. For GPS III, the need for further increases in M-code power will probably lead to a spot beam of about 1,000 kilometers (621.4 miles).

Though all specifics of the GPS III concept are still to be determined, the United States intends to continue to provide and improve on a worldwide continuously available, precise, navigation signal that is free to all of the world. GPS III will undoubtedly continue in that tradition and provide a yet more robust and more accurate system of positioning on a global scale.

See also **Satellites (Communications and Navigation)**.

Additional Reading

Francisco, S.G. "GPS Operational Control Segment," in B.W. Parkinson, J.J. Spilker, P. Axelrad, and P. Enge, eds., *Global Positioning System: Theory and Applications*, Vol. I, AIAA, Washington, DC, 1996.

Guier, W.H., and G.C. Weiffenbach: *John Hopkins APL Tech. Dig.* **18** (2), 178–181 (1997).

Klobuchar, J.A.: "Ionospheric effects on GPS," in B.W. Parkinson, J.J. Spilker, P. Axelrad, and P. Enge eds., *Global Positioning System: Theory and Applications*, Vol. I, AIAA, Washington, DC, 1996.

Leick, A.: *GPS Satellite Surveying*, 3rd Edition, John Wiley & Sons, Inc., New York, NY, 2003.

Misra, P., and P. Enge.: *Global Position System: Signals, Measurements, and Performance*, Ganga-Jamuna Press, Lincoln, MA, 2001, pp. 284–287.

Parkinson, B.W.: "Introduction and heritage of NAVSTAR, the Global Positioning System," In B.W. Parkinson, J.J. Spilker, P. Axelrad, and P. Enge (eds), *Global Positioning System: Theory and Applications*, Vol. I, AIAA, Washington, DC, 1996.

Parkinson, B.W.: "GPS Error Analysis," In B.W. Parkinson, J.J. Spilker, P. Axelrad, and P. Enge, eds., *Global Positioning System: Theory and Applications*, Vol. I, AIAA, Washington, DC, 1996.

Piscane, V.L.: *John Hopkins APL Tech. Dig.*, **19** (1), 4–10 (1998).

Reaser, R.: U.S. Air Force Joint Program Office 2002. http://www.ccit.edu.tw/ccchang/Gps_modernization_ppt.pdf.

Roddy, D.: *Satellite Communications*, The McGraw-Hill Companies, Inc., New York, NY, 2006.

Rycroft, M.J.: *Satellite Navigation Systems: Policy, Commercial and Technical Interaction (Space Studies)*, Vol. 8, Springer-Verlag New York, LLC, New York, NY, 2003.

Sobel, D.: *Longitude: The True Story of a Lone Genius Who Solved the Greatest Scientific Problem of His Time*, Walker, New York, NY, 1995.

Spilker, J.J.: "GPS Signal Structure and Theoretical Performance," in B.W. Parkinson, J.J. Spilker, P. Axelrad, and P. Enge, eds., *Global Positioning System: Theory and Applications*, Vol. I, AIAA, Washington, DC, 1996.

Spilker, J.J.: "Fundamentals of Signal Tracking Theory," in B.W. Parkinson, J.J. Spilker, P. Axelrad, and P. Enge, eds., *Global Positioning System: Theory and Applications*, Vol. I, AIAA, Washington, DC, 1996.

Taylor, G., and G. Blewitt: *Intelligent Positioning–GIS-GPS Unification*, John Wiley & Sons, Inc., Hoboken, NJ, 2006.

Tsui, J. Bao-Yen: *Fundamentals of Global Positioning System Receivers: A Software Approach*, 2nd Edition, John Wiley & Sons, Inc., New York, NY, 2004.

U.S. Air Force. Navstar GPS Space Segment/Navigation User Interfaces, ICD-GPS-200 C, 1997. http://gps.losangeles.af.mil/gpsarchives/1000-public/1300-LIB/documents/Other_Data/icdgps200c_irn1thru4.pdf

U.S. Coast Guard. Nationwide DGPS Status Report. 2001. http://www.navcen.uscg.gov/dgps/ndgps/default.htm.

U.S. DOD and DOT. Federal RadioNavigation Plan, 1999. http://avnwww.jccbi.gov/icasc/PDF/frp1999.pdf.

U.S. Federal Aviation Administration. WAAS, 2002. http://gps.faa.gov/Programs/WAAS/waas.htm.

U.S. Federal Aviation Administration. LAAS, 2002. http://gps.faa.gov/Programs/LAAS/laas.htm.

BRADFORD PARKINSON
JAMES SPIKER
GABRIEL ELKAIM
Stanford University, Stanford, CA

GLOBAL PRECIPITATION MEASUREMENT (GPM). See **Earth Observing System (EOS)**; and **Space Science Missions: Earth**.

GLOBAL SYSTEM FOR MOBILE COMMUNICATION. See **Wireless Communications Applications**.

GLOBAL TEMPERATURE. Global surface temperatures in 1999 fell back from the record setting high level of 1998, which was the warmest year in the period of instrumental data, report researchers at the NASA Goddard Institute for Space Studies who analyze data collected from several thousand meteorological stations around the world. But 1999 was still one of the warmest years of the century, as shown in Figure 1.

Although global temperature fluctuates considerably from year to year due to chaotic variability of the atmosphere and ocean, there has been a long-term global warming trend underway since the early 1960s, as illustrated in Figure 1. The 1999 data are consistent with a continuation of that warming trend, with 1999 being approximately the sixth warmest year in the record. The ranking of years is approximate because of incomplete global coverage of measurement stations and small errors in the measurements.

Most parts of the world were warmer than normal, i.e., warmer than the 30-year period 1951–1980, as illustrated in Figure 2. It was particularly warm across most of North America (except the West Coast) and most of

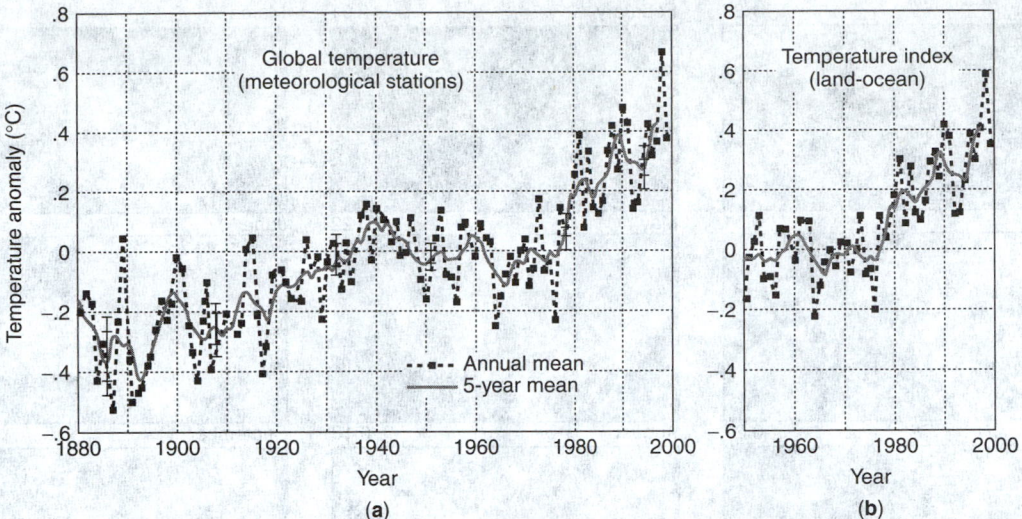

Fig. 1. (a) Near-global annual-mean surface air temperature change, based on meteorological station network, (b) global land-ocean surface temperature index, which combines sea surface temperature measurements for ocean areas with surface air temperature measurements at meteorological stations.

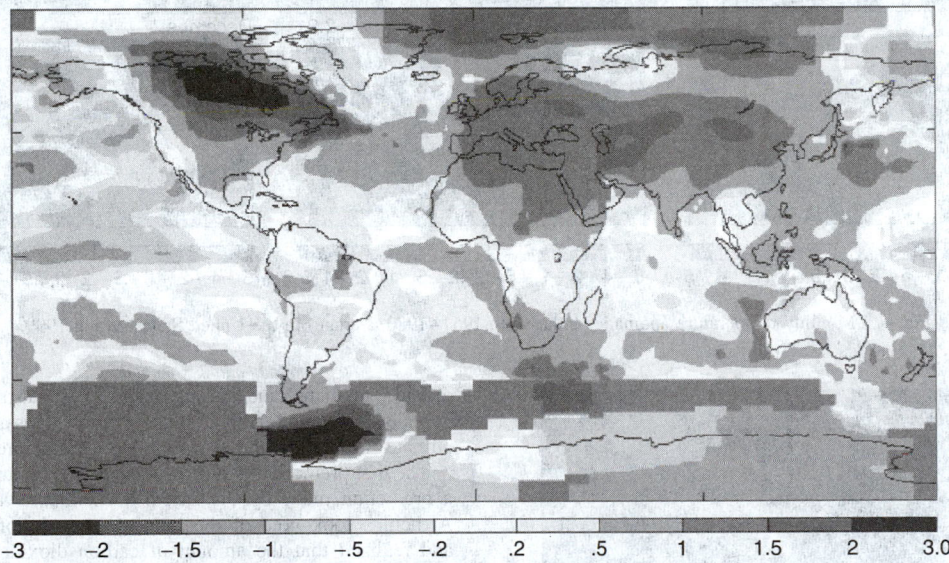

Fig. 2.   Surface temperature anomaly for the 1999 calendar year derived from measurements at several thousand meteorological stations and satellite measurements of the ocean surface. (Hansen et al.; Reynolds and Smith.)

Eurasia. However, the tropical Pacific Ocean was cool due to a strong La Niña. During a La Niña the near equatorial region is cooled by upwelling of cool water from the deep ocean. See also **La Niña**.

The temperature in the United States was also warm, about 0.7 °C above the 1951–1980 average. See Fig. 3. 1999 was approximately the 10th warmest year of the century. The warmest years in the United States occurred during the dust bowl era, with 1934 being the warmest year.

Our analyzed temperature, in the United States and the rest of the world, includes corrections for urban effects on the record. Nearby rural stations are used to adjust the long-term trends at urban stations, as described by [Hansen et al. 1999].

The temperature anomalies fluctuate substantially from month to month, as illustrated for the United States in Figure 4.

February and November were both exceptionally warm in the United States. Averaged over the year, most of the United States was warm in 1999 (See Figure 5), except the West Coast and Florida.

These maps for the United States illustrate that even with the level of warmth that occurred in the United States in 1999, the local warming trend is less than natural year-to-year fluctuations of monthly mean temperature. Thus for any given location in the United States there are generally at least a few months in the year that are cooler than normal. But the

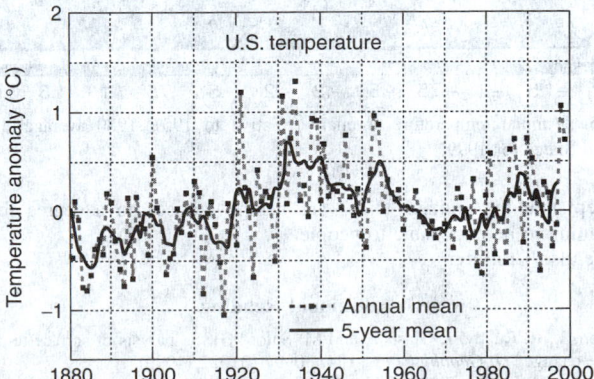

Fig. 3.   Temperature anomaly (deviation from the 1951–1980 average) for calendar years for the contiguous United States.

overall tendency toward warming is enough that it is beginning to effect the probability of a month or a season being warmer than normal. In our discussion of 1998 temperatures (Reynolds and Smith) discussed this

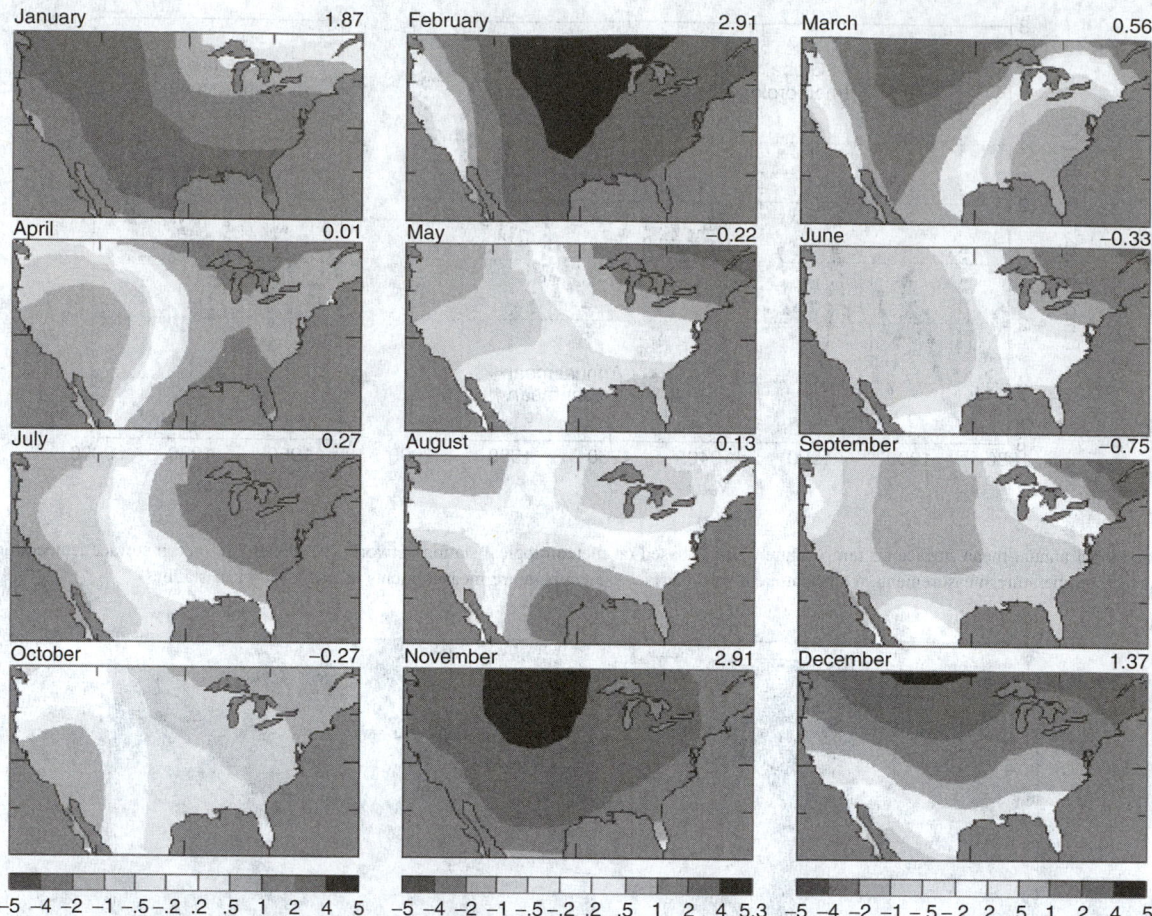

Fig. 4.   Monthly temperature anomalies (relative to 1951–1980 average) in the United States during 1999.

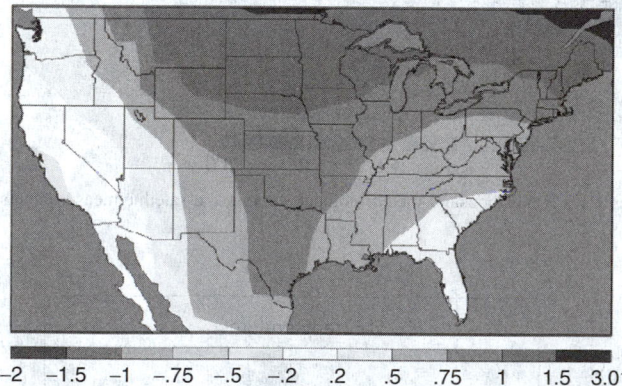

Fig. 5.   Annual temperature anomaly (relative to 1951–1980 average) in the United States for 1999.

concept that the climate "dice" are being "loaded" to a degree that is beginning to be noticeable to people.

See also **Climate**.

Additional Reading

Hansen, J., R. Ruedy, J. Glascoe, and M. Sato: "GISS analysis of surface temperature change," *J. Geophys. Res.*, **104**, 30997–31022 (1999).

Reynolds, R.W. and T.M. Smith: "Improved Global Sea Surface Temperature Analyses," *J. Climate*, **7**, 929–948 (1994).

Goddard Institute for Space Studies, New York, NY

GLOBAL WARMING. The Earth's climate is predicted to change because human activities are altering the chemical composition of the atmosphere through the buildup of greenhouse gases, primarily carbon dioxide, methane, and nitrous oxide. The heat-trapping property of these gases is undisputed. Although uncertainty exists about exactly how Earth's climate responds to these gases, global temperatures are rising.

The first prediction of climate change due to human activities began with a prediction made by the Swedish chemist, Svante Arrhenius, in 1896. Arrhenius took note of the industrial revolution then getting underway and realized that the amount of carbon dioxide being released into the atmosphere was increasing. Moreover, he believed carbon dioxide concentrations would continue to increase as the world's consumption of fossil fuels, particularly coal, increased ever more rapidly. His understanding of the role of carbon dioxide in heating Earth, even at that early date, led him to predict that if atmospheric carbon dioxide doubled, Earth would become several degrees warmer. Arrhenius was referring to a potential modification of what is now called the greenhouse gas effect. However, little attention was paid to what must have been seen to be a rather far-out prediction that had no apparent consequence for people living at that time.

Our Changing Atmosphere

Energy from the sun drives the Earth's weather and climate, and heats the Earth's surface; in turn, the Earth radiates energy back into space. Atmospheric greenhouse gases (water vapor, carbon dioxide, and other gases) trap some of the outgoing energy, retaining heat somewhat like the glass panels of a greenhouse.

Without this natural "greenhouse effect," temperatures would be much lower than they are now, and life as known today would not be possible. Instead, thanks to greenhouse gases, the Earth's average temperature is a more hospitable 60 °F. However, problems may arise when the atmospheric concentration of greenhouse gases increases. Since the beginning of the industrial revolution, atmospheric concentrations of carbon dioxide have increased nearly 30%, methane concentrations have more than doubled, and nitrous oxide concentrations have risen by about 15%. These increases have enhanced the heat-trapping capability of the earth's atmosphere. Sulfate aerosols, a common air pollutant, cool the atmosphere by reflecting light back into space; however, sulfates are short-lived in the atmosphere and vary regionally.

Why are greenhouse gas concentrations increasing? Scientists generally believe that the combustion of fossil fuels and other human activities are the primary reason for the increased concentration of carbon dioxide. Plant respiration and the decomposition of organic matter release more than 10 times the CO_2 released by human activities; but these releases have generally been in balance during the centuries leading up to the industrial revolution with carbon dioxide absorbed by terrestrial vegetation and the oceans.

What has changed in the last few hundred years is the additional release of carbon dioxide by human activities. Fossil fuels burned to run cars and trucks, heat homes and businesses, and power factories are responsible for about 98% of U.S. carbon dioxide emissions, 24% of methane emissions, and 18% of nitrous oxide emissions. Increased agriculture, deforestation, landfills, industrial production, and mining also contribute a significant share of emissions. In 1997, the United States emitted about one-fifth of total global greenhouse gases.

Estimating future emissions is difficult, because it depends on demographic, economic, technological, policy, and institutional developments. Several emissions scenarios have been developed based on differing projections of these underlying factors. For example, by 2100, in the absence of emissions control policies, carbon dioxide concentrations are projected to be 30–150% higher than today's levels.

Changing Climate

Global mean surface temperatures have increased 0.5–1.0 °F since the late 19th century. The 20th century's 10 warmest years all occurred in the last 15 years of the century. Of these, 1998 was the warmest year on record. The snow cover in the Northern Hemisphere and floating ice in the Arctic Ocean have decreased. Globally, sea level has risen 4–8 inches over the past century. Worldwide precipitation over land has increased by about 1%. The frequency of extreme rainfall events has increased throughout much of the United States. See Fig. 1.

Increasing concentrations of greenhouse gases are likely to accelerate the rate of climate change. Scientists expect that the average global surface temperature could rise 1–4.5 °F (0.6–2.5 °C) in the next fifty years, and 2.2–10 °F (1.4–5.8 °C) in the next century, with significant regional variation. Evaporation will increase as the climate warms, which will increase average global precipitation. Soil moisture is likely to decline in many regions, and intense rainstorms are likely to become more frequent. Sea level is likely to rise two feet along most of the U.S. coast.

Calculations of climate change for specific areas are much less reliable than global ones, and it is unclear whether regional climate will become more variable.

Trends

Temperature. Global temperatures are rising. Observations collected over the last century suggest that the average land surface temperature has risen 0.45–0.6 °C (0.8–1.0 °F) in the last century. The surface of the ocean has also been warming at a similar rate. Studies that combine land and sea measurements have generally estimated that global temperatures have warmed 0.3–0.6 °C (0.5–1.0 °F) in the last century. About two-thirds of this warming took place between 1900 and 1940. Global temperatures declined slightly from the 1940s through the 1970s; but have risen more rapidly during the last 25 years than in the period before 1940.

Surface temperatures are not rising uniformly. Night-time low temperatures are rising on average about twice as rapidly as daytime highs. The winters in areas between 50 and 70° North Latitude (the latitude of Canada and Alaska) are warming relatively fast, while summer temperatures show little trend. Urban areas are warming somewhat more rapidly than rural areas, because of both the changes in land cover and the consumption of energy that take place in densely developed areas (a feature known as the "urban heat island" effect).

In the United States, temperatures in the last 50 years have cooled in the East while warming in the West. Over the last 100 years, the pattern is similar, except that New England is warmer than 100 years ago because it warmed more in the first half of the 20th century by more than it cooled in the second half. This pattern of warming and cooling may be part of a worldwide pattern: while most of the Earth has warmed, the regions that are downwind from major sources of sulfur dioxide emissions have generally cooled.

Precipitation. Precipitation has increased by about 1% over the world's continents in the last century. High latitude areas are tending to see more significant increases in rainfall, while precipitation has actually declined in many tropical areas.

In North America, precipitation has increased significantly. Precipitation in the United States has increased by an average of about 5% in the last century. Along the northern tier states and in Southern Canada, rainfall has increased 10–15%. Much of the increase in rainfall has been taking place between September and November. Rainfall is also tending to be more concentrated in heavy downpours, according to studies by the National Oceanic and Atmospheric Administration NOAA. At the beginning of the 20th century, only 9% of the nation experienced a storm each year in which more than two inches of precipitation fell in a 24-hour period. In recent decades, such a severe storm has occurred each year over close to 11% of the nation.

Sea Level. Sea level has risen worldwide approximately 15–20 cm (6–8 inches) in the last century. Approximately 2–5 cm (1–2 inches) of the rise has resulted from the melting of mountain glaciers. Another 2–7 cm has resulted from the expansion of ocean water that resulted from warmer

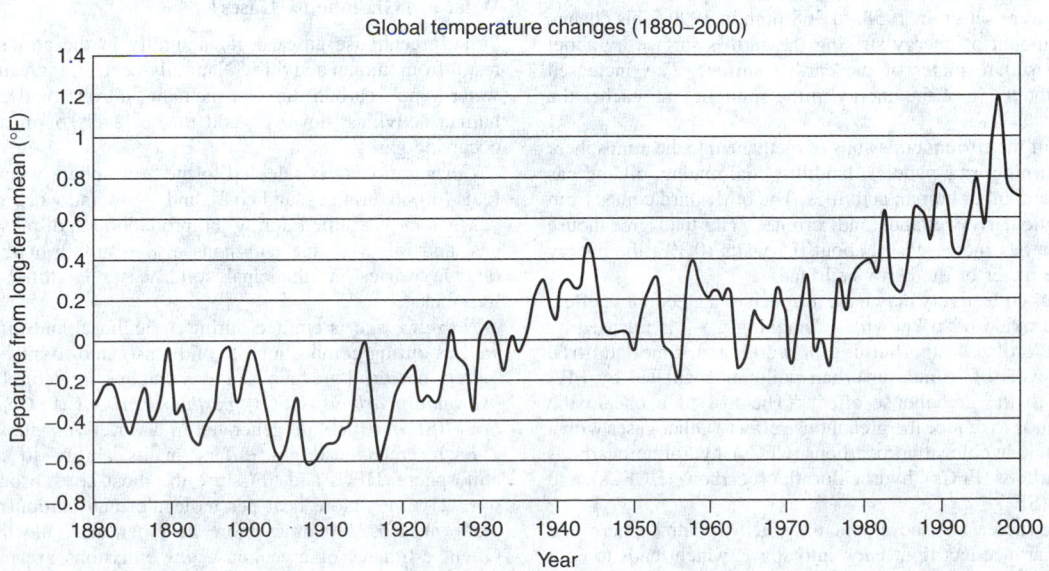

Fig. 1. *Source*: U.S. National Climatic Date Center, 2001.

ocean temperatures. The pumping of groundwater and melting of the polar ice sheets may have also added water to the oceans.

Along most of the U.S. coast, sea level has been rising 2.5–3.0 mm/yr (10–12 inches per century). Nevertheless, the rate varies from about one cm per year (three feet per century) along the Louisiana Coast, to a drop of several millimeters per year (a few inches per decade) in parts of Alaska. The rapid rate in Louisiana resulted from the settling of newly created land formed by the sediments that washed down the Mississippi River. In Galveston, the removal of groundwater led the land above the water table to sink. In areas that were covered by glaciers during the last Ice Age, by contrast, the land is rising because of the removal of the weight of the ice, which had previously compressed the land downward. As a result, the sea is dropping relative to these coasts.

Atmospheric Change

Humanity has been adding gases to the atmosphere that tend to warm the earth, known as "greenhouse gases." People are also adding small particles and droplets called aerosols that reflect light back into space and tend to cause some areas to cool. In the coming decades, humans are likely to continue to change our atmosphere. Because the greenhouse gases that warm the earth stay in the atmosphere longer than the aerosols that cool the earth, the earth's average temperature is likely to continue to warm.

Past. The temperatures of the Earth and any other planet depends mainly on (1) the amount of sunlight received, (2) the amount of sunlight reflected into space, and (3) the extent to which the atmosphere retains heat. Over the last two million years, changes in the timing and amount of sunlight striking the earth has been responsible for inducing ice ages during which temperatures were about $5\,°C$ ($9\,°F$) colder than today, and interglacial warm periods during which temperatures have been approximately the temperature of today. During the 20th and 21st century, however, the other two factors may be more important.

The water vapor and carbon dioxide found naturally in the atmosphere keep the Earth warmer than it would otherwise be. The clear atmosphere allows sunlight to penetrate to the Earth's surface and warm it. The surface releases this energy as infrared radiation, which is absorbed by water vapor and CO_2 in the atmosphere. This mechanism is commonly known as the "greenhouse effect." Without the greenhouse effect, the earth would be about $33\,°C$ ($60\,°F$) colder than it is currently.

Humanity is altering the energy balance of the planet by adding gases that absorb infrared radiation to the atmosphere, and thereby strengthening the greenhouse effect. The chief "greenhouse gases" are CO_2, methane, and nitrous oxide. Whenever oil, coal, gas, or wood are burned, carbon dioxide is released into the atmosphere. Approximately half of the CO_2 that is released is soon absorbed by the oceans or by increased plant photosynthesis. The other half remains in the atmosphere for many decades. As a result, the atmospheric concentration of CO_2 is increasing. The average concentration of carbon dioxide has increased from around 275 parts per million before the industrial revolution, to 315 ppm when precise monitoring stations were set up in 1958, to 368 ppm in 1999. This change has increased the amount of energy striking the earth's surface by about 1.5 watts for every square meter of the earth's surface. This increased energy is equal to about 1% of the energy in the sunlight that reaches the earth's surface.

About two thirds of the current emissions of methane into the atmosphere result from cattle farming, rice paddies, landfills, coal mining, oil and gas production, and several other human activities. The other third comes from natural sources, particularly wetlands and termites. The total greenhouse effect from methane has increased by about 0.5 watts (0.3%) the energy striking each square meter of the earth's surface.

Several other gases collectively may have as much of a greenhouse effect as methane. Nitrous oxide (also known as "laughing gas") is released by the use of nitrogen fertilizers, the burning of wood, and some industrial processes. Higher levels of ozone, an urban pollutant regulated by EPA NAAQS, also add to the greenhouse effect. (The loss of ozone in the upper atmosphere tends to reduce the greenhouse effect.) Other gases with a greenhouse effect include chlorofluorocarbons (CFCs), hydrofluorocarbons (HFCs), perfluorocarbons (PFCs), hydrochlorofluorocarbons (HCFCs), and sulfur hexafluoride (SF_6).

Humanity is also increasing the extent to which the atmosphere and the surface of the Earth reflect light back into space, which tends to cool the earth. Changes in land use can change the reflectivity of the earth's surface; tropical deforestation appears to be the most important change, but has only reduced the amount of sunlight absorbed by about 0.1 watt per square meter. Changes in the reflectivity of the atmosphere, however, appear to have an impact that could be as great as the impact of carbon dioxide, albeit in the opposite direction. Most importantly, society is adding very fine particles and droplets known as aerosols. See also **Aerosol**.

The most important aerosols are sulfates. Power plants that burn coal, as well as copper, lead, and zinc smelters, release sulfur dioxide, which reacts with water vapor in the atmosphere to form sulfates. Sulfates currently reflect enough light back into space to reduce the amount of energy striking the Earth's surface by somewhere between 0.5 and 1.5 watts per square meter. Unlike CO_2, methane, and other greenhouse gases, which remain in the atmosphere for decades or longer, most of the sulfates are removed by precipitation within a few weeks of being emitted. As a result, sulfates tend to be concentrated in the areas immediately downwind of major industrial areas. The ability of sulfates to scatter light also causes visibility problems in the Grand Canyon and other scenic vistas. Sulfates also cause acid rain. See also **Acid Rain**.

Future. The extent and speed at which humanity changes the climate will depend to a large extent on the rate at which society adds additional greenhouse gases to the atmosphere. Until a few decades ago, energy consumption grew at about the same rate as the Gross National Product.

Emissions

Once, all climate changes occurred naturally. However, during the Industrial Revolution, people began altering the climate and environment through changing agricultural and industrial practices. Before the Industrial Revolution, human activity released very few gases into the atmosphere, but now through population growth, fossil fuel burning, and deforestation, the mixture of gases in the atmosphere are being affected.

International Emissions. Greenhouse gases are global in their effect upon the atmosphere. The primary greenhouse gases, unlike many local air pollutants like carbon monoxide, oxides of nitrogen, and volatile organic compounds, are considered stock pollutants. A stock air pollutant is one that has a long lifetime in the atmosphere, and therefore can accumulate over time. Stock air pollutants are also generally well mixed in the atmosphere. As a consequence of this mixing, the impact a greenhouse gas has on the atmosphere is mostly independent of where it was emitted. These characteristics of greenhouse gases imply that they should be addressed on a global (i.e., international) scale. See also **Pollution (Air)**.

Anthropogenic emissions of greenhouse gases occur in every country of the world. These emissions result from many of the industrial, transportation, agricultural, and other activities that take place in each country. Countries that are signatories to the United Nations Framework Convention on Climate Change (UNFCCC) are committed to reporting their anthropogenic emissions of greenhouse gases to the Secretariat of the convention.

What are Greenhouse Gases?

Some greenhouse gases occur naturally in the atmosphere, while others result from human activities. Naturally occurring greenhouse gases include water vapor, carbon dioxide, methane, nitrous oxide, and ozone. Certain human activities, however, add to the levels of most of these naturally occurring gases:

Carbon dioxide is released to the atmosphere when solid waste, fossil fuels (oil, natural gas, and coal), and wood and wood products are burned.

Methane is emitted during the production and transport of coal, natural gas, and oil. Methane emissions also result from the decomposition of organic wastes in municipal solid waste landfills, and the raising of livestock.

Nitrous oxide is emitted during agricultural and industrial activities, as well as during combustion of solid waste and fossil fuels.

Very powerful greenhouse gases that are not naturally occurring include *hydrofluorocarbons* (HFCs), *perfluorocarbons* (PFCs), and *sulfur hexafluoride* (SF_6), which are generated in a variety of industrial processes.

Each greenhouse gas differs in its ability to absorb heat in the atmosphere. HFCs and PFCs are the most heat-absorbent. Methane traps over 21 times more heat per molecule than carbon dioxide, and nitrous oxide absorbs 270 times more heat per molecule than carbon dioxide. Often, estimates of greenhouse gas emissions are presented in units of millions of metric tons of carbon equivalents (MMTCE), which weights each gas by its GWP value, or *Global Warming Potential*.

What are Emissions Inventories?

An emission inventory is an accounting of the amount of air pollutants discharged into the atmosphere. It is generally characterized by the following factors:

The chemical or physical identity of the pollutants included.
The geographic area covered.
The institutional entities covered.
The time period over which emissions are estimated.
The types of activities that cause emissions.

Emission inventories are developed for a variety of purposes. Inventories of natural and anthropogenic emissions are used by scientists as inputs to air quality models, by policy makers to develop strategies and policies or track progress of standards, and by facilities and regulatory agencies to establish compliance records with allowable emission rates. A well constructed inventory should include enough documentation and other data to allow readers to understand the underlying assumptions and to reconstruct the calculations for each of the estimates included.

What are Sinks?

A sink is a reservoir that uptakes a chemical element or compound from another part of its cycle. For example, soil and trees tend to act as natural sinks for carbon. Each year hundreds of billions of tons of carbon in the form of CO_2 are absorbed by oceans, soils, and trees.

Impacts of Global Warming

Rising global temperatures are expected to raise sea level, and change precipitation and other local climate conditions. Changing regional climate could alter forests, crop yields, and water supplies. It could also threaten human health, and harm birds, fish, and many types of ecosystems. Deserts may expand into existing rangelands, and the character of some of our National Parks may be permanently altered.

Most of the United States is expected to warm, although sulfates may limit warming in some areas. Scientists currently are unable to determine which parts of the United States will become wetter or drier, but there is likely to be an overall trend toward increased precipitation and evaporation, more intense rainstorms, and drier soils.

Unfortunately, many of the potentially most important impacts depend upon whether rainfall increases or decreases, which can not be reliably projected for specific areas.

Health. Throughout the world, the prevalence of particular diseases and other threats to human health depend largely on local climate. Extreme temperatures can directly cause the loss of life. Moreover, several serious diseases only appear in warm areas. Finally, warm temperatures can increase air and water pollution, which in turn harm human health.

The most direct effect of climate change would be the impacts of hotter temperatures themselves. Extremely hot temperatures increase the number of people who die on a given day for many reasons: People with heart problems are vulnerable because one's cardiovascular system must work harder to keep the body cool during hot weather. Heat exhaustion and some respiratory problems increase.

Higher air temperatures also increase the concentration of ozone at ground level. The natural layer of ozone in the upper atmosphere blocks harmful ultraviolet radiation from reaching the earth's surface; but in the lower atmosphere, ozone is a harmful pollutant. Ozone damages lung tissue, and causes particular problems for people with asthma and other lung diseases. Even modest exposure to ozone can cause healthy individuals to experience chest pains, nausea, and pulmonary congestion. In much of the nation, a warming of four degrees (°F) could increase ozone concentrations by about 5%.

Statistics on mortality and hospital admissions show that death rates increase during extremely hot days, particularly among very old and very young people living in cities. In July 1995, a heat wave killed more than 700 people in the Chicago area alone. Studies based on these types of statistics estimate that in Atlanta, for example, even a warming of about two degrees (°F) would increase heat-related deaths from 78 today to anywhere from 96 to 247 people per year. If people are able to install air conditioning and otherwise acclimatize themselves to the hotter temperatures, the lower estimate is more likely.

Warmer temperatures may decrease the number of people who die each year from cold weather. However, in the United States, only 1000 people die from the cold each year, while twice that many die from the heat. Moreover, of the ten states with the greatest number of cold-related deaths, Alaska and Illinois are the only northern states. For the most part, cold-related deaths occur during occasional cold spells in areas with mild winters where people prepare less for the cold, or during extreme events like the severe snow storm that struck Colorado in November of 1997. Global warming is unlikely to reduce either of these situations. Finally, deaths due to the heat are more sensitive to temperature changes than deaths due to the cold; the difference between $-20\,°F$ and $-15\,°F$, for example, has a much smaller impact than an increase from $95\,°F$ to $100\,°F$.

Global warming may also increase the risk of some infectious diseases, particularly those diseases that only appear in warm areas. Diseases that are spread by mosquitoes and other insects could become more prevalent if warmer temperatures enabled those insects to become established farther north; such "vector-borne" diseases include malaria, dengue fever, yellow fever, and encephalitis. Some scientists believe that algal blooms could occur more frequently as temperatures warm, particularly in areas with polluted waters, in which case diseases such a cholera that tend to accompany algal blooms could become more frequent.

In spite of these risks, increased mortality is not an inevitable consequence of global warming. Malaria, for example, is rare in the United States even in warmer regions where the mosquito that transmits the disease is found, because this nation has the ability to rapidly identify and contain outbreaks when they appear. Heat-related deaths can be prevented by emergency measures to move vulnerable people to air-conditioned buildings, and by reducing the emissions of photochemical oxidants which cause ground-level ozone. Many of the impacts of climate change on health could be avoided through the maintenance of strong public health programs to monitor, quarantine, and treat the spread of infectious diseases and respond to other health emergencies as they occur. Although air-conditioning and public health programs may impose additional costs on the public and private sectors, they would often be preferable to the impacts on human health that would otherwise occur.

Water Resources. Changing climate is expected to increase both evaporation and precipitation in most areas of the United States. In those areas where evaporation increases more than precipitation, soil will become drier, lake levels will drop, and rivers will carry less water.

Lower river flows and lower lake levels could impair navigation, hydroelectric power generation, and water quality, and reduce the supplies of water available for agricultural, residential, and industrial uses. Some areas may experience both increased flooding during winter and spring, as well as lower supplies during summer. In California's Central Valley, for example, melting snow provides much of the summer water supply; warmer temperatures would cause the snow to melt earlier, and thus, reduce summer supplies even if rainfall increased during the spring. More generally, the tendency for rainfall to be more concentrated in large storms as temperatures rise would tend to increase river flooding, without increasing the amount of water available.

Navigation. Climate change could impair navigation by changing average water levels in rivers and lakes, increasing the frequency of both floods during which navigation is hazardous and droughts during which passage is difficult, and necessitating changes in navigational infrastructure. On the other hand, warmer temperatures could extend the ice-free season.

Hydropower. Changes in the flows of rivers would have a direct impact on the amount of hydropower generated, because hydropower production decreases with lower flows. Because of the ambiguous projections of changes in future river flow, studies of the impacts of climate change show ambiguous effects on hydropower production. As a general rule, however, a 1% decrease in runoff produces a greater than 1% decrease in hydropower production. Not only does less water run through the turbines, but the lower reservoir levels reduce the water pressure and hence the power produced by a given amount of water. In the Colorado River's lower basin, for example, a 10% decrease in runoff reduces power production 36%.

Water Supply and Demand. In some parts of the Western United States, the most widely discussed potential impact of climate change is the impact on water supply and demand. The potential changes in water supplies would result directly from the changes in runoff and the levels of rivers, lakes, and aquifers. In the eastern half of the United States, consumptive withdrawals are usually only a fraction of total runoff; hence these withdrawals are unlikely to be threatened except during severe droughts or where nonconsumptive uses such as navigation or environmental quality take precedence. In the west, by contrast, where many rivers are already fully allocated, a decline in runoff would translate directly to a reduced supply of water.

Environmental Quality and Recreation. Decreased river flows and higher temperatures could harm the water quality of the nation's rivers, bays, and lakes. In areas where river flows decrease, pollution concentrations will rise because there will be less water to dilute the pollutants; to keep pollution concentrations from increasing sewage treatment plants and other water pollution controls will have to be upgraded, thus could cost billions of dollars per year. Increased frequency of severe rainstorms could increase the amount of chemicals that run off from farms, lawns, and streets into the nations rivers, lakes, and bays.

The amount of dissolved oxygen in the water could also be reduced, effectively suffocating the fish in some areas. Higher water temperatures decrease the solubility of oxygen in water. Moreover, warmer water hastens the rate at which organic pollutants degrade; this degradation exerts a "biochemical oxygen demand" (BOD). The combined effect of lower oxygen solubility and higher BOD will be to reduce the availability of dissolved oxygen, which is critical to the health of aquatic organisms. One study estimated that throughout the Southeastern United States, the warmer water temperatures would push dissolved oxygen levels in most rivers to below the 5 ppm necessary to sustain most fish.

Flood Control. Although the impacts of sea level rise and associated coastal flooding have been more widely discussed, global climate change could also change the frequency and severity of inland flooding, particularly along rivers. General circulation models suggest that some regions of the United States may have more rainfall during the wet season, which would increase river and Lake levels. Moreover, increased flooding could occur even in areas that do not become wetter: (1) earlier snowmelt could worsen spring flooding while diminishing summer water availability; (2) some climate models suggest wetter winters and drier summers; (3) the need to ensure summer/drought water supplies could lead water managers to keep reservoir levels higher, thereby limiting the capacity for additional water retention during unexpected wet spells; (4) warm areas generally have a more intense hydrologic cycle and thus more rain in a severe storm; and, (5) many areas may receive more intense rainfall. With the exception of earlier snowmelt, however, it is also possible that these processes will become more benign as climate changes, at least in some areas.

Forests. The projected 2 °C (3.6 °F) warming could shift the ideal range for many North American forest species by about 300 km (200 mi.) to the north. If the climate changes slowly enough, warmer temperatures may enable the trees to colonize north into areas that are currently too cold, at about the same rate as southern areas became too hot and dry for the species to survive. If the earth warms 2 °C (3.6 °F) in 100 years, however, the species would have to migrate about two miles every year.

Trees whose seeds are spread by birds may be able to spread at that rate. But neither trees whose seeds are carried by the wind, nor such nut-bearing trees such as oaks, are likely to spread by more than a few hundred feet per year. Poor soils may also limit the rate at which tree species can spread north. Thus, the range over which a particular species is found may tend to be squeezed as southern areas become inhospitably hot. The net result is that some forests may tend to have a less diverse mix of tree species.

Several other impacts associated with changing climate further complicate the picture. On the positive side, CO_2 has a beneficial fertilization effect on plants, and also enables plants to use water more efficiently. These effects might enable some species to resist the adverse effects of warmer temperatures or drier soils. On the negative side, forest fires are likely to become more frequent and severe if soils become drier. Changes in pest populations could further increase the stress on forests. Managed forests may tend to be less vulnerable than unmanaged forests, because the managers will be able to shift to tree species appropriate for the warmer climate.

Perhaps the most important complicating factor is uncertainty whether particular regions will become wetter or drier. If climate becomes wetter, then forests are likely to expand toward rangelands and other areas that are dry today; if climate becomes drier, then forests will retreat away from those areas. Because of these fundamental uncertainties, existing studies of the impact of climate change have ambiguous results.

Rangelands. Grasslands comprise a large portion of the United States west of the 100th meridian. Although these areas receive too little rainfall

to sustain a forest, the pioneers that settled the western frontier of the United States found dense foliage for grazing sheep and cattle. Today, a large portion of these "rangelands" is administered by the Federal Bureau of Land Management, which leases grazing rights to private ranchers. Much of the private land is irrigated for agriculture.

The impacts of climate change on grasslands has not been studied in the same detail as the implication for forests. Nevertheless, the existing research suggests a number of likely outcomes. Perhaps most importantly, climate change could harm grazing activities on both federal and private lands. Availability of water in these areas is often the single most important factor determining the value of land for grazing. The decline in western water availability suggested by several studies would seriously decrease the economic viability of grazing on these lands. Because grazing on the open range accounts for a small and declining fraction of U.S. cattle, national beef production would not be seriously impaired.

Changing climate is also likely to alter both the geographical extent and the plant composition of rangelands. If a drier climate causes some areas of the Southeast or Midwest to lose their ability to sustain a forest, the terrain in those areas may come to resemble the landscape of the open range. A wetter climate, by contrast, might enable forests to grow in areas that are now grasslands, while also enabling range grasses to grow in areas that are deserts today. Within existing rangelands, elevated levels of carbon dioxide may induce a shift from grasses toward shrubs and other woody plants.

Deserts. Relatively little research has been conducted on the possible impact of climate change on deserts. Scientists have focused on a different question: Are deserts likely to expand?

Studies that focus on what is likely to occur over the next century, however, do not indicate whether deserts are more likely to expand or contract. Studies using the biogeographic models have estimated that desert shrublands could increase by as much as 185% or decrease by as much as 56%. See also **Biome**.

Non-tidal Wetlands. Among the most biologically productive lands, wetlands cover approximately 4–6% of the Earth's land surface. The high productivity results from the essential characteristic of a wetland: an area that is flooded part of the time but not all of the time. This flooding ensures that the wetlands have ample supplies of water, minerals, or both. In addition to the high biological productivity, wetlands are important habitat for birds, fish, and other species. Wetlands are also important cleansing mechanisms for preventing pollutants from farms and other activities from running off and polluting rivers, lakes, and streams. See also **Wetlands**.

The impact of climate change on non-tidal wetlands is uncertain, because it depends on changes in the amount of rainfall, as well as when it occurs, which scientists are unable to forecast. In those areas where the climate becomes drier, drought will tend to lower water tables more than the lowering that occurs during droughts today. The resulting effect on prairie potholes would be similar to what happens today when farmers drain the wetlands. (See the EPA Wetlands Office site for more information.) http://www.epa.gov/OWOW/wetlands/. The open water ponds, which are critical habitat for waterfowl, would be replaced by relatively damp land, although some form of wetland vegetation would remain.

Coastal Zones. Sea level is rising more rapidly along the U.S. coast than worldwide. Studies by EPA and others have estimated that along the Gulf and Atlantic coasts, a one foot (30 cm) rise in sea level is likely by 2050 and could occur as soon as 2025. In the next century, a two foot rise is most likely, but a four foot rise is possible; and sea level will probably continue to rise for several centuries, even if global temperatures were to stop rising a few decades hence. See also **Ocean**.

Rising sea level inundates wetlands and other low-lying lands, erodes beaches, intensifies flooding, and increases the salinity of rivers, bays, and groundwater tables. Some of these effects may be further compounded by other effects of changing climate. Measures that people take to protect private property from rising sea level may have adverse effects on the environment and on public uses of beaches and waterways. Federal, state, and local governments are already starting to take measures to prepare for the consequences of rising sea level.

Agriculture. Developing the ability to estimate the impacts of climate change on agriculture confidently is critically important. If ever achieved, it could provide the global information needed to help farmers develop their own long-range responses to climate change. Unfortunately, we are a long way from having such a capability, and it may take a decade or more to substantially improve the resolution and accuracy of the Global Climate Models (GCMs) and evaluate the implications for agriculture.

Many studies have examined the likely impacts of climate change on agriculture both in the United States and abroad over the last couple of decades. Although the focus is on the impacts on U.S. agriculture resources, it is important to note that the global situation appears to be much less reassuring. Developing countries are likely to have considerably more difficulty adapting to climate change due to many factors, such as less developed technology and less available capital. In addition, global climate change will clearly impact U.S. agriculture exports, imports and market prices. Although very important, such global considerations are largely beyond the scope of this discussion.

Fisheries. Global warming could have many impacts on fish and other aquatic species. Some bodies of water may become too warm for the fish that currently inhabit those areas; but warmer temperatures may also enable fish in cold ocean waters to grow more rapidly. Global warming may also change the chemical composition of the water that fish inhabit: the amount of oxygen in the water may decline, while pollution and salinity levels may increase. Loss of wetlands could diminish habitat and alter the availability of food for some fish species. Scientists have examined the implications for three types of fisheries: (1) inland freshwater fisheries found in non-tidal rivers, lakes, and streams; (2) coastal fisheries, which extend from tidal freshwater rivers, to estuaries, to coastal ocean fisheries; and (3) deep ocean fisheries.

Inland Fisheries. Higher water temperatures may have the most important implications for inland fisheries. Like plants and birds, most species of fish tolerate, and many require, winter cooling and summer warming by tens of degrees. Nevertheless, most fish have limits to how hot or cold the water can be before they must either find more hospitable temperatures or die. Again like plants and birds, most freshwater fish are found throughout a fairly large region where the water is rarely too hot or too cold for them to live.

Scientists are not yet certain whether the overall level of fishing will increase or decrease in these waters. Because warmer temperatures often promote biological activity, fisheries biologists generally believe that the increase in warm water species would more than offset the decline of cold water fish in the Great Lakes, and possibly in other freshwater fisheries as well. But the total amount of fish that can be caught is not the only important consideration.

Coastal Fisheries. Wetland loss, salinity changes, and higher temperatures are all likely to affect finfish and shellfish in the coastal zone. The most vulnerable species are those that either reproduce in coastal wetlands, spend their entire lifetimes in an estuary, or both.

Coastal marshes are the primary nursery grounds for crab, shrimp, menhaden, and several other important fish. Most of the reproduction occurs in the part of the wetlands that are within about 50–100 feet of the open water, because fish find it more difficult to access parts of the marsh farther from the open water. As sea level rises and inundates wetlands, the initial effect in some areas is to increase the amount of channels through the marsh, which increases the total area of marsh to which the fish have access. Hence, sea level rise initially tends to increase the production of these species. As sea level continues to rise, however, the loss of marsh accelerates; and eventually most or all of the wetlands in an area are replaced by open bodies of water. Thus, in the long run, an accelerated rise in sea level would decrease production of these species.

A number of marine species that are not important for commercial fishing are also vulnerable to the inundation and erosion of coastal habitat. The horseshoe crab lays its eggs on sandy estuarine beaches of Delaware Bay and other beaches in the mid-Atlantic; officials in that region are already concerned that this species is being threatened by loss of sandy beaches. The horseshoe crab and the birds that rely on a plentiful supply of their eggs could be further endangered as rising sea level increases the erosion of these beaches, especially if shores are armored. In Hawaii, seals and turtles nest on small sandy islands and sand spits, which could be eroded by rising sea level.

Estuarine species are also vulnerable to other implications of climate change. Oyster predators such as MSX and oyster drill require salty water; hence the increased salinity from rising sea level or increased drought could lower oyster harvests, which have already decline by more than 90% in the last several decades in many estuaries. Warmer temperatures could leave estuaries too warm for some of the current inhabitants: In

Apalachicola Bay (Florida), for example, a warming of 7°F could cause several species to flee the bay on hot days for the cooler waters of the Gulf of Mexico, making them more vulnerable to predators; secile species such as clams and oysters would not have this option. The lower dissolved oxygen content resulting from warmer water could also lead to fish kills in some estuaries.

Approximately 50% of ocean fish, including shrimp, menhaden, flounder, sea trout, croker and red drum spend most of their lives in the ocean, but spawn in estuaries. As climate warms, these fish might be able to migrate north, as long as an estuarine environment remains available. The loss of wetlands from rising sea level may substantially diminish the critical coastal habits of shrimp and other fish that rely on the marsh. Moreover, scientists are unsure whether these fish would be able to shift to more northern estuaries. Some species tend to automatically seek the estuary in which they were born; as the water warms they may therefore tend to return to the same estuary even if it was inhospitably warm, rather than seek cooler waters. On the Pacific Coast, the small number of estuaries would further increase the difficulty of finding a cooler estuary if, for example, global warming were to leave San Francisco Bay too warm for the oceangoing fish that spawn there. See also **Ocean Resources (Living)**.

Ocean Fisheries. Scientists generally expect fish on the high seas to be less affected by global warming than coastal and inland fisheries. The year-to-year variations in climate conditions appear to be much greater than the change expected from greenhouse gases over the next century. Nevertheless, oceanographers are currently unable to rule out the possibility that global warming may exacerbate El Nino and other causes of fluctuation. See also **El Niño**.

Warmer temperatures are likely to enhance fishing in many areas. Overall biological activity is greater at higher temperatures, more food is available, fish grow faster, and they reproduce at a younger age. The expected increase in fisheries from warmer temperatures may be partly offset, however, by a decline in the upward flow of deep ocean water to the surface (upwelling). These upward flows bring nutrients to the upper layers of the ocean, increasing the growth of the aquatic plants that form the base of the marine food chain.

Birds. Climate change is likely to have both direct and indirect effects on birds. Higher temperatures can directly alter their life cycles. The loss of wetlands, beaches, and other habitat could have an equally important indirect effect, by making some regions less hospitable to birds than those regions are today.

As temperatures warm, birds will tend to inhabit more northerly areas (in the Northern Hemisphere). Data collected by the National Audubon Society's Christmas Bird Count show that during years with warmer temperatures, the majority of the bird species do not have to fly as far south for the winter. Warmer temperatures also allow birds to spend their summers farther north.

Warmer temperatures can also affect how birds respond to the change in seasons. Several types of birds that fly north to Michigan during spring now arrive two or three weeks earlier than in 1960. Scientists at the British Trust for Ornithology have found that 20 of 65 species of birds are laying their eggs an average of 9 days earlier today than in 1971. The earlier nesting appears to result in part because plants are flowering and growing leaves sooner, which in turn causes earlier availability of the insects that these birds eat.

Scientists do not know whether birds will benefit from these changes. Earlier nesting means that birds will be a week or so older when the time come to migrate south, which may improve their odds of survive their first winters. The changing climate, however, may impair the extent to which a bird's life cycle is synchronized with its food supply. While birds can adjust to warmer temperatures by flying to more northern areas in any given year, the vegetation upon which they (or the insects they eat) rely may take decades or longer to adjust.

National Parks and Public Lands. Approximately 30% of the nation's land is owned by the public. In the 11 contiguous states west of the 100th meridian, approximately 50% is owned by the federal government, including 80% of Nevada. About two-thirds of the land in Alaska is owned by the federal government. Even the east and Midwest, however, have large areas of publicly owned land, particularly Louisiana (20%), New Hampshire (13%), Florida (10%), Michigan (10%) and Virginia (7%). Although most of these public lands are owned by the federal government, many states also have large parks and state forests; and most coastal wetlands below mean high water are owned by the state.

The diversity of National Parks and other public lands mirrors the diversity of the nation from which these lands are drawn. As a result, global warming will have the same types of impacts on these lands as occur in areas that are not owned by the government. Sea level rise will tend to erode and inundate the beaches of the National Seashores and the wetlands of various National Wildlife Refuges and National Parks in coastal areas. Regional climate change combined with the fertilizing effect of CO_2 in the atmosphere will have the same effect on forests within National Parks and National Forests as occur in other forests. The intensification of evaporation and precipitation will tend to increase the frequency during which Wild and Scenic Rivers experience either extreme floods or extremely low flows of water.

State Impacts. See http://www.epa.gov/globalwarming/impacts/stateimp/index.html

International Impacts. The effects of global warming and a changing climate will not be felt equally across our planet. Regional climate changes will likely be very different from changes in the global average. Differences from region to region could be in both the magnitude and rate of climate change. Furthermore, not all things, whether they be natural ecosystems or human settlements, are equally sensitive to changes in climate. And finally, nations (and indeed regions within nations) vary in their ability to cope and adapt to global warming and a changing climate.

With that said, some nations will likely experience more adverse effects than others, while other nations may benefit more than others. Poorer nations are generally more vulnerable to the consequences of global warming. These nations tend to be more dependent on climate-sensitive sectors, such as subsistence agriculture, and lack the resources to buffer themselves against the changes that global warming may bring. The Intergovernmental Panel on Climate Change (IPCC) has identified Africa as "the continent most vulnerable to the impacts of projected changes because widespread poverty limits adaptation capabilities."

For more detailed information on how global warming may impact different regions across the globe, see the IPCC's special report, edited by R.T. Watson, M.C. Zinyowera, and R.H. Moss.

Actions

Today, action is occurring at every level to reduce, to avoid, and to better understand the risks associated with climate change. Many cities and states across the country have prepared greenhouse gas inventories; and many are actively pursuing programs and policies that will result in greenhouse gas emission reductions.

At the national level, the U.S. Global Change Research Program (USGCRP) coordinates the world's most extensive research effort on climate change. In addition, EPA and other federal agencies are actively engaging the private sector, states, and localities in partnerships based on a win–win philosophy and aimed at addressing the challenge of global warming while, at the same time, strengthening the economy.

At the global level, countries around the world have expressed a firm commitment to strengthening international responses to the risks of climate change. The U.S. is working to strengthen international action and broaden participation under the auspices of the Framework Convention on Climate Change.

International. Climate change is a global problem requiring action from the entire international community. Countries from around the world are working together to share technologies, experience, resources and talent to lower net greenhouse gas emissions and reduce the threat of global climate change. The United States participates in and supports several international efforts designed to help countries to address climate change.

One important strategy for reducing global greenhouse gas emissions is developing and sharing climate-friendly technologies, commonly referred to as *Technology Cooperation.* See also http://www.epa.gov/globalwarming/actions/international/techcoop/index.html. These efforts can occur between nations, private entities, and organizations around the world. The United States participates in various bilateral and multilateral technology cooperation initiatives that aim to encourage the use of technologies that will reduce greenhouse gases. Through the U.S. Initiative on Joint Implementation, organizations in the United States and other countries have been encouraged to implement projects that reduce, avoid, or sequester greenhouse gas emissions. In addition, the *U.S. Countries Studies Program* has provided developing countries

and countries with economies in transition with funding and technical assistance to support greenhouse gas inventories, mitigation assessments, vulnerability and adaptation assessments and national action plans for addressing climate change. See also http://www.epa.gov/globalwarming/actions/international/countrystudies/index.html. Finally, international efforts are underway to establish guidelines for *Land Use, Land Use Change, and Forestry* practices that reduce greenhouse gas emissions and increase carbon sinks. See also http://www.epa.gov/global-warming/actions/international/landuse/index.html.

As countries continue to grow and develop, international cooperation will become increasingly important as the global community searches for ways to meet the climate change challenge efficiently and effectively. The key to successful cooperation is finding activities that will help all countries achieve their economic, environmental, and developmental goals in a climate-friendly manner.

National Level. In response to mounting concern over the potential risks posed by global warming, the Framework Convention on Climate Change was opened for signature in Rio di Janeiro at the United Nations Conference on Environment and Development (also referred to as the "Earth Summit") in June 1992. More than 150 nations, including the United States, signed the treaty, which entered into force less than two years later, on March 21, 1994. It has now been ratified by more than 155 nations.

The U.S. was the fourth nation overall, and the first industrialized nation, to ratify this landmark accord. Under this treaty, the world's industrialized nations pledged to establish policies and measures that reduce emissions of the greenhouse gases that are changing the Earth's climate.

See also **Arrhenius, Svante**; **Atmosphere (Earth)**; **Climate**; and **Earth**; **Global Change**; and **Global Temperature**.

Additional Reading

Adler, T.: "Climate Change May Make Insects Winners," *Science News*, **145**(15), 230 (1994).

Boersma, P.D.: "Population Trends of the Galapagos Penguin: Impacts of El Nino and La Nina," *Condor*, **100**, 245–253. (1998).

Cameron, G.C. and D. Scheel: "A GIS Model of the Effects of Global Climate Change on Mammals," *Geocarto International*, **4**, 19–32 (1993).

Dobson, A., A. Jolly, and D. Rubenstein: "The Greenhouse Effect and Biological Diversity," *Trends Ecol. Evol.*, **4**(3), 64–68 (1989).

Docherty, M.: "Does Man Play Dice with the Environment? Insect-tree Interactions and Environmental Change," *Antenna*, **20**(3), 105–109 (1996).

Graham, R.W. and E.C. Grimm: "Effects of Global Climate Change on the Patterns of Terrestrial Biological Communities," *Trends Ecol. Evol.*, **5**(9), 289–292 (1990).

Gunn, A. and T. Skogland: "Responses of Caribou and Reindeer to Global Warming," *Ecological Studies*, **124**, 189–200 (1997).

Harvey, D.D.: *Global Warming; The Hard Science*, Prentice-Hall, Inc., Upper Saddle River, NJ, 1999.

Hocking, C., C. Sneider, and L. Bergman: *Global Warming and the Greenhouse Effect*, University of California, Berkeley, Hall of Science, Berkeley, CA, 1999.

Hoffmann, A.A. and M.W. Blows: "Evolutionary Genetics and Climate Change: Will Animals Adapt to Global Warming?" In *Biotic Interactions and Global Change*, P.M. Kareiva, J.G. Kingsolver, and R.B. Huey, eds., Sinauer Associates, Sunderland, MA, 1993.

Hughes, J.M. and K.A. Evans: "Global Warming and Pest Risk Assessment," *Aspects of Applied Biology*, **45**, 339–342 (1996).

Huntley, B.: "Plant Species' Response to Climate Change: Implications for the Conservation of European Birds," *Ibis*, **137**(Sup. 1), S127–S138 (1995).

Jutro, P.R.: "Biological Diversity, Ecology, and Global Climate Change," *Environmental Health Perspectives*, **96**, 167–170 (1991).

La Roe, T.: "The Effects of Global Climate Change on Fish and Wildlife Resources," *Trans. North Am. Wildl. Nat. Resour. Conf.*, **56**, 171–176 (1991).

Larson, D.L.: Effects of Climate on Numbers of Northern Prairie Wetlands," *Climatic Change*, **30**, 169–180 (1995).

Lester, R.T. and J.P. Myers: "Double Jeopardy for Migrating Wildlife," In R.L. Wyman, ed., *Global Climate Change and Life on Earth*, Chapman and Hall, New York, NY, 1991, pp. 119–133.

Malcolm, J.R. and A. Markham: *Climate Change Threats to National Parks and Protected Areas in the United States and Canada*, World Wildlife Fund, Washington, DC, 1997.

Parmesan, C.: "Climate and Species Range," *Nature*, **382**, 765–766 (1996).

Philander, G.S.: *Is the Temperature Rising?: The Uncertain Science of Global Warming*, Princeton University Press, New York, NY, 2000.

Reid, W.V. and M.C. Trexler: "Responding to Potential Impacts of Climate Change on U.S. Coastal Biodiversity," *Coastal Management*, **20**(2), 117–142 (1992).

Shvarts, E.A., S.V. Pushkaryov, V.G. Krever, and M.A. Ostrovsky: "Geography of Mammal Diversity and Searching for Ways to Predict Global Changes in Biodiversity," *Journal of Biogeography*, **22**, 907–914 (1995).

Staff: Environmental Protection Agency, "Ecological Impacts from Climate Change: An Economic Analysis of Freshwater Recreational Fishing," *U.S. Environmental Protection Agency Policy*, Planning and Evaluation Report #220-R-95-004, Washington, DC, 1995.

Tramer, E.J.: "Global Warming: An Imminent Threat to Birds?" *Living Bird*, **11**(2), 8–12 (1992).

Veit, R., et al.: "Apex Marine Predator Declines 90% in Association with Changing Oceanic Climate," *Global Change Biology*, **3**, 23–28 (1997).

Watson R.T., M.C. Zinyowera, and R.H. Moss, eds., *The regional Impacts of Climate Change—An assessment of values ability*, Cambridge University Press, 1988.

GLOBAR (or Globar Lamp). A ceramic rod consisting largely of silicon carbide (carborundum) which has some electrical conductivity at room temperature and which can be heated to an almost white heat in air without rapid deterioration. It radiates almost like a black body. Globars are used as a radiation source like the Nernst glower in infrared spectrometers.

They have the advantage over Nernst glowers of not requiring a secondary heat source for starting and in being more rugged. However, they cannot be made as small as Nernst glowers and, in general, some sort of cooling device, such as a water jacket, is necessary.

GLOBULINS. Proteins that are insoluble in water, but that dissolve readily in aqueous salt solutions. The term globulins is applied to certain subgroups of the plasma proteins. See also **Antibody**; and **Blood**.

GLOMERATE. The textural term, proposed by R.M. Field, for a sedimentary rock with a coarse and poorly graded texture, when the origin of the shape of the larger constituents has either been undetermined or is indeterminable.

GLORY. Small, faintly colored rings of light surrounding the antisolar point, seen when looking down at a water cloud. Having a radius of only a few degrees, the glory often surrounds an airplane's shadow cast on a cloud or a mountain climber's shadow cast on fog in a valley. (The shadow of the observer plays no role in the phenomenon other than as easy way of quickly finding the antisolar point.)The glory is not as easily described by simple theory as is the corona. Nevertheless, some similarities hold: the angular size of a particular ring is approximately inversely proportional to drop size. The result is that glories are formed by droplets with radii smaller than about 25 μm (the rings from larger droplets are washed out by the angular width of the sun). Similarly, a broad droplet distribution will destroy the glory.

GLORY OBSERVATORY MISSION.. See **Earth Observing System (EOS)**.

GLOSSITIS. An inflammation of the tongue resulting from nutritional deficiencies or bacterial infections. Taste buds disappear and the tongue becomes smooth and shiny. The condition may indicate pernicious anemia and vitamin B deficiencies.

GLOSSMETER. An instrument for measuring the ratio of the light regularly or specularly reflected from a surface, to the total light reflected.

GLOW WORM (*Insecta, Coleoptera*). Wingless females of certain beetles. They resemble larvae throughout life and are luminous.

Glow worm also refers to the larvae of the firefly.

GLUCOSE. See **Carbohydrates**; **Starch**; and **Sweeteners**.

GLUTAMINE. See **Amino Acids**.

GLUTEN. See **Starch**.

GLYCEROL. [CAS: 56-81-5] Glycerol, propanetriol, glycyl alcohol, "glycerine," $CH_2OH \cdot CHOH \cdot CH_2OH$, is a colorless, viscous liquid, of sweetish taste, odorless, boiling point 290°C. Glycerol reacts (1) with phosphorus pentachloride to form glyceryl trichloride, $CH_2Cl \cdot CHCl \cdot CH_2Cl$, (2) with acids to form esters, e.g., glycerol monoacetate $CH_2OH \cdot$

$CHOH \cdot CH_2OOCCH_3$, glycerol diacetate $C_3H_5(OH)(OCOCH_3)_2$, glycerol triacetate (triacetin), $CH_2OOCCH_3 \cdot CHOOCCH_3 \cdot CH_2OOCCH_3$, glycerol mononitrates (alpha, $CH_2OH \cdot CHOH \cdot CH_2ONO_2$; beta, $CH_2OH \cdot CHONO_2 \cdot CH_2OH$), glycerol dinitrates (1, 2, $CH_2OH \cdot CHONO_2 \cdot CH_2ONO_2$; 1, 3, $CH_2ONO_2 \cdot CHOH \cdot CH_2ONO_2$), glyceryl trinitrate ("nitroglycerine"), $CH_2ONO_2 \cdot CHONO_2 \cdot CH_2ONO_2$, glyceryl tristearate (tristearin), $CH_2OOCC_{17}H_{35} \cdot CHOO\text{-}CC_{17}H_{35} \cdot CH_2OOCC_{17}H_{35}$, indirectly, glycerol monophosphates (alpha, $CH_2OH \cdot CHOH \cdot CH_2OPO(OH)_2$, beta, $CH_2OH \cdot CHOPO(OH)_2 \cdot CH_2OH$, (3) with oxidizing agents, e.g., dilute nitric acid, to form glyceric acid, $CH_2OH \cdot CHOH \cdot COOH$, tartaric acid, $COOH \cdot CHOH \cdot COOH$, mesoxalic acid, $COOH \cdot CO \cdot COOH$, (4) with phosphorus plus iodine, to form allyl iodide, $CH_2 : CHCH_2I$, which with hydrogen iodide yields propylene, $CH_2 : CHCH_3$, and then iso-propyl iodide, CH_3CHICH_3, (5) with sodium or sodium hydroxide to form alcoholates, (6) with sodium hydrogen sulfate or phosphorus pentoxide heated, to form acrolein, $CH_2 : CHCHO$. Glycide alcohol is obtained by treatment of glycerol alphamonochlorohydrin

$$CH_2OH \cdot \underset{\underset{O}{\rfloor}}{CH} \cdot CH_2$$

$CH_2OH \cdot CHOH \cdot CH_2Cl$, which is made by reaction of hypochlorous acid and allyl alcohol with barium hydroxide. With hydrogen chloride, glycide alcohol yields epichlorohydrin

$$CH_2Cl \cdot \underset{\underset{O}{\rfloor}}{CH} \cdot CH_2$$

Glycerol may be detected by the characteristic odor of acrolein, found on heating with potassium bisulfate.

Glycerol is used (1) in the manufacture of high explosives, e.g., glyceryl trinitrate ("nitroglycerin"), which is the main component of dynamite, (2) in antifreeze solutions, especially for automobile radiators, (3) to maintain a moist condition in fruits and tobacco, (4) in cosmetics and skin preparations, and (5) to prepare glycerol phosphoric acid, used in medicine, and "boroglyceride" used as a preservative. See Table 1.

TABLE 1. CHARACTERISTICS OF WATER SOLUTIONS OF GLYCEROL

% Glycerol by Weight	Specific Gravity (15.6°C/60°F)	Freezing Point, °C
20	1.049	−5.0
40	1.103	−15.6
60	1.158	−34.0

GLYCOL. [CAS: 107-21-1] A dihydric alcohol (i.e., a compound containing two alcoholic hydroxyl groups). The chemical properties are represented by those of the simplest members of the class, ethylene glycol, 1,2-ethanediol, $CH_2OH \cdot CH_2OH$, which is a colorless, viscous liquid, of sweetish taste, odorless, boiling point 197°C, miscible in all proportions with water or alcohol, slightly soluble in ether. Like ethyl alcohol, ethylene glycol is often called by the class name.

Glycol reacts (1) with sodium to form sodium glycol, $CH_2OH \cdot CH_2ONa$, and disodium glycol, $CH_2ONa \cdot CH_2ONa$; (2) with phosphorus pentachloride to form ethylene dichloride, $CH_2Cl \cdot CH_2Cl$ (3) with carboxy acids to form mono- and disubstituted esters, e.g., glycol monoacetate, $CH_2OH \cdot CH_2OOCCH_3$, glycol diacetate, $CH_3COOCH_2 \cdot CH_2OOCCH_3$; (4) with nitric acid (with sulfuric acid), to form glycol mononitrate, $CH_2OH \cdot CH_2ONO_2$, glycol dinitrate, $CH_2ONO_2 \cdot CH_2ONO_2$; (5) with hydrogen chloride, heated, to form glycol chlorohydrin (ethylene chlorohydrin, $CH_2OH \cdot CHCl$); (6) upon regulated oxidation to form glycollic aldehyde, $CH_2OH \cdot CHO$, glyoxal, $CHO \cdot CHO$, glycollic acid, $CH_2OH \cdot COOH$, glyoxalic acid, $CHO \cdot COOH$, oxalic acid, $COOH \cdot COOH$.

Glycol is made by reaction of ethylene and chlorine or hypochlorous acid to form ethylene dichloride or ethylene chlorohydrin, respectively, followed by treatment of either of these with sodium carbonate solution heated under pressure. Glycol is also formed when ethylene is treated with potassium permanganate.

Glycol is used (1) in antifreeze solutions, especially for automobile radiators; (2) in the preparation of ethers and esters, especially nitrate for explosive; (3) as a solvent substitute for glycerol.

See Table 1.

TABLE 1. CHARACTERISTICS OF WATER SOLUTIONS OF GLYCOL

% Glycerol by Weight	Specific Gravity (15.6°C/60°F)	Freezing Point, °C
17	1.026	−6.7
32.5	1.048	−17.8
44	1.063	−28.9

GLYCOLYSIS. A series of about 10 enzyme-stimulated reactions in which glucose is broken down into pyruvic acid in cell respiration. No oxygen is needed for glycolysis and it is used as the sole energy source for anaerobic organisms. In aerobic metabolism, however, the pyruvic acid is then taken through the tricarboxylic acid (TCA) cycle, and the balance of the energy is extracted. It appears that glycolysis can take place free in the cytoplasm, but that the tricarboxylic acid cycle must take place within the mitochondria of the cell.

Glycolysis was defined in the late 1920s by Otto Warburg as "the splitting of carbohydrate into lactic acid." This type of lactic acid fermentation was well known to Berzelius, Liebig, Pasteur, and Claude Bernard in the mid-1800s, as was also alcoholic fermentation. Various kinds of carbohydrates may serve as substrates for glycolysis. It is remarkable that although glycolysis is the sum of a very large number of consecutive intermediate compounds, enzymes, and coenzymes, knowledge of these components and their sequences was acquired many years ago. For most animal cells studied, the biochemical sequence from glucose may be summarized as shown in Table 1.

TABLE 1. EMBDEN-MEYERHOF PATHWAY

Step	Product	By way of
	Glucose (start)	
1	D-Glucose	Glucokinase, ATP, Mg^{2+}, insulin: anti-insulin regulators
2	D-Glucose-6-phosphate	Phosphoglucoisomerase
3	D-Fructose-6-phosphate	Phosphofructokinase, ATP, Mg^{2+}
4	D-Fructose-1,6-diphosphate	Fructaldose
5	D-Glyceraldehyde-3-phosphate	Glyceraldehyde-3-phosphate dehydrogenase, DPN, $HOPO_3^{2-}$
6	1,3-Diphospho-D-glycerate	3-Phosphoglycerate kinase, ADP, Mg^{2+}
7	3-Phospho-D-glycerate	Phosphoglycerate mutase, Mg^{2+}
8	2-Phospho-D-glycerate	Enolase, Mg^{2+}
9	Phosphoenolpyruvate	Pyruvate kinase, ADP, Mg^{2+}
10	Pyruvate	Pyruvate reductase = lactate dehydrogenase, $DPNH_2$
11	L-Lactate	

The splitting of sugar to lactic acid is thus, briefly, the shifting of hydrogen by means of the nicotinamide moiety of diphosphopyridine nucleotide (also termed nicotinamide adenine dinucleotide). Nicotinamide in DPN takes away two atoms of hydrogen from phosphorylated carbohydrate, and after dephosphorylation gives back two hydrogens (in $DPNH_2$) to pyruvic acid.

The biochemical importance of the foregoing sequence in glycolysis is at least twofold. (1) Each one of the intermediate compounds formed leads to one or more important possible side reactions. These, in turn, lead to innumerable reactions indispensable to life processes, including respiration. (2) In the entire sequence, and also in some of its parts, comparatively large amounts of free energy are made available — up to a maximum of 28,000 cal/mole lactate formed under common *in vivo* conditions from one-half mole of glucose. This free energy available is considerably larger than the approximately 9,000 calories free energy available from hydrolysis

of the high-energy ATP to ADP and inorganic phosphate, although much smaller than the free energy of combustion of a mole of lactate to carbon dioxide and water, some 332,000 calories. Whereas the free and heat energies of combustion of lactate are nearly equal, lactic acid fermentation from glucose represents an instance of the relatively rare situation in which the free energy liberated is considerably greater (about 50%) than the heat energy liberated, owing to the large entropy change involved in the formation of the additional carbonyl (=C=O) bond in two lactates derived from one glucose molecule.

The foregoing reaction sequence, commonly called the Embden-Meyerhof pathway after its initial investigators, was in due course worked out in greater detail by Warburg. This pathway is also common to ethyl alcohol fermentation down to the pyruvate stage, which then branches off (via carboxylase) to form acetaldehyde and finally (via alcohol dehydrogenase, $DPNH_2$) to ethanol. Alcoholic fermentation is sometimes erroneously referred to as glycolysis. Ordinary respiration, by this same reasoning, could be called glycolysis, since it too shares the common pathway down to pyruvate. Just as lactate fermentation is the most common fermentation met with in animal cells, so alcoholic fermentation is the most common fermentation met with in plant cells, a distinction most easily observed under anaerobic conditions.

See also **Carbohydrates**.

GLYCOSIDES (STEROID). In plants, steroids occur as glycosides, as acyl glycosides, as esters, and in the free form. Many of the steroid glycosides are important drugs or starting materials for the partial synthesis of drugs.

Sterolins

Although the sterols are largely in the free and esterified form, plants contain significant quantities of sterolins (steryl glycosides and acyl glycosides). As a rule, the 3-hydroxyl group of the sterol is linked to glucose or some other common sugar to form a heteroside, but other hydroxyl groups in the sterol molecule and higher saccharides may be involved. The two most common sterol aglycones in higher plants are sitosterol [CAS: 83-46-5] (previously called β-sitosterol) and stigmasterol [CAS: 83-48-7], shown in Fig. 1. Cholesterol [CAS: 57-88-5] is usually present in very small amounts. See also **Cholesterol**. The biosynthesis of sitosterol differs from that of cholesterol in the alkylation of a precursor at C-24 and involves the successive introduction of two methyl groups. Stigmasterol is formed by the dehydrogenation of sitosterol.

Cholesterol and the C_{29} sterols are used by plants as the starting materials for the biosynthesis of other steroids with the same number of carbons, such as the insect-molting hormones and the steroidal sapogenins and alkaloids or they are converted to progesterone and other steroids with a lower number of carbons. Fig. 1 shows the structure of one representative of the insect-molting hormones, ponasterone A, which has been isolated from plants in the form of its 3-glucoside, ponasteroside A.

Steroidal Saponins and Glycoalkaloids

Certain plants have the ability to hydroxylate cholesterol stereospecifically at C-26 or C-27. A glycoside of cholesterol with a glucose residue attached to this terminal hydroxyl group and chacotriose (2L-rhamnoses + D-glucose) attached to its 3-hydroxyl group is a biogenetic precursor of the saponin, dioscin, shown in Fig. 1. Steroidal saponins are glycosides of spiroketals, which form spontaneously, when the terminal sugar in an analogous 16-hydroxy-22-keto-steroid is enzymatically removed. While the configuration at C-22 is the same in all natural sapogenins, the orientation of the methyl group at C-25 depends on the position of the terminal hydroxyl group in the sterol precursor. In the D- or isosapogenin series the methyl group is α-oriented (equatorial), as in diosgenin (see Fig. 1), whereas in the L-, normal, or neosapogenin series it is β-oriented (axial).

Sapogenins are widely distributed in monocots belonging to the genera *Yucca, Trillium, Chlorogalum, Smilax, Nolina, Agapanthus, Agave, Manfreda,* and *Dioscorea,* and in dicots belonging to the genera *Digitalis, Solanum, Lycopersicon,* and *Cestrum.* Diosgenin (Fig. 1) from Mexican barbasco root (*Dioscorea* tubers) and hecogenin (Fig. 1) from wastes of African sisal fibers (*Agave sisalana*) are important starting materials for the commercial preparation of synthetic hormones.

The glycoalkaloids are nitrogen analogs of the saponins, occurring in the Solanaceae. Two of their aglycones are shown in Fig. 2. Tomatidine

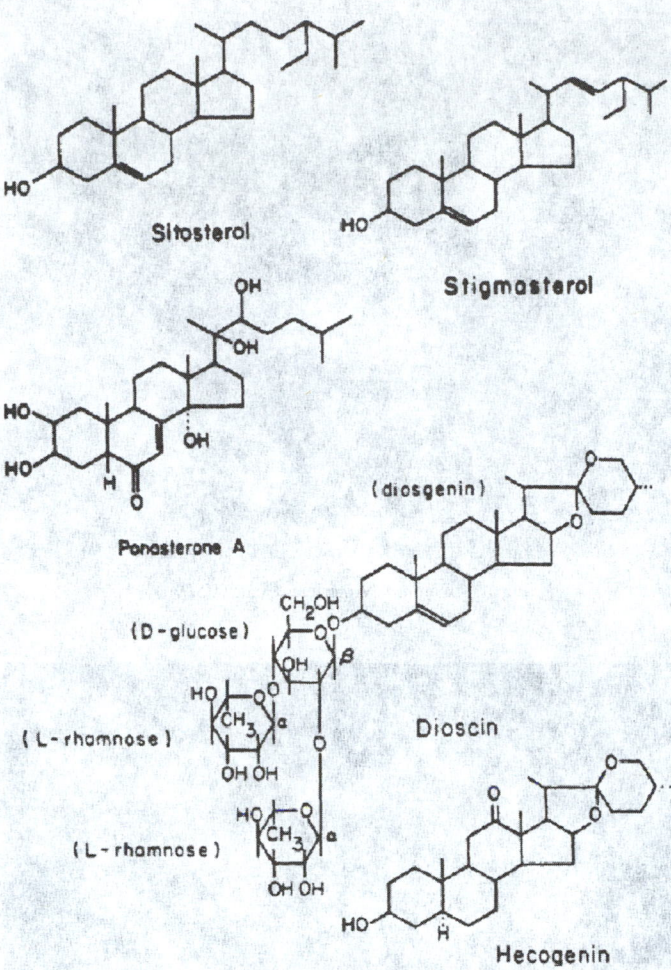

Fig. 1. Steroidal saponins and related compounds.

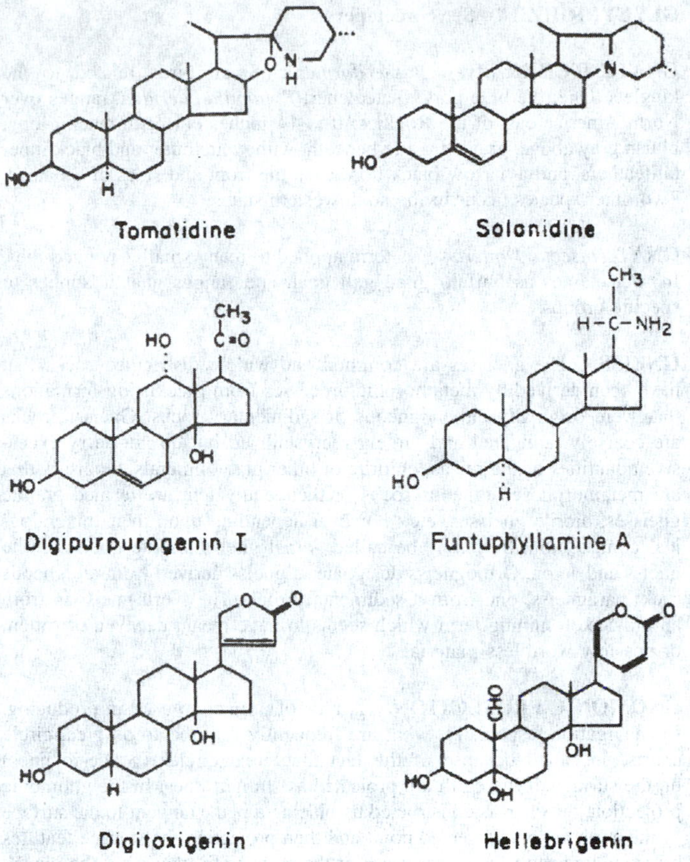

Fig. 2. Representative glycoalkaloids occurring in Solanaceae.

occurs in the form of a glycoside, tomatine, in tomato vines and may also be used for the partial synthesis of steroid hormones. Solanidine is found in potatoes and other *Solanum* species in the form of various glycosides, the solanines and chaconines. For instance α-chaconine differs from dioscin (Fig. 1) only in the nature of the aglycone. The glycoalkaloids are likewise synthesized by plants from cholesterol.

Saponins and glycoalkaloids are characterized by their surface activity and hemolytic effect as well as their ability to form complexes with cholesterol and similar sterols. The best-known cholesterol-precipitating agent is digitonin, a saponin in *Digitalis*. While ingested saponins are nontoxic to warmblooded animals, the glycoalkaloids are toxic. Tomatine and other glycoalkaloids have antifungal and cytostatic activity.

Pregnane Derivatives

The degradation of cholesterol in plants and animals produces pregnenolone (δ^5-pregnen-3β-ol-20-one), which is oxidized to progesterone [CAS: 57-83-] (Δ^4-pregnene-3,20-dione). Various plants contain neutral pregnane derivatives with C and D rings in cis-fusion in the form of glycosides. They have been called digitanol glycosides, because they were first isolated from *Digitalis* plants, and contain the rare hexoses otherwise only found in the cardiac glycosides. Fig. 2 shows an example of the digitanols, digipurpurogenin I, which occurs in *Digitalis purpurea* as digipurpurin, a glycoside in which three molecules of D-digitoxose are attached to the 3-hydroxyl group.

Plants belonging to the Apocynaceae and Buxaceae aminate the pregnane derivatives to produce alkaloids, which may be present as either glycosides or esters. Only one example of Apocynaceae alkaloid is shown in Fig. 2, funtuphyllamine A, which was isolated from *Funtumia africana*. The alkaloids of kurchi bark (*Holarrhena antidysenterica*) have many interesting pharmacological properties and one of them, conessine, is used as an amebicide and as a starting material for the partial synthesis of aldosterone.

Cardiac Glycosides

About eleven plant families are known to elaborate cardiac glycosides. Their genins have either 23 (cardenolides) or 24 (bufadienolides) carbon atoms and their sugars are not found elsewhere in nature. Cardenolides and bufadienolides have not been found together in the same genus. Both types of genins are synthesized in plants from a C_{21} steroid, usually progesterone, and contain a 14β-hydroxyl group.

Fig. 2 gives one example of a cardenolide, digitoxigenin, and one representative of the bufadienolides, hellebrigenin. *Digitalis* plants contain three cardenolides: digitoxigenin [CAS: 143-62-4], gitoxigenin [CAS: 545-26-6] (16β-hydroxy digitoxigenin), and digoxigenin [CAS: 1672-46-4](12β-hydroxy-digitoxigenin). These genins are combined with 2 molecules of digitoxose, 1 molecule of acetyldigitoxose, and 1 molecule of glucose to form the lanatosides (digilanides) A, B, and C, respectively. When the acetyl group is removed by mild alkaline hydrolysis, one obtains the corresponding purpurea glycosides (desacetyllanatosides or desacetyldigilanides). Enzymatic removal of the glucose unit, on the other hand, gives the acetyl derivatives of the digitoxose triosides. Combination of both hydrolytic procedures yields the digitoxose triosides digitoxin [CAS: 71-63-6], gitoxin [CAS: 4562-36-1], and digoxin [CAS: 20830-75-5].

The bufadienolides occur in plants as well as animals, but only in plants are they in the form of glycosides. Their 3-hydroxyl group is attached to glucose, rhamnose [CAS: 3615-41-6], or thevetose. Hellebrigenin (Fig. 2), which is also known as bufotalidin, occurs in the rhizomes of the Christmas rose and other *Helleborus* species in the form of a rhamnoside. Bufadienolides have so far been found in plants in only two families, the buttercup and the lily family.

Crude leaf preparations of *Digitalis* have been in medical use since 1785. Pure cardiac glycosides are now available. These preparations in injectable tinctures or powdered leaf tablets are used extensively for the treatment of congestive heart failure. They increase the force of the heart muscle and the power of systolic contraction, apparently by inhibiting the active transport of K^+ and Na^+ ions through cell membranes.

See also **Steroids**.

GLYCYRRHIZIN.   See **Sweeteners.**

GNATCATCHER *(Aves, Passeriformes).*   Small birds related to the kinglets. One, the blue-gray gnatcatcher, *Polioptila caerulea,* ranges over North America east of the Rockies. It is $4\frac{1}{2}$ inches (11 centimeters) long, bluish gray above, grayish white beneath, with white outer and black inner tailfeathers, and a narrow black border on the front and sides of the head. Two other species occur in the southwestern states.

GNAT *(Insecta, Diptera).*   A term applied to many small 2-winged flies. In such names as buffalo gnat, gall gnat, and fungus gnat it applies to specific groups.

GNEISS.   The gneisses are common and widely distributed rocks that have been derived by metamorphic processes from preexisting formations that were originally either igneous or sedimentary rocks. Gneissic rocks are coarsely laminated and largely recrystallized but do not carry excessive quantities of the micas, chlorite or other platy minerals. Gneisses that are metamorphosed igneous rocks or their equivalent are termed granite gneisses, diorite gneisses, etc.; however depending upon their mineralogical composition, they may be called garnet gneiss, biotite gneiss, albite gneiss and so on. Orthogneiss designates a gneiss derived from an igneous rock; paragneiss, one from a sedimentary rock. The word gneiss is from an old Saxon mining term which seems to have meant decayed or rotten, or possibly worthless material.

GNOMONIC PROJECTION.   A type of projection used in producing, for navigation, especially, what are frequently referred to as great-circle charts, so called because of the fact that great circles (geodesic lines) on the surface of the earth are projected as straight lines. In the gnomonic projection, the chart is constructed by placing a plane tangent to the surface of the earth at some selected point and then projecting the surface features by extending radii from the center of the earth until they meet the plane.

In the gnomonic projection, the distortion of both shape and size is very severe except for a very limited area immediately about the point of tangency with the earth. The great value of the charts lies in the fact that the shortest distance, even between very widely separated points, will be projected as a straight line. A series of charts are available on this type of projection for all the principal cruising areas of the world, and they are of immense value to navigators for determining at a glance whether or not the following of the shortest course between two points (great-circle course) is practicable. See also **Great-Circle Course.**

GNU.   See **Antelope.**

GOATS AND SHEEP *(Mammalia, Artiodactyla).*   The goats and sheep *(Caprines)* comprise a significant group in the order *Artiodactyla* (eventoed hoofed mammals). Because of the general familiarity with the domesticated goat, this description will start with that animal. As is true of the dog, sheep (discussed later in this description), and several other domesticated animals, the ancestry of the "farm yard" goat is not entirely clear. See Fig. 1. No creatures exactly like or even very closely resembling the domesticated goat exist in the wild today. All known wild species may be described as being exaggerated forms and these appear among the Tahrs, Markhors, Ibexes, and Turs. Some authorities believe that the domesticated goat is a descendant of *Capra aegagrus* and that these animals were Persian in origin. Variations in the domesticated goat now are generally identified in terms of the country from which they originally came—thus, Swiss goats; Nubians (from Egypt and north Africa); Indian goats; and Israeli and Syrian goats, etc.

Goats are, of course, very important commercially. They can produce very large quantities of milk. A Great Britain saanen (Swiss) goat is on record of having produced 6,400 pounds (2,903 kilograms) of milk in 365 days of lactation. In the United States, a saanen produced 4,900 pounds (2,223 kilograms) of milk, representing 150 pounds (68 kilograms) of butterfat, in 305 days of lactation. See Fig. 2. In California, a Nubian goat produced just under 4,250 pounds of milk, representing 185 pounds (84 kilograms) of butterfat, during a similar period. Where there are extremes of temperature (tropical or arctic), the milk from goats is considered superior to that from cows. The milk, pure white in color, is easily digested and is used for some infants and invalids,

Fig. 1.   Common American Goat. (*USDA.*)

Fig. 2.   Purebred Saanen buck goat. (*USDA.*)

as well as by people who are allergic to cow's milk. The curds are smaller, more soluble, and the fat globules are finer and more easily assimilated making homogenization usually unnecessary. Of course, cheese from goat milk is made on a high-tonnage basis, particularly in Europe.

The flesh of the goat is edible and, in particular, that of the young kids. The hair is used (mohair) and the skin is used for leather.

Goats produce a litter of two, although triplets are fairly common. The female is sometimes referred to as the "nanny" or doe and comes in heat once every three weeks. The gestation period is from 21 to 22 weeks. The life span of the goat ranges from 8 to 12 years.

Some authorities believe that the best breeding goats are Swiss. A majority of the French and German goats stem from Swiss stock, as do the goats in Scandinavia and the Netherlands where the goat is held in high esteem. See Fig. 3. The Maltese goat is considered to have blood strains of eastern goats.

Nubians are large goats with short legs, lop ears, "Roman" noses, and are short-haired. They are partially colored or spotted. Syrian goats have long hair and large lop ears, colored black with or without patches of white. Most goats found in Great Britain are fairly small with short legs, long hair, and gray in color. The breeding of fine goats, with the importation of excellent Swiss specimens, commenced in earnest in the United States in about 1910.

Fig. 3.   French Alpine buck goat. (*USDA.*)

TABLE 1. GENERAL ORGANIZATION OF THE GOATS AND SHEEP CAPRINES

GAZELLE-GOATS (*Saiginae*)	TRUE GOATS (*Caprinae*)
The Chiru (*Panthalops*)	Domesticates (*C. hircus*)
The Saiga (*Saiga*)	Markhors (*C. falconeri*)
	The Tur (*C. caucasica*)
	Ibexes (*C. ibex*)
ROCK-GOATS (*Rupicaprinae*)	Tahrs (*Hemitragus*)
The Goral (*Naemorhedus*)	
Serows (*Capricornis*)	
Chamois (*Rupicapra*)	SHEEP (*Ovinae*)
Rocky Mountain Goats (*Orlamnos*)	The Aoudad (*Ammotragus*)
	— Maned or Barbary Sheep
	The Bharal (*Pseudois*)
OX-GOATS (*Oriborinae*)	True Sheep (*Ovis*)
Takins (*Budorcas*)	— Argalis (*O. ammon*)
Muskox (*Oribos*)	— Mouflon (*O. musimon*)

Organization of the caprines is shown in Table 1. The following paragraphs of brief description follow the order of that table.

Gazelle-goats are not to be confused with the goat-gazelles which are described under **Antelope**. At one time, gazelle-goats were placed formally with the gazelles, but now are considered to be typical members of the caprines. The Chiru is of moderate size with long horns, ringed in the basal half, and is fawn-gray color with white underneath. The animal is somewhat sheep-shaped and lives on the high plateaus of Tibet. The male may be 30 inches (76 centimeters) at the shoulders with horns as long as the animal is tall. They weigh about 120 pounds (54.5 kilograms). The speed of the Chiru is faster than that of a dog or wolf, but not as fast as some antelopes. The males may have a harem of from 10 to 20 females at mating time and often, fierce battles take place among competing males. Mating occurs in the autumn and the fawns are born in May. The Chiru is held sacred by many Tibetans — they do not eat the flesh, but it has been reported as quite good. The Saiga is a rather ugly-appearing beast and considered somewhat clumsy. The animal is small and is found on the steppes of western Asia and eastern Europe. Its most conspicuous feature is the peculiarly swollen face with nostrils that point straight downward.

The Rock-goats are widely distributed over the Northern hemisphere. They like to climb and dwell in rocky country. Their performance in climbing and walking along narrow ledges and precipices has been described by some authorities as unbelievable. The Goral is fairly small, dull olive brown with backwardly curving horns, and prefers mountainsides, ranging from the Himalayas to Amuria and Korea. The Serow is widely distributed in eastern and southeastern Asia, inhabiting hilly or mountainous country, sometimes at an altitude of 12,000 feet (3,658 meters). Unofficially, they are sometimes called "goat-antelopes." The Chamois is mainly native to the barren mountain ranges of Europe — the Alps, Carpathians, and the Caucasus Mountains, preferring the edges of the tree line where it can scurry for protection when in danger. The animal, originally of Switzerland, is about the size of a male deer. The herds are usually small. The chamois has two small horns between the ears. The horns turn backward and are sharply pointed. The ears are long, alert, and tapered to a point. The tail is about 4 to 5 inches (10 to 12.5 centimeters) in length, and the face, back, and tail have black and white markings. The coat is chestnut brown in summer, turning to gray in the winter. The animal is timid and is protected from hunters by law in many localities. At one time, the widely used chamois skin was derived from this animal, but the product used today, if not synthetic, usually comes from kid, sheep, or buckskin. However, even today, for high performance, the original chamois skin is preferred particularly for drying off expensively decorated surfaces. The Rocky Mountain Goat is well distributed through the Rocky Mountains of the United States — from Alaska southward through Canada and into Montana. Unlike the gorals, these goats do not descend to the tree areas, but prefer remaining in the barren, rocky areas at all times. Their diet is stunted growth, mosses, and lichens.

Of all the caprines, the Ox-goats are the least goat-like in appearance and are considered carryovers from very early species. The Takin is a moderately large animal of heavy build, with strong curved horns. The animal ranges in mountainous country from the eastern Himalayas through Assam and northern Burma to eastern Tibet, and Szechwan, Kansu, and Shensi provinces in the People's Republic of China. They are not truly mountain animals, but prefer giant bamboo forests and thick woodlands.

The Muskox is about two-thirds as large as the American bison and is clothed with long shaggy hair. It has a thick coating of underwool. The horns are broad at the base, but become rapidly narrower as they curve downward from the forehead over the sides of the head. The more slender tips turn abruptly upward. The muskox lives on the treeless Arctic tundras and snowfields. The animals are hunted by the Eskimos for their hides and flesh. The animal population has shrunk in recent years and is no longer spread across northern Canada from Labrador to Alaska as it once was, but now is found in the vicinity of Hudson Bay and the Mackenzie River. Another herd is found on the islands to the north — from Banks Island in the west, eastward to Greenland. The muskox has huge feet and widely splayed hoofs; the legs are short and stout.

The muskox travels in herds of from 20 to 50 animals. When attacked by wolves or other predators, the animals form rings around the attacker(s) with their sharp horns pointing toward the center of the ring. This method of protection is considered quite effective. The muskox diets on scrub grasses, stunted growth, lichens, and mosses. The muskox is not to be confused with the musk deer; the latter is well known for the highly aromatic material it secretes, which is used as a fixative in perfumes.

Members of the true-goats, other than the domesticated varieties previously described, include the Markhors which are rather magnificent animals, standing proudly upright, with a heavy mane and beautifully twisted horns. The horns appear something like a twisted or spiral candle. There are several variations of the Tur. These animals inhabit the Caucasus Mountains. The Tur is of a rich-brown coloration, with short hair and a forwardly brushed beard and huge horns. There are several variations of the Ibex and they are distributed widely in locations of a mountainous nature. A Siberian ibex is shown in Fig. 4.

In briefly describing the subfamily of sheep, it should be noted that, as with the goat, it is difficult to look into the wildlife known today and to find what would seem to be a "wild" sheep of the type known so well by way of the millions of domesticated animals; or, in fact, to identify what appears to be an ancestor of the domestic sheep. However, zoologically, as indicated by the accompanying table, there are three broad classes of sheep. The Aoudad (also Udad), also known as the maned or Barbary sheep, is the only indigenous sheep of Africa and is found around the Sahara. The animals are powerful, with large and very thick horns, and are avid rock-climbers. See Fig. 5.

The true sheep have horns that resemble those of domestic breeds. The Argalis is found in east central Asia. The Mouflon is a wild sheep of Europe and inhabits the areas around the Mediterranean, including Corsica and Sardinia. They are reddish in coloration and quite distinguished, with large sweeping horns.

A group of wild sheep occurs in North America, with two distinct species. One of these is the Canadian, Rocky Mountain, or Bighorn sheep,

Fig. 4.   Siberian ibex. (*New York Zoological Society.*)

Fig. 5.   Udad (Barbary wild sheep). (*New York Zoological Society.*)

Fig. 6.   Bighorn sheep.

which is found from British Columbia southward to Lower California and as far east as the western Mexican mainland. Close relatives of this sheep are found in eastern Siberia. Another species, Dall's sheep, is closely related to the Bighorn. See Fig. 6.

The Sha is an Asiatic sheep with a very wide range. It is found from Iran into India and northward through Tibet. The name more widely used

Fig. 7.   Registered Rambouillet ram, 15 months old. (*American Rambouillet Sheep Breeders' Association.*)

Fig. 8.   Columbia yearling ram. (*The Columbia Sheep Breeders' Association of America.*)

is *urial*, the term *sha* applying to a large variety of sheep ranging from northern Tibet to Afghanistan. The sheep inhabits areas up to altitudes of 14,000 feet (4,267 meters).

Two of many domesticated breeds of sheep are shown in Figs. 7 and 8.

Additional Reading

Considine, D.M.: *Foods and Food Production Encyclopedia*, Van Nostrand Reinhold, New York, NY, 1982.

Haynes, N.: *Keeping Livestock Healthy: A Veterinary Guide to Horses, Cattle, Pigs, Goats and Sheep*, Storey Communications, Inc., Pownal, VT, 1994.

Maijala, K.: *Genetic Resources of Pig, Sheep and Goat*, Elsevier Science, New York, NY, 1991.

Smith, B.: *Large Animal Internal Medicine: Diseases of Horses, Cattle, Sheep, and Goats*, Mosby-Year Book, Inc., St. Louis, MO, 1996.

Web References

American Cheviot Sheep Society: http://members.aol.com/culhamef/bcheviots/cheviot.htm

American Dairy Goat Association–ADGA: http://www.adga.org/

American Dorper Sheep Breeders' Society: http://showcase.netins.net/web/sam/adsbs.htm

American Finnsheep Breeders' Association: http://www.finnsheep.org/

American Rambouillet Sheep Breeders Association: http://www.rambouilletsheep.org/

American Sheep Industry Association: http://www.sheepusa.org/

Blackbelly Barbados Sheep Association International: http://www.blackbellysheep. org/

Greater Wisconsin Rambouillet Association: http://www.applehollow.com/ rambouilletsheep/

Icelandic Sheep Breeders of North America: http://www.isbona.com/

Jacob Sheep Breeders Association: http://www.jacobsheep.com/jsba.htm

Katahdin Hair Sheep International: http://www.KHSI.org/

Montadale Sheep Breeders' Association: http://www.countrylovin.com/msba/index. html

Navajo-Churro Sheep Association: http://www.navajo-churrosheep.com/

North American Barbados Blackbelly Sheep Registry: http://members.aol.com/ barbadoshp/

North American Romanov Sheep Association: http://www.netwalk.com/%7Enarsa

The American Miniature Brecknock Hill Cheviot Sheep Breeders Association: http://myweb.accessone.com/~brighton/

U.S. Targhee Sheep Association: http://www.ttc-cmc.net/%7Eschuldt/

GOBIES (Osteichthyes). Of the suborder *Gobioidea* and family *Gobiidae*, there are numerous species of gobioid fishes. They are characterized by a sucker which is present under the forward part of the body, and by two dorsal fins. Some of the over 400 species are quite small, ranging from $\frac{1}{2}$-inch (13 millimeters) long up to about 4 inches (10 centimeters). Most gobies are quite colorful. *Pandaka pygmaea*, a freshwater fish found in the Philippines, is considered by some authorities to be the smallest vertebrate animal in terms of length. The species *Eviota* found in the Indo-Pacific is also quite small. Even the freshwater gobies spawn in saline waters and advantage of this fact is taken by fisheries in the Philippines to capture extremely large schools of the genus *Paragobiodon*. A fermented paste (bagoong) is made from the tiny fish (ipon). Several species of gobies develop a symbiotic relationship with other creatures, such as crabs, burrowing worms, and notably shrimp, wherein the gobies share shelters with their hosts, but also serve to warn their hosts of impending danger. The *Elecatinus oceanops* is a small 2-inch (5-centimeter) neon goby that is noted for its ability to clean parasites from larger fishes. Several species of gobies are favorites among tropical-fish fanciers. There are about 15 species of eel gobies (*Taenioididae*). They are elongated with a maximum length of just over a foot (0.3 meter). They are found in tropical Indo-Pacific waters. The loach goby (*Rhyacichthys aspro*) is the only member of the family *Rhyacichthyidae*, growing to a length of about 9 inches (23 centimeters) and found in large streams and rivers of the Philippines and Indonesia. Were it not for its spiny first dorsal fin, this species would be difficult to distinguish from the homalopterid loaches. See also **Loaches (Osteichthyes)**.

GODDARD, ROBERT, H. (1882–1945). Dr. Robert Hutchings Goddard, considered by many to be the father of modern rocketry, was a physicist of great insight who had a genius for invention. Along with Konstantin Tsiolkovsky of Russia and Hermann Oberth of Germany, Goddard envisioned the exploration of space.

Born in Worcester, Massachusetts, Goddard graduated from Worcester Polytechnic Institute (WPI) in 1908 and remained at his alma mater as a physics instructor. That same year he began graduate work in physics at Clark University and received his Master's and Doctorate degrees in 1910 and 1911, respectively. In 1912, Goddard became a research fellow at Princeton University and in 1913 developed the mathematical theory of rocket propulsion. The following year he joined the faculty at Clark University. In 1915, he proved that rocket engines could produce thrust in a vacuum, therefore making space flight a practical goal. Goddard became a full professor at Clark in 1919.

Goddard first obtained public notoriety in 1907 when he fired a powder rocket in the basement of the physics building at WPI. School officials then took an immediate interest in Goddard's work and, to their credit, did not expel him for the incident.

In 1914, Goddard received two U.S. patents, one for a rocket that used liquid fuel, the other for a two- or three-stage rocket using solid fuel. At his own expense, he began to make systematic studies about propulsion provided by various types of gunpowder. This work resulted in his classic study in 1916 requesting funds from the Smithsonian Institution to continue his research. This was published along with his subsequent research in the *Smithsonian Miscellaneous Publication No. 2540* (January 1920) entitled "A Method of Reaching Extreme Altitudes." In this treatise, he detailed a search for methods of raising weather-recording instruments higher than sounding balloons.

In this 1920 publication, Goddard outlined the possibility of a rocket reaching the moon and exploding a load of flash powder on its surface to mark the rocket's arrival. The bulk of his scientific report to the Smithsonian was a dry explanation of how he used the $5000 grant in his research. The press picked up Goddard's proposal about a rocket flight to the moon and sparked a journalistic controversy concerning the feasibility of such a concept. Goddard was widely ridiculed, causing him to deeply resent the press corps, a view that he held for the rest of his life.

By 1926, Goddard had constructed and successfully tested the first liquid-fueled rocket. The flight of Goddard's rocket on March 16, 1926, at Auburn, Massachusetts, was a historic feat. It was one of Goddard's "firsts" in the now booming field of rocket propulsion in military missilery and the scientific exploration of space. See Fig. 1.

Fig. 1. The first liquid-fueled flight lasted only 4.2 seconds, reached an altitude of merely 41 feet, and landed just 184 feet from its launch pad. However, this modest accomplishment marked the beginning of the space age. (Photograph courtesy *NASA Goddard Space Flight Center*.)

Goddard moved his experiments to Roswell, New Mexico, as his rockets got bigger. By the mid 1930s, Goddard's rockets had broken the sound barrier [1191 km per hour (741 mph)] and flown to heights of up to 2.7 km (1.7 miles). See Fig. 2.

Goddard's greatest engineering contributions were made during his work in the 1920s and 1930s. By 1927 he had received a total of $10,000 from the Smithsonian Institution, and through the personal efforts of Charles Lindbergh, he subsequently received financial support from the Daniel and Florence Guggenheim Foundation. Progress on all of his work appeared in "Liquid Propellant Rocket Development," which was published by the Smithsonian in 1936.

Goddard's work was virtually ignored in the United States and made little impression upon government officials. Ironically, his rocket designs (including gyroscopic control, steering by means of vanes in the jet stream of the rocket motor, gimbal steering, power-driven fuel pumps) shared many similarities with those developed by German rocket engineers during the 1930s and World War II. Thus, many people assumed that the Germans had used copies of Goddard's work in their development of the V-2 rocket. However, Goddard's secrecy had prevented the Germans from learning much about his work, and the similarity of design was mostly coincidence. It was not until after the war that Goddard's work was widely published.

From 1917 to 1918, Goddard developed the basis for a rocket-propelled weapon, now known as the bazooka. Using a music stand as his launching

Fig. 2.   Construction of the rocket used in the flight of April 19, 1932. (Photograph courtesy *NASA Goddard Space Flight Center*.)

platform, he successfully demonstrated the idea at the Aberdeen Proving Grounds in Maryland before representatives of the U.S. Armed Services. Dr. Clarence N. Hickman, a young Ph.D. from Clark University, worked with Goddard in 1918 and continued his research that produced the World War II bazooka. In World War II, Goddard was assigned by the U.S. Navy to develop practical jet-assisted takeoff and liquid propellant rocket motors capable of variable thrust. He was successful in both ventures.

In memory of this brilliant scientist, a major space and earth science laboratory, NASA's Goddard Space Flight Center in Greenbelt, Maryland, was established on May 1, 1959. Later that year on September 16, Congress authorized the issuance of a gold medal in honor of Professor Robert H. Goddard.

Goddard died on August 10, 1945. At the time of his death, Goddard held 214 patents in rocketry.

Historic Firsts

Robert H. Goddard's basic contributions to missilery and space flight is a lengthy list. As such, it is an eloquent testimonial to his lifetime of work in establishing and demonstrating the fundamental principles of rocket propulsion.

1912: First to explore mathematically the practicality of using rocket propulsion to reach high altitudes and even the moon.

1914: First to receive a U.S. patent for the idea of a multistage rocket.

1915: First to prove, by actual static test, that a rocket will work in a vacuum.

1926: First to develop and shoot a liquid fuel rocket using a mixture of gasoline and liquid oxygen.

1929: First to shoot a scientific payload (barometer and camera) in a rocket flight.

1932: First to use vanes in the rocket motor blast for guidance.

1932: First to develop a gyro control apparatus for rocket flight.

1935: First to launch a liquid-propellant rocket that attained a speed greater than the speed of sound (700 mph).

1937: First to successfully launch a rocket with a motor pivoted on gimbals under the influence of a gyro mechanism.

Web Reference

"Robert Goddard, father of the space age," Clark University: http://140.232.1.5/god_dardfolder/goddard.html

GOETHITE.   The mineral goethite is a hydroxide of iron corresponding to the formula FeO(OH) crystallizing in the orthorhombic system. It occurs in prisms, but is often found in foliated or other massive forms. When observable it shows one good cleavage parallel to the prism; fracture, uneven; hardness, 5–5.5; specific gravity, 3.3–4.3; luster, adamantine to dull; color; yellowish, reddish, brownish to nearly black; translucent to opaque. It is found associated with hematite and limonite, being perhaps in part an alteration product of the latter mineral. Goethite is used as an ore of iron. There are many European localities, including Bohemia, Saxony, Westphalia, and Cornwall. In the United States it is found in the hematite mines of the Lake Superior region and in Colorado. This mineral was named in honor of the German poet Johannes Wolfgang von Goethe.

GOITER.   See **Iodine (In Biological Systems)**; and **Thyroid Gland**.

GOITRE.   See **Iodine (In Biological Systems); Nutritional Science (The History)**; and **Thyroid Gland**.

GOLAY PNEUMATIC CELL.   A small transparent cell containing gas used to detect radiation. A very thin film within the cell absorbs incident radiation, which increases the cell temperature and pressure. Changes in pressure are recorded as indications of the amount of incident radiation.

GOLD.   [CAS: 7440-57-5] Chemical element symbol Au (from Latin aurum), at. no. 79, at. wt. 196.967, periodic table group 11 (transition metals), mp 1,064.43 °C, bp approximately 3080 °C, density 19.32 g/cm³ (20 °C). Elemental gold has a face-centered cubic crystal structure.

Gold is a yellow metal, soft, and extremely malleable. The purity of gold (sometimes referred to as "fineness") is expressed in karats. Pure gold is 24 karat. See also **Radioactivity**. In terms of cosmic abundance, in the estimate of Harold C. Urey (1952), using silicon as a base with a figure of 10,000, gold was ranked number 79 among the elements, with an abundance figure of 0.0015. In terms of abundance in seawater, gold is ranked number 59 among the elements, with an estimated content of 38 pounds per cubic mile (4 kilograms per cubic kilometer) of seawater.

Electronic configuration is

$$1s^2 2s^2 2p^6 3s^2 3p^6 3d^{10} 4s^2 4p^6 4d^{10} 4f^{14} 5s^2 5p^6 5d^{10} 6s^1.$$

First ionization potential is 9.223 eV; second 19.95 eV. Oxidation potentials: $Au \rightarrow Au^{1+}$, $E° = -1.68$ V: $Au \rightarrow Au^{3+}$, $E° = -1.50$ V. Other important physical properties of gold are given under Chemical Elements.

Gold is one of the most ancient metals. Gold jewelry and ornaments made as early as 3500 B.C. have been discovered at Ur in Mesopotamia. During the period from 3000 to 2000 B.C., lead cupellation was used to purify gold and most modern jewelry techniques were developed during that time.

Occurrence and Processing

Gold is found chiefly as the free metal scattered through gravel (placer gold) or disseminated in veins of quartz (vein gold). Small quantities also are found in lead and copper sulfide ores. Nuggets of native gold, varying in size from that of a tiny pebble to a mass weighing as much as 248 pounds (112.5 kilograms), have been found. In a combined state, gold occurs in sylvanite, a telluride of gold and silver, $(Au, Ag)Te_2$, a rich ore found in Colorado. The bulk of the gold ores contain very little gold (about 5 to 15 grams/metric ton). Some of the richest ores found in Africa contain from 20 to 30 grams/metric ton. Almost all countries produce some gold. The leader, by far, is the Republic of South Africa, followed by Russia and Canada. Far behind, other producers include the United States, Australia, Ghana, and Zimbabwe. See also **Mineralogy**.

The treatment of gold ores involves: (1) grinding, amalgamation, and/or cyanidation of those ores containing coarse free gold, and (2) the very fine grinding, flotation, roasting, and amalgamation and/or cyanidation of those ores containing gold telluride or sulfide. These processes produce an impure gold metal containing considerable silver and some copper plus other base metals. The impure gold is purified by melting and oxidizing the base metals or by melting and chlorinating (Miller process) which removes the base metals and silver. The silver-containing oxidized gold is purified by the electrolysis of gold chloride solutions containing an HCl solution (Wohlwill process). In the latter process, the anode is the alloy (gold-silver) and the cathode is pure gold. The gold deposits then on the cathode and the silver forms silver chloride and remains as a deposit about the anode.

Throughout early mining history, it was believed that ores, such as placer gold, resulted from mechanical weathering, wind, and water erosion of the veins of ore. However, since the early 1800s, geologists have found that biological processes also play a role in shaping some mineral deposits. Watterson (U.S. Geological Survey), in the early 1980s, made a serendipitous observation that gold solutions are lethal to many soil bacteria. Thin coats of gold tend to condense around the bacterial spores, clogging the narrow pores in their cell walls, through which nutrients enter. Watterson's findings were confirmed by inspection of placer gold particles in an Alaskan stream. Masses of gilded cells were found as the result of this biological process in connection with Pedomicrobia and related bacteria. Stephen Mann (Univ. of Bath), who specializes in biomineralization, observes that many bacteria can become encased in mineral coatings under favorable conditions. It should be noted that some mining firms are using bacteria to assist in extracting metals from low-grade ores. Further detail is given in the Rennie reference listed.

Uses of Gold

The monetary aspects of gold have long dominated commercial interest in the metal. Gold through history has provided a common base from which the value of materials and services can be measured. Gold probably became a medium of exchange as early as 3400 B.C.

Jewelry is the largest commercial user of gold, accounting for nearly 65% of the total consumption. Most jewelry is made by the "lost wax process," a casting method that dates to 3000 B.C. or earlier. Usually these jewelry products employ karat golds which contain 10 and 14 karats, and less commonly 18 karat, of gold (41.7, 58.3 and 75.0 weight percent of gold, respectively). These gold alloys are of two general types. Red, yellow, and green golds are basically alloys of gold, copper, and silver. A wide variety of color shades can be produced by varying composition within this ternary alloy system, with reddish hues provided by high copper to silver ratios, and pale green tint when silver is predominant. These alloys almost always contain minor amounts of zinc and deoxidizers or grain refiners to facilitate fabrication. The second widely used class is the white karat golds, which are produced in two basic alloy types. These are the original gold-nickel-zinc-copper (18 karat) and the gold-copper-nickel-zinc (10 and 14 karats) alloys, and the more recent gold-palladium-silver-copper, and gold-copper-nickel-palladium-silver alloys which are usually 14- and 10-karat alloys. The pink golds are derived from the system gold-silver-copper-nickel-zinc. These are essentially red golds, which are "whitened" by the addition of silver, nickel, and zinc.

Considerable brazing is done by jewelry manufacturers and the solders that are used may be of a lower karat content than the alloy being brazed. Usually they contain much more silver and zinc than the alloys themselves.

The use of gold in the electrical, electronic, and other industrial fields has grown considerably in recent years, estimated at about 25%. The electrical and thermal conductivity, resistance to oxidation, and ease of being electroplated make gold an excellent coating for electrical contacts. See Fig. 1. This has been particularly true in metallized ceramics for use in microelectronics and other electronic components. Here gold does not migrate into the ceramic as does silver. Gold is widely used as a conductor in thin and thick film circuitry. It is also useful as bonding wire for integrated circuit electrical connections and mechanical packaging of semiconductor chips (die bonding).

Fig. 1.   Electrolytically deposited gold crystals. (*Bausch & Lomb.*)

Gold is used extensively in many industrial solders and brazing alloys. These range from the low-melting eutectics of gold with germanium, silicon, and tin to gold-copper, gold-nickel, and gold-palladium-nickel alloys. The latter brazing materials have the ability to withstand long use at high temperatures and are particularly applicable to jet engine fabrication.

Gold is also used in dentistry. This application has declined in recent years; however, it still accounts for about 7% of gold consumption. Gold alloys, such as gold-silver-copper with varying amounts of platinum and palladium, are used for restorations and for bridges, inlays, and partial dentures. These are cast with much more precision than jewelry, and have, in fact, replaced wrought gold wire in many of these dental appliances. Gold wire is now used principally in orthodontic and prosthetic appliances. These are complex alloys containing gold, platinum, palladium, silver, copper, nickel, and zinc.

Some of the minor commercial uses of gold are among the most interesting. Gold is used to produce a very beautiful ruby glass. When an oxidizing glass is melted with a gold salt, the gold dissolves, forming colorless ions. If reducing agents like Sn, Sb, Bi, Pb, Se, or Te are present, the glass will become red after heating at temperatures between 600–700°C, as a result of the precipitation of minute particles of gold. Gold films deposited on glass by evaporation are superior to other metals for reflectivity in the infrared. Mirrors thus coated have application in spectroscopy and space science. Thin films applied to plate glass give adequate transmission of light combined with good infrared reflectivity, reducing the overheating of office windows during hot weather. Gold is extremely malleable. It can be rolled and beaten into foil less than 5 millionths of an inch (0.00013 millimeter) thick. Such foil has been used for indoor and outdoor decoration for centuries. One of the most conspicuous examples is the gold leaf dome, an architectural highlight in many important structures.

Chemistry of Gold

Gold has a $5d^{10}6s^1$ electron configuration, like the similar ones at lower levels of copper and silver, and thus the d electrons can take part in bonding. However, for gold the $+3$ oxidation state is the most stable, and the $+1$ state next to it in stability, so that Au^{3+} as well as Au^+ are found both in simple compounds and in complexes. As with copper and silver, the bonds in most gold compounds, including the oxides, are largely covalent. In most of its compounds gold is univalent or trivalent. While a few compounds are known in which it is divalent, some of these are considered to consist of Au(I) and Au(III), rather than Au(II). Thus, the compound with cesium and chlorine, $CsAuCl_3$, is black and diamagnetic, and so contains both Au(I) and Au(III). A similar compound with cesium, silver, and chlorine, $Cs_2AuAgCl_6$, yields $[AuCl_4]$ and $[AgCl_2]^-$ ions on hydrolysis. However, the sulfide, AuS, probably contains divalent gold.

Gold does not combine directly with oxygen. Gold(I) oxide, Au_2O, formed by heating AuOH to 200°C, is very easily reduced to gold. It is essentially covalent. Gold(I) hydroxide, AuOH, is prepared from a gold(I) solution by the addition of potassium hydroxide solution in theoretical amounts. It forms a deep-blue "solution" believed to be a colloidal sol. It dissolves in excess alkali to form aurates(I), such as $KAu(OH)_2$. Gold(III) oxide, Au_2O_3, is formed by heating $Au(OH)_3$ at 100°C in the presence of a dehydrating agent. Like Au_2O, it is easily reduced to gold. It dissolves in hydrochloric, hydrobromic, and hydriodic acids, forming the haloauric acids, $HAuX_4$. It also dissolves in excess of alkali hydroxide, forming an aurate, containing the ion $[Au(OH)_4]^-$. Gold(III) hydroxide, $Au(OH)_3$, is precipitated by the addition of potassium hydroxide solution in equivalent amount, to a solution of chloroauric acid (obtained by dissolution of gold in aqua regia). It is insoluble in H_2O, gives many of the reactions of Au_2O_3, and may be a hydrous form of that compound. Gold(II) oxide, AuO, formed by the action of potassium bicarbonate upon solutions of chloroauric acid, is believed, as stated above, to consist of gold(I) and gold(III), based on properties of other divalent gold compounds.

Gold does not react directly with fluorine, but dissolves in bromine trifluoride, BrF_3, to form BrF_2AuF_4, which loses BrF_3 at 120°C to give gold(III) trifluoride, AuF_3 which decomposes into the elements at about 500°C. Water decomposes AuF_3, into hydrogen fluoride and $Au(OH)_3$. The chlorides, on the other hand, are the most important of the gold salts. Gold(I) chloride, AuCl, may be produced by heating gold(III) chloride, $AuCl_3$, in air at 170°C; it is hydrolyzed by H_2O, to $AuCl_3$ and gold. Gold(III) chloride, $AuCl_3$, is formed directly from the elements at 200°C; unlike AuCl, it is soluble in H_2O forming initially $H[AuCl_3(OH)]$, which then undergoes further hydrolysis. With hydrochloric acid, $AuCl_3$ forms tetrachloroauric(III) acid. $[AuCl_4]$, of which many salts are known. Gold(I) bromide, AuBr, is formed by continued heating of bromoauric(III) acid above 100°C. Like the AuCl, it readily undergoes hydrolysis. Gold(III) bromide is formed by the action of bromine water upon gold. The equivalence of its three Au-Br bonds have been proved by a tracer technique with radioactive bromine. With hydrobromic acid it forms $H[AuBr_4]$. Gold(I) iodide is prepared from the elements at $50\,°C$, or by the slow decomposition of AuI_3 at room temperature. It decomposes on heating above $120\,°C$. It dissolves in potassium iodide, KI, solution, forming $KAuI_2$, which then decomposes to gold and $KAuI_4$. Gold(III) iodide, obtained by evaporation of a 1 : 1 hydriodic acid solution of $AuCl_3$, is unstable, decomposing, when dry or when heated with H_2O, into the elements. It dissolves in hydriodic acid as $H[AuI_4]$. The gold(I) halides are the least soluble of the univalent halides except for silver iodide. The solubility product constants are AuI, 1.6×10^{-23}; $AuBr$, 5.0×10^{-17}; AgI, 8.30×10^{17}; $AuCl$, 2.0×10^{-13}; and $AgBr$, 4.27×10^{-13}.

There are many gold complexes. The gold(I) and gold(III) halocomplexes, involving the groups $[AuX_2]^-$ and $[AuX_4]^-$ have already been discussed. Apparently, there are no other gold(I) halocomplexes than the chloro-compound. There are also fluorocomplexes of the form $M[AuF_4]$ formed by fluorination of $M[AuCl_4]$ where M is an alkali metal or ammonium. Due to the polar character of the AuF bond, they are readily hydrolyzed. Many hexachloroaurates, such as $Cs_2M[AuCl_6]$, are known.

Gold forms complexes with ammonia much less readily than do copper and silver. A few ammonia complexes of gold(III), such as $KAuCl_4 \cdot 3NH_3$, have been prepared. Gold(I) halides react more readily. AuCl forms $[Au(NH_3)_2]Cl$, while AuBr and AuI react, but only with anhydrous ammonia, to form $[Au(NH_3)_2]Br$ and $[Au(NH_3)_6]$. Gold(I) cyanide dissolves in excess cyanide to form the very stable ion $[Au(CN)_2]^-$, $K_{inst} = 10^{-38.3}$. This complex is so stable that gold metal dissolves in potassium cyanide solution in the presence of air. This is of importance in the separation of gold from its ores; while $Au(CN)_3$ reacts to form $[Au(CN)_4]^-$, $K_{inst} = 10^{-56}$. Treatment of salts of this ion with sulfites, gold(III) forms such complexes as $K_5[Au(SO_3)_4] \cdot 5H_2O$ and $Na_5[Au(SO_3)_4] \cdot 14H_2O$. In these complexes, the sulfite group is monodentate and is attached to the gold atom through the sulfur atom (really an aurisulfonate ion); however, a bidentate compound is also known.

Gold(III) chloride or tetrachloroaurates(III) also form thiosulfate complexes, especially in the presence of NaI, of the form $Na_3[Au(S_2O_3)_2]$, in which the gold is monovalent.

Gold forms thiocyanate complexes $M[Au(SCN)_2]$ and $M[Au(SCN)_4]$.

A striking difference between gold and copper or silver is the fact that its oxyacid compounds do not exist in stable form, and few have been isolated. Among the few that are known are the gold(III) orthoarsenite, $AuAsO_3 \cdot H_2O$, the gold(III) selenate, $Au_2(SeO_4)_3$, and the gold(III) iodate, $Au(IO_3)_3$. Nevertheless a number of complexes of oxyacids are known, including $M[Au(NO_3)_4] \cdot 2H_2O$ (M = H_3O^+, NH_4^+, K^+, Rb^+), $Mg[Au(CH_3CO_2)_4]$.

Gold is unique among the coinage metals in forming true (i.e., sigma-bonded) stable organometallics. The action of methyllithium on $AuBr_3$ in ether at $-65\,°C$ produces a solution of $(CH_3)_3Au$, which begins to decompose at $-35\,°C$ into gold, ethane, and methane. The presence of benzylamine or ethylenediamine, however, stabilizes the solution up to room temperature. Triethylgold is less stable than trimethylgold. The action of a hydrogen halide on a trialkylgold or the action of an alkyl Grignard reagent in pyridine on gold(III) halides produces dialkylgold halides, which are much more stable. Appropriate methathetical reactions of these produce the corresponding cyanides, sulfates, etc. These are all covalent compounds, as attested by the solubility of the sulfates, $(R_2Au)_2SO_4$, in benzene and chloroform. The melting points of a few dialkylgold compounds are: $(CH_3)_2AuBr$, $68\,°C$; $(C_2H_5)_2AuCl$, $48\,°C$; $(C_2H_5)_2AuBr$, $58\,°C$; $(C_2H_5)_2AuCN$, $103-105\,°C$; $(n\text{-}C_3H_7)_2AuCN$. $94-95\,°C$; $(i\text{-}C_3H_7)_2AuCN$, $88-90\,°C$; $(i\text{-}C_5H_{11})_2AuCN$. $70\,°C$; $(C_6H_5CH_2)_2AuCN$, $100\,°C$ decomposes; $(C_6H_5CH_2CH_2)_2AuBr$, $112.5\,°C$. The n-propyl chloride and bromide, and the n-butyl, i-butyl, and i-amyl bromides are liquid at room temperature.

The dialkylgold halides are dimeric, having the planar structure:

The cyanides, on the other hand are tetrameric, having the structure shown below.

Additional Reading

Brady, G.S.S., H.R. Clauser, and J.A. Vaccari: *Materials Handbook*, 14th Edition, The McGraw-Hill Companies, Inc., New York, NY, 1996.

Davis, J.R.: *Metals Handbook*, 2nd Edition, ASM International, Materials Park, OH, 1998.

Gasparrini, C.: *Gold and Other Precious Metals: From Ore to Market*, Springer-Verlag, Inc., New York, NY, 1993.

Greener, E.H.: "Dental Materials," *Encyclopedia of Materials Science and Engineering*, MIT Press, Cambridge, MA, 1986.

Krebs, R.E.: *The History and Use of Our Earth's Chemical Elements: A Reference Guide*, Greenwood Publishing Group, Inc., Westport, CT, 1998.

Lagowski, J.J.: *MacMillan Encyclopedia of Chemistry*, Vol. 1, Macmillan Library Reference, New York, NY, 1997.

Lechtman, H.: "Pre-Columbian Surface Metallurgy," *Sci. Amer.*, **53** (June 1984).

Lide, D.R.: *CRC Handbook of Chemistry and Physics*, 88th Edition, CRC Press, LLC., Boca Raton, Fl, 2007.

Meyer, C.: "Ore Metals Through Geologic History," *Science*, **227**, 1421–1428 (1985).

Parker, P.: *McGraw-Hill Encyclopedia of Chemistry*, 2nd Edition, The McGraw-Hill Companies, Inc., New York, NY, 1993.

Rennie, J.: "Bug in a Gilded Cage: All That Glitters is Sometimes Bacterial," *Sci. Amer.*, **27** (September 1992).

Schmidbauer, H.: *Gold: Progress in Chemistry, Biochemistry, and Technology*, John Wiley & Sons, Inc., New York, NY, 1998.

Stwertka, A.: *A Guide to the Elements*, Oxford University Press, Inc., New York, NY, 1998.

DONALD A. CORRIGAN, Handy & Harman, Fairfield, CT

GOLDBEATER'S SKIN HYGROMETER. A hygrometer using gold-beater's skin as the sensitive element.

Variations of the physical dimensions of the skin caused by its hygroscopic character indicate atmospheric relative humidity. The length of a piece of goldbeater's skin changes between 5% and 7% for a change in humidity from 0% to 100%. The time constant of response becomes extremely long both at low ambient temperatures and at very high and very low relative humidity.

Note: Goldbeater's skin is the prepared outside membrane of the large intestine of an ox; it is used in goldbeating to separate the leaves of the metal.

AMS

GOLDEN-EYE. 1. *Insecta, Neuroptera*. The lace-wing, adult of the aphis-lion. These insects are small and delicate, with large many-veined wings of yellowish or green color and shining eyes. They have a disagreeable odor.

2. *Aves, Anseriformes*. A North American and European duck, *Bucephala*. See also **Eagle**.

GOLD NUMBER. When certain colloids (hydrophilic), such as gelatine, are added to a gold sol, the gold sol is strongly protected against the flocculating action of electrolytes. This protective action on red gold sols may be measured by utilizing the color change red to blue which indicates the first stage of coagulation. The "gold number" as defined by Zsigmondy is the weight in milligrams of protective colloid which is just sufficient to prevent the change from red to blue in 10 cm^3 of a standard gold sol (0.0053 to 0.0058 percent Au) after the addition of 1 cm^3 of a 10 percent sodium chloride solution.

GOLDSCHMIDT REDUCTION PROCESS. Reaction of oxides of various metals with aluminum to yield aluminum oxide and the free metal. This reaction has been used to produce certain metals, e.g., chromium and zirconium, from oxide ores; and it is also used in welding (iron oxide plus aluminum giving metallic iron and aluminum oxide, plus considerable heat). (Thermite process.)

A method of producing formates by heating sodium hydroxide with carbon monoxide under pressure.

A process for recovery of tin, by treatment of scrap tinplate with dry chlorine, better known as the Goldschmidt detinning process.

GOLGI BODY. See Cell (Biology).

GOLGI, CAMILLO (1843–1926). Golgi studied medicine at the University of Pavia under Mantegazza, Bizzozero and Oehl. After graduating in 1865 he continued to work in Pavia at the Hospital of St. Matteo. Golgi himself stated that Bizzozero greatly influenced him and his methods of scientific research; at that time most of his investigations were concerned with the nervous system, i.e., insanity, neurology, and the lymphatics of the brain.

Golgi shared the Nobel Prize in Medicine in 1906 with Santiago Ramón y Cajal for their work on the structure of the nervous system. One of his important contributions was a brain tissue staining technique and the discovery of the very high-density packing of neurons in brain tissue. However, the work of greatest importance that Golgi carried out was a revolutionary method of staining individual nerve and cell structures, which is referred to as the "black reaction". This method uses a weak solution of silver nitrate and is particularly valuable in tracing the processes and most delicate ramifications of cells. Golgi himself was extremely modest and reticent about his work and it is not known when exactly he made this invention. All through his life, however, he continued to work on these lines, modifying and improving this technique.

See also **Cell (Biology)**; and **Central and Peripheral Nervous Systems**.

J. M. I.

GONADS. Both the female sex gland (*ovary*) and the male sex gland (*testis*) are referred to by the general word, *gonads*. Not only are the gonads the fundamental organs of reproduction, but they also produce several hormones. The two testes are made up of tissues that specialize in producing the male germ cells and tissues that manufacture the male hormone. The two ovaries provide the egg (*ovum*) and several hormones that are involved in the regulation of sexual function. Because the ovaries and testes produce hormones, they are considered endocrine glands. See also **Endocrine System**. Collectively, these male and female hormones are called gonadal hormones.

The hormones produced by the ovaries are called female hormones. The name female or male hormone does not imply that these substances are produced exclusively by either sex, but that they are produced predominantly by one sex. Thus, certain structures in males, especially the adrenals, can and do produce female hormones that are excreted in the urine. Women also produce male hormones and in some instances where the balance is disturbed by disease the effects of overproduction of male hormone become evident. In such instances, women develop signs of masculinity.

Sex hormones act primarily upon the reproductive system, which in both men and women is made up of the gonads and the accessory or secondary sex organs. The proper development and functioning of the accessory sex organs are dependent upon the production of sex hormones. In women, the accessory sex organs are the breasts, the womb (uterus), Fallopian tubes, vagina, vulva, and clitoris; each serves a particular function in the complex process of reproduction. In men, the secondary or accessory sex organs are represented by a series of tubes or ducts that convey the germ cells from the testes through the penis to the outside of the body, plus several glands located at different points. These glands are the prostate glands, the seminal vesicles, and Cowper's glands. Again, each performs a particular function, and each is dependent on the male hormone for its proper functioning. Castration results in a decrease in size of all these structures, and eventually they cease to function. The effect is the result of removing the source of male hormone.

Secondary Sex Characteristics

Male and female hormones are poured into the blood like all other hormones. They exert different actions on different parts of the body, imparting qualities that are typical of each sex. Thus, the distribution of hair on the body, particularly pubic hair, varies greatly. In women, pubic hair is limited above by a horizontal line, and the hair may grow in a triangular zone. In men, the growth of pubic hair may extend from the navel to the anus. The hair on other parts of the body is more abundant in men. The female voice is high pitched, and the larynx is less developed than in the male. Other qualities, such as breast development, shape of pelvis, and distribution of fat are also different in the sexes as a result of different sex hormone production.

Both male and female sex hormones belong to the group of substances called *steroids* to which also belong the hormones produced by the adrenal cortex. See also **Steroids**. Pregnant women excrete large quantities of certain sex hormones in their urine. In the past, urine from pregnant women was used as a source of a female sex hormone (*estrone*). Pregnant mares also eliminate large amounts of estrone in their urine. Sex hormones produced by animals are identical with those produced by humans. Urine obtained from postmenopausal women has been utilized on an industrial scale to obtain a hormone that stimulates the gonads. It is termed *human menopausal gonadotropin*. Also check *hormones* and *sex* in alphabetical index.

The Testes

Production of sperm cells takes place in the testes. The testes are two oval-shaped organs located outside the abdominal cavity below the penis, and held by a pouch called the *scrotum*. In addition to the reproduction function, the testes produce male sex hormones, which are secreted into the bloodstream. Rarely is an individual born having both testes and ovaries. When such occurs this is *true hermaphroditism*.

Before a boy is born, the testes are present within the abdominal cavity where they have been formed and descend gradually until, by the time of birth, they make their exit through a passage called the *inguinal canal* and have become localized in the scrotum.

For the testes to function effectively, they must be at a lower temperature than that of the abdomen. When the temperature increases, the testes do not produce mature spermatozoa. Because they are located within the scrotum outside the abdominal cavity, the testes are kept at a temperature a few degrees lower than that of the body. When the outside temperature is lowered, the spermatic cord that is attached to the testes and the scrotum draws upward, keeping the testes close to the body and allowing them to be warmed by the body's heat. The reverse occurs when the outside temperature is raised.

The surface of the testes is covered by a layer of fibrous tissue called the *tunica vaginalis*. The internal structure of the testes is divided into sections separated by thin membranes. Within each section are long, thin, tube-like strands, called the *seminiferous tubules*. It is within these tubules that the spermatozoa are produced. In the spaces or interstices that exist between the tubules are the interstitial cells which produce the male hormone. If a section of the testes is observed with a powerful microscope, a number of circular structures representing cross sections of the tubules can be seen. Within the circular structures are seen the spermatozoa at different stages of development. Toward the center of the tubules are seen the mature spermatozoa with complete heads and tails.

During the maturing process, the spermatozoa pass into multiple small tubes (*vasa efferentia*) which lead to the *epididymis*. The epididymis is a long, thin duct (*ductus* or *vas deferens*). Upward in its course toward the abdomen, the vas deferens is joined by the resticular arteries, veins, lymphatics, and nerves to form a thick tube, the *spermatic cord*. The spermatic cord, containing the vas deferens and other vessels, passes into the abdomen through the inguinal canal, and descends by the side of the urinary bladder to the prostate, through which it passes to reach the urethra. It is there joined by the small duct of the *seminal vesicles*. For each testis, there is one spermatic cord, one vas deferens, and one seminal vesicle.

The seminal vesicles are two pouches located between the bladder and the rectum, although not connected to either. The lower ends of the two seminal vesicles unite to form two short ducts that serve to carry the spermatic fluid to the large duct in the penis (urethra) and outside the body. These are the *ejaculatory ducts*, which are two small ducts that penetrate the prostate. From this point, both the semen and the urine share the same passage, the remaining portion of the urethra.

The prostate is an organ located at the base of the bladder; it completely surrounds the portion of the urethra that leads from the bladder. The prostate is an accessory organ of reproduction, containing numerous glands that produce the *prostatic fluid*, an important component of the *semen*. The secretion is produced at a low, but constant rate, and is poured into the urethra in small amounts; small quantities escape into the urine. Sexual stimulation accelerates production of prostatic fluid. During ejaculation, the prostatic fluid is delivered in larger quantities and is mixed with the seminal plasma to form the semen. In addition to serving as a housing and transporting vehicle for the sperm, the prostatic fluid appears to be necessary to maintain viable spermatozoa in the vagina, possibly by protecting the sperm from the acid condition of the vagina.

A single ejaculation may contain over a quarter of a billion spermatozoa. If fertilization does not occur, all of these cells die; if fertilization does occur, only one spermatozoon will survive; it will fertilize the egg. Occasionally, two ova may be produced within a short period of time and two spermatozoa will fertilize them, producing *fraternal twins*. *Identical twins* develop from a single ovum. Fraternal twins may be of different sexes, but identical twins are of the same sex and look alike. The sperm cells that swim in the semen are microscopic. Their propulsion is by movements of their tails. When sperm are deposited in the vagina during sexual intercourse, they move gradually upward toward the womb. The fatality rate of the sperm is high, but the chances of one arriving alive in the womb are usually good. The life span of a sperm cell is not precisely known, but it is believed that the sperm has the ability to penetrate and fertilize an ovum for only about 48 hours. The energy necessary for maintenance and propulsion of spermatozoa is derived mostly from the various types of nourishment present in the seminal plasma.

The Penis. In sexual intercourse, the penis serves to convey the semen into the vagina of the female. The shape of the penis varies greatly depending on whether it is flaccid or erect. In the flaccid state, the penis is cylindrical, but when erect, it assumes a triangular shape in cross section. The organ consists of three cylindrical masses of erectile tissue held together by fibrous tissue and covered by skin. Two of the cylindrical bodies lie side by side, and the third, which holds the urethra, is located underneath the other two. The lower cylinder ends in a cone-shaped body (the *glans*), which constitutes the free end of the penis; in the center of the glans is the opening of the urethra. The skin that covers the penis is thin and has no hairs except near the root of the organ, but possesses numerous glands that produce secretion.

The glans of the penis is covered by a circular fold of skin called the *prepuce*. In many instances, the prepuce, or foreskin, may cover the entire glans, obstructing the passage of urine. Under these conditions, the secretion of the skin glands accumulates, creating a constant source of irritation and infection. Therefore, surgical removal of the foreskin (*circumcision*) may be desirable as a prophylactic measure, and is usually performed shortly after birth. The operation was performed in ancient Egypt before it was introduced among the Hebrews. Today, it is practiced among the Jews and Mohammedans as a religious rite. However, it is practiced widely as a hygienic measure by peoples of all continents.

Erection is necessary for normal transmission of semen into the body of the female. Sexual stimulus, either mental or physical, sets off a series of reactions that culminate in erection. The sexual stimulus received by the nervous system causes a flow of blood, from the arteries that lead to the penis and within the penis, to the many vessels and cavities of the erectile tissue. This occurs at a faster rate than the blood flows from the penis via the veins. The penis becomes engorged with blood, thus becoming firm and erect. The organ returns to its original flaccid state when the process is reversed after erection.

Male Sex Hormones. The male sex hormone is produced after complete development of the testes. At puberty, the secondary sexual characteristics make their appearance rapidly. In normal boys, signs of puberty may appear at any age between 10 and 17 years. The average onset is 12 to 13 years. A related problem in the development of sexual characteristics in boys is *cryptorchidism*, or undescended testes.

When the output of male hormone is less than normal, a condition known as *hypogonadism* develops. A patient who is of adolescent age or younger may develop symptoms characterized by effeminate traits and retarded development of the sexual organs. In men who have attained maturity, the signs of *androgen* (male sex hormone) deficiency are less conspicuous. The most common events are reduction in prostatic size, diminished growth of the beard and body hair, the appearance of fine wrinkles around the eyes, and a pasty, sallow complexion. Also, semen volume is reduced.

Klinefelter's syndrome is a common form of hypogonadism. Feminine characteristics and infertility may exist. Patients with this condition are often tall with disproportionately long lower extremities. Mental retardation and psychopathic behavior are not uncommon, and men with this syndrome are often poorly adapted socially. In 1956, it was discovered that Klinefelter's syndrome is the result of a genetically determined defect. Treatment for patients with this condition must be closely supervised by a physician, as the use of hormones is usually involved.

In the male, with age, sexual activity declines gradually. The climacteric (*change of life*) is not as conspicuous as it is in women, and the age at which it occurs varies over a wider range. At the time of the male climacteric, sexual activity declines to a lower level.

Tumors of the testes are uncommon. The greatest incidence occurs in men in their twenties and thirties. The most common testicular tumor is called seminoma. Generally, this tumor is relatively slow growing and responds well to radiotherapy.

See also **Pituitary Gland**.

The Ovaries

Located on each side of the womb, the ovaries are two almond-shaped organs. Each is about the size of a walnut. The ovaries, unlike the testes, produce several hormones. Although different, they are grouped under the term *female sex hormones*. These substances regulate various functions of the body, but their major duty is regulation of the female reproductive system. Two chemically determined types of ovarian hormones are (1) the estrogenic steroids or *estrogens* (*estradiol, estrone*, etc.), and (2) the *progestagens* (*progesterone*, etc.). Within recent years, it has been possible to produce these hormones synthetically.

The control that ovarian hormones exert upon the reproductive system is not limited to the accessory or secondary sex organs, i.e., the womb, Fallopian tubes, vagina, vulva, and clitoris. In an indirect sense, the ovaries themselves are affected by their own secretions, since a reciprocal ovary-pituitary relationship is of importance in the regulation of the ovaries. The maturation of the eggs, ovulation, and other changes that occur in the ovaries are dependent, then, to some degree, on the hormones from the ovaries. See also **Pituitary Gland**.

The sexual cycle in women is well-regulated as long as the production and secretion of both the gonadotrophic hormones of the pituitary gland and the sex hormones from the ovaries are normal. This occurs most of the time, but occasionally the pituitary gland, the ovaries, or both may vary in their production of hormones. When the pituitary gland becomes underactive as a result of disease, the production of all the pituitary hormones is affected.

The two ovaries establish contact with the uterus by means of the two Fallopian tubes, which convey the egg cells from the ovaries to the womb. The womb (uterus) is a muscular organ with great capacity for expansion. The inside of the womb is hollow and the walls are covered by a mucous membrane known as the *endometrium*. Here, the fertilized ovum develops into a baby.

The hollow portion of the female reproductive system constitutes a continuous structure, so that the ovaries, tubes, and womb may be regarded as a unit. The uterus forms the center of this unit, and is located in the pelvic cavity between the urinary bladder and the rectum, and the tubes form a passageway to the ovaries, which are located on each side of the uterus.

The female reproductive system does not produce a fluid corresponding to the male seminal fluid. Under the influence of sexual stimulation, however, the walls of the vagina secrete fluids, which serve as lubricants that facilitate intercourse.

The egg cells or ova are periodically produced in the ovaries at intervals of approximately 4 weeks. At the end of each 4-week period, one egg reaches maturity and passes into one of the Fallopian tubes. The egg descends gradually and remains viable for a short while. Following intercourse, the sperm cells swim toward the tubes, in one of which fertilization may take place. Since neither the male nor the female reproductive cells live long, successful fertilization can occur only during a short period of time each month. This period of maximum fertility in women can be ascertained by various means, including temperature measurements.

If the egg is fertilized by the sperm, the fertilized ovum enters the uterus and becomes attached to the uterine wall where the child develops. Ordinarily, only one egg is produced each month, although more than one egg may be produced and, in some cases, may lead to multiple birth. If pregnancy occurs, usually no eggs are produced until after the child is born, or pregnancy is interrupted.

The maturing of the egg is a continuous process regulated by the endocrine system. Within the ovary, there is a layer of cells called the *germinal epithelium.* Here, the potential egg begins its existence and continues to develop until a *primary follicle* is formed around it, which is a clump of cells isolated from the main layer. The central cell of the clump is the egg, the remaining cells forming a ring around the egg. During a lifetime, each ovary forms between 200,000 and 400,000 follicles. Of all these potential eggs, only a few develop into mature eggs; most of them degenerate at the follicle stage. Those follicles that do not degenerate increase in size; meanwhile the egg cell itself enlarges until the original size is doubled. The one-ring layer of cells around the egg then multiplies and forms several layers. Fluid begins to accumulate in little pools which merge and form larger ones until one large pool is formed with the egg inside of it.

Other changes occur in the areas adjacent to the follicle. As the follicle matures, it moves toward the surface of the ovary; when the maturation process is complete, the follicle protrudes from the surface of the ovary. At this time ovulation occurs. The follicle bursts and the egg, with its fluid, is expelled from the surface of the ovary, leaving a cavity. Consequently, the adult woman who has ovulated many times possesses ovaries that have a pitted appearance. See also **Gamete**; and **Pregnancy (or Cyesis; Gestation)**.

The Uterus. Commonly known as the womb, this is a pear-shaped organ the size of a small fist and is located in the pelvic cavity of the female. The uterus is the organ that receives the fertilized egg from the Fallopian tube and provides the necessary nourishment and protection of the fetus during the various stages of pregnancy, and expels the developed child by the action of its muscular walls. The walls of the uterus are elastic, allowing for distention during pregnancy and return to the original thickness after childbirth.

The cavity of the womb is lined with the endometrium, a mucous membrane. The endometrium is not of the same thickness and consistency all the time, but varies considerably during the menstrual cycle. During menstruation, the endometrium disintegrates and is expelled with the menstrual blood, but a new endometrium lining begins to form immediately following each menstruation. The womb possesses two parts called the "body" (*fundus*) and the "neck" (*cervix*). The cervix is below the fundus and connects with the vagina at a right angle. The position of the womb is not always the same. In general, the long axis of the womb extends from front to back and slightly downward. The neck of the womb is then pointed toward the rectum and meets the vagina at a right angle. The urinary bladder lies in front and the rectum in the back of the womb.

The cervix, or neck of the womb, is an important organ that has numerous functions in the reproductive system. During pregnancy, the cervix protects the fetus, and during childbirth it distends to permit passage of the child. The cervix may be the origin of a variety of disorders and the site of numerous infections.

The Vagina. During sexual intercourse, the vagina receives the male sperm cells. The organ is made up of muscular tissue that possesses a considerable degree of elasticity. This permits distention without tearing when the child passes from the womb to the exterior of the body. The vagina is located between the urinary bladder and the rectum, although it is not directly connected to either. The vagina serves as a passageway between the opening of the vulva and the opening of the cervix.

In the adult woman, the size of the vagina varies but the average length is approximately 3 inches (7.5 centimeters). When the woman is in a standing position, the direction of the vagina is backward and upward, forming almost a right angle with the long axis of the uterus. The outer opening of the vagina is surrounded by a mucous membrane called the *hymen*. In the virgin woman, the hymen covers a considerable area of the vaginal opening; in rare instances, it may cover it entirely (*imperforate hymen*) causing retention of the menstrual flow. The hymen varies considerably in shape, but in general is semicircular. If the hymen is intact at the incident of first intercourse, it is usually ruptured at that time, although not always; sometimes it does not tear, but merely stretches. Consequently, absence of a hymen or a ruptured hymen should not be construed to mean that a woman is not a virgin.

The lining of the vagina secretes a fluid that is acid in nature and serves as a cleanser and lubricant. In an acid environment only certain types of bacteria can live, most of which are harmless and even helpful. The vaginal lining is smooth only in women that have borne children or after the menopause in childless women. In the young woman, the lining forms a series of folds.

The Vulva. Vulva is a collective name applied to the external female organs of reproduction and includes the mons pubis, labia majora, labia minora, clitoris, vestibular bulbs, vestibule, Bartholin's glands, Skene's glands, and hymen. The urethra, which is part of the urinary system, is often regarded as a structure of the vulva.

The *mons pubis* is located on top of the pubic bone just above the genital organs. This is a pad of fatty tissue covering the underlying bone. It forms an inverted triangular area covered with hair in the adult woman. The sides of the triangular area are delimited by the groins. From the top of the triangle, the mons pubis bends gradually downward and backward, dividing in the center to form two distinct sides that eventually, toward the perineum, become indistinguishable from the labia majora. The mons pubis contains many erogenous nerve endings which, when stimulated, add to the female's excitement.

Labia majora means "major lips," and as the name indicates they are two large folds of tissue located around the vaginal opening. When the woman is in the erect position, the labia majora conceal most of the other external organs of reproduction. Extending downward they gradually decrease in thickness until they disappear into the region of the perineum. The perineum is the area between the vulva and the anus. When the labia majora are pulled aside, the remainder of the female organs of reproduction becomes visible.

Within the labia majora lie the *labia minora*, which means "minor lips." These are folds of skin, which form an angle. The area bounded by this angle is called the vestibule, and within this area is located the opening of the vagina. The labia minora have an abundance of erogenous nerve endings. When stimulated during sexual excitement, the labia minora thicken two to three times their normal size.

The *clitoris*, which is located at the apex of the triangular area delimited by the labia minora, is a relatively small organ made up of erectile tissue. Erectile tissue becomes firm and engorged with blood in response to stimulation. The clitoris in the female and the penis in the male are somewhat similar in structure and response. The clitoris is covered by a fold of skin known as the prepuce; the tip of the clitoris is called the glans.

The opening of the urethra and the vagina are located in the vestibule. The urethral opening and openings of the Skene's glands lie just below the clitoris. Below these lies the opening of the vagina. Skene's glands secrete an alkaline substance that reduces the acidity of the vagina. The Bartholin's glands are located in the lower portion of the vestibule and are not normally conspicuous, but become prominent when inflamed and infected. Bartholin's glands produce a drop or so of mucous secretion, which at one time was thought to serve as a lubricant during sexual intercourse. However, this secretion is insufficient for that purpose.

Diseases and Disorders of Female Reproduction Organs

Ovarian Tumors. The diseases not related to the endocrine system that affect the ovaries comprise a large number, of which tumor formation is the most important. Tumor does not necessarily imply cancer and actually most ovarian tumors are not cancers.

Most ovarian tumors develop without presenting symptoms, except those that produce hormones. Eventually, pain is caused by the tumor pressing against neighboring organs, tension of the tumor mass, rupture, or infection. When a positive diagnosis of tumor has been made, surgical exploration becomes necessary in almost every case. Abdominal exploration is necessary to secure a complete diagnosis and to remove the tumor. All ovarian tumors may be dangerous if not removed, because it is almost impossible to determine which will or will not develop into a cancer. The extensive growth of tumors of the ovary can be prevented only by early discovery and removal. Therefore, periodic pelvic examinations are extremely important in the early detection of cancer. Also, a pelvic examination is of great importance for early detection of cancer of the cervix. An examination of the cervix by a physician is a simple procedure. A procedure known as the "Pap" test is a cytologic examination and was developed chiefly by the late Dr. George N. Papanicolaou. The test involves the microscopic examination of cells collected from the vagina. These are cells shed from the uterus into the vagina as a part of the normal life process. If microscopic examination of the smear reveals any abnormal cells, bits of tissue are taken from the cervix for further microscopic study.

Infection and Tumors of the Fallopian Tubes. Infections of the Fallopian tubes frequently cause permanent sterility. The Fallopian tubes are attacked most often by the organisms causing gonorrhea, infections produced during childbirth, tuberculosis, and a variety of systemic infections. These infections may be acute or chronic. In some instances, they may involve the entire reproductive system. Tumors may develop in the Fallopian tubes, usually as a secondary growth which originated in some other organ of the body. Tumors of the Fallopian tubes are relatively rare.

Tubal Pregnancy. The Fallopian tube is at times the site of an abnormal type of pregnancy, called tubal pregnancy. In these cases, the embryo fails to descend into the womb and develops instead in the Fallopian tube. As the fertilized egg grows within the tube, the tension increases, and the tube may rupture, causing death of the fetus. Once the existence of tubal pregnancy has been established, surgical intervention to remove the tube and the embryo is usually required. Often, there may be no symptoms of tubal pregnancy prior to rupture. This condition endangers the patient because hemorrhage is imminent in nearly every case. Tubal pregnancy is not the only form of abnormal pregnancy that takes place outside the womb, but it is perhaps the most common abnormal type. Other types include abdominal and ovarian pregnancies.

Retrodisplacement of the Uterus. The uterus is held in place by the floor of the pelvis and a series of tough bands of tissue (ligaments). Thus, the womb is not rigidly fixed in one position, but is movable. Abnormal displacements may occur when the position of the womb changes beyond certain limits. The uterus can turn backward, causing retrodisplacement. The most common cause is childbirth. During labor there is often considerable stretching of the supports that keep the womb in place. To avoid displacement, the physician instructs the mother to lie on her abdomen or side during convalescence. Once the condition has been discovered, the physician institutes treatment. This generally consists of bringing the uterus to a normal position by manual manipulation and maintaining it in a normal position by some mechanical support. Such supports vary in design and shape and are called *pessaries*, usually consisting of a flexible ring made of rubber or plastic.

Prolapse. At childbirth, the stretching of the uterine supports may cause both retrodisplacement and *prolapse* of the uterus. In the latter condition, the womb falls from the normal position and the cervix pushes far into the vagina. Severe prolapse can cause the womb to push the cervix through the vagina. Complications ensue, usually associated with ulcerations of the cervix as a result of irritation produced by continuous contact with the clothing of the patient. The pressure exerted by the prolapsed womb upon the urinary bladder causes an inability to retain urine. Frequently, incontinence is the complaint that induces the patient to consult the physician. Prolapse is corrected with pessaries and by surgical means. The restoration of the normal position of the womb does not necessarily involve loss of reproductive function.

Endometriosis. The lining (endometrium) of the womb sometimes behaves abnormally and grows not only on the walls of the womb, but within the walls, or on adjacent pelvic organs, causing a condition known as *endometriosis*. The patient with this condition may suffer irregularities in the menstrual cycle. Menstruation is often painful and copious. The manner in which bits of lining are transported from the womb and lodge in other parts of the body is not fully understood. Apparently, they can be transported by way of the Fallopian tubes, the blood, and the lymph. External endometriosis may necessitate surgical treatment. The results are satisfactory in most cases.

Uterine Tumors. The uterus is one of the most frequent sites of tumor formation, being second only to the breast. Tumors develop in nearly any part of the organ. Tumors of the fundus are of many types, but most common are fibroids (*leiomyomata*) of the uterus which develop from muscle tissue. The patient may have a group of small fibroids for many years and suffer no ill effects. However, the size of the tumors varies, sometimes reaching large proportions. Treatment of patients who have fibroids varies according to type and size of the tumors. If small and cause no symptoms, no treatment may be deemed necessary. Others that may endanger health usually are removed surgically. There are many other forms of tumors that can grow in the fundus and that can arise from any of its component tissues. In their early stages, many of these growths can be treated successfully either surgically or radiologically. Some, such as choriocarcinoma, respond to chemotherapy.

Cervical Cancer. When cancer develops in the cervix, it is at first confined to this organ, but, depending on the type of growth, spreads at different rates to the adjacent organs. In the early stages of the disease, there are no specific symptoms except perhaps irregular bleeding and discharge. The patient may delay examination until she is sure that the bleeding will

not disappear. After such delay, the cancer may have advanced beyond hope of cure. Any unusual bleeding or discharge, other irregularities in the menstrual cycle, periods in which there is profuse bleeding, and the recurrence of a period after several months without periods should be recognized as danger signals. Cervical cancer rarely appears in women under age 20; sometimes before age 30; but most commonly in women around 45 years of age.

See also **Cancer and Oncology**.

Trichomonas vaginalis, a parasitic protozoan, may infect the vagina, producing an irritative discharge. *Monilasis*, a fungus infection caused by *Candida albicans*, may affect the vaginal wall causing a white discharge and white patches. Nonspecific infections, caused by a number of bacteria, may be present in the vagina. Bacterial infections can usually be controlled by administration of one of the antibiotic drugs. Vaginal tumors are relatively uncommon. The most common type is called "inclusion cyst," which in most instances is not serious.

Vulvitis. Inflammation of the vulva may be caused by a number of factors. Since the external portion of the vulva is covered by skin, many skin conditions, such as eczema, ringworm, crysipelas, contact dermatitis, etc., may occur. Acute vulvitis occurs in children and obese women because of constant irritation. Vulvitis occurring in diabetic patients is caused by increased sugar content of the urine, which produces irritation and provides a favorable environment for the growth of yeasts and fungi.

Premenstrual Syndrome. Complex signs and symptoms of the premenstrual syndrome occur during the second half of the menstrual cycle. In most cases, the clinical features promptly cease with the onset of the menstrual flow, and a symptom-free period follows. Symptoms include bloating, edema, emotional lability, headache, changes in appetite or craving for specific foods, breast swelling and tenderness, constipation, and decreased ability to concentrate mentally. The syndrome was recognized as early as the 1930s, at which time the cause was attributed to excess estrogen. This hypothesis, along with newer numerous explanations, have not been professionally accepted. There is general agreement, however, that there may be a relationship between premenstrual syndrome and ovarian function. That relationship may include a delayed effect of sex steroids on neurotransmitter turnover within the hypothalamic centers that modulate reproductive and other hormones, which may induce symptoms of premenstrual syndrome and even affect the centers controlling mood and behavior. Statistics and studies pertaining to the syndrome generally have been unsatisfactory in terms of pointing a pathway to research. No endocrine or physiologic markers to distinguish women with the syndrome from unaffected women have so far been identified. Based largely upon unproven concepts, a variety of treatments, a few with reported success, have been used. These include administration of vitamin B$_6$ and progesterone supplementation. Another procedure is the administration of an agonist of gonadotropin-releasing hormone, sometimes referred to as the reversible "medical ovariectomy."

Dysmenorrhea. This complaint consists of moderate to severe lower abdominal cramping and back pain during menses, sometimes with nausea, vomiting, and other symptoms. For years, patients with those symptoms were considered to have a psychological and not an organic disorder. Not until systematic, scholarly studies were undertaken did the role of prostaglandins in mediating most of the symptoms associated with dysmenorrhea become evident. Women are now successfully treated with drugs that affect prostaglandin synthesis. Until more experience was gained with premenstrual syndrome, it was often misdiagnosed as dysmenorrhea.

Toxic Shock Syndrome. This syndrome (TSS) was first noted in 1978. By the end of 1980, nearly a thousand patients had been identified in the United States. 99% of cases seen in women, and 98% of cases noted as occurring during menstruation in women using tampons. In most studies, *Staphylococcus aureus* has been isolated from vaginal cultures of more than 90% of menstruating women with TSS, but found in only 10% of otherwise well, menstruating women. The very small number of cases of TSS noted in males or nonmenstruating females has been associated with focal staphylococcal infections.

Onset of the illness usually occurs on the third or fourth day of menstruation. Symptoms involve multiple organ systems. In addition to having a sore throat, TSS patients may also develop a strawberry tongue, resembling scarlet fever. Recurrences tend to be less severe than the initial episode, indicating that immunity may provide partial protection.

Administration of a beta-lactamase-resistant penicillin or a cephalosporin during an episode of TSS reduces the likelihood of recurrence.

In making a differential diagnosis of TSS, the symptoms are highly suggestive of scarlet fever, but prominent hypotension and the lack of bacteriologic evidence of a Group A streptococcal infection eliminates this diagnosis.

In addition to antimicrobial therapy, hypotension and shock should be immediately treated employing vigorous fluid replacement and possibly supplemental use of catecholamines. Most patients recover in one or two weeks. Mortality has been reported as high as 10%.

A CDC (Centers for Disease Control) survey of 285 women showed that those who used tampons were at 33 times greater risk of contracting TSS than non-tampon users. The studies also showed that the risk for tampon users range from 5 to 80 times higher, depending on the type of tampon used. Present knowledge suggests that the higher the absorbency of a tampon, the higher the risk. There is the assumption that extra-absorbant tampons may create a better environment for the bacteria responsible for TSS. It is reported that suppliers of tampons are extensively researching TSS, out of which studies a safe, highly absorbent tampon can be developed.

Venereal Diseases. These are described elsewhere in this encyclopedia. Check alphabetical index.

Menopause

Cessation of menstruation marks the commencement of the *menopause*. This is a period when there are numerous biochemical and hormonal changes in the body, the symptoms of which vary widely from one woman to the next. In addition to physical changes, there are frequently accompanying psychological features in many cases. Some women experience no symptoms, whereas others require varying degrees of medical assistance in making the adjustments. Physiologically, as the result of ovarian failure, the amount of estrogens produced declines. The average age of ovarian failure is 48 years (statistic for the United States). Some women become amenorrheic in their earlier forties; others continue to menstruate and ovulate regularly into their fifties. The functions of estrogen still are not fully understood, but a number of the signs of menopause are associated with estrogen deficiency. These include vascular symptoms (among these are "hot flashes") in the shorter term and, extended over a period of time, consequences of estrogen deficiency may include an acceleration of atherogenesis (see also **Arteries and Veins (Vascular System)**); osteoporosis (see also **Bone**); and urethral and vaginal atrophy. See also **Geriatrics**; and **Gerontology**.

Because the wide range of symptoms and their degree of severity, there is no universal approach to treating them. Several years ago, estrogen replacement therapy was welcomed by patients and physicians alike as an excellent pathway to alleviating many of the problems of menopause. Several studies were published in the mid- and late-1970s linking estrogens to increased incidence of endometrial (lining of the uterus) carcinoma. Studies have since convinced many physicians that there are some risks in estrogen therapy — risks that must be weighed against the specific symptoms and needs of the patient. Thus, the present situation is one of using estrogens with the utmost of discretion. However, in the case of younger women who have lost their ovaries surgically, some physicians suggest that they should have full estrogen replacement until the time of a natural menopause, at which time the continuation of the therapy must be reevaluated.

Additional Reading

Anderson, M.C., F. Sharp, and J.A. Jordan: *Integrated Colposcopy: For Colposcopists, Histopathologists and Cytologists*, 2nd Edition, Chapman & Hall, New York, NY, 1996.

Barron, W.B., M.D. Lindheimer, and J.M. Davison: "Medical Disorders During Pregnancy," in 3rd Edition, *Mosby-Year Book*, St. Louis, MO, 2000.

Bellino, O.F.L.: *Biology of Menopause*, Springer-Verlag, Inc., New York, NY, 2000.

Bergdoll, M.S. and P.J. Chesney: *Toxic Shock Syndrome*, CRC Press, LLC., Boca Raton, FL, 1991.

Berkow, R. and M.H. Beers: *The Merck Manual*, 17th Edition, Merck & Company, Inc., Whitehouse Station, NJ, 1999.

Blackledge, G.R.P., J.A. Jordan, and H.M. Singleton: *Textbook of Gynecologic Oncology*, W.B. Saunders Company, *Philadelphia, PA*, 1991.

Bulletti, C. et al.: *Uterus: Endometrium and Myometrium*, New York Academy of Sciences," New York, NY, 1997.

Burnett, A.L. et al.: "Nitric Oxide: A Physiologic Mediator of Penile Erection," *Science*, 401 (July 17, 1992).

Buttino, L., Jr., M.I Evans, and B.M. Sibai: *Principles and Practice of Medical Therapy in Pregnancy*, 3rd Edition, Appleton & Lange, Old Tappan, NJ, 1998.

Chamberlain, G. and F. Broughton-Pipkin: *Clinical Physiology in Obstetrics*, 3rd Edition, Blackwell Science, Inc., Malden, MA, 1998.

Chard, T. and J.G. Grudzinskas: *The Uterus*, Cambridge University Press, New York, NY, 1994.

Disaia, P.J., M. Doherty, W.T. Creasman, et al.: *Clinical Gynecologic Oncology*, 5th Edition, Mosby-Year Book, Inc., St Louis, MO, 1997.

Dulbecco, R.: *Encyclopedia of Human Biology*, Academic Press, Inc., San Diego, CA, 1997.

Evans, M.I. and K.A. Ginsburg: *Obstetrics & Gynecology Pretest Self Assessment and Review*, 9th Edition, The McGraw-Hill Companies, Inc., New York, NY, 2000.

Frederiksen, M.A.: *Obstetrics and Gynecology (Rypins' Review)*, Lippincott Williams & Wilkins, Philadelphia, PA, 1999.

Genazzani, A.R., F. Petraglia, and R.H Purdy: *The Brain: Source and Target fro Sex Steroid Hormones*, Parthenon Publishing Group, New York, NY, 1996.

Hoskins, W.J., C.A. Perez, and R.C. Young: *Principles and Practice of Gynecologic Oncology*, in 3rd Edition, Lippincott Williams & Wilkins, Philadelphia, PA, 2000.

Jaffe, R. and S.L. Warsof: *Color Doppler Imaging in Obstetrics and Gynecology*, The McGraw-Hill Companies, Inc., New York, NY, 1992.

Kruman, R.J., E.J. Wilkinson, and H.J. Norris: *Tumors of the Cervix, Vagina, and Vulva*, American Registry of Pathology, Chicago IL, 1992.

Reece, E.A. and J.C. Hobbins: *Medicine of the Fetus and Mother*, 2nd Edition, Lippincott Williams & Wilkins, Philadelphia, PA, 1998.

Russell, P. and A. Farnsworth: *Surgical Pathology of the Ovaries*, 2nd Edition, Churchill Livingstone, Inc., Philadelphia, PA, 1997.

Shepherd, J.H. and J.M. Monaghan: *Clinical Gynaecological Oncology*, 2nd Edition, Blackwell Science Inc., Malden, MA, 1991.

Spitzer, A.R.: *Intensive Care of the Fetus and Neonate*, 3rd Edition, Mosby-Year Book, Inc., St. Louis, MO, 1996.

Thompson, J.D. and R.W. Te Linde: *Te Linde's Operative Gynecology*, 8th Edition, Lippincott Williams & Wilkins, Philadelphia, PA, 1996.

GONIOMETER. 1. An instrument for measuring the angles between the reflecting surfaces of a crystal or a prism. Parallel rays from a collimator, impinging upon the polished surfaces, are reflected in different directions. Two methods may be used. In one, the crystal or prism is held stationary. The angle between the reflected beams from the two faces, received in succession by a telescope moving around a graduated circle, is measured on the circle; the angle between the two faces is then $\frac{1}{2}$ of this. See Fig. 1. In the other method, the telescope is clamped in some convenient position and the crystal or prism is rotated so that first one and then the other face reflects light into it; the angle between the faces is the supplement of the angle through which the prism mounting is turned. An ordinary spectrometer may be used for the purpose. An instrument similar in geometrical principle, but employing x-rays instead of light and an ionization chamber instead of a telescope, is used for measuring angles between the atomic planes within crystals.

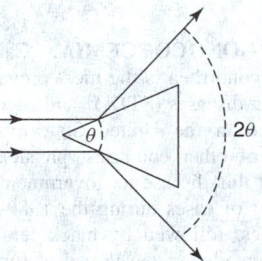

Fig. 1. Angle between reflected rays is twice the angle between prism faces.

2. For approximate measurement of interfacial angles on larger crystals, a simpler instrument, known as a *contact goniometer*, can be used. This instrument consists of a protractor with a movable arm attached to its base at a point exactly perpendicular to its 90° reading. The crystal is held between the base of the protractor and the movable arm with the interfacial plane surfaces making parallel contact with the protractor base and the arm. The corresponding external angle is then read directly from the protractor. In using this instrument, it is required that it be held perpendicular to the interfacial crystal planes being measured. The angle desired will be the

internal angle, which as in the example of the preceding paragraph, will be the supplement of the measured *external* angle.

3. Sometimes in the use of a loop antenna for directional purposes it is convenient or impossible to rotate the loop. This is especially true for transmitting loops where the size becomes appreciable in order to improve the radiation efficiency. To overcome this difficulty the goniometer is used. As shown in Fig. 2, the instrument consists of crossed stationary coils feeding fixed, crossed (90°) loops with the coupling to the moving coil proportional to the cosine of the angle of rotation. The movable coil is fed from the transmitter. The amount of energy transferred from the rotating coil to the fixed coils and hence to their antennas is determined by the position of the moving coil with respect to the others. The effect as far as the resultant field pattern of the antennas is concerned is exactly the same as if a single-loop antenna had been physically rotated. An extension of this principle is used with two sets of crossed loops in many radio range systems. A further advantage of the goniometer is that the antennas may be connected through a transmission line and thus it is not necessary that the operating position be near the antenna. It may be used equally well for reception.

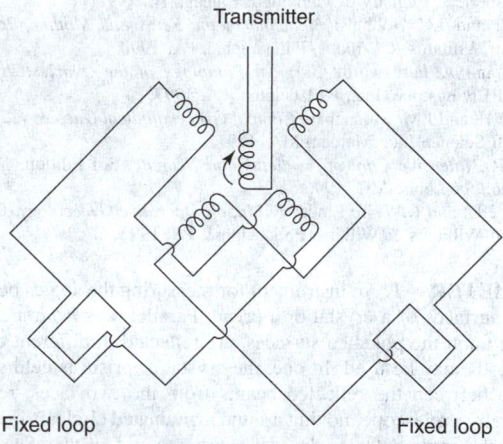

Fig. 2.   Goniometer (radio).

GO/NO-GO DETECTOR.   An instrument that has only two stable states of indication, and which therefore will give full response to any stimulus capable of actuating it. For example, a common fuse is a go/no-go detector, since either it is intact, or it is burned out. An ammeter, however, can respond continuously to the same current.

Go/no-go detectors are widely used in automated sorting and inspecting machines.

GONORRHEA AND GONOCOCCEMIA.   Caused by the diplococcus Neisseria gonorrhoea, gonorrhea is the most prevalent and widespread of the sexually transmitted diseases (STDs)[1] and accounts for approximately 461,000 cases reported in the United States in 1992. In the United States and a number of other countries, physicians and health centers are mandated to report this disease to government agencies. By race and sex, the largest number of cases during the 1980s and early 1990s were reported in black males, followed by black females, white males, and white females. These details are developed in further detail by Arel and Holmes (see reference listed). States accounting for over 20,000 cases during 1992 include California, Florida, Georgia, Illinois, New York, North Carolina, Ohio, and Texas. States reporting fewer than 100 cases in 1992 included Maine, Nebraska, New Hampshire, North Dakota, Vermont, and Wyoming. Generally, incidence of gonorrhea follows the normal population distribution, with high concentrations of cases occurring in major cities.

[1] Other STD's include syphilis, chlamydia, herpes simplex virus, cytomegalovirus, trichomonas vaginalis, bacterial vaginosis, and AIDS. See alphabetical index.

Nature of Infection and Symptoms

The infecting diplococcus is a fragile and fastidious organism that usually invades the transitional and columnar epithelial surfaces of the genitourinary tract, rectum, and conjunctivae. Stratified squamous epithelium is much more resistant to the organism. It invades the mucosal cells and, after penetration, colonizes the subepithelial tissues.

Five types of the organism have been identified, but only two are virulent — those that have hairlike appendages (pili) projecting from their surface, enabling the cocci to attach to the body cells.

Direct contact between persons, usually of a sexual nature, is required for the transmission of gonorrhea. Approximately 90% of cases occur in persons under their mid-30s, and, of these, 25% occurs in the teenage bracket. Statistics have shown that persons who regularly engage in fleeting sexual relationships with numerous partners run the greatest risk of contracting the disease.

At one time, it was reasonably well accepted that many females essentially were reservoirs of the disease without being aware of having the disease — because of lack of symptoms, i.e., the disease was spread mainly by asymptomatic females to males who almost always became symptomatic. It was believed that one female could infect several males within a short or long period prior to her awareness of disease in her body. Because symptom awareness in males is more vivid than in females, there is some strength to this early observation. However, it is now also realized that, although most males who develop urethral gonococcal infection recognize it and seek medical attention shortly after symptoms develop, there are also some males who can be infected for extended periods. Thus, they are available to infect females over extended periods without experiencing the usual vivid symptoms. Or, the symptoms may be so mild that medical attention is not sought. Thus, some members of both sexes from a practical standpoint can serve as carriers of the disease, spreading infection to dozens or even scores of sexual partners if they are very sexually promiscuous. A few thousand individuals in these categories can create a pandemic or epidemic situation.

The symptoms of gonorrhea in heterosexual males are anterior urethritis with a purulent urethral exudate and dysuria (difficult and/or painful urination). Incubation requires 3 to 4 days, but can be as many as 14 days or more. Without treatment, the course of gonorrhea and its complications can include epididymitis (testicular infection), prostatitis (prostate gland infection), infection of the paraurethral glands, and sometimes urethral stricture. There is also always the possibility of the development of disseminated gonococcal disease, described later. In homosexual males, there is gonococcal infection of the urethra, as well as infection of the anal canal (30–55% of cases) and infection of the pharynx (throat) in some 21% of the cases. Anal infections are frequently asymptomatic, but may include pain (upon defecation), a feeling of rectal fullness, and rectal discharge. Pharyngeal infection can lead to acute exudative pharyngitis.

The symptoms of gonorrhea in heterosexual females are quite versatile. As previously mentioned, they may go unnoticed (asymptomatic) for a long period. The infection may be associated with vaginal discharge, discomfort in the area of the lower abdomen, as well as abnormal uterine bleeding. The anal canal may be infected by vaginal secretion. Other symptoms may include dysuria, pyuria (pus in urine), and less frequently, hematuria (blood in urine). Untreated, the course of gonorrhea and its complications may include abscess of Bartholin's glands (vestibular glands in vagina), acute pelvic inflammatory disease, conjunctivitis, and disseminated gonococcal disease.

Gonococci may infect an infant during birth, notably in the eyes. Gonococcal ophthalmia neonatorum is prevented by the administration of silver nitrate solution to the infant's eyes, but this procedure alone does not guarantee full and permanent protection of the infant. Some mothers may have an asymptomatic gonococcal infection during pregnancy. Thus, hematogenous gonococcal arthritis may occur in the neonate by infection of the anogenital, oropharyngeal, or umbilical area during birth. Thus the need to screen pregnant women, particularly in instances where multiple sexual relationships may be suspected, for possible gonococcal infection during the prenatal period.

Gonorrhea is also known, infrequently, among prepubertal children, sometimes the result of sexual molestation or precocious childhood sexual activity. The characteristics of the disease are consistent with that of adults.

Diagnosis. The definitive diagnosis of gonorrhea is contingent on the recovery of gonococci from the patient. Gram's-stained smears from the urethra or freshly cleansed cervix can be used for tentative diagnosis. When typical Gram-negative diplococci are seen within three or more polymorphonuclear leukocytes, the degree of certainty is 90% in males and females, although sensitivity is only about 65%. Definitive confirmation requires selective culture media. These include Thayer-Martin and transglow media. Cultures are obtained from all clinically infected sites, whether or not local symptoms are present. In women, a culture of the endocervix is an effective screening test. These tests are positive in from 80–90% of infected females. In instances where fellatio is practiced, pharyngeal cultures should also be obtained. Cultures of material from the female urethra are usually omitted. In men, urethral cultures are paramount. In the case of homosexual males, rectal and pharyngeal cultures are also made.

Treatment. Persons to be treated for gonorrhea should be screened for evidence of syphilis because this will alter the course of treatment.

As pointed out by Handsfield and associated authors (see reference listed), "The proportion of isolates of Neisseria gonorrhoeae in the United States that had absolute or relative resistance to the penicillins or tetracyclines rose greatly in the 1980s, especially in the second half of the decade." This degree of resistance continued, alerting the U.S. Public Health Service to administrate cefriaxone intramuscularly. An oral version (cefixime) is now available. Based upon a randomized, unblinded multicenter study of 209 men and 124 women, with uncomplicated gonorrhea, it is now believed that a single oral dose of cefixime (400 to 800 mg) is as effective as the regimen of cefriaxone (250 mg) given intramuscularly.

Individuals with a recent known exposure to gonorrhea should receive the same treatment used for the established disease.

Disseminated Gonococcal Disease. Also called the arthritis-dermatitis syndrome, disseminated gonococcal infection (gonococcemia) is the most common cause of infectious arthritis in young adults. This condition develops as the result of gonococci invading the bloodstream. Many more cases are seen in women than in men. Symptoms include tender, pustular skin lesions (5 to 25 per patient) of a distinctive and repelling appearance, ranging from 5 to 15 millimeters ($\frac{1}{5} - \frac{3}{5}$-inch) in diameter. They often have a necrotic center. In the later phase of gonococcemia (a week or more after onset), purulent arthritis involving one or two joints will appear, with gonococci present in the synovial fluid in over 50% of the cases. The pattern of disease development varies from one patient to the next. Sexually active persons who display skin rashes and acute arthritis at the same time should be considered arthritis-dermatitis syndrome suspects because relatively few other diseases mimic this condition. High-risk diseases, such as meningitis and endocarditis, although infrequent, may develop as a consequence of gonococcemia. Gonococcal meningitis and endocarditis require prolonged intravenous penicillin therapy with accompanying supportive measures. For treatment of endocarditis in patients allergic to penicillin G, a procedure to desensitize the patient to the antibiotic may be required. Chloramphenicol can be effective against gonococcal meningitis.

Additional Reading

Arel, S.O. and K.K. Holmes: "Sexually Transmitted Diseases in the AIDS Era," *Sci. Amer.*, 62 (February 1991).

Baxter, R.A.: "Sexually-Transmitted Diseases and Rape," *N. Eng. J. Med.*, 1141 (October 18, 1990).

Beck-Sague, C.M. et al.: "Laboratory Diagnosis of Sexually Transmitted Diseases in Facilities within the United States," *Sex Transm Dis*, **23**, 342–349 (1996).

Cates, W., Jr. and A.R. Hinman: "Sexually-Transmitted Diseases in the 1990s," *N. Eng. J. Med.*, 1368 (November 7, 1991).

Endersbe, J.K.: *Sexually Transmitted Diseases: What are They?* Capstone Press, Mankato, MN, 1999.

Fox, K.K. et al.: "Antimicrobial Resistance in Neisseria Gonorrhoeae in the United States, 1988–1994: The Emergence of Decreased Susceptibility to the Fluoroquinolones," *J. Infect Dis.*, **175**, 1396–1403 (1997).

Handsfield, H.H. et al.: "Asymptomatic Gonorrhea in Men," *New Engl. J. Med.*, **290**, 117 (1974).

Handsfield, H.H.: *Color Atlas and Synopsis of Sexually Transmitted Diseases*, 2nd Edition, The McGraw-Hill Companies, Inc., New York, NY, 2000.

Holmes, K.K., P.F. Sparling, S.M. Lemon, and Per A. Mardh: *Sexually Transmitted Diseases*, 3rd Edition, The McGraw-Hill Companies, Inc., New York, NY, 1999.

Hunter, H. et al.: "A Comparison of Single-Dose Cefixime with Ceftriaxone as Treatment for Uncomplicated Gonorrhea," *N. Eng. J. Med.*, 1337 (November 7, 1991).

Kilmarx, P.H. et al.: "Inter-city Spread of Gonococci with Decreased Susceptibility to Fluoroquinolones: A Unique Focus in the United States," *J. Inf Dis*, **1777**, 677–682 (1998).

Knapp, J.S., K.K. Fox, D.L. Trees, and W.L. Whittington: "Fluoroquinolone Resistance in Neisseria Gonorrhoeae," *Emer Infect Dis*, **3**, 33–39 (1997).

Knapp, J.S. et al.: "Molecular Epidemiology, in 1994, of Neisseria Gonorrhoeae in Manila and Cebu City, Republic of the Philippines," *Sex Transm Dis*, **24**, 2–7 (1997).

Knapp, J.S. et al.: "Antimicrobial Susceptibilities of Strains of Neisseria Gonorrhoeae in Bangkok, Thailand: 1994–1995," *Sex Transm Dis*, **24**, 142–148 (1997).

MMS: "Morbidity and Mortality Weekly Report," in *Massachusetts Medical Society*, Waltham, Massachusetts (published weekly).

Schwarz, D., J.S. Knapp, R.J. Rice, and J. Paavonen: "Pelvic Inflammatory Disease," in Morse, S.A., et al.: *Atlas of Sexually Transmitted Diseases and AIDS*, 133–147, 2nd Edition, Mosby-Wolfe, London, UK, 1996.

Trees, D.L., A.L. Sandul, W.L. Whittington, and J.S. Knapp: "Identification of Novel Mutation Patterns in the parC Gene of Ciprofloxacin-Resistant Isolates of Neisseria Gonorrhoeae," *Antimicrob Agents Chemother*, **42**, 2103–2105 (1998).

Woods, S.G.: *Everything You Need to Know about Std: Sexually Transmitted Disease*, Rosen Publishing Group, Inc., New York, NY, 2000.

Web Reference

Centers for Disease Control and Prevention: http://www.cdc.gov/health/diseases.htm

GOODYEAR, CHARLES (1800–1860). Born in Wodurn, MA, Goodyear was the first to realize the potentialities of natural rubber. Frustrated by its lack of stability to temperature and other weaknesses in the uncured state, he experimented with additives such as magnesium and sulfur. The discovery of vulcanization was not accidental, as is often stated, but the result of intelligent trials and correct evaluation of their results. Though Goodyear's patents were contested by Hancock in England, he well merits the credit for making rubber usable in countless ways and helping to make the automobile possible.

See also **Vulcanization**.

GOOSE. See **Poultry**; and **Waterfowl**.

GOPHER. See **Squirrels and Other Sciuromorphs**.

GORGE WIND. See **Winds and Air Movement**.

GORILLA. See **Anthropoids**.

GOSSAN. This term is applied to the decomposed upper parts of mineral veins and ore deposits. It usually consists chiefly of hydrated iron oxide resulting from the weathering of pyrite, chalcopyrite, etc. Gossans have been important sources for the release of the relatively insoluble precious metals and gems which are washed away to form placer deposits. Many valuable gold ore bodies have been traced to their source by means of their derived placers. Also, secondary enriched sulfide ores of copper have been discovered beneath gossans that were originally prospected for the more precious metals.

GOUGE. A term used to designate soft or clay-like material between the sides of a mineral vein or ore deposit and the wall rock; also (structural geology), a layer of finely comminuted material between the walls of a fault.

GOURDS. See **Curcurbitaceae**.

GOUT. A syndrome made up of a number of physical and chemical factors. These include: abnormally high levels of uric acid [CAS: 69-93-2] (hyperuricemia) in the blood, usually symptomatic of gout, but which can occur from a few other causes; attacks of acute arthritis, with the presence of deposits of uric acid salts in and within the region of joints and tendons as well as in the kidney parenchyma; and formation of uric acid stones in the urinary tract and renal collecting system. The latter condition may lead to occasional kidney failure. The patient with acute gout is usually debilitated for a period, the length of time depending upon promptness of treatment and response to therapy. In Europe and America, the incidence of gout is about 3 cases per 1000 population. The disease occurs in males about ten times more often than in females. In the latter, the disease rarely occurs before menopause. The Maoris of New Zealand are particularly prone to gout, with the disease found in about one out of every ten males. It is estimated that many gouty people go undiagnosed unless the complications of serious joint and renal changes occur. Although the genetics of the disease are not accurately known, experience has shown familial connections.

Gout is a manifestation of faulty purine metabolism. Uric acid is a product of the purines. See also **Purines**. These occur in all tissues and are characteristic constituents of the nucleoproteins. Nucleoproteins are found in the nuclei and cytoplasm of all living tissues, plant and animal. In the breakdown of nucleoproteins, nucleic acids are released, and purines are located in these portions of the nucleoproteins. The purines have as their end product, in humans, an oxidized purine, namely, uric acid, In addition, there is a pathway for the formation of uric acid which does not involve purines, but which does involve glycine and other simple products.

Normally, the excretion of uric acid by the kidney keeps pace with its formation from purines of the food, purine metabolism of the tissues, and synthesis of uric acid. An elevation of serum uric acid may occur if the kidney cannot eliminate it at a normal rate, or if the rate of tissue breakdown is accelerated. This is usually accompanied by rises in other nitrogenous constituents of the blood, e.g., urea, and may not always be associated with gout. In gout, the only nitrogenous constituent of serum that rises characteristically is uric acid. As a result of this increased amount in blood, uric acid precipitates out in various locations of the body. The onset of the first attack is usually a very severe pain in the joint of a finger or toe. The joint becomes red, swollen, and extremely tender. Other joints are sometimes affected and frequently more than one finger or toe is involved. In severe cases, knob-like deformities around the affected joints appear, due to the deposition of uric acid to form "tophi."

Other conditions that may cause hyperuricemia sometimes interfere with an accurate diagnosis, particularly in the milder cases. These include individuals who have drastically lowered their carbohydrate intake in connection with dieting, rheumatic fever, rheumatoid arthritis, septic arthritis, cellulitis, and bursitis. Pseudogout sometimes is seen in older people and is easily distinguished by x-ray examination, which shows calcification of tissues, often involving the knee joint.

Gout therapy frequently involves the use of colchicine, which is reasonably specific to acute gouty arthritis. See also **Alkaloids**. Oral doses (sometimes intravenous) are given frequently for several hours until vomiting or diarrhea is induced. Within 12 to 24 hours, considerable improvement usually will be noted. Colchicine is then resumed in small dosages at less frequent intervals. The physician is aware of possible gastrointestinal side effects of this drug. Other drugs used include indomethacin and phenylbutazone. When the rare patient does not respond to any of these drugs, parenteral glucocorticoids may be used.

To inhibit *interval gout*, the plasma urate concentration will be maintained at proper levels. This is usually done with colchicine therapy. In the treatment of *chronic gout*, several objectives must be met: (1) further precipitation of monosodium urate crystals in tissues must be prevented; (2) dissolution of crystalline deposits already formed must occur; and (3) the function of affected joints must be restored. The drug probenecid acts effectively in most patients as a uricosuric agent. See also Diuretic and **Diuretic and Uricosuric Agents**. Other drugs of this type are salicylic acid (sodium salicylate), sulfinpyrazone, allopurinol and benzbromarone (see Table 1). Sometimes probenecid and colchicine are administered in combination. This type of therapy, although sometimes prolonged, usually is successful. It is uncommon to have to surgically remove the tophaceous deposits. Dietary procedures appear to be of little avail, probably because uric acid can be synthesized from generously available very small molecules.

Uricosuric Agents

Sodium Salicylate. The uricosuric properties of sodium salicylate (Table 2) [CAS: 54-21-7] were noted before 1890, and its use continued through 1950. As late as 1955, sodium salicylate was used for the long-term treatment of gout. For adequate uricosuric activity, however, salicylate

TABLE 1. URICOSURIC AGENTS

Generic Name	Trade Name	Structure
Salicylic acid		
Probenecid	Benemid	
Sulfinpyrazone	Anturane	
Allopurinol	Zyloprim	
Benzbromarone	Desuric, Minuric, Narcaricin	

must be administered in doses greater than 5 g/day, often resulting in serious side effects, so that its usage has gradually declined.

Probenecid. Probenecid (Table 1) [CAS: 57-66-9] was developed as a result of a search for a compound that would depress the renal tubular secretion of penicillin at a time when the supply of penicillin was limited. Recognition of the uricosuric properties of probenecid resulted from prior experience with the uricosuric effects of the related compound carinamide

(1)

in normal subjects and in gouty subjects. Carinamide had been introduced as an agent for increasing penicillin blood levels by blocking its rapid excretion through the kidney. Its biological half-life was relatively short, and the search for compounds with a longer half-life that would not have to be administered so frequently led to probenecid.

In a study of a series of *N*-dialkylsulfamoylbenzoates(2),

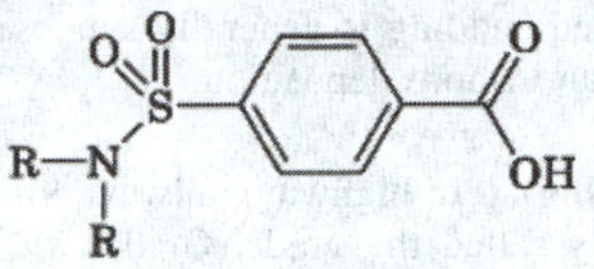

(2) R = H, Methyl, Ethyl, Propyl

Beyer found that as the length of the *N*-alkyl groups increased, the renal clearance of the compounds decreased. This most likely results from the enhanced lipid solubility imparted by the longer alkyl groups, which would account for their complete back diffusion in acidic urine. Optimal activity was found in probenecid, the *N*-dipropyl derivative. The structure-activity relationship of probenecid congeners and that of other uricosuric agents has been reviewed in detail by Gutman.

Normally, a high percentage of the uric acid filtered by the glomerulus is reabsorbed by an active transport process in the proximal tubule. It is now clear that the human proximal tubule also secretes uric acid, as does the proximal tubule of many lower animals. Small doses of probenecid depresses the excretion of uric acid by blocking tubular secretion, whereas high doses lead to greatly enhanced excretion of uric acid by depressing proximal reabsorbtion of uric acid.

Probenecid is completely absorbed after oral administration; peak plasma levels are reached in 2–4 h. The half-life of the drug in plasma for most patients is 6–12 h. The drug is 85–95% bound to plasma proteins. The small unbound portion is filtered at the glomerulus; a much larger portion is actively secreted by the proximal tubule. The high lipid solubility of the undissociated form results in virtually complete reabsortion by back diffusion unless the urine is markedly alkaline.

Probenecid is insoluble in water, but the sodium salt is freely soluble. In the treatment of chronic gout, a single daily dose of 250 mg is given for 1 week, followed by 500 mg administered twice daily. A daily dose of up to 2 g may be required.

Sulfinpyrazone. (Table 1), [CAS: 57-96-5]. Despite the therapeutic efficacy of phenylbutazone (3) [CAS: 50-33-9]

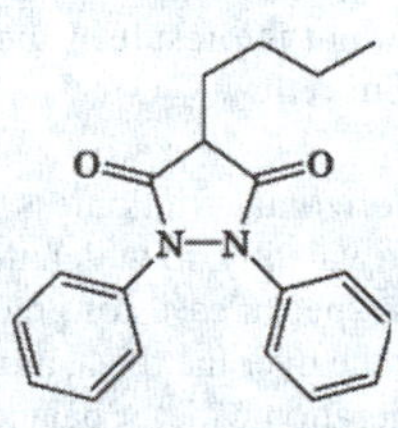

(3)

as an anti-inflammatory and uricosuric agent, its side effects were severe enough to preclude its continuous use in the treatment of chronic gout. Evaluation of several chemical congeners indicated that the phenylthioethyl analog of phenylbutazone

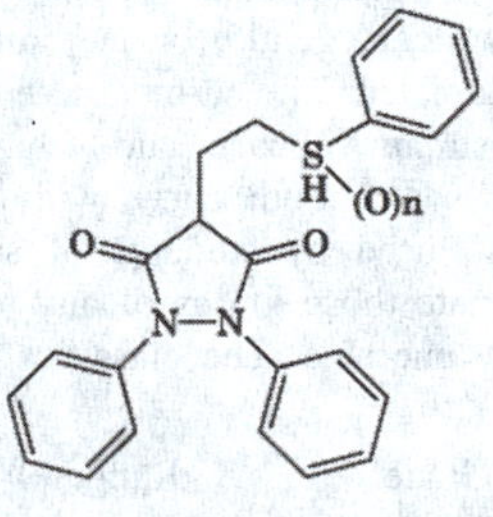

(4) n = 0
(5) n = 1
(6) n = 2

had promising anti-inflammatory and uricosuric activity. A metabolite, the sulfoxide pyrazone (5), exhibited enhanced uricosuric activity. Interestingly, the corresponding sulfone (6) does not appear to be a metabolite. Sulfinpyrazone lacks the clinically striking anti-inflammatory and analgesic properties of phenylbutazone (4).

Sulfinpyrazone is a strong acid ($pK_a = 2.8$) and readily forms soluble salts. Evaluation of a number of congeners indicated that a low pK_a and polar side chain substituents favor uricosuric activity) and increase the rate of renal excretion. The inverse relationship between uricosuric potency and pK_a has also been confirmed in a number of 2-substituted analogs of probenecid (7)

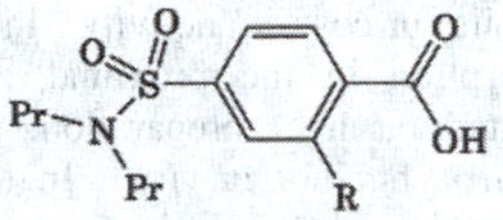

(7) R = OH, Cl, NO₂

(probenecid R = H; Table 1). All three compounds were considerably stronger acids than probenecid. Evaluation in the *Cebus albifrons* monkey indicated that these compounds were about 10 times as potent as probenecid when compared on the basis of concentration of drug in plasma.

In small doses, as seen with other uricosuric agents, sulfinpyrazone may reduce the excretion of uric acid, presumably by inhibiting secretion but not tubular reabsorbtion. Its uricosuric action is additive to that of probenecid and phenylbutazone but antagonizes that of the salicylates. Sulfinpyrazone can displace to an unusual degree other organic anions that are bound extensively to plasma protein (e.g., sulfonamides and salicylates), thus altering their tissue distribution and renal excretion. Depending on concomitant medication, this may be a clinical asset or liability.

For the treatment of chronic gout, the initial dosage is 100–200 mg/day. After the first week the dose may be increased up to 400 mg/day until a satisfactory lowering of plasma uric acid is achieved.

Allopurinol. (Table 1), [CAS: 315-30-0]. Allopurinol (8)

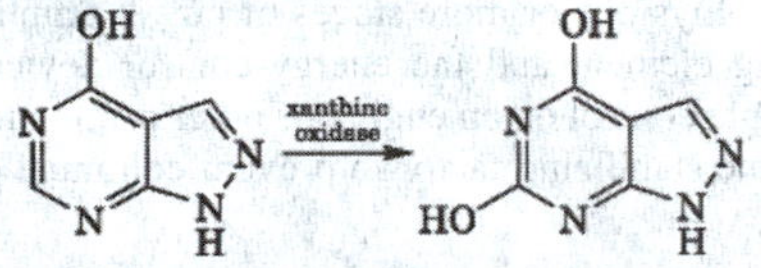

(8)   **(9)**

does not reduce serum uric acid levels by increasing renal uric acid excretion; instead it lowers plasma urate levels by inhibiting the final steps in uric acid biosynthesis.

Uric acid in humans is formed primarily by xanthine oxidase-catalyzed oxidation of hypoxanthine [CAS: 68-94-0] and xanthine [CAS: 69089-6] to uric acid. Allopurinol (8) and its primary metabolite, alloxanthine (9) [CAS: 2465-59-0], are inhibitors of xanthine oxidase.

Inhibition of the last two steps in uric acid biosynthesis by blocking xanthine oxidase reduces the plasma concentration and urinary excretion of uric acid and increases the plasma levels and renal excretion of the more soluble oxypurine precursors. Normally, in humans the urinary purine content is almost solely uric acid; treatment with allopurinol results in the urinary excretion of hypoxanthine, xanthine, and uric acid, each with its independent solubility. Lowering the uric acid concentration in plasma below its limit of solubility facilitates the dissolution of uric acid deposits. The effectiveness of allopurinol in the treatment of gout and hyperuricemia that results from hematogical disorders and antineoplastic therapy has been demonstrated.

For the control of hyperuricemia in gout, an initial daily dose of 100 mg is increased weekly at intervals by 100 mg. The usual daily maintenance dose for adults is 300 mg.

Benzbromarone. (Table 1), [CAS: 3562-84-3]. Benzbromarone (10)

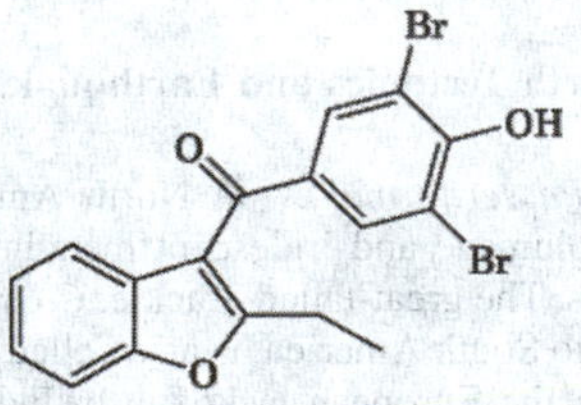

(10)

is a benzofuran derivative that has been reported to lower serum urate levels in animals and human studies. In normal and hyperuricaemic subjects, benzbromarone reduced serum uric acid levels by one-third to one-half. In comparison with other urate-lowering drugs, 80 mg of micronized or 100 mg of nonmicronized benzbromarone had equal urate-lowering activity to 1–1.5 g of probenecid or 400–800 mg of sulfinpyrazone.

The mechanism of the urate-lowering activity of benzbromarone appears to be attributable to its uricosuric activity. In rats, benzobromarone inhibited urate reabsorption in the proximal tubules when given at 10 mg/kg i.v. In isolated rat liver preparation, benzobromarone inhibits xanthine oxidase *in vitro* but not *in vivo* . In humans, this compound only weakly inhibits xanthine oxidase and no increase in urinary excretion of xanthine or hypoxanthine was observed. After oral administration, about 50% of benzbromarone is absorbed. The drug undergoes extensive dehalogenation in the liver and is excreted mainly in the bile and feces. For control of gout the usual therapeutic dose is 100–200 mg daily. Benzbromarone has few side effects and is usually well tolerated.

Additional Reading

Emmerson, B. and E. Cochrane: *Getting Rid of Gout*, 2nd Edition, Oxford University Press, Inc., New York, NY, 2003.

Harkness, R. et al: *Purine and Pyrimidine Metabolism in Man: Chemotherapy, ATP Depletion and Gout*, Kluwer Academic Publishers, New York, NY, 1991.

Porter R. and G. Rousseau: *Gout*, Yale University Press, New Haven, CT, 2000.

GOVERNOR. An automatic controller for maintaining the rotative speed of a machine. The governor senses the speed, compares the measured value with the desired value, and acts to correct any error between these two values — most often by adjusting the flow of energy to the machine. The two major types of governors are: (1) designs wherein the speed-sensing element operates an energy-metering device directly; and (2) a design which employs one or more stages of power amplification between the speed-sensing element and the energy-control device. The first type usually gives stable control on an engine or other prime mover. The second type requires some stabilizing factor to prevent continual oscillation of the speed (*hunting*).

GRAB BUCKET. A grab bucket is an apparatus able to pick up a load of bulk material by "biting" into the surface of the material. The particular usefulness of the grab bucket is that it may be lowered from the end of a boom onto the surface of the material to be moved, where it is operated to bite into this material, picking up a load, which can then be raised and deposited where wanted. Figure 1 shows a grab bucket in open and closed positions.

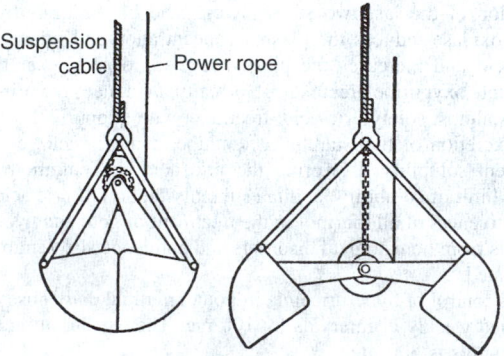

Fig. 1.   Grab bucket.

GRABEN.   See **Earth Tectonics and Earthquakes**.

GRACKLE (*Aves, Passeriformes*).   In North America, several species of birds with black plumage and iridescent metallic luster, related to the orioles and blackbirds. The great-tailed grackle, *Cassidix mexicanus*, which ranges from Texas into South America, is also called the jackdaw. It should not be confused with the European jackdaw. In India the hill mynas and related species are called grackles.

GRADED BEDDING.   A geological term denoting a type of bedding or stratification characterized by a cyclic or rhythmic deposition of coarse to fine sediments. Graded bedding is generally supposed to be characteristic of offshore rather than inshore deposition.

GRADE (Engineering).   In highway, railway, or municipal engineering, the slope of a line is called the grade. Grades are usually expressed as percentages preceded by a plus or minus sign. As an example, a +2% grade indicates a rise of 2 feet in every 100 feet (2 meters in every 100 meters) measured horizontally in the direction of travel; a −2% grade indicates a drop of 2 feet in every 100 feet (2 meters in every 100 meters). A curve known as a vertical curve is used to make the transition at a point of change in the grade of a highway or railroad. A second-degree parabola is used because it is the only curve in which the rate of change of slope is constant. The length is a function of the difference of the connected grades and the allowable rate of change of slope of the parabola per hundred feet measured horizontally. In the case of highways, the length of a vertical curve, at a point where the grade changes from plus to minus (at the crest of a hill), is governed by the safe sight distance.

GRADIENT CURRENT.   In oceanography, a current determined by the condition that the horizontal pressure gradient due to the (hydrostatic) distribution of mass balances the Coriolis force due to the earth's rotation. The gradient current corresponds to the geostrophic wind in meteorology. In practice, the distribution of density is determined by measurements of salinity and temperature at a series of depths in a number of positions. From this the geopotential topography of any isobaric surface relative to any other isobaric surface may be computed and the horizontal pressure gradient may be expressed by the geopotential slope of the isobaric surface. In this way relative isobaric surface currents are obtained, corresponding to thermal wind in meteorology. If one isobaric surface is known to be level, the absolute geopotential topography of any other surface may be computed by reference to this, and hence absolute gradient currents are obtained. Where no isobaric surface is known to be level, the total gradient current will consist of the relative gradient current, due to the distribution of density, and the slope current, due to that portion of the inclination of the isobaric surfaces that is not the result of the distribution of density.

See also **Geostrophic Current**.

GRADIENT FLOW.   Horizontal frictionless flow in which isobars and streamlines coincide; or equivalently, in which the tangential acceleration is everywhere zero. Important special cases of gradient flow, in which two of the normal forces predominate over the third, are: (1) *Cyclostrophic flow*, in which the centripetal acceleration exactly balances the horizontal pressure force; (2) *Geostrophic flow*, where the Coriolis force exactly balances the horizontal pressure force; (3) *Inertial flow*, which is flow in the absence of external forces; in meteorology, frictionless flow in a geopotential surface in which there is no pressure gradient, so that centripetal and Coriolis accelerations must be equal and opposite.

GRADIENT (Geology).   The term is applied to streams to refer to the slope of their beds, as steep, gentle, or in terms of so many feet per mile or meters per kilometer. The term is synonymous with *grade* as used in engineering. A stream valley is said to have become graded when its longitudinal profile is a smooth curve without waterfalls or rapids. The term grade is also used by students of sedimentary rocks, in a textural sense, to designate those grains of any sediment or sedimentary rock which are of the same size. The classification of grade-sizes is as follows:

Name of Grade		Range of Diameters
Pebbles		Greater than 10 mm
Gravel		10 mm to 2 mm
Sand	Very Coarse	2 mm to 1 mm
	Coarse	1 mm to 0.5 mm
	Medium	0.5 mm to 0.25 mm
	Fine	0.25 mm to 0.1 mm
Silt		0.1 mm to 0.01 mm
Clay		Less than 0.01 mm

GRADIENT (Mathematics). A vector obtained by the application of the vector differential operator del (∇) to a scalar point function. In rectangular coordinates, it is

$$\text{grad } \phi = \nabla \phi = i\frac{\partial \phi}{\partial x} + \mathbf{j}\frac{\partial \phi}{\partial y} + \mathbf{k}\frac{\partial \phi}{\partial z}$$

where $\mathbf{i}$, $\mathbf{j}$, $\mathbf{k}$ are unit vectors. It expresses, both in magnitude and direction, the greatest space rate of change of the scalar ϕ. At any point, P, it is normal to the surface $\phi(x, y, z) = $ constant, which passes through P.

GRADIENT WIND. See **Winds and Air Movement**.

GRAFTING AND BUDDING. Grafting is the process of inserting a part of one plant into another in such manner that the two unite and the inserted piece continues to grow. The part which is inserted is called the scion, the plant into which it is inserted is the stock. Budding is a similar process in which the part inserted consists of a bud with some of the bark adjoining it.

This process is possible because of the cambium cells. The successful union of the two pieces is caused by the formation of callus tissue by the cambium cells. Callus tissue is composed of a mass of parenchyma cells that fill in or grow over wounds, thus repairing the injury. In graft unions, the cells of the callus tissue soon begin maturing into cells of various types, as xylem and phloem cells, while others become typical cambium cells joining the cambium layer of stock and scion. In grafting, the cambium layers of the two parts are to be brought as closely together as is possible.

There are several methods of grafting. A very common method is known as cleft grafting. In this method a small twig having several buds is removed from the plant which is selected as desirable. The lower end of this twig is cut wedge-shaped. A branch of the plant used as stock is cut off, and a vertical cut made in the end. Into this cut the prepared scion is inserted in such a position that its cambium layer and that of the stock come together. To prevent drying of the tissues the entire cut surface is covered with a prepared wax. Usually, several scions are inserted in a branch of the stock. When union has taken place and the scion has started to grow, all but one may be cut off.

Another method is whip grafting, which is used when the stock is too small for successful cleft grafting. In whip grafting, both stock and scion are cut in a long oblique cut. In the cut surface of each a vertical cut is made. They are then fitted together so that the parts of one slide into and against those of the other, with the cambium of one in contact with that of the other. The two parts are then bound firmly together and the whole covered with wax.

In budding, a small bit of bark bearing a bud is removed from the selected plant. Usually, little wood is taken with this. In the stem of the stock, a T-shaped cut is made in the bark and the flaps so formed loosened. The prepared bud is inserted under the flaps, which are then pressed down over it and bound tightly in place to insure contact between the two cambium layers. Wax is used here also to prevent loss of water.

In modern horticulture, grafting is a very important practice. Many plants, for instance, do not come true when grown from seed. It becomes necessary, therefore, to propagate such desirable plants vegetatively. This may be done in two ways. One is by means of cuttings, pieces of the plant that are rooted and grown into new plants. The other method is grafting, which is now done on an immense scale. Vegetative propagation must be used also in those plants that do not bear seed, as seedless oranges and seedless grapes.

Commonly, the stock used in such cases is not a mature plant but a seedling. This is often chosen for its hardness or its resistance to diseases and pests. The seedlings are allowed to grow until their roots are well established. The graft is then inserted at the base of the stem. As soon as union has taken place and the scion has started to grow, the shoot of the stock is cut off, so that all substances absorbed by the root are sent into the scion. Grafting of this sort is used in producing nursery stock for rubber plantations, as well as nearly all common fruit trees.

Successful grafting can only take place between plants that are of the same kind or closely related. Others fail entirely to develop any union between the two parts. In nearly all cases, the nature of the scion is constant after grafting, so that one can be sure of the product that will result. Because of this, it is possible to graft several different scions on a single stock. Now infrequently one sees an apple tree bearing many different kinds of apples maturing at different times of the year. Dwarf apple and pear trees are produced by budding, using quince as stock. Grafting also hastens the time of fruiting, grafted plants coming into bearing earlier than those growing from seed.

Bridge grafting is done for a very different reason. Often, trees are completely girdled at the surface of the ground by rodents, especially during the winter months. Damage of this sort is fatal to the trees unless quickly corrected. Correction is done by bridge grafting. This is done by trimming the edges of the girdled region and inserting small twigs across the gap in the bark in such a way that the cambium region of the strips is in contact with that of the tree in which it is inserted. Long sloping ends greatly increase the probability of such contact. These "bridges" unite with the damaged tissues and allow movement of materials to occur. Gradually the damaged and new tissues fill in the gap, and the damage is repaired.

See also **Budding**.

GRAHAM LAW. The rates of diffusion of two gases are inversely proportional to the square roots of their densities.

GRAHAM, THOMAS (1805–1869). Born in Scotland, Graham is famous for his basic studies in diffusion that led to the development of colloid chemistry. He was the first to observe a marked difference in the rate of passage of certain types of substances through a parchment membrane. Those that readily crystallize, like sugar, pass rapidly through the membrane, but gelatinous types are "slow in the extreme." Graham designated the later, which comprise albumin, starch, gums, etc., as colloids and their solutions as colloidal solutions. The former, which he called crystalloids, form "true" or molecularly, dispersed solutions.

See also **Colloid Chemistry**; and **Colloid System**.

GRAIN. (1) The smallest unit of mass in the avoidupois system; 1 grain = 0.0648 gram; one ounce contains 437.5 grains. (2) Any cereal plant, as wheat, corn, barley, etc., (3) Crystalline particles of metals. (4) The dehaired side of a skin of hide.

GRAIN BOUNDARY. The surface separating two regions of a solid in which the crystal axes are differently oriented. It has been shown that such a boundary may be thought of as built up of an array, or network of dislocations, whose spacing depends on the tilt θ of the axes across the surface. The energy (per unit area) of a grain boundary is given by

$$E/E_m = (\theta/\theta_m)\{1 - \ln(\theta/\theta_m)\}$$

where E_m and θ_m are parameters depending on the material.

Grain boundary relaxation is a source of internal friction in solids due to the motion of grain boundaries under stress.

GRAINS. See **Grasses**

GRAIN SIZE. In metallurgy, it is common practice to call the crystals of a polycrystalline metal its grains. The grain or crystal size of metals is determined by microscopic examination of a suitably prepared section. There are two principal standards of grain size in use in the United States. Both are standards of the American Society for Testing and Materials.

For most nonferrous alloys, particularly brass and bronze and other alloys having homogeneous grain structures with twin bands, a set of ten photomicrographs having average grain diameters ranging from 0.010 to 0.200 millimeter are used for direct comparison with microstructures at a magnification of 75 times.

The A.S.T.M. standard grain size chart for steels covers about the same range of average grain diameters but the comparison is made at 100 times magnification and the grain size is expressed by numbers from 1 to 8. The following single equation relates the grain size number to the grain sizes:

$$n = 2^{N-1}$$

where N is the grain size number and n the number of grains per square inch. In general, grain sizes 1 to 3 are considered coarse, 4 to 6 intermediate, and 7 to 8 fine. The grain size of steel can also be judged from a clean fracture if the steel can be fractured without appreciable plastic deformation because the fracture surface mirrors the grain structure. This is possible with most heat-treated machine steels and tool steels, but low-carbon steels are often too tough to break with a crystalline fracture. A series of standard fractures is available for direct visual comparison, and the numbering system for these standards coincides with that of the charts used for microscopic determination of grain size.

The grain size of metals is related to many important properties. In general, fine grain size is an indication of relatively high strength, hardness, and toughness while coarse grain indicates softness and plasticity. However, the hardenability of steels by heat treatment is highest for coarse grain steel. Coarse grain size is usually desirable for creep strength at elevated temperatures.

In the case of sheet and strip for drawing or stamping, coarse grain may give a rough surface. On the other hand, metal with too fine a grain size may lack plasticity and crack in the dies; therefore, a compromise must be reached.

The grain size of castings is generally much coarser than that of wrought products such as rod or sheet. In the case of steel castings the original coarse structure may be refined by heat treatment. This is not possible in the case of most nonferrous alloys because they do not undergo a change in type of crystal structure on heating or cooling.

In the case of hot-rolled or forged metals, the finishing temperature has an important influence on grain size. A high finish-forging temperature, for example, will permit grain growth after recrystallization. In the case of metals finished by cold-working processes, the final annealing temperature establishes the grain size. A high annealing temperature results in coarse grain size.

GRAIN-STORAGE INSECTS. Attack on stored grain varies from region to region. This damage in the United States is divided into four regions as shown by map. Damage is heaviest in the southern region, where long summers and high temperatures permit development of many insect generations during the year. In colder climates, stored-grain insects are generally fewer and less troublesome. However, a large infestation can heat up the grain and cause it to remain active, even in cold weather.

Treatment

In the United States, the rice weevil, the red flour beetle, the lesser grain borer, the saw-toothed grain beetle, and the granary weevil are among the most destructive insect pests of stored grains. In many other areas of the world, the khapra beetle is also very destructive.

Precautionary measures for reducing insect populations in storage areas are simple and straightforward, but must be observed if infestations are to be avoided. Fundamental rules include: (1) Clean the storage bin and area around it; (2) spray bins before storing grain in them; and (3) treat and inspect the grain regularly. When a spray mixture is used to treat bins, only one day's supply should be prepared at one time.

After the bins and storage area have been thoroughly cleaned and sprayed, further steps can be taken to protect against infestation, including: (1) applying insecticide to the grain as it goes into the bin; (2) applying a surface dressing to the grain after it is in the bin; or (3) fumigating the grain to disinfect it. Such protection will last for about one season. Because of the warmer temperatures in region 4 (see Fig. 1), sprays or dusts are less effective than in the other regions.

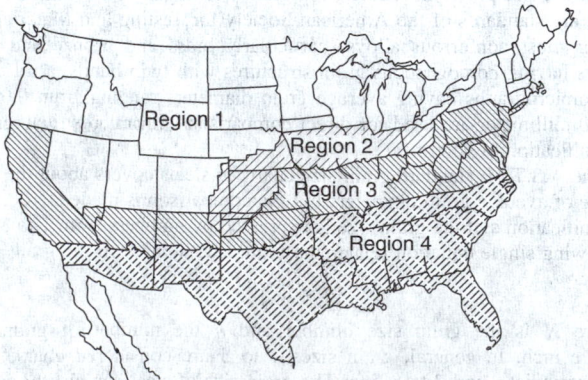

Fig. 1. The map shows, by regions, the degrees to which farm-stored grain in the United States is subjected to insect attack: *Region 1*: Little if any damage occurs to grain on the farm during the first season's storage. *Region 2*: Insects may be troublesome during the first season. *Region 3*: Insects are troublesome every year. *Region 4*: Insects are a serious problem through the storage period.

To get rid of an infestation that is already established, fumigation is almost always necessary. Fumigants are sold under various trade names with ingredients listed on the label.

Fumigants should be applied only by a trained operator wearing a gas mask and equipped with a fresh canister. Before fumigating, grain surface should be level to ensure even distribution of the fumigant. Any crust on the surface of the grain should be broken up. An assistant always should be present during the fumigating procedure. Fumigation should occur within 2 weeks after binning the grain if installation is in region 4; within 6 weeks for regions 2 and 3; and only when required by inspection in region 1. Samples of grain from the center of the bin should be taken once per month for insect inspection. The samples should be sifted through a screen with mesh large enough to let most kinds of insects fall through, but small enough to hold back the grain. Most stored-grain insects are smaller than the grain. Fumigate at once if even only one granary weevil, rice weevil, or lesser grain borer is present. Methyl bromide has been used to fumigate farm-type bins of wheat and corn (maize).[1]

Important Pests

Confused flour beetle (Tribolium confusum, Duval*).* A very common insect of grain storage areas. This insect is also a pest in grocery stores and warehouses and can be a serious pest in flour mills. The beetles are small, about $\frac{1}{7}$-inch (3–3.5 millimeters) long, with elongated bodies and a reddish-brown coloration. Large numbers of them will appear whenever the stored product is slightly disturbed. Mixed with the adult beetles may be found the brownish-white, flat, 6-legged larvae that feed on the inside of the grain kernels. The larvae sometimes are called "barn bugs." In addition to all kinds of grain products, these insects like anything of a starchy nature, including beans, baking powder, peas, dried plant roots, dried fruits, nuts, chocolate, certain drugs, snuff, cayenne pepper, and many other substances. It is interesting to note that they are pests on insect collections. These insects occur worldwide and were first noted in the United States in 1893. The beetle is found more frequently in the northern United States than in the southern states.

Red flour beetle (Tribolium castaneum, Herbst*).* This beetle is closely allied with the confused flour beetle, but occurs more in the southern climes than in the north. It is seldom encountered north of latitude 41°N. Under the best circumstances, from four to five generations of both of these beetles will take place per year.

Saw-toothed grain beetle (Oryzaephilus surinamensis, Linne*).* The feeding habits are much the same as those for the confused flour and red beetle. It can penetrate packages that would seem to be tightly sealed. Often these beetles will follow the damage of other insects because it cannot successfully devour sound seeds. Distribution is worldwide. Only the adult stage overwinters in unheated structures. Normally there are from four to six generations per year. Under the best of conditions, the entire life cycle of this insect could occur in less than one month.

Granary weevil (Sitophilus granarius, Linne*).* The granary and the rice weevil are considered by some experts as the most destructive of all grain insects. They are true weevils. If undisturbed these weevils can cause almost complete destruction of grain stored in elevators, ships, and on the farm. A telltale of infested grain is a rise of the surface temperature as well as wetness. Sometimes, sprouting of seed will be noted. The beetles have prominent snouts, which are an advantage in feeding upon grain. The larvae prefer the interior of the kernels. Substances particularly attractive to the granary weevil include buckwheat, barley, maize (corn), macaroni, oats, kaffir seed, and wheat. The weevil is distributed widely throughout the world, but less abundant in tropical and semitropical areas.

The weevil overwinters as adult or larva. The adult can withstand subzero temperatures for many hours. The adult weevil is dark brown or nearly black and has ridged wing-covers and a long snout extending downward from the front of the head. Length is about $\frac{1}{16}$-inch (1.5 millimeters). The female weevil deposits her eggs (from 300 to 400 small and white) in small cavities found in the grain kernels. Legless, soft, fleshy, white grubs are hatched within a few days and immediately commence feeding on the interior of the kernel. When fully grown, the larvae are about $\frac{1}{8}$-inch (3 millimeters) long. The full life cycle ranges from 4 to 7 weeks. The adults can go for long periods without food, if necessary, and

---

[1] This application is well described in Report 929, U.S. Department of Agriculture, Washington, DC, (revised periodically).

it has been estimated that they can live over 2 years on a starvation diet. There are four to five generations of granary weevils per year.

Rice weevil (Sitophilus oryza, Linne). This beetle is very similar in construction and habits to the granary beetle. A major difference is that the rice weevil has well- developed wings and can fly, and frequently does so, particularly under warm conditions. The granary weevil's wing covers are grown together, keeping it from flying.

Mealyworms (Tenebrio molitor, Linne, yellow; and *T. obscurus,* Fabricius, dark colored). These worms have shiny bodies, yellow-to-brown in color, with smooth coats. They have some resemblance to wireworms of black beetles. They are relatively large, about 1 inch (2.5 centimeters) in length. They are found in dark, damp locations where grain or other attractive substances have been stored for a long period. In addition to grain and grain products, mealyworms enjoy feathers, dead insects, and scraps from meat packing operations. Native to Europe, mealyworms are distributed worldwide. As adults the two species are much alike and often are difficult to distinguish. From about 1 to 3 weeks are required for eggs to hatch. The eggs are placed in stored food substances. One female may lay as many as 250–1000 eggs.

Cadelle (Tenebroides mauritanicus, Linne). If not present in overabundance, the cadelle beetle can help in controlling the population of other damaging beetles, because the adults often kill and feed upon other insects. However, in this regard, they are not considered to be predaceous insects. On balance, the cadelle is a serious pest of grain bins and like storage places. This insect is notably damaging in flour mills, not only consuming grain, but also destroying flour sacs, cloth used in machinery, cardboard containers, etc. The insect may overwinter as an adult or larva, but not as a pupa. The black adult beetle ranges in length from $\frac{1}{3}$ to $\frac{1}{2}$ inch (8 to 12 millimeters). A single female can lay up to 1300 eggs in cracks and crevices. Hatching occurs within 1 to 2 weeks. The resulting larvae are an off-white color, have prominent black heads, some black spots, and two hooks at the rear of the body (a bit like an earwig). When fully developed the larvae are about 2/3-inch (17 millimeters) in length. These larvae have the additional bad habit of boring into wood as may be found in grain bins, ships holds, etc. Tunnels in wood make excellent hiding places, where the pupal stage is passed. The full development period of the cadelle is considerably longer than most of the other grain pests, ranging from 7 to 14 months, although the average time span is 2 to 3 months. The cadelle adult has been known to live as long as 3.5 years.

Lesser grain borer (Rhyzopertha dominica, Fabricius). Also known as the Australian wheat weevil, the insect is widely distributed throughout the southern and midwestern United States. It is rarely found in the northern states. Adult beetles are brown-to-black, cylindrical in shape, about $\frac{1}{8}$-inch (3 millimeters) long by $\frac{1}{4}$-inch (6 millimeters) wide. The larvae appear as grubs and assume a curved posture. They are about $\frac{1}{10}$ inch (2.5 millimeters) long. The habits of this insect are similar to the other insects described, but they have a wider range of attractive feeding substances. In addition to grain, they like seeds, certain drugs, dry roots, cork, wood, and paper boxes. But, it thrives in wheat and is one of the most common of the wheat pests. The *larger grain borer (Dinoderus truncatus,* Horn) is similar to the lesser grain borer in most respects, with exception that it is a bit larger, and prefers corn to wheat. It does not occur widely in the United States, with the exception of a few locations in the southern states.

Angoumois grain moth (Sitotroga cerealella, Olivier). A buff-colored and delicate adult moth having a wingspread of from $\frac{1}{2}$ to $\frac{2}{3}$ inch (12 to 18 millimeters) was first found to be a damaging insect on wheat, corn (maize), and other grains in France (Province of Angoumois) in about 1736. It occurs in many parts of the world, including all of the United States. The other stages of the insect are seldom seen because the larvae and pupae habitate the internals of seeds and the eggs are extremely tiny. The fully grown larva is about $\frac{1}{5}$ inch (5 millimeters) in length. The eggs are deposited by the female moths in the hundreds on grain in the shock or on the heads in the field. Only 1 to 4 weeks is required for hatching, at which time the larvae burrow into the kernel. The full life cycle is about 5 weeks. The larvae may overwinter in the grain. Thus, the insect goes with the grain into storage and adults may appear from time to time while the grain is in storage. Reproduction can continue during storage. In mild latitudes, there are about two generations per year, whereas in southern climates, there may be as many as six generations per year. The insect not only destroys corn in the crib, but also damages ripening grain in the field before storage.

Mediterranean flour moth (Ephestia or Anagasta künniella, Zeller). At one time this was a very serious pest in flour milling operations. Fumigating procedures have largely brought the insect under control. Conveyors and chutes that carry flour may be webbed over when the small caterpillars are present. A telltale is a number of small gray moths that will be present in infested structures. Although the insect prefers flour, it will also feed upon breakfast cereals, maize, bran, and whole grain wheat. It will also feed on pollen in beehives. Although widely distributed in the United States and Canada (first reported in 1889), the insect is also found in many other regions of the world. The life cycle requires from 9 to 10 weeks. The eggs are laid in crevices, cracks, undisturbed accumulations of flour, etc. The eggs hatch within less than a week, after which the caterpillars spin silken threads to form small tubes, in which they live and feed. The web spinning and the clogging of machinery that results is the principal damage caused by the insect.

Indian meal moth (Plodia interpunctella, Hübner). In addition to feeding on grain, this insect (native to Europe) feeds on breakfast cereals, soyabean, nuts, seeds, dried roots, dead insects, powdered milk, beehive pollen, and soybean. The insect is also a pest in museums where it attacks specimens. The Indian meal moth is also a serious pest in confection factories. Distribution is throughout the United States and many other regions of the world. All phases of the life cycle (4 to 6 weeks) can be present at the same time, with exception of unheated structures during winter, under which conditions the insect winters over as a larva. As with the Mediterranean flour moth, a principal damaging aspect of this moth is its web spinning and its binding together dirt with larvae excreta in the nearness of processed foods and processing machinery that is subject to clogging and jamming by the webs.

The *flour mite* is described under **Mite**.

See also **Khapra Beetle**.

Additional Reading

Subramanyam, B. and D. Hagstrum: *Integrated Management of Insects in Stored Products,* Marcel Dekker, Inc., New York, NY, 1999.

White, N. et al.: *Stored-Grain Ecosystems,* Marcel Dekker, Inc., New York, NY, 1995.

Williams, D. et al.: http://muextension.missouri.edu/xplor/extcirc/ec0960.htm *Grain Storage Management: A Guide for Keeping Your Grain in Top Condition* (abstract only), University of Missouri, Columbia, MO, 1999.

GRAM. See **Units and Standards**.

GRAM-ATOM. That quantity of an element having a mass in grams numerically equal to the atomic weight. One gram-atom contains the Avogadro number of atoms.

GRAM-CHARLIER SERIES. This series attempts to represent frequency functions in statistics by an expansion, resembling a Taylor series, in terms of derivatives of the normal (Gaussian) distribution.

$$F(x) = \sum_{k=0}^{\infty} c_k e^{-x^2/2} H_k(x)$$

where the constants c_k depend on the frequency function represented over the interval $[-\infty, \infty]$ and the $H_k(x)$ are the Hermite polynomials. The Gram-Charlier series is similar to the Edgeworth series, and indeed the two are identical for infinite series; their difference arises in regard to the stoppage point when a finite number of terms only is taken as an approximation, in which case Edgeworth's form is probably preferable. See also **Edgeworth Series**.

GRAM-EQUIVALENT. The gram-atomic weight of an element (or formula weight of a radical) divided by its valence. In the case of multivalent substances there will be more than one value for the gram-equivalent, viz., Fe(II) = 27.92 grams, Fe(III) = 18.61 grams, and the proper value for the particular reaction must be chosen.

GRAM-MOLECULAR WEIGHT. That amount of a pure substance having a weight in grams numerically equal to the molecular weight. One gram-molecular weight contains the Avogadro number of molecules. It is also designated as the mole or mol.

GRAM STAIN. A method of staining micoorganisms which enables such organisms to be classified into two main groups, those which retain the stain being described as *Gram-positive*; and those from which the stain is decolorized being described as *Gram-negative*. The organisms are first stained with either gentian violet or its analogue, crystal violet, and then treated with the solution of iodine. An organic solvent, usually alcohol, is then applied which washes out the stain from the Gram-negative organisms, leaving Gram-positive organisms with the violet stain unaffected. A counterstain of some contrasting color is then applied to demonstrate the Gram-negative organisms. Gram-positive organisms include staphylococci, streptococci, pneumococci; among the Gram-negative organisms are gonococci, meningococci, *Bacillus coli*, and the salmonellae.

GRANDFATHER CLOCK.   See **Clocks**.

GRAND MAL.   See **Seizure (Neurological)**.

GRANITE.   This name is applied to a common and widely occurring group of deep-seated igneous rocks consisting of orthoclase, plagioclase, quartz, hornblende, biotite, muscovite and minor accessories such as magnetite, garnet, zircon and apatite. Rarely, a pyroxene is present. Ordinary granite always carries a small amount of plagioclase, but when this is absent the rock is then referred to as an alkali-granite. An increasing proportion of plagioclase feldspar causes granite to pass into granodiorite. A rock consisting of equal proportions of orthoclase and plagioclase plus quartz may be considered a quartz monzonite. A granite containing both muscovite and biotite micas is called a binary granite.

The word granite comes from the Latin *granum*, a grain, in reference to the grained structure of such a crystalline rock.

Granite occurs as stock-like masses and as batholiths often associated with mountain ranges and frequently of great extent. Granite has been intruded into the crust of the earth during all geologic periods, except perhaps the most recent; much of it is of pre-Cambrian age. Granite is widely distributed throughout the earth.

Graphic granite is a coarsely crystalline variety of granite or pegmatite composed almost entirely of quartz and feldspar that have intergrown in such a manner as to simulate Semitic or cuneiform characters.

GRANITOID.   A textural term derived from granite and signifying the relatively uniform and coarse grain of batholithic rocks, such as granite, syenite, anorthosite, etc. In a typical granitoid rock, each species of mineral occurs as a single generation; the silicates crystallize first, and any surplus of free silica crystallizes last in the form of quartz, or is finally driven off with the surplus water to form quartz veins.

GRANULITE (also Leptite).   This is a general term for a group of rocks that vary considerably in composition but for the most part seem to be derived by metamorphic processes from quartz-feldspar rocks. The classic locality for granulite is in Saxony, where there occurs a granular gneiss of quartz and feldspar plus such accessory minerals as pyroxene and garnet, with occasionally small quantities of kyanite, spinel and similar minerals. The Saxon granulites have a decided banded structure and seem to resemble injection gneisses. It appears reasonable to suppose that these and other granulites may have been derived from sedimentary formations severely altered by igneous processes. Leptite is a term used in the Scandinavian countries for fine-grained granulites that originally were rhyolitic tuffs and lavas.

Other than in Saxony and Scandinavia, these rocks are found in the northern highlands of Scotland, India, West Africa, and Canada.

GRANULOCYTES.   See **Blood**.

GRAPE-LEAF FOLDER (*Insecta, Lepidoptera*).   A moth, *Desmia funeralis*, whose larva eats the leaves of grapevines and lives in a fold fastened with silk. It is not an important pest.

GRAPE-LEAF SKELETONIZER (*Insecta, Lepidoptera*).   A moth, *Harrisina americana*, whose larvae, working in groups, destroy the soft tissues of the grape leaf, leaving the network of veins. It is rarely an important pest.

GRAPE PHYLLOXERA (*Insecta, Homoptera*).   A sucking insect related to the plant lice aphids and scale insects. The many species make up a subfamily, which, with the adelgids, constitutes the family *Phylloxeridae*.

They differ from the aphids in that all females lay eggs and form the scales in their more complex structure, including the four wings of the winged stages.

The most important phylloxerid is a species (*Phylloxera vitifoliae*, Fitch) which attacks grapevines, working on the leaves and roots. It once threatened to ruin the vineyards of France and has destroyed millions of acres of vines. The use of roots of certain American grapes, which are not seriously harmed by the pest, has greatly lessened the danger from its attack. Tender varieties are grafted onto the resistant roots.

GRAPES AND WINES.   Of the family *Vitaceae* (grape family), grapes are climbing plants of numerous species that have been cultivated for centuries for their fruits and the various products obtainable from them.

Climbing in grapes is made possible by tendrils, modified stems that coil tightly around any suitable support. These tendrils are usually interpreted as terminal portions of the stem, which have been pushed to one side by the more rapid growth of an axillary bud. The leaves of grapes are simple, palmately lobed and alternate, with small stipules. The stems elongate rapidly and are of a coarse porous nature; the internodes of young stems are frequently hollow, the nodes solid. The flowers are borne in compact panicles. Each flower is small and inconspicuous. The calyx is a mere rim around the tip of the pedicel; the corolla five-parted and greenish. When the flower opens, the petals, united at their tips but free at the base, are forced away from the base of the flower and drop off. There are five stamens and a single pistil. The fruit is a 2-celled berry. See Fig. 1.

Fig. 1.   Grapes ready to harvest. (*U.S. Department of Agriculture.*)

Commercial grapes are largely derived from three species, *Vitis vinifera*, the wine grape of Europe, a native of Asia, *Vitis labrusca*, the northern fox grape of eastern North America, and *Vitis rotundifolia*, the southern fox grape. Many varieties and hybrids of these exist, as well as hybrids with other wild species. In commercial vineyards, grapevines are variously pruned to increase yield and improve quality. Pruning cuts are made through the nodes, to prevent the leaving of hollow internodes in which disease might gain entrance to the plant. Propagation of the grape is mainly by means of stem cuttings, a method used in Europe for centuries.

Grapes are used as a table fruit, as raisins when dried, and for making wine. Fewer than a dozen important varieties of grapes are grown for table grapes. Most of the sweet juice produced in North America is from the *Concord*. Only a few varieties are used for canning. *Concord* grapes are used extensively for juice and also for jams, jellies, puddings, and pies. Table grapes, such as *Emperor, Thompson Seedless, Tokay, Cardinal, Ribier*, and others are mostly eaten out of hand, but are also used in salads, fruit cups, pies, puddings, cakes, stewed fruit, and as meat accompaniments. See Fig. 2. Dried or raisin grapes are mainly *Thompson Seedless* (also known as *Sultanina*), *Black Corinth*, and *Muscat of Alexandria*. A variety closely related to *Thompson Seedless* is important and dominates the raisin vineyards of Greece, Iran, and Turkey. Remarkably few grapes are well suited for wine as well as fresh (table) use or raisin production. Worldwide, the *Muscat* grape is considered a triple-purpose grape. There are numerous subvarieties of the *Muscat*, but all possess the characteristic *Muscat* odor and flavor. For wines, the *Muscat* is used principally in making sweet, fortified wines. In California, the *Thompson Seedless* grape plays the three roles and some production is used in making wines. Wines from this grape, however, tend to be rather neutral and bland and thus are mainly used for blending purposes.

Fig. 3.   Raisins drying in a California field are protected from insects through the use of treated paper. (*U.S. Department of Agriculture.*)

Fig. 2.   Close-up of table variety Steuben grapes. (*U.S. Department of Agriculture.*)

Fig. 4.   Raisins entering a California processing plant first go over a shaker to remove stems and foreign materials prior to final cleaning, inspecting, and packaging. (*U.S. Department of Agriculture.*)

Raisins. These are either sun-dried or artificially dried grapes. Because of the risk of rainfall occurring during the drying season, artificial drying has become increasingly popular among growers. The principal problem is the requirement for additional energy. When this process is used, the fresh grapes go through a hot caustic solution, which removes the waxy coating (bloom) and makes tiny cracks in the skins. The grapes are then spread onto long, shallow wooden trays. In the case of golden seedless raisins, grapes for processing are transferred to a chamber where they are exposed to sulfur dioxide for about five hours. This treatment prevents darkening during drying. The grapes are then transferred to dehydrating tunnels where they are exposed to warm, dry air for about 18 hours. Raisins require a residual moisture content because most consumers do not like a thoroughly dry or crispy raisin. Residual moisture encourages mold and yeast growth. Protection can be obtained by dipping the fruit

in weak solutions of potassium sorbate, thus leaving a fine coating of the antimicrobial agent on the fruit pieces. See Figs. 3 and 4.

Wine Grapes. Among the better known wine grapes are *Cabernet-Sauvignon, Chardonnay, Chenin Blanc, Gamay, Grenache, Grignolino, Gutadel, Müller-Thurgau, Pinot Noir, Riesling, Sauvignon Blanc, Sémillon, Silvaner, Trollinger*, and *Zinfandel*. As indicated by their names, most of the famous wine-variety grapes were originated in Europe, notably in France, Germany, Italy, Spain, and Austria.

Scores of varieties of grapes are used for wine production. Some of the more important varieties are listed in Table 1.

TABLE 1. PRINCIPAL WINE GRAPES OF THE WORLD

ALEATICO. Native Italian grape. Produces red wine with a Muscat flavor.

ALICANTE. Another name for Grenache grape (See this list).

ALIGOTE. White grape extensively grown in France for production of White Burgundies.

ARAMON. Productive red grape grown in France and California. Quality of red wine is marginal.

AURORE. A French-American hybrid grape, originally designated Seibel 5279, the parentage of which includes *Vitis linecumii*, *V. rupestris*, and *V. vinifera*. As observed by Cobb et al. (1978), although identification of flavor components from many studies on grape juices and wines from California and Europe have appeared, very little information is available on volatile components of native North American species and hybrids. Considerable advancement has been made in this direction by Cobb et al. See reference listed.

BARBERA. Source of red wine produced in Italy; to a lesser extent in California.

BLANC-FUMÉ. Local French name for Sauvignon Blanc grape (See this list).

BOAL. A Sherry wine grape cultivated in Madeira. Also spelled Bual.

BONARDA. Native Italian grape. Produces red wines.

BOUCHET. Local French name for Cabernet-Sauvignon grape (See this list).

BOUSCHET. A hybrid (named after Henry Bouschet), very productive grape for producing high-volume wines. Found in Algeria, California, France.

BRACCHETO. A native Italian red grape.

CABERNET-SAUVIGNON. Renowned red grape used in production of superb Clarets of Bordeaux. In addition to France, the variety is cultivated in Australia, California, Chile, and South Africa, among other countries. The Ruby Cabernet extensively planted in California is a cross of the Carignane and the Cabernet-Sauvignon. In the Saint Émilion district of France, the Cabernet Franc is the principal variety.

CARIGNANE. A productive wine grape used for ordinary red table wines. Cultivated in Algeria, California, France, Israel, and Spain.

CATAWBA. A light-red grape native to North America and used in making Ohio wines and New York State Champagnes. First found in the Carolinas, vineyards were later concentrated in New York and Ohio. It is of the Vitis lubrusca family. This species is also cultivated in Canada.

CÉPA or CÉPAGE. A prefix used for varieties of grapes that have been grown from vine stock that has been transferred to a new area. Cépa means individual vine or vine stock. Thus one finds in parts of Spain, grapes with the names Cépa Chablis; Cepa Médoc; Cépa Borgona, etc.

CHARDONNAY. Renowned white grape used in production of superb White Burgundies (Chablis, Montrachet, Pouilly-Fuissé, etc.) in France. It is also the white grape used in production of Champagnes. The grape is also cultivated in Alsace and California. Although the term Pinot is sometimes used in connection with this variety, such as Pinot Chardonnay (considered by some authorities as the best American white table wine), botanists have not established a true relationship between the Pinot Noir and Pinot Blanc, among others. In recent years, a French hybrid of the Chardonnay has been cultivated in Canada.

CHASSELAS. A white and sometimes pink grape cultivated in Alsace, Australia, France, Germany, and Switzerland. It is known as the Gutadel in Germany. The variety has not done well in California. In Europe, it is also a table grape. The Chasselas produces wines of medium quality.

CHENIN-BLANC. Very highly regarded white grape and sometimes referred to as the Pineau de la Loire. In addition to France, the variety is successfully cultivated in northern California (Napa, San Benito, Santa Clara, and Sonoma countries, in particular). White wines made from this variety are the predominant wines in several of the French provinces where it is grown. The variety is sometimes referred to as the Pinot Blanc, in error.

CINSAULT. A high-quality grape that yields deeply colored red wines. It is also used in production of rosé wines. Primarily cultivated in France.

CLAIRETTE. A white wine grape, cultivated mainly in southern France. It is grown in California, but not extensively.

COLUMBARD. A white wine grape of good quality, cultivated principally in France (Cognac district). Also cultivated in California, where it is sometimes called the French Columbard. Wine from the Columbard is sometimes blended with other California wines, such as Chablis and in some California Champagne. The grape is also well suited for distillation.

CONCORD. A blue-black grape native to North America and used primarily in making Kosher-type wines, unfermented grape juice, and jellies. The wine is also used in New York State Burgundies and Ports. The concord is also grown in Canada. The grape is of Vitis labrusca. This grape is of much current interest as a source of food colorant. See Calvi and Francis (1978) reference.

CORTESE. A native Italian white grape. Quality is generally considered superior.

CORVINO. A native Italian red grape. Wine production and distribution is essentially limited to northern Italy.

CROATINA. A native Italian red grape. Cultivated mainly in Italy.

DELAWARE. A native North American pink grape that produces white juice. It is cultivated in New York State and Ohio as well as in Canada for making table wines. It is one of the most widely planted of the native North American varieties.

DIANA. A native North American grape. Produces red wines. It is cultivated in the eastern United States and Canada.

DUCHESS. A native North American white grape grown in the eastern United States and Canada for making medium-quality white wines.

DURIFF. A red wine grape grown mainly in France, which somewhat resembles the Syrah (See this list). Some botanists observe that the Petit Sirah grown in California may be, in actuality, a Duriff grape.

ELBING. Although of less than superior quality, this variety is highly productive and grown in Alsace, California, Germany, and Luxembourg, mainly for production of less-expensive sparkling wines. The variety is no longer considered a legal grape in Alsace.

ELDERBERRY. Not of the genus *Vitis*, but rather of *Sambucus*. Can produce wine, but much added sugar is required. Elderberry juice no longer can be used to color light-red wines and Ports.

ELVIRA. A native North American grape. Used for production of white wines. It is sometimes referred to as the "Missouri Riesling" but is not related. Cultivation is mainly in the eastern United States and Canada, with a concentration of vineyards in the Finger Lakes region of New York State.

ERBALUCE. Native Italian white wine grape.

FOLLE BLANCHE. Mainly grown in France for production of white wines. The variety is cultivated in California and is sometimes used in California "Chablis" and California Champagnes.

FRIULARO. A native Italian red wine grape. Mainly cultivated and distributed in and near Venice.

FURMINT. The well-known white grape cultivated in Hungary and the basis of Tokay wine and other Hungarian wines. The variety is also cultivated in Rumania.

GALEGO DOURADO. A white wine grape cultivated mainly in Portugal.

GAMAY. A highly regarded red wine grape and dominant in the Beaujolais country of France. The Gamay grape planted in California is considered of an inferior quality, although it is considerably more productive.

GARGANEGA. A native Italian white grape, sometimes used as a blend with other white wines and the principal constituent of Soave wine.

GEWÜRZTRAMINER. A pink wine grape derived from the Traminer and cultivated in Alsace, Germany, and Italy.

GRENACHE. Also called the Alicante, this is a red grape cultivated in California, France, Germany, Israel, and Spain. It is used in sweet and heavy dessert wines, in some vin rosés, and California Ports. A white variety is much less widely grown.

GRIGNOLINO. A native Italian red wine grape of highly regarded quality. The color of the wine is somewhat different from the usual reds, having a crimson coloration. Some Grignolino grapes have been planted in southern California where they are used for producing vin rosés.

GROLLEAU. Also commonly called the Groslot, this is a red wine grape of medium quality, but quite productive. It is cultivated mainly in France for production of less-expensive wines.

GROPELLO. A native Italian wine grape used in lesser-known red wines.

GROS PLANT. Another name for the Folle Blanche grape (See this list).

GUTADEL (Weisser Gutadel). This vine requires sites that are well sheltered against winds and with a rich, deep humus soil, found most readily in the Baden region of Germany. Ripening period falls between that of the Müller-Thurgau and the Silvaner. The wine is light, pleasing, and agreeable. The soft, sweet Gutadel grape is also appreciated as a dessert grape. Of vineyard areas in Germany, this grape represents 1.4% of total.

HANEPOOT. A variety of grape grown in South Africa for production of South African Sherries.

IONA. A native North American grape. Although grape is reddish-purple, it produces white wines. Principal vineyards are near the Finger Lakes and along the Hudson River in New York State.

ISLAND BELLE. A native North American grape of rather poor quality, but planted in Washington State along the Pacific Coast. Wine produced from the variety has what is known as a foxy flavor.

IVES. A native North American grape of relatively poor quality. It is used in some New York State Burgundies. Wine from this variety are considered rather coarse and with a foxy flavor.

JAMES. A native North American grape of the Muscadine family and sometimes used for making local wines in the southeastern states.

JOHANNISBERG RIESLING. A number of untrue Rieslings, such as the Franken Riesling and Grey Riesling are planted in California. Johannisberg Riesling is used to indicate a true Riesling grape. One of German's most famous vineyards is located in Johannisberg. See also Riesling this list.

TABLE 1. (*Continued*)

KADARKA. A native Hungarian red wine grape. Plantings are extensive. The Zinfandel is no longer regarded as identical with the Kadarka. The Kadarka is used in production of a number of red wines in Hungary.

KERNER. A relatively recent development, out of the Trollinger and the Riesling vine. The grape grows in all soil conditions. Favored regions in Germany include the Württemberg, Rhenish Palatinate, and Francoia areas. The wine is lively, pleasing, Rieslinglike, with a light muscat bouquet. Of vineyard areas in Germany, this grape represents 2% of total.

KNIPPERLÉ. A rather poor-quality white wine grape grown in Alsace. Most wine made from it is for local consumption.

MALBEC. A highly regard red wine grape, found mainly in the Bordeaux district of France, but also planted in Australia, Chile, and Israel. The wines from this grape are considered well-balanced. Also called Cot or Pressac.

MALMSEY. See Malvasia (this list).

MALVASIA. A native of Greece and considered of ancient origin. This white grape is now found in several parts of the world, including California, France, Madeira, Portugal, and South Africa. In France, the grape is called Malvoisie. The grape is used for producing medium-quality table wines and some sparkling wines.

MARATHEFTIKA. A native of Cyprus and used for producing local red wines.

MATARÓ. See Mourvedre (this list).

MAVRODAPHNE. Widely planted in eastern Europe and the Balkans, it is the basis of many red wines produced in these regions, including various sweet wines and Ports.

MAVRON. A native of Cyprus and used for producing local red wines.

MELON. The preferred name for this white wine grape is Melon, although it is locally known in the Loire Valley region of France as the Muscadet. Wines have a muscat flavor.

MERLOT. A well-regarded red wine grape, somewhat comparable in quality with the Cabernets in Bordeaux region. Merlot wines tend to be somewhat more mellow than the Cabernets. The Merlot is also cultivated in California, Chile, Italy, and Switzerland.

MEUNIER. Related to the Pinot Noir, the Meunier is highly regarded, but not quite on same level as the Pinot Noir. The grape is planted in the Burgundy and Champagne districts of France as well as in California. The Meunier is sometimes confused with the Pinot Noir in California.

MISSION. Although botanists consider this variety as originating in Europe, it has been raised in Mexico for a number of centuries. It is now planted in California and is used principally for production of Angelica, a marginal wine. The grape is not highly regarded by the experts.

MOLINARA. An exceptional red wine grape native to Italy.

MOORE'S DIAMOND. A native North American grape planted in the eastern United States and Canada, and used for producing a tart, pale wine. Vineyards are concentrated in the Finger Lakes district of New York State.

MORIO-MUSCAT. This grape is a crossing of Silvaner and Weisser Burgunder (Pinot Blanc). It ripens fairly early and gives a very good yield. The vine grows particularly well in the Rhenish Palatinate and the Rheinhessen wine-producing regions of Germany. Its wine has a strong muscat bouquet, which can become very potent in very ripe wine. This grape represents 3.1% of the total vineyard areas in Germany.

MOURASTEL. A red wine grape grown in parts of California and used mainly in making common red wines.

MOURVEDRE. A red wine grape extensively planted in California and about of equal quality with the Carignane. The grape is quite productive and is regarded as of French or Spanish origin. There are some plantings in France. It is used mainly for producing common red wines. Also called Mataró.

MÜLLER-THURGAU. A Geisenheimer cultivation, produced in 1862 by the Swiss cultivator, Prof. Müller, from the kanton of Thurgau. The Müller-Thurgau is thought to be a cross of Riesling and Silvaner. The grape ripens early and brings a good yield of mild, well-balanced, forthcoming wine with a delicate muscat bouquet and taste. The vine is found mainly in the Franconia, Rheinhessen, Baden, and Nahe wine-producing regions of Germany. Geisenheim, where the grape was developed, is an important wine-producing area in the Rheingau. Of vineyard areas in Germany, this grape represents 27.2% of total.

MUSCADELLE. A white wine grape cultivated principally in the Bordeaux district of France. It is sometimes planted in with the vines of the Sémillon and Sauvignon Blanc. The grape provides a Muscat flavor to finished wine. The variety also has been planted in South Africa.

MUSCADET. See Melon (this list).

MUSCADINE. A native North American grape, found in the southeastern United States. The wine has a characteristic flavor. Considerable sugar must be added to the juice prior to fermentation. The grape is also widely used for jellies, candies, etc.

MUSCAT. Numerous subvarieties of this grape exist, ranging in color from yellow to blue-black. Thus a variety of wines and uses are made of it, including sweet red dessert wines and use as a blend with some Sauternes.

The muscat is also popular as a table and raisin grape. Plantings are widespread, including Alsace, Austria, California, Cyprus, France, Greece, Hungary, Israel, Italy, Portugal, Spain, and Tunisia, as well as other Mediterranean countries.

NEBBIOLO. A native Italian red wine grape.

NEGRARA. A native Italian red wine grape and highly regarded for making excellent red wines.

NIAGARA. A native North American white grape used for making sweetish table wines of a golden color. Principal vineyards are in the Finger Lakes district of New York State. The Niagara grape is also grown in Canada on the Niagara Peninsula.

OPTHALMA. A native red wine grape of Cyprus.

PALOMINO. A variety of grape cultivated mainly in Spain for making Sherry. The grape is also planted in South Africa.

PEDRO XIMÉNEZ. A variety of grape cultivated mainly in Spain for making Sherry. Resulting wine is quite sweet and is a main contributor of sweetness to Spanish Sherries.

PETIT SIRAH. A red wine grape variety with extensive plantings in California. It is related to the Syrah of Hermitage (See this list), but is much more productive. Some authorities, however, believe that the Petit Sirah is actually the Duriff (See this list). It is used essentially for producing common red wines.

PINEAU DE LA LOIRE. A well-regarded white grape and the basis for some of the better white wines of France. It is an ingredient of the better California Champagnes. Proper name is Chenin Blanc (See this list).

PINOT BLANC. Planted mainly in Alsace, France, Germany, and Italy. Yields a white wine of good quality.

PINOT CHARDONNAY. See Chardonnay (this list).

PINOT GRIS. Related to other members of Pinot varieties, it is called the Ruländer in Germany. Sometimes it is incorrectly referred to as Tokay. The rose-gray grapes yield white wines. Plantings are rather widespread, including Alsace, California, France, Germany, Hungary, Italy, Luxembourg, and Rumania.

PINOT NOIR. Regarded by most authorities as one of the superior red wine grapes. It is the basis for excellent red wines and is also used in Champagnes. Plantings are widespread and, in addition to France, are found in Alsace, Australia, California, Canada (hybrid is used), Hungary, and Italy.

PORTUGIESER (Blauer). This blue grape did not originate in Portugal, but was introduced into Germany around 1800 from the Danube region. The grape is deep blue and the vine is modest in its demands of site and soil. It grows mainly in the Ahr, Rhenish Palatinate, Württemberg, and Rheinhessen wine-producing regions of Germany. The grape ripens early. Yields a pleasant "little wine" (Carafe wine); light, agreeable, mild. The wine is red. Of vineyard areas in Germany, this grape represents 4.9% of total.

PROSECCO. A native Italian white wine grape that grows north of Venice. The variety yields a number of sparkling and semisparkling wines of good quality.

RARA-NJAGRA. A red wine grape grown in the U.S.S.R.

REFOSCO. A native Italian red wine grape of fair quality used in making common red table wines for local consumption. Some years ago, the variety was planted in California.

RIESLING. Considered by many authorities as the noblest white wine grape known. Small, insignificant-looking berries, very late ripening; finds favorable growing conditions in all German regions, particularly in the Mosel-Saar-Ruwer region, Rheingau, Rhenish Palatinate, and the Nahe and Mittelrhein regions. Riesling wines are racy, usually of high quality, and delicately fragrant. Not to be confused with other vine species, such as the Welsch or Italian Riesling. However, the species has been extensively transplanted and is now found in Australia, Austria, California, Chile, Luxembourg, Rumania, South Africa, and Switzerland. Of vineyard areas in Germany, this grape represents 21.4% of total.

RIVANER. A white wine grape, representing a crossing of Riesling with Sylvaner; and with Müller-Thurgau. The variety is found principally in Luxembourg.

RONDINELLA. A native Italian red grape wine of principal interest locally.

ROUSSANE. A white wine grape found mainly in France and capable of yielding fine quality wines.

RULÄNDER. German variety of the Grauer Burgunder (Pinot Gris). The grape is of medium-size, heavy, and strong. The vine prefers a rich, deep soil. Ripens relatively early, but may extend late into the season. The species favors the growing conditions found in the Baden, Rhenish Palatinate, Rheinhessen, and Hessische Bergstrasse wine-producing regions of Germany. The wine is fiery, full-bodied and of uniquely delicate bouquet. Its Spätlese and Auslese belong to the range of German high quality wines. Of vineyard areas in Germany, the Ruländer represents 3.7% of the total.

SACY. A white wine grape found principally in France.

(*continued*)

TABLE 1. (*Continued*)

SAN GIOVETO. A highly regarded native Italian red wine grape and the most important variety cultivated in the Chianti country of Italy. Some plantings have been made in California.

SAPARVI. A red wine grape grown in the U.S.S.R.

SAUVIGNON BLANC. Sometimes only the word Sauvignon is used to identify this outstanding white wine grape. Some authorities believe that this variety is only second in quality to the Chardonnay or true Riesling. It is extensively planted in the Graves region of France. The variety is also planted in California, Chile, South Africa, and the U.S.S.R.

SCHEUREBE. A relatively new breeding cross between Silvaner and Riesling. The grape ripens late. Grows well in Rheinhessen, Rhenish Palatinate, and Franconia regions of Germany. Produces full-bodied, flowery wines of Riesling character. Its bouquet is strongly aromatic, reminiscent of black currants. Of vineyard areas in Germany, this grape represents 2.7% of total.

SCHIAVA. A native Italian red wine grape as well as an excellent table grape. Wines produced from this variety are highly regarded.

SÉMILLON. A highly regarded white wine grape. It is grown in the southwestern part of France and is often planted along with Sauvignon Blanc. The wine from this grape blends well with wines that have a hint of sweetness. The Sémillon is also planted in Australia, California, Chile, and Israel.

SERCIAL. A high-quality white grape used in producing dry wines of Madeira.

SILVANER (Sylvaner). A well-regarded, productive white wine grape that originated either in Austria or Germany. Although most extensively planted in Germany, where it represents 17.2% of total vineyard area, the Silvaner is also found in Austria, California, and Chile. In Germany, the Silvaner is grown predominantly in Rheinhessen, Rhenish Palatinate, the Nahe and Franconia regions. The grape is of medium-size, very juicy, producing a pleasant, mild wine with a pleasing low-acid content.

SPÄTBURGUNDER (Blauer). German variety of the Pinot Noir and has been cultivated in Germany for over 500 years. The small, blue grapes require deep, fertile soil and grows best in the Ahr, Baden, and Württemberg wine-producing areas of Germany. Ripens fairly early, but may extend late into season. Its deep-red wine ranks as Germany's best red wine, with a velvety taste, and a bouquet reminiscent of bitter almonds. Of vineyard areas in Germany, this grape represents 3.5% of total.

STEIN. A variety cultivated in South Africa for producing South African Sherry.

SYRAH. A very high-quality red wine grape and is the red variety used in the production of Hermitage, renowned wine of the Rhône Valley. The variety also has been transplanted in Australia and California. However, regarding the California plantings, see Petit Sirah (this list).

THOMPSON SEEDLESS. Essentially a table and raisin grape with extensive plantings in California. It is capable of yielding a bland and rather neutral wine used for blending with other wines.

TINTA. A term used to describe a family of red wine grapes in Spain. These include Tinta Alvarelhão, Tinta Carvalha, Tinta Madeira, etc. These grapes are used mainly for Ports and a few red table wines. Plantings in California have been made of Tinto Cão and Tinta Madeira, which yield good quality California Ports.

TRAMINER. A familiar white wine grape with a characteristic aroma that is transferred to the wines which it yields. The wines have been described as soft with a hint of sweetness. Found principally in Alsace, Germany's Rhine Valley, and Italy, the grapes also are grown in Austria, Australia, California, Luxembourg, and Rumania.

TREBBIANO. A medium-quality white wine group, of greater importance in Italy than in France. In France, the variety is referred to as the Ugni Blanc. The variety also has been planted in California.

TROJA. A very productive and deeply colored, native Italian red wine grape.

TROLLINGER (Blauer). A large, sweet, reddish-blue grape. The vine favors a rich soil, but will grow on poor soil, if not too dry. Plantings in Germany are almost exclusively in the Württemberg region. The grape ripens late. The wine tastes fresh, racy, fruity, and is usually of a light-red color. Of vineyard areas in Germany, this grape represents 2.2% of total.

UGNI BLANC. See Trebbiano (this list).

UGRETTA. A native Italian red wine grape.

VELTLINER. Mainly important in Austria, this is a white wine grape of good quality. The variety also has been planted in California.

VERDELHO. A superior grape variety cultivated on Madeira and used mainly in producing fortified wines.

VERDISO. An exceptionally fine white wine grape well known for dry white wines made in northern Italy.

VERDOT. A highly regarded red wine grape of France (Bordeaux district). It is often grown with Cabernets, Merlot, and Malbec. Wines are high in tannin.

VESPOLINA. A native Italian red wine grape.

VIOGNIER. A white wine grape, grown in the Rhone Valley of France, capable of yielding wines of fine quality.

VITIS LABRUSCA. The grapes originally used by the wine industry in the northeastern United States were of *Vitis labrusca* (See Catawba, Concord, Delaware, Diana, Duchess, Elvira, Iona, Ives, Moore's Diamond, Niagara, this list). In recent years, cultivars of the original labruscans and of the European *Vitis vinifera* have been developed. The pure labruscans are high in fruity flavors. Changes in nonvolatile acids and other chemical constituents of New York State grapes and wines during maturation and fermentation have been investigated by Kluba and Mattick (1978). See reference listed.

WHITE PINOT. A term sometimes used in California when referring to Chenin Blanc, which is seen in this list.

WÜRZBURGER PERLE. A white wine grape representing a cross between the Gewurztraminer and the Müller-Thurgau, and cultivated principally in Germany.

ZINFANDEL. An extensively planted red wine grape in California. It is quite productive. The exact origin of this grape has not been successfully traced. The wine yielded is of good quality and with a characteristic flavor of its own, identified as a "bramble" flavor (suggestive of wild blackberries or dewberries) by some tasters.

Additional grapes planted in California and not included in this table are: Reds or blacks-Aramon, Royalty, Rubired, Ruby Claret, St. Macaire, Salvador, Souzao, and Valdepeñas; Whites-Burger and Flora.

Inclusion of all varieties and subvarieties of local interest would require a list many times longer. Differences in language tend to complicate the problem of sorting out the various wine grape varieties. In some cases, the French name may have become the common international designation for a variety; in other cases, the German or Spanish names. Sometimes, these designations are used interchangeably.

Depending upon the variety of grape, the water content of a ripe berry will range between 70 and 80%. Most of this water is contained in the *pulp* of the berry, that is, the fleshy and juicy part. But, there are also liquid and some semiliquid components in the *skins* (peels, husks, or hulls) and in the *stems*; these liquids are freed when the total mass of berries is subjected to considerable pressure (squeezing force). Further, in any crushing, macerating, or pressing operation applied to a mass of berries, there is an inevitable mixing of both solid and liquid components—so that a reasonably complete separation of liquid (juice) components from the grape requires more than one crushing or squeezing operation. The purest juice (from the pulp) is obtained from the first squeezing action. This is known as *free-run juice*. Many of the traditional hydraulically operated basket presses have been replaced by roller-type crushers, Garolla blade-type crushers, or disintegrators.

The grape juice and/or the mass of crushed grapes on the way to wine production is referred to as *must*. The grape pressings (skins, seeds, etc.) after the juice has been fully extracted are known as *marc* or *pomace*. The antiseptic and antioxidant properties of sulfur dioxide are used effectively in the treatment of musts prior to fermentation and later in the winemaking process. Many winemakers prefer compressed SO_2 gas, but sulfurous acid or sodium or potassium metabisulfite may be used. These essentially sterilize the must, which can be later reinoculated with a specially selected yeast culture.

Remarkably few grapes are well suited for wine as well as fresh (table) use or raisin production. Worldwide, the Muscat grape is considered a triple-purpose grape. There are numerous subvarieties of the Muscat, but all possess the characteristic Muscat odor and flavor. For wines, the Muscat is used principally in making sweet, fortified wines. In California, the Thompson seedless grape plays the three roles and some production is used in making wines. Wines from this grape tend to be rather neutral and bland and thus are mainly used for blending purposes.

Leading wine-producing countries include France, Italy, Germany, the United States, Spain, Portugal, Greece, Austria, Russia, and some of the countries formerly of the Soviet block. Argentina and South Africa are also notable wine producers. In France, wine production is divided into 6 major regions: Bordeaux, Champagne, Chablis, Burgundy, Loire, Rhone, and Alsace. Italy is divided into 18 regions, the most important of which are Puglia (located in the extreme southeastern part of the country), Sicily, Veneto, in northeastern Italy, Emilia, in northern Italy, and Piedmont, in the extreme northwest of the country. There are approximately 11 major wine-producing areas in Germany, including the Ahr (south of Bonn), the Mosel-Saar-Ruwer region, and the Mittelrhein, Rheingau, Nahe, Rheinhessen,

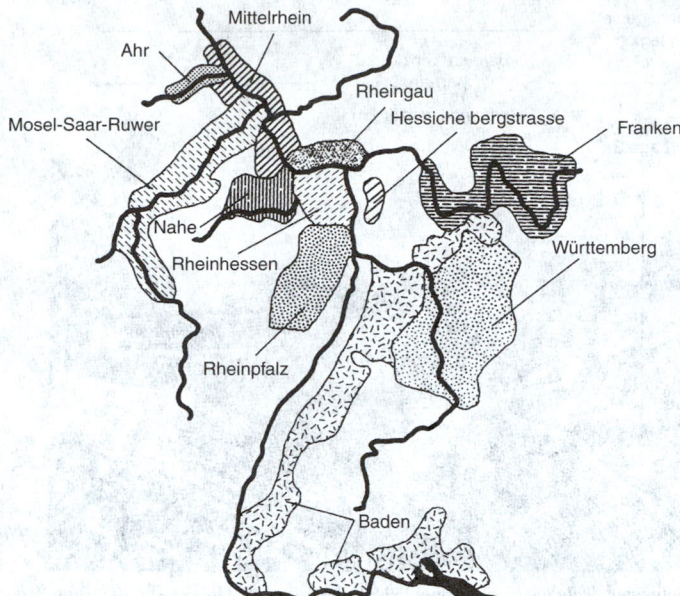

Fig. 5. The eleven major wine-producing regions of western Germany. The seven major rivers shown greatly affect the microclimates of these regions.

Labels on figure: Ahr, Mittelrhein, Rheingau, Hessische bergstrasse, Franken, Mosel-Saar-Ruwer, Nahe, Rheinhessen, Württemberg, Rheinpfalz, Baden

Rheimpfalz, Würtemberg, Hessiche Bergstrasse, and Baden districts. See Fig. 5.

In the United States, about 93% of the wine produced comes from California; the other 7% is produced in eastern states, notably New York and Ohio. In California, over 75% is produced in the San Joaquin Valley, and most of the remainder from the Napa Valley.

The Sherry district of Spain is located in the southwestern corner of the country, close to the Portuguese border. Spanish table wines are produced in a number of districts, including Malaga, Alicante, Valencia, and Tarragona, all along the Mediterranean shore. In Argentina, vineyards are located along the foothills of the Andes, near the Chilean border. Sometimes not fully appreciated, the wine industry of South Africa is some three centuries or more old. Most of the vineyards are located east and northeast of Cape Town. There is one stretch of land that enjoys a climate much like that of the Mediterranean countries.

Grape-Growing Conditions. The character and quality of a wine depend upon the kind of vine, the natural setting of the vineyards, and the human effort in the care of the vines in the vineyard and in the production and storage of the wine. The natural prerequisites for the vine to grow and thrive are the local climate, the location and topography of the vineyard sites, and the kinds of soil. Variations of all of these factors contribute to the diversities of wine and are reflected in the differences of their chemical compositions. Sometimes, the differences are so subtle that they cannot be revealed by visual examination of chemical data. Although beyond the scope of this volume, in an effort to improve upon ways and means for classifying wines, vineyard pattern recognition techniques have been developed.

Concept of Microclimate. While it is safe to say that no two vineyards are exactly alike, there are some major wine-producing areas of the world, such as some of the large grape-producing areas of California, where there is relatively more uniformity among vineyard environments (climate, soil, moisture, etc.) than will be found in areas where there are numerous rivers and tributaries and wide variations in topography (numerous hills and valleys) as well as variations in forestation that alter air circulation patterns. In the former situation, one is dealing essentially with what might be termed a macroclimate, whereas in the latter situation, the concept of microclimate applying to essentially individual vineyards predominates. Closely coupled with climatic variations are soil variations, which, again, are more likely to be large in hilly or mountainous terrain and less so in flat valleys or plateaus.

In lieu of extensive scientific examination, viticulture has developed as the result of several hundreds of years of trial and error. Grape growers have found which varieties do best in given locations and how to tend the vines to maximum advantage in a given location. Likewise, the winemakers

(frequently also the growers) have learned how best to process grapes from a given location into acceptable, if not always superior wines. Similarly, over the years, discriminating consumers of wine have learned to associate the origin of a wine with quality, always allowing, of course, for the overall reputation of the winery and appreciating the fact that, for many wines, there are excellent growing seasons (vintage years) in a given location, as well as average and poorer years.

In recognition of geographical variances, France pioneered the concept of the "Appellation D'Origine," that is, a wine's name in geographical terms. In practice, this can reduce in terms of a whole district, such as Bordeaux or Burgundy; sometimes in terms of a river valley, such as Loire (*Vins de la Loire*); sometimes in terms of township, such as Vosne-Romanée; sometimes in terms of an estate, such as Château d'Yquem; and sometimes, in the extreme, a single vineyard or grouping of vineyards, such as Richebourg Appellations like these are applied to nearly all of the famous French wines, which are marked A.O.C. (Appellation Controlée), which stand for the registering and controlling organization, "Institut National d'Appellations d'Origine des Vins et Eaux-de-Vie." With lesser wines, there are varying degrees of association with geographic location. The entire system is somewhat complex and beyond the scope of this volume. The principal French wine-producing districts were described briefly earlier in this entry.

Microclimates of German Vineyards. Possibly nowhere in the world of grape-growing is the concept of microclimate more important than in Germany. The importance of rivers, valleys, hills, and mountains has previously been mentioned and is well exemplified by the topography of Germany's wine-producing regions. As previously shown by Fig. 5, seven of the most important rivers that affect climatic conditions in Germany essentially determine wine-growing areas from which characteristic types of wine are formed. The 11 classified regions, as previously described briefly are noted in Fig. 5. At a latitude of approximately 50°N, the Germany wine-growing country is the most northerly of the major wine-growing regions of the world. For over 1000 years, German viticulture has flourished in regions most favorable to the vine. Considering the variety of wines produced, the number of grape species grown seems limited. However, some experts point out that the type of vine is the third factor, with soil and climate the most important factors. In recent years, German viticultural research has examined several hundred descendants of the original European vine that have appeared in recent centuries in order to find the sites and conditions best suited to each variety. Within the last century, new species, such as the Müller-Thurgau, Scheurebe, and Morio-Muskat, have been developed. About 87% of the grapes cultivated in Germany are white species, with the remaining 13% devoted to red species, such as Blauer (Blue) Portugieser, Blauer (Blue) Spätburgunder (Pinot Noir), and Blauer (Blue) Trollinger.

For relating a wine with its microclimate, the 11-district classification is far from sufficient. For accurate labeling purposes, it is desirable to get as close to identifying a specific vineyard as may be practical. The microclimate of a vineyard depends upon several factors: Whether it faces south or east; the gradient of its incline; the intensity of the sun's reflection from the surface of a river; the proximity of sheltering forest or mountain peak; altitude; and soil moisture. On steep inclines, the soil is frequently slatey; where the incline gently slopes down to its base, there is fertile alluvial land; other areas show lime deposits or volcanic rocks; all of which naturally influence the taste of the wine. German oenologists have observed that, separated by a distance of only a few hundred meters, wine of world acclaim may grow-or nothing more than gorse bushes.

Thus, the 11 German wine-growing regions have been divided into 130 general sites and approximately 2600 individual sites. This breaks down as follows:

1. Anbaugebiete (specified regions).
2. Bereiche (district). Bereiche are part-areas within the Anbaugebiete, where conditions of growth are largely similar, so that wines growing there show similar aspects of quality. A Bereiche embraces a fairly large number of wine-growing communities.
3. Name of wine-growing communities or towns/villages. These are identical with the political community.
4. Names of Lagen (sites) entered into the Register of Vineyards. This is the smallest geographical unit, i.e., the vineyard site (Weinbergslage). The minimum size of a site is 5 hectares (2.47 acres).

The foregoing terms are illustrated in Fig. 6. The special significance of geographical detail in terms of the consumer gave rise to the publication

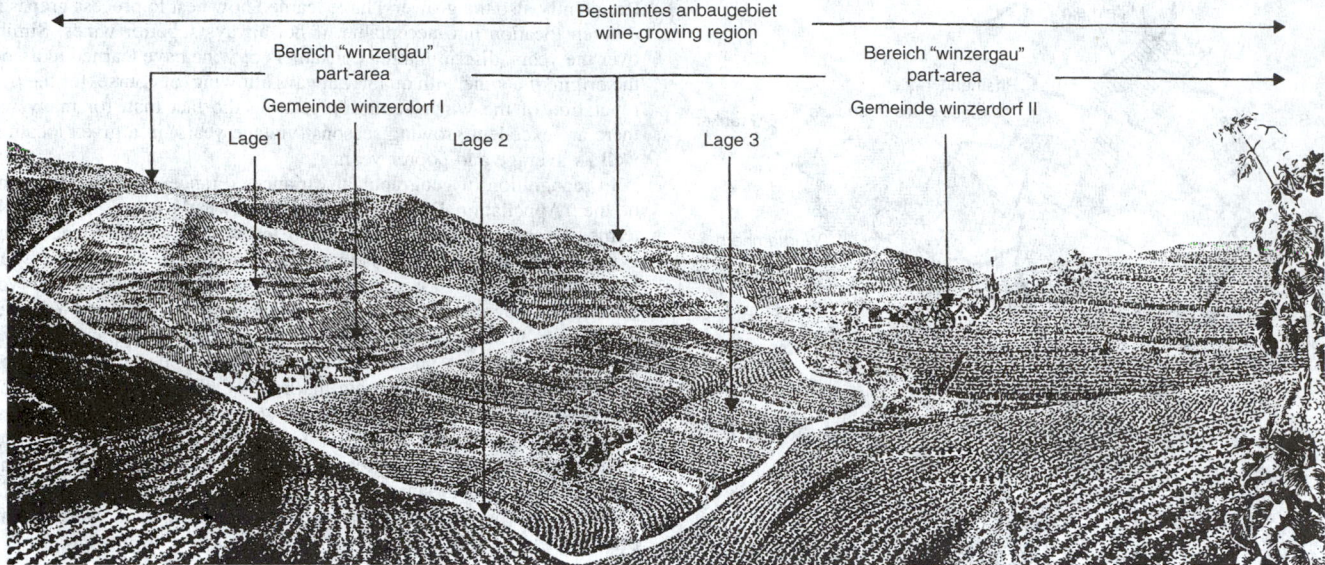

Fig. 6.   Illustration of German system for classification of vineyards in terms of microclimates. *Anbaugebiet* = region; *Bereich* = district; *Lage* = site. (*German Wine Atlas.*)

of the "German Wine Atlas and Vineyard Register," which contains a complete compilation of all names and sites recorded.

Soil Conditions. It is interesting to note the number of terms that have been coined to reflect the effect of soil conditions on various wines. Among these terms are: *Barro*, a Spanish word for describing flavor that derives from clay soil and is used especially as regards the vineyards in the Sherry country of Spain. The finest vines for Sherry are grown on chalky, white soils, known as *albariz*. Coarse, heavy wines result from grapes cultivated on clay. The poorest of all soils for the vine are sandy (*arena*). The term *bodenton* (English), Bodengeschmack (German), goût de terroir (French) is used to describe a disagreeable and unmistakable flavor that results in wines prepared from grapes (certain varieties) when grown on heavy clay or alluvial soils.

Weather Abnormalities. Possibly the greatest concern of the vintner is fear of heavy frost, particularly at certain stages of the growing season. Inasmuch as many of the highest-quality table and sparkling wines are produced in northern regions which represent the climatic limitation for vines, this fear is universal throughout the northern wine-producing regions of France, Germany, Switzerland, Spain, Austria, among others. However, even in regions such as parts of California, the vintner is not free of this worry. Severe frost damage, for example, was suffered in April 1964 in the Napa Valley of California. In France and Germany, there is a period of approximately 6 weeks (1 April to about 15 May) when frost represents a major threat. This is particularly true of German vineyards located in the Moselle and Saar Valleys. Thus, celebration of Ice Saints Day (4 days from 12 to 15 May) can be joyful if frost has not appeared because in these areas, frosts are essentially unknown after 15 May. However, there are always exceptions. On the 28th of May, 1961, the vines of the Pouilly-Fumé grape were severely damaged in the Loire region of France, the latest known killing frost in French weather records. Severe frost damage can carry into poor yields in the following year as well. As in California, some of the European vintners now use smudge-pots, fans, and stoves in some of their most frost-susceptible vineyards — and with considerable success.

Hail is also a major hazard of the vine. Hailstorms are not infrequent and, if heavy, can destroy a crop, with damage extending into the following year. During the season when the grapes are ripening, even a light hailstorm can be damaging. Slightly bruised berries, particularly for red wine, can impart the faintest hint of rot in an otherwise fine wine. The French refer to this as *hail taste*.

Infrequently, grape vineyards may be invaded by the sucking insect phylloxeran, which attacks the roots of the grapevine. See also **Phylloxeran (Phylloxerid)**. In the last century, millions of acres of vineyards in France were destroyed, but since that time more resistant grapevines have been found. However, in the summer of 1992, there was the scare of possible excessive damage to vines by this cause in California's Napa Valley. It is too early to forecast how serious the present threat may be.

Wine Production

One of the continuing and fundamental problems of winemaking is the lack of uniformity of the grapes used, from one season to the next — a lack of consistency which in some years is responsible for truly exceptional and great wines and, in other years, wines that are only passable to good. Two important factors are sugar content and acidity. When these factors are purposely altered after the grapes are picked by way of adding sugar, water, or acid, the process is referred to as *amelioration*. While practiced in some wine-producing regions, the practice is frowned upon and is outlawed in some regions. Where permitted, amelioration is strictly regulated. When water is the only additive, the term *gallisation* is used. The term *chaptalisation* (French) refers to the addition of sugar.

It has been observed since ancient times that grape juice at ordinary temperatures does not retain its freshness, but instead commences to turn to wine, that is, the juice (or must) exhibits the aromatic characteristics of alcohol, but when retained for a still longer period it turns into vinegar. Centuries ago, the Latin word *fermentum*, from *fervere* (to boil) was first used to describe the bubbling nature of the process, and thus the word *fermentation* became a part of the language. A few centuries of research have gone into the process, particularly as applied in winemaking. Considering the total time span of winemaking, the addition of yeast cultures purposely by the winemaker to the fermentation vats is a relatively recent action.

From whence did the yeast organisms come that made wine possible during all of those earlier centuries? Yeast occurs naturally on grapes, but there are numerous species — some desirable and many more undesirable for making the best wines. The species most favorable to the winemaker is *Saccharomyces cervisae* var. *ellipsoideus*. Numerous molds are found on green grapes and, as grapes ripen, so-called wild yeasts appear. These yeasts can cause many problems, including off-flavors, off-colors, spoilage, and a host of problems that sometimes are difficult to trace. Poor or unacceptable wines are sometimes referred to as greasy or ropy, wines that become cloudy and pour like oil. Other wines may be flat or bitter. Although Pasteur offered a depth of understanding to the entire winemaking process, his recommendation for overcoming the problems associated with undesired microorganisms was a short-cut solution, namely, *pasteurization*. Over the years, winemakers have regarded pasteurizing with mixed feelings. While pasteurization kills many undesirable microorganisms, rendering wine stable and suitable for long-term storage, it also eliminates or interferes with the possibilities of improving the product during normal aging. Pasteurization is widely used for common table and dessert wines produced in high volume for early consumption. In contrast, winemakers who target to superior and excellent wines regard pasteurization with much caution. The longer, more painstaking procedure for overcoming contamination by wild yeasts involves sulfiting, the addition of specially selected yeast cultures to the sterilized must, and the careful manipulation of all process variables which favor the type and degree of fermentation desired for any given kind of wine.

Must fermentation occurs in three stages: (1) an initial slow stage during which the yeast cells are multiplying; (2) a very vigorous stage, accompanied by bubbling and a marked rise in temperature; and (3) quiet fermentation that can proceed for quite a long time at a lower and lower rate. The main fermentation stages (1 and 2) take place in a variety of vessels, ranging from concrete vats (not often glass-lined) or in wooden tanks (oak, redwood), and ranging from 10,000 to 60,000 gallons (380–2,280 hectoliters) and more. While some continuous fermenting systems have been built, by and large fermenting remains a batch operation. Fermenting may range from 2 to 20 days, depending upon numerous variables. With alcohol-tolerant yeasts, fermentation proceeds rapidly to completion, producing from 10 to 12.5% alcohol by volume. When the sugar content exceeds 23%, this may inhibit fermentation rate as well as full completion of fermentation. At total acidities of less than 1% (pH greater than 3), alcohol fermentation is not inhibited. Yeasts require a number of amino acids, but fortunately these are present in most grapes in ample amounts. Some winemakers will sometimes add nitrogen-bearing substances in small quantities as yeast food.

Temperature is quite critical to the fermenting process. Each winemaker may have opinions as to which temperature is best for any given type of wine. For white wines, the optimum temperature ranges between 50 and 60 °F (10 and 15.6 °C); for sherry, the optimum is about 80 °F (26.7 °C); for red wines, about 85 °F (29.4 °C); for wines from Pinot Noir grapes, 70–80 °F (21.1–26.7 °C); and for Cabernet-Sauvignon grapes, 70 °F (21.1 °C). For some wines, retardation of fermentation commences at about 85 °F (29.4 °C), and for all must fermentations the action is greatly weakened with a temperature rise to 95 °F (35 °C); above 100–105 °F (37.8–40.5 °C), fermentation essentially ceases. At temperatures above 90 °F (32.2 °C), it is likely that wine flavor and bouquet will be injured.

The end of fermentation is signaled by a clearing of the liquid, by a vinous taste and aroma, and by a drop in temperature, and can be confirmed by checking sugar residual. It is interesting to note that fermentation can be halted as the result of a temperature too high or too low. In this case, the condition is referred to as a "stuck wine." If a batch is stuck at a low temperature, warming will usually cause fermentation to resume. In the case of a batch stuck because of high temperature, cooling alone may not suffice. The addition of small quantities of ammonium phosphate will usually help to restart fermentation.

To date, no substitute for time has been found in the transformation of the green wine (after drawn off the fermenters) into an acceptable product. Considerable settling of finely divided solid particles and colloidal materials is required, the subtle and slow chemical reactions involving aldehydes, esters, etc. that enter into the ultimate bouquet of a wine — all are time-related events, much more critical with some wines than others. There is a requirement for all wines for a minimum of clarification, stabilizing, and settling that occurs at the winery prior to containerizing for the market; there is the additional aging that goes on once a wine has reached the market. Popular writers interested in viniculture tend to overemphasize the aging aspects of wine, considering that 90% or more of the wine produced is for the mass market and relatively early consumption. For example, most of the common table wines in Spain are consumed when less than two years old. It has been shown that common California red wines can be adequately matured in a year or less. In contrast, a fine Cabernet requires up to a minimum of three years in wood. Such wines should be further aged a year in the bottle at the winery prior to labeling and releasing for sale. Such wines usually will continue to improve for a period of from 5 to 15 years in the bottle.

One authority estimates that about 75% of all wine produced is as good when about two years old as it is likely to be and deterioration is likely to commence after three years. Wine recommended for consumption within three to five years includes: Vin Rosé (California, France, etc.); most California white wines, with the exception of a select few prepared from Chardonnay, Chenin Blanc, Pinot Blanc, Sauvignon Blanc, and Johannesberg Riesling; most white Burgundies, with the exception of those from the excellent vineyards in good vintage years; and nearly all Italian wines, except a few select red wines.

Fining agents that bring about clarification of wine include gelatin, casein, tannin, and bentonite. Fining is most efficiently accomplished in relatively small vessels, including barrels. Because of so many variables

involved, a careful laboratory examination of the wine is made prior to selection and determination of the amount of fining agent to be used.

The French use the term *maderise'* to describe a wine that is overage (past its prime condition), which has become partially oxidized, and which has frequently acquired a brownish tinge, and an aroma and flavor remindful of Madeira (not desirable except in a Madeira wine). The term is more commonly applied to white and *rosé* wines.

Other terms used in winemaking include: *Racking* — the drawing off of the clear portion of a young wine from one vessel and transferring it to another vessel. In this process, the lees and sediment formed during the prior storage period are separated. To hasten the total process, more rackings are required. Winemakers have found that refrigeration helps to hurry the aging process. *Binning* involves laying away bottled wine for aging. Always with table and sparkling wines, the bottles should be stored on their side so that the wine is in constant contact with the cork. The wine should be stored at a cool temperature. There is a relationship between the size of the container and the time of aging. A half-bottle will be ready earlier than a full bottle. *Blending* is widely used in connection with high-volume wines and where year-to-year quality is important to consumer acceptance. *Filtering* is commonly practiced in connection with high-volume wines and with most other wines with relatively few exceptions. Many winemakers prefer a lighter filtration so that the wine will not take on what is known as a character of *numbness*, that is, removal of some constituents that help prior to their ultimately becoming sediment upon aging in the bottle. It is a well accepted fact that discriminating wine consumers do not look upon sediment, particularly in certain wines, such as old red wines, as a defect, but rather as a natural result of proper aging. Sediment in white and rosé wines is in the form of colorless crystals of cream of tartar, which is tasteless and harmless and often disappears when the wine is slightly warmed. The sediment in red wines is of larger amount and complexity, made up of pigments, small quantities of mineral salts and tannins, all of which can be removed by careful decanting.

Fortification signifies a wine that contains more alcohol than is obtainable through natural fermentation. Fortified wine is not grape juice to which alcohol has been added (known as *mistelle*). Port is a fortified wine, to which about half-way through the fermentation, juice is drawn off and put into vessels that contain high-proof grape brandy of a predetermined volume. Sherry is also fortified with high-proof brandy. Not regarded as fortification, but effective in adding a few percentage points of alcohol to wine, some winemakers use alcohol-tolerant yeasts. *Brandy* is made by distilling wine.

Additional Reading

Cox, J.: *From Vines to Wines: The Complete Guide to Growing Grapes and Making Your Own Wine*, Storey Communications, Inc., Pownal, VT, 1999.

Robinson, J.: *Jancis Robinson's Guide to Wine Grapes: A Unique A-Z Reference to Grape Varieties*, Oxford University Press, Inc., New York, NY, 1996.

Web Reference

New York Wine and Grape Foundation: http://www.nywine.com/, 1999.

GRAPE SUGAR.   See **Carbohydrates**.

GRAPH COMPONENT.   A component of a graph *G* is a nonseparable maximal connected subgraph. The decomposition of a graph into components is unique.

GRAPHITE.   An allotropic form of carbon, graphite occurs in nature and also is produced artificially. Graphite crystallizes in the hexagonal system, often in the form of scales or plates, or in large foliated masses. Graphite has a perfect basal cleavage, is soft (hardness between 0.5–1 on the Mohs scale — similar to talc), and feels greasy to the touch. Specific gravity 2–2.2, black to steel gray, lustrous metallic appearance, very opaque. Graphite finds many uses: (1) in the manufacture of "lead" pencils, graphite (the marking medium) is mixed with clay as a bonder, the amount of clay used determining the hardness of the pencil lead; (2) in the manufacture of self-lubricative metals in which graphite is mixed with copper, lead, and tin, after which the mix is sintered and subjected to powder metallurgy techniques to form alloys which will hold relatively large volumes of lubricating oil over long periods of use; (3) in the construction of heat-resistance structures, such as rocket casings and chemical process equipment, allowing operating temperatures up to 3,000 °C and greater; (4) in the manufacture of corrosion-resistant apparatus for chemical processing; (5) in the manufacture of packings where the lubricative and corrosion-resistant characteristics of graphite are advantageous; (6) in the production of electrodes for electric furnaces and electrolysis equipment; and (7) a special pyrolytic graphite, with excellent electrical and thermal conductivity properties, good tensile strength at temperatures up to about 2,800 °C, and impervious to gases and liquids, finds use in various electrical apparatus and, when mixed with boron, makes an effective nuclear radiation shield. Graphite slows the flow of neutrons without capturing them.

Graphite in Composites. Graphite has been used in composite materials of construction for a number of years, notably pioneered in structures for aircraft. See also **Airplane**. The use of composite materials based upon graphite (carbon) fibers, fiber glass, numerous plastics (including epoxies), and ceramic fibers, among others, has received zealous attention in the materials community in the last half of the 1980s. Carbon-carbon (C/C) composites emerged from requirements of the aerospace field and their numerous advantages are now being extended to a variety of industrial and transportation equipment applications, including the automotive field. As observed by Klein (Nov. 1986 reference), not only can C/C withstand the heat generated at the nose cone and leading edge of space vehicles, C/C has endured such conditions mission after mission. The temperature capabilities of C/C extend to over 3300 °C (5972 °F), and C/C composites are twenty times stronger than conventional graphite, yet are 30% lighter, with a density of about 85 lb/ft3 (1.38 g/cc). C/C can endure higher temperatures for longer periods of time than other ablative materials. It also resists thermal shock, permitting rapid transition from −158 °C (−250 °F) in the cold of space to nearly 1650 °C (3002 °F) during reentry, well beyond the capabilities of metals and ceramics.

The C/C nose cone is made by a two-dimensional layup. In a first step, graphite cloth, preimpregnated with phenolic resin, is laid in a mold and cured. The part is trimmed, then pyrolyzed, driving off gases and moisture as the phenolic resin converts to graphite. At this point, the relatively soft composite is impregnated with furfuryl alcohol and pyrolyzed three additional times, each step increasing the density, strength, and modulus. A ceramic coating of silica and alumina is applied in the form of a powder that is finer than the pores in a human hand. To prevent the C/C from oxidizing, a coating of silicon carbide is caused to form on the top two layers of the laminate. Because the SiC is brittle and susceptible to craze-cracking, additional protection is provided by impregnating the surface with tetraethylorthosilicate, which is cured, leaving a silicon dioxide residue throughout the coating, further reducing the area of exposed carbon. C/C is stiff and resists buckling, maintaining its aerodynamic shape over a wide temperature range. The composite has long fatigue life when subjected to thermal cycling. Numerous other, similar techniques are used to make C/C composites for a variety of applications, an excellent example of which is racing car brake disks. See Figs. 1 and 2.

Sources of Graphite. Graphite is formed during the metallurgical operations of producing pig iron, cast iron, malleable cast iron, and some special die steels and has a marked effect upon the characteristics of these materials. See also **Iron Metals, Alloys, and Steels**. The effects may be positive or negative. When present in cast iron in excessive amounts, or in the form of large interlocking flakes or films, graphite reduces the tensile strength.

Graphite is a rather widely distributed mineral and is found in a variety of rocks. It occurs in marbles, gneisses or schists; granites and other igneous rocks often carry graphite. It has been noted in pegmatites. It is likely that graphite has been formed by different processes, by magmatic separation of the graphite as an original constituent or as the result of assimilation of carbonaceous rocks, by pneumatolytic action, or by the metamorphism of sedimentary rocks that contained original carbonaceous matter. Well-known localities are in Siberia, on the Island of Ceylon, which is the chief producing district at present; England, Madagascar, Mexico, and Canada. In the United States it is found in the Adirondack region of New York State, in Massachusetts, Rhode Island, Pennsylvania, Alabama, New Mexico, and Montana. Natural graphite sometimes is referred to as plumbago, black lead, and Flanders stone.

Graphite is made artificially by heating coke to a very high temperature, usually in an electric furnace. To prevent oxidation, the coke is covered with a layer of sand.

The German mineralogist, A.G. Werner, devised the name graphite from the Greek meaning *to write*, with reference to its use in pencils.

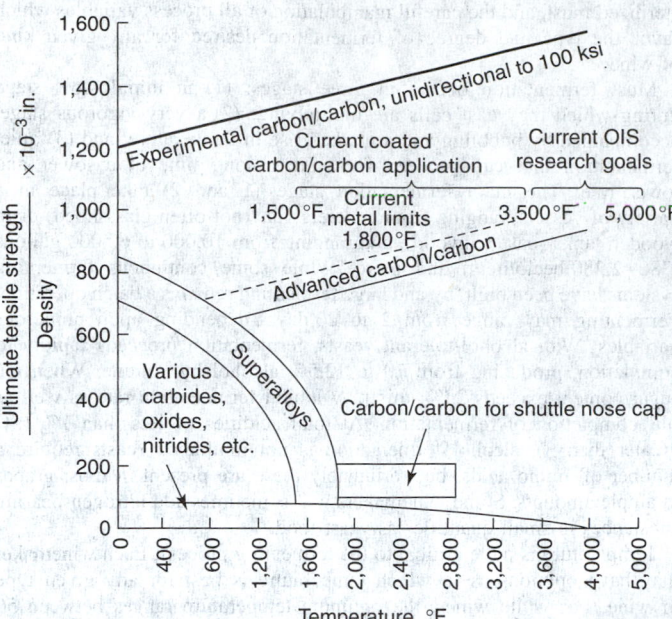

Fig. 1. Carbon-carbon retention of strength at high temperatures. (*LTV Aerospace and Defense.*)

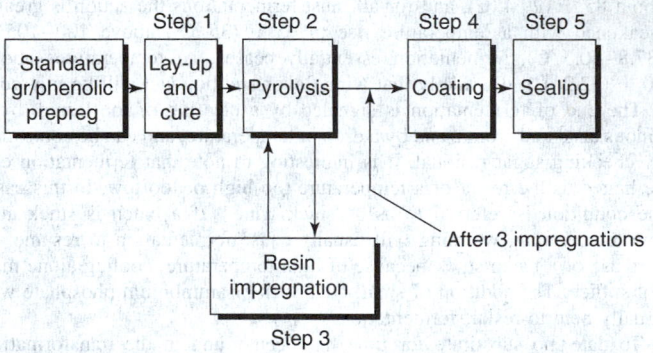

Fig. 2. Processing C/C composites. Graphite cloth is impregnated with furfuryl alcohol and pyrolyzed three or more times, each time increasing part density, strength, and modulus. Next, the part is packed with ceramic powder and fired at 1650 °C to form a silicon carbide coating on the top two layers of the laminate, to prevent oxidation. (*LTV Aerospace and Defense.*)

For a comparison of the characteristics and crystalline structure of graphite and diamond, see also **Carbon**; and **Diamond**.

Additional Reading

Arsenault, R., R. Bhaget, and S. Fishman: *Mechanisms and Mechanics of Composites Fracture*, ASM International, Materials Park, OH, 1993.

Clemmer, C.R. and T.P. Beebe, Jr.: "Graphite: A Mimic for DNA and Other Biomolecules in Scanning Tunneling Microscope Studies," *Science*, 640 (February 8, 1991).

Gutowski, T.G.: *Advanced Composites Manufacturing*, John Wiley & Sons, Inc., New York, NY, 1997.

Harper, C.A.: *Handbook of Plastics, Elastomers, and Composites*, The McGraw-Hill Companies, Inc., New York, NY, 1996.

Inagaki, M.: *New Carbons: Control of Structure and Functions*, Elsevier Science, New York, NY, 2000.

Klein, A.J.: "Composites that Fight Fatigue," *Adv. Mat. And Processes*, 33–36 (February 1986).

Pierson, H.O.: *Handbook of Carbon, Graphite, Diamond, and Fullerenes: Properties, Processing, and Applications*, Noyes Publications, New York, NY, 1994.

Rabe, J.P. and S. Buchholz: "Commensurability and Mobility in Two-Dimensional Molecular Patterns on Graphite," *Science*, 424 (July 26, 1991).

Rohatgi, P., Editor: *Friction, Lubrication, and Wear Technologies for Advanced Composite Materials*, ASM International, Materials Park, OH, 1993.

Staff: *Advanced Synthesis of Engineered Structural Materials*, ASM International, Materials Park, OH, 1993.

Upadhya, K., Editor: *Processing, Fabrication and Application of Advanced Composites*, ASM International, Materials Park, OH, 1993.

Utsumi, W. and T. Yagi: "Light-Transparent Phase Formed by Room-Temperature Compression of Graphite," *Science*, 1542 (June 14, 1991).

Woishnis, W.A.: *Engineering Plastics and Composites*, 2nd Edition, ASM International, Materials Park, OH, 1993.

Yoshimura, S. and R.P. Chang: *Supercarbon: Synthesis, Properties and Applications*, Springer-Verlag, Inc., New York, NY, 2000.

GRAPH (Mathematics). Generally, a curve or surface on which the locus of a function is shown on a series of coordinates which are set at right angles to each other.

Graph (Complete). A complete graph G is a linear graph in which every two distinct vertices are endpoints of an edge in G. Figure 1 is a complete graph with four vertices. The total number N of distinct labeled trees in a complete graph containing ν vertices is $N = \nu^{\nu-2}$, a result due to Caylet. Thus, this example has 16 trees.

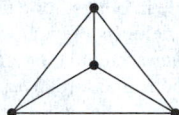

Fig. 1.   Complete graph with four vertices.

Graph (Connected). A graph is connected if there exists a path between any two vertices. Stated in another way, any two distinct vertices β_1 and β_2 are the terminal vertices of some path.

Graph (Directed). See also **Digraph**.

Graph (Dual). The linear graph G_2 is the dual of the linear graph G_1 if the conditions enumerated below are satisfied:

1. The edges of G_1 and G_2 are in one-to-one correspondence.

2. If H_1 is any subgraph of G_1 and H_2 is the complement of the corresponding subgraph in G_2.

$$r_2 = R_2 - n_1$$

where r_2 is the graph rank of H_2, R_2 is the rank of G_2 and n_1 is the nullity of H_1.

It follows easily from this definition that rank G_1 = nullity G_2 and rank G_2 = nullity G_1. Furthermore if G_2 is the dual of G_1, G_1 is the dual of G_2.

Two extremely useful and significant results are that the dual of a nonseparable graph is nonseparable and that a linear graph is planar if and only if it possesses a dual.

The usual geometric procedure for finding the dual of a planar graph G involves three steps:

1.   Choose a set of fundamental circuits. See also **Circuits, Fundamental (Mathematics)**.

2.   Put a node in each such circuit and a node outside the graph.

3.   Connect any two nodes that are on opposite sides of a branch by a line segment.

The resulting graph is the dual of G. These rules are illustrated in Fig. 2, in which the dual appears dotted.

Graph (Finite). A finite graph contains only a finite number of line segments and vertices.

Graph (Homeomorphic). Two graphs G and G' are homeomorphic if there exists a one-to-one bicontinuous mapping between the two pointsets defined by G and G'. Refer to description of planar graph.

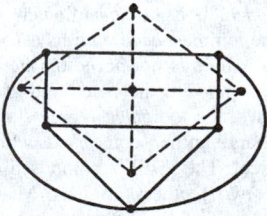

Fig. 2.   Dual graph (dotted).

Graph (Infinite). Graph containing an infinite number of line segments and vertices. Such graphs have many interesting mathematical properties.

Graph (Isomorphic). Two graphs G and G' are said to be isomorphic if there exists a one-to-one transformation which maps the vertices of G onto the vertices of G' and the edges of G onto the edges of G' in such a way as to preserve incidence relationships. Thus, if vertex B and edge ε are incident in G, the respective images β' and ε' are incident in G'. The one-to-one transformation is an isomorphism of G with G'.

Graph (Linear). A collection of edges no two of which have a point in common that is not a vertex. The words linear-complex and 1-complex are frequently used alternatives. As defined here, a graph is an abstract graph devoid of any geometric significance. It is true, however, that a graph can be interpreted as a configuration in three-dimensional Euclidean space.

Graph (Nonoriented). A linear graph in which the elements have not been assigned an orientation is said to be nonoriented. A graph of this type also is called *ordinary*.

Graph (Nonseparable). A graph of which every subgraph has at least two vertices in common with its complement.

Graph (Nullity). The nullity μ of a graph G possessing ν vertices, e edges and P maximal connected subgraphs is

$$\mu = e - \nu + P \geq 0$$

Graph (Oriented). A linear graph is oriented when an orientation has been assigned to each of its elements. By long-standing convention the phrase "oriented graph" is applied only to graphs which possess at most one directed segment between any two vertices. (For the more general case in which parallel edges are permitted see **Digraph**.)

Graph (Planar). A linear graph G can be viewed from either a geometric or a topological standpoint. In the first, it is considered a collection of edges, no two of which have a point in common that is not a vertex. In the latter, it is thought of as defining a set of points in three dimensions, whose members are the points which make up the edges of the graph. This point set is the topological graph G^* corresponding to the linear graph G. G is said to be planar if G^* can be mapped on a plane by a one-to-one continuous transformation in such a way that no two image edges have a point in common that is not the image of a vertex in G.

It has been shown by Kuratowski that a linear graph is planar if and only if it does not contain either of the two graphs shown in Fig. 3 as subgraphs.

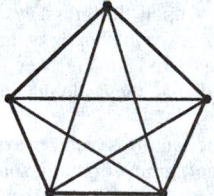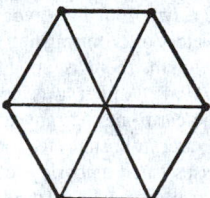

Fig. 3.   Kuratowski graphs.

Graph (Separable). A connected graph is separable if it contains at least one subgraph which has only one vertex in common with its complement. Otherwise the graph is nonseparable.

GRAPH RANK. The rank of a graph G is $\nu - P$ where ν is the number of vertices and P the number of maximal connected subgraphs of G.

GRAS. In the United States, the acronym for "generally recognized as safe," used for designating foods and materials used in food products with regard to their impact upon human health. During recent years, there has been a gradual erosion of the list of GRAS substances as the result of research efforts on the part of various government regulatory bodies, in Canada, France, Germany, the United Kingdom, etc., as well as in the United States, and also on the part of various industry self-regulating bodies. Research activities have been directed essentially in terms of determining and confirming possible carcinogenic qualities of such substances. Some GRAS substances have been eliminated and there is a trend toward lowering the levels of usage generally recognized as safe. The parts per million (ppm) levels range considerably from one type of food substance to the next. For a number of years, at periodic intervals, the Institute of Food Technologists (U.S.) has reported summaries of current progress in the consideration of flavoring ingredients under the Food Additives Amendment (U.S.). These summaries appear in *Food Technology* magazine. Lists of GRAS substances are also obtainable from the U.S. Food and Drug Administration, Washington, DC, and from its counterparts of other governments in many major countries.

GRASHOF NUMBER.

A nondimensional parameter appearing in the theory of flows caused by free convection. It is

$$G = \frac{\alpha \theta g d^3}{\nu^2}$$

where θ is the temperature difference producing the convection, α is the coefficient of thermal expansion of the fluid, d is the length scale of the system, and ν is the kinematic viscosity. Flows without large density changes caused by the temperature differences are dynamically similar if the Grashof and Prandtl numbers are equal. Similar nondimensional numbers include the Froude number, the Mach number, and the Reynolds number.

GRASSES.

Of all plant families, the grass family (*Gramineae*) is one of the most important economically. With the many thousands of species of grasses, this is one of the largest families in the plant kingdom. Members of the grass family were probably among the first plants to be cultivated by humans. Grasses are found just about everywhere plants can grow, ranging from the polar regions to the tropics and to the upper limits of vegetation on mountains.

Most grasses are herbaceous plants of low stature. A few, notably the Bamboos, become woody plants of great height, and a small number are of clambering or trailing habit. The cereals, and many other grasses, are annuals, completing their growth in a single growing season; others are perennial plants. Some of the former are winter annuals, plants which start growth in one season, remain dormant over winter, and complete growth and fruit in the following season. Winter wheat is an example.

Among the earliest records of the grasses are those of the Old Testament, all of which emphasize the importance of the grasses to populations thousands of years ago. *Genesis* 1:12: "And the earth brought forth grass... whose seed was in itself, after its kind; and God saw that it was good." *Deuteronomy* 11:15: "And I will seed grass in thy fields for the cattle, that thou mayest eat and be full." *Proverbs* 19:12: "The king's wrath is as the roaring of a lion; but his favor is as dew upon the grass." *Isaiah* 15:6: "For the waters of Nimrim shall be desolate; for the hay is withered away, the grass faileth, there is no green thing." And, in the New Testament, *Revelation* 9:4 "And it was commanded that they should not hurt the grass of the earth, neither any green thing...."

Grasses are important to food production in several ways. (1) The cereal grasses, such as barley, corn (maize), grain sorghum, some millets, oats, rice, rye, and wheat, furnish the cereal grains, are basic foodstuffs and frequently are the sources of important food byproducts, such as edible oils. Cereals also become part of feedstuffs for livestock. (2) The *forage grasses*, such as the bluegrasses, the bromegrasses, the fescues, the ryegrasses, timothy, and wheatgrasses, among many others, along with a number of legumes, comprise pasturage, fodder, green feed, hay, and silage for consumption by livestock, and are the basic ingredients for processed feedstuffs consumed by livestock of many kinds, including beef and dairy cattle, sheep, and poultry. (3) The grasses also aid in the production of field food crops by playing an important role in soil conservation. Grasses are highly effective in reducing erosion and runoff. It is generally agreed among experts that a mat of grass and grass roots has no equal in holding soil. The establishment of grass waterways is an accepted procedure in many areas for routing excessive rainfall.

Botany of the Grasses

The characteristic growth of the principal elements of a representative grass plant is shown in Fig. 1.

The *root system* of a grass plant is made up entirely of fine fibrous roots, which enlarge but little, remaining about the same diameter throughout their length. These roots are mainly adventitious, arising from the lowermost nodes of the stem. The roots of many grasses penetrate deeply into the ground, thus reaching supplies of moisture which enable the plant to live in dry regions where surface moisture may be rare.

The *stems* of grasses, frequently called *culms*, are cylindrical and in most genera hollow except in the region of the nodes, where solid plugs occur. When young, the stem is solid, but as growth continues the central portion fails to keep pace with the outer and gradually becomes hollow. Maize (corn) is an exception, the stems being permanently solid in the plant. In most grasses, the stem grows erect, but frequently falls over during the growing season, because of climatic disturbances or lack of suitable nutrient sources to give it strength. Such fallen stems do not remain flat, but gradually become erect through renewed growth in the nodal regions. The cause of such a growth is not definitely known. The upward bend, negative geotropism, may be produced by auxin, which accumulates in the lower half of the node and stimulates overgrowth in that region. In many species of grass the lowermost nodes normally give rise to a number of buds which develop into lateral branches which give the plant a tufted appearance. Such basal branches are known as tillers or stools, and the habit of forming them as tillering or stooling. It is a valuable property of many cereals, and undesirable in others, for example, corn, where it causes a considerable reduction in yield. In a few grasses, the basal portion of the stem becomes enlarged by an accumulation of reserve food material, the plant being known as a bulbous grass. Many grasses develop underground stems known as rhizomes, from the nodes of which erect branch stems may develop, as well as numerous adventitious roots. These rhizomes may be short and the erect branches numerous, producing a tufted grass, or they may be long and wide spreading, as in the case of witch grass, *Agropyron repens*, also called quack grass. Due to the readiness with which the joints of the rhizomes of the latter grass strike root and develop to erect stems, it becomes a pestiferous weed. Eradication by chopping up the rhizome with a hoe only serves to increase its numbers, each joint or node producing a new plant. Only by preventing the green tops from forming can the plant be controlled and eliminated, or of course by complete removal of the entire underground rhizome. In some grasses the stem grows out over the surface of the ground, being then known as a stolon. Rhizomes and stolons form an effective way of propagating the plant, and in many species insure considerable dispersal over a limited area.

The leaves of grasses are composed of two parts, a basal sheath which enwraps the stem, and a flat elongate blade. The veins of the leaf are all parallel to one another, with few inconspicuous interconnecting veinlets. The blades of grasses grow from the bases, so that the apical portion is older and the cells of the basal portion retain for some time the ability to divide and increase. Because of this property grasses can be mowed by machines or cropped by animals, the upper portions of the blades being removed and the basal portion growing to renew the blade. Each node bears a single leaf, which is often reduced to a small scale, especially in the lowermost nodes, and in modified stems, such as rhizomes. At the junction of the sheath with the blade there occurs in many grasses a distinct structure called the ligule. This appears on the stem side of the leaf, and is a membranous or cartilaginous fringe or ring.

The inflorescence, in grasses, is composed of large numbers of groups of flowers, called spikelets, attached to the main stem or rachis. These spikelets are variously arranged. If they grow directly from the main stem and the latter is unbranched, the inflorescence is said to be a spike. If the main stem produces many branches, which in turn branch, the resulting inflorescence is a panicle. The nature of the branches, whether long or short, spreading or appressed, determines the nature of the panicle. In other grasses the inflorescence is a raceme, the spikelets being borne on short unbranched lateral branches. See Fig. 2.

The individual spikelet of a grass is composed of a short axis called a rachilla from which arise a series of opposite overlapping bracts. The two lowermost bracts are called glumes; these are empty, that is, have no flowers formed in their axils. The next bract above the glumes is the lemma, in the axil of which is borne a flower. In many grasses, each spikelet contains several lemmas, each with its associated flower. Opposite the lemma is the palea, which is not borne on the rachilla, but on a short pedicel, or flowerstalk. Opposite the palea and at the base of the ovary appear two minute scales, the lodicules. Three stamens, each with a long slender filament and a large anther, come next, while a single pistil grows at the apex of the pedicel. The pistil is composed of a 1-celled, 1-seeded ovary, two styles and two feathery stigmas. Many variations from the typical spikelet described occur in different species, the number of parts being increased, or parts being completely absent. In many species of grass,

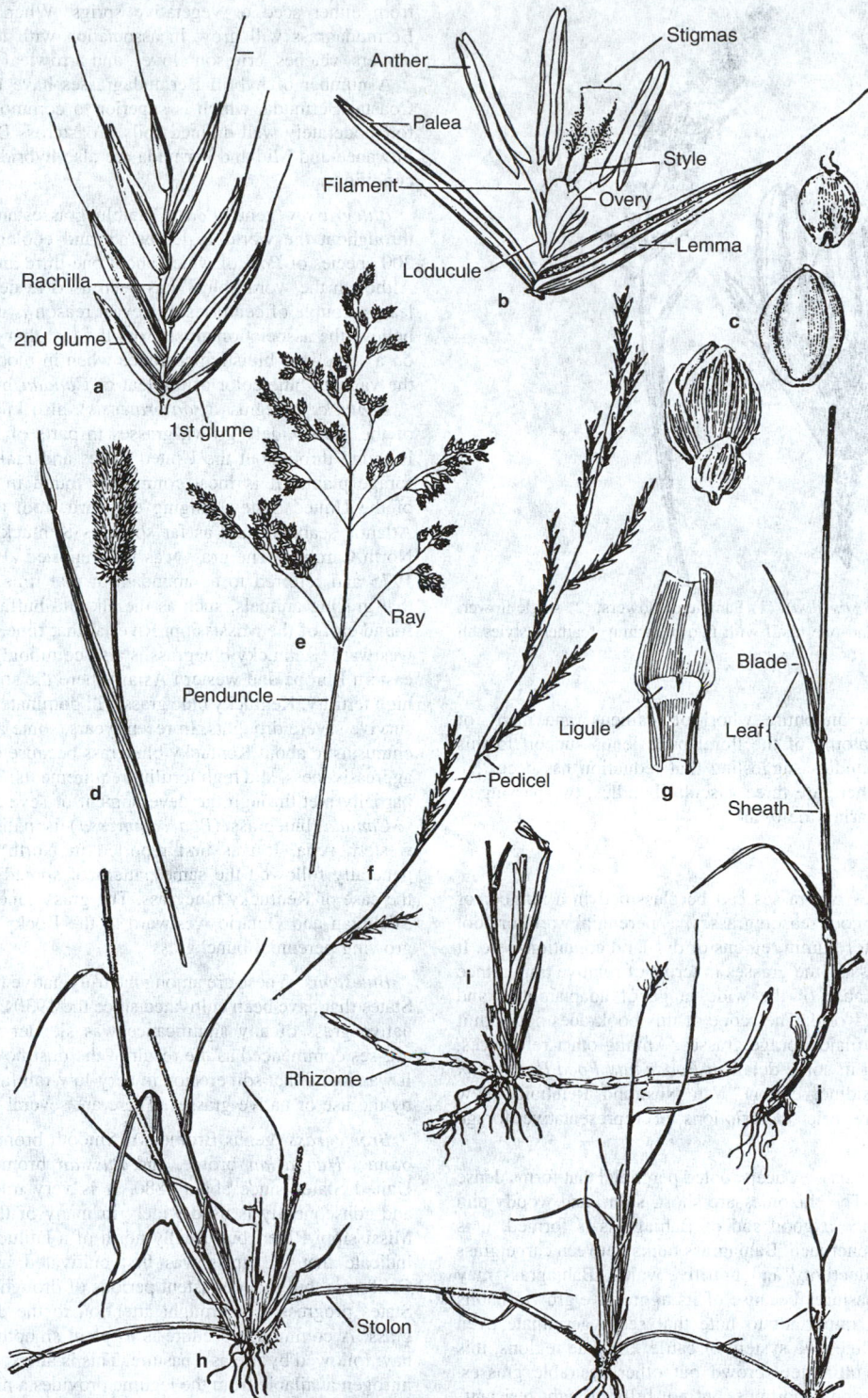

Fig. 1.   Characteristic growth of the parts of a representative grass plant: (**a**) Flowers in a spikelet arranged on a central axis enclosed in two empty glumes or bracts; (**b**) the different parts of a grass flower; (**c**) the developed fruit or seed (a caryopsia). This is shown successively enclosed in the outer glumes, with the lemma and paleas both closely adhering and free; (**d**) spikelets arranged in a terminal spike; (**e**) spikelets arranged in a panicle; (**f**) spikelets in a raceme; (**g**) a ligule, at the junction of the leaf blade and leaf sheath; (**h, i, j**) means of propagating or spreading—stolon, rhizome, and bulb, respectively. (*USDA diagram.*)

conspicuous prolongations on the glumes or the lemmas are noted—these are the awns.

Pollination in grasses is almost entirely by wind, the light dry pollen being scattered from the open anthers, often in conspicuous clouds. Grass pollen is a particularly common cause of hay fever.

The fruit of grasses is one-seeded, dry and indehiscent, that is, does not split open at maturity to liberate the seed. The ovary wall, or pericarp, is attached to the seedcoat. Within the latter is an abundant starchy endosperm. Such a fruit is known as a grain or a karyopsis.

Considerable speculation has been advanced as to the probable origin of grasses, whether they are primitive monocotyledonous plants from which others such as lilies may have developed, or whether they are reduced plants. To many the available evidence indicates reduction from lily-like ancestors, a reduction in which two of the three pistil lobes of the ancestral

2   1

Fig. 2.   A grass (redtop, *Agrostis alba*): (1) Panicle of flowers; (2) single flower, consisting of three stamens and one pistil with two branching feathery styles all enclosed by scales.

form have been lost, also an entire whorl of stamens, and many of the perianth parts. The anatomy of the floral parts lends support to this conception; the vascular bundles suggesting that reduction has occurred. For example, in the pistil there are three vascular bundles, two passing to the styles, and the third bearing the ovule.

The Forage Grasses

The forage or pasture types of grasses can be classified in a number of ways—as annual warm or cool season grasses; as perennial warm or cool season grasses; as grasses for humid regions or dry-land conditions; etc. It is extremely difficult to classify the grasses in terms of relative importance (quantity grown, etc.) because of the wide range of adaptabilities and preferences throughout the world. The scope of this book does not permit detailed descriptions of all major forage grasses. Among other references, these grasses are described in some detail in *Foods and Food Production Encyclopedia*, (D.M. Considine, editor), Van Nostrand Reinhold, New York, 1982. Following are brief descriptions of representative forage grasses.

Bahiagrass (genus *Paspalum*). A deep-rooted perennial that forms dense beds even on sandy soils. The rhizomes are short, stout, and woody and reach out horizontally. Once a good sod of Bahiagrass is formed, it is difficult for other plants to encroach. Bahiagrass ranks between carpetgrass and Bermudagrass in productivity and nutritive value. Bahiagrass may become a pest in certain pastures because of its aggressive growth habits and prolific seeding. It is important to note that seeds germinate even after passage through the digestive system of cattle. In some regions, this has caused Bahiagrass to ultimately crowd out other desirable grasses. Bahiagrass is suitable to range conditions, but not fully drought resistant.

Bermudagrass (Cynodon dactylon). This grass is commonly found in tropical and subtropical regions of the world. Because Bermudagrass grows so widely in India, it was believed for a long time that the species originated there. However, recent research indicates a much greater diversity of types found in Africa, and thus Africa is now considered by many authorities as the original source of Bermudagrass. In the United States, Bermudagrass is found mainly in the southern portions, ranging from southern California eastward to the North Carolina coast. The Midland variety ranges a bit further north, particularly east of the Mississippi River, where it is found in Kentucky, West Virginia, and northward to southern New England.

Bermudagrass has been an important pasture cover since the early 1800s. It is believed that it was first introduced to Savannah, Georgia as early as 1751. Common Bermudagrass is a fast-spreading grass that can be used effectively to prevent soil erosion. Common Bermudagrass is established from either seed or vegetative sprigs. When grazed closely, common Bermudagrass will grow in association with lespedeza, improved white clovers, vetches, crimson clover, and arrowleaf clover.

A number of hybrid Bermudagrasses have been developed, including Coastal Bermuda, which is superior to common bermuda and is adapted for moderately well-drained soils. Coastcross Bermuda is another hybrid. Suwanee and Midland bermuda are also hybrids, developed for particular conditions.

Bluegrasses (genus *Poa*). The bluegrasses are found widely distributed throughout the world in temperate and cooler regions. There are some 200 species of *Poa*, of which about one-third are native to North America. Although the word "blue" has been used to describe these grasses for at least a couple of centuries, the exact reason is unknown. Some authorities believe the association arose from the fact that some of these grasses take on a somewhat bluish appearance when in bloom. Others attribute this to the vaguely blue color of the leaf of *Canada* bluegrass.

Kentucky bluegrass (*Poa pratensis*), also known as *June* grass, is one of the most widely grown grasses in parts of North America. The grass is found throughout the United States and ranks as one of the important forage plants. It is most commonly found in the northeastern quadrant of the United States, ranging eastward from the eastern Dakotas to the Atlantic seaboard and as far south as Kentucky, Tennessee, and western North Carolina. The grass was first reported at Grassy Lick, Kentucky in 1775 and referred to as abundant at that time. Some authorities believe that grazing animals, such as the elk and buffalo, which were commonly found east of the Mississippi River at that time, helped to spread the grass westward. Kentucky bluegrass is also commonly found in the meadows of eastern Europe and western Asia. Where the soil pH is 5 or higher and of high fertility, Kentucky blue grass will dominate other plants. The grass can survive severe droughts. In recent years, some authorities have grown less enthusiastic about Kentucky bluegrass because of its low midseason yield, aggressiveness, and high fertility requirements. These objections have been partially met through the development of several new varieties.

Canada bluegrass (*Poa compressa*) is native to eastern Europe and western Asia. It was first reported in North America about 1792 and generally followed the same pattern of spread across the continent as in the case of Kentucky bluegrass. The grass generally ranges from northern Michigan and Ontario westward to the Rocky Mountains. It is an erect-growing perennial bunchgrass.

Bluestems. These are among the truly native forage grasses of the United States that have been cultivated since the 1930s. Prior to that time, the only native grass of any significance was slender wheatgrass. Use of native grasses commenced as the result of the dust bowl conditions of the 1930s. It was found that soil erosion in very-low-rainfall areas could be controlled by the use of native grasses. There are several bluestems.

Bromegrass (genus *Bromous*). Smooth brome, also known as *Austrian* brome, *Hungarian* brome, and *Russian* brome, has been grown in the United States since about 1880. It is very tolerant of heat and drought and consequently is used widely in many of the dry regions west of the Mississippi River, but usually north of a latitude of about 36°N. Records indicate that the grass was first cultivated in the west and widely in California, but that persistent periods of drought in the midwestern United States progressively brought attention to the desirable properties of this grass. A common procedure is to plant smooth brome with a legume for hay, followed by use as a pasture. This is an excellent combination because nitrogen available from the legume provides a nitrogen supply for the grass for several years.

Brome grass may be described as an extremely hardy perennial that grows to a height of 3 to 4 feet (0.9 to 1.2 meters). The root system is highly branched and sometimes reaches a depth of 6 to 8 feet (1.8 to 2.4 meters).

Varieties of brome, in addition to smooth brome, include field bromegrass, cheat bromegrass, nodding brome, and fescuegrass. See Fig. 3.

Buffalograss. This is highly regarded as a range pasture plant. The grass has numerous qualities that are attractive to stockmen—very palatable and nutritious when green in summer, but also retains a good feeding value when dried and cured for winter feeding. It tolerates heavy grazing.

Carpetgrass (Axonopus affinis). This is a low-growing, creeping perennial that makes a dense sod. Native to Central America and the West Indies, the grass was introduced into the United States in the early 1830s

Fig. 3. Bromer mountain bromegrass, a development of the Washington Agriculture Experiment Station. (*USDA and Soil Conservation Service.*)

and first reported in the New Orleans area. The grass is well suited to sandy or sandy loam soils. It is a prolific seeder and does not require high fertility. The grass does not do well in swampy areas. In the United States, carpetgrass is found mainly in the southeastern coastal area. The grass will tolerate close grazing, but is not as nutritious and productive as many other pasture plants.

Dallisgrass (genus *Paspalum*). Also known as *watergrass*, this is a fast-growing, rather stout perennial primarily utilized for pasture in the southeastern United States, ranging as far west as Texas. Dallisgrass is native to South America, ranging from Brazil to Argentina. It is believed that the grass was accidentally introduced into the United States in the mid-1800s. It is not suitable for hay production.

Fescues (genus *Festuca*). Made up of both annuals and perennials, there are some one hundred or more species of fescue. The growth habit may be creeping or erect. Of the species, tall fescue (*Festuca arundinacea*) is one of the more important forage grasses in the western, northwestern, and southeastern United States. It is also widely used in other grassland regions throughout the world. Tall fescue is a deep-rooted, strongly tufted, winter-hardy perennial with broad basal leaves that are dark green, coarse, and flat. It will tolerate a high water table and may be used in areas too low and wet for other pasture plants.

A few years after tall fescue was introduced to New Zealand from Europe (early 1800s), livestock that grazed on the grass for extensive periods were noted to develop a lameness, a condition called *fescue foot*. The situation became widespread and serious and a program was undertaken to eradicate the grass from that country. Although exacting conclusions may not have been drawn, some authorities believe that it was a peculiar grouping of circumstances rather than the qualities of the grass.

For example, where fescue foot was observed the areas usually were wet, low, and swampy with extensive deficiencies of minerals.

Foxtail. This grass is commonly used in Europe for pasture and hay and is well adapted to wet lands. Records indicate that it was first used in the mid-1700s. In the United States, it performs well in the northwestern states, including Alaska. It prefers a cool, moist climate and does not resist high-temperature and drought conditions. There is a superficial resemblance of foxtail with timothy. The palatability of the grass, both as pasture and hay, is very good. Varieties of foxtail include meadow, creeping, and reed foxtail.

Grama Grasses. Two species of the grama grasses are of significance in the Great Plains regions of the United States—a bunch form and perennial known as *sideoats grama* and a more drought-resistant form known as *blue grama*. Grama grasses are palatable and retain their flavor and nutrition well into the winter months. However, grama grasses are not suitable for hay. There are a number of important native varieties of grama which are cultivated for forage locally.

Johnsongrass (Sorghum halapense). Not commonly considered a cultivated grass, but more often as a weed by some food crop growers, nevertheless Johnson grass is an important hay grass in the southeastern United States. This grass also can be used as an effective soil-conserving crop. It requires relatively fertile and loose soil and does not endure close grazing.

Lovegrasses. The principal attractions of the lovegrasses are their toleration of low fertility and sandy soils. These grasses produce abundant quantities of seed, which germinate readily. In the United States, one native and three introduced species occur. Native to the central southern Great Plains is *sand lovegrass*. The value of the grass was not formally recognized until the late 1930s, after the dust bowl period.

Millet Grasses. These grasses, of several species, offer the advantage of only requiring 60 to 70 days from seeding to maturity. Some authorities have found that the foxtail millets (not to be confused with foxtail grass) exceed all other crops in their efficient use of water.

Napiergrass (Pennisetum purpureum). A grass native to equatorial Africa and introduced into the United States in 1913. This grass is adapted to the Gulf coastal region from Texas to and including all of Florida. It also does well in southern California. The grass will grow on almost any soil that will support ordinary food crops. The useful area of the grass can be extended northward if planted on rather fertile soils.

Natalgrass (Tricholaena rosea Nees). A grass native to South Africa and introduced into the United States in the late 1860s. It is also known as *Hawaiian redtop* and *Australian redtop*. First attention was brought to the grass because of its ornamental potential. The grass is suited to well-drained, poor, sandy soils. An outstanding advantage of Natalgrass is its resistance to attack by nematodes. The grass often succeeds as a forage crop in areas where no other forage grass can grow. It can be cut for hay.

Oatgrass. Tall oatgrass, at one time, was very important as a forage grass in Europe. It was introduced into the United States in the early 1800s. Of secondary importance, the grass is found mainly in the northwestern United States. It is not drought or heat resistant.

Orchardgrass (Cactylis glomerata L.). This grass is native to western and central Europe, but has been under cultivation in the United States since 1760. It is a cool season perennial that grows in clumps producing an open stand. It makes excellent hay. It is tolerant of partial shade and grows well in mixtures with white clover. However, the grass is highly susceptible to a number of diseases. The flowering culms of the plant reach a height of 2 to 4 feet (0.6 to 1.2 meters). The importance of this grass in North America has increased manyfold since the early 1930s. In some states, such as Virginia, Kentucky, and Tennessee, this is the major forage grass. It is frequently part of a mixture, particularly with red clover or alfalfa for hay. Throughout the United States, in terms of quantity, orchardgrass probably is exceeded only by smooth bromegrass, timothy, and Kentucky bluegrass, although reliable Figures are difficult to obtain. Persistence of the grass under continuous grazing is limited. Rotational grazing is the best practice for orchardgrass.

The use of orchardgrass in the British Isles has increased considerably during the last couple of decades. Orchardgrass possesses much versatility, being adapted for harvest for hay or silage as well as for grazing. Much orchardgrass seed is produced in Oregon, Washington, and California. High applications of nitrogen can increase seed production by a factor of 100%. Shattering is a problem in seed processing. There are numerous varieties of orchardgrass. The *Akaroa* variety was released for use in western Washington in 1951 and for use in California in 1952. This grass has long been popular in New Zealand. It is well adapted to all of the Pacific coastal states, but must be irrigated in California.

Redtop (Agrostis alba L.*).* Of the same genus as the bentgrasses, redtop at one time (until 1940s) was second only to Kentucky bluegrass as an important forage and pasture grass in North America. Since that time, redtop has been significantly displaced by a number of other grasses and grass-legume mixtures. In addition to forage uses, redtop finds application for lawns, recreational areas, highway plantings, etc. Redtop is most common in the northeastern quadrant of the United States. Most frequently, redtop is sown with legumes and other grasses.

Reed Canarygrass (Phalaris arundinacea). This is an important grass, not only as a hay and silage crop, but also for use in soil conservation programs. The grass will frequently produce good yields of forage from soils that are too wet or poorly drained for other grasses and legumes. Variations of reed canarygrass include ribbongrass and Hardinggrass.

Ryegrasses (genus Lolium). These are hardy winter annual bunch grasses with glossy, dark-green foliage. Ryegrass furnishes grazing in the late fall, winter, and spring. The grass is most often used in mixtures of small grain and annual clover. Ryegrass adds to nutritive value when grown with wheat for silage. It will extend the grazing period in the late spring. Ryegrass does best when heavily fertilized, especially with nitrogen. The greatest concentrations of ryegrass are found in the Gulf coast states, as well as Georgia, South Carolina, and parts of North Carolina. Ryegrass is not extensively used in Florida.

Saint Augustinegrass (Stenotaphrum secundatum). This grass is native to the West Indies, and possibly to Australia and southern Mexico. The grass is also found in South Africa. It was introduced into France and Italy from Africa and probably introduced into the United States from Cuba. This grass is also called saltgrass, sheepgrass, and jointgrass. The grass does well in most kinds of soil, but requires a lot of moisture. It is notably well adapted to mucky soils and partially shaded areas.

Sudangrass (genus Sorgos). The grass sorghums include a number of varieties, one of the most important being Sudangrass. This is an excellent annual grass and used extensively in the United States, with exception of the far north and southeastern states. In these areas, because of frost or disease problems, Sudangrass is essentially replaced by pearl millet. For clarity, it should be pointed out that there are many kinds of sorghum — grain sorghum, forage sorghum, sirup sorghum, grass sorghum, and broomcorn.

Timothy (Phelum pratense). At one time, timothy was the most important and widely used of the many forage grasses. The existence of timothy dates back to antiquity. The grass is native to most of Europe, eastward through Siberia and north to a latitude of 70 °N. The grass also occurs naturally in the Caucasus region and in Algeria. In the New England states, timothy is sometimes called herdgrass. Timothy grows best in a cool and humid climate. Although it may survive in some hot humid or hot dry climates, it does not yield well. Best results are achieved when the plant is grown on clay or silt loam soils that are fairly well drained. Timothy roots are shallow and fibrous. Timothy is a bunch grass with erect culms, ranging from 20 to 40 inches (51 to 102 centimeters) in height. It produces a dense, cylindrical, spike-like inflorescence (the head). See Fig. 4.

Although it is grown alone, more often timothy is sown in mixtures with legumes, such as medium-red or Alsike clover. Principal regions for plantings in the United States are in the northeastern quadrant. Timothy is grown mainly for hay. Improved varieties of timothy have been developed in recent years.

Wheatgrasses (genus Agropyron, tribe *Hordeae).* At one time, the wheatgrasses were considered to be in the same genus as wheat. The common name stems from the fact that the seed heads resemble those of wheat. The wheatgrasses are widely distributed through the temperature regions of the world. Of the 150 known species, about 30 are native to North America. Most species originated in eastern Europe and western Asia in desert or steppe soils and in climates ranging from semihumid to arid. A few species are confined to South America. In North America, most of the wheatgrasses are found in the northwestern quadrant, including British Columbia. They

Fig. 4.   Close-up of timothy in heading stage. (*USDA and Soil Conservation Service.*)

range eastward as far as Minnesota and Ontario and as far south as northern Texas. These are cool-season grasses and are highly valued where suited as important sources of very nutritious early-season forage. They also are highly regarded for control of wind and water erosion. Some authorities have estimated that natural wheatgrasses in the United States are found on 300 million or more acres (120 million hectares). Species introduced into the United States include Crested wheatgrass (*Agropyron desertorum*), a hardy, drought-resistant bunchgrass native to eastern Russia, western Siberia, and central Asia. Fairway was introduced from Canada. *Siberian* wheatgrass, a drought-resistant bunchgrass, was introduced from the U.S.S.R. in 1934. There were many other introductions.

Wildrye Grasses. These grasses are closely related to the wheatgrasses, differing mainly by the fact that the wildryes have two spikelets at each rachis node. Among the wildrye grasses in North America, some are native, but several have been introduced and are now used in grassland agriculture in the United States and Canada. *Russian* wildrye is particularly well adapted to the northern Great Plains region. This grass does well in several of the Canadian provinces. This variety is drought-resistant. Other varieties include *Canada* wildrye, *Virginia* wildrye, *Basin* wildrye, and *Beardless* wildrye, among others.

The *cereal grasses* are described in separate alphabetical entries. See also **Oats**; **Rye**; and **Wheat**.

Additional Reading

Brown, L.: *Grasses: An Identification Guide*, Houghton Mifflin Company, New York, NY, 1992.

Darke, R.: *The Color Encyclopedia of Ornamental Grasses: Sedges, Rushes, Restios, Cat-Tails, and Selected Bamboos*, Timber Press, Inc., Portland, OR, 1999.

Knobel, E., Revised by M. Faust: *Field Guide to the Grasses, Sedges and Rushes of the United States*, Dover Publications, Inc., Mineola, NY, 1981.

GRASSHOPPER *(Insecta, Orthoptera).* Also known as locusts and of many species of the family *Locustidae*, grasshoppers have been known

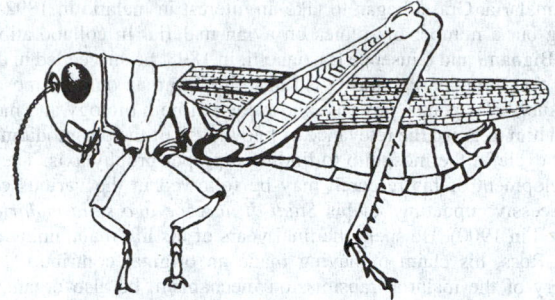

Fig. 1.   Grasshopper, also known as locust. (*USDA.*)

since ancient times and associated with devastating crop losses and resulting famines. See Fig. 1. Practically no plant, cultivated or wild, is immune from attack by one or several species of grasshopper. These insects occur worldwide. In the United States, serious outbreaks of grasshopper seldom develop east of the Mississippi River, but they are not uncommon in the western two-thirds of the country. Grasshoppers often severely damage range grasses. Their feeding is one of the main reasons for loss of productive grasslands in many of the western states.

When range grass is scarce and outbreaks are severe, grasshoppers often migrate into and severely damage the foliage of alfalfa, clover, corn (maize), small grains, potato, and fruit trees. In fruit orchards, grasshoppers sometimes fully strip the leaves and may kill young trees. Both insecticides and cultural practices can be effective and must be used for effective grasshopper control.

Many species of grasshopper winter in the egg stage. The eggs are laid in masses that are found from $\frac{1}{2}$ to 3 inches (about 2 to 8 centimeters) below the soil surface. Each mass will have from 20 to 120 elongated eggs, held together securely by cement. One female may deposit from 8 to 25 egg masses. The eggs usually are deposited in uncultivated ground, often in alfalfa, clover, and stubble fields. The egg-laying procedure varies from one species of grasshopper to the next. The pellucid grasshopper prefers sod land and heavy soil. The migratory grasshopper prefers crop land. Other species prefer uncultivated ground, as previously mentioned.

The red-legged grasshopper (*Melanoplus femur-rubrum*, De Geer) is a small species, ranging up to 1 inch (2.5 centimeters) in length when fully developed. The insect is severely destructive of legumes, notably soyabean. The color is a brown-red. The hind tibiae are a pinkish-red with black spines.

The migratory grasshopper (*Melanoplus bilituralus*, Walker) is the most destructive and widespread of all species. The insect has a great ability to survive in dry and waste lands. This insect is also about 1 inch (2.5 centimeters) long when fully grown. The term migratory is used in describing the species because the partially developed nymphs normally travel or migrate from their breeding ground to find more attractive vegetation. See Fig. 2. The adults also may fly for many miles in search of more attractive feeding areas.

The clear-winged grasshopper (*Camnula pellucida*, Scudder). In terms of damage, this insect is only second to the migratory grasshopper. It occurs throughout the United States, but is most common in the west and it seems to prefer relatively high elevations. It is well adapted to survive heat and drought. The hind wings are nearly transparent.

The differential grasshopper (*Melanoplus differentiallis*, Thomas). This insect prefers cultivated areas and does not survive long dry periods. In such times, they will be found only near ditches and irrigated areas. The insect ranges from $1\frac{1}{2}$ to $1\frac{3}{4}$ inch (about 3.5 to 4.5 centimeters) in length and of a brown-green color with yellow underparts. The differential grasshopper is a severe destroyer of corn (maize).

The two striped grasshopper (*Melanoplus bivittalus*, Say). This is a strong species and is of an olive-green color with yellow stripes on each side. The species is frequently found in clover fields.

The Carolina grasshopper (*Dissosteira carolina*, Linne). One of the largest of the grasshoppers, attaining a length of about 2 inches (5 centimeters). One of the most commonly observed species, although somewhat less destructive of crops.

Fig. 2.   Grasshoppers sometimes gather in swarms and migrate hundreds of miles (kilometers). (*USDA.*)

Closely allied to the grasshopper are the **Cicada**; **Katydid**; and **Locust**; see separate entries under these headings.

Cultural Practices

Grasshoppers, particularly those that lay their eggs in fields planted to crops, may be controlled to some extent by tillage and seeding operations. Cultural operations do not eliminate the need for insecticides, but they reduce the amount of chemicals needed.

Tillage. Working the soil kills grasshoppers in several ways. It can bury their eggs so deep that young grasshoppers do not hatch. It can bring the eggs to the surface where they are destroyed by drying of sun and wind. Tillage also discourages egg-laying, preventing dispersal of the pests and forcing grasshoppers scattered over a field to concentrate in a smaller area. Proper tillage before eggs have hatched often gives excellent control of threatening grain-stubble infestations. Fall tillage is preferable, but spring tillage can be effective. Tillage immediately after harvest will make the soil less attractive to egg-laying and will assist in destroying eggs already laid.

Shallow cultivation is less effective than moldboard plowing, but it will destroy many of the eggs by exposing them to sun and wind. The one-way disk is the best implement for this operation. The duck-foot cultivator, the single or double harrow, and the one-way harrow, also are satisfactory. Blade tillers used in stubble-mulch farming are less effective than the others. Shallow cultivation is most effective during dry weather.

Grasshopper-infested grain stubble that is to be summer-fallowed should be worked before the eggs hatch. If tillage is delayed until after the young grasshoppers appear, it still may be useful in preventing the insect from moving to nearby crops. This tillage can be accomplished by cultivating a guard strip 3 rods (about 5 meters) wide around the entire field. If the strip is kept cleanly fallowed, the young grasshoppers can usually be held within the field for a week or two. There may be time to complete tillage operations before they escape. Tillage done after the establishment of the guard strip should start next to the strip and extend until only a small block of unworked stubble remains in the center of the field. The grasshoppers will then be concentrated in this small area. Here they can be killed with insecticide at much less cost than would be required for spraying the entire field. Large tracts of sod or idle land should not be plowed or shallow-tilled for control of grasshoppers unless the land is intended for seeding or summer-fallow. Cultivation ruins such land for pasture and makes it subject to soil blowing.

Aircraft are frequently used to spread control chemicals in connection with grasshopper infestations.

Seeding. In years when grasshoppers are abundant, small grains may be planted on fall- or spring-tilled land, or on clean summer-fallowed land.

Few grasshoppers emerge from such land. A grain drill should not be used on heavily infested, unworked stubble. This will destroy only a few eggs by the seeding process. When the eggs hatch, the field will swarm with young grasshoppers. Then, immediate spraying of the entire field will be required to save the crop.

Early spring seeding is important in reducing grasshopper damage. These crops make considerable growth before grasshoppers hatch. Thus, they withstand a longer period of feeding than late-seeded crops and also provide a better opportunity to kill the grasshoppers with chemicals.

When small grains are ripening, flying grasshoppers frequently congregate in late-seeded crops that are still green and succulent. Such crops are often severely damaged before the grasshoppers are noticed. Well advanced crops are much less attractive to the pests. Barley, oats, and wheat that have headed can withstand considerable defoliation without serious reduction in yield of grain.

Regrassing Field Margins. Weedy field margins, including roadsides and fence rows, contain more grasshopper eggs than other habitats. Replacing broad-leaved weeds with perennial grasses greatly reduces the number of grasshoppers in such locations. Crested wheatgrass can be used for this purpose. It is easily and quickly established and is less attractive for egg-laying than native grasses. Elimination of weeds and prevention of soil erosion are additional benefits of grassed field margins.

Immune Crops. Some of the sorghums, such as sorgo and kafir, after reaching a height of 8 to 10 inches (20 to 25 centimeters), are practically immune to grasshopper attack. They can be planted rather late in the season to provide valuable feed for livestock.

Irrigation. When alfalfa and other legumes are irrigated, large numbers of grasshoppers are sometimes driven to ditchbanks and other dry places. Here, they can be killed with sprays at very low cost. Flooding hay meadows where grasshopper eggs have recently hatched will destroy many young grasshoppers.

Additional Reading

Chapman, R. and A. Joern: *Biology of Grasshoppers*, John Wiley & Sons, Inc., New York, NY, 1990.
Gangwere, S. et al.: *Bionomics of Grasshoppers, Katydids and Their Kin*, CAB International, New York, NY, 1997.
Preston-Mafham, K. and R. Preston-Mafham: *Grasshoppers and Mantids of the World*, Blandford Press, UK, 1998.

GRASSI, GIOVANNI BATTISTA (1854–1925). Giovanni Grassi was an Italian zoologist and parasitologist who established the role of Anopheles mosquitoes in transmitting malaria parasites. Born on 27 May 1854 at Rovellasca, in the Italian province of Como, Giovanni Battista Grassi (or Battista Grassi, as he preferred to sign his name), graduated in medicine from Pavia University in 1878. While still a student he carried out fundamental studies on helminth parasitosis, among other things describing, together with Corrado Parona, the *Ancylostoma duodenalis* epidemic affecting workers building the Gotthard Tunnel. By the time he graduated, Grassi was already a world renowned parasitologist who was arguing with Rudolph Leuckart over the meaning of the specificity of the parasites' development cycle and of host–parasite interactions. He embarked on a study tour to meet the great German morphologists and in Karl Gegenbauer's laboratory he investigated the relationship between morphological development and evolution in vertebrates and arthropods. Then, in 1883, he was offered the chair of zoology at Catania University. There he resumed the study of entozoan biology and, in 1896, was called to the chair of Comparative Anatomy and Physiology of Rome University. In Rome he carried out studies on the mechanism of malaria transmission, on phylloxera and the life cycle of the moray eel. He also dedicated himself to developing and assessing ways and means of combating malarial infection. His relationship with evolutionary theory was fraught with problems. Initially a moderate Darwinian evolutionist, he later judged the problem of the origin of species as insoluble, openly displaying sympathy for German vitalist morphology. In his later years he embraced the chemical–materialist theories of life.

Grassi was the greatest Italian zoologist and made an essential contribution to the description of the development cycle of many parasites. In his research he followed a comparative approach based on the concept of the biological specificity of parasitism. The heuristic value of this approach was exemplary in studies on the transmission mechanism of human malaria. Grassi began to take an interest in malaria in 1892–1893, carrying out a number of studies on avian malaria. In collaboration with Amico Bignami and Giuseppe Bastianelli, in 1898, he succeeded in demonstrating that the malaria parasite is transmitted to man only by mosquitoes of the *Anopheles* genus. His vast knowledge of both biology and medicine allowed him to grasp the relevance of knowledge of the plasmodium development cycle in the mosquito to finding a proper prophylaxis. The stages of development of his research may be followed in the various editions and successive updatings of his *Studi di uno zoologo sulla malaria* (first published in 1900). He spent the final years of his life maintaining against Ronald Ross, his claim of having made an original contribution to the discovery of the malaria transmission mechanism; he also organized the antimalaria campaign in the Rome rural area. His image as a doctor and benefactor became a popular legend on the Roman Plain and at his death he asked to be buried at Fiumicino (near Rome). He was accepted as a member of the Accademia Nazionale dei Lincei, as well as of many prestigious foreign academies. For his work on malaria, he was awarded an *honoris causa* degree from Leipzig University. See also **Malaria**; and **Ross, Ronald (1857–1932).**

Additional Reading

Alippi C.M.: *Grassi, Giovanni Battista*, Dizionario biografico degli italiani, Istituto dell'Enciclopedia Italiana, vol. 58, pp. 630–640, 2002.
Corbellini, G., and L. Merzagora: "Battista Grassi e la malaria," In Ca-panna, E., ed., *Battista Grassi. Catalogo del Museo Grassi, Parassitologia*, **38**(Suppl. 1): 23–36, 1996.
Grassi, B.: "The Transmission of Human Malaria,"*Nature*, **113**, 304–307; 353; 458 (1924).
Neghme, A.: "An Appraisal of Giovan Battista Grassi: His Work in Biology and Parasitology," *Experimental Parasitology*, **15**(3), 260–278 (1964).

GIBERTO CORBELLNI, University 'La Sapienza', Rome, Italy

GRATICULE.   1. A graticule is a reticle composed of lines ruled on a transparent plate, instead of the usual fine threads or wires.

2. By extension, the pattern of lines representing parallels of latitude and meridians of longitude on a map or chart is known as the graticule of the chart. A person familiar with the various types of map projection can usually tell by examination of the graticule the type of projection that was used in constructing the sheet.

GRATING.   Any framework or latticework, consisting of a regular arrangement of bars, rods, or other long, narrow objects with interstices between them. A diffraction grating consists of rulings upon the surface of a light-transmitting or light-reflecting substance; it is used for the production of spectra.

GRAVEL.   An unconsolidated, natural accumulation of rounded rock fragments resulting from erosion, consisting predominantly of particles larger than sand (diameter greater than 2 millimeters; $\frac{1}{12}$ inch), such as boulders, cobbles, pebbles, granules, or any combination of these fragments; the unconsolidated equivalent of conglomerate. In the United Kingdom, the range of 2–10 millimeters has been specified.

Gravel is also a popularly used term for loose accumulation of rock fragments, such as detrital sediment associated especially with streams or beaches, composed predominantly of more or less rounded pebbles and small stones, and mixed with sand that may compose 50–70% of the total mass.

Gravel is also a term for rock or mineral particles having a diameter in the range of 2–50 millimeters. In the United States, the term is used for rounded rock or mineral soil particles having a diameter in the range of 2–75 millimeters $\frac{1}{6}$ to 3 inches); formerly the term applied to fragments having diameters ranging from 1–2 millimeters.

See also **Ocean Resources (Mineral)**.

GRAVE'S DISEASE.   See **Thyroid Gland**.

GRAVIMETRIC ANALYSIS.   A type of quantitative analysis involving precipitation of a compound that can be weighed and analyzed after drying. It is also used in determining specific gravity.

See also **Specific Gravity**.

GRAVIMETRY.   See **Earth**.

GRAVITATION.

During the early 1990s, there was an increased interest shown by theoretical physicists in their views on the nature of gravity, which have been widely held since Einstein's proposals of three-fourths of a century ago. A number of interesting new experiments have been proposed.

Although such experiments could have a major, fundamental (but not necessarily practical) bearing on our understanding of natural forces, and even though the proposed experiments carry relatively modest costs, the national support for such experiments in the United States, as well as other leading nations worldwide, has been less than overwhelming.

Thus, the exact timing of the proposed gravity-related experiments will depend upon the priorities for science projects as established by government planners.

Newton's Gravity. Gravitation is a phenomenon characterized by the mutual attraction of any two physical bodies.[1] This universal character of the gravitational force was first recognized by Sir Isaac Newton who also gave its quantitative expression. For point masses or spherical bodies, a simple expression results:

$$F = \frac{GM_1M_2}{R_2} \qquad (1)$$

In addition to the masses M_1, M_2 of the two bodies and their distance apart R, the force depends only on a constant $G = 6.670 \times 10^{-8}$ dyne cm^2 gm^{-2} which is independent of all properties of the particular bodies involved. The same force law describes the motion of the planets around the sun, of the moon around the earth, as well as the falling of an apple to the earth. A body moving under an inverse square law as given in Equation (1) satisfies the three laws established by Kepler for the motion of the planets around the sun:

1. The planets move in elliptical orbits with the sun at one focus (the general orbit is a conic section). Fig. 1.

2. The radius vector sweeps out equal areas in equal times.

3. The square of the period of revolution is proportional to the cube of the semi-major axis: $a^3 = (2\pi)^{-2} GM \odot T^2$. Here $M\odot$ is the mass of the sun and T is the period of the planet.

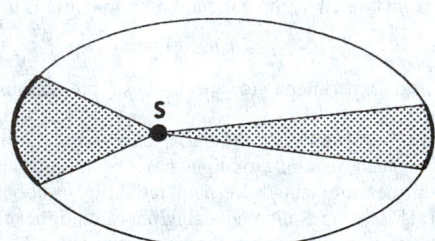

Fig. 1. An elliptical orbit for a planet around the sun. The shaded areas indicate equal areas swept out in equal times at different parts of the orbit. Clearly, the speed of the planet varies with its position in its orbit.

These results together with a detailed analysis of anomalies in the motion of the moon established the correctness of the Newtonian theory of gravitation.

The *weight* of a body of mass M on the earth is the force with which it is attracted to the center of the earth. On the surface of the earth the weight is given by

$$W = Mg$$

where the *acceleration due to gravity* is obtained from Equation (1):

$$g = \frac{GM_E}{R_E{}^2} = 980.665 \text{ cm/sec}^2$$

$$= 32.174 \text{ ft/sec}^2$$

All freely falling bodies near the surface of the earth are accelerated at the same rate g. It is for this reason that Galileo found that both light and

[1] Einstein's general relativity theory is essentially the modern statement of gravity and reference to the entry on **Relativity and Relativity Theory** is also suggested, where the topic is approached from a somewhat different direction and viewpoint.

heavy objects take the same time to reach the ground when dropped from the Leaning Tower of Pisa.

An astronaut is said to be in a state of *weightlessness* when in orbit. Strictly speaking, the body still has weight for the earth's gravity still acts on it. Otherwise the astronaut would fly off into outer space. However, when in free fall, the local effects of the gravitational field are eliminated for the astronaut. Objects that are released fall together with him and hence remain in his vicinity unlike the situation on the ground. Therefore, the organs of the body respond as though the gravitational field was absent, and this gives the sensation of weightlessness.

Precise determination of the Newtonian gravitational constant G has been attempted by many investigators, both in the field and in laboratories. Because of deficiencies associated with instruments in the past, the geophysically determined values did not have the accuracy to match that obtained in laboratories. A.T. Hsui (University of Illinois) reports that the geophysically determined Newtonian gravitational constant is consistently larger than the laboratory value by 1 to 2% on the basis of gravity measurements in Australian mines. This discrepancy may have strong implications for the physics of gravitation. To test whether similar results can be observed in a different geological environment, gravity measurements in a Michigan borehole have been examined. Although these results cannot be taken as conclusive, owing to the large uncertainties involved in mass determination on a geophysical scale, these measurements are generally consistent with those of the Australian experiment. The Michigan test site is known as State Burch #1–20 borehole and is located near the eastern shore of Lake Michigan (44°10′N; 86°6′W).

Gravitational Field. According to Newtonian theory, the sun exerts the gravitational force directly on the earth without an intervening medium for transmitting that force. The behavior of such forces is called "action at a distance." To overcome the conceptual difficulty of a force acting directly over large distances, one assumes that a *gravitational field* fills all space. The force acting on any mass is determined by the gravitational field in its neighborhood. Thus, at the point P a distance R from the center of the earth, the gravitational field has the magnitude

$$\mathscr{G} = \frac{GM_E}{R^2}$$

and magnitude of the force on a mass M at P is simply $F = M\mathscr{G}$. Note that the field is to exist at P even in the absence of the mass M.

It is sometimes convenient to introduce the gravitational potential which determines the field through its gradient. For a spherical earth, it is defined as

$$\phi = \frac{GM_E}{R}, \quad \mathscr{G} = -\text{grad}\phi$$

In general ϕ will satisfy Poisson's equation

$$\frac{\partial^2\phi}{\partial x^2} + \frac{\partial^2\phi}{\partial y^2} + \frac{\partial^2\phi}{\partial z^2} = 4\pi\rho$$

ρ is the density of matter. The potential energy of a mass M, in the field is simply expressed in terms of ϕ,

$$V = M\phi$$

Although one can introduce the gravitational field, it is an auxiliary concept in Newtonian theory for the field has no independent dynamical behavior as is true of the electromagnetic field (e.g., electromagnetic waves). At any time, the Newtonian gravitational field is determined by the configuration of masses at that instant and does not depend on previous history or state of motion. Thus if the sun were to vanish, the gravitational force on the earth would immediately be removed. This property may be thought of in terms of an infinite velocity of propagation for the gravitational field. Letting the velocity of light become infinite in Maxwell's equations eliminates all independent dynamical behavior for the electromagnetic field. In that case there could be no radio or television. The special theory of relativity, which is based on the velocity of light in vacuum being the maximum velocity for the transmission of energy, implies that Newton's theory requires modification.

Principle of Equivalence. The mass of a body may be measured either by weighing $W = Mg$ (*gravitational mass*) or by observing its motion under a known applied force using Newton's second law of motion $F = MA$ (*inertial mass*). The equality of these two differently defined masses has been measured by R.H. Dicke to an accuracy of 1×10^{-11}

improving an earlier measurement by Eötvös. It is this equality which distinguishes the gravitational force from all other forces in giving all bodies the same acceleration. The discussion of weightlessness pointed out that local effects of the gravitational field are eliminated for an observer in free fall precisely because all bodies fall at the same rate. It follows that the gravitational field *measured* by an observer will depend on his state of motion. In a sense *there is an equivalence between a gravitational field down and an acceleration up for the observer.* However, the equivalence is not complete, for real gravitational fields converge on their sources so that two particles released at the same time will drift closer together as they fall. On the other hand, acceleration fields have no effect on the separation of particles moving on parallel paths. See Fig. 2. In a curved space, initially parallel geodesics—the "straight lines"—do not maintain a constant separation (e.g., great circles on a sphere). Thus, the gravitational field may have its explanation in the geometry of a curved space-time.

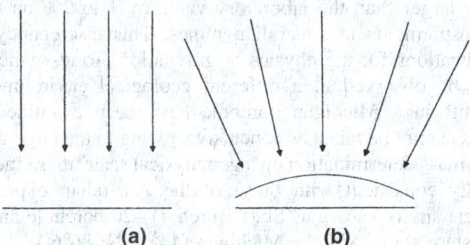

Fig. 2. (a) The paths of particles released in an acceleration field (the acceleration is up, the apparent force is down); (b) the paths of particles released in a gravitational field showing convergence toward the source.

Red Shift. According to the quantum theory, a photon of frequency ν has an energy $h\nu$ (h is Planck's constant), and by the relation $E = mc^2$, this quantum has a mass $m = h\nu/c^2$. To lift a mass m a height H requires expenditure of the energy mgH. Therefore, a photon emitted at the surface of the earth arrives at the height H with the energy

$$h\nu - (h\nu/c^2)gH = h\nu\left(1 - \frac{gH}{c^2}\right) = h\nu'$$

At the surface of the earth, the frequency shift amounts to

$$\frac{\Delta\nu}{\nu} = 1.1 \times 10^{-16} H \, (H \text{ in meters})$$

This shift was measured by Pound and Rebka using the Mössbauer effect in good agreement with the prediction. As time standards are determined by frequency, it follows that if the same photon were emitted at the height H, it would be measured to have the frequency ν, not ν'. Therefore, an observer at H must conclude that his clock is running faster than the same clock would run on the surface of the earth in the ratio $\Delta T/T = -\Delta\nu/\nu$. See Fig. 3.

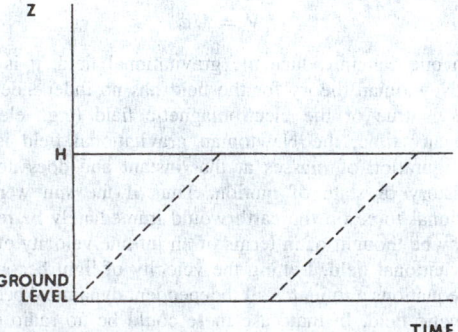

Fig. 3. Photons are emitted on the ground and are received at the height H. Between the two dotted lines representing the beginning and end of a pulse, the same number of oscillations, n, are received at H as are emitted at the ground level. Because of the red shift, the interval t' between oscillations at H is greater than the interval t between oscillations on the ground. Therefore, the time measured at H for the reception of the n oscillations is greater than the time required for their emission on the ground: $nt' > nt$. This result implies that clocks run faster at H than on the ground.

Einstein's Theory of Gravitation. Albert Einstein assumed that gravitation is a physical effect produced by the curvature of a four-dimensional space-time. The generalization of Newton's gravitational potential is the metric tensor g_ν in terms of which the four-dimensional distance, and hence the geometry of space-time, is determined:

$$ds^2 = \sum_{\mu,\nu=1}^{4} g_{\mu\nu} \, dx^\mu \, dx^\nu$$

The curvature of space-time is defined in terms of a four index tensor $R^\mu_{\nu\rho\sigma}$, the curvature tensor. The vanishing of the curvature tensor means that no real gravitational field is present. The field equations are ten linear combinations of the curvature components which are of the second order in the derivatives of the metric tensor and are a generalization of Poisson's equation [Equation (2)]. Symbolically these equations are written

$$G^\nu = 8\pi\kappa T^\nu$$

where T^ν is a symmetric tensor which describes the distribution of matter and energy throughout space-time and $\kappa = G/c^2$. In a weak field static approximation, these equations contain Newton's theory of gravitation with the Newtonian gravitational potential. Given by $2\phi = 1 - g_{44}$.

The metric tensor outside a static spherically symmetric mass distribution is given by the Schwarzschild solution:

$$ds^2 = \left(1 - \frac{2\kappa m}{r}\right) dt^2 - \left(1 - \frac{2\kappa m}{r}\right)^{-1} dr^2 - r^2 \, d\theta^2 - r^2 \sin^2\theta \, d\phi^2$$

This geometry exhibits the red shift described above and in addition shows three other effects:

1 The bending of a ray of light passing near the sun's edge by

$$\delta\theta = 1.75''$$

2 The precession of the perihelion of Mercury by

$$\delta\phi = 43''.03/\text{century}$$

3 The retardation of signals passing near the sun; for a radar pulse reflected from Mercury, this amounts to a maximum time delay

$$\Delta t = 1.6 \times 10^{-4} \text{ sec}$$

Observations and experiments to check these predictions are still in progress.

Since one can see stars near the sun's edge only during an eclipse, the optical data on the bending of light have been slow and difficult to obtain and such measurements have poor reliability—about 10–25%. A group under H. Hill set up equipment using photomultiplier tubes sensitive to a narrow spectral range so that the solar background can be filtered out. As a result, measurements at a fixed site can be made continuously as the sun moves into and out of a selected field of stars. Therefore, much improved accuracy is possible. Using radio frequency measurements, Shapiro observed the angular position of two sources, 3C279 and 3C273, which have an angular separation of about 10°. The latter source acts as the reference, as 3C279 is occulted by the sun each year on October 8. Results gave agreement with predicted value within 20%.

Shapiro also reevaluated the optical data with regard to the solar system and established new data, using radar ranging. In both cases, he found agreement with the predicted value for the perihelion precession of Mercury within 3%. By combining the data, the error can be reduced to 1%. As another test of Einstein's theory of general relativity, Shapiro suggested measuring the retardation of radar echo signals from Mercury when the planet moves into a position of superior conjunction. The gravitational field of the sun, as represented by the Schwartzschild solution, not only produces a bending of the ray, but also affects the time of flight of the signal. Therefore, the time delay between the transmission of a radar pulse to Mercury and the reception of the reflected signal will depend not only on the relative positions of the earth and Mercury in their respective orbits, but also on whether the radar signals pass near the sun. See Fig. 4. Measurements have given agreement within 5%.

Gravitational Collapse. The gravitational force between any two masses is attractive. Therefore, given a quantity of matter, under action of gravity alone it will become as compact as possible. In the planets, the compaction process is stopped by the electrical forces, which act

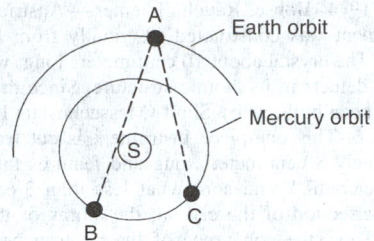

Fig. 4. Conditions for testing Einstein's theory of general relativity, S = sun.

(a)   (b)

Fig. 5. (a) A circular arrangement of dust particles before a gravitational wave arrives; (b) the same particles after a passage of a wave consisting of one mode. The second mode would produce the same effect, rotated at 45°.

between atoms and molecules in close range. The pressure in the sun, however, is much too great to be supported by such solid body forces. The tremendous pressure is balanced primarily by the counterpressure of electromagnetic radiation, which is produced by the nuclear processes at the sun's center. Stars in which the nuclear processes have ended undergo a further contraction, which is stopped by the pressure of free electrons at the densities associated with white dwarfs. This pressure, which occurs because electrons obey the Pauli exclusion principle, is capable of supporting up to 1.4 solar masses within a volume of 10^{-4} to 10^{-8} of the solar volume. Objects that are more massive continue the crush. Neutrons become the most stable particles in the interior and the contraction is stopped by repulsive nuclear forces when a neutron occupies only about 10^{-39} cubic centimeter, the nuclear volume. If the resulting neutron star is one solar mass, its radius is just 10 kilometers and its volume 10^{-15} the sun's volume. Objects with more than about 1.2 solar masses cannot be stable as neutron stars. They continue to contract. Beyond this point, the situation is confused by the abundance of exotic elementary particles, but there is no theoretical evidence that the contraction can be stopped.

One might have hoped that Einstein's theory of gravitation would contain a short-range repulsion that would stop this endless contraction. However, the opposite is the case. First of all, all forms of energy contribute to the attractive mass in general relativity, and secondly, the fact that matter determines the geometry indicates that there should be peculiarities in the space when the body is highly collapsed. There are several general theorems, particularly by Penrose and Hawkins, whose general conclusion seems to be that, as long as the energy density remains everywhere positive, collapse is inevitable. This does not mean that collapse actually occurs in nature. As a very massive star proceeds through the various stages indicated in the foregoing paragraph. It may become unstable and throw off enough mass through an explosive process, such as a supernova, that it may settle down at a planetary size, or as a white dwarf, or as a neutron star. There is evidence for the existence of these objects. A pulsar is considered to be a rapidly rotating neutron star. And, thus it is unlikely that everything continues to collapse. But there are many very massive stars and, in the absence of more information, it is not unreasonable to rule out the possibility that some indeed go through an indefinite collapse or that some may have already done so.

What physical effects result from the collapse? It was pointed out (Eq. 4) that at the Schwarzschild radius, the escape velocity from a point mass is the velocity of light. Thus, no signal can escape from a body which has collapsed below R_s. This result can be deduced from the Schwarzschild solution of the Einstein equations. As a result, knowledge of events is limited at the Schwarzschild radius; the surface $r = R_s$ is an *absolute event horizon*. Because no light or other signal can be received from a source that has collapsed below its Schwarzschild radius, it has been called a *black hole*.

A neutron star of one solar mass has a radius of 10 kilometers, while $R_s = 3$ kilometers; a neutron star of 10 solar masses will have a radius of 30 kilometers. Thus, there is observational evidence for the existence of objects that are very nearly black holes. See also **Black Hole**; and **Cosmology**.

Gravitational Waves. Einstein's field equations require that the gravitational field have a finite velocity of propagation—the same as that for light. Therefore, the gravitational field has independent dynamical degrees of freedom which permit gravitational waves to exist in two states of polarization. These states are wholly transverse, i.e., the waves act on matter only in planes which are orthogonal to the direction of propagation. In passing through matter, one state produces oscillations such that

there is a compression followed by elongation along one axis and a corresponding elongation followed by compression along the perpendicular axis. See Fig. 5. For a periodic wave this process repeats at the frequency of the wave. The other state of polarization has the same effect along axes rotated by 45°. This character for the modes is caused by the tensor nature of the potentials g_{uv} which limits the lowest order of gravitational waves to quadrupole radiation. A crude estimate of the energy radiated by the earth-sun system per year amounts to 10^{16} ergs (about 10^6 kWh). Radiating at this rate, the earth has lost about 10^{-15} of its available mechanical energy since its formation possibly some 5×10^9 years ago. Presumably there are stronger sources of gravitational waves available in the universe.

Experiments to detect gravitation radiation were begun in 1958 by Weber. For a detector, Weber used an aluminum cylinder suspended in the earth's gravitational field. An incident gravitational wave sets up transverse oscillations in the cylinder. These oscillations are transformed into electrical signals by piezoelectric crystals bonded to the surface of the cylinder. The apparatus is acoustically insulated from outside interferences.

The initial detection program used principally two identical cylinders, 153 centimeters long and 66 centimeters in diameter. These were located at the University of Maryland and the Argonne National Laboratory, respectively, some 100 kilometers apart. The electronic recording system was narrowly tuned to 1660 Hz, which has an acoustic half-wavelength of 153 centimeters in aluminum. Thermal oscillations are randomly generated and one would not expect correlation between the outputs of two detectors 100 kilometers apart. Therefore, Weber looked for coincidences in the output signals of the two detectors. The observation technique was to record each signal separately at its own location and, at the same time, to transmit the Argonne signal to Maryland where it could be compared directly with the Maryland signal. Coincidences of a certain pulse height were then marked. The coincidence rate due to random fluctuations was correlated with the observed rate by careful statistical analysis. Weber concluded that there is a "significant coincidence rate of about one every two days."

Both cylinders were lined up in an east-west direction. Therefore, some directional information was available by studying the change in coincidence rate as the earth rotated on its axis. The information was not very precise, as the two-cylinder array was a broad-beam detector and there was twelve-hour symmetry in orientation because the earth does not absorb much energy. Nonetheless, there was a definite indication that a source of radiation lies in the direction of the center of the galaxy.

Gravitational Wave Antennas. As described further in the entry on **Quantum Mechanics**, the principles of quantum mechanics were introduced in the 1920s. Among other guidelines, they state that when the property of an electron or other microparticle is measured, the state of that particle will inevitably be disturbed—and disturbed in some unpredictable fashion. It follows that the more accurate the measurement, the greater and more unpredictable will be the disturbance (Heisenberg uncertainty principle). These ground rules contribute to the complexity of designing antennas and detectors used in gravity-wave research.

Typically, gravity-wave detectors are made of aluminum, sapphire, or silicon bars that weigh as little as 10 kilograms and up to several hundred kilograms. With an instrumental ability of measuring end-to-end vibrations with the accuracy required (10^{-19} centimeter), the device will behave quantum mechanically. Scientists in Russia and California have proposed a quantum nondemolition (QND) method to circumvent the effects of the Heisenberg uncertainty principle. It has been proposed that instead of measuring the position of a 10-ton bar (visualized for future experiments), the momentum of the bar would be measured. The bar would purposely be set in motion so that the effects of a passing gravity wave on the bar's momentum could be detected.

Michelson, Price, and Taber (High Energy Physics Laboratory. Stanford University) reported in mid-1987 on a network of second-generation low-temperature gravitational radiation detectors. These detectors, sensitive to mechanical strains of order 10^{-18}, are possible because of a variety of technical innovations that have been made in cryogenics, low-noise superconducting instrumentation, and vibration isolation techniques. Another five orders of magnitude improvement in energy sensitivity of resonant-mass detectors is possible before the linear amplifier quantum limit is encountered. The interaction of a gravitational wave with a resonant-mass detector and the signal-to-noise analysis and detector optimization for linear transducer readouts are all now well understood. Such an analysis shows that even a relatively large energy flux of gravitational radiation, expected from some astrophysical sources, couples very weakly to a detector. By considering the signal and all the relevant detector noise sources, one can understand the fundamental sensitivity and bandwidth limitations of resonant-mass detectors. For example, a high-Q antenna resonance does not lead to a narrow detection bandwidth.

Research groups developing resonant-mass gravitational radiation detectors are listed in Table 1.

Sources of Gravity Waves. The types of signal, frequency, and strength from various astrophysical sources have been estimated by Jeffries, et al. (see reference):

Source	Characteristics
Stellar binary	Periodic signal; 1 MHz or lower; strength, 10^{-21}.
Neutron-star binary	Quasiperiodic signal; sweeps up to 1 kHz; strength, 10^{-22}
Accreting neutron star	Periodic signal; 200–800 Hz; strength 3×10^{-27}.
Type II supernova	Impulsive signal; 1 kHz; strength, 10^{-21}.
Vibrating black hole	Damped sinusoidal signal; 10 kHz for one solar mass, 10 Hz for 1000 solar masses; strength unknown.
Galaxy formation (by cosmic strings)	Noisy signal; broad band, 1 cycle/year 300 Hz; strength, 10^{-14} to 10^{-24}.

Neutron Interferometer. Prior to the mid-1970s, little tangible experimentation occurred that would permit the establishment of a good relationship between quantum mechanics and the general theory of relativity (the modern theory of gravitation). For one thing, there is a vast gap of scale between quantum theory and the general theory of relativity, with quantum mechanics concerned with particles at the atomic scale of 10^{-8} centimeter, whereas the effects of gravity appear significant only in terms of a stellar or cosmic scale. Among ways to narrow this gap and to learn more about gravity is the neutron interferometer. As early as 1964, Bonse and Hart (Cornell University) constructed an X-ray interferometer, but it was not felt at that time that an instrument of this type would work in the case of neutron beams. Thus, the first neutron interferometer was not

constructed until 1974 (Bonse, Rauch, Triemer—Austrian Nuclear Institute). The instrument was constructed essentially from a single, perfect crystal of silicon. The crystal about 10 centimeters long, was free of dislocations and other defects in its atomic structure. Since then, other similar instruments have been built, as by Shull (Massachusetts Institute of Technology). See Fig. 6. This one-piece instrument is cut from a cylindrical crystal approximately 8 centimeters long and features three ears that are about 0.5 centimeter thick and somewhat less than 3 centimeters apart. Because of the perfection of the crystal, the atoms of the three ears all line up exactly. Thus, the coherence of the neutron beam entering the instrument is not disturbed.

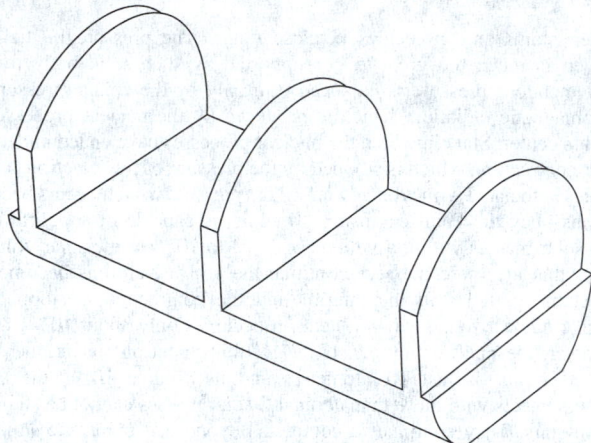

Fig. 6.   Typical neutron interferometer. The instrument is constructed from a single perfect crystal of silicon. The ears are each 0.5 centimeter thick.

Scattering of the neutron beams does not occur from the surface of the ears; rather, they are scattered by the planes of atoms in the crystal. The behavior of neutron beams in the interferometer is somewhat complex and is well explained by Greenberger/Overhauser (1980).

In assessing the use of the neutron interferometer as a means of detecting gravitational effects, it is important to note that although neutron waves have much in common with light waves and water waves (reinforcement and cancellation when exactly in or out of phase, respectively), there are some basic differences. The neutron possesses both mass and a magnetic moment; the photon does not. Thus, a neutron is affected by a magnetic field and can be caused to rotate, whereas such a field has no effect on a photon. It follows that the characteristic of the neutron wave is such that it will be affected much more strongly by gravity than will a light wave, the measurable gravity-photon interactions of which can be observed only on a cosmic scale. The shorter neutron wave length (10^{-8} centimeter)

TABLE 1.   RESEARCH GROUPS DEVELOPING RESONANT-MASS GRAVITATIONAL RADIATION DETECTORS

Institute of Physics, Academia Sinica, Beijing: Al bar and low-frequency tuning fork at room temperature. Piezoelectric transducers with field-effect transistor amplifiers.

Louisiana State University: Al bar at 4 K. Inductive superconducting transducer with SQUID (superconducting quantum interference device) amplifier and parametric transducer.

Moscow State University: Ultrahigh-Q sapphire bars and quantum nondemolition methods.

Stanford University: Al bars at 4 K. Inductive superconducting transducer with SQUID amplifier.

University of Maryland: Al bars at 4 K and 300 K. Inductive superconducting transducer and SQUID amplifier.

University of Rome: Al bars at 4 K. Electrostatic transducer.

University of Tokyo: Disk antenna for low-frequency monochromatic waves. Microwave parametric transducer.

University of Western Australia: Niobium bars at 4 K. Microwave parametric transducer.

Zhongshan University, Guangzhou: Al bar and low-frequency tuning fork at room temperature. Piezoelectric transducers with junction field-effect transistor amplifiers.

California Institute of Technology: Two evacuated pipes that stretch 40 meters down two hallways. Laser beam is directed by mirrors and optical filters into a vacuum tank. The tank contains a beam splitter, or partially reflecting mirror, that divides light equally between the two pipes. Mirrors mounted on freely suspended masses at each end of the pipes reflect the light. The light beams bounce back and forth the length of the laboratory approximately 10,000 times. Resulting interference is observed. A passing gravitational wave would slightly alter distance between one or both pairs of masses and thereby change the interference. Apparatus is sensitive to changes as small as 3×10^{-16} meter, or $\frac{1}{3}$ diameter of a proton, lasting for as little as one millisecond.

Massachusetts Institute of Technology: As of 1987, under construction is a 1.5 meter and a 5 meter interferometer.

compared with the longer light wave (10^{-5} centimeter) permits resolution of effects on a smaller scale.

In 1975, Coella, Overhauser, and Werner conducted an experiment (termed COW for the initials of the investigators) to measure the effect of the earth's gravity on the phase of the neutron wave. As pointed out by Greenberger/Overhauser (1980), "it was already known experimentally that the neutron falls in the earth's gravitation field as any other massive particle does. That fall, however, is strictly Galilean, or classical. The question is whether one can observe an effect of gravity on the wave nature of the neutron. The way to do this is through an interference effect, for which the neutron interferometer is ideally suited (provided the effect is large enough to detect)." It is interesting to note that it has been estimated that the force of gravity at the earth's surface is derived from some 10^{52} protons and neutrons of which the earth is comprised. Also, it has been established that the electric repulsion between two protons is 10^{36} times greater than their gravitational attraction. And, two protons at an atomic distance of 10^{-8} from each other have an electric force on each other that is some 10^{16} times greater than the gravitational force exerted on them by the entire earth. Thus, the investigators had the task of proving that such a weak gravitational force could produce measurable effects in the neutron interferometer.

The neutron wave, as previously mentioned, maintains its coherency over the full 10-centimeter length of the instrument crystal. During this distance, the wave oscillates 10^9 times. It was possible with the instrument to observe 100 additional oscillations—these extra oscillations attributed to gravity effects. As the scientists pointed out, "As weak as gravity is, it has a measurable effect on the wave function because the neutron wave is coherent on a macroscopic scale."

The experimental data obtained agreed precisely with the amount predicted by the Schrödinger equation. In their explanation of the experiment, the scientists describe why it is believed that the measurement is due to gravitational force and is not a manifestation of the time difference or red shift effect described by Einstein in 1916. That is, in the case of this experiment, the difference between the time on a clock moving along with one beam and the time on a clock moving along with the other beam. Since the COW experiment, a number of other sophisticated experiments have been conducted with the neutron interferometer. Their complexity is beyond the scope of this encyclopedia, but details can be found in some of the references listed.

Gravity Lens. It is currently believed that the comparatively weak forces of gravity waves require a cosmic scale to observe their effects. What was believed to be twin quasars were photographed in the early 1950s, using the 1.2-meter Schmidt telescope on Palomar Mountain (California). In these early views, the image of the bodies appeared fused because of the motion of the earth's atmosphere. Scientists have observed that had the telescope been above the earth's atmosphere, it could have resolved objects 60 times closer together than the twins. But, until March of 1979, these bodies were considered twins. Subsequent research involving the 2.1-meter telescope at the Kitt Peak National Observatory and the 2.3 meter telescope of the University of Arizona yielded spectral information that was strikingly similar for both bodies. A red shift 1.4 was measured for each body and this, coupled with the similarity of spectral data puzzled the astronomers. The spectral and velocity measurements were further confirmed, using a multiple-mirror telescope of the Smithsonian Astrophysical Observatory and the University of Arizona. Later, data were gathered by the National Radio Astronomy Observator's Very Large Array (near Socorro, New Mexico). A computer-generated display on a cathode-ray tube of one image of a quasar whose radiation has been deflected to form two images by a gravitational lens is shown in Fig. 7. It is believed that an elliptical galaxy is acting as a gravitational lens. As pointed out by Chaffee (1980), "Eight months of theoretical work and intensive investigation with the largest optical and radio telescopes has demonstrated that these "twin" quasars are not two distinct objects at all. Rather, they are a single object whose light has been split into two images by the gravitational field of a galaxy between the quasar and our galaxy; a kind of optical illusion on a cosmic scale." Several technical objections were raised concerning the conclusion that a gravitational lens is involved, most of which have since been resolved.

Additional Reading

Abbott, L.: "The Mystery of the Cosmological Constant," *Sci. Amer.*, 106 (May 1988).

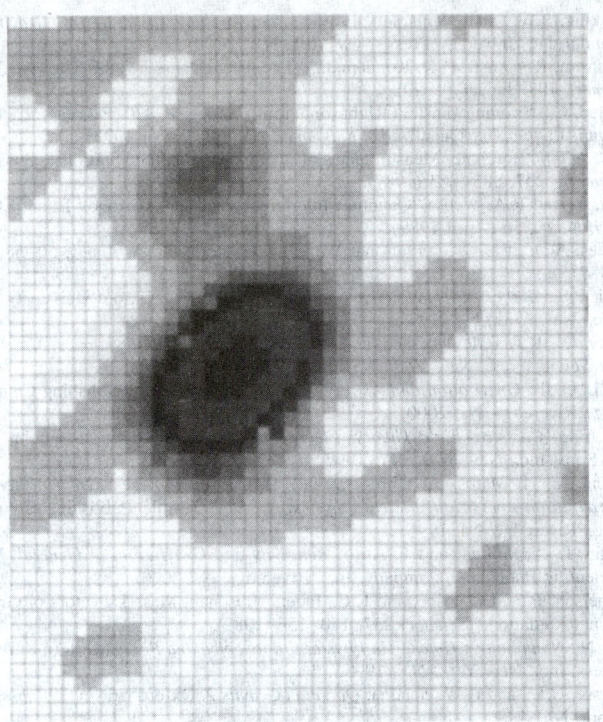

Fig. 7. Twin quasars 0957 + 561 A, B. Reasonable facsimile of cathode-ray tube image. Increased shades of gray indicate increased intensity of radio waves (wavelength = 6 centimeters).

Abramovici, A. et al.: "LIGO: The Laser Interferometer Gravitational-Wave Observatory," *Science*, 325 (April 17, 1992).

Adar, R.K.: "A Flaw in a Universal Mirror," *Sci. Amer.*, 50 (February 1988).

Blair, D.G.: *The Detection of Gravitational Waves*, Cambridge University Press, New York, NY, 1991.

Blandford, R.D. et al.: "Gravitational Lens Optics," *Science*, 824 (August 25, 1989).

Boslough, J.: "Searching for the Secrets of Gravity," *Nat'l. Geographic*, 562 (May 1989).

Brush, S.G.: "Prediction and Theory Evaluation: The Case of Light Bending," *Science*, 1124 (December 1, 1989).

Chaffee, F.H., Jr.: "The Discovery of a Gravitational Lens," *Sci. Amer.*, **243**(5), 70–78 (1980).

Ciufolini, I. and J.A. Wheeler: *Gravitation and Intertia*, Princeton University Press, Princeton, NJ, 1995.

Cohen, I.B.: "Newton's Discovery of Gravity," in *Scientific Genius and Creativity*, W.H. Freeman, New York, NY, 1987.

Davies, P.: *The New Physics*, Cambridge University Press, New York, NY, 1992.

Dhurandhar, S. and T. Padmanabhan: *Gravitation and Cosmology*, Vol. 211, Kluwer Academic Publishers, Norwell, MA, 1997.

Einstein, A.: "On the Generalized Theory of Gravitation (A Classic Nobel Laureate Paper)," *Sci. Amer.*, (April 1950).

Einstein, Albert: "On the Generalized Theory of Gravitation (April 1950)," in *A classic reference in The Laureaters' Anthology*, 1–5, Scientific American, Inc., New York, NY, 1990.

Ellis, P.J. and Y.C. Tang: *Trends in Theoretical Physics*, Addison-Wesley Longman, Inc., Reading, MA, 1990.

Falomir, H., R.E. Gomboa, and F.A. Schaposnki: *Trends in Theoretical Physics*, American Institute of Physics, College Park, MD, 1998.

Gibbons, A.: "Putting Einstein to the Test—In Space," *Science*, 939 (November 15, 1991).

Glick, T.F.: *The Comparative Reception of Relativity*, Kluwer Academic Publishers, Norwell, MA, 1987.

Goldman, T., R.J. Hughes, and M.M. Nieto: "Gravity and Antimatter," *Sci. Amer.*, 48 (March 1988). A Classic Reference.

Greenberger, D.M. and A.W. Overhauser: "The Role of Gravity in Quantum Theory," *Sci. Amer.*, **242**(5), 66–76 (1980).

Hakim, R.: *An Introduction to Relativistic Gravitation*, Cambridge University Press, New York, NY, 1998.

Hamilton, D.P.: "Gazing Through a Gravitational Lens," *Science*, 1662 (December 21, 1990).

Hamilton, D.P.: "LIGO In Limbo," *Science*, 635 (May 3, 1991).

Hawking, S.W. and W. Israel, Editors: *General Relativity*, Cambridge University Press, New York, NY, 1980.

Hawking, S.W. and W. Israel, Editors: *Three Hundred Years of Gravitation*, Cambridge University Press, New York, NY, 1992.

Hegstrom, R.A. and D.K. Kondepudi: "The Handedness of the Universe," *Sci. Amer.*, 108 (January 1990).

Holden, C.: "Proving Einstein Right (or Wrong)," *Science*, 870 (February 22, 1991).

Horgan, J.: "Gravity Quantized?" *Sci. Amer.*, 18 (September 1992).

Hsui, A.T.: "Borehole Measurement of the Newtonian Gravitational Constant," *Science*, **237**, 881–883 (1987).

Imry, Y. and R.A. Webb: "Quantum Interference and the Abaronov-Bohm Effect," *Sci. Amer.*, 56 (April 1989).

Jeffries, A.D. et al.: "Gravitational Wave Observatories," *Sci. Amer.*, 50–58 (June 1987).

Joshi, P.S.: *Global Aspects in Gravitation and Cosmology*, Oxford University Press, Inc., New York, NY, 1994.

Lahav, O., R.J. Terlevich, and E. Terlevich: *Gravitational Dynamics*, Cambridge University Press, New York, NY, 1996.

Liu, L. et al.: *Gravitation and Astrophysics*, World Scientific Publishing Company, Inc., Riveredge, NJ, 2000.

Low, F.E.: *Classical Field Theory: Electromagnetism and Gravitation*, John Wiley & Sons, Inc., New York, NY, 1997.

Marcia, A., T. Matos, O. Obregon, and H. Quevedo: *Recent Developments in Gravitation and Mathematical Physics*, World Scientific Publishing Company, Inc., Riveredge, NJ, 1996.

Martin, J., A. Molina, and F. Atrio: *Relativity and Gravitation in General*, World Scientific Publishing Company, Inc., Riveredge, NJ, 2000.

Michelson, P.F., J.C. Price, and R.C. Taber: "Resonant-Mass Detectors of Gravitational Radiation," *Science*, 237, 150–156 (1987).

Pais, N.: *Niels Bohr's Times — In Physics, Philosophy, and Policy*, Oxford University Press, Inc., New York, NY, 1993.

Penrose, C.J.I. and D.W. Sciama: *Quantum Gravity 2*, Oxford University Press, New York, NY, 1981.

Peterson, I.: "Antimatter Takes a Free Gravitational Fall," *Sci. News*, 135 (March 2, 1991).

Pool, R.: "'Fifth Force' Update: More Tests Needed," *Science*, 1499 (December 10, 1988).

Pool, R.: "Closing In on Einstein's Special Relativity Theory," *Science*, 1207 (November 30, 1990).

Rembielinski, J. and D.G. Seiler: *Particles, Fields, and Gravitation*, American Institute of Physics, College Park, MD, 1999.

Ruthen, R.: "Waves are Waves," *Sci. Amer.*, 21 (August 1991).

Ruthen, R.: "Catching the Wave," *Sci. Amer.*, 90 (March 1992).

Tourrenc, P.: *Relativity and Gravitation*, Cambridge University Press, New York, NY, 1997.

Turner, E.L.: "Gravitational Lenses," *Sci. Amer.*, 54 (July 1988).

Vogt, R.E.: "The U.S. Laser Interferometer Gravitational-Wave Observatory (LIGO) Project," *Proceedings of the Sixth Marcel Grossmann Meeting on General Relativity*, 91–97, Kyoto, Japan (June 1991).

Weber, J.: *General Relativity and Gravitational Waves*, John Wiley & Sons, Inc., New York, NY, 1961.

Wheeler, J.A.: *A Journey Into Gravity and Space-Time*, W.H. Freeman Company, New York, NY, 1990.

Will, C.M.: "General Relativity at 75: How Right was Einstein?" *Science*, 770 (November 9, 1990).

Will, C.M.: *Theory and Experiment in Gravitational Physics*, Cambridge University Press, New York, NY, 1993.

Zee, A.: *Einstein's Universe: Gravity at Work and Play*, Oxford University Press, Inc., New York, NY, 2001.

Zichichi, A.L., N. Sanchez, and V. De Sabbata: *Gravitation and Modern Cosmology: The Cosmological Constant Problem*, Kluwer Academic Publishers, Norwell, MA, 1991.

GRAVITY AND MICROGRAVITY. Gravity is such an accepted part of our lives that we rarely think about it, even though it affects everything we do. Any time we drop or throw something and watch it fall to the ground, we see gravity in action. Although gravity is a universal force, there are times when it is not desirable to conduct scientific research under its full influence. In these cases, scientists perform their experiments in microgravity—a condition in which the effects of gravity are greatly reduced, sometimes described as "weightlessness."

Any object in freefall experiences microgravity conditions, which occur when the object falls toward the Earth with an acceleration equal to that due to gravity alone [approximately 9.8 meters per second squared (m/s^2), or 1 g at Earth's surface].

Brief periods of microgravity can be achieved on Earth by dropping objects from tall structures. Longer periods are created through the use of airplanes, rockets, and spacecraft. The microgravity environment associated with the space shuttle is a result of the spacecraft being in orbit, which is a state of continuous freefall around the Earth.

Newton offered a thought experiment to explain how an object could stay in orbit while falling toward the Earth. He imagined a cannon at the top of a tall mountain that fired cannonballs. Each cannonball was acted upon by two forces: the force from the explosion and the force of gravity. The combination of the two forces would cause the cannonballs to travel in an arc. If the cannonballs were fired with more and more energy, they would hit the ground farther and farther away from the cannon. If the cannonball was fired with enough energy, it would fall entirely around the Earth and return to its starting point, completing an orbit. See also **Gravitation**.

Instead of being fired from a cannon atop a mountain, a spacecraft is launched in a trajectory that arcs above the Earth. When a particular speed and altitude are attained, the craft's falling path will be parallel to the curvature of the Earth, and a microgravity environment is established.

This microgravity environment gives researchers a unique opportunity to study the fundamental states of matter, solids, liquids, and gases, and the forces that affect them. In microgravity, researchers can isolate and study the influence of gravity on physical processes, as well as phenomena that are normally masked by gravity and thus difficult, if not impossible, to study on Earth.

Microgravity Research

Working in partnership with the scientific community and commercial industry, NASA's Microgravity Research Program strives to increase understanding of the effects of gravity on biological, chemical and physical systems.

Using both space flight and ground-based experiments, researchers throughout the nation, as well as international partners, are working together to benefit economic, social and industrial aspects of life for the United States and the entire Earth. Ten U.S. universities, designated by NASA as "Commercial Space Centers," share these space advancements with U.S. industry to create new commercial products, applications and processes.

Under the NASA Headquarters' Office of Life and Microgravity Sciences and Application, the Microgravity Research Program supports NASA's strategic plan in the Human Exploration and Development of Space Enterprise.

Microgravity research has been performed by NASA for more than 25 years. The term "microgravity" literally means a state of very little gravity. The prefix "micro" comes from the Greek word mikros, meaning "small." In metric terms, the prefix means "one part in a million" (0.000001). Gravity dominates everything on Earth, from the way life has developed to the way materials interact. But aboard a spacecraft orbiting the Earth, the effects of gravity are barely felt. In this "microgravity environment," scientists can conduct experiments that are all but impossible to perform on Earth. In this virtual absence of gravity as we know it, space flight gives scientists a unique opportunity to study the states of matter (solids, liquids and gases), and the forces and processes that affect them.

Marshall Space Flight Center in Huntsville, Ala. is the lead center for NASA's Microgravity Research Program. The program manages Microgravity Science and Applications Project Offices at the Lewis Research Center in Cleveland, Ohio, the Jet Propulsion Laboratory in Pasadena, Calif., and also project offices at the Marshall Center.

Under the project offices, the Microgravity Research Program is divided into nine major areas: five science disciplines, three research infrastructure programs and the Space Product Development Office.

The science disciplines include Biotechnology, Fluid Physics, Materials Science, Combustion Science and Fundamental Physics. The infrastructure activities include Acceleration Measurement, Advanced Technology and the Glovebox Flight Programs.

Marshall Center manages the Biotechnology program and Material Science program as well as the Glovebox Flight program and the Space Products Development office. Lewis Research Center manages the Fluid Physics, Combustion Science and Acceleration Measurement programs. The Jet Propulsion Laboratory manages the Fundamental Physics and the Advanced Technology Development program.

Microgravity Biotechnology

Biotechnology is the application of engineering and technology to life sciences research. NASA partnerships with private industry and academia in space-based, biotechnological research are helping ensure that scientific advances in the field continue to produce technical innovations for improved health care on Earth. Biotechnology research in space focuses on protein crystal growth, (1) growing organic crystals with thousands of atoms and (2) on cell/tissue culturing, the study of how cells interact in a low-gravity or low-shear environment.

Pure, precisely ordered protein crystals of sufficient size and uniformity for X-ray analysis are in demand by the pharmaceutical industry as tools for research. Structural information gained from protein crystals can provide a better understanding of the role of a given protein in the body's immune system. Protein crystal research could ultimately aid in the development of more effective drugs and life-saving treatments for many diseases.

Since the mid-1980s, NASA has sponsored protein crystal growth experiments to learn about the effects of space on the growth process and to refine techniques for obtaining the highest quality crystals in space and on the ground. The result is that generally, protein crystals produced in space are larger and more precisely ordered than those produced on Earth. These improvements are important to scientists who analyze a crystal's three-dimensional structure (the key to understanding a protein's activity) and possibly develop new and more effective medicines. Knowledge of the molecular structure of the antibody may lead to development of treatments and vaccines to cure the disease that causes pneumonia and severe upper respiratory infection in nearly four million children, ages 1 to 5 in the United States annually.

The other focus of biotechnology in microgravity is cell and tissue culturing experiments. Located at Johnson Space Flight Center in Houston, Texas, the goal of this research is to grow cells on a tissue in near-weightlessness, that otherwise is unachievable on Earth.

The medical benefit of microgravity tissue and culture engineering may lead to new research models in cellular and molecular biology. These studies also are developing new tissues for potential transplant operations.

Biotechnology research results have provided significant advances in the understanding of many diseases including AIDs, heart disease, cancer, diabetes, and hepatitis. See also **Bioprocess Engineering (Biotechnology)**.

Microgravity Fluid Physics

Everyone has practical experience with fluids, i.e., liquids and gases, and knows how a fluid will behave under "normal" circumstances. Steam rises from the surface of a hot spring or a boiling pot, and water spilled on a tabletop runs over, then off, the surface. Gravity drives much of the fluid behavior we are accustomed to on Earth.

Many of our intuitive expectations do not hold up in microgravity, though, because other forces such as surface tension control fluid behavior. Surface tension causes drops of any liquid to form almost perfect spheres when the influence of gravity is absent. On Earth, gravity distorts the shape when liquid is resting on or attached to a surface. Although these differences in fluid behavior often present engineers and astronauts with practical problems, they also offer scientists unique opportunities to explore different aspects of the physics of fluids.

Research conducted in microgravity is increasing our understanding of fluid physics to provide a foundation for predicting, controlling and improving a vast range of technological processes. The behavior of fluids is at the heart of many phenomena in materials science, biotechnology and combustion science. Surface tension-driven flows, for example, affect some techniques of semiconductor crystal growth, welding, and the spread of flames on liquids. The dynamics of liquid drops are an important aspect of chemical process technologies and meteorology.

Results from microgravity Fluid physics research will lead to better understanding of the effects of miniaturization of electronic materials. Advances in the field will lead to even smaller and more efficient electronic devices with reduced costs for the consumer.

Microgravity Materials Science

Materials science investigates the relationships between the structure, properties, and processing of materials. Structure is the arrangement of the atoms in the material. Properties include physical, chemical, electronic, thermal and magnetic characteristics. Processing is the method by which materials are formed. They can be solidified, evaporated and condensed, or dissolved and then separated from a solution. NASA's materials science microgravity program uses the unique characteristics of the microgravity space environment to study these fundamental relationships in materials solidification and crystal growth.

In the production of electronic materials, crystals have achieved far greater value as conductors than they ever had as gemstones. Pioneering research is leading to next-generation commercial crystal products.

Material science also has a focus on the production of alloys and composites. High-strength metals are needed in the aviation, aerospace, power generation and propulsion industries. Processing these materials in space helps researchers understand how to make better materials on Earth and is allowing scientists to create new metal alloys. Alloys are mixtures of metals or metals and nonmetals. When combined, they can produce materials with improved strength or better resistance to corrosion.

On Earth, when a melted alloy solidifies, it forms pine-tree-shaped crystals called dendrites. These dendrites play a very important role in determining the properties of the alloy and its subsequent usefulness. Gravity causes fluid flows in the alloy, leading to the formation of irregular dendrites that weaken the alloy or metal structure. This type of processing is so complex that it is difficult to measure and predict, and even more difficult to control. In space, gravity-related phenomena such as convection are reduced, thus simplifying the process for study. See also **Dendrite**.

Ceramics and glass experiments also are part of the Material Science program. Optical engineering is being revolutionized by new glasses, crystals and other materials that surpass conventional substances in quality. However, production of these superior materials is difficult. Some glasses have chemical mixes that react with their containers. Others are extremely sensitive to contamination levels from impurities of even a few parts per billion. For example, certain fluoride glasses are of great interest for their infrared transmission properties. These glasses can be made on Earth, but trace contaminants from processing containers have prevented them from reaching their highest potential.

Containerless processing, in which a sample is suspended and manipulated without touching contaminating containers, is an attractive solution to these problems. Containerless processing on massive samples can only be done in the microgravity environment of space where the forces used for suspending and manipulating the samples are not overwhelmed by gravity.

Microgravity Materials results will contribute to future models of industrial and manufacturing processes. This will lead to new, stronger, lighter alloys with never-seen-before properties.

Microgravity Combustion Science

NASA's combustion research program focuses on understanding the important processes of ignition, flame spreading and flame extinction during combustion in low gravity. Research is directed at gaining basic knowledge of combustion processes, as well as addressing issues of fire safety in space.

Scientists are interested in the physical characteristics of flame, such as size and shape, and the role of soot formation in combustion. Investigations also study air flows and the transfer of heat and mass in fuel vapors, liquid pools, paper and metal solids.

Since the physical and chemical mechanisms that cause flames to spread on Earth are strongly influenced by gravity, researchers are finding out flames behave very differently in the low-gravity of an orbiting spacecraft. It is well known that material flammability and flame growth are strongly affected by the environment, including oxygen content, pressure and air flow. However, the effects of these conditions in the microgravity environment are largely unknown. Scientists want to understand combustion to improve efficiency of our fuel-driven machines and to evaluate potential fire hazards aboard spacecraft. Combustion research will lead to more efficient fuels, better fire safety and a cleaner environment.

Microgravity Fundamental Physics

Fundamental physics researchers use the low-gravity environment of space to test basic scientific theories not possible in the gravity environment on Earth in fields such as thermophysical measurements, atomic physics and relativistic physics.

This research is important because it seeks to uncover principles that govern the behavior of the physical world, such as the influence of heat energy, new forms of matter and low-temperature physics.

Fundamental physics research in microgravity is driving the development of new technologies that will advance scientific knowledge and improve life on Earth. The benefits of this research can be seen in improvements in ultra-sensitive detectors of temperature and magnetic fields, as well as valves that can function at temperatures close to absolute zero.

Also, scientists have discovered the transition between different forms of matter, whether magnetic or non-magnetic, solid or liquid, and the similarities between the different systems. Theories resulting from studies of superfluid helium in microgravity can help to understand many other systems. Scientists can use these theories to better understand the formation of weather systems such as tornadoes and hurricanes, how water seeps through soil, and how cracks propagate in metals.

New understanding of nature's processes have made much exploratory surgery unnecessary with the advent of magnetic resonance imagers (MRIs). Another application of this research is liquefied gases. Liquid oxygen is used to supply breathing gas in hospitals, and helps fuel the powerful rockets that have made human exploration of space possible.

Acceleration Measurement Program

The Acceleration Measurement Program is an operational space flight infrastructure program to measure and track accelerations with a recording system. Acceleration is the force that pushes the passengers in a car against the side opposite of a turn. Acceleration measurement systems serve a wide variety of microgravity science and technology experiments. These systems can measure and record low-gravity accelerations at as many as three experiment sites simultaneously. They can be mounted on or near an experiment to measure the accelerations experienced by the experiment. Understanding the interactions of accelerations contributes to improved microgravity research.

Units mounted on spacecraft exteriors historically have had the capability for remote commanding from the ground and downlinking mission data. Units mounted inside recently have been modified to incorporate this capability. The data are displayed at NASA's Lewis Telescience Support Center in Cleveland, Ohio or NASA's Marshall Payload Operations Control Center (POCC) in Huntsville, Ala.

Advanced Technology Development Program

The Advanced Technology Development Program was developed by NASA's Microgravity Science and Applications Division in response to the challenges researchers face when defining experiment requirements and designing associated hardware. The program provides efficient, cost-effective, state-of-the-art technological support for microgravity science investigations. The Advanced Technology infrastructure program enables new types of scientific investigations and gives researchers capable, high-quality experimental hardware to overcome existing technology-based constraints. The goal is to investigate and develop high-risk microgravity research technologies before they are needed on the critical development path for actual flight hardware.

Historically, Advanced Technology projects have encompassed a broad range of activities. Project funding includes the development of diagnostic instrumentation and measurement techniques, observational instrumentation and data recording methods, acceleration characterization and control techniques, and advancements in methodologies associated with hardware design technology.

Glovebox Flight Program

The Glovebox Flight Program is a microgravity infrastructure program that provides facilities for performing investigations not requiring large, specialized equipment. The glovebox offers scientists the capability to conduct experiments, test science procedures, and develop new technologies in microgravity. The facility enables crew members to handle, transfer and manipulate experiment hardware and materials that are not approved for use in the open Spacelab.

By providing hardware, development and investigation integration services, the program allows researchers to concentrate on scientific objectives, rather than facility development and vehicle and mission issues. The Glovebox Flight Program lowers cost and allows for quicker and easier access to space for proof-of-concept demonstrations and new investigations. The program office is developing laboratory support equipment items for the International Space Station.

Space Product Development Office

The Space Product Development office examines the opportunities for space commerce, offering a full range of support capabilities to demonstrate the commercial value of space. To ensure continued growth of U.S. industry, the office initiates and guides pilot projects in an effort to eliminate barriers to viable space commercialization.

The Space Product Development Office manages projects such as a new advance in insulation called "Aerogel," a NASA flight project aimed at revolutionizing window insulation and light emitting diodes, a project originally aimed at developing plant growth facilities for space flight and today is funding state-of-the-art cancer-fighting treatments.

Ground-Based Research

A major challenge facing NASA's space-based microgravity program is to conduct scientifically significant and productive research through the wisest possible use of space. To achieve this, NASA uses a ground-based research program to assess whether scientific investigations are worthy of a space flight opportunity. These studies are then refined for the ultimate experimental test during a space mission.

To create low-gravity environments on Earth for research, free-fall facilities are used in a variety of ways. Releasing experiment samples from tall drop towers provides about four seconds of microgravity. Research aircraft can expose experiments to about 30 seconds of low-gravity while the aircraft approaches the top of a steep climb and begins a sharp descent. This parabolic curve is generally repeated about 40 times during each flight, which is used primarily to perform experiments requiring short times for experimental equipment tests or for crew training.

Low-cost sounding rockets, such as Space Processing Applications Rockets, also have parabolic flight paths. They ascend and then descend, rather than proceeding into orbit around the Earth. Sounding rocket flights provide five to seven minutes of low-gravity. Although these periods of microgravity are brief, the test facilities are beneficial both for space flight preparation and for some actual microgravity research.

Shuttle-Mir Microgravity

The Microgravity Research Program Office manages the development and integration of microgravity science experiments of the Shuttle-Mir program. Both United States and international microgravity science partners used the facilities aboard Mir to conduct investigations in fluid physics, combustion, biotechnology and materials science. The microgravity facilities aboard the Mir space station included furnaces, glovebox and a system to isolate experiments from the station's vibration environment.

The Future of Microgravity Science; The International Space Station

The NASA Microgravity Research Program is evolving to take maximum advantage of the upcoming International Space Station. The Space Station will permit long-duration microgravity experiments in an environment otherwise more similar to Earth-based laboratories minus the gravity.

Rather than experiments being limited to a week or two, as they are aboard the Shuttle, Space Station microgravity experiments will stretch over long periods of time. This longer duration will greatly increase the number and types of materials that can be processed to full term. This will be a great advantage to experiments in areas such as solution and vapor crystal growth, which require 15 to 30 days of continuous growth to produce crystals of the desired size.

On the Space Station, with crew members to observe experiments and with equipment for analyzing samples in orbit, it will not be necessary to return all specimens to Earth for analysis before running the next experiment. This will allow researchers to conduct experiments in a series which builds on prior results without waiting years for another flight opportunity.

Future space research will stress both scientific and commercial goals. Products will include crystals, metals, ceramics, glasses, and biological materials. Processes will include solidification of metals and alloys, as well as transporting fluids and chemicals in microgravity. As research in these areas develops, the benefits will become increasingly apparent on Earth: new materials, more efficient use of fuel resources, new medicines, advanced computers and lasers and better communications. Like space, opportunities offered by microgravity science are vast, and only beginning to be explored. See also **Space Stations**.

NASA's Microgravity Research Program Office at Marshall Center is responsible for the definition and development of microgravity science and space product development projects planned for the International Space Station.

See also **Microgravity and Materials Processing**.

Additional Reading

Barlow, P.W. and D. Moore: *Life Sciences: Microgravity*, Elsevier Science, New York, NY, 1999.

El-Genk, M.S.: *Space Technology and Applications International Forum 2000: Conference on International Space Station Utilization; Conference on Thermophysics in Microgravity; Conference on Enabling Technology and Required Science*, American Institute of Physics, College Park, MD, 2000.

El-Genk, M.S.: *Space Technology and Applications International Forum: Conference on International Space Station Utilization; Conference on Global Virtual Presence; Conference on Applications of Thermophysics in Microgravity and Breakthrough Propulsion Physics*, American Institute of Physics, College Park, MD, 2000.

Moore, D., P. Bie, and H. Oser: *Biological and Medical Research in Space: An Overview of Life Sciences Research in Microgravity*, Springer-Verlag Inc., New York, NY, 1996.

Staff: National Research Council; Committee on Physical Science; Space Studies Board: *Microgravity Research in Support of Technologies for the Human Exploration and Development of Space and Planetary Bodies*, National Academy Press, New York, NY, 2000.

Zee, A.: *Einstein's Universe: Gravity at Work and Play*, Oxford University Press, Inc., New York, NY, 2001.

Web References

Jet Propulsion Laboratory. http://www.jpl.nasa.gov/

Marshall Space Flight Center. http://www.msfc.nasa.gov/

GRAVITY PROBE B (GP-B) MISSION. See **Space Science Missions: Universe**.

GRAVITY RECOVERY AND CLIMATE EXPERIMENT (GRACE). See **Earth System Science Pathfinder (ESSP) Program**; and **Space Science Missions: Earth**.

GRAVITY WAVE. A wave disturbance in which buoyancy (or reduced gravity) acts as the restoring force on parcels displaced from hydrostatic equilibrium. Also called *gravitational wave*.

There is a direct oscillatory conversion between potential and kinetic energy in the wave motion. Pure gravity waves are stable for fluid systems that have static stability. This static stability may be 1) concentrated in an interface or 2) continuously distributed along the axis of gravity. The following remarks apply to the two types, respectively.

1. A wave generated at an interface is similar to a surface wave, having maximum amplitude at the interface. A plane gravity wave is characteristically composed of a pair of waves, the two moving in opposite directions with equal speed relative to the fluid itself. In the case where the upper fluid has zero density, the interface is a free surface and the two gravity waves move with speeds

$$c = U \pm \left[\frac{gL}{2\pi} \tanh \left(\frac{2\pi H}{L} \right) \right]^{1/2},$$

where U is the current speed of fluid, g the acceleration of gravity, L the wavelength, and H the depth of the fluid. For deep-water waves (or Stokesian waves or short waves), $H \gg L$ and the wave speed reduces to

$$c = U \pm \left(\frac{gL}{2\pi} \right)^{1/2}.$$

For shallow-water waves (or Lagrangian waves or long waves), $H \gg L$, and

$$c = U \pm (gH)^{1/2}.$$

All waves of consequence on the ocean surface or interfaces are gravity waves, for the surface tension of the water becomes negligible at wavelengths of greater than a few centimeters.

2. Heterogeneous fluids, such as the atmosphere, have static stability arising from a stratification in which the environmental lapse rate is less than the process lapse rate. The atmosphere can support short internal gravity waves and long external gravity waves. The short waves (of the order of 10 kilometers (6 miles)) have been associated, for example, with lee waves and billow waves. Such waves have vertical accelerations that cannot be neglected in the vertical equation of perturbation motion. The long gravity waves, moving relative to the atmosphere with speed $\pm (gH)^{1/2}$, where H is the height of the corresponding homogeneous atmosphere, have

small vertical accelerations and are therefore consistent with the quasi-hydrostatic approximation. In neither type of gravity wave, however, is the horizontal divergence negligible. For meteorological purposes in which neither type is desired as a solution, for example, numerical forecasting, they may be eliminated by some restriction on the magnitude of the horizontal divergence. The above discussion is based upon the method of small perturbations. In certain special cases of water waves, for example, the Gerstner wave or the solitary wave, a theory of finite-amplitude disturbances exists. See also **Shear-Gravity Wave**.

Additional Reading

Gill, A.E.: *Atmosphere–Ocean Dynamics*, Academic Press, New York, NY, 1982.

AMS

GRAVITY WIND. See **Winds and Air Movement**.

GRAY, ASA (1810–1888). Asa Gray was an American botanist and botanical taxonomist who collaborated in producing the *Flora of North America* and became the leading promoter of Darwin's theory of evolution by natural selection in the United States.

Gray studied medicine at Fairfield, in western New York, before practicing briefly and gaining his MD in 1831. However, a passion for collecting plants was already established. Part-time teaching and library positions continued for a period of five years, during which Gray's growing expertise in botany, led John Torrey, then the leading American botanist, to invite him to collaborate on *A Flora of North America*. In this ambitious project Torrey and Grey intended systematically to describe for the first time all the plants of North America. See also **Plant Sciences (The History)**; and **Taxonomy (The History)**.

A professorship at the new University of Michigan in 1838 included a year in Europe: purchasing library books and plant specimens and meeting European colleagues. In 1842 Gray joined Harvard University as Fisher professor of Natural History, albeit to concentrate on botany. He remained at Harvard, continuing to nurture the botanical garden until his death, although he retired from teaching in 1873.

In the *Flora* Gray revised the Linnaean taxonomy for American plants and followed the classification scheme of A. L. Jussieu and A. P. de Candolle which relied on the anatomy of the fruit rather than on the gross morphology. Only two volumes (1838–1843) had been published when pressure of work compelled the authors to abandon the project. In addition to his Harvard appointment, the westward expansion in America had yielded an impossibly large number of new specimens. Gray's *A Manual of the Botany of the Northern United States* (1848) quickly became a classic, passing through five editions during his lifetime. This general volume did much to popularize botany. Gray also wrote books for school children (1858, 1872). See also **de Candolle, Augustin-Pyramus (1778–1841)**; and **de Jussieu, Antoine-Laurent (1748–1836)**.

In the early 1850s, Gray met Charles Darwin and began a direct correspondence with him on the geographical distribution of plants in addition to his regular communications with Joseph D. Hooker. Darwin suggested to Gray in 1857 the direction in which his work was proceeding. In turn Gray used Darwin's ideas in analyzing the specimens arriving in America from expeditions to Japan. Gray considered that plants from eastern Asia and eastern North America were derived from a common stock pushed south by Pleistocene glaciation. Gray supported the underlying basis for Darwin's theory of evolution by natural selection before publication of *Origin of Species* in 1859 and engaged in public debate with his anti-evolutionary colleague at Harvard, Louis Agassiz. See also **Agassiz, Jean Louis Rodolphe (1807–1873)**; and **Hooker, Joseph Dalton (1817–1911)**.

Gray reviewed *Origin*, acted as Darwin's agent in securing royalties on the American edition (there was no copyright agreement between Britain and the USA at this time) and following his own thinking urged Darwin along with Charles Lyell to accommodate religious objections. Gray argued for a continued belief in the argument from design and therefore a role for God. Darwin ultimately rejected this theological interpretation, but he and Gray continued to correspond and Darwin dedicated his 1877 work on flowering plants to Gray. Gray's stance did not prevent his being elected to the National Academy of Sciences in 1863. He continued to defend his theism writing anonymous articles against what he perceived to be the extremes for and against Darwinism: the agnosticism of Thomas H. Huxley and the religious opponents of natural selection. These were collected into a volume called *Darwiniana* in 1876 with the help of a like-minded clergyman, George F. Wright. See also **Huxley, Thomas Henry (1887–1975)**; and **Lyell, Sir Charles (1797–1875)**.

Additional Reading

Dupree, A.H.: *Asa Gray, American Botanist, Friend of Darwin*, Johns Hopkins Paperbacks, Baltimore, MD, 1988.

Gray, A.: *Botany for Young People and Common Schools: How Plants Grow, a Simple Introduction to Structural Botany with a Popular Flora, or an Arrangement and Description of Common Plants, both Wild and Cultivated Illustrated by 500 Wood Engravings*, American Book Company, New York, NY, 1858.

Gray, A.: *Gray's School and Field Botany: Consisting of 'Lessons in Botany' and 'Field, Forest, and Garden Botany'*, Ivison, Blakeman, Taylor, New York, NY, 1872.

Gray, J.L.: *The Letters of Asa Gray*, 2 Vols, Macmillan, London, UK, 1893.

Torrey, J., and A. Gray: *A Flora of North America*, 2 Vols, John Wiley & Putnam, New York, NY, 1838-1843.

H. J. POWER, Wellcome Trust Centre for the History of Medicine at UCL, London, UK

GRAY BODY. A radiator whose spectral emissivity is constant throughout the spectrum, being in a constant ratio to that of a black body at the same temperature.

GRAYLING (*Osteichthyes*). Of the order *Isospondyli*, family *Thymallidae*, the grayling is highly regarded both as a sport and food fish. Graylings occur in the Northern Hemisphere and are found in cold lakes and streams, both in North America and Eurasia. There are several species. The European species is *Thymallus thymallus* and the American species is *T. arcticus*. At one time, graylings were found in Michigan, but they are now considered extinct in that area. Availability is limited in the United States, Montana being an exception. Graylings range in length from 12 to 16 inches (30 to 41 centimeters) and weigh from 1 to 2 pounds (0.5 to 1 kilogram). All graylings are freshwater fish.

GRAY MATTER. Grayish-brown color matter, especially of neural tissue in the brain and spinal cord. Such matter contains nerve-cell bodies as well as nerve fibers. See also **Central and Peripheral Nervous Systems**.

GRAY SCALE. A series of achromatic tones ranging from black to white. A gray scale may be divided into three or more steps but 10 is a common number of divisions. A gray scale is sometimes included with the subject when making a color photograph so that measurements of its densities on the separation negatives or tripack will give the density range of that stage in the reproduction. A gray scale is helpful in controlling the processing stages in the analysis and synthesis of a color photograph.

GRAYWACKE (or Grauwacke). This term is of British origin and is not used extensively outside of Western Europe. As originally defined graywacke designates hard, dark-colored, coarse sandstones and grits having an argillaceous matrix or cement and occurring among the lower Paleozoic formations of Wales, England. Many typical graywackes are similar to basic arkoses, the dark color being due to a preponderance of the ferric minerals and plagioclase feldspar.

GREASE. A lubricating agent of higher viscosity than oils, consisting originally of a calcium or sodium soap jelly emulsified with mineral oil. Greases are employed where heavy pressures exist, where oil drip from the bearings is undesirable, and where the motion of the contacting surfaces is discontinuous so that it is difficult to maintain a separating film in the bearing. Grease-lubricated bearings have greater frictional characteristics at the beginning of operation, causing a temperature rise which tends to melt the grease and give the effect of an oil-lubricated bearing.

The principal categories of greases are: (1) calcium soap greases; (2) sodium soap greases; (3) complex soap greases—combinations of soaps and fatty acids used to impart high-temperature properties and moisture resistance. A low-molecular-weight soap can be used as a binding agent between the oil and soap in place of water; (4) lithium soap greases—excellent as multipurpose greases; (5) extreme-pressure greases, usually containing some form of sulfur, phosphorus, or other reactive agent—particularly suited to uses where there are sudden shock loads or continuous high pressures, as in steel rollingmill bearings; (6) nonsoap greases—exemplified by organically modified clays which hold the lubricating oil both by absorption and adsorption. Such greases are often used in high-temperature applications because they actually have no melting point; (7) asphalt-base greases—blends of asphaltic materials with lubricating oil, enabling a wide range of consistencies; and (8) filler-type greases—frequently calcium-base greases that contain solid materials having unctuous properties. The filler essentially serves as a cushion for absorbing impacts. Calcium and sodium base greases are most commonly used; sodium base greases have higher melting point than calcium base greases but are not resistant to the action of water. Graphite, either by itself or mixed with grease, is also employed as a lubricant. Gear greases consist of rosin oil, thickened with lime and mixed with mineral oil, with some percentage of water. The special-purpose greases often contain glycerol and sorbitan esters. They are used, for example, for low temperature conditions. See also **Lubricant**.

Standard methods for testing greases are published by the American Society for Testing and Materials, Philadelphia, Pennsylvania.

GREASE ICE. A thin skin of frazil crystals coagulated on the sea surface having a dark, greasy appearance. Also called *ice fat*, or *lard ice*. See also **Ice**.

It precedes the development of shuga. See also **Shuga**.

GREAT CATS. See **Cats**.

GREAT-CIRCLE COURSE. The shortest distance between any two points on the surface of a sphere is a great circle. For all practical purposes of navigation, the earth may be considered a sphere, and hence, the shortest course that a vessel may follow between any two ports is a great-circle course.

The great-circle course between two ports is frequently impractical for a ship to follow because of the fact that it may lead across land or into dangerous waters. For example, the great-circle course between two points that are in the same latitude but are separated by 180° of longitude will lead across a pole of the earth. Before deciding whether or not the great circle is practical, it is necessary to compute the course, computing a sufficient number of points so that the track may be plotted on a chart. Such computation is laborious, and to avoid the necessity of doing the computing, a great-circle chart may be used. On such a chart, any great circle appears as a straight line, and all that is necessary for the purpose of studying a great-circle course is to draw a straight line between the two points on the chart and examine it.

Even when the great-circle course does not lead the ship into danger, it is a very difficult course to follow because it makes a different angle with each successive meridian and requires the helmsman to continually change his course. To avoid this difficulty, as well as to avoid dangers, and yet to still approximate as closely as practical the shortest distance between the ports, the composite course is the type almost universally followed by vessels and aircraft on long-distance flights.

See also **Course**; **Gnomonic Projection**; and **Navigation**.

GREAT OBSERVATORY PROGRAM. To grasp the wonders of the cosmos, and understand its infinite variety and splendor, we must collect and analyze radiation emitted by phenomena throughout the entire electromagnetic (EM) spectrum. Towards that end, NASA proposed the concept of Great Observatories, a series of four space-borne observatories designed to conduct astronomical studies over many different wavelengths (visible, gamma rays, X-rays, and infrared). An important aspect of the Great Observatory program was to overlap the operations phases of the missions to enable astronomers to make contemporaneous observations of an object at different spectral wavelengths.

Hubble Space Telescope. The first element of the program—and arguably the best known—is the Hubble Space Telescope (HST). The Hubble telescope was deployed by a NASA Space Shuttle in 1990. A subsequent Shuttle mission in 1993 serviced HST and recovered its full capability. A second successful servicing mission took place in 1997. Subsequent servicing missions have added additional capabilities to HST,

Fig. 1. Hubble get revitalized in new servicing mission. (*Image courtesy of European Space Agency*).

Fig. 2. Hubble in Orbit. (*ESA*).

which observes the Universe at ultraviolet, visual, and near-infrared wavelengths. See Fig.1.

Since its preliminary inception, HST was designed to be a different type of mission for NASA—a long term space-based observatory. From its position 380 miles (612 kilometers) above Earth's surface, the Hubble Space Telescope has contributed enormously to astronomy. See Fig. 2 and 3. It has expanded our understanding of star birth, star death, and galaxy evolution, and has helped move black holes from scientific theory to fact. Credited with thousands of images and the subject of thousands of research papers, the space telescope is helping astronomers answer a wide range of intriguing questions about the origin and evolution of the universe. See also Hubble Space Telescope (HST).

Compton Gamma Ray Observatory. The Compton Gamma Ray Observatory (CGRO) was the second of NASA's Great Observatories. Compton, at 17 tons (17,273 kilograms), was the heaviest astrophysical payload ever

flown at the time of its launch on April 5, 1991, aboard the space shuttle Atlantis. This mission collected data on some of the most violent physical processes in the Universe, characterized by their extremely high energies. See Fig.4.

Compton had four instruments that covered an unprecedented six decades of the electromagnetic spectrum, from 30 keV to 30 GeV. In order of increasing spectral energy coverage, these instruments were the Burst And Transient Source Experiment (BATSE), the Oriented Scintillation Spectrometer Experiment (OSSE), the Imaging Compton Telescope (COMPTEL), and the Energetic Gamma Ray Experiment Telescope (EGRET). For each of the instruments, an improvement in sensitivity of better than a factor of ten was realized over previous missions. See Fig.5.

The Observatory was named in honor of Dr. Arthur Holly Compton, who won the Nobel prize in physics for work on scattering of high-energy

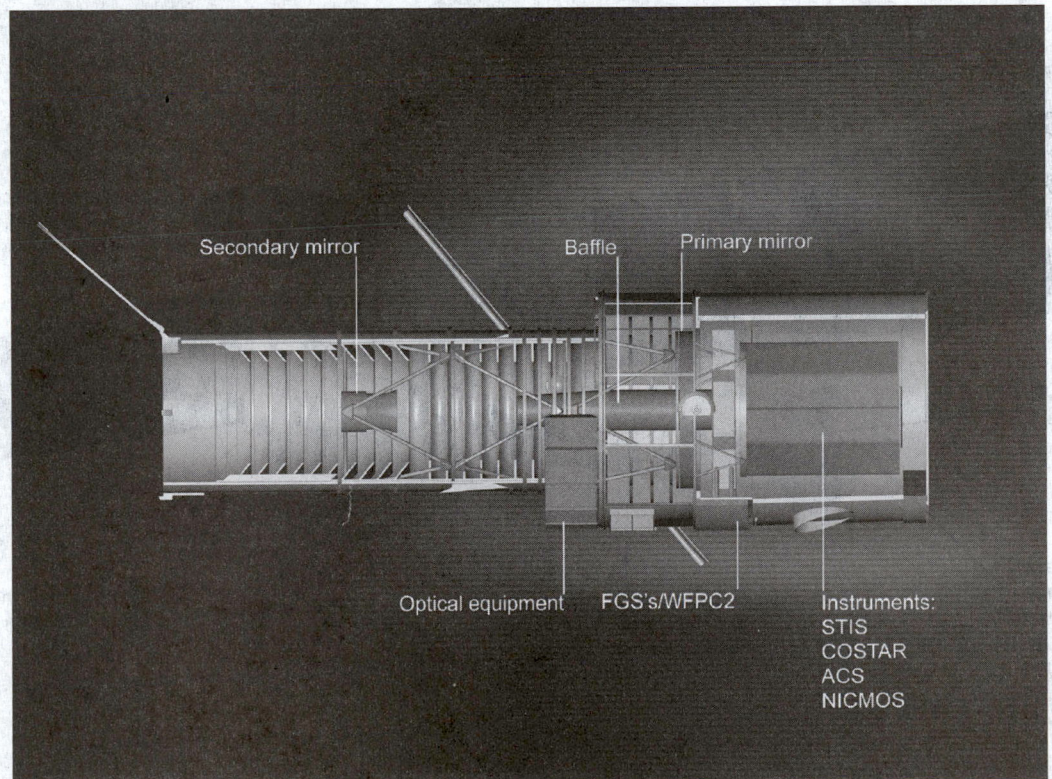

Fig. 3. Spacecraft illustration (*ESA*).

Fig. 4.   CGRO in orbit (*NASA*).

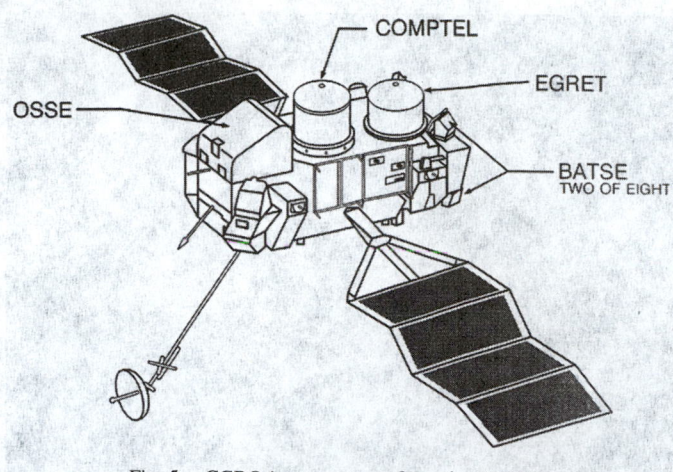

Fig. 5.   CGRO instrument configuration (*NASA*).

photons by electrons—a process which is central to the gamma-ray detection techniques of all four instruments.

Compton was safely deorbited and re-entered the Earth's atmosphere on June 4, 2000. See also **Compton Gamma-Ray Observatory (CGRO)**.

Chandra X-Ray Observatory. The third member of the Great Observatory family, the Chandra X-Ray Observatory (CXO), was deployed from a Space Shuttle and boosted into a high-Earth orbit in July 1999. This observatory is observing such objects as black holes, quasars, and high-temperature gases throughout the x-ray portion of the EM spectrum. See Fig.6.

Chandra detects and images X-ray sources that are billions of light years away. The mirrors on Chandra are the largest, most precisely shaped and aligned, and smoothest mirrors ever constructed. If the surface of Earth

was as smooth as the Chandra mirrors, the highest mountain would be less than six feet (1.8 meters) tall! The images Chandra makes are twenty-five times sharper than the best previous X-ray telescope. This focusing power is equivalent to the ability to read a newspaper at a distance of half a mile. Chandra's improved sensitivity is making possible more detailed studies of black holes, supernovas, and dark matter. Chandra will increase our understanding of the origin, evolution, and destiny of the universe. See also **Chandra X-Ray Observatory**.

Spitzer Space Telescope. The Spitzer Space Telescope represents the fourth and final element in NASA's Great Observatory program. Spitzer fills in an important gap in wavelength coverage not available from the ground—the thermal infrared.

The Spitzer Space Telescope was launched into space by a Delta rocket on August 25, 2003. Spitzer obtains images and spectra by detecting the infrared energy, or heat, radiated by objects in space between wavelengths of 3 and 180 microns (1 micron is one-millionth of a meter). Most of

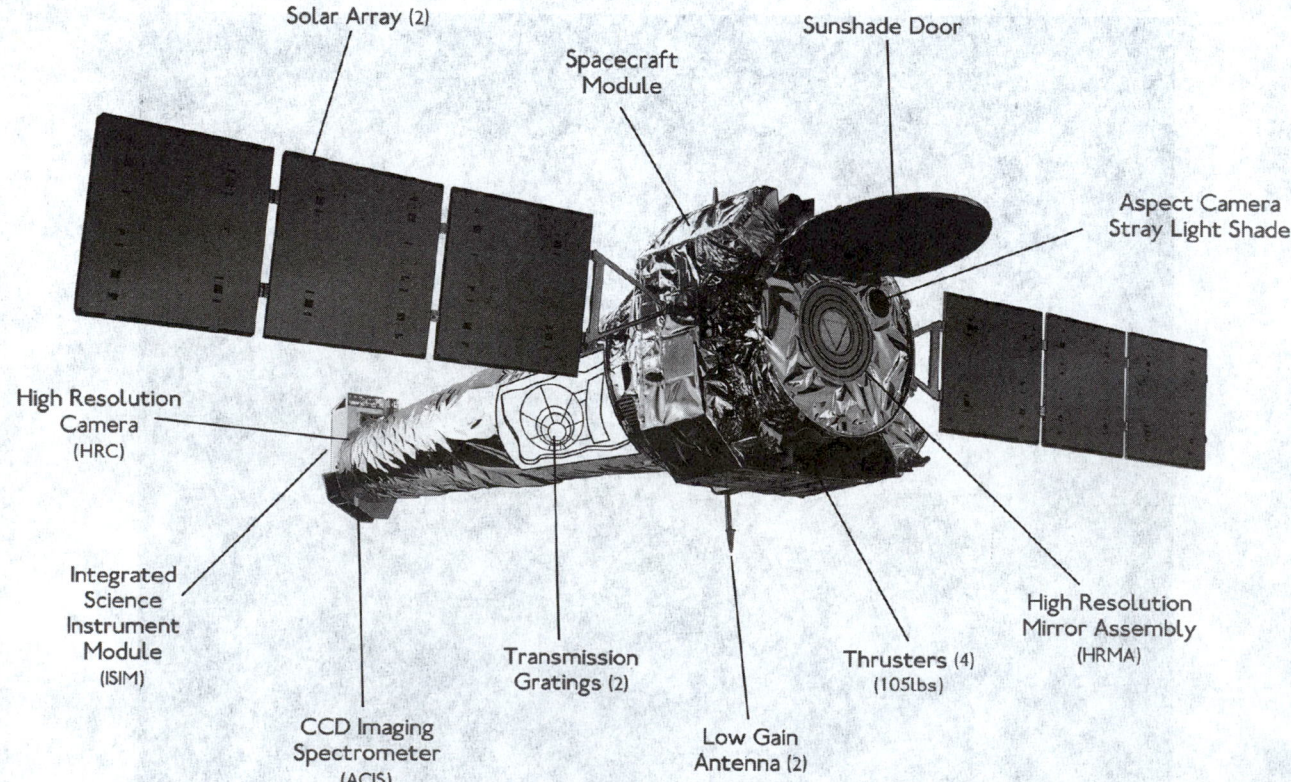

Fig. 6.   Labeled illustration of the Chandra X-Ray Observatory (*NASA/CXC/SAO*).

this infrared radiation is blocked by the Earth's atmosphere and cannot be observed from the ground.

Consisting of a 0.85-meter (2.8 foot) telescope and three cryogenically-cooled science instruments, Spitzer is the largest infrared telescope ever launched into space. See Fig.7. Its highly sensitive instruments give us a unique view of the Universe and allow us to peer into regions of space which are hidden from optical telescopes. Many areas of space are filled with vast, dense clouds of gas and dust which block our view. Infrared light, however, can penetrate these clouds, allowing us to peer into regions of star formation, the centers of galaxies, and into newly forming planetary systems. Infrared also brings us information about the cooler objects in space, such as smaller stars which are too dim to be detected by their visible light, extrasolar planets, and giant molecular clouds. Also, many molecules in space, including organic molecules, have their unique signatures in the infrared. See also **Spitzer Space Telescope (SST)**.

Fig. 7.   Artist's conception of Spitzer in its heliocentric orbit (*NASA/JPL-Caltech*).

Web References

Chandra X-Ray Observatory: http://chandra.harvard.edu/
Compton Gamma Ray Observatory: http://cossc.gsfc.nasa.gov/docs/cgro/index.html
Hubble Image Archive: http://www.spacetelescope.org/images/index.html
Hubble Site: http://hubblesite.org/
Spitzer Space Telescope: http://www.spitzer.caltech.edu/index.shtml

GREAT RED SPOT.   See **Jupiter**.

GREAT WHITE SHARK.   See **Sharks**.

GREBE (*Aves, Podicepediformes, Podicipedidae*).   This order of birds has a long geological history. They evolved in the Northern Hemisphere but now inhabit all continents except the Antarctic. They are from thrush to duck size; the length is 20–78 centimeters (8–31 inches), and the weight is 120–1500 grams (4–53 ounces). There are 17 to 21 cervical vertebrae. Some thoracic vertebrae are fused. The legs are positioned far back on the trunk. The tarsus is laterally compressed with a sharp front edge; on the back a double row of horny sawteeth is found, which is not known in any other group of birds. The lobed membrane along one side of the toes is 1 centimeter wide (0.4 inch). The claw of the mid-toe resembles a fingernail and is somewhat comb-like at the tip; possibly this is used to clean the plumage. Tail feathers are small and soft (unlike most birds), and so these birds appear tailess. See Fig. 1.

There are four genera with nine species: (a) Grebes (*Podiceps*) with six species; (b) Pied-Billed Grebes (*Podilymbus*) with two species; (c) the Running Grebes have only one species, the Western Grebe (*Aechmophorus occidentalis*); (d) Titicaca Grebes have only one species (*Centropelma micropterum*).

Grebes move on land only when they have no other choice, such as when building nests, in order to incubate, or to get from one open water hole to another in severe frost.

Fig. 1.   Grebe (Pied-billed grebe). (Sketch by Glenn D. Considine.)

They are all excellent divers, although they dive neither as deeply nor for as long as the loons, generally for less than half a minute and less then 7 meters (23 feet) deep. They live in still, fresh water and are seen at sea only outside the breeding season. The felt-like, thick, silky-soft contour feathers protect the underside against the water.

The nests are built of rotting plants, they float, and they are anchored to reeds or branches. A clutch consists of at least three eggs; and incubating birds always cover the eggs when leaving the nest. The eggs are at first snow-white and covered with chalky calcium carbonate, but soon they become chocolate brown on the wet plants on which they lie. The downy young are generally colorfully marked and striped, often producing a clown-like effect; right after hatching they move under the wings into the fur-like back plumage of whichever parent happens to be on the nest. Thus protected, they swim and dive with their parents weeks before they can dive themselves. In the breeding season of the second year of their lives, they usually resemble their parents. See also **Podicipediformes**.

GREEN FLASH.   A flash of green light seen on or (seemingly) adjacent to the upper rim of the low sun (at either sunrise or sunset). The green flash is a mirage, but the image formed in this case is of a portion of the sun rather than of an earthbound object. In addition to the displacement and distortion that is characteristic of mirages, there is also significant dispersion. The upper edge of the low sun normally has a thin green rim (occasionally blue) that is too narrow to be seen by the naked eye unless the rest of the sun is obstructed, say, by the horizon. It is often asserted that the green flash is seen in this way: a mere transient view of the green rim between obscuration by the rest of the sun and obstruction by the horizon. Yet such a sequence produces a singularly poor flash. Rather, the remarkable flashes always seem to involve multiple and magnified images of the green rim. Indeed, the presence of such multiple images of a small portion of the sun is a good indicator of a forthcoming flash. The optical signature of multiple images is a serrated edge to the sun. The refraction that displaces the image of the low sun up from the position it would occupy in the absence of an atmosphere does so more strongly for shorter wavelengths. This leads to a red rim on the bottom of the sun and a blue or green rim on the top.

GREEN FUNCTION.   The name of George Green (1793–1841), an English mathematician, is attached to several different mathematical results and not always consistently by different writers. The relation, called Green's theorem by some, is called Green's equation by others. It has seemed useful to collect all of these results in one item. The names given are chosen in accordance with what seems to be the most prevalent usage, but they are uniquely determined only by the accompanying equations.

1. Green function. A symmetric kernel $G(x, z)$ used to convert a Sturm-Liouville equation and its boundary conditions into an integral equation. It is defined to have the properties: (a) continuity over the range $a < x < b$ and with continuous derivatives of orders up to $(n - 2)$ where n is the order of the differential equation; (b) its derivative of order $(n - 1)$ is discontinuous at a point z within the range (a, b); (c) it satisfies the differential equation everywhere except at $x = z$.

2. Green formula. In the general theory of the nth-order linear differential operator, the linear differential operator, L, and its adjoint, $\overline{L}$ are of interest. Then the homogeneous equation $L(u) = 0$ is adjoint to $\overline{L}(v) = 0$ and Green's formula is

$$\int_a^b [vL(u) - uL(v)]\, dx = [P(u, v)]_a^b$$

where the left-hand side is the Lagrange identity. The right-hand side is a bilinear form in the $2n$ quantities $u(a), u'(a), \ldots, u^{(n-1)}(a)$; $u(b), u'(b), \ldots, u^{(n-1)}(b)$; $v(a), \ldots, v^{(n-1)}(a)$; $v(b), \ldots, v^{(n-1)}(b)$. Its determinant does not vanish and $P(u, v)$ is called the bilinear concomitant.

3. Green theorem. In vector analysis, there are several relations between single and multiple integrals. If u, v are scalar functions, and S indicates a double and τ a triple integral, the Gauss theorem in vector form is

$$\int_{\phi} \nabla u \cdot \nabla v \, d\tau + \int_{\phi} u \nabla^2 v \, d\tau = \int_{S} u \nabla v \cdot dS$$

On exchanging u and v and subtracting the result from this equation, the Green theorem results:

$$\int_{\phi} (u \nabla^2 v - v \nabla^2 u) \, d\tau = \int_{S} (u \nabla v - v \nabla u) \cdot dS$$

These relations, which correspond to integration by parts in scalar calculus, are also known as Gauss theorems for the divergence theorem.

See also **Divergence (Mathematics)**.

GREENHOUSE EFFECT. The water vapor and carbon dioxide found naturally in the atmosphere keep the Earth warmer than it would otherwise be. The clear atmosphere allows sunlight to penetrate to the Earth's surface and warm it. The surface releases this energy as infrared radiation, which is absorbed by water vapor and CO_2 in the atmosphere. This mechanism is commonly known as the "greenhouse effect." Without the greenhouse effect, the earth would be about $33\,°C$ ($60\,°F$) colder than it is currently. See also **Climate**; and **Global Warming**.

Humanity is altering the energy balance of the planet by adding gases that absorb infrared radiation to the atmosphere, and thereby strengthening the greenhouse effect. The chief "greenhouse gases" are CO_2, methane, and nitrous oxide. Whenever oil, coal, gas, or wood are burned, carbon dioxide is released into the atmosphere. Approximately half of the CO_2 that is released is soon absorbed by the oceans or by increased plant photosynthesis. The other half remains in the atmosphere for many decades. As a result, the atmospheric concentration of CO_2 is increasing. The average concentration of carbon dioxide has increased from around 275 parts per million before the industrial revolution, to 315 ppm when precise monitoring stations were set up in 1958, to 368 ppm in 1999. This change has increased the amount of energy striking the earth's surface by about 1.5 watts for every square meter of the earth's surface. This increased energy is equal to about 1% of the energy in the sunlight that reaches the earth's surface. See also **Greenhouse Gases**.

About two thirds of the current emissions of methane into the atmosphere result from cattle farming, rice paddies, landfills, coal mining, oil and gas production, and several other human activities. The other third comes from natural sources, particularly wetlands and termites. The total greenhouse effect from methane has increased by about 0.5 watts (0.3%) the energy striking each square meter of the earth's surface.

Several other gases collectively may have as much of a greenhouse effect as methane. Nitrous oxide (also known as "laughing gas") is released by the use of nitrogen fertilizers, the burning of wood, and some industrial processes. Higher levels of ozone, an urban pollutant regulated by EPA NAAQS, also add to the greenhouse effect. (The loss of ozone in the upper atmosphere tends to reduce the greenhouse effect.) Other gases with a greenhouse effect include chlorofluorocarbons (CFCs), hydrofluorocarbons (HFCs), perfluorocarbons (PFCs), hydrochlorofluorocarbons (HCFCs), and sulfur hexafluoride (SF_6).

Additional Reading

Dobson, A., A. Jolly, and D. Rubenstein: "The Greenhouse Effect and Biological Diversity," *Trends Ecol. Evol.*, **4**(3), 64–68 (1989).
Hardy, J.T.: *Climate Change: Causes, Effects, and Solutions*, John Wiley & Sons, Inc., New York, NY, 2003.
Hocking, C., C. Sneider, and L. Bergman: *Global Warming and the Greenhouse Effect*, University of California, Berkeley, Hall of Science, Berkeley, CA, 1999.
Long, D.: *Global Warming*, Facts on File, Inc., New York, NY, 2003.

GREENHOUSE GASES. Greenhouse gases are global in their effect upon the atmosphere. The primary greenhouse gases, unlike many local air pollutants like carbon monoxide, oxides of nitrogen, and volatile organic compounds, are considered stock pollutants. A stock air pollutant is one that has a long lifetime in the atmosphere, and therefore can accumulate over time. Stock air pollutants are also generally well mixed in the atmosphere. As a consequence of this mixing, the impact a greenhouse gas has on the atmosphere is mostly independent of where it was emitted. These characteristics of greenhouse gases imply that they should be addressed on a global (i.e., international) scale. See also **Pollution (Air)**.

Anthropogenic emissions of greenhouse gases occur in every country of the world. These emissions result from many of the industrial, transportation, agricultural, and other activities that take place in each country. Countries that are signatories to the United Nations Framework Convention on Climate Change (UNFCCC) are committed to reporting their anthropogenic emissions of greenhouse gases to the Secretariat of the convention.

What are Greenhouse Gases?

Some greenhouse gases occur naturally in the atmosphere, while others result from human activities. Naturally occurring greenhouse gases include water vapor, carbon dioxide, methane, nitrous oxide, and ozone. Certain human activities, however, add to the levels of most of these naturally occurring gases:

Carbon dioxide is released to the atmosphere when solid waste, fossil fuels (oil, natural gas, and coal), and wood and wood products are burned.

Methane is emitted during the production and transport of coal, natural gas, and oil. Methane emissions also result from the decomposition of organic wastes in municipal solid waste landfills, and the raising of livestock.

Nitrous oxide is emitted during agricultural and industrial activities, as well as during combustion of solid waste and fossil fuels.

Very powerful greenhouse gases that are not naturally occurring include *hydrofluorocarbons* (HFCs), *perfluorocarbons* (PFCs), and *sulfur hexafluoride* (SF_6), which are generated in a variety of industrial processes.

Each greenhouse gas differs in its ability to absorb heat in the atmosphere. HFCs and PFCs are the most heat-absorbent. Methane traps over 21 times more heat per molecule than carbon dioxide, and nitrous oxide absorbs 270 times more heat per molecule than carbon dioxide. Often, estimates of greenhouse gas emissions are presented in units of millions of metric tons of carbon equivalents (MMTCE), which weights each gas by its GWP value, or *Global Warming Potential*.

See also **Greenhouse Effect**.

GREENOCKITE. The mineral greenockite is cadmium sulfide, CdS, and is used as an ore of that metal. It is found rarely in hexagonal crystals, sometimes as earthy coatings on other minerals. Its hardness is 3–3.5; specific gravity, 4.9–5.0; luster, adamantine to earthy; color, yellow to yellowish-orange; subtransparent. It is found in Scotland, Bohemia, and France; also, in the United States, at Franklin Furnace. New Jersey; and Marion County, Arkansas, where it occurs as a yellow coloring matter in smithsonite; and in Mono County, California. It was named for Lord Greenock.

GREEN REVOLUTION. A popular term used mainly in the 1965–1975 period to describe the results of technology transfer to the growing of certain crops in some of the developing countries, such as India, Mexico, Pakistan, and the Philippines, this new technology increasing yields beyond the expectations of many experts. However, enthusiasm for the green revolution has been tempered somewhat in recent years. Generally credited with these productivity improvements is the work done by Borlaug and his associates on wheat genetics at the International Maize and Wheat Improvement Center (CIMMYT) in Mexico. Originally sponsored by the Rockefeller Foundation, the Center developed HYVs (high-yielding varieties) of wheat. Some of the current semidwarf HYVs, of course, are the offspring of varieties developed from similar ancestors in other breeding programs. The relatively short and stiff stalk of the semidwarfs means that they respond to improved cultural practices through increased yields rather than through increased plant growth, which would also result in lodging (falling over of the plant). The semidwarf varieties in use of the early 1980s, while considered by some to be revolutionary in their impact, are the product of a long developmental process. Semidwarf wheats were noticed in Japan in the 1800s.

In 1946, S.C. Salmon, a U.S. Department of Agriculture scientist acting as agricultural advisor to the occupation army in Japan, noticed *Norin 10* growing at the Morioka Branch Research Station in northern Honshu. The stems were short, but produced many full-sized heads. Salmon brought 16 varieties of this plant back to the United States. They were grown in a detention nursery for a year and then made available to breeders in seven locations. Although *Norin 10* was not satisfactory for direct use in the United States, it was useful for breeding. O.A. Vogel, a U.S. Department of Agriculture scientist stationed at Washington State University, was the first to recognize its worth and to use it in a breeding program as early as 1949.

In the interim, word about the short-strawed germ plasm had reached Borlaug in Mexico. His breeding efforts had run into a yield plateau because of lodging under high levels of nitrogen fertilization. Introduction of the *Norin 10* genes led to the development of a number of Mexican dwarf and semidwarf bread varieties of wheat. International diffusion of these varieties began very quickly at the experimental level and India and Pakistan were the first countries to be substantially involved.

The first Mexican wheats arrived in India in 1962 by way of the international nursery system. They became of immediate interest to M.S. Swaminathan of the Indian Agricultural Research Institute (IARI) in the spring of 1963. Borlaug, at the request of IARI, toured wheat areas in India and, upon his return to Mexico, he sent 100 kilograms of each of four varieties and small samples of over 600 other selections. The material was grown and studied at seven locations during the 1963–1964 season, as a part of the All-India Coordinated Wheat Trials. In 1965, two varieties, *Lerma Rojo* and *Sonora 64*, were released for general cultivation.

In another undertaking, in the spring of 1962, Borlaug gave some of the improved seeds to two trainees from Pakistan. The seeds were subsequently planted at the Agricultural Research Institute near Lyallpur. Borlaug visited Lyallpur in the spring of 1963 and later sent 203 kilograms of experimental Mexican seed to Pakistan. In the spring of 1964, Borlaug again visited Pakistan and soon secured government and foundation support for the varieties. Pakistan purchased several hundred tons of Mexican seed for planting during the 1965–1966 and 1967–1968 seasons.

The Mexican varieties proved remarkably adapted to India and Pakistan—for several reasons: (1) They had been bred in Mexico with alternate generations in different climatic and day-length regimes, primarily in order to get two generations each year. A valuable side-effects of this system was to establish a good degree of insensitivity to photoperiod. (2) Selection for disease resistance had also been practiced and the stocks introduced were found to show a remarkable level of resistance under the conditions in India and Pakistan. (3) The original stocks incorporated diversity. They had not been bred to pure line standards and there remained in them a reservoir of genetic potential that Indian wheat breeders were quick to exploit.

By the mid-to-latter 1970s, the process of varietal change had gone through four stages in India. A large percentage of plantings in India, Pakistan, Afghanistan, and Nepal, among other less-developed countries, is planted to varieties of Mexican origin. Exceptional increases in yield were obtained.

A number of improved varieties of corn (maize) also came out of the outstanding research done in Mexico. Dr. Borlaug, one of the leading researchers in an international crop improvement program, received the Nobel Peace Prize in 1970 in recognition for his efforts.

Research into the genetics of rice with an objective of improving yields also occurred during the green revolution period. The activities of the International Rice Research Institute (IRRI), established in 1962 in Los Banos, Philippines, and of the Indian Council of Agricultural Research, are particularly well known. Since the inception of those programs, numerous new varieties have been introduced. The rice situation was well summarized by K.L. Bachman of the Food and Agriculture Organization (United Nations): "The most important factor influencing the adoption of the new strains was their potential to give much higher yields than traditional and improved local varieties. With the new varieties, it now paid to apply more fertilizers and pesticides and to devote more time and money to improved cultural practices; with the older varieties, it was risky to use even modest amounts of fertilizers owing to the danger of lodging, particularly in the wet season."

As evidence of the successes achieved during the green revolution, Pakistan's 1971 wheat production was up 76% from its 1961–1965 average; Latin American corn (maize) production was up more than 50%; the Indian wheat crop of 1971 was almost double that of six years earlier; and Pakistan's 1974 rice crop set an all-time record.

Progress from improved varieties has, in some instances, reduced the nutritional level of people in farming areas because more emphasis was placed on wheat, rice, and corn (maize), and the production of food legumes was lowered. Recent years have indicated that much more than improved crop variety is required to improve the food status of many of the underdeveloped countries. Better means of storage are needed as well; it has been estimated that 15% of all the rice and other cereal crops raised in the Orient is destroyed by rats, either in the field or in storage. Better means of distribution and processing are also required.

An excellent summary of the green revolution era of agriculture is contained in Report No. 95, "Development and Spread of High-yielding Varieties of Wheat and Rice in the Less Developed Countries," by D.G. Dalrymple, U.S. Department of Agriculture, Washington, DC, 1976.

See also **Plant Breeding**.

Additional Reading

Coleman, D.: *Ecopolitics: Building a Green Society*, Rutgers University Press, Piscataway, NJ, 1994.
Perkins, J.: *Geopolitics and the Green Revolution: Wheat, Genes, and the Cold War*, Oxford University Press, Inc., New York, NY, 1997.
Sale, K. and E. Foner: *The Green Revolution: The American Environmental Movement, 1962–1992*, Hill & Wang, Inc., New York, NY, 1993.

GREEN RIM.   See **Meteorology**.

GREENSTONE.   Greenstone is an old field term for more or less altered basalts and dolerites, which, because of the development of chlorite, or perhaps hornblende or epidote, develop a characteristic green color. Many diabases and epidiorites have been called greenstones.

GREEN THUNDERSTORM.   Any thunderstorm that is perceived by observers to be green. The perceptually dominant wavelength of light from green thunderstorms ranges from blue- green to yellow-green. The purity of the color is generally low and the physical mechanism that causes the green appearance is not understood. Although green clouds often occur in conjunction with severe weather, there is no evidence to support anecdotal attributions of the cause of this green to specific characteristics of severe storms, such as hail or tornadoes.

Additional Reading

Bohren, C.F., and A.B. Fraser: "Green Thunderstorms," *Bull. Amer. Meteor. Soc.*, **74**, 2185–2193 (1993).

AMS

GREGALE.   See **Winds and Air Movement**.

GREGARIOUSNESS.   An association of animals of the same species; it may be of benefit to the individual but is not essential. The incidental grouping of animals, as in the swarms of maggots in a dead body, is not an association of this type. But the grouping of caterpillars of certain moths, even though the group originates in a like manner by the deposition of eggs in a mass, must be regarded as a gregarious association because the maintenance of the group is due to the behavior of the individuals. They are free to scatter but do not.

Herds of grazing animals cooperate for the common defense and such animals as the killer whale and the wolves are able to attack large animals by hunting in groups, but in all such cases the individual is able to subsist without the assistance of his fellows.

GREGORIAN TELESCOPE.   A reflecting telescope with a concave secondary mirror, located extrafocally, that reflects the light through an opening in the primary mirror and forms a real image behind the primary mirror. See also **Telescope**.

GREGORY FORMULA.   A formula for the numerical evaluation of an integral. It is obtained from the Newton formula for interpolation and may be written

$$\int_a^b f(x)\,dx = h\left[\frac{y_0}{2}y_1 + y_2 + \cdots + y_{n-1} + \frac{y_n}{2}\right]$$
$$- \frac{h}{12}(\Delta y_{n-1} - \Delta y_0) - \frac{h}{24}(\Delta^2 y_{n-2} + \Delta^2 y_0)$$
$$- \frac{19h}{720}(\Delta^3 y_{n-3} - \Delta^3 y_0) - \frac{3h}{160}(\Delta^4 y_{n-4} + \Delta^4 y_0) - \cdots$$

where h is the interval between equally spaced values of the independent variable x and the quantities $\Delta^m y_k$ are finite differences. Gregory's formula is equivalent to the trapezoidal rule, with correction terms in these differences.

GREISEN. An old German petrological term originally proposed by Werner for an igneous rock of granitic or aplitic texture composed principally of quartz, alkali feldspar, the fluorine-rich micas, and sometimes containing topaz. Greisens are pneumatolytically altered granites which are closely associated with the development of the tin ore mineral cassiterite.

GRIBBLE (*Crustacea, Isopoda*). A small marine crustacean, *Limnoria lignorum*, which bores into submerged timbers. It is a source of serious damage to docks and piling.

GRIFFITH CRACK THEORY. A theory relating to the brittle fracture of solids. The observed strength of ordinary window glass is less than one-hundredth of its theoretical strength. This discrepancy led Griffith to postulate that the low observed strength was due to the presence of small cracks or flaws in the glass. Because the ends of cracks have the ability to act as stress raisers, Griffith assumed that the theoretical strength was obtained at the ends of a crack, even though the average stress was still far below the theoretical strength. Fracture, according to this concept, occurs when the stress at the ends of the cracks exceeds the theoretical stress. When this occurs, the crack expands catastrophically. With the aid of the additional assumption that the strain energy released by the spreading of a crack is converted into the energy of the surfaces created by the fracture, it is possible to derive the following equation

$$S_n = \left(\frac{\sigma E}{2c}\right)^{1/2}$$

where S_n is the average applied stress necessary to make a crack spread, σ is the specific surface energy, $2c$ is the crack length, and E is Young's modulus.

GRIGNARD REACTIONS. Very important to the synthesis of numerous organic compounds, both in the laboratory and on a large scale in industry, is a two-step reaction involving the use of organo-magnesium halides. These reactions were studied intensively by Victor Grignard during the early 1900s and for this work he was awarded the Nobel Prize in Chemistry in 1912. The reactions are universally referred to as Grignard reactions and the many magnesium compounds required by the reactions are known as Grignard reagents. Grignard's work stemmed from a discovery by Barbier in 1899 that dimethylheptenol could be prepared by reacting methyl iodide, dimethylheptenone, and magnesium in ethyl ether. In studying the mechanics of Barbier's reaction, Grignard found that the reaction proceeds in two steps: (1) the reaction of magnesium [CAS: 7439-95-4] and an alkyl halide to form the corresponding alkyl magnesium halides; and (2) the reaction of the alkyl magnesium halide with a compound containing a carbonyl group to form a new carbon-carbon bond. Through subsequent years of experience, researchers have learned that nearly all alkyl and aryl halides react with magnesium to form Grignard reagents. However, the aryl and vinyl derivatives are, with more difficulty, achieved. In the mid-1950s, Normant and Ramsden showed that some of the less reactive halides, such as vinyl chloride [CAS: 75-01-4] and chlorobenzene [CAS: 108-90-7], will form a Grignard reagent with comparative ease if tetrahydrofuran is used as the solvent. See Table 1.

Because of the importance of the Grignard reaction techniques, they have received much study and numerous proposals have been made concerning the detailed mechanics involved. Originally, Grignard represented a Grignard reagent by RMgX, where R is the alkyl or aryl radical and X is the halide. Thus, magnesium ethyl bromide, a Grignard reagent, would appear in Grignard's symbolism as C_2H_2MgBr. Two of the main factors which make Grignard reagents so important are: (1) the many kinds of reagents that can be formulated, considering the substitution possibilities of the R and the X in the formula; and (2) the variety of reactions in which the Grignard reagents participate to yield numerous kinds of compounds. This versatility is demonstrated partially by Table 1.

The general sequence of the reactions is now embodied in the following generic forms, where RX = an organic halide (most typically a chloride or bromide, although fluorides can be induced to react); S = a coordinating

TABLE 1. REACTIONS OF GRIGNARD REAGENTS

Grignard reagents react with	To yield
H_2O, alcohols, primary or secondary amines	Hydrocarbons
Oxygen	Alcohols and phenols
CO_2	Carboxylic acids
Nitriles	Ketones
Metal halides	Organometallic compounds
NH_3	Hydrocarbons
γ-Lactones	Glycols
Acid esters	Tertiary alcohols (except formic acid which yields secondary alcohols or aldehydes)
Aldehydes	Secondary alcohols (except formaldehyde which yields primary alcohols)
Carboxylic acids	Tertiary alcohols
Acid halides	Tertiary alcohols or ketones
Ketones	Tertiary alcohols
Hydrogen halides	Hydrocarbons
Sulfur	Mercaptans

solvent (such as an ether or an amine); and AZ = a substrate with an electronegative group, Z:

$$RX + Mg + nS \rightarrow RMgX \cdot S_n$$

$$RMgX \cdot S_n + AZ \rightarrow RAZMgX \cdot S_n$$

$$RAZMgX \cdot S_n \rightarrow RA + ZMgX \cdot S_n$$

The heterolysis of AZ is dependent on the substrate and does not always occur. The final isolation of the product usually involves a hydrolysis step.

The development of improved industrial procedures, including the substitution of tetrahydrofuran [CAS: 109-99-9] (THF) for diethyl ether [CAS: 60-29-7] and the demonstration that the less reactive, but significantly less expensive, vinyl and aryl chlorides could be successfully used, has greatly expanded the commercial possibilities of this reaction. In the flavor, fragrance, pharmaceutical, and fine chemical industries, its use can generally be regarded as routine. Tens of thousands of metric tons of Grignard reagents are produced annually for captive use or merchant sale.

The great value of the Grignard reaction to the synthetic chemist is its general applicability as a building block for an impressive range of structures and functional groups. The Grignard reagent can act both as a prototypical carbon nucleophile that can undergo addition and substitution reactions and as a strong base that can deprotonate acidic substrates, resulting in the conjugate base or in some cases elimination reactions. Grignard reagents react with most functional groups containing polar multiple bonds (e.g., ketones, nitriles, sulfones, and imines), highly strained rings (epoxides), acidic hydrogens (e.g., alkynes), and certain highly polar single bonds (e.g., carbon–halogen and metal–halogen).

Preparation of Grignard Reagents

A Grignard reagent is prepared by first adding magnesium and a partial charge of solvent to the reactor, followed by the addition of RX, in the remaining solvent, to the reaction flask.

The most critical aspect of the solvent is that is must be dry (less than 0.02 wt % of H_2O) and free of O_2.

Other considerations for the solvent are the solubility of the Grignard reagent and the temperatures required for initiation and adventitious reactions of the Grignard with the solvent. Based on these three considerations, the best general solvent for the preparation of a Grignard reagent is THF. However, other solvents that are commonly used are diethyl ether, methyl t-butyl ether, di-n-butyl ether, glycol diethers, toluene, dioxane (R_2Mg), and hexane.

A surface coating resulting from the oxidation or hydration of the metal surface is the principal problem encountered for the magnesium reaction component. Fortunately, there are dozens of methods to remove the inert coating, thus activating the magnesium. For industrial use, the best method is using freshly chipped Mg turnings with a small quantity of the desired Grignard added to the reactor before addition of RX.

Just as for Mg and the solvent, the organic halide must be dry (less than 0.02 wt % of H_2O) and free of O_2. The relative reactivity of the halogens is reflected in the rate of disappearance of Mg, which follows the general order I > Br > Cl > F. Unfortunately, the rate of disappearance

Mg of does not always correlate with the formation of active Grignard. Typically, the more reactive the RX is, the higher the probability of forming a homocoupled product. Therefore, when choosing X, the rate of reactivity, product selectivity, and cost must be taken into account.

There are several common alternative methods for making Grignard reagents. Metal-exchange reactions are straightforward and MgR$_2$ can easily be prepared by this route.

Hydromagnesation reactions allow for the economical preparation of a Grignard reagent from an olefin.

Industrial Manufacturing Process

In spite of its industrial use for many years, the commercial-scale production of Grignard reagents has not been extensively described. The only practically important method is the batch method described by Grignard in 1900, namely formation of the Grignard reagent, reaction with a substrate, followed by hydrolysis of the reaction mixture.

The equipment can usually be constructed of carbon steel except for the hydrolysis vessel, which is usually glass-lined to avoid corrosion by aqueous acids. All vessels must be supplied with an inert gas (nitrogen or argon) for purging and blanketing and are vented to release off-gases. It is imperative that the reaction vessel be protected with a rupture disk.

Analysis of Grignard Reagents

There are three potential problems that may occur during Grignard reagent preparation: oxidation by O$_2$, hydrolysis by H$_2$O, or homocoupling during the addition of alkyl or aryl halide. All three of these reactions decrease the active Grignard reagent while maintaining the same equivalents of base. Consequently, the concentration of a Grignard reagent should not be assumed, based on the reactants. The disadvantages of not analyzing the Grignard reagent are improper stoichiometry, potentially deleterious side reactions, highly exothermic quenching processes, phase splits, waste disposal, and cost problems. The analytical technique must be able to differentiate between active Grignard and total basicity. Many methods are available to measure the active Grignard, ranging from titration to electrophilic quenching followed by gc analysis.

Health and Safety Factors

The hazards associated with the manufacture, transport, and use of Grignard reagents are related to the flammability of the solvents employed and the exothermic reactions involved in their preparation and use.

Because of their high reactivity, there is little meaningful information on the health hazards of Grignard reagents *per se*. Rather, consideration needs to be given to the reagents employed, including the solvents and the products (or by-products) of the reaction. Some starting materials, such as organic halides (notably methyl bromide and vinyl chloride), are particularly toxic.

Commercial use of a Grignard reagent in the United States requires that it appear on the Environmental Protection Agency (EPA) list of Chemical Substances in Commerce. A corresponding registration exists for the European community and for Japan.

Because they are classified as flammable liquids, Grignard reagents in the United States must be packaged in drums or other suitable containers bearing a red U.S. Department of Transportation label.

Reactions and Applications of Grignard Reagents

Reactions and applications of Grignard reagents include asymmetric syntheses using Grignard reagents, Grignard reactions with inorganic chlorides, Grignard reagents as bases, metal-assisted modified Grignard reactions, intramolecular Grignard reactions, Grignards as methacrylate polymerization catalysts, and Grignard reagents as supports for the Ziegler-Natta process.

Additional Reading

Kharasch, M. and O. Reinmuth: *Grignard Reactions of Nonmetallic Substances*, Prentice-Hall, Inc., New York, NY, 1954.
Okubo, M. and K. Matsuo: *Rev. Heteroatom Chem.*, **10**, 213 (1994).
Raston, C. and G. Salem: *Chem. Met.–Carbon Bond*, **4**, 159 (1987).
Silverman, G.S. and P.E. Rakita, eds.: *The Grignard Reagent Handbook*, Marcel Dekker, Inc., New York, NY, 1996.

GRILLAGE. A grillage is a system of timber or steel beams; it is used under columns to spread the loads over a comparatively large area. Timber grillages, consisting of layers of wooden beams, laid at right angles to each other, are generally used for temporary construction, although there are instances in which they have been enclosed in concrete for permanent construction. If this grillage is used for permanent foundation it should be either entirely submerged or creosoted to withstand deterioration.

The steel grillage consists of one or more layers or tiers of beams encased in concrete. If there are two or more tiers the beams in one tier are laid at right angles to those in the next tier. The individual beams in each tier are held in place by rods and pipe separators, cast iron separators or steel diaphragms. Since the concrete-encased steel grillage has more resistance to bending than the ordinary reinforced concrete spread footing it can be used to distribute heavy column loads over large areas.

GRINDING AND POLISHING AGENTS. These materials are comprised of abrasives in some form. An abrasive is a hard substance that, in particulate form, is capable of effecting a physical change in a surface, ranging from the removal of a thin film of tarnish to the cutting of heavy metal cross sections and cutting stone. Abrasive action can be negative, as in the case of grit in lubrication oil that will cause engine wear; or it can be positive, as used in scores of different abrasive products, ranging from sandpaper to grinding wheels.

The two principal categories of abrasives are: (1) natural abrasives, such as quartz [CAS: 14808-60-7], emery [CAS: 12415-34-8], corundum [CAS: 1344-28-1], garnet, tripoli [CAS: 1317-95-9], diatomaceous earth (diatomite) [CAS: 61790-53-2], pumice [CAS: 1332-09-8], and diamond [CAS: 7782-40-3]. (2) synthetic abrasives, such as fused alumina [CAS: 1344-28-1], silicon carbide [CAS: 409-21-2], boron nitride [CAS: 10043-11-5], metallic abrasives, and synthetic diamond. Quartz, emery, garnet, and corundum were used in prehistoric times. Natural diamond was first used in India as an abrasive in about 800 B.C. The first grinding machines were developed in France in about 1300 A.D. Early shellac-bonded abrasives were developed in India about 1825. The first cylindrical grinding machine was made in the United States in 1860. Vitrified bonded abrasives were developed in the United States in 1872. Fused alumina and silicon were developed in 1901. Resinoid bonded abrasives were developed by L. Baekeland in Belgium in 1923. Metal bonded diamond wheels appeared in 1936, and synthetic diamond abrasives and cubic boron nitride were developed in the United States during 1955 and 1957, respectively.

The general properties of natural and synthetic abrasives are given in Table 1. Both natural and synthetic abrasives must be crushed to small particle size before bonding to cloth or paper for mounting on various tools. See Table 2.

The Nature of the Grinding Process has been the object of much research. It appears to be a mixture of chemical and physical processes. Chips are cut from the metal by the sharp abrasive points which are heated in the process to the melting temperature of the metal. At this temperature chemical reactions take place which involve the abrasive, the metal, and the surrounding atmosphere. These reactions cause a dulling of the abrasive points necessitating a wearing of the wheel structure to expose new ones. The reactions also have a beneficial effect in preventing the rewelding of the chips to the base metal and their adhesion to the wheel structure. The latter is termed "loading." The detrimental reactions involving the abrasive are minimized by the control of its purity. The beneficial reactions can be augmented by the incorporation of chemical aids either into the wheel structure or into a fluid applied to the point of contact. Substances commonly used for this purpose are organic and inorganic sulfides and chlorides.

Polishing appears to be a quasi-chemical process in which the metal (or other material) is removed in particles approaching molecular size. See also **Abrasives**.

GRIT. An old term for coarse-grained sandstones whose components are angular or "gritty." There is a tendency to use it for any coarse-grained sandstone without regard to the angularity of the fragments.

GROSS-AUSTAUSCH. The exchange of airmass properties and the associated momentum and energy transports produced on a worldwide scale by the migratory large-scale disturbances of middle latitudes. When the atmospheric circulation is regarded as a large-scale turbulence process, the cyclones and anticyclones are considered to be eddies superposed on the average zonal wind currents. The mixing length, that is, the average

TABLE 1. GENERAL PROPERTIES OF ABRASIVES

Abrasive	Composition	Trade names and synonyms	Knoop hardness	Melting point C	Specific gravity
Quartz	SiO_2	Sand, flint	820	1,700	2.65
Emery	imp. Al_2O_3		2,000	1,900	4.00
Corundum	imp. Al_2O_3		2,000	2,050	3.95
Garnet			1,360	1,200[a]	4.25
Tripoli	98% SiO_2	Rottenstone	820	1,700	2.50
Diatomite	89% SiO_2	Diatomaceous earth Kieselguhr	820	1,700	2.50
Pumice	70% SiO_2 + Oxides	Pumicite, black ash			2.50
Diamond	C'		6,500	1,000[a]	3.51
Fused alumina	93–97% Al_2O_3	Alundum, Aloxite, Lionite	2000	2,050	3.95
Silicon carbide	SiC	Carborundum, Crystolon	2,450	2,400?	3.20
Cubic boron nitride	BN	Borazon	4,700	2,000[a]	

[a] decomposes

Source: Norton Company.

TABLE 2. AVERAGE PARTICLE SIZE OF ABRASIVE GRAIN USED IN GRINDING WHEELS

Grit size	Inches	Micrometers
8	.1817	4,620
10	.1366	3,460
12	.1003	2,550
14	.0830	2,100
16	.0655	1,660
20	.0528	1,340
24	.0408	1,035
30	.0365	930
36	.0280	710
46	.0200	508
54	.0170	430
60	.0160	406
70	.0131	328
80	.0105	266
90	.0085	216
100	.0068	173
120	.0056	142
150	.0048	122
180	.0034	86
220	.0026	66
240	.00248	63
280	.00175	44
320	.00128	32
400	.00090	23
500	.00065	16
600	.00033	8

distance over which these traveling eddies maintain the characteristics of their original environment, has a value near 10^8 cm. The coefficient of turbulent mass exchange or the exchange coefficient, which is a measure of the intensity of the large-scale exchange processes and hence of the intensity of the general circulation, has a magnitude of between 10^8 and 10^6 gm cm^{-1} s^{-1} (compared with a value of about 10^2 gm cm^{-1} s^{1} for small-scale turbulence). On the basis of results obtained so far, there is considerable doubt whether turbulence concepts can be applied meaningfully to the large-scale features of the atmosphere.

AMS

GROSS THRUST. The total thrust of a jet engine without deduction of the drag due to the momentum of the incoming air (ram drag). The gross thrust is equal to the product of the mass rate of fluid flow and the velocity of the fluid relative to the nozzle, plus the product of the nozzle exit area and the difference between the exhaust pressure and ambient pressure. See also **Airplane**.

GROSS WEIGHT. The total weight of an aircraft, rocket, etc., as loaded; specifically, the total weight with full crew, full tanks, payload, etc. Also called *take-off weight*.

GROUND-EFFECT MACHINE. Sometimes also referred to as air-cushion vehicle or hover craft, the ground-effect machine essentially "traps" a volume of air between itself and the ground or water beneath it. Depending upon the design, the vehicle can be lifted from a fraction of an inch (centimeter) up to several feet (meters) above the underlying surface, with sustaining pressures, or the equivalent in lifting force, of some 36 psi (2.4 atmospheres) or more. Normally, operational economy requires that the machine be kept as close to the surface over which it is to travel as may be possible.

The ground-effect principle has been employed in vehicles for traveling over water and land, for industrial conveyors, and for industrial towing vehicles.

GROUND (Electrical). A ground is a conductor connected to earth, or a large conductor whose potential is taken as zero (e.g., the steel frame of a car). A ground may be an undesirable, inadvertent, or accidental path taken by an electrical current in its effort to reach ground potential; or it may be the deliberate provision of conductors well connected to the ground by means of plates buried therein, or similar device.

There is always the possibility that, during the life of an insulated conductor, the insulation may be punctured or broken down and a ground occurs. Usually, a ground develops rapidly into a low-resistance path through which currents of damaging magnitude may flow. Insulation may be damaged in many ways — by the effect of moisture, or chemical vapors, by age, heat, abrasion, breaking, or crushing. Two-wire dc systems are permanently grounded on one side of the line, three-wire dc systems permanently grounded on the neutral wire. The same applies to two- or three-wire single-phase ac systems. The common grounding point of station three-phase lines is the generator neutral.

The grounding system of the ac generating station fulfills two distinct functions. The first is the grounding of noncurrent-carrying parts, the second is the furnishing of a ground connection for generator or transformer neutral to provide for the operation of a ground protection system. A common ground bus is employed, to which are connected the frames of all electric machines, the cases of instruments, transformers, circuit breakers, the secondaries of current and potential transformers, the switchboard ground bus, conduits, insulator bases, building structural steel, etc. Thus, if the grounding system is effective, a zero, or earth, potential will be established on all metal parts which might otherwise be dangerous in case a ground developed. To the common ground bus is also connected the fault bus, when used.

Connection to or insulation from grounds are also very important to the successful operation of various instrumentation, data processing, and telemetry systems. See also **Common-Mode Rejection Ratio**; **Common-Mode Voltage**; and **Electrical Ground Fault Circuit Interrupters (GFCI)**.

GROUND MORAINE. When a valley glacier melts completely away the debris carried on or within it is dropped upon the valley floor, forming a deposit called ground moraine. The ground moraine from the melting of the great Pleistocene ice sheets is usually spoken of as till.

GROUNDNUT OIL. See **Vegetable Oils (Edible)**.

GROUND PEARL (*Insecta, Homoptera*). The iridescent covering secreted by some of the scale insects that live on the roots of plants. Used as ornaments.

GROUND SPEED. The speed of an airborne object relative to the earth's surface. It is the magnitude of the vector sum of the object's velocity with respect to the air and the wind velocity, or, expressed in a different manner, the algebraic sum of the aircraft's airspeed and the wind factor.

GROUND STATE. The lowest energy level of an atom or molecule, corresponding to the most stable configuration of the atoms and electrons. At the relatively cold temperatures encountered in the atmosphere, most species normally exist in their ground states. Excitation to higher energy levels usually occurs via the absorption of radiation.

GROUND STREAMER. An upward advancing column of high ionization (a streamer or arc) that typically ascends from a point on the earth's surface toward a descending stepped leader. The ground streamer usually joins the stepped leader about 50 meters (164 feet) above the ground, after which the upward propagating light and current of the return stroke begin. Ground streamers occur because of the very high electric field intensities that build up directly below the descending, charged stepped leader. Often, more than one ground streamer starts up from the general area under a descending leader, but usually only one makes contact with the leader.

GROUND SUPPORT EQUIPMENT (GSE). That equipment on the ground, including all implements, tools, and devices (mobile or fixed), required to inspect, test, adjust, calibrate, appraise, gage, measure, repair, overhaul, assemble, disassemble, transport, safeguard, record, store, or otherwise function in support of a rocket, space vehicle, or the like, either in the research and development phase or in an operational phase, or in support of the guidance system used with the missile, vehicle, or the like.

The GSE is not considered to include land or buildings; nor does it include the guidance-station equipment itself, but it does include the test and checkout equipment required for operation of the guidance-station equipment.

GROUND SWELL. Swell as it passes through shallow water; it is characterized by a marked increase in height in water shallower than one-tenth wave length. To the seaman, ground swell is an indication of shoal water; to the shore-dweller, it is often an indication of approaching bad weather.

GROUNDWATER. At varying depths below the surface of the earth, depending upon wet or dry seasons, underground structures, and other natural and unnatural factors, is a zone which is saturated with water most of which comes from rain which has penetrated the ground. The upper surface of this saturated zone is called the water table, and the water itself, the groundwater or the subsurface water. The region above the upper surface of the water table is called the zone of aeration or vadose zone.

There is a lower limit to the saturated zone as well as an upper limit. Little groundwater exists at depths below 2,000–3,000 feet (610–914 meters). Deep down in the earth's crust the pressure must be so great that all pores in the rocks are completely closed; thus at depths of several miles below the surface there could exist no zone of saturation.

The groundwater moves through the rocks and unconsolidated materials of the earth near the surface, constantly seeping into streams and lakes to maintain these bodies of water between rains. If this seepage is sufficiently strong on hillsides or elsewhere springs may result. A well is simply an opening dug deep enough to encounter the zone of saturation.

In certain cases, the groundwater will flow through porous tilted beds called aquifers, from higher to lower localities, establishing a "head." This is sometimes sufficiently great to cause the water to flow out under pressure and rise above the surface of the ground, when the aquifer is penetrated by a drill. Such a source of water is called an artesian well (see Fig. 1), from Artois, France, a classic locality for such waters. Artesian conditions exist along much of the Atlantic Coastal Plain of the United States and in North and South Dakota, Nebraska, Kansas, Illinois, Indiana, Missouri, and Arkansas. Since the supply of underground water is largely dependent upon structure, the geology of water supply is one of the most important economic phases of the earth sciences. From the point of view of their origin, groundwaters are classified as juvenile, connate, and meteoric. Juvenile waters are of volcanic or magmatic origin, hence original. Connate waters are those in which the sediments were originally deposited. Meteoric waters are those of atmospheric origin.

All pure water, and most of all of the underground waters are of meteoric or surface-water origin. See also **Hydrology**; and **Wastes and Pollution**.

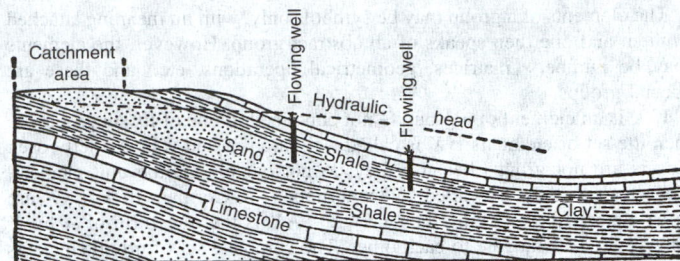

Fig. 1.   Ground cross section showing flowing artesian wells in a monocline.

GROUND WAVE. A radio wave that propagates by means of interaction with the earth's surface, as opposed to free-space or sky wave propagation. At low frequencies, ground waves can propagate thousands of kilometers following the earth's curvature.

Ground wave propagation is particularly important on the LF and MF portion of the radio spectrum. Ground wave propagation is used to provide relatively local coverage, especially by radio broadcast stations that require to cover a particular locality.

Ground wave radio signal propagation is ideal for relatively short distance propagation on these frequencies during the daytime. Sky-wave ionospheric propagation is not possible during the day because of the attenuation of the signals on these frequencies caused by the D region in the ionosphere. In view of this stations need to rely on the ground-wave propagation to achieve their coverage.

A ground wave signal is made up from a number of constituents. If the antennas are in the line of sight then there will be a direct wave as well as a reflected signal. As the names suggest the direct signal is one that travels directly between the two antenna and is not affected by the locality. There will also be a reflected signal as the transmission will be reflected by a number of objects including the earth's surface and any hills, or large buildings. That may be present.

In addition to this there is surface wave. This tends to follow the curvature of the earth and enables coverage to be achieved beyond the horizon. It is the sum of all these components that is known as the ground wave. See also **Surface Wave**.

Additional Reading

Barton, D.K., A.I. Leonov, S.A. Leonov, I.A. Morozov, and P.C. Hamilton: *Radar Technology Encyclopedia*, Artech House, Inc., Norwood, MA, 1997.

GROUP. A set of elements, finite or infinite in number, satisfying the following conditions: (1) There is a defined operation by which to each ordered pair of elements A and B in the group G there is associated an element C of G, denoted by $C = AB$, and called the product of A and B. (2) For this operation the associative law holds: $(AB)C = A(BC) = ABC$ for any three elements A, B, C of G. There exists: (3) a unit element E in G such that $EA = A$ for every element A of G, and (4) to each element A of G a reciprocal (or inverse) element A^{-1} of G such that $A^{-1}A = E$.

It must be understood that product, as defined in (1), is a convenient word to use for the result of combining two or more elements in a group but the law of combination is not confined to multiplication. For example, let the group elements be the integers $0, \pm 1, \pm 2, \ldots$ and let the combination law be addition, then the product of any two elements is their algebraic sum. These integers, regarded as elements of a group, will be seen to satisfy the requirements (1)–(4).

Infinite groups are discrete if the elements are denumerable; continuous, if they contain a nondenumerable infinity of elements. A finite group containing n elements is of order n. If $m < n$ elements satisfy the requirements of (1)–(4), they form a subgroup. Every group contains at least two subgroups: the unit element and the group itself.

The elements of a group may be symbols only, with no meaning attached to them and one then speaks of an abstract group. However, the elements may be numbers, matrices, geometrical operations, etc., and these are special groups.

If X is an element of a group **G** not contained in one of its subgroups **H**, then the set of elements **H**X is called a right coset and X**H** is a left coset. Cosets are not groups because they do not contain E, the unit element. Nevertheless, they are called "Nebengruppen" in German. If A, B, X are three elements of a group, then $B = X^{-1}AX$ is the transform of A by X and A, B are conjugate to each other. The complete set of group elements conjugate among themselves is a class of the group.

If **H** is a subgroup of the group **G** and X is an element of **G**, but not necessarily contained in **H**, then X^{-1}**H**X is also a subgroup of **G** and a conjugate subgroup to **H**. If **H** and **H**$' = X^{-1}$**H**X are conjugate then these two subgroups are invariant if **H** = **H**$'$. It is also called a normal subgroup or a normal divisor.

Suppose **H** is an invariant subgroup of a group **G** and that **H**X, **H**Y, ... are its cosets. The elements of **H** can be considered collectively as the unit element of another group and the various cosets as the remaining elements. It is called the quotient or factor group and is often designated by **G/H**. The multiplication properties of this group are similar to those of **G**.

Given a group **G**$'$ of order m with elements $A_1, A_2, \ldots, A_m$ and a second group **G**$''$ of order n with elements $B_1, B_2, \ldots, B_n$ such that every element of **G**$'$ commutes with every element of **G**$''$, then the mn element $A_i B_j$ for a group **G** = **G**$' \times$ **G**$''$ is of order mn and is called the direct product of **G**$'$ and **G**$''$. (See also **Lie Group**.)

Many other types of groups have been studied. They are of interest in geometry, differential equations, topology, and other branches of mathematics. In physics and chemistry, groups are used in the study of quantum mechanics; molecular, crystal, and nuclear structure; electrical circuits, etc.

GROUPERS. See **Bass**.

GROUP VELOCITY (Wave Train). The velocity of propagation of an interference pattern between two or more wave trains traveling in the same direction with different speeds. It may be quite different from the velocity of any one of the component wave trains. If there are more than two components, the character (waveform) of the resultant wave changes as the "group" progresses, so that the group velocity becomes ambiguous. For two components, the analysis is fairly simple.

To illustrate, first suppose for the moment that the wave train A of shorter wavelength λ is standing still, and the other, B of wavelength $\lambda + \Delta\lambda$ is moving past it in the positive direction (see Fig. 1). For example, let $\lambda = 1$ cm and $\lambda + \Delta\lambda = 1.1$ cm, and let the velocity Δv of the train B relative to the (stationary) train A be $+3$ cm per sec. As often as B moves forward 0.1 cm, the coincidence or beat maximum X moves backward 1 cm; consequently, X moves with respect to A with the velocity -30 cm per sec, which is -10 times, or, in general $\lambda/\Delta\lambda$ times, the velocity Δv with which B moves. (The analogy to a vernier should be quite apparent.) Now suppose that an additional velocity v is imposed upon both wave trains, so that now A moves with velocity v and B with velocity $v + \Delta v$. If $v = +100$ cm per sec, A moves with this velocity, B moves 103 cm per sec, but X moves only $100 - 30 = 70$ cm per sec. That is, the velocity of the interference maximum X is $u = v - \lambda \cdot \Delta v/\Delta\lambda$. This is the group velocity, usually written

$$u = v - \lambda \frac{dv}{d\lambda}$$

In the case of media in which there is dispersion, v is a function of λ; where there is no dispersion, $u = v$, since $(dv/d\lambda) dv$ is then zero.

Take the case of sodium light traveling through carbon bisulfide. This light has two close components with respective wavelengths 5,890 Å and 5,896 Å (in air). The refractive index for the 5.890 Å component being about 1.64, the velocity v of this component in CS_2 is about 1.83×10^{10} cm per sec. Now the dispersion of CS_2 in this part of the spectrum is such that $dv/d\lambda$ is readily computed to be 3.81×10^{13} cm per sec per cm, while the wavelength λ in CS_2 is 3,590 Å or 3.59×10^{-5} cm. Hence the group velocity u is 1.83×10^{10} cm per sec -3.59×10^{-5} cm $\times$ 3.81×10^{13} cm per see per cm $= 1.69 \times 10^{10}$ cm per sec.

Michelson, using the same revolving-mirror method as in measuring the speed of light in vacuo, actually obtained this velocity in carbon bisulfide, showing that it is the group velocity which this method really measures.

GROUSE (*Aves, Galliformes*). Game birds with compact rounded bodies and legs feathered to the feet. The closely related ptarmigans have both legs and feet feathered. Grouse are birds of the northern hemisphere. The ptarmigans, including the red grouse of the British Isles and the willow grouse, are found at high altitudes and in the north. Most of these birds have white plumage in the winter. Grouse vary in habits, some frequenting woodlands and others open ground.

The blackcock is the same as the heathcock. It is a large grouse (*Tetrao tetrix*) of Europe, named for its glossy black feathers. It is sometimes called black grouse. The hen is gray with mixed darker colors. She is called gray hen or heath hen.

The sage grouse (*Centrocercus urophasianus*) is the largest grouse in North America. It measures about 2 feet (0.6 meter) in length, largely comprised of tail. The male weighs from 6 to 8 pounds (3 to $3\frac{1}{2}$ kilograms). All of the male grouse have air sacs at the neck, some as large as golf balls and brightly colored.

The ruffled grouse (*Bonasa umbellus*) has plumage of a rich-brown coloration. The birds nest on the ground with 11 to 12 eggs at incubation time. Hatching requires 21 days.

The prairie chicken (*Tympanuchus cupido*) is of a pale-brown color and is found from Canada to Texas. The eastern heath hen is extinct in the United States.

Grouse are well known for their courtship dance. During this dance, the colored air sacs are inflated and feathers stand straight up to encase most of the fowl's body. The dance occurs just before daylight when the males of the field gather to be chosen for mates. As the males go into the dance, they are about 6 feet (1.8 meters) apart and start shuffling their feet, dancing back and forth, making loud, deep, pumping-like noises all during the dance. The females, attracted by these maneuvers, gather around to ultimately select their choice of the brightest, strongest male for a mate. Once the selection has been made, the female immediately starts to build a nest. The female incubates the eggs. The young remain in the nest about one week after hatching, after which time the young poults follow the female in a covey.

The capercaillie is a large woodland grouse of Scandinavian stock and is found in northern and central Europe and Asia. The male measures about 3 feet (0.9 meter) in length, averaging about 1 foot (0.3 meter) longer than the females. The species (*Tetrao urogallus*) is also known as the capercally, capercailzie, wood-grouse, and cock-of-the-walk. These birds are very shy and are clever in avoiding hunters. However, the birds tend to enter a hypnotic state during the courtship dance and are comparatively easy to capture during such display maneuvers.

The characteristics and habits of most all grouse are much alike. Different coloring and slight variations are visible, but mainly all are about the same. See also **Galliformes**; and **Ptarmigan**.

GROWING DEGREE DAY (GDD). A heat index that relates the development of plants, insects, and disease organisms to environmental air temperature. GDD is calculated by subtracting a base temperature from the daily mean temperature and GDD values less than zero are set to zero. The summation over time is related to development of plants, insects, and disease organisms. The reference temperature (base temperature) below which development either slows or stops is species dependent. For example, cool season plants (canning pea, spring wheat, etc.): base temperature is 40 °F (5 °C); warm season plants (sweet corn, green bean, etc.): base temperature is 50 °F (10 °C); and very warm season plants (cotton, okra, etc.): base temperature is 60 °F (15 °C). See also **Climate**; and **Degree Days**.

GROWING SEASON. In agriculture, the growing season is the period of each year when crops can be grown. It is usually determined by climate

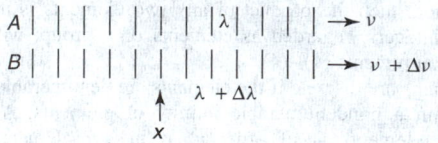

Fig. 1. Two sets of waves traveling at different velocities. Resultant maximum is at X.

and crop selection. Depending on the location, temperature, daylight hours (photoperiod), and rainfall, may all be critical environmental factors. This is an important concept in agricultural climatology, but it suffers greatly from vagueness and complexity. The growing season is highly variable due to plant varieties as related to temperature sensitivity. Currently, the most common measure of this period, "the average length of growing season," is defined as the number of days between the average dates of the last killing frost in spring and the first killing frost of autumn. The lack of a positive, practical definition for (and means of determining) a "killing" frost seriously limits the scientific usefulness of this measure. To provide some economic significance, the effective growing season is defined as the length of growing season that prevails in 80% of the years. Another measure, the frost-free season, is defined as the interval between the last and first occurrences of 32 °F (0 °C) temperatures in spring and fall. This may be observed exactly, but its relationship to the local microclimate is variable and nonspecific, and it does not consider differences in types of vegetation. Still a fourth measure, the vegetative period or vegetation season, attempts to allow for the greater microclimatic temperature range and for the general growth retardation by cold temperatures, and is defined as the summer period confined between occurrences of 42 °F (6 °C) or 41 °F or 43 °F temperatures. At best, any of the above is an index of growing season length, rather than a direct measure of it. Basically, the growing season (and "killing frost") should be defined biologically rather than meteorologically and should consider the detailed microclimate, plant resistance to frost, growth rate versus temperature, and probably other factors.

GROWTH.   Increase in size and complexity. Growth of living structures depends upon increase in the number of cells or in the bulk of cells and intercellular material. It is based on the process of intussusception through which materials received as food become an integral part of the structures already present. Accretional growth is of very limited occurrence in living things and is not independent of intussusception.

(Over decades of traditional biological and medical research, a large fund of essentially qualitative information concerning the growth process has been accumulated, a condensation of which appears in the following paragraphs. There are high expectations that much more will be learned during the next few years as the result of intense studies directed at the gene and molecular level. See also **Genetics and Gene Science**; and **Molecular Biology**.)

Most animals exhibit determinate growth; that is, they increase in size until they approximate a limit characteristic of their kind. A few mature within rather wide limits according to the amount of food available. In the adult body, the capacity of various tissues to continue their growth varies, but in all cases tissues worn away in the course of normal life have the power of renewal. Some, such as the bone-producing cells of vertebrates, are capable of becoming active for the restoration of damaged structures. These aspects of growth are closely associated with regeneration.

The rate of growth in different parts of the body also varies, as also does the rate of total growth at different periods of life. Most mammals increase in size rapidly during early life and gradually slow down as maturity is approached; whereas man grows rapidly during infancy, slowly during childhood, rapidly again during youth, and more slowly toward the completion of his size. In the human body, the nervous system most rapidly approaches its maximum size, and the reproductive system lags until the onset of maturity. Some of the glandular tissues increase rapidly before maturity and then decrease in bulk. The balance of all these processes when normal food is available results in the gradual process of general growth, and the attainment of stability in adult life is a result of their correlation with external factors. Although no one factor is wholly responsible for growth, hormones of the pituitary and thyroid glands are of great importance in its regulation in vertebrates. Deficiency of either gland may result in dwarfing, and pituitary excess sometimes causes human beings to attain unusual height. Heights of more than 7 feet (2.1 meters) are probably due in all cases to such abnormality.

Plant growth is indeterminate. In the higher plants, primary growth is confined to the tips of stems and roots, secondary growth to cambium layers, which produce wood and bark. The cambiums and the undifferentiated tissues at the tips of stems and roots are called meristems. Meristem cells divide rapidly and some of them finally become the mature cells of the plant. Each cell starts to grow, like an animal cell, by adding more

protoplasm but finally increases tremendously in size by taking up a quantity of water to form a large central vacuole. Tissues are differentiated by the accumulation of excess food (cellulose, lignin, suberin) on the outside of each cell in the form of a cell wall. Certain columns of cells thicken their side walls, digest their end walls, and then die, leaving long tubes (vessels) which conduct water. Other cells die from an excess accumulation of impervious wall material and become fibers or cork cells. Others remain alive for a season or two and manufacture, transport, or store food, much more food than the plant can ever use. Some few cells become concerned with the isolation of meristems in reproductive organs (ovules, seeds). These isolated meristems produce the cells of new plants. The life of a plant need never terminate. There is no adult stage as in animals. Propagation may serve to keep a single set of meristems in action continuously.

GROWTH CURVE.   1. An activity curve in which the activity increases with time, or that portion of an activity curve showing such an increase. 2. A theoretical or experimental curve showing, as a function of time, the number of atoms, or the mass, or the activity of a nuclide being produced in a radioactive transformation or in an induced nuclear reaction. See also **Logistic Curve**.

GRUB.   The larva of certain insects, usually of beetles and flies. The term *worm* is sometimes applied to a grub. The grub is frequently the most damaging stage in the life cycle of an insect.

GRUIFORMES *(Aves).*   The cranes and their relatives form this order of wading and swimming birds. Hardly any other order among birds has so little uniformity. Cranes cover a wide variety of forms, such as the common moorhens and coots, the long-legged cranes, the heavy bustards, and the peculiar seriemas. Even in appearance the various families do not resemble each other very much.

All cranes are covered with down and able to run about when newly born. The length is 10–150 centimeters (4–59 inches), and the weight is 5 grams to 16 kilograms (2 ounces to 35 pounds). The cranes are characterized by the absence of horny ridges in the beak, of ramicorn over the nostril sheath, of elongated patellae, of a crop, and of fully developed toe-webbings.

There are eleven families: The Rails *(Rallidae)*; The Stilt Rails *(Mesitornithidae)*; Sun Bitterns *(Eurypygidae)*; Finfoots *(Heliornithidae)*; Kagus *(Rhynochetidae)*; Cranes *(Gruidae)*; Limpkins *(Aramidae)*; Trumpeters *(Psophiidae)*; Bustards *(Otididae)*; Seriemas *(Cariamidae)*; and Buttonquails *(Turnicidae)*.

Rails, cranes, bustards, and buttonquails all inhabit northern parts of both the Old and New Worlds. The rest of the families are confined to warm regions: stilt rails in Madagascar, limpkins in Central and South America, trumpeters, sunbittern, and seriemas in South America and Central and South Africa, and the kagus in New Caledonia. See also **Rails, Coots, and Cranes**.

GRUNION.   See **Silversides**.

GRUNTS *(Osteichthyes).*   Of the order *Percomorphi*, suborder *Percoidea*, family *Pomadasyidae*, grunts are named after sounds which they produce, much the same way that croakers, somewhat related, received their name for acoustic reasons. In the grunt, the noise stems from sharp pharyngeal teeth which, when ground together and assisted by a nearby air bladder acting as a resonator, create deep vibrations. The sounds can be picked up underwater by a hydrophone and can also be heard when the fish is taken out of water. In appearance, the grunts look quite a lot like snappers. They favor tropical marine waters.

The grunts include white grunts *(Haemulon plumieri)* and French grunts *(H. flavolineatum)* both of which inhabit American Atlantic waters. The latter is considered a beautiful fish. The porkfish *(Anisotremus virginicus)* is also quite a spectacular fish in this group. The *Anisotremus davidsoni* is the only western species and may be described as a dull silver fish that attains a length of about 20 inches (51 centimeters). Grunt-like fishes in Indo-Australian waters are tropical marine varieties, sometimes called sweetlips.

See also **Fishes**.

GRUS (the crane). A southern constellation located between Tucana and Piscis Australis.

GUANACO. See **Camels and Llamas.**

GUANIDINE. Guanidine, [CAS: 113-00-8] or carbamidine or iminourea, $(NH_2)C=NH$ is formed (1) by heating ammonium thiocyanate [CAS: 1762-95-4] to $180\,°C$, (2) by ammonolysis of orthocarbonates, $C(OC_2H_5)_4 + 3NH_3 \rightarrow (NH_2)_2C=NH + 4C_2H_5OH$, (3) by ammonolysis of chloropicrin [CAS: 76006-2], $Cl_3CNO_2 + 7NH_3 \rightarrow NH_2)_2 C=NH + 3NH_4Cl + N_2 + 3H_2O$, (4) by ammonolysis of cyanogen chloride [CAS: 506-77-4], $ClCN + NH_3 \rightarrow ClC(NH_2)=NH \rightarrow HN=C=NH \rightarrow (NH_2)_2 C=NH$.

Guanidine forms salts with acids, e.g., guanidine nitrate, $HNC(NH_2)_2 \cdot HNO_3$. By heating at $120\,°C$ for several hours, a mixture of ammonium thiocyanate and dicyanodiamide [CAS: 461-58-5], guanidine thiocyanate [CAS: 593-84-0] solution is obtained by extracting with water. Treating guanidine with a mixture of nitric and sulfuric acids forms nitroguanidine

which is reduced by zinc and acetic acid to aminoguanidine

By treating aminoguanidine (1) with dilute acid or alkali, there is obtained, first, semicarbazide, finally hydrazine; (2) with nitrous acid, diazoguanidine

which is decomposed by alkali into alkali azide (e.g., NaN_3) plus cyanamide ($H_2N \cdot CN$) plus water.

In the Pauling theory of its structure, guanidine is a resonance compound of the molecular structure cited $[(NH_2)_2C=NH]$ and two ionic structures in which the nitrogen of the imino group gains an electron lost by one of the amino groups.

The monoalkyl- and N,N-dialkyl guanidines are somewhat weaker bases than guanidine, because resonance of the double bond to the substituted $-NH_2$ group is restricted by the fact that carbon is more electronegative than hydrogen, and renders more difficult the acquisition of a positive charge by an adjacent nitrogen atom. This effect is still more marked with the N,N'-dialkyl guanidines, while, in contrast, the N,N',N'-trialkyl guanidines are essentially as strong bases as guanidine.

Table 1 lists seven representative substituted guanidines.

GUAR GUM. See **Food Additives** and **Gums and Mucilages.**

GUAVA TREES. Of the family *Myrtaceae* (myrtle family), there are some 150 species of guava trees and shrubs, including *Psidium guajava* and *P. cattleyanum*. These plants are indigenous to tropical America. It is recorded that the guava was one of the favorite foods of the Aztecan and Incan Indians. In South American countries, the fruit is called the *guayaba*. These plants have oblong, short-petioled leaves, and white flowers. The fruits are aromatic and slightly acid. The seedy pulp is used for making guava jelly and as a blending agent by ice cream manufacturers. Significant quantities of the fruits are also consumed fresh. The fruit is very rich in ascorbic acid (vitamin C), having about ten times the quantity contained in an average orange. The guava is also an excellent source of vitamin

TABLE 1. GUANIDINE

Guanidine	Formula	Melting point °C
1. Guanidine		
2. 1,3-diphenylguanidine		147
3. 1,1,3,3-tetraphenylguanidine		130
4. 1,2,3-triphenylguanidine		144
5. 1,1,3-triphenylguanidine		131
6. Guanylurea		105
7. Aminoguanidine decomposes		

B_1. The tree has been widely introduced into tropical areas throughout the world, including Florida, Hawaii, and southern California.

GUIANA CURRENT. An ocean current flowing northwestward along the northern coast of South America (the Guianas).

The Guiana current is an extension of the south equatorial current (flowing west across the ocean between the equator and $20\,°S$), which crosses the equator and approaches the coast of South America. Eventually, it is joined by part of the north equatorial current and becomes, successively, the Caribbean current and the Florida current.

GUIDANCE. The process of directing the movements of an aeronautical vehicle or space vehicle, with particular reference to the selection of a flight path. In preset guidance a predetermined path is set into the guidance mechanism and not altered, in inertial guidance accelerations are measured and integrated within the craft, in command guidance the craft responds to information received from an outside source.

Beam-rider guidance utilizes a beam; terrestrial-reference guidance, some influence of the earth; celestial guidance, the celestial bodies and particularly the stars; and homing guidance the information is in response to transmissions from the craft, in *semiactive homing guidance* the transmissions are from a source other than the craft, and in *passive homing guidance* natural radiations from the destination are utilized.

Midcourse guidance extends from the end of the launching phase to an arbitrary point enroute and terminal guidance extends from this point to the destination.

GUILLEMIN, ROGER CHARLES LOUIS (1924–Present). Roger Guillemin was a French-born American endocrinologist and Nobel Prize winner who discovered the releasing hormone of hypothalamus.

Guillemin was born at Dijon, France on 11 January 1924, the son of a machine tool maker. Following his BSc (1942) from the University of Dijon, he studied medicine at the University of Lyon, where he qualified in 1949. Research with Hans Selye at the Institute for Experimental Medicine at the French-speaking University of Montreal led to a doctorate in 1953. From 1953 to 1970 he was Professor of Neuroendocrinology at the Baylor College of Medicine, Houston, Texas, where he and Andrew Schally began a search for the hypothalamic hormone postulated by the English endocrinologist Geoffrey Harris. They soon became bitter rivals. Since 1970 he has been at the University of California, San Diego, where he is also a member of the Salk Institute, La Jolla. He became an American citizen in 1963. He is married to a musician and has six children. In 1974 he was made a Chevalier Legion d'Honneur as well as a member of the National Academy of Sciences. See also **Schally, Andrew Victor (1926–Present)**.

Guillemin's life's work has been concerned with the discovery, identification and chemical synthesis of the peptide hormones of the brain. In 1969, after working with half a ton of sheep's' brains, he succeeded in isolating a milligram of hormone. Produced in the hypothalamus, this stimulates or inhibits the release of pituitary hormones which, in turn, control the activities of other glands. Guillemin identified the corticotrophin- and thyrotrophin-releasing factors (CRF and TRF) in 1981. He also identified somatostatin, an inhibiting release hormone in 1972. Guillemin and Schally shared the Nobel Prize in Physiology or Medicine with Rosalyn Yalow in 1977 for the isolation of the peptide hormones of the hypothalamus. Since 1977 Guillemin has worked on the isolation and synthesis of endomorphins. See also **Central and Peripheral Nervous Systems**; and **Yalow, Rosalyn Sussman (1921–Present)**.

Additional Reading

Guillemin, R.: "Peptides in the Brain: the New Endocrinology of the Neuron," In: Lindsten, J.: *Nobel Lectures Physiology or Medicine 1971–1980*, pp. 357–397, World Scientific Publishing Company, Inc., Hackensack, NJ, 1992.

Locke, W., A.V. Schally, T.K. Setze and J.C. Oehler: *Hypothalamus and Pituitary in Health and Disease*, Charles C. Thomas Publisher, Ltd., Springfield, IL, 1972.

McCann, S.M.: "Saga of the Discovery of Hypothalamic Releasing and Inhibiting Hormones," In: McCann, S.M.: *Endocrinology: People and Ideas*: American Psychological Society, Bethesda, MD, 1988.

Sherman, L.: "Roger Charles Louis Guillemin," In: Fox, D.M., M. Meldrum, and I. Rezak: *Nobel Laureates in Medicine and Physiology: A Biographical Dictionary*, Garland, New York, NY, 1990, pp. 219–224.

W. H. BROCK, University of Leicester, Leicester, UK

GUINEA FOWL.   See **Pheasant**.

GUINEA PIG.   See **Rodentia**.

GUINEA WORM (*Nemathelminthes, Nematoda*).   A large roundworm, *Dracunculus* (*Filaria*) *medinensis*, parasitic in man. It sometimes reaches a length of more than a yard. The worm lives in the superficial tissues, especially of the legs, forming an abscess open to the surface, and can be removed by gradual traction on the end of the worm exposed in this opening.

GULF STREAM.   As the North Equatorial Current in the Atlantic Ocean moves westward, it is deflected, first, by the continental land mass and, second, by the Coriolis effect. This intensification, turning clockwise in the Northern Hemisphere, results in a warm, powerful current known as the *Gulf Stream*. Originating in the Gulf of Mexico, the stream passes through the Straits of Florida, and flows northeast parallel to the U.S. coastline. Finally, it slows down and spreads out to become the North Atlantic Drift, an eastward movement of warm water that is responsible for the warmth of Western Europe commonly attributed to the Gulf Stream. Presently, the ocean thermal differences existing along the Gulf Stream and similar ocean currents are being considered as energy sources for solar sea power stations. See also **Irminger Current**; **Ocean**; and **Ocean Resources (Energy)**.

GULF STREAM COUNTERCURRENT.   A density ocean current flowing southwestward in the vicinity of Cape Hatteras and skirting the Bahamas. It flows at a depth of approximately 6,000–9,000 feet (1,830–2,745 meters) and at a rate of about 8 miles (12.8 kilometers) a day.

GULL.   See **Petrels and Albatrosses**; and **Waders, Shorebirds and Gulls**.

GUM ARABIC.   See **Acacia Trees**; and **Food Additives**.

GUM RESINS.   See **Resins (Natural)**.

GUMS AND MUCILAGES.   Natural gums and mucilages are carbohydrate polymers of high molecular weight obtained from plants. They can be dispersed in cold water to give viscous or mucilaginous solutions that normally do not gel. They are composed of acidic and/or neutral monosaccharide building units joined by glycosidic bonds. The acid groups ($-CO_2H$, $-SO_3H$) are usually present as salts of calcium, magnesium, sodium, and potassium; in certain cases substituents such as acetyl (karaya gum) and methyl groups (mesquite gum) may be present as well. Pyruvic acid residues, linked as ketals, are present in several cases (such as agar). The properties of several gums are described in Table 1.

Gums are of particular importance in the food processing field where they perform at least three functions—emulsifying, stabilizing, and thickening. A few also function as gelling agents, bodying agents, foam enhancers, and suspension agents. Gum guiac also serves as an antioxidant and preservative.

Sources of Gums. Gums and mucilages may be found either in the *intracellular parts* of plants or as *extracellular exudates*. Those found within plant cells represent storage material in seeds and roots. They also serve as a water reservoir and as protection for germinating seed. The polysaccharides found as extracellular exudates of higher plants appear to be produced as a result of injury caused by mechanical means or by insects. It has not been well established whether the exudates are formed at the site of the injury, or whether they are generated elsewhere and then transported to the injured area.

The true exudates, such as gum arabic and the East African and Indian gums are picked by hand. Seldom are commercial samples pure. This is a serious disadvantage in product control. They are classified according to grade, which, in turn, depends upon color and contamination with foreign bodies, such as wood and bark. The exudates are processed simply by grinding, their only prior treatment being sorting and sometimes bleaching under the sun. In some cases, they are purified by extraction with water and precipitated by alcohol.

Gums and mucilages present in roots, tubers and seaweeds are usually extracted with hot water, dried, and marketed as a powder. Those gums found on the inner side of the seed coat as vitreous layers (e.g., locust bean, guar bean, etc.) are best obtained by a suitable milling process. This first removes the seed coat and then makes use of the fact that the gum layer is very hard and tough as compared with the seed endosperm. The intracellular gums and mucilages can be purified by precipitation with alcohol from aqueous solution as in the case of the plant gum exudates, or by a process such as acetylation. In a similar way, the bacterial polysaccharides can be precipitated from the cell-free culture fluid with alcohol, or as the salt of a quaternary ammonium compound where acidic groups are present.

Characteristics of Gums. The extracellular plant gums and mucilages (gum arabic, karaya gum, and tragacanth, for example) generally have a more complex structure than the intracellular types. They are made up of a number of different sugar-building units linked together by a variety of glycosidic bonds. They possess a central core or nucleus composed mainly of D-galactose and D-glucuronic acid units joined by glycosidic bonds, which are relatively stable to hydrolysis by acids. To this central nucleus are attached as side chains those sugar units, which are removed by mild acid hydrolysis. Thus, in the case of gum arabic, the acid-resistant portion of the molecule is composed of D-glucuronic acid and D-galactose; to this nucleus are attached units of L-arabinose, L-rhamnose, and D-galactopyranosyl $(1 \rightarrow 3)$ L-arabinose.

The neutral mucilages and gums, such as mannans, glactomannans, and glucomannans extracted from seed and roots, have a relatively simple structure. The kinds of building units are fewer and the molecules are much less branched. The galactomannans are usually composed of a backbone of linear chains of D-mannose units jointed by 1,6-glycosidic bonds, to which are attached at regular intervals side chains of D-galactose residues. The glucomannans are essentially linear polymers united by 1,4-linkages.

TABLE 1. GUMS AND MUCILAGES — PROPERTIES AND APPLICATIONS

Acacia gum (arabic gum)

The dried water-soluble exudate from stems of *Acacia senegal* or related species. Thin flakes, powder, granules, or angular fragments; color white to yellowish white; almost odorless, mucilaginous taste. Completely soluble in hot and cold water, yielding a viscous solution of mucilage; insoluble in alcohol. Aqueous solution is acid to litmus. Produced in the Sudan, Nigeria, and other parts of west Africa. Used in adhesives, inks, textile printing, cosmetics; as a thickening agent and colloidal stabilizer in confectionery and other food products.

Alginic acid ($C_6H_8O_6)_n$

White to yellow powder, possessing marked hydrophilic colloidal properties for suspending, thickening, emulsifying, and stabilizing. Insoluble in organic solvents; slowly soluble in alkaline solutions. Used in food industry as thickener and emulsifier; as a protective colloid; in tooth paste, cosmetics, pharmaceuticals, textile sizing, coatings; as a waterproofing agent for concrete; in boiler water treatment; in oil-well drilling muds; in storage of gasoline as a solid.

Agar

Thin, translucent, membranous pieces or pale bluff powder. Strongly hydrophilic — absorbs 20 times its weight of cold water with swelling; forms strong gels at about 40 °C. Agar (sometimes called agar-agar) is a phycocolloid derived from red algae, such as *Gelidium* and *Gracilaria*. It is a polysaccharide mixture of agarose and agaropectin. Agar is used as a culture medium in microbiology and bacteriology; as an antistaling agent in bakery products; in confectionery; in meats and poultry; as a gelation agent; in desserts and beverages; as a protective colloid in ice cream; in pet foods, health foods; as a laxative, in pharmaceuticals; for making dental impressions; as a laboratory reagent; in photographic emulsions.

Calcium alginate

White or cream-colored powder, or filaments, grains, or granules. Slight odor and taste. Insoluble in water; insoluble in acids, but soluble in alkaline solutions. It is used in pharmaceutical products; as a food additive; as a thickening agent and stabilizer in ice cream, cheese products, canned fruits, and sausage casings also used in synthetic fibers.

Carrageenan

A yellowish to colorless, coarse to fine powder, practically odorless, but with a mucilaginous taste. Moderately soluble (1 gram in 100 milliliters of water at 27 °C), forming a viscous, clear, or slightly opalescent solution which flows readily. Carrageenan disperses in water more readily if first moistened with alcohol, glycerin, or a saturated solution of sucrose in water. Carrageenan is a hydrocolloid consisting mainly of a sulfated polysaccharide, the dominant hexose units of which are galactose and anhydrogalactose. It is a two-component, polyanionic colloid. The *kappa* and *lambda* components occur in varying proportions and degrees of polymerization and are associated with ammonium, calcium, potassium, or sodium ions, or with a combination of these four. Varying proportions alter the physical qualities of the substance. Carrageenan is obtained by extraction with water of members of the *Gigartinaceae* and *Solieriaceae* families of the class *Rhodophyceae* (red seaweed). The seaweed is also called Irish Moss and is prevalent off the coasts of Canada, New England, and New Jersey, but is found in other parts of the world. Carageenan is used as an emulsifier in food products, especially chocolate milk; in toothpastes, cosmetics, pharmaceuticals; as a protective colloid; and as a stabilizing aid in ice cream (0.02%).

Guar gum

Yellowish-white powder. Dispersible in hot or cold water. It possesses 5–8 times the thickening power of starch. Reduces friction drag of water on metals. Guar gum is obtained from the ground endosperms of *Cyanopsis tetragonoloba*, which is cultivated in Pakistan and used there as a livestock feed. The water-soluble portion of the flour (85%) is called *guaran* and consists of 35% galactose, 63% mannose, probably combined in a polysaccharide, and $5\frac{7}{8}$% protein. Guar gum is used in paper manufacture; cosmetics; pharmaceuticals; as an interior coating of fire-bose nozzles; as a fracturing aid in oil wells, in textiles, printing, polishing; as a thickener and emulsifier in food products.

Guiac gum

Moderate yellow-brown powder, becoming olive brown upon exposure to air. Odor is balsamic. Taste is slightly acrid. Dissolves incompletely but readily in alcohol, ether, chloroform, and in solutions of alkalies. Slightly soluble in carbon disulfide and benzene. Occurs as irregular masses enclosing fragments of vegetable tissues, or in large, nearly homogenous masses. Source is resin of the wood of *Guajacum officinale*, principally found in Central America.

Karaya gum

A pale yellow to pinkish brown, translucent, and horny gum with a slightly acetous odor and a mucilaginous and slightly acetous taste. In powdered form it is light gray to pinkish gray. Karaya gum is insoluble in alcohol, but swells in water to form a gel. Karaya gum is obtained as a dried gummy exudate from *Sterculia urens* and other species of *Sterculiaceae* family, or from *Cochlospermum gossypium*. It occurs in tears of variable size or in broken irregular pieces having a somewhat crystalline appearance. The properties depend upon freshness and time of storage. Viscosity greatly decreases over a 6-month period. The gum is used in pharmaceuticals, textile coatings, ice cream and other food products, adhesives; as a protective colloid, stabilizer, thickener, and emulsifier.

Locust bean gum (carob-bean gum)

White to yellowish-white, nearly odorless powder. It is dispersible in either hot or cold water, forming a sol, having a pH between 5.4 and 7.0, which may be converted to a gel by the addition of small amounts of sodium borate. It has a molecular weight of about 310,000. The gum swells in water, but viscosity increases when heated. Insoluble in organic solvents. The gum is extracted from the ground endosperms of *Ceratonia siliqua* of the *Leguminosae* family. The gum is used in foods as a stabilizer, thickener, and emulsifier; in packaging material, cosmetics, sizing and finishes for textiles, pharmaceuticals, paints.

Potassium alginate

Occurs in filamentous, grainy, granular, and powdered forms. It is colorless or slightly yellow and may have a slight characteristic odor and taste. Slowly soluble in water, forming a viscous solution; insoluble in alcohol. The gum is used as a thickening agent and stabilizer in dairy products, canned fruits, and sausage casings. It is variously used as an emulsifier.

Sodium alginate

A colorless or slightly yellow solid occurring in filamentous, granular, and powdered form. Forms a viscous colloidal solution with water, insoluble in alcohol, ether, and chloroform. It is extracted from brown seaweeds. The gum is used as a thickener, stabilizer, and emulsifier in foods, especially ice cream. Also used in boiler compounds, pharmaceuticals, textile printing, cement compositions, paper coatings, and in some water-base paints.

Tragacanth gum

Dull white, translucent plates or yellowish powder. Soluble in alkaline solutions, aqueous hydrogen peroxide solution; strongly hydrophilic; insoluble in alcohol. One gram in 50 milliliters of water swells to form a smooth, stiff, opalescent mucilage free from cellular fragments. It is obtained as a dried gummy exudate from *Astragalus gummifer*, or other Asiatic species of *Astragalus* (*Leguminosae* family). The gum is used in pharmaceutical emulsions, adhesives, leather dressings, textile printing and sizing, dyes, food products (notably ice cream and desserts), toothpastes; for coating soap chips and powders; and in hair wave preparations.

See separate entry on **Xanthan Gum.**

The algal polysaccharides resembled the relatively simplified structures of the neutral mucilages, as in the case of carrageenan. A wider spectrum of structures is found in the bacterial gums, which are generally of the highly branched type exuded by higher plants.

Food processing and other industrial applications of gums and mucilages take advantage of their physical properties, especially the viscosity and colloidal nature. They are substances of high molecular weight. For example, gum arabic has a molecular weight of 250,000 to 300,000. The gums and mucilages that possess relatively linear molecules, such as gum tragacanth, form more viscous solutions than the more spherically shaped gums, such as gum arabic, when at the same concentration. Consequently, for some applications, the gums with linear molecules are more economic to use. Due also to the elongated molecular shape of the seed gums and mucilages, the viscosity of their aqueous solutions varies widely with concentration. They exhibit structure viscosity. In contrast, the gums and mucilages of more spherical shape, i.e., the exudates, give solutions whose viscosities do not depend so much upon concentration.

Gums and mucilages influence each other. Mixing of two gums of the same viscosity may result in a mixture with a different viscosity. The viscosity of solutions of gums and the mucilages is dependent upon the pH, especially for those containing acid groups. In certain cases, the viscosity decreases upon standing as the result of enzymatic breakdown of the molecules. The molecules can undergo large changes in shape and size under the osmotic influence of opposing ions. Some of them, such as carrageenan from Irish Moss, can be fractionated by dilute salt solutions (potassium chloride) and the poly-β-glucosan from barley grain may be precipitated with ammonium sulfate. Gum arabic shows the phenomenon of coacervation when mixed with gelatin. See also **Coacervation**.

The specific uses of gums are wide and diverse. By way of a few examples, seaweed gums (e.g., carrageenan) and seed mucilages (guar gum) are used as stabilizers in dairy products, such as ice cream and certain cheeses. They are used in confectionery, in making jams, jellies, and in stabilizing citrus oil emulsions and salad dressings. They have been used as fixatives for 2,3-butanedione in the baking industry. Outside the food field gums and mucilages find scores of applications.

In 1974, the Northern Regional Research Center (Peoria, Illinois) of the U.S. Department of Agriculture and the Kelco Company were joint recipients of the Institute of Food Technologists award for the development and commercialization of xanthan gum. See also **Xanthan Gum**. This gum differs by virtue of its production by pure-culture fermentation of a carbohydrate as contrasted with refining a naturally occurring substance.

See list of references under **Colloidal Systems**. A particularly good reference covering the physical properties and procedures for testing various gums and mucilages is 'Food Chemicals Codex,' published by the National Academy of Sciences, Washington, DC, (revised periodically).

Additional Reading

Nussinovitch, A.: *Hydrocolloid Applications: Gum Technology in the Food and Other Industries*, Chapman and Hall, New York, NY, 1999.
Satterlee, D.: www.herbaldave.com/Herbs/Phytochemicals/Classes/Gums+Muculage.htm, 2000.
Williams, P. and G. Phillips: *Gums and Stabilizers for the Food Industry*, Springer-Verlag, Inc., New York, NY, 1998.

GUM TREES. See **Eucalyptus Trees**.

GUPPY. See **Viviparous Topminnows**.

GURNARDS (*Osteichthyes*). Of the suborder *Dactylopteroidea*, family *Dactylopteridae*, these are tropical marine fishes frequently called flying gurnards because of their apparent ability to propel themselves out of water. There is little documentation available to indicate their abilities at flight as in the instances of the true flying fishes. See also **Hatchet Fishes**.

The *flying gurnards* are characterized by greatly developed pectoral fins, the rear portion of which has become a large, wing-like structure. The front of the pectoral fins is short. There are two long individual spines in front of the first dorsal fin. The body is elongate with firmly attached scales. The pre-opercle gill cover has strong spines; the opercle has no spine. The jaws bear small teeth. Flying gurnards are very similar to sea robins. See also **Sea Robbins**. However, they differ from them in the arrangement of the skull bones. The snout is short and very steep. The top of the skull is flat. The gill openings are very small. One species is found in the Atlantic Ocean and Mediterranean Sea, while three species are found in the Indo-Pacific region. Flying gurnards prefer warm-to-subtropical seas.

Juvenile flying gurnards, with small pectoral fins, and adults, with wing-like pectoral fins, are so different that the young were once classed in another genus. In older studies, the flying gurnards were often confused with the flying fishes. See entry on **Flying Fishes**. The chief enemies of flying gurnards are sea breams and mackerel; but while in the air they are fed upon by frigate birds, gulls, white-tailed sea eagles, procellariids, and tropical birds.

Many travelers have reported flying gurnard schools some 13 to 16 feet (4 to 5 meters) in the air for flights extending up to 300 feet (90 meters). This spectacle is repeated continuously. One group flies out of the water, leans forward, and then disappears again into the sea, while a second group has already shot into the air; then comes a third, and so forth. When flying gurnards leap out of the water at night, they glow with a phosphorescent light. When the sea is calm, the rushing sound of their beating pectoral fins can be heard, as well as the whistling sound of the air shooting through the gill openings.

Can flying gurnards actually fly? Most marine researchers say not. They claim that these fishes could never lift themselves out of the sea and fly over it because of the armored, spiny skull, the heavy body with its thick

scales, and the caudal fin with its two small tips. Further detailed research is required to form a definite conclusion in the opinion of other authorities.

GUSSET PLATE. A gusset plate is a flat plate connecting two or more structural members where they meet at a joint. See Fig. 1. Stress is transferred between the members through the gusset plate by riveted, bolted, or welded connections. A gusset plate should be of a shape giving a minimum waste of material, and which can be fabricated in the shop with minimum amount of labor. For this reason it should be cut with straight edges. The thickness of a gusset plate should be sufficient to give bearing value, so that the material or the rivet will not be crushed. Minimum thicknesses of gusset plates are usually $\frac{1}{4}$-inch (6 millimeters) for inside protected structures and $\frac{3}{8}$-inch (9 millimeters) for outside exposed structures. The area between rivet holes should be great enough to transmit the stress from one member to another. Examples of gusset plates are to be found in all types of welded and riveted steel structures, and in gussets which strengthen and make the joints in the rib structure of an airplane wing.

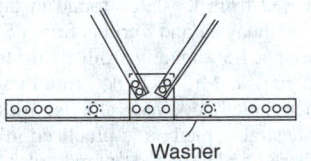

Washer

Fig. 1. Gusset plate at joint.

GUST. See **Winds and Air Movement**.

GUST FRONT. Thunderstorms and some showers are accompanied by small-area but frequently intense rain and sometimes hail. The precipitation originates well up in the cumulonimbus clouds and cascades earthward, accompanied by a downdraft of cold air which arrives at the earth's surface significantly colder than environing air. The temperature difference may be as much as 27 °F (15 °C). See Fig. 1.

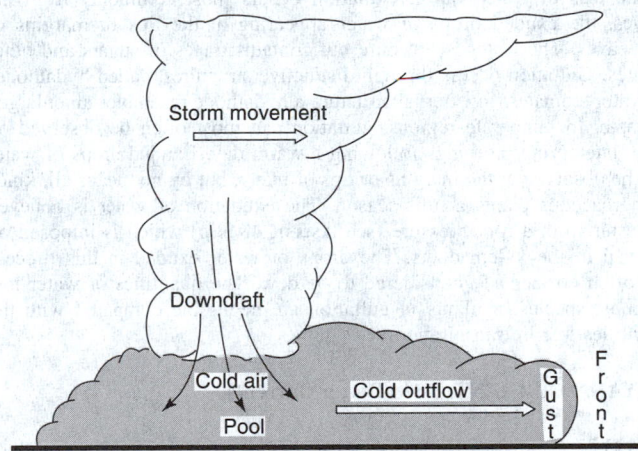

Fig. 1. Schematic cross section showing the mechanics of a gust front. Vertical and lateral dimensions are not to scale.

The cold air accumulates under the downdraft and forms a pool of air which is heavier (more dense) than the environing air by reason of its lower temperature. Very quickly the cold air begins to flow away from the area of accumulation under the influence of gravity, that is, a gravity-induced flow of a heavier fluid into a region of lighter and less dense fluid. The leading edge of the outflowing cold air becomes a *gust front*, along which there is a wind shift, often vigorous, and a temperature drop.

Gust fronts tend to be most vigorous near the cold air source region and diminish as they move outward and away. The most intense wind shift is usually on the side toward which the storm is moving. Gust fronts have been observed as many as 15 miles (24 kilometers) from the parent storm.

Gust fronts associated with a line of thunderstorms tend to form a common front and move as a *squall line*, triggering new thunderstorms as the front moves.

See references listed at ends of entries on **Climate**; and **Meteorology**.

PETER E. KRAGHT, Consulting Metereologist, Mabank, TX

GUTENBERG, BENO (1889–1960). Gutenberg was a German-born United States geophysicist who contributed to the geophysics field during the 20th century. He used the velocities of seismic waves to calculate the depth of the Earth's core at 2,900 km/ 1,812 mm. The boundary between Earth's mantle and Earth's core is called the Gutenberg discontinuity after him. Gutenberg suggested in 1948 the existence of a low-velocity zone below the Earth's surface in which seismic waves travel more slowly. This zone is now known to be the asthenosphere.

See also **Astenosphere**; and **Earth**.

J. M. I.

GUTTA PERCHA (*Palaquium Gutta,* and related species; *Sapotaceae*). Gutta percha is prepared from the latex found in the stem and leaves of certain trees native in Malaysia and various South Sea Islands. To obtain the latex, which does not flow readily from living trees, the tree may be felled and a series of rings cut in the bark. From these the latex oozes and may be gathered. Such a method is naturally very destructive to continued production. A more desirable method is practiced in plantations of today. Fresh leaves are gathered and chopped up and crushed. The crushed mass is then boiled in water and the gum removed and pressed into blocks.

In South America a related tree, *Mimusops Balata* (*Sapotaceae*) yields a similar gum of somewhat inferior quality. This tree is usually tapped by cutting a row of zigzag gashes that connect one with another. Down these the latex flows, to be gathered in a cup at the bottom, and later coagulated in trays.

Gutta percha is a yellowish or brownish somewhat leathery solid containing up to 90% of a hydrocarbon gutta. On heating, it becomes plastic and is very resistant to water.

GUTTATION. The loss of liquid water from intact plants is called guttation. This process should not be confused with transpiration, which is the loss of water vapor. Guttation occurs most commonly from the leaves, the exuded drops of water appearing at the tips or margins of the leaves. The water is not pure but contains traces of sugars and other solutes. Guttation occurs through distinctive structures, called hydathodes or water stomates. In external structure a hydathode resembles an enlarged stomate. In temperate regions, guttation can most often be observed on cool, late spring mornings following a warm day. Exuded drops of water can be observed at the margins or tips of many, but by no means all, kinds of herbaceous plants at this season. The exudation of water is believed to result from a root pressure (see **Ascent of Sap**) which is imposed on the sap in the xylem ducts. The drops of water exuded in this process are often erroneously considered to be dew. The quantities of water lost by most species of plants in guttation are negligible compared with the quantities lost in transpiration.

GUYANA CURRENT. See **Ocean Currents**.

GUYOT. A guyot is a flat-topped seamount rising from the ocean floor like a volcano but planed off on top and covered by appreciable water depth. Guyots show evidence of having been above the surface with gradual subsidence through stages from fringed reefed mountain, coral atoll, and finally a flat topped submerged mountain. Guyots are very commonly found in the Pacific Ocean, and are considered to be extinct volcanoes. The Emperor Seamounts are an excellent example of an entire volcanic chain undergoing this process and contain many guyots among their older examples. See also **Seamounts**.

Guyots were first identified by Harry Hess[1] who collected data using echo-sounding equipment on a ship he commanded during World War

II. The data showed the configuration of the seafloor where he saw that some undersea mountains had flat tops. Calling these guyots (after the 19th century geographer Arnold Henry Guyot) Hess postulated they were once volcanic islands that were beheaded by wave action yet they are now deep under sea level. This idea was used to help bolster the theory of plate tectonics.

GYMNOPHIONA (*Caecilians*). The order *Gymnophiona* (also sometimes called *Apoda* or *Caecilia*) is a group of highly specialized tropical fossorial (burrowing) amphibians. Because most are burrowers, they are rarely seen and, until the last two decades, little was known of their biology. They have no widely used common names and are generally called caecilians. They are found in tropical South America, Africa, the Seychelles, the Indian subcontinent and parts of southeast Asia, but are absent from Madagascar and Australasia.

Caecilians share the characteristics of *Lissamphibia* including a pulmocutaneous circulatory system associated with respiratory gas exchange through the skin, toxin-secreting granular glands in the skin, pedicellate teeth, very reduced straight ribs, and vertebrae composed predominantly of membrane bone without a cartilage precursor. Caecilians resemble earthworms superficially, having long, apparently segmented, bodies, tiny eyes and no tympana, no limbs or girdles, and a short or absent postcloacal tail. See also **Annelida**; and **Amphibia**.

The skin is smooth and contains mucous and granular glands. It is arranged in annuli (rings) separated by annular grooves which range from two per segment in primitive caecilians to one per segment in advanced forms. In primitive caecilians, these grooves contain rows of thin bony disc-like dermal scales, homologous to the dermal scales of fishes. Such scales are absent in other living amphibians and in various advanced caecilians, having apparently been lost more than once. See also **Fishes**; and **Osteichthyes(Bony Fishes)**.

The skull is highly modified, apparently as a consequence of the role of the head in burrowing. It is fairly compact and cylindrical, and several elements fuse during development. The orbit is tiny and ahead of it is the tentacular fossa for a sensory tentacle, this opening sometimes being larger than the orbit, and sometimes confluent with it. The skull roof is solid, without temporal fossae, except in the primitive Rhinotrematidae. The palate is dominated by a broad parasphenoid and the occiput is largely made up of a single composite os basale. Advanced caecilians have a long retroarticular process on the mandible. This is associated with a unique jaw-closing mechanism in which the jaw is closed by a downward and backward pull on the retroarticular process by the muscularis interhyoideus, which arises from the lower trunk wall. The dentition is unusual in that there is a double row of teeth on upper and lower jaws. The upper tooth rows are on the premaxillae and maxillopalatine outer edges, while the inner tooth row is on the vomers and inner maxillopalatine ramus. On the lower jaws, the outer tooth row is on the dentary and the inner tooth row is on the splenial. These tooth rows interlock, the dentary row between upper rows, to give a crimping bite. The teeth are pedicellate as in most lissamphibians, but vary greatly in crown structure. Some are bicuspid — the typical lissamphibian condition — but others are monocuspid. They are frequently curved medially and have sharp cutting edges. In viviparous forms, the fetus has a distinctive dentition in which the crowns are spatulate with 5–16 tiny spiky cusps per tooth.

The eyes are reduced but are always functional as light-sensitive structures. The lens varies from functional to absent and some of the musculature may be lost, but parts of the optical system are modified to support a sensor that is characteristic of the group — the sensory tentacle, found in a pit just ahead of the orbit. The sensory tentacle is olfactory and can be extended ahead of the animal. Its retractor muscles and innervation are modified from those found in the orbits of other vertebrates and it is lubricated by the Harderian gland, squeezed by the levator bulbi muscle. In scolecomorphids, the association is so close that the eye is embedded in the tentacle and can be protruded with the tentacle.

The trunk is long, 95–285 trunk vertebrae, and the postcloacal tail varies from 12 vertebrae in length in primitive forms to absent in most advanced forms in which the cloaca is terminal. The ribs are small and double-headed. There is no trace of pectoral or pelvic girdles, or limbs, in any modern caecilian. The fundamental locomotor mechanism appears to

[1] Harry Hammond Hess (1906–1969) was an American geologist. He is best known for his theories on sea floor spreading, specifically work on relationships between island arcs, seafloor gravity anomalies, and serpentinized peridotite, suggesting that the convection of the Earth's mantle was the driving force behind

this process. This work provided a conceptual base for the development of the theory of plate tectonics.

be internal concertina-crawling in burrows and lateral undulation on the surface. Aquatic typhlonectines use only lateral undulation. The internal concertina-crawling is, in part, powered hydrostatically by vertical body muscles squeezing body fluid and causing the body to become rigid and elongate. A crossed-helix array of tendons turns this pressure into a powerful forward thrust of the head, necessary for pushing through soil.

Respiration may be by buccopharyngeal, cutaneous and/or pulmonary gas exchange. Caecilians vary between possession of symmetrical lungs, asymmetrical lungs, one lung or no lungs. Buccopharyngeal breathing is certainly fundamental for the group but the long body does provide a more favorable surface–volume ratio for cutaneous exchange than in most amphibians.

Fertilization is internal, the male having an intromittent organ comprising an eversible cloaca. Primitive caecilians lay large yolky eggs in a burrow in wet ground near streams, sometimes guarding the eggs until they hatch. The larvae are aquatic and retain a lateral-line system, one or more pairs of gill slits but do not have external gills. In advanced caecilians, viviparity predominates, with up to 20 young developing in the mother's oviducts. The fetuses respire by large external gills which are appressed to the vascular oviduct wall of the mother. When their yolk sacs are exhausted, they continue to feed by rasping at secretory tissue in the mother's oviduct wall, which produces a milk-like nutrient. The fetal dentition is an adaptation for this. They metamorphose before birth, losing the gills and larval teeth and acquiring the adult dentition.

Diversity and Lifestyles

In the late 1990s, caecilian diversity was estimated at 164 species. Most species grow to 300 mm (12 inches) but some reach only 50 mm (2 inches) and one species of *Caecilia* reaches 1500 mm (59 inches). Most live in wet soil or leaf-litter and are opportunistic carnivores, feeding on worms, arthropods and small vertebrates. Caecilians have skin toxins but are recorded as prey for snakes and birds. Five families are recognized, four being small families, all more primitive than the Caeciliidae, and all endemic to specific continents.

The Rhinatrematidae are small burrowing forms from South America. Where known, they have aquatic larvae. The Ichthyophiidae and Uraeotyphlidae are small–medium burrowers from India (both families), Sri Lanka and southeast Asia (ichthyophiids). They lay large yolky eggs and have aquatic larvae. The Scolecomorphidae is a small family of medium-sized viviparous burrowers from subSaharan Africa. It is monogeneric, comprising the highly derived genus *Scolecomorphus*, which has a very modified skull suggestive of a different burrowing mechanism.

The Caeciliidae is a large and diverse family found across tropical South America, Africa, the Seychelles and the Indian subcontinent. It contains about two-thirds of the known species and is currently divided into three subfamilies: the Caeciliinae for a few South American genera with enlarged upper marginal dentition; the Typhlonectinae for a group of specialized aquatic genera, also from South America (see below); and the Dermophinae which is a large, rather miscellaneous, grade of genera of obscure relationships. The family diversity extends to patterns of reproduction, including both oviparous and viviparous forms. Members of the subfamily Typhlonectinae are large (700 mm (28 inches)) aquatic caeciliids. The posterior trunk is laterally compressed for swimming. Despite their aquatic lifestyle, they do not have free-living larvae. They are viviparous, with the young developing in the oviduct and respiring through unique large membranous external gills that are resorbed before birth. Typhlonectines are believed to be secondarily aquatic and derived from burrowing caeciliid ancestors.

Fossil Record and Phylogeny

Morphological analysis of the relationships of living amphibians suggests that caecilians are slightly more distantly related to frogs and salamanders than the latter groups are to each other, whereas molecular analyses indicate that caecilians are closer relatives of salamanders. They might have diverged from other amphibians at any time between the late Permian and the end of the Triassic. The earliest fossil form is *Eocaecilia* from the Lower Jurassic Kayenta Formation of Arizona. This was already a long-bodied burrower, with a caecilian-type skull bearing a tentacular fossa. However, it retained girdles and small limbs, pleurocentrum-dominated endochondral vertebrae and several primitive skull elements lost in modern forms. The only later fossils described to date are isolated caeciliid-type vertebrae from the Mid-Cretaceous of the Sudan and the Palaeocene of Brazil and Bolivia. All our understanding of caecilian phylogeny comes from analyses of the modern taxa. See also **Amphibia**; and **Salamanders**.

Diversification of the caecilian crown group appears to have largely taken place in Gondwana during the Mesozoic. The presence of caecilians on the Seychelles—a Gondwanan platelet isolated since the Cretaceous—is the most obvious testimony of this. The most primitive living family is the South American Rhinatrematidae, which retains temporal fossae between the parietals and squamosals like other lissamphibians but unlike other caecilians. It is followed by an Asian clade comprising the Ichthyophiidae and Uraeotyphlidae which share with higher caecilians the possession of a solidly roofed skull. All three families retain short tails and dermal scales and all are oviparous with aquatic larvae. The higher caecilians, the scolecomorphids and caeciliids, have gained a long retroarticular process on the mandible and lost or fused the pterygoid in the palate. They have all lost the tail and most have lost the scales; most are viviparous though some caecilids remain oviparous. See also **Mesozoic**.

It appears that the South American rhinotrematids, the Asian ichthyophiids and uraeotyphlids, and the African scolecomorphids are the relicts of a first radiation within Gondwana, while the Caeciliidae represent a second radiation, overlapping all the more primitive families in range. Both radiations took place before the separation of South America, Africa, the Seychelles and India as evidenced by their present distributions. The southeast Asian ichthyophiids are assumed to represent range extensions from India after it fused with the rest of Asia in the late Cretaceous or early Tertiary.

Additional Reading

Billo, R., and M.H. Wake: "Tentacle Development in *Dermophis mexicanus* (Amphibia, Gymnophiona) with an Hypothesis of Tentacle Origin," *Journal of Morphology*, **192**, 101–111 (1987).

Bruce, R.C., L.D. Houck, and R.G. Jaeger: *Biology of Plethodontid Salamanders*, Springer-Verlag New York, LLC, New York, NY, 2000.

Duellman, W.E., and L. Trueb: *Biology of Amphibians*, McGraw-Hill, New York, NY, 1986.

Evans, S.E., A.R. Milner, and C. Werner: "Sirenid Salamanders and a Gymnophionan Amphibian from the Cretaceous of the Sudan," *Palaeontology*, **39**, 77–95 (1996).

Hedges, S.B., R.L. Nussbaum, and L.R. Maxson: "Caecilian Phylogeny and Biogeography Inferred from Mitochondrial DNA Sequences of the 12 S rRNA and 16 S rRNA Gene," *Herpetological Monographs*, **7**, 64–76 (1993).

Jenkins, F.A., and D.M. Walsh: "An Early Jurassic Caecilian with Limbs," *Nature*, **365**, 246–250 (1993).

Lever, C.: *Naturalized Reptiles and Amphibians of the World*, Oxford University Press, New York, NY, 2003.

O'Reilly, J.C., R.A. Nussbaum, and D. Boone: "Vertebrate with Protrusible Eyes," *Nature*, **382**, 33 (1996).

O'Reilly, J.C., D.A. Ritter, and D.R. Carrier: "Hydrostatic Locomotion in a Limbless Tetrapod," *Nature*, **386**, 269–272 (1997).

Summers, A.P., and J.C. O'Reilly: "A Comparative Study of Locomotion in the Caecilians *Dermophis mexicanus* and *Typhlonectes natans* (Amphibia: Gymnophiona)," *Zoological Journal of the Linnean Society*, **121**, 65–76 (1997).

Wake, M.H.: "Fetal Maintenance and its Evolutionary Significance in the Amphibia: Gymnophiona," *Journal of Herpetology*, **11**, 379–386 (1977).

Wake, M.H.: "The Comparative Morphology and Evolution of the Eyes of Caecilians (Amphibia, Gymnophiona)," *Zoomorphology*, **105**, 277–295 (1985).

Zug, G.R., L.J. Vitt, and J. Caldwell: *Herpetology: An Introductory Biology of Amphibians and Reptiles*, 2nd Edition, Elsevier Science & Technology Books, New York, NY, 2001.

ANDREW R. MILNER, Birkbeck College, London, UK

GYMNOSPERMS. The characteristic feature of the gymnosperms is the occurrence of the ovule on the surface of the scale that bears it, and not surrounded by an ovary wall. In most gymnosperms, the reproductive bodies are borne in cones. The gymnosperms are the most primitive of seed plants. Arising early in geological time, these plants became abundant and widespread in the Carboniferous period. From that period to the present, gymnosperms have decreased in numbers, many groups becoming entirely extinct. There remain some 500 species, occurring in nearly all parts of the world, but attaining their greatest development in the temperate zones. They often form a dominant forest tree.

The gymnosperms are woody plants. The majority of them are trees, often attaining immense size, as exemplified by the Giant Sequoias of California. See also **Giant Sequoia**. However, a few are low shrubby plants, and a very small number of vine-like species still exist. Nearly all gymnosperms are plants of xerophytic habit, that is, fitted to survive in regions in which water is not abundant. Some, like the Welwitschia of the arid deserts of southwestern Africa, live in regions where the annual rainfall is less than 0.5 inch (12 millimeters).

GYMNOSPORE. A sexual reproductive cell which is naked and capable of active locomotion by amoeboid movement or by cilia or flagella.

GYMNOTID EELS *(Osteichthyes)*. These eels, along with knifefishes and the electric eel, are members of the order *Ostariophysi* (which includes characins, minnows, and catfishes), and the family *Gymnotidae*. They are not true eels. Characteristics of the gymnotids include: (1) diminutive beady eyes; (2) no true dorsal fin with fin rays; (3) presence of a long, undulating anal fin, extending the greater length of the fish; (4) thin cylindrical body sometimes resembling a ribbon; and (5) a thin, often pointed tail. The long tail, of course, accounts for the extreme ability of the gymnotids to move in all directions speedily and easily. Gymnotids essentially are habitants of Central and South American waters, southward at least to Paraguay. There are probably less than 50 species of gymnotids, of which there are four convenient groups: (1) the *Rhamphichthys rostratus*, a food fish that may attain a length up to about $4\frac{1}{2}$ feet (1.4 meters); (2) the knifefishes (*stenarchids*), some of which are sought by tropical fish hobbyists; (3) other knifefishes, including the banded knifefish (*Gymnotus carapo*); and (4) *Electrophorus electricus*, the well known electric "eel."

The electric organs of the electric eel are so powerful that it appears to have no enemies other than people. As an air breather, the fish must surface about every 15 minutes. Rather than lungs or truly functioning gills, this fish has a unique tissue lining in its mouth, which permits obtaining oxygen directly from air. Thus the fish can be left out of water for many hours as long as moisture is provided to keep the special tissue moist. Advantage has been taken of this fact by experimental biologists.

Electrically, the fish is positive toward the head; negative toward the tail—just the opposite of the electrical profile of the electric catfish. Authorities have recorded outputs as high as 650 volts, but the average is about 350 volts for a 3-foot-long eel. The ability to generate voltage levels off with age, but amperage increases slightly. The electric eel possesses a combination of battery power. The principal battery occupies most of the body of the fish and creates the highest voltage. Discharge of this battery takes the form of a train of waves of about 0.002- second duration each. The train may consist of six or more waves, each varying some in time interval and voltage.

Because the amperage is low (0.5 to 0.75 amperes), a shock from an electric eel is not necessarily lethal, depending of course upon the size and physical characteristics of the victim.

Although electric eels have been kept in captivity, they have not been bred. Apparently because of a protective antibiotic exuded by the electric eel, they survive best in water that is not frequently changed. The electric eel has effective eyes when young, but these tend to become cloudy with age and it is theorized that this may be due to the effects of electrical discharges by other eels. Thus, the older electric eels must use their electrical form of detection to find potential sources of nourishment. The electric eel is found in the Amazon River and tributaries.

GYNANDROMORPH. An abnormal individual whose body shows the characteristics of the two sexes in different parts. Not synonymous with hermaphrodite although this term is sometimes applied to these abnormalities. It is due to abnormalities in the distribution of the chromosomes, especially the sex chromosome, in cell division during development.

Gynandromorphs are fairly common among the insects, where they are often of the bilateral type. Such individuals have one side of the body male and the other female, with a sharp boundary in the median line. Mosaic gynandromorphs present an irregular distribution of the sexual characters.

GYNECOLOGY. The study, diagnosis and treatment of diseases and disorders of the female genital organs.

GYPSUM. The mineral gypsum is hydrous calcium sulfate, $CaSO_4 \cdot 2H_2O$. It occurs as flattened monoclinic crystals, often twinned, transparent cleavable masses, called selenite, or silky and fibrous, called satin spar; it may also be granular or quite compact. It is a soft mineral, hardness 2; has two good cleavages, which yield rhombic plates whose angles are 66° and 114°. Its specific gravity is 2.31–2.33; luster, vitreous to silky or pearly; color, colorless to white and gray, may be tinted red, yellow, blue, brown, etc., by impurities; transparent to opaque. A very fine-grained white or lightly tinted variety of gypsum is called alabaster, and prized for ornamental work of various sorts.

Gypsum is a very common mineral, thick and extensive beds of which are associated with sedimentary rocks. The largest deposits known occur in strata of Permian age. Besides being a result of deposition in sea and lake waters, gypsum has been deposited by hot springs, from volcanic vapors, and by sulfate solutions in veins. Notable localities for gypsum are in Greece, the Czech Republic and Slovakia, Austria, Saxony, Bavaria, Italy, France, Spain, England and Mexico. In the United States, well-known localities are at Lockport, New York; the Mammoth Cave, Kentucky; Ellsworth, Ohio; Grand Rapids, Michigan; Hermosa, South Dakota; Wayne County, Utah; and San Bernardino County, California. In Canada, the Provinces of New Brunswick and Nova Scotia have large gypsum deposits. Because the gypsum from the quarries of the Montmartre district of Paris has long furnished burnt gypsum used for various purposes, this material has been called plaster of Paris.

Often, there is confusion between the mineral gypsum, $CaSO_4 \cdot 2H_2O$, and the useful product of partial dehydration, $CaSO_4 \cdot 1/2H_2O$. See Table 1. There are numerous commercial products based upon gypsum. *Plaster*, made from gypsum, is widely used for the economical fabrication of building products. Importantly, the setting time of gypsum plaster can be carefully controlled through the addition of fractional percentages of *accelerators* (typically water-soluble salts, such as K_2SO_4, or finely-ground gypsum) and *retarders*, which frequently are modified organic substances, such as glue, casein, blood, hair, and hoof meal; or citric, boric, and phosphoric acids and their salts. Accelerators are believed to function by providing additional nuclei for crystallization, whereas retarders are believed to provide protective colloids or insoluble salts which block water access to the plaster particle. A controlled rate of reaction can be obtained by incorporating a combination of retarders and accelerators in the gypsum plaster mix.

Wallboard (Sheetrock) is a large single user of gypsum. The product usually consists of a core of gypsum sandwiched between two layers of paper. Characteristics of the product include fire resistance, dimensional stability, low cost, and easy workability. Wallboard conventionally measures $\frac{1}{2}$ inch (1.3 centimeters) thick, 48 inches (1.2 meters) wide, and 8 to 20 feet (2.4 to 6 meters) in length. In manufacture, foamed plaster slurry is mixed and discharged on a moving web of paper. The edges of the bottom paper are scored and folded so that the slurry is completely contained between that sheet and the top paper, which is laid on the slurry. The paper surfaces not only provide strength and paintability to the finished board, but also form a continuous mold within which the gypsum is cast. The board machine operates continuously. Within five minutes after forming, the gypsum is sufficiently hard to be cut, after which the sheets are dried further before storage and shipment. Fibers may be added to provide crack resistance and additional fire resistance. Water-repellent chemicals may be added to the board core or to the paper surface. Also, decorative and functional finishes may be factory-applied.

Industrial plasters of a gypsum base include dental plasters, used in making tooth impressions, orthopedic plasters for immobilizing broken bones, pottery plasters, oil-well cements, permeable plasters for casting nonferrous metals, art and statuary casting, lamp bases, patching and grouting compounds, insulating-brick production, and pattern and model making for the aircraft and automotive industries. Water-reducing additives and reinforcing resins and cements may be added to achieve a compressive strength of over 15,000 pounds per square inch (1021 atmospheres).

Portland cement also consumes large quantities of gypsum. About 5% of gypsum is added to the cement clinker before grinding. Addition of gypsum aids in increasing the early strength of the cement and prevents undesirable false set.

Agriculturally, gypsum serves as a soil conditioner, providing a source of available calcium and sulfate, assisting the retention of organic nitrogen, without the addition of acidity or alkalinity to the soil. Gypsum is widely used in areas where the soils are deficient in sulfur. Gypsum also has been used in mixed fertilizers and animal feeds.

Terra alba or dead-burned, fine white gypsum is used as a paper filler, in plastics, and as an extender for titanium dioxide. Pharmaceutically pure gypsum can be added to bread and other bakery products, finds use in

TABLE 1. TERMINOLOGY AND PROPERTIES OF CALCIUM SULFATE-WATER COMPOUNDS

Chemical Formula	Designations Commonly Used	Properties
$CaSO_4 \cdot 2H_2O$	Calcium sulfate dihydrate; rock gypsum; chemical gypsum; alabaster (white fine-grained); selenite (translucent platey); satin spar (fibrous); land plaster (pulverized gypsum)	All forms (natural, synthetic, and recrystallized) are thermodynamically and crystallographically equivalent. Habit may be needles, plates, or prisms.
$CaSO_4 \cdot 1/2H_2O$	Calcium sulfate hemihydrate; calcined gypsum; stucco; plaster of Paris; molding plaster; gypsum plaster; chemical hemihydrate.	Alpha and beta types exist, depending upon conditions of calcination. Alpha type is more stable, crystalline, of lower energy. Beta type is less stable, disordered, of higher energy.
$CaSO_4$	Anhydrite	
I	Anhydrite 1: high-temperature anhydrite.	Produced by high-temperature ($>1,000\,°C$) calcining. Contains free CaO.
II	Anhydrite II: insoluble anhydrite; inactive anhydrite; dead-burned gypsum; chemical anhydrite; mineral anhydrite.	Produced by calcining at $250-1,000\,°C$. Relatively inert. Reactivity depends upon calcining-time-temperature relationship and particle size.
III	Anhydrite III: soluble anhydrite; active anhydrite; dehydrated hemihydrate.	Produced by low-temperature ($175-250\,°C$) dehydration of hemihydrate. Reacts vigorously with water and moist air to form hemihydrate.

Source: United States Gypsum Company. Des Plaines, Illinois.

beer production, and as a pharmaceutical-tablet diluent. In Japan, calcium sulfate is used in making *tofu*, a soyabean curd.

Gypsum may be a potential source of sulfur and sulfuric acid. Some European plants make Portland cement and sulfuric acid from gypsum or anhydrite. In the Muller-Kuhne process, gypsum is mixed with clay and silica in quantities necessary to make cement, along with coke to reduce $CaSO_4$ to CaO. In equipment similar to that for portland-cement manufacture, the SO_2 is driven off and converted to sulfuric acid by the contact process.

Additional Reading

Coburn, A. et al.: *Gypsum Plaster: Its Manufacture & Use*, Intermediate Technology Publications, London, UK, 1989.
Gerhartz, W. (Editor): *Benzyl Alcohol to Calcium Sulfate*, Vol. 4, John Wiley & Sons, Inc., New York, NY, 1985.
Staff: *Cement, Lime, Gypsum*, American Society for Testing & Materials, West Conshohocken, PA, 2000.

Web References

Natural Resources Canada, Minerals and Metals Sector: http://nrcan.gc.ca/mms/efab/mmsd/minerals/gypsum.htm, 1999.
Technology Information, Forecasting & Assessment Council (TIFAC): http://www.tifac.org.in/offer/tlbo/rep/TMS149.htm

GYPSY MOTH *(Insecta, Lepidoptera)*. A moth, *Lymantria (Porthetria) dispar*, introduced from Europe and now a serious pest in the northeastern United States. The caterpillars are able to defoliate shade and forest trees and also attack apple trees and sometimes the conifers. The damage and control are the same as in the case of the brown-tail moth.

The female moth does not fly. It measures about 2 inches (5 centimeters) from wing tip to wing tip, has black markings on the wings, and is creamy white in color. Usually from 300 to 500 eggs are deposited on the underside of a branch, in the bark of a tree, or along tree roots where they are hidden from view. The larvae feed on leaves and can cause serious damage. After the caterpillars transform to pupae, they soon emerge as adult insects, requiring a period of about ten days. Several insects help to control the population of the gypsy moth, but nevertheless effective means of eradicating the insect are under intense investigation. One approach under study is that of destroying the reproductivity of the insect.

GYRE. See **Ocean Resources (Energy)**.

GYRES. Oceanic current systems of planetary scale driven by the global wind system. The subtropical gyres are driven by the trade winds and by the westerlies of the temperate regions, the subpolar gyres by the westerlies and the polar easterlies. Gyres consist of a narrow, swift-flowing western boundary current, an eastward-flowing zonal current, a broad and slow-moving eastern boundary current, and a westward flowing zonal current.

Eight gyres are distinguished in the World Ocean: In the Atlantic, the Brazil, South Atlantic, Benguela, and South Equatorial Currents form the subtropical gyre of the Southern Hemisphere; the Gulf Stream, Azores, Canary, and North Equatorial Currents form the subtropical gyre in the Northern Hemisphere; the Labrador, North Atlantic, Irminger, and East Greenland Currents form the subpolar gyre. In the Pacific, the East Australian, South Pacific, Peru/Chile, and South Equatorial Currents form the subtropical gyre of the Southern Hemisphere; the Kuroshio, North Pacific, California, and North Equatorial Currents form the subtropical gyre of the Northern Hemisphere; the Oyashio, Aleutian, California, and Alaskan Currents and the Alaskan Stream form the subpolar gyre; a second subpolar gyre exists in the Bering Sea. In the Indian Ocean, the Agulhas, South Indian, West Australian, and South Equatorial Currents form the only subtropical gyre. See also **Ocean Currents**; **Ocean Resources (Energy)**; and **Winds and Air Movement**.

GYROMAGNETIC RATIO. Two important uses of this term are:
1. The ratio of the magnetic moment of a system to its angular momentum.
2. The ratio of moment of momentum to magnetic moment. An electron traveling around a circular orbit f times per second generates a magnetic moment equal to the product of the orbit area and the equivalent current:

$$\mu_0 = ef\pi r^2/c$$

Since the charge is negative, the mechanical angular momentum is in the opposite direction and has the magnitude

$$L_0 = O\pi fmr^2$$

yielding the gyromagnetic ratio, for orbital motion

$$G_0 = \frac{\mu_0}{L_0} = \frac{e}{2mc}$$

The factor c disappears throughout when mksa units are used. For an electron spinning about its own center, the quantum-theory values of magnetic moment and mechanical angular momentum yield

$$G_s = 2G_0 = e/mc$$

twice that for orbital motion, leading to a g factor that has a magnitude of 2. Similarly, nuclear gyromagnetic ratios are ratios of magnetic moment and angular momentum for atomic nuclei.

GYROSCOPE. A heavy symmetrical disk free to rotate about an axis which itself is confined within a framework that is free to rotate about one axis or two. The two qualities of a gyroscope that account for its usefulness are: (1) the axis of a free gyroscope will remain fixed with respect to space, provided that no external forces act upon it; and (2) a gyroscope can be made to deliver a torque (or a signal) proportional to the angular velocity about a perpendicular axis. Both qualities stem from the principle of conservation of angular momentum, which may be stated as follows: in any system of particles, the total angular momentum of the system relative to any point fixed in space remains constant, provided no external forces act on the system.

Gyroscopes are frequently spoken of as having one or two degrees of freedom, or as being *free gyroscopes*. This terminology is confusing because it results from the conventional use of the number of degrees of freedom of the vector of angular momentum rather than from the actual degrees of rotational freedom. Figure 1a shows diagrammatically the mounting of what is commonly called a *single-degree-of-freedom*, or "rate," gyroscope. Although there are obviously two rotational axes involved, in its use it is a single-degree-of-freedom system. Figure 1b illustrates the gimballing arrangement for what is sometimes called a *two-degree-of-freedom* gyroscope. As can be seen, a gyro wheel so mounted has three degrees of rotational freedom, except when all three axes are in the same plane. When the measurements of motion are made only from two coordinate axes, or when the outer axes lie in the same plane, this arrangement is frequently called a two-degree-of-freedom gyroscope. A free gyroscope is defined as one wherein the wheel has three degrees of rotational freedom and is unconstrained with respect to rotation. Although the wheel illustrated in Fig. 1b fulfills this definition as long as the axes are not aligned, a wheel so mounted as to be capable of rotation about five intersecting axes has three degrees of rotational freedom, whatever the direction of the axes.

Precession. The phenomenon of gyroscopic precession is explained readily by Newton's law of motion for rotation, which may be stated: The time rate of change of angular momentum about any given axis is equal to the torque applied about the given axis. When a torque is applied about the input axis of the gyroscope illustrated in Fig. 2 and the speed of the wheel is held constant, the angular momentum of the rotor may be changed only by rotating the projection of the spin axis with respect to the input axis. That is, the rate of rotation of the spin axis about the output axis is proportional to the applied torque. This may be stated in equation as

$$T = I\omega_r\Omega$$

where T = torque

I = inertia of the gyroscope rotor about the spin axis

ω_r = rotor speed

Ω = angular velocity about the axis

The rule for determination of the direction of precession about the output axis is: Precession is always in such direction as to align the direction of rotation of the rotor with the direction of rotation of the applied torque. This is illustrated in Fig. 2, which indicates the direction of precession about the output axis as a result of the applied torque. The output axis (or axis of precession) is always at right angles to the input axis.

Gyroscopic precession differs from angular acceleration about a fixed axis in that it is theoretically possible for the fixed axis acceleration to continue indefinitely, whereas the precessional response to torque has a well-defined limit. The limit is reached when the spin axis is turned sufficiently to align itself with the torque axis. No further precessional response to torque input is possible when this condition has been reached, because all the angular momentum of the system is already about the input axis.

Gyroscopes are used to provide fixed reference directions for compasses on ships and aircraft. They also are used in space vehicle stabilization systems. One type of mass flowmeter is based upon the gyroscopic principle.

Up to this point, this entry has dealt with the gyroscope from the standpoint of its basic principles. The construction of a practical device for a given purpose, however, introduces a number of other considerations. One of these is *drift*, i.e., departure of the motion from the theoretical, and may be caused by unwanted torques due to friction in rotor suspensions or mass shifts in the rotor itself, magnetic effects, and various other causes.

A method widely used to eliminate friction at rotor suspensions is to eliminate them entirely by floating the rotor (and its driving motor) in a viscous, high-density liquid, such as one of the fluorocarbons. This method does have the disadvantage that most of these liquids polymerize over a period of time due to the heat generated. Moreover, these systems require close temperature control to avoid convection currents due to temperature differences in the fluid.

An alternative solution is to retain the bearings, but change them from the ordinary mechanical type to "gas bearings," in which the shaft is actually supported by high-pressure gas. Helium, air, and hydrogen have been used for the purpose. Still another solution is to support the rotor in a high vacuum by an electric field (this is the *electrostatic gyro*), or by a magnetic field. The latter type has been developed effectively by cooling it to the extremely low temperatures at which the rotor becomes superconductive, so that the external magnetic field generates in it currents great enough to produce a "counter" electromagnetic field in it to balance the external field. Because of the low temperature used, this type is called the *cryogenic gyroscope*.

It should be noted that other rotating objects, which are free to precess, exhibit gyroscopic properties. They range from spinning tops to such particles as electrons, atoms and molecules at one end of the size scale, and astronomical bodies, such as satellites and planets, at the other.

Moreover, gyroscope devices are not limited to the basic mechanical type. An example of a quite different kind is the laser gyroscope, developed as an inertial sensor. It consists of a solid quartz block, into which holes are drilled to provide paths for the laser beam. Thin-film mirrors are sealed onto the unit. Laser energy is transmitted clockwise and counterclockwise simultaneously—at rest, they are the same frequency. But when an input rate is present, an output signal is generated that is proportional to that input rate, that does not require a rotating mass as in conventional gyroscopes.

Gyroscope using Fiber Optics. The gyroscope consists of a coil of fiber-optic cable and a 1-inch (2.5-centimeter) square chip containing a laser, beam splitters, a modulator, detectors, and data-processing circuits. The sensor, developed by Hughes Aircraft Company for NASA's Jet Propulsion Laboratory, detects motion by sensing changes in the path of light going in and out of the fiber-optic coil.

Additional Reading

Bao, Min-Hang: *Micro Mechanical Transducers: Pressure Sensors, Accelerometers, and Gyroscopes*, Elsevier Science, New York, NY, 2000.
Burns, W.K.: *Optical Fiber Rotation Sensing*, Academic Press, Inc., San Diego, CA, 1993.

GYROSCOPIC EFFECT. See **Helicopters and V/STOL Craft**.

GYROSYN COMPASS. See **Compass (Navigation)**.

H

HAAR. See **Meteorology**.

HABER-BOSCH PROCESS. See **Ammonia**.

HABER, FRITZ (1868–1934). Born in Breslau, Germany, Haber's great contribution to chemistry for which he was awarded the Nobel Prize in 1918, was his development (with Bosch) of a workable method for synthesizing ammonia by the water-gas reaction from hot coke, air, and steam; the gas mixture obtained includes nitrogen from the air, as well as hydrogen from the steam. It was the first successful attempt to "fix" atmospheric nitrogen in an industrial process. This discovery was developed to production scale in approximately 1912; it enabled Germany to manufacture an independent supply of explosives for World War I.

HABIT. As used by the mineralogist, this term denotes the sum of the external characteristics of a mineral. It is also, but more rarely, applied to rocks.

HABITAT. See **Ecology**.

HABIT PLANE. Many phenomena, such as twinning and martensite transformations, occur in metals where plate-like structures develop inside crystals. The crystallographic plane or planes of the parent phase parallel to the sides of these plates are called the habit plane or planes of the phenomena.

HABOOB. A strong wind and sandstorm or duststorm in northern and central Sudan, especially around Khartoum, where the average number is about 24 a year. (There are numerous variant spellings, including habbub, habub, haboub, hubbob, hubbub.) The name comes from the Arabic word habb, meaning "wind." Haboobs are most frequent from May through September, especially in June, but they have occurred in every month except November. Their average duration is three hours; they are most severe in April and May when the soil is driest. They may approach from any direction, but most commonly from the north in winter and from the south, southeast, or east in summer. The average maximum wind velocity is over 13 m s^{-1} (30 mph) and a speed of 28 m s^{-1} (62 mph) has been recorded. The sand and dust form a dense whirling wall that may be 1,000 meters (3,000 ft) high; it is often preceded by isolated dust whirls. During these storms, enormous quantities of sand are deposited.

Haboobs usually occur after a few days of rising temperature and falling pressure.

See also **Duststorm**; and **Sandstorm**.

Additional Reading

Sutton, L.J.: "Haboobs," *Quart. J. Roy. Meteor. Soc.*, **51**, 25–30 (1925).

AMS

HACHURE. A short line drawn parallel to the slope as a means of illustrating topography on a map.

HACKBERRY AND ZELKOVA TREES. These trees are members of *Ulmaceae* (elm family). The hackberry tree (*Celtis occidentalis*) is found in the United States, principally in the eastern part—from the coast west to Indiana. Other concentrations are found in Colorado and New Mexico. The hackberry is a medium-to-large tree, on the average attaining a height between 50 and 100 feet (15 to 30 meters). As shown in Table 1, some specimens attain greater heights. The twig is red-brown, having leafy scars that are small and oval. The bud is small, approximately $\frac{1}{8}$ inch (0.3 centimeter) long, pointed, and somewhat flattened. The leaf is from 4 to 6 inches (10 to 15 centimeters) long. It is individual, alternate, and simple. It features sharp teeth with deep veins. The fruit is a drupe, small, and deep purple in color. It ripens in September and October. The taste is bitter. The bark is an ash gray, rough, near wart-like as the tree grows old. The wood is heavy, compact, and pale-yellow in color. It weighs approximately 40 pounds per cubic foot (643 kilograms per cubic meter). Other species of hackberry include the Georgia hackberry (*Celtis tenuifolia*); the Lindheimer hackberry (*C. lindheimer*); and the netleaf hackberry (*C. reticulata*).

Zelkova trees are ornamental trees of attractive habit and handsome foliage. They are deciduous with alternate leaves and polygamous flowers. They have a 1-Seeded drupe. Five species are found in Crete, the Caucasus, and eastern Asia. *Zelkova serrata* is an important timber tree, having very durable wood and considered one of the best of building materials in Japan. Young wood is yellowish-white; old wood is a dark brown and known for a beautiful grain. Zelkovas appear much as small-leaved elms, but sometimes take on a shrubby appearance. Small greenish flowers and the fruits are inconspicuous. *Z. serrata* and *Z. Davidii* are also found in North America. They are hardy and can withstand northern climates. The *Z. ulmoides* is less hardy and usually is not found north of Massachusetts. Because of its upright stems, *Z. Davidii* makes an excellent shrub. *Z. serrata* can attain

TABLE 1. RECORD HACKBERRY TREES IN THE UNITED STATES[1]

Specimen	Circumference[2]		Height		Spread		Location
	Inches	Centimeters	Feet	Meters	Feet	Meters	
Common hackberry (1993) (*Celtis accidentalis*)	235	597	94	28.7	88	26.8	Illinois
Georgia hackberry (1999) (*Celtis tenuifolia*)	22	56	30	9.1	25	7.6	Georgia
Lindheimer hackberry (1975) (*Celtis lindheimeri*)	72	183	43	13.1	46	14	Texas
Netleaf hackberry (1988) (*Celtis reticulata*)	180	457	69	21	75	22.9	New Mexico
Spiny hackberry (1996) (*Celtis pallida*)	31	79	22	6.7	24	7.3	Arizona

[1]From the "National Register of Big Trees," American Forests (by permission).
[2]At 4.5 feet (1.4 meters).

a height up to 100 feet (30 meters), featuring a broad, round-topped head and slender branches.

HACKLY. A term used by mineralogists to describe a jagged fracture.

HADDOCK. See **Codfishes**.

HADLEY CELL. A direct thermally driven and zonally symmetric circulation under the strong influence of the earth's rotation, first proposed by George Hadley in 1735 as an explanation for the trade winds. See also **Winds and Air Movement**.

It consists of the equatorward movement of the trade winds between about latitude 30° and the equator in each hemisphere, with rising wind components near the equator, poleward flow aloft, and, finally, descending components at about latitude 30° again. In a dishpan experiment, a Hadley cell is any direct thermally driven vertical cell of the approximate scale of the dishpan. See also **Trade Inversion**.

AMS

HADRONS. These are subatomic particles, the strong interactions of which are manifested by the forces that hold neutrons and protons together in the atomic nucleus. Hadrons include the proton, the neutron, and pion, among others. These particles show signs of an inner structure, i.e., they are made up of other particles, which has led over a period of the last several years to consider the hadrons as combinations of constituents known as *quarks*. See also **Quarks**; and **Particles (Subatomic)**.

HAECKEL, ERNST HEINRICH PHILIPP AUGUST (1834–1919). Ernst Haeckel was a German zoologist who elaborated a complex view of evolution based on the concept that ontogeny recapitulates phylogeny and a whole worldview called Monism.

The younger son of a civil servant, Haeckel was born in Potsdam, Prussia. He studied medicine at Würzburg, where his teacher Rudolph Virchow introduced him to the cell theory. He also studied in Berlin where he was one of Johannes Müller's last disciples. Müller was his greatest scientific inspiration and Haeckel appreciated his teaching in marine zoology and morphology more than the physiological work. Having decided to embark on an academic career in zoology, he was approached by the comparative anatomist Carl Gegenbaur to move to the University of Jena as a zoologist but before going, Haeckel went to southern Italy in 1859 to study marine invertebrates, especially Radiolarians. When Haeckel returned to his academic career the following year, he read the German translation of Darwin's *Origin of Species* and declared himself a convinced supporter of evolution. From 1861, when he obtained his *Habilitation*, to his death he lived and worked in Jena collaborating with Gegenbaur to reform morphology in the light of evolutionary principles. In 1866, he published his major theoretical work, *Die generelle Morphologie der Organismen*, and two years later the first edition of his most successful scientific book, *Die Natürliche Schöpfungsgeschichte (The History of Creation)*. Haeckel's evolutionary view was a combination of Lamarckian and Darwinian concepts: he believed in spontaneous generation as the origin of life—an aspect not considered by Darwin—and in common descent. He also believed that natural selection applied to groups—types, or as he called them, *Phyla*—rather than individuals. His original contribution to evolutionary theory was the so-called *biogenetic law*. This is summarized in his famous aphorism, *Ontogeny recapitulates Phylogeny*; that is, the history of the individual (ontogeny) from cell to embryo to adult recapitulates the history of the group or *Phylum (Phylogeny)* to which that individual belongs. This law could explain the entire history of life on earth, starting from simple primordial creatures called *Monera (Moners)* and moving to increasingly more complex forms culminating with man. This history of life could be visualized in the form of evolutionary trees, thus leading to an evolutionary reworking of taxonomy with the absence or presence of a true intestine as a key feature (this was what he called *Gastraea Theory*). The entire phylogenetic tree of living forms could be integrated by hypothetical forms whenever a loop was found. Morphology would be revised in the light of evolutionary thought so that the teleology of the old idealistic morphology was replaced by natural selection as the mechanical cause of the changes of forms in time and not only in space. See also **Virchow, Rudolf Carl (1821–1902)**.

Haeckel showed great interest in human evolution. He postulated a hypothetical form intermediate between apes and humans, which he called

Phytecanthropus alalus, Speechless Ape-Man. Because of this, he studied the connection between languages and evolution, first under the influence of Slavist August Schleicher and then his cousin Wilhelm Bleek, who had moved to Cape Town, South Africa. Bleek had studied the bushmen (Koi-Koi) and believed that their language corresponded to the original language of humankind. Haeckel based on that a culturally based anthropology opposed to the merely physical anthropology of his former teacher and then rival Virchow. Haeckel divided humankind into races, which could even be considered separate species, on the connection between human morphology and human languages.

Haeckel saw evolution as a founding element in a comprehensive worldview he called *Monism* that was based on the eternity, unity and development of substance and force. He believed that evolution was a directional process leading to progress: its development would lead to the final unity of all natural processes and of all human activities and provided a connection between science and religion. He strongly opposed revealed religions, above all Roman Catholicism, and had what was close to a pantheistic view of nature and religion.

Haeckel was a good painter with strong artistic inclinations. Inspired by Goethe, he saw a connection between science and art in a unified vision and expressed his interpretation of this connection in his *Kunstformen der Natur* of 1899–1904. He also tried to establish a direct link between what is true (science), good (ethics) and beautiful (aesthetics), all this with an originally biological foundation. In 1899, his book *Die Welträthsel (The Riddle of the Universe)* exposed such views and became a true bestseller translated into several languages. Haeckel dedicated most of the last twenty years of his life to spread his monistic philosophy and became the best-known representative of evolutionary and, broadly speaking, scientific views in the world. In 1904, he wrote *Die Lebenswunder (The Wonders of Life)*, which was less successful and supported such controversial measures as Spartan selection and the suppression of mentally challenged people, which reflected widespread fears of degeneration in the West. Politically, he always refused to be connected with any party, although he had become an admirer and supporter of Bismarck during the process of formation of the German *Reich* in the early 1870s. He was also involved with the foundation of a Phyletic Museum in Jena and once he saw that the museum had not become the shrine of his ideas as he had projected it, he turned his own residence, Villa Medusa in Jena, into such a shrine. During World War I, he expressed extremely chauvinistic views and saw the war on the one side as the end of progress and on the other as the absurd fight between the sister races of Germany and Britain. He insisted on the eternity of the universe and the ephemeral existence of individual life. In 1917, he published his last scientific work, *Kristallseelen,* related to Otto Lehmann's studies on liquid crystals, which seemed to him the vindication of his views of the continuity of substance.

Additional Reading

Briedbach, O.: *Visions of Nature: The Art and Science of Ernst Haeckel*, Prestel Publishing, New York, NY, and Prestel Germany, 2006.
Di Gregorio, M.A.: *From Here to Eternity: Ernst Haeckel and Scientific Faith*, Vandenhoeck & Ruprecht, Göttingen, Germany, 2005.
Krausse, E.: *Ernst Haeckel*, Teubner, Leipzig, Germany, 1987.

MARIO A. DI GREGORI, University of L'Aquila, Italy

HAFNIUM. [CAS: 7440-58-6] Chemical element symbol Hf, at. no. 72, at. wt. 178.49, periodic table group 4, mp 2207–2247 °C, bp 4601–4603 °C, density 13.3 g/cm^3. The alpha form of elemental hafnium has a close-packed hexagonal crystal structure; the beta form, a body-centered cubic structure. Metallic hafnium, like zirconium, exhibits passivity in air due to formation of adherent coatings of oxide or nitride. Urbain reported evidence of the element in 1911, but hafnium was not fully identified until 1923 by D. Coster and G.C. de Hevesy. The remarkable similarity between hafnium and zirconium accounts mainly for its late isolation, as compared with the majority of elements. In terms of abundance, there is an average of about 4 ppm hafnium in the earth's crust. The element occurs with zirconium in certain varieties of zircon, including malacon, cyrtolite, and alvite. One mineral found in Scandinavia, thortveitite, contains more hafnium than zirconium. Pegmatite, monazite, baddeleyite, and zerkelite also contain hafnium. First ionization potential 5.5 eV. Oxidation potentials $Hf + H_2O \rightarrow HfO^{2+} + 2H + 4e^-$, 1.68 V; $Hf + 4OH^- \rightarrow HfO(OH)_2 + H_2O + 4e^-$, 2.60 V. Electron configuration $1s^2 2s^2 2p^6 3s^2 3d^{10} 4s^2 4p^6 4d^{10} 4f^{14} 5s^2 5p^6 5d^2 6s^2$. Ionic radius Hf^{+4}, 0.75 Å. Other important physical properties of hafnium are given under **Chemical Elements**.

Hafnium usually is extracted from ores along with zirconium. In one process, zircon sand is broken down by carbiding or carbonitriding, followed by chlorination. The mixture formed is dissolved with a complexing agent, after which it is introduced into a liquid-liquid extraction process. The final product is $HfCl_4$. Fractional crystallization of the fluorides of hafnium and zirconium also is practiced. Metallic hafnium is made by the Kroll process in which the $HfCl_4$ is reduced in an inert atmosphere by magnesium. The hafnium sponge and magnesium chloride resulting is vacuum-distilled to accomplish the final separation. In a modified Kroll process, sodium or sodium amalgam may be used. The latter requires less rigid temperature and pressure control during processing, costs less, and introduces fewer impurities into the process. For further purification of hafnium metal, a number of methods have been used, including electrorefining, arc and induction melting, zone refining, and the hot-wire or van Arkelde Boer process.

Uses. Compared with most metals, the annual production of hafnium is low. Mainly produced in the United States, France, and Russia, the combined production is in the range of 100 metric tons annually, or less. Several uses have been found for hafnium: (1) as a control material in water-cooled nuclear reactors. Also hafnium is an effective flux-depressor in a reactor for absorbing neutrons to decrease the peaks in neutron flux; (2) as a filament in gas-filled incandescent light bulbs; (3) as an alloying ingredient to add strength to tungsten and molybdenum filaments and electrodes used in high-pressure discharge tubes; (4) as a cathode in x-rays tubes; (5) as a getter material in vacuum tubes and systems; (6) as a minor alloying ingredient in nichrome heating elements where hafnium appears to significantly increase the lifespan of the elements; and (7) usually with zirconium, as an ingredient of several alloys.

Chemistry and Compounds. Hafnium metal dissolves in HCl (warm) and slowly in H_2SO_4, more rapidly if fluoride ion F^- is present, forming compounds of HfO^{2+}, or fluoro complexes in the latter case. The metal resists the attack of weak acids and their salts.

Due to its $5d^2 6s^2$ electron configuration, hafnium forms tetravalent compounds readily, although the Hf^{4+} ion does not exist as such in aqueous solution except at very low pH values, the common cation being HfO^{2+} (or $Hf(OH)_2^{2+}$) and many of the tetravalent compounds are partly covalent. There are also less stable Hf(III) compounds. There is close similarity in chemical properties to those of zirconium due to the similar outer electron configuration ($4d^2 5s^2$ for zirconium) and the almost identical ionic radii (Zr^{4+} is 0.80 Å) the relatively low value for Hf^{4+} being due to the Lanthanide contraction.

With improved means to separate the compounds of these two elements, future research will yield more details of specific hafnium compounds. The methods of separation used effectively include ion exchange techniques, a particularly effective one using a column of silica gel, with a solution of the tetrachlorides in methanol as feed and a 1.9 N HCl solution as eluant for zirconium. Separations also have been accomplished through the distillation of the phosphorus oxychloride addition products.

See list of references at end of entry on **Chemical Elements**.

HAGFISHES (*Agnatha*). A jawless fish of the family *Myxinidae*, is an aggressive scavenger usually averaging less than 30 inches (76 centimeters) in length. The hagfish is characterized by the primitive features of jawless fishes—no scales, no sympathetic nervous system, a cartilage skeleton, and single nostril. The hagfish is elongate, rather wormlike, and blind. Because the fish can exude large quantities of a slimy mucus, it is sometimes called a "slime eel." Among species of hagfishes are the Japanese *Paramyxine*, the *Eptatretus*, the Atlantic *Myxine glutinosa*, and the Pacific *Heptatretus stouti*. The latter species has been used in medical research, particularly in studies of the hag heart (no heart nerves or sympathetic nerves). Generally, hagfishes prefer cold to temperate marine waters from shallow levels down to about 3,000 feet (900 meters). They cannot tolerate fresh or brackish waters.

See also **Cyclostomata**; and **Fishes**.

HAIDINGER FRINGES. Optical interference fringes seen with thick, flat plates near normal incidence. The fringes of the Fabry-Perot interferometer are of this type. They are also known as constant angle or constant deviation fringes.

HAIL. Precipitation in the form of balls or irregular lumps of ice, always produced by convective clouds, nearly always cumulonimbus. An individual unit of hail is called a hailstone. By convention, hail has a diameter of 5 millimeters ($\frac{1}{4}$ inch) or more, while smaller particles of similar origin, formerly called small hail, may be classed as either ice pellets or snow pellets. Thunderstorms that are characterized by strong updrafts, large liquid water contents, large cloud-drop sizes, and great vertical height are favorable to hail formation. The destructive effects of hailstorms upon plant and animal life, buildings and property, and aircraft in flight render them a prime object of weather modification studies. In aviation weather observations, hail is encoded A. See also **Hailstone**; and **Precipitation and Hydrometeors**.

HAILSTONE. A single unit of hail, ranging in size from that of a pea to that of a grapefruit (i.e., from 5 millimeters ($\frac{1}{4}$ inch) to more than 15 centimeters (6 inches) in diameter). Hailstones may be spheroidal, conical, or generally irregular in shape. The spheroidal stones often exhibit a layered internal structure, with layers of ice containing many air bubbles alternating with layers of relatively clear ice. These probably correspond to dry growth and wet growth and are called rime and glaze, respectively. The conical stones fall with their bases downward without much tumbling and are often smaller and not as layered. Irregular hailstones often have a lobate structure and are not composed of smaller hailstones frozen together. Hailstones grow by accretion of supercooled water drops and sometimes also by accretion of minor amounts of small ice particles. Large hail may contain liquid water and be spongy (an intimate mixture of ice and water) in some regions; it is usually solid ice with density greater than 0.8 g cm^{-3}. Small hail may be indistinguishable from large graupel (snow pellets) except for the convention that hail must be larger than 5 millimeters ($\frac{1}{4}$ inch) in diameter. The density of small hail can be much less than 0.8 g cm^{-3} if they are dry; if partly melted such hailstones become spongy. The largest recorded hailstone in the United States fell in a hailstorm in Coffeyville, Kansas on 3 September 1970. It weighed 766 grams (27 ounces or 1.7 lbs), had a longest dimension about 15 centimeters (6 inches), and had protrusions (lobes) several centimeters ($\frac{3}{4}$ inch) long on one side that formed as it grew. See also **Precipitation and Hydrometeors**.

HAIR. There are several kinds of hair on the human body. The appearance depends on age and body location. The so-called *lanugo* is that hair which develops on the unborn child. Usually, it is shed before birth, or within the first few months after birth. The lanugo is immediately replaced by secondary hair, which is fine and soft and is often called "baby hair." The coarser hair of later life is called *tertiary hair*. Hairs are continually lost from all parts of the body throughout life, and up to a certain age; those that replace them often are coarser than their predecessors.

There are about 125,000 hairs on the scalp of the average person. Darker persons usually have fewer scalp hairs than blondes. Scalp hair usually grows from 3 to 5 inches (7.5 to 12.5 centimeters) per year and, if permitted, can become as long as 2 to 3 feet (0.6 to 0.9 meter), or even longer.

The hairs of the body originate from hair follicles embedded in the skin. The lower part of the follicle extends into the dermis where it is supplied with blood vessels. Generally, only one hair grows from a single follicle. That part of the hair beneath the surface of the skin is termed the *root*, while that part extending outward from the skin is called the *shaft*. The sebaceous glands of the skin have their openings in the hair follicles. These glands secrete a substance (sebum) which is responsible for the oily appearance of the skin or scalp. Persons with oily skin possess overactive sebaceous glands. When the hair follicle becomes plugged, the sebum collects within it, turns dark at the surface, and becomes a "blackhead."

Minute muscles (*erectors pilorum*) are connected to the hair follicle. When these muscles contract, they temporarily displace the entire follicle, causing the hair to "stand on end." The skin surrounding the hair is also elevated by the contraction of these muscles, giving the skin a prickled appearance, sometimes called "goose pimples." Contraction of the muscles also exerts pressure on the sebaceous glands, causing the emission of extra amounts of sebum. Thus, this set of reactions aids in protecting the body from sudden cold, the hairs forming better insulation when standing erect, and the sebum coats the skin with a further barrier against the cold.

The partial or complete absence of hair from the body is called *alopecia*.

The use of hair analysis and examination in the forensic sciences has been known for a number of years. Regarding the analysis of Sir Isaac Newton's hair for mercury, see **Mercury**.

HAIR HYGROMETER.   A hygrometer that measures relative humidity by means of the variation in length of a strand of human hair. The length variation of a properly treated hair is 2%–2.5% when the humidity changes from 0%–100%. The hair hygrometer is considered to be a satisfactory instrument in situations where extreme and very low humidities are seldom or never found. The rate of response is very dependent on air temperature; the lag time increases with decreasing temperature. For air temperatures between 0° (32°F) and 30°C (86°F) and relative humidities between 20% and 80%, a good hair hygrometer should indicate 90% of a sudden change in humidity within about three minutes. See also **Hygrometry and Psychrometry.**

HAIRSTREAK *(Insecta, Lepidoptera).*   Small butterflies, those of the temperate zone dull-colored and those of the tropics often brilliant. The hind wings of most species bear hair-like tails. With the coppers and blues they make up the family *Lycaenidae.*

HAIRTAIL.   See **Cutlassfishes.**

HAIRWORM *(Nematomorpha).*   or *Gordiacea*; formerly placed in the phylum *Nemathelminthes* with the *Nematoda*). Long slender round worms of small size, which live as parasites in the bodies of invertebrates, chiefly insects.

HAKE *(Osteichthyes).*   Of the family *Merluccidae*, the hakes are closely related to the codfishes, but have a special systematic position due to their unusual distribution. The family has just one genus, *Merluccius.* The slender body, skull structure, and the large-toothed mouth give this carnivorous fish a gar-like appearance. There are two dorsal fins and one long anal fin, which is almost the mirror image of the second dorsal fin in shape, size, and position.

The hake (*Merluccius merluccius*) is found in the northeastern Atlantic Ocean off the western and southwestern coasts of Europe, along the continental shelf. The northern border of the distribution is formed where branches of the Gulf Stream meet masses of polar waters. This is also the northern limit of the *American hake* or *silver-hake* (*Merluccius bilnearis*). See Fig. 1. Living in deep water has enabled the hakes to penetrate the tropical Atlantic Ocean and inhabit oceanic regions in the southern hemisphere with temperate to subtropical conditions. This accounts for the large South Atlantic populations of *stockfish* (*Merluccius capensis*) off southwestern Africa; and *Merluccius hubbsi* from the coasts of southern Brazil and Argentina. There are also Pacific Ocean species: *Merluccius gayi* and *M. productus*, off the western coasts of North and South America. Their presence has been explained by a presumed migration around Cape Horn. The New Zealand species, *Merluccius australis*, may also have come by this route.

Hakes can be over 3 feet (1 meter) in length, but there are small- and medium-sized species as well. They are predators, feeding chiefly on herring and other schooling fishes. The European hake seeks its prey at night in the upper water levels. During the day, it is less active and stays near the floor, at which time it can be caught easily, even with a dragnet. This species spawn in spring, apparently without preferred spawning sites. The floating eggs then drift within the hake distribution region. The commercial importance of hake has increased since the early 1960s.

The *Cape hake* in South African waters is a whitefish (*M. capensis*) and it appears that there are two distinct populations on the trawling grounds. About half of the population attains sexual maturity at an age of 3 to 4 years. Peak spawning occurs during spring and early summer. Preliminary studies indicate that in the case of the Cape hake, diurnal vertical migration is much less pronounced than in the case of the European cod. The diet of the adult hake is comprised of rattails, maasbanker, and squid, and cannibalism is quite common.

See also **Fishes.**

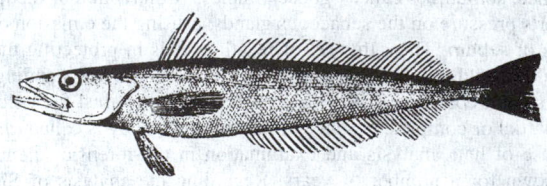

Fig. 1.   Silver hake.

HALDANE, JOHN SCOTT1860–1936.   Haldane read medicine at Edinburgh, graduating in 1884. Shortly after the publication of his first research in 1887, he moved to Oxford as demonstrator in physiology working with his uncle, John Burdon Sanderson, Waynflete Professor of Physiology.

Haldane's career was characterized by his moving between the laboratory and situations in life where the respiratory system can be compromised, for example coal mines, deep-sea diving, high altitude, ships and submarines. His 1896 report on the cause of death in coal-mining disasters highlighted the importance of carbon monoxide and laboratory studies showed that this gas binds with the hemoglobin, therefore reducing the body's ability to take up oxygen. Experiments with mice illustrated the benefits of a hyperbaric environment. Alone and then with Joseph Barcoft he developed a readily usable clinical apparatus to determine blood gas content. Later, in the 1900s, Haldane worked out the system of stage decompression for divers with the bends.

His work with J. G. Priestly (1905), and subsequently extended during his career, elucidated the mechanism by which the amount of carbon dioxide in the arterial blood traveling to the respiratory center of the midbrain controls pulmonary ventilation. Later shown to be too simple a model, nevertheless the basic principle that respiration (in most circumstances) is controlled by carbon dioxide rather than oxygen level was correct. See also **Priestley, Joseph (1733–1804)**.

Haldane's research did much to raise the status of respiratory physiology and continued the rising prominence of the Oxford school of physiology. He was elected a Fellow of the Royal Society in 1897. He directed a laboratory first at Doncaster and then at Birmingham run under the auspices of the coal-mining industry and after 1921 spent a great deal of time in Birmingham, although he remained a fellow of New College, Oxford until his death. See also **Physiology (The History)**.

Additional Reading

Gillispie, C.C.: *Dictionary of Scientific Biography*, Charles Scribner's Sons, New York, NY, 1972.
Phillips, J.L.: *The Bends: Compressed Air in the History of Science, Diving and Engineering*, Yale University Press, New Haven, CT, 1988.
Sturdy, S.: "Biology as Social Theory: John Scott Haldane and Physiological Regulation," *British Journal for the History of Science*, **21**, 315–340 (1988).
West, J.B., and I.B. Weiner: *High Life: A History of High-Altitude Physiology and Medicine*, Oxford University Press, New York, NY, 1998.

HELEN J. POWER, Wellcome Trust Centre for the History of Medicine at UCL, London, UK

HALF-ADDER.   A circuit having two output points, *S* and *C*, representing sum without carry and carry, and two input points, *A* and *B*, representing addend and augend, such that the output is related to the input according to the following table:

Input		Output	
A	*B*	*S*	*C*
0	0	0	0
0	1	1	0
1	0	1	0
1	1	0	1

Two half-adders and an Inclusive-OR circuit, properly connected, can provide a Full-Adder having two inputs (augend and addend) and a carry input which produces a sum output (without carry) and a carry output.

HALF-CELL.   An electrochemical system consisting of a single electrode and an electrolytic solution, with usually a (reversible) ionization process in progress between electrode and electrolyte. See also **Galvanic Cell.**

HALF-LIFE (Biological).   The time of survival of half the individual members of an unstable system. The half-life $t_{1/2}$ of the system is related to the decay constant λ and the mean life τ by the relation:

$$t_{1/2} = \frac{\ln 2}{\lambda} = \frac{0.693}{\lambda} = 0.693\tau$$

The term half-life is most commonly applied to systems of radionuclides but may also be applied to other systems that decay.

The biological half-life of a substance is the time in which a living tissue, organ or individual eliminates, through biological processes, one-half of a given amount of a substance which has been introduced into it. The effective half-life is a term usually applied to a radioactive substance in a biological organism. It is defined in terms of the half-life of the radioactive substance itself, and its biological half-life in the organism, by the following expression:

$$\text{effective half-life} = \frac{\text{radioactive half-life} \times \text{biological half-life}}{\text{radioactive half-life} + \text{biological half-life}}$$

HALF-LIFE (Elements).   See **Chemical Elements**.

HALF-SILVERED SURFACE.   A surface coated with a metallic film of such thickness that it transmits approximately half of the light falling on it at normal incidence and reflects approximately half.

HALF-THICKNESS (Absorber).   The thickness of a particular absorber that will reduce the intensity of a beam of radiation to one-half its initial value. If the absorption is exponential, the half-thickness is related to the linear or mass absorption coefficient and the mean free path as follows:

$$d_{1/2} = \frac{\ln 2}{\mu} = \frac{0.693}{\mu} = 0.693l$$

where $d_{1/2}$ is the half-thickness, μ is the absorption coefficient and l is the mean free path.

HALFWIDTH OF A SPECTRAL LINE.   The intensity within a spectral line may be expressed as $I(x)$, where x is a measure of wavelength, frequency or wave number, and where $I(x)\, dx$ is a measure of the contribution to the intensity between x and $x + dx$. The halfwidth of the line is the halfwidth of the function $I(x)$.

HALIBUT.   See **Flatfishes**.

HALIDES.   A compound made up of a halogen (astatine, bromine, chlorine, fluorine, or iodine) and another element or radical may be termed a *halide*. Fundamentally, there are three classes: (1) the *ionic* (saline) halides, (2) the *covalent* (acid) halides, and (3) the *complex* halides. The ionic halides are most sharply characterized by the halides of the alkali and alkaline earth metals, plus those of certain Lanthanide and Actinide metals. They form ionic or semi-ionic crystals in the solid state, have high boiling points and melting points, and are soluble in polar solvents. Their bonding is electrovalent, varying in degree with the difference between the electronegativities of the halogen and the metal. Potassium iodide and silver fluoride are ionic, but silver iodide is essentially covalent. The fluorides exhibit a primarily ionic character for most of the metals, but the other halogens form fewer ionic compounds. The degree of ionicity varies down as well as across the periodic table.

The covalent (acid) halides have low boiling and melting points, are soluble in nonpolar solvents and insoluble in polar solvents, although they often react with the latter. The degree of covalence generally is greatest for the nonmetals. For a given nonmetal, the boiling point depends upon both the number of atoms of the halogen with which it is combined and the symmetry of the molecule. For example, the boiling points of bromine(I) fluoride, bromine(III) trifluoride, and bromine(V) pentafluoride, BrF, BrF$_3$ and BrF$_5$, are 20, 135, and 40.5 °C, respectively.

The complex halides are very numerous, because of the readiness with which halide ions form coordination compounds with metals. In general, stability of these complexes depends upon the size and electronic structure of the metal ion — the smaller cations form their more stable compounds with the smaller halide ions, notably with fluoride. With larger cations the order of stability is that of the ability to be polarized of the halide, i.e., decreasing from iodide to fluoride. The more electronegative transition elements form especially stable complexes; e.g., those of palladium, platinum, etc., PdCl$_4{}^{2-}$, PtF$_6{}^{2-}$, etc. The most common halo complexes have four or six halogen ions coordinated with the cation, although such complexes as those of copper, gold and mercury, e.g., CuI$_2{}^-$, AuCl$_2{}^-$, HgCl$_3{}^-$, etc., are notable exceptions.

See also **Bromine**; **Carbon**; **Chlorine**; **Chlorinated Organics**; **Fluorine**; **Halogen Group**; and **Iodine**.

HALITE (Rock Salt).   The mineral halite (rock salt) is naturally occurring sodium chloride, NaCl, common salt. It is isometric with cubic habit and cleavage. It is brittle; hardness, 2.5; specific gravity, 2.168; luster, vitreous; colorless when pure, but usually white, yellow, red, or blue. It is soluble in water. Halite occurs interbedded with sedimentary rocks in all parts of the world and in all but the very oldest rocks. It frequently occurs in association with anhydrite and gypsum. In the United States this type of "salt beds" has been exploited in Michigan, New York, Ohio, and Pennsylvania. Louisiana produces salt from great subsurface dome-shaped masses, often 2,000–4,000 feet thick. The salt domes of the Gulf Coastal Plain are particularly important as subsurface structures, on the flanks of which are apt to occur large and important pools of petroleum. Poland, Saxony, Austria, and France possess well-known deposits of salt, as well as the former U.S.S.R., England, Algeria, India, and China. Salt is chiefly used in cooking and as a preservative; in the manufacture of soda ash for the glass industry; and as a source of many sodium compounds. It derives its name from the halogen group of elements to which chlorine belongs.

See also **Sodium Chloride**.

HALL, CHARLES MARTIN (1863–1914).   A native of Ohio, Hall invented a method of reducing aluminum oxide in molten cryolite by electrochemical means. This discovery made possible the large-scale production of metallic aluminum and resulted in formation of the Aluminum Company of America. The process requires high electric power input. Hall is generally considered the founder of the aluminum industry.

HALL EFFECT AND QUANTIZED HALL EFFECT.   In 1879, Edwin H. Hall (Johns Hopkins University), discovered that if a strip of gold leaf, carrying an electric current longitudinally, was placed in a magnetic field with the plane of the strip perpendicular to the direction of the field, the points directly opposite each other on the edges of the strip acquired different electric potentials; and that if such points were joined through a sensitive galvanometer, a feeble current would be indicated. In other words, the equipotential lines, ordinarily running across at right angles to the edges, were skewed into an oblique position, and the electric lines of flow in the plane of the strip were deflected to one side.

If one looks along the strip in the direction of the current, with the magnetic field directed downward, then, with strips of antimony, cobalt, zinc, or iron, the electric potential drop is toward the right and the effect is said to be positive. With gold, silver, platinum, nickel, bismuth, copper, and aluminum, it is toward the left, and the effect is called negative. The transverse electric potential gradient per unit magnetic field intensity per unit current density is called the "Hall coefficient" for the metal in question. Thus, the Hall coefficient R_H is defined as

$$R_H = \frac{E_y}{j_x H_z}$$

where E_y is the electric field developed in the y direction when a current of current density j_x flows in the x direction through a magnetic field H_z in the z direction. According to the free electron theory of metals, the Hall coefficient should be given by

$$R_H = \frac{B}{ne}$$

where N is the number of free electrons per unit volume, of charge e (in esu), and c is the velocity of light. The observed result that for some metals the carriers would seem to have positive charges is explained by the band theory of solids. In a nearly filled band, the wave functions of the electrons near the top of the band are so modified that it is the holes in the band that behave like particles. Since a hole represents the absence of negative charge, it behaves as if positively charged. The Hall angle is the ratio of E_y (defined above) to the field E_x, generating the current in the magnetic field H_z. The Hall mobility is the mobility of the electrons or holes in a semiconductor as measured by the Hall effect.

A number of transducers utilize the Hall effect. Shown in Fig. 1 is a direct-current oscilloscope probe based on the effect. A steady direct current I_c is applied to one axis of the Hall generator and a magnetic field B, proportional to the current through the conductor, is applied to a second axis. An output voltage V_c is taken across the third axis of the Hall generator. The output voltage can be calculated from:

$$V_c = \frac{10^{-5} R_H}{t} I_c B$$

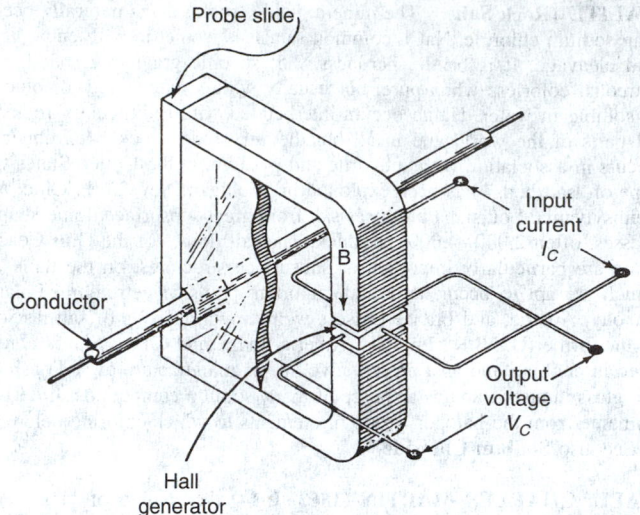

Fig. 1. Direct-current oscilloscope probe based on Hall effect.

where V_c = Hall voltage, volts
      R_H = Hall coefficient, cm^3 coulomb
      t = thickness, cm
      B = magnetic field density, kilogauss

Exploring the Complexities of the Hall Effect

Over the intervening century since Hall's discovery and notably since the advent of semiconductor technology, the Hall effect has inspired research. A number of related effects have been observed. One of these, for example, is the widely studied galvanomagnetic effect, referred to as *transfer magnetoresistance*. By shorting the Hall field or by choosing a disk geometry so that such a field does not exist, one obtains a "magnetoresistance" (more strictly, a *magnetoconductivity*) which does not saturate. This is called the Corbino magnetoresistance or Corbino effect. There are several thermal effects in a magnetic field that can produce transverse voltages or temperature gradients. These result from the velocity separation of charge carriers by the Lorentz force — the energetic ones going to one side, the slower ones going to the other. Temperature gradients are produced, and also electric fields. In the Righi-Leduc effect, a longitudinal temperature gradient produces a transverse temperature gradient (thermal analog of the Hall effect). In the Nernst effect, it produces a transverse electric field. In the Ettingshausen effect, a longitudinal electric current produces a transverse temperature gradient. This latter effect, if large, can disturb the Hall field, since the potential probes and leads are seldom made of the same material as the specimen. Therefore, the Ettingshausen temperature gradient can produce a thermoelectric voltage, which adds to the Hall voltage.

Analysis of Hall-effect data has been one of the most widely used techniques for studying conduction mechanisms in solids, especially semiconductors. For the single-carrier case, one readily obtains carrier concentrations and mobilities, and it is usually of interest to study these as functions of temperature. This can supply information on the predominant charge-carrier scattering mechanisms and on activation energies, i.e., the energies necessary to excite carriers from impurity levels into the conduction band. Where two or more carriers are present, the analysis becomes more complex, but much more information can be obtained from studies of the temperature and magnetic field dependencies.

Unlike, for example, the magnetoresistance, the Hall effect is a first-order phenomenon. A weak magnetic field, it depends linearly on the magnetic field intensity and it does not vanish in isotropic solids if all the carriers have essentially the same velocity or if the scattering is characterized by a relaxation time which is independent of the carrier energy. As previously indicated, the Hall effect forms the basis of a number of devices used in isolating circuits, transducers, multipliers, converters, rectifiers, and gaussmeters (for measurement of magnetic fields). The fundamental component of such devices is a slab of material (often called a "Hall generator") possessing favorable Hall characteristics.

The Quantized Hall Effect

In 1980, Klaus von Klitzing[1] (High Magnetic Field Laboratory, Max Planck Institute), made some unusual findings while studying the Hall effect in devices in which the electrons free to carry current are confined within a *thin layer* of material. The researcher found that by cooling an experimental device to within a degree of absolute zero and by placing the device in a *very strong* magnetic field, the behavior of the ordinary resistance and the Hall resistance differed dramatically from that expected of a traditional Hall-effect device. Instead of increasing steadily and linearly as the strength of the magnetic field was increased, the Hall resistance increased in a *series of plateaus*. There were intervals observed in which the Hall resistance did not vary at all when the strength of the magnetic field was varied. Between the plateaus, the Hall resistance increased smoothly with increasing magnetic field. It was also found that during the same intervals of magnetic field strength during which the Hall resistance exhibited plateaus, the voltage drop parallel to the current was noted to disappear completely (no electrical resistance in sample and current flows without dissipating any energy).

The vanishing electrical resistance and the plateaus in the Hall resistance are remarkable phenomena. It is even more remarkable, as pointed out by Halperin (See reference), that, on each plateau, the value of the Hall resistance satisfies a remarkably simple condition. That is, the reciprocal of the Hall resistance is equal to an integer multiplied by the square of the charge on the electron and divided by Planck's constant (the fundamental constant of quantum mechanics). Each plateau is characterized by a different integer. Essentially, in such a system, the Hall resistance is reduced to the formula

$$R_H = \frac{1}{Nce}$$

where n = density of electrons (per square meter) in the sample. If the two-dimensional system is connected to an external reservoir of electrons and the magnetic field B is allowed to vary, then the density of electrons in the layer will vary with B in such a way as to minimize the combined energy of the layer and reservoir.

The degree of precision of the quantized Hall effect has amazed even the experts. Measured values of the Hall resistance at various integer plateaus are accurate to about one part in six million. The effect can be used to construct a laboratory standard of electrical resistance that is much more accurate than the standard resistors currently in use. Authorities also observe that, if the quantized Hall effect is combined with a new calibration of an absolute resistance standard, it should be able to yield an improved measurement of the fundamental dimensionless constant of quantum electrodynamics, the fine-structure constant α.

In his original experiment, von Klitzing used a silicon field-effect transistor (MOSFET) of exceptional quality and of the type used on integrated circuit chips. In the device, electrons are trapped in a so-called inversion layer near the surface of a silicon crystal that is covered with a film of insulating silicon oxide, on top of which is deposited a metal "gate electrode," used to control the density of conduction electrons in the inversion layer.

A somewhat similar Hall effect phenomenon, known as the *fractional quantized Hall effect*, was observed at the National Magnet Laboratory (Cambridge, Massachusetts) by Tsui, Störmer, and Gossard (AT&T Bell Laboratories) a couple of years after von Klitzing's finding. The fractional quantized Hall effect was first noted in a heterojunction (an interface of crystals made of two different semiconducting materials). As pointed out by Halperin, in a heterojunction, electrons from one semiconductor are attracted to more energetically favorable locations in the other semiconductor. The positive charge thereby created in the "donor" semiconductor provides a force attracting the electrons back, however, and they become trapped in a thin layer at the interface of the two crystals.

Additional Reading

Abrikosov, A. and R.A. Silverman: *Quantum Field Theory*, Dover Publications, Inc., Mineola, NY, 1975.

Ando, T. et al.: *Mesoscopic Physics and Electronics*, Springer-Verlag, Inc., New York, NY, 1998.

Cerdeira, H.A., B. Kramer, and G. Schon: *Quantum Dynamics of Submicron Structures*, Kluwer Academic Publishers, Norwell, MA, 1995.

Chakraborty, T. and P. Pietilainen: *The Quantum Hall Effects: Integral and Fractional*, Vol. 85, Springer-Verlag, Inc., New York, NY, 1995.

[1] Nobel Prize winner for Physics, 1985.

Eisenstein, J.F. and H.L. Stormer: "The Fractional Quantum Hall Effect," *Science*, **1510** (June 22, 1990).

Ellis, P.J. and Y.C. Tang, Editors: *Trends in Theoretical Physics*, Addison-Wesley Longman, Inc., Reading, MA, 1990.

Falomir, H., R.E. Gomboa, and F.A. Schaposnki: *Trends in Theoretical Physics*, American Institute of Physics, College Park, MD, 1998.

Fisk, Z. et al.: *Physical Phenomena at High Magnetic Fields II*, World Scientific Publishing Company, Inc., Riveredge, NJ, 1996.

Halperin, B.I.: "Theory of the Quantized Hall Conductance," *Helvetica Physica Acta*, **56**(4603), 1241–1246 (June 17, 1983).

Halperin, B.I.: "The Quantized Hall Effect," *Sci. Amer.*, 52–60 (April 1986).

Halperin, B.I.: "The 1985 Nobel Prize in Physics," *Science*, **231**, 820–822 (1986).

Janssen, M. and J. Hajdu: *Introduction to the Theory of the Integer Quantum Hall Effect*, John Wiley & Sons, Inc., New York, NY, 1994.

Nicholas, R.J., Ed.: "Proceedings of the Fifth Int. Conf. on Electrical Properties of Two-Dimensional Systems," in *Surface Science*, 142 (1984).

Pinczuk, A. and S. Das Sarma: *Perspectives in Quantum Hall Effects*, John Wiley & Sons, Inc., New York, NY, 1996.

Schwartzschild, B.: "Von Klitzing Wins Nobel Prize for Quantum Hall Effect," *Physics Today*, **38**(12), 17–20 (December 1985).

Semenoff, G.W. and L. Vinet: *Particles and Fields*, Springer-Verlag, Inc., New York, NY, 2000.

von Klitzing, K., G. Dorada, and M. Pepper: *Phys. Rev. Lett.*, **45**, 494 (1980).

HÄLLEFLINTA. A Swedish term for hard, dense, metamorphic rocks composed chiefly of microscopic crystals of quartz and feldspar with occasional phenocrysts. Accessory minerals may be hornblende, chlorite, hematite or magnetite. The texture and composition of hälleflinta suggests that it is the metamorphosed equivalent of acid lava flows or tuffs.

HALLER, ALBRECHT VON (1708–1777). Haller was a Swiss physiologist, anatomist and writer who developed the principle of irritability.

Born into a well-established Swiss family, Haller grew up in the city of Bern. His education ranged broadly both in content and in style. Educated, in turn, by a former pastor, in the local school, and by his stepuncle, Haller eventually began to write poetry and thought of becoming a physician. He studied medicine, botany and anatomy at Tübingen and then moved on to Leiden, where he studied with, among others, Hermann Boerhaave. After receiving his medical degree at the age of 18, Haller toured Europe studying at various locations including London, Oxford, Paris and Strasbourg. In 1728 he ended up in Basel, where he spent several months learning mathematics from Johann Bernoulli. Simultaneously, Haller spent his free time hiking in the Alps collecting botanical specimens. From 1728 to 1729 he lectured in anatomy in Basel but then returned to Bern to practice medicine, continue his anatomical studies, collect plants for his herbarium, and tutor students.

None of this truly counted as a career, however, and Haller also continued to look for more suitable employment. He found it in 1736 when he was chosen as the first Professor of Anatomy, Surgery and Medicine at the newly established University of Göttingen. Surprisingly, he gave up his professorship and returned to Bern in 1753 in order to accept a political post and a position running the Bern salt works. Married three times, eight of his children survived to adulthood. He suffered from a number of ailments including gout, eye pain and stomach distress. An overindulgence in tea led to many sleepless nights, something he counteracted with opium, to which he became addicted.

Throughout his life he produced both scientific and literary works and worked hard to improve his local community. He published a work of poetry in 1732, *Die Alpen*, which earned him some early fame. These poems received their inspiration from his alpine excursions to collect plants. Towards the end of his life he returned to literature and wrote several philosophical romances. A religious man, he strove to reconcile faith and science and to defend Christianity against atheism. He became well known during his lifetime and was given membership in many scientific societies throughout Europe as well as numerous honors and awards. He also served as president of the Göttingen Royal Society of Sciences. In addition to his work in medicine, embryology, physiology and botany, he compiled lengthy bibliographies on several different subjects, contributed a number of articles to the *Encyclopédie* of Diderot and d'Alembert, and wrote innumerable book reviews for various journals, especially the *Göttingische Anzeigen von gelehrten Sachen*, a journal he also edited from 1747 to 1753.

Haller's physiological and anatomical studies employed numerous observations and experiments. Arguably his most important discovery, and a major blow against the idea that the body was simply a machine (iatromechanics, an idea that prevailed during this period), was that of the property of irritability. Francis Glisson had put this idea forth first in the seventeenth century but Haller experimentally demonstrated its veracity. Experiments with the gallbladder showed that the organ discharged more bile when irritated rather than simply maintaining a constant flow of bile. This indicated that the gallbladder responded to the needs of the body and that matter had the power to move and regulate itself. Haller applied this understanding of irritability to the heart in order to explain cardiac activity. Haller also differentiated the concept of irritability from that of sensibility. While irritability implied a physical response to certain conditions, sensible tissues sent a message to the brain for processing; the key examples here were nerve endings which sent messages to the brain but which did not necessitate any particular action. See also **Glisson, Francis (1597–1677)**; and **Physiology (The History)**.

Haller expressed some caution concerning his experiments. It did seem clear to him that irritability was innate in tissues; evidence for this came from the involuntary muscular contractions in both live animals and in dead muscle tissue. He did not want, however, to imply that bodily tissues were simply part of a mechanical model. Instead, he felt that matter essentially was passive and that a God was required to have motion. Haller claimed that his physiology was an *anatomia animata* and that he wanted to examine and explain the behavior of matter without attempting to go beyond the realm of the senses. Nonetheless, his ideas were taken up eagerly by eighteenth century materialists and atheists such as Julien Offray de La Mettrie. La Mettrie even praised Haller in the dedication of his book, *L'homme-machine*; Haller was identified, somewhat ironically and certainly against his wishes, as a spiritual father to materialism. Haller responded to this with a public letter disavowing any association with La Mettrie in particular and materialism in general. See also **La Mettrie, Julien Offray de (1709–1751)**.

Haller also became embroiled in a debate over embryology with the Count de Buffon. In this area, as in his studies of physiology and anatomy, Haller was also an experimentalist. Haller initially took his cue in this subject from his former teacher Boerhaave who had argued in favor of spermaticist preformation. Boerhaave had taught that the embryo could be found within the male semen and that this seed developed into a complete organism within the female. The work of Abraham Trembley and Charles Bonnet on the regeneration of polyps, however, caused Haller to convert, temporarily, to epigenesis. He realigned himself with the preformationist camp, this time in favor of ovist preformation, after conducting experiments on chicken eggs, particularly on the formation of the bones, heart and lungs of the embryo. In addition to his own experiments, Haller was also reacting to the publication of Buffon's theory of generation in 1749 and to Buffon's *Histoire Naturelle* more generally. Haller wrote a preface to the second volume of the German edition of Buffon's massive work in which he both summarized and criticized Buffon's views. Buffon had put forward a two-semen theory that was intended to account for the fact that children often resembled both of their parents. Haller completely dismissed the notion of female semen on the basis of his own anatomical observations. More crucially, Haller objected to the lack of any possible organization in the coming together of the semen. He did not believe Buffon's theory could ever result in complex organisms. Haller continued his research up until his death in 1777. See also **Boerhaave, Herman (1668–1738)**; **Bonnet, Charles (1720–1793)**; **Buffon, Georges Louis (1707–1788)**; and **Trembley, Abraham (1710–1784)**.

H. J. Power, Wellcome Trust Centre for the History of Medicine at UCL, London, UK

HALLETT–MOSSOP PROCESS. One of the mechanisms thought to be responsible for secondary ice production, when ice crystal concentrations in clouds well in excess ($\times 10\,000$) of the ice nucleus concentration are found. Also called *rime splintering*.

Ice particles are produced in the range of temperature $-3°$ ($27°F$) to $-8°C$ ($18°F$) (with a maximum at $-4°C$ ($25°F$) as graupel grows by accretion provided the cloud droplet spectrum contains appropriate numbers of droplets smaller than $12\ \mu m$ and greater than $25\ \mu m$. About 50 splinters are produced per milligram of accreted ice.

HALLEY, EDMOND (1656–1742). Halley was an English astronomer and mathematician. He studied the planetary orbits and predicted the return in 1758 of a comet. Now named Halley's Comet the probably the most famous of all comets. He also did studies on magnetic variations.

Halley was instrumental in helping Newton publish his most notable book, *Principia Mathematica Philosophiae Naturalis* (1687).

See also **Halley's Comet**.

J. M. I.

HALLEY'S COMET. This is probably the most famous of all the comets. It is the brightest periodic comet, and so was the first to have its return predicted. In 1705, Edmund Halley (whose name almost certainly rhymed with valley) computed the orbit of the great comet that he and others observed in 1682, and found the elements to be almost identical with those that he derived for the prominent comets observed by Kepler and Longomontanus in 1607 and by Peter Apian in 1531. He noted that the intervals between the three dates were not quite identical, and correctly attributed the difference to gravitational perturbations of the comet's motion by the planets. Celestial mechanics had not yet advanced to the stage at which planetary perturbations could be readily evaluated, either to prove the aforementioned conclusion about the slight inequality in the intervals of the comet's return, or to enable a completely accurate prediction of the next return after 1682. Nonetheless Halley's predicted time for the next return, late 1758 or early 1759, did partially allow for the effect of Jupiter, the most important of the perturbing planets. The comet was actually first Seen again on Christmas night of 1758. By that time, with the aid of improved mathematical methods, perihelion passage had been computed beforehand by Clairaut, Lalande, and Madame Lepaute, and predicted to be within a month of mid-April, 1759. The comet actually passed perihelion on March 13 of that year. With each successive return since 1682, Halley's comet has had its position determined with methods of increasing precision. This, coupled with increasingly sophisticated computational procedures, has led to increasingly accurate predictions for subsequent returns: in 1835 it passed perihelion within a few days of the predicted time, and for the return of 1910 the agreement was better still. A 1971 ephemeris for the 1986 return made some allowance for nongravitational forces (See also **Comet**), namely, the partially known effects of asymmetric emission of materials from the nucleus. This ephemeris apparently predicted the February 9, 1986 perihelion correctly within $1\frac{1}{2}$ hours.

Following the apparition of 1910, it was possible to compute the dates of perihelion passage backward for many centuries. Examination of ancient records — which, prior to the fourteenth century, are mainly Chinese — has enabled observations of Halley's comet to be identified with confidence as far back as 87 B.C. This is no small achievement, in view of the planetary perturbation problem on the one hand, and, on the other, of the fact that many comets are as bright as or brighter than Halley's have been recorded.

At its return in 1910, Halley's comet was first picked up by Wolf at Heidelberg, on September 11, 1909, at 5×10^8 km from the sun, intermediate between the distances of Mars and Jupiter. It approached the sun in the evening sky as a telescopic object, passing within the earth's orbit on the far side from the sun, and passed perihelion, at 0.59 au, on April 20. It emerged into the morning sky in the first weeks of May as a beautiful naked-eye object for those possessing a dark sky. See Figs. 1 and 2. It then turned eastward and passed between the earth and sun; according to computation the head of the comet actually transited the solar disk on May 19 (universal time), although it was undetectable in doing so. Around this time, experienced observers with good skies traced the tail to a distance of 120° from the below-horizon head. The earth grazed the comet's tail and probably passed through it, and patent-medicine vendors advertised concoctions for warding off the effects of the comet's tail, which had, however, no detectable terrestrial effects. A few days later the head of the comet was visible to the naked eye in the evening sky, despite interference by bright moonlight (except for a total lunar eclipse on May 23). It remained a naked-eye object, for experienced observers with good skies, throughout most of June. It was followed photographically, at that epoch already a more sensitive method than visual detection, until July 1, 1911, when it was 8.3×10^8 km from the sun or slightly more distant than Jupiter. It should have reached aphelion, more distant than the planet Neptune, in 1949.

Return of the Comet in 1986. On its next return, the comet was seen on October 16, 1982, almost $3\frac{1}{2}$ years before perihelion, by observers using a charge-coupled device on the Mt. Palomar 200-inch telescope. The comet was magnitude 24.2 and 1.6×10^9 km from the sun, or slightly more distant than the planet Saturn. It was within 8 seconds of arc of its predicted position.

Fig. 1. Visual aspect of Halley's comet on the morning of May 12, 1910. The Square of Pegasus is on the left; the planet Venus is on the right. The comet tail is about 32 degrees long. The observer was in Mexico.

At perihelion in 1986, Halley's comet was almost directly behind the sun. See Fig. 3. As had been foreseen, this circumstance made the 1985/86 return the poorest for visual observation in many centuries. This was especially true for dwellers of the north temperate zone, since, as the comet moved out from behind the sun, it moved for about six weeks in the skies of the far south. Experienced observers agree that, especially for northern observers, the past two decades have provided several comets that were visually more impressive than Halley's 1985/86 display; certainly comet Bennet (1969) in April, 1970 far outperformed the recent Halley. Nevertheless, even northern observers saw comet Halley with the naked eye, in dark skies. Moreover, from within northern cities the comet was readily visible in binoculars and small telescopes in December and January of 1985/86, and again in mid-March, late April, and the first days of May. The far-southern history occurred from the second half of March through most of April. The urban appearance of the comet was much like the central bulge of the Andromeda nebula, which, as was the comet at its brightest, is a naked-eye object in reasonably dark skies. A naked-eye tail was visible in March and April of 1986 (on dark skies), estimates of length ranging from a few degrees to a few tens of degrees.

Undoubtedly the most important thing about the recent visit of Halley's comet to the inner solar system is the fact that it was met by no less than five spacecraft, sent by the European Space Agency (ESA), Russia, and Japan. The ESA flyby, called Giotto, made much the closest approach, penetrating deeply within the comet's head and coming within about 500 km of the solid nucleus on March 14, 1986. An excellent view was also made by the aging *Pioneer Venus* spacecraft.

Since Giotto was launched in the direction of the Earth's motion around the sun, while Halley's comet moves in the opposite sense, the spacecraft met the comet at the terrific relative velocity of 68 km sec^{-1}. Consequently, the imaging optics were badly sand-blasted by cometary

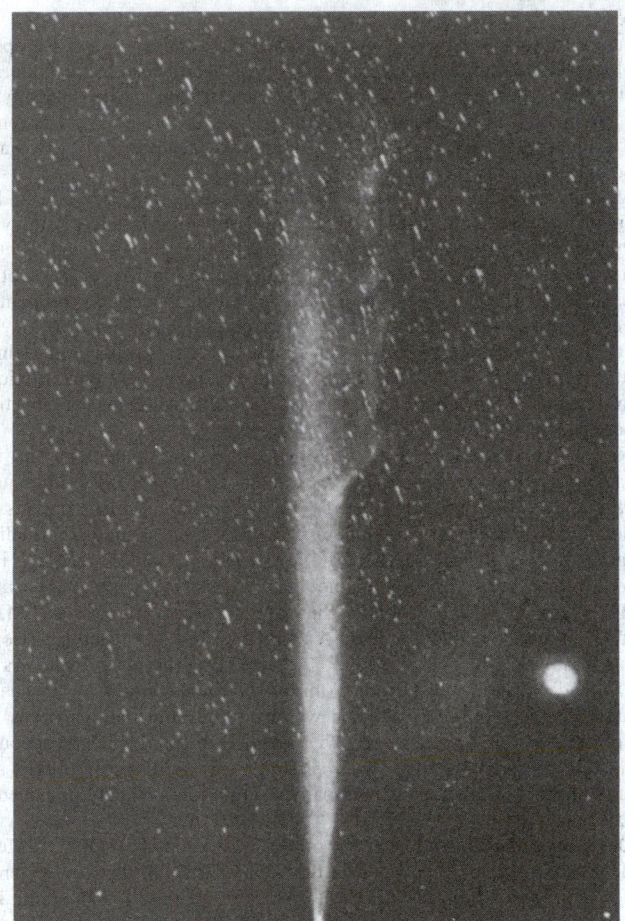

Fig. 2. Lowell Observatory photograph of Halley's comet on May 15, 1910. Venus, its image greatly enlarged by overexposure, is on the right.

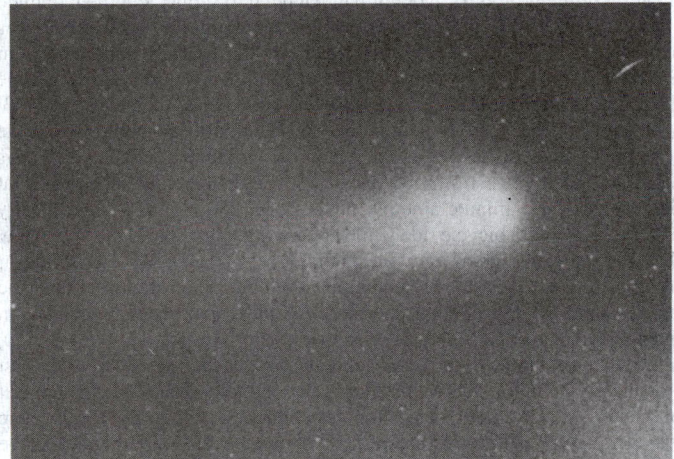

Fig. 3. Warner and Swasey Observatory photograph of Halley's comet on March 8, 1986.

dust long before closest approach to the nucleus, and the best view of the nucleus was transmitted at 11,000 km separation. The nucleus appeared oblong, about 15 × 7 km in size. Outgassing and dust emission were occurring at only a few areas on the sunlit portion of the nucleus, whose albedo is estimated at 2–4% (close to indirect ground-base estimates). See Fig. 4.

Both Giotto and the Russian flybys were instrumented to analyze the comet dust, which proved to be rich in H, C, N, and O. Hence it appears not unlikely that comet dust includes organic materials. The flybys also greatly strengthened the evidence that a major and probably principal molecule

Fig. 4. This picture is a composite of 60 of the images of Comet Halley that were taken by the Halley Multicolour Camera during the GIOTTO flyby of the nucleus. The very successful GIOTTO mission was ESA's first interplanetary mission. The spacecraft flew to within 600 km (375 miles) of the nucleus at 00:03:02 UT on 14 March, 1986. The Halley Multicolour Camera (HMC) was built by an international team led by the Max Planck Institut für Aeronomie, FRG. West Germany provided the electronics, structure and mechanisms, France the optics, Italy the optical baffle and deflecting mirror, and Belgium the optical simulator. The United States' participation was directed by NASA and performed by Ball Aerospace Systems Division, which provided preliminary engineering design, baffle design, program management services and image processing of the camera data, and is participating in the analysis of the images.

In this picture the sun is to the left and north is up. The bright areas are the source regions for the active dust jets. The dark night side of the nucleus is silhouetted against light scattered from dust which lies on the far side of the nucleus. Sunlight is illuminating the nucleus from an angle of 107 degrees from the viewing direction, and the sunrise terminator can be clearly Seen in the picture. The complex surface structure can be Seen at the foot of the dust jets. The bright spot in the night side of the nucleus is caused by the top of a "hill" which extends up into the morning sunlight. The surface has very low reflectivity (about 4 percent), suggesting that surface is covered by a dark mantle which is porous and traps the light. The surface is a good insulator, separating the surface at a temperature (in sunlight) of 300 K to 400 K (100F to 250F) from the icy core with a temperature of 50 to 100 K (−400F to −300F). The active regions are associated with cracks in the surface mantle caused by thermal stress. The long dimension of the nucleus is about 15 km (9.25 miles) and the short dimension about 8 km (5 miles). The images were taken at distances of 20,000 km to 4000 km (12,500 to 2500 miles) from the nucleus at a relative velocity of 68 km/sec (153,000 miles per hour). The resolution varies from 800 meters ($\frac{1}{2}$ mile) at the lower right to 100 meters (330 feet) at the foot of the bright jet. The nucleus rotates about an axis perpendicular to the long dimension with a period of about 54 hours, and there is some evidence for a nutation period of about 7.3 days.

Image processing by Ball Aerospace. HMC data are copyright 1986 by the Max Planck Institut, West Germany. This reproduction is by permission of Ball Aerospace Systems Division, Boulder, Colorado (Harold J. Reitsema).

present in the nucleus is H_2O. The "dirty snowball" model of the cometary nucleus, proposed by Fred Whipple in the 1950s, is thus vindicated.

See also **Stardust** mission.

References

Halley's summary of his comet work was published in *Synopsis Astronomiae Cometicae*, Oxford 1705, and *Tabulae Astronomicae*, London 1749. See also Shapley and Howarth, *A Source Book in Astronomy*. Contemporary English-language accounts of the 1910 apparition of Halley's comet are given in the journals *The Observatory*, Vol. 33, and *Popular Astronomy*, Vol. 18. All the securely identified perihelion passages of the comet are listed in Marsden's *Catalogue of Cometary Orbits* (Smithsonian Astrophysical Observatory, 5th ed., 1970). The 1971 calculation of the 1985–86 ephemeris is by Brady and Carpenter, published in The Astronomical Journal, **76**, 728 ff. Calculations by Yeomans, leading to the date of the 2061 return, are discussed in *The Astronomical Journal*, **82**, 435 ff.; this paper also lists in detail the early basic papers. The spacecraft Giotto was named for the painter Giotto; See also "Giotto's Portrait of Halley's Comet," by R.J.M. Olson, *Sci., Amer.*, **240**(5), 160–170 (1979).

Additional Reading

Anisimov, S. et al.: *Thermophysical Aspects of Meteoroid Protection in Halley's Comet Project Vega*, Vol. 2, Gordon and Breach Publishing Group, Newark, NJ, 1989.

Balsiger, H.K., H. Fechtig, and J. Geiss: "A Close Look at Halley's Comet," *Sci. Amer.*, 96 (September 1988).

Belton, M.J.S.: "P/Halley: The Quintessential Comet," *Science*, 1229 (December 13, 1985).

Brandt, J.C. and M.B. Niedner, Jr.: "The Structure of Comet Tails," *Sci. Amer.*, 49 (January 1986).

Cowen, R.: "Frozen Relics of the Early Solar System," *Science News*, 248 (April 21, 1990).

Etter, R. and S. Schneider: *Halley's Comet: Memories of 1910*, Abbeville Press, New York, NY, 1985.

Gore, R.: "Much More Than Met the Eye: Halley's Comet '86," *Nat'l Geographic*, 758 (December 1986).

Mason, J.W.: *Comet Halley: Investigations, Results, Interpretations*, Vol. 1, Prentice-Hall, Inc., Upper Saddle River, NJ, 1990.

Pollard, S. and N. Calder: *Comets: Speculation and Discovery*, Dover Publications, Mineola, NY, 1994.

Rahe, J.H., M. Neugebauer, and R.L. Newburn, Jr.: *Comets in the Post-Halley Era*, Kluwer Academic Publishers, Norwell, MA, 1991.

Reitsman, H.J., W.A. Delamare, and F.L. Shipple: "Active Polar Region on the Nucleus of Comet Halley," *Science*, 198 (January 13, 1989).

Sagan, C. and A. Druyan: *Comet*, Ballantine Books, Inc., New York, NY, 1997.

Schaaf, F.: *Comet of the Century: From Halley to Hale-Boop*, Springer-Verlag Inc., New York, NY, 1996.

Smyth, W.H. et al.: "Analysis of the Pioneer-Venus Lyman-Alpha Image of the Hydrogen Coma of Comet P/Halley," *Science*, 1008 (August 30, 1991).

C. BRUCE STEPHENSON, Case Western University, Cleveland, OH

HALLUCINOGENIC DRUGS (Hallucinogens).

Hallucinogen are one of the oldest classes of drugs used by humanity. Although they have been used by people in ritual settings for thousands of years, they remain contemporary drugs that have their own distinct subculture and usage today. According to the 1996 National Household Survey on Drug Abuse, hallucinogens are the second most frequently used illicit drug in the United States after marijuana, with 10% of the adult population (about 2.1 million people) having tried them at least once. This means that more Americans have used hallucinogens than have used cocaine, heroin or other illegal drugs. These statistics are similar to those found in other countries. Thus, research on the effects, pharmacology and history of hallucinogens, while neglected in large part since the 1960s, remains a pertinent area of investigation.

Widespread Use of Hallucinogenic Drugs Throughout Human History

Ancient cultures, almost exclusively in the New World, identified hallucinogen-containing plants and incorporated their unusual effects into religious and mystical rituals. Mescaline is one of the oldest known, with archaeological evidence of use dating back some 3000 years. It is found in high concentrations in the peyote cactus and is ingested by chewing the "buttons" of this cactus.

Although Spanish colonialists attempted to eradicate ceremonial use of peyote, the practice spread beyond Mexico in the late nineteenth century to native American tribes of Central and North America. At that point, writers of European descent were introduced to peyote, including the poet William Butler Yeats, who was impressed with the "intense pleasure" that it produced. In the following century, Aldous Huxley described his experience with mescaline in *The Doors of Perception*. The Huichol tribe in Mexico is perhaps the most famous for its modern-day use of peyote, involving long journeys to gather the plants for religious purposes. In the United States, it is illegal to possess or grow peyote for recreational use. Members of the Native American Church, however, do have the legal right to use peyote in religious ceremonies, following many years of court battles.

Psilocybin-containing mushrooms furnish another hallucinogen and have been used in the indigenous cultures of Central America since ancient times, with ceremonial mushroom-shaped stones dating to 1000 BC. Psychedelic mushrooms first appear in recorded history when European explorers came into contact with Central American tribes in the sixteenth century. In their writings there is the description of participants at the coronation of Montezuma II consuming "inebriating mushrooms" and having heavenly visions. These fungi were known by the name "Teonanacatl", which translates to "divine mushroom."

More recently, in 1955 R. Gordon Wasson (then vice president of J.P. Morgan bank) and his wife popularized knowledge of this native psilocybin culture after they experienced the mushrooms themselves through contact with the curandera (healing woman) Maria Sabina in Mexico. Scientific interest followed, leading to the infamous "Good Friday" experiment in which divinity students in Boston were given either 30 mg of psilocybin or a small amount of vitamin B combined with amphetamine in a double-blind study held in a chapel. Not surprisingly, only those students who received the active hallucinogen reported religious or mystical experiences.

Native use of hallucinogens does not always originate with the use of a single plant. One notable example is ayahuasca (also known as *yage*), an admixture of two plants, *Psychotria viridius* (which contains the hallucinogen DMT) and *Banisteriopsis caapi* (which contains harmaline, a monoamine oxidase (MAO) inhibitor) that are boiled together into a foul-tasting brew. The harmaline is a necessary component because DMT is not active orally unless the naturally occurring enzyme MAO in the gut that metabolizes DMT can be blocked from acting. Use of ayahuasca was first recorded in the 1850s by European explorers in Brazil and Peru. Participants in Amazonian rituals ingest ayahuasca in a structured religious ceremony in which the ensuing hallucinations are interpreted as visions from god. This practice inspired modern accounts of ayahuasca use, including *The Yage Letters*, by William Burroughs and Allen Ginsburg. Today, an ayahuasca-based religion is widely practiced in Brazil.

One important structural variant of DMT is 5-methoxy-DMT (5MeO-DMT). This compound is found in the bark of trees in the *Virola* genus, which is ground up by tribes in South America who use it as a snuff. The powder is forcefully blown into the nostrils of a tribe member through long, narrow tubes made from bones. Intense hallucinations result, typically followed by a stupor.

No account of the history of hallucinogen use would be complete without mention of the prototypic modern hallucinogen, lysergic acid diethylamide (LSD; the S stands for *Sauer*, the German word for acid). LSD was synthesized in 1938 by Albert Hofmann at the Sandoz Laboratories in Switzerland when he developed a series of compounds to strengthen uterine contractions during labor. While working with these chemicals in 1943, Hofmann noticed peculiar sensations. He decided to go home, where he fell into a state of "extreme activity of the imagination," Several days later, intrigued by his strange responses, he deliberately ingested what he thought was a minute amount of the drug (250 μg) and subsequently experienced a wild bicycle ride while attempting to go home. We now know that 250 μg (microgram) is a substantial dose of LSD. He later wrote of his experience, "There surged upon me an uninterrupted stream of fantastic images of extraordinary plasticity and vividness and accompanied by an intense, kaleidoscope-like play of colors."

LSD was not the first hallucinogen to be produced synthetically (the first was mescaline in the 1880s), but it was the one to make the biggest impact on modern society. Long before hippies embraced LSD in the 1960s, there was an active underground use of LSD by intellectual suburbanites in the 1950s who were interested in its effects on the mind and on creativity. Access to LSD at that time was legal and it often came from psychiatrists who were investigating its behavioral effects in humans. LSD remains the most easily accessible street hallucinogen and it is often sold as other, more exotic psychedelics.

Finally, MDMA ("Ecstasy") has become a drug of choice at "raves," all-night dance parties featuring electronic music and colorful light shows. MDMA was synthesized in the early part of the twentieth century, but only gained popularity in the 1980s for its ability to induce emotional openness and its stimulant-like properties.

Different Classes of Hallucinogenic Drugs

The classic hallucinogens are generally divided on the basis of chemical structure into two families: tryptamines (indolealkylamines) and phenethylamines. Tryptamines contain an indole ring, which is a six-membered ring attached to a five-membered ring containing nitrogen. Phenethylamines are compounds containing a phenyl ring attached to an ethyl group ending in a nitrogenous amine group.

Tryptamines

LSD. LSD is a tryptamine, but it has a much more complicated chemical structure than most tryptamine hallucinogens and is better classified as an ergoline. The chemical base for synthesis of LSD comes from a black fungus (*Claviceps purpurea*) that grows on rye, known as ergot. LSD is one of the most potent drugs known and its effects can be felt after an oral dose of only 25–50 μg (0.025–0.050 mg). The dose of LSD found on the street typically ranges from 50 to 100 μg.

The half-life of LSD (the amount of time it takes for one-half of the dose of the drug to be metabolized in the body) is approximately 3 h. Since it takes about 3–4 half-lives before most drugs cease to have noticeable effects, the duration of action of LSD will range from 9 to 12 h in most people.

Psilocybin Mushrooms. Psilocybin is a tryptamine hallucinogen with a phosphoryloxy substitution at the 4-position of the indole nucleus (4-phosphoryloxy-N,N-dimethyltryptamine). Hydrolysis of psilocybin in the body strips off the phosphoric acid and leaves psilocin, which is about 1.4 times as potent as a hallucinogen as the parent compound. When the ratio of the biological potency of these two drugs is compared, it is the same as the ratio as that of the molecular masses of the drugs.

Psilocybin is present in certain mushrooms genera, including *Psilocybe*, *Stropharia* and *Panaeolus*. Mushrooms comprise about 90% water, but there is on average approximately 2–3 mg of psilocybin per gram of dry weight mushroom. The potency of psilocybin is about 1/100 of that of LSD, making the typical dose of psilocybin between 5 and 15 mg. Since the half-life of the active metabolite psilocin is approximately 1.5 h, the psilocybin experience will generally last approximately 4–6 h.

Psilocybin mushrooms are available on the street either fresh or dried. Common lore holds that fresh hallucinogenic mushrooms will turn blue when crushed; while this is true for some species, it is not a reliable identifying characteristic and cannot be used for determining whether a mushroom is safe to consume.

DMT and Related Compounds. Dimethyltryptamine (DMT) is another psychedelic tryptamine. It is present chemically in a variety of plants (including *Psychotria viridis*), as well as in minute quantities in the human body. This led to the speculation that perhaps DMT could induce psychosis endogenously, but this has never been shown to be true. DMT is biologically active in pharmacological doses and a seemingly endless series of substitutions on the basic DMT structure are also hallucinogenic.

As mentioned above, DMTs are not active by ingestion without inhibition of the enzyme MAO in the gut. Use of DMT is therefore typically by smoking the chemical in powder form or through injection. A dose of 50–100 mg is sufficient for hallucinogenic effects when DMT is smoked, but as little as 4 mg are needed when it is used intravenously. Both methods are extremely rapid in delivering DMT to the brain (smoking is actually faster because of a more direct blood route) and the effects subside within 60 min.

Phenethylamines

Mescaline. Chemically, mescaline is 3,4,5-trimethoxybenzene ethanamine, or 3,4,5-trimethoxyphenethylamine. The most familiar source of mescaline is the peyote cactus, *Lophophora williamsii* but it is also found in significant quantities in the San Pedro cactus, *Trichocereus pachanoi*. An average human dose is approximately 400 mg, making it 10 000 times less potent than LSD. Mescaline is a long-lasting hallucinogen, persisting for 12–15 hours. A notable side effect of mescaline is an early and intense nausea, often accompanied by vomiting. This response is so typical that it has been incorporated into the ritual use of the drug as a symbolic cleansing phase.

DOM and Related Compounds. In the mid-1960s, structure–activity relationship investigations led to the synthesis of a new phenethylamine hallucinogen, DOM (2,5-dimethoxy-4-methylamphetamine). (It should be noted here that the name amphetamine is simply chemical shorthand for its structure: α-methylphenethylamine.) An effective dose of DOM ranges from 3 to 10 mg orally. An extensive family of DOM derivatives have been synthesized, including DOB (substituting bromine at the 4-position) and DOET (substituting an ethyl group at the 4-position). These drugs have a long onset time (up to 2 h) and their effects can persist for 15–20 h. The unusually long duration is related to the chemical structure. The presence of an α-methyl group on the phenethylamine physically prevents easy enzymatic degradation of the drug, extending the time the drug acts in the body.

In contrast, a related series of phenethylamine hallucinogens without the α-methyl group are short-acting (2–6 h) because the metabolizing enzyme has easy access for biological degradation. Recently, one of these, 2-CB (2,5-dimethoxy-4-bromophenethylamine), a derivative of DOB, has become a popular street drug, especially among college students.

MDMA. In addition to the traditional phenethylamine psychedelics, MDMA (3,4-methylenedioxymethamphetamine) and MDA (3,4-methylene dioxyamphetamine) form a class of phenethylamines known as the hallucinogenic amphetamines. This term is something of a misnomer, since MDMA is characterized as having the ability to increase emotional warmth and stimulate the desire for movement rather than to produce hallucinations. However, users do note that some of the physical and psychological effects of MDMA do bear a similarity to those of true hallucinogens. A typical dose of MDMA, which lasts 2–4 h, is 125 mg. It is not uncommon for users to 'boost' with another dose of the drug when the effects of the initial dose begin to subside.

Conditions of Use that Influence the Behavioral Effect of Hallucinogens

Hallucinogenic compounds are usually ingested orally, while some DMT-based hallucinogens are not active orally by themselves and are typically smoked. When taken by mouth, the effects of hallucinogens usually begin within 30–60 min. The duration of these effects is determined by the chemical structure of the drug and can range from a few hours to more than a day. Acute tolerance develops rapidly, so that ingesting the same dose of the same hallucinogen a short period after the first dose will not reproduce the same level of response. Tolerance to a single dose of most hallucinogens persists for several days and limits the abuse potential of these drugs.

More than most other classes of drugs, the response a person has to hallucinogens depends as much on psychological factors as on pharmacological factors such as chemical structure or dose. The common term for someone experiencing the effects of a hallucinogen is that of "tripping." The initial effects are somatic and include sympathomimetic signs like increased heart rate and respiration, dilated pupils, and the sensation of butterflies in the stomach. Within an hour after the onset of physical responses, many (but not all) individuals begin to experience visual hallucinations. The ethnobotanist Richard Schultes of Harvard University, for example, spent decades identifying Amazonian hallucinogens but claimed that he never hallucinated when ingesting the plants with local tribespeople.

When visual hallucinations do occur, they can take the form of moving patterns and shapes, often in colorful interplays. Walls may seem to breathe and the appearance of everyday objects can take on amusing or upsetting qualities. In contrast to the effects of marijuana, which are often described as being "dream-like," hallucinogens may impart a "super-real" quality to existence. Senses may seem to merge, leading to the experience of synaesthesia, where smells may be "heard" or colors may be "tasted." Auditory hallucinations are also possible, despite clinical lore suggesting that such effects can only occur with schizophrenia. See also **Schizophrenia**.

During and after the peak of the hallucinatory phase, pronounced psychological changes of an intensely personal or mystical nature can begin to occur. Indeed, it is these unique effects that characterize hallucinogens as a class, perhaps more so than hallucinations themselves. These responses can include changes in meaning or insight, alterations in the ability to concentrate, development of selective attention, misperceptions of time, and a sense of losing control that can lead to a depersonalization or an "anxious passivity."

As the dose of any hallucinogen is increased, there will initially be a quantitative increase in the effects of the experience. Above a certain dose range for each individual psychedelic, the hallucinogenic response may progressively take on a qualitatively different character. At such higher doses, an individual's connection to the real world may begin to erode significantly, such that ego boundaries may disintegrate, one's thoughts may seem to merge profoundly with the universe, and hallucinations may be so profuse that it is difficult to see one's surroundings.

It is extremely uncommon for any other class of drugs to produce effects that are so profound that a single exposure will lead an individual to claim that her or his "life has been changed"–whether positively or

2382   HALLUCINOGENIC DRUGS (Hallucinogens)

negatively. As a means of encapsulating such psychological changes in an understandable manner, many religious tendencies have been founded in response to the psychedelic experience. Shamanistic rituals involving peyote or psilocybin mushrooms have been conducted since prehistory, with adherents ingesting these plants to produce visions of gods or ancestors with knowledge of the future.

In modern times, recreational users of hallucinogens often view their consumption of these drugs as enhancing personal or spiritual growth. It is this belief that led to early suggestions that hallucinogens had great potential as aids in psychotherapeutic contexts. Despite the fact that decades of encouraging therapeutic work with these compounds often went unpublished, there has been renewed interest in and approval of research studies into the usefulness of hallucinogens for such purposes, both in the United States and elsewhere. Applications under investigation include administration of LSD to treat drug abuse or administration of MDMA to ease psychological distress in terminal cancer patients.

As the acute effects of hallucinogens subside, after several hours, users can experience a variety of emotions ranging from feeling at peace to feeling unsettled and suspicious. Most "bad trips" typically happen after the peak of drug effects. This is because once frank hallucinations have stopped, some people can begin to think that they no longer are under the influence of a drug. Yet if unusual psychological responses continue, such people may be unprepared and may feel anxious. It is worth noting that emergency-room visits are rare in hallucinogen users because direct physical distress is unlikely.

One concept that has aided in the process of calming someone during a bad drug experience is that of "set and setting." This phrase suggests that the response a person will have to hallucinogens is based on the "set" of the mind (e.g. contented or agitated) and the setting in which the person consumes the drug (e.g. comfortable living room with friends or alone in an unfamiliar situation). The incidence of negative reactions is obviously reduced under conditions where people can be made to feel at ease with both the mindset and the circumstances in which they take hallucinogens.

In contrast to other psychedelics, MDMA does not typically induce hallucinations. Rather, MDMA has effects that remind users of both the sensory-enhancing qualities of hallucinogens and the energizing properties of amphetamines. More profoundly, MDMA often creates an emotional state of deep trust and lowered defenses. Therapists applied this quality of this class of drugs, also known as *empathogens* or *entactogens*, as a means of accessing difficult psychological problems in their patients or in couples counseling.

Interactions with other Drugs. Given that hallucinogens of all classes have virtually no physiologically toxic effects in typical doses, users have often been somewhat cavalier in combining hallucinogens with other drugs to alter the quality of the "high" produced. While adverse reactions are unusual, there are some interactions that should be mentioned.

The most important of these is that acute or chronic administration of an MAO inhibitor can have life-threatening effects when combined with MDMA (or any other amphetamine). The idea behind using an MAO inhibitor with a psychedelic is usually based on an awareness of the pharmacology of ayahuasca, where an MAO inhibitor can activate a tryptamine hallucinogen with relative safety. The combination of an MAO inhibitor and MDMA is not safe, and there are numerous reports in the medical literature of a dangerous increase in blood pressure known as a hypertensive crisis when these drugs are taken together. Individuals should be severely cautioned against exposing themselves to this sort of risk.

Other antidepressants can have profound psychological effects on the human hallucinogenic response that do not seem to involve adverse physiological effects. Specifically, prolonged use (3 weeks or longer) of serotonin-reuptake inhibitors (such as fluoxetine (Prozac)) or MAO inhibitors (such as phenelzine (Nardil)) will significantly reduce or eliminate the hallucinogenic response to the classic hallucinogens. In contrast, prolonged use of lithium or tricyclic antidepressants (such as desipramine (Norpramine)) will produce an exacerbation of the hallucinogenic response, often to a very unpleasant degree.

However, acute (single-dose) administration of antidepressants does not appear to have the same effect. Acute administration of serotonin-reuptake inhibitors or MAO inhibitors, for example, seems to potentiate the response to LSD. This difference in effect, based on duration of antidepressant administration, seems to be related to adaptive changes in serotonin levels and neurotransmission.

Pharmacology of Hallucinogens

The pharmacology of the classic tryptamine and phenethylamine hallucinogens has been studied by scientists since the beginning of the twentieth century, but it is only since the 1970s that we have had a clearer picture of the distinct mechanisms of action of these chemicals on the brain. The initial focus of these efforts was the neurochemical serotonin (also known as 5-HT, 5-hydroxytryptamine). This is because serotonin itself is a tryptamine and contains an indole ring, as do all tryptamine hallucinogens. Research has subsequently shown that serotonergic receptors are the primary binding sites of both tryptaminergic and phenethylamine hallucinogens.

Although phenethylamine hallucinogens chemically resemble the neurotransmitter noradrenaline, many years of investigation have demonstrated that this neurochemical system is not directly responsible for their effects. However, noradrenergic systems do become hyperresponsive to sensory stimulation after any hallucinogen has been administered. This suggests that noradrenaline may contribute to hallucinogen-induced alterations in the processing of sensory information by the brain. Similarly, dopaminergic systems may play a role in the characteristic responses to some hallucinogens, both directly through activation of dopamine receptors and indirectly through modulation by serotonergic systems, but they are not responsible for the psychedelic effects of these drugs per se. See also **Adrenaline and Noradrenaline**; and **Neurotransmitters**.

The serotonergic systems of the brain are so complex that there are currently 14 known serotonin receptor subtypes. Of these, only a few appear to play a significant role in the hallucinogenic response. The most critical receptor is thought to be the 5-HT$_{2A}$ site. This is based on the fact that there is a high correlation between binding of both tryptamine and phenethylamine hallucinogens at the 5-HT$_{2A}$ receptor and their ability to produce hallucinatory effects in humans. Tryptamine hallucinogens also have high binding affinity at 5-HT$_{1A}$ receptors, which may be important for certain aspects of their hallucinatory response that distinguish them from phenethylamine hallucinogens. Additionally, hallucinogens have high binding affinity to the 5-HT$_{2C}$ receptor, as well as to two more recently identified receptors subtypes, 5-HT$_6$ and 5-HT$_7$, that have not been fully characterized.

Investigations of the pharmacological mechanism of action of hallucinogens have typically been in animals. In addition to biochemical studies of receptors, observational investigations of behaviors following the administration of hallucinogens have also been conducted. One of the most frequently used techniques is that of drug discrimination. In this operant conditioning method, an animal is trained to press a bar on one side of a test cage when it is given a hallucinogen, and a bar on the other side of the test cage when it is given saline. The discrimination between the internal sensations (known as *interoceptive cues*) evoked by a drug or by saline can be learned reliably after repeated training sessions. It is then possible to test whether the animal will press the hallucinogen-associated bar when it is injected with a challenge compound that may have hallucinogenic effects. If it does, the challenge drug is said to demonstrate stimulus generalization with the training drug, presumably because the animal has received similar interoceptive stimulation.

For instance, it has been shown that when a rat is trained to discriminate LSD, it will later identify the hallucinogens mescaline or psilocybin as being more like LSD than like saline in this test. Generalizations such as this hold true between all hallucinogens in a variety of training and testing conditions in both rats and in humans. In fact, there is a near-perfect correlation between the ED$_{50}$ (the median effective dose) of hallucinogens in drug discrimination in rats and their hallucinogenic potency in humans. Additionally, there is a near-perfect correlation between behavioral potency and the affinity of a hallucinogen to bind at the 5-HT$_{2A}$ receptor. Drug discrimination has also been used to show that pretreatment of a rat with 5-HT$_2$ antagonists will block the ability of the rat to identify the interoceptive cue of a hallucinogen, causing the rat to press the saline bar. Impressively, a recent clinical study has shown that 5-HT$_2$ antagonists can prevent humans from feeling any of the effects of a behaviorally-active dose of psilocybin. Taken together, these data form the basis of the theory that the 5-HT$_{2A}$ receptor is critical in the hallucinogenic response. See also **Pavlov, Ivan Petrovitch (1849–1936)**.

Hallucinogens also evoke distinct behaviors in animals. These behaviors have been used to suggest selective activation of serotonin receptors. Both tryptamine and phenethylamine hallucinogens will cause a rat to shake its head or its full body (a behavior known as "wet dog shakes") if given in sufficient doses. Experiments with serotonin antagonists have shown that these behaviors are directly related to activation of the 5-HT$_{2A}$ receptor. In contrast, most tryptamine hallucinogens (but only a few phenethylamine hallucinogens) produce the "serotonin syndrome," a behavioral complex involving flat body posture, movement of the front paws in a "piano playing" motion, splayed rear legs and a tail standing straight up ("Straub tail"). Again, through serotonin antagonist studies, it has been shown that this syndrome is related to stimulation of 5-HT$_{1A}$ receptors. If a novel hallucinogen is thought to have a particular receptor binding profile, watching for these two types of behaviors after administration of the drug can be indicative of where the drug might be acting in the brain.

Another behavioral test that has been shown to be sensitive to hallucinogens is the Behavior Pattern Monitor (BPM). The BPM is a large test chamber that can localize rat movement using photobeams. When rats are given LSD or other hallucinogens and placed in the BPM, there is a decrease in their natural tendency to explore an unfamiliar test chamber. This is interpreted as an increase in neophobia. Hallucinogens also increase the tendency of rats to avoid the open space in the center of the BPM, interpreted as an increase in the response to adversive or threatening stimuli. Both of these effects are blocked when rats are pretreated with 5-HT$_2$ antagonists. The behavioral patterns produced by hallucinogens are distinct from those produced by other drugs.

All of the behavioral methods detailed above have been used to demonstrate that tolerance develops rapidly to repeated doses of hallucinogens. Thus, the initial administration of a hallucinogen will produce the expected behavior, but giving the same drug again at the same dose a day later will produce less of that behavior. Continued administration will further reduce and then eliminate the behavior. Cross-tolerance has also been demonstrated in these tests, such that once an animal is tolerant to one hallucinogen, it will also be tolerant to other hallucinogens at comparable doses of efficacy. It has been shown that behavioral tolerance is the result of a decrease in (downregulation of) serotonin receptors following continual drug stimulation. These data support the idea that there is a common neurochemical mechanism of action responsible for the behaviors at the receptor level.

Mechanistically, MDMA is different from other hallucinogens in that its effects do not seem to rely on direct activation at serotonergic receptors (although there is rather poor affinity for 5-HT$_{2A}$ and 5-HT$_{1B}$ receptors). Instead, MDMA acts dualistically as a serotonin-reuptake inhibitor (like the antidepressant fluoxetine (Prozac)) and as a serotonin releaser (like fenfluramine (Pondimin), the weight-reduction drug). It also has the ability to inhibit MAO. All of these actions cause the increase of serotonin in the synapse, thereby increasing serotonergic neurotransmission. This can result in the well-known effect of hyperthermia. In rats, the increase in core body temperature due to MDMA can be exacerbated by enclosed spaces and a lack of hydration–exactly the conditions found at "raves." It is debatable, however, whether the human deaths attributed to MDMA use in clubs are the direct result of drug ingestion or rather can be attributed to club conditions or mixing of classes of drugs.

Interestingly enough, animals will self-administer nearly every drug that is used recreationally or is abused in humans, but this is not the case with hallucinogens. It is unclear why this is so, and it suggests that the explanation for human hallucinogen use may involve more than simple reinforcement theories or associated neural tracts.

Negative Aspects of Hallucinogen Use

It should be stated first that hallucinogens are like any other psychoactive drug in that they can produce unpleasant or disturbing psychological effects that may be difficult for a user to overcome while under the influence of the drug. Intoxication with any drug can similarly cause irrational and highly risky behavior. Although the stereotyped "bad trip" that can result from hallucinogen use is certainly an undesirable and unique condition, it is seen much less frequently among users of psychedelics than are the drunken rages seen in alcohol users or the manic craving for drugs seen in cocaine users. See also **Alcoholism**.

Many myths that have no factual basis have been promoted over the years regarding hallucinogens. There is no known lasting physical or neural damage from hallucinogen use, nor is there any evidence of physical dependence, a withdrawal syndrome or addiction to these drugs. One would have to consume massive amounts of a hallucinogen to overdose and die. The threat of damage to sex chromosomes is nonexistent. Hallucinogens do not remain in or destroy the cerebrospinal fluid, and physical side effects

like stomach cramps are probably not from a poor-quality drug but are instead related to the intrinsic properties of hallucinogens acting on the mesenteric serotonergic system.

One of the most famous negative attributes related to hallucinogen use is that of "flashbacks," as well as the more recent "post hallucinogen perceptual disorder." These states clearly exist in some individuals who have taken hallucinogens, but the frequency of these reactions in the population of users is known to be extremely rare. There is no evidence that these responses are based on chemical insult to the brain and it is probable that these after-effects are psychological phenomena not unlike the persistence of memories about events with significant emotional valence (such as becoming engaged or going to war). Additionally, hallucinogens should not be confused with THC (tetrahydrocannabinol; the psychoactive component of marijuana) in that they are not fat-soluble and therefore are not able to be harbored in or released from adipose tissue long after use. There is no evidence that THC can induce flashbacks.

Another negative concern is that MDMA is a neurotoxin. Part of this association may be from a confusion of MDMA with the "designer drug" 1-methyl-4-phenyl-1,2,3,6-tetrahydropyridine (MPTP) (a derivative of the opioid meperidine), which did indeed cause dopaminergic degeneration and subsequent Parkinson-like immobility. In contrast, MDMA can induce a persistent reduction of serotonergic nerve terminals when given in rodent studies. Whether this effect constitutes neurotoxicity is controversial. Careful analysis reveals that the doses used in such rats studies are comparable to 3–4 times the doses typically consumed by humans. Since it is not uncommon for users to take much larger quantities of MDMA when "boosting," it is possible that alterations in serotonin systems may occur with very frequent, high-dose use of the drug. Recent clinical studies have suggested that MDMA users have deficits in performance on a selected number of cognitive tasks (such as certain tests of memory). It is difficult to know whether these data represent behavioral toxicity since the reported effects are present only in individuals who have taken extremely high doses of MDMA for extended periods, there is a lack of baseline measurement prior to MDMA use, and the contributions of other illicit drugs also consumed are not taken into account. See also **Parkinson's Disease**.

Hallucinogen-Like Effects from Nonhallucinogenic Drugs

Other classes of drugs may also induce effects that are similar in some fashion to those of hallucinogens. Although these other drugs have certain effects that resemble hallucinogens, they elicit their own distinct physical and psychological responses based on their unique pharmacology. Drugs with certain effects that may resemble those of hallucinogens include:

- Cannabinoids such as marijuana
- Dissociative anesthetics like PCP (phencyclidine), ketamine, or the cough suppressant dextromethorphan, which act as NMDA (N-methyl-d-aspartate) channel blockers
- Deliriants such as the anticholinergics scopolomine or atropine
- Kappa and/or sigma opioid receptor agonists such as pentazocine
- Inhalants such as nitrous oxide that act nonspecifically at opioid receptors
- Muscimol, a GABA agonist found in the mushroom *Amanita muscaria*

In addition, hallucinations can be induced without drugs, under conditions such as high fever, discontinuation of chronic drug use, extreme physical or psychological stimulation, oxygen deprivation and psychotic states. What is fascinating is that the nervous system has so many different mechanisms to produce a hallucinatory response accompanied by unusual mental states.

Conclusion

Hallucinogens are a class of drugs that produce a complex set of psychological and physical effects that have been experienced by humans throughout the ages. Many of these hallucinogens occur naturally in plants, while others have been synthesized in laboratories, based on two primary chemical structures, tryptamines and phenethylamines. Only in the latter part of the twentieth century have scientists been able to identify the precise sites within the serotonergic systems of the brain that are responsible for the unique responses of hallucinogens. Given that these drugs are widely used recreationally and may have therapeutic applications, continued investigation into the neurochemical and behavioral responses they produce will continue.

Note: This is a US Government work and is in the public domain in the United States of America.

Additional Reading

Aghajanian, G.K., and G.J. Marek: "Serotonin and Hallucinogens," *Neuropsychopharmacology*, **21**(2)(suppl.), S16–S23 (1999).

Halpern, J.H., and H.G. Pope Jr.: "Do Hallucinogens Cause Residual Neuropsychological Toxicity?" *Drug and Alcohol Dependence*, **53**(3), 247–256 (1999).

Laing, R., J.A. Siegel: *Hallucinogens: A Forensic Handbook*, Elsevier Science & Technology Books, New York, NY, 2003.

Metzner, R.: "Hallucinogenic Drugs and Plants in Psychotherapy and Shamanism," *Journal of Psychoactive Drugs*, **30**(4), 333–341 (1998).

Nichols, D.E.: "Role of Serotonergic Neurons and 5-HT Receptors in the Action of Hallucinogens," In: Baumgarten, H.G., and M. Göthert: *Handbook of Experimental Pharmacology, Serotoninergic Neurons and 5-HT Receptors in the CNS*, pp. 563–585, Springer-Verlag, Heidelberg, Germany, 1997.

Shulgin, A.: *PIHKAL (Phenethylamines I Have Known and Loved)*, Transform Press, Berkeley, CA, 1991.

Shulgin, A.: *TIHKAL (Tryptamines I Have Known and Loved)*, Transform Press, Berkeley, CA, 1991.

Staff, Heffter Research Institute: *Research at the Frontiers of the Mind*, www.heffter.org

Vollenweider, F.X.: "Advances and Pathophysiological Models of Hallucinogenic Drug Actions in Humans," *Pharmacopsychiatry*, **31**(2)(suppl.), S92–S103 (1998).

Winter, J.C., D.J. Fiorella, D.M. Timineri, et al.: "Serotonergic Receptor Subtypes and Hallucinogen-induced Stimulus Control," *Pharmacology, Biochemistry and Behavior*, **64**(2), 283–293 (1999).

KATHERINE R.BONSON, National Institute of Mental Health, Bethesda, MD

HALO.   Any of a family of colored or whitish rings, arcs, pillars or spots of light that appear in the sky and are explained by the reflection or refraction of light by ice crystals. They are usually found in the vicinity of the light source, the most important of which are the sun and moon, but may arise from artificial lights if seen, say, through an ice fog. Halos exhibiting some prismatic coloration are explained, at least in part, by the refraction of light by the ice crystals. However, the color is usually fairly pale, being best on a red edge next to the light. The exceptions, in having very good colors, are the circumhorizontal and circumzenithal arcs, the positions of which are not determined by the minimum angle of refraction. Halos that are white, or show the same color as the light source itself, are explained by the reflection of light off the crystal faces. Whether explained by reflection or refraction, the pattern that emerges depends upon the crystal type, crystal orientation (actually the probability of various orientations within a population of crystals), and the elevation angle of the light (sun). With such a rich range of possibilities, a large variety of halos are theoretically possible and over 50 different halo phenomena have been documented photographically. Some halos predicted theoretically have yet to be reported, others that have been reported have yet to be explained. The most common halo is the halo of 22°. Other frequently seen halos are the parhelia of the halo of 22°, the sun pillar, the 22° tangent arcs, the circumzenithal arc, the halo of 46°, and the parhelic circle. On very rare occasions an observer's sky will be filled with a display with 10, 20, or more different halos, usually persisting for only a few minutes. Much supernatural lore has been prompted by observations of old of such events. Halos must be distinguished from optical phenomena arising from water drops, such as the rainbow, corona, and glory.

Halo of 22°

A halo in the form of a circle, or portion of a circle, with an angular radius of about 22° about a light source, such as the sun or moon. This is the most common of all halos. The sky is darker just to the inside of the halo than it is to the outside. The halo exhibits a pale coloration from a reddish tint on the inside fading to a bluish white on the outside. The 22° halo is explained by the refraction of light that enters one prism face and leaves by the second prism face beyond, thus being refracted by a prism with an effective angle of 60°. The angle of minimum deviation for an ice prism of this prism angle is about 22°, so such light does not appear inside the halo, accounting for the darker region there. The minimum angle of deviation varies slightly with wavelength, with the longer wavelengths being deviated least. This causes the reddish inner edge, outside which the additional contributions from light of increasingly shorter wavelengths decrease the color purity. The orientation that a crystal must have to contribute light to the halo depends on both the elevation angle of the light source and the portion of the halo in question, but the probability that a crystal will have a particular orientation depends upon its type and size. Consequently, it is frequently the case that for a given population of crystals and sun height, only a portion of the halo will be seen, while a change in one or the other might

enable the full circle to form. It is not the case, despite being frequently asserted, that a view of the full circle requires crystals to have random orientations.

Halo of 46°

A halo in the form of a circle, or portion of a circle, with an angular radius of about 46° about a light source, such as the sun or moon. The coloration is reddish on the inner edge to bluish on the outer edge. This halo is much less common that the halo of 22°. The 46° halo is explained by the refraction of light passing through the 90° prism formed between the side and basal faces of a hexagonal ice crystal. The minimum angle of deviation for this ice prism is about 46°. Closely associated with this halo are the 46° infralateral arcs and the 46° supralateral arcs. In particular, the shape of the 46° supralateral arc often follows the uppermost parts of the 46° halo so closely that the two are almost impossible to distinguish. In fact, upon examination, a large fraction of the halos commonly interpreted as being 46° halos turn out to be pieces of 46° supralateral arcs. Which of the two halos (46° halo or 46° supralateral arc) is more frequent has not been settled.

See also **Atmospheric Optical Phenomena**; and **Sun Pillar**.

HALOCARBONS.   A collective term for the group of partially halogenated organic species, including the chlorofluorocarbons, hydrochlorofluorocarbons, hydrofluorocarbons, halons, methyl chloride, methyl bromide, etc. These compounds, some of which are naturally occurring (largely through production in the oceans) and some of which are of anthropogenic origin are the major source of the halogens; fluorine (F) [CAS: 7782-41-4], chlorine (Cl) [CAS: 7782-50-5], bromine (Br) [CAS: 7726-95-6], iodine (I) [CAS: 7553-56-2], and astatine (At) [CAS: 7440-68-8] to the stratosphere. See also **Ozone**; **Ozone Depletion (Science of)**; and **Stratospheric Ozone**. The lower members of the various homologous series are used as refrigerants, propellant gases, fire-extinguishing agents, and blowing agents for urethane foams. When polymerized, they yield plastics characterized by extreme chemical resistance, high electrical resistivity, and good heat resistance.

See also **Fluorocarbon**.

HALOCARBONS (Ozone Depletion).   See **Oxygen**.

HALOGENATED COMPOUNDS.   See **Chlorinated Organics**; and **Organic Chemistry**.

HALOGEN GROUP.   The elements of group 17 (formerly 7a) of the periodic classification sometimes are referred to as the Halogen Group. The individual elements commonly are called *halogens*. In order of increasing atomic number, they are fluorine, chlorine, bromine, iodine, and astatine. The elements of this group are characterized by the presence of seven electrons in an outer shell, and hence have the ability to gain an electron to form negative ions with a completed octet of valence electrons. The halogens present striking similarities of chemical behavior, all being very reactive and, in particular, readily form substitution compounds with numerous organic compounds. Although these elements also have other valences all have a −1 valence in common.

HALON.   See **Ozone Depletion (Science of)**.

HAMILTON, ALICE (1869–1970).   The first American physician to devote her life to the practice of industrial medicine. In studying the lead industries in Illinois, she discovered and ameliorated lead poisoning among bathtub enamelers in Chicago. She wrote about phossy jaw, which occurred among American matchmakers who used white or yellow phosphorus. She studied the effects of carbon monoxide among steel workers, the toxicity of nitroglycerin among munitions makers during World War I, the symptoms of hatters exposed to mercury in Danbury, Connecticut, and the "dead fingers" syndrome of workers utilizing the early jackhammers. She also described the toxic effects to the blood-forming cells from benzol, and the neurologic and psychological responses of workers in the viscose rayon industry. In 1919, Hamilton was appointed assistant professor of industrial medicine at Harvard Medical School. The first woman on the Harvard faculty, she gave occupational medicine respectability as an academic pursuit.

HAMILTONIAN (or Hamiltonian Function of a System).   Generally denoted by the symbol H, the Hamiltonian is defined by the equation

$$H(q_k, p_k, t) = -L(q_k, p_k, t) + \sum_{l=1}^{3n} p_l \dot{q}_l(q_k, p_k, t)$$

L is the Lagrangian function of the system, expressed as a function of the coordinates, momenta and time. $\dot{q}_l$ stands for the generalized velocities, also expressed as functions of the coordinates, momenta and time, where q are the coordinates of position, p, those of momentum, and the dot means the derivative with respect to time. n is the number of particles of the system. If the time does not occur explicitly, the system is called conservative, and H is identical with the total energy of the system.

HAMMER FORGING.   See **Forging**; and **Iron Metals, Alloys, and Steels**.

HAMMERHEAD SHARKS.   See **Sharks**.

HAMSTER.   See **Rodentia**.

HAND.   The terminal portion of the pectoral appendage of mammals, developed for grasping and in some species largely freed from locomotor uses. Although many mammals and other animals have grasping appendages similar in form to a hand, these are scientifically not considered to be so, and have other varying names, including paws. Using the term hand is merely a scientific usage of anthropomorphization, to distinguish the terminations of the front paws from the hind ones. True hands appear only in the mammalian order of primates.

The hands (*med./lat.*: manus, pl. manūs) are the two intricate, prehensile, multi-fingered body parts normally located at the end of each arm (medically: "terminating each anterior limb/appendage") of a human or other primate. They are our chief organs for physically manipulating the environment, from the roughest to the finest motor skills (wielding a club; threading a needle), and since the fingertips contain some of the densest areas of nerve endings on the human body, they are also our richest source of tactile feedback from our environment, so that our sense of touch is intimately associated with our hands. Like other paired organs (eyes, ears, legs), each hand is dominantly controlled by the opposing brain hemisphere, and thus *handedness*, or preferred hand choice for single-handed activities such as writing with a pen, reflects a significant individual trait. The human hand is the most versatile grasping organ in the animal kingdom.

Human Anatomy of the Hand

The skeletal structure of the human hand is subdivided into three segments: the **carpus** or wrist bones; the **metacarpus** or bones of the palm; and the **phalanges** or bones of the digits (Fig. 1).

The Carpus (Ossa Carpi)

The **carpal bones,** eight in number, are arranged in two rows. Those of the proximal row, from the radial to the ulnar side, are named the **navicular, lunate, triangular, and pisiform;** those of the distal row, in the same order, are named the **greater multangular, lesser multangular, capitate, and hamate.**

Common Characteristics of the Carpal Bones. Each bone (excepting the pisiform) presents six surfaces. Of these the *volar* or *anterior* and the *dorsal* or *posterior surfaces* are rough, for ligamentous attachment; the dorsal surfaces being the broader, except in the navicular and lunate. The *superior* or *proximal,* and *inferior* or *distal surfaces* are articular, the superior generally convex, the inferior concave; the *medial* and *lateral surfaces* are also articular where they are in contact with contiguous bones, otherwise they are rough and tuberculated. The structure in all is similar, viz., cancellous tissue enclosed in a layer of compact bone.

Bones of the Proximal Row (Upper Row)

The Navicular Bone. (Os naviculare manus; scaphoid bone) is the largest bone of the proximal row, and has received its name from its fancied resemblance to a boat. It is situated at the radial side of the carpus, its long axis being from above downward, lateralward, and forward. The navicular articulates with *five* bones: the radius proximally, greater and lesser multangulars distally, and capitate and lunate medially.

The Lunate Bone. (*Os lunatum; semilunar bone*) may be distinguished by its deep concavity and crescentic outline. It is situated in the center of the proximal row of the carpus, between the navicular and triangular. The lunate articulates with *five* bones: the radius proximally, capitate and hamate distally, navicular laterally, and triangular medially.

The Triangular Bone. (*Os triquetum; cuneiform bone*) may be distinguished by its pyramidal shape, and by an oval isolated facet for articulation with the pisiform bone. It is situated at the upper and ulnar side of the carpus. The triangular articulates with *three* bones: the lunate laterally, the pisiform in front, the hamate distally; and with the triangular articular disk which separates it from the lower end of the ulna.

The Pisiform Bone. (*Os pisiforme*) may be known by its small size, and by its presenting a single articular facet. It is situated on a plane anterior to the other carpal bones and is spheroidal in form. The pisiform articulates with *one* bone, the triangular.

Bones of the Distal Row

The Greater Multangular Bone. (*Os multangulum majus; trapezium*) may be distinguished by a deep groove on its volar surface. It is situated at the radial side of the carpus, between the navicular and the first metacarpal bone. The greater multangular articulates with *four* bones: the navicular proximally, the first metacarpal distally, and the lesser multangular and second metacarpal medially.

The Lesser Multangular Bone. (*Os multangulum minus; trapezoid bone*) is the smallest bone in the distal row. It may be known by its wedge-shaped form, the broad end of the wedge constituting the dorsal, the narrow end the volar surface; and by its having four articular facets touching each other, and separated by sharp edges. The lesser multangular articulates with *four* bones: the navicular proximally, second metacarpal distally, greater multangular laterally, and capitate medially.

The Capitate Bone. (*Os capitatum; os magnum*) is the largest of the carpal bones, and occupies the center of the wrist. It presents, above, a rounded portion or head, which is received into the concavity formed by the navicular and lunate; a constricted portion or neck; and below this, the body. The capitate articulates with *seven* bones: the navicular and lunate proximally, the second, third, and fourth metacarpals distally, the lesser multangular on the radial side, and the hamate on the ulnar side.

The Hamate Bone. (*Os hamatum; unciform bone*) may be readily distinguished by its wedge-shaped form, and the hook-like process which projects from its volar surface. It is situated at the medial and lower angle of the carpus, with its base downward, resting on the fourth and fifth metacarpal bones, and its apex directed upward and lateralward. The hamate articulates with *five* bones: the lunate proximally, the fourth and fifth metacarpals distally, the triangular medially, the capitate laterally.

The Metacarpus

The metacarpus consists of five cylindrical bones which are numbered from the lateral side (*ossa metacarpalia I-V*); each consists of a body and two extremities.

Common Characteristics of the Metacarpal Bones

The Body. (*Corpus; shaft*). The body is prismoid in form, and curved, so as to be convex in the longitudinal direction behind, concave in front. It presents three surfaces: medial, lateral, and dorsal. The **medial** and **lateral surfaces** are concave, for the attachment of the Interossei, and separated from one another by a prominent anterior ridge. The **dorsal surface** presents in its distal two-thirds a smooth, triangular, flattened area which is covered in the fresh state, by the tendons of the Extensor muscles. This surface is bounded by two lines, which commence in small tubercles situated on either side of the digital extremity, and, passing upward, converge and meet some distance above the center of the bone and form a ridge which runs along the rest of the dorsal surface to the carpal extremity. This ridge separates two sloping surfaces for the attachment of the Interossei dorsales. To the tubercles on the digital extremities are attached the collateral ligaments of the metacarpophalangeal joints.

The Base or Carpal Extremity. (*Basis*) is of a cuboidal form, and broader behind than in front: it articulates with the carpus, and with the adjoining metacarpal bones; its **dorsal** and **volar surfaces** are rough, for the attachment of ligaments.

The Head or Digital Extremity. (*Capitulum*) presents an oblong surface markedly convex from before backward, less so transversely, and flattened from side to side; it articulates with the proximal phalanx. It is broader,

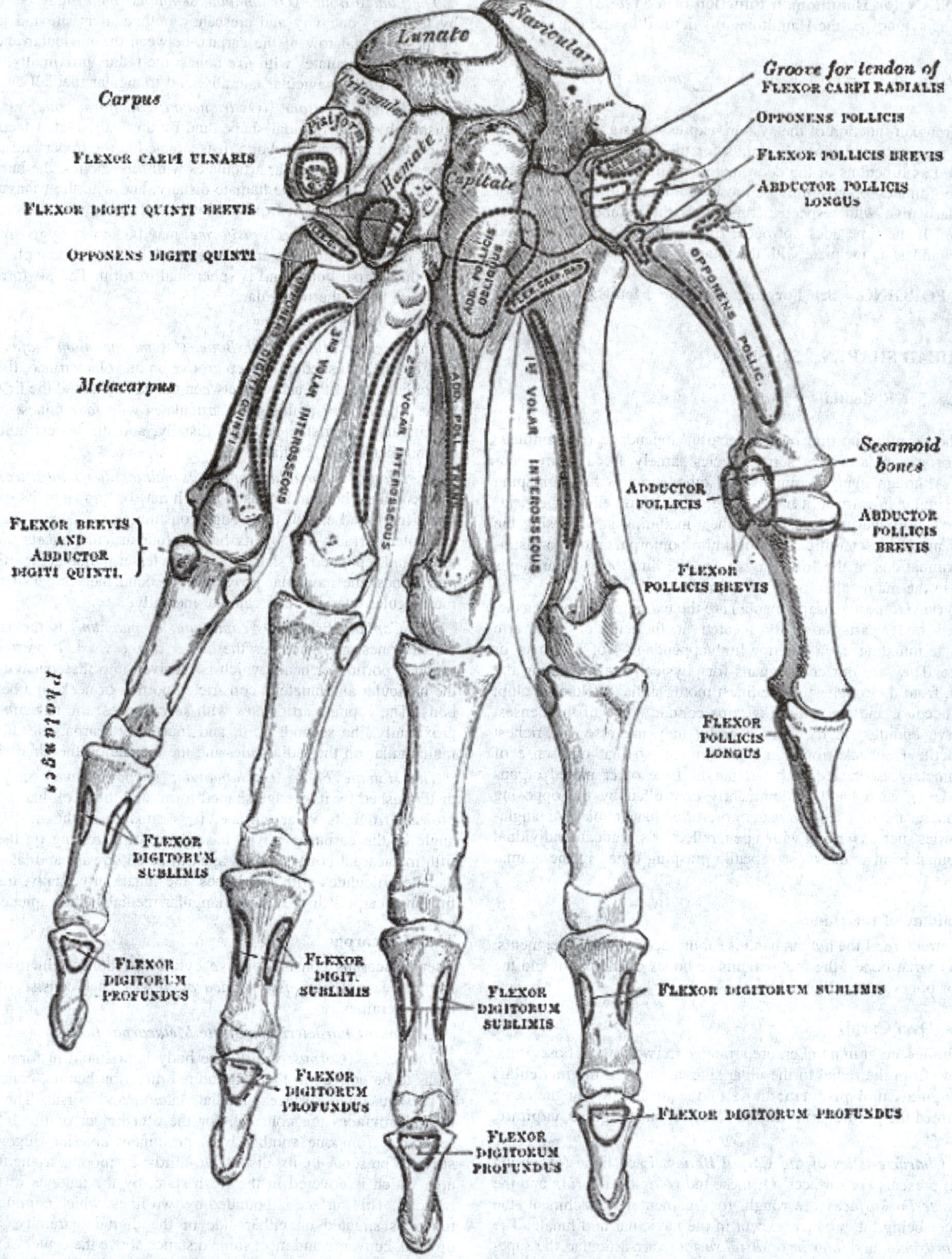

Fig. 1.   Bones of the left hand, Volar surface (from Gray's *Anatomy of the Human Body*, 1918 publication).

and extends farther upward, on the volar than on the dorsal aspect, and is longer in the antero-posterior than in the transverse diameter. On either side of the head is a tubercle for the attachment of the collateral ligament of the metacarpophalangeal joint. The **dorsal surface,** broad and flat, supports the Extensor tendons; the **volar surface** is grooved in the middle line for the passage of the Flexor tendons, and marked on either side by an articular eminence continuous with the terminal articular surface.

Characteristics of Individual Metacarpal Bones. The First Metacarpal Bone (*os metacarpale I; metacarpal bone of the* **thumb**) is shorter and stouter than the others, diverges to a greater degree from the carpus, and its volar surface is directed toward the palm.

The Second Metacarpal Bone (*os metacarpale II; metacarpal bone of the index finger, pointer finger, or forefinger*) is the longest, and its base the largest, of the four remaining bones.

The Third Metacarpal Bone (*os metacarpale III; metacarpal bone of the middle finger*) is a little smaller than the second.

The Fourth Metacarpal Bone (*os metacarpale IV; metacarpal bone of the ring finger*) is shorter and smaller than the third.

The Fifth Metacarpal Bone (*os metacarpale V; metacarpal bone of the little finger or "pinky"*).

Articulations. Besides their phalangeal articulations, the metacarpal bones articulate as follows: the first with the greater multangular; the second with the greater multangular, lesser multangular, capitate and third metacarpal; the third with the capitate and second and fourth metacarpals; the fourth with the capitate, hamate, and third and fifth metacarpals; and the fifth with the hamate and fourth metacarpal.

The Phalanges (Phalanges Digitorum Manus)

The phalanges are fourteen in number, three for each finger, and two for the thumb. Each consists of a body and two extremities. The **body** tapers from above downward, is convex posteriorly, concave in front from above downward, flat from side to side; its sides are marked by rough which give attachment to the fibrous sheaths of the Flexor tendons. The **proximal extremities** of the bones of the first row present oval, concave articular surfaces, broader from side to side than from before backward. The proximal extremity of each of the bones of the second and third rows presents a double concavity separated by a median ridge. The **distal extremities** are smaller than the proximal, and each ends in two condyles separated by a shallow groove; the articular surface extends farther on the volar than on the dorsal surface, a condition best marked in the bones of the first row.

The **ungual phalanges** are convex on their dorsal and flat on their volar surfaces; they are recognized by their small size, and by a roughened, elevated surface of a horseshoe form on the volar surface of the distal extremity of each which serves to support the sensitive pulp of the finger.

Articulations. In the four fingers the phalanges of the first row articulate with those of the second row and with the metacarpals; the phalanges of the second row with those of the first and third rows, and the ungual phalanges with those of the second row. In the thumb, which has only two phalanges, the first phalanx articulates by its proximal extremity with the metacarpal bone and by its distal with the ungual phalanx.

See also **Carpal Tunnel Syndrome**.

Robotic Hands

Since the serious introduction of robotics to industry in the early 1970s, designers have devoted much research to studies of the human hand, with the target of duplicating the manipulative skills of the end effectors (robotic hands and fingers). Scientists at the Department of Energy, Massachusetts Institute of Technology, have been studying the explicit manipulation of the human hand, particularly in the development of robots for handling radioactive materials by telerobotics, in which a robot is controlled remotely by human operators. Essentially, robotic hands currently are inferior to the human hand in terms of what may be called a "sense of touch." Common ground is being explored by robot scientists and neurosurgeons.

Outstanding work also is being conducted at the Johns Hopkins Applied Physics Laboratory, which by December 1992 had developed a tactual simulator consisting of a 20×20 arrangement of 400 pins spaced as little as 0.4 mm apart, each controlled by its own microprocessor. The array is reported to vibrate up to 400 times a second at varying amplitudes. In some instances, it is believed that the array matches or exceeds the sensory capabilities of the skin. It is interesting to note, however, that the device will sense an area of touch of about 64 square millimeters, roughly the area of one human fingertip. Like humans, robots require more than one finger in most instances to accomplish their assigned tasks. Robotic manipulators are described further in entry on *Robot and Robotics*.

Web Reference

Gray's Anatomy of the Human Body Online: http://www.bartleby.com/107/

HANDEDNESS (Right- and Left-).

Defined in terms of the motion of a screw. A *right-handed* screw, when rotated in the sense of Fig. 1(a) (counterclockwise looking down at the page), will move out of the page; when rotated in the sense of Fig. 1(b), a right-handed screw will move into the page. A *left-handed* screw will move into the page in (a) and out of

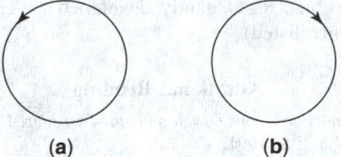

Fig. 1.   Screw rotations.

the page in (b). The mirror image of a right-handed screw is left-handed and vice versa.

The vector product is also defined in terms of a right-handed screw. Thus $\mathbf{A} \times \mathbf{B} = \mathbf{C}$ where the magnitude of $\mathbf{C}$ is $|A||B| \sin \theta$, and the direction of $\mathbf{C}$ is given by the direction of progression of a right-handed screw rotating in the sense of rotating $\mathbf{A}$ into $\mathbf{B}$ through the smaller angle (θ). $\mathbf{C}$ is thus a vector pointing into the page in Fig. 2.

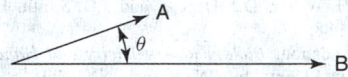

Fig. 2.   Vector product.

Coordinate systems are also classed as right or left handed. In Fig. 3, coordinate system (a) is right-handed, since rotation of the unit vector $\mathbf{i}$ into the unit vector $\mathbf{j}$ would make a right-handed screw progress in the direction of $\mathbf{k}$, $\mathbf{i} \times \mathbf{j} = \mathbf{k}$ and also $\mathbf{j} \times \mathbf{k} = \mathbf{i}$, $\mathbf{k} \times \mathbf{i} = \mathbf{j}$. Thus in a right-handed coordinate system, the above cyclic relations among the unit vectors hold. Coordinate system (b), on the other hand, is left-handed, i.e., it would take a left-handed screw to carry the unit vectors into each other in cyclic order of vector multiplication.

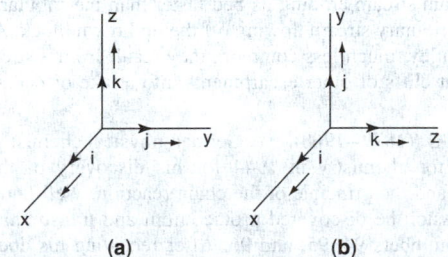

Fig. 3.   Coordinate system.

Circular polarization of electromagnetic waves is described as right-handed or left-handed depending on whether the direction of rotation of the electric vector and the direction of progression of the electromagnetic wave are related to a right-handed or a left-handed screw.

Handedness is evident throughout many natural phenomena. For example, twining vines, as they grow, wind from left to right (Trumpet honeysuckle, *Lonicera sempervirens*) or from right to left (blindweed, *Convolvulus arvensis*). Although snail shells (*Liguus virgineus*) usually are right-handed, some left-handed varieties are known, presumably because of mutations. The bacterium *Bacillus subtilis* also is asymmetric with respect to left and right. Normally, the bacterium forms right-handed spiral colonies, but, when heated, the colonies change to left-handedness.

The broad concept of left and right is widely manifested in the isomeric compounds of chemistry. This effect was discovered by Louis Pasteur in 1848 in connection with his work with tartaric acid. Pasteur found that there are two versions of tartaric acid, even though they are of identical chemical composition. Dextro- and levotartaric acids became a primary example of optical isomerism. See also **Isomerism**; and **Tartaric Acid**.

Humans are seldom ambidextrous. Most people are right-handed, a fact that does not indicate any relationship to races or cultures. Reliable information on the possible genetics of this characteristic preference remains to be developed.

An elementary particle, such as an electron or proton, is spherically symmetrical (i.e., it is *achiral*). However, once in motion, the spinning particle becomes chiral (i.e., it acquires the characteristic of handedness (chirality)). Some researchers have proclaimed that the universe is asymmetric with respect to chirality.

The foregoing subject is elegantly developed by Hegstrom and Kondepudi (See reference listed).

Additional Reading

Bishop, D.V.M.: *Handedness and Developmental Disorder*, Lippincott Williams & Wilkins, Philadelphia, PA, 1991.

Coren, S.: *Left-Handedness: Behavioral Implications and Anomalies*, Elsevier Science, New York, NY, 1990.

Gardner, M., Editor: *The Ambidextrous Universe: Mirror Asymmetry and Time-Reversed Worlds*, 2nd Edition, Charles Scribner's Sons, New York, NY, 1979.

Hegstrom, R.A., J.P. Chamberlain, K. Seto, and R.G. Watson: *Amer. J. of Physics*, **56**, 12, 1086 (December 1988).

Hegstrom, R.A. and D.K. Kondepudi: "The Handedness of the Universe," *Sci. Amer.*, 108 (January 1990).

Kondepudi, D.K. and G.W. Nelson: "Weak Neutral Currents and Origin of Biomolecular Chirality," *Nature*, **314**, 6010, 438 (April 14, 1985).

Kondepudi, D.: "Parity Violation and the Origin of Biomolecular Chirality," in *Entropy, Information, and Evolution: New Perspectives in Physical and Biological Evolution* (Bruce H. Weber, D.J. Depew, and J.D. Smith, Editors), MIT Press, Cambridge, MA, 1988.

Marsh, J., CIBA Foundation: *Biological Asymmetry and Handedness–Symposium No. 162*, John Wiley & Sons, Inc., New York, NY, 1991.

Ruthen, R.: "Quantum Pinball: A Device That Can't Tell Left from Right," *Sci. Amer.*, 36 (November 1991).

HANGING VALLEY. Under normal conditions a tributary stream enters the main stream at grade, that is, at the same level. Under certain circumstances the tributary valley may be at a greater elevation than the main valley into which the tributary stream will plunge, forming a waterfall. In such cases the tributary valley is called a hanging valley, and the stream in it is said to be out of adjustment with the main stream.

Hanging valleys originate in the following ways: by glacial action, the main glacier cutting down its valley faster than a tributary glacier; by river action, the main stream eroding its bed faster than the tributary stream; by faulting, the tributary stream flowing off the upthrown block. A fourth type of hanging valley, much less common, may result from a stream plunging over wave-cut cliffs or other escarpments into a lake or ocean basin.

HANN, OTTO (1879–1968). A German physical chemist who won the Nobel Prize for chemistry in 1944 for his discovery of the fission of heavy nuclei and the principle of the chain reaction. Well-known for work on nuclear fission he discovered protactinium and transuranium elements with atomic numbers 94, 95, and 96. After receiving his doctorate at the University of Munich, he worked in Canada before returning to Europe.

HAPLOID. See **Cell (Biology)**.

HAPTEN. See **Immune System and Immunology**.

HARDENABILITY OF STEEL. The hardenability of steel refers to the ease with which it can be hardened rather than the maximum hardness value attainable. For example, a 1-inch diameter bar of a certain 0.20% carbon alloy steel can be hardened to 50 Rockwell "C" in the center by quenching in oil. A similar bar of plain carbon steel requires a drastic quench in brine to attain the same hardness, and therefore, has a lower hardenability. Neither bar can be quenched to a greater hardness because 50 Rockwell "C" is the maximum attainable for a 0.20% carbon steel. A 0.40% carbon steel can be hardened to a maximum of about 60 Rockwell "C" and the maximum for high-carbon steel is about 65 Rockwell "C."

Of the several methods for determining the relative hardenability of steels, the Jominy test is the most widely used. A cylindrical specimen 1 inch (2.5 centimeters) in diameter and about 3 inches (7.6 centimeters) long is heated to the hardening temperature and quenched in a special fixture which holds the specimen in a vertical position and directs a stream of water on the bottom surface. The stream takes an "umbrella" shape and does not wet the sides. Cooling occurs progressively from the bottom to the top of the cylinder and the cooling rate at any distance from the bottom is known and reproducible from one sample to another. The hardness along the length of a quenched Jominy bar decreases from bottom to top. The distance from the bottom, expressed in sixteenths of an inch, to the point where the hardness is 50 Rockwell "C" is one method of reporting the hardenability.

One of the principal functions of alloying elements in steel, such as manganese, chromium, nickel, molybdenum, etc., is to increase the hardenability. Whereas prodigious amounts of expensive alloys were formerly used to insure full hardening, especially in medium and heavy sections, wartime shortages focused attention on the use of as little alloy as possible within the hardenability requirements. A large number of steels were developed containing relatively small additions of a number of elements, and a number of these steels have continued in use.

HARDEN, ARTHUR (1865–1940). Arthur Harden was an English biochemist known for his study of the alcoholic fermentation of sugars.

Harden, the son of a Manchester businessman, was educated at Owens College (the forerunner of the University of Manchester) and at Erlangen in Germany, where he received a PhD in synthetic organic chemistry in 1888. This was followed by nearly a decade as a lecturer at Manchester, until in 1897 he became head of the chemistry department at the British Institute of Preventive Medicine (later the Lister Institute). In 1912 he took up a chair of biochemistry (only the second to be created in Britain) at the University of London. See also **Biochemistry (The History)**.

Although Harden did some work on bacterial chemistry, his research was devoted mainly to the alcoholic fermentation of sugars (his *Alcoholic Fermentations* appeared in 1911). In 1897 Eduard Buchner discovered that fermentation would occur in the presence of a cell-free yeast extract thought to contain an enzyme called zymase. Harden showed that zymase was actually a mixture of enzymes, in which the larger protein molecules required a small nonprotein coenzyme to activate them (later named cozymase by Hans von Euler-Chelpin). This, together with his finding that sugar phosphates are key intermediates in fermentation, was an important step in biochemical development, and in 1929 he and von Euler-Chelpin shared the Nobel Prize for chemistry. See also **Buchner, Eduard (1860–1917)**; and **Euler-Chelpin, Hans Karl August Simon von (1873–1964)**.

Harden was known as a reserved, cheerful, but above all fair, man. His interest in the history of science was manifested in several publications, and he edited the *Biochemical Journal* for many years. He was knighted in 1936.

Additional Reading

Farber, E.: *Nobel Prize Winners in Chemistry 1901–1961*, Abelard-Schuman, London, UK, 1963.

Harden, A.: *Alcoholic Fermentations*, (revised 1914, 1923, 1932), Longmans, Green and Company, London, UK, 1911.

James, L.K.: *Nobel Laureates in Chemistry 1901–1992*, American Chemical Society and Chemical Heritage Foundation, Washington, DC, 1993.

McMurray, E.J.: *Notable Twentieth Century Scientists*, Gale Research, New York, NY, 1995.

Wasson, T.: *Nobel Prize Winners*, HW Wilson Company, New York, NY, 1987.

KATHERINE D. WATSON, University of Oxford, Oxford, UK

HARDENING OF METALS. There are three principal methods of hardening metals and alloys: cold working (see also **Cold-Worked Metal**) by plastic deformation, precipitation hardening, and quench hardening as applied to steel. The last two methods involve heating and cooling operations. A pure metal may also be hardened through the addition of alloying elements. When a solid solution is formed it is normally harder than the pure metal. If additional phases are formed by alloying, these may also be harder than the pure metal and contribute to the hardness of the metal.

HARDENING (Precipitation). See **Precipitation Hardening**.

HARD FACING. Deposition of a hard wear-resistant alloy on a metal surface. The material to be deposited is generally in the form of a welding rod and may be applied by gas or arc welding. Such surfaces are usually finished by grinding.

While hard facing or hard surfacing is usually a maintenance operation, it is also used in new production. The surfacing material may be cemented carbides, nonferrous Stellite-type alloys, or iron-base alloys with alloying additions such as chromium, tungsten, manganese, silicon, nickel, and carbon. While hard facing is most often applied to steel, cast iron and some of the nonferrous alloys such as Monel metal can also be coated. Typical applications are metal-working dies, oil well drilling tools, excavating equipment, shafting, and rolling mill rolls.

HARD FREEZE. See **Freeze**.

HARDNESS. The significance of this term as applied to solids has various interpretations. Commonly, it refers to the resistance of the substance to surface abrasion, so that of two solids, the one that will scratch the other, as diamond scratches glass, is the harder. Again, it may denote rigidity, or lack of plasticity, or even strength; in some cases a combination of several such properties. The original Mohs' Scale of Hardness is delineated in Table 1 and further described under *Mineralogy*.

TABLE 1. HARDNESS SCALES

Moh's Scale	Ridgway's Extension of Mohs' Scale	Metal Equivalent	Others
1. Talc			
2. Gypsum			
			2.5. Finger Nail
3. Calcite			
4. Fluorite			
5. Apatite			
			5.5. Window Glass
6. Feldspar (Orthoclase)	6. Orthoclase or Periclase		
			6.5. Steel (Knife Blade; File)
7. Quartz	7. Vitreous Pure Silica		
8. Topaz	8. Quartz	8. Stellite	
9. Corundum or Sapphire	9. Garnet		
	10. Topaz		
	11. Fused Zirconia	11. Tantalum Carbide	
	12. Fused Alumina	12. Tungsten Carbide	
	13. Silicon Carbide		
	14. Boron Carbide		
10. Diamond	15. Diamond		

1. In the above scales each abrasive is capable of scratching all others above it in each scale and may be scratched by all abrasives below it.
2. The gap between 9 and 10 in the original Mohs' scale is much greater than that between 1 and 9 in the same scale.
3. Various Additional hardness scales have been devised by different investigators; in general, different materials maintain the same order of hardness in all these scales.

In metallurgy and engineering, hardness is determined by methods based on resistance to penetration by an indenter of greater hardness than the material being tested. Aluminum, copper, lead, magnesium, tin, and their alloys, as well as plastics are generally indented by hardened steel balls ranging in size in the various tests from $\frac{1}{16}$ inch to 10 millimeters in diameter. The same methods may be used for soft steels and irons, but for heat-treated steels and all other alloys that develop high hardness special diamond indenters, or in some cases sintered tungsten carbide balls, are used. In all of the technological tests, the indenters are impressed into the test material under carefully regulated loads; thus, the relative size of the resulting indentation becomes a measure of hardness. (See Table 2.) The operating principles of the instruments most widely used in this country follow:

Brinell. The **Brinell scale** characterises the indentation hardness of materials through the scale of penetration of an indenter, loaded on a material test-piece. It is one of several definitions of hardness in materials science.

Brinelling refers to surface fatigue caused by repeated impact or overloading. It is a common cause of roller bearing failures, and loss of preload in bolted joints when a hardened washer is not used. Engineers will use the Brinell hardness of materials in their calculations to avoid

TABLE 2. TYPICAL HARDNESS VALUES

Material	Brinell 500 kg	3,000 kg	Rockwell	Vickers 50 kg
Aluminum, annealed	23		H 45	25
Magnesium alloy	63		B 21	63
Armco iron	66	73	B 31	71
Yellow brass, annealed	72	82	B 40	77
Copper, cold rolled	99	83	B 55	110
Mild steel, annealed	107	117	B 70	123
Aluminum alloy, 24st	130	144	B 78	146
Stainless steel, annealed	121	145	B 80	153
Yellow brass, cold rolled	174	178	B 91	189
Ni-Moly steel, quenched in water, tempered at 1200 °F (649 °C)		241	C 23	255
Same, 1000 °F (538 °C)		293	C 31	310
Same, 800 °F (427 °C)		363	C 38	380
High-speed tool steel		684	C 62	740

this mode of failure. Fretting corrosion can cause a similar-looking kind of damage and is called false brinelling even though the mechanism is different.

Proposed by a Swedish engineer Johan August Brinell (1849–1925) in 1900, it was the first widely used and standardised hardness test in engineering and metallurgy. The large size of indentation and possible damage to test-piece limits its usefulness.

The typical test uses a 10-millimeter (0.4 in) diameter hardened steel ball as an indenter. A sintered tungsten-carbide ball is also coming into use, especially for testing hard metals. The indentation is measured and hardness calculated as:

$$\text{BHN} = \frac{2P}{\pi D(D - \sqrt{(D^2 - d^2)})}$$

where:

P = applied force (kgf)
D = diameter of indenter (mm)
d = diameter of indentation (mm)

The load applied is generally 500 kilograms (1,102 pounds) for soft metals and 3,000 kilograms (6,614 pounds) for steels and hard metals. Brinell hardness is equal to the load (kilogram(pounds) divided by the surface area square millimeter (square inch) of the impression made in the test material. Tables are available for direct conversion to hardness from the diameter of the indentation as measured with a calibrated magnifier after removal of the piece from the testing machine.

Accuracy of Brinell Hardness Test. When commercial apparatus, as ordinarily used for making the Brinell test, is employed, and the test is carried out with ordinary care and precaution, it is reliable within an error of five Brinell units above or below the actual hardness. In other words, if the hardness of two pieces of metal is tested, and the difference on the Brinell scale is more than ten hardness units, it is certain that there is an absolute difference in the hardness of the pieces tested. With regard to the conditions under which the tests should be made, it may be stated that the pressure should be gradually applied for two minutes or more, and the pressure should be kept on the test piece for a period of at least five minutes.

Rockwell. Indenter is $\frac{1}{16}$-, $\frac{1}{8}$-, or $\frac{1}{4}$-inch-diameter (1.6, 3.2, or 6.4 millimeter) steel ball or a conical diamond having an apex angle of 120° and a slightly rounded point. The various scales used are designated by letters. Rockwell "B," for example, indicates a 100-kilogram (220.5-pound) load on a $\frac{1}{16}$-inch (1.6 millimeter) diameter ball. Rockwell "C" indicates a 150-kilogram (331 pound) load on the diamond indenter. Rockwell "30T" designates a load of 30 kilograms (66 pounds) on a $\frac{1}{16}$-inch (1.6 millimeter) diameter ball. An instrument of higher sensitivity known as the Rockwell Superficial Tester is used for loads of 15, 30, and 45 kilograms (33, 66, and 99 pounds). The size of the indentation is measured by a dial gauge as the final depth minus a small preliminary penetration produced by a minor preload of 10-kilograms (220-pounds). The Rockwell hardness values are arbitrary numbers having an inverse relationship to the depth of the indentation.

The **Rockwell scale** characterizès the indentation hardness of materials through the depth of penetration of an indenter, loaded on a material sample and compared to the penetration in some reference material. It is one of several definitions of hardness in materials science. It involves the application of a minor load followed by a major load, and then noting the hardness value directly from a dial. Its chief advantage is its ability to display hardness values directly, thus obviating tedious calculations involved in other hardness measurement techniques. Also, the relatively simple and inexpensive set-up enables its installation in college laboratories.

There are several alternative scales, the most commonly used being the "B", and "C" scales. Both express hardness as an arbitrary dimensionless number. The B-scale is used for softer materials (such as aluminum, brass, and softer steels). It employs a hardened steel ball as the indenter and a 100 kg weight to obtain a value expressed as "HRB". The C-scale, for harder materials, uses a diamond cone, known as a *Brale indenter* and a 150 kg weight to obtain a value expressed as "HRC".

It is typically used in engineering and metallurgy and is most common in the USA. Its commercial popularity arises from its speed, reliability, robustness, resolution and small area of indentation.

Knoop. The Knoop hardness test is a microhardness test - a test for mechanical hardness used particularly for very brittle materials or thin sheets, where only a small indentation may be made for testing purposes. A pyramidal diamond point is pressed into the polished surface of the test material with a known force, for a specified dwell time, and the resulting indentation is measured using a microscope. The Knoop hardness *HK* or *KHN* is then given by the formula:

$$HK = \frac{\text{load(kgf)}}{\text{impression area(mm}^2)} = \frac{P}{C_p L^2}$$

where:

L = length of indentation along its long axis
C_p = correction factor related to the shape of the indenter, usually about 0.07
P = load

HK values are typically in the range from 100 to 1000, when specified in the conventional units of kgf· mm^{-2}. SI units (pascals) are sometimes used instead: 1 kgf · mm^{-2} =9.80665 MPa.

The test was developed by F. Knoop and colleagues at the National Bureau of Standards (now NIST) of the USA in 1939, and is defined by the ASTM D-1474 standard.

The advantages of the test are that only a very small sample of material is required, and that it is valid for a wide range of test forces. The main disadvantages are the difficulty of using a microscope to measure the indentation (with an accuracy of 0.5 micrometre), and the time needed to prepare the sample and apply the indenter.

Vickers. Also known as Diamond Pyramid Hardness. The Vickers hardness test was developed in the early 1920s and uses a pyramid-shaped indenter made of diamond. It is based on the principle that impressions made by this indenter are geometrically similar regardless of load. Accordingly, loads of various size are applied, depending on the hardness of the material to be measured. The Vickers Hardness (HV) is then determined from the formula

$$HV = 1.854 \frac{F}{D^2}$$

where F = applied load, kg; and D = the mean of the two diagonals of the impression made by the indenter, in mm. The Vickers test can be used for all metals and has one of the widest scales among hardness tests.

Loads may vary from 1 to 120 kilograms (2.2 to 264.5 pounds) with 10, 30, and 50 kilograms (22, 66 and 110 pounds) in common use. Hardness is equal to load kilograms (pounds) divided by surface area square millimeter (square inch) of the permanent indentation. It is determined directly from optical measurements of the diagonals of the indentation, which appear square at the surface of the metal.

Tukon. A highly sensitive instrument for determining micro-hardness under very light loads down to 25 grams (0.88 ounce). The tukon hardness test uses a Knoop diamond indenter or Vickers square-base pyramid indenter. The small indentations are measured at high magnifications up to 1,000 times. The indenter is a diamond pyramid that makes an elongated impression, one diagonal being 7 times the other in length.

Eberbach. Also used for very light loads. Consists of a spring-loaded, Vickers-type diamond pyramid indenter arranged for use on a metallurgical microscope.

Scleroscope. Depends on the height of rebound of a diamond-tipped body falling under the force of gravity from a fixed height. The instrument is relatively small and is portable. One type reads directly on a graduated dial.

While there is overlapping in the field of useful application of the various hardness tests, each has certain special qualifications. The Brinell test makes a large indentation, giving an average hardness value for several grains even in rather coarse-grained metals; however, it cannot be used on small or thin specimens. The various Rockwell tests are widely used, especially for rapid production inspection of parts. The Vickers test, which originated in England, is less rapid than the Rockwell but has the advantage of a single scale covering the hardness of all metals from lead to the hardest tool materials. The Tukon test makes it possible to determine the hardness of very thin sheets and of thin metallic coatings such as chromium plate, or zinc on galvanized steel. The Scleroscope test is used principally on heavy forgings or castings that cannot be placed in an indentation-type instrument, or for field tests where a portable instrument is required. See also **Sclerometer**.

HARDNESS (Mineral).   See **Mineralogy**.

HARDNESS (Water).   See **Feedwater (Boiler)**.

HARDPAN.   The term which prospectors and miners give to the subsurface or basal layers of placer deposits in which the gold-bearing gravels have been cemented and hardened. The same term is also used to designate till or boulder clay which has been cemented by limonite.

HARDWOODS.   See **Wood (or Timber; Lumber)**.

HARE.   See **Rabbits and Hares**.

HARELIP.   A congenital deformity in which there is a failure of fusion of the maxillary and median nasal processes, resulting in a cleft in the upper lip. This is part of the same defect associated with cleft palate. In some cases, the harelip may be double, in which case there is a division on either side of the mid-line of the lip. Correction of the deformity must be by surgery, and this is best accomplished at a very early age. Usually, correction of the lip is done first, thus enabling the infant to suck. Surgery on the palate normally is undertaken just as soon as there is sufficient tissue to cover over the bony palate after repair. This usually occurs at the age of 18 to 24 months, well before abnormal speech habits are formed. In some cases, the operation must be performed in several stages. Cosmetically and functionally, the surgical results usually are excellent.

HARLEQUIN BUG (*Insecta, Hemiptera*).   Of the family *Pentatomidae* (stink bug), also known as the "fire bug" and the "calico black" bug, this insect has a major economic impact on cabbage and cruciferous crops in the southern United States. When not controlled, an entire crop can be destroyed. The bug kills plants by sucking sap from the underground portions of the plant. In addition to cabbage, the harlequin is injurious to Brussels sprouts, collard, cauliflower, horseradish, kohlrabi, mustard, radish, and turnip. If these preferred sources of nourishment are not immediately available to the insect, it will attack asparagus, bean, eggplant, okra, potato, and tomato. The insect also can damage numerous other garden crops, certain weeds, small fruit trees, and field crops.

The harlequin bug is colorful, with red and black spots, of a shield shape, and about $\frac{3}{8}$ inch (9 to 10 millimeters) long. The nymphs, somewhat smaller, have a similar appearance see Fig. 1. All stages of the insect may be found from the early to the late months of the year. In the United States, the pest is found in the southern portion in all areas from the Atlantic to the Pacific coasts. The insect is believed to be native to Mexico. In more northern areas, the bug winters as an adult. During the first warm spell of spring, eggs are deposited on the underside of leaves. They appear something like tiny white beer kegs, standing on end and usually in a double row. There are two black bands around each "keg." Hatching of the eggs occurs within 1 to 4 weeks, depending upon temperature.

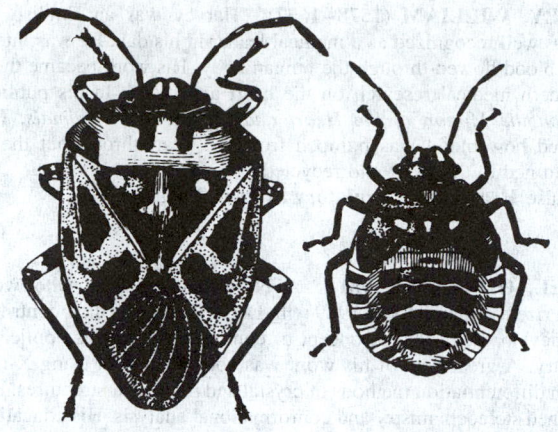

Fig. 1.   Harlequin bug. Adult at left; nymph at right. (*USDA.*)

Immediately, the nymphs commence feeding and destroying target plants. During a period ranging from 4 to 9 weeks, the insect passes through five instars, after which it is ready to mate and lay eggs for the next generation. Normally there are three to four generations per year.

In addition to being controlled by chemicals, populations can be controlled by destroying weeds that attract the insects. These include *Amaranthus* and wild mustard. Advantage of the insect's preference for certain crops, such as kale, mustard, radish, and turnip, can be taken by planting a small area to such plants either very early in the season or after harvest. Large concentrations of the insects thus can be killed with relatively small amounts of insecticide. The debris from such decoy crops should be burned.

HARMATTAN.   See **Winds and Air Movement**.

HARMONIC.   A sinusoidal frequency component of a waveform. The harmonic has a frequency that is an integral multiple of the fundamental frequency. The frequency of the second harmonic will be double that of the fundamental frequency (first harmonic).

Harmonic distortion is nonlinear distortion characterized by the appearance in the output of harmonics other than the fundamental component when the input wave is sinusoidal. Harmonic distortion is sometimes called amplitude distortion.

HARMONIC ANALYSIS.   Not only is it possible to combine two or more simple harmonic motions of different period, amplitude, and phase to form a complex motion, but there are also means of analyzing the resultant motion, when the latter is given, to find its component harmonics. For example, if the wave form of such a complex tone as that produced

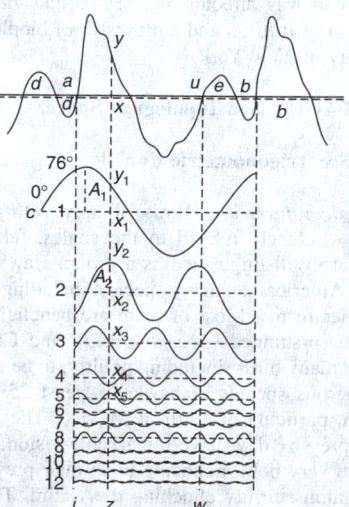

Fig. 1.   Records of a complex sound and twelve of its components.

by a bell or a saxophone is accurately graphed by means of a phonodeik the equation of the vibratory motion can be deduced in such form as to show the separate components. Fourier showed that the same analysis is possible for any periodic motion, however complicated. The equation, called Fourier's series, may be written

$$y = a \sin 2\pi nt + b \cos 2\pi nt + c \sin 4\pi nt + d \cos 4\pi nt$$
$$+ e \sin 6\pi nt + f \cos 6\pi nt + \cdots$$

in which y is the displacement of the vibrating particle and t is the time. The fundamental frequency n and the constants a, b, c, d, etc., must be calculated from the given wave form or the data from which it is plotted. There is a type of instrument, called a "harmonic analyzer," which automatically computes the coefficients; or it may be done mathematically, though the process is very laborious. Figure 1 shows the wave form and the twelve components of complex tone, analyzed by Professor D.C. Miller.

HARMONIC MEAN.   See **Average**.

HARMONIC MOTION.   A distinct type of periodic motion, or vibration, characteristic of elastic bodies; illustrated by a bird-cage bobbing up and down at the end of a spiral spring, or (approximately) by the piston of the steam engine. It may be either simple, with only one frequency and amplitude, or made up of two or more simple components and consequently of more complex character. The essential feature of simple harmonic motion is that, with its range extending to equal distances on both sides of an equilibrium position or origin, the acceleration is always toward the origin and directly proportional to the distance from it. With elastic vibrations this is easily seen to follow from Hooke's law, since the force tending to restore the deformed body to equilibrium is proportional to the deformation. See also **Elasticity**. The motion is called "harmonic" undoubtedly because the vibrations of bodies emitting musical sounds are of this character. Any simple harmonic motion may be represented by the equation.

$$y = a \cos(2\pi nt + \phi)$$

in which y is the distance at time, t, a is the amplitude, n is the frequency or number of vibrations per unit time, and ϕ is the phase constant, such that when $t = 0$, $y = a \cos \phi$.

It is interesting to note the relationship between harmonic and circular motion. If a peg is inserted in the face of a circular disk or wheel and the latter uniformly rotated, the motion of the peg, as viewed with the wheel Seen edgewise, is simple harmonic. In fact, uniform circular motion is made up of two simple harmonic components of the same period and amplitude at right angles, one being a quarter-period ahead of the other in phase. If the two harmonic components have a phase difference other than a quarter-period, the resultant in general is motion in an ellipse; while if they have unequal periods, the path is one of a class of more or less complicated loci called "Lissajous' curves."

HARMONIC OPERATION.   *Impeded* harmonic operation is constrained magnetization or forced magnetization. It is the type of operation that takes place in a magnetic amplifier in which the impedance of the control circuit and any circuit closely coupled to it is so great as to substantially prevent the flow of all harmonic currents in such circuits.

Unimpeded harmonic operation is natural magnetization or free magnetization. It is the type of operation that takes place in a magnetic amplifier in which the impedance of the control circuit or any circuit closely coupled to it is so small as to permit substantially unimpeded flow of all harmonic currents in such circuit.

See also **Amplifier**.

HARMONIC PROGRESSION.   See **Progression**.

HARMONIC SYNTHESIZER.   A machine which combines elementary harmonic constituents into a single periodic function. A machine performing the opposite function is called a harmonic analyzer.

HARMOTOME.   The mineral harmotome is a zeolite, composition approximately (Ba, K) (Al, Si)$_2$Si$_6$O$_{16}$ · 6H$_2$O; it is monoclinic but often forms double twins giving the effect of a square prism. It is a brittle mineral; hardness, 4.5; specific gravity, 2.41–2.50; luster, vitreous; color, white to gray or perhaps yellow, red or brown; white streak; translucent.

Harmotome like other zeolites is found in cavities in basalts and similar rocks, sometimes in trachytes or in gneisses, occasionally as a gangue mineral in veins of metallic minerals. Some well-known localities are in Bavaria; the Harz Mountains; Norway; and Scotland. Harmotome occurs in the United States with stilbite, near Port Arthur, Lake Superior. The name harmotome comes from the Greek meaning joint and to cut, referring to the division of the pyramid formed by the prismatic faces of the mineral when in the twinned position.

HARPY. See **Eagle**.

HARRIER. See **Eagle**.

HARTEBEEST. See **Antelope**.

HARTLEY. In information theory, a unit of logarithmic measures of information equal to the decision content of a set of ten mutually exclusive events expressed by the logarithm with the base ten. For example, the decision content of a set of 8 characters equals $\log_{10} 8$, or 0.903 Hartley. Synonymous with information content decimal unit. (*American National Dictionary for Information Processing*.) See also **Shannon**.

HARTLEY OSCILLATOR. See **Oscillator**.

HARTLEY PRINCIPLES (Transmission). The amount of information that can be transmitted is proportional to the width of the frequency range, and the time it is available. Information content is equated to the total number of code elements, multiplied by the logarithm of the number of possible values a code element may assume. Information content is independent of how the code elements are grouped. By quantizing, the continuous magnitude-time function used in ordinary telephony may be transmitted by a succession of code symbols such as are employed in telegraphy. To obtain the maximum rate of transmission of information, the signal elements need to be spaced uniformly.

Time-Frequency Duality. As implied by the Fourier integral, a time function cannot be confined within a small region on the time scale when the steady-state transmission characteristic is confined to a narrow range on the time scale. For example, it is well known that, if a telegraph dot is made narrower and narrower, its corresponding significant-frequency spectrum becomes broader and broader until, in the limit when the dot becomes an impulse, its significant-frequency spectrum is of infinite intent.

HARTMANN TEST. Hartmann devised various optical tests, including the following: (1) Hartmann test for telescope mirrors. For a perfect mirror, light from all points on the mirror should come to the same focus. By covering the mirror with a screen, in which regularly spaced holes have been cut, and then permitting the reflected light to strike a photographic plate placed near the focus, the failure of dots on the plate to be regularly spaced indicates a fault of the mirror. (2) Hartmann test for spectrometers. Light is passed through different parts of the entrance slit. Any change in the spectrum as different parts of the slit are used indicates a fault of the instrument. A "Hartmann diaphragm" is one device for using only one part of the entrance slit at a time.

HARTREE-FOCK APPROXIMATION. Also called Hartree-Fock-Slater approximation. A method for the solution of a many electron problem, e.g., that which arises in considering the band theory of solids or an atom with more than one electron. The antisymmetric wave function for the N-electron system is expanded as a linear combination of determinants of order N, having as elements one electron wave functions. This procedure introduces exchange terms in the Hamiltonian, of the form:

$$e^2 \int \left[\frac{\psi_i(r_1)\psi_j^*(r_2)}{r_{12}} d\tau_2 \right] \psi_j(r_1)$$

where r_{12} is the separation of the points defined by the vectors r_1 and r_2.

HARVESTMAN (*Arachnida, Phalangida*). Spider-like animals, most species with small oval bodies and extremely long slender legs. Those with shorter legs are more easily confused with the true spiders but all may be recognized by the segmented abdomen. Daddy longlegs.

HARVEY, WILLIAM (1578–1657). Harvey was an English doctor who was well recognized as a medical leader in his day. He was interested in how blood flowed through the human body. His work became the basis for modern medical research on the heart and blood. In his publication, *Essay on the Motion of the Heart and the Blood of Animals*, Harvey explained how blood was pumped from the heart throughout the body, then returned to the heart and recycled.

See also **Heart and Circulatory System (Physiology)**.

J. M. I.

HASSEL, ODD (1897–1981). A Norwegian chemist who won the Nobel Prize for chemistry in 1969 with Derek Barton for their contributions to the development of the concept of conformation and its application in chemistry. A great deal of his work was concerned with using X-ray and electron differentiation methods of crystal and molecular structures. He also researched stereochemistry and conformational analysis. His education and teaching career were in his homeland.

HASSIUM. See **Chemical Elements**; and **Chemical Elements: The History of the Origin**.

HASTELLOY. See **Nickel**.

HATCHET FISHES (*Osteichthyes*). Of the order *Isospondyli*, family *Sternoptychidae*, hatchet fishes are small, rarely exceeding $3\frac{1}{2}$ inches (9 centimeters) in length. They are silvery and are so named because of their hatchet-head appearance. They possess photophores (light organs) on their sides and undersurfaces. The genus *Argyroplecus* features telescopic eyes, which are aimed in an upward direction. They are a food source for tuna, but are not nearly so abundant as their relatives, the bristlemouths. Some species of hatchet fishes have been favorites among tropical-fish fanciers.

The flying hatchet fishes of South America of the order *Ostariophysi*, family *Characidae* are the only fishes credited with performing true flight. See also **Characids (Osteichthyes)**.

Hatchet fishes are fully adapted to a life near the water surface. Like speedboats, the front of which rises off the water at high speed, these fishes can also rise off the water surface. They literally fly several yards (meters) through the air, after a starting movement of a few yards (meters). The initiation of the movement has been seen in nature, but the actual flying is difficult to observe. Laboratory investigations, however, indicate that the process is accompanied by a humming sound.

HAUPTMAN, HERBERT A. (1917–). An American biophysicist who won the Nobel prize for chemistry in 1985 along with Jerome Karle for their outstanding achievements in the development of direct methods for the determination of crystal structures. Hauptman's work involved developing equations that allow determination of phase information from X-ray crystallography intensity patterns. The use of computers permitted use of the equations to determine the conformation of thousands of chemicals. Hauptman was director of research and vice president of the Medical Foundation of Buffalo and a professor of biophysics in Buffalo at the State University of New York.

HAUSDORFF SPACE. See **Topological Space**.

HAVERSINE. See **Trigonometric Function**.

HAWK (*Aves, Falconiformes*). Birds of prey with hooked beaks and large curved claws, closely related to the eagles, falcons, harriers, and others and not sharply distinguished as a group. Hawks are found on all continents. North America has many species, including buzzards, harriers, goshawks and other forms. Most of them are beneficial as destroyers of vermin but the sharp-shinned (*Acceptor vela*), and Cooper (*A. cooperi*) hawks destroy too many birds, including poultry, to be regarded as friends.

There are numerous species of hawks, at least 25 of these occurring in North America, particularly north of Mexico. The hawks are swift in flight, seek their prey by day, have remarkable vision, and eat only what they kill. They are very bold, pouncing upon their prey in a rapid swoop, using claws and talons, firmly clinching the victim. The hawk prefers to take its victim to a private location for consumption.

In the African rain forest, three species are known as darters.

The osprey is a large bird of prey of almost worldwide distribution. It is a skillful fisher and is known in North America as the fish hawk, Pandion haliaëtus.

Buzzards are birds of prey of several species that belong to the genus *Buteo*. The North American representatives are commonly called hawks, as Swanson's hawk. The same may be said of the nearly related rough-legged buzzards; American representatives of the genus are the rough-legged hawks. The name is incorrectly, although commonly, applied to the turkey buzzard, which is a vulture. See also **Falconiformes**.

Size Dimorphism. It has been known since medieval times among hawk and falcon fanciers that the male hawk is considerably smaller than the female, whereas the norm for birds in general is the reverse. In a 1985 paper, E. Temples (University of California, Davis) points out that in addition to size dimorphism among these birds of prey, there is a correlation between the type of diet and degree of dimorphism. The degree of size difference between females and their mates increases as diet moves through carrion, insects, fish, and mammals to birds. Could it be postulated that the faster the prey moves, the greater will be the size dimorphism in the pursuers? I. Newton (Institute of Terrestrial Ecology, Monks Wood, England) observes that the striking link between diet and size dimorphism possibly may be misleading. Because so much of the biology of birds correlates with their diet in some way, there are many factors that will also correlate with size dimorphism, but not be the cause of it. Mueller (University of North Carolina), in reviewing the various size dimorphism hypotheses, has suggested that a behavioral explanation may be the most reasonable, namely, the female dominance hypothesis. First advanced in the 1960s by T. Cade (Cornell University) and later by S. Smith (Mount Holyoke College), the concept involves some form of protection mechanism for the females, which enhances pair-bonding and pair-bond maintenance.

In nonpredatory birds, where the male usually is physically and behaviorally dominant to the female, courtship often involves something of a role reversal, where the male offers food and essentially amuses the female in his efforts to attract her. Some authorities argue that because the predatory bird is equipped with dangerous talons, beaks, and killer instincts, such social interactions during courtship would be potentially hazardous to the females. Thus, the larger, better equipped female could better protect herself during contact with the male. It is observed that if female predatory birds chose mates smaller than themselves, then the reverse dimorphism among them would emerge. Others suggest that as small males are better equipped for hunting, then female choice for that skill would produce the observed pattern of size dimorphism. Or, if small body size were important in aerial competition for females, this also could produce the size relationships as observed.

Temeles, in proposing a prey vulnerability hypothesis, pointed out that the greater agility of prey not only narrows the potential size range that a predator can exploit, but it also reduces the likely success during each hunting exploit. In essence, this determines the amount of energy captured per amount of energy put out in hunting. In field tests with raptors, Temeles did indeed find that hunting success varied according to the nature of the prey—in the following approximate order: invertebrates, 82%; fish, 58%; mammals, 23%; and birds, 13%.

It can be further postulated that the larger female may be better suited for capturing some prey, the male for others. Perhaps the size dimorphism is not a matter of competition between sexes, but rather a balancing of advantages in insuring the success of the species as a whole. It is expected that many years will be required to produce a reasonably complete answer for the size dimorphism question.

Additional Reading

Clark, W. and L. White (Editor): *Field Guide to Hawks: North America*, Houghton Mifflin Company, New York, NY, 1998.

Dunne, P. et al.: *Hawks in Flight: The Flight Identification of North American Migrant Raptors*, Houghton Mifflin Company, New York, NY, 1989.

Johnsgard, P.: *Hawks, Eagles and Falcons of North America: Biology and Natural History*, Smithsonian Institution Press, Washington, DC, 1990.

HAWKING, STEPHEN WILLIAM (1942–). Hawking is a British theoretical physicist. Interestingly, Hawking was born on the 300th anniversary of Galileo's death and his reputation is equal or exceeds that of Albert Einstein. Moreover, he now holds the prestigious post of Lucasian professor of mathematics; a post that once was held by Sir Issac Newton.

Hawking graduated from Oxford in 1962 and then went to Cambridge. At Cambridge he became interested in black holes as first proposed by Robert Oppenheimer and in events in which the laws of physics seem to break down called "space-time singularities". He received his Ph.D. in 1966 in the field of cosmology. His work on black holes has almost become synonymous with his name. He made the startling discovery that black holes emit subatomic particles, radiation, now known as Hawking radiation.

Hawking has worked to find ways to link relativity with quantum mechanics. He has shown "black hole evaporation". In 1983, he published, along with Hartle, the "no boundary theory" proposing that the universe is infinite with no beginning or end.

Hawking's name is known in the general public because of the publication of his book *A Brief History of Time: From the Big Bang to Black Holes*, which he wrote about his theories of the cosmos in lay terms. With this book, Hawking proved himself a scientific educator for the common man.

Hawking's accomplishments are even more noteworthy when one realizes that Hawking has amytropic lateral sclerosis and is confined to a wheelchair and is able to move only a few of his fingers and he uses a voice synthesizer to speak.

See also **Black Hole**; and **Cosmology**.

Additional Reading

Boslough, J.: *Stephen Hawking's Universe: An Introduction to the Most Remarkable Scientist of Our Time*, William Morrow & Company, New York, NY, 1989.

Cole, R.: *Stephen Hawking: Solving the Mysteries of the Universe*, Streck-Vaughn Company, Austin, TX, 1998.

Filkin, D.: *Stephen Hawking's Universe: The Cosmos Explained*, Basic Books, Boulder, CO, 1998.

Hawking, S.W.: *Hawking on the Big Bang and Black Holes*, World Scientific Publishing Company, Inc., Riveredge, NJ, 1993.

J. M. I.

HAWK MOTH (*Insecta, Lepidoptera*). Large moths composing the family *Sphingidae*, one of the largest of the order. These moths have a long, rather stout body projecting beyond the narrow wings. The front wings are much longer than the hinder pair, and because of their limited surface they are vibrated rapidly in flight. The moths have long tongues and visit deep-throated flowers. From their habit of hovering as they probe the flower for nectar they are also called hummingbird moths. Another common name is sphinx moth.

HAWORTH, SIR WALTER N. (1883–1950). An English chemist who received the Nobel Prize in chemistry in 1937 along with Paul Karrer. He recommended the name *ascorbic acid* and synthesized vitamin C. He accomplished much work on carbohydrate structure and developed a substitute for blood plasma using carbohydrates. During World War II, he developed gaseous diffusion separation on uranium isotopes. He received his Ph.D. in Manchester, England.

HAWTHORN TREES AND SHRUBS.   See Rose Family.

HAY BRIDGE.   See Bridge Circuits (Electrical/Electronic).

HAY FEVER.   See Allergy.

HAZE. Particles suspended in air, reducing visibility by scattering light; often a mixture of aerosols and photochemical smog. Many aerosols increase in size with increasing relative humidity due to deliquescence, drastically decreasing visibility. On Köhler curve plots of saturation relative humidity versus aerosol particle radius, equilibrium haze particles are to the left of the peak, while growing cloud droplets are to the right. Many haze formations are caused by the presence of an abundance of condensation nuclei which may grow in size, due to a variety of causes, and become mist, fog, or cloud. Distinction is sometimes drawn between dry haze and damp haze, largely on the basis of differences in optical effects produced by the smaller particles (dry haze) and larger particles (damp haze), which develop from slow condensation upon the hygroscopic haze particles. Dry haze particles, with diameters of the order of 0.1 μm, are small enough to scatter shorter wavelengths of light preferentially though not according to the inverse fourth-power law of Rayleigh scattering. Such haze particles produce a bluish color when the haze is viewed against a dark background, for dispersion allows only the slightly bluish scattered light to reach the eye. The same type of haze, when viewed against a light background, appears as a yellowish veil, for here the principal effect is the removal of the bluer components from the light originating in the distant light-colored background. Haze may be distinguished by this same effect from mist, which yields only a gray obscuration, since the particle sizes are too large to yield appreciable differential scattering of various wavelengths. See also **Precipitation and Hydrometeors**.

HAZELNUT SHRUBS. Of the family *Corylaceae*, genus *Corylus*, hazelnut shrubs (rarely trees) are deciduous, and characterized by male catkins that hang from the tree during most of the winter months, and by their edible and tasty fruit, an ovoid nut in a toothed container, simply known as the hazelnut of commerce. The term *filbert* is generally reserved for use with reference to the fruits of two European hazelnut plants, *Corylus avellana pontica* and *C. maxima*. The American hazelnut (*C. americana*) is a shrub ranging from 3 to 8 feet (0.9 to 2.4 meters) in height. It is commonly found in thickets and hedgerows. The leaves are narrow, heart-shaped or sometimes ovate, with abrupt points. They are of a lackluster dark green color and from 3 to 5 inches (7.6 to 12.7 centimeters) in length. The stems are short. The staminate catkins are from 3 to 4 inches (7.6 to 10.1 centimeters) in length. This shrub ranges from Maine westward to Alberta and Kansas and southward to Florida. The record American hazelnut growing in the United States and selected in 1989 is located in Mississippi. As compiled by American Forests, this specimen has a circumference (at 4.5 feet (1.4 m) above ground level) of 12 inches (30.5 cm) a height of 34 feet (10.4 m), and a spread of 24 feet (7.3 m). The beaked hazelnut (*C. rostrata*) ranges from 3 to 8 feet (0.9 to 2.4 meters) in height and commonly occurs along the road in thickets, ranging throughout Canada from Quebec westward to the Pacific slopes and south into the United States to Missouri, Michigan, and Ohio, and Delaware in the east. It is found in the mountains as far south as northern Georgia. The fruit is edible and sweet and in the form of an ovoid nut. The nut is enclosed in a bristly cup which has a beak-like termination, hence the name.

The California hazelnut (*C. cornuta* or *californica*) is well known for its velvety leaves and makes an attractive garden shrub. For purple coloration

Fig. 1. Filberts (hazelnuts). Upper left, note nuts in husk; upper right, shell opened to expose kernel. (*U.S. Dept. of Agriculture photo.*)

in gardens, the purple hazel (*C. maxima purpurea*) is sometimes used. The Turkish hazel (*C. colurna*) can be classified as a tree, in that it can attain a height up to 75 feet (22.5 meters), but in most respects it is similar to the lesser hazelnut shrubs. Another species is the corkscrew hazel or Harry Lauder's walking stick, a shrub which attains a height up to 10 feet (3 meters). It makes an attractive shrub, particularly in winter months when the catkins are on display.

The record hazelnut (California hazelnut, *Corylus cornuta var. californica*) and Hazelnut (*Corylus americana*) growing in the United States and selected by the American Forests are listed in Table 1.

The familiar filbert (hazelnut) of commerce is from transplantation of European varieties several decades (1860) ago, mainly in Oregon and Washington, as well as in a few other western parts of the United States. European species now growing in these areas are *Corylus avellana* and *C. maxima*. Varieties of these include Barcelona, Daviana, and Du Chilly. A grouping of filberts is shown in Fig. 1.

H-DISPLAY. In radar, a B-display modified to include indication of angle of elevation. The target appears as two closely spaced blips, which approximate a short bright line, the slope of which is in proportion to the sine of the angle of elevation. Also called *H-scan, H-scope*, or *H-indicator*.

HEADHYDROSTATIC. See **Hydrostatic Pressure**

HEADACHE. Head pain is a symptom and not a disease. Headache is one of the most common symptoms of a disorder, not only of the nervous system, but of other parts of the body as well. Consequently, discovery of the primary cause of headache is often difficult. The degree of pain associated with headache does not necessarily correlate with seriousness of a cause, a violent headache sometimes being associated with a relatively minor injury. Diagnosis of headache complaint can be facilitated by providing accurate information to the physician—events occurring before the headache, such as emotional stress, exertion, eating, and so on; the time of day or night when headache usually occurs; and other symptoms that may accompany headache, such as nausea, flashes of light, ringing in the ears, rapid or slow onset of the headache, as well as how the headache usually ceases.

The large veins (venous sinuses) and their tributaries that drain the surface of the brain are sensitive to pain, as are the arteries. The brain substance itself apparently is not sensitive to pain, but the coverings of the brain are. The sinuses, teeth, ears, and muscles in the area of the head may be affected so that pain from them, at first local, later covers a wider area.

At least eight pain mechanisms have been identified as causative factors in headache: (1) dilation of the cranial arteries; (2) pulling or traction upon pain-sensitive intracranial structures; (3) traction on and dilation of intracranial blood vessels; (4) inflammation of structures within the skull; (5) contraction of skeletal muscles over the head and neck; (6) spread of pain from stimulation elsewhere in the head; (7) pain from allergenic reaction; and (8) mentally produced (*psychogenic*) pain. The majority of headaches for which medical attention is sought arise either from dilation of the cranial arteries or contraction of the muscles of the head and neck, or by combinations of these factors. Fortunately, headaches of this type arise from conditions that usually are easy to correct.

Vascular Headache

This term is applied to a type of headache caused by dilation of the cranial arteries. It is associated with general infections, migraine headaches, or those resulting from taking certain drugs; and is largely responsible for so-called hunger and hangover headaches. The headaches of suddenly increased blood pressure are in this group, as well as headaches that follow

TABLE 1. RECORD HAZELNUT IN THE UNITED STATES[1]

| Specimen | Circumference[2] | | Height | | Spread | | Location |
	Inches	Centimeters	Feet	Meters	Feet	Meters	
California hazelnut (1989) (*Corylus cornuta var. califonica*)	66	168	50	15.2	42	12.8	Oregon
Hazelnut (1997) (*Corylus americana*)	30	76	27	8.2	21	6.4	Maryland

[1] From the "National Register of Big Trees," American Forests (by permission).
[2] At 4.5 feet (1.4 meters).

convulsive seizures or head injury. Headaches of this type usually have a throbbing quality, but this may not be present if the headache is prolonged.

Treatment of vascular headache is generally directed to the underlying cause. The inhalation of high concentrations of oxygen is particularly helpful to persons whose headaches are caused by lack of oxygen. Headaches caused by traction or pressure on intracranial structures are associated with expanding intracranial masses, with brain tumors, abscesses, and hematomas, as examples. Such headaches are aggravated by coughing or straining and are not relieved by drugs which constrict the arteries. Headache associated with brain tumor may be intermittent and mild to moderate in severity and usually does not interfere with sleep.

The headache produced by a hematoma (swelling or tumor filled with blood) is dull, steady, and felt throughout the head. The pain from brain abscess is similar to that of tumor. However, the abscess must be of sufficient size to cause traction before pain is felt.

Headaches caused by traction upon and dilation of the intracranial vessels are typified by the headache that frequently follows lumbar spinal puncture. Despite precautions, at times there may be slow leakage of the spinal fluid through the hole made by the needle. This results in headaches that are ordinarily mild, but can be severe. Once the headache develops, bed rest is about all that is required. The condition heals spontaneously.

Headaches resulting from inflammation of cranial structures are experienced if the patient has any infection within the skull, such as meningitis or encephalitis. Such a headache also occurs as a result of the inflammation that follows brain hemorrhages. These headaches may be intense and require narcotics.

Migraine Headaches

Headaches of this type have been reported since ancient times. Migraine has been termed one of the most common complaints of civilized people. The onset of migraine headaches usually occurs between the ages of 12 to 25, but they can begin at any age. Persons who perform mental work are more likely to be affected than blue collar workers. Also, urban dwellers seem to be more affected than people in rural areas. Often, the migraine victim will be an ambitious, hard driving, meticulous, and exceptionally intelligent individual.

Women who suffer attacks of migraine usually do not have any episodes during pregnancy; and the attacks may disappear entirely after the change of life. The disease may disappear in men and women at all ages, but most frequently attacks cease at around 50 years of age, when the elasticity of the blood vessels has diminished, so that the dilation previously described in the etiology of migraine has decreased.

An outstanding feature of migraine headache, thus differentiating it from other types, is that it affects one side of the head. Other distinctions are the periodic recurrence. There is some evidence that migraine may be hereditary. In most instances, the headaches occur about once every two weeks. In women, it may be associated with the menstrual period. Attacks in some persons, however, do not show this regularity, with headaches being separated by months or even years. A migraine headache may last from a few hours to more than a week. In any individual, the characteristic pain, accompanying symptoms, and length of time are usually about the same for each attack. Some sufferers can generally predict such experiences.

The typical migraine headache commences in the temple, eyeball, or forehead, and soon spreads to include either the left- or right-half of the head. The pain may involve the face and neck and sometimes the arms. Sometimes the headache is preceded by disturbances of vision (dullness of vision, blinding flashes of light, sensitivity to light or sound, or dizziness). As the attack begins, the patient may notice a blind spot, that is, several words in a printed sentence may not be seen. This spot, in rare instances, increases in size until vision in one field is fully gone. The patient may regain the ability to see in the later stages of the attack, but he may still be troubled with dazzling white flashes of light. During the attack, the victim's face usually is pale and sallow and the skin may be sweaty and clammy. The arms and legs may feel cool to the patient, even though there may be fever. Nausea and violent vomiting often mark the climax of the attack. After the attack has run its course, if there has been no vomiting, the patient usually feels relaxed and relieved. The patient may be filled with energy and tend to be overactive, although a dull headache may persist for a day or two.

Many migraine sufferers report that attacks seem to occur in relation to periods of let-down or of exhilaration. Many have noted that their headaches commence on weekends, the first day of a holiday, or on days of planned social engagements or travel. Often, on the eve of onset, the victim may be in high spirits, with an unusually increased appetite. However, on the following morning, the victim may arise with a very depressed or melancholic attitude. The victim may become restless, irritable, and confused, with an inability to concentrate on routine tasks, or to make decisions.

Early Studies of Migraine. Many theories over the years have been offered as to the cause of migraine headaches. Such headaches have been associated with distention of the cranial arteries in the scalp as an immediate cause, but the cause of such distensions has not been well understood. Some authorities have attempted to develop a relationship between the personality traits of migraine patients and those having high blood pressure. For some time, an approximate relationship between children of a migraine-prone parent was considered. Some researchers have attributed migraine headaches to food allergies, asthma, eyestrain, and imbalances of the endocrine system.

Traditional Treatment of Migraine. Until quite recently, physicians have utilized the following procedures:

Generally, the patient should be left alone in a quiet darkened room because most migraine patients are extremely sensitive to light and odors. An ice bag on the head and hot water bottle at the feet may provide some relief from pain. In some patients, sitting in an upright position rather than lying down reduces the intensity of the pain. Ergotamine tartrate, to be prescribed only by a physician, has been found helpful in terminating the headache in many instances, if given at the beginning of an attack. Inhalation of 100% oxygen may alleviate pain. Strong drugs should not be taken for a migraine attack unless prescribed by a physician, for it is too easy for migraine sufferers to develop a drug habit. The victim should make every effort to determine the factors associated with attacks and to avoid wherever possible such factors. Avoidance of fatigue, late hours, strain, and worry tend to reduce frequency or severity of migraine attacks. In most persons, physical or mental tension is often the immediate cause of an attack.

Breakthrough in Understanding and Treatment. Migraine attacks may persist from 4 to 72 hours and sometimes are preceded by transient focal neurologic symptoms. The conventional treatment as just described has been directed at acute symptomatic relief and sometimes involves prophylactic treatment to reduce the frequency of attacks. It is generally agreed, however, that the efficacy of such treatments is rather poor.

A hypothesis was developed in the 1960s that serotonin [a phenolic amine (5-hydroxytryptamine) and powerful vasoconstrictor found in the blood serum] is important in the pathogenesis of migraine. This finding created much attention because it possibly could become the first drug capable of preventing or reducing the intensity and frequency of migraine. Subsequent tests, however, were disappointing, and only recently has the interest in serotonin been revived.

Recently, a new drug, sumatriptan, was introduced and is reported to be effective in the treatment of migraine and cluster headaches. It is reported that this *design drug* was developed for a particular subpopulation of 5-hydroxytryptamine receptors. The question still posed is where are these receptors? Are they vascular or neural?

In the past, many researchers have attributed migraine to vasodilation, but evidence to this effect has been unconvincing. In fact, the exact cause of migraine-induced pain still remains uncertain, but a central pain mechanism is strongly suggested because so many migraine sufferers also note upper- and lower-limb pain concurrent with head pain. This has focused attention on the probable neural component of the illness. This led researchers to tentatively associate migraine (and cluster headache) with abnormal serotoninergic transmission, possibly at different loci. This could be the central factor to the variable symptoms of both migraine and cluster headaches.

In 1991, The Subcutaneous Sumatriptan International Study Group reported on a study of 639 patients with migraine attacks in a randomized, double-blind, placebo-controlled parallel-group clinical trial in which sumatriptan was tested. Conclusions reported: "We conclude that a single 6-mg dose of sumatriptan given subcutaneously is a highly effective, rapid-acting, and well-tolerated treatment for migraine attacks. The administration of a second dose 60 minutes later to patients not responding well to an initial dose affords little Additional benefit."

Cluster Headaches. These headaches are characterized by recurrent, unilateral attacks of great intensity and brief duration. They may be accompanied by local signs and symptoms of autonomic nervous system dysfunction. The attacks occur in series lasting weeks or months, thus the name "cluster" headaches. The pain is reported as severe and reaches a maximum within a comparatively short period of time. The cluster headache syndrome occurs with unusual periodicity and, for that reason, sometimes is referred to as the "alarm clock" headache. As explained by Rankin, "The autonomic symptoms are bilateral but are more severe on the same side as the pain. The hypothalamus may be an activation site in this disorder. The posterior hypothalamus contains cells that regulate autonomic functions, and the anterior hypothalamus contains cells (the suprachiasmatic nuclei) that serve as the principal circadian pacemaker in mammals. Activation of both is necessary to explain the symptoms of cluster headache. The pacemaker is modulated by a 5-hydroxytryptamine-mediated (serotoninergic) system." Thus sumatriptan appears to be effective in the treatment of cluster headaches as well as migraines. With this recent knowledge of the biological mechanism, new designer drugs may be developed to treat both migraine and cluster headaches for patients who do not respond fully to sumatriptan. A study group known as The Sumatriptan Cluster Headache Study Group conducted a clinical trial in 1991 (similar to the previously mentioned trial involving migraine sufferers) and concluded: "Sumatriptan is an effective and well-tolerated treatment of acute attacks of cluster headache."

Additional Reading

Diamond, S.: *Migraine Headache Prevention and Management*, Marcel Dekker, New York, NY, 1990.

Ekbom, K. et al.: *Treatment of Acute Cluster Headache with Sumatriptan*, Department of Neurology, Söder Hospital, Stockholm, Sweden. [Abstract published in *New Eng. J. Med.*, 322 (August 1, 1991).]

Evans, R.W. and N.T. Mathew: *Handbook of Headache*, Lippincott Williams & Wilkins, Philadelphia, PA, 2000.

Ferrari, M.D. et al.: *Treatment of Migraine Attacks with Sumatriptan*, Department of Neurology, University Hospital, Leiden, the Netherlands. [Abstract published in *New Eng. J. Med.*, 316 (August 1, 1991).]

Jay, G.W.: *The Headache Handbook: Diagnosis and Treatment*, CRC Press, LLC., Boca Raton, FL, 1998.

Kittredge, M.: *Headaches*, Chelsea House Publishers, Broomall, PA, 2000.

Lance, J.W. and P.J. Goadsby: *Mechanism and Management of Headache*, Butterworth-Heinemann, Inc., Woburn, MA, 1998.

Lance, J.W.: *Migraine and Other Headaches*, Simon & Schuster Inc., New York, NY, 1999.

Moe, B.: *Everything You Need to Know about Migraines and Other Headaches*, Rosen Publishing Group, Inc., New York, NY, 2000.

Mongini, F.: *Headache and Facial Pain*, Thieme Medical Publishers, Inc., New York, NY, 1999.

Olesen, J., P. Tfelt-Hansen, and K.M. Welch: *The Headaches*, Lippincott Williams & Wilkins, Philadelphia, PA, 1999.

Raskin, N.H.: "Serotonin Receptors and Headache," *New Eng. J. Med.*, 353 (August 1, 1991).

Robbins, L.D.: *Management of Headache and Headache Medications*, Springer-Verlag, Inc., New York, NY, 2000.

Spierings, E.L.H.: *Headache*, Butterworth-Heinemann, Inc., Woburn, MA, 1998.

Siberstein, S.D. and T.D. Rozen: *An Atlas of Headaches: The Encyclopedia of Visual Medicine Series*, Parthenon Publishing Group, New York, NY, 2001.

Swerdlow, B.: *Whiplash and Related Headaches*, CRC Press, LLC., Boca Raton, FL, 1998.

HEADER. Any pipe, conduit, duct, or channel, which acts as a central point of distribution of a fluid flow to several branch lines, is a header.

HEADING. The direction of the forward end of the keel of a ship (either airborne or seaborne) is known as the heading of the ship. Unless a qualifying adjective is used with the term heading, it means direction with reference to true north. Compass heading, or magnetic heading, may be converted to heading by applying the compass corrections.

See also **Compass (Navigation)**; **Course**; and **Navigation**.

HEADWIND. See **Jet Streams**; and **Winds and Air Movement**.

HEAD (Zoology). The region of a bilaterally symmetrical animal body lying at the front end in relation to the ordinary direction of locomotion, or, in bipedal vertebrates like humans and some of the birds, at the highest level.

The development of a head is indicated in animals that are without sharply separated body regions, such as the flatworms. This process of cephalization is closely correlated with bilateral symmetry. The portion of a bilateral animal that goes first inevitably is the first to encounter new sources of stimuli, and shows some concentration of sense organs. Usually the chief nerve center, a cerebral ganglion or brain, also develops here. The concentration of sense organs and nervous control in the head remains characteristic of the region throughout the animal kingdom and in most groups is accompanied by the location of the mouth in the head, together with associated structures for securing food.

HEARING AND THE EAR. The role of the sense organ of hearing (the ear) is to code acoustic disturbances into neural signals suitable for transmission to the brain. The study of this process necessarily involves anatomy and physiology of the ear, the nature of auditory pathways and central nervous system activity in hearing, properties of acoustic signals that elicit auditory responses, and observed phenomena of auditory behavior. These aspects serve to define and delineate areas for investigations of hearing.

The truly phenomenal aspects of hearing can be observed in such behavior as localization of sounds, speech perception and particularly the understanding of one voice in the noisy environment of many, and the recognition of acoustic events that only last a few milliseconds. These and other behavioral phenomena remain to be fully accounted for in theories of hearing.[1]

Structure and Function of the Human Ear

What the human ear can do in processing auditory signals has been established for several years and in rather exquisite quantitative detail. How sounds are conducted to the inner ear is relatively well understood, as is the manner in which signals move from the cochlea to the brain along the eighth cranial nerve. The *total hearing process*, however, continues to elude researchers because of the extreme complexity of the ear's transducer, the *cochlea*. About the size of a pea and containing the organ of Corti, the cochlea incorporates well over a million essential moving parts. These are *hair cells* which, with remarkable subtlety, combine mechanical, hydrodynamic, electrical, and biochemical phenomena in their processing and measuring of incoming acoustic signals. They do this with amazing sensitivity and excellent frequent discrimination. Studies of the inner ear in recent years have taken advantage of advanced technology, including scanning electron microscopy. As mentioned by Hudspeth (University of California School of Medicine), a central goal of current auditory research is the elucidation of the cellular and molecular bases for the active process in the organ of Corti. If the present models are correct, the contribution of this active process must occur every few microseconds or tens of microseconds to facilitate high-frequency hearing. To date, these have been demonstrated to occur on a time scale of seconds to minutes, not of microseconds. This is but one gap that hopefully will be ultimately explained in further biophysical studies of the hair cells.

General Structure of the Ear. Traditionally, a description of the ear is based upon three regions: (1) external, (2) middle, and (3) inner ear. From

---

[1] Theories pertaining to the human hearing process date back a century and a half. Various hypotheses have been proposed. In the early years of study, two contradictory concepts were proposed. First, Seebeck (1843) suggested that the pitch of complex tones composed of higher harmonics (integral multiples of the fundamental frequency) corresponds to that of a pure tone whose frequency equals that of the fundamental frequency, and that the pitch does not change, even when the fundamental frequency (missing fundamental) is removed. The perceived pitch is assumed to be related to the temporal structure of the auditory stimulus (*periodicity* or *virtual* pitch). In the second theory, Helmnoltz (1862) suggested a systematic spatial representation of pure tones in the auditory system according to their frequency (*tonotopic* organization). This was supported by later invasive physiological methods. Helmholtz suggested that frequency information is encoded as *place* information and that perceived pitch is related to the *place* of cortical excitation.

In 1956 Licklider attempted to unify the two hypotheses (*place* versus *periodicity* pitch).

The refinement of biomagnetic measurements made it possible to test Licklider's unified hypothesis in human hearing systems. Further theoretical developments are beyond the scope of this article, but are elegantly described in the Pantev reference listed. See also **Acoustics**.

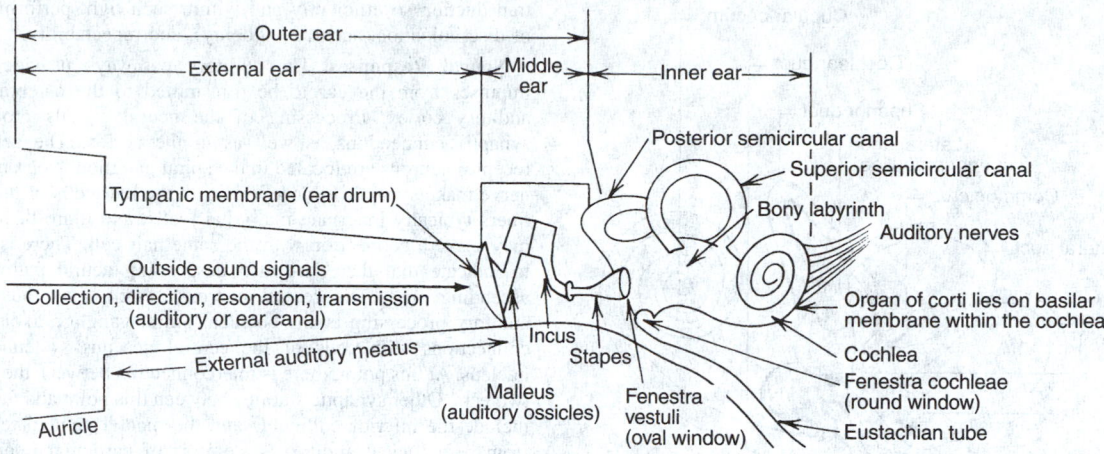

Fig. 1.   Highly schematic representation of human auditory system.

a functional standpoint, however, the ear may be divided into the outer and inner regions, as indicated in the highly schematic diagram (Fig. 1).

External Ear. This includes the auricle (that part of the ear which can be seen) as well as the *auditory* or *ear canal*, which extends to the *tympanic membrane* (eardrum). The outer ear performs the process of transforming acoustic energy into mechanical energy. The space between the auricle and the ear drum is called the *external auditory meatus*. The meatus is an irregularly shaped tube approximately 27 mm long (adult), with a diameter of about 7 mm, and terminated by the tympanic membrane. The ear canal is an acoustic resonator. Frequencies in the range of 3 to 4 thousand Hz are increased in pressure at the eardrum, as compared with the pressure at the entrance to the canal. The eardrum is in a protected position at the end of the canal and thus humidity and temperature conditions at the drum are relatively independent of those external to the ear.

Middle Ear. This is an irregularly shaped, air-filled space in the pestrous portion of the temporal bone. The three auditory ossicles of the middle ear, (a) the *malleus*, (b) the *incus*, and (c) the *stapes*, provide mechanical linkage between the tympanic membrane and the *fenestra vestuli*, an opening in the vestibule of the inner ear, commonly referred to as the oval window. The auditory ossicles are shown greatly enlarged in Fig. 2. The handle of the malleus attaches to the tympanic membrane, and the footplate of the stapes attaches to the oval window. Two important functions are provided by the middle ear. The first is to amplify and deliver sound vibrations from the drum to the inner ear, and the second is that of protecting the inner ear from very loud sounds. The amplification of sound waves is accomplished by apparent lever action of the ossicles that produces a greater force at the oval window than the force at the drum, and because of the gain in force that results from the relationship between the larger drum area to the smaller stapedial footplate area. The area of the drum is approximately 25 times that of the oval window. The amplification gain of these two factors is approximately 25 dB. The effectiveness of the middle ear action in increasing hearing sensitivity is evidenced in middle ear pathologies where the ossicular chain is disrupted. A hearing loss of 25 dB or more occurs. The second function of the middle ear, that of protecting the inner ear from loud sounds, is accomplished by reflex action of the middle ear musculature, the tensor tympani, and the stapedius. The action of the muscles is to retract the eardrum, draw

the stapes away from the oval window, and change ossicle vibrations in such a way as to decrease the transmitted pressure. Latency of muscle contraction and possible muscle fatigue limit protection of the inner ear by these mechanisms. Middle ear air pressure is equalized by virtue of the Eustachian tube, which connects the middle ear and the nasopharynx. The pressure equalization is necessary for normal ear drum movement.

Inner Ear. This is a system of cavities in the dense petrous portion of the temporal bone. One of the cavities is the cochlea, a bony labyrinth that is approximately 35 mm in length, coiled around a central core for $2\frac{3}{4}$ turns.

The Cochlea. Hudspeth, a contemporary researcher in the field, describes the mammalian cochlea as an extraordinarily complex structure that operates in a manner that is fundamentally simple. Nevertheless, the details of the cochlea are still rather poorly understood. Sound made up of a pattern of pressure changes at the eardrum is mechanically conducted through the chain of bones within the middle ear. The stapes, the last of the three bones, is mounted like a piston in contact with fluid within the cochlea. As the stapes moves back and forth in response to stimulation, pressure changes are transmitted into the cochlear fluids. The cochlea is comprised of three fluid-filled chambers, two bony and one membraneous. See Figs. 3 and 4. These chambers are separated from one another by two elastic partitions, which are helically coiled, one top another, about a common axis.

When the stapes compresses the fluid within one chamber (basilar membrane), one of the partitions between the cochlear chambers is

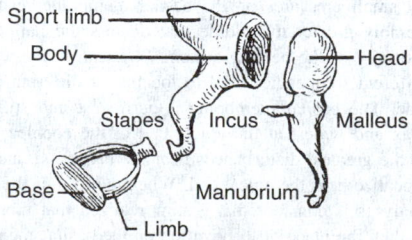

Fig. 2.   Greatly enlarged sketches of the malleus, incus, and stapes. (*Anatomy of the Ear, Grace Hewitt.*)

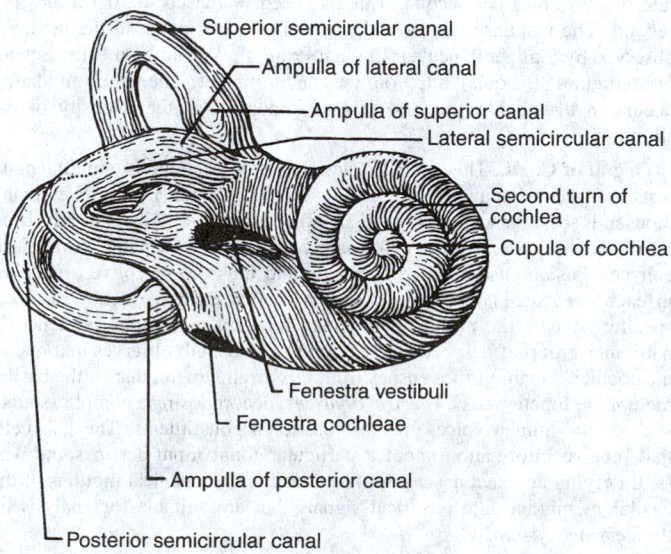

Fig. 3.   The bony labyrinth. (*Anatomy of the Ear, Grace Hewitt.*)

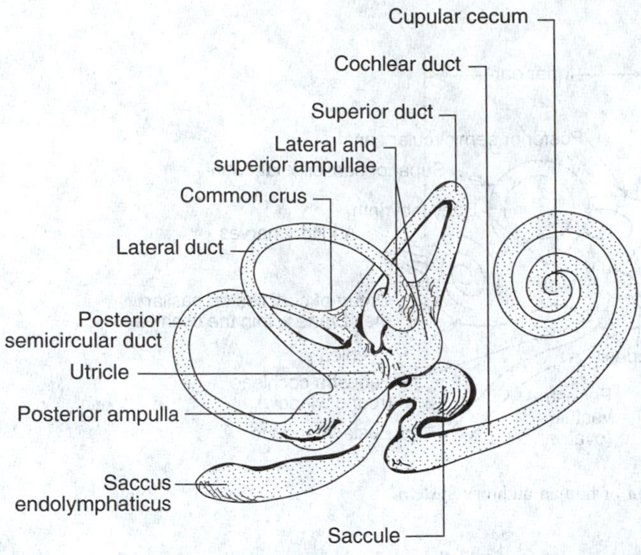
Cupular cecum
Cochlear duct
Superior duct
Lateral and
superior ampullae
Common crus
Lateral duct
Posterior
semicircular duct
Utricle
Posterior ampulla
Saccus
endolymphaticus
Saccule

Fig. 4. The membranous labyrinth. (*Anatomy of the Ear, Grace Hewitt.*)

deflected. It has been found that even when stimulated with a simple sound, such as a pure tone, nevertheless the basilar membrane moves in a complex fashion. As explained by Hudspeth, because the dimensions and mechanical properties of the membrane vary from its base to its apex, the membrane does not act like a homogeneous string on a plucked musical instrument. Rather, the basilar membrane develops a traveling wave in a region along its length that depends upon the stimulus frequency. It has been determined that low frequencies (down to 20 Hz in humans) excite motions near the apex of the cochlea. High frequencies (up to 20 kHz in humans) deflect the basal parts of the partition.

For persons with profound sensorineural hearing loss in both ears that cannot be helped by hearing aids, cochlear implants may be useful. Initially announced and approved for use in the United States in 1985, these implants have been used in a comparatively few thousand patients thus far, but estimates indicate that one-quarter million people could benefit from them. A wire electrode that is 1 millimeter in diameter at its widest point is "threaded" into the cochlea to electrically stimulate the auditory nerves. Deafness of this type is caused by damage to the 12,000+ sensory hair cells that line the normal cochlea. Since 1985, the implants have been improved considerably. A 22-channel implant was the first to be approved for use in children in 1990. Tests have shown speech comprehension improvement of from 5 to 160%. In original devices, signals were sent at a maximum rate of 300 times per second. This has been increased to 1000 times per second. The response ranges widely from one individual to the next. As observed by Skinner (Cochlear Implant Program, Washington Univ. School of Medicine, St. Louis), http://oto.wustl.edu/implant, where the stimulation occurs in the cochlea may be just as important as the intensity of the stimulation.

Organ of Corti. This helical structure, which is about 34 mm in length, rests on the basilar membrane in humans. This organ incorporates many thousands of hair cells as well as other types of cells—it has been estimated to have about 16,000 hair cells, in four parallel rows. Each hair cell has about a hundred sterocilia and thus the receptive organelles in each ear exceed 1 million. It is known that each frequency moves a specific zone of the basilar membrane; thus any given tone influences a particular group of hair cells most strongly. Hudspeth observes that one of the cochlea's main virtues ensues from this arrangement, that is, the basilar membrane functions as a *spectral analyzer*, decomposing a complex sound, such as the human voice, into its pure tonal constituents. The hair cells that receive information about a particular tonal input act in some way (still defying an exact description) to transduce mechanical motions of the basilar membrane into electrical signals that are suitable for analysis by the nervous system.

A number of models of the organ of Corti have been constructed in an effort to explain how the hair cells and basilar membrane accomplish transduction so efficiently and within such tight performance parameters of discrimination, accuracy (fidelity), and repeatability.[2]

Neural Responses. The auditory pathways provide for the neural impulses from the ear to be transmitted to the cerebral centers of the auditory cortex. Processing of the neural signals probably occurs at synaptic connections as well as in the cortex. The cell bodies of the receptor neurons are located in the spiral ganglion. Neurons of the auditory nerve make synaptic connections with the hair cells of the cochlea. Nerve fibers typically innervate many hair cells, and more than one nerve fiber may make a connection with the same hair cell. There is recent evidence to indicate that there are also descending neural pathways as well as ascending ones. The central nervous system may thus be involved in auditory processing at the cochlea. Spiral ganglion axons make synaptic connections with cells of the central nervous system at the cochlear nucleus. At this point, there is interconnection between the pathways for the two ears. Other synaptic stations between this point and the auditory cortex include the inferior colliculus and the medial geniculate body. Evidence from pathological auditory systems is of particular interest with respect to the auditory pathways. An impaired cochlea, for example, may result in a better than normal response to small amplitude changes in a sound. A lesion of the eighth cranial nerve is frequently manifested by a rapid decrease in the ability to respond under sustained stimulation. The ability to process speech is markedly affected when there is an involvement of the lower central nervous system. Cortical involvement does not affect usual speech or pure tone inputs.

Diseases and Disorders of the Auditory System

Common earache may arise from many causes and occurs in numerous forms. The most frequent cause of pain, aside from mechanical injuries, arises from some kind of bacterial infection. A physician should be consulted when an earache persists over several hours.

Otitis Media—Acute. An infection of the middle ear. Normally, the middle ear is sterile. The problem occurs rather commonly in early childhood, but the incidence decreases with increasing age. The disease takes a number of forms. When bacteria ascend from the nose and throat to the middle ear, the condition is referred to as *purulent otitis media.* The predominant symptoms are pain, fever, and often, diminished hearing. Perforation of the tympanic membrane (see Fig. 1) and otorrhea (discharge) may occur. One key to diagnosis is a bulging tympanic membrane with accompanying obscuration of the bony landmarks. The most frequent cause of purulent otitis media is pneumococcus, followed in order by *Haemophilus influenzae*. Anaerobic bacteria, although prominent in the normal flora of the upper respiratory tract, rarely cause acute otitis. Staphylococci are rarely involved. *H. influenzae* is a common cause in young children and is Seen in about one-third of the patients between 5 and 9 years of age. A number of other microorganisms can be involved (streptococci, *Neisseria catarrhalis*, and *S. epidermis*, among others).

In past years, acute suppurative mastoiditis and, less frequently, meningitis, sometimes followed acute otitis media. With current antibiotic therapy, these are rare occurrences. Therapy includes pain relievers (analgesics),

[2] One end of the hair cells rests on the basilar membrane; the other ends of the hair cells are the cilia, very fine hairlike processes, which make contact with the tectorial membrane, a membrane that overlaps the organ of Corti and that functionally behaves as if it were hinged at the cochlear wall. There are three rows of outer, and one row of inner, hair cells along most of the length of the basilar membrane. When vibrations are introduced into the inner and cause displacement of the basilar membrane, a shearing of the action of the cilia occurs that results in neural activity. It is assumed that amplifications occurs in the inner ear in that small pressures on the basilar membrane result in a shearing force of considerably greater magnitude that distorts the hair cells. The result is increased sensitivity of the hearing system. Physical properties of the cochlea are such that different frequencies tend to localize at different points along the basilar membrane. The basilar membrane is narrowest and stiffest at the basal end, and most lax and widest at the apical end of the cochlea. High-frequency sounds result in the greatest disturbances near the basal end, and low-frequency sounds tend to localize near the apical end. When the role of the cochlea in pitch and loudness analyses is considered it is now realized that more is involved in pitch perception than the place of localization on the basilar membrane, although the particular neural fibers involved are probably relevant. Loudness is probably related to the total number of neural impulses per unit time.

decongestants, and antibiotics. Ampicillin is frequently the drug of choice for children, and penicillin for adults. Where ampicillin-resistant strains of *H. influenzae* are encountered and where there is allergic response to penicillin, combinations of erythromycin and sulfisoxazole or trimethoprim and sulfamethoxazole are used. The latter drug alone is sometimes used for the chemoprophylaxis of recurrent otitis in children.

Otitis Media — Chronic. This condition usually results from neglected or recurrent acute otitis media and is seen in all age groups. Pain and fever may be absent, with hearing loss and foul discharge being the major symptoms. The tympanic membrane will be perforated. A number of microorganisms (staphylococci, streptococci, *Pseudomonas aeruginosa*, and enteric gram-negative bacilli, among others) may be cultured from the discharge. Antibiotics generally are ineffective. Surgery may be required in advanced cases. In some tropical and developing countries, where clostridia may be introduced with dirty cloths used for removing ear drainage, otogenous tetanus may develop. Otitis media can be a predisposing cause of bacterial meningitis and also may follow a measles infection.

During recent years, there has been considerable interest in the way otitis media occurring during early life may produce lasting developmental impairment. In the past, disorders of speech, language, cognition, and behavior have sometimes been attributed to early otitis media. The mechanism presumed involved deficits in conductive hearing and corresponding "auditory deprivation" during supposedly critical periods of brain development. As pointed out by Hubbard et al., the question has both practical and public health significance because, at one time or other, otitis media affects a large proportion of children. A study reported in 1985 involving nearly 50 children supported the hypothesis that early, long-standing otitis media may result in impairment of hearing and of speech, but no support was found for the hypothesis that cognitive, language, and psychosocial development are adversely affected.

Serous Otitis Media. Also called *secretory otitis*, this condition is characterized by the collection of fluid in the middle ear. This fluid may be either clear (serous) or gluelike (mucous). The predominant symptom is impaired hearing, which varies from a slight to almost total loss. Children who have serous otitis media may be subject to frequent upper respiratory infections and often have enlarged lymphoid tissue in the nasopharynx. If there is an underlying allergy or infection, appropriate antihistamines, antibiotics, or sulfonamides may be administered. Draining the fluid through an incision in the eardrum may relieve the condition. When there are repeated attacks, tiny plastic tubes can be inserted into the middle ear to provide adequate aeration, a procedure that requires a hospital environment. These tubes may be left in place for 3 to 4 months. Many cases of severely impaired hearing in adults can be attributed to middle ear infections in childhood. In infants and children, the Eustachian tube is shorter and more nearly horizontal than in adults, thus making the tube more likely to be an avenue of infection.

Otitis Externa. This disorder originates from the same causes as all middle ear infections, but it differs in the type of inflammation and the changes that occur in the tissues. A head cold may precede the infection. The attack of inflammation is sudden and causes congestion in the linings of the ear spaces, Eustachian tube, and mastoid cells. The ear itself fills with fluid, which gradually becomes puslike. Pain is the main symptom and can be severe, radiating, and throbbing. In children, early symptoms may include refusal to eat, nausea and vomiting, rolling the head, or tugging at the ear. Temperature generally runs high. A ringing sensation and dizziness may be present. Hearing is impaired as long as pus remains in the middle ear. If the condition is left untreated, after several days the eardrum ruptures spontaneously. For as long as three weeks, fluid Seeps through the canal and then subsides. The parts of the middle ear are so intricate and delicate that infection spreads easily. Pain resulting from movement of the external ear assists in distinguishing otitis externa from otitis media. Topical therapy includes polymyxin B and neomycin, usually with excellent results. Where true cellulitis of the external ear develops (infrequently), systemic antibiotics and possibly debridement of infected cartilage may be indicated. Very rare neurological complications of this condition can be life-threatening and require parenteral therapy with tobramycin and carbenicillin, as well as surgical debridement.

Aero-Otitis Media. In this disorder, the structures of the middle ear are affected by changes of pressure that occur during airplane flights. In milder cases, there is a sensation of stuffiness in the ears, with a slight inflammation of the eardrum, and perhaps some minor hearing impairment. Excruciating pain and hemorrhages in the tympanic membrane may occur in more severe cases. Although the condition still may occur among sensitive individuals, pressurized aircraft cabins have greatly alleviated the problem. If one senses this developing, chewing gum or moving the lower jaw with the mouth open will usually prevent it by opening the Eustachian tube, which will equalize the pressure. The problem is more common with persons who have upper respiratory infection or severe nasal allergy.

Mastoiditis. The middle ear is generally involved when there is an infection of the mastoid process of the temporal bone. The acute form of this disease (*acute mastoiditis*) has been practically eliminated since antibiotic drugs became available to combat middle ear infections.

The inflammation in mastoiditis involves the lining of the mastoid cells. The infection may enter the bone, which becomes soft and decayed. The causes of mastoiditis include respiratory infection, abnormal anatomy of the ear in infants and children, improper channels for ear drainage, and lowered resistance to infection. Mastoiditis may occur as a secondary infection to various diseases. The predominating symptom is pain, which may be either continuous or intermittent. If the patient is not treated, the intense pain could persist for 6 or more days, which may not be true for middle ear infection. Also unlike middle ear infection, mastoiditis is characterized by a definite, localized tenderness over the mastoid process.

In *chronic mastoiditis*, which now occurs more often than the acute type, drainage from the ear (*otorrhea*) is the principal symptom. Fever may or may not be present. If acute mastoiditis should occur, the physician may perform a *mastoidectomy*. In this operation, the infected mastoid cells are removed through an incision in the area behind the ear, or in the external auditory meatus.

Punctured Eardrum. The most common cause of a punctured eardrum is the insertion of a sharp object into the ear. Violent explosions near the ear may cause the drum to tear or rupture. Decreased air pressure during or after descent from high altitudes, severe sneezing, diving, and increased pressure, frequently are responsible for damaged membranes. Sometimes, diagnosis is difficult. The pain accompanying a puncture is sharp and intermittent. Blood may ooze from the injury, but this is not positive proof of a drum tear, because the same symptom may be present in a skull fracture. Dizziness, ringing sounds, and headaches also are significant symptoms. A tear in the eardrum may heal without treatment within a period of a few weeks, but there may be aftereffects which may not be noticed, even for as long as a year. A grafting operation known as *tympanoplasty* can be employed in cases in which the tear does not close.

Growth on the Eardrum. Following rupture or perforation of the eardrum, small chalky (lime) deposits may form at the site of healing as a result of repeated attacks of middle ear infection. If they form from a healed perforation, they mark the path of least resistance for a future rupture. It is the general opinion of physicians that such deposits do not affect normal hearing. There is no successful way of removing the chalk deposits without injuring the eardrum seriously or depressing the hearing. Hence, it is rarely attempted.

Boils or Furuncles. When present in the external ear, these often produce severe pain because the skin in this region normally adheres closely to the underlying cartilage and bone. If infection is allowed to persist, perforations of the eardrum may occur. Through them, infection may spread to the middle ear, the inner ear, or the mastoid area. An x-ray will assist in determining the nature of any secondary complication.

Fungus Infection. *Otomycosis* is a fungus infection of the outer ear and canal. The inside of the ear appears dirty and crusty, and fluid seeps out continually. When the crusts and scales are removed, the skin beneath is raw and bleeds easily. Itching causes much discomfort. Pain is usually present because of the swelling of the canal; hearing may be impaired. Treatment is by specific solutions and ointments. Home remedies are not recommended.

Tinnitus. Most persons, at one time or another, experience this disorder, a sensation of ear noise, which is more noticeable in a quiet environment. Such sounds may seem to be in the head rather than the ear, and may affect one or both ears. The symptom is associated with many conditions, including middle ear infection, Ménière's syndrome, exposure to intense noise, circulatory diseases, otosclerosis, and neuritis of the auditory nerve. The symptom also may be caused by excessive amounts of coffee, tobacco, or alcohol. Quinine, certain antibiotics, or large doses of aspirin also may produce tinnitus. Such sounds occur most often in persons between ages 50 and 70. The reason for the sensation has not been established. In as much as the symptom could be an early warning of hearing damage, it should be investigated.

Cauliflower Ear. Known as *hematoma of the auricle*, this disorder has long been recognized as the badge of the prizefighter. It is caused by injury to the external ear. A hard blow may cause bleeding below the skin. If this accumulation of blood remains for sometime, it becomes fibrous tissue and eventually will be converted into a bone-like or cartilaginous substance. Thus, the ear will be deformed by this irregular mass of extra tissue. For prevention, the blood should be removed before it clots. Plastic surgery also is used for restoration of affected ears.

Congenital Malformations. These occur rather frequently, but generally they are not gross enough to impair hearing. They may be unsightly. Absence of the lobe or the outer rim of the ear (*helix*), large protruding ears, and irregular shapes are among the more common malformations. Plastic surgery can restore most of these conditions to normal appearance. Occasionally, a congenital defect, such as an obstruction in the canal, may have to be removed before hearing improves. In rare instances, the ears may be displaced on the head, and in some extreme cases when the lower jaw is grossly misshapen, they may even be fused together (*synotia* or *otocephaly*).

Vestibular Disturbances. The semicircular canals of the inner ear are partially responsible for adjusting the body to changes in motion. The rate of these changes normally allows sufficient time for the canals to maintain body equilibrium. Rapid, irregular, and continuous waves of motions, when they persist over a period of time, may interfere with the vestibular apparatus of the ear and the result is **motion sickness**. This unpleasant condition may be encountered at sea, in the air, while riding in an automobile, on an elevator, etc. The personal reactions to motion sickness are highly individualistic. Recovery is rapid, once the cause is avoided. A number of oral drugs, such as dimenhydrinate (Dramamine®), meclizine, cyclizine, or promethazine, are used as preventive measures—taken an hour prior to boarding a boat, car, etc. Drowsiness may be a side-effect, thus the drugs should not be taken by persons operating automobiles or other vehicles and dangerous machinery.

Ménière's Syndrome. Prosper Ménière described this malady in 1861 and correctly attributed its origin to the inner ear. Its characteristic symptoms are sudden severe episodes of *vertigo* (dizziness), tinnitus, and fluctuating hearing loss. The term syndrome continues to be used because the exact causes of the disorder have not been fully established. Persons in the middle age group are more commonly affected by the syndrome. The vertigo associated with an attack may be so severe that the simplest activities become impossible. Usually, the patient has a sensation that objects are whirling about. The same type of dizziness occurs with certain cardiovascular disorders and middle ear infections. Attacks may last for minutes or weeks. The tinnitus, usually a roaring noise, sometimes persists between attacks. Nausea and vomiting are also usual symptoms.

The course of the syndrome is unpredictable. Remissions of up to several years often occur. About two-thirds of the patients improve or recover regardless of treatment. No single form of therapy has been fully successful. Certain drugs, such as Dramamine®, often help control the vertigo. Sedatives or tranquilizers are occasionally helpful. If the condition is disabling and unilateral, the diseased parts of the labyrinth may be surgically removed. The procedure stops the vertigo, but balance is impaired and hearing loss in the affected ear is total. Ultrasonic radiation has been used to irradiate the labyrinth with the objective of destroying the diseased portions. For relief of severe vertigo, some surgeons recommend the Tack operation to drain the saccule, which contains endolymph. A tack, a small pointed piece of metal, is placed through the footplate into the sac, thus allowing drainage. According to one theory, this syndrome is related to an imbalance of pressure between the perilymph and the endolymph. Another innovation has been the use of surgical instruments maintained at temperatures as low as $-140\,°C$. With these instruments a surgical procedure should be less likely to damage the cochlea.

Vestibular Neuronitis. This is a comparatively common syndrome, the manifestations of which are vertigo, vomiting, and imbalance. Some authorities believe the disorder results from irritation of the vestibular portion of the eighth cranial nerve. Although vestibular neuronitis resembles Ménière's syndrome in many respects, there are no audiologic symptoms, and in particular no hearing loss. The disease is benign and there is no specific treatment.

The Ear and the Nature of Sound

Sound as a physical phenomenon is described in considerable detail in the article on **Acoustics**. Sound involves a disturbance in the air that is a forward and backward, rarefaction and compression, movement of air parcels. The unit of force usually used in acoustics is the dyne. Sound pressure is frequently expressed in dynes per square centimeter. Intensities of sounds are usually measured on a decibel scale, a logarithmic ratio scale. The tremendous loudness range of the ear is exemplified by the fact that the most intense sound that can be tolerated is a million, million times greater in intensity than a sound that is just audible. This is a range of approximately 120 dB. The frequency range of hearing is frequently given as 16 to 20,000 Hz. The ear is most sensitive in the middle frequency range of 1,000 to 6,000 Hz. In terms of discrimination of frequency and intensity, it is possible for about 1,400 pitches and 280 intensity levels to be distinguished.

Hearing Loss and Deafness

Deafness means nearly complete or total loss of hearing. There are two types: (1) congenital, and (2) acquired. In the congenital type, the person is born deaf or later becomes deaf because of an inborn defect. Hard of hearing is a term that applies to those who lose some of the ability to hear later in life, but who have learned how to speak before the loss occurred.

Causes of deafness are many. Some conditions which may cause deafness or milder hearing difficulties include: (1) temporary or chronic infections in one or both ears; (2) secondary complications of disease elsewhere in the body; (3) direct damage or defect in some part of the hearing system; (4) aging; (5) occlusion of the auditory canal; (6) aero-otitis media; (7) Ménières syndrome; (8) ostosclerosis; (9) noise; and (10) certain toxic drugs. Side effects of the loop diuretics, ethacrynic acid and furosemide, include transient hearing impairment. Complete deafness has been reported after intravenous administration of ethacrynic acid and a permanent hearing deficit after chronic use.

Conductive deafness results when sound waves are not transmitted properly through the outer and the middle ear. If the damage is to the inner ear or the nerve pathway to the brain, a *sensorineural* (also called *nerve* or *perceptive*) *deafness* occurs. The latter type is generally a greater handicap and usually cannot be reversed. In *mixed hearing loss*, there are elements of both conductive and sensorineural types of loss. Some deafness is caused by a disorder in the central nervous system.

Otosclerosis. Usually first detected during early adulthood, *otosclerosis* can cause a conductive type of hearing loss. Bony growths form just inside the inner ear where the middle ear's stirrup (*stapes*) enters it. Eventually, the footplate of the stapes becomes anchored and no longer conducts sound waves to the inner ear. About 10% of the population is affected to some extent in this way, although they may have no hearing loss for many years. Experience indicates that the disorder may become arrested at any stage. Heredity appears to be an important factor. Middle ear infections are not a cause. The disorder occurs about twice as often in females as in males.

Noise. Individuals vary in terms of susceptibility to noise-induced hearing loss. If sufficiently exposed to intense noise for extended periods, all persons are considered as candidates for loss of hearing. Any noise in excess of 85 dB is considered damaging. Frequently, the hearing loss will be accompanied by a high-frequency tinnitus. Noise-induced hearing loss usually is first noted at about 4 kHz, progressively moving into the lower frequencies with continued exposure. The alterations in the inner ear caused by external noise are not well understood. Recovery from the hearing loss is not to be expected. Avoidance of noise or wearing protective devices, such as ear protectors and plugs, is recommended. The best form of prevention is that of taking measures to reduce the amount of noise radiation that escapes from heavy industrial equipment, vehicles, etc.

Even with an increased awareness of the adverse effects of noise, the environment continues to become noisier. Overamplified music and noisy vehicles, particular favorites of young people, have been implicated in the cause of hearing loss. One of the frustrating effects of noise is the masking of speech. For example, if the speaker and listener are separated by 5 feet (1.5 meters), the levels of noise that will barely permit reliable word intelligibility are 50 decibels for normal conversation; 57 dB for raised speech; 63 dB for very loud speech; and 69 dB for shouting. As shown in Table 1, these levels are approached or exceeded in several day-to-day industrial and commercial activities.

Instrumental Methods for Measuring Sound and Hearing

The most common measurement of hearing function is the pure-tone audiogram in which a frequency from 125 to 8,000 Hz is plotted against hearing loss in decibels. The audiogram displays the ability of the ear to

TABLE 1. NOISE LEVELS FOR VARIOUS SOURCES AND LOCATIONS

Description of Noise	Noise Level (dB)
Threshold of hearing	0
Rustle of leaves in gentle breeze	10
Quiet whisper (distance of 5 feet)	10
Average whisper (distance of 4 feet)	20
House in country (average situation)	30
House in city (average situation)	40
Apartment (average situation)	40
Hotel	42
Theater (between performances)	42
Small retail establishment	52
Commercial garage	55
Medium-size office	58
Residential street	58
Restaurant	60+
Medium-size retail establishment	62
Factory or warehouse office	63
Large retail establishment	63
Ordinary conversation (distance of 3 feet)	65
Large office	65
Traffic on busy street	68
Factory (light-to-medium work)	78
Riveter (distance of 35 feet)	97
Hammer blows on steel plate (distance of 2 feet)	114
Threshold of pain	130

Based upon original data by H.F. Olson ("Acoustical Engineering," Van Nostrand Reinhold, New York, 1957).

hear a pure sine-wave tone at a given frequency compared with a "normal" ear. The unit of loudness is the decibel, defined as $10 \times \log_{10}(P_1/P_2)$, where P_1 is the power of the sound being applied and P_2 is the just-audible power required at the given frequency for the "normal" ear to hear. The standard audiometer contains a frequency-selection knob, an attenuator calibrated in 5-dB increments, a key that connects the output of the instrument to the earphones is placed on the subject's head. The procedure is to increase the amplitude slowly while depressing the key in short pulses until the subject reports that the sound can just be detected.

In addition to pure tones, speech sounds are also used as test signals. Using +9 dB (referred to 0.0002 dyne/square centimeter) as a 0-dB threshold level, it is possible to determine the extent of the hearing loss for speech using specially selected two-syllable words having approximately equal stress on each syllable (called "spondaic" words). The equipment used for this measurement consists of a microphone, audio amplifier, and a pair of headsets, the system having a float frequency response between 125 Hz and 8 kHz. Sensitivity, or gain, of the amplifier is controlled by a step attenuator calibrated in 1-dB steps, and the output is arranged to go into either ear separately, or both ears simultaneously.

In the von Békésy pure-tone audiometer, the amplitude control is run up and down by a motor while the subject operates a key. The amplitude is slowly increased until the subject hears the sound, which reverses the motor. The frequency is similarly increased slowly and automatically. The resulting curve is somewhat sawtooth in form and more accurately brackets the threshold values.

In designing and using audiometers, great care must be given to the elimination of background noise and hum. If more than one tone is presented at a time, "masking effects" may occur, giving different results than would be obtained with each sound separately.

Sound-level meters are widely used throughout industry in an effort to stay within legislatively prescribed limitations. Noise-level dosimeters, which automatically compute cumulative noise exposures (for example, the exposure of a worker to noise over an 8-hour workday) are also available. Allowable noise limits in the United States are monitored for compliance by OSHA (Occupation Safety and Health Administration). These limitations are subject to change from time to time as experience is gained.

Hearing Devices

In addition to portable personal hearing aids, which have been available for many years, a few researchers are taking a different approach to the problem with the target of developing implantable prostheses for delivering electrical stimuli directly to the auditory nerves. Such devices would be applicable to individuals whose hearing loss is the result of damage to the hair cells of the inner ear. In one design (experimental), an 8-channel, bipolar solid-state device would deliver stimuli at eight different frequencies to separate groups of auditory-nerve fibers in the cochlea. Eight closely spaced pairs of electrical contacts are distributed along the length of the implanted device. The many problems remaining to be solved with such endeavors are well outlined by Loeb.

See also **Voice and Sound Production**.

Additional Reading

Berlin, C.I. and B.J. Keats: *Genetics and Hearing Loss*, Singular Publishing Group, Inc., San Diego, CA, 2000.

Blachman, N.: *Noise and Its Effect on Communication*, Krieger Publishing Company, Melbourne, FL, 1982.

Bohnke, F.: *Cochlear Implants*, S. Karger Publishers, Inc., Farmington, CT, 2000.

Borg, E. and S.A. Counter: "The Middle-Ear Muscles," *Sci. Amer.*, 74 (August 1989).

Brownell, W.E. et al.: "Evoked Mechanical Responses of Isolated Cochlear Outer Hair Cells," *Science*, **227**, 194–196 (1985).

Canalis, R.F. and P.R. Lambert: *The Ear: Comprehensive Otology*, Lippincott Williams & Wilkins, Philadelphia, PA, 1999.

Corwin, J.T. and D.A. Cotanche: "Regeneration of Sensory Hair Cells After Acoustic Trauma," *Science*, 1772 (June 24, 1988).

Dooling, R.J. and S.H. Huklse, Editor: *The Comparative Psychology of Audition: Perceiving Complex Sounds*, Lawrence Erlbaum Associates, Inc., Hillsdale, NJ, 1989.

Erickson, D.: "Electronic Earful: Cochlear Implants Sound Better All the Time," *Sci. Amer.*, 132 (November 1990).

Gelfand, S.: *Hearing: An Introduction to Psychological and Physiological Acoustics*, Marcel Dekker, Inc., New York, NY, 1998.

Henderson, D., J.M. Berstein, B. Maurizio, et al.: *Immunologic Diseases of the Ear*, New York Academy of Sciences, New York, NY, 1997.

Hubbard, T.W. et al.: "Consequences of Unremitting Middle-Ear Disease in Early Life," *N. Eng. J. Med.*, **312**(24), 1529–1534 (June 13, 1985).

Huckabee, M.L. and C.A. Pelletier: *Management of Adult Neurogenic Dysphagia*, Singular Publishing Group, Inc., San Diego, CA, 2000.

Hudspeth, A.J.: "The Cellular Basis of Hearing: The Biophysics of Hair Cells," *Science*, **230**, 745–752 (1985).

Jahn, A.F. and J. Santos-Sacchi: *Physiology of the Ear*, Singular Publishing Group, Inc., San Diego, CA, 2000.

Kryter, K.D.: *The Effects of Noise on Man*, 2nd Ed., Academic Press, Orlando, Florida, 1985.

Licklider, J.C.R.: in *Information Theory* (C. Cherry, Editor), Butterworths, London, 1956.

Loeb, G.E.: "The Functional Replacement of the Ear," *Sci. Amer.*, 104–111 (February 1985).

Myers, D.G.: *A Quiet World: Living with Hearing Loss*, Yale University Press, New Haven, CT, 2000.

Northern, J.: *Hearing in Children*, Lippincott Williams & Wilkins, Philadelphia, PA, 2000.

Pantev, C., M. Hoke, B. Lütkenhöner, and K. Lehnertz: "Tonotopic Organization of the Auditory Cortex: Pitch Versus Frequency Representation," *Science*, 486 (October 27, 1989).

Romani, G.L., S.J. Williamson, and L. Kaufman: *Science*, **216**, 1339, (1983).

Seebeck, A.: *Ann. Phys. Chem.*, **53**, 417 (1843). (A classic reference.)

Suter, A.H.: "Noise Wars," *Technology Review (MIT)*, 42 (November–December 1990).

Tos, M.: *Surgical Solutions for Conductive Hearing Loss*, Thieme Medical Publishers, Inc., New York, NY, 2000.

von Helmholtz, H.: *Die Lehre von den Tonempfindungen als physiologische Grundlage für die Theorie der Musik*, Vieweg, Braunschweig, 1862. (A classic reference.)

Waltzman, S.B. and J.L. Cohen: *Cochlear Implants*, Thieme Medical Publishers, Inc., New York, NY, 1999.

Yost, W.A.: *Fundamentals of Hearing: An Introduction*, 4th Edition, Academic Press, Inc., San Diego, CA, 2000.

Web References

Better Hearing Institute: http://www.betterhearing.org/
Clarion Cochlear Implant: http://www.cochlearimplant.com/
Hearing Disorders: http://healthlink.mcw.edu/hearing-disorders/
Hearing Loss. Health factsheets from BUPA: http://hcd2.bupa.co.uk/fact_sheets/Mosby_factsheets/Hearing_Loss.html

HEARING (Fishes). See **Fishes**.

HEARING ORGANS. See **Sensory Organs**.

HEART AND CIRCULATORY SYSTEM (Physiology).

The circulatory or cardiovascular system of the human body is comprised of the heart and the blood vessels (arteries, veins, and capillaries). These organs are highly interdependent. The study, diagnosis, and treatment of diseases and disorders of this system fall under the general classification of *cardiovascular medicine*.

Several other articles in this encyclopedia relate to the heart and circulatory system and include:

Aneurysm	Cerebrovascular Diseases	Endocarditis
Angiography	Collateral Circulation	Hemorrhage
Anticoagulants	Congestive Heart Failure	Hypertension
Aorta	Diastole	Hypotension
Arrhythmias	Diuretics	Ischemic Heart Disease
Arteries and Veins	Echocardiography	Pulse
Blood Pressure	Electrocardiography	Sphygmomanometer

Heart

The heart is the muscular organ that pumps blood through various conduits to and from all parts of the body. Depending upon the size of the adult individual, the human heart weighs somewhat less than three-quarters of a pound (about 340 grams). The organ essentially is a hollow muscle capable of contraction like other muscles. A contraction of the heart is referred to in general terms as a *heartbeat*. The rate of the heartbeats can be changed by two different sets of nerves: (1) The accelerating nerves are connected to the spinal cord and are a part of the sympathetic nervous system; (2) the *vagus nerve* depresses the rate and is connected to the brain stem. The beating of the heart commences long before birth and continues as long as life continues. Beats occur at the rate of 70–80 times per minute in adults, but may increase to 100 beats per minute during exertion, or in the presence of emotional disturbance. During a 70-year life span, it is estimated that the heart beats some 3 billion times, an average of about 42 million beats per year. Each contraction of the heart moves slightly more than 2 fluid ounces (~59 cubic centimeters) out into the arteries, providing a change of blood over the body about once every minute. During a lifetime of 70 years, a total of 250 million quarts (~236.5 million liters) of blood are moved, almost enough to fill a large football stadium. There are only a little over 6 quarts (~5.7 liters) of blood in the average human body, so that this blood requires not only rapid circulation, but also a fine adjustment of controls to assure the proper and effective distribution required by the body.

The highly schematic diagram of the heart given in Fig. 1 indicates the principal components of the heart structure. The heart is divided into four chambers—two auricles, referred to as the right and the left auricle; and two ventricles, referred to as the right and the left ventricle. The flow of blood through these chambers is controlled by four valves, as numbered in the diagram: (1) the tricuspid valve; (2) the mitral valve; (3) the pulmonic valve; and (4) the aortic valve.

Blood coming from over the body through the large veins (venae cavae) enters the right auricle at A. This blood has been partially depleted of its oxygen. As the lower, thick-muscled ventricles expand, this blood enters the right ventricle through the tricuspid valve. Then, the ventricle contracts and forces the blood into the pulmonary artery toward the capillaries of the lungs and is prevented from running back into the heart by the closure of the pulmonic valve. In the meantime, the purified blood in the left auricle has just arrived from the lungs through the pulmonary veins, at B. From here it passes into the thick-walled left ventricle through the mitral valve. When the right ventricle forces blood out into the pulmonary artery, the left ventricle at the same time contracts and sends blood out into the arteries of the body, passing through the aortic valve into the aorta. The auricles thus act as collecting chambers, while the ventricles serve as pumps. The right side of the heart collects the blood and forces it through the lungs; while the left side collects it from the lungs and forces it through the body as a whole. The four valves between the various chambers of the heart prevent the blood from flowing backward and maintain the pressure between heartbeats because of the closed system that results.

In order that blood can be moved forward in an orderly manner, it is important that the heart muscles expand and contract at just the right time and that all the valves open and close completely at the proper time during the cycle. This control is accomplished by a special structure known as the *sino-auricular node*. This is the pacemaker of the heart. It is not entirely dependent upon the general nervous system, and it has been known to function for some time after breathing has ceased. Sudden changes in temperature, unusual nervous stimuli, fright, a sense of impending danger, or a happy thought can affect this heart center and, thereby cause speeding or slowing of the heart action. All warm-blooded animals have such a fine adjustment that acceleration or retardation may occur within $\frac{1}{100}$ th second.

Traditionally, the accepted model of the heart's function was derived mainly from work done near the end of the 19th Century by Otto Frank (Germany) and Ernest H. Starling (England) who postulated that the energy imparted to the blood by the contraction of a ventricle, independent of any control by nerves or hormones, is proportional to the length of the ventricular muscle fibers at the end of the preceding diastole. It was assumed that once systolic contraction was complete, the subsequent diastolic filling becomes a passive function of venous pressure, which stretches the relaxed muscle of the ventricle wall. The Frank-Starling concept is that the energy expended in contraction has no essential role in the diastolic filling of the ventricles.

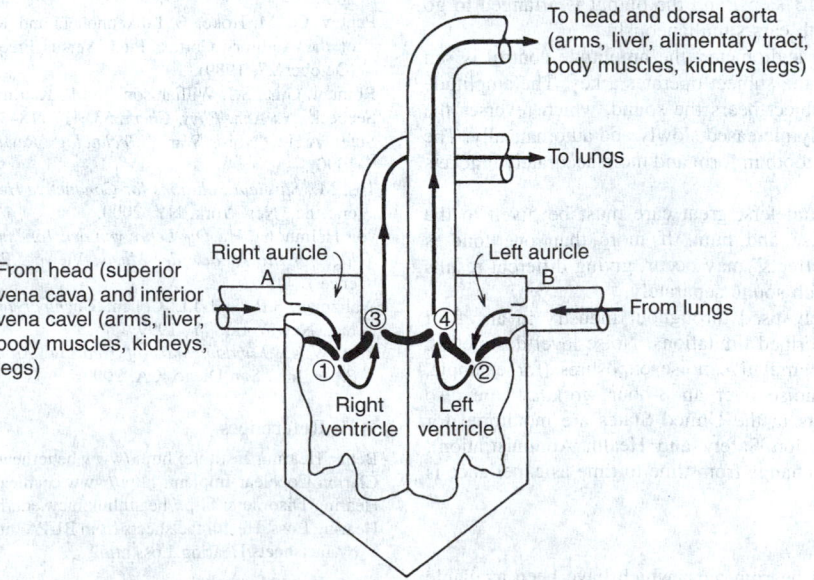

Fig. 1. Highly schematic diagram of major components of human heart: (**a**) Entrance of blood from venae cavae to right auricle; (**b**) entrance of blood from lungs; (1) tricuspid valve; (2) mitral valve; (3) pulmonic valve; (4) aortic valve. Diagram is not to scale.

In 1986, reporting on research of the early 1980s, researchers T.F. Robinson, S.M. Factor, and E.H. Sonnenblick (Albert Einstein College of Medicine, Yeshiva University) describe a new model of the heart, suggesting that some energy from each contraction is stored within the muscle to provide the power for a suction that aids filling. The effect appears to be amplified by the motion of the heart as a whole. They point out that the Frank-Starling model does not reflect the dynamic interplay between systole and diastole. The researchers concede that vast improvements in the instrumentation for accurately measuring the parameters of the heart's performance were not available at the time the original model was proposed. Thus, in the earlier model, the mechanism by which the heart is filled is relatively static. In the new model, the dynamic relation between systole and diastole is critical to the proper function of the heart. The systolic contraction provides much of the energy that drives the process of diastolic expansion. The researchers explain that this energy is stored and recovered in two ways: (1) By the gross motion of the heart itself (when the heart contracts, it propels blood upward and thus, by Newton's law of action and reaction, it propels itself downward within the body). Recoil stretches the great elastic vessels and connective tissue that hold the heart in place. Subsequently, as the heart relaxes, it springs upward, thus meeting the inflow of blood head on. Thus, the velocity of the blood with respect to the heart is raised and assists in powering the filling process. (2) The energy of systole is stored in the deformation of the heart itself. In the new model, the systolic contraction compresses the elastic elements of the heart and its muscle fibers, so that, without any external filling, there is a natural propensity for the ventricles to expand — an expansion that creates a negative pressure (suction) that pulls blood into the ventricles from the atria. In summary, the Frank-Starling model is of a static pressure pump; the new model is of a dynamic suction pump.

Networking in the Heart. The sino-auricular node lies in the wall of the right auricle, embedded within the muscular tissue. A heavy partition extends between the left and right side of the heart, so that there is no direct connection between them except for a group of structures consisting of the auriculo-ventricular node, the *common bundle* and its left and right branches. The auriculo-ventricular node transmits impulses from the common bundle, also known as the *bundle of His*, thence to the two branches, and from there to a network of muscle fibers which covers the inside of each ventricle. The network extends to the outer covering of the heart and is called the *Purkinje system*. This system assures an almost instantaneous response of the muscles of the ventricles once the impulse has passed into it. Although the heartbeat is not entirely independent of the general nervous system, it may carry on for some time without the ordinary nerve impulses. This is illustrated by the fact that the heart of a rabbit, for example, may continue to beat long after the animal has died. This automaticity of the heartbeat allows for cardiac transplantation.

The normal beating of the heart is associated with the production of bioelectric currents in the organ. Although these currents are not strong, they are carried to the surface of the body where they may be measured by a sensitive instrument, the electrocardiograph. The beat of a normal heart shows a characteristic pattern of electrical responses. See also **Electrocardiography**.

There are thousands of small muscle fibers interwoven to make up the walls of the heart. The organ also has its own circulatory system to provide the muscle with nourishment. The whole structure is sheathed with a tough sac, the *pericardium*, containing a small amount of fluid. This provides for lubrication of the rapidly moving heart.

The 70 or 80 normal heartbeats per minute do not allow much time between the expansion and contraction of the four heart chambers. The period of relaxation of the muscles, during which the heart fills, is about equal to that of contraction, when it empties. This period of relaxation permits the heart to recover fully from its work period. The contraction of the heart is called *systole*; the relaxation is called *diastole*.

Valving in Heart and Circulatory System. To make the blood move in only one direction, there not only are the valves inside the heart, but also valves in the veins. In addition, in the small veins there is a constricting type of valve, which helps adjust the rate of blood flow and the distribution of blood between the several organs in accordance with need. The capillaries act as the final speed control, by being so small that only one or two rows of blood cells may pass through at a time. Here the speed of the flow is so reduced that time is allowed for rebalancing the mineral content of the area, the exchange of oxygen for carbon dioxide, and soluble food for waste materials.

Heart as a Hormone Source. Since the publication of "Essay on the Motion of the Heart and the Blood in Animals" by William Harvey in 1628, the heart has essentially been considered a *pump*, albeit a complex one. John Peters (Yale University School of Medicine), as early as 1935, speculated about a mechanism that may be located in or near the heart to "sense the fullness of the blood-stream," in essence, some biochemical substance that fine-tunes the regulation of blood volume. Later, during the 1950s and 1960s, some researchers established the overall properties of what they called a "natriuresis" (derived from excretion of sodium and diuresis, excretion of water) hormone. The hormone was also referred to as the "third factor," third because of the other two established regulators of blood pressure and blood volume, namely, (1) the hormone aldosterone and (2) the process of filtering the blood by the kidneys. In the mid-1950s, while studying the constituents of heart-muscle cells, J.D. Jamieson and G.E. Palade (Yale) noted unexplained dense bodies in the cells. In 1974, a research team of M. Cantin and J. Genest (Clinical Research Institute of Montreal) and colleagues noted a similarity between the unknown dense bodies and the storage granules seen in the endocrine (hormone-secreting) cells, such as the pancreas and the anterior pituitary gland. In 1976, Pierre-Yves Hatt (University of Paris) and colleagues, by way of experimentation with laboratory animals, noted a relationship between the number of granules in the atrial cardiocytes (heart-muscle cells) and the amount of sodium in the animal diet. As reported by Cantin and Genest (1986), the breakthrough came in 1981 when A.J. deBold, H. Sonnenberg and associates (Queen's University, Ontario, Canada) injected homogenized rat atria into rats and observed a rapid, massive, and brief diuresis and natriuresis. Thus, they concluded that the atria contained a factor that promotes these effects and named it atrial natriuretic factor (ANF). The heart-produced substance is classified as a hormone and subsequently it has been found that ANF exerts its effects on the blood vessels, the kidneys, the adrenal glands, and on a large number of regulatory regions in the brain.

Blood Transporting Vessels

The aggregate length of conduits required to transport blood throughout the human body would be measured in terms of miles or kilometers. The beat of the heart forces a temporarily increased amount of blood into the arteries. The arterial walls are elastic and expand to accommodate this larger volume of blood. Between beats, the walls gradually contract, forcing the blood through the capillaries at an approximately constant rate. In this manner, the arteries act as a reservoir that prevents the blood from flowing to the tissues in gushes. See also **Arteries and Veins (Vascular System)**.

Blood passing from the heart through the lungs has only about one-sixth of the pressure of the blood as it is forced out over the body through the *aorta*. See Fig. 2. The pressure is still sufficient, however, to cause flow through the multitude of capillaries in the walls of the lungs. The lungs are composed of innumerable small sacs that have a supply of changing air. In the lung or pulmonary capillaries, the blood releases carbon dioxide and takes up oxygen.

The blood continues to flow back through the pulmonary veins and into the left auricle for distribution over the body. The loss of carbon dioxide and the assimilation of oxygen are accompanied by a change of color in the blood, from a dark to a bright red.

Although the liver does not have a special connection with the heart, it acts as a storage organ for blood. Blood is carried to the liver from the stomach and intestinal tract by the portal vein and from the rest of the body by the hepatic artery. It has been estimated that the liver and portal vein drainage system may hold as much as one-third of all the blood in the body. When the body is inactive and requires a smaller amount of blood, the liver and portal vein system relieves the remainder of the system by holding a large part of the excess. Some impurities are removed in the liver and excreted into the digestive tract. The hepatic vein returns the blood from the liver to the larger *vena cava* and heart for distribution over the circulatory system.

The blood supply of the heart itself is by way of special *coronary* arteries. These are necessary to supply the thick heart muscles with the large amounts of food and oxygen necessary for their continuous activity. The walls of the blood vessels themselves contain small canals through which blood is transported to nourish the cells of these tissues.

In addition to its function in the transportation of materials throughout the body, the circulatory system is important in temperature regulation. This arises by virtue of the ability of the muscular walls of the blood

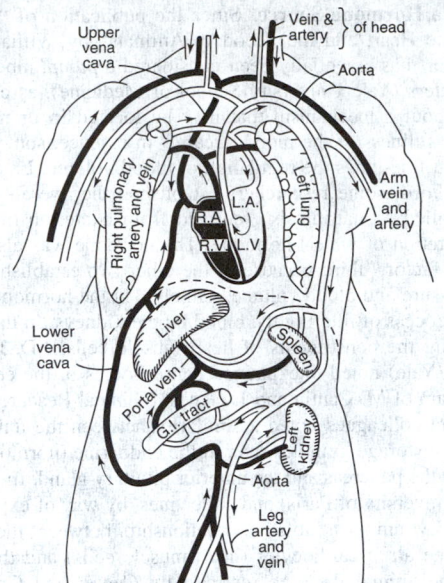

Fig. 2.   Highly schematic representation of circulatory system of human body: R.A. = right auricle; L.A. = left auricle; R.V. = right ventricle; L.V. = left ventricle; G.I. = gastrointestinal tract.

vessels to expand or contract, thereby changing the diameter of the vessels. When the capillaries in the skin are expanded or dilated, a larger amount of blood flows through them. If the temperature outside of the body is below body temperature, the blood in these capillaries is cooled. This cooled blood is then transported to the interior of the body where it is able to counterbalance any tendency toward a rise in temperature. On a cold day, these surface capillaries will be constricted so that the blood will not lose undue amounts of heat to the atmosphere.

The size of the various blood vessels thus varies automatically with the particular needs of the body. Drugs which cause a constriction of the blood vessels (*vasoconstrictors*) bring about a rise in blood pressure even though blood content remains fixed. By contrast, *vasodilators* generally bring about a reduction in blood pressure. Physiological changes in the sizes of the blood vessels are in part under the control of vasodilators and vasoconstrictors produced naturally in the body, and partially under the control of the nervous system. Sometimes, a substance that causes a constriction of the blood vessels in one tissue may dilate the vessels in another. The hormone secreted by the medulla of the adrenal glands is one example of a natural vasoconstrictor that aids in regulating the blood pressure in the body.

Cardiac Disorders and Diseases

Generally, three conditions are symptomatic of heart disease: (1) *myocardial ischemia* (a decrease in blood supply to the heart muscle); (2) disturbances in *cardiac rhythm*; and (3) disorders in the *pumping efficiency* of the heart, as may be manifested by increased filling pressure, which causes upstream venous circulation, or decreased systolic pumping, which results in an inadequate circulation of blood to organs that are located downstream of the heart. Common symptoms of heart problems include chest pain, palpitation, syncope (loss of consciousness), dyspnea (labored breathing), and edema (accumulation of fluid). These conditions, of course, are not exclusive to heart conditions.

Except in emergencies, when time is of the essence, milder symptoms of heart problems will be methodically diagnosed through the use of a number of instrumental techniques. The well-established cornerstone of heart diagnosis remains the electrocardiogram, preferably using twelve leads. Although the interpretation of electrocardiograms has been computerized to a degree and has been found useful in studies of mass populations, the input of an experienced cardiologist is considered mandatory in the analysis of specific patients with possible heart problems. During recent years, the two-step exercise procedure has largely been replaced by treadmill exercise. Ambulatory electrocardiographic measurements also have been emphasized in recent years. See also **Electrocardiography**.

The use of ultrasound in a technique known as echocardiography has been available since the late 1960s and is growing in acceptance, being

of particular value in the diagnosis of such conditions as pericardial effusion, mitral valve prolapse, and left atrial tumors, among others. See also **Echocardiography**.

Another relatively new diagnostic tool is *isotope imaging*. Examples include radionuclide angiocardiography, using radioactive technetium, myocardial scanning with techetium pyrophosphate, and myocardial perfusion scanning with radioactive thallium Positron emission tomography (PET) is also used as a diagnostic tool in heart disorders.

Invasive procedures are still required in the diagnosis of many cardiac problems. These include cardiac catheterization, angiocardiography, coronary arteriography, intracardiac electrophysiological studies, and myocardial biopsy. These methods generally are limited to situations of an advanced, more serious nature and where other diagnostic procedures do not suffice. Principal limiting factors in their use are the risks generally attendant to invasive procedures, patient discomfort, and cost.

The major cardiac disorders and diseases are described in separate entries in this encyclopedia.

Valvular Heart Disease. The function of the valves of the heart has previously been described in this entry. See Fig. 1. At one time, most heart valvular damage was ascribed to rheumatic fever. See also **Rheumatic Fever**. It has since been established that there are over twenty forms of nonrheumatic valvular diseases.

In *rheumatic heart disease*, there is fibrotic scarring of the valvular tissue, which ultimately produces *stenosis* or *regurgitation*. In nearly all cases, some stenosis is present. With exception of rare congenital causes, *stenosis of the mitral valve* is usually considered of rheumatic origin. Stenosis is defined as the narrowing or contraction of a passage or opening. Regurgitation is the abnormal backward progression of fluids; in the case of the heart, the backward return of blood through the valves of the heart. Stenosis adds an extra load on the heart because of increased pressure required to overcome resistance to flow; regurgitation reduces the efficiency of the heart as a pump.

In *nonrheumatic mitral regurgitation*, there is the *floppy valve syndrome*, a dysfunction related to coronary artery disease as well. In floppy valve syndrome, in what is described as an idiopathic pathologic process, there is a loss of fibrous and elastic tissue; this is sometimes called *myxomatous degeneration*. Mitral regurgitation also may result from rupture of the papillary muscle (*papillae* are conical projections from the walls of the cardiac ventricles attached to the cusps of the atrioventricular valves by the *chordae tendineae*). Rupture may occur as the result of infarction or ischemia and thus contributes to mitral regurgitation.

Aortic valve stenosis in adults (particularly the elderly) is considered of nonrheumatic origin and results from a gradual but progressive degenerative thickening and calcification of the leaflets in the valves. The disease process is considered to be somewhat like that occurring in atherogenesis. See also **Arteries and Veins (Vascular System)**.

Aortic regurgitation is also generally considered a nonrheumatic disorder and frequently occurs as a secondary manifestation of other diseases (syphilis, ankylosing spondylitis, aortic dissection, aortic aneurysm, and inherited diseases that affect connective tissue).

Prevention and therapy in valvular heart disease include the long-term administration of prophylactic antibiotics to decrease the possibilities of a return of rheumatic fever for persons who previously have had acute rheumatic fever. The length of time during which such prophylaxis should be given is debatable among authorities. In persons with rheumatic heart disease featuring aortic regurgitation or a bicuspid aortal valve, most specialists suggest the administration of antibiotics to prevent the development of bacterial endocarditis after dental and surgical procedures. In valvular disease, the physician will be aware of the risk of systemic embolism which sometimes develops in connection with rheumatic heart disease. Long-term administration of anticoagulants may be indicated in such cases. Cardiac arrhythmias arising from valvular disease will be handled as described in the entry on **Arrhythmias (Cardiac)**.

Surgery is frequently indicated in valvular heart disease. This may range from repair of malfunctioning parts to valve replacement. Over the years, over three dozen designs of *artificial (prosthetic) valves* have been used. Designs of preference in recent years have included the Starr-Edwards, the Smeloff-Cutter, and the Björk-Shiley valves. These valves are considered to have ample durability. The principal problems sometimes involved include thrombus formation and embolism, and thus long-term anticoagulant therapy is usually indicated.

In the United States, the natural tissue valve preferred is the porcine aortic valve. In Europe, some valves are configured from dura mater (outermost membrane of the brain and spinal cord), pericardium (membrane enclosing the heart), and fascia lata (wide, dense sheath of the thigh muscles). The valves from pig hearts make excellent replacements for human heart valves. They are durable, resistant to infection, and not readily rejected by the human body. There has been a shortage of valves of the proper size from this source. The valves are taken from pigs of various sizes, with most of the animals weighing less than 80 pounds (36 kilograms). Since most pigs in the United States are slaughtered at around 200 pounds (90.7 kilograms), the supply of hearts from small pigs is limited. Also, only about one of every ten valves is suitable for placement in the human heart. The cost of raising pigs strictly for their heart valves has proved prohibitive. A number of countries slaughter pigs weighing less than 80 pounds (36 kilograms) and the hearts from these pigs can be obtained rather inexpensively at slaughterhouses. However, they have the potential of introducing exotic diseases of swine into the United States. Such diseases as African swine fever, hog cholera, foot and mouth disease, and swine vesicular disease could devastate the pork industry in the United States. To prevent the introduction of such diseases, scientists at the U.S. Department of Agriculture (Plum Island, New York) have developed a method for inactivating these viruses. They have found that glutaraldehyde, a substance used to stabilize pig heart valves prior to their transplantation in humans, will kill the viruses associated with these diseases. Nevertheless, great care must be exercised in making certain that all porcine valves are fully free of such viruses prior to surgery.

Cardiomyopathies. Dysfunctions of the heart muscle (*myocardium*) that are *not* related to coronary atherosclerosis, hypertension, or valvular problems, fall into four categories which when considered as a group are called *cardiomyopathies*. From the standpoint of hemodynamics (study of movements of the blood), these categories (Goodwin, 1970) are: (1) *congestive*; (2) *hypertrophic*; (3) *restrictive*; and (4) *obliterative*.

In *congestive cardiomyopathy*, the contractility of the heart muscle is subnormal. Common symptoms include dyspnea (labored or difficult breathing) and fatigue. Often pulmonary congestion accompanies the disorder. There is often mild elevation of blood pressure. Cardiac enlargement is common. This condition must be differentiated from acute myocarditis. The usual course of congestive cardiomyopathy is to congestive heart failure, ultimately the cause of death of persons with the condition. The prognosis is variable. Therapy includes salt reduction, digitalis glycosides, and diuretics.

For many years, it has been observed that congestive cardiomyopathy is frequently seen in alcoholics. The term *alcoholic cardiomyopathy* now frequently appears in the literature. Alcohol has not been definitely identified as the cause; possibly the malnutrition usually associated with alcoholism may be the major contributor. In the midwestern United States and Canada in the 1960s, there was an epidemic of cardiomyopathy, but this was ultimately traced to cobalt toxicity derived from an additive used in making the beer consumed in the region.

In recent years, there has been considerable rethinking as regards the possible connection between cardiomyopathy and coronary artery disease; in the past, the presence of cardiomyopathy by definition ruled out coronary artery disease.

A common cause of congestive cardiomyopathy in certain regions of South America is **Chaga's Disease**. **Hypertrophic cardiomyopathy** has been known for many years, but possibly well defined for the first time by Teare (1958), who termed the disorder "asymmetrical hypertrophy of the heart." Hypertrophy is an increase in the volume of a tissue or organ caused entirely by enlargement of existing cells. Asymmetry refers to the disproportionate hypertrophy of the left ventricle which effectively reduces the size of the left ventricular chamber. The result is obstruction to left ventricular outflow. In recent years, new names have been given to the disease—*muscular subaortic stenosis*; and *idiopathic hypertrophic subaortic stenosis*. Symptoms include angina, syncope, palpitations, and congestive heart failure. See also **Congestive Heart Failure (CHF)**. Although the symptoms of the disease worsen with time, the process may be slow—a span of years. In some cases, however, sudden death may occur, particularly in children and men with a family history of this condition. About 15% of cases are treatable by surgery. Drug therapy is not universal, but is directed toward the profile of symptoms presented.

In *restrictive cardiomyopathy*, the myocardium loses its resilience and becomes rigid—conditions which offer resistance to ventricular filling and elevate cardiac filling pressures. The condition tends to mimic constrictive pericarditis. Symptoms are those of congestive heart failure. There are no fixed therapies for this disease that have proven effective. Some authorities believe that removal of excess iron in the body by phlebotomy (incision of a vein) may provide some relief.

In *obliterative cardiomyopathy*, there is a massive fibrosis (formation of fibrous tissue) of the endocardium. This reduces the size of the ventricular cavities. Although the disease, of unknown etiology, is frequently seen in eastern Africa, it is seldom encountered in Europe and the Western world.

Pericarditis. Inflammation of the membrane enclosing the heart (*pericardium*) may take three fundamental forms, all of which are generally termed *pericarditis*. *Acute pericarditis* is usually associated with a viral infection. There is chest pain, which increases with inspiration (contrast with myocardial infarction), a low-grade fever, and sometimes tachycardia. The physician will listen for the sounds of a characteristic pericardial friction rub. Where a bacterial infection is diagnosed, antibiotics will be used; for neoplasms, radiation or chemotherapy may be indicated. In *pericardial effusion*, fluids accumulate in the pericardial cavity. Echocardiography is commonly used in diagnosis. In acute forms of cardiac tamponade (compression of heart due to collection of fluid in pericardium), as may arise from an injury, an aortic dissection, or rupture of an aortic aneurysm, prompt surgery may be indicated. In a less severe situation, pericardiocentesis (puncture and aspiration) may be used. In *constrictive pericarditis*, diastolic filling of the heart is impeded, the results of which are an increase in venous pressure and reduced cardiac output. At one time, this condition was almost exclusively attributed to a tuberculous lesion. A majority of cases are classified as idiopathic, but some are related to radiation exposure, to rheumatoid arthritis, or uremia. Surgical removal of the pericardium is sometime indicated.

Sudden Cardiac Death. This term applies to the unexpected cessation of breathing and circulation when the hearts stops pumping, usually caused by an underlying heart disease, such as atherosclerosis of the coronary arteries. If the patient's breathing and circulation are not restored within a few minutes, permanent biological death, precipitated by irreversible brain damage, will result. It is estimated that between 20% and 30% of sudden cardiac deaths result from myocardial infarction; the remainder (statistics not yet reliable) is divided between myocardial ischemia and primary rhythm disturbance. Provided exceptionally effective emergency measures are applied (difficult in many situations), the long-term prognosis for attacks resulting from myocardial infarction or ischemia are good; they are poor in the case of a primary rhythm disturbance. About 25% of heart attacks can be classified as out-of-hospital sudden cardiac deaths, instances in which coronary heart disease has precipitated the attack with very little warning, often no warning whatsoever. Currently, fewer than 5% of sudden cardiac death patients are successfully resuscitated. The persons in the United States with coronary heart disease run into the several millions, of which 1.5 million (approximately) suffer heart attacks each year. It is estimated that 75% of these persons are admitted to a hospital in time (warnings noted hours, weeks, or months in advance), of which 80% are discharged, but usually having to follow some therapeutic regimen. On the other hand, 25% of the 1.5 million persons suffer sudden cardiac death outside of a hospital. Approximately 95% of these attacks are fatal (some 600,000 deaths per year).

In large communities, or exceptionally progressive smaller communities, some progress has been made in getting persons to a hospital barely in time to effect treatment. Emergency medical technician (EMT) teams have been formed. They have been trained for handling cardiac emergencies. The immediate treatment is cardiopulmonary resuscitation (CPR), which is a repeated series of mouth-to-mouth respirations and chest compressions that circulates a small amount of oxygenated blood to the brain, heart, and other vital organs. This is followed by specific medical treatment, designed to restore normal circulation and respiration. This usually includes the insertion of a breathing tube into the trachea, delivery of drugs, and defibrillation. The latter applies an electric shock across the victim's chest to depolarize all heart cells simultaneously and thus reset, so to speak, the pacemaking nodes of the heart. Defibrillation thus interrupts chaotic twitching of the heart muscle. Seattle, Miami, Los Angeles, and Columbus (Ohio) pioneered the EMT program. The concept, however, was first applied in Belfast (U.K.) in the late 1960s.

Coronary Artery Bypass Surgery and Percutaneous Coronary Angioplasty are discussed in article on **Ischemic Heart Disease**.

Congenital Disorders and Anomalies

Most congenital disorders of the circulatory system appear in the embryo as the result of some defect in development, usually between the fifth and eight week of pregnancy. An infection in the mother during pregnancy, or rubella (German measles), may be responsible for the abnormality. In some cases, the heart may be located in the right side of the body, although this seldom causes any difficulty and may not be noticed immediately. More serious defects are those which involve the size and development of the chambers of the heart, its valves, and connecting vessels. In some patients, such congenital defects may manifest themselves only after many years, and cause nothing more than a slight discomfort in breathing. In other instances, the defects may be such as to inhibit seriously the flow of blood through the heart and lungs.

In one of the malformations (*patent ductus arteriosus*), a small duct connecting the aorta and the pulmonary artery fails to close at birth. Since the pressure is higher in the aorta, blood will flow from this vessel to the pulmonary artery and back to the lungs, from which it had just come. This means that even when the lungs are working at full capacity, all of the oxygenated blood is not being circulated to the body. Difficulty in breathing and palpitation are outstanding symptoms. Once it is discovered, this defect can be repaired surgically by tying or dividing and sewing the open ends of the duct.

If defects exist which allow a mixing of arterial and venous blood, the patient frequently has a bluish or *cyanotic* appearance. This condition, if not corrected, may limit the life of the patient to a relatively few years. Best known of the cyanotic congenital heart defects are those that are found in "blue babies." One of the most common conditions causing blue babies is really a combination of four malformations (*tetralogy of Fallot*). In this disorder, the prenatal partition (*septum*) between the two pumping chambers (*ventricles*) of the heart has failed to close at birth. In addition, the major artery (*aorta*) leading from the heart is slightly out of place, and the artery leading from the heart to the lungs is constricted. The right ventricle, therefore, not only must pump blood through the lungs, but also must work directly against pressure from the left, so that the ventricle becomes enlarged because of the extra work. Blood that has been through the lungs becomes mixed with that which has not. An increase in the number of red blood cells may occur to compensate for the circulatory insufficiency. The child's fingers may be club-shaped and there may be a failure on the part of the child to develop physically in a normal manner. Breathlessness is common.

At one time, the treatment of blue babies was limited and consisted mainly in preventing infection and overactivity of the patient. The span of life was short. Now, in a special surgical procedure, one of the arteries — the *aorta, common carotid, subclavian,* or *innominate* — is connected to the pulmonary artery. There is then an increase of the blood flow to the lungs sufficient to permit the patient maximum activity without placing undue strain on the heart. This operation, when needed, is performed during the very early years of childhood. At a later date, the individual can be fully corrected with a second operation, utilizing the heart-lung machine.

Congenital Anomalies. Each of the four valves of the heart may have congenital anomalies. The *tricuspid valve* may have a deformity of the leaflets, known as *Ebstein's malformation of the tricuspid valve.* Or there may be *tricuspid atresia*, in which the valve never forms, preventing the normal flow of blood from the right auricle into the right ventricle. Instead, it flows from the right auricle into the left auricle through a hole in the wall between the two upper chambers of the heart. The *pulmonary valve* cusps are partially fused in some individuals and prevent the proper flow of blood, *pulmonary stenosis*. This condition can be caused by narrowing of the orifice leading to the valve or by fusion of the leaves of the valve itself. In the normal heart, the systolic pressure is the same on both sides of the valve. If the pressure is found to be lower in the pulmonary artery than in the right ventricle, the physician knows that *pulmonary stenosis* exists. The mitral valve may have *atresia, incompetence,* or *stenosis*, although isolated cases of these conditions are rare. The aortic valve in the heart may have a congenital narrowing of the orifice or fusion of the cusps, known as *aortic stenosis*. Most of these abnormalities of heart valves can be corrected surgically.

The most common congenital malformation, occurring as a single lesion, is *ventricular septal defect*, in which there is a hole in the wall between the left and right ventricles. Following diagnosis, this abnormality can be corrected surgically by sewing a patch composed of a tough, resilient plastic material over the opening. A hole between the two auricles, *atrial septal defect*, allows blood to flow from the left side of the heart as the result of pressure differences. This defect can be corrected by directly suturing the edges of the defect.

A more complicated group of defects occurs when there is a hole between both the upper chambers (*atria*) and lower chambers (*ventricula*) with malformed intervening tissue and one or both valves between the atria and ventricula. These most difficult lesions can be corrected with the use of the heart-lung machine and require the use of a patch and sometimes a prosthetic valve.

In some cases, the oxygenated blood from the lungs returns partially or totally to the right side of the heart instead of draining into the left auricle. This type of malformation, *anomalous drainage of pulmonary veins*, is characterized by an abnormal condition — the same amount of oxygen being present in all the chambers of the heart, the pulmonary artery, and the aorta. This condition can be corrected by various surgical procedures in which the anomalous drainage is redirected into the correct left auricle.

In *coarctation of the aorta*, another rather common genital heart defect, the main artery leaving the heart is constricted to such an extent that the flow of blood to all parts of the body is restricted. When the diagnosis of this condition has been confirmed, the constriction can be removed surgically and the ends of the aorta reunited or the defect bridged with a synthetic vessel, thus allowing the blood to flow freely.

Cardiac Transplantation

Since the first human heart transplantation was accomplished by Christian Barnard, a South African surgeon, in December 1967, the practical feasibility of the procedure for extending life in patients with obviously terminal heart disease has been under severe scrutiny, not only by the medical professions, but by government regulators and the lay public. The gamut of technical, social, and economic pros and cons has been discussed extensively but not fully resolved. Moratoriums by governments and by hospital groups have been invoked and revoked. Aside from economic restraints, decisions to use or not use the procedure largely rest with the medical professionals and their institutions (hospital facilities, etc.) and, of course, with the patient.

In assuming that there will be improvements both in postoperative survival and quality of life, some authorities estimate that the number of *technically justifiable* (unrelated to socioeconomic factors) cardiac transplantations in the United States will not exceed 1000 to 5000 per year. This figure will be affected largely by the guidelines used by the medical profession in selecting candidate patients. One set of guidelines suggests that the prospective candidate not be over 50 years old and have no significant systemic disease other than very advanced cardiac malfunction. It is apparent that a major controlling factor will be the availability of donor hearts.

In 1984, it was estimated that the number of persons with irreversible brain death and identified as suitable allograft donors (of heart and other organs) does not exceed 2000 per year.

As pointed out by Austen and Cosimi, important improvements in cardiac transplantation have resulted from several factors: (1) better definition of criteria for selection of appropriate patients; (2) refinement of the use of antilymphocyte serum and T-lymphocyte monitoring for management of immunosuppression (organ rejection), (3) improved myocardial preservation due to perfection of effective cardioplegic techniques, and (4) the use of the fungal metabolite, *cyclosporine*. With cyclosporine, the episodes of rejection are less dangerous and easier to treat. Hospital stays, hence costs, have been reduced for those patients receiving this powerful and specific immunosuppressive drug.

It is generally felt that cardiac transplantation should be restricted to highly specialized medical centers. From a technical standpoint, the procedure is now viewed with cautious optimism.

Heart-Lung Transplantation. This comparatively new and highly selective procedure has thus far been used in the United States for a very limited number of patients who had terminal and irreversible pulmonary hypertension and were near death. Survival rate has exceeded 50%, but the number of procedures is so small that it is difficult to forecast future survival statistics. In a number of cases, after surgery, the short-term improvement has been excellent. Lung function has been restored to near-normal levels. Improvements in dyspnea, pulmonary parenchymal function, gas exchange, pulmonary vascular function, and cardiac function have been observed.

In this procedure, the lungs and heart are transplanted as a single unit. This simplifies the transplantation procedure. Cyclosporine is used to suppress graft rejection and appears to do so without the toxicity of conventional agents. A principal toxic effect of cyclosporine is impaired renal function in nearly all patients. Although renal function usually returns toward normal after the drug dose is reduced, some patients have required dialysis therapy, and have had a prolonged moderate impairment of renal function.

Heart-lung transplantation ultimately may be used in patients with other forms of lung disease, including obstructive lung disease, restrictive lung disease, and cystic fibrosis. Some authorities also stress that this procedure may lead to increased knowledge concerning the pathophysiology of diseases of the lung and pulmonary vasculature.

Cardiopulmonary Bypass Technology. This has been a key not only to heart transplantations, but to all procedures that involve "open heart" surgery. This technology (the so-called heart-lung machine) permits surgeons to operate on the heart for long periods of time in a dry, bloodless field, under direct vision. A pump draws blood from the vena cava, through tubes connected to these veins before they enter the heart. The blood is pumped under controlled pressure and flows to an "artificial lung" usually a plastic, membranous structure, where it is allowed to contact a steady stream of oxygen. The oxygenated blood is then pumped through another tube into the arterial system. The oxygen content, temperature, degree of alkalinity or acidity, rate of flow, and pressure, among other instrumental variables, must be carefully regulated throughout the entire surgical procedure. Checks on the circulation in the extremities are made continuously during the bypass procedure to prevent death of any tissues because of inadequate blood supply.

Artificial Heart

The concept of an artificial (prosthetic) heart dates back as early as 1812, when Julien-Jean Céesar La Gallois observed, "if one could substitute for the heart a kind of injection (of arterial blood), one would succeed easily in maintaining alive indefinitely any part of the body." Mechanical perfusion experiments with heart and lung organs date back to about 1880, exemplified by the work of Henry Martin. Martin's work prepared the foundation for modern cardiopulmonary bypass technology.

From a bioengineering standpoint, it is interesting to note that the total power output of the human heart is about 2.5 watts, of which 80% is required by the left ventricle (the output side of the heart which pumps blood into the arteries and ultimately to the capillaries). The pressure parameters are well established. See the entry on **Hypertension (High Blood Pressure)**, which gives diastolic and systolic pressures.

By 1950, well over 30 artificial heart-lung designs had been proposed. Well known is the work of Lindbergh (Charles) and Carrel in the 1930s in connection with their perfusion pump, then reported by the news media as a "robot heart." Prominent in the search for an artificial heart has been the concept of a device that will be implanted in the human body and take over all heart functions. Lindbergh and Carrel added an interesting new dimension to this objective, as suggested by the following quotation from one of their publications: "We can perhaps dream of removing diseased organs from the body and placing them in the Lindbergh pump as the patients are placed in a hospital. There [the organs] could be treated far more energetically than within the organism and, if cured, replanted in the patient."

Working essentially with these criteria in mind, a number of teams have been researching and experimenting with artificial hearts. These include work at the Cleveland Clinic, dating back to the 1950s. In 1957, these researchers were able to keep dogs alive for about 1.5 hours with a plastic polyvinyl chloride heart energized by compressed air. It should be noted that these experiments were conducted at a time before attempts at human heart transplantation had been made, and when open-heart surgery was in the early pioneering stage. Research on artificial heart valves had just commenced. Progressively, these and other workers refined their designs and selection of materials of construction as well as various energy supplies, including electrically driven apparatus. Nuclear power as a source was considered. By the mid-1960s, researchers at the Cleveland Clinic were able to keep calves alive for 1.5 days with an artificial heart.

It should be mentioned that in England in 1928, Dale and Schuster built a pump with the objective of temporarily bypassing the heart during heart surgery. Dodrill (General Motors Corporation), in 1952, developed a mechanical heart which was used for nearly an hour during human heart surgery. Jarvik stresses, however, that open-heart surgery as it is known today requires the heart-lung machine, not simply a pump to replace the heart.

In 1969, for the first time, an artificial heart was installed in a human being. This artifact was designed by Liotta and Hall (Texas Heart Institute) and sustained life for about 64 hours, during which time a natural heart was being sought for transplantation.

In the early 1980s, Jarvik, a present-generation pioneer in the field, listed at least six criteria for what may be termed "the total artificial heart". They are: (1) *small size*, to fit into the existing human cardiac cavity; (2) *work output* ample to provide all needs supplied by a natural heart; (3) *variable output* in accordance with the changing rate of body requirements (range from rest to vigorous exercise); (4) *gentle handling of blood* to avoid hemolysis (disintegration of the elements of the blood); (5) *ease of sterilization*; and (6) *durability*. There are, of course, numerous other criteria, certainly one of which is economics.

The current consensus in the medical profession today targets the prosthetic heart as a means for maintaining life in a patient who is waiting for a suitable donor heart. This may be a period of days or several weeks. The Jarvik heart and other recently conceived designs have been used successfully for this purpose.

The technical principles of the prosthetic heart, namely, those of a pump, are not complex, but the detailed engineering in selecting materials, means of connection to human tissue, size, durability, resistance to rejection, strength, etc. are indeed very complex and unfortunately beyond the scope of this encyclopedia.

Additional Reading

Akutsu, T. and H. Koyanagi: *Heart Replacement: Artificial Heart 6*, Springer-Verlag, Inc., New York, NY, 1998.

Alpert, J.S., J.E. Dalen, and S.H. Rahimtoola: *Valvular Heart Disease*, Lippincott Williams & Wilkins, Philadelphia, PA, 1999.

Austen, W.G. and A.B. Cosini: "Heart Transplantation," *N. Eng. J. Med.*, **311**(22), 1436–1438 (November 29, 1984).

Bashore, T.M., Ed.: *Invasive Cardiology: Principles and Techniques*, B.C. Decker, Philadelphia, PA, 1990.

Braunwald, E.: *Cardiomyopathies, Myocarditis, and Pericardial Disease*, Mosby-Year Book, Inc., St. Louis, MO, 1996.

Cantin, M. and J. Genest: "The Heart and the Atrial Natriuretic Factor," *Endocrine Reviews*, **6**(2) 107–127 (Spring 1985).

Cantin, M. and J. Genest: "The Heart as an Endocrine Gland," *Sci. Amer.*, 76–81 (February 1986).

Carabello, B. and G. Vetrovec: *Invasive Cardiology: Current Diagnostic and Therapeutic Issues*, Futura Publishing Company, Inc., Aramonk, NY, 1980.

Chatterjee, K. et al., Eds.: *Cardiology: An Illustrated Text/Reference*, J.B. Lippincott/Gower Medical, Philadelphia, PA, 1991.

Chou, Te-Chuan, L.S. Ramaiah, and R. Zorab: *Electrocardiography in Clinical Practice*, W. B. Saunders, Philadelphia, PA, 1996.

Cohn, P.F.: *Silent Myocardial Ischemia and Infarction*, 4th Edition, Marcel Dekker, Inc., New York, NY, 2000.

De Bold, A.J.: "Atrial Natriuretic Factor: A Hormone Produced by the Heart," *Science*, **230**, 767–770 (1985).

Dorman, C. and B. Katzir: *Cognitive Effects of Early Brain Injury*, Johns Hopkins University Press, Baltimore, MD, 1994.

Goldstein, S. et al.: *Sudden Cardiac Death*, Futura Publishing Company, Inc., Armonk, NY, 1994.

Goodwin, J.F.: "Congestive and Hypertrophic Cardiomyopathies," *Lancet*, **1**, 731 (1970).

Hosenpud, J.D. et al.: *Cardiac Transplantation: A Manual for Health Care Professionals*, Springer-Verlag, Inc., New York, NY, 1991.

Jarvik, R.K. et al.: "Criteria for Human Total Artificial Heart Implantation Based on Steady State Animal Data," *Trans. Amer. Soc. Artif. Int. Organs*, **23**, 535–542 (1977).

Jarvik, R.K.: "The Total Artificial Heart," *Sci. Amer.*, **244**, 1, 74–80 (1981).

Jelliffe, R.W.: *Fundamentals of Electrocardiography*, Springer-Verlag, Inc., New York, NY, 1990.

Julian, D.: *Diseases of the Heart*, W. B. Saunders Company, Darien, IL, 1996.

Luderitz, B.: *History of the Disorders of Cardiac Rhythm*, Futura Publishing Company, Inc., Armonk, NY, 1998.

Mehler, R.E. and L. Sompayrac: *How the Circulatory System Works*, Blackwell Science, Inc., Malden, MA, 2000.

Otto, C.M. and R. Zorab: *Valvular Heart Disease*, W.B. Saunders Company, Philadelphia, PA, 1999.

Potparic, O. and J. Gibson: *A Dictionary of Congenital Malformations and Disorders*, Parthenon Publishing Group, New York, NY, 1997.

Rajskina, M.E. and D.P. Zipes: *Ventricular Fibrillation and Sudden Coronary Death*, Kluwer Academic Publishers, Norwell, MA, 1999.

Robinson, T.F., S.M. Factor, and E.H. Sonnenblick: "The Heart as a Suction Pump," *Sci. Amer.*, 84–91 (June 1986).

Rose, E.A. and L.W. Stevenson: *Management of End-Stage Heart Disease*, Lippincott Williams & Wilkins, Philadelphia, PA, 1996.

Rowlands, D.: *Emergency Cardiology*, Butterworth-Heinemann Inc., Woburn, MA, 1989.

Sekiguchi, M. and P.J. Richardson: *Prognosis and Treatment of Cardiomyopathies and Myocarditis*, Columbia University Press, New York, NY, 1995.

Spodick, D.H.: *The Pericardium: A Comprehensive Textbook*, Vol. 27, Marcel Dekker, Inc., New York, NY, 1996.

Sutton, G.: *Clinical Cardiology: An Illustrated Text*, Chapman and Hall, New York, NY, 1999.

Teare, D.: "Asymmetrical Hypertrophy of the Heart in Young Adults," *Br. Heart J.*, **20**, 1 (1958).

Vogel, S.: *Vital Circuits: On Pumps, Pipes, and the Workings of Circulatory System*, Oxford University Press, Inc., New York, NY, 1992.

Wilson, G.N. and W.C. Cooley: *Preventive Management of Children with Congenital Anomalies and Syndromes*, Cambridge University Press, New York, NY, 2000.

HEART ATTACK. See **Coronary Artery Disease (CAD); Coronary Thrombosis**.

HEARTBURN. See **Esophagus**.

HEART FAILURE (Congestive). See **Congestive Heart Failure**.

HEARTWORM DISEASE (Dirofilariasis). This is a serious and potentially fatal disease in dogs. It is caused by a worm (*Dirofilaria immitis*) that is found in the animal's heart and large adjacent vessels. The female worm is 6–14 inches (15.2–35.6 centimeters) long and about $\frac{1}{8}$ inch (3 millimeters) wide. The male is smaller. One dog may have as many as 300 worms. Adult heartworms live in the animal up to 5 years and during that period the female produces millions of young *microfilariae*. These microfilariae live in the bloodstream, mainly in the small blood vessels. They cannot grow to adults without passing through an intermediate host (a mosquito). As many as 30 species of mosquito can serve as host. The microfilariae develop for 10–30 days in the mosquito and then enter the saliva of the insect. At this point, the organisms are *infective larvae* because at this stage of development they will grow to adults when they enter a dog. The mosquito bites the dog, mostly on the abdomen where the haircoat is thinnest.

Adult worms cause disease by clogging the heart and major blood vessels leading from the heart. They interfere with the valve action. By clogging the main blood vessels, the blood supply to other organs of the body is reduced, particularly the lungs, liver, and kidneys. Most dogs infected with heartworms do not show external signs of the disease. When symptoms develop after some period of infection, these will include soft dry chronic cough, shortness of breath, weakness, nervousness, listlessness, and loss of stamina. These features are noticed particularly after exercise. The microfilariae circulate throughout the body, but remain mainly in the small blood vessels, which they tend to clog. Ultimately there is destruction of lung and kidney tissue.

An arsenical drug is used in treatment and usually requires a hospital environment. The treatment requires injections of the drug over a period of 2–3 days. About 6 weeks after the adult worms have been eradicated, further injections of other drugs are required to eradicate the microfilariae. Some veterinarians prefer to eradicate the microfilariae first. To prevent heartworms, many veterinarians recommend the use of diethylcarbamazine citrate (e.g., Filarbits®) in the pet's diet during the mosquito season.

HEAT. The agency whose addition to or removal from a physical system is the cause of thermal changes of various types. These include rise and fall of temperature, changes in length and volume, changes of physical states, such as melting, evaporations, etc.

During the eighteenth century heat was assumed to be a subtle fluid called *caloric*, filling the interstices between the ultimate particles of matter and, under conditions of isolation from the surroundings, known to satisfy a conservation law. The production of heat by friction as well as its disappearance during the performance of external mechanical work established its essential physical nature as another form of *energy* and led to the overthrow of the caloric theory. Nevertheless, we still speak of the *flow* of heat as though it were a fluid and have retained the methods of measuring the *quantity of heat* originally devised by the upholders of the caloric view.

Our direct knowledge of heat is provided by the sensation of hotness and coldness when we come in contact with various physical bodies. It is possible to arrange a set of bodies in a sequence such that A feels hotter than B, B hotter than C, etc. We say that A has a higher *temperature* than B, B a higher one than C, and so on. Of course our sensations are qualitative and are considerably influenced by the thermal conductivity of the body we touch. Thus, on a frosty morning, the head of an ax being metal feels considerably colder than the wooden handle though the two are presumably at the same temperature. To obtain a continuous and reproducible physical scale of temperature, various types of thermometers have been devised of which the mercury-in-glass or colored-alcohol-in-glass are familiar examples. The two temperature scales in common use are the Fahrenheit scale and the Celsius scale. The first assigns values of 32° and 212° to the normal freezing and boiling points of pure water, respectively, and divides this interval into 180 equal subintervals or degrees. The Celsius, formerly called the Centigrade scale assigns the respective values of 0° and 100° to the above fixed points; the standard interval is then divided into 100 equal degrees.

Temperature changes are produced by the addition or subtraction of heat from a body. Thus, temperature may be regarded as a measure of the concentration or *intensity* of heat. In general, the more heat we add to a given body the more its temperature rises.

Measurement of Heat. Since heat is imponderable and not directly observable, it is necessary to measure the size of a given quantity of heat by its effect on another body. If this effect is the production of a rise in temperature from some initial temperature, t_1, to a final temperature, t, then the rise $(t - t_1)$ is found to vary inversely with the mass of the test body. It is thus natural, following the calorists, to regard the quantity of heat, say Q, as determined by the product of m and $(t - t_1)$. Thus we say

$$Q \text{ is proportional to } m \times (t - t_1)$$

To make this statement into an equation we write

$$Q = \text{constant} \times m \times (t - t_1) \tag{1}$$

where the constant of proportionality depends on the substance, being large for some materials and small for others. This constant for water, for example, is about 33 times as great as for lead; water is said therefore to have a greater *heat capacity* than lead. Notice that the constant in Equation (1) actually gives the numerical value of Q, which is required to warm a unit mass of the substance through a temperature interval of exactly 1°. This constant is accordingly called the *specific heat capacity* (usually abbreviated to *specific heat*) and is indicated by c. Since it is found that the value of the specific heat, particularly for gases, but in principle for all materials, depends on the conditions under which the heat is absorbed, this must be indicated. We thus have c_p and c_r, for example, for the two important cases of absorption at constant pressure and constant volume, respectively. Since the former characterizes the common laboratory case of working under atmospheric pressure, we accordingly rewrite Equation (1) as

$$Q_p = c_p m (t - t_1) \tag{2}$$

Q_p now measures the heat absorbed under constant pressure, and c_p is the constant pressure specific heat. Since the right side of Equation (2) contains *three* quantities, a mere choice of a mass unit and a degree unit is insufficient to establish a unit of heat. It is necessary to select some substance as a standard reference body and assign an arbitrary value of, say c_p equal to unity for it. Water is the universal choice for this standard body due not only to its cheapness and ease of purification, but also to its large heat capacity.

With the selection of water as the standard with $c_p = 1$, the left side of Equation (2) clearly becomes of unit value when m and $(t - t_1)$ are each of unit value. In the English system, we accordingly have the *British thermal-unit* (or Btu) as the heat required to warm 1 pound of pure water through an interval of 1 °F. In the metric system, the corresponding unit is the *calorie*, the heat required to warm 1 gram of water 1 °C. A large unit or *kilocalorie* corresponding to 1,000 ordinary calories is also frequently used in scientific work.

Specific Heats. Use of Equation (2) reveals that the values of c_p obtained experimentally depend on the temperature interval used, indicating a dependence of c_p on temperature. Thus, if c_p for water were actually uniform throughout the 0 to 100 °C range, a mass of water at 100 °C mixed with an equal mass at 10 °C would give a final mixture at exactly 50 °C. The actual value is near 50.05°; this difference although small, indicates the need to specify the calorie at some particular temperature. For this purpose, we suppose a system of mass m is warmed from t to $t + \Delta t$ by the addition at constant pressure of an increment of heat ΔQ_p. Then Equation (2) becomes

$$\Delta Q_p = m\bar{c}_p \Delta t \qquad (3)$$

where now $\bar{c}_p$ is an average value of c_p over this interval. Then we define the *instantaneous* heat capacity, c_p at t by the following relation

$$c_p \frac{1}{m} \lim_{\Delta t \to 0} \frac{\Delta Q_p}{\Delta t} = \frac{1}{m} \frac{dQ_p}{dt}$$

i.e., the heat absorbed per unit mass per degree as the interval becomes smaller and smaller without limit. This leads to the differential form of Equation (3)

$$dQ_p = mc_p \, dt \qquad (4)$$

where dQ_p is the differential heat absorption which produces a differential temperature rise dt in a body of mass m and specific heat c_p.

The standard or 15° calorie is now defined as the rate of absorption of heat per gram per degree at 15 °C and in practice is essentially the same as the average calorie over the 1° interval from 14.5 to 15.5 °C.

If a mass m of water is warmed from t_1 to t, the integral of Equation (4) gives for the total heat absorbed in 15° calories

$$Q_p = \int_{t_1}^{t} dQ_p = m \int_{t_1}^{t} c_p \, dt = m \left[\int_{0°}^{t} c_p \, dt - \int_{0°}^{t} c_p \, dt \right] \qquad (5)$$

where the integral of c_p over the range t_1 to t has been written as the difference of two integrals from a common lower limit of 0 °C. If, therefore, we evaluate an integral of the type $\int_0^t c_p \, dt$ with t varying in 1° steps and arrange these in a table, the right side of Equation (5) may be evaluated by merely subtracting appropriate entries.

In Fig. 1, the value of c_p in 15° calories per gram per degree is plotted graphically from 0 to 100 °C, and the integrals on the right of Equation (5) are represented by appropriate areas under the c_p curve. Thus the integral from 0° to t is hatched with lines sloping up to the right, while that from 0° to t_1 has the lines sloping up to the left. The value of Q_p is then the singly hatched area.

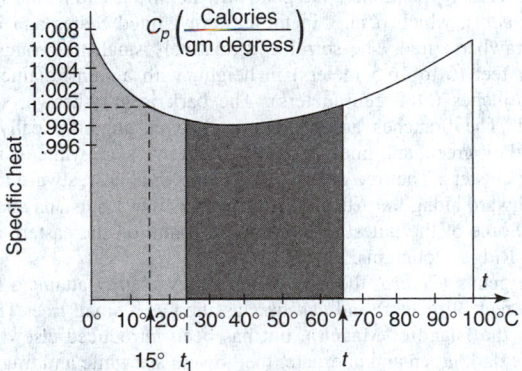

Fig. 1.   Specific heat of water versus temperature.

With heat quantities measured in 15° calories, from the observed rise or fall of temperature in known masses of water, the specific heats of various substances — the heats absorbed on melting solids to liquids (heats of fusion), the heats absorbed on passage from the liquid to the vapor state (heats of vaporization), the heats evolved on combination of various substances, and the heats absorbed or evolved in chemical changes — are at once determinable (see also **Calorimetry**). For the present purpose, Table 1 gives the values of the constant pressure heat capacities of a few typical substances, variations with temperature being disregarded. Notice that c_p, although expressed in terms of calories per gram per degree, is in fact independent of the system of units since water is the reference body

TABLE 1. APPROXIMATE CONSTANT-PRESSURE SPECIFIC HEAT OF SELECTED MATERIALS

Substance	State	c_p (cal/g deg)
Water	Vapor	0.48
Water	Liquid	1.00
Water	Solid	0.50
Ethyl alcohol	Liquid	.54
Hydrogen	Gas	3.44
Air	Gas	.24
Aluminum	Solid	.22
Iron	Solid	.11
Lead	Solid	.03

in all systems. Thus the specific heat of water in the English system would be 1 Btu per pound per degree Fahrenheit.

The Mechanical Nature of Heat. The conservation of heat *per se* is observed only for systems involving the performance of no mechanical or electrical work. Count Rumford (ca. 1800) was the first to establish this fact in his famous cannon-boring experiments carried out in the arsenal of the Duchy of Bavaria in Munich. He observed that when his drills became dull, heat was produced in great quantities limited only by the amount of work done against friction. He concluded that the large scale mechanical energy used in overcoming friction could only be converted into the motions of the ultimate particles of matter, a motion not directly observable but detected by our senses as heat. His results were confirmed and extended by the later work of Joule and Helmholtz, in particular, and also provided a more reliable value for the so-called *mechanical equivalent of heat*. This is taken as the amount of mechanical (or electrical) energy which when converted into heat is equivalent to exactly 1 calorie. The presently accepted value for this important constant is 4.185 joules per 15° calorie. Here the joule is the work performed when power is expended at the rate of 1 watt for 1 second. Thus an ordinary 100-watt lamp bulb converts 100 joules of electrical energy to thermal each second; this amounts to 100/4.185 or about 24 calories.

As a result of experiments such as these and a host of others, we are forced to recognize that heat is merely another form of the universal quantity *energy*. Its transformation always occurs at the rate of 4.185 joules per calorie whether heat goes into external work or work is dissipated through friction into heat.

See also entries that follow; and **Thermodynamics**.

Additional Reading

Butterworth, D. and C.F. Mascone: "Heat Transfer Heads Into 21st Century," *Chem. Eng. Progress*, 30 (September 1991).
Incropera, F.P. and D.P. Dewitt: *Introduction to Heat and Mass Transfer*, 3rd Edition, John Wiley & Sons, Inc., New York, NY, 2000.
Rolle, K.C.: *Heat and Mass Transfer*, Prentice-Hall, Inc., Upper Saddle River, NJ, 1999.
Rosenberg, R.: *Companion to Chemical Thermodynamics*, 6th Edition, John Wiley and Sons, Inc., New York, NY, 2000.
Sandler, S.I.: *Chemical and Engineering Thermodynamics*, 3rd Edition, John Wiley & Sons, Inc., New York, NY, 1998.
Smith, J.M.M. and H. Van Ness: *Introduction to Chemical Engineering Thermodynamics*, The McGraw-Hill Companies, Inc., New York, NY, 2000.
Welty, J.R. et al.: *Fundamentals of Momentum, Heat, and Mass Transfer*, 4th Edition, John Wiley & Sons, Inc., New York, NY, 2000.

HEAT (Atomic).   See **Atomic Heat**.

HEAT BALANCE (Distillation).   See **Distillation**.

HEAT BALANCE (Planet).   The equilibrium that exists on the average between the radiation received by a planet and its atmosphere from the sun and that emitted by the planet and atmosphere. That the equilibrium does exist in the mean is demonstrated by the observed long-term constancy of the earth's surface temperature. On the average, regions of the earth nearer the equator than about 35° latitude receive more energy from the sun than they are able to radiate, whereas latitudes higher than 35° received less. The excess of heat is carried from low latitudes to higher latitudes by atmospheric and oceanic circulations and is re-radiated there.

HEAT BALANCE (Process).   A heat balance is a method of accounting for all heat units in a process or change during which heat is transferred.

Examples of cases where heat balances might be undertaken are as follows. (1) Determining the nature and the magnitude of the various losses that occur when fuel is burned in a steam boiler furnace. (2) Accounting for all heat units during the operation of a prime mover, such as a Diesel engine or a steam turbine. (3) Determining the distribution of heat in a static heating device, such as a water heater supplied with steam.

Heat balance work is based upon the first law of thermodynamics, a statement of which is: Energy may not be created or destroyed, but may be converted from one form to another. The significance of this law applied to the heat balance is that the total energy may be accounted for by straight addition, hence striking a heat balance resembles bookkeeping, with heat supplied on the credit side of the ledger, and various heats usefully employed on the debit side. One way of showing a heat balance is a tabular form; another shows the heat as a stream, properly branched and subdivided to indicate the distribution of heat. Briefly, a heat balance might be said to be the bookkeeping by which heat supplied is shown to be equal to the sum of heat utilized and lost.

(It should be added that the above statement of the First Law, while adequate for many engineering calculations, is subject to modification in accordance with the principle of mass-energy equivalence.)

HEAT CAPACITY.

The amount of heat necessary, to raise the temperature of a system, entity, or substance by one degree of temperature. It is most frequently expressed in calories per degree centigrade or Btu per degree Fahrenheit. If the mass of a substance is specified, then certain derived values of the heat capacity can be obtained, such as the atomic heat, molar heat, or specific heat.

HEAT CAPACITY EQUATION (Einstein).

A quantum relationship for the heat capacity at constant volume of an element of the form:

$$C_v = 3R \left(\frac{hv}{kT}\right)^2 \left(\frac{e^{hv/kT}}{(e^{hv/kT} - 1)^2}\right)$$

in which C_v is the heat capacity at constant volume for one gram-atom of an element, R is the gas constant, h is Planck's constant, k is the Boltzmann constant, v is the characteristic frequency of oscillation of the atoms of the element, T is the absolute temperature, and e is the natural logarithmic base.

The Einstein equation was the first approximation to a quantum theoretical explanation of the variation of specific heat with temperature. It was later replaced by the Debye theory of specific heat and its modifications.

HEAT CONSERVATION.

See **Insulation (Thermal)**.

HEAT CONTENT.

See **Enthalpy**.

HEAT DEGREE DAY.

See **Climate**.

HEAT ENGINE.

As used in thermodynamics the term denotes a thermodynamic system, e.g., a sample of gas, carried through a cyclic process in such a way that a closed path is traced out on a pressure-volume $(P - V)$ diagram, and positive work is done by the system. If Q_1 is the positive amount of heat energy absorbed by the system, Q_2 the positive amount of heat energy rejected by the system and W the net amount of work done by the system, then the first law of thermodynamics (conservation of energy) gives $W = Q_1 - Q_2$. The efficiency of the engine is defined as

$$\eta = \frac{W}{Q_1} = 1 - \frac{Q_2}{Q_1}$$

For an engine following a reversible Carnot cycle (Carnot engine), the efficiency is given by $\eta = (T_1 - T_2)/T_1$ where T_1 is the Kelvin temperature of the reservoir at which Q_1 is absorbed, and T_2 is the Kelvin temperature at which Q_2 is rejected. The second law of thermodynamics states that no engine working between these same two temperatures can have a greater efficiency than that of the Carnot engine.

A thermodynamic engine run backwards becomes a *refrigerator*. Thus a positive amount of heat Q_2 is absorbed at a low temperature, work W is done, and positive heat Q_1 is rejected at a higher temperature. The first law now gives $Q_1 = W + Q_2$. The ratio Q_2/W is known as the *coefficient of performance* of the refrigerator. See also **Solar Energy**.

HEATER (Hysteresis).

See **Hysteresis Heater**.

HEAT EXCHANGER.

A vessel in which an outgoing hot liquid or vapor transfers a large part of its heat to an incoming cool liquid; in the case of vapors, the latent heat of condensation is thus utilized to heat the entering liquid. The shell-and –tube type is widely used; here the hot liquid or vapor is contained in the shell while the cool liquid passes through the tubes, which are usually arranged in coils for maximum contact with the heat source. Heat exchangers are used in many chemical operations, e.g., evaporation and pulp manufacture, as well as to produce steam from the heat developed in nuclear reactors for power generation.

See also **Evaporation**; and **Heat Transfer**.

HEATHER SHRUBS AND TREES.

The heather or heath family (*Ericaceae*) is comprised of a number of genera, many species, hybrids, clones, and cultivars. Three of the main genera are: *Arbutus*, small to large evergreen trees, of which the strawberry tree (not to be confused with the fruit-bearing plant of the rose family) and the madrona tree are examples; *Clethra*, a small genus of deciduous or evergreen trees or shrubs, of which the Lily-of-the-valley clethra is representative; and *Rhododendron*, evergreen or deciduous shrubs (usually) or trees, of which azaleas and rhododendrons are members.

The strawberry tree (*Arbutus unedo*) is characterized by small whitish flowers occurring in clusters, a small fruit, about $\frac{1}{2}$ inch (1.2 centimeters) in diameter, which appears something like a strawberry, and narrow, oval leaves of medium length. This tree, which can reach a height of 40 feet (12 meters), does well in southwestern Ireland and the Mediterranean region. However, the tree can withstand somewhat colder climes and is found in parts of Britain and North America. Some authorities describe the fruit more as a roughened cherry than as a strawberry. Flowering occurs in late autumn, a definite attraction to the gardener. The A. andrachne is the strawberry tree of the eastern Mediterranean region. It is a slightly smaller tree, generally attaining a height of about 30 to 35 feet (9 to 10.5 meters). The flowers are an off-white and occur in broad clusters. Flowering occurs in the spring. The fruit is similar to the A. unedo.

The strawberry tree of western and southwestern North America is the A. menziestii, or, as commonly termed, the *madrona* or *madrone* tree. As shown in Table 1, there are closely related, localized species, such as the A. arizonica and A. texana. As noted, these trees are capable of achieving excellent heights under favorable conditions. The leaves are of medium length, oval, with dark green coloration above and a bluish-white color underneath. The tree usually flowers late in the spring. The fruit is a pea-sized berry. The tree is also found in Europe, where it may achieve a height of 50 to 55 feet (15 to 16.5 meters).

Not previously mentioned, the genus *Oxydendrum* claims the sorrel tree (*O. arboreum*), which occurs in the eastern United States and is related to the strawberry tree. The sorrel is a relatively small tree, ranging from 20 to 55 feet (6 to 16.5 meters) in height, with a trunk diameter up to about 20 inches (50.8 centimeters). The bark is gray-brown, somewhat furrowed. The branches are pendulous. Leaves are elliptically shaped, pointed, dark green, and finely toothed. The flowers are white and occur in drooping clusters. The tree occurs from Pennsylvania westward to Indiana and southward along the Alleghany Mountains into Louisiana and western Florida. Some of the tallest specimens are found on the eastern slopes of the Blue Ridge Mountains.

Of the genus *Clethra*, the Lily-of-the-valley clethra attains a height of about 30 feet (9 meters) and can be classified as a small tree. The tree is found on the Island of Madeira, but has been introduced elsewhere. The leaves are dark green and alternate; the flowers are white and fragrant. The tree is sensitive to climate and soil, requiring a rich and acid mix.

It is interesting to note that prior to 1820, rhododendrons other than the European and American varieties were unknown. In particular, the R. maximum of the eastern United States was best known then. In that year, the *Rhododendron arboreum* was brought out of the Himalayas. During the intervening years, numerous species from Asia have been found and propagated.

The R. maximum, also sometimes referred to as the Great Laurel or rose bay, is a shrub/tree that can attain a height of 40 feet (12 meters) under favorable conditions, with a trunk diameter approaching a foot (0.3 meter). The bark is gray-brown, smooth, with minor scaling. The leaves are evergreen and lustrous, quite large — from 4 to 8 inches (10.1 to 20.3 centimeters) in length. The flowers occur in large clusters and may be described as pale pink, with spots of coloration in the upper part of the throat. In nature, the plant is found, usually in damp woodsy areas or along

TABLE 1. RECORD MADRONE AND RELATED TREES IN THE UNITED STATES[1]
(Heather or Heath Family)

Specimen	Circumference[2]		Height		Spread		Location
	Inches	Centimeters	Feet	Meters	Feet	Meters	
Clethra or Cinnamon tree (1995) (*Clethra acuminata*)	10	25	33	10.1	12	3.7	North Carolina
Clethra or Cinnamon tree (1997) (*Clethra acuminata*)	9	23	29	8.8	10	3.04	North Carolina
MADRONES							
Arizona madrone (1997) (*Arbutus arizonica*)	176	447	42	12.8	46	14	Arizona
Pacific madrone (1997) (*Arbutus menziesii*)	418	1062	80	24.4	121	36.9	California
Texas madrone (1999) (*Arbutus texana*)	175	445	26	7.9	50	15.2	New Mexico
RHODODENDRONS							
Catawba rhododendron (1991) (*Rhododendron catawbiense*)	10	25	26	7.9	11	3.4	North Carolina
Catawba rhododendron (1995) (*Rhododendron catawbiense*)	17	43	19	5.8	17	5.2	North Carolina
Pacific rhododendron (1976) (*Rhododendron macrophyllum*)	20	51	33	10.1	20	6.1	California
Rosebay rhododendron (1981) (*Rhododendron maximum*)	25	64	40	12.2	22	6.7	South Carolina

[1] From the "National Register of Big Trees," American Forests (by permission).
[2] At 4.5 Feet (1.4 meters).

streams, from Nova Scotia westward through Quebec and Ontario to Ohio and Lake Erie. As they range southward, they become more numerous, notably through the Allegheny Mountains and in the southeastern states as far as Georgia.

Other species of rhododendrons occurring in the mountains and woods of the United States, notably east of the Rocky Mountains include: *R. viscosum*, also known as clammy azalea or white swamp honeysuckle; *R. nudiflorum*, also known as Pinxter flower; *R. arborescens* or the smooth azalea; *R. canescens* or mountain azalea; *R. calendulaceum* or flame azalea; *R. canadense* or rhodora; *R. catawbiense* or rose bay; *R. lapponicum* or Lapland rose bay; and *R. hispida* or rose acacia of the southeastern United States. Rhododendrons from Asia include the previously mentioned *R. arboreum*. This plant is found in the Himalayas in the Khasia Hills, Sri Lanka and from Kashmir to Bhutan. It displays bell-shaped flowers of dark red color. The height ranges from 30 to 40 feet (9 to 12 meters). The *R. barbatum* is found in Bhutan, Nepal, and Sikkim and is capable of growing to a height of 40 feet. The flowering is similar to that of *R. arboreum*. The *R. calophytum*, with similar flowers but ranging from white to rose pink and with a characteristic maroon blotch, is found in western China. It can attain a height of about 35 feet (10.5 meters). Also from the Bhutan, Nepal, and Sikkim regions is the *R. falconeri*, with purple-blotched white flowers, and reaching a height of about 30 feet (9 meters). The *R. giganteum* of Yunnan is well named because it ranges in height between 40 and 80 feet (12 and 24 meters). The flowers are of a deep rose-crimson color. Also of Yunnan and of southeastern Tibet and upper Burma is the *R. sinogrande* which can rise to a height of about 45 feet (13.5 meters), displaying white/yellow flowers.

In a brief description of the heather family, certainly the common heather shrub (*Culluna vulgaris*) should not be omitted. This is a small, straggly shrub ranging from about 6 to 16 inches (15.2 to 40.6 centimeters) in height. It was introduced into America from Europe. The leaves are small, gray-green, perhaps $\frac{1}{8}$ inch (0.3 centimeters) in length and overlap the branches. Tiny bell-shaped flowers, white or of a deep pink color, occur as spikes. The Scottish heather counterpart is the *Erica cineria*. Another heather, the *Erica Tetralix*, was introduced along with the aforementioned two species into Nantucket Island, at the same time the Scots pine was introduced, during the years 1875–1877.

HEAT INDEX. The heat index (sometimes called the *apparent temperature*) is a measure of the contribution that high temperature and high humidity (expressed as relative humidity (RH)) make in reducing the body's ability to cool itself. The human body normally cools itself by perspiration, or sweating, in which the water in the sweat evaporates and carries heat away from the body. However, when the relative humidity is high, the evaporation rate of water is reduced. This means heat is removed from the body at a lower rate, causing it to retain more heat than it would in dry air. Measurements have been taken based on subjective descriptions of how hot subjects feel for a given temperature and humidity, allowing an index to be made which corresponds a temperature and humidity combination to a higher temperature in dry air.

Table 1 may be used to estimate the heat index. The heat index (HI) is an accurate measure of how hot it really feels when the affects of humidity are added to high temperature.

To reduce the risk of the danger from high humidity and high temperature wear light and loose fitting clothing and a large hat. Stay in the shade as much as possible, avoid alcoholic beverages, and drink water frequently. It is usually best to perform strenuous outdoor activity early in the morning or during the evening hours when the heat index is expected to be relatively low.

Heat Index/Heat Disorders

When the heat index is between 80°–90°F (27°–32°C) fatigue is possible with prolonged exposure and/or physical activity. When the heat index is between 90°–105°F (32°–41°C) sunstroke, heat cramps or heat exhaustion are possible with prolonged exposure and/or physical activity. When the index is between 105°–130°F (41°–54°C) sunstroke, heat cramps, or heat exhaustion is likely and heatstroke is possible. Heat indices of 130° (54°C) or higher will result in heatstroke or sunstroke quickly.

In Canada, the similar humidex is used in place of the heat index. The name is sort of a meld of the words "humidity" and "index". Sometimes Canadian governmental agencies will invent their own words for things to make translation easier between the two official languages. The humidex was introduced in 1965 by what is now the Meteorological Service of Canada. The formula currently used to determine the humidex has been in use since 1979.

The Weather Service makes the following assessment of how unpleasant to how dire humidex ratings can seem: Less than 29°C (85°F): no discomfort 30°C to 39°C (86°F to 103°F): some discomfort 40°C to

TABLE 1. HEAT INDEX °F (°C)

TEMP.	40	45	50	55	60	65	70	75	80	85	90	95	100
					Relative Humidity (%)								
110 (43)	136 (58)	143 (61.7)	152 (66.7)	161 (71.7)	171 (77.2)	182 (83.3)							
109 (42.8)	133 (56)	140 (60)	148 (64.4)	157 (69.4)	167 (75)	177 (80.5)	188 (86.7)						
108 (42.2)	130 (54.4)	137 (58.3)	144 (62.2)	153 (67.2)	162 (72.2)	172 (77.8)	182 (83.3)	193 (89.4)					
107 (41.7)	127 (52.8)	134 (56.7)	141 (60.6)	149 (65)	157 (69.4)	167 (75)	177 (80.5)	188 (86.7)	199 (92.8)				
106 (41.1)	124 (51.1)	130 (54.4)	137 (58.3)	145 (62.8)	153 (67.2)	162 (72.2)	172 (77.8)	182 (83.3)	193 (89.4)				
105 (40.6)	121 (49.4)	127 (52.8)	134 (56.7)	141 (60.6)	149 (65)	157 (69.4)	166 (74.4)	176 (80)	187 (86.1)	198 (92.2)			
104 (40)	119 (48.3)	124 (51.1)	131 (55)	137 (58.3)	145 (62.8)	153 (67.2)	161 (71.7)	171 (77.2)	181 (82.8)	191 (88.3)			
103 (39.4)	116 (46.7)	122 (50)	127 (52.8)	134 (56.7)	141 (60.6)	148 (64.4)	157 (69.4)	165 (73.9)	175 (79.4)	185 (85)	195 (90.6)		
102 (38.9)	114 (45.6)	119 (48.3)	124 (51.1)	130 (54.4)	137 (58.3)	144 (62.2)	152 (66.7)	160 (71.1)	169 (76.1)	179 (81.7)	189 (87.2)	199 (92.8)	
101 (38.3)	112 (44.4)	116 (46.7)	121 (49.4)	127 (52.8)	133 (56.1)	140 (60)	147 (63.9)	155 (68.3)	164 (73.3)	173 (78.3)	182 (83.3)	192 (88.9)	
100 (37.8)	109 (42.8)	114 (45.6)	118 (47.8)	124 (51.1)	129 (53.9)	136 (57.8)	143 (61.7)	150 (65.6)	158 (70)	167 (75)	176 (80)	185 (85)	195 (90.6)
99 (37.2)	107 (41.7)	111 (43.9)	115 (46.1)	120 (48.9)	126 (52.2)	132 (55.6)	138 (58.9)	145 (62.8)	153 (67.2)	161 (71.7)	170 (76.7)	179 (81.7)	188 (86.7)
98 (36.7)	105 (40.6)	109 (42.8)	113 (45)	117 (47.2)	123 (50.6)	128 (53.3)	134 (56.7)	141 (60.6)	148 (64.4)	155 (68.3)	164 (73.3)	172 (77.8)	181 (82.8)
97 (36.1)	103 (39.4)	106 (41.1)	110 (43.3)	114 (45.6)	119 (48.3)	125 (51.7)	130 (54.4)	136 (57.8)	143 (61.7)	150 (65.6)	158 (70)	166 (74.4)	174 (78.9)
96 (35.6)	101 (38.3)	104 (40)	108 (42.2)	112 (44.4)	116 (46.7)	121 (49.4)	126 (52.2)	132 (55.6)	138 (58.9)	145 (62.8)	152 (66.7)	160 (71.1)	168 (75.6)
95 (35)	99 (37.2)	102 (38.9)	105 (40.6)	109 (42.8)	113 (45)	118 (47.8)	123 (50.6)	128 (53.3)	134 (56.7)	140 (60)	147 (63.9)	154 (67.8)	161 (71.7)
94 (34.4)	97 (36.1)	100 (37.8)	103 (39.4)	106 (41.1)	110 (43.3)	114 (45.6)	119 (48.3)	124 (51.1)	129 (53.9)	135 (57.2)	141 (60.6)	148 (64.4)	155 (68.3)
93 (33.9)	95 (35)	98 (36.7)	101 (38.3)	104 (40)	107 (41.7)	111 (43.9)	116 (46.7)	120 (48.9)	125 (51.7)	130 (54.4)	136 (57.8)	142 (61.1)	149 (65)
92 (33.3)	94 (34.4)	96 (35.6)	99 (37.2)	101 (38.3)	105 (40.6)	108 (42.2)	112 (44.4)	116 (46.7)	121 (49.4)	126 (52.2)	131 (55)	137 (58.3)	143 (61.7)
91 (32.8)	92 (33.3)	94 (34.4)	97 (36.1)	99 (37.2)	102 (38.9)	105 (40.6)	109 (42.8)	113 (45)	117 (47.2)	122 (50)	126 (52.2)	132 (55.6)	137 (58.3)
90 (32.2)	91 (32.8)	92 (33.3)	95 (35)	97 (36.1)	100 (37.8)	103 (39.4)	106 (41.1)	109 (42.8)	113 (45)	117 (47.2)	122 (50)	127 (52.8)	132 (55.6)
89 (31.7)	89 (31.7)	91 (32.8)	93 (33.9)	95 (35)	97 (36.1)	100 (37.8)	103 (39.4)	106 (41.1)	110 (43.3)	113 (45)	117 (47.2)	122 (50)	126 (52.2)
88 (31.1)	88 (31.1)	89 (31.7)	91 (32.8)	93 (33.9)	95 (35)	98 (36.7)	100 (37.8)	103 (39.4)	106 (41.1)	110 (43.3)	113 (45)	117 (47.2)	121 (49.4)
87 (30.6)	87 (30.6)	88 (31.1)	89 (31.7)	91 (32.8)	93 (33.9)	95 (35)	98 (36.7)	100 (37.8)	103 (39.4)	106 (41.1)	109 (42.8)	113 (45)	116 (46.7)
86 (30)	85 (29.4)	87 (30.6)	88 (31.1)	89 (31.7)	91 (32.8)	93 (33.9)	95 (35)	97 (36.1)	100 (37.8)	102 (38.9)	105 (40.6)	108 (42.2)	112 (44.4)

TABLE 1. (*Continued*)

TEMP.	40	45	50	55	60	65	70	75	80	85	90	95	100
					Relative Humidity (%)								
85 (29.4)	84 (28.9)	85 (29.4)	86 (30)	88 (31.1)	89 (31.7)	91 (32.8)	93 (33.9)	95 (35)	97 (36.1)	99 (37.2)	102 (38.9)	104 (40)	107 (41.7)
84 (28.9)	83 (28.3)	84 (28.9)	85 (29.4)	86 (30)	88 (31.1)	89 (31.7)	90 (32.2)	92 (33.3)	94 (34.4)	96 (35.6)	98 (36.7)	100 (37.8)	103 (39.4)
83 (28.3)	82 (27.7)	83 (28.3)	84 (28.9)	85 (29.4)	86 (30)	87 (30.6)	88 (31.1)	90 (32.2)	91 (32.8)	93 (33.9)	95 (35)	97 (36.1)	99 (37.2)
82 (27.7)	81 (27.2)	82 (27.7)	83 (28.3)	84 (28.9)	84 (28.9)	85 (29.4)	86 (30)	88 (31.1)	89 (31.7)	90 (32.2)	91 (32.8)	93 (33.9)	95 (35)
81 (27.2)	81 (27.2)	81 (27.2)	82 (27.7)	82 (27.7)	83 (28.3)	84 (28.9)	85 (29.4)	85 (29.4)	86 (30)	87 (30.6)	88 (31.1)	90 (32.2)	91 (32.8)
80 (26.7)	80 (26.7)	80 (26.7)	81 (27.2)	81 (27.2)	82 (27.7)	82 (27.7)	83 (28.3)	84 (28.9)	84 (28.9)	85 (29.4)	86 (30)	86 (30)	87 (30.6)

45 °C (104 °F to 113 °F): great discomfort; avoid exertion Above 45 °C (113 °F): dangerous Above 54 °C (129 °F): heat stroke imminent

Web References

Humidex Temperature and Relative Humidity Charts: http://www.msc-smc.ec.gc.ca/cd/brochures/humidex_table_e.cfm

National Weather Service; Meteorological Calculator: http://www.srh.noaa.gov/tsa/met_calc.html

HEATING (Geothermal). See **Geothermal Energy**.

HEATING (Solar). See **Solar Energy**.

HEAT LIGHTNING. See **Lightning**.

HEAT OF COMBUSTION. See **Calorimetry**; and **Combustion**.

HEAT PUMP. A system involving a compressor, heat exchangers, a refrigerant, and a flow restriction that can be used to supply or remove heat. In its cooling cycle, the traditional heat pump operates very much like a conventional air conditioner. Although the principle of the heat pump has been known for decades, it has received renewed interest in recent years in connection with the search for more energy efficient systems. The fundamentals of the heat pump for both its heating and cooling cycles are described briefly in the caption for Fig. 1.

The heat pump has been well established as an efficient user of electrical power. However, heat pumps are not universally applicable to year-round heating/cooling applications if they are to operate at maximum efficiency and thus outperform separate heating and cooling equipment. Comparisons can be made by determining the coefficient of performance (COP) of a given heat pump for a given application. The COP is a ratio, namely, of the amount of heat required by the condenser (exchanger A in diagram b), or Q_c, divided by the quantity of electrical energy consumed to power the pump, or W. That is, $COP = Q_c/E$.

Assuming that the average efficiency of an electric generating and distributing utility is 30%, the use of a heat pump can bring the overall heating efficiency up to 75%. (See "Design Improvements" later in this article.) It is interesting to note that a heat pump can yield more thermal energy than the electrical energy it consumes when operating under certain conditions. This in no way defies the law of energy conservation because the heat pump picks up increments of thermal energy from the evaporator when used in the heating cycle. As pointed out in diagrams a and b, when on the heating cycle, a heat pump system, by virtue of absorbing heat from already cold ambient air, dumps air that is below ambient temperature to the atmosphere. And, when on the cooling cycle, a heat pump system, by virtue of absorbing heat from already warm room air, dumps air that is above ambient temperature to the atmosphere.

For a given capacity heat pump, the volume flow rate of refrigerant vapor through the compressor is approximately constant. It will be noted from diagrams a and b that, in either the heating cycle or cooling cycle, the medium taken into the compressor is in the vapor phase at a comparatively low temperature and pressure. Also, the medium exiting the compressor is in the vapor phase at a comparatively high temperature and pressure. It is evident from diagram a that the temperature of the cold ambient air in its effect on heat exchanger A determines the temperature of the medium exiting the exchanger and thus entering the compressor. Thus, the colder the ambient air, the lower will be the vapor pressure and the density of the medium. These conditions reduce the effective mass of the medium moving through the compressor. This decreased mass flow rate lowers the thermal capacity of the medium, thus reducing the quantity of heat energy it can transfer from the compressor to heat exchanger B and thence to heat the conditioned space. Depending upon the base capacity of the unit, the point may be reached where the ambient temperature is just too low to permit the heat pump to heat the conditioned space adequately. In the present state of the art, this leaves the designer two options—either use a unit with greater capacity (larger initial investment, etc.) or arrange to furnish for auxiliary heating during abnormally cold periods when such conditions may persist. Depending upon the form of the auxiliary energy and the efficiency of the auxiliary system, some or all of the advantages of the heat pump (cost, normal operating efficiency, etc.) may be negated. In general, it has been a practice to size the heat pump to the summer's cooling requirements and permit the winter heating to fall where it may, with dependence upon auxiliary heating. In southern climes, this may be acceptable because relatively little if any heating may be required of the unit during the winter season. In northern climes, designing to the summer cooling load will normally lead to the requirement for auxiliary heating.

Design Improvements. As has been pointed out over the years since the introduction of the heat pump, there has been a wide disparity of opinion among experts between the ideal heat pump performance and the actual performance of the equipment. Some of the shortfall of performance, particularly in older installed equipment, arises from an emphasis on minimizing manufacturing costs (thus initial costs) and consumer prices. For example, heat exchangers, condensers, and evaporators can be made of a relatively small size to keep initial costs down. But small components have limited capacity to transfer heat between refrigerant and air. To achieve a high rate of heat transfer with a small heat exchanger requires a rather large temperature difference between the refrigerant and the air. Thus, in the heating mode, the refrigerant in the condenser must be much warmer than the indoor air, and the refrigerant in the evaporator must be much colder than the outside (ambient) air. In some designs, to maintain these exaggerated temperature differences, the compressor literally must work overtime. The expected result is that the cost coefficient of performance is lower than it would be with larger, more expensive heat exchangers.

To improve heat performance, several concepts have been proposed in recent years. These include the following. (1) Use of a volume of water to store and provide low-temperature heat—with some ice forming in the evaporator and deliberately used. The concept has been called the

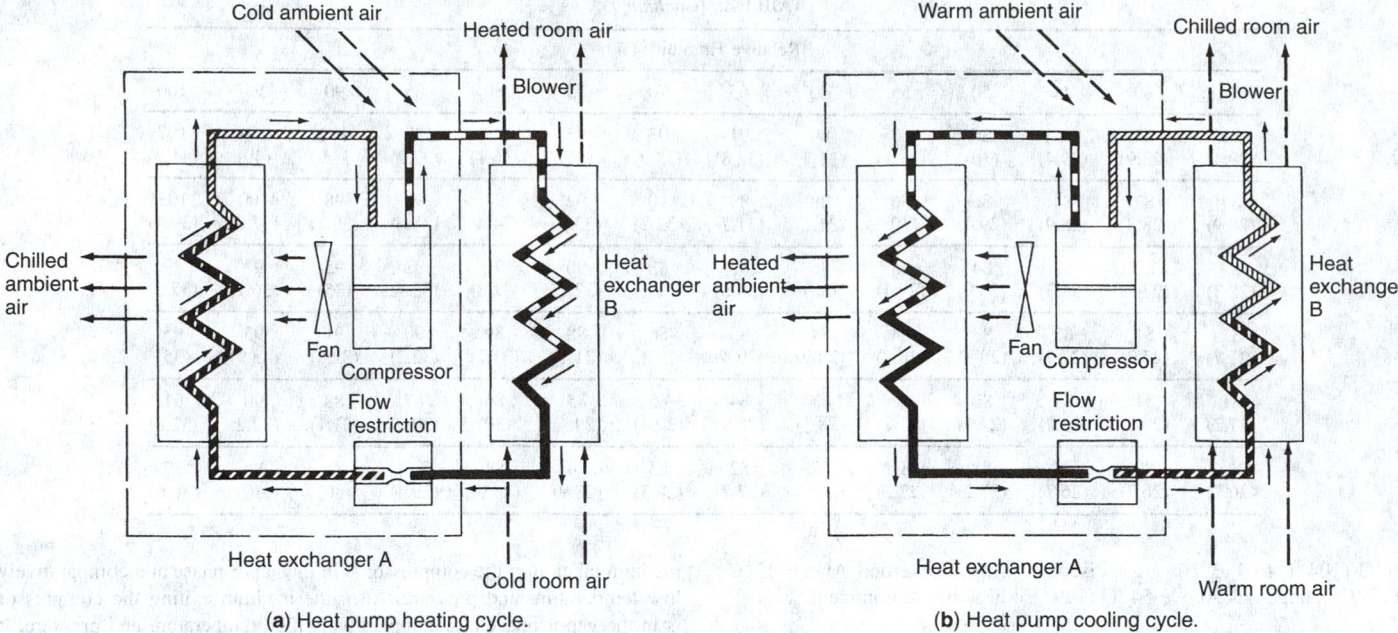

(a) Heat pump heating cycle.

(b) Heat pump cooling cycle.

Fig. 1.   Operating cycles of traditional heat pump. Although the heat pump, which can be used for both heating and cooling, is a relatively simple concept, its operation is sometimes not fully understood from the first reading of a description. Rather than begin with the usual comparison of a heat pump with an air conditioning unit and making reference to the terms condenser and evaporator, the present description commences with the heating cycle and refers to heat exchangers.

(a) There are four principal elements of equipment — a compressor, a flow restriction, and two heat exchangers, A and B. These are represented very schematically in the diagram. The heat-exchange medium (a refrigerant liquid, such as Freon O) exists the compressor in the vapor phase at a high temperature and pressure. It passes to heat exchanger B, where it is gradually cooled by the cold room air, which it in turn warms. The medium exits heat exchanger B in the liquid phase at moderate temperature, but still under high pressure, then proceeds to a flow restriction which effects a pressure drop. The medium exits the restriction as mixed liquid and vapor phase at a much lower temperature and pressure. (The lower temperature is the result of cooling caused by expansion.) This liquid/vapor mixture enters heat exchanger A, where it absorbs some heat from cold ambient air, making the ambient air just a bit colder in the vicinity of the unit. The medium exits the exchanger A as a vapor at low temperature and pressure, and returns to the compressor, where the cycle begins anew.

(b) By using a simple arrangement of valves, the flows can be reversed to make the heat pump a means for chilling room air rather than warming it. The heat-exchange medium exits the compressor in the vapor phase at high temperature and pressure. It passes to heat exchanger A, from which it exits in the liquid phase at moderate temperature and high pressure, then passes through the flow restriction, from which it exits as a mixed liquid and vapor at a low temperature and pressure. Again, the cooling is the result of expansion of the vapor. This mixed liquid and vapor phase enters heat exchanger B, where it is gradually warmed by warm room air, which it in turn cools. The medium exits heat exchanger B in the vapor phase at low temperature and pressure, and thence returns to the compressor, where the cooling cycle begins anew.

It will be noted that during the heating cycle the medium actually extracts heat from already cold ambient air as it passes through heat exchanger A. In contrast, during the cooling cycle, the medium actually adds heat to already warm ambient air as it passes through heat exchanger A.

It will be evident that, when operating in the cooling cycle, the heat pump operates like a conventional air conditioner. Heat exchanger A is the evaporator and heat exchanger B is the condenser. These units reverse their roles for the heating cycle.

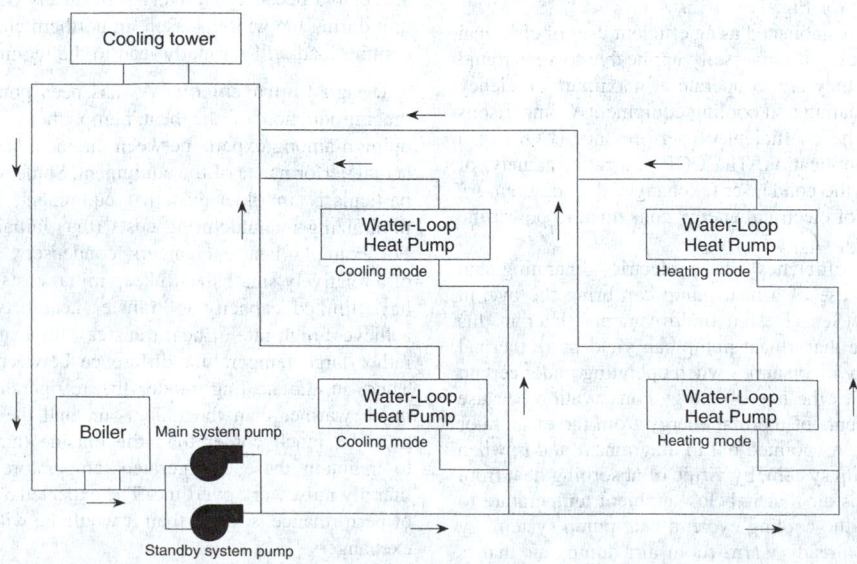

Fig. 2.   A typical water-loop heat pump system, consisting of a water circulation loop (a two-pipe supply and return system), a series of heat pumps that use the water as a heat source or sink to perform heating or cooling, and a boiler and a cooling tower that operate as required to keep the temperature of the circulating water within an optimum temperature range. Heat pumps operating in the cooling mode add heat to the loop; those operating in the heating mode extract heat from it. (*Electric Power Research Institute.*)

"annual cycle energy system." (2) Use of a heat pump to supplement a solar collector system. (3) Combined use of a thermal storage system and heat pump to partially solve the problems of oversize heat pumps for cooling, particularly in northern climes. (4) A system for varying the capacity of the system by throttling down large heat pumps when their full capacity is too great for either heating or cooling requirements. (5) Use of high-efficiency natural gas-fired heat pumps as the energy source.

Enhanced Water-loop Heat Pump. Research by EPRI (Electric Power Research Institute) and others in a cooperative effort have developed an enhanced water-loop heat pump for heating and cooling large and medium-size commercial buildings. Claims for the new system include improved energy efficiency through inherent heat recovery, low first cost, zoning flexibility, simple control, and reduced space requirements.

The typical water-loop heat pump system (WLHP) is simple in concept, consisting of a pipe loop for circulating water and a *series* of heat pumps (one in each thermal zone) that use the piped water as a heat source or sink. The system also requires a means of removing heat from the pipe loop (typically a cooling tower) and a means of adding heat (typically a boiler).

The cooling tower and the boiler operate as necessary to keep the temperature of the water in the loop within a 60–90 °F (15.6–32.2 °C) range, allowing use of uninsulated piping, which significantly reduces installed costs. Because each heat pump can perform both heating and cooling, it is possible to use a two-pipe system rather than the usual four-pipe system, further cutting distribution system costs.

The WLHP efficiency is evident when a building requires simultaneous heating and cooling needs. See Fig. 2. The system has been tested in an office building in the northeastern part of the United States for over 3 years. As a test installation, the system has been subjected to severe research scrutiny. A number of options also have been designed into the basic system. (See EPRI reference listed.)

Additional Reading

Bevington, R. and A.H. Rosenfield: "Energy for Buildings and Homes," *Sci. Amer.*, 76 (September 1990).

Bose, J. and I. Fischer: *Geothermal Heat Pumps: Introductory Guide*, Ground Source Heat Pump Publications, Stillwater, OK, 1997.

Cooper, W.B., R. Lee, and R. Featherstone: *Warm Air Heating for Climate Control*, 4th Edition, Simon & Schuster, Inc., New York, NY, 2000.

Culp, A.W.: *Principles of Energy Conversion*, 2nd Edition, McGraw-Hill Companies, Inc., New York, NY, 1991.

EPRI: *WLHP Information Brochure*, Electric Power Research Institute, Palo Alto, California 1992. http://www.epri.com/

Herold, K.E., R. Radermacher, and S.A. Klein: *Adsorption Chillers and Heat Pumps*, CRC Press, LLC., Boca Raton, FL, 1995.

Kane, D., A. Morrison, C. Ireland, et al.: *Operating Experiences with Commercial Ground-Source Heat Pump Systems*, American Society of Heating, Refrigerating & Air Conditioning, Atlanta, GA, 1998.

Lucas, C.E.: "Refrigeration System Stability Linked to Compressor and Process Characteristics," *Chem. Eng. Progress*, 37 (November 1989).

Maili, A.: "Heat Pump for Distillation Columns," *Chem. Eng. Progress*, 60 (June 1990).

Rezendes, V.S.: *Geothermal Energy: Barriers to the Use of Geothermal Heat Pumps*, DIANE Publishing Company, Collingdale, PA, 1994.

Staff: "Handbook of Fundamentals," *Amer. Soc. Heating, Refrigeration and Air Conditioning Engineers*, New York, NY. (Revised periodically.)

Walker, G. et al.: *The Stirling Alternative: Power Systems, Refrigerants and Heat Pumps*, Gordon & Breach Publishing Group, Newark, NJ, 1994.

Woodson, R.D.: *Troubleshooting and Repairing Heat Pumps*, The McGraw-Hill Companies, Inc., New York, NY, 1995.

Web References

American Society of Heating, Refrigerating & Air Conditioning: http://www.ashrae.org/

The International Ground Source Heat Pump Association: Stillwater, OK, http://www.igshpa.okstate.edu/Technology/InfoPackets/WhatsIGSHPA.html

HEAT SINK. 1. In thermodynamic theory, a means by which heat is stored, or is dissipated or transferred from the system under consideration.

2. A place toward which the heat moves in a system.

3. A material capable of absorbing heat; a device utilizing such a material and used as a thermal protection device on a spacecraft or reentry vehicle.

4. In nuclear propulsion, any thermodynamic device, such as a radiator or condenser, that is designed to absorb the excess heat energy of the working fluid. Also called *heat dump*.

HEAT STORAGE (Solar).   See **Solar Energy**.

HEAT STRESS, EXHAUSTION, AND STROKE. With ever-increasing emphasis on exercise and sports activities, there is a growing awareness of the effects of heat and exercise stress. The effects of heat stress can be serious and sometimes life threatening.

Weakness, mental fogginess, incapacity for work, and irritability are characteristics of heat exhaustion. These symptoms also appear in dehydration, alcoholism, and periods of insufficient rest and sleep. Heat exhaustion is produced in some persons when they are confined to an uncomfortably warm environment for an extensive period, during which time body fluids and salt may be depleted. The usual immediate symptoms are subnormal body temperature, clammy skin, gastric muscular spasms, and, less frequently, vomiting and diarrhea. Some authorities suggest that this syndrome may be the result of a sharp curtailment of heat production within the body, with a corresponding suppression of other functions of the adrenal cortex. Incidences of heat exhaustion can be prevented in many instances by curtailing vigorous physical exercise (as in the case of military training exercises in hot, dry or hot, humid areas) when the temperature exceeds 100 °F (38 °C). Another preventive measure is the scheduling of frequent rest periods in cooler locations where this is practical. See Table 1.

Heatstroke is a much more serious manifestation of similar factors and usually occurs when the body is subjected to very high temperatures and high humidities over relatively long periods. Epidemics of heatstroke occur in metropolitan areas during a heat wave, causing the deaths of hundreds of persons, particularly the elderly. Many such deaths go unreported and are not always identified with the cause. Poorly ventilated areas, such as barracks, sauna baths, and crowded facilities, as found in old nursing homes for the aged, aggravate the underlying conditions.

Heatstroke frequently may be manifested quite precipitously, with delirium, impaired senses, seizures, and coma. Heatstroke victims may be found with body temperatures as high as 104 °F (40 °C). Sweating is not always present. There may be extreme tachycardia, circulation may fail, and pulmonary edema and shock may result. Dehydration is often present. Most patients have vomiting and diarrhea.

Many deaths result from heatstroke because persons are not found and advised in time to initiate treatment. The initial step in treatment is removal of the person from the causative hot, humid environment. This should be followed by immersion in ice baths, application of ice packs, or sponging with alcohol in a relatively cool environment where there is good movement of air. Preferably with a rectal temperature sensor, the body temperature should be carefully monitored so that the patient will not be overcooled or allowed to reaccumulate risky heat loads. Phenothiazines may be administered to prevent excessive shivering or seizures, both of which conditions tend to increase body temperature. Water and electrolyte deficiencies should be corrected. The patient should be checked for possible renal failure and disseminated intravascular coagulation.

Information furnished for parts of this article by the Hughston Sports Medicine Foundation, Inc., Columbus, Georgia, is gratefully acknowledged.

TABLE 1. HEAT STRESS CHECKLIST

Recognizing Symptoms of Heat Stress

Heat Exhaustion
  Dizziness, lightheadedness, fainting
  Fatigue
  Headache
  General weakness
  Gastrointestinal discomfort, nausea, and vomiting
  Pale, moist skin
  Heat cramps
    Commonly encountered in sports activities, such as football. Result from muscular tightening and spasm occurring during or after prolonged exercise in a hot environment. Exquisitely painful, usually involving larger muscles of the calf and thigh. Abdominal or stomach muscles also can be affected.

(Continued)

TABLE 1. (Continued)

Heatstroke
  Malfunction of the central nervous system
  Aggressiveness or irritability
  Restlessness or delirium
  Confusion or disorientation
  Muscular incoordination
  Incoherent speech
  Seizures
  Hot, flushed, dry skin
    Heat stress occurs when the heat produced by the body and heat
    transmitted to the body from a hot environment exceeds the body's ability
    to dispose of the heat. The human body is designed to function within a
    narrow internal temperature range. To maintain its operating temperature,
    the body must rid itself of the large amount of heat that working muscles
    produce. Heavy sports equipment, as encountered in football and other
    sports, increase muscular work and hence heat production. Such excess
    heat can cause death in 15 to 20 minutes if allowed to accumulate. Heat
    from muscles is transferred to the blood and thence removed in two ways:

  1. Amount of blood flow to the skin increases. The "hot" blood can passively
     transfer heat to the environment, or
  2. It can be used to evaporate sweat.

Passive heat transfer requires that the air temperature be lower than the skin
temperature. If outside temperature is higher, the body will gain rather than lose
heat. In hot weather, sweat evaporation is the main method of heat loss. To be
effective, this method of cooling requires that (1) there must be adequate blood
flow to transport heat to the skin, (2) the skin must be exposed to the air, and
(3) the air must be able to absorb additional moisture (determined by humidity).
Death from heatstroke has occurred at temperatures below 75°F (23.9°C) when
humidity is 95%. Sweating is ineffective as a cooling medium when the sweat
cannot evaporate. Sweat production during exercise in hot conditions can exceed
1 to 2 quarts (liters) per hour. With continued exercise, blood volume (50%
water) drops and thus cannot furnish critical body needs. It is the inadequate
supply of blood to the brain that produces the warning signals — headache,
dizziness, nausea, and eventually unconsciousness and seizures.

High-risk Factors for Heat Stress

Alcohol consumption, including recovery period from excessive drinking.
Age, particularly the preteens and over 50 years.
Excess body weight loss (over 5% of body weight during a short period of
minutes to a few hours).
Prior tendency toward heat stress problems.
Inadequate sleep.
Excessive body fat and decreased aerobic conditioning.
No prior heat acclimatization.
Recent fever or gastrointestinal illness.

Preventive Measures to Avoid Heat Stress

Adequate hydration
  Thirst, alone, is an insufficient warning of dehydration. Prior to increased
  and vigorous exercise, one should practice "forced" drinking of water. As
  a general rule, athletes (football players, for example) should consume 12
  to 20 ounces (0.4 to 0.6 l) of cool water or other dilute fluid prior to
  exercise and a minimum of 8 ounces (0.3 l) for every 15 to 20 minutes of
  active play. Players and coaches should monitor fluid consumption to
  assure that it is adequate.
Adequate body cooling
  Clothing should be worn that assists the body's cooling mechanism.
  Changing sweat-soaked clothing improves sweat evaporation.
  Loose-fitting jerseys and shirts permit air to reach the skin. Low-cut socks
  are helpful. Head gear should be removed as often as practical because
  the blood supply to the head is high.
Fluid electrolytes
  Normally, cool water is the best fluid replacement. However, specially
  prepared drinks for athletes are available to assure adequate replacement
  of sodium and other ions needed by the body. Salt usually is not
  necessary. Some drinks also incorporate glucose as an energy supplement,
  but some authorities observe that glucose tends to slow water absorption
  from the gut and may inhibit the body's normal cooling mechanisms.
Acclimatization
  Vigorous exercise over long periods should be approached in a step-like
  manner to permit the body to adjust to an exercise program, rather than
  expose the body suddenly to thermal shock.
Drug avoidance
  Heat intolerance is increased by medications, such as antihistamines,
  anticholinergics, beta-blockers, diuretics, thyroid preparations, and
  antidepressants.

HEAT TRANSFER.   Although there are three generally accepted methods for transferring heat from one medium to another, or from one locale to
another within a given medium, it is uncommon for one method to act unilaterally. Particularly where convection may predominate, some conduction
of heat will be involved. In conduction, heat must diffuse through material
substances; in convection, heat is essentially carried from one locale to
another by actual movement of the transport medium; in radiation, heat
transfer involves radiant wave energy.

Conduction. From a microscopic standpoint, thermal conduction refers
to energy being handed down from one atom or molecule to the next
one. In a liquid or gas, these particles change their position continuously
even without visible movement and they transport energy also in this
way. From a macroscopic or continuum viewpoint, thermal conduction
is quantitatively described by Fourier's equation, which states that the
heat flux q per unit time and unit area through an area element arbitrarily
located in the medium is proportional to the drop in temperature, $-\text{grad }T$, per unit length in the direction normal to the area and to a transport
property k characteristic of the medium and called *thermal conductivity*:

$$q = -k \text{ grad } T \qquad (1)$$

Predictions for the value of the thermal conductivity k can be made
from considerations of the atomic structure. Accurate values, however,
require experimentation in which the heat flux q and the temperature
gradient, grad T, are measured and these values are inserted into Fourier's
equation. Thermal conductivity values for a number of media over a
large temperature range are shown in Fig. 1. Metals have the largest
conductivities and, among these, pure metals have larger values than alloys.
Gases, in contrast, have very low heat conductivity values. Electrically
nonconducting solids and liquids are arranged in between. The low thermal
conductivity of air is utilized in the development of thermally insulating
materials. Such materials, like cork or glass fiber, consist of a solid
substance with a very large number of small spaces filled by air. The
thermal transport occurs then essentially through the air spaces, and
the solid structure only supplies the framework that prevents convective
currents. It will be noted that the thermal conductivities indicated in Fig. 1
(at ambient temperature) extend through five powers of 10. This range is
still small when compared with the range for the electric conductivity of

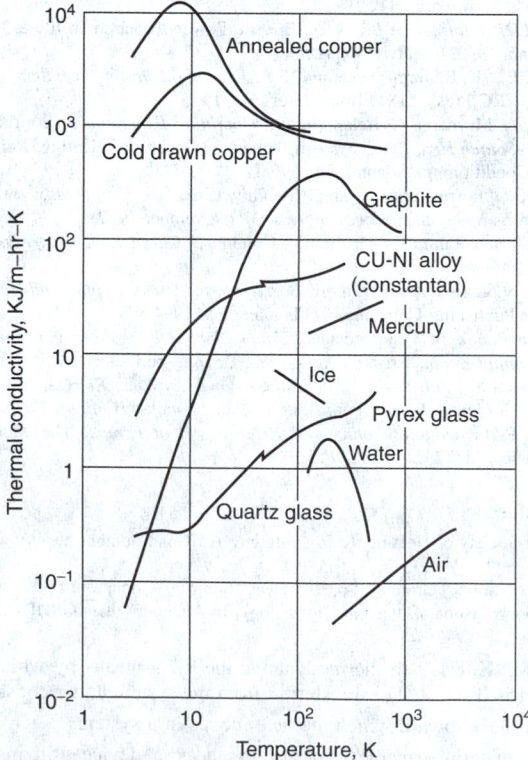

Fig. 1.   Thermal conductivity values for a wide range of substances and over a
temperature range of 1 to 10^4 K.

various substances, where electric conductors have values that are larger by 25 powers of 10 than electric insulators. As a consequence, it is much easier to channel electricity along a desired path than to do so with heat, a fact that accounts for the difficulty in accurate experimentation in the field of heat transfer.

Fourier's equation can be used together with a statement on energy conservation to derive a differential equation describing the temperature field in a medium. Fourier was the first person to develop this equation and to devise means for its solution. In vector notation, this equation is:

$$\rho c = \frac{\partial \mathbf{T}}{\partial t} = \nabla(k\nabla \mathbf{T}) \tag{2}$$

where ρ is the density, c is the specific heat, t is time, and ∇ is the Nabla (vector differential) operator. The temperature field in a substance can either change in time (unsteady state), or it can be independent of time (steady state, $\partial \mathbf{T}/\partial t = 0$). For a steady-state situation, the temperature field depends primarily on the geometry of the body involved and on the boundary conditions. The simplest case of a steady-state temperature field is a plane wall with temperatures that are uniform on each surface, but different at the two surfaces. The temperature in the wall then changes linearly in the direction of the surface normal as long as the variation of the thermal conductivity in the temperature range involved can be neglected. For an unsteady process, the capacity of the medium to store energy enters the energy conservation equation; correspondingly, the specific heat of the material and its density become factors for the conduction process, as well as the thermal conductivity. A combination of these properties, defined as the ratio of the thermal conductivity to the product of specific heat and density, called *thermal diffusivity* $(k/\rho c)$, then determines how fast existing temperature differences in a medium equalizes in time. It is found that metals and gases have thermal diffusivity values approximately equal in magnitude and are considerably higher than thermal diffusivities of liquid and solid nonconductors. This means that temperature differences equalize much faster in metals and gases than in other substances.

Various other physical processes lead in their mathematical description to equations of the same form as Eq. (2), especially in its steady-state form. Such processes include the conduction of electricity in a conductor, or the shape of a thin membrane stretched over a curved boundary. This situation has led to the development of analogies (electric analogy, soap film analogy) to heat conduction processes, which are useful because they often offer the advantages of simpler experimentation.

Convection. When energy is transported by convection in fluids, conduction usually takes care of the transport of heat from one stream tube to another and is the dominating mode of transfer near solid walls. Convection transports heat along the stream lines and is dominating in the main body of the fluid where the velocities are large. In many situations, the flow is turbulent; this means that unsteady mixing motions are superimposed on the mean flow. These mixing motions contribute also to a transport of heat between stream tubes, a process which can be described by an "effective" conductivity which often has values by several powers of ten larger than the actual conductivity of the fluid.

Movement of the fluid may be generated by means external to the heat transfer process, as by fans, blowers, or pumps. It may also be created by density differences connected with the heat transfer process itself. The first mode is called *forced convection*; the second one *natural* or *free convection*. Convection heat transfer may also be classified as heat transfer in *duct flow*, or in *internal flow* (over cylinders, spheres, air foils, and similar objects). In the case of external flow, the heat transfer process is essentially concentrated in a thin fluid layer surrounding the object (boundary layer).

Of special interest in such heat transfer processes is the knowledge of the heat flux from the surface of a solid object exposed to the flow. This heat flux q_w per unit area and time is conventionally described by Newton's equation:

$$q_w = h(T_w - T_f) \tag{3}$$

where T_w is the surface temperature and T_f is a characteristic temperature in the fluid. This equation defining the heat transfer coefficient h is convenient, because, in many situations, the heat flux is at least approximately proportional to the temperature difference $T_w - T_f$. Information on the heat transfer coefficients can be obtained by a solution of the Navier-Stokes equation describing the flow of a viscous fluid and the related energy equation, or they are found by experimentation. Computers enhance the ability to study heat transfer analytically at least for laminar flow, whereas in turbulent flow the bulk of the information is determined experimentally.

Experimentation is difficult because of the large number of parameters involved. Dimensional analysis has been applied to reduce the number of influencing parameters, and relations for convective heat transfer are correspondingly presented in many handbooks as relations between dimensionless parameters. Such an analysis demonstrates that heat transfer in forced flow can be described by a relation of the form

$$Nu = f(Re, Pr) \tag{4}$$

in which the Nusselt number Nu is a dimensionless parameter hL/k, containing the heat transfer coefficient h, the Reynolds number describes essentially the nature of the flow, and the Prandtl number $Pr = c_p\mu/k$ can be considered a dimensionless transport property characterizing the fluid involved. L and V are an arbitrarily selected characteristic length and velocity, respectively; ρ denotes the density, μ the viscosity, and C_p the specific heat of the fluid at constant pressure. See also **Reynolds Number**.

Convection is frequently thought of in terms of space heating and industrial heat-exchange processes. It should be pointed out that convection plays a cosmic role (in the sun's photosphere, for example), and a very large role in connection with the atmosphere of the earth and some other planetary bodies. For example, when normal convective transport is inadequate, temperature inversions occur and create smog hazards over large cities. See also **Atmosphere (Earth)**.

Attempts to develop a theory for convection date back at least to the 1790s when Thompson (Count Rumford) introduced the concept of heat convection. Very little theoretical work was undertaken, however, until the early 1900s, when Bénard (France) undertook experimental investigations. Modern convection physics stems from the work of Lord Rayleigh, who first published on the subject in 1916. In current times, advanced convection research studies have been undertaken by Velarde and Normand (1980), among others. See reference listed. See also **Boiler (Steam Generator)**; and **Heat**.

Radiation. In the transfer of energy from one location to another in the form of photons (electromagnetic waves), usually a multiplicity of wavelengths is involved. In vacuum, all waves regardless of their wavelength move with the same speed (2.9977×10^8 meters per second). In various substances, the wave velocity c changes somewhat with wavelength, and the ratio of the wave velocity in vacuum to the velocity in a substance is equal to the optical refraction index. Air and generally all gases have refractive indices that differ from one only in the fourth decimal. Their wave velocity is therefore practically equal to that in vacuum. See also **Waves and Wave Mechanics**.

Prévost's principle states that the amount of energy emitted by a volume element within a radiating substance is completely independent of its surroundings. Whether the volume element increases or decreases its temperature by the process of radiation depends upon whether it absorbs more foreign radiation than it emits or vice versa. One refers to thermal radiation when the emission of photons is thermally excited, i.e., when the substance within the volume element is nearly in thermodynamic equilibrium. For such radiation, Kirchhoff was able to derive a number of relations by consideration of a system of media in thermodynamic equilibrium. If jv indicates the coefficient of emission, i.e., the radiative flux at the frequency v^* emitted per unit volume into a unit solid angle, and η is the coefficient of absorption at the same frequency, i.e., the fraction of the intensity of a radiant beam that is absorbed per unit path length, then one of these relations states

$$c^2\frac{jv}{\eta_v} = f(T, v) \tag{5}$$

with c denoting the wave velocity. According to this relation, the combination of parameters on the left-hand side of Eq. (5) is a function of temperature T and frequency v of the radiation only, but does not depend upon the substance under consideration. Kirchhoff's law can also be expressed in parameters that refer to the interface of two media (1 and 2). It then takes the form

$$c^2\frac{i_v}{\alpha_v} = f(T, v) \tag{6}$$

in which i_v is the monochromatic intensity of the radiative flux at frequency v originating in medium 2 and traveling through the interface into medium 1 per unit solid angle and area normal to the direction of the radiant beam. α_v is the monochromatic absorptance or absorptivity, i.e., that fraction of a radiant beam approaching the interface in the medium 1 in the opposite direction that is absorbed in medium 2. The wave velocity in medium 1 is c. Kirchhoff's law states that the combination of the parameters on the left-hand side of Eq. (6) is again a function of temperature and frequency only, but does not depend upon the nature of the medium. A medium which absorbs all the radiation traveling into it through an interface ($\alpha_v = 1$) is called a *blackbody*. The intensity of radiation emitted by an arbitrary medium is, according to Eq. (6), in the following way related to the intensity of radiation i_{bv} emitted by a black body at the same temperature and frequency:

$$\frac{i_v}{v} i_{bv} \tag{7}$$

See also **Planck Radiation Formula**.

The amount of heat transferred by radiation can be determined by use of the *Stefan-Boltzmann Law*

$$Q = bA(T_1^4 - T_2^4) \tag{8}$$

where Q is the amount of heat transferred per unit time, b is a constant, A is the area of the radiating surface, T_1 is the absolute temperature of the radiating body and T_2 is the absolute temperature of the receiving body. Various correction factors are introduced into the formula to account for the shape of the bodies, their thermal radiation characteristics and the properties of the media through which the radiant rays must pass while traveling from radiator to absorber. The thermal radiation characteristics are its emissivity, a measure of its ability to radiate at a given temperature, its absorptivity, a measure of its ability to absorb heat and its reflectivity, which measures its ability to reflect without absorbing.

Radiant energy travels in a straight line. Therefore to transmit it to an object out of sight of the radiator requires a reflector, such as a furnace wall, to deflect the rays to their objective.

It is possible to set up controlled laboratory radiation between simple plane surfaces and determine therefrom accurate coefficients to incorporate into radiation equations. However, the radiation of heat from furnace gases, consisting of non-luminous gases, luminous carbon particles in flame, ash globules, etc., to the walls and tubes of a steam generator in commercial operation at variable load, is another matter. Here, empirical data which are gathered and interpreted from field tests on similar equipment, must still be resorted to however great the designer's urge to go back to basic laws of heat transfer.

Radiant heat transfer in furnaces is roughly proportioned to the difference in the fourth power of the absolute temperatures of the radiating and receiving surfaces. The water wall surface is approximately at boiler saturation temperature, while the superheater surface varies from this to somewhat above the temperature of the steam at the superheater outlet. However, the mean radiating temperature of the furnace gases is usually over 1204 °C. The fourth power of the receiving surface temperature is thus seen to be small compared to the fourth power of the transmitting surface temperature; consequently the latter controls the transmittance, and boiler tube temperature does not need to be considered a variable to be accounted for.

Figure 2 shows some of the arrangements in which radiant heat-absorbing surface is disposed. It may be used to illustrate another of the difficulties that beset the designer in following a rational or semi-rational form of radiation analysis. Projected radiant surface is one thing; actual radiant energy receiving surface may be quite a different area. For example, suppose the tubes of case (a) to be separated and spaced l_1 inches on centers. The *projected* areas of cases (a) and (c) would then be the same, but it seems obvious that re-radiation from the wall causes more of a (c) tube to receive radiant energy than is the case with an (a) tube. Also, if δ is a factor correcting projected area to *equivalent* absorbing surface, what value should be assigned to it in the case of a bank of tubes which may receive by re-radiation some radiant energy deep in the tube bank? Here δ

has a minimum value of 1, but some investigators have derived expressions which indicate that δ may have a magnitude of 3 or more.

Industrial Heat Transfer Equipment

Some of the more common cases of industrial heat transfer are:

1. Radiation from fuel beds and luminous gases to absorptive surfaces such as boilers, cylinder walls, etc.
2. Radiation from heat generators such as drying lamps.
3. Convection of heat out of combustion regions.
4. Convection of heat from hot surfaces under either free or forced convection.
5. Conduction of heat through the tubes of boilers, heaters, heat exchangers, condensers, etc.
6. Conduction in walls, pipe covering, and other so-called "heat insulators."
7. Conduction of heat through the plates of plate-type heat exchangers and regenerators.

Heat exchangers perform many functions within a manufacturing facility. Often they are given special names, even though they remain fundamentally heat exchangers. These include:

Chiller — a device which cools fluids to temperature below those obtainable with ordinary cooling water by using the vaporization of a refrigerant. The fluid to be cooled is routed through the tubes while the low-boiling refrigerant vaporizes from a pool of liquid in the shell.

Partial Condensers — Many overhead vapors from distillation columns in petroleum-refinery services are a mixture of light and heavy hydrocarbons and noncondensable gases, i.e., gases that are not condensed at the outlet temperature and pressure of the condenser (air, hydrogen sulfide, methane, and other light ends). These vapors are routed through the shell side while water is used as the cooling medium on the tube side of the unit. Condensation on the shell side begins at the saturation temperature of the heavy components and continues over a decreasing temperature range until part of the lighter components are condensed. Part of the existing liquid is sent back to the tower as reflux, while the remainder is further refined or passes to the trim cooler and storage.

Trim Cooler — This unit condenses the last remaining light-end vapors and cools the liquid to the ultimate storage temperature (often about 100 °F: 38 °C) by using cooling water. This cooling usually is not conducted in the main condenser because it would reduce column pressure.

Thermosiphon Reboiler — Flow of the vaporizing fluid depends upon the difference in static head between the column of liquid flowing from the tower to the reboiler and the partially vaporized column of liquid returning from the exchanger to the tower.

Reboilers — These exchangers operate in conjunction with a distillation tower to vaporize enough liquid to assure vaporization of the overhead product. A hot process stream of steam may be used as the heating medium. Most reboilers are shell-and-tube exchangers located at the base of the tower. The vaporizing fluid is routed through the shell side of the exchanger.

Forced-circulation Reboiler — A pump is used to provide more positive circulation than available with the thermosiphon effect, e.g., in the vaporization of viscous fluids.

Vapor Heat Exchanger — Units of this type preheat a cool stream of process fluid by using heat from partially condensing vapor. The objective is to conserve heat and eliminate the requirement for a separate preheater.

Air-cooled Exchanger — As used in the petroleum industry, air-cooled exchangers normally comprise two headers joined by a horizontal bank of finned tubes. Usually two motor-driven fans located above (induced draft) or below (forced draft) the tubes are used to circulate the air over the finned surface.

Superheater — A unit of this type heats vapor above the saturation temperature.

Waste-heat Boiler — A unit of this type generates steam and is similar to a regular steam generator except that hot gas or liquid produced by a chemical reaction (often combustion) is the heating medium.

Types of Heat Exchangers. In terms of heat exchange for recovering and recycling thermal energy, the shell-and-tube heat exchanger of the type

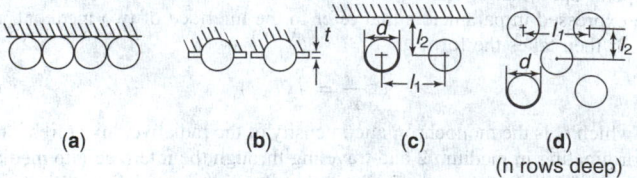

(a)	(b)	(c)	(d)
			(n rows deep)

Fig. 2.   Arrangements of radiant heat-absorbing surface.

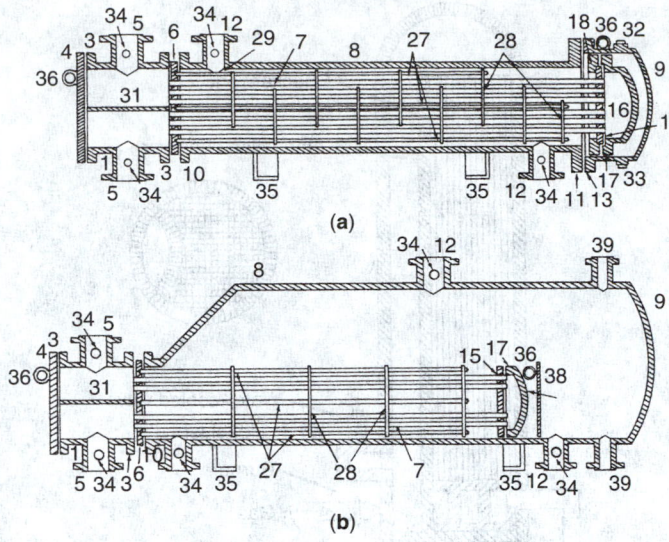

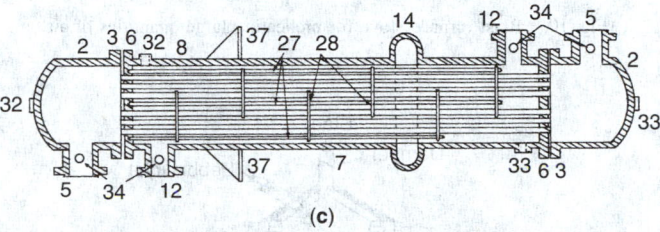

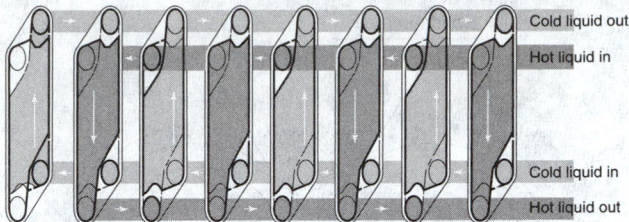

Fig. 4. Principle of plate-and-frame heat exchanger. (*After Carlson.*)

Increasingly, they are finding use in condensing and boiling applications, where their compact size and thinner material requirements for wetted parts offer advantages over other types of heat-exchange designs.

Less commonly used are the heat-transfer configurations shown in Figs. 5 through 9.

Heat Storage

It is often necessary to store heat in rather large quantities in specially designed apparatus. Hot water, of course, is one of the easiest forms in

Fig. 3. Three common types of shell-and-tube heat exchangers: (**a**) Type AES, internal-floating-head exchanger (with floating-head backing device); (**b**) Type AKT, kettle-type floating-head reboiler; and (**c**) Type BEM, fixed-tube sheet exchanger. (*Sketches adapted from specification diagrams from the Standards of Tubular Exchange Manufacturers Association.*) Legend: (1) Stationary head, channel, (2) stationary head bonnet, (3) stationary-head flange, channel, or bonnet, (4) channel cover, (5) stationary-head nozzle, (11) stationary tube sheet, (12) tubes, (13) shell, (9) shell cover, (10) shell flange, stationary-head end, (11) shell flange, rear-head end, (12) shell nozzle, (13) shell cover flange, (14) expansion joint, (15) floating tube sheet, (16) floating-head cover, (17) floating-head flange, (18) floating-head backing device, (19) split shear ring, (20) slip-on backing flange, (21) floating-head cover, external, (22) floating tube sheet skirt, (23) packing-box flange, (24) packing, (25) packing follower ring, (26) lantern ring, (27) tie rods and spacers, (28) transverse baffle, (29) impingement baffle, (30) longitudinal baffle, (31) pass partition, (32) vent connection, (33) drain connection, (34) instrument connection, (35) support saddle, (36) lifting lug, (37) support bracket, (38) weir, and (39) liquid-level connection.

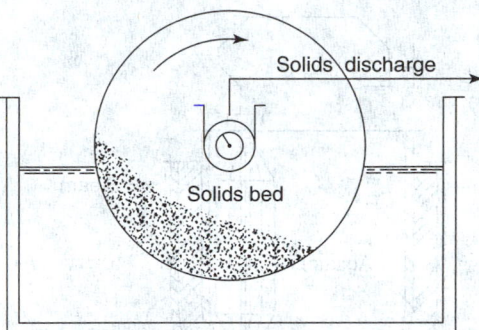

Fig. 5. Plain rotating shell used for both heating and cooling. For high-range heating, tempered combustion gases may be used instead of water.

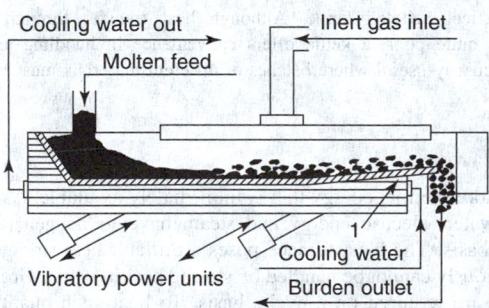

Fig. 6. Vibrating-type heat-transfer equipment for batch solidification. Sometimes referred to as a *caster*, the machine is used widely in a number of industries. After cooling and solidification, intense vibratory action shatters cake into lumps.

shown in Fig. 3 for many decades has been the most common type. It can be used with liquid on both sides, gas on both sides, or liquid on one side and gas on the other side. The most common requirement is for liquid-liquid exchangers. Heat exchanges may be used strictly for processing purposes—that is, materials need to be heated (or cooled) prior to entering some processing application, such as reacting, distilling, vaporizing, and the like. Or heat exchangers may be used simply for recovering the energy from hot fluids for use elsewhere.

Although immensely improved over the years from the standpoint of design efficiency, resistance to corrosion, and ease of maintenance, among other objectives, the fundamental design has remained unchanged. However, within the past few years, the plate-and-frame heat exchanger has been introduced. See Fig. 4. This type of exchanger consists of a frame that carries a series of closely spaced metal plates that have been pressed, with a corrugated trough pattern. The plates, which are clamped between a fixed head and movable follower, have corner ports to permit the passage of process and service liquids. There are elastomeric gaskets around the ports and plate edges to avoid leakage. The plates are grouped into passes within the heat exchanger. The product and service fluids flow countercurrent to each other between the parallel passages in each pass. Initially, plate-and-frame exchangers were used for liquid-liquid thermal exchange purposes.

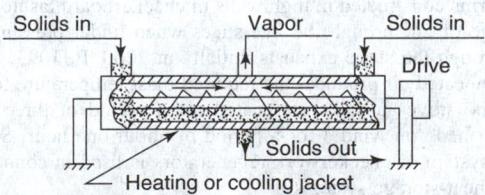

Fig. 7. Tank equipped with mixing ribbon spirals provides considerable agitation and is useful for melting or cooking dry powdered solids. Heat-transfer efficiency is only moderate because of the relatively deep beds of solid particles.

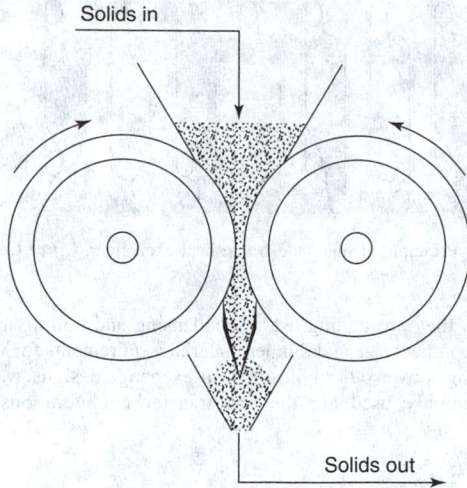

Fig. 8. Double drum. Scraping knives may be engaged continuously or intermittently, depending upon the nature of the heated product.

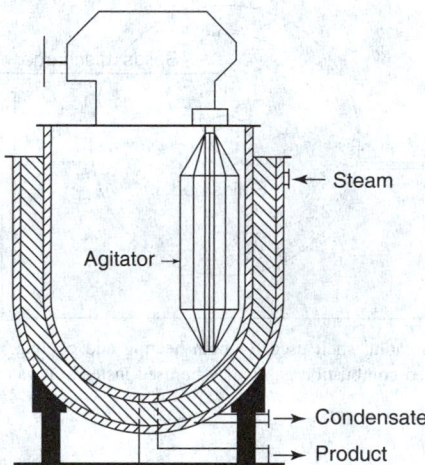

Fig. 9. Vertical agitated kettle. Although heat transfer through the jacket normally is quite poor, a kettle offers convenience in handling and cleaning and is particularly useful where batches of different materials must be processed frequently.

which to store thermal energy that is immediately available. As contrasted with hot water, electric energy and steam have to be generated on an as-needed basis. The blast furnace poses a difficult heat storage problem, which obviously cannot be handled by storing heat in water. Great amounts of hot gas are required on a cyclic basis. To heat such quantities of air on a continuous, as-required basis would be quite impractical with the present stage of the art. The solution used involves several stoves that are quite large, often over 100 feet (30 meters) in height and about 25 feet (7.5 meters) in diameter. The blast temperature of approximately 1,000 °F (538 °C) is accomplished by preheating the stove checkerwork to a much higher temperature. Checkerwork is comprised of refractory material forms constructed in high walls in checkerboard fashion to permit free passage of air through the interstices when under pressure. The gas passing through the stove exhausts initially at 2000 °F (1093 °C). Mixing this with unheated air produces the required blast temperature for the blast furnace. The stoves usually are heated for a period of three hours and exhaust (termed "on wind") for a period of about one hour. See Fig. 10. A similar system of checkerwork regenerators is used in connection with glass-tank heat-storage systems.

Flowing streams of pebbles also have been used in the chemical industry for removing heat from gases. Pebbles and stones are also used in some solar energy storage systems. See Fig. 11. See also **Solar Energy**.

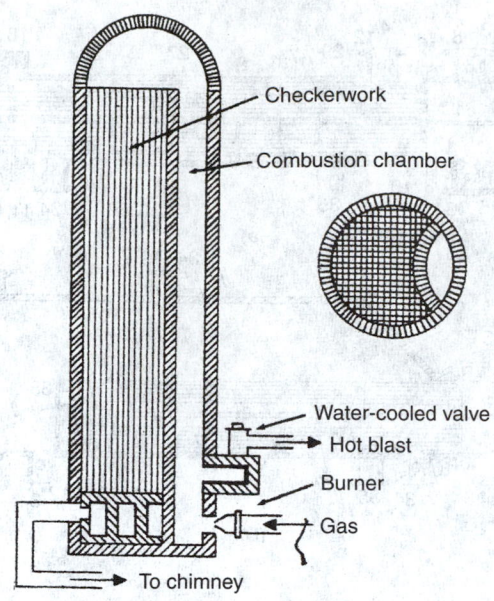

Fig. 10. Blast furnace stove for preheating large quantities of air.

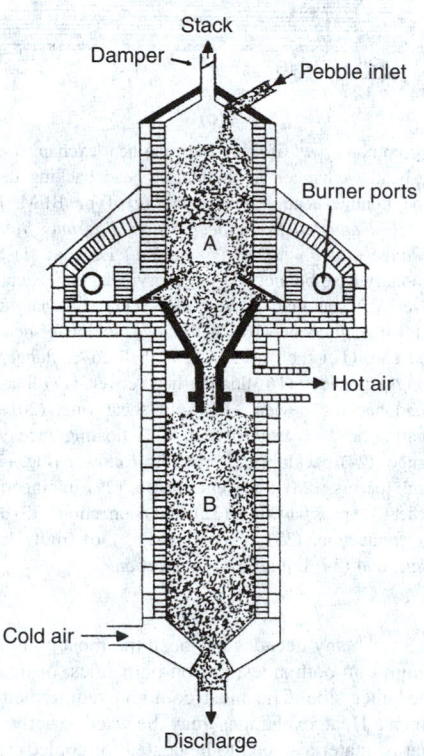

Fig. 11. Pebble heater for heating steam to temperatures impractical in metallic units. Also used for heating air, hydrogen, methane, and other gases for processing purposes. In reverse, a pebble heater may be used to recover heat from hot gases. The pebbles are heated in top chamber A by direct contact with combustion gases and passed through a throat to lower chamber B, where heat is transferred to cool gases. The two chambers are maintained at the same temperature so that there will be no gas flow between them. An average cycle on the pebbles is 30–50 minutes.

Additional Reading

Arpaci, V.S. et al.: *Introduction to Heat Transfer*, Prentice-Hall, Inc., Upper Saddle River, NJ, 2000.

Bejan, A. and J.S. Jones: *Heat Transfer*, John Wiley & Sons, Inc., New York, NY, 1999.

Butterworth, D. and C.F. Mascone: "Heat Transfer Heads Into the 21st Century," *Chem. Eng. Progress*, 30 (September 1991).

Carlson, J.A.: "Understand the Capabilities of Plate-and-Frame Heat Exchangers," *Chem. Eng. Progress*, 26 (July 1992).

Corsi, R.: "Specify Bayonet Heat Exchangers Properly," *Chem. Eng. Progress*, 32 (July 1992).

Ganapathy, V.: "Heat-Recovery Boilers: The Options," *Chem. Eng. Progress*, 59 (February 1992).

Incropera, F.P. and D.P. Dewitt: *Introduction to Heat and Mass Transfer*, 3rd Edition, John Wiley & Sons, Inc., New York, NY, 2000.

Incropera, F.P. and D.P. Dewitt: *Fundamentals of Heat and Mass Transfer*, 4th Edition, John Wiley & Sons, Inc., New York, NY, 2000.

Janna, W.S.: *Engineering Heat Transfer*, 2nd Edition, John Wiley & Sons, Inc., New York, NY, 1999.

Kakapc, S. and H. Liu: *Heat Exchanges: Selection, Rating, and Thermal Design*, CRC Press, LLC., Boca Raton, FL, 1997.

Kakapc, S., A.E. Bergles, F. Mayinger, and H. Yuncu: *Heat Transfer Enhancement of Heat Exchangers*, Kluwer Academic Publishers, Norwell, MA, 1999.

McKetta, J.J.: *Heat Transfer Design Methods*, Marcel Dekker, Inc., New York, NY, 1992.

Mukherjee, R.: "Use Double Segmental Baffles in Shell-and-Tube Heat Exchangers," *Chem. Eng. Progress*, 47 (November 1992).

Perry, R.H., D.W. Green, and J.O. Maloney: *Perry's Chemical Engineers' Handbook*, 7th Edition, McGraw-Hill Companies, Inc., New York, NY, 1997.

Reddy, J.N. and D.K. Gartling: *The Finite Element Method in Heat Transfer and Fluid Dynamics*, CRC Press, LLC., Boca Raton, FL, 2000.

Someah, K.: "On-Line Tube Cleaning: The Basics," *Chem. Eng. Progress*, 39 (July 1992).

Velarde, M.G. and C. Normand: "Convection," *Sci. Amer.*, **243**, 1, 92–108 (1980).

Welty, J.R. et al.: *Fundamentals of Momentum, Heat, and Mass Transfer*, 4th Edition, John Wiley & Sons, Inc., New York, NY, 2000.

Wood, R.M. et al.: "A New Option for Heat Exchanger Network Design," *Chem. Eng. Progress*, 38 (September 1991).

Yokell, S.A.: *A Working Guide to Shell-and-Tube Heat Exchangers*, The McGraw-Hill Companies, Inc., New York, NY, 1990.

HEAT TRANSFER (Nusselt Number).   See **Nusselt Number**.

HEAT TREATING.   Heating and cooling of metals to effect changes in properties. Annealing and normalizing are generally for the purpose of softening or improving the grain structure. Patenting is also a softening process in which cold drawn carbon-steel wire is heated above its critical temperature range followed by cooling to below this range in a molten lead or molten salt bath, with subsequent cooling to room temperature.

While heat treating includes the softening treatments, it most often implies hardening and strengthening. In the case of steels this requires heating to above the critical temperature range followed by rapid cooling (quenching) in oil, water, or brine, except in the case of special grades which harden on cooling in air. This is followed by tempering, a low-temperature reheating treatment which reduces the internal stresses caused by the hardening treatment. Tempering may be carried to a high enough temperature to reduce somewhat the extreme hardness of the as-quenched steel and increase the toughness and ductility, depending on the requirements of the part. See also **Iron Metals, Alloys and steels**.

Another important form of heat treatment for hardening is precipitation hardening. See also **Annealing**; **Carbonitriding**; **Carburizing**; **Case Hardening**; and **Nitriding**. See Table 1.

Thermal or heat treating has been an inherent part of metalworking and fabricating for well over a century, and, in fact, some aspects of the topic date back to ancient times. Within the last few decades, heat treating has become quite sophisticated through the incorporation of modern computing and modeling techniques. Process modeling can be used as a scheduling tool to optimize throughput of a continuous furnace, for example. Much more instrumentation has been added to heat-treating processes, allowing better control over a larger number of variables that affect final product quality.

As of the early 1990s, one of the most interesting new processes is the use of solar energy as a direct heat source for surface hardening, cladding, and other surface modifications. An impressive demonstration project has been established at the Solar Energy Research Institute (SERI) in Golden, Colorado. Its heliostat has an area of 31.8 square meters (342 sq ft) and features an ultraviolet (UV)–enhanced aluminum coating on its front surface. The primary concentrator consists of 23 hexagonal facets, each of which is a spherical mirror ground to a 14.6-meter (48-ft) radius of curvature and is aluminum coated. At the target, 94% of the energy falls inside a 100 millimeter (4-in–diameter) circle. The beam has a Gaussian shape, with a peak flux of 2.5 MW/m², without a secondary concentrator.

TABLE 1. PRINCIPAL HEAT-TREATING PROCESSES

Carburizing

Pack and gas carburizing create a diffused carbon case. Base metals are low-carbon steels and low-carbon alloy steels. Process temperature range is 815–980 °C (1500–2000 °F).

Liquid carburizing creates a diffused carbon (possibly nitrogen) case. Base metals are low-carbon steels and low-carbon alloy steels. Process temperature range is 815–980 °C (1500–1800 °F).

Vacuum carburizing creates a diffused carbon case. Base metals are low-carbon steels and low-carbon alloy steels. Process temperature range is 815–1090 °C (1500–2000 °F).

Nitriding

Gas nitriding creates a diffused nitrogen (nitrogen compounds) case. Base metals are alloy steels, nitriding steels, and stainless steels. Process temperature range is 480–590 °C (900–1100 °F).

Salt nitriding creates a diffused nitrogen (nitrogen compounds) case. Base metals are ferrous metals, including cast irons. Process temperature range is 510–565 °C (950–1050 °F).

Ion nitriding creates a diffused nitrogen (nitrogen compounds) case. Base metals are alloy steels, nitriding steels, and stainless steels. Process temperature range is 340–565 °C (650–1050 °F).

Carbonitriding

Gas carbonitriding creates a diffused carbon and nitrogen case. Base metals are low-carbon steels, low-carbon alloy steels, and stainless steels. Process temperature range is 760–870 °C (1400–1600 °F).

Liquid (cyaniding) creates a diffused carbon and nitrogen case. Base metals are low-carbon steels. Process temperature range is 760–870 °C (1400–1600 °F).

Ferric nitrocarburizing creates a diffused carbon and nitrogen case. Base metals are low-carbon steels. Process temperature range is 565–675 °C (1050–1250 °F).

Aluminizing

Creates a diffused aluminum case. Base metals are low-carbon steels. Process temperature range is 870–980 °C (1600–1800 °F).

Siliconizing (chemical vapor deposition).

Creates a diffused silicon case. Base metals are low-carbon steels. Process temperature range is 925–1040 °C (1700–1900 °F).

Chromizing (chemical vapor deposition).

Creates a diffused chromium case. Base metals are low- and high-carbon steels. Process temperature range is 980–1090 °C (1800–2000 °F).

Titanium carbide

Creates a diffused carbon, titanium, and TIC case. Base metals are alloy and tool steels. Process temperature range is 900–1010 °C (1650–1850 °F).

Boriding

Creates a diffused born (boron compounds) case. Base metals are alloy and tool steels; cobalt- and nickel-base alloys. Process temperature range is 400–1150 °C (750–2100 °F).

TABLE 2. HIGH-FLUX SOLAR FACILITIES WORLDWIDE

Location	Total Power kW	Peak Flux MW/m²
Alburquerque, New Mexico		
(Central receiver test facility)	5000	2.4
Furnace	22	3.0
Atlanta, Georgia		
Furnace	1.3	9.5
Golden, Colorado	10	2.5
White Stands, New Mexico	30	3.6
Odello, France		
Horizontal furnace	1000	16.0
Vertical furnace	6.5	15.0
Rehovot, Israel		
(Central receiver test facility)	2900	—
Furnace	16	11.0
Uzbek, Russia	1000	17.0

Source: Solar Energy Research Institute, Golden, Colorado.

A solar facility requires a major resource of direct normal radiation. Thus, in the United States, a facility of this type is limited to locations in Arizona, Colorado, Nevada, New Mexico, and Utah, or in a nearby region of one of these bordering states. Obviously, the facility cannot operate at night or during periods of dense cloud cover. Even with these limitations, design calculations show that a solar furnace can compete economically with laser and arc-lamp sources. The case for solar-furnace technology becomes even more attractive for materials-processing applications in space.

Currently, high-flux solar facilities are comparatively few, as shown in Table 2.

Additional Reading

Coffey, J.A.: "Supercell Carburizing," *Adv. Mat. & Proc.*, 81 (September 1989).

Conybear, J.G.: "Advanced Controls Offer Heat-Treating Flexibility," *Adv. Mat. & Proc.*, 38 (October 1989).

Conybear, J.G.: "Gas-Fired Vacuum Furnace Speeds Ion Nitriding," *Adv. Mat. & Proc.*, 87 (September 1991).

Dekumbis, R.: "Surface Treatment of Materials by Lasers," *Chem. Eng. Progress*, 23 (December 1987).

Doak, K.W.: "Furnaces Focus on New Processes, New Materials," *Adv. Mat & Proc.*, 84 (September 1989).

Holm, T.: "Synthetic Heat-Treating Atmospheres," *Adv. Mat. & Proc.*, 45 (October 1989).

Jones, L.E. and E.D. Jamieson: "Computers Tackle Heat-Treating Problems," *Adv. Mat. & Proc.*, 33 (March 1990).

Krauss, G.: "Thermal Processing of Steel," *Adv. Mat & Proc.*, 57 (January 1990).

Moerdijk, I.W.: "Polymer Quenchants," *Adv. Mat. & Proc.*, 19 (March 1990).

Persampieri, D., A. San Roman, and P.D. Hilton: "Process Modeling for Improved Heat Treating," *Adv. Mat. & Proc.*, 19 (March 1991).

Smidt, F.A.: "Surface Modification," *Adv. Mat. & Proc.*, 61 (January 1990).

Staff: *ASM Handbook: Heat Treating*, Vol. 4, ASM International, Materials Park, OH, 1991.

Staff: *Metal Casting and Heat Treating Industry*, DIANE Publishing Company, Collingdale, PA, 1996.

Stanley, J.T., C.L. Fields, and J.R. Pitts: "Surface Treating (Solar)," *Adv. Mat. & Proc.*, 16 (December 1990).

Totten, G.E.: "Polymer Quenchants: The Basics," *Adv. Mat. & Proc.*, 51 (March 1990).

HEAT UNITS. See **Units and Standards**.

HEAT WAVE. A period of abnormally and uncomfortably hot and usually humid weather. Also called *hot wave* , and *warm wave* . To be a heat wave such a period should last at least one day, but conventionally it lasts from several days to several weeks. In 1900, A. T. Burrows more rigidly defined a *hot wave* as a spell of three or more days on each of which the maximum shade temperature reaches or exceeds $90\,°F$ ($32.2\,°C$). More realistically, the comfort criteria for any one region are dependent upon the normal conditions of that region. In the eastern United States, heat waves generally build up with southerly winds on the western flank of an anticyclone centered over the southeastern states, the air being warmed by passage over a land surface heated by the sun.

AMS

HEAVISIDE LAYER. See **Ionosphere**.

HEAVY HYDROGEN. See **Deuteron**.

HEAVY WATER. Water in which the hydrogen of the water molecule consists entirely of the heavy hydrogen isotope of mass 2 (deuterium). Written D_2O. Density, 1.1076 at 20 degrees C. It is used as a moderator in certain types of nuclear reactors. The term is sometimes applied to water whose deuterium content is greater than natural water. See also **Nuclear Power Technology**.

HEAVY WATER REACTOR. See **Nuclear Power Technology**.

HEEL. The prominence at the posterior end of the foot. It is based on the projection of one bone, the calcaneum, behind the articulation of the bones of the lower leg. In the long-footed mammals, both the hoofed species and the clawed forms that walk on the toes, the heel is well above the ground at the apex of the angular joint known as the hock or hough. In plentigrade species it rests on the ground.

HEILIGENSCHEIN. A diffuse bright region surrounding the shadow an observer's head casts on a irregular surface. It is most apparent when the sun is low in the sky and when the surface is dew-covered. The explanation of the heiligenschein varies depending upon whether it is seen over a dry or a dew- covered surface. When an observer's shadow is cast on a dry, irregular surface (such as gravel or vegetation), each irregularity near the antisolar point covers its own shadow. In other directions, the average brightness results from a mixture of sunlit and shaded surfaces. The lower the sun in the sky, the longer the shadows and so the greater the contrast with the brighter region near the antisolar point. While evident over virtually any irregular surface, the ready appearance of the heiligenschein in sunlit wooded areas when seen from an airplane has spawned the epithet, the "hot spot in the forest." Observations of the "hot spot in the forest." are undoubtedly all the more striking when the plane is high enough that its own shadow (the umbra portion) has vanished. The presence of dew on some species of grass greatly enhances the heiligenschein. Dewdrops held off the surface of the leaf by small hairs focus sunlight on the leaf where it is diffusely reflected. The drop, acting in a manner similar to the lens in a lighthouse, then collects a large fraction of this diffusely reflected light that would have otherwise gone in other directions and sends it back toward the source and the observer. The heiligenschein is occasionally called Cellini's halo, after Benvenuto Cellini who described its behavior in his memoirs of 1562. He even pointed out that "it appears to the greatest advantage when the grass is moist with dew," but felt its appearance bespoke "the wondrous ways of [God's] providence toward me." Shades of this interpretation are also found in the name heiligenschein: It is German for "the light of the holy one." See also **Dew**.

HEISENBERG FORCE. The Heisenberg force is a phenomenologically postulated force between two nucleons derivable from a potential in which there appears an operator that exchanges the spins and positions of the two particles.

HEISENBERG REPRESENTATION. Representation of the equations of motion in quantum mechanics and in quantized field theory where the vector describing the state is treated as constant and the time dependence is transferred to the operators that operate on this state vector. This may be represented in Hilbert space by keeping the state vector constant and allowing the axes to rotate with time as the motion of the system develops. Matrices representing operators referred to these axes are thus time dependent and obey the Heisenberg equation of motion. The theory developed in this representation is therefore called matrix mechanics.

HEISENBERG, WERNER KARL (1901–1976). Heisenberg was a German physicist who earned his Ph.D. from the University of Munich, and then became Assistant to Max Born at the University of Göttingen, and in 1924 he gained the *venia legendi* at that University. From 1924 until 1925 he worked, with a Rockefeller Grant, with Niels Bohr, at the University of Copenhagen, returning for the summer of 1925 to Göttingen.

In 1926 he was appointed Lecturer in Theoretical Physics at the University of Copenhagen under Niels Bohr and in 1927, when he was only 26, he was appointed Professor of Theoretical Physics at the University of Leipzig, and worked in the field of quantum mechanics.

In 1941 he was appointed Professor of Physics at the University of Berlin and Director of the Kaiser Wilhelm.

At the end of the Second World War he, and other German physicists, were taken prisoner by American troops and sent to England, but in 1946 he returned to Germany and reorganized, with his colleagues, the Institute for Physics at Göttingen. This Institute was, in 1948, renamed the Max Planck Institute for Physics.

Heisenberg's name will always be associated with his theory of quantum mechanics, published in 1925, when he was only 23 years old. For this theory and the applications of it, which resulted especially in the discovery of allotropic forms of hydrogen, Heisenberg was awarded the Nobel Prize for Physics for 1932.

His new theory was based only on what can be observed, that is to say, on the radiation emitted by the atom. We cannot, he said, always assign to an electron a position in space at a given time, nor follow it in its orbit, so that we cannot assume that the planetary orbits postulated by Niels Bohr actually exist. Mechanical quantities, such as position, velocity, etc., should be represented, not by ordinary numbers, but by abstract mathematical structures called "matrices" and he formulated his new theory in terms of matrix equations.

See also **Field Theory**; **Heisenberg Force**; **Heisenberg Representation**; and **Quantum Mechanics**.

J. M. I.

HELIARC WELDING.   See **Welding**.

HELICAL GEARING.   For high pitch-line velocities and heavy loads, some form of "twisted tooth" gear is generally used. Two important types are helical gears and double-helical or *herringbone gears*. Both helical and herringbone gears are essentially spur gears with teeth twisted across the face in the form of a helix about the axis of rotation.

When spur gear teeth engage, the contact extends across the entire tooth on a line parallel to the axis of rotation, and may result in noise and shock at high speeds. In helical gear engagement, contact begins at one end of the entering tooth and gradually extends along a diagonal line across the tooth face as the gears rotate. The nature of the contact is such that with sufficient face width, two or more teeth are in contact and are carrying the load at all times. Helical gears are therefore used for transmission ratios as high as 10:1, and at pitchline velocities up to 2,000 feet (610 meters) per minute for commercially cut units. Herringbone gear sets of special design have been successfully operated at pitch-line speeds of 12,000 feet (3,660 meters) per minute or above.

Tooth elements of helical gears are similar to those of spur gears. See Fig. 1. The *helix angle H* of the tooth is measured between the line tangent to the tooth helix at the pitch circle and the shaft axis. In any pair, the gears have teeth with mating right-hand and left-hand helices. The usual method of tooth measurement is by diametral pitch P_d, which corresponds to circular pitch P_c in the diametral plane, perpendicular to the axis of rotation. By using standard pitches, the pitch diameters (and therefore the center distance of helical gear sets) can be given in commonly used fractions or integers; consequently, a spur gear set of a certain size can be replaced directly by a similar helical gear set. Actual tooth thickness depends upon the pitch and the size of the helix angle; if the circular pitch P_c be held constant, the actual tooth thickness measured perpendicular to its elements will decrease as the helix angle H is increased. A different cutter is required for every change in helix angle, although the pitch may remain constant. To eliminate an extensive variety of cutters, commercially available helical gears are made in several standard helix angles, among which are 7°30′, 15°, and 23°.

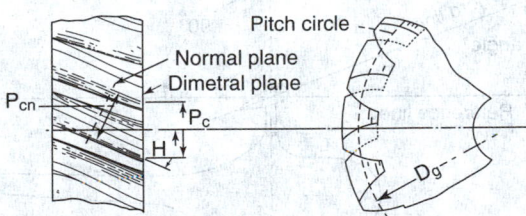

Fig. 1.   Tooth elements of helical gear.

By using a standard pitch in a plane normal to the tooth helix, the pitch diameter of a helical gear can be varied to suit a particular center distance by changing the helix angle. In this method of tooth measurement, the normal diameter pitch P_n corresponds to a normal circular pitch P_{cn} in the normal plane. Helical gear teeth designed with normal diametral pitches may be cut with standard spur gear cutters or hobs.

The pitch diameter D_g of a helical gear, based upon normal pitch P_n, is given by

$$D_g = N_g / P_n \cos H$$

where N_g is the number of teeth in the gear. The power transmitting capacity of helical and herringbone gears may be found by methods analogous to those used for spur gearing.

End thrust inherent in single helical gears can be eliminated by the use of herringbone gears that consist virtually of two integral single helical gears of opposite hand, which absorb the axial thrust within the gear. Herringbone gears are used for hoisting and mining machinery, rolling mills, sugar mill and lumber machinery, turbine and compressor drives.

HELICAL SCANNING.   A scanning technique in airborne Doppler radar that makes use of both the aircraft motion and scanning to the side

of the aircraft to survey the volume of atmosphere surrounding the flight track. Typically, the radar antenna is located in the tail of the airplane and rotates continuously about an axis aligned with the airplane. The pointing direction may be normal to the flight track or, as common for dual-beam airborne radars, may include pointing angles both forward and rearward of the direction normal to the flight track. The fore and aft directions provide measurements for dual-Doppler analysis that can be used to determine the three-dimensional velocity vector of the echoes.

AMS

HELICITY.   One-half the scalar product of the velocity and vorticity vectors. It is a conserved quantity if the flow is inviscid and homogeneous in density, but is not conserved in more general viscous flows with buoyancy effects. The concept is useful in understanding severe convective storms and tornadoes, since in strong updrafts the velocity and vorticity vectors tend to be aligned, yielding high helicity. Three-dimensional turbulence containing a nonzero mean value of helicity may develop an inertial decay range, but the development is slowed by helicity. The reluctance of helical turbulence to cascade into an inertial range means that small-scale atmospheric flows with high helicity are less unstable and more predictable than small-scale flows with low helicity.

Storm-relative Environmental Helicity (SREH)

A measure of the streamwise vorticity within the inflow environment of a convective storm. It is calculated by multiplying the storm-relative inflow velocity vector by the streamwise vorticity and integrating this quantity over the inflow depth. Geometrically, the storm-relative environmental helicity is represented by the area on a hodograph swept out by the storm-relative wind vectors between specified levels (typically the surface and 3 km to represent the primary storm inflow). It is thought to be a measure of the tendency of a supercell to rotate.

HELICOPTERS AND V/STOL CRAFT.   Although both fixed- and rotary-winged aircraft use airfoils to produce lift, in fixed-wing craft the wings can move no faster than the fuselage to produce lift and in order to fly; the whole aircraft must maintain considerable forward speed at all times. In the helicopter the wings (called rotor blades) are rotated at high speed, and there is no relationship between blade speed and fuselage speed. The helicopter, employing one or more horizontal rotors to give both lift and translation, can rise and descend vertically from the ground, hover over a spot on the ground, and fly backward and sideward as well as forward. With these flight characteristics, the helicopter does not require a prepared runway or landing area. A clearing about the size of a tennis court is adequate for landing, even though the surface be rough or uneven.

V/STOL aircraft are of more conventional lines. V/STOL is an abbreviation for a vertical or short take-off and landing. VTOL signifies vertical takeoff and landing.

Rotary-Wing Aerodynamics

The aerodynamics of rotary-wing and fixed-wing aircraft are basically the same. See also **Aerodynamics and Aerostatics**. Both types of aircraft employ airfoils to produce lift; and both are subjected to identical fundamental forces of lift, drag, thrust, and gravity. It is true, however, that the flight characteristics of the helicopter differ widely from those of the fixed-wing craft.

Lift.   Weight and lift are closely associated inasmuch as weight tends to pull the helicopter down and lift acts to hold it up. The similarity between the fixed-wing airplane and the helicopter is apparent; both are heavier than air, and both are sustained in flight by reaction of airflow over airfoils. The helicopter's airfoils are rotor blades, which are turned at high speed. In a fixed-wing aircraft, if the angle of attack is increased, lift is increased until the stalling angle is reached; and for a given angle of attack, the greater the speed, the greater the lift. The helicopter rotates the rotor blades at high speed in rpm, to establish high-speed airflow over the airfoils. It is normal for the tip speed of the rotor blades to be as much as 350 miles (563 kilometers) per hour when the speed of the fuselage is zero. This explains why the helicopter does not require forward speed to produce lift, and why it can hover or fly backward, sideward, or forward.

Airflow.   During normal operation conditions, the direction of airflow is from the top down through the main rotor system. As the blades are rotated with a positive angle of attack, they, in effect, screw upward into

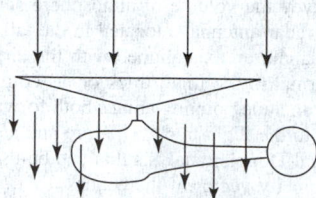

Fig. 1.   Downwash through the rotor system.

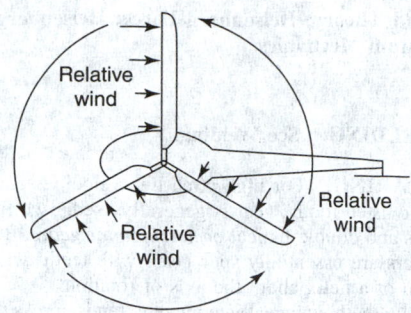

Fig. 5.   Direction of relative wind.

the air; thus a downwash of air (Fig. 1) is established through the rotor system. Notice that the leading edge of each blade bites into air throughout the complete cycle of rotation, forcing the air downward.

At the root of the blade, airflow is slightly more than zero, but the velocity progressively increases throughout the length of the blade and at the tip may be 350 miles (563 kilometers) per hour or higher. It is the blade velocity that determines the resultant strength and direction of the relative wind at a positive angle of attack. The helicopter changes the angle of attack by varying the pitch of the main rotor blades. In a helicopter, the relative wind is developed throughout the complete cycle of 360 degrees by rotation of the rotor system, and it usually varies considerably. This variation is dependent upon flight conditions. See Figs. 2 through 6.

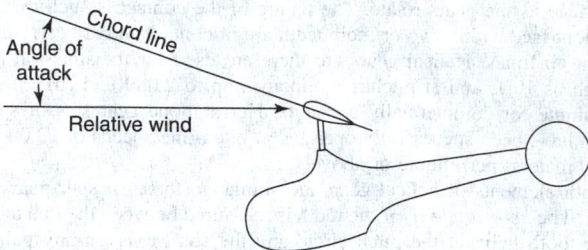

Fig. 6.   Rotor-blade angle of attack.

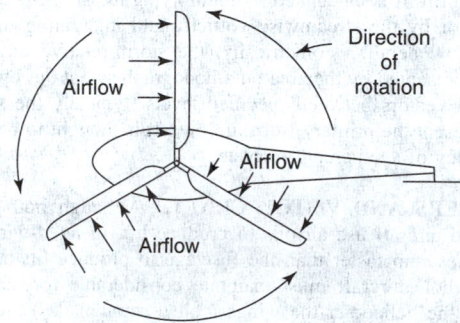

Fig. 2.   Airflow in the rotor system.

Angle of Incidence. This is the angle formed by the chord of the airfoil and the longitudinal axis of the aircraft. The conventional airplane's angle of incidence is built into the aircraft by the designer and in most aircraft cannot be changed. In the case of the helicopter, however, the pilot continually changes the angle of incidence during flight by increasing or decreasing the pitch of the main rotor blades. See Fig. 7.

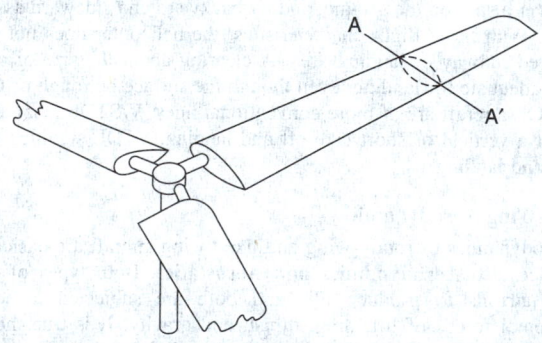

Fig. 3.   High-speed blade section.

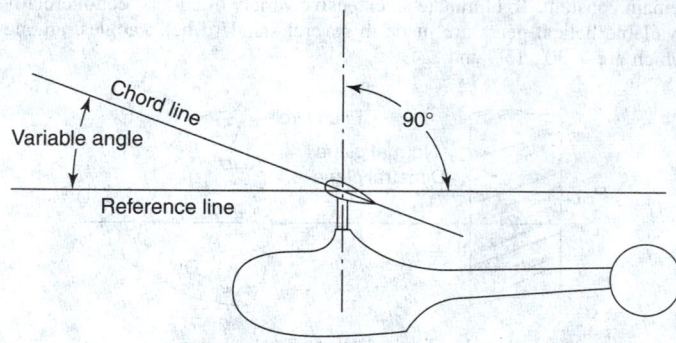

Fig. 7.   Rotor-blade angle of incidence.

Airfoil Section. The type of wing used on conventional airplanes varies considerably; the airfoils may be symmetrical or unsymmetrical, usually dependent upon some specific requirements. The unsymmetrical airfoil may be efficient for an airplane wing, but it has one disadvantage that makes it unsatisfactory for use as a rotor blade. It is normal for the center of pressure to "walk" forward and rearward as the angle of attack is changed. The center of pressure is the imaginary point on the airfoil where all the aerodynamic forces are considered to be concentrated. See Fig. 8.

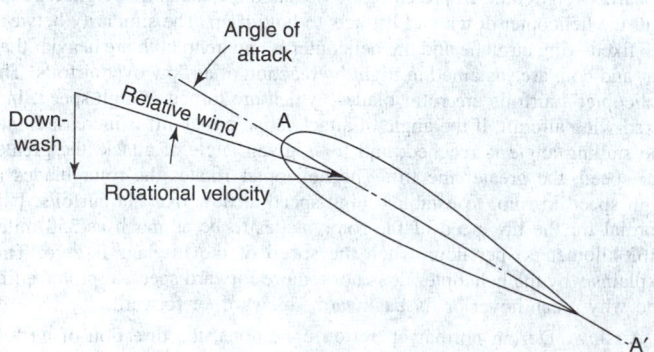

Fig. 4.   Relative wind components.

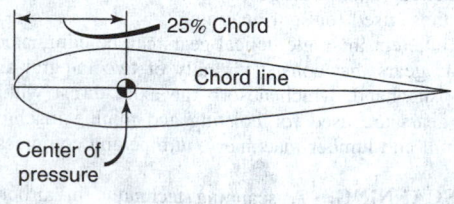

Fig. 8.   Airfoils.

The airfoil section used for rotor blades is symmetrical, having equal camber above and below the chord line. Normally the greatest thickness of the blade is at a point about one-fourth of the way back from the leading edge. It is at this point that the center of pressure is located. There are several reasons for using the symmetrical airfoil for rotor blades. (1) There is a restricted migration of the center of pressure on a symmetrical airfoil when the angle of attack is changed; (2) the lift-drag ratio is very good even though the velocity of the blade varies from root to tip; and (3) the symmetrical airfoils permit ease of construction. If the center of pressure were permitted to travel during angle of attack variations, pitching moments would be introduced into the rotor system; this condition would set up violent vibrations. Good lift-drag ratio throughout a wide range of velocities is important because it is necessary to have the lift forces spread over a wide area in order to equalize stresses. Usually a slight twist is built into the blade to help equalize these forces.

Thrust and Drag. As weight and lift are closely associated, so are thrust and drag. Thrust moves the helicopter in a designated direction and drag tends to hold it back. The helicopter develops both lift and thrust in the main rotor system. In vertical ascent, thrust acts upward in a vertical direction; drag, the opposing force, acts vertically downward. Lift sustains the weight of the helicopter; and excess thrust is available to give translation or vertical acceleration. During vertical ascent, drag is considerably increased by the downwash of the main rotor system striking the fuselage. Thrust must be sufficient to overcome both drag and downwash. The force representing the total reaction of the airfoils with the air is divided into two components: (1) Lift, and (2) thrust. However, drag is a separate force from weight, as shown in Fig. 9.

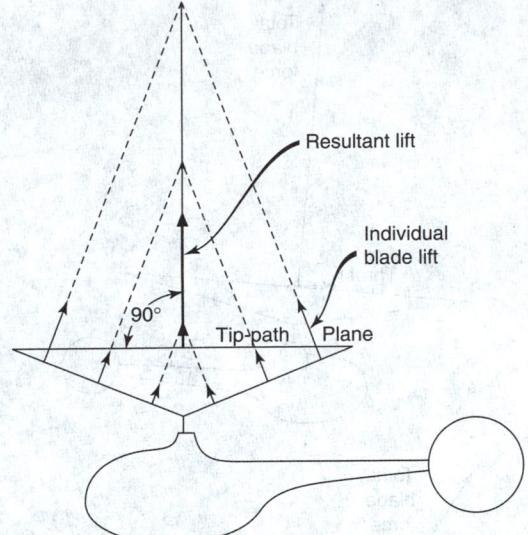

Fig. 10.   Direction of resultant lift.

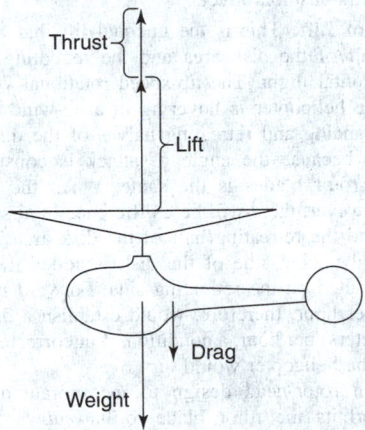

Fig. 9.   Forces in vertical ascent.

At all times, the lift forces of the rotor system are perpendicular to the tip-path plane. The tip-path plane is the imaginary plane described by the tips of the blades in making a cycle of rotation. The lift on the individual blade is perpendicular to the airfoil, but the resultant lift developed by the several blades is perpendicular to the tip-path plane. See Fig. 10. Lift increases in magnitude from root to tip of blade because of increase in velocity.

In vertical flight, the tip-path plane is horizontal; and in forward, backward, or sideward flight, the plane of rotation is tilted off the horizontal, thus inducing thrust in the direction of inclination. For example, to establish forward flight, resultant lift is inclined forward. See Fig. 11. Total force, being tilted off the vertical, acts both upward and forward; therefore it can be resolved into two components. One component is lift; the other is thrust. Likewise, flight may be established sideward, or in any horizontal direction, by tilting the tip-path plane in the direction of desired flight. Also, the rate of movement or speed depends upon the degree of tilt of the resultant lift force. Note the magnitude of thrust at the two speeds shown in Fig. 12.

Torque. Torque effect is displayed in a helicopter by the turning of the fuselage in the opposite direction to the rotation of the main rotor system. This reaction is in accord with Newton's third law of motion (to every action there is an equal and opposite reaction). The engine is the initiating force that drives the rotor system in a counterclockwise direction, and the

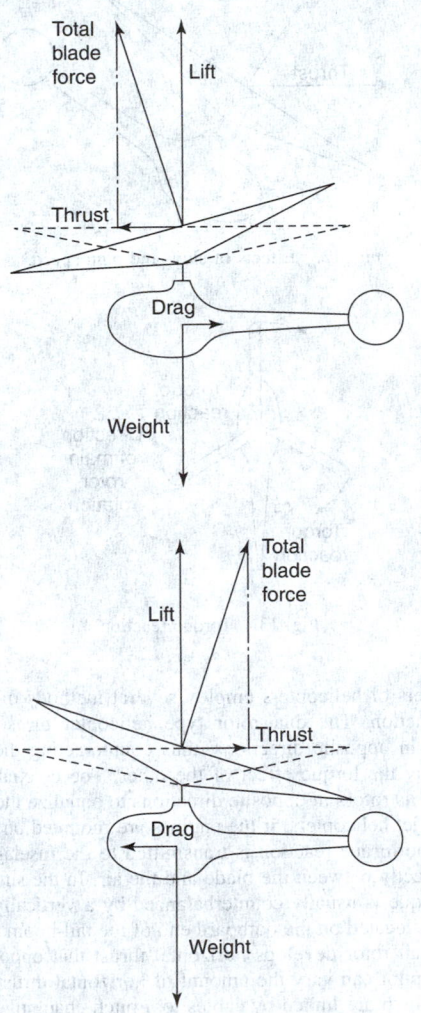

Fig. 11.   Forces in forward and rear flight.

reaction to this driving force would cause the fuselage of the helicopter to rotate with an equal force in a clockwise direction. See Fig. 13. Torque is of real concern to both the pilot and designer. Adequate means must be provided not only to counteract torque, but also for positive control over its effect during flight.

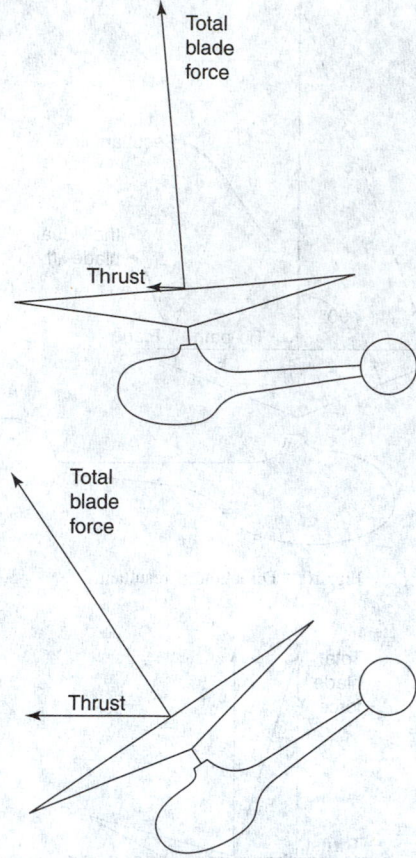

Fig. 12.   Effects of slow and high speed.

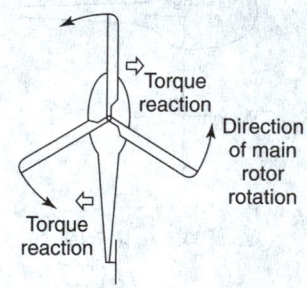

Fig. 13.   Torque reaction.

The designers of helicopters employ several methods of compensating for torque reaction. The dual-rotor type helicopter turns the two main rotor systems in opposite directions, thus counteracting the torque effect of one rotor by the torque effect of the other. The coaxial configuration likewise turns its rotors in opposite directions to equalize the torque effect. In the case of jet helicopters, if the engines are mounted on the tips of the rotor blades, no torque reaction is transmitted to the fuselage because the reaction is directly between the blade and the air. In the single main rotor helicopter, torque is usually counterbalanced by a vertically-mounted tail rotor, which is located on the outboard end of the tail-boom extension. See Fig. 14. The tail rotor develops horizontal thrust that opposes the torque reaction. The pilot can vary the amount of horizontal thrust by activating foot pedals, which are linked by cables to a pitch changing mechanism in the tail rotor system.

Ground Cushion. Also called ground effect, this is a volume of packed air built up between the rotor blades and the ground when the helicopter hovers near the ground. The downward flow of air strikes the ground and is partially trapped under the main rotor system. The air packs because it cannot escape as rapidly as the downward flow; therefore a cushion of slightly compressed air is established. The packed air is denser, thus increasing the efficiency of both the engine and the rotor system. The

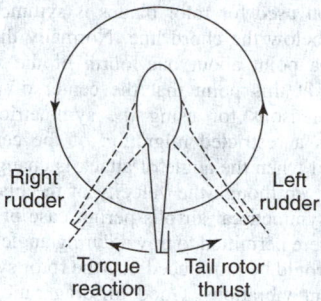

Fig. 14.   Torque correction.

ground cushion is effective to a height of approximately one-half the rotor diameter; above this height, the air cannot be effectively trapped: Also, the ground-cushioning effect is lost at airspeeds in excess of ten miles per hour.

Translational Lift. This is the additional lift developed by a helicopter in horizontal flight. This lift becomes noticeably effective at an airspeed of 10 to 15 miles (16 to 24 kilometers) per hour and it continues to increase in magnitude as speed is increased. As horizontal flight is progressively induced, a higher inflow of air is established through the rotor disk, and greater lift is produced because of increased rotor efficiency. However, when a speed of from 45 to 50 miles (72 to 80 kilometers) per hour is reached, translational lift is canceled by fuselage drag. When hovering from 6 to 8 feet (1.8 to 2.4 meters) above the ground, the helicopter is aided by the ground-cushion effect.

Dissymmetry of Lift. This is the unequal lift that develops between the advancing half of the disk area and the retreating half of the disk area during horizontal flight. The tip-speed rotational velocity is usually constant when the helicopter is hovering in a no-wind condition. Lift is equal on the advancing and retreating halves of the disk area when the craft is hovering because the angle of attack is constant and velocity airflow over the rotor blades is the same. When the helicopter enters forward flight, however, there will be a difference in airspeed between the advancing half and the retreating half of the disk area. To the rotational velocity on the advancing side of the disk is added the forward speed; the latter is subtracted on the retreating side. Forward flight of 50 miles (80 kilometers) per hour, therefore, would establish a differential of 100 miles (161 kilometers) per hour, a condition, if uncorrected, would develop unequal lift and the helicopter would turn over.

It is normal in rotor-head design to incorporate a flapping hinge, a device that permits the rotor blade to flap upward. Under normal operational conditions, the high-speed rotation of the rotor system develops a centrifugal force of approximately 20,000 pounds (88,930 N) on each blade. Centrifugal force holds the blades in a horizontal plane, but lift will cause the blade to rise vertically. The rotor blade will take the resultant position between centrifugal force and lift. It is normal for the rotor blades to take this coned-up attitude. See Fig. 15. The centrifugal force will be constant throughout the complete cycle of 360 degrees. Lift will vary between the advancing and retreating portion of the disk area in forward flight because of difference in air flow velocity. During forward flight, the advancing blade will flap higher because it has greater lift, and the retreating blade will flap to a lower angle because it has less lift. As the rotor blade flaps up, the effective lift area is lessened; and vice versa. On the retreating half of the disk area, however, reduced airspeed

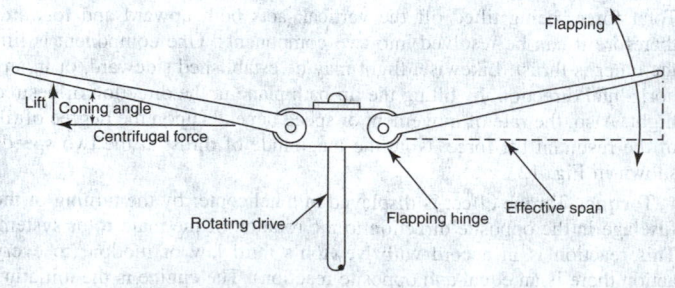

Fig. 15.   Flapping hinge device.

developed less lift. Therefore the retreating blade will assume a more horizontal attitude. Experience has proved that blades free to flap will assume a position that develops symmetry of lift between the advancing and retreating portions of the disk area.

Gyroscopic Precession. This is the innate quality of all rotating bodies by which application of a force perpendicular to the plane of rotation will produce a maximum displacement of the plane approximately 90 degrees later in the direction of rotation. See Fig. 16. Thus, if a downward force is applied to the right side of a rotating disk, gyroscopic precession will cause the disk plane to tilt to the front, provided the disk is turning from right to left. Maximum resulting displacement occurs approximately 90 degrees further in the direction of turning, but speed of rotation, weight, and diameter of the disk, and friction are factors which determine the actual displacement of the system. The main rotor system of a helicopter displays the phenomenon of gyroscopic precession. This is known as the *gyroscopic effect* for helicopters. The applied force is introduced by pitch change on the main rotor system. As the pilot moves the cyclic stick forward, it causes the control plane to tilt forward, thus introducing an equal but opposite pitch change at points 180 degrees apart in the cycle of rotation. Thus, if a linkage were not provided to take care of precession, the helicopter would fly 90 degrees out of phase. Forward stick movement would cause the craft to fly to the left. Thus, it is common practice to set the cyclic pitch change back approximately 90 degrees in the cycle of rotation. Various types of linkages are used.

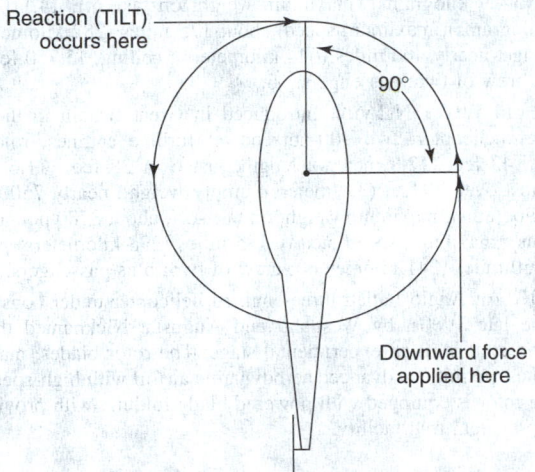

Fig. 16.   Gyroscopic precession.

Autorotation. This is the process of producing lift with rotor blades that freely rotate because of the developed aerodynamic forces resulting from the flow of air up through the rotor system. Under power-off conditions, the helicopter will descend; thus the flow of air will be established upward through the system. The rotor is automatically disengaged from the engine by a free-wheeling device and the necessary power required to overcome parasitic and induced drag of the rotor blades is obtained from the potential energy due to the helicopter's weight and height above ground. This potential energy is converted into kinetic energy, which is used to drive the rotor system during descent.

During autorotation, it is essential that the pitch angle of the rotor blades be reduced materially. The change in direction of airflow through the rotor causes a change in the direction of the relative wind, which greatly increases the angle of attack at which the rotor blades are operating. If the pitch were not reduced, the blade would stall for much the same reason that a conventional airplane's wing stalls when the nose of the aircraft is pulled up too high. When the pitch angle of the blade is low and the angle of attack is large, the resultant lift force lies ahead of the axis of rotation of the blades, tending to keep the blades turning in their normal direction. See Figs. 17 and 18. If, on the other hand, the pitch angle remains high, drag is increased and the resultant lift force lies behind the axis of rotation, tending to slow and stop the rotor. Autorotation is an emergency procedure that permits the helicopter to make a safe landing in case of engine failure. It is necessary to maintain the speed of the rotor at sufficient

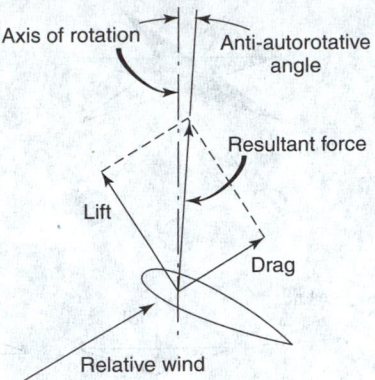

Fig. 17.   Low-pitch angle.

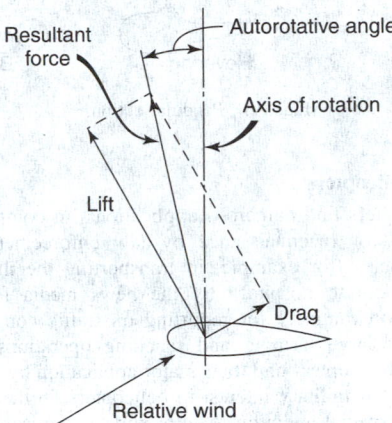

Fig. 18.   High-pitch angle.

rpm to provide not only adequate airflow over the rotor blades, but also the required centrifugal force to hold the blades in an extended attitude. Otherwise the blades would fold up and the helicopter would tumble out of control.

Pendular Action. The fuselage of the helicopter is suspended from the drive shaft that mounts the main rotor head. Because the fuselage is bulky and suspended from a single point of attachment, it is free to oscillate laterally and longitudinally much in the same fashion as a freely-swinging pendulum. As the rotor system introduces horizontal translation, the fuselage is dragged in the direction of induced flight. During established forward flight, the fuselage will assume a nose-low attitude. In effect, as the tip-path plane is inclined forward, resultant lift is inclined from the vertical, thus introducing thrust. The main drive shaft of the helicopter will have a tendency to align itself with the inclined resultant lift force. See Fig. 19.

Other factors that are important to the design of a helicopter include (1) resonance; and (2) weight and balance. Generally, sympathetic resonance has been well overcome by controlling design features of gear boxes and other mechanisms. Ground resonance always has been a knotty problem. This is a self-excited vibration that develops when the landing gear repeatedly strikes the ground, thus unseating the center of mass of the main rotor system. The pounding effect of the landing gear is prone to occur during take-off and landing when the helicopter is from 87 to 93% airborne. The aircraft, being light on the landing gear, bounces from one wheel to another in rapid succession, setting up a pendular oscillation of the fuselage. The succession of shocks is transmitted to the main rotor system, and the main rotor blades straddling the pounding wheel are forced to change their angular relationship. This condition unbalances the main rotor system, which in turn transmits the shock back to the landing gear. To control this potentially damaging condition, various dampening devices are used to control the unbalancing of the rotor system, and helicopter pilots are trained to avoid critical maneuvers conducive to agitating ground resonance.

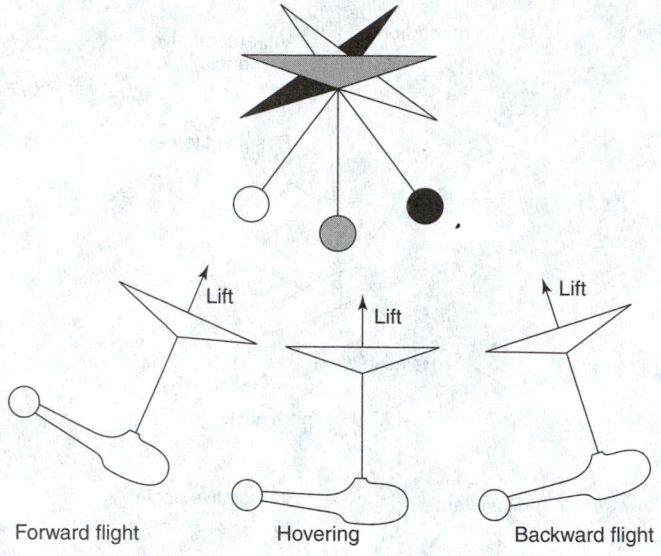

Fig. 19.   Pendular action.

Operational Helicopters

Although helicopters find numerous applications–in commercial aviation for commuting, for reconnaissance by law enforcement and civilian emergency agencies (for example, in transporting the injured from the site of an accident to hospital), by the news media for coverage of special events and routinely for reporting on traffic congestion, and by the operators of large farming and ranching operations, among other important uses–helicopters find their major application by the military. As just one example of military interest in helicopters, in the late 1950s, the U.S. Navy paid special attention to improving its antisubmarine warfare capabilities. The integration of carrier-based and land-based fixed wing aircraft with helicopters and surface units proved to be a giant step in the right direction. The amphibious assault mission made great strides with the concept of vertical assault, which employed helicopters to speed materials and personnel from shipboard to points ashore. Thus, the 1960s witnessed the passing of the lighter-than-air vehicles and flying boats from the Navy's inventory. While the importance of the helicopter has not diminished, much attention in recent years has also been given to VTOL craft. The role of the helicopter in the Vietnam War hardly requires Additional emphasis here. It was the principal tool used during that war for moving Army equipment and personnel.

The first American helicopter with metal rotor blades (the Sikorsky S-52) was originally flown in two-seat form on February 12, 1947. It was powered by a 178 hp Franklin engine and was claimed to have been the first helicopter to have performed a loop. In later developments for the Army and Marine Corps, the S-52 spawned several models, some of which are exhibited today at the U.S. Army Aviation Museum, Fort Rucker, Alabama. Other firms active early in the helicopter field included Bell Aircraft and Hughes Aircraft, among others.

An abridged list follows of operational helicopters, most of which are periodically modified, introduced during the last twenty years.

Sikorsky S-65 (heavy assault helicopter). The "Sky Crane." The HH-52 amphibious helicopter for search and rescue missions. The S-70 (U.S. Army UH-60A), an 11-seat troop transport also used for medical evacuation, battlefield command and control, and as a reconnaissance aircraft. Lifting capacity of cargo hook, 8000 pounds (3600 kg). The S-76 designed for civil aviation, carrying 12 passengers for a distance of about 460 mi (740 km), with a model especially adapted for use in offshore rig support operations, having auxiliary fuel tanks which permit the craft to transport 8 passengers for more than 690 mi (1110 km).

Bell UH-1 (also called "Huey" or "Iroquois"). Introduced as a multipurpose craft in the 1970s. Maximum weight on takeoff, 10,000 pounds (4536 kg), maximum speed of 121 mi (195 km) per hour, range of about 300 mi (483 km), ceiling of 11,500 ft (3505 m), requiring a crew of one and carrying 14 passengers. The OH-58A ("Kiowa"), introduced as a multipurpose craft in the late 1960s. Rotor blade diameter, 35.4 ft (10.8 m), maximum weight on takeoff, 3000 pounds (1361 kg), range of 350 mi

(563 km), ceiling, 19,000 ft (5791 m), crew of 2, passengers 2. The AH-1 ("Cobra"), a slim-bodied gunship with a fighterlike cockpit seating a gunner in front and pilot above and behind. The craft accommodates diverse forms of armament. Rotor blade diameter, 44 ft (13.4 m), maximum weight on takeoff, 10,000 pounds (4536 kg), maximum speed 210 mi (338 km) per hour, range 360 mi (579 km), ceiling 10,550 ft (3216 m), combat crew of 2.

Dornier Do 132. Introduced in Germany in 1971. A passenger helicopter, one 720-horsepower engine; rotor blade diameter, 35.1 feet (10.7) meters); length, nearly 25 feet (7.6 meters); height, just over 9 feet (2.7 meters); empty weight, about 1500 pounds (680 kilograms); maximum weight on take-off, about 3635 pounds (1649 kilograms); maximum speed, just over 140 miles (225 kilometers) per hour; range, 275 miles (442 kilometers); crew of 1; passengers, 4.

Augusta A-106. Introduced in Italy in late 1960s as an antisubmarine helicopter; one 350-horsepower turbine engine; rotor blade diameter, just over 31 feet (9.4 meters); height, just over 8 feet (2.4 meters); length, nearly 29 feet. (8.8 meters); empty weight, about 1520 pounds (689 kilograms); maximum weight on take-off, about 3085 pounds (1399 kilograms); maximum speed, 110 miles (177 kilometers) per hour; range, 460 miles (740 kilometers); crew of 1 (total).

Aerospatiale-Westland SA-300. Introduced in France, operational in 1970; transport helicopter; two 1320-horsepower turbine engines; rotor blade diameter, just over 49 feet (14.9 meters); length, just over 46 feet (14 meters); height, nearly 14 feet (4.3 meters); empty weight, about 7560 pounds (3429 kilograms); maximum weight on take-off, 14,110 pounds (6400 kilograms); maximum speed, about 175 miles (282 kilometers) per hour; range, nearly 400 miles (644 kilometers); ceiling, 15,750 feet (4801 meters); crew of two; passengers, 16.

Westland WG 13N Lynx. Introduced in Great Britain in the 1970s; passenger helicopter; two 900-horsepower turbine engines; rotor blade diameter, 42 feet (12.8 meters); length, just over 38 feet (11.6 meters); height, just over 11 feet (3.4 meters); empty weight, nearly 7500 pounds (3402 kilograms); maximum weight on take-off, about 8780 pounds (3983 kilograms); maximum speed, nearly 185 miles (298 kilometers) per hour; range, 150 miles (241 kilometers); crew of two; plus passengers.

EH101. An Anglo-Italian three-engined helicopter under construction, as of the late 1980s, by Westland and Augusta. Nicknamed the "Iron Bird," the craft is in an experimental stage. The rotor blades, made from composites, permit an advanced aerodynamic airfoil with high-speed blade end. The rotor is equipped with powered blade folding, with provision for a backup manual fold facility.

V/STOL Craft

The concept of taking off and landing vertically inspired early efforts on helicopters just described. Attempts during the early 1900s included: (1) The Gyroplane, designed and built in Europe in 1907 by Bréguet and Richet, managed to lift only a few feet off the ground. The engines were inadequate. Following these experiments, Bréguet decided to concentrate, and successfully, on the design of more conventional airplanes. (2) Frenchman Paul Cornu, while aboard his craft, succeeded in lifting a fragile design off the ground about one foot for a period of about 20 seconds in 1907. (3) Igor Sikorsky, in 1909, constructed a helicopter prototype using an Anzani engine, but the latter proved inadequate to lift the craft off the ground. For several years thereafter, Sikorsky concentrated successfully on the design of seaplanes. (4) In Denmark, Ellehammer, a designer of prior fixed-winged airplanes, managed to lift his helicopter design prototype a few inches above the ground in 1916. This craft was equipped with two coaxial rotors. (5) Frenchman Etienne Oemichen succeeded with his design to take off vertically for a flight of a few hundred yards in mid-1924; (6) The Marquis Raul Patteras Pescara (Spain) in the same year managed a flight of 2,415 feet (736 meters). Pescara continued with other models, but ceased activity upon the successful demonstration of Juan de la Cierva's autogiro which, at that time, appeared to provide the answers being sought from helicopters; (7) In 1930, Italian D'Ascanio's design, piloted by Marinello Nelli, broke the prior distance record, achieving a flight of some 3,535 feet (1077 meters), and at a record height of 59 feet (16 meters) above ground. But probably success for the first practical helicopter designs should go to the Germans with the development of the Focke-Wulf FW-61, which appeared in 1936; and one year later when Igor Sikorsky built and flew the VS-300.

Autogiro. This craft was developed in Spain and flown in 1923. The autogiro consists of a wingless fuselage mounting a pylon which contains a rotating head to which are affixed three or four balanced blades of airfoil section resembling a propeller configuration. The blades are rotated by the action of the relative wind and thus are self-rotating, hence the name of the craft. The autogiro is equipped with a regular power plant and propeller that produce the necessary forward thrust to get the craft in motion and keep it in motion when in flight. The angle of the rotor to the fuselage is controlled by the pilot, and this takes the place of the normal control surfaces of the conventional airplane. The blades rotate at speeds that give an average air velocity over them considerably in excess of the autogiro's airspeed, and so the autogiro may be flown at speeds lower than the stalling speed of the airfoil section. This is an advantage, in that it permits the autogiro to land in small fields, with short landing runs, to descend almost vertically, and to approach "hovering" flight. Although autogiro development essentially has been abandoned, primarily because of the successes of later helicopter designs, much of the information gained in autogiro experiments has proven of much value to helicopter pioneers and later VTOL designers.

During the past few decades, a number of V/STOL have been proposed and built, either experimentally or in production. Principal configurations include the following. (1) Compound or convertible aircraft—essentially helicopters with propellers for horizontal flight. At take-off, the engines power the rotors; at altitude, the power is transferred to the propellers. A serious disadvantage of this concept is the substantial drag presented by the rotors during horizontal flight. (2) Vertiplanes—large, propeller-equipped airplanes, with a jet engine for effecting vertical take-off. In appearance, past designs of these craft look like an essentially conventional fixed-wing airplane. (3) Tilt-Prop or Tilt-Wing Airplanes—essentially airplanes with wings and jets or propellers, either of which can be tilted by 90 degrees during take-off and landing. (4) Bidirectional Jet Aircraft—jet-propelled aircraft in which the jet engine(s) can be directed downward for vertical take-off and turned toward the rear for horizontal flight when at altitude.

The AV-8B V/STOL. Developed for the United States Marine Corps, this aircraft is a single-seat, single-turbofan-powered craft for close air support and interdiction missions. The craft is powered by a Rolls Royce Pegasus 11 vectored thrust turbofan, with 21,500 pounds of thrust (95,600 N) without afterburning. The length is 46.3 feet (14.1 meters); height is 11.6 feet (3.5 meters); the wingspan is 30.3 feet (9.2 meters); and the wing area is 230 square feet (21.4 square meters). Take-off distance is 0–1100 feet (0–305 meters). The first flight of the YAV-8B prototype occurred in November 1978. Combat range is 600+ nautical miles (1112 kilometers); ferry range is 2460 nautical miles (4558 kilometers). See Fig. 20. The aircraft can carry armament up to 9200 pounds (4175 kilograms) on several

Fig. 20.   U.S. Marine Corps' AV-8B V/STOL aircraft for close air support and interdiction missions. (*McDonnell Aircraft Company; McDonnell Douglas Corporation.*)

stations: two 30 millimeter cannons; laser or TV-guided weapons; as well as up to four 300-gallon (1136-liter) fuel tanks. The Marine Corps V/STOL close air support uses three types of land sites as well as a variety of ships. Closest to the battle area would be a number of V/STOL forward sites, accommodating two to four AV-8Bs. These sites would normally have fuel and ordnance for turnaround operations only. If a STO (short take-off) strip is not available, the aircraft would operate in the VTOL (vertical take-off and landing) mode. Located about 50 miles (80 kilometers) from the battle-field would be a V/STOL facility with at least a 600-foot (182-meter) strip, providing six to ten aircraft turnaround, support, and maintenance. The V/STOL main base would have at least a 1500-foot (457-meter) strip and all the logistics and support assets for prolonged AV-8B operations.

The AV-8B's vertical takeoff and landing ability is derived from four exhaust nozzles—two on either side of the aircraft—positioned around the plane's center of gravity. These nozzles can be rotated from the full-aft position, for forward flight, to a full-down position for vertical operations. Within the engine, some rotating parts turn clockwise while others turn counterclockwise. This is necessary to prevent gyroscopic effects, which, if all parts rotate in the same direction, can make the aircraft difficult or impossible to control during hover, and during transition from hover to forward flight or from forward flight to hover.

All that is required for take-off or landing is an amphibious assault ship, or a clearing large enough for a 72-foot (22-meter) square aluminum mat, a section of two-lane road, or even a damaged airfield. As with the trend in the design and construction of modern aircraft, much stress is being given to composites and graphite epoxy materials. See Fig. 21.

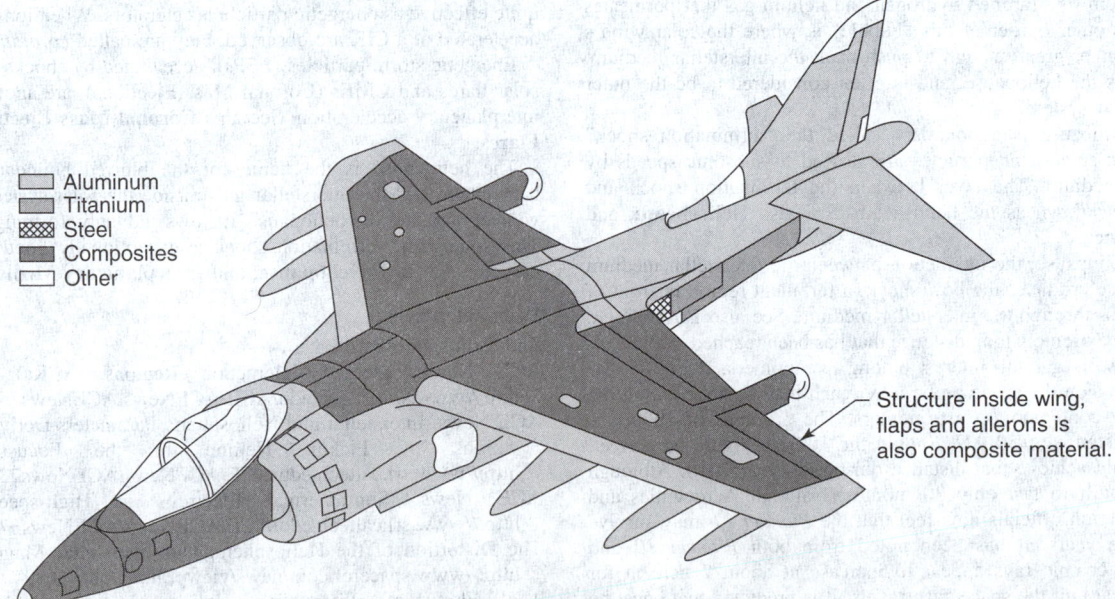

Aluminum
Titanium
Steel
Composites
Other

Structure inside wing, flaps and ailerons is also composite material.

Fig. 21.   Types of materials used in the AV-8B V/STOL aircraft. As a percent of structural weight: aluminum, 48.4%; titanium, 8.5%; steel, 14.5%; composites, 23.3%; other materials 5.3%. Pylons not included in these Figures. Approximately 1186 pounds (540 kg) of graphite epoxy are used in the structure. (*McDonnell Aircraft Company; McDonnell Douglas Corporation.*)

VTOL Craft

For many years, aircraft designers have been seeking a vehicle that would fly like a rotary wing aircraft in a hover and fly like a fixed wing aircraft in up-and-away flight. The "X-Wing," currently under development by NASA (National Aeronautics and Space Administration), DARPA (Defense Advanced Research Projects Agency), and Sikorsky Aircraft in a joint program, is the most promising concept as of the late 1980s for performing both the traditional airplane and helicopter roles. The unique aspect of the X-wing is that it derives its lift for both hover and forward flight from the same airfoil. In August 1986, the X-wing prototype was delivered by Sikorsky to NASA's Dryden facility at Edwards Air Force Base for flight testing.

As pointed out by Brahney, the most challenging part of X-wing aerodynamics is the conversion between helicopter and fixed wing modes of flight. The retreating side of the rotor disk must continue to support its share of total system lift, even as its airflow is reversing direction. As the rotor/wing slows, the azimuthal segment with the leading edge blowing grows until it encompasses the entire left side of the disk. Thus, on the left side of the disk, what was the leading edge in helicopter flight becomes the trailing edge in the fixed wing mode.

Air is fed out of the compressor into a plenum. The inner wall of the plenum has two rows of 24 valves. The top row supplies the leading edges; the bottom row feeds the trailing edges. The valves and plenum do not rotate with the rotor system. What does rotate is a series of 8 receiver ducts (four for leading edges and four for trailing edges) inside the plenum which receive the valve-modulated air from the plenum and feed it to the individual blades. This rather complex system is described in detail in the Brahney (1986) reference listed.

Additional Reading

Brahney, J.H.: "75 Years of Naval Aviation," *Aerospace Engineering*, 34–39 (May 1987).

Leishman, J.: *Principles of Helicopter Aerodynamics*, Cambridge University Press, New York, NY, 2000.

McCormick, B.: *Aerodynamics of V/Stol Flight*, Dover Publications, Inc., Mineola, NY, 1999.

Prouty, R.: *Helicopter Performance, Stability and Control*, Krieger Publishing Company, Melbourne, FL, 1995.

HELIOPAUSE. This is defined as that distant location in outer space at which the sun's influence ends and is replaced by the interstellar medium. For a number of years, astronomers have estimated that the heliopause occurs between 50 and 100 astronomical units (AU) from Earth. [One AU 5 approximately 93,000,000 miles (150 million km)]. Some experts have revised this estimate to between 85 and 100 AU.

The solar wind blows a "bubble" known as the heliosphere in the interstellar medium (the rarefied hydrogen and helium gas that permeates the galaxy). The outer border of this "bubble" is where the solar wind's strength is no longer great enough to push back the interstellar medium. This is known as the heliopause, and is often considered to be the outer border of the solar system.

Inside the heliopause is a boundary called the "termination shock" where supersonic solar wind particles are slowed to subsonic speeds by the interstellar medium. The layer between the termination shock and the heliopause is known as the heliosheath. See also **Heliosheath**; and **Termination Shock**.

Outside the heliopause, the interaction between the interstellar medium and the heliopause produces the bow shock, a turbulent region in front of the Sun's progress through the interstellar medium. See also **Bow Shock**.

Because of the extremely long distance that has been reached by *Pioneer 10*, the heliopause is gaining interest among astronomers. If the 85 AU estimate is close, then the space probe may reach that distance within the lifetime of many contemporary astronomers. The probe as of 1990 was estimated at a distance of 50 AU. Should the 100 AU figure be correct, *Pioneer 10* should achieve that distance during the year 2010. Although the probe was built to last only 30 months, National Aeronautics and Space Administration officials now feel that the *Pioneer 11* may survive for several more years. It has been noted from both *Pioneer 10* and *Pioneer 11* that cosmic rays appear to increase at about 2 percent for every astronomical unit the spacecraft travels. It is predicted that once the heliopause is reached, the cosmic radiation will be constant with time.

The current mission of the Voyager 1 and 2 spacecraft is to find and study the termination shock, heliosheath, and heliopause. Voyager 1 crossed the termination shock and entered the heliosheath in December 2004, at 94 AU. It is expected that Voyager 1 will reach the heliopause in about 2015. Voyager 2 could cross the termination shock between 2008 and 2010 and reach the heliopause about 10 years later.

See also **Interplanetary Medium**; and **Voyager Missions to Jupiter and Saturn**.

Web Reference

Voyager Interstellar Mission: http://voyager.jpl.nasa.gov/Proposal-2005/VgrProp05.pdf

HELIOSEISMOLOGY. See **Meteorology**.

HELIOSHEATH. The "heliosheath" is the zone between the termination shock and the heliopause at the outer border of the solar system. It lies along the edge of the heliosphere, a "bubble" caused by solar winds. See also **Heliopause**; **Solar Wind**; and **Termination Shock**.

The heliosheath's distance from the Sun is approximately 80 to 100 astronomical units (AU). The current mission of the Voyager 1 and Voyager 2 space probes includes studying the heliosheath. See also **Astronomical Unit**; **Interplanetary Medium**; and **Voyager Missions to Jupiter and Saturn**.

In May 2005, NASA announced that Voyager 1 had crossed the termination shock and entered the heliosheath in December 2004, at a distance of 94 AU.

See http://www.nasa.gov/vision/universe/solarsystem/voyager_agu.html.

HELIOSPHERE. "Helio-" means having to do with the Sun. The heliosphere is the immense magnetic bubble containing our solar system, solar wind, and the entire solar magnetic field. It extends well beyond the orbit of Pluto. While the density of particles in the heliosphere is very low (it's a much better vacuum than is created in a laboratory), it is full of particles of interest to heliospheric scientists. See Fig. 1. below for a diagram of the heliosphere.

The solar wind near our Sun's surface contains alternating streams of high and low speed. These streams *corotate* with the Sun, that is, they rotate along with it. The high-speed streams originate in coronal holes and extend toward the solar poles; the low-speed streams come from near the Sun's equator. There are compositional differences between the high and low speed streams of the solar wind. See also **Coronal Holes**; and **Solar Wind**.

With increasing distance from the Sun, the high-speed streams overtake the slower plasma, producing Corotating Interaction Regions (CIRs) on their leading edges. CIRs are bounded by two shocks at the front and rear edges called the *forward* and *reverse* shocks. At these shocks, the density, pressure, and magnetic field strength are all higher. These regions are quite effective as energetic particle accelerators. When ions that have been accelerated at a CIR are observed, they are called *corotating ion events*.

Energetic storm particles (ESPs), accelerated by shocks associated with solar flares and CMEs (Coronal Mass Ejections), are another example of interplanetary acceleration. See also **Coronal Mass Ejections**; and **Solar Flares**.

The heliopause is the name for the blurred boundary between the heliosphere and the interstellar gas outside the solar system. As the solar wind approaches the heliopause, it slows suddenly, forming a shock wave. This solar wind termination shock is exceptionally good at accelerating particles. See also **Heliopause**; and **Interplanetary Medium**.

Web References

The Heliosphere in the News

ACE News: Merged Interaction Regions (MIRs) at 1 AU: http://www.srl.caltech.edu/ACE/ACENews/ACENews75.html

ACE News: Interstellar and Heliospheric Parameters Derived from Observations of Pickup Helium in the Focusing Cone: http://www.srl.caltech.edu/ACE/ACENews/ACENews73.html

ACE News: Suprathermal Electrons in High-speed Streams: http://www.srl.caltech.edu/ACE/ACENews/ACENews67.html

The Distortion of the Heliosphere: Our Interstellar Magnetic Compass: http://www.spaceref.com/news/viewpr.html?pid=16394

The Flow of Interstellar Helium in the Solar System: http://www.spaceref.com/news/viewpr.html?pid=15256

Ulysses, Fifteen Years and Going Strong: http://www.spaceref.com/news/viewpr.html?pid=17993

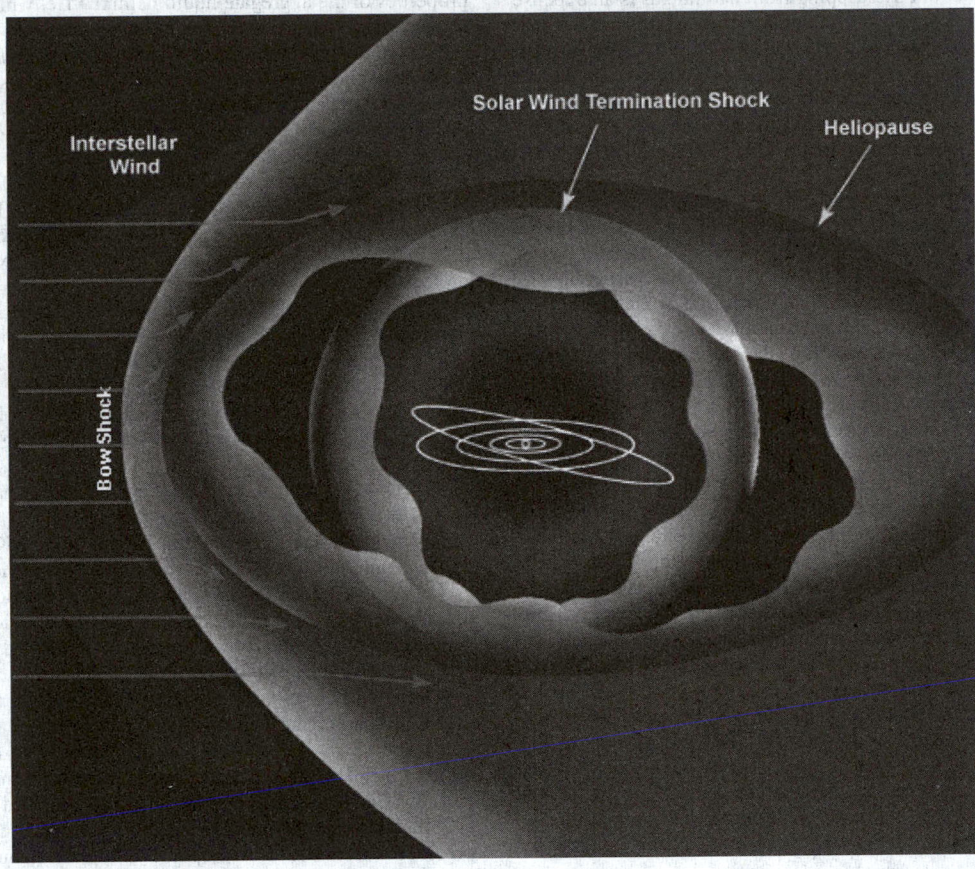

Fig. 1.   Diagram of heliosphere. (*NASA*)

Ulysses Mission Extended: http://www.spaceflightnow.com/news/n0402/
    14ulysses/
Voyager Finds Three Surprises Near Our Solar System's Edge:
    http://www.physorg.com/news6788.html
Voyager Spared NASA's Budget Ax: http://dsc.discovery.com/news/briefs
    /20051219/voyager_spa.html

HELIOSTAT.   An arrangement of mirrors, driven by clockwork, used to
reflect a beam of sunlight in a fixed direction as the sun moves across the
sky. The heliostat is used in the control of some astronomical instruments
as well as in some solar energy systems for tracking the sun. See also
Solar Energy.

HELIOTROPE.   See **Bloodstone**.

HELIOTROPIC WIND.   See **Winds and Air Movement**.

HELIUM.   [CAS: 7440-59-7] Chemical element symbol He, at. no. 2, at.
wt. 4.0026, periodic table group 18 (inert or noble gases), mp $-272.2\,^\circ$C
(20 atmospheres), bp $-268.93\,^\circ$C (4.2144K), specific gravity 0.124 at
4.2144 K. The element has no triple point and can be solidified only by
applying high pressure to the liquid phase. Described later, liquid helium
undergoes a change in its physical properties at 2.178 K, known as the
lambda point. Solid helium has a close-packed hexagonal crystal structure
(subject to further study and confirmation). At standard conditions, helium
is a colorless, tasteless, odorless gas. There are two natural isotopes ^{3}He
and ^{4}He, with ^{4}He being slightly less than 100% abundant. The boiling
point is 3.2 K for ^{3}He. Radioactive ^{5}He and ^{6}He have extremely short
half-lives. See also **Radioactivity**. The first ionization potential for helium
is 24.58 eV; second, 54.14 eV. Other physical properties of helium are
described under **Chemical Elements**.

Like the other rare gases, helium exhibits negative chemical properties
with ordinary materials under normal conditions. Under the influence of
electric glow discharge or electron bombardment, helium forms compounds
with tungsten and other metals, as well as with iodine, sulfur, and
phosphorus. In a vacuum electric discharge tube helium shows green to
canary-yellow glow. Discovered first in the vapors surrounding the sun
by Lockyer in 1868, through the yellow spectral line near the two yellow
lines of sodium, then by Ramsay in 1895 in the mineral clevite.

Helium occurs (1) in minerals of uranium and thorium, such as
clevites, pitchblende, carnotite, monazite, and also in beryl, (2) in mineral
waters (1 part He per thousand of water, in some Iceland waters), (3) in
volcanic gases, (4) especially in certain natural gases of the United States.
The first discovery of this kind was made in Kansas.

Uses.   Industrially, helium is used: to provide an inert gaseous shield for
arc welding, for growing transistor crystals, in the production of titanium
and zirconium, to fill the space between optical lenses in instruments, as the
carrier gas in some chromatographic apparatus, as a liquid bath for masers
and cryotrons, as a refrigerant for furnishing the low temperature required
for superconducting electrical equipment, in lasers, as a diluent gas in deep-
sea diving applications, as a heat-transfer medium in gas-cooled nuclear
reactors, and as a leak-detecting medium for testing pressure and vacuum
equipment. Now, to a rather limited extent, helium is used as a lifting
gas for airships and for balloons used in meteorological investigations.
Helium is used in aerospace programs in several ways, including its use
in propellant tanks as a compressed gas which expands and takes the
place of fuel as the fuel is consumed, in ground-support equipment, and
in communication satellites for providing the low temperature required for
sensitive electronic systems. In medicine, helium sometimes is mixed with
oxygen for patients with certain respiratory ailments and also it is mixed
with certain anesthetics to reduce the hazards of forming an explosive
mixture with air.

Future possible uses of helium have a direct influence on the conservation of helium resources. The known resources are not large in comparison
with most other raw materials, and most authorities are of the opinion
that conservation measures should be continued. However, most helium
demand projections have proved overly optimistic and consequently have
lessened the pressure for conservation. Natural gas streams, of which He is

but a minor constituent, are produced commercially for sale and consumption as fuel. To separate helium by stripping from natural gas is an expense that is not attractive to the producers of natural gas. Thus, it is evident that helium conservation, under current supply/demand conditions, must stem from government regulation. Even though stripping helium from natural gas is costly, later costs to recover helium lost to the atmosphere from burning He-bearing natural gas would be many times greater.

The history of helium conservation measures dates back to 1925 when the U.S. Congress passed the Helium Act of 1925. Congress amended the act in 1960 to provide for stripping natural gas of its helium, for purchase of the separated helium by the government, and for its long-term storage. In 1971, after about 28 billion cubic feet had been stored (in a federally owned gas field called Cliffside near Amarillo, Texas), the purchase program was terminated by the government, an action that, as reported by Hammel et al., unleashed several lawsuits and not a little acrimony. As of the present, most of the litigation has been concluded, much of the He that could have been saved has been wasted to the atmosphere, and the gas fields supplying the He are almost depleted. However, in the meantime, a new and rich source of He has been discovered in southwestern Wyoming that could ensure adequate supplies for many decades if an appropriate new federal policy on He were developed and implemented. The new field, first explored by Mobil in 1960, led to an initial estimate of 3 to 15 billion cubic feet of He in the Tip Top drilling unit of that field. However, the natural gas from that unit was initially judged unfit for sale as a natural gas fuel. The borehole was cemented shut and the well abandoned. In the early 1980s, it was established by Additional drilling that the amount of He recoverable from the Wyoming field is at least 200 billion cubic feet. Left untouched this would represent an excellent long-term helium reserve. More than 90% of this field lies within federal land boundaries.

With increasing incentives because of natural gas pricing (Natural Gas Policy Act of 1978), several firms have plans to drill and develop nearly 250 deep wells, from which about 2.8 billion cubic feet of acidic gas per day would be produced. If private developers elect not to conserve the helium from this project (Riley Ridge Natural Gas Project), it is estimated that about 5 billion cubic feet of He would be vented to the atmosphere each year.

Unexpected Uses for Helium. Hammel (1984) points out that new uses for He continue to emerge. These include a 49-meter-diameter He-filled sphere that has been proposed as a lighter-than-air hoist capable of moving loads in excess of 90 tons. Another use is in superconducting magnets for imaging with nuclear magnetic resonance. Proposals for the use of large amounts of He continue to be made in the national security area.

Origin of Helium. Helium has a geologic occurrence and distribution unique among the elements. It is a product of radioactive disintegration of uranium and thorium within the earth's mantle and crust. But it flows to the surface at a rate less than that of its generation, because most of it is driven into crystal structures of rock minerals until released by alpha radiation damage near radioactive concentrations. Mobile helium rising through the crust may then be trapped, along with other gases, beneath relatively impermeable barriers. Nitrogen is almost always associated with helium in natural gases, although this has not been fully explained. Also, carbon dioxide is abundant in some helium-rich gas mixtures.

Liquefaction of Helium. This was accomplished by Onnes in 1908 in Leiden, and Keosom in 1926 succeeded in solidifying helium in the same laboratory. Relatively recently, helium has been solidified at room temperature. The melting pressure at $24\,^{\circ}C$ is 115 kilobars, in complete agreement with the Simon equation. Besson and Pinceaux (1979) developed an original apparatus for the experiment, which allowed loading of the cell at room temperature. Diamond anvil cells were used in the procedure.

Liquid Helium II. Upon cooling, ^{4}He liquefies at atmospheric pressure at 4.216 K to form an essentially normal liquid, liquid helium I. On further cooling to the lambda-point, 2.178 K at one atmosphere, a change occurs to liquid helium II. The latter has a very low viscosity (hence the name "superfluid") and a very high thermal conductivity, which produce such phenomena as the creeping of a film over the edge of the container, and the fountain effect, in which the liquid sprays out of a capillary. Superfluidity is commonly explained in terms of a two-fluid theory. Thus, London and Tirza attribute the properties of helium II to a mathematical peculiarity in the distribution function of Bose-Einstein statistics, whereby below the λ-point, a finite fraction of the atoms fall into a ground state of zero thermal energy. In this state they would have the properties of a superfluid.

However, this theory has not yielded good quantitative predictions of the properties of the aggregate liquid helium. ^{3}He, which follows Fermi-Dirac statistics, does not have a superfluid state.

Landau treats liquid helium by an approach similar to that of the Debye theory of solids. The longitudinal and transverse sound waves, which are the elementary excitations of that theory of solids, correspond in the case of liquid helium to phonons and rotons. The *phonons* are the longitudinal sound waves, while the *rotons* are another type of elementary excitation postulated by Landau to represent the rotational motion of the liquid, because a liquid cannot support transverse waves. The specific heat can be expressed as the sum of contributions from phonons and rotons. Landau derived expressions for these that fit the data and experiments quite closely up to 1.6 K.

Feynman developed wave functions to provide an atomistic interpretation of Landau's spectrum of elementary excitations.

The complexity of the helium II problem is apparent at once when one attempts to extend the equations of classical hydrodynamics to this two-component system, in which each component has its own density and velocity. Khalatnikov derived such equations by ignoring terms of second order.

Still another area of investigation has been that of the properties of ^{3}He-^{4}He mixtures. As stated above, ^{3}He exhibits no λ-transition and no superfluidity. It has a critical temperature of 3.35 K and a boiling point of 3.2 K, against values of 5.2 K and 4.216 K for ^{4}He.

The most abundant helium atoms, ^{4}He, are bosons, but the ^{3}He atoms are fermions. This has a consequence that liquid ^{3}He does not show superfluidity—a property very probably connected with the Bose-Einstein statistics obeyed by the ^{4}He atoms.

Donnelly and associated researchers (University of Oregon) has observed that in the future liquid helium rather than air may be used in a much down-scaled wind tunnel, perhaps with experiments conducted within the space of an average room versus current, very large wind tunnels. It is envisioned that a tunnel could be filled with superfluid liquid helium, taking advantage of the liquid's absence of viscosity and friction as previously described. Based upon quantum-mechanical factors, a source of heat in the tunnel could cause extremely fast currents to flow in the liquid.

Peterson (reference listed) reported in early 1991 that researchers at Harvard University made what is considered a remarkable prediction regarding the energy-level transitions that occur in a helium atom. The agreement between theoretical calculations and experimental results show that computational methods for constructing a model of a two-electron atom can work, thus bridging the gap between theory and practice.

Chemistry. The most striking properties of helium are: its emission as the positively charged (+2) alpha particles in radioactive changes, its formation in radioactive change by uranium-radium and thorium-containing substances, emitting alpha particles, later losing the charge to become helium, and its production artificially by bombardment of lithium or boron with high-velocity protons or alpha rays.

Unlike the other inert gases, helium gives little evidence of compound formation with organic substances. Like neon, but unlike the others, it forms no hydrate. However, it forms compounds much more readily under excitation, due apparently to unpairing of its $1s$ electrons and promoting of one of them to the $2s$ state. The 460 kcal/g-atom of energy is readily obtained by electric discharge or electron bombardment. Under such conditions the helium molecule-ion, He_2^+, with a pair of bonding electrons ($1s$) and a single antibonding electron ($1s$), is formed, as are combinations of the type of HeH^+ and HeH_2^+. In a mercury discharge tube, the compound $HgHe_{10}$ has been found, and with various metallic electrodes corresponding helides, such as the compounds of tungsten, platinum, iron, palladium, bismuth, etc., e.g., WHe_2, Pt_3He, $FeHe$, $PdHe$, $BiHe$, etc., have been formed.

Additional Reading

Andronikashvili, E.L. and A.K. Ishkhneli: *Reflections on Liquid Helium*, American Institute of Physics, College Park, MD, 1998.

Besson, J.M., and J.P. Pinceaux: "Melting of Helium at Room Temperature and High Pressure," *Science*, **206**, 1073–1075 (1979).

Donnelly, R.J.: "Superfluid Turbulence (Helium)," *Sci. Amer.*, 100 (November 1988).

Epstein, A.W.: "Cool Breeze: A Helium Superwind for Wind-Tunnel Experiments," *Sci. Amer.*, 30 (May 1990).

Greenwood, N.N. and A. Earnshaw: *Chemistry of the Elements*, 2nd Edition, Butterworth-Heinemann, Inc., Woburn, MA, 1997.

Hammel, E.F., M.C. Krupka, and K.D. Williamson, Jr.: "The Continuing U.S. Helium Saga," *Science*, **223**, 789–792 (1984).

Krebs, R.E.: *The History and Use of Our Earth's Chemical Elements: A Reference*, Greenwood Publishing Group, Inc., Westport, CT, 1998.

Lide, D.R.: *CRC Handbook of Chemistry and Physics*, 88th Edition, CRC Press, LLC., Boca Raton, FL, 2007.

Parker, P.: *McGraw-Hill Encyclopedia of Chemistry*, 2nd Edition, The McGraw-Hill Companies, Inc., New York, NY, 1993.

Peterson, I.: "Helium Theory Gets High-Precision Test," *Science News*, 86 (February 9, 1991).

Vollhardt, D. and P. Wolfle: *The Superfluid Phases of Helium 3*, Taylor and Francis, Inc., New York, NY, 1990.

Volovik, G.E.: *Exotic Properties of Superfluid Helium Three*, World Scientific Publishing Company, Inc., Riveredge, NJ, 1992.

HELIUM LEAK DETECTION. See **Mass Spectrometry**.

HELIX. A space curve traced on a cylinder or conical surface in such a way that all elements of the surface are cut at a constant angle. A circular helix lies on a right-circular cylindrical surface. In parametric form, its equation is $x = a\cos\theta$, $y = a\sin\theta$, $z = b\theta$ where a, b are constants and θ is the parameter. The thread of a screw is often a circular helix.

See also **Conical Surface**.

HELIX FEEDER. See **Feeder (Volumetric)**.

HELLBENDER *(Amphibia, Urodela)*. A large aquatic salamander, *Cryptobranchus allegheniensis*, of the Mississippi river system. It reaches a length of 18 inches (46 cm) and has a flattened head and body, short legs, and a compressed tail. The gills are concealed, but otherwise it resembles the mudpuppy.

HELLGRAMMITE *(Insecta, Neuroptera)*. The large aquatic larva of the dobson fly, *Corydalus*. It lives in running water and is an excellent bait for bass.

HELMHOLTZ EQUATION. An equation of the form

$$n_1 y_1 \tan\theta_1 = n_2 y_2 \tan\theta_2$$

expressing the relation between the linear and the angular magnification at a spherical refracting interface. y_1, y_2 are linear dimensions of object and image, θ_1, θ_2 the angles made by focal rays and axis at object and image points and n_1, n_2 are refractive indices of object and image space. Also called Lagrange-Helmholtz equation. (See, however, the **Abbe Sine Condition**.) A spherical surface cannot satisfy both these equations for finite angles. Hence a spherical surface can never make a perfect image.

HELMHOLTZ FREE ENERGY. A thermodynamic function of state that, in a reversible isothermal process, increases with work done on the system. Also called *Helmholtz function*, and *work function* .

In typical notation the Helmholtz free energy is

$$F = U - TS,$$

where F is the Helmholtz free energy (sometimes designated as A), U is the internal energy, T is temperature, and S is the entropy. By use of the first law of thermodynamics for reversible processes, the rate of change of the Helmholtz free energy is given by

$$\frac{dF}{dt} = -S\frac{dT}{dt} - \frac{dW}{dt},$$

where W is the work done by the system.

HELMHOLTZ FUNCTION. See **Thermodynamics**.

HELMHOLTZ, HERMANN VON (1821–1894). Helmholtz was a German physician and scientist who made contributions to physiology, optics, acoustics, mathematics, meteorology, and electrodynamics. His early professional life was spent as a surgeon in the Prussian army.

Helmholtz is well known for publishing the world's first treatise on the principle of energy conservation and for his trichromatic theory of color vision. He proved that any color of the spectrum could be matched by various amounts of three primary colors that were red, green, and blue.

He is credited with two important inventions, the ophthalmoscope to study the eye's interior, and the ophthalmometer to measure the eye's accommodation to changing conditions.

In 1871, Helmholtz became professor of physics at the University of Berlin. Almost until his death, his main focus during his latter years was on electrodynamics.

See also **Equation of State**; **Free Energy**; **Helmholtz Equation**; **Helmholtz Resonator**; **Helmholtz Theorem**; and **Thermodynamics**.

J. M. I.

HELMHOLTZ INSTABILITY. The hydrodynamic instability arising from a shear, or discontinuity, in current speed at the interface between two fluids in two-dimensional motion. Also called *shearing instability* .

The perturbation gains kinetic energy at the expense of that of the basic currents. According to the theory of small perturbations, waves of all wavelengths on such an interface are unstable, their rate of growth being $e^{\mu t}$ with μ given by

$$\mu = \frac{\pi}{\lambda}|U - U'|,$$

where λ is the wavelength and U and U' the current speeds of the two fluids. Such waves are called Helmholtz waves or shear waves, and move with a phase speed c equal to the mean of the current speeds,

$$c = \frac{1}{2}(U + U').$$

With an assumed density difference in the fluids, gravity waves may also be generated. The combination of these effects yields a critical wavelength λ_c,

$$\lambda_c = \frac{2\pi}{g}\left(\frac{\rho\rho'}{\rho^2 - \rho'^2}\right)(U - U')^2,$$

where and ' are the densities of the lower and upper fluids, respectively. Waves shorter than the critical are unstable, longer waves, stable. This analysis has been applied to billow clouds; however, the critical wavelength is considered too small (of the order of a few kilometers) for this sort of instability to be the explanation for the growth of cyclonic disturbances on atmospheric fronts.

AMS

HELMHOLTZ RESONATOR. An enclosure communicating with the external medium through an opening of small cross-sectional area. Such a device resonates at a single frequency dependent on the geometry of the resonator.

HELMHOLTZ'S THEOREM. The statement that if F is a vector field satisfying certain quite general mathematical conditions, then F is the sum of two vectors, one of which is irrotational (has no vorticity), the other solenoidal (has no divergence). Thus, the horizontal velocity field, for example, may be expressed by

$$\mathbf{v} = \nabla_H\chi + \mathbf{k}\times\nabla_H\psi = \mathbf{v}_\chi + \mathbf{v}_\psi,$$

where $\mathbf{v}_\chi$ is irrotational, that is, $\nabla_H\times\mathbf{v}_\chi = 0$; $\mathbf{v}_\psi$ is solenoidal, that is, $\nabla_H\cdot\mathbf{v}_\psi = 0$; $\mathbf{k}$ is a unit vector directed vertically; and X and ψ are scalar functions that may be computed from the given wind field.

Additional Reading

Salby, M.L., R. Dmowska, and R.A. Pielke: *Fundamentals of Atmospheric Physics*, Elsevier Science & Technology Books, New York, NY, 1996.

AMS

HELMOHOLTZ WAVE. An unstable wave in a system of two homogeneous fluids with a velocity discontinuity at the interface. See also **Helmholtz Instability**.

HELMINTHOLOGY. A biological science dealing with the worms, more particularly parasitic flatworms and roundworms. Since many worms are parasitic, the term parasitology is more commonly used. The study of roundworms is important in agriculture and has resulted in the science of nematology (see also **Nematodes**), which is properly a subsidiary of helminthology.

HELM WIND. See **Winds and Air Movement**.

HEMATITE. The mineral hematite, ferric oxide, Fe_2O_3, occurs as thick or thin tabular rhombohedral forms, sometimes in pyramids but rarely in hexagonal prisms. It also assumes botryoidal, columnar and lamellar shapes, and may be granular or compact. Its hardness is 5.6; specific gravity, 5.26; luster, metallic to earthy or dull; color, dark gray to black; earthy forms may be different shades of red; streak, red to red-brown; translucent (in very thin flakes) to opaque. Hematite with a metallic luster is called specular iron.

It is a widely distributed and common mineral, found in igneous, sedimentary and metamorphic rocks as beds and veins, having probably been formed in many different ways under very different conditions. Beautifully crystallized hematite has been found in the Urals of the former U.S.S.R.; Rumania; Switzerland; the Island of Elba; Alsace, France; Cumberland, England. Extremely rich, large hematite ore bodies have been found and are being worked in Minas Gerais, Brazil; Cerro de Mercado, Durango, Mexico; Quebec and Labrador in Canada. The hematite ore deposits that lie along the southern and northwestern sides of Lake Superior in Michigan, Wisconsin and Minnesota have been worked to near depletion. Extensive beds of hematite are found throughout the Appalachian region from New York to Alabama, being mined near Birmingham in the latter state. Hematite occurs in quantity in Nova Scotia and Newfoundland. It is the most important ore of iron, and has other industrial uses in paint manufacture and polishing compounds. The name hematite is derived from the Greek word meaning blood.

HEMATOLOGY. That branch of medicine having to do with the study of the blood, the blood-forming tissues and the diseases of the blood.

HEMATOMA. An accumulation of free blood in the body tissues forming a localized mass. This usually follows an injury in which rupture of blood vessels takes place. See also **Brain (Injury)**; and **Cerebrovascular Diseases**.

HEMATOPOIESIS. See **Blood**.

HEMATURIA. The presence of blood in the urine. This condition is found in certain forms of nephritis and with injury, tumors, stones, or calculi in the urinary tract. It is also seen in scurvy and in some cases of severe sepsis.

HEME. See **Cytochromes**.

HEMI- AND HOLOCELLULOSE. See **Pulp (Wood) Production and Processing**.

HEMICELLULOSE. Hemicellulose [CAS: 9034-32-6] is the least utilized component of the biomass triad comprising cellulose, lignin, and hemicellulose. The term was originated by Schulze (1) and is used here to distinguish the noncellulosic polysaccharides of plant cell walls from those that are not part of the wall structure. Confusion arises because other hemicellulose definitions based on solvent extraction are often used in the literature. The term polyose is used in Europe to describe these noncellulosic polysaccharides from wood, whereas hemicellulose is used to describe the alkaline extracts from commercial pulps. See also **Cellulose**; and **Lignin**.

Pure hemicellulose components are seldom extracted directly from their source. Extracts are a mixture of polysaccharides, lignin, and lignin—hemicellulose complexes (by chemical linkages and possibly physical interactions) characteristic of their origin and the solvent employed. Hemicellulose has a lower degree of polymerization (DP) than cellulose (about 200 vs more than 10,000) and its lower limits have not been clearly defined. The extract may contain two or more polymers of similar composition but different structures (polydiversity) or of different distributions and amounts of branching or bonding in otherwise similar molecules (polydispersity). If a single polymer is present, it may exhibit a spectrum of molecular weights (polymolecularity) which may exhibit a Gaussian or biased distribution. A pure hemicellulose component is one where polydiversity has been avoided and a degree of heterogeneity has been attained compatible with end use application.

The most common hemicellulose in angiosperms is composed of D-xylose [CAS: 58-86-6] arranged in a linear manner. D-Mannose [CAS: 3458-28-4] is derived from a glucomannan which is the most common

hemicellulose in most gymnosperms. Both contain other sugars and exist in a variety of configurations and molecular weights. This article concentrates primarily on the components of tracheids and fibers of arborescent plants.

Techniques for the isolation of hemicellulose depend on the intended end use and whether it occurs in soluble waste material or is part of a solid matrix. Isolation is more difficult from solids as diminution of particle size and removal of undesired encrustants such as lignin is necessary to increase accessibility and destroy lignin hemicellulose bonds. Delignification techniques, except for those using ethanolamine, employ oxidants. Peroxides, peroxyacetic acid, and chlorine dioxide have been used but the most common reagents are chlorine and acidified sodium chlorite.

Delignification extracts varying amounts of hemicellulose. Low reaction temperatures and (where possible) high salt concentrations minimize losses and concomitant chemical degradations such as oxidation and the effects of pH. Pectic substances and easily soluble arabinans, arabinogalactans, galactoglucomannans, xylans, and compression wood galactans are found in waste chlorite liquors. Carbonyl groups (excepting carboxyls) are frequently reduced with a suitable reagent before alkaline extractions are attempted to minimize β-elimination reactions. The use of an inert atmosphere during alkaline extraction prevents oxidation by oxygen.

Extraction of hemicellulose is a complex process that alters or degrades hemicellulose in some manner. Alkaline reagents that break hydrogen bonds are the most effective solvents but they de-esterify and initiate β-elimination reactions. Polar solvents such as DMSO and dimethylformamide are more specific and are used to extract partially acetylated polymers from milled wood or holocellulose.

The separation of the polysaccharide components utilizes their different solubilities, polar groups, extents of branching, molecular weights, and molecular flexibilities and may be accomplished batchwise or with easily automated column techniques such as column or high performance liquid chromatography. These procedures have been summarized in several reviews.

The increasing sophistication of analytical techniques coupled with suitable fractionation procedures has made the heterogeneity of hemicellulose components increasingly apparent. These techniques common to polymer chemistry include gas chromatography—mass spectroscopy, and proton and ^{13}C-nuclear magnetic resonance spectroscopy. Molecular studies employ viscometry, osmometry, x-ray techniques, light scattering, and chromatographic and centrifugal techniques, as well as the use of optical rotatory dispersion and circular dichroism.

Pulping

The complex behavior of hemicellulose during pulping has been reviewed. When hemicellulose and lignin dissolve with the help of chemical transformations, fresh cellulosic surfaces are created and competition for deposition in these spaces arises between the dissolved components. Hemicellulose degradations also occur and are related to pH (acid hydrolysis, β-elimination reactions, redox reactions, etc.), and pyrolytic effects to an extent dependent upon the time, temperature, and liquor composition of the cook. These reactions are rendered more complex because the cell wall controls the diffusion of the reactants and products into and out of the fiber so that hemicellulose may not be able to react or diffuse out of the fiber before the cook is completed. Under pulping conditions, the formation and cleavage of lignin-hemicellulose bonds and possible carbohydrate—carbohydrate bonds occurs complicating the nature of the product. The extent to which these competing reactions is accomplished is reflected in product composition and end use quality and is the subject of much empirical research.

Those pulps with about 15% hemicellulose are usually used for paper manufacture, whereas those with 5% or less are used where a high cellulose content is required. The proportion of glucomannan is slightly greater in sulfite pulps, whereas the quantity of xylan is somewhat greater in kraft and soda pulps. These proportions can be altered slightly by changes in the cooking schedule.

Suitable pretreatment of wood before pulping alters the behavior of hemicellulose significantly. Saponification of the acetyl groups of softwood before sulfite cooking results in glucomannan retention in the final product. Those treatments that limit the peeling reaction during alkaline pulping processes (reductions with $NaBH_4$, H_2S, or oxidations with chlorite, polysulfide, anthraquinone, etc), can result in polysaccharide retention. The pretreatment of wood with mineral acid or liberated acids of wood at elevated temperatures (i.e., 170°C for 30 min) diminishes the DP of hemicellulose components sufficiently that they will be consumed mostly during a subsequent alkaline cook. The resulting pulp behaves more like cotton cellulose in many industrial applications.

The Effect of Hemicellulose in Commercial Products

Hemicellulose components have an effect on the properties of products in which they are present. In the case of viscose manufacture much of the hemicellulose which remains in the product has only a marginal effect on strength properties and brightness and no effect on heat stability. It does contribute to swelling in yarn, and that remaining in the spent viscose liquor is harmful to filter presses. Increased clogging of spinnerets, low color index, and decreased yarn strength can also result when resin, cations, and hemicellulose are together in the steeping liquor.

Applications

Hemicellulose and hemicellulose-like polysaccharides are beneficial components of foodstuffs because of their interactions between water and water-insoluble components. Endogenous polysaccharides are responsible for processing characteristics as well as texture and mouthfeel. Other properties such as gel formation, swelling of dough, fermentation, and optical properties are achieved using suitable treatments with enzymes and chemicals.

Besides inherent usefulness as a naturally occurring component of some manufactured products, hemicellulose can be utilized either as a polymer or as the source of chemical intermediates. The former use is complicated since the mixture of polysaccharides and lignin in extracts requires special treatment if one component is to be isolated. As a result, the naturally occurring gums and mucilages are at a competitive advantage in the marketplace.

Larch arabinogalactan is easy to isolate and requires limited purification for many uses. The mixture of sugars, oligosaccharides, degraded hemicellulose, and lignin found in the liquors and condensates from Asplund-like and prehydrolysis pulping processes is used for binding and extending animal fodder. The sugars and oligosaccharides in waste sulfite liquor can be used for furfural production and the growth of yeast. Pectin is used in the food industry, and apart from specialty uses finds limited application in the paper industry.

Derivatives of hemicellulose components have properties similar to the cellulosic equivalents but modified by the effects of their lower molecular weight, more extensive branching, labile constituents, and more heterogeneous nature. Acetates, ethers, carboxymethylxylan, and xylan–poly(sodium acrylate) have been prepared. See also **Diet**; and **Dietary Fiber**.

Additional Reading

Aspinall, G.O., in G.O. Aspinall, ed., *The Polysaccharides*, Vol. 2, Academic Press Inc., New York, NY, 1983, Chapt. 1.

Clayton, D. et al.: "Chemistry of Alkaline Pulping," in *Pulp and Paper Manufacture*, 3rd ed., Alkaline Pulping, Vol. 5, The Joint Textbook Committee of the Paper Industry, TAPPI, CPPA, Technology Park, Atlanta, GA, 1989.

Fengel, D., and G. Wegener, in D. Fengel and G. Wegener, *Wood, Ultrastructure, Reactions*, DeGruyter, Berlin, 1983, Chapt. 5.

Gatenholm, P. and M. Tenkanen: *Hemicelluloses Science and Technology*, American Chemical Society, Washington, DC, 2003.

Schulze, E., and E. Steiger, *Berichte*, **20**, 290–294 (1887); **23**, 3110–3113 (1890).

Shimizu, K.: in N.S. Hon and N. Shiraishi, eds., *Wood and Cellulosic Chemistry*, Marcel Dekker, Inc., New York, NY, 1991, Chapt. 5.

Thompson, N.S.: in I.S. Goldstein ed., *Organic Chemicals from Biomass*, CRC press, Boca Raton, Fla., 1981.

Whistler, R.L. and C.-C. Chen: in M. Lewin and I.S. Goldstein, eds., *Wood Structure and Composition*, International Fiber Science and Technology Series, Vol. 11, Marcel Dekker, Inc., New York, NY, 1991, Chapt. 7.

Wilkie, K.C.B.: *Chemtech*, 306–319 (1983); *Adv. Carbohydr. Chem. Biochem.*, **36**, 215–264 (1979).

HEMICHORDATA. A subphylum of the phylum *Chordata* containing only a few primitive marine animals without common names. The genus *Balanoglossus* has lent its name to the forms most commonly seen, although some belong to other genera. They are worm-like animals that live in mud and sand at the bottom of the ocean. The central nervous system is dorsal in this group but it remains partly or wholly at the surface. The notochord is limited to the anterior part of the body and is sometimes connected with the alimentary tract. Gill slits vary from one to many pairs. The group is also commonly named Enteropneusta and rarely Adelochorda.

There are two orders:

Order *Balanoglossida*. Worm-like animals with many gill slits and with a fleshy proboscis before the mouth. *Balanoglossus* and related forms.

Order *Pterobranchia* (*Cephalodisca*). Sessile animals, some solitary and some colonial. One pair of gill slits. A proboscis and branching tentacles lie before the mouth and the intestine is U-shaped. *Cephalodiscus* and *Rhabdopleura*.

HEMICOLLOID. A colloid composed of particles of small size, i.e., ranging from 0.005 to 0.0025 micrometer in length.

HEMIHEDRITY. A term describing crystal symmetry operations, to indicate that only half of a symmetrical structure undergoes modification. For example, if, in truncating a cube, the process is carried out symmetrically on four out of the eight solid angles, the resulting structure exhibits hemihedral symmetry.

HEMIMETABOLA. A division of the insects characterized by incomplete metamorphosis. The immature insect differs conspicuously from the adult in form and is adapted to an entirely different mode of life; in this the group resembles the *Holometabola*. The young have compound eyes, however, and the wings develop externally as in the *Paurometabola*. The group includes the three orders, *Plecoptera*, *Ephemerida*, and *Odonata*, all with aquatic larvae that are called naiads.

HEMIMORPHITE. This mineral is zinc silicate, $Zn_4Si_2O_7(OH)_2 \cdot 2H_2O$, occurring in tabular and prismatic orthorhombic crystals, although often in massive and fibrous forms. There is a perfect cleavage parallel to the prism; it is brittle with a subconchoidal fracture; hardness, 4.5–5; specific gravity, 3.40–3.50; luster, vitreous; color, white, tending to translucent. Hemimorphite differs from willemite, also a zinc silicate, in that the former contains considerable water which may be driven off when heated to a high temperature.

There are many localities for hemimorphite in Europe, fine specimens having come from Saxony; Sardinia; Cumberland, Alston Moor and Derbyshire, England. It is found in Siberia, Algeria, and Mexico. In the United States, hemimorphite has been found at Sterling Hill, New Jersey; in Lehigh County, Pennsylvania, and in Virginia, Missouri, Montana, Colorado, Utah, New Mexico, and Nevada.

The mineral is so named because of the tendency to form doubly terminated crystals showing a different grouping of faces at either end. The name is derived from the Greek words meaning half and form.

HEMIPLEGIA. Loss of voluntary movement on one side of the body, commonly resulting from damage to the cerebral cortex on the opposite side of the body, or to the nervous pathways leading from it. Transient hemiplegias occur in epilepsy and hysteria but the majority are persistent and are due to hemorrhage or sometimes tumor compressing or destroying the cerebral cortex and associated tracts of nerve fibers.

HEMIPODE. See **Rails, Coots, and Cranes**.

HEMIPTERA. Many hemiptera, an order of insects containing about 21,000 species, are of economic importance. They have piercing and sucking mouth parts and live on the blood or juices of animals or the sap of plants. The wings, when present, are usually distinctive. The basal half is thicker than the terminal, and the tips overlap partially so that the margins of the wings form an X on the back. Metamorphosis is usually gradual. The chinch bug and bedbug are species of economic importance.

Bugs of several families are aquatic and some forms live on the surface of the water, supported by the surface film. The swimming forms are the water boatmen, back swimmers, and giant water bugs and the water striders skate on the surface. One of the last, *Halobates*, is the only marine insect known. Shore forms include the toad bugs. On dry land the order is represented in almost every possible habitat. The main families of

Hemiptera include:

Belostomatidae	Giant water bugs
Cimicidae	Bed bugs
Coreidae	Squash bug
Corixidae	Water boatmen
Gerridae (also called *Hydrobatidae*)	Water striders
Lygaeidae	Chinch bugs
Miridae (also called *Capsidae*)	Leaf bugs
Nabidae	Damsel bugs
Nepidae	Water scorpions
Notonectidae	Back swimmers
Pentatomidae	Stink bugs
Phymatidae	Ambush bugs
Reduviidae	Assassin or kissing bugs
Tingidae	Lace bugs

HEMISPHERIC MODEL. See **Meteorology.**

HEMITROPIC. A term used by mineralogists for a crystal that appears to be composed of two halves of the same crystal turned partly around.

HEMLOCK TREES. Members of the family Pinaceae (pine family), these trees are of the genus Tsuga. The trees are sometimes referred to as hemlock spruces or hemlock firs. The hemlocks are evergreen trees, broadly conical. They are well known for their immunity to disease, with the exception of normal decay with age. The trees are quite tolerant of shade. The principal species not listed in Table 1 include:

Black hemlock	Tsuga martensiana
Canadian hemlock	*T. canadensis*
Low weeper form	*T.c.* 'Pendula'
Formosan hemlock	*T. formosana*
Himalayan hemlock	*T. dumosa*
Northern Japanese hemlock	*T. diversifolia*
Southern Japanese hemlock	*T. sieboldii*

The Canadian or Eastern hemlock normally attains a height between 50 and 80 feet (15 to 24 meters), but under favorable conditions can approach 100 feet. This species sometimes has several stems which form a spreading tree. The foliage may be described as feathery or plumelike. The bark is a dull brown. The tree is commonly found in swamps, ravines, rocky woods, and the mountain slopes of cold areas. It is found in some of the eastern mountains up to an altitude of about 2,000 feet (600 meters). The natural range of the tree is from Labrador, Newfoundland, and Nova Scotia westward to Michigan and Minnesota and southward to Delaware and Maryland and on the mountain slopes as far south as Georgia and Alabama. It is found throughout most of New England and is particularly common in the central portions of Maine. Commercially the tree is often called hemlock spruce in the northern states; and spruce pine in the southern states. Timber from the tree is used, but is not considered a high-grade wood. It is of uneven texture, tending to splinter easily. Major uses are for pulp wood, and boxes and crating. In the green condition, Eastern hemlock wood has a moisture content of 111% and weighs 50 pounds per cubic foot (801 kilograms per cubic meter). When air-dried to 12% moisture content, the weight is 28 pounds per cubic foot (448.5 kilograms per cubic meter) or 1,000 board-feet (2.36 cubic meter) weigh 2,330 pounds (1057 kilograms). The compressive or crushing strength of the dry wood, parallel to the grain, is 5,410 pounds per square inch (37.3 MPa); the tensile strength perpendicular to the grain in the green wood is 230 pounds per square inch (1.6 MPa).

The tree makes an excellent hedge and is often used for this purpose in landscaping.

The Carolina hemlock is essentially exclusive to the Allegheny Mountains. It is a smaller tree, but has many of the characteristics of the Eastern hemlock. Normal height is about 50–60 feet (15 to 18 meters), but can grow higher under favorable conditions. The tree prefers dry, rocky mountain soil as found in Virginia, North and South Carolina, Tennessee, and Georgia.

The Western hemlock is considered the master tree of the genus and can attain a height well in excess of 150 feet (45 meters). The branches are slender and pendulous. The crown is narrow and pyramidal. The needles are dark green. This tree ranges widely from central California northward to Oregon, Washington, and British Columbia on into southern Alaska and eastward to the Rocky Mountains, mainly in Idaho and Montana. Along with the mountain pine, Douglas fir, white fir, and Engelmann's spruce, the Western hemlock makes up a significant portion of the forests of western United States and Canada. The tree is a major source of timber and commercially may be called West Coast hemlock, hemlock spruce, Prince Albert fir, gray fir, Western hemlock fir, or Alaskan pine. The wood has a slight pinkish tinge, is moderately soft, straight-grained, and nonresinous. Select grades are free of knots and suitable for preferred construction uses. However, although the wood is easy to work, it does not plane smoothly. Unfortunately, the wood has frequent dark streaks from heart rot, particularly common in the older trees. In the green condition, Western hemlock wood has a moisture content of 74% and weighs 41 pounds per cubic foot (657 kilograms per cubic meter). When air-dried to 12% moisture content, the weight is 29 pounds per cubic foot (465 kilograms per cubic meter); or 2,420 pounds (1098 kilograms) for 1,000 board-feet (2.36 cubic meter). The compression or crushing strength parallel to the grain is 2,990 pounds per square inch (20.6 MPa) for the green wood; 6,210 pounds per square inch (42.8 MPa) for the dried wood. The tensile strength perpendicular to the grain for the green wood is 310 pounds per square foot (1513 kilograms per square meter) and about the same for the dried wood. Commercially, hemlock timber is commonly mixed with Douglas fir. The bark of the Western hemlock contains 22% tannin. However, most hemlock-bark extract is obtained from the Eastern hemlock.

As will be noted by the names given in the prior list of hemlock species, the hemlocks also occur in Asia at similar latitudes. See also **Conifers**.

HEMOCHROMATOSIS. See **Anemias**; and **Liver.**

HEMOCYANIN. An oxygen-absorbing substance in the plasma of the blood of the crayfish and many other arthropods. It is a clear material in the blood, but turns blue when removed and allowed to stand for a time. It serves a purpose similar to hemoglobin such as is found in higher forms of animal life.

TABLE 1. RECORD HEMLOCK TREES IN THE UNITED STATES[1]

Specimen	Circumference[2]		Height		Spread		Location
	Inches	Centimeters	Feet	Meters	Feet	Meters	
Carolina hemlock (1999) (*Tsuga caroliniana*)	161	409	99	30.2	50	15.2	North Carolina
Eastern hemlock (1995) (*Tsuga canadensis*)	202	513	165	50.1	38	11.6	Tennessee
Mountain hemlock (1999) (*Tsuga mertensiana*)	320	813	105	32	33	10.1	Oregon
Western hemlock (1999) (*Tsuga heterophylla*)	338	859	194	59.1	51	15.5	Washington

[1] From the "National Register of Big Trees," American Forests (by permission).
[2] At 4.5 feet (1.4 meters).

HEMOGLOBIN. The main function of the hemoglobin molecule is oxygen transport. The hemoglobin molecules from each species of organism that has been examined differ in the sequence of amino acids in their polypeptide chains unless they are very closely related. Chimpanzee and human hemoglobins are apparently identical. Sometimes two or more different kinds of hemoglobin are found simultaneously in the same organism. These structural variations may give rise to differences in the physiological properties that help to determine the efficiency of oxygen transport by the blood from lungs or gills to the tissues. Hemoglobin also plays an important role in carbon dioxide transport. See also **Blood**.

Vertebrate hemoglobins are usually composed of four polypeptide chains of two types, called α and β. The molecules can, therefore, be described as $\alpha_2\beta_2$. An iron porphyrin moiety, *heme*, is associated with each chain. Evidence indicates that combination of the heme with oxygen results in structural changes in the protein to which it is bound. Studies of single crystals of horse and human hemoglobins by x-ray diffraction show that removal of oxygen from the iron atoms of the four hemes results in a separation of the β-chains from one another; the relative positions of the α-chains do not appear to change. Although the molecular basis is not fully understood, the consequences are important. It is certain that any change in the mutual relationships of the polypeptide chains will alter the environment of many amino acid residues. These environmental changes are probably responsible for the degree of oxygenation to the oxygen pressure; and the dependence of the oxygenation upon pH and upon carbon dioxide concentration.

Mutations which alter the amino acid sequence can occur in either the α or the β-chain of the adult. However, most mutations are deleterious and changes in the α-chain would be more severely selected against in a process of natural selection because any change in the α-chain would affect the sensitive fetus, whereas changes in the β-chain would affect only the adult. This means that evolution tends to favor changes in the β-chain over changes in the α-chain. These considerations indicate that molecular adaptation of hemoglobin, at least in mammals, may involve changes more in the β-chain than in the α-chain.

Hemoglobins can be dissociated into their α- and β-subunits. Not only are hemoglobins capable of dissociating into their polypeptide subunits, but certain hemoglobins are also capable of polymerization. Many reptiles and amphibians and certain mice possess hemoglobins which polymerize to form double molecules $(\alpha_2\beta_2)_2$ and sometimes triple or quadruple molecules. Many hemoglobins from invertebrate animals have very large molecular weights and are composed of a large number of subunits — as many as 180 in some species. The nature of the forces holding these large aggregates together is under study.

The amino acid sequences of hemoglobins have been extensively altered by mutation during evolution. Data on the amino acid sequences of the chains from a variety of mammalian and other vertebrate hemoglobins show that the sequence can be varied extensively without drastic change in function. There appears to exist a hierarchy in the functional importance of different parts of a protein. Substitutions in different segments of a polypeptide chain may, according to the type and position of the substitution, exhibit a spectrum of effects, ranging from detectable to catastrophic. For example, the single substitution of valine for glutamic acid in the 6th position of the β-chain in human sickle cell hemoglobin results in a large decrease in the solubility of deoxygenated hemoglobin within the red cells. The hemoglobin, by forming a gel, distorts the red cell shape ("sickle") in such a way that flow through the capillaries is retarded. Such drastic consequences do not result if the substitution is lysine rather than glutamic acid (hemoglobin C). Histidine in position 63 of the human β-chain has an essential role stabilizing the ferrous state of the heme iron. Substitution by tyrosine (in hemoglobins "M") results in the loss of this stability because the ferric iron can form a strong linkage with the —OH group of tyrosine. Such a substitution results in a complete loss of capacity to combine reversibly with oxygen.

The foregoing are radical substitutions. Most effective substitutions appear to be relatively conservative and do not drastically affect the oxygen transport function. Therefore, the number of differences between homologous chains appears to be related not to functional differences, but to the time which has elapsed since the chains diverged from a hypothetical polypeptide ancestor. The mean number of differences between the hemoglobin chains of man, horse, pig, rabbit, and cattle is approximately 11. The common ancestor of these mammals may have existed some 80 million years ago. Thus, approximately 11 effective mutations per chain occurred in 80 million years, or 1 substitution per chain in 7 million years. Zuckerkandl and Pauling, using standard probability theory, have used this figure to estimate the time at which the different human hemoglobin chains (α, β, γ, and δ) are believed to have arisen by gene duplication. These estimates are shown in Table 1.

TABLE 1. DIVERGENCE OF HEMOGLOBIN CHAINS WITH TIME

Type of Chain Divergence	Number of Differences	Estimated Time since Divergence
β-δ	10	35 million years
β-γ	37	150 million years
β-α	76	380 million years
(α-β)-myoglobin	~135	650 million years

Estimates like these indicate that hemoglobins are very old and that it may be possible to find relatives of vertebrate hemoglobins in invertebrate animals. They also suggest that the gene duplication believed to be responsible for the divergence of the α- and β-chains took place in the Devonian period at the time of the appearance of early amphibians and the dominance of fish.

The suggested relationship between numbers of differences and evolutionary time is not wholly secure. It assumes uniformity in the rate of effective amino acid substitution, but this rate may be neither uniform with time, nor uniform in different parts of the polypeptide chain. Differences in the rate of effective substitution along the polypeptide chain may be due not only to restrictions imposed by the required tertiary structure, but also to differences in the rate at which various parts of the DNA or the gene mutate. The evolution of hemoglobin may be contrasted with that of cytochrome c in which approximately 50% of the molecule appears to have remained invariant during the time yeast and man have evolved.

Additional Reading

Chang, T.: *Blood Substitutes and Oxygen Carriers*, Marcel Dekker, Inc., New York, NY, 1992.

Honig, G. and J. Adams: *Human Hemoglobin Genetics*, Springer-Verlag, New York, Inc., New York, NY, 1986.

Perutz, M.: *Science Is Not a Quiet Life: Unravelling the Atomic Mechanism of Haemoglobin*, Imperial College Press, London, UK, 1999.

HEMOGLOBINURIA. The presence of hemoglobin in the urine. This occurs when red cells of the blood are destroyed at such a rate that the hemoglobin set free cannot be disposed of by the normal processes, but appears unchanged in the urine. Myohemoglobin, the pigment of muscle cells, may similarly appear in urine, especially after extensive crush injury, from mismatched blood transfusion, from allergy to the bean, *vicia faba* and in certain rare conditions, e.g., after exposure to cold in certain persons whose blood contains a hemolytic agent active only when the blood is cooled (Donath-Landsteiner reaction), in certain otherwise normal persons after exercise (march hemoglobinuria) and in a rare type of hemolytic anemia (Marchiafava-Micheli anemia) in which the hemoglobinuria occurs only in sleep. See also **Kidney and Urinary Tract**.

HEMOLYMPH. The blood of higher invertebrates, consisting of a clear plasma and white cells but without red cells. Respiratory pigments are dissolved in the plasma. It contains a lower percentage of water than the blood of more primitive forms.

HEMOLYSIS. See **Blood**.

HEMOLYTIC ANEMIAS. See **Anemias**.

HEMOPHILIA. Hemophilia is a hereditary bleeding disorder in which the blood fails to coagulate. This is caused by a deficiency of one of two plasma clotting factors, factor VIII or factor IX. The clinical symptoms of this X-linked recessive disease vary depending on the level of activity of the affected factor. Patients may be asymptomatic, except in the face of trauma or surgery, or suffer from mild to severe, often spontaneous, hemorrhage.

History and Royal Pedigrees

Hemophilia was one of the first inherited diseases to be described and was already documented in the Talmud about 2000 years ago as a severe and sometimes fatal hemorrhagic tendency in males occurring between the Euphrates and the Tigris rivers. Even at that early time, knowledge about the hereditary nature of the disease lead to rabbinical advice that if the first two sons died as a result of circumcision, it should not be performed on a third or fourth son. The term hemophilia was apparently first used by Schönlein in 1828, although it was originally transiently named hemorrhaphilia. The year 1893 marked the beginning of an understanding about the nature of the defect, when Wright noted a prolonged clotting time of blood from patients with hemophilia. It was not until 1947 that hemophilia was recognized as consisting of different types, when Pavlovsky noted that the clotting defect in certain hemophilic plasma samples could be corrected by adding plasma from certain other patients with hemophilia. Although the genetics of the disease were elucidated in the early 1800, complete characterization of the underlying molecular bases with cloning and sequencing of the genes was achieved only recently.

Queen Victoria (1819–1901) was undoubtedly the most famous carrier of hemophilia in history. Together with her husband, Prince Albert of Saxe-Coburg, she had nine children (five daughters and four sons) and 40 grandchildren; at the time of her death 37 great grandchildren were alive. Through marriages of her grandchildren with members of other European royal families, she established royal lines with Germany, Russia, Spain, Sweden and others, which made her the "grandmother of Europe." Queen Victoria had inherited the defective gene from her mother, who was a member of the Coburg royalty, and spread it to many of her descendants. Two of Victoria's daughters were carriers, and one son had hemophilia.

Most notably, the hemophilia trait was introduced into the Romanov family by the marriage of Alice, a carrier-granddaughter of Queen Victoria, with Nikolaus, who became the tsar of Russia in 1894. They had only one son, Alexis, who was severely affected by the disease. Another granddaughter, Victoria Eugenie, introduced the disease into the Spanish monarchy when she married Alfonso XIII, the king of Spain. Three of their five sons had hemophilia.

Introduction

The X-linked recessive bleeding disorders hemophilia A and hemophilia B result from the lack of functional clotting factor VIII or IX respectively, in the circulation. Both factors are critical components of the middle stages of the blood coagulation cascade. After being activated factor XI (factor XIa) or factor VIIa and tissue factor, factor IXa will form a complex with its cofactor, factor VIIIa, an enzyme cofactor that has no procoagulant activity of its own. This complex in the presence of calcium and phospholipids will then activate factor X to factor Xa (Fig. 1), a reaction that is accelerated more than 10 000-fold by factor VIIIa. The roles of factors VIII and IX as parts of a single critical complex in the coagulation process explain why the two diseases, although caused by molecular defects in two different proteins, are clinically indistinguishable. See also **Blood Clotting: General Pathway**.

The genes for both factors are located on the X-chromosome, which explains their X-linked recessive pattern of inheritance. The gene for factor VIII is mutated in hemophilia A, and hemophilia B is caused by mutations in the gene for factor IX. The disease can be characterized either by the complete absence of clotting factor protein, or by varying levels of a partly or totally dysfunctional protein. The former are classified as cross-reactive material (CRM)-negative variants, while the latter are CRM-positive variants. Only about one-third of patients with hemophilia B have detectable antigen in their plasma.

The incidence of hemophilia is approximately 20 cases per 100 000 males (one case per 10 000 of the whole population). Hemophilia A accounts for about 85% of the cases and hemophilia B for the remainder. Apparently, there is no difference in incidence between different races and geographical areas. Since it is an X-linked recessive disorder, hemophilia occurs almost exclusively in males. Females are affected only in the rare case of extreme lyonization of one of the X-alleles or, theoretically, as offspring of a hemophilic father and a carrier mother. The latter, however, has not been observed in humans, but is well known in hemophilic dog colonies.

The clinical diagnosis of severe hemophilia is often suspected by a lifelong history of hemarthrosis and, unless the disease is caused by a novel mutation in the individual, a family history of bleeding events. The finding of a prolonged activated partial thromboplastin time together with an isolated reduction of factor VIII or factor IX activity confirms the diagnosis. The patient's plasma will be unable to correct the coagulation time when added to plasma known to lack factor VIII or factor IX activity. With the exception of a special CRM-positive form of hemophilia B (hemophilia Bm, characterized by a prolonged ox brain prothrombin time), the prothrombin time, thrombin time and bleeding time are all within the normal range.

The physiological tissue of production of factor VIII has not yet been definitively established. It has been suggested that the liver is the primary site of synthesis, but it is likely that other cell types, such as fibroblasts, lymphocytes and vascular endothelial cells, produce functional factor VIII as well. The role of the liver as one important site of synthesis is demonstrated by the detection of factor VIII in previously antigen-negative patients with hemophilia A after liver transplantation. However, severe failure of liver function does not lead to factor VIII deficiency, which suggests that other tissues also play an important role in synthesis. For factor IX the liver has conclusively been established as the physiological site of production, from which it is secreted into the bloodstream. In the attempt to find other tissues that would be easier targets for gene therapy approaches, it was found that a variety of tissues, such as fibroblasts, myoblasts and myocytes, can also produce active factor IX. However, these tissues do not perform the posttranslational modifications nearly as well as hepatocytes.

Factor VIII

Factor VIII is encoded by a gene consisting of 26 exons spanning about 186 kilobases (kb). Constituting about 0.1% of the entire X-chromosome, it is one of the largest human genes known. The 26 exons vary in size from 69 to 3106 nucleotides, adding up to 9 kb, whereas the introns account for 177 kb. Elucidation of the complete coding sequence has permitted the deduction of the primary amino acid sequence of the protein. It consists of a C-terminal light chain of 80 000 Da in molecular weight and an N-terminal heavy chain of 90 000–200 000 Da. Different functional domains (three A domains, one B domain and two C domains) can be recognized within the protein. The order of these domains is A-1, A-2, B, A-3, C-1, C-2. The A domains have a marked homology with one another and with the copper-binding protein ceruloplasmin, which suggests that they may be involved in metal-ion binding. The C domains show homology to discoidins, which can bind glycoconjugates and phospholipids. The precise function of the B domain is still unknown, and its deletion has no effect on procoagulant activity. It is coded for by only one exon (exon 14) and shows no homology with other known proteins.

The factor VIII protein undergoes a series of intracellular modifications in its biosynthetic pathway before it is secreted into the circulation as a functional protein. Cleavage at two sites within the B domain occurs within the Golgi apparatus and removes the B domain, generating a light and a heavy chain. Upon secretion into the circulation, the association of these two chains is greatly promoted by von Willebrand factor (VWF), which stabilizes the factor VIII protein. People lacking functional VWF also have defective factor VIII function. Binding of VWF to factor VIII also protects it from activation by factor Xa (but not by thrombin) and from inactivation by activated protein C.

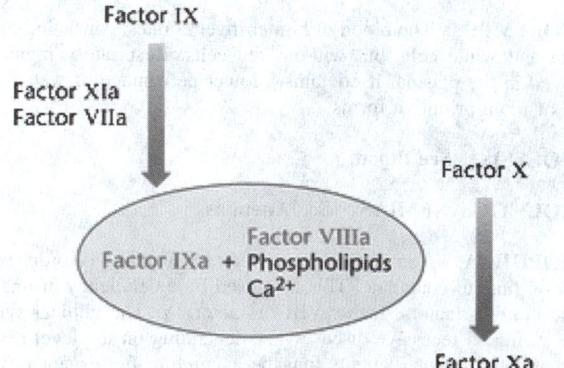

Fig. 1. Middle stages of the coagulation cascade. Factors beside the arrows denote the activators in the respective reaction

Factor IX

The factor IX gene was first cloned by two independent groups in 1982, and the complete sequence of the gene was first established in 1985. The gene is about 34 kb long and encompasses eight exons and seven introns. It encodes a protein that consists of 415 amino acids and has a molecular weight of 57 000 Da. The mature protein has several distinct structural domains that are each encoded by one or two specific exons. To become functional, the protein has to undergo a series of posttranslational modifications. The most important of these is the γ-carboxylation of Glu residues within the Gla domain near the N-terminus, a process that requires vitamin K. Processing of these residues is essential for calcium-dependent binding to phospholipids of factor IX. This modification can be inhibited by anticoagulants such as warfarin. After secretion into the blood, the factor IX protein circulates as an inactive single-chain glycoprotein. Upon activation by factor XIa or factor VIIa, one domain of the protein, the activation peptide, is cleaved out of the protein chain. This generates an N-terminal light chain and a C-terminal heavy chain that are linked by a disulfide bond. The activation peptide, which consists of 35 amino acids, remains noncovalently bound to factor IXa. The active site of the protein, a serine residue in the heavy chain, is activated by the binding of its adjacent residue to the free N-terminus of the heavy chain. Factor X is activated by factor IXa by cleavage of a peptide bond by this active site.

Clinical Manifestations

The clinical manifestations of hemophilia span a wide spectrum and differ considerably among individuals affected by the disease. Within a kindred, whose affected subjects all have the same underlying genetic defect, the clinical manifestations are, however, largely constant.

The frequency and severity of bleeding episodes depend mainly on the plasma levels of factor VIII or factor IX activity. Since these levels vary by 50–150% even in healthy people, a reference standard consisting of pooled plasma from healthy donors is assumed to have 100% factor activity. Patients with activity levels lower than 1% of normal suffer from severe hemophilia and usually present with an average of two to four bleeding episodes per month that require substitution therapy, although the intervals between hemorrhages vary. The hallmark of severe forms of hemophilia is spontaneous bleeding primarily into weight-bearing joints or muscle tissue. In these cases often no antecedent trauma can be recognized for a specific bleeding episode. Patients with levels of 5–15% are considered mild hemophiliacs and may not be diagnosed with the disease until adult life. In these less severely affected patients, bleeding often occurs only in response to a traumatic event, although it may be delayed until several days after the trauma. They may also present with bleeding after surgery but only occasionally experience spontaneous bleeding episodes. Patients whose levels are between 1% and 5% have moderate disease with severity of symptoms falling between the two extremes.

Depending on the underlying molecular defect, hemophilia can present with varying severity of symptoms, and thus the initial diagnosis may be made in almost any age group. Interestingly, even in newborns with severe disease, vaginal delivery is usually uneventful. Sometimes the first clinical symptoms of the disease are seen during circumcision, although more often affected infants suffer from easy bruising and frequent hematomas. Since patients with hemophilia have intact platelet function, major hemorrhage from minor cuts or abrasions is not a typical finding. The hallmark of severe hemophilia is spontaneous and repeated hemorrhage into joints, which is responsible for the often crippling arthropathy seen in these patients. Hemarthrosis stimulates chronic inflammation and hypervascularization of the synovial membrane of the affected joint, which may predispose to further bleeding episodes–a vicious cycle that perpetuates the sequelae of the disease. The joints most frequently involved are, in decreasing order of frequency, the knees, elbows, ankles, shoulders and hips. The recurrent bleeding episodes finally lead to chronic arthritis with intraarticular destruction that cause progressive deterioration of joint function. Thus, hemarthrosis with its sequelae plays an important role in morbidity in these patients.

The most life-threatening events in patients with hemophilia, however, are intracranial bleeds. They are often caused by a minor trauma that may eventually be recognized only in retrospect. Any neurological symptoms such as an unusual headache should thus be treated immediately with factor concentrates before any further diagnostic measures such as computed tomography are taken. Peripheral neurological symptoms such as nerve paralysis can also develop if dissecting hematomas lead to nerve

compression. Hematuria, gross or microscopic, is a very common finding in hemophiliacs, and almost every patient will experience it at some point in life. Often no causative lesion within the urinary tract can be identified, although the sources are usually one of the kidneys or the urinary bladder. Serious complications such as a renal colic can ensue if a blood clot stops urinary flow through the ureter. Sometimes patients experience hemorrhages after surgical procedures, such as tooth extractions. See also **Computed Tomography (CT)**.

Molecular Genetics of the Hemophilia's

The molecular bases of hemophilia A and B are mutations in the genes for factor VIII and IX, respectively. The genes for both factors are located near the terminus of the long arm of the X-chromosome. Both forms are caused by many different mutations, a fact that makes this disease quite heterogeneous at the molecular level. Hemophilia is inherited in an X-chromosomal recessive pattern, although about one-third of all cases of both types are the result of novel mutational events. Due to this mode of inheritance, a few generalizations can be made about the hemophilia's: (1) only males are affected by the disease (only in rare cases of extreme lionization of one X-chromosome allele are females affected), (2) male offspring of a female carrier will have a 50% chance of being affected, while female offspring will be a carrier in 50% of the cases and normal in the remaining 50% and (3) female offspring of a patient with hemophilia will be obligatory carriers, whereas male offspring will be normal.

Because of its hereditary nature, genetic counseling based on carrier detection and prenatal diagnosis can be offered to relatives of patients. In general, female carriers have about 50% normal factor activity in plasma, although owing to the great variation of normal levels carrier detection by this method is inaccurate. Assessment of the molecular status is the most reliable method. Due to the large size of the factor VIII gene, sequencing, although available, is expensive. Since polymorphisms exist in and near the factor VIII gene, a mutant allele can be tracked within a family by linkage analysis using restriction fragment length polymorphisms (RFLPs). If the mutation creates or abolishes a recognition site for a restriction enzyme, diagnosis can be made simply by enzyme digestion and agarose gel electrophoresis. For factor IX, where the gene is much smaller, sequencing of the coding region and identification of the mutation by direct sequencing is usually feasible. See also **Agarose Gel Electrophoresis**.

The molecular bases of both forms of hemophilia have been elucidated in many different affected kindred's by identifying the underlying genetic defects. A considerable amount of information about the function of the different structural domains of the proteins has been gained from mutational analysis. Owing to the size of the factor VIII gene, characterization of mutations has been slower than for factor IX.

In both the factor VIII and factor IX genes, all types of mutations such as deletions, insertions and point mutations have been described. Gross gene deletions account for only 5–10% of all mutations in hemophilia. They cause severe disease and are characterized by the complete lack of any detectable antigen and activity in plasma. Other deletional mutations that affect only a part of the entire gene may lead to unstable or truncated proteins, and most of them are also characterized by a severe phenotype. About 45% of all cases of severe hemophilia A are characterized by recurrent deoxyribonucleic acid (DNA) inversions that generally occur between a region of intron 22 and one of two homologous copies of this region located telomeric to the factor VIII gene. Intrachromosomal homologous recombination between these sites leads to an inversion of all intervening DNA sequences and to the destruction of the normal factor VIII gene. Deletions or insertions of only one nucleotide, leading to a shift in the normal trinucleotide reading frame, cause the formation of a premature stop codon further downstream of the actual mutational site; these also often present as severe hemophilia. Missense mutations, exchanging one amino acid for another, can cause a variety of symptoms and are largely responsible for the broad range of clinical severity of hemophilia.

Although there are no specific regions within either of the two genes that show an accumulation of defects, the CpG dinucleotide has clearly been identified as a mutational hotspot. This can be explained by the methylation of the cytosine residue and the subsequent deamination of methylcytosine to thymine in the presence of mutagens, or even spontaneously.

Treatment

Factor Replacement Therapy. The mainstay of therapy for both types of hemophilia is replacement of the missing factor by purified factor concentrates, either plasma-derived or recombinant. The cost of this therapy is high and with current regimen is estimated at about $50,000–100,000 yearly per patient depending on the severity of the disease and the weight of the patient. These figures explain why replacement therapy is, for many patients, instituted only in response to bleeding episodes rather than prophylactically. Hemophilic arthropathy, however, can be prevented by lifelong prophylactic treatment starting at the age of 1–2 years. Based on the observation that chronic arthropathy is much less frequent and less severe in moderate hemophilia compared with that in patients with severe disease, minimum trough levels greater than 1% of normal should be maintained. Several clinical studies have shown that initiation of prophylactic treatment early in life is important to reduce the risk of developing a target joint and also to prevent serious bleeds in other tissues. This form of treatment substantially reduces the number of joint bleeds and the ensuing arthropathic sequelae in almost all patients. Although the cost of prophylactic treatment is high and in fact prohibitive in many countries, studies have shown that about 16% of patients treated on an on-demand basis consume similar factor levels per year as the average patient on prophylactic treatment.

For patients suffering from chronic arthropathy, conservative orthopedic and physical therapy in addition to factor replacement is of paramount importance. Remarkable improvements have been noted after several weeks of physical therapy for muscle building and joint stabilization. Patients with chronic synovitis should be treated both with factor replacement and physical therapy. In these patients surgical synovectomy is an alternative if the conservative approach fails. If all other therapeutic measures have failed, joint replacement is the ultimate treatment option. Total hip and knee replacement have been performed with excellent results, but should be reserved only for patients in whom no other alternative exists.

The regimen and dosing, as well as the choice of the replacement product, depend mainly on the severity and location of the bleeding episode. When calculating the amount of factor activity to be infused to reach a certain level of factor activity in the patient's plasma, it has to be considered that only up to 50% of infused factor VIII or factor IX is recovered as a result of redistribution into the extravascular space. Since the half-life of factor VIII is about 8–12 h and that of factor IX 18–24 h, about half of the initial dose of concentrate has to be administered at these time points to maintain therapeutic levels. In hemophilia B a factor IX activity level of about 25% should be reached in response to an episode of hemarthosis or superficial hematoma. After major bleeding events, such as intracranial or gastrointestinal bleeds, a level of up to 100% should be achieved.

Patients with hemophilia can undergo both emergency and elective surgery if certain precautions are taken. Optimally, consultation between the surgeon, hematologist and coagulation laboratory should be maintained. Generally, the patient's missing factor has to be replaced immediately before the procedure and should then be kept at a slightly lower level for about 10–14 days. Before institution of clotting factor replacement therapy, the presence of an inhibitor to the infused factor has to be ruled out in order to ensure a reliable effect. For major surgery the targeted clotting factor activity level should be brought to about 60–80% of normal and be kept at 30–40% over the following days.

Complications. A serious complication of replacement therapy has been the infection with different viruses, notably hepatitis B and C viruses and human immunodeficiency virus (HIV), from products made from pooled plasma. In the early 1980s about 70% of severe hemophiliac patients were infected with HIV, the cause of acquired immune deficiency syndrome. Since then, the development of methods for inactivating viruses has improved the safety of transfused products to the point where the threat of infection with these viruses has essentially been eliminated. Recombinant factor VIII, produced without human plasma, has been available for several years now, and the first recombinant factor IX is currently in clinical trials. These preparations have added a measure of safety for patients receiving replacement therapy. See also **Hepatitis Viruses**; and **Human Immunodeficiency Viruses (HIV)**

Gene Therapy. Because of the genetic nature of hemophilia, gene therapy would be a useful alternative to current treatment, since it would provide continuous production of the deficient clotting factor. Several features of hemophilia make gene therapy approaches attractive. First, tissue-specific expression is not necessarily required, since a number of different cell types have been shown to produce functionally active factor VIII and factor IX. A second advantage is that precise regulation of transgene expression is not required since levels greater than 1% of normal are clinically beneficial, and even levels up to 200% are not associated with any ill effects. In addition, small (mice) and large (dogs) animal models exist for both hemophilia A and B, which helps substantially to establish an experimental basis for gene therapy. On the other hand, the large size as well as the complex structure of factor VIII, and the crucial posttranslational modifications of factor IX, are problems that need to be solved. The majority of gene therapy studies on hemophilia have used viral vectors as the vehicle to introduce the coding sequences of the gene into the target cells, most commonly retroviral, adenoviral and adeno-associated viral vectors. In virtually all experimental studies to date, low levels or short duration of transgene expression in immunocompetent animals have been major obstacles that will have to be addressed. Different promising approaches, such as modification to the vector system or modulation of the host immune system, are currently underway and may have the potential for eventual human application. See also **Human Gene Therapy**.

Additional Reading

Berntorp, E., K. Hoots, and L. Christine: *Textbook of Hemophilia*, Blackwell Publishers, Malden, MA, 2004.

Hoffman, R., L.E. Silberstein, E.J. Benz, S.J. Shattil, and B. Furie: *Hematology: Basic Principles and Practice*, 4th Edition, Elsevier Health Sciences, New York, NY, 2004.

Jones, P.: *Living with Haemophilia*, 5th Edition, Oxford University Press, New York, NY, 2002.

Ljung, R.C.R.: "Can Hemophilic Arthropathy be Prevented?" *British Journal of Haematology*, **101**, 215–219 (1998).

Ludlam, C.A.: "Treatment of Hemophilia," *British Journal of Haematology*, **101** (supplement 1), 13–14 (1998).

Miller, J.M., and S.K. Miller: "Vignette of Medical History: Queen Victoria and Hemophilia," *Maryland Medical Journal*, **42**(6), 581–583 (1993).

Potts, D.M., and W.T. Potts: *Queen Victoria's Gene: Haemophilia and the Royal Family*, Sutton Publishing, Gloucestershire, UK, 1999.

Staff, Icon Health Publications: *Hemophilia—a Medical Dictionary, Bibliography, and Annotated Research Guide to Internet References*, ICON Health Publications, San Diego, CA, 2004.

Staff, Icon Health Publications: *Official Patient's Source Book on Hemophilia*, ICON Health Publications, San Diego, CA, 2005.

Walter, J., and K.A. High: "Gene Therapy for the Hemophilia's," *Advances of Veterinary Medicine*, **40**, 119–134 (1997).

JOHANNES WALTER, Kaiser Franz Josef Spital, Vienna, Austria

KATHERINE A. HIGH, Children's Hospital of Philadelphia, Philadelphia, PA

HEMOPTYSIS. The bringing up of blood from the larynx, trachea, bronchi or lungs. The commonest cause is pulmonary tuberculosis; carcinoma of the bronchus is a frequent cause; it may also occur in any chronic bronchial or pulmonary disease and certain varieties of heart disease, especially mitral stenosis, in which the pulmonary blood pressure is persistently raised; and in aneurysm of the aorta. Other diseases which may predispose hemoptysis include polyarteritis nodosa (a subacute or chronic, remittent, disseminated vascular disease characterized by focal necrotizing inflammation of the walls of medium- and small-sized arteries and arterioles); Weil's disease (a severe form of leptospirosis); and wool sorter's disease. See also **Leptospirosis**.

HEMORRHAGE. Bleeding. Escape of blood from the vessels. Anemias caused by sudden blood loss as in traumatic injury are generally normocytic, that is, the cells are of normal size, but reduced in number. When the blood is lost over a longer period of time, as from bleeding hemorrhoids, peptic ulcer, in hookworm disease, and in excessive menstrual bleeding (menorrhagia), a microcytic anemia may result. Following hemorrhage, body fluids seep into the blood which restore it to its former volume; consequently, dilution of the blood occurs, and anemia may result. It may require some time for the body to manufacture the necessary red cells and other substances necessary to return the blood to normal. The symptoms of such a blood-loss anemia include a general weakness, dizziness, and faintness. In more severe cases, there may be vomiting and a great thirst, the heart rate may be rapid, and the breathing weak and shallow.

The first step in the treatment of persons with a posthemorrhagic anemia is to stop the loss of blood. Blood transfusions may be given to return the blood to its proper volume before excessive dilution occurs. In milder hemorrhages, however, the body may be able to restore the lost blood without transfusion. This is often accomplished by ample rest and a good diet, including adequate amounts of iron and protein necessary for red cell building.

Hemophilia is a rather rare hemorrhagic disease. See also **Hemophilia**. Other conditions exist in which unusually large amounts of blood may be lost. In many such cases, bleeding may take place into the skin, as in a bruise. This symptom is referred to as *purpura*.

Essential thrombocytopenic purpura is a disease characterized by hemorrhage, and caused by a deficiency in the number of blood platelets. The spleen may be responsible for this disease by destroying the blood platelets. Corticosteroid therapy helps control the bleeding, and in most patients, is regarded as a desirable precaution prior to removal of the spleen.

Purpura and excessive bleeding may occur in persons suffering from deficiency of vitamins C and K. Some newborn infants contract a hemorrhagic disease, which once was frequently fatal; the victims now recover rapidly when treated with vitamin K. Purpura may occur in persons receiving antitoxin treatments, or as a symptom of snakebite poisoning, or with some types of food poisoning. The taking of certain drugs may bring about abnormal bleeding. Purpura is occasionally a symptom of such varied conditions as meningitis, scarlet fever, severe measles, chronic kidney disease, endocrine disorders, liver disease, macrocytic anemias, allergies, typhus fever, and a specific bacterial heart disease. The symptom disappears in each case when the primary cause is removed.

Additional Reading

Kase, C. and L. Caplan: *Intracerebral Hemorrhage*, Butterworth-Heinemann, Inc., Woburn, MA, 1994.

Sarwar, M. and S. Batnitzky: *Imaging of Non-Traumatic and Hemorrhagic Disorders of the Central Nervous System*, Kluwer Academic Publishers, Norwell, MA, 1989.

Weir, B.: *Subarachnoid Hemorrhage: Causes and Cures*, Vol. 52, Oxford University Press, Inc., New York, NY, 1998.

HEMORRHAGIC FEVER VIRUSES.   Viral hemorrhagic fever (VHF) is a clinical definition, incorporating several diseases that present, in their severe forms, with a common haemorrhagic syndrome; however, their clinical pictures, as well as their epidemiological characteristics and aetiological agents, are different. Some have been known for centuries, such as yellow fever in Africa [Monath, 1998] or hemorrhagic fever with renal syndrome (HFRS) in China [Schmaljohn and Hjelle, 1997], whereas others have been recognized during the last decades. All are zoonoses, and recent ecological perturbations induced by the growing human population are the main causes of their outbreaks or spread.

Hemorrhagic fever viruses belong to different families (Table 1). All have a single-stranded ribonucleic acid (RNA) genome. The genome of the *Flaviviridae* is a positive-sense RNA. Members of this family include the yellow fever and dengue viruses, which are transmitted by mosquitoes, and viruses of the tick-borne encephalitis complex, which may be responsible for haemorrhagic fevers like Omsk hemorrhagic fever or Kyasanur Forest disease. The other families, *Arenaviridae, Bunyaviridae* and *Filoviridae*, have negative-sense RNA. The *Arenaviridae* and *Bunyaviridae* genomes are segmented, with two segments for the *Arenaviridae* and three segments for the *Bunyaviridae*. The *Filoviridae* are nonsegmented and closely related to two other families which share the same genomic organization, the *Paramyxoviridae* and the *Rhabdoviridae*, with which they form the *Mononegavirales* order. See also **Flaviviruses**.

DHF, dengue hemorrhagic fever; DSS, dengue shock syndrome; HF, haemorrhagic fever; HFRS, hemorrhagic fever with renal syndrome; HPS, hantavirus pulmonary syndrome.

The viruses have names taken from places where they first caused recognized outbreaks of disease: towns, like Junin, Marburg, Guanarito, Puumala; rivers, like Hantaan, Machupo, Ebola; or regions, like Congo, Crimea or Rift Valley.

Viruses responsible for hemorrhagic fevers have numerous closely related neighbors in their families that are mild pathogens, and others that are nonpathogenic for humans. An exception is the *Filoviridae* family: the two known members, Marburg and Ebola, are highly pathogenic for primates. This means that the animal (or vegetal) genera that harbor the *Filoviridae* have never been correctly studied from the virological point of view. When the natural hosts of Ebola or Marburg are identified, other filoviruses will probably be recognized. There is as yet no clue to explain what, in the structure and mechanism of these different viruses, leads to the common vascular disorders that have given the name to the hemorrhagic fevers they induce. See also **Ebola Hemorrhagic Fever**; **Filoviruses**; and **Marburg Hemorrhagic Fever**.

TABLE 1. CLASSIFICATION OF THE MAIN HEMORRHAGIC FEVER VIRUSES

Family	Genus	Virus/ disease	Transmission[a]
Flaviviridae	*Flavivirus*	Yellow fever	Mosquitoes
		Dengue 1–4/DHF-DSS	Mosquitoes
		Kyasanur Forest disease	Ticks
		Omsk HF	Ticks
Bunyaviridae	*Phlebovirus*	Rift Valley fever	Mosquitoes
	Nairovirus	Crimean–Congo HF	Ticks
	Hantavirus	Hantaan/HFRS	Rodents (mice)
		Seoul/HFRS	Rodents (rats)
		Belgrade/HFRS	Rodents
		Puumala/nephropathia epidemica[b]	Rodents (bank vole)
		Sin Nombre/HPS[b]	Rodents (deer mice)
Arenaviridae	*Arenavirus*	Junin/Argentinean HF	Rodents
		Machupo/Bolivian HF	Rodents
		Guanarito/Venezuelan HF	Rodents
		Sabia/Brazilian HF	Rodents
		Lassa/Lassa fever	Rodents
Filoviridae	*Filovirus*	Marburg/HF	Unknown
		Ebola/HF	Unknown

[a] Arthropod vector for arboviral transmission, and animal reservoir if nonarboviral.
[b] These are not HF viruses/diseases, but viruses are in the same *Hantavirus* genus as Hantaan.

Epidemiology

Since these viruses are all zoonotic agents, their epidemiology is linked to the relations of their hosts with humans. Another major determinant is the ease of human-to-human transmission. Viruses belonging to different families may have similar epidemiology, like the rodent-borne hantaviruses and arenaviruses, whereas closely related viruses like Yellow fever virus (YFV) and Omsk hemorrhagic fever virus (OHFV) have completely different epidemiological patterns because of the biology of their vectors: mosquitoes for YFV and ticks for OHFV. Outbreaks occur when human behavior or modifications in the environment favor human contact with viral reservoirs or the increase of viral populations. Comparable reservoirs lead to comparable gross epidemiological patterns; however, differences appear with a finer analysis. See also **Yellow Fever**.

Rodent-Borne Viruses. The natural hosts of the hantaviruses and the arenaviruses are rodents. Once infected, these animals remain chronic carriers of the viruses, and may shed infectious particles in their urine, faeces and saliva throughout their lives. These viruses are good models of the adaptation of the viruses to their hosts: if each viral species may infect different rodents, it generally has only one reservoir, the rodent species able to become a chronic carrier [Bowen, et al.: 1997]. The other species are only transient carriers. There is a difference between arenavirus and hantavirus transmission in rodent populations: antenatal or perinatal infection of the rodents is necessary to induce chronic carriage of the arenaviruses, whereas hantaviruses are mainly transmitted horizontally by bite. The prevalence is higher for these hantaviruses among adult males than other members of a troop. The specificity of the hosts explains why the rodent-borne diseases are limited to specific geographical regions where these animals live. The hantavirus pulmonary syndrome (HPS) has been recognized only in the Americas because it is linked to viruses that infect rodents of the Sigmodontinae subfamily, which is limited to the New World. Furthermore, even in endemic regions, human cases appear in foci. The majority of the rodent species spend their whole lives in a very small area. Thus, the probability of transmission of the viruses they harbor to other populations of the same species living further away is very low. For example, the hantavirus Belgrade is responsible for HFRS in several European countries. Puumala virus-infected cases always appear in the same forested areas where activities bring humans into contact with the reservoir, a woodland rodent–the bank vole. See also **Arenaviruses** and **Hantaviruses**.

Humans are infected by the respiratory route through aerosols contaminated by the urine or faeces of the rodents or by direct contact of mucous membranes or skin abrasions with excrement or blood of the animals. Populations at risk and incidence rate vary with the behavior of the rodents. The reservoirs of *Lassa virus* (*Mastomys natalensis*) and *Machupo virus* (*Calomys callosus*) enter human habitations. All age groups and both genders may be infected. Furthermore, *Mastomys* (rats and mice) are hunted and eaten in Sierra Leone and Liberia. This is a high risk for infection, which can be prevented by education. Other rodents are field-adapted, like *C. musculinus*, the reservoir of *Junin virus*, or *Apodemus agrarius*, the reservoir of *Hantaan virus*; male agricultural workers form the majority of cases.

There is no human-to-human transmission of the hantaviruses, except for some cases of HPS associated with the Argentinean strain, Andes. Outbreaks among humans are directly related to the abundance of infected animals. There are cycles: seasonal or annual. In cold climates, as, for example, in Scandinavia, the cycle of the bank vole is mainly seasonal, with little difference between years. The renal syndrome in humans appears endemic. In temperate countries like France, the dynamics of the rodent population follows a 3-year cycle. Every third year there is an outbreak with about 200 cases, whereas only 50 cases are diagnosed between these peak years. Besides these regular cycles, ecological events may induce the abnormal multiplication of the rodents; for example, the rainy winter of 1992 in the mountains and deserts of New Mexico, Nevada and Colorado. The exceptional humidity favored a particularly abundant crop of pine kernels, which are a major diet for the deer mouse *Peromyscus maniculatus*. The population density of the animals multiplied 10-fold between 1992 and 1993. These animals are the reservoir of a hantavirus now called *Sin Nombre* ("no name" in Spanish) and they infected enough people to allow the epidemiologists to recognize a cluster of cases of an unknown disease, now called *hantavirus pulmonary syndrome*.

In addition to natural phenomena, outbreaks may be linked to changes in agricultural patterns: in South America, four hemorrhagic fevers due to arenaviruses have been identified since 1950. *Junin virus* causes Argentine hemorrhagic fever (AHF), which was recognized in the pampas west of Buenos Aires in 1953, and caused epidemics every year until the use of a vaccine in 1993. The cultivation of large areas of maize supported huge populations of *C. laucha* and *C musculinus*, because these species were better adapted to these new linear habitats than other rodent species. Beside this effect, the increased need for agricultural workers during the harvest period multiplied contacts between these rodents and humans. Today, mechanization has placed the operators of agricultural machinery in the front line: combine harvesters not only produce clouds of infectious dust, they also create an aerosol of infectious blood when they accidentally crush the animals. Human infections with two other arenaviruses responsible for hemorrhagic fevers have been associated with deforestation of the Amazon region: Machupo, which appeared in Bolivia in 1959, and Guanarito in Venezuela in 1989. The cause of the emergence of the Bolivian disease is not clear, but it is limited to a small part of the Beni region. The infected villages were at the limit between the pampas and the Amazon forest where subsistence crops were raised in areas called *chacos*, cut out from the forest by a "slash and burn" method. In 1989 the same human behavior was clearly the cause of an epidemic in Venezuela due to an arenavirus, now called *Guanarito virus*. The first 15 cases belonged to a rural community that had started to clear a forested region in the centre of the Portugesa state. One reservoir is the rat *Zygodontomis brevicauda*. More than 100 additional cases were subsequently diagnosed in the same area.

Another cause of increased contact between humans and rodents is war: soldiers living in the field, and more particularly in trenches, have contact with animals rarely encountered, such as nocturnal rodents. In Asia HFRS due to the *Hantaan virus* existed for centuries. Western countries first became interested in this illness during the Korean War, when more than 2000 United Nations troops suffered from it between 1951 and 1953. The European form of hantaviral illness due to *Puumala virus* was described in Sweden in 1934 as the "nephropathia epidemica", but medical records of French and English military physicians reported a similar disease called *néphrite de guerre* or *trench nephritis* during the 1914–1918 war. The largest epidemic ever reported, with about 4000 cases, occurred among Russian and German troops fighting during 1942 in Lapland. More recently, about 700 cases were reported among Bosnian and Republica Serbska soldiers and displaced populations around Sarajevo,

Tuzla and Bihac during 1994–1996. Several soldiers from the Italian, French, English and Canadian troops were also infected.

Mosquito-Borne Viruses. There are three main hemorrhagic fever viruses transmitted by mosquitoes: YFV, *Dengue virus* and Rift Valley fever virus (RVFV). These modes of transmission allow large and often unpredictable epidemics. In Ethiopia, where the disease had not been diagnosed before, 600 000 cases of yellow fever were reported in 1960–1962, and millions of dengue infections occur every year in tropical countries. Hopefully, the percentage of the hemorrhagic form is less than 1/1000 for this disease. The mosquito represents only one partner in the natural cycles of these arthropod-borne viruses (arboviruses): one or several vertebrate hosts are necessary to maintain the viruses in nature, even if the female mosquito may infect its eggs by so-called *transovarian transmission*. The natural cycles of *Dengue virus* and YFV were similar during previous centuries. Monkeys and mosquitoes were the main actors in tropical forests. Today, dengue is a worldwide problem in urban areas of the tropics [Gubler and Clark, 1995]. Humans are now the vertebrate partners, with *Aedes aegypti* as the main arthropod partner, in this new urban cycle. Urban outbreaks of yellow fever were reported in Central and North America during the eighteenth and nineteenth centuries, but the problem has been solved by mosquito control. A sylvatic cycle of yellow fever still exists between monkeys and hemagogus mosquitoes in South American rainforests, and the reinvasion of towns of these regions by the *Ae. aegypti* mosquito is a threat of recurrence of urban yellow fever. In Africa, there is permanent circulation of the virus in forested areas, and outbreaks occur regularly in rural savannah and towns. There is no direct human-to-human transmission for these two viruses. See also **Arthropod-Borne Viruses**; **Dengue Fever and Dengue Hemorrhagic Fever** and **Yellow Fever**.

If yellow fever and dengue epidemics seem unpredictable, abundance of water has been clearly associated with outbreaks of RVFV. This virus exists everywhere in Africa south of the Sahara. The natural cycle is not completely described. Several mosquitoes and other hematophagous diptera have been found to be infected; however, no vertebrate wildlife species has been clearly demonstrated to be the second partner of the cycle. Transovarian transmission in mosquitoes of the *Aedes* genus plays an important role, and natural or artificial flooding induced by heavy rain or dams allows millions of infected mosquito eggs to hatch at the same time. A large epidemic responsible for approximately 400 human cases in late 1997 and early 1998 in Kenya and Southern Somalia occurred after unusual rains. Freshwater areas created by dams not only provide breeding areas for mosquitoes but also attract humans and their herds together in new population centers. These two factors probably explain two other epidemics of Rift Valley fever in Africa: one in 1977 in Egypt and the other in 1987 in Mauritania. After the construction of the Aswan dam, there were major losses of cattle and, of the 200 000 people infected, 600 died. In 1987 another epidemic followed the damming of the Senegal River in Mauritania, with a total of approximately 200 cases. If Rift Valley fever outbreaks are triggered by an abundance of infected mosquitoes which transmit the virus to various vertebrates, the majority of human cases are due to direct transmission from infected cattle to humans during abortion or slaughtering. Human-to-human transmission has never been clearly demonstrated. See also **Rift Valley Fever**.

Tick-Borne Viruses. Tick-borne viruses have little opportunity of causing large human epidemics. Mosquitoes fly to their blood meals, whereas ticks wait on a vegetative support until an animal passes by. The main tick-borne hemorrhagic fever virus is Crimean–Congo haemorrhagic fever virus (CCHFV) [Hoogstraal, 1979]. Several tick species have been involved in its cycle; the majority belong to the *Hyalomma* genus and live in arid countries in Asia, Europe and Africa. This large distribution may be related to the fact that some tick larvae and nymphs feed on birds, and may be transported in this manner. In the major outbreak of Crimean–Congo hemorrhagic fever reported in 1944–1945 among Soviet soldiers and peasants in the Crimean Oblast, who were bitten by ticks during the harvest, this type of infection was generally responsible only for sporadic cases. The outbreaks among humans are linked to the high infectious potential of blood from patients or infected cattle. The majority of cases arise in hospitals or slaughter houses.

Nosocomial Transmission. For some hemorrhagic fever viruses, other than CCHFV, the transmission from patients or infected animals to healthy people is also the main cause of outbreaks. The best examples are the filoviruses. The natural cycles of *Marburg virus* and *Ebola virus* are not known, but appear, at least for Ebola, to be linked to tropical rainforests [Le Guenno and Galabru, 1997]. The number of human cases directly infected from the unknown reservoirs(s) is three for Marburg and fewer than 10 for Ebola [Feldmann and Klenk, 1996]. Nearly 1000 other Ebola cases reported since 1967 were due to contact with sick people or primates, or their blood and excreta. The outbreaks in Nzara, Makokou and Booue were limited to 30–60 cases because the patients were mainly treated in villages by their families. The large Ebola outbreaks in Yambuku, Zaire and Maridi, Sudan in 1976, and Kikwit, Zaire in 1995 were responsible for nearly 300 cases each, and each outbreak was amplified by the introduction of infected patients into hospitals with poor hygiene. The nosocomial transmission of *Lassa virus* was the cause of the first outbreaks described in 1969 and 1970. However, it is now well-established that the majority of human infections have occurred as a result of contact with rodents or their urine. See also **Ebola Hemorrhagic Fever Lassa Fever (African Hemorrhagic Fever)** and **Marburg Hemorrhagic Fever**.

The hanta viruses, except the Argentinean Andes strain, and the flaviviruses, dengue and YFV, are not transmissible directly from person to person. For the other viruses, the risk varies, but is mainly due to contact with blood. Direct close contact with patients or with soiled instruments or clothes seems the most dangerous. Airborne transmission of these viruses directly from the upper respiratory tract of patients to others is doubtful; however, there is a real risk of aerogenous contamination if droplets of infected liquid are dispersed in air.

Pathogenesis

The pathogenesis of VHF is not clearly understood, and may vary among different VHF viruses. Even simple data, such as the existence of intravascular coagulation or the level of thrombopenia in patients, are often not available because the majority of cases appear in tropical countries with hospitals of low technical capabilities. However, hypotheses may by drawn from animal models or *in vitro* studies. Macrophages and endothelial cells appear to be the most important target for the replication of these viruses [Lewis, et al.: 1989], except for YFV. Both produce tumor necrosis factor

α (TNFα), interleukin 6 (IL-6) and numerous other cytokines when they are stimulated (Figure 1).

Shock. The main symptom and cause of death in these diseases is hypovolaemic shock, linked to leakage of water and low molecular weight components of the plasma at capillary level. If viral antigens are abundant in the endothelial cells, viral replication does not cause direct gross damage to these cells, and the leakage cannot be explained by vascular lesions. Only swelling of endothelial cells and mild oedema of vessel walls have been regularly reported. The default in vascular tightness could be induced by the action of cytokines like TNF α and IL-6 on the endothelial cells [Feldmann, et al.: 1996].

Hemorrhage. The second biological feature common to these diseases is the hemorrhagic syndrome. Thrombopenia and/or platelet dysfunction have been reported in different cases and models. The infection of endothelial cells probably modifies their membranes and favors the aggregation of platelets, which leads to disseminated intravascular coagulation (Figure 1). Another hypothesis, particularly for dengue, is an inhibition of platelet production at the medullary level due to a direct or a cytokine-mediated action of the viruses on the megakaryocytes. Normal numbers of platelets, but defaults in their function, have been reported in Lassa infections.

Other Manifestations. These vary with the virus. If the pathogenesis results from direct viral cytopathology in some cases, as in Rift Valley fever or yellow fever hepatitis, the majority of these viruses act through unsuitable immune response and triggering of the inflammatory cascade. These viruses are pantropic but obvious lesions of the various organs are rarely found. Cytokines, coagulation factors and complement factors are involved, but these complex pathways have been only partially studied in different models for some viruses. Among the cytokines, interferon has been studied in two cases, the Rift Valley fever and Junín infections. In Rift Valley fever, clearance of the virus and recovery seem linked to an early production of interferon and antibodies, whereas in Junín-associated AHF, high levels of interferon are found in severely ill patients and those with terminal illness. See also **Interferons** and **Lymphokines**.

The TNF α is produced by infected macrophages and endothelial cells. Its key position within a network of mediators could explain the similarity

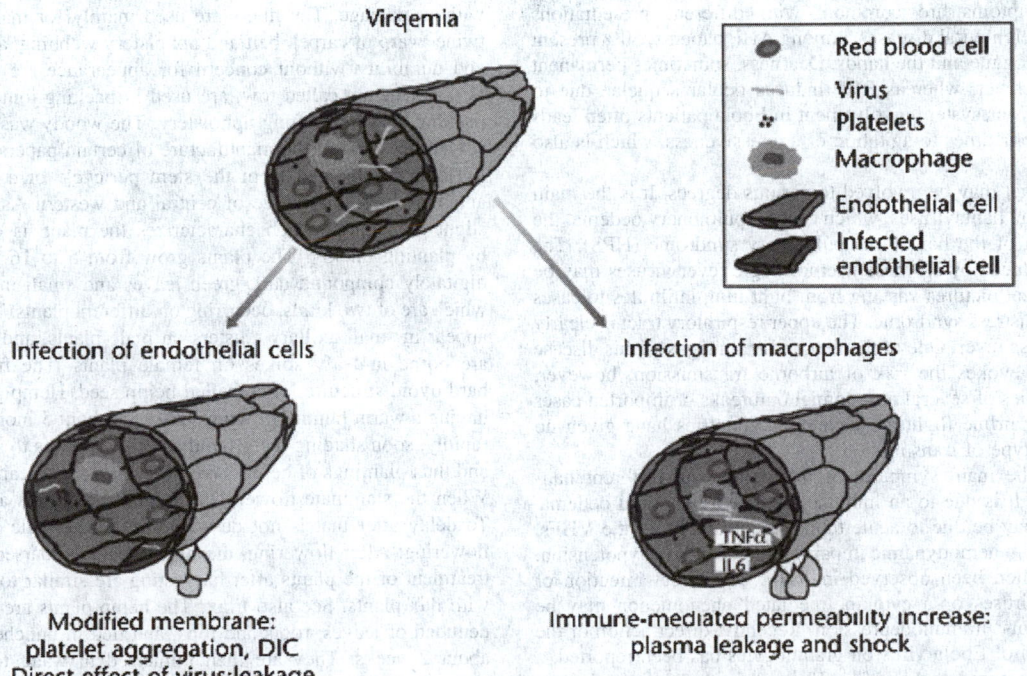

Fig. 1. Hypothetical mechanism of intravascular coagulation and shock induced by the hemorrhagic fever viruses. Viruses in the bloodstream may infect macrophages and endothelial cells. Stimulated cells produce tumor necrosis factor α (TNFα) and interleukin 6 (IL-6), known to induce capillary permeability. These molecules trigger the production of other cytokines, leading to the systemic inflammatory response syndrome. The endothelial cells are not destroyed by viral infection; however, their membrane is modified and favors the aggregation of platelets, leading to disseminated intravascular coagulation (DIC) and hemorrhages

of severe cases of hemorrhagic fever due to different viruses. It is also an explanation of their similarity with other viral, bacterial or parasitic acute infections that lead from a systemic inflammatory response syndrome to septic shock. [Reinhart, et al.: 1996].

Clinical Features

The Common Syndrome. Hemorrhagic fever is characterized by an incubation period of between 3 days and 3 weeks. The shortest incubation period is for RVFV, sometimes limited to 48 h, whereas incubation periods over 1 month have been reported for hantavirus. This period is followed by a gradual or sudden onset of systemic symptoms. The main symptom is a high fever, generally accompanied by headache, myalgia, arthralgia and nausea. The diseases may be confused with influenza or malaria during the first 2–3 days, and mild forms exist, limited to these early symptoms. In severe cases, this period is followed by a general deterioration in health, during which bleeding may occur. Superficial bleeding reveals itself through skin lesions, such as petecchiae and purpura. The serious cases present with epistaxis, melena and hematemesis; however, even in lethal forms, spontaneous hemorrhages may be absent. Hemorrhages are more prominent in CCHFV, filoviruses, RVFV and YFV infections than in the arenavirus, hantavirus and dengue cases. They are rarely a direct cause of death, except when intractable gastrointestinal bleeding (seen in yellow fever or Crimean–Congo hemorrhagic fever) or cerebral hemorrhages occur. See also **Influenza** and **Malaria**.

Other Cardiovascular, Digestive, Neurological, Respiratory and Renal Complications. If hepatitis is frank in yellow fever, Rift Valley fever and sometimes in Crimean–Congo hemorrhagic fever, it is not an obvious symptom for the other hemorrhagic fevers. Increases in alanine aminotransferase and aspartate aminotransferase without jaundice are observed in Lassa, Ebola or Marburg infections, but only disseminated necrotic foci may be detected in the liver at autopsy.

Vomiting and diarrhea are frequent and non specific symptoms in acute infections. Nearly all the hemorrhagic fevers may be accompanied by these signs. However, the involvement of the gastrointestinal tract is the main sign of Ebola and Marburg and an important one in CCHF infections. About 70% of the Ebola patients present with diarrhea. See also **Ebola Hemorrhagic Fever** and **Marburg Hemorrhagic Fever**.

Neurological symptoms are common, with different presentations according to the different viral diseases. Among AHF patients, 50% present with tremors of the tongue and the hands. Deafness, sometimes permanent is caused by Lassa fever, whereas RVF induces ocular sequelae due to retinitis. Central nervous system involvement in Ebola patients often leads to prostration, but sometimes to agitation or aggressiveness, which is also reported in AHF.

The respiratory tract may be involved to various degrees. It is the main target of the American hantaviruses, which cause a pulmonary oedema, the principal component of the hantavirus pulmonary syndrome (HPS). The capillary leakage induced by all these haemorrhagic fever viruses may be responsible for clinical pictures varying from light lung infiltrates to cases of adult respiratory distress syndrome. The upper respiratory tract is clearly involved only in Lassa fever. One of the first and main signs of this disease is pharyngitis. This evokes the risk of airborne transmission; however, epidemiological studies of several nosocomial outbreaks or imported cases through commercial airline flights to developed countries have given no data to support this type of transmission.

Renal failure is the main symptom of infections with different hantaviruses in Eurasia. It is due to an immune-mediated interstitial oedema. In yellow fever, it may be due to acute tubular necrosis. In all the VHFs, the kidneys suffer from hemodynamic impairment induced by hypotension.

Bacteremia has often been observed in VHF. The direct infection of leucocytes by the viruses or a cytokine-mediated phenomenon may be responsible for patients' immunodepression. Recently, direct action of the secreted glycoprotein of Ebola virus on granulocytes has been reported.

The lethality varies from less than 1% in Puumala or dengue infections to more than 80% with the Zaire strain of Ebola. The clinical picture of these patients is often hypotensive shock with hemorrhages, acute respiratory distress syndrome and/or multiorgan failure.

Additional Reading

Bowen, M., C. Peters, and S. Nichol: "Phylogenetic Analysis of the *Arenaviridae*: Patterns of Virus Evolution and Evidence for Cospeciation between Arenaviruses and Their Rodent Hosts," *Molecular Phylogenetics and Evolution*, 8(3), 301–316 (1997).

Feldmann, H., and H. Klenk: "Marburg and Ebola Viruses," *Advances in Virus Research*, 47, 1–52 (1996).

Feldmann, H., H. Bugany, F. Mahner, et al.: "Filovirus-induced Endothelial Leakage Triggered by Infected Monocytes/macrophages," *Journal of Virology*, 70, 2208–2214 (1996).

Gubler, D., and G. Clark: "Dengue/dengue Hemorrhagic Fever: The Emergence of a Global Health Problem," *Emerging Infectious Diseases*, 1(2), 55–57 (1995).

Hoogstraa, 1 H.: "The Epidemiology of Tick-borne Crimean–Congo Hemorrhagic Fever in Asia, Europe and Africa," *Journal of Medical Entomology*, 15, 307–417 (1979).

Le Guenno, B., and J. Galabru: "Ebola Virus," *Bulletin de l'Institut Pasteur*, 95, 73–83 (1997).

Lewis, R., J. Morrill, P. Jahrling, and T. Cosgriff: "Replication of Hemorrhagic Fever Viruses in Monocytic Cells," *Reviews of Infectious Diseases*, 11 (supplement 4), S736–742 (1989).

Monath, T: "Yellow Fever," In: Monath, T: *The Arboviruses, Epidemiology and Ecology*, pp. 139–231. CRC Press, LLC, Boca Raton, FL, 1988.

Porterfield, J.: *Exotic Viral Infections*, Chapman & Hall, Inc., New York, NY, (1995).

Reinhart, K., C. Wiegand-Lohnert, F. Grimminger, et al.: "Assessment of the Safety and Efficacy of the Monoclonal Anti-tumor Necrosis Factor Antibody-Fragment, MAK 195F, in Patients with Sepsis and Septic Shock: a Multicenter, Randomized, Placebo-controlled, Dose-ranging Study," *Critical Care Medicine*, 24(5), 733–742 (1996).

Schmaljohn, C., and B. Hjelle: "Hantaviruses: a Global Disease Problem," *Emerging Infectious Diseases*, 3, 95–104 (1997).

Staff, Novartis Foundation Symposium: *New Treatment Strategies for Dengue and Other Flaviviral Diseases*, John Wiley & Sons, Inc., Hoboken, NJ, 2006.

BERNARD LE GUENNO, Institut Pasteur, Paris, France

HEMORRHOIDS. See **Arterial and Venous Disorders**.

HEMOSTASIS. See **Blood Clotting: General Pathway**.

HEMOTOXIN. See **Snake**.

HEMP. The fibers of the hemp plant, *Cannabis sativa*, of the family *Cannabinaceae* (hemp family) are coarse and rather harsh and much less pliable than flax fibers. They are dark colored and not easily bleached without damage. The fibers are used mainly for making rope and coarse twine, warp of carpet, belt and upholstery webbing, and wherever strength and durability without concern for appearance are of importance. Short fibers of hemp, called tow, are used in packing joints in pipes, for pump packing, and for stuffing upholstery. The woody waste from hemp fiber is sometimes used in the manufacture of certain papers. See also **Rope**.

Hemp is obtained from the stem pericycle of a tall hollow-stemmed annual, which is a native of central and western Asia. In cultivation, the slight branching, which characterizes the plant, is considerably reduced by planting thickly. The plants grow from 5 to 16 feet high. They have digitately compound dark green leaves and small inconspicuous flowers, which are of two kinds, occurring on different plants. The staminate flowers appear in small axillary clusters on male plants, and the pistillate flowers are borne in leafy spikes on female plants. The fruit, an achene, is a hard ovoid structure, often called hemp seed. Hemp grows best in regions having a warm humid growing season of about 5 months. The plants grow rapidly, soon shading the ground so effectively as to suppress other plants, and thus plantings of hemp have been used as a means to eradicate weeds. When the staminate flowers are mature, the plants are ready for harvest. To delay after that is not desirable, since the male plants die soon after flowering. After flowering, the fibers become coarser. Harvesting and the treatment of the plants after harvesting are similar to the procedures used with flax plants. See also **Flax**. The hemp plants are cut off or pulled up, denuded of leaves, roots and tops, and tied in bunches and left to dry for about 2 weeks. They are then immersed in water to ret. In retting, the intercellular substance of the stems is acted upon by bacteria and softened so that the fibers are readily cleaned of surrounding tissues. Scutching removes the woody tissue, after which the rough hemp fibers are hackled, or drawn over coarse combs that pull out the fibers.

In recent years, the cultivation of hemp has been subject to controls because marijuana is prepared from the dried leaves and flowers of the plant, which then are smoked in the form of cigarettes as a narcotic. The species *Cannabis indica* is usually used in this connection. See also **Marijuana**.

Additional Reading

Bosca, I. and M. Karus: *The Cultivation of Hemp: Botany, Varieties, Cultivation and Harvesting*, Hemptech, Sebastopol, CA, 1998.

Ranalli, P. (Editor): *Advances in Hemp Research*, Haworth Press, Inc., Binghamton, NY, 1998.

Robinson, R. and R. Nelson: *The Great Book of Hemp: The Complete Guide to the Environmental, Commercial, & Medicinal Uses of the World's Most Extraordinary Plant*, Inner Traditions International, Ltd., Rochester, VT, 1995.

Web References

The Industrial Hemp Information Network. http://hemptech.com/
North American Industrial Hemp Council. www.NAIHC.org

HENCH, PHILIP SHOWALTER (1896–1965). Hench studied medicine at the University of Pittsburgh, graduating in 1920. He was long associated with the Mayo Clinic, in Rochester, Minnesota, becoming Head of the Section for Rheumatic Diseases in 1926. Between 1928 and 1929 he studied in Freiburg and in Munich.

From 1926 Hench became immersed in the study of rheumatic disorders. In April 1929 he observed that the condition of a patient with rheumatoid arthritis was dramatically improved during a bout of jaundice. He had also noticed a similar remission of the disease in patients who were pregnant, or had undergone general anesthesia. Hench began to look for a common factor that might explain these improvements, in what was a severely debilitating disease. This was the search for what Hench called *anti-rheumatic factor X*. Hench believed that such a factor might be found within the adrenal cortex. See also **Rheumatoid Arthritis**.

Coincidentally, in 1929, Hench's colleague Edward Kendall began to separate and characterize the many hormones of the adrenal cortex. In September 1948 one of these extracts, compound E, was administered by Hench to patients with rheumatoid arthritis. Their condition was transformed; a formerly bedridden patient even attempted to dance. Hench named compound E cortisone. It was hailed a wonder drug, akin to insulin, but it was not the panacea it was hoped to be. For this research, Hench and Kendall shared the 1950 Nobel Prize for Medicine or Physiology with Tadeus Reichstein. See also **Kendall, Edward Calvin (1886–1972)**.

Additional Reading

Fox, D.M., M. Meldrum, and I. Rezak: *Nobel Laureates in Medicine or Physiology*, Garland, New York, NY, 1990, pp. 231–237.

Glyn, J.: "The discovery and early use of cortisone," *Journal of the Royal Society of Medicine*, **91**, 513–517 (1998).

CHAIRE E. J. HERRICK, London, UK

HENDERSON, LAWRENCE JOSEPH (1878–1942). Lawrence Henderson was an American biochemist, philosopher and sociologist concerned with the relationship between organisms and the environment.

Henderson entered Harvard at the age of 16, an institution with which he remained linked for the whole of his career. In 1898 he entered Harvard Medical School, receiving his MD in 1902. He never practiced medicine, preferring to study biochemistry. He spent two years in the laboratory of the biochemist Franz Hofmeister in Strasbourg, and then studied physical chemistry with T. W. Richards at Harvard. In 1905 he was appointed a lecturer in biochemistry at Harvard. See also **Biochemistry (The History)**.

Henderson's career was devoted to the study of the organization of organisms, the universe and society. He wrote on philosophy and sociology, as well as on science. He became interested in the fitness of the inorganic environment to support organic life. For Henderson there was a harmonious order about the biological and social universes, an idea that he elaborated in two main works: *The Fitness of the Environment* (1913), and *The Order of Nature* (1917). In 1919, in what was perhaps his greatest contribution to biochemistry, he described the blood as a physicochemical system which could be defined by mathematics.

He also applied scientific thought to the study of society, and in particular, human relations. In 1927 he helped to found the Fatigue Laboratory in the Graduate School of Business Administration at Harvard, which provided an opportunity to explore the relationship between individuals and their environment. Henderson was also instrumental in introducing the teaching of history of science at Harvard.

Additional Reading

Gillispie, C.C.: *Dictionary of Scientific Biography* (1970–1980), Vol. 6, Charles Scribner's Sons, New York, NY, pp. 260–262.

Henderson, L.J.: *Blood: A Study in General Physiology*, Yale University Press, New Haven, CT, 1928.

CHAIRE E. J. HERRICK, London UK

HENLE, FRIEDRICH GUSTAV JAKOB (1809–1885). Friedrich Henle was a German anatomist and pathologist who provided intellectual background to the rise of the germ theory.

Henle was born at Fürth, Germany on 19 July 1809. After studying at Bonn and Heidelberg, Henle worked in Berlin as Johannes Müller's assistant. While in Berlin, Henle became acquainted with Matthias Jacob Schleiden and Theodor Schwann who were developing the cell theory, and Henle was influenced by their work. He was the first to describe the structure of human epithelium (1837). See also **Schleiden, Matthias Jacob (1804–1881); Schwann, Theodor Ambrose Hubert (1810–1882)**.

In 1840 Henle was appointed Professor of Anatomy at Zurich and, four years later, was called to Heidelberg. In 1852 he became Professor of Anatomy at Göttingen, where he remained until his death. Henle was the author of more than eighty items among which were two influential pathology texts, *Pathologische Untersuchungen* (1840) and *Handbuch der rationellen Pathologie* (two volumes 1846, 1853), and a widely used and pioneering text on systematic anatomy, *Handbuch der systematischen Anatomie des Menschen* (three volumes, 1855–1871). Henle is credited with several anatomical discoveries.

In the early decades of the nineteenth century, physicians believed that any disease could be caused by a variety of different and unrelated factors. Henle was critical of this way of thinking: "Only in medicine can the same effect flow from the most varied possible sources ... This is no more scientific than if a physicist were to teach that the fall of a body depended on the removal of a board or a beam, from the breaking of a rope or a cable, or from the existence of an opening, and so forth." At the time there was little empirical evidence for the germ theory of disease, but Henle became interested in the possibility that many diseases could be due to living agents that invaded the body. Henle followed Augustino Bassi's investigations of muscardine and Johann Lucas Schönlein's research on favus, and he reviewed these and other related discoveries in *Zeitschrift der rationelle Medicin* , a periodical that he co founded and co edited. In his pathological writings, Henle proposed classifying diseases by cause, and he recognized three broad classes: those due to miasms (e.g. malaria), those brought on by a specific contagion (e.g. syphilis and scabies), and those due either to a contagion or to a miasm (e.g. fevers). While he himself contributed no experimental results to the germ theory, Henle's interest in microorganisms was a source of inspiration to early researchers.

In connection with his discussion of infectious diseases, Henle proposed empirical tests for determining whether any specific organism was the cause of a particular disease. These tests included what we now know as the isolation, cultivation and reinoculation of the purported causal agent. There is no evidence that Henle actually attempted to apply these tests in his own research −indeed he remarked that doing so would probably be impossible. Nevertheless, because he identified these possible tests and because Robert Koch was Henle's student at Göttingen, it has been widely supposed that Koch's famous criteria for demonstrating causality, the so-called *Koch's postulates* , were derived from Henle. However, as Koch himself admitted, he adopted the postulates from Edwin Klebs and, in this respect; he seems not to have been influenced by Henle. See also **Bacteriology (The History): and Koch, Heinrich Hermann Robert (1843–1910)**.

K. CODELL CARTER, Bringham Young University, Provo, UT

HENNA SHRUB. Of the family *Lythraceae*, the *Lawsonia inermis* is a small shrub native to Africa and Asia and cultivated in tropical countries. Well known since the time of the early Egyptians as a red dye for hair, nails, hoofs of animals, etc., the leaves of this plant are powdered and made into a paste, which is then used as a dyeing medium. The small flowers of the plant are inconspicuous, but fragrant.

HENRY (abbr h). The unit of electrical inductance; the inductance of a closed circuit in which an electromotive force of 1 volt is produced when the electric current in the circuit varies uniformly at the rate of 1 ampere per second.

HENRY, JOSEPH (1797–1878). Henry was an American experimental physicist. He worked most of his life to understand the relationships between electricity, magnetism, light, and heat. He is recognized for his contribution in finding that an electromotive force can be induced in a circuit by a changing magnetic field (known as mutual induction). He also discovered that any electrical circuit in which there is a varying current induces an electromotive force in itself (known as self-induction). The international unit of inductance, the "henry," is named after him.

See also **Units and Standards**.

J. M. I.

HENRY'S LAW.   The concentration of a chemical species in a liquid solution is proportional to its gas- phase partial pressure. The constant of proportionality is called the *Henry's law constant* and provides a measure of the solubility of a particular compound. For molecules that form ions in solution, Henry's law is modified to take account of the increased solubility as a function of pH.

HEPARIN.   A complex organic acid (mucopolysaccharide) present in mammalian tissue; a strong inhibitor of blood coagulation. Precise chemical formula has not been fully established, but the formula $(C_{12}H_{16}NS_2Na_3)_{20}$, with a molecular weight of 12,000, has been suggested for sodium heparinate. The drug is derived from animal livers or lungs. Heparin is used in deep venous thrombosis therapy. It is also used in rodenticides which cause internal hemorrhaging. Pets exposed to such poisons must receive immediate treatment with the administration of vitamin K, also sometimes called the antihemorrhagic vitamin. See also **Anticoagulants**; and **Vitamin K**.

HEPATITIS.   See **Liver**; and **Virus**.

HEPATITIS VIRUSES.   Viral hepatitis is a major public health problem throughout the world, affecting many hundreds of millions of people. Hepatitis is a cause of considerable morbidity and mortality in the human population both from acute infection and chronic sequelae, which include, with hepatitis B and hepatitis C infection, chronic liver damage, chronic active hepatitis, cirrhosis and primary liver cancer.

Classification

The hepatitis viruses include a range of unrelated human pathogens:

- *Hepatitis A virus* (HAV) is a small unenveloped isometric ribonucleic acid (RNA) virus which shares many of the characteristics of the picornavirus family. This virus has been classified in the genus *Hepatovirus* within the family *Picornaviridae* and is the cause of hepatitis A, previously known as infectious or epidemic hepatitis, transmitted by the faecal–oral route.
- *Hepatitis B virus* (HBV) is a member of the family *Hepadnaviridae*, double-stranded deoxyribonucleic acid (DNA) viruses which replicate by reverse transcription. The virus is endemic in the human population and hyperendemic in many parts of the world; it is transmitted essentially by blood-to-blood contact and by the sexual route. Natural hepadnavirus infections also occur in other mammals, including woodchucks, beechy ground squirrels and ducks.
- *Hepatitis C virus* (HCV) is an enveloped single-stranded RNA virus which is related distantly to flaviviruses. Seroprevalence studies confirm the importance of the parenteral route of transmission and transmission by blood and blood products, but in many carriers and patients the origin of the infection has not been identified, although intravenous drug abuse appears to be the most common factor in Western countries. Several genotypes and subtypes have been described. Infection with this virus is common in many countries, and it is associated with chronic liver disease and also with primary liver cancer.
- *Hepatitis δD virus* (HDV) is an unusual virus with single-stranded circular RNA and a number of similarities to certain plant viral satellites and viroids. It requires hepadnavirus helper functions for propagation and it is an important and common cause of acute and severe chronic liver damage in some regions of the world. The modes of transmission are similar to those of HBV.
- *Hepatitis E virus* (HEV) is an enterically transmitted, nonenveloped, single-stranded RNA virus, which shares many biophysical and biochemical features with caliciviruses. However, HEV open reading frame (ORF) 1 sequences bear a closer similarity to those of *Rubella virus* and plant furoviruses. HEV is an important cause of large epidemics of acute hepatitis in the subcontinent of India, central and southeast Asia,

the Middle East, parts of Africa and elsewhere. Infection with this virus causes high mortality during the third trimester of pregnancy.
- The GB viruses and *Hepatitis G virus* (GBV-A, B, C and GBV-C/HGV): Two viruses, GBV-A and GBV-B, were identified in infectious plasma of tamarins inoculated with serum from a patient (G.B.) with acute hepatitis. A third virus, GBV-C, was isolated subsequently from a human hepatitis specimen. The following year another virus, named *Hepatitis G virus* (HGV), was described as a new transfusion-transmitted agent; however, HGV represents an independent isolate of GBV-C. There is no evidence that these viruses cause acute hepatitis or chronic liver disease in humans.

Phylogenetic studies showed that the GBV viruses are distinct viruses and they are not genotypes of HCV. The three GB viruses and HCV share limited overall amino acid sequence homology. The genomic organization of the GBV viruses shows that they are single-stranded RNA viruses with local regions of sequence identity with various flaviviruses.

Hepatitis A

Outbreaks of jaundice have been described frequently for many centuries and the term "infectious hepatitis" was coined in 1912 to describe the epidemic form of the disease. HAV is spread by the faecal–oral route and continues to be endemic throughout the world and hyperendemic in areas with poor standards of sanitation and hygiene. The seroprevalence of antibodies to HAV has declined since World War II in many countries and infection results most commonly from person-to-person contact, but large epidemics do occur. For example, an outbreak of hepatitis A associated with the consumption of clams in Shanghai in 1988 resulted in almost 300,000 cases. The subject was reviewed by [Koff 1998].

Structure of Hepatitis A Virus.

HAV is a small, nonenveloped spherical particle with cubic symmetry, measuring on average 27 nm in diameter (Fig. 1), typical of picornaviruses. Sequence comparisons with other picornaviruses showed limited homology with the enteroviruses and rhinoviruses, although the structure and genome organization is typical of the *Picornaviridae*. HAV is now classified in the *Hepatovirus* genus.

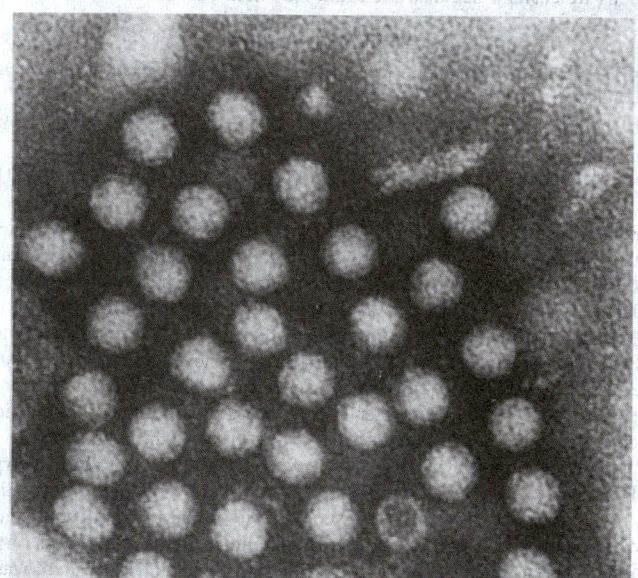

Fig. 1.   Electron micrograph of *Hepatitis A virus* in a faecal extract, ×400000. From a series by Thornton and Zuckerman.

The buoyant density of HAV is $1.33-1.34$ g mL^{-1} in caesium chloride, and the sedimentation coefficient is $156-160$ S. The virus is stable at low pH, and is resistant to degradation by environmental conditions.

There is one human serotype of HAV. Seven genotypes have been identified from wild-type strains, but all have a highly conserved single immunodominant epitope which is responsible for generating neutralizing antibodies. Simian strains are distinct genetically and have phenotypic differences from human strains. Infectivity studies using the simian strains suggest that these may constitute naturally attenuated variants.

Organization and Replication of Hepatitis A Virus.   The HAV genome comprises about 7500 nucleotides of positive-sense linear, single-stranded

RNA which is polyadenylated at the 3′-end and has a polypeptide (VPg) attached to the 5′-end. A single, large ORF occupies most of the genome and encodes a polyprotein with a theoretical molecular mass of M_r 252000. An untranslated region of around 735 nt precedes the ORF. Secondary structure within this region of the genome may be important for efficient translation of the RNA. There is also a short untranslated region at the 3′-end of the HAV genome. There are four capsomeric structural proteins and seven nonstructural proteins. The virus replicates in the cytoplasm of hepatocytes, sharing the features of replication of the picornaviruses.

Epidemiology. Hepatitis A (infectious or epidemic hepatitis) occurs endemically in all parts of the world, with frequent reports of minor and major outbreaks. The exact incidence is difficult to estimate because of the high proportion of subclinical infections and infections without jaundice, differences in surveillance and differing patterns of disease. The degree of underreporting is very high.

HAV enters the body by ingestion. The virus then spreads, probably by the bloodstream, to the liver, a target organ, where it replicates in the hepatocytes. Large numbers of virus particles are detectable in faeces during the incubation period, beginning as early as 10–14 days after exposure and continuing, in general, until peak elevation of serum aminotransferases. Virus is also detected in faeces early in the acute phase of illness but relatively infrequently after the onset of clinical jaundice. Immunoglobulin G (IgG) antibody to HAV that persists is also detectable late in the incubation period, coinciding approximately with the onset of biochemical evidence of liver damage. The virus does not persist and chronic excretion of HAV does not occur. There is no evidence of progression to chronic liver disease.

The mode of transmission of HAV is by the faecal–oral route, most commonly by person-to-person contact, and infection occurs readily under conditions of poor sanitation and overcrowding. Common-source outbreaks are initiated most frequently by faecal contamination of water and food, but water-borne transmission is not a major factor in maintaining this infection in industrialized communities. On the other hand, many food-borne outbreaks have been reported. This can be attributed to the shedding of large quantities of virus in the faeces during the incubation period of the illness in infected food handlers; the source of the outbreak can often be traced to uncooked food or food that has been handled after cooking. Although hepatitis A remains endemic and common in the developed countries, the infection occurs mainly in small clusters, often with only a few identified cases.

Hepatitis A is recognized as an important travel-related infection in travelers from low prevalence areas to endemic countries. Two other sectors of the population are at an increased risk of infection with HAV: those who undertake oral–anal sexual practices and homosexuality and injecting drug users. The latter group is at risk because of a combination of poor personal hygiene, faecal contamination of injection equipment which is often shared, the use of water drawn from toilet pans to dissolve drugs and possible contamination of illicit drugs, which are transported in the intestine after swallowing or carriage in the rectum.

Hepatitis A is rarely transmitted by blood transfusion, although transmission by inadequately inactivated and treated blood coagulation products has been reported, as have cases in patients with cancer treated with lymphokine-activated killer cells and interleukin 2 prepared with tissue culture medium supplemented with pooled human serum.

The incubation period of hepatitis A is 3–5 weeks, with a mean of 28 days. Subclinical and anicteric cases are common and, although the disease has in general a low mortality rate, patients may be incapacitated for many weeks. There is no evidence of progression to chronic liver damage.

Clinical Features and Laboratory Diagnosis. Inapparent or subclinical infections and infection without jaundice are common with all the different hepatitis viruses. The clinical picture ranges from an asymptomatic infection to a mild anicteric illness, to acute disease with jaundice, to severe prolonged jaundice, to fulminant hepatitis.

Differences between the clinical syndromes of acute hepatitis A, acute hepatitis B and other types of viral hepatitis become apparent on analysis of large numbers of well-documented cases, but these differences are not reliable for the diagnosis of individual patients with jaundice.

The following description of the acute disease applies to all types of viral hepatitis. Prodromal nonspecific symptoms such as fever, chills, headache, fatigue, malaise and aches and pains are followed a few days later by anorexia, nausea, vomiting and right upper quadrant abdominal pain, followed by the passage of dark urine and clay coloured stools. Jaundice of the sclera and the skin develop. With the appearance of jaundice, there is usually a rapid subjective improvement of symptoms. The jaundice usually deepens for a few days and persists for 1–2 weeks. The faeces then darken and the jaundice diminishes over a period of about 2 weeks. Convalescence may be prolonged. See also **Jaundice** and **Liver**.

HAV and HEV do not persist in the liver and there is no evidence of progression to chronic liver damage. HBV, with or without its satellite HDV, and HCV, may be associated with persistent infection, a prolonged carrier state and progression to chronic liver disease, which may be severe. There is an aetiological association between HBV and HCV and hepatocellular carcinoma. GBV-C virus tends to cause persistent infection.

Various serological diagnostic tests are available for markers of infection with each of the hepatitis viruses, mostly based on enzyme-linked immunosorbent assays (ELISAs) and various nucleic acid amplification techniques.

Prevention and Control. In areas of high prevalence, most children are infected early in life and such infections are generally asymptomatic. Less than 10% of cases of acute hepatitis A in children up to the age of 6 years are icteric, but this increases to 40–50% in the 6–14 age group and to 70–80% in adults. Of 115,551 cases of hepatitis A in the USA between 1983 and 1987, only 9% of the cases, but more than 70% of the fatalities, were in those aged over 49.

Patients with chronic liver disease, especially if visiting an endemic area, should be immunized against hepatitis A. Patients with chronic blood-clotting disorders should be immunized. In some developing countries, the incidence of clinical hepatitis A is increasing as improvements in socioeconomic conditions result in infection later in life and strategies for immunization are yet to be developed and agreed.

Since faecal shedding of the virus is at its highest during the late incubation period and the prodromal phase of the illness, strict isolation of patients is not a useful control measure. Spread of hepatitis A is reduced by simple hygienic measures and the sanitary disposal of excreta.

Passive Immunization. Normal human immunoglobulin, containing at least 100 IU mL^{-1} of antihepatitis A antibody, given intramuscularly before exposure to the virus or early during the incubation period will prevent or attenuate a clinical illness. The dosage should be at least 2 IU of antihepatitis A antibody per kilogram body weight but in special cases, such as in pregnancy or in patients with liver disease, that dosage may be doubled. Immunoglobulin does not always prevent infection and excretion of HAV, and inapparent or subclinical hepatitis may develop. The efficacy of passive immunization is based on the presence of hepatitis A antibody in the immunoglobulin, and the minimum titre of antibody required for protection is believed to be about 10 IU L^{-1}.

Immunoglobulin is used most commonly for close personal contacts of patients with hepatitis A and for those exposed to contaminated food. Immunoglobulin has also been used effectively for controlling outbreaks in institutions such as homes for the mentally handicapped and in nursery schools. Prophylaxis with immunoglobulin is recommended for persons without hepatitis A antibody visiting highly endemic areas. After a period of 6 months the administration of immunoglobulin for travelers needs to be repeated, unless it has been demonstrated that the recipient had developed hepatitis A antibodies. Active immunization for travelers is strongly recommended, and it is the most common vaccine-preventable infection in travelers.

Active Immunization. In areas of high prevalence, most children have antibodies to HAV by the age of 3 years and such infections are generally asymptomatic. Infections acquired later in life are of increasing clinical severity. It is important, therefore, to protect those at risk because of personal contact or because of travel to highly endemic areas.

Other groups at risk of hepatitis A infection include staff and residents of institutions for the mentally handicapped, those in day care centers for children, sexually active male homosexuals, intravenous narcotic drug abusers, food handlers, sewage workers, health care workers, military personnel, and certain low socioeconomic groups in defined community settings.

In some developing countries, the incidence of clinical hepatitis A is increasing as improvements in socioeconomic conditions result in infection later in life; protection by immunization would be prudent but strategies are yet to be agreed. Global control of hepatitis A will require universal immunization of infants and will become possible when hepatitis A vaccine is combined in a polyvalent form with other childhood vaccines, such as diphtheria, pertussis, tetanus, measles, rubella, mumps and hepatitis B vaccines.

Killed Hepatitis A Vaccines. These vaccines are prepared from virus grown in tissue culture and inactivated with formalin. The first such vaccine was licensed in 1992 and several preparations are available, including a combined hepatitis A and B vaccine. These vaccines are highly immunogenic and provide long-term protection against infection.

Hepatitis B

Hepatitis B was originally referred to as "serum hepatitis," and HBV is an important and frequent cause of acute and chronic infection of the liver in many countries.

More than a third of the world's population has been infected with HBV, and the World Health Organization estimates that it results in 1–2 million deaths every year.

The clinical features of acute infection resemble those of the other viral hepatitides. The virus persists in approximately 5–10% of immunocompetent adults, and in as many as 90% of infants infected perinatally. Persistent carriage of HBV, defined by the presence of hepatitis B surface antigen (HBsAg) in the serum for more than 6 months, has been estimated to affect about 350 million people worldwide. Long-term continuing virus replication may lead to progression to chronic liver disease, cirrhosis and hepatocellular carcinoma.

Hepatitis B was recently reviewed by Harrison, et al.: [2004].

Structure and Organization of Hepatitis B Virus. The hepatitis B virion is a 42 nm particle comprising an electron-dense core (nucleocapsid), 27 nm in diameter, surrounded by an outer envelope of HBsAg embedded in membranous lipid derived from the host cell (Fig. 2). The surface antigen, originally referred to as Australia antigen, is produced in excess by the infected hepatocytes and is secreted in the form of 22 nm particles and tubular structures of the same diameter. See also **Antigen**.

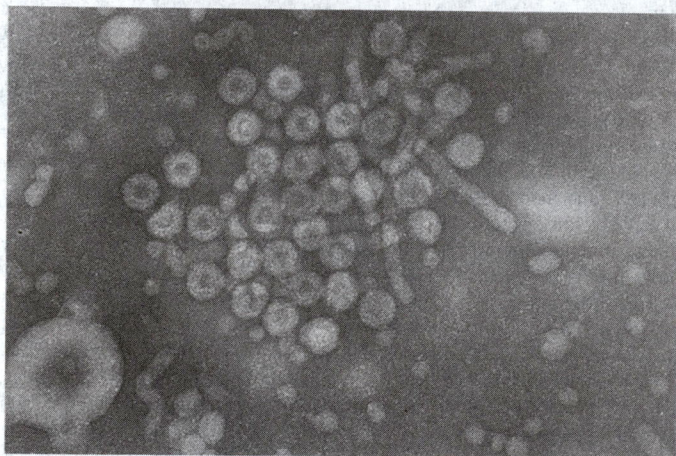

Fig. 2. Electron micrograph of whole serum showing the three morphological forms of hepatitis B: small pleomorphic surface antigen particles; tubular forms and the complete double-shelled hepatitis B virions, ×200000. From a series by Zuckerman et al.

The 22 nm particles are composed of the major surface protein in both nonglycosylated (p24) and glycosylated (gp27) form in approximately equimolar amounts, together with a minority component of the so-called middle proteins (gp33 and gp36), which contain the pre-S2 domain, a glycosylated 55 amino acid *N*-terminal extension. The surface of the virion has a similar composition but also contains the large surface proteins (gp39 and gp42), which include both the pre-S1 and pre-S2 regions. These large surface proteins are not found in the 22 nm spherical particles (but may be present in the tubular forms in highly viraemic individuals) and their detection in serum correlates with viraemia. The domain which binds to the specific HBV receptor on the hepatocyte is believed to reside within the pre-S1 region.

There is a variation in the epitopes presented on the surface of the virions and subviral 22 nm particles so that there are several subtypes of HBV, which differ in their geographical distribution. All isolates of the virus share a common epitope, *a*, a domain of the major surface protein, which

is believed to protrude as a double loop from the surface of the particle. Two other pairs of mutually exclusive antigenic determinants, *d* or *y* and *w* or *r*, are also present on the major surface protein. These variations have been correlated with single-nucleotide changes in the surface ORF that lead to variation in single amino acids in the protein. Four principal subtypes of HBV are recognized: *adw*, *adr*, *ayw* and *ayr*. Subtype *adws* predominates in northern Europe, the Americas and Australasia and also is found in Africa and Asia. Subtype *ayw* is found in the Mediterranean region, eastern Europe, northern and western Africa, the near East and the Indian subcontinent. In the Far East, *adr* predominates but the rarer *ayr* may occasionally be found in Japan and Papua New Guinea.

The nucleocapsid of the virion consists of the viral genome surrounded by the core antigen (HBcAg). The genome, which is approximately 3.2 kb in length, has an unusual structure and is composed of two linear strands of DNA held in a circular configuration by base pairing at the 5′-ends. One of the strands is incomplete and the 3′-end is associated with a DNA polymerase molecule, which is able to complete that strand in the presence of deoxynucleoside triphosphates (Fig. 3).

The genomes of many isolates of HBV have been cloned and the complete nucleotide sequences determined. Analysis of the coding potential of the genome reveals four ORFs, which are conserved between all of these isolates, but there is some variation in sequence of up to 17% of nucleotides.

The first ORF encodes the various forms of the surface protein and contains three in-frame methionine codons, which are used for initiation of translation. A second promoter is located upstream of the pre-S1 initiation codon. This directs the synthesis of a 2.4-kb messenger RNA (mRNA), which is coterminal with the other surface messages and is translated to yield the large (pre-S1) surface proteins.

The core ORF also has two in-phase initiation codons. The "precore" region is highly conserved, has the properties of a signal sequence and is responsible for the secretion of HBeAg.

The third ORF, which is the largest and overlaps the other three, encodes the viral polymerase. This protein appears to be another translation product of the 3.5 kb RNA, and is synthesized apparently following internal initiation of the ribosome. The *N*-terminal domain is believed to be the protein primer for minus-strand synthesis. There is then a spacer region followed by the (RNA- and DNA-dependent) DNA polymerase.

The fourth ORF was designated "x" because the function of its small gene product was not known initially. However, x has now been demonstrated to be a transcriptional transactivator, and may enhance the expression of other viral proteins.

Surface Antigen Mutants. Production of antibodies to the group antigenic determinant *a* mediates cross-protection against all subtypes, as has been demonstrated by challenge with a second subtype of the virus following recovery from an initial experimental infection. The epitope *a* is located in the region of amino acids 124–148 of the major surface protein, and appears to have a double-loop conformation. See also **Epitopes**.

A monoclonal antibody, which recognizes a region within this *a* epitope is capable of neutralizing the infectivity of HBV for chimpanzees, and competitive inhibition assays using the same monoclonal antibody demonstrate that equivalent antibodies are present in the sera of subjects immunized with either plasma-derived or recombinant hepatitis B vaccine.

During a study of the immunogenicity and efficacy of hepatitis B vaccines in Italy, a number of individuals who had apparently mounted a successful immune response and became antisurface antibody (anti-HBs)-positive, later became infected with HBV. These cases were characterized by the coexistence of noncomplexed anti-HBs and HBsAg, and there were other markers of hepatitis B infection.

Furthermore, analysis of the antigen using monoclonal antibodies suggested that the *a* epitope was either absent or masked by antibody. Subsequent sequence analysis of the virus from one of these cases revealed a mutation in the nucleotide sequence encoding the *a* epitope, the consequence of which was a substitution of arginine for glycine at amino acid position 145.

There is now considerable evidence for a wide geographical distribution of the point mutation in HBV from guanosine to adenosine at position 587, resulting in an amino acid substitution at position 145 from glycine to arginine in the highly antigenic group determinant *a* of the surface antigen. This is a stable mutation which has been found in viral isolates from children and adults and it has been described in Italy, Singapore, Japan, Brunei, Taiwan, India, the USA and elsewhere and from liver transplant

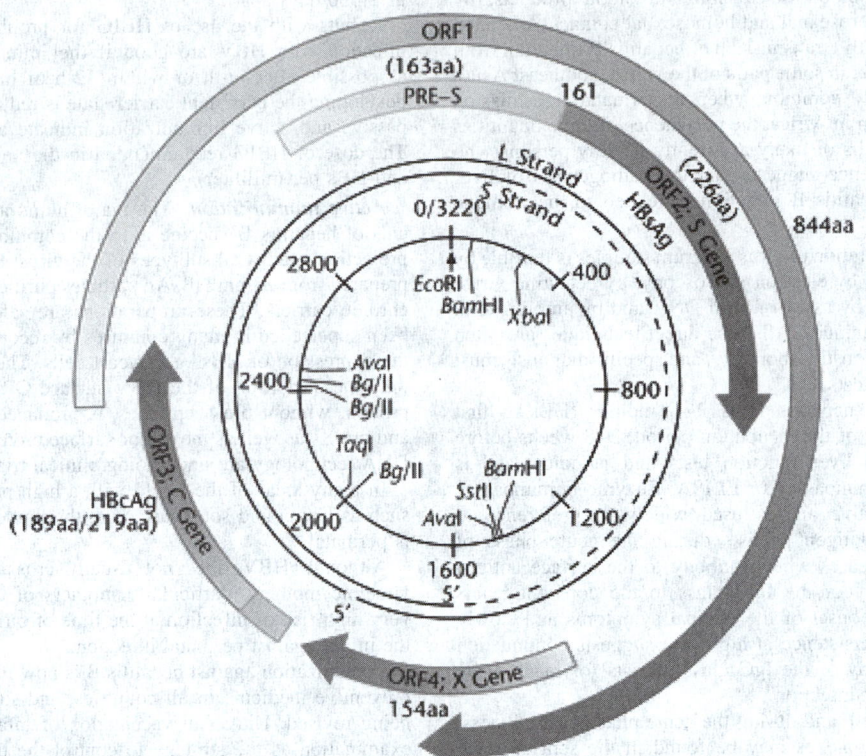

Fig. 3.   The genomic structure of *Hepatitis B virus*. aa, amino acids; HBcAg, hepatitis B core antigen; HBsAg, hepatitis B surface antigen; ORF, open reading frame.

recipients with hepatitis B, in the USA, Germany and the UK, who had been treated with specific hepatitis B immunoglobulin or humanized hepatitis B monoclonal antibody, and in patients with chronic hepatitis in Japan and elsewhere. Other point mutations and substitutions have also been described. The 145 mutation appears to be the most common.

The region in which this mutation occurs is an important virus epitope to which vaccine-induced neutralizing antibody binds, as discussed above, and the mutant virus is not neutralized by antibody to this specificity. The 145 variant virus can replicate as a competent virus, implying that the amino acid substitution does not alter the attachment of the virus to the liver cell. Variants of HBV with altered antigenicity of the envelope protein show that HBV is not as antigenically singular as believed previously and that humoral escape mutation can occur *in vivo*. This finding gives rise to two causes for concern: failure to detect HBsAg may lead to transmission through donated blood or organs; and HBV may infect individuals who are anti-HBs-positive after immunization [reviewed by Oon, et al.: 1995 and Zuckerman, 2000].

Mathematical modeling suggests that HBV variants may become dominant over the current wild-type virus in 50–100 years.

Pre-Core Mutants. In 1988, a report was published on the nucleotide sequence of the genome of a strain of HBV cloned from the serum of a naturally infected chimpanzee. A surprising feature was a point mutation in the penultimate codon of the pre-core region, which changed the tryptophan codon (TGG) to an amber termination codon (TAG). The nucleotide sequence of the HBV pre-core region from a number of anti-HBe-positive Greek patients was investigated by direct sequencing polymerase chain reaction (PCR)-amplified HBV DNA from serum. An identical mutation of the penultimate codon of the pre-core region to a termination codon was found in seven of eight anti-HBe-positive patients who were positive for HBV DNA in serum by hybridization. In most cases, there was an additional mutation in the preceding codon.

Similar variants were found by amplification of HBV DNA from serum from anti-HBe-positive patients in Italy and Greece. These variants are not confined to the Mediterranean region; the same nonsense mutation (without a second mutation in the adjacent codon) has been observed in patients from Japan and other countries, along with rarer examples of defective pre-core regions caused by frameshifts or loss of the initiation codon for the pre-core region.

In many cases, pre-core variants have been described in patients with severe chronic liver disease and who may have failed to respond to therapy with interferon. This observation has raised the question of whether the variants are more pathogenic than the wild-type virus.

Replication of Hepatitis B Virus. HBV and the recognized animal hepadnaviruses are unique among animal DNA viruses in that they replicate through an RNA intermediate [reviewed by Birkenmeyer, 2004].

When the virus infects the hepatocyte, the viral DNA is uncoated and converted into a covalently closed circular form in the nucleus, which is the template for transcription of the viral RNAs. There are at least four viral promoters and all of the RNAs are 3'-coterminal, being polyadenylated in response to a signal in the core ORF. The largest RNAs are longer than the genome of the virus, and while some act as mRNAs for the synthesis of e antigen, core antigen and the viral polymerase, a subset are intermediates for the synthesis of progeny genomes. Binding of the polymerase to secondary structure at the 5'-end of the pregenome leads to packaging into immature viral cores in the cytoplasm of the hepatocytes.

The *N*-terminal domain of the viral polymerase acts as the primer for minus-strand DNA synthesis, and, following synthesis of a four nucleotide nascent strand, translocates to a complementary four base sequence near the 3'-end of the RNA template. This protein remains covalently linked to the 5'-end of that strand in the mature virus. Minus-strand synthesis follows by reverse transcription of the pregenome by the polymerase. The remaining oligoribonucleotide is at the position of the direct repeat, DR1, and is believed to translocate to the other copy of the direct repeat, DR2, on the minus strand and to prime synthesis of the plus strand. The minus strand has a short terminal redundancy of approximately eight nucleotides, which permits circularization of the genome as the plus strand is synthesized. The cores are then coated with the surface antigen to form mature virus particles.

Epidemiology. Although various body fluids (blood, saliva, menstrual and vaginal discharges, serous exudates, seminal fluid and breast milk) have been implicated in the spread of infection, infectivity appears to be especially related to blood and to body fluids contaminated with blood. The epidemiological propensities of this infection are therefore wide;

they include infection by inadequately sterilized syringes and instruments, transmission by unscreened blood transfusion and blood products, by close contact and by both heterosexual and homosexual contact. Antenatal (rarely) and perinatal (frequently) transmission of hepatitis B infection from mother to child may take place; in some parts of the world (southeast Asia), perinatal transmission is very common, whereas perinatal transmission does not appear to be common in Africa. the persistence of large quantities of surface antigen in liver cells of many apparently healthy persons who are carriers. Additional evidence suggests that the pathogenesis of liver damage in the course of hepatitis B infection is related to the immune response by the host.

Diagnosis. Direct demonstration of virus in serum samples is feasible by visualizing the virus particles by electron microscopy, by detecting virus-associated DNA polymerase, by assay of viral DNA and by amplification of viral DNA by various techniques. All these direct techniques are often impractical in the general diagnostic laboratory, and specific diagnosis must therefore rely on serological tests.

The incubation period of hepatitis B is 2–6 months. HBsAg first appears during the late stages of the incubation period, 2–8 weeks before the appearance of abnormal liver function tests and jaundice, and is easily detectable by radioimmunoassay or ELISA. Enzyme immunoassay is specific and highly sensitive and is used widely in preference to radioisotope methods. The antigen persists during the acute phase of the disease and sharply decreases when antibody to the surface antigen becomes detectable. Antibody of the IgM class to the core antigen is found in the serum after the onset of the clinical symptoms and slowly declines after recovery. Its persistence at high titre suggests continuation of the infection. Core antibody of the IgG class persists for many years and provides evidence of past infection.

During the incubation period, and during the acute phase of the illness, surface antigen–antibody complexes may be found in the sera of some patients.

It should be noted that transmission of the infection may result from accidental inoculation of minute amounts of blood, or fluid contaminated with blood, during medical, surgical and dental procedures; immunization with inadequately sterilized syringes and needles; intravenous and percutaneous drug abuse; tattooing; ear and nose piercing; acupuncture; laboratory accidents and accidental inoculation with razors and similar objects that have been contaminated with blood. Additional factors may be important for the transmission of hepatitis B infection in the tropics; these include traditional tattooing and scarification, bloodletting, ritual circumcision and repeated biting by blood-sucking arthropod vectors. Investigation of the role that biting insects may play in the spread of hepatitis B has yielded conflicting results. HBsAg has been detected in several species of mosquito and in bed bugs that were either trapped in the wild or fed experimentally on infected blood, but no convincing evidence of replication of the virus in insects has been obtained. Mechanical transmission of the infection, however, is a possibility, but does not appear to be an important route of transmission of HBV.

Immune Responses. Antibody and cell-mediated immune responses to various types of antigen are induced during the infection; however, not all these are protective and, in some instances, may cause autoimmune phenomena that contribute to disease pathogenesis. The immune response to infection with HBV is directed towards at least four antigens: hepatitis B surface antigen, the core antigen, the e antigen and the X region. The view that hepatitis B exerts its damaging effect on hepatocytes by direct cytopathic changes is inconsistent with Immune complexes have been found by electron microscopy in the sera of all patients with fulminant hepatitis, but are seen only infrequently in nonfulminant infection. Immune complexes are also important in the pathogenesis of other disease syndromes characterized by severe damage of blood vessels (for example, polyarthritis nodosa, some forms of chronic glomerulonephritis and infantile papular acrodermatitis). See also **Electron Microscope**.

Passive Immunization. Hepatitis B immunoglobulin (HBIG) is prepared specifically from pooled plasma with high titre of hepatitis B surface antibody and may confer temporary passive immunity under certain defined conditions. The major indication for the administration of HBIG is a single acute exposure to HBV, such as occurs when blood containing surface antigen is inoculated, ingested or splashed on to mucous membranes and the conjunctiva. It should be administered as early as possible after exposure and preferably within 48 h, usually 3 mL (containing 200 IU of anti-HBs per milliliter) in adults. It should not be administered 7 days or more after

exposure. It is generally recommended that two doses of HBIG should be given 30 days apart.

Results with the use of HBIG for prophylaxis in neonates at risk of infection with HBV are good if the immunoglobulin is given as soon as possible after birth or within 12 h of birth, and the risk of the baby developing the persistent carrier state is reduced by about 70%. Combined passive and active immunization indicate an efficacy approaching 90%. The dose of HBIG recommended in the newborn is 1–2 mL (200 IU of anti-HBs per milliliter).

Active Immunization. The major humoral antibody response of recipients of hepatitis B vaccine is to the common *a* epitope, with consequent protection against all subtypes of the virus. First-generation vaccines were prepared from 22 nm HBsAg particles purified from plasma donations from chronic carriers. These preparations are safe and immunogenic but have been superseded in many countries by recombinant vaccines produced by the expression of HBsAg in yeast cells. The expression plasmid contains only the 3′-portion of the HBV surface ORF and only the major surface protein, without pre-S epitopes, is produced. Vaccines containing pre-S2 and pre-S1 as well as the major surface proteins expressed by recombinant DNA technology are undergoing clinical trial.

In many areas of the world with a high prevalence of HBsAg carriage, such as China and southeast Asia, the predominant route of transmission is perinatal.

Although HBV does not usually cross the placenta, the infants of viraemic mothers, particularly mothers of certain ethnic groups, have a very high risk of infection at the time of birth, and immunization protects the infant against perinatal infection.

Immunization against hepatitis B is now recognized as a high priority in preventive medicine in all countries, and strategies for immunization are being revised. Universal vaccination of infants and adolescents is under examination as the strategy to control the transmission of this infection. More than 165 countries now offer hepatitis B vaccine to all children, including the USA, Canada, Italy, France and most western European countries [reviewed by Kane, 1996].

However, immunization against hepatitis B is at present recommended in a number of countries with a low prevalence of hepatitis B only to groups, which are at an increased risk of acquiring this infection. These groups include individuals requiring repeated transfusions of blood or blood products, tissue and organ graft recipients, prolonged inpatient treatment, patients who require frequent tissue penetration or need repeated circulatory access, patients with natural or acquired immune deficiency and patients with malignant disease. Viral hepatitis is an occupational hazard among health care personnel and the staff of institutions for the mentally handicapped, and in some semi closed institutions such as prisons.

High rates of infection with hepatitis B occur in narcotic drug addicts and intravenous drug abusers, sexually active male homosexuals and prostitutes. Individuals working in highly endemic areas are, however, at an increased risk of infection and should be immunized.

Young infants, children and susceptible persons (including travelers) living in certain tropical and subtropical areas where present socioeconomic conditions are poor and the prevalence of hepatitis B is high should also be immunized. It should be noted that in about 30% of patients with hepatitis B the mode of infection is not known and this is, therefore, a powerful argument for universal immunization.

Site of Injection for Vaccination and Antibody Response. Hepatitis B vaccination should be given in the upper arm or the anterolateral aspect of the thigh and not in the buttock. There are over 100 reports of unexpectedly low antibody seroconversion rates after hepatitis B vaccination using injection into the buttock.

Apart from the site of injection there are several other factors that are associated with a poor, or no, antibody response to currently licensed vaccines. Indeed, all studies of antibody response to plasma-derived hepatitis B vaccines and hepatitis B vaccines prepared by recombinant DNA technology have shown that between 5 and 10% or more of healthy immunocompetent subjects do not mount an antibody response (anti-HBs) to the surface antigen component (HBsAg) present in these preparations (nonresponders), or that they respond poorly (hyporesponders). The exact proportion depends partly on the definition of nonresponsiveness or hyporesponsiveness, generally less than 10 IU L^{-1} or 100 IU L^{-1}, respectively, against an international antibody standard.

Hepatitis B and Hepatocellular Carcinoma. Hepatocellular carcinoma (HCC) is one of the 10 most common cancers in the world and one of the most prevalent cancers, particularly among males, in southeast Asia, the western Pacific, sub-Saharan Africa and in a number of other developing regions. Epidemiological data from case–control and cohort studies and laboratory investigations indicate that there is a consistent and specific causal association between infection with HBV and HCC and that up to 80% of such cancers are attributable to this virus. HBV is thus second only to tobacco (cigarette smoking) among the known human carcinogens.

The propensity of HBV for persistence following perinatal infection from a carrier mother, particularly among certain ethnic groups (for example, the Chinese), or infection in early life, and in 10–15% following acute infection with HBV in adults, is an essential factor in the development of HCC because of the relentless progression to chronic liver disease, including chronic active hepatitis and cirrhosis, in approximately 25% of carriers. The increased risk of developing HCC has been estimated to be 100-fold higher in chronic HBV carriers compared with noninfected populations. The highest incidence of HCC is observed in patients with cirrhosis, although this tumor also occurs in those with chronic hepatitis and in asymptomatic carriers.

In addition to the epidemiological evidence, another important link is provided by the related animal hepadnaviruses, and in particular the *Woodchuck hepatitis virus* (WHV). Woodchucks infected chronically in the wild with WHV frequently develop liver cancer, and virtually all animals infected experimentally during the neonatal period develop liver tumors. However, the mechanisms of hepatocarcinogenesis of HBV remain largely unknown, although one key event is essential, namely the integration of viral DNA into the host cell chromosomal DNA, where it may activate or suppress cellular genes involved in cell growth and proliferation. Integration is known to occur at random and is likely to mediate rearrangement of both cellular and viral DNA, often leading to novel *trans*-activation and alteration of functional cell proteins and cellular gene expression.

It follows, therefore, that if integration of viral DNA can be prevented then cell transformation and neoplastic changes will not occur. This crucial step has been described by Chang, et al.: [1997], providing at the same time evidence for a direct causal relationship between HBV and HCC. Immunization against hepatitis B in Taiwan reduced carriage of hepatitis B in children aged 6 years from about 10% between 1981 and 1986 to between 0.9 and 0.8% between 1990 and 1994. This highly significant reduction in the prevalence of carriage of hepatitis B surface antigen was accompanied by a sharp decline in the average annual incidence of HCC in children and adolescents. Similar observations have been described since from Hong Kong, Singapore and the Gambia.

Antiviral Therapy for Chronic Hepatitis B: Interferon α. Several preparations of interferon α have been licensed for treatment of hepatitis B. These preparations consist either of a mixture of interferon species derived from virus-stimulated Namalwa cells or a single-component preparation produced by recombinant DNA techniques. Pegylated interferon (PEG interferon) is now in use. See also **Interferons**.

About 35–40% of patients with HBeAg respond to treatment with 2.5–5 million IU two or three times weekly for 4–6 months. Patients with anti-HBe with core antigen in the hepatocytes and chronic active hepatitis due to HBV core mutants may have severe liver disease. Approximately 10–25% have long-term responses to treatment with 9–10 million IU three times weekly for 6–12 months.

Nucleoside Analogues. Several nucleoside analogues have been licensed for treatment, including famciclovir, lamivudine and adefovir dipivoxil. These drugs are given orally. Combination therapy is also being used. See also **Antiviral Drugs**.

Hepatitis C

Attempts to clone the agent of parenterally transmitted non-A, non-B hepatitis were made, from a plasma known to contain high titres of the agent, by experimental transmission to nonhuman primates. Because it was not known whether the genome was DNA or RNA, a denaturation step was included before the synthesis of complementary DNA (cDNA) so that either DNA or RNA could serve as a template. The resultant cDNA was then inserted into the bacteriophage expression vector λ gt11 and the libraries screened using serum from a patient with chronic non-A, non-B hepatitis.

This led to the detection of a clone, which was found to bind to antibodies present in the sera of several individuals infected with non-A, non-B hepatitis. This clone was used as a probe to detect a larger, overlapping clone in the same library. It was possible to demonstrate that these sequences hybridized to a positive-sense RNA molecule of about 10000 nt, that was present in the livers of infected chimpanzees but not in uninfected controls. Homologous sequences were not detected in the chimpanzee or human genomes. By employing a "walking" technique, the newly detected overlapping clones were used as hybridization probes to detect further virus-specific clones in the library. Thus, clones covering the entire viral genome were assembled and the complete nucleotide sequence determined [reviewed by McGarvey, et al.: 1998].

Structure of Hepatitis C Virus. The genome of HCV comprises about 10,000 nt of positive-sense RNA and lacks a 3′-poly(A) tract; it has a genome organization similar to that of *Flaviviridae* and is considered the prototype of a third genus of this family. All of these genomes contain a single large ORF, which is translated to yield a polyprotein (of around 3,000 amino acids in the case of HCV) from which the viral proteins are derived by post-translational cleavage and other modifications. See also **Flaviviruses**.

The amino acid sequence of the nucleocapsid protein is highly conserved among different isolates of HCV. The next domain in the polyprotein also has a signal sequence at its *C*-terminus. The product is a glycoprotein which is probably found in the viral envelope and is variably termed E1/S or gp35. The third domain may be cleaved by a protease within the viral polyprotein to yield what is probably a second surface glycoprotein, E2/NS1 or gp70. These proteins are the focus of considerable interest because of their potential use for tests for the direct detection of viral proteins and for HCV vaccines. Nucleotide sequencing reveals that both domains contain hypervariable regions.

The nonstructural region of the HCV genome is divided into regions NS2–5. In the flaviviruses, NS3 has two functional domains: a protease, which is involved in cleavage of the nonstructural region of the polyprotein; and a helicase, which is presumably involved in RNA replication. Motifs within this region of the HCV genome have homology to the appropriate consensus sequences, suggesting similar functions. NS5 seems to be the replicase and contains the Gly-Asp-Asp motif common to viral RNA-dependent RNA polymerases. The HCV protease is a major target for developing specific antiviral drugs. The HCV NS5 is cleaved to yield NS5a and NS5b. NS5b is likely to be the HCV replicase and NS5a may also be involved in genome replication.

HCV comprises a family of highly related but nevertheless several distinct genotypes and various subtypes with differing geographical distribution. The C, NS3 and NS4 domains are the most highly conserved regions of the genome, and therefore these proteins are the most suitable for use as capture antigens for broadly reactive tests for antibodies to HCV.

The degree of divergence apparent within the viral envelope proteins implies the absence of a broad cross-neutralizing antibody response to infection by viruses of different genotypes. In addition, there is considerable sequence heterogeneity among almost all HCV isolates in the *N*-terminal region of E2/NS1, suggesting that this region may be under strong immune selection. Sequence changes within this region may occur during the evolution of disease in individual patients and may play an important role in progression to chronicity.

Epidemiology. Infection with HCV occurs throughout the world. Many of the seroprevalence data are based on blood donors, who represent a selected population. The prevalence of antibodies to HCV in blood donors varies from 0.02 to 1.25% in different countries. Higher rates have been found in southern Italy, Spain, Central Europe, Japan and parts of the Middle East, with as many as 19% in Egyptian blood donors. Until screening of blood donors was introduced, hepatitis C accounted for the vast majority of non-A, non-B post-transfusion hepatitis. However, it is clear that, while blood transfusion and the transfusion of blood products are efficient routes of transmission of HCV, these account for a small proportion of cases of acute clinical hepatitis in a number of countries (with the exception of patients with hemophilia). Current data indicate that in 50% or more of patients in industrialized countries, the source of infection cannot be identified, although transmission by contact with blood and contaminated materials is likely to be important; 35% of patients have a history of intravenous drug misuse; household contact and sexual exposure do not appear to be major factors in the epidemiology of this common infection and occupational exposure in the healthcare setting accounts for about 2% of cases. Transmission of HCV from mother to infant occurs in about 10% of viraemic mothers and the risk appears to be related to the level of viraemia. It should be noted, however, that information on the natural history of hepatitis C is limited because the onset of the infection is often unrecognized and the early course of the disease is indolent and protracted in most patients.

Clinical Features. Most acute infections are asymptomatic: less than 30% of patients with acute infections have nonspecific symptoms and some develop mild jaundice. Fulminant hepatitis has been described. Extrahepatic manifestations include mixed cryoglobulinemia, membranous proliferative glomerulonephritis and porphyria cutanea tarda.

About 80% of patients do not clear the virus by 6 months and develop chronic hepatitis. The majority have fluctuating abnormal alanine transaminase levels, but some 30% have normal levels. Histological examination of liver biopsies from asymptomatic HCV-carriers (blood donors) reveals that none has normal histology and that up to 70% have chronic active hepatitis and/or cirrhosis. The rate of progression of chronic hepatitis is highly variable. Whether the virus is cytopathic or whether there is an immunopathological element remains unclear. The presence of antibodies to specific antigen components is variable and may or may not reflect viraemia, and in the case of interferon treatment there is a correlation between response and loss of specific antibodies to the E2 component.

Detection and monitoring of viraemia are important for management and treatment, and sensitive techniques are available for the measurement of HCV RNA. The identification of specific types and subtypes is important, with observations suggesting an association between response to interferon and particular genotypes, and the possibility that different types may differ in their pathogenicity.

Chronic hepatitis C infection leads to cirrhosis within two decades of the onset of infection in at least 20% of patients. Chronic infection is also associated with an increased risk of hepatocellular carcinoma, which occurs on a background of inflammation and regeneration related to chronic hepatitis over three or more decades. The risk of developing HCC is estimated at 1–5% after 20 years, but this varies considerably in different areas of the world. It develops more commonly in men than in women.

Control and Treatment. A vaccine against HCV is not available. There is evidence that alcohol and hepatitis C may aggravate synergistically hepatic damage. Alcohol restriction is essential and abstinence from alcohol is strongly recommended.

Interferon α is indicated for treatment of patients with chronic hepatitis C who have circulating HCV RNA, elevated serum aminotransferases and histological evidence of liver damage. Approximately 50% of patients will respond initially to 3 million IU three times weekly, but many will relapse after stopping treatment for 6 months. Between 15 and 25% of patients will have a sustained virological response and, usually, histological improvement. PEG interferon is more effective. See also **Interferons**.

Treatment is now prolonged to 12 months. Combination therapy with ribavirin, a synthetic guanosine nucleoside analogue, along with PEG interferon is more effective than interferon alone. Triple therapy with amantadine and other drugs is being evaluated.

Hepatitis D

δ Hepatitis was first recognized following detection of a novel protein, δ antigen, by immunofluorescent staining in the nuclei of hepatocytes from patients with chronic active hepatitis B. HDV is now known to require an HBV helper function for its transmission. HDV is coated with HBsAg, which is needed for release from the host hepatocyte and for entry in the next round of infection.

Two forms of δ hepatitis infection are known. In the first, a susceptible individual is coinfected with HBV and HDV, often leading to a more severe form of acute hepatitis caused by HBV. In the second, an individual chronically infected with HBV becomes superinfected with HDV. This may cause a second episode of clinical hepatitis and accelerate the course of the chronic liver disease, or cause overt disease in asymptomatic carriers of HBV. HDV is cytopathic and hepatitis δ antigen (HDAg) may be directly cytotoxic. δ Hepatitis is common in the Mediterranean region, parts of eastern Europe, the Middle East, Africa and South America. It has been estimated that 5% of HBsAg carrier's worldwide (approximately 18 million people) are infected with HDV. In areas of low prevalence of HBV, those at risk of hepatitis B, particularly intravenous drug abusers, are also at risk of HDV infection.

Properties and Structural Organization of Hepatitis δ Virus. HDV is approximately 36 nm in diameter, with an RNA genome associated with HDAg, surrounded by an outer coat of HBsAg. The genome is a closed-circular RNA molecule of 1679 nucleotides, resembling those of

the satellite viroids and virusoids of plants, and seems to be replicated by the host RNA polymerase II with autocatalytic cleavage and circularization of the progeny genomes via *trans*-esterification reactions. Consensus sequences of viroids that are believed to be involved in these processes also are conserved in HDV.

Unlike the plant viroids, HDV codes for a protein, HDAg. This is encoded in an ORF in the antigenomic RNA, but four other ORFs which are also present in the genome do not appear to be used.

Other Features of Hepatitis D. Laboratory diagnosis in acute infection is based on specific serological tests for anti-HDV IgM or HDV RNA or HDAg in serum. Acute infection is usually self limited and markers of HDV infection often disappear within a few weeks.

Superinfection with HDV in chronic hepatitis B may lead to suppression of HBV markers during the acute phase. Chronic infection with HDV (and HBV) is the usual outcome in nonfulminant disease. Outbreaks of severe hepatitis with high mortality have been reported in Indians of the Amazon Basin and in areas of Central Africa.

Antibody to δ hepatitis has been found in most countries, commonly among intravenous drug abusers, patients with hemophilia and those requiring treatment by blood and blood products. A high prevalence of infection has been found in Italy and the countries bordering the Mediterranean, eastern Europe and particularly Romania; the former Soviet Union; South America and particularly the Amazon Basin, Venezuela, Columbia (hepatitis de Sierra Nevada de Santa Marta), Brazil (Labrea black fever) and Peru; and parts of Africa, particularly West Africa.

The ratio of clinical to subclinical cases of HDV and superinfection is not known; however, the general severity of both forms of infection suggests that most cases are clinically significant. A low persistence of infection occurs in 1–3% of acute infections and about 80% or higher in superinfection of chronic HBV carriers. The mortality rate is high, particularly in the case of superinfection, ranging from 2 to 20% [reviewed by Hadziyannis, 1997].

Prevention and Control. Prevention and control measures for HDV are similar to those for hepatitis B. Immunization against hepatitis B protects against HDV. The difficulty is protection against superinfection of the many millions of established carriers of hepatitis B. Studies are in progress to determine whether specific immunization against HDV based on HDAg is feasible.

Treatment with interferon α at doses of 3–10 million IU three times weekly for 6 months results in biochemical and virological improvement, but many patients relapse when treatment is stopped.

Hepatitis E

Epidemics of enterically transmitted non-A, non-B hepatitis in the Indian subcontinent were first reported in 1980, but outbreaks involving tens of thousands of cases have also been documented in the former USSR, southeast Asia, northern and eastern Africa, and Mexico. Infection has been reported in returning travelers. The incubation period is slightly longer than for hepatitis A, with a mean of 6 weeks. The epidemiological features of the infection resemble those of hepatitis A. The highest attack rates are found in young adults, and high mortality rates of 20–39% have been reported in women infected during the third trimester of pregnancy [reviewed by Tam, et al.: 1998].

Properties of Hepatitis E Virus. Physicochemical studies have shown that the virus is very labile and sensitive to freeze-thawing, cesium chloride and pelleting by ultracentrifugation. The sedimentation coefficient is 183 S with defective particles at 165 S, and the buoyant density is 1.18 g cm^{-1} in a potassium tartrate/glycerol gradient. Morphologically, the virus is spherical and unenveloped, measuring 27–34 nm in diameter, with spikes and indentations visible on the surface of the particle.

Confirmation that the virus has been propagated in cell culture is awaited. All these properties suggest that HEV is similar to the caliciviruses.

However, HEV resembles most closely the sequences of *Rubella virus* and a plant virus, *Beet necrotic yellow vein virus*. It has, therefore, been proposed that these three viruses should be placed in separate but related families. The virus was cloned in 1991 and the entire 7.5-kb sequence is known. The genome is a single-stranded, positive-sense polyadenylated RNA molecule, with three overlapping ORFs.

HEV sequences have been classified into four major genotypes: most infections in several countries in Asia and Africa are caused by genotype I, the majority of infections in Mexico and Nigeria are caused by genotype II and isolated cases of infection with genotype III and IV have been described in the USA, European countries, Argentina, Taiwan, China and Japan.

Man is the natural host of HEV. A number of nonhuman primates such as chimpanzees, cynomolgus monkeys, rhesus monkeys, pigtail monkeys, owl monkeys, tamarins and African green monkeys are susceptible to natural (and experimental) infection with human strains of HEV. Swine strains have been identified and are able to cross the species barrier and infect humans. In endemic areas, antibodies to HEV acquired naturally have been found in 42–67% of domestic farm animals: cows, sheep and goats. In addition, there is evidence of widespread HEV or HEV-like infection in rodents in the USA, raising the possibility of reservoirs of HEV infection in industrialized countries.

Serological Tests and Laboratory Diagnosis. All HEV strains examined to date comprise a single serotype. Major epitopes are present near the carboxyl ends of ORF 2 and ORF 3, and the epitopes contained in ORF 2 are more conserved (90.5%) than those contained in ORF 3 (73.5%). Serological tests for anti-HEV based on expressed ORF 2 sequences are more sensitive for detecting IgM and IgG anti-HEV than tests based on antigens containing ORF 3 sequences. In addition, proteins expressed from ORF 2 measure antibodies that correlate with protection against hepatitis E.

Diagnostic tests are based on ELISA, and followed by Western blot. PCR is used for the detection of HEV RNA in serum and faeces; immunofluorescent antibody blocking assays to detect antibody to HEV in serum and liver; and immune electron microscopy to visualize viral particles in faeces.

Epidemiology. HEV is spread by the faecal–oral route. Consumption of drinking water contaminated with faecal material has led to epidemics, and the ingestion of raw or uncooked shellfish has caused sporadic infections and epidemics in endemic areas. The highest prevalence of infection occurs in regions with low standards of sanitation and nonchlorinated drinking water. The incubation period is 2–9 weeks, with an average of 6 weeks. Zoonotic spread of HEV is not excluded, and indeed appears likely particularly from swine and possibly rodents. For example, although hepatitis E is not endemic in the USA and other developed countries, anti-HEV has been found in a significant proportion (up to 28% in some areas) of healthy persons in these countries. Subclinical infection might be the explanation. Infection in town dwellers might be caused by rodents. The prevalence of anti-HEV in blood donors (a highly selected sector of the population) in Central Europe and North America is 1.4–2.5%, in South Africa 1.4%, Thailand 2.8%, Saudi Arabia 9.5% and 24% in Egypt.

At the same time it should be noted that the prevalence of anti-HEV in endemic regions is 3–26%, which is much lower than expected, although HEV infections account for more than 50% of acute sporadic hepatitis in some highly endemic areas.

Virus is excreted from the liver via the bile duct into the intestine and faeces. Viraemia and shedding of HEV in the faeces reach a peak during the incubation period, and excretion of the faeces continues for up to 14 days after the onset of jaundice. The quantity of virus in the faeces is small, which is consistent with the low rate of secondary spread by person-to-person contact. There is no evidence for sexual transmission or for transmission by transfusion.

Clinical Features. The clinical spectrum of infection with hepatitis E is similar to infection caused by other hepatitis viruses, and includes subclinical and anicteric infections, acute hepatitis with jaundice and fulminant hepatitis. Hepatitis E does not progress to chronic liver disease and there is no evidence of persistent infection. As with other forms of viral hepatitis, hepatitis E is more likely to be asymptomatic, subclinical and anicteric in young children, but severe overt HEV in children has been reported. In general, typical symptoms include fever, chills, anorexia and nausea, vomiting, joint pain, epigastric pain, dark urine, clay-coloured stools and jaundice. Cholestatic features are common.

The preicteric phase lasts for 1–10 days and icteric phase is 15–40 days. Seroconversion is followed by clearing of the virus, return to normal of liver enzyme elevation and resolution of the disease. Recurrence of hepatitis E has not been reported.

Infection with hepatitis E is associated with a relatively high mortality of 1–4% of patients admitted to hospital from the general population. Fulminant hepatitis in pregnancy may lead to a mortality rate of 20% during the third trimester. Premature delivery with high infant mortality of up to 33% has been observed [reviewed by Zuckerman, 2003].

Control and Prevention. Outbreaks are more common in countries with a hot climate, and have been reported from Algeria, Bangladesh, Borneo, China, Egypt, Ethiopia, Greece, India, Indonesia, Iran, the Ivory Coast, Jordan, Libya, Mexico, Myanmar (Burma), Nepal, Nigeria, Pakistan, Southern Russia, Somalia, Sudan and the Gambia. Most outbreaks have occurred following heavy rain and flooding and contamination of drinking water, contamination of well water and untreated sewage gaining access into city water treatment plants. Food-borne outbreaks have been associated with raw or uncooked shellfish. Therefore, the provision of safe (and chlorinated) drinking water and safe disposal of sanitary waste are essential, including safeguarding the water supply from animal waste from farm animals.

Smaller outbreaks and sporadic cases have been reported from many countries from south east Asia, central Asia, the Middle East, northern and western Africa, Mexico and also Italy and Spain. Sporadic cases have been reported from many other countries including Taiwan, Japan, the USA, South America and many countries in Europe among returning travelers and also among those who have not undertaken travel outside their own country.

In highly endemic areas, boiling is a good way of treating water, and should be available both for drinking and for brushing teeth. Bottled water or water in sealed cans of well-known brand names should be used for drinking. All bottles must be sealed and opened in the presence of the consumer.

Raw or uncooked shellfish must be avoided, and the other usual elementary food hygiene precautions are recommended. These include not eating uncooked fruit or vegetables that are not peeled or prepared by the consumer.

Pregnant women traveling to countries where outbreaks have been reported and countries where HEV is endemic should be counseled and the importance of the precautions outlined above must be stressed.

Immunoglobulin and vaccines: passive protection with immunoglobulin prepared from pooled plasma obtained from blood donors does not afford protection, and the efficacy of immunoglobulin prophylaxis with globulin prepared from donors living in endemic countries has not been demonstrated. Several recombinant and subunit HEV vaccines are under development, and one such preparation is under clinical trial [reviewed by Emmerson and Purcell, 2003].

GBV-C/HGV (Hepatitis G)

The recent isolation of three viruses, *GB virus A* (GBV-A), *GB virus B* (GBV-B) and *GB virus C* (GBV-C), from marmosets inoculated initially with a serum from a patient with jaundice was followed by the isolation of another strain of GBV-C from a patient with post-transfusion hepatitis. This isolate was named HGV. GBV-A and GBV-B appear to be animal viruses. Studies including phylogenetic analysis of genomic sequences showed that GBV-A, B and C are not genotypes of HCV, and that GBV-A and GBV-C are closely related. GBV-A/C and GBV-B and the hepatitis C viruses are members of distinct viral groups. The organization of the genes of the GBV-A, B and C genomes shows that they are related to other positive-strand RNA viruses, with local regions of sequence identity with various flaviviruses.

Serological reagents were prepared with recombinant antigens and limited testing for antibodies and by reverse transcriptase-PCR (RT-PCR) for specific RNA was carried out in groups of patients, blood donors and other selected individuals, patients with non-A, B, C, D, E hepatitis, multitransfused patients, intravenous drug addicts and other populations with a high incidence of viral hepatitis. Preliminary studies indicated the presence of antibody to each of the GB viruses from 3% to as many as 14%.

The blood-borne nature of GBV-C/HGV has been clearly demonstrated and there is evidence that it persists. The association of the virus with liver damage is illustrated by raised alanine transaminase levels and detectable HGV RNA in a number of patients; however, between 40 and 90% of viraemic subjects have normal alanine transaminase levels. In a significant proportion of patients there is coinfection with HBV and/or HCV. The primary manifestations, if any, of GBV-C/HGV infection may be extrahepatic, and liver damage may be the result of coinfection with other hepatitis viruses–or an, as yet unidentified, hepatotropic agent. The development of sensitive and specific serological tests for GBV-C/HGV, the application of modern virological techniques and histological studies, where appropriate, will determine the clinical significance of this newly identified virus [reviewed by Mushahwar and Zuckerman, 1998].

Additional Reading

Birkenmeyer, L.: "Hepatitis B Virus: Life Cycle and Morphogenesis," In: Mushahwar, I.K.: *Viral Hepatitis: Molecular Biology, Diagnosis, Epidemiology and Control. Perspectives in Medical Virology*, Vol. 10, Elsevier, Amsterdam, The Netherlands. 2004, pp. 109–125.

Chang, M.H., C.J. Chen, M.S. Lai, et al.: "Universal Hepatitis B Vaccination in Taiwan and the Incidence of Hepatocellular Carcinoma," *New England Journal of Medicine*, **336**, 1855–1859 (1997).

Emmerson, S.U., and R.H. Purcell: "Hepatitis E Virus," *Reviews in Medical Virology*, **13**, 145–154 (2003).

Harrison, T.J., and A.J. Zuckerman: *The Molecular Medicine of Viral Hepatitis*, John Wiley & Sons, Inc., New York, NY, 1996.

Hollinger, F.B., and J.R. Ticehurst: "Hepatitis A Virus," In: Knipe, D.M., P.M. Howley, M.A. Martin, D.E. Griffin, and R.A. Lamb: *Fields Virology*, 5th Edition, Vol. 1, Lippincott-Raven, Philadelphia, PA, 2006.

Hydziyannis, S.J.: "Hepatitis Delta Review," *Journal of Gastroenterology and Hepatology*, **12**, 289–298 (1997).

Kane, M.: "Global Strategies for the Control of Hepatitis B," In: Zuckerman, A.J.: *Prevention of Hepatitis B in the Newborn, Children and Adolescents*, pp. 87–96. Royal College of Physicians, London, UK, 1996.

Koff, R.S.: "Hepatitis A," *Lancet*, **351**, 1643–1649 (1998).

Leary, T.P., A.S. Muerhoff, J.N. Simons, et al.: "Sequence and Genomic Organization of GBV-C: a Novel Member of the Flaviviridae Associated with Human Non-A–E Hepatitis," *Journal of Medical Virology*, **48**, 60–67 (1996).

Lau, J.Y.: *Hepatitis B and D Protocols: Detection, Genotypes, and Characterization*, Vol. 1, Springer-Verlag New York, LLC, New York, NY, 2004.

Lee, W.M.: "Hepatitis B," *Clinics in Liver Disease*, Vol. 3, no. 2, W.B. Saunders, Philadelphia, PA, 1999, pp. 1–432.

McGarvey, M.I., M. Houghton, and A.J. Weiner: "Hepatitis C Virus: Structure and Molecular Virology," In: Zuckerman, A.J. and H.C. Thomas: *Viral Hepatitis*, 2nd Edition, Churchill Livingstone, Edinburgh, UK, 1998, pp. 253–270.

Mushawhar, I., and J.N. Zukerman: "The Clinical Significance of GBV-C," *Journal of Medical Virology*, **56**, 1–3 (1998).

Mushawhar, I.K.: *Viral Hepatitis: Molecular Biology, Diagnosis, Epidemiology and Control. Perspectives in Viral Virology*, Vol. 10, Elsevier, Amsterdam, The Netherlands. 2004, pp. 1–264.

Oon, C.J., G.K. Lim, Z. Ye, et al.: "Molecular Epidemiology of Hepatitis B Virus Vaccine Variants in Singapore," *Vaccine*, **13**, 699–702 (1995).

Schinazi, R.F., J.-P., Sommadossi, and C.M. Rice: *Frontiers in Viral Hepatitis*, Elsevier Science & Technology Books, New York, NY, 2003.

Staff, Icon Health Publications: *Hepatitis B Virus: A Medical Dictionary, Bibliography, and Annotated Research Guide to Internet References*, ICON Health Publications, San Diego, CA, 2004.

Staff, Icon Health Publications: *Hepatitis C Virus: A Medical Dictionary, Bibliography, and Annotated Research Guide to Internet References*, ICON Health Publications, San Diego, CA, 2004.

Tam, A.W., D.W. Bradley, K. Krawczynski, E.E. Mast, and P.O. Yarborough: "Hepatitis E Virus," In: Zuckerman, A.J. and H.C. Thomas: *Viral Hepatitis*, 2nd Edition, pp. 395–416, Churchill Livingstone, Edinburgh, UK, 1998.

Tan, Seng-Lai: "Hepatitis C Viruses," *Genomes and Molecular Biology*, Taylor & Francis, Inc., Philadelphia, PA, 2006.

Thomas, H.C., S. Lemon, and A.J. Zuckerman: *Viral Hepatitis*, 3rd Edition, Blackwell Publishers, Oxford, UK, 2005.

Zuckerman, A.J.: "Effect of Hepatitis B Virus Mutants on Efficacy of Vaccination," *Lancet*, **355**, 1382–1384 (2000).

Zuckerman, J.N.: "Hepatitis E and the traveler," *Travel Medicine and Infectious Disease*, **1**, 73–76 (2003).

Web References

Centers for Disease Control and Prevention: http://www.cdc.gov/ncidod/diseases/hepatitis/

Hepatitis Foundation: http://www.hepfi.org/

Hepatitis B Foundation: http://www.hepb.org/

Hepatitis Magazine: http://www.hepatitismag.com/

MayoClinic.com: http://www.mayoclinic.com/health/search/search

World Health Organization: http://www.who.int/topics/hepatitis/en/

ARIE J. ZUCKERMAN, University of London, London, UK

HERBICIDES. Herbicides are important for many reasons. Cultivation of plants for economic or ornamental purposes entails an incessant struggle against losses from pests. Weeds not only reduce yields by competing for sunlight, water, and nutrients, but they also reduce the quality of products and overgrow adjacent areas and bodies of water. Some of them actually produce phytotoxins (allelochemicals) that reduce crop growth. Uncontrolled weed infestations drastically reduce crop yields and decrease crop, turf, timber, and forage quality. For example, the post-harvest presence of weed seeds reduces crop quality, ie, cocklebur in soybeans, wild mustard in canola, and red rice and Northern jointvetch in rice. Weeds also serve

as alternative hosts for crop-infesting fungi and harbor insect pests such as whiteflies. Furthermore, certain weeds, such as nightshade, produce toxins that can have severe health consequences for both livestock and humans.

The broadest definition of "herbicides" includes all agents that destroy or inhibit plant growth. Thus, an herbicidal agent may be animal, ie, a home-gardener with a hoe or a grazing herbivore; vegetable, ie, a parasitic weed or one plant species competing successfully with another; or mineral, ie, chemicals with herbicidal activity. The definition of a weed as "a plant growing where it is not wanted" is convenient, although perhaps not scientific. It focuses on one of the basic problems of weed control, ie, selectively killing weeds without crop damage. Whether a plant is considered a weed depends entirely on the circumstances.

Weeds can be controlled by crop rotation, mowing, tilling the soil, and crowding out by crop competition. Extensive infestations of certain single weed species can be controlled by biological methods. Insects, herbivores, or diseases destroy certain weeds. However, these techniques have the disadvantage that weed seeds remain dormant in the soil and are unaffected. Cultural practices are important, but the use of chemicals for weed control has been adopted globally.

Pest-control chemicals, ie, pesticides, have contributed significantly to agricultural productivity in the United States and often provide the farmer's first line of defense against pests. The term "pesticide" includes all classes of chemicals used against insects, weeds, plant pathogens, rodents, algae, snails, and other pests. Legally, it also includes growth regulators. The term "herbicide" refers specifically to weedkillers. See also **Pesticides**.

Modern agriculture demands that herbicides and other crop-protection chemicals be integrated into a production system that includes the development of pest-resistant and high yielding crop varieties, crop management, plant nutrition, and mechanization of farming methods and pest-control techniques. In this system, chemical control is an important component. Pesticides have been stated to increase production of crops, livestock, and forest products by 25% and thus contribute to the stability of food prices.

The widespread introduction of chemicals for weed control (herbicides) brought about major changes in agriculture affecting not only the economics of farming, but also the communities that were founded and based on crop production. Populations shifted from rural areas as labor demands decreased. The changes came about initially in the United States where this technology developed rapidly but parallel developments took place in Europe. Chemical control of pests was widely adopted by large-scale agricultural systems in areas throughout the world. There were many benefits from the use of chemicals. Not only was there increased potential for food production at lower cost, but also there was potential for conservation of soil resources through reduced tillage. The reduction in tillage that was made possible by the use of herbicides has resulted in a dramatic reduction in soil erosion. The application and development of the technology required sophisticated users. Industrial research and development received increasing support, as did the efforts of counterparts in government and academia. The quantity of herbicides used grew throughout the last decades of the twentieth century, and the industry that supported this growth flourished. The major companies became multinational corporations.

The chemical inputs were expensive. Adverse environmental effects and other problems gradually offset some of the benefits. Regulatory agencies both national and international have called for more stringent regulations on the types and amounts of chemicals that could be used. There was little initial understanding of the implications of the widespread use of chemicals in the environment, but the growth of this field of science soon paralleled progress in pesticide research. Increased costs of safety tests and the introduction of government-mandated requirements to reduce pesticide use made some industries reluctant to continue investing in development of new pesticides. It had become very costly to introduce new herbicides. The fact that farmers were already treating large acreages successfully meant that the market had become extremely competitive and there was a general reduction of effort by major companies, many of whom have separated or divested themselves of their agrochemical departments.

Reductions in the use of herbicides have been driven to some extent by regulation, but more significant changes in the patterns of herbicide use are due to progress in the applications of biotechnology to agriculture. The understanding of metabolic processes in plants, modes of action of herbicides, and plant genetics coupled with the ability to manipulate genes and facilitate their expression in plants are major factors in these changes. Industrial research emphasized the potential of biotechnology, and industry invested heavily. Some developments, such as herbicide-resistant crops and plants incorporating insecticides, are currently approved and widely adopted. Such new directions are the current focus of the major North American and European chemical industries whereas outside Europe and the United States, manufacturing plants and new industries have originated to satisfy the needs for herbicides.

The knowledge that chemicals could kill plants or render soils sterile has existed since ancient times. The use of selective herbicides that could kill weeds without damage to crops growing in the same cultivated area is a twentieth-century development that has brought about major changes in agriculture and agricultural communities.

Sulfuric acid, sodium chlorate, arsenic compounds, copper sulfate, and other inorganic compounds have been used as weed killers since the early twentieth century. Until the introduction of synthetic organic chemicals, weed control in fields and turf depended on inorganic compounds and various combinations of surface tillage, mowing, chopping, hand weeding, scorching, and burning of unwanted plants. Those time-honored but highly inefficient and labor-intensive methods were essential to agriculture because weeds successfully compete with crop plants for water, sunlight, and nutrients. Early in the twentieth century, sodium chlorate was used to control deep-rooted perennial weeds in noncrop areas. Borates also found use for control of weeds in specific locations. The introduction of synthetic organic herbicides that acted selectively against broad-leaved weeds changed the situation irreversibly. The first organic chemical herbicide to be introduced was 4,6-dinitro-o-cresol [CAS: 534-52-1] (DNOC) in 1932.

DNOC was used initially as an insecticide, and the selective herbicidal properties of this and related compounds were discovered later. This was followed by the introduction in the 1940s of the substituted phenoxy acids, and in 1951 of the substituted ureas and uracils. The triazine family of herbicides appeared in 1955, and the bipyridiniums in 1960. Chemicals of many other classes rapidly entered the herbicide market and their usage in major crops expanded rapidly.

Herbicides can be classified as selective and nonselective. Selective herbicides, like 2,4-D (2,4-dichlorophenoxyacetic acid), metolachlor [CAS: 51218-45-2], and EPTC [CAS: 759-94-4], are more effective against some types of plants than others, e.g., broadleaved plants vs grasses. Glyphosate [CAS: 1071-83-6] is representative of the nonselective herbicides used for total vegetable control.

The classes of herbicidally active toxophores are limited in number.

Herbicides are also sometimes classified according to mode of action, selectivity, registered uses, and toxicity.

Modes of Herbicide Action

Modes of herbicide action include: Photosystem I inhibitors; Photosystem II inhibitors; Bleaching herbicides; Chlorophyll biosynthesis inhibitors; Lipid and wax synthesis inhibitors; Inducers of damage to antioxidants systems; Herbicidal inhibition of enzymes; Amino acid and nucleotide biosynthesis inhibitors; Cell division inhibitors; and Plant growth regulator synthesis and function inhibitors.

Environmental Fate of Herbicides

Beyond modes of action and structure–activity relationships, developers of new herbicides must also consider uptake by plants, translocation within the plant, and possible deactivation of herbicides by contact with soil. Some of these problematic factors can be addressed as part of the QSAR studies and during the screening process. Considerable attention is also being paid to the use of safeners which protect the crop from herbicides that specifically target the weeds usually associated with that crop. Environmental protection and pesticide regulation concerns are the driving forces in the current efforts toward minimizing application rates, optimizing delivery through improved formulations and application equipment, and increasing target specificity. These research and development efforts include other important and related areas of interest to chemists, eg, the fate and detection of herbicides in the soil and ground and surface water.

The fate of herbicides in the environment is influenced by many chemical, biological, and physical factors. The principal transport and dissipation pathways include sorption to organic and mineral soil and sediment constituents; transport to groundwater in the solution phase by mass flow and/or diffusion; transport to surface water in either the solution or sorbed phases; loss to the atmosphere through volatilization, with redeposition at a later time and location; transformation or mineralization by biological, chemical, or photochemical processes; and uptake by plant

or animal species. These processes do not operate as isolated systems, but occur simultaneously and involve significant interaction and feedback. Although the environmental fates of most herbicides are controlled primarily by one or two of the outlined processes, all of these factors influence the fate to some extent.

Continued concern is expressed over the potential contamination of surface and groundwaters by agricultural chemicals. Herbicides have received much of this attention, due to their widespread use and the large total volume applied. However, this perceived threat to groundwater resources appears to be largely unfounded. A survey of private wells and public water well supplies in the United States has revealed that contain herbicides at levels that would affect human or animal health. In addition, those sources that are contaminated can usually be attributed to point rather than nonpoint sources. A point source of contamination is readily located and thus more easily controlled and remediated, and is generally associated with industrial sources or municipal wastewater plants, although agricultural sources such as herbicide equipment rinsing stations also could be point sources. A nonpoint contamination source is one in which the exact source is unknown. They are typically diffuse, often of large areal extent, and are generally of agricultural origin. Nonpoint sources are generally treated by modifications in agricultural management practices. Typical modifications would include the use of alternative herbicide formulations, the splitting of the herbicide application in time, or the installation of vegetative buffer strips to trap runoff.

A re-evaluation of the water quality problem has revealed that surface water resources, rather than groundwater resources, are at higher risk of contamination from agricultural chemicals.

The public health implications of drinking water contamination by herbicides are unclear. The levels that have been detected in groundwater are generally in the part per billion (ppb) or part per trillion (ppt) range and are below estimated acute toxicity levels. However, the long-term health effects of this exposure are generally unknown. Several studies have demonstrated that the mortality from some types of cancer is significantly higher in rural residents of many corn belt states. This trend is particularly evident in a study from Kansas involving 2,4-D exposure; however, factors other than 2,4-D exposure are also being considered. The U.S. Environmental Protection Agency (EPA) developed (ca 1993) a classification scheme in an attempt to further evaluate the carcinogenic potential of herbicides and pesticides. In this system, chemicals are placed in one of five groups, A–E, according to their carcinogenic potential, ranging from definite (A) human carcinogens to no evidence of carcinogenicity for humans (E). The principal difference between these groups is the amount of accumulated evidence demonstrating carcinogenic potential.

This classification scheme is used in part in the determination and calculation of health advisory (HA) drinking water levels or carcinogenic risk estimates. The majority of herbicides in use in the United States (ca 1993) for which HAs have been issued fall into Group D, with a smaller percentage falling into Group C. This would indicate that there are insufficient data to classify the carcinogenic potential of many herbicides. This does not imply that chemical companies are not adequately testing herbicides. To the contrary, exhaustive toxilogical testing of a potential herbicide is required by the U.S. EPA before registration. The lack of data does indicate, however, that further testing will be required before the carcinogenic potential of many herbicides is known. Based on available HAs and the U.S. EPA classification scheme, acifluorfen, alachlor, amitrole, haloxyfop–methyl, lactofen, and oxadiazon have been listed as B2 carcinogens.

Since 1984, dramatic technical advances have been made in the analysis of trace organic chemicals in the environment. Indeed, these advances have been largely responsible for the increased public and governmental awareness of the wide distribution of herbicides in the environment. The ability to detect herbicides at ppb and ppt levels has resulted in the discovery of trace herbicide residues in many unexpected and unwanted areas. The realization that herbicides are being transported throughout the environment, albeit at extremely low levels, has caused much public and governmental concern. However, the public health implications remain unclear.

An increased emphasis has been placed on the first step in the environmental sampling process, that of obtaining a representative, uncontaminated sample. If this is to be accomplished, consideration must be made of such factors as sample size and location. After the sample has been obtained, it must be stored in such a way as to minimize degradation. This generally consists of refrigeration, possibly preceded by some type of drying.

Preparation of soil–sediment of water samples for herbicide analysis generally has consisted of solvent extraction of the sample, followed by cleanup of the extract through liquid–liquid or column chromatography, and finally, concentration through evaporation. This complex but necessary series of procedures is time-consuming and is responsible for the high cost of herbicide analyses. The advent of solid-phase extraction techniques in which the sample is simultaneously cleaned up and concentrated has condensed these steps and thus greatly simplified sample preparation.

Traditionally, herbicides have been analyzed by gas chromatography (gc) or spectrophotometric methods. The method of choice when accuracy and sensitivity are of the utmost importance is gc, especially when combined with mass spectrometry. However, several other methods are used for routine monitoring or screening purposes. High pressure liquid chromatography (hplc) provides detection limits that nearly rival gc and require significantly less sample preparation and cleanup. Advances in the 1980s have made thin-layer chromatography (tlc) a valuable tool in herbicide analysis. The combination of high performance tlc plates and scanning densitometers allows quantitative results to be obtained at detection limits that nearly rival hplc. Significant advances have been made in stationary phases for both hplc and tlc systems, including reverse-phase options. These have proven to be invaluable for herbicide analysis. Another analytical tool that has received much attention and shows great promise for routine analysis is enzyme immunoassay (eia). This technique offers the advantages of a low cost analysis, few interferences, high specificity and sensitivity, and a minimal amount of sample preparation.

A mobility ranking based on soil thin-layer chromatography (stlc) is used to classify the herbicide leaching potential of various herbicides. The rankings range from I (immobile) to V (very mobile) with intermediate categories of II (low mobility), III (intermediate), and IV (mobile). This method is widely used and has been accepted for submission of leaching data for herbicide registration purposes by the U.S. EPA.

Herbicide Groups

Herbicides can be grouped according to common structural features. Sometimes the assignment is arbitrary when there are a multitude of functional groups, eg, acifluorfen which is a diphenyl ether (phenoxy compound) as well as a trifluoromethyl compound.

Phenoxyalkanoics. The phenoxyalkanoic herbicide grouping is composed of two subgroups, the phenoxyacetic acids and the phenoxypropionic acids. The phenoxyacetic acid herbicides include some of the first commercially successful herbicides, eg, 2,4-D. They continue to be widely used for foliar control of broadleaf weeds. The more heavily functionalized phenoxypropionic acid herbicides are relatively new compared to the phenoxyacetic acids and are used primarily for selective control of grassy weeds in broadleaf crops.

The phenoxyalkanoic herbicides are acidic in nature and thus subject to some degree of ionization. The extent to which the herbicide ionizes is controlled by the acid dissociation constant (K_a) of the herbicide in question and the soil solution pH. The leaching potential is significantly influenced by these reactions.

Considerable research has been conducted on the breakdown of phenoxyacetic acids in soil. The decomposition of 2,4-D appears to be primarily a microbial process that occurs rapidly in surface soils under aerobic conditions and decreases with depth. MCPA is also degraded by microbial means although at a slower rate than 2,4-D. MCPA has also been shown to photodecompose rapidly, losing of the initial herbicide application in six days of exposure. 2,4-D has been shown to be completely degraded in aerated solutions of hydrogen peroxide containing Fe^{3+}. This reaction can be further accelerated by irradiation with visible light containing a small ultraviolet (uv) component. Other related materials include bifenox [CAS: 42576-02-3] and (4-(4-chloro-2-methyl-phenoxy)-butanic acid) MCPB [CAS: 94-81-5].

The phenoxypropionic acids, eg, haloxyfop–methyl and fluazifop–butyl (Table 1) bear a variety of other functional groups and are not strongly sorbed to soils of widely varying constituents. Thus the leaching potential of these compounds is significant. Both compounds are rapidly hydrolyzed to the parent free acids in soil and gradually decomposed by microbial means. Diclofop–methyl (Table 1) is hydrolyzed in hydroalcoholic solutions in the presence of montmorillonite. Finally, mecoprop [CAS:

7085-19-0] was degraded by several microbial communities but not by the individual members, indicating a co-metabolic relationship. Other phenoxypropionic acid herbicides include dichlorprop [CAS: 120-36-5] and quizalofop–ethyl [CAS: 76578-14-8].

Considerable concern has been raised over the carcinogenic potential of 2,4-D. However, the World Health Organization (WHO) has evaluated the environmental health aspects of this chemical and concluded that 2,4-D posed an insignificant threat to the environment. They did indicate, however, that only limited data on toxicology in humans are available. An HA has been issued for MCPA. It was found in 4 of 18 SW samples analyzed and in none of 118 GW samples, and has been placed in Group D for carcinogenic potential.

Bipyridiniums. The bipyridinium herbicides, paraquat and diquat, are nonselective contact herbicides and crop desiccants. Diquat is also used as a general aquatic herbicide. Bipyridinium herbicides are organic cations and are retained in the soil complex via cation exchange. They are strongly sorbed to most soils and are not readily desorbed. Both paraquat and diquat are not readily leached.

Benzonitrile, Acetic Acid, and Phthalic Compounds. Benzonitrile herbicides are generally used for pre-emergence and post-emergence control of broadleaf weeds. Dichlobenil [CAS: 1194-65-6] also controls grass weeds and dichlobenil, endothall [CAS: 145-73-3], and fenac [CAS: 85-34-7] are used as aquatic herbicides. Most benzonitriles are selective in their control. Benzonitrile herbicides are acidic in nature, thus their environmental fate is influenced by changes in soil pH. Sorption of these herbicides is expected to increase with decreasing pH. This is the case with dicamba, which is minimally sorbed at near neutral pH but demonstrates a dramatic increase in sorption as the soil pH decreases. Endothall is also only minimally sorbed at a near neutral pH.

Benzonitrile herbicides tend to possess a high leaching potential; dichlobenil is an exception, due to its stronger sorption. The benzonitrile herbicides are also prone to volatilization losses and off-site deposition.

Benzonitrile herbicides are readily degraded by soil microbes. Dicamba is rapidly degraded in soil and water samples by microbial means, and is metabolized by several species of soil bacteria. DCPA degrades very rapidly under optimum conditions (25 °C) and slower at lower temperatures. Bromoxynil can be degraded by microbial or photochemical means. The photodegradation of bromoxynil is highly pH-dependent, a decrease in degradation occurring at lower pH values. Endothall, which is widely used as an aquatic herbicide, is rapidly dissipated in water and dissipates only slightly slower in soil. Dichlobenil is apparently degraded by a combination of microbial and photochemical processes. Finally, fenac is slowly degraded in water and soil by microbial processes.

Dinitroanilines and Derivatives. Dinitroaniline herbicides are used principally for the selective, pre-emergence control of annual grasses and broadleaved weeds. They have little or no post-emergence activity. Oryzalin is used for selective weed control in flooded rice culture. In general, dinitroaniline herbicides are extremely prone to volatilization losses. For this reason, they should always be incorporated in the soil immediately after application; oryzalin and pendimethalin are exceptions to this statement. Dinitroaniline herbicides are nonionic and retained in soil primarily by hydrogen bonding to soil organic matter, or possibly through hydrophobic or van der Waals forces. The uptake of nonionic herbicides has been described as a chemical partition of the herbicide into soil organic matter. The reactions are governed, to a large extent, by the herbicide's polarity.

Acid Amides. The principal use of acid amide herbicides is the selective control of seedling grass and certain broadleaved weeds. The majority of acid amide herbicides are applied pre-emergence or pre-plant incorporated, except for propanil which is applied post-emergence. In general, the acid amide herbicides are not considered subject to large volatilization losses. However, under ideal conditions, eg, high soil moisture and low soil sorption, volatilization may be significant.

Acid amide herbicides are nonionic and moderately retained by soils. Acetochlor [CAS: 34256-82-1] is sorbed more than either alachlor or metolachlor, which are similarly sorbed by a variety of soils. Sorption of all the herbicides is well correlated to soil organic matter content. In a field lysimeter study, metolachlor has been found to be more mobile and persistent than alachlor; diphenamid [CAS: 957-51-7] and napropamide [CAS: 15299-99-2] have been found to be more readily leached.

Phenylcarbamates. Phenylcarbamate herbicides represent one of two subgroups of carbamate herbicides, the phenylcarbamates and the thiocarbamates. The carbamate herbicides are used, in general, for the selective pre-emergence control of grass and broadleaved weeds. Exceptions would include barban, desmedipham, and phenmedipham, which are applied post-emergence.

Phenylcarbamate herbicides are nonionic and, in general, readily leached in soils. One notable exception is chlorpropham which is strongly sorbed to soils. Movement of karbutilate [CAS: 4849-32-5] has been studied in several Texas rangeland soils and found to be greater in a loamy sand soil than in a clay loam soil, but more persistent in the clay loam soil. The phosphonate fosamine–ammonium [CAS: 25954-13-6] readily degrades in soils, having a half-life of one week in the field and 10 days in the laboratory; applications of fosamine–ammonium to the soil do not adversely affect soil microbial populations. Finally, a bacterial strain has been isolated that can utilize chlorpropham as a sole carbon source. Other phenylcarbamate herbicides include desmedipham [CAS: 13684-56-5] and phenmedipham [CAS: 13684-63-4].

Thiocarbamates. Thiocarbamate herbicides are nonionic. Diallate and triallate were strongly sorbed to both cation- and anion-exchange resins but minimally to kaolinite or montmorillonite. This behavior suggests a physical, rather than ionic mechanism of attraction. The mobility of the thiocarbamate herbicides increases with increasing water solubility. The ranking of five thiocarbamate herbicides, in terms of leaching depth, is molinate [CAS: 2212-67-1] >EPTC >vernolate [CAS: 1929-77-7] >pebulate [CAS: 1114-71-2] >cycloate. Thiobencarb [CAS: 408-27-5] has been found to be relatively immobile in soil columns under saturated flow conditions.

Triazines. Triazine herbicides are one of several herbicide groups that are heterocyclic nitrogen derivatives. Triazine herbicides include the chloro-, methylthio-, and methoxytriazines. They are used for the selective pre-emergence control and early post-emergence control of seedling grass and broadleaved weeds in cropland. In addition, some of the triazines, particularly atrazine, prometon [CAS: 1610-18-0], and simazine [CAS: 122-34-9], are used for the nonselective control of vegetation in noncropland. Simazine may be used for selective control of aquatic weeds.

Triazine herbicides are not readily volatilized. However, given ideal circumstances, volatilization losses may be significant. The tendency to volatilize varies among herbicides and is highly dependent on soil type and moisture conditions.

Pyridines and Pyridazinones. Pyridine herbicides are auxin-type herbicides generally used for selective control of broadleaved weeds in cropland, rangelands, and noncroplands. The pyridazinones are used primarily for the selective pre- and post-emergence control of seedling grass and broadleaved weeds in cotton and sugarbeets. The pyridines are slightly acidic in nature and the pyridazinones, slightly basic.

Pyridine herbicides are not strongly sorbed to soils and are readily leached. The mobility of fluroxypyr [CAS: 69377-81-7] has been found to decrease with increasing incubation time; this is attributed to entrapment of the herbicide within the soil organic matter.

Pyridazinone herbicides tend to be strongly sorbed in soils and do not leach readily. Norflurazon sorption increases as organic matter and clay contents increase, and it is subject to degradation through photolysis but only minimally through volatilization processes. Pyrazon [CAS: 1698-60-8] sorption also has been shown to increase, and mobility to decrease, with increasing soil organic matter contents. The degradation of pyrazon appears to be a microbially mediated process directly related to soil organic matter content. Difenzoquat [CAS: 43222-43-6] also is a pyridazinone herbicide.

Sulfonylureas. Sulfonylurea herbicides are a relatively new class of herbicides generally used for selective pre- and post-emergence control of broadleaved weeds in croplands. In general, the sulfonylureas are applied in significantly lower amounts than most herbicides, and they tend to be more active against broadleaved species than grasses. Sulfometuron–methyl [CAS: 74222-97-2] is used for broad-spectrum selective or nonselective weed control in noncroplands.

Sulfonylurea herbicides are weak acids and, in general, are not strongly sorbed to soils. Sorption of chlorsulfuron and metsulfuron–methyl is inversely related to soil pH and is positively correlated to soil organic matter.

Imidazoles. Imidazole herbicides are generally used for selective pre- and post-emergence control of grass and broadleaved weeds in croplands.

Buthidazole [CAS: 55511-98-6] and imazapyr are used for broad-spectrum, nonselective weed control in noncroplands. Imidazole herbicides are amphoteric, possessing both acidic and basic functional groups. A notable exception is buthidazole which is nonionic in nature. At typical soil pH values, most of the imidazole herbicides exist as anions

Ureas and Uracils. Urea herbicides are generally used for selective pre-emergence and early post-emergence control of seedling grass and broadleaved weeds. Uracil herbicides are generally used for selective control of annual and perennial weed control in certain crops and for general weed control in noncrop areas. Bromacil, linuron [CAS: 330-55-2], and tebuthiuron [CAS: 34014-18-1] are used for the nonselective control of weeds in noncropland. Bromacil is also used in citrus crops, and linuron is used in sorghum and corn crops. Urea herbicides are nonionic and generally of low water solubility. The uracils are ionic herbicides that are not strongly sorbed to soils and readily leach.

Urea and uracil herbicides tend to be persistent in soils and may carry over from one season to the next. However, there is significant variation between compounds. Bromacil is debrominated under anaerobic conditions but does not undergo further transformation, linuron is degraded in a field soil and does not accumulate or cause carryover problems, and terbacil [CAS: 5902-51-2] is slowly degraded in a Russian soil by microbial means. The half-lives for this breakdown range from 76 to 2,475 days and are affected by several factors including moisture and temperature. Finally, tebuthiuron applied to rangeland has been shown to be phytotoxic after 615 days, and the estimated time for total dissipation of the herbicide is from 2.9 to 7.2 years.

Aliphatic–Carboxylics. There are only two herbicides present in this class, trichloroacetate [CAS: 76-03-9] (TCA) and dalapon [CAS: 75-99-0]. These are used primarily for the selective control of annual and perennial grass weeds in cropland and noncropland. Dalapon is also used as a selective aquatic herbicide. Dalapon and TCA are acidic in nature and are not strongly sorbed by soils. They are reported to be rapidly degraded in both soil and water by microbial processes. However, the breakdown of TCA occurs very slowly when incubated at $14-15\,^{\circ}$C in acidic soils. iming not only accelerates this degradation but also increases the numbers of TCA-degrading bacteria.

Metal Organics and Inorganics. The metal organic herbicides are arsenicals used for the selective, post-emergence control of grass and broadleaved weeds in cropland and noncroplands. These herbicides are particularly useful for weed control in cotton and turf crops. Cacodylic acid is a contact herbicide used for nonselective weed control in cropland and noncropland. Ammonium sulfamate [CAS: 7773-06-0] (AMS) is an inorganic herbicide used for control of woody plants and herbaceous perennials.

Arsenical herbicides are salts of methylarsonic acid, eg, calcium salt of methylarsonic acid [CAS: 5902-95-4] (CMA), and are thus freely soluble in water. They are strongly sorbed to soils and not readily leached. The sorption of DSMA is greater on clay soils than on sandy soils. In addition, the amount sorbed is greater on kaolinite than on montmorillonite or vermiculite, indicating possible retention by exposed hydroxyl groups. Sorption of MSMA is also significantly higher on clay soils than on sandy soils, and MSMA is essentially immobile in field studies and not expected to leach. AMS is not retained in soils and is susceptible to leaching losses. Cacodylic acid and MSMA are both degraded in field soils and do not accumulate with repeated application. MSMA is degraded at a faster rate under flooded soil conditions than in soils at a moisture content less than field capacity. Finally, MSMA appears to be degraded, at least partially, by soil microbes.

Amino Acid Analogues. Amino acid analogue herbicides also control a large variety of weeds. Diethatyl [CAS: 38725-95-0] is used for selective, pre-emergence control of grass and broadleaved weeds. Flamprop [CAS: 58667-63-3] is used to control the growth of wild oats in wheat.

Glyphosate is zwitterionic and thus can be sorbed as an anion, cation, or zwitterion. Although the amount of glyphosate sorbed decreases with increasing soil pH, at the pH of typical agricultural soils glyphosate is strongly sorbed relatively immobile. The mobility classification varies from immobile on an acidic sandy clay loam soil to low mobility on an alkaline clay loam soil. The increase in mobility with increasing pH arises from a decrease in sorption.

Registration of Herbicides

A herbicide that promises to be commercially successful must be officially approved or registered with the EPA before it can be used or sold in the United States. Labeling and marketing of pesticides in interstate commerce are regulated in the United States by the Federal Insecticide, Fungicide, and Rodenticide Act (FIFRA), as amended, which is administered by the U.S. EPA. The most significant changes to FIFRA took place with the passage of the Food Quality Protection Act (FQPA) of 1996. Most states have similar laws. Federal registration does not remove the requirement for a state registration. Safety tests evaluate hazards to human, the environment, and nontarget species, and acute and chronic toxicity data are obtained. Methods of residue analysis must be devised and validated. If residues might occur on foodstuffs, a tolerance or exemption therefrom must be obtained. Tolerance has been defined as the maximum concentration of pesticide residue that is permitted in or on food at a specified stage in the harvesting, storage, transport marketing, or preparation of the food, up to a final point of consumption, and the concentration is expressed in parts by weight of the pesticide residue per million parts of the food (ppm).

At the present time, many governments mandate reductions in pesticide usage. This may be achieved to some degree by using more effective chemicals (lower rates of application) and by improved application technology (formulation and precision agriculture). It was suggested that reductions in application rates were driven by discovery rather than by regulation. The combination of selectivity with improved efficacy helps to meet environmental objectives, as exemplified by the steady decrease in application rates of new classes of herbicides introduced between 1954 and 1981 (atrazine 1959, alachlor 1967, acifluorfen 1975, chlorsulfuron 1979, imazaquin 1981, as representative members of the classes of triazine, chloracetanilide, phenoxy acid, and imidazolinone, respectively).

Another factor in reduction of pesticide use is the adoption of integrated pest management practices. The federal government of the United States is committed to the concept of IPM, and compatibility of new pest control chemicals or technologies with IPM is an important factor in regulatory approval. In Europe, integrated crop management (ICM), a similar concept, has developed.

Innovative Weed Management Agents

Adoption by the agricultural community requires that an innovative weed management agent must be an effective control of the target species, be cost-effective, and be practical to employ. It must not interfere with crop production practices such as crop rotation or the use of other pesticides. Additionally, new weed-control agents cannot pose a significant threat to human health or the environment. Considerable costs are incurred in the development, registration, production, and marketing of weed control agents. These costs require that an herbicide have sufficient long-term market viability and market niche potential to justify these costs in time and money. The need for safe and effective methods of crop production in an environment that contains competitive weeds is becoming increasingly critical.

Additional Reading

Cobb, A.H., and R.C. Kirkwood: *Herbicides and Their Mechanisms of Action*, CRC Press LLC., Boca Raton, FL, 2000.

Draber, W. and T. Fujita, eds.: *Rational Approaches to Structure, Activity, and Ecotoxicology of Agrochemicals*, CRC Press, Boca Raton, FL, 1992.

Grover, R. and A.J. Cessna, eds.: *Environmental Chemistry of Herbicides*, Vols. 1 and 2, CRC Press, Boca Raton, FL, 1991.

Hakansson, S.: *Weeds and Weed Management on Arable Land: An Ecological Approach*, CAB International, New York, NY, 2003.

Herbicide Handbook, 7th Edition, Weed Science Society of America, Champaign, IL, 1994.

Liebman, M., C.L. Mohler, and C.P. Staver: *Ecological Management of Agricultural Weeds*, Cambridge University Press, New York, NY, 2001.

Marrs, T.C., and B. Ballantyne: *Pesticide Toxicology and International Regulation*, John Wiley & Sons, Inc., New York, NY, 2003.

Monaco, T.J., S.C. Weller, S.C. Weller, and F.M. Ashton: *Weed Science: Principles and Practices*, 4th Edition, John Wiley & Sons, Inc., New York, NY, 2002.

U.S. EPA, *Manual of Chemical Methods for Pesticides and Devices*, 2nd Edition, Assoc. Off. Anal. Chem., Arlington, VA, 1992.

Whitehead, R. ed.: *U.K. Pesticide Guide*, 1995, CAB International, Wallingford, Oxon, UK.

HERBS.   See **Composite Family**.

HERCULES.   A large constellation lying between Lyra and Corona Borealis. Hercules contains no strikingly bright stars, and hence is somewhat difficult to locate. Once found, however, it is a fertile field for a small telescope. In 1934, this constellation received considerable notice because of the brilliant nova that appeared in it just before Christmas. Perhaps the most interesting object within it is a remarkable star cluster, which was first noted by Halley, in 1714. Although this cluster can be distinguished as such in a telescope of only 2-inch aperture, it requires a telescope larger than 6 inches to appreciate the magnificence of the object. (See map accompanying entry on **Constellations**.)

HEREDITARY MECHANICS.   The field of mechanics involving boundary conditions extending over continuous intervals of space and time and demanding integrals for their representation. For example, in the application of stress to a deformable elastic medium, the final strain at any instant depends not only on the stress at that instant but on the whole previous stress to which the medium has been exposed. Analytically,

$$\delta(t) = kX(t) + \int_{t0}^{t} \theta(t, \tau)X(\tau)d\tau$$

where δ is the final strain at time t, $X(t)$ is the instantaneous stress at time t and the integral represents the effect of the stress heredity of the system. The quantity $\theta(t, \tau)$ is called the coefficient of heredity. The above equation may be considered an integral equation for the evaluation of X when δ is known.

HEREDITY.   The transmission of developmental potentialities from one generation of living things to the next and following generations through the natural process of reproduction. The materials of the parent bodies from which a new individual develops are its actual heritage. During its own embryonic development, the potentialities of this heritage are expressed in the structural characteristics of the new body, normally like those of the parents or those of a more remote generation of ancestors. This fact leads to the statement that the organism inherits certain characters; although this may not be precisely true, the interpretation is permissible for ordinary purposes of description.

Genetics is that branch of biology that deals with the phenomena of heredity and the variations between parents and offspring. See also **Genetics and Gene Science (Classical)**.

Work of Mendel.   The first steps in genetics were taken by plant hybridizers of the eighteenth and nineteenth centuries, chiefly in Europe, and culminated in the experiments of Gregor Johann Mendel, a monk at Brno, Czechoslovakia, then Brünn in Austria. Mendel's results were published in 1866 and lay almost unnoticed until 1990, when they were corroborated by three scientists in the birth of modern genetics. The published report of Mendel's work repeated the significant observations of his predecessors and added a simple mathematical analysis that had not been previously expressed. As a result of the importance of this work, the term Mendelian heredity is applied to the established fundamentals with which subsequent discoveries have been correlated.

Mendelian heredity depends on three fundamental concepts: (1) The organism is a mosaic of unit characters capable of separate hereditary transmission. (2) A unit character may mask a related unit character completely when the potentialities for the development of both are present in the same individual. This principle is called dominance, and the masked character is said to be recessive. (3) Unit characters may be segregated during reproduction, regardless of the combinations in which they have been associated.

To these concepts, later scientists in the field added that the association of different related unit characters in one individual may result in the development of both in different parts of the body, in a mosaic inheritance, or in an intermediate condition.

Some characters, particularly of a quantitative nature, are due to multiple genes. Such characters must be studied by statistical methods. They were the foundation of another attempt to formulate laws of inheritance made by Sir Francis Galton, from which we retain the law of ancestral inheritance and the law of filial regression. The former indicates that each parent contributes one-quarter of the total heritage of the individual, each grandparent one-sixteenth, and so on in a rapidly diminishing percentage. The law indicates the great reduction of the possibility of a hereditary character reappearing after a lapse of generations. Filial regression is the tendency of extreme parents to produce offspring less extreme than themselves. Thus tall parents beget tall children, but usually shorter than themselves. Galton studied human inheritance and in addition to his mathematical analyses, so necessary in this field, took the initial steps in proposing deliberate control, which led to the science of eugenics.

Modern gene science has added vast amounts of quantitative information, strengthening the recognition that hereditary potentialities are resident in the chromosomes of body cells and that definitely located genes within these chromosomes are the determiners through which specific unit characters are brought to expression. The behavior of chromosomes has been found to be in harmony with the transmission of characters by Mendelian heredity. Since nothing was known of chromosomes during Mendel's life, this correlation had to await further advances in cytology.

Mendel's chief contributions were derived from the study of garden peas, in which he observed seven pairs of unit characters, all similar in behavior. He noted, for example, that seed colors included two unit characters, yellow and green. When he crossed parent plants of the two strains the resulting hybrid seeds were entirely yellow, indicating the dominance of this color over green. He then inbred the hybrids, and in their offspring both yellow and green seeds appeared in the ratio of three yellow to one green. Related unit characters of this kind are said to be alleles or allelomorphs. It is now known that their genes occupy the same position in the paired chromosomes of the cells, while only one can be represented in the single chromosome of a germ cell. Since each parent contributes one chromosome to each pair in its offspring, it may also contribute one gene of an allelic pair. The one parent plant contributed a gene for yellow, the other for green, and through dominance the offspring were yellow. Segregation, however, enabled these hybrids to transmit either yellow or green during their reproduction, and through random fertilization all possible combinations of these determiners were established. The characters are commonly represented by symbols, using a capital letter for the dominant and a small letter for the related recessive, as Y and y for yellow and green, respectively. For the pair of characters mentioned, in diagram (1) is representative:

Parental generation (P):	YY	yy
Germ cells:	Y	y
Hybrids of first filial generation (F_1):	Yy	
Gametes of F_1 generation	Y	y

	Y	y
Y	YY	Yy
y	Yy	yy

and their combinations in the F2 generation, in a Punnett square:

The YY and yy individuals in this diagram are homozygous, and the Yy individuals are heterozygous. Since all YY and Yy individuals look alike, due to the dominance of Y, they belong to the same phenotype, but since their hereditary potentialities are different they belong to different genotypes. The yy individuals from hybrid parents are known as extracted recessives. There are twice as many heterozygotes as homozygotes of either kind in this 3:1 ratio because similar individuals in this category result from reciprocal combinations of genes, half of the individuals receiving the dominant from one parent and half from the other. Examples of this kind, involving only one pair of allelic characters, are known as monohybrids.

Additional complexity arises in dihybrids, trihybrids, and polyhybrids of still more characters through the free reassortment of the unrelated pairs of alleles. Thus peas from smooth yellow seeds crossed with others from wrinkled green seeds, a dihybrid combination, produce only yellow smooth seeds in the F1 generation. But when inbred, these plants give rise in the F2 generation to the four possible combinations: smooth yellow, smooth green, wrinkled yellow, and wrinkled green, in the ratio 9:3:3:1. The reason is evident in diagram (2).

In this diagram, each pair of symbols above and at the left side represents the contribution of one parent in one of its germ cells, and in the small squares the possible combinations from the two parents are shown. Dominance prevails as in the monohybrid.

	SY	Sy	sY	sy
SY	SY SY	Sy SY	sY SY	sy SY
Sy	SY Sy	Sy Sy	sY Sy	sy Sy
sY	SY sY	Sy sY	sY sY	sy sY
sy	SY sy	Sy sy	sY sy	sy sy

In a trihybrid, free reassortment results in an F2 ratio of 27:9:9:9: 3:3:3:1. The number of phenotypes is always a power of two indicated by the number of pairs of alleles under consideration. See also **Cell (Biology)**.

Studies of Fruit Fly. The study of heredity in animals has shown that these principles are applicable in that kingdom as well as in plants, but relatively few animals are sufficiently prolific to demonstrate complex ratios. The fruit fly, Drosophila melanogaster, has been the most productive of all genetic subjects, whereas man and the domestic animals yield very limited Mendelian data. See also **Fruit fly**.

Modern genetics, largely from studies of the fruit fly, has disclosed many principles as corollaries of simple Mendelian heredity. The more important are as follows:

Multiple alleles: More than two unit characters may be related to each other as alleles. In such cases only two of the series may be present in any one individual, and dominance is in a graded series, as may be determined by experimental results.

Multiple genes: More than one gene may be necessary for the production of a single unit character. If two genes are essential for its appearance and either alone is incapable of expression, they are said to be complementary. If one expresses itself alone, a gene that modifies this expression is supplementary. If two are capable of producing the same effect whether present singly or in combination, so that the resulting character is absent only from homozygous recessives, they are said to be duplicate genes. In all cases, recombination of the genes during reproduction follows the same course as in simple Mendelian heredity, but the resulting phenotypic ratios differ because fewer unit characters are involved.

Lethal genes: Some genes completely inhibit development or modify it in such a way that the individual dies. They also modify the usual ratios of associated characters.

Linkage: Some characters, although not allelic, are inherited in definite groups; they are said to be linked. Modern genetics shows that linkage is due to the presence of genes for the linked characters in the same chromosomes.

Crossing over: Linkage relations are sometimes interrupted in a limited number of individuals, permitting some reassortment of normally grouped characters. This change is due to the breaking of paired chromosomes in synapsis and the reunion of their fragments in new combinations to form similar chromosomes, sometimes with new combinations of genes.

Translocation: This change is a shifting of the relations of genes in the chromosomes, due to looping, fusion, and rupture, or to the attachment of fragments to other chromosomes. It may result in the duplication of genes within a chromosome or in a change in the serial arrangement of the included genes.

The inheritance of sex has also been shown in many cases to depend on a simple chromosomal mechanism. Males of many species have an X chromosome without a synaptic mate or with a Y chromosome mate that is evidently abortive. The females of such species have two X chromosomes. In the formation of germ cells all eggs receive an X chromosome while half of the sperm cells receive an X chromosome and half a Y or none. Random combination of these cells restores the XX combination in one-half and X or XY in the other, thus producing half females and half males. Other investigations have shown that the quantitative balance between the sex and other chromosomes is the active factor in conditioning the differentiation of the sexes.

This disclosure also explains the phenomenon of sex linkage. Genes lying in the sex chromosomes, mostly in the X chromosomes but a few in the Y, are inevitably transmitted and expressed in some definite relation with sex; hence they are said to be sex linked. Such characters need have no active sexual role.

Plant and Animal Breeding. The findings of genetics have been of value in plant and animal breeding. Although the improvement of cultivated plants and domestic animals by selection preceded by many years the formulation of scientific principles of heredity, the discovery of these principles has made possible much more precise and efficient procedure in the establishment of useful strains. Hybridization and selection together are the chief means of improvement. Applied by scientists they have brought about many modifications of living things and have disclosed many facts concerning heredity. Corn (maize) has been studied in detail and subjected to many experiments, both practical and purely scientific. Tomatoes, radishes, various cereals, and flowers of many species have also commanded attention. More has been done with plants than with animals because the domestic animals are less amenable to experiment. From the practical point of view plants are more satisfactory subjects because desirable hybrid strains may often be propagated by cuttings, grafting, and other asexual methods which avoid the segregation that is inevitable in sexual processes. Only rigid selection can establish desired hybrid combinations in plants or animals that must be produced sexually.

The study of human heredity depends on studies of families. Genealogical records have furnished a large amount of valuable material and the records of public institutions have been equally useful to the geneticist. Such records are not to be compared with scientifically assembled experimental data, but they leave no doubt that the principles of heredity worked out in the study of other organisms are also applicable to humans.

The clearest evidences of human heredity are found in the behavior of simple structural defects, such as the appearance of extra digits (polydactylism), the fusion of bones in the digits (symphalangism), and shortness of the fingers (brachydactylism). These defects are transmitted as Mendelian unit characters allelic to normal structure. Red-green color-blindness (vision) is one of the most striking examples of inheritance in man. It is a sex-linked recessive allele of normal vision. Both X chromosomes of the female must carry the gene for the defect if she is to be color-blind, whereas the male would be color-blind if he receives such a gene in his one X chromosome. Females may be heterozygous carriers of the defect, with normal vision; males are either strictly normal or defective. In this type of inheritance the male always receives the genes for his characters from his mother; therefore a carrier mother may have some color-blind sons. A color-blind man and a genotypically normal woman cannot produce color-blind children, but all of their daughters are carriers. On the other hand, a color-blind woman and a normal man will produce carrier daughters and color-blind sons. Hemophilia is inherited in a like manner, except that the recessive genes for hemophilia are lethal in the homozygous condition in the female. See also **Hemophilia**.

Pigmentation of the skin is controlled by multiple-factor inheritance. Since variations of skin color within a race are not always readily identified and may be partly environmental, knowledge of skin color inheritance has had to come mainly from study of black-white marriages. When a black person without white ancestry marries a white person without black ancestry, their children are typically intermediate in color, or mulattoes. Children from the marriage of a typical mulatto to another typical mulatto may vary in skin color from the black of the black grandparent to the light color of the white grandparent. It has been estimated that the color differences in blacks and whites are controlled by from two to four pairs of alleles. It is possible for a white-skinned person of black-white ancestry to have all the genes of the white genotype. Children from such a person married to a white or similar near-white should be all-white. Children from the marriage between two near-whites are seldom much darker than their parents, and some would have light skin color. If a near-white marries a white, their children are usually no darker than their near-white parents; there is no well-established evidence that a very dark or black child could be born to them.

Albinism is a rare inherited condition in which the skin, hair, and eyes lack the melanin pigment normally present. It results from a biochemical deficiency in which specialized skin cells called melanocytes are unable to synthesize melanin from the amino acid tyrosine. See also **Albinism**.

There is evidence that some allergies may be inherited. Inherited weaknesses in the tissues may make it easier for some antigens to enter the body of certain persons. See also **Allergy**.

Diseases of genetic origin are discussed in the entry on **Genetics and Gene Science**.

Extranuclear Inheritance. The existence of cytoplasmic genes was suggested as long ago as 1909 when the first examples of non-Mendelian inheritance were described by Correns and Baur. However, the demonstration that chloroplasts, mitochondria, and the kinetoplasts of trypanosomes contain specific DNA of their own came as a surprise to most biologists. It is now recognized that organelle DNAs are present in the cell in small amounts, perhaps $1-10\%$ of the total cellular DNA. Organelle DNAs are also distinct entities, as indicated by average nucleotide compositions different from nuclear DNA. All organelle DNAs examined thus far consist of covalently closed circles and exhibit autonomous replication. Although the functions of such organelle DNAs remain largely unknown, it appears that ribosomal RNAs and most if not all tRNAs of chloroplasts and mitochondria are transcribed from the corresponding DNAs. Specific proteins either coded by organelle genes or synthesized with the organelle have been more difficult to identify. The importance to the cell of organelle DNA and the resultant extranuclear inheritance of genes present in this DNA is illustrated by the petite mutants of yeasts. These mutants contain an altered mitochondrial DNA which results in lack of mitochondrial respiratory function. Thus, to survive, petite mutants must utilize an alternative source of energy such as anaerobic fermentation of carbohydrates. Genetic analysis has established that inheritance patterns of the defect are consistent with cytoplasmic inheritance.

Additional Reading

Bateson, W.: *Mendel's Principles of Heredity: A Defence*, Genetics Heritage Press, Placitas, NM, 1996.

Cummings, M.: *Human Heredity: Principles and Issues*, Brooks/Cole Publishing Company, Pacific Grove, CA, 1999.

Gayon, J.: *Darwinism's Struggle for Survival: Heredity and the Hypothesis of Natural Selection*, Cambridge University Press, New York, NY, 1998.

Haymer, D.: *Modern Genetics Laboratory: Heredity and Molecular Biology*, Kendall/Hunt Publishing Company, Dubuque, IA, 1999.

Knight, J. and R. McClenaghan: *Encyclopedia of Genetics: Aggression–Heredity and Environment*, Vol. 1, Salem Press, Hackensack, NJ, 1999.

Santrock, J.: *Evolution and Heredity*, The McGraw-Hill Companies, Inc., New York, NY, 1999.

Singer, S.: *Human Genetics: An Introduction to the Principles of Heredity*, W. H. Freeman Company, New York, NY, 1998.

Sternberg, R.J. and E. Grigorenko: *Intelligence, Heredity, and Environment*, Cambridge University Press, New York, NY, 1996.

ANN C. DeBALDO, University of South Florida, Tampa, FL

HERMAPHRODITE. An animal with functional reproductive organs of both sexes.

HERMAPHRODITISM. A condition characterized by the presence of both ovarian and testicular tissue. Because of overactivity of the adrenal glands, excessive hormones can be produced. In some patients with overactive adrenals, there is an excessive development of fat, accompanied by sexual disturbances. The symptoms vary according to age and sex. If the disease develops during fetal life and the child is a female, a form of hermaphroditism, or dual sexuality, may result, in which the clitoris is enlarged and resembles the penis. Other signs of masculinization accompany this condition. Sometimes a true hermaphrodite may appear to be a normal female, but who is found at surgery to possess testes in the groin region. Only about a dozen cases of true hermaphroditism in the human race have been reported. This term signifies the presence of all of the functioning genital organs of both sexes in one individual. The reported cases were claimed to have both testicles and ovaries present. However, the ability to impregnate as well as to conceive has never been reported in one individual.

Many cases of pseudo-hermaphroditism have been seen. In this condition the genital organs, internal or external, do not conform either totally or in part with the sexual glands (testicles or ovaries) present. In the male hermaphrodite, testicles are present but may be abdominal in position. The penis is small and more nearly resembles a large clitoris; the scrotum is divided by a cleft resembling the female labia with a small short vagina. Uterus and tubes are not present.

The female hermaphrodite has a large clitoris more like a small penis, rudimentary vagina, a uterus and ovaries. Various in-between stages may be present, giving a very bizarre picture where the sex can only be determined by microscopic study of sex characteristics shown by the nuclei of the tissue cells. Such cells may be examined by biopsy of the skin, and a definite decision as to sex given with accuracy of a high degree. Where biopsy is not desired or facilities are not available, the nuclei of epithelial cells scraped from the inside of the mouth, or even of polymorphonuclear leucocytes in the blood will furnish a slightly less reliable answer. Such sexing should be done as soon as possible after birth in any infant in whom the identity of the sex organs appears dubious; by this means mistakes in naming and upbringing can be avoided. Where such a decision as to sex is not made in very early life, it is probably wise to bring up the child according to the sex that seems most apparent and defer final decision until the onset of puberty, when the development of sex consciousness may reveal psychological orientation to one sex or the other.

Dewald et al. (Mayo Clinic and Mayo Foundation), using chromosome heteromorphisms and blood cell types as genetic markers, demonstrated chimerism in a chi46, XX/46,XY true hermaphrodite. The pattern of inheritance of the chromosome heteromorphisms indicated that this individual was probably conceived by the fertilization, by two different spermatozoa, of an ovum and the second meiotic division polar body derived from the ovum and subsequent fusion of the two zygotes. A chimera may be defined as an individual with two or more genetic cell types resulting from the fusion of different zygotes. As described by Dewald et al., "Chimeras can be readily classified as whole-body or partial chimeras according to their mode of origin. Partial chimeras can arise by placental cross-fertilization between dizygotic twins, maternal-fetal transplacental exchange, transfusions, or grafting. Because of lack of suitable studies, the origin of wholebody chimeras is less clear. Theoretically, they can arise by: (1) early fusion of different embryos; (2) fertilization of an ovum and any polar body by two different sperm and subsequent fusion of the zygotes; (3) fertilization of a haploid ovum or polar body and subsequent fusion with a diploid polar body or ovum; or (4) fusion of a diploid sperm with an embryo." Most reported chimeras have sexual abnormalities, such as clitoral hypertrophy or true hermaphroditism. More detail will be found in Dewald, G., et al.: "Origin of chi46,XX/46,XY Chimerism in Human True Hermaphrodite," *Science*, **207**, 321–323 (1980).

HERMITE EQUATION. A second-order differential equation

$$y'' - 2xy' + 2ny = 0$$

where n is a constant. The Hermite polynomials (see also **Generating Function**) are solutions. The equation occurs in the quantum mechanical problem of the harmonic oscillator. (See also **Weber Equation**, from which the Hermite equation can be obtained by a change of variable.)

HERNIA. At one time commonly called rupture, a hernia is an abnormal protrusion of a part or organ through the containing wall of its cavity. In common usage, the term hernia usually applies to the abdominal cavity and implies a covering or sac over the protrusion. There are two main classes of hernias: (1) *congenital* hernia, in which the sac was present before birth; and (2) *acquired* hernia, in which the sac is formed after birth and pushes through an opening in the muscle wall that failed to close at birth, or that was formed following an incision. A large percentage of acquired hernias result from injury or strain, such as those hernias which occur when a person lifts a heavy object. Hernias may occur in the groin, the navel, the membrane separating the abdominal and chest cavities (diaphragm), in surgical incisions, and elsewhere. All herniation takes place through a normal opening, or through an opening that should have been eliminated at some period of development, or through an opening which had closed and then reopened in later life.

The hernial sac has a mouth, a neck, and a body. The mouth connects with the abdominal cavity and is called the hernial ring; the body is the pouch or sac that projects outside the abdominal wall; and the neck connects the mouth and body of the sac.

The contents of the sac might be any of the abdominal organs, in whole or part; loops of the intestine are commonly found in hernias. The sac and its contents are subject to injury that can lead to serious complications. The skin surface is vulnerable to blows, falls, pressure, irritation from binders or trusses, or may become inflamed, infected, or abscessed. From within, the contents of the sac are prone to strangulation when the blood supply is cut off by a narrow or constricted hernial ring; gangrene may set in if treatment is not sought promptly.

Hernias are considered *reducible* or *irreducible*. Reduction may be spontaneous; for example, sac contents may return unaided to the abdominal

cavity when the patient lies flat. If the patient remains untreated, however, a reducible hernia may become irreducible, that is, the contents of the sac can no longer be returned to the abdominal cavity. Irreducibility may be caused by increased size of the hernia, by formation of adhesions, or by development of a small or constricted hernial ring. Hernias of enormous size, hanging down to the knees, have been reported. An irreducible hernia is a constant source of danger.

Hernias occurring in the groin are either *inguinal* hernias or *femoral* hernias. Inguinal hernias account for about 92% of all hernias. Superficially, inguinal and femoral hernias look alike because the bulge is in the groin. However, they differ anatomically. Inguinal hernias slip through the normal openings for the passage of nerves or organs of the reproductive system. Femoral hernias occur through the passageway for nerves and vessels to the thigh.

Normally, the deep and shallow layers of muscles and ligaments on the abdominal wall protect these normal openings against herniation. With rise of intra-abdominal tension, as by straining, coughing, or lifting, the muscles contract and flatten like a shutter in a normal situation. But if the muscles and/or other protective structures of these openings are weak, the shutter action fails and an increase of intra-abdominal tension may push part of the abdominal organs through the opening into the preformed sac, and thus a hernia is begun. Successive incidents of tension increase the size of the sac by forcing Additional intra-abdominal tissue into it.

Hernias of the navel are called *umbilical* hernias. The navel is an opening that should close in the process of development. After birth, it is a scar formed of interlaced muscle fibers of the contracted umbilical ring. Sometimes a defect occurring before birth prevents its closing, and the baby is born with a hernia, or may soon acquire one. In adults between 25 and 40 years of age, obesity and pregnancy are the most common predisposing causes of this form of hernia.

Hiatus hernia occurs over the diaphragm. See also **Esophagus**.

Obese women are the most frequent subjects of hernia in the site of a surgical incision. Some incisional hernias are caused by failure of the layers of deep muscle and fascia to knit firmly after surgery. Blood clot, infection, exudate, and swelling in the line of incision, as well as increased intra-abdominal tension, also are factors favoring herniation. The neck of the incisional hernia is a firm ring of scar tissue. Because of the large hernial ring, these hernias are difficult to control by a truss. Large incisional hernias may cause invalidism unless surgical relief is obtained.

The treatment of a hernia patient can be accomplished by a mechanical device (truss), or surgery. Most authorities agree that a truss is a makeshift, which is acceptable only when surgery would be hazardous. Improved techniques have made possible the surgical repair of hernias, which not many decades ago would have been irreparable. It is often necessary to close the opening with a fascial graft, or an inert foreign material, such as polypropylene mesh.

Additional Reading

Bendavid, R.: *Abdominal Wall Hernias: Principles and Management*, Springer-Verlag, Inc., New York, NY, 2001.

Chevrel, Jean-Paul, and E. Goldstein: *Hernia and Surgery of the Abdominal Wall*, Springer-Verlag, Inc., New York, NY, 1997.

Darzi, A.: *Laparoscopic Inguinal Hernia Repair*, Mosby-Year Book Inc., St. Louis, MO, 1994.

Dunn, C.C. and D. Menzies: *Hernia Repair: The Laparoscopic Approach*, Blackwell Science, Inc., Malden, MA, 1995.

Kavic, M.: *Laparoscopic Hernia Repair*, Gordon & Breach Publishing Group, Newark, NJ, 1997.

Maddern, G.J.: *Abdominal Wall Hernias*, W.B. Saunders Company, Philadelphia, PA, 1996.

Nyhus, L. and R. Condon: *Hernia*, Lippincott-Raven Publishers, Philadelphia, PA, 1994.

Postacchini, F.: *Lumbar Disc Herniation*, Springer-Verlag, Inc., New York, NY, 1999.

Schumpelick, V. and A.N. Kingsnorth: *Incisional Hernia*, Springer-Verlag, Inc., New York, NY, 1999.

Skandalakis, L.J. et al.: *Modern Hernia Repair: The Embryological and Anatomical Basis for Surgery*, Parthenon Publishing Group, New York, NY, 1995.

HEROIN. See **Alkaloids**

HERON (*Aves, Ciconiiformes*). Long-legged wading birds (*Aves*) with a sharp slender beak and when adult with plumes or a crest. They live chiefly on fish.

Herons are found throughout the world. The most widely known North American species are the great blue heron, *Ardea herodias*, the green heron, *Butorides virescens*, and the egret, *Egretta*. The last is a white bird which bears beautiful plumes known as aigrettes during the breeding season. It was once threatened with extinction through the use of these plumes as ornaments for hats, but the remaining birds are adequately protected. See Fig. 1.

Fig. 1.   Egret. (*National Audubon Society; Grant M. Haist.*)

These birds are remarkable for the down that they produce. It is exceptionally light and fluffy and grows all over the breast, rump, and flanks. The birds roost in tall trees during the day and feed mostly at night. They have long legs and long beaks. Sometimes they will reach a height of about 20 inches (51 centimeters). Some species are gray with black on head and neck. However, there is a wide variation both in size and coloration. See also **Ciconiiformes**.

HERPES SIMPLEX VIRUS DISEASES.   Four major herpesviruses cause infections in humans: (1) Herpes simplex; (2) varicella-zoster; (3) cytomegalovirus; and (4) Epstein-Barr virus. These are among the most widespread of all human pathogens and, characteristically, tend to follow cycles of dormancy and activity within an individual, such cycles often extending over long periods. A long span of dormancy may be interrupted by a flareup resulting from unusual physical or psychological stress. There is no known effective treatment for achieving their full eradication. Incidence of herpesvirus infections tends to occur more frequently in immunosuppressed patients. See also **Immune System and Immunology**. These viruses are described further in the entry on **Virus**. See also **Cancer Research**.

There are two types of herpes simplex virus (HSV), with multiple strains of each type. Type 1 infects mucous membranes of the oral cavity, perioral skin, eyes, and skin above the waist. Type 2 usually causes a genital infection, an infection which is the second most common venereal disease found in the United States and a number of other countries. It has been estimated that between 30 and 90% of young adults carry antibody to one or both types of herpes simplex virus. Most infections from Type 1 HSV occur during childhood, but may occur any time during life. Type 2 usually does not appear before puberty and the commencement of sexual activity. The incidence of Type 2 antibody peaks by age 35 years. Type 2 antibody can be found in 20–35% of the general population. Certain occupational groups, such as health professionals and prostitutes, are at greater risk for HSV infection — simply because of the greater number of possible contacts with the virus. However, the infection is found in all segments of society.

Both types of herpes simplex virus appear to be spread by close contact between infected and susceptible individuals. Incubation period ranges from 2 to 20 days. Even in persons with antibody to the virus, second infections are observed. Virus may be shed from the oral or genital mucous membranes. Where there is adequate immunity, the shed virus may produce either subclinical infection or observable clinical disease. In persons who are highly immunosuppressed (as in cases of persons who have received organ transplants), the infection may be widespread and not heal for many months.

The most serious infection of Type 1 HSV is herpes keratitis, which can lead to destruction of the cornea. Other primary infections of Type 1 include stomatitis, pharyngitis, tracheobronchitis, and dermatitis. Type 2 usually involves vulvovaginitis or balanitis. Some clinical studies have shown that, in addition to causing oral, ocular, and genital lesions, HSV infections may involve visceral sites, such as the throat, lungs, esophagus, brain, meninges, liver, spleen, and pancreas. It has been observed that organ transplantation and the widespread use of cancer chemotherapy have increased the frequency of herpes simplex visceral infection.

The neonate is seriously threatened in cases of maternal genital infection. Such infections can be fatal in about half of the cases of the newborn. Studies have shown that even when the mother is asymptomatic, there is a risk to the infant. Some authorities have suggested that the incidence of neonatal Type 2 disease could be lowered by performing a Cesarean section in symptomatic women.

Herpes simplex virus obtained from infected sites grows easily. The virus also can be demonstrated in tissues by fluorescent antibody staining. A fourfold rise in complement-fixing antibody titer supports the diagnosis, but is not itself diagnostic. Intranuclear inclusion bodies may be observed in tissues from patients who are infected with herpes simplex as well as in tissue from patients infected with varicella-zoster virus or cytomegalovirus.

Therapy depends on the site involved. For herpes keratitis, vidarabine, trifluridine, and idoxuridine are licensed. Topical acyclovir is effective for initial genital herpes and for localized mucocutaneous lesions in immunocompromised patients. Intravenous acyclovir has been approved for herpes simplex infection in immunocompromised patients and for initial genital herpes infection in immunocompetent patients that is sufficiently severe to require hospitalization.

Every effort should be made to deter transmission of the virus. Medical personnel and others close to infected patients, such as sexual partners, should avoid direct contact with lesions. Asymptomatic shedding, unfortunately, limits the effectiveness of effort to prevent spread.

See also **Herpesviruses (Human)**.

Additional Reading

Berns, K. and R. Whitley: "Latency by Herpes Simplex Viruses," *Journal: Intervirology*, Vol. 32, S. Karger, AG Basel, Switzerland, 1991.

Brown, M.: *Herpes Simplex Virus Protocols*, Vol. 10, Humana Press, Totowa, NJ, 1997.

Calisher, C.H.: *Immunity to and Prevention of Herpes Zoster*, Springer-Verlag, Inc., New York, NY, 2001.

Hajjar, D.P. and S.M. Schwartz: *Role of Herpes Viruses in Atherogenesis*, Gordon & Breach Publishing Group, Newark, NJ, 1999.

Hutto, S.C., and G.B. Scott: *Diagnosis of Congenital and Perinatal Infections: A Concise Guide*, Springer-Verlag New York, LLC, New York, NY, 2006.

Kirchner, H.: *Immunobiology of Infection with Herpes Simplex Virus*, S. Karger, AG, Basel, Switzerland, 1982.

Posner, T.N.: *Herpes Simplex*, Routledge, New York, NY, 1998.

Stanberry, L.R.: *Genital and Neonatal Herpes*, John Wiley & Sons., Inc., New York, NY, 1997.

Stanberry, L.R.: *Understanding Herpes*, University Press of Mississippi, Jackson, MS, 1998.

Studahl, M., P. Cinque, and T. Bergstrom: *Herpes Simplex Viruses*, Taylor & Francis, Inc., Philadelphia, PA, 2006.

Web References

Centers for Disease Control and Prevention: http://www.cdc.gov/std/Herpes/ STD Fact-Herpes.htm

E medicine: http://www.emedicine.com/MED/topic1006.htm

Medline Plus; National Institutes of Health: http://www.nlm.nih.gov/medlineplus/ herpessimplex.html;   and   http://www.nlm.nih.gov/medlineplus/ency/article/ 001324.htm

National Institutes of Health: http://www.niaid.nih.gov/factsheets/stdherp.htm

HERPESVIRUSES (HUMAN).   In nature, herpesviruses infect both vertebrate and nonvertebrate species, and over 100 have been at least partially characterized. Only eight of these have been isolated routinely from humans. They are known as the human herpesviruses and are *Herpes simplex virus type 1* (HSV-1), *Herpes simplex virus type 2* (HSV-2), *Varicella-zoster virus* (VZV), *Cytomegalovirus* (CMV), *Epstein–Barr virus* (EBV), *Human herpesvirus 6* (HHV-6), *Human herpesvirus 7* (HHV-7) and, most recently, *Kaposi sarcoma-associated herpesvirus* (KSHV or HHV-8). See also **Herpes Simplex Virus Diseases**.

General Biological Properties

The human herpesviruses share four significant biological properties [Roizman and Pellett, 2001]:

- All code for unique enzymes that process nucleic acids. These enzymes are structurally diverse and provide unique sites for inhibition by antiviral agents.
- Synthesis and assembly of viral deoxyribonucleic acid (DNA) is initiated in the nucleus. Assembly of the capsid is also initiated in the nucleus.
- Release of progeny virus from the infected cell is accompanied by cell death.
- All establish latent infection indicative of tissue tropism of each member of this family [Straus, 2000].

Structure

Membership in the family *Herpesviridae* is based on the structure of the virion. These viruses contain double-stranded DNA (dsDNA), which is located at the central core. Herpesvirus DNA varies in molecular weight from approximately 80 to 150 million, or 120 to 250 kb, depending on the virus. This DNA core is surrounded by a capsid, which consists of 162 capsomers, arranged in icosapentahedral symmetry. The capsid is approximately 100–110 nm in diameter. Tightly adherent to the outside of the capsid is the tegument, which appears to consist of amorphous protein material. Loosely surrounding the capsid and tegument is a lipid bilayer envelope derived from host cell membranes. The envelope consists of cellular polyamines and lipids, and viral glycoproteins. These glycoproteins confer distinctive properties to each virus and provide unique antigens to which the host immune system is capable of responding [Roizman and Pellett, 2001].

Herpesviruses can be divided into six groups arbitrarily classified A to F, according to genomic sequence arrangement [Roizman and Pellett, 2001]. Unique structures are demonstrable for those herpesviruses that infect humans (groups C, D and E). In group C genomes (EBV and KSHV), a number of terminal and internal reiterations divide the genome into several well-delineated domains. The group D genomes (VZV) have sequences from one terminus repeated in an inverted orientation internally [Kost and Straus, 1996]. Thus, the DNA extracted from these virions consists of two equal molar populations. For group E viral genomes (HSV and CMV), the genomes are divided into unique long and short sequences (U_L and U_S), whereby both termini are repeated in an inverted orientation at the junction of U_L and U_S. Thus, the genomes can form four equimolar populations that differ in the relative orientation of the U_L and U_S segments [Whitley et al., 1998].

Classification

The grouping of herpesviruses into subfamilies serves the purpose of identifying evolutionary relatedness as well as summarizing unique properties of each member [Roizman and Pellett, 2001]. Members of the *Alphaherpesvirinae* have an extremely short replicative cycle (hours), destroy the host cell and replicate in a wide variety of host tissues. Since these viruses have a propensity to replicate in neuronal tissue, latent infection is established in sensory nerve ganglia. This subfamily consists of HSV-1 and HSV-2 and VZV. In contrast to the alphaherpesviruses, members of the *Betaherpesvirinae* have a restricted host range. Their reproductive life cycle is long (days), with infection progressing slowly in cell culture. A characteristic of these viruses is their ability to form multinucleated giant cells exemplified by human HSV infection. These viruses establish latent infection in secretory glands, cells of the reticuloendothelial system and kidneys. Other members of this subfamily include HHV-6 and HHV-7. Members of the *Gammaherpesvirinae* have the most limited host range and replicate in lymphoblastoid cells, causing lytic infections in targeted cells. Latent virus has been demonstrated in lymphoid tissue. EBV is a member of this subfamily, as is KSHV.

Replication and Latency

Replication of all herpesviruses is a multistep process. Following the onset of infection, DNA is uncoated and transported to the nucleus of the host cell. Transcription of immediate early genes follows, allowing encoding of regulatory proteins. Expression of immediate early gene products is followed by the synthesis of proteins encoded by early and, subsequently, late genes. Assembly of the viral core and capsid takes place within the nucleus. This is followed by envelopment at the nuclear membrane and transport from the nucleus through the endoplasmic reticulum and Golgi,

possibly with deenvelopment into the cytoplasm and reenvelopment with tegument added at Golgi or similar membranes [Gershon and Silverstein, 2002]. Mature virions are transported to the outer membrane of the host cell inside vesicles. Release of progeny virus is accompanied by cell death. Replication for all herpesviruses is considered inefficient, with a high ratio of noninfectious to infectious viral particles.

Latency is established in specific host cells, and the latent viral genome may be either extrachromosomal or integrated into host cell DNA. HSV-1, HSV-2 and VZV establish latency in the dorsal root ganglia. EBV can maintain latency within B lymphocytes and salivary glands. CMV, HHV-6 and HHV-7 probably establish latency in endothelial cells and monocytes and macrophages. KSHV has an unknown site of latency. Latent virus may be reactivated and enter a replicative cycle at any point in time. The reactivation of latent virus to produce recurrent infections is a well-recognized biological phenomenon, but not one that is understood from a biochemical or genetic standpoint. Stimuli that are associated with the reactivation of latent HSV include stress, menstruation and exposure to ultraviolet light. Precisely how these factors interact at the level of the ganglia remains to be defined. Reactivation of herpesviruses may be clinically asymptomatic, or may produce life-threatening disease.

Diagnosis

The definitive diagnosis of a herpesvirus infection usually requires either isolation of virus or detection of viral gene products. For virus isolation, swabs of clinical specimens or other body fluids can be inoculated into susceptible cell lines and observed for the development of characteristic cytopathic effects. This technique is most useful for the diagnosis of infection due to HSV-1 and HSV-2 or VZV because of relatively short replicative cycles. HHV-6, HHV-7 and KSHV have unique growth characteristics that make identification in cell culture systems difficult.

Newer and more rapid diagnostic techniques involve the detection of viral gene products. This can be done by applying fluorescence antibody directed against immediate early or late gene products to tissue cultures after 24–72 h of incubation. A positive result is the appearance of intranuclear fluorescence. Alternatively, fluorescence antibodies may be applied directly to cell monolayers or scrapings of clinical lesions, with intranuclear fluorescence again indicating a positive result.

Polymerase chain reaction (PCR) amplification of viral DNA has proven of value in the diagnosis of some herpesvirus diseases, especially HSV infections of the central nervous system when applied to cerebrospinal fluid. This tool has been used to study the natural history of genital HSV infections as well as to identify new herpesviruses (i.e. KSHV).

In addition to new tests for virus gene products and viral DNA, improved serological assays are also becoming available, particularly the application of immunoblot technology to distinguish HSV-1 from HSV-2 infections. However, these tests are only useful for making a diagnosis in retrospect.

Herpes Simplex Viruses

Of all the herpesviruses, HSV-1 and HSV-2 are the most closely related, having nearly 70% genomic homology. These two viruses can be distinguished most reliably by DNA composition; however, differences in antigen expression and biological properties also serve as methods for differentiation.

A critical factor for transmission of HSV, regardless of virus type, is the requirement for intimate contact between a person who is shedding virus and a susceptible host. After inoculation on to the skin or mucous membrane and an incubation period of 4–6 days, HSV replicates in epithelial cells [Whitley and Roizman, 2001]. As replication continues, cell lysis and local inflammation ensue, resulting in characteristic vesicles on an erythematous base. Regional lymphatics and lymph nodes become involved; viraemia and visceral dissemination may develop, depending upon the immunological competence of the host. In all hosts, the virus generally ascends the peripheral sensory nerves to reach the dorsal root ganglia.

Latency is established when HSV reaches the dorsal root ganglia after anterograde transmission via sensory nerve pathways. In its latent form, intracellular HSV DNA cannot be detected routinely unless specific molecular probes are utilized. Reactivation and replication of HSV within neural tissue is followed by retrograde axonal spread of the virus back to other mucosal and skin surfaces via the peripheral sensory nerves. Virus replicates further in epithelial cells, reproducing the lesions of the initial infection until infection is contained through host systemic and mucosal immunity.

Mucocutaneous infections are the most common clinical manifestations of HSV-1 and HSV-2. Gingivostomatitis, usually caused by HSV-1, occurs most frequently in children under 5 years of age [Whitley, 2001]. It is characterized by fever, sore throat, pharyngeal oedema and erythema, followed by the development of vesicular or ulcerative lesions on the oral and pharyngeal mucosa. Recurrent HSV-1 infections of the oropharynx are most frequently manifested as herpes simplex labialis (cold sores), and usually appear on the vermillion border of the lip. Intraoral manifestations of recurrent disease are uncommon in the normal host but do occur frequently in immunocompromised individuals.

Genital herpes is most frequently caused by HSV-2 but an ever-increasing number of cases are attributed to HSV-1 [Stanberry et al., 1999]. Primary infection in women usually involves the vulva, vagina and cervix. In men, initial infection is often associated with lesions on the glans penis, prepuce or penile shaft. In individuals of either gender, primary disease is associated with fever, malaise, anorexia and bilateral inguinal adenopathy. Women frequently have dysuria and urinary retention due to urethral involvement. It is estimated that as many as 10% of individuals will develop an aseptic meningitis with primary infection. Sacral radiculomyelitis may occur in both men and women, resulting in neuralgias, urinary retention or obstipation. The complete healing of primary infection may take several weeks. The first episode of genital infection is less severe in individuals who have had previous HSV infections at other sites, such as herpes simplex labialis.

Recurrent genital infections in either men or women can be particularly distressing. The frequency of recurrence varies significantly from one individual to another. Approximately, one-third of individuals with genital herpes have virtually no recurrences, one-third have approximately three recurrences per year and another third more than three per year. Seroepidemiological studies found that between 35 and 65% of individuals in the United States have antibodies to HSV-2 and that seroprevalence is dependent upon the number of sexual partners. A follow-up study indicated an increase of nearly one-third a decade later. By applying PCR to genital swabs from women with a history of recurrent genital herpes, viral DNA can be detected in the absence of culture proof of infection, suggesting chronicity as opposed to a recurrence due to reactivation of a latent infection.

Other HSV skin infections occur. Common among healthcare workers are lesions on abraded skin of the fingers, known as herpetic whitlow. Similarly, wrestlers, because of physical contact, may develop disseminated cutaneous lesions known as herpes gladiatorum.

Herpes simplex keratitis is usually caused by HSV-1 and is accompanied by conjunctivitis in many cases. It is considered the most common infectious cause of blindness in the United States. The characteristic lesions of HSV keratoconjunctivitis are dendritic ulcers, and are best detected by fluorescein staining. Deep stromal involvement occurs and may result in visual impairment unless prompt antiviral therapy is instituted.

Neonatal HSV infection is estimated to occur in approximately 1 in 3000 deliveries in the United States annually [Whitley, 2004a]. Approximately 70% of cases are caused by HSV-2 and usually result from contact of the fetus with infected maternal genital secretions at the time of delivery. Manifestations of neonatal HSV infection can be divided into three categories: (1) skin, eye and mouth disease, (2) encephalitis and (3) disseminated infection. As the name implies, skin, eye and mouth disease consists of cutaneous lesions and does not involve other organ systems. Involvement of the central nervous system may occur with encephalitis or disseminated infection, and generally results in diffuse encephalitis. The cerebrospinal fluid formula characteristically reveals an elevated protein and a mononuclear pleocytosis. Disseminated infection involves multiple organ systems and can produce disseminated intravascular coagulation, haemorrhagic pneumonitis, encephalitis and cutaneous lesions. Diagnosis can be particularly difficult in the absence of skin lesions. The mortality rate for each disease classification varies from zero for skin, eye and mouth disease to 15% for encephalitis and 50% for neonates with disseminated infection. In addition to the high mortality associated with these infections, morbidity is significant, in that children with encephalitis or disseminated disease develop normally in only approximately 40% of cases, even with the administration of appropriate antiviral therapy.

Herpes simplex encephalitis is characterized by haemorrhagic necrosis of the inferiomedial portion of the temporal lobe [Whitley, 2004a]. Disease begins unilaterally, and then spreads to the contralateral temporal lobe. It is the most common cause of focal, sporadic encephalitis in the United States

today, and occurs in approximately 1 in 150000 individuals. Most cases are caused by HSV-1. The actual pathogenesis of HSV encephalitis requires further clarification, although it has been speculated that primary or recurrent virus can reach the temporal lobe by ascending neural pathways, such as the trigeminal tracts or the olfactory nerves.

Clinical manifestations of HSV encephalitis include headache, fever, altered consciousness and abnormalities of speech and behavior. Focal seizures may also occur. The cerebrospinal fluid formula for these patients is variable, but usually consists of a pleocytosis, with both polymorphonuclear leucocytes and monocytes present. The protein concentration is characteristically elevated and glucose is usually normal. Historically, a definitive diagnosis could only be achieved by brain biopsy, as other pathogens may produce a clinically similar illness. However, the application of PCR for detection of viral DNA in cerebrospinal fluid has replaced brain biopsy as the standard for diagnosis. The mortality and morbidity are high, even when appropriate antiviral therapy is administered. At present, the mortality rate is approximately 30%, and 70% of survivors have significant neurological sequelae. *Simian herpesvirus B*, a nonhuman primate member of the *alphaherpesviruses* along with HSV and VZV, can also cause potentially fatal encephalitis in humans who are infected by monkey bites or, as in one recent case, by a monkey spitting into the eye of a laboratory worker.

HSV infections in the immunocompromised host are clinically more severe, may be progressive and require more time for healing. Manifestations of HSV infections in this patient population include pneumonitis, oesophagitis, hepatitis, colitis and disseminated cutaneous disease. Individuals suffering from human immunodeficiency virus (HIV) infection may have extensive perineal or orofacial ulcerations. HSV infections are also noted to be of increased severity in individuals who are burned.

Varicella-Zoster Virus

VZV is one of the most common viruses encountered by humans. It is usually transmitted by airborne routes (droplet spread), with initial replication in the oropharynx. In the susceptible or seronegative individual, replication of virus in the oropharynx leads to primary viraemia, with subsequent development of a vesicular rash. The replication of VZV *in vitro* is similar to that for HSV, although the period of replication is somewhat prolonged.

Varicella, or chickenpox, is the manifestation of primary VZV infection. This infection occurs most commonly in young children of preschool age and has a characteristic disseminated vesicular rash, which appears after an incubation period of 14–17 days. The rash begins on the face and trunk and spreads to the extremities. The lesions of chickenpox are initially vesicles that become pustular, crusted and then scabbed prior to healing. The average duration of lesion formation is 3–5 days in the normal child; however, it is longer in adolescents, adults and in the immunocompromised host. Because of 'waves' of viraemia with VZV, all stages of lesion development can be seen simultaneously, in contrast to the synchronous progression of all lesions characteristic of smallpox. Serious complications of chickenpox in the nonimmunocompromised child are rare, but secondary bacterial infection (such as group A streptococcus) can be problematic. At the time of primary infection, VZV establishes latency in dorsal root ganglia. See also **Chickenpox (Varicella)**.

Herpes zoster, or shingles, is the recurrent form of VZV, and is the consequence of reactivation of latent virus, manifesting as a localized vesicular rash with a dermatomal distribution. Herpes zoster is not transmitted from one individual to another; however, spread of virus from the vesicles of herpes zoster may lead to the development of *varicella* (chickenpox) in a susceptible host, such as from an elderly person to a young grandchild. Individuals over the age of 50 experience zoster at a frequency of approximately 1%. The rash initially appears within the dermatome as erythema, but is soon followed by the development of vesicles. Some individuals will have coalescence of vesicles into bullous lesions. New vesicles may form for 5–7 days, then evolve through the sequence of healing described for the lesions of *varicella*. The average time to heal for individuals with shingles ranges from 10 to 21 days, depending upon the age and immune status of the individual.

Characteristic of herpes zoster is the appearance of both acute neuritis and postherpetic neuralgia. Acute neuritis is present in most individuals with localized zoster, the exception being young children. Postherpetic neuralgia will develop in as many as 50% of individuals over 60 years of age. The treatment of postherpetic neuralgia can be problematic for individual patients.

VZV infections in the immunocompromised host are associated with a higher incidence of visceral disease. Immunocompromised children, particularly those with acute lymphoblastic leukaemia, are at increased risk for progressive disease, specifically resulting in pneumonitis and/or hepatitis. One of three immunocompromised children suffers from visceral disease, with a mortality of 15% in the absence of antiviral therapy.

Herpes zoster in the immunocompromised host may be associated with cutaneous dissemination and visceral complications. In the absence of antiviral therapy, as many as 25% of individuals with lymphoproliferative malignancies will have cutaneous dissemination and 10% will develop visceral complications, with an overall mortality rate of approximately 8%.

Human Herpesvirus 6 and 7

HHV-6 and HHV-7 have recently been isolated, and the cellular/tissue reservoir and their mode of transmission are not well understood at the present time. Loss of transplacental antibodies, followed by acquisition of antibodies early in life, implies horizontal transmission within the home environment. By the age of 5 years, antibodies to both of these viruses are present in virtually 100% of the population. The high prevalence of antibodies early in life would implicate transmission from oropharyngeal secretions; indeed, saliva is a frequent source of virus in adults. HHV-6 exists as types A and B. The type A variant was the original isolate retrieved from an immunocompromised host. Both the type B variant of HHV-6 and HHV-7 are associated with exanthem subitum, or roseola. This illness is characterized by 3–5 days of fever, followed by the appearance of a maculopapular 'slapped cheek' rash. In addition, there has been an association between HHV-6 and rejection of transplanted kidneys, fulminant hepatitis and infections of the central nervous system [Black and Pellett, 1999].

Kaposi Sarcoma-Associated Herpesvirus

A new herpesvirus (KSHV or HHV-8) has been associated with Kaposi sarcoma and, together with EBV, AIDS-related lymphomas of organ cavities [Chang, et al., 1994; Roizman, 1995]. The DNA of this virus is partially homologous to the DNA of EBV and *Herpesvirus saimiri*, a nonhuman primate member of the *gammaherpesviruses* with tropism for T cells.

Control of Herpesvirus Infection

Prevention. Only one vaccine is licensed for the prevention of a herpesvirus infection; it is directed against VZV. This live, attenuated vaccine is licensed for use in the normal child, and not in immunocompromised individuals. Experimental vaccines for HSV-1 and HSV-2, EBV and CMV are in various stages of clinical trials but initial studies have not been fruitful. Passive immunization with immune or hyperimmune serum, including monoclonal antibodies, has been used either to prevent infection or as an adjunct to therapy. The administration of VZV immune globulin to the immunocompromised child exposed to this virus is routinely used to prevent, or at least attenuate, chickenpox in these high-risk individuals. More recently, CMV immune globulin has been utilized along with antiviral drugs to treat life-threatening infection in immunocompromised patients, or as immune prophylaxis for CMV-seronegative organ transplant recipients.

Treatment. Infections due to HSV, VZV and, to a lesser extent, CMV are the most amenable to therapy with antiviral drugs [Whitley, 2004a]. Aciclovir (Acyclovir in North America; trade name Zovirax®) has proved useful for the management of specific infections caused by HSV and VZV. It is the usual treatment for mucocutaneous HSV infections in the immunocompromised host, herpes simplex encephalitis, neonatal HSV infections and VZV infections in the immunocompromised host. Although both valaciclovir and famciclovir may offer pharmacokinetic and clinical advantages, intravenous administration of aciclovir is preferred for therapy against life-threatening disease. Immunocompromised individuals with mucocutaneous HSV infections that are not life-threatening may be given oral aciclovir. Caution must be exercised when aciclovir is used intravenously, since it may crystallize in the renal tubules when given too rapidly or to dehydrated patients. Topical trifluridine, iododeoxyuridine or vidarabine is effective as therapy for herpes simplex keratitis. See also **Antiviral Drugs**.

The two prodrugs licensed for the treatment of herpes zoster in the elderly are valaciclovir, the prodrug of aciclovir, and famciclovir, the prodrug of penciclovir. Both provide high plasma levels of the parent compounds and offer added efficacy, as well as decreased dosing frequency, in the management of shingles.

Ganciclovir, foscarnet and cidofovir are licensed for the treatment of CMV retinitis in immunocompromised individuals. Treatment with ganciclovir is associated with potential haematological toxicity, notably neutropenia and thrombocytopenia. Dose reductions are required if evidence of toxicity appears. Foscarnet is associated with electrolyte imbalances, particularly hypocalcaemia. Cidofovir is a phosphonate analogue, which can be administered every other week on maintenance regimens but is associated with significant nephrotoxicity. An antisense oligonucleotide compound known as fomivirsen was licensed in 1998 as intraocular therapy for CMV retinitis, the first such antisense compound licensed as specific therapy against any infectious agents.

There is at present no established therapy for infection due to EBV, HHV-6, HHV-7 or KSHV, although ganciclovir has been used experimentally for EBV and HHV-6 disease.

Additional Reading

Arvin, A. M: "Varicella-zoster Virus," In: Howley R.M., and D.M. Knipe: *Fields Virology*, Lippincott Williams & Wilkins, Philadelphia, PA, 2001, pp. 2731–2770.
Arvin, A.M., B. Roizman, R. Witley, P.S. Moore, and G. Campandielli-Flume: *Human Herpesviruses: Biology, Therapy, and Immunoprophylaxis*, Cambridge University Press, New York, NY, 2006.
Becker, Y., A.L. Aurelian: *Herpesviruses, the Immune System, and AIDS*, Springer-Verlag New York, LLC, New York, NY, 2004.
Black, J.B., and P.E. Pellett: "Human Herpesvirus 7," *Reviews in Medical Virology*, **9**, 245–262 (1999).
Chang, Y., E. Cesarman, M.S. Pessin, et al.: "Identification of Herpesvirus-like DNA Sequences in AIDS-associated Kaposi's Sarcoma," *Science*, **266**, 1865–1869 (1994).
Gershon, A.A., and S.J. Silverstein: "Varicella-zoster Virus," In: Richman, D.D., R.J. Whitley, and F.G. Hayden: *Clinical Virology*, ASM Press, Washington, DC, 2002, pp. 413–432.
Kost, R.G., and S.E. Straus: "Postherpetic Neuralgia: Pathogenesis, Treatment, and Prevention," *New England Journal of Medicine*, **335**, 32–42 (1996).
Krueger, G., and D.V. Ablashi: *Human Herpesvirus-6: General Virology, Epidemiology and Clinical Pathology*, 2nd Edition, Elsevier Science & Technology Books, New York, NY, 2006.
Minarovits, J., E. Gonczol, and T. Valyi-Nagy: *Latency Strategies of Herpesviruses*, Springer-Verlag New York, LLC, New York, NY, 2006.
Roizman, B.R.: "New Viral Footprints in Kaposi's Sarcoma," *New England Journal of Medicine*, **332**, 1227–1228 (1995).
Roizman, B., and P.E. Pellett: "Herpesviridae," In: Knipe, D.M., P.M. Howley, D. Griffin, et al.: *Fields Virology*, 4th Edition, Lippincott Williams & Wilkins, Philadelphia, PA, 2001, pp. 2381–2397.
Stanberry, L.R., A. Cunningham, G. Mertz, et al.: "New Developments in the Epidemiology, Natural History and Management of Genital Herpes," *Antiviral Research*, **42**, 1–14 (1999).
Stanberry, L.R.: *Understanding Herpes*, 2nd Edition, University Press of Mississippi, Jackson, MS, 2006.
Straus, S.: "Introduction to Herpesviridae," In: Mandell, G.L., J.E. Bennett, and R. Dolin: *Principles and Practice of Infectious Diseases*, 5th Edition, pp. 1557–1564, Churchill Livingstone, New York, NY, 2000.
Whitley, R.J.: "Herpes Simplex Virus," In: Knipe, D.M., P.M. Howley, D. Griffin, et al.: *Fields Virology*, 4th Edition, Lippincott Williams & Wilkins, Philadelphia, PA, 2001, pp. 2461–2509.
Whitley, R.J.: "Antiviral Therapy (non-aids)," In: Goldman, L., and J.C. Bennett: *Cecil Textbook of Medicine*, 22nd Edition, Saunders, Philadelphia, PA, 2004 pp. 1960–1967.
Whitley, R.J. "Herpes Simplex Virus," In: Scheld, W.M., R.J. Whitley, and C.M. Marra: *Infections of the Central Nervous System*, 2nd Edition, Lippincott Williams & Wilkins, Philadelphia, PA, 2004 pp. 123–144.
Whitley, R.J., and B. Roizman: "Herpes Simplex Viruses," *Lancet*, **357**, 1513–1518 (2001).
Whitley, R.J., D.W. Kimberlin, and B. Roizman: "Herpes Simplex Virus," *Clinical Infectious Diseases*, **26**, 541–555 (1998).

RICHARD J. WHITLEY, University of Alabama at Birmingham, Birmingham, AL

HERPETOLOGY. The study of amphibians and reptiles, often mistakenly believed to be a study of reptiles only and snakes in particular. See also **Snakes**.

HERPOLHODE. The curve along which the cone traced out by the angular velocity vector intersects the invariable plane tangent to the momental ellipsoid and perpendicular to the angular momentum vector, in the case of a rotating rigid body not subject to any external torque. The concept is useful in studying the dynamics of a rigid body.

HERRINGBONE BEAR.   See **Helical Gearing**.

HERRING (*Osteichthyes*). Of the order *Clupeiformes*, herring are a characteristic fish group of the oceans, but also include a number of species that inhabit tropical fresh water. They form schools and are found near shores as well as in the open sea. Many of them are migratory. Herrings can be distinguished from other species by several characteristics — there are no rayed canals on the gill cover bones; lateral line pores are absent; there are keel scales along the medial line of the belly. Noteworthy skull characteristics include a suprabranchial organ with unknown function which joins the fourth and fifth gill arches; there is little dentition in the mouth, since most species feed on plankton; and there are no teeth on the parasphenoid (a bone at the base of the skull).

Commercially important herring are of the suborder *Clupeoidei* and most statistics report the catch of clupeoids without distinguishing specific subtypes. About 25 herring genera, with some 100 species, live in the sea. Until recent times, herring was considered the most important commercial fish catch.

Atlantic Herring. This fish (*Clupea harengus*), shown in Fig. 1, is one of the most important of commercial fishes in the northeastern Atlantic Ocean. Originally, it was presumed that the entire herring population in the northeastern Atlantic Ocean was a unified group, inhabiting the region from the Arctic Ocean to the English Channel. From here, this group presumably migrated extensively to the north and south (and back) during the course of a year. But in recent years structural differences in herring caught in different regions have been noted. Based upon these and other studies, herring researchers met in Copenhagen in 1956 and proposed the following classification:

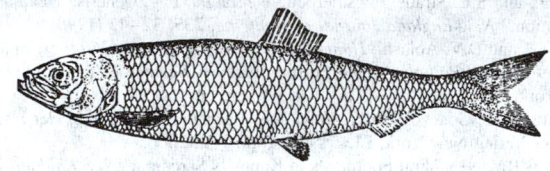

Fig. 1.   Atlantic herring.

Class A Herring — Known as the *Atlanto-scandian herring*, which inhabit the open Atlantic Ocean and which spawn on the Atlantic coasts of northern Europe in mid-winter, spring and possibly in early summer. These fishes are of appreciable size and are characterized by an intermediate number of vertebrae (57 or more).

Class B Herring — Known as *shelf herring*, which inhabit the North Sea on the shelf west of the British Isles and in the transition zone between the North and Baltic Seas. These fishes spawn between August and January along the coasts. They reach a smaller size and have an intermediate number of vertebrae.

Class C Herring — Distributed inside coastal waters of the North Sea, in the traditional area between the North and Baltic Seas, and in the Baltic Sea. They spawn in shallow water during winter and spring. Body size and vertebral number are smaller than in the Class B herring.

Class D Herring — Found in the most northeastern part of the Atlantic Ocean and which, at one time, were classified with the Pacific herring.

The eggs of the Atlantic herring have a diameter of about 1 to 2 millimeters. In general, the winter-spring spawners have a relatively lower fertility and larger eggs, while the opposite is true of the summer-fall spawners. Winter-spring spawners lay from 22,000 to 40,000 eggs, whereas the summer-fall spawners lay from 48,000 up to 70,000 eggs. Freshly hatched larvae are found in tremendous masses on the spawning grounds and vicinity. They are transparent and very slender. Even when the fish is only about 0.8 inch (2 centimeters) in length, the distinguishing characteristics of the herring can be observed.

The availability of suitable plankton is the most important determinant for the development and ultimate survival of the larvae. The prey must be as close as 0.2 inch (0.5 centimeter) from the herring larvae in order for it to be perceived and eaten. The larvae feed only on moving organisms.

Summer and fall spawners of the North Sea reach sexual maturity in the third or fourth year, at which time they have a length of about 9.5 inches (24 centimeters). Life expectancy of the summer-fall spawners is from 12 to 16 years, while late-winter spawners of the Norwegian coast may live from 23 to 25 years.

Herrings have been found in schools ranging from hundreds to thousands of individuals, in all sizes from young to sexually mature adults. The individuals within a particular school are generally of equal size and age. It is not known how long they remain together. Studies have indicated that herring in the North Sea spend the day at the floor. With the beginning of dusk, they ascend to depths of some 100 to 165 feet (30 to 50 meters) into a warmer-temperature zone. Light plays an important role in this movement. During darkness, they seem to remain scattered at the higher level, but with the onset of dawn, they collect and return to the floor. In recent years, herring behavior at darkness has been studied closely. Soviet researchers have observed daily activities from a submarine and report that the Atlanto-scandian herring spends the night motionless at the surface of the water in an oblique position, as if sleeping. They become active shortly before dawn and begin their move to greater depths. Other researchers believe that schooling behavior ceases at night.

In recent years, the behavior of commercial fishes has been studied intensively with a view toward developing better fishing methods. One finding of these studies has been that vision plays a significant role in herring. Experiments have shown that herring do not avoid plastic sheets if they are transparent. Herring in which the eyes were covered could not perceive a net, while those with sight did detect the obstruction. During the day, a wall of air bubbles can act as an obstacle to herring, but they will swim right through it at night. Herring can also detect noises and vibrations, to which they respond with fright behavior. This has been confirmed with echolocation tracking. When a ship moves toward a herring school, the school sinks to a depth of many meters. Fright can also be induced in an aquarium by tapping on the wall.

During a few months of the year, herring can be found in certain regions in tremendous quantities, while at other times these same areas are completely devoid of the fishes. On the other hand, herring can be caught in some places throughout the year, but the catch varies from year to year. Marking studies have shown that Atlanto-scandian herring migrate between feeding grounds off Iceland and spawning grounds on the Norwegian coast. It has been established that this major herring population has 3 major growth and development areas, i.e., in the Norwegian fjords, the Barents Sea, and in the southern and eastern parts of the ocean off northern Europe. Young herring, which have developed in the fjords, migrate to the sea at an age of 2 to 3 years, where they meet those herring that have been developing there.

Herring feed on plankton, which is not simply filtered, but selected — a phenomenon found by stomach content investigations and aquarium studies. Some researchers indicate that herring select food visually and then again test it in the mouth. Materials that are useless or of bad taste are immediately rejected.

Three stocks of herring are taken in Icelandic waters. Two of them are of Icelandic origin and the third of Norwegian origin. The Newfoundland herring fishery is entirely coastal, particularly concentrated in the west and south coast regions. The fish are caught during winter and spring with gillnets and purse seines. The industry in Newfoundland has increased by 200% to 300% during the past 20 years.

Pacific Herring. This fish (*Clupea pallastii*) inhabits the coasts of the northern Pacific Ocean from the Bering Strait to Korea and in the Arctic Sea to the mouth of the Lena River. On the North American coast, its distribution extends from California to Nome, Alaska. Herring in the White Sea and from Cape Kanin to the Kara Sea are very similar to the Pacific herring. This species differs from the Atlantic herring by the smaller number of vertebrae, among other features. Generally, the eggs are laid in brackish water on plants. Pacific herring form spawning groups whose distribution is limited to very specific narrow zones. They apparently do not migrate to a great extent. On the Asiatic coast, 10 spawning groups are known in the region from Korea to the Sea of Okhotsk. Of these, the *Hokkaido-Sakhalin herring* is commercially the most important. The principal herring fisheries on the Pacific coast of the United States are in the bays and channels of southeastern and central Alaska. A general downward trend of the herring catch in Alaska, as contrasted with British Columbia, commenced in the early 1950s.

Sprat. This fish (genus *Sprattus*) is closely related to the herring. Six well-defined species have been identified. The majority are found in the southern hemisphere. Length ranges up to about 8 inches (20 centimeters). Coloration of the best known species (*Sprattus sprattus*) resembles herring and is iridescent. These fishes are found in the Northern Hemisphere on the European coast from Tromsö to the Baltic Sea and the Bay of Biscay, as well as in the Mediterranean and in the bordering waters of the Black Sea. Sprats do not undertake long migrations like herring. They generally stay near the coast and in river mouths. In the Baltic Sea, the sprat is found in water with low salt content. The sprat apparently avoids areas far from the coast. It reaches sexual maturity at an age of 2 to 3 years. Spawning takes place some distance from the coast and occurs in the North Sea from April to July; in the Kattegat and Skagerrak from May to June; in the Baltic Sea from May to August. Sprats, like herring, are commercially important fishes.

Sardines. Also related to sprats and herrings and of large commercial importance are the *Sardinops* sardines, of which there are at least 5 significant species: (1) *Pacific sardine* (*Sardinops cerulean*); (2) *South American sardine* (*S. sagax*); (3) *Japanese sardine* (*S. melanosticta*); (4) *Australian sardine* (*S. neo-pilchardus*); and (5) *South African sardine* (*S. ocellata*). Commercially, the latter species is of the least commercial importance.

The *Pacific sardine* is of large commercial importance for the United States and is found on the east coast of the Pacific Ocean from Baja California to British Columbia. The species lives in schools near the surface of the water and spawns between January and June, chiefly in March and April. Spawning takes place on the high seas off Baja California and southern California, as much as 300 nautical miles (556 kilometers) from shore. The larvae hatch 3 to 4 days after spawning and they migrate to the coast at a length of 3 to 5 inches (7.5 to 12.5 centimeters). They are caught in great masses and used as bait for tuna. At a length of about 6.5+ inches (17 centimeters), they leave the feeding grounds off the coast and meet the adults swimming on the open sea. Sexual maturity is attained at a length between 6.5 and 9.8 inches (17 and 25 centimeters), which occurs at an age of 2 to 3 years. The species can reach an age of 13 years. In California waters, the sardine catch has decreased dramatically since the late 1930s.

Sardine Ecology. The California sardine fishery has become the classic example of an ecologically complex community modified by an intensive fishery. A simple matter of overfishing might, on theoretical grounds, be overcome by abstention from fishing for an appropriate period. Apparently an ecologically related and evidently competitive species, the anchovy, has occupied the gap left by the exploitation of the sardines (the gap was evidently increased by harvesting during a period of conditions unfavorable for reproductive success). In the absence of a similar market for anchovies, a reduced technology for processing sardines, and legislative restrictions on harvesting anchovies, the situation reached the stage where a sardine fishery of nearly any magnitude further decreased the stock. Possibly this imbalance could be redressed by an unpredictable alteration in natural conditions in favor of the sardine, but this does not appear likely. It is interesting to note that the Pacific sardine supported the largest fishery in the Western Hemisphere in the early 1930s (exceeding over 1 billion pounds; 0.45 billion kilograms) taken from California waters, as compared with lower catches of just a few million pounds annually in recent years. From an economic standpoint, much of the loss of production of California and Oregon sardines has been compensated for by large increases in menhaden catches in the south Atlantic Ocean and off the Gulf states.

The *South American sardine* and the *South African sardine* have increased in production since World War II. The South African sardine, sometimes referred to as the *pilchard* (*Sardinops ocellata*), has a wide geographical distribution and is known from St. Lucia Bay (north of Durban) to Bahia dos Tigres on the Angolan coast. The main commercial concentrations are limited to the Walvis Bay region, the waters of St. Helena Bay, and the area between Cape Point and Cape Agulhas. The species is normally found within 25 miles (40 kilometers) of the coastline, but occasionally schools have been reported up to 80 miles (129 kilometers) offshore.

The South African pilchard is a fast-growing fish and reaches sexual maturity at the age of about 2.5 years, by which time it attains a length of some 8.25 inches (21 centimeters). The main spawning seasons are spring and early summer. Spawning occurs offshore and three main grounds have been identified—those off Walvis Bay; near St. Helena Bay; and east of Cape Point. The pilchard is a filter-feeder, its diet consisting of both phytoplankton and zooplankton. Tagging experiments have established that there is periodically an influx of pilchards from the Walvis region into Cape waters.

The *Japanese sardine* is a warm-water species, which attains a length of about 11.5 inches (29 centimeters). Distribution is chiefly in a temperature of from 59 to 79 °F (15 to 26 °C). Spawning grounds are off the south coast of Japan and Korea at some distance from the coast. Japanese sardines spawn from December to May on the high seas at a water temperature of 55 to 68 °F (13 to 20 °C). The number of eggs varies between 27,000 and 84,000. After spawning, a migration to the north takes place on the far eastern coast, where a large fishing industry has developed. These sardines feed chiefly on plankton. The annual catch (mainly by Japanese, Russian, and Korean fishers) varies considerably from year to year.

True Sardine. Only one species belongs to the genus of the true sardines, the *Sardina pilchardus*, also called *pilchard*, and not to be confused with the South African fish previously described. See Fig. 2. The true sardine reaches a length of about 11.8 inches (30 centimeters), but is generally from 9 to 9.8 inches (23 to 25 centimeters) long. Commonly, the larger sizes are called pilchards, while the smaller fishes are called sardines, the latter ranging between 5 and 6 inches (13 and 16 centimeters) in length. Distribution is on the coasts of west and southwest Europe and north Africa, from southern Ireland, the southern part of the North Sea, and the Kattegat in the north to Madeira and the Canary islands in the south. Distribution also includes the northern parts of the Mediterranean and bordering waters. There are two subspecies. The spawning period of the pilchard is rather extended. Off the Iberian peninsula, spawning takes place from February to March; in the North Sea, from July to August; off the coast of west Britany, November to June; in the Mediterranean, September to May; and in the Black Sea, July and August.

Fig. 2.   True sardine.

The distribution of the true sardine is approximately limited by the 68 °F (20 °C) isotherm. Until 1930, the Strait of Dover was apparently the northern limit for pilchards. Those found in Norwegian waters and in the Kattegat only occurred in small numbers. Since that time, the northern population has increased significantly. In the late 1930s, large quantities of sardine eggs were reported off the East Frisian Islands. In late 1940s, the first large quantities were found in the area near Amrum Island (in the North Frisians). Climatic changes are probably responsible for the extension of the northern limit.

Adult pilchards feed primarily on zooplankton. The catch of pilchards has been increasing progressively over a number of years and is very important to Portugal, Morocco, Spain, France, and the former Yugoslav Republics, most of the catch being used in production of canned sardine oil.

Shad. This fish of the genus *Alosa* is also related to the herring. Shads have a compressed upper body and the keel scales form a sharp keel. The teeth in the jaw are either small or absent altogether, and in adults there are no vomerine teeth. There are four species in the north Atlantic Ocean, Mediterranean, and in the northern Pacific Ocean. These species migrate into fresh water. The best known species in Europe is the shad (*Alosa alosa*). The length of the shad exceeds 27.5 inches (70 centimeters) and the jaw protrudes forward. The scales are not as lightly attached as in the herring. Distribution is on the European coast from Norway to the Iberian Peninsula, and along the north African coast to Morocco, as well as the western part of the Baltic Sea and the Mediterranean. In March, shad migrate from the sea to spawn in the rivers well upstream. Earlier, they were found in the Neckar River, Germany. During recent years, shad have disappeared from much of their original habitat, largely as the result of pollution.

The *American shad* (*A. sapidissima*) is found mainly in the Atlantic Ocean, from the Gulf of St. Lawrence to Florida. In 1871, the species was introduced to the Sacramento and Columbia Rivers in the western United States and, by 1876, shad were caught off Vancouver island. Since that time, the species has spread along the entire coast of the Pacific Ocean from southern California to Alaska and Kamchatka. It is a prevalent fish in the California rivers. Since the earliest settlements in North America, the American shad has been an important and valuable commercial fish, while the Alabama species has enjoyed much regional acclaim. Over the years, however, the shad population has decreased.

See also **Fishes**.

Additional Reading

Douglass, R.: *Atlantic Coast Fishes*, Houghton Mifflin Company, New York, NY, 1999.

Lythgoe, J. and G. Lythgoe: *Fishes of the Sea: The North Atlantic and Mediterranean*, MIT Press, Cambridge, MA, 1992.

Schul'man, G. and R. Love: *Advances in Marine Biology: Vol. 36 the Biochemical Ecology of Marine Fishes*, Academic Press, Inc., San Diego, CA, 1999.

HERSCHBACH, DUDLEY R. (1932–). Awarded the Nobel Prize in chemistry in 1986 for work reporting that the energies of reactions of crossed molecular beams of isolated alkali metal atoms and alkyl halide molecules appeared mostly as vibrational excited states of products. This method of studying all types of chemical reactions led to a more detailed knowledge of reaction processes. His doctorate was awarded from Harvard in 1958.

HERSCHEL, F.W. (1738–1822). Herschel was a German musician who supported himself by teaching, performing, conducting, and composing music. His interest in music led him to read Robert Smith's books *Harmonics and A Compleat System of Opticks in Four Books*. Smith's books changed Herschel's life because from Smith's work, Herschel learned how to build telescopes.

Herschel and his sister, Caroline Lucretia Herschel, began building telescopes, which were superior to most of their time. Herschel began searching the heavens and cataloging stars. In March 1781, he discovered Uranus and became instantly famous. In the same year, he correctly calculated the rotation time of Mars. In 1887, he discovered the moons of Uranus and the moons of Saturn. The Royal Observatory acknowledged Herschel's superior telescopes and the King George III awarded him an annual grant to do his building. By 1789 Herschel had built the world's largest telescope for his time.

See also **Telescope (Astronomical-Optical)**; and **Uranus**.

J. M. I.

HERSCHEL SPACE OBSERVATORY. See **Space Science Missions: Universe**.

HERSHEY, ALFRED DAY 1908–1997. Hershey was an American bacteriologist who demonstrated that deoxyribonucleic acid (DNA) is foundational for cellular reproduction. Born in 1908 in Oswosso, Michigan, Hershey was a prominent twentieth-century bacteriologist. He received his PhD from Michigan State University in 1934 and took a teaching position at Washington University in St Louis. From 1950 until the end of his career, he worked in the Genetic Research Unit at Cold Spring Harbor on Long Island, New York.

Hershey's primary research interest was the bacteriophage, a virus that attacks bacteria and whose only function is its own reproduction. He was awarded the 1969 Nobel Prize in Physiology or Medicine — along with fellow bacteriologists Max Delbrück and Salvador Luria — for investigating the genetic make-up and reproduction of viruses, which furthered the understanding of viruses and viral diseases. See also **Bacteriology (The History)**; **Bacteriophage**; and **Delbruck, Max Ludwig Henning (1906–1981)**.

Hershey, a renowned experimentalist, shared in this recognition primarily for his 1952 'blender experiment', in which he, and his assistant Martha Chase, sought to identify the genetically active material in the bacteriophage. Putting different markers on protein and deoxyribonucleic acid (DNA) in a phage, they allowed it to attack a bacterial culture. The culture was then mixed in a Waring blender to rupture the bacterial cell walls and finally centrifuged to separate the components. Hershey and Chase found that the DNA was actively replicating and thus concluded that it, and not protein, is foundational for cellular self-reproduction. This experiment paved the way for the discovery of DNA's double-helix structure in 1953. Hershey's research also was foundational for the discovery of vaccines for viral diseases such as polio. See also **Virology (The History)**.

C. Christopher Smith Indiana University, Bloomington, IN

HERTWIG, WILHELM AUGUST OSCAR (1849–1922). Wilhelm Hertwig was a German cell biologist who was the first to observe the fusion of male and female nuclei during fertilization.

Oscar Hertwig was the older son of the chemist Carl Hertwig and Elise Trapp. In 1868, together with his younger brother Richard (1850–1932), he went to Jena to study with Ernst Haeckel. Hertwig stayed in Jena until 1888, being appointed an associate professor in anatomy (*Extraordinarius*) in 1878 and full professor (*Ordinarius*) in 1881. Hertwig left Jena for Berlin in 1888 when he took the newly created chair of cytology and embryology and became director of the new Anatomical-Biological Institute. He married Marie Gesenius in 1884. Both of their two children, Günther (born 1888), and Paula (born 1889), would later also become biologists and collaborate with their father on his studies of the biological effects of radiation. See also **Haeckel, Ernst Heinrich Philipp August**.

In his doctoral thesis Hertwig investigated the development and structure of elastic tissue in the cartilaginous matrix. His results lent further support to the protoplasmic theory of the cell and also gave Hertwig the necessary technical skills for his future work in cytology. Turning his attention to the problem of fertilization, Hertwig was lucky to find the right research material during a trip to the Mediterranean together with Ernst Haeckel and his brother Richard in 1875. The eggs of the sea urchin *Toxopneustes lividus* were transparent enough to observe the fate of both the egg and the sperm nucleus before and after fertilization. Hertwig observed the presence of the egg nucleus before and immediately after the entry of the sperm and 5 to 10 minutes later he detected a single nucleus. He therefore inferred that the new nucleus of the zygote was a product of the fusion of the two gametic nuclei. Furthermore, he also observed that just one spermatozoon fertilizes the egg. His observations disproved previous theories of fertilization, such as G. W. Bischoff's contact theory (sperm acts as "ferment") or Auerbach's idea that the nucleus is formed anew from a mixture of gametic materials. Hertwig also advocated the genetic primacy of the nucleus, but rejected August Weismann's theory of cellular differentiation through the selective loss of genetic material (*idioplasm*) during development. He asserted the genetic equivalence of all body cells and in 1890 also established the equivalence of spermatogenesis with oogenesis.

Oscar Hertwig also collaborated with his brother Richard. In a series of papers during the early 1880s they investigated the problem of the origin of the third germ-layer (mesoderm) and the central body cavity of higher organisms. In their coelom theory they claimed that the body cavity of higher organisms is the product of a secondary invagination of the endoderm. The coelom theory was subsequently widely discussed among students of phylogeny, even though it contributed relatively little to experimental embryology.

Hertwig was immensely critical of Darwinism and attacked the concept of natural selection. His own studies of cells in development led him to a more cooperative view of the origin of complex characters and organizations. In his later years he also expanded this notion of cooperation between the cells within an organism to include the social and political organization of states and societies.

Additional Reading

Olby, R.: "Oscar Hertwig," In: Gillispie, C.C.: *Dictionary of Scientific Biography*, Charles Scribner's Sons, New York, NY, 1981.

Weindling, P.J.: *Darwinism and Social Darwinism in Imperial Germany: The Contribution of the Cell Biologist Oscar Hertwig (1849–1922)*, Gustav Fischer, Stuttgart, Germany, 1991.

M. D. Laubichler Princeton University, Princeton, NJ

HERTZ. The standard unit of measurement for frequency in cycles/second. Prefixes include kilo (10^3), mega (10^6), and giga (10^9), abbreviated kHz, MHz, and GHz; replaces cycles per second in modern usage.

HERTZ, GUSTAV LUDWIG (1887–1975). Hertz was a German physicist. He attended the Johanneum School in Hamburg before commencing his university education at Göttingen in 1906; he subsequently studied at the Universities of Munich and Berlin, graduating in 1911. He earned a Ph.D. from the University of Berlin.

Hertz's early researches, for his thesis, involved studies on the infrared absorption of carbon dioxide in relation to pressure and partial pressure. Together with J. Franck he began his studies on electron impact in

1913 and before his mobilization, he spent much patient work on the study and measurement of ionization potentials in various gases. He later demonstrated the quantitative relations between the series of spectral lines and the energy losses of electrons in collision with atoms corresponding to the stationary energy states of the atoms. His results were in perfect agreement with Bohr's theory of atomic structure, which included the application of Planck's quantum theory.

Hertz published many papers, alone, with Franck, and with Kloppers, on the quantitative exchange of energy between electrons and atoms, and on the measurement of ionization potentials. He also is the author of some papers concerning the separation of isotopes.

In 1925, Hertz shared the Nobel Prize in Physics with James Franck.

See also **Bohr Theory of Atomic Spectra**; and **Planck Law**.

J. M. I.

HERZBERG, GERHARD (1904–1999). A German-born physicist who won the Nobel Prize for chemistry in 1971, for his work on the composition of molecules. His research involved the spectroscopy of atoms and molecules and their excitation behavior. He became a Canadian citizen and was the director of the Division of Pure Physics of the National Research Council of Canada.

HESSIAN. A functional determinant, related to the Jacobian and defined for six variables by the equation

$$H(F) = \frac{\partial(u, v, w)}{\partial(x, y, z)} = \begin{vmatrix} F_{xx} & F_{xy} & F_{xz} \\ F_{xy} & F_{yy} & F_{yz} \\ F_{xz} & F_{yz} & F_{zz} \end{vmatrix}$$

where u, v, w are differential coefficients of another function, $F(x, y, z)$ and

$$u = \partial F/\partial x = F_x, v = F_y, w = F_z; \ \partial^2 F/\partial x^2 = F_{xx}, \text{etc.}$$

It can be generalized for any number of variables. The Hessian of two binary quantics is a covariant; hence, it is useful in studying the invariants of algebraic functions. As an example, the Hessian of a quadratic is its discriminant.

See also **Determinant**.

HESSIAN FLY *(Insecta, Diptera).* One of the worst pests of wheat. It is a small two-winged fly *Mayetiola (Phytophaga) destructor*, a member of the gall-gnat family, which was introduced into the United States in the Revolutionary period. The larva lives between the base of a leaf and the stem of the wheat plant and either kills or weakens the plant so that no grain develops. Other cereals are attacked to some extent.

Fall plowing and burning stubble aid in destroying many insects. The most effective means of avoiding damage to winter wheat is to sow late enough to avoid the attack of most of the adults. They live no more than ten days and the date of emergence is known for various regions; hence, late planting subjects the crop only to the light infestation due to the eggs deposited by the relatively few flies which emerge late. Phorate is an effective chemical control.

HESSITE. A mineral telluride of silver, Ag_2Te, with some gold, crystallizing in the monoclinic system at normal temperatures; isometric system above 149.5°F (65.3°C). Crystalline form not obvious at normal temperatures. Hardness, 2–3; specific gravity, 8.24–8.45; color, gray with metallic luster; opaque. Named after G.H. Hess (1802–1850).

HETEROCYCLIC COMPOUNDS. See **Compound (Chemical)**; and **Organic Chemistry**.

HETERODYNE. This term is used in communications terminology as an adjective or a verb, but in either case it concerns the beating together in an electrical circuit of two frequencies to produce new frequencies which are the sum or difference of the original ones. When two voltages of different frequencies are applied simultaneously to a circuit containing a nonlinear impedance, for example, one in which the signal current varies as the square or higher power of the input signal voltage, the output of the circuit will contain new frequencies, among them one equal to the sum and another equal to the difference of the applied frequencies. Either one or both of these may be selected by properly tuning or filtering the output.

HETEROGAMY. The occurrence or union of male and female gametes of different size and structure; anisogamy. The alternation of two sexual generations, one true sexual, the other parthenogenetic.

HETEROGENEOUS. (Latin "different kinds"). Any mixture or solution comprised of two or more substances regardless of whether they are uniformly dispersed. Common examples are such diverse materials as air (a mixture of 20% oxygen and 80% nitrogen), milk, marble, paint, gasoline, blood and mayonnaise. In all such cases, the mixtures can be separated mechanically into their components. "Homogenized" milk is a heterogeneous as regular milk and the term is, strictly speaking, a misnomer.

See also **Homogeneous**.

HETEROGENEOUS CHEMISTRY. A wide-ranging subject that consists of chemistry involving two phases, usually one or more gaseous reactants, and a condensed phase substrate where the reaction occurs, either liquid or solid. Generally the substrate facilitates the reaction and many heterogeneous reactions do not proceed in the gas phase. Reactants include free radicals as well as closed-shell molecules. Often, a reactant will hydrolyze (react with H_2O) heterogeneously; in this case a major constituent of the substrate is a reactant. Common substrates are fog and rain droplets as well as aerosol in the lower atmosphere and sulfuric acid aerosol (*see* **stratospheric sulfate layer**) and water ice particles in the upper atmosphere. See also **Clouds and Cloud Formation Polar Stratospheric Clouds (PSC)**.

AMS

HETEROMOLYBDATES. A large group of complex molybdenum salts and acids in which the anion contains oxygen atoms and from 2 to 18 hexavalent molybdenum atoms, as well as one or more other metal or nonmetal atoms (phosphorus, arsenic, iron, and tellurium). The latter are referred to as hetero atoms, and any of approximately 35 elements may be present in this manner. Example: $Na_3PMo_{12}O_{40}$, sodium phospho-12-molybdate. The molecular weights of these compounds range up to 3000. The acids and most of the salts are very soluble in water, and the acids and some salts are soluble in organic solvents.

HETEROMORPHOSIS. Deviation from normal form. Malformation or deformity and also less extreme departures incidental to slightly different conditions in the animal or its environment.

HETEROPOLYACIDS. Acids derived from two or more other acids, under such conditions that the negative radicals of the individual acids retain their structural identity within the complex radical or molecule formed. The term heteropolyacids is usually restricted to complex acids in which both radicals are derived from oxides, such as phosphomolybdic acid.

HETEROSPHERE. The upper portion of a two-part division of the atmosphere according to the general homogeneity of atmospheric composition; the layer above the homosphere. The heterosphere is characterized by variation in composition and mean molecular weight of constituent gases. This region starts at 80 and 100 kilometers above the earth, and therefore closely coincides with the ionosphere and the thermosphere.

HETEROSPORY. The production of two distinct types of spores by a plant, in contrast to homospory, which is the production of only one type of spore. The two kinds of spores produced in heterospory are known as microspores and megaspores. The microspores are very small and grow into the male gametophyte. The megapores are much larger and form the female gametophyte. All of the Seed plants have heterospory and a few of the minor subphyla of vascular plants do also. The *Lycopsida* is one of these subphyla. See also **Lycopsida**.

HETEROZYGOUS. Bearing two allelic genes of a different nature. The opposite of homozygous, which means bearing allelic genes of the same kind. For instance, if a person is homozygous for the recessive gene for albinism, the person will bear two such genes, represented as *aa*. The person will be an albino. A person who is heterozygous, however, will bear one gene for normal pigmentation and one gene for albinism, represented as *Aa*. A person can also be homozygous for the dominant gene, *AA*. Both heterozygous and homozygous dominant persons will have normal pigmentation.

HEULANDITE. The mineral heulandite is a monoclinic zeolite whose crystals are often quite suggestive of orthorhombic forms. Its chemical composition is probably $(Na, Ca)_{4-6}Al_6(Al, Si)_4Si_{26}O_{72} \cdot 24H_2O$; strontium may be present. Heulandite has one good cleavage; is brittle with a conchoidal fracture; hardness, 3.4–4; specific gravity, 2.18–2.22; luster, vitreous to pearly; color, white to gray, red or brown; streak, white; transparent to translucent. Occurs chiefly in cavities in basaltic rocks with other zeolites, but may be found in granites, pegmatites, gneisses, and schists. Famous localities are in Iceland, India, the Harz Mountains, Italy, Switzerland, Scotland, Nova Scotia; and in the United States at Bergen Hill and West Paterson, New Jersey. This mineral was named for the English mineralogist Heuland.

HEURISTIC. In artificial intelligence, a rule of thumb, generally based on expert experience or common sense rather than an underlying theory or mathematical model, that can be incorporated in a knowledge base and used to guide a problem-solving process. See also **Artificial Intelligence**.

Most procedures used by human weather forecasters are heuristic, as are many pattern-recognition techniques in radar and satellite meteorology.

HEVESY, GEORG DE (1885–1966). A Hungarian chemist who won the Nobel Prize in chemistry in 1943 for his work on the use of isotopes as tracers in the study of chemical processes. He discovered the element hafnium in 1923. One of his interesting projects involved the calculation of the percentages of chemical elements in the universe. He also was involved in research using radioactive lead and phosphorus traces. His work included the separation of isotopes by physical means. His Ph.D. was granted at Freiburg in 1908.

HEXACTINELLIDA. The glass sponges, constituting a class of the phylum *Porifera*. The spicules of the skeleton are silicious and of six-rayed form. Many of the species have a large central cavity, resulting in a tubular or vase-like form, and when freed of organic matter appear to be made of spun glass. These sponges are found in deep water in the ocean. Venus' flower basket, *Euplectella*, and the glass-rope sponge, *Hyalonema*, are the most common examples.

HEXADECIMAL NUMBER. In computer design, the hexadecimal (radix 16) numbering system is used as a convenient method for representing large binary numbers, which often consist of long strings of zeros and ones. The latter are difficult to handle in a computer. Each hexadecimal digit stands for four binary digits.

Hexadecimal notation calls for the use of 16 symbols to represent 16 number values. Inasmuch as the decimal system provides only 10 number symbols (0 to 9), six Additional marks thus are needed to represent the remaining values. The letters A, B, C, D, E, and F are used for this purpose. As shown in Table 1, the list of hexadecimal symbols is comprised of 0, 1, 2, 3, 4, 5, 6, 7, 8, 9, A, B, C, D, E, and F, in ascending sequence. From the table, note that upon reaching decimal 16, the hexadecimal symbols are used up and hence a "1 carry" must be placed in front of each hexadecimal symbol during its second cycle, i.e., from decimal 16 to decimal 31.

Binary numbers are converted to hexadecimal notation simply by dividing the numbers into groups of four binary digits, commencing from the right, and replacing each group by the corresponding hexadecimal symbol. Where the left-hand group is incomplete, zeros are filled in as required. This is illustrated by the following example.

$$111110011011010011 = 0011/1110/0110/1101/0011$$

$$= 3\ E6\ D3$$

$$= (3E6D3)_{16}$$

Hexadecimal numbers are best understood in terms of expansion in powers of 16. In the case of hexadecimal number 2CA.B6, for example, when decimals are substituted for hexadecimal symbols, it is evaluated as

$$2 \times 16^2 + 12 \times 16^1 + 10 \times 16^0 + 11 \times 16^{-1} + 6 \times 16^{-2}$$

$$= 2 \times 256 + 12 \times 16 + 10 \times 1 + 11/16 + 6/256$$

$$= 512 + 192 + 10 + 0.6875 + 0.0234375$$

$$= 714 + 0.7109375$$

$$= (714.7109375)_{10}$$

TABLE 1. COMPARISON OF DECIMAL, HEXADECIMAL, AND BINARY NOTATION

Decimal	Hexadecimal	Binary
0	0	0000
1	1	0001
2	2	0010
3	3	0011
4	4	0100
5	5	0101
6	6	0110
7	7	0111
8	8	1000
9	9	1001
10	A	1010
11	B	1011
12	C	1100
13	D	1101
14	E	1110
15	F	1111
16	10	10000
17	11	10001
18	12	10010
19	13	10011
20	14	10100
21	15	10101
22	16	10110
23	17	10111
24	18	11000
25	19	11001
26	1A	11010
27	1B	11011
28	1C	11100
29	1D	11101
30	1E	11110
31	1F	11111

HEXAGONAL COLUMN. One of the many forms in which ice crystals are found in the atmosphere. This particular crystal habit of ice is characterized by hexagonal cross section in a plane perpendicular to the long direction (principal axis, optic axis, or c axis) of the columns. It differs from that found in hexagonal platelets only in that environmental conditions have favored growth along the principal axis rather than perpendicular to that axis. Growth by vapor deposition at temperatures of from $-3°$ ($27°F$) to about $-8°C$ ($18°F$) and also at lower temperature below $-25°C$ ($-13°F$) leads to growth of columnar crystals, though other crystal features (needles, scrolls) also appear in this temperature interval, depending on the degree of water vapor supersaturation and the crystal fall velocity. See also **Column (Meteorological)**.

HEXAGONAL PLATELET. A small ice crystal of hexagonal tabular form. The distance across the hexagonal facet of the crystal may be as large as 1–2 mm (0.04–0.08 in) and the ratio of thickness to diameter as small as 1/100. This crystal form usually grows by vapor deposition at temperatures between $-8°$ ($18°F$) and $-25°C$ ($-13°F$). At temperatures between $-12°$ ($10°F$) and $-16°C$ ($3°F$), as crystals grow and attain a higher fall velocity, provided that the saturation ratio approaches water saturation, the corners sprout and form dendritic (treelike) side arms. See also **Dendrite**.

HEXAMINE. [CAS: 109-97-0] $(CH_2)_6N_4$, formula weight 140.19, white crystalline solid, mp $280°C$, decomposes at higher temperatures. Also known as hexamethylenetetramine, methenamine, and urotropine, the compound is soluble in H_2O and only very slightly soluble in alcohol or ether. Although used to some extent in medicine as an internal antiseptic, the primary use of hexamine is in the manufacture of synthetic resins where the compound is a substitute for formalin (aqueous solution of paraformaldehyde) and its NaOH catalyst. Hexamine also is used as an accelerator for rubber.

On a commercial scale, hexamine is manufactured from anhydrous NH_3 and a 45% solution of methanol-free formaldehyde. These raw materials, plus recycle mother liquor, are charged continuously at carefully controlled rates to a high-velocity reactor. The reaction is exothermic. The reactor effluent is discharged into a vacuum evaporator which also serves as a crystallizer. The hexamine crystals then are washed, dried, and screened.

Average yield of the process is about 96% conversion of ingredients to produce hexamine.

HEXAMITOSIS. Hexamitosis is a disease of chickens, turkeys, quail, and pheasants in which there is an infectious catarrhal enteritis in the duodenum and small intestine. The disease, caused by the protozoan *Hexamita meleagridis*, occurs in the United States, the United Kingdom, South America, and parts of Europe. It is primarily a disease of young birds under 10 weeks old in which mortality may be as high as 80%. The infected animals develop diarrhea, lose weight rapidly, appear listless and weak, and may eventually die. Hexamitosis is transmitted through contaminated food

TABLE 1. HEXAMITOSIS ANTIPROTOZOAL AGENTS

Structure number	Compound name	CAS Registry Number	Molecular formula	Structure
(1)	Penicillin	[1406-05-9]		
(2)	Enheptin	[121-66-4]	$C_3H_3N_3O_2S$	
(3)	Streptomycin	[57-92-1]	$C_{21}H_{39}N_7O_{12}$	
(4)	Ronidazole	[7681-76-7]	$C_6H_8N_4O_4$	
(5)	Dimetridazole	[551-92-8]	$C_5H_7N_3O_2$	
(6)	Acetylenheptin	[140-40-9]	$C_5H_5N_3O_3S$	
(7)	Nithiazide	[139-94-6]	$C_6H_8N_4O_3S$	
(8)	Ipronidazole	[14885-29-1]	$C_7H_{11}N_3O_2$	

and water. Carrier birds are adults that have survived earlier infections. Hot weather and overcrowding may exacerbate the severity of the outbreak.

There is no effective treatment for hexamitosis. Penicillin (1), oxytetracycline (See also **Amebiasis (Amebic Dysentery)**, chlorotetracycline (See also **Anaplasmosis**), Enheptin (2, 2-amino-5-nitrothiazole [121-66-4]), and streptomycin (3) sulfate [CAS: 3810-74-0] were found to have limited value. Hexamitosis in carrier pigeons caused by *Hexamita columbae* was treated successfully with ronidazole (4, Dugro) (Table 1). An experimental infection of nude mice with *Hexamita muris* was treated with dimetridazole (5, 1,2-dimethyl-5-nitro-1*H*-imidazole [CAS: 557-92-8]), metronidazole (See also **Amebiasis (Amebic Dysentery)**), tinidazole, and acranil (See also **Giardiasis.**). All compounds lowered or suppressed the fecal discharge of cysts, but the latter reappeared when the 1–3 week treatment terminated. In a related study, dimetridazole controlled the clinical disease in mice but did not eliminate the infection. See also **Antiparasitic Agents, Antiprotozoals**.

Additional Reading

Herenda, D.C., and D.A. Franco: *Poultry Diseases and Meat Hygiene: A Color Atlas*, Iowa State Press, Ames, IA, 1996.

Kaufmann, J.: *Parasitic Infections of Domestic Animals: A Diagnostic Manual*, Birkhauser Verlag, Cambridge, MA, 2001.

Kierzenbaum, F.: *Parasitic Infections and the Immune System*, Elsevier Science & Technology Books, New York, NY, 1993.

Staff: *Infectious and Parasitic Diseases of Wild Mammals*, 3rd Ed., Iowa State Press, Ames, IA. 2001.

Wakelin, D.: *Immunity to Parasites: How Parasitic Infections are Controlled*, 2nd Ed., Cambridge University Press, New York, NY, 1996.

DANIEL L.KLAYMAN Walter Reed Army Institute of Research

HEXAPODA.   Synonymous with Insecta.

HEXOSE MONOPHOSPHATE OXIDATIVE PATHWAY.   See **Carbohydrates**.

HEYROVSKY, JAROSLAV (1890–1967).   A Czechoslovakian physiochemist who won the Nobel Prize for chemistry in 1959, for his discovery and development of the polarographic and oscillo-polarographic methods of analysis. Although his Ph.D. was from the University of Prague, he later studied in London.

HIATUS HERNIA.   See **Esophagus**.

HIBERNATING SPACECRAFT.   A spacecraft maintaining an orbit without using propellant power and without maintaining orientation within the orbit, but with inherent power capability. A hibernating spacecraft could be in an orbit around the sun for months or years before power is triggered from a station on earth at an opportune time.

HICKORY AND WINGNUT TREES.   Of the family *Juglandaceae* (walnut family), hickory trees are of the genus Carya and are one of the most, if not the most distinctly North American tree. They are relatively unknown on the other continents. One authority aptly describes the hickories as walnuts with greater height and grace. The principal species are indicated in Table 1. It should be appreciated that the tree dimensions given in the table represent record specimens and that the average tree, most likely growing under somewhat more adverse conditions, will not attain such dimensions.

TABLE 1. RECORD HICKORY TREES IN THE UNITED STATES[1]

Specimen	Circumference[2]		Height		Spread		Location
	Inches	Centimeters	Feet	Meters	Feet	Meters	
Bitternut hickory (1999) (*Carya cordiformis*)	182	462	120	36.6	100	30.5	Tennessee
Black hickory (1996) (*Carya texana*)	128	325	139	42.4	84	25.6	Texas
Carolina hickory (1997) (*Carya ovata var. australis*)	94	239	150	45.7	65	19.8	Alabama
Mackernut hickory (1989) (*Carya tomentosa*)	140	356	156	47.5	70	21.3	Mississippi
Nutmeg hickory (1985) (*Carya myristiciformis*)	132	335	145	44.2	80	24.4	Alabama
Pecan hickory (1999) (*Carya illinoensis*)	257	653	91	27.7	120	36.6	Texas
Pignut hickory (1997) (*Carya glabra var glabra*)	200	508	105	32	125	38.1	Georgia
Red hickory (1997) (*Carya glabra var adorata*)	84	213	140	42.7	62	18.9	Tennessee
Sand hickory (1997) (*Carya pallida*)	143	363	94	28.7	86	26.2	New Jersey
Sand hickory (1998) (*Carya pallida*)	138	351	104	31.7	68	20.7	Maryland
Scrub hickory (1992) (*Carya floridana*)	62	158	47	14.3	48	14.6	Florida
Shagbark hickory (1984) (*Carya ovata var. ovata*)	132	335	153	46.6	56	17.1	South Carolina
Shagbark hickory (1994) (*Carya ovata var. ovata*)	144	366	132	40.2	109	33.2	Kentucky
Shellbark hickory (1994) (*Carya laciniosa*)	167	424	139	42.4	80	24.8	Kentucky
Water hickory (1993) (*Carya aquatica*)	228	579	101	30.8	85	25.9	Florida
Water hickory (1996) (*Carya aquatica*)	190	483	148	45.1	49	14.9	South Carolina

[1]From the "National Register of Big Trees," American Forests (by permission).
[2]At 4.5 feet (1.4 meters).

The bitternut or swamp hickory (*C. cordiformis*) has a light-brown or gray-brown thin bark. The fissures are shallow. The leaves are compound. The leaflets are of a deep yellow-green color, somewhat lighter underneath. The ovoid fruit is about an inch long and is contained in a thin husk. The tree prefers a rich woodsy environment, but will tolerate a variety of soils. The tree ranges from southern Maine and western Quebec westward to the Great Lakes and Minnesota and south through Nebraska, Kansas, Oklahoma, and Texas. It ranges eastward and south to Florida. The tree is found commonly only in southern New England, with only occasional representation in Vermont and New Hampshire. The tree attains its greatest height in the mountains of the Carolinas. As compared with other hickories, the wood is considered inferior.

The nutmeg hickory (*C. myristicaeformis*) prefers alluvial soil. It is found in the southeastern states and westward through Arkansas. The fruit is a little over an inch long and contained within a thin husk. The shell is very hard; the kernel is not edible. The wood is strong, hard, and is of a light-brown color.

The mockernut or bigbud hickory (*C. tomentosa*) occurs in southeastern Canada and the eastern United States. It tends to be a very tall tree with a round head. The dark-green leaves are long, with from five to nine toothed and pointed leaflets. The male catkins are from 3 to 5 inches (7.6 to 12.7 centimeters) long.

The water hickory or bitter pecan (*C. aquatica*) is generally a tree of the coastal plain, ranging from Virginia southward to Florida and then westward into Texas. The tree is capable of attaining great heights and is generally quite slender. The bark is an ashen gray, thin, and often quite shaggy on older trees. The leaves are compound. Leaflets have sharp points, of a deep yellow-green color, with slightly lighter coloration underneath. The wood is considered inferior as compared with other hickory species.

The shagbark or shellbark hickory (*C. ovata*) is a tree of stature and beauty and of great utility. It is capable of attaining great height, as evident from the Table 1. It is valued for its wood and nuts. See Fig. 1. The bark is a pale-brown/gray and very shredded and shaggy — hence the name. Often, the bark will hang loosely in strips of a foot or more in length. The branches are pendulous and the foliage is a deep green. The leaves are large, from 4 to 6 inches (10.1 to 15.2 centimeters) in length. The staminate catkins occur in clusters of three and are green. The fruit may be described as globular in shape, with a very thick husk. The nut is white, thin-shelled and the kernel is sweet. This is the most important of the hickory nuts marketed, not of course including the pecans to be described shortly.

The shagbark hickory ranges from the Saint Lawrence River valley southward into Maine and generally following the Appalachian Mountains into the southeastern states. The tree ranges westward through Michigan and Minnesota and southward into Kansas, Oklahoma, and Texas. The tree does very well in certain parts of New England and particularly well in the Piedmont region of North Carolina. The wood is well known for its hardness, density, toughness, and close-grain. It is of a pale-brown color and remains the preferred wood for quality tool and implement handles and other heavy-duty applications. In the green state, the wood has a moisture content of 57% and a weight per cubic foot of 63 pounds per cubic foot (1009 kilograms per cubic meter). After air-drying to 12% moisture content, the weight per cubic foot is 51 pounds (817 kilograms per cubic meter) and 1,000 board-feet (2.36 cubic meter) of nominal sizes weigh 4,250 pounds (1927 kilograms). Crushing strength of the green wood when compression is applied parallel to the grain is 4,570 pounds per square inch (31.5 MPa); of the dry wood, 8,970 psi (61.9 MPa). The wood has 30% greater strength than white oak and double the shock resistance.

The species *C. illinoensis* is well known for its production of pecans. See Fig. 2. The tree ranges through much of the eastern United States. In particular, these trees are extensively cultivated in the southern states for their nut crop. The trees have huge branches and are capable of attaining a height of 100 feet (30 meters) or more. The head is rounded. There are 11 to 17 toothed, pointed leaflets. The tree was introduced to Europe many years ago and does well in the central and southern parts of France. For top production of pecans, the tree requires hot summers.

Wingnut trees are of the genus *Pterocarya*, deciduous, with large alternate, pinnate leaves. The leaflets are toothed. There are unisexual flowers in separate catkins appearing on the same tree. The trees bear small, winged nuts which occur on long hanging spikes. The trees are fast-growing and not too particular about soil. The Caucasian wingnut tree (*Pterocarya*) occurs in the Caucasus. It can attain a height of 100 feet

Fig. 1.   Shellbark hickory nut. (*USDA photo.*)

Fig. 2.   Pecans. (*USDA photo.*)

(30 meters), is broad and spreading, with large oblong leaves from 8 to 14 inches (20.3 to 35.5 centimeters) in length. The *P. stenoptera* occurs in China and also can attain a height of 100 feet (30 meters). The *P. x. rehdrena* is a hybrid of the two aforementioned species and is a shorter (up to 40 feet (12 meters) in height) broad-domed tree.

HIGH. In meteorology, an area of high pressure, referring to a maximum of atmospheric pressure in two dimensions (closed isobars) in the synoptic surface chart, or a maximum of height (closed contours) in the constant-pressure chart. Since a high is, on the synoptic chart, always associated with anticyclonic circulation, the term is used interchangeably with anticyclone.

See also **Low**.

HIGH ENERGY TRANSIENT EXPLORER (HETE-2) MISSION. See **Space Science Missions: Universe**.

HIGH FIDELITY. The quality of a sound reproducing system such that the acoustical characteristics of the reproduced sounds (usually musical) match as closely as possible the characteristics of the original sounds when made under their normal conditions. Thus a high fidelity reproduction of a symphonic work should sound the same to the listener as if he were present in a concert auditorium, listening to the orchestra directly, even though the sounds used in the recording were actually transcribed in a recording studio with extremely artificial acoustical characteristics.

HIGH-G ACCELEROMETER. See **Acceleration**.

HIGH-LIFT DEVICES. See **Aerodynamics and Aerostatics**.

HIGH LIMITING CONTROL. See **Control Action**.

HIGH-PASS FILTER. See **Filter (Communications System)**.

HIGH-PRESSURE TECHNOLOGY. See **Pressure**.

HIGH-RESOLUTION INFRARED RADIATION SOUNDER (HIRS). See **Nimbus Satellite Program**.

HIGH-RESOLUTION INFRARED RADIOMETER (HRIR). See **Nimbus Satellite Program**.

HIGH-SPEED CIRCUIT-SWITCED DATA (HSCSD). See **Wireless Communications Applications**.

HIGH TECHNOLOGY. A buzz term of the 1980s and early 1990s used mainly by the lay media to identify relatively new, complex, and sophisticated scientific and industrial pursuits, such as late-generation computers, electronic components, gene science, medical research, lasers and advanced optics, automation, communication systems, etc. It is frequently abbreviated as *high tech*. Use of the term was intended to connote a marked distinction between these more recent activities that are aligned with the information needs of society and the long-established, but more prosaic and heavy industries that are geared mainly to the material needs of society. In the case of high tech, the field is predominantly white collar, intellectual, clean, and nonpolluting, versus the older industries with their large blue-collar workforce. Many of the glowing promises of high tech have been much slower in reaching fruition than initially forecast. In retrospect, it is interesting to note that each generation of the past could have made its claims to high tech, as witness the light bulb, the telephone, radio and television, the automobile, the airplane, etc., which in its day probably were more revolutionary than most of the high-tech claims of the current generation. The high tech of today is based upon the high tech of yesterday.

HIGH-TEMPERATURE RESEARCH. See **Solar Energy**.

HIGH VACUUM. The condition in a gas-filled space at pressure less than 10^{-3} torr. The term high vacuum has frequently been defined as a pressure less than some upper limit. High vacuum (and similar vacuum terms) should not be defined as a pressure but rather as the condition or state in a gas-filled space at pressures less than some upper limit or within specified limits. The following classification of degrees of high vacuum has been proposed:

Condition	Pressure Range
high vacuum	10^{-3} to 10^{-6} torr
very high vacuum	10^{-6} to 10^{-9} torr
ultrahigh vacuum	10^{-9} torr and below

HIGHWAY BANKING. See **Superelevation**.

HILDEBRAND, JOEL (1891–1983). One of the most distinguished American chemists and teacher. Born in New Jersey, he obtained his doctorate in chemistry and physics from the University of Pennsylvania. After studying abroad under Nernst and van't Hoff, he became professor of Chemistry at the University of California, Berkeley, in 1913 where he remained until retirement. He made many important contributions to physical chemistry, particularly in the area of nonelectrolyte solutions; his treatise on this subject is a recognized classic and his textbook on the principles of chemistry established a new standard of excellence. He also made important contributions to the thermodynamics of vaporization of liquids. He proposed the use of helium in deep-sea diving equipment, which has become accepted practice. A gifted teacher and lecturer, he continued his constructive research to the end of his long life. He was unusually active in outdoor sports such as swimming, skiing, and hiking. Among his numerous awards were the Nichols and William Gibbs medals and the Priestley medal.

Additional Reading

Hildebrand, J., R.L. Scott, and J.M. Prausnitz: *Regular and Related Solutions: The Solubility of Gases, Liquids and Solids*, John Wiley & Sons, Inc., New York, NY, 1979.
Hildebrand, J.: *Principles of Chemistry*, The Macmillan Company, New York, NY, 1949.

HILDEBRAND RULE. The entropy of vaporization, i.e., the ratio of the heat of vaporization to the temperature at which it occurs, is a constant for many substances if it is determined at the same molal concentration of vapor for each substance.

HILL, ARCHIBALD VIVIAN (1886–1977). Archibald Hill was an English physiologist noted for his work on the thermodynamics of muscle fibres. Hill, the son of a Bristol timber merchant, read mathematics at Trinity College, Cambridge. On the advice of his tutor, Walter Morley Fletcher, he turned his mathematical ability (third wrangler in 1907) in the direction of physiology. First as a recipient of the George Henry Lewes Studentship in Physiology and then as a research fellow of Trinity College, Hill studied the processes of heat production in muscle. His early work, much of which consisted of developing new experimental techniques and improving equipment, helped found the speciality of biophysics. Hill's contributions during World War I (anti-aircraft research) and II (radar) provide additional evidence of his technical prowess. After the end of World War I, as a newly elected Fellow of the Royal Society (1918), Hill moved to the University of Manchester as Professor of Physiology. After winning the Nobel Prize in Physiology or Medicine in 1922, he moved to University College London (UCL), succeeding Ernest Starling in the physiology chair. Although he gained the Foulerton Research Professorship from the Royal Society in 1926, he remained at UCL until retirement. See also **Physiology (The History)**.

Hill proved to be something of a scientific administrator, serving as biological and foreign secretaries of the Royal Society and secretary and foreign secretary of the Physiological Society. In addition to his own scientific research Hill was committed to internationalism and the promotion of political neutrality within science. His commitment to such ideals was manifest through the Academic Assistance Council (later called the Society for the Protection of Science and Learning), an organization he founded and helped to run, and by personally finding bench space in laboratories (including his own) for scientific refugees from Europe in the 1930s.

Hill began the work that led to his Nobel Prize (shared with Otto Meyerhof) in 1910. Initially he wanted to understand the relationship between heat production during muscle contraction, the appearance of lactic acid and its subsequent breakdown into carbon dioxide and water. He determined that heat was generated during contraction and recovery of the muscle fibres. With his colleague William Hartree, Hill made thermal measurements of muscle and found that the heat was directly related to the biochemical reactions involved in producing lactic acid and its dissipation, through an unidentified compound, to the known products of water and carbon dioxide. To Hill's thermodynamic approach the German Otto Meyerhof added his own biochemical studies, hypothesizing that glycogen fuelled the production of heat and its regeneration accounted for the heat produced during recovery. Although using different methodologies both Hill and Meyerhof were attempting to explain some of the fundamental physiology taking place within living cells. See also **Meyerhof, Otto Fritz (1884–1951)**.

Following his success in elucidating the thermodynamics of muscle fibres, Hill turned his attention in the later 1920s and 1930s to the much smaller amount of heat generated by nerve fibres and the underlying physiology of nerve impulses, which he approached by determining the relevant physico-chemical equations. Just before World War II and again afterwards he returned to the study of muscle heat, revising earlier ideas with the benefit of hindsight and new research.

H. J. POWER Wellcome Trust Centre for the History of Medicine at UCL, London, UK

HINODE (SOLAR-B).   See **Space Science Missions: Sun**.

HINSBERG REACTION.
Reaction of primary and secondary amines with sulfonyl halides to give sulfonamides; because the products from primary amines are soluble in alkali and those from secondary amines are not, and since tertiary amines do not react, this method is useful for the separation and identification of amines.

HINSHELWOOD, CYRIL NORMAN (1897–1967).
Cyril Hinshelwood was a British physical chemist who made notable contributions to chemical kinetics.

Hinshelwood was born in London in 1897. In World War I he worked as a chemist in an explosives factory and this stimulated his interest in the rates and mechanisms of chemical change. He entered Balliol College, Oxford in 1919, became a Research Fellow in 1920, was Fellow and Tutor of Trinity College (1921–1937), and Dr Lee's Professor of Chemistry (1937–1964).

His pioneering work on gas reactions in the 1920s, and their interpretation in terms of kinetic theory, led to his best-known book *The Kinetics of Chemical Change in Gaseous Systems* (1926), revised as *The Kinetics of Chemical Change* (1940).

His studies of "unimolecular" gas reactions laid the foundations for much later work. He also studied chain reactions, particularly the hydrogen–oxygen reaction. Related work was done by the Russian N. N. Semenov, with whom Hinshelwood shared a Nobel Prize for Chemistry in 1956. He also worked on heterogeneous catalysis and the kinetics of reactions in solution, but from the mid-1930s he was increasingly interested in the kinetics of bacterial growth. He saw the bacterial cell as a complex system of reactions, to be dealt with through chemical kinetics. This work led to *The Chemical Kinetics of the Bacterial Cell* (1946) and *Growth, Function, and Regulation in Bacterial Cells* (1966, with A. C. R. Dean). Some of his ideas were initially much criticized by biologists, but later became accepted.

Hinshelwood was President of the Chemical Society (1946–1948) and of the Royal Society (1955–1960). He was knighted in 1948 and admitted to the Order of Merit in 1960. A fine linguist, with a good knowledge of the classics, he was President of the Classical Association in 1959. He was also a talented painter.

Additional Reading
Dean, A.C.R., and C.N. Hinshelwood: *Growth, Function, and Regulation in Bacterial Cells*, Clarendon Press, Oxford, UK, 1966.
Thompson, H.W.: "Cyril Norman Hinshelwood 1897–1967," *Biographical Memoirs of Fellows of the Royal Society*, **19**, 374–431 (1973).

JOHN SHORTER, University of Hull, Hull, UK

HIP.
The joint at the attachment of the human thigh to the body. Also, the adjacent portion of the thigh where it merges with the buttocks and less commonly the corresponding part of the leg in various animals.

Congenital dislocation of the hip, caused by improper development during the fetal life, is thought to be a heritable condition. Females are

much more likely to be afflicted than males, and the dislocation may be of one or both hips. The condition is often difficult to diagnose before the child begins to walk, although it is during infancy that treatment is most useful. The first symptom may be a more pronounced rotation of the femur than in normal infants. When only one hip is affected, the creases in the infant's thighs may not be symmetrical. Upon starting to walk, the individual may develop a limp and marked lordosis (forward curvature of the spine). Later, there is a shortening of the thigh and a wide space between the thighs when the child stands with feet together. Usually there is no pain associated with the dislocation until adulthood and that is usually a low-back pain resulting from the lordosis. Treatment involves a long tedious procedure employing casts and weights to gradually correct the dislocation. Surgical treatment may be required if soft tissues have developed in the space to which the head of the femur is to be restored.

Accidental dislocation of the hip, as with other bones, causes pain and limitation of movement. Nerves also may be severely injured. In hip dislocations, the major motor nerve may be paralyzed; and if nutrient-supplying blood vessels are torn, the head of the thighbone may become necrotic, soft, and die, or osteoarthritis may develop.

A total hip replacement is one of the most successful orthopedic procedures. Approximately 150,000 hip replacements are performed annually in the United States. A major goal of such procedures is the relief of pain. What is referred to as a total hip replacement may have a limited life, depending largely upon the kind of exercise and work that are customary for a given patient. Breakage and wear of the artificial joint theoretically are potential problems, but the most significant problem is that of the prosthesis loosening.

Forces generated in a total hip replacement during normal walking are about 1.5 to 3 times body weight. With activities, such as running and jumping, these forces can reach 4 to 6 times body weight. An artificial hip replacement joint is shown in Fig. 1. Ultimately, most patients can participate in numerous activities, including walking, bicycling, golf, hiking, swimming, rowing, and cross-country skiing. High-impact activities, however, should be avoided, and these include running, jumping, heavy lifting, and contact sports.

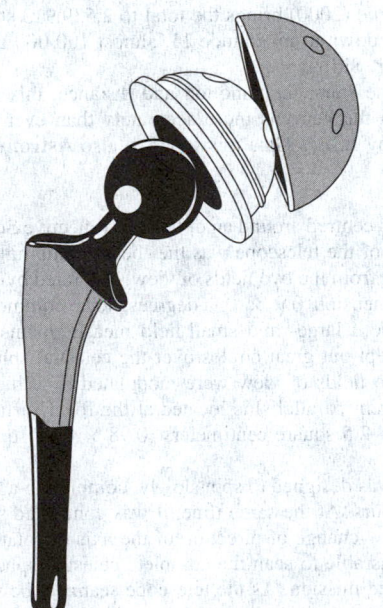

Fig. 1.   An artificial replacement hip joint.

HIPPARCOS MISSION.
Hipparcos is the acronym for high precision parallax collecting satellite, which points out the main astrophysical objective, but recalls also Hipparchus, the Greek astronomer who discovered precession and is the author of the first star catalog. See also **Star Catalogues**. The satellite was launched by the European Space Agency (ESA) on 8 August 1989 23:25 UT aboard an Ariane-4 launch vehicle from Kourou, French Guiana. A geostationary orbit was aimed at, but due to the failure of the apogee boost motor, the final orbit was very elongated; the perigee was at an altitude of 500 km (311 miles), and the

apogee was at 36,500 km (22,680 miles). The period was 10 hours and 40 minutes. Communications with Earth were secured by three stations in Odenwald (Germany), Perth (Australia), and Goldstone (USA). They ensured direct visibility that covered 97% of the orbit and 93% of the useful observing time. Satellite control and pretreatment of the data were provided by the ESA Operation Center (ESOC in Darmstadt, Germany). http://www.esa.int/spacecraftops/.

Because of the difficulties that arose from the change from the nominal to the actual orbit, the operations started only at the end of November 1989. They stopped in March 1993, after the failure of several onboard gyroscopes. Taking into account several interruptions, the total useful data collected represents an accumulated 37 months of observations. However, instead of quasi-continuous 24 hour a day observations anticipated for the nominal mission, only 7 to 9 hours and sometimes less observation time per orbital revolution could be achieved because occultation times by Earth, passage through radiation belts which induces strong Cerenkov radiations in the optics, and illumination by the Moon when it was near a field of view that produced a noise that masked the signal had to be excluded.

Hipparcos, pinpointed the positions of more than one hundred thousand stars with high precision, and more than one million stars with lesser precision. Hipparcos turned slowly on its axis and repeatedly scanned right around the sky on different slants. It measured angles between widely separated stars, and recorded their brightness, which were often variable from one visit to the next. Each star selected for study was visited about 100 times during four years.

Hipparcos sent million million bits of information, radioed from Hipparcos to ground stations in Germany, Australia, and the United States, went into the biggest computation in the history of astronomy.

Hipparcos confirmed Einstein's prediction of the effect of gravity on starlight. The mission discovered that the Milky Way is changing shape. Its data also helped predict the impact of Comet Shoemaker-Levy with Jupiter in 1994.

Calculations from observations by the main instrument generated the Hipparcos Catalogue of 118,218 stars charted with the highest precision. An auxiliary star mapper pinpointed many more stars with lesser but still unprecedented accuracy, in the Tycho Catalogue of 1,058,332 stars. The Tycho 2 Catalogue (2000) brings the total to 2,539,913 stars, and includes 99% of all stars down to magnitude 11, almost 100 000 times fainter than the brightest star, Sirius.

Pinpointing the stars, their motion and distance, this totally European mission mapped the heavens more accurately than ever before. Its work will be refined by ESA's Gaia mission. See also **Astrometry**.

Spacecraft

The payload was centred around an optical all-reflective Schmidt telescope. A novel feature of the telescope was the 'beam-combining' mirror, which brought the light from the two fields of view, separated by about 58 degrees and each of dimension 0.9×0.9 degrees, to a common focal surface. This achieved both large- and small-field measurements simultaneously. The satellite swept out great circles over the celestial sphere, and the star images from two fields of view were modulated by a highly regular grid of 2688 transparent parallel slits located at the focal surface and covering an area of 2.5×2.5 square centimeters (0.98×0.98 square inches). See Fig. 1.

The satellite was designed to spin slowly, completing a full revolution in just over two hours. At the same time, it was controlled so that there was a continuous slow change of direction of the axis of rotation. In this way, the telescope was able to scan the complete celestial sphere several times during its planned mission. As the telescope scanned the sky, the starlight was modulated by the slit system, and the modulated light was sampled by an image-dissector-tube detector, at a frequency of 1200 Hz. At any one time, some four or five of the selected (or programme) stars were present in the combined fields of view. The detector had a small sensitive area which covered an area of about 38 arcsec in diameter (projected on the sky). The detector could only follow the path of one star at a time. However, with rapid computer control, it could be switched to all the programme stars for short intervals of time during their passage across the field, which took about 20 seconds.

The telescope was continually determining the relative (along-scan) positions of the programme stars which appeared first in the preceding field of view and then in the following field of view due to the rotation of the satellite. In this way, astronomers obtained several comparisons

Fig. 1.   Schmidt telescope.

with different stars. As the scans also overlapped 'sideways' when the satellite axis of rotation changed on each sweep of the sky, the stars appeared again, but this time compared with other stars. In this way, a dense net of measurements of the relative angular separations of the stars was progressively built up.

In addition to the main instrument (designed to measure about one hundred thousand stars down to about 12 mag), the payload included star mappers whose function was to provide data allowing precise real-time satellite attitude determination (a task performed on board the satellite). The star mappers consisted of two sets of four slits. Each slit was set at different inclinations with respect to the scanning direction, so that the satellite attitude could be derived from the detector signals as the star images moved across the grid. The modulated light signal was converted into photon counts by the two photomultiplier tubes.

The digitised photon counts from the main detector were sent to the ground. Along with relevant attitude information from the satellite's star mappers and other house-keeping data, the relative phases of the star images present within the combined fields of view were derived. The data processing was carried out on the ground. A full analysis of the data collected during the mission lifetime led to the final catalogue of star positions, parallaxes, and proper motions.

Principle of Hipparcos

The principle of Hipparcos is sketched in Fig. 2. The main characteristic is that two fields of view are focused on a single focal surface. The optical axes from the center of each field are combined by two glued half-mirrors called a beam combiner whose angle sets the angular reference γ, known as the basic angle. On the focal surface is a grid composed of slits parallel to the intersection of the mirrors, a fundamental direction that we shall call vertical. The satellite revolves slowly around an axis parallel to this direction, and the light of the star is modulated by the grid. A photoelectric receiver registers the resulting signal. On both sides of the main grid are two systems of vertical and chevron slits, called star-mappers. Another photoelectric system registers the light crossing a star-mapper that provides data for determining the satellite's attitude. These data are also used for astrometry in the framework of the Tycho program. Note that, except in part for the star-mapper, the observations are one-dimensional, they amount to determining the transit time of star images through the grid. To cover the whole sky, it is necessary to modify the satellite's attitude in a predetermined way.

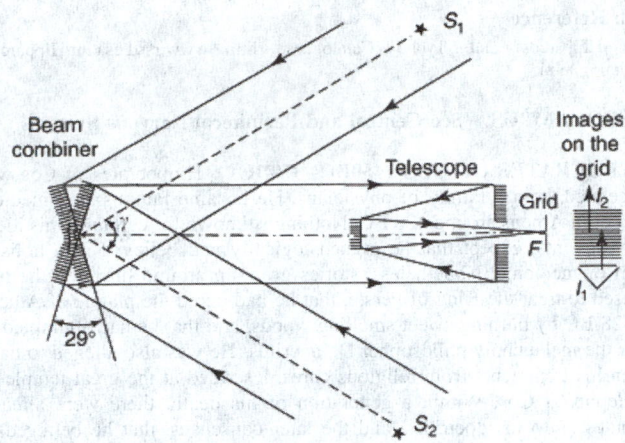

Fig. 2.   Principle of Hipparcos showing the motion of the images I_1 and I_2 of the stars S_1 and S_2 in different fields of view.

Description of the Hipparcos Payload

Construction of the payload followed the principles described before. The basic angle is $\gamma = 58°31.25'$. The beam combiner was produced by cutting a 29-cm (11 inches) Schmidt corrected mirror into two halves. The refractive Schmidt configuration was chosen to have a large astrometrically good field of view across more than 1.3° in diameter. The space structure of the dual telescope and its baffles that protect it from stray solar light is shown in Fig. 3. The light paths from the two fields to the optical block on which the grids are engraved are shown in the figure. The equivalent focal distance is 140 cm (55 inches). A star produces a diffraction pattern on the grid that is elongated along the vertical direction. The along-track dimensions range between 0.5 and 0.7 seconds of arc in the sky, depending on the color of the star. The grids are engraved on the front side of the optical block which is curved so that it matches the focal surface of the telescope. The main grid consists of 2688 regularly spaced slits whose period is 8.20 μm (1.208' in the sky); the transparent width is 3.13 μm (0.46' in the sky). It covers two fields of view in the sky of 0.9° × 0.9°. The vertical extension of the star-mapper grids is limited to 0.7°. Each grid is composed of four 0.9' wide slits that have pseudorandom separations respectively of a, $3a$, and $2a$ ($a = 5.675'$ in the sky). Only the preceding star-mapper was operated. The second, which was redundant, was actually never used.

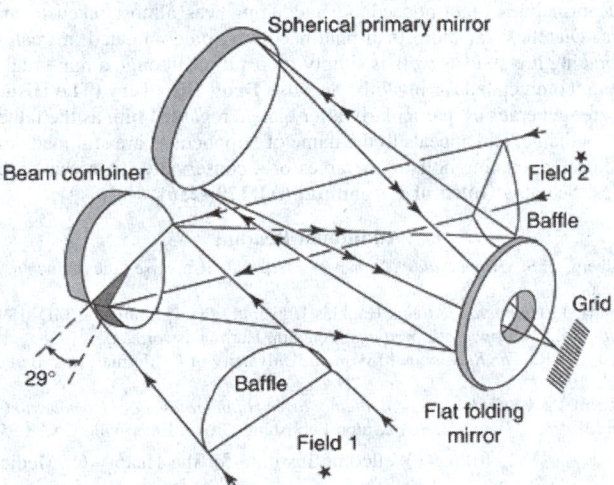

Fig. 3.   Space configuration of the Hipparcos optics.

The receiving systems were quite different for the main grid and for the star-mapper. The rear side of the main grid was curved to serve as a field lens of the optical system used to image the full grid on the entrance of an image dissector. This special kind of photomultiplier produces an electronic image on the back wall of the tube. A small hole leaves the way open only to electrons coming from a tiny portion of the electronic image. A set of magnetic deflectors can be controlled to shift any point of the electronic image on the hole. As a consequence, only the light coming from a 30' radius circle in the whole focal image is recorded. The way this is used will be described later. In contrast, all of the light that enters the star-mapper grids is transmitted to a dichroic mirror that splits the light into two wavelength ranges that correspond roughly to the B and V filters of the Johnson UBV photometric system, called here B_T and V_T. Then each channel is directed toward a different photomultiplier. The photoelectrons are recorded at a rate of 1200 Hz for the main grid and 600 Hz for each of the star-mapper channels. More details on the Hipparcos instrumentation and operations are given in Volume 2 of the Hipparcos and Tycho Catalogue [ESA Ref.].

Scanning the Sky. Because observations are made along a scan in a narrow band 0.9° wide, it is necessary to modify the attitude of the satellite so that all of the sky is scanned as homogeneously as possible and all of the stars are observed for roughly the same amount of time. Various scanning laws can do this, but there are additional constraints. First, and this is the most important condition, the angle between the observed fields of view and the Sun must be at least 45° to minimize stray light. However, the inclination of the scan with respect to the ecliptic should be as small as possible. The parallactic deviation is parallel to the plane of the ecliptic. One wishes to maximize its projection along track because this quantity is used to determine parallaxes. So the inclination chosen is at the limit of acceptance by the first condition. Finally, the attitude should change slowly so that there are overlaps between successive scans.

Nominal Scanning Law. As a compromise between these conditions, the following nominal scanning law was adopted. The satellite rotates in 2 hours 8 minutes allowing 19 seconds for each star to cross the main grid. The rotational axis circles the direction of the Sun in 57 days, keeping an angular distance of 43° from the Sun. Figure 4 shows the motion of the axis of rotation in one year and the part of the sky scanned in 70 days.

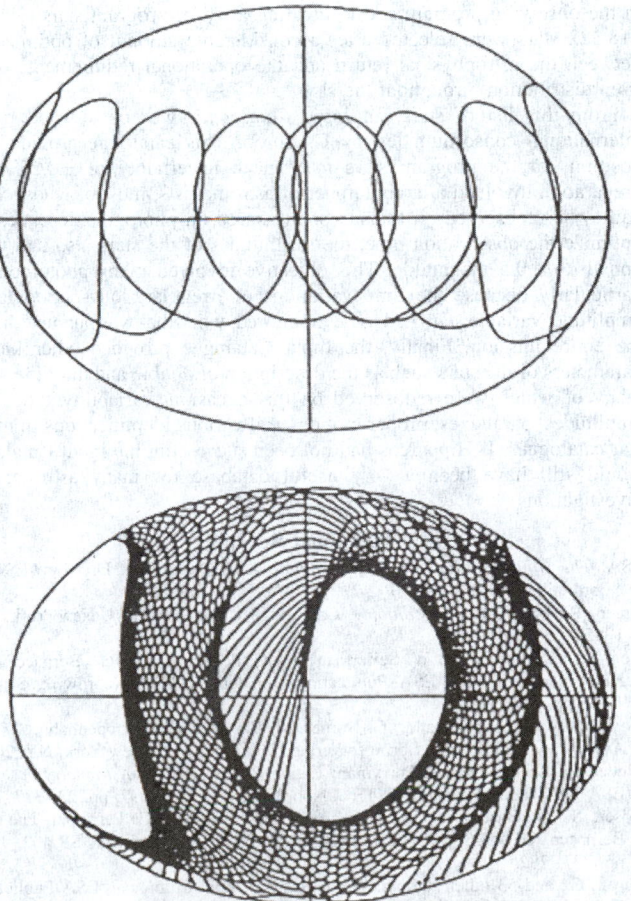

Fig. 4.   Hipparcos sky scanning law in ecliptic coordinates. *Above*: Motion of the satellite axis in one year. *Below*: Part of the sky scanned in 70 consecutive days.

Left to itself, the rotational axis would drift rapidly due to the various torques that are applied to the satellite (gravitation, radiative pressure, reaction of the gyroscopes, etc.). To follow the scanning law demands active attitude control that is realized by six gas-jet thrusters using compressed nitrogen. The satellite attitude is monitored onboard (see earlier section). When it deviates by 10 minutes of arc from the nominal scanning, gasjets were actuated to reverse the natural attitude drift. In practice, this happened four to six times an hour during the observation conditions. When the satellite was in Earth's shadow or in radiation belts, attitude control was sometimes very bad, special scanning law recovery procedures had to be applied, and observations were not possible during these maneuvers.

Onboard Attitude Determination. The attitude had to be known continuously during the observations at an accuracy better than one second of arc. The onboard attitude determination used five rate-integrating gyroscopes calibrated in real time using star-mapper data from star crossings. When, a star image crosses a star-mapper slit at a time t, one may write an equation stating that the star lies on the projection of the slit in the sky. Knowing the celestial coordinates of the star corrected for stellar and orbital aberrations, the equation sets one condition on the parameters describing the attitude at time t. If the evolution of the attitude between two gas-jet actuations is smooth, a few equations of condition are sufficient to calibrate the gyroscope drifts and therefore to know the attitude with the desired precision.

The Input Catalogue. As just stated, the necessary real-time orbit was determined on the basis of rough knowledge of the position of some stars. Actually, there was also the same need in the operations of the main grid: to shift the image of a star on the hole of the image dissector, it is necessary to know the position of the star as well as the attitude of the satellite. In addition, because the main grid is periodic, there is no other method to know on what slit the image of a star lies than to know the position significantly better than the grid period. Finally, if all of the observable (up to magnitude 12.5) stars were to be observed, the time allocated to each star would have been insufficient. This led to a limit of the number of stars in the observing program and to creating a list of program stars. Finally, 118,322 stars were selected after a considerable amount of optimization between the astrophysical return and the operational requirements of an even distribution throughout the sky.

From this list of stars, an *Input Catalogue* was constructed by an international consortium led by C. Turon. This catalogue provided the positions of the program stars to a mean uncertainty of $\pm0.25'$. The preparation involved a large number of astrometrists, most of whose work had to be reobserved on transits or measured on photographic plates. To optimize the observation time, the magnitudes of the stars also had to be known to ±0.5 magnitude. This objective involved many photometrists, particularly because the time variations of irregular, long-period large-amplitude, variable stars had to be monitored, work that was pursued during the entire mission. Finally, the Input Catalogue provided other known parameters of the stars such as the description of double and multiple stars, many of which were reobserved on this occasion, variability types and amplitudes, parallaxes, proper motions and various identifications in major star catalogues. If Hipparcos had not been successful, the Input Catalogue would still have been a very useful database for many astronomical investigations.

Additional Reading

ESA: *The Hipparcos and Tycho Catalogues*, ESA Publication Division, ESTEC, Noordwijk, SP-1200, 1997.

Green, R.M.: *Spherical Astronomy*, Cambridge University Press, New York, NY, 1985.

Høg, E., U. Bastian, and W. Seifert: in M.A.C. Perryman and P.L. Bernacca eds., *Hipparcos Venice'97*, ESA Publication Division, ESTEC, Noordwijk, SP-402, 1997, pp. 783–788.

Kovalevsky, J., I Appenzeller, G. Borner, M. Harwit, and R. Kippenhahn: *Modern Astrometry*, 2nd Edition, Springer-Verlag New York, LLC, New York, NY, 2001.

Lindegren, L., and M.A.C. Perryman: *Future Possibilities for Astrometry in Space*. ESA Publication Division, ESTEC, Noordwijk, SP-379, 1997, pp. 23–32.

Röser, S., U. Bastian, et al.: *Hipparcos Venice'97*, M.A.C. Perryman and P.L. Bernacca (eds), ESA Publication Division, ESTEC, Noordwijk, SP-402, 1997, pp. 777–782.

Turon, C., and 53 other authors: *The Hipparcos Input Catalogue*, ESA Publication Division, ESTEC, Noordwijk, SP-1136, 1992.

Yoshisawa, M., K. Sato, et al.: *Hipparcos Venice'97*, M.A.C. Perryman, and P.L. Bernacca (eds), ESA Publication Division, ESTEC, Noordwijk, SP-402, 1997, pp. 795–797.

Web Reference

The Hipparcos and Tycho Catalogues: http://www.rssd.esa.int/Hipparcos/catalog.html

HIPPOCAMPUS. See **Central and Peripheral Nervous Systems**.

HIPPOCRATES OF COS (460BC–370BC). Hippocrates of Cos was descended from a family of physicians. He became famous as a medical teacher in Athens around 420 BC. Nothing is known for certain of his ideas or treatments, except that he argued logically and believed in a holistic form of medicine. Nonetheless, stories grew up around him, that he had refused to treat the King of Persia, that he had cured the plague of Athens in 428 BC by burning sweet-smelling woods, and that he had attempted to treat the melancholy philosopher Democritus. He was also alleged to have taken his treatments from religious cures described at the great temple of Asclepius at Cos. Within a generation of his death, there were already disputes as to his doctrines, and the later consensus, that he believed in a theory of the four humours and strongly advocated prognosis, is based on weak foundations. By 250 BC many writings had become associated with his name, most written in his lifetime, but not certainly by him. Most defined health and illness in terms of balance or imbalance of bodily fluids (or humours) capable of being controlled by the doctor. Others thought in terms of harmful fluxes, or gases produced by the residues of undigested food in the body. What these fluids, residues or competing forces might be was a matter for considerable debate; some, like the author of *Ancient Medicine*, believed in a multiplicity of forces, others, like the author of *Sacred Disease*, in the dangerous effects of bile or phlegm.

Theorizing about the body (known almost entirely through surface anatomy and observation) might take place in public as well as private, and the doctor was expected to defend his own decisions publicly against competitors, who included root-cutters, herbalists and exorcists. Although many medical writers of the Hippocratic Corpus (some 60 or so treatises in all) were religious in outlook, they endeavoured to exclude god or the gods from direct intervention in health and sickness. They also developed a professional ethic, based on efficacy – the *Oath* is anomalous in its moral orientation.

Close observation lies at the base of much Hippocratic medicine, whether of patients, groups of patients (as in the so-called *Constitutions* in some books of the *Epidemics*) or the environment. Signs and symptoms tended to be ascribed to individual mixtures of humours, although some treatises, often associated with doctors from the peninsula of Cnidus, talk in terms of more specific disease types.

Complex surgery was rare, although the Hippocratic writers made sound observations about the consequences of head wounds, and understood the basic principles of orthopaedics. Bandaging was almost taken to an art form. Dietetics, far more than pharmacology, predominated in treatment, embracing life style as well as simply foodstuffs, although a thin gruel was often recommended for invalids. See also **Drug Discovery (The History)**.

Later generations, particularly after Galen, regarded him as the father of true medicine, and appeals to the name of Hippocrates are still made today by those advocating holistic therapies or a conservative stance in medical ethics. See also **Galen of Pergamum (AD129–216)**.

Additional Reading

Goldberg, H.S.: *Hippocrates: Father of Medicine*, iUniverse, Inc., Lincoln, NE, 2006.

Jouanna, J.: *Hippocrates*, Johns Hopkins University Press, Baltimore, MD, 1999.

Lloyd, G.E.R.: *Hippocratic Writings*, Penguin, Harmondsworth, UK, 1978.

Lloyd, G.E.R.: *The Revolution of Wisdom*, University of California Press, Berkeley, CA, 1989.

Strathern, P.: *A Brief History of Medicine: From Hippocrates' Four Humours to Crick and Watson's Double Helix*, Avalon Publishing Group, Emeryville CA, 2005.

V. NUTTON, Wellcome Institute for the History Of Medicine, London, UK

HIPPOPOTAMUS (*Mammalia, Artiodactyla*). The group of *Hippotamines* is one of the smaller in the order *Artiodactyla* (even-toed hoofed animals). There are two extant species: (1) the greater Hippopotamus (*Hippopotamus*); and (2) the Pigmy Hippopotamus (*Choeropsis*).

The greater hippopotamus is a large, rather commonly occurring animal of the rivers of tropical Africa. The body is large, barrel-shaped, bulky, with short strong legs, and a very broad muzzle. The beast may attain a length of about 14 feet (4.2 meters) and a height of about 4 feet

(1.2 meters). Four tons is usually the figure quoted for the larger specimens. They are almost hairless, and of a gray-black coloration with white underneath. Hippopotamuses are largely aquatic in habits and live entirely on vegetation, including both water plants and terrestrial species. They have been known to do extensive damage to crops in their roving. The animals feed mostly at night. The greater hippopotamus is the largest of the living nonruminating even-toed mammals. They are fast swimmers and can run about as fast as humans on land. They can issue loud grunts and bellows. Multiple births are rather uncommon. The gestation period is about eight months. The baby animal is known as a calf. See Fig. 1.

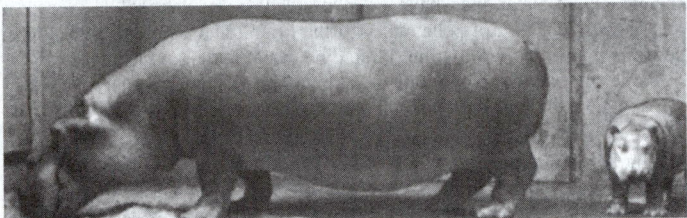

Fig. 1.   Female hippopotamus and baby. (*A.M. Winchester.*)

In some regions, the hippopotamus has been a staple food among tribesmen who harpoon the animal from small canoes. The manner in which the animal is prepared for consumption by some tribesmen is rather offensive to most people. After the body of the animal is dragged to the shore of the river, it is allowed to "ripen" under the hot tropical sun for a few days. The animal is then ripped open and the natives tear apart the softened carcass, gorging themselves on the rotten flesh.

Normally, the hippopotamus has a reasonably good disposition, but is proprietary concerning staked out stretches of the river. Usually, the animal will move out of the way of boats, floating just beneath the surface, but watching the passerby with use of the periscopic eyes, which are just above the water line. However, the animal has been known to attack boats for unknown reasons, stomping and chewing the boat and occupants. Considering the size of the creature's mouth, the bite of a hippopotamus is no less than ghastly and usually terminal. The animal also is unpredictable when encountered on land, particularly at night.

The pigmy hippopotamus attains the size of a large pig. It has a large mouth equipped with fang-like teeth. For habitat, the pigmy hippopotamus prefers small lakes and rivers, ponds and stagnant pools and seldom wanders far from its aquatic habitat. The animal cannot afford to stay out of water very long because the skin is equipped with very large pores and, unless kept moist, the skin cracks easily. Apparently, the animals use these pores for absorbing water into their system, rather than taking all of their liquid input by mouth.

Naturalists always have suspected a relationship between the *Hippopotamines* and the *Suines*, but fossil remains have failed to yield evidence for this connection. And thus they remain in a separate classification. It is interesting to note that the Romans regarded them as pigs of a very special nature.

HIRSUTISM. Abnormal growth of hair, particularly on the face of women. Although not well understood, causative factors appear to include a predisposition to the condition by inheritance, variations in endocrine activity, and imbalance of the metabolic processes. The condition usually does not appear until middle age. Treatment essentially is of a cosmetic nature.

Hirsutism is one of the principal features of polycystic ovary syndrome. Control of this hirsutism is extremely difficult. Best results have been obtained with therapy directed to suppressing adrenal and ovarian functions. Hirsutism is also Seen in Cushing's syndrome (hyperfunction of adrenal cortex). See also **Androgens**.

HIRUDINEA. The leeches, a class of segmented worms (phylum *Annelida*), well known for their habit of sucking blood. Marine and freshwater species are known, and in the moist tropical forests terrestrial species occur. They often attach themselves to bathers.

The members of this class are distinguished from other annelids by the following characteristics. (1) The body is relatively short, usually with 32 segments. (2) The external segments are annuli, numbering from 2 to 14 to each metamere. (3) Each end of the body bears a sucker. (4) The mouth is usually provided with three toothed plates or jaws. (5) The alimentary tract is provided with an enormous pouched crop in which blood is stored prior to digestion. (6) The anus opens dorsally to the posterior sucker. (7) The coelom is partially obliterated by a peculiar mesenchymal tissue. (8) At the anterior end of the ventral nerve cord, several ganglia (ganglion) are fused to form a large mass.

Leeches were once extensively used in medicine for letting blood and are still of minor importance for this purpose. Otherwise they are of no importance to man save as an occasional annoyance. They eat small aquatic animals as well as the blood of vertebrates, and some species are entirely predacious.

Two orders are recognized:

Order *Rhynchobdellida*. With a protrusible proboscis, colorless blood, and no jaws. Marine and freshwater.

Order *Gnathobdellida*. With jaws and red blood. No proboscis. Freshwater and terrestrial. The medicinal leech belongs to this order. It is native to Europe but is naturalized in ponds and streams of the eastern United States.

HISTAMINE AND HISTAMINE ANTAGONISTS. Histamine is an important protein involved in many allergic reactions. Allergies are caused by an immune response to a normally innocuous substance (i.e. pollen, dust) that comes in contact with lymphocytes specific for that substance, or antigen.

The history of histamine [CAS: 51-45-6], $C_5H_9N_3$, and the development of antihistamines have been reviewed in [Drugs of Today (1986) and the Journal of Allergy & Clinical Immunology]. Histamine was the first to be characterized of a series of biogenic amines that are released in the inflammatory process. As early as 1910, it was shown that histamine caused constriction of isolated guinea pig ileum and, subsequently, it was found that histamine induced a shock-like syndrome. In 1927 the presence of histamine in normal tissues was demonstrated. Attempts to reduce histamine manifestations led to the report, in 1933, that certain phenolic ethers inhibited histamine action. Toxicity precluded clinical use. In 1942 phenbenzamine [CAS: 961-71-7] (Antergan), $C_{17}H_{22}N_2$, was the first antihistamine to be successfully used in humans.

In 1966, the name H_1 was proposed for receptors blocked by the at that time known antihistamines. It was also speculated that the other actions of histamine were likely to be mediated by other histamine receptors. The existence of the H_2 receptor was accepted in 1972 and the H_3 receptor was recognized in rat brain in 1983. H_3 receptors in the brain appear to be involved in the feedback control of both histamine synthesis and release, whereas release of various other neurotransmitters, eg, serotinin (5-HT), dopamine [CAS: 51-61-6], noradrenaline [CAS: 51-41-2], and acetylcholine [CAS: 51-84-3], is also modulated. H_3 receptor effects have also been demonstrated in various peripheral tissues and H_3 agonists and antagonists are undergoing intensive study for therapeutic applications. See also **Neuroregulators**.

Histamine Synthesis, Metabolism, and Distribution

Histamine 2-(4-imidazolyl)ethylamine, is formed by decarboxylation of histidine by the enzyme L-histidine decarboxylase. Most histamine is stored preformed in cytoplasmic granules of mast cells and basophils. In humans mast cells are found in the loose connective tissue of all organs, especially around blood and lymphatic vessels and nerves. These cells are most abundant in the organs expressing allergic diseases: the skin, respiratory tract, and gastrointestinal tract.

Histamine release is mainly caused by cross-linking of immunoglobulin E on the mast cell surface by antigens. Basophil degranulation is caused mainly by histamine-releasing factors produced by inflammatory cells, such as neutrophils, platelets, and eosinophils. After its release, histamine diffuses rapidly into the blood stream and surrounding tissues.

Metabolism of histamine occurs via two principal enzymatic pathways. Most (50 to 70%) histamine is metabolized to N-methylhistamine by N-methyltransferase, and some is metabolized further by monoamine oxidase to N-methylimidazoleacetic acid and excreted in the urine. The remaining 30 to 40% of histamine is metabolized to imidazoleacetic acid by diamine oxidase, also called histaminase. Only 2 to 3% of histamine is excreted unchanged in the urine.

There is evidence that histamine functions as a neurotransmitter or a neuromodulator in the brain. In the brain, histamine is related to functions such as the regulation of neuroendocrine and cardiovascular systems, thermoregulation, the circadian rhythm of sleep-wakefulness, behavior, vestibular function, cerebral vascular regulation, and antinociception and analgesia.

Histamine is present in sympathetic nerves and has a distribution within the heart that parallels that of norepinephrine. A physiological role for cardiac histamine as a modulator of sympathetic responses is highly plausible.

Histamine Receptors

The actions of histamine are mediated through at least three distinct receptors defined pharmacologically by the actions of the respective agonists and antagonists.

The H_1 Receptor and Its Ligands. The H_1 receptor mediates most of the important histamine effects in allergic diseases. These include smooth muscle contraction, increased vascular permeability, pruritus, prostaglandin generation, decreased atrioventricular node conduction time with resultant tachycardia, activation of vagal reflexes, and increased cyclic guanosine monophosphate (cGMP) production.

H_1 Receptor Agonists. In the histamine molecule there are two principal structural elements: an imidazole moiety and an ethylamine side chain. Only the N_π;-position is absolutely necessary for H_1 agonism. The imidazole ring can be replaced, e.g., 2-pyridylethylamine, 2-thiazolylethylamine, or substituted at the 2-position. 2-Methylhistamine is often used as a selective H_1 agonist; however, larger substituents are not allowed unless a phenyl ring is used. 2-Phenylhistamine analogues appear to be very selective H_1 receptor agonists.

H_1 Receptor Antagonists. The classical H_1 receptor antagonists are reversible, competitive, dose-dependent inhibitors of the action of histamine on H_1 receptors. Histamine H_1 antagonists are usually divided into two classes: the first-generation or classical H_1 antagonists and the second-generation H_1 antagonists. The main distinction between the first- and second-generation drugs is the absence of sedative and anticholinergic side effects in the latter.

The classical histamine H_1 receptor antagonists are structurally very similar, all being substituted ethylamines. The classical H_1 receptor antagonists can be subdivided into six classes: aminoalkylethers (diphenhydramine, $C_{17}H_{21}NO$), ethylenediamines (tripelennamine, $C_{16}H_{18}N_3$), alkylamines (chlorpheniramine, $C_{16}H_{19}ClN_2$), piperazines (hydroxyzine, $C_{21}H_{27}ClN_2O_2$), phenothiazines (promethazine, $C_{17}H_{20}N_2S$), piperidines (cyproheptadine, $C_{21}H_{21}N$). See Table 1.

The prototype aminoalkylether or ethanolamine is diphenhydramine (Benadryl). Clemastine [CAS: 15686-51-8], $C_{21}H_{26}ClNO$, differs from it in having one phenyl ring chlorinated, a methylated benzhydryl carbon, and a three-carbon amino side chain, partly incorporated into a pyrrolidinyl ring system. Setastine [CAS: 64294-95-7], $C_{22}H_{28}ClNO$, which has been launched in Hungary, differs from clemastine only in the type of heterocyclic ring.

The prototype of a pure ethylenediamine is tripelennamine; antazoline [CAS: 91-75-8], $C_{17}H_{19}N_3$, belongs to the same family of compounds. Several well-known alkylamines in addition to chlorpheniramine are known. Dexchlorpheniramine maleate [CAS: 2438-32-6] and triprolidine monohydrochloride monohydrate [CAS: 6138-79-0], an alkenyl derivative, are two examples.

Replacement of the oxygen bridge atom of diphenhydramine by nitrogen and incorporation of the two nitrogens into a piperazine ring leads to cyclizines. Hydroxyzine is a well-known piperazine derivative. Cinnarizine [CAS: 298-57-7], $C_{26}H_{28}N_2$, and oxatomide [CAS: 60607-34-3], $C_{27}H_{30}N_4O$, both structurally related to cyclizine [CAS: 82-92-8], differ only by having respectively a cinnamyl or a propylbenzimidazolone N-substituent instead of a methyl group. Oxatomide also has mast cell stabilizing actions.

There are two types of tricyclic H_1 antagonists: the phenothiazine and piperidine derivatives. Promethazine is the prototype molecule of the phenothiazine derivatives. Mequitazine [CAS: 29216-28-2], $C_{20}H_{22}N_2S$, a quinuclidinylmethyl derivative, is less sedative than promethazine. Cyproheptadine and pizotifen [CAS: 5189-11-7], $C_{19}H_{21}NS\cdot C_4H_6O_5$, are piperidine derivatives. Though antihistamines, they are mainly used because of their antiserotonergic properties. Azatadine maleate [CAS: 3978-86-7], $C_{20}H_{22}N_2$, is a pyridyl variant of cyproheptadine. The compound mianserin hydrochloride [CAS: 21535-47-7], $C_{18}H_{21}ClN_2$, may structurally be regarded as a double-bridged analogue of phenbenzamine.

Ketotifen [CAS: 34580-14-8], $C_{19}H_{19}NOS$, is also a piperidine derivative; like oxatomide, it has mast cell stabilizing actions besides its antihistamine properties.

Second-Generation H_1 Receptor Antagonists. Because of undesirable side effects caused by classical H_1 receptor antagonists, drugs having enhanced clinical effectiveness and a reduced side-effect profile have been sought. The main progress has been in the development of antihistamines that, unlike the classical H_1 antagonists, do not cause sedation. Thus the term second-generation antihistamines usually refers to those nonsedating antihistamines. The reduced ability to penetrate the central nervous system (CNS) can generally be explained by physicochemical properties.

Several antihistamines have been derived from classical H_1-receptor antagonists, but do not penetrate into the brain. They include acrivastine, cetirizine, ebastine, epinastine, loratadine, pibaxizine, and terfenadine.

Some of the second-generation H_1 antagonists have nonclassical structures, e.g., astemizole, azelastine, emedastine, levocabastine, and mizolastine.

The H_2 Receptor and Its Ligands. The discovery of H2 receptors and antagonists occurred in 1972.

The H_2 receptor mediates effects, through an increase in cyclic adenosine monophosphate (cAMP), such as gastric acid secretion; relaxation of airway smooth muscle and of pulmonary vessels; increased lower airway mucus secretion; esophageal contraction; inhibition of basophil, but not mast cell histamine release; inhibition of neutrophil activation; and induction of suppressor T cells. There is no evidence that the H_2 receptor causes significant modulation of lung function in the healthy human subject or in the asthmatic. In the brain, the H2 receptors are distributed widely and mainly in association with neurones.

Combined H_1/H_2 receptor stimulation by histamine is responsible for vasodilation-related symptoms, such as hypotension, flushing, and headache, as well as for tachycardia stimulated indirectly through vasodilation and catecholamine secretion.

Structural requirements of histamine as an H_2 agonist are considered to be the protonated side-chain nitrogen atom and the ability of the imidazole amidine system to undergo a tautomeric shift. 4-Methylhistamine is often used as a selective H_2 agonist. Larger substituents are not allowed. Methylation of the amine group is allowed, but leads to nonselective analogues. H_2 agonists are divided in three chemical classes, i.e., analogues of histamine, dimaprit, and impromidine.

The prototype examples of H_2 antagonists are cimetidine, famotidine, and ranitidine.

The H_3 Receptor and Its Ligands. The first evidence for the existence of a third histamine receptor subtype was published in 1983 and great advances have been made since then. The H_3 receptor has been reported to modulate the release of a variety of neurotransmitters and can be regarded as a general regulatory mechanism.

In contrast to the development of selective agonists for the H_1 and H_2 receptor, potent agonists for the H_3 receptor can be obtained by simple modification of the histamine molecule. Histamine itself is already a rather potent agonist of the H_3 receptor, although it is of course not very selective. Its high affinity for the H_3 receptor in comparison with its affinity for the H_1 and H_2 receptor is striking. Modification of the histamine molecule is usually not well tolerated. Substitution or replacement of the imidazole moiety is not beneficial for H_3 receptor activity. As for the H_2 receptor, both nitrogen atoms of the imidazole ring appear to be essential for agonistic activity. See Table 4.

H_3 receptor antagonists include betahistine, clobenpropit, and thioperamide.

Uses of Histamine Receptor Ligands

Allergic Diseases. H_1-receptor antagonists are used for the symptomatic treatment of several allergic diseases where histamine release form mast cells is induced via immunological or nonimmunological mechanisms.

Allergic Seasonal or Perennial Rhinoconjunctivitis. Histamine can cause all pathologic features of allergic rhinitis, with the exception of late-phase inflammatory reactions. Pruritus is caused by stimulation of H_1 receptors on sensory nerve endings; prostaglandins may also contribute. See also **Prostaglandins**. Sneezing, like pruritus, is an H_1-mediated neural reflex and can also be mediated by eicosanoids. Mucosal edema, which manifests as nasal obstruction, can be caused by H_1 stimulation as well as by eicosanoids and kinins. Increased vascular permeability is H_1 mediated. Nasal mucus secretion can be mediated by histamine both directly and indirectly through muscarinic discharge and by eicosanoids. Late-phase reactions, which manifest as nasal congestion and hyperirritability, are not mediated by histamine, but rather by inflammatory and chemotactic factors such as eicosanoids. In view of the excellent response to H_1 antagonists experienced by most patients with allergic rhinitis, histamine is a likely primary mediator of this disease. In numerous studies in this indication, H_1-receptor antagonists have proven to be extremely useful in ameliorating sneezing, nasal discharge, itchy nose and eyes, tearing, and eye redness. However, they are rather ineffective in relieving nasal congestion. Hence, decongestants such as pseudoephedrine or phenylpropanolamine have been added to H_1 antihistamines in order to provide relief of congestion.

The action of histamine on the skin is manifested by the classical Lewis triple response: edema (wheal), erythema (owing to direct vasodilation, redness) which spreads because of axon reflex (flare), and pruritus or pain. Histamine, acting through its H_1 receptor, can mediate all three pathologic components of urticaria. The mechanism by which H_1 receptors mediate pruritus is indirect and involves stimulation of sensory nerve endings. Histamine-induced cutaneous vasodilation and flushing are probably partially mediated by neurohormones; other vasoactive mediators, eg, bradykinin and prostaglandins, can contribute. The partial failure of H_1 antihistamines in the treatment of urticaria may result from the involvement of other mediators or from the presence of H_2 as well as H_1 receptors in the skin. Direct vasodilation seems to involve H_2 receptors. H_3 receptors have not yet been found in human skin. Nevertheless, H_1 antihistamines remain the primary therapy for urticaria; these reduce pruritus and the number, size, and duration of urticarial lesions.

In asthma, bronchospasm and mucosal edema can be caused by H_1 receptor stimulation. H_2 and possibly H_1 activation may be minor causes of mucus secretion. However, other mediators, such as leukotrienes, prostaglandins, bradykinin, and PAF, may be more important in asthma than histamine. Airway inflammation is not stimulated by histamine but can be caused by inflammatory factors, such as chemotactic factors. Mucus glycoprotein secretion can be induced by H_2 receptor activation, whereas increased movement of interstitial fluid into the airway lumen can be mediated by H_1 receptors. Clinical trials using classical H_1 antagonists for the treatment of asthma have yielded disappointing results. The drug concentrations achieved in the lungs with the available H_1 antagonists may not be high enough for an effective H_1 receptor blockade. H_1 antihistamines provide some relief from seasonal or chronic asthma when taken over weeks or months by patients with mild asthma, but these certainly are not drugs of first choice. High dose nonsedating H_1 antagonists, however, deserve further study as potential agents in the treatment of asthma.

All the symptoms of anaphylaxis can be reproduced by histamine. Vascular permeability manifests as angioedema, and laryngeal and intestinal edema; vasodilation leads to flushing and headache; smooth muscle contraction results in wheezing, abdominal cramping, and diarrhea. Tachycardia, often in the form of palpitations, can be caused by decreased atrioventricular node conduction time and indirectly by histamine-induced vasodilation and resultant catecholamine secretion. Reduced peripheral vascular resistance is responsible for syncope. Mucus secretion manifests as rhinorrhea and bronchorrhea. In acute anaphylaxis, treatment of first choice is epinephrine; H_1 receptor antagonists are a useful adjunctive treatment for control of pruritus, rhinorrhea, and some other symptoms.

The mechanism of itching associated with atopic dermatitis remains unknown, but histamine is almost certainly involved to some extent as histamine concentrations are increased in the skin and in the plasma of patients with this disorder. Second-generation H_1 receptor antagonists, unlike first-generation H_1 receptor antagonists, have not been uniformly found to be effective in relieving itching in atopic dermatitis, which may be related to the absence of a sedative effect.

Clinical Efficacy and Side-Effects. It is evident from the mechanism of action of antihistamines and the etiology of allergic diseases that antihistamines in no sense achieve a cure of the patient's allergy. After the administration of a therapeutic dose, a temporal blockade of the effects of histamine is obtained. Whereas classical antihistamines needed at least twice daily administration, for most of the more recently introduced agents administration once daily is sufficient.

Nevertheless, although the nonsedating H_1 antihistamines have substantially improved the acceptability and clinical efficacy of this class of compounds, these do not provide complete relief; eye disease responds less well than nasal disease, of the rhinitis symptoms nasal congestion responds poorly, breakthrough symptoms occur at high pollen counts, and only some 70% of patients report excellent to good treatment responses. Considerable research therefore still continues in the H_1 antihistamine field. New antihistamines are continually being introduced.

The classical H_1 antihistamines are not very specific, and several compounds have some degree of anticholinergic activity. Anticholinergic effects can present side effects such as dry mouth, blurred vision, and urine retention, whereas the anticholinergic action of some antihistamines is probably the reason for effectiveness in motion sickness. Interactions in the brain with noradrenergic, serotonergic, and dopaminergic uptake systems may play a role in behavioral effects of H_1 antihistamines.

At therapeutic doses, the classical H_1-receptor antagonists generally produce sedation. This usually unwanted effect is probably caused by the H_1-receptor blockade in the CNS.

The second-generation H_1 antihistamines generally present few side effects and, in particular, are considered not to cause sedation, mainly because of reduced ability to penetrate the CNS. However, terfenadine and astemizole have been associated with prolongation of the QT-interval in the electrocardiogram (ECG) and ventricular arrhythmias, generally at higher than therapeutic plasma levels.

Tropical Application. Azelastine and levocabastine have been developed for topical application. The topical antihistamines address the preference of some patients for a local treatment and allow administration of drug directly to the site required. The advantage of this therapeutic approach is likely to be in the speed of onset of symptom relief. In contrast to earlier reports of sensitization with older antihistamines locally applied to the skin, sensitization has not been reported with local application to the nose or eyes.

H_2 Agonists and Antagonists. H_2-antagonists inhibit histamine-induced gastric acid secretion. They are used widely in the treatment of peptic ulcer disease and esophageal reflux.

H_3 Agonists and Antagonists. No clear therapeutic indications have been reported for H_3 receptor ligands, yet with their use insights in the role of histamine H_3 receptors in various (patho)physiological processes have been obtained. Interesting options for therapeutic application of H_3 agonists could be in asthmatic diseases, gastrointestinal disorders, and in the regulation of sleep/wakefulness patterns.

Additional Reading

Arrang, J.-M., M. Garbarg, and J.-C. Schwartz: *Nature*, **302**, 832 (1983).
Ash, A.S.F., and H.O. Schild: *Brit. J. Pharmacol.*, **27**, 427 (1966).
Black, J.W., and co-workers: *Nature*, **236**, 385 (1972).
Emanuel, M.B. *Drugs of Today*, **22**(1), 39 (1986).
Fourneau, E., and D. Bovet: *Arch. Intl. Pharmacodyn. Ther.*, **46**, 178 (1933).
Leurs, R., and H. Timmerman: *Progress in Drug Research*, **39**, 127 (1992).
Leurs, R. and H. Timmerman: *Histamine H3 Receptor: A Target for New Drugs*, Elsevier Science, New York, NY, NY, 1998.
Middleton, E. Jr. et al.: *Principles and Practice*, 3rd Edition, The C. V. Mosby Co., St. Louis, Mo., 1988.
Norman, P.S.: *J. All. Clin. Immunol.*, **76**, 366 (1985).
Uvnäs, B. ed.: *Histamine and Histamine Antagonists*, Springer-Verlag, New York, NY, 1991.

HISTIDINE. See **Amino Acids**.

HISTOGRAM. A histogram is a graphical representation of a grouped frequency distribution. Rectangles are formed by using the class interval as the base and the frequency of the class as the height. Equal areas represent equal frequencies. See Fig. 1.

HISTOLOGY. The science that deals with the minute structure of living things: microscopic morphology. The study of the structure and functions of cells is the special province of cytology, leaving the study of special forms of cells and their association in tissues and organs as the field of histology, but histology necessarily includes much cytological matter.

The science is made up of two subordinate fields, general and special histology. In the former are considered the specialization of cells in the multicellular body and the characteristics and classification of the tissues in which they are grouped. The details of minute structure of the organs and organ systems are the materials of the latter. This field of histology is necessarily extensive and detailed, even in the study of a single species.

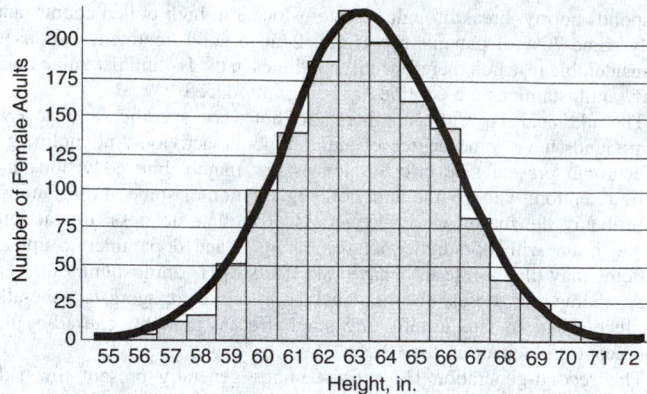

Fig. 1. Histogram showing heights of female adults prepared from a survey of approximately 1400 women. Normal distribution curve is fitted to the histogram. This familiar bell-shaped curve typifies numerous empirical distributions found in biology.

Histology recognizes five principal kinds of animal tissues, epithelium, nervous tissue, mesenchymal (connective and supporting) tissues, muscular tissue, and vascular tissue. All organs are made up of these components.

The tissues of plants are not so easily separable, but plant histologists recognize epithelial tissue, vascular tissue, supporting tissue, and parenchymous tissue. Plant tissue is studied more by its location than by the particular kinds of tissue.

HISTOMONIASIS.   Histomoniasis (histomonas, enterohepatitis, blackhead disease) is primarily an affliction of chickens and turkeys, but it also affects wild populations of peafowl, guinea fowl, pheasant, grouse, quail, and partridge. The disease occurs throughout the world and extensive outbreaks have been responsible for enormous losses of domesticated fowl; however, good sanitation reduces the number of affected animals. Because chickens are carriers of the disease, they must be separated from turkeys, which are more readily infected. Young turkeys from 3 to 12 weeks of age are particularly susceptible to the disease and may die within three days of the appearance of the first symptoms. Mortality decreases with age. Diseased birds look weak and droopy, may have diarrhea due to enterohepatitis, the head may become darkened, and death may ensue.

Histomoniasis is caused by the protozoan, *Histomonas meleagridis*, found in the liver and cecum of birds. The protozoans from the cecum are more infective than those from the liver. Transmission primarily occurs by ingestion of the eggs of the cecal worm, *Heterakis gallinarium*, which carries the protozoan and can survive several years in the soil.

A favored treatment for histomoniasis is dimetridazole [CAS: 551-92-8]. See also **Hexamitosis**. Other useful antiprotozoal agents include the thiazoles, Enheptin [CAS: 121-66-4], acetylenheptin (**1**,2-acetamido-5-nitrothiazole, aminitrozole, Enheptin-A), and nithiazide [**2**, 1-ethyl-3-(5-nitro-2-thiazolyl)urea, Hepzide], which are administered continuously into the feed for prevention and suppression. Furazolidone [CAS: 67-45-8] and ipronidazole (**3**, Ipropran) have also been used. The latter has been evaluated, in addition, as a growth promotant. In an *in vitro* study, metronidazole [CAS: 443-48-1], monensin [CAS: 17090-79-8], and tinidazole [CAS: 19387-91-8] showed good anti-histomonal activity. Arsenicals that have shown activity are 4-nitrophenylarsonic acid and carbarsone [CAS: 121-59-4]. The drugs mentioned above, in order to prevent relapses, typically require continuous administration in the drinking water of the poultry until a few days prior to their slaughter, and are responsible for reduced egg production. See also **Amebiasis (Amebic Dysentery)**; **Antiparasitic Agents, Antiprotozoals**; **Balantidiasis**; and **Coccidiosis**.

Additional Reading

Herenda, D.C., and D.A. Franco: *Poultry Diseases and Meat Hygiene: A Color Atlas*, Iowa State Press, Ames, IA, 1996.
Kaufmann, J.: *Parasitic Infections of Domestic Animals: A Diagnostic Manual*, Birkhauser Verlag, Cambridge, MA, 2001.
Wakelin, D.: *Immunity to Parasites: How Parasitic Infections are Controlled*, 2nd Edition, Cambridge University Press, New York, NY, 1996.

Web Reference

Blackhead: http://www.michigan.gov/dnr/0,1607,7-153-10370_12150_12220-26481−,00.html.
Histomoniasis, The Merck Veterinary Manual: http://www.merckvetmanual.com/mvm/index.jsp?cfile=htm/bc/203000.htm.

DANIEL L. KLAYMAN, Walter Reed Army Institute of Research

HISTONES.   Basic proteins which occur in the nuclei of both plant and animal cells. They are less basic than the protamines, having isoelectric points at about pH11. Some investigators restrict the term *histone* to only those basic proteins anatomically and chemically associated with DNA (deoxyribonucleic acids). The close associations of histones with DNA led to the hypothesis that histones might play a role in the control of genetic expression at the cellular level. Advances in molecular biology have permitted more detailed mechanisms for such control to be proposed. Histones, by blocking some areas of the DNA molecule, may permit only part of the DNA base sequences to act as templates for the formation of messenger RNA. Thus histones, by controlling messenger RNA formation, may ultimately control protein biosynthesis within the cell. Or, the primary role of histone may be structural, histone being essential for stabilizing the DNA helix, for the integration of DNA strands into more complex chromosomal structures, and for fixing and maintaining during cell division

TABLE 1. HISTOMONIASIS ANTIPROTOZOAL AGENTS

Structure number	Compound name	CAS Registry Number	Molecular formula	Structure
(**1**)	Acetylenheptin	[140-40-9]	$C_5H_5N_3O_3S$	
(**2**)	Nithiazide	[139-94-6]	$C_6H_8N_4O_3S$	
(**3**)	Ipronidazole	[14885-29-1]	$C_7H_{11}N_3O_2$	

chromosomal changes occurring during differentiation and development. The foregoing two concepts are not mutually exclusive, i.e., histone may fix chromosomal structure in a specific configuration in which the position of the histone molecules also limit RNA formation. The possibility that histones play a role in genetic mechanisms suggests the possibility that histone changes may initiate or accompany early cellular changes, leading to the formation of tumors. Further investigation is needed to ascertain whether or not tumor histones differ from those of corresponding normal tissues. See also **Cell (Biology)**.

HISTOPLASMOSIS.   Histoplasmosis is an infectious disease caused by inhaling the spores of a fungus called *Histoplasma capsulatum* which has large (8 to 20 micrometers diameter) spores with thick capsules. Histoplasmosis is not contagious; it cannot be transmitted from an infected person or animal to someone else.

Histoplasmosis primarily affects a person's lungs, and its symptoms vary greatly. The vast majority of infected people are asymptomatic (have no apparent ill effects), or they experience symptoms so mild they do not seek medical attention and may not even realize that their illness was histoplasmosis. If symptoms do occur, they will usually start within 3 to 17 days after exposure, with an average of 10 days. Histoplasmosis can appear as a mild, flu-like respiratory illness and has a combination of symptoms, including malaise (a general ill feeling), fever, chest pain, dry or nonproductive cough, headache, loss of appetite, shortness of breath, joint and muscle pains, chills, and hoarseness. A chest X-ray can reveal distinct markings on an infected person's lungs.

Chronic lung disease due to histoplasmosis resembles tuberculosis and can worsen over months or years. Special antifungal medications are needed to arrest the disease. The most severe and rarest form of this disease is disseminated histoplasmosis, which involves spreading of the fungus to other organs outside the lungs. Disseminated histoplasmosis is fatal if untreated, but death can also occur in some patients even when medical treatment is received. People with weakened immune systems are at the greatest risk for developing severe and disseminated histoplasmosis. Included in this high-risk group are persons with acquired immunodeficiency syndrome (AIDS) or cancer and persons receiving cancer chemotherapy; high-dose, long-term steroid therapy; or other immunosuppressive drugs.

Impaired vision and even blindness develop in some people because of a rare condition called "presumed ocular histoplasmosis." The factors causing this condition are poorly understood. Results of laboratory tests suggest that presumed ocular histoplasmosis is associated with hypersensitivity to *H. capsulatum* and not from direct exposure of the eyes to the microorganism. What delayed events convert the condition from asymptomatic to symptomatic are also unknown.

How is Histoplasmosis Diagnosed?

Histoplasmosis can be diagnosed by identifying *H. capsulatum* in clinical samples of a symptomatic person's tissues or secretions, testing the patient's blood serum for antibodies to the microorganism, and testing urine, serum, or other body fluids for *H. capsulatum* antigen. On occasion, diagnosis may require a transbronchial biopsy.

How is Histoplasmosis Tested?

Culturing of *H. capsulatum*. Culturing clinical specimens is a standard method of microbial identification, but the culturing process for isolating *H. capsulatum* is costly and time-consuming. To complicate matters, positive results are seldom obtained during the acute stage of the illness, except from clinical specimens from patients with disseminated histoplasmosis. Research advances in polymerase chain reaction (PCR) technology suggest that a laboratory method may soon be available that will allow direct identification of pathogenic fungi in clinical samples without the need for culturing them.

Serologic Tests. Most cases of histoplasmosis are diagnosed serologically. Because of their convenience, availability, and utility, the most widely accepted serologic tests are the immunodiffusion test and the complement-fixation test. Serologic test results are useful when positive. However, sometimes test results are negative even when a person is sick with histoplasmosis, a situation that arises especially in patients with weakened immune systems.

The immunodiffusion test qualitatively measures precipitating antibodies (H and M precipitin lines or bands) to concentrated histoplasmin. While this test is more specific for histoplasmosis (i.e., a person who is not infected with *H. capsulatum* is unlikely to have a positive test result) than the complement-fixation test, it is less sensitive (i.e., someone who is acutely infected can have a negative test result). Because the H band of the immunodiffusion test is usually present for only 4–6 weeks after exposure, it indicates active infection. The M band is observed more frequently, appears soon after infection, and may persist up to three years after a patient recovers.

The complement-fixation test, which measures antibodies to the intact yeast form and mycelial (histoplasmin) antigen, is more sensitive but less specific than the immunodiffusion test. Complement-fixing antibodies may appear in 3–6 weeks (sometimes as early as 2 weeks) following infection by *H. capsulatum*, and repeated tests will give positive results for months. The results of complement-fixation tests are of greatest diagnostic usefulness when both acute and convalescent serum specimens can be obtained. A high titer (1:32 or higher) or a fourfold increase is indicative of active histoplasmosis. Lower titers (1:8 or 1:16), although less specific, may also provide presumptive evidence of infection, but they can also be measured in the serum of healthy persons from regions where histoplasmosis is endemic. Antibody titers will gradually decline and eventually disappear months to years after a patient recovers.

Detection of *H. Capsulatum* Antigen. A radioimmunoassay method can be used to measure *H. capsulatum* polysaccharide antigen (HPA) levels in samples of a patient's urine, serum, and other body fluids. The test appears to meet the important need for a rapid and accurate method for early diagnosis of disseminated histoplasmosis, especially in patients with AIDS. HPA is detected in body fluid samples of most patients with disseminated infection and in the urine and serum of 25 to 50% of those with less severe infections.

Histoplasmin skin test. A person can learn from a histoplasmin skin test whether he or she has been previously infected by *H. capsulatum*. This test, similar to a tuberculin skin test, is available at many physicians' offices and medical clinics. A histoplasmin skin test becomes positive 2–4 weeks after a person is infected by *H. capsulatum*, and repeated tests will usually give positive results for the rest of the person's life. A previous infection by *H. capsulatum* can provide partial protection against ill effects if a person is reinfected. Since a positive skin test does not mean that a person is completely protected against ill effects, appropriate exposure precautions should be taken regardless of a worker's skin-test status. Furthermore, while histoplasmin skin test information is useful to epidemiologists, a positive skin test does not help diagnose acute histoplasmosis, unless a previous skin test is known to have been negative.

Where are *H. Capsulatum* Spores Found?

H. capsulatum grows in soils throughout the world. In the United States, the fungus is endemic and the proportion of people infected by *H. capsulatum* is higher in central and eastern states, especially along the valleys of the Ohio, Mississippi, and St. Lawrence rivers, and the Rio Grande. The fungus seems to grow best in soils having a high nitrogen content, especially those enriched with bird manure or bat droppings. The organism can be carried on the wings, feet, and beaks of birds and infect soil under roosting sites or manure accumulations inside or outside buildings. Active and inactive roosts of blackbirds (e.g., starlings, grackles, red-winged blackbirds, and cowbirds) have been found heavily contaminated by *H. capsulatum*. Therefore, the soil in a stand of trees where blackbirds have roosted for three or more years should be suspected of being contaminated by the fungus. Habitats of pigeons and bats, and poultry houses with dirt floors have also been found contaminated by *H. capsulatum*.

On the other hand, fresh bird droppings on surfaces such as sidewalks and windowsills have not been shown to present a health risk for histoplasmosis because birds themselves do not appear to be infected by *H. capsulatum*. Rather, bird manure is primarily a nutrient source for the growth of *H. capsulatum* already present in soil. Unlike birds, bats can become infected with *H. capsulatum* and consequently can excrete the organism in their droppings.

To learn whether soil or droppings are contaminated with *H. capsulatum* spores, samples must be collected and cultured. The culturing process involves inoculating mice with small portions of a sample, sacrificing the mice after 4 weeks, and streaking agar plates with portions of each mouse's liver and spleen. Then for four more weeks, the plates are watched for the growth of *H. capsulatum*. Enough samples must be collected so that small but highly contaminated areas are not overlooked. On several occasions, *H. capsulatum* has not been recovered from any of the samples collected from material believed responsible for causing illness in people diagnosed from the results of clinical tests as having histoplasmosis.

Until a less expensive and more rapid method is available, testing field samples for *H. capsulatum* will be impractical in most situations. Consequently, when thorough testing is not done, the safest approach is to assume that the soil in regions where *H. capsulatum* is endemic and any accumulations of bat droppings or bird manure are contaminated with *H. capsulatum* and to take appropriate exposure precautions.

Who is at Risk for Exposure to *H. Capsulatum* Spores?

Anyone working at a job or present near activities where material contaminated with *H. capsulatum* becomes airborne can develop histoplasmosis if enough spores are inhaled. After an exposure, how ill a person becomes varies greatly and most likely depends on the number of spores inhaled and a person's age and susceptibility to the disease. The number of inhaled spores needed to cause disease is unknown. Infants, young children, and older persons, in particular those with chronic lung disease, are at increased risk for developing symptomatic histoplasmosis.

The U.S. Public Health Service (USPHS) and the Infectious Diseases Society of America (IDSA) have jointly published guidelines for the prevention of opportunistic infections in persons infected with the human immunodeficiency virus (HIV). The USPHS/IDSA Prevention of Opportunistic Infections Working Group recommended that HIV-infected persons "should avoid activities known to be associated with increased risk (e.g., cleaning chicken coops, disturbing soil beneath bird-roosting sites, and exploring caves)." HIV-infected persons should consult their health care provider about appropriate exposure precautions that should be taken for any activity with a risk of exposure to *H. capsulatum*.

Below is a partial list of occupations and hobbies with risks for exposure to *H. capsulatum* spores. Appropriate exposure precautions, should be taken by these people and others whenever contaminated soil, bat droppings, or bird manure are disturbed.

- Bridge inspector or painter.
- Chimney cleaner.
- Construction worker.
- Demolition worker.
- Farmer.
- Gardener.
- Heating and air-conditioning system installer or service person.
- Microbiology laboratory worker.
- Pest control worker.
- Restorer of historic or abandoned buildings.
- Roofer.
- Spelunker (cave explorer).

If someone who engages in these activities develops flulike symptoms days or even weeks after disturbing material that might be contaminated with *H. capsulatum*, and the illness worsens rather than subsides after a few days, medical care should be sought and the health care provider informed about the exposure.

See also **Fungal Disease**.

Additional Reading

Benenson, A.S.: *Control of Communicable Diseases Manual*, 16th ed., American Public Health Association, Washington, DC, 1995, pp. 237–240.

Bowman, B.H.: "Designing a PCR/probe Detection System for Pathogenic Fungi," *Clin. Immunol. Newsletter*, **12**, 65–69 (1992).

Check, W.A.: "Molecular Techniques Shed Light on Fungal Genetics," *Am. Soc. Microbiol. News.*, **60**, 593–596 (1994).

Davies, S.F.: "Histoplasmosis: Update 1989," *Semin. Respir. Infections*, **5**(2), 93–104 (1990).

Deepe, G.S.: "The Immune Response to Histoplasma Capsulatum: Unearthing its Secrets," *J. Lab. Clin. Med.*, **123**, 201–205 (1994).

George, R.B. and R.L. Penn: "Histoplasmosis," in: G.A. Sarosi and S.F. Davies, *Fungal Diseases of the Lung*, Harcourt Brace Jovanovich, Orlando, FL, 1986, pp. 69–85.

Greenfield, R.A.: "Pulmonary Infections Due to Higher Bacteria and Fungi in the Immunocompromised host," *Semin. Respir. Med.*, **10**, 68–77 (1989).

Hajjeh, R.A.: "Disseminated Histoplasmosis in Persons Infected with Human Immunodeficiency Virus," *Clin. Infectious Dis.*, **21**(Suppl. 1), S108–S110 (1995).

Johnson, P.C and G.A. Sarosi: "Histoplasmosis," *Semin. Respir. Med.*, **9**(2), 145–151 (1987).

Larsh, H.W.: "Histoplasmosis," in A.F DiSalvo, ed., *Occupational Mycoses*, Lea and Febiger, Philadelphia, PA, 1983, pp. 29–41.

Mitchell, T.G.: "Systemic Mycoses," in W.K. Joklik, H.P. Willett, D.B. Amos, and C.M. Wifert, eds., *Zinsser Microbiology*, 20th ed., Appleton and Lange, Norwalk, CT, 1992, pp. 1091–1112.

Newell, F.W.: *Ophthalmology Principles and Concepts*, 7th ed., Mosby Year Book, Inc., St. Louis, MO, 1992, p. 439.

Schwarz, J.: "Histoplasmosis of the Eye," in *Histoplasmosis*, Praeger Publishers, New York, NY, 1981, pp. 317–350.

Selik, R.M. and J.M. Ward: "Effect of the Human Immunodeficiency Virus Epidemic on Mortality from Opportunistic Infections in the United States in 1993," *J. Infect. Dis.*, **176**, 632–636 (1997).

Stobierski, M.G., C.J. Hospedales, W.N. Hall, et al.: "Outbreak of Histoplasmosis Among Employees in a Paper Factory—Michigan, 1993," *J. Clin. Microbiol.*, **34**(5), 1220–1223 (1996).

Walsh, T.J., T.G. Mitchell, and D.H. Larone: "Histoplasma, Blastomyces, Coccidioides, and Other Dimorphic Fungi Causing Systemic Mycoses," in P.R. Murray: *Manual of Clinical Microbiology*, 6th ed., American Society for Microbiology Press, Washington, DC, 1995, pp. 749–764.

Wheat J., M.L.V. French, R.B. Kohler, et al.: "The Diagnostic Laboratory Tests for Histoplasmosis," *Ann. Int. Med.*, **97**(5), 680–685 (1982).

Wheat, L.J., P.A. Connolly-Stringfield, R.L. Baker, et al.: "Disseminated Histoplasmosis in the Acquired Immune Deficiency Syndrome: Clinical Findings, Diagnosis and Treatment, and Review of the Literature," *J. Lab. Clin. Med.*, **69**(6), 361–374 (1990).

HISTORICAL CLIMATE.   See **Climate**.

HISTOSOLS.   See **Soil**.

HITCHINGS, GEORGE HERBERT (1905–1998).   George Hitchings was an American pharmacologist who, together with his long-standing collaborator Gertrude Elion, introduced many new drugs, including 6-mercaptopurine and azathioprine, and developed a rational approach to drug design.

Unlike many pharmaceutical scientists who synthesized naturally occurring compounds by simple empirical methods, Hitchings believed that identifying the difference between healthy and diseased cells would make it possible to design drugs that selectively destroyed diseased cells, while leaving healthy ones intact. He applied this approach to cancer, and to viral and autoimmune diseases. See also **Drug Discovery (The History)**.

Hitchings was the son of a naval architect, whose death when he was twelve created a profound impression upon him, and drove him towards a career in medicine. He received a bachelor's, followed by a master's degree in chemistry from Washington University in 1927–1928. He went on to do a PhD in biological chemistry at Harvard, which he obtained in 1933. His doctoral thesis was on the metabolism of nucleic acids, at a time when few showed any interest in these substances that were later found to play a crucial role in cell division (for nucleic acids –natural polymers in which bases are attached to a sugar phosphate backbone –are what DNA is made of).

After nine years as a teaching fellow, first at Harvard, then at Western Reserve University, in 1942 he took up a position at the Burroughs Wellcome laboratories in Tuckahoe, New York. He rose from chief biochemist in 1946, to associate director in 1955, research director of the Chemotherapeutic Division in 1963, and director in 1968. Having grasped the significance of the discovery of the nucleic acid composition of DNA, and of DNA's role in cell division, Hitchings resumed the work he had begun as a graduate student and conceived the idea of nucleic acid analogues to inhibit DNA synthesis in cancer cells. He and Gertrude Elion, who had joined him at Burroughs Wellcome, synthesized analogues of the two bases, purines and pyrimidines, that compose nucleic acids, and found that these killed cancer cells during division. In 1951, Elion synthesized the purine analogue 6-mercaptopurine (6MP), which clinical trials run by the Sloan-Kettering Institute showed to be effective in treating childhood leukemia. Later, 6MP and another purine analogue, thioguanine, also synthesized by Elion, were used in the treatment of other acute leukemia's. See also **Elion, Gertrude Belle**

Further investigation showed 6MP to inhibit the production of antibodies in rabbits. After researchers at the Tufts and Harvard Medical Schools had succeeded in preventing the rejection of transplanted kidneys in dogs by using 6MP, Hitchings and Elion developed a derivative of 6MP, azathioprine, better known under the name Imuran, which was to play an important part not only in organ transplants, but also in the treatment of autoimmune diseases, such as rheumatoid arthritis. See also **Autoimmune Disease**; **Rheumatoid Arthritis**; and **Transplantation**.

Among other drugs to emerge out of Hitchings' laboratory were antiviral drugs such as Zovirax, which became a Burroughs Wellcome best-seller, and was effective against herpes, shingles and chicken pox. The first drug to treat AIDS, azidothymidine (better known as AZT), was also discovered there.

During his 35-year long career, Hitchings accumulated 85 patents, but also many awards. In 1988 he was awarded the Nobel Prize in Physiology or Medicine, together with Gertrude Elion and the British pharmacologist James Black. He received the Albert Schweitzer International Prize for Medicine, after being elected a member of the National Academy of Sciences in 1977. See also **Black, James Whyte (1924–present)**.

Additional Reading

George, K.H.: "George H. Hitchings (1905–): American Pharmacologist," In: Murray, E.J.: *Notable Twentieth-Century Scientists*, Vol. 1, Gale Group, Detroit, MI, 1995, pp. 933–934.

Hitchings, G.H.: "Relevance of Basic Research to Pharmaceutical Invention," *Trends in Pharmacological Sciences*, **1**, 167–168 (1980).

Hitchings, G.H.: "Rational Design of Anticancer Drugs: Here, Imminent, or Illusive?," In: Cheng, Y-C: *The Development of Target-Oriented Anticancer Drugs*, Raven Press, New York, NY, 1983, pp. 227–238.

Hitchings, G.H.: In: *American Men and Women of Science*, Vol. 3, R.R. Bowker, New Providence, NJ, 1998, pp. 874–875.

VIVIANE QUIRKE, The Royal Institution of Great Britain, London, UK

HITTORF, JOHANN WILHELM (1824–1914). Hittorf was a German physicist who computed the electricity-carrying capacity of charged atoms and molecules (ions). His computation was significant for understanding electrochemical reactions. He also discovered new properties of cathode rays.

See also **Cathode Dark Space**; **Cathode Ray**; and **Hittorf Principle**.

J. M. I.

HITTORF PRINCIPLE. An application of the Paschen law. The Hittorf principle states that discharge between electrodes in gas at a given pressure will not always occur between the closest points of the electrodes if the distance between these points corresponds to a point to the left of the minimum of the ignition potential curve.

HIV (Human Immunodeficiency Virus). See **Acquired Immune Deficiency Syndrome (AIDS)**; and **Immune System and Immunology**.

HIVES. See **Urticaria**.

H LINES. A contour along which the electromagnetic field strength is constant with respect to some reference plane.

HOARFROST. See **Precipitation and Hydrometeors**.

HOATZINS (*Aves, Galliformes*). A very strange bird, which at first sight looks like a small curassow, belonging to the suborder *Opisthocomi*. There is only one species, the Hoatzin (*Opisthocomus hoazin*). The size is approximately that of a crow, the length is 60 centimeters ($23\frac{1}{2}$ inches) and the weight is about 800 grams (28 ounces). See Fig. 1.

The hoatzins lay claim to a special position among all birds: first, as a result of their specialized diet; and second, because of the ability of their young, while they are still very undeveloped, to climb about a network of branches on all fours with the aid of the primary wing feathers, which have talons. Young hoatzins have particularly long and movable first and second digits; each has a strong claw which retrogresses later. The oldest bird so far known, *Archaeopteryx* from the Jurassic period, also had such flexible fingers with claws. We assume that it, too, used them to climb around in trees. Therefore, young hoatzins look primitive when they move around in the branches like reptiles. They not only climb, but also swim and dive with all fours when danger impels them to drop into the water.

Old hoatzins, in contrast, avoid the water and almost never touch the ground. Yet they, too, give the impression of being "primitive" when they flit about in tree branches or awkwardly fly short distances. But that has to do with their diet, in which their crop plays a peculiar part. Hoatzins primarily eat leaves of various arum types that they pick or from which

Fig. 1.   Hoatzin. (*Sketch by Glenn D. Considine.*)

they tear off large pieces with their beaks. They form the pieces into a ball in their mouths and swallow these large chunks. The leaves are ground into a fine mash in their huge crops, which are extremely muscular, with horny ridges, and are divided into several sections. The mash then passes through the small gizzard and the short intestine. The crop is fifty times as large as the gizzard and represents 13% of the entire weight of the bird. In no other bird is the crop comparatively as large. See also **Galliformes**.

HOB. A milling cutter with form-type teeth of helicoidal shape and with profiles such that conjugate surfaces on cylindrical parts may be machined by rotating the work and the hob at a constant velocity ratio. Hobs are extensively used for cutting spur gears, and hobbing is the only really precise method of cutting heavy-duty worm wheels. Two types of gear hobs are commonly used; the radial or infeed type, and the tapered or tangential feed hob. The latter is superior, particularly for hobbing worm gears with high helix angles and high pressure angle. Hobbing processes are also used for spline cutting, and for generating ratchet teeth. See also **Worm Gearing**.

The term hobbing is also used to designate a method of die sinking, in which a hardened master punch, a duplicate of the part to be formed, is pressed into an unheated die blank so that the shape of the hob is reproduced in the die impression. This method of producing die cavities is simpler than die sinking by cutting away the material, since it is considerably easier to machine the surface of the hob than to machine the die cavity. It is also advantageous in the production of multiple die cavities, since a single hob can be used for a series of duplicate dies. The process is also referred to as "hubbing."

HOBBING. See **Worm Gearing**.

HOCK (or Hough). The joint at the attachment of the foot and the leg in animals that walk on the toes (digitigrade or unguligrade), commonly applied to domestic animals. It corresponds to the ankle joint of other species. Also the back of the human knee.

HODGKIN, ALAN LLOYD(1914–1998). Alan Hodgkin was an English physiologist and biophysicist who made fundamental discoveries about the mechanisms of nervous system activity and the conduction of nerve impulses. Born in Banbury, into a Quaker family, Hodgkin read natural sciences at Trinity College, Cambridge, and apart from short research breaks and war service working on airborne radar research, spent his entire career in the Physiological Laboratory, Cambridge. Much of his research was in collaboration with his former student Sir Andrew Huxley and the two shared the 1963 Nobel Prize for Physiology or Medicine with the Australian Sir John Carew Eccles. See also **Eccles, John Carew (1903–1997)**; and **Huxley, Andrew Fielding (1917–)**.

A great deal of their work was done at the Marine Biological Association's Laboratory in Plymouth, studying the basic mechanisms by which nerve fibres transmit information, using the large diameter giant nerve fibres of the squid. The size of the fibre allowed them to insert fine glass capillary tubes and microelectrodes into the nerve. In 1939 they first measured the electrical potential of the nerve fibre both at rest and when conducting an action potential. It was not until after World War II, however, that they could continue this work. After working for six months on the physiological problems of high altitude flying, Hodgkin moved in early 1940 to airborne radar research, where he learned much, especially in electronics, that was to be invaluable to his postwar research.

In 1945 he resumed experiments with Huxley. They showed that the inside of the nerve cell axon was electrically negative, and the outside positive, during the resting potential. During the conduction of a nervous impulse, this membrane potential reversed, the inside becoming positive and the outside negative. Over the next two decades they undertook detailed experiments from which they described, in physico-chemical and mathematical terms, the mechanisms by which nerves conduct impulses, by the complex and complementary movement of electrically charged particles, sodium and potassium ions, across the nerve membrane. It was these sequential movements that produced the short-lasting electrical pulse of the nerve conduction.

Hodgkin also analysed the ionic processes that occur in the light-sensitive receptor cells of the retina, when illuminated. Working with a number of collaborators he was involved in elucidating many of the complex biochemical and biophysical mechanisms that followed the absorption of one photon of light by a single rod or cone cell of the retina. In 1948 he was elected a Fellow of the Royal Society, and in 1970 he became President of the Royal Society for five years, and received many honours and awards, including the Order of Merit in 1973.

Additional Reading

Hodgkin, A.L.: *Chance and Design: Reminiscences of Science in Peace and War*, Cambridge University Press, Cambridge, UK, 1992.

Weismann, S.: "Sir Alan Lloyd Hodgkin," In: Fox, D.M., M. Meldrum, and I. Rezak: (eds) *Nobel Laureates in Medicine or Physiology: A Biographical Dictionary*, Garland, New York, NY, 1990, pp. 255–260.

E. M. TANSEY Welcome Institute for the History of Medicine, London, UK

HODGKIN, DOROTHY MARY CROWFOOT (1919–1994).

Dorothy Hodgkin was British chemist and crystallographer who used X-ray crystal analysis to determine the structure of complex biological molecules.

Hodgkin was born in Cairo–then under British rule–where her archaeologist father worked for the Ministry of Education. Self-reliant and confident, she was raised and educated in England, and became interested in chemistry and crystals at an early age. During her final year at Somerville College, Oxford, she specialized in crystallography, taking a first-class honors degree in 1932. She then went to Cambridge to work with John D. Bernal, and completed her PhD on the chemistry and crystallography of sterols in 1937. In the same year she married historian Thomas Hodgkin, with whom she shared a happy and supportive relationship; they had three children.

In 1934 Hodgkin was persuaded to return to Somerville as a research fellow, and remained there for the rest of her career, becoming a college fellow and chemistry tutor in 1936, university lecturer (1946) and reader (1955), and finally Wolfson Research Professor of the Royal Society (1960–1977). In 1947 she became the third woman to be elected to the Royal Society, and in 1965 she was awarded the Order of Merit–the first woman since Florence Nightingale to be so honored.

Hodgkin began her crystallographic studies of large biological molecules at Cambridge, where Bernal took the first X-ray photograph of a protein crystal, pepsin, in 1934. She had a remarkable ability to interpret X-ray diffraction patterns, and found early success with the determination of the structure of cholesterol iodide (1943). In 1942 Oxford scientists purified penicillin and Hodgkin began working on its structure, a task she completed in 1949. She turned next to vitamin B_{12} (lack of which causes pernicious anaemia), a much larger and more complicated molecule, taking the first X-ray photographs of it in 1948. At the beginning she used punch-card machines to calculate electron density maps, but electronic computers introduced during the 1950s expedited the work and she finished it in 1957. In 1964 she became the third woman to win a Nobel Prize in Chemistry, in recognition of this work.

Hodgkin's third major scientific achievement arose from a project that began in 1935, when she took the first X-ray photograph of insulin. She was fascinated by the molecule, which contains nearly 1000 atoms, but was unable to determine its three-dimensional structure until 1969, when modern computers could perform the necessary calculations. Although she never wrote a book, her original papers on this and other topics are reprinted in her *Collected Works* (1994).

Hodgkin, who suffered from rheumatoid arthritis for most of her life, was a tireless campaigner for world peace and traveled widely in pursuit of this goal; she served as president of the Pugwash Conference on Science and World Affairs from 1975 to 1988. She was a kind and sympathetic person who established contacts with crystallographers all over the world; consequently her laboratory was a friendly place with a strongly international character.

Additional Reading

Dodson G, J.P. Glusker, and D. Sayre: *Structural Studies on Molecules of Biological Interest: A Volume in Honor of Professor Dorothy Hodgkin*, Clarendon Press, Oxford, UK, 1981.

Dodson, G., J.P. Glusker, S. Ramaseshan, and K. Venkatesan: *The Collected Works of Dorothy Crowfoot Hodgkin*, 3 vols. Interline Publishing, Bangalore, India, 1994.

Dodson, G: "Dorothy Hodgkin 1910–1994," *Structure*, **2**, 891–893 (1994).

Ferry, G.: *Dorothy Hodgkin: A Life*, Granta Books, London, UK, 1998.

McGrayne, S.B.: "Dorothy Crowfoot Hodgkin," *Nobel Prize Women in Science: Their Lives, Struggles, and Momentous Discoveries*, Birch Lane Press, New York, NY, 1993, pp. 225–254.

KATHERINE D. WATSON, University of Oxford, Oxford, UK

HODGKIN'S DISEASE.

A malady characterized by a painless localized enlargement of lymph nodes, usually beginning in one side of the neck, but occasionally in the axillary or inguinal-femoral region. On examination, the mass is found to be a discrete, rubbery, painless lymphadenopathy, frequently surrounded by enlarged lymph nodes. Some patients have an intermittent evening fever alternating with afebrile periods sometimes lasting days or weeks. Pruritis is usually general and when severe is a characteristic symptom. Profound anemia develops in some cases at the onset of the disease, but more commonly is seen during the course.

The cause of the disease is unknown, but is probably of viral origin because high titers to Epstein-Barr virus are found in the sera of victims. Patients with Hodgkin's disease have a defect in delayed hypersensitivity and, in general, the more advanced the clinical extent of the disease, the more complete is the loss of immunological reaction. Because of this defect in cellular immunity, Hodgkin's patients are particularly susceptible to viral and bacterial infections.

Diagnosis is by examination of excised lymph tissue; histology shows destruction of nodal architecture with proliferation of abnormal reticulum cells and the development of characteristic Reed-Sternberg giant cells. The reticulum cells are monocytic macrophages having surface receptors for crystalline fragments of immunoglobulin and a tendency to ingest IgC. Four variants are distinguished by histology: *lymphocyte predominant*, in which the infiltrate consists mainly of small lymphocytes with a small number of histocytes; the *nodular sclerosing form*, in which tumor nodules are separated by collagenous connective tissue; the *mixed cellular form*, presenting a diffuse architectural replacement with the infiltrate containing conspicuous granulocytes and plentiful Reed-Sternberg cells; and the *lymphocyte-depleted form*, in which there are Seen large numbers of H.D.-2 atypical reticulum cells, often pleomorphic, with bizarre mitoses and Reed-Sternberg cells.

Surgical excision, irradiation, and chemotherapy all have a place in the treatment of patients with Hodgkin's disease. Surgical excision can be used when the condition is localized, followed sometimes by local irradiation and/or chemotherapy. X-irradiation alone is valuable for localized disease. Massive doses in the early stages can produce dramatic results, including the rapid disappearance of masses and long remissions. Nitrogen mustard has been beneficial in patients with disseminated disease. Other drugs that have been used include chlorambucil, cyclophosphamide, and vinblastine sulfate.

Hodgkin's disease has a bimodal, age-specific incidence rate in the United States and northern Europe. There is a high rate between the ages of 15 and 34 and after the age of 50 years. The first age mode appears to be absent in Japan. Hodgkin's disease in children under 10 years of age is Seen much more frequently in some of the developing countries of Latin America and the Middle East than in the United States and is found in boys from 8 to 10 times more frequently than in girls. In the United States, the incidence of Hodgkin's disease is about 30 cases per million population per year, but the incidence varies widely by sex, age, and socioeconomic status.

The clinical course of the disease can be extremely variable. In addition, almost all patients receive treatment that may profoundly affect the course of the disease. Sometimes the treatment results in apparent cure, and sometimes it produces complications that become difficult to separate from the disease itself. However, in time, nearly all patients with untreated or uncontrollable Hodgkin's disease develop increasingly severe systemic symptoms. High continuous fever, drenching night sweats, malaise, fatigue, anorexia, and weight loss characterize the terminal picture.

Hodgkin's disease no longer can be considered inevitably fatal. No matter what the stage of disease, patients now have the potential for cure, although the probability of cure ranges between 25 and 90%.

See also **Immune System and Immunology**.

Additional Reading

Canellos, G.P. et al.: "Chemotherapy of Advanced Hodgkin's Disease with MOPP, ABVD, or MOPP Alternating with ABVD," *N. Eng. J. Med.*, 1478 (November 19, 1992).

Dana, B.W.: *Malignant Lymphomas, Including Hodgkin's Disease: Diagnosis, Management, and Special Problems*, Vol. 66, Kluwer Academic Publishers, Norwell, MA, 1993.

Davis, T.H. et al.: "Hodgkin's Disease, Lymphomatoid Papulosis, and Cutaneous T-Cell Lymphoma Derived from a Common T-Cell Clone," *N. Eng. J. Med.*, 1115 (April 23, 1992).

Hancock, S.L., R.S. Cox, and I.R. McDougall: "Thyroid Diseases after Treatment of Hodgkin's Disease," *N. Eng. J. Med.*, 599 (August 29, 1991).

Jarrett, R.F.: *Etiology of Hodgkin's Disease*, Vol. 280, Plenum Publishing Corporation, New York, NY, 1996.

Mason, D.Y. and N.L. Harris: *Human Lymphoma: Clinical Implications of the Real Classification*, Springer-Verlag, Inc., New York, NY, 1999.

Melby, J.C. and A.L. Vickery, Jr.: "A 27-Year-Old Woman with Hodgkin's Disease and an Adrenal Mass," *N. Eng. J. Med.*, 400 (February 7, 1991).

Urba, W.J. and D.L. Longo: "Medical Progress: Hodgkin's Disease," *N. Eng. J. Med.*, 678 (March 5, 1992).

Wotherspoon, A.C.C.: *Lymphoma*, Cold Spring Harbor Laboratory Press, Cold Spring Harbor, NY, 1999.

Web Reference

http://www.leukemia.org/ The Leukemia & Lymphoma Society.

R.C. VICKERY, M.D., D.Sc., Blanton/Dade City, FL

HODOGRAPH. In general (mathematics), the locus of one end of a variable vector as the other end remains fixed. A common hodograph in meteorology represents the vertical distribution of the horizontal wind.

HOFFMAN, ROALD (1937–). A Polish-born chemist who won the Nobel Prize for chemistry with Kenichi Fukui in 1981. His work involved applying the theories of quantum mechanics to predict the course of chemical reactions.

HOFMANN, AUGUST WILHELM (1818–1892). A German organic chemist who studied under Liebig. While professor of chemistry at the Royal College of Chemistry in London, he did original research on coal-tar derivatives that later led him into a study of organic dyes. Perkin, who first synthesized the dye mauveine in England, was a student of Hofmann. When the latter returned to Germany he continued his work in the field of dyes, which became the basis of German leadership in synthetic dye manufacture that continued until World War I.

HOFMANN DEGRADATION. Formation of an olefin and a tertiary amine by pyrolysis of a quaternary ammonium hydroxide; useful for the preparation of some cyclic olefins and for opening nitrogen-containing ring compounds.

HOFMANN ISONITRILE SYNTHESIS. Formation of isonitriles by the reaction of primary amines with chloroform in the presence of an alkali; the odor of the isocyanide is a test for a primary amine.

HOFMANN RULE. When a quaternary ammonium hydroxide containing different primary alkyl radicals is decomposed, the least-substituted olefin is formed preferentially.

HOFMANN'S REACTION. Reaction used for preparation of a primary amine from an amide by treatment with a halogen (usually bromine) and caustic soda. The resulting amine has one fewer carbon atom than the amide used.

HOGBACK. Ridge-like topographic features, the result of the differential erosion of highly tilted hard and soft strata. The steeper, or dip-slope, side is developed on the harder or less soluble formation, while the gentler slope is developed on the opposite side, on the softer rocks.

HOHLRAUM. In radiation thermodynamics, a cavity whose walls are in radiative equilibrium with the radiant energy within the cavity. This idealized cavity can be approximated in practice by making a small perforation in the wall of a hollow container of any opaque material. The radiation escaping through such a perforation will be a good approximation to black-body radiation at the temperature of the interior of the container.

HOIST. Any device for lifting materials, weights, articles, etc., may be called a hoist. Hoists often compose a part of other apparatus whose purpose may extend to movement of material other than vertically. For example, the bridge crane incorporates within it a hoist for vertical lift. The energy required for lifting is derived ultimately from a number of various sources. For example, in the hoisting field one finds such varied power sources as compressed air, internal combustion engines, hydraulic power, steam and electric power. The pneumatic drives may be either a direct lift supplied by air acting on a piston connected directly to the load, or it may be employed in compressed air engines, whose crankshaft is geared to the hoisting apparatus. In the internal combustion engine type hoist, the gasoline engine is generally used for the light-capacity hoist, and the Diesel engine for heavier hoists. It has the advantage over other drives for portable service, such as locomotive cranes, and power shovels.

The essential parts of a hoist are a rope or chain, which is wrapped around a drum or drive sheave. A hook, grapnel magnet, or other device for handling the load is attached to the free end. The rotation of the drum winds up the rope, thus shortening the distance between the drum and the load. If the drum is fixed in position over the load, naturally the load must be hoisted. To drive the drum, one of the power supplies just mentioned is connected with the drum through a suitable speed-reducing, torque-increasing mechanism. A gear train is often used. These component parts when supplied with a brake controlling the speed during lowering of weights, are the essential elements of all hoists except the direct-acting.

HOLLERITH. Pertaining to a widely used system of encoding alphanumeric information onto cards (described by American National Standard ANSI X3.26–1970). The term Hollerith cards is synonymous with punch cards. Such cards were first used in 1890 for the United States Census and were named after Herman Hollerith, their originator. See also **Calculators**.

HOLLYHOCK. See **Malvaceae (Mallow Family)**.

HOLLY, ROBERT WILLIAM (1922–1993). Robert Holly was an American biochemist and Nobel Prize winner who developed techniques for determining the structure of nucleic.

Holley, the son of teachers, was born at Urbana, Illinois on 28 January 1922. He began the study of chemistry at the University of Illinois in 1938 and obtained a BA in 1942. He then did graduate work on penicillin at Cornell University, where he obtained his PhD in organic chemistry in 1947. After a brief period of teaching at the State College of Washington, Seattle, he returned to Cornell in 1948 as Professor of Organic Chemistry, switching to biochemistry in 1957. During this time he also worked for the U. S. Plant, Soil and Nutrition Laboratory at Geneva, NY. In 1964 Cornell gave him a laboratory for molecular biology. In 1968 he became a resident fellow at the Salk Institute, La Jolla while also acting as adjunct professor at the neighboring University of California, San Diego.

In 1955, inspired by Watson and Crick's work on DNA, Holley began to work on protein synthesis. Following a decade's research, he succeeded in isolating a low molecular weight nucleic acid that he identified as pure alanine transfer RNA (tRNA). He then spent five years determining its sequence of amino acids, announcing the result in 1965. He deduced a mechanism whereby tRNA bound itself to a matching area of messenger RNA (mRNA). This process aligned amino acids so that a series of nucleotides in DNA was translated into a series of amino acids that constituted specific proteins. In 1968 Holley shared the Nobel Prize in Physiology or Medicine with H. G. Khorana and Marshall Nirenberg for his determination of the complete nucleotide sequence of the alanine transfer ribonucleic acids. His later research at the Salk Institute involved a search for factors that affected cell division and the growth of cells. He succeeded in sequencing a number of growth factors. His work proved important for pharmaceutical companies who were searching for anti-cancer drugs. See also **Genetics and Gene Science (Classical)**; and **Nirenberg, Marshall Warren (1927–Present)**.

Additional Reading

Holley, R.: "An Alanine-dependent, Ribonuclease-inhibited Conversion of AMP to ATP, and its Possible Relationship to Protein Synthesis," *Journal of the American Chemical Society*, **79**, 658–662 (1957).

Holley, R.: "Structure of RNA," *Science*, **147**, 1462–1465 (1965).

Holley, R.: "The Nucleotide Sequence of a Nucleic Acid," *Scientific American*, **214**, 30–40 (1966).

Holley, R.: "Alanine Transfer RNA," *Nobel Lectures Physiology or Medicine 1963–1970*, Elsevier, Amsterdam, Netherlands. 1972, pp. 319–340

Magill, F.N.: *The Nobel Prize Winners: Physiology or Medicine, Vol. 2: 1944–1969*, Salem Press, Englewood Cliffs, NJ, 1991, pp. 1007–1017

W. H. Brock, University of Leicester, Leicester, UK

HOLLY TREES AND SHRUBS. Of the family *Aquifoliaceae* (holly family), genus *Ilex*, there are numerous species of hollies and many hybrids and cultivars, making both nomenclature and generalization difficult. The plants may be deciduous or evergreen. They often are spiny with leathery leaves. The flowers frequently are white, usually polygamous. They bear small fruit and can withstand full sun or partial shade. The hollies tend to be more resistant to pests than most plants. Some of the important varieties include:

American holly	*Ilex opaca*
Azorean holly	*I. perado*
Chinese holly	*I. pernyi*
Dahoon holly	*I. cassine*
English holly	*I. aquifolium*
Highclere hybrid holly	*I. × altacierensis*
Longstalk holly	*I. pedunculosa*
Posshmhaw holly	*I. decidua*
Tarajo holly	*I. latifolia*

Depending upon height, the American holly may be considered a shrub or a tree. The plant can range from about 15 to 30 feet (4.5 to 9 meters) in height, although as shown in Table 1, under favorable conditions, the plant can develop into a very sizeable tree. The foliage may be described

TABLE 1. RECORD HOLLY TREES IN THE UNITED STATES[1]

Specimen	Circumference[2] Inches	Circumference[2] Centimeters	Height Feet	Height Meters	Spread Feet	Spread Meters	Location
DAHOON HOLLY							
Ilex cassine (1984)	55	140	40	12.2	36	11	Florida
Ilex cassine (1995)	32	81	68	20.7	31	9.4	Florida
Ilex cassine (1995)	54	137	42	12.8	36	11	Florida
Ilex cassine (1994)	34	86	61	18.6	34	10.4	Florida
Myrtle dahoon holly (1998) (*Ilex myrtifolia*)	77	196	55	16.8	35	10.7	Florida
GALLBERRY HOLLY							
Large gallberry holly (1999) (*Iled coriacea*)	10	25	27	8.2	12	3.7	Virginia
HOLLY							
American holly (1999) (*Ilex apaca*)	125	318	76	23.2	48	14.6	Alabama
Carolina holly (1995) (*Ilex ambigua*)	18	46	27	8.2	30	9.1	Florida
Georgia holly (1998) (*Ilex longipes*)	5	13	12	3.7	12	3.7	Tennessee
Sarvis holly (1999) (*Ilex amelanchier*)	8	20	14	4.3	18	5.5	Virginia
MOUNTAIN HOLLY							
Nemopanthus callinus (1982)	13	33	20	6.1	10	3.04	Michigan
POSSUMHAW HOLLY							
Ilex decidua (1995)	36	91	42	12.8	52	15.8	South Carolina
WINTERBERRY HOLLY							
Common winterberry holly (1994) (*Ilex verticillata*)	24	62	13	3.96	16	4.87	Virginia
Common winterberry holly (1994) (*Ilex verticillata*)	21	53.3	13	3.96	16	4.9	Virginia
Mountain winterberry holly (1995) (*Ilex montana*)	38	97	30	9.1	30	9.1	New York
Smooth winterberry holly (1999) (*Ilex laevigata*)	8	20	18	5.5	21	6.4	Virginia
YAUPON HOLLY							
Ilex vomitoria	19	48	32	9.8	21	6.4	Florida

[1] From the "National Register of Big Trees," American Forests (by permission).
[2] At 4.5 feet (1.4 meters).

as being of a bronze-green or olive-green color. The leaves are glossy and quite spiny, but less so than the English species. The leaves are from 2 to 3 inches (5 to 7.6 centimeters) in length. The fruit is a scarlet red, sometimes (in the 'Xanthocarpa') a bright yellow, and a little over 1/4 inch (0.6 centimeters) in diameter. It is berrylike on short stems and often clings to the plant throughout most of the winter months. With proper care, there are numerous areas in the United States where the plant does quite well. The natural occurrence generally follows the coastal regions. Sheltered locations are preferred. The plant ranges from Massachusetts southward into Florida and westward to the Mississippi Valley south of lower Indiana and Illinois.

As the name suggests, Azorean holly is found on the Azores and also on the Canary Islands. This species is a small evergreen tree with dark green foliage and deep red berries. *I. pernyi* is found in central and westward China and is a narrow tree of pyramidal form that can rise to a height of about 30 feet (9 meters). The leaves are small, leathery, and of a lustrous dark green color. The tree bears clusters of small red berries.

The Dahoon holly is found in the southeastern United States and is characterized by a somewhat heavier trunk than found on most hollies. The leaves are evergreen and narrow, about 2 to 4 inches (5 to 10 centimeters) in length. Their color is dark green. The flowers and fruit are similar to the American holly. The plant ranges from the southern part of Virginia to Florida and along the Gulf coast west to Louisiana.

English holly occurs naturally in western Asia, northern Africa, and southern Europe, as well as the British Isles from which it derives its name. However, several forms of this holly are not so hardy in England as they may be on the continent. Because of the numerous hybrids, a great variety of leaf and fruit colorations, as well as other characteristics of the shrubs and trees, is obtainable. Some of the more important varieties include: Perry's weeping silver holly ('Argenteomarginata Pendula'), which has silver foliage and lots of berries; the silver milk-boy ('Argenteo-Medico Picta'), which grows to a height of about 30 feet (9 meters) and has dark green, spiny leaves with cream-colored spots in their central portion; the golden queen ('Aurea Regina'), with yellow-edged dark green leaves; and the silver hedgehog holly ('Ferrox Argentea'), which has leaves featuring white spines and margins, and ranging up to 15 feet (4.5 meters). Other varieties include the 'Bacciflavia,' the 'Crispa,' the 'Elegantissima,' the 'Ferox,' and the 'Hastata.'

The highclere hybrid hollies are known for their vigor. They have quite large evergreen leaves. They are known for their toleration of industrial and seaside environments. They range in height from small bushes to trees of 50 feet (15 meters). These hybrids were obtained by crossing the English holly with the Azorean holly. Some of the more important varieties include: 'Camellifolia,' a tree of conical contour, characterized by a purple bark, almost spineless evergreen leaves that are purple when young, later turning a dark green; the 'Golden King,' which has green leaves with yellow edges, nearly spineless; the 'J.C. van Tol,' almost spineless leaves of dark-green color and produces large quantities of berries; the 'Lawsoniana,' which has large leaves with yellow borders and marbleized centers; the 'Purple Shaft,' which is known for its vigor and large quantities of berries; and the 'Silver Sentinel,' which has mottled leaves that are flat and almost spineless.

The longstalk holly is found in Japan. It ranges from a shrub to a small tree of about 30 feet (9 meters) in height. The plant has evergreen leaves and small red fruits.

The Possumhaw holly is found in the southeastern United States. It is also sometimes referred to as the swamp holly. Normally, it is a small shrub or tree, but as shown by the Table 1, the plant can attain very respectable dimensions under favorable conditions. The leaves are a lustrous deep green, deciduous, and from $1\frac{1}{2}$ to 3 inches (7.6 centimeters) in length. The flowers are similar to those of the American holly. Its natural range is between the Atlantic coast and the Appalachian Mountains south of Virginia and into western Florida and westward to Arkansas, Missouri, and Texas.

The Tarajo holly is found in Japan. This species can attain a height of 60 feet (18 meters) or more and features the largest leaves of any holly. The leaves are evergreen of a dark green color, yellow underneath. The fruit occurs in large numbers of orange-red clusters.

HOLMIUM. [CAS: 7440-60-0] Chemical element symbol Ho, at. no. 67, at. wt. 164.93, tenth in the Lanthanide Series in the periodic table, mp 1,474 °C, bp 2700 °C, density 8.795 g/cm3 (20 °C). Elemental holmium has a close-packed hexagonal crystal structure at 25 °C. The pure holmium is silver-gray in color, slow to tarnish or oxidize at room temperature in normal atmospheres. Even at relatively high temperatures, the metal is slow to oxidize. Under a vacuum of about 10 torr, holmium will react when hot with water vapor, CO_2, NH_3, and hydrocarbons. Holmium is soft and can be worked by conventional equipment. There is one natural isotope of holmium, ^{165}Ho, and 18 artificial isotopes have been produced. The natural isotope is not radioactive. In terms of abundance, holmium is present on the average of 1.2 ppm in the earth's crust, ranking ahead of bismuth, antimony, cadmium, and mercury in potential availability. The element was first identified by P.T. Cleve and J.L. Soret in 1879. The metal has a low acute-toxicity rating. Electronic configuration $1s^2 2s^2 2p^6 3s^2 3p^6 3d^{10} 4s^2 4p^6 4d^{10} 4f^{10} 5s^2 5p^6 5d^1 6s^2$. Ionic radius Ho^{3+} 0.901 Å. Metallic radius 1.766 Å. First ionization potential is 5.43 eV; second 13.9 eV. Other important physical properties of holmium are given under **Rare-Earth Elements and Metals**.

Holmium occurs in apatite, xenotime, and yttrium and heavy rare-earth minerals. The element of a purity of 99.9% can be obtained through organic ion-exchange techniques. Supplies of holmium are available commercially as the result of yttrium production. To date, the applications for holmium have been very limited. When added to orthoferrites, it has shown promise for use in electronic circuits. Uses in semiconductors, lasers, thermoelectric devices and phosphors currently are being studied.

See references listed at ends of entries on **Chemical Elements**; and **Rare-Earth Elements and Metals**.

K. A. Gschneidner, Jr., and B. Evans, Iowa State University, Ames, IA

HOLOCRYSTALLINE. The term applied by petrologists to igneous rocks composed entirely of crystals; in contradistinction to igneous rocks which are partly or entirely composed of natural glass, such as obsidian.

HOLOGRAPHY. The technique of holography is similar to photography in many respects, yet it is fundamentally different. With photography, one generally records, by means of lens and film, the two-dimensional irradiance distribution in the image of an object. With holography, one records not the optically formed image of an object, but the object wave itself. This wave is recorded (frequently on photographic film) in such a way that a subsequent illumination of this record, called a *hologram*, reconstructs the original object wave. A visual observation of this reconstructed wavefront then yields a view of the object that is practically indiscernible from the original, including three-dimensional parallax effects. The process was discovered by Gabor[1] (England) in 1948. It was then identified as a two-step method of optical imagery. During the past couple of decades, holography has become widely known and a limited number of practical uses for it have been developed. This later progress is attributed to the general availability of the laser, with the outstanding temporal and spatial coherence of its light. Much of the work in the laser to was carried out by Upatnieks and Leith at the University of Michigan during the early 1960s.

With reference to Fig. 1, one starts with a single, monochromatic beam of light that has originated from a very small source. This single beam is split into two components, one of which is directed toward the object and the other to a suitable recording medium, most commonly a photographic emulsion. The component that is incident on the object is scattered by it, and this scattered radiation, now called the object wave, impinges on the recording medium. The wave that proceeds directly to the recording medium is called the *reference wave*. Since the object and reference waves originate from the same source, they are mutually coherent and form a stable interference pattern when they meet at the recording medium. The detailed record of this interference pattern constitutes the hologram.

Types of Holograms

When the hologram is illuminated with a beam similar to the original reference wave, it modulates the phase and/or amplitude of the illuminating wave in such a way that the transmitted wave divided into three separate components, one of which exactly duplicates the original object wave.

If the two interfering beams are traveling in substantially the same direction, the recording of the interference pattern is said to be a *Gabor hologram* or *in-line hologram*. If the two interfering beams arrive at the recording medium from substantially different directions, the recording is a *Leith-Upatnieks* or *off-axis hologram*. If the two interfering beams are traveling in essentially opposite directions, the recorded hologram is said to be a *Lippmann* or *reflection hologram*, first invented by Denisyuk.

[1] For which he received the Nobel Prize in physics (1971).

Electromagnetic radiation is most commonly used, although acoustic radiation can be used. The most common electromagnetic radiation employed is light, but holograms have also been recorded successfully with electron beams, x-radiation, and microwaves.

Holograms can be classified by the way they diffract light. In an *amplitude hologram*, the varying irradiance distribution of the interference pattern is recorded as a density variation of the recording medium. In this type of hologram, the illuminating wave is always partially absorbed, i.e., the illuminating wave is *amplitude-modulated*. In the *phase hologram*, a *phase modulation* is imposed on the illuminating beam which, in turn, results in diffraction of the light. Phase modulation occurs when the optical path (thickness × index) varies with position. A phase hologram results from either relief-image or index variation, or both.

Either phase or amplitude holograms can be classified further as *Fresnel holograms* or as *Fraunhofer holograms*. Generally speaking, if the object is reasonably close to the recording medium, say just a few hologram or object diameters distant, the field at the hologram plane is the Fresnel diffraction pattern of the object. A hologram recorded in this manner is termed a *Fresnel hologram*.

If the object and hologram are separated by many object or hologram diameters, the field at the hologram due to the object alone is the Fraunhofer diffraction pattern of the object. A hologram recorded in this manner is termed a *Fraunhofer hologram*.

Any of these hologram types may be recorded as either a *thick* or a *thin* hologram. A thin hologram is one for which the thickness of the recording medium is thin compared to the space between the recorded interference fringes. A thick or volume hologram is one in which the thickness of the recording medium is of the order of or greater than the spacing of the recorded fringes.

Conceptually, the simplest form of an off-axis hologram is one for which the object is just a single, infinitely distant point so that the object wave at the recording medium is a plane wave. If the reference wave is also plane, and incident on the recording medium at an angle to the object wave, the hologram will consist of a series of Young's interference fringes. These recorded fringes are equally spaced straight lines running perpendicular to the plane of incidence. Since the hologram consists of a series of alternating clear and opaque strips, it is in the form of a diffraction grating. When the hologram is illuminated with a plane wave, the transmitted light consists of a zero-order wave traveling in the direction of the illuminating wave, plus two first-order waves. The higher diffracted orders are generally missing or very weak, inasmuch as the irradiance distribution of a two-beam interference pattern is sinusoidal. As long as the recording is essentially linear (irradiance proportional to final amplitude transmittance), the hologram will be a diffraction grating varying sinusoidally in amplitude transmittance, and only the first diffracted orders will be observed. One of these first-order waves will be traveling in the same direction as the object wave. This is the reconstructed wave.

Holographic Recording

The recording of a hologram and the subsequent reconstruction is shown in Fig. 1. In Fig. 1(a), the laser beam is first expanded and then divided by a mirror, which directs part of the beam directly onto the photographic plate; the rest of the light is reflected from the object. After processing, the hologram plate may be replaced in its original position (Fig. 1(b)), and the object removed. The light diffracted by the hologram forms, in part, the same wavefront that was originally scattered by the object. A viewer looking through the hologram will See an undistorted view of the object, just as if it were still present.

In addition to the *virtual* or *primary image*, a real, or *conjugate image* will be formed on the observer's side of the hologram. This image will appear unsharp and highly distorted, and it will also be inverted in depth, i.e., reversed front to back, as shown in Fig. 1(b). However, a distortion-free real image can be formed by changing the position of the illuminating beam so that all of the rays of the reference beam are reversed in direction. In this way, an undistorted, real, three-dimensional image of the object scene appears in front of the hologram, as shown in Fig. 1(c).

Holograms may be recorded with diverging, parallel, or converging reference beams. If care is taken to maintain the recording geometry during reconstruction, it is possible to form holograms with an arbitrary reference beam, the only requirement being that it be coherent with the object beam.

Color holograms can be produced by recording three separate holograms on a single photographic plate, each in a different color. Subsequent illumination with a three-color beam yields three separate wavefronts, one in each of the three colors representing the portion of the object corresponding to that color.

Holograms also can be made that can be viewed in reflection. This is done by allowing the reference and object beams to enter the recording medium from opposite sides. The fringes formed are planes lying approximately parallel to the plane of the hologram. When such a hologram is illuminated by a beam similar to the reference wave, a reflected wave is formed which exactly duplicates the object wave. The image is viewed in reflected light. This type of hologram can be illuminated with white

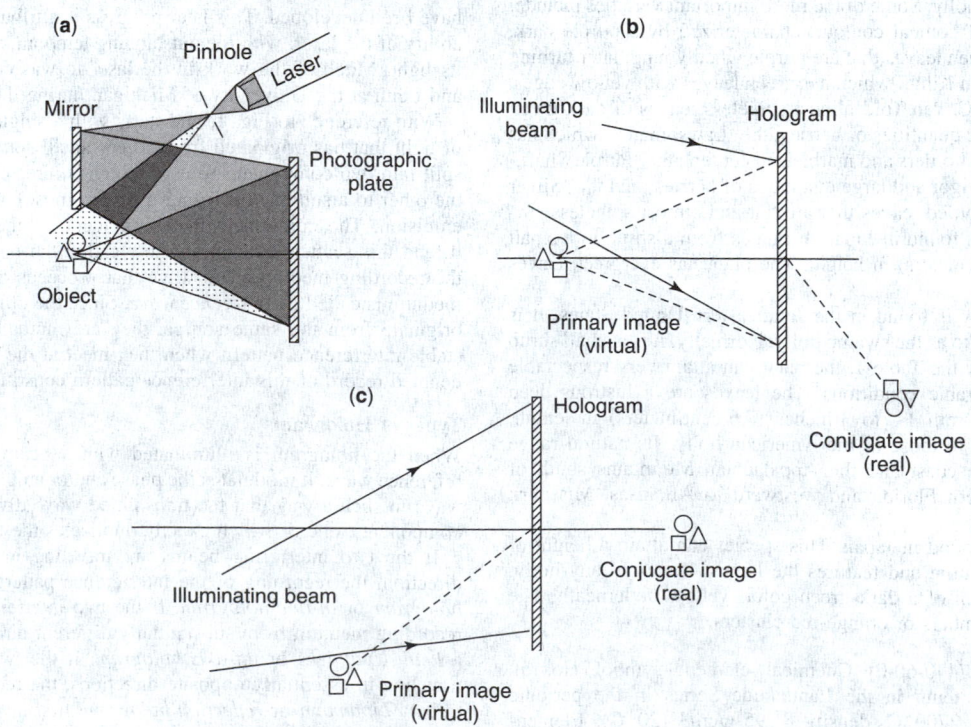

Fig. 1.   A typical holographic arrangement: (**a**) Recording the hologram; (**b**) reconstructing the primary object wave; (**c**) reconstructing an undistorted conjugate wave.

light. The interference planes filter the light by acting as a $\lambda/2$ multilayer interference filter, in the same way as in Lippmann color photography.

One of the most striking aspects of the modern hologram is the three-dimensional image that it is capable of producing. The three-dimensional image indicates that there is a large amount of information contained in a single hologram—much more than is contained in a conventional photograph of the same size. Because of the many perspectives available, the hologram is well suited to display purposes. With a hologram, one can present all of the observable characteristics of a three-dimensional object clearly and concisely. Complex molecular or anatomical structure can be simply presented with a single holographic image, with little chance of error or misinterpretation on the part of the viewer. Thus holograms may reduce the number of conventional drawings or photographs to illustrate a single object. It has been proposed that the use of holograms in textbooks would be an aid to readers, particularly in fields where three dimensions are important. Holograms can be made to be viewed with a small penlight and a colored filter.

Applications of Holography

Early applications of holography, essentially prior to the early 1980s, were more of a novel than scientific nature. In 1984, Chang pointed out that holography is more than simply a curiosity related to three-dimensional photography. It is a technology involving the precise structure of light waves, with advanced implications for solutions to engineering problems. Engineering applications of holography utilize the interference patterns created by superimposing holographic images from a target made under slightly different conditions. The patterns can be examined visually or the data can be digitized for computer analysis. In realtime holography, the object is viewed through a hologram of itself. Double-exposure holography involves recording the interference patterns obtained from the same target before and after distortion. Time-average holograms, employed for vibration analysis for example, are made by exposing the plate while the object is driven in resonance.

As described by Chang, some of the more recently developed applications for holography include: (1) Analysis of dimensional instability when an object is stressed. Small distortions resulting from the applications of forces, changes in environment, and other factors yield interference fringes in the superimposed images equivalent to strain contour lines. The

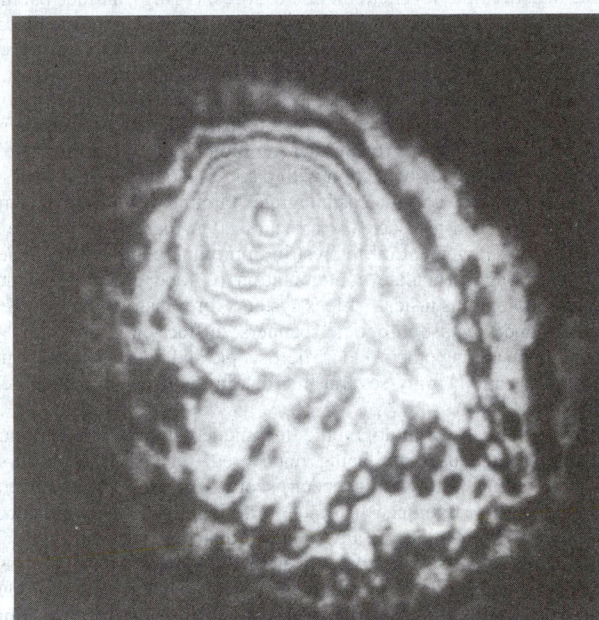

Fig. 2. Holographic interferometry used here to determine delamination in the bonding between skin and honeycomb structure of a composite helicopter rotor blade. (*After M. Chang, Newport Corp.*)

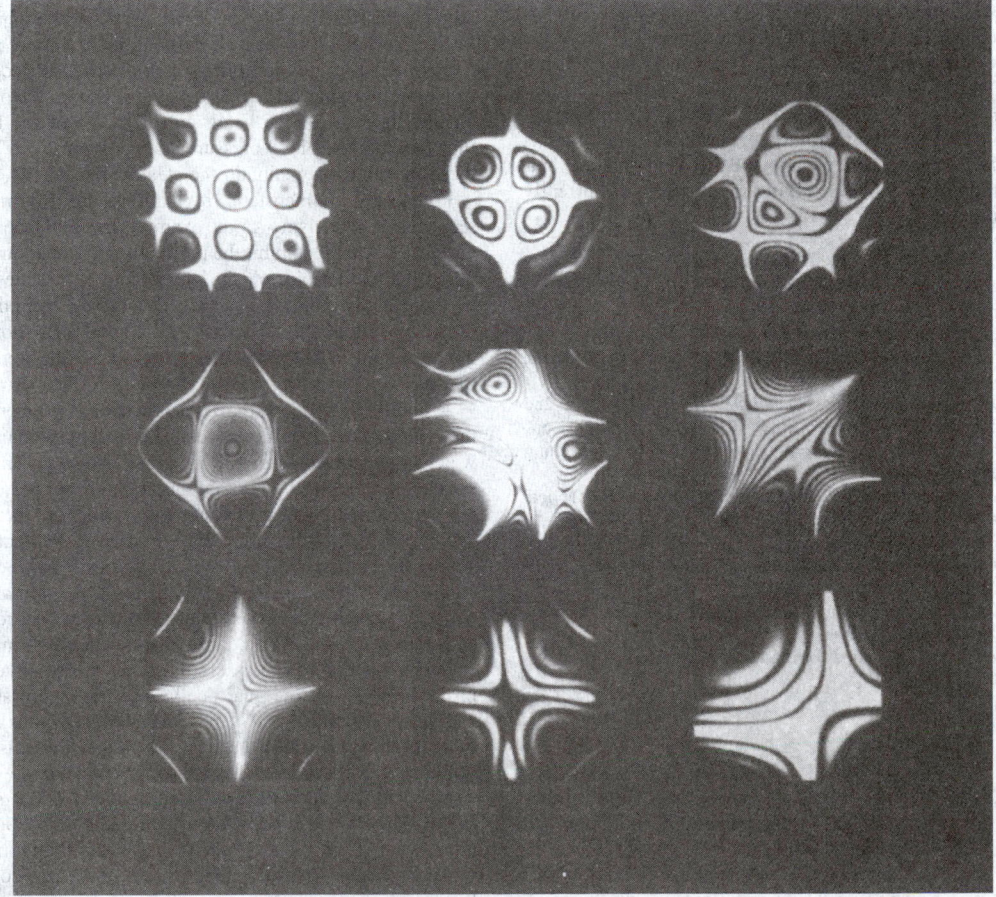

Fig. 3. Examples of resonant modes of plate as obtained through time-averaged holographic interferometry. (*After M. Chang, Newport Corp.*)

approach has been used in connection with thermally induced changes in large-dish antennas at very low temperatures, for observing the performance of miniature gyroscopes, and for examining deformations due to pressure in a vessel or pipe. (2) Checking for cracks in welding by seeking fringe discontinuity across a seam. (3) Determining voids in layered objects, including the inspection of composite aircraft components, clutch plate facings, multiple-layer circuit boards, tires, O-rings, antique paintings, nuclear fuel rods, and detecting the delamination in the composite blades of a helicopter, among others. See Fig. 2. (4) Vibration analysis of such components and subsystems as turbine blades, loudspeakers, rocket castings, and automobile engines. See Fig. 3. (5) Studies for flow visualization in connection with air foils, plasmas, and combustion flames — as an alternate to Schlieren photography. (6) Studies of biological and crystal growth. (7) Image processing in connection with pattern recognition in robotics and in finding defects in parts, such as integrated circuits. (8) Monitoring materials properties, such as index of refraction. (9) Analysis of particle size distribution and movement, of interest in improving combustion efficiency and for checking particulate contamination of food and drug products. (10) Use of holograms instead of lens systems for transforming light beams and images and thus use in optical scanners, diffraction gratings, and pattern generators, among others. (11) Microscopic and interferometric studies where objects may be only several microns in diameter — without depth-of-field limitations.

As the potential for holographic techniques become better known, more sophisticated uses are being uncovered. One such application is picosecond holographic-grating spectroscopy. As reported by Wiersma and Duppen (1987 reference listed), interfering light waves produce an optical interference pattern in any medium that interacts with light. This modulation of some physical parameter of the system acts as a classical holographic grating for optical radiation. When such a grating is produced through interaction of pulsed light waves with an optical transition, a transient grating is formed whose decay is a measure of the relaxation time of the excited state. Transient gratings can be formed in real space or in frequency space, depending on the time ordering of the interfering light waves. The two gratings are related by a space-time transformation and contain complementary information on the optical dynamics of the system. The status of a grating can be probed by a delayed third pulse, which diffracts off this grating in a direction determined by the wave vector difference of the interfering light beams. This generalized concept of a transient grating can be used to interpret many picosecond-pulse optical experiments on condensed-phase systems. In their paper, Wiersma and Duppen illustrate some low-temperature experiments. The impact of nonlinear photon-interference spectroscopy on the field of transient-grating and more generally on four-wave mixing spectroscopy is currently significant and is expanding.

Additional Reading

Brcic, V.: *Application of Holography and Hologram Interferometry to Photoelasticity*, Springer-Verlag Inc., New York, NY, 1975.

Chang, M.: "Holography," *Instrumentation Technology*, 39–41 (March 1984).

Fournier, J.M., H. Weber, T. Asakura, et al.: *Holography: The First 50 Years*, Springer-Verlag, Inc., New York, NY, 2000.

Hariharan, P.: *Optical Holography: Principles, Techniques, and Applications*, 2nd Edition, Cambridge University Press, New York, NY, 1996.

Mehta, P.C. and V.V. Rampal: *Lasers and Holography*, World Scientific Publishing Company, Inc., Riveredge, NJ, 1993.

Staff: Optical Society of America *1996 Technical Digest of Holography*, Optical Society of America, Washington, DC, 1996.

Tonomura, A.: *Electron Holography*, Springer-Verlag, Inc., New York, NY, 1999.

Vacca, J.: *Holograms and Holography: Design Techniques and Commercial Applications*, Charles River Media, Rockland, MA, 2000.

Voelkl, E., L.F. Allard, D.C. Joy: *Introduction to Electron Holography*, Perseus Publishing Company, Boulder, CO, 1999.

Wiersma, D.A. and K. Duppen: "Piosecond Holographic-Grating Spectroscopy," *Science*, **237**, 1147–1153 (1987).

Web Reference

http://www.holo.com/

HOLOHEDRAL CRYSTAL. A crystal in which the full number of faces are developed, corresponding to the maximum and complete symmetry of the system. See also **Mineralogy**.

HOLOTHUROIDEA. The sea cucumbers, a class of the phylum *Echinodermata*.

These animals differ from other echinoderms in several particulars: (1) The principal axis is elongated and the animal rests on its side. (2) The body wall is soft because of the reduction of the calcareous ossicles. (3) A branching respiratory tree extends from the alimentary tract into the body cavity.

Sea cucumbers are used as food in the Oriental region. They are dried for the market and in this form are called trepang or bêche-de-mer.

The class includes five orders:

Order *Aspidochirota*. Tropical species with shield-shaped tentacles. In shallow water.

Order *Elasipoda*. Benthonic species of deep water.

Order *Dendrochirota*. Shallow water species with branching tentacles.

Order *Molpadonia*. Burrowing species. Tentacles unbranched or slightly branched.

Order *Apida*. (*Synaptida, Paractinopoda*.) Burrowing species without respiratory trees.

HOLOTYPE. A term used by biologists and paleontologists to mean the specimen to which all others should ultimately refer to determine the species. The holotype does not necessarily have to be the originally described species (type) and frequently is not.

HOMEOSTASIS. Maintenance of the steady state. As applied to living organisms, this refers to the many adjustments constantly being made to keep the organism in a rather constant environment internally in spite of the fact that there may be many variations in external environment. Life within individual cells can continue only within a rather narrow range of conditions, and each form of life possesses many self-regulating systems whereby it can maintain a favorable internal environment in spite of the great variations in its surroundings. When a person goes from bright sunlight into a dark room the eyes undergo certain changes as a result of automatic internal adjustments which permit the eyes to function in spite of the greatly reduced light intensity. A person living in arctic regions of the north and persons on a tropical beach have an internal temperature that does not vary more than a fraction of a degree. Internal thermostatic adjustments regulate the body temperature to keep it at such a constant level. The human brain must have a blood supply at a constant pressure; a slight drop in pressure brings a "blackout" and too great a pressure will cause the bursting of capillaries and a "stroke." Homeostatic mechanisms change the beat of the heart and the force of the blood to the head and thus regulate the blood pressure at a constant level. We sometimes say that these mechanisms maintain the steady state.

Frequently, homeostatic regulations are by means of the negative feedback mechanism. As an example the male hormones, androgens, of vertebrate animals inhibit the production of gonadotropin from the pituitary gland. Gonadotropin, on the other hand, stimulates androgen production by the testes. In this way, the level of androgens in the blood is kept within rather close tolerances. A castrated animal will show a sudden rise in the gonadotropin in the blood and urine due to the removal of the androgens that usually serve as a control. See Fig. 1. See also **Hormones**.

A good example of homeostasis in plants concerns the maintenance of the water balance in the leaves. The leaf must maintain a steady state of water concentration in spite of great variations in the amount of water available in the soil and variations in the humidity and temperature of the air, which affect the loss of water from the leaves through transpiration. The leaves have tiny stomata, which admit air to the leaf. This air is needed to supply the carbon dioxide for photosynthesis, but it can also carry moisture from the leaf. The guard cells surrounding the stomata minimize the loss of water by opening and closing the stomata in accordance with the amount of water in the leaf. The guard cells have a tough inner portion that, as the cells swell with turgor pressure, becomes more convex and opens the stoma in between. When the guard cells lose turgor pressure, the inner portions become less convex and the stoma is closed, thus preventing the loss of more water when the water level drops in the cells. The guard cells also function according to the usage of carbon dioxide during photosynthesis. As the carbon dioxide level drops, some of the starch is converted to sugar. This increase of solutes within the cell causes the cell to absorb more water from the surrounding cells and the stoma is opened. At night when no carbon dioxide is being used, the sugar concentration is low and the stoma is closed. The concentration of carbonic acid, from the carbon dioxide in the cells, is the factor that activates the starch-splitting enzymes.

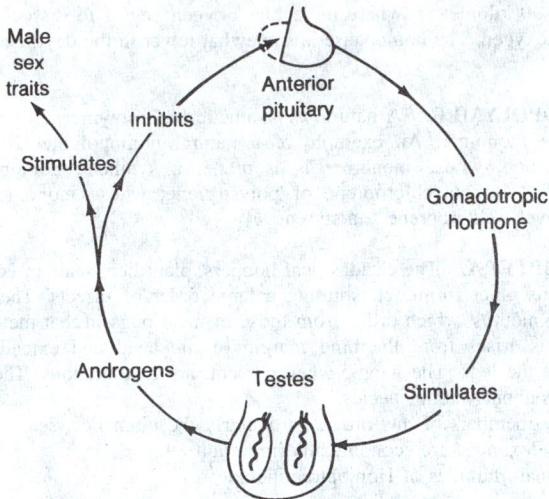

Fig. 1.   Homeostatic regulation of male hormone. (*A.M. Winchester.*)

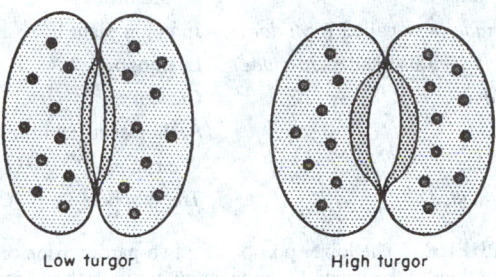

Fig. 2.   Guard cell closing. (*A.M. Winchester.*)

With low carbonic acid, there is an inactivity of the enzymes, with high acid, the enzymes are most active. See Fig. 2.

There can also be homeostasis of a genetic nature. The balance of genes within a gene pool is homeostatically regulated. Suppose a certain harmful gene is continually being added to a population through mutation. As the gene is expressed it will reduce the reproductive potential of the individual and there will be a gradual elimination of the gene from the population. Soon the input through mutation is exactly balanced by the outgo through genetic death and the gene remains at a stable level in the population. If we increase the mutation rate, say, by radiation, then the concentration in the gene pool will increase. If we reduce the rate of elimination by medical means, we can also increase the gene pool.

Ecological homeostasis concerns the balance of nature. When certain plants that serve as food for herbivores increase, the number of herbivores increases. This, in turn, results in an increase of the predators. A balance is established that will vary according to variations in environmental factors which cause an increase or decrease of any one of the organisms in the complicated food web.

HOMEOTYPE. A term used by biologists and paleontologists for a specimen that has been identified by an authority by comparing it with the type.

HOMOCENTRIC RAYS. Rays having the same focal point. (It may be at infinity; in other words, the rays may be parallel.)

HOMOCLINE. Group of strata that dip in one and the same direction. Never a complete structure and usually representing the limb of an anticline or syncline.

HOMODYNE RECEPTION. In this system, used in connection with radio reception for suppressed-carrier systems of radiotelephony, the receiver generates a voltage that has the original carrier frequency. This is combined with the incoming signal. The term zero-beat reception is also used.

HOMOGENEOUS. (Latin, "the same kind"). This term, in its strict sense, describes the chemical constitution of a compound or element. A compound is homogeneous since it is composed of one and only one group of atoms represented by a formula. For example, pure water is homogeneous because it contains no other substance than is indicated by its formula, H_2O. Homogeneity is a characteristic property of compounds and elements (collectively called substances) as opposed to mixtures. The term is often loosely used to describe a mixture or solution composed of two or more compounds or elements that are uniformly dispersed in each other. Actually, no solution or mixture can be homogeneous; the situation is more accurately described by the phrase "uniformly dispersed." Thus so-called homogenized milk is not truly homogeneous; it is a mixture in which the fat particles have been mechanically reduced to a size that permits uniform dispersion and consequent stability.

See also **Heterogeneous**.

HOMOGENEOUS ATMOSPHERE. 1.   A hypothetical atmosphere in which the density is constant with height. The lapse rate of temperature in such an atmosphere is known as the autoconvective lapse rate and is equal to g/R (or approximately 3.4 °C per 100 meters) where g is the acceleration of gravity and R is the gas constant for air. A homogeneous atmosphere has a finite total thickness which is given by $RdTv/g$, where Rd is the gas constant for dry air and Tv is the virtual temperature (K) at the surface. For a surface temperature of 273 K, the vertical extent of the homogeneous atmosphere on the earth is approximately 8000 meters. At the top of such an atmosphere both the pressure and absolute temperature vanish.

2.   With respect to radio propagation, an atmosphere, which has a constant index of refraction, or one in which, radio waves travel in straight lines at constant speed. Free space is the ideal homogeneous atmosphere in this sense.

3.   Same as adiabatic atmosphere.

HOMOGENEOUS (Mathematics). The term has several meanings. Inhomogeneous is the opposite of homogeneous.

A function $f(x_1, x_2, \ldots, x_n)$ is homogeneous in all of its variables if, for any parameter t, $f(tx_1, tx_2, \ldots, tx_n) = t^n f(x_1, x_2, \ldots, x_n)$. The exponent n is the degree or order of the function. The behavior of such a function is known as the Euler theorem on homogeneous functions.

The term is used with two meanings for a differential equation:

1. A first-order equation, $y' = M(x, y)/N(x, y)$ is a homogeneous equation if M, N are homogeneous functions of the same degree.
2. The general equation, $f(x, y, y', y'', \ldots) = 0$ is homogeneous and linear if f is a homogeneous function of y and all its derivatives. If the right-hand side equals a function of x, the independent variable, it is still linear but now inhomogeneous.

An integral equation, a boundary condition, or a system of simultaneous linear algebraic equations can also be homogeneous or inhomogeneous in a similar way.

HOMOGENIZING. A process for reducing the size of particles in a liquid and useful in the preparation of numerous food substances, including milk, ice cream, salad dressings, various fruit juices, flavor concentrates, infant foods, among others. The homogenization of milk is described in some detail in entry on **Milk and Dairy Products**.

A reduction of particle or globule size in a mixture of two immiscible liquids makes an emulsion possible. If an emulsifying agent is present, a more stable emulsion can be produced and coalescence of the dispersed phase is prevented. The homogenizer is also used to produce dispersions by reducing the particle size in solid-in-liquid mixtures. As in the preparation of an emulsion, a dispersing agent is needed to maintain a homogeneous mixture.

Typically, a homogenizer consists of a high-pressure, positive-displacement pump and an adjustable orifice. The pump is a piston or plunger type, usually consisting of three plungers, although some homogenizers are made with five or even seven plungers. The cylinder for each plunger has an inlet and discharge valve. The plunger pump must push the product through the homogenizing valve (adjustable orifice). For two-stage homogenization, two valves are arranged in series.

A typical homogenizing valve consists of a seat and plug of very hard abrasion-resistant materials (alloys such as Stellite are used). The seating surfaces must be lapped smooth and be parallel. In operation, the plug is spring-loaded against the seat. Spring compression is adjusted so that when the product flows, energy in the form of pressure is required to lift the plug. Although many products can be homogenized at pressures below 3000 pounds per square inch (204 atmospheres), machines are made to develop pressures in excess of 8000 pounds per square inch (544 atmospheres). In another design, a valve uses a compressed cone of stainless-steel wire inserted into a socket, the product being homogenized by flowing between the wires.

A number of theories have been proposed as to what actually breaks up the particles in the homogenizer. (1) As the product enters the area between the lapped surfaces, it is suddenly accelerated to velocities as high as 30,000 feet per minute (9,144 meters per minute) at a pressure of 5,000 pounds per square inch (340 atmospheres). When acceleration is this sudden, the particle (especially the liquid particle) is stretched or elongated to the point of breaking. (2) At this high velocity, there are shear forces between layers of liquids under flow that break up particles. (3) Cavitation may be the major cause of homogenization. When the pressure energy is converted into velocity energy, the vapor pressure of the product exceeds product pressure, resulting in the formation of vapor cavities, which collapse upon leaving the valve at higher pressures. This collapsing, or implosion, of cavitation exerts tremendous force, breaking up the particles. Most homogenizers are designed to incorporate one or more of the foregoing principles.

HOMOIOTHERMY. Warm-bloodedness. The maintenance of a body temperature above that of the environment is common among animals; hence the usual terms warm-blooded and cold-blooded are inaccurate. Cold-blooded forms are those whose body temperature fluctuates with that of the surrounding air or water, so that the animal's activity is directly conditioned by external temperatures. They are more accurately described as poikilothermal. In contrast, homoiothermal animals tend to maintain a constant body temperature in spite of external fluctuations. Fluctuations are normal, although the human body usually maintains a constant temperature.

Only birds and mammals are homoiothermal. Both regulate the body temperature by producing excess heat and by regulating its radiation from the surface. Regulation is accomplished by nervous control of the blood vessels near the surface, by insulating vestiture, and by the evaporation of water from the body. When the surrounding air is warm, the blood flows more freely near the surface of the body and more heat is radiated, but when the air is cold, less blood reaches the surface and the heat is conserved. In air too warm to permit adequate radiation, the animal reduces its activity, exposes as much surface as possible, and either sweats or pants. The evaporation of water either from the mouth or from the sweat glands absorbs heat from the underlying tissues. Vestiture plays a passive role as an insulating coat, but it is capable of some regulation, especially in the birds. The erection of the feathers provides a thicker and looser covering of high insulating value, and their depression results in less interference with radiation.

Homoiothermy is one of the highest adaptations of living things, since it provides for the maintenance of optimum conditions for the vital processes of the body. Through it the animal becomes virtually independent of one of the most important of the fluctuating environmental conditions.

HOMOLOGOUS SERIES. Two organic compounds are said to be homologous if their molecular formulas differ by CH_2, or a multiple of CH_2. For example, the alkane series has the general formula, C_nH_{2n+2}, its first three members being methane, CH_4, ethane, C_2H_6, and propane, C_3H_8.

HOMOLOGY. Fundamental structural relationship, based on similarity of embryological development and evolutionary history. The antithesis of analogy, which is superficial likeness based on adaptation for similar uses.

The anterior appendages of terrestrial vertebrates, for example, are regarded as fundamentally similar structures, derived from the pentadactyl appendage; yet they include the wings of birds, flippers of aquatic mammals, and a great variety of less extreme adaptations, including the legs of animals and the arms of man. In contrast, the wings of birds and of insects are broad thin structures used for flight, but in structure and origin they show no resemblance beyond this point and so are analogous.

HOMOPAUSE. The top of the homosphere, or the level of transition between it and the heterosphere. The homopause probably lies between

80 and 90 kilometers, where molecular oxygen begins to dissociate into atomic oxygen. The homopause is somewhat lower in the daytime than at night.

HOMOPOLYMER. A natural or synthetic high polymer derived from a single monomer. An example of a natural homopolymer is rubber hydrocarbon, whose monomer is isoprene; a synthetic homopolymer is typified by polychloroprene or polystryrene, whose monomers are, respectively, chloroprene and styrene.

HOMOPTERA. The cicadas, leaf hoppers, plant lice, scale insects, and numerous other forms, constituting a large order of insects. They have sucking mouths, which differ from those of most bugs in that the slender proboscis arises from the hind margin of the head and extends back between the legs. The wings, when present, are membranous. The order includes about 16,000 species.

Many members of this order, particularly the plant lice, scale insects, and phylloxerans, are economically important.

The main families of Homoptera include:

Aleyrodidae	White flies
Aphidae	Aphids or plant lice
Cercopidae	Spittle bugs
Chermidae (also called *Psyllidae*)	Jumping plant lice
Cicadellidae (also called *Jassidae*)	Leafhoppers
Cicadidae	Cicadas
Coccidae	Scale insects and mealybugs
Fulgoridae	Plant hoppers
Membracidae	Tree hoppers

HOMOSPHERE. The lower portion of a two-part division of the atmosphere according to the general homogeneity of atmospheric composition; opposed to the heterosphere. The homosphere is the region in which there is no gross change in atmospheric composition, that is, all of the atmosphere from the earth's surface to about 80 or 100 kilometers (50 or 62 miles). See also **Earth**.

HONEY. Raw, unprocessed honey is a thick, viscous, high-density, very sweet, hygroscopic liquid that is formed by honeybees from the nectar of flowers and, to a limited extent, from the juices of fruits and honeydew. Honey is available commercially as a liquid, as crystallized honey, as comb honey, as chunk honey, and as powdered honey. Honey contains a large percentage of simple sugars, as well as essential oils of the flowers from which it is derived, plus about 20% water. The flavor of honey depends upon the flowers from which the nectar is derived, upon manufacturing conditions if it is processed, upon the season and climate during which it is gathered and stored by the honeybees, and upon its age. Under appropriate conditions, honey is one of the most storable of foods and can be kept for many years, particularly in a frozen state.

In food processing, honey is frequently used because of its qualities as a humectant and a source of reducing sugars. Bakers and candy makers prefer honey for these reasons plus the fact that it promotes caramelization and aids in obtaining uniform browning of baked goods, as well as providing clarity to glazes. In addition to bakery products and confections, honey is used in the manufacture of breakfast foods, snacks, sauces, and syrups, as well as a sweetener and bodying agent in some canned fruits, jams, jellies, and spreads. Honey is a common ingredient of graham crackers, where it blends with the dark whole-wheat flours, as it also does with whole-wheat breads.

Physical and Chemical Properties of Honey

Although a seemingly simple substance, honey is relatively complex and requires some rather exacting conditions when it is processed. Important factors include: (1) moisture content which, if excessive, causes the honey to ferment over a period of time; (2) the tendency of the glucose to crystallize out of the liquid phase, a process known in the trade as *granulation*; and (3) the presence of nitrogeneous substances, which even in very small amounts cause the honey to darken with age. Because water content is important to ultimate quality of the honey, including the possibility of fermentation, the water content is strictly regulated by most countries. In the United States, U.S. Grade A (Fancy) and Grade B (Choice) cannot contain over 18.6% water. Grade C (Standard) for reprocessing may contain up to 20% water. Honey with greater amounts of water is Grade D (Substandard).

Beeswax. This is a commercial byproduct of honey production. The wax represents approximately 1.9% of the weight of honey produced. In the United States, beeswax production ranges between 3 and 5 million pounds (1.4 and 2.3 million kilograms) per year. Freshly made wax is of a light yellow color, but becomes brown with age. However, the wax may be bleached by sunlight or with acids. Beeswax is made up mainly of a complex long-chain ester, myricil palmitate, $C_{15}H_{31}COOC_{30}H_{61}$, and cerotic acid, $C_{25}H_{51}COOH$; specific gravity, 0.965–0.969; mp 63 °C. The wax is easily colored with dyes and finds numerous uses, as in polishes, candles, leather dressings, adhesives, cosmetics, and molded articles.

Mead (Honey Wine). This is one of the most ancient of fermented products and wines. Although regarded by some people as a curiosity in modern times, there is some demand for mead in various parts of the world. It is usually prepared and consumed on a regional basis.

Honey and Infant Botulism. Honey has been implicated in about one-third of the cases of infant botulism. Consequently, some physicians recommend that honey not be fed to infants under one year of age.

See also list of references at end of article on **Honeybees**.

Additional Reading

Style, S.: *Honey: From Hive to Honeypot: A Celebration of Bees and Their Bounty*, DIANE Publishing Company, Collingdale, PA, 1998.

HONEYBEES.

There was little scientific knowledge of honeybees until the 1850s. Even though European scientists of the 1700s and early 1800s developed an understanding of the biological aspects of honey and wax production by bees, this knowledge did not contribute in a major way to the practical aspects of beekeeping. It was not until 1852 that an American minister (L.L. Langstroth) discovered what became known as "bee space." This concept led to development of the first practical, movable-frame hives. This breakthrough, coupled with other important equipment developments, such as the centrifugal honey extractor (commercialized in 1870), the bee smoker, bee escape, and queen excluder, transformed beekeeping during the latter half of the 1800s from a minor activity on the part of many farmers to a serious business capable of centralization and full-time management. Prior to that time, there was little if any organization in terms of producing and marketing honey.

It is estimated that 60–70% of the honey produced in the United States comes from 1200 to 1500 fulltime beekeepers who operate 35–40% of the nation's nearly 5 million bee colonies. Production of honey in the world is estimated at 683,000 metric tons, of which the United States accounts for about 114,000 metric tons per year. Other large honey producing countries include Russia, China, Mexico, Uruguay, Canada, and Turkey. The principal honey producing states in the United States are Florida, North Dakota, Minnesota, California, and Wisconsin. However, honey production is found in practically all of the states to some degree.

A full-time beekeeper maintains a minimum of 400 hives, but the average is about 1,200 hives. Large operators will maintain 20,000 or more hives. The optimal hive density for good honey yield usually is insufficient for efficient crop pollination. Thus crop growers compensate beekeepers for reduced honey crops resulting from maintaining relatively high hive density and also for the costs of moving colonies from one farm to the next during pollinating season. One of the specialty sectors of beekeeping is that of raising bees and queens for sale to beekeepers and growers of crops. This activity is concentrated in the Gulf states and southeastern states.

Strains and Hybrids

The most common strain of bees kept in North America are of Italian origin. These bees, yellow to brown in color, are industrious and relatively unexcitable. The Caucasian strain, even more gentle than the Italian, is also kept. This honeybee is gray to black in coloration. An import from Japan in the late 1970s, the species *Osmia cornifrons*, is under intensive study by the U.S. Department of Agriculture. It is believed that the species may be of particular value for solving pollination problems for small-farm fruit tree growers. The usual honeybee suffers from diseases and predator pests; they swarm; they sometimes abscond; they require considerable attention; and they sting. The Japanese honeybee, about two-thirds the size of the

Italian or Caucasian bees, has few of the foregoing disadvantages. They produce no honey or beeswax, they are more active at cooler temperatures, they rarely range more than 300 feet (90 meters) from their nests, they are docile, and their sting is about like that of a mosquito bite. Although they produce no honey, they appear to be quite ideal for pollinating.

Biological Aspects. Bees are of the subgroup *Aculeata* within the order of *Hymenoptera* (*Insecta*). *Aculeata* also includes wasps and ants, these insects all incorporating the sting, which was developed in the course of evolution from the ovipositor apparatus. This modification involved a change of function, for it no longer serves for egg-laying; rather it is employed as an effective weapon of defense or as an injection cannula for the paralysis of prey. In view of the origin of the device, it is obvious that only female *Hymenoptera* can sting. The sting consists of several reciprocally movable chitinous elements. Into it there open the ducts of two glands, one of which produces poison. In the stinging *Hymenoptera* the eggs are ejected from the opening of the genital chamber at the base of the ovipositor.

Bees are members of the superfamily *Apoidea* and include, in addition to the universally known honeybee, more than 20,000 other species. They are distributed almost worldwide, and in the north their range extends well beyond the Arctic Circle. The smallest bees measure barely 2 millimeters, while the largest approach 4 centimeters in body length. They are among the most economically important animals, particularly since they play a critical role in the reproduction of numerous cultivated plants. Moreover, the study of their life history has charms all its own, for they have evolved fascinating forms of social interaction.

Bees are classified into six families, of which two, the *Halictidae* and the *Apidae*, have evolved social species. In addition to the honeybee, there are plasterer bees, mining bees or burrowing bees, mason bees, leaf-cutter bees, carpenter bees, and bumblebees.

At present, four species of *true honeybees* (tribe *Apini*) are known, all of them native to tropical southeastern Asia. But the so-called domesticated honeybee (*Apis mellifera*) has been distributed by humans all over the world. Before cane and beet sugar came into use, honey was the primary sweetener for foods. For that reason, the bees were most highly prized for their honey and wax. But, in more recent years, they have become even more prized for their beneficial activity in pollinating flowering crops.

Social Structure of Honeybee Colonies

All honeybees (See Fig. 1) build their combs of pure wax, which can be produced only by the workers. The combs are hung up vertically either in the open or in enclosed areas, and on both sides they have geometrically perfect hexagonal cells, a shape which minimizes the construction material required. The cells for the worker brood and those for the storage of honey and pollen are similar, while those for the males (drones) are larger. The queens are raised in special vertically hanging chambers, as shown by Fig. 2. Only the workers have the organs and instincts necessary for the activities of construction and foraging. The queen, somewhat larger than the workers, has as her only task the laying of eggs. Fertilized eggs give rise to workers or queens, depending upon the food given to the larvae, while the drones come from unfertilized eggs. The time at which reproductive individuals are to be produced is determined by the workers, who then prepare the appropriate cells and food. All the larvae are fed and cared for continually during their development. A queen can lay as many as 2000 eggs in a single day. In her 4 to 5 years of life, she produces about 2 million eggs. More than 80,000 bees can live in a colony.

The tasks of the workers are manifold. They supply the colony with food, guard the nest, and build the combs. They also keep the combs clean, for these are used several times for the brood. By fanning with their wings, the workers cool the nest, and by their muscular activity they warm

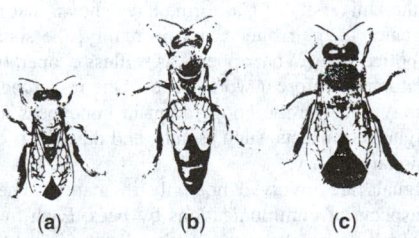

Fig. 1. Relative sizes of honeybees: (**a**) worker, (**b**) queen, (**c**) drone. (*USDA.*)

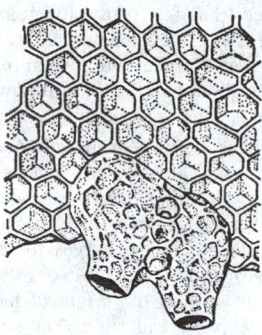

Fig. 2. Section of a honeybee comb. Drone cells are above and two queen cells are shown at the lower edge.

it. Thus, they ensure that the temperature in the brood area stays close to 35 °C (95 °F). Although every worker bee is, should special circumstances demand it, capable of performing any of these tasks, ordinarily there is an orderly division of labor, corresponding to the age of the workers. In their first days as workers, they act as janitors, keeping the combs clean. Next, after the pharyngeal gland in the head has matured, they devote themselves to the larvae, feeding them first with the secretion of this gland and later with pollen and honey as well.

The food given by worker honeybees to the young larvae during the first 3 days of their existence and to the larvae of queens until they are fully developed is known as *royal jelly*. This is a thick, white liquid formed in the stomach of the worker by partial digestion of honey and pollen and is apparently a highly concentrated food. Queen cells are supplied with the material in excess of the needs of the larvae. If conditions within the colony deprive queen larvae of this abundance, they fail to become large and, in some cases, they may revert to development as intermediate forms between queen and worker. Such individuals may, however, have the instincts of queens and so may mate and lay fertile eggs. The change from royal jelly to a less concentrated food in the case of worker larvae apparently is responsible for the development of worker bees, since both queens and workers may develop from identical eggs.

At about the tenth day, the workers fly out briefly for the first time and become acquainted with the surroundings of the hive. In the following days, the pharyngeal gland becomes reduced and the wax glands begin to function. Now the worker bee becomes a construction worker, in addition to having responsibility for the food stores and carrying away refuse. Finally, when the worker approaches the age of 20 days, it goes out into the open more and more often. At first, it takes over guard duty at the flight hole; then it forages with great industry for pollen and nectar until the end of its life, in summer, after 4 to 5 weeks of labor.

Bee colonies propagate by swarms. In the early summer, shortly before one or more queens emerge from the queen cells, the old queen leaves the hive with about half its population. She first gathers her court into a cluster near the hive and then follows the advance guard, which has found a new nest site. The young queens which have matured in the old hive fly out repeatedly until they have mated, each with several drones. The supply of sperm thus accumulated must last the queen a lifetime. They mate in flight, at assembly places where the drones often congregate in great numbers. There the drones fling themselves on every female that flies past and has the appropriate scent signal. The inseminated queen returns to the nest and stings to death any rivals that may still be present. The drones too meet their fate in late summer. The workers drive them out of the nest with bites and stings, after which they are left to starve.

Although, traditionally, honeybees have been admired for their altruism, researchers at the University of California have shown that they project an aspect of allegiance to their queen and, in reality, the social structure is more that of a police state. What appears as selfless cooperation is enforced by a special platoon of enforcer worker bees. One researcher has observed that coercion may underlie the cooperation in honeybees as well as with other highly organized insects, such as ants, and that this is consistent with modern evolutionary theory.

Chemical signals are involved not only in mating. They also play a role in many aspects of communications by bees. Each queen secretes a substance by which the workers are continually reassured that the colony is not without a mother. Scent signals spread the alarm when the bees are in danger and also serve to identify food sources and the entrance to the hive.

Foraging for Nectar

Bees store up honey collected from various sources. Blossom honey is the thickened nectar of millions of flowers, which has passed through the stomachs of many bees and has been altered by glandular secretions. Bees also collect the sweet secretions of aphids that stick to leaves. With this, they form the leaf honey or pine honey, which is considered a delicacy. Several hundred kinds of plants produce nectar, which the bees also use for honey, but only a few kinds are common enough, or produce enough nectar, to be considered as major sources. The best sources of nectar for producing surplus honey vary from place to place. Some plants that are major nectar sources in the United States include: Alfalfa, aster, buckwheat, catclaw, citrus fruit, clover, cotton, fireweed, goldenrod, holly, horsemint, locust, mesquite, palmetto, tulip tree, tupelo, sage, sourwood, star thistle, sweet clover, sumac, and willow. The varying qualities of these sources is reflected in the color and flavor of the raw honey taken from the hive.

Since a good colony can store as much as 1 kilogram (2.2 pounds) of honey per day, the number of foraging flights undertaken in a day is astronomical. This efficiency depends upon two special achievements: (1) The bees can orient themselves; and (2) they can communicate with one another. Considering orientation, each bee must cover the distance between hive and collecting place as quickly and accurately as possible. This phenomenon was studied by Karl von Frisch and his co-workers for over 50 years. It was found that the bee uses all its senses—color leads it to the bright flowers; sense of smell enables it to distinguish different species of flowers. On the flight out and back, it not only observes conspicuous landmarks, but also uses the sun as a compass by keeping the flight path at the proper angles to the direction of the sun. Even the fact that the sun seemingly moves across the sky does not confuse the bee. Its sense of time permits it to take the time of day into account in its flights, correctly altering the setting of the sun compass. This sense of time also makes it possible for the bee to go at any time of the day to the sort of flower that is producing nectar then. Even if the sun is covered by clouds, the forager is not helpless. The direction of polarization of the light waves that penetrate from a patch of blue sky is dependent upon the position of the sun. Since the eyes of the honeybee, unlike human eyes, can measure this direction of polarization, the bee can orient to that just as well as to the sun itself.

Honeybee Communications

Because of their highly organized social structure, it is no surprise that the systematics of procedure should extend from the hive to their activities in the field when the bees are seeking nectar. Once a good source of nectar is found, this is effectively conveyed to other bees of the colony. Over many decades, naturalists have postulated the existence of such communications, but the mechanism escaped understanding. It was established that scout bees communicate to worker bees, not only in the hive, but also while in flight.

For a number of years, it was widely believed that the bees used aerial "dances" to convey their findings, but in recent years that theory has been challenged vigorously by another school of researchers.

Aerial Dance Theory. For what is considered one of the most exciting revelations of biology, Karl von Frisch received a 1973 Nobel Prize for describing how various kinds of aerial dances used by bees are a primary means of bee-to-bee communications. For example, he described waggle dances and round dances, as indicated very schematically in Fig. 3. Details are beyond the scope of this article. Although these performances of honeybees appear to border on the miraculous, they have been observed and documented by Dr. von Frisch and the supporters of his concept.

For example, Moffett (see reference listed) reported in 1990 findings based upon the use of an electronic robot bee in Germany. Construction of the electronic bee research robot is elegantly diagrammed in the Moffett reference. Inasmuch as the beehive is completely dark, the bees cannot depend upon visual perception. It was believed for many years that honeybees were incapable of detecting sounds. It now appears that vibrations from the bee robot are picked up by the honeybee's antennae. Hence, in essence, the researchers can "talk" to the bees by simulating the varying wing vibrations, which carry the information that the insect requires to locate a foraging target. Although many details remain unclear, investigators observe, "The angle between the dance direction and the

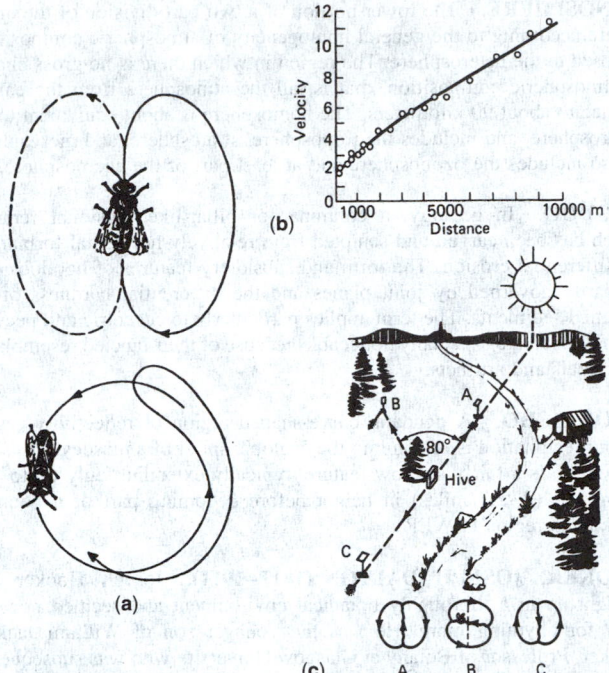

Fig. 3. (a) Dances with which honeybees communicate. The "waggle" dance (top) is performed by long-distance foragers to communicate location of food sources. Both direction and distance are communicated. Direction is illustrated in (c). Distance is related to the speed or frequency with which individual dances are repeated. The round dance, indicated by the lower diagram, announces to the bees in the hive that a rich harvest has been located. Foragers repeat this dance each time they return to the hive until a more important new source is located. It is important to realize that the dance is normally done in the pitch darkness of a closed beehive, so that the dance is invisible to its comrades, who depend upon their senses of touch (vibrations) and smell to follow the tortuous course of the dance.

(b) Velocity of the waggle dance (ordinate) as a function of the distance (abscissa) between feeding place and hive.

(c) Relationship between flight angle with respect to sun and dance angle in the hive with respect to gravity. The latter is diagrammed for the three feeding sites shown (A, B, and C).

vertical is known to signal the direction from the hive to food in relation to the sun. The bee waggles her abdomen and quivers her wings, indicating the distance. The intensity of the dance plus samples offered and the lingering odors on the bee's body suggest the type of food and its quality."

Also, having built another bee robot, researchers at Occidental College and the University of California (Santa Barbara), plus Additional investigators at Princeton University, among others, refute the "aerial dance" theory. To date, they have been unable to gather evidence in support of the theory and conclude that honeybees depend largely on their sense of odor when searching for nectar. Thus, in the community of biologists, as of the mid-1990s, there remained a dichotomy of opinion.

Cognitive Maps. Research into honeybee behavior continues apace with major emphasis on bee navigation and flower recognition. Particular concern involves the so-called invertebrate-vertebrate dichotomy of navigation. As reported by J.L. Gould (Princeton University), the higher vertebrates use landmarks as a part of an overall map of an area so that a selection of routes (in contrast with a series of route-specific steps) may be used. In other words, the higher vertebrates have considerable flexibility in their navigation. Traditionally, it has been assumed that invertebrates were limited to navigating step by step from one specific landmark to the next. Recent extensive experiments by Gould tend to invalidate former assumptions and indicate that bees can store (in some fashion not fully understood) broad, map-like information, thus enhancing their navigational capabilities. In his research, Gould used individually marked honeybees (*Apis mellifera ligustica*).

Flower Recognition. In another series of studies, Gould reported that bees are able to learn to distinguish between flowers with different shapes or patterns. Earlier studies suggested that bees remember only isolated features, such as spatial frequency and line angles, rather than the photographic search images that are characteristic of vertebrates. New information indicates that bees can store flower patterns as a low-resolution eidetic image or photograph. Several decades ago, M. Hertz had suggested that bees spontaneously prefer highly dissected patterns (shapes with a high ratio of edge to area, or high spatial frequency) and that only the crudest sort of learned discrimination was possible. It was later found, however, that *shape learning* could be considerably more subtle, i.e., the memory process could be more sophisticated and complex than previously thought, but that nevertheless no photograph-like (eidetic) images were involved. Thus, two distinct hypotheses developed — the *isolated-feature* or *parameter hypothesis* and the *eidetic image* or *picture hypothesis*. Although not fully resolved, recent research tends to favor the latter hypothesis. Gould has suggested that the limiting factor is the resolution of the eidetic storage in the brain of the bee.

Other factors enter into how the bee discriminates pollen sources. For example, bees prefer to land on and learn to recognize most quickly violet-colored food sources. This, apparently, is an innate bias, which must be accommodated by the insect's information storage system.

Interesting investigations also have been conducted among other foraging insects. A.C. Lewis (University of Colorado) has studied the memory constraints and flower choice in the cabbage butterfly (*Pieris rapae*). Decades ago, Darwin hypothesized that *flower constancy* in insects that feed on nectar results from the need to learn how to extract nectar from a flower of a given species. In experiments conducted by Lewis, it was found that the cabbage butterfly shows a flower constancy by continuing to visit flower species with which it had experience. The time required by individuals to find the source of nectar in flowers decreased with successive attempts, the performance following a learning curve. Learning to extract nectar from a second species interfered with the ability to extract nectar from the first. Insects that switch species thus experience a cost in time to learn.

Honeybee studies generally have involved the domestic honeybee (*Apis mellifera*), whose many subspecies have been distributed by human intervention throughout the world.

Other Honeybee Subspecies. The Indian subspecies (*A. mellifera derana*) closely parallels the domestic honeybee in structure and habits. Subspecies *A. mellifera dorsata* and *A. mellifera florea*, both of which live in southeastern Asia, represent a somewhat lower level of performance. Each of these species builds only one comb and hangs it in the open on a branch. These subspecies are not important commercially in terms of honey production.

"Africanized" Bees

The common honeybee of Europe and North and South America commonly is considered of Italian (European) origin. Although they are comparatively poor honey producers, they are known for their good temperament. In 1956, in an effort to breed a better honey producer, Brazilian entomologists imported 46 South African queen bees, which now are known to be of a much different temperament in terms of their interaction with society in general and thus developed the term "killer bee." Unfortunately, 26 queens escaped from the Brazilian laboratory and since that time have been spreading at an asymptotic pace northward into Central and North America.

Vigorous preventive measures were taken by Mexican officials and the U.S. Department of Agriculture. Migration of the "Africanized" bees was slowed, but not stopped. The first arrival of the unwanted bees was reported from a location near Hidalgo, Texas, in October 1990.

Entomologists predict that the bees ultimately will spread throughout the southern states of the United States and most likely will penetrate as far north as Pennsylvania in the east, Iowa in the Midwest, and northern California in the west. It is believed that this rate of progression will depend largely on the pattern of winter temperatures experienced as the bees travel northward. Responsibility for monitoring the progress and for taking preventive and planning measures to protect the American honey-producing and crop-pollinating industries rests with the U.S. Department of Agriculture (USDA) Laboratory located in Baton Rouge, Louisiana.

Hybridization of the African with the established European honeybees is unclear and difficult to forecast. Today, 80 percent of honey production in Argentina is credited to the "Africanized" bees, including hybridized bees.

Even though the Africanized honeybee has caused more deaths to humans than have the traditional European types, it is reported that the sting of the Africanized bee is no more serious than that from the European bee. What is different is the aggressiveness during swarming and the attendant "attacks" on people and animals in areas and situations where precautions normally would not be required, as, for example, in urban areas. A major concern is the probability that African bees may take over hives in the South and will affect traveling beekeepers in controlling their bees. Traveling beekeepers transport their hives around the countryside to pollinate an estimated $10 billion worth of crops each year.

Because of hybridization that is occurring in some of the transition zones as the African bees progress northward, the threat may be lessened. But, as one agricultural expert observes, this may require 25 years to develop.

Studies in honeybee genetics have been spurred in an effort to develop more productive countermeasures than those currently known.

Additional Reading

Bailey, L.: *Honey Bee Pathology*, Academic Press, Inc., San Diego, CA, 1991.

Cowen, R.: "Bumblebee Energy," *Science News*, 215 (October 6, 1990).

Darwin, C.: *The Effects of Cross- and Self-Fertilization in the Animal Kingdom*, classic reference, Murray, London, 1876.

Doherty, J.: "The Hobby that Challenges You to Think Like a Bee," *Smithsonian*, 62 (July 1987).

Dyer, F.C. and J.L. Gould: "Honey Bee Orientation: A Backup System for Cloudy Days," *Science*, **214**, 1041–1042 (1981).

Gould, J.L.: *The Biology of Learning* (P. Marler and H. Terrace, Eds.), Springer-Verlag, Berlin, 1984.

Gould, J.L.: "How Bees Remember Flower Shapes," *Science*, **227**, 1492–1494 (1985).

Gould, J.L.: "The Locale Map of Honey Bees: Do Insects Have Cognitive Maps?" *Science*, **232**, 861–863 (1986).

Heinrich, B.: "The Regulation of Temperature in the Honeybee Swarm," *Sci. Amer.*, **244**(6), 146–160 (June 1981).

Hepburn, H.R. and S.E. Radloff: *Honeybees of Africa*, Springer-Verlag, Inc., New York, NY, 1998.

Horgan, J.: "Do Bees Think?" *Sci. Amer.*, 36 (May 1989).

Horgan, J.: "Disco-Bee (Robot)," *Sci. Amer.*, 31 (June 1989).

Horgan, J.: "Bee Police," *Sci. Amer.*, 28 (March 1990).

Horgan, J.: "Stinging Criticism," *Sci. Amer.*, 29 (November 1990).

Hubbell, S.: *A Book of Bes: And How to Keep Them*, Houghton Mifflin Company, New York, NY, 1998.

Lewis, A.C.: "Memory Constraints and Flower Choice in Pieris rapae," *Science*, **232**, 863–865 (1986).

Moffett, M.W.: "Dance of the Electronic Bee," *Nat'l Geographic*, 134 (January 1990).

Needham, G.R. et al., Editors: *Africanized Honey Bees and Bee Mites*, Wiley, New York, NY, 1988.

Real, L.A.: "Animal Choice Behavior and the Evolution of Cognitive Architecture," *Science*, 980 (August 30, 1991).

Rinderer, T.E. et al.: "Hybridization Between European and Africanized Honey Bees in the Neotropical Yucatan Peninsula," *Science*, 3039 (July 19, 1991).

Robinson, G.E. et al.: "Hormonal and Genetic Control of Behavior Integration in Honey Bee Colonies," *Science*, 109 (October 6, 1989).

Rouibik, D.W.: *Ecology and Natural History of Tropical Bees*, Cambridge Univ. Press, New York, NY, 1989.

Spivak, M., D.J.C. Gletcher, and D. Michael: *The African Honey Bee*, Westview, Boulder, CO, 1991.

Staff: "Bee Mites Buzz Off," *Nat'l. Food Review*, 43, U.S. Dept. of Agriculture, Washington, DC, (October–December 1989).

Staff: "Myth of the Killer Bees," *Technology Review (MIT)*, 80 (February/March 1991).

Staff: "Africanized Bees Reach U.S.; Prepare to Settle," *Nat'l. Geographic, Geographica section* (April 1991).

Von Frisch, K. and M. Lindauer: *Naturwissenschaften*, **41**, 245 (1954).

Von Frisch, K.: *The Dance Language and Orientation of Bees*, Harvard University Press, Cambridge, MA, 1993.

Weiss, R.: "The African Advantage," *Science News*, 328 (May 26, 1990).

Winston, M.L.: *The Biology of the Honey Bee*, Harvard University Press, Cambridge, MA, 1991.

Winston, M.L.: *From Where I Sit: Essays on Bees, Beekeeping, and Science*, Cornell University Press, Ithaca, NY, 1998.

HONEY BUZZARD.   See **Eagle**.

HONEYDEW. A sweet secretion produced by aphids, which, when abundant on trees, sometimes cause spotting on the leaves and anything below the tree like a heavy dew. Honeydew is freely sought by ants and is sometimes gathered by bees, but it makes a very inferior honey.

HONOSPHERE.   The lower portion of a two-part division of the atmosphere according to the general homogeneity of atmospheric composition; opposed to the heterosphere. The region in which there is no gross change in atmospheric composition, that is, all the atmosphere from the earth's surface to about 90 kilometers. The homosphere is about equivalent to the neutrosphere, and includes the troposphere, stratosphere, and mesosphere; it also includes the ozonosphere and at least part of the chemosphere.

HOODOO.   In geology, a columnar or pillar-like erosional remnant which has been carved and sculpted from relatively horizontal formations by differential erosion. The form and subsidiary features of hoodoos may be partly governed by joint planes and the differential hardness of the stratified sediments. The term applies particularly to eccentric and peculiar forms, which are especially noticeable because of their fancied resemblance to animals and artifacts.

HOOK ECHO.   A pendant, curve-shaped region of reflectivity caused when precipitation is drawn into the cyclonic spiral of a mesocyclone. The hook echo is a fairly shallow feature, typically extending only up to 3–4 kilometers (2–2.5 miles) in height before becoming part of a bounded weak echo region (BWER).

HOOKER, JOSEPH DALTON (1817–1911).   Joseph Hooker was brought up in a thoroughly botanical environment and seemed never to wish for anything more. He was the younger son of William Jackson Hooker, Professor of Botany at Glasgow University who was subsequently Director of Kew Gardens outside London. Educated at Glasgow, Hooker graduated Doctor of Medicine in 1839. Soon afterwards he was offered a place as assistant-surgeon and naturalist on James Clark Ross's exploring voyage to Antarctica. The *Erebus* and *Terror* departed in 1839, returning in 1843. Hooker collected plants extensively, bringing back material for very substantial scientific monographs: the *Flora Antarctica* (1844–1847), *New Zealand* (1852–1854), and *Tasmania* (1855–1860). These were prefaced by thoughtful introductory essays in which Hooker discussed the species question, variability and geographical distribution. He also provided instructions for colonial botanists, seeing Kew Gardens as the hub of a global supply system.

In 1855, after a succession of unfulfilling jobs, he was appointed Assistant-Director at Kew, gradually taking on more administration until he succeeded as Director on his father's death in 1865. His career was thereafter notable for consolidating Kew's position as a preeminent center of colonial botany, masterminding the plantation economy of the British Empire and encouraging high-level taxonomic research. He published prolifically in scientific journals, ranging over the entire field of systematics, geography, morphology and paleontology. At this point he also began a lifelong undertaking with George Bentham to reexamine the whole botanical kingdom for an authoritative *Genera Plantarum*. See also **Paleontology (The History)** and **Taxonomy (The History)**.

Before his appointment to Kew, Hooker made an important expedition to India, 1848–1851, exploring the Himalayan Mountains for rhododendrons and other rare species, traveling partly with Thomas Thomson, with whom he wrote an unfinished *Flora Indica* (1855), and partly with Andrew Campbell, with whom he was briefly imprisoned in Sikkim. He published several fine illustrated works depicting the plants he found, and a less successful *Himalayan Journals* (1854). Although his time was afterwards increasingly occupied with Kew, he subsequently traveled Syria (1860), where he investigated the Cedars of Lebanon; Morocco (1871); and the Rocky Mountains in Colorado (1877), accompanied by Asa Gray. In 1872 he was involved in an unpleasant dispute with A. S. Ayrton over the running of Kew Gardens. See also **Gray, Asa (1810–1888)**.

Much of Hooker's personal and public life was tied up with that of Charles Darwin. He had met Darwin briefly before the Antarctic voyage and became close friends with him soon after his return. The two corresponded for nearly forty years. Darwin depended on him for botanical advice and used to say that Hooker's letters were like a "jam-pot" to him. Early on, Darwin revealed his theory of natural selection to Hooker and Hooker played a key part, with Charles Lyell, in bringing Darwin's and Alfred Russel Wallace's views to the Linnean Society in 1858 and afterwards championing them vigorously. At the British Association in 1866 he delivered an address that marked the general acceptance of Darwin's views by British scientists. At the end of his life Darwin bequeathed a sum of money to enable Hooker to start the *Index*

Kewensis (1882–1895). See also **Darwin, Charles (1809–1882) Lyell, Sir Charles (1797–1875)** and **Wallace, Alfred Russel (1823–1912)**.

Additional Reading

Desmond, R.: *Sir J. D. Hooker, Traveler and Plant Collector*, Antique Collectors' Club and Royal Botanic Gardens Kew, Woodbridge, Suffolk, UK, 1999.

Desmond, R.: *Sir Joseph Dalton Hooker*, Antique Collectors' Club, Woodbridge, Suffolk, UK, 2008.

Web Reference

Kew Gardens: http://www.rbgkew.org.uk/

JANET BROWNE, Wellcome Institute for the History of Medicine, London, UK

HOOKE, ROBERT (1635–1702). Robert Hooke was an English instrument-maker, experimentalist and natural philosopher who made key contributions in a wide range of areas including physiology, geology and mechanics.

Born on the Isle of Wight, Hooke showed early aptitude with the design of mechanical toys. At Westminster School he learnt mathematics and geometry, and at Christ Church, Oxford, he joined a remarkable group of natural philosophers working before the Restoration on physiological and physical topics. Much of Hooke's career was driven by financial uncertainty. As an employee, working firstly with Robert Boyle and then as curator of experiments at the new Royal Society from 1662 to 1677, Hooke's status as a professional in a society of gentlemen natural philosophers was problematic. He was continually concerned to establish his credibility, and his defenses of ownership of his ideas were often vitriolic. See also **Boyle, Robert (1627–1691)**.

Hooke's disputes with Newton over light, mechanics, and the theory of planetary motion, in particular, have dominated assessments of his place in the history of science. Contrasts are often drawn between the single-minded genius and the dispersed interests of the technician: "Hooke never achieved the highest status as a scientist, since he was not a theorist but a practitioner, who had not advanced far in mathematics, and tended to work by intuitive understanding rather than sustained thought" (Vickers 1987, *English Science: Bacon to Newton*, pp. 99–100). Other recent historians, suspicious of such neat dichotomies and the evaluations they support, have emphasized the intellectual coherence of Hooke's dauntingly diverse practical and theoretical projects.

Having published a pamphlet on capillary action in 1661, Hooke embarked on an ambitious series of experiments with the microscope on plants, on insect anatomy, and on combustion and respiration. He demonstrated that the function of respiration is simply to bring a constant supply of air to the lungs, not to cool or to pump. Key members of the Royal Society, including Christopher Wren (with whom Hooke would work closely as surveyor and architect in rebuilding London after the fire of 1666), were closely involved in the preparation of Hooke's *Micrographia*. With its magnificent plates of subjects like moss, mites and flies, and the blue mould, it showed critics of the new Society the relevance of the experimental method for the understanding of life. Hooke's discussion of the structure of cork included a use of the term "cell" from which the modern biological usage descends. When Samuel Pepys bought his copy of this "most ingenious book" on 20 January 1665, he "sat up till 2-a-clock in [his] chamber, reading of Mr Hooke's Microscopicall Observations." See also **Plant Sciences (The History)**.

The microscope supported the central claim of the mechanical philosophers that the subvisible world was composed of textured parts like those in the macroscopic realm. After the Fall, our cognitive limitations require us to supplement our senses with external aids to penetrate to the reality behind appearances. Apparently mysterious microphysical effects, which had traditionally been attributed to intrinsic, occult powers, might now, argued Hooke, be seen to be "perform'd by the small *Machines* of Nature, which are not to be discern'd without these helps, seeming the meer products of *Motion, Figure*, and *Magnitude*."

While officially Hooke upheld the mechanists' claim that all natural change is by contact action between material bodies in motion, he did in practice often allow matter to have intrinsically active powers such as gravity, congruity or incongruity, or sympathy. His studies of vibration and attraction owed much to music theory and to empirical traditions in natural magic. Hooke continued working on light, cosmology, earthquakes and fossils, cartography and meteorology. Often collaborating with a network of craftsmen and entrepreneurs, he designed and tested an array of instruments, including barometers and thermometers, clocks and marine chronometers, and telescopes. See also **Paleontology (The History)**.

Hooke always saw knowledge of nature as a sure way to uphold religious orthodoxy, and scriptural exegesis plays a significant part, for example, in his outstanding work in geology and Earth history. He consistently ridiculed reductionist approaches to life and mind. The theory of memory that he outlined in the *Lectures of Light* of 1682 (published in 1705), for example, was based on an executive soul which can radiate its attention out to the "material and bulky" ideas splayed on the coils of memory in the brain.

Additional Reading

Bennett, J.A.: "Robert Hooke as Mechanic and Natural Philosopher," *Notes and Records of the Royal Society of London*, **35**, 33–48 (1980).

Frank, R.: *Harvey and the Oxford Physiologists*, University of California Press, Berkeley, CA, 1980.

Hunter, M., and S. Schaffer: *Robert Hooke: New Studies*, Boydell Press, Woodbridge, UK, 1989.

Kassler, J.C.: *Inner Music: Hobbes, Hooke, and North on Internal Character as Musical Activity*, Athlone Press, London, UK, 1995.

McKie, D.: "Fire and the *Flamma Vitalis*: Boyle, Hooke, and Mayow," In: Underwood, E.: *Science, Medicine, and History*, 1, Oxford University Press, Oxford, UK, 1953, pp. 469–488.

Vickers, B.: *English Science: Bacon to Newton*, Cambridge University Press, Cambridge, UK, 1987.

JOHN SUTTON, Macquarie University, Sydney, New South Wales, Australia

HOOKE'S LAW. When a load is applied to any elastic body so that the body is deformed or strained, then the resulting stress (the tendency of the body to resume its normal condition) is proportional to the strain. Stress is measured in units of force per unit area; strain is the extent of the deformation. For example, when a bar of metal is subjected to a stretching load, the extent of the increase in length of the bar is directly proportional to the force per unit area, i.e., to the stretching load or stress. In general, Hooke's law applies only up to a certain stress called the *yield strength*.

HOOKWORM. See **Dermatitis and Dermatosis**.

HOOPOE. See **Kingfishers and Other Coraciiformes**.

HOPKINS, FREDERICK GOWLAND (1861–1947). Frederick Hopkins was a British biochemist who made important contributions to the discovery of vitamins and was a pioneer of dynamic biochemistry.

Brought up in southern England by his widowed mother, Hopkins left school at 16 to become first a clerk in a London office, and then an assistant in a commercial analytical laboratory. He took night classes in chemistry and his excellent examination performance led to his being invited to enter a prestigious forensic laboratory. At age 27, aided by an inheritance, he became a medical student in a London hospital and, after graduation, worked and published with Archibald Garrod on genetic abnormalities of metabolism. See also **Garrod, Archibald Edward (1857–1936)**.

In 1898 he was invited to the Department of Physiology at Cambridge to develop the teaching of chemical physiology, but his appointment also involved long hours of tutoring students in anatomy. In 1902 (together with S. W. Cole) he reported the isolation of the amino acid tryptophan and in 1906 (with E. G. Willcock) published results indicating that it was nutritionally essential for mice—the first demonstration of this kind for any specific organic molecule.

Hopkins then began to grow young rats on purified diets and concluded that they still lacked some unknown trace nutrient(s). In 1910 he had a breakdown from overwork, but was finally, at almost 50, given a position that removed his tutoring duties so that he could concentrate on research. In 1912 he published what has come to be regarded as a classic paper (*Journal of Physiology* 44: 425–460) demonstrating the value of small quantities of milk in providing "accessory food factors" needed by growing rats, and showing that their effect was not just on the appetites of the animals. In 1929 he would share a Nobel Prize with Christiaan Eijkman for the discovery of vitamins. However, it has been commonly concluded that really it was the full breadth of his work that had earned him this recognition. See also **Eijkman, Christiaan (1858–1930)**; and **Nutritional Science (The History)**.

In 1913 Hopkins gave an invited address to the British Association for the Advancement of Science that set out his ideas as to the future of biochemistry as a basic science, independent of its direct assistance to medical practice. He argued that intermediary metabolism was capable of study because it involved simple substances undergoing comprehensible reactions. He viewed the living cell as an organized, highly differentiated system of interdependent processes in dynamic equilibrium.

In the following year he was appointed as Cambridge's first Professor of Biochemistry, and began to develop a department active in both teaching and research. His own research concentrated on the function of glutathione in oxidative systems in cells, but this turned out to be less significant than he had previously thought, and his most important contribution was probably giving encouragement and finding finance for the independent workers in his large group. He also spent much time with the activities of the Medical Research Council, became President of the Royal Society in 1931, and received many honors including a knighthood and the Order of Merit. He was repeatedly described as the most modest and gentle of men, but with an unexpected element of determination. See also **Biochemistry (The History)**.

Additional Reading

Gillispie, C.C.: *Dictionary of Scientific Biography (1970–1980)*, Charles Scribner's Sons, New York, NY.

Hopkins, F.G. "Accessory Food Factors," *Journal of Physiology*, **44**, 425–460, (1912).

Kamminga, H., and M.W. Weatherall: "The Making of a Biochemist I: Frederick Gowland Hopkins' Construction of Dynamic Biochemistry," *Medical History*, **40**, 269–292 (1996).

Needham, J.: *Hopkins and Biochemistry*, Heffer and Sons, Cambridge, UK., 1949.

K. J. CARPENTER, University of California, Berkeley, CA

HOPPE-SEYLER, ERNST FELIX IMMANUEL (1825–1895). Born Felix Hoppe at Freiburg, Germany, he was orphaned at the age of nine, and brought up by his brother-in-law, Dr Seyler, whose name he added to his own in 1864. He trained in medicine at the Universities of Halle, Leipzig and Berlin, and completed his MD in 1851. In 1856, after finding medical practice uncongenial, he joined R. Virchow's Pathological Institute in Berlin as Head of Clinical Chemistry. From 1861 to 1872 he taught physiological chemistry at the University of Tübingen before assuming a separate chair of physiological chemistry at the University of Strasbourg, where he spent the remainder of his life. In 1858 he published a handbook of analytical tests to be used in physiological and pathological chemistry which remained in standard use until the end of the nineteenth century. Using the new technique of absorption spectroscopy, he made many innovations in the chemical study of blood and urine, including the isolation of the coloring matters of the blood. His investigation of the reactions between hemoglobin, oxygen and carbon monoxide showed that physiological oxidations were affected in the tissues and not in the blood itself. See also **Virchow, Rudolf Carl (1821–1902)**.

Perhaps Hoppe-Seyler's greatest achievement was to argue for the separation of physiological chemistry from medicine and to make biochemistry an independent academic discipline. To this end, in 1877, he founded the *Zeitschrift für physiologische Chemie* (Magazine for Physiological Chemistry). In 1881 he summarized all that was then known of biochemistry in *Physiologische Chemie*, which is over a thousand pages in length. He died while studying gases dissolved in Lake Constance. See also **Biochemistry (The History)**.

Additional Reading

Fruton, J.S.: Hoppe-Seyler, In: Gillispie, C.C.: *Dictionary of Scientific Biography*, Vol.6, Charles Scribner's Sons, New York, NY, 1972, pp. 504–506.

W. H. BROCK, University of Leicester, Leicester, UK

HOPS. See **Mulberry Family**.

HORIZON (Astronomical). Also called sensible horizon; real horizon. The plane that passes through the observer's eye and is perpendicular to the zenith at that point; or, the intersection of that plane with the celestial sphere (i.e., a great circle on the celestial sphere equidistant from the observer's zenith and nadir). It is the projection of a horizontal plane in every direction from the point of orientation. See also **Celestial Sphere and Astronomical Triangle**.

HORIZON (Celestial). Also called rational horizon; geometrical horizon; true horizon. The plane, through the center of the earth, perpendicular to a radius of the earth that passes through the point of observation on the earth's surface; or, the intersection of that plane with the celestial sphere.

In astronomy, the term horizon is used to describe the great circle cut out on the celestial sphere by a plane perpendicular to the direction of gravity. If this plane is tangent to the surface of the earth, the horizon so described is the *apparent horizon*; if the plane passes through the center of the earth, we have the *geocentric horizon*.

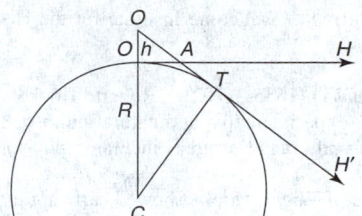

Fig. 1.   Visible and astronomical horizons.

The difference in direction between the visible horizon (as the line where earth and sky meet) and the apparent astronomical horizon is known as the dip of the horizon. In Fig. 1, $O'H'$ represents the direction of the visible horizon from an observer at a station O' elevated above the surface of the earth by an amount h. OH represents the direction of the horizon for the observer on the surface of the earth at O. The angle HAH' is the dip of the horizon, and may be given, very approximately, by this relation: the dip of the horizon (expressed in minutes of arc) is equal to the square root of the height of the observer above the surface of the earth (expressed in feet). The distance $O'T$ from the observer to the visible horizon is approximately as follows: the distance of the visible horizon (expressed in miles) is given by the square root of $\frac{3}{2}$ the height of the observer above the surface of the earth (expressed in feet). This distance is frequently very much increased by an effect known as looming of the horizon, produced by refraction of light in heated (or cooled) layers of the air near the surface. Both the expressions for the dip and distance of the horizon are applicable only when the point of observation of the visible horizon (the point T) is actually on the surface of the earth. See also **Celestial Sphere and Astronomical Triangle**.

HORIZON (Geographic). Also called apparent horizon, local horizon, and visible horizon, the distant line along which earth and sky appear to meet. In both popular usage and weather observing, this is the usual conception of horizon. Nearby prominences are said to obscure the horizon and are not considered to be a part of it. For observational reference, the minimum desirable horizon distance should be of the order of three miles.

Sea-level horizon, also called ideal horizon, sensible horizon, sea horizon, visible horizon, and apparent horizon, is the apparent junction of the sky and the sea-level surface of the earth; the horizon that is actually observed at sea. This type of horizon is used as the reference for establishing times of sunrise and sunset.

In these definitions of horizon, the zenith is considered to be at right angles to the horizon.

HORIZONTAL COORDINATE SYSTEM (Astronomy). A system of spherical coordinates on the celestial sphere, which uses the horizon as a fundamental plane. Planes perpendicular to the horizon cut out great circles on the celestial sphere known as vertical circles. The fundamental direction selected in the fundamental plane is true south. The azimuth of a point on the celestial sphere is the angular distance, measured in the plane of the horizon, from the true south direction to the point of intersection of the vertical circle through the object with the horizon. There are several different methods for expressing azimuth, but the astronomical method is to measure azimuth from the south through the west through 360°. The altitude of a point on the celestial sphere is the angular distance, measured along the vertical circle through the point, from the plane of the horizon to the point.

The horizontal system of spherical coordinates is frequently referred to as the altazimuth system.

See also **Celestial Sphere and Astronomical Triangle**.

HORMONES. During the past few years, the classical concept of hormone has been undergoing revision and expansion. A half-century ago, when only a few hormones were reasonably well understood, it was generally believed that there were a comparatively few very important complex biological substances generated by a few glands that stimulate and inhibit principal body functions. Investigators have learned, particularly since the early 1980s, that the creation of hormones is not an exclusive process for the endocrine glands, that receptors for any given hormone are not usually simply concentrated in relatively limited locations of the body, but are found sometimes where they were least suspected. For example, a few years ago when J. Roth and J. Kova (National Institutes of Health) confirmed insulin receptors in the human brain, they sought and found receptors in the testes and liver. In the case of insulin, this had led investigators to explore the possibility of insulin synthesis occurring, not just in the pancreas, but elsewhere in the body.

Researchers have also commenced studies of other life forms, insects and annelida for example, in a search for hormone receptors and, indeed, have found material similar to insulin and receptors that are reminiscent of human insulin receptors. That hormone production is not limited to the endocrine glands has been suspected for a number of years, but such cases were considered the exception rather than the rule. Modern investigators no longer accept this hypothesis. Suspected earlier, but not well confirmed until the mid-1980s, the human heart, once simply considered as a pump, is now known to create at least one hormone, *atrial natriuretic factor* (ANF). A few years earlier, researchers found that the human brain synthesizes important substances, such as endorphins and enkephalins.

Thus, the study and indeed the definition of hormones is departing rapidly from the former exclusive association with endocrinology. The field is becoming broader and more complex and the number of previously unidentified substances playing some form of hormonal role is increasing. More hormones are being found and in more locations in the living process, both human and other life forms. The field no longer is limited to a comparatively exclusive few sophisticated substances.

In connection with a new hypothesis for hormones, Roth has suggested that cell hormones and neurotransmitters began as what cell biologists term *tissue factor*—substances that stimulate cells to grow or come together or otherwise react biochemically. Only when animals evolved to have extreme cell differentiation and cellular organization did glands evolve to overproduce these hormones so the animals could use them in more clever and sophisticated ways. This theory would explain why many mammalian hormones are also tissue factors. As examples, insulin and glucagon, in addition to playing roles as hormones, also act locally as tissue factors on cells within the pancreas. Exocrine and endocrine functions overlap—there is no difference between exocrine and endocrine functions at the level of unicellular organisms. Roth points out that such a hypothesis would explain the finding that many classical messenger molecules, such as prostaglandins, nerve growth factor, and the hormonal substances are found in exocrine fluids such as saliva, intestinal secretions, milk, and semen.

Investigations are also being conducted apace in what might be called hormone genetics. It has been found, for example, that guinea pigs produce two different insulins, one type made in the pancreas, the other type synthesized in the brain and other organs. Thus, there are differences in gene expression. New findings may explain why cancer cells sometimes secrete hormones that cause severe metabolic disturbances. Lung cancers, for example, are prone to produce vasopressin, a cause of water retention. Perhaps the tumor-generated vasopressin is not normal vasopressin, but a slightly different hormone that has escaped detection by radioimmunoassays. There may be numerous other instances where investigators are seeking one of the better understood hormones, when a somewhat different, uncataloged hormone should be the target. Admittedly, the thoughtful speculation involved in the reevaluation of hormone science will require considerably more research and proof.

In studies of the immune system, investigators have reported on the finding of a heretofore unknown immunoregulatory hormone, 1,25-dihydroxyvitamin D_3. The substance was found to be effective in suppressing interleukin-2.

By applying recombinant DNA techniques, a group of researchers has produced two human fertility hormones, human chorionic gonadotropin (hCG) and human luteinizing hormone (hLH). This is one of the first examples in which recombinant DNA techniques have been used to produce molecules that are a combination of proteins and carbohydrates in mammalian cells. The two hormones are similarly structured, consisting of two polypeptide chains that are put together inside cells and processed. It has been suggested that the hormones will be useful in the treatment of infertility because they can induce both ovulation and sperm production. Past hormone treatments have involved extracts from pituitaries, urine, or placentas which do not yield a pure product.

Hormone Science in the Traditional Sense

In animals, hormones are organic compounds (usually of considerable complexity and even after years of research not fully understood) that are secreted by endocrine (ductless) glands, such as the adrenal gland, the thyroid and parathyroid glands, the pituitary gland, and the gonads, among others. See also **Endocrine System**. Hormones are sometimes commonly called by the names of the glands that secrete them. Thus, there are adrenal cortical hormones, thyroid and parathyroid hormones, etc. Hormones are regulators of physiological processes within the body, exerting control over such processes as metabolism, growth, reproduction, molting, pigmentation, and electrolytic and osmotic balance, among other processes. Apparently, hormones achieve these objectives chemically and electrically, although the mechanisms are not fully understood and, in fact, the mechanisms may vary from one situation to the next. At one time, hormones were loosely called "chemical messengers" because they are transported from point to point within the organism and thus effect actions at distances from the region where they are made. If one visualizes secreting glands as sensors of a type detecting need for correction of some physiological process, then the hormones might be visualized as both the transmitters or carriers of this information and the initiators of actions as well. The conventional concept is that cells have receptors on their surfaces, which sense the presence of specific hormones. At one time, it was firmly believed that hormones, particularly polypeptide hormones, such as insulin, prolactin, and growth hormone, all of which are large charged molecules, could not penetrate through the cell's membrane and actually enter the cell. This belief has since been altered because researchers have shown that insulin, for example, can enter into the cell. Referring to this process as "internalization," one investigator in 1978 suggested that the internalization of polypeptide hormones will be one of the most active topics in cell biology for a number of years.

Research has shown that hormones and/or their receptors may be degraded. As gross examples of this type of situation, it is known that many obese people with high concentrations of insulin in their blood also have normal concentrations of blood sugar. Why doesn't the insulin decrease the blood sugar concentration in these cases? It is well known that pregnant women produce much angiotensin II, which normally increases blood pressure, but these women usually do not have hypertension. What alterations in the hormone-cell mechanism provide this result? There are also instances where males have tumors that secrete large quantities of a hormone that stimulates the production of testosterone, and yet there is no evidence of abnormal amounts of testosterone. At least two questions can be posed: Do certain hormones lose their effectiveness with time? Or, are there changes in target receptor cells? Research has indicated that there may be a relationship between concentration of hormones and the surface receptors which bind them, such receptors being inactive for a time or possibly disappearing from the cell surface altogether. A number of investigators have observed that a better understanding of the manner in which hormones affect their own and other receptors possibly may result in new ways to treat certain diseases, including insulin-resistant diabetes.

In recent years, it has been shown that a wide variety of receptors are regulated by hormones. Some receptors are sensitive to only one hormone; this appears to be the case with insulin. Others appear to be regulated not only by one hormone, but others as well. For example, it has been shown that the receptors for TRH (thyrotropin-releasing hormone) are not exclusively regulated by TRH, but also by other hormones. Receptors for gonadotropins (pituitary hormones that act on the gonads) appear to be regulated by hormones in addition to gonadotropin. There are numerous other instances of this kind.

Hormones display not only great variations in function, but also in their chemical nature, of which there is a great diversity. Some are steroids, such as estrogen, progesterone, cortisone, etc., while others are amino acids (thyroxine), polypeptides (vasopressin), low-molecular-weight proteins, and conjugated proteins. Amino acid and steroid hormones have been isolated and many, including insulin, have been synthesized. Other types are prepared directly from the endocrine organs of animals.

Hormones produced by one species usually show similar activity in other species. The hormones showing greatest species specificity are proteins or conjugated proteins.

Hormones are markedly affected by deficiencies or excesses of the various vitamins and other dietary essentials.

Because of the great complexity of a number of the natural hormones, conventional approaches of organic synthesis which have been used so successfully over the years in connection with many drugs have not proved viable to date with some of the hormones. Insulin is an example. Presently, millions of diabetics still depend upon animal insulin as extracted from the pancreatic glands of slaughtered pigs. If diabetes mellitus continues to become more prevalent, as it has over the past several years, natural sources may not be sufficient.

Recombinant DNA technology was applied to the problem the first time a few decades ago. One group has been successful in inducing the bacterium *Escherichia coli* to manufacture and secrete rat proinsulin, an immediate precursor of rat insulin that incorporates insulin itself. Research like this is an important step toward the objective of developing bacterium-based industrial systems that can replace animal and human tissues as the source of medically useful proteins, such as insulin, growth hormone, and clotting factor.

Classes of Hormones

Hormones may be grouped into two distinct types: (1) *direct-acting*; and (2) *stimulating*-substances that stimulate other organs to produce their own characteristic hormones. The latter group is sometimes called the *tropic hormones*. See summary of hormones in Table 1.

Thyroid Hormones. These are compounds of the amino acid *thyronine*. They are present in the free form only to a slight extent, existing chiefly as constituents of the protein thyroglobulin. The most important of these acids in terms of hormone action are the 3,5,3'-tri, and the 3,5,3',5'-tetraiodocompounds, *triiodothyronine* and *thyroxin*, the structures of which are given in Part 1 and Part 2 of Table 1. See also **Thyroid Gland**. The action of thyroid hormones is to accelerate cellular reactions and to increase the metabolic rate and oxygen consumption of tissues. They effect this action by stimulating many of the enzyme systems, not only the glucose oxidation system and the cytochrome chain for dehydrogenating the coenzyme NADPH, but other processes, such as the synthesis of proteins from amino acids. Their effects are clearly apparent in the pathological changes in the organism caused by their excess or deficiency. The thyrotrophic hormone and other biochemical interactions with the thyroid gland are discussed later in this entry.

Parathyroid Hormones. The influence of the parathyroid glands on the regulation of calcium concentrations in the blood of mammals was first recognized by MacCullum and Voegtlin in 1909.

More recently, several groups of investigators have succeeded in purifying and partially identifying the structure of the hormone, variously called *parathormone* and *parathyroid hormone*. This is a single chain peptide hormone with a molecular weight of about 8,000. A second parathyroid hormone, *calcitonin*, was postulated by Copp (1961). Subsequent research has indicated that this hormone is actually the hormone produced by the thyroid gland. However, a parathyroid calcitonin may exist in certain species.

The more classical function of parathyroid hormone is concerned with its control of the maintenance of constant circulating calcium levels. Its action is on (1) the kidney, where it increases the phosphate in the urine, (2) the skeletal system, where it causes calcium resorption from bone, and (3) the digestive system, where it accelerates (stimulates) calcium absorption into the blood. The hormone and gland exhibit characteristics of feedback control; when the concentration of calcium ions in the blood falls, the secretion of the hormone increases, and when their concentration rises, the secretion of hormone decreases. See also **Parathyroid Glands**.

Adrenal Cortical Hormones. The adrenal gland is made up of two parts, the medulla and the cortex, each of which secretes characteristic hormones. The hormones of the adrenal medulla are the catecholamines, epinephrine (adrenalin) and norepinephrine (noradrenalin), which are closely related chemically, differing only in that epinephrine has an added methyl group. See Table 1. In fact, animal experiments have established a metabolic pathway for the biosynthesis of both compounds from the amino acid phenylalanine, which involves enzymatic oxidation and decarboxylation reactions. It is also to be noted that the isomeric form of norepinephrine is most important; the natural D-form (which

incidentally, is levorotatory) has many times the activity of the synthetic isomer. Epinephrine has a pronounced action upon the circulatory system, increasing both blood pressure and pulse rate, and hence the cardiac output by its direct action upon the heart muscle, and especially because it causes constriction of the arterioles. However, its effects upon smooth muscles vary; it relaxes the muscles of the digestive system, but contracts the pyloric sphincter.

Norepinephrine does not affect the cardiac output, although it does raise the blood pressure by constricting the arterioles. Its muscular effects are less pronounced. Both epinephrine and norepinephrine release free fatty acids from adipose tissue, so raising its level in the blood. This effect is due to the action of the hormones in accelerating enzymatic reactions whereby the esters of the fatty acids are hydrolyzed. The third type of action of epinephrine is its effect upon the carbohydrate metabolism, notably the acceleration of the hydrolysis of glycogen in muscular tissue and the liver. It raises the glucose level in the blood, and the rate of glucose oxidation, with resulting increase in oxygen utilization, carbon dioxide production, and body temperature. See also **Adrenal Glands**.

The hormones of the adrenal cortex are steroids. See also **Steroids**. Among them there are a number of hormones with androgenic activity, such as adrenosterone and 17α-hydroxyprogesterone, which are discussed under the sex hormones later in this entry. In all, over ten steroids have been identified in the adrenal cortex, including seven of characteristic cortical activity. These are corticosterone, from which the others are named, 17α-hydroxyl-11-dehydrocorticosterone (cortisone), 17α-hydroxycorticosterone (cortisol or hydrocortisone), and 18-oxocorticosterone (aldosterone). Only two hormones, cortisol and corticosterone, are normally released in fairly large quantities, and another, aldosterone, deserves mention because of its somewhat different effects, even though it is released to a far lesser extent.

All of these hormones are synthesized from cholesterol in the adrenal cortex, by an extended series of reactions, which include many related compounds. Although these hormones have widespread effects throughout the organism, their primary mechanism is not known, so that many of the effects may be indirect. Much of the knowledge of their action arises from studies of insufficiency or hyperactivity of the adrenal cortex, which produces a wide variety of pathological conditions. See Table 1.

It is generally considered that aldosterone, and to some extent the other hormones, have a regulatory effect upon the metabolism of electrolytes and water, particularly upon the concentration of the ions of the alkali metals in intracellular fluids. Administration of steroids also increases the concentration of calcium ions in those fluids. However, all three of these hormones have a number of other effects, roughly in the order of potency—cortisol, corticosterone, aldosterone. They produce changes in the metabolism of carbohydrates, proteins, and fats.

For the carbohydrates alone, three major effects are evident—increase in the rate of formation of glucose, increase in the rate of release of glucose from the liver, and increase in the rate of utilization of glucose. These hormones affect the digestive system, increasing the secretion of hydrochloric acid, pepsinogen, and trypsinogen. They prevent inflammatory responses to bacterial or even chemical stimuli; they counteract anaphylactic shock, and other effects of hypersensitivity. Obviously, these properties have led to their widespread therapeutic use.

There are relationships between the adrenal cortical hormones and the thyroid and pituitary glands. Depression of the function of the adrenals produces thyroid deficiency, whereas administration of thyroxine stimulates the ACTH-adrenal cortical mechanism.

Pituitary Hormones. The hormones of the hypophysis (pituitary gland) are quite numerous, being secreted variously in three parts of the gland—the neurohypophysis (posterior lobe), the adenohypophysis (anterior lobe), and the *pars intermedia*, which connects the other two.

The chief hormones of the neurohypophysis are the polypeptides oxytocin and vasopressin. The hormone characteristic of the *pars intermedia* is the melanocyte-stimulating hormone. It is usually spoken of in the plural, since in most mammals both alpha and beta forms are known. The structures of the first two are shown in the Table 1. See also **Diabetes Insipidus**.

The most prominent effect of oxytocin is the contraction of smooth muscle, especially of the uterus. It also has a major effect upon the muscles about the breast, and so stimulates the ejection of milk in lactating animals. It has a definite stimulating effect upon the muscles of the ureter, urinary bladder, intestine, and gall bladder.

TABLE 1. REPRESENTATIVE HUMAN HORMONES

Hormone Common Names, (Synonyms), Structure and Production Site	Principal Physiological Functions	Interrelationships with Vitamins
Adrenocorticotropic Hormone (ACTH) [CAS: 9061-27-2] (Adrenocorticotropin; corticotropic hormone) Straight-chain, simple, polypeptide, 39 amino acids, no S — S bridges. (See text, Fig. 1.) Molecular Weight ~4500 Production Site: Anterior pituitary	Maintenance of adrenal cortex Promotes secretion of steroids, oxidative phosphorylation in adrenal cortex Mobilizes and increases oxidation of free fatty acid in adipose tissue Increases gluconeogenesis in liver; increases cyclic adenosine monophosphate (AMP) in adrenal cortex Decreases urea formation in liver	Ascorbic acid: depleted in adrenal cortex on stimulation by ACTH Biotin and vitamin A: adrenocortical insufficiency noted in biotin and vitamin A deficiency Niacin: production of reduced nicotinamide adenine dinucleotide (phosphate) (NADPH) by ACTH via cyclic adenosine monophosphate (AMP) Niacin and pantothenic acid: synergistic with ACTH in steroid hormone synthesis Vitamin D: antagonized directly by ACTH via cortisol action
Aldosterone [CAS: 52-39-1] (Aldocortin; electrocortin; mineralocorticoid; 18-oxocorticosterone) Molecular Weight 360.4 Production Site: Adrenal cortex	Maintenance of normal electrolyte blood balances Prolongs survival of adrenalectomized animals Accelerates gluconeogenesis Regulates kidney function	Ascorbic acid: adrenal cortex depleted of ascorbic acid on production of aldosterone Biotin: prolongs life in adrenalectomized rats Niacin: nicotinamide adenine dinucleotide (phosphate) (NADPH) involved in synthesis of aldosterone
Cortisol [CAS: 50-23-7] (Hydrocortisone, 17-hydroxycorticosterone) Molecular Weight 362.5 Production Site: Adrenal cortex	Increases (1) protein catabolism (excepting liver) gluconeogenesis; (2) carbohydrate anabolism (liver); (3) blood sugar; (4) glucose absorption; (5) brain excitation; (6) spread of infections; (7) urinary glucose and nitrogen; (8) stress tolerance; (9) lactation; (10) water diuresis Regulates general adaptation syndrome, water balance, blood pressure, and hormone release. Decreases (1) fat anabolism; (2) growth rate; (3) inflammation; (4) eosinophils; (5) lymphocytes; (6) antigen sensitivity; (7) respiratory quotient; (8) ketosis; (9) wound healing; (10) skin pigmentation; (11) RBC hemolysis	Ascorbic acid: may be required for steroid hormone biosynthesis; depleted from adrenal cortex on cortical secretion Biotin: adrenocortical insufficiency noted in biotin deficiency Folic acid and pantothenic acids maintain secretions of steroids by adrenal cortex Niacin: nicotinamide adenine dinucleotide (phosphate) (NADPH) required for steroid hormone biosynthesis Vitamin A: deficiency causes cortical necrosis Vitamin D: action antagonized by cortisol by reducing calcium absorption in intestine
Epinephrine [CAS: 51-43-4] (Adrenaline, adrenin, suprarenin, vasotonin, vasoconstrictine, adrenamine, levorenine) Molecular Weight 183.2 Production Site: Adrenal medulla and chromaffin cells in gut	Blood circulation: increases blood pressure; peripheral vasodilator; increases heart output and rate; flow increases in brain, liver, and skeletal muscle Central nervous system: causes restlessness, anxiety Kidney: reduces glomerular filtration rate Lung, intestine, genital system: inhibited motility Metabolic effects: increases oxygen consumption, temperature, basal metabolic rate, gluconeogenesis Pituitary effects: stimulates production and release of ACTH and corticoids	Ascorbic acid: maintains reduced state of epinephrine Ascorbic acid, folic acid, and vitamins B_6, and B_{12} are cofactors in synthesis of epinephrine from phenylalanine

(continued)

TABLE 1. (*Continued*)

Estradiol [CAS: 50-28-2] (Female hormone; dihydrotheelin; dihydrofollicular hormone dihydrofolliculin) Molecular Weight 272.4 Production Sites: Ovarian follicles; tests; corpus luteum; adrenal cortex; placenta	Regulates menstrual cycle, female sex behavior Maintains secondary sex characteristics Affects antibody properties Induces estrus, uterine hypertrophy, vaginal cornification; potentiate sand stimulates calcitonin secretion	Folic acid: involved in mitotic effect of estradiol Niacin, diphosphopyridine nucleotide (DPN), triphosphopyridine nucleotide (TPN): involved in increased respiration and in cholesterol precursor synthesis Pyridixine: competes as cofactor with estrogen sulfate in kynurenine aminotransferase activity Vitamin D: synergistic in calcium metabolism with estradiol Vitamin E: involved in follotropin production or release
Follicle-Stimulating Hormone (FSH) (Follotropin, luteoantine, thylakentrin, Prolan A, gonadotropin 1, gametogenic hormone, follicle ripening hormone, gametokinetic hormone) Structure: Not fully definitized. Production Site: Anterior pituitary.	Female: stimulates ovarian follicles to grow and to develop, forming multiple layers and antra Male: stimulates seminferous tubules; stimulates spermatogenesis	Ascorbic acid: depletion in ovary due to follicle-stimulating hormone and luteinizing hormone action Vitamin E: required to maintenance of membranes in sex organs
Glucagon (HGF) (Hyperglycemic-glycogenolyltic factor; glucagon; HG-factor) Structure: Polypeptide, 29 amino acids (structure determined). No S—S bridges Molecular Weight ~3500 Production Site: Alpha cells in pancreas.	Increases: blood sugar; blood K^+, oxygen consumption, liver glycogenolysis, gluconeogenesis, nitrogen and salt excretion Decreases: liver glycogen, protein formation, gastric juice, fatty acid synthesis	Ascorbic acid: depletion of adrenal ascorbic acid by glucagon
Insulin (no synonyms) Structure: 51 amino acids. Known and synthesized. 3 S—S bridges. (See text, Fig. 4) Molecular Weight 5,734 (monomer); 12,000–48,000 (polymer), depending upon pH. Production Site: Beta cells of islets of pancreas.	Regulates carbohydrate and fat metabolism, especially glucose and fat oxidations Stimulates amino acid and glucose transport into cells and protein synthesis	Ascorbic acid: acts similarly to alloxan (i.e., antagonist)
Luteinizing Hormone (LH) (Luteotropin, ISCH) Structure: Globular glycoprotein with S—S bridges. Molecular Weight 26,000 Production Site: Anterior pituitary.	Female: promotes estrogen and progesterone secretion, ovulation; maintains ovarian tissues Male: stimulates Leydig cells to secrete testosterone; gametogenic with follotropin (FSH)	Ascorbic acid: ovarian depletion on LH stimulation Vitamin E: involved in spermatogenesis
Melanocyte-stimulating Hormone (MSH) (Melanotropin, chromatophorotropic hormone; pigmentation hormone) Structure: Polypeptide; purified, synthesized; alpha and beta forms; straight chains. Molecular Weight: 1500 (alpha) 2100–2600 (beta) Production Site: Intermediate lobe of pituitary.	Mammals: exerts small effect on skin pigmentation (protection from sunlight not fully proved) Expands or contracts pigments in various chromatophores Expands melanophore pigments with color changes in amphibia (adaptation to environment) Lower vertebrates: increases sensitivity to light; decreases dark adaptation time.	Ascorbic acid: adrenal cortex depleted on ACTH and MSH activity Vitamin A: MSH decreases dark adaptation time.
Norepinephrine [CAS: 149-95-1] (Arterenol; noradrenaline; levarterenol) Molecular Weight 169.2 Production Site: Adrenal medulla; adrenergic nerve endings; chromaffin cells.	Blood circulation: increase blood pressure; peripheral vasoconstrictor without change or slight decrease in output and heart rate. No flow increase in brain, liver, or muscle Central nervous system effects: adrenergic transmitter agent at synapses; no brain excitation Kidney: decreases glomerular filtration rate Lung, intestine, genital system: inhibited Metabolic effects: weak epinephrine effect	Ascorbic acid: protects against oxidation of norepinephrine Ascorbic acid, folic acid, and vitamin B_6 are cofactors in synthesis of norepinephrine from phenylalanine

TABLE 1. (*Continued*)

Oxytocin [CAS: 50-56-6] (Oxytocic hormone; pitocin; uteracon; α-hypophamine) Cys—Tyr—Ile—Gln—Asn—Cys—Pro—Leu—Gly NH$_2$ Molecular Weight 1007 Production Site: Hypothalamus.	Uterine contraction, milk ejection, facilitates sperm ascent in female tract Decreases membrane potential of myometrium, basic metabolic rate, and liver glycogen Stimulates oviposition in hen, releases luteinizing hormone (LH) Increases blood sugar and urinary sodium and potassium	Findings on interrelationships with vitamins are not extensive
Parathyroid Hormone (PTH) (Parathormone) Structure: Simple polypeptide (83 amino acids), sequence determined; straight chain; No S—S bridges. Production Site: Parathyroid glands.	Increases blood calcium, kidney calcium reabsorption, phosphate excretion, and blood citrate level Mobilizes calcium and phosphate from bone Activates calcium and phosphate absorption from the gastrointestinal tract (for which vitamin D is required) Increases osteoclast formation	Vitamin D: synergistic with PTH in maintenance of serum calcium
Progesterone [CAS: 57-83-0] (Progestin, luteosterone) 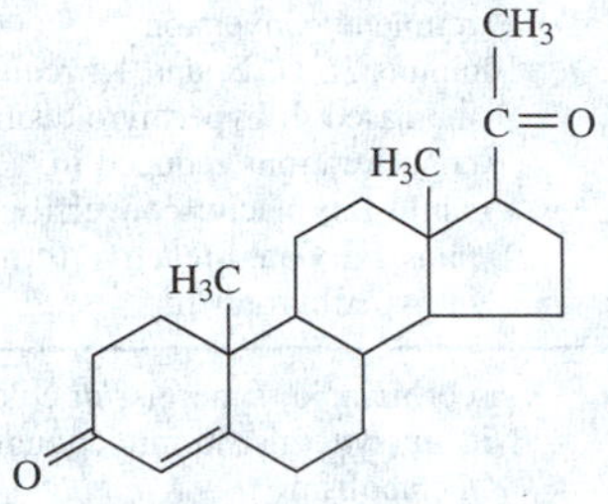 Molecular Weight 314.5 Production Sites: Ovary (follicles, corpus luteum); testicles; adrenal cortex; placenta	In low concentrations: prepares uterus for blastocyst implantation; promotes ovulation and mammary gland development; regulates female sex accessory organs; weak corticosteroid properties; precursor to sex hormones In high concentrations: maintains pregnancy; represses ovulation and sex activity; inhibits vaginal cornification and parturition; decreases myometrial excitation	Ascorbic acid: depleted from adrenal cortex or ovary on progesterone formation Niacin: diphosphopyridine nucleotide (DPN) involved in progesterone synthesis
Prolactin LTH (Lactogenic hormone; lactogen; galactin; mammotropin) Structure: Single-chain protein, 205 amino acids Molecular Weight 23,000–25,000 Production Site: Anterior pituitary	Initiates lactation Develops mammary glands in female Increases weight and growth (similar to somatotrophin in some species) Participates in nidation of zygote Protein anabolism (some species) Growth and secretion of crop gland (birds) Luteotropic (only in mouse, rat) Promotes maternal behavior	Not fully determined. Generally participates with other substances having growth action

Relaxin
  (Releasin, cervilaxin)
Structure: Polypeptide (4 peptides with activity have been isolated); about 30–40 amino acids in each peptide
  Molecular Weight 4000–5000
Production Site: Corpus luteum in pregnancy
	Enlarges birth canal in preparation for parturition
	Separation of symphysis pubis, loss of rigidity in pelvic bones
	Decreases uterine motility
	Maintains pregnancy
	Increases sensitivity to oxytocin; release oxytocin
	Stimulates mammary gland
	Stimulates inhibition of water in uterus
	Inhibits uterine contraction
		Ascorbic acid: maintains mucoprotein ground substance in connective tissue, affected by relaxin

Somatotropin (STH) (Growth hormone, GH; somatotrophic hormone; hypophyseal growth hormone) Structure: Known and synthesized; coiled, unbranched protein; 188 amino acid residues; 2 S—S bridges Molecular Weight 21,500 Production Site: Anterior pituitary	Promotes general growth of organism Promotes skeletal growth, protein anabolism, fat metabolism, carbohydrate metabolism, water, and salt metabolism	Relates with all vitamins in connection with growth actions

(continued)

TABLE 1. *(Continued)*

Testosterone [CAS: 58-22-0] (17 beta-hydroxy-4-androsten-3-one) 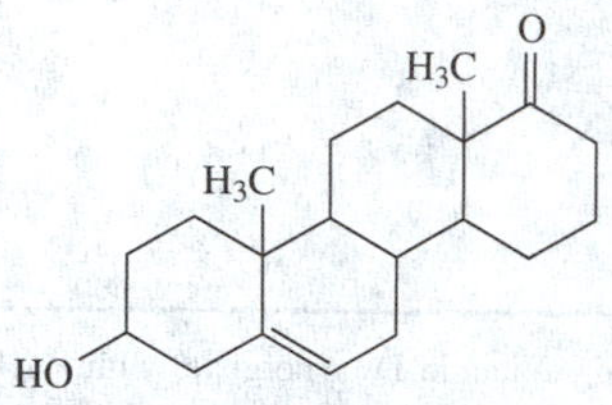 Molecular Weight 288.4 Production Sites: Interstitial cells of ovary and testis; adrenal cortex; embryonic placenta	Controls secondary male sex characteristics Maintains functional competence of male reproductive ducts and glands Increases protein anabolism; maintains spermatogenesis; inhibits follotropin Increases male sex behavior; increases closure of epiphyseal plates	Ascorbic acid, folic acid, vitamins A and E are synergists with testosterone for maturation of germ cells and increased anabolic activity
Thyroid-stimulating Hormone (TSH) (Thyrotrophic hormone, thyrotropin) Structure: Glycoprotein (300 amino acids) Molecular Weight 26,000–30,000 Production Site: S^2 type cell, anterior pituitary	Regulates body temperature via thyroxine Maintains thyroid gland and its secretory activity (colloid discharge) Maintains iodine uptake by thyroid gland Promotes differentiation in embryo during development via thyroxine Stimulates coupling of diodotyrosine to form thyroxine	Ascorbic acid, thiamine, riboflavin, and vitamin B_{12}: requirements increase in hyperthyroidism; tissue concentrations reduced Vitamin A: massive doses of vitamin A inhibit secretion of TSH; thyroid hormones required for carotene and retimene conversions Vitamins A, D, E, and K: requirements increased in hyperthyroidism; tissue concentrations reduced in Vitamin B_6, niacin: conversion to phosphorylated reactive forms impaired in hyperthyroidism
Thyroxine (T_4) [CAS: 7488-70-2] (3,5,3′,5′ tetraiodothyronine) 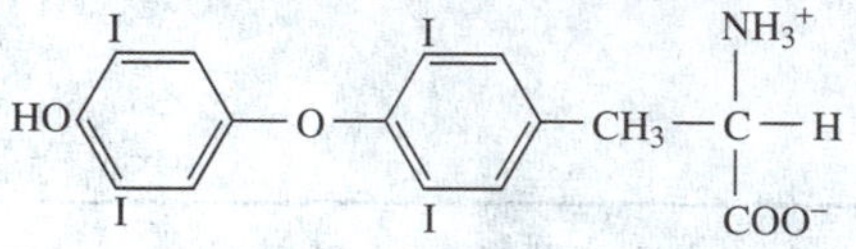 Molecular Weight 776.9 Production Site: Thyroid gland	Regulates growth, differentiation, oxidative metabolism, electrolytic balance Increases carbohydrate metabolism, calorigenesis, protein anabolism, basal metabolic rate, oxygen consumption, fat catabolism, fertility Sensitizes nervous system	Ascorbic acid: synergist in cold survival Niacin: synergist in mitochondrial metabolism Vitamin A: T_4 is required for vitamin A synthesis in liver Vitamin B_{12}: T_4 aids in B_{12} absorption B complex vitamins: deficiencies develop in hyperthyroidism
Vasopressin (Arginine vasopressin; antidiuretic hormone; ADH; pitressin; tonephin; vasophysin) Cys – Tyr – Phe – Glu NH₂ – Asp NH₂ – Cys - Pro - Leu - Gly NH₂ Vasopressin Molecular Weight 1084 (arginine-vasopressin) Production Site: Hypothalamus	Elevates blood pressure (mammals) (reverse effect in birds) Decreases kidney blood flow Antidiuretic, releases ACTH Increases sodium chloride and urea excretion Regulates water balance Stimulates contraction of smooth muscles Increases renal tubular water reabsorption Releases anterior pituitary hormones	Not fully determined

The most prominent effect of vasopressin is upon the kidneys, where it stimulates the resorption of water in the tubules (which by repeated release and absorption concentrate the urine). It also constricts the coronary arteries, raises the blood pressure, and exhibits the effect of oxytocin upon smooth muscles, but generally to a lesser degree.

The action of the melanocyte-stimulating hormones has been established by studies of animals, in which they cause dispersal of certain black pigments from the cells that contain them, with resulting darkening of the skin.

The adenohypophysis is the part of the gland in which the tropic hormones are secreted. They include the adrenocorticotropic hormone (ACTH), the thyrotropic hormone (TSH), and somatotropin, as well as three hormones with pronounced effects upon the gonads: the hormone prolactin, the follicle-stimulating hormone (FSH) and the luteinizing or interstitial cell stimulating hormone (LH or ISCH).

ACTH. Adrenocorticotropin (ACTH) in humans is a polypeptide containing a sequence of 39 amino acids, although work with animal forms

of it and with degradation products of the human form have shown that not all of them are essential to the activity of the hormone. This sequence for the human ACTH is shown in Fig. 1.

The primary function of ACTH is the stimulation of the adrenal cortex to produce its hormones, which have already been discussed. This is evident from the therapeutic effect of administration of ACTH, which is closely

Ser—Tyr—Ser—Met—Glu—His—Phe—Arg—Tyr—Gly—Lys—Pro—
Val—Gly—Lys—Lys—Arg—Arg—Pro—Val—Lys—Val—Tyr—Pro—
NH₂
|
Asp—Ala—Gly—Glu—Asp—Glu—Ser—Ala—Glu—Ala—Phe—Pro—
Leu—Glu—Phe

Fig. 1. Amino acid sequence of human adrenocorticotropin (ACTH). Abbreviations of amino acids will be found in entry on **Amino Acids**.

similar to that of these hormones, so that if the action of only one of them is sought, its administration is preferable. Moreover, ACTH stimulates secretion of the androgenic substances mentioned as produced by the adrenal cortex.

Thyrotropic Hormone. This hormone (TSH) stimulates the development of the thyroid and controls its secretion. Although purified preparations of it have been obtained, they consist of a mixture of proteins of high mean molecular weight (about 30,000). Some of their amino acids have been determined, as well as their carbohydrates, but the structures have not been elucidated.

Growth Hormone. Somatotropin is the growth hormone. Purified preparations of extracts of it from the human adenohypophysis have been crystalized. They are known to be proteins, of mean molecular weight 21,000, and containing a single polypeptide chain. This hormone differs from the others of its group in not acting primarily upon the other endocrine glands, but in controlling the gain in body weight and the rate of skeletal growth. The growth abnormalities, such as dwarfism and giantism, have been shown to result from its hypo- and hypersecretion. In addition to its effect upon growth and anabolism generally, it has been found to affect the kidneys and pancreas, and to influence glucose, galactose, and lipid metabolism. See also **Pituitary Gland**.

Gonadotropic Hormones. These include follicle stimulating hormones (FSH), luteinizing or interstitial cell stimulating hormone (LH or ISCH), and prolactin. Their structures are not known; the molecular weight of human LH is about 26,000, that of human FSH is about 30,000, and that of human prolactin is uncertain. They are proteins, with variable amounts of carbohydrates. FSH induces the growth of Graafian follicles in the ovary and the production of spermatozoa in the testis. LH stimulates the final development of the ovarian follicles, the appearance of estrus, and the change of the follicles to corpora lutea. In the male, it stimulates the secretion of testosterone. Since these effects are due to the effect of this hormone upon interstitial cells, it is also called ISCH. Prolactin stimulates lactation after birth, acts with estrogen to promote the growth of the mammary gland, and influences the activity of the corpora lutea.

Male Hormones. The androgenic hormones produced in the testes (and adrenal gland) have a widespread effect upon the development of secondary sexual characteristics (musculature, facial hair, larynx, etc.), as well as upon the sexual organs and responses themselves. They also promote anabolism to a marked degree by their effect upon nitrogen and calcium metabolism. The structure of testosterone is shown in the Table 1. See also **Gonads**.

Female Hormones. Closely related to the male androgenic hormones, and probably synthesized from them in the female organism, are the estrogenic hormones which are produced principally in the ovary. Although β-estradiol is the normally secreted ovarian hormone, a number of other estrogenic substances have been isolated from urine and from animal studies. They include α-estradiol, estriol, and estrone. The structures of these hormones are given in Fig. 2.

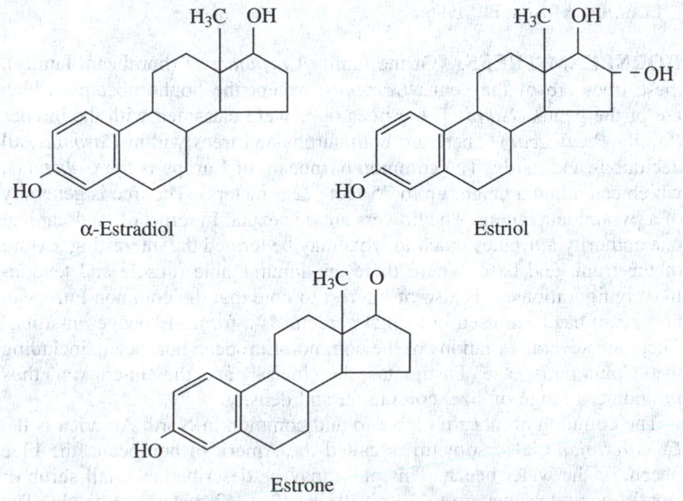

These hormones are important in both the menstrual cycle and the reproductive cycle, and of course play an important role in oral contraceptives (the "pill"). They induce growth of the vaginal epithelium, secretion of mucus by the glands of the cervix, and initiate the growth of the endometrium, which is taken over by progesterone (from the corpus luteum) later in the cycle. They activate the proliferation of the mammary gland during pregnancy. As the androgens do for the male, the estrogens bring about the secondary sexual characteristics of the female. They have a number of effects upon metabolism, notably that of calcium and phosphorus, and of lipids and proteins. A number of other estrogens, some made synthetically and others obtained from animals, are known.

The corpus luteum produces two hormones, progesterone and relaxin. The structures of these hormones are shown in Part 1 and Part 2 of table.

Progesterone acts to complete the proliferation of the endometrium, which was initiated by the estrogenic hormones, and to prepare it for the ovum. In pregnancy the continued action of progesterone is necessary. It aids the growth of the breasts and has a definite effect against ovulation. It is also the biosynthetic precursor of some of the estrogenic hormones. Relaxin has been shown to have a relaxing effect on the cartilaginous junction of the public bones in preparation for parturition. See also **Embryo**.

Feedback in Hormone Control Systems

Not only do the hormones initiate or stimulate biological processes, both directly and by bringing about production of other hormones in other glands, but they also act to maintain the organism in a steady state, or *homeostasis*. Thus the gonadotropic hormones from the hypophysis stimulate the testes, but the resulting production there of androgens like testosterone, inhibits the action of the hypophysis in producing the gonadotropic hormones. The complicated cycle of adjustment in the human female is shown in the cycle illustrated in Fig. 3.

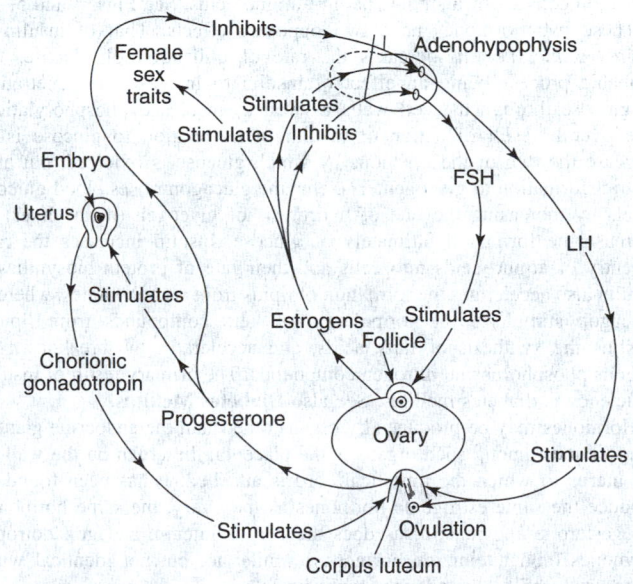

Fig. 3.   Cycle of hormone adjustment in human female.

As shown in the figure, the regulation of the ovarian hormones in the human female involves both positive and negative feedback. The follicle-stimulating hormone (FSH) from the adenohypophysis stimulates the Graafian follicles, which thus produce estrogens. These not only inhibit FSH production through negative feedback, but also stimulate the adenohypophysis to increase its production of luteinizing hormone (LH) through positive feedback. This hormone in turn brings about ovulation from the Graafian follicle. After the ova are discharged, the LH stimulates the empty follicle, now the corpus luteum, to produce progesterone.

This hormone brings about the changes in the reproductive organs required for the development of the embryo. Then the progesterone partly inhibits the adenohypophysis from producing further LH, an example of negative feedback; as a result, there is no further ovulation. The progesterone also acts as a positive feedback and stimulates the production of FSH.

H_3C OH

HO

α-Estradiol

H_3C OH

$-OH$

HO

Estriol

H_3C O

HO

Estrone

Fig. 2.   Major ovarian hormones.

NH2 S S NH2 NH2 NH2
Gly·Ileu·Val·Glu·Glu·Cy·Cy·Ala·Ser·Val·Cy·Ser·Leu·Tyr·Glu·Leu·Glu·Asp·Tyr·Cy·Asp
S S
NH2 NH2 S S
Phe·Val·Asp·Glu·His·Leu·Cy·Gly·Ser·His·Leu·Val·Glu·Ala·Leu·Tyr·Leu·Val·Cy·Gly·Glu·Arg·Gly·Phe·Phe·Tyr·Thr·Pro·Lys·Ala

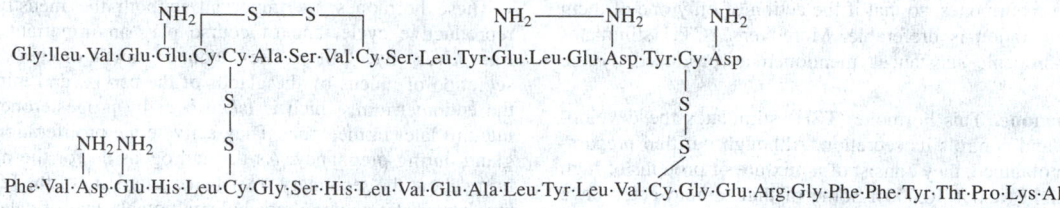

Fig. 4. Primary structure of bovine insulin.

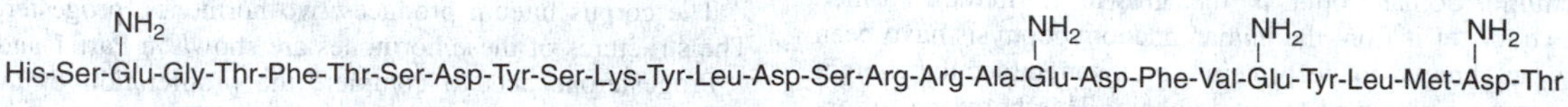

NH2 NH2 NH2 NH2
His-Ser-Glu-Gly-Thr-Phe-Thr-Ser-Asp-Tyr-Ser-Lys-Tyr-Leu-Asp-Ser-Arg-Arg-Ala-Glu-Asp-Phe-Val-Glu-Tyr-Leu-Met-Asp-Thr

Fig. 5. Glucagon.

When pregnancy intervenes, a new feedback mechanism must be introduced, or the embryo would be expelled, by the shedding of the lining of the uterus in menstruation. Here the placenta (chorion) of the embryo itself produces hormones, as already noted. Its LH stimulates continuing production of progesterone from the corpus luteum, thus preventing menstruation and stimulating the continuing development of the uterus as needed by the growing embryo. The extra progesterone also inhibits further ovulation in spite of the presence of the gonadotropin from the placenta (chorion).

Pancreas and Nonendocrine Hormone Sources. In addition to producing hormones, the pancreas also generates digestive fluids (*pancreatic juice*). It is the hormone function that makes the pancreas a part of the endocrine system. See also **Endocrine System**; and **Pancreas**. The pancreas secretes *insulin* and *glucagon*, both hormones. The structure of glucagon consists of a single chain of amino acids. See Figs. 4 and 5.

These two hormones have two opposing effects. That of insulin is *hypoglycemic*, i.e., it increases the rate of utilization of glucose, the probable process being an effect of insulin to increase the penetration of glucose through the cell walls as well as increased phosphorylation. The overall result of action of insulin in its relation to glucose is to increase the rate of the reactions by which glucose is oxidized, but also its transformation to glycogen. The enzyme glucagon raises blood glucose levels by increasing the rate of hydrolysis of glycogen (in the liver) to increase the formation ultimately of glucose. Insulin increases the rate of entry of amino acids into cells and their rate of protein biosynthesis. Insulin also accelerates the formation of lipids from carbohydrates, whereas glucagon stimulates the formulation of keto compounds from lipids, inhibits the synthesis of fatty acids, and accelerates the breakdown of various phosphorus and nitrogen compounds. The primary result of insulin deficiency is diabetes mellitus. See also **Diabetes Mellitus**.

Hormones may be produced by organs other than the endocrine glands. Conspicuous among such organs is the placenta, the organ on the wall of the uterus to which the umbilical cord is attached. It has been found to produce the same estrogenic hormones as the ovary, the same hormones (progesterone and relaxin) as does the corpus luteum, and gonadotropic hormones (and luteinizing hormones) similar to, but not identical with, those produced by the adenohypophysis.

Other hormones which do not originate in endocrine glands are the cholecystokinin of the intestine, and the enterogastrone and gastrin of the stomach. The first is produced by the upper intestinal mucosa and causes the gall bladder to contract; the enterogastrone is produced in the same tissue and inhibits gastric motility and secretion; it also excites secretion of digestive fluids, principally hydrochloric acid.

In addition to the entries covering specific endocrine glands, See also **Endocrine System**; **Central and Peripheral Nervous Systems**; and **Steroids**.

In plants, a *plant hormone* or "phytohormone" is an organic compound produced by the plant, controlling growth and other functions at sites remote from where the hormone is produced. Plant hormones also act in very minute amounts. Plant hormones include the auxins, gibberellins, and kinetins. These are described in the entries on **Gibberellic Acid and Gibberellin Plant Growth Hormones**; and **Plant Growth Modification and Regulation**. Plant hormones are also mentioned in a number of specific plant-related entries.

Additional Reading

Bengtsson, Bengt-Ake: *Growth Hormone*, Kluwer Academic Publishers, Norwell, MA, 1999.

Buckingham, J.C., G. Gillies, and A.M. Cowell: *Stress, Stress Hormones and the Immune System*, John Wiley & Sons, Inc., New York, NY, 1997.

Conn, P.M. and W.F. Crowley, Jr.: "Gonadotropin-Releasing Hormone and Its Analogues," *N. Eng. J. Med.*, 93 (January 10, 1991).

Copp, D.H.: "Parathyroids, Calcitonin, and Control of Plasma Calcium," *Recent Progr. Hormone Res.*, **20**, 59–77 (1964).

Erickson, D.: "Human Growth Hormone," *Sci. Amer.*, 164 (September 1990).

Erickson, D.: "Hormone Derivatives May Combat PMS and Epilepsy," *Sci. Amer.*, 124 (May 1991).

Evans, R.M.: "The Steroid and Thyroid Hormone Receptor Superfamily," *Science*, 889 (May 13, 1988).

Fackelmann, K.A.: "High-Pressure Hormone," *Sci. News*, 344 (December 1, 1990).

Hadley, M.E.: *Endocrinology*, Prentice-Hall, Inc., Upper Saddle River, NJ, 2000.

Harvey, P.W., A. Cockburn, and K.C. Rush: *Endocrine and Hormonal Toxicology*, John Wiley & Sons, Inc., New York, NY, 1999.

Kanellis, A.K. et al.: *Biology and Biotechnology of the Plant Hormone Ethylene II*, Kluwer Academic Publishers, Norwell, MA, 1999.

Kritchevsky, D. and D. Heber: *Dietary Fats, Lipids, Hormones, and Tumorigenesis: New Horizons in Basic Research*, Vol. 299, Kluwer Academic Publishers, Norwell, MA, 1996.

Kutsky, J.F.: *Handbook of Vitamins and Hormones*, Van Nostrand Reinhold, New York, NY, 1973.

Litwack, G. and T. Begley: *Vitamins and Hormones*, Vol. 61, Academic Press, Inc., San Diego, CA, 2000.

Marx, J.: "How Peptide Hormones Get Ready for Work," *Science*, 779 (May 10, 1991).

Norman, A.W. and G. Litwack: *Hormones*, Academic Press, Inc., San Diego, CA, 1997.

Seifer, D.B. and E.A. Kennard: *Menopause: Endocrinology and Management*, Vol. 18, Humana Press, Totowa, NJ, 1999.

Smith, R.G. and M.O. Thorner: *Human Growth Hormone: Research and Clinical Practice*, Vol. 19, Humana Press, Totowa, NJ, 1999.

Strauss, J. et al.: *Molecular Biology in Reproductive Medicine*, Parthenon Publishing Group, New York, NY, 1999.

Timiras, P.S., A. Vernadakis, and W.B. Quay: *Hormones and Aging*, CRC Press, LLC., Boca Raton, FL, 1996.

HORNBEAM TREES. Of the family *Carpinaceae* (hornbeam family), these trees are of the genus *Carpinus*, except the hophornbeams, which are of the genus *Ostrya*. These trees once were classified with the birches (family *Betulaceae*). There are both shrubs and trees within *Carpinus*, all deciduous and hardy. The common hornbeam of Europe is the *C. betulus*, which can attain a height up to 75 feet (22.5 meters). The tree is generally of a pyramidal contour. The flowers are unisexual. In terms of landscaping, one authority attributes much to what may be termed the interesting texture of the trunk and base, where there are innumerable muscle and tendon-like configurations. It is also of interest to note that the common European hornbeam has been used in formal gardens as a form of "hedge-on-stilts." There are several variations of the common European hornbeam, including the 'Columnaris,' the 'Fastigiata,' the 'Incisa,' and the 'Intertexta,' thus providing a range of size, coloration, and density.

The common hornbeam native to and common in North America is the *C. caroliniana*, also sometimes called the American hornbeam, the blue beech, or the water beech. This plant may be described as a tall shrub or small tree and ranges up to 30 or 40 feet (9 to 12 meters) in height. The current champion tree selected by American Forests is shown in Table 1.

TABLE 1. RECORD HORNBEAM TREES IN THE UNITED STATES[1]

Specimen	Circumference[2]		Height		Spread		Location
	Inches	Centimeters	Feet	Meters	Fee	Meters	
American hornbeam (1983) (*Carpinus caroliniana*)	95	241	69	21	56	17.1	New York
Chisos hophornbeam (1983) (*Ostrya chisosensis*)	28	181	32	9.8	24	7.3	Texas
Eastern hophornbeam (1991) (*Ostrya virginiana*)	115	292	74	22.6	111	33.8	Michigan
Knowlton hophornbeam (1996) (*Ostrya knowltonii*)	39	99	44	13.4	29	8.8	Arizona

[1] From the "National Register of Big Trees," American Forests (by permission).
[2] At 4.5 feet (1.4 meters).

The branches of the American hornbeam are slender and extended nearly horizontally from the trunk. The leaf is narrow, ovate, sharp-pointed and a dull light green, lighter color underneath. The staminate catkins are about 11/2 inches (3.8 centimeters) long. The bark is scaly and gray-brown. This tree ranges from Nova Scotia and Quebec west to the Great Lakes and south through Nebraska, Kansas, and Oklahoma, to Texas in the southwest and Florida in the southeast. The tree is quite common in New England.

The Japanese hornbeam (*C. japonica*) reaches a height of about 50 feet (15 meters) and is wide-spreading. It has male catkins from 1 to 2 inches (2.5 to 5 centimeters) in length. A more ornamental Japanese hornbeam is the *C. laxiflora*. Other species include the *C. orientalis*, a bushy shrub of southeastern Europe and Asia Minor; and the *C. turczaninowil*, a thin, spindly hornbeam.

Closely related to the hornbeams are the hophornbeams. See Fig. 1. Generally, these trees are medium-to-large in size and are deciduous. Several excellent specimens selected by American Forests are shown in Table 1. They are also characterized by drooping male catkins and upright female catkins. Their fruit is in the form of a nutlet contained in a husk. The hophornbeam of southern Europe and Asia Minor is the *Ostrya carpinifolia*, which can attain a height up to 65 feet (19.5 meters). The common hophornbeam in America is the American hornbeam (*O. virginiana*), sometimes also called the ironwood or leverwood. Generally, this is a fairly small tree, ranging from 30 to 45 feet (9 to 13.5 meters) in height, but under favorable conditions the tree can do much better. The bark is gray-brown, scaly, and has perpendicular scoring. The leaves are from 3 to 4 inches (7.6 to 10 centimeters) in length, narrow, ovate, double-toothed, sharp-pointed, and of a dull light-green color. The staminate flowers normally occur in three drooping catkins. The tree prefers a dry soil and open woods. It ranges from Nova Scotia and New Brunswick southward along the Saint Lawrence and Lower Ottawa Rivers, westward to Lake Huron, and northwest to Minnesota and the Dakotas, thence southward as far as Kansas and Nebraska. In the east, it is found in the Alleghany Mountains and south into Florida. The tree also is found in eastern Texas. Thus, the tree has a broad climatic range.

The *Ostrya knowltoni* is found on the southern slopes of the Colorado River canyon in Arizona and northward to Flagstaff. It is abundant to the 6,000 to 7,000-foot (1800 to 2100-meter) level. The height of the tree ranges from 25 to 40 feet (7.5 to 12 meters). The trunk is about 15 inches (38 centimeters) in diameter. The bark is light brown and scaly, with bright orange underneath. The leaf is small, 11/2 inches to 2 inches (3.8 to 5 centimeters) long, ovate, soft, hairy above and smooth underneath.

The Catalina ironwood (*Lyonothamnus Gray*) or Lyon tree is of interest because it is so rare and because of its unusual location. The tree grows on the canyon slopes of the steep shores of Santa Catalina Island, just off the coast of southern California. The tree was discovered by William Lyon, a young forester, in 1884 and thus so named. It is postulated that the tree was on the island at the time when the sea level was many hundreds of feet lower than it is today. There are a number of species of the tree, particularly on the eastern side of the island. They are also found on Santa Rosa and Santa Cruz Islands nearby. Geologists postulate that at one time these islands were a connected land mass. However, the tree leaf differs from one location to the next. The tree has compound foliage on Santa Rosa and Santa Cruz, whereas this is not the case on Santa Catalina.

Fig. 1. Eastern hophornbeam tree (*Ostrya virginiana*).

If the trees were numerous, the wood would be of considerable economic value because it is very strong and tough, weighing about 50 pounds per cubic foot (801 kilograms per cubic meter). It is believed that the Canalino Indians once used the wood for handles and shaft wood. The trees grow straight and tall, varying greatly in size, some towering up to 60 feet (18 meters) in height, with trunks of about 11/2 feet (0.5 meter) in diameter. They are found at an elevation between 500 and 2,000 feet (150 and 600 meters). The flowers are small and lacy, in groups on branches that are somewhat flat. They bloom in June and July.

The leaf has a simple-bladed fern-like foliage, projecting an aura of antiquity. It is blade-shaped with teeth coarsely cut. The Seed is oblong and light-brown. The bark is dark brown, but appears to weather to a lighter color. It is often tattered, with the underbark showing through. The tree is difficult to propagate from cuttings or Seeds. Root sprouts thus far have proved most effective.

HORNBILL. See **Kingfishers and Other Coraciiformes**.

HORNBLENDE. The mineral hornblende is a complex silicate. It is probably an isomorphous mixture of three molecules, a calcium-iron-magnesium silicate, an aluminum-iron-magnesium silicate and an iron-magnesium silicate. A general formula is

$$(Ca, Na, K)_{2-3}(Mg, Fe^{2+}, Fe^{3+}, Al)_5(Al, Si)_8O_{22}(OH)_2.$$

Manganese and alkalies are sometimes present as is also titanium. It is monoclinic, with prismatic crystals, often pseudo-hexagonal. Bladed, fibrous, columnar, granular and compact massive varieties also are common. It has a perfect prismatic cleavage; hardness, 5–6; specific gravity, 3.02–3.27; color, green, greenish-brown, brown and black; luster, vitreous to silky; transparent to opaque.

Hornblende is a common constituent of many of the igneous rocks such as granite, syenite, diorite, or gabbro, of gneisses and schists and is the principal mineral of the amphibolites. Hornblende alters easily to chlorite and epidote. A variety of hornblende that contains little (less than 5%) of iron oxides is gray to white in color and named edenite, from its locality in Edenville, New York. Very dark brown to black hornblendes, which contain titanium, ordinarily are called basaltic hornblende from the fact that they are usually a constituent of basalts and similar rocks.

Well-known localities for hornblende are in The Czech Republic and Slovakia, Mount Vesuvius, Italy, Norway, Sweden, and, in the United States, in Massachusetts, New Hampshire, and New York. Black hornblende is found in Renfrew County, Canada. The word hornblende is derived from the German horn, and blende, to blind or dazzle. The term blende was often used to refer to a brilliant nonmetallic luster, i.e., zinc blende.

See also terms listed under **Mineralogy**.

ELMER B. ROWLEY, Union College, Schenectady, NY

HORNBLENDITE. A coarse-grained rock related to gabbro that consists almost wholly of hornblende. Olivine being present, this rock may grade into a hornblende-peridotite (cortlandtite). Hornblendite is a rare rock type and of relatively little importance.

HORNED TOAD (*Reptilia, Sauria; Phrynosoma*). Small spiny lizards of the southwestern states and Mexico. They have short broad bodies and short tails, hence the confusion of terms in the common name. Horned lizard is a better term. See Fig. 1.

Fig. 1. Horned toad. (*A.M. Winchester.*)

Horned toads are desert animals and are capable of living for incredibly long periods without food or water. They cannot, however, survive for the long periods of years as has sometimes been claimed.

HORN (Electromagnetic). Horn radiators are used to obtain directional radiation characteristics that cannot be obtained as conveniently with simple antennae. As such directors, they are used both with conventional antennae and with waveguides, but in either case they serve to direct the radiation in a pattern from the open end of the horn in a manner determined by the dimensions of the horn. The important dimensions are the horn opening (in terms of wavelength of the radiation) and the flare angle. While theoretically an infinitely long horn will give a radiation pattern whose angle conforms to that of the horn, those of practical length do not confine the beam to quite this degree. For example, a horn with an angle of 15° may give a radiation pattern that spreads 23°. The common horns may be divided into three classes, sectoral, pyramidal, and biconical. See Fig. 1. The sectoral horn has two sides that are parallel and the other two flared. The pyramidal horn has all sides flared. The conical horn is really a pyramidal horn with a circular cross section. The biconical horn consists of two cones with their vertices coinciding or adjacent to one another. The first two types are used where a singly directed beam of radiation is desired, the exact pattern in both vertical and horizontal planes being determined by the dimensions. The biconical horn gives a uniform pattern in a plane perpendicular to the axis and highly directional in any plane containing the axis.

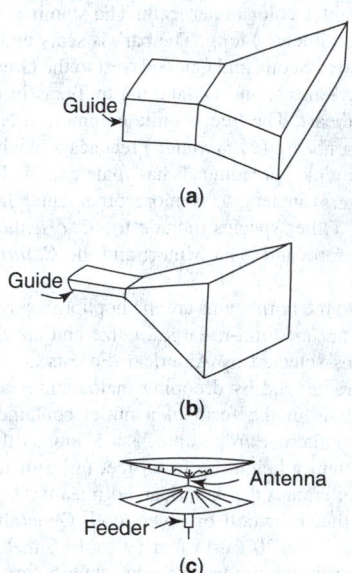

Fig. 1. Types of electromagnetic horns: (**a**) sectoral; (**b**) pyramidal; (**c**) biconical.

Sectoral and pyramidal horns may be excited by more or less conventional antennae or by waveguides. In the former case, a short section of waveguide is attached to the end of the horn, and this is excited by the antennae. In the latter case, the horn is really a flared extension of the guide and may be looked upon as an impedance-transforming section for matching the impedance of the guide to that of free space. Biconical horns are excited by a variety of antenna arrangements in the space between vertices of the horns. Because of space considerations, horns are not feasible except at ultra-high frequencies, but in the microwave region they are widely used as radiators.

HORNER, JOHANN FRIEDRICH (1831–1886). Johann Friedrich Horner was a Swiss ophthalmologist who received his schooling at the University of Zurich, and following a period of compulsory military service, he commenced the study of medicine in Zurich in 1849 under Karl Ewald Hasse (1810–1902) and Ernst Hasse. He obtained his doctorate in 1854 with a highly praised thesis on the subject of spinal curvature.

Following graduation he went abroad, as was usual at the time to do additional studies. He came to Vienna, where he became familiar with several specialties, among them the diseases of the eye. Eduard Jaeger Ritter von Jaxtthal (1818–1884) taught him the use of the ophthalmoscope, which had been invented a few years earlier by Hermann Helmholtz (1821–1894). Through Eduard Jæger he came into contact with the famous Albrecht von Graefe in Berlin, to whom Horner became his assistant.

Horner returned to Zurich in 1856, where he was habilitated and associated to the University of Zurich's ophthalmological clinic, then a part of the department of surgery. The eye clinic soon became independent, however, and was built up by Horner, who became its director and in 1862 was appointed professor extraordinary. Parallel to his educational duties he carried on a busy private practice, attracting patients from near and far. His total number of patients treated has been estimated to be some 100,000. Horner became a full professor of ophthalmology in 1873.

Horner was loved as a physician caring about his patients, for whom he had an impressive memory of persons. He took advantage of the great experience gained from his private practice when teaching. He was highly appreciated as a clinical teacher with a great talent for rhetoric's. His many-sided presentation with the patients in the midst — where diseases of the eye were related to other disorders of the body — mirrored his own broad knowledge. He was a competent surgeon who carried out some 2,000 cataract/glaucoma procedures — with a decline in complications as he was the first to introduce antiseptical methods in the treatment of eye diseases. Horner published some 40 papers illustrating his interests in the various diseases of the eye. He left many of his observations to his students for publication in no less than 28 doctoral dissertations.

Besides being an important physician, Horner was also a great organizer. He founded his own ophthalmological clinic, Hottinghof, and contributed to the establishment of a children's hospital by his planning and fund collecting. In 1867 he was one of the driving forces fighting the epidemic of cholera.

In 1876 he established that a man with a red-green colour blindness transmitted this anomaly to his male grandchildren through his daughter who was not color blind, similar to hemophilia, i.e., sex linked transmission.

HORNET (*Insecta, Hymenoptera*).   A name loosely applied to many of the larger wasps, particularly to the species which build paper nests and have a severe sting. See Fig. 1.

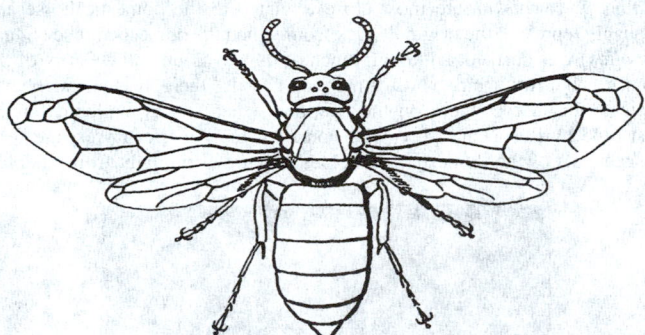

Fig. 1.   Hornet, a member of the wasp family (*Vespidae*). (*USDA diagram.*)

The true hornet is a European wasp (*Vespa crabro*). In America, the term may refer to any form of large stinging wasp that makes paper nests. Hornets are all social. In the southern United States, a smaller species (*V. carolina*) goes by the name of hornet. The white-faced hornet is the common American hornet (*V. maculata*).

The hornet is usually yellow and black in coloration and has a pugnacious spirit. Its sting is severe. The nest is usually pear-shaped, but sometimes round, and suspends from a branch of a tree or roof of a building. The nest consists of horizontal cones all facing downward. A small hole is left in the side of the nest as an entrance. The nest may accommodate from a few hundred to over 5,000 hornets.

These insects live much as honeybees, with similar work habits. Their food is the nectar of flowers. Adult hornets prefer carbohydrate food sources; the young prefer protein foods from caterpillars. Hornets are susceptible to bacterial and fungus disease and have numerous insect enemies, factors that keep their numbers under control. Among their worst enemies are other hornets, which pillage the food they store.

HORNFELS.   A more or less general term applied to fine-grained, massive, and frequently speckled rock, the result of contact metamorphism developed in slates by granitic intrusions.

HORN FLY (*Insecta, Diptera*).   This insect is most irritating and injurious to cattle, but also will attack goat, horse, and sheep. On occasions, the fly will also be a pest on dogs. The species *Haemotobia* or *Siphona irritans* (Linne) pierces the animal's skin and sucks blood. The associated pain and irritation trouble the animals during resting and feeding periods and, as a result, the cattle lose weight, milk cows have a lower milk yield, and the general health of the animals deteriorates, particularly when the irritation continues over a long period. The horn fly is about half the size of the common house fly, but appears very much like it. Although not fully proven, some scientists believe that the horn fly may carry anthrax disease.

The insect was first noted in the United States in Philadelphia in 1887, but has since spread throughout the continental United States and also to Hawaii.

Control is usually by spraying the animals along their back and flanks with one of several formulations, including methoxychlor, ronnel, or toxaphene. Treatment of dairy cows should be confined to handrubbing a suitable formulation around the neck area, thus avoiding possible contamination of milk. Pyrethrins and allethrin also have proved effective. Where practical, housing the animals in darkened structures equipped with entrance curtains that help to remove the flies can be helpful.

The fly maggots depend largely on animal dung for their food. Thus, cleanliness about shelters frequented by the animals cannot be overstressed.

HORN (Substance).   A hard translucent material formed by the development of epidermal cells containing a substance known as keratin. The outer layers of the skin are keratinized and the nails, claws and hoofs of mammals are formed of similar material. Horn is also developed in large amounts in the appendages of the head which go by the same name. Horns may be bony cores sheathed in horn or solid bony growths. The former occur in cattle and the latter in deer. The median horn or horns borne on the head of rhinoceroses are quite unlike true horns. They are formed of aggregated hair-like components firmly based on roughened areas of the underlying bones.

HORN-TAIL (*Insecta, Hymenoptera*).   Large sawflies (woodwasps) whose larvae bore in the trunks of trees. The adults have a cylindrical body and in the female sex a short strong ovipositor which is the source of the name horn-tail. With this organ holes are drilled into the wood of the tree for the deposition of the eggs.

HORNWORM (*Insecta, Lepidoptera*).   Closely related species of this worm are known as tobacco worms and tomato hornworms. The adult moths do not injure plants, but their larvae are quite damaging.

Southern or tobacco hornworm (*Protoparce sexta*, Johanssen). See Fig. 1. This insect is found in most of the United States and ranges southward into South America.

Fig. 1.   Tobacco hornworm moth. (*USDA.*)

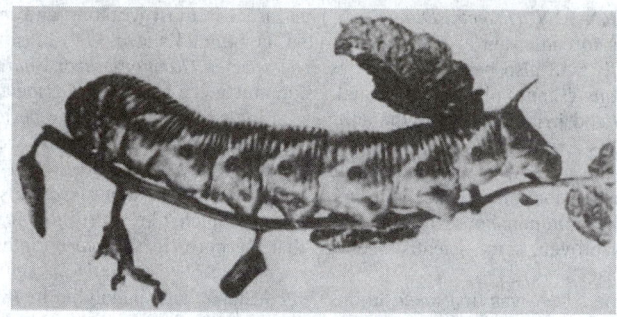

Fig. 2.   Tomato Hornworm larva. (*USDA.*)

Northern or tomato hornworm (*Protoparce quinquemaculata*, Haworth). See Fig. 2. This insect also is found throughout the United States and ranges northward into Canada.

In habit and damage, the two species are strikingly similar and, in fact, both species may be found attacking the same crop. In addition to tobacco and tomato, the insect is injurious to eggplant, pepper, and tomato. The worms also feed on a number of weeds often associated with these crops.

The worms are green with diagonal lines on their sides. They have a prominent horn (red horn on tobacco hornworm; black horn on tomato hornworm) on the rear end. They range up to 4 inches (10 centimeters) in length. Damage is caused by their eating of foliage and fruit. While widely distributed, infestations are usually localized. Because the worms are so large, handpicking is comparatively easy. Control chemicals used are carbaryl, endosulfan, and toxaphene. A natural enemy is the braconid wasp *Apanteles congregatus* (Say).

HOROLOGIUM.   A southern constellation situated near Eridanus.

HORSE FLY (*Insecta, Diptera*).   These insects are of special species (*Tabanua* and *Chrysops* spp.) and are irritating and injurious to horses, mules, cattle, hogs, deer, and other wild animals. On occasion, they are pests to humans. The flies look something like bees and are of a tan to brown coloration. The wings are faintly spotted. During spring and summer, the adult flies severely irritate the aforementioned animals and hover closely about the head, neck, and forequarters of the host, waiting for an opportunity to strike and lay their eggs. The bite is painful because the mouth parts of the insect are very sharp. See Fig. 1. The fly will suck blood from the animal's neck or back for several minutes. The animal may twitch and run about in an unmanageable fashion. Some scientists have estimated that in regions where the horse fly is abundant, an animal in the field may lose as much as 3 ounces (about 90 grams) of blood per day to horse fly inflictions. Some species of horse fly carry such diseases as tularaemia, Calabar swellings (filariasis), and el debab, a disease of camels and horses that occurs in Algeria. Evidence also indicates that these flies are carriers of swamp fever, surra, and anthrax.

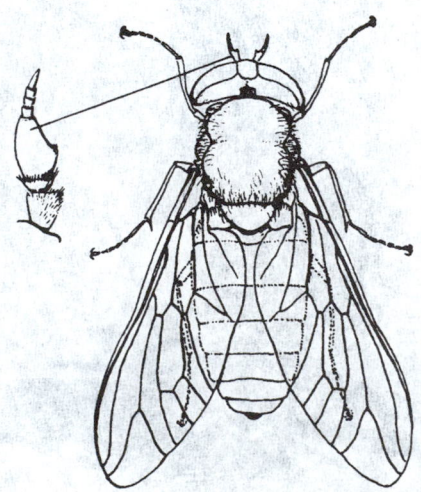

Fig. 1.   Horse fly showing detail of biting structure. (*USDA.*)

Control measures usually are directed against the maggots. The insect usually winters as a fully-grown larva in some wet area near a stream. The maggot is pointed at both ends and is about 2 inches (5 cm) in length. They are fully grown by late spring, after which they pupate until early summer in dried mud. Adult flies commence to appear in early summer. Depending upon species, there are one or two generations per year.

Preventive measures, such as draining wet places and taking precautions similar to those for mosquito control, are effective. Spraying of the animals with allethrin, incorporating piperonyl butoxide, can be effective. Broadcasting a granular insecticide over areas frequented by animals also is effective.

HORSE LATITUDES.   The belts of latitude over the oceans at approximately 30°–35°N and S where winds are predominantly calm or very light and weather is hot and dry. These latitudes mark the normal axis of the subtropical highs, and move north and south by about 5° following the sun. The two calm belts are known as the calms of Cancer and calms of Capricorn in the Northern and Southern Hemispheres, respectively; in the North Atlantic Ocean, these are the latitudes of the Sargasso Sea. The name is believed to have originated in the days of sailing ships, when the voyage across the Atlantic in those latitudes was often prolonged by calms or baffling winds so that water ran short, and ships carrying horses to the West Indies found it necessary to throw the horses overboard. See also **Winds and Air Movement**.

HORSEPOWER.   See **Units and Standards**.

HORSES, ASSES, AND ZEBRAS (*Mammalia, Perissodactyla*).   With exception of Grevy's Zebra, all horses, asses, and zebras are placed in the genus *Equus* (Equines).

The horse is a hoofed animal with a single toe on each foot, encased in a massive hoof. The teeth are very high-crowned grinding structures. Two wild species of central Asia, the Tarpan and Prezewalski's horse (*Equus prezewalski*), are most closely related to the domestic horse and probably represent the original stock from which the domestic horse (*Equus caballus*) was derived. Although much of the developmental history of the horses is known from North American fossils, there is no evidence to indicate that horses were on the continent when it was explored for the first time by Europeans. The wild horses of the American West are feral descendents of the domesticated horses imported originally from Spain.

Fig. 1.   Prezewalski's horse. (*A.M. Winchester.*)

Prezewalski's horse is believed to be the only true ancestor of the domestic horses that are found in the wild today. This is a small, quite heavyset animal, reddish-brown in color. See Fig. 1. Many years ago, it commonly roamed the plains of Eurasia. As nomads settled on the steppes, their domesticated stock won out in competition for water and pasture. The last Prezewalski's horse to be captured occurred in 1947 and was moved to Askanya-Nova, a reserve in the Ukraine. The horse was named for a Russian colonel who first reported its existence in the 1870s. About a thousand of this species remain, all born in captivity in various zoos and wildlife parks. As of 1992, plans were underway to reintroduce the horse into the wild on its former Mongolian native ranges. It is interesting to note that Prezewalski's horse has a few characteristics that are unique to the species. The horse has a distinctively short mane and no forelock. Genetically, the horse has 66 chromosomes instead of 64, as found in domestic horses and feral horses, such as mustangs.

TABLE 1. COMMON HORSE AND PONY BREEDS WORLDWIDE

Horse Breeds

AFRICA
Egyptian Arabian, Libyan Berber, Berber, Fulani horse, Nigerian horse, Basuto pony

ARGENTINA
Criollo

AUSTRALIA AND NEW ZEALAND
Brumby, Wales horse, New Zealand pony

AUSTRIA
Lipizzan

BELGIUM
Belgian warm-blooded horse, Brabant, Ardenner

BRAZIL
Crioulo, campolino, mangalarga

CANADA
Royal Canadian mounted police horse, Stable Island pony, Canadian cutting horse

CHINA
China pony

CZECH REPUBLIC AND SLOVAKIA
Kladrub

DENMARK
Fjord horse, Frederiksborger, Knabstniper, Jutländer

FINLAND
Finnish universal, Finnish draft horse

FRANCE
French thoroughbred, half-bred trotting horse, Norman trotter, Norman horse, Anglo-Arabian Camargue horse, Ardenne horse, trait du nord, Bretonne, Percheron, Boullonnais, Poitevine

GERMANY
East Prussian horse, Hannoveraner, Oldenburger, Holsteiner, Dülmener, Württemberger, Rhinish, Schlesiger, Pinzgauer

GREAT BRITAIN
Exmoor pony, new forest pony, Dartmoor pony, fell pony, dales pony, Welsh mountain pony, Welsh pony, highland pony, Shetland pony, Welsh cob, English thoroughbred, Shire, Clydesdale, Suffolk punch

GREECE
Peneia pony, Pindos pony, Skyros pony

HAITI
Haiti pony

HUNGARY
Nonius, furioso, lipizzan, Arabian (shagya)

ICELAND
Icelandic pony

INDIA
Kathiawar pony, Marwar pony

INDONESIA
Sumba pony, sandalwood pony, Sumbawa pony, Java pony, Timor pony, Batak pony, Bali pony

IRAN
Persian Arabian, darashoori (schiras horse), jaf, tchenarani, Turkomane, Polo ponies, Turkmen pony, British pony

IRELAND
Irish Clydesdale

ITALY
Salemer, Kalabrier, Aveligneser

MEXICO
Mexican horse

NETHERLANDS
Frisian horse, gelderse, Groninger

NORWAY
Fjord horse, Döle horse, Gudbrandsdaler, Döle trotting horse

PERU
Criollo (costeno), morochuco

POLAND
Huzuler, komik, Sokólsker, Mazure, Poznan horse Arabian, Anglo-Arabian

PORTUGAL
Lusitanian

RUSSIA
Viatka, zemaitnika (petschora), tori, Bukhyonii horse, Kirghiz horse, karabagh, lokai, yomud, akhal-tekkiner, Arabian, Métis, Orlov trotting horse, Russian-American trotting horse, Vladimir

SPAIN
Sorreia, Andalusian, Arabian

SWEDEN
Gotlander (Skogruss), Swedish warm-blooded horse

SWITZERLAND
Freiburger, Einsiedler horse, Anglo-Norman, Holsteiner

TABLE 1. (Continued)

TIBET
Tibetan horse

TURKEY
Anatolian pony, Karakabeyer

UNITED STATES
American thoroughbred, quarter horse, Kentucky saddle horse, Tennessee walking horse, Missouri fox-trotting horse, Morgan, standardbred trotting horse, Spanish mustang (the wild horse of Wyoming), galiceno, Chincoteague pony, Assateague pony, American Cleveland-bay, American hackney, Welsh and Shetland ponies, American pony (P.O.A. = Pony of America), pinto, appaloosa, palomino, albino, American Percheron, American Belgian

VENEZUELA
Llanero (prairie) horse

Donkey Breeds

Poitou ass
Puli ass
Spanish giant ass
Gascogne ass
Savoy ass
Sicilian ass
Macedonian ass
Maskat ass

The tarpan is a wild horse of the steppes of central Asia. This species has been regarded as feral rather than a natural species, but this interpretation has been disputed. It is in any case closely related to the domestic horse and may be an ancestral form. The western American mustang, a spirited, agile horse, is exemplary of interbreeding in the wild and is believed to be descended from stock introduced by Spanish explorers. The mustang also is known as the Indian pony or bronco. A worldwide classification of common horse and pony breeds is given in Table 1.

Classification of Horses. Under domestication, many varieties of horses have been developed for riding, driving, draft animals, and other uses. They also have been crossed with the domestic ass to produce mules for various uses, and have been hybridized experimentally with other species. Some authorities believe that the horse has no equal in its capacity and adaptability to withstand extreme climatic conditions and the various uses to which it has been placed by humans.

There are several ways to classify horses, as, for example, the draft horse (for working and pulling heavy loads); the light horse (light loading and riding); and the pony (a small horse, normally not over 14 hands high). A hand is considered four inches (10 centimeters) in breadth; thus $4 \times 14 = 56 = 4$ feet, 8 inches, or (1.42 meters). Horses also are commonly designated by color. Bay is considered the more or less standard color and this is reddish-brown. Other colors of horses include brown, gray, chestnut (a particular reddish-brown), and roan (various — reddish-brown with a significant sprinkling of gray or white). Pinto signifies marked with spots of two or more colors; Palomino designates a golden or brownish-gray horse with an ivory or silvery-white mane and tail.

There are numerous breeds of horses, often classified generally as light, coach, and heavy breeds. The coach breeds originated in England and include the Cleveland Bay, the Yorkshire Coach, and the Hackney. The latter is possibly the most popular and is $14\frac{1}{2}$ to $15\frac{1}{2}$ hands high, quite strong, with high carriage of the tail, and usually of a dark color. Heavy breeds include the Clydesdale, developed in Scotland and named for the Clyde River water. The height is 16 to $16\frac{1}{2}$ hands high; color is dark brown or bay; noted for its heavy fetlock (tuft of hair on back of leg just above hoof), and high action. The Belgian is another heavy breed, originated in Belgium, with a height of from 16 to 17 hands, weight from 1,800 to 2,200 pounds (816 to 997 kilograms); color chestnut or sorrel (particular reddish-brown shade); flaxen mane and tail. The Percheron is a famous draft breed that originated in Normandy. The head is small; the contour is Arabian; height is from 16 to 17 hands; weight from 1,900 to 2,100 pounds (861 to 953 kilograms); heavy, but active and supple: excellent for agricultural jobs.

Thoroughbred race horses were originally developed by crossing English with Turkish and Arabic horses. Many special racing and riding breeds have been developed, much too numerous for description here. Commonly known breeds include the Tennessee walking horse which has a distinctive gait — the running walk; the American saddle horse; the American quarter

Fig. 2. White mare and black colt of a domesticated horse found mainly in Germany and Austria. At an early age, the solid-black colt changes its pigmentation and becomes a white adult.

horse, particularly adapted for running short distances; and numerous others.

One breed of domesticated horse found in Germany and Austria is shown in Fig. 2. These horses are born totally raven black, but within a comparatively short time turn nearly pure white.

Asses

One breed of wild asses occurs in Asia and two breeds occur in Africa. However, all are believed to have been derived from the Nubian Wild Ass. Possibly the best known is the Onager (*Equus onager*), of a rust color, and very horse-like in appearance. The animal is found in the more arid regions of India and throughout Iran and Afghanistan. Traveling in small herds, the animals appear to prefer a semi-desert habitat. Possibly, the Kiang (*E. hemionus*) accounts for the greatest population of wild asses today. This animal is found in Tibet and Mongolia and is considerably larger than the Onager. In winter, the animal has a coat of dark, shaggy hair and a white and red coloration in summer. Another breed or two, small in stature and in number, can be found in the semiarid parts of Mongolia. The asses found in Africa are usually of a gray tone, with white muzzles and white under sides. The Nubian is characterized by distinct black shoulder stripes. The African asses prefer mountain and desert terrain. A domesticated ass is referred to as a donkey.

The Quagga (couagga) is now an extinct species. It was a South African animal (*E. quagga*) related to the zebras and asses. It was reddish-brown above, blending to white on the legs, and marked with dark brown stripes on the head, neck, and fore part of the body. The last known specimen died in 1875 at the Berlin Zoo, although hopefully some wild specimens still may be roaming in the secluded areas of South Africa. The animal was overharvested for use as native labor and food.

A mule is a hybrid between the domestic horse and ass, produced by mating a mare and a jack. The reciprocal cross of stallion and she ass is called a hinny. Mules have large ears, small hoofs, and tufted tail characteristic of the ass and the stature of the horse. They are strong and hardy, resistant to disease and generally adverse conditions. Since they are infertile, mules are always bred by crossing the two species. Some authorities claim that hinnys are smaller and lacking in the qualities desired in mules; while others claim that both forms fall within the range of variation to be expected in the hybrid.

Zebras

The Zebra (*E. burchelli*,...) is related to the asses and the quagga and is distinguished by the complete or nearly complete transverse striping of the body and legs. The stripes vary in the several species from white to yellow brown, alternating with dark brown to black. They range throughout much of Africa south of the Sahara Desert, but are concentrated in the south. Subspecies include: The Common Zebra (characterized by a V-shaped junction-pattern occurring in the middle of the sides); the Damaraland Zebra (Zaire, Zambia, Botswana regions); the East African

Zebra (Rhodesia, Abyssinia, the Sudan); and Selou's Zebra (central and southeast Africa). General characteristics of zebras include: grazing like horses; traveling in big herds; mixing with other animals; a main dietary item for lions; shy and nervous; can be quite pugnacious in self-defense, kicking vigorously with either front or hind feet and inflict severe bites; coloration and striping puts them at disadvantage when viewed against a green backdrop, but is an advantage in tall grass and open plains; frequently harbor intestinal parasites which are believed to aid in their digestion. See Fig. 3.

Fig. 3.   Chapman's zebra (*Equus antiquorium*).

Grevy's Zebra (*Dolichohippus*) is not classified with *Equus*. However, it appears much as a horse, with large head and big ears. The animal prefers desert scrub foliage and requires a minimum of water. It is found in Somalia. Certain anatomical distinctions set it in its own genus.

Other members of the order of *Perissodactyla* are listed under **Perisso-dactyla**.

Additional Reading

England, G. and W. Allen: *Allen's Fertility and Obstetrics in the Horse*, Blackwell Science, Inc., Malden, MA, 1996.

Fredericks, A. and G. Ellis: *Zebras*, Lerner Publishing Group, North Minneapolis, MN, 2000.

Gould, E. and G. McKay: *Encyclopedia of Mammals*, 2nd Edition, Academic Press, Inc., San Diego, CA, 1998.

Grzimek, B.: *Grzimek's Encyclopedia of Mammals*, 2nd Edition, The McGraw-Hill Companies, Inc., New York, NY, 1990.

Haines, G.: *The American Paint Horse*, The University of Oklahoma Press, Norman, OK, 1991.

Hawcroft, T.: *A A-Z of Horse Diseases and Health Problems: Signs, Diagnoses, Causes, Treatment*, Howell Books, New York, NY, 1990.

Macdonald, D.: *Encyclopedia of Mammals*, Barnes & Noble Books, New York, NY, 1999.

Price, S.: *The American Quarter Horse: An Introduction to Selection, Care, and Enjoyment*, The Lyons Press, New York, NY, 1999.

Staff: *Encyclopedia of Mammals*, Marshall Cavendish Inc., Tarrytown, NY, 1997.

HORST. See **Fault**.

HOST.   An animal used as a source of food by a parasite. The parasite may live on the surface of the body or within it and may be harmless or harmful, but in all cases the host is the source of its food.

HOT SHORTNESS.   Brittleness of metals in the hot working temperature range.

HOT SPRINGS.   Springs have been important in the social, cultural, and economic history of humankind. They have been an important source of water for humans. Our ancestors did not have drilling equipment, so in the absence of surface water, they had to rely on springs. Many early towns in Greece, Egypt, Mesopotamia, India, and China developed around springs. Many towns in the United States have taken their names from springs: Steamboat Springs, Colorado; Glenwood Springs, Colorado; Springfield, Missouri; and Rock Springs, Wyoming, are a few examples.

For centuries, hot springs and mineral springs were considered of medicinal or therapeutic value. Bottled "springwater" is still considered by many people of higher quality than ordinary tap water. This belief is at least partially responsible for the current high-volume sale of bottled water. Unfortunately, this association of springwater with purity is not based on fact. Springwater can contain higher concentrations of dissolved solids than local public water supplies, and springs become easily contaminated. Springs become contaminated because they are open, unprotected, and accessible to human beings and animals. Contamination can also occur from surface water flow or flooding into spring collection systems.

More than 3,400 public water supply systems in the United States obtain part or all of their drinking water from springs. These systems provide drinking water for more than 7 million people. However, the average number of people served by a single public supply system using springs as a water source is small. Springs are an important component of the drinking water supply in many states. Public water supplies that use springs are more numerous in the western United States. Wyoming has 80 springs that are used for public water supply. A box used to collect water from springs is shown in Fig. 1.

A spring is a source of water that flows naturally from an aquifer or soil onto the land or into a body of surface water. Its occurrence depends on the nature and relationship of rocks (especially permeable and impermeable strata), on the position of the water table, and on topography. See also **Aquifers**.

Springs are generally classified as gravity springs and artesian springs. Thermal springs are typically considered a type of artesian spring. The vast majority of springs are gravity springs. Gravity springs are created by water that moves downgradient (downhill) and emerges at the surface. Two gravity springs, contact and fault springs are shown in Fig. 2. Artesian (nongravitational) springs occur when the potentiometric level of the groundwater system is above the land surface and water flows at the land surface under pressure either at the aquifer outcrop or from fractures or faults. See also **Artesian Water**.

Springs are replenished by precipitation that recharges aquifers. The precipitation seeps into the soil and enters fractures, joints, bedding planes, or pore spaces in sand aquifers and sedimentary rocks. Springs occur when water flows through aquifers and discharges at the land surface through faults, fractures, or by flow along an impermeable layer. They can also occur where water flows from large orifices that result when the water dissolves carbonate rock (karst) and enlarges fractures or joints to create a passage (Fig. 2).

Springwater is sometimes forced to the surface along a fault from deep sources by thermal pressure gradients. Springs are associated with volcanism and fractures that extend to great depths in the earth's crust. Springs are a common feature in some hydrogeologic settings. For example, springs are common in mountains and in watersheds that are underlain by fractured rock aquifers.

Springs can be regional (long flow paths) or local (short flow paths). Local springs are comparatively small, can have low flow, and are typically from shallow aquifers. The discharge from these springs often fluctuates either seasonally or in greater cycles, sometimes in response to local precipitation. Local aquifers are quickly recharged, and water movement through them is comparatively rapid, resulting in low water mineralization. Springs supported by local aquifers are more likely than regional flow springs to stop flowing periodically. Regional springs more typically have large discharges. Regional springs typically have nearly constant discharge and are more mineralized than local springs. Regional springs rarely stop flowing even during long droughts.

The discharge rate of a spring is a function of three main variables: 1. hydraulic conductivity (permeability) of the aquifer; 2. area contributing recharge to the aquifer; and 3. quantity of recharge.

Springs are also classified based on hydrogeologic characteristics. The Water Resources Division of the U.S. Geological Survey recognizes eight principal types of springs based on hydrogeologic characteristics:

1. An artesian spring is a release of pressurized water from a confined aquifer at the aquifer outcrop or through an opening in the confining unit.
2. A depression spring is formed when the water table intersects a steeply sloping land surface. This type of spring is sensitive to seasonal fluctuations in groundwater storage and frequently disappears during dry periods.

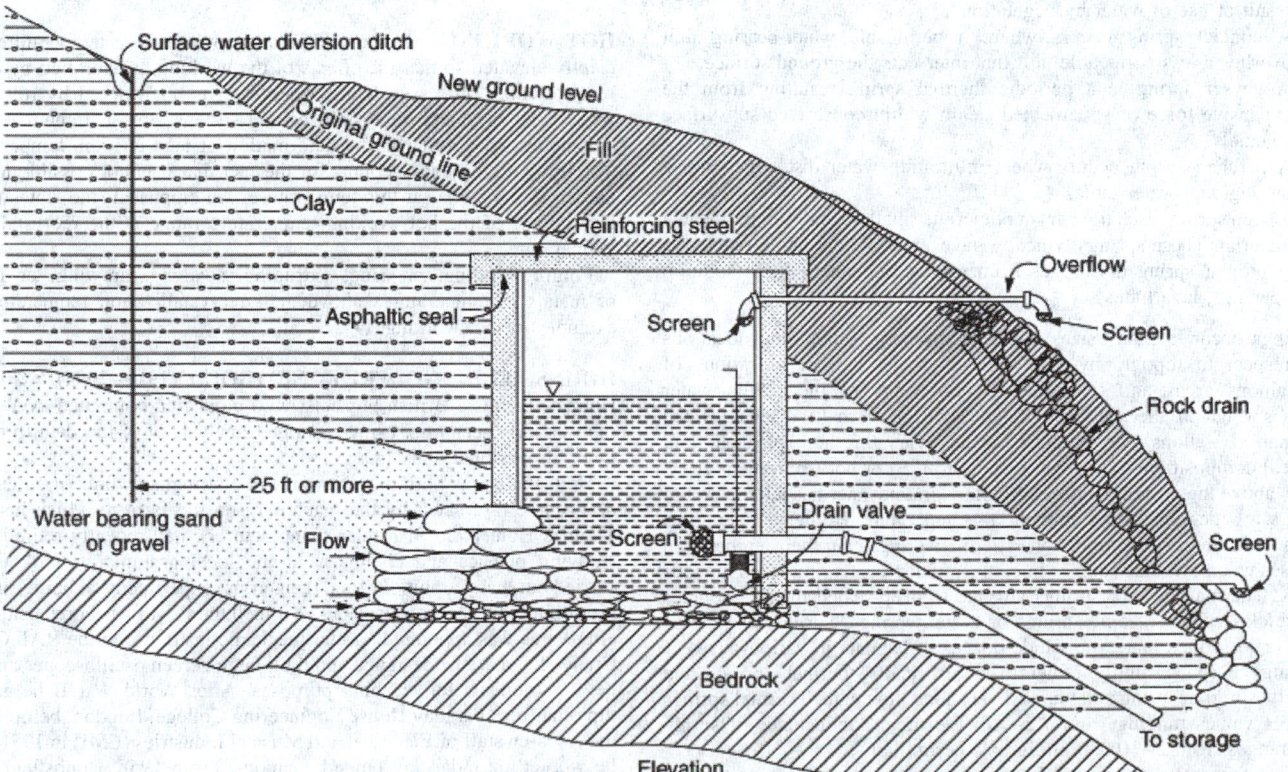

Fig. 1.   Diagram of a spring collection box. (*courtesy of the EPA.*)

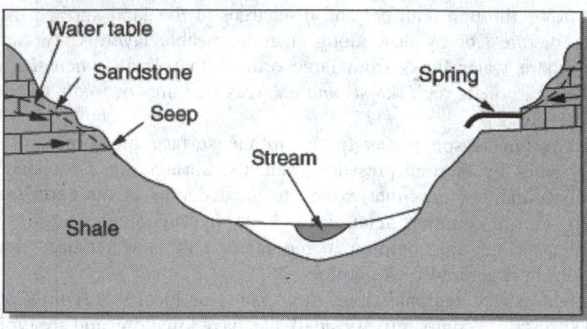

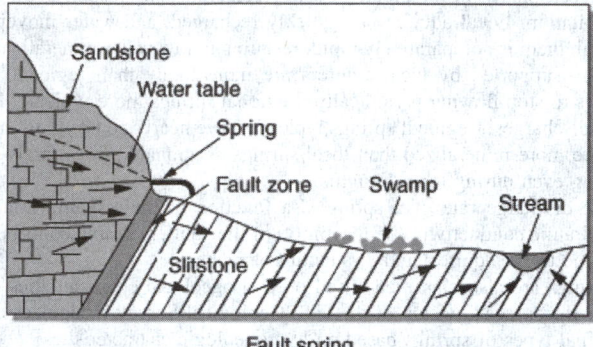

Fig. 2.   Contact and fault springs. (*courtesy of the American Geological Institute.*)

3. A fracture spring (fault spring) is formed in fractured or jointed rocks. Water movement is through fractures, and springs form where the fractures intersect the land surface. This type of spring is particularly sensitive to seasonal fluctuations in groundwater storage and frequently disappears due to reduced flow during dry periods. Daily fluctuations of discharge of small springs are commonly the result of use of water by vegetation.

4. A contact spring occurs where a permeable water-bearing unit overlies a less permeable unit that intersects the ground surface.

5. A geyser spring is a periodic thermal spring resulting from the expansive force of superheated steam within constricted subsurface channels.

6. A perched spring occurs where infiltrating water discharges above the regional water table.

7. A seep spring discharges from numerous small openings in permeable material. These springs typically have very low discharge rates.

8. A tubular spring discharges from rounded channels (karst solution openings, lava tubes).

Springs occur in many sizes, types of discharge points, and locations with respect to topography. They occur in the highest elevations of mountainous areas and on valley floors. Many of the public water supply springs in the western United States are small; discharges are less than 5 gallons per minute. Springs vary in their physical and chemical compositions. They can be cold (50 °F) or hot (more than 10 °C (50 °F) above the air temperature). Shallow groundwater has a temperature within a few degrees of the ambient air temperature. Higher temperatures usually indicate deeper circulation. Thermal springs gain their temperature increases when the water comes in contact with recently emplaced igneous masses. Thermal and hot springs are due to deep-seated thermal sources and are classed as volcanic springs or fissure springs. Springs may occur singly or in groups that can include dozens of habitats in various sizes.

Springs may be highly mineralized, especially thermal springs and sometimes region springs that have a very long flow path. Thermal springs have pH values ranging from 7.2–7.6. Electrical conductance of spring water ranges from 2–10,000 milligrams per million.

A list of question that could be asked to evaluate spring conditions is shown below.

- Are there standing water and surface drainage around the spring?
- Is there deeply rooted vegetation in the spring collection area?
- Are there roots in the collection pits?
- Is there a lock-tight fence around the collection area?
- Is there a diversion channel capable of diverting surface water away from the collection area?
- Are there privies, septic systems, underground storage tanks, or barnyards near the spring?
- Is the collection chamber properly constructed (lockable)?

Additional Reading

Bassington, R.; *Field Hydrogeology*, 2nd Edition, John Wiley & Sons, Inc., New York, NY, 2001.

Davis, S.N. and R.J.M. De Wiest: *Hydrogeology*, Krieger Publishing Company, Melbourne, FL, 1991.

Fetter, C.W.: *Applied Hydrogeology*, Prentice-Hall, Inc., Upper Saddle River, NJ. 2000.

LaMoreaux, P.E. and J.T. Tanner: *Springs and Bottled Waters of the World*, Springer-Verlag New York, LLC, New York, NY. 2002.

Meinzer, O.E.: *Outline of Ground Water Hydrology*, U.S. Geology Survey Water Supply Paper 494. **71**, p. 1923.

Moore, J.E. et al.: *Groundwater-a primer*, American Geological Institute, Arlington, VA. 1995.

Moore, J.E.: *Field Hydrogeology*, Lewis, Boca Raton, FL, 2002.

Moore, J.E. et al.: *A Field Guide to Evaluate Springs*, U.S. EPA, Washington, DC, 2004.

Sanders, L.L.: *A Manual of Field Hydrogeology*, Prentice-Hall, Inc., Upper Saddle River, NJ, 1998.

Staff: *Investigation of Hydrogeologic Mapping to Delineate Protection Zones Around Springs-Report of Two Case Studies*, U.S. Environmental Protection Agency, Washington, DC, 1997.

Staff: *Guidance Manual for Conducting Sanitary Surveys of Public Water Systems*, U.S. Environmental Protection Agency, Washington, DC, 1999.

Weight, W.D., and J.L. Sonderegger: *Manual of Applied Field Hydrogeology*, The McGraw-Hill Companies, Inc., New York, NY, 2001.

JOHN E. MOORE, USGS (Retired), Denver, CO

HOTWELL.   A hotwell is a tank or container in which heated liquid collects. An example is the hotwell attached to and made part of a steam condenser of the surface type. As the steam is condensed, the condensate drops to the bottom of the condenser shell and flows into the hotwell, from which it is pumped.

HOT WORKING.   Plastic deformation of metals at temperatures sufficiently elevated so that the effects of the working are nullified by concurrent softening processes. Thus, when steel is hot worked by rolling at a white heat, the metal recrystallizes and softens almost immediately after it is deformed. Similarly, the deformation of lead at room temperature is also hot working and accounts for the fact that it is not possible to work-harden this material at this temperature. An empirical rule states that the lower limit of the hot-working temperature range is the recrystallization temperature.

Forging, rolling, pressing, extruding, swaging, drawing, or forming of metals at temperatures above their recrystallization temperatures are examples of hot working.

HOUNSFIELD, GODFREY NEWBOLD (1919–2004).   Godfrey Hounsfield was a British physicist who pioneered computed axial tomography (CAT) scanning, using computer analysis of X-ray data to produce three-dimensional body images.

Hounsfield was born near Newark, in Nottinghamshire. At an early age he became fascinated by the machinery on the farm to which his father, Thomas Hounsfield (a former manufacturer), had recently retired. While studying physics and chemistry at the local grammar school, Godfrey carried out many ambitious (and dangerous) experiments at home, and built himself a radio receiver. In 1939 he joined the Royal Air Force, and shortly afterwards was appointed a radar mechanic instructor at the RAF College, Cranwell, where he designed and built large-screen oscilloscopes and other new instruments for teaching purposes. After World War II he gained a diploma from Faraday House Engineering College, London, before joining the research staff of Electrical and Musical Industries (EMI) in 1951, where he worked on radar and guided weapons. From 1958, Hounsfield led the team that built the first British solid-state electronic computer, the EMI-DEC 1100, in the process making important innovations in the construction of transistors. He also carried out pioneering work on the use of thin film devices for high capacity information storage.

During this period he acquired an interest in the use of computers for pattern recognition, and in 1967 he began working on computer methods for the analysis of information produced by X-ray examination of the human body. In the USA, Allan Cormack had made some progress in this direction, but it was Hounsfield who, in 1972, produced the first computerized axial tomography (CAT) scanner. This instrument rotates a beam of X-rays around the patient, while an array of detectors measures the attenuation of the transmitted X-rays. A computer then analyses the resulting pattern, using the data to construct cross-sectional images of the relevant organs, which can be viewed on a computer screen. It is thus possible to build up accurate three-dimensional images of internal body parts by collating a sequence of two-dimensional scans—a technique that proved particularly valuable for investigating the brain. A series of improvements to the original prototype enabled Hounsfield to increase the precision of the instrument, while reducing the amount of radiation to which the patient was exposed, and by 1975 whole-body scanners were commercially available. Thereafter, Hounsfield continued to work on this and other body scanning techniques, including magnetic resonance imaging (MRI). In 1979, he and Cormack shared the Nobel Prize in Physiology or Medicine for the development of CAT scanning, and in 1981 Hounsfield was knighted. See also **Computed Tomography; Cormack, Allan Macleod;** and **Magnetic Resonance Imaging (MRI)**.

Additional Reading

Sherby, L.S., and W. Odelberg: *The Who's Who of Nobel Prize Winners 1901–2000*, 4th Edition, Greenwood Publishing Group, Inc., Westport, CT, 2001.

MICHAEL A. SUTTON, University of Northumbria, Newcastle upon Tyne, UK

HOUR ANGLE. In reference to a celestial object, the spherical coordinate, in the equatorial system of coordinates, measured in the plane of the celestial equator from the local meridian, in the direction of apparent rotation of the celestial sphere, to the intersection of the hour circle through the object with the equator. Since time and hour angle are practically synonymous (e.g., the hour angle of the mean sun is local mean time), the determination of hour angle is vitally necessary for the determination of local time and, hence, longitude.

At sea, hour angle is determined by measuring the altitude of the object by means of the sextant, reducing the observed altitude to true geocentric, and solving the astronomical triangle. For the solution of the triangle, both the declination of the object and the latitude of the observer must be known. The declination may be immediately obtained from the tabulated coordinates of the object, but the latitude can be obtained only by some previous observation. If the ship is in motion, the latitude must be obtained by dead reckoning from the previously determined position.

See also **Celestial Sphere and Astronomical Triangle**.

HOUR CIRCLE. See **Celestial Sphere and Astronomical Triangle**.

HOURGLASS. See **Clocks**.

HOUSEFLY (*Insecta, Diptera*). A true fly, *Musca domestica*, well known for its habit of frequenting houses and alighting on all kinds of food. Since it also visits filth of any kind, it is an important carrier of disease, especially typhoid fever, and has been the object of public health crusades for many years. With the improvement of sanitation, the danger has been lessened, although it has not been entirely eliminated.

The housefly breeds in horse manure and in various kinds of decaying organic matter. Proper disposal of such wastes is an important measure in the control of the insect.

The housefly is found worldwide except at high altitudes. Hair covers most of the head of the housefly. Its jaws work horizontally and the stubby snout has a piercing stylet. It does not bite, but pierces and sucks its food. The eyes are large, covering about three-fourths of the facial area. The eyes of a fly are made up of approximately 4000 six-sided facets. All facets together frame the total object of view. However, the fly does not focus for a sharp image, nor does it have the ability to close its eyes. The vision is believed to be reasonably sharp for distances of 2 to 3 feet (up to 1 meter) or less.

Several generations of flies may be born in one season. They live for only a few weeks, but their eggs live on in fertile debris until spring.

The housefly is a principal agent for transmitting many diseases in areas where it is not carefully controlled. The legs are hairy and well designed to carry filth. Infection may be transferred by the piercing proboscis of the fly, or the fly may vomit its own food, leaving a trail of infection. As many as several million microorganisms may be found in the intestines of a housefly. The flying speed of the housefly is approximately 5 miles (8 kilometers) per hour.

HUBBLE, EDWIN POWELL (1889–1953). Hubble was an American pioneer in the study of extragalactic astronomy. He is acknowledged and remembered for his dramatic discovery that other galaxies exist in the Universe beyond the Milky Way and for observing that the universe is expanding at a constant rate.

He entered the University of Chicago and majored in astronomy and mathematics. After graduation, he received a Rhodes scholarship and studied Spanish and law at Oxford University. He passed his bar examine but only briefly practiced law. He taught physics and Spanish at the high school level in Indiana for a year' then went back to the University of Chicago to again study astronomy. Soon, thereafter, he became an assistant at Yerkes Observatory in Williams Bay, Wisconsin. At Yerkes, Hubble began a program of nebular photography. He soon made his first discovery finding a nebula that had changed over a few years. His Ph.D. thesis was, "Photographic Investigations of Faint Nebulae."

Hubble did some service for the U.S. Army between 1917 and 1919 and then went to work at the Mount Wilson Observatory in California. From this mountain top in California, Hubble discovered that the Universe is expanding. He was able to make this discovery with the aid of new technology, specifically, a 100-inch telescope, the Hooker Telescope. Hubble was able to show that what appears as fuzzy patches of light in less powerful telescopes are other islands in space far beyond the Milky Way. Hubble discovered a family of stars called the Cepheids. These stars vary in a regular way with a repeating pattern of brightening and dimming. The Cepheids are in the Andromeda nebula. Hubble estimated their distances and confirmed it was outside the Milky Way. His discovery showed the universe was much bigger than had been previously thought. Hubble developed estimating distances of galaxies based on their luminosity and velocity–distance relationships. From his research he composed the classification scheme for the structure of galaxies that we use today, and Hubble's Law and Hubble's Constant which suggests the theory of the expanding universe.

Hubble is also remembered for being given the honor to be the first person to use the 200-inch reflecting telescope at Mount Palomar Observatory in 1949.

The 1990s new technology gave us Hubble Space Telescope, named in honor of Edwin Hubble. This powerful telescope produces images from deep space. Now there are photographic views that support Hubble's expanding universe theory and confirms that it has been expanding for some 15 billion years.

See also **Cosmology**; **Galaxy**; **Hubble Space Telescope (HST)**; and **Red Shift**.

J. M. I.

HUBBLE'S LAW. See **Cosmology**.

HUBBLE SPACE TELESCOPE (HST). The *Hubble Space Telescope (HST)* is the first major infrared-optical-ultraviolet telescope to be placed into orbit around the earth. The telescope is named after the American astronomer Edwin P. Hubble who, in the 1920's, found galaxies beyond our Milky Way, and discovered that the universe is uniformly expanding. Located high above Earth's obscuring atmosphere the *Hubble Space Telescope* has provided the clearest views of the universe yet obtained in optical astronomy.

The heart of the telescope is the primary mirror, which is 94.5 inches (2.4 meters) in diameter. It is the smoothest optical mirror ever polished, with surface tolerance of one-millionth of an inch. It is made of fused silica glass, and weighs about 1,800 pounds. See Fig. 1.

Outside the blurring effects of Earth's turbulent atmosphere, the telescope can resolve astronomical objects with an angular size of 0.05 arc second. Depending on atmospheric observing conditions, this is ten to twenty times better than typical resolution with large ground-based telescopes.

The *Hubble Space Telescope* can detect objects as faint as 31st magnitude, which is slightly better than the sensitivity of much larger Earth-based telescopes. The *HST* has been used to probe the limits of the

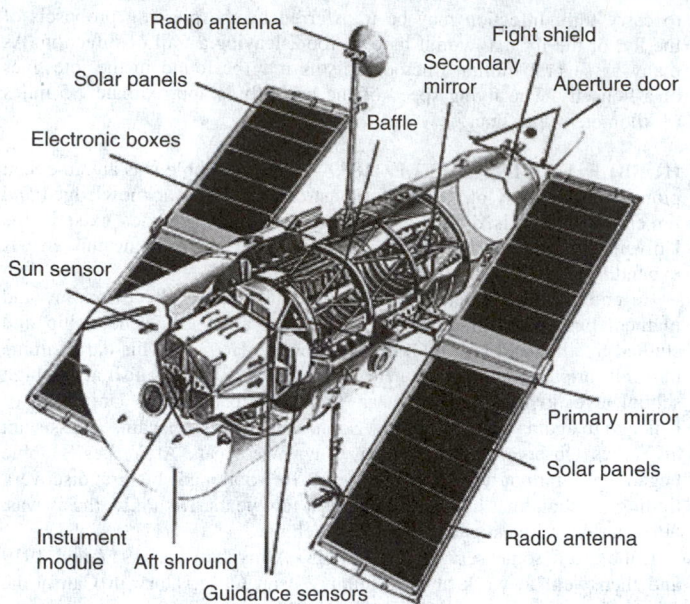

Fig. 1.   Detailed illustration of the Hubble Space Telescope. (*Courtesy of Ball Aerospace & Technologies Corp.*)

visible universe and uncover never-before-seen objects near the horizon of the universe. Because it is outside of our atmosphere, the *HST* can view astronomical objects across a broad swath of the electromagnetic spectrum, from ultraviolet light through visible and on to near-infrared wavelengths. The *HST* can also see faint objects near bright objects. This is a significant capability and an important requirement for researchers studying the environments around stars and the glowing nuclei of active galaxies.

The crystal-clear vision of the HST has triggered a revolution in optical astronomy. It has revealed a whole new level of detail and complexity in a variety of celestial phenomena, from nearby stars to galaxies near the limits of the observable universe. This has provided key new insights into the structure and evolution of our universe across a broad scale.

History

The *Hubble Space Telescope* was launched by the space shuttle *Discovery* on April 24, 1990. See Fig. 2. The original *HST* was equipped with

five science instruments: the Wide-Field Planetary Camera (WFPC), the Faint Object Camera, the Faint Object Spectrograph, the High-Resolution Spectrograph, and the High-Speed Photometer. In addition, the HST was fitted with three fine guidance sensors used for pointing and precision astrometry, or the measure of angles on the sky. See also **Astrometry**.

After the HST was launched, scientists discovered that its primary mirror was misshapen due to a fabrication error. This resulted in spherical aberration: the blurring of starlight because the telescope could not bring all the light to a single focal point. Using image-processing techniques, scientists were able to do significant research with the *HST* until an optical repair could be developed.

In December 1993, the first HST servicing mission carried replacement instruments and supplemental optics aboard the space shuttle *Endeavor* to restore the telescope to full optical performance. A corrective optical device, called the *Corrective Optics Space Telescope Axial Replacement* (*COSTAR*), was installed, which required removing the High Speed Photometer, so that it could improve the sharpness of the first-generation instruments. The original Wide-Field Planetary Camera (WFPC) was replaced with a second camera, the Wide-Field Planetary Camera2 (WFPC-2), which has a built-in correction for the aberration in the primary mirror.

In March 1997, the space shuttle *Discovery* returned to the Hubble Space Telescope for a second servicing mission. See Figs. 3 and 4. Two advanced instruments, the Near Infrared Camera and Multi-Object Spectrometer (NICMOS) and the Space Telescope Imaging Spectrograph (STIS) were installed to replace two first-generation instruments. The astronauts also replaced or enhanced several electronic subsystems, and patched unexpected tears in the *HST's* shiny, aluminized, thermal insulation blankets, which gives the telescope its distinctive foil-wrapped appearance. Figures 5–11 show images taken by each of the instruments discussed.

Fig. 5. Images of change in gas shell around Nova Cygni as taken by the European Space Agency's Faint Object Camera (FOC) and NASA's Corrective Optics Space Telescope Axial Replacement (COSTAR).

Fig. 6. A series of eight snapshots, the first of which was taken by the Wide-Field Planetary Camera (WFPC) in July 1993, the remainder of which was taken after the servicing mission by Wide-Field Planetary Camera 2 (WFPC-2).

Fig. 7. Black and white images taken in near-ultraviolet light with the Wide-Field Planetary Camera 2 (WFPC-2).

Fig. 8. An infrared view of the moon, ring, and clouds of Jupiter taken by the Near Infrared Camera and Multi-Object Spectrometer (NICMOS).

Fig. 9. An infrared view of Saturn taken by the Near Infrared Camera and Multi-Object Spectrometer (NICMOS) on January 4, 1998.

Fig. 2.   Hubble Space Telescope (HST) being deployed during the STS 31 flight. (*NASA & NSSDC.*)

Fig. 3. Hubble Space Telescope (HST) being refurbished during the STS 61 flight. Astronauts Story Musgrave and Jeffrey Hoffman are Seen during the last of the five EVAs. Australia's west coast can be Seen in the background. (*NASA & NSSDC.*)

Fig. 4. Hubble Space Telescope (HST) during the STS 61 flight. The new solar arrays are Seen here from the aft flight deck, backlit against the black background of space. (*NASA & NSSDC.*)

Fig. 10. Clear images of Saturn's ultraviolet aurora taken by the Space Telescope Imaging Spectrograph (STIS).

Fig. 11. STIS chemically analyzes the ring around Supernova 1987A.

In December 1999 a 3rd servicing mission replaced a number of subsystems on Hubble, but added no new instruments. About a month before the mission a critical gyroscope had failed bringing the total to four out of six inoperative gyros. This had left the telescope incapable of precision pointing to doing science. The December mission restored Hubble to having six fully functioning gyroscopes. The telescope's main computer was upgrade from a 1960s computer with 48 kilobytes of memory, to an Intel 486 microprocessor. Other electronics were replaced, to improve Hubble's redundancy.

Two more *HST* servicing missions are planned for the years 2001 and 2003. The Advanced Camera for Surveys (ACS) will be installed in late 2001. It will yield even sharper pictures and a wider field of view than the present cameras. a new set of high-efficiency solar panels will be added to the telescope. A news set of thermal blankets, to insulate the telescope from the heat and cold of space, will be overlaid on the tattered original blankets. The astronauts will resemble wallpaper hangers as they unfurl the rolls of new insulation.

During the 2003 servicing mission the Wide-Field Planetary Camera 3 (WFPC-3) and Cosmic Origins Spectrograph (COS) will be installed. The telescope's science operations are expected to end in 2010. The telescope will then either be de-orbited to disintegrate in the atmosphere, or it will be captured and returned to earth intact inside the shuttle's cargo bay.

Hubble Operations

The *HST* is controlled at the Goddard Space Flight Center in Greenbelt, Maryland, USA. The Space Telescope Science Institute (STScI) directs the science mission at the Johns Hopkins University in Baltimore, Maryland.

Images and other data collected by Hubble are stored onboard the telescope in solid-state data recorders. Depending on the importance of the observation they delay in reaching the astronomers can be from a few hours to a day or two Sometimes near-real-time images, needed for repointing the telescope, can be transmitted to astronomers in a matter of minutes.

All the information from HST is transmitted to a ground station in White Sands, New Mexico. The information is then relay via a domestic communications satellite to the Goddard Space Flight Center, and then via landline to the Space Telescope Science Institute (STScI) in Baltimore MD. All data are archive on optical disk. The present archive contains 6 terrabytes of Hubble data. The STScI sends the results to astronomers around the globe via the Internet or data tape.

HST research and funding engages a significant fraction of the worldwide community of professional astronomers. Astronomers compete annually for observation time on the *Hubble Space Telescope*. The oversubscription is typically 4-to-1. Observing proposals are submitted to peer review committees of astronomer experts. The STScI director makes the final acceptance, and can use his own discretionary time for special programs. Accepted proposals must be meticulously planned and scheduled by experts at STScI to maximize the telescope's efficiency.

The telescope is not pointed by direct remote control but instead automatically carries out a series of preprogrammed commands over the course of a day.

A date "pipeline" assembled and maintained by STScI ensures that all observations are stored on optical disk for archival research. The data are sent to research astronomers for analysis, and then made available to astronomers worldwide one year after the observation. To date, Hubble has looked at over 13,000 celestial targets and stored 4.6 gigabytes of data onto large optical disks. The telescope has made nearly one quarter million

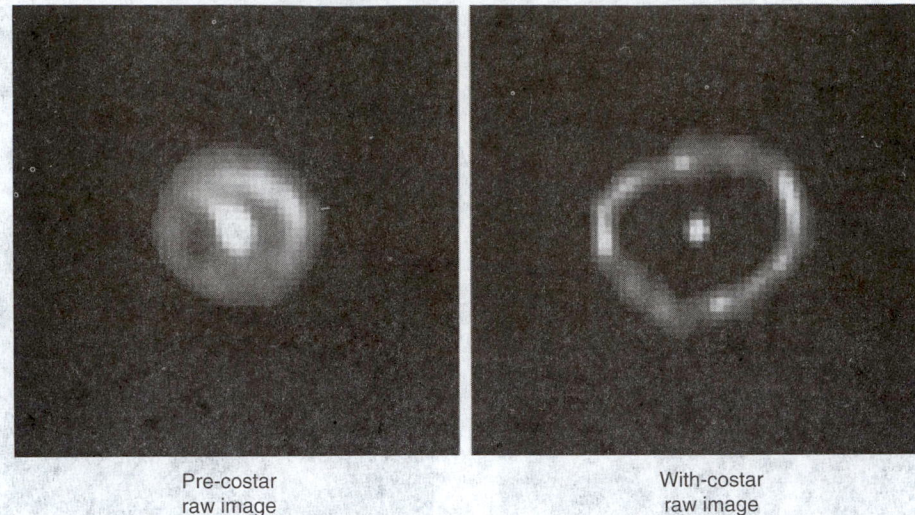

Pre-costar
raw image

With-costar
raw image

Fig. 5. The European Space Agency's ESA Faint Object Camera (FOC) utilizing the corrective optics provided by NASA's Corrective Optics Space Telescope Axial Replacement (COSTAR) has given astronomers their best look yet at a rapidly ballooning bubble of gas blasted off a star. The shell surrounds Nova Cygni 1992, which erupted on February 19, 1992. A nova is a thermonuclear explosion that occurs on the surface of a white dwarf star in a double star system. The new HST image (right) reveals an elliptical and slightly lumpy ring-like structure. The ring is the edge of a bubble of hot gas blasted into space by the nova. The shell is so thin that the FOC does not resolve its true thickness, even with HST's restored vision. An HST image taken May 31, 1993 (left), 467 days after the explosion, provided the first glimpse of the ring and a mysterious bar-like structure. But the image interpretation was severely hampered by HST's optical aberration, that scattered light from the central star, which contaminated the ring's image. A comparison of the pre and post COSTAR/FOC images reveals that the ring has evolved in the seven months that have elapsed between the two observations. The ring has expanded from a diameter of approximately 74 to 96 billion miles. The bar-like structure Seen in the earlier HST image has disappeared. These changes might confirm theories that the bar was produced by a dense layer of gas thrown off in the orbital plane of the double star system. The gas has subsequently grown more tenuous and so the bar has faded. The ring has also grown noticeably more oblong since the earlier image. This suggests that the hot gas is escaping more rapidly above and below the system's orbital plane. As the gas continues escaping, the ring should grow increasingly egg-shaped in the coming years. HST's new improved sensitivity and high resolution provides a unique opportunity to understand the novae by resolving the effects of the explosion long before they can be resolved in ground-based telescopes. Nova Cygni is 10,430 light years away (as measured directly from the ring's diameter), and is located in the summer constellation Cygnus the Swan. (*Credit: F. Paresce, R. Jedrzejewski (STScI), NASA/ESA.*)

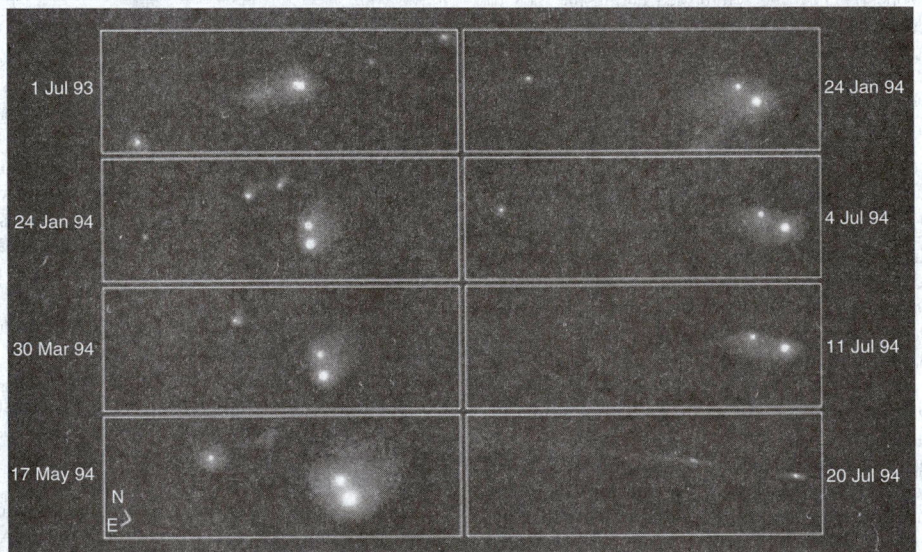

Fig. 6. This series of eight NASA Hubble Space Telescope "snapshots" shows the evolution of the P-Q complex, also called the "gang of four" region, of comet P/Shoemaker-Levy 9. The eight individual frames chronicle changes in the comet during the 12 months before colliding with Jupiter. The sequence shows that the relative separations of the various cometary fragments, thought to range in size from about 500 meters to almost 4 km (2.5 miles) across, changed dramatically over this period. The apparent separation of Q1 and Q2 was only about 1100 kilometers (680 miles) on July 1, 1993 and increased to 28,000 kilometers (17,400 miles) by July 20, 1994. The P-Q complex demonstrates that further fragmentation occurred after the breakup of the parent body in July 1992. Fragments Q1 and Q2 were probably together at some point in a single body. However, it is not clear how P1 and P2, and the P and Q objects are related. Between January 24 and March 30, 1994, the P2 nucleus broke-up into two separate fragments, one of which disappeared by late June. (It might be present in the mid-May image.) The P1 nucleus had a "streaked" appearance on January 24, 1994 and then became a barely discernible "puff" through mid-May. It was not detected in subsequent observations. Throughout the period, most nuclei were within a 4000 kilometer-wide (2500 miles) spherical cloud of dust, called a coma. However, shortly before impact, the coma around each nucleus became highly elongated along the comet's travel path due to "stretching" by Jupiter's rapidly increasing gravity. This stretching is dramatic in the image of the Q-complex taken on July 20, 1994, just 10 hours before collision. Despite the coma's changes, HST images show that the core of each nucleus always remained concentrated. This shows that the nuclei were probably not catastrophically fragmenting, at least not up to 10 hours before impact. (*Credit Dr. H.A. Weaver, Mr. T.E. Smith and Mr. K.B. Jones (STSci), and NASA.*)

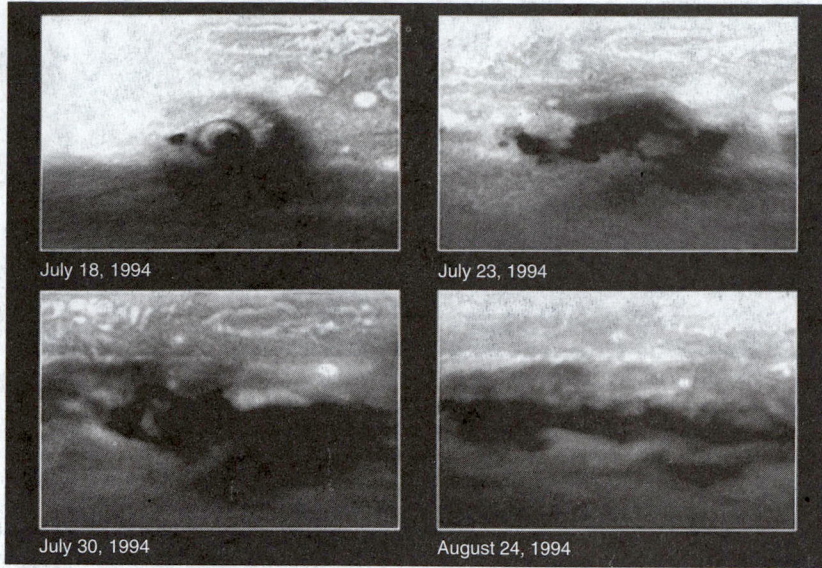

Fig. 7. This series of snapshots, taken with NASA's Hubble Space Telescope, shows evolution of the comet P/Shoemaker-Levy 9 impact region called the D/G complex. This feature was produced by two nuclei of comet P/Shoemaker-Levy 9 that collided with Jupiter on July 17 and 18, 1994, respectively, and was later modified again by the impact of the S fragment on July 21, 1994.

Upper Left: This first image was taken about 90 minutes after the G impact on July 18, 1994. Nearly all of the structure in this image was created by the impact of fragment G, although a small dark spot to the left was the remainder of small fragment D that collided one day earlier. The explosion of the nucleus in Jupiter's atmosphere created the unique ring structure, which may be analogous to a "sonic boom" on earth. Though this structure is best Seen for the G impact, it is not unique. Hubble reveals similar rings around several other fresh impact sites. They are all clear evidence for coherent outward motion of this wave phenomena.

Upper Right: This second image, obtained on July 23, shows that the Jovian winds have swept the material into a striking "curly-cue" structure.

Lower Left, Lower Right: The structure Seen in earlier views has disappeared rapidly in the images taken on July 30 and August 24, respectively. Almost all of the changes between the images are due to Jupiter's east-west winds that play a key role in the dispersing of the dark material. Hubble Space Telescope's high resolution will allow astronomers to continue to trace the impact debris as it is transported by the Jovian winds. This information promises to advance current understanding of the physics of Jupiter's atmosphere. (*Credit: H. Hammel, MIT and NASA*).

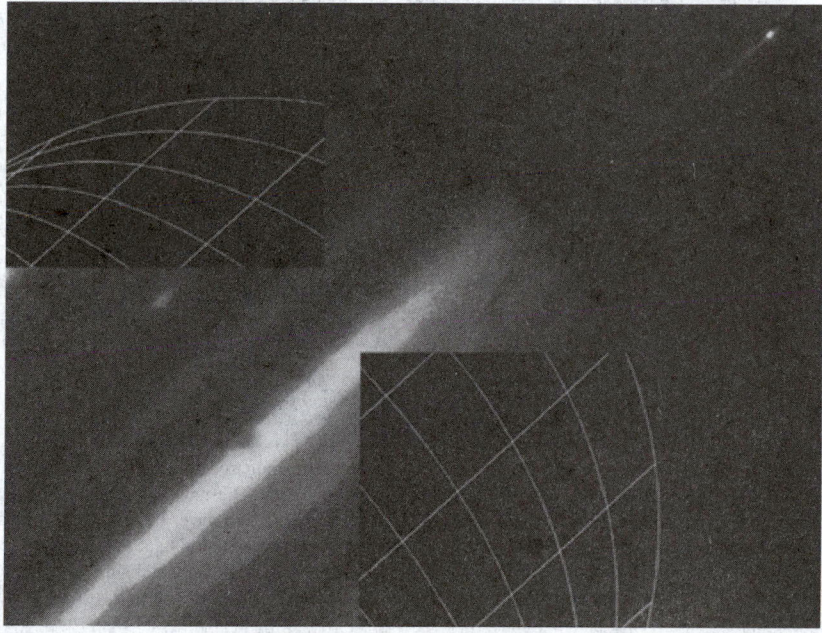

Fig. 8. Probing Jupiter's atmosphere for the first time, NASA Hubble Space Telescope's new Near Infrared Camera and Multi-Object spectrometer (NICMOS) provides a sharp glimpse of the planet's ring, moon, and high-altitude clouds. The presence of methane in Jupiter's hydrogen- and helium-rich atmosphere has allowed NICMOS to plumb Jupiter's atmosphere, revealing bands of high-altitude clouds. Visible light observations cannot provide a clear view of these high clouds because the underlying clouds reflect so much visible light that the higher level clouds are indistinguishable from the lower layer. The methane gas between the main cloud deck and the high clouds absorbs the reflected infrared light, allowing those clouds that are above most of the atmosphere to appear bright. Scientists will use NICMOS to study the high altitude portion of Jupiter's atmosphere to study clouds at lower levels. They will then analyze those images along with visible light information to compile a clearer picture of the planet's weather. Clouds at different levels tell unique stories. On Earth, for example, ice crystal (cirrus) clouds are found at high altitudes while water (cumulus) clouds are at lower levels. Besides showing details of the planet's high-altitude clouds, NICMOS also provides a clear view of the ring and the moon, Metis. Jupiter's ring plane, Seen nearly edge-on, is visible as a faint line on the upper right portion of the NICMOS image. Metis can be Seen in the ring plane (the bright circle on the ring's outer edge). The moon is 25 miles wide and about 80,000 miles from Jupiter. Because of the near-infrared camera's narrow field of view, this image is a mosaic constructed from three individual images taken September 17, 1997. The color intensity was adjusted to accentuate the high-altitude clouds. The dark circle on the disk of Jupiter (center of image) is an artifact of the imaging system. (*Credits: Reta Beebe (New Mexico State University), and NASA*.)

Fig. 9.   This image is courtesy of the new Near Infrared Camera and Multi-Object Spectrometer (NICMOS), which has taken its first peek at Saturn. The false-color image—taken January 4, 1998—shows the planet's reflected infrared light. This view provides detailed information on the clouds and hazes in Saturn's atmosphere. The cloud particles are believed to be ammonia ice crystals. Most of the Northern Hemisphere that is visible above the rings is relatively clear. The dark region around the South Pole at the bottom indicates a big hole in the main cloud layer. Most of the Southern Hemisphere (the lower part of Saturn) is quite hazy. These layers are aligned with latitude lines, due to Saturn's east-west winds. The rings cast their shadow onto Saturn. The bright line Seen within this shadow is sunlight through the Cassini Division, the separation between the two bright rings. It is best observed on the left side, just above three' rings. This view is possible due to a rare geometry during the observation. The next time this is observable from Earth will be in 2006. An accurate investigation of the ring's shadow also shows sunlight shining through the Encke Gap, a thin division very close to the outer edge of the ring system. Two of Saturn's satellites were recorded, Dione on the lower left and Tethys on the upper right. Tethys is just ending its transit across the disk of Saturn. They appear in different colors, yellow and green, indicating different conditions on their icy surfaces. Wavelengths: A color image consists of three exposures (or three film layers). For visible true-color images, the wavelengths of these three exposures are 0.4, 0.5, and 0.6 micrometers for blue, green, and red light, respectively. This Saturn image was taken at longer infrared wavelengths of 1.0, 1.8, and 2.1 micrometers, displayed as blue, green, and red. Reflected sunlight is Seen at all these wavelengths, since Saturn's own heat glows only at wavelengths above 4 micrometers. (*Credit: Erich Karkoschka (University of Arizona), and NASA.*)

Fig. 10.   This is the first image of Saturn's ultraviolet aurora taken by the Space Telescope Imaging spectrograph (STIS) on board the Hubble Space Telescope in October 1997, when Saturn was a distance of 810 million miles (1.3 billion kilometers) from Earth. The new instrument, used as a camera, provides more than ten times the sensitivity of previous Hubble instruments in the ultraviolet. STIS images reveal exquisite detail never before Seen in the spectacular auroral curtains of light that encircle Saturn's North and South Poles and rise more than a thousand miles above the cloud tops. Saturn's auroral displays are caused by an energetic wind from the Sun that sweeps over the planet, much like the Earth's aurora that is occasionally Seen in the nighttime sky and similar to the phenomenon that causes fluorescent lamps to glow. But unlike the Earth, Saturn's aurora is only Seen in ultraviolet light that is invisible from the Earth's surface, hence the aurora can only be observed from space. New Hubble images reveal ripples and overall patterns that evolve slowly, appearing generally fixed in our view and independent of planet rotation. At the same time, the curtains show local brightening that often follow the rotation of the planet and exhibit rapid variations on time scales of minutes. These variations and regularities indicate that the aurora is primarily shaped and powered by a continual tug-of-war between Saturn's magnetic field and the flow of charged particles from the Sun. Study of the aurora on Saturn had its beginnings in 1979, when the Pioneer 11 spacecraft observed a far-ultraviolet brightening on Saturn's poles. The Saturn flybys of the *Voyager 1* and *2* spacecraft in the early 1980's provided a basic description of the aurora and mapped for the first time the planet's enormous magnetic field that guides energetic electrons into the atmosphere near the north and south poles. The first images of Saturn's aurora were provided in 1994–1995 by the Hubble Space Telescope's Wide Field Planetary Camera 2 (WFPC2). Much greater ultraviolet sensitivity of the new STIS instrument allows the workings of Saturn's magnetosphere and upper atmosphere to be studied in much greater detail. These Hubble aurora investigations provide a framework that will ultimately complement the in situ measurements of Saturn's magnetic field and charged particles by NASA/ESA's Casini spacecraft, now en route to its rendezvous with Saturn early in the next decade. Two STIS imaging modes have been used to discriminate between ultraviolet emissions predominately from hydrogen atoms (shown in red) and emissions due to molecular hydrogen (shown in blue). Hence the bright red aurora features are dominated by atomic hydrogen, while the white traces within them map the more tightly confined regions of molecular hydrogen emissions. The southern aurora is Seen lower right, the northern at upper left. (*Credit: J.T. Trauger (Jet Propulsion Laboratory) and NASA.*)

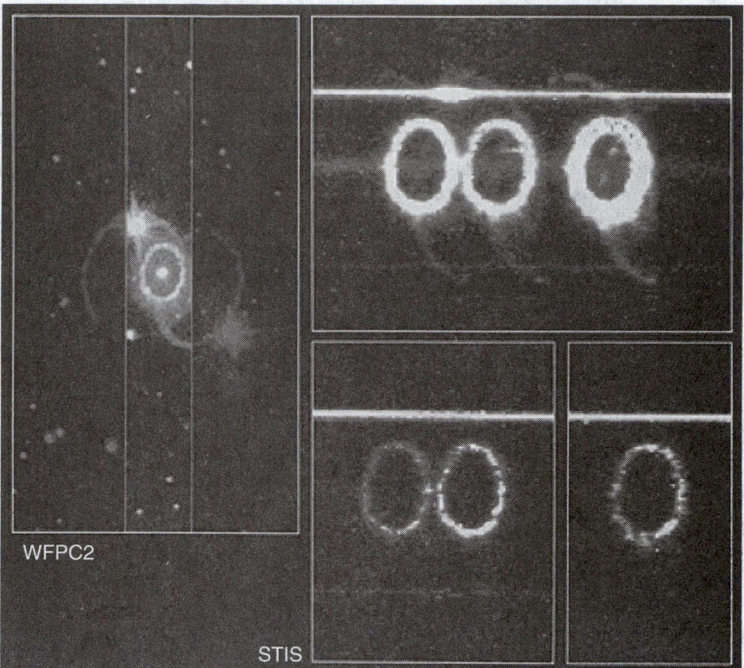

Fig. 11. These images from the Hubble's Space Telescope Imaging Spectrograph (STIS) provide a new and unprecedented look at one of the most unique and complex structures in the universe — a light-year wide ring of glowing gas around supernova 1987A, the nearest stellar explosion in 400 years, which occurred in February 1987. The STIS long-slit spectrograph viewed the entire ring system, dissecting its light and producing a detailed image of the ring in each of its component colors. Each color represents light from specific elements in the ring's gasses, including oxygen (single green ring), nitrogen and hydrogen (triple orange rings), and sulfur (double red rings). The ratio of the ring's brightness in different colors, emanating from the same elements, gives a measure of the concentration of the gasses. The light from different elements also identifies gasses at different temperatures. By dismantling the ring into its different puzzle pieces — its component elements — astronomers hope to put together a picture of stellar process and physics which created the ring. The ring formed 30,000 years before the star exploded and so is a fossil record of the final stages of the star's existence. The light from the supernova heated the gas in the ring so that it now glows at temperatures from 5,000 to 25,000 Kelvin. Supernova 1987A is located 167,000 light-years away from Earth in the Large Megellanic Cloud. (*Credit: STIS images — George Sonneborn (GSFC) and NASA, WFPC image — Jason Pun (NOAO) and SINS Collaboration.*)

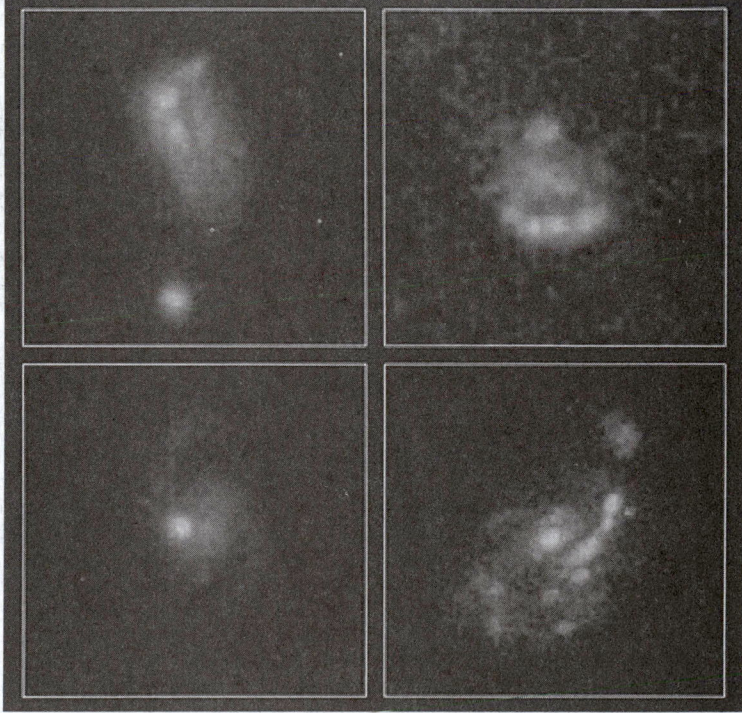

Fig. 12. This is a NASA Hubble Space Telescope image of a variety of galaxies with irregular and peculiar shapes. These galaxies are so far away that they are seen when the universe was a fraction of its current age. The bright blue regions indicate a rapid episode of star formation. Hubble reveals that these objects once far outnumbered large galaxies like our Milky Way, but have faded or self-destructed by today. This image is part of a serendipitous sky survey that has been conducted over the past three years by Professor Richard Griffiths and colleagues at the Johns Hopkins University, Baltimore, MD, with a team of astronomers in the United States and Britain. The survey is one of the key projects for Hubble. Over the past three years the deep survey has uncovered a bizarre variety of shapes and structures in distant galaxies, which previously appeared as fuzzy blobs from ground-based telescopes. (*Credit: Richard Griffiths (JHU), The Medium Deep Survey Team, and NASA.*)

exposures, approximately half of these are of astronomical targets and the rest are calibration exposures.

Hubble Science

Hubble Space Telescope has made dramatic inroads into a broad range of astronomical frontiers. Astronomers used *Hubble Space Telescope* to look out into the universe over distances exceeding 12 billion light-years. Because the starlight harvested from remote objects began its journey toward earth billions of years ago, *Hubble Space Telescope* (as well as all large telescopes) looks further back into time the farther away it looks into space. *HST* has seen back to a time when the universe was only about five percent of its present age.

Hubble Space Telescope's deepest views of the universe, made with its visible and infrared cameras, are collectively called the *Hubble Deep Field*. These "long-exposures" of the universe have revealed galaxies that existed when the universe was less than 1 billion years old. The *Hubble Deep Field* also uncovered hundreds of galaxies at various stages of evolution, and strung along a corridor of billions of light-years. The high resolution of the *HST* enables astronomers to actually see the shapes of galaxies in the distant past and to study how they have evolved over time. See Fig. 12 for images from deep-sky survey that show irregular and peculiarly shaped galaxies, and Figs. 13 and 14 for *Hubble's* deepest views of the universe.

Another key project for the *Hubble Space Telescope* has been to make precise distance measurements for calculating the rate of expansion of the universe. Determining the exact value of this rate is fundamental to calculating the age of the universe. To find out if the universe was expanding at a faster rate long ago, the *Hubble Space Telescope* peered halfway across the universe to find ancient exploding stars called supernova. These stars can be used to calculate vast astronomical distances. *HST* observations, as well as other observations done with ground-based telescopes, show that the universe has not decelerated, and therefore will likely expand forever. This realization challenges fundamental models of the Big Bang theory. See also **Cosmology**. Contrary to simple Big Bang models, the observational results strongly imply there is not enough mass in the universe to provide sufficient gravity to slow or halt its expansion.

Black Holes

The *Hubble Space Telescope* has provided convincing evidence of the existence of super-massive black holes that are millions or even a billion times more massive than our sun. *HST* observations of both active and quiescent galaxies, which pour out prodigious amounts of energy, have shown that super-massive black holes are commonly found at the hub of a galaxy. This suggests that they may be intimately linked to the birth and evolution of a galaxy.

The *Hubble Space Telescope* uncovered the existence of super-massive black holes by measuring the speed of gas and stars trapped in the gravitational field of a black hole. This allowed for a direct calculation of the mass of the black hole. The measurements show that there is far more mass than can be accounted for by starlight at a galaxy's core. This unseen mass is locked away inside the black hole.

The ability of the *HST* to discern faint objects near bright objects allowed for definitive observations that show the true nature of quasars, which are compact powerhouses of light that reside largely at the outer reaches of the universe. The *Hubble Space Telescope* conclusively showed that quasars dwell in the cores of galaxies, which means they are powered by super-massive black holes which are swallowing material at a furious rate–about the equivalent of one solar mass per year.

The *HST* played a key role in helping astronomers resolve questions regarding the nature of mysterious gamma-ray bursts. Gamma-ray bursts are powerful blasts, which come from random directions in the universe about once per day. The *Hubble Space Telescope* found host galaxies associated with some of these blasts. This places the bursts at cosmological distances rather then being localized phenomena within our galaxy. The *HST* also showed that the blasts occur among the young stars in the spiral arms of a host galaxy. This favors neutron star collisions or neutron star—black hole collisions. Some of these galaxies observed are so far away the explosions behind them are the most powerful events happening since the big bang itself. See also **Black Hole**.

Stellar Environments

The *Hubble Space Telescope* has unveiled a wide variety of shapes, structures and fireworks that accompany the birth and death of stars. *HST*

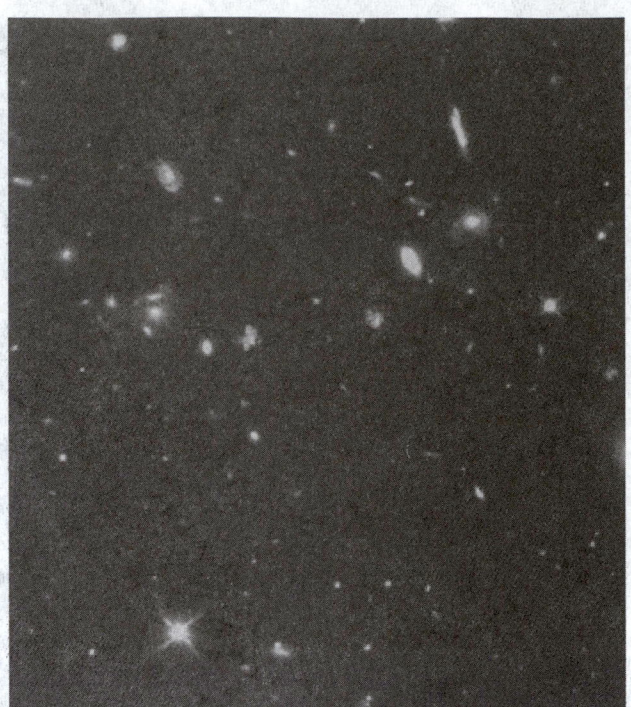

Fig. 13.   One of the deepest images of the sky taken to date with NASA's Hubble Space Telescope reveals a population of faint blue galaxies which turn out to be the most common class of objects in the universe. Their distances are estimated to range from three to eight billion light-years, meaning that they were abundant when the universe was a fraction of its present age, but are rare or harder to find today because they have faded or self-destructed. This picture, in combination with a series of images from the Hubble Space Telescope Medium Deep Survey that covers a larger area of sky, is allowing astronomers to solve the longstanding "faint blue galaxy mystery" by showing the true nature of these dim and remote objects. Deciphering the formation and evolution of these blue dwarf galaxies may provide new values to understanding the process of galaxy evolution, including the formation of our Milky Way Galaxy. Hubble's high resolution shows that most of these faint galaxies do not resemble elliptical and well-defined spiral galaxies that are common in the present universe. Instead, they have a wide variety of shapes suggesting that galaxy collisions and other interactions were more common in the past. The galaxies are blue because they are undergoing episodes of intense star-formation, which produce a lot of young, hot, and blue stars. This picture is a true-color image made from separate exposures taken in blue, green, and far-red light with the Wide Field and Planetary Camera 2. It required a total of 48 orbits around the Earth (amounting to roughly one day of exposure time) to make the observation and detect objects about four billion times fainter than the unaided eye can See (30th magnitude). The image resolution is about 0.06 arc seconds. The image covers a relatively small area of sky — only one-tenth the diameter of the full moon — in the constellation Hercules. (*Credits: Rogier Windhorst and Simon Driver (Arizona State University), Bill Keel (University of Alabama), and NASA.*)

images have provided a clear look at pancake-shaped disks of dust and gas swirling around and feeding embryonic stars. Besides helping build the star, the disks are also the prerequisite for condensing planets. The *HST* images also show blowtorch-like jets of hot gas streaming deep within the disks. These jets are an "exhaust product" of star formation. In dramatic images, *HST* has shown the effects of very massive young stars on their surrounding nebulae. The astronomical equivalent of a hurricane, the intense flow of visible and ultraviolet radiation from an exceptionally massive young star eats into surrounding clouds of cold hydrogen gas, laced with dust. This helps trigger a firestorm of starbirth in the neighborhood around the star.

The *Hubble Space Telescope* has produced a dazzling array of images of colorful shells of gas blasted into space by dying stars. These intricate structures are "fossil evidence" which show that the final stages of a star's life are more complex than once thought. An aging star sheds its outer layers of gas through stellar winds. Late in a star's life, these winds become more like a gale, and consequently sculpt strikingly complex shapes as they

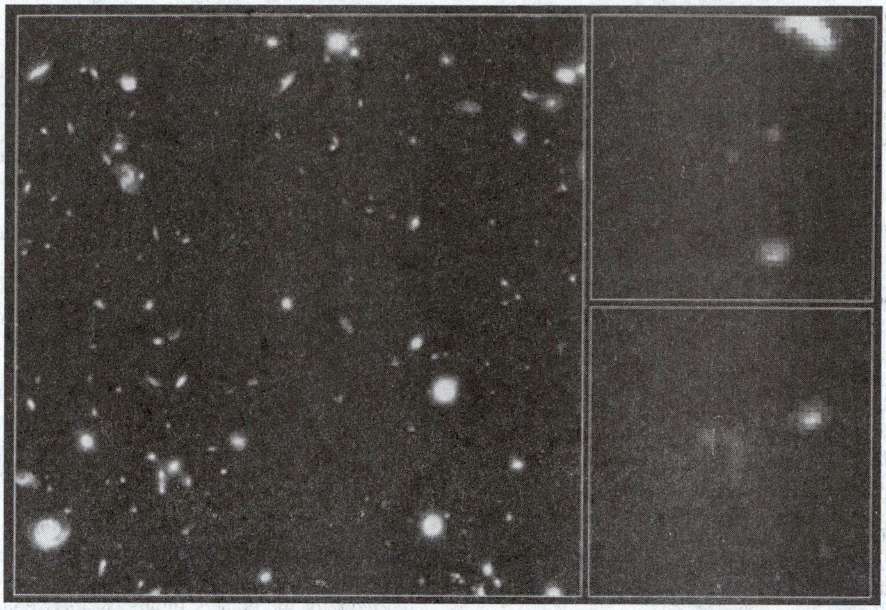

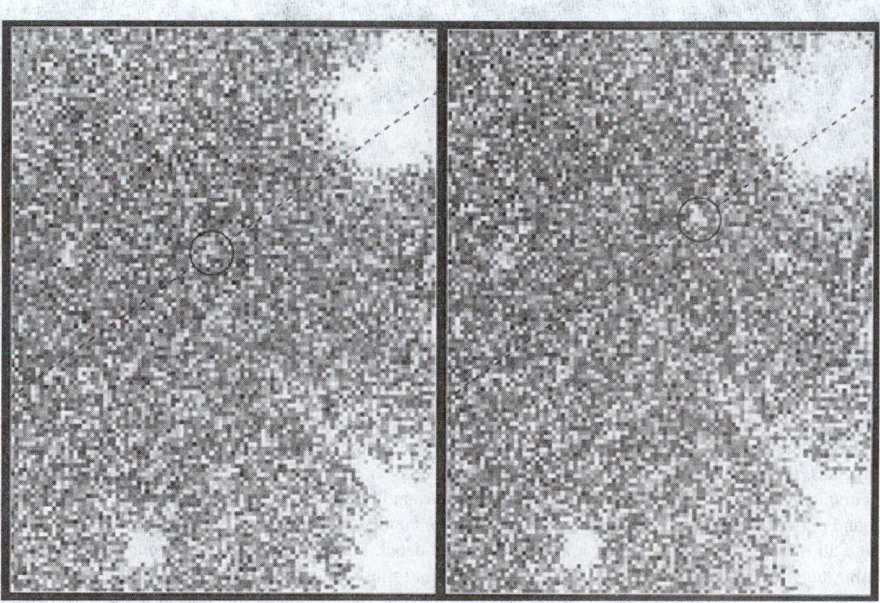

Fig. 14.   (Left)—A NASA Hubble Space Telescope view of the faintest galaxies ever Seen in the universe, taken in infrared light with the Near Infrared Camera and Multi-Object Spectrometer (NICMOS). The picture contains over 300 galaxies having spiral, elliptical, and irregular shapes. Though most of these galaxies were first Seen in 1995 when Hubble was used to take a visible-light deep exposure of the same field, NICMOS uncovers many new objects. Most of these objects are too small and faint to be apparent in the full field NICMOS view. Some of the reddest and faintest of the newly detected objects may be over 12 billion light-years away, as derived from a standard model of the universe. However, a powerful new generation of telescopes will be needed to confirm the suspected distance of these objects. The field of view is 2 million light-years across, at its maximum. Yet, on a cosmic scale, it represents only a thin pencil beam look across the universe. The area of sky is merely 1/100th the apparent diameter on the full moon.

   (Right)—Two close-up NICMOS views of candidate objects which may be over 12 billion light-years away. Each candidate is centered in the frame. The reddish color may mean that all of the starlight has been stretched to infrared wavelengths by the universe's expansion. Alternative explanations are that the objects are closer to us, but that the light has been reddened by dust scattering. A new generation of telescope will be needed to make follow-up observations capable of establishing true distance. The image was taken in January 1998 and required an exposure time of 36 hours to detect objects down to 30th magnitude. Hubble was aimed in the direction of the constellation Ursa Major, in a region just above the handle of the Big Dipper. The color corresponds to blue (0.45 microns), green (1.1 microns), and red (1.6 microns). (*Credit: Roger I. Thompson (University of Arizona), and NASA.*)

Fig. 15.   This is sample data from NASA's Hubble Space Telescope that illustrates the detection of comets in the Kuiper Belt, a region of space beyond the orbit of the planet Neptune. This pair of images, taken with the Wide Field Planetary Camera 2 (WFPC2), shows one of the candidate Kuiper Belt objects found with Hubble. Believed to be an icy comet nucleus several miles across, the object is so distant and faint that Hubble's search is the equivalent of finding the proverbial needle-in-the-haystack. Each photo is a 5-hour exposure of a piece of sky carefully selected such that it is nearly devoid of background stars and galaxies that could mask the elusive comet. The left image, taken on August 22, 1994, shows the candidate comet object (inside circle) embedded in the background. The right picture, taken of the same region one hour forty-five minutes later shows the object has apparently moved in the predicted direction and at the predicted rate of motion for a Kuiper belt member. The dotted line on the images is a possible orbit that this Kuiper belt comet is following. A star (lower right corner) and a galaxy (upper right corner) provide a static background reference. In addition, other objects in the picture have not moved during this time, indicating they are outside our solar system. Through this search technique astronomers have identified 29 candidate comet nuclei belonging to an estimated population of 200 million particles orbiting the edge of our solar system. The Kuiper Belt was theorized 40 years ago, and its larger members detected several years ago. However, Hubble has found the underlying population of normal comet-sized bodies. (*Credit: A. Cochran (University of Texas) and NASA.*)

plow into slower-moving material that was ejected earlier in the star's life. The most dramatic star-death observation for the *Hubble Space Telescope* has been to track the expanding wave of debris from the explosion of supernova 1987A. *HST* observations show that debris from the supernova blast is slamming into a ring of material around the dying star. The crash has allowed scientists to probe the structure around the supernova and uncover new clues about the final years of the progenitor star. For *HST* images of comets, galaxies, nebula, and stars including nova and super nova. See Figs. 15–34.

Comets

Fig. 15. Hubble detects comet nucleus at fringe of solar system.
Fig. 16. Hubble Sees material ejected from Comet Hale-Bopp.

Galaxies

Fig. 17. Hubble reveals stellar fireworks accompanying galaxy collision.
Fig. 18. Blue stragglers in globular cluster 47 Tucanae.
Fig. 19. Hubble spies globular cluster in neighboring galaxy.
Fig. 20. Turbulent cauldron of starbirth in nearby active galaxy.
Fig. 21. Distance measurements to a type 1A supernova-bearing galaxy.
Fig. 22. Hot white dwarf shines in young star cluster.
Fig. 23. Images of a galaxy in the Fornax cluster of galaxies.
Fig. 24. Close-up view of star formation in Antennae galaxy.

Nebula

Fig. 25. Hubble Sees supersonic exhaust from Nebula.
Fig. 26. Super-sharp view of the doomed star Eta Carinae.
Fig. 27. Hubble probes the great Orion Nebula.
Fig. 28. Bubble Nebula.

Nova

Fig. 29. Hubble pinpoints distant Supernovae.
Fig. 30. Hubble finds mysterious ring structure around Supernova 1987A.

Stars

Fig. 31. Nearby massive star cluster yields insights into early universe.
Fig. 32. Hubble uncovers brilliant star in Milky Way's core.
Fig. 33. HST captures first direct image of a star.
Fig. 34. The Sagittarius star cloud in a brilliant display.

Planets

Although space probes have offered a close look at every planet in the solar system, excluding Pluto, the *Hubble Space Telescope* has provided additional, remarkable observations which track changes on the planets. Many of the details seen by *HST* are not visible with ground-based telescopes. *HST* provided invaluable weather information about the atmosphere on Mars in support of the July 1997 *Mars Pathfinder* landing during which NASA sent a rover-vehicle to Mars. Similar observations of changes in Jupiter's atmosphere helped target observations by the Jupiter-orbiting Galileo probe. The *Hubble Space Telescope* also provided detailed pictures showing the effects of the collision of comet Shoemaker-Levy with Jupiter in 1994. To see figures provided by the *Hubble Space Telescope* of the planets Jupiter, Mars, Neptune, Saturn, and Uranus. See Figs. 35–46.

Jupiter

Fig. 35. Hubble provides a complete view of Jupiter's Auroras.
Fig. 36. Jupiter's upper atmospheric winds revealed in ultraviolet images.

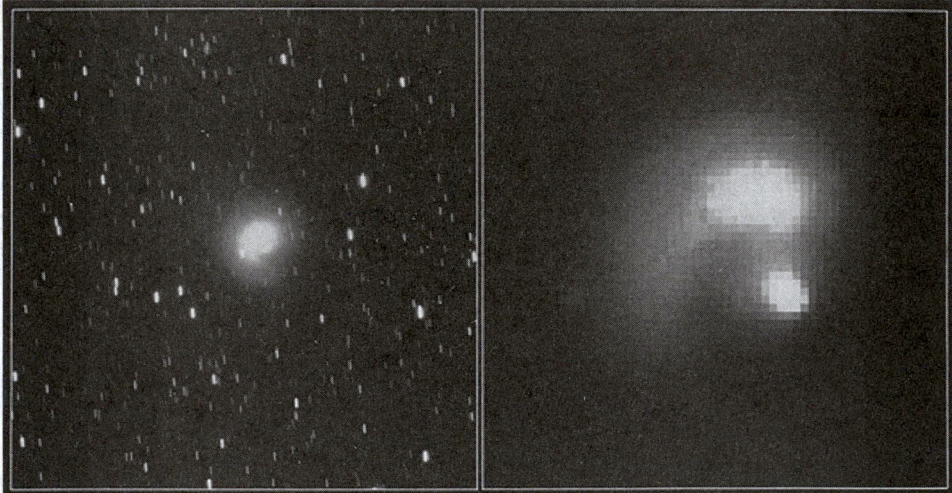

Fig. 16.   These NASA Hubble Space Telescope pictures of comet Hale-Bopp show a remarkable "pinwheel" pattern and a blob of free-flying debris near the nucleus. The bright clump of light along the spiral (above the nucleus, which is near the center of the frame) may be a piece of the comet's icy crust that was ejected into space by a combination of ice evaporation and the comet's rotation, and which then disintegrated into a bright cloud of particles. Although the "blob" is about 3.5 times fainter than the brightest portion at the nucleus, the lump appears brighter because it covers a larger area. The debris follows a spiral pattern outward because the solid nucleus is rotating like a lawn sprinkler, completing a single rotation about once per week. Ground-based observations conducted over the past two months have documented at least two separate episodes of jet and pinwheel formation and fading. By coincidence, the first Hubble images of Hale-Bopp, taken on September 26, 1995, immediately followed one of these outbursts and allow researchers to examine it at unprecedented detail. For the first time they See a clear separation between the nucleus and some of the debris being shed. By putting together information from the Hubble images and those taken during the recent outburst using the 82 cm telescope of the Teide Observatory (Tenerife, Canary Islands, Spain), astronomers find that the debris is moving away from the nucleus at a speed (projected on the sky) of about 68 miles per hour (109 kilometers per hour). The Hubble observations will be used to determine if Hale-Bopp is really a giant comet or rather a more moderate-sized object whose current activity is driven by outgassing from a very volatile ice which will "burn out" over the next year. Comet Hale-Bopp was discovered on July 23, 1995 by amateur astronomers Alan Hale and Thomas Bopp. Though this comet is still well outside the orbit of Jupiter (almost 600 million miles, or one billion kilometers from Earth) it looks surprisingly bright, fueling predictions that it could become the brightest comet of the century in early 1997. The full-field picture of the left, taken with the Wide Field Planetary Camera 2 (in WF mode), shows the comet against a stellar backdrop in the constellation Sagittarius. The stars are streaked due to a combination of Hubble's orbital motion and its tracking of the nucleus, which is now falling toward the Sun at 33,800 miles per hour (54,000 km/hr). In the close-up picture on the right, the stars have been subtracted through image processing. Each picture element is nearly 300 miles (480 km) across at the comet's distance. In this false color scale the faintest regions are black, the brightest regions are white, and intermediate intensities are represented by different levels of red. Even more detailed Hubble images will be taken with the Planetary Camera in late October to follow the further evolution of the spiral, look for more outbursts, place limits on the size of the nucleus, and use spectroscopy to study the enigmatic comet's chemical composition. (*Credit: H.A. Weaver (Applied Research Corp.), P.D. Feldman (John Hopkins University), and NASA.*)

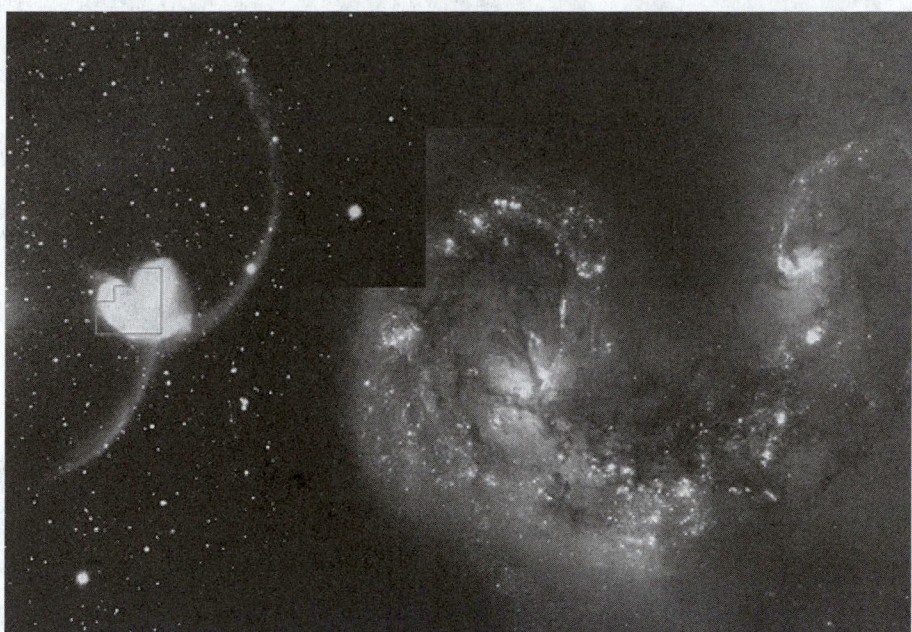

Fig. 17.   This Hubble Space Telescope image provides a detailed look at a brilliant "fireworks show" at the center of a collision between two galaxies. Hubble has uncovered over 1,000 bright, young star clusters bursting to life as a result of the head-on wreck.

Left-A ground-based telescopic view of the Antennae galaxies (known formally as NGC 4038/4039) — so named because a pair of long tails of luminous matter, formed by the gravitational tidal forces of their encounter, resembles an insect's antennae. The galaxies are located 63 million light-years away in the southern constellation Corvus.

Right-The respective cores of the twin galaxies are the orange blobs, left and right of image center, crisscrossed by filaments of dark dust. A wide band of chaotic dust, called the overlap region, stretches between the cores of the two galaxies. The sweeping spiral-like patterns, traced by bright blue star clusters, show the result of a firestorm of star birth activity which was triggered by the collision. This natural-color image is a composite of four separately filtered images taken with the Wide Field Planetary Camera 2 (WFPC2), on January 20, 1996. Resolution is 15 light-years per pixel (picture element). (*Credit: Brad Whitmore (STScI), and NASA.*)

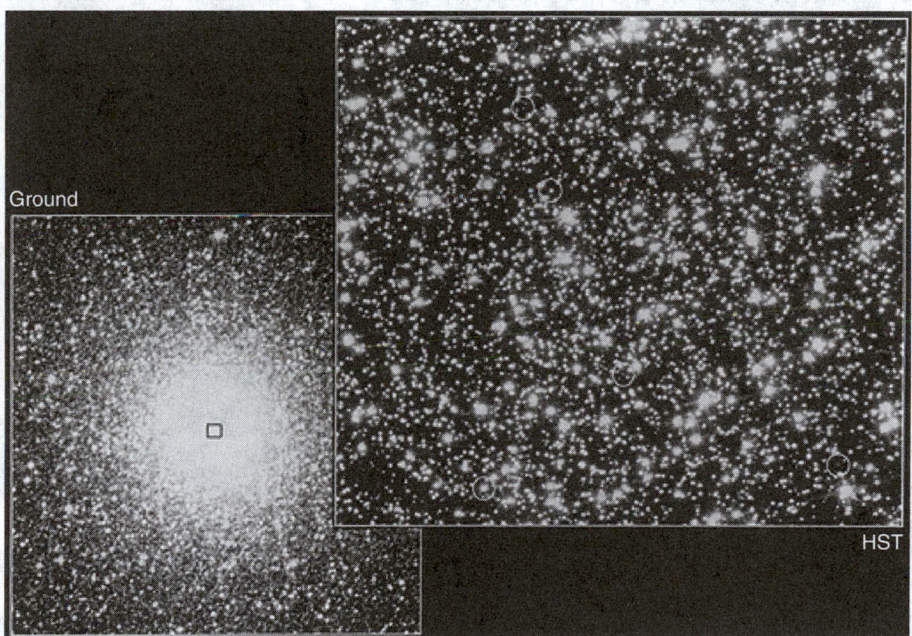

Fig. 18.   The core of globular cluster 47 Tucanae is home to many blue stragglers, rejuvenated stars that glow with the blue light of young stars. A ground-based telescope image (on the left) shows the entire crowded core of 47 Tucanae, located 15,000 light-years away in the constellation Tucana. Peering into the heart of the globular cluster's bright core, the Hubble Space Telescope's Wide Field Planetary Camera 2 separated the dense clump of stars into many individual stars (image on right). Some of these stars shine with the light of old stars; others with the blue light of blue stragglers. The yellow circles in the Hubble telescope image highlight several of the cluster's blue stragglers. Analysis for this observation centered on one massive blue straggler. Astronomers theorize that blue stragglers are formed either by the slow merger of stars in a double-star system or by the collision of two unrelated stars. For the blue straggler in 47 Tucanae, astronomers favor the slow merger scenario. This image is a 3-color composite of archival Hubble Wide Field Planetary Camera 2 images in the ultraviolet (blue), blue (green), and violet (red) filters. Color Tables were assigned and scaled so that the red giant stars appear orange, main-sequence stars are white/green, and blue stragglers are appropriately blue. The ultraviolet images were taken on October 25, 1995, and the blue and violet images were taken on September 1, 1995. (*Credits: Rex Saffer (Villanova University) and Dave Zurek (STScI), and NASA.*)

Fig. 19. Hubble Space Telescope has captured a view of a globular cluster called G1, a large, bright ball of light in the center of the photograph consisting of at least 300,000 old stars. G1, also known as Mayall II, orbits the Andromeda galaxy (M31), the nearest major spiral galaxy to our Milky Way. Located 130,000 light-years from Andromeda's nucleus, G1 is the brightest globular cluster in the Local Group of galaxies. The Local Group consists of about 20 nearby galaxies, including the Milky Way. The crisp image is comparable to ground-based telescope views of similar clusters orbiting the Milky Way. The Andromeda cluster, however, is nearly 100 times farther away. A glimpse into the cluster's finer details allows astronomers to See its fainter helium-burning stars whose temperatures and brightnesses show that this cluster in Andromeda and the oldest Milky Way clusters have approximately the same age. These clusters were formed shortly after the beginning of the universe, providing astronomers with a record of the earliest era of galaxy formation. During the next two years, astronomers will use Hubble to study about 20 more globular clusters in Andromeda. The color picture was assembled from separate images taken in visible and near-infrared wavelengths taken in July of 1994. (*Credits: Michael Rich, Kenneth Mighell, and James D. Neill (Columbia University), and Wendy Freedman (Carnegie Observatories), and NASA.*)

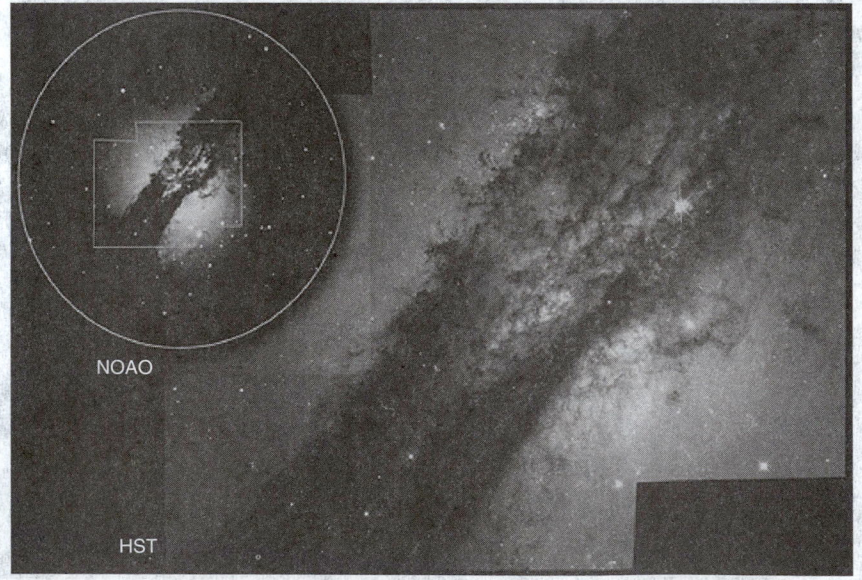

Fig. 20. NASA's Hubble Space Telescope offers a stunning unprecedented close-up view of a turbulent firestorm of starbirth along a nearly edge-on dust disk girdling Centaurus A, the nearest active galaxy to Earth. A ground-based telescopic view (upper left insert) shows that the dust lane girdles the entire elliptical galaxy. This lane has long been considered the dust remnant of a smaller spiral galaxy that merged with the large elliptical galaxy. The spiral galaxy deposited its gas and dust into the elliptical galaxy, and the shock of the collision compressed interstellar gas, precipitating a flurry of star formation. Resembling looming storm clouds, dark filaments of dust mixed with cold hydrogen gas are silhouetted against the incandescent yellow-orange glow from hot gas and stars behind it. Brilliant clusters of young blue stars lie along the edge of the dark dust rift. Outside the rift the sky is filled with the soft hazy glow of the galaxy's much older resident population of red giant and red dwarf stars. The dusty disk is tilted nearly edge-on, its inclination estimated to be only 10 or 20 degrees from out line-of-sight. The dust lane has not yet had enough time since the recent merger to settle down into a flat disk. At this oblique angle, bends and warps in the dust lane cause us to See a rippled "washboard" structure. The picture is a mosaic of two Hubble Space Telescope images taken with the Wide Field Planetary Camera 2, on August 1, 1997 and January 10, 1998. The approximately natural color is assembled from images taken in blue, green and red light. Details as small as seven light-years across can be resolved. The blue color is due to the light from extremely hot, newborn stars. The reddish-yellow color is due in part to hot gas, in part to older stars in the elliptical galaxy and in part to scattering of blue light by dust — the same effect that produces brilliant orange sunsets on Earth. (*Credit: E.J. Schreier, (STScI) and NASA.*)

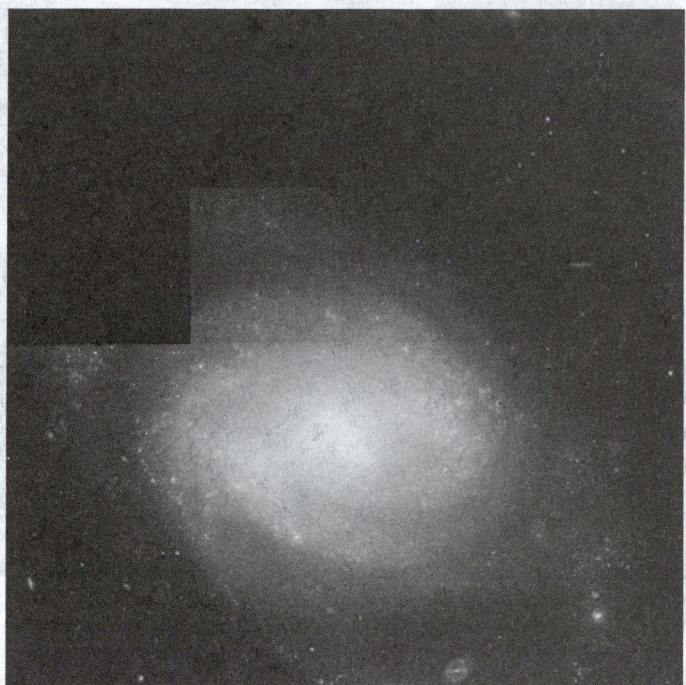

Fig. 21. This Hubble Space Telescope image shows NGC 4639, a spiral galaxy located 78 million light-years away in the Virgo cluster of galaxies. The blue dots in the galaxy's outlying regions indicate the presence of young stars. Among them are young, bright stars called Cepheids, which are used as reliable milepost markers to obtain accurate distances to nearby galaxies. Astronomers measure the brightness of Cepheids to calculate the distance to a galaxy. Allan Sandage's team used Cepheids to measure the distance to NGC 4639, the farthest galaxy to which Cepheid distance has been calculated. After using Cepheids to calculate the distance to NGC 4639, the team compared the results to the peak brightness measurements of SN 1990N, a type Ia supernova located in the galaxy. Then they compared those numbers with the peak brightness of supernovae similarly calibrated in nearby galaxies. The team then determined that type Ia supernovae are reliable secondary distance markers, and can be used to determine distances to galaxies several hundred times farther away than Cepheids. An accurate value for the Hubble Constant depends on Cepheids and secondary distance methods. The color image was made from separate exposures taken in the visible and near-infrared regions of the spectrum with the Wide Field Planetary Camera 2. (*Credits: A. Sandage (Carnegie Observatories), A. Saha (Space Telescope Science Institute), G.A. Tammann, and L. Labhardt (Astronomical Institute, University of Basel), F.D. Macchetto and N. Panagia (Space Telescope Science Institute/European Space Agency), and NASA.*)

Fig. 22. A dazzling "jewel-box" collection of over 20,000 stars can be seen in crystal clarity in this NASA Hubble Space Telescope image, taken with the Wide Field Planetary Camera 2 (WFPC-2). The young (40 million year old) cluster, called NGC 1818, is 164,000 light-years away in the Large Magellanic Cloud (LMC), a satellite galaxy of our Milky Way. The LMC, a site of vigorous current star formation, is an ideal nearby laboratory for studying stellar evolution. The circled star is a young white dwarf star, which has only very recently formed following the burnout of a red giant. Based on this observation astronomers conclude that the red giant progenitor star was 7.6 times the mass of our Sun. Previously, astronomers have estimated that stars anywhere from 6 to 10 solar masses would not just quietly fade away as white dwarfs but abruptly self-destruct in torrential explosions. Hubble can easily resolve the star in the crowded cluster, and detect its intense blue-white glow from a sizzling surface temperature of 50,000 degrees Fahrenheit. (*Credits: Rebecca Elson and Richard Sword (Cambridge UK), and NASA. [Original WFPC-2 image courtesy J. Westpal (Caltech).]*)

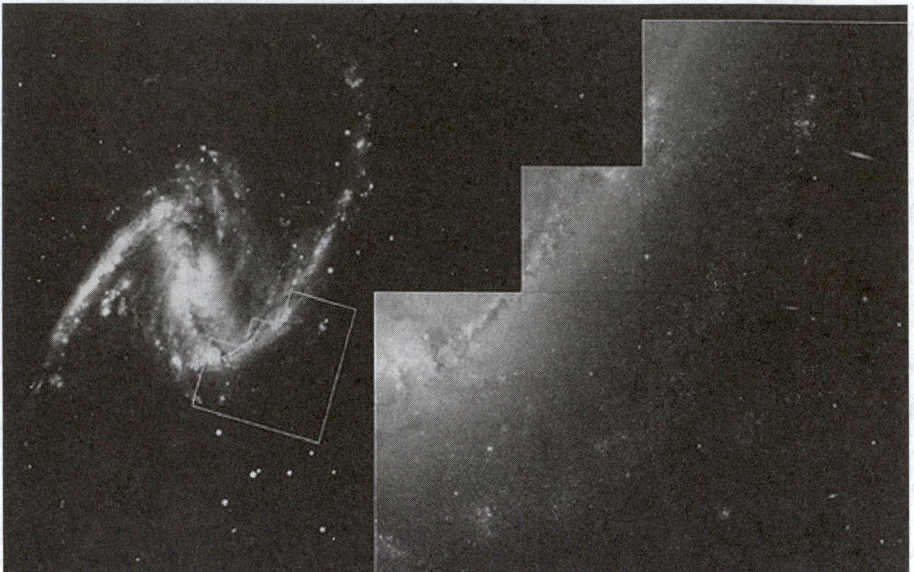

Fig. 23. This color image from the Hubble Space Telescope shows a region in NGC 1365, a barred spiral galaxy located in a cluster of galaxies called Fornax. A barred spiral galaxy is characterized by a "bar" of stars, dust and gas across its center. The black and white photograph from a ground-based telescope shows the entire galaxy, which is visible from the Southern Hemisphere. Members of the Key Project team, who have been measuring the distance to the Fornax cluster, have estimated it to be 60 million light-years from Earth. The team arrived at their preliminary estimate by using Cepheids, bright, young stars that are used as milepost markers to calculate distances to nearby galaxies. The line of small blue dots in the color image shows the formation of stars in the galaxy's spiral arms, making them ideal targets for the discovery of Cepheids. The group has discovered about 50 Cepheids in the galaxy. The team also has used the Fornax cluster to calibrate and compare many secondary distance methods. Cepheids are accurate distance markers for nearby galaxies, but astronomers need secondary methods to measure distances to faraway galaxies. An accurate value for the Hubble Constant is dependent on reliable secondary distance methods. (*Credit: W. Freedman (Carnegie Observatories), the Hubble Space Telescope Key Project team, and NASA.*)

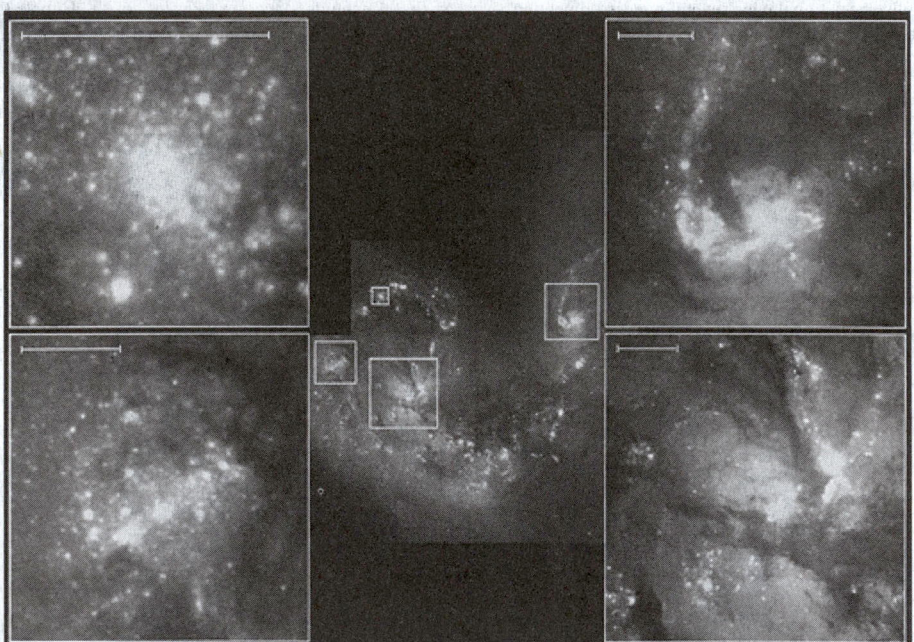

Fig. 24. These four close-up views are taken from a head-on collision between two spiral galaxies called the Antennae Galaxies, Seen at image center. The scale bar at the top of each image is 1,500 light-years across.

Left Images-The collision triggers the birth of new stars in brilliant blue star clusters, the brightest of which contains roughly a million stars. The star clusters are blue because they are very young, the youngest being only a few million years old, a mere blink of the eye on the astronomical time scale.

Right Images — These close-up views of the cores of each galaxy show entrapped dust and gas funneled into the center. The nucleus of NGC 4038 (lower right) is obscured by dust, which dims and reddens starlight by scattering the shorter, bluer wavelengths. This is also the reason the young star clusters in the dusty regions appear red instead of blue. This natural-color image is a composite of four separately filtered images taken with the Wide Field Planetary Camera 2 (WFPC2), on January 20, 1996. Resolution is 15 light-years per pixel (picture element). (*Credit: Brad Whitmore (STScI), and NASA.*)

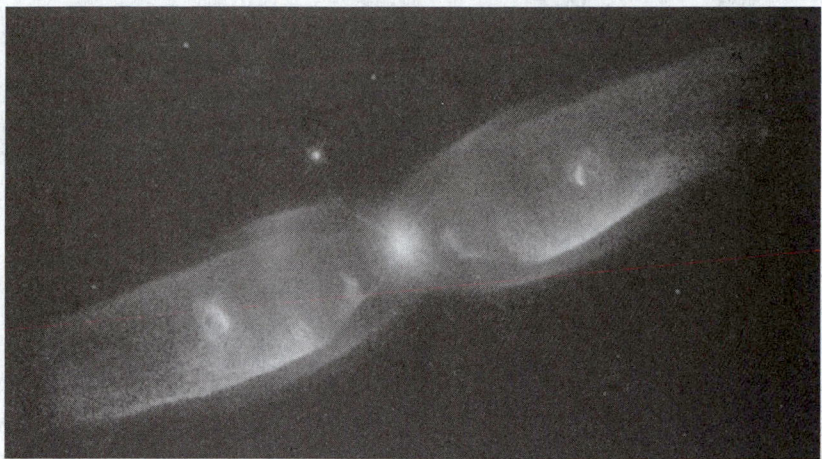

Fig. 25. M2-9 is a striking example of a "butterfly" or a bipolar planetary nebula. Another more revealing name might be the "Twin Jet Nebula". If the nebula is sliced across the star, each side of it appears much like a pair of exhausts from jet engines. Indeed, because of the nebula's shape and the measured velocity of the gas, in excess of 200 miles per second, astronomers believe that the description as a super-super-sonic jet exhaust is quite apt. Ground-based studies have shown that the nebula's size increases with time, suggesting that the stellar outburst that formed the lobes occurred just 1,200 years ago. The central star in M2-9 is known to be one of a very close pair, which orbit one another at perilously close distances. It is even possible that one star is being engulfed by the other. Astronomers suspect the gravity of one star pulls weakly bound gas from the surface of the other and flings it into a thin, dense disk, which surrounds both stars and extends well into space. The disk can actually be seen in shorter exposure images obtained with the Hubble telescope. It measures approximately 10 times the diameter of Pluto's orbit. Models of the type that are used to design jet engines ("hydrodynamics") show that such a disk can successfully account for the jet-exhaust-like appearance of M2-9. The high-speed wind from one of the stars rams into the surrounding disk, which serves as a nozzle. The wind is deflected in a perpendicular direction and forms the pair of jets that we See in the nebula's image. This is much the same process that takes place in a jet engine. The burning and expanding gases are deflected by the engine walls through a nozzle to form long, collimated jets of hot air at high speeds. M2-9 is 2,100 light-years away in the constellation Ophiucus. The observation was taken August 2, 1997 by the Hubble telescope's Wide Field and Planetary Camera 2. In this image, neutral oxygen is shown in red, once-ionized nitrogen in green, and twice-ionized oxygen in blue. (*Credits: Bruce Balick (University of Washington), Vincent Icke (Leiden University, The Netherlands), Garrelt Mellema (Stockholm University), and NASA.*)

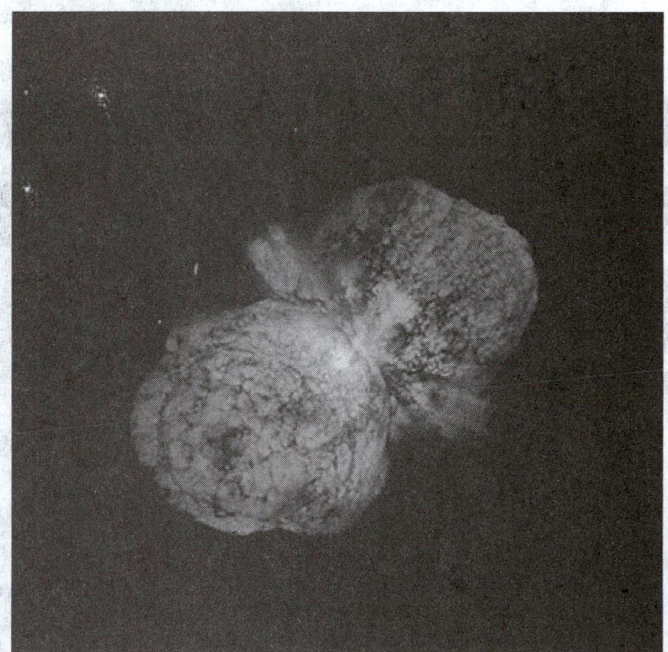

Fig. 26. A huge, billowing pair of gas and dust clouds is captured in this stunning NASA Hubble Space Telescope image of the supermassive star Eta Carinae. Using a combination of image processing techniques (dithering, subsampling and deconvolution), astronomers created one of the highest resolution images of an extended object ever produced by Hubble Space Telescope. The resulting picture reveals astonishing detail. Even though Eta Carinea is more than 8,000 light-years away, structures only 10 billion miles across (about the diameter of our solar system) can be distinguished. Dust lanes, tiny condensations, and strange radial streaks all appear with unprecedented clarity. Eta Carinae was observed by Hubble in September 1995 with the Wide Field Planetary Camera 2 (WFPC2). Images taken through red and near-ultraviolet filters were subsequently combined to produce the color image shown. A sequence of eight exposures was necessary to cover the object's huge dynamic range: the outer ejecta blobs are 100,000 times fainter than the brilliant central star. Eta Carinae was the site of a giant outburst about 150 years ago, when it became one of the brightest stars in the southern sky. Though the star released as much visible light as a supernova explosion, it survived the outburst. Somehow, the explosion produced two polar lobes and a large thin equatorial disk, all moving outward at about 1.5 million miles per hour. The new observation shows that excess violet light escapes along the equatorial plane between the bipolar lobes. Apparently there is relatively little dusty debris between the lobes down by the star; most of the blue light is able to escape. The lobes, on the other hand, contain large amounts of dust which preferentially absorb blue light, causing the lobes to appear reddish. Estimated to be 100 times more massive than our Sun, Eta Carinae may be one of the most massive stars in our Galaxy. It radiates about five million times more power than our Sun. The star remains one of the great mysteries of stellar astronomy, and the new Hubble images raise further puzzles. Eventually, this star's outburst may provide unique clues to other, more modest stellar bipolar explosions and to hydrodynamic flows from stars in general. (*Credit: Jon Morse (University of Colorado), and NASA.*)

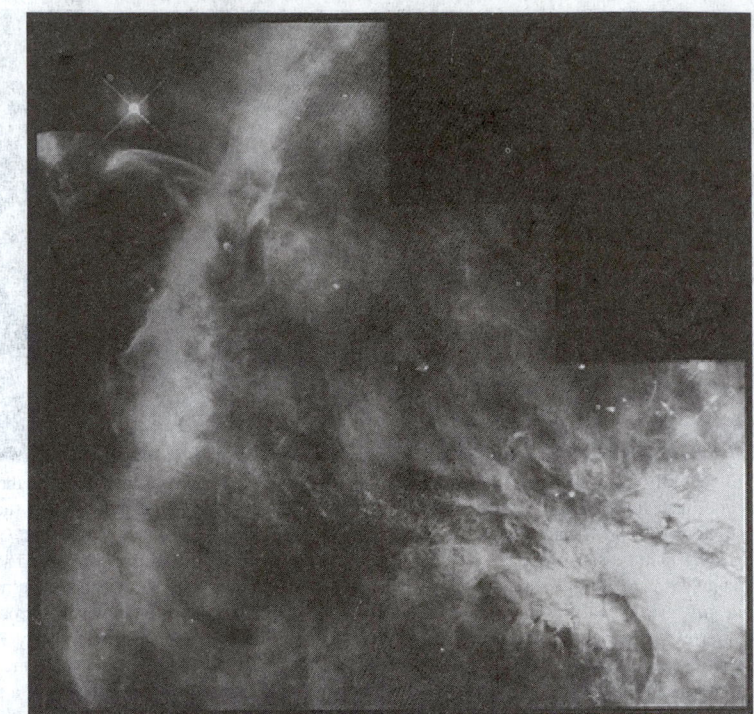

Fig. 27.   This is one of the nearest regions of very recent star formation (300,000 years ago). The nebula is a giant gas cloud illuminated by the brightest of the young hot stars at the top of the picture. Many of the fainter young stars are surrounded by disks of dust and gas that are slightly more than twice the diameter of the Solar System. The great plume of gas in the lower left in this picture is the result of the ejection of material from a recently formed star. The brightest portions are "hills" on the surface of the nebula, and the long bright bar is where Earth observers look along a long "wall" on a gaseous surface. The diagonal length of the image is 1.6 light-years. Red light depicts emission in Nitrogen; green is Hydrogen; and blue is Oxygen. The Orion Nebula star-birth region is 1,500 light-years away, in the direction of the constellation Orion the Hunter. The image was taken on 29 December 1993 with the HST's Wide Field Planetary Camera 2. (*Credit: C.R. O'Dell/Rice University and NASA.*)

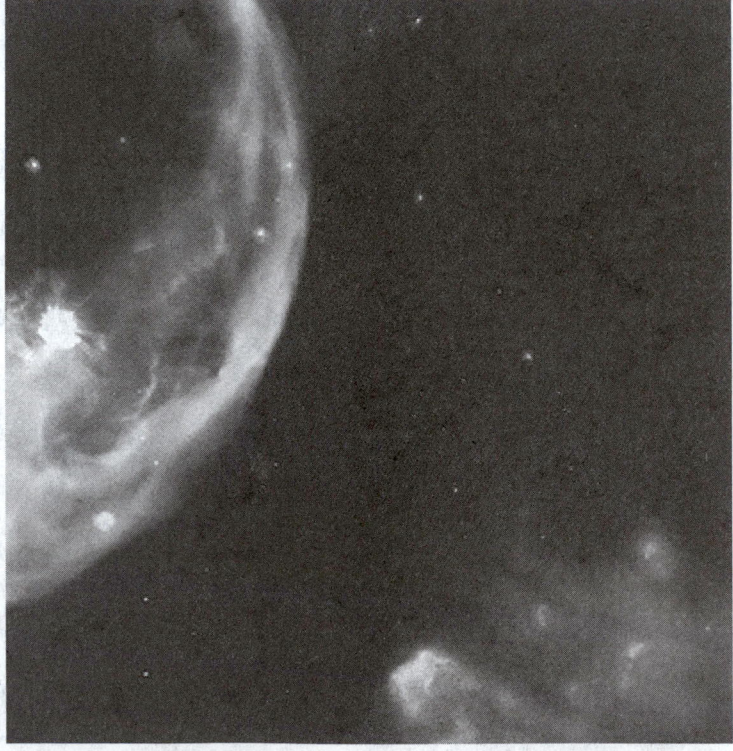

Fig. 28.   This NASA Hubble Space Telescope image reveals an expanding shell of glowing gas surrounding a hot, massive star in our Milky Way Galaxy. This shell is being shaped by strong stellar winds of material and radiation produced by the bright star at the left, which is 10 to 20 times more massive than our Sun. These fierce winds are sculpting the surrounding materials—composed of gas and dust—into the curve-shaped bubble. Astronomers have dubbed it the Bubble Nebula (NGC 7635). The nebula is 10 light-years across, more than twice the distance from Earth to the nearest star. Only part of the bubble is visible in this image. The glowing gas in the lower right-hand corner is a dense region of material that is getting blasted by radiation from the Bubble Nebula's massive star. The radiation is eating into the gas, creating finger-like features. This interaction also heats up the gas, causing it to glow. Scientists study the Bubble Nebula to understand how hot stars interact with the surrounding material. (*Credit: Hubble Heritage Team, AURA/STScI/NASA.*)

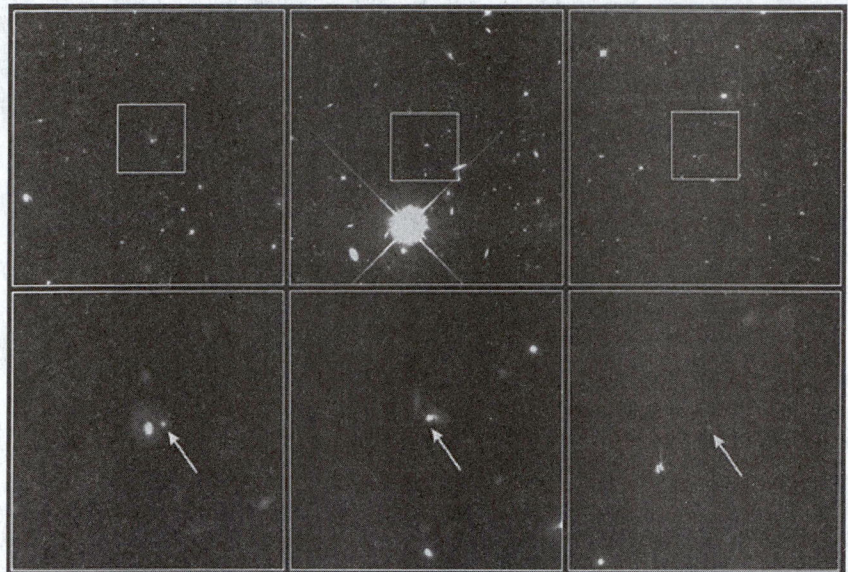

Fig. 29. These Hubble Space Telescope images pinpoint three distant supernovae, which exploded and died billions of years ago. Scientist are using these faraway light sources to estimate if the universe was expanding at a faster rate long ago and is now slowing down. Images of SN 1997cj are in the left hand column; SN 1997ce, in the middle; and SN 1997ck, on the right. All images were taken by the Hubble telescope's Wide Field and Planetary Camera 2. The images in the top row are wider views of the supernovae. The supernovae were discovered in April 1997 in a ground-based survey at the Canada-France-Hawaii telescope on Mauna Kea, Hawaii. Once the supernovae were discovered, the Hubble telescope was used to distinguish the supernovae from the light of their host galaxies. A series of Hubble telescope images were taken in May and June 1997 as the supernovae faded. Six Hubble telescope observations spanning five weeks were taken for each supernova. This time series enabled scientist to measure the brightness and create a light curve. Scientists then used the light curve to make an accurate estimate of the distances to the supernovae. Scientists combined the estimated distance with the measured velocity of the supernova's host galaxy to determine the expansion rate of the universe in the past (5 to 7 billion years ago) and compare it with the current rate. These supernovae belong to a class called Type Ia, which are considered reliable distance indicators. Looking at great distances also means looking back in time because of the finite velocity of light. SN 1997ck exploded when the universe was half its present age. It is the most distant supernova ever discovered (at a redshift of 0.97), erupting 7.7 billion years ago. The two other supernovae exploded about 5 billion years ago. SN 1997ce has a redshift of 0.44; SN 1997cj, 0.50. SN 1997ck is in the constellation Hercules, SN 1997ce is in Lynx, just north of Gemini, and SN 1997cj is in Ursa Major, near the Hubble Deep Field. (*Credits: Peter Garnavich (Harvard-Smithsonian Center for Astrophysics), the High-z Supernova Search Team, and NASA.*)

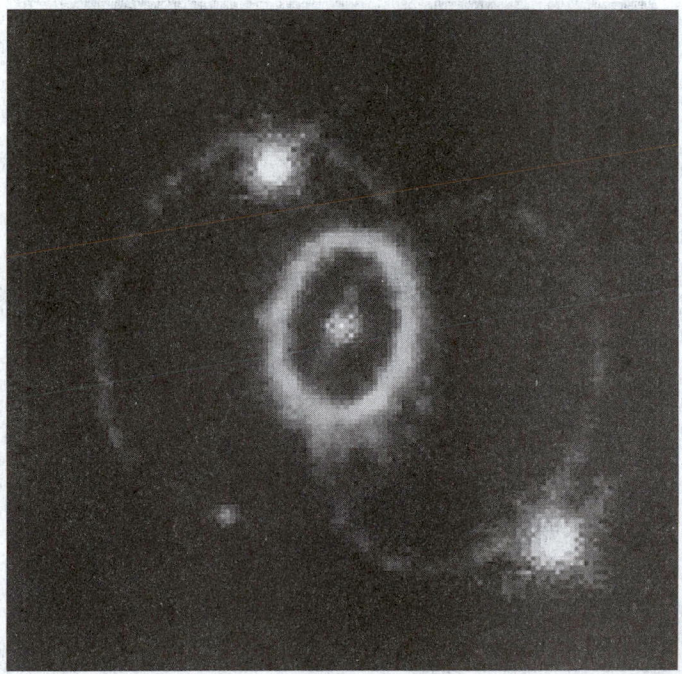

Fig. 30. This striking NASA Hubble Space Telescope picture shows three rings of glowing gas encircling the site of supernova 1987A, a star which exploded in February 1987. Though all of the rings appear inclined to our view (so that they appear to intersect) they are probably in three different planes. The small bright ring (first identified by HST in August 1990) lies in a plane containing the supernova, the two larger rings lie in front and behind it. The rings are a surprise because astronomers expected to See, instead, an hourglass shaped bubble of gas being blown into space by the supernova's progenitor star (based on previous HST observations, and images at lower resolution taken at ground-based observatories). One possibility is that the two rings are being "painted" on the invisible hourglass by a high-energy beam of radiation that is sweeping across the gas, like a searchlight sweeping across clouds. The source of the radiation might be a previously unknown stellar remnant that is a binary companion to a star that exploded in 1987. The supernova is 169,000 light years away, and lies in the dwarf galaxy called the large Magellanic Cloud, which can be Seen from the southern hemisphere. The image was taken in visible light (hydrogen-alpha emission), with the Wide Field Planetary Camera 2, in February 1994. (*Credit: Dr. Christopher Burrows, ESA/STSci and NASA.*)

Fig. 31. A NASA Hubble Space Telescope "family portrait" of young, ultra-bright stars nested in their embryonic cloud of glowing gases. The celestial maternity ward, called N81, is located 200,000 light-years away in the Small Magellanic Cloud (SMC), a small irregular satellite galaxy of our Milky Way. Hubble's exquisite resolution allows astronomers to pinpoint 50 separate stars tightly packed in the nebula's core within a 10 light-year diameter—slightly more than twice the distance between earth and the nearest star to our sun. The closest pair of stars is only 1/3 of a light-year apart (0.3 arc seconds in the sky). This furious rate of mass loss from these super-hot stars is evident in the Hubble picture that reveals dramatic shapes sculpted in the nebula's wall of glowing gases by violent stellar winds and shock waves. A pair of bright stars in the center of the nebula is pouring out most of the ultraviolet radiation to make the nebula glow. Just above them, a small dark knot is all that's left of the cold cloud of molecular hydrogen and dust the stars were born from. Dark absorption lanes of residual dust trisect the nebula. The nebula offers a unique opportunity for a close-up glimpse at the "firestorm" accompanying the birth of extremely massive stars, each blazing with the brilliance of 300,000 of our suns. Such galactic fireworks were much more common billions of years ago in the early universe, when most star formation took place. The "natural color" view was assembled from separate images taken with the Wide Field Planetary Camera 2, in ultraviolet light and two narrow emission lines of ionized Hydrogen (H-alpha, H-beta). The picture was taken on September 4, 1997. (*Credits: Mohammad Heydari (Paris Observatory, France), NASA/ESA.*)

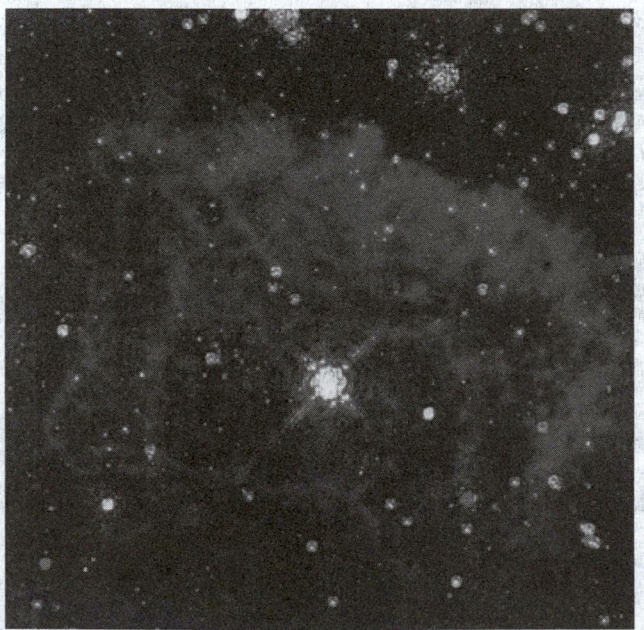

Fig. 32. One of the intrinsically brightest stars in our galaxy appears as the bright white dot in the center of this image taken with NASA's Hubble Space Telescope. Hubble's Near Infrared Camera and Multi-Object Spectrometer (NICMOS) was needed to take the picture, because the star is hidden at the galactic center, behind obscuring dust. NICMOS' infrared vision penetrated the dust to reveal the star, which is glowing with the radiance of 10 million suns. The image also shows one of the most massive stellar eruptions ever Seen in space. The radiant star has enough raw power to blow off two expanding shells (magenta) of gas equal to the mass of several of our suns. The largest shell is so big (4 light-years) it would stretch nearly all the way from our Sun to the next nearest star. The outbursts Seen by Hubble are estimated to be only 4,000 and 6,000 years old, respectively. Despite such a tremendous mass loss, astronomers estimate the extraordinary star may presently be 100 times more massive than our Sun, and may have started with as much as 200 solar masses of material, but it is violently shedding much of its mass. The star is 25,000 light-years away in the direction of the constellation Sagittarius. Despite its great distance, the star would be visible to the naked eye as a modest 4th magnitude object if it were not for the dust between it and the Earth. This false-colored image is a composite of two separately filtered images taken with the NICMOS, on September 13, 1997. The field of view is 4.8 light-years across, at the star's distance of 25,000 light-years. Resolution is 0.075 arc seconds per pixel (picture element). (*Credit: Don F. Figer (UCLA), and NASA.*)

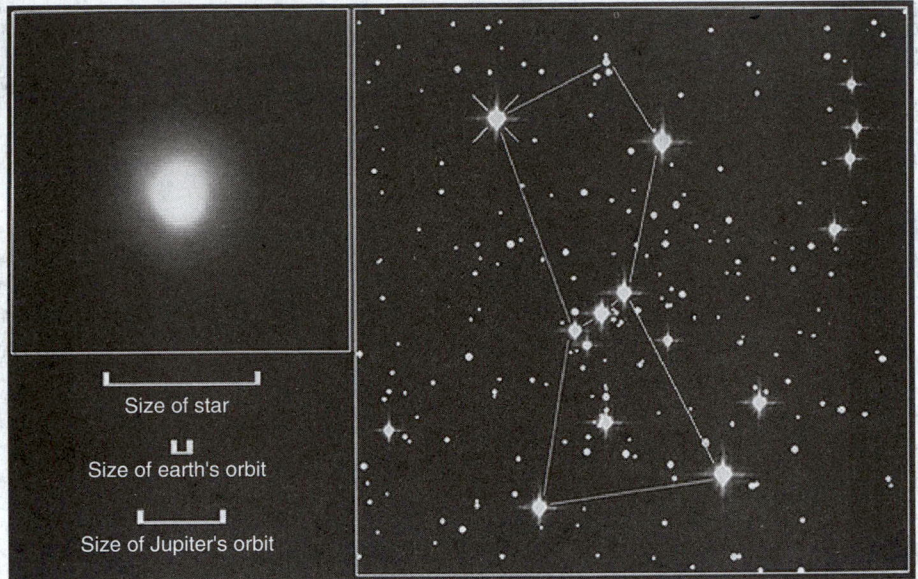

Size of star

Size of earth's orbit

Size of Jupiter's orbit

Fig. 33. This is the first direct image of a star other than the Sun, made with NASA's Hubble Space Telescope. Called Alpha Orionis, or Betelgeuse, it is a red supergiant star marking the shoulder of the winter constellation Orion the Hunter (diagram at right). The Hubble image reveals a huge ultraviolet atmosphere with a mysterious hot spot on the stellar behemoth's surface. The enormous bright spot, more than ten times the diameter of Earth, is at least 2,000 Kelvin degrees hotter than the surface of the star. The image suggests that a totally new physical phenomenon may be affecting the atmospheres of some stars. Follow-up observations will be needed to help astronomers understand whether the spot is linked to oscillations previously detected in the giant star, or whether it moves systematically across the star's surface under the grip of powerful magnetic fields. The observations were made by Andrea Dupree of the Harvard-Smithsonian Center for Astrophysics in Cambridge, MA, and Ronald Gilliland of the Space Telescope Science Institute in Baltimore, MD, who announced their discovery at the 187th meeting of the American Astronomical Society in San Antonio, Texas. The image was taken in ultraviolet light with the Faint Object Camera on March 3, 1995. Hubble can resolve the star even though the apparent size is 20,000 times smaller than the width of the full Moon — roughly equivalent to being able to resolve a car's headlights at a distance of 6,000 miles. Betelgeuse is so huge that, if it replaced the Sun at the center of our Solar System, its outer atmosphere would extend past the obit of Jupiter (scale at lower left). (Credit: Andrea Dupree (Harvard-Smithsonian CfA), Ronald Gilliland (STScI), NASA and ESA).)

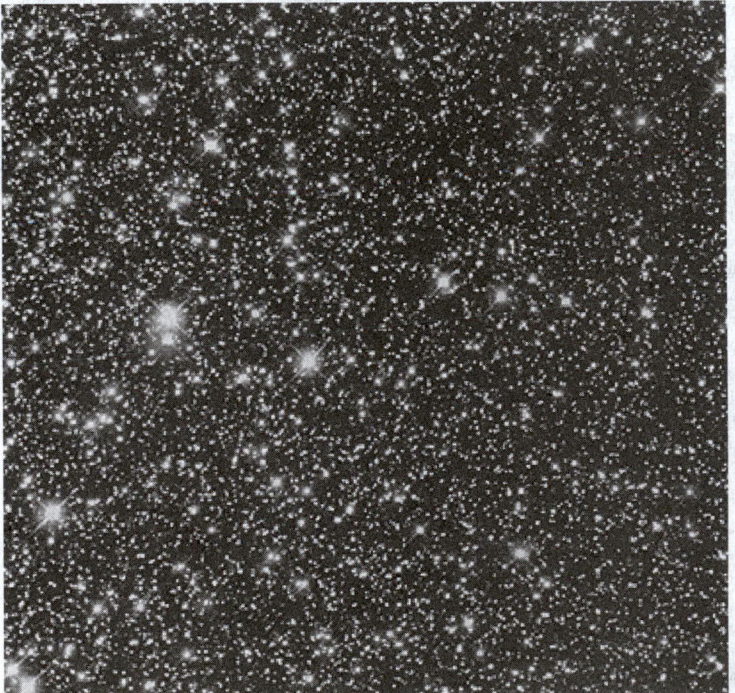

Fig. 34. NASA's Hubble Space Telescope has given us a key hole view toward the heart of our Milky Way Galaxy, where a dazzling array of stars reside. Most of the view of our galaxy is obscured by dust. Hubble peered into the Sagittarius Star Cloud, a narrow, dust-free region, providing this spectacular glimpse of a treasure chest full of stars. Some of these gems are among the oldest inhabitants of our galaxy. By studying the older stars that pack our Milky Way's hub, scientists can learn more about the evolution of our galaxy. Many of the brighter stars in this image show vivid colors. A star's color reveals its temperature, one of its most "vital statistics." Knowing a star's temperature and the power of the star's radiation allow scientists to make conclusions about its age and mass. Most blue stars are young and hot, up to ten times hotter than our Sun. They consume their fuel much faster and live shorter lives than our Sun. Red stars come in two flavors: small stars and "red giants." Smaller red stars generally have a temperature about half of our Sun, consuming their fuel slowly and thus, live the longest. "Red giant" stars are at the end of their lives because they have exhausted their fuel. Although many "red giant" stars may have been ordinary stars like our Sun, as they die the y swell up in size, become much cooler, and are much more luminous than they were during the majority of their stellar life. (Credit: Hubble Heritage Team (AURA/STScI/Nasa).)

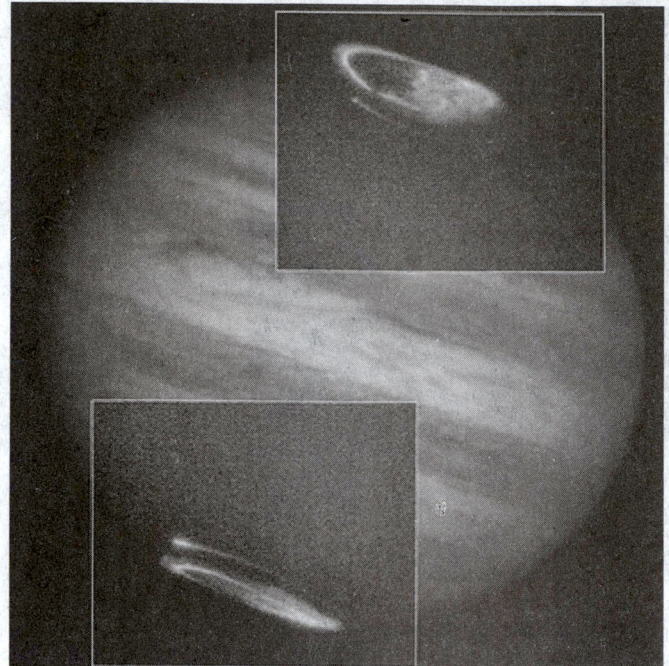

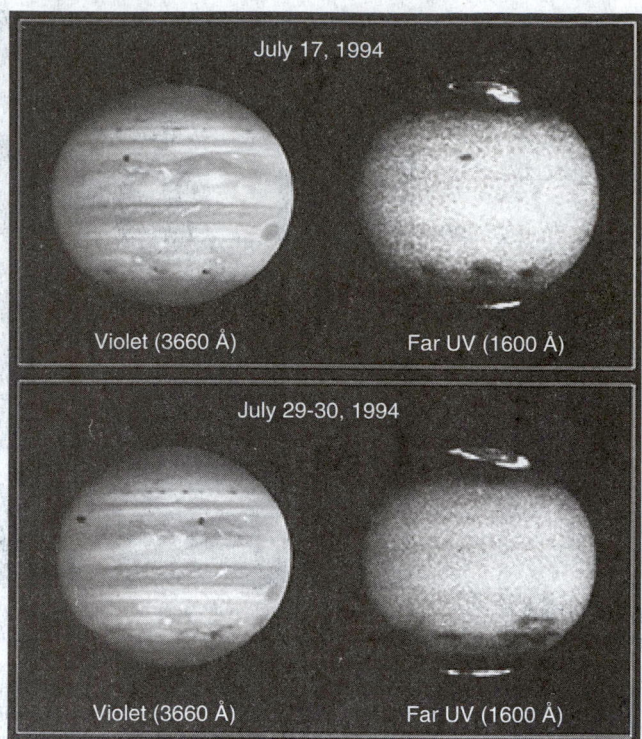

July 17, 1994

Violet (3660 Å)   Far UV (1600 Å)

July 29-30, 1994

Violet (3660 Å)   Far UV (1600 Å)

Fig. 35.   NASA's Hubble Space Telescope has captured a complete view of Jupiter's northern and southern auroras. Images taken in ultraviolet light by the Space Telescope Imaging Spectrograph (STIS) show both auroras, the oval-shaped objects in the inset photos. While the Hubble telescope has obtained images of Jupiter's northern and southern lights since 1990, the new STIS instrument is 10 times more sensitive than earlier cameras. This allows for short exposures, reducing the blurring of the image caused by Jupiter's rotation and providing two to five times higher resolution than earlier cameras. The resolution in these images is sufficient to show the "curtain" of auroral light extending several hundred miles above Jupiter's limb (edge). Images of Earth's auroral curtains, taken from the space shuttle, have a similar appearance. Jupiter's auroral images are superimposed on a Wide Field Planetary Camera 2 image of the entire planet. The auroras are brilliant curtains of light in Jupiter's upper atmosphere. Jovian auroral storms, like Earth's, develop when electrically charged particles trapped in the magnetic field surrounding the planet spiral inward at high energies toward the north and south magnetic poles. When these particles hit the upper atmosphere, they excite atoms and molecules there, causing them to glow (the same process acting in street lights). The electrons that strike Earth's atmosphere come from the sun, and the auroral lights remain concentrated above the night sky in response to the "solar wind," as Earth rotates underneath. Earth's auroras exhibit storms that extend to lower latitudes in response to solar activity, which can be easily Seen from the northern U.S. But Jupiter's auroras are caused by particles spewed out by volcanoes on Io, one of Jupiter's moons. These charged particles are then magnetically trapped and begin to rotate with Jupiter, producing ovals of auroral light centered on Jupiter's magnetic poles in both the day and night skies. Scientists are comparing the Hubble telescope images with measurements taken by NASA's Galileo spacecraft of Jupiter's magnetic field and co-rotating charged particles. They believe the data will help them understand the production of Jupiter's auroras. Both auroras clearly show vapor trails of light left by Io. These vapor trails are the white, comet-shaped streaks just outside both auroral ovals. These streaks are not part of the auroral ovals. They are caused when an invisible electrical current of charged particles (equal to about 1 million amperes), ejected from Io, flow along Jupiter's magnetic field lines to the planets' north and south magnetic poles. The brightness part of both emissions (on the left in both images) pinpoints where Io's magnetic field lines leave its footprint on the planet. The trail of light following both emissions extends to the right all the way to Jupiter's edge and represents the most sensitive detection of ultraviolet emissions from Jupiter to date. These emissions are related to magnetically trapped ions and electrons that are carried by Jupiter's magnetic field along Io's orbital path and some of these charged particles continue to be driven down into Jupiter's atmosphere for several hours after Io has passed by. The images were taken Sept. 20, 1997. The artificial colors used here have been constructed by combining images taken in two different ultraviolet band passes, with one ultraviolet color presented as blue and the other as red. In this color representation, the planet's reflected sunlight appears brown, while the auroral emissions appear white or shades of blue or red. (*Credits: John Clarke (University of Michigan) and NASA.*)

Fig. 36.   These four NASA Hubble Space Telescope images of Jupiter, as Seen in visible (violet) and far-ultraviolet (UV) wavelengths, show the remarkable spreading of the clouds of smoke and dust thrown into the atmosphere after the impacts of the fragments of comet P/Shoemaker-Levy 9. These dark regions provide the only information ever obtained on the wind direction and speed in Jupiter's upper atmosphere. TOP Three impact sites appear as dark smudges lined up along Jupiter's Southern Hemisphere (from left to right, sites C, A, and E). This pair of images was obtained on 17 July, several hours after the E impact. These 3 impact sites appear strikingly darker in the far-ultraviolet images to the right. This is because the smoke and dust rising from the fireballs absorbs UV light more strongly than violet light, so that the clouds appear both darker and larger in the UV images. Apparently, the fireball and plume threw large amounts of material completely above the atmosphere. This material diffused back down through the atmosphere with the smaller and lighter particles suspended at high altitudes. BOTTOM Hubble's view of the same hemisphere of Jupiter 12-13 days later shows that the smoke and dust have now been spread mainly in the east/west direction by the prevailing winds at the altitude where the dark material is suspended or "floating" in the atmosphere. HST shows that winds in Jupiter's upper atmosphere carry the high altitude smoke and dust in different directions than in the lower atmosphere. For example, the UV image shows a fainter cloud near 45 deg. South latitude, which does not appear in the violet image. The fainter cloud may be due to high altitude material drifting with the upper atmospheric winds to the north away from the polar regions. However, in the left-hand impact regions the clouds being observed are lower in the atmosphere, where there is apparently no such northerly wind. The violet images show the Great Red Spot, on the eastern (right) limb, one of Jupiter's moons crossing in front of the planet in the northern hemisphere (and its shadow on Jupiter's clouds on the left-hand side in the lower image), and the dark clouds above 3 of the impact sites near 45 deg. South latitude. In addition, Jupiter's polar aurora can also be Seen in the far-ultraviolet images near both northern and southern poles. (*Credit: J.T. Clarke, G.E. Ballester (University of Michigan), and J.T. Trauger (Jet Propulsion Laboratory), and NASA.*)

Mars

Fig. 37. Hubble's look at Mars shows canyon duststorm.

Fig. 38. Four views of Mars in northern summer.

Fig. 39. Image of seasonal changes in Mars' north polar ice cap.

Neptune

Fig. 40. Images of weather on opposite hemispheres of Neptune.

Saturn

Fig. 41. View of Saturn's rings edge-on.

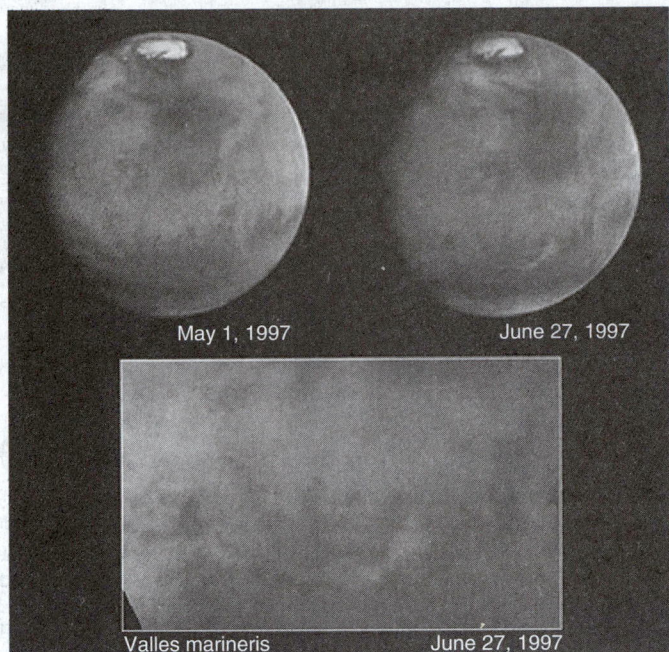

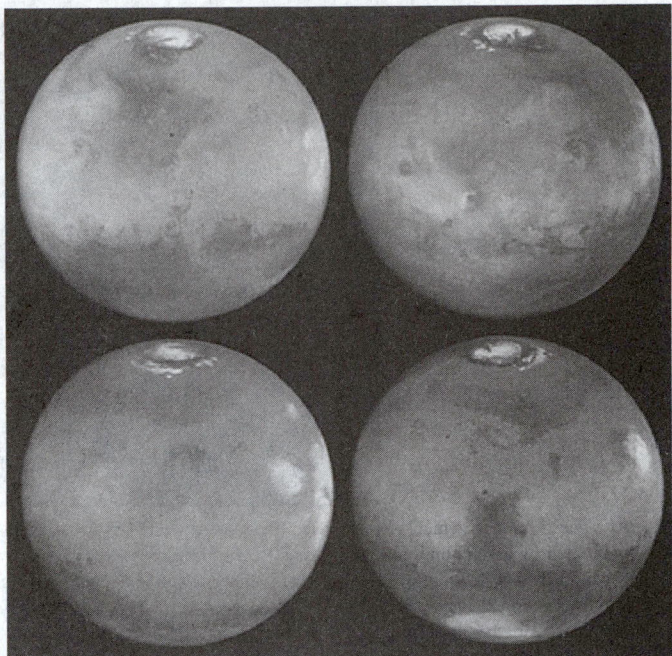

Fig. 37. Hubble Space Telescope images of Mars taken on June 27, 1997, reveal a significant dust storm which fills much of the Valles Marineris canyon system and extends into Xanthe Terra, about 600 miles (1.000 kilometers) south of the Pathfinder landing site. It is difficult to predict the evolution of this storm and whether it will affect the Pathfinder observations. The pictures were taken in order to monitor the site in Ares Vallis where the Pathfinder spacecraft would land on July 4. The two images of Mars at the top of the figure are Hubble observations from June 27 (right) and May 17 (left). Visual comparison of these images clearly shows the dust storm between 5 and 7 o'clock and about 2/3 of the way from the center of the southern edge of the June image. (*Credit: Steve Lee (University of Colorado), Mike Wolff and Phil James (University of Toledo), and NASA.*)

Fig. 42. Hubble views Saturn ring-plane crossing.
Fig. 43. View of Saturn in natural colors.

Uranus
Fig. 44. Hubble tracks clouds on Uranus showing the planet's rotation.
Fig. 45. Hubble captures detailed image of Uranus' atmosphere.
Fig. 46. Hubble spots northern hemispheric clouds on Uranus.

Other remarkable solar system findings include:

- In 1995, *Hubble Space Telescope* astronomers viewed Saturn's rings edge-on, an alignment that occurs only once every 24 years. They discovered clouds of debris from shattered moonlets near the edge of Saturn's F ring.
- *Hubble Space Telescope* is tracking weather changes on Uranus as the sun begins to shine in the planet's Southern Hemisphere.
- *Hubble Space Telescope* has followed the rapid disappearance and reoccurrence of Neptune's great dark spot (first discovered by the Voyager-2 spacecraft in 1989).
- Hubble Space Telescope resolved surface features on the planet Pluto, which is a challenging task equivalent to discerning the pattern on a soccer ball 40 miles away.
- In July 2000 Hubble saw the breakup of Comet Linear, whose nucleus mysteriously vanished following the comet's passage around the sun. Hubble resolved an armada of at least a dozen "mini-comets" which were left behind from the breakup. They are the primordial building blocks of comets.
- Hubble's ultraviolet sensitivity makes it ideas for watching auroras on the giant planets Jupiter and Saturn. The immense magnetic fields of these planets trap solar cosmic radiation that then excites gasses in the planets' upper atmospheres to glow as aurorae, just like the aurorae that appear above earth's polar regions.
- Hubble's infrared eyes found evidence for a fresh impact crater on the surface of a black and cold comet nucleus wandering from the Kupier

Fig. 38. Four faces of Mars as Seen on March 30, 1997 are presented in this montage of NASA Hubble Space Telescope images. Proceeding in the order upper-left, upper-right, lower-left, lower-right, Mars has rotated about ninety degrees between each successive time step. For example the Tharsis volcanoes, which are Seen (between 7:30 and 9 o'clock positions) in mid-morning in the upper-left view, are Seen near the late afternoon edge of the planet (about 3 o'clock position) in the lower-left image. All of these color images are composed of individual red (673 nanometers), green (502 nm), and blue (410 nm) Planetary Camera exposures.

Upper Left-This view is centered on Ares Valles, where Pathfinder was to land on July 4, 1997; the Valles Marineris canyon system stretches to the west across the lower left portion of the planet, while the bright, orangish desert of Arabia Planitia is to the east. The bright polar water-ice cap, surrounded by a dark ring of sand dunes, is obvious in the north; since it is northern summer and the pole is tilted toward us, the residual north polar cap is Seen in its entirety in all four images. Acidalia Planitia, the prominent dark area fanning southward from the polar region, is thought to have a surface covered with dark sand. Numerous "dark wind streaks" are visible to the south of Acidalia, resulting from wind-blown sand streaming out of the interiors of craters.

Upper Right-The Tharsis volcanoes and associated clouds are prominent in the western half of this view. Olympus Mons, spanning 340 miles (550 km) across its base and reaching an elevation of 16 miles (25 km), extends through the cloud deck near the western limb, while (from the south) Arsia Mons, Pavonis Mons, and Ascraeus Mons are to the west of center. Valles Marineris stretches to the east, and the Pathfinder landing site is shrouded in clouds near the afternoon limb.

Lower Left-This relatively featureless sector of Mars stretches from the Elysium volcanic region in the west to the Tharsis volcanoes (shrouded by the bright clouds near the afternoon limb) in the east. The group of three dark specks just left of center are all that remain of Cerberus, a very prominent dark region during the Viking and Mariner 9 missions. This is an example of the remarkable large scale changes which can occur on Mars due to windblown dust: the former dark area has now been covered by a layer of bright dust, masking the underlying material.

Lower Right-The dark Syrtis Major region dominates this image. Syrtis Major is one of the most prominent dark features on Mars, and has been visible since ground-based observers first peered at Mars through telescopes. The bright cloud at 3 o'clock is associated with Elysium Mons. The bright bluish-white feature near the southern limb of the planet is Hellas, a 1,200 mile (2,000 km) diameter impact basin formed by the collision of a large body with Mars long ago. Hellas is covered with dry ice frost and clouds during this season (winter in the south). (*Credits: Steve Lee (University of Colorado), Todd Clancy (Space Science Inst., Boulder, CO), Phil James (University of Toledo), and NASA.*)

belt, a reservoir of comets just beyond Pluto's orbit. Only 40 miles across, the nucleus is too small for Hubble to see the crater directly. The crater was inferred from the presence of fresh surface ices excavated by the impact that made the crater.

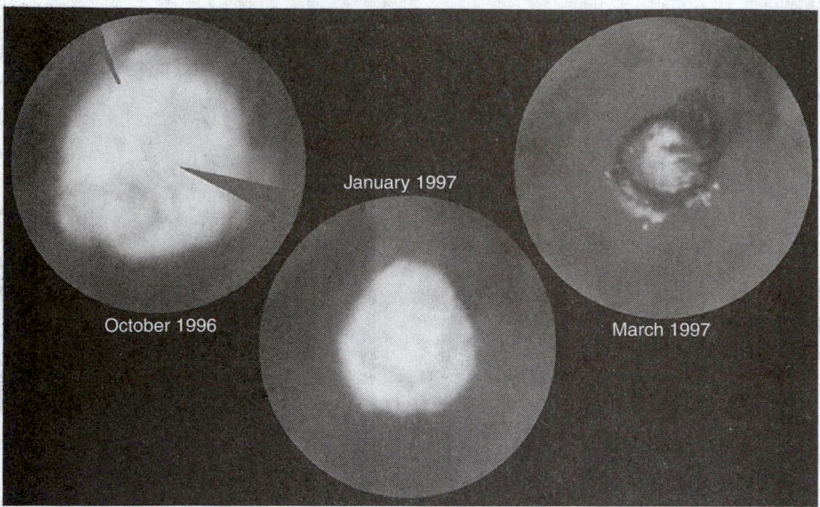

Fig. 39.   These images, which seem to have been taken while NASA's Hubble Space Telescope (HST) was looking directly down on the Martian North Pole, were actually created by assembling mosaics of three sets of images of Mars taken by HST in October, 1996 and in January and March, 1997 and projecting them to appear as they would if Seen from above the pole. This first mosaic is a view which could not actually be Seen in nature because at this season a portion of the pole would have actually been in shadow; the last view, taken near the summer solstice, would correspond to the Midnight Sun on Earth with the pole fully illuminated all day. The resulting polar maps begin at 50 degrees N latitude and are oriented with 0 degrees longitude at the 12 o'clock position. This series of pictures captures the seasonal retreat of Mars' north polar cap. October 1996 (early spring in the Northern Hemisphere): In this map, assembled from images obtained between October 8 and 15, the cap extends down to 60 degrees N latitude, nearly its maximum winter extent. (The notches are areas where Hubble data were not available.) A thin, comma shaped cloud of dust can be Seen as a salmon-colored crescent at the 7 o'clock position. The cap is actually fairly circular about the geographic pole at this season; the bluish "knobs" where the cap Seems to extend further are actually clouds that occurred near the edges of the three separate sets of images used to make the mosaic. January 1997 (mid-spring): Increased warming as a spring progresses in the Northern Hemisphere has sublimated the carbon dioxide ice and frost below 70 degrees N latitude. The faint darker circle inside the cap boundary marks the location of circumpolar sand dunes (See March 1997 map); these dark dunes are warmed more by solar heating than are the brighter surroundings, so the surface frost sublimates from the dunes earlier than from the neighboring areas. Particularly evident is the marked hexagonal shape of the polar cap at this season, noted previously by HST in 1995 and Mariner 9 in 1972; this may be due to topography, which isn't well know, or to wave structure in the circulation. This map was assembled from WFPC2 images obtained between December 30, 1996 and January 4, 1997. March 30, 1997 (early summer): The cap has fully retreated to its remnant core of water-ice. This residual cap is actually almost cut into two by a large, horn-shaped canyon called Chasma Borealis which is cut deeply into the polar terrain. The HST images also reveal a curious layered terrain which is evidence of past climatic changes on Mars. The sublimation of all of the carbon dioxide has exposed the ring of dark sand dunes which encircle the North Polar Cap. Outliers of ice persist south of the polar sand sea (between the 3 o'clock and 9 o'clock positions). The bright circular features at 3, 6, and 9 o'clock are ice-filled craters. All images were taken with the Wide Field Planetary Camera 2. The color is constructed from images taken in red (673 nm), blue (410 nm) and green (502 nm) light. The resolution at the North Pole ranges from about 115 km/pixel in October '96 to about 45 km/pixel in March 1997. (*Credits: Phil James (University of Toledo), Todd Clancy (Space Science Inst., Boulder, CO), Steve Lee (University of Colorado), and NASA.*)

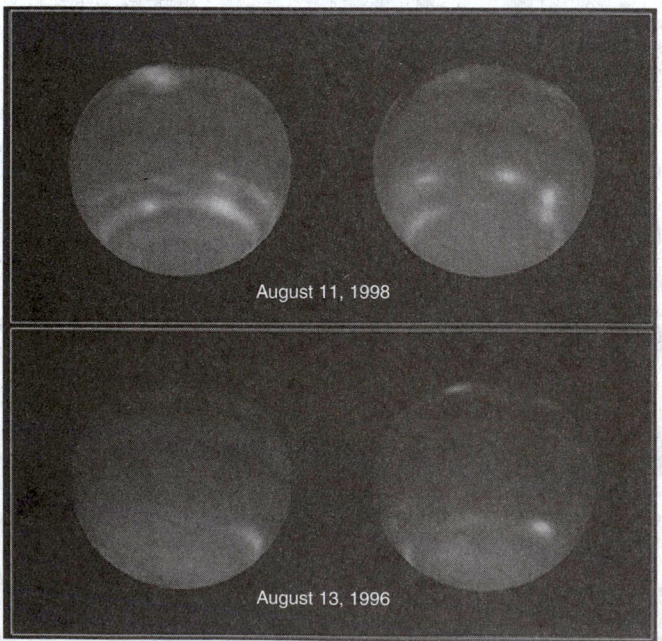

Fig. 40.   These views of Neptune, as Seen through the Hubble Space Telescope, are helping planetary scientists gain some insight into the weird and wild weather that is a hallmark of the eighth planet from the Sun. On Neptune, winds blow at 900 miles per hour and huge storms — some the size of Earth itself — come and go with regularity. How, precisely, Neptune's weather is driven is a mystery since the Sun, which drives the Earth's weather, is 900 times dimmer there than on Earth. The bottom images show Neptune's Hubble portrait August 13, 1996. The top images taken August 11, 1998 help illustrate the dynamic weather features that dominate the planet. (*Credit: L. Sromovsky (University of Wisconsin) and NASA.*)

December 1994

May 1995

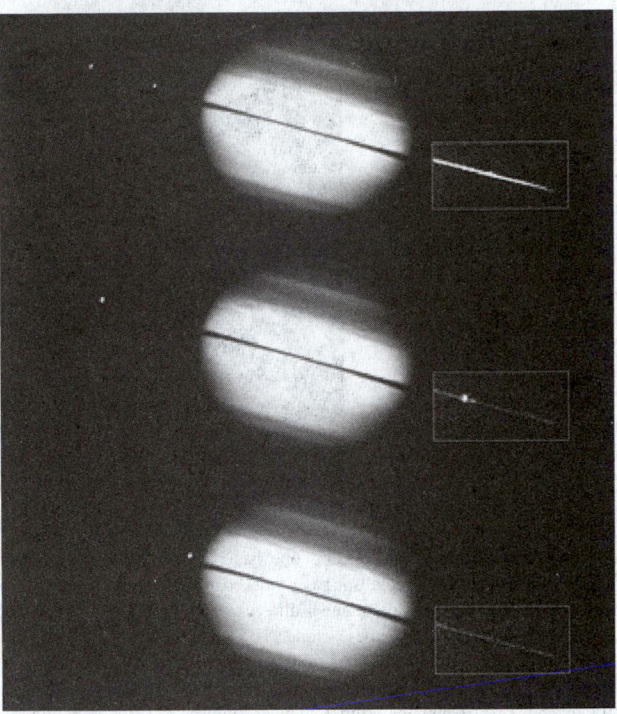

Fig. 41. In one of nature's most dramatic examples of "now-you -See-them, now-you-don't", NASA's Hubble Space Telescope captured Saturn on May 22, 1995 as the planet's magnificent ring system turned edge-on. This ring-plane crossing occurs approximately every 15 years when the Earth passes through Saturn's ring plane. For comparison, the top picture was taken by Hubble on December 1, 1994 and shows the rings in a more familiar configuration for Earth observers. The bottom picture was taken shortly before the ring plane crossing. The rings do not disappear completely because the edge of the rings reflects sunlight. The dark band across the middle of Saturn is the shadow of the rings cast on the planet (the Sun is almost 3 degrees above the ring plane). The bright stripe directly above the ring shadow is caused by sunlight reflected off the rings onto Saturn's atmosphere. Two of Saturn's icy moons are visible as tiny starlike objects in or near the ring plane. They are, from left to right, Tethys (slightly above the ring plane) and Dione. This observation will be used to determine the time of ring-plane crossing and the thickness of the main rings and to search for as yet undiscovered satellites. Knowledge of the exact time of ring-plane crossing will lead to an improved determination of the rate at which Saturn "wobbles" about its axis (polar precession). Both pictures were taken with Hubble's Wide Field Planetary Camera 2. The top image was taken in visible light. Saturn's disk appears different in the bottom image because a narrowband filter (which only lets through light that is not absorbed by methane gas in Saturn's atmosphere) was used to reduce the bright glare of the planet. Though Saturn is approximately 900 million miles away, Hubble can See details as small as 450 miles across. (Credits: Reta Beebe (New Mexico State University), D. Gilmore, L. Bergeron (STScI) and NASA.)

- Hubble followed the eruptions of volcanoes on Io the innermost of the Galilean satellites orbiting Jupiter.

RAY VILLARD, Space Telescope Science Institute, Baltimore, MD

HUBER, ROBERT (1937–). Awarded the Nobel Prize for chemistry in 1988, along with Johann Deisenhofer and Hartmut Michel, for work that revealed the three-dimensional structure of closely linked proteins that are essential to photosynthesis. Doctorate awarded in 1963 by Technical University of Munich, Germany.

HUE. The attribute of color perception that determines whether it is red, yellow, green, blue, purple, or the like. White, black, and gray are not considered hues.

HUFFMAN CODING. See **Data Compression**.

HUFFYUV. See **Data Compression**.

Fig. 42. This sequence of image from NASA's Hubble Space Telescope documents a rare astronomical alignment–Saturn's magnificent ring system turned edge-on. This occurs when the Earth passes through Saturn's ring plane, as it does approximately every 15 years. These pictures were taken with Hubble's Wide Field Planetary Camera 2 on 22 May 1995, when Saturn was at a distance of 919 million miles (1.5 billion kilometers) from Earth. At Saturn, Hubble can See details as small as 450 miles (725 km) across. In each image, the dark band across Saturn is the ring shadow cast by the Sun which is still 2.7 degrees above Saturn's ring plane. The box around the western portion of the rings (to the right of Saturn) in each image indicates the area in which the faint light from the rings has been multiplied through image processing (by a factor of 25) to make the rings more visible.

TOP—This image was taken while the earth was above the lit face of the rings. The moons Tethys and Dione are visible to the east (left) of Saturn; Janus is the bright spot near the center of the ring portion in the box, and Pandora is faintly visible just inside the left edge of this box. Saturn's atmosphere shows remarkable detail: multiple banding in both the northern and southern hemisphere, wispy structure at the north edge of the equatorial zone, and a bright area above the ring shadow that is caused by sunlight scattering off the rings onto the atmosphere. There is evidence of a faint polar haze over the North Pole of Saturn and a fainter haze over the south.

CENTER—This image was taken close to the time of ring-plane crossing. The rings are 75% fainter than in the top image, though they do not disappear completely because the vertical face of the rings still reflects sunlight when the rings are edge-on. Rhea is visible to the east of Saturn, Enceladus is the bright satellite in the rings to the west, and Janus is the fainter blip to its right. Pandora is just to the left of Enceladus, but is not visible because Enceladus is too bright. An oval-shaped atmospheric feature has just rotated into view (near the eastern limb, at the northern edge of the equatorial zone), and appears to be a local circulation pattern that is not penetrated by the bright clouds that are deflected around it.

BOTTOM—This image was taken approximately 96 minutes (one Hubble orbit) after the center image. The rings are 10% brighter than they were in that image. Rhea is visible just off the eastern limb of Saturn, and casts a shadow on the south face of Saturn. During this exposure, the Earth and Sun were on opposite sides of Saturn's ring plane (they remained in this configuration until 10 August 1995). The atmospheric circulation pattern has rotated to just past the center of the planet's disk, and is followed by more wispy structure in the bright band of clouds, reminiscent of the structure Seen during the Saturn storm observed in 1990. These images will be used to determine the time of ring-plane crossing and the thickness of the main rings and to search for as yet undiscovered satellites. Knowledge of the exact time of ring-plane crossing will lead to an improved determination of the rate at which Saturn "wobbles" about its axis (polar procession). (Credits: Amanda S. Bosh (Lowell Observatory), Andrew S. Rivkin (University of Arizona/LPL), the HST High Speed Photometer Instrument Team (R.C. Bless, PI), and NASA.)

Fig. 43. NASA's Hubble Space Telescope has provided images of Saturn in many colors, from black-and-white, to orange, to blue, green, and red. But in this picture, image processing specialists have worked to provide a crisp, extremely accurate view of Saturn, which highlights the planet's pastel colors. Bands of subtle color-yellows, browns, grays-distinguish differences in the clouds over Saturn, the second largest planet in the solar system. Saturn's high-altitude clouds are made of colorless ammonia ice. Above these clouds is a layer of haze or smog, produced when ultraviolet light from the Sun shines on methane gas. The smog contributes to the planet's subtle color variations. One of Saturn's moons, Enceladus, is Seen casting a shadow on the giant planet as it passes just above the ring system. The flattened disk swirling around Saturn is the planet's most recognizable feature, and this image displays it in sharp detail. This is the planet's ring system, consisting mostly of chunks of water ice. Although it appears as if the disk is composed of only a few rings, it actually consists of tens of thousands of thin "ringlets." This picture also shows the two classic divisions in the ring system. The narrow Encke Gap is nearest to the disk's outer edge; the Cassini division is the wide gap near the center. (*Credit: Hubble Heritage Team (AURA/STScI/NASA).*)

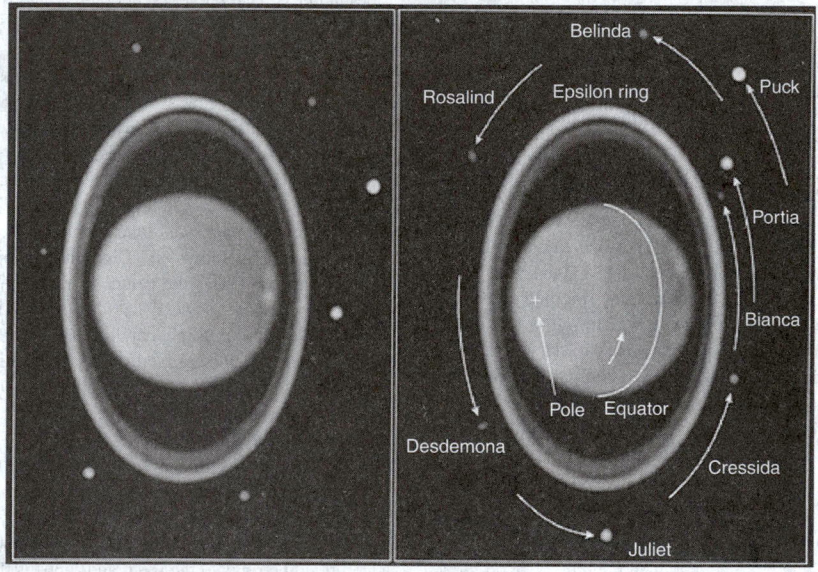

Fig. 44. The image on the right, taken 90 minutes after the left-hand image, shows the rotation of Uranus. Each image is a composite of three near-infrared images. They are called false-color images because the human eye cannot detect infrared light. Therefore, colors corresponding to visible light were assigned to the images. (The wavelengths for the "blue", "green", and "red" exposures are 1.1, 1.6, and 1.9 micrometers, respectively). At visible and near-infrared light, sunlight is reflected from hazes and clouds in the atmosphere of Uranus. However, at near-infrared light, absorption by gases in the Uranian atmosphere limits the view to different altitudes, causing intense contrasts and colors. In these images, the blue exposure probes the deepest atmospheric levels. A blue color indicates clear atmospheric conditions, prevalent at mid-latitudes near the center of the disk. The green exposure is sensitive to absorption by methane gas, indicating a clear atmosphere; but in hazy atmospheric regions, the green color is seen because sunlight is reflected back before it is absorbed. The green color around the South Pole (marked by "+") shows a strong local haze. The red exposure reveals absorption by hydrogen, the most abundant gas in the atmosphere of Uranus. Most sunlight shows patches of haze high in the atmosphere. A red color near the limb (edge) of the disk indicates the presence of a high-altitude haze. The purple color to the right of the equator also suggests haze high in the atmosphere with a clear atmosphere below. The five clouds visible near the right limb rotated counterclockwise during the time between both images. They reach high into the atmosphere, as indicated by their red color. Features of such high contrast have never been seen before on Uranus. The clouds are almost as large as continents on Earth, such as Europe. Another cloud (which barely can be seen) rotated along the path shown by the white arrow. It is located at lower altitudes, as indicated by its green color. The rings of Uranus are extremely faint in visible light but quite prominent in the near infrared. The brightest ring, the epsilon ring, has a variable width around its circumference. Its widest and thus brightest part is at the top in this image. Two fainter, inner rings are visible next to the epsilon ring. Eight of the 10 small Uranus satellites, discovered by Voyager 2, can be seen in both images. Their sizes range from about 25 miles (40 kilometers) for Bianca to 100 miles (150 kilometers) for Puck. The smallest of these satellites have not been detected since the departure of Voyager 2 from Uranus in 1986. These eight satellites revolve around Uranus in less than a day. The inner ones are faster than the outer ones. Their motion in the 90 minutes between both images is marked in the right panel. The area outside the rings was slightly enhanced in brightness to improve the visibility of these faint satellites. (*Credit: Erich Karkoschka (University of Arizona), and NASA.*)

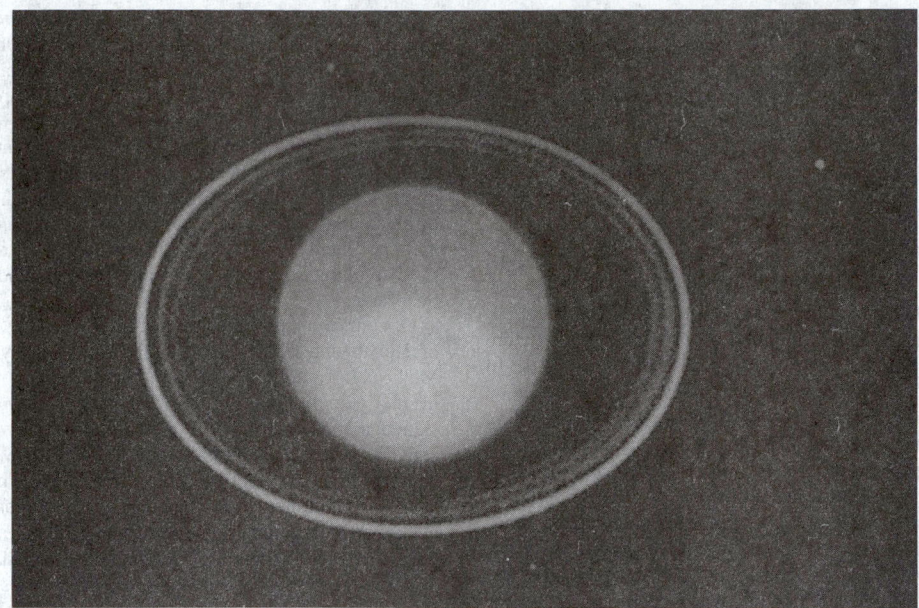

Fig. 45. Hubble Space Telescope has peered deep into Uranus' atmosphere to See clear and hazy layers created by a mixture of gases. Using infrared filters, Hubble captured detailed features of three layers of Uranus' atmosphere. Hubble's images are different from the ones taken by the Voyager 2 spacecraft, which flew by Uranus 10 years ago. Those images–not taken in infrared light–showed a greenish-blue disk with very little detail. The infrared image allows astronomers to probe the structure of Uranus' atmosphere, which consists of mostly hydrogen with traces of methane. The red around the planet's edge represents a very thin haze at a high altitude. The haze is so thin that it can only be Seen by looking at the edges of the disk, and is similar to looking at the edge of a soap bubble. The yellow near the bottom of Uranus is another hazy layer. The deepest layer, the blue near the top of Uranus, shows a clearer atmosphere. Image processing has been used to brighten the rings around Uranus so that astronomers can study their structure. In reality, the rings are as dark as black lava or charcoal. This false color picture was assembled from several exposures taken July 3, 1995 by the Wide Field Planetary Camera 2. (*Credit: Erich Karkoschka (University of Arizona Lunar & Planetary Lab) and NASA.*)

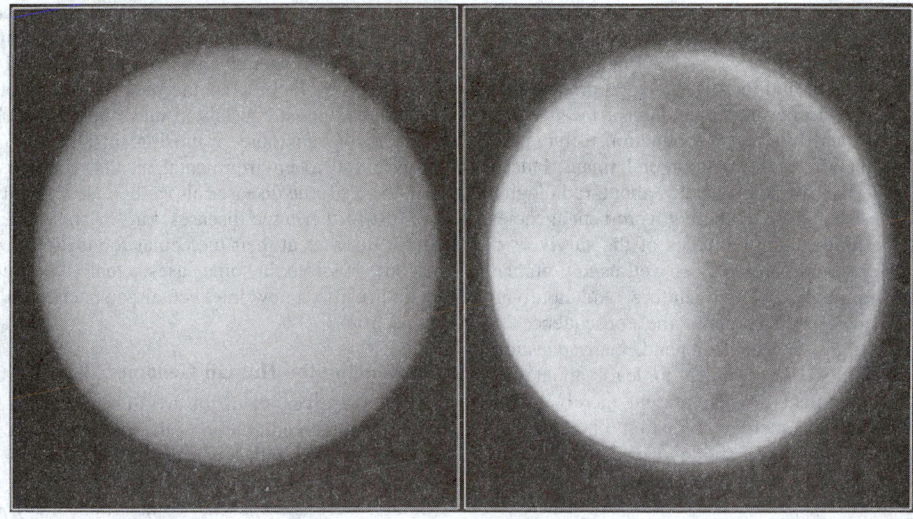

Fig. 46. Using visible light, astronomers for the first time this century have detected clouds in the Northern Hemisphere of Uranus. The newest images, taken July 31 and August 1, 1997 with NASA Hubble Space Telescope's Wide Field and Planetary Camera 2, show banded structure and multiple clouds. Using these images, Dr. Heidi Hammel (Massachusetts Institute of Technology) and colleagues Wes Lockwood (Lowell Observatory) and Kathy Rages (NASA Ames Research Center) plan to measure the wind speeds in the Northern Hemisphere for the first time. Uranus is sometimes called the "sideways" planet, because its rotation axis is tipped more than 90 degrees from the planet's orbit around the Sun. The "year" on Uranus lasts 84 Earth years, which creates extremely long seasons. Winter in the Northern Hemisphere has lasted for nearly 20 years. Uranus has also been called bland and boring, because no clouds have been detectable in ground-based images of the planet. Even to the cameras of the Voyager spacecraft in 1986, Uranus presented a nearly uniform blank disk, and discrete clouds were detectable only in the Southern Hemisphere. Voyager flew over the planet's cloud tops near the dead of northern winter (when the Northern Hemisphere was completely shrouded in darkness). Spring has finally come to the Northern Hemisphere of Uranus. The newest images, both the visible-wavelength ones described here and those taken a few days earlier with the Near Infrared and Multi-Object Spectrometer (NICMOS) by Erich Karkoschka (University of Arizona), show a planet with banded structure and detectable clouds. Two images are show here. The "aqua" image (on the left) is taken at 5,470 Angstroms, which is near the human eye's peak response to wavelength. Color has been added to the image to show what a person on a spacecraft near Uranus might See. Little structure is evident at this wavelength, though with image-processing techniques, a small cloud can be Seen near the planet's northern limb (rightmost edge). The "red" image (on the right) is taken at 6,190 Angstroms, and is sensitive to absorption by methane molecules in the planet's atmosphere. The banded structure of Uranus is evident, and the small cloud near the northern limb is now visible. Scientists are expecting that the discrete clouds and banded structure may become even more pronounced as Uranus continues in its slow pace around the sun. "Some parts of Uranus haven't Seen the Sun in decades," says Dr. Hammel, "and historical records suggest that we may See the development of more banded structure and patchy clouds as the planet's year progresses." Some scientists have speculated that the winds of Uranus are not symmetric around the planet's equator, but no clouds were visible to test those theories. The new data will provide the opportunity to measure the northern winds. Hammel and colleagues expect to have results soon. (*Credit: Heidi Hammel (Massachusetts Institute of Technology), and NASA.*)

HUMAN BIOCLIMATOLOGY. A major branch of bioclimatology that deals with effects of climate upon man. Currently, its major emphasis is on 1) the heat balance of the human body under different conditions of air temperature, humidity, and wind; 2) the effects of radiation, especially nuclear and ultraviolet, on genetics and general health; 3) the effects of atmospheric conditions and of types of changes of weather and climate on human health, vigor, and disease; and recently, 4) the effects of electrical conditions, including the atmospheric potential gradient and longwave radiations.

AMS

HUMAN GENOME PROJECT (The). The Human Genome Initiative is a worldwide research effort that has the goal of analyzing the structure of human DNA and determining the location of all human genes. In parallel with this effort, the DNA of a set of model organisms will be studied to provide the comparative information necessary for understanding the functioning of the human genome. The information generated by the human genome project is expected to be the source book for biomedical science in the 21st century. See Fig. 1. It will have a profound impact on and expedite progress in a variety of biological fields, including those such as developmental biology and neurobiology, where scientists are just beginning to understand the underlying molecular mechanisms. The analysis and interpretation of the information will occupy scientists for many years to come. Thus, the maximal benefit of the human genome project will only be achieved if it is surrounded by research efforts that are focussed on understanding and taking advantage of the human genetic information.

To Know Ourselves: The Human Genome Project 1983–1998

The biosciences research community is now embarked on a program whose boldness, even audacity, has prompted comparisons with such visionary efforts as the Apollo space program and the Manhattan project. That life scientists should conceive such an ambitious project is not remarkable; what is surprising — at least at first blush — is that the project should trace its roots to the Department of Energy.

For close to a half-century, the DOE and its governmental predecessors have been charged with pursuing a deeper understanding of the potential health risks posed by energy use and by energy-production technologies with special interest focused on the effects of radiation on humans. Indeed, it is fair to say that most of what we know today about radiological health hazards stems from studies supported by these government agencies. Among these investigations are long-standing studies of the survivors of the atomic bombings of Hiroshima and Nagasaki, as well as any number of experimental studies using animals, cells in culture, and nonliving systems. Much has been learned, especially about the consequences of exposure to high doses of radiation. On the other hand, many questions remain unanswered; in particular, we have much to learn about how low doses produce their insidious effects. When present merely in low but significant amounts, toxic agents such as radiation or mutagenic chemicals work their mischief in the most subtle ways, altering only slightly the genetic instructions in our cells. The consequences can be heritable mutations too slight to produce discernible effects in a generation or two but in their persistence and irreversibility, deeply troublesome nonetheless.

Until recently, science offered little hope for detecting at first hand these tiny changes to the DNA that encodes our genetic program. Needed was a tool that could detect a change in one "word" of the program, among perhaps a hundred million. Then, in 1984, at a meeting convened jointly by the DOE and the International Commission for Protection Against Environmental Mutagens and Carcinogens, the question was first seriously asked: Can and should we sequence the human genome? That is, can we develop the technology to obtain a word-by-word copy of the entire genetic script for an "average" human being, and thus to establish a benchmark for detecting the elusive mutagenic effects of radiation and cancer-causing toxins? Answering such a question was not simple. Workshops were convened in 1985 and 1986; the issue was studied by a DOE advisory group, by the Congressional Office of Technology Assessment, and by the National Academy of Sciences; and the matter was debated publicly and privately among biologists themselves. In the end, however, a consensus emerged that we should make a start. For a timeline of major events in the U.S. Human Genome Project see Table 1.

Adding impetus to the DOE's earliest interest in the human genome was the Department's stewardship of the national laboratories, with their demonstrated ability to conduct large multidisciplinary projects — just the sort of effort that would be needed to develop and implement the technological know-how needed for the Human Genome Project. Biological research programs already in place at the national labs benefited from the contributions of engineers, physicists, chemists, computer scientists, and mathematicians, working together in teams. Thus, with the infrastructure in place and with a particular interest in the ultimate results, the Department of Energy, in 1986, was the first federal agency to announce and to fund an initiative to pursue a detailed understanding of the human genome. See Table 1.

Of course, interest was not restricted to the DOE. The National Institutes of Health, the Cold Spring Harbor Laboratory and the Howard Hughes Medical Institute had also sponsored workshops. In 1988 the NIH joined in the pursuit, and in the fall of that year, the DOE and the NIH signed a memorandum of understanding that laid the foundation for a concerted interagency effort. See Table 1. The basis for this community-wide excitement is not hard to comprehend. The first impulse behind the DOE's commitment was only one of many reasons for coveting a deeper insight into the human genetic script. Defective genes directly account for an estimated 4000 hereditary human diseases; maladies such as Huntington disease and cystic fibrosis. In some such cases, a single misplaced letter among three billion can have lethal consequences. For most of us, though, even greater interest focuses on the far more common ailments in, which altered genes influence but do not prescribe. Heart disease, many cancers, and some psychiatric disorders, for example, can emerge from complicated interplays of environmental factors and genetic misinformation.

The first steps in the Human Genome Project are to develop the needed technologies, then to "map" and "sequence" the genome. But in a sense, these well-publicized efforts aim only to provide the raw material for the next, longer strides. The ultimate goal is to exploit those resources for a truly profound molecular-level understanding of how we develop from embryo to adult, what makes us work, and what causes things to go wrong. The benefits to be reaped stretch the imagination. In the offing is a new era of molecular medicine characterized not by treating symptoms, but rather by looking to the deepest causes of disease. Rapid and more accurate diagnostic tests will make possible earlier treatment for countless maladies. Even more promising, insights into genetic susceptibilities to disease and to environmental insults, coupled with preventive therapies, will thwart some diseases altogether. New, highly targeted pharmaceuticals, not just for heritable diseases, but for communicable ailments as well, will attack diseases at their molecular foundations. And even gene therapy will become possible, in some cases actually "fixing" genetic errors. All of this in addition to a new intellectual perspective on who we are and where we came from.

Introducing the Human Genome

For all the diversity of the world's five and a half billion people, full of creativity and contradictions, the machinery of every human mind and body is built and run with fewer than 100,000 kinds of protein molecules. And for each of these proteins, we can imagine a single corresponding gene (though there is sometimes some redundancy) whose job it is to ensure an adequate and timely supply. In a material sense, then, all of the subtlety of our species, all of our art and science, is ultimately accounted for by a surprisingly small set of discrete genetic instructions. More surprising still, the differences between two unrelated individuals, between the man next door and Mozart, may reflect a mere handful of differences in their genomic recipes — perhaps one altered word in five hundred. We are far more alike than we are different. At the same time, there is room for near-infinite variety.

It is no overstatement to say that to decode our 30,000 genes in some fundamental way would be an epochal step toward unraveling the manifold mysteries of life.

The human genome is the full complement of genetic material in a human cell. (Despite five and a half billion variations on a theme, the differences from one genome to the next are minute; hence, we hear about the human genome — as if there were only one.) The genome, in turn, is distributed among 23 sets of chromosomes, which, in each of us, have been replicated and re-replicated since the fusion of sperm and egg that marked our conception. The source of our personal uniqueness, our full genome, is therefore preserved in each of our body's several trillion cells. At a more

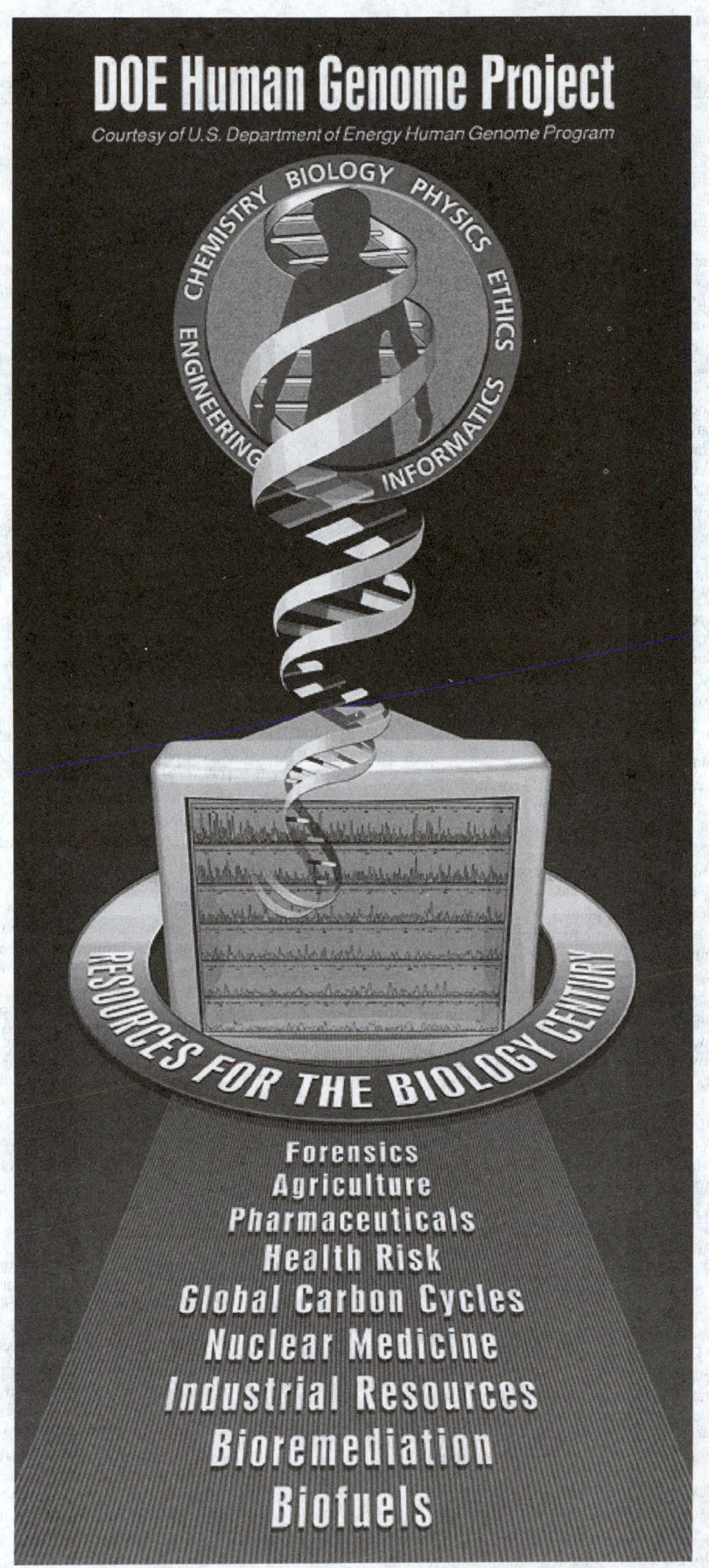

Fig. 1. Illustration of the DOE Human Genome Project. (Courtesy of U.S. Department of Energy Human Genome Project.)

TABLE 1. MAJOR EVENTS IN THE U.S. HUMAN GENOME PROJECT AND RELATED PROJECTS[a]

Date	Event
1983	• The Los Alamos National Laboratory, a Department of Energy Laboratory (LANL) and Lawrence Berkeley National Laboratory, a Department of Energy Laboratory (LLNL) begin production of DNA clone (cosmid) libraries representing single chromosomes.
1984	• DOE OHER and ICPEMC cosponsor Alta, Utah, conference highlighting the growing role of recombinant DNA technologies. See http://www.ornl.gov/hgmis/project/alta.html for more information on this conference. The Office of Technology Assessment (OTA) incorporates Alta proceedings into report acknowledging value of human genome reference sequence.
1985	• Robert Sinsheimer holds meeting on human genome sequencing at University of California, Santa Cruz.
	• At OHER Charles DeLisi and David A. Smith (Director of the DOE Human Genome Program) commission the first Santa Fe conference to assess the feasibility of a Human Genome Initiative. See http://www.ornl.gov/hgmis/publicat/hgn/v11n3/05delisi.html for a perspective by Charles DeLisi, HGP Pioneer. See http://www.ornl.gov/TechResources/Human_Genome/publicat/hgn/v7n3/02smithr.html Evolution of a Vision: Genome Project Origins, Present and Future Challenges, and Far-Reaching Benefits by David A. Smith.
1986	• Following the Santa Fe conference, DOE OHER announces Human Genome Initiative. With $5.3 million, pilot projects begin at DOE national laboratories to develop critical resources and technologies.
1987	• Congressionally chartered DOE advisory committee, HERAC, recommends a 15-year, multidisciplinary, scientific, and technological undertaking to map and sequence the human genome. DOE designates multidisciplinary human genome centers. See http://www.ornl.gov/hgmis/project/herac2.html for the Human Genome Initiative Office of Health and Environmental Research report.
	• NIH NIGMS begins funding of genome projects.
1988	• Reports by congressional OTA and NAS NRC committees recommend concerted genome research program.
	• HUGO founded by scientists to coordinate efforts internationally. See http://www.gene.ucl.ac.uk/hugo/
	• First annual Cold Spring Harbor Laboratory meeting on human genome mapping and sequencing.
	• DOE and NIH sign MOU outlining plans for cooperation on genome research. See http://www.ornl.gov/hgmis/publicat/hgn/v2n1/03memo.html for more information on this memorandum of understanding.
	• Telomere (chromosome end) sequence having implications for aging and cancer research is identified at LANL.
1989	• DNA STSs recommended to correlate diverse types of DNA clones. See http://www.ornl.gov/hgmis/publicat/hgn/v2n3/01stsnew.html Human Genome News, 2(3) (September 1990); "STS-New Strategy May Provide Common Link for Mapping."
	• DOE and NIH establish Joint ELSI Working Group. See http://www.ornl.gov/hgmis/publicat/hgn/v2n1/05elsi.html Human Genome News, 2(1) (May 1990); "NIH-DOE Joint Working Group on Ethical, Legal, and Social Issues Established."
1990	• DOE and NIH present joint 5-year U.S. HGP plan to Congress. The 15-year project formally begins. See http://www.ornl.gov/hgmis/publicat/hgn/v2n1/04five.html Human Genome News, 2(1) (May 1990); "Five-year Plan Goes to Capitol Hill." Projects begun to mark gene sites on chromosome maps as sites of mRNA expression.
	• Research and development begun for efficient production of more stable, large-insert BACs.
1991	• Human chromosome mapping data repository, GDB, established. See http://www.gdb.org/ The Genome Database. Established at Johns Hopkins University in Baltimore, Maryland, USA in 1990, the Genome Database (GDB) is the official central repository for genomic mapping data resulting from the Human Genome Initiative.
1992	• Low-resolution genetic linkage map of entire human genome published.
	• Guidelines for data release and resource sharing announced by DOE and NIH. See http://www.ornl.gov/hgmis/publicat/hgn/v4n5/04share.html Human Genome News, 4(5) (January 1993); "NIH, DOE Guidelines Encourage Sharing of Data, Resources."

Date	Event
1993	• International IMAGE Consortium established to coordinate efficient mapping and sequencing of gene-representing cDNAs. See http://www.ornl.gov/hgmis/publicat/hgn/v6n6/3image.html Human Genome News, 6(6) (Mar.–Apr. 1995); 3 "IMAGE Characterizes cDNA Clones."
	The Scientist 13(4); 17 (Feb. 15, 1999) Hot Papers In Genomics: G. Lennon, C. Auffray, M. Polymeropoulos, M.B. Soares, "The I.M.A.G.E. Consortium: An Integrated Molecular Analysis of Genomes and Their Expression," Genomics, 33, 151–152, 1996. (Cited in more than 290 papers since publication). See http://www.ornl.gov/meetings/wccs/hot1_990215.html
	• DOE-NIH ELSI Working Group's Task Force on Genetic and Insurance Information releases recommendations. See http://www.ornl.gov/hgmis/publicat/hgn/v5n2/01task.html Human Genome News, 5(2) (July 1993); "Insurance Task Force Makes Recommendations."
	• DOE and NIH revise 5-year goals [Science 262, 43–46 (Oct. 1, 1993)].
	• French Généthon provides mega-YACs to the genome community.
	• IOM releases U.S. HGP-funded report, "Assessing Genetic Risks." See http://www.ornl.gov/hgmis/publicat/hgn/v5n4/iomreprt.html Human Genome News, 5(4) (November 1993): "IOM Issues Report on Genetic Testing."
	• LBNL implements novel transposon-mediated chromosome-sequencing system.
	• GRAIL sequence-interpretation service provides Internet access at ORNL. See http://genome.ornl.gov/ Computational Biology at ORNL.
1994	• Genetic-mapping 5-year goal achieved 1 year ahead of schedule. See http://www.ornl.gov/hgmis/publicat/hgn/V6N4/MAPGOALS.HTML Human Genome News, 6(4), 1 (Nov. 1994); "Genetic Map Goal Met Ahead of Schedule."
	• Completion of second-generation DNA clone libraries representing each human chromosome by LLNL and LBNL.
	• Genetic Privacy Act, first U.S. HGP legislative product, proposed to regulate collection, analysis, storage, and use of DNA samples and genetic information obtained from them; endorsed by ELSI Working Group. See http://www.ornl.gov/hgmis/publicat/hgn/v6n6/4genetic.html Human Genome News, 6(6), 4 (Mar.–Apr. 1995); "Genetic Privacy Act Introduced."
	• DOE MGP launched; spin-off of HGP. See http://www.ornl.gov/hgmis/publicat/hgn/V6N3/OHER.HTML Human Genome News, 6(3), 7 (September 1994); "OHER Launches Microbial Genome Initiative."
	• LLNL chromosome paints commercialized. See http://www.ornl.gov/hgmis/publicat/hgn/v7n5/01collab.html Human Genome News, 7(5), 1 (January–March 1996); "Collaborations Multiply Research, Commercial Benefits."
	• SBH technologies from ANL commercialized.
	• DOE HGP Information Web site activated for public and researchers. See http://www.ornl.gov/hgmis/ Human Genome Project Information.
1995	• LANL and LLNL announce high-resolution physical maps of chromosome 16 and chromosome 19, respectively. See http://www.ornl.gov/hgmis/publicat/hgn/v6n5/2safchrm.html Human Genome News, 6(5), 2 (Jan.–Feb. 1995); "High-Resolution Physical Maps of Chromosomes 16 and 19 Completed."
	• Moderate-resolution maps of chromosomes 3, 11, 12, and 22 maps published. See http://www.ornl.gov/hgmis/publicat/hgn/v6n5/14chrom2.html Human Genome News, 6(5), 14 (Jan.–Feb. 1995); "Groups Publish Detailed Chromosome 22 Map."
	• Physical map with over 15,000 STS markers published. See http://www.ornl.gov/hgmis/publicat/hgn/v7n5/05detail.html Human Genome News, 7(5) (January–March 1996); "Detailed Human Physical Map Published by Whitehead-MIT."
	• First (nonviral) whole genome sequenced (for the bacterium Haemophilus influenzae). See http://www.ornl.gov/hgmis/publicat/hgn/v7n1/05microb.html Human Genome News, 7(1) (May–June 1995); "Two Bacterial Genomes Sequenced."

TABLE 1. (Continued)

Date	Event

- Sequence of smallest bacterium, *Mycoplasma genitalium*, completed; provides a model of the minimum number of genes needed for independent existence. See http://www.ornl.gov/hgmis/publicat/hgn/v7n1/05microb.html *Human Genome News*, 7(1) (May–June 1995).
- EEOC guidelines extend ADA employment protection to cover discrimination based on genetic information related to illness, disease, or other conditions. See http://www.ornl.gov/hgmis/publicat/hgn/v7n2/4eeocada.html *Human Genome News*, 7(2), 4 (July–August 1995); "New EEOC Guidelines Clarify Disability."

1996
- *Methanococcus jannaschii* genome sequenced; confirms existence of third major branch of life on earth. See http://www.ornl.gov/hgmis/publicat/hgn/v8n1/01archae.html *Human Genome News*, 8(1) (July–September 1996); "Third Branch of Life Confirmed."
- DOE initiates 6 pilot projects on BAC end sequencing. See http://www.ornl.gov/hgmis/publicat/hgn/v8n1/08bacend.html *Human Genome News*, 8(1) (July–September 1996); "BAC End-Sequencing Projects Initiated."
- Health Care Portability and Accountability Act prohibits use of genetic information in certain health-insurance eligibility decisions, requires DHHS to enforce health-information privacy provisions. See http://www.ornl.gov/hgmis/publicat/hgn/v8n3/01fear.html *Human Genome News*, 8(3 & 4) (January–June 1997); "Fear of Genetic Discrimination Drives Legislative Interest."
- HGP Participants Agree on Sequencing Data Release Policies Bermuda Conference. See http://www.ornl.gov/hgmis/research/bermuda.html Summary of the Report of the Second International Strategy Meeting on Human Genome Sequencing (Bermuda, 27th February–2nd March, 1997) as reported by HUGO.
- DOE and NCHGR issue guidelines on use of human subjects for large-scale sequencing projects. See http://www.ornl.gov/hgmis/publicat/hgn/v8n1/08humans.html *Human Genome News*, 8(1) (July–September 1996); DOE, NCHGR Issue Human Subject Guidelines."
- *Saccharomyces cerevisiae* (yeast) genome sequence completed by international consortium. See http://www.ornl.gov/hgmis/publicat/hgn/v8n3/08yeast.html *Human Genome News*, 8(3 & 4) (January–June 1997); "Yeast Genome Directory."
- Sequence of the human T-cell receptor region completed. See http://www.ornl.gov/hgmis/publicat/hgn/v8n1/09immune.html *Human Genome News*, 8(1) (July–September 1996); Immune System Genes Reveal Surprises."
- Wellcome Trust sponsors large-scale sequencing strategy meeting for international coordination of human genome sequencing. See http://www.ornl.gov/hgmis/publicat/hgn/v7n6/19intern.html *Human Genome News*, 7(6) (April–June 1996); International Large-scale Sequencing Meeting."

1997
- NIH NCHGR becomes National Human Genome Research Institute (NHGRI). See http://www.ornl.gov/hgmis/publicat/hgn/v8n3/08nchgr.html *Human Genome News*, 8(3 & 4) (January–June 1997); "NCHGR Becomes NIH Institute."
- *Escherichia coli* genome sequence completed. See http://www.ornl.gov/hgmis/publicat/hgn/v9n1/07ecoli.html *Human Genome News*, 9(1–2) (January 1998); "Complete *E. coli* Genome Sequence in Public Database."
- Second large-scale sequencing strategy meeting held in Bermuda. See http://www.gene.ucl.ac.uk/hugo/bermuda2.htm "Summary of the Report of the Second International Strategy Meeting on Human Genome Sequencing Bermuda, 27th February–2nd March 1997."
- High-resolution physical maps of chromosomes X and 7 completed. See http://www.ornl.gov/hgmis/publicat/hgn/v8n3/08xmap.html *Human Genome News*, 8(3 & 4) (January–June 1997); "Chromosome X Map Completed," and http://www.ornl.gov/hgmis/publicat/hgn/v9n1/05mile.html *Human Genome News*, 9(1–2) (January 1998); "Team Completes HGP Milestone: Human Chromosome 7 Map."
- DOE-NIH Task Force on Genetic Testing releases final report and recommendations. See http://www.med.jhu.edu/tfgtelsi/promoting/ "Promoting Safe and Effective Genetic Testing in the United States: Principles and Recommendations, Task Force on Genetic Testing."

- DOE forms Joint Genome Institute for implementing high-throughput activities at DOE human genome centers, initially in sequencing and functional genomics. See http://www.ornl.gov/hgmis/publicat/hgn/v8n2/01doe.html *Human Genome News*, 8(2) (October–December 1996); "DOE Merges Genome Center Sequencing Efforts."
- UNESCO See http://www.unesco.org/ adopts Universal Declaration on the Human Genome and Human Rights. See http://www.unesco.org/human_rights/hrbc.htm

1998
- Hospital for Sick Children, Toronto, Ontario, to continue GDB data collection, curation. See http://www.ornl.gov/hgmis/publicat/hgn/v10n1/21gdb.html
- *Caenorhabditis elegans* genome sequence completed. See http://www.ornl.gov/hgmis/publicat/hgn/v10n1/08celeg.html
- DOE and NIH reveal new five-year plan for HGP, predict project completion by 2003. See http://www.ornl.gov/hgmis/publicat/hgn/v10n1/03goal.html *Human Genome News*, 10(1–2) (February 1999); "Five-Year Research Goals of the U.S. Human Genome Project: October 1, 1998 to September 30, 2003."
- JGI exceeds sequencing goal, achieves 20 Mb for FY 1998. See http://www.ornl.gov/hgmis/publicat/hgn/v10n1/01jgi.html *Human Genome News*, 10(1–2) (February 1999); "DOE Joint Genome Institute Exceeds DNA Sequencing Goal."
- GeneMap'98 containing 30,000 markers released. See http://www.ornl.gov/hgmis/publicat/hgn/v10n1/07genmap.html *Human Genome News*, 10(1–2) (February 1999); "GeneMap 98 Doubles Density of 1996 Map."
- Incyte Pharmaceuticals announces plans to sequence human genome in 2 years. See http://www.ornl.gov/hgmis/publicat/hgn/v10n1/07incyte.html *Human Genome News*, 10(1–2) (February 1999); "Second Private Human Genome Sequencing Project Under Way."
- *Mycobacterium tuberculosis* bacterium sequenced. See http://www.ornl.gov/hgmis/publicat/hgn/v9n3/16tigr.html#tb *Human Genome News*, 9(3) (July 1998); "Microbial Genome News."
- Celera Genomics formed to sequence much of human genome in 3 years using HGP-generated resources. See http://www.ornl.gov/hgmis/publicat/hgn/v9n3/01venter.html *Human Genome News*, 9(3) (July 1998); "Private-Sector Sequencing Planned." In May, J. Craig Venter [The Institute for Genomic Research (TIGR)] announced plans to form a new company with Perkin-Elmer's Applied Biosystems Division (PE-ABD) to sequence a large portion of the human genome in 3 years for $300 million.
- DOE funds production BAC end sequencing projects. See http://www.ornl.gov/hgmis/publicat/hgn/v10n1/04bacend.html *Human Genome News*, 10(1–2) (February 1999); "BAC End Sequencing Speeds Large and Small Projects."
- Largest-ever ELSI meeting attended by over 800 from diverse disciplines and sponsored by DOE; Whitehead Institute; and the American Society of Law, Medicine, and Ethics. See http://www.ornl.gov/hgmis/publicat/hgn/v10n1/13white.html *Human Genome News*, 10(1–2) (February 1999); "The Human Genome Project: Science, Law, and Social Change in the 21st Century: Reports from Cambridge Symposium."
- Human Genome Project passes midpoint. See http://www.ornl.gov/hgmis/project/midpt.html

1999
- First Human Chromosome Completely Sequenced! On December 1, researchers in the Human Genome Project announced the complete sequencing of the DNA making up human chromosome 22. See http://www.ornl.gov/hgmis/project/chr22.html
- Joint Genome Institute sequencing facility opens in Walnut Creek, CA. See http://www.ornl.gov/hgmis/publicat/hgn/v10n3/03richard.html *Human Genome News*, 10(3–4) (October 1999); "Richardson Attends Facility Opening."
- Major Drug Firms Create Public SNP Consortium. See http://www.ornl.gov/hgmis/publicat/hgn/v10n3/13snp.html *Human Genome News*, 10(3–4) (October 1999); "Major Drug Firms Create Public SNP Resource

(continued)

TABLE 1. (*Continued*)

Date	Event
	• The Billion Base Pair Celebration November 23, 1999. Bruce Alberts, President, National Academy of Sciences and early planner of the Genome Project; Francis Collins, Director, NHGRI; Secretary of HHS, Donna Shalala; Secretary of DOE, Bill Richardson. See http://www.ornl.gov/hgmis/graphics/video/nhgri112399.ram (Total Running Time: 01:09:45; Bandwidth: 146 Kbps).
	• HGP advances goal for obtaining a draft sequence of the entire human genome from 2001 to 2000. See http://www.ornl.gov/hgmis/project/update.html
2000	• HGP leaders and President Clinton announce the completion of a "working draft" DNA sequence of the human genome. See http://www.ornl.gov/hgmis/project/clinton1.html White House Press Conference: The Human Genome Project, June 26, 2000 (Total Running Time: 00:41:23; Bandwidth: 33 Kbps) See http://www.ornl.gov/hgmis/graphics/video/whpc062600.ram
	• International research consortium publishes chromosome 21 genome, the smallest human chromosome and the fifth to be completed. See http://hgp.gsc.riken.go.jp/chr21/
	• DOE researchers announce completion of chromosomes 5, 16, and 19 draft sequence. See http://www.ornl.gov/hgmis/project/51619jgi.html "Researchers Decode Three Human Chromosomes: Information May Lead to Treatments for Kidney Disease, Diabetes, and Prostate and Colorectal Cancer."
	• International collaborators publish genome of fruit fly *Drosophila melanogaster*, the largest organism sequenced to date. See http://www.ornl.gov/hgmis/archive/articles/drosophila.html
	• President Clinton signs executive order prohibiting federal departments and agencies from using genetic information in hiring or promoting workers. See http://www.ornl.gov/hgmis/elsi/legislat.html#clinton
2001	• **Publication of Initial Working Draft Sequence February 12, 2001** Special issues of *Science* (Feb. 16, 2001) and *Nature* (Feb. 15, 2001) contain the working draft of the human genome sequence. Nature papers include initial analysis of the descriptions of the sequence generated by the publicly sponsored Human Genome Project, while Science publications focus on the draft sequence reported by the private company, Celera Genomics.

Links for more information are:

Nature: http://www.nature.com/genomics/

Science: http://www.sciencemag.org/content/vol291/issue5507/index.shtml

Press releases on First Analysis of Genome Sequence: http://www.ornl.gov/hgmis/resource/media.html#releases

• Pieter de Jong's team (now at the Oakland Children's Hospital, Oakland, CA) was a major provider of the BAC libraries used in the sequencing of the human and several other genomes. See http://www.ornl.gov/meetings/bacpac/

---

a Acronyms

ADA — Americans with Disabilities Act.
ANL — Argonne National Laboratory, a Department of Energy Laboratory.
BAC — bacterial artificial chromosome.
cDNA — complementary deoxyribonucleic acid.
DHHS — Department of Health and Human Services at National Institutes of Health (NIH).
DNA — deoxyribonucleic acid.
DOE — Department of Energy.
EEOC — Equal Employment Opportunity Commission.
ELSI — ethical, legal, and social issues.
FY — federal fiscal year (October 1 to September 30).
GDB — Genome Database.
GRAIL — Gene Recognition and Analysis Internet Link.
HERAC — Health and Environmental Research Advisory Committee.
HGI — Human Genome Initiative.
HGP — Human Genome Project, Human Genome Program.
HUGO — Human Genome Organisation.
ICPEMC — International Commission for Protection Against Environmental Mutagens and Carcinogens.
IMAGE — Integrated Molecular Analysis of Gene Expression.
IOM — Institute of Medicine
JGI — the Department of Energy's Joint Genome Institute in Walnut Creek, California.

The JGI houses the DOE's production sequencing facility.
LANL — Los Alamos National Laboratory, a Department of Energy Laboratory.
LBNL — Lawrence Berkeley National Laboratory, a Department of Energy Laboratory.
LLNL — Lawrence Livermore National Laboratory, a Department of Energy Laboratory.
MGP — Microbial Genome Project.
MOU — memorandum of understanding.
mRNA — messenger ribonucleic acid.
NAS — National Academy of Sciences.
NCHGR — National Center for Human Genome Research at National Institutes of Health (NIH).
NHGRI — National Human Genome Research Institute at National Institutes of Health (NIH).
NIGMS — National Institute of General Medical Sciences at National Institutes of Health (NIH).
NIH — National Institutes of Health.
NRC — National Research Council.
OBER — Office of Biological and Environmental Research, U.S. Department of Energy (formerly Office of Health and Environmental Research).
OHER — Office of Health and Environmental Research, U.S. Department of Energy (now Office of Biological and Environmental Research).
ORNL — Oak Ridge National Laboratory, a Department of Energy Laboratory.
OTA — Office of Technology Assessment.
R&D — research and development.
SBH — Sequencing by hybridization.
STS — sequence tagged site.
UNESCO — United Nations Educational, Scientific, and Cultural Organization.
YAC — yeast artificial chromosome.

---

basic level, the genome is DNA, deoxyribonucleic acid, a natural polymer built up of repeating **nucleotides**, each consisting of a simple sugar, a phosphate group, and one of four nitrogenous bases. The hierarchy of structure from chromosome to nucleotide is shown in Figure. 2. In the chromosomes, two DNA strands are twisted together into an entwined spiral — the famous double helix — held together by weak bonds between complementary bases, adenine (A) in one strand to thymine (T) in the other, and cytosine to guanine (C-G). In the language of molecular genetics, each of these linkages constitutes a **base pair**. All told, if we count only one of each pair of chromosomes, the human genome comprises about three billion base pairs.

The specificity of these base-pair linkages underlies all that is wonderful about DNA. First, replication becomes straightforward. Unzipping the double helix provides unambiguous templates for the synthesis of daughter molecules: One helix begets two with near-perfect fidelity. Second, by a similar template-based process, depicted in Figure 3, a means is also available for producing a DNA-like messenger to the cell cytoplasm. There, this **messenger RNA**, the faithful complement of a particular DNA segment, directs the synthesis of a particular protein. Many subtleties are entailed in the synthesis of proteins, but in a schematic sense, the process is elegantly simple.

Every **protein** is made up of one or more polypeptide chains, each a series of (typically) several hundred molecules known as **amino acids**, linked by so-called peptide bonds. Remarkably, only 20 different kinds of amino acids suffice as the building blocks for all human proteins. The synthesis of a protein chain, then, is simply a matter of specifying a particular sequence of amino acids. This is the role of the messenger RNA. (The same nitrogenous bases are at work in RNA as in DNA, except that uracil takes the place of the DNA base thymine.) Each linear sequence of three bases (both in RNA and in DNA) corresponds uniquely to a single amino acid. The RNA sequence AAU thus dictates that the amino acid asparagine should be added to a polypeptide chain, GCA specifies alanine — and so on. A segment of the chromosomal DNA that directs the synthesis of a single type of protein constitutes a single **gene**.

In 1990 the Department of Energy and the National Institutes of Health developed a joint research plan for their genome programs, outlining specific goals for the ensuing five years. Three years later emboldened by progress that was on track or even ahead of schedule, the two agencies put forth an updated five-year plan. Improvements in technology, together with the experience of three years, allowed an even more ambitious prospect. For more information on these plans refer to Table 1.

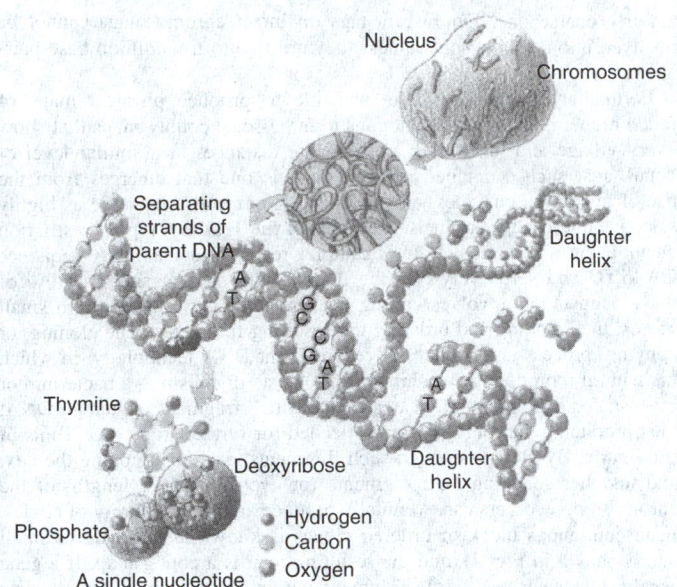

Fig. 2.   Apart from reproductive gametes, each cell of the human body contains 23 pairs of chromosomes, each a packet of compressed and entwined DNA. Every strand of the DNA is a huge natural polymer of repeating nucleotide units, each of which comprises a phosphate group, a sugar (deoxyribose), and a base (either adenine, thymine, cytosine, or guanine). Every strand thus embodies a code of four characters (A's, T's, C's, and G's), the recipe for the machinery of human life. In its normal state, DNA takes the form of a highly regular double-stranded helix, the strands of which are linked by hydrogen bonds between adenine and thymine (ADT) and between cytosine and guanine (CDG). Each such linkage is said to constitute a base pair; some three billion base pairs constitute the human genome. It is the specificity of these base-pair linkages that underlies the mechanism of DNA replication illustrated here. Each strand of the double helix serves as a template for the synthesis of a new strand, the nucleotide sequence of, which is strictly determined. Replication thus produces twin daughter helices, each an exact replica of its sole parent. (*U.S. Department of Energy.*)

In broad terms, the revised plan includes goals for genetic and physical mapping of the genome, DNA sequencing, identifying and locating genes, and pursuing further developments in technology and informatics. In addition, the plan emphasizes the continuing importance of the ethical, legal, and social implications of genome research, and it underscores the critical roles of scientific training, technology transfer, and public access to research data and materials. Most of the goals focus on the human genome, but the importance of continuing research on widely studied "model organisms" is also explicitly recognized.

Among the scientific goals of human genome research, several are especially notable, as they provide clear milestones for future progress. In reciting them, however, it is important to note an underlying assumption of adequate research support. Such support is obviously crucial if the joint plan is to succeed. Some of the central goals for 1993–1998 follow:

- Complete a genetic linkage map at a resolution of two to five centimorgans by 1995—As discussed in **Exploring the Genomic Landscape**, this goal was far surpassed by the fall of 1994.
- Complete a physical map at a resolution of 100 kilobases by 1998—This implies a genome map with 30,000 "signposts," separated by an average of 100,000 base pairs. Further, each signpost will be a sequence-tagged site, a stretch of DNA with a unique and well-defined DNA sequence. Such a map will greatly facilitate "production sequencing" of the entire genome. By the end of 1995, molecular biologists were halfway to this goal: A physical map was announced with 15,000 sequence-tagged signposts. Physical mapping is discussed in the **Exploring the Genomic Landscape**.
- By 1998 develop the capacity to sequence 50 million base pairs per year in long continuous segments—Adequate fiscal investment and continuing progress beyond 1998 should then produce a fully sequenced human genome by the year 2005 or earlier. Sequencing is discussed in the **Exploring the Genomic Landscape**.
- Develop efficient methods for identifying and locating known genes on physical maps or sequenced DNA—The goals here are less quantifiable, but the aim is central to the Human Genome Project: to home in on and ultimately to understand the most important human genes, namely, the ones responsible for serious diseases and those crucial for healthy development and normal functions.

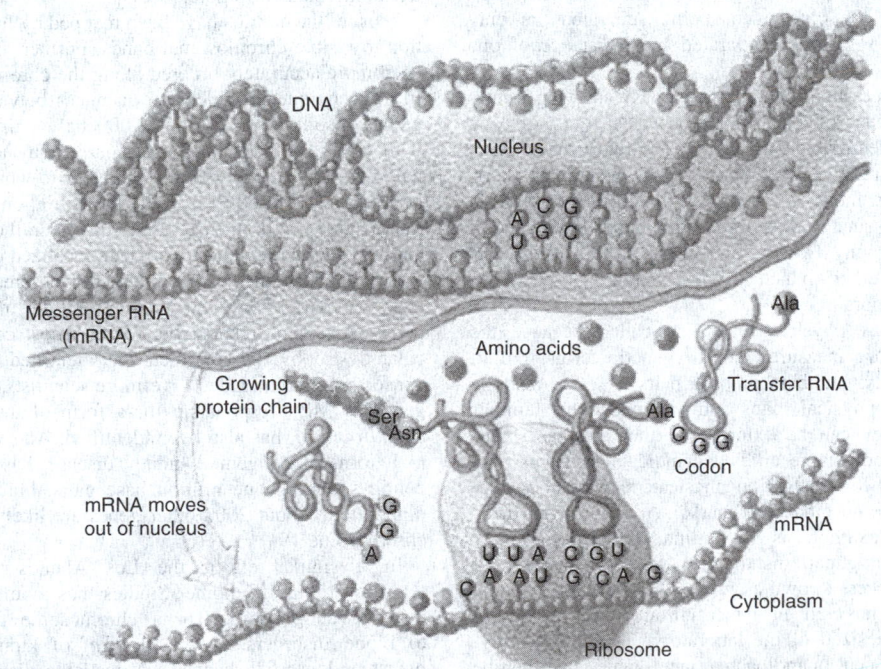

Fig. 3.   In the cell nucleus, RNA is produced by transcription, in much the same way that DNA replicates itself. RNA, however, substitutes the sugar ribose for deoxyribose and the base uracil for thymine, and is usually single-stranded. One form of RNA, messenger RNA or mRNA, conveys the DNA recipe for protein synthesis to the cell cytoplasm. There, bound temporarily to a cytoplasmic particle known as a ribosome, each three-base codon of the mRNA links to a specific form of transfer RNA (tRNA) containing the complementary three-base sequence. This tRNA, in turn, transfers a single amino acid to a growing protein chain. Each codon thus unambiguously directs the addition of one amino acid to the protein. On the other hand, the same amino acid can be added by different codons; in this illustration, the mRNA sequences GCA and GCC are both specifying the addition of the amino acid alanine (Ala). (*U.S. Department of Energy.*)

- Pursue technological developments in areas such as automation and robotics — A continuing emphasis on technological advance is critical. Innovative technologies, such as those described in **Beyond Biology**, are the necessary underpinnings of future large-scale sequencing efforts.
- Continue the development of database tools and software for managing and interpreting genome data — This is the area of informatics, discussed in **Beyond Biology**. The challenge is not so much the volume of data, but rather the need to mount a system compatible with researchers around the world, and one that will allow scientists to contribute new data and to freely interrogate the existing databases. The ultimate measure of success will be the ease with which biologists can fruitfully use the information produced by the genome project.
- Continue to explore the ethical, legal, and social implications of genome research — Much emphasis continues to be placed on issues of privacy and the fair use of genetic information. New goals focus on defining additional pertinent issues and developing policy responses to them, disseminating policy options regarding genetic testing services, fostering greater acceptance of human genetic variation, and enhancing public and professional education that is sensitive to sociocultural and psychological issues. This side of the genome project is discussed in **Ethical, Legal, and Social Implications**.

Exploring the Genomic Landscape

Mapping the Terrain. One of the central goals of the Human Genome Project is to produce a detailed "map" of the human genome. But, just as there are topographic maps and political maps and highway maps of the United States, so there are different kinds of genome maps, the variety of which is suggested in Figure 4. Genomic geography. One type, a genetic linkage map, is based on careful analyses of human inheritance patterns. It indicates for each chromosome the whereabouts of genes or other "heritable markers," with distances measured in centimorgans, a measure of recombination frequency. During the formation of sperm and egg cells, a process of genetic recombination — or "crossing over" — occurs in which pieces of genetic material are swapped between paired chromosomes. This process of chromosomal scrambling accounts for the differences invariably seen even in siblings (apart from identical twins). Logically, the closer two genes are to each other on a single chromosome, the less likely they are to get split up during genetic recombination. When they are close enough that the chances of being separated are only one in a hundred, they are said to be separated by a distance of one centimorgan.

The role of human pedigrees now becomes clear. By studying family trees and tracing the inheritance of diseases and physical traits, or even unique segments of DNA identifiable only in the laboratory, geneticists can begin to pin down the relative positions of these genetic markers. By the end of 1994, a comprehensive map was available (see Table 1) that included more than 5800 such markers, including genes implicated in cystic fibrosis, myotonic dystrophy, Huntington disease, Tay-Sachs disease, several cancers, and many other maladies. The average gap between markers was about 0.7 centimorgan.

Other maps are known as physical maps, so called because the distances between features are measured not in genetic terms, but in "real" physical units, typically, numbers of base pairs. A close analogy can thus be drawn between physical maps and the road maps familiar to us all. Indeed, the analogy can be extended further. Just as small-scale road maps may show only large cities and indicate distances only between major features, so a low-resolution physical map includes only a relative sprinkling of chromosomal landmarks. A well-known low-resolution physical map, for example, is the familiar chromosomal map, showing the distinctive staining patterns that can be seen in the light microscope. Further, by a process known as *in situ* hybridization, specific segments of DNA can be targeted in intact chromosomes by using complementary strands synthesized in the laboratory. These laboratory-made "probes" carry a fluorescent or radioactive label, which can then be detected and thus pinpointed on a specific region of the chromosome. Of particular interest are probes known as cDNA (for complementary DNA), which are synthesized by using molecules of messenger RNA as templates. These molecules of cDNA thus hybridize to "expressed" chromosomal regions — regions that directly dictate the synthesis of proteins. However, a physical map that depended only on in situ hybridization would be a fairly coarse one. Fluorescent tags on intact chromosomes cannot be resolved into separate spots unless they are two to five million base pairs apart.

Fortunately, means are also available to produce physical maps of much higher resolution — analogous to large-scale county maps that show every village and farm road, and indicate distances at a similar level of detail. Just such a detailed physical map is one that emerges from the use of restriction enzymes — DNA-cleaving enzymes that serve as highly selective microscopic scalpels. See **Tools of the Trade**. A typical restriction enzyme known as EcoRI, for example, recognizes the DNA sequence GAATTC and selectively cuts the double helix at that site. One use of these handy tools involves cutting up a selected chromosome into small pieces, then cloning and ordering the resulting fragments. The cloning, or copying, process is a product of recombinant DNA technology, in which the natural reproductive machinery of a "host" organism — a bacterium or a yeast, for example — replicates a "parasitic" fragment of human DNA, thus producing the multiple copies needed for further study. See **Tools of the Trade**. By cloning enough such fragments, each overlapping the next and together spanning long segments (or even the entire length) of the chromosome, workers can eventually produce an ordered library of clones. Each contiguous block of ordered clones is known as a **contig** (a small one is shown in Fig. 4), and the resulting map is a contig map. If a gene can be localized to a single fragment within a contig map, its physical location is thereby accurately pinned down. Further, these conveniently sized clones become resources for further studies by researchers around the world — as well as the natural starting points for systematic sequencing efforts.

Two giant steps: Chromosomes 16 and 19. One of the signal achievements of the DOE genome effort so far is the successful physical mapping of chromosomes 16 and 19. The high-resolution chromosome 19 map, constructed at the Lawrence Livermore National Laboratory, is based on restriction fragments cloned in cosmids, synthetic cloning "vectors" modeled after bacteria-infecting viruses known as *bacteriophages*. Like a phage, a cosmid hijacks the cellular machinery of a bacterium to mass-produce its own genetic material, together with any "foreign" human DNA that has been smuggled into it. The foundation of the chromosome 19 map is a large set of cosmid contigs that were assembled by automated analysis of overlapping but unordered restriction fragments. These contigs span an estimated 54 million base pairs, more than 95 percent of the chromosome, excluding the centromere.

Most of the contigs have been mapped by fluorescence in situ hybridization to visible chromosomal bands. Further, more than 200 cosmids have been more accurately ordered along the chromosome by a high-resolution FISH technique in which the distances between cosmids are determined with a resolution of about 50,000 base pairs. This ordered FISH map, with cosmid reference points separated by an average of 230,000 base pairs, provides the essential framework to which other cosmid contigs can be anchored. Moreover, the EcoRI restriction sites have been mapped on more than 45 million base pairs of the overall cosmid map. Over 450 genes and genetic markers have also been localized on this map, of which nearly 300 have been incorporated into the ordered map. Figure 5 an emerging gene map shows the locations of the mapped genes. Among these genes is the one responsible for the most common form of adult muscular dystrophy (DM), which was identified in 1992 by an international consortium that included Livermore scientists. A second important disease gene (COMP), responsible for a form of dwarfism known as *pseudoachondroplasia*, has also been identified. And yet another gene, one linked to a form of congenital kidney disease, has been localized to a single contig spanning one million base pairs, but has not yet been precisely pinpointed. About 2000 other genes are likely to be found eventually on chromosome 19.

In a similar effort, the Los Alamos National Laboratory Center for Human Genome Studies has completed a highly integrated map of chromosome 16, a chromosome that contains genes linked to blood disorders, a second form of kidney disease, leukemia, and breast and prostate cancers. A readable display of this integrated map covers a sheet of paper more than 15 feet long; a portion of it, much reduced and showing only some of its central features. See http://www.ornl.gov/hgmis/publicat/tko/04d_img.html "Mapping chromosome 16." The framework for the Los Alamos effort is yet another kind of map, a "cytogenetic breakpoint map" based on 78 lines of cultured cells, each a hybrid that contains mouse chromosomes and a fragment

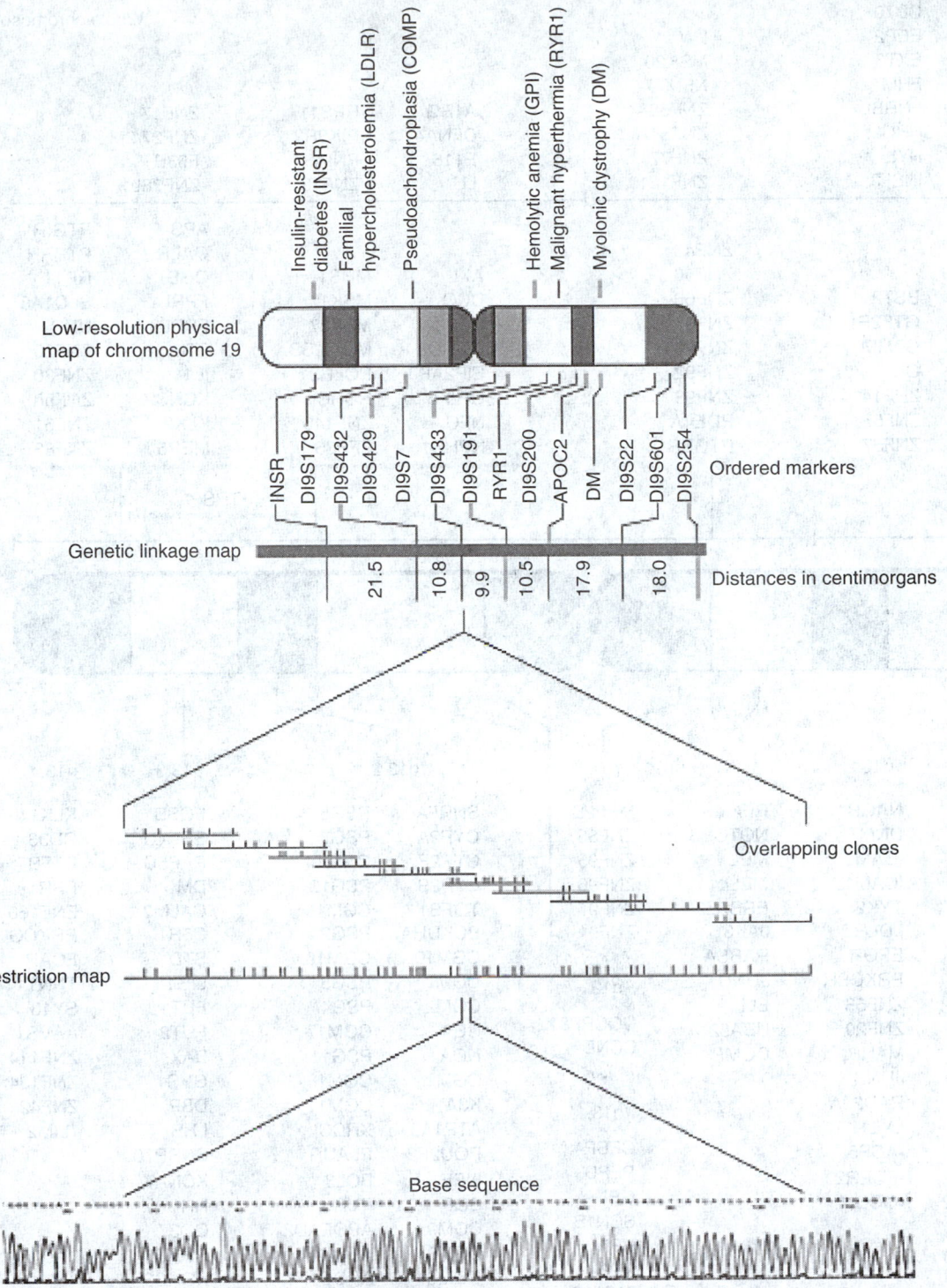

Fig. 4. Genomic geography. The human genome can be mapped in a number of ways. The familiar and reproducible banding pattern of the chromosomes constitutes one kind of physical map, and in many cases, the positions of genes or other heritable markers have been localized to one band or another. More useful are genetic linkage maps, on which the relative positions of markers have been established by studying how frequently the markers are separated during a natural process of chromosomal shuffling called genetic recombination. The cryptically coded ordered markers near the top of this figure are physically mapped to specific regions of chromosome 19; some of them also constitute a low-resolution genetic linkage map. (Hundreds of genes and other markers have been mapped on chromosome 19; only a few are indicated here. See Fig. 5 an emerging gene map for a display of mapped genes.) A higher-resolution physical map might describe, as shown here, the cutting sites (the short vertical lines) for certain DNA-cleaving enzymes. The overlapping fragments that allow such a map to be constructed are then the resources for obtaining the ultimate physical map, the base-pair sequence for the human genome. At the bottom of this figure is an example of output from an automatic sequencing machine. (*U.S. Department of Energy.*)

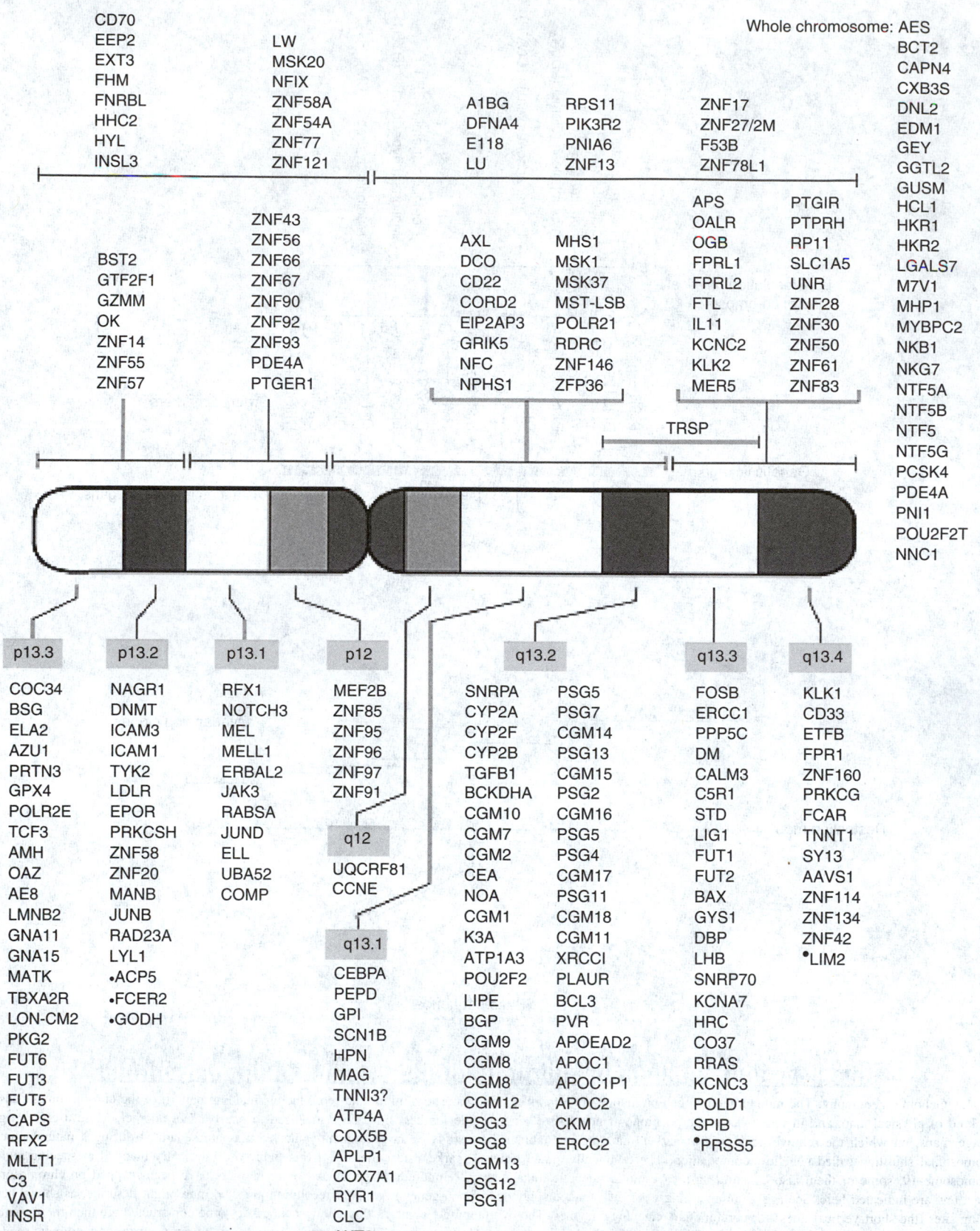

Fig. 5. An emerging gene map. More than 250 genes have already been mapped to chromosome 19. Those listed on the lower half of this illustration have been assigned to specific cosmids and (except for those marked with asterisks) have been ordered on the Livermore physical map. Their positions are therefore known with far greater accuracy than shown here. The genes listed above the chromosome have been mapped to larger regions of the chromosome — or merely localized to chromosome 19 generally — and have not yet been assigned to cosmids in the Livermore database. The text mentions several of the most important genes mapped so far. Others include INSR, which codes for an insulin receptor and is involved in adult-onset diabetes; LDLR, a gene for a low-density lipoprotein receptor involved in hypercholesterolemia; and ERCC2, a DNA repair gene implicated in one form of xeroderma pigmentosum. (*U.S. Department of Energy.*)

of human chromosome 16. Natural breakpoints in chromosome 16 are thus identified, leading to a breakpoint map that divides the chromosome into segments whose lengths average 1.1 million base pairs. Anchored to this framework are a low-resolution contig map based on YAC clones and a high-resolution contig map based largely on cosmids (for more on YACs, yeast artificial chromosomes. See **Tools of the Trade**. The low-resolution map, comprising 700 YACs from a library constructed by the Centre d'Etude du Polymorphisme Humain (CEPH), provides practically complete coverage of the chromosome, except the highly repetitive DNA in the centromere region. The high-resolution map comprises some 4000 cosmid clones, assembled into about 500 contigs covering 60 percent of the chromosome. In addition, it includes 250 smaller YAC clones that have been merged with the cosmid contig map. The cosmid contig map is an especially important step forward, since it is a "sequence-ready" map. It is based on bacterial clones that are ideal substrates for DNA sequencing, and further, these clones have been restriction mapped to allow identification of a minimum set of overlapping clones for a large-scale sequencing effort.

The high- and low-resolution maps have been tied together by sequence-tagged sites (STSs), short but unique stretches of DNA sequence. They have also been integrated into the breakpoint map, and with genetic maps developed at the Adelaide Children's Hospital and by CEPH. The integrated map also includes a transcription map of 1000 sequenced **exons** (expressed fragments of genes) and more than 600 other markers developed at other laboratories around the world.

Sequencing the Genome. Ultimately, though, these physical maps and the clones they point to are mere stepping stones to the most visible goal of the genome project, the string of three billion characters—A's, T's, C's, and G's—representing the sequence of base pairs that defines our species. Included, of course, would be the sequence for every gene, as well as the sequences for stretches of DNA whose functions we don't yet know (but which may be involved in such little-understood processes as orchestrating gene expression in different parts of our bodies, at different times of our lives). Should anyone undertake to print it all out, the result would fill several hundred volumes the size of a big-city phone book.

Only the barest start has been made in taking this dramatic step in the Human Genome Project. Several hundred million base pairs have been sequenced and archived in databases, but the great majority of these are from short "sequence tags" on cloned fragments. Only about 30 million base pairs of human DNA (roughly one percent of the total) have been sequenced in longer stretches, the longest being about 685,000 base pairs long. Even more daunting is the realization that we will eventually need to sequence many parts of the genome many times, thus to reveal differences that indicate various forms of the same gene.

Hence, as with so many human enterprises, the challenge of sequencing the genome is largely one of doing the job cheaper and faster. At the beginning of the project, the cost of sequencing a single base pair was between $2 and $10, and one researcher could produce between 20,000 and 50,000 base pairs of continuous, accurate sequence in a year. Sequencing the genome by the year 2005 would therefore likely cost $10–20 billion and require a dedicated cadre of at least 5000 workers. Clearly, a major effort in technology development was called for—an effort that would drive the cost well below $1 per base pair and that would allow automation of the sequencing process. From the beginning, therefore, the DOE has emphasized programs to pave the way for expeditious and economical sequencing efforts—programs to develop new technologies, including new cloning vectors, and to establish suitable resources for sequencing, including clone libraries and libraries of expressed sequences.

Efforts to develop new cloning vectors have been especially productive. YACs remain a classic tool for cloning large fragments of human DNA, but they are not perfect. Some regions of the genome, for example, resist cloning in YACs, and others are prone to rearrangement. New vectors such as bacterial artificial chromosomes (BACs), P1 phages, and P1-derived artificial cloning systems (PACs) have thus been devised to address these problems. These new approaches are critical for ensuring that the entire genome can be faithfully represented in clone libraries, without the danger of deletions, rearrangements, or spurious insertions.

Marked progress is also evident in the development of sequencing technologies, though all of those in widespread current use are still based on methods developed in 1977 by Allan Maxam and Walter Gilbert and by Frederick Sanger and his coworkers. See **Tools of the Trade**. Both of these methods rely on gel-based electrophoresis systems to separate DNA fragments, and recent advances in commercial systems include increasing the number of gel lanes, decreasing run times, and enhancing the accuracy of base identification. As a result of such improvements, a standard sequencing machine can now turn out raw, unverified sequences of 50,000 to 75,000 bases per day.

Equally important to the sequencing goals of the genome project is a rational system for organizing and distributing the material to be sequenced. The DOE's commitment to such resources dates back to 1984, when it organized the National Laboratory Gene Library Project. Based on cell- and chromosome-sorting technologies developed at Livermore and Los Alamos, libraries of clones were established for each of the human chromosomes, and the individual clones are widely available for mapping and for isolating genes. These clones were invaluable in such notable "gene hunts" as the successful searches for the cystic fibrosis and Huntington disease genes. More recently, as more efficient vectors have become available, complete human DNA libraries have been established using BACs, PACs, and YACs.

Another critical resource is being assembled in an effort known as I.M.A.G.E. (Integrated Molecular Analysis of Genomes and their Expression), cofounded by the Livermore Human Genome Center. The aim is a master set of mapped and sequenced human cDNA, representing the expressed parts of the human genome. By early 1996, I.M.A.G.E. had distributed over 250,000 partial and complete cDNA clones, most of them with one or both ends sequenced to provide unique identifiers. These identifiers, expressed sequence tags (ESTs), are usually 300–500 base pairs each. Twenty-five hundred genes have also been newly mapped as part of this coordinated effort.

Shotguns and Transposons. Such advances as these, in both technology development and the assembly of resource libraries, have brought much nearer the day when "production sequencing" can begin. A great deal of variety remains, however, in the approaches available to sequencing the human genome, and it is not yet clear which will prove the most efficient and most cost-effective way to read long stretches of DNA over the next decade. One of the available choices, for example, is between "shotgun" and "directed" strategies. Another is the degree of redundancy—that is, how many times must a given strand be sequenced to ensure acceptable confidence in the result?

Shotgun sequencing derives its name from the randomly generated DNA fragments that are the objects of scrutiny. Many copies of a single large clone are broken into pieces of perhaps 1500 base pairs, either by restriction enzymes or by physical shearing. Each fragment is then separately cloned, and a convenient portion of it sequenced. A computational assembly process then compares the terminal sequences of the many fragments and, by finding overlaps that indicate neighboring fragments, constructs an ordered library for the parent clone. The members of this ordered library can then be sequenced from end to end to yield a complete sequence for the parent. The statistics involved in taking this approach require that many copies of the original clone be randomly fragmented, if no gaps are to be tolerated in the final sequence. A benefit is that the final sequence is highly reliable; the main disadvantage is that the same sequence must be done many times (in the many overlapping fragments). Nevertheless, shotgun sequencing has been the primary means for generating most of the genomic sequence data in public DNA databases. This includes the longest contiguous fragment of sequenced human DNA, from the human T-cell receptor beta region, of about 685,000 base pairs—a product of DOE-supported work at the University of Washington.

The shotgun strategy is also being used at the Genome Therapeutics Corporation and The Institute for Genomic Research (TIGR), as part of the DOE-supported Microbial Genome Initiative. Genome Therapeutics has sequenced 1.8 million base pairs of *Methanobacterium thermoautotrophicum*, a bacterium important in energy production and bioremediation, and TIGR has successfully sequenced the complete genomes of three free-living bacteria, *Haemophilus influenzae* (1,830,137 base pairs; an effort supported mostly by private funds), *Mycoplasma genitalium* (580,070 base pairs), and *Methanococcus jannaschii* (1,739,933 base pairs).

The alternative to shotgun sequencing is a directed approach, in which one seeks to sequence the target clone from end to end with a minimum of duplication. The essence of this approach is embodied in a technique known as primer walking. Starting at one end of a single large fragment, one replicates a stretch of DNA—say, 400 base pairs long—that can be sequenced in one run. With the sequence for this first segment in hand, the next stretch of DNA, just overlapping the first, is then tackled in the same way. In principle, one can thus "walk" the entire length of the original clone. Unfortunately, this conceptually simple approach has been historically beset with disadvantages, mainly the expense and inconvenience of custom-synthesizing a primer as the necessary starting point for each sequencing step. The widely automated Sanger sequencing method involves a DNA replication step that must be "primed" by a DNA fragment that is complementary to 15 to 20 base pairs of the strand to be sequenced. See **Tools of the Trade**. Until recently, making these primers was an expensive and time-consuming business, but recent innovations have made primer walking, and similar directed strategies, more and more economically feasible.

One way to deal with the primer bottleneck, for example, is to use sets of very short fragments to prime the next sequencing step. As an illustration, the four nucleotides (A, T, C, and G) can be ordered in more than 68 billion ways to create an 18-base primer, an imposing set of possibilities. But it is eminently practical to create a library of the 4096 possible 6-base primers. Three of these "6-mers" can be matched to the end of the fragment to be sequenced, thus serving as an 18-base primer. This modular primer technology, developed at the Brookhaven National Laboratory, is currently being applied to *Borrelia burgdorferi*, the organism that causes Lyme disease; a 34,000-base-pair fragment has already been sequenced.

Another directed approach uses a naturally occurring genetic element called a transposon, which insinuates itself more or less randomly in longer DNA strands. This predilection for random insertion and the fact that the transposon's DNA sequence is well known are the keys to the sequencing strategy depicted schematically in Figure 6 (Taking a directed approach.) The largest clones are broken into smaller subclones (each of about 3000 base pairs), which then become the targets of the transposons. Multiple copies of each subclone are exposed to the transposons, and reaction conditions are controlled to yield, on average, a single insertion in each 3000-base-pair strand. The individual strands are then analyzed to yield, for each, the approximate position of the inserted transposon. By mapping these positions, a "minimum tiling path" can be determined for each subclone—that is, a set of strands can be identified whose transposon insertions are roughly 300 base pairs apart. In this set of strands, the region around each transposon is then sequenced, using the inserted transposons as starting points. The known transposon sequence allows a single primer to be used for sequencing the full set of overlapping regions.

At the Lawrence Berkeley National Laboratory, this technique has been used to sequence over 1.5 million base pairs of DNA on human chromosomes 5 and 20, as well as over three million base pairs from the fruit fly *Drosophila melanogaster*. On chromosome 5, interest focuses on a region of three million base pairs that is rich in growth factor and receptor genes; whereas, on chromosome 20, Berkeley researchers are interested in a region of about two million base pairs that is implicated in 15 to 20 percent of all primary breast carcinomas.

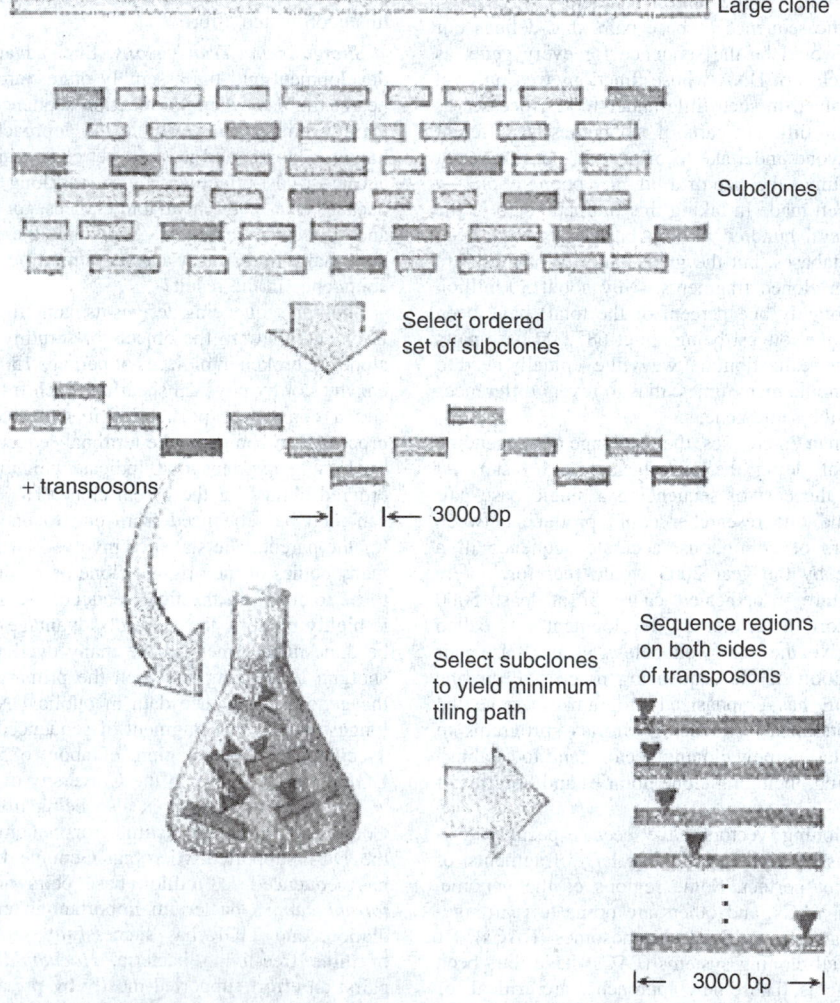

Fig. 6. *Taking a directed approach.* One directed sequencing strategy exploits a naturally occurring genetic element known as a transposon. The starting point is an ordered set of subclones, each about 3000 base pairs long, derived from a much larger clone (say, a YAC). For each subclone, a preparation is then made in which transposons insert themselves randomly into the subclone—on average, one transposon in each 3000-base-pair strand. The positions of the transposons are mapped, and a set of strands is selected such that the insertion points are about 300 base pairs apart. Sequencing then proceeds in both directions from the transposon insertion points, using the known transposon sequence as a primer. The full set of overlapping regions yields the sequence for the entire subclone, and the sequences of the full set of subclones yield the sequence for the larger original clone. (*U.S. Department of Energy.*)

Researchers supported by the DOE at the University of Utah are also pursuing the use of directed sequencing. In addition, they have developed a methodology for "multiplex" DNA sequencing, which offers a way of increasing throughput with either shotgun or directed approaches. By attaching a unique identifying sequence to each sequencing sample in a mixture of, say, 50 such samples, the entire mixture can be analyzed in a single electrophoresis lane. The 50 samples can be resolved sequentially by probing, first, for bands containing the first identifier, then for bands containing the second, and so forth. In a similar way, multiplexing can also be used for mapping. The Utah group is now able to map almost 5000 transposons in a single experiment, and they are using multiplexing in concert with a directed sequencing strategy to sequence the 1.8 million base pairs of the thermophilic microbe *Pyrococcus furiosus* and two important regions of human chromosome 17.

The completed physical maps of chromosomes 16 and 19, with their extensive coverage in many different kinds of cloning vectors, are especially ripe for large-scale sequencing. Los Alamos scientists have therefore begun sequencing chromosome 16, focusing special effort on locating the estimated 3000 expressed genes on that chromosome and using those sites as starting points for directed genomic sequencing. A region of 60,000 base pairs has already been sequenced around the adult polycystic kidney gene, and good starts have been made in mapping other genes. Interestingly, even random sequencing has led to the identification of gene DNA in over 15 percent of the samples, confirming the apparent high density of genes on this chromosome. Between chromosome 16 and the short arm of chromosome 5, another Los Alamos target, the genome center there has produced almost two million base pairs of human DNA sequence.

A parallel effort is under way at Livermore on chromosome 19 and other targeted genomic regions. Using a shotgun approach, researchers there have completed over 1.3 million bases of genomic sequence. Initially, they are attacking two major regions of chromosome 19: one of about two million base pairs, containing several genes involved in DNA repair and replication, and another of approximately one million base pairs, containing a kidney disease gene. The Livermore scientists are making use of the I.M.A.G.E. cDNA resource to sequence the cDNA from these regions, along with the associated segments of the genome. In addition, Livermore scientists have targeted DNA repair gene regions throughout the genome and, in many cases, have done comparative sequencing of these genes in other species, especially the mouse. Such comparative sequencing has identified conserved sequence elements that might act as regulatory regions for these genes and has also assisted in the identification of gene function. See **Mouse Genetics**.

How Good is Good Enough? The goal of most sequencing to date has been to guarantee an error rate below 1 in 10,000, sometimes even 1 in 100,000. However, the difference between one human being and another is more like one base pair in five hundred, so most researchers now agree that one error in a thousand is a more reasonable standard. To assure a higher level of confidence, and perhaps to uncover important individual differences, the most biologically or medically important regions would still be sequenced more exhaustively, but using this lowered standard would greatly reduce the cost of acquiring sequence data for the bulk of human DNA.

With this philosophy in mind, Los Alamos scientists have begun a project to determine the cost and throughput of a low-redundancy sequencing strategy known as sample sequencing (SASE, or "sassy"). Clones are selected from the high-resolution Los Alamos cosmid map, then physically broken into 3000-base-pair subclones—much as in other sequencing approaches. In contrast to, say, shotgun sequencing, though, only a small random set of the subclones is then selected for sequencing. Sequence fragments already known—end sequences, sequence-tagged sites, and so forth—are used as the starting points. The result is sequence coverage for about 70 percent of the original cosmid clone, enough to allow identification of genes and ESTs, thus pinpointing the most critical targets for later, more thorough sequencing efforts. Further, the SASE-derived sequences provide enough information for researchers elsewhere to pursue just such comprehensive efforts, using whole genomic DNA. In addition, the cost of SASE sequencing is only one-tenth the cost of obtaining a complete sequence, and a genomic region can be "sampled" ten times as fast.

As the first major target of SASE analysis, Los Alamos scientists chose a cosmid contig of four million base pairs at the end (the telomere) of the short arm of chromosome 16. By early 1996, over 1.4 million base pairs had been sequenced, and a gene, EST, or suspected coding region had been located on every cosmid sampled.

In addition, Los Alamos is building on the SASE effort by using SASE sequence data as the basis for an efficient primer walking strategy for detailed genomic sequencing. The first application of this strategy, to a telomeric region on the long arm of chromosome 7, proved to be as efficient as typical shotgun sequencing, but it required only two- to threefold redundancy to produce a complete sequence, in contrast to the seven- to tenfold redundancy required in shotgun approaches. The resulting 230,000-base-pair sequence is the second-longest stretch of contiguous human DNA sequence ever produced.

In a sense, though, even a complete genome sequence—the ultimate physical map—is only a start in understanding the human genome. The deepest mystery is how the potential of 100,000 genes is regulated and controlled, how blood cells and brain cells are able to perform their very different functions with the same genetic program, and how these and countless other cell types arise in the first place from an single undifferentiated egg cell. A first step toward solving these subtle mysteries, though, is a more complete physical picture of the master molecules that lie at the heart of it all.

Tools of the Trade

Over the next decade, as molecular biologists tackle the task of sequencing the human genome on a massive scale, any number of innovations can be expected in mapping and sequencing technologies. But several of the central tools of molecular genetics are likely to stay with us, much improved perhaps, but not fundamentally different. One such tool is the class of DNA-cutting proteins known as **restriction enzymes**. These enzymes, the first of which were discovered in the late 1960s, cleave double-stranded DNA molecules at specific recognition sites, usually four or six nucleotides long. For example, a restriction enzyme called *Eco* RI recognizes the single-strand sequence GAATTC and invariably cuts the double helix as shown in Figure 7.

When digested with a particular restriction enzyme, then, identical segments of human DNA yield identical sets of restriction fragments. On the other hand, DNA from the same genomic region of two different people, with their subtly different genomic sequences, can yield dissimilar sets of fragments, which then produce different patterns when sorted according to size.

This leads directly to discussion of a second essential tool of modern molecular genetics, gel electrophoresis, for it is by electrophoresis that DNA fragments of different sizes are most often separated. In classical gel electrophoresis, electrically charged macromolecules are caused to migrate through a polymeric gel under the influence of an imposed static electric field. In time the molecules sort themselves by size, since the smaller ones move more rapidly through the gel than do larger ones. In 1984 a further advance was made with the invention of pulsed-field gel electrophoresis, in which the strength and direction of the applied field is varied rapidly, thus allowing DNA strands of more than 50,000 base pairs to be separated.

A third necessary tool is some means of DNA "amplification." The classic example is the cloning vector, which may be circular DNA molecules derived from bacteria or from bacteriophages (viruslike parasites of bacteria), or artificial chromosomes constructed from yeast or bacterial genomic DNA. The characteristic all these vectors share is that fragments of "foreign" DNA can be inserted into them, whereby the inserted DNA is replicated along with the rest of the vector as the host reproduces itself. A yeast artificial chromosome, or YAC, for instance, is constructed by assembling the essential functional parts of a natural yeast chromosome—DNA sequences that initiate replication, sequences that mark the ends of the chromosomes, and sequences required for chromosome separation during cell division—then splicing in a fragment of human DNA. This engineered chromosome is then reinserted into a yeast cell, which reproduces the YAC during cell division, as if it were part of the yeast's normal complement of chromosomes. The result is a colony of yeast cells, each containing a copy, or clone, of the same fragment of human DNA. One of the important achievements of the Human Genome Project has been to establish several libraries of such cloned fragments, using several different vectors (bacterial artificial chromosomes, P1 phages, and P1-derived cloning systems), that cover the entire human genome.

Another way of amplifying DNA is the polymerase chain reaction, or PCR. This enzymatic replication technique requires that initiators, or PCR

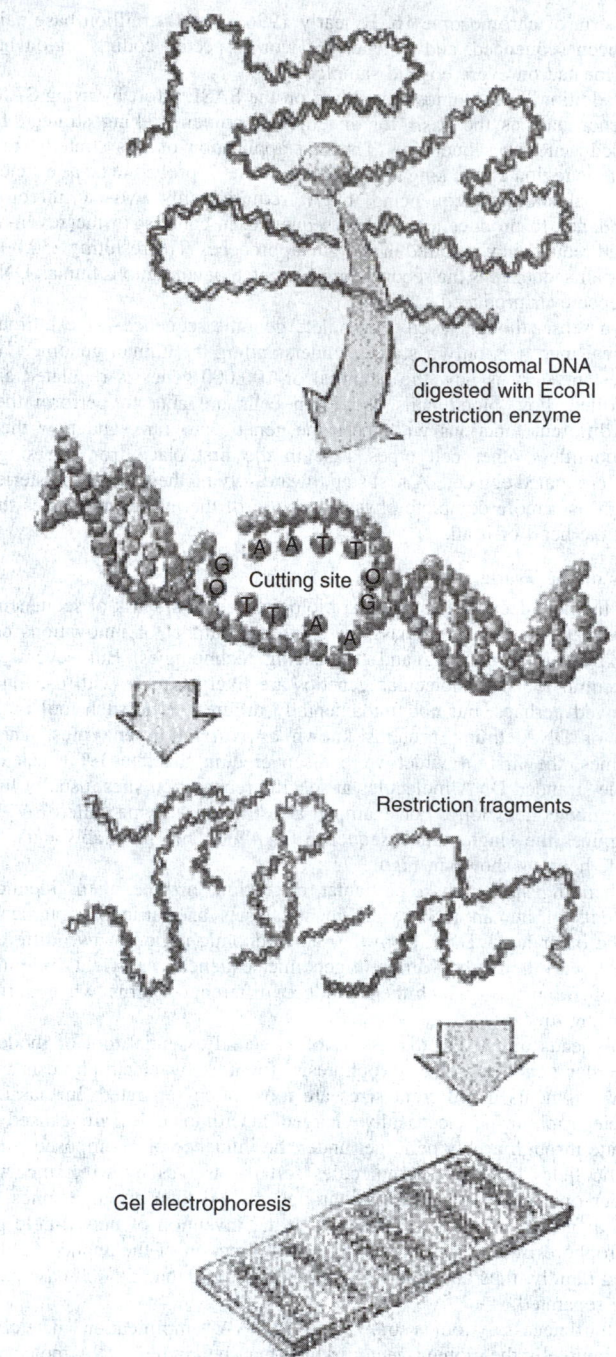

Chromosomal DNA
digested with EcoRI
restriction enzyme

Cutting site

Restriction fragments

Gel electrophoresis

Fig. 7.   Digesting DNA. Isolated from various bacteria, restriction enzymes serve as microscopic scalpels that cut DNA molecules at specific sites. The enzyme EcoRI, for example, cuts double-stranded DNA only where it finds the sequence GAATTC. The resulting fragments can then be separated by gel electrophoresis. The electrophoresis pattern itself can be of interest, since variations in the pattern from a given chromosomal region can sometimes be associated with variations in genetic traits, including susceptibilities to certain diseases. Knowledge of the cutting sites also yields a kind of physical map known as a restriction map. (*U.S. Department of Energy.*)

primers, be attached as short complementary strands at the ends of the separated DNA fragments to be replicated. An enzyme then completes the synthesis of the complementary strands, thus doubling the amount of DNA originally present. Again and again, the strands can be separated and the polymerase reaction repeated—so effectively, in fact, that DNA can be amplified by 100,000-fold in less than three hours. As with cloning vectors, the result is a large collection of copies of the original DNA fragment.

When a clone library can be ordered—that is, when the relative positions on the human chromosomes can be established for all the fragments—one then has the perfect resource for achieving the project's central goal, sequencing the human genome. How the sequencing is actually done can be illustrated by the most popular method in current use, the Sanger procedure, which is depicted schematically in Figure 8. The first step is to prime each identical DNA strand in a preparation of cloned fragments. The preparation is then divided into four portions, each of which contains a different reaction-terminating nucleotide, together with the usual reagents for replication. In one batch, the replication reaction always produces complementary strands that end with A; in another, with G; and so on. Gel electrophoresis is used to sift the resulting products according to size, allowing one to infer the exact nucleotide sequence for the original DNA strand.

Mouse Genetics

The human genome is not so very different from that of chimpanzees or mice, and it even shares many common elements with the genome of the lowly fruit fly. Obviously, the differences are critical, but so are the similarities. In particular, genetic experiments on other organisms can illuminate much that we could not otherwise learn about homologous human genes—that is, genes that are basically the same in the two species.

In some cases, the connection between a newly identified human gene and a known health disorder can be quickly established. More often, however, clear links between cloned genes and human hereditary diseases or disease susceptibilities are extremely elusive. Diseases that are modified by other genetic predispositions, for example, or by environment, diet, and lifestyle can be exceedingly difficult to trace in human families. The same holds for very rare diseases and for genetic factors contributing to birth defects and other developmental disorders. By contrast, disorders such as these can sometimes be followed relatively easily in animal systems, where uniform genetic backgrounds and controlled breeding schemes can be used to avoid the variability that often confounds human population studies. As a consequence, researchers looking for clues to the causes of many complex health problems are focusing more and more attention on model animal systems.

Among such systems, which range in complexity from yeast and bacteria to mammals, the most prominent is the mouse. Because of its small size, high fertility rate, and experimental manipulability, the mouse offers great promise in studying the genetic causes and pathological progress of ailments, as well as understanding the genetic role in disease susceptibility. In pursuing such studies, the DOE is exploiting several resources, among them the experimental mouse genetics facility at the Oak Ridge National Laboratory. Initially established for genetic risk assessment and toxicology studies, the Oak Ridge facility is one of the world's largest. Mutant strains there express a variety of inherited developmental and health disorders, ranging from dwarfism and limb deformities to sickle cell anemia, atherosclerosis, and unusual susceptibilities to cancer.

Most of these existing mutant strains have arisen from random alterations of genes, caused by the same processes that occur naturally in all living populations. However, other, more directed means of gene alteration are also available. So-called transgenic methods, which have been developed and refined over the past 15 years, allow DNA sequences engineered in the laboratory to be introduced directly into the genomes of mouse embryos. The embryos are subsequently transferred to a foster mother, where they develop into mice carrying specifically designed alterations in a particular gene. The differences in form, basic health, fertility, and longevity produced by these "designer mutations" then allow researchers to study the effects of genetic defects that can mimic those found in human patients. The payoff can be clues that aid in the design of drugs and other treatments for the human diseases.

The Human Genome Center at Berkeley is using mice for similar purposes. In vivo libraries of overlapping human genome fragments (each 100,000 to 1,000,000 base pairs long) are being propagated in transgenic mice. The region of chromosome 21 responsible for Down syndrome, for example, is now almost fully represented in a panel of transgenic mice. Such libraries have several uses. For example, the precise biochemical means by which identified genes produce their effects can be studied in detail, and new genes can be recognized by analyzing the effects of particular genome fragments on the transgenic animals. In such ways, the promise of the massive effort to map and sequence the human genome can be translated into the kind of biological knowledge coveted by pharmaceutical designers and medical researchers.

Adding to the potential value of mutant mice as models for human genetic disease is growing evidence of similarities between mouse and

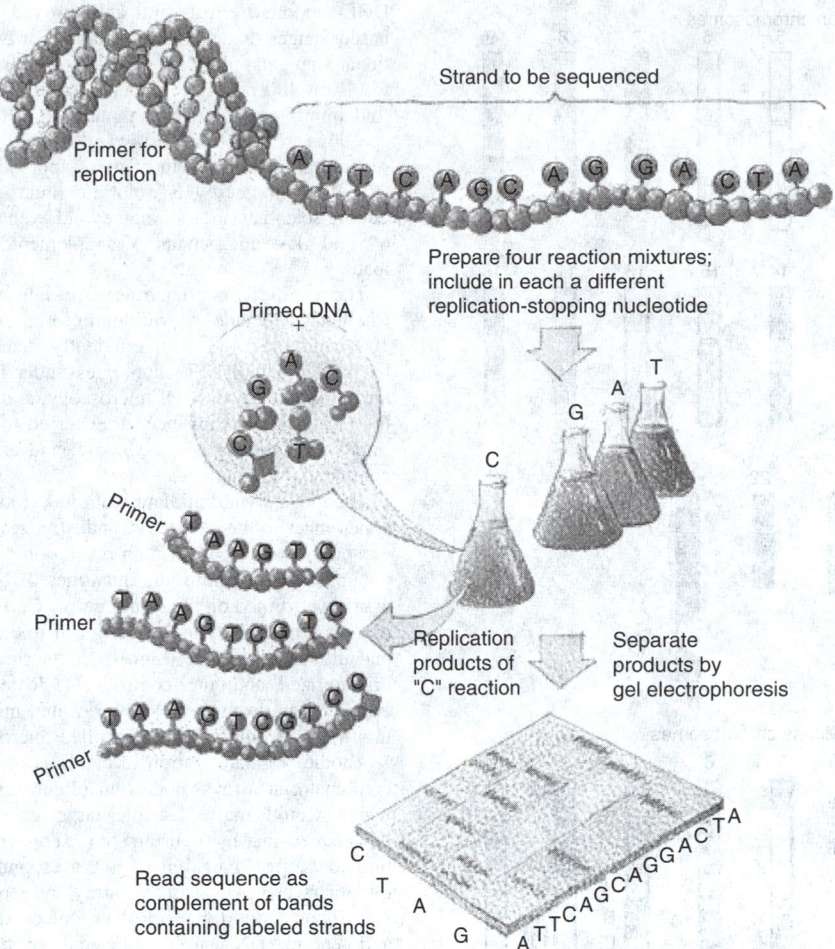

Fig. 8.   In the much-automated Sanger sequencing method, the single-stranded DNA to be sequenced is "primed" for replication with a short complementary strand at one end. This preparation is then divided into four batches, and each is treated with a different replication-halting nucleotide (depicted here with a diamond shape), together with the four "usual" nucleotides. Each replication reaction then proceeds until a reaction-terminating nucleotide is incorporated into the growing strand, whereupon replication stops. Thus, the "C" reaction produces new strands that terminate at positions corresponding to the G's in the strand being sequenced. (Note that when long strands are being sequenced the concentration of the reaction-terminating nucleotide must be carefully chosen, so that a "normal" C is usually paired with a G; otherwise, replication would typically stop with the first or second G) Gel electrophoresis — one lane per reaction mixture — is then used to separate the replication products, from which the sequence of the original single strand can be inferred. (*U.S. Department of Energy.*)

human genes. Indeed, practically every human gene appears to have a counterpart in the mouse genome. Furthermore, the related mouse and human genes often share very similar DNA sequences and the same basic biological function. If we imagine that the 23 pairs of human chromosomes were shattered into smaller blocks — to yield a total of, say, 150 pieces, ranging in size from very small bits containing just a few genes to whole chromosome arms — those pieces could be reassembled to produce a serviceable model of the mouse genome. This mouse genome jigsaw puzzle is shown to the right. Thanks to this mouse-human genomic homology, a newly located gene on a human chromosome can often lead to a confident prediction of where a closely related gene will be found in the mouse, and vice versa. See Fig. 9.

Thus, a crippling heritable muscle disorder in mice maps to a location on the mouse X chromosome that is closely analogous to the map location for the X-linked human Duchenne muscular dystrophy gene (DMD). Indeed, we now know that these two similar diseases are caused by the mouse and human versions of the same gene. Although mutations in the mouse *mdx* gene produce a muscle disease that is less severe than the heartbreaking, fatal disease resulting from the DMD mutation in humans, the two genes produce proteins that function in very similar ways and that are clearly required for normal muscle development and function in the corresponding species. Likewise, the discovery of a mouse gene associated with pigmentation, reproductive, and blood cell defects was the crucial key to uncovering the basis for a human disease known as the *piebald trait*. Owing to such close human-mouse relationships as these, together with the benefits of transgenic technologies, the mouse offers enormous potential

in identifying new human genes, deciphering their complex functions, and even treating genetic diseases.

Beyond Biology

Instrumentation and Informatics. From the start, it has been clear that the Human Genome Project would require advanced instrumentation and automation if its mapping and sequencing goals were to be met. And here, especially, the DOE's engineering infrastructure and tradition of instrumentation development have been crucial contributors to the international effort. Significant DOE resources have been committed to innovations in instrumentation, ranging from straightforward applications of automation to improve the speed and efficiency of conventional laboratory protocols to the development of technologies on the cutting edge — technologies that might potentially increase mapping and sequencing efficiencies by orders of magnitude.

On the first of these fronts, genome researchers are seeing significant improvements in the rate, efficiency, and economy of large-scale mapping and sequencing efforts as a result of improved laboratory automation tools. In many cases, commercial robots have simply been mechanically reconfigured and reprogrammed to perform repetitive tasks, including the replication of large clone libraries, the pooling of libraries as a prelude to various assays, and the arraying of clone libraries for hybridization studies. In other cases, custom-designed instruments have proved more efficient. A notable illustration is the world's fastest cell and chromosome sorter, developed at Livermore and now being commercialized, which is used to sort human chromosomes for chromosome-specific libraries.

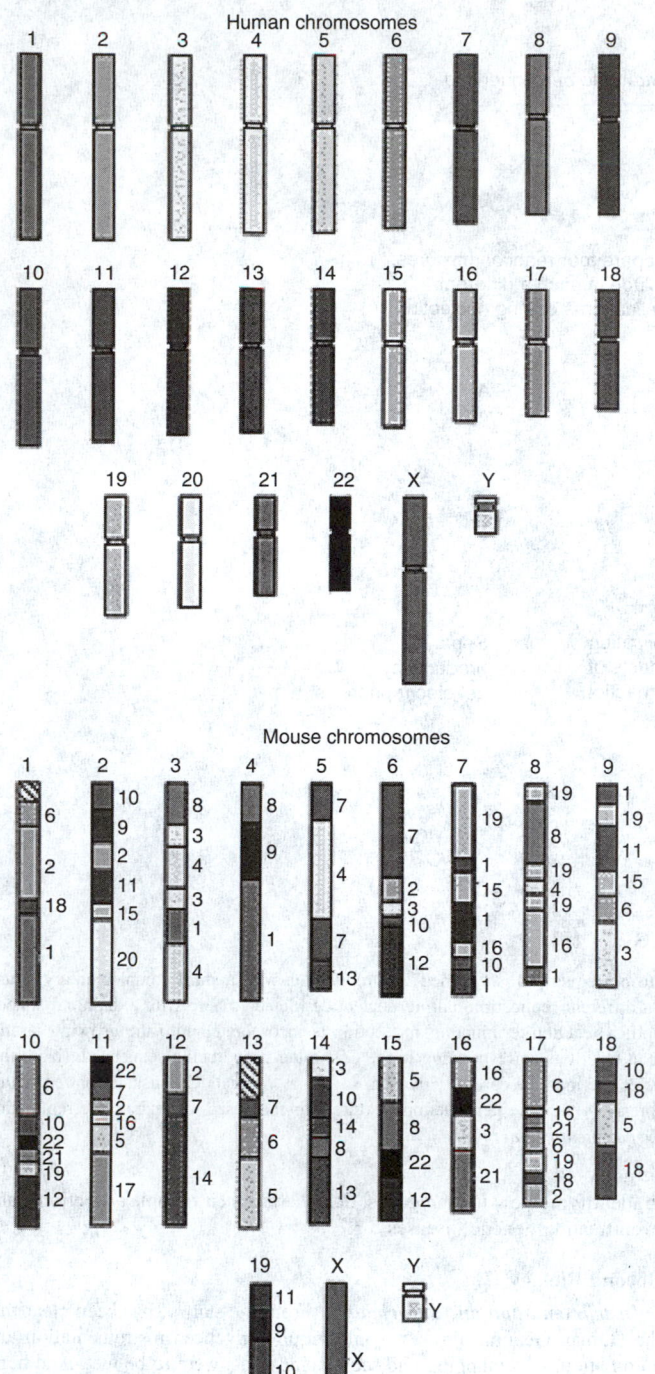

Fig. 9.   The genetic similarity (or homology) of superficially dissimilar species is amply demonstrated here. The full complement of human chromosomes can be cut, schematically at least, into about 150 pieces (only about 100 are large enough to appear in this illustration), then reassembled into a reasonable approximation of the mouse genome. The colors of the mouse chromosomes and the numbers alongside indicate the human chromosomes containing homologous segments. This piecewise similarity between the mouse and human genomes means that insights into mouse genetics are likely to illuminate human genetics as well. (*U.S. Department of Energy.*)

Other examples include a high-speed, robotics-compatible thermal cycler developed at Berkeley, which greatly accelerates PCR amplifications, and instruments developed at Utah for automated hybridization in multiplex sequencing schemes.

Beyond "mere" automation are efforts aimed at more fundamental enhancements of established techniques. In particular, a number of

DOE-supported efforts aim at improved versions of the automated gel-based Sanger sequencing technique. For example, in place of the conventional slab gels, ultrathin gels, less than 0.1 millimeter thick, can be used to obtain 400 bases of sequence from each lane in an hour's run, a five-fold improvement in throughput over conventional systems. Even faster speedups are seen when arrays of 0.1-millimeter capillaries are used as the separation medium. Both of these approaches exploit higher electric field strengths to increase DNA mobility and to reduce analysis times. And Livermore scientists are looking beyond even capillaries, to sequencing arrays of rigid glass microchannels, supplemented by automated gel and sample loading.

The capillary approach is especially ripe for further development. Challenges include providing uniform excitation over arrays of 50 to 100 capillaries and then efficiently detecting the fluorescence emitted by labeled samples. Technologies under investigation include fiber-optic arrays, scanning confocal microscopy, and cooled CCD cameras. Some of this effort has already been transferred to the private sector, and tenfold improvements in speed, economy, and efficiency are projected in future commercial instruments.

The move toward miniaturization is afoot elsewhere as well. Building on experiences in the electronics industry, several DOE-supported groups are exploring ways to adapt high-resolution photolithographic methods to the manipulation of minuscule quantities of biological reagents, followed by assays performed on the same "chip." Current thrusts of this "nanotechnology" approach include the design of microscopic electrophoresis systems and ultrasmall-volume, high-speed thermal cycling systems for PCR. A miniaturized, computer-controlled PCR device under development at Livermore operates on 9-volt batteries and might ultimately lead to arrays of thousands of individually controlled micro-PCR chambers.

Another miniaturization effort aims at the fabrication of high-density combinatorial arrays of custom **oligomers** (short chains of nucleotides), which would make feasible large-scale hybridization assays, including sequencing by hybridization. This innovative technique uses short oligomers that pair up with corresponding sequences of DNA. The oligomers are placed on an array by a process similar to that of making silicon chips for electronics. Successful matches between oligomers and genomic DNA are then detected by fluorescence, and the application of sophisticated statistical analyses reassembles the target sequence. This same technology has already been used for genetic screening and cDNA fingerprinting. Similar approaches can be envisioned to understand differences in patterns of gene expression: Which genes are active (which are producing mRNA) in which cells? Which are active at different times during an organism's development? Which are active, or inactive, in disease?

Sequencing by hybridization is only one of several forward-looking ideas for revolutionizing sequencing technology. In spite of continuing improvements to sequencers based on the classic methods, it is nonetheless desirable to explore altogether new approaches, with an eye to simplifying sample preparation, reducing measurement times, increasing the length of the strands that can be analyzed in a single run, and facilitating interpretation of the results. Over the course of the past few years, several alternative approaches to direct sequencing have been explored, including atomic-resolution molecular scanning, single-molecule detection of individual bases, and mass spectrometry of DNA fragments.

All of these alternatives look promising in the long term, but mass spectrometry has perhaps demonstrated the greatest near-term potential. Mass spectrometry measures the masses of ionized DNA fragments by recording their time-of-flight in vacuum. It would therefore replace traditional gel electrophoresis as the last step in a conventional sequencing scheme. Routine application of this technique still lies in the future, but fragments of up to 500 bases have been analyzed, and practical systems based on high-resolution mass separations of DNA fragments of fewer than 100 bases are currently being developed at several universities and national laboratories.

Another innovative sequencing method is under investigation at Los Alamos. If each of the four bases (A, T, C, G) in a single strand of DNA receives a different fluorescent label, then the bases are enzymatically detached, one at a time. The characteristic fluorescence is detected by a laser system, thereby yielding the sequence, base by base. This approach is beset by major technical challenges, and direct sequencing has not yet been achieved. But the potential benefits are great, and much of the instrumentation for sensitive detection of fluorescence signals has already proved useful for molecular sizing in mapping applications.

Dealing with the Data. Among the less visible challenges of the Human Genome Project is the daunting prospect of coping with all the data that success implies. Appropriate information systems are needed not only during data acquisition, but also for sophisticated data analysis and for the management and public distribution of unprecedented quantities of biological information. Further, because much of the challenge is interpreting genomic data and making the results available for scientific and technological applications, the challenge extends not just to the Human Genome Project, but also to the microbial genome program and to public- and private-sector programs focused on areas such as health effects, structural biology, and environmental remediation. Efforts in all these areas are the mandate of the DOE genome informatics program, whose products are already widely used in genome laboratories, general molecular biology and medical laboratories, biotechnology companies, and biopharmaceutical companies around the world.

The roles of laboratory data acquisition and management systems include the construction of genetic and physical maps, DNA sequencing, and gene expression analysis. These systems typically comprise databases for tracking biological materials and experimental procedures, software for controlling robots or other automated systems, and software for acquiring laboratory data and presenting it in useful form. Among such systems are physical mapping databases developed at Livermore and Los Alamos, robot control software developed at Berkeley and Livermore, and DNA sequence assembly software developed at the University of Arizona. These systems are the keys to efficient, cost-effective data production in both DOE laboratories and the many other laboratories that use them.

The interpretation of map and sequence data is the job of data analysis systems. These systems typically include task-specific computational engines, together with graphics and user-friendly interfaces that invite their use by biologists and other non-computer scientists. The genome informatics program is the world leader in developing automated systems for identifying genes in DNA sequence data from humans and other organisms, supporting efforts at Oak Ridge National Laboratory and elsewhere. The Oak Ridge-developed GRAIL system, is a world-standard gene identification tool. In 1995 alone, more than 180 million base pairs of DNA were analyzed with GRAIL.

A third area of informatics reflects, in a sense, the ultimate product of the Human Genome Project—information readily available to the scientific and lay communities. Public resource databases must provide data and interpretive analyses to a worldwide research and development community. As this community of researchers expands and as the quantity of data grows, the challenges of maintaining accessible and useful databases likewise increase. For example, it is critical to develop scientific databases that "interoperate," sharing data and protocols so that users can expect answers to complex questions that demand information from geographically distributed data resources. As the genome project continues to provide data that interlink structural and functional biochemistry, molecular, cellular, and developmental biology, physiology and medicine, and environmental science, such interoperable databases will be the critical resources for both research and technology development. The DOE genome informatics program is crucial to the multiagency effort to develop just such databases. Systems now in place include the Genome Database of human genome map data at Johns Hopkins University, the Genome Sequence DataBase at the National Center for Genome Resources in Santa Fe, and the Molecular Structure Database at Brookhaven National Laboratory.

Ethical, Legal, and Social Implications

The Human Genome Project is rich with promise, but also fraught with social implications. We expect to learn the underlying causes of thousands of genetic diseases, including sickle cell anemia, Tay-Sachs disease, Huntington disease, myotonic dystrophy, cystic fibrosis, and many forms of cancer—and thus to predict the likelihood of their occurrence in any individual. Likewise, genetic information might be used to predict sensitivities to various industrial or environmental agents. The dangers of misuse and the potential threats to personal privacy are not to be taken lightly.

In recognition of these important issues, both the DOE and the National Institutes of Health devote a portion of their resources to studies of the ethical, legal, and social implications (ELSI) of human genome research. Perhaps the most critical of social issues are the questions of privacy and fair use of genetic information. Most observers agree that personal knowledge of genetic susceptibility can be expected to serve us well,

opening the door to more accurate diagnoses, preventive intervention, intensified screening, lifestyle changes, and early and effective treatment. But such knowledge has another side, too: the risk of anxiety, unwelcome changes in personal relationships, and the danger of stigmatization. Consider, for example, the impact of information that is likely to be incomplete and indeterminate (say, an indication of a 25 percent increase in the risk of cancer). And further, if handled carelessly, genetic information could threaten us with discrimination by potential employers and insurers. Other issues are perhaps less immediate than these personal concerns, but they are no less challenging. How, for example, are the "products" of the Human Genome Project to be patented and commercialized? How are the judicial, medical, and educational communities—not to mention the public at large—to be effectively educated about genetic research and its implications?

To confront all these issues, the NIH-DOE Joint Working Group on Ethical, Legal, and Social Implications of Human Genome Research was created in 1990 to coordinate ELSI policy and research between the two agencies. See Table 1. One focus of DOE activity has been to foster educational programs aimed both at private citizens and at policy-makers and educators. Fruits of these efforts include radio and television documentaries, high school curricula and other educational material, and science museum displays. In addition, the DOE has concentrated on issues associated with privacy and the confidentiality of genetic information, on workplace and commercialization issues (especially screening for susceptibilities to environmental or workplace agents), and on the implications of research findings regarding the interactions among multiple genes and environmental influences.

Whereas the issues raised by modern genome research are among the most challenging we face, they are not unprecedented. Issues of privacy, knotty questions of how knowledge is to be commercialized, problems of dealing with probabilistic risks, and the imperatives of education have all been confronted before. As usual, defensible perspectives and reasonable arguments, even precious rights, exist on opposing sides of every issue. It is a balance that must be sought. Accordingly, further study is needed, as well as continuing efforts to promote public awareness and understanding, as we strive to define policies for the intelligent use of the profound knowledge we seek about ourselves.

The Age of Discovery was the age of da Gama, Columbus, and Magellan, an era when European civilization reached out to the Far East and thus filled many of the voids in its map of the world. But in a larger sense, we have never ceased from our exploration and discovery. Science has been unstinting over the ages in its efforts to complete our intellectual picture of the universe. In this century, our explorations have extended from the subatomic to the cosmic, as we have mapped the heavens to their farthest reaches and charted the properties of the most fleeting elementary particles. Nor have we neglected to look inward, seeking, as it were, to define the topography of the human body. Beginning with the first modern anatomical studies in the sixteenth century, we have added dramatically to our picture of human anatomy, physiology, and biochemistry. The Human Genome Project is thus the next stage in an epic voyage of discovery—a voyage that will bring us to a profound understanding of human biology.

In an important way, though, the genome project is very different from many of our exploratory adventures. It is spurred by a conviction of practical value, a certainty that human benefits will follow in the wake of success. The product of the Human Genome Project will be an enormously rich biological database, the key to tracking down every human gene—and thus to unveiling, and eventually to subverting, the causes of thousands of human diseases. The sequence of our genome will ultimately allow us to unlock the secrets of life's processes, the biochemical underpinnings of our senses and our memory, our development and our aging, our similarities and our differences.

It has further been said that the Human Genome Project is *guaranteed* to succeed: Its goal is nothing more assuming than a sequence of three billion characters. And we have a very good idea of how to read those characters. Unlike perilous voyages or searches for unknown subatomic particles, this venture is assured of its goal. But beyond a detailed picture of human DNA, no one can predict the form success will take. The genome project itself offers no promises of cancer cures or quick fixes for Alzheimer's disease, no detailed understanding of genius or schizophrenia. But if we are ever to uncover the mysteries of carcinogenesis, if we are ever to know how biochemistry contributes to mental illness and dementia, if we ever hope to really understand the processes of growth and development, we must first have a detailed map of the genetic landscape. That's what the Human Genome Project promises. In a way, it's a rather prosaic step, but what lies beyond is breathtaking.

Editor's Note: The staff of this encyclopedia is most appreciative of the cooperation extended by the DOE, for furnishing the information "To Know Ourselves" prepared at the request of the U.S. Department of Energy, by the Office of Health and Environmental Research, as an overview of the Human Genome Project.

The Human Genome Project 1998–Present

The Human Genome Project has successfully completed all the major goals from the first five-year plan 1991–1995 and revised second five-year plan covering the period 1993–98. A third and final plan was developed (October 1, 1998 to September 30, 2003) during a series of DOE and NIH workshops. Some 18 countries participate in the worldwide effort, with significant contributions from the Sanger Center in the United Kingdom and research centers in Germany, France, and Japan. See Table 1 (Major Events in the U.S. Human Genome Project.)

Final five-year Research Goals of the U.S. Human Genome Project

Human DNA Sequence.

- Finish the complete human genome sequence by the end of 2003.
- Finish one-third of the human DNA sequence by the end of 2001.
- Achieve coverage of at least 90% of the genome in a working draft based on mapped clones by the end of 2001.
- Make the sequence totally and freely accessible.

Sequencing Technology.

- Continue to increase the throughput and reduce the cost of current sequencing technology.
- Support research on novel technologies that can lead to significant improvements in sequencing technology.
- Develop effective methods for the advanced development and introduction of new sequencing technologies into the sequencing process.

Human Genome Sequence Variation.

- Develop technologies for rapid, large-scale identification and/or scoring of single nucleotide polymorphisms and other DNA sequence variants.
- Identify common variants in the coding regions of the majority of identified genes during this five-year period.
- Create a SNP map of at least 100,000 markers.
- Develop the intellectual foundations for studies of sequence variation.
- Create public resources of DNA samples and cell lines.

Functional Genomics Technology.

- Generate sets of full-length cDNA clones and sequences that represent human genes and model organisms.
- Support research on methods for studying functions of nonprotein-coding sequences.
- Develop technology for comprehensive analysis of gene expression.
- Improve methods for genome-wide mutagenesis.
- Develop technology for large-scale protein analyses.

Comparative Genomics.

- Complete the sequence of the roundworm *C. elegans* genome by 1998. Complete the sequence of the fruitfly *Drosophila* genome by 2002.
- Develop an integrated physical and genetic map for the mouse, generate additional mouse cDNA resources, and complete the sequence of the mouse genome by 2008.
- Identify other useful model organisms and support appropriate genomic studies.

Ethical, Legal, and Social Issues.

- Examine issues surrounding the completion of the human DNA sequence and the study of human genetic variation.
- Examine issues raised by the integration of genetic technologies and information into health care and public health activities.
- Examine issues raised by the integration of knowledge about genomics and gene-environment interactions in non-clinical settings.
- Explore how new genetic knowledge may interact with a variety of philosophical theological, and ethical perspectives.

- Explore how racial, ethnic, and socioeconomic factors affect the use, understanding, and interpretation of genetic information; the use of genetic services; and the development of policy.

Bioinformatics and Computational Biology.

- Improve content and utility of databases.
- Develop better tools for data generation, capture, and annotation.
- Develop and improve tools and databases for comprehensive functional studies.
- Develop and improve tools for representing and analyzing sequence similarity and variation.
- Create mechanisms to support effective approaches for producing robust, exportable software that can be widely shared.

Training and Manpower.

- Nurture the training of scientists skilled in genomics research.
- Encourage the establishment of academic career paths for genomic scientists.
- Increase the number of scholars who are knowledgeable in both genomic and genetic sciences and in ethics, law, or the social sciences.

The Human Genome Project (HGP) is fulfilling its promise as the single most important project in biology and the biomedical sciences–one that will permanently change biology and medicine. With the recent completion of the genome sequences of several microorganisms, including *Escherichia coli, Saccharomyces cerevisiae*, and the completion of the sequence of the metazoan *Caenorhabditis elegans*, the door has opened wide on the era of whole genome science. The ability to analyze entire genomes is accelerating gene discovery and revolutionizing the breadth and depth of biological questions that can be addressed in model organisms. These exciting successes confirm the view that acquisition of a comprehensive, high-quality human genome sequence will have unprecedented impact and long-lasting value for basic biology, biomedical research, biotechnology, and health care. The transition to sequence-based biology will spur continued progress in understanding gene-environment interactions and in development of highly accurate DNA-based medical diagnostics and therapeutics.

The flagship endeavor of the HGP, Human DNA sequencing, is entering its decisive phase. It will be the project's central focus during the next 5 years. While partial subsets of the DNA sequence, such as expressed sequence tags (ESTs), have proven enormously valuable, experience with simpler organisms confirms that there can be no substitute for the complete genome sequence. In order to move vigorously toward this goal, the crucial task ahead is building sustainable capacity for producing publicly available DNA sequence. The full and incisive use of the human sequence, including comparisons to other vertebrate genomes, will require further increases in sustainable capacity at high accuracy and lower costs. Thus, a high-priority commitment to develop and deploy new and improved sequencing technologies must also be made.

In September of 1999, international leaders of Human Genome Project (HGP) sequencing confirmed a plan to complete a rough draft of the human genome by next summer, a year ahead of schedule. This accelerated pace is made possible by the commercialization of a new generation of automated capillary DNA sequencing machines and by BAC mapping resources generated from DOE-sponsored clone projects.

On June 26, 2000 President Clinton and HGP leaders Craig Venter (head of Celera Genomics), and Francis Collins (director, NIH National Human Genome Research Institute) announced the completion of a "working draft" DNA sequence of the human genome. See Table 1.

On February 12, 2001 scientists from the public Human Genome Project and the private company Celera Genomics published the long-awaited details of the "working-draft" DNA sequence achieved less than a year before. Although the draft is filled with mysteries, the first panoramic view of the human genetic landscape has revealed a wealth of information and some early surprises. Papers describing research observations in the journals, *Nature* (Feb. 15, 2001) http://www.nature.com/genomics/ and *Science* (Feb. 16, 2001) http://www.sciencemag.org/content/vol291/issue5507/index.shtml are freely accessible via the Web.

Although clearly not a Holy Grail or Rosetta Stone for deciphering all of biology two early metaphors commonly used to describe the coveted prize the sequence is a magnificent and unprecedented resource that will serve as a basis for research and discovery throughout this century and beyond.

It will have diverse practical applications and a profound impact upon how we view our place and ourselves in the tapestry of life around us.

One insight already gleaned from the sequence is that, even on the molecular level, we are more than the sum of our 35,000 or so genes. Surprisingly, this newly estimated number of genes is only one-third as great as previously thought and is only twice as many as those of a tiny transparent worm, although the numbers may be revised as more computational and experimental analyses are performed. At once humbled and intrigued by this finding, scientists suggest that the genetic key to human complexity lies not in the number of genes but in how gene parts are used to build different products in a process called *alternative splicing*. Other sources of added complexity are the thousands of post-translational chemical modifications made to proteins and the repertoire of regulatory mechanisms controlling these processes.

The draft encompasses 90% of the human genome's euchromatic portion, which contains the most genes. In constructing the working draft, the 16 genome sequencing centers produced over 22.1 billion bases of raw sequence data, comprising overlapping fragments totaling 3.9 billion bases and providing sevenfold coverage (sequenced seven times) of the human genome. Over 30% is high-quality, finished sequence, with eight- to tenfold coverage, 99.99% accuracy, and few gaps.

The entire working draft will be finished to high quality by 2003. Coincidentally, that year also will be the 50th anniversary of Watson and Crick's publication of DNA structure that launched the era of molecular genetics. See www.nature.com/genomics/human/watson-crick. Much will remain to be deciphered even then. Some highlights from *Nature*, *Science*, and The Wellcome Trust is described below.

What Does the Draft Human Genome Sequence Tell Us?

- The human genome contains 3164.7 million chemical nucleotide bases (A, C, T, and G).
- The average gene consists of 3000 bases, but sizes vary greatly, with the largest known human gene being dystrophin at 2.4 million bases.
- The total number of genes is estimated at 30,000 to 35,000 much lower than previous estimates of 80,000 to 140,000 that had been based on extrapolations from gene-rich areas as opposed to a composite of gene-rich and gene-poor areas.
- Almost all (99.9%) nucleotide bases are exactly the same in all people.
- The functions are unknown for over 50% of discovered genes.

The Wheat from the Chaff.

- Less than 2% of the genome codes for proteins.
- Repeated sequences that do not code for proteins ("junk DNA") make up at least 50% of the human genome.
- Repetitive sequences are thought to have no direct functions, but they shed light on chromosome structure and dynamics. Over time, these repeats reshape the genome by rearranging it, creating entirely new genes, and modifying and reshuffling existing genes.
- During the past 50 million years, a dramatic decrease seems to have occurred in the rate of accumulation of repeats in the human genome.

How It's Arranged.

- The human genome's gene-dense "urban centers" are predominantly composed of the DNA building blocks G and C.
- In contrast, the gene-poor "deserts" are rich in the DNA building blocks A and T. GC- and AT-rich regions usually can be seen through a microscope as light and dark bands on chromosomes.
- Genes appear to be concentrated in random areas along the genome, with vast expanses of noncoding DNA between.
- Stretches of up to 30,000 C and G bases repeating over and over often occur adjacent to gene-rich areas, forming a barrier between the genes and the "junk DNA." These CpG islands are believed to help regulate gene activity.
- Chromosome 1 has the most genes (2968), and the Y chromosome has the fewest (231).

How the Human Compares with Other Organisms.

- Unlike the human's seemingly random distribution of gene-rich areas, many other organisms' genomes are more uniform, with genes evenly spaced throughout.
- Humans have on average three times as many kinds of proteins as the fly or worm because of mRNA transcript "alternative splicing" and chemical modifications to the proteins. This process can yield different protein products from the same gene.
- Humans share most of the same protein families with worms, flies, and plants, but the number of gene family members has expanded in humans, especially in proteins involved in development and immunity.
- The human genome has a much greater portion (50%) of repeat sequences than the mustard weed (11%), the worm (7%), and the fly (3%).
- Although humans appear to have stopped accumulating repeated DNA over 50 million years ago, there seems to be no such decline in rodents. This may account for some of the fundamental differences between hominids and rodents, although gene estimates are similar in these species.
- Scientists have proposed many theories to explain evolutionary contrasts between humans and other organisms, including those of life span, litter sizes, inbreeding, and genetic drift.

Variations and Mutations.

- Scientists have identified about 1.4 million locations where single-base DNA differences (SNPs) occur in humans. This information promises to evolutionize the processes of finding chromosomal locations for disease-associated sequences and tracing human history.
- The ratio of germline (sperm or egg cell) mutations is 2 : 1 in males vs females. Researchers point to several reasons for the higher mutation rate in the male germline, including the greater number of cell divisions required for sperm formation than for eggs.

Applications and Future Challenges

Deriving meaningful knowledge from the DNA sequence will define research through the coming decades to inform our understanding of biological systems. This enormous task will require the expertise and creativity of tens of thousands of scientists from varied disciplines in both the public and private sectors worldwide.

The draft sequence already is having an impact on finding genes associated with disease. Over 30 genes have been pinpointed and associated with breast cancer, muscle disease, deafness, and blindness. Additionally, finding the DNA sequences underlying such common diseases as cardiovascular disease, diabetes, arthritis, and cancers is being aided by the human variation maps (SNPs) generated in the HGP in cooperation with the private sector. These genes and SNPs provide focused targets for the development of effective new therapies.

One of the greatest impacts of having the sequence may well be in enabling an entirely new approach to biological research. In the past, researchers studied one or a few genes at a time. With whole-genome sequences and new high-throughput technologies, they can approach questions systematically and on a grand scale. They can study all the genes in a genome, for example, or all the transcripts in a particular tissue or organ or tumor, or how tens of thousands of genes and proteins work together in interconnected networks to orchestrate the chemistry of life.

The Next Step: Functional Genomics

The words of Winston Churchill, spoken in 1942 after 3 years of war, capture well the HGP era: "Now this is not the end. It is not even the beginning of the end. But it is, perhaps, the end of the beginning."

The avalanche of genome data grows daily. The new challenge will be to use this vast reservoir of data to explore how DNA and proteins work with each other and the environment to create complex, dynamic living systems. Systematic studies of function on a grand scale-functional genomics-will be the focus of biological explorations in this century and beyond. These explorations will encompass studies in transcriptomics, proteomics, structural genomics, new experimental methodologies, and comparative genomics.

- **Transcriptomics** involves large-scale analysis of messenger RNAs transcribed from active genes to follow when, where, and under what conditions genes are expressed.
- Studying protein expression and function—or proteomics—can bring researchers closer to what's actually happening in the cell than gene-expression studies. This capability has applications to drug design.
- **Structural genomics** initiatives are being launched worldwide to generate the 3-D structures of one or more proteins from each protein family, thus offering clues to function and biological targets for drug design.

- Experimental methods for understanding the function of DNA sequences and the proteins they encode include knockout studies to inactivate genes in living organisms and monitor any changes that could reveal their functions.
- **Comparative genomics**— analyzing DNA sequence patterns of humans and well-studied model organisms side-by-side-has become one of the most powerful strategies for identifying human genes and interpreting their function.

Potential Benefits of Human Genome Project Research

Rapid progress in genome science and a glimpse into its potential applications have spurred observers to predict that biology will be the foremost science of the 21st century. Technology and resources generated by the Human Genome Project and other genomics research are already having a major impact on research across the life sciences. The potential for commercial development of genomics research presents U.S. industry with a wealth of opportunities, and sales of DNA-based products and technologies in the biotechnology industry are projected to exceed $45 billion by 2009 (Consulting Resources Corporation Newsletter, Spring 1999). Some current and potential applications of genome research include:

Molecular Medicine

- *improved diagnosis of disease*
- *earlier detection of genetic predispositions to disease rational drug design*
- *gene therapy and control systems for drugs*
- *pharmacogenomics "custom drugs"*

Technology and resources promoted by the Human Genome Project are starting to have profound impacts on biomedical research and promise to revolutionize the wider spectrum of biological research and clinical medicine. Increasingly detailed genome maps have aided researchers seeking genes associated with dozens of genetic conditions, including myotonic dystrophy, fragile X syndrome, neurofibromatosis types 1 and 2, inherited colon cancer, Alzheimer's disease, and familial breast cancer.

On the horizon is a new era of molecular medicine characterized less by treating symptoms and more by looking to the most fundamental causes of disease. Rapid and more specific diagnostic tests will make possible earlier treatment of countless maladies. Medical researchers also will be able to devise novel therapeutic regimens based on new classes of drugs, immunotherapy techniques, avoidance of environmental conditions that may trigger disease, and possible augmentation or even replacement of defective genes through gene therapy.

For more information, see http://www.ornl.gov/hgmis/medicine/medicine.html "Medicine and the New Genetics" and "Fast Forward to 2020: What to Expect in Molecular Medicine" an article for online magazine TNTY Futures. http://www.ornl.gov/hgmis/medicine/tnty.html

Microbial Genomics

- *new energy sources (biofuels)*
- *environmental monitoring to detect pollutants*
- *protection from biological and chemical warfare*
- *safe, efficient toxic waste cleanup*
- *understanding disease vulnerabilities and revealing drug targets*

In 1994, taking advantage of new capabilities developed by the genome project, DOE initiated the Microbial Genome Program to sequence the genomes of bacteria useful in energy production, environmental remediation, toxic waste reduction, and industrial processing.

Despite our reliance on the inhabitants of the microbial world, we know little of their number or their nature: estimates are that less than 0.01% of all microbes have been cultivated and characterized. Programs like the DOE Microbial Genome Program help lay a foundation for knowledge that will ultimately benefit human health and the environment. The economy will benefit from further industrial applications of microbial capabilities.

Information gleaned from the characterization of complete genomes in MGP will lead to insights into the development of such new energy-related biotechnologies as photosynthetic systems, microbial systems that function in extreme environments, and organisms that can metabolize readily available renewable resources and waste material with equal facility. Expected benefits also include development of diverse new products, processes, and test methods that will open the door to a cleaner

environment. Biomanufacturing will use nontoxic chemicals and enzymes to reduce the cost and improve the efficiency of industrial processes. Already, microbial enzymes are being used to bleach paper pulp, stone wash denim, remove lipstick from glassware, break down starch in brewing, and coagulate milk protein for cheese production. In the health arena, microbial sequences may help researchers find new human genes and shed light on the disease-producing properties of pathogens.

Microbial genomics will also help pharmaceutical researchers gain a better understanding of how pathogenic microbes cause disease. Sequencing these microbes will help reveal vulnerabilities and identify new drug targets.

Gaining a deeper understanding of the microbial world also will provide insights into the strategies and limits of life on this planet. Data generated in this young program already have helped scientists identify the minimum number of genes necessary for life and confirm the existence of a third major kingdom of life. Additionally, the new genetic techniques now allow us to establish more precisely the diversity of microorganisms and identify those critical to maintaining or restoring the function and integrity of large and small ecosystems; this knowledge also can be useful in monitoring and predicting environmental change. Finally, studies on microbial communities provide models for understanding biological interactions and evolutionary history.

For more information, see: Microbial Genome Program: http://www.ornl.gov/microbialgenomes/ and Microbe World http://www.microbeworld.org/.

Risk Assessment

- *assess health damage and risks caused by radiation exposure, including low-dose exposures*
- *assess health damage and risks caused by exposure to mutagenic chemicals and cancer-causing toxins*
- *reduce the likelihood of heritable mutations*

Understanding the human genome will have an enormous impact on the ability to assess risks posed to individuals by exposure to toxic agents. Scientists know that genetic differences make some people more susceptible and others more resistant to such agents. Far more work must be done to determine the genetic basis of such variability. This knowledge will directly address DOE's long-term mission to understand the effects of low-level exposures to radiation and other energy-related agents, especially in terms of cancer risk.

Bioarchaeology, Anthropology, Evolution, and Human Migration

- *study evolution through germline mutations in lineages*
- *study migration of different population groups based on female genetic inheritance*
- *study mutations on the Y chromosome to trace lineage and migration of males*
- *compare breakpoints in the evolution of mutations with ages of populations and historical events*

Understanding genomics will help us understand human evolution and the common biology we share with all of life. Comparative genomics between humans and other organisms such as mice already has led to similar genes associated with diseases and traits. Further comparative studies will help determine the yet-unknown function of thousands of other genes.

Comparing the DNA sequences of entire genomes of different microbes will provide new insights about relationships among the three kingdoms of life: archaebacteria, eukaryotes, and prokaryotes.

For more information, see http://www.wellcome.ac.uk/en/old/AWTpubNWSwlkBCKw17.html *Wellcome News Issue* 17 Q4 1998.

DNA Forensics (Identification)

- *identify potential suspects whose DNA may match evidence left at crime scenes*
- *exonerate persons wrongly accused of crimes*
- *identify crime and catastrophe victims*
- *establish paternity and other family relationships*
- *identify endangered and protected species as an aid to wildlife officials (could be used for prosecuting poachers)*
- *detect bacteria and other organisms that may pollute air, water, soil, and food*

- *match organ donors with recipients in transplant programs*
- *determine pedigree for seed or livestock breeds*
- *authenticate consumables such as caviar and wine*

Any type of organism can be identified by examination of DNA sequences unique to that species. Identifying individuals is less precise at this time, although when DNA sequencing technologies progress further, direct characterization of very large DNA segments, and possibly even whole genomes, will become feasible and practical and will allow precise individual identification.

To identify individuals, forensic scientists scan about 10 DNA regions that vary from person to person and use the data to create a DNA profile of that individual (sometimes called a DNA fingerprint). There is an extremely small chance that another person has the same DNA profile for a particular set of regions.

For more information, see the DNA Forensics site: http://www.ornl.gov/hgmis/elsi/forensics.html

Agriculture, Livestock Breeding, and Bioprocessing

- *disease-, insect-, and drought-resistant crops*
- *healthier, more productive, disease-resistant farm animals*
- *more nutritious produce*
- *biopesticides*
- *edible vaccines incorporated into food products*
- *new environmental cleanup uses for plants like tobacco*

Understanding plant and animal genomes will allow us to create stronger, more disease-resistant plants and animals—reducing the costs of agriculture and providing consumers with more nutritious, pesticide-free foods. Already growers are using bioengineered seeds to grow insect- and drought-resistant crops that require little or no pesticide. Farmers have been able to increase outputs and reduce waste because their crops and herds are healthier.

Alternate uses for crops such as tobacco have been found. One researcher has genetically engineered tobacco plants in his laboratory to produce a bacterial enzyme that breaks down explosives such as TNT and dinitroglycerin. Waste that would take centuries to break down in the soil can be cleaned up by simply growing these special plants in the polluted area.

For more information, see the Access Excellence Website's Biotech Applied http://www.accessexcellence.org/AB/IWT/

Challenges for the Future: What We Still Do not Know

- Gene number, exact locations, and functions
- Gene regulation
- DNA sequence organization
- Chromosomal structure and organization
- Noncoding DNA types, amount, distribution, information content, and functions
- Coordination of gene expression, protein synthesis, and post-translational events
- Interaction of proteins in complex molecular machines
- Predicted vs experimentally determined gene function
- Evolutionary conservation among organisms
- Protein conservation (structure and function)
- Proteomes (total protein content and function) in organisms
- Correlation of SNPs (single-base DNA variations among individuals) with health and disease
- Disease-susceptibility prediction based on gene sequence variation
- Genes involved in complex traits and multigene diseases
- Complex systems biology including microbial consortia useful for environmental restoration
- Developmental genetics, genomics

Projects Beyond the Human Genome Project

Post-sequencing projects are well under way worldwide. See **Genomes to Life**. These explorations will result in a more comprehensive, new, and profound understanding of complex living systems, with applications to human health, energy, global climate change, and environmental cleanup, among others.

Genomes to Life

The Genomes to Life program began with a 1999 charge by Martha Krebs, former director of DOE's Office of Science, that the BER advisory committee define DOE's potential scientific roles after the Human Genome Project is completed. The current program plan (roadmap) document (April 2001) see http://doegenomestolife.org/roadmap/GTLcontents.pdf was prepared in response to recommendations set forth in the resulting report, "Bringing the Genome to Life" (August 2000). See http://doegenomestolife.org/history/genome-to-life-rpt.html. The FY 2002 budget for GTL is $19.5 million.

Building on the successes of the Human Genome Project, DOE has initiated an ambitious program to achieve the most far-reaching of all biological goals: a fundamental, comprehensive, and systematic understanding of life.

The DOE's Genomes to Life program will make important contributions in the quest to venture beyond characterizing such individual life components as genes and other DNA sequences toward a more comprehensive, integrated view of biology at a whole-systems level. The DOE offices of Biological and Environmental Research and Advanced Scientific Computing Research have formed a strategic alliance to meet this grand challenge.

The plan for the 10-year program is to use DNA sequences from microbes and higher organisms, including humans, as starting points for systematically tackling questions about the essential processes of living systems. Advanced technological and computational resources will help to identify and understand the underlying mechanisms that enable organisms to develop, survive, carry out their normal functions, and reproduce under myriad environmental conditions.

This approach ultimately will foster an integrated and predictive understanding of biological systems and offer insights into how both microbial and human cells respond to environmental changes. In a related program, the Microbial Cell Project (MCP) provides the ultimate test of GTL findings. The MCP program takes a whole-genome approach to understanding the function and regulation of genes for a single living system and the pathways in which the protein products interact. See **Microbial Cell Project**. The applications of this level of knowledge will be extraordinary and will help DOE fulfill its broad missions in energy, environmental remediation, and the protection of human health.

Specific Genomes to Life Goals

- Identify the protein machines that carry out critical life functions.
- Characterize the gene regulatory networks that control these machines.
- Explore the functional repertoire of complex microbial communities in their natural environments to provide a foundation for understanding and using their remarkably diverse capabilities to address DOE missions.
- Develop the computational capabilities to integrate and understand these data and begin to model complex biological systems.

Microbial Cell Project

The whole-system approach taken by the Microbial Cell Project is complementary to that proposed for the Genomes to Life program, which seeks to develop a genome-based understanding of the components and processes of living systems.

During the last decade, scientists have amassed millions of DNA sequences containing the complete genetic instructions for a growing list of microbes and viruses. These DNA sequences offer a virtual "parts list" for life in its simplest form, but scientists do not know what many of the parts do. Furthermore, DNA sequences provide little information on how the parts work together to orchestrate the chemistry of life. (By analogy, a pile of automobile parts would tell us very little about the complex function of an automobile.) In biology, the whole is much greater than the sum of the parts, and understanding this complexity is the exciting challenge science now faces. Revolutionary breakthroughs in genome sequencing, new methods of protein characterization, and access to powerful supercomputers now position scientists to begin to understand the complex pathways that give a microbial cell its life. The Microbial Cell Project (MCP) is an exciting new initiative that will address these challenges. The project builds on previous research sponsored by the Office of Science, including the Microbial Genome Program (MGP), itself a spinoff of the Human Genome Program.

The Microbial Cell Project will support core missions of the Department of Energy and is consistent with several Office of Science strategic goals.

One of DOE's missions is to help ensure that the United States continues to have access to sources of affordable and environmentally friendly energy (Goal 1, "Science for Clean and Affordable Energy"). While physical sciences have been the backbone of energy research, new concepts in the biological sciences will shape our energy future by providing ways to use living organisms to produce energy and clean the environment (Goal 2: "Energy Impacts on People and the Biosphere").

Microbes have evolved for 3.8 billion years and have colonized almost every environment on Earth. In the process, they have developed an astonishingly diverse collection of capabilities that will help DOE meet its challenges in toxic waste cleanup, energy production, global climate change, and biotechnology (Goal 3: "Building Blocks from Atoms to Life"). All of this will take place in the context of outstanding science that has characterized DOE since its inception (Goal 5: "Scientific and Operational Excellence").

To embark upon this journey will require the development of new technologies, analytical tools, and modeling capabilities. In addition to working with academic, nonprofit, and industrial partners, DOE will take advantage of the scientific talents available in its national laboratories. These talents include high-throughput genomic DNA sequencing, microbial biochemistry and physiology, imaging, and structural biology. National user facilities such as synchrotrons will play important roles, as will capabilities in high-performance computing. Interdisciplinary collaborations among biologists, chemists, physicists, engineers, and computer experts will be critical to this effort.

In the MCP, scientists will begin to write a comprehensive "owner's manual" for a microbial cell. Microbial cells have internal organization and complex control systems that allow them to respond to their environment. They can work as miniature chemistry laboratories, making unique products and carrying out specialized functions. Ultimately, understanding the complex functioning of a single microbial cell will enable science to go far beyond just exploiting the beneficial capabilities of microbes to meet DOE's missions. The knowledge gained will apply to cells in all living things. Thus the MCP represents a first step in moving from cataloguing molecular parts to constructing an integrative view of life at the level of a whole organism—microbe, plant, or animal.

A glossary of terms used to describe all aspects of the Human Genome Project would include:

Acquired Genetic Mutation. See **Somatic Cell Genetic Mutation**.

Additive Genetic Effects. When the combined effects of alleles at different loci are equal to the sum of their individual effects. See **Anticipation**; and **Complex Trait**.

Adenine (A). A nitrogenous base, one member of the base pair AT (adenine-thymine). See **Base Pair** and **Nucleotide**.

Affected Relative Pair. Individuals related by blood, each of whom is affected with the same trait. Examples are affected sibling, cousin, and avuncular pairs. See **Avuncular Relationship**.

Aggregation Technique. A technique used in model organism studies in which embryos at the 8-cell stage of development are pushed together to yield a single embryo (used as an alternative to microinjection). See **Model Organisms**.

Allele. Alternative form of a genetic locus; a single allele for each locus is inherited from each parent (e.g., at a locus for eye color the allele might result in blue or brown eyes). See **Locus**; and **Gene Expression**.

Allogeneic. Variation in alleles among members of the same species.

Alternative Splicing. Different ways of combining a gene's exons to make variants of the complete protein.

Amino Acid. Any of a class of 20 molecules that are combined to form proteins in living things. The sequence of amino acids in a protein and hence protein function are determined by the genetic code. See also **Amino Acids**.

Amplification. An increase in the number of copies of a specific DNA fragment; can be in vivo or in vitro. See **Cloning** and **Polymerase Chain Reaction**.

Animal Model. See **Model Organisms**.

Annotation. Adding pertinent information such as gene coded for, amino acid sequence, or other commentary to the database entry of raw sequence of DNA bases. See **Bioinformatics**.

Anticipation. Each generation of offspring has increased severity of a genetic disorder; e.g., a grandchild may have earlier onset and more severe symptoms than the parent, who had earlier onset than the grandparent. See **Additive Genetic Effects**; and **Complex Trait**.

Antisense. Nucleic acid that has a sequence exactly opposite to an mRNA molecule made by the body; binds to the mRNA molecule to prevent a protein from being made. See **Transcription**.

Apoptosis. Programmed cell death, the body's normal method of disposing of damaged, unwanted, or unneeded cells. See **Cell**.

Arrayed Library. Individual primary recombinant clones (hosted in phage, cosmid, YAC, or other vector) that are placed in two-dimensional arrays in microtiter dishes. Each primary clone can be identified by the identity of the plate and the clone location (row and column) on that plate. Arrayed libraries of clones can be used for many applications, including screening for a specific gene or genomic region of interest. See **Library**; **Genomic Library**; and **Gene Chip Technology**.

Assembly. Putting sequenced fragments of DNA into their correct chromosomal positions.

Autoradiography. A technique that uses X-ray film to visualize radioactively labeled molecules or fragments of molecules; used in analyzing length and number of DNA fragments after they are separated by gel electrophoresis.

Autosomal Dominant. A gene on one of the non-sex chromosomes that is always expressed, even if only one copy is present. The chance of passing the gene to offspring is 50% for each pregnancy. See **Autosome**; **Dominant**; and **Gene**.

Autosome. A chromosome not involved in sex determination. The diploid human genome consists of a total of 46 chromosomes: 22 pairs of autosomes, and 1 pair of sex chromosomes (the X and Y chromosomes). See **Sex Chromosome**.

Avuncular Relationship. The genetic relationship between nieces and nephews and their aunts and uncles.

Backcross. A cross between an animal that is heterozygous for alleles obtained from two parental strains and a second animal from one of those parental strains. Also used to describe the breeding protocol of an outcross followed by a backcross. See **Model Organisms**.

Bacterial Artificial Chromosome (BAC). A vector used to clone DNA fragments (100- to 300-kb insert size; average, 150 kb) in *Escherichia coli* cells. Based on naturally occurring F-factor plasmid found in the bacterium E. coli. See **Cloning Vector**.

Bacteriophage. See **Phage**.

Base. One of the molecules that form DNA and RNA molecules. See **Nucleotide**; **Base Pair**; and **Base Sequence**.

Base Pair (bp). Two nitrogenous bases (adenine and thymine or guanine and cytosine) held together by weak bonds. Two strands of DNA are held together in the shape of a double helix by the bonds between base pairs.

Base Sequence. The order of nucleotide bases in a DNA molecule; determines structure of proteins encoded by that DNA.

Base Sequence Analysis. A method, sometimes automated, for determining the base sequence.

Behavioral Genetics. The study of genes that may influence behavior.

Bioinformatics. The science of managing and analyzing biological data using advanced computing techniques. Especially important in analyzing genomic research data. See **Informatics**.

Bioremediation. The use of biological organisms such as plants or microbes to aid in removing hazardous substances from an area.

Biotechnology. A set of biological techniques developed through basic research and now applied to research and product development. In particular, biotechnology refers to the use by industry of recombinant DNA, cell fusion, and new bioprocessing techniques.

Birth Defect. Any harmful trait, physical or biochemical, present at birth, whether a result of a genetic mutation or some other nongenetic factor. See **Congenital**; **Gene**; **Mutation**; and **Syndrome**.

BLAST. A computer program that identifies homologous (similar) genes in different organisms, such as human, fruit fly, or nematode.

Candidate Gene. A gene located in a chromosome region suspected of being involved in a disease. See **Positional Cloning**; and **Protein**.

Capillary Array. Gel-filled silica capillaries used to separate fragments for DNA sequencing. The small diameter of the capillaries permit the application of higher electric fields, providing high speed, high throughput separations that are significantly faster than traditional slab gels.

Carcinogen. Something that causes cancer to occur by causing changes in a cell's DNA. See **Mutagene**.

Carrier. An individual who possesses an unexpressed, recessive trait.

cDNA Library. A collection of DNA sequences that code for genes. The sequences are generated in the laboratory from mRNA sequences. See **Messenger RNA**.

Cell. The basic unit of any living organism that carries on the biochemical processes of life. See **Genome**; and **Nucleus**.

Centimorgan (cM). A unit of measure of recombination frequency. One centimorgan is equal to a 1% chance that a marker at one genetic locus will be separated from a marker at a second locus due to crossing over in a single generation. In human beings, one centimorgan is equivalent, on average, to one million base pairs. See **Megabase**.

Centromere. A specialized chromosome region to which spindle fibers attach during cell division.

Chimera (pl. chimaera). An organism that contains cells or tissues with a different genotype. These can be mutated cells of the host organism or cells from a different organism or species.

Chimeraplasty. An experimental targeted repair process in which a desirable sequence of DNA is combined with RNA to form a chimeraplast. These molecules bind selectively to the target DNA. Once bound, the chimeraplast activates a naturally occurring gene-correcting mechanism. Does not use viral or other conventional gene-delivery vectors. See Gene Therapy; and Cloning Vector.

Chloroplast Chromosome. Circular DNA found in the photosynthesizing organelle (chloroplast) of plants instead of the cell nucleus where most genetic material is located.

Chromomere. One of the serially aligned beads or granules of a eukaryotic chromosome, resulting from local coiling of a continuous DNA thread.

Chromosomal Deletion. The loss of part of a chromosome's DNA.

Chromosomal Inversion. Chromosome segments that have been turned 180 degrees. The gene sequence for the segment is reversed with respect to the rest of the chromosome.

Chromosome. The self-replicating genetic structure of cells containing the cellular DNA that bears in its nucleotide sequence the linear array of genes. In prokaryotes, chromosomal DNA is circular, and the entire genome is carried on one chromosome. Eukaryotic genomes consist of a number of chromosomes whose DNA is associated with different kinds of proteins.

Chromosome Painting. Attachment of certain fluorescent dyes to targeted parts of the chromosome. Used as a diagnositic for particular diseases, e.g. types of leukemia.

Chromosome Region p. A designation for the short arm of a chromosome.

Chromosome Region q. A designation for the long arm of a chromosome.

Clone. An exact copy made of biological material such as a DNA segment (e.g., a gene or other region), a whole cell, or a complete organism.

Clone Bank. See **Genomic Library**.

Cloning. Using specialized DNA technology to produce multiple, exact copies of a single gene or other segment of DNA to obtain enough material for further study. This process, used by researchers in the Human Genome Project, is referred to as cloning DNA. The resulting cloned (copied) collections of DNA molecules are called clone libraries. A second type of cloning exploits the natural process of cell division to make many copies of an entire cell. The genetic makeup of these cloned cells, called a cell line, is identical to the original cell. A third type of cloning produces complete, genetically identical animals such as the famous Scottish sheep, Dolly. See **Cloning Vector**.

Cloning Vector. DNA molecule originating from a virus, a plasmid, or the cell of a higher organism into which another DNA fragment of appropriate size can be integrated without loss of the vector's capacity for self-replication. Vectors introduce foreign DNA into host cells, where the DNA can be reproduced in large quantities. Examples are plasmids, cosmids, and yeast artificial chromosomes; vectors are often recombinant molecules containing DNA sequences from several sources.

Code. See **Genetic Code**.

Codominance. Situation in which two different alleles for a genetic trait are both expressed. See Autosomal Dominant; and Recessive Gene.

Codon. A triplet of nucleotides [three nucleic acid units (residues) in a row] that code for an amino acid (triplet code) or a termination signal. See **Genetic Code**; and **Messenger RNA (mRNA)**.

Coisogenic or Congenic. Nearly identical strains of an organism; they vary at only a single locus.

Comparative Genomics. The study of human genetics by comparisons with model organisms such as mice, the fruit fly, and the bacterium *E. coli.*

Complementary DNA (cDNA). A single-stranded DNA that is complementary to a strand of RNA. The DNA synthesized *in vitro* by an enzyme known as reverse transcriptase. It is a DNA copy of mRNA (messenger RNA) and this "rebukes" the Central Dogma. See **Messenger RNA (mRNA)**.

Complementary Sequence. Nucleic acid base sequence that can form a double-stranded structure with another DNA fragment by following base-pairing rules (A pairs with T and C with G). The complementary sequence to GTAC for example, is CATG.

Complex Trait. Trait that has a genetic component that does not follow strict Mendelian inheritance. May involve the interaction of two or more genes or gene-environment interactions. See **Mendelian Inheritance**; and **Additive Genetic Effects**.

Computational Biology. See **Bioinformatics**.

Confidentiality. In genetics, the expectation that genetic material and the information gained from testing that material will not be available without the donor's consent.

Congenital. Any trait present at birth, whether the result of a genetic or nongenetic factor. See **Birth Defect**.

Conserved Sequence. A base sequence in a DNA molecule (or an amino acid sequence in a protein) that has remained essentially unchanged throughout evolution.

Constitutive Ablation. Gene expression that results in cell death.

Contig. Group of cloned (copied) pieces of DNA representing overlapping regions of a particular chromosome.

Contig Map. A map depicting the relative order of a linked library of overlapping clones representing a complete chromosomal segment.

Cosmid. Artificially constructed cloning vector containing the cos gene of phage lambda. Cosmids can be packaged in lambda phage particles for infection into *E. coli*; this permits cloning of larger DNA fragments (up to 45kb) than can be introduced into bacterial hosts in plasmid vectors.

Crossing Over. The breaking during meiosis of one maternal and one paternal chromosome, the exchange of corresponding sections of DNA, and the rejoining of the chromosomes. This process can result in an exchange of alleles between chromosomes. See **Recombination**.

Cytogenetics. The study of the physical appearance of chromosomes. See **Karyotype**.

Cytological Band. An area of the chromosome that stains differently from areas around it. See **Cytological Map**.

Cytological Map. A type of chromosome map whereby genes are located on the basis of cytological findings obtained with the aid of chromosome mutations.

Cytoplasmic Trait. A genetic characteristic in which the genes are found outside the nucleus, in chloroplasts or mitochondria. Results in offspring inheriting genetic material from only one parent.

Cytosine (C). A nitrogenous base, one member of the base pair GC (guanine and cytosine) in DNA. See **Base Pair**; and **Nucleotide**.

Data Warehouse. A collection of databases, data tables, and mechanisms to access the data on a single subject.

Deletion. A loss of part of the DNA from a chromosome; can lead to a disease or abnormality. See **Chromosome**; and **Mutation**.

Deletion Map. A description of a specific chromosome that uses defined mutations—specific deleted areas in the genome—as 'biochemical signposts,' or markers for specific areas.

Deoxyribonucleotide. See **Nucleotide**.

Deoxyribose. A type of sugar that is one component of DNA (deoxyribonucleic acid).

Diploid. A full set of genetic material consisting of paired chromosomes, one from each parental set. Most animal cells except the gametes have a diploid set of chromosomes. The diploid human genome has 46 chromosomes. See **Haploid**.

Directed Evolution. A laboratory process used on isolated molecules or microbes to cause mutations and identify subsequent adaptations to novel environments.

Directed Mutagenesis. Alteration of DNA at a specific site and its reinsertion into an organism to study any effects of the change.

Directed Sequencing. Successively sequencing DNA from adjacent stretches of chromosome.

Disease-associated Genes. Alleles carrying particular DNA sequences associated with the presence of disease.

DNA (deoxyribonucleic acid). The molecule that encodes genetic information. DNA is a double-stranded molecule held together by weak bonds between pairs of nucleotides. The four nucleotides in DNA contain the bases adenine (A), guanine (G), cytosine (C), and thymine (T). In nature, base pairs form only between A and T and between G and C; thus the base sequence of each single strand can be deduced from that of its partner.

DNA Bank. A service that stores DNA extracted from blood samples or other human tissue.

DNA Probe. Also called gene probe or genetic probe. Short, specific (complementary to desired gene) artificially produced segments of DNA used to combine with and detect the presence of specific genes (or shorter DNA segments) within a chromosome. If a DNA probe of known composition and length is mingled with pieces of DNA (genes) from a chromosome, the probe will cling to its exact counterpart in the "chromosomal DNA pieces" (genes), forming a stable double-stranded hybrid. The presence of this (now) "labeled" probe is detected visually or with the aid of another detection instrument. See **Probe**.

DNA Profiling. Invented in 1985 by Alec Jeffreys, it is a technique used by forensic (i.e., crime-solving) chemists to match biological evidence (e.g., a blood stain) from a crime scene to the person (e.g., the assailant) involved in that particular crime.

DNA Repair Genes. Genes encoding proteins that correct errors in DNA sequencing.

DNA Replication. The use of existing DNA as a template for the synthesis of new.

DNA Strands. In humans and other eukaryotes, replication occurs in the cell nucleus.

DNA Sequence. The relative order of base pairs, whether in a DNA fragment, gene, chromosome, or an entire genome. See **Base Sequence Analysis**.

Domain. A discrete portion of a protein with its own function. The combination of domains in a single protein determines its overall function.

Dominant Allele. Discovered by Gregor Mendal in the 1860's, it is a gene that produces the same phenotype when it is heterozygous as it does when t is homozygous (i.e., trait, or protein, is expressed even if only one copy of the gene is present in the genome). See **Gene**; and **Genome**.

Double Helix. The twisted-ladder shape that two linear strands of DNA assume when complementary nucleotides on opposing strands bond together. This structure was first put forward by Watson and Crick in 1953.

Draft Sequence. The sequence generated by the HGP as of June 2000 that, while incomplete, offers a virtual road map to an estimated 95% of all human genes. Draft sequence data are mostly in the form of 10,000 base pair-sized fragments whose approximate chromosomal locations are known. See **Sequencing**; **Finished DNA Sequence**; and **Working Draft DNA Sequence**.

Electrophoresis. A method of separating large molecules (such as DNA fragments or proteins) from a mixture of similar molecules. An electric current is passed through a medium containing the mixture, and each kind of molecule travels through the medium at a different rate, depending on its electrical charge and size. Agarose and acrylamide gels are the media commonly used for electrophoresis of proteins and nucleic acids.

Electroporation. A process using high-voltage current to make cell membranes permeable to allow the introduction of new DNA; commonly used in recombinant DNA technology. See **Transfection**.

Embryonic Stem (ES) Cells. An embryonic cell that can replicate indefinitely, transform into other types of cells, and serve as a continuous source of new cells.

Endonuclease. A class of enzymes capable of hydrolyzing (breaking) the interior phosphodiester bonds of DNA or RNA chains. As opposed to cleavage (by exonucleases) at the terminal bonds (ends) of a chain. See **Exonuclease**; and **Restriction Enzyme**.

Enzyme. An organic protein-based catalyst that is not itself used up in the reaction. It is naturally produced by living cells to catalyze biochemical reactions. Each enzyme is highly specific with regard to the type of chemical reaction that it catalyzes, and to the substances called *substrates* upon which it acts.

Epistasis. One gene interferes with or prevents the expression of another gene located at a different locus.

Escherichia coli. Common bacterium that has been studied intensively by geneticists because of its small genome size, normal lack of pathogenicity, and ease of growth in the laboratory.

Eugenics. First formulated by Francis Galton, who was a contemporary of Gregor Mendel in the 19th century. Eugenics is the study of improving a species by artificial selection and usually refers to the selective breeding of humans.

Eukaryote. Cell or organism with membrane-bound, structurally discrete nucleus and other well-developed subcellular compartments. Eukaryotes include all organisms except viruses, bacteria, and bluegreen algae. See **Prokaryote**; and **Chromosome**.

Evolutionarily Conserved. See **Conserved Sequence**.

Exogenous DNA. DNA originating outside an organism that has been introduced into the organism.

Exon. The protein-coding DNA sequence of a gene. See **Intron**.

Exonuclease. An enzyme that cleaves nucleotides sequentially from free ends of a linear nucleic acid substrate.

Expressed Gene. See **Gene Expression**.

Expressed Sequence Tag (EST). A short strand of DNA that is a part of a cDNA molecule and can act as identifier of a gene. Used in locating and mapping genes. See **cDNA**; and **Sequence Tagged Site**.

Filial Generation (F1, F2). Each generation of offspring in a breeding program, designated F1, F2, etc.

Fingerprinting. In genetics, the identification of multiple specific alleles on a person's DNA to produce a unique identifier for that person. See **Forensics**.

Finished DNA Sequence. High-quality, low error, gap-free DNA sequence of the human genome. Achieving this ultimate 2003 HGP goal requires additional sequencing to close gaps, reduce ambiguities, and allow for only a single error every 10,000 bases, the agreed-upon standard for HGP finished sequence. See **Sequencing**; and **Draft Sequence**.

Flow Cytometry. Analysis of biological material by detection of the light-absorbing or fluorescing properties of cells or subcellular fractions (i.e., chromosomes) passing in a narrow stream through a laser beam. An absorbance or fluorescence profile of the sample is produced. Automated sorting devices, used to fractionate samples, sort successive droplets of the analyzed stream into different fractions depending on the fluorescence emitted by each droplet.

Flow Karyotyping. Use of flow cytometry to analyze and separate chromosomes according to their DNA content.

Fluorescence in Situ Hybridization (FISH). A physical mapping approach that uses fluorescein tags to detect hybridization of probes with metaphase chromosomes and with the less-condensed somatic interphase chromatin.

Forensics. The use of DNA for identification. Some examples of DNA use are to establish paternity in child support cases; establish the presence of a suspect at a crime scene, and identify accident victims.

Fraternal Twin. Siblings born at the same time as the result of fertilization of two ova by two sperm. They share the same genetic relationship to each other as any other siblings. See **Identical Twin**.

Full Gene Sequence. The complete order of bases in a gene. This order determines which protein a gene will produce.

Functional Genomics. The study of genes, their resulting proteins, and the role played by the proteins the body's biochemical processes.

Gamete. Mature male or female reproductive cell (sperm or ovum) with a haploid set of chromosomes (23 for humans).

GC-rich Area. Many DNA sequences carry long stretches of repeated G and C, which often indicate a gene-rich region.

Gel Electrophoresis. See **Electrophoresis**.

Gene. The fundamental physical and functional unit of heredity. A gene is an ordered sequence of nucleotides located in a particular position on a particular chromosome that encodes a specific functional product (i.e., a protein or RNA molecule). See **Gene Expression**.

Gene Amplification. The copying of segments (e.g., genes) within the DNA or RNA molecule. This can be done by man (e.g., polymerase chain reaction), can be caused by certain chemical carcinogens (e.g., phorbol ester), or occur naturally (e.g., in procaryotes and certain lower eucaryotes). The five primary techniques that are used by man to perform gene amplification are: 1) Polymerase Chain Reaction (PCR), 2) Ligase Chain Reaction (LCR), 3) Self-sustained Sequence Replication (SSR), 4) Q-beta Replicase Technique, and 5) Strand Displacement Amplification (SDA).

Gene Chip Technology. Development of cDNA microarrays from a large number of genes. Used to monitor and measure changes in gene expression for each gene represented on the chip.

Gene Expression. The process by which a gene's coded information is converted into the structures present and operating in the cell. Expressed genes include those that are transcribed into mRNA and then translated into protein and those that are transcribed into RNA but not translated into protein (e.g., transfer and ribosomal RNAs).

Gene Family. Group of closely related genes that make similar products.

Gene Library. See **Genomic Library**.

Gene Mapping. Determination of the relative positions of genes on a DNA molecule (chromosome or plasmid) and of the distance, in linkage units or physical units, between them.

Gene Pool. All the variations of genes in a species. See **Allele**; **Gene**; and **Polymorphism**.

Gene Prediction. Predictions of possible genes made by a computer program based on how well a stretch of DNA sequence matches known gene sequences.

Gene Product. The biochemical material, either RNA or protein, resulting from expression of a gene. The amount of gene product is used to measure how active a gene is; abnormal amounts can be correlated with disease-causing alleles.

Gene Therapy. An experimental procedure aimed at replacing, manipulating, or supplementing nonfunctional or misfunctioning genes with healthy genes. See **Gene**; **Inherit**; **Somatic Cell Gene Therapy**; and **Germ Line Gene Therapy**.

Gene Transfer. Incorporation of new DNA into and organism's cells, usually by a vector such as a modified virus. Used in gene therapy. See **Mutation**; **Gene Therapy**; and **Vector**.

Genetic Code. The set of triplet code words in DNA coding for all of the amino acids. There are more than 20 different amino acids and only four bases (adenine, thymine, cytosine, and guanine). The mRNA code is a triplet code, that is, each successive "frame" of three nucleotides (sometimes called a codon) of the mRNA corresponds to one amino acid of the protein. This rule of correspondence in the genetic code.

Genetic Counseling. Provides patients and their families with education and information about genetic-related conditions and helps them make informed decisions.

Genetic Discrimination. Prejudice against those who have or are likely to develop an inherited disorder.

Genetic Engineering. The selective, deliberate alteration of genes (genetic material) by man. This term as come to have a very broad meaning including the manipulation and alteration of the genetic material (constitution) of an organism in such a way as to allow it to produce endogenous proteins with properties different from those of the normal, or to produce entirely different (foreign) proteins altogether. Other words applicable to the same process are gene splicing, gene manipulation, or recombinant DNA technology.

Genetic Map. A diagram showing the relative sequence and position of the specific genes along a chromosome molecule.

Genetic Marker. A gene or other identifiable portion of DNA whose inheritance can be followed. See **Chromosome**; **DNA**; **Gene**; and **Inherit**.

Genetic Mosaic. An organism in which different cells contain different genetic sequence. This can be the result of a mutation during development or fusion of embryos at an early developmental stage.

Genetic Polymorphism. Difference in DNA sequence among individuals, groups, or populations (e.g., genes for blue eyes versus brown eyes).

Genetic Predisposition. Susceptibility to a genetic disease. May or may not result in actual development of the disease.

Genetic Screening. Testing a group of people to identify individuals at high risk of having or passing on a specific genetic disorder.

Genetic Testing. Analyzing an individual's genetic material to determine predisposition to a particular health condition or to confirm a diagnosis of genetic disease.

Genetics. The branch of biology concerned with heredity, it was literally invented by Gregor Mendel in the 19th century. It is the study of the manner in which genes operate and are transmitted from parents to offspring. It involves the study of the mechanism of gene action — the manner in which the genetic material (DNA) affects physiological reactions within the cell.

Genome. The entire hereditary material (which was proven by Oswald Avery in 1944 to be DNA) in a cell. All the genetic material in the chromosomes of a particular organism; its size is generally given as its total number of base pairs.

Genome Project. Research and technology-development effort aimed at mapping and sequencing the genome of human beings and certain model organisms. See **Human Genome Initiative**.

Genomic Library. A collection of clones made from a set of randomly generated overlapping DNA fragments that represent the entire genome of an organism. See **Library**; and **Arrayed Library**.

Genomic Sequence. See **DNA**.

Genomics. The scientific study of genes and their role in an organism's structure, growth, health, disease (an/or resistance to disease, etc.).

Genotype. The total genetic, or hereditary, constitution of an organism, as distinguished from its physical appearance (its phenotype).

Germ Cell. Sperm and egg cells and their precursors. Germ cells are haploid and have only one set of chromosomes (23 in all), while all other cells have two copies (46 in all).

Germ Line. The continuation of a set of genetic information from one generation to the next. See **Inherit**.

Germ Line Gene Therapy. An experimental process of inserting genes into germ cells or fertilized eggs to cause a genetic change that can be passed on to offspring. May be used to alleviate effects associated with a genetic disease. See **Genomics**; and **Somatic Cell Gene Therapy**.

Guanine (G). A nitrogenous base, one member of the base pair GC (guanine and cytosine) in DNA. See **Base Pair**; and **Nucleotide**.

Gyandromorph. Organisms that have both male and female cells and therefore express both male and female characteristics.

Haploid. A single set of chromosomes (half the full set of genetic material) present in the egg and sperm cells of animals and in the egg and pollen cells of plants. Human beings have 23 chromosomes in their reproductive cells. See **Diploid**.

Haplotype. A way of denoting the collective genotype of a number of closely linked loci on a chromosome.

Hemizygous. Having only one copy of a particular gene. For example, in humans, males are hemizygous for genes found on the Y chromosome.

Hereditary Cancer. Cancer that occurs due to the inheritance of an altered gene within a family. See **Sporadic Cancer**.

Heterozygosity. The presence of different alleles at one or more loci on homologous chromosomes.

Highly Conserved Sequence. DNA sequence that is very similar across several different types of organisms. See **Gene**; and **Mutation**.

High-throughput Sequencing. A fast method of determining the order of bases in DNA. See **Sequencing**.

Homeobox. A short stretch of nucleotides whose base sequence is virtually identical in all the genes that contain it. Homeoboxes have been found in many organisms from fruit flies to human beings. In the fruit fly, a homeobox appears to determine when particular groups of genes are expressed during development.

Homolog. A member of a chromosome pair in diploid organisms or a gene that has the same origin and functions in two or more species.

Homologous Chromosome. Chromosome containing the same linear gene sequences as another, each derived from one parent.

Homologous Recombination. Swapping of DNA fragments between paired chromosomes.

Homology. Similarity in DNA or protein sequences between individuals of the same species or among different species.

Homozygote. An organism in which the corresponding genes (alleles) on the two genomes are identical. An organism which possesses an identical pair of alleles in regard to a given (genetic) characteristic.

Human Artificial Chromosome (HAC). Chromosomes that have been synthesized (made) from chemicals that are identical to chromosomes within human cells. A vector used to hold large DNA fragments. See **Chromosome**; and **DNA**.

Human Genome Initiative. Collective name for several projects begun in 1986 by DOE to create an ordered set of DNA segments from known chromosomal locations, develop new computational methods for analyzing genetic map and DNA sequence data, and develop new techniques and instruments for detecting and analyzing DNA. This DOE initiative is now known as the Human Genome Program. The joint national effort, led by DOE and NIH, is known as the Human Genome Project.

Human Genome Project (HGP). Formerly titled Human Genome Initiative.

Hybrid. The offspring of genetically different parents.

Hybridization. The process of joining two complementary strands of DNA or one each of DNA and RNA to form a double-stranded molecule.

Identical Twin. Twins produced by the division of a single zygote; both have identical genotypes. See **Fraternal Twin**.

Immunotherapy. Using the immune system to treat disease, for example, in the development of vaccines. May also refer to the therapy of diseases caused by the immune system.

Imprinting. A phenomenon in which the disease phenotype depends on which parent passed on the disease gene. For instance, both Prader-Willi and Angelman syndromes are inherited when the same part of chromosome 15 is missing. When the father's complement of 15 is missing, the child has Prader-Willi, but when the mother's complement of 15 is missing, the child has Angelman syndrome.

In Situ Hybridization. Use of a DNA or RNA probe to detect the presence of the complementary DNA sequence in cloned bacterial or cultured eukaryotic cells.

In vitro. Studies performed outside a living organism such as in a laboratory.

In vivo. The testing of a substance or experimentation in (using) a living, whole organism.

Independent Assortment. During meiosis each of the two copies of a gene is distributed to the germ cells independently of the distribution of other genes. See **Linkage**.

Informatics. See **Bioinformatics**.

Inherit. In genetics, to receive genetic material from parents through biological processes.

Insertion. A chromosome abnormality in which a piece of DNA is incorporated into a gene and thereby disrupts the gene's normal function. See **Chromosome**; **DNA**; **Gene**; and **Mutation**.

Interference. One crossover event inhibits the chances of another crossover event. Also known as positive interference. Negative interference increases the chance of a second crossover. See **Crossing Over**.

Interphase. The period in the cell cycle when DNA is replicated in the nucleus; followed by mitosis.

Intron. DNA sequence that interrupts the protein-coding sequence of a gene; an intron is transcribed into RNA but is cut out of the message before it is translated into protein. See **Exon**.

Isoenzyme. An enzyme performing the same function as another enzyme but having a different set of amino acids. The two enzymes may function at different speeds.

Junk DNA. Stretches of DNA that do not code for genes; most of the genome consists of so-called *junk DNA* which may have regulatory and other functions. Also called *non-coding DNA*.

Karyotype. A photomicrograph of an individual's chromosomes arranged in a standard format showing the number, size, and shape of each chromosome type; used in low-resolution physical mapping to correlate gross chromosomal abnormalities with the characteristics of specific diseases.

Kilobase (kb). Unit of length for DNA fragments equal to 1000 nucleotides.

Knockout. Deactivation of specific genes; used in laboratory organisms to study gene function. See **Gene**; **Locus**; and **Model Organisms**.

Library. An unordered collection of clones (i.e., cloned DNA from a particular organism) whose relationship to each other can be established by physical mapping. See **Genomic Library**; and **Arrayed Library**.

Linkage. A phenomenon discovered by Thomas Hunt Morgan in the early 1900's via his experiments with fruit flies. The proximity of two or more markers (e.g., genes, RFLP markers) on a chromosome. The closer the markers, the lower the probability that they will be separated during DNA repair or replication processes (binary fission in prokaryotes, mitosis or meiosis in eukaryotes), and hence the greater the probability that they will be inherited together.

Linkage Disequilibrium. Where alleles occur together more often than can be accounted for by chance. Indicates that the two alleles are physically close on the DNA strand. See **Mendelian Inheritance**.

Linkage Map. A map of the relative positions of genetic loci on a chromosome, determined on the basis of how often the loci are inherited together. Distance is measured in centimorgans (cM).

Localize. Determination of the original position (locus) of a gene or other marker on a chromosome.

Locus (pl. loci). The position on a chromosome of a gene or other chromosome marker; also, the DNA at that position. The use of locus is sometimes restricted to mean expressed DNA regions. See **Gene Expression**.

Long-Range Restriction Mapping. Restriction enzymes are proteins that cut DNA at precise locations. Restriction maps depict the chromosomal positions of restriction-enzyme cutting sites. These are used as biochemical "signposts," or markers of specific areas along the chromosomes. The map will detail the positions where the DNA molecule is cut by particular restriction enzymes.

Macrorestriction Map. Map depicting the order of and distance between sites at which restriction enzymes cleave chromosomes.

Mapping Population. The group of related organisms used in constructing a genetic map.

Megabase (Mb). Unit of length for DNA fragments equal to 1 million nucleotides and roughly equal to 1 cM. See **Centimorgan**.

Meiosis. The process of two consecutive cell divisions in the diploid progenitors of sex cells. Meiosis results in four rather than two daughter cells, each with a haploid set of chromosomes. See **Mitosis**.

Mendelian Inheritance. One method in which genetic traits are passed from parents to offspring. Named for Gregor Mendel, who first studied and recognized the existence of genes and this method of inheritance. See **Autosomal Dominant**; **Recessive Gene**; and **Sex-linked**.

Messenger RNA (mRNA). Messenger ribonucleic acid. The intermediary molecule between DNA and ribosomes (in a cell) which synthesize (i.e., make) those proteins coded for by the cell's DNA. Upon receiving the "message" encoded in the DNA, the messenger RNA passes through the ribosomes like a reel of punched paper passes through an old player piano giving the ribosomes the specifications for making the coded-for proteins. See **Genetic Code**.

Metaphase. A stage in mitosis or meiosis during which the chromosomes are aligned along the equatorial plane of the cell.

Microarray. Sets of miniaturized chemical reaction areas that may also be used to test DNA fragments, antibodies, or proteins.

Microbial Genetics. The study of genes and gene function in bacteria, archaea, and other microorganisms. Often used in research in the fields of bioremediation, alternative energy, and disease prevention. See **Model Organisms**; and **Bioremediation**.

Microinjection. A technique for introducing a solution of DNA into a cell using a fine microcapillary pipet.

Micronuclei. Chromosome fragments that are not incorporated into the nucleus at cell division.

Mitochondrial DNA. The genetic material found in mitochondria, the organelles that generate energy for the cell. Not inherited in the same fashion as nucleic DNA. See **Cell**; **DNA**; **Genome**; and **Nucleus**.

Mitosis. The process of nuclear division in cells that produces daughter cells that are genetically identical to each other and to the parent cell. See **Meiosis**.

Model Organisms. A laboratory animal or other organism useful for research.

Modeling. The use of statistical analysis, computer analysis, or model organisms to predict outcomes of research.

Molecular Biology. A term coined by Vannevar Bush during the 1940's that eventually came to mean the study and manipulation of molecules that constitute, or interact with, cells. Molecular biology as a distinct scientific discipline originated largely as a result of a decision to provide "support for the application of new physical and chemical techniques to biology" during the 1930's by Warren Weaver, director of the biology (funding) program at America's Rockefeller Foundation.

Molecular Farming. The development of transgenic animals to produce human proteins for medical use.

Molecular Genetics. The science dealing with the study of the nature and biochemistry of the genetic material.

Molecular Medicine. The treatment of injury or disease at the molecular level. Examples include the use of DNA-based diagnostic tests or medicine derived from DNA sequence information.

Monogenic Disorder. A disorder caused by mutation of a single gene. See **Mutation**; and **Polygenic Disorder**.

Monosomy. Possessing only one copy of a particular chromosome instead of the normal two copies. See **Cell**; **Chromosome**; **Gene Expression**; and **Trisomy**.

Morbid Map. A diagram showing the chromosomal location of genes associated with disease.

Multifactorial or Multigenic Disorder. See **Polygenic Disorder**.

Multiplexing. A laboratory approach that performs multiple sets of reactions in parallel (simultaneously); greatly increasing speed and throughput.

Murine. Organism in the genus Mus. A rat or mouse.

Mutagen. An agent that causes a permanent genetic change in a cell. Does not include changes occurring during normal genetic recombination.

Mutagenicity. The capacity of a chemical or physical agent to cause permanent genetic alterations. See **Somatic Cell Genetic Mutation**.

Mutation. From the Latin term *mutare*, meaning "to change." Any change that alters the sequence of the nucleotide bases in the genetic material (DNA) of an organism or cell; with alteration occurring either by displacement, addition, deletion, cross-linking, or other destruction. The alteration to the DNA sequence would alter its meaning, that is, its ability to produce the normal amount or normal kind of protein, so the (organism or cell) is itself altered. Such an altered organism is called a *mutant*. See **Polymorphism**.

Nitrogenous Base. A nitrogen-containing molecule having the chemical properties of a base. DNA contains the nitrogenous bases adenine (A), guanine (G), cytosine (C), and thymine (T). See **DNA**.

Northern Blot. A gel-based laboratory procedure that locates mRNA sequences on a gel that are complementary to a piece of DNA used as a probe. See **DNA**; and **Library**.

Nuclear Transfer. A laboratory procedure in which a cell's nucleus is removed and placed into an oocyte with its own nucleus removed so the genetic information from the donor nucleus controls the resulting cell. Such cells can be induced to form embryos. This process was used to create the cloned sheep "Dolly". See **Cloning**.

Nucleic Acid. A nucleotide polymer. A large, chain-like molecule containing phosphate groups, sugar groups, and purine and pyrimidine bases; two types are ribonucleic acid (RNA) and dexyribonucleic acid (DNA). The bases involved are adenine, guanine, cytosine, and thymine (uracil in RNA). See **DNA**.

Nucleolar Organizing Region. A part of the chromosome containing rRNA genes.

Nucleotide. A subunit of DNA or RNA consisting of a nitrogenous base (adenine, guanine, thymine, or cytosine in DNA; adenine, guanine, uracil, or cytosine in RNA), a phosphate molecule, and a sugar molecule (deoxyribose in DNA and ribose in RNA). Thousands of nucleotides are linked to form a DNA or RNA molecule. See **DNA**; **Base Pair**; and **RNA**.

Nucleus. The usually spherical body with each living cell that contains its hereditary biological material (e.g., metabolism, growth, and reproduction). The nucleus is a highly differentiated, relatively large organelle lying in the cytoplasm of the cell. The nucleus is surrounded by a (nuclear) membrane, which is quite similar to the plasma (cell) membrane; except the nuclear membrane contains holes or pores.

Oligogenic. A phenotypic trait produced by two or more genes working together. See **Polygenic Disorder**.

Oligonucleotide. A molecule usually composed of 25 or fewer nucleotides; used as a DNA synthesis primer. Oligonucleotides (also called, simply "oligos") are used as synthetic (i.e., man-made) genes, DNA probes, and in site-directed mutagenesis. See Nucleotide.

Oncogene. A gene, one or more forms of, which is associated with cancer. Many oncogenes are involved, directly or indirectly, in controlling the rate of cell growth.

Open Reading Frame (ORF). The sequence of DNA or RNA located between the start-code sequence (initiation codon) and the stop-code sequence (termination codon).

Operon. A gene unit consisting of one or more genes that specify a polypeptide and an operator unit that regulates the structural gene, that is, the production of messenger RNA (mRNA) and hence, ultimately, of a number of proteins. Generally an operon is defined as a group of functionally related structural genes mapping (that is, being) close to each other.

Overlapping Clones. See: **Genomic Library**.

P1-derived Artificial Chromosome (PAC). One type of vector used to *clone DNA fragments (100- to 300-kb insert size; average, 150 kb) in Escherichia* coli cells. Based on bacteriophage (a virus) P1 genome. See **Cloning Vector**.

Patent. In genetics, conferring the right or title to genes, gene variations, or identifiable portions of sequenced genetic material to an individual or organization. See Gene.

Pedigree. A family tree diagram that shows how a particular genetic trait or disease has been inherited. See **Inherit**.

Penetrance. The probability of a gene or genetic trait being expressed. "Complete" penetrance means the gene or genes for a trait are expressed in all the population who have the genes. "Incomplete" penetrance means the genetic trait is expressed in only part of the population. The percent penetrance also may change with the age range of the population.

Peptide. Two or more amino acids joined by a bond called a "peptide bond." An oligomer component of a polypeptide. A dipeptide, for example, consists of two (di) amino acids joined together by a peptide bond or linkage. By analogy, this structure would correspond to two joined links of a chain. See **Polypeptide**.

Phage. A virus for which the natural host is a bacterial cell. A virus that attacks bacteria is known as a bacteriophage. Bacteriophages are frequently used as vectors for carrying (foreign) DNA into cells by genetic engineers.

Pharmacogenomics. The study of the interaction of an individual's genetic makeup and response to a drug.

Phenocopy. A trait not caused by inheritance of a gene but appears to be identical to a genetic trait.

Phenotype. The outward appearance (structure) or other characteristics of an organism or the presence of a disease that may or may not be genetic. See **Genotype**.

Physical Map. A map of the locations of identifiable landmarks on DNA (e.g., restriction-enzyme cutting sites, genes), regardless of inheritance. Distance is measured in base pairs. For the human genome, the lowest-resolution physical map is the banding patterns on the 24 different chromosomes; the highest-resolution map is the complete nucleotide sequence of the chromosomes.

Plasmid. Autonomously replicating extra-chromosomal circular DNA molecules, distinct from the normal bacterial genome and nonessential for cell survival under nonselective conditions. Some plasmids are capable of integrating into the host genome. A number of artificially constructed plasmids are used as cloning vectors.

Pleiotropy. One gene that causes many different physical traits such as multiple disease symptoms.

Pluripotency. The potential of a cell to develop into more than one type of mature cell, depending on environment.

Polygenic Disorder. Genetic disorder resulting from the combined action of alleles of more than one gene (e.g., heart disease, diabetes, and some cancers). Although such disorders are inherited, they depend on the simultaneous presence of several alleles; thus the hereditary patterns usually are more complex than those of single-gene disorders. See **Single-gene Disorder**.

Polymerase Chain Reaction (PCR). Developed in 1984 and 1985 by Kary B. Mullis, Randall K. Saiki, Stephen J. Scharf, Fred A. Faloona, Glenn Horn, Henry A. Erlich, and Norman Arnheim. A method for amplifying a DNA base sequence using a heat-stable polymerase and two 20-base primers, one complementary to the (+) strand at one end of the sequence to be amplified and one complementary to the (−) strand at the other end. Because the newly synthesized DNA strands can subsequently serve as additional templates for the same primer sequences, successive rounds of primer annealing, strand elongation, and dissociation produce rapid and highly specific amplification of the desired sequence. PCR also can be used to detect the existence of the defined sequence in a DNA sample.

Polymerase, DNA or RNA. Enzyme that catalyzes the synthesis of nucleic acids on pre-existing nucleic acid templates, assembling RNA from ribonucleotides or DNA from deoxyribonucleotides.

Polymorphism. Difference in DNA sequence among individuals that may underlie differences in health. Genetic variations occurring in more than 1% of a population would be considered useful polymorphisms for genetic linkage analysis. See **Mutation**.

Polypeptide. A protein or part of a protein made of a chain of amino acids joined by a peptide bond. Synonymous with protein. Via the synthesis (of this "chain") performed by ribosomes, each polypeptide (protein) in nature is the ultimate expression product of a gene.

Population Genetics. The study of variation in genes among a group of individuals.

Positional Cloning. A technique used to identify genes, usually those that are associated with diseases, based on their location on a chromosome.

Premature Chromosome Condensation (PCC). A method of studying chromosomes in the interphase stage of the cell cycle.

Primer. Short preexisting polynucleotide chain to which new deoxyribonucleotides can be added by DNA polymerase.

Probe. Single-stranded DNA or RNA molecules of specific base sequence, labeled either radioactively or immunologically, that are used to detect the complementary base sequence by hybridization.

Prokaryote. Cell or organism lacking a membrane-bound, structurally discrete nucleus and other subcellular compartments. Bacteria are examples of prokaryotes. See **Chromosome**; and **Eukaryote**.

Promoter. A DNA site to which RNA polymerase will bind and initiate transcription.

Pronucleus. The nucleus of a sperm or egg prior to fertilization. See **Nucleus**; and **Transgenic**.

Protein. From the Greek word *proteios*, which means "the first" or "the most important." A large molecule composed of one or more chains of amino acids in a specific order; the order is determined by the base sequence of nucleotides in the gene that codes for the protein. Proteins are required for the structure, function, and regulation of the body's cells, tissues, and organs; and each protein has unique functions. Examples are hormones, enzymes, and antibodies.

Proteome. Proteins expressed by a cell or organ at a particular time and under specific conditions.

Proteomics. The study of the full set of proteins encoded by a genome.

Pseudogene. A sequence of DNA similar to a gene but nonfunctional; probably the remnant of a once-functional gene that accumulated mutations.

Purine. A nitrogen-containing, double-ring, basic compound that occurs in nucleic acids. The purines in DNA and RNA are adenine and guanine. See **Base Pair**.

Pyrimidine. A nitrogen-containing, single-ring, basic compound that occurs in nucleic acids. The pyrimidines in DNA are cytosine and thymine, in RNA, cytosine and uracil. See Base Pair.

Radiation Hybrid. A hybrid cell containing small fragments of irradiated human chromosomes. Maps of irradiation sites on chromosomes for the human, rat, mouse, and other genomes provide important markers, allowing the construction of very precise STS maps indispensable to studying multifactorial diseases. See **Sequence Tagged Site**.

Recessive Allele. Discovered by Gregor Mendel in the 1960's, this refers to an allelic gene whose existence is obscured in the phenotype of heterozygote by the dominant allele.

Recessive Gene. A gene which will be expressed only if there are 2 identical copies or, for a male, if one copy is present on the X chromosome.

Reciprocal Translocation. When a pair of chromosomes exchange exactly the same length and area of DNA. Results in a shuffling of genes.

Recombinant Clone. Clone containing recombinant DNA molecules.

Recombinant DNA Molecules. A combination of DNA molecules of different origin that are joined using recombinant DNA technologies.

Recombinant DNA Technology. Procedure used to join together DNA segments in a cell-free system (an environment outside a cell or organism). Under appropriate conditions, a recombinant DNA molecule can enter a cell and replicate there, either autonomously or after it has become integrated into a cellular chromosome.

Recombination. The process by which progeny derive a combination of genes different from that of either parent. In higher organisms, this can occur by crossing over. See **Crossing Over**; and **Mutation**.

Regulatory Region or Sequence. A DNA base sequence that controls gene expression.

Repetitive DNA. Sequences of varying lengths that occur in multiple copies in the genome; it represents much of the human genome.

Replication (of DNA). Reproduction of a DNA molecule (inside a cell). This process can be viewed as occurring in stages, in which the first stage consists of an enzyme "unwinding" the double helix of the DNA molecule at a replication origin, forming a replication fork. At the replication fork, the two separated (DNA) strands serve as templates for new DNA synthesis. The new DNA synthesis is accomplished on each strand via enzymes known as DNA polymerase, which travel along each (single) strand making a second complementary strand by catalyzing the addition of DNA bases (to the new, growing strands). The end result is two new double helices (DNA molecules), each of which has one chain from the original DNA molecule and one chain that was newly synthesized by the DNA polymerase enzymes.

Resolution. Degree of molecular detail on a physical map of DNA, ranging from low to high.

Restriction Enzyme, Endonuclease. A protein that recognizes specific, short nucleotide sequences and cuts DNA at those sites. Bacteria contain over 400 such enzymes that recognize and cut more than 100 different DNA sequences.

Restriction Fragment Length Polymorphism (RFLP). Variation between individuals in DNA fragment sizes cut by specific restriction enzymes; polymorphic sequences that result in RFLPs are used as markers on both physical maps and genetic linkage maps. RFLPs usually are caused by mutation at a cutting site. See **Marker**; and **Polymorphism**.

Restriction-enzyme Cutting Site. A specific nucleotide sequence of DNA at which a particular restriction enzyme cuts the DNA. Some sites occur frequently in DNA (e.g., every several hundred base pairs); others much less frequently (rare-cutter; e.g., every 10,000 base pairs).

Retroviral Infection. The presence of retroviral vectors, such as some viruses, which use their recombinant DNA to insert their genetic material into the chromosomes of the host's cells. The virus is then propagated by the host cell.

Reverse Transcriptase. Also known as RNA-directed DNA polymerases. An enzyme used by retroviruses to form a complementary DNA sequence (cDNA) from their RNA. The resulting DNA is then inserted into the chromosome of the host cell.

Ribose. The five-carbon sugar that serves as a component of RNA. See **Ribonucleic Acid**; and **Deoxyribose**.

Ribosomal RNA (rRNA). A class of RNA found in the ribosomes of cells.

Ribosomes. Small cellular components composed of specialized ribosomal RNA and protein; site of protein synthesis. See **RNA**.

Risk Communication. In genetics, a process in which a genetic counselor or other medical professional interprets genetic test results and advises patients of the consequences for them and their offspring.

RNA (Ribonucleic Acid). A chemical found in the nucleus and cytoplasm of cells; it plays an important role in protein synthesis and other chemical activities of the cell. The structure of RNA is similar to that of DNA. There are several classes of RNA molecules, including messenger RNA, transfer RNA, ribosomal RNA, and other small RNAs, each serving a different purpose.

Sanger Sequencing. A widely used method of determining the order of bases in DNA. See **Sequencing**; and **Shotgun Sequencing**.

Satellite. A chromosomal segment that branches off from the rest of the chromosome but is still connected by a thin filament or stalk.

Scaffold. In genomic mapping, a series of contigs that are in the right order but not necessarily connected in one continuous stretch of sequence.

Segregation. The normal biological process whereby the two pieces of a chromosome pair are separated during meiosis and randomly distributed to the germ cells.

Sequence Assembly. A process whereby the order of multiple sequenced DNA fragments is determined.

Sequence Tagged Site (STS). Short (200 to 500 base pairs) DNA sequence that has a single occurrence in the human genome and whose location and base sequence are known. Detectable by polymerase chain reaction, STSs are useful for localizing and orienting the mapping and sequence data reported from many different laboratories and serve as landmarks on the developing physical map of the human genome. Expressed sequence tags (ESTs) are STSs derived from cDNAs.

Sequencing. Determination of the order of nucleotides (base sequences) in a DNA or RNA molecule or the order of amino acids in a protein.

Sequencing Technology. The instrumentation and procedures used to determine the order of nucleotides in DNA.

Sex Chromosome. The X or Y chromosome in human beings that determines the sex of an individual. Females have two X chromosomes in diploid cells; males have an X and a Y chromosome. The sex chromosomes comprise the 23rd chromosome pair in a karyotype. See **Autosome**.

Sex-linked. Traits or diseases associated with the X or Y chromosome; generally seen in males. See **Gene**; **Mutation**; and **Sex Chromosome**.

Shotgun Method. Sequencing method that involves randomly sequenced cloned pieces of the genome, with no foreknowledge of where the piece originally came from. This can be contrasted with "directed" strategies, in which pieces of DNA from known chromosomal locations are sequenced. Because there are advantages to both strategies, researchers use both random (or shotgun) and directed strategies in combination to sequence the human genome. See **Library**; and **Genomic Library**.

Single Nucleotide Polymorphism (SNP). DNA sequence variations that occur when a single nucleotide (A, T, C, or G) in the genome sequence is altered. See **Mutation**; **Polymorphism**; and **Single-gene Disorder**.

Single-gene Disorder. Hereditary disorder caused by a mutant allele of a single gene (e.g., Duchenne muscular dystrophy, retinoblastoma, sickle cell disease). See **Polygenic Disorders**.

Somatic Cell. Any cell in the body except gametes and their precursors. See **Gamete**.

Somatic Cell Gene Therapy. Incorporating new genetic material into cells for therapeutic purposes. The new genetic material cannot be passed to offspring. See **Gene Therapy**.

Somatic Cell Genetic Mutation. A change in the genetic structure that is neither inherited nor passed to offspring. Also called *acquired mutations*. See **Germ Line Genetic Mutation**.

Southern Blotting. Transfer by absorption of DNA fragments separated in electrophoretic gels to membrane filters for detection of specific base sequences by radio-labeled complementary probes.

Spectral Karyotype (SKY). A graphic of all an organism's chromosomes, each labeled with a different color. Useful for identifying chromosomal abnormalities. See **Chromosome**.

Splice Site. Location in the DNA sequence where RNA removes the noncoding areas to form a continuous gene transcript for translation into a protein.

Sporadic Cancer. Cancer that occurs randomly and is not inherited from parents. Caused by DNA changes in one cell that grows and divides, spreading throughout the body. See **Hereditary Cancer**.

Stem Cell. Undifferentiated, primitive cells in the bone marrow that have the ability both to multiply and to differentiate into specific blood cells.

Structural Genomics. The effort to determine the 3D structures of large numbers of proteins using both experimental techniques and computer simulation.

Substitution. In genetics, a type of mutation due to replacement of one nucleotide in a DNA sequence by another nucleotide or replacement of one amino acid in a protein by another amino acid. See **Mutation**.

Suppressor Gene. A gene that can suppress the action of another gene.

Syndrome. The group or recognizable pattern of symptoms or abnormalities that indicate a particular trait or disease.

Syngeneic. Genetically identical members of the same species.

Synteny. Genes occurring in the same order on chromosomes of different species. See **Linkage**; and **Conserved Sequence**.

Tandem Repeat Sequences. Multiple copies of the same base sequence on a chromosome; used as markers in physical mapping. See Physical Map.

Targeted Mutagenesis. Deliberate change in the genetic structure directed at a specific site on the chromosome. Used in research to determine the targeted region's function. See Mutation; and Polymorphism.

Technology Transfer. The process of transferring scientific findings from research laboratories to the commercial sector.

Telomerase. The enzyme that directs the replication of telomeres.

Telomere. The end of a chromosome. This specialized structure is involved in the replication and stability of linear DNA molecules. See **DNA Replication**.

Teratogenic. Substances such as chemicals or radiation that cause abnormal development of a embryo. See **Mutagen**.

Thymine (T). A nitrogenous base, one member of the base pair AT (adenine-thymine). See **Base Pair**; and **Nucleotide**.

Toxicogenomics. The study of how genomes respond to environmental stressors or toxicants. Combines genome-wide mRNA expression profiling with protein expression patterns using bioinformatics to understand the role of gene-environment interactions in disease and dysfunction.

Transcription. The synthesis of an RNA copy from a sequence of DNA (a gene); the first step in gene expression. See **Translation**.

Transcription Factor. A protein that binds to regulatory regions and helps control gene expression.

Transcriptome. The full complement of activated genes, mRNAs, or transcripts in a particular tissue at a particular time.

Transfection. The introduction of foreign DNA into a host cell. See **Cloning Vector**; and **Gene Therapy**.

Transfer RNA (tRNA). A class of RNA having structures with triplet nucleotide sequences that are complementary to the triplet nucleotide coding sequences of mRNA. The role of tRNAs in protein synthesis is to bond with amino acids and transfer them to the ribosomes, where proteins are assembled according to the genetic code carried by mRNA.

Transformation. A process by which the genetic material carried by an individual cell is altered by incorporation of exogenous DNA into its genome.

Transgenic. An experimentally produced organism in which DNA has been artificially introduced and incorporated into the organism's germ line. See **Cell**; **DNA**; **Gene**; **Nucleus**; and **Germ Line**.

Translation. The process in which the genetic code carried by mRNA directs the synthesis of proteins from amino acids. See **Transcription**.

Translocation. A mutation in which a large segment of one chromosome breaks off and attaches to another chromosome. See **Mutation**.

Transposable Element. A class of DNA sequences that can move from one chromosomal site to another.

Trisomy. Possessing three copies of a particular chromosome instead of the normal two copies. See **Cell**; **Gene**; **Gene Expression**; and **Chromosome**.

Uracil. A nitrogenous base normally found in RNA but not DNA; uracil is capable of forming a base pair with adenine. See **Base Pair**; and **Nucleotide**.

Virus. A noncellular biological entity that can reproduce only within a host cell. Viruses consist of nucleic acid covered by protein; some animal viruses are also surrounded by membrane. Inside the infected cell, the virus uses the synthetic capability of the host to produce progeny virus. See **Cloning Vector**.

Western Blot. A technique used to identify and locate proteins based on their ability to bind to specific antibodies. See **DNA; Northern Blot; Protein; RNA**; and Southern Blotting.

Wild Type. The form of an organism that occurs most frequently in nature.

X Chromosome. One of the two sex chromosomes, X and Y. See **Sex Chromosome**.

Xenograft. Tissue or organs from an individual of one species transplanted into or grafted onto an organism of another species, genus, or family. A common example is the use of pig heart valves in humans.

Y Chromosome. One of the two sex chromosomes, X and Y. See Sex Chromosome.

Yeast Artificial Chromosome (YAC). Constructed from yeast DNA, it is a vector used to clone large DNA fragments. See **Cloning Vector**; and **Cosmid**.

Zinc-finger Protein. A secondary feature of some proteins containing a zinc atom; a DNA-binding protein.

Editor's Note: The staff of this encyclopedia is most appreciative of the cooperation extended by the DOE, for furnishing the glossary.

See also **Bioprocess Engineering (Biotechnology); Carcinogens; Cell (Biology); Cloning (Mammals); Cloning (The Story of Dolly the Sheep); Genetics and Gene Science (Classical); Heredity; Proteins; Sex-Limited Inheritance**; and **Sex-Linked Inheritance**.

The World Wide Web offers the easiest path to current news about the Human Genome Project. Good places to start include the following:

Web References

Brookhaven National Laboratory: http://www.bnl.gov/world/
Cold Spring Harbor Laboratory: http://www.cshl.org/
DOE Human Genome Program: http://www.er.doe.gov/production/ober/hug_top.html
"Evolution of a Vision—Part I" by David Smith: http://www.ornl.gov/break hgmis/publicat/hgn/v7n3/02smithr.html
"Evolution of a Vision—Part II" by Francis S. Collins: http://www.ornl.gov/hgmis/publicat/hgn/v7n3/03collin.html
Genome Therapeutics Corporation: http://www.genomecorp.com/
Human Genome Management Information System at Oak Ridge National Laboratory: http://www.ornl.gov/TechResources/Human_Genome/home.html
Human Genome News: Full text of all issues is available online: http://www.ornl.gov/hgmis/publicat/hgn/hgn.html
"Introducing the Human Genome Project: Its Relevance, Triumphs, and Challenges" by Ari Patrinos and Daniel W. Drell: http://www.ornl.gov/TechResources/Human_Genome/publicat/judges/drell.html
Lawrence Berkeley National Laboratory Human Genome Center: http://www-hgc.lbl.gov/GenomeHome.html
Lawrence Livermore National Laboratory: http://www.llnl.gov/
Livermore Human Genome Center: http://www-bio.llnl.gov/bbrp/genome/genome.html
Los Alamos National Laboratory Center for Human Genome http://www-ls.lanl.gov/index.html
National Academy of Sciences: http://www.nas.edu/
National Institutes of Health: http://www.nih.gov/
NIH National Center for Human Genome Research: http://www.nhgri.nih.gov/
The Genome Database at Johns Hopkins University School of Medicine: http://gdbwww.gdb.org/
The Howard Hughes Medical Institute: http://www.hhmi.org/intro/
The Institute for Genomic Research—TIGR: http://www.tigr.org/cet/
The National Center for Genome Resources: http://www.ncgr.org/

HUMAN HERPESVIRUSES (HHV).
See **Antiviral Drugs; Herpes Simples Virus Diseases; Herpesviruses (Human)**; and **Sexually Transmitted Diseases**.

HUMAN IMMUNODEFICIENCY VIRUSES (HIV).
Human immunodeficiency viruses (HIVs) are members of the retroviral family *Retroviridae*, which also includes mammalian and avian type C viruses, bovine leukaemia viruses (including human T-cell leukaemia/lymphoma viruses), and other lentiviruses (including bovine, feline and simian immunodeficiency viruses (SIVs)). There are two main types of HIV (Figure 1):

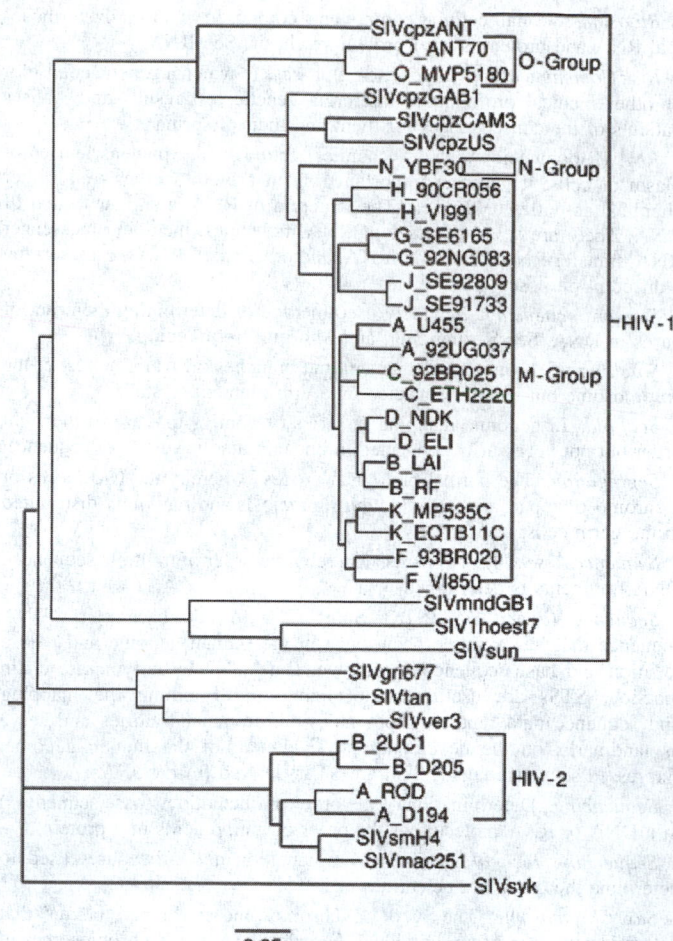

Fig. 1. Estimated phylogenetic (evolutionary) relationships among human immunodeficiency virus types 1 (HIV-1), types 2 (HIV-2) and simian immunodeficiency viruses (SIVs), including the three main groups of HIV-1 (groups O, N and M) as well as all the major subtypes of HIV-1 within group M (subtype A–K). The phylogeny is based on amino acid sequences from the Pol protein. (Courtesy of David Robertson)

Human immunodeficiency virus 1 (HIV-1), which is the causative agent of the majority of acquired immune deficiency syndrome (AIDS) cases [Blattner et al., 1988; Weiss, 1993], and *Human immunodeficiency virus 2* (HIV-2), which causes a slowly developing immune deficiency syndrome. HIV-1 and HIV-2 are closely related to, indeed are interrelated with, simian immunodeficiency viruses isolated from a variety of monkeys and chimpanzees (Figure 1). This article will focus mainly on HIV-1. See also **Acquired Immune Deficiency Syndrome (AIDS)**.

HIV-1 Origins

The most recent estimates of evolutionary relationships suggest that HIV-1 originated via a host shift from chimpanzees to humans [Gao et al.,1999]. While the phylogenetic evidence suggested this many years ago, researchers still favored the 'HIV from an as yet unidentified monkey species' hypothesis because of the high seroprevalence of SIV in monkey species [Sharp et al., 1994]. However, recent evidence from more extensive sequencing of SIVs from chimpanzees suggests that SIVs from chimpanzees and HIVs share a most recent common ancestor. The mechanism of this host shift has been hotly debated [Cohen, 2000; Hillis, 2000], but conventional wisdom suggests transfer via increased human–chimpanzee contact involving blood products, such as butchering of chimpanzees. The probable geographic location of this transfer is in central Africa. Estimates of the timing of the origin of HIV-1 suggest somewhere around 1930 ± 15 years [Korber et al., 2000].

Subtypes and Global Distribution of HIV-1

HIV-1 variants have been classified into three groups: the main group ('M'), the outliers ('O'), and the recently described 'N' group (Figure 1).

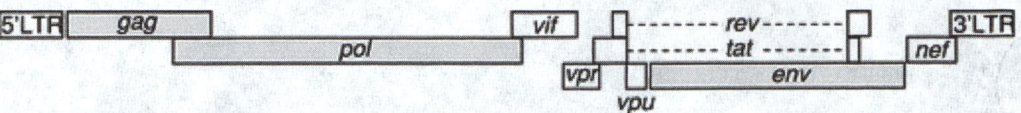

Fig. 2.   Genome organization of the human immunodeficiency virus 1 (HIV-1). The shaded regions are the structural genes common to all primate lentiviruses (*gag*, *pol* and *env*). In addition, there are essential regulatory genes (*tat* and *rev*) and nonessential genes (*nef*, *vif*, *vpu* and *vpr*). LTR, long terminal repeat.

Within the M group, distinct lineages or 'subtypes' have been described, based on the phylogenetic clustering of viral sequences [McCutchan, 1999]. The greatest diversity of subtypes is in sub-Saharan Africa, where nearly all subtypes are found. Outside this region, subtypes are highly geographically structured. For example, subtype B is found predominately in North America. Recently, researchers are able to sequence more easily complete genomes of HIV-1. The phylogenetic analysis of complete genome sequence and comparison of different regions of the genome has led to the discovery of recombinant forms of HIV-1 sequences [Robertson et al., 1995]. Because of recombination, one gene region in the virus can be from one subtype and another region from another subtype. The discovery of circulating recombinant forms has led researchers to require complete genome sequences for classification purposes. Furthermore, there is currently great effort being expended on characterizing the diversity of HIV-1 and monitoring these circulating recombinant forms. In geographic regions, such as sub-Saharan Africa where so many distinct subtypes are cocirculating, the potential for a significant recombination event is high. These subtypes typically differ by 30% of their amino acids in the envelope gene (*env*) (see Structures) and 15% in their *gag* gene. Thus, recombination offers an opportunity for generating an incredible amount of variation upon which evolutionary forces can act.

Structures

The HIV genomes are characteristic of all retroviruses, in that they are molecules of single-stranded RNA coding for at least four characteristic genes: 5'-*gag-pol-env*-3' (that is, the gag, protease, polymerase, and envelope) (Figure 2). The *gag* gene encodes the precursor for virion capsid proteins, which function to package genomic viral ribonucleic acid (RNA) into virions and may be involved in the early stages of replication. The *pol* gene encodes the precursor for several virion enzymes, including protease, reverse transcriptase and integrase. The *env* gene encodes the precursor for envelope glycoprotein. These proteins are on the surface of HIV particles and bind to CD4 receptors (T-cell proteins that participate in recognizing MHC−peptide complexes of target cells). HIVs are also characterized by long terminal repeats (LTRs) on both ends of the genome. Genome sizes of HIV-1 are typically of the order of 9200 bases in length.

Replication

The replication cycle of HIV consists of seven main events (Figure 3) [Coffin, 1999]. The first is entry. Entry of the virion into the host cell is mediated by the envelope protein. The HIV has a number of unique adaptations to the envelope protein that make it distinct from other retroviruses. The envelope is made up of variable and conserved regions. Sequence analysis of the variable regions suggests that there is diversifying selection occurring at these regions, generating the variability needed by the virus to escape the host immune system. The principal receptor cell for HIV in the host is the cell-surface molecule CD4. However, unlike other retroviruses, HIV infection also requires a coreceptor. Two principal coreceptors, the chemokine receptors, are used by HIV depending on the strain of virus. M-tropic viruses use a coreceptor known as CCR5, whereas T-tropic viruses use a different coreceptor, CXCR4. Sequence variation in the host chemokine receptors can influence the speed of the establishment of infection and disease progression towards AIDS after HIV-1 infection [Martin et al., 1998].

The second step in the replication process is that of DNA synthesis. Deoxyribonucleic acid (DNA) synthesis occurs when the single-stranded RNA form is converted to double-stranded DNA, catalysed by the enzyme reverse transcriptase. This complex reaction has two noteworthy aspects that make HIV such a difficult virus to combat. First, because the reverse transcriptase enzyme can jump from one template to another combined with the dimeric nature of the protein, the recombination rate is high in HIV: of the order of one crossover per genome per replication cycle [Temin, 1991]. Not only does this allow the generation of intact DNA copies of damaged RNA, it also provides the virus with a mechanism for generating genetic

diversity [Coffin, 1999]. Thus, recombination alone during replication can generate significant genetic variation upon which natural selection can act (for example, to escape drug therapies; see **Antiviral Agents**). This, coupled with the second important aspect of DNA synthesis, the high error rate, makes obvious the importance of genetic diversity to the success of this virus. RNA viruses typically have a much higher error rate than their DNA counterparts [Pathak and Temin, 1990]. Indeed, the estimated error rate for HIV is about 3×10^{-5} errors per base per replication cycle [Mansky and Temin, 1995], compared with substitution rates of viral hosts of the order of 10^{-9}. See also **Baltimore, David**; **Temin, Howard Martin (1934−1994)**.

The third step of HIV replication is that of integration of the viral DNA into the cellular DNA [Coffin, 1999]. This process involves the formation of a preintegration complex, which then directs the integration of the DNA into chromosomal DNA. The integration then leads to the covalent bonding of the viral DNA to cellular DNA. It is this integration step that makes retroviruses unique among infectious agents by allowing some members to cause malignancies, to acquire cellular sequences and express them as oncogenes, and to become incorporated into the germline as endogenous proviruses [Coffin, 1999].

Transcription of the DNA occurs through the normal transcriptional machinery of the cell, as the cell sees the integrated provirus as though it were a normal gene [Coffin, 1999]. To complete the replication cycle, the provirus must be copied into both genome and messenger RNA (mRNA) molecules by being modified through RNA processing and transport. An important component of the processing includes RNA splicing to various genes found in complex retroviruses. Thus, the RNA splicing allows for an array of genes and gene products, some functions of which are still unknown [Coffin, 1999]. Once the mRNAs are made, protein synthesis occurs at two sites in the cell; the envelope gene is translated on the membrane-bound polyribosomes and the rest are translated on free polyribosomes.

The final step in the replication cycle is the assembly of the internal viral components (i.e., RNA, and the Gag and Gag-Pro-Pol precursor proteins) and their release from the cell via *budding*. The final step in budding is the release of virions from the infected cell. Now the virions are free, but immature. Protease-mediated cleavage transforms the virions from their immature, noninfectious form into active forms [Coffin, 1999]. Thus, researchers early on in drug development targeted the protease inhibitors as likely candidates for effective drug therapy (see below, **Antiviral Agents**).

Epidemiology

While HIV has been known for less than 20 years, and has been in existence for perhaps less than 70 or 80 years, it has quickly become the leading cause of death in Africa and the fourth leading cause of death worldwide. Over 16 million people have died from AIDS since the epidemic began, and today an estimated 34.3 million people are infected with the HIV virus. Women are becoming increasingly infected with HIV, with approximately 46% of the adult infections. The AIDS epidemic has left behind a cumulative total of 13.2 million orphans, defined as those having lost their mother before reaching the age of 15. The overwhelming majority of people infected with HIV (some 95% of the global total) are living in the developing world. See also **Acquired Immune Deficiency Syndrome (AIDS)**.

HIV is spread from one infected person to another through blood-to-blood contact or sexual contact. There are several body fluids that have been shown to spread HIV, including blood, semen, vaginal fluid, breast milk, and other body fluids containing blood. Infection occurs when an infected individual transfers, via one of the body fluids, the HIV virus to an uninfected individual. The common modes of transfer are via sexual intercourse (heterosexual or homosexual; anal and vaginal), intravenous drug use with a shared, infected needle; and from HIV-infected women to their babies *in utero*, during birth, or through breast-feeding after birth. Infection has also occurred through the use of contaminated blood transfusions and through inadvertent exposure to contaminated blood by

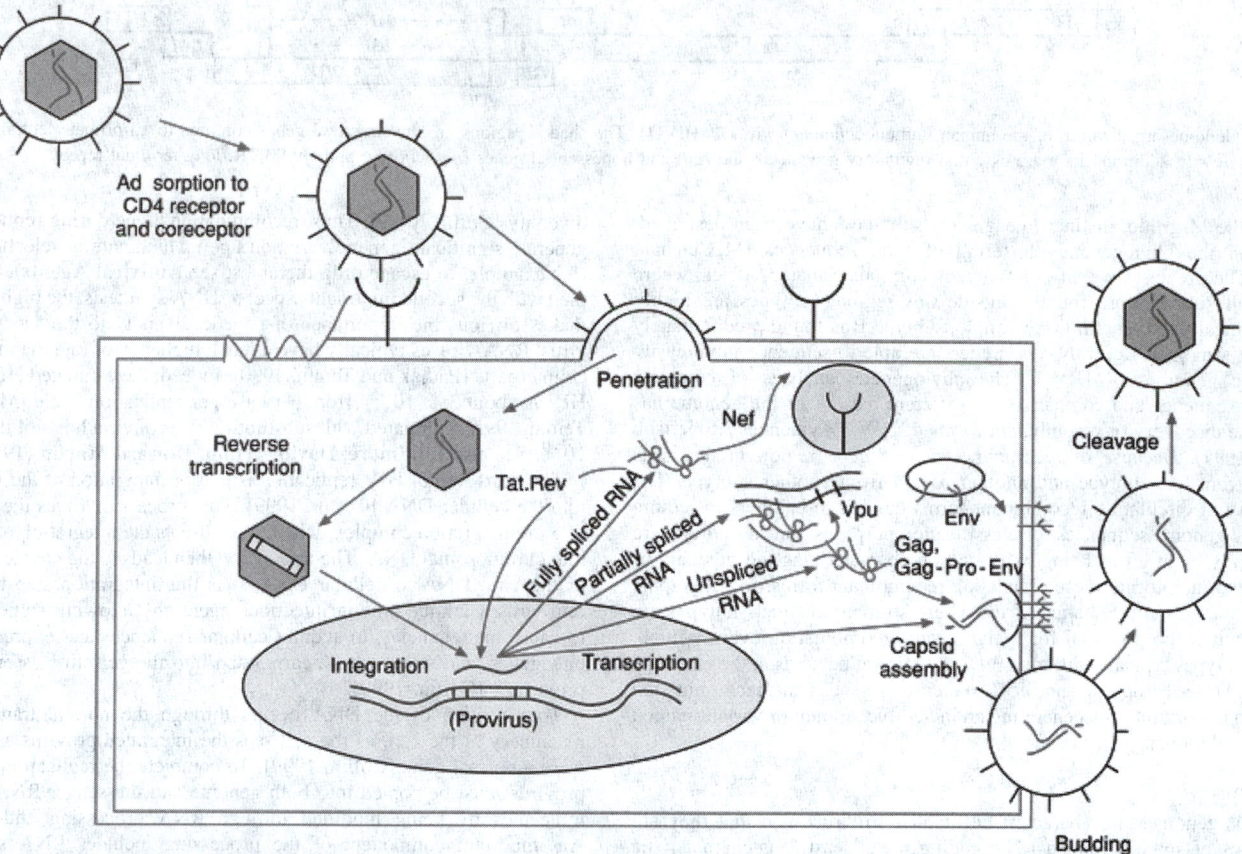

Fig. 3. The replication cycle of the human immunodeficiency virus (HIV) (after Coffin, 1999) showing the seven main steps in HIV replication: (1) viral entry, (2) DNA synthesis, (3) integration, (4) transcription, (5) RNA processing and transport, (6) protein synthesis, and (7) assembly and budding.

healthcare workers. By far the best preventative measures in curtailing the HIV epidemic have been through education leading to prevention measures.

Pathogenesis

HIV infection and subsequent progression to AIDS is characterized by three primary stages (Figure 4) [Pantaleo et al., 1993]. The first stage is primary infection. This stage can be clinically asymptomatic or characterized by influenza-like symptoms. During this stage, high levels of viral particles are present in the plasma (viral load) and CD4+ T-cell counts go down. With this primary infection, an immune response is initiated, resulting in a decrease of viral load and a leveling off of CD4+ counts. After this reduction in viral replication, most patients enter the second stage of infection, clinical latency. This period is characterized by continual viral replication, albeit at reduced levels, and adaptation of the viral population to escape the host immune response. Eventually most patients experience a gradual deterioration of their immune system, coupled with an increased occurrence of opportunistic infections. This deterioration ushers in the third stage, that of AIDS. This stage is characterized by an increase in viral loads and a decrease in CD4+ counts, typically to less than 200 cells mL^{-1}. Many studies have attempted to characterize the role of genetic variation of the virus in the course of pathogenesis [Viscidi, 1999]. However, to date, there is still ample debate about the relationship between genetic diversity and disease progression. It is clear, however, that the advent of drug therapies has greatly altered this relationship (see below, **Antiviral Agents**). Likewise, it is clear that the host's genotype appears to play a significant role in determining the length of the clinical latency period (reviewed in Crandall et al., 1999b).

Antiviral Agents

The development of antiviral agents has been a very effective approach to limiting the replication ability and viral load of HIV infection in patients. When antiviral agents were first developed, it was hoped that they might be a cure for this disease by eliminating replicating virus from the infected patient. These early drug therapies consisted of reverse

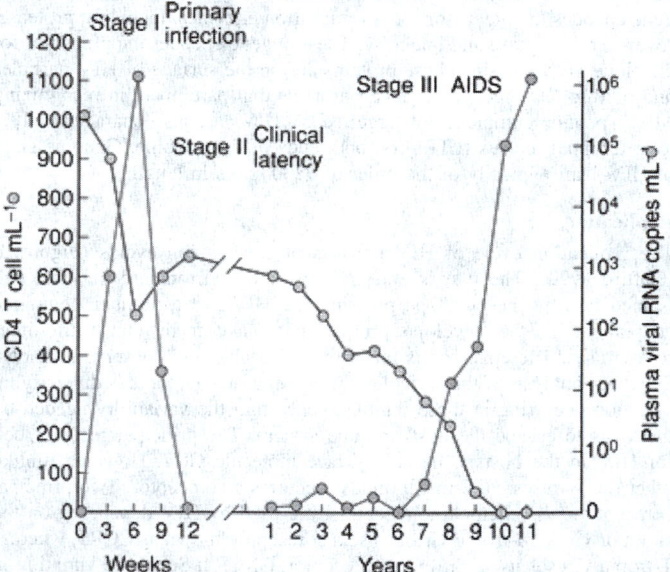

Fig. 4. The typical course of infection and disease progression in human immunodeficiency virus 1 (HIV-1)-infected patients showing the change in CD4+ counts and viral load over time (from Viscidi, 1999). Highlighted are the three main stages of disease progression: (1) primary infection followed by an immune response by the host, leading to (2) a period of clinical latency during which the virus eventually evolves a way to get around the host immune system (and even drug therapy), leading to (3) a clinical diagnoses of AIDS.

transcriptase inhibitors such as azidothymidine (AZT). The object of these drugs was to target reverse transcriptase and thereby interfere with the replication process (see above, **Replication**). However, the virus clearly had an ability to evolve drug resistance to the antiviral agents. The drug approach then escalated to a combination of antiviral drugs, known as highly active antiretroviral therapy or HAART. HAART was initially claimed as a 'cure' for AIDS [Wain-Hobson, 1997]. However, researchers once again underestimated the ability of the virus to persist and evolve, and shortly evidence began building of the ability of the virus to escape even the triple drug therapy of HAART [Chun et al., 1997]. The evolutionary potential (number of surviving new variants) of the virus is enhanced by the difficulty of the drug regimen and patients wavering from this regimen [Chun et al., 1999]. Furthermore, as drug therapies select for HIV variants with drug-resistant mutations [Crandall et al., 1999a], the incidence of drug-resistant mutations in the general population of HIV-infected patients will increase [Kozal et al., 1996]. Thus, the gamble on the evolution of drug resistance appears to be lost [Leigh Brown and Richman, 1997], as the extraordinary ability for HIV to generate variation (see above, **Replication**) allows too much potential for evolutionary change. Researchers are now turning to vaccine development as the best hope for halting (or slowing) this rampaging virus. See also **Antiviral Drugs**.

Antiviral agents used to treat HIV-1 infection can be broadly classified into four categories: (1) nucleoside inhibitors of reverse transcriptase; (2) nonnucleoside inhibitors of reverse transcriptase; (3) protease inhibitors; and (4) foscarnet and nucleotides [Shankarappa, 1999]. Nucleoside inhibitors of reverse transcriptase (including AZT or zidovudine (ZDV), lamivudine (LMV or 3TC), stavudine (SVD or d4 T), didanosine (ddI) and dideoxycytodine (ddC)) cause a termination of the nucleic acid addition during the replication process following their incorporation by the reverse transcriptase. Nonnucleoside inhibitors (including nevirapine (NVP) and delvridine (DLV)) act by binding to a nonsubstrate binding hydrophobic pocket within the polymerase subunit. Thus, unlike the nucleoside inhibitors, the nonnucleoside inhibitors do not require metabolic activation inside an infected cell to be functional. Protease inhibitors target the protease gene, resulting in a block of the maturation process of the HIV-1 viral particle. Protease inhibitors have an advantage over the reverse transcriptase inhibitors, in that they prevent infection of new cells (as the reverse transcriptase inhibitors), and they also suppress viral replication in chronically infected cells. The fourth category, foscarnet and nucleotides, is used to a limited extent and will not be discussed further.

As discussed above, the amazing ability of HIV-1 to generate mutational changes through point mutation (errors in the replication process), and bring these mutations together through recombination, provides a wealth of variants upon which natural selection can act to bring into high frequency mutations that confer drug resistance. Researchers have documented that there is lower fitness relative to wild-type associated with drug-resistant variants in a drug-free environment [Zennou et al., 1998]. However, once a patient begins drug therapy, the environment to which the viral population is exposed changes dramatically. These variants now confer a greater fitness on the associated viral particles, and drug-resistant variants are swept by natural selection into high frequency in the population, allowing HIV to escape drug therapy. The early hope for drug therapy as a cure was based on the observation that viral loads dropped below detectable levels. However, even with 'undetectable levels' of viral load (<50 copies mL^{-1}), sensitive molecular biological assays could still detect viral DNA in patients. Presumably, virus was 'hiding' in tissues that served as 'latent' viral reservoirs [Siliciano, 1999]. These protected populations of virus could not be reached by drug therapy. Furthermore, these populations appeared to be undergoing low levels of replication, producing variants that eventually evolved drug resistance. Thus, it seems only a matter of time before drug therapy will fail in individual patients. This notion of protected viral reservoirs is further supported by the observation that, in patients who abandon drug therapy, the viral populations rebound and appear to be genetically similar to the populations at the initiation of drug therapy [Imamichi et al., 2001]. Drug therapy is further complicated by the extreme expense associated with the drug treatments (estimated at US$11 000 annually) and the stringent and complicated schedule of drugs. In sub-Saharan Africa, where the epidemic is raging, the annual household expenditure on health care is of the order of US$8 per year. Furthermore, in trials for tuberculosis therapy, the drug therapy completion rate was 25% [Senior, 1999]. Here the regimen consisted of one or two tablets per day for only 6 months. Thus, the prospect for successful adherence to the difficult drug regimen for HIV-1 is not good, even if drugs were provided with no cost to patients. While drug therapy has been extremely successful in extending lives and quality of lives for patients infected with HIV-1, and in reducing the risk of transmission in mother–infant cases [Kalish et al.,

1999], it has failed to produce a cure for AIDS. Therefore, concentrated effort has now been focused on the development of HIV vaccines.

Prospects for Vaccine Development

The development of an HIV vaccine has been frustratingly slow due to the enormously complex nature of vaccine development in general and the added complications imposed by HIV itself. Significant progress has been made over the last 15 years in understanding the basic components of the replication cycle (above), genome structure and variability (above), and immunology of HIV, and how these components relate to vaccine development. Most of the vaccine development has focused on the envelope region and consists of a variety of approaches, including recombinant proteins, synthetic peptides, recombinant viral vectors, recombinant bacterial vectors, recombinant particles, and whole-killed and live-attenuated HIV. The latter two have not progressed into clinical trials due to an unfavorable benefit: risk ratio, further supported by experimental evidence from humans and simian models [Greenough, et al., 1999; Baba, et al., 2000]. However, it is now clear that the vaccine approach is our best strategy for combating AIDS. Therefore, significant effort and resources are being targeted toward the development of an HIV vaccine.

Not the least of the hurdles in the development of an effective HIV vaccine is the clinical trial process to demonstrate efficacy against the virus. Vaccine trials against HIV are composed of three phases. Phase I and II trials are performed on small numbers of volunteers at relatively low risk for HIV-1 infection. These trials attempt to establish a record of safety and provide valuable immunogenicity data concerning the vaccine. Phase III trials, on the other hand, are large-scale trials on populations of high incidence and high risk for HIV-1 infection. The first phase I trial of an HIV-1 candidate vaccine was held in 1987. Since then, over 60 phase I and II trials have taken place for over 30 candidate vaccines. Of these numerous candidates, only two, developed by the company VaxGen, have progressed to phase III trials. There has been significant debate about the requirements to move from phase II to phase III [Boily, et al., 1999], which perhaps explains why only two vaccines have made it to phase III trials at this point. This includes the lack of a validated correlate of protective immunity against HIV-1 (i.e. a reference standard of natural immunity on which to base our search for protective correlates) [Castillo, et al., 2000]. These vaccines are based on the gp120 of a laboratory strain of subtype B being tested in the United States and a gp120 of subtype B and E from Thailand being tested in Thailand. The results from these Phase III trials are expected in late 2002. The hope is that the vaccine will provide some effectiveness against HIV infection by decreasing the odds of infection by only, say, 30%. Ideally, an HIV vaccine should elicit both neutralizing antibody responses and strong cellular immunity. It should induce broadly cross neutralizing antibodies against primary isolates from divergent clades (A to K; Figure 1) and against viruses of both phenotypes, lymphotropic, syncytium-inducing viruses, which use the CXCR4 coreceptor for viral entry, as well as macrophage-tropic, nonsyncytium-inducing viruses, which utilize the chemokine CCR5 receptor [Klein, 1999]. Once positive data are obtained from a phase III trial, then phase IV trials can begin to provide additional data on effectiveness and population specificity. Remember, these first two phase III vaccines are developed from a laboratory strain of subtype B and a native B/E mixture. Thus, the strains involved in the sub-Saharan epidemic are genetically very distinct from those used to develop these vaccines. It is unclear how this extensive genetic diversity will affect the efficacy of vaccines [Zolla-Pazner, et al., 1999]. The near future holds great potential for vaccine development and this area of research will attract greater and greater attention with the relative failure of antiviral agents to produce a cure for AIDS.

Additional Reading

Baba, T.W., V. Liska, R. Hofmann-Lehmann, et al.: "Human Neutralizing Monoclonal Antibodies of the IgG1 Subtype Protect Against Mucosal Simian-Human Immunodeficiency Virus Infection," *Nature Medicine*, **6**, 200–206 (2000).

Blattner, W., R.C. Gallo, and H.M. Temin: "HIV Causes AIDS," *Science*, **241**, 515–516 (1988).

Boily, M.C., B.R. Masse, K. Desai, M. Alary, and R.M. Anderson: "Some Important Issues in the Planning of Phase III HIV Vaccine Efficacy Trials," *Vaccine*, **17**, 989–1004 (1999).

Castillo, R.C., S. Arango-Jaramillo, R. John, et al.: "Resistance to Human Immunodeficiency Virus Type 1 *in vitro* as a Surrogate of Vaccine-induced Protective Immunity," *Journal of Infectious Diseases*, **181**, 897–903 (2000).

Chun, T.W., L. Stuyver, S.B. Mizell, et al.: "Presence of an Inducible HIV-1 Latent Reservoir during Highly Active Antiretroviral Therapy," *Proceedings of the National Academy of Sciences of the USA*, **94**, 13193–13197 (1997).

Chun, T.W., R.T. Davey, D. Engel, H.C. Lane, and A.S. Fauci: "AIDS: Re-emergence of HIV after Stopping Therapy," *Nature*, **401**, 874–875 (1999).

Coffin, J.M., S.H. Hughes, and H.E. Varmus: *Retroviruses*, Cold Spring Harbor Laboratory Press, Cold Spring Harbor, NY, (1997).

Coffin, J.M.: "Molecular Biology of HIV," In: Crandall, K.A.: *The Evolution of HIV*, pp. 3–40, Johns Hopkins University Press, Baltimore, MD, (1999).

Cohen, J.: "Vaccine Theory of AIDS Origins Disputed at Royal Society," *Science*, **289**, 1850–1851 (2000).

Crandall, K.A., C.R. Kelsey, H. Imamichi, and N.P. Salzman: "Parallel Evolution of Drug Resistance in HIV: Failure of Nonsynonymous/synonymous Substitution Rate Ratio to Detect Selection," *Molecular Biology and Evolution*, **16**, 372–382 (1999).

Crandall, K.A., D. Vasco, D. Posada, and H. Imamichi: "Advances in Understanding the Evolution of HIV," *AIDS*, **13**, S39–S47 (1999).

Crandall, K.A.: *The Evolution of HIV*, Johns Hopkins University Press, Baltimore, MD, (1999).

Gao, F., E. Bailes, D.L. Robertson, et al.: "Origin of HIV-1 in the Chimpanzee *Pan troglodytes troglodytes*," *Nature*, **397**, 436–441 (1999).

Greenough, T.C., J.L. Sullivan, and R.C. Desrosiers: "Declining CD4 T-cell Counts in a Person Infected with Nef-deleted HIV-1," *New England Journal of Medicine*, **340**, 236–237 (1999).

Hillis, D.M.: "How to Resolve the Debate on the Origin of AIDS," *Science*, **289**, 1877–1878 (2000).

Imamichi, H., K.A. Crandall, V. Natarajan, et al.: "The Viral Quasispecies of HIV-1 Rebounding Following Discontinuation of HAART are Similar to the Viral Quasispecies Present Prior to the Initiation of Therapy," *Journal of Infectious Disease*, **183**, 36–50 (2001).

Kalish, M.L., D.M. Thea, and R.W. Steketee: "Perinatal HIV infection," In: Crandall, K.A.: *The Evolution of HIV*, Johns Hopkins University Press, Baltimore, MD, 1999, pp. 390–431.

Klein, M.: "AIDS and HIV Vaccines," *Vaccine*, **17**, S65–S70 (1999).

Korber, B., M. Muldoon, J. Theiler, et al.: "Timing the Ancestor of the HIV-1 Pandemic Strains," *Science*, **288**, 1789–1796 (2000).

Kozal, M.J., N. Shah, N. Shen, et al.: "Extensive Polymorphisms Observed in HIV-1 Clade B Protease Gene Using High-density Oligonucleotide Arrays," *Nature Medicine*, **2**, 753–759 (1996).

Leigh Brown, A.J., and D.D. Richman: "HIV-1: Gambling on the Evolution of Drug Resistance?" *Nature Medicine*, **3**, 268–271 (1997).

Mansky, L.M., and H.M. Temin: "Lower *in Vivo* Mutation Rate of Human Immunodeficiency Virus Type 1 than that Predicted from the Fidelity of Purified Reverse Transcriptase," *Journal of Virology*, **69**, 5087–5094 (1995).

Martin, M.P., M. Dean, M.W. Smith, et al.: "Genetic Acceleration of AIDS Progression by a Promoter Variant of *CCR5*," *Science*, **282**, 1907–1911 (1998).

McCutchan, F.E.: "Global Diversity in Human Immunodeficiency Viruses," In: Crandall, K.A.: *The Evolution of HIV*, Johns Hopkins University Press, Baltimore, MD, 1999, pp. 41–101.

Pantaleo, G., C. Graziosi, and A.S. Fauci: "New Concepts in the Immunopathogenesis of Human Immunodeficiency Virus Infection," *New England Journal of Medicine*, **328**, 327–335 (1993).

Pathak, V.K., and H.M. Temin: "Broad Spectrum of In *Vivo* forward Mutations, Hypermutations, and Mutational Hotspots in a Retroviral Shuttle Vector after a Single Replication Cycle: Substitutions, Frameshifts and Hypermutations," *Proceedings of the National Academy of Sciences of the USA*, **87**, 6019–6023 (1990).

Robertson, D.L., P.M. Sharp, F.E. McCutchan, and B.H. Hahn: "Recombination in HIV-1," *Nature*, **374**, 124–126 (1995).

Senior, K.: "Ups and Downs on the Road to an HIV Vaccine," *Molecular Medicine Today*, **5**, 141–142 (1999).

Shankarappa, R.: "Evolution of HIV-1 Resistance to Antiviral Agents," In: Crandall, K.A.: *The Evolution of HIV*, Johns Hopkins University Press, Baltimore, MD, 1999, pp. 469–490.

Sharp, P.M., D.L. Robertson, F. Gao, and B.H. Hahn: "Origins and Diversity of Human Immunodeficiency Viruses," *AIDS*, **8**, S27–S42 (1994).

Siliciano, R.F.: "Latency and Reservoirs for HIV-1," *AIDS*, **13**, S49–S58 (1999).

Temin, H.M.: "Sex and Recombination in Retroviruses," *Trends in Genetics*, **7**, 71–74 (1991).

Viscidi, R.P.: "HIV Evolution and Disease Progression via Longitudinal Studies," In: Crandall, K.A.: *The Evolution of HIV*, Johns Hopkins University Press, Baltimore, MD, 1999, pp. 346–389.

Wain-Hobson, S.: "Down or out in Blood and Lymph?" *Nature*, **387**, 123–124 (1997).

Weiss, R.A.: "How Does HIV Cause AIDS?" *Science*, **260**, 1273–1279 (1993).

Zennou, V., F. Mammano, S. Paulous, D. Mathez, and F. Clavel: "Loss of Viral Fitness Associated with Multiple Gag and Gag-Pol Processing Defects in Human Immunodeficiency Virus type 1 Variants Selected for Resistance to Protease Inhibitors *in Vivo*," *Journal of Virology*, **72**, 3300–3306 (1998).

Zolla-Pazner, S., M.K. Gomy, and P.N. Nyambi: "The Implications of Antigenic Diversity for Vaccine Development," *Immunology Letters*, **66**, 159–164 (1999).

Web References

Center for AIDS Prevention Studies at the University of California San Francisco: http://www.caps.ucsf.edu/siteindex.php

Center for Disease Control and Prevention Information Network: http://www.cdc npin.org/scripts/index.asp

Journal of the American Medical Association's website for HIV: http://pubs.ama-assn.org/

Los Alamos National Laboratory HIV Database: http://www.hiv.lanl.gov/content/index

National Institutes of Health, National Institute of Allergy and Infectious Diseases, Division of Acquired Immunodeficiency Syndrome: http://www3.niaid.nih.gov/research/daids.htm

The Body: An AIDS and HIV Information Resource: http://www.thebody.com/index.shtml

UNAIDS–Joint United Nations Programme on HIV/AIDS: http://www.unaids.org/en/default.asp

KEITH A. CRANDALL, Brigham Young University, Provo, UT

HUMAN PAPILLOMA VIRUS.   See **Dermatitis and Dermatosis**; and **Wart** *(Verrucae)*.

HUMBOLDT, ALEXANDER VON (1769–1859).   (Friedrich Wilhelm Heinrich) Alexander, Freiherr von (Baron) Humboldt was born in Berlin, into an independently wealthy family of minor nobility. His godparents included the later Prussian King Friedrich Wilhelm II. Whereas outside Germany the name "Humboldt" is associated first and foremost with Alexander, within Germany, it more often refers to Alexander's elder brother Wilhelm (1767–1835), statesman and philologist. Humboldt Sr. died when Alexander was only nine years old, and his education was directed by his reputedly severe mother. His university education proceeded from Frankfurt an der Oder (1787–1788), to Berlin (1789), where he was taught botany by Carl Ludwig Willdenow, and Göttingen (1789–1790), where he came under the influence of Johann Friedrich von Blumenbach. Additionally, Humboldt studied at the Hamburg Business Academy (1790–1791) and the Freiberg Mining Academy (1791–1792), where he followed the lectures by Abraham Gottlob Werner. From 1792 till the death of his mother in 1796, Humboldt held various positions in the mining industry of Prussia and Franconia, when he invented safety devices to protect miners against gas explosions. His inheritance made it possible to fulfill his desire to travel abroad, and he now left for Paris. When Napoleon excluded him from the military and scientific Expedition to Egypt, Humboldt, together with the botanist Aimé–Bonpland, set out via Madrid on what became a classic journey of exploration of the Americas (1799–1804), visiting the regions that today are Venezuela, Cuba, Colombia, Ecuador, Peru, Mexico and the USA. This journey became a trend- and standard-setting scientific exploration, characterized by the exemplary use of precision instruments for the determination of latitude, longitude, altitude, barometric pressure, humidity and various other physical features, including the blueness of the sky. Humboldt's studies and illustrations of the distribution of vegetation were particularly innovative. Upon his return to Europe, Humboldt perfected the use of isolines to depict temperature variations across the surface of the earth, and inspired colleagues, among them the cartographer Heinrich Berghaus, to construct global distribution maps, not only of climate and vegetation, but also of ocean currents, rainfall, magnetism and various other features of physical geography. His scientific style, referred to in recent years as "Humboldtian science," became widely admired and imitated in the course of the nineteenth century. See also **Blumenbach, Johann Friedrich (1752–1840)**.

The American journey divides Humboldt's very many publications into three groups. First, there were his early writings, from before 1799, most famously a work on organic electricity, *Versuche über die gereizte Muskel- und Nervenfaser*. Second, there were the many and sometimes voluminous publications that resulted from his exploration of the equatorial Americas. Most of these were written during Humboldt's Parisian period (1808–1826), when he lived as an independently wealthy, private scholar. The production of this American *oeuvre* was a major undertaking that exhausted Humboldt's personal fortune. The thirty-volume work carried the collective title *Voyage aux Regions équinoxiales du Nouveau Continent*. It dealt with botany, plant geography, zoology, physical geography, political economy, and included such classics as the *Essai politique sur le royaume de la Nouvelle-Espagne* and the *Relation historique du voyage*. Third, there were the publications of Humboldt's Berlin period

(1827–1859) during which he was employed as a royal chamberlain at the Prussian Court. Among these were the accounts of his relatively short and hasty Siberian journey (1829), but also and most famously his *Kosmos*, a work that constituted both the summary of most of Humboldt's lifelong interests and an holistic digest of the scientific study of celestial and terrestrial phenomena. On the coat-tails of its success, a new, third German edition of *Ansichten der Natur* (Opinions of nature) (first published in 1808) was produced. See also **Plant Sciences (The History)**.

Humboldt was a sociable man with a sharp tongue. He enjoyed pouring forth his vast and detailed scientific knowledge, in private meetings as well as at Parisian soirées. He was fair-minded and generally gave his support to disadvantaged groups, becoming known for his anti-slavery stance as well as his philosemitism. Yet there was much ambiguity in his allegiances. For example, on the one hand, he paid homage to and loyally served royalty ranging from the Spanish Crown to the Prussian Court and the Russian Tsar; on the other hand, he simultaneously gave cautious support to revolutionary causes in Latin America as well as in Europe. His fame has proved lasting and has taken on extraordinary proportions, with over a thousand places in the world being named after him and the secondary literature on him (books and articles) surpassing ten thousand. Recently, his holistic, geobotanical work has been interpreted as part of the foundations of environmentalism and green politics. See also **Biogeography (The History)**.

NICOLAAS A. RUPKE Göttingen University, Göttingen, Germany

HUMBOLDT CURRENT (also called Peru Current). The cold ocean current flowing north along the coasts of Chile and Peru. It is one of the swiftest of ocean currents. The Peru current originates where part of the water that flows toward the east across the subantarctic Pacific Ocean is deflected toward the north as it approaches South America. The northern limit of the current can be placed a little south of the equator, where the flow turns toward the west, joining the south equatorial current.

The southern portion of the Humboldt current is sometimes called the *Chile current*.

HUMECTANTS AND MOISTURE-RETAINING AGENTS. Substances that have affinity for water, with stabilizing action on the water content of a material, are called *humectants* or moisture-retaining agents. Ideally, a humectant maintains within a rather narrow range the moisture content caused by humidity fluctuations. These materials are widely used in certain food products, as well as tobacco, and in recent years have taken on increasing importance in the case of intermediate-moisture foods. Traditionally, humectants have been used to retain moisture in foods like coconut and marshmallows which otherwise would quickly dry and become tasteless. For example, flaked coconut is kept moist in the container by adding glycerine and glyceryl monostearate.

Among the most commonly used humectant are glycerine, potassium polymetaphosphate, propylene glycol, sodium chloride, sorbitol, sucrose, and triacetin. Also, phosphates are added to the pickling solutions used to treat cured meats, such as ham, bacon, corned beef, etc., by soaking or injection. Their principal purpose is for moisture binding to reduce the loss of fluids during curing and cooking.

During the last few years, important research has gone into the addition of multiple humectants and water to food systems. Studies have shown that a hysteresis effect may occur with certain humectants, i.e., a different rate of moisture absorption than the rate for moisture desorption. Multiple humectants tend to compensate these hysteresis effects, giving uniform rates in both directions.

HUME-ROTHERY RULES. When alloy systems form distinct phases, it is found that the ratio of the number of valence electrons to the number of atoms is characteristic of the phase (e.g., β, γ-, ε-) whatever the actual elements making up the alloy. Thus, both $Na_{31}Pb_8$ and Ni_5Zn_{21} are γ-structures, with the electron-atom ratio 21:13. The rules are explained by the tendency to form a structure in which all the Brillouin zones are nearly full, or else entirely empty.

HUMID CLIMATE. See **Climate**.

HUMIDITY. Generally, some measure of water-vapor content of air. *Absolute humidity* is the ratio of the mass of water vapor present to the volume occupied by the mixture; that is, the density of the water vapor

component. The percentage of water vapor in the total composition of the air may be determined by passing a measured quantity of air through a tube containing an absorbing substance that removes all the vapor, and which can be weighed before and after the absorption.

Absolute humidity is usually expressed in grams of water vapor per cubic meter or, in engineering practice, in grains per cubic foot. Because this measure of atmospheric humidity is not conservative with respect to adiabatic expansion or compression, it is not commonly used by meteorologists. As occasionally used in air-conditioning practice, absolute humidity refers to the number of grains of water vapor per pound of moist air, which is dimensionally identical with the specific humidity (defined below).

Critical humidity is the point at which the partial pressure of water vapor in the atmosphere is equal to the saturation vapor pressure. Condensation on suitable nuclei will occur when the humidity reaches or exceeds this value.

Relative humidity is the ratio of the actual vapor pressure of the air, at any temperature, to the maximum of saturation vapor pressure at the same temperature. It expresses the vapor content as a fraction or percentage of the concentration necessary to render the vapor saturated at the given temperature. At the dew point, the relative humidity is 100%. A rise of temperature without the addition of more vapor reduces the relative humidity (but not the absolute humidity), while a fall of temperature increases it and may bring about saturation. Relative humidity is measured by the hygrometer.

Specific humidity is the (dimensionless) ratio of the mass of water vapor to the total mass of the system. It may be approximated by the mixing ratio for many purposes:

$$q = \frac{w}{1+w}$$

where q is the specific humidity and w the mixing ratio.

See also **Psychrometric Chart**.

HUMIDITY INDEX. See **Meteorology**.

HUMMOCKED ICE. See **Pressure Ice**.

HUNTER, JOHN (1728–1793). The younger brother and one-time protégé of the anatomist William Hunter, John Hunter received a rather desultory education, though he was always interested in natural history. In 1748 he joined his brother in London, where he was inspired by anatomy, and also learned surgery at St Bartholomew's and St George's Hospitals. He was sent to Oxford University in 1755, with the idea of becoming a physician and gaining polish, but the experiment failed and he returned to London after a few months. Here he busied himself with studies of congenital hernia, the structure and function of the lymphatic system and comparative anatomy. His health giving way, he joined the army as a surgeon and spent two years in France and Portugal, gaining experience he put to good use in his work on the treatment of gunshot wounds.

On his discharge from the army in 1763, he established a surgical practice in London and in 1768 was elected to the surgical staff of St George's Hospital. He had already begun giving lectures in surgery, and the post at St George's allowed him to take a steady stream of apprentices and pupils, of whom the most famous was Edward Jenner. His surgical practice also flourishing. He was able to acquire a large estate just west of London as it was (now Earl's Court) and, in 1783, a fine house in Leicester Square, where he established a museum, gallery and lecture room. The estate in Earl's Court had room for a large menagerie of exotic animals which he used for experimentation and dissection. See also **Jenner, Edward (1749–1823)**.

Hunter was a restless man with an insatiable curiosity. He was more interested in physiology (which he called the *animal economy*) than in operative surgery, and as much concerned with disease processes as disease consequences. As a surgeon, his most innovative contribution was in the treatment of arterial aneurysms by ligation, his understanding of the development of collateral circulation resting on animal experiments. He was also a pioneer in what later was called *endocrinology*, in transplantation surgery and in the scientific study of dentistry. His *The Natural History of Human Teeth* (1771) and a supplemental volume of 1778 provided a detailed, well-illustrated exposition of the anatomy, pathology and development of the human jaw and teeth; Hunter coined the terms cuspids, bicuspids, molars and incisors, and showed that transplantation of teeth is surgically possible.

Hunter was always especially interested in the blood, where he believed the vital principle resided. His investigations of inflammation, contained in his last book, *A Treatise on the Blood, Inflammation, and Gun-shot Wounds* (1794), were suggestive, though difficult and idiosyncratic. Rather more accessible was his *Treatise on the Venereal Disease* (1786), the title of which reflects Hunter's belief that gonorrhea and syphilis were simply stages of a single disease process. This mistaken belief was substantiated by his famous experiment whereby he gave an unfortunate experimental subject (sometimes said without conclusive evidence to have been himself) both diseases through a single inoculation of pus taken from a syphilitic chancre.

Hunter's extensive museum reflected the breadth of his interests in the natural world. He invested substantial amounts of time and money in its formation, and it was acquired, through a Parliamentary grant, for the Royal College of Surgeons after Hunter's death. The College received its Royal Charter in 1800, and the Hunterian legacy was perpetuated there through the museum as well as a Professorship and an annual Oration. He himself suffered from angina for a number of years, and died suddenly after being provoked to anger at a staff meeting at St George's Hospital. See also **Physiology (The History)**.

W. F. Bynum Wellcome Trust Centre for the History of Medicine at UCL, London, UK

HUNTING. The tendency of a rotating mechanism that normally should operate at constant speed to pulsate in speed above and below the normal point, is known as hunting. It may occur in prime movers controlled by governors which are too isosynchronous, or in electric apparatus where rotating and stationary parts are electrically coupled. The nature of such coupling is essentially elastic, and may, under certain circumstances, lead to hunting action on the part of the rotor. Governors which hunt must be corrected by the use of dash pots or other damping devices, and the introduction of the governor characteristic of a slight amount of speed regulation.

See also **Governor**.

HURRICANE-FORCE WIND. See **Winds and Air Movement**.

HUXLEY, ANDREW FIELDING (1917–Present). Andrew Huxley was an English physiologist and biophysicist who made fundamental discoveries about the basic mechanisms of the nervous system, and how muscle fibers contract.

The grandson of Thomas Henry Huxley (1825–1895) and half-brother of Aldous and Julian Huxley, Andrew Huxley was born in London. He read natural sciences at Trinity College, Cambridge and in the Physiological Laboratory, Cambridge, graduating in 1938. Whilst an undergraduate his tutor had been (Sir) Alan Hodgkin, and in 1939 he began research with Hodgkin, examining the physico-chemical basis of the nerve action potential, using the large diameter nerve fibre of the squid. During World War II (1940–1945) he worked on operational anti-aircraft research. He returned to Cambridge to Trinity College and the Physiological Laboratory in 1945, remaining there until 1960 when he moved to University College London as Jodrell Professor of Physiology (1960–1969) and later as Royal Society Research Professor (1969–1983). In 1984 he moved back to Cambridge as Master of Trinity College until 1989. See also **Hodgkin, Alan Lloyd (1914–1998); Huxley, Julian Sorrell (1887–1975);** and **Huxley, Thomas Henry (1825–1895)**.

With Hodgkin, Huxley studied the membrane potential of the nerve fiber, analyzing in precise chemical and physical detail the mechanisms by which nerve fibers transmit electrical messages. They showed that the selective permeability of the nerve membrane to charged particles, or ions, was responsible for the conduction of nervous activity along the nerve fiber. They demonstrated that at rest the membrane was selectively permeable to potassium ions, but when activated became selectively permeable to sodium ions. For this work, carried out mainly during the late 1940s and 1950s, Hodgkin and Huxley were awarded the 1963 Nobel Prize in Physiology or Medicine, which they shared with the Australian physiologist (Sir) John Eccles. In 1950 Huxley changed research direction, to muscle physiology, and he devised a special interference microscope with which to study and measure the contraction and relaxation of different components of the muscle cell fiber. From his observations, he postulated that between the thick and thin filaments of the muscle fiber, composed of the proteins actin and myosin, were "bridges." These generated a force in the direction of shortening, as the fibers slid over each other to overlap. This "sliding filament" theory is still accepted to account for the functional and morphological changes that occur during muscular contraction. See also **Eccles, John Carew (1903–1997)**; and **Neurochemistry (The History)**.

Additional Reading

Weismann, S.: "Sir Andrew Fielding Huxley," In: Fox, D.M., M. Meldrum, and I. Rezak: *Nobel Laureates in Medicine or Physiology: a Biographical Dictionary*, Garland, New York, NY, 1990, pp. 284–288.

E. M. Tansey, Wellcome Institute for the History of Medicine, London, UK

HUXLEY, JULIAN SORRELL (1887–1975). Julian Huxley was a British biologist and science popularizer who influenced the development of evolutionary theory, championed the idea that science should benefit society and served as first director-general of UNESCO (United Nations Educational, Scientific and Cultural Organization); http://portal.unesco.org/en/ev.php-URL ID=29008&URL DO=DO TOPIC &URL SECTION=201.html.

Julian Huxley was a scientist of many talents and prodigious intellectual output. His life consisted of a chain of intense periods — stints of brilliant laboratory and field research, episodes of self-doubt and clinical depression, intervals of writing prodigious synthetic works, and periods of public administration and global politicking. He was part of an elite intellectual family descending from his famous grandfathers, T. H. Huxley (the biologist sometimes called *Darwin's Bulldog*) and Thomas Arnold, famous headmaster of Rugby School. Huxley's family included his aunt, the novelist Mrs Humphrey Ward, his brother, the writer Aldous Huxley, and his half-brother, Andrew Huxley, who won the Nobel Prize for his research in physiology. While the focus of Julian Huxley's professional activities was never fixed for long, the motivation and interests underlying his work remained fairly constant. He sought to develop grand syntheses in biology, to create a religion of evolutionary humanism based on biology, and to bring these efforts to fruition through popularization and liberal political action.

As a research biologist, Huxley covered an extraordinary range of topics, although in general his laboratory work was related to the development of individual organisms and his fieldwork concerned evolution — in particular, the evolution of ritualized behavior exhibited by birds. In addition to dozens of specialized articles, Huxley wrote three major scientific books in which he attempted to synthesize broad ranges of biological findings concerning relative growth, embryology and evolution. Although Huxley's work on evolutionary theory is typically cited as his greatest contribution to biology (he coined many terms and phrases used by contemporary evolutionary biologists including "cline" and "the evolutionary synthesis,") the bulk of his original research was devoted to issues of development.

Huxley conducted his research as a Lecturer and Demonstrator at Oxford (1909–1912), founding member of the Department of Biology at Rice Institute (now University) in Houston, Texas (1913–1916), as Fellow and Senior Demonstrator at Oxford (1919–1925) and at London University as Professor of Zoology at King's College (1925–1927). He surprised his colleagues at King's College by resigning his professorship to devote full attention to writing *The Science of Life* with HG Wells and Wells' son.

As a science popularizer, Huxley wrote a large number of essays and books about biology and about the relation of biological knowledge to areas of general humanistic concern, including religion and ethics. He was a regular on the BBC and even advanced the popularization of science by directing and reforming the London zoo for several years (1935–1942).

Huxley's public ambitions were not limited to popularization and he had a keen interest in promoting the use of scientific knowledge to improve the human condition. This led him to advance evolutionary humanism and to participate in the eugenics movement, a movement based on the premise that scientific knowledge about heredity would make it possible to guide the future course of human evolution in directions that would "improve" the human species. Huxley's interest in seeing science used to improve the human condition also led him to insist that the United Nations' Education and Cultural Organization should incorporate science, and he was credited with putting the 'S' in UNESCO.

Julian Huxley did not fit the traditional mould of a research scientist working in isolation on narrow and esoteric problems. It might be said that he was as much a statesman of science as a research scientist.

Additional Reading

Huxley, J.S.: *Problems of Relative Growth*, Methuen, London, UK, 1932.

Huxley, J.S.: *The Uniqueness of Man*, Chatto &; Windus, London, UK; American title, *Man Stands Alone*, Harper, New York and London, 1941

Huxley, J.S.: *Elements of Experimental Embryology*, Cambridge University Press, Cambridge, UK, 1963.

Waters, C.K., and A. Van Helden: *Julian Huxley: Biologist and Statesman of Science*, Rice University Press, Houston, TX, 1992.

C. KENNETH WATERS, Minnesota Center for Philosophy of Science, University of Minnesota, MN

HUXLEY, THOMAS HENRY (1825–1895). Thomas Huxley was an English biologist and leading man of science who played a prominent role in promoting Darwinism. Huxley was the son of a schoolmaster in Ealing, the youngest survivor of seven. He always considered himself completely like his mother, Rachel Withers, both mentally and physically, including the hot temper and obstinacy that later earned him a well-deserved reputation as a controversialist. He had little early education and in 1841 became apprenticed as a surgeon to his brother-in-law Dr J. G. Scott. This was followed by a spell at Charing Cross Hospital, where he was greatly influenced by Henry Wharton Jones' lectures on physiology. He showed great talent and graduated MB in 1845, winning a medal for anatomy and physiology. He then applied for an appointment in the Royal Navy, sailing with Captain Owen Stanley as assistant-surgeon on HMS Rattlesnake from 1846 to 1850. The voyage, which surveyed the coastline of Australia, allowed him to make extended and very accurate examinations of marine invertebrates. He sent two path-breaking studies back to England, one dealing with the structure of hydrozoa (1847), the other on the affinities of medusae ("jellyfish" 1849). At this time he rejected the concept of an abstract morphological type and searched instead for resemblances in the embryological stages. His abilities were quickly recognized and on his return Huxley was elected FRS.

Much of Huxley's all-consuming passion for the professionalization of science dates from this time when he could not find adequately paid employment. He took a succession of small lecturing positions, including natural history at the Royal School of Mines and naturalist to the Geological Survey, and published prolifically in comparative anatomy and paleontology. In 1855 he estimated he could afford to marry his Australian fiancée, Henrietta Heathorn. By 1858 he was especially interested in the embryological development of the vertebrate skull, delivering the Royal Society's Croonian lecture on this subject.

His life took a significant turn in 1859 with publication of Darwin's *Origin of Species*. By then the two men were close friends although Huxley had been mostly skeptical about the possibility of transmutation. He was deeply impressed by Darwin's *Origin*, telling Darwin he would defend it from all attacks. But he regarded the theory as unproven until evidence of increasing sterility between closely related organisms could be produced. This proviso ran through all of his most famous articles defending evolution; and it was discussed at length in letters between the two over many subsequent decades. Even so, Huxley threw himself vigorously into the ensuing turbulence over the *Origin*, becoming the main protagonist for evolutionary theory. He fearlessly defended the naturalistic, reformist principles behind Darwin's work, seeing the theory as a step in the right direction towards a completely new kind of British science, free from church and state interference and divested of its hierarchical, class-based structure. His motives were mixed. Huxley certainly considered Darwinism as a key part of his own reformist program. He also used Darwinism as personal ammunition in his long-standing argument with Richard Owen. And the controversies resulted in gratifying fame for himself. Yet he truly believed in Darwin's achievement and was undoubtedly the major force in getting the theory accepted. He published three significant reviews in the first six months, declaring in the *Westminster Review* that the *Origin of Species* was a "Whitworth gun in the armory of liberalism" and in *Macmillan's Magazine* that "Extinguished theologians lie about the cradle of every science as the strangled snakes beside that of Hercules." Later, in the 1870s, Huxley aptly described himself as "Darwin's Bulldog." See also **Owen, Richard (1804–1892)**.

His part in the British Association meeting at Oxford in 1860 was notorious, not so much for what was said (few of the participants remembered the precise words) but for the dramatic symbolism of a public clash between church and science. The clash took place between Huxley and Samuel Wilberforce, Bishop of Oxford, at the end of the session on 30 June 1860. Neither was a scheduled speaker. Wilberforce reiterated the

substance of a long review of the *Origin* he had just composed, including many points drawn from Richard Owen. At the end he apparently asked whether Huxley claimed descent from apes on his mother's or father's side. Huxley had already argued bitterly with Owen over the structure of the brain, both during the meeting, and for years beforehand. Owen claimed the human brain was distinctively characterized by the hippocampus minor, a fold in the ventricle floor. Huxley claimed the structure could also be seen in ape brains. In replying to Wilberforce at Oxford, Huxley was therefore also continuing his disagreement with Owen. He said something to the effect that he would rather have an ape for an ancestor than a man who twisted the truth merely to make an impact. The crowd erupted, several other speakers tried to make themselves heard, and the meeting dissolved. Both sides apparently felt their man had won, although most modern historical accounts paint Huxley as a victor for science. The meeting has rightly been seen as a turning point in the public relations between science and religion in Victorian times.

Huxley's career advanced rapidly through the following decades. He wrote an influential little book on the anatomical relationship of apes and humankind, *Man's Place in Nature* (1863). He also became deeply involved in delivering lectures for working men and in educational reform, and he edited several scientific journals, taking an active part in the foundation of *Nature* in 1869. He served on several royal commissions and accepted many scientific appointments and honors, most notably Hunterian Professor at the Royal College of Surgeons and Fullerian Professor at the Royal Institution. In 1872, when the School of Mines moved from Jermyn Street to South Kensington, in London, he was able to create a new form of laboratory-based biological instruction. His students and demonstrators included W. Thisleton Dyer, Michael Foster, Ray Lankester and H. G. Wells, who all spoke highly of the training they received. With Huxley's determined intervention in the committee structure of the examination boards, this emphasis on personal observation and experiment spread throughout the educational system of Britain. He thoroughly relished participating in the administration of science and from 1864 had been the leading figure in a private dining club, the X club, which by virtue of the positions of its members, more or less controlled Victorian science.

Between 1870 and 1895 Huxley published a number of philosophical essays, including a sketch of David Hume (1879) whose form of practical skepticism deeply attracted him. By now Huxley was a professed agnostic, a word he coined to describe himself in 1869. He never enjoyed good health and in 1871 was forced to take a long holiday, arranged and financially subsidized by his scientific friends. For health reasons he gave up his public commitments in 1885 and in 1890 he left London for Eastbourne. His Romanes lecture on evolution and ethics was delivered in Oxford in 1893. He died in Eastbourne in 1895, and was buried in Finchley.

Additional Reading

Desmond, A.: *Huxley: From Devil's Disciple to Evolution's High Priest*, Addison-Wesley, Reading, MA, 1997.

Huxley, T.H.: *Select Works of Thomas H. Huxley*, Reprint, Kessinger Publishing Company, Whitefish, MT, 2007.

Mitchell, P.C.: *Thomas Henry Huxley: A Sketch of His Life and Work*, Kessinger Publishing Company, Whitefish, MT, 2006.

JANET BROWNE, Wellcome Institute for the History of Medicine, London, UK

HUYGENS, CHRISTIAN (1629–1695). Huygens was a Dutch physicist and astronomer whose areas of achievement included astronomy, invention, mathematics, physics, and technology.

Reading the works of Descartes had a great impact on his views of science. With his brother, he developed a better way to grind lenses. He built an improved telescope and studied Saturn. In 1655, Huygens discovered Saturn's largest moon, Titan, and he also discovered its rings.

Huygen had a mechanistic view of nature. Although not all of his ideas were correct, he was correct in his belief that light propagates as waves and Huygens' principle is still used today in describing the wave properties of light.

Huygen is also known for developing a weight-driven pendulum clock in 1656.

See also **Cassini Mission to Saturn**; **Eyepiece**; **Pendulum Clock**; **Saturn**; and **Interference (Wave)**.

J. M. I.

HUYGENS PRINCIPLE.   A very general principle applying to all forms of wave motion which states that every point on the instantaneous position of an advancing phase front (wave front) may be regarded as a source of secondary spherical wavelets. The position of the phase front a moment later is then determined as the envelope of all the secondary wavelets (ad infinitum). This principle, stated by Dutch physicist Christian Huygens (1629–1695), is extremely useful in understanding effects due to refraction, reflection, diffraction, and scattering of all types of radiation, including sonic radiation as well as electromagnetic radiation and applying even to ocean-wave propagation.

HUYGENS WAVELETS.   The assemblage of secondary waves asserted by Huygens to be set up at each instant at all points on the advancing surface of a wave, or phase front. Many phenomena of wave optics can be neatly explained on this assumption (Huygens principle) of the continual creation of new wavelets and the subsequent destructive or constructive interference between the wavelets to set up the next-imagined state of the advancing wave front.

HYADES.   An open, V-shaped, moving cluster of stars in the constellation of Taurus. References to the Hyades are to be found in all the ancient literatures, Virgil referring to them as the "rainy Hyades." The group is exceedingly rich in double stars, which, even with a small telescope and low magnifying power, present a beautiful appearance.

The Hyades form one of the best known of the so-called moving star clusters. The brightest star of the Hyades, Aldebaran, is not a member of the cluster, but has an independent motion through space and just happens to be in its present position at this time.

HYATT, JOHN WESLEY (1837–1920).   Hyatt is generally credited as being the father of the plastics industry. In 1869, he and his brother patented a mixture of cellulose nitrate and camphor which could be molded and hardened. Its first commercial use was for billiard balls. The TM "Celluloid" was the first ever applied to a synthetic plastic product; it flammability hazard limits its use.

HYBRID (Biological).   An organism produced by parents belonging to different species or to different strains of the same species. A hybrid combines characteristics derived from the two parent stocks and in some cases is more desirable than either. Beauty of flowers, productivity of various plants, and appearance and hardiness of animals have been enhanced by controlled hybridization.

When a hybrid is once secured its propagation is hampered by the fact that the diverse hereditary characters are reassorted in hereditary transmission by sexual reproduction. Hybrids are often infertile but even when they are capable of producing offspring they rarely breed true. The mule is the only animal hybrid of great value, and it is produced always by parents of the two species, horse and ass. Plant hybrids are not subject to this limitation, for they can usually be propagated by bulbs, cuttings, or grafts. Plants produced in this fashion are sometimes referred to as cultivars. See also **Plant Breeding**.

HYDANTOIN PROCESS.   See **Amino Acids**.

HYDATID DISEASE.   Also referred to as *echinococcosis*, this is an infection with *Echinococcus granulosus* or *E. multiocularis*, which are cestodes. These worms live in the intestines of dogs and wolves, whose feces include infective eggs. Such material may inadvertently find its way to a substance that is ingested by sheep, cattle, or humans. Infection with *E. granulosus* is most commonly found in regions where sheep and cattle are produced as, for example, in the western United States, parts of Canada, and Alaska. Upon ingestion of eggs, oncospheres are carried by the bloodstream to the liver, lungs, and other organs. These cause the development of cysts, often with neurologic symptoms. The cysts may grow to a diameter of 6 inches (15 centimeters), and contain many worms.

Mice and small animals are intermediate hosts for *E. multiocularis*, which resides in the intestines of foxes and dogs. These cestodes are found mainly in the Northern Hemisphere—Europe, Canada, Alaska, and north-central United States. They produce extensive alveolar hydatid cysts, frequently resulting in jaundice.

Frequently, the therapy for hydatid disease involves surgical excision of cysts. Cryosurgery is frequently used. Some success has been reported with the drug mebendazole in the treatment of the disease. See also **Tapeworm**.

HYDATOGENESIS.   A term used by petrologists to designate the process by which rocks are formed from highly aqueous solutions. Some petrologists limit the use of the term to rocks which have been deposited from water-rich magmatic solutions.

HYDRATE.   Excluding the loose usages in which the term hydrate indicates merely the presence of water or of its elements in 2:1 ratio, as in carbohydrate, the term hydrate denotes the appearance of water in compounds. There are a number of ways in which water may appear in stoichiometric proportions in compounds. Moreover, these ways may be described from more than one point of view. A somewhat systematic approach is to view these compounds from the point of view of the extent of integration of the water, or its elements, into the compound.

The term "water of constitution" is a somewhat old usage, applied to compounds in which no H_2O groupings appear in the structure of the compound, but the compound may undergo reaction, usually reversible, in which water is one of the products. Magnesium hydroxide and sulfuric acid could thus be said to have "water of constitution," even though it appears in their structure as hydroxyl groups, or hydroxyl groups and hydrogen atoms (protons).

The term "cationic water" may be used to describe the situation in which water appears in coordination compounds apparently joined to cations by covalent bonds. However, the fact that a number of such compounds exhibit "hydrate isomerism" is evidence for cationic bonding, as well as it is for the existence of other forms of these compounds in which the presence of water is due to electrostatic attractions or crystal stability requirements.

The term "anionic water" describes the situation in which water is joined to anions through covalent bonds, or more frequently, through hydrogen bonds. The type case is copper(II) sulfate pentahydrate, where the cation has a coordination number of four and presumably the fifth molecule of H_2O is bound to the sulfate ion (as well as to other H_2O molecules) by hydrogen bonds.

The term "lattice water" is commonly applied to cases in which the water molecules are occupying definite positions in the crystal lattice but are apparently not coordinated with either cations or anions. Again, clear-cut cases are those in which the compound is so highly hydrated that both lattice water and "ion water" are present.

The water in crystals may, however, be present in other than definite lattice positions. For example, the water molecules may be found in holes in the lattices, or they may occupy random positions in the lattices. The latter situation is often found in ion exchange resins where loss of water, up to a certain point, does not materially change the lattice structure.

Finally, in essentially noncrystalline materials, such as hydrous precipitates and colloidal gels, the water present is at the limiting case of being a hydrate, in which virtually no bonding, in the chemical sense, exists.

HYDRA (the serpent).   A southern constellation that forms the outline of a serpent.

HYDRATION.   *Hydration* can be defined as the process by which water is added to produce a *hydrate*. The process is reversible, that is, water can be reextracted from the hydrate. A second meaning of *hydration* involves a *hydration reaction*. This is where water is permanently and chemically combined with a reactant in a way that it can no longer be reextracted. The second meaning of hydration will not be considered here.

Hydration to form a hydrate can occur when water (*solvent*) interacts with *solute* molecules to form a solution, which involves *hydrophobic hydration*. Hydrophobic hydration is the main focus here.

Hydration of Hydrophobic Molecules

Hydration of nonpolar molecules is usually associated with hydrophobic effects in which water cavities or hydration shells are formed. This is accompanied by a loss in entropy and aggregation of the nonpolar (hydrophobic) molecules. Hence, ordering of the water hydration shell around hydrophobic molecules has been attributed to "clathrate" behavior in which the water hydration shell is dominated by pentagons compared to bulk liquid water. However, thermodynamic studies suggest that although there are larger numbers of pentagons in the solvation shell compared to the bulk, the hydration shell also contains significant numbers of hexagons and larger polygons. It was suggested that the existence of larger polygons in the hydration shell indicates that the clathrate analogy is too simple to explain water organization along hydrophobic surfaces.

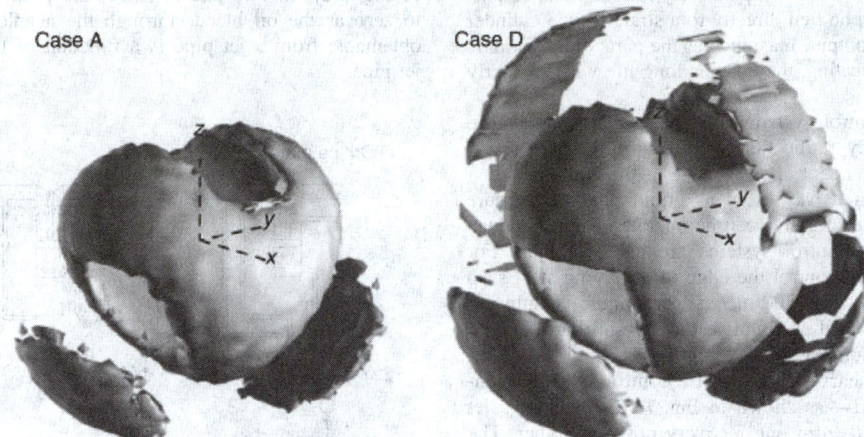

Fig. 1.   The hydration shell structure around methane: Case A, before hydrate formation (180 bar, 18 °C (64 °F); case D, after hydrate formation (180 bar, 4 °C (39 °F).

The water hydration shell around nonpolar molecules, such as methane, has been studied before, during and after gas clathrate hydrate formation using neutron diffraction with H/D isotopic substitution coupled with empirical potential structure refinement (EPSR) computer simulations. Gas clathrate hydrates are crystalline inclusion compounds that are formed when water and nonpolar molecules, such as methane, come into contact at high pressure and low temperature. Figure 1 illustrates the expansion of the water hydration shell and increased ordering of the second hydration shell around methane after clathrate hydrate formation.

Hydrophobic behavior has been observed for a concentrated solution of 0.86 mole fraction of tertiary butanol in water. This behavior is generally seen in more dilute solutions. A solution of tertiary butanol in water presents a classic example of an ampliphile–water solution. Neutron diffraction with H/D isotopic substitution coupled with EPSR computer simulations was used to examine the intermolecular correlations between alcohol and water molecules. The results revealed that "water pockets" are created within the concentrated alcohol–water solution, which reduce the degree of alcohol–alcohol direct hydrogen bonding. The pockets were found, on average, to contain between two and three water molecules. These water molecules exhibit a strong tendency to interact with the polar hydroxyl groups on the alcohol. As a result, the nonpolar groups became more tightly packed (i.e., exhibiting hydrophobic association) compared to that in the pure liquid alcohol. In the case of more dilute solutions of tertiary butanol in water (0.06 mole fraction of alcohol), the structural data do not support the conventional view that hydrophobic processes are dominated by perturbations of the first hydration shell. Conversely, significant changes are observed in the second hydration shell of water, with a "compression and tightening" of water molecules in the second shell.

A two-moment information theory was developed to describe the hydrophobic effect. This theory was shown to have a clear connection with the molecular principles of statistical thermodynamics. It was suggested that this theory could also be applied to describe hydrophobic effects on biopolymer structure in aqueous solution.

Hydration of Ions

Concentrated solutions of sodium hydroxide in water have been studied using neutron diffraction with H/D isotopic substitution and EPSR computer simulations. The solute was found to affect the tetrahedral network of hydrogen-bonded water molecules (cf. when high pressure is applied to pure water). The results also indicated that there was a competition between the hydrogen-bonding interactions and Coulomb forces in determining the orientation of water molecules within the cation solvation shell.

The effect of adding sodium chloride as a salting-out agent to a dilute solution of tertiary butanol in water (0.02 mole fraction of alcohol) has also been studied using neutron diffraction. Contrary to previous understanding on salting-out, these measurements show that an anion bridge is formed between the polar ends of nearby alcohol molecules. Hence, there is significant enhancement of polar–polar interactions of the alcohol hydroxyl

groups. As a result of the anion bridge formation, further hydrophobic interactions occur between the nonpolar/nonpolar groups.

Conclusion

Hydration of nonpolar molecules and salts and hydrophobic hydration have been studied in detail over the last couple of decades using computer simulation and theory. Recently, microscopic tools such as neutron diffraction with H/D isotopic substitution have been applied to study hydration of nonpolar molecules and salts. Neutron diffraction measurements coupled with molecular simulation provide a powerful tool to verify existing models of hydration of nonpolar molecules and salts, as well as combinations of these species.

This article originally appeared in the *Water Encyclopedia*, 5 Vols. Lehr, J. H., and J. Keeley, Editors, John Wiley & Sons, Inc, Hoboken, NJ.2005.

Additional Reading

Botti, A. et al.: "The Microscopic Structure of Concentrated NaOH Solutions," *J. Chem. Phys.*, **120**, 10154–10162 (2004).

Bowron, D.T., A.K. Soper, and J.L. Finney: "Temperature Dependence of the Structure of a 0.06 Mole Fraction Tertiary Butanol–water Solution," *J. Chem. Phys.*, **114**, 6203–6219 (2001).

Bowron, D.T. and J.L. Finney: "Structure of a Salt–amphiphile–water Solution and the Mechanism of Salting Out," *J. Chem. Phys.*, **118**, 8357–8372 (2003).

Buchanan, P. et al.: "*In situ* Neutron Diffraction Studies of Methane Hydrate Formation and Decomposition," *J. Chem. Eng. Data*, **48**, 778–782 (2003).

Bowron, D.T. and S.D. Moreno: "The Structure of a Concentrated Aqueous Solution of Tertiary Butanol: Water Pockets and Resulting Perturbations," *J. Chem. Phys.*, **117**, 3753–3761 (2002).

Franks, F.: *Water: A Comprehensive Treatise*, Vol. 2, Plenum, New York, NY, 1973, pp. 1–48.

Head-Gordon, T.: "Is Water Structure around Hydrophobic Groups Clathrate-like?" *Proc. Natl. Acad. Sci.*, **92**, 8308–8312 (1995).

Hummer, G. et al.: "An Information Theory Model of Hydrophobic Interactions," *Proc. Natl. Acad. Sci.*, **93**, 8951–8955 (1996).

Koh, C.A. et al.: "Water Ordering Around Methane During Hydrate Formation," *J. Chem. Phys.*, **113**, 6390–6397 (2000).

Koh, C.A.: "Towards a Fundamental Understanding of Natural Gas Hydrates," *Chem. Soc. Rev.*, **31**, 157–167 (2002).

Koh, C.A. and E.D. Sloan: "Clathrate Hydrates," in *Encyclopedia of Water*, John Wiley& Sons, Hoboken, NJ, 2005.

ANN KOH CAROLYN, Colorado School of Mines, Golden, CO

HYDRAULIC CONTROLLER. A device that uses a liquid control medium to provide an output signal, which is a function of an input error signal. Aside from the use of a liquid controlling medium, hydraulic controllers are similar in operating principle to electric, electronic, and pneumatic controllers. In fact, there are striking similarities between hydraulic control and pneumatic control. Because a liquid control medium is essentially incompressible, there is an excellent speed of response between controller and final actuating element. Hydraulic control systems also are characterized by high power gain inasmuch as liquids can be converted readily to high pressures or flows through the use of various types of pumps. The final actuators are comparatively simple; most outputs are two hydraulic lines that can be tied directly to a straight-type cylinder to provide a linear mechanical output. Inasmuch as the parts of a hydraulic system are essentially self-lubricating, they have a long life when properly designed.

Limitations of hydraulic control systems include special maintenance problems in connection with hydraulic fluids—fire hazard and leakage, and somewhat higher cost, dependent upon the size of the equipment.

Hydraulic controllers are extensively used as liquid pipeline-pressure controllers where a pipeline control valve can be operated against sudden pressure surges. Edge-guiding control systems are also common. For example, a hydraulic system can control the edge of a moving steel strip (typical strip velocity of 1,000 feet; 300 meters per minute) to plus or minus $\frac{1}{64}$-inch (0.4 millimeter) and accomplish this by shifting a coil of steel weighing up to 50,000 pounds (22,680 kilograms).

The hydraulic relay is the heart of a hydraulic control system. Commonly, a jet-pipe valve is used—as shown in Fig. 1. By pivoting a jet pipe, a fluid jet can be directed from one recovery port to another. The fluid energy is converted entirely into a velocity head as it leaves the jet-pipe tip and then is reconverted into a pressure head as it is recovered by the recovery ports. The relationship between jet-pipe motion and recovery pressure is shown in Fig. 2. Although the jet pipe can be used at higher pressures, most applications are less than 800 psi (~54 atmospheres). The proportional operation of the jet pipe makes it useful in proportional-speed floating systems (integral control) as indicated in Fig. 3(a). Position feedback can be provided by rebalancing the jet pipe from the work cylinder

as shown in Fig. 3(b). A proportional-plus-reset arrangement is shown in Fig. 3(c). In this last instance, the proportional feedback is reduced to zero as the oil bleeds through the needle valve. The hydraulic flow obtainable from a jet pipe is a function of the pressure drop across the jet pipe.

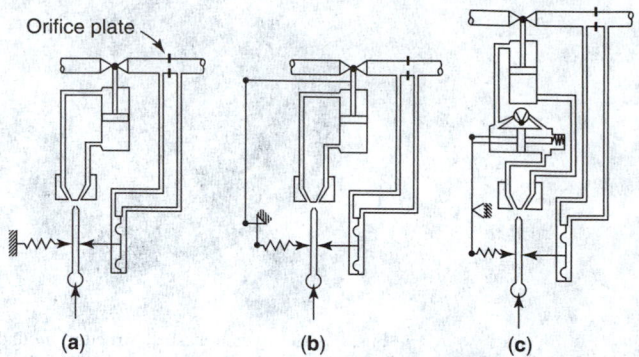

Fig. 3.   Hydraulic controllers: (a) proportional speed floating control; (b) proportional position control; (c) proportional plus reset control.

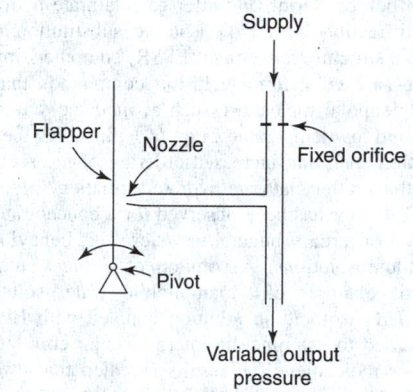

Fig. 4.   Single flapper valve.

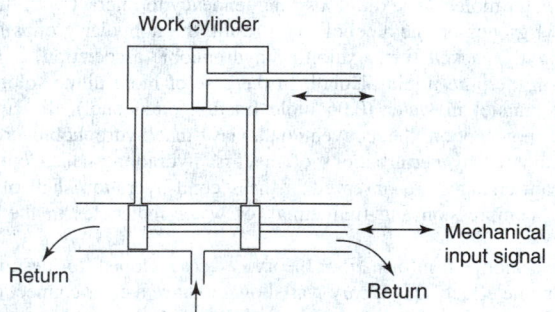

Fig. 5.   Spool valve.

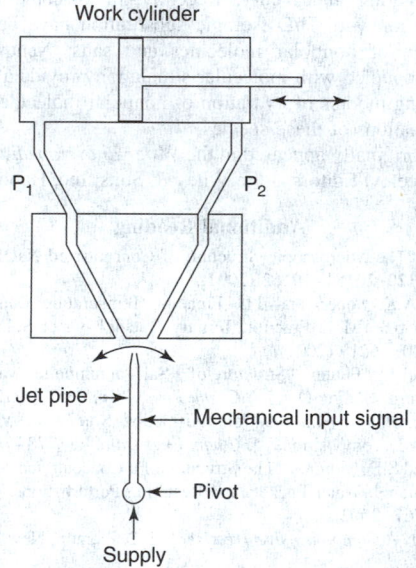

Fig. 1.   Jet-pipe valve used in hydraulic control system.

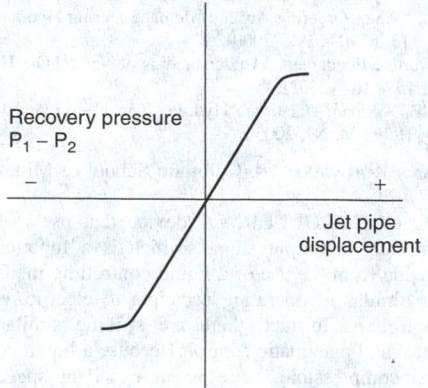

Fig. 2.   Jet-pipe motion and recovery pressure relationship.

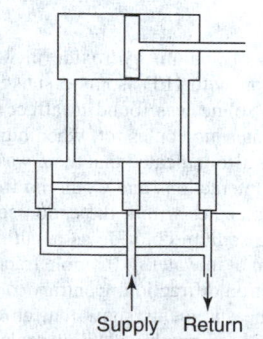

Fig. 6.   Four-way spool valve.

Flapper valves of the type shown in Fig. 4 also are used. The spool valve, shown in Fig. 5, when used as a hydraulic relay usually is constructed in either a three-way or a four-way valve-porting arrangement. See Fig. 6. The mechanical displacement of the spools allows the hydraulic-pressure supply to be ported in a fashion that will displace the work cylinder in either direction, depending upon the spool displacement.

HYDRAULIC FLUID. A liquid or mixture of liquids designed to transfer pressure from one point to another in a system on the basis of Pascal's law, i.e., pressure on a confined liquid transmitted equally in all directions. For industrial use, such fluids are based on paraffinic and cycloparaffinic petroleum fractions, usually with added antioxidant and viscosity index improvers. Flame-resistant types include additives such as phosphate esters or emulsions of water and ethylene glycol. The brake fluids used in auto are composed of (1) a lubricant (polypropylene glycol of 1000–2000 mw, a castor oil derivative or a synthetic polymeric mixture of monobutyl ethers of oxyethylene and oxypropylene glycols); (2) a solvent blend (mixture of glycol ethers); and (3) additives for corrosive resistance buffering, etc., bp 375–550 °F. The composition and performance characteristics are specified by the Society of Automotive Engineers: http://www.sae.org/.

HYDRAULIC RADIUS. The theory of hydraulics indicates that the ratio of the frictional area to the volume of the liquid stream is an important dimension governing the friction loss. The hydraulic radius, which expresses this fact, is the cross-sectional area of flow divided by the wetted perimeter of a cross section of the conduit. The hydraulic radius of a circular pipe flowing full of water is one-fourth of the diameter. The hydraulic radius of an open canal is the cross-sectional area of the stream divided by the wetted perimeter of the cross section.

HYDRAULICS. Hydraulics is the dynamics of liquids (hydrodynamics), especially applied to the practical problems of engineering. Although this general definition is entirely correct, in common usage hydraulics is the study of water at rest or in motion. This conception of hydraulics is used in this article. The mechanics of fluids (liquids and gases) in general is termed fluid mechanics. A basic proposition of hydraulics is that water is incompressible. While this condition is not completely met in fact, the compressibility of water is so small as to be negligible for practically all propositions of hydraulics. The viscosity of water varies with the temperature and is one reason for change of conditions of water flow in pipes with changing temperature. The unit weight of fresh water is usually taken as 62.4 pounds per cubic foot ($\sim$1000 kg/cu meter).

The science of hydraulics is divisible into hydrostatics and hydrokinetics. Hydrostatics is the hydrodynamics of liquids considered apart from their motion: hydrokinetics is the hydrodynamics of moving, especially flowing, liquids. Among the subjects included in any study of hydrostatics are the following: (1) the pressure on a submerged area of any shape or inclination, (2) the measurement of pressure on water at rest by manometers or pressure gauges, (3) buoyancy and flotation. Practical application of (1) is to be found in problems associated with water gates, large valves, pressure against dams, tanks, hydraulic presses, etc.

Hydrokinetics includes a great many different phases of hydraulics. Most of these will be found treated in specialized articles, references to which are given below. The flow of fluids supplies many cases of the application of hydraulic science. These include flows of steady, uniform, unsteady and nonuniform types, and the friction losses occasioned thereby, in closed or open conduits. The measurement of flows and the discharges under given conditions are part of this phase of hydraulics. Also, there is to be considered the flow of water through openings, such as orifices, nozzles, and weirs. The flow of water in pipe lines offers a great many problems in addition to friction: the discharge through different sections of branching and looping pipes, siphons, fittings, valves, etc., is included. Measurement of discharge of large amounts of water, as in stream and river flow, offers problems different from those met in closed conduits. Furthermore, the forces occasioned by deviated flows of water, as met in hydraulic turbines, the pump, and other hydraulic machinery, are fit subjects to be included in any study of hydromechanics. See also **Fluid and Fluid Flow**.

Additional Reading

Morris, H. and J. Wiggert: *Applied Hydraulics in Engineering*, 2nd Edition, John Wiley and Sons, Inc., New York, NY, 1972.

Parmley, R.O.: *Hydraulics Field Manual*, The McGraw-Hill Companies, Inc., New York, NY, 2000.

Smith, P. (Editor): *Applying Research to Hydraulic Practice*, American Society of Civil Engineers, Reston, VA, 1982.

Staff: United States Army Material Command, *Engineering Design Handbook: Hydraulic Fluids*, University Press of the Pacific, San Jose, CA, 2000.

Stutman, P.: *Applied Marine Hydraulics*, Cornell Maritime Press, Centreville, MD, 1988.

Zipparro, V.J. and H. Hasen: *Davis' Handbook of Applied Hydraulics*, The McGraw-Hill Companies, Inc., New York, NY, 1992.

HYDRAZINE. [CAS: 302-01-2] $H_2N \cdot NH_2$, formula weight 32.04, colorless, fuming liquid, mp 1 °C, bp 113 °C, sp gr 1.011, decomposes when heated above 350 °C at atmospheric pressure into N_2 and NH_2, also decomposes in presence of a catalyst (e.g., platinum) into N_2 and NH_3. Hydrazine burns when ignited in air with a violet-colored flame. The compound is soluble in all proportions with H_2O and is soluble in alcohol. Hydrazine forms a hydrate with one molecule of H_2O. Upon moderate heating or in a vacuum, the hydrate yields hydrazine and H_2O. Hydrazine is a base slightly weaker than NH_4OH.

Hydrazine is a tonnage chemical with numerous uses, including that of a propellant for rockets, yielding exhaust products at a high temperature and of a low molecular weight; use as a strong reducing agent in the manufacture of various chemicals; and as a blowing agent for foamed rubber. The compound reacts with citric acid to form *Continazin*, an antituberculan drug.

Although the earlier processes for the commercial production of hydrazine used urea as a raw material, modern processes employ direct ammonia oxidation. In one such process, reactions occur in two steps:

$$NH_3 + NaOCl \longrightarrow NH_2Cl + NaOH \tag{1}$$

$$NH_3 + NH_2Cl + NaOH \longrightarrow H_2N \cdot NH_2 + NaCl + H_2O. \tag{2}$$

High-grade hypochlorite is required for Step 1. Special agents, such as gelatin, ethylenediamine tetracetic acid, glue, high alcohols, and formaldehyde, are required to inhibit undesirable side reactions that would reduce the hydrazine yield through formation of ammonium chloride and N_2. In another hydrazine process, chlorine, NH_3, and H_2SO_4, along with methylethyl ketone, are used as the charge. The products of this process include hydrazine hydrate, hydrazine sulfate, ketazine, and dialkyldiazacyclopropane. Hydrazine also is used as a start-up ingredient in the preparation of cooling water for nuclear reactors where it is desired to keep the oxygen content of the water to an absolute minimum and thus decrease corrosion. Oxygen reacts with hydrazine. $H_2N \cdot NH_2 + O_2 \rightarrow N_2 + 2H_2O$. When no oxygen is present in the water, the hydrazine acts as a sink for dissolved oxygen that may enter later, by maintaining metal oxides at their lower oxidation states.

Hydrazine forms two series of salts: (1) hydrazinium (1+) chloride, $H_2NNH_3^+Cl^-$, nitrate, $H_2NNH_3^+NO_3^-$, hemisulfate, $(H_2NNH_3^+)_2 SO_4^{2-}$, (2) hydrazinium (2+) chloride, $H_3NNH_3^{2+}(Cl^-)_2$, dinitrate, $H_3NNH_3^{2+}(NO_3^-)_2$, hydrogen sulfate, $H_3NNH_3^{2+}(HSO_4^-)_2$, all soluble in H_2O. This last is produced when hydrogen azide reacts with concentrated H_2SO_4. It is very hygroscopic and decomposes in aqueous solution to give the slightly soluble monosulfate and H_2SO_4. The monosulfate and difluoride, which have been thought to have the structures $N_2H_5^+HSO_4^-$ and $N_2H_5^+HF_2^-$ in the solids, have been shown in fact to be $N_2H_6^{2+}SO_4^{2-}$ and $N_2H_6^{2+}(F^-)_2$. Hydrazinium azide, $N_2H_5^+N_3^-$, is a soluble solid.

In the laboratory, hydrazine can be prepared by converting one-half of a given amount of NH_3 into chloramine, NH_2Cl, by sodium hypochlorite solution in the presence of a colloid and heating. The remaining one-half of the NH_3 reacts with chloramine to form hydrazine. The product is then cooled to 0 °C and H_2SO_4 added in amount to react with the hydrazine to form hydrazine sulfate, $N_2H_6SO_4$, insoluble solid. Hydrazine hemisulfate, $(N_2H_5)_2SO_4$, is soluble in H_2O. It can also be made by the reaction of NH_3 and hydroxylamine-*O*-sulfonic acid.

Phenylhydrazine [CAS: 100-63-0] is a colorless liquid, slightly soluble in H_2O, miscible in all proportions with alcohol or ether, forms salts with acids, e.g., phenylhydrazine hydrochloride or phenylhydrazinium chloride, $C_6H_5NHNH_3Cl$, is a powerful reducing agent, with alkaline copper(II) salt solution (Fehling's solution) yields copper(I) oxide precipitate, reacts with carbonyl group of aldehydes or ketones yielding phenylhydrazones, white solids, of definite melting point and utilized in identification of aldehydes and ketones, e.g., acetaldehyde phenylhydrazone, $CH_3CH:NNHC_6H_5$.

Phenylhydrazine, as hydrochloride solution plus sodium acetate, reacts with polyhydroxy aldehydes or ketones yielding *osazones* or diphenylhydrazones, yellow solids, of definite melting point and utilized in identification of sugars, e.g., phenyl-d-glucosazone, CH_2OH $(CHOH)_3C$: $(NNHC_6H_5)CH:(NNHC_6H_5)$ plus aniline $C_6H_5NH_2$ plus NH_3.

Attention should be given to the difference between osazones and osones. An *osone* is formed by reaction of an osazone with HCl, e.g., glucosone, $CH_2OH(CHOH)_3CO \cdot CHO$.

1,1-Diphenylhydrazine is made by reduction of diphenylnitrosamine,, by zinc plus acetic acid, the nitrosamine being formed by reaction of diphenylamine,, and nitrous acid.

Tetraphenylhydrazine is a white solid, soluble in chloroform, acetone, benzene, or toluene, and upon standing is changed into triphenylamine plus azobenzene. In solution, tetraphenylhydrazine dissociates into nitrogen diphenyl, $(C_6H_5)_2N\cdot$, free radical, which in toluene at $90\,°C$ reacts with nitric oxide, NO. Tetraphenylhydrazine is formed by oxidation of diphenylamine,, by lead dioxide.

Hydrazine reacts with ketones to form *azines*.

HYDRAZOIC ACID.   [CAS: 7782-79-8] HN_3, formula weight 43.03, colorless, odorous, poisonous liquid, mp $-80\,°C$, bp $37\,°C$, explodes with marked violence. Also known as azoimide and hydronitric acid, the compound is miscible in all proportions with H_2O, alcohol, and ether. Hydrazoic acid reacts (1) with metals, e.g., magnesium, aluminum, zinc, iron, to form azides or hydrazoates (or trinitrides), (2) with heavy metal salt solutions to form insoluble azides, e.g., silver azide AgN_3, mercury(I) azide HgN_3, lead azide PbN_6. Silver, mercury(I), and copper(I) azides decompose in the light to form nitrogen plus the metal. (3) It reacts with NH_4OH to form ammonium azide $NH_4 \cdot N_3$, (4) with hydrazine to form hydrazine azide $N_2H_4 \cdot HN_3$, (5) with sodium hypochlorite plus acetic acid to form chlorazide ClN_3, explosive, (6) with sodium amalgam to form NH_3 with some hydrazine, (7) with potassium permanganate to form nitrogen and H_2O.

Hydrazoic acid is formed (1) by reaction of sodium nitrate with molten sodamide, (2) by reaction of nitrous oxide with molten sodamide, (3) by reaction of nitrous acid and hydrazinium ion $(N_2H_5{}^+)$, (4) by oxidation of hydrazinium salts, (5) by reaction of ethyl nitrite with NaOH solution and acidifying. See also **Azides**.

HYDRAZONES.   The products of the reaction between an aldehyde or a ketone with phenylhydrazine are termed *hydrazones*. Sometimes the compounds are referred to as phenylhydrazones.

$$CH_3\cdot CHO \; + \; C_6H_5\cdot NH\cdot NH_2 \longrightarrow CH_3\cdot CH:N\cdot NH\cdot C_6H_5 + H_2O$$
(acetaldehyde)   (phenylhydrazine)   (acetaldehyde hydrazone)

$$C_6H_5CHO \; + C_6H_5\cdot NH\cdot NH_2 \longrightarrow C_6H_5\cdot CH:N\cdot NH\cdot C_6H_5 + H_2O$$
(benzaldehyde)   (benzylidenehydrazone)

$$(CH_3)_2CO + C_6H_5\cdot NH\cdot NH_2 \longrightarrow (CH_3)_2C:N\cdot NH\cdot C_6H_5 + H_2O$$
(acetone)   (acetone hydrazone)

$$C_6H_5\cdot CO\cdot CH_3 + C_6H_5\cdot NH\cdot NH_2 \longrightarrow (C_6H_5)(CH_3)C:N\cdot NH\cdot C_6H_5 + H_2O$$
(acetophenone)   (acetophenonehydrazone)

Several of the hydrazones may be decomposed by strong acids whereupon the original aldehyde or ketone is regenerated, along with the formation of a phenylhydrazine salt. When reduced, hydrazones yield primary amines.

HYDRIDES.   Hydrides are compounds that contain hydrogen in a reduced or electron-rich state. Hydrides may be either simple binary compounds or complex ones. In the former, the negative hydrogen is bonded ionically or covalently to a metal, or is present as a solid solution in the metal lattice. In the latter, which comprise a large group of chemical compounds, complex hydridic anions such as BH_4^-, AlH_4^-, and derivatives of these, exist.

Commercial applications of hydrides have become important and some of these compounds have become industrial chemicals manufactured and used on a large scale.

Simple (Binary) Hydrides

The ionic or saline hydrides contain metal cations and negatively charged hydrogen ions. They crystallize in the cubic lattice similar to the corresponding metal halide, and when pure, are white solids. When dissolved in molten salts or hydroxides and electrolyzed, hydrogen gas is liberated at the anode. Their densities are greater than those of the parent metal, and their formation is exothermic. All are strong bases.

Sodium hydride finds commercial usage in organic synthesis in condensation and alkylation reactions.

Calcium hydride is a convenient portable source of hydrogen gas, which results from its reaction with water.

In all hydrides, hydrogen is bound to an atom of lower electronegativity ($X_H = 2.1$) than itself. In covalent hydrides, the hydrogen–metal bond is effected through a common electron pair. Beryllium and magnesium hydrides are included in this group and are polymeric materials, as is aluminum hydride. The simple hydrides of silicon, germanium, tin, and arsenic are gaseous or easily volatile compounds.

Beryllium Hydride [CAS: 7787-52-2], BeH_2 is an amorphous, colorless, highly toxic polymeric solid (H = 18.3%) that is stable to water but hydrolyzed by acid. It is insoluble in organic solvents but reacts with tertiary amines at $160\,°C$ to form stable adducts, eg, $(R_3N \cdot BeH_2)_2$. Beryllium hydride was formerly of interest as a rocket fuel and as a moderator for nuclear reactors. Toxicity has been a serious barrier to its commercialization.

Magnesium Hydride [CAS: 7693-27-8], MgH_2, is a gray powder of about 97% purity which is insoluble in inert organic solvents. It is easily oxidized, and when heated to about $280\,°C$, dissociates without melting. When prepared by direct reaction of the elements, magnesium hydride is stable in air and only mildly reactive with water. However, when it is obtained by pyrolysis of diethylmagnesium or by reaction of diethylmagnesium and $LiAlH_4$, it is very reactive with both air and water. This difference in reactivity mainly results from the much finer particle size of the product obtained by the pyrolysis route. This high reactivity allows reaction with 1-alkenes to form dialkyl magnesium compounds used in the production of Ziegler catalysts. Other uses for magnesium hydride include hydrogen storage and as a drying agent.

Aluminum Hydride [CAS: 7784-21-6] is a relatively unstable polymeric covalent hydride that received considerable attention in the mid-1960s because of its potential as a high energy additive to solid rocket propellants. The projected uses, including aluminum plating, never materialized, and in spite of intense research and development, commercial manufacture has not been undertaken. The synthetic methods developed were costly.

Silane [CAS: 7803-62-5], SiH_4 is a colorless gas that is spontaneously flammable in air and slowly decomposed by water; in the presence of aqueous alkali it is completely hydrolyzed to form hydrogen and silicates. It is manufactured on a commercial scale and sold as a compressed gas in cylinders. Silane, pure or doped, is used to prepare semiconducting silicon by thermal decomposition at $>600\,°C$. Gaseous dopants such as germane, arsine, or diborane may be added to the silane at very low concentrations in the epitaxial growing of semiconducting silicon for the electronics industry. Higher silanes, eg, Si_2H_6 and Si_3H_8, are known but are less stable than SiH_4. These are analogues of lower saturated hydrocarbons. See also **Silicon**.

Germane [CAS: 7782-65-2], GeH_4 is a colorless gas, spontaneously flammable in air. It is manufactured in small amounts and is available as a compressed gas in cylinders. Germane is used primarily to produce high purity germanium metal or epitaxial deposits of germanium on substrates for electronics by thermal decomposition at about $350\,°C$. See also **Germanium**.

Stannane [CAS: 2406-52-2], SxH_4 is a colorless poisonous gas that decomposes rapidly at room temperature. A large number of organostannanes, eg, R_3SnH and R_2SnH_2, are known, and their properties as organic reducing agents have been extensively investigated. Tributyltin hydride [CAS: 688-73-3] is used frequently to dehalogenate organic compounds.

Arsine [CAS: 7784-42-1], AsH_3 is a highly toxic colorless gas, made in small amounts as a dopant for silicon in the electronics industry by the reaction.

Other Simple Hydrides. Potassium hydride is commercially available. It is a stronger base than NaH and is used to make the strong reducing agent $KBH(C_2H_5)_3$ and super bases RNHK and ROK. Strontium and barium hydrides resemble calcium hydride in properties and reactivity. They have no significant commercial applications.

Complex Hydrides

The complex hydrides are a large group of compounds in which hydrogen is combined in fixed proportions with two other constituents, generally metallic elements. These compounds have the general formula $M(M'H_4)_n$, where n is the valence of M, and M' is a trivalent Group 3 (IIIA) element such as boron, aluminum, or gallium. In the BH_4- and AlH_4- anions, the hydrogen atoms are arranged tetrahedrally around the boron or aluminum and retain significant hydride or electron-rich character. For this reason, the complex hydrides have achieved significant and broad use as reducing agents in many different areas of chemistry. Lithium, sodium, and potassium borohydrides, lithium aluminum hydride, and sodium dihydrobis(2-methoxyethoxy)aluminate are commercially available, but many others are known, such as the corresponding alkaline-earth and other metal, as well as quaternary ammonium and phosphonium complex hydrides. Sodium borohydride in particular and lithium aluminum hydride are the most important commercially. In addition, compounds have been prepared in which from one to three hydrogen atoms have been replaced by other groups.

The alkali metal borohydrides are the most important complex hydrides. They are ionic, white, crystalline, high melting solids that are sensitive to moisture but not to oxygen. Group 3 (IIIA) and transition-metal borohydrides, on the other hand, are covalently bonded and are either liquids or sublimable solids. The alkaline-earth borohydrides are intermediate between these two extremes, and display some covalent character. They include lithium borohydride, $LiBH_4$, and sodium borohydride [CAS: 16940-66-2], $NaBH_4$.

Complete hydrolysis of $NaBH_4$ produces 2.37 L hydrogen (STP) per gram of borohydride; similarly, addition of acid to a cold aqueous solution liberates the theoretical amount of hydrogen. The inorganic reductions of $NaBH_4$ are numerous and varied. Sodium borohydride reacts with boron halides to form diborane, B_2H_6, which is more conveniently handled as the monomer BH_3 complexed with an ether, sulfide, or amine. Sodium borohydride is used extensively for the reduction of organic compounds. Sodium borohydride is manufactured from sodium hydride and trimethyl borate in a mineral oil medium at about 275°C. Sodium borohydride is classified as a flammable solid. It is available as powder, caplets, and granules and as a 12% solution in caustic soda. The principal uses of $NaBH_4$ are in synthesis of pharmaceuticals and fine organic chemicals; removal of trace impurities from bulk organic chemicals; wood-pulp bleaching, clay leaching, and vat-dye reductions; and removal and recovery of trace metals from plant effluents.

Potassium borohydride [CAS: 13762-51-1] was formerly used in color reversal development of photographic film and was preferred over sodium borohydride because of its much lower hygroscopicity. Because other borohydrides are made from sodium borohydride, they are correspondingly more expensive. Generally their reducing properties are not sufficiently different to warrant the added cost Zinc borohydride, $Zn(BH_4)_2$, however, has found many applications in stereoselective reductions.

The principal uses of $NaBH_4$ are in synthesis of pharmaceuticals and fine organic chemicals; removal of trace impurities from bulk organic chemicals; wood-pulp bleaching, clay leaching, and vat-dye reductions; and removal and recovery of trace metals from plant effluents. In pharmaceutical applications, the selectivity of sodium borohydride is ideally suited for conversion of high value intermediates, such as steroids, in multistep syntheses. It is used in the manufacture of a broad spectrum of products such as analgesics, antiarthritics, antibiotics, prostaglandins, and central nervous system suppressants.

Modification of the BH_4^- anion has provided derivatives of widely differing reducing properties. Alkoxyborohydrides, such as sodium trimethyoxyborohydride, $NaBH(OCH_3)_3$, exhibit enhanced reducing power but are less selective and more sensitive to decomposition by water. Sodium cyanoborohydride [CAS: 25895-60-7], $NaBH_3CN$, on the other hand, shows weakened reducing properties and is unique among the complex hydrides because it is stable in acidic aqueous solutions to a pH of about 3.

Sodium or tetramethylammonium triacetoxyborohydride has become the reagent of choice for diastereoselective reduction of β-hydroxyketones to antidiols. Trialkylborohydrides, e.g., alkali metal tri-sec-butylboro hydrides, show outstanding stereoselectivity in ketone reductions.

Aluminohydrides

In general, the aluminohydrides are more active and powerful reducing agents than the corresponding borohydrides. They decompose vigorously with water. Reaction also occurs with alcohols, although more moderately, providing a route to substituted derivatives.

Freshly prepared lithium aluminum hydride is a white crystalline solid that tends to become gray during storage, although very little loss in purity occurs. Although lithium aluminum hydride is best known as a nucleophilic reagent for organic reductions, it converts many metal halides to the corresponding hydride, e.g., Ge, As, Sn, Sb, and Si. Commercial manufacture of $LiAlH_4$ uses the original synthetic method, i.e., addition of a diethyl ether solution of aluminum chloride to a slurry of lithium hydride.

Sodium aluminum hydride can be prepared from NaH, but direct synthesis from the elements is more economical.

The few known derivatives of the aluminohydrides are principally alkoxy substitutions, including the trimethoxy, $LiAlH(OCH_3)_3$, triethoxy. $LiAlH(OC_2H_5)_3$, and tri-t-butoxy aluminohydrides, $LiAlH(O-t-C_4H_9)_3$.

Metallic Hydrides

A number of metal alloys are very useful for safely storing large volumes of hydrogen because these easily dissolve hydrogen at relatively low temperatures and pressures, forming interstitial hydrides. The hydrogen is subsequently released by applying heat and lowering the pressure. Many metals and binary and ternary alloys have been thoroughly studied for this application. The Sandrock and Thomas reference contains an exhaustive compilation of alloys screened for hydrogen storage. An on-line version is available at http://hydpark.ca.sandia.gov.

Hydrogen-storage alloys are commercially available from several companies in the United States, Japan, and Europe. A commercial use has been developed in rechargeable nickel–metal hydride batteries which are superior to nickel–cadmium batteries by virtue of improved capacity and elimination of the toxic metal cadmium. See also **Batteries: Secondary Cells**. Other uses are expected to develop in nonpolluting internal combustion engines and fuel cells, heat pumps and refrigerators, and electric utility peak-load shaving. See also **Fuel Cells**.

Transition-metal hydrides of titanium and zirconium have commercial use besides those related to hydrogen storage. The following paragraphs describe these two hydrides in more detail.

Titanium Hydride, [CAS: 7704-98-5], TiH_2, is a brittle, metallic-gray solid, density 3.8 g/cm^3, which produces 448 mL H_2 at STP per gram TiH_2. Titanium hydride powder is stable at room temperature and inert to water and most chemical reagents. Titanium hydride is used as a source for Ti powder, alloys, and coatings; as a getter in vacuum systems and electronic tubes; as a sealer of metals; and as a hydrogen source.

Zirconium Hydride, [CAS: 7704-99-6], ZrH_2, is a brittle, metallic-gray solid that is stable in air and water, and has a density of 5.6 g/cm^3. The chemical properties of ZrH_2 closely resemble those of titanium hydride. Commercial uses are as a getter in the manufacture of vacuum tubes and other systems; as a hydrogen source for foaming metals; as a hydrogen reservoir; for the introduction of zirconium into powdered alloys; for metal–ceramic and metal–metal bonding; as a moderator in nuclear reactors; and as a source of Zr metal powder and alloys.

Activated rare-earth metals react directly with hydrogen even at room temperature. These metals are activated by heating to 300°C in H_2, followed by cooling under H_2. Lanthanum dihydride [CAS: 13823-36-4], and lathanum trihydride [CAS: 13864-01-2] are both known compositions. Cerium hydride [CAS: 13569-50-1], CeH_2, and hydrides of higher hydrogen content ($CeH_{<3}$) have been made and studied. See also **Cerium**. Both cerium and lanthanum hydrides are black pyrophoric solids. Hydrogen content can be varied with hydrogen pressure and temperature over the metal hydride. Cerium and lanthanum hydrides are very reactive chemically and ignite spontaneously in air. They react with N_2 at 20°C and with water at 0°C. The most important uses of cerium and lanthanum hydrides are in the hydrogen-storage alloys $LaNi_5$ and $CeMg_2$. There is some interest in these materials as hydrogenation catalysts. The other rare earth metals (praseodymium, samarium, europium, gadolinium, and yttrium) react with hydrogen to form hydrides of varying composition. See also **Rare-Earth Elements and Metals**.

Group 5 (VB) Hydrides are formed from the metals, preferably by heating the metals in powder form in a hydrogen atmosphere up to 1000°C. Trace impurities (oxides, nitrides) in the metal prevent complete hydriding. The hydrides are brittle powders that can be handled in air. Heating to above 400°C initiates hydrogen evolution; complete hydrogen removal is usually obtained at 700°C under vacuum. Tantalum and niobium hydrides are superconductors at <10K. These hydrides are manufactured in small amounts mainly for research and development work in powder metallurgy. See also **Powder Metallurgy**.

Health and Safety Factors

In general, hydrides react exothermically with water, resulting in the generation of hydrogen. This hydrolysis reaction is accelerated by acids or heat and, in some instances, by catalysts. Because the flammable gas hydrogen is formed, a potential fire hazard may result unless adequate ventilation is provided. Ingestion of hydrides must be avoided because hydrolysis to form hydrogen could result in gas embolism.

Another aspect of the hydrolysis of hydrides is the alkalinity that results, especially from alkali metal and alkaline-earth hydrides. This alkalinity can cause chemical burns in skin and other tissues. Hydrolysis considerations obviously demand that hydrides be kept away from contact with acids.

Although there is little toxicity information published on hydrides, a threshold limit value (TLV) for lithium hydride in air of 25 $\mu g/m^3$ has been established. More extensive data are available for sodium borohydride in the powder and solution forms. The acute oral LD_{50} of $NaBH_4$ is 50–100 mg/kg for $NaBH_4$ and 500–1000 mg/kg for the solution. The acute dermal LD_{50} (on dry skin) is 4–8 g/kg for $NaBH_4$ and 100–500 mg/kg for the solution. The reaction or decomposition by-product sodium metaborate is slightly toxic orally (LD_{50} is 2000–4000 mg/kg) and nontoxic dermally.

Additional Reading

Adams, R.M. and A.R. Siedle: *Boron, Metallo-Boron Compounds and Boranes*, Wiley-Interscience, New York, NY, 1964, Chapt. 6, pp. 373–506.

Dedieu, A.: *Transition Metal Hydrides*, John Wiley and Sons, Inc., New York, NY, 1991.

Dedina, J. and D. Tsalev: *Hydride Generation Atomic Absorption Spectrometry*, John Wiley and Sons, Inc., New York, NY, 1995.

James, B.D. and M.G.H. Wallbridge: *Progr. in Inorganic Chemistry*, Vol. 11, Wiley-Interscience, New York, NY, 1970, pp. 99–231.

Mueller, W.M., J.P. Blackledge, and G.G. Libovitz: *Metal Hydrides*, Academic Press, Inc., New York, NY, 1968, Chapt. 12, pp. 546–674.

Sandrock, G., and G. Thomas: *Compilation of IEA/DOE/SNL Hydride Databases*, IEA Technical Report IEA/H2/A12-97, Sept. 1997.

Sastri, M.V.C., B. Viswanathan, and S.S. Murthy: *Metal Hydrides: Fundamentals and Applications*, Springer-Verlag, Inc., New York, NY, 1998.

Shrilrain, E. and S. Amoretty: *Thermophysical Properties of Lithium Hydride, Deuteride, and Tritide and of Their Solutions with Lithium*, Springer Verlag-New York, New York, NY, 1987.

Sodium Borohydride Digest, 3rd Edition, Morton International, 1995.

Walker, E.R.H.: *Chem. Soc. Rev.*, **5**, 23 (1976).

HYDROBORATION.   The reaction of diboranes either with alkenes (olefins) to form trialkylboron compounds or with acetylene to yield alkenylboranes. Much research has been devoted to developing these reactions, the products of which are called organoboranes. They are useful in many complex organic syntheses, including prostaglandins and insect pheeromones.

See also **Borane**; **Carborane**; and **Organoborane**.

HYDROCEPHALUS.   A condition characterized by abnormally large amounts of cerebrospinal fluid around or within the brain, usually associated with enlargement of the cerebral ventricles. See also **Meningitis**.

HYDROCHLORIC ACID.   [CAS: 7647-01-0] HCl (hydrogen chloride gas) in aqueous solution, colorless when pure. Commercial grades of HCl (also known as muriatic acid) generally are marketed in three concentrations: (1) 18° Bé (sp gr 1.1417 at 15.6 °C, 27.92% HCl); (2) 20° Bé (sp gr 1.160, 31.45% HCl); and (3) 22° Bé (sp gr 1.1789, 35.21% HCl). Frequently the commercial grades are slightly yellow because of impurities, notably dissolved iron. Fuming hydrochloric acid contains about 37% HCl, with a sp gr 1.194. Reagent grade hydrochloric acid usually is of this latter high strength, and is perfectly clear and colorless. The maximum limits set on impurities commonly are: NH_4 0.003%; arsenic 0.000001%; free chlorine 0.0001%; heavy metals, such as lead 0.001%; iron 0.00002%; sulfates 0.0001%; sulfites 0.0001%; and residue after ignition 0.0005%. A mixture of three parts HCl and one part is HNO_3 known as *aqua regia*, a powerful solvent and oxidizing agent which will dissolve materials that may be unaffected by either acid alone. Gold and platinum are soluble in aqua regia.

Hydrochloric acid is a very-high-tonnage chemical, finding major uses in (1) the cleaning and preparation of metals prior to application of coatings, (2) the recovery of zinc from galvanized iron scrap, (3) the production of numerous chlorides, and (4) production of chlorine. At one time, HCl was extensively used as a source of both hydrogen and chlorine by way of electrolysis. This process was made obsolete many years ago when the chlor-alkali process (electrolysis of sodium chloride brines) was introduced for the production of chlorine. In recent years, however, the production of byproduct HCl, resulting from chlorination of numerous organic compounds, has increased. In some of these instances, the installation of a HCl electrolysis plant may be economically feasible. For industrial consumption anhydrous HCl gas also is available in steel cylinders under a pressure of 1,000 psi (68 atmospheres). Hydrochloric acid forms a constant-boiling solution with H_2O (20.22% HCl) which has a bp 108.58 °C (760 mm Hg).

Dilute HCl reacts (1) with many hydroxides, e.g., NaOH, to yield the corresponding chloride, e.g., sodium chloride, solution, (2) with many ordinary oxides, e.g., magnesium oxide, to yield the corresponding chloride, e.g., magnesium chloride, solution, (3) with many carbonates, e.g., calcium carbonate, to yield the corresponding chloride, e.g., calcium chloride solution plus CO_2, (4) with many sulfides, e.g., ferrous sulfide, to yield the corresponding chloride, e.g., ferrous chloride, solution plus H_2S, (5) with many metals, e.g., zinc (but not copper) to yield the corresponding chloride, e.g., zinc chloride, solution plus hydrogen gas, (6) with some special oxides, e.g., lead or manganese dioxide, to yield lead or manganese chloride plus chlorine gas, (7) with solution of some salts, e.g., silver nitrate, to yield the corresponding chloride, silver chloride, precipitate. Higher strengths of hydrochloric acid usually react similarly to the dilute. Hydrochloric acid sometimes reacts as a reducing acid, e.g., (6) above.

All metallic chlorides, except silver chloride and mercurous chloride, are soluble in H_2O, but lead chloride, cuprous chloride and thallium chloride are only slightly soluble. Metallic chlorides when heated melt, and volatilize or decompose, e.g., sodium chloride, mp 804 °C; calcium, strontium, barium chloride volatilize at red heat; magnesium chloride crystals yield magnesium oxide residue and hydrogen chloride; cupric chloride yields cuprous chloride and chlorine. See also **Chlorine**; **Chlorinated Organics**; **Halides**; **Hypochlorites**; and **Sodium Chloride**.

Hydrogen Chloride.   This is a colorless gas, heavier than air, density 1.639 g/l at standard conditions. The gas is poisonous and quickly causes suffocation. Formula weight 36.47, mp −111 °C, bp −85 °C, critical pressure 83 atm, critical temperature 51.3 °C. The gas is very soluble in H_2O, accounting for the high concentrations of hydrochloric acid obtainable. Although hydrogen chloride gas may be used directly in some industrial operations, normally it is generated for the purpose of dissolving in H_2O to form hydrochloric acid. The most common route to HCl is by reacting sodium chloride with H_2SO_4. This is a two-step, exothermic reaction: (1) $NaCl + H_2SO_4 \rightarrow NaHSO_4 + HCl$, and (2) $NaCl + NaHSO_4 \rightarrow Na_2SO_4 + HCl$. Preparation of hydrochloric acid from the gas involves an absorption tower where the gas meets a fine spray of H_2O. Ratio controllers are used to assure maximum yield of the acid of desired concentration. These controls are easily adjusted for obtaining different concentrations. In most chlorinations of organic compounds, only half of the chlorine is used to substitute for hydrogen atoms, the remaining chlorine forming HCl. Frequently, this byproduct HCl is recycled or recovered.

Additional Reading

Behrens, D.: *DECHEMA Corrosion Handbook: Corrosive Agents and Their Interaction with Materials, Vol. 5, Aliphatic Amines, Alkaline Earth Chlorides, Alkaline Earth Hydroxides, Fluorine, Hydrogen Fluoride and Hydrofluoric Acid Hydrochloric Acid*, John Wiley and Sons, Inc., New York, NY, 1989.

Considine, D.M. and G.D. Considine: *Van Nostrand Reinhold Encyclopedia of Chemistry*, 4th Edition, Van Nostrand Reinhold Company, Inc., New York, NY, 1984.

Lewis, R.J. and N.I. Sax: *Sax's Dangerous Properties of Industrial Materials*, 10th Edition, John Wiley & Sons, Inc., New York, NY, 1999.

Wu, T. and T. Young: *Enthalpies of Dilution of Aqueous Electrolytes: Sulfuric Acid, Hydrochloric Acid, and Lithium Chloride*, National Bureau of Standards, National Engineering Lab, Washington, DC, 1979.

Web References

Clinical Center of the United States Government National Institutes of Health. http://www.cc.nih.gov/cp/about_clin_path/hcl.html

HYDROCHLOROFLUOROCARBONS (HCFCS).   A collection of partially chlorinated and fluorinated hydrocarbons (mostly methanes and ethanes), used as refrigerants, foam-blowing agents, and solvents. These species have been developed as replacements for the now-banned chlorofluorocarbons.

Examples include HCFC-141b (CH_3CFCl_2) and HCFC-142b (CH_3CF_2Cl). These compounds are less harmful to the ozone layer than the chlorofluorocarbons, due to their shorter atmospheric lifetimes. However, because their ozone depletion potentials are nonzero, these compounds will only be in use temporarily and will ultimately be phased out of production.

See also **Ozone**; and **Ozone Depletion (Science of)**.

HYDROCOLLOID. A hydrophilic colloidal material used largely in food products as emulsifying, thickening, and gelling agents. They readily absorb water, thus increasing viscosity and imparting smoothness and body texture to the product, even in concentrations of less than 1%. Natural types are plant exudates (gum arabic), seaweed extracts (agar), plant seed gums or mucilages (guar gum), cereal gums (starches), fermentation gums (dextran), and animal products (gelatin). Semisynthetic types are modified celluloses and modified starches. Completely synthetic types are also available, e.g., polyvinylpyrolidone. Most are carbohydrate polymers, but a few such as gelatin and casein are proteins.

HYDRODEALKYLATION (HDA). A type of hydrogenation used in petroleum refining in which heat and pressure in the presence of hydrogen are used to remose methyl or larger alkyl groups from hydrocarbon molecules, or to change the position of such groups. The process is used to upgrade products of low value, such as heavy reformate fractions, naphthenic crudes, or recycle stocks from catalytic cracking. Also toluene and pyrolysis gasoline are converted to benzene, and methyl naphthalenes to naphthalene, by this process.

See also **Hydrogenation**.

HYDRODEIK. A form of psychrometer with wet- and dry-bulb thermometers mounted on opposite edges of a specially designed graph of the psychrometric tables. It is so arranged that the intersections of two curves determined by the wet- and dry-bulb readings yield the relative humidity, dewpoint, and absolute humidity.

HYDRODYNAMIC PRESSURE. The difference between the pressure and the hydrostatic pressure. This concept is useful chiefly in problems of the steady flow of an incompressible fluid in which the hydrostatic pressure is constant for a given elevation (as when the fluid is bounded above by a rigid plate), so that the external force field (gravity) may be eliminated from the problem. If p^* is the hydrodynamic pressure, the density, and V the speed, Bernoulli's equation gives

$$p^* + \frac{1}{2}\rho V^2 = \text{constant along a streamline.}$$

See also **Static Pressure**.

HYDRODYNAMICS. The study of fluid motion. "Fluid" here refers ambiguously to liquids and gases. Although "classical" hydrodynamics was primarily concerned with incompressible fluids, the term aerodynamics has been reserved for such a specialized aspect of compressible fluid flow that most of meteorological dynamics is best included under the general heading of "hydrodynamics." W. and J. Bjerknes refer to the hydrodynamics of compressible fluids as physical hydrodynamics.

AMS

HYDROELECTRIC POWER. In a hydroelectric power plant, advantage is taken of the gravitational energy available from water flowing from a higher level to a lower level. In seeking a lower level, the flowing water is directed to exit through a hydraulic turbine, which in turn drives an electric generator. A substantial portion of the world's electric power is derived from hydro facilities, particularly in some countries. In general, however, the hydro percentage of total power generation has been declining over the past several years, notably for economic reasons. The technology of hydro power has been developed to a very advanced state.

Classification of Hydroelectric Plants

Low-, Medium-, and High-Head Facilities. There is no definite line of demarcation between high, medium, and low hydraulic heads. Generally, a plant with a head of more than 500 feet (152 m) can be considered a high-head development; a plant with a head of 50 feet (15 m) is definitely in the low-head class. See Figs. 1 and 2.

Briefly, the characteristics of the low-head plant are: vertical, reaction type, runners using large volumes of water and requiring large water

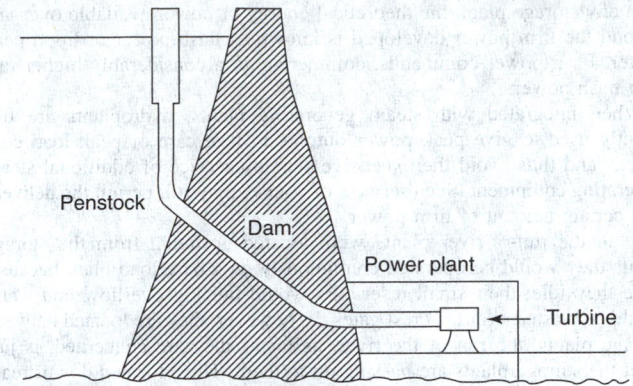

Fig. 1. River plant (high head). Typified by the Hoover Dam and Niagara River installations.

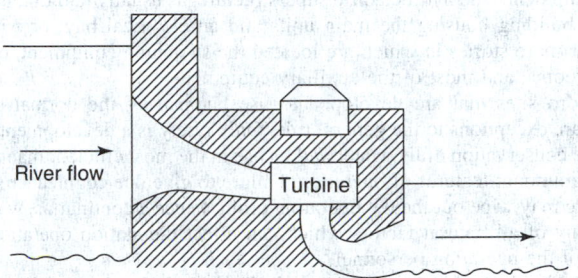

Fig. 2. River plant (low head). Typified by the Bonneville and St. Lawrence Waterway installations.

passages. Substructure is both extensive and expensive, and intake works are large and complicated. Large diameter generators are made necessary by the low rotational speeds. Characteristics of the high-head plant are: horizontal impulse turbines, small volumes of water at high pressures, plant at some distance from the dam. The advantage of smaller and simpler substructure is offset by the presence of a long water conduit, or penstock, between dam and plant. The turbines are high-speed and allow smaller generator diameter. The high-speed is accounted for by the high heads used. Inherently, the impulse turbine has a low characteristic speed.

Impounded Volume. The possible hydroelectric development sites along the flow of a stream are of two types, namely, those suitable for run-of-the-river plants (See Fig. 3) and those offering natural impounding basins for storage plants. In general, the run-of-the-river plant is cheaper than the storage plant of equal capacity, but it suffers seasonal variation of output more or less proportional to the variation of stream flow.

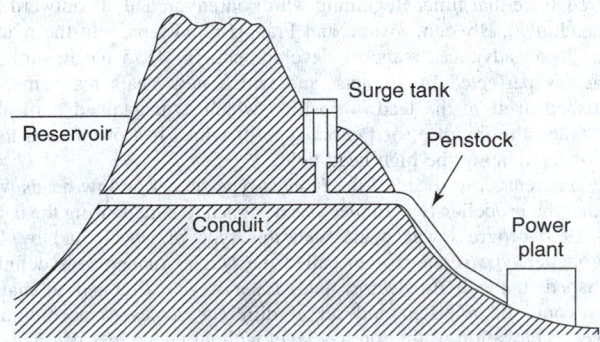

Fig. 3. Mountain reservoir plant.

Storage plants give a greater proportion of firm power, which can be delivered day by day on a regular schedule. This firm power is in more or less direct ratio to the degree of regulation of the flow of the stream and this in turn is a function of the impounded volume. Complete regulation of stream flow is rarely possible or practical, although 80–90% regulation is not infrequent.

In any storage plant, the theoretical energy or power available over and beyond the firm power developed is known as flash power or flood peak power. Firm power commands, commercially, a considerably higher rate than flash power.

When integrated with steam generating plants, hydroplants are frequently used to give peak power outputs to take care of peak load conditions and thus avoid the expensive standby service of additional steam generating equipment. Such service, of course, may still permit the delivery of a certain amount of firm power.

If all the run-of-river plants were located upstream from the storage plants they would be operated continuously on a base load plan, because, were they idle, their small reservoirs would quickly overflow and water would be wasted over the crest gates. If, however, they are located between storage plants, the run of the river, as far as they are concerned, is just what the storage plants are passing on to them. Thus, located downstream from a storage plant, a run-of-river plant will produce an increase in output when the storage plant increases its output.

In the hydroelectric plant the turbines and generators are the main items of equipment. The hydroelectric superstructure, as usually laid out, has one large building housing the main units and an electrical bay, or wing, of one or more stories in which are located the switching equipment, offices, storerooms, and most of the auxiliary equipment.

Hydro sites that are developed to use but part of the normal stream flow are exceptions to the general rule. Only rarely is a development made where conservation of the water and its use in the most efficient manner are not paramount features of operation. Failure to give due cognizance to this feature may wipe out the net operating profit; hence a continuous, watchful scrutiny of all natural factors which can affect the station operation is a duty of the operating personnel.

A hydraulic turbine suffers loss of efficiency at heads above or below the designed value because of shock losses. At the correct head there will be one point of best efficiency, somewhere between 80–95% of full load. When a number of units are installed in a plant, and when steam reserve is available, it is generally possible to operate the units near the point of best efficiency. There are four faults of operation and maintenance, which can reduce the maximum energy production of a plant:

1. Waste of water over spillways.
2. Improper distribution of the load between the station units.
3. Water leakage through valves, gates, dam or flow line.
4. Wear on moving parts, especially corrosion or erosion of the runner.

The relative simplicity of hydroelectric equipment makes hydraulic efficiency of the turbine the principal consideration.

Combined hydro and steam power plants, designed for use as pumped-storage plants are described later.

Hydraulic Turbines

The fundamentals of the turbine were incorporated into the wheels built before the turn of the nineteenth century, but its principal development has occurred since that time. Beginning with Fourenyon and his outward flow turbine, Jonval, Boyden, Swain, and Francis rapidly brought the reaction turbine to an advanced stage of development. By 1875 the inward flow turbine, as perfected by Francis, and which now bears his name, had established itself in the lead, a position which it maintained until about 1900, when the impulse, or Pelton, type of wheel had progressed to the point of dominating the high head field.

The inherent slow speed of the Francis-type runner on low heads was a fault that the propeller-type runner was designed to cure. During the decade 1910–1920 progress was made with this type of wheel, and by 1920 the propeller-type runner, often called the Nagler runner, was definitely established in the hydroelectric field. Later it was arranged so that the blades could be adjusted and set at different angles to accommodate changes in elevation of the forebay level without undue loss of efficiency. The success of the propeller-type turbine encouraged American adoption of the Kaplan turbine, on which the blade adjustment is performed automatically, being under the same control as the turbine gates.

As between impulse and reaction types, the action in the impulse turbine is easiest to understand. There is no difficulty in visualizing the transformation of pressure head into velocity head at the nozzle, nor of understanding the push, or impulse, that is given to the buckets by the stream of water. The jet is directed upon the rotor tangentially, and hence this type is also called the tangential turbine. The velocity of the jet of water is only slightly less

than the free spouting velocity under the effective head h. Impulse buckets are divided into two halves by a "splitter" and the axial thrusts which would otherwise have to be borne by special bearings are equalized.

The essential difference between the impulse and reaction types is that in the former the entire energy received by the wheel is in the velocity form, while in the latter it may be partially in the velocity form, but is also, in a large measure, still in the pressure form. The reaction of conversion of residual pressure into velocity in the runner is the source of much of the torque delivered to the reaction turbine. If the turbine were blocked stationary and had its gates opened, the water would issue from the turbine as from a nozzle. Now, by removing the blocking, let these nozzles begin to rotate, and the absolute velocity of water leaving them is found to be diminishing, the energy having been absorbed by the runner. At the best speed the final velocity will be just sufficient to enable the water to clear the runner. At this time, the wheel may be absorbing from 90–95% of the energy that the water had in the pressure form just before reaching the turbine gates.

In a Francis turbine, the water flows inward, then downward and into the draft tube. See Fig. 4.

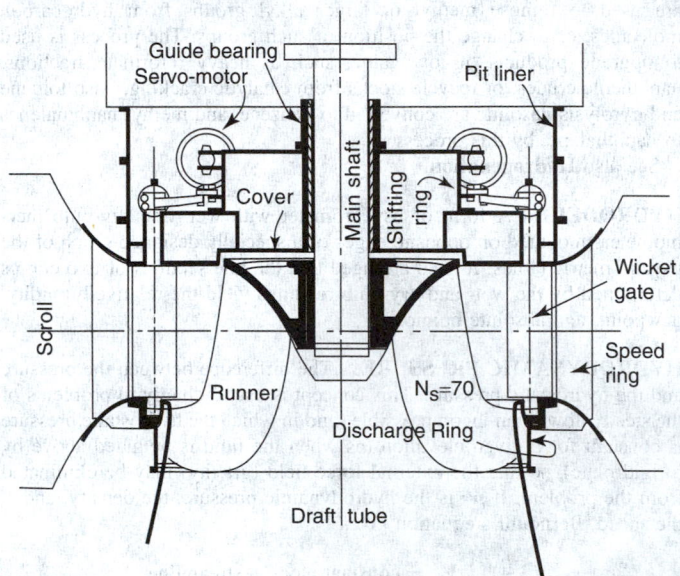

Fig. 4.   Cross section showing component parts of a Francis turbine.

A convenient classification of hydraulic turbines is:

1. *Reaction Turbine*   (Water under pressure is only partially converted into velocity before it enters the turbine runner.)

   (a) Francis Turbine

   (b) Propeller Turbine

     — 1 Fixed-blade

     — 2 Adjustable-blade (*Example*: Kaplan turbine)

     — 3 Axial-flow (*Example*: Dariaz turbine)

     — 4 Diagonal-flow

2. *Impulse Turbine* (*Example*:   (Water under pressure is entirely Pelton Wheel)   converted into velocity before it enters the turbine runner.)

The Francis turbine is rarely a horizontal shaft machine, except in small sizes and where it is desired to avoid the expense of excavation for a vertical setting. The standard runner consists of two crowns between which the buckets or blades are placed. It is best adapted to vertical setting. In order to pass the large discharges possible in a high specific speed wheel, the buckets are curved downward. Some axial flow action is present in runners of high specific speed. Water is admitted to the runner through guide vanes and gates.

Loss of efficiency at part load is sometimes a serious fault as, for instance, where only one or two units are installed in an isolated plant.

The feature of the Kaplan turbine is that the blade angles and gates are adjusted simultaneously by the governor mechanism so that the blades are always in the position best suited for full utilization of the flow, through the reduction of eddying and shock losses. The result is that the efficiency at part load holds up remarkably well.

Conveying the water from the penstock and directing the proper amount of it correctly against the runner requires first, a scroll case; second, a speed ring; and third, turbine gates.

The scroll case for medium and high head development is circular in form. In plan, it leads from the penstock and wraps, in spiral form, around the speed ring. The cross section of the spiral at any point should be such that the water flows with uniform velocity. This leads to the spiral form, because the water is being delivered to the turbine uniformly around the entire circumference.

The speed ring is that part of the turbine which joins the discharge ring with the turbine cover and pit liner. The ribs between the top and bottom portions must be strong enough to support the dead weight above the casing, consisting of concrete, generator, and turbine rotative parts; hence the speed ring is a very important part of the turbine.

Inside the speed ring, and rigidly bolted to it, is the inlet gate mechanism. The mechanism is operated by the governor, which, by opening or closing the gates, can maintain a control of speed under variable load. Gates are of the guide vane type and, while various types of gates have been used, the wicket gate is in general use at the present time. Its principal advantage is its efficiency. Shock losses at part gate opening are reduced to a minimum in the wicket gate. It is not particularly tight and has many wearing parts, most of which are bronze bushed and grease lubricated.

The Pelton wheel is either a solid or open disk, to the rim of which are attached buckets upon which a jet of water is played from a stationary nozzle. A horizontal shaft is the usual arrangement, but vertical shaft units have also been put into operation. The advantage of using the vertical arrangement is that more than one jet can be played on the buckets; this is obtained, however, at the expense of some loss of efficiency. The Pelton wheel is overhung on the bearing and often, for additional capacity, two wheels are overhung on the same generator. Variable power demand is met by decreasing the amount of water in the jet, by deflecting the jet from the buckets, or both. Some turbines of this type have a relief jet, which opens as the main jet closes. Afterwards, a dash pot slowly closes the relief jet, slowly enough to prevent a large pressure rise in the penstock. The same is also accomplished by deflecting the jet from the wheel upon loss of load, then slowly closing the valve controlling the jet.

Draft Tube. Hydraulic turbines frequently discharge the water with considerably more velocity than would be economical from the efficiency viewpoint, were it not possible to recover a great deal of that energy by the proper use of a diffusing chamber at the outlet. The diffusing chamber or tube is known as the draft tube, and there are a variety of types. However, the main objective is to convert the velocity head residing in the water leaving the turbine into pressure head. If this can be done efficiently, the turbine can be set somewhat below normal tailwater level.

The greater the specific speed of a turbine runner the higher will be the velocity of the water discharged into the draft tube, and the more important the recovery of this velocity by draft tube design. The draft tube is to take the water from the turbine at a point where the pressure is considerably less than atmospheric, and, by efficiently reducing the velocity, convert it into pressure head so that it can emerge smoothly into the tailrace at atmospheric pressure. By "efficiently" is meant without shock or whirl loss. Not all the velocity head can be recovered, for the water must be given to the tailrace at normal tailrace velocity to prevent its backing up into the turbine. Also, whatever friction loss occurs in the draft tube adds to this reduction of useful head.

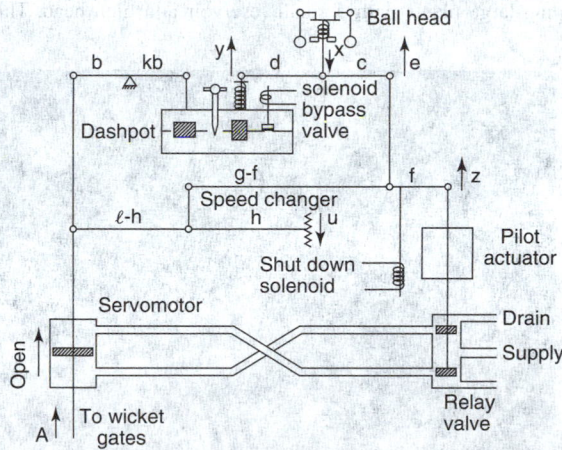

Fig. 6.   Typical hydro governor system.

Fig. 7.   The first hydroelectric development in Newfoundland was built at Petty Harbour, and electricity was transmitted to St. John's for the first time on April 19, 1900. The original equipment consisted of a Pelton wheel turbine, which was connected to a 250kVA General Electric generator. In 1907, a second Pelton wheel turbine and a generator similar in size to the original were installed. To meet increased demand and to provide for further expansion of the system, the building was extended in 1914 and a Voith turbine and a 500 kVA Westinghouse generator were installed. (*Courtesy, Newfoundland Light & Power Co. Limited.*)

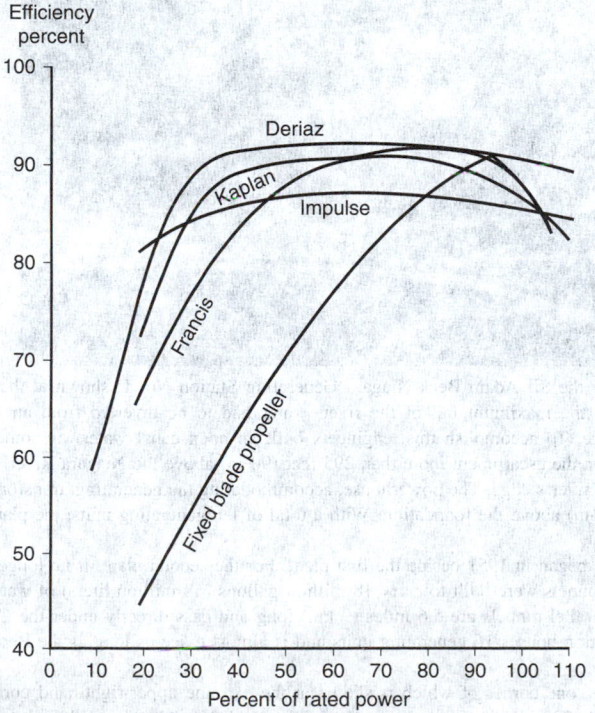

Fig. 5.   Hydraulic turbine efficiency curves.

Typical turbine efficiency curves are shown in Fig. 5. A schematic diagram of a typical hydro governor system is shown in Fig. 6.

Pumped-Storage Plants

Growing emphasis over the past couple of decades has been placed upon the use of special hydro plants as a means of storing energy in the form of a head of water — pumped into an upper reservoir during off-peak hours. The history of pumped-storage plants dates back to the late 1920s when several plants were first installed in Europe. The Rocky River plant of Connecticut Light and Power Company was the first to be built in the United States during that early period. From the viewpoint of plant location, there are three categories of pumped-storage installations:

1. *Combined with conventional hydro plant.* Plants of this type are used in locations suitable for conventional hydro plants, but where rainfall or water availability and system demand are out of phase. For example, in Switzerland, demand is highest in winter when water is scarce and lowest in summer where there is an abundance of water. The available energy in summer can be used to fill the reservoirs and store the available water for later use in meeting the winter demand.

2. *Pure pumped storage.* The advantages of this type of plant are its flexibility of location, in that the upper reservoir need have no source of water other than what is pumped into it, and the possibility of developing large plants with a small reservoir and high head. This type of plant is commonly used in steam-based systems, which lack the many advantages of available hydro generation. As well as providing fast and reliable peaking power, pumped-storage units have the added advantage of smoothing the weekly load curve and enabling more of the efficient base-loaded steam plants to be operated continuously.

3. *Pumped storage with diversion.* This situation arises when available water must be shared between power generation and irrigation use. Water that must be pumped to a higher reservoir to feed an irrigation canal can be used as a source of peaking power if allowed to run back down through a pump-turbine.

Design Configurations. The three fundamental configurations include: (1) separate pump and turbine on the same shaft; (2) reversible pump-turbines; and (3) axial-flow units.

The turbines used in the first class are Francis type for heads in the range 100 to 1,000 feet (30.5 to 305 meters) and Pelton wheels for heads up to 3,000 feet (914 meters). These units are usually mounted on a horizontal shaft with a clutch (hydraulic or friction) between the motor-generator and the pump. The turbine is usually rigidly connected to the shaft and is dewatered, using compressed air during pumping. Sometimes small impulse turbines are installed on the shaft for starting and braking.

Reversible pump-turbines are of radial or mixed flow type — Francis or Deriaz — and have been designed to operate at a wide range of heads.

Fig. 8.   Called the Queenston-Chippaway development when construction began in 1917, the Sir Adam Beck-Niagara Generating Station No. 1, shown at the right of the view, was for many years the largest hydroelectric plant in the world. To utilize the maximum fall of the river, water had to be diverted from an intake 2 miles (3.2 km) above the Horseshoe Falls to the plant at the base of the Niagara Gorge. To accomplish this, engineers built an open canal waterway some $12\frac{1}{2}$ miles (20 km) long from Chippaway across country to a triangular basin called a *forebay* on the escarpment more than 295 feet (90 m) above the Niagara River. From the forebay, giant *penstocks* (tubes) carry the water to the powerhouse below, located on the river's edge. The powerhouse, accommodating the generating, transforming, and controlling equipment, rises more than halfway up the cliff to a height of 180 feet (55 m) above the foundation. With a total of ten generating units, the plant has an installed capacity of 414,650 kW.

Construction of the Sir Adam Beck Generating Station No. 2, shown at the left of view, began in 1951 beside the first plant. For the second plant, it no longer was feasible to interrupt surface traffic to build another open canal. Instead, two underground tunnels were built to carry 18 million gallons (68 million liters) of water per minute from Chippaway to the forebay. With a finished diameter of 46 feet (14 m), the parallel tunnels are 5.6 miles (9 km) long and pass directly under the City of Niagara Falls, Ontario at a depth of 331 feet (101 m). Opened in 1954, the Beck No. 2 station houses 16 generating units and is almost twice as long as the Beck No. 1 station. The more recent plant has an installed capacity of 1,223,600 kW.

To accommodate for peak loads on the Beck No. 2 station, a pumped-storage reservoir, one corner of which is shown at the extreme upper-right-hand corner of view, was created. A separate pumping-generating station, containing 6 generators, adds 176,700 kW to the total capacity of the Beck No. 2 station. When the pumps are reversed, they act as turbogenerators. (*Ontario Hydro.*)

Axial-flow units are designed to operate at low heads, around 20 feet (6 meters), and have adjustable blades similar to those of a Kaplan turbine. The bulb type is used in Europe and the tube type has been designed and built in the United States. Both can operate as turbines or as pumps in both directions of flow by reversing the pitch of the blades. A 9-megawatt bulb unit was installed at Saint Malo, France, and in the Rance tidal project near Saint Malo, there are twenty-four 10-megawatt bulb units. These are described in entry on **Tidal Energy**.

Hydroelectric Power in the United States

In the 1930s, hydroelectric power furnished almost 40% of the electric energy needs of the United States. Since the 1950s, hydropower has grown less rapidly than other forms of electricity generation. Total hydropower generated by hydro plants in the early 1990s represents 10–13% of the nation's needs. Total output is estimated at about 70 GW (gigawatts). Construction of new hydro facilities has decreased for several reasons, notable of which are very high construction costs and environmental factors. Although pure water is the fuel entering a hydro plant and essentially pure water exits the facility and there is no release of carbon dioxide, nitrogen oxides, sulfur dioxide, and so on into the atmosphere, objections have been raised pertaining to the amount of land (and aesthetics) that must be sacrificed for a hydro plant and the large dam and reservoir needed. Another objection is the manner in which some water species are threatened. Another important factor, of course, is that many of the excellent sites are already in the hydropower network.

Large hydro installations in the United States include Grand Coulee (Washington), 6.5 GW; John Day (Oregon), 2.2 GW; and Chief Joseph (Washington), 2.1 GW. Only Grand Coulee ranks among the top ten largest installed capacity hydro plants in the world. Throughout the United States, there are over 40 hydro plants. The U.S. Corps of Engineers has estimated that the ultimate potential for hydro power in the United States is over a half-million GW, with the possible development of over 10,000 different sites. Realistically, however, hydro power probably will not exceed 75 to 105 GW by the year 2020. Approximately 85% of the potential for new plant sites are located in the western states.

Hydroelectric Power in Canada

In contrast with the United States, 57 percent of Canada's electric power is furnished by hydro plants. Approximately one-half of this energy comes from hydro plants in Quebec. Less than one-third of Canada's power is generated by burning fossil fuels. British Colombia, Ontario, Newfoundland, and Manitoba also have extensive hydro facilities.

Historically, Canada had depended upon hydro plants for many years. See Fig. 7. Nevertheless, planners in Canada expect that hydropower as a percentage of total power generated will decline gradually over the next several years because of several factors:

1. Most of the better sites have been used.
2. The growth rates of real fossil fuel prices were negative between 1950 and 1973, which favored the construction of thermal, including nuclear, facilities during that period.
3. A planned development of nuclear energy as an alternative source for future energy demand.

Canada has two hydro plants that are among the ten largest hydro facilities in the world: La Grande 2, Quebec (5.3 GW); and Churchill Falls,

Newfoundland (5.2 GW). The development of power, jointly by Canada and the U.S., along the St. Lawrence River is exemplary of international cooperation. The total installed capacity, involving 3 dams and 16 miles (26 km) of dykes, utilize the drop in water level between Lake Ontario and the powerhouses 125 miles (201 km) downstream. The main dam and powerhouses form a continuous structure some 3300 feet (1006 m) long. Generators, totaling 32 in number with a total capacity of 1.8 GW are not housed in conventional structures, but are protected by removable hatch covers. Generators on the Canadian side of the river feed into Ontario Hydro's grid system. Their capacity totals more than 0.9 GW, equal to the

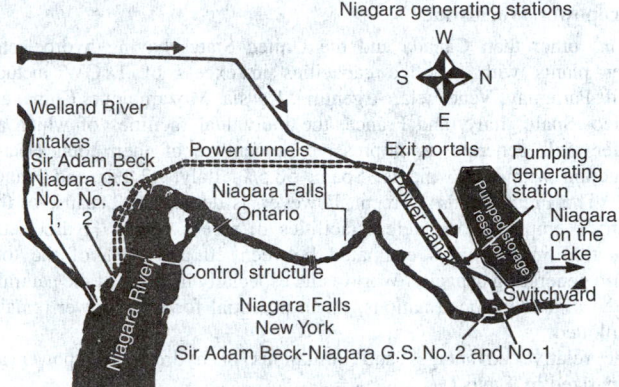

Fig. 10. Detail of site of the Sir Adam Beck Niagara generating stations Nos. 1 and 2. (*Ontario Hydro.*)

TABLE 1. RELATIVE ADVANTAGES AND LIMITATIONS OF HYDRO-ELECTRIC POWER INSTALLATIONS

ADVANTAGES

- Continuous low-cost power production except when droughts occur.
- Low maintenance costs.
- No consumption of irreplaceable fossil fuel. a
- No air pollution.
- Reservoir lakes can be used for recreation in majority, but not in all cases.
- Reservoirs can provide considerable, but not complete flood protection to downstream areas.
- Reservoirs are capable of storing large quantities of water for long periods of time, but not indefinitely.
- Downstream flow can be managed to aid in water-quality control and to level out the extremes of winter versus summer stream conditions.
- Ground-water reserves are increased by recharging from the reservoir.

LIMITATIONS

- High initial cost of construction.
- Recreational facilities can be adversely affected in reservoirs where draw down in the dry season lowers the water level.
- Flood protection can best be provided by an *empty* reservoir, while power production is best from a full reservoir. A *full* reservoir cannot retain a major flood; an empty reservoir generates no power. The compromise then, is to retain enough water in a reservoir to insure continuous power generation, but leave a margin of free board to take the major surges out of a sudden torrential rain storm.
- Loss of land suitable for agriculture.
- Power production may be curtailed or even discontinued in time of drought.
- Original stream valley is inundated.
- Some water is lost by evaporation from the reservoir surface.
- In coastal areas, such as Oregon and Washington, the construction of dams prohibits the upstream migration of anadromous fish, such as the Pacific Salmon, unless some arrangements, such as a "fish ladder" is provided.

Source: Battele Memorial Institute, Columbus, Ohio.

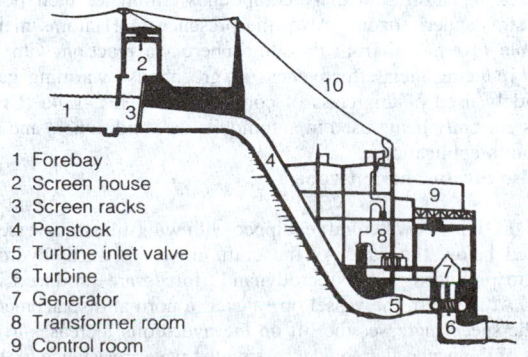

1 Forebay
2 Screen house
3 Screen racks
4 Penstock
5 Turbine inlet valve
6 Turbine
7 Generator
8 Transformer room
9 Control room
10 Transmission line

Fig. 9. Sectional view of Sir Adam Beck generating station. (*Ontario Hydro.*)

needs of about 600,000 homes. The Canadian power station (Robert H. Saunders) is located at Cornwall, Ontario. The main U.S. counterpart is located at the Robert Moses Power Dam. Flooding of the huge headpond area called for vast removal of property. Homes and even cemeteries were relocated. Ontario Hydro built new shopping centers, schools, churches, roads, sidewalks, waterworks, sewage treatment plants, and recreation areas for the 6500 persons displaced. Only farm families and cottage owners were involved on the sparsely populated American side of the project. Both Canadian and U.S. stations were opened in 1958. The headpond area is estimated at 100 sq miles (259,000 sq km) and the watershed area at nearly 300,000 sq miles (777,000 sq km). See Figs. 8, 9, and 10.

Hydropower Worldwide

Nations other than Canada and the United States having hydroelectric power plants with installed capacities in excess of 1 GW include: Brazil/ Paraguay, Venezuela, Argentina, Russia, Mozambique, China, and Mexico, Spain, Italy, and France, the individual facilities of which are smaller, still generate an impressive percentage of their total electric production with hydro plants: Spain, 58.5%; Italy, 42.2%; and France, 32%. (The French hydro percent, however, is decreasing because of that country's emphasis on nuclear facilities in recent years.) Hydroelectric production in Russia is estimated between 20 and 22% of the total electric generating capacity. Worldwide, especially in some of the naturally favored underdeveloped nations, much potential for hydropower remains unexploited.

The relative advantages and limitations of hydroelectric power are summarized in Table 1.

Additional Reading

In addition to specific references indicated below, statistics on hydroelectric power are continuously updated and available from such organizations as: U.S. Federal Power Commission, U.S. Army Corps of Engineers, Bureau of Reclamation, U.S. Department of the Interior — all in Washington, DC, Also, Edison Electric Institute, New York; and Electric Power Research Institute, Palo Alto, California.

In Canada, Additional information can be obtained from Energy, Mines and Resources, Canada, Information Services, Ottawa, Ontario; and from Ontario Hydro, Corporate Communications, Toronto, Ontario.

HYDROFLUORIC ACID. [CAS: 7664-34-3] HF (hydrogen fluoride gas) in aqueous solution, colorless when pure, fuming (dependent on concentration), highly corrosive, extremely reactive, available commercially in 30, 52, 60, and 80% HF concentrations. There is a maximum constant boiling point 111 °C (750 torr) at 43% HF (distillate) for mixtures of HF and water. Because HF attacks glass and many other container materials, the laboratory HF reagent is packaged in polyethylene bottles or carboys. Larger containers for industrial use usually are steel drums or tanks with a polyethylene lining. Anhydrous HF is available in tank cars of 22- or 42-ton (20- or 38-metric ton) capacity, as well as in steel cylinders of 100- or 200-pound (45- or 90-kilogram) capacity.

The formula weight of HF is 20.01 (calculated). However, its apparent molecular weight ranges widely with temperature and pressure. The molecular weight of saturated HF vapor at 19.51 °C is 78.24; at 100 °C, the value is 49.08. Because of strong hydrogen bonding between molecules, significant polymerization occurs, thus resulting in marked departures from ideal behavior, both in the gaseous and liquid phases. The polymerization mechanism has not been fully determined, but both ring- and chain-type structures have been suggested. Of interest is the comparison of the high boiling point of HF (+19.5 °C) with the boiling points of other acids in the halogen series: HCl, −85 °C; HBr, −65 °C; HI, −36 °C. Because of this polymerization, the formula H_2F_2 often has been used for hydrogen fluoride, although the polymers in the gas appear to be chiefly $(HF)_6$. Evidence of the stability of these hydrogen bonds is furnished by the existence of the hydrogen fluoride ion (HF_2^-) in ionic crystals and acid fluoride solutions.

Liquid hydrogen fluoride is one of the three binary hydrides (the others are H_2O and NH_3) which are self-ionized and highly associated (Trouton constant 26.6, bp 19.54 °C, mp −83.7 °C). It has a dielectric constant of 83.6 at 0 °C and is an excellent ionizing solvent. Because of its very high acidity, most oxygen-containing substances are protonated in solution in HF, forming substituted oxonium ions or oxonium ion itself by solvolysis. It has a very low viscosity — 0.256 centipoises at 0 °C. Its surface tension is also exceptionally low. Its density at the boiling point is 0.991 g/ml.

The effects of hydrogen fluoride on glass were observed by A.S. Marggraf in 1764. In 1771, Scheele established that a new acid had been discovered. In 1814, Davy showed that the acid contained a newly found element, fluorine. Fluorine was not isolated until 1886 by Moissan. The first anhydrous HF was not prepared until 1856 by Frémy. The first commercial shipment of HF (anhydrous acid) was not made until 1936.

Because of the strong affinity of HF for H_2O, there is no known chemical substance that can be used for drying it. HF immediately reacts on complexes with the drying agents. Thus, the compound can be dehydrated only by electrolysis. Even though HF is highly reactive, it has the characteristics of a weak acid, due to the extensive polymerization of the HF.

Anhydrous hydrogen fluoride is prepared commercially by reacting calcium fluoride (acid-grade fluorspar) with concentrated H_2SO_4 in a heated reactor. The presence of silica in the calcium fluoride is highly objectionable inasmuch as each pound of silica present will consume 2.6 lb of CaF_2 to form silicon tetrafluoride. When the latter compound is absorbed with HF in H_2O, fluosilicic acid is formed, representing a further loss of net HF produced: (1) $CaF_2 + H_2SO_4 \rightarrow CaSO_4 + 2HF$; (2) $4HF + SiO_2 \rightarrow SiF_4 + 2H_2O$; (3) $2HF + SiF_4 \rightarrow H_2SiF_6$. After reacting at a temperature of $200 - 250$ °C, the HF is treated to remove dust and H_2SO_4 fumes and then condensed as 99% HF.

Uses: Principal uses for HF include: (1) the production of aluminum fluoride and synthetic cryolite required for aluminum production, (2) the production of fluorinated organics of several types and for several applications, including aerosol propellants, special-purpose solvents, refrigerants, and plastics (polytetrafluoroethylene, polyvinylidene fluoride, polychlorotrifluoroethylene), (3) in the formulation of atomic-energy feed materials, (4) as an alkylation catalyst in petroleum processing, (5) as a pickling acid in stainless-steel and nonferrous metals manufacture, (6) as an agent for etching and polishing glass, (7) as a reactant in several organic syntheses, (8) in the manufacture of elemental fluorine, and (9) as a starting material for the preparation of fluorides and fluoborates.

Although the manufacture of atmosphere-damaging chlorofluorocarbons in the past has represented a high-tonnage requirement for hydrofluoric acid, the newer, less-polluting hydrochlorofluorocarbons will require an even larger supply of hydrofluoric acid. A new manufacturing facility was opened in Coahuila, Mexico, near Corpus Christi, Texas, in 1992.

Additional Reading

Lewis, R.J. and N.I. Sax: *Sax'x Dangerous Properties of Industrial Materials*, 10th Edition, John Wiley & Sons, Inc., New York, NY, 1999.

Lide, D.R.: *CRC Handbook of Chemistry and Physics*, 88th Edition, CRC Press, LLC, Boca Raton, FL, 2007.

Meyers, R.A.: *Handbook of Chemicals Production Processes*, The McGraw-Hill, Companies, Inc., New York, NY, 1986.

Staff: "World-Wide hazardous Chemicals and Pollutants," *Forum for Scientific Excellence*, American Chemical Society, Washington, DC, 1990.

Staff: "Hydrofluoric Acid (Anhydrous and Aqueous)," *Chemical Safety Data Sheet*, Manufacturing Chemists Association, Inc., Washington, DC. (Updated periodically.)

Welch, J.T., Editor: *Selective Fluorination in Organic and Bioorganic Chemistry*, American Chemical Society, Washington, DC, 1991.

HYDROFLUOROCARBONS (HFCS). A collection of partially fluorinated hydrocarbons in use or under development as replacement compounds for the chlorofluorocarbons (CFCs).

The lack of chlorine in these compounds eliminates their potential to destroy stratospheric ozone. Also, the presence of H atoms makes them susceptible to removal from the atmosphere via reaction with hydroxyl radicals, reducing their effectiveness as greenhouse warming gases. The most widely used of this class of compounds is HFC-134a (CF_3CFH_2), which is currently being used in automobile air conditioners and domestic refrigeration applications.

See also **Stratospheric Ozone**.

HYDROFOIL. A watercraft equipped with wing-like transverse surfaces suspended below the hull. As the craft moves forward, hydrodynamic forces are produced, just as aerodynamic forces are produced in air, to give lift. At the start, the vessel operates as a normal displacement vessel, but as the speed increases the lift on the hydrofoils increases to raise the craft out of the water, thereby decreasing the water resistance to that of the hydrofoils alone. Higher speeds can be obtained, since the water resistance for the required lift is considerably less for the gear still submerged in the water. To be most effective, propulsive units, such as engine-propeller combinations, or jet engines operating well above the water, are necessary.

Hydrofoil craft are of a general class of designs collectively referred to as *interface vehicles*.

HYDROFORMING. The use of hydrogen in the presence of heat, pressure, and catalysts (usually platinum) to convert olefinic hydrocarbons to branched chain paraffins (isomerization) to yield high-octane gasoline. Catforming and similar terms are often used in the same sense.

HYDROGEN. [CAS: 133-74-0] Chemical element symbol H, at. no. 1, at. wt. 1,008, periodic table group 1, mp $-259.14\,°C$, bp $-252.87\,°C$, density 0.089 (solid at 4.2K), 0.071 (liquid at 20.4K), sp gr 0.0696 (air = 10,000). Solid hydrogen has a hexagonal crystal structure. Hydrogen at standard conditions is a colorless, odorless, tasteless gas, suffocating, but not toxic. Hydrogen occurs chiefly combined with oxygen in H_2O, with carbon in hydrocarbons, with carbon and oxygen, and with carbon and several other elements, including oxygen, nitrogen, sulfur, phosphorus, and most metals in a vast variety of hundreds of thousands of organic compounds. See also **Organic Chemistry**. Hydrogen is considered by some scientists as the primordial substance from which all other elements in the universe were developed. In terms of cosmic abundance, with a rating of silicon = 10,000, it has been estimated that the figure for hydrogen is about 3.5×10^8, this figure compared with that of carbon = 80,000, nitrogen = 160,000, and oxygen = 220,000. For further comparison, the figure for gold is 0.0015 and for uranium it is 0.0002. In terms of abundance of the chemical elements in seawater, hydrogen ranks second (behind oxygen) with an estimated 510 million tons per cubic mile ($\sim$109 million metric tons per cubic kilometer). Hydrogen ranks eleventh in terms of content in igneous rocks in the earth's crust, the estimate of average content being 0.13%. Although free hydrogen escaped from the earth's lower atmosphere, some of the planets appear to have significant amounts, including the atmospheres of Jupiter, Saturn, and Uranus. At an altitude of 1,000 miles (1609 kilometers) above the surface of the earth, there is a greater abundance of hydrogen atoms than of nitrogen or oxygen atoms.

Hydrogen was first identified by Cavendish in 1766. The element was named by Lavoisier in 1783. However, it was not until 1931 that a second isotope of hydrogen (deuterium) with a mass number 2 was discovered by Urey. In 1934, Rutherford, Oliphant, and Harteck prepared a third isotope (tritium) with a mass number 3. Normal hydrogen (protium) and deuterium are stable, whereas tritium is radioactive, with a half-life of 12.26 years. Tritium emits a negative electron to form ^{-3}H. It is estimated that the isotopic abundance of ^1H (protium) in natural occurring hydrogen is 99.9851% and on the basis of carbon = 12 (atomic weight scale), protium has a mass of 1.007825 amu. The isotopic abundance of ^2H (deuterium) is estimated at 0.0149% with a mass of 2.014101 amu. The artificially prepared ^3H (tritium), $^9Be +^2 H \rightarrow 2^4He +^3 H$, has a mass of 3.01605 amu. Heavy water is deuterium oxide, ^2H_2O, usually written D_2O. Deuterium and deuterium oxide gained prominence largely because of their excellent properties as moderators in nuclear reactors. The ionization potential of hydrogen is 13.59765 ± 0.00022 eV. Other physical properties of hydrogen are given under **Chemical Elements**. See also **Deuteron**; and **Deuterium**.

When ignited, hydrogen burns in air with a pale blue to colorless, nonluminous flame, yielding H_2O. When mixed with air, the flammability limit is 4–74% hydrogen. When mixed with oxygen, the flammability limit is 4–94% hydrogen. Care always must be exercised where there may be hydrogen mixtures with air or oxygen because violent explosions may occur. In sunlight or magnesium light, hydrogen combines with chlorine with violent release of energy, forming hydrogen chloride HCl. When hydrogen is heated with sodium, calcium, and several other metals, the corresponding hydride is formed. In the presence of a catalyst, hydrogen reacts with nitrogen to form ammonia NH_3. Upon heating sulfur in the presence of hydrogen, hydrogen sulfide, H_2S, is formed. At elevated temperatures, hydrogen will reduce many of the metal oxides to the metal, notably copper, iron, nickel, tin, and lead. The oxides of zinc, aluminum, and magnesium are not so reduced. Hydrogen reacts with unsaturated organic compounds in most cases to form saturated compounds. For example, in the presence of a catalyst, hydrogen will add to oleic acid $C_{17}H_{33}COOH$ to form stearic acid $C_{17}H_{35}COOH$. See also **Hydrogenation**.

Production of Hydrogen. For chemical and petroleum processes, hydrogen is an extremely high-tonnage and one of the most fundamental raw materials. Sources of hydrogen and processes for producing it are described in entry on **Hydrogen (Fuel)**.

Uses. In terms of consumption, NH_3 is by far the largest user of hydrogen. Petroleum refining processes and methanol synthesis are the next largest consumers. Hydrogen needs for these uses are almost always fulfilled by hydrogen-generation capacity on the premises. What might be termed commodity hydrogen is shipped from hydrogen plants to various users. Some of the more important uses include the hydrogenation of numerous organic compounds, such as vegetable and animal oils, the oxyhydrogen and atomic-hydrogen welding applications, the reduction of several metallic oxides, such as iron, copper, nickel, cobalt, tungsten, and molybdenum, and the use of liquid hydrogen as a rocket fuel. See also **Ammonia**; **Hydrogenation**; **Methyl Alcohol**; **Petrochemicals**; and **Synthesis Gas**. For the potential role as a fuel. See also **Hydrogen (Fuel)**.

Ortho- and Para-Hydrogen. On the basis of nuclear spin, two forms of hydrogen are known: *ortho-hydrogen*, in which the two nuclei in the H_2 molecule have parallel spins, and *para-hydrogen*, in which the nuclear spins are antiparallel. At ordinary temperatures (and above) ortho-hydrogen is present to the extent of about 75%; at lower temperatures, the ortho changes to para-hydrogen, until at very low temperatures, as that of liquid hydrogen, the para form is present to the extent of 99.7%. There is some difference in properties between the two, notably in thermal conductivity.

The transition from ortho- to para-hydrogen releases heat in amount of 168 cal/g. The heat of vaporization of liquid hydrogen is 107 cal/g. Thus, more than ample heat is released to revaporize liquid hydrogen. Knowledge of the existence of the ortho-para transition and the development of catalysts to equilibrate the liquid during liquefaction essentially have made possible the very large-scale manufacture, use, and storage of liquid hydrogen.

Below $-220\,°C$ the specific heat of hydrogen is that of a monoatomic gas like helium (He). Practically pure para-hydrogen may be obtained by adsorption of ordinary hydrogen, which is three-fourths ortho and one-fourth para, on charcoal at about $-225\,°C$. The mp of para-hydrogen is $0.13\,°C$ lower (ortho-hydrogen $0.04\,°C$ higher) than ordinary hydrogen, and the bp at 60 mm pressure is $0.13\,°C$ lower (ortho-hydrogen $0.04\,°C$ higher) than ordinary hydrogen. Para-hydrogen reverts slowly to ordinary hydrogen, but immediately in the presence of platinized asbestos.

Atomic Hydrogen. At high temperatures, the loss of heat from a glowing wire in hydrogen is larger than expected on regular assumptions. This is believed to be due to dissociation of ordinary hydrogen into atomic hydrogen (H). See Table 1.

TABLE 1. DISSOCIATION OF HYDROGEN

Temperature, °C	Pressure	
	At 760 mm	At 1 mm
1730	0.33%	8.7%
2230	3.1	57.5
2730	34	99.3

When hydrogen is passed through an electric arc between tungsten poles, a considerable transformation into atomic hydrogen occurs, and when a stream of this gas strikes a surface a large evolution of heat takes place through recombination to ordinary hydrogen. This atomic hydrogen flame is of temperature sufficiently high to melt tungsten (mp $3,370\,°C$). The half-life of the hydrogen atom is one-third second at 0.5 mm pressure. This reaction is endothermic, values of 98–105 kcal per mole having been reported for it. It is an active reducing agent, reducing many metallic oxides and halides to the free metals, and forming hydrides with many nonmetals. The energy of its exothermic recombination is utilized, in combination with the energy released by the oxidation of the H_2 formed, by atmospheric oxygen, in the oxyhydrogen welding process.

Ionization. The ionization potential of hydrogen is $13.59765 \pm .00022$ eV, and the ionization process (in the case of protium) yields an electron and a free proton. The electric field of the proton is strong, due to its small radius, so that it readily combines with polarizable atoms. Thus, in aqueous solution, it shares an unshared pair of electrons of the oxygen atom of H_2O to form H_3O^+, the hydronium ion; with NH_3 it forms NH_4^+, the ammonium ion; with phosphine it forms the phosphonium ion, PH_4^+, etc. The hydrogen atom can also add an electron, to form the hydride anion, H^-, this potential (electron affinity) being only about 0.7 eV. Hydride ions have been shown (by electrolysis, crystal structure, etc.) to exist in the hydrides which hydrogen forms with the alkali metals and some of the other metals on the left side of the periodic table. While most other hydrogen compounds are essentially covalent, the binary compounds with the halogens and some of the other elements on the right side of the periodic table exhibit a considerable degree of ionicity, varying considerably in the same group.

The hydrogen atoms in many compounds tend to be shared between the electronegative atom or group to which they are attached and similar groups on other molecules. These hydrogen bonds increase the intermolecular forces and boiling points of hydrogen fluoride, water, organic acids and alcohols, etc. A descriptive explanation of the process is the positive polarity of the H atom that is attached to the electronegative atom or group, which gives it an effective coordination number of 2, so that it can attract an unshared electron pair of a fluorine, oxygen, nitrogen, atom of another molecule. The atom having the unshared pair must be negatively polarized or easily polarizable. For example, tertiary arsines form stronger hydrogen bonds with phenols than do tertiary phosphines.

A number of hydrogen compounds ionize to yield solvated protons, i.e., $2H_2O \rightleftharpoons OH_3^+ + OH^-$, and $2NH_3(\text{liq.}) \rightleftharpoons NH_4^+ + NH_2^-$. Moreover, many hydrogen compounds, when dissolved in such solvents, ionize more or less completely to give solvated protons and anions. In the case of polybasic acids, ionization constants are reported for each step in this dissociation.

Hydrides. See section on hydrides in entry on **Hydrogen (Fuel)**; and separate entry on **Hydride**.

Water and Acids. The properties of the most prevailing hydrogen-bearing compound, water, are given under **Water**. The characteristics of acids are attributed essentially to the presence of hydrogen ions. These topics are treated under **Acids and Bases**; and **pH (Hydrogen Ion Concentration)**.

Hydrogen Under Extreme Pressure

Interest in the possible existence of hydrogen as a metal is spurred by the prospect that hydrogen may be able to conduct electricity with zero resistance near room temperature and that, because of the tremendous concentration of energy, as contrasted with liquid hydrogen, it could serve as a rocket fuel and high explosive.

In 1989, Mao and Hemley (Geophysical Laboratory, Carnegie Institution of Washington) reported on an investigation of the insulator-metal transformation in solid hydrogen at high pressure. Much earlier, theoretical calculations made by Wigner and Huntington (1935) revealed that the transition may occur in the 250-to-400 GPa (2.5-to-40 megabar) range. With the high-pressure research tools (diamond anvil cell) available today, this transition point of hydrogen has become a primary target for some researchers.

Mao and Hemley (see reference listed) reported that direct optical observations of solid hydrogen at the aforementioned pressure range and at 77 K indicated that the hydrogen sample appeared nearly opaque and that optical data were consistent with a band-overlap mechanism of metallization. These findings were later challenged by Silvera (reference listed) to the effect that "Visual darkening of a sample is not sufficient evidence of metallization, just as a lustery metallic reflection is not. Good examples are the semiconductors germanium and silicon, which as thin films, are metallic in appearance. Darkening of a sample can arise from any number of physical mechanisms that cause absorption throughout the visible spectrum." Further justification, however, was given by Mao and Hemley.

In mid-1991, Badding, Hemley, and Mao reported on studies of the high-pressure chemistry of hydrogen in metals and made specific studies of the reaction between iron and hydrogen at sudden pressure-induced expansion at 3.5 gigapascals of iron samples immersed in fluid hydrogen. The investigators mention numerous specific areas of interest that may be addressed with a better understanding of the behavior of hydrogen with metallic environments, including the hydrogen degradation of ferrous metals.

Additional Reading

Badding, J.V., R.J. Hemley, and H.K. Mao: "High-Pressure Chemistry of Hydrogen in Metals: In Situ Study of Iron Hydride," *Science*, 421 (July 26, 1991).

Considine, D.M. and G.D. Considine: *Van Nostrand Reinhold Encyclopedia of Chemistry*, 4th Edition, Van Nostran Reinhold Company, Inc., New York, NY, 1984.

Crawford, M.: "Accelerator Eyes for Warhead Tritium," *Science*, 469 (January 27, 1989).

Giacobbe, F.G., Iaquaniello, and O. Loiacono: "Increase Hydrogen Production," *Hydrocarbon Processing*, 69 (March 1992).

Greenwood, N.N. and A. Earnshaw: *Chemistry of the Elements*, 2nd Edition, Butterworth-Heinemann, Inc., Woburn MA, 1997.

Krebs, R.E.: *The History and Use of Our Earth's Chemical Elements: A Reference Guide*, Greenwood Publishing Group, Inc., Westport, CT, 1998.

Lide, D.R.: *CRC Handbook of Chemistry and Physics*, 88th Edition, CRC Press, LLC., Boca Raton, FL, 2007.

Mao, H.K. and R.J. Hemley: "Optical Studies of Hydrogen Atoms Above 200 Gigapascals: Evidence for Metallization by Band Overlap," *Science*, 1462 (June 23, 1989).

Parker, P.: *McGraw-Hill Encyclopedia of Chemistry*, 2nd Edition, The McGraw-Hill Companies, Inc., New York, NY, 1993.

Peterson, I.: "Squeezing Hydrogen to Molecular Metal," *Science News*, 164 (March 17, 1990).

Pool, R.: "The Chase Continues for Metallic Hydrogen," *Science*, 1545 (March 30, 1990).

Ross, P. and R. Ruthen: "Hard Pressed: Squeezed Hydrogen Forms Metal with Superconducting Potential," *Sci. Amer.*, 26 (November 1989).

Silvera, I.F.: "Evidence for Band Overlap Metallization of Hydrogen — (Technical Comments)," *Science*, 863 (February 16, 1990).

HYDROGENATION. In its simplest interpretation, to hydrogenate is to add hydrogen. There are scores of examples where hydrogenation is used as a unit process throughout the chemical and process industries. Generally, the process is associated with relatively high pressure, elevated temperature, and the presence of a catalyst.

Nickel, prepared in finely divided form by reduction of nickel oxide in a stream of hydrogen gas at about 300°C, was introduced by Sabatier (1897) as a catalyst for the reaction of hydrogen with unsaturated organic substances to be conducted at about 175°C. Nickel proved to be one of the most successful catalysts for such reactions. The unsaturated organic substances that are hydrogenated are usually those containing a double bond, but those containing a triple bond also may be hydrogenated. Platinum black, palladium black, copper metal, copper oxide (Adkin catalyst), nickel oxide, aluminum, and other materials have subsequently been developed as hydrogenation catalysts. Temperatures and pressures have been increased in many instances to improve yields of desired product. The hydrogenation of methyl ester to fatty alcohol and methanol for example, occurs at about 3,000 psig (204 atmospheres) and 290–315°C. In the hydrotreating of liquid hydrocarbon fuels to improve quality, the reaction may take place in fixed-bed reactors at pressures ranging from 100 to 3,000 (7 to 204 atmospheres) psig. Many hydrogenation processes are of a proprietary nature, with numerous combinations of catalysts, temperature, and pressure possible.

Among the better known products of hydrogenation are hydrogenated vegetable and fish oils, which may be hardened or solidified by catalytic hydrogenation. Some of these oils can be partially hydrogenated to clarify and deodorize them. Fatty oils, such as oleic acid, may be converted into stearic acid by hydrogenation. Through hydrogenation, peanut oil, cottonseed oil, and coconut oil can be converted to materials that taste, appear, and smell like lard; or by varying the process, they can be made to resemble tallow. Most synthetic shortenings are comprised of hydrogenated oils. Usually, hydrogenated oils will have higher melting points and lower iodine values than the natural untreated oils.

Hydrogenation of Coal and Crudes. The interest in hydrogenation has been greatly intensified since the early- and mid-1970s in connection with the synthesis of new types of fuels to augment the world energy supplies. Basically, however, the hydrogenation of coal is not a new concept, but dates back at least a half-century to the time when manufactured gas (artificial, illuminating, producer, water gas, etc.) was used prior to the more general availability of low-cost, cleaner natural gas. In 1927, a White Paper was published discussing the processes then available for production of oil from coal. One of the first large-scale applications of the Fischer-Tropsch process for the production of oil from coal was that of the South African Coal, Oil and Gas Corporation's plant in Sasolburg, Republic of

South Africa, constructed in the mid-1950s and expanded and improved several times during the interim.

Similarly, sour crudes, heavy residuums, and other petroleum-base starting materials can be hydrogenated, sometimes coupled with other processes, to sweeten, reduce viscosity, and otherwise improve the materials for better use as fuels. See also **Coal**; **Hydrotreating**; and **Petroleum**.

HYDROGENATION (Vegetable Oils).   See **Vegetable Oils (Edible)**.

HYDROGEN CYANIDE.

[CAS: 74-90-8] HCN, formula weight 27.03, colorless gas with characteristic odor, very poisonous, mp $-14\,°C$, bp $26\,°C$, critical temperature $183.5\,°C$, critical pressure 50 atmospheres, density 0.20 g/cm^3, sp gr 0.697 ($18\,°C$). There are two isomeric forms: (1) HCN which forms cyanides, (2) HNC (inferred from its derivatives) which forms isocyanides. Hydrogen cyanide is soluble in H_2O, or alcohol, or ether in all proportions. The compound usually is marketed as an aqueous solution containing $2-10\%$ (weight) HCN. For many process uses, it is frequently more convenient to generate HCN as needed and thus avoid storage and handling problems. HCN burns with a red-blue flame, yielding CO_2, nitrogen, and H_2O. Aqueous solutions of HCN decompose slowly, yielding ammonium formate: $HCN + 2H_2O \rightarrow HCOONH_4$. Decomposition is slowed by storage in dark locations. Peaches, apricots, bitter almonds, cherries, and plums contain some HCN derivatives in their kernels, frequently in combination with glucose and benzaldehyde as a glucoside (amygdalin). The bitter almond fragrance of HCN and its derivatives sometimes can be detected in such kernels.

Production. Hydrogen cyanide can be prepared from a mixture of NH_3, methane, and air by partial combustion in the presence of a platinum catalyst:

$$HN_3 + CH_4 + 1.5\ O_2 + 6\ N_2 \longrightarrow HCN + 3\ H_2O + 6N_2$$

The process is carried out at about $900-1,000\,°C$; yield ranges from $55-60\%$. In another process, methane (contained in natural gas) is reacted with $NH3$ over a platinum catalyst at from $1,200-1,300\,°C$, the reaction requiring considerable heat input. In still another process, a mixture of methane and propane is reacted with NH_3: $C_3H_8 + 3NH_3 \rightarrow 3HCN + 7H_2$; or $CH_4 + NH_3 \rightarrow HCN + 3H_2$. An electrically heated fluidized bed reactor is used. Reaction temperature is approximately $1,510\,°C$.

The high-tonnage uses of HCN are in the preparation of numerous chemical products and intermediates for organic syntheses. As a gas, HCN sometimes is applied as a disinfectant; or cellulosic disks impregnated with HCN may be used. In ore processing and metal treating, cyanides are widely used.

Hydrogen cyanide reacts with hydrogen at $140\,°C$ in the presence of a catalyst, e.g., platinum black, to form methyl amine CH_3NH_2. When burned in air, it produces a pale violet flame; when heated with dilute sulfuric acid, it forms formamide $HCONH_2$ and ammonium formate $HCOONH_4$; when exposed to sunlight with chlorine it forms cyanogen chloride CNCl, plus hydrogen chloride. An important reaction of hydrogen cyanide is that with aldehydes or ketones, whereby cyanhydrins are formed, e.g., acetaldehyde cyanhydrin $CH_3CHOH \cdot CH$, and the resulting cyanhydrins are readily converted into alpha-hydroxy acids, e.g., alpha-hydroxypropionic acid $CH_3 \cdot CHOH \cdot COOH$.

Metallic cyanides are (1) soluble, e.g., sodium cyanide NaCN, potassium cyanide KCN, calcium cyanide $Ca(CN)_2$, mercuric cyanide $Hg(CN)_2$, aurous cyanide AuCN, (2) insoluble, e.g., silver cyanide AgCN, cuprous cyanide CuCN, (3) complex, (a) decomposed by dilute H_2SO_4 and not affected by dilute NaOH, e.g., sodium silver cyanide $NaAg(CN)_2$ solution, sodium cuprous cyanide $NaCu(CN)_2$ colorless solution. (b) changed only to acid by dilute H_2SO_4 and reactive with dilute NaOH, e.g., potassium hexacyanoferrate(II) $K_4Fe(CN)_6$ yields, with dilute H_2SO_4, hexacyanoferric(II) acid, cupric hexacyanoferrate(II) $Cu_2Fe(CN)_6$ yields, with dilute NaOH, cupric hydroxide.

Sodium cyanide solution dissolves certain metals (1) with absorption of oxygen, e.g., gold, silver, mercury, lead, and (2) with evolution of hydrogen, e.g., copper, nickel, iron, zinc, aluminum, magnesium; and solid sodium cyanide, when heated with certain oxides, e.g., lead monoxide PbO, stannic oxide SnO_2, yields the metal of the oxide, e.g., lead, tin, respectively, and sodium cyanate NaCNO. Two classes of esters are known, cyanides or nitriles, and isocyanides, isonitriles or carbylamines, the latter being very poisonous and of marked nauseating odor.

Methyl cyanide CH_3CN, bp $82\,°C$, formed by reaction of (1) methyl iodide and potassium cyanide, (2) acetamide and phosphorus pentoxide. Methyl isocyanide CH_3NC, bp $60\,°C$, formed by reaction (1) of methyl iodide and silver cyanide, (2) of methylamine, chloroform and NaOH solution warmed. Ethyl isocyanide C_2H_5NC, bp $78\,°C$. Phenyl isocyanide C_6H_5NC, bp $78\,°C$ at 40 torr pressure.

Oxidation of cyanide ion (e.g., by copper(II) gives cyanogen or oxalonitrile NCCN, poisonous colorless gas, bp $-21\,°C$. This reacts with organic compounds and bases like a halogen, for example, disproportionating in aqueous alkali to cyanide and cyanate. In aqueous acid, hydrolysis to oxalamide and ultimately oxalic acid takes place. Oxidation of cyanides by oxygen donors (e.g., lead monoxide or dioxide, manganese dioxide or dichromate) a little below red heat produces cyanates.

HYDROGEN FLUORIDE.   See **Fluorine**.

HYDROGEN (Fuel).

Because of the wide use of hydrogen in the processing industries and for the hydrogenation of various oils and fats in the food and related industries, hydrogen has become much better understood during the past several decades. For many years, hydrogen has served as a specialized fuel for certain applications, such as oxyhydrogen cutting and welding torches. But generally, until the late 1960s, the possible role of hydrogen as a major energy source fuel was rarely discussed. The word *hydrogen* took on a negative connotation with the development of the hydrogen bomb, as it also did some years ago when the hydrogen-filled dirigible Hindenburg exploded as it moved toward its mooring mast in Lakewood, New Jersey in 1937.

The probable future of hydrogen in the world's energy system was the subject of prophecy over one hundred years ago. In 1874, Jules Verne wrote: "I believe that water will one day be employed as a fuel; that hydrogen or oxygen, which constitute it, used singly or together, will furnish an inexhaustible source of heat and light." And, in the early 1900s, Britain's Lord Haldane said: "It is axiomatic that the exhaustion of our coal and oil fields is a matter of centuries only. ... As it has often been assumed that their exhaustion would lead to the collapse of industrial civilization, I may perhaps be pardoned if I give some of the reasons which led me to doubt this proposition." Haldane envisioned networks of windmills generating the electricity needed to separate hydrogen from water. The hydrogen would then be liquefied and stored underground.

Some readily apparent advantages of hydrogen, both as a direct and an indirect fuel (discussed later) have been extrapolated into terms of a future hydrogen economy. As the result of continuing and concentrated research and development in the energy field, many experts see a hydrogen energy economy gradually emerging.

Like other energy proposals (and there have been many in the past decade or so) three factors will likely determine the pace of hydrogen energy technology: (1) the manner in which, step-by-step, hydrogen-oriented systems and subsystems will compete economically and environmentally with other energy source, conservation, and utilization proposals; (2) the pace of technological advancement in related fields, such as nuclear engineering, upon which hydrogen systems may depend; and (3) the pace of unilateral efforts on behalf of hydrogen-oriented systems, including the refinement of current planning-purpose data and opinions into actual operating information relating to hydrogen generation, transportation, conversion and/or end-utilization, and safety. Without the funding of a series of "crash programs," unilateral developments probably will be relatively slow. Most likely, the information bank for hydrogen systems will stem from an increasing awareness of the energy characteristics of hydrogen and the progressive use of hydrogen subsystems in situations where they are eminently superior.

The present concept of a hydrogen fuel economy includes a primary energy source, such as a nuclear fission or fusion reactor, a geothermal source, or a solar-powered source, with hydrogen being produced as the portable energy carrier. See Fig. 1. Thermal energy from nuclear sources would be used to generate electricity that would then be used to electrolyze water for the production of hydrogen and oxygen. The hydrogen would be distributed by pipeline to distant points of use, with storage provided by underground gas storage, or by liquefaction and refrigerated storage.

Fuel-related Background of Hydrogen. Although the abundant hydrogen isotope *protium* is the simplest known atom, it forms two diatomic molecules, namely, *ortho-hydrogen*, in which the two atomic nuclei spin in the same direction; and *para-hydrogen*, in which the nuclei spin in opposite

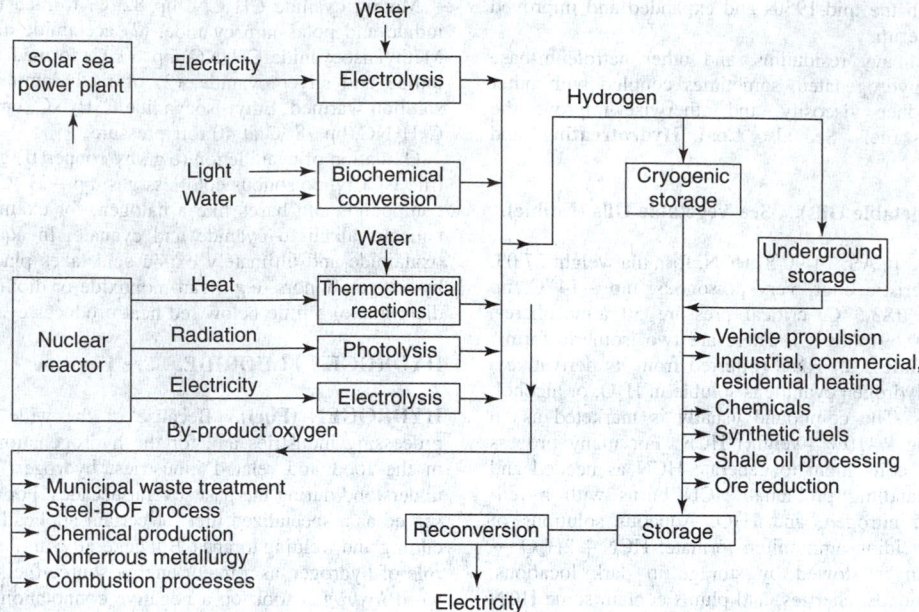

Fig. 1.   Major elements of a hydrogen fuel economy.

directions. While the equilibrium composition of hydrogen gas is 75% ortho at ambient temperature, it changes to 99.8% para in the liquid state. The transition from ortho- to para-hydrogen is exothermic (168 cal/gram), so that the heat released is more than enough to revaporize liquid hydrogen (heat of vaporization 107 cal/gram). Recognition of the existence of the ortho-para transition and the development of catalysts to equilibrate the liquid during liquefaction have made possible the large-scale production, use, and storage of liquid hydrogen.

Hydrogen molecules dissociate to atoms endothermally at high temperatures (heat of dissociation about 103 cal/gram mole), in an electric arc, or by irradiation. This property is used to effect atomic-hydrogen arc welding, in which hydrogen gas is dissociated by an electric arc between two tungsten electrodes, the hydrogen atoms recombining at the metal surface to provide the heat required for welding.

Pertinent properties of hydrogen are given in Table 1.

Actual and potential uses for hydrogen can be predicted by inspection of its properties. Its low density, 7% that of air, plus its high thermal conductivity, 6.7 times that of air, have led to its use as a coolant in large rotating electrical equipment. The low density reduces windage friction losses to less than 10% those with air, while its high thermal conductivity and heat capacity permit more efficient heat transfer, the result being an overall increase in generator efficiency of as much as 1%.

The high heats of reaction of hydrogen with oxygen or fluorine, plus the low molecular weights of the product gases, have made hydrogen a prime fuel for rocket propulsion, since rocket thrust increases directly with the temperature and inversely with the molecular weight of the exhaust gases. Liquid hydrogen and oxygen were used in the second- and third-stage Saturn engines in the Apollo moon flights. The low atomic weight of hydrogen has made it the preferred propellant for nuclear rockets, in which nuclear emission provides heat for exhausting hydrogen gas at high temperatures.

Some studies have indicated that the cost of transporting and distributing hydrogen by pipeline may be less than the cost of transporting and distributing electric power. Presumably existing natural gas pipelines and distribution systems can be adapted to the use of hydrogen. Although hydrogen has a net heating value of only 275 Btus per cubic foot (2448 Calories per cubic meter) as compared with 913 Btus per cubic foot (8126 Calories per cubic meter) for methane, the lower density and viscosity of hydrogen make it possible for a pipeline to deliver about the same amount of thermal energy as with methane, at a somewhat greater compression cost. The thermal energy in hydrogen can be utilized more efficiently in home heating than natural gas, because hydrogen can be burned in nonvented heaters, with no loss of heat, since its only primary combustion product is water. By using flameless catalytic heaters, nitrogen oxide

TABLE 1. FUEL PROPERTIES OF HYDROGEN

Melting point, K	13.96
Heat of fusion at 14.0 K, calories/gram	14.0
Boiling point at 1 atmosphere, K	20.39
Heat of vaporization at 20.4 K, calories/gram	107
Density, grams/cubic centimeter	
Solid at 4.2 K	0.089
Liquid at 20.4 K	0.071
Critical temperature, K	33.3
Critical pressure, atmosphere absolute	12.8
Critical volume, cubic centimeters/mole	65.0
Critical density, grams/cubic centimeter	0.031
Heat of transition, ortho to para at 20.4 K, calories/gram	168
Specific heat (At constant pressure C_p, calories/gram)	
Liquid at 17.2 K	1.93
Solid at 13.4 K	0.63
0–200°C	3.44
Specific heat (At constant volume C_p (0–200°C) calories/gram	2.46
Specific heat: Ratio C_p/C_r (0–200°C)	1.40
Gas density, 0°C and 1 atmosphere, grams/liter	0.0899
Gas specific gravity (Air = 1.0)	0.0695
Gas thermal conductivity, 25°C, (cal)(cm)/(s)(cm^2)(°C)	0.00044
Gas viscosity, 25°C and 1 atmosphere, centipoise	0.0089
Coefficient of thermal expansion per °C	0.00356
Heat of combustion at 25°C, kcal/gram mole	
Gross	63.3174
Net	57.7976
Energy release upon combustion, calories/gram	29.000
calories/cubic centimeter	2.050
joule/gram	1.21×10^5
Flame temperature, K	2.483
Autoignition temperature, K	858
Heat of formation of HF at 25°C, kcal/gram mole ΔH	−64.2
Flammability limit, percent	
In oxygen	4 to 94
In air	4 to 74

formation can be eliminated. However, oxygen depletion of closed spaces will still present a hazard.

One advantage of hydrogen as a source of thermal energy, as compared with electricity, is that it can be stored for later use—it is a commodity with weight and volume. Electricity, although it can be converted into chemical energy in batteries, essentially is a form of energy that must be used as it is generated. Hydrogen, like natural gas or substitute natural gases, may be stored and transported as a refrigerated liquid, or stored as a gas under pressure in underground systems. Hydrogen also may be stored as a metallic hydride.

Categories of Energy-related Hydrogen Uses. The probable functions of hydrogen in future energy technology may be put into two major categories: (1) *direct functions*, in which hydrogen serves as a fuel, that is, as the source of heat, power, and light without prior conversion to some other energy form; and (2) *indirect functions*, in which hydrogen is an important component of the total energy system, but before the end-use of that energy, the hydrogen is involved in some conversion, possibly chemically, used in the creation of a synthetic fuel, such as substitute natural gas, or possibly converted into electrical energy which becomes the final end-energy used. One of the major indirect or secondary roles proposed for hydrogen is that of an energy transporter, wherein in one scheme, other forms of energy would be consumed to generate hydrogen which then would be pipelined and stored at distant points available for another conversion step—for example, converted into electrical energy as needed.

Hydrogen as Energy Source for Motive Power

Aside from their relatively low costs until the mid-1970s and continuing into the 1980s, the hydrocarbon fuels, notably gasoline and kerosene, have offered convenience in handling and transportability for use in connection with powered vehicles. And, during the past decade, the political factors that arise from the striking geographic imbalance between petroleum resources and petroleum consumption in most regions of the world have provided ample incentives to strike out for alternative sources of vehicular power.

Hydrogen, when cost competitive, can provide many of the advantages of petroleum liquids and offer the additional attraction of decreasing air pollution.

Because of its low density, the net storage volume required would be at least as much as for gasoline. The storage tank must be maintained at a temperature of $-423\,°F$ $(-253\,°C)$, which is the boiling point of hydrogen at atmospheric pressure. This would require insulation that would increase the overall size of the storage container. Vaporization losses from the storage tank, amounting to perhaps 2% or more per day, must be vented so that no ignition of the vented hydrogen gas can occur, and no accumulation of explosive hydrogen-air mixtures are possible. The lower explosive limit of hydrogen in air is 4%, so adequate ventilation must be provided. Fortunately, hydrogen gas, being the lightest gas with a specific gravity of 0.07 referred to air, will rise and diffuse rapidly and thus can be easily dispersed. Service stations for dispensing liquid hydrogen will require more expensive storage and pumping facilities than required for gasoline.

The estimated weights and volumes expressed in Table 2 are relative to the same energy content of gasoline. Relative weight includes that of containers. The data indicate that magnesium hydride would be at a 4.6 weight disadvantage and thus require four times the tankage in comparison with the use of gasoline in a conventional automobile. New hydrides, as described later, may change this. This would be equivalent to 450 pounds (204 kilograms) of added vehicle weight and 60 more gallons (227 liters) ($2 \times 2 \times 2$ feet storage; 0.2 cubic meter) over that required for a vehicle with a 20-gallon (76 liters) gasoline tank. Bursting upon collision for liquid storage can be overcome by using containers capable of withstanding 30 Gs, which are presently available.

If a designer were to elect the option of using hydrogen in the gaseous phase at 2000 psi (136 atmospheres), this would require a metal container weighing some 30 times and requiring a volume of some 24 times that required for an energy equivalent volume of a hydrocarbon fuel. Also important in the total energy equation is the additional energy required to compress hydrogen (gaseous phase) or to liquefy it.

It is most likely that the first major use of liquid hydrogen as an energy source for motive power will be jet aircraft, largely because of the excellent weight advantage and the less serious nature of the boil-off loss and

TABLE 2. SOME HYDROGEN STORAGE OPTIONS FOR VEHICLES

Storage System	Relative System Weight[a]	Relative Contained Volume[a]
Gaseous phase, 2,000 psi (136 atmospheres)	~30.0	~24.0
Solid (as magnesium hydride with 40% porosity)	4.6	4.0
Liquid phase at 37 °R	2.4	3.8

[a] Relative to gasoline, as unity for same energy content.

distribution problems as compared with other forms of transportation. City buses and long-haul motor trucks, already equipped mainly with hydride hydrogen power, have been tested in the United States and Germany, among other countries. These may follow as candidates wherein refueling may be effected through replacement of entire storage tanks (dewars). Because the private motorcar presents the most crucial logistics problems, including the small-capacity fuel system, concern with safety, boil-off loss of fuel even when vehicle is not in use, and the education and acceptance involving millions of users, it probably will follow rather than lead the use of hydrogen in other modes of transportation. However, during the last few years, a few firms have offered hydrogen-powered private motor vehicles, set up to switch from hydrocarbon fuel to hydrogen and vice versa, but at a cost that is not competitive with mass-marketed vehicles.

Metal Hydrides. For a number of years many scientists and advanced planners have considered the possible use of metal hydrides to store hydrogen at atmospheric or reasonable pressures and at relatively low temperatures (comparable to current metal temperatures in some conventional engines). The fact that hydrogen will form hydrides with most metals has been known for many years, during which time, a number of these hydrides have been formed and tested. Until comparatively recently, magnesium hydride and a hydride of a rare-earth metal plus nickel, such as LaNi$_5$, appeared to be best suited as hydrogen storage media.

The hydride-forming reaction is exothermic and reversible: Metal + Hydrogen $\rightleftharpoons$ Metal Hydride + Heat. Thus, when it is desired to call for the separation of hydrogen from the hydride, heat (of decomposition) is required. As may be expected, the heat of decomposition is roughly proportional to the stability of the particular hydride. It is thus evident that for a metal hydride to serve as an efficient and viable means for storing hydrogen, it should be capable of decomposition at a relatively low temperature—say 300 °C or lower. At the same time, the hydride must be reasonably stable and, of course, not require a high hydrogen pressure to manufacture it. It is further evident that the metal portion of the hydride must be comparatively inexpensive and thus common and readily available in the quantities that may be required. Metal hydride storage, operating as it does in terms of hydrogen as a battery operates in terms of electricity, must be capable of easy and efficient replenishing or "recharging" cycles. In every respect, the hydride storage element must be as safe as current vehicular fuel systems.

Researchers have investigated a large number of known binary hydrides, i.e., compounds that contain one metal and hydrogen. Investigators now regard magnesium hydride (MgH$_2$) as a borderline possibility. This binary hydride evolves hydrogen at a pressure of one atmosphere and requires a decomposition temperature of 289 °C.

In comparatively recent research, much has been learned concerning the manner in which hydride compounds hold hydrogen. It has been known for a long time, of course, that the metal portion of the hydride should be comprised of tiny particles so that there is a large surface area available for reaction. In searching for reasons why hydrides permit such a high density of hydrogen, Reilly/Sandrock (1980) have observed that it is possible to pack more hydrogen into a metal hydride than into the same volume of liquid hydrogen. When the subject metal is first exposed to diatomic hydrogen (H$_2$), the hydrogen atoms are adsorbed onto the surface of the metal. Immediately, some of the hydrogen is dissociated into monoatomic hydrogen (H). This permits the monoatomic hydrogen to penetrate deeply into the crystal lattice of the metal and to occupy what are known as *interstitial sites*. Investigators have found that these sites must have a critical minimum volume if they are to easily receive the hydrogen atom. Upon increasing the pressure of the hydrogen applied, the metal reaches a saturated phase—the metal hydride phase. It has been found that under certain conditions and with certain metals, the number of hydrogen atoms contained in the crystal will range from 2 to 3 times the number of metal atoms.

The most recent experimentation with metal hydrides has involved multiple-metal hydrides. It has been known for some time that hydrogen reacts with alloy metal combinations. Considerable research has gone forth in connection with ternary hydrides (2 metals + hydrogen), and one of the most promising of these compounds as of the early 1980s is iron-titanium hydride (FeTiH$_x$), where x may range from 1 to 2. Reilly/Sandrock report that the hydrogen storage capacity by weight percent of this ternary hydride is 1.75 and by volume (grams per milliliter) is 0.096. The energy density by weight is 593 calories per gram; and the energy density by volume is 3254 calories per milliliter. Thus, this ternary hydride has a higher hydrogen-storage capacity than an equal volume of liquid or gaseous hydrogen (at 100 atmospheres). Another promising intermetallic hydride is lanthanum-pentanickel hydride (LaNi$_5$H$_x$), although it is more costly to produce. In this hydride, x may range from 1 to 6. Both the iron-titanium hydride and the lanthanum-pentanickel hydride have low temperatures of formation and decomposition, contributing to easy charging and discharging at ambient temperature.

In connection with hydrogen engine design, it is assumed that the heat of decomposition required by the hydride can be furnished from the inevitable waste heat generated by any engine. See also **Hydride**.

Hydrogen as a Heating Fuel

The routine use of hydrogen as a heating fuel for industry and commercial-residential installations entails even greater complications and would appear to be much more dependent upon the overall economic and technical aspects of a so-called hydrogen fuel economy. From many standpoints, assuming availability, hydrogen can be an excellent fuel for almost any heating application. Hydrogen can be used in the home for cooking and heating (and even lighting) and likewise in commerce and industry. Compared with natural gas, hydrogen burns with a faster, hotter flame. Hydrogen-air mixtures are flammable over wider limits of mixtures. Hydrogen burns without producing noxious exhaust products, allowing unvented appliances except where water vapor and resulting increased humidity may be objectionable. In winter, the additional humidity can, in fact, be highly desirable. But, in humid locations in summer, the water vapor produced could be objectionable. Adequate ventilation must be provided to prevent depletion of oxygen in closed spaces.

But, generally because of the absence of hazards from carbon monoxide and other fumes, large savings could be achieved from the elimination or at least simplification of flues. Some experts suggest that not only construction costs could be lowered as the result of clean burning, but that an increase of some 30% in the efficiency of a gas-fired home heating system could be achieved. The concept of peripherally placed unflued devices, particularly through the use of catalytic "flameless" heaters, could ultimately lead to a serious revision of the widely accepted central heating concept. By maintaining the temperature of a catalytic bed as low as 100°C, the production of nitrogen oxides would be virtually eliminated.

Because hydrogen burns with a hotter flame, some design features of heating apparatus would require change. The energy content per unit mass of liquid hydrogen is about 2.75 times greater than that of hydrocarbon fuels. On the other hand, there are only 325 Btus per standard cubic foot (2893 Calories per standard cubic meter) of hydrogen as compared with about 1,000 Btus per standard cubic foot (8900 Calories per cubic meter) of natural gas, thus dictating further design changes. The ignition energy of hydrogen is about 0.02 millijoules, which is less than 7% that of natural gas, a major factor in making low-temperature catalytic burners possible; also a major factor in designing for safe operation.

Despite the numerous advantages of hydrogen as a direct heating fuel, particularly in the home, the application of hydrogen must be viewed in terms of the total energy concept of an exclusively hydrogen-supplied (all-hydrogen home) installation. Where the direct use of hydrogen for heating is large, the economy will be most favorable. If a substantial amount of the hydrogen must be converted into electrical energy, as by a fuel cell, then economic justification becomes more difficult.

Lighting in the all-hydrogen home may be accomplished by condolu-minescence, a cold process. A phosphor is spread on the inside of a tube similar to the conventional fluorescent lamp. Upon coming in contact with the phosphor, small amounts of hydrogen combine with the oxygen in the air to excite bright luminescence in the phosphor.

Conversion of burners and other design aspects of heating systems and appliances to pure hydrogen, or to a hydrogen-enriched natural or substitute natural gas supply, while costly and inconvenient, is certainly not in the economically insurmountable category. Similar alterations over the years were made in the United States when communities switched from manufactured gas (about 50% hydrogen) to natural gas. Such switchovers are even more recent in European communities.

As more hydrogen becomes available for transportation use and as more hydrogen is pipelined regionally or transcontinentally (depending largely on the demand placed upon the supply of hydrogen for industrial uses), it may be that the hydrogen content of community gas supplies will be progressively enriched (in a periodic, stepwise manner because of switchover problems) and thus contribute in a gradual manner to less pollution and to the conservation of natural gas.

Hydrogen as an Energy Transporter

With the possible use of hydrogen as a source of motive power in the transportation field, the *non*chemical interest in hydrogen in the total energy picture is directed to the use of hydrogen as a means or mode of storing and transporting energy. It is in this area that hydrogen directly confronts the past ever-increasing trend toward a fully electrical energy economy. Undeniably, hydrogen energy has a major starting advantage over electrical energy, namely, hydrogen is a storable energy form. Investigations are showing that hydrogen in pipelines may cost less to transport than electricity flowing over long power lines. Thus, hydrogen may play an important future role simply as a mode of storage and transport, even though source and terminal energy conversions may be required.

Electrical power plants are most efficient when operated at constant output at full-rate load. Because of wide fluctuations in consumer load (daily and seasonally), generating rates require constant adjustment. Communication systems and some emergency systems employ batteries for interim storage of electrical energy, but these applications are minuscule when compared with the total electrical generating and distribution system. The principal means of large-scale storage is the use of pumped storage, i.e., in essence a reversible hydroelectric station wherein electrical energy is temporarily converted to a hydraulic head by pumping water to an elevated reservoir. Unfortunately, the topography has to be suitable for such an installation and thus this approach is limited to comparatively few power-generating sites. See also **Hydroelectric Power**.

The high-voltage cables required to transmit electricity from generating stations to load centers are costly. The cost of going to underground cables for transmitting bulk current ranges from 9 to 20 times that of overhead configurations. The effective use of cryogenic superconducting cables may lower underground costs considerably, but much research remains to be completed before this is possible.

Because of the tremendous volumes of fuel that can be moved in pipelines, the construction, maintenance, and operating costs of a buried pipeline are much less in terms of a percentage of the total product moved. Pipeline operations have been profitable even at the relatively low price ranges for liquid and gaseous fuels prevailing prior to the price rises of the 1970s and 1980s. Pipeline technology, of course, is well established—with several hundred thousand miles of trunklines installed and operating in the United States. These lines transport nearly 23 trillion cubic feet (0.65 trillion cubic meters) of gas. Typical pipelines range from 600 to 1,000 miles (965–1609 kilometers) in length and are up to 48 inches (1.2 meters) in diameter. Line pressures may range from 600 to 800 psi (41–54 atmospheres) but go up to 1,000 psi (68 atmospheres). A representative 36-inch (0.9 meter) pipeline will carry a gaseous fuel with the equivalent of 37,500 billion Btus (9450 billion Calories) per hour. The electrical energy equivalent would be 11,000 megawatts. By comparison, this is ten times the energy-carrying capacity of a single-circuit 500-kilovolt overhead transmission line.

The figures for pipeline transportation of pure hydrogen are not quite so attractive, but nevertheless the comparison with electric transmission costs remains highly significant.

One study shows that the pipeline transmission costs for hydrogen will range from 30% to 50% more than for natural gas. Conversion of an existing natural gas line to hydrogen service is estimated to require a rise of compressor capacity by a factor of 3.8 and compressor horsepower by 5.5.

Obviously, hydrogen transmission costs represent but one part of a total system. Should the costs of generating hydrogen in the first place, and the subsequent conversion of hydrogen into electricity at the terminal end of the system remain excessively high, then the savings in energy transportation costs, of course, become academic.

Sources of Hydrogen

The major source of chemical hydrogen over the past several decades has been natural gas. In strictly terms of chemical needs, where economic factors are favorable, natural gas has served this need well. Obviously, in terms of total energy conservation, where hydrogen is looked to as a means of conserving fossil-fuel sources, a much less costly and much more abundant hydrogen-containing raw material must be sought. The logical candidate is water. Particularly in areas of the world where hydrocarbons are not readily available, reasonably large water electrolysis installations have been made, notably in locations with low electricity costs.

In addition to electrolysis, the principal means under consideration for deriving hydrogen from water is that of thermochemical splitting. The waste heat and high temperature available from certain types of nuclear reactors would effect a series of chemical reactions, still much in the research phase, to free hydrogen and oxygen from water. Additional proposals have included the use of ultraviolet radiation from the plasma of a fusion reactor for the direct photolysis of water vapor (Department of Energy) and the use of some forms of algae, under the stimulation of light, to convert hydrogen ions to hydrogen gas by a complex chain of biochemical reactions (Case Western Reserve University).

Electrolysis. Because of years of operating experience, electrolysis is possibly an order of magnitude ahead of other proposals from a technological standpoint. Although simple in concept, electrolysis is costly—hence the research efforts to find other ways of splitting water carry a high incentive. Nevertheless, this side of one or more breakthroughs in other areas, most likely electrolysis operations will continue to serve as the basis for costs in extending the use of hydrogen in the relatively near term.

As of the early 1980s, industrial electrolyzers ranged in size from 500 standard cubic feet (14.2 cubic meters) of hydrogen production per day, consuming 3 kilowatts of electricity, to more than 40 million standard cubic feet (~1.1 million cubic meters) of hydrogen per day, consuming 240,000 kilowatts. Most common installations are from 10,000 to 500,000 standard cubic feet (283–14,160 cubic meters) of hydrogen per day. Two factors generally characterize an electrolyzer installation: (1) access to comparatively low-cost electricity, as found in some areas served by hydroelectric installations; and (2) need for the oxygen which accompanies the production of the hydrogen. Industrial electrolyzers usually operate at efficiencies of about 60% to 70%. Some high-pressure prototype models have reached 85%. It has been pointed out (D.P. Gregory, Institute of Gas Technology) that, in theory, electrolyzers can approach a maximum electrical efficiency of nearly 120% as the result of the ideal unit absorbing ambient heat and also converting this energy into hydrogen. A reasonable, practical target for an improved electrolyzer appears to be around 100%. Thus, the production of electrolytic hydrogen would be limited only by the efficiency of electric current generation, namely, between 35% and 45%. An estimate has been made (E.C. Tanner, Princeton University; R. Huse, Public Service Electric & Gas Co.) that the overall conversion efficiency of electricity-to-hydrogen-to-electricity will approximate 38%. The theoretical power required to produce hydrogen from water is 79 kilowatts per 1,000 cubic feet (~28 cubic meters) of hydrogen gas. One of the largest electrolyzers operating commercially is that of Cominco, Limited (British Columbia). This is a 90-megawatt installation that produces approximately 36 tons (32.4 metric tons) of hydrogen gas per day for use in ammonia synthesis. Other large plants are located in Norway and Egypt.

Two main types of electrolyzers are in commercial use: (1) Tank cells with monopolar electrodes. Porous diaphragms separate the alternate cathodes and anodes to prevent gas mixing. The anodes and cathodes are connected in parallel to keep the required voltage at approximately 2 volts and to permit high current densities. This arrangement requires a large floor area; (2) bipolar electrodes, connected in series and suitably insulated. The electrodes are cathodic on one side; anodic on the other side. This arrangement requires less floor space, is more complex, and requires high voltages.

High pressure can increase efficiency and this concept has been under development for many years. A commercial electrolyzer (Lurgi) is available which operates at a pressure of 30 atmospheres and 90 °C,

requiring 300 amperes of electric current at 217 volts. In the mid-1960s, bipolar cells of porous nickel electrodes were developed which operate at current densities of 800 and 1600 amperes per square foot (0.09 square meter).

In the mid-1960s, electric-high-temperature, vapor-phase electrolysis (General Electric Co.) was developed. In this process, the electrolyte is solid, porous zirconia, which contains dopants. Operating temperature ranges from 500° to 800 °C. A modification of the process is under development which will produce only hydrogen by consuming byproduct oxygen.

Among electrolyzer design improvements that may occur are better electrodes which may result as a spinoff from fuel-cell work. There are indications that electrode improvement could cut the costs of electrolytic hydrogen by about 20% to 25%. Electrolysis looms high in consideration of utilization of ocean thermal gradients and thus these two technologies are closely interacting.

Thermochemical Splitting. The major objective is to find one or more series of chemical reactions that will result in the satisfactory separation of hydrogen (and oxygen) from water. Considerable work has been going forth at the Nuclear Research Center, Julich, Federal Republic of Germany, where much attention has been given to sulfur- and chlorine-base thermochemical cycles. Other researchers (Institute of Gas Technology; General Electric Co.; European Atomic Energy Community) have been probing various combinations of at least 56 chemical elements, including over 700 different compounds that may show promise in various schemes for a closed-water-splitting cycle. It is understood that approximately 20 promising schemes have emerged, mainly centered in chlorine compounds. The most frequent flaw encountered among prospective reactions is the large amount of free energy required to force one or possibly two of the series of reactions; and the appearance of reactions that produce stable compounds incapable of regeneration.

Some of these reactions would rely upon a nuclear reactor as a heat source and would not have to await the emergence of a practical, operating fusion reactor. One sequence of reactions, in particular, is of interest:

$$CaBr_2 + 2H_2O \longrightarrow Ca(OH)_2 + 2\ HBr$$

$$Hg + 2HBr \longrightarrow HgBr_2 + H_2$$

$$HgBr_2 + Ca(OH)_2 \longrightarrow CaBr_2 + HgO + H_2O$$

$$HgO \longrightarrow Hg + \tfrac{1}{2}\ O_2$$

A drawback of this sequence is its use of highly corrosive hydrogen bromide. The scheme also requires a large inventory of mercury.

Of major concern to investigators in the thermochemical splitting schemes is the availability of appropriate materials of construction. Heat exchangers between the nuclear side and the chemical side must withstand both corrosion and radioactive contamination. The conventional nickel-chromium alloys are capable up to about 1050K; exotic, but available alloys, up to about 1400 K. Above these temperatures, ceramics and new alloys may have to be used. Considerable materials research along these lines is going forth at the Los Alamos Scientific Laboratory.

Conventional Hydrogen Uses. Even before its serious consideration in the fuel economy, the demand for hydrogen grew at a rate of about 15% annually since World War II. About 3 trillion standard cubic feet (~85 million cubic meters) of hydrogen (8 million tons; 7.2 million metric tons) were produced in the United States in 1970. Not including energy applications, the chemical requirements for hydrogen are expected to increase by about 7% per year through the year 2000. Among demands for hydrogen include petroleum refining, plastics, elastomers, increased desulfurization of fuel oils, increased use in iron ore reduction, aerospace uses, and hydrogen/air fuel cells. About 42% of the hydrogen produced now is consumed in ammonia production; about 38% is used in petroleum refining. The other large consumers are metallurgical and food processing.

In terms of presently nonconventional fuels that will require increasing quantities of hydrogen as new processes develop, it is estimated that (1) synthetic crude oil from coal will require 6,500 standard cubic feet (184 cubic meters) of hydrogen per barrel of oil; (2) 1,300 standard cubic feet (37 cubic meters) of hydrogen will be required per barrel of oil from shale; and (3) 1,500 standard cubic feet (42 cubic meters) of hydrogen will be required for every 1,000 standard cubic feet (~28 cubic meters) of synthetic pipeline gas produced from the gasification of coal. Petroleum refining use of hydrogen is expected to increase to 610 standard cubic feet (~17.3 cubic meters) per barrel of crude refined. Direct iron ore reduction use of hydrogen is expected to increase to 20,000 standard cubic feet (566 cubic meters) per ton (0.9 metric ton) of iron. If there were not other hydrogen sources available, the hydrogen needs could be met by using approximately 10% of the natural gas production.

Additional Reading

Considine, D.M.: *Energy Technology Handbook*, McGraw-Hill Companies, Inc., New York, NY, 1973. A classic reference.

Gregory, D.P.: "The Hydrogen Economy," *Sci. Amer.*, **228**, 1, 13–21 (1973).

Norbeck, J.: *Hydrogen Fuel for Surface Transportation*, Society of Automotive Engineers, Warrendale, PA, 1996.

Peavey, M.: *Fuel from Water: Energy Independence with Hydrogen*, Merit Products, Bloomington, CA, 1988.

Peschka, W.: *Liquid Hydrogen-Fuel of the Future*, Springer Verlag-New York, New York, NY, 1992.

Reilly, J.J. and G.D. Sandrock: "Hydrogen Storage in Metal Hydrides," *Sci. Amer.*, 118–129 (February 1980).

HYDROGEN PEROXIDE. [CAS: 7722-84-1] H_2O_2, formula weight 34.02, in pure, anhydrous form is a viscous, colorless liquid, sp gr 1.44, mp $-0.89\,°C$, bp $151.4\,°C$. Hydrogen peroxide is soluble in H_2O in all proportions, soluble in alcohol, or ether, but not in hydrocarbons. Reagent, chemically pure (CP) grade H_2O_2 is a solution of 90% H_2O_2 and 10% H_2O, sp gr 1.39. This concentration contains 42% active oxygen by weight. One volume yields 410 volumes of oxygen. Hydrogen peroxide solutions are high-tonnage chemicals and are supplied commercially in several strengths, ranging from 3–35% H_2O_2 by weight. Commercial grades for oxidation and bleaching normally contain 27.5–35% H_2O_2.

To reduce the tendency of H_2O_2 solutions to decompose, storage must be at comparatively low temperatures and in light-tight containers. Often, an organic material, such as acetanalide, will retard degradation. H_2O_2 has been used as an oxidizer in liquid bipropellant systems, or as a monopropellant through controlled catalytic decomposition, in supplying oxygen to various fuel mixtures for rockets and torpedoes. Low-concentration (normally 3% H_2O_2) solutions have been used for many years as antiseptics in medical applications. Bleaching is a primary outlet for H_2O_2, particularly in connection with cotton, wool, groundwood pulp—as well as hair-bleaching formulations. The compound is used as a source of gas in foaming rubber plastics. The highly reactive H_2O_2 molecule readily participates in oxidation, epoxidation, and hydroxylation reactions and is frequently used in an intermediate capacity in chemical syntheses. In restoring old paintings, H_2O_2 has been used to convert black PbS tarnish into the original white lead sulfate.

Use in Food Processing. Within recent years, there has been increased interest in the use of H_2O_2 as a bactericidal and sporicidal agent in aseptic systems used for sterilizing food processing equipment and packaging materials. Several factors affect the success of such use. At low concentration, H_2O_2 may be regarded as bactericidal, but not highly sporicidal. The latter requires concentrations of up to 35% H_2O_2. Elevated solution temperature also increases effectiveness. Hydrogen peroxide solutions can be applied at a temperature up to $95\,°C$ because of their excellent thermal stability. Such treatment must be followed by hot-air heating at about $125\,°C$ in order to dissipate H_2O_2 residuals, which must be ≤ 0.1 ppm H_2O_2. Some inorganic salts, notably cupric salts, increase bactericidal activity. Treatment with H_2O_2 and ultrasonic radiation has been shown to produce a synergistic effect on the destruction of bacterial spores. Similarly, the combination of ultraviolet radiation and H_2O_2 appears to be synergistic. In one system, a UV-irradiated solution of H_2O_2 is used. The resistance of spores varies with species. In general, the resistance of clostridial spores to H_2O_2 is lower than that of spores of bacilli. Further details are given in the Stevenson/Shafer reference listed.

Researchers have found that alkaline hydrogen peroxide renders plant fibers more digestible by ruminants and thus suggests that a number of alternative feed sources, including crop residues and other cellulosic plant biomass, may be used in animal production. Researchers at the University of Illinois and the U.S. Department of Agriculture have treated lignocellulosic residues (wheat straw, corncobs, and cornstalks) with dilute alkaline H_2O_2 solutions and have found that the fermentability of such substances increases and that the byproducts produced may be considered an acceptable energy source for the ruminant animal. Details are given in the Kerley, et al. reference listed.

Industrial Production of Hydrogen Peroxide. The traditional process for manufacturing H_2O_2 has been the electrolysis of aqueous solutions of $KHSO_4$, H_2SO_4, or NH_4HSO_4. In recent years, chemical auto-oxidation processes have grown in favor, largely because of energy costs. In these processes, the feedstock may be an alkylated quinone, alkylated anthraquinone, and hydroquinone solvents, together with hydrogen, air or oxygen, H_2O, and a nickel, palladium, or platinum catalyst. The process yields a 15–75% solution of H_2O_2 in H_2O, depending upon adjustment of process concentrations and conditions to provide desired concentration. The yield for this type of process is about 90% of theoretical. The process proceeds essentially in two steps. In the first step, anthraquinone contained in a solvent is hydrogenated at a temperature of about $40\,°C$ and a pressure of 1–3 atmospheres. The anthraquinone is reduced to hydroquinone (*p*-dihydroxybenzene):

$$C_6H_4 : (CO)_2 : C_6H_2 \longrightarrow C_6H_4 : (COH)_2 : C_6H_3R$$

R is a radical such as ethyl or tertiary butyl. In the second step, the hydroquinone solution is oxidized with air or oxygen: $C_6H_4 : (COH)_2 : C_6H_3R + O_2 \rightarrow C_6H_4 : (CO)_2 : C_6H_3R + H_2O_2$. In theory, the process consumes only hydrogen, atmospheric oxygen, and H_2O. A solvent must be used that will minimize side reactions during hydrogenation while also dissolving both the hydrogenated and oxidized forms of the organic compound. Solvents referred to in this connection are benzene-methylcyclohexanol mixtures and primary and secondary nonyl alcohols. Very tight purity precautions are required because any impurities in the H_2O_2 cause spontaneous catalytic decomposition of the product. As the result of these necessary precautions, the resulting H_2O_2 is one of the purest of commercial chemicals.

The process is highly corrosive. At one time, enameled steel vessels were standard for H_2O_2 processing. Aluminum, once properly passified through pickling and treatment after fabrication, has been found satisfactory.

Hydrogen peroxide reacts (1) with alkalis to form peroxides, (2) with potassium iodide solution, in presence of ferrous sulfate, to liberate iodine. This reaction serves to indicate the presence of as small an amount as 1 part by weight of hydrogen peroxide in 25,000,000 parts of H_2O, (3) with lead sulfide PbS, brown solid, to form lead sulfate $PbSO_4$, white solid, and sometimes used to brighten the lead pigment of darkened oil paintings, (4) with lead dioxide to form lead oxide, (5) with sulfites, especially in alkaline solution, to form sulfates, (6) with nitrites to form nitrates, (7) with arsenites for form arsenates, (8) with ferrous compounds to form ferric, (9) with chromic compounds to form chromates (See also **Chromium**), (10) with permanganates in acid solution to form manganous compounds plus oxygen of twice the volume available from the hydrogen peroxide, (11) with dichromates in acid solution cold to form perchromic acid, blue solution, more soluble in ether than in acid, (12) with titanic salt solutions to form pertitanic acid, yellow solution, (13) with colored organic materials, e.g., litmus, indigo, to destroy the color, and thus used for bleaching hair, silk, feathers, straw, ivory, teeth, bones, gelatin, flour. When hydrogen peroxide solution is treated with finely divided platinum or other substances, or comes in contact with rough surfaces, e.g., ground glass, oxygen is evolved (water also formed).

In the laboratory, hydrogen peroxide is prepared from barium peroxide by treatment with ice-cold dilute acid; when H_2SO_4 is used barium sulfate insoluble may be separated by filtration. Other peroxides, e.g., sodium peroxide, react similarly with acids to form hydrogen peroxide plus the salt corresponding to the peroxide and acid used. Hydrogen peroxide is formed when ether is exposed to sunlight, when a hydrogen-oxygen flame impinges on ice, and when H_2O in a quartz vessel is exposed to ultraviolet light.

Additional Reading

Kent, J.A.: *Reigel's Handbook of Industrial Chemistry*, Chapman & Hall, New York, NY, 1992.

Kerley, M.S. et al.: "Alkaline Hydrogen Peroxide Treatment Unlocks Energy in Agricultural By-Products," *Science*, **230**, 820–822 (1985).

Lide, D.R.: *CRC Handbook of Chemistry and Physics*, 88th Edition, CRC Press, LLC., Boca Raton, FL, 2007.

Stevenson, K.E. and B.D. Shafer: "Bacterial Spore Resistance to Hydrogen Peroxide," *Food Technology*, **37**(11), 111–114 (1983).

HYDROGEN POLYSULFIDES. See **Sulfur Compounds**.

HYDROGEN SCALE. 1. A thermometric scale. (See also **Temperature.**)

2. Since there is no reliable method for determining the absolute potential of a single electrode, electrode potentials are measured against a reference electrode whose potential is arbitrarily taken as zero. The arbitrary zero in general use is the potential of a reversible hydrogen electrode, with gas at 1 atmosphere pressure, in a solution of hydrogen ions of unit activity, or other electrodes calibrated against the hydrogen electrode.

HYDROGEN SULFIDE. [CAS: 7783-06-4]. H_2S, formula weight 34.08, colorless, odorous gas, mp $-82.9\,°C$, bp $-59.6\,°C$, sp gr 1.1895 (air $= 1$). The gas must be handled carefully because of (1) its toxic properties (particularly dangerous because it may paralyze the olfactory nerves), and (2) its explosive tendencies (low ignition temperature of $260\,°C$ and wide flammability range from 4.3 to 44% by volume in air). Hydrogen sulfide liberates considerable heat upon burning (6,230 calories/liter at $15.6\,°C$). The gas is produced by acid hydrolysis of many sulfides and by water hydrolysis of those elements higher in the hydrogen scale.

An aqueous solution of hydrogen sulfide is termed hydrosulfuric acid, which undergoes slow atmospheric oxidation to sulfur. The acid is a strong reducing agent, usually with the separation of sulfur, e.g., with nitric acid (nitric oxide formed), with concentrated $H_2SO_4(SO_2$ is formed), with permanganate (manganous ion formed in the presence of acid), dichromate (chromic ion formed in the presence of acid).

Fluorine, chlorine, bromine, and iodine react with H_2S to form the corresponding halogen acid. Metal sulfides are formed when H_2S is passed into solutions of the heavy metals, such as Ag, Pb, Cu, and Mn. This reaction is responsible for the tarnishing of Ag and is the basis for the separation of these metals in classical wet qualitative analytical methods. Hydrogen sulfide reacts with many organic compounds.

The gas results from the decomposition of metal sulfides and albuminous matter and is found in the areas of mineral springs, sewers, and in some mines where it is referred to as "stink damp." H_2S also is a byproduct of several industrial processes, including synthetic rubber, viscose rayon, petroleum refining, dyeing, and leather-treating operations. In the laboratory, H_2S usually is prepared by treating a sulfide with an acid, such as iron pyrites and HCl, or by heating thioacetamide $CH_3C(:S)NH_2$. Three processes are used industrially to produce H_2S in large quantities: (1) treating a sulfide with an acid, $2NaHS+H_2SO_4 \rightarrow 2H_2S+Na_2SO_4$, (2) reacting sulfur with an alkali, $4S+2NaOH+2H_2O \rightarrow 2H_2S+Na_2S_2O_3$, and (3) directly reacting sulfur with hydrogen, $S+H_2 \rightarrow H_2S$. Large quantities of byproduct H_2S usually are converted into elemental sulfur or H_2SO_4.

Industrial uses for H_2S include: (1) the preparation of sulfides, such as sodium sulfide and sodium hydrosulfide; (2) the production of sulfur-bearing organic compounds, such as thiophenes, mercaptans, and organic sulfides; (3) the removal of Cu, Cd, and Ti from spent catalysts where the gas acts as a precipitant; (4) the formulation of extreme-pressure lubricants; and (5) the preparation of rare-earth phosphors used in color TV tubes.

See also **Coal.**

HYDROGRAPH. By graphing the discharge of a stream as ordinate against time sequence as the abscissa, a hydrograph of the stream flow is obtained. The hydrograph proves to be an important source of information in the design of sewerage and water supply systems and in the design of hydroelectric power projects. The reliability of the information it contains increases as the period of time over which the hydrograph extends is lengthened. Hydrographs extending over periods of less than 10 years are liable to be deceptive in the information they convey regarding maximum and minimum flows. The United States Geological Survey water supply papers form a valuable and important reference source for data upon which hydrographs are constructed.

Runoff and *stream flow* are synonymous. Surface runoff, interflow, and groundwater flow in varying proportions make up the total runoff in stream channels.

The *direct runoff* is that runoff which enters the stream promptly after rainfall or melting of snow. It is equal to the *surface runoff*, the *prompt subsurface runoff*, plus the *channel precipitation* that falls directly on the water surfaces of lakes and streams. Surface runoff is commonly represented in the form of a hydrograph similar to that shown in Fig. 1.

Interflow is that part of the precipitation which infiltrates the surface soil, moves laterally through the upper soil horizons as ephemeral, shallow, perched groundwater above the main groundwater level, and reaches the

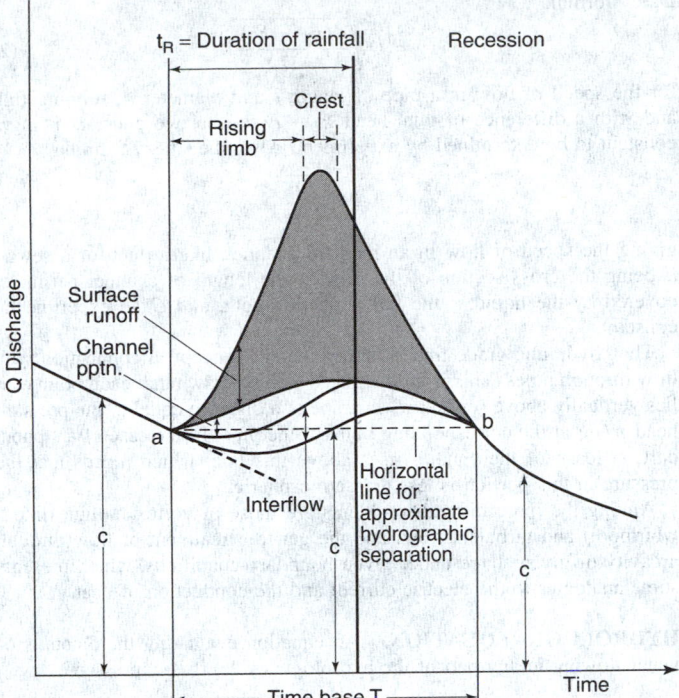

Fig. 1.   Hydrograph parts and flow contributions: (**a**) and (**b**) are reference points; (**c**) is groundwater flow.

stream before it reaches the water table. This lateral movement results from the presence of relatively impervious horizons near the surface.

See also **Drainage Systems**; and **Hydrology**.

HYDROID. One of the two forms of individuals in the coelenterates. The polyp. This form is a tubular or sac-like individual whose body wall is composed of two cellular layers separated by a thin mesogloea. The latter contains some cells derived from the other layers but is not developed as a third cellular layer. The hydroid is usually attached to the stalk of a colony or directly to a supporting surface. Its cavity opens at the free end of the body, and the mouth is surrounded by a circlet of slender tentacles except in specialized individuals found in some colonial species. The other form of coelenterate individual is the medusa.

HYDROKINETICS. The flowing of liquids is due to the three principal causes: pressure difference, gravity, and inertia. Bernoulli's law expresses an ideal condition fulfilled by the three components of "head" corresponding to these three causes. The value of this head (whether constant or not) is, at a given point (x, y, z) of the liquid,

$$e + \frac{p}{\rho g} + \frac{v^2}{2g} = F(x, y, z) \tag{1}$$

The terms of this expression represent lengths, usually given in centimeters or feet. The assumption of constant density requires that the product of the speed of flow by the cross section of any conserved portion of the stream shall be constant and that the streamlines (paths of the moving particles) therefore converge as the speed increases. If one could assume that the function F is really constant, or if it were possible to obtain F as a known function of the coordinates of the moving particle, then all hydrokinetic problems could be solved by applying suitable mathematics to the equation which would thus develop from (1).

Various attempts have been made to do this. Useful formulae result from assuming F constant (Bernoulli's law) and applying the equation to special cases. But when such formulas are tested, the calculated results are found to be in error, in every case indicating that appreciable energy has been lost in friction. While some improvement is obtained by introducing a friction factor, it has on the whole been found more satisfactory to employ empirical formulas adapted to each type of problem. Thus, we have the

Darcy formula,

$$v = D\sqrt{\frac{d(F_1 - F_2)}{l}} \qquad (2)$$

for the speed of flow in a pipe of length l and diameter d, running full, and with a difference of total head $F_1 - F_2$ at the two ends; D being a constant to be determined by experiment. Also, the Chézy formula,

$$v = C\sqrt{\frac{as}{u}} \qquad (3)$$

giving the speed of flow in an inclined channel, like a ditch or a sewer; a being the cross section of the flow, u the length of channel perimeter covered by the liquid, s the fall per unit length, and C an experimental constant.

The "hydraulic grade line" is a convenient concept in connection with flow through pipes. This is an imaginary line so drawn that each point of it lies vertically above (or below) the pipe at a distance equal to the pressure head $p/\rho g$ at the corresponding point of the pipe. In the case of a siphon, part, at least, of the conduit rises above this line, which means that the pressure in this portion is less than atmospheric.

Among the more difficult problems are those of vortex motion (like a whirlpool) and turbulent flow; and the general treatment of flow through a cavity of given shape under given boundary conditions, which presents some analogies to the electric current and the conduction of heat.

HYDROLOGIC EQUATION. An equation evaluating the amounts of water flowing in any part of the hydrologic cycle. It is expressed as

$$I - O = \Delta S,$$

where I is inflow into system during a defined period, O is outflow from a system during a defined period, and ΔS is change in storage in the system during the period.

HYDROLOGY. The science, or study, of water, especially in relation to its occurrence in streams, lakes, underground structures, and as snow. The study of glaciers, their origin and geological effects is usually included under the heading of Glaciology. The term hydrology is derived from the Greek meaning water, and reason, hence the science of water, including its discovery, uses, control and conservation. Since water ranks first of all the natural resources, the science of hydrology is of great practical importance. The basis of hydrology is the hydrologic cycle. All terrestrial (fresh) waters are derived from the great oceanic reservoirs through evaporation and precipitation. See also **Glacier**; and **Glaciology**.

Hydrology is one of several scientific disciplines that will play a major role in the "Global Change Research Program" (GCRP) http://www.usgcrp.gov/, announced by the U.S. National Aeronautics and Space Administration (NASA) in 1992 as a cooperative effort with other nations to use platforms and satellites in space to provide images and measurements in a concerted effort to understand better how Earth is changing, particularly as the result of human activities on Earth.

The Hydrologic Cycle. Also known as the water cycle, this is the never-ending circulation of water and water vapor over the entire earth. This circulation penetrates the three parts of the total earth system: the atmosphere (gaseous envelope above the hydrosphere), the hydrosphere (water covering the surface of the earth), and the lithosphere (solid rock beneath the hydrosphere). Solar energy and gravity provide the energy for the circulation.

Water is evaporated from the oceans and the land, with the former providing the largest amounts. The evaporated water is carried into the atmosphere, usually drifting tens to hundreds of miles before being returned to the earth as rain, snow, hail, or sleet. This precipitated water may be intercepted by plants, may run over the ground surface and into streams, may infiltrate into the ground, or fall back into the oceans. A considerable part of the water intercepted and transpired by plants and the surface runoff returns to the air by evaporation. The infiltrated water may seep down to deeper zones of the earth, forming groundwater storage which may later flow out to streams as runoff and finally evaporate into the atmosphere to complete the hydrologic cycle. Thus, the main processes involved in the hydrologic cycle are evaporation, precipitation, interception, transpiration, infiltration, seepage, storage, and runoff.

Magnitude of the Hydrologic Cycle. Each year, approximately 96,000 cubic miles (4×10^5 cubic kilometers or 4×10^{20} grams) of water are

evaporated from the earth's surface. Of this amount, the oceans account for 84.4%, and inland water bodies and wet soils providing the remaining 15.6%. Most of the inland evaporation occurs into relatively dry air masses. Much of the water evaporated from the oceans is transported by maritime air masses (which can hold considerably more water vapor than continental air masses) to the continents, where total precipitation amounts to 24,000 cubic miles/year (100,000 cubic kilometers/year). This amount of water would cover the entire state of Texas (267,339 square miles; 692,408 square kilometers) to a depth of 475 feet (144.8 meters). Of the 24,000 cubic miles of water precipitated, 9,000 cubic miles (37.5%) returns to the sea as runoff to balance the excess precipitation over evaporation inland.

E. Reichel calculated that the mean annual precipitation for the entire world is 34 inches (86.4 centimeters), which is balanced by a comparable amount of evaporation. It is estimated that 97% of all the water in the world or over one quadrillion (10^{15}) acre-feet ($1,234 \times 10^{15}$ cubic meters) is contained within the oceans. If the earth were a uniform sphere, this volume of water would cover the earth to a depth of 800 feet (243.8 meters), as estimated by A. Wolman.

The total volume of fresh water on the earth is estimated at 33 trillion acre-feet (4.1×10^{15} cubic meters; or 4.1×10^{21} gallons) distributed, as estimated by V.T. Chow, as follows:

Polar ice and glaciers	75%
Groundwater between 2,500 and 12,500 feet (762 and 3,810 meters)	14
Groundwater between the surface and a depth of 2,500 feet (762 meters)	11
Lakes	0.3
Soil moisture	0.06
Atmosphere	0.035
Streams	0.03

The foregoing figures are stationary estimates of distribution. Huge amounts of water pass through the atmosphere while the water content is relatively small at any given instant.

The average annual precipitation over the continental United States would amount to 30 inches (76.2 centimeters) if it were spread evenly. In actuality, the precipitation ranges from a few inches in the arid southwest to over 100 inches (254 centimeters) in parts of the Pacific Northwest. Although the 17 western states contain 60% of the land area, they receive only 25% of the total precipitation. The 30 inches of water for the United States represents 4,800,000,000 acre-feet/year or 4,300 billion gallons/day. Of this amount, 21.5 inches (54.6 centimeters), or 71.7% is returned to the atmosphere by processes of evapotranspiration. The remaining 8.5 inches (21.6 centimeters) (28.3%) becomes surface and groundwater runoff into the oceans. The foregoing estimates were made by C.J. Robinove.

Mount Waialeale, Hawaii (Kauai) receives the most rain of any location in the world, averaging 460 inches (1.168 centimeters). There is also a wide range of precipitation over Canada, from about 10 inches (25.4 centimeters) in parts of Yukon to nearly 70 inches (178 centimeters) in Halifax, Nova Scotia. St. John's, Newfoundland also receives over 60 inches (152 centimeters) of precipitation.

Function of the Hydrologic Cycle. If the atmosphere and the earth were considered as separate entities, radiation and conduction fail to provide balanced heat budgets, because the earth's surface has a net gain and the free atmosphere a net loss. The link between the gain and loss is the hydrologic cycle.

Some of the heat absorbed by the earth's surface is expended in evaporation, and therefore is transferred to latent heat, which is later realized as sensible heat and released to the atmosphere when the vapor condenses to clouds. Evaporation is high where relatively cool air sweeps over warmer oceans. The highest evaporation values found in the northern hemisphere occur in the Atlantic and Pacific trade wind belts south of 30°N. High values also occur over the northwestern Pacific and North Atlantic oceans during winter when cold, dry continental air masses move over warmer waters.

The average life of water vapor molecules in air varies from an hour to several days. Latent heat is usually liberated far from the regions where evaporation occurred. This is particularly true of evaporation in the trade wind belts, which supply much of the vapor that eventually precipitates in middle and high latitudes. Thus, the circulation of water is a key part of heat transfer from low to high latitudes and from oceans to continents.

Return of Water to the Oceans. Although there is a relatively uniform pattern of evaporation in the various latitudinal belts of the ocean, there is a marked regional imbalance in the return flow of water to the oceans. The explanation lies in the concentration of major rivers (Amazon, Mississippi, Congo, Niger, St. Lawrence, Danube, Po, Nile, and Rhine) which drain into the Atlantic Ocean and its marginal seas (Gulf of Mexico, Black Sea). In contrast, the Pacific has only a limited number of major discharge outlets (Yangtze, Hwang-Ho, Yukon, Columbia, and Colorado).

The mean annual discharges of the world's major rivers are summarized by Livingstone (U.S. Geological Survey Professional Paper 440-G) in Table 1. The information in Table 2 also provides further evidence that the Atlantic not only drains the largest portion of the earth's land surface, but has the highest proportion of land area drained to ocean area.

TABLE 1. ESTIMATED RUNOFF OF MAJOR RIVERS OF THE WORLD

River	Cubic Feet/Second (Thousands)	Cubic Meters/Second (Hundreds)
Rivers discharging into Atlantic Ocean		
Eastern North America		
Mississippi	620	175.5
St. Lawrence	500	141.5
South Atlantic slope	325	92.0
North Atlantic slope	210	59.4
	1655	468.4
Europe		
Danube	225	63.7
Rhine	76	21.5
Rhone	59	16.7
Dnieper	59	16.7
Elbe	24	6.8
Garonne	24	6.8
Don	24	6.8
	491	139.0
South America		
Amazon	3600	1018.8
Orinoco	600	169.8
Parana	526	148.9
Uruguay	136	38.5
	4862	1376.0
Africa		
Congo	1600	452.8
Niger	326	92.3
Orange and Zambezi	352	99.6
Nile	100	28.3
	2378	673.0
TOTAL for Atlantic Ocean	9386	2656.4
Rivers discharging into Pacific Ocean		
Columbia	345	97.6
Colorado	23	6.5
Yukon	180	50.9
Australia	354	100.2
Japan and Korea	225	63.7
Middle latitude Asian rivers	2250	636.8
TOTAL for Pacific Ocean	3377	955.7
GRAND TOTAL		
Atlantic and Pacific Oceans	12763	3612.1

Basic Principles of Hydrology

Since the eighteenth century, the development of hydrologic principles has been aimed at refinement of the understanding of each distinct phase of the hydrologic cycle and of the relationship between the phases. A few of the more important principles are listed as follows, not necessarily in order of importance or discovery:

1. The recognition that groundwater moves from points of high pressure to points of low pressure (down gradient) and that gradients are often, but not exclusively, related to rock type and structure.
2. The fact that the velocity of flowing water, on the surface or underground, is governed by the differences in pressure head, or slope, and the resistance of the confining channel or of the aquifer.
3. The knowledge that water is capable of dissolving and carrying large amounts of mineral matter that changes composition as the water comes in contact with various types of potential solutes.
4. The geologic recognition that water transports and deposits vast quantities of solid rock waste and is a major agent in the modification of land forms and in chemical alterations underground.
5. The fact that natural (underground) or artificial (surface) storage of water modifies the regimen of water in an area by changing the time of flow.

With these basic principles in mind, scientists, geographers, and engineers, who practice in the field of hydrology, are constantly attempting to refine two areas of knowledge: (1) an inventory or description of the water resources of the world—the amount of water in storage, rates and volumes of precipitation, recharge and discharge, the quantitative availability and suitability of water for use and the effects of water in terms of floods and droughts; and (2) a full understanding of water in all of its properties and cycles.

Hydrology of Coastal Terrain

In coastal districts, the fresh water in the water table migrates slowly downhill to the sea. Because of their different densities, the fresh water and salt water do not generally mix, except in the ocean where the tides, waves, and currents do the mixing. In the aquifers in coastal districts, the less dense fresh water tends to float on the more dense fresh water sometimes like an iceberg. The shape of the fresh water lens on a sandy island, assuming that fresh water is being replenished by rainfall, is shown in Fig. 1. The relationship between the thickness of the fresh water body (a) and the depth of the lowest part of the fresh water body below sea level (b) is:

$$b/a = \frac{\text{Specific gravity of fresh water}}{\text{Specific gravity of seawater}} = 40/42$$

Thus, for every foot the fresh water stands above sea level, the surface of the salt water lies some forty times as many feet below sea level. These figures, of course, only approximate the condition and depend upon the salinity of the seawater and the purity of the fresh water. The flow lines, i.e., the paths of water movement, for the fresh water contained within the lens are indicated in Fig. 2. Both the lens and the underlying salt water will rise and fall with the tide unless there is a barrier between the underground water and the sea. The time of the peaks and troughs of the fluctuations becomes later as traced inland, just as the time of high and low tide becomes progressively later as it is traced up the tidal portion of a river. The time between the peaks and troughs will remain the same, while the time lag will be constant for a given well.

Some mixing of the fresh and salt water does occur at the interface. Usually this is negligible, but it can be appreciable when favored by certain conditions. This produces a brackish water zone, which may be quite thick.

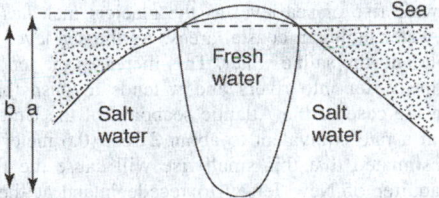

Fig. 1. Characteristic shape of the freshwater lens in islands made of uniformly permeable materials in humid areas.

TABLE 2. OCEANIC AND LAND-DRAINAGE AREAS (IN MILLIONS OF SQUARE MILES AND MILLIONS OF SQUARE KILOMETERS)

	Area		Land Area Drained		Percent of Total	Percent of Area
Ocean	Square Miles	Square Kilometers	Square	Square Kilometers	Land Area miles	Drained to Ocean Area
Atlantic	37.8	98	25.9	67	45.3	68.5
Indian	25.3	65.5	6.6	17	11.5	26.1
Antarctic	12.4	32	5.4	14	9.4	43.5
Pacific	63.7	165	6.9	18	12.1	10.8
Interior Drainage	—	—	12.4	32	21.7	—
Total	139.2	360.5	57.2	148	100.0	

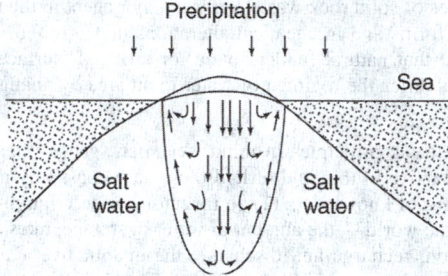

Fig. 2.   Flow lines within the freshwater lens on the island shown in Fig. 1. The lens is recharged by infiltration of rain and snow into the ground, but loses water by diffusion and by flow into the sea.

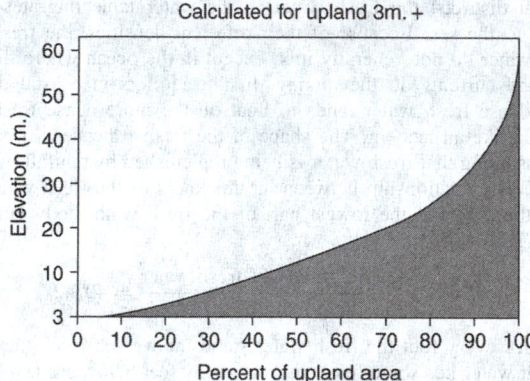

Fig. 3.   Hypsometric curve for the Town of Falmouth, Massachusetts, located on Cape Cod. From the curve, one can read the percentage of upland area of the town that lies below any given elevation. Almost all Falmouth upland is less than 60 meters (197 feet) high, and about 50% of the town is less than 15 meters (49 feet) high. A hypsometric curve presents the distribution of the area of a given land (or sea-floor) surface area with respect to elevation. (*After Giese and Aubrey.*)

This zone occurs where there are considerable fluctuations in the level of the interface due to tidal action or irregular heavy rains. Thus, a strong development of a brackish zone is found in the basalt aquifers along the coast of Oahu in Hawaii. It is also increased by pumping the wells in these regions. Like the fresh water, the brackish water lens moves slowly downslope.

The worldwide rise of sea level as the glaciers melt has an important effect on the water tables in coastal areas. As the sea level rises, so does the water table and the saline water. This increases the tendency of tides to sweep saline water into rivers and it tends to push the fresh water shoreward. In the case of the Atlantic seaboard of the United States, the sea is rising at a rate equivalent to about 2 feet (0.6 meter) per century. It has been estimated that this small rise will cause the fresh water in the artesian aquifer of New Jersey to recede inland at the rate of 1 to 4 miles (1.6 to 6.4 kilometers) per century, depending on the dip of the aquifer.

The thin layer of fresh water underlain by salt water means that great care must be taken in exploiting the fresh water. The usual method of exploitation is by well from which the water flows or is pumped. When the water is taken from a well, the surface of the water table is lowered close to the well by an amount depending on the output of the well and the porosity of the aquifer, as well as other factors. As mentioned before, for every foot of lowering of the surface of the water table, the fresh-saltwater boundary moves 40 feet (12 meters) nearer the surface. Thus, it does not take a great output of water to cause the bottom of the fresh water layer to rise to the bottom of the well. Thereafter, the water produced by the well will be saline. By limiting the production, contamination can be prevented.

Particular care must be taken in the case of artesian basins in coastal regions. The quantity of fresh water that is stored is finite, and the amount of recharge is limited. Overpumping will cause influx of saline water, as has occurred in the Savannah area of Georgia and South Carolina. Only restricted pumping or artificial recharge will prevent the eventual salinization of such a productive fresh water source.

Coastal Preservation. Of hydrological concern, the preservation of ocean shorelines is of specific interest to ocean science and engineering. Giese and Aubrey (see reference listed) estimate that by the year 2025 the amount of upland lost in Massachusetts (for example) as the result of relative sea-level rise will range from 3000 to 10,000 acres (1215 to 4050 hectares). Now mostly occupied by private residences and commercial structures, "upland" refers to terrain landward of wetland that has not been altered appreciably by coastal processes (i.e., waves and tides). Giese and Aubrey observe, "Past studies of coastal upland retreat have concentrated on shore erosion and have neglected *passive submergence*, probably because such losses have been considered to be relatively small. Such is not the case, however, even at present rates of relative sea-level rise. In addition, when we consider the possible importance of measuring this loss due to relative sea-level rise, the importance of measuring this loss becomes obvious. Thus, we set out to quantify the passive retreat of upland within the coastal communities of Massachusetts."

Hypsometric curves are a convenient method for illustrating the community upland retreat rates. See Fig. 3. Calculations by Giese and Aubrey show that, by approximately the year 2025, using past sea-level rise numbers, the total upland loss for Massachusetts will be 2950 acres (1190 hectares), resulting from a sea-level rise of 0.45 feet (0.14 meters). This could triple with a sea-level rise of 1.57 feet (0.48 meters).

Within a geologic period of increasing sea level, ocean beach properties in many areas of the world are threatened. Most research has been directed toward the development of a rationale for future seashore land development programs. One example of this is the University of Florida Sea Grant Programs. This has led to the establishment of the Florida Coastal Construction Control Line, a line that delineates the 100-year coastal hazard zone within which the state has construction-permitting jurisdiction. Evidence of past erosion of beaches by sea-level rise are attested by the numerous barrier islands formed along oceanic coastlines. It is estimated that these islands were formed about 6000 to 7000 years ago during a period of similar sea-level rise. How can technology reduce the effects of sea-level rise? Some past concepts have failed; only a few have succeeded.

One concept, "beach nourishment," is the placement of large quantities of quality sand on the beach to advance the shoreline seaward. Costs are high, ranging up to $6 million/mile ($3.7 million/kilometer). Thus, the longevity of the nourishment is a very important economic factor.

One geologist, whose opinions have been challenged, has observed that nourished beaches erode ten times faster than natural beaches. Experience in terms of measured effectiveness is inadequate. For example, it has been observed that a beach that has been nourished may fail in its specific effects on a given beach, but that the nourishment may be passed along shore (downdrift) and assist in a small way toward nourishing beaches in adjacent areas. Dean (see reference listed) has observed, "An example is the Port Canaveral, Florida beach nourishment project (1974) in which 2.4 million cubic yards (1.8 million cubic meters) were placed over 2.1 miles (3.4 kilometers) of beach immediately downdrift of the port entrance. Recent surveys indicate nearly every grain of sand placed, although it has been transported downdrift (southward), can be accounted for. Findings like this are now being included in the preparation of cost/benefit analysis procedures for future beach nourishing projects."

Waves breaking across the surf zone are a primary function for cross-shore and longshore sediment transport. Also, of course, waves represent a significant destructive force in terms of any object located along the shore, as witnessed by hurricanes.

Dean (University of Florida), who has pioneered in the field of beach erosion, observes, "Future research agendas must include the dynamics of the surf zone and recognition of our poor understanding of this region. Although relevant knowledge has increased many fold in the last few decades, there is still much to be learned prior to development of rational design capabilities. Obvious questions include the rate of longshore and cross-shore transport under given weather conditions, the relative roles of bed load and suspended load transport, the cause of rip currents, and the mechanics of longshore bar formation."

Deltas and Estuaries. Deltas represent a special situation where fresh-waters from rivers meet freshwater or seawater reservoirs, such as lakes or the oceans. See also **Estuary**; and **Delta**. As pointed out by D.J. Stanley (Mediterranean Basin Project, U.S. National Museum of Natural History), "The northeastern margin of the Nile delta, including Lake Manzala, Port Said, and the northern Suez Canal, has subsided rapidly at rates of up to 0.5 centimeter per year since some 7500 years ago. This subsidence has diverted at least four major distributaries of the Nile River into this region. The combined effects of continued subsidence and sea-level rise may flood a large part of the northern delta plain by as much as 1 meter by the year 2100. The impact of continued subsidence, now occurring when sediment input along the coast has been sharply reduced because of the Aswan High Dam, is likely to be substantial, particularly in the Port Said area and as far inland as south of Lake Manzala." The aforementioned areas, with a population estimated at over 1 million, may be susceptible to flooding because it lies over one of the more rapidly subsiding parts of the delta. Continued scientific studies are needed to precisely measure present subsidence and to better determine its possible effects and thus possibly take remedial actions. In late 1990, Stanley observed, "Unless Egypt moves promptly with new coastal protection measures, the sea may advance inland by as much as 30 kilometers (18.6 miles) within the next 100 years."

Possibly the most intensely studied of the world's estuaries is Lake Michigan's Green Bay, located in the heart of North America. As pointed out by P.L. Smith and associated researchers at the University of Wisconsin Sea Grant Institute, "Green Bay is best characterized as an estuary since it functions as a nutrient trap, has exceptionally high biological productivity, and because of the thermal and chemical differences between the water of its tributaries and that of Lake Michigan. The bay's mixing process is driven by a strong wind-induced seiche coupled with a small lunar tide." Rather than one of hydrological concern, the problem with restoration activities of the estuary, which commenced in earnest in 1969, has been one of overexploitation and biological pollution. The studies and actions taken to date are elegantly described by Smith and co-researchers. See reference listed.

Hydrology of Limestone Terrain

The primary factor in the hydrology of limestone terrains is the solubility of carbonate rocks in aqueous solutions, which leads to underground networks of pipes and channels known as *karst*. Limestone terrains are defined as regions where carbonate rock formations extend from near the surface to below the water table. The soil, and the material directly beneath, strongly influence the hydrology of these terrains. An impermeable cover overlying a thick limestone sequence can cause most of the precipitation to pass out of the area as surface runoff and the effect of the limestone is minimized. A pervious soil horizon, or erosion of the impermeable cover, will facilitate infiltration to the limestone substrate.

The hydrology of a carbonate region can range from a karst area with many interconnecting passages and caves, readily absorbing surface waters and transmitting these waters fairly rapidly from source area to discharge or storage areas, to a situation where a limestone formation of low permeability acts as an aquiclude and the terrain has high surface runoff and little available groundwater. The variability of the hydrology is as significant as the unique hydrology of a fully developed karst terrain and makes the concept of an average hydrology rather hypothetical. This variability can be demonstrated by comparing an area in Texas where the Edwards limestone absorbs an estimated 156,000 acre-feet (192.4×10^6 cubic meters)/day from surface stream flow, with a chalky portion of the Cooper Marl in South Carolina, which is used as an unlined public water supply conduit for Charleston. The permeability of carbonate rocks varies greatly as a function of its purity, ratio of calcium/magnesium, texture, structure, and history.

The water table in limestone terrains that have undergone extensive karstification fluctuates greatly, rising in rapid response to precipitation input and falling due to rapid flow of water through solution passages to the discharge zones. Because there may be perched flow or storage on impermeable layers and along some bedding planes, some hydrologists question the validity of the water table concept in limestone terrains.

See also **Groundwater**.

Hydrology of Semiarid Regions

The semiarid regions of the earth's surface occur as transition zones between the arid deserts and the subhumid belts. Water movement will shape the landscape, according to its geology and past topography, and will work in conjunction with wind erosion, solar insolation, temperature changes, and soils (stable or in movement), as well as with the vegetation and the animals that live thereon, to produce an ecological balance of all factors, either in a temporary or a permanent sense.

In defining arid and semiarid areas, Peveril Meigs used only three factors: humidity, season of precipitation, and temperature. An extremely arid region is defined as an area with at least one entirely rainless month per year; arid regions are defined as regions where precipitation is less than potential evaporation. A typical semiarid region is defined as one in which precipitation occurs in a cold winter, with the coldest month in the 0–10°C (32–50°F) average range and the hottest month in the 20–30°C (68–86°F) range. These conditions are typical of Mediterranean semiarid climate, occurring in Morocco, Algeria, Lebanon, northern Iran and also on the western coast of the United States around latitude 35°N.

Arid and semiarid lands account for over one-third of the land surface of the earth, while cultivated lands account for but one-tenth of the whole. The greatest belt of arid and semiarid regions extends across North Africa as the Sahara, through the Arabian Peninsula with the "Empty Quarter" of extreme aridity, into the Salt Desert of Iran, and the Takla Makan of Central Asia. In North America, the Great American Desert falls generally within this classification.

Precipitation in the semiarid regions is restricted and kept low by the inability of moisture-bearing winds to penetrate into, and cool down within, such regions. Zones of high pressure may prevent the entry of winds, and the great desert areas are mainly associated with this meteorological phenomenon. Such winds as do enter arid and semiarid regions may have had no opportunity of acquiring moisture by passage over oceans or sea, or they may have been forced to lose their moisture in passing over high mountains, as in the "rainshadow" deserts of Imperial Valley, California and the Jordan-Syrian steppe. Again, lack of orogenic effects within the regions, combined with high heat reflected from the ground, may prevent cooling of the incoming winds so that no moisture condenses to form clouds of precipitation, as in the coastal deserts of Chile, southern California, Morocco, and western Australia.

Almost all precipitation in the semiarid and arid regions occurs as rain, except in higher altitudes. Dew and even hoarfrost are also of importance, and are due to the great differences between day and night temperatures. Where infiltration conditions are good, as over coastal sand dunes, or suitable vegetation exists, such dew may make a permanent addition to the useful water resources of the area because of this type of moisture intake by certain vegetation.

Wind-wells have been reported in use in some areas of the Crimea and the south of France, wherein airborne moisture condenses on rapidly cooling stones. It has been reported that the Byzantines had irrigated vines by planting them at the base of an octahedron of open stones, the upper pyramid above ground surface to condense moisture and the lower inverted pyramid leading the condensate down to the vine root.

High evaporation and high transpiration in vegetated areas is the dominant hydrological characteristic of the semiarid and arid zones. Both transpiration and evaporation are high because abundant heat energy is supplied to change the limited amounts of liquid water into water vapor, either directly or through the biological processes. In this way, the heat balance of the area is maintained.

The inability of surface waters to maintain themselves against evaporation has a long-term effect in that it permits the formation of basins of inland drainage. Any basin of closed drainage will cease to exist if the average annual storage of surface water exceeds evaporation from its central lake system. This is because the lake will rise and spread each year till it overtops the lowest point of the encircling water divide over which it will discharge to the ocean level and also cut its way down so as to reduce the size of the lake. Basins of closed drainage are characteristic of the semiarid lands of the world. The ability of the Nile, the Euphrates-Tigris, the Indus, the Colorado, and similar rivers, to keep open their basins is due to the fact that the amount of incoming surface waters (originating in nonarid regions) exceeds the evaporation losses.

Precipitation is never pure water, but contains salts and gases in solution. The salts are dissociated into cations, mainly calcium, magnesium, sodium, and potassium, while the anions are bicarbonate, chloride, and sulfate. Carbon dioxide is the principal dissolved gas. These elements in solution in precipitation may be of marine or terrestrial origin. It has been estimated that the annual precipitation of sea salts to be about three kilograms per hectare for the drier steppe regions south of the Sahara; as two kilograms per hectare in the Kalahari; and one kilogram per hectare for parts of Iraq and Iran. Full evaporation of the water which carries these salts will result in their deposition more or less where they fell; surface runoff will concentrate them in the central evaporating pans in basins of closed drainage, while infiltration to the aquifers may be with water which already is far from pure. Thus, in Syria, the chemical composition of the precipitation may be altered from a starting figure of about 20 ppm to concentrations of 100 to 200 ppm by evapotranspiration and leaching of precipitates in the zones of precipitation. Thus, D.J. Burdon and S. Mazloum report that the recharge waters to aquifers in Syria may contain from 50 to 200 ppm of total soluble salts.

In some aquifers, such as the Fars Formation of Iraq-Syria (of lagoonal facies), such soluble salts will be very abundant, while in other aquifers, such as the continental arkositic sandstones of the Sahara and Arabia, soluble minerals are almost completely absent. When the amount of groundwater flowing through the aquifer is large, such soluble salts tend to be removed and the aquifer flushed and cleaned out; likewise, fast-moving groundwater will flush an aquifer quicker than slow-moving water. Since the amount and often the rate of movement of groundwater in the semiarid zone tend to be small, mineralization by dissolution of the aquifer tends to be high.

At the point of natural discharge from aquifers, springs or marshy ground appears. If the spring is large, a perennial river carries off the discharge, and an oasis is formed, or else a large city, such as Damascus (fed by the Barada River flowing mainly from Ain Figeh) comes into existence. If the discharge is small or diffuse, a saline marsh tends to form, of which one of the greatest is the Qatarra Depression in Egypt, the probable discharge zone for the sandstone aquifer of the Western Desert of Egypt.

Serious studies and efforts have been underway for several ways to improve the overall hydrological conditions of semiarid zones, including: (1) surface management—directed to making use of the water before it is lost by evaporation—increasing transportation through useful vegetation; (2) control of storage of surface runoff—often intensive and short-lived—possibly spreading the water over large areas by diverting it from the wadi or stream bed and controlling it behind earth banks in such a way that its flow velocity is never sufficient to erode the retaining structures; (3) control of aquifers—use of proper techniques in connection with extraction from galleries, wells, and bore holes; (4) underground storage of groundwater borrowing some of the successful techniques used in storing surplus natural gas and oil underground; (5) possibly using weather modification techniques and through the construction of dams and other holding means to allow large surfaces of water to form upwind of the semiarid area under consideration (Would introduction of the Mediterranean Sea to the Qatarra depression increase precipitation along the Alexandrian coast?); and (6) desalting of brackish waters.

Hydrology of Volcanic Terrain

Volcanic terrains are made of rocks erupted from volcanos and intrusive rocks that congealed below the surface. The eruptives fall into three main categories:

1. Basalt, a dark-colored rock low in silica and high in ferromagnesian minerals;
2. Rhyolite, a light-colored rock high in silica and low in ferromagnesian minerals; and
3. A whole series of rocks of intermediate composition between basalt and rhyolite, such as andesite, dacite, latite, and trachyte.

The silica-rich magmas commonly are more explosive and tend to form steep cones close to their vents. Such vents include Mt. Lassen, California, Mt. Hood, Oregon, and Mt. Ranier, Washington. The basalt, being more fluid, forms plains, such as the Snake River Plain of Idaho and the Columbia River plateau of Washington, Oregon, and California. Basalts form the high islands of the Central Pacific, of which Mauna Loa and Kilauea volcanos on Hawaii are well known. If poured out molten, they are *flows*; if blown out, they are *pyroclastics*; if solidified in cracks or other voids in the crust, they are *intrusives*; if deposited as fragments in a vent, they are *throat breccias*.

The hydrology of volcanic terrain depends largely upon the permeability of the rocks present. Of the nearly 70 first-magnitude springs in the United States that discharge more than 100 cubic feet/second (2.83 cubic meters/second), 36 of these issue from basalt. Big Springs, Idaho, near Yellowstone Park, discharges about 180 cubic feet/second (5.1 cubic meters/second) from spherolitic obsidian at the terminus of a blocky silica-rich lava flow in an ancient caldera. Several other large springs issue in the same area from silicious lavas, presumably filling ancient valleys.

During the 1950s, wells were developed that were sufficient to irrigate large areas of pineapple on the island of Lanai, Hawaii. Also, large quantities of water have been developed in the Hawaiian Islands by tunnels penetrating dike complexes. A large development of groundwater from basalt has been done by way of wells, tunnels, and shafts on Oahu.

Glossary

An abridged glossary of hydrologic terms would include:

Absorption—the process by which, substances in gaseous, liquid or solid form dissolve or mix with other substances. See also **Absorption**.

Adsorption—adherence of gas molecules, ions, or molecules in solution to the surface of solids. See also **Adsorption**.

Adsorption isotherm—a graphical representation of the relationship between the bulk activity of adsorbate and the amount adsorbed at constant temperature.

Advection—is the transport of a conserved scalar quantity that is transported in a vector field. A good example to have in mind would be the transport of pollution in a river: the motion of the water carries the polluted water downstream.

Air-space-ratio—is the ratio of (a) the volume of water that can be drained from a saturated soil or rock under the action of force of gravity to (b) the total volume of voids.

Anisotropy—is the condition of having different properties in different directions.

Anisotropic mass—a mass having different properties in different directions at any given point.

Aquiclude—a hydrogeologic unit which, although porous and capable of storing water, does not transmit it at rates sufficient to furnish an appreciable supply for a well or spring.

Aquifer—a geologic formation, group of formations, or part of a formation capable of yielding a significant amount of ground water to wells or springs. Any saturated zone created by uranium or thorium recovery operations would not be considered an aquifer unless the zone is or potentially is (1) hydraulically interconnected to a natural aquifer, (2) capable of discharge to surface water, or (3) reasonably accessible because of migration beyond the vertical projection of the boundary of the land transferred for long-term government ownership and care. See also **Aquifers**.

Aquifer system—a body of permeable and poorly permeable material that functions regionally as a water-yielding unit; it comprises two or more permeable beds separated at least locally by confining beds that impede ground-water movement but do not greatly affect the regional hydraulic continuity of the system; includes both saturated and unsaturated parts of permeable material.

Aquifer test—a test to determine hydrologic properties of the aquifer involving the withdrawal of measured quantities of water from or addition of water to a well and the measurement of resulting changes in head in the aquifer both during and after the period of discharge or additions.

Aquifuge—a hydrogeologic unit which has no interconnected openings and, hence cannot store or transmit water.

Aquitard—a confining bed that retards but does not prevent the flow of water to or from an adjacent aquifer; a leaky confining bed. It does not readily yield water to wells or springs, but may serve as a storage unit for ground water.

Artesian well—a well deriving its water from an artesian or confined aquifer.

Base flow—that part of the stream discharge that is not attributable to direct runoff from precipitation or melting snow; it is usually sustained by ground-water discharge.

Baseline monitoring—is the establishment and operation of a designed surveillance system for continuous or periodic measurements and recording of existing and changing conditions that will be compared with future observations.

Breakthrough curve—is a plot of relative concentration versus time, where relative concentration is defined as C/Co with C as the concentration at a point in the ground-water flow domain, and Co as the source concentration.

Buildup—the vertical distance the water table or potentiometric surface is raised, or the increase of the pressure head due to the addition of water.

Capillary action—is the movement of water in the interstices of a porous medium due to capillary forces.

Capillary fringe—is the lower subdivision of the unsaturated zone immediately above the water table in which the interstices are filled with water under pressure less than that of the atmosphere, being continuous with the water below the water table but held above it by capillary forces.

Capillary head—the potential, expressed in head of water that causes the water to flow by capillary action.

Capillary potential—the scalar quantity that represents the work required to move a unit mass of water from the soil to a chosen reference location and energy state.

Capillary pressure—is the difference in pressure across the interface between two immiscible fluid phases jointly occupying the interstices of a porous medium caused by interfacial tension between the two phases.

Capillary rise—is the height above a free water surface to which water will rise by capillary action.

Cascading water—In reference to wells, ground water which trickles or pours through cracks or perforations down the casing or uncased borehole above the water level in the well.

Cation exchange capacity—is the sum total of exchangeable cations that a porous medium can absorb. Expressed in moles of ion charge per kilogram (per pound) of soil (or of other exchanges such as clay).

Concentration gradient—is the change in solute concentration per unit distance in solute. Concentration gradients cause Fickian diffusion (spreading) of solutes from regions of highest to regions of lowest concentrations. In slow moving ground water, this is the dominant mixing process.

Confined aquifer—an aquifer bounded above and below by confining units of distinctly lower permeability than that of the aquifer itself. See also **Aquifers**.

Confining unit—a hydrogeologic unit of impermeable or distinctly less permeable material bounding one or more aquifers and is a general term that replaces aquitard, aquifuge, aquiclude. See also **Aquifers**.

Contamination—the addition to water of any substance or property preventing the use or reducing the usability of the water. Sometimes considered synonymous with pollution. **Contaminant plume** - An elongated body of ground water containing contaminants, emanating and migrating from a point source within a hydrogeologic unit(s).

Convection—is the transfer of heat by currents within a fluid. It may arise from temperature differences either within the fluid or between the fluid and its boundary, which would affect density. Other sources of density variations, such as variable salinity, or from the application of an external motive force are also often causes. It is one of the three primary mechanisms of heat transfer, the others being conduction and radiation. Convection occurs in atmospheres, oceans, and planetary mantles. See also **Advection**; **Convection**; and **Atmosphere (Earth)**.

Darcey's law—an empirical law which states that the velocity of flow through d porous medium is directly proportional to the hydraulic gradient assuming that the flow is laminar and inertia can be neglected (after Darcy, 1856).

Deep percolation—is the drainage of soil water downward by gravity below the maximum effective depth of the root zone toward storage in subsurface strata.

Discharge area—an area in which ground water is discharged to the land surface, surface water, or atmosphere.

Differential water capacity—is the absolute value of the rate of change of water content with soil water pressure. The water capacity at a given water content will depend on the particular desorption or adsorption curve employed. Distinction should be made between volumetric and specific water capacity.

Diffusion—is the process whereby ionic or molecular constituents move under the influence of their kinetic activity in the direction of their concentration gradient. See also **Diffusion**.

Diffusivity, soil water—is he hydraulic conductivity divided by the differential water capacity (care being taken to be consistent with units), or the flux of water per unit gradient of moisture content in the absence of other force fields.

Dispersivity—is a geometric property of a porous medium which determines the dispersion characteristics of the medium by relating the components of pore velocity to the dispersion coefficient.

Drainage well—(1) a well installed to drain surface water, storm water, or treated waste water into underground strata. (2) a water well constructed to remove subsurface water or to reduce a hydrogeologic unit's potentiometric surface.

Drawdown—is the vertical distance the water elevation is lowered or the reduction of the pressure head due to the removal of water.

Evapotranspiration—is the combined loss of water from a given area by evaporation from the land and transpiration from plants. See also **Evapotranspiration**.

Fickiam diffusion—is the spreading of solutes from regions of highest to regions of lower concentrations caused by the concentration gradient. In slow moving ground water, this is the dominant mixing process.

Flow line—the general path that a particle of water follows under laminar flow conditions. See also **Laminar Flow**.

Flow net—a graphical representation of flow lines and equipotential lines for two-dimensional, steady-state ground-water flow.

Flow path—is the subsurface course a water molecule or solute would follow in a given ground-water velocity field.

Free water—See **Gravitational water**.

Fresh water—Water that contains less than 1,000 milligrams per liter (mg/L) of dissolved solids; generally more than 500 mg/L is undesirable for drinking and many industrial uses. See also **Water**.

Gaining stream—is a stream or reach of a stream whose flow is being increased by inflow of groundwater.

Gravitational head—the component of total hydraulic head related to the position of a given mass of water relative to an arbitrary datum.

Gravitational water—is water which moves into, through, or out of the soil or rock mass under the influence of gravity.

Groundwater—is water located beneath the ground surface in soil pore spaces and in the fractures of geologic formations. A formation of rock/soil is called an aquifer when it can yield a useable quantity of water. The depth at which soil pore spaces become saturated with water is called the water table. Groundwater is recharged from, and eventually flows to, the surface naturally; natural discharge often occurs at springs and seeps and can form oases or wetlands. See also **Groundwater**.

Ground-water barrier—Rock or artificial material which has a relatively low permeability and which occurs below the land surface where it impedes the movement of ground water and consequently causes a pronounced difference in the potentiometric surface on opposite sides of it.

Ground-water basin—a general term used to define a ground-water flow system that has defined boundaries and may include permeable materials that are capable of storing or furnishing a significant water supply, the basin includes both the surface area and the permeable materials beneath it.

Ground-water travel time — is the time required for a unit volume of ground water to travel between two locations. The travel time is the length of the flow path divided by the velocity, where velocity is the average groundwater flux passing through the cross-sectional area of the geologic medium through which flow occurs, perpendicular to the flow direction, divided by the effective porosity along the flow path. If discrete segments of the flow path have different hydrologic properties the total travel time will be the sum of the travel times for each discrete segment.

Head, static — is the height above a standard datum of the surface of a column of water (or other liquid) that can be supported by the static pressure at a given point. The static head is the sum of the elevation head and the pressure head.

Head, total — the total head of a liquid at a given point is the sum of three components: (a) the elevation head, which is equal to the elevation of the point above a datum, (b) the pressure head, which is the height of a column of static water that can be supported by the static pressure at the point, and (c) the velocity head, which is the height to which the kinetic energy of the liquid is capable of lifting the liquid.

Hydraulic gradient — is the change in static head per unit of distance in a given direction. If not specified, the direction generally is understood to be that of the maximum rate of decrease in head.

Hydraulic head — is the height above a datum plane (such as sea level) of the column of water that can be supported by the hydraulic pressure at a given point in a ground water system. For a well, the hydraulic head is equal to the distance between the water level in the well and the datum plane.

Hydrochemical facies — are distinct zones that have cation and anion concentrations of diagnostic chemical character of water solutions in hydrologic systems which is describable within defined composition categories.

Hydrogeologic unit — is any soil or rock unit or zone which by virtue of its hydraulic properties has a distinct influence on the storage or movement of ground water.

Hydrologic properties — Those properties of a rock that govern the entrance of water and the capacity to hold, transmit, and deliver water, such as porosity, effective porosity, specific retention, permeability, and the directions of maximum and minimum permeabilities.

Impermeable — a characteristic of some geologic material that limits its ability to transmit significant quantities of water under the head differences ordinarily found in the subsurface.

Infiltration rate — the rate at which a soil or rock under specified conditions absorbs falling rain, melting snow, or surface water expressed in depth of water per unit time.

Irrigation return flow — the part of artificially applied water that is not consumed by evapotranspiration and that migrates to an aquifer or surface water body. See also **Evapotranspiration**.

Laminar flow — Flow in which the head loss is proportional to the first power of the velocity. See also **Laminar Flow**.

Leaching — is the removal of materials in solution from soil, rock, or waste. See also **Leaching**.

Leakage — is the flow of water from one hydrogeologic unit to another. The leakage may be natural, as through semi-impervious confining layer, or human-made, as through an uncased well.

Lysimeter — a device for measuring percolation and leaching losses from a column of soil under controlled conditions. See also **Lysimeters**.

Matric potential — the energy required to extract water from a porous medium to overcome the capillary and adsorptive forces.

Moisture content — the ratio, expressed as a percentage, of either (a) the weight of water to the weight of solid particles expressed as moisture weight percentage or (b) the volume of water to the volume of solid particles expressed as moisture volume percentage in a given volume of porous medium. See **Water content** .

Moisture equivalent — the percentage of water retained in a soil sample 1 cm thick after it has been saturated and subjected to a centrifugal force 1000 times gravity for 30 minutes. Centrifuge moisture equivalent is the water content of a soil after it has been saturated with water and then subjected for 1 hour to a force equal to 1000 times that of gravity.

Moisture tension — the equivalent negative pressure of water in an unsaturated porous medium equal to the pressure that must be applied to the medium to bring the water to hydraulic equilibrium through a porous permeable material with a pool of water of the same composition.

Moisture volume percentage — is the ratio of the volume of water in a soil to the total bulk volume of the soil.

Moisture weight percentage — is the moisture content expressed as a percentage of the oven-dry weight of a soil.

Peclet number — a relationship between the advective and diffusive components of solute transport expressed as the ratio of the product of the average interstitial velocity, times the characteristic length, divided by the coefficient of molecular diffusion; small values indicate diffusion dominance, large values indicate advection dominance.

Percent saturation — the ratio, expressed as a percentage, of (a) the volume of water to (b) the total volume of intergranular space (voids) in a given porous medium.

Percolation — is the downward flow of water in saturated or nearly saturated porous medium at hydraulic gradients of the order of 1.0 or less.

Piezometer — a devise used to measure ground-water pressure head at a point in the subsurface.

Pollutant — Shall include, but not be limited to, any element, substance, compound, or mixture, including disease-causing agents, which after release into the environment and upon exposure, ingestion, inhalation, or assimilation into any organism, either directly from the environment or indirectly by ingestion through food chains, will or may reasonably be anticipated to cause death, disease, behavioral abnormalities, cancer, genetic mutation, physiological malfunctions (including malfunctions in reproduction) or physical deformations, in such organisms or their offspring.

Pollution — Specific impairment of water quality by agricultural, domestic, or industrial wastes (including thermal and atomic wastes), to a degree that has an adverse effect upon any beneficial use of water. See also **Water Pollution**.

Saline water — Water that generally is considered unsuitable for human consumption or for irrigation because of its high content of dissolved solids. Commonly expressed as milligrams per liter (mg/L) of dissolved solids, with 35,000 mg/L defined as equivalent to sea water, slightly saline as 1,000 –3,000 mg/L, moderately saline as 3,000 –10,000 mg/L, very saline as 10,000 –35,000 mg/L, and brine has more than 35,000 mg/L.

Seep — is a wet place, where a liquid, usually groundwater, has oozed from the ground to the surface. Seeps are usually not flowing, with the liquid sourced only from underground.

The term seep may also refer to the movement of liquid hydrocarbons to the surface and between geological layers. Oil may rise from the sea floor to sea surface and can form tar balls that can wash up on beaches.

Site characterization — a general term applied to the investigation activities at a specific location that examines natural phenomena and human-induced conditions important to the resolution of environmental, safety and water-resource issues.

Soil-water pressure — the pressure (positive or negative), in relation to the external gas pressure on the soil water, to which a solution identical in composition with the soil water must be subjected in order to be in equilibrium through a porous permeable wall with the soil water.

Solubility — the total amount of solute species that will remain indefinitely in a solution maintained at constant temperature and pressure in contact with the solid crystals from which the solutes were derived.

Solute — the substance present in a solution in the smaller amount. For convenience, water is generally considered the solvent even in concentrated solutions with water molecules in the minority.

Solute transport — the net flux of solute through a hydrogeologic unit controlled by the flow of subsurface water and transport mechanisms.

Solution — a homogeneous mixture of two or more components. In ideal solutions, the movement of molecules in charged species are independent of each other. In aqueous solutions charged species interact even at very low concentrations, decreasing the activity of the solutes.

Sorption — a general term used to encompass the process of absorption and adsorption.

Specific discharge — the rate of discharge of ground water per unit area of a porous medium measured at right angle to the direction of flow.

Specific retention — the ratio of the Volume of water which the porous medium, after being saturated, will retain against the pull of gravity to the volume of the porous medium.

Specific storage — the volume of water released from or taken into storage per unit volume of the porous medium per unit change in head.

Specific yield — the ratio of the volume of water which the porous medium after being saturated, will yield by gravity to the volume of the porous medium.

Tensiometer — a device used to measure the moisture tension in the unsaturated zone.

Transmissivity — the rate at which, water of the prevailing kinematic viscosity is transmitted through a unit width of the aquifer under a unit hydraulic gradient. It is equal to an integration of the hydraulic conductivities across the saturated part of the aquifer perpendicular to the flow paths.

Turbulent flow — the flow condition in which inertial forces predominate over viscous forces and in which head loss is not linearly related to velocity.

Unsaturated zone — the zone between the land surface and the deepest water table which includes the capillary fringe. Generally, water in this zone is under less than atmospheric pressure, and some of the voids may contain air or other gases at atmospheric pressure. Beneath flooded areas or in perched water bodies the water pressure locally may be greater than atmospheric.

Volatiles — substances with relatively large vapor pressures. Many organic substances are almost insoluble in water so that they occur primarily in a gas phase in contact with water, even though their vapor pressure may be very small.

Additional Reading

Baird, A. and R. Wilby: *ECO-Hydrology*, Routledge, New York, NY, 1999.

Bates, P.D. and S. Lane: *High Resolution Flow Modelling in Hydrology and Geomorphology: Advances in Hydrological Process*, John Wiley & Sons, Inc., New York, NY, 1999.

Bloschi, G.: *Scale and Scaling in Hydrology: A Framework of Thinking Analysis*, John Wiley & Sons, Inc., New York, NY, 2000.

Brown, A.C. and A. McLachlan: *Ecology of Sandy Shores*, Elsevier, Amsterdam, 1990.

Brutsaert, W.H.: *Hydrology: An Introduction*, Cambridge University Press, New York, NY, 2005.

Burdon, D.J. and S. Mazloum: "Some Chemical Types of Groundwater from Syria," in *Salinity Problems in the Arid Zones, UNESCO-Arid Zone Research XIV*, pp. 73–90, Paris, 1959.

Burdon, D.J.: "Hydrology, Semiarid Regions," in the *Encyclopedia of Geochemistry and Environmental Sciences*, Vol. IVA (R.W. Fairbridge, editor), John Wiley & Sons, Inc., New York, NY, 1982.

Chin, D.A.: *Water-Resources Engineering*, Prentice-Hall, Inc., Upper Saddle River, NJ, 1999.

Czaya, E.: *Rivers of the World*, Van Nostrand Reinhold, New York, NY, 1982.

Dean, R.G.: "Managing Sand and Preserving Shorelines," *Oceanus*, 49 (Fall 1988).

Dingman, S.L.: *Physical Hydrology*, 2nd Edition, Prentice-Hall, Inc., Upper Saddle River, NJ, 2001.

Giese, G.S. and D.G. Aubrey: "Losing Coastal Upland to Relative Sea-Level Rise: 3 Scenarios for Massachusetts," *Oceanus*, 16 (Fall 1987).

Goff, J.C. and B.P.J. Williams: *Fluid Flow in Sedimentary Basins and Aquifers*, Blackwell Scientific, Palo Alto, CA, 1987.

Gore, R. et al.: "Between Monterey Tides," *Nat'l. Geographic*, 2 (February 1990).

Grant, S.: *Contaminant Hydrology: Cold Region Modeling*, Lewis Publishers, Boca Raton, FL2000.

Hamilton, D.P.: "Death of the Nile Delta?" *Science*, 1084 (November 23, 1990).

Hauck, G.F.W.: "The Roman Aqueduct of Nîmes," *Sci. Amer.*, 98 (March 1989).

Herschy, R.W., and R.W. Fairbridge: *Encyclopedia of Hydrology and Water Resources*, Springer-Verlag New York, LLC, New York, NY, 2006.

Hey, R.D., J.C. Bathurst, and C.R. Thorne: *Gravel-Bed Rivers*, John Wiley & Sons, Inc., New York, NY, 1982.

Kirby, C. and H. Wheater: *Hydrology in a Changing Environment*, John Wiley & Sons, Inc., New York, NY, 2000.

Lowenstein, F.: "The Rising Tide," *Technology Review (MIT)*, 17 (July 1988).

Marshall, C.P. and R.W. Fairbridge: *Encyclopedia of Geochemistry*, Kluwer Academic Publishers, Norwell, MA, 1999.

Moore, J.E.: *Glossary of Hydrology*, American Geological Institute, Alexandria, VA, 2000.

Parlange, M.B. et al.: *Vadose Zone Hydrology: Cutting across Disciplines*, Oxford University Press, Inc., New York, NY, 1999.

Patra, K.C.: *Hydrology and Water Resources Engineering*, CRC Press, LLC, Boca Raton, FL, 2000.

Seymour, R.J.: *Nearshore Sediment Transport*, Plenum, New York, NY, 1989.

Schultz, G.A. and E.T. Engman: *Remote Sensing in Hydrology and Water Management*, Springer-Verlag, Inc., New York, NY, 2000.

Singhal, B.B. and R.P. Gupta: *Applied Hydrology of Fractured Rocks*, Chapman & Hall, New York, NY, 1999.

Smith, P.L. et al.: "Estuary Rehabilitation: The Green Bay Story," *Oceanus*, 12 (Fall 1988).

Stanley, D.J.: "Subsidence in the Northeastern Nile Delta: Rapid Rates, Possible Causes, and Consequences," *Science*, 407 (April 22, 1988).

Thorne, C.R., R.D. Hey, and J.C. Bathurst: *Sediment Transport in Gravel-Bed Rivers*, John Wiley & Sons, Inc., New York, NY, 2002.

Todd, D.K.: *Groundwater Hydrology*, 3rd Edition, John Wiley & Sons, Inc., Hoboken, NJ, 2004.

Van Dam, J.C.: *Impacts of Climate Change and Climate Variability on Hydrological Regimes*, Cambridge University Press, New York, NY, 2003.

Yeh, G.T.: *Computational Subsurface Hydrology: Fluid Flows*, Kluwer Academic Publishers, Norwell, MA, 1999.

HYDROLYMPH. The watery body fluid or blood of lower invertebrates. It carries nutriment to organs and tissues and removes waste; it has no respiratory function generally, though may contain proteins able to function as oxygen carriers.

HYDROLYSIS. A chemical reaction in which water reacts with another substance to form two or more substances. This involves ionization of the water molecules as well as splitting of the compound hydrolyzed, e.g., $CH_3COOC_2H_5 + H \cdot OH \rightarrow CH_3COOH + C_2H_5OH$. Examples are: conversion of starch to glucose by water in the presence of suitable catalysts; or the conversion of sucrose (cane sugar) to glucose and fructose by reaction with water in the presence of an enzyme or acid catalyst; or conversion of natural fats into fatty acids and glycerin by reaction with water, as occurs in one stage of soap manufacturing; or the reaction of the ions of a dissolved salt to form various products, such as acids, complex ions, etc. See also **Cellulose Ester Plastics (Organic)**; **Organic Chemistry**; and **Starch**.

HYDROMAGNETIC EQUATIONS. The time-dependent equations which describe the behavior of a plasma in a magnetic field, assuming that the plasma is a compressible fluid and the plasma pressure P is a scalar. These equations are:

$$\rho \frac{d\mathbf{V}}{dt} = \mathbf{j} \times \mathbf{B} + q\mathbf{E} - \nabla P + \rho\mathbf{g} \tag{1}$$

$$\nabla \cdot (\rho\mathbf{V}) = -\frac{\partial \rho}{\partial t} \tag{2}$$

$$\mathbf{E} + \frac{\mathbf{V}}{c} \times \mathbf{B} = \frac{1}{\sigma}(c\mathbf{j} - q\mathbf{V}) \tag{3}$$

$$\frac{1}{P}\frac{dP}{dt} = \frac{\gamma}{\rho}\frac{d\rho}{dt} \tag{4}$$

$$\nabla \times \mathbf{B} = 4\pi\mathbf{j} + \frac{1}{c}\frac{\partial \mathbf{E}}{\partial t} \tag{5}$$

$$\nabla \cdot \mathbf{B} = 0 \tag{6}$$

$$\nabla \times \mathbf{E} = -\frac{1}{c}\frac{\partial \mathbf{B}}{\partial t} \tag{7}$$

$$\nabla \cdot \mathbf{E} = 4\pi q \tag{8}$$

The first equation is a force equation including gravitational forces. The second equation is a statement of mass conservation, while the third is analogous to Ohm's law. Number four is a statement of the adiabatic condition of the motion where γ is the ratio of specific heats of the plasma. The next four equations are the familiar Maxwell equations with no distinction made for **B** and **H** and **D** and **E** because all currents and charges are treated explicitly. The electromagnetic quantities are given in mixed Gaussian units and the conductivity σ in esu.

HYDROMETEOR. Any product of condensation or deposition of atmospheric water vapor, whether formed in the free atmosphere or at the earth's surface; also, any water particle blown by the wind from the earth's surface. Hydrometeors may be classified in a number of different ways, of which the following is one example: 1) liquid or solid water particles formed and remaining suspended in the air, for example, damp (high relative humidity) haze, cloud, fog, ice fog, and mist; 2) liquid precipitation, for example, drizzle and rain; 3) freezing precipitation, for example, freezing drizzle and freezing rain; 4) solid (frozen) precipitation, for example, snow, hail, ice pellets, snow pellets (soft hail, graupel), snow grains, and ice crystals; 5) falling particles that evaporate before reaching the ground, for example, virga; 6) liquid or solid water particles lifted by the wind from the earth's surface, for example, drifting snow, blowing snow, and blowing spray. See also **Precipitation and Hydrometeors**.

HYDROMETEOROLOGY.   1. Study of the atmospheric and terrestrial phases of the hydrological cycle with emphasis on the interrelationship between them. 2. Meteorology plus hydrology. Many countries use the word in this sense to name the official service charged with the dual responsibility of weather and hydrologic functions. 3. (Rare.) That branch of meteorology that deals with the hydrometeors.

HYDROMETER.   An instrument used for measuring the specific gravity of a liquid. It is usually made of glass and consists of a cylindrical stem and a bulb weighted with mercury or shot to make it float upright. The liquid is poured into a tall jar, and the hydrometer is gently lowered into the liquid until it floats freely.

The point where the surface of the liquid touches the stem of the hydrometer is noted. Hydrometers usually contain a paper scale inside the stem, so that the specific gravity (or density) can be read directly in grams per cubic centimeter.

In light liquids like kerosene, gasoline, and alcohol, the hydrometer must sink deeper to displace its weight of liquid than in heavy liquids like brine, milk, and acids. In fact, it is usual to have two separate instruments, one for heavy liquids, on which the mark 1.000 for water is near the top, and one for light liquids, on which the mark 1.000 is near the bottom of the stem.

The function of the hydrometer is based on Archimedes principle that a solid suspended in a liquid will be buoyed up by a force equal to the weight of the liquid displaced. Thus, the lower the density of the substance, the lower the hydrometer will sink.

See also **Specific Gravity**.

HYDRONIUM ION.   An ion found in water and all its solutions, which has the formula H_3O^+ and which consists of a proton combined with a water molecule. It has been established that hydrogen ions do not exist free in aqueous solution, but are present as hydronium ions. Formation of such ions is statistically rare, resulting from the interaction of water molecules in a ratio of 1 to 556 million.

HYDROPHILIC.   Having a strong tendency to bind or absorb water, which results in swelling and formation of reversible gels. This property is characteristic of carbohydrates, such as algin, vegetable gums, pectins, starches, and of complex proteins, such as gelatin and collagen. See also **Colloid System**; and **Detergents**.

HYDROPHOBIC.   Antagonistic to water; incapable of dissolving in water. This property is characteristic of oils, fats, waxes, and many resins, as well as of finely divided powders, such as carbon black and magnesium carbonate. Some interesting concepts are explored in "The Hydrophobic Effect and the Organization of Living Matter," by C. Tanford, *Science*, **200**, 1012–1018 (1978). See also **Colloid System**.

HYDROPHONE.   A transducer that responds to water-borne sound waves and, if of electroacoustic design, produces equivalent electric waves as output. Types of hydrophones include:

Line Hydrophone. A directional hydrophone consisting of a single straight line element, or an array of contiguous or spaced electroacoustic transducing elements disposed on a straight line, or the acoustic equivalent of such an array.

Split Hydrophone. A directional hydrophone in which electroacoustic transducing elements are so divided and arranged that each division may induce a separate electromotive force between its own electric terminals.

Directional Hydrophone. A hydrophone the response of which varies significantly with the direction of incoming sound.

HYDROPHYTES.   Sometimes called water plants, these plants can grow only where there is an abundance of water, essentially growing in water or saturated soil. Those hydrophytes that are flowering plants are probably those that have reverted to an aquatic habitat. The reverting land plants may first have become marsh plants and then gradually developed into definite hydrophytes.

An aqueous environment presents conditions far more constant than an aerial one does. In the tropics, such conditions permit the plants to grow throughout the year. In colder regions there is a definite winter period when growth must cease. Many hydrophytes of temperate regions merely sink to the bottom and remain dormant during the winter. Others accumulate food reserves in rhizomes, which remain rooted in the bottom and renew growth in the spring. Still others form winter buds, consisting of large apical buds surrounded by many closely packed leaves containing much reserve food material. A few hydrophytes form small tubers.

The stems of hydrophytes contain a very small amount of vascular tissue, since support is largely afforded by the water, and conduction is not a great problem. In many of these plants the stem is very porous so that the plant floats in the water. The leaves of hydrophytes are of two types. Submerged leaves are thin and of various shapes; some, like eel grass leaves, are long and ribbon-like; others, like bladderworts, are finely dissected; while others are reduced to awl-shaped structures of small size. Floating leaves are usually large, undivided, and with stomata on the upper surface.

Reproduction in hydrophytes occurs both asexually and sexually. The flowers of nearly all hydrophytes are wind and insect pollinated, apparently a hangover from the time when they lived on land. A few have become modified to such an extent that pollination takes place on the surface of the water, the pollen floating about thereon and eventually reaching the stigma. A small number of hydrophytes are pollinated under water.

Nearly all algae and many fungi are hydrophytes; so also are some of the higher plants. In the flowering plants there are many water plants, such as the water lilies, bladderwort, eel grass and pondweeds. Many are very interesting plants; several are aquarium plants, serving to oxygenate the water; few are of any economic value. See also **Algae**; and **Fungus**.

HYDROPONICS.   The soilless culture of plants. In this technique, plants are grown with their roots immersed in a solution containing the necessary mineral salts or rooted in a sand medium that is kept moistened with such a solution. In one version of the method, the plants are supported in a matrix of peat, excelsior or some similar material on a wire screen with their roots dipping into the solution below. Aeration of the solution must also be provided if the best results are to be obtained. In another method, the plants are rooted in a medium of sand, gravel, or some similar material contained in a shallow tank into which the solution is automatically pumped at suitable intervals. Between pumpings, the solution gradually drains back into a reservoir tank.

The elements known to be necessary in chemically detectable amounts for the development of plants are carbon, oxygen, hydrogen, nitrogen, phosphorus, sulfur, potassium, magnesium, calcium, iron, manganese, boron, copper, zinc, and perhaps molybdenum. The first three of these elements are obtained by the plant from atmospheric gases or from water absorbed from the soil. The others are all absorbed in the form of mineral salts from the soil. Of the elements absorbed as salts the iron, manganese, boron, copper, zinc, and molybdenum are required in relatively minute quantities and are often called micronutrient elements. The principal elements that must be provided in the form of dissolved salts in hydroponic techniques, therefore, are nitrogen, phosphorus, sulfur, potassium, calcium, and magnesium.

Numerous solutions have been devised for use in the solution or sand culture of plants on both large and small scales. One solution which has been widely and successfully used for such purposes is made as follows: To each liter of water (preferably distilled or rain) add 1 M solution of the following salts as indicated: 1 cubic centimeter KH_2PO_4, 5 cubic centimeters KNO_3, 5 cubic centimeters $Ca(NO_3)_2$, and 2 cubic centimeters $MgSO_4$. To this solution then add 1 cubic centimeter per liter of a solution of micronutrients made as follows: 2.5 grams H_3BO_3, 1.8 grams $MnCl_2 \cdot 4H_2O$, 0.1 gram $ZnCl_2$, 0.05 gram $CuCl_2 \cdot 2H_2O$, and 0.075 gram MoO_3 per liter of distilled water. Also add to each liter of the solution made as described first, 1 cubic centimeter of a 0.5% solution of iron tartrate. The solution must be replaced with a fresh one at suitable intervals, and it is often necessary to add more of the iron solution between replacements.

Crop yields of at least some kinds of plants fully equal to those obtained on fertile soils can be obtained by hydroponic methods. The raising of crops by this method, however, probably will prove to be economically sound only for certain intensive types of agriculture or under certain special conditions. Some greenhouse floricultural and horticultural crops are now being grown successfully by this method. In regions where there is no soil, or where the soil is extremely infertile, but in which the climate is suitable to the development of plants, it Seems likely that hydroponic techniques may prove useful. They have been used with some success, for example, on some of the coral islands of the Pacific Ocean.

See also **Aquaculture**.

HYDROQUINONES. Hydroquinones [CAS: 123-31-9] are dihydroxy aromatic compounds with the two groups in positions corresponding to *ortho* or *para* substitution in the benzene ring. They are closely related to the quinones from which they can be obtained by reduction. Thus *o*-dihydroxybenzene (catechol) can be obtained from *o*-benzoquinone, and hydroquinone (*p*-dihydroxybenzene or quinol) from *p*-benzoquinone. Resorcinol (*m-d*ihydroxybenzene) [CAS: 108-46-3] is not properly a hydroquinone since the corresponding *meta* quinone is not known to exist. Homologs of hydroquinone are usually named after the parent hydrocarbon. Thus toluhydroquinone is 2,5-dihydroxy-1- methylbenzene and naphthohydroquinone is 1,4-dihydroxy-naphthalene. Unlike many of the quinones, the ring systems are fully aromatic and undergo substitution reactions common to phenols and other benzene derivatives. However they are easily oxidized by some reagents to the less stable quinones and degradation products frequently result. Thus treatment of hydroquinone with nitric acid yields oxalic acid while halogenation with sulfuryl chloride results in a mixture of chlorohydroquinones, quinone, quinone chlorides, and tetrachloro-*p*-benzoquinone. The formation of side and degradation products can be minimized if the molecule is protected against oxidation by acetylating or benzoylating at least one of the hydroxyl groups. For example 2-nitro-hydroquinone can be prepared in good yield by the nitration of monobenzoyl hydroquinone followed by hydrolysis. Concentrated sulfuric acid gives hydroquinone-2.5-disulfonic acid directly, and tertiary amyl groups can be introduced into the ring in the 2 and 5 positions by treatment with amylene in the presence of sulfuric acid. The hydroxyl groups are weakly acidic and can readily be converted to ethers by treatment with alkyl halides or sulfates in the presence of alkali. A diacetate is formed on treatment with acetic anhydride. The most characteristic reaction of hydroquinones is their reversible oxidation to quinones.

Catechol, or 1,2-dihydroxybenzene [CAS: 120-80-9], was first prepared by the dry distillation of catechin obtained from *Mimosa catechu*. It can also be formed by the hydrolysis of its methyl ether, guaiacol, which is a constituent of beechwood tar. It is prepared synthetically by fusing phenol-*o*-sulfonic acid with sodium hydroxide, or treating *o*-chlorophenol with aqueous alkali in the presence of copper at a high temperature and pressure. It crystallizes from benzene in colorless monoclinic plates which melt at 105 °C. The lead salt can be oxidized to *o*-benzoquinone by a solution of iodine in chloroform. The ethers of catechol are of considerable importance and can be derived from a number of naturally occurring substances. The methylene ether of protocatechualdehyde is known as piperonal, and is closely related to various natural products including piperine, safrole, and isosafrole, from which it can be derived. These compounds have been used for the synthesis of pyrethrin synergists. *Vanillin*, the principal flavoring constituent of vanilla, is the 3-methyl ether of protocatechualdehyde.

Hydroquinone is found in nature combined in the glycoside arbutin, from which it can be released by hydrolysis with emulsin or dilute sulfuric acid. It is prepared commercially from *p*-benzoquinone by reduction with sulfur dioxide. It is a dimorphic solid with the stable form melting at 170.5 °C. Hydroquinone is one of a number of compounds that possess the property of forming molecular compounds with gases such as hydrogen sulfide, sulfur dioxide, krypton, xenon, etc. These are known as clathrate compounds, and their existence is due to the entrapment of atoms or molecules of the gas in the crystal lattice of the hydroquinone. Three moles of hydroquinone can entrap one mole of gas, which is firmly held but which is liberated when the clathrate is dissolved in water. The most important commercial use of hydroquinone is for the development of photographic film. Its effectiveness is dependent on its ability to reduce the silver subhalide formed on exposure of the film to light to metallic silver. It gives films of high density and it is often necessary to reduce the harshness of contrast by using it in combination with other developers such as metol or paramidophenol. Hydroquinone and its derivatives are effective antioxidants for the preservation of fats, oils, and rubber. It has also been used as a short-stopping agent for controlling polymerization in the production of synthetic rubber of the butadienestyrene type.

H. P. BURCHFIELD, Gulf South Research Institute, New Iberia, Louisiana

HYDROSPHERE. The hydrosphere [Greek *hydor* water and *sphera* sphere] refers to the water on or surrounding the surface of the globe, as distinguished from those of the lithosphere (the solid upper crust of the earth) and the atmosphere (the air surrounding the earth). More specifically, the hydrosphere includes the region that includes all the earth's liquid water, frozen and floating ice, water in the upper layer of soil, and the small amounts of water vapor in the earth's atmosphere. The hydrosphere is the major setting for the earth's hydrologic cycle. See also **Hydrology**.

Origin of Water on Earth

The most significant feature of the earth, in contrast to our neighboring planets, is the liquid water that covers more than two-thirds of the planet's surface. This water came about during the early days of the formation of the earth, when the earth's surface cooled down and the oxygen and hydroxides contained in the accreted material, diffused toward the surface. These gases then cooled and condensed to form the earth's oceans. It is believed that since then, there has been little loss or gain in the overall quantity of the hydrosphere, despite minor fluctuations such as gain from continued degassing and infalling comets and loss in the upper layers of the atmosphere caused by ultraviolet light breaking up water molecules. See also **Earth**; and **Ocean**.

Distribution

The earth's water has six major reservoirs in which water resides. These include the oceans, the atmosphere (split into two reservoirs, one over the land and one over the oceans), surface water (including water in lakes, streams, and the water held in the soil), groundwater (water held in the pore spaces of rocks below the surface), and snow and ice. The locations of some major reservoirs on earth are shown in Fig. 1.

The approximate contribution of the different components of the reservoirs to the hydrosphere, the annual recycled volumes, and the average replacement periods are shown in Table in Table 1.

Table 1 highlights the enormous disparity between the huge volume of salt water and the tiny fraction of freshwater and, in addition, the long residence time of polar ice and groundwater, as opposed to the brief period for which water remains in the atmosphere. Some 96.5% of the total volume of the world's water, it is estimated, exists in the oceans and only 2.5% as freshwater. Of this freshwater, nearly 70% occurs in the ice sheets and glaciers in the Antarctic, Greenland and in mountainous areas, whereas a little less than 30%, it is calculated, is stored as groundwater in the world's aquifers.

Water moves through the reservoirs by a variety of processes, at different rates, and for unique residence times within any reservoir. This flow of water constitutes the Earth's hydrologic cycle. A brief summary of the major processes involved in this movement, along with the amount of water transferred per unit time are shown in Fig. 2.

Biochemistry of the Hydrosphere

The quality of natural water in the various reservoirs of the hydrosphere depends on a number of interrelated factors. These factors include geology, climate, topography, biological processes, land use, and the time for which the water has been in residence. Table 2 gives a comparison of major elements in selected reservoirs.

Rainwater has a low concentration of nutrients compared to the other reservoirs because it originates as evaporated water vapor and also has a relatively short residence time in the atmosphere. Even so, it is never pure. The major constituents originate from dissolution of aerosol particles, which are formed from natural processes, such as evaporation of sea spray or human activities, such as burning of fossil fuels. Naturally rain water has a slightly acid pH (about 5.5). This results from the formation of mild carbonic acid, when rainwater reacts with atmospheric carbon dioxide:

$$CO_2 + H_2O \rightarrow H_2CO_3$$

In areas of high emission of sulfur dioxide or nitrogen oxide gases from industrial activities or fossil fuel burning, hydrolysis of rainwater may result in more acidic rain and a pH as low as 4. See also **Acid Rain**.

River waters have an intermediate concentration of ions compared to that of rainwater and oceans. The main factor controlling the composition or river water is the weathering reaction between rainfall and rocks through

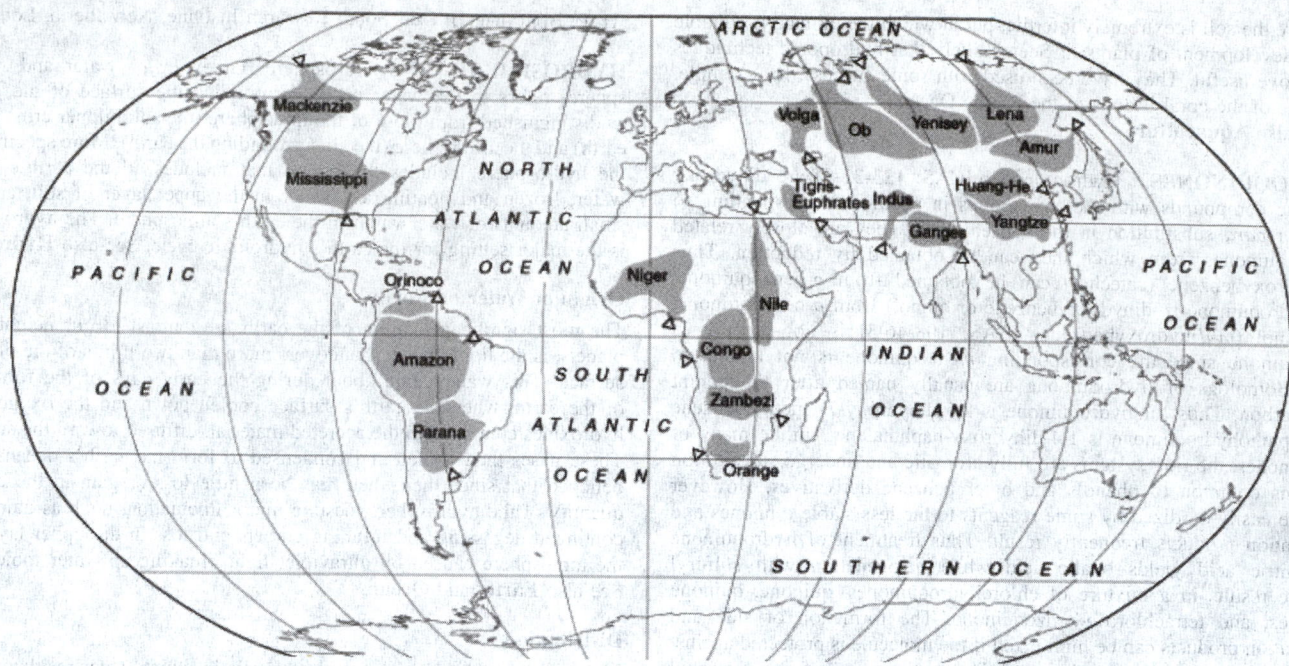

Fig. 1.   The locations of some major global water reservoirs: oceans and surface water drainage basins. (*Ernst Ref.*).

TABLE 1. THE DISTRIBUTION OF WATER ACROSS THE GLOBE[a]

Location	Volume 10^3 km^3	% of Total Volume in Hydrosphere	% of Freshwater	Volume Recycled Annually km^3	Renewal Period, Years
Ocean	1,338,000	96.5	—	505,000	2,500
Groundwater (gravity and capillary)	23,400[b]	1.7		16,700	1,400
Predominantly fresh groundwater	10,530	0.76	30.1		
Soil moisture	16.5	0.001	0.05	16,500	1
Glaciers and permanent snow cover	24,064	1.74	68.7		
Antarctica	21,600	1.56	61.7		
Greenland	2,340	0.17	6.68	2,477	9,700
Arctic Islands	83.5	0.006	0.24		
Mountainous regions	40.6	0.003	0.12	25	1,600
Ground ice (permafrost)	300	0.022	0.86	30	10,000
Water in lakes	176.4	0.013	—	10,376	17
Fresh	91.0	0.007	0.26		
Salt	85.4	0.006	—		
Marshes and swamps	11.5	0.0008	0.03	2294	5
River water	2.12	0.0002	0.006	43,000	16 days
Biological water	1.12	0.0001	0.003		—
Water in the atmosphere	12.9	0.001	0.04	600,000	8 days
Total volume in the hydrosphere	1,386,000	100	—		
Total freshwater	35,029.2	2.53	100		

[a]Shiklomanov, Ref.
[b]Excluding groundwater in the Antarctic estimated at 2 million km^3, including predominantly freshwater of about 1 million km^3.

which this water passes. An example is that of calcite in limestone, which reacts with carbonic acid of rainfall, as

$$CaCO_3 + H_2CO_3 \rightarrow Ca^{2+} + 2HCO_3{}^-$$

Lakes also have an intermediate concentration of ions compared to those of river and seawater. Lake waters constitute a reservoir of freshwater, and their composition depends on four factors: the hydrology (e.g., the relative importance of groundwater or surface water inputs, evaporation),

the surrounding geology (e.g., carbonate rocks or granite), temperature-driven circulation patterns, and anthropogenic factors (e.g., acid rain, agricultural fertilizers). In some instances, evaporation of water from lakes formed in closed basins may result in a high concentration of salts, as opposed to areas of high rainfall. See also **Lakes**.

Sea and ocean waters are dominated by sodium and chloride ions, followed by sulfate and magnesium. Surface sea water is alkaline at an average pH of about 8. Seawater tends to have a more or less uniform

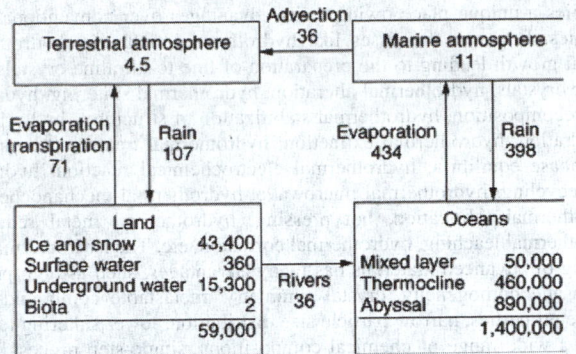

Fig. 2. Estimates of global water reservoirs (in 10^{15} kg and 10^{15} kg/yr) global water cycle fluxes. (*Chahine Ref.*)

TABLE 2. CHEMISTRY OF SOME HYDROSPHERIC COMPONENTS (IN PARTS PER MILLION — PPM)

Major Element	Average Seawater	Average Natural River Water	Average Rainwater
Chloride (Cl^-)	19,000	5.75	3.79
Sodium (Na^+)	10,500	5.15	1.98
Sulphate (SO_4^{2-})	2,700	8.25	0.58
Magnesium (Mg^{2+})	1,350	3.35	0.27
Calcium (Ca^{2+})	410	13.4	0.09
Potassium (K^+)	390	1.3	0.3
Bicarbonate (HCO_3^-)	142	52	0.12
Bromide (Br^-)	67	0.02	
Strontium (Sr^{2+})	8	0.03	
Silica (SiO_2)	6.4	10.4	—
Boron (B)	4.5	0.01	
Fluoride (F^-)	1.3	0.1	

composition in the major elements. But concentrations of minor constituents, including trace and heavy metals and nutrients, vary with depth and location, resulting in marked differences in biological productivity. Organisms living on the surface of the seawater are also involved in changes in its composition via removal of nutrients and breakdown of organic matter at different depths. See also **Ocean Water**.

Groundwater composition is the result of the rock type in which it is confined (e.g., limy is to calcium as argillaceous is to silica); the chemical processes of dissolution, hydrolysis, oxidation-reduction; and biological processes. Moreover, anthropogenic contaminants such as excess fertilizers and heavy metals may also affect the composition of groundwater. See also **Groundwater**.

Ice is a pure solid and has thus only few impurities in its structure, but particulate matter and gases may be trapped within it. Analysis of successively trapped gases or other anthropogenic substances such as carbon dioxide in polar ice caps, has been used to study consecutive changes in the atmospheric composition of past times.

See also **Hydrology**; and **Polar Research**.

Additional Reading

Chahine, M.T.: "The Hydrological Cycle and its Influence on Climate," *Nature*, **359**, 373–380 (1992).
Ernst, W.G.: *Earth Systems, Processes and Issues*, Cambridge University Press, Cambridge, UK, 2000.
Shiklomanov, I.A.: *World Water Resources at the Beginning of the 21st Century*, Cambridge University Press, Cambridge, UK, 2004.

HYDROSTATIC MODEL.   An atmospheric model in which the hydrostatic approximation replaces the vertical momentum equation. This implies that vertical acceleration is negligible compared to vertical pressure gradients and vertical buoyancy forces, a good approximation for synoptic and subsynoptic scales of motion. Hydrostatic models have been successfully applied with horizontal resolutions as small as about 10 kilometers (6 miles), resolving even some mesoscale circulations. Global and regional weather prediction models have traditionally been hydrostatic models.

HYDROSTATIC PRESSURE.   The pressure created by a superimposed layer of a liquid is hydrostatic pressure. The intensity of hydrostatic pressure is commonly expressed as pounds per square inch. A head of 2.31 feet of fresh water creates a hydrostatic pressure of 1 pound per square inch. At a given depth of immersion in water, the pressure acts with equal intensity in all directions, that is, hydrostatic pressure is not directional in effect. Hydrostatic pressures are measured by means of pressure gauges of the Bourden tube type, or manometers of the U-tube type. Hydrostatic pressure sometimes is referred to as hydrostatic head.

In compressible flow, the hydrostatic pressure must be defined more carefully. A suitable definition is in terms of the Helmholtz free energy, $A = U - TS$,

$$p = -\left(\frac{\partial A}{\partial(1/\rho)}\right)_T$$

HYDROSTATICS.   This branch of mechanics has to do with the equilibrium of liquids and the laws relating to liquid pressure. A study of these laws makes it clear that the components of pressure in a liquid fall naturally into two classes, according to the way in which they are produced. These are (1) pressures due to forces applied externally, as by the atmosphere or by the piston of a pump, and (2) pressures due to causes operating throughout the body of liquid, such as gravity or inertia.

Pascal's law applied only to the first class, and states that any pressure in an enclosed liquid, originating in forces applied at its boundary, is communicated with unaltered intensity to all parts of the liquid. A familiar illustration of this fundamental law is the hydraulic press, which consists of two communicating cylinders, usually of different diameter, fitted with pistons, the force acting upon one piston and the force exerted by the other being in proportion to their areas.

The pressure in an enclosed liquid due to its own weight, on the other hand, increases uniformly with the depth below its highest point, and is equal to the product of the depth by the weight per unit volume. For fresh water, the pressure at depth h feet is $62.4\ h$ pounds per square foot. The total pressure of water against a submerged plane are area is equal to the intensity of pressure on the center of gravity times the area.

Problems of flotation, draft, and buoyant stability always involve the density of the liquid and the volume and shape of the floating object. A floating body of mass m in a liquid of density ρ, will float with a volume v submerged, v determined by the relationship $v = (m/\rho)$. That is, a floating body displaces a volume of liquid having the same weight as the body.

The buoyancy may be said to be the force that is equivalent to the weight of the liquid displaced by the submerged portion of the floating object. Buoyancy and weight do not, in general, act in the same vertical line. The weight acts at the center of gravity of the floating object, the buoyancy at the center of gravity of the displaced liquid, called the center of buoyancy. The relative positions of the buoyancy and the weight when a floating object is disturbed from an upright floating position, determines whether it is stable or unstable flotation. If the vertical drawn through the center of buoyancy passes above the center of gravity of the body, there is a righting moment, and the body is stable. Whereas, if it passes below the center of gravity, it is unstable in that the buoyancy tends to tip the object still further. The intersection of the line of buoyancy with the axis of symmetry of the floating body is the metacenter, and the distance from the metacenter to the center of gravity is the metacentric height. The latter is used to measure the stability of a hull. Another case of flotation is illustrated by the balance of the hydrometer. It is apparent from the above that with a given weight, the volume of immersion varies inversely with the density of the liquid. In other words, a floating body rides higher in a denser liquid. This fact is put to use in the hydrometer, which has a given weight and which is immersed in fluids to measure their density. The hydrometer is calibrated to read the volume submerged directly in terms of density of the liquid.

An important general principle of hydrostatics is that which determines the free liquid surface in equilibrium. The direction of the surface at any point is perpendicular to the resultant of all forces acting upon a particle at that point. Thus, if only gravity is acting, the surface is horizontal or "level"; but if there are capillary forces, or if the external pressure is not uniform, the surface is inclined. An interesting case is that of a liquid rotating uniformly in a cylindrical tub; the surface then assumes the form of a paraboloid of revolution, symmetrical about the vertical axis.

Hydrostatic Equation. The form assumed by the vertical component of the vector equation of fluid motion when Coriolis, earth curvature, frictional, and vertical acceleration terms are considered negligible compared with those involving the vertical pressure force and the force of gravity. Thus,

$$\frac{\partial p}{\partial z} = -\rho g$$

where p is the pressure, ρ the density, g the acceleration of gravity, and z the geometric height.

Hydrostatic Equilibrium. The state of a fluid whose surfaces of constant pressure and constant mass (or density) coincide and are horizontal throughout. Complete balance exists between the force of gravity and the pressure force. The relation between the pressure and the geometric height is given by the hydrostatic equation. The analysis of atmospheric stability has been developed most completely for an atmosphere in hydrostatic equilibrium.

Hydrostatic Pressure. The pressure created by a superimposed layer of liquid is hydrostatic pressure. See also **Hydrostatic Pressure**.

HYDROTHERMAL PROCESSING.

Hydrothermal processing of advanced materials has becoming popular among scientists and technologists of different disciplines, for the last 20 years. The term hydrothermal is purely of geological origin. It was first used by the British Geologist, Sir Roderick Murchison (1792–1871), to describe the action of water at elevated temperature and pressure, in bringing about changes in the earth's crust leading to the formation of various rocks and minerals. It is well known that the largest single crystal formed in Nature (beryl crystal of >1000 g) and some of the largest quantities of single crystals created by humans in one experimental run (quartz crystals of several thousands of gram) are both of hydrothermal origin.

The term advanced material refers to a chemical substance, whether organic or inorganic in composition possessing the desired physical and chemical properties. In current context, the term materials processing is used in a very broad sense to cover all sets of technologies and processes of materials preparation to meet the demand from a wide range of industrial sectors. Obviously, it refers to the preparation of materials with a desired application potential. Among various technologies available today in advanced materials processing, the hydrothermal technique

occupies a unique place owing to its advantages over conventional technologies. It covers processes like hydrothermal synthesis, hydrothermal crystal growth leading to the preparation of fine-to-ultrafine crystals, bulk single crystals, hydrothermal alteration, hydrothermal sintering, hydrothermal decomposition, hydrothermal stabilization of structures, hydrothermal dehydration, hydrothermal extraction, hydrothermal treatment, hydrothermal phase equilibria, hydrothermal electrochemical reaction, hydrothermal recycling, hydrothermal microwave, hydrothermal-mechanochemical, hydrothermal fabrication, hot pressing, hydrothermal metal reduction, hydrothermal leaching, hydrothermal corrosion, etc. The hydrothermal processing of advanced materials has many advantages, such as high product purity and homogeneity, crystal symmetry, metastable compounds with unique properties, narrow particle size distribution, lower sintering temperature, a wide range of chemical compositions, single-step process, dense sintered powders, submicron particles with narrow size distribution, simple equipment, lower energy requirements, fast reaction times, growth of crystals with polymorphic modifications, growth of crystals with very low solubility, and a host of others.

Hydrothermal processing can be defined as any heterogeneous reaction in the presence of aqueous solvents or mineralizers under high pressure and temperature conditions to dissolve and recrystallize (recover) materials that are relatively insoluble under ordinary conditions. (Byrappa and Yoshimura, 2001) define *hydrothermal* as any heterogeneous chemical reaction in the presence of a solvent (whether aqueous or nonaqueous) above room temperature and at a pressure >1 atm in a closed system. However, there is still some confusion with regard to the very usage of the term hydrothermal. For example, chemists prefer to use a term, viz, *solvothermal*, meaning any chemical reaction in the presence of a nonaqueous solvent or solvent in supercritical or near supercritical conditions. Similarly, there are several other terms like *glycothermal, alcothermal, ammonothermal,* etc.

Earth is a blue planet of the universe where water is an essential component. Circulation of water and other components such as entropy (energy) are driven by water vapor and heat (either external or internal). Water has a very important role in the formation of material or transformation of materials in Nature and the hydrothermal circulation has always been assisted by the bacterial activity. Hydrothermal processing as such is a part of the solution processing. It can be described as a superheated aqueous solution processing. Figure 1 shows the pressure–temperature map of various materials processing techniques. The solution processing is located

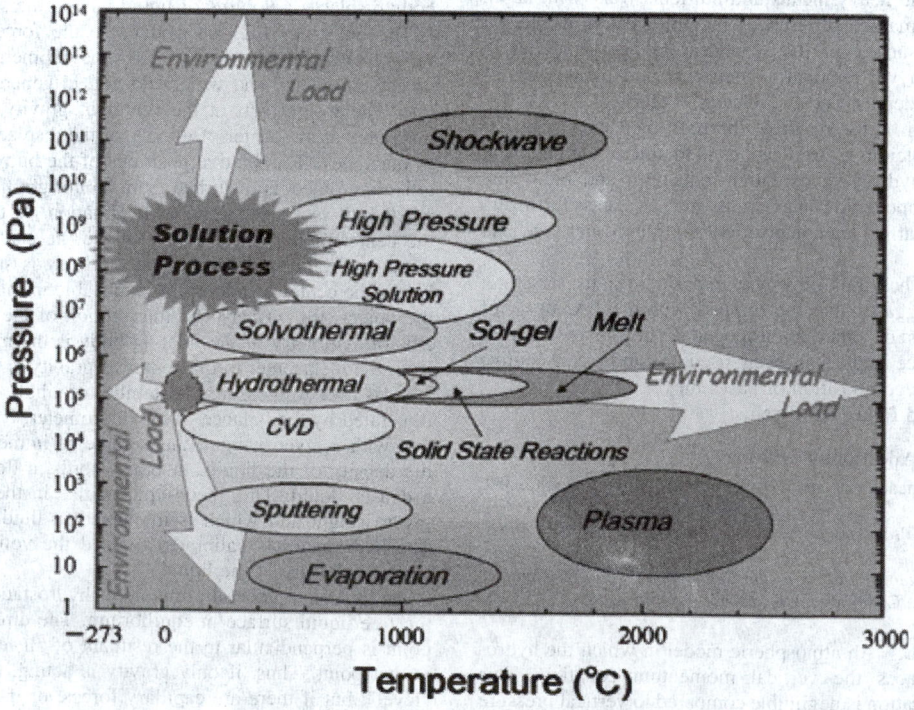

Fig. 1.   Pressure–temperature map of materials processing techniques.

in the pressure–temperature range characteristic for conditions of life on earth and the hydrothermal processing method becomes a part of this solution processing. All other processing routes are connected with increasing temperature and/or increasing (or decreasing) pressure. Therefore, they are environmentally stressed. Thus hydrothermal processing can be considered as environmentally benign. In the last decade, the hydrothermal technique has offered several advantages, e.g., homogeneous precipitation using metal chelates under hydrothermal conditions, decomposition of hazardous and/or refractory chemical substances; monomerization of high polymers, e.g., polyethylene terephthalate; and a host of other environmental engineering and chemical engineering issues dealing with recycling of rubbers and plastics instead of burning. The solvation properties of supercritical solvents are being extensively used for detoxifying organic and pharmaceutical wastes and also to replace toxic solvents commonly used for chemical synthesis. Similarly, it is used to remove caffeine and other food-related compounds selectively. In fact, the food and nutrition experts in recent years are using a new term *hydrothermal cooking*. These unique properties take the hydrothermal technique altogether in a new direction for the twenty-first century and one can forecast a slow emergence of a new branch of science and technology for sustained human development. The important topics of technology in the twenty-first century are predicted to be the balance of environmental and resource and / or energy problems. This has led to the development of a new concept related to the processing of advanced materials, viz, *industrial ecology*: science of sustainability.

During the twenty-first century, hydrothermal technology on the whole will not be limited to crystal growth, or leaching of metals, but it is going to take a very broad shape covering several interdisciplinary branches of science. Therefore, it has to be viewed from a different perspective. Further, the growing interest to enhance hydrothermal reaction kinetics using microwave, ultrasonic, mechanical, and electrochemical reactions will be distinct. Also, duration of the experiments is being reduced by at least two to four orders of magnitude, which in turn makes the technique more economic. With an ever-increasing demand for composite nanostructures, the hydrothermal technique offers a unique method for coating of various compounds on metals polymers and ceramics, as well as fabrication of powders or bulk ceramic bodies.

From the above statements, it is clear that hydrothermal processing of advanced materials is a highly interdisciplinary subject and the technique is popularly used by physicists, chemists, ceramists, hydrometallurgists, materials scientists, engineers, biologists, geologists, technologists, etc.

The first successful commercial application of hydrothermal technology began with mineral extraction or ore beneficiation in the nineteenth century. The use of sodium hydroxide [CAS: 1310-73-2] to leach bauxite [CAS: 1318-16-7] was invented in 1892 by Karl Josef Bayer (1871–1908) as a process for obtaining pure aluminum hydroxide [CAS: 21645-51-2], which can be converted to pure aluminum oxide Al_2O_3 [CAS: 1344-28-1] suitable for processing the metal. Even today, 90 million tons of bauxite ore is treated annually by this process. Similarly, ilmenite [CAS: 12168-52-4], wolframite, cassiterite [CAS: 18282-10-5], laterites, a host of uranium ores, sulfides of gold, copper [CAS: 22205-45-4], nickel [CAS: 12137-12-1], zinc [CAS: 1314-98-3], arsenic [CAS: 12612-21-4], antimony [CAS: 1315-04-4], etc, are treated by this process to extract the metal.

This process is easy to achieve and the leaching can be carried out in a few minutes at ~330 °C (~626 °F) and 25,000 kPa.

Further, the importance of the hydrothermal technique for the synthesis of inorganic compounds in a commercial way was realized soon after the synthesis of large single crystals of quartz by Nacken (1946) and zeolites by Barrer (1948) during 1940s. The sudden demand for large size quartz crystals during World War II, forced many laboratories in Europe and North America to grow large crystals. Subsequently, the first synthesis of zeolite that did not have a natural counterpart was carried out by Barrer, in (1948) and this opened an altogether new field of science, viz, *molecular sieve technology*. The success in the growth of quartz crystals has provided further stimulii for hydrothermal crystal growth. See also **Molecular Sieve**.

In any hydrothermal system or reaction, the role played by the solvent under the action of temperature and pressure is very important. It has been interpreted in various ways by many workers. Yoshimura and Suda (1994) described the action of the hydrothermal fluid on solid substances under elevated pressure and temperature conditions and it is represented in Table 1.

Through proper interpretation of the above listed processes, one can easily develop the required hydrothermal processing using a suitable

TABLE 1. ACTION OF HYDROTHERMAL FLUID (HIGH TEMPERATURE–HIGH PRESSURE AQUEOUS SOLUTION / VAPOR) ON SOLID-STATE MATERIALS

Classified	Action	Application
Transfer	Transfer of kinetic	Erosion, machining abrasion, HIP forming, etc.
Medium	Energy, heat, and pressure	
Adsorbate	Adsorption–desorption at the surface	Dispersion, surface diffusion, catalyst, crystallization, sintering, ion exchange, etc.
Solvent	Dissolution–precipitation	Synthesis, growth, purification, extraction, modification, degradation, etching, corrosion, etc.
Reactant	Reaction	Formation–decomposition (hydrates, hydroxides, oxides), corrosion, etc.

solvent to increase the solubility of the desired compound. Water is the most important solvent and it was popularly used as a hydrothermal mineralizer in all the earlier experiments. However, several compounds do not show high solubility for water even at supercritical temperature. Hence, the size of the crystals or minerals obtained in all the earlier hydrothermal experiments of the nineteenth century did not exceed thousandths or hundredths of a millimeter. Similarly, the time required for the processing of materials has longer durations. Therefore, the search for other suitable mineralizers began in the nineteenth century itself. A variety of aqueous and nonaqueous solutions were tried to suit the preparation of a particular compound. The knowledge acquired through the use of several new mineralizers has helped to implement this hydrothermal technique as an effective one in preparative chemistry.

Natural Hydrothermal Systems

The beginning of hydrothermal research is firmly associated with the study of natural systems by earth scientists, who were interested in understanding the genesis of various rocks, minerals, and ore deposits through laboratory simulations of the conditions existing in the earth's crust. During the nineteenth century, much of the hydrothermal work was confined to mineral synthesis, particularly silicate minerals. In fact, some even included this hydrothermal technology as a part of *silicate technology*. More recently, Japanese researchers are discussing a new concept, viz, *geothermal reactor*. The principles of a geothermal reactor include the direct use of geothermal energy as a heat source or driving force for chemical reactions. It helps to produce hydrothermal synthesis of minerals and a host of inorganic materials, extraction of useful chemical elements contained in crustal materials, such as basalt, and use them as raw materials for hydrothermal synthesis. Thus, the concept of a geothermal reactor leads to the construction of a high temperature and pressure autoclave underground. This has several advantages over conventional autoclave technology. The main disadvantage of the geothermal reactor is that the flow characteristics of a high temperature slurry accompanied by a chemical reaction must be well understood for controlling the reaction. The volumetric capacity of the geothermal reactor is very large compared to that of a usual autoclave, and the operation must be continuous. The merit of the geothermal reactor and its operational cost can be realized only if the target material is to be developed in large quantities.

The spectacular nature of the submarine hydrothermal ecosystem with features, such as "*black smokers*", "*white smokers*", and peculiar ecosystems that are independent of sunlight as a source of reducing power, has focused much interest on hydrothermal processes for the explanation of an array of geochemical processes and phenomena. See also **Ocean**. The submarine hydrothermal systems reveal that most primitive organisms found in modern environments are thermophiles (eg, *archea*). Many scientists believe in a "redox neutral" in the primitive atmosphere. The most important aspect is the possibility that iron vapor and reduced carbon liberated from impacting objects like meteorites would leave the ocean reducing for a long period. In addition, those submarine hydrothermal systems are the only environments where primitive life would have been protected against postulated meteoritic impacts and partial vaporization of the ocean. The presence of supercritical fluids, such as H_2O, CO_2, or CH_4, are the main constituents of any hydrothermal system. They serve as excellent solvents of organic compounds and would probably be of great potential for several of the chemical reactions eventually leading to the origin of life. Further, the pressure and temperature gradients existing in natural hydrothermal systems have a dramatic effect on the properties of the hydrothermal fluids.

There are several reports on the laboratory hydrothermal synthesis of amino acids to justify the origin of life on earth. It is appropriate to mention that the organic synthesis on the whole is not new, and in fact, it began in the nineteenth century. However, the organic synthesis under hydrothermal conditions with reference to the origin of life began in a systematic way only during the 1980s, after the discovery of hydrothermal activity in the deep sea on a Galapagos spreading rich in thermophile organisms in 1977. These workers synthesized amino acids in a temperature range of 150–275 °C (302–527 °F) from aqueous solutions containing KCN, NH_3, HCHO, CO_2, H_2, O_2, CaC_2, NaCN, and NH_4HCO_3 solutions. Recently, based on thermodynamic calculations, Amend and Shock (1988) showed that the autotrophic synthesis of all 20 protein-forming amino acids was energetically favored in hot (100 °C (212 °F), and moderately reduced submarine hydrothermal solutions relative to their synthesis in cold (18 °C (64 °F), oxidized, surface sea water. Although these studies do not support the challenge raised by several others over life in submarine hydrothermal ecosystems, they definitely have set a new trend in hydrothermal research related to biological science. See also **Hydrothermal Vents**.

Physical Chemistry of Hydrothermal Processing of Advanced Materials

Physical chemistry of hydrothermal processing of materials is perhaps the least known aspect in the literature. The Nobel symposium organized by the Royal Swedish Academy of Sciences in 1978, followed by the First International Symposium on hydrothermal reactions organized by the Tokyo Institute of Technology in 1982, helped in setting a new trend in hydrothermal technology by attracting physical chemists in large numbers. Hydrothermal physical chemistry today has enriched our knowledge greatly through a proper understanding of hydrothermal solution chemistry. The behavior of the solvent under hydrothermal conditions dealing with aspects, such as structure at critical, supercritical, and subcritical conditions; dielectric constants; pH variation; viscosity; coefficient of expansion; and density, are to be understood with respect to the pressure and temperature. Similarly, the thermodynamic studies yield rich information on the behavior of solutions with varying pressure–temperature conditions. Some of the commonly studied aspects are solubility, stability, yield, dissolution–precipitation reactions, etc., under hydrothermal conditions. Hydrothermal crystallization is only one of the areas where our fundamental understanding of hydrothermal kinetics is lacking due to the absence of data related to the intermediate phases forming in solution. Thus our fundamental understanding of hydrothermal crystallization kinetics is at an early stage, although the importance of the kinetics of crystallization studies was realized with the commercialization of the synthesis of zeolites during the 1950s and 1960s. In the absence of predictive models, we must empirically define the fundamental role of temperature, pressure, precursor, and time on the crystallization kinetics of various compounds. Insight into this would enable us to understand how to control the formation of solution species, solid phases, and the rate of their formation. In recent years, the thermochemical modeling of chemical reactions under hydrothermal conditions is becoming very popular. Thermochemical computation data helps in intelligent engineering of the hydrothermal processing of advanced materials. The modeling can be successfully applied to very complex aqueous electrolyte and non-aqueous systems over wide ranges of temperature and concentration, and is widely used in both industry and academia. For example, OLI Systems Inc. provides the software for such thermochemical modeling, and using such a package of aqueous systems can be studied within the temperature range −50–300 °C (−58–572 °F), with pressure ranging from 0 to 1500 bar, and concentrations from 0–30 mm molal ionic strength. For the non-aqueous systems, the temperature range covered is from 0 to 1200 °C (32 to 2192 °F) and the pressure is from 0 to 1500 bar with species concentration from 0 to 1.0 mol fraction.

A key limitation to the conventional hydrothermal method has been the need for time-consuming empirical trial and error methods as a mean for process development. Currently, the research is focused on the development of an overall rational engineering-based approach that will speed up process development. The rational approach involves the following four steps:

1. Compute thermodynamic equilibria as a function of chemical processing variables.

2. Generate equilibrium diagrams to map the process variable space for the phases of interest.

3. Design hydrothermal experiments to test and validate the computed diagrams.

4. Utilize the processing variables to explore opportunities for control of reaction and crystallization kinetics.

Such a rational approach has been used quite successfully to predict optimal synthesis conditions for controlling phase purity, particle size, size distribution, and particle morphology of PZT, hydroxyapatite (HAp), and other related systems. The software algorithm considers the standard-state properties of all system species as well as a comprehensive activity coefficient model for the solute species.

By using such a modeling approach, theoretical stability field diagrams (also popularly known as the yield diagrams) are constructed to get 100% yield.

Apparatus Used in Hydrothermal Processing of Materials

Materials processing under hydrothermal conditions requires a pressure vessel capable of containing highly corrosive solvent at high temperature and pressure. Hydrothermal experimental investigators require facilities that must operate routinely and reliably under extreme pressure–temperature conditions. Often, they face a variety of difficulties, and even some peculiar problems pertaining to design, procedure, and analysis. Designing a suitable or ideal hydrothermal apparatus popularly known as an autoclave, reactor, pressure vessel, or high pressure bomb, is a difficult task and perhaps impossible to define, because each project has different objectives and tolerances. However, an ideal hydrothermal autoclave should have the following characteristics:

1. Inertness to acids, bases, and oxidizing agents.

2. Easy to assemble and dissemble.

3. A sufficient length to obtain a desired temperature gradient.

4. Leak-proof with unlimited capabilities to the required temperature and pressure.

5. Rugged enough to bear high pressure and temperature experiments for a long time, so that no machining or treatment is needed after each experimental run.

Keeping these requirements in mind, the autoclave fabrication is done using a thick glass cylinder, thick quartz cylinder, high strength alloys [such as 300 series (austenitic) stainless steel], iron, nickel, cobalt-based super alloys, and titanium and its alloys. Here, it is impossible to describe all of the autoclaves design and working principles. Instead, it is prefered to describe only a few selected and commonly used autoclaves in the hydrothermal processing of materials.

While selecting a suitable autoclave, the first and foremost parameter is the experimental temperature and pressure conditions and their corrosion resistance in that pressure–temperature range in a given solvent or hydrothermal fluid. If the reaction takes place directly in the vessel, the corrosion resistance is of course a prime factor in the choice of autoclave material. In some of the experiments, the autoclaves do not insist on any lining, liners, or cans. For example, the growth of quartz can be carried out in low carbon steel autoclaves. The low carbon steel is corrosion resistant in systems containing silica and NaOH, because, the relatively insoluble NaFe–silicate forms and protectively coats the ground vessel. In contrast, the materials processing from the aqueous phosphoric acid media or other highly corrosive media like extreme pH conditions, then it requires a Teflon lining, beakers, platinum tubes, or lining to protect the autoclave body from the highly corrosive media. Therefore, the corrosion resistance of any metal under hydrothermal conditions is very important. For example, turbine engineers have long known that boiler water with pH >7 is less corrosive than slightly acidic water, especially for alloys containing silicon.

Hydrothermal autoclaves can be used for a variety of applications, such as materials syntheses, crystal growth, phase equilibria study, hydrothermal alteration, reduction, structure stabilization, etc. See also **Silica: Synthetic Quartz Crystals**. There are several new autoclave designs commercially available, which are popularly known as the stirred autoclaves or reactors. These reactors have special features, e.g., the reactor contents can be continuously stirred at different rates, the fluids can be withdrawn while running the hydrothermal experiment, and the desired gas can be supplied externally into the reactors. Such features help greatly to withdraw the fluids from time to time and are subject to various analytical techniques to determine the intermediate phases, which facilitate the understanding of the hydrothermal reaction mechanism for a given material preparation.

There are many more reactor designs for special purposes such as rocking autoclaves, PVT apparatus, multichamber autoclaves, fluid sampling autoclaves, microautoclaves, autoclaves for visual examination, hydrothermal hot pressing, vertical autoclaves, continuous flow reactors, hydrothermal electrochemical autoclaves, autoclaves for solubility measurements, autoclaves for kinetic study, pendulum autoclave, horizontal autoclaves for controlled diffusion study, etc.

Safety and maintenance of the autoclaves is the prime factor to keep in mind in carrying out experiments under hydrothermal conditions. It is estimated that for a 100-cm^3 vessel at 20,000 psi, the stored energy is ~15,000 foot-1b. The hydrothermal solutions — either acidic or alkaline — at high temperatures are hazardous to humans, if the autoclave explodes. Therefore, the vessels should have rupture disks calibrated to burst above a given pressure. Such rupture disks are commercially available for various ranges of bursting pressure. The most important arrangement is that provision should be made for venting the live volatiles out in the event of rupture. Proper shielding of the autoclave should be given to divert the corrosive volatiles away from personnel. In the case of a large autoclave, the vessels are to be placed in a pit with proper shielding.

Conclusions

Hydrothermal processing of advanced materials is an important branch of science and technology owing to its advantages over conventional technologies in terms of materials purity, quality, and performance, as well as its being environmentally friendly, since it consumes lesser energy and the reactions are carried out in a closed system. Hydrothermal processing covers a very broad range of processing techniques that are in use for a great variety of inorganic and organic materials preparation. To summarize, the recent trends in hydrothermal processing of materials covers the study of solvent chemistry, solubility, and solvent–solute interaction, complexation. Modeling of crystallization mechanisms and intelligent engineering of materials; soft hydrothermal technology for materials synthesis and processing; reduction in pressure and temperature; cost effective, environmentally benign processing; use of organic precursors; simple autoclave design; possibility of visual observation; synthesis of new materials; hydrothermal electrochemical or film growth; hydrothermal synthesis of diamond; hydrothermal preparation of whiskers; hydrothermal synthesis of life-forming organics; hydrothermal microwave; hydrothermal mechanochemical; hydrothermal sonochemical; hydrothermal recycling; hydrothermal with selective gas systems; hydrothermal nanotechnology; continuous processing of materials (flow reactors, closed system, solvothermal recycling); hydrothermal treatment (alteration, extraction, etc.); hydrothermal crystal synthesis and growth of bulk crystals; hydrothermal preparation of fine crystals with controlled size, shape, and composition; hydrothermal sintering with or without reactions; hydrothermal etching; hydrothermal corrosion; etc. Thus, hydrothermal technology has moved significantly from geology to technology.

Additional Reading

Barrer, R.M.: *Hydrothermal Chemistry of Zeolites*, Academic Press, London, UK, 1982.

Byrappa, K., and M. Yoshimura: *Handbook of Hydrothermal Technology*, Noyes Publications, Park Ridge, NJ, 2001.

Feng, S.H., J.S. Chen, and Z. Shi: *Hydrothermal Reactions and Techniques*, World Scientific Publishing Company, Inc., River Edge, NJ, 2003.

Habashi, F.: *A Textbook of Hydrometallurgy*, Librairie Universitaire du Quebec, Quebec, Canada, 1993.

Habashi, F.: Recent Advances in Pressure Leaching Technology, Proceeding of the First International Conference on Solvothermal Reactions, Takamatsu, Japan, Dec. 5–7, 1994 pp. 13–16.

Kuhn, M.: *Reactive Flow Modeling in Hydrothermal Systems*, Springer-Verlag New York, LLC, Boca Raton, FL, 2004.

Nakatsuka, K.: "Geothermal Reactor–Concept and Perspectives," T. Moriyoshi, ed., *Proceedings of the Second International Conference on Solvothermal Reactions*, Takamatsu, Japan, Dec. 18–20, 1996, p. 150.

Wickman, F.E., and D. Rickard: Chemistry and Geochemistry of Solutions at High Temperatures and Pressures, Proceedings of the Nobel Symposium in Physics and Chemical Earth, Vols. 13/14, 1981, p. 562.

Yoshimura, M, and H. Suda: "Hydrothermal Processing of Hydroxyapatite: Past, Present, and Future," in P.W. Brown and B. Constanz, eds., *Hydroxyapatite and Related Materials*, CRC Press, LLC, Boca Raton, FL, 1994, pp. 45–72.

HYDROTHERMAL VENTS. Hot springs on the ocean floor are called *hydrothermal vents*. The most numerous and spectacular hydrothermal vents are found along world's mid-ocean ridges. The heat source for these springs is the magma (molten rock) beneath the volcanic ridge system. Geothermal activity beneath 2,000 to 5,000 meters (6,562 to 16, 404 feet) of seawater is markedly different than on land because of the high pressure at the bottom of the ocean. As seawater descends into the region of partly molten rock beneath the mid-ocean ridge, it heats up to 300-400 °C (572–752 °F) and becomes extremely corrosive. This hot fluid is capable of dissolving the surrounding basaltic rock and leaching out metals and other elements. This 300-400 °C (572–752 °F) fluid is also very buoyant and begins rising rapidly back to the surface, and eventually reenters the ocean at hydrothermal vents. See also **Mid-Ocean Ridges**.

ALVIN, an ONR-research submersible (a small submarine) operated by Woods Hole Oceanographic Institute, made an amazing discover in 1977. While diving nearly 8,000 feet (2,400 meters) on the East Pacific Rise near the Pacific Ocean's Galapagos Islands, the submersible and its three passengers happened upon a hydrothermal vent, the first ever seen by humans! Since then, they have been found in the Atlantic, Indian, and most recently, the Arctic Ocean. Completely isolated from the world of light, whole communities of organisms (creatures) live in places where warm water flows from chimneys in the ocean floor. These vents are found in some of the deepest places in the ocean, far beyond the reach of normal submarines or divers. Since then, they have been found in the Atlantic, Indian, and most recently, the Arctic Ocean.

The most spectacular kind of hydrothermal vents are called "black smokers", where a steady stream of "smoke" gushes from a chimney-like structure. The "smoke" consists of tiny metallic sulfide particles that precipitate out of the hot vent fluid as it mixes with the cold seawater. Plumes from such vents can be traced in the ocean for hundreds of meters upwards and hundreds of kilometers horizontally. The chimneys are made out of sulfide minerals that precipitate out of the vent fluid and can grow 10's of meters high. Many large ore deposits now found on land were formed at hydrothermal vents millions or even billions of years ago. Black smokers are an example of focused vents, in which almost all the vent fluid comes out of one small pipe. See also **Black Smoker**.

Sometimes the hot fluids rising from depth are mixed with cold seawater and spread out before they emerge back onto the seafloor. These are called diffuse vents and are usually only a few tens of degrees above the near freezing deep ocean water. Diffuse vent areas have warm water exiting the seafloor over a large area and consequently do not build sulfide chimneys. However, they still contain high levels of hydrogen sulfide and other compounds that specialized microbes can use for energy. This is the basis for an ecosystem that is largely independent of the sun and gives rise to the specialized vent animals such as large tubeworms and clams. The relatively low temperature allows the animals to remain immersed in the nutrient rich water and allows the diffuse vent sites to develop into complex ecosystems. Often chimneys with focused, high-temperature venting are surrounded by areas of diffuse, low-temperature venting. See also **Ocean**.

Web References

Hydrothermal Vents: http://www.ocean.udel.edu/deepsea/level-2/geology/vents.html
NOVA: http://www.pbs.org/wgbh/nova/abyss/life/extremes.html
Vents Program: http://www.pmel.noaa.gov/vents/

HYDROTREATING. A specialized kind of hydrogenation in which the quality of liquid hydrocarbon streams is improved by subjecting them to mild or severe conditions of hydrogen pressure in the presence of a catalyst. The objective is to convert undesirable material in the feedstock to either desired materials or easily disposed byproducts, on a highly selective basis. As of the early 1980s about 45% of the crude oil refined in the United States is hydrotreated. Some applications of hydrotreating include: (1) improvement of the burning quality of jet fuels, kerosines, and diesel fuels; (2) purification of light aromatic byproducts from pyrolysis operations; (3) pretreatment of naphtha feeds for catalytic reforming units; (4) reduction in sulfur content of residual fuel oils; (5) pretreatment of catalytic cracking feeds and cycle oils by removal of metals, sulfur, nitrogen, and reduction of polycyclic aromatics;

(6) desulfurization of distillate fuels; (7) upgrading of lubricating oil quality; and (8) improvement of color, odor, and storage stability of various fuels.

Some of the specific reactions involved include: (1) hydrogenation of monoaromatics to naphthenes to improve burning quality of certain fuels; (2) removal of nitrogen as ammonia from its organic combinations; (3) removal of oxygen from its organic combinations as water; (4) hydrogenation of polycyclic aromatics so that only one aromatic ring remains in the molecule; (5) hydrogenation of diolefins and olefins to paraffins or naphthenes; (6) removal of sulfur from its organic combinations in various types of sulfur compounds by hydrodesulfurization to form hydrogen sulfide; and (7) decomposition and removal of organometals, such as arsenic compounds in naphthas, by retention of these metals on the catalyst. Vanadium and nickel also can be removed.

In the hydrotreating process shown by the Fig. 1, the liquid feed is preheated by exchange with the reactor effluent. It is then heated to the desired reactor-inlet temperature in a fired heater. At this point, recycle hydrogen joins the feedstock. An excess of hydrogen is used to suppress accumulation of deactivating carbonaceous deposits on the catalyst. Fresh makeup hydrogen enters the process to maintain a sufficient supply and also pressure on the system. Cooled effluent from the reactor goes to a separator vessel at which point the recycle or net hydrogen is removed. The liquid then goes to a stripper or stabilizer where hydrogen, hydrogen sulfide, ammonia, water, and light hydrocarbons dissolved in the separator liquid are removed. The stabilized hydro-treated liquid, free of dissolved, unwanted contaminants, is routed to subsequent processing or to product fuel blending.

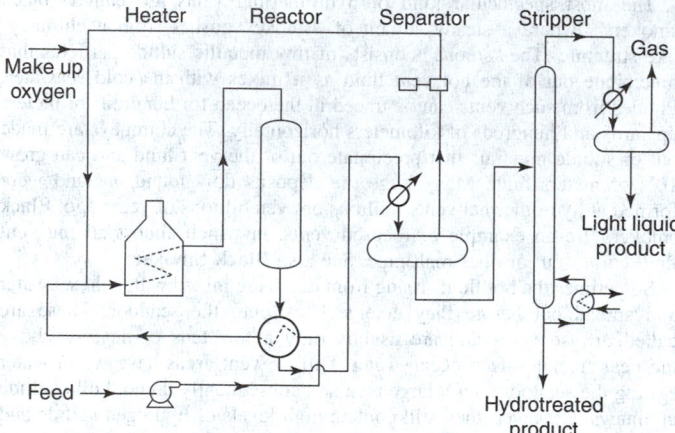

Fig. 1. Representative hydrotreating unit used in petroleum industry. (*UOP Inc.*)

It is interesting to note that there are over 25 proprietary versions of this basic process. Numerous modifications are required, depending upon the nature of the feedstock and desired end products.

Technical Staff, UOP Inc., Des Plaines, Illinois

HYDROXPROLINE.   See Amino Acids.

HYDROXYACETIC ACID.   See Hydroxy Dicarboxylic Acids.

HYDROXYLAMINE. [CAS: 7803-49-8] H_2NOH, formula weight 33.02, white, odorless solid, mp 33 °C, bp 56 °C (22 mm pressure), explosive, soluble in all proportions in H_2O or alcohol. Hydroxylamine is: (1) A weak base forming with acids soluble salts that decompose more or less violently when heated, e.g., hydroxylamine hydrochloride (hydroxylammonium chloride, $H_2NOH \cdot HCl$), mp 151 °C, nitrate $H_2NOH \cdot HNO_3$, hemisulfate $H_2NOH \cdot \frac{1}{2}H_2SO_4$. Dihydroxylamine oxalate and trihydroxylamine phosphate are insoluble in H_2O. Hydroxylamine hydrochloride is soluble in alcohol. (2) A weak acid forming with bases soluble salts, e.g., sodium hydroxylamite H_2NONa. Hydroxylamine salt solution is a powerful reducing agent, more especially in alkaline than in acid solution, for example, cupric salt solutions changed to cuprous oxide, silver salt solutions to silver, mercuric chloride solution to mercurous chloride, ferric salt

solutions (in acid) to ferrous. Ferrous hydroxide in sodium hydroxide is, however, oxidized by hydroxylamine to ferric hydroxide plus NH_3.

Hydroxylamine reacts with carbonyl group $=CO$ of aldehydes, ketones or quinones, yielding *oximes*, white solids, of definite melting point and used in identification of aldehydes and ketones, e.g., acetaldehyde oxime $CH_3CH:NOH$:

Beta-phenylhydroxylamine, *N*-phenylhydroxylamine, is a white solid, slightly soluble in water, very soluble in alcohol or ether, forms salts with acids, e.g., beta-phenylhydroxylamine hydrochloride $C_6H_5 NHOH \cdot HCl$, upon exposure to air the water solution forms azobenzene $C_6H_5N:NC_6H_5$. Beta-phenylhydroxylamine reacts (1) with oxidizing agents, such as chromic acid or ferric chloride, to form nitrosobenzene C_6H_5NO, (2) with reducing agents, such as tin plus hydrochloric acid, to form aniline $C_6H_5NH_2$, (3) with alkaline cupric salt solution (Fehling's solution) at room temperature to form cuprous oxide, (4) with ammonio-silver salt solution (Tollen's solution) at room temperature to form silver, (5) in the presence of hydrochloric acid to form paraminophenol $HO \cdot C_6H_4 \cdot NH_2(1,4)$.

Beta-phenylhydroxylamine is formed by reduction of nitrobenzene (1) by zinc and calcium chloride or ammonium chloride solution, (2) by electrolysis in acetic acid plus sodium acetate solution.

Diphenylhydroxylamine is prepared by reaction of nitrosobenzene and phenylmagnesium bromide in anhydrous ether, followed by treatment with H_2O (magnesium hydroxybromide also formed).

When hydroxylamine reacts with aldehydes, the resulting compounds are termed *aldoximes* as, for example, acetaldoxime. $CH_3 \cdot CHO + H_2NOH \rightarrow CH_3 \cdot CH:N \cdot OH(\text{acetaldoxime}) + H_2O$. Hydroxylamine reactions with ketones produce *ketoximes*. $(CH_3)_2CO + H_2NOH \rightarrow (CH_3)_2C:N \cdot OH(\text{dimethylketoxime}) + H_2O$.

The lower aldoximes are essentially odorless, volatile liquids, and miscible with H_2O in all proportions. The higher members are only slightly soluble. Ketoximes have similar properties.

HYDROZOA.   A class of the phylum *Coelenterata* composed chiefly of small animals without common names. Many species are colonial and the colonies of a few, such as the Portuguese man-of-war, are quite large.

The class differs from the other coelenterates in the occurrence of both hydroid polyps and medusae in the same species, usually in alternating generations. In many colonies, additional specialization occurs among the polyps for the performance of different functions; the gonozooids, gastrozooids, and dactylozooids of the siphonophores are such individuals. Hydrozoan medusae differ from jellyfishes and are called medusoids. In some cases, they are specialized forms, which remain attached to the colony and show no resemblance to medusae, but the free-swimming forms differ from medusae only in details of structure. The medusoids are sexual reproductive individuals.

Relatively few species of hydrozoans live in fresh water. *Hydra*, the most widely known genus, includes a number of species without a medusa stage. The polyps are solitary and carry on both asexual and sexual reproduction. Several freshwater medusae for which no polyp stage has been discovered are also known from lakes in Europe, Africa, and the Americas. The marine species are numerous.

The class can be conveniently classified as follows:

Order 1. *Hydroidea*. Fixed zoophyte stage.
  Suborder a. *Anthomedusae* (*Athecata*). Polyps and reproductive zooids not protected.
  Suborder b. *Leptomedusae* (*Thecata*). Polyps protected by hydrothecae and reproductive zooids by gonotheca.
Order 2. *Hydrocorallina*. Massive skeleton of calcium carbonate secreted from coenosare — Hydroid corals.
Order 3. *Trachylinae*. No fixed zoophyte stage, all members being locomotive medusae.
  Suborder a. *Trachymedusae*. Tentacles from margin of umbrella and gonads develop in connection in the radial canals.
  Suborder b. *Narcomedusae*. Tentacles from ex-umbrella away from margin and gonads develop in connection with the manabrium.
Order 4. *Siphonophora*. Pelagic forms; colony usually exhibits polymorphism of its zooids.

HYENA (*Mammalia, Carnivora*).   A large animal slightly resembling a wolf, but more closely related to the civets. The Aard-Wolf (*Protelinae*) and the Hyena (*Hyaeninae*) make up a small, special grouping in the order of *Carnivora*. The aard-wolf is believed by some authorities to bridge the gap between the Hyaenines and the Viverrines. See also **Viverrines**. Some years ago, investigators believed the aard-wolf was a type of civet and belonged with the Viverrines. This animal is difficult to describe, appearing something like a clumsy dog, with a rather fox-like face, with wooly hair along the back in a form of a permanently erected crest, giving it something of a skunk-like appearance. The aard-wolf differs from the hyena, in that it has five toes on the forefeet; four on the hind feet. Because the animal's teeth are widely set and reduced in number and size, the principal diet is comprised of insects or very decomposed meat or newly born animals. The aard-wolf is found uncommonly in eastern Africa (north of the Kalahari Desert) and on the west coast of Africa as far north as Angola.

At one time, hyenas ranged over much of Europe in climes south of Scotland and the Scandinavian countries and reached eastward through eastern Europe and central Asia. They are found today in most parts of Africa south of the great deserts.

The Striped Hyena (*Hyaena*) is found in Africa as described, as well as in fairly large numbers in northern India. It is also characterized by a crest along its back and by striping on the flanks, with cross-stripes on the legs. The animal is sturdily built, with massive head and somewhat disproportionately long front legs.

The Spotted Hyena (*Crocuta*) is larger than the striped hyena and its limbs are better proportioned. The animal is extremely powerful and can crack large bones, including those of the elephant and hippopotamus. It can put up a good fight with the Big Cats. The striped hyena is considered of very poor disposition, sometimes described as sneaky and generally unpleasant. However, except when in danger, it is described as cowardly. In nature, the animal is extremely dirty, carrying around a most offensive odor. Surprisingly, however, the animal is reported to make an excellent, docile, and trustworthy pet if taken young and trained.

The hyenas are known for their bloodcurdling howling and noise like an insane laugh which usually occurs at the end of a barking streak. On flat ground, their speed exceeds that of a horse.

HYETAL EQUATOR.   See **Meteorology**.

HYETAL REGION.   See **Meteorology**.

HYGRISTOR.   An electric humidity sensor element often used in radiosonde equipment that relies on changes in the resistance of a humidity-sensitive component.

A carbon humidity element is used in the United States. It comprises a polystyrene slide or strip with two metal electrodes along the long edges sprayed with a mixture of carbon particles and a cellulose binder. The binder changes its volume with relative humidity in such a way that it separates the carbon particles from each other as the humidity increases, thus increasing the resistance between the electrodes. The relative humidity is calibrated as a function of resistance.

HYGROMETRY AND PSYCHROMETRY.   These are instrumental methods for measuring humidity. Humidity can be expressed in a variety of

different forms: wet bulb temperature; percent relative humidity (% RH); vapor pressure; mixing ratio; dew/frost point; grains per pound; grams per kilogram; and parts per million, among others. These parameters can be measured by a number of different instruments, each capable of accurate measurement under certain conditions and within specific limitations.

Definition of Humidity.   Unless one is routinely working with humidity measurements, there is a tendency to overlook the fact that humidity is water gas, behaving in accordance with the ideal gas laws. One of the easiest ways to put humidity in its proper perspective is through application of Dalton's law of partial pressures to the most commonly encountered gas—*air*.

Dalton's law states that the total pressure P_m exerted by a mixture of gases or vapors is the sum of the pressure of each gas if it were to occupy the same volume by itself. The pressure of each individual gas is called its *partial pressure*. The total pressure of an air-water gas mixture, containing oxygen, nitrogen, and water, is equal to the sum of the partial pressures of each gas:

$$P_m = P_{N_2} + P_{O_2} + P_{H_2O} + \cdots$$

Therefore, the partial pressure of water vapor in air is directly related to the measurement of humidity. This vapor pressure varies from 1.22×10^{-3} mb (millibar) of mercury (0.122 Pascal) at the $-75\,°C$ frost point of "bone dry" arctic or industrial dry air—to 1.013×10^3 mb of mercury (0.1013×10^6 Pascal) at the $100\,°C$ dew point of saturated hot air in a product drier. This is a change of almost a million to one over the span of interest in industrial humidity measurement.

The ideal humidity instrument would be a linear, wide-range pressure gage, specific to water vapor and employing a primary or fundamental measuring method. Such an instrument, although physically possible, would be cumbersome. Most humidity measurements are made by some secondary instrument which is responsive to humidity-related phenomena.

Humidity Parameters.   The humidity parameters most often encountered in scientific and industrial applications are given in Table 1. In addition to these common parameters, numerous other formats exist for use in narrow applications or specific technologies. However, most of these are variations of the parameters listed.

The psychrometric chart provides a quick means for converting from one humidity format to another because dew point, relative humidity, ambient temperature, and wet bulb temperature can be conveniently related to each other on a single sheet of paper. The psychrometric chart has long been the basic tool of air conditioning engineers. A chart of this type is given in the entry entitled **Psychrometric Chart**. Psychrometric charts are available for higher temperatures and humidities and are quite useful in drier and condensation system design. Charts are also available for lower temperatures, but these tend to be less useful because wet bulb measurements are difficult to make with any accuracy at temperatures below $-7\,°C$.

Wet Bulb/Dry Bulb Measurements

Psychrometry has long been a popular method for monitoring humidity, primarily due to its simplicity and inherent low cost. A typical industrial

TABLE 1. HUMIDITY MEASUREMENT METHODS

Parameter	Description	Units	Typical Applications
Wet bulb temperature	Minimum temperature reached by a wetted thermometer in an airstream	°F or °C	High temperature driers, air conditioning, meteorology, test chambers
Percent relative humidity	The ratio of the actual vapor pressure to the saturation vapor pressure, with respect to water, at the prevailing dry bulb temperature	0–100%	Monitoring conditioning rooms, test chambers, pharmaceutical and foodpackaging
Dew/frost point	Dew point is the temperature to which the air must be cooled to achieve saturation. If the temperature is below 32 °F, it is called the frost point	°F or °C	Heat treating, annealing atmospheres, drier control, instrument air monitoring, meteorological/environmental measurements
Volume or mass ratio	Parts per million (ppm) by volume is the ratio of the partial pressure of the water vapor to the partial pressure of the dry carrier gas. PPM by weight is identical to ppm by volume, but the ratio changes according to the molecular weight of the carrier gas.	ppm_V, ppm_W	Used primarily to insure dryness of industrial process gases such as air, nitrogen, oxygen, methane, hydrogen, etc.

psychrometer consists of a pair of matched electrical thermometers, one of which is maintained in a wetted condition. Water evaporation cools the wetted thermometer, resulting in a measurable difference between it and the ambient, or dry bulb measurement. When the wet bulb reaches its maximum temperature depression, the humidity is determined by comparing the wet bulb/dry bulb temperatures on a psychrometric chart. In a properly designed psychrometer, both sensors are aspirated at an airstream rate between 4 and 10 meters per second for proper cooling of the wet bulb, and both are thermally shielded to minimize errors from radiation.

A properly designed and utilized psychrometer, such as the Assman laboratory type, is capable of providing accurate data. However, very few industrial psychrometers meet these criteria and are limited to applications where low cost and moderate accuracy are the underlying requirements. The psychrometer does have certain inherent advantages: (1) The psychrometer is capable of highest accuracy near 100% RH. From an accuracy standpoint, it is superior to most other humidity sensors near saturation. Since the dry bulb and wet bulb sensors can be connected differentially, this allows the wet bulb depression (which approaches zero as the relative humidity approaches 100%) to be measured with a minimum of error. (2) Although large errors can occur if the wet bulb becomes contaminated or improperly fitted, the simplicity of the device affords easy repair at minimum cost. (3) The psychrometer can be used at ambient temperature above 100 °C, and the wet bulb measurement is usable up to 100 °C.

Major limitations of the psychrometer include: (1) As relative humidity drops below about 20% RH, the problem of cooling the wet bulb to its full depression becomes difficult. The result is impaired accuracy below 20% RH, and few psychrometers work at all below 10% RH. (2) Wet bulb measurement at temperatures below 0 °C are difficult to obtain with any high degree of confidence. Automatic water feeds are not feasible, because of freezing. (3) Because a wet bulb psychrometer is a source of moisture, it can only be used in environments where added water vapor from the psychrometer exhaust is not a significant component of the total volume. (4) Generally speaking, psychrometers cannot be used in small, closed volumes.

Percent Relative Humidity

Percent relative humidity is the best known and perhaps the most widely used method for expressing the water vapor content of air. Percent relative humidity is defined as the ratio of the prevailing water vapor pressure e_a to the water vapor pressure if the air were saturated, e_s, multiplied by 100:

$$\% \, \text{RH} = (e_a/e_s) \times 100$$

The term "percent relative humidity" appears to be derived from the invention of the hair hygrometer in the 17th Century. The hair hygrometer operates on the principle that many organic filaments, such as hair, goldbeater's skin, and even nylon, change length as a nearly linear function of the *ratio* of *prevailing water vapor pressure* to the *saturation vapor pressure*.

Basically, percent relative humidity is an indicator of the water vapor saturation deficit of the gas mixture, rather than an indicator of sorption, desorption, comfort, or evaporation. A measurement of % RH without a corresponding measurement of dry bulb temperature is not of particular value, since the water vapor content cannot be determined from % RH alone.

Sensors for Measuring % RH. Over the years, devices other than the simple hair hygrometer have evolved which permit a direct measurement of % RH. These devices are, for the most part, electrochemical sensors that offer a degree of ruggedness, compactness, and remote electronic readout ability not afforded by hair devices.

Two widely used electronic % RH sensors are the Dunmore element and the Pope cell. The Dunmore sensor employees a bifilar-wound inert wire grid on an insulative substrate which is coated with a lithium chloride solution of a controlled concentration. The hygroscopic nature of this salt causes it to take up water vapor from the surrounding atmosphere. The ac resistance of the sensor is an indication of the prevailing % RH. Dunmore cells are excellent RH sensors, but, because of the characteristics of lithium chloride, are usually designed to cover a narrow range of interest. For example, a single sensor may cover from 40 to 60% RH and the sensor output is usable only in that range. See Fig. 1(a).

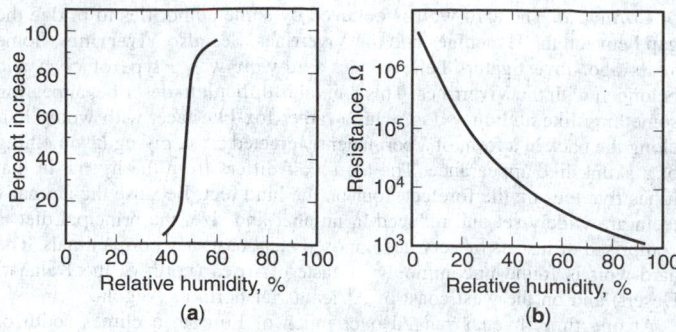

Fig. 1. Resistance characteristics of typical Dunmore and Pope sensors. (**a**) Dunmore sensors are limited to a narrow range of humidity. This sensor operates between 40 and 60% RH. (**b**) Pope sensors operate over a wide humidity range, but output impedance of the sensor varies from 1000 ohms (100% RH) to several megohms (10% RH), which complicates readout circuitry.

Wide-range Dunmore sensors can be made with a cluster of narrow range sensors in a common housing, mated with an electrical matching network. This arrangement, however, usually results in a rather bulky sensor.

The Pope cell employs a similar bifilar conductive grid on an insulative substrate. In this sensor, the substrate is made from polystyrene, which has been treated in a prescribed fashion with sulfuric acid. This results in sulfonation of the longer-chain polystyrene molecules. Because the sulfate radical (SO_4) is highly mobile in the presence of hydrogen ions (available from the water molecule in vapor form), the $(SO_4)^{2-}$ ions can detach and take on H^+ ions, thereby altering the surface resistivity of the sensor as a function of humidity.

In both the Dunmore and Pope sensors, the element is arranged in an ac-excited Wheatstone bridge so that only alternating current flows through the grid. Direct current excitation of either the Dunmore or Pope elements polarizes the sensor, eventually causing loss of calibration.

The Pope sensor has one significant advantage over the Dunmore sensor in that the Pope unit is a wide-range sensor, typically covering 15% RH to 99% RH in a single element. See Fig. 1(b). Considerable attention must be given to readout circuitry for the Pope sensor because the resistance varies in a nonlinear fashion from 1000 ohms to several megohms.

Dew Point Hygrometry

Dew point measurements are widely used in scientific and industrial applications when precise measurement of water vapor pressure is needed. Dew point, the temperature at which water condensate begins to form on a surface, can be accurately measured from −75 °C to +100 °C across the entire range of humidity with a condensation (chilled mirror) hygrometer.

Three types of instruments have received wide acceptance in dew point measurements: (1) the saturated salt dew point sensor; (2) the condensation-type hygrometer; and (3) the aluminum oxide sensor. Many other instruments are used in specialized applications, including pressure ratio devices, dewcups, and fog chambers. The latter are manually operated.

Saturated Salt Dew Point Sensors. The saturated salt (lithium chloride) dew point sensor is widely used because of its inherent simplicity, ruggedness, and low cost. Both the United States and Canadian government weather services use this type of sensor for most official groundbased humidity measurements. However, some of these are being converted to the more accurate condensation hygrometers.

The principle of the saturated salt dew point sensor is based on the relationship that the vapor pressure of water is reduced in the presence of a salt. When water vapor in the air condenses on a soluble salt, it forms a saturated layer on the surface of the salt. This saturated layer has a lower vapor pressure than water vapor in the surrounding air. If the salt is heated, its vapor pressure increases to a point where it matches the water vapor pressure in the surrounding air and the evaporation/condensation process reaches equilibrium. The temperature at which equilibrium is reached is directly related to the dew point.

A saturated salt sensor is constructed with an absorbent fabric bobbin covered with a bifilar winding of inert electrodes and coated with a dilute solution of lithium chloride. Lithium chloride (LiCl) is often used as the saturating salt because of its hygroscopic nature, which permits application in relative humidities between 11 and 100%.

An alternating current is passed through the winding and salt solution, causing resistive heating. As the bobbin heats, water evaporates into the surrounding air from the diluted LiCl solution. The rate of evaporation is determined by the vapor pressure of water in the surrounding air. When the bobbin begins to dry out, due to evaporation of water, resistance of the salt solution increases. With less current through the winding, because of increased resistance, the bobbin cools and water begins to condense, forming a saturated solution on the bobbin surface. Eventually, equilibrium is reached and the bobbin neither takes on nor loses any water.

Properly used, a saturated salt sensor is accurate to $\pm 1\,°C$ between dew point temperatures of -12 and $+38\,°C$. Outside these limits, small errors may occur as a result of the multiple hydration characteristics of lithium chloride, which may produce ambiguous results at $41\,°C$, $-12\,°C$, and $-34\,°C$ dew points. Maximum errors at these ambiguity points are 1.4, 1.6, and 3.4 °C, respectively, but actual errors encountered in typical applications are usually less.

Applications. The saturated salt sensor has certain advantages over other electrical humidity sensors, such as % RH instruments. Because the salt sensor operates as a current carrier saturated with Li and Cl ions, addition of contaminating ions has little effect on its behavior compared to a typical RH sensor, which operates "starved" of ions and is easily contaminated. A properly designed saturated salt sensor is not easily contaminated since, from an ionic standpoint, it can be considered precontaminated.

If a saturated salt sensor does become contaminated, it can be washed with an ordinary sudsy ammonia solution, rinsed and recharged with lithium chloride. It is seldom necessary to discard a saturated salt sensor if proper maintenance procedures are observed.

Limitations of saturated salt sensors include: (1) relatively slow response time; and (2) a lower limit to the measurement range imposed by the nature of lithium chloride. The sensor cannot be used to measure dew points when the vapor pressure of water is below the saturation vapor pressure of lithium chloride, which occurs at about 11% RH. In certain gases, ambient temperatures can be reduced, increasing the RH to above 11%; but the extra effort needed to cool the gas usually warrants selection of a different type of sensor. Fortunately, a large number of scientific and industrial measurements fall above this limitation and are readily handled by the sensor.

Condensation-Type Hygrometers. The condensation-type dew point hygrometer is one of the most accurate and reliable of sensors for humidity measurements, and has the widest range. These features are achieved, however, through increased complexity and cost. In the condensation-type hygrometer, a surface is cooled (either thermoelectrically, mechanically, or chemically) until dew or frost begins to condense out. The condensate surface is maintained electronically in vapor pressure equilibrium with the surrounding gas, while surface condensation is detected by optical, electrical, or nuclear techniques. See Fig. 2. The surface temperature is then the dew point temperature, by definition.

The largest source of error in a condensation hygrometer stems from the difficulty in measuring condensate surface temperature accurately. Typical industrial versions of the instrument are accurate to $\pm 0.2\,°C$ over very wide temperature spans. Laboratory models offer accuracies up to $\pm 0.1\,°C$.

Wide span and minimal errors are two main features. A properly designed condensation hygrometer can measure dew points from $100\,°C$ down to frost points of $-75\,°C$.

Response time of a condensation dew point hygrometer is usually specified in terms of its cooling/heating rate, typically 1.5°/second, making it considerably faster than a saturated salt dew point sensor and nearly as fast as most electrical % RH sensors. Perhaps the most significant feature of the condensation hygrometer is its fundamental measuring technique, which essentially renders the instrument self-calibrating. For calibration, it is only necessary to manually override the surface-cooling control loop, causing the surface to heat, and witness that the instrument recools to the same dew point when the loop is closed. Assuming that the surface temperature measuring system is calibrated, this is a reasonable and valid check on the instrument's performance.

Because of its fundamental nature and superior accuracy and repeatability, this kind of instrument is widely used as a secondary standard (National Bureau of Standards) for calibrating other lower level humidity instruments.

The inert construction of the condensation hygrometer makes it virtually indestructible. Although the instrument can become contaminated, it is easy to wash and return to service without impairment of performance or calibration.

The condensation (chilled mirror) hygrometer measures dew/frost temperature. Unfortunately, many applications require measurement of % RH, water vapor in parts per million, or some other humidity parameter. In such cases, the user must decide whether to employ the fundamental, high accuracy condensation hygrometer and convert the dew/frost point measurement to the desired parameter, or use lower level instrumentation to measure these parameters directly. In recent years, microprocessors have been developed which can be incorporated in the design of a condensation hygrometer, resulting in instrumentation that can offer accurate measurements of humidity in terms of almost any humidity parameter.

Electrolytic Hygrometer. A typical electrolytic hygrometer utilizes a cell coated with a thin film of phosphorous pentoxide (P_2O_5), which absorbs water from the sample gas. See Fig. 3. The cell has a bifilar winding of inert electrodes on a fluorinated hydrocarbon capillary. Direct current applied to the electrodes dissociates the water, which is absorbed by the P_2O_5, into hydrogen and oxygen. Two electrons are required to electrolyze each water molecule and thus the current in the cell represents the number of molecules dissociated. A further calculation, based on flow rate, temperature and current, yields the parts per million concentration of water vapor.

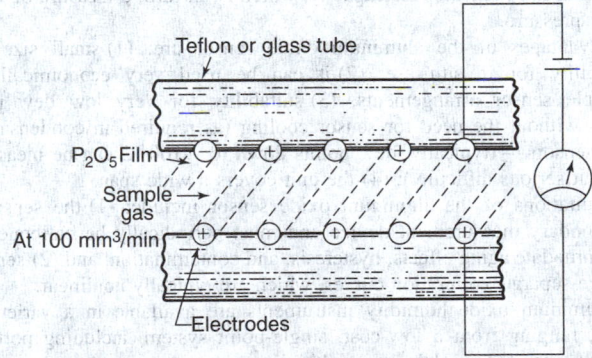

Fig. 3. An electrolytic hygrometer dissociates water, absorbed by P_2O_5, into hydrogen and oxygen by electrolysis. Since two electrons are required to electrolyze a molecule of water, the amount of current used by the hygrometer relates to parts per million of water vapor.

In order to obtain accurate data, the flow rate of the sample gas through the cell must be known and constant. Since the ppm calculation is partially based on flow, an error in the flow rate causes a direct error in measurement.

A typical sampling system for insuring constant flow is shown in Fig. 4. Constant pressure is maintained within the cell. Sample gas enters the inlet, passes through a stainless steel filter, and enters a stainless steel manifold block. It is very important that all components prior to the sensor be made of an inert material, such as stainless steel, to minimize contamination.

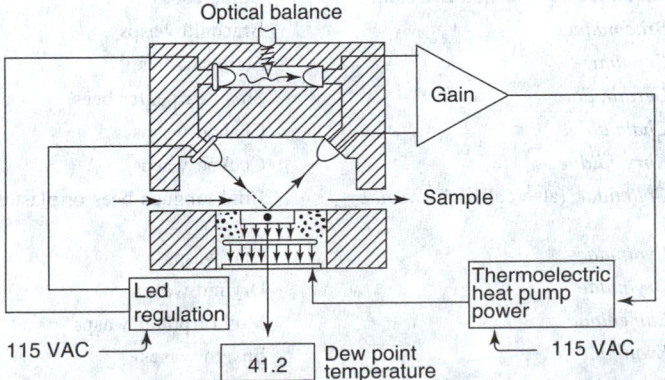

Fig. 2. Dew is detected in a condensation hygrometer by cooling a surface until water begins to condense. Condensation is detected optically or electronically. The signal is fed into a control circuit which maintains the surface temperature at the precise dew point.

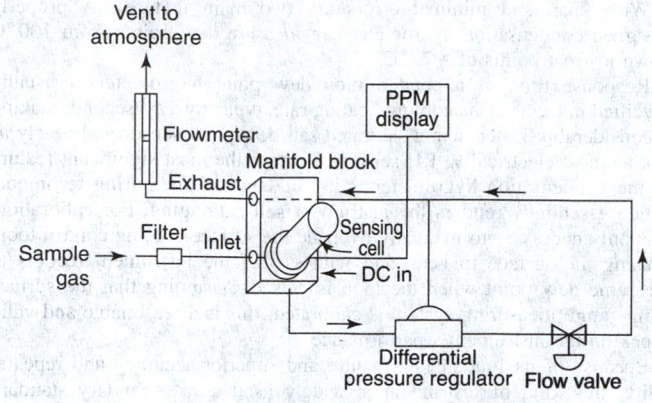

Fig. 4. Calculation of the water vapor content in an electrolytic hygrometer is dependent on precise control of the flow rate. This arrangement controls the sample pressure across the cell, ensuring correct flow regardless of input pressure fluctuations.

After passing through the sensor, the sample gas pressure is controlled by a differential pressure regulator which compares pressure of the gas leaving the sensor with the pressure of the gas venting to atmosphere through a preset valve and flowmeter. In this way, constant flow is maintained even though there may be nominal pressure fluctuations at the inlet port.

A typical electrolytic hygrometer can cover a span from 0 to 2000 ppm with an accuracy of $\pm 5\%$ of the reading, more than adequate for most industrial applications. The sensor is suitable for most inert elemental gases and organic and inorganic gas compounds that do not react with P_2O_5.

Electrolytic hygrometers cannot be exposed to high water vapor levels for any long period of time because this results in a high usage rate for the P_2O_5 and high cell currents.

Aluminum Oxide Moisture Sensor. This type of sensor is a capacitor, formed by depositing a layer of porous aluminum oxide onto a conductive substrate, and then coating the oxide with a thin film of gold. The conductive base and the gold layer become the capacitor's electrodes. Water vapor penetrates the gold layer and is absorbed by the porous oxidation layer. The number of water molecules absorbed determines the electrical impedance of the capacity, which is, in turn, a measure of water vapor pressure.

Advantages of the aluminum oxide sensor are: (1) small size and suitability for in situ use; (2) it can be used very economically in multiple sensor arrangements; (3) suitability for very low dew point levels without the need for sensor cooling (as required in condensation-type sensors—(typically, dew points down to $-100\,°C$ can be measured without serious difficulty); (4) the unit covers a wide span.

Limitations of the aluminum oxide sensor include: (1) the sensor is a secondary measurement device and must periodically be calibrated to accommodate aging effects, hysteresis, and contamination; and (2) sensors require separate calibration curves, which are typically nonlinear.

Aluminum oxide humidity instruments are available in a variety of types, ranging from a low-cost, single-point system, including portable battery operated models, to multipoint microprocessor based systems with capability to compute and display humidity information in different parameters, such as dew point, % RH, etc.

The aluminum oxide sensor is also used for moisture measurements in liquids (hydrocarbons). Because of its low power usage, it is suitable for use in explosion proof installations. These sensors are frequently used in petrochemical applications where low dew points are to be monitored on line and where the reduced accuracies and other limitations are acceptable. The advantages of the sensor must be weighted against the fact that accuracy is lower than with any of the fundamental measurement sensor types. As a secondary measurement device, it can provide reliable data only if kept in calibration and if damage due to incompatible contaminants is avoided.

PIETER R. WIEDERHOLD, General Eastern Instruments Corp., Watertown, MA

HYGROSCOPIC. 1. Pertaining to a marked ability to accelerate the condensation of water vapor. In meteorology, this term is applied principally to those condensation nuclei composed of salts that yield aqueous solutions of a very low equilibrium vapor pressure compared with that of pure water at the same temperature. Condensation on hygroscopic nuclei may begin at a relative humidity much lower than 100% (about 75% for sodium chloride); while on so-called nonhygroscopic nuclei, which merely furnish sufficiently large (by molecular standards) wettable surfaces, relative humidities of nearly 100% are required.

2. Descriptive of a substance, the physical characteristics of which are appreciably altered by effects of water vapor. The hygroscopicity of certain materials has been advantageously utilized in humidity measurement and control devices; for example, the hair element of a hair hygrometer.

HYMENOPTERA. One of the large orders of insects, including ants, bees, wasps, sawflies, and many species without common names. The mouth is formed for biting or for biting and sucking and the wings, when present, are four in number and membranous. Metamorphosis is complete. The order includes plant-eating, parasitic, and predacious species, and in the ants and bees displays some of the finest examples of social organization. The order includes about 70,000 species. See Fig. 1.

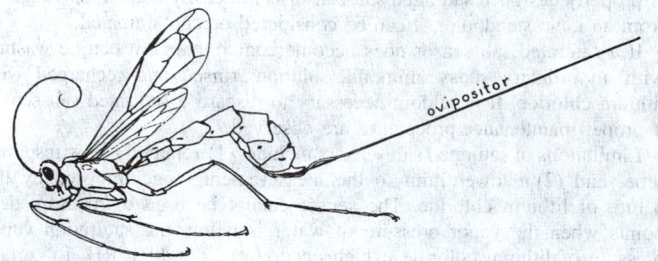

Fig. 1. The degrees of specialization represented by members of *Hymenoptera* are exemplified by this ichneumon wasp, which incorporates a greatly extended proboscis for placing eggs in the bodies of other insects, notably caterpillars. (*USDA.*)

Owing to its extent and diversity this division of the insects includes many species of economic importance. Some of the sawflies and gall wasps are harmful to plants and, on the other hand, the fig insects are beneficial and the galls produced by some gall wasps are of commercial value. Many parasitic species are of undoubted value in holding in check important insect pests. Ants are sometimes very troublesome and the large carpenter bee sometimes damages wood in construction. The most important single species is the honeybee, which is of great value as a producer of honey and wax and in the cross pollination of fruit trees.

The main families of Hymenoptera include:

Andrenidae (also called *Halictidae*)	Mining bees, sweat bees
Apidae	Honeybees
Anthophoridae	Anthophorid bees
Bombidae (also called *Bremidae*)	Bumblebees
Braconidae	Braconid wasps
Cephidae	Stem sawflies
Ceratinidae	Small carpenter bees
Chalcididae	Chalcid wasps
Chrysididae	Cuckoo wasps
Colletidae (also called *Hylaeidae*)	Bifid-tongued bees or plaster bees
Cynipidae	Gall wasps
Dryinidae	Dryinid wasps
Eumenidae	Mud or potter wasps
Evaniidae	Ensign wasps
Formicidae	Ants
Ichneumonidae	Ichneumon wasps
Megachilidae	Leaf-cutting bees and mason bees

Multilidae	Velvet ants
Nomadidae	Cuckoo bees
Pompilidae (also called *Psammocharidae*)	Spider asps
Proctotrupidae (also called *Serphiodea*)	Egg-parasite wasps
Prosopidae (also called *Hylaeidae*)	Obtuse-tongued bees or wasp-like bees
Scoliidae	Vespoid digger wasps
Siricidae	Horn-tails
Sphecidae	Digger wasps, mud-daubers, thread-waisted wasps
Tenthredinidae	Sawflies
Vespidae	Social wasps, paper-nest wasps, hornets, yellow jackets
Xylocopidae	Large carpenter bees

HYPABYSSAL. A general term sometimes used by structural geologists and petrologists to designate those igneous rocks such as sills and dikes which have congealed under less pressure than the plutonic or deep-seated rocks, but under greater pressure than the effusive rocks (lavas).

HYPERACTIVITY (Children). Professionally, this disorder of some children is termed *Attention Deficit-Hyperactivity Disorder* (ADHD). The disorder may be described in lay terms as an *excessively rambunctious behavior*. ADHD affects, in varying degrees, an estimated 5 million children in the United States.

ADHD is poorly understood. At one time, the disorder was considered to be a purely psychological problem. Today, ADHD is considered a physical disorder, and some experts believe it may be inherited. It has been proposed, but not proved, that ADHD children lack certain neurotransmitters (chemical "messengers" that transmit signals within the brain). This insufficiency may have a genetic base. As with most scientifically unknown situations, there is a tendency to suspect nearly anything within reason as a cause. Thus, diet, lead poisoning, food additives, allergies, and other factors have been suspected but not proved to date as a cause of ADHD. One factor that has been well established is that ADHD is not associated with brain damage or impaired intelligence. About half of ADHD children outgrow the disorder.

Some clinics that specialize in their attempts to treat ADHD use a dual approach involving medication and modification of behavior and environment. Medications are selected to control the child's hyperactivity. Parental control of the ADHD child is important. Some authorities suggest:

1. *Set limits* — Establish a system of rewards and punishment that is consistent with the child's behavior. Consistency of attention and approach to the child is very important.
2. *Encourage a sense of responsibility* — An ADHD child should not be removed from responsibility simply because of his or her actions or reactions.
3. *Monitor educational needs* — If the child falls behind in such subjects as reading and mathematics, remedial education classes should be encouraged wherever available. Testing can reveal the need for special education. Close cooperation between teacher, parent, and child must be given a very high priority.

Only after careful analysis by one or more physicians should stimulants, on the one hand, or tranquilizers, on the other hand, be considered.

In diagnosing ADHD, one or more of the following characteristics will be determined:

1. What sets ADHD children apart from other feisty and inattentive children with boundless energy is the intensity and persistence of ADHD behavior. The child acts much younger than his or her chronological age.
2. Easy distraction and a very short attention span, sometimes with extreme mood swings. Although there is a pattern of going from one project to another in many children, the ADHD child in most cases will do this persistently. However, in rarer cases, an ADHD child may become deeply absorbed in certain pursuits.

3. Hyperactivity may become evident at an early age, with feeding and sleeping problems and unexplained crying. Drumming fingers, shuffling feet, and the inability to sit still are commonly manifested.
4. The ADHD child is impulsive, acting on the spur of the moment. Although there are exceptions, untidiness and risk-taking behavior are common.
5. Attention-demanding behavior. The ADHD child desires to be center stage. Actions may include virtual nonstop talking, whining, badgering, teasing, and bossing of other children. But, in some cases, the ADHD child may be "cold" emotionally and quite unresponsive to affection or discipline.

ADHD is a serious but manageable condition when parents, teachers, and friends are aware of the disorder. With proper supervision of a physician knowledgeable and experienced in handling ADHD cases and a supportive home environment, the ADHD child can enjoy the many positive aspects of childhood.

Web Reference

Centers for Disease Control and Prevention. http://www.cdc.gov/health/diseases.htm

HYPERBOLA. A conic section obtained by a plane cutting both nappes of a right-circular conical surface. It is the locus of a point which moves so that the difference of its distances from two foci is a constant. Its eccentricity is greater than unity.

The standard equation may be taken as $x^2/a^2 - y^2/b^2 = 1$. The curve is a central conic for it is symmetric about both the X- and Y-axes when placed in this standard position and the coordinate origin is its center. The transverse axis, coincident with the X-axis, is of length $2a$; the conjugate axis, along the Y-axis, has length $2b(b < a)$. The distance from the center of the hyperbola to either focus is $\sqrt{a^2 + b^2}$ the eccentricity, $e = \sqrt{a^2 + b^2}a$; the length of the latus rectum is $2b^2/a$; the equations for the directrices are $x = \pm a/e$, the same as for the ellipse. The distance from any point on the hyperbola to a focus is a focal radius and the differences between any two focal radii equals $2a$. The lines $y = \pm bx/a$ are asymptotes to the hyperbola. If the length of the transverse axis becomes equal to that of the conjugate axis ($a = b$), the curve is an equilateral or rectangular hyperbola. In this case, the asymptotes are perpendicular to each other. If the coordinate axes are rotated so that they coincide with the asymptotes, the equation for the rectangular hyperbola becomes $xy = a^2/2$, a form which is familiar to students of physical chemistry as Boyle's law.

The polar equation of the hyperbola is $r = a(e^2 - 1)/(e \cos \theta - 1)$ and its parametric equations are $x = a \cosh u$, $y = b \sinh u$ or $x = a \sec \phi$, $y = b \tan \phi$. Its evolute is similar to that of the ellipse $X^{2/3} - Y^{2/3} = 1$, where $X = ax/e^2$, $Y = by/e^2$.

With reference to Fig. 1, the shaded area $= ab \log_e(x/a + y/b)$. In an equilateral hyperbola, $a = b$, in which case the shaded area $= a^2 \log((x + y)/a) = a^2 \log(a/(x - y)) = a^2 \sin h^{-1}(y/a)$, or $a^2 \cos^{-1}(x/a)$.

See also **Conic Section**.

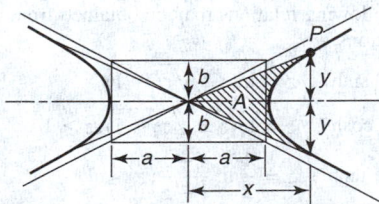

Fig. 1. Major parameters of hyperbola.

HYPERBOLIC FUNCTION. Combinations of $e^{\pm z}$ with properties similar to those of the trigonometric functions. They are defined by:

$$\sinh z = (e^z - e^{-z})/2 = z + \frac{z^3}{3!} + \frac{z^5}{5!} + \cdots$$

$$\cosh z = (e^z - e^{-z})/2 = 1 + \frac{z^2}{2!} + \frac{z^4}{4!} + \cdots$$

$$\tanh z = \sinh z/\cosh z; \coth z = 1/\tanh z$$

$$\operatorname{sech} z = 1/\cosh z; \operatorname{csch} z = 1/\sinh z$$

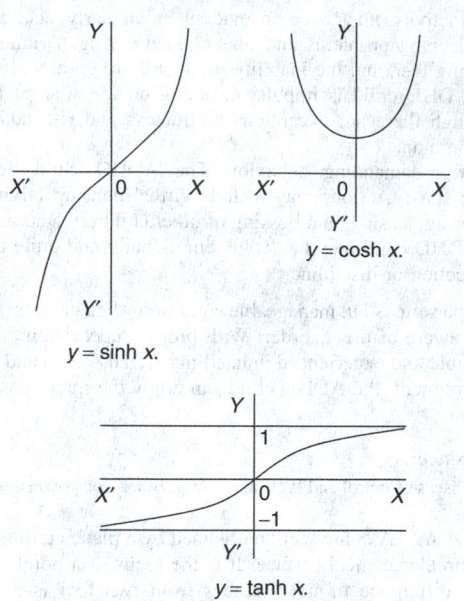

$y = \cosh x.$

$y = \sinh x.$

$y = \tanh x.$

Fig. 1.   Major hyperbolic functions.

If n is a positive integer, $i = \sqrt{-1}$, $u = n\pi i$; $\sinh u = \tanh u = 0$; $\cosh u = (-1)^n$, $\sinh (z + u) = (-1)^n \sinh z$;

$$\cosh (z + u) = (-1)^n \cosh u.$$

For real z, the hyperbolic functions are related to the hyperbola in the same way that the trigonometric functions are related to a circle. If $x^2 \pm y^2 = a^2$, with a plus sign, is the equation for a circle of radius a, or, with a minus sign, for a rectangular hyperbola, the parametric equations are $x = a \cos \phi$, $y = a \sin \phi$ and $x = a\cosh z$, $y = a\sinh z$, respectively. The equations $x = a \sec \phi$, $y = a \tan \phi$ also apply to the hyperbola, and ϕ is the same angle as that for the circle. Comparison of these results shows that $\cosh z = \sec \phi$, $\sinh z = \tan \phi$, with $-\pi/2 < \phi < \pi/2$. Further relations are obtained from the equations defining the hyperbolic functions: $\operatorname{sech} z = \cos \phi$, $\cosh z = \cot\phi$, $\tanh z = \sin \phi$, $\coth z = \csc\phi$. The equation $\sinh z = \tan \phi$ determines ϕ as a function of z. It is called the gudermannian of z, and thus $\phi = \tan^{-1} \sinh z = \operatorname{gd} z$. See Fig. 1.

Again, with real z, hyperbolic and circular (trigonometric) functions are related as follows: $\sinh iz = i \sin z$; $\cosh iz = i \cos z$; $\tanh iz = i \tan z$. Additional formulas, similar to those familiar from trigonometry, are: $\cosh^2 z - \sinh^2 z = 1$; $1 - \tanh^2 z = \operatorname{sech}^2 z$; $\cosh^2 z + \sinh^2 z = \cosh 2z$; $2\sinh z\cosh z = \sinh 2z$.

The inverse hyperbolic functions are also denoted in a manner similar to that for the inverse trigonometric functions. Thus, if $y = \sinh z$, the inverse function is the angle whose hyperbolic sine is y, or $z = \sinh^{-1} y = \operatorname{arcsinh} y$. The following relations may be obtained from the definitions of the various functions:

$$\sinh^{-1} z = \ln(z + \sqrt{z^2 + 1})$$

$$\cosh^{-1} z = \ln(z \pm \sqrt{z^2 + 1}); z \geq 1$$

$$\tanh^{-1} z = \frac{1}{2} \ln \frac{1 + z}{1 - z}; z^2 < 1$$

$$\coth^{-1} z = \frac{1}{2} \ln \frac{z + 1}{z - 1}; z^2 > 1$$

$$\operatorname{sech}^{-1} z = \ln \frac{1 \pm \sqrt{1 - z^2}}{z}; 0 < z \leq 1$$

$$\operatorname{csch}^{-1} z = \frac{1 \pm \sqrt{1 + z^2}}{z}$$

HYPERBOLIC POINT.   (Sometimes called *neutral point*). A singular point in a streamline field that constitutes the intersection of a convergence line and divergence line. It is analogous to a col in the field of a single-valued scalar quantity.

See also **Neutral Point**.

HYPERBOLIC SPIRAL.   A transcendental plane curve, also known as a reciprocal spiral, with polar equation $r\theta = a$ and thus inverse to Archimedes' spiral. It begins at an infinite point from the pole, but as it winds around it never reaches the pole. It has an asymptote $y = a$. Its equation can also be taken as $xt = a \cos t$, $yt = a \sin t$, where t is a parameter. See Fig. 1.

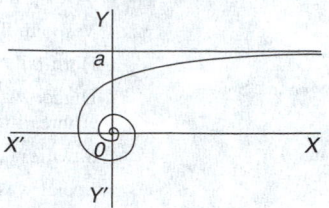

Fig. 1.   Hyperbolic spiral.

HYPERBOLOID.   A central quadric surface with one or two negative terms in its equation. If there is only one, so that $x^2/a^2 + y^2/b^2 - z^2/c^2 = 1$, the surface is a hyperboloid on one sheet. It is given this name because any point on the surface may be reached from any other point on the surface. A plane parallel to the XY-plane cuts out an ellipse, but if the sections are parallel to the XZ- or YZ-planes the results are hyperbolas. When $a = b$, the sections by planes $z =$ constant are circles and the surfaces can be generated by revolving the hyperbola, $x^2/a^2 - z^2/c^2 = 1$ about its conjugate axis, the Z-axis.

If there are two negative terms in the equation, $x^2/a^2 - y^2/b^2 - z^2/c^2 = 1$, the surface is a hyperboloid of two sheets, separated into two parts symmetrically located above and below the planes $x =$ constant. Traces parallel to the XY- and XZ-planes are hyperbolas and traces parallel to the YZ-planes are ellipses, provided $x > a$. When $b = c$, the sections by planes $x =$ constant are circles and a surface of revolution results when the hyperbola $x^2/a^2 - y^2/c^2 = 1$ is rotated about its X- or transverse axis.

See also **Quadric Surface**.

HYPERCONJUGATION.   The description of the properties of a molecule in terms of resonance structures in which an atom or group is not joined by any sort of bond to the atom to which it is ordinarily considered linked. Also called no-bond resonance. The hypothesis of hyperconjugation has been advanced to interpret some properties of substances containing but 1 double bond by analogy with those of substances containing conjugated double bonds. Consider a substance with a terminal structure $H_3C-CH=CH-....$ One of the possible resonating structures of this group is

$$H_3 \equiv C - CH = CH \longrightarrow H_3 = C = CH - CH -$$

the dotted line indicating two unpaired electrons with opposite spins.

HYPEREUTECTIC ALLOY.   An alloy with a composition falling on the right of the eutectic point of a binary phase diagram that freezes with a structure containing some eutectic.

HYPERFINE STRUCTURE.   In general, a set of very closely spaced lines in atomic spectra or other kinds of spectra. There may be many causes of hyperfine structure: (1) for a single atomic species or nuclide, the occurrence of spectral lines as doublets, triplets, etc., due to the interaction, or coupling, of the total angular momentum of the orbital electrons with the nuclear spin and associated magnetic moment; (2) for an element consisting of several isotopes, the occurrence of components for each spectral line that is observable under high resolution, each isotope contributing one or more components. This type of hyperfine structure is often called isotope structure to differentiate it from the first type of hyperfine structure discussed above. See also **Atomic Spectra**.

HYPERGEOMETRIC DISTRIBUTION.   A distribution of a discrete random variable generally associated with sampling from a finite population without replacement. The frequency of r "successes" and $n - r$

"failures" in a sample of n so drawn from a population of N in which there are N_p "successes" and N_q "failures" $(p + q = 1)$ is

$$\frac{1}{N^n} \binom{x}{r} (N_p)^{[r]} (N_q)^{[n-r]}$$

where $N^{[r]} = N(N-1)\ldots(N-r+1)$. As N tends to infinity the distribution tends to the ordinary binomial form. The distribution derives its name from the fact that the probability generating function may be put in the form of a hypergeometric series.

HYPERONS. These are subatomic particles that are more massive than nucleons (protons and neutrons). *Strangeness* is a property of elementary particles found useful in classifying hyperons. Each particle is assigned a strangeness quantum number S which is related to the electric charge Q, the isospin number T, and the baryon number B by the formula $Q = T + (S + B)/2$. ($T = \frac{1}{2}$ for a proton and $-\frac{1}{2}$ for a neutron; other particles may have $T = 0$ or $T = 1$, depending on the type.) Strangeness is conserved in reactions involving the strong interaction. The selection rules resulting from strangeness conservation are important in understanding why some reactions take place much more slowly than others. See also **Particles (Subatomic)**.

HYPEROPIA (Farsightedness). Hyperopia is the ability of the eye to clearly see objects at a distance but not up close. For instance, a person who is farsighted can see to drive, but needs corrective lenses for reading.

Hyperopia is an inherited condition that occurs when the cornea is too flat or the distance from the cornea to the retina is too short. When this happens, the light rays coming from an object strike the retina before coming to sharp focus, or the image is theoretically focused at an imaginary point behind the retina. The result is a blurred image when trying to focus on something that is up close, but distance vision remains sharp.

Children who are farsighted can sometimes compensate without corrective lenses because of the strength and agility of their natural lenses. With a high degree of hyperopia, however, they may exhibit nonvisual symptoms such as headaches and lack of interest in reading. As the eye gets older, it loses some of its ability to commodate(focus), and eventually, most farsighted individuals need corrective lenses.

The usual treatment for hyperopia is prescription eyeglasses with convex lenses that curve outward, or contact lenses that counteract the distortion created by corneas that are too flat in shape. A convex lens moves the image of a distant object forward onto the retina, thereby bringing it into proper focus.

Refractive eye surgery, which steepens the cornea, has recently become an option for the correction of farsightedness. The most popular of those procedures is Laser In-Situ Keratomileusis (LASIK), which uses an Excimer laser to reshape the cornea. Other procedures that show promise for the surgical correction of hyperopia include Implantable Contact Lenses (ICLs) that fit between the iris and the natural lens of the eye, and Clear Lens Extraction (CLE) during which the eye's natural lens is replaced by a plastic intraocular prescription lens. Another procedure that shows promise for the treatment of hyperopia is Laser Thermal Keratoplasty (LTK), which uses a holmium laser to shrink the peripheral area of the cornea in order to steepen the shape. Refractive eye surgery is usually not recommended for people under 18 years of age. See also **Laser In-Situ Keratomileusis (LASIK)**; **Refractive Eye Surgery**; and **Vision and the Eye**.

Vision Rx, Inc., Elmsford, NY

HYPERSENSITIVITY: ANAPHYLACTIC (TYPE I). A significant portion of the human population suffers an allergic reaction sometime during their lifetime. People experience adverse immunologically mediated reactions to allergens from a wide variety of sources, including animal dander, chemical additives, drugs, foods, fungi, insect stings and pollens. The allergic reactions occur when an individual encounters an allergen for the second time. The severity of the reaction can range from mildly irritating to life threatening. No matter what the source of the allergen is, or the severity of the reaction, immunoglobulin E (IgE) mediates all type I immediate hypersensitivity reactions. This immunoglobulin is typically produced in response to a pathogen, but in sensitized individuals it is produced to an otherwise innocuous foreign substance. Most allergic reactions are not life threatening. However, certain allergies such as to food, insect stings and drugs are more likely than others to result in anaphylaxis. See also **Allergy**.

Mechanism

The development of a type I hypersensitivity reaction to a specific allergen involves a series of interactions between antigen-presenting cells (APCs), thymus derived (T) cells and bursa derived (B) cells. On initial contact with the allergen, APCs present small peptide fragments (T-cell epitopes) in conjunction with major histocompatibility complex (MHC) class II molecules to T cells. T cells bearing the appropriate T-cell receptor (TCR) specific for the peptide and MHC class II, bind to the peptide–MHC complex on the APC, leading to further cognate interactions that result in the generation of a 'second' signal, T-cell proliferation and cytokine generation that promote an IgE response (T_H2-like lymphocyte activation). These T cells subsequently interact with B cells bearing appropriate allergen-specific receptors and induce B cells to produce allergen-specific IgE. The allergen-specific IgE then binds to surface receptors of mast cells, basophils, macrophages and other APCs, enabling the immune system to respond to the next encounter with the specific allergen (B-cell epitope) (Figure 1).

T cells play a crucial role in the initial sensitization of an individual to an allergen. The majority of T cells can be divided into two distinct phenotypes, CD4+ cells that are associated with recognition of endocytosed or phagocytosed foreign antigens (e.g. bacteria, allergens, etc.), and CD8+ cells that respond to intracellular pathogens (i.e. viruses). Based on their cytokine expression profiles, CD4+ helper cells are subdivided into T_H1, T_H2 cells and a recently described T_H0 cell type. There is increasing evidence linking allergic diseases with elevated levels of T_H2 cells, known to secrete interleukin 4 (IL-4), IL-5, IL-6, IL-10 and IL-13, and stimulate eosinophilia. T_H1 cells secrete IL-2, interferon γ (IFNγ) and tumor necrosis factor β (TNFβ) and mediate macrophage activation. Cytokine secretion by T-cell lines of allergic individuals has also revealed T-cell clones with T_H0 phenotypes. Thus, the T_H2 and T_H0 cells promote IgE antibody production by B cells, which is critical for the generation of type I hypersensitivity responses.

For B cells to produce a functional immunoglobulin molecule, two separate DNA rearrangement events must occur. VDJ (variable, diversity and joining regions) recombination determines the composition of the antigen-binding site. Thus, it is this region of the immunoglobulin that is responsible for recognition of a specific allergen. The second DNA rearrangement involves isotype switching. With respect to type I hypersensitivity, this rearrangement places a VDJ region proximal to an IgE constant region. It is this portion of the immunoglobulin that interacts with mast cells and basophils, the cells responsible for producing the clinical symptoms of allergy.

The mast cell plasma membrane contains receptors for the Fc portion of IgE. This receptor binds IgE that recognizes the allergen. Mast cell activation is accomplished through the binding of an allergen simultaneously to more than one molecule of IgE. This crosslinking of at least two surface-bound IgE molecules brings the receptor proteins into close association with one another in the plane of the mast cell membrane. Kinases associated with these receptors become activated as a result of this proximity, initiating the second-messenger cascade, which results in mast cell degranulation. The clinical symptoms of allergy and anaphylaxis are primarily due to the release of histamine and a large variety of other bioactive compounds from mast cells and basophils. These cells contain numerous secretory granules, in which these substances are stored at extremely high concentrations. Activation of the mast cell results in the fusion of these granules with the cell-surface membrane, leading to the exocytosis of the granule contents and the induction of allergic symptoms

Histamine, also known as vasoactive or biogenic amine, acts by binding to a variety of different target cell receptors. Histamine receptors on vascular endothelial cells, when occupied by histamine, cause these cells to synthesize vascular smooth-muscle cell relaxants that cause vasodilation and vascular cell leakage, resulting in many of the clinical symptoms associated with an allergic reaction. In addition, histamine can also cause intestinal smooth-muscle cells to spasm, resulting in an increased peristalsis causing the gastrointestinal distress associated with some allergic reactions. See also **Histamine and Histamine Antagonists**.

Examples of Type I Hypersensitivities

Most of the severe allergic reactions that result in anaphylaxis are caused by insect stings, foods and drugs. The allergens responsible for the clinical symptoms of patients allergic to these substances have been identified. Identification and purification of allergens has been essential for the

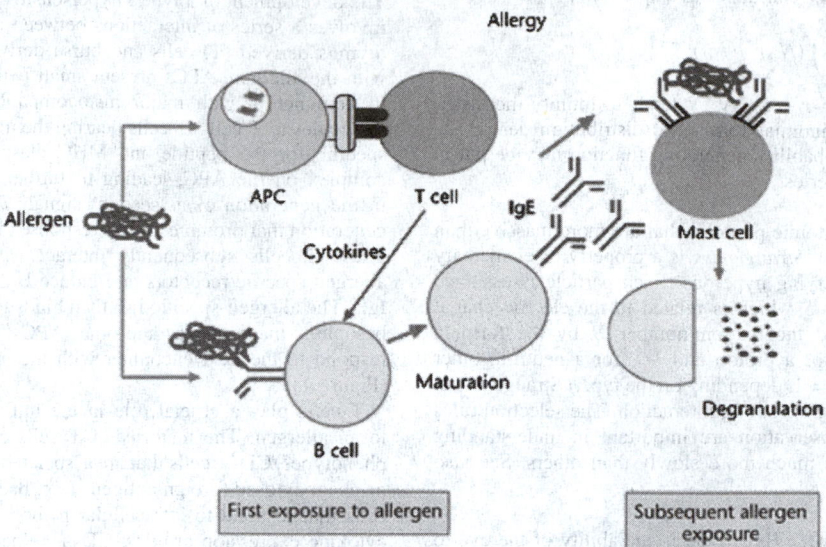

Fig. 1. Interaction of the immune system with an allergen. On first exposure to an allergen, antigen-presenting cells (APCs) will take up, degrade, and then present allergen fragments (T-cell epitopes) to T cells. In a susceptible individual a T_H2-type response will take place where the T cell will secrete cytokines and other mediators that cause the B cell to proliferate and produce immunoglobulin E (IgE). The IgE will be bound by mast cells through the IgE-binding receptor. On subsequent allergen exposures, the allergen will be bound by the mast cells causing a degranulation, or exocytosis, of its granular contents. The granular contents contain the chemicals and mediators (histamine) that cause the clinical symptoms of allergy.

structural and immunological studies necessary to understand how these molecules stimulate IgE antibody formation. Most allergens are low-molecular-weight proteins or glycoproteins readily soluble in aqueous solution. These properties seemingly allow rapid penetration of the allergen at mucosal membranes, facilitating the immediate symptoms observed in allergic patients. For example, insect sting hypersensitivity is due to the allergenicity of the major components of insect venom. These components are melittin, phospholipase A_2 and hyaluronidase. Allergic reactions to single stings can range from large local reactions at the site of the sting, which include diffuse and extensive cellulitis, to a more generalized urticaria and anaphylaxis. Knowledge of the proteins and their IgE-binding sites are critical to successful treatment of allergic disease and anaphylaxis. In the case of insect sting hypersensitivity, knowledge of the proteins and their IgE-binding sites has led to effective treatments for this disease (see the section on treatment below). See also **Allergens**.

Food allergy has commonly been defined as an adverse reaction to foods. However, a true allergic reaction is an aberrant immune response and should be distinguished from food intolerance. Ingested food is digested by salivary enzymes followed by gastric pepsin, hydrochloric acid, pancreatic enzymes, intestinal peptidase and hydrolysis by lysozymes. The larger peptide fragments that find their way into the interstitial space are susceptible to phagocytosis by macrophages. Undigested protein fragments gaining access to gut-associated lymphoid tissue lead to the synthesis of secretory IgA molecules, which bind to and diminish migration of antigen to enterocyte surface membranes where they are endocytosed. It is believed that despite these protective measures small amounts of the larger protein fragments (resistant to digestion) manage to escape the protective measures and become accessible to lymphoid cells, initiating an immune response. Some of the reasons thought to be responsible for the development of a food hypersensitivity include downregulation of local gut immune response, abundance of a certain ingested food protein, higher permeability of infant intestine to antigenic molecules, or early exposure of an infant to antigen prior to immune system development. Upon ingestion, an allergen rapidly crosses the gastrointestinal barrier, where a cascade of events orchestrated by peripheral lymphoid cells cause an immune response leading to type I, IgE-mediated food hypersensitivity.

The consumer marketplace reflects widespread interest and concern about adverse reactions to certain foods and food additives. A recent consumer survey indicated that 30% of the people interviewed reported that they or some family member had an allergy to a food product. This survey also found that 22% avoided foods on the mere possibility that the food may contain an allergen. While the public perception of food allergies is high, true type I hypersensitivity reactions occur in about 6% of children and 1–2% of adults. Peanuts, fish, tree nuts, wheat and shellfish account for the majority of food hypersensitivity reactions in adults, while peanuts, milk and eggs cause over 80% of food hypersensitivity reactions in children. The food allergens are typically low-molecular weight compounds less than 70 kDa. These molecules are usually resistant to proteases, heat and denaturants, allowing them to resist degradation during food preparation and digestion.

Peanuts are one of the most allergenic foods and sensitive individuals may experience symptoms ranging from hives to complete circulatory shutdown (anaphylaxis). Many cases of fatal anaphylaxis have been reported. Most published studies have reported sensitivity to peanuts in children, but many adults are similarly affected. Typically, peanut sensitivity appears early in life and often persists indefinitely. In general, peanuts appear in the diet in easily recognizable forms such as peanut butter powder, peanut butter chips, peanut flours and pressed deflavored peanuts. However, the increasing use of new peanut products, particularly as protein extenders in processed foods, has significantly increased the risk to the peanut-sensitive individual. The latest recommendation for introduction of peanuts into the diet suggests delaying the addition of this food until 3 years of age to prevent sensitization.

The peanut (*Arachis hypogaea*) is an annual plant belonging to the family Leguminosae (legumes) and is native to South America. Several peanut varieties are grown in the United States, including Virginia, Spanish and runners. Some of the best characterized food allergens are those involved in hypersensitivity reactions to peanuts. Various studies over recent years have examined the nature and location of the multiple allergens in peanuts. These showed that the allergenic portion of the peanut is in the protein of the cotyledon and does not differ significantly between different varieties of peanuts. There are three major peanut allergens, all of which are classified as seed storage proteins. Similar to all other food allergens, peanut allergens have multiple IgE-binding epitopes. IgE-binding epitopes are those regions on the allergen that specifically interact with the immunoglobulin. These regions can be either a continuous amino acid sequence composed of 6–15 amino acids or a specific structure composed of different regions of the allergen. Interestingly, many of the

IgE-binding epitopes identified in foods are linear peptides as opposed to the conformational epitopes found in many aeroallergens.

Treatment

The sequence of events necessary to produce allergen-specific IgE, and the resulting clinical symptoms from a second exposure to the allergen, suggests two strategies for treatment. The first strategy involves minimizing the clinical symptoms of an allergic reaction, while the second strategy involves immunotherapeutic approaches designed to decrease the initial immune response that produced allergen-specific IgE.

Approaches designed to minimize the clinical symptoms of an allergy include avoidance of the allergen and use of prescription drugs designed to counteract the effects of inflammatory mediators released during an allergic attack. Avoidance of a particular allergen is not always practical or possible. For example, amongst the most widely used legumes, used in the preparation of a variety of foods throughout the world and as protein extenders in developing countries, are peanuts. It is increasingly difficult for the public to avoid an abundantly utilized and often disguised food source such as peanuts, leading to sensitization in genetically predisposed individuals or accidental ingestion, anaphylaxis and possible death in peanut allergic patients.

Several medications are available that are designed to minimize the symptoms of an allergic response. Antihistamines are one of the most common class of medications used. These drugs act as histamine receptor antagonists, effectively blocking the binding of histamine to its receptor and inhibiting the clinical symptoms associated with histamine-activated cells. Other drugs include ketotifin, corticosteroids, inhibitors of prostaglandin synthase and antileucotrienes. Unfortunately, these drugs are limited in their effectiveness and can lead to unwanted side effects, such as drowsiness. While some of these medications are able to alleviate symptoms associated with an allergic reaction (sneezing, watery eyes, runny nose), none of them have been shown to be effective in treating the underlying conditions associated with the onset of an immediate hypersensitivity reaction.

One of the most common prescription drugs used to counteract a severe allergic reaction is the sympathomimetic drug adrenaline. Adrenaline (epinephrine) is typically given subcutaneously or intramuscularly and has a rapid onset and short duration of action. It helps to maintain blood pressure, reduces the effects of released mediators, and inhibits the release of additional mediators of anaphylaxis. Patients with a history of allergic anaphylaxis usually have EpiPens (registered trademark of Schwarz Pharma) prescribed to them, which allow self-administration of adrenaline after they have come into contact with an allergen.

Approaches designed to modulate the immune response to a particular allergen fall into two categories. The first category includes strategies designed to inhibit interaction of mediators important to IgE production, while the second category includes strategies designed to alter T-cell activation. Strategies that fall into the first category include the use of allergen-specific, soluble TCRs to inhibit allergen presentation to T cells, the use of peptides that compete with allergen-specific peptides for TCR binding; and the use of antibodies that would interfere with cytokine interactions that promote B cell production of IgE. Approaches that fall into the second category include induction of T-cell anergy by use of allergen-specific peptides (T-cell epitopes) and modulation of T_H cell development by alteration of T-cell epitope structure to favour T_H1 cytokine response or prevent T-cell proliferation. For any of these approaches to be successful, it must be specific to the allergen that the individual's immune system recognizes.

Immunotherapy designed to desensitize individuals to a specific allergen was first introduced in 1911 and involved the injection of increasing amounts of nonstandardized extracts from the source of the allergen. More recently, the introduction of standardized extracts has made it possible to increase the efficacy of immunotherapy, particularly in the case of insect sting hypersensitivity. Guidelines set by the Canadian Society of Allergy and Clinical Immunology indicate that immunotherapy regimens with the venom of insects is >95% protective in patients with previous anaphylactic reactions to insect venoms. For any immunotherapeutic approach to be successful, patients must be treated with high doses of the allergen responsible for the reaction. Dosage is usually determined by the patient's history and positive skin test results. Duration of the allergen injection treatments is usually 4–5 years. See also **Venoms**.

Even though allergen immunotherapy is successful for some types of allergic disease, there are some allergies that are difficult to treat

with this approach. For example, initial trials of immunotherapy to food allergens has shown an unacceptable safety/efficacy ratio, prompting investigators to discontinue this form of treatment for most food allergic patients. Unfortunately, the only effective treatment currently available for food hypersensitivity is food avoidance. As described above, allergen immunotherapy is an effective therapeutic modality to prevent further anaphylactic episodes for patients with insect sting hypersensitivity because it can downregulate the specific IgE response and the cellular response to allergens. Since food allergy can be a significant health problem due to the potential severity of the allergic reaction, the chronicity of the allergic sensitivity, and the ubiquity of disguised food products in processed foods, novel immunotherapeutic approaches have to be developed to deal with this sometimes life-threatening allergic disease.

Additional Reading

Fireman, P.: *Atlas of Allergies and Clinical Immunology*, Elsevier Health Sciences, New York, NY, 2005.
Golden, D.B.: "Stinging Insect Allergy," *American Family of Physician*, **15**(12), 2541–2546 (2003).
Herrick, C.A., L. Xu, A.N. McKenzie, R.E. Tigelaar, K. Bottomly: "IL-13 is Necessary, Not Simply Sufficient, for Epicutaneously Induced Th2 Responses to Soluble Protein Antigen," *Journal of Immunology*, **170**(5), 2488–2495 (2003).
Huang, S.K.: "Molecular Modulation of Allergic Responses," *Journal of Allergy and Clinical Immunology*, **102**, 887–892 (1998).
Kirman, J., and G. Le Gros: "Which is the True Regulator of TH2 Cell Development in Allergic Immune Responses?" *Clinical and Experimental Allergy*, **28**, 908–910 (1998).
Lieberman, P.L., and M.S. Blaiss: *Atlas of Allergic Diseases*, 2nd Edition, Springer-Verlag New York, LLC, New York, NY, 2005.
Maleki, S.J., A.W. Burks, and R.M. Helm: *Food Allergy*, ASM Press, Washington, DC, 2006.
Rolland, J., and R. O'Hehir: "Immunotherapy of Allergy: Anergy, Deletion, and Immune Deviation," *Current Opinion in Immunology*, **10**, 640–645 (1998).
Spiegelberg, H.L., K. Takabayashi, L. Beck, and E. Raz: "DNA-based Vaccines for Allergic Disease," *Expert Reviews in Vaccines*, **1**(2), 169–177 (2002).
Staff, Springhouse Publishing Company: *Handbook of Allergic Disorders*, Lippincott Williams & Wilkins, Philadelphia, PA, 2002.
Stanley, J.S., and G.A. Bannon: "Molecular Mechanisms of Food Allergy," In: Zempleni, J., and H. Daniel: *Molecular Nutrition*, CAB International, New York, NY, 2003, pp. 369–379.
Sturm, G., B. Kranke, C. Rudolph, and W. Aberer: "Rush Hymenoptera Venom Immunotherapy: a Safe and Practical Protocol for High-risk Patients," *Journal of Allergy and Clinical Immunology*, **110**(6), 928–933 (2002).

GARY A. BANNON, Monsanto Company, St Louis, MO

HYPERSONIC FLOW. In aerodynamics, flow of a fluid over a body at speeds much greater than the speed of sound and in which the shock waves start at a finite distance from the surface of the body.

HYPERSORPTION. Process in which activated carbon selectively absorbs the less-volatile components from a gaseous mix, while the more-volatile components pass on unaffected. Particularly applicable to separations of low-boiling mixtures such as hydrogen and methane, ethane from natural gas, ethylene from refinery gas, etc.

HYPERSTHENE. The mineral hypersthene is an orthorhombic pyroxene, chemically a ferromagnesian silicate, differing from enstatite in that the iron content is considerable (FeO being greater than 15%). A general formula is $(Mg, Fe)SiO_3$. It is usually found as a massive mineral, whose crystals tend to be prismatic or tabular in habit. It has a distinct prismatic cleavage; fracture, uneven; brittle; hardness, 5–6; specific gravity, 3.42–3.84; luster, pearly to somewhat metallic; color, brownish-green, brown, greenish-black to grayish-black; streak, grayish-brown; translucent to opaque. Hypersthene is often associated with labradorite in gabbro and norite and in extrusive rocks like andesite. It is occasionally encountered in meteorites. Hypersthene is associated with pyrrhotite in Bavaria, with labradorite on the Isle St. Paul, Labrador. It is also found in Montmorency County, Quebec; and in the United States in the rocks of the Cortlandt series in the Hudson River Valley, and the andesites of Colorado and northern California. Superb crystals of exceptional size and quality have been found growing into and within the almandinepyrope garnets at Gore Mountain, North River, New York. The rarity of hypersthene in crystal form makes this occurrence noteworthy. The word hypersthene comes from the Greek words meaning *strong* or *tough*.

See also **Pyroxene**.

HYPERTENSION (High Blood Pressure). Commonly regarded and treated as a disorder in itself, high blood pressure may be more accurately described as a major symptom of a complex of disorders. Not all of these disorders are present in one person. There is a variety of patterns of these underlying disorders which readily explains what was a puzzle for many years—namely, the manner in which different people with the symptom of hypertension react differently to various drug therapies. Considering the numerous body systems—circulatory, nervous, endocrine, excretory, among others—that interact in different ways to produce a universal symptom (hypertension), it would indeed be surprising if all persons with high blood pressure did react precisely in the same manner to therapy.

Considering the heart as a pump, (1) the *systole* is the period of the heart's contraction, or the contraction itself—the systolic pressure represents the highest arterial blood pressure; and (2) the *diastole* is the period of the heart's dilation—the diastolic pressure represents the lowest arterial blood pressure that occurs between the pulse waves. The arterial blood pressure in humans is usually measured in the arm at the brachial artery, preferably with the patient seated or lying down with the arm slightly flexed and at heart level. In a thorough examination for hypertension, multiple readings will be made—both arms and legs. With the aid of a stethoscope, the examiner will determine both systolic and diastolic pressure with the *sphygmomanometer*, an instrument whose pressure scale is calibrated in millimeters of mercury. See also **Blood Pressure**; **Manometer**; and **Sphygmomanometer**.

Statistically, over decades, expected ranges for these pressures in healthy persons have been established. These ranges have become established standards against which individual readings are compared and from which a diagnosis of high blood pressure (*hypertension*) or low blood pressure (*hypotension*) is made. The pressures in the arteries and the veins were first measured as early as 1733 by Stephen Hales, who used a rather crude measurement technique in making determinations of these pressures in a mare. By 1828, Hales had developed a method using a U-tube manometer, the prototype of current instruments.

The statistical averages for blood pressure of healthy adults are:

Systolic   110–120 mm mercury
Diastolic   65–80 mm mercury

Some observers have reported that the systolic pressure is higher in men than in women. The normal upper limits of systolic pressure are

140 mm mercury (men)

130 mm mercury (women)

In a conservative approach to hypertension, treatment is indicated as follows:

Age	Systolic Pressure
Under 35	Greater than 140 mm mercury
35–59	Greater than 150 mm mercury
60+	Greater than 160 mm mercury

According to Koch-Wester (1973), *hypertension* exists when the systolic pressure exceeds 150 mm mercury and the diastolic pressure is greater than 90 mm mercury. *Borderline* hypertension has been defined as the intermittent elevation of systolic or diastolic pressure above the accepted normal value for a person's age and sex.

Hypertension has been variously estimated to affect between 20 and 35 million persons in the United States alone and thus it is considered the most common of the chronic disorders. Of the millions of people affected, it is estimated that, as of the early 1980s, only about 50% of cases have been diagnosed and are known to the individuals. Of the remaining 50% of cases, only about half are being treated. Hypertension is considered a major health problem because the disorder predisposes individuals to debilitating and often fatal diseases, the major categories of which are heart attack and heart diseases, stroke, and kidney failure. The risk of these consequences is *greatly reduced* with proper therapy for lowering blood pressure.

The foregoing is exemplified by a study made in Framingham, Massachusetts several years ago, which took into account five risk factors: (1) glucose intolerance (indicative of diabetes), (2) cholesterol level, (3) cigarette smoking, (4) left ventricular hypertrophy (increase in volume of a tissue or organ produced entirely by enlargement of existing cells), and (5) hypertension. See Table 1.

Primary and Secondary Hypertension. Traditionally, authorities in the field have made a distinction between two forms of hypertension. *Primary* or *essential* hypertension is a disorder of unknown etiology. This form accounts for approximately 90% of all cases of hypertension. In *secondary* hypertension, a cause for the disorder can be identified. The causes of secondary hypertension are many and include: various drugs, such as amphetamines, oral contraceptives, estrogens, steroids, and thyroid hormones; increased intracranial pressure; certain tumors, such as pheochromocytoma; primary aldosteronism (abnormal aldosterone secretion by adrenal cortex, causing excessive loads of potassium and muscular weakness); several renal diseases, such as chronic pyelonephritis, diabetic nephropathy, glomerulonephritis, gout, polycystic disease, vasculitis, and renovascular hypertension, among others. Hypertension is also associated with toxemia of pregnancy, acute pulmonary edema, acute myocardial infarction, dissecting aortic aneurysm, and cerebral hemorrhage.

Primary (Essential) Hypertension

Although considerable progress has been made during the past few years in understanding the complex root causes which contribute to primary hypertension, much of this information has stemmed from observing the actions of various drugs used in the therapy of hypertension.

The Renin-Angiotensin-Aldosterone System. In the early 1900s, Tigerstedt and Bergman suggested that renin,[1] a proteolytic enzyme elicited by ischemia of the kidneys or by diminished pulse pressure, played an important role in blood pressure homeostasis and in the pathogenesis of hypertension. In a revival of interest in the role of renin, which commenced during the 1960s, considerable new information has been gained, with an increasing implication of the kidneys in hypertension. Additional renin-like enzymes have been identified and their possible functions are being researched. In a rather complex pathway, renin cleaves renin substrate to yield *angiotensin I*, a decapeptide and apparently quite inactive physiologically. Through further enzymatic action during its passage through the pulmonary circulation, angiotensin I is cleaved to produce *angiotensin II*. This substance is the most potent vasoconstrictor known, causing constriction of the arterioles (small arteries that branch to form the capillaries—the smallest of blood vessels at the sites where the blood exchanges nutrients and waste products with the tissues).

Angiotensin also stimulates adrenocortical production of aldosterone, which promotes the reabsorption of sodium and water by the renal tubules. The resulting augmentation of the fluid content of the circulatory system elevates the blood pressure. The action of the angiotensins also involves the nervous system, releasing the neurotransmitter norepinephrine by the nerve terminals of the sympathetic nervous system and by the adrenal medulla (inner portion of adrenal gland). This action potentiates the action of the norepinephrine, which causes increased blood pressure as the result of constriction of the arterioles. This complex is known as the *renin-angiotensin-aldosterone system*. When this system is functioning normally, a decrease of pressure of the blood flowing through the kidneys will stimulate renin release, while an increase of that pressure "signals" the kidney to halt the release of renin as well as angiotensin, an action accomplished by a feedback mechanism. Thus, any increase in blood pressure will be transient. But, when the feedback apparatus dysfunctions, chronic hypertension may result.

Statistics show that about 90% of persons with primary hypertension have elevated levels of renin in the blood, although they usually do not exhibit symptoms of kidney damage. One researcher has found that persons with high plasma renin activity run the highest risk of heart attack, stroke, and kidney failure. The reverse situation holds for those persons with low plasma renin activity.

Some researchers have learned that angiotensin II must combine with specific receptors on target organs before its effects can be produced. Most peptides are angiotensin antagonists and thus, by binding with the receptors, prevent angiotensin from binding. These antagonist substances have been synthesized by several investigators (Cleveland Clinic Foundation; Washington University Medical School). One of these antagonists is the octapeptide called Saralasin® (first synthesized by Norwich Pharmacal Company). It has been found that intravenous injection of this drug will lower the blood pressure to near-normal levels if the cause of hypertension

[1] Not to be confused with rennin, an enzyme secreted by the glands of the stomach which cases curdling of milk.

TABLE 1. EFFECT OF VARIOUS RISK FACTORS (INCLUDING HYPERTENSION) ON OCCURRENCE OF CARDIOVASCULAR DISEASE (Within eight years in a 45-year-old male)

Risk Factor Effects	Glucose Intolerance	Cholesterol Level[1]	Smoke Cigarettes	Left Ventricular Hypertrophy	Probable Cases per Thousand at a Systolic Blood Pressure of (Millimeters Mercury):			
					105	135	165	195
ONE RISK FACTOR PRESENT								
None present	No	Low	No	No	22	35	54	84
Glucose intolerance (GI)	Yes	Low	No	No	39	61	95	143
High cholesterol (HC)	No	High	No	No	44	68	105	158
Smoking (SM)	No	Low	Yes	No	38	59	91	138
Left ventricular hypertrophy (LVH)	No	Low	No	Yes	60	93	141	208
COMBINED MULTIPLE RISK FACTORS PRESENT								
GI + HC	Yes	High	No	No	145	214	304	411
GI + HC + SM	Yes	High	Yes	No	229	323	433	550
GI + HC + SM + LVH	Yes	High	Yes	Yes	460	577	686	778

Notes: The data are based upon a follow-up of patients in the Framingham Study. Framingham males in the study had the following characteristics at the age of 45 years:
Average systolic blood pressure, 131 millimeters mercury; average serum cholesterol level, 234 milligrams/100 milliliters; 0.7% have definite left ventricular hypertrophy as shown by an electrocardiogram; 3.9% have glucose intolerance. Considering these average values, the probability of having cardiovascular disease within 8 years is 75/1000.
[1]Low cholesterol level is considered 185 milligrams/100 milliliters; high cholesterol level, 335 milligrams./100 milliliters.

is renin. This kind of information has proved helpful in determining the most effective drug therapy for primary hypertension.

Other investigators have found that the destruction of a portion of the mid-brain (*subnucleus medialis*) will negate the blood pressure response, probably because the area is involved in the control of peripheral resistance to blood flow. Apparently angiotensin II has the ability to cross the blood-brain barrier. More recently, some researchers have suggested that angiotensin is synthesized in the brain. Much of the experimentation to date has been carried out with laboratory dogs.

Involvement of the nervous system in hypertension has attracted much interest in recent years and ultimately may prove or disprove the association of stress with hypertension. Other investigators have found that prostaglandin E_2 tends to decrease blood pressure by countering the angiotensin-induced constriction of the blood vessels.

Additional Reading

Bonna, K.H. et al.: "Effect of Eicosapentaenoic and Docosahexaenoic Acids on Blood Pressure in Hypertension: A Population-Based Intervention Trial from the Tromose Study," *N. Eng. J. Med.*, 795 (March 22, 1990).

Bulpitt, C.J.: *Epidemiology of Hypertension*, Elsevier Science, New York, NY, 2000.

Calhoun, D.A. and S. Oparil: "Current Concepts—Treatment of Hypertensive Crisis," *N. Eng. J. Med.*, 1177 (October 25, 1990).

Dominiczak, A.F., F. Soubrier, and J.M. Connell: *Molecular Genetics of Hypertension*, Academic Press Inc., San Diego, CA, 2000.

Fackelmann, K.A.: "High-Pressure Hormone," *Science News*, 344 (December 1, 1990).

Grimm, R.H., Jr., et al.: "The Influence of Oral Potassium Chloride on Blood Pressure in Hypertensive Men on a Low-Sodium Diet," *N. Eng. J. Med.*, 569 (March 1, 1990).

Grobbee, D.E. et al.: "Coffee, Caffeine, and Cardiovascular Disease in Men," *New Eng. J. Med.*, 1026 (October 11, 1990).

Guyton, A.C.: "Blood Pressure Control—Special Role of the Kidneys and Body Fluids," *Science*, 1813 (June 28, 1991).

Kem, D.C. and R.D. Brown: "Renin—From Beginning to End," *N. Eng. J. Med.*, 1136 (October 18, 1990).

Kilgour, F.G.: *William Harvey, in Scientific Genius and Creativity*, W.H. Freeman, New York, NY, 1987.

Koch-Weser, J.: "Correlation of Pathophysiology and Pharmacotherapy in Primary Hypertension," *Am. J. Cardiol.*, **32**, 499 (1973).

Leyva, F., and A.J. Coats: *Hypertension and Coexisting Disease*, Blackwell Science, Inc., Malden, MA, 2000.

Manger, W.M. and R.W. Gifford, Jr.: *100 Questions and Answers about Hypertension*, Blackwell Science, Inc., Malden, MA, 2000.

O'Rourke, M. and M. Safar: *The Arterial System in Hypertension*, Kluwer Academic Publishers, New York, NY, 1993.

Panza, J.A. et al.: "Abnormal Endothelium-Dependent Vascular Relaxation in Patients with Essential Hypertension," *New Eng. J. Med.*, 22 (July 5, 1990).

Rowan, R.: *Control High Blood Pressure without Drugs: A Complete Hypertension Handbook*, Simon & Schuster Trade, New York, NY, 2001.

Salisbury, D.: "Hypertension: A Discriminating Disease," *Technology Review (MIT)*, 22 (October 1990).

Schwab, S. et al.: *1998 Year Book of Nephrology, Hypertension and Mineral Metabolism*, Mosby-Year Book, Inc., St. Louis, MO, 1998.

Staff: *Diabetes Mellitus and Hypertension*, Mosby-Year Book, Inc., St. Louis, MO, 2000.

Sytkowski, P.A., W.B. Kannel, and R.B. D'Agostino: "Changes in Risk Factors and the Decline in Mortality from Cardiovascular Disease—The Framingham Heart Study," *N. Eng. J. Med.*, 1635 (June 7, 1990).

Vane, J.R., E.E. Anggard, and R.M. Botting: "Regulatory Functions of the Vascular Endothelium," *N. Eng. J. Med.*, 27 (July 5, 1990).

Vinson, G. and D. Anderson: *Adrenal Glands, Vascular System and Hypertension*, Blackwell Science, Inc., Malden, MA, 1997.

HYPERTEXT TRANSFER PROTOCOL (HTTP).   See **Transmission Control Protocol (TCP)/Internet Protocol (IP) Suite**; and **Wireless Communications Applications**.

HYPERVENTILATION.   Overbreathing. A respiratory-minute volume, or pulmonary ventilation, that is greater than normal. Hyperventilation often results in an abnormal loss of carbon dioxide from the lungs and blood, which may lead to dizziness, confusion, and muscular cramps.

HYPERVENTILATION SYNDROME.   The syndrome of blurring of vision, (feeling of) tingling of the extremities, faintness, and dizziness, which may progress to unconsciousness, and convulsions, caused by reduction of the normal carbon dioxide tension of the human body, due to increased pulmonary ventilation.

HYPOBARIC (CONTROLLED-ATMOSPHERE) SYSTEMS.   Sensitive materials, notably fresh foods, normally cannot withstand long periods of transportation and storage prior to consumption. Over the years, much of the effort extended toward offering produce in marketplaces far distant from the source was concentrated on reducing the time for delivery. Thus, the extensive use of air express and air freight. Conventional refrigeration systems for trucks and railway cars also were especially adapted for use during transport. But even with all of these improvements in technology, certain transporting feats (as, for example, shipping midwestern pork to the California and even Hawaii markets) were difficult to achieve. The controlled-atmosphere hypobaric concept has greatly extended the potential for distant shipping of delicate, perishable materials (not necessarily limited to foodstuffs).

In 1964, the Institute of Food Technologists annual award was given in recognition of the development of a controlled-atmosphere storage system. Essentially, the process was designed to reduce the rate of deterioration of certain fruits and vegetables in refrigerated storage by reducing the oxygen level, increasing the carbon dioxide level, and maintaining the relative humidity close to 100%. In an initial design, the conditions were created by using a home-furnace size catalytic generator that burned natural gas or propane gas to create the atmosphere that essentially halts the natural respiration of the stored products. Later, the gas generator was replaced by cryogenic liquefied gases, allowing additional flexibility and the creation of any desired gas mixture. Atmospheres can be tailored to particular perishables. For example, an atmosphere of 15–20% carbon dioxide and 80–85% nitrogen is optimal for strawberries. For iceberg lettuce, an atmosphere of 8–10% oxygen, less than 10% carbon dioxide, with the remainder nitrogen is used. As of the early 1980s, the system has been installed on over ten thousand rail cars and 7000 sea vans.

In 1979, another IFT award was given in recognition of a hypobaric transport and storage system for fresh meats and meat products. Hypobarics is defined as a precisely controlled combination of low pressure, low temperature, high humidity, and ventilation which, when properly applied, extends up to six times the length of time a perishable commodity remains fresh. This makes possible the shipment of perishable items by way of relatively low-cost surface transportation to distant points. In developing the concept, it was observed that refrigerated storage of fruits in closed containers will result in accumulation of gases generated by the fruit, i.e., ethylene and carbon dioxide, an atmosphere which hastens ripening and spoilage. Although ventilation of fruit containers can prevent accumulation of the gases, the gases are not removed from within the product itself—with no prevention of accumulation of gases within the cells of the fruit. The researchers made the supposition that by drawing a partial vacuum on a closed vessel containing the fruit, the low pressure would increase the diffusivity of the gases, thus promoting release and removal of the gases. At the same time, a reduction of pressure would reduce the oxygen concentration, thus retarding respiration and attendant spoilage. Combined with refrigeration, this would decelerate the metabolic processes, not only of the fruit, but also of any bacteria present. Humidification of the chamber would prevent any drying of the fruit. After testing the concept on bananas and other perishables, the system was patented.

Generally the storage temperature for meats is about $-1\,°C$, and up to 10 or $12\,°C$ for various fruits and vegetables. In all cases, the relative humidity is controlled at about 95%. Pressure ranges between 10 and 80 millimeters of mercury. Lower pressures are maintained for meats and seafoods; somewhat higher pressures for fruits and vegetables.

HYPOCAPNIA. Deficiency of carbon dioxide in the blood and body tissues, which may result in dizziness, confusion, and muscular cramps.

HYPOCHLORITES. When chlorine is reacted with an alkali, a hypochlorite is formed. These compounds are very high-tonnage chemicals for sanitizing and bleaching purposes. Commercial sodium hypochlorite NaClO usually is available in two strengths (1) the familiar household liquid bleach which contains about 5.25% (weight) NaClO, and (2) commercial bleach which contains about 13% (weight) NaClO. The latter compound sometimes is referred to as 15% bleach because the chlorine content is approximately 150 grams/liter of available chlorine. The term "liquid chlorine" usually refers to a solution of NaClO (up to 10%) used in the swimming-pool trade. "Dry chlorine" is part of the registered trademark of a proprietary calcium hypochlorite product containing 70% available chlorine. See also **Bleaching Agents**.

Sodium hypochlorite normally is manufactured in batches by diluting caustic soda to the proper starting concentration. This is approximately 6.8% NaOH for the 5.25% bleach; and about 18.5% NaOH for the 15% bleach. After cooling the caustic soda solution, chlorine gas is added through a sparger pipe until the desired concentration is reached. This usually is determined by making a series of titration analyses. Bleaching powder $CaOCl_2$ is made by passing chlorine gas over slaked lime. This was the first type of chlorine bleaching agent made and dates back to 1799. The product usually contains about 30% available chlorine. Over the

years, it was used extensively in the bleaching of textiles and for sanitizing even though the compound is unstable and difficult to use. The original bleaching powder largely has been replaced by an improved calcium hypochlorite product, which contains about 70% available chlorine. The compound essentially is a calcium hypochlorite dihydrate and, in one process, is made by chlorinating a slurry of lime and caustic soda. The crystals that precipitate out are mixed with calcium chloride and chlorinated lime. When warmed, the calcium hypochlorite dihydrate precipitates, with sodium chloride remaining in solution. After filtering, the cake is dried, granulated, sized, and packaged. In addition to use in swimming pools, products of this type are used widely for water purification, algae control, and sanitation. On a very high-tonnage basis, calcium hypochlorite $Ca(ClO)_2 \cdot 4H_2O$ is used for pulp bleaching in the paper industry. Bleach liquor containing from 20–40% available chlorine may be produced in batches or continuously. In a continuous system, the flow of chlorine is controlled by making frequent (or continuous) measurements of oxidation-reduction potential.

A common means of detecting hypochlorites is the production of a blue color (caused by free iodine) with starch iodide paper by hypochlorites in weakly alkaline solution. Silver nitrate also precipitates part of the hypochlorite in solutions as white silver chloride.

Hypochlorous Acid. [CAS: 7790-92-3] This compound, HOCl, is prepared by the reaction of (1) chlorine monoxide Cl_2O with H_2O, (2) sodium hypochlorite and an acid, excess acid yielding chlorine and oxygen, and (3) chlorine with mercuric oxide suspended in water, mercuric chloride being formed simultaneously. Hypochlorous acid is a yellow solution of characteristic odor. It decomposes upon standing, the rate depending upon (1) concentration, (2) exposure to light, (3) presence of a catalyst (cobaltous hydroxide, for example, promotes the evolution of oxygen), and (4) acidity or alkalinity. Hypochlorous acid is a powerful oxidizing agent and sometimes used as a bleaching agent for organic colors.

Perchloric Acid. [CAS: 7601-90-3] This compound, $HClO_4$, is a colorless, fuming, oily liquid, miscible with H_2O, volatile under diminished pressure. A maximum constant-boiling solution ($203\,°C$, 760 millimeters Hg) results when the concentration of $HClO_4$ reaches 73% in H_2O. Cold dilute perchloric acid reacts with such metals as zinc and iron, yielding hydrogen gas and the corresponding perchlorate in solution. It is stable from the point of view of oxidation and reduction (except that iodine is oxidized to periodic acid, with liberation of chlorine, ferrous salt solutions to ferric, titanous salt solutions to titanic). Concentrated hot perchloric acid, on the other hand, is a powerful oxidizing agent, exploding violently in contact with charcoal, paper, or alcohol; causes serious wounds in contact with the skin. Prepared by distilling ammonium perchlorate with HNO_3 and HCl.

Metallic perchlorates are soluble in water, except that potassium perchlorate is slightly soluble. Potassium perchlorate is, however, insoluble in alcohol containing perchloric acid, a property made use of in the qualitative recognition and quantitative estimation of potassium in salt solutions. Perchlorates, when heated, evolve oxygen and leave the chloride as a residue. Potassium perchlorate decomposes at $400\,°C$.

HYPOCYCLOID. A special case of a cyclic curve, thus a higher plane curve and, in particular, the case where a circle of radius r rolls around inside a fixed circle of radius R. Its parametric equations are

$$x = (R - r)\cos\phi + r\cos\frac{(R - r)\phi}{r}$$

$$y = (R - r)\sin\phi - r\sin\frac{(R - r)\phi}{r}$$

Reference to the corresponding equations for the epicycloid, where the circle rolls around the outside of the fixed circle, will show that the hypocycloid (R, r) is identical with the hypocycloid $(R, R - r)$ or the epicycloid $(R, r - R)$.

Considerations similar to those used for the epicycloid will also show that the curve may or may not repeat itself and that it will produce cusps when its generating point touches the fixed circle. The special case is that in which $R = 4r$ has four cusps, and is called the asteroid.

See also **Asteroid (Mathematics)**; and **Curve (Higher Plane)**.

HYPODERMIS. The cellular layer of the integument (integumentary system), in the invertebrates, which secretes the outer cuticula.

HYPOEUTECTIC ALLOY.   An alloy to the left of the eutectic point in a binary phase diagram that freezes with a structure containing some eutectic.

HYPOFLUORITE.   Any compound containing the group—OF. The simple anion FO^- is unknown. A number of covalent hypofluorites are known, including such compounds with carbon, oxygen, nitrogen, sulfur, chlorine and arsenic (uncertain), CF_3OF, CF_3COOF, C_2F_5COOF, NO_2OF, OF_2, O_2F_2, O_3F_2, SF_5OF, FSO_2OF, ClO_3OF and possibly AsF_4OF. These are all powerful fluorinating agents. They react violently with water yielding OF_2 as one product. The oxygen fluorides O_3F_2 and O_2F_2 decompose about $-158\,°C$ and $-100\,°C$, respectively, the former into the latter and the latter into the elements. Nitryl and perchloryl hypofluorites (fluorine nitrate and fluorine perchlorate) easily detonate. The perfluoracyl hypofluorites are much more stable but may also decompose violently. The others appear to be stable.

HYPOGENE.   Originated by the geologist Charles Lyell for all igneous rocks that assumed their form, fabric, and texture at great depths beneath the surface of the lithosphere.

HYPOIODOUS ACID AND HYPOIODITES.   Hypoiodous acid (HOI) is a greenish-yellow solution, of characteristic odor. It is unstable, and cannot be distilled unchanged.

Prepared by reaction (1) of iodine and mercuric oxide (see also **Mercury**) suspension in water, mercuric iodide being simultaneously formed, (2) of sodium hypoiodite and an acid, excess acid yielding iodine.

Sodium hydroxide solution reacts with iodine to form iodide and hypoiodite, the latter decomposing in a few hours at ordinary temperatures to form iodide and iodate.

HYPONITROUS ACID AND HYPONITRITES.   Hyponitrous acid [CAS: 10024-97-2] $H_2N_2O_2$ is a white solid, explosive even at as low a temperature as $0\,°C$, soluble in water, more soluble in ether, can thus be extracted from water solution by ether and the latter evaporated, water solution decomposes quickly into nitrous oxide plus water. Hyponitrous acid is nonreactive with hydriodic acid (a strong reducing agent), but reactive with permanganic acid (a strong oxidizing agent) to form nitrous or nitric acid.

Prepared (1) by reaction of silver hyponitrite $Ag_2N_2O_2$ and hydrogen chloride in anhydrous ether, an evaporation of the resulting solution, (2) by reaction of hydroxylamine H_2NOH plus nitrous acid HONO.

Sodium hyponitrite $Na_2N_2O_2$ is formed (1) by reaction of sodium nitrate or nitrite solution with sodium amalgam (sodium dissolved in mercury), after which acetic acid is added to neutralize the alkali. Sodium stannite ferrous hydroxide, or electrolytic reduction with mercury cathode may also be utilized, (2) by reaction of hydroxylamine sulfonic acid and sodium hydroxide. Silver hyponitrite is formed by reaction of silver nitrate solution and sodium hyponitrite.

HYPOPHOSPHORIC ACID AND HYPOPHOSPHATES.   Hypophosphoric acid [CAS: 7803-60-3] (H_2PO_3 or $H_4P_2O_6$) is a solid, melting point, $55\,°C$, decomposing in solution to form phosphorous plus phosphoric acids. Hypophosphoric acid is used in solution and is a reducing agent, but only with strong oxidizing agents, such as potassium permanganate; and the acid is unaffected by zinc and dilute sulfuric acid (distinction from phosphorous acid). Dehydration of hypophosphoric acid does not yield phosphorus tetroxide; hydration of phosphorus tetroxide does not yield hypophosphoric acid but phosphorous plus phosphoric acids.

Hypophosphoric acid is formed by reaction (1) of yellow phosphorous and potassium permanganate in sodium hydroxide medium, (2) of red phosphorus and calcium hypochlorite solution, (3) also one of the products of slow oxidation at ordinary temperatures of phosphorus in moist air.

There are recorded the following sodium hypophosphates: Na_2PO_3 (or $Na_4P_2O_6$), $NaHPO_3$ (or $Na_2H_2P_2O_6$), $Na_3H(PO_3)_2$ (or $Na_3HP_2O_6$), and $(NaH_3PO_3)_2$ (or $NaH_3P_2O_6$). There is evidence in support of each of the formulas H_2PO_3, $H_4P_2O_6$ for hypophosphoric acid.

Ester: Dimethyl hypophosphate $(CH_3)_2PO_3$ or $(CH_3O)_2PO$. See also **Phosphorus**.

HYPOPHOSPHOROUS ACID AND HYPOPHOSPHITES.   Hypophosphorous acid [CAS: 6303-21-5] (H_3PO_2, or $H \cdot PO_2H_2$) is a colorless liquid, melting point $26.5\,°C$, density 1.493.

Hypophosphorous acid is miscible with water in all proportions and a commercial strength is 30% H_3PO_2. Hypophosphites are used in medicine.

Hypophosphorous acid is a powerful reducing agent, e.g., with copper sulfate forms cuprous hydride Cu_2H_2, brown precipitate, which evolves hydrogen gas and leaves copper on warming; with silver nitrate yields finely divided silver; with sulfurous acid yields sulfur and some hydrogen sulfide; with sulfuric acid yields sulfurous acid, which reacts as above; forms manganous immediately with permanganate.

Hypophosphorous acid is formed by reaction of barium hypophosphite and sulfuric acid, and filtering off barium sulfate. By evaporation of the solution in vacuum at $80\,°C$, and then cooling to $0\,°C$, hypophosphorous acid crystallizes.

Sodium hypophosphite $NaPO_2H_2$, the only sodium hypophosphite, is formed (1) by reaction of yellow phosphorus and sodium hydroxide solution (phosphine simultaneously formed), (2) by reaction of hypophosphorous acid and sodium hydroxide, and evaporating. Sodium hypophosphite, upon heating, yields sodium phosphate and sodium phosphide. Common tests for the hypophosphites are as follows:

1. Zinc reduces dilute sulfuric acid solution of hypophosphites to phosphine recognizable by odor (difference from phosphates).

2. Barium chloride produces no precipitate (difference from phosphites). See also **Phosphorus**.

HYPOPLASIA.   Defective or insufficient development of any tissue. Thymic hypoplasia, also known as DiGeorge's syndrome, results from embryopathy of third and fourth pharyngeal pouch area. There are deficiencies of cell-mediated immunity (CMI) and impaired antibodies. Attendant features of the condition are hypoparathyroidism, abnormal feces, and cardiovascular abnormalities. See also **Immune System and Immunology**.

HYPOPROTHROMBINEMIA.   Lack of adequate amounts of, prothrombin in the blood, resulting in tendency to hemorrhage from impairment of the clotting mechanism.

HYPOSULFUROUS ACID AND HYPOSULFITES.   Hyposulfurous acid $H_2S_2O_4$ is a yellow solution rapidly oxidized in air to sulfurous acid and then to sulfuric acid. Commercially known as hydrosulfurous acid and its salts as hydrosulfites (but not to be confused with "hypo" which is sodium thiosulfate).

Hyposulfurous acid is a powerful reducing agent, e.g., with copper sulfate forms cuprous hydride Cu_2H_2, brown precipitate, which evolves hydrogen gas and leaves copper on warning, with silver nitrate yields finely divided silver, with permanganate yields manganous compounds. Hyposulfurous acid is formed by reaction of sodium hyposulfite and an acid.

Sodium hyposulfite, sodium hydrosulfite $Na_2S_2O_4 \cdot 2H_2O$ is formed (1) by reaction of zinc and sulfurous acid (or sodium hydrogen sulfite), yielding zinc hyposulfite and then converted by sodium chloride into sodium hyposulfite, (2) by electrolysis of sodium hydrogen sulfite and then addition of sodium chloride.

Sodium hyposulfite is used to bleach sugar, indigo, wood pulp. With moist hydrogen sulfide, sulfur is precipitated and sodium thiosulfate simultaneously formed.

HYPOTENSION.   When the systolic arterial pressure is consistently below 100 millimeters of mercury, low blood pressure (hypotension) is said to exist. Many healthy individuals have a blood pressure that is somewhat below average. A moderately low value is usually considered conducive to longer life. When no cause for the low pressure can be found, the condition is referred to as *essential hypotension*. There often are no significant symptoms.

In *orthostatic* or *postural hypotension*, the regulatory mechanism does not function properly so that a person with this condition may suffer unconsciousness simply in changing from a reclining or sitting position to a standing position—as the result of the action causing an abnormal drop in blood pressure. Some normal individuals may from time to time experience a slight giddiness when standing up quickly, but the severe changes in postural hypotension are such that they should be called to the attention of a physician.

Frequently, unrelated diseases, largely degenerative in nature, may cause hypotension as a secondary symptom. Such conditions include acute fevers,

Addison's disease, heart failure, hypothyroidism, malnutrition, hyperinsulinism, and anemia. Sometimes associated with transient hypotension are internal hemorrhage, shock, fainting, and anesthesia. In most situations of this type, the blood pressure returns to normal upon removal of the original causative condition. In most instances, hypotension is of major significance only when the blood pressure falls below that required to produce adequate filtration through the kidneys.

See also **Blood Pressure**; **Heart and Circulatory System (Physiology)**; **Hypertension (High Blood Pressure)**; and **Shock Syndrome**.

HYPOTHERMIA AND COLD-RELATED INJURIES.   The human body contains water in and around its cells. When water inside the cells freezes, the cells burst and die. When water outside the cells freezes, water is drawn out of the cells, causing damage. Lower temperatures also make the blood thicker (viscous) and the blood vessels narrower, resulting in poor circulation and tissue damage.

Several factors affect risks for cold weather injuries. Most of these are self-evident, but still can be overlooked. They include: (1) inappropriate clothing, (2) inactivity, (3) age of the person exposed, (7) lack of customization to cold weather, (8) prior cold injury, (9) cardiovascular disease, (10) use of alcohol, (11) a concurrent accident which may temporarily superexclude attention to a less serious condition, (9) victim is in a state of shock, (10) high altitude, (11) lack of sleep, (12) poor nutrition, (12) lack of fitness, and (13) dehydration. Different levels of injury occur, depending on extremes of weather, the time exposed, and how well the body is protected.

Chillblains.   The mildest form of cold-related injury. This occurs with repeated exposure of bare skin to weather ranging from 32° to 60 °F (−1° to 15.6 °C). The skin swells, turns red, and itches.

Frostbite.   This a serious cold injury. Severity depends on temperature, wind, and duration of exposure. Superficial frostbite involves only the skin. A waxy appearance is common, and blisters appear in 1 to 3 days, followed by generalized swelling of the affected area. Deep frostbite damages not only affect the skin, but also deeper tissues and even bone. The nose, ears, fingers, toes, penis, buttocks, and chin are the parts of the body most commonly frostbitten. Thorough rewarming can limit the extent. Severe cold injuries resemble burns and usually are treated in a similar manner.

Hypothermia.   Considered a major emergency, the victim may be found semiconscious or unconscious, often some distance from shelter. Cardiac arrest may have occurred, in which case cardiopulmonary resuscitation measures should be applied for a very long time. This occurs when the body core temperature is below 85 °F (29 °C). Extreme measures must be taken to improve the airway and ventilation. If transfer to hospital cannot be made immediately, the victim's hands and forearms should be immersed in water maintained at about 113° to 118 °F (45° to 48 °C) and controlled by a thermometer if possible. As a guide, the water should feel uncomfortable but bearably hot to the rescuer's elbow. A conscious victim should be given hot drinks. At hospital, some authorities prefer whole-body immersion in hot water at 113° to 118 °F (45° to 48 °C). Other authorities indicate that nothing will succeed if the victim's rectal temperature continues to fall after rescue and that whole-body immersion may be counterproductive. Core temperature can be raised by gastric lavage with warm water containing dextrose. For more detail in treatment of hypothermia, reference to the *Merck Manual* (frequently updated) is suggested.

Trench Foot or Immersion Foot.   This injury occurs after prolonged exposure to wet, cold weather, usually ranging from 32° to 50 °F (0 °C to 10 °C). See also **Trench Foot**.

HYPOTHESIS.   A tentative assumption, usually based upon some reasonable concept, made in order to generate interest in obtaining proof and to consider the consequences of the assumption.

HYPOVENTILATION.   A respiratory-minute volume, or pulmonary ventilation, that is less than normal. Also called *underbreathing*.

HYPOXAEMIA.   The condition of reduction of the normal oxygen tension in the blood. Also called *anoxaemia*.

HYPOXIA.   Oxygen want or deficiency; any state wherein a physiologically inadequate amount of oxygen is available to, or utilized by, tissue without respect to cause or degree.

HYPSOMETER.   Literally, an instrument for measuring height; specifically, an instrument for measuring atmospheric pressure by determining the boiling point of a liquid at the station. The relationship between the boiling point of a liquid and atmospheric pressure is given by the Clapeyron–Clausius equation. The sensitivity of the hypsometer increases with decreasing pressure, making it more useful for high-altitude work. Consequently, hypsometers are frequently used for height estimation.

The instrument consists of a cylindrical vessel in which the liquid, usually water, is boiled, surmounted by a jacketed column, in the outer partitions of which the vapour circulates, while in the central one a thermometer is placed. To deduce the height of the station from the observed boiling point, it is necessary to know the relation existing between the boiling point and pressure, and also between the pressure and height of the atmosphere.

HYRAXES *(Hyracoidea)*.   A very small group of *Mammalia*, hyraxes are small animals and are of two genera: Dassies *(Procavia)* and Tree-Hyraxes *(Dendrohyrax)*. These rabbit-shaped animals are popularly termed Coneys, a term used in the Bible. Classification of these animals has been a problem for zoologists over the years, finally solved by creating a separate small group. At one time, they were considered to be rodents closely associated with guinea-pigs. At another time, they were classified with the Pachyderms, at a time when elephants, rhinoceroses, and hippopotamuses were all grouped together. In the mean time, all of the aforementioned mammals have been reclassified, Pachyderm being an obsolete term.

Hyraxes are nocturnal in nature and are considered highly aggressive and essentially mean, attempting to bite anything that gets close to them. These animals also are quite noisy and can issue a number of different sounds, including whistles and screams. The fur contrasts in color on their mid-backs. The fur is thick and coarse. They are good jumpers. Vacuum cups in their padded feet enable them to cling to vertical surfaces. They have dagger-like teeth. The Dassies are found in Africa south of the Sahara Desert. The Coney mentioned in the Bible is found in the Sinai Peninsula, Palestine, and Syria. They are rock dwellers and live in fur-lined nests. The Tree-Hyraxes prefer a mountain habitat and frequently are found at relatively high altitudes—7,000 to 10,000 feet (2100 to 3000 meters). They are omnivorous. They prefer closed-canopy forests in the central and west-central regions of Africa. See Fig. 1.

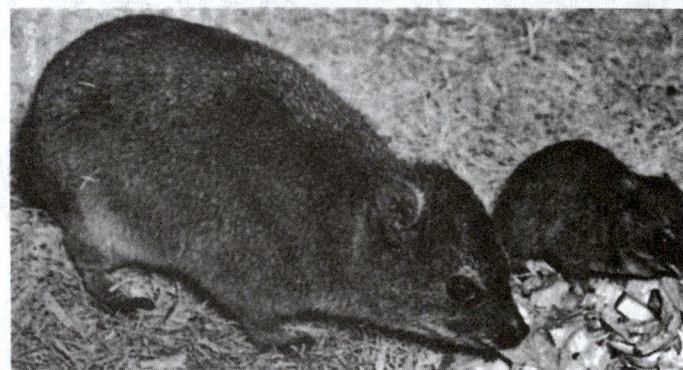

Fig. 1.   Adult cape hyrax with young. *(New York Zoological Society.)*

HYSTERECTOMY.   Total or partial removal of the uterus.

HYSTERESIS.   In general, the phenomenon exhibited by a system whose state depends on its previous history. This term usually refers to magnetic hysteresis, of importance in alternating-current machinery. When a ferromagnetic material such as iron is placed in a magnetic field, a certain amount of energy is involved in bringing about its magnetization. If the field is a rapidly alternating one, the material may become noticeably warm. It appears that the repeated changes of orientation in whatever it is within the substance that responds to the reversals of field are opposed by something like viscous friction.

A quantitative study of the process indicates that, as the field intensity H increases, the magnetic induction B also increases in a manner characteristic of the substance. This is conveniently represented by a graph,

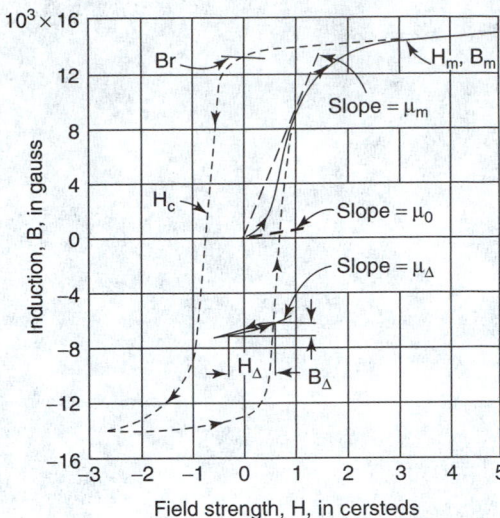

Fig. 1.   Hysteresis loop (dotted). Some important magnetic quantities are shown.

which is called the magnetization curve. See Fig. 1. Its initial slope is the initial permeability (μ_0). If H is carried to some maximum value H_m and then reduced (to $-H_m$), B follows the dotted hysteresis curve. B does not fall off as it was built up (solid line); the residual induction B_r is the induction remaining when H has been reduced to zero; the reverse H needed to reduce B to zero is called the coercive force (H_c). From this point the cycle proceeds to describe the closed curve shown by the dotted lines, which is called the hysteresis loop. The initial portion (solid line) is not retraced. The amount of energy converted into heat is proportional to the area of the cycle.

Electric hysteresis is a somewhat analogous phenomenon exhibited by dielectrics in the electric field and gives rise to heating in capacitors.

Some solids exhibit what is called elastic hysteresis, in which the variables corresponding to H and B in the magnetic case are the stress and the strain or deformation. Elastic bodies such as metals operating at stresses below the proportional limit also undergo hysteresis.

Hysteresis energy is that energy used per cycle of operation to overcome the effect of hysteresis.

HYSTERESIS DISTORTION.
The distortion of voltage and/or current waveforms in circuits containing magnetic components, which is caused by the nonlinear hysteresis effect.

HYSTERESIS HEATER.
An induction device in which a charge or a muffle about the charge is heated principally by hysteresis losses due to a magnetic flux which is produced in it. A distinction should be made between hysteresis heating and the enhanced induction heating in a magnetic charge.

HYSTERESIS (Instrument).
With reference to industrial and scientific instruments, the Scientific Apparatus Makers Association defines hysteresis as:

1. When used as a performance specification, the maximum difference for the same input between the upscale and downscale output values during a full range traverse in each direction. See Fig. 1(c). This is a common usage definition, which includes hysteretic error and dead band. That portion of the difference dependent on the history of prior excursion is hysteretic error, while that portion due to dead band may be determined by a conventional dead band test.

2. When describing a physical property, that property of an element evidenced by the dependence of the value of the output, for a given excursion of the input, upon the history of prior excursions and the direction of the current traverse. Some reversal of the output will occur on any small reversal of the input if a device exhibits hysteretic error without dead band.

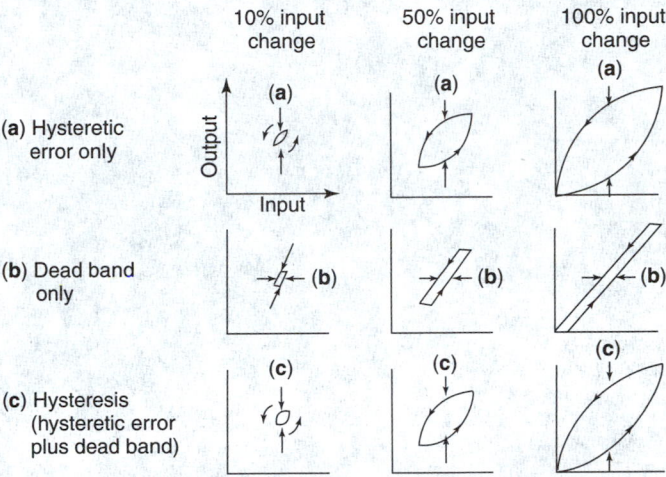

Fig. 1.   Hysteretic error, dead band, and hysteresis.

Hysteretic Error. That portion of hysteresis due to energy absorption in the elements of a measuring instrument. It is obtained by subtracting the value of dead band from the corresponding value of hysteresis for a given input. See Fig. 1(a). The energy absorbed is conceived as produced by molecular friction and appears as heat in dynamic cycling when cyclic mechanical force is applied to a spring or cyclic magnetizing force to a magnetic material.

See also **Backlash**; and **Core Loss**.

HYTHERGRAPH.   See **Meteorology**.

I

IATROGENIC. Caused by medical action.

IBEX. See **Goats and Sheep**.

IBIS (*Aves, Ciconiiformes*). Long-legged wading birds related to the storks. Some of the 28 species occur in all continents, but mainly are found in tropical climates. There are two main groupings: (1) the ibises with a slender, curved bill and naked black head and neck; and (2) the spoonbills which have a bill shaped somewhat like a large spatula. The birds reach a length of about $2\frac{1}{2}$ feet (0.75 meter). They are much like the heron in many respects, but do not have powdered down. The ibises and spoonbills love water and fly and glide in unison. Groups of spoonbills often fly in a V-formation. The most beautiful ibis is *Eudocimus ruber* of scarlet coloration with black primaries. This species is found in South America. The white ibis (*E. albus*) is found in Mexico and parts of South America. The ibis roosts in trees; the spoonbills prefer the ground. There are from 3 to 5 brown eggs marked with gray. The incubation period is 21 days. One species known as the sacred ibis was well known in the days of ancient Egypt. See also **Ciconiiformes**.

ICE. The solid form of water. All commonly occurring forms of ice are crystalline, although large single crystals are relatively rare except in glaciers. At a pressure of one atmosphere, ice melts at 0°C by definition of the centigrade temperature scale. On the other hand, ice does not invariably form in liquid water cooled to 0°C, because of supercooling and the absence of ice nuclei. Ice is found in the atmosphere in such forms as ice crystals, snow, hail, and ice pellets; and on the earth's surface in such forms as hoarfrost, rime, and glaze, and the following:

Anchor ice is ice attached to the bed of streams, lakes, and shallow seas, irrespective of its nature of formation. On clear, cold nights in relatively still water, it may form directly on submerged objects. It also develops in supercooled water of turbulence sufficient to maintain uniform temperature at all depths. When the water temperature increases to above 0°C, the ice rises to the surface, often carrying with it the object on which it had accumulated. (Cf. ground ice, defined below.)

Droxtal is a tiny ice particle, about 10 to 20 microns in diameter, formed by direct freezing of supercooled water droplets at temperatures below −30°C. The term is coined by combining the words "drop" and "crystal."

Fossil ice is ice that was formed in the geologic past, found in regions of permafrost or where present-day temperatures are not low enough to have formed it.

Frazil crystals are ice crystals that form in supercooled water too turbulent to permit coagulation into smooth sheet ice. This is most common in swiftly flowing streams, but is also found in a turbulent sea (cf. lolly ice, defined below). It may accumulate as anchor ice on submerged objects obstructing the water flow.

Glacier ice is any ice that is or was once a part of a glacier. It has been consolidated from firn (i.e., old snow that has become granular and compacted) by further melting and refreezing, and by static pressure. It may be found in the sea as icebergs.

Ground ice is a body of clear ice in frozen ground. It is most commonly found in more-or-less permanently frozen ground, and may be of sufficient age to be termed fossil ice.

Lolly ice is salt water frazil.

Pack ice is ice covering more than half the visible sea surface; no open water whatever is visible in unbroken pack ice, such as that which sometimes covers the central Arctic Ocean.

Sludge is a dense accumulation of frazil or lolly ice; an early stage in the freezing of a body of water. The sea surface becomes thick and soupy and sometimes greasy in appearance. Sludge depth seldom exceeds 1 foot (0.3 meter).

Molecular Forms of Ice. The H_2O molecules in an ice crystal are much further apart than the molecules in liquid water and thus the density of ice is less than that of liquid water, permitting ice to float atop liquid water. In the ice crystal, the molecules are joined by highly directional, obtuse-angled hydrogen bonds, forming a regular hexagonal design. As ice is warmed and passes through the freezing point, the characteristic *rigid* but open structure of the ice crystal gives way, thus allowing H_2O molecules to crowd into the former "open spaces."

Through the application of pressure (2000 atmospheres and higher), water molecules can be forced to assume various deformed patterns (as compared with ordinary ice). For many years, eight solid forms of water have been so produced, designated ice II through ice IX (water is designated ice I). These high-pressure forms of solid water exist at specific temperature-pressure domains in the extended phase diagram of water. Upon release of the applied high pressure, these ice structures revert to common ice or water depending upon the temperature.

Still another form of solid water was proposed several years ago by Holzapfel (a German physicist). This form, designated ice X, was described by Holzapfel as a very dense crystal structure that would not be made up of well-defined molecules linked by hydrogen bonds. Rather, each oxygen atom would be surrounded by a tight cubic array of nearest-neighbor oxygen atoms, and a hydrogen atom would be located halfway between each pair of oxygen atoms such that a hydrogen atom would be associated no more with one oxygen atom than with the other. It was predicted that this form of solid water would exist at a pressure greater than 350,000 atmospheres. At the time of this prediction, equipment for producing pressure that high was not available.

With the development of the diamond-anvil high-pressure cell, researchers at the Argonne National Laboratory, in a series of experiments, subjected water to high pressures ranging between 300,000 and 670,000 atmospheres. Through the use of Brillouin-scattering spectroscopy, an anomaly was noted at a pressure of 440,000 atmospheres. The researchers tentatively observe that this "is the tenth known solid phase of H_2O and ... is probably the predicted symmetric ice. As such it would be the first nonmolecular structure for H_2O." (In Brillouin-scattering spectroscopy, the compressibility of a sample is ascertained indirectly by measuring the reflection of laser light from highly directional sound waves in the sample.)

See also **Clouds and Cloud Formation**; **Glacier**; **Polar Research**; **Precipitation and Hydrometeors**; and **Water**.

Additional Reading

Fairbridge, R.W., Ed.: *Encyclopedia of Geochemistry and Environmental Sciences*, John Wiley & Sons, Inc., New York, NY, 1972.

Hochheimer, H.D. and R.D. Etters: *Frontiers of High Pressure Research*, Kluwer Academic Publishers, Norwell, MA, 1991.

Jayaraman, A.: "Diamond Anvil Cell and High-Pressure Physical Investigations," *Rev. of Modern Physics*, **55**(1), 65–108 (January 1983).

Jayaraman, A.: "The Diamond-Anvil High-Pressure Cell," *Sci. Amer.*, **250**(4), 54–62 (April 1984).

Kukla, G. and J. Gavin: "Summer Ice and Carbon Dioxide," *Science*, **214**, 497–503 (1981).

Lide, D.R.: *CRC Handbook of Chemistry and Physics*, 88th Edition, CRC Press, Inc., Boca Raton, FL, 2007.

Oerlemans, J. and C.J. Van Der Veen: *Ice Sheets and Climate*, Kluwer Academic Publishers, Norwell, MA, 1984.

Peluer, W.R.: *Ice in the Climate System*, Springer-Verlag, Inc., New York, NY, 1994.

ICE AGE.   See Ocean.

ICE AGES.   Ice ages are periods of the Earth's history characterized by glacier advance and cold climate. The response of the Earth's systems to ice ages includes extinctions and migrations of the biosphere, expansion and contraction of oceans and lakes, and the waxing and waning of glaciers. The interactions among these systems produce a complex environmental chronology. The date assigned to the beginning of the current (Pleistocene) ice age depends upon the geographical region and the Earth system that is being studied. An age of 2,500,000 years is preferred by oceanographers, but an age of 1,620,000 years is supported by paleontologists. The first Pleistocene glaciers began growing 30,000,000 years ago in Antarctica.

Over the past 500,000 years, ice age climate has followed regular 100,000 year cycles of extreme cold (glacial), followed by a rapid transition to intense warmth (interglacial), then by gradual cooling. The glacial and interglacial extremes have lasted about 10,000 years. If this trend continues, the current interglacial (the Holocene) should be nearing its end, as is supported by evidence for global cooling that began about 3,000 years ago. After the end of the Holocene, climate will continue to cool and glaciers to expand, reaching a full-glacial maximum about 90,000 years in the future.

Several factors interact to produce the climate oscillations of the ice age, including changes in the Earth's orbit, atmosphere, oceans and vegetation. Each of these factors has a characteristic magnitude, periodicity, and level of interaction with the other climate factors. It is generally accepted that changes in the Earth's orbit (the Milankovitch theory) produce the primary timing signal for major glacial–interglacial cycles of 100,000 years duration. The three elements of the Earth's orbit are eccentricity, tilt, and precession. They influence the distribution of incoming solar energy for each season of the year and for each latitude of the Earth. Summer insolation at latitudes of 45–65° N is particularly important for the growth and melting of continental glaciers. The solar radiation changes at these latitudes are amplified by other climatic factors such as ocean circulation and terrestrial biomass, which influence the concentration of greenhouse gases (carbon dioxide and methane) in the atmosphere, which in turn control global temperature.

Evidence for the Pleistocene Ice Ages

During the late nineteenth and early twentieth centuries, European and American scientists strove to explain their natural surroundings, including unusual features such as the bones of extinct animals and curious geological formations. The oddities included geological features such as boulders of unusual composition (erratics) and polished bedrock (glacial scour). As written discussions of these anomalies progressed, it became clear that the Earth's environment had been very different in the not-too-distant past. Nineteenth century scientists such as Louis Agassiz (1840) later studied the effects of contemporary glaciers in the Alps, and used this knowledge to interpret the unusual features as the results of ancient glaciers. Another oddity discussed by nineteenth-century scientists was the remains of polar animals and extinct organisms in central and southern Europe. Eventually, these too were given an ice age explanation and, in 1948, the International Geological Congress defined the beginning of the Pleistocene using the earliest occurrence of two North Atlantic marine organisms in Mediterranean sediments. See also **Agassiz, Jean Louis Rodolphe (1807–1873)**.

Evidence for the ice age steadily accumulated during the early twentieth century, but the details of its chronology remained obscured by the nature of glacial deposits. With each successive expansion of the continental glaciers, the deposits left by the previous glaciation were eroded to bedrock. Based on the available evidence, scientists came to recognize four (or five) glaciations during the Pleistocene ice age. In southern Europe these were named the 'Gunz, Mindel, Riss and Würm'; and in the America Midwest the 'Nebraskan, Kansan, Illinoian and Wisconsinan'. However, none of the glacial deposits were independently dated, and no single stratigraphic succession included all four glaciations in one place. Then, in the mid-twentieth century, the discovery of two kinds of uninterrupted records revolutionized the study of ice ages.

The first, and subsequently best developed, uninterrupted record of Pleistocene glaciations is ocean sediment. Ocean cores were first examined in detail in the 1940s by Cesare Emiliani, who studied the oxygen isotopes

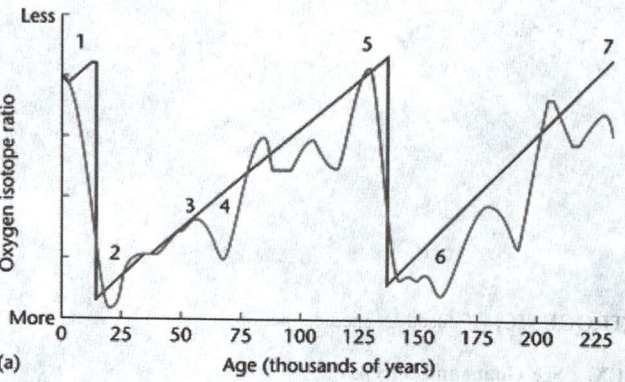

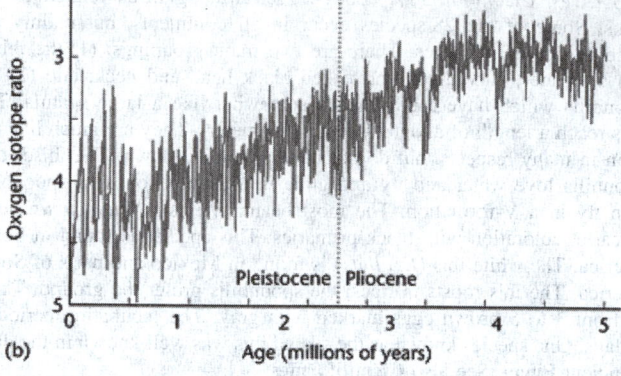

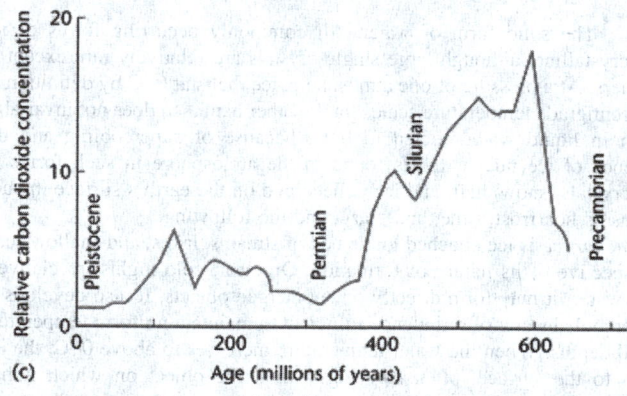

Fig. 1.   Ice age variations on three time scales. **(a)** Oxygen isotope stages as proposed by Cesare Emiliani (1950). Interglacials are odd-numbered stages (blue), and have lower $^{18}O/^{16}O$ ratios. Glacials are even-numbered (red) and have higher $^{18}O/^{16}O$ ratios. Stages 3 and 4 are intermediate, but follow Emiliani's original publication, as does the orientation of the y-axis. The saw-toothed pattern of the glacial–interglacial cycles is emphasized by the blue line. **(b)** Oxygen isotope variations of the past 5,000,000 years as recorded in DSDP 846, data provided by Ralf Tiedman. Note the long-term increase of the $^{18}O/^{16}O$ ratios, the beginning of the Pleistocene ice age, and 100 000 year periodicity of the last 500 000 years. **(c)** Older ice ages plotted on a graph of atmospheric carbon dioxide concentration relative to a modern value of '1'. Data after Berner (1994).

in the shells of marine organisms (foraminifera) in a core from the western Pacific Ocean. Emiliani showed that, during glacials, the ratio of oxygen isotope 18 (^{18}O) to ^{16}O was greater, while during interglacials the proportion was smaller (Fig. 1). Ultimately, this variation was shown to result from the preferential incorporation of the lighter ^{16}O in the continental ice sheets during glaciations. The depletion of the ^{16}O in the oceans required the build-up of massive ice sheets over 3 km thick and the lowering of global sea level by 100–125 m (328–410 ft). These sea-level changes have left a record of raised beaches in coastal regions undergoing rapid geological uplift.

Emiliani labeled the sedimentary intervals with low $^{18}O/^{16}O$ ratios with odd numbers, starting with '1' for the current (Holocene) interglacial, and with even numbers starting with '2' for the last full glacial. This simple nomenclature has been extended to over 100 glacial–interglacial cycles, with chronological control provided by palaeomagnetism and dated volcanic ashes.

The effects of glacial–interglacial cycles on ocean sediment are not limited to oxygen isotopes. During glaciations, the amount of chalk (calcium carbonate) in ocean sediments is less, and the kinds of preserved marine organisms are cold-loving. Also, the amount of iceberg debris is much greater in ocean sediments during glacials. The iceberg-transported particles vary from sand to boulder sized. One of the primary indications for the beginning of the Pleistocene ice age is the presence of these iceberg 'drop stones' in ocean cores from the North Pacific and North Atlantic dated to 2,500,000 years.

A second important source of uninterrupted ice age data comes from wind-blown silt (loess) of Eastern Europe and China. These deposits were first studied by George Kukla in the 1970s, who gathered various geological and paleontological data from alternating light and dark layers in the brick-clay quarries throughout Czechoslovakia. The silts are derived from the various tributaries of the Danube River that drain the northern Alps. During glaciations, these tributaries carry huge amounts of rock-flour produced by the Alpine glaciers. The strong ice age winds blow the rock-flour eastward to produce thick, light-colored layers of loess, containing the remains of polar and boreal animals. These alternate with dark-colored layers produced by forest or grassland soils during interglacials. The hardened silts provided an excellent paleomagnetic signal for dating the deposits. See also **Geological Time: Dating Techniques**.

More recent research by Kukla and various Chinese scientists have produced an equally impressive record of ice age climate for China preserved in loess. There, loesses are produced by wind erosion of the deserts of central Asia, blown onto the loess plateau of Northern China. The primary climate indicator in the Chinese loess record is paleomagnetism. Interglacial soils have a much stronger paleomagnetic signal than glacial loesses, probably due to soil-forming processes. Chinese researchers have produced a simple chronology, with 'S' denoting interglacial soils and 'L' representing glacial loesses. The chronology starts with 'S1' for the Holocene, 'L1' for the last full glacial, and so on downward for at least 20 cycles. See also **Paleoclimatology**.

The marine and loess records provided the initial convincing evidence for a continuous record of the Pleistocene ice age. They were soon joined by an incredibly detailed record preserved in the polar and alpine glaciers. In its upper layers, glacier ice can be annually dated by the alternation of lighter (winter) and darker (summer, partially melted) layers. These layers include seasonal accumulations of wind-blown fossils and isotopes. The growing ice crystals also trap air–a record of past atmospheric composition. The oldest high-mountain glaciers studied so far are over 20,000 years in age, but the oldest polar ice cores span two glacial–interglacial cycles. The trapped air bubbles from the ice cores contain much less greenhouse gas (carbon dioxide and methane) than does the modern atmosphere. This suggests that greenhouse gases may have been the ultimate cause of climatic change during the Pleistocene ice age. Orbital changes, amplified by marine and terrestrial feedbacks, altered the atmospheric composition, and the greenhouse gases caused global cooling or warming.

Evidence for ice age climate for temperate regions comes from lake sediments and cave deposits. Ancient lakes provide a diversity of sedimentary, isotopic, plant and animal data, which can be dated by paleomagnetism, isotopes and volcanic ashes. Pollen and seed analyses of lake sediment indicate plant migrations by hundreds of kilometers north–south and hundreds of meters in elevation in response to glacial–interglacial cycles. They also record general aridity during the glaciations. The Sahara Desert was much more extensive during the last glaciation, and the Amazon and Congo rainforests were replaced by savannah and grassland. The longest lake records show that progressive transformations of terrestrial vegetation are superimposed on the regular glacial–interglacial plant migrations. In Europe, this includes the local extinction of forest species, and in western North America it entails the origination and expansion of cold desert species such as sagebrush and saltbush. The progressive changes of the Pleistocene ice age are more clear-cut in the lake records of terrestrial vegetation than in the marine, loess and ice core data.

In western North America, central Asia, and other regions of internal drainage, the Pleistocene climate oscillations are associated with high lake levels (pluvials). In regions of winter rainfall, the maximum lake depths coincide with glacial climate, but in regions of monsoonal (summer) rainfall, lakes are fullest during the early Holocene. The expansion of lakes and associated rivers permits the migration of aquatic animals such as snails and fish. Coincident biogeographical patterns of several species have been used to document these formerly continuous waterways.

The effects of the ice age on terrestrial animals is no less profound than its effects on aquatic animals and on the vegetation. Of particular importance are the effects of sea level on land bridges connecting the continents. Lowering of the sea surface by 125 m (410 ft) exposes extensive areas of ocean floor between Siberia and Alaska, between Central and South America, and between many coastal islands and the mainland. Similarly, lowering of vegetation zones permits forested vegetation to spread between isolated mountain ranges. The exposure of intercontinental land bridges allowed migrations of land animals, including the migration of humans from Asia to Australia and from Asia to America during the last glaciation. It has been argued that this human migration had profound effects on the native animals of these continents. It has been proposed that, in North America alone, hunting by the recent immigrants led to 'Pleistocene overkill'–the demise of dozens of genera of large animals (>44 kg, 100 lb adult body weight), including mammoths, mastodons and giant ground sloths.

Older Ice Ages

Throughout the 4,600,000,000 years of Earth history, the planet has experienced many ice ages, with the Pleistocene ice age being merely the most recent. Geological evidence exists for four such ice ages over the last 1,000,000,000 years. The oldest of these, approximately 700,000,000 years ago, may have been the most severe. During this Precambrian ice age, the Earth may have been an ice-covered 'snowball planet', with glaciers forming at sea level within 12° of the palaeo-equator. At least two advances at this time, each of several million years' duration, are recorded in sedimentary rocks of what is currently Namibia, Africa. During each event, large changes of carbon isotope ratios in ocean sediment indicate a dramatic lowering of atmospheric carbon dioxide, which presumably led to global cooling and glacier expansion (Fig. 1c). See also **Terrestrialization (Precambrian–Devonian)**.

Evidence for two Paleozoic ice ages comes from glacier debris deposited in ancient deep-sea sediments by icebergs. These ice-rafted deposits are most extensive during the Ordovician–Silurian (440,000,000 years ago), the Carboniferous–Permian (300,000,000 years ago), and the Pleistocene. Additional geological evidence for the Paleozoic ice ages comes from glacial deposits (tillites) overlying ancient bedrock surfaces that have been polished and grooved by advancing glaciers. Unlike the Pleistocene ice ages, the Paleozoic ice ages were most extensive on the land surrounding the South Pole, on the ancient supercontinent of Gondwanaland.

Geological evidence for the Ordovician–Silurian ice age is preserved in what is now the Sahara Desert and eastern South America. The orientation of glacier grooves indicates the flow of ice was outward from the center of the (current) African continent. By the time of the Carboniferous–Permian ice age, paleomagnetic data show that the position of the continents relative to the South Pole had shifted. The tillites and grooves indicate that the glaciers were centered in what is today South Africa and Southern Australia. The Carboniferous–Permian ice age lasted 60,000,000 years, and sedimentary cycles from North America indicate that it consisted of many individual glacial advances, each of about 400,000 years' duration.

The three pre-Pleistocene ice ages have lasted about 10,000,000 years, and have been interspersed 150,000,000–250,000,000 years apart. Geographical and atmospheric similarities exist among these ice ages and the Pleistocene. Each is characterized by land masses near the poles: Asia and North America during the Pleistocene, Gondwanaland during the Carboniferous–Permian and Ordovician–Silurian ice ages, and the supercontinent Rodinia during the Precambrian ice age. This geography would have affected oceanic and atmospheric circulation, and provided a stable platform for glaciers. It has recently been proposed that climate has alternated between hothouse and icehouse as the concentration of greenhouse gases has fluctuated. The plate-tectonic mechanisms are thought to be primarily responsible for the variation in greenhouse gases, with volcanoes releasing carbon dioxide to the atmosphere and the production and burial of carbonate rocks removing it.

See also **Ocean**.

Additional Reading

Agassiz, L.: *Études sur les glaciers*, Jent et Gassmann, Neuchâtel, France, 1840.

Berner, R.A.: "GEOCARB II: A Revised Model of Atmospheric CO_2 over Phanerozoic time," *American Journal of Science*, **294**, 56–91 (1994).

Broecker, W.S.: *The Glacial World According to Wally*, Eldigo Press, Lamont Doherty Earth Observatory, Columbia University, Palisades, NY, 1995.

Crowley, J., and G.R. North: "Paleoclimatology," *Oxford Monographs Geology Geophysics 16*, Oxford University Press, New York, 1991.

Emiliani, C.: "Pleistocene Temperatures," *Journal of Geology*, **63**, 538–578 (1950).

Frakes, L.A., and J.E. Francis: "A Guide to Phanerozoic Cold Polar Climates from High-latitude Ice Rafting in the Cretaceous," *Nature*, **33**, 547–549 (1988).

Gillespie, A.R., S.C. Porter, and B.F. Atwater: *Quaternary Period in the United States*, Elsevier Science & Technology Books, New York, NY, 2003.

Hecht, A.D.: *Paleoclimate Analysis and Modeling*, John Wiley & Sons Inc., New York, NY, 1995.

Hoffman, P.F., A.J. Kaufman, G.P. Halverson, and D.P. Schrag: "A Neoproterozoic Snowball Earth," *Science*, **281**, 1342–1346 (1998).

Imbrie, J., and K.P. Imbrie: *Ice Ages, Solving the Mystery*, Harvard University Press, Cambridge, MA, 2005.

Kukla, G.J.: "Pleistocene Land–sea Correlations. I. Europe," *Earth Science Reviews*, **13**, 307–374 (1977).

MacDougall, D.: *Frozen Earth: The Once and Future Story of Ice Ages*, University of California Press, Berkeley, CA, 2006.

Martin, P.S.: "The Discovery of America," *Science*, **179**, 969–974 (1973).

Martin, P.S., and R.G. Klein: *Quaternary Extinctions, A Prehistoric Revolution*, University of Arizona Press, Tucson, AZ, 1984.

Prothero, D.R.: *The Eocene–Oligocene Transition: Paradise Lost*, Columbia University Press, New York, NY, 1994.

Rampino, M.R., J.E. Sanders, W.S. Newman, and L.K. Königsson: *Climate History, Periodicity, and Predictability*, Van Nostrand Reinhold, New York, NY, 1987.

OWEN DAVIS, University of Arizona, Tucson, AZ

ICEBERG. A large mass of floating or stranded ice that has broken away from a glacier; usually more than 5 meters (16 feet) above sea level. The unmodified term "iceberg" usually refers to the irregular masses of ice formed by the calving of glaciers along an orographically rough coast, whereas tabular icebergs and ice islands are calved from an ice shelf, and floebergs are formed from sea ice. In decreasing size, they are classified as: ice island (few thousand square meters to 500 square meters (5,382 square feet) in area); tabular iceberg; iceberg; bergy bit (less than 5 m above sea level, between 1 and 200 square meters in area (11 and 2,153 square feet); and growler (less than 1 meter (3.3 feet) above sea level, about 20 square meters (237 square feet) in area. See also **Polar Research**; **Tabular Iceberg**; and **Water Resources**.

ICE, CLOUDS, AND LAND ELEVATION SATELLITE (ICESat) MISSION. See **Earth Observing System (EOS)**; and **Space Science Missions: Earth**.

ICE CRYSTAL. Any one of a number of macroscopic, crystalline forms in which ice appears, including hexagonal columns, hexagonal platelets, dendritic crystals, ice needles, and combinations of these forms. The crystal lattice of ice is hexagonal in its symmetry under most atmospheric conditions. Varying conditions of temperature and vapor pressure can lead to growth of crystalline forms in which the simple hexagonal pattern is present in widely different habits (a thin hexagonal plate or a long thin hexagonal column). In many ice crystals, trigonal symmetry can be observed, suggesting an influence of a cubic symmetry. The principal axis (c axis) of a single crystal of ice is perpendicular to the axis of hexagonal symmetry. Planes perpendicular to this axis are called basal planes (a axes related to the prism facets) and present a hexagonal cross section. Ice is anisotropic in both its optical and electrical properties and has a high dielectric constant (even higher than water) resulting from its water dipole structure. The electrical relaxation time for water is much shorter than for ice (10^9 Hz compared with 10^4 Hz), resulting from a chain reaction requirement for molecules to relax through defects in the ice lattice. In the free air, ice crystals compose cirrus-type clouds, and near the ground

they form the hydrometeor called, remarkably enough, "ice crystals" (or ice prisms). They are one constituent of ice fog, the other constituent being droxtals. On terrestrial objects the ice crystal is the elemental unit of hoarfrost in all of its various forms. Ice crystals that form in slightly supercooled water are termed frazil. Ice originating as frozen water (e.g., hail, graupel, and lake ice) still has hexagonal symmetry but lacks any external hexagonal form. Analysis of their sections (0.5 mm) in polarized light reveals different crystal shapes and orientations, depending on the freezing and any annealing and subsequent recrystallization process. See also **Precipitation and Hydrometeors**.

ICE-CRYSTAL CLOUD. See **Clouds and Cloud Formation**.

ICE DAY. In climatology, a day on which the maximum air temperature in a thermometer shelter does not rise above $0\,°C$ ($32\,°F$), and ice on the surface of water does not thaw. This term is not used in the United States, but is used in the United Kingdom, throughout most of Europe, and probably in many other parts of the world. See also **Climate**.

ICE FISHES. For many years, the extremely cold-water ice fishes of the polar oceans were a biological oddity and mystery. How did a relatively few species of fishes evolve and adapt to such cold temperatures? Norwegian fisherman noted that the gills of the *Chaenocephalus aceratus* (Fig. 1) had white rather than red gills and consider the creatures bloodless, or at least devoid of hemoglobin in their blood. This particular species prompted some early research work, but a concerted effort to investigate so-called ice fishes did not commence until the 1970s and 1980s.

Fig. 1. Ice fish (*Chaenocephalus aceratus.*)

As early as 1899, a few explorers commenced the first serious expedition of the Antarctic, the coldest marine habitat on Earth. At that time, the renowned zoologist, Nicholai Hanson, found to his surprise that large numbers of fishes survived the rigorous cold waters. Hanson collected numerous examples of previously unknown species. Later research was directed mainly to the region of McMurdo Sound. See Fig. 2. The first average yearly temperatures of McMurdo Sound were determined by J.L. Littlepage of Stanford University and found to be $-1.87\,°C$, and it was estimated that, during the summer season, less than 1% of solar radiation received at the water surface penetrated through the ice surface. Later researchers found, however, that these cold temperatures were not the major threat to the ice fishes, but rather the most dangerous factor in their survival was ice crystal formations under the surface of the water that could penetrate the flesh of the fishes. Subsequent research has demonstrated how damaging such penetrations are to the survival of the ice fishes. Arthur DeVries (University of Illinois) explored the expanse of McMurdo Sound over a period of some 20 years. DeVries and fellow researchers targeted their studies on two assumptions pertaining to the adaptation of the ice fishes: (1) the possible ability of the fishes to produce powerful antifreeze compounds; and (2) the possible development of "neutral" buoyancy or weightlessness, thus sparing the fishes from having to expend critical energy on flotation.

Relatively recent research in the laboratory, particularly involving the Antarctic species, *Dissostichus mawsoni*, has revealed that the blood of this fish contains a mixture of eight glycopeptides that differ from each other only in length. These substances, according to C.A. Knight (National

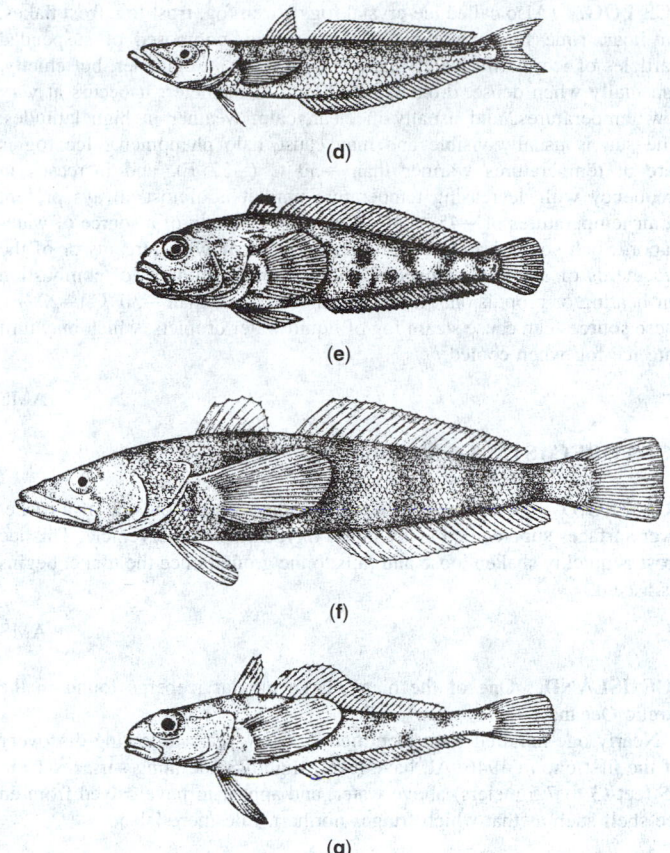

(d)

(e)

(f)

(g)

Fig. 3.   (*continued*)

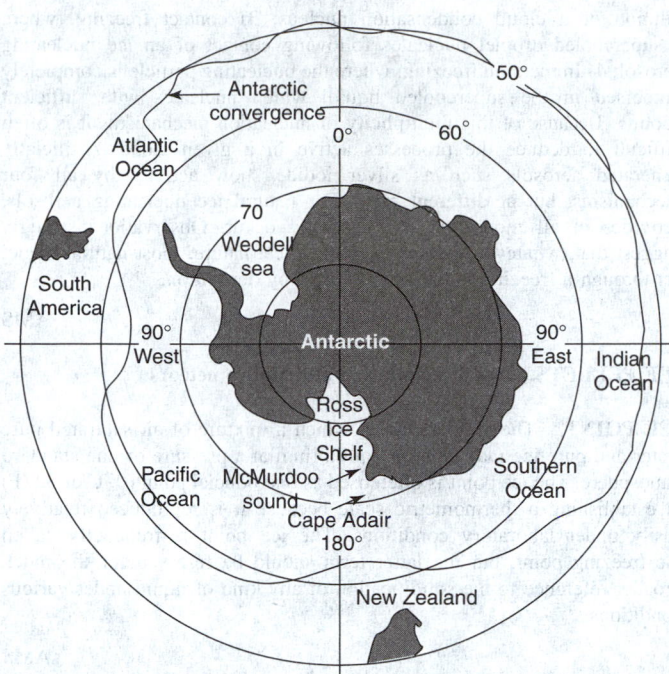

Fig. 2.   Most research directed toward Antarctic ice fishes has concerned the species found in and near McMurdo Sound.

Center for Atmospheric Research) and Arthur DeVries, "produce a novel kind of antifreeze that prevent ice crystal growth but do not substantially lower the equilibrium freezing point of the water (or solution) in which they are dissolved."

The notothenioids, including the Antarctic cod, appear to dominate other species found in the cold waters of Antarctica. They live, feed, and reproduce near the sea floor. Different species are found at varying depths, ranging from about 10 meters to 500 meters. Some of these species are shown in Fig. 3.

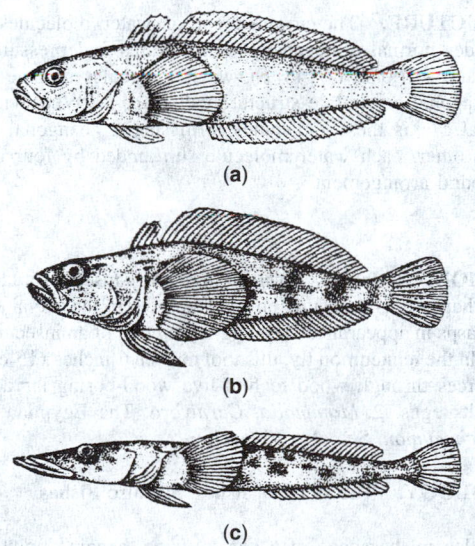

(a)

(b)

(c)

Fig. 3.   Relative size and appearance of species of Antarctic icefishes (notothenioids). In some respects, these fishes resemble warmwater perches. The largest species (e) is 127 centimeters (50 inches) long and weighs 28 kilograms (12.7 pounds). Species (a), (b), (c), (d), and (e) live within 100 meters (328 feet) below the ocean surface and top crust of ice. Species (f) lives about 400 meters (1,310 feet) below the surface, and species (g) lives about 500 meters (1,640 feet) below the surface. (**a**) *Pagothenia borchgrevinki*. (**b**) *Trematomus nicolai*. (**c**) *Gymnodraco acuticeps*. (**d**) *Pleuragramma antarcticum*. (**e**) *Trematomus bernacchii*. (**f**) *Dissostichus mawsoni*. (**g**) *Trematomus loennbergii*.

Suggestions have been made that some 80 million years ago, in connection with the separation of Antarctica from Australia and as the result of tectonic events of that period, a thermal barrier caused by accompanying extreme environmental changes halted the migration of warmer-water species into the Antarctic Ocean. But there was sufficient time for the ice fishes to adapt and evolve. Evidence that warm-water species at one time were present in the region is provided by numerous fossils that have been found on the land mass of Antarctica.

Similar conditions exist in Arctic waters, but they are not quite so extreme, thus explaining why most research has been directed to the Antarctic species.

Additional Reading

DeVries, A.L.: "Biological Antifreeze Agents in Coldwater Fishes," *Comparative Biochemistry and Physiology*, **73A**(4), 627 (1982).

Eastman, J.T.: "The Evolution of Neutrally Buoyant Antarctic Fishes: Their Specialization and Potential Interactions in the Antarctic Marine Food Web," in *Antarctic Nutrient Cycles and Food Webs (Proceedings of the Fourth SCAR Symposium on Antarctic Biology)* (R.W. Siegfried, P.R. Condy and R.M. Laws, Editors), Springer-Verlag, Inc., New York, NY, 1985.

Eastman, J.T. and A.L. DeVries: "Antarctic Fishes," *Sci. Amer.*, 106 (November 1986).

Eschmeyer, W.N., C.J. Ferraris, M.D. Hoang and D.J. Long: *Catalog of Fishes*, California Academy of Sciences, San Francisco, CA, 1998.

Knight, C.A. and A.L. DeVries: "Melting Inhibition and Superheating of Ice by an Antifreeze Glycopeptide," *Science*, 505 (August 4, 1989).

Kock, Karl-Hermann: *Antarctic Fish and Fisheries*, Cambridge University Press, New York, NY, 1992.

Maresca, B. and B. Tota: *Biology of Antarctic Fish*, Springer-Verlag, Inc., New York, NY, 1991.

Paxton, J.R. and W.N. Eschmeyer: *Encyclopedia of Fishes*, 2nd Edition, Academic Press, Inc., San Diego, CA, 1998.

Waller, G.: *Sea Life: A Complete Guide to the Marine Environment*, Smithsonian Institution Press, Washington, DC, 1996.

ICE FLOW.   See **Glacier**.

ICE FOG.   (Also called ice-crystal fog, frozen fog, frost fog, frost flakes, air hoar, rime fog, pogonip.) A type of fog, composed of suspended particles of ice, partly ice crystals 20 to 100 μm in diameter, but chiefly, especially when dense, droxtals 12–20 μm in diameter. It occurs at very low temperatures, and usually in clear, calm weather in high latitudes. The sun is usually visible and may cause halo phenomena. Ice fog is rare at temperatures warmer than −30 °C (−22 °F), and increases in frequency with decreasing temperature until it is almost always present at air temperatures of −45 °C (−49 °F) in the vicinity of a source of water vapor. Such sources are the open water of fast-flowing streams or of the sea, herds of animals, volcanoes, and especially products of combustion for heating or propulsion. At temperatures warmer than −30 °C (−22 °F), these sources can cause steam fog of liquid water droplets, which may turn into ice fog when cooled.

AMS

ICE FORECAST.   See **Meteorology**.

ICE FROST.   A thickness of ice that gathers on the outside of a rocket over surfaces supercooled as by liquid oxygen inside the vehicle. This ice frost is quickly shaken loose and falls to the ground once the rocket begins its ascent.

AMS

ICE ISLAND.   One of the many, large tubular icebergs found in the Arctic Ocean.

Nearly one hundred were identified in a few years following discovery of the first one in 1946. All have level, slightly undulating surfaces 10 to 25 feet (3 to 7.5 meters) above water, and appear to have calved from an ice shelf such as that which fringes northern Ellesmere Island.

ICELANDIC LOW.   See **Meteorology**.

ICE MULTIPLICATION.   A process from which more ice particles are produced from existing ice crystals in clouds. Sometimes known as ice enhancement.

The process is inferred from the observation that ice particle concentration often exceeds that of ice nuclei, sometimes by several orders of magnitude. Currently the following mechanisms are thought to be responsible for the ice multiplication phenomenon: 1) mechanical fracturing of ice crystals during evaporation; 2) shattering or partial fragmentation of large drops during freezing; and 3) ice splinter formation during the riming of ice particles (Hallett–Mossop process). See also **Hallett-Mossop Process**.

ICE NUCLEI COUNTER.   (Abbreviated IN counter). Any of several devices for counting atmospheric particles that serve as heterogeneous ice nuclei that are suited by composition to catalyze the formation of ice crystals in the atmosphere.

The devices operate on varied principles, include means to cool and moisturize the air within a chamber or over a nucleus-collection filter, and are intended to measure the number concentration of ice nuclei that form ice crystals as a function of temperatures that would occur in subzero tropospheric clouds (thus, 0° (32 °F) to near −40 °C (−40 °F). The product is the nucleation activity temperature spectrum. Some, but by no means all, types of ice nuclei counters are designed to attempt to replicate supersaturations that occur where ice crystals form in natural tropospheric clouds.

AMS

ICE NUCLEUS.   Any particle that serves as a nucleus leading to the formation of ice crystals without regard to the particular physical processes involved in the nucleation. The process is referred to as heterogeneous nucleation, as opposed to homogeneous nucleation, which depends on the formation of an ice particle large enough to grow by random motion of water molecules alone. Four processes are generally distinguished: 1) deposition (sorption; previously called sublimation), where the ice phase forms directly from water vapor; 2) condensation freezing, where the ice phase forms in a supercooled solution following growth and

dilution of a cloud condensation nucleus; 3) contact freezing, where a supercooled droplet nucleates following contact of an ice nucleating aerosol; 4) immersion freezing, where the nucleating particle is completely immersed in the supercooled liquid, which nucleates with sufficient cooling. Because of this multiplicity of nucleation mechanisms it is often difficult to deduce the processes active in a given cloud. Artificially generated aerosols such as silver iodide show activity by all four mechanisms, but at different rates. For natural ice nucleating aerosols, activities in all modes do not generally occur. Observations strongly suggest that, whatever their physico-chemical nature, most natural nuclei act through a freezing process rather than by deposition.

AMS

ICE PELLETS.   See **Precipitation and Hydrometeors**.

ICE POINT.   The temperature at which a mixture of air-saturated pure water and pure ice may exist in equilibrium at a pressure of one standard atmosphere. The ice point is often used as one fiducial point (0 °C or 32 °F) in establishing a thermometric scale because it is reproduced relatively easily under laboratory conditions. The ice point is frequently called the freezing point, but the latter term should be reserved for the much broader reference to the solidification of any kind of liquid under various conditions.

AMS

ICE-RAFTING.   The transporting of rock and other minerals, of a wide variety of sizes, on or within icebergs, ice floes, river drift, or other forms of floating ice. The term *ice-rafted* is especially said of till that is deposited by the melting of floating ice and that is found widely distributed in marine sediments. The amount of material so moved can be tremendous. For example, Ruddiman and McIntyre (1977) estimated that the total ice-rafted input from 125,000 to 10,000 years B.P. in the North Atlantic, exclusive of continental shelves, was 1.4×10^{18} grams of sand. Data from several eastern North Atlantic cores suggest that the total amount of noncarbonate detritus, including silt and clay, may be higher by a factor of roughly 7.1 than the sand input, or 10×10^{18} grams. Although some of this material may have been windblown or deposited from suspension in surface waters, scientists believe that the bulk was ice-rafted. B.P. = Before Present.

ICE STRUCTURE.   The arrangement of water molecules in an ice crystal. Under normal atmospheric temperatures and pressures between 0° (32 °F) and −100 °C (−148 °F), water molecules arrange themselves into an hexagonal crystalline structure called *ice-lh*. When viewed along the principal c axis these molecules form spatial hexagonal rings lying above each other, each water molecule surrounded by four others, in a near tetrahedral arrangement.

AMS

ICHNEUMON.   1. *Insecta, Hymenoptera*. Any insect of a large number of species that live as parasites on other insects in the larval stage. They resemble wasps in appearance. The egg-laying tool of an insect is called the ovipositor. In the ichneumon fly, this tool may be 6 inches (15 centimeters) long. It pierces through wood to find live wood-boring larvae on which to deposit its eggs. 2. *Mammalia, Carnivora*. The Egyptian mongoose, *Herpestes ichneumon*. See also **Viverrines**.

ICHTHYOLOGY.   The study of fishes. See also **Fishes**.

ICICLE.   Ice in the shape of a narrow cone, hanging point downward from a roof, fence, cliffside, etc. An icicle is formed when above-freezing water, for example, snowmelt or groundwater, runs or drips into subfreezing air. The water freezes as it drips or runs, forming a narrow cone pointed downward and growing in both length and width, widest at its top. Most icicles are found hanging from the edges of heated, snow-topped roofs, with any water that has not frozen in its downward traverse forming ice on the surfaces below.

AMS

ICING. 1. In general, any deposit or coating of ice on an object, caused by the impingement and freezing of liquid (usually supercooled) hydrometeors; to be distinguished from hoarfrost in that the latter results from the deposition of water vapor. The two basic types of icing are rime and glaze. See also **Glaze**; and **Rime**.

2. [Also known as flood icing, flooding ice, aufeis (German), naled (Russian).] A mass or sheet of ice formed during the winter by successive freezing of sheets of water that may seep from the ground, from a river, or from a spring.

AMS

ICONOSCOPE. An early form of camera tube used in television, developed by Dr. Vladimir K. Zworykin. This tube operates under the *storage* principle, with photoemission being accumulated as charge at each image point. The charge that has accumulated is removed each picture period, as the picture elements are scanned. This results in a photocurrent, which has been increased by an amount equal to the number of picture elements.

See also **Television (TV)**.

-IC. A suffix, used in naming inorganic compounds, that indicates that the central element is present in its highest oxidation state. Thus in ferric chloride ($FeCl_3$) the iron has an oxidation number of +3, equivalent to its valence: in an ionized state it would have three positive charges (Fe^{3+}). (A recommended change in this system of nomenclature is to use the common name of the element (iron) together with a Roman numeral showing the oxidation number; thus, ferric chloride would be iron (III) chloride.)

IDEAL GAS LAW. An "ideal gas" would, if kept at a constant temperature, behave as respects volume and pressure in strict accord with Boyle's law. If now the temperature is also allowed to vary, we must combine the law of Charles (or of Gay Lussac) with Boyle's law, yielding the Boyle-Charles law:

$$pv = p_0 v_0 (1 + at), \tag{1}$$

in which $p_0 v_0$ is the value of the pressure-volume product pv when the temperature t is zero, a is the coefficient of expansion of the gas, practically the same for all gases, and in the ideal case equal to the reciprocal of the absolute temperature of the scale zero. If the centigrade scale is used, the value of a is approximately 1/273.2 per degree. Substituting this, Equation (1) may be written

$$pv = \frac{p_0 v_0 (t + 273.2°)}{273.2°} \tag{2}$$

which is one expression for the ideal gas law.

The factor $t + 273.2°$ will be recognized as the absolute temperature T of the gas. And since the ideal gas obeys Boyle's law, the product $p_0 v_0$ is constant however p_0 and v_0 may vary between themselves. We may thus denote the coefficient $p_0 v_0 / 273.2°$ by a single constant symbol, say R, and the ideal gas equation then takes the usual form

$$pv = RT \tag{3}$$

The value of R depends, of course, upon the quantity of gas used, since at any pressure p_0 it is proportional to the volume v_0. For 1 gram of air, R equals about 2,868,000 gcm^2/sec^2 deg. At the zero of temperature and at any given pressure p_0, the gram molecular weights, or moles, of all pure gases have equal volumes. (This follows from Avogadro's law.) Hence if one mole of any pure gas is used, R will always have the same value, in c.g.s. units about 8.316×10^7 g cm^2/sec^2 deg; which is called the "ideal gas constant." Many physical formulas involve a quantity which may be regarded as the ideal gas constant per molecule, that is, the above molar gas constant divided by the number of molecules in a mole, 6.025×10^{23}, giving 1.3803×10^{-16} g cm^2/sec^2 deg. This is the "Boltzmann constant."

Since actual gases, even those with the smallest molecules, hydrogen and helium, do not obey the ideal gas law exactly, various empirical characteristic equations have been devised to represent their behavior.

See also **Avogadro Law**; **Boyle-Charles Law**; **Boyle's Law**; **Characteristic Equation**; and **Combustion**.

IDEAL SYSTEM. A thermodynamic system is called an ideal system when the chemical potentials of all the components are of the form

$$\mu_i = \mu_i(T, p) + RT \ln x_i \tag{1}$$

where $\mu_i(T, p)$ is a function only of the variables absolute temperature, T, and pressure, p. The x_i are the mole fractions of the components.

Systems for which μ_i has this form possess remarkably simple properties. Moreover, mixtures of perfect gases (i.e., gases under conditions which can be approximated with sufficient accuracy by the ideal gas law) and very dilute solutions have these properties.

According to Equation (2), a system is called ideal if the chemical potential of component i varies linearly with the logarithm of the mole fraction of i, with a slope RT. This linear relation need not necessarily extend over the whole concentration range, so that the quantity $\mu_i(T, p)$ is, in general, the value of μ_i extrapolated to $x_i = 1$ at constant T, p. If the system is ideal in a concentration range which extends to $x_i = 1$, then

$$\mu_i(T, p) = \mu_i^0(T, p) \tag{2}$$

where μ_i^0 is the chemical potential of the pure component i. They are two important cases: (1) The mixture is ideal for all values of x_i and for all i. It is then called a *perfect mixture* and Equation (2) is verified for all i. (2) The mixture is ideal when all components but one (index 1) are present in very small amount. Such systems are called *ideal dilute solutions*. Then Equation (2) is only valid for component 1.

Different kinds of ideal systems are distinguished by the form of $\mu_i(T, p)$. In a mixture of perfect gases, $\mu_i(T, p)$ varies logarithmically with pressure, while for a liquid or solid solution, one can, to a first approximation, regard μ_i as independent of pressure.

See also **Perfect Gas**.

-IDE. A suffix, used in naming compounds composed of two elements; in such names the first (electropositive) element retains its name without change, while the second (electronegative) bears the suffix-*ide* as a modification of the elemental name. Examples: sodium hydroxide, magnesium chloride, hydrogen sulfide, etc. Similarly, oxygen is modified to oxide, fluorine to fluoride, phosphorus to phosphide, and carbon to carbide.

IDIOPATHIC. In medical usage, descriptive of a primary or singular, unassociated disorder or condition—as contrasted with diseases and disorders which may arise from other contributory factors, or which may cause complications, leading to other disorders, infections, etc.

I-DISPLAY. In radar, a display in which a target appears as a complete circle when the radar antenna is correctly pointed at it and in which the radius of the circle is proportional to target distance. When not correctly pointing at the target, the circle reduces to a segment of a circle, the segment length being inversely proportional to the magnitude of the pointing error and its angular position being reciprocal to the direction of pointing error. Also called *I-scan*, *I-scope*, or *I-indicator*.

IDOE PROGRAM. A coordinated, international ocean exploration effort is not unprecedented-the international Decade of Ocean Exploration (IDOE), 1971–1980, was established by the Marine Sciences Act of 1966 and motivated both by anticipated discoveries of useful and important marine resources and by scientific curiosity. Questions about the health of the world's oceans led scientists to argue for systematic baseline surveys that were not possible from randomly spaced observations. The IDOE program recognized that exploration of the ocean required a sustained global effort with international participation, and justification for the program included issues of clear international interest. More information was necessary to describe the ability of the oceans to provide food for an expanding world population, to protect the United States and other nations from maritime threats to world order, to assuage the deterioration of water quality and waterfronts in coastal cities, to support expanded ocean shipping, and to locate new supplies of seabed oil, gas, and minerals. The objective of IDOE was to "achieve more comprehensive knowledge of ocean characteristics and their changes and more profound understanding of oceanic processes for the purpose of effective utilization of the ocean and its resources" (National Academy of Sciences, 1969). More specifically, it was expected that the program would help increase the yield from ocean resources, improve predictions of and responses to natural phenomena, and protect or improve the quality of the marine environment. IDOE was a great success-it provided observational databases on the physics, geochemistry, paleoceanography, biology, and geophysics of the ocean which has fueled hypothesis-driven research for decades. See also **Ocean**.

IGNEOUS ROCK. Igneous rocks are rocks which have solidified (congealed) with, or without, crystallization from hot natural solutions such as magma or lava. Igneous rocks are classified by their texture, fabric, chemical (mineral) composition, and their field relationship or mode of occurrence. See Fig. 1. Under mode of occurrence igneous rocks are classified as intrusive (plutonic) or extrusive (effusive). The intrusive rocks are classified according to the shape and size of the intrusive body and its relation to the other formations which it intrudes. Typical intrusives are batholiths, laccoliths, sills and dikes. The extrusive types are called lavas. Over 700 species of igneous rocks have been described, the bulk of which are intrusives.

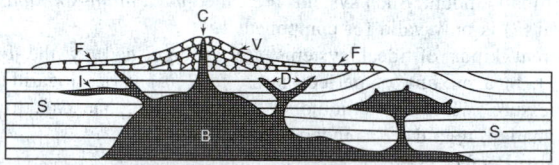

Fig. 1.   Diagrammatic structure section illustrating modes of occurrence of igneous rocks: (S) strata; (B) batholith of plutonic rock; (L) laccolith; (D) dikes; (I) intrusive sheet or sill; (V) volcano; (N) neck of volcano; (F) lava flow; (C) crater.

An excellent reference is "The Evolution of the Igneous Rocks," by H.S. Yoder, Jr., Princeton Univ. Press, Princeton, New Jersey, 1979.

IGNITRON.   An electron tube of the mercury-arc type having a special starting principle. The tube consists of a mercury pool, to serve as cathode, and an anode for the main part of the circuit, and an auxiliary electrode, the igniter, which dips into the mercury pool. For rectification of alternating current, where no provision is made for keeping the arc alive from cycle to cycle, or for control purposes, where the tube may be alternately turned on and off under the influence of auxiliary equipment, the arc must be restarted at intervals. In the ignitron, this is accomplished by the igniter, a rough-surfaced material which will not be "wet" by the mercury. The resultant points of contact between the igniter rod and the mercury will carry very high current densities if a pulse having only a nominal current value is passed through it. This tube has the advantages of the ordinary mercury-arc tube, plus the feature of an easily controlled starting mechanism not involving any moving parts. See also **Electron Tube**.

IGUANIDS.   Of the class *Reptilia*, subclass *Lepidosauria*, order *Squamata* (scaly reptiles), suborder *Sauria* (lizards), infraorder *Gekkota*, according to the classification of Grzimek (1972), the iguanid families include some of the most striking lizards in North and South America. Their counterparts in the Old World are classified as agamids. See also **Agamids**.

Iguanids enjoy wide adaption to a variety of habitats, ranging from deserts, steppes, rain forests, high mountains, and seacoasts. Among the better-known iguanids is the marine iguana, which is found in vast numbers in the Galapagos Islands. However, because of slaughter, their

numbers have diminished from some of these islands, threatening their survival. Some iguanids possess unusual structural features, such as helmets (casques) on the head or crests along the back. These features participate during threat behavior. There are over 50 genera of iguanids, with over 700 species. As indicated on the classification table in the entry on **Lizards**, there are 5 subfamilies. Particular genera are listed for each family. In nontechnical terms, these include the spiny lizards, the side-blotched lizard, the banded rock lizard, the earless lizards, the horned lizards, the spiny-tailed iguanids, the crested keeled lizards, the smooth-throated lizards, the common iguanas, the rhinoceros iguana, the marine iguana, the chuckwalla, the basilisks, and the anoles. See Fig. 1.

Common Iguana. Frequently mentioned and of striking appearance, the common iguana (*Iguana iguana*) frequents the Caribbean and the northern one-third of South America. Closely related species, such as the black iguana and the West Indian iguana, occupy the same general habitat. But they are also found elsewhere as, for example, the desert iguana and chuckwalla, which are found as far north as the southwestern United States; the Fijian iguana found in Polynesia; and the land iguana of the Galapagos Islands, among others. The common iguana prefers forests along the banks of rivers. As shown in Fig. 2, the common iguana can achieve a length of up to 2.2 meters, but only 45 centimeters is taken up by the head and trunk. The animal has a permanently visible dewlap. Mature common iguanas prefer a diet of vegetable matter, although the young consume small animals. When pursued, the animal may jump from heights of 6 meters, frequently escaping by diving into water. The flesh and eggs of the animal are palatable, and thus it is hunted intensively. The eggs (up to 30) are buried by the female and require about 2 months to hatch. The common iguana's defense is its muscular tail, which can whiplash the enemy. In an encounter, a dog is no match for the iguana. It is interesting to note the reaching ability of adult iguanas, which makes it possible for them to consume foliage that may be up to 20 meters above ground level. The common iguana is powerful, stoutly built, and of a dusky gray or olive-green coloration.

The marine iguana (*Amblyrhynchus*) is unique among lizards because it qualifies as a marine animal. As pointed out by W. Kästle, the animal lives on the algae and seaweed that can be found on rocks above and below the surface of the water. Excess salt that is consumed is excreted by means of glands in the nasal cavities. Charles Darwin in 1835 studied the behavior of this lizard. Darwin reported, "When in the water this lizard swims with perfect ease and quickness, by a serpentine movement of its body and flattened tail — the legs being motionless and closely collapsed on its dies... The nature of the lizard's foot, as well as the structure of its tail and feet, and the fact of its having been seen voluntarily swimming out at sea, absolutely prove its aquatic habits; yet there is in this respect one strange anomaly, namely, that when frightened it will not enter the water."

Other Iguanidae. In addition to the subfamily *Iguanidae* just described, there are six other subfamilies, namely *Sceloporinae*, *Tropidurinae*, *Basiliscinae*, *Anolinae*, *Agamidae*, and *Chamaeleontidae*.

The basilisks (*Basiliscus*) are of particular interest. These animals range up to 80 centimeters in length, one-third of which is taken up by the head and trunk. Adults have a narrow casque on the head and a tail crest along the back. Legs are long. These animals feed on small animals and fruit. A frequent location is on a branch that overhangs a body of water. They are excellent swimmers and divers and often hide on the bottom, underwater. Basilisks can run on land, but also on the surface of water — the latter at speeds up to 12 kilometers/hour. One species (*B. vittatus*) has been reported as crossing a lake some 400 meters wide. In this kind of feat, the animals use only their hind legs. They do not sink because of their widened toes, where a border of skin contacts the surface of the water at brief intervals because of rapid leg movement. Since they move so fast, the animals are protected from both land and water predators. Common basilisks are shown in Fig. 1(l).

Anolinae. This subfamily of the *Iguanidae* includes 300 forms, thus making the anoles the largest genus in the *Iguanidae*. The anoles are found in tropical and subtropical America. In the West Indies alone, over a hundred species are known. They populate over three thousand islands. One particularly interesting species is the so-called "false chameleon," which by classification is *not* a chameleon, but rather is an anole.

Fig. 1.   Iguanids and their various characteristics (**a**) When threatening, the male fence lizard flattens its body, so as to display the dark blue belly; (**b**) The hind legs of the genus *Uma* have combs of scales on their toes to increase their effectiveness as digging tools; (**c**) Threatening males of the genus *Uma* flatten their bodies at an angle to exhibit the black spots on their sides; (**d**) Collared lizard in full flight runs on hind legs alone; (**e**) Collared lizard (top) at high temperatures lifts its body off the ground; a male (below) tries to impress an opponent by flattening body and expanding dewlap; (**f**) Deep folds of skin at the sides of neck of *Pica* can be expanded to form pointed appendages of vivid orange when the creature is excited (**g**) A high-altitude iguanid of the southern Andes (*Phymaturus*) makes its body, covered with small scales, very flat when sunbathing; (**h**) A genus *Hoplocercus* iguanid has a very compact body and spiny tail; (**i**) An old male rhinoceros iguana (*Cyclura cornuta*) has well-developed rolls of fat at back of head; (**j**) Duel between male marine iguanas; (**k**) Complicated lever mechanism formed by the hyoid bone expands the anole's dewlap; (**l**) Basilisks (top to bottom), Common basilisk (*Basiliscus basiliscus*), double-breasted basilisk (*B. plumifrons*), and banded basilisk (*B. vittatus*); (**m**) A genus *Corytophanes* iguanid (top) in normal posture and (below) during a threat display; (**n**) A long-legged lizard (*Polychrus*) preparing to jump down.

Fig. 2.   Iguana. (*A.M. Winchester.*)

Additional Reading

Coborn, J.: *Green Iguanas and Other Iguanids*, Chelsea House Publishers, Broomall, PA, 1998.

Rogner, M. and J. Hackworth: *Lizards: Husbandry and Reproduction in the Vivarium: Geckoes, Flap-Footed Lizards, Agamas, Chameleons, and Iguanas*, Vol. 1, Krieger Publishing Company, Melbourne, FL, 1997.

Schettino, L. Rodriguez and A. Coy Otero: *The Iguanid Lizards of Cuba*, University Press of Florida, Gainsville, FL, 1999.

ILEITIS.   See **Colitis and Other Inflammatory Bowel Diseases**.

ILEOCECAL VALVE.   See **Digestive System (Human)**.

ILLUMINATION. When an object is exposed (irradiated) by electromagnetic radiation, it is said to be illuminated. Illumination not only applies to visible light to which the eye, as a detector, is sensitive, but to other portions of the electromagnetic spectrum—for example, illumination is used in connection with television and radar. In this article, illumination is confined to visible light and to artificial rather than natural light sources, such as the sun. The design and application of light sources is the function of illumination engineering, also commonly called lighting engineering.

Illuminance Units and Law. The lighting (illumination) of a surface is the luminous flux it receives per unit area. The three most commonly used units are: (1) the foot-candle (fc) = 1 lumen per square foot; (2) the lux (lx) = 1 lumen per square meter; and (3) the phot (ph) = 1 lumen per square centimeter. The International System (SI) unit is the lux, which equals 0.0929 foot-candles; one foot-candle equals 10.76 lux.

If the luminous flux is uniform over the surface, the illumination is the quotient of the total flux divided by the area. If not, the illumination of a point on a surface is defined as the quotient of the flux incident on an infinitesimal element of surface containing the point, divided by the area of that element.

Illumination obeys a cosine law, like the radiation from a surface and for similar reasons. That is, if the illumination is I_0 for zero angle of incidence, then for any other angle θ it is $I = I_0 \cos \theta$. Since for perpendicular incidence the illumination from a concentrated source of luminous intensity L at distance r is $I_0 = kL/r^2$, the illumination at incidence angle θ is

$$I = k\frac{L}{r^2} \cos \theta$$

The value of the constant k depends on the units of I, L, and r; for example, if I is in foot-candles, L in candles, and r in ft., $k = 1$ (lumen per candle). See Table 1.

TABLE 1. RELATIVE LUMINANCE OF MAJOR LIGHT SOURCES

Type of Light Source	Luminance, Candles per Sq. Centimeter
Fluorescent lamp	0.6–0
Tungsten filament:	
vacuum; 10 lumens/W	200
gas-filled; 20 lumens/W	1200
projector lamp—750 W; 26 lumens/W	7500
Mercury vapor (clear; 400 W)	970
Metal halide (clear; 400 W)	810
High-pressure sodium (400 W)	780
Candle flame (bright spot)	1
Welsbach mantle (bright spot)	6.2
Acetylene flame (Mees burner)	10.5

Light Sources

Since antiquity, people have used crude sources of flame-producing substances to cause light and thus cope with darkness. The earliest lamps burned animal or vegetable oil. It is reported that the Roman elite preferred beeswax, but, for the masses of their citizenry, the more available and less expensive oils were widely used. It is interesting to note that candles are not mentioned in the early literature until the first century AD. The earliest candles were made from tallow produced from animal fats. Candles remained quite competitive with other sources of light up to the end of the nineteenth century. During the early 1800s, candle manufacturers developed mechanized means for making candles and thus cut production costs. The greatest boon to the candle manufacturer, however, occurred with the discovery of oil in the American fields, which made large quantities of paraffin wax available at comparatively low prices. Concurrent with that development, large quantities of inexpensive paraffin oil also became available and was shipped worldwide for use in lamps, accounting in some measure for the first large fortunes made as the result of oil field development. As contrasted with animal oils, paraffin oil had little odor, flowed readily up the wick, and was relatively inexpensive.

As pointed out by G.G. Roberts (reference listed), "The first public demonstration of gas lights was in Paul Mall (London) in 1807 and helped celebrate King George III's birthday." The light of the gas flame is due to particles of carbon that are made luminous by the heat of the flame. Improvement of gas lighting came about with the introduction of gas mantles, consisting of a web of carbonized cotton supporting oxides of thorium and cerium. Much later, Auer von Welsbach (Vienna) invented a glowing mantle made of lanthanum and zirconium.

A major shortcoming of gas lighting was the heat produced when they were installed in confined indoor places, such as meeting halls.

One of the earliest attempts to utilize electricity as a source of light was the rather impractical arc lamp developed by Jablochkoff in England. The lamp consisted of two parallel carbon rods separated by a thin layer of plaster of Paris. Only a few experimental installations were made in 1878. Sixteen of the Jablochkoff "candles" were installed in London's Billingsgate fish market, and several were used to illuminate a soccer match in Sheffield. Charles Brush of Cleveland, Ohio, generally is credited for designing a more practical electric arc lamp, the first of which was installed for street lighting in 1877. See also **Arc Lamp.**

A major breakthrough in electrical artificial lighting occurred when Edison invented the incandescent lamp at his laboratory in Menlo Park, New Jersey, in 1879. Although the lamp has been improved innumerable times and is continuing to compete with other important electrical sources of light, the fundamental principles, as outlined in Edison's original patent (Fig. 1), remain intact. Concurrent with Edison's work was that of Swan, an inventor in Northeast England. Swan's lamp was quite similar, but used a carbonized cotton filament instead of the carbonized bamboo filament used by Edison. Principal improvements in incandescent lamps over the years have included the use of a tungsten filament, the introduction of small traces of halogen gas into the lamp enclosure, several redesigns of the shape of the bulb, improved glass enclosures, and superior sockets. Even in view of the availability of fluorescent and other basically different principles, the modern incandescent lamps continue to be sold worldwide in terms of billions of units each year.

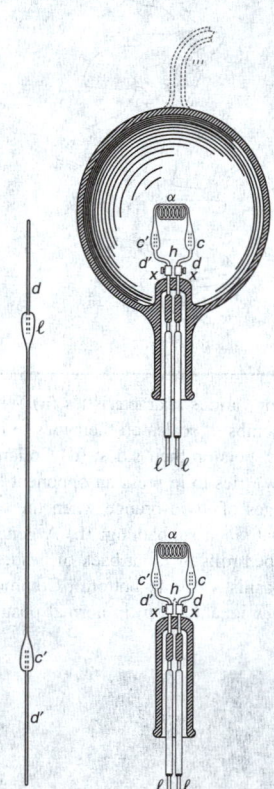

Fig. 1.   Patent Drawing (U.S. Patent No. 223,898; January 27, 1880) of Edison's first successful incandescent lamp.

Fluorescent Lamps. These lamps provide an efficient way of generally lighting building interiors. Designers are working on shorter, fatter lamps operating at high frequencies, with new cathode designs that emit greater amounts of light per foot of length. The use of fluorescent lamps for residential lighting has increased at a good rate in recent years and helps to reduce the relative amount of power required to light homes and apartments.

Fluorescent tubes use a gas discharge with a substantial ultraviolet component. This ultraviolet light excites electrons in fluorescent centers on the walls of the tube. The electrons drop immediately from the highly excited state to an intermediate state with a lower energy. From this state, they finally drop down to the original ground energy level with the emission of visible light.

During the past several years, fluorescent lamp developments have concentrated on improvements in color rendering and refinements in light output and depreciation characteristics. U-shaped 40-watt designs, while not generally as economical as straight 40-watt lamps for conventional installations, are frequently used where luminaire dimensions are limited to 2 × 2-foot (0.6 × 0.6-meter) square sections. Where dimming is required, a 40-watt rapid-start system is the frequent choice. Solid-state control devices have expanded the dimming range and reduced wiring and operational problems.

More highly loaded, high-output lamps are good choices where the overall number of luminaires must be kept to a minimum for a given lighting level and in large areas of relatively high illumination levels. Fluorescent lamps become more efficient as lamp length and current loading increases, and often high-output lamps will provide lower cost of light than 40-watt lamps in office, store, school, and other commercial installations.

Fluctuating ambient temperatures reduce light output of fluorescent lamps unless provision is made to keep the bulb wall temperature near 100 °F (37.8 °C) corresponding to an ambient temperature of about 77 °F (25 °C). Above these values, lamp watts and light output drop gradually (roughly 25% per 40 °F). Too-cold operation is more critical. Light output drops about 75% per 40 °F while lamp watts drop only about 25%

and the lamp consumes almost full power while emitting very little light.

Indoor luminaires typically operate the lamp too warm unless some provision has been made for returning air through the luminaire or otherwise controlling the temperature of the air surrounding the lamp. The life of fluorescent lamps as well as some of the newer high intensity discharge sources is affected by the number of times the lamp is turned on and off. As fluorescent lamp life ratings have increased, this effect has become less important.

In 1990, a double twin tube fluorescent lamp (Fig. 2) was introduced. These 18- and 26-watt light sources are designed to replace 75- and 100-watt conventional incandescent lamps, respectively, thus providing energy savings. The new lamps produce 69 lumens per watt, which is approximately 4 times that of the incandescent lamps that they are designed to replace. The new lamps are claimed to have a lifetime of 10,000 hours, which is up to 13 times longer than that of the incandescent bulbs.

In 1992, a new 15-watt, soft white, compact fluorescent bulb (Fig. 3) was introduced. This bulb was designed for "bare-built" fixtures and lamps, down lights, and ceiling fixtures.

Compact fluorescent tubes cost considerably more (from 10× to 20×) than conventional incandescent bulbs, but the incentives to purchase them are energy savings and much longer bulb life. The latter factor can be very important to both household and commercial users in terms of replacing lamps in hard-to-reach places. In early 1991, a western U.S. utility presented gratis over 800,000 compact fluorescent lamps in an energy savings promotion. A few other utilities have encouraged customers to lease the new bulbs by adding several cents per bulb to their monthly light bill. It is interesting to note that many years ago, when electrification of the U.S. was in an early stage, some utilities furnished light bulbs free of charge in an effort to encourage more and better lighting and higher light bills.

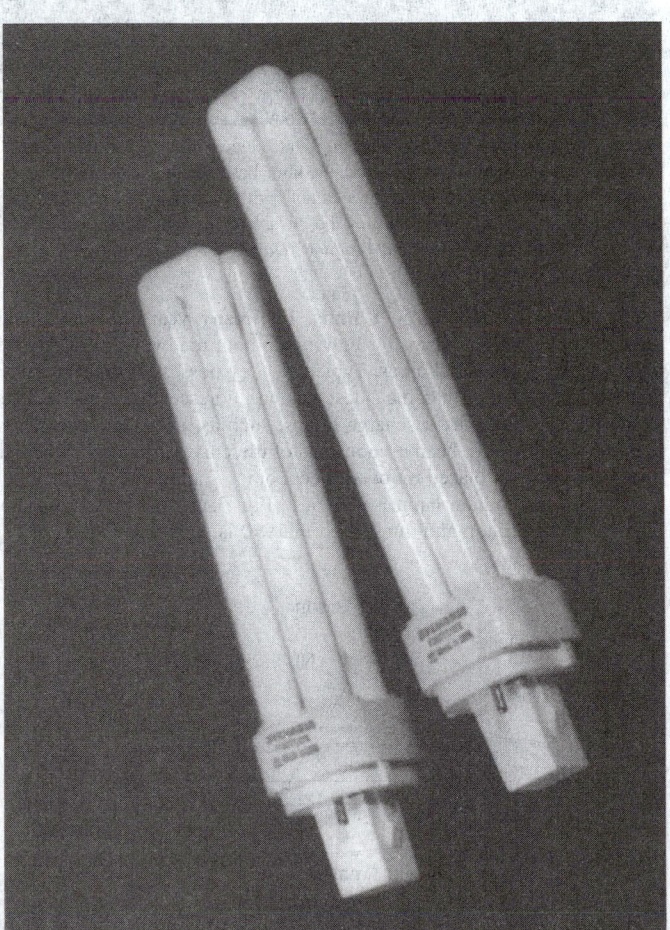

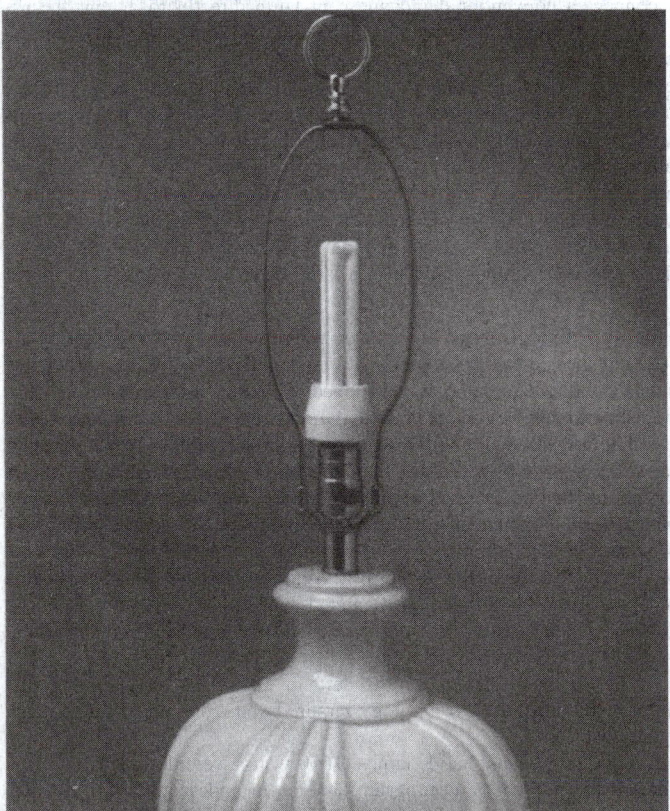

Fig. 2.   (a) An 18-watt (left) and a 26-watt (right) double twin tube fluorescent lamp. (b) Bulb installed in a "bare" fixture. (*GTE Electrical Products, Danvers, Massachusetts.*)

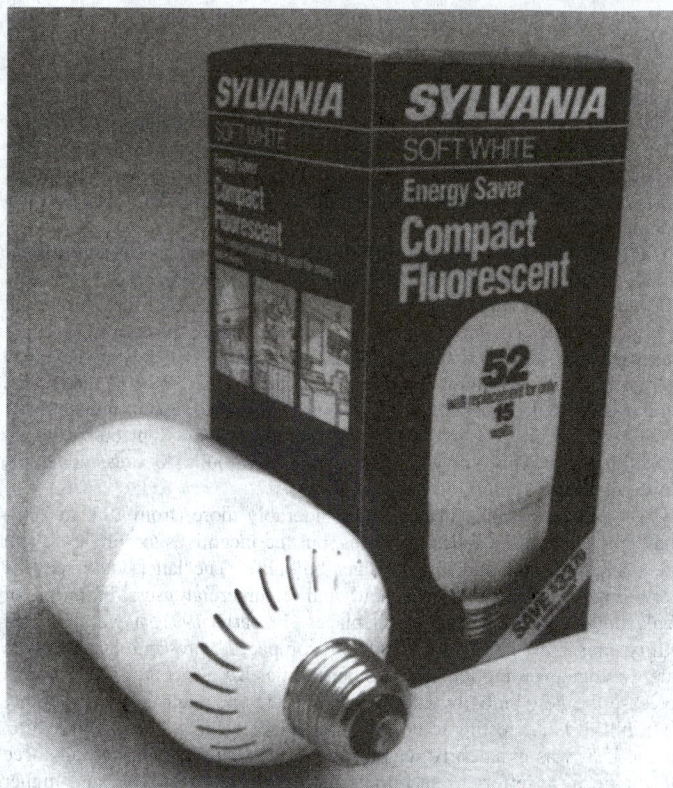

Fig. 3.   A 15-watt, soft white, compact fluorescent lamp. (*GTE Electrical Products, Danvers, Massachusetts.*)

Fig. 4.   Low-wattage metal halide lamps introduced in 1992. Available in 50-, 70-, and 100-watt units, the lamps fit medium base sockets and can be used in open bottom fixtures. The units are available with a clear or coated bulb. Initial lumens are 3300, 5500, and 8500 for the clear lamps; 2800, 4800, and 8000 for the coated bulbs. (*GTE Electrical Products, Danvers, Massachusetts.*)

For many years, a serious deterrent to the wider acceptance of fluorescent lamps has been the need for a ballast. It always has been possible to connect a conventional incandescent lamp directly to a light fixture. Fluorescent lamps have required suitable circuitry and control gear (ballast components) for starting their operation. Mainly over the past decade, developments in electronics have enabled smaller and more efficient ballasts. As noted previously in Figs. 2 and 3, fluorescent lamps now can be fitted directly into conventional lamp sockets. By incorporating ballasts by way of silicon chip technology, this will stimulate even greater flat, compact, and geometric luminaire designs. Dimming can be made easy, and flickering (sometimes encountered in some countries that have a 50 Hz power supply) can be eliminated. Also, controls can be added that will automatically turn a lamp back on in instances of power failures.

Metal Halide Lamps. In an era when ways of saving energy are receiving the increasing attention of consumers, the very poor efficiency of the incandescent lamp is mentioned with increasing repetition. It is interesting to note that many years ago the British physicist, Lord Rayleigh, described a theoretical solution—namely, to coat a bulb with a thin film that reflects heat back at the filament while transmitting visible light. Converting theory to practice has been extremely difficult and had to await the development of new materials processing from the technology of electronics fabrication, such as chemical vapor deposition. The goal has been one of forming a uniform film (a few microns in thickness) that could withstand the temperatures created in high-performance light bulbs. Multilayer metal oxide films now are used in a number of commercially available halogen bulbs.

The clear bulb shown at the left in Fig. 4 is indicative of the complex manufacturing technology required and, consequently, the much higher cost.

Initially, large metal halide lamps (1500 watts) were designed to replace standard incandescent and tungsten halogen incandescent lamps for such applications as athletic fields, large stadiums, and arenas. Besides higher efficacy (over 3 times more light per watt than incandescent), the sources have excellent color characteristics for color television pickup.

High-Intensity Discharge Lamps. For many years, mercury lamps were the conventional means for lighting outdoor roadways, railway yards, and other large industrial areas. Typically, they possessed superior optical control characteristics than incandescent and fluorescent systems. In recent years, the high-intensity discharge lamps, such as metal halide and high-pressure sodium, have been used instead of mercury, particularly where the emphasis has been placed on overall economy. The operating principle of a high-pressure sodium lamp is shown in Fig. 5. A high-pressure sodium lamp of recent design and introduced in 1990 is shown in Fig. 6.

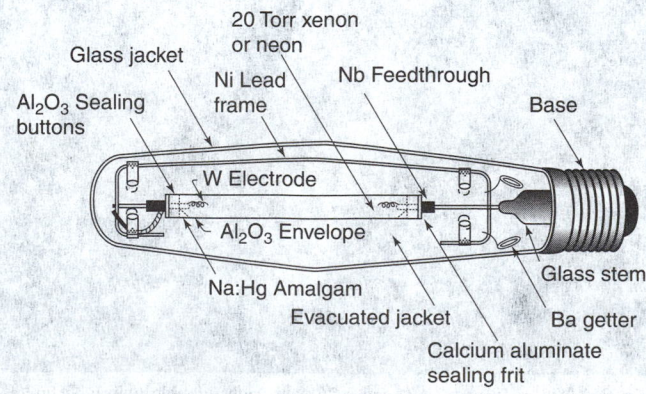

Fig. 5.   Construction of a representative high-pressure sodium lamp.

on instantly, with no warm-up period. Cost of the lamp is estimated to lie between 10× and 20× that of a traditional incandescent lamp of equivalent light output.

Instrument and Data Display Illuminators

Prior to the use of the cathode-ray tube (CRT) for information displays, miniature incandescent lamps were the predominant light source. The fact that the lamps could be made small, bright, and in a variety of colors gave the display designer a considerable range of choice. Although still used and preferred for some display situations, the incandescent lamp was markedly impacted by the light-emitting diode (LED), somewhat later by liquid-crystal displays (LCD) and, to a lesser extent, but increasing in acceptance, the plasma display. Check alphabetical index for descriptions of CRTs, LEDs, and LCDs.

Plasma Displays. Plasma displays are bright, easy to view, and relatively reliable, but they are expensive and difficult to drive. Quite recently, refinements have enlarged pixel formats and frame sizes, decreased the number of drive components, reduced power consumption, and lowered costs. More gray-scale displays are now available and better color capabilities are emerging. Some computer makers have recently opted for plasma technology for flat-panel displays.

Plasma displays are constructed by sandwiching a neon-argon gas mixture between two sealed glass plates. Parallel electrodes are deposited on the inner side of each plate, one set of electrodes running at right angles to the other set facing it. The electrode cross points form the pixel matrix. Either an ac or dc voltage can be applied to the electrode cross points. Current flows, causing the gas nearby to discharge a red-orange glow, the characteristic color of the plasma display thus far. The glow is visible through the coverplate because the electrodes on it are transparent.

In a dc plasma display, the electrodes are in direct contact with the neon-argon gas. The system is inherently simple, but material transfer between electrodes and gas tends to degrade their discharge brightness over time. The dc plasma display is a refresh technology; thus there is no extended memory characteristic when voltage is removed. No dc memory displays are currently available. The selected dots of a row remain on only for the scan time of that row. Each row's duty cycle can be defined as $1/n$, where n is the number of rows in the panel. A Ac refresh technology operates on the same principle.

In an ac plasma display, the electrodes are covered with a dielectric material, glass, through which they are capacitively coupled to the neon-argon gas. The polarity of the voltage across the cell is alternated, causing an alternating current to flow. This current flow and the glassy electrodes extend the ac panel's life, with no diminution in brightness over time. The A.C. plasma display exists in both memory and refresh modes. Ac memory plasma uses a select-and-sustain system of electrodes, whereby a cell once selected remains on until it is turned off. Thus, a selected dot in a row of the pixel matrix remains on continuously during subsequent addressing of the panel. The result is a high selected-dot duty cycle and a brilliant display.

An ac memory plasma unit is constructed from thick, flat glass and precision spacers, thus it is bulkier and more expensive than either the ac or the dc refresh technology. However, it does offer better brightness and is capable of yielding comparatively large and high-density arrays.

An ac refresh system is a lighter in weight than ac memory. It uses thin-glass, thick-film processing and is adequately bright up to about 400 to 500 scan lines. It is lighter, more portable, and more uniform in appearance than ac memory plasmas, and less costly.

Many of the new plasma products are graphics rather than alphanumeric displays. As users require increasingly more information on the screen, the size of the screens has enlarged. One firm now offers a 1728 × 1280 pixel display, equivalent to an entire newspaper display, or four $8\frac{1}{2}$ × 11 (in.) pages. As of 1990, the largest known plasma display constructed is a 2048 × 2048 pixel ac memory unit with a 4.95 ft (1.5 m) diagonal. The main activity in graphics displays lies in the 640 × 400 pixel product, which essentially has replaced the 640 × 200 version within the last few years.

A contemporary ac plasma display is shown in Fig. 7.

Characteristics of Light Sources

Color. This is one of the most difficult characteristics to evaluate. The reason is that the colored appearance that a source gives an object or a space

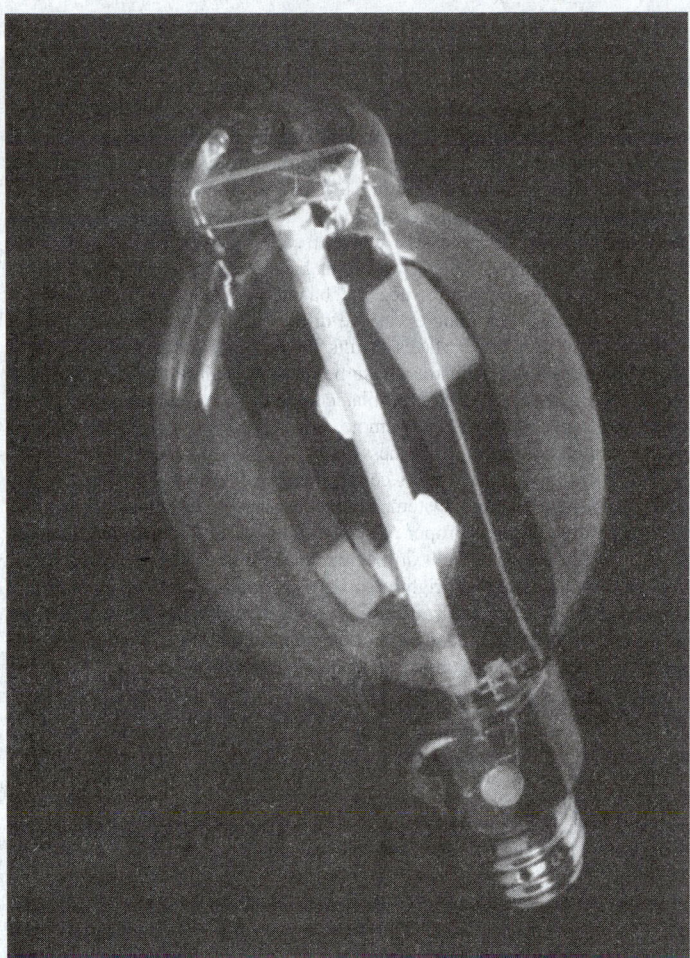

Fig. 6. A 760-watt high-pressure sodium lamp introduced in 1990. The lamp has a high efficiency of 145 lumens per watt, which represents a significant design improvement over prior lamps. The lamp features a reduced-sized arc tube, thus contributing to its compact size. The lamp finds use in roadway lighting, parking lot illumination, and high bay industrial warehouses. Rate life of the lamp is proclaimed to be 16,000 hours, thus reducing replacement maintenance costs. (*GTE Electrical Products, Danvers, Massachusetts.*)

With reference to Fig. 5, the rare gas fill promotes easier starting of the discharge. As the lamp warms up, mercury develops sufficient vapor pressure to enter into the discharge. Finally, sufficient sodium enters the vapor phase, resulting in a 3700 °C plasma temperature and spectral emission from 569 to 2205 nm with a broad maximum around the 589-nm Na D line. The cold spot temperature near the electrode is approximately 700 °C, and controls the sodium vapor pressure to approximately 100 torr. The envelope stabilizes the plasma and reaches a peak temperature of about 1200 °C. A fundamental property requirement for a lamp envelope material is that it have a wide band gap for transparency in the visible wavelength range and for electrical insulation to isolate the electrodes. It is desirable, but not essential, that the material have a cubic crystal structure to limit birefringent light scattering at grain boundaries.

A so-called "E-lamp" currently under development was described at a June 1992 meeting of the Edison Electric Institute. The developers claim that the new lamp will have a long life and feature the light intensity of incandescent bulbs, but with the energy efficiency of fluorescent lamps. The E-lamp is reported to consist of a magnetic coil that generates a high-frequency radio signal. When a sealed glass globe containing the same gas mixture used in a conventional fluorescent lamp interacts with this signal, the gas is converted into plasma. The plasma, in turn, emits non-visible light, which strikes the phosphor coating on the inside of the glass, which then glows with visible light. It is claimed that the E-lamp, with the same lighting output, will use one-fourth of the energy and will produce less heat than a standard 75-watt incandescent lamp. The E-lamp will switch

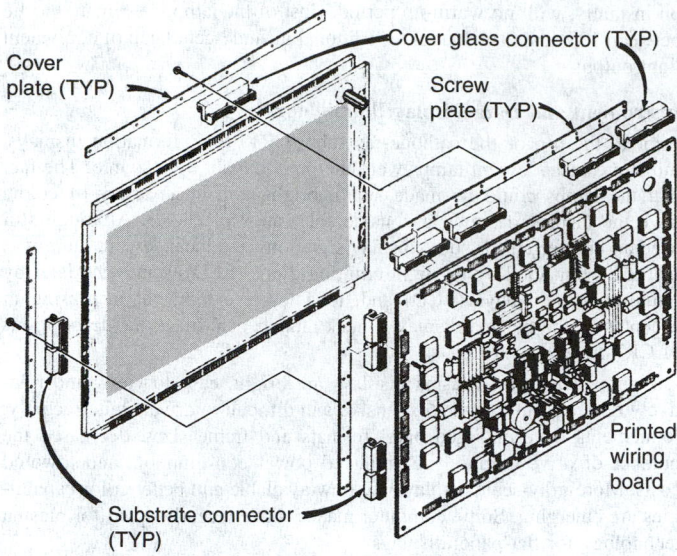

Fig. 7. Exploded view of typical as plasma display. Available in seven sizes, ranging from 192 rows × 320 columns to 400 rows × 640 columns. To operate the display, the external video controller must provide a continuous dot clock signal, horizontal and vertical synchronization signals (HSYNC and VSYNC), and serial video data in either monochrome or color (red, green, and blue). The video signal interface circuit is also resident on the plasma display circuit board. (*AT&T Technologies, Allentown, Pennsylvania.*)

is subjective, strongly influenced by the viewer. Because the technical terms used to describe the color characteristics of light sources have rather limited use outside scientific circles and often are highly mathematical, the specifier of lighting systems may find it difficult to describe what is wanted in precise terms. In this regard, two terms are helpful.

Chromaticity. Apparent color temperature (sometimes called correlated color temperature) and color rendering index (written as R_a in color literature) are these terms. Chromaticity is the measure of a light source's "warmth" or "coolness," expressed in the Kelvin temperature scale (K). It describes the appearance an object would have if it were heated to incandescence—the point of emitting light—then to higher temperatures where the appearance changes from ruddy red through a range of warm colors to white, then finally to blue-white. See Fig. 8.

Chromaticity provides no information about how well a source will render objects. Natural light in general has excellent color rendering. Light is present throughout the visible spectrum, although some colors may be slightly distorted by weather and cloud conditions, time of day, season, and latitude. All electric sources, however, distort colors in some way inasmuch as their emission of the different colors is out of balance. An incandescent lamp, for example, emits little blue and green light relative to red and so tends to mute or gray down the "cool" colors, such as blue.

A measure of this distortion characteristic or how well a light source renders colors is the color-rendering index. Essentially, this is a number which compares a given source against a "perfect" or reference source on a 0–100 scale. The system is limited because the comparison is meaningful only if the two sources being compared have the same chromaticity. Thus, it would not be meaningful to compare the color-rendering index of an incandescent lamp against a cool-white fluorescent lamp, because the chromaticity of the incandescent is 2,900 K, versus the cool-white fluorescent at 4,200 K. A comparison could be made between cool-white, $R_a = 66$, and deluxe cool white fluorescent, $R_a = 89$, however, because both have the same chromaticity. This can also be qualitatively determined by examining the spectral power distribution curve of these sources. Lamps with emission throughout the spectrum, especially in the red regions, tend to render all colors well. From a design viewpoint, if color-rendering is important, one approach might be to select a chromaticity range for its warmth or coolness. Then, find a source with the highest color-rendering

index in that range, being aware that high color-rendering lamps have lower luminous efficacies, so there may have to be a trade-off against illumination level.

Intensities of Illumination. People are more comfortable in an environment in which luminaire brightness is controlled. Excessive quantities of raw light entering the eye interfere with the ability to see. A broad classification of recommended lighting levels is given in Table 2. Adequate lighting is needed not only on the task, but also on the areas immediately surrounding it. Excessive contrast can cause ocular fatigue if the eye has to constantly adapt to different brightnesses. Contrast is defined in terms of brightness ratio, i.e., the balance of reflected light between adjacent surfaces. Reducing excessive differences in reflectivity of two or more contiguous work surfaces results in a more comfortable brightness ratio. Sometimes this adjustment can be made by painting areas peripheral to the work. There should be similar contrast control between the proximate work area and the more remote areas. Table 3 on p. 1906 provides recommended maximum brightness ratios.

The reflectivity of a factory's ceiling, walls, and floor contributes to the utilization of the lighting system. The appearance of these surfaces affects the visual environment. Proper painting enhances both. An all-white room would afford maximum light utilization and brightness, an all-black room, minimum utilization and brightness. Few occupants, however, would feel comfortable in either environment. Room colors should be selected for optimum light utilization and environmental acceptability. Modern practice suggests the reflectances shown in Table 4 on p. 1906.

Streamlining Light Source Selection. Tens of interacting variables are encountered in choosing the most optimum light sources for given applications. The process, however, can be orderly, as suggested by Fig. 9 on p. 1906.

Lighting Engineering. Among several factors of lighting system design that affect worker productivity are:

1. *Adequate lighting* of working areas, tools, and complex machinery, the latter often having complex contours and shapes and that must be viewed from varying distances;
2. *Elimination of glare, shadows, and unnecessary reflection*; and
3. *Color.*

Thus, the designer responsible for the lighting of a facility must not only select the best possible lamps and other light sources, but also must pay close attention to the design and placement of the luminaires to be specified. Each situation has its particular differences, and thus it is not always wise to allow all decisions to be made by an electrical equipment contractor. Lighting consultants and architects who have had extensive experience with factory and office lighting are available to assist at a cost that can be retrieved quickly because of more productive and satisfied workers.

The illuminance for a number of industrial areas is given in Table 5 on p. 1906. A number of excellent references are available that can assist in making the right lighting decisions. See **Additional Reading** list.

In the case of office lighting, the bulk of the cost of lighting an office is for electricity. The bulk of the cost of an office operation is for people. The cost of the lamps is 2–3% of the cost of the light and $\frac{2}{200}$ of 1% of the total cost of the office; yet lamp efficiency can have a significant effect on energy use. Lamp color and light output have a significant effect on worker performance.

Light Glare. Glare produced by lighting or outside light should be eliminated wherever possible. Where instrument settings must be read and parts manipulated, glare can cause very serious errors. There are two types of glare: (1) direct and (2) reflected. Glare is unwanted light in the field of view. It may cause worker annoyance, discomfort, and even loss of visual performance. Glare occurs when luminance within the visual field is substantially greater than the amount of luminance to which the eyes are adapted. Direct glare often results from the luminaire not shielding the lamp from view. See Fig. 10 on p. 1907. Direct glare may be exceedingly severe with lamps of high lumen and/or luminance

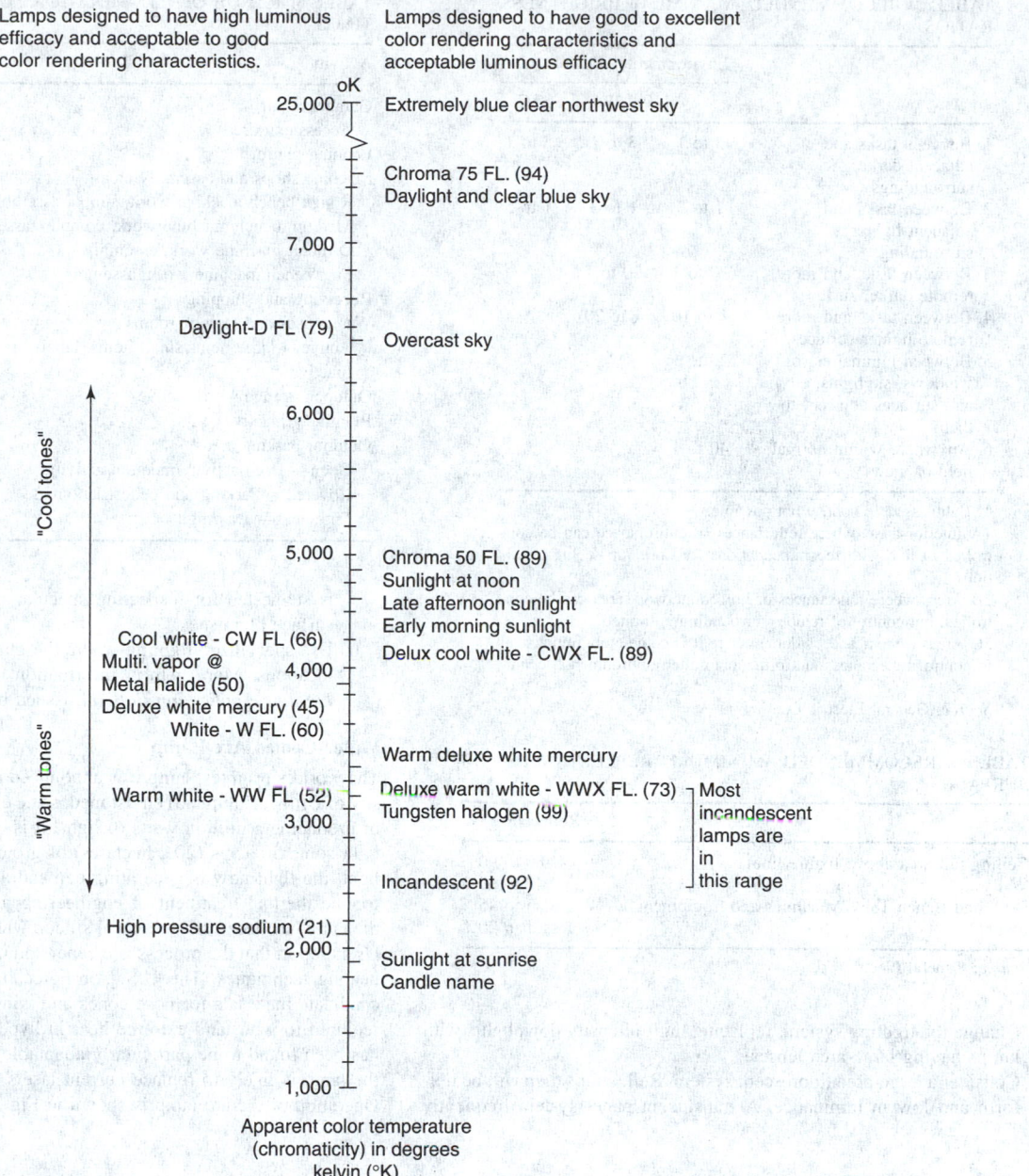

Fig. 8. Color characteristics of light sources. Numbers in parentheses denote the lamp's color rendering index, R_a.

TABLE 2. MINIMUM RECOMMENDED LIGHT-
ING LEVELS[a]

Seeing Task	Footcandles	Hectolux
Casual	30	3.2
Rough	50	5.4
Medium	100	11
Fine	500	54
Extra Fine	1,000	110

[a]Recommendations based upon young adults with normal
eyes and minimum lighting on the task at any time.
Source: General Electric Co.

values. For example, a clear 250-watt high-pressure sodium lamp that emits 27,500 lumens from a cigarette-size arc tube will be much brighter than a 400-watt metal-halide lamp that emits 36,000 lumens from a melon-size, phosphor-coated lamp envelope. Portions of some specular (mirrored) reflectors can reflect an overly bright image of the lamp. See Fig. 11 on p. 1907. Lens-enclosed luminaires may eliminate bare-lamp brightness, but can themselves cause direct glare at certain angles. Fluorescent luminaires with their relatively low brightness lamps usually will be less glaring. Avoiding direct glare is simply a matter of choosing the right luminaire and its correct location.

Reflected glare is somewhat more insidious. Images of the lamps or the luminaires are reflected from the task. See Fig. 12 on p. 1907. These "veiling reflections" can be severe and may cause errors during product machining, fabrication, assembly, or inspection, or at an office desk when handling paperwork.

Means used to reduce or eliminate reflected glare include:

1. Either change position of offending luminaire or the position of the task.
2. Use a dull or matte finish on surfaces surrounding the task.
3. Replace clear lamps with diffuse-coated ones.

TABLE 3. RECOMMENDED MAXIMUM BRIGHTNESS RATIOS

	Environmental Classification		
	1A	2B	3C
1. Between tasks and adjacent darker surroundings	3 to 1	3 to 1	5 to 1
2. Between tasks and adjacent lighter surroundings	1 to 3	1 to 3	1 to 5
3. Between tasks and more remote darker surfaces	10 to 1	20 to 1	*
4. Between tasks and more remote lighter surfaces	1 to 10	1 to 20	*
5. Between luminaires (or windows, skylights, etc.) and surfaces adjacent to them	20 to 1	*	*
6. Anywhere within normal field of view	40 to 1	*	*

*Brightness ratio control not possible.

1A Interior areas where reflectances of entire space can be controlled in line with recommendations for optimum seeing conditions.

2B Areas where reflectances of immediate work area can be controlled, but control of remote surrounding is limited.

3C Areas (indoor and outdoor) where it is completely impractical to control reflectances and difficult to alter environmental conditions.

Source: General Electric Co.

TABLE 4. RECOMMENDED MINIMUM REFLECTANCES OF SURFACES

	Reflectance
Ceiling (all area above fixture line)	80 to 90%
Walls	40 to 60%
Desk and Bench Tops, Machines and Equipment	25 to 45%
Floors	Not less than 20%

Source: General Electric Co.

4. Change the lighting system, replacing high-intensity downlights with lamps having large-area lenses.
5. Consider a large-area fluorescent system. Reflections then will be uniform and low in luminance. A translucent panel system frequently

TABLE 5. RANGE OF ILLUMINATION FOR SOME INDUSTRIAL SETTINGS

Activity	Foot-candles Required
Garage, repair	50–100
Access/exit areas	10–20
Loading platform	20
Machine shops and assembly areas	
Rough bench/machine work, simple assembly	20–50
Medium bench/machine work, complex assembly	50–100
Difficult machine work, assembly	100–200
Fine bench/machine work, assembly	200–500
Receiving and shipping	20–50
Warehouses and storage rooms	
Active — large items/small items, labels	15–30
Inactive	5
Outdoor storage yards, active	15–30
Relatively inactive	5
Outdoor parking areas	
Open — High activity/medium activity	2–1
Covered — Parking and pedestrian areas	5
Entrances — Day/night	50–5

is suggested for inspecting specular metal and plastic products for surface blemishes.

6. Use specialized lighting where the glare problem is limited to a few locations. More helpful information on this topic can be found in *Lighting Application Bulletin*, issued by GE Lighting.

Water-Cooled Arc Lamp

The world's brightest lamp was announced in 1990. G.G. Albach (University of British Columbia) envisioned in the early 1980s an arc lamp capable of producing a million watts of light. It is claimed that the new lamp can light some 50 acres (20.2 hectares) of ground. Rather than as a source of light, the light now is generating demand as a tremendous source of heat for the thermal treatment of engineering materials. A 300-kW model of the lamp has been used to anneal silicon wafers in the electronics industry. Users report that the process is cleaner and more controlled than traditional heating techniques. The U.S. Air Force has also used the lamp to test candidate materials for nose cones and wing leading edges, areas that are required to withstand extreme heat in hypersonic aircraft. The lamp also has been found to be particularly adaptable for selective hardening. Thus, the lamp is likely to replace current lasers systems for surface treatment. Operation of the arc lamp is shown in Fig. 13.

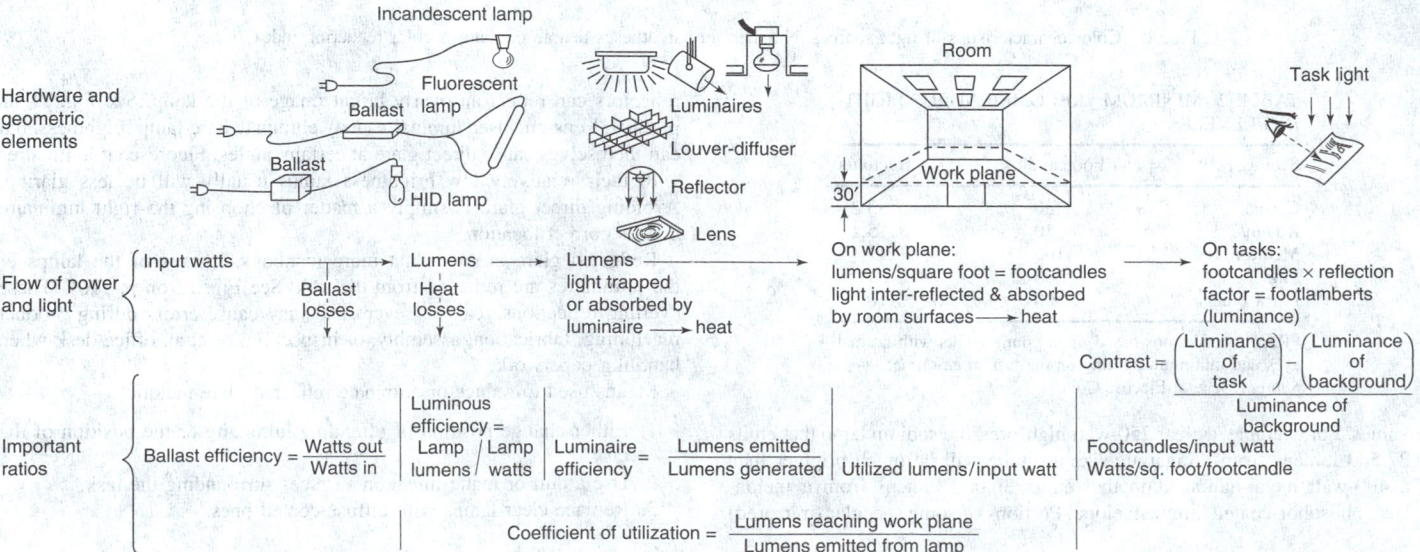

Fig. 9.   Major factors in design and selection of an overall lighting system. (*GE Lighting.*)

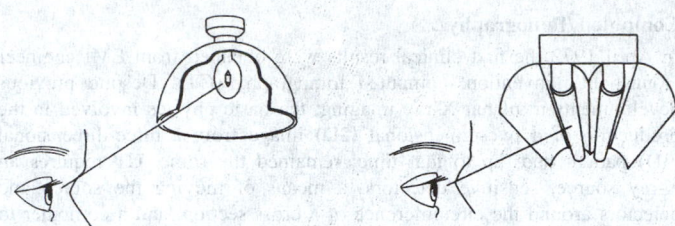

Fig. 10. Examples of direct glare where the eye of a person at a working position can see the lamp directly. (*GE Lighting*.)

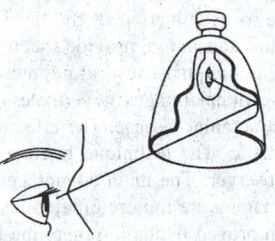

Fig. 11. Example of reflected glare from internal reflections within a luminaire. (*GE Lighting*.)

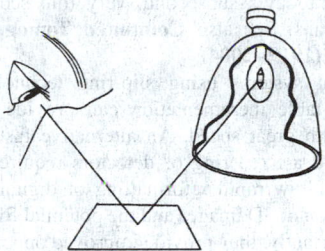

Fig. 12. Example of veiling reflection, where light from the luminaire is reflected back to the visual path of the worker. (*GE Lighting*.)

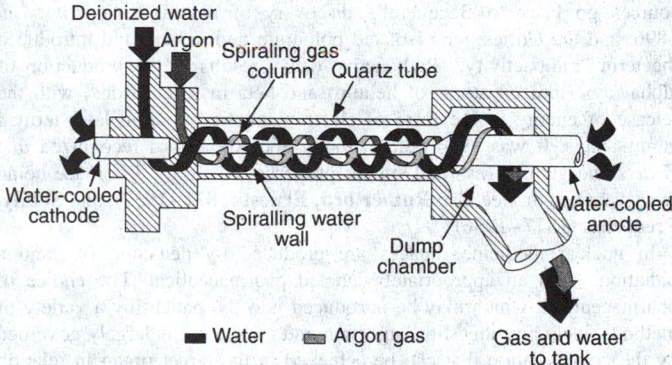

Fig. 13. Sectional view of a 300-kW arc lamp. The arc temperature is between 12,000 and 13,000 °C. The Vortek lamp utilizes a spiraling wall of water to cool a quartz containment tube and to wash away debris from its tungsten electrode. It is estimated that the arc temperature is about one-half the temperature of the surface of the sun.

Additional Reading

Baker, H.: "A Bright Idea," *Advanced Materials & Processes*, 8 (May 1989).
Brou, P., et al.: "The Color of Things," *Sci. Amer.*, 84 (September 1986).
Buckwald, J.D.: *The Rise of the Wave Theory of Light: Optical Theory and Experiment in the Early Nineteenth Century*, University of Chicago Press, Chicago, IL, 1990.
Cherfas, J.: "Skeptics and Visionaries Examine Energy Savings (Fluorescent Bulbs)," *Science*, 154 (January 11, 1991).
Corcoran, E.: "Body Heat: QWIPs Offer a New Way to See in the Dark," *Sci. Amer.*, 123 (October 1991).
Greenberg, D.P.: "Light Reflection Models for Computer Graphics," *Science*, 166 (April 14, 1989).

Hamilton, D.P.: "Efficient Bulb Sees (Most of) The Light," *Science*, 1084 (November 22, 1990).
Hurlburt, A.C. and T.A. Poggio: "Synthesizing a Color Algorithm from Examples," *Science*, 482 (January 29, 1988).
Kubel, E.J., Jr.: "Surface Treatment with a High-Intensity Arc Lamp," *Advanced Materials & Processes*, 37 (September 1990).
Lieberman, K. et al.: "A Light Source Smaller Than the Optical Wavelength," *Science*, 59 (January 5, 1990).
Lindsey, J., et al.: *Applied Illumination Engineering*, Prentice Hall, Inc., Upper Saddle River, NJ, 1997.
Peterson, I.: "Putting a Far Finer Point on Visible Light," *Science News*, 7 (January 6, 1990).
Peterson, I.: "Bubble Light in the Blink of an Eye," *Science News*, 292 (May 11, 1991).
Roberts, G.G.: "The Bridge of Technology (Light Sources)," *Review (The Univ. of Wales)*, 57 (Spring 1989).
Ross, P.E.: "A Million Watts of Light: World's Most Powerful Light," *Sci. Amer.*, 138 (November 1990).
Schivelbusch, W. and A. Davies (Editor): *Disenchanted Night—The Industrialization of Light in the Nineteenth Century*, University of California Press, Berkeley, CA, 1995.
Walker, J.: "The Colors Seen in the Sky Offer Lessons in Optical Scattering," *Sci. Amer.*, 102 (January 1989).
Williams, D.: "Let There Be Lights!" *Case Alumnus*, 2 (Spring/Summer 1992).

ILLUVIATION. See **Soil**.

ILMENITE. A mineral oxide of iron and titanium, $FeTiO_3$. Magnesium and manganous manganese may replace ferrous iron to form a complete isomorphous series between ilmenite, and its magnesium-manganese end members, geikielite and pyrophanite. It crystallizes in the rhombohedral division of the hexagonal system; hardness, 5–6; specific gravity, 4.72; brittle, with uneven to conchoidal fracture. Crystals tabular, rarely rhombohedral, also massive, lamellar, granular. Color, iron black; opaque, with metallic to dull luster.

Ilmenite occurs as a common accessory mineral in both igneous and metamorphic rocks, and as heavy concentrations in certain black beach sands with magnetite, rutile, and zircon. Also found in pegmatites and as vein deposits. Valuable deposits are found in Norway; Sweden; Mexico; Finland; Ilmen Mountains, former U.S.S.R.; Canada; England; Brazil; and Italy. Brazil and India are rich in beach sand deposits. United States localities include California, Idaho, Colorado, Wyoming, Arkansas, Kentucky, Pennsylvania, Massachusetts, Connecticut, Orange County and the Adirondack Mountain Deposits in New York, and as beach sands in Florida north of St. Augustine.

Named after the Ilmen Mountains, former U.S.S.R.

IMAGE. A two-dimensional array of data with values at each element of the array related to an intensity or a color. An image is typically defined as the result of some type of image collection system; however, it could be the representation in two dimensions of any data by intensity or color.

IMAGE COMPRESSION. See **Data Compression**.

IMAGE DISSECTOR CAMERA SYSTEM (IDCS). See **Nimbus Satellite Program**.

IMAGE ENHANCEMENT. The process of changing the display levels in an image to highlight particular information in the image. This includes, but is not limited to, contrast improvement, edge enhancement, spatial filtering, noise suppression, image smoothing, and image sharpening. The result of this process is an enhanced image.

See also **Photography and Imagery**.

IMAGE FILES. See **File Types**.

IMAGE PROCESSING. The use of automated or manual techniques to provide the means of assessing, preprocessing, extracting features, classifying, identifying, and displaying the original or processed imagery for subjective evaluation, interpretation, and further interaction with the data.

IMAGER FOR MAGNETOPAUSE-TO-AURORA GLOBAL EX-PLORATION (IMAGE) MISSION.   See **Space Science Missions: Sun**.

IMAGINARY NUMBER.   See **Complex variable**.

IMAGING (MEDICAL): AN OVERVIEW.   The spectrum of diagnostic imaging now extends from morphological display to functional analysis, with an increasingly wide range of diagnostic and therapeutic possibilities. It is no longer necessary or even desirablein some instances for *in vivo* biological events to be recorded pictorially. The "image" may be entirely numerical, graphic or spectral. Increasingly, the analogue image is derived from, or replaced by, digital data. See also **Digital Processing and Information Technology in Imaging**.

In vivo biological imaging was born of four major discoveries at the turn of the nineteenth century. While the discovery of X-rays by Roentgen in November 1895 overshadowed the others in the public mind, the discoveries of radioactivity by Becquerel, the electron by J. J. Thomson and the splitting of the spectral lines of light using magnetic fields by Zeeman and Lorentz were all of seminal importance. Radioactivity led to new fields of both diagnosis and treatment. An understanding of the electron exposed the entire panoply of twentieth century science to medical diagnosis. The Zeeman effect proved to be a powerful tool in the unraveling of atomic structure and decisive for the later discovery of electron spin, the basis of nuclear magnetic resonance. The vacuum tube, electricity and photography were all catalysts in the observation and recording of these developments. The available phenomena that can now be detected or propagated through tissue include electromagnetic and particulate radiations, electric and magnetic fields and ultrasound.

X-Rays

Roentgen's serendipitous discovery of a penetrating new ray (X) emanating from a Lenard cathode-ray tube, by the chance observation of fluorescing crystals nearby, changed medical diagnosis for ever. J. J. Thomson considered, in 1897, that the negatively charged cathode ray was particulate and that the particle had a charge, later termed the electron. More significantly, the same particle could be found in different gases and different materials—indeed, a constituent of all matter. The duality of the electron in terms of its alternative waveform was later explored by Thomson's son, George. See also **Roentgen, Wilhelm Conrad (1845–1923)**.

X-rays employed in diagnostic imaging are polychromatic, i.e. they consist of a high-energy spectrum of electromagnetic radiation produced by accelerating electrons from the negative cathode. When an X-ray has more than 15 eV of energy it is referred to as ionizing radiation. On entering biological tissue, X-ray photons are attenuated by (1) the effective number of electrons, i.e. the mass per unit volume, and (2) the effective atomic number of the materials they encounter. Attenuation is due primarily tophotoelectric absorption of the lower energies in the spectrum and Comptonscattering of the higher energies. The photoelectric effect is highly dependenton atomic number (z^3). Added contrast agents usually contain high-atomic number elements, e.g. iodine, while lead provides a protective barrier. The final recorded image represents the interaction of emerging transmitted primary or secondary photons with an appropriate detector.

X-rays became the dominant imaging tool in the first half of the twentieth century, influenced by the improved care of both civilian and military injured. Between the two World Wars, film replaced glass plates, intensifying screens improved, exposures were measured in seconds and fractions of seconds rather than minutes and clinical applications became more numerous. Tomography, the ability to image "slices" of tissue, became a practical diagnostic tool in the 1930s. Until the 1950s most developments had been concerned with the links between the radiation source and the detector, i.e. film or fluorescent screen, but it then became possible to exploit the relationship between the detector and the observer. Image intensification madedark-adaptation unnecessary and allowed the observer to exercise normal visual acuity and contrast discrimination. The introduction of percutaneous catheterization of blood vessels, by Seldinger in 1953, rapid filming devices and safer contrast media enabled rapid advances to be made. The addition of closed circuit television brought imaging into full daylight and provided the opportunities and environment for detailed interventional procedures under imaging control. See also **X-Ray**.

Computed Tomography

In April 1972 the first clinical results were obtained from EMI engineer Hounsfield's invention, computed tomography (CT). Despite previous developments in planar X-ray imaging, the basic physics involved in the production of a two-dimensional (2D) image from a three-dimensional (3D) patient had, up to that time, remained the same. CT requires an X-ray source, sensitive detectors, a means of moving the source and detectors around the circumference of a cross-section, and a computer to calculate the X-ray transmission measurements across the section. A grey-scale picture resembling a radiograph can then be produced by digital-to-analogue conversion using a standardized scale of Hounsfield units. The original "EMI scanner" required a water bath for reference, thereby limiting the technique to examination of the head. The geometry of that first system, rotation and translation, provided sections 1–2 cm in thickness on an 80×80 matrix and required several minutes to acquire and process the data. The cross-sectional images nevertheless revealed, for the first time, the brain and intracranial disease. The effect on diagnostic medicine and surgery was dramatic. The technique liberated, as it were, the brain of both patient and observer. The main advantages of CT compared with conventional X-ray imaging are a more efficient use of the transmitted X-rays, a significantly improved dynamic range, the ability to separate very thin cross-sections of tissue, and the opportunity to express, manipulate and quantify the entire content of an *in vivo* image. Modern scanners now employ thousands of highly efficient detectors and a variety of geometries, which have led to today's subsecond, very thin section (1 mm), high-resolution body scans. See also **Computed Tomography**; **Hounsfield, Godfrey Newbold (1919–2004)**.

Helical or spiral systems using slip-ring technology together with continuous patient-table incrementation can provide many sections for a given volume with great speed. An alternative fast system employing an electron gun and a fixed ring of detectors requires no moving parts. Such technologies allow rapid reformatting of digital data in alternative planes, reconstruction of 3D images and the potential for virtual endoscopy. Perhaps more importantly, they provide improved opportunities of studying function in addition to morphology.

Radio isotope Imaging

The roots of nuclear medicine using emergent gamma rays from radioactive sources go back to Becquerel's discovery of natural radioactivity in 1896, and the Curies, who isolated polonium and radium and introduced the term "radioactivity." Radioactive decay results in the production of alpha rays (ionized atoms of helium) and beta rays (electrons) with the release of energy in the form of electromagnetic radiation later termed gamma rays. It was 1914 before Soddy and Rutherford recognized that X-rays and gamma rays had similar properties, the only difference being their wavelength. See also **Rutherford, Ernest (1871–1937)**; and **Soddy, Frederick (1877–1956)**.

In nuclear medicine, images are produced by detection of gamma radiation, from an appropriately labeled pharmaceutical. The choice of pharmaceutical, which may be introduced into the patient by a variety of methods, including ingestion, injection and inhalation, is largely governed by the concentration that can be achieved in the target organ in relation to adjacenttissue. The principal objective is to achieve a high chemical resolution to enable evaluation of the target organ's functional capacity. Practical clinical scanning using radionuclides began only in the early 1960s, when most studies were performed using technetium-99 m agents. Conventional radionuclide imaging is performed with a gamma camera, which consists of a large scintillation crystal viewed by an array of photomultiplier tubes. The digital data so acquired can be displayed as an analogue image or subjected to computer processing. Single photon emission tomography (SPECT), by rotation of the gamma camera and the detector array, can provide the data necessary to reconstruct a sectional image.

In positron emission tomography (PET), isotopes that emit positive electrons, or positrons (emitted by beta decay), are attached to analogues of naturally occurring substances which can be introduced into a patient. The high-energy gamma rays, produced when the positrons collide with electrons in the organ of interest, are measured by a coincidence counting system. Isotopes that emit positrons have short half-lives (2–100 min), enabling many biochemical events taking place within the body to be measured. PET has permitted major advances to be made in the assessment of cardiac and cerebral perfusion, tumor metabolism and receptor ligand systems. See also **Positron Emission Tomography (PET)**.

Ultrasound

High-frequency mechanical vibrations produced by a piezo crystal using an electric field were appreciated by Pierre Curie 15 years before Roentgen's discovery of X-rays and were used by Langevin in the early 1900s to detect submarines. Ultrasound in the frequency range of 2–10 mHz can be employed in transmission or echo modes. Changes in acoustic impedance result in reflection or back scattering. Detected echoes can be displayed in one-dimensional format (A-mode or motion M-mode) or a two-dimensional format (B-mode). Current clinical practice relies on the detection and analysis of reflected echoes using a transducer which acts as both transmitter and receptor. Ultrasound does not use ionizing radiation, is highly mobile and is relatively inexpensive. It is, however, operator dependent. Probes of differing frequencies are available for selection by the operator. The selection is made according to the ability of the transducer to scan, with optimum resolution, at a specific tissue depth. The higher the frequency, the less the tissue is penetrated. At 10 mHz, for example, the focus is limited to 4 centimeter (1.5 inch), a depth suitable for examination of the breast. Real-time ultrasound is now a routine imaging procedure in a variety of clinical circumstances, ranging from evaluation of the fetus in pregnancy to intraoperative and laparoscopic diagnosis and the guidance of tissue sampling procedures. Endoscopic and intravacular probes are also currently employed. The application of the Doppler effect, i.e. the change in sound produced by a shift in frequency, can, because it is both directional and quantitative, be used to display and to measure blood flow. See also **Curie, Pierre (1859–1906)**; and **Ultrasound**.

Magnetic Resonance Imaging

The fact that certain atomic nuclei might behave like magnets by virtue of their spin and associated electric charge had been postulated by Pauli in 1924, and the possibility of a resonance method for detecting such magnetic moments was discussed by several workers in the 1930s. The first demonstrations of nuclear magnetic resonance (NMR) were achieved simultaneously by Bloch and Purcell in 1946. The phenomenon, whereby nuclei in a magnetic field are caused to resonate by the application of a specific radiofrequency (RF) radiation, became an important laboratory tool. When the source of the RF radiation is switched off, the nuclei continue to resonate, emitting an RF radiation which can be detected. Only after some 40 years did the technique become a nonionizing imaging facility for medicine. The concept of magnetic resonance (MR) was introduced in 1973 by Lauterbur, who argued that, if a gradient magnetic field were to be applied to a structured object, each nucleus would respond with its own frequency, determined by its position. Gradient fields in three orthogonal directions enable sufficient digital information to be acquired for the production of a sectional image.

Not all nuclei resonate. The most suitable and abundant nucleus in the body is hydrogen, occurring in both water and fat. The term "proton imaging" is often used to describe routine MR imaging. The first live human proton images were produced in 1976 by Mansfield and Maudsley. Echo planar MR is a rapid (60-ms) method of data acquisition used to obtain multiple signals within encoded spatial information.

MR provides the facilities to image in any plane, manipulate contrast, measure body composition and study biological phenomena *in vivo*, all without ionizing radiation. The multiplicity of MR parameters, including proton density, relaxation times, paramagnetism, chemical shift, diffusion, perfusion and flow, which can all be influenced by the type and timing of the RF pulse sequence, has provided unprecedented opportunities to explore morphology, pathology, physiology and biochemistry. Blood flow, for example, can be quantified in terms of velocity, acceleration and turbulence, and also be used to record the distribution of blood vessels without addition of contrast agents (magnetic resonance angiography, MRA). See also **Magnetic Resonance Imaging (MRI)**; and **Nuclear Magnetic Resonance (NMR)**.

Fundamental Limits

All medical images are representations of the distribution of some property of the human body which shows the structure and/or function of the organ or tissue under investigation. The imaging process has two components: signal detection and display. The performance of imaging systems can be assessed in quantitative terms that describe the transfer characteristics and the system noise properties. The quality of the detected image depends on the design of the imaging system, the efficiency of data capture, the available display facilities and the complexity of the diagnostic task. The interpretation of image data is presently dependent on the human observer, whose performance can be assessed by appropriate psychophysical methods, for example the application of receiver operating characteristic (ROC) curves.

The most important limiting factor in the use of ionizing radiation is radiation dose. The roentgen (R) as a measure of the quantity of X-radiation, i.e. exposure, was introduced in 1928 and defined as the charge ofions produced per unit mass of air, based on measurements with air-filled ionization chambers. The current definition of the roentgen applies to both X- and gamma radiation. The unit of absorbed dose, i.e. the energy deposited per unit mass of any material is the rad (*r*adiation *a*bsorbed *d*ose), equal to 0.01 J kg^{-1}. The rad was introduced in 1953 and later replaced in 1975, from the International System of Units (SI), by the gray (Gy), equal to 1 J kg^{-1} (1 Gy = 100 rad = 100 cGy). The dose equivalent (H) is the product of the absorbed dose and a quality factor which characterizes the specific radiation. H was expressed in rems, now replaced by the SI unit, the sievert (Sv). The unit of radioactivity, i.e. the number of atoms undergoing radioactive decay per unit time, the curie (Ci), has also been superseded by the SI unit, the becquerel (Bq); 1 Bq is approximately equal to 2.7×10^{-11} Ci.

Manmade radiation constitutes only 15% of the collective dose equivalent; the remaining 85% is derived from natural background radiation. Of the manmade radiation received, 90% is provided by X-rays and 4% by nuclear medicine techniques. The biological effects of ionizing radiation can be stochastic or deterministic. The principal stochastic effects are genetic and carcinogenic, whereas the deterministic effects are somatic, varying according to the severity and duration of exposure and the radiosensitivityof the tissue irradiated. Hematopoietic tissue, the lens of the eye and the gonads are particularly vulnerable. Conventional diagnostic imaging withionizing radiation should not give rise to a somatic hazard. Any theoretical long-term genetic or carcinogenic risk, sometimes extrapolated from atom bomb survivors exposed to very large doses of ionizing (1–10 Gy) and other radiations, must be weighed against the known benefits resulting from clinical diagnosis. For example, a dose of 1 mGy to the breast during mammography may, after a 10-year period, be calculated to cause one excess breast cancer death per year per 6 000 000 women examined–a risk equivalent to smoking an eighth of a cigarette or simply existing for 3 min at the age of 60. The risks from ionizing radiation indiagnostic imaging have diminished significantly in recent years by the application of strict codes of practice. These require justification for the procedure, optimization of protection based on the principle that all dosesshould be as low as reasonably achievable (ALARA), better quality assuranceand better education. The effective dose from a chest X-ray may only be 0.04 mSv but CT can provide much higher doses. The maximum permissible dose for an operator is 5 mSv whole body effective dose equivalent, i.e. a tenth of the annual dose limit (50 mSv) for an occupationally exposed person. A major piece of legislation in the European Union, the Council Directive "Health protection of individuals against the dangers ofionizing radiation in relation to medical exposures" ((MED) 97/43), was adopted in national legislation from May 2000. Since the largest fraction ofimaging procedures worldwide is carried out with ionizing radiation, any reduction in patient and operator dose is clearly desirable.

Ultrasound is a nonionizing radiation and generally considered to be noninvasive. There is, however, a wide range of exposure conditions from conventional 2D imaging scanners. Attenuation by intervening tissue may be significantly reduced, for example during examination of the potentially sensitive ovary by the transvaginal approach. Power levels in duplex and 3D Doppler flow imaging are significantly higher than those used in conventional ultrasound imaging. Some pulsed Doppler transducers may emit relativelyhigh intensity waves, 100 mW cm^{-2} or even higher. Concern has also been expressed about the safety of ultrasound contrast agents and the effect they may have on cavitation in tissue.

There is no known evidence of detrimental effects, short- or long-term, to humans subjected to clinical MR imaging techniques, provided certain guidelines are adopted. The limits are set by the potential adverse effects of the static magnetic field, the induced currents due to applied gradient fields, i.e. low frequency time varying magnetic fields, the heating effects from RF pulses on vulnerable tissues, e.g. the lens of the eye, and acoustic noise.

The unit of magnetic flux density is the tesla (T). Most conventional MR large-bore systems for body imaging operate at field strengths at or below 2.5 T but limbs can be imaged at 4.0 T. A designated Controlled Area should contain the 0.5 mT (5 gauss) magnetic field contour, as some physical hazards can arise in strong magnetic fields, e.g. cardiac pacemakers may malfunction in static fields above 0.5 mT, metallic implants and operative clips may be displaced and the projectile effect of ferromagnetic material can be dangerous. Nonferromagnetic patient monitoring and resuscitation equipment are required in an Inner Controlled Area containing the 3 mT (30 gauss) magnetic field contour. Gradient fields used to localize RF signals in three dimensions may, for reasons of speed, have to be switched rapidly. Gradient field strengths are usually about 10 mT m^{-1}. In echo planar MRimaging 20–30 mT may be used. Exposure to RF fields should be such that whole body temperature rise is limited to 0.5–1.0 °C (33–33.8 °F). Hearing protection is advisable when acoustic noise exposure rises to 85 dB or above.

Contrast agents to highlight both structural and functional change inbody systems during X-ray procedures have been explored since 1896. The introduction of contrast agents for MR and US has provided further opportunities to influence diagnosis and clinical management. It is estimated that some 100 million patients per year will be subjected to vascular administration of contrast media. Approximately 1 in 40,000 patients die from adversereactions in these circumstances.

Future Prospects

The future prospects for diagnostic imaging include improvements in the demonstration of normal and disordered function. The molecular diagnosis of gene mutations and the targeting of enzymes for cancer therapy by nuclear medicine or MR imaging techniques will, for example, have profound effects on medicine. There is an increasingly important role for imaging in the planning, guidance, monitoring and control of gene therapy, interventional radiological procedures and minimally invasive therapy (MIT). It is estimated that in 10 years time 70% of surgical procedures will be conducted by MIT, with the advantages of reduction in pain, morbidity and hospital stay. Real-time images and the full perception of 3D space are the keys to the future of MIT.

The advent of MR-guided therapy employing novel magnet designs and instrumentation with position-sensing devices can allow the operator's hands to remain in the imaging field without any of the hazards of ionizing radiation.

Imaging development in the 1980s and 1990s was dominated by new technologies and improvements in existing hardware. From 2000 and beyond, developments will be influenced more by increased computing speeds and advances in software. Potential applications are likely to include reasoning with uncertainty or "fuzzy logic," allowing a computer to proceed with incomplete information, object-orientated modeling and, perhaps most importantly, with the development of neural networks, image understanding by machine. Future hospital design will have to take into account changes in patient handling, shorter hospital stays and the need, for both financial and practicalreasons, to locate imaging and treatment facilities centrally.

Multimedia systems incorporating X-rays, microthin CT and MR sectionstogether with color photography provide the facility, electronically, to dissect and reassemble anatomical specimens. The technology for teleconferencing, telemedicine and the transmission of digital images in local or distant locations, static or mobile, is presently available. With sufficient virtual reality information to enable the operator to feel physically present at a remote site, telesurgery under imaging control should become a real possibility. The opportunities for anatomical, medical and surgical trainingby these means are clear.

Only twice in history, in Ancient Greece and the Renaissance, have systematic attempts been made to approximate an image to reality. Perhaps medical imaging of the twenty-first century will come to be regarded as a third such period.

IAN ISHERWOOD, University of Manchester, Manchester, UK

IMBRICATE STRUCTURE.   The type of compound, low-angle (almost horizontal) thrust faults, which produce mechanical piles similar in arrangement to slates on a roof. This type of compound faulting consists of the lower thrust plane or sole, the intervening multiple thrusts (imbricate) and the overlying thrust plane or thrust proper. Imbricate structure is particularly characteristic of the North West Highlands of Scotland.

IMIDES.   An imide may be defined as a compound that has the divalent radical NH combined with two acid radicals. The definition implies that the acid from which an imide is derived must be a dibasic acid, such as oxalic acid [CAS: 144-62-7], HOOCCOOH, or succinic acid [CAS: 110-15-6], HOOCCH$_2$CH$_2$COOH. The derivatives of these two acids illustrate the relationship between amides and imides.

Phthalimide [CAS: 85-41-6], C$_6$H$_4$(CO)$_2$NH, is an imide of commercial and industrial importance, forming a number of interesting derivatives. With alcoholic potash, phthalimide forms a potassium derivative, C$_6$H$_4$(CO)$_2$ NK, which, when reacted with ethyl iodide (or other alkyl halides), yields ethylphthalimide [CAS: 5022-29-7], C$_6$H$_4$(CO)$_2$ N C$_2$H$_5$. The latter product, when hydrolyzed with an acid or alkali, further yields ethylamine [CAS: 75-04-7] C$_2$H$_7$N. Such reaction chains are useful in the preparation of certain primary amines and their derivatives.

IMINO COMPOUNDS.   Imino compounds are organic compounds containing the imino group ⟩NH e.g., dimethylamine, (CH$_3$)$_2$NH, dibenzamide, (C$_6$H$_5$CO)$_2$NH, succinimide,

NH, purrole (C$_4$H$_4$NH), and uric acid,

IMMUNE SYSTEM AND IMMUNOLOGY.   The word *immunity* is derived from the Latin *immunis* (free of). The term originally referred to the ability of the body to resist invasion by pathogenic organisms, but has now been expanded to include specific reactions to antigens (Ags) in general, and to include reactions observed in the emerging field of tumor immunology. See also **Antigen**.

Immunity is derived from the *immune system* which, when functioning properly, protects the organism from infection. Failures of the immune system produce some of the most challenging and serious diseases that a physician can meet in the patient population.

The immune system consists of a number of lymphoid organs, including the thymus, lymph nodes, spleen, and tonsils. It also includes aggregates of lymphoid tissue in nonlymphoid organs, such as Peyer's patches in the intestines and clusters of lymphoid tissue dispersed throughout the connective and epithelial tissues of the body. The immunologically active cells of the immune system comprise the various classes of lymphocytes. A number of cells, however, including monocytes (macrophages) and polymorphonuclear leucocytes play important accessory roles. The stem cells from which the lymphocytes arise are derived from the yolk sac and the fetal liver, later some stem cells originate from the bone marrow and differentiate into lymphocytes in the primary lymphoid organs.

The function of the immune system is the preservation of the body's integrity against antigens recognized by the lymphocytes as foreign, e.g., surface structures of microorganisms, tissue transplants, or a wide variety of chemicals. Specifically, antigens include such structurally diverse substances as proteins, polysaccharides, nucleic acids, and lipids. Large, rigid proteins are the most antigenic, and the more insoluble the foreign material, the more antigenic it appears.

The various antigens are recognized by lymphocytes, which have a memory and specificity, an ability to increase the number of antigen-specific lymphocytes following the antigenic stimulus, and an ability to distinguish between self and nonself. The production of immunoglobulins (Igs — see later) by the immune response is under the control of genes which are located in the same chromosome as, and very close to, another group of genes which control the production of the histocompatibility antigens (HLA). These are antigens that identify as self the cells they are on and differentiate from cells of other individuals. These two groups of genes form the major histocompatibility complex (MHC) which plays a crucial role in the immune system. An intriguing aspect of the genetic control of immunoglobulin synthesis is the diversity of the product; plasma cells can make antibodies that react with more than a million antigens.

As previously indicated, the primary cells involved in the immune response are lymphocytes which have a centrally located round nucleus, lack specific granules, and have a basophilic cytoplasm containing free ribosomes. The (thymus-dependent) T-lymphocytes are involved in cell mediated reactions and also interact with B-lymphocytes (see later) to regulate the production of antibody. The B cells differentiate into the antibody-producing plasma cells. There is growing evidence that neither T nor B cells constitute a homogeneous population, but actually consist of a number of subgroups which can be differentiated from each other by their surface markers and by their function.

Thymus-dependent antigens are those in which antibody production requires thymus-derived (T) cell participation, i.e., serum proteins. Thymus-independent antigens do not require this participation, i.e., polysaccharides, such as endotoxins.

Antigen is any substance capable of generating an immune response that is reacting with T and B cells to induce the formation of antibodies and sensitized lymphocytes and then reacting with those antibodies and cells once they are formed. The basis for the general immunogenicity of proteins is not known, but is probably related to their unique and stable configuration. Antigens invoke immune responses by the host which include the production of *antibodies* (Abs) possessing a specificity for the antigen which is determined by the latter's structure. See also **Antibody**. Antibodies belong to a group of *immunoglobulins* (Igs) which bind with the antigens to form complexes in which the two components are held together by weak hydrogen bonds, van der Waals forces, or ionic bonds, but not by covalent bonding. An antibody that binds a given antigen will also bind antigens having similar structural configurations. This is referred to as *cross-reactivity*. The extent to which this binding occurs indicates the measure of similarity of the two antigens. Most antigens have several antigenic determinants or Ab binding sites and the antibody response to any antigen is thus the sum of responses to each individual determinant.

In addition to antigens per se, two other types of substance are recognized by the immune system: (1) *haptens*, which are molecules capable of reacting with antibodies, but which are unable to stimulate their production unless coupled to a carrier — an immunogenic substance that is usually a protein or a synthetic polypeptide; and (2) *adjuvants*, which enhance the immune response to an antigen. These include, but are not limited to, aluminum salts, bacterial endotoxins, *Bacillus Calmette-Guérin* (BCG), *Bordetella pertussis*, and mycobacteria.

The immune response to antigens can proceed through several paths, the two major ones being *humoral* and *cellular*. Most immune responses involve both pathways. Humoral immunity is mediated by antigen-specific antibodies, which circulate through the body; they act at a site distant from that of their production; it can be transferred from one person to another by serum transfusion. Cell-mediated immunity is mediated by specifically sensitized cells, which release mediators in the vicinity of the antigen. This form of immunity can be transferred from one person to another by cell transfer. Cell-mediated immune responses usually take longer to develop and are responsible for resistance to many infectious agents and tumors as well as for some drug allergies, rejection of foreign organ grafts, and some autoimmune diseases.

The level of immune response to antigens varies. A *primary response* is seen to antigens the body has never before encountered. In this, the first

antibody to be produced is IgM (described later), the serum concentration of which peaks after a lag of several days and then decreases. IgG production shows a longer lag time, but its concentration remains elevated for a more extended period. *Secondary responses* are more rapid and of greater intensity; they differ markedly from primary responses in having a shorter lag period, higher serum antibody levels with earlier and more pronounced emphasis upon IgG production. These secondary responses occur when the immune system faces a previously encountered antigen and illustrates the basic characteristic of immune response — *memory*. For this reason, secondary responses are also known as *anamnestic*. These memorized responses are derived from two of the major cell types which work together to produce immune responses — T and B lymphocytes. The former are thymus-dependent, are responsible for cell-mediated immune responses and for providing help for most antibody responses. B cells depend upon another central lymphoid organ (in birds, this has been identified as the Bursa of Fabricus). There does not appear to be a bursa equivalent in mammals, however, even though it was originally thought that such animals had some gut-associated primary lymphoid organ. Current evidence suggests that stem cells can differentiate into B cells in the bone marrow and in the peripheral lymphoid organs themselves. B lymphocytes have immunoglobulin on their surface and it can be shown that the cell itself has produced this. When stimulated by antigens, B cells become *plasma cells* and produce antibodies specific for that antigen. T and B cells cannot be distinguished morphologically, but may be differentiated, such as by theta antigen on T cells and surface immunoglobulin on B cells. Unlike T cells, B cells tend to migrate.

The spleen, lymph nodes, tonsils, and gut-associated lymphoid tissue comprise the secondary or peripheral system wherein T and B lymphocytes undergo terminal differentiation in response to antigen stimulation.

The third major type of cell involved in the immune process is the *macrophage*. The B and T lymphocytes are rarely phagocytic — this is the function of the macrophage. After specific recognition of the invading antigen by the lymphocytes, the macrophage acts nonspecifically, and it moves by *chemotaxis* (movement along a chemical concentration of increasing gradient) to the site of the immune response. The macrophage ingests and eventually digests the antigenic inert particle or the living or dead microorganism responsible for engendering the immune response.

These major cell types may be subdivided into populations that interact by means of soluble mediators called *lymphokines* or *monokines*. Lymphokines are products of activated lymphocytes that exert regulatory effects upon other cells of the immune system. Monokines are products of activated microphages. The lymphokines or lymphocyte mediators are soluble substances produced by lymphocytes that help to amplify and regulate a variety of immune responses. They are not immunoglobulins. There are clearly many different lymphokines and they are classified on the basis of the target cells they affect. It is likely that they have more than one biological function. Lymphokines are generally synthesized and secreted by sensitized lymphocytes stimulated by a specific antigen; they are not stored in a preformed state. Most are proteins or glycoproteins, and while it has been shown that protein synthesis inhibitors prevent their formation, their exact structures are undetermined. Some play a role in humoral reactions and others in cell-mediated responses. The soluble mediators allow such activities as help, suppression, or cytotoxicity to be manifested by the target cells. See also **Lymphokines**.

A given antigen induces proliferation and differentiation of clones of cells capable of producing antibodies in response to that antigen; a process called *clonal selection*. Each stimulated cell produces antibodies of only one specificity. Thus, the immense heterogenicity of antibodies to a given antigen results from a great diversity of responsive cells.

As previously indicated, antibodies are immunoglobulins (Igs), produced by B-cell-derived plasma cells during a humoral response to antigen. They are synthesized on polyribosomes attached to the rough endoplasmic reticulum of plasma cells upon inoculation of antigen into the host, and they are specific for that antigen. They have two functions — specific recognition of the antigen and effector functions, such as agglutination or lysis of bacteria, complement fixation, or opsonization.

Immunoglobulins are high-molecular-weight glycoproteins having symmetrical four-polypeptide chain structures composed of two heavy and two light chains held in configuration by disulfide linkages. Based upon serological characteristics of the heavy chains, five distinct classes or isotypes have been found — IgG, IgA, IgM, IgD, and IgE. Subclasses of

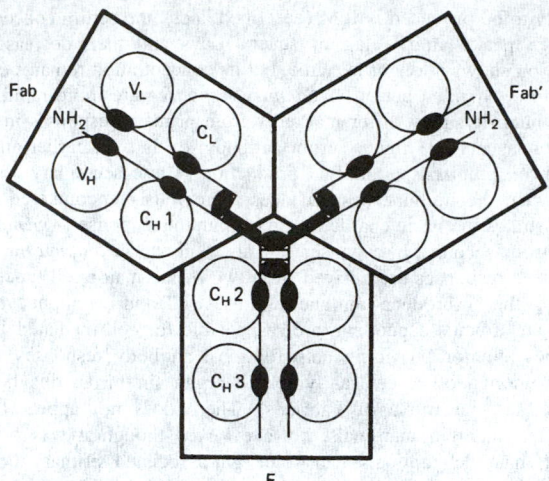

Fig. 1. Hypothetical model of human IgG along the lines proposed by Poljak and others. Note heavy solid bars, which indicate hinge region (at center of configuration). Solid ellipses indicate S–S linkages.

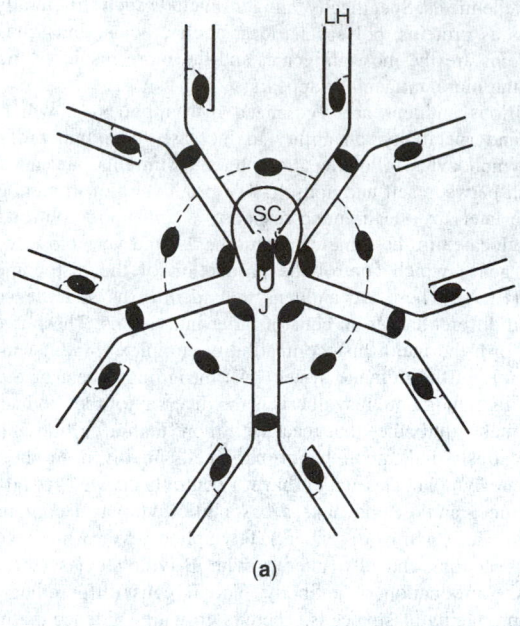

(a)

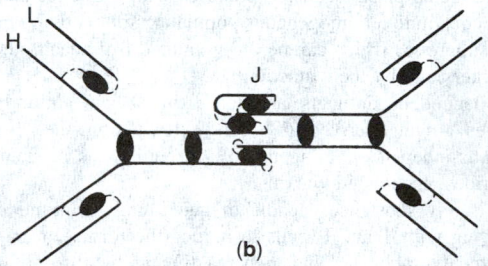

(b)

Fig. 2. Hypothetical models utilizing the clasp configuration proposed by Koshland, et al. (1975). (a) Pentameric IgM; (b) dimeric IgA. L = light chain; H = heavy chain; J = J chain. Solid ellipses indicate S–S linkages.

these isotypes have also been found. Bonds between chains—heavy-heavy or heavy-light—are by disulfide bridges with the number and positions of these being characteristic of different classes and subclasses of immunoglobulins. See Fig. 1. In a given immunoglobulin molecule, all of the heavy chains are identical and, although there are two types of light chains—kappa and lambda—each of which may be associated with any heavy chain, both of the light chains in a given Ig are identical. Polymerization of the basic Ig configuration can occur, but IgG, IgD, and IgE occur only as monomers. IgA may be either monomeric or dimeric; IgM is pentameric. See Fig. 2.

In addition to the basic configurations of the immunoglobulin molecule, the light and heavy chains have variable and constant regions where some variability in amino acid sequences occurs. The variable regions of the antibody molecule contain the structures responsible for the antigenic specificity of the Ig. These are found in the amino terminal of the heavy and light polypeptide chains—the Fab portion obtained as a cleavage fragment following treatment with pepsin or papain. Since each antigen-binding region is composed of the variable region of one light and one heavy chain, an immunoglobulin molecule has two identical antigen binding sites. The antigen-binding site of every Ig molecule is structurally unique. Therefore, each Ig has unique antigenic determinants not shared by any other Ig molecule. These unique determinants are called the *idiotypes* of the particular immunoglobulin. On the other hand, an *allotype* is an antigenic specificity representing any one of a number of possible allelic markers present on a significant percentage of Ig molecules in an individual which are recognizable as foreign by another individual lacking that allotype. Allotypic determinants distinguish between the Igs of a particular isotype and are genetically determined by Mendelian laws in a manner similar to those that determine the ABO blood group. *Isotypic* determinants are found in the sera of normal individuals and differentiate the various classes and subclasses of heavy and light chains.

The F$_c$ portion of the immunoglobulin molecule identified by papain cleavage is responsible for biological activity other than antigen binding, e.g., complement fixation, transplacental transfer, binding to cells such as macrophages and granulocytes, and the rate of synthesis and catabolism of the Ig molecule.

IgG and IgM are present in the highest concentrations throughout the body and compose the major portion of a systemic humoral response to antigenic challenge. IgM exists as a pentamer of five immunoglobulin molecules bound together by a protein molecule called a J chain. IgM has a molecular weight of approximately 900,000 and a sedimentation coefficient of 19S, and is thus referred to as 19S Ig. It is the main immunoglobulin of early humoral response, will fix complement and is not cytophilic. IgG is the major component of secondary humoral response, and its ability to diffuse through tissues makes it indispensable for host defense. It has a molecular weight of 150,000 and a sedimentation coefficient of 7S. Eighty percent of serum immunoglobulin is IgG and several subclasses of this

immunoglobulin exist which have different functions, i.e., complement fixation, binding to macrophages or passage across the placenta.

IgA is the primary secretory immunoglobulin and occurs in tears, nasal and intestinal secretions, saliva, and bile. It is present as a 7S IgA, it is an 11S dimer with the monomeric structures joined by a J chain and linked to a glycoprotein called secretory component. Secretory IgA plays a role in initial protection against external pathogens, aggregates bacteria, and neutralizes viruses.

IgE, with a molecular weight of 190,000, has a short half-life and occurs at low concentrations in the serum. It is cytophilic and binds strongly to most cells and basophils. Plasma cells producing IgE are primarily found in gastrointestinal mucosa and along the respiratory tract. This immunoglobulin is responsible for immediate *hypersensitivity reactions*, since crosslinking by antigens of two IgE molecules causes the release of mediators such as histamine, SRS-A (Slow Reacting Substance of Anaphylaxis), and ECF-A (Eosinophil Chemotactic Factor) which are responsible for the immediate hypersensitivity reactions. Hence, IgE plays a definitive role in allergies and may have a part in providing resistance to parasites and in protection of mucosal surfaces.

IgD occurs in extremely low concentrations in serum and has a short serum half-life. It has a molecular weight of 170,000–200,000, is monomeric, and fixes complement. Otherwise its role is unknown; it may act as a cell surface receptor on B lymphocytes.

Associated with humoral responses involving antibodies is *complement*. The *complement system* is a complex system of 18 plasma proteins circulating in inactive form in the extracellular liquid. It is the principal humoral effector of immunologically induced inflammation. It plays a crucial role, both in immunologically induced and nonspecific resistance to infection and in the pathogenesis of tissue injury. The products of complement activity regulate a number of biological events including the release of mediators from mast cells, which increases vascular permeability. Activation of complement involves a series of steps, each one generating a new active component that, in turn, activates the next component and so on. There are two major pathways of complement activation—the classical pathway and the alternate or properdin pathway. The classical pathway is initiated by the binding of antigen to antibody (two molecules of IgG or one molecule of IgM). The alternate pathway does not require the presence of antigen-specific antibody, but can be activated by endotoxin, bacterial cell-wall polysaccharides, and aggregated immunoglobulins. The ability to be initiated before an immune response can occur makes the alternate pathway a first line of defense against microorganisms. The ultimate result of complement activation is lysis of the invading cell. Mediators produced during the activation sequence play major roles by allowing *immunoadherence, opsonization*, or *chemotaxis* to occur, but others, such as *anaphylatoxins* may be harmful to the host. The importance of complement to normal host defense is illustrated by many pathologic conditions and high susceptibility to infections seen in persons with congenital complement deficiencies.

Additional Reading

Abbas, A.K. and A.H. Lichtman: *Basic Immunology: The Function of the Immune System*, W.B. Saunders Company, Philadelphia, PA, 2001.

Angell, M.: "A Dual Approach to the AIDS Epidemic," *N. Eng. J. Med.*, 1498 (May 23, 1991).

Austin, K.F., S.J. Burakoff and T.B. Storm: *Therapeutic Immunology*, 2nd Edition, Blackwell Science, Inc., Malden, MA, 2000.

Balter, M.: "East Europe: A Chance to Stop HIV," *Science*, 1964 (December 24, 1993).

Bayer, R.: "Public Health Policy and the AIDS Epidemic," *N. Eng. J. Med.*, 1500 (May 23, 1991).

Chapel, H. and M. Haeney: *Essentials of Clinical Immunology*, Blackwell Science, Inc., Malden, MA, 1993.

Coe-Clough, R. and J.A. Roth: *Understanding Immunology*, Mosby-Year Book, Inc., St. Louis, MO, 1998.

Cohen, J.: "Aids Vaccine Research," *Science*, 1820 (December 17, 1993).

Cohen, J.: "T Cell Shift: Key to AIDS Therapy?" *Science*, 175 (October 8, 1993).

Cohen, I. and L.A. Segal: *Design Principles for the Immune System and Other Distributed Autonomous Systems*, Oxford University Press, Inc., New York, NY, 2001.

DasGupta, D.: *Artificial Immune Systems and Their Applications*, Springer-Verlag, Inc., New York, NY, 1998.

Descotes, J. and C. Bernard: *An Introduction to Immunotoxicology*, Taylor & Francis, Inc., Philadelphia, PA, 1999.

Diamond, J.: "The Mysterious Origin of AIDS," *Natural History*, 24 (September 1992).

Fauci, A.S.: "Multifactorial Nature of Human Immunodeficiency Virus Disease: Implications for Therapy," *Science*, 1011 (November 12, 1993).

Frank, M.M. and J.E. Volanakis: *The Human Complement System in Health Disease*, Marvel Dekker, Inc., New York, NY, 1998.

Greene, W.C.: "AIDS and the Immune System," *Sci. Amer.*, 98 (September 1993).

Janeway, C.A., Jr.: "How the Immune System Recognizes Invaders," *Sci. Amer.*, 772 (September 1993).

Jenkins, M.K.: "The Role of Cell Division in the Induction of Clonal Anergy," *Immunology Today*, 69 (February 1992).

Kuby, J.: *Immunology*, 3rd Edition, W.H. Freeman Company, New York, NY, 1999.

Leffell, M.S., M.R. Rose and A.D. Donnenberg: *Handbook of Human Immunology*, CRC Press, LLC, Boca Raton, FL, 1997.

Lichtenstein, L.M.: "Allergy and the Immune System," *Sci. Amer.*, 116 (September 1993).

Marrack, P. and J.W. Kappler: "How the Immune System Recognizes the Body," *Sci. Amer.*, 80 (September 1993).

Marx, J.: "Cell Communication Failure Leads to Immune Disorder," *Science*, 896 (February 12, 1993).

Nossal, G.J.V.: "Life, Death and the Immune System," *Sci. Amer.*, 52 (September 1993).

Passwater, R.A. and S. Davis: *Beta-Carotene and Other Carotenoids: The Antioxidant Family That Protects against Cnacer and Heart Disease and Strengthens the Immune System*, Keats Publishing, Inc., Chicago, IL, 1999.

Patterson, P.H., C. Kordon and Y. Christen: *Neuro-Immune Interactions in Neurologic and Psychiatric Disorders*, Springer-Verlag, Inc., New York, NY, 1999.

Paul, W.E.: "Infectious Diseases and the Immune System," *Sci. Amer.*, 90 (September 1993).

Paul, W. (Editor): *Fundamental Immunology*, Lippincott-Raven Publishers, Philadelphia, PA, 1998.

Roitt, I.M.: *Roitt's Essential Immunology*, Blackwell Science, Inc. Malden, MA, 1997.

Rothwell, N.J.: *Immune Responses in the Nervous System*, Oxford University Press, Inc., New York, NY, 1998.

Schwartz, R.H.: "A Cell Culture Model for T Lymphocyte Clonal Anergy," *Science*, 1349 (June 15, 1990).

Schwartz, R.H.: "Costimulation of T Lymphocytes: The Role of CD28, CTLA-4 and B7/BBi in Interleukin-2 Production and Immunotherapy," *Cell*, 1065 (December 24, 1992).

Schwartz, R.H.: "Immunologic Tolerance," in *Fundamental Immunology*, 3rd Edition, Raven Press, New York, NY, 1993.

Schwartz, R.H.: "T Cell Anergy," *Sci. Amer.*, 62 (August 1993).

Sheehan, C.: *Clinical Immunology: Principles and Laboratory Diagnosis*, 2nd Edition, Lippincott Williams & Wilkins, Philadelphia, PA, 1997.

Staff: *Signaling and Gene Expression in the Immune System*, Vol. 64, Cold Spring Harbor Laboratory, Cold Spring, NY, 2000.

Steinman, L.: "Autoimmune Disease," *Sci. Amer.*, 106 (September 1993).

Stine, G.: *AIDS: Update 2000*, Prentice Hall, Inc., Upper Saddle River, NJ, 1999.

Storad, C.: *Inside AIDS: HIV Attacks the Immune System*, The Lerner Publishing Group, North Minneapolis, MN, 1998.

Toufu, Z., et al.: "Genotypic and Phenotypic Characterization of HIV-1 in Patients with Primary Infection," *Science*, 1179 (August 27, 1993).

Weissman, I.L. and M.D. Cooper: "How the Immune System Develops," *Sci. Amer.*, 64 (September 1993).

Wigzell, H.: "The Immune System as a Therapeutic Agent," *Sci. Amer.*, 126 (September 1993).

ANN C. DeBALDO, Ph.D., College of Public Health,
University of South Florida, Tampa, FL

IMMUNOASSAY. Immunoassay is the term used to denote any technique employed for the quantification of an analyte that depends upon the reaction between an antigen and its complementary antibody. The nature of this reaction defines the unique characteristics of an assay. An antigen is so called because of its ability to induce an immune response when injected into an animal, in which it is treated as a foreign species, the production of an antibody being a natural defense mechanism. Small molecules, known as haptens, will not inherently produce an immune response, being removed from the body by other means; however, if a hapten is coupled to a protein then it is possible to produce antibodies that will bind to portions of the hapten molecule. See also **Antibody**; and **Antigen**.

Immunoassay therefore depends on a natural phenomenon that generates a biological recognition molecule–the antibody. An immunoassay may also be designed for the quantification of an antibody–the antigen then providing the recognition molecule. It is the nature of the reaction between the antigen and its complementary antibody that defines the unique characteristics of an assay.

IMMUNODEFICIENCY. Immunodeficiency is a condition caused by one or more immune system defects and is characterized clinically by increased susceptibility to infections with consequent severe, acute, recurrent or chronic disease. An immunodeficiency disorder should be considered in anyone with infections that are unusually frequent, severe and resistant; without a symptom-free interval; from an unusual organism; or with unexpected or severe complications. Immunodeficiencies may be either primary or secondary.

Primary Immunodeficiencies

Primary immunodeficiencies include a variety of disorders that render patients more susceptible to infections. If left untreated, these infections may be fatal. The disorders constitute a spectrum of more than 80 innate defects in the body's immune system. Primary immunodeficiencies generally are considered to be relatively uncommon. There may be as many as 500,000 cases in the United States, of which about 50,000 cases are diagnosed each year.

The primary immunodeficiencies are classified into four main groups depending on which component of the immune system is deficient: B cells, T cells, phagocytic cells, or the complement cascade. See also **Lymphocytes**.

Of the primary immunodeficiencies, B cell-associated antibody defect-spredominate; selective IgA deficiency (usually asymptomatic) may occur in 1in 400 people. Excluding asymptomatic IgA deficiency, B-cell defects still account for 50% of the primary immunodeficiencies, T-cell defects for about 30%, phagocytic deficiencies for 18%, and complement deficiencies for 2%. T-cell defects include several disorders with associated B-cell

TABLE 1. PRIMARY IMMUNODEFICIENCY DISORDERS[a]

Disorder	Associated findings
B-cell (antibody) deficiencies	
X-linked agammaglobulinaemia (Bruton)	B lymphocytes absent
Hyperimmunoglobulin M syndrome (XL)	Neutropenia, lymphadenopathy; absence of CD40 ligand on T cells
Immunoglobulin (Ig) A deficiency	Autoimmunity; respiratory or food allergy; respiratory infection
IgG subclass deficiencies	IgA deficiency; recurrent respiratory infections
Common variable immunodeficiency	Autoimmunity; B cells present
Transient hypogammaglobulinemia of infancy	Prematurity; recurrent infections
T-cell (cellular) deficiencies	
Predominant T-cell deficiency	
DiGeorge anomaly	Hypocalcaemia, peculiar facies, aortic arch and heart abnormalities
Chronic mucocutaneous candidiasis	Endocrinopathies; bacterial infections
Nezelof syndrome	Bronchiectasis; increased IgE levels
Nucleoside phosphorylase deficiency (AR)	Severe viral infection; decreased T cells
Combined T- and B-cell deficiencies	
Severe combined immunodeficiency (AR or XL)	Failure to thrive, diarrhea
Adenosine deaminase deficiency (AR)	Skeletal abnormalities
Bare lymphocyte syndrome	Absence of HLA antigens on lymphocytes
Ataxia–telangiectasia (AR)	Dermatitis, neurological deterioration
Wiskott–Aldrich syndrome (XL)	Eczema, thrombocytopenia
Short-limbed dwarfism	Cartilage–hair hypoplasia
XL lymphoproliferative syndrome	Epstein–Barr virus infection
Phagocytic disorders	
Defects of cell movement	
Hyperimmunoglobulinemia E syndrome	Eczema, dermatitis
Leucocyte adhesion defects type I (AR)	Prolonged attachment of umbilical cord, leucocytosis, periodontitis
Defects of microbicidal activity	
Chronic granulomatous disease (XL or AR)	Lymphadenopathy, perianal abscesses
Myeloperoxidase deficiency (AR)	Often asymptomatic
Chediak–Higashi syndrome (AR)	Oculocutaneous albinism, giant granules in leucocytes, neuropathy
Complement disorders	
C1, C4, C2 deficiency	Systemic lupus erythematosus-like syndrome, glomerulonephritis
C3 deficiency	Pyogenic infections
C5	Bacterial infections
C6, C7, C8, C9	*Neisseria* infection
C1 inhibitor deficiency (AD)	Angioedema, systemic lupus erythematosus
Factor H deficiency (ACD)	Hemolytic–uremic syndrome, glomerulonephritis
Factor D deficiency (ACD)	Pyogenic infections
Properin deficiency (XL)	*Neisseria* infections

[a] ACD, autosomal codominant; AD, autosomal dominant; AR, autosomal recessive; Ig, immunoglobulin; XL, X linked.

(antibody) defects, which is understandable since B and T cells originate from a common precursor stem cell and, in addition, T cells influence B-cell function. Phagocytic diseases include disorders in which the primary defect is one of cell movement (chemotaxis) and those in which the primary defect is one of microbicidal activity. A classification of primary immunodeficiencies is shown in Table 1.

Secondary Immunodeficiencies

Secondary immunodeficiency is an impairment of the immune system resulting from an infection, medications or malignancy in a previously normal person. The impairment is often reversible if the underlying condition or illness resolves. Secondary immunodeficiencies are considerably more common than primary immunodeficiencies and occur in many hospitalized patients. Nearly every prolonged serious illness interferes with the immune system to some degree. A classification of the secondary immunodeficiencies is shown in Table 2.

Diagnosis of Immunodeficiency

The most common manifestation of immunodeficiency is frequent infections, usually beginning with recurrent respiratory infections. Most immunodeficient patients eventually develop severe bacterial infections that persist, recur or lead to complications (e.g. sinusitis, chronic otitis and bronchitis often follow repeated episodes of sore throat). Bronchitis may progress to pneumonia, bronchiectasis and respiratory failure, the most common

cause of death in these patients. Infections with opportunistic organisms (e.g. *Pneumocystis carinii* or *Cytomegalovirus*) may occur, particularly in patients with T-cell deficiencies.

Infection of the skin and mucous membranes also is common. Resistant thrush (oral candidial infection) may be the first sign of T-cell immunodeficiency. Oral ulcers and periodontitis also are noted, particularly in granulocytic disorders. Other common symptoms include diarrhea, malabsorption and failure to thrive. The diarrhea may be noninfectious or associated with *Giardia lamblia, Rotavirus, Cytomegalovirus* or *Cryptosporidium*. In some patients, the diarrhea may be exudative with loss of serum proteins and lymphocytes. Less common manifestations of immunodeficiency include hematological abnormalities (autoimmune hemolytic anaemia, leucopenia, thrombocytopenia), autoimmune disorder (vasculitis, arthritis, endocrinopathies)and central nervous system disease (chronic encephalitis, slow development, seizures).

Associated History. If there is a family history of early death, similar disease, autoimmune illness, allergy, early malignancy or consanguinity, then a pedigree chart will help to identify a hereditary pattern.

The age of onset also may help in diagnosis. Patients with T-cell disorders usually present at under 6 months of age. Onset of illness at around6 months of age, when transplacentally acquired maternal antibodies have disappeared, suggests congenital antibody deficiency. Patients with phagocytic disorders often develop symptoms in infancy but sometimes are not diagnosed until later in childhood.

TABLE 2. SECONDARY IMMUNODEFICIENCY DISORDERS

Predisposing factors	Specific factors
Premature and newborn infants	Physiological immunodeficiency due to immaturity of the immune system
Hereditary and metabolic diseases	Chromosome abnormalities (e.g. Down syndrome)
	Uremia
	Diabetes mellitus
	Malnutrition, vitamin and mineral deficiency
	Protein-losing enteropathies
	Nephrotic syndrome
	Myotonic dystrophy
	Sickle cell disease
Immunosuppressive agents	Radiation
	Immunosuppressive drugs, corticosteroids
	Antilymphocyte or antithymocyte globulin
	Anti-T-cell monoclonal antibodies
Infectious	Congenital infections (rubella)
	Viruses (measles, varicella, human immunodeficiency virus, cytomegalovirus, Epstein–Barr virus)
	Acute bacterial disease
	Severe mycobacterial or fungal disease
Infiltrative and hematological	Histiocytosis
	Sarcoidosis
	Hodgkin disease and lymphoma
	Leukemia, myeloma
	Agranulocytosis and aplastic anaemia
Surgery and trauma	Burns
	Splenectomy
	Anesthesia
Miscellaneous	Systemic lupus erythematosus
	Chronic active hepatitis
	Alcoholic cirrhosis
	Ageing
	Anticonvulsant drugs
Graft-versus-host disease	

Physical Findings. Patients with immunodeficiency often appear chronically ill, with pallor, malaise, malnutrition and a distended abdomen. Rashes, vesicles, pyoderma, eczema, petechiae, alopecia or telangiectasia may appear on the skin, and conjunctivitis is common. Cervical lymph nodes and adenoid and tonsillar tissue are typically absent in some B- or T-cell immunodeficiencies, despite a history of recurrent throat infections. However, the lymph nodes may be enlarged and draining in patients with phagocytic defects. The tympanic membranes often are scarred or perforated, and the nostrils may be excoriated and crusted, indicative of purulent nasal discharge. Postnasal drip and nocturnal cough suggest chronic sinusitis. Rales and wheezes are often present on auscultation of the lungs. The liver and spleen are frequently enlarged. Muscle mass and fat deposits of the buttocks are diminished. In infants, there may be excoriation around the anus as a result of chronic diarrhea or a candidial nappy rash. Neurological examination may reveal delayed developmental milestones or ataxia.

A characteristic constellation of findings permits a tentative clinical diagnosis in a number of immunodeficiency syndromes: newborns with DiGeorge anomaly typically exhibit infections, hypocalcaemia, peculiar facies and congenital heart disease; boys with Wiskott–Aldrich syndrome typically suffer from pyogenic infections, eczema and bleeding manifestations; children with ataxia–telangiectasia develop recurrent sinopulmonary infections, ataxia and telangiectasia in early childhood; and patients with hyper immunoglobulin E syndrome develop severe eczema and severe pulmonary infections.

Laboratory Screening Studies. Screening tests for immunodeficiency can be performed in most offices and include a complete blood count (CBC) with differential and platelet count; determination of IgG, IgM and IgA levels; assessment of specific antibody function; and evaluation for infections with appropriate cultures.

The CBC will establish the presence of anaemia, thrombocytopenia, lymphopenia, neutropenia or leucocytosis. A total lymphocyte count of less than 1500 per pL is suggestive of T-cell immunodeficiency. The peripheral blood smear should be examined for the presence of Howell–Jolly bodies and other unusual red blood cell forms suggestive of asplenia or poor splenic function. The granulocytes may show morphological abnormalities (e.g. granules of the Chediak–Higashi syndrome).

Immunoglobulin levels also are part of the initial screen, and values must be interpreted with care because of marked alterations with age; all infants aged 2–6 months are hypogammaglobulinemic by adult standards. A total IgG level of less than 400 mg dL with normal screening functional antibody test results usually excludes antibody deficiency. A total immunoglobulin level of less than 200 mg dL^{-1} usually indicates significant antibody deficiency. IgM function may be estimated by isoagglutinin titres (antiA and/or antiB). Antibodies to these and certain bacterial polysaccharides are selectively deficient in certain disorder (e.g. Wiskott–Aldrich syndrome, IgG2 subclass deficiency). In the immunized patient, specific antibody titres to *Hemophilus influenzae* type B, hepatitis B, rubella virus, tetanus or diphtheria antigens can be used to estimate IgG function. An adequate antibody level to one or more of these antigens or a higher-titre antibody response following reimmunization is evidence against antibody deficiency. Finally, screening should include a search for chronic infection. The sedimentation rate often is raised, usually in proportion to the degree of infection. Appropriate radiographs (chest, sinus) and cultures should be obtained.

If the results of all these screening tests are normal, immunodeficiency (particularly antibody deficiency) can usually be excluded. However, if chronic infection is documented, if the history is unusually suspicious or if the results of screening tests are positive, advanced tests must be performed.

Tests for B-cell (antibody) Deficiency. If immunoglobulin levels are very low (total less than 200 mg dL^{-1}), a diagnosis of antibody deficiency is established and other procedures are indicated only to define the exact illness and to identify other immunological defects. If immunoglobulin levels and preexisting antibody titres are low but not absent, the antibody responses to one or more standardized antigens should be assessed. Antibody titres are obtained before and 3–4 weeks after immunization with tetanus toxoid or diphtheria vaccine to test for protein antigen responsiveness, or after immunization with pneumococcal or meningococcal vaccine to test for polysaccharide antigen responsiveness. Responsiveness to polysaccharide antigens, however, does not usually occur until 2 years of age. An inadequate response (less than a 4-fold rise in titre) is suggestive of antibody deficiency regardless of total immunoglobulin levels.

If immunoglobulin levels are low, B-cell enumeration is performed by assessing the percentage of lymphocytes reacting with fluoresceinated antibodies to B cell-specific antigens (e.g. CD19, CD20) as assessed by flow cytometry. Normally, 10–20% of peripheral blood lymphocytes are B cells.

Next, serum levels of IgG subclasses should be obtained. IgG subclass determinations are indicated if IgG levels are normal or near normal but antibody function is deficient. Selective deficiencies of one of the four subclasses may be present. High and low levels of IgD and IgE may occur in some antibody deficiency syndromes. IgE levels may be high in chemotactic disorders, partial T-cell immunodeficiencies, allergic disorders and parasitism. See also **Allergy**.

Other laboratory tests for B-cell deficiencies are indicated in certain circumstances. A lymph node biopsy (sometimes preceded by immunization in the adjacent extremity) is indicated in the presence of lymphadenopathy to exclude malignancy or infection. If rapid IgG catabolism or IgG loss through the skin or gastrointestinal tract is suspected, an IgG survival study may be indicated. In illnesses in which the genetic defect has been identified, the mutant gene or mutant gene product can be identified (e.g. Bruton tyrosine kinase gene in X-linked agammaglobulinaemia; CD40 ligand in hyper-IgM syndrome) by special laboratory testing.

Tests for T-cell Deficiency. Profound and prolonged lymphopenia usually suggests a T-cell immunodeficiency. Chest radiography is a useful screening test in infants. An absent thymic shadow in the newborn period suggests T-cell deficiency, particularly if the radiograph is obtained before the onset of infection or other stress that may shrink the thymus. See also **Thymus Gland.**

Delayed hypersensitivity skin tests are valuable screening tests after the age of 2 years. The following antigens are used: mumps, *Candida*(1:100), fluid tetanus toxoid (1:10) and *Trichophyton*. Nearly all adults and most immunized infants and children will react to one or more of these antigens with erythema and induration (greater than 5 mm) at 48 h. The presence of one or more positive delayed skin test results generally confirms an intact T-cell system.

The most valuable advanced test in cellular immunodeficiency is T-cell and T-subset (helper/inducer and suppressor/cytotoxic) enumeration, usually performed by flow cytometry using T cell-specific monoclonal murine antibodies. A T helper cell (CD4) count lower than 500 cells per pL is highly suggestive of a T-cell immunodeficiency, and a CD4 count below 200 cells perpL indicates a profound T-cell immunodeficiency. The ratio of CD4/CD8 (helper/suppressor) cells should be greater than 1.0; reversal of this ratio also suggests T-cell immunodeficiency (e.g. in acquired immune deficiency syndrome (AIDS), a decline in the CD4/CD8 ratio indicates progressive immunological impairment). Monoclonal antibodies also are available to identify activated cells (human leucocyte antigen (HLA)-DR, CD25), natural killer cells (CD16 and CD56) and immature T-cell (thymocyte) antigens (CD1). See also **Acquired Immune Deficiency Syndrome (AIDS)**.

Another useful advanced test measures the ability of the patient's lymphocytes to proliferate and enlarge (transform) when cultured in the presence of mitogens, irradiated allogeneic mononuclear cells (in the mixed leucocyte reaction) or antigens to which the patient has been exposed previously. With these stimuli, normal lymphocytes undergo rapid division, which can be assessed morphologically or by uptake of radioactive thymidine into dividing cells. Patients with T-cell immunodeficiency have low or absent proliferative responses in proportion to the degree of immune impairment. See also **Lymphocytes**.

Special tests also assess lymphokine production after mitogen or antigen stimulation. Different types of cytotoxicity (natural killer, antibody-dependent or cytotoxic T cell) are measured using different tumor cell or virus-infected target cells. In some forms of combined immunodeficiency, enzymes of the purine pathway (adenosine deaminase, nucleoside phosphorylase) are deficient and can be assayed using the patient's erythrocytes. HLA typing can be valuable for assessing the presence of two populations of cells (chimerism) due to cells from the mother or from a blood transfusion, and for excluding deficiencies of HLA antigens (bare lymphocyte syndrome).

Tests for Phagocytic Cell Deficiency. An investigation is indicated when a patient with a convincing history of immunodeficiency has normal B- and T-cell immunity. Recurrent staphylococcal infections, perianal abscesses and delayed umbilical cord detachmentwith marked leucocytosis are suggestive of a phagocytic defect.

Initial screening with a CBC may reveal neutropenia but serial blood counts (at twice-weekly intervals) may be necessary to rule out cyclic neutropenia. Other testing may include determination of IgE concentration, which is raised in the hyperIgE syndrome. A nitroblue tetrazolium (NBT) dye reduction test will test for chronic granulomatous disease (CGD), the most common phagocytic disorder. The NBT test is based on the increased oxidative burst activity of granulocytes following activation with reduction of colorless NBT to blue formazan due to the release of oxygen radicals. This color change can be assessed visually, microscopically or by spectrophotometry. Carriers of the X-linked form of CGD will have a partial response. Granulocytes can also be tested for the presence or absence of myeloperoxidase by special staining techniques.

A chemotactic abnormality can be assessed by an *in vitro* assay in which migration of granulocytes or monocytes is measured, using a special chemotactic chamber (Boyden) or an agarose plate; cell movement toward a chemoattractant (complement fragments, chemotactic peptide) is assessed.

Next, phagocytosis is tested by measuring uptake of latex particles or bacteria by isolated granulocytes or monocytes. Microbial killing is thenassessed by mixing the patient's granulocytes in fresh serum with a known number of live bacteria, followed by serial quantitative bacterial assays over a 2-h period.

Other specialized tests define phagocytic defects: assays of granulocyte mobilization after administering corticosteroids, adrenaline (epinephrine) or endotoxin; assays for granulocyte oxidant products (hydrogen peroxide, superoxide); and assays for specific granulocyte adhesions proteins suchas CD11/CD18.

Tests for Complement Deficiency. A complement abnormality is screened by measuring the total serum complement activity (CH50) and serum C3 and C4 levels. Low levels of any of these should be followed by titration of the classical and alternative complement pathways and the measurement of individual complement components. Monospecific antisera or sensitized erythrocytes and solutions that contain allcomponents except for the one to be assessed are used to measure complement components.

Antisera also are available to measure complement control proteins. Deficiency of C1 inhibitor is associated with hereditary angioedema, and deficiency of factor I (C3 inhibitor) is associated with C3 deficiency with C3hypercatabolism. Assays of serum opsonic activity, serum chemotactic activity or serum bactericidal activity can be used to test complement function indirectly.

Infections in Immunodeficient Patients

Recurrent infections are the primary feature of immunodeficiency diseases. Opportunistic infections are caused by organisms often present in theenvironment that usually do not cause significant infections in patients with intact immune systems, such as *Pneumocystis* and some fungi. These opportunistic infections usually occur in patients with primary T-cell deficiencies or AIDS. However, the more common pathogens, such as *Staphylococcus*, are also opportunistic as they may result in life-threatening infections in these immunodeficient patients.

The type of infection may suggest the nature of the immunodeficiency. Infections with major Gram-positive organisms (pneumococci, streptococci) are noted in antibody (B cell) immunodeficiencies. Severe infections from viruses, fungi and other opportunistic organisms are common in cellular (T cell) immunodeficiencies. Recurrent staphylococcal and Gram-negative infections are common in phagocytic deficiencies. Recurrent *Neisseria* infection is characteristic in patients with complement component deficiencies. Table 3 lists pathogens associated with specific immunodeficiency diseases.

Treatment of Immunodeficiency

General management of patients with immunodeficiency requires an extraordinary amount of care to maintain optimal health and nutrition, to manage infections, to prevent emotional problems related to the illness, and to cope with costs. Patients should be protected from unnecessary exposure to infection, should sleep in their own beds, and preferably have their own rooms. Killed vaccines should be given regularly if there is evidence of someantibody function. The teeth should be kept in good repair.

Antibiotics are life saving for treating infections; selection and dosage are identical to those used normally. However, because immunodeficient-patients may succumb rapidly to infection, fever and other manifestations of infection are assumed to be secondary to bacterial infection, and antibiotic treatment is begun immediately. Throat, blood or other cultures are obtained before most therapies; these are especially useful subsequently when the infection does not respond to the initial antibiotic and when the infectious organism is unusual.

Continuous prophylactic antibiotics often are beneficial, particularly when there is the risk of sudden overwhelming infection (e.g. Wiskott–Aldrich syndrome, asplenic syndromes); when other forms of immune therapy are unavailable (e.g. in phagocytic disorders) or insufficient (e.g. recurrent infection in agammaglobulinaemia despite immunoglobulin (IG) therapy); and when there is a high risk for a specific infection (e.g. *P. carinii* in cellular immunodeficiency disorders). Prophylactic therapy with antifungal agents, such as itraconazole, have been used in some patients with phagocytic disorders (e.g. CGD and hyperimmunoglobulin E syndrome).

Antivirals, including amantadine or rimantidine for influenza, aciclovir for herpes infection (including varicella zoster) and ribavirin for respiratory syncytial virus, may be life saving in immunodeficient patients with viral infections.

IG is effective replacement therapy in most forms of antibody deficiency. The largest intramuscular dose at one site is 10 mL in adults and 5 mL in children; accordingly, multiple injections at various sites may be necessary. High doses of intravenous IG ($400–800$ mg kg^{-1} month^{-1}) can be given and are beneficial to some antibody-deficient patients not responding well

TABLE 3. INFECTIONS IN IMMUNODEFICIENT PATIENTS

Disorder	Pathogens
Phagocytic disorders	
Congenital or cyclic neutropenia	Bacteria, Gram-positive and Gram-negative
Leucocyte adhesion deficiency	Fungus (*Candida*, *Aspergillus*)
Chediak–Higashi syndrome	*Staphylococcus aureus* is predominant pathogen with increased susceptibility to other bacteria and fungi as in neutropenia
Hyperimmunoglobulin E syndrome	*S. aureus* is predominant pathogen
	Other Gram-positive and Gram-negative bacteria
	Fungi (*Candida albicans*, *Aspergillus*)
Chronic granulomatous disease	Bacteria, catalase-positive including *S. aureus*, *Serratia marcescens*, *Salmonella*, *Burkholdia cepacia*, *Nocardia*
	Fungus (*Aspergillus*)
Myeloperoxidase deficiency	*C. albicans*
B-cell (antibody) disorder	
X-linked agammaglobulinaemia (Bruton)	Bacteria, Gram-positive and Gram-negative
	Viruses (echo, coxsackie, adenovirus)
	Parasites (*Giardia*)
Common variable immunodeficiency	Similar to Bruton disease, plus *Cryptosporidium*
Hyperimmunoglobulin M syndrome	Similar to Bruton disease, plus opportunistic organisms
Immunoglobulin A deficiency	Common viral and bacterial respiratory pathogens
	Giardia
Immunoglobulin G subclass deficiency	Encapsulated bacteria (*Streptococcus pneumoniae*, *Haemophilus influenzae*) .
Transient hypogammaglobulinemia of infancy	Common bacterial respiratory pathogens
T-cell and severe combined immune deficiency disorders	
Di George syndrome (highly variable with regard to degree of immunodeficiency) and severe combined immune deficiency (often more severe than DiGeorge syndrome)	*Candida* species (thrush common)
	Bacteria (common and opportunistic) including *S. pneumoniae*, *Pseudomonas aeruginosa*, *Mycobacteria* species
	Viruses (*Herpes simplex virus*, *Cytomegalovirus*, *Varicella-zoster virus*)
	Protozoa (*Pneumocystis carinii*, *Crytosporidium*)
Wiskott–Aldrich syndrome	Bacteria (encapsulated) including *H. influenzae*, *S. pneumoniae*
	Candida species
	Viruses (*Cytomegalovirus*, *Herpes simplex virus*, *Varicella-zoster virus*, *Epstein–Barr virus*, virus causing molluscum contagiosum)
Ataxia telangiectasia	Bacteria (more common pathogens)
Chronic mucocutaneous candidiasis	Fungi (*Candida*, *Histoplasma*)
	Bacteria, (*S. aureus*, *S. pneumoniae*, *H. influenzae*, *Nocardia*)
	Viruses (*Herpes simplex virus*, *Varicella-zoster virus*)
	Giardia
Complement disorders	
C1, C4, C2, C3	Bacteria (encapsulated) including, *S. pneumoniae*, *H. influenzae*
C5, C6, C7, C8, C9	*Neisseria meningitidis* (meningitis)
	Disseminated gonococcal infection
Properidin	*Neisseria meningitidis*

to conventional doses, particularly those with chronic lung disease. The aim with high-dose intravenous IG is to keep IgG trough levels in the normal range (i.e. greater than 500 mg dL^{-1}). Slow subcutaneous infusions of 10% IG given at weekly intervals has also been needed to deliver high-dose immunoglobulin therapy (i.e. more than 400 mg kg^{-1} month^{-1}1) in patients with adverse reactions to intravenous infusions or poor venous access.

Other therapies, including immunologically enhancing drugs (isoprinosine), biological agents (transfer factor, interleukins) and hormones (thymic), have been of limited value in treating cellular or phagocytic immunodeficiencies, although interferon γ is the only cytokine specifically approved for treatment of a primary immunodeficiency (e.g. CGD). Enzyme replacement with bovine adenosine deaminase conjugated to polyethylene glycol (PEG-ADA) has benefited patients with adenosine deaminase deficiency. See also **Interferons**.

Stem cell transplantation can often achieve complete correction of immunodeficiency. In severe combined immune deficiency and its variants, bonemarrow transplantation from an HLA-identical, mixed leucocyte culture-matched sibling has resulted in restored immunity in over 300 cases. In patientswith intact or partial cellular immunodeficiency (e.g. Wiskott–Aldrich syndrome), prior immunosuppression must be given to ensure engraftment. When a matched sibling donor is unavailable, haploidentical (half-matched) bone marrow from a parent can be used. Under these circumstances, mature T lymphocytes that will cause graft-versushost disease must be removed from the parenteral marrow before its administration. Alternatively, bone marrow from a matched but unrelated person identified through the International Bone Marrow Transplant Registry can be used. It is also possible to harvest CD34 stem cells from the peripheral blood of donors by leucopheresis followed by isolation of CD34 cells. Umbilical cord blood can also be used as a source of stem cells, from a HLA-matched sibling or banked HLA-compatible cord blood.

Fetal thymus transplants, thymic epithelial cell transplants and fetal liver transplants succeed occasionally, particularly fetal thymus transplants in DiGeorge anomaly.

Additional Reading

Chapel, H., M. Haeney, S. Misbah, and N. Snowden: *Essentials of Clinical Immunology*, 5th Edition, Blackwell Publishing Ltd., Malden, MA, 2006.

Magee, D.J., F.A. Oski, D. Ginsburg, S.H. Orkin, and A.T. Look: *Hematology of Infancy and Childhood*, 6th Edition, Elsevier Science, New York, NY, 2003.

Mak, T.W., and M.E. Saunders: *Elsevier Science & Technology Books*, New York, NY, 2005.

Rose, N.R., and I.R. Mackay: *The Autoimmune Diseases*, 4th Edition, Elsevier Science & Technology Books, New York, NY, 2006.

Stiehm, E.R., J.A. Winkelstein, and H.D. Ochs: *Immunologic Disorders in Infants and Children*, 5th Edition, Elsevier Health Sciences, New York, NY, 2004.

Todd, I., and G. Spickett: *Immunology*, 5th Edition, Blackwell Publishers, Malden, MA. 2005.

Virella, G.: *Medical Immunology*, 6th Edition, CRC Press, LLC, Boca Raton, FL, 2007.

ROBERT L. ROBERTS
RICHARD E. STIEHM
University of California at Los Angeles, Los Angeles, CA

IMMUNOLOGY (THE HISTORY).

Immunity to disease has long been recognized, but only during the late nineteenth century, thanks to Louis Pasteur, Robert Koch, Elie Metchnikoff and Paul Ehrlich, did immunology become a true science able to explain mechanisms and to develop preventive vaccines. Originally medically oriented (1880–1910), the field passed through a stage of chemical domination (1910–1960) and then became biologically and, once again, medically oriented.

Early Theories of Acquired Immunity

From the earliest of humankind's social organizations, epidemic disease must have been a frequent visitor; both the annals of the oldest dynasties of Egypt as well as the Babylonian Epic of Gilgamesh record visitations of disease and pestilence. However, most early societies and even many modern primitive peoples have held that both humans and nature are ruled by the magical influences of spirits and demons, or the mystical influences of the gods. It was thus natural to assume that disease represents a punishment for some infraction of a tribal taboo, or some sin against the gods. The pantheons of many cultures contain a god of disease, and throughout the Old Testament, God frequently smites those who trespass against Him, often employing pestilential disease. Even in Greek legend, the sun god, Phoebus Apollo, was held to have caused the plague of Thebes as punishment for the misdeeds of Oedipus Rex, and Apollo is supposed to have rained plague arrows on the Greek army before Troy, because Agamemnon had abducted the daughter of Apollo's priest.

While the cause of these various epidemics might be unknown, it couldnot fail to be noticed that those who had survived a disease might be spared further involvement on its return. Thucydides described this phenomenon in his history of the plague of Athens of 430 BC, when he wrote: "Yet it was with those who had recovered from the disease that the sick and the dyingfound most compassion. These knew what it was from experience, and had now no fear for themselves; for the same man was never attacked twice–never at least fatally." This "plague" was most probably not due to *Pasteurella pestis*, but the plague of Justinian some thousand years later was more likely to have been bubonic plague, and of this Procopius said, "At a later time it [the plague] came back; then those who dwelt round aboutthis land, whom formerly it had afflicted most sorely, it did not touch at all." In time, this resistance to reinfection came to be known by the term immunity, from the Latin *immunitas*, which in ancient Rome originally described the exemption of an individual from service or duty to the state.

The idea that disease might originate from a vengeful deity carries with it an implicit theory of immunity. If disease be considered as punishment for sin, then being spared during a raging epidemic (i.e. natural immunity) would automatically be viewed as the inevitable result of having led apious life. But a significant change occurred in early Christian times. Now, not only does God punish the sins of man with disease, but He might also employ disease to cleanse man of his sins. If disease can be viewed as an expiation and purgative, then the recovery from a deadly plague would notonly imply that one's sins had been minor but also that, once cleansed of these sins, one would not merit further punishment when the plague returned (i.e., acquired immunity).

It has only been during the last millennium that explicit theories of acquired immunity were advanced; these were invariably imaginative, and were all eminently consistent with the then-prevailing notion of disease pathogenesis. Since smallpox was one of the earliest diseases to be identified clinically, and because the lifelong immunity that it conferred could hardly escape notice, it is not surprising that most early theories of immunity would be formulated in terms of this disease.

Expulsion Theories of Acquired Immunity. The tenth-century Islamic physician Rhazes differentiated smallpox from measles and other exanthematous diseases for the first time. He recognized that recovery from smallpox infection provides lasting immunity, and advanced the first explicit theory of acquired immunity. Rhazes followed the Hippocratic tradition which held that disease is due to quantitative imbalances among the four humors, or to their fermentation. He claimed that smallpox is due to a fermentation of the blood, which helps to dispel the "excess moisture" that he thought was present in the blood of the young. Thus, the pustules that form on the skin during this disease and break to release fluid provided the presumed exit through which the body expels the excess moisture contained in the blood. He compared the maturation of an individual to the fermentation of wine from grape juice, wherein the blood, like grape juice, matures by progressively losing excess moisture; he even suggested that smallpox disease itself might assist in this normal process!

This theory seemed to explain well all that was known about smallpox: almost everyone is affected, especially during youth (since then the blood is most moist). Further, the disease is seldom seen in adults and almost never in old age (because by then the normal ageing process would have sufficiently dried the blood, so that it no longer could support the infection). Finally, a single infection induces lasting immunity, since recurrence of the disease is impossible because the initial attack would have expelled all of the "excess moisture" which the theory required as a prerequisite for the disease process. It is interesting that Rhazes presented the smallpox of the tenth century as an almost benign childhood disease, and even as a salutary phenomenon.

In his 1546 book *On Contagion*, Girolamo Fracastoro claimed that all disease is caused by small seeds or germs (*seminaria*) which may spread from person to person, each of which possesses a specific affinity for a given plant or animal, and for a given organ or humor. Fracastoro claimed that the germ of smallpox has an affinity for and causes the fermentation only of that trace of menstrual blood contaminant which he supposed taints all mammalian young *in utero*. When a (young) person was infected, then the menstrual contaminant would ferment, rise to the surface beneath the skin in the form of pustules, and be expelled when the pustules break. "This ebullition is a kind of purification of the blood ... That is why almost all of us suffer from this malady, ... and this fever is of itself seldom fatal (sic!), but is rather a purgation ... the malady usually does not recur because the infection has already been secreted in the previous attack."

Depletion Theories of Acquired Immunity. The introduction of variolation (the use of material from a diseased smallpox victim to immunize normal recipients) early in the eighteenth century led to renewed interest in the nature of acquired immunity. The inoculation of pustular fluid as a preventive appears to have been employed extensively in the folk medicine of many cultures. While it was condemned on both religious and medical grounds, it did attain a degree of acceptance, especially in England, thanks to the example set in 1722 by the Prince and Princess of Wales in permitting their children to be inoculated. Inoculation proved especially popular during periods of smallpox epidemic, when the casemortality rate often reached 15–20%; the rate of disfigurement was even higher. In contrast, inoculation protected well against reinfection, caused little or no facial scarring, and was accompanied by at most a 2–3% death rate. See also **Poxviruses**; and **Vaccination**.

When the practice of inoculation was given currency in the pages of the *Philosophical Transactions of the Royal Society*, some were led to speculate on its meaning. In 1721, the New England divine Cotton Mather advanced a theory of acquired immunity which held that some unidentified substrate in the blood is depleted either forcefully during natural infection or more benignly following inoculation; the absence of this material thenceforth prevents development of the disease a second time. It was in the context of three-quarters of a century of smallpox inoculation that Edward Jenner published in 1798 his report on a safer and even more efficacious vaccine against smallpox, derived from cowpox pustules. Jenner seems never to have speculated on why his vaccine caused immunity, perhaps influenced by the earlier advice of his teacher John Hunter: "Why think? Why not try the experiment?" See also **Jenner, Edward (1749–1823)**; and **Hunter, John (1728–1793)**.

There developed, during the seventeenth and eighteenth centuries, a most interesting concept of disease pathogenesis and thus of disease immunity, that of the innate seed. Humans (and animals) were thought to be born with the seeds or ovule for every different disease to which they were subject, each of which could be "fertilized" specifically by the appropriate contagious agent to produce the given disease. Here was not only a concept of specific aetiology, but in the depletion of the seeds of a given diseaseit offered a plausible explanation of acquired immunity that is at once specific and lasting.

The idea of an immunity based upon the depletion of some type of substance required by the disease process itself was repeated often during the eighteenth century. Thus, one finds statements in the contemporary literature such as one in 1755, "I lately tried this experiment [inoculation] upon myself, ... and it had no effect upon my blood, as it had been sufficiently defecated 15 years before." Again, susceptibility to smallpox was likened to a body which a single spark might set afire, but which thenceforth has become "incombustible" although surrounded by flames, and thus immune to further infection.

The Origins and Research Program of Early Immunology

Preventive Immunization. The 1870s saw increasing acceptance of the germ theory of disease and the identification of specific bacterial agents, thanks to the efforts of Louis Pasteur, Robert Koch, and others. The newer concepts of disease pathogenesis overthrew all earlier concepts of the mechanism of immunity. Immunology as a science was born in the laboratory of Louis Pasteur, whose earlier work on the agents responsible for certain diseases in the French silkworm and wine industries had convinced him that each disease is the reproducible result of an infection by a specific microorganism. In collaboration with Emile Roux, Pasteur devised techniques for the attenuation of cultures of virulent bacteria, working initially with the organism responsible for the disease chicken cholera. In one of those happy instances of serendipityin science, it was discovered that chickens that had recovered from a mild attack of chicken cholera induced by an attenuated strain were thenceforth protected from challenge with more lethal strains. See also **Koch, Heinrich Hermann Robert (1843–1910)**; **Pasteur, Louis (1822–1895)**; and **Roux, Pierre Paul Emile (1853–1933)**.

This 1880 report was the first generalization on Edward Jenner's use of cowpox vaccine to protect against smallpox, and opened up an entirely new research program of prophylactic immunization. Pasteur was quick to seize upon these possibilities, as his subsequent work on anthrax, rabies, and other diseases amply testifies. Over the next quarter-century, as new pathogens for different diseases were reported with increasing frequency, scientists throughout the world sought to develop the corresponding vaccines, using Pasteurian approaches. See also **Vaccination**; and **Vaccines**.

Cellular Immunity. The second significant addition to the immunological research program of the nineteenth century came in 1884 with Elie (Ilya) Metchnikoff's cellular theory of immunity. Based upon Darwinian evolutionary principles, Metchnikoff proposed that the primitive intracellular digestive functions of lower animals had persisted in the capacity of the mobile phagocytes of higher forms to ingest and digest foreign substances. Thus, he suggested that the phagocytic cell is the primary element in natural immunity (the first line of defense against infection), and critical also for acquired immunity. Another notable contribution of the phagocytic theory was to the field of general pathology. Most believed at the time that inflammation was a damaging component of the disease process itself. Metchnikoff disagreed, and suggested that the inflammatory response has in fact evolved to protect the organism. See also **Metchnikoff, Elie (Ilya) (1845–1916)**.

Serotherapy. In 1888, Emile Roux and Alexandre Yersin demonstrated that a soluble toxin could be isolated from the supernatants of cultures of the diphtheriaorganism; the toxin alone produces all of the symptoms of typical diphtheria in experimental animals. It did not take long for Emil Behring and Shibasaburo Kitasato to exploit this observation. They reported in 1890 that animals immunized with diphtheria and tetanus toxins produce something in their blood that can neutralize or destroy the toxin, thus preventing disease. Indeed, the antitoxic sera from experimental animals could be transferred to normal animals and even to children, to confer immunity and even to abort ongoing infection. Soon the more general and noncommittal term "antibody" was used to describe this new class of protective substances, while the material responsible for generating these antibodies came to be known as the "antigen." See also **Antibody**; **Antigen**; **Kitasato, Shibasaburo (1852–1931)**; **Von Behring, Emil Adolf (1854–1917)**; and **Yersin, Alexandre (1863–1943)**.

The cure of children with active diphtheria by means of these protective serums (called *serotherapy*) stimulated an explosion of laboratoryand clinical experimentation and of high expectation for its application toother diseases. In recognition of this important discovery, Behring received the first Nobel Prize in Physiology or Medicine, in 1901.

Cytotoxic Antibodies and Autoimmunity. The next significant area that occupied early immunologists emerged from the demonstration by Jules Bordet in 1898 that erythrocytes could be destroyed (hemolysed) by specific antibodies operating in conjunction with the serum factor complement. For the first time, the cells and tissues of the immunized host itself were seen possibly to be at risk by an "aberrant" immune response against self components. Soon, scientists everywhere beganto immunize experimental animals with suspensions or extracts of almost every tissue or organ in the body, in an attempt to find cytotoxic antibodies that might be responsible for one or another local disease. While it was quickly discovered that xenoantibodies (formed in another species) and isoantibodies (formed in another member of the same species) were readily formed and might be cytotoxic against the target tissue or organ, autoantibodies (formed in the host itself) were rarely produced and seemed never to cause disease. This led Paul Ehrlich to formulate his famous dictum of *horror autotoxicus*, which held that for reasons unknown, an individual is unable to mount a destructive immune response against self-constituents. But in1904, Julius Donath and Karl Landsteiner reported the observation of a trueautoimmune disease, paroxysmal cold hemoglobinuria, the first of many such observations to come. See also **Bordet, Jules Jean Baptiste Vincent (1870–1961)**; **Ehrlich, Paul (1854–1915)**; and **Landsteiiner, Karl (1868–1943)**.

Serology and Immunodiagnosis. Starting in 1896 with the discovery of bacterial agglutination, it was quickly recognized that bacteria could be identified and differentiated with the use of appropriate specific antisera. Even the serum of patients could be tested for their ability to agglutinate a given organism, thus determining whether prior exposure to that organism had occurred. The discovery of the antigen–antibody precipitin reaction extended this approach even further, to include the assay of antigens and antibodies in systems involving bacterial products, or even nonbacterial agents. When Bordet showed that immune antigen–antibody complexes fix complement, and that the degree of such fixation might be measured, a new approach was opened to the diagnosis of disease. This approach was exploited by August von Wassermann and his colleagues in the development of a serodiagnostic complement fixation test for syphilis, and soon complement fixation was adapted to many other systems. Not only could prior exposure to a given pathogen be assessed, but even the course of a disease might be followed serologically. See also **Immunoassay**; and **Syphilis**.

Allergy and Immunopathology. A seminal discovery was made in 1902 by physiologists Paul Portier and Charles Richet. Until then, the immune response had been viewed as a protective mechanism against exogenous pathogens and toxins; the work of those searching for cytotoxic antibodies had done little to alter this view. These investigators now demonstrated that even bland substances might cause severe systemic shock-like symptoms and even death, when injected into individuals previously exposed to the same material. They termed this phenomenonanaphylaxis, in an attempt to distinguish it from the usual prophylactic results expected of the immune system. Shortly thereafter, Maurice Arthus demonstrated that bland antigens can cause local necrotizing lesions when they react with specific antibody in the skin of test animals, the so-called *Arthus phenomenon*. The floodgates were now open. In 1906, Clemens von Pirquet and Bela Schick demonstrated that the pathogenesis of so-called *serum sickness* depends upon an antibody response by the host to the injection of large quantities of foreign protein antigens, such as accompanies the administration of horse antidiphtheria toxin according to Behring's serotherapeutic doctrine. Then it was demonstrated that two of the significant curses of humankind, hay fever and asthma, also belong to this group of specific antibody-mediated diseases. Here was a system that had presumably evolved for defensive functions that somehow "went astray" to produce a variety of pathological conditions. This teleological view of immunity was so deeply ingrained that for over half a century the mechanisms of allergy were treated as quite separate from those of immunity. Only in recent years has immunopathology been incorporated into the broader context of immunological phenomena. See also **Allergy; Asthma; Richet, Charles (1850–1935);** and **Von Behring, Emil Adolf (1854–1917)**.

Immunohematology. The initial demonstration that antibodies directed against erythrocytes might mediate their agglutination and hemolysis focused early attention on these cells as antigens. It was soon found that many animal and even human sera contain "natural" isoantibodies capable of agglutinating the erythrocytes of certain other members of the same species. In 1901, Karl Landsteiner showed that humans could be divided into several groups, depending on the presence in their sera of agglutinins specific for the erythrocytes of other humans. These observations served as the basis for the ABO system of blood groups, important for the typing of blood for transfusion; for this discovery, Landsteiner was awarded the Nobel Prize in Physiology or Medicine for 1930. Landsteiner followed up this observation in the 1920s by discovering with Philip Levine the M, N and P blood groups, and in 1940 he and Alexander Wiener discovered the rhesus factor, important in both blood transfusions and as the principal contributor to the transplacental disease of the newborn, erythroblastosis fetalis. Since that time, many other minor erythrocyte antigens have been identified, and immunohematology has contributed significantly to theoretical immunology, to forensic medicine, and to anthropological studies of racial relationships and mass migrations. See also **Landsteiner, Karl (1868–1943)**.

Ehrlich's Side-chain Theory. The term antibody was originally a noncommittal one, employed to designate whatever it was in immune serum that had the capacity to neutralize toxins and pathogenic bacteria. Behring's work on serotherapy made it apparent that antibody must be a discrete substance somehow formed within the immune host; the mechanism of its formation thus became a valid topic for speculation and study. Early on, it was suggested that antigen itself carried the information for antibody specificity by somehow being incorporated into the antibody molecule. This theory could not long survive the quantitative studies that showed that much more antibody was formed than could be accounted for by the quantity of antigen injected, and that antibody formation, once started, would continue for long periods without further administration of antigen. See also **Antigen**.

In 1897, Paul Ehrlich advanced a comprehensive theory of antibody formation. He suggested that antibodies are naturally occurring products of the body that function as specific receptors on the surface of the cell, there to fulfill normal physiological functions. Ehrlich believed that the specificity for antigen depends upon the presence of certain stereochemical configurations (side-chains) whose complementarity to analogous structures on the antigen permit specific interaction. The selection of these antibody receptors by the appropriate antigen would stimulate the cell to produce more of the same, appearing in the blood as circulating antibody.

Ehrlich's imaginative theory held great sway for many years. Few were troubled at the time by any hint that the potential size of the immunological repertoire of antibody specificities presented any problem, since the only antibodies known in the mid-1890s were thought to be antitoxins directed against a rather limited number of human and animal pathogens. However, two observations were made over the succeeding decades to cast doubt on Ehrlich's theory. The first was the increasing demonstration that antibodies could be produced against a wide variety of naturally occurring and even innocuous animal and plant substances, including many to which the host species would normally never be exposed. Then, in the second decade of the twentieth century came the observations that showed that antibodies can be formed against almost any artificial chemical capable of being coupled to a protein carrier. These observations made it appear unreasonable that evolution would have endowed the individual with the capacity to make specific antibodies against so great a number of foreign and even nonthreatening substances, and Ehrlich's theory was soon consigned to status of historical footnote.

The Cellular Versus Humoral Controversy. It often happens that the times of greatest advance in a field are those punctuated by strenuous disputes between two opposing schools; at such times, each side is stimulated to devise experiments designed to uphold its own position and to challenge the opposing view. Perhaps no dispute in immunology lasted so long, or had such important consequences for the future development of the field, as did that between the proponents of a cellular theory of immunity and those who argued that all immunity was based on the action of humoral elements. As we saw above, Ilya Metchnikoff based his theory of the importance of cells in mediating immunity on observations that even marine invertebrates possess macrophages capable of ingesting and destroying both foreign bodies or invading bacteria, or at least of walling them off by the formation of giant cells and granulomatous reactions. He suggested that the vertebrate phagocytic cells perform a similar protective function, and in fact are the most important contributors to both natural and acquired immunity. Pasteur invited Metchnikoff to join him at the newly constructed Pasteur Institute in Paris, where Metchnikoff and a succession of distinguished students spent the next 28 years working productively and imaginatively to verify and to extend the cellular (phagocytic) theory of immunity.

Metchnikoff's cellular theory quickly excited opposition. First, Richard Pfeiffer described a phenomenon in which circulating antibody (even that passively transferred to a normal recipient) would cause the specific lysis of cholera vibrios injected into the peritoneal cavity of immune guinea-pigs. Then, Behring and Kitasato showed that immunity to diphtheria and tetanus is clearly mediated by circulating antibody rather than by phagocytic cells. As time went on, circulating antibodies were found for most of the new pathogenic organisms that were rapidly being discovered. Finally, Metchnikoff's own student Bordet described the lysis of erythrocytes by humoral antibody and complement. See also **Kitasato, Shibasaburo (1852–1931);** and **Von Behring, Emil Adolf (1854–1917)**.

Metchnikoff and his students were quick to respond to these strong attacks against the phagocytic theory. In paper after paper, they showed that there is often no relationship between the bactericidal powers of the blood and host resistance to infection against a given organism. Rather, species resistance can often be directly correlated with the ability of its phagocytes to ingest the pathogen, as in the case of anthrax. Metchnikoff also showed that the creation of a macrophage-rich peritoneal exudate would protect the host against intraperitoneal injection of otherwise lethal doses of different bacterial pathogens. But the tide had obviously turned against the phagocytic theory during the 1890s, and Metchnikoff's last-ditch attempt to reestablish the importance of phagocytosis with the publication in 1901 of his famous book on *Immunity in Infectious Diseases* came too late. The book was widely admired for its scholarship, but made few converts among the unbelievers. See also **Metchnikoff, Elie (Ilya) (1845–1916)**.

Two attempts were made later to mediate the cellular–humoral dispute. In 1908, the Swedish Academy conferred the Nobel Prize in Physiology or Medicine jointly to Metchnikoff, the champion of cellularism, and to Ehrlich, the then-leading exponent of humoralist doctrines. Somewhat earlier, Sir Almroth Wright and S. R. Douglas had attempted to rationalize the differences between these two schools by their extensive work on the process of opsonization (Greek *opsonein*, to render palatable). These investigators claimed that both humoral and cellular factors were equally important and interdependent, in that humoral antibody appears to interact specifically with its target microorganism to render it more susceptible to phagocytosis by macrophages. Wright's espousal of this doctrine became so popular in England that Bernard Shaw used it as the subject of his play *The Doctor's Dilemma*. In his *Preface on Doctors*, Shaw summarized Wright's approach in an otherwise scathing castigation of the medical profession: "Sir Almroth Wright, following up one of Metchnikoff's most suggestive biological romances, discovered that the white corpuscles or phagocytes, which attack and devour disease germs for us, do their work only when we butter the disease germs appetizingly for them with a natural sauce which Sir Almroth named opsonin." But Wright's approach soon fell out of favor, and his efforts to revivify the cellular theory of immunity had little long-lasting effect. See also **Wright, Almroth Edward (1861–1947)**.

Transplantation and Immunogenetics. Transplantation research has been pursued since the turn of the century. At the outset, however, those who worked in the area were not immunologists, but rather surgeons, oncologists, biologists and geneticists, with little connection to contemporary immunology. For centuries, surgeons had dreamed of replacing missing or defective tissues and organs, but all attempts failed, save for the occasional success with corneal grafts. Then, at the end of the nineteenth century, it was shown that tumors could be passaged in experimental animals, but the grafts invariably failed due to some sort of immunity. If the secret of tumor graft rejection could be discovered, then humans might be induced to reject their tumors as well. The experiments on tumor rejection in animals generally employed normal tissues such as skin as a control, and it was quickly established that it was the host, and not some peculiarity of cancer tissues, that accounted for rejection. See also **Transplantation**.

In not much more than a decade, the general rules of graft rejection had been worked out: (1) transplantation into a foreign (xenogeneic) species invariably fails; (2) grafts to unrelated members of the same (allogeneic) species usually fail; (3) autografts almost invariably succeed; (4) in anallogeneic recipient, there is a primary take of the first graft, and then a delayed rejection; (5) rejection of a second graft from the same donor is accelerated; (6) the closer the 'blood relationship' between donor and recipient, the more likely is successful transplantation; and (7) these rules apply to normal as well as tumor tissues. Even early in the century, rejection was viewed as an active response on the part of the host's immune system; indeed, the term "transplantation immunity" was coined at that time.

Studies with inbred mice showed the genetic nature of donor–recipient incompatibility, and that this was not inherited as "a single Mendelizing factor." It was a similar interest in the tumor problem that stimulated Clarence Little to found the Jackson Laboratories in Bar Harbor, Maine in 1929, where George Snell "invented" the congenic mouse in the mid-1930s, and helped to define the major histocompatibility complex (MHC), the genetic material that defines the antigens responsible for tissue incompatibility among individuals. It is interesting that all of this immunological activity on the part of the surgeons, tumor biologists, and geneticists went substantially unnoticed by the immunologists of the day.

Immunology in Transition, 1912–1950s

The Fate of the Early Immunological Research Program. During the thirty years after 1880, the young field of immunology had organized itself predominantly in terms of the major areas of interest outlined above. While most of its practitioners might not yet have called themselves "immunologists," the beginnings of institutionalization of the discipline were evident. An institute devoted to immunological research had been established for Paul Ehrlich, and departments and services dedicated to the discipline had been formed within many of the leading research institutions around the world. Immunology occupied a prominent place at international congresses of medicine or hygiene, and an "invisible college" provided the means for informal exchange among researchers. The *Annales de l'Institut Pasteur* had long been devoting much space to immunological reports, 1908 saw the founding of the *Zeitschrift für Immunitätsforschung*, and the *American Journal of Immunology* first appeared in 1916. Finally, the discipline was more formally recognized by the founding of the American Association of Immunologists in 1913.

What was the fate of the components of the early immunological research program? The great victories of preventive immunization had been with chicken cholera, anthrax, rabies, plague, and several other important diseases. But it was proving impossible to prepare efficacious vaccines against such newly discovered pathogens as the tubercle and leprous bacilli, the cholera vibrio, the spirochete of syphilis and the large group of Gram-positive organisms, to say nothing of the growing number of deadly viruses and parasites. By 1910, the great early promise of Pasteurian immunization was no longer being fulfilled; new successes would thenceforth be few and far between. Research and development in this area soon left the "classical" immunology laboratory, and was taken over by bacteriologists, virologists and parasitologists, interested more in the agents themselves than in immunological mechanisms.

As we saw above, the study of the role of cells in immunology went into decline early in the twentieth century, under the attack of the humoralists. No similar techniques which made antibodies visible and measurable (such as cellular agglutination, the antigen–antibody precipitin reaction, immune

hemolysis, and the ability to transfer antibody passively from one animal to another) existed in the field of cell studies. The cell was still considered something of a mystery, whereas antibodies and their specific protective functions could be comprehended readily.

The techniques of serotherapy suffered a fate similar to that of preventive immunization. Following the demonstration of the efficacy of horse antidiphtheria and antitetanus sera in the treatment of these diseases, no further famous victories were recorded in this area. Interest in this approach waned, since there are so few other significant diseases that are caused by exotoxins and thus immediately amenable to this approach. Later, when passive transfer of antibody would be employed, it would be by hematologists using human gamma globulin to prevent erythroblastosis fetalis, or by pediatricians employing convalescent sera to deal with poliomyelitis or by clinicians to fight hepatitis virus.

The investigation of cytotoxic antibodies and autoimmunity was short-lived. Almost all attempts to demonstrate that anti-tissue and anti-organ antibodies play a role in the pathogenesis of disease proved fruitless, and immunologists did not recognize that the autoimmune pathogenesis of paroxysmal cold hemoglobinuria might represent the tip of an autoimmune disease iceberg. By 1912, immune cytotoxic phenomena interested few immunologists, and was investigated primarily by ophthalmologists interested in the disease sympathetic ophthalmia and in a putative autoimmune disease of the lens.

The case of serodiagnosis represents a more typical example of disciplinary differentiation. This approach to the diagnosis of infectious disease was central to the research as well as the clinical interests of the new immunologists. Syphilis remained the mainstay of diagnostic laboratories, and the techniques were improved and applied to other diseases throughout the period under discussion. In due course, however, serodiagnosis was standardized and became quite routine; immunologists interested in basic mechanisms quickly lost interest. The field was taken over by classical bacteriologists, and soon those who devoted themselves to serodiagnosis began to call themselves "serologists" and worked principally in hospital diagnostic laboratories rather than in those devoted to basic immunological research. See also **Immunoassay**.

Shortly after the discovery of anaphylaxis, it and its related diseases also became an intimate concern of immunologists interested in basic mechanisms. They wished to determine the nature of the antibodies responsible for these phenomena, and how these antibodies cause disease. Here was a paradox; a system that supposedly had evolved to protect the body from infection was now shown capable of causing disease. But answers proved to be elusive, and the immunologists soon deserted the field to others. Those who chose to pursue these interests were predominantly clinicians interested in hay fever and asthma, and clinical allergy soon became an accepted medical subspecialty. The study of anaphylactic and related mechanisms was also of great interest to physiologists, and it was they who discovered the important role of histamine and other pharmacological agents in the various allergic diseases. See also **Histamine**.

It was the surgeons, among those who had studied tissue transplantation, who left the field first, because tissue and organ grafting seemed to present insuperable problems. The tumor biologists also were unable to transfer the knowledge gained from animal experiments to humans, and they soon turned their attention to other aspects of the cancer problem. The geneticists who had worked on transplantation biology made a more subtle transition. They had viewed the inbred mouse as the perfect tool to study susceptibility to cancer formation, with possible therapeutic implications. Where as the early work to plot the histocompatibility loci in mice was aimed at understanding the immune response to tumors, it soon lost its oncological and even immunological basis, and became a study in pure genetics. Of primary interest now was the size and polymorphism of the histocompatibility genes, and the nature of their protein products. Only after World War II, when their applicability to tissue typing and histocompatibility matching was established, would the investigations return to the realm of transplantation immunobiology.

The Rise to Dominance of Immunochemistry. From the point of view of the medically oriented research scientist, immunology had lost most of its fascination. Those portions of the field that had not passed into the hands of new disciplines underwent a major shift in the direction of chemical approaches. The seeds of this new interest in the chemistry of antigens and antibodies can be traced directly to the work of Paul Ehrlich. Ehrlich's side-chain theory of antibody formation had pictured antigen, antibody and complement as chemical molecules, and their combining sites as stereochemical structures whose close fit with one another would account for their specificity.

It is common practice to trace the paternity of the field of immunochemistry to the physical chemist Svante Arrhenius; he had coined the term "immunochemistry" in 1907, in a book with that title. Arrhenius became interested in diphtheria toxin–antitoxin interactions, and proposed that they are reversible, like the interactions of weak acids and weak bases. However, Arrhenius's contributions could not be adequately tested at the time, and thus exercised little influence on subsequent events. The most significant impetus for a new direction was probably the discovery in 1906 that protein antigens could be chemically altered to change their immunological specificity, by the addition of simple chemical structures (called *haptens*). A powerful new tool was now available with which to dissect the nature of immunological specificity and the structure of the combining site on antibody. This approach was exploited most effectively by Karl Landsteiner who, from 1917 until his death in 1944, almost single-handedly defined the new discipline of immunochemistry. Yet another insight into the chemistry of antigens and antibodies was afforded by organic chemist Michael Heidelberger, who demonstrated that antibodies might be formed against polysaccharide as well as against protein antigens. Heidelberger studied the immunological responses to the polysaccharides of different strains of the pneumococcus. He not only developed effective vaccines against pneumococcal pneumonia, but simultaneously developed many quantitative techniques that helped to establish immunology as a more exact science. See also **Antigen**; and **Arrhenius, Svante August (1859–1906)**.

It is clear that during this period, the biology of antibody formation and the biological consequences of the antigen–antibody interaction took a distant second place to interest in the chemical nature of antigens and antibodies.

Chemical Theories of Antibody Formation. The theories of immunity advanced by zoologist Ilya Metchnikoff and medical researcher Paul Ehrlich were biological, and based upon Darwinian evolutionary principles. But the investigators who dominated immunology after World War I were interested in antibodies and antigens as chemical molecules, and in the origin of immunological specificity. Thus, new theories of antibody formation were not slow to appear. Now, however, the emphasis was no longer on antibody function, but rather on antibody structure. The paradox to be explained involved the ability of the vertebrate host to produce so many different specific antibodies. It seemed highly improbable that evolution could have accounted for the spontaneous production of so many different antibodies, the great majority of which were directed against bland and even artificial antigens of no obvious evolutionary selective force.

The new chemical theories advanced to explain antibody formation and specificity were quite Lamarckian in nature; whereas the molecules of the biologist generally have an evolutionary history, those of the chemist do not. The first of the new theories to be advanced, in 1930, was instructive, in contrast to Ehrlich's idea that antigen selects among preformed substances. It proposed that only the antigen can contain the information necessary for antibody formation. It acts as a template against which a nascent protein is constructed, composed of a unique sequence of amino acids. Here was a ready explanation not only for the impressively large repertoire of different antibodies, but also for how so fine a specificity could be imparted to the antibody molecule. The theory was later modified by such prominent theoreticians as Linus Pauling and Macfarlane Burnet, but continued to emphasize that the information required for specificity was derived from the antigen. See also **Burnet, Frank Macfarlane (1899–1985)**; **Lamarck, Jean–Baptiste Pierre Antoine de Monet de (1744–1829)**; and **Pauling, Linus Carl (1901–1994)**.

There was a problem with these chemical theories; while they appeared to satisfy the requirements for chemical structure, they did not explain such biological phenomena as the continued production of antibody in the apparent absence of antigen, or why a second exposure to the same antigen should result in an enhanced booster response. Even more perplexing was the failure of these theories to explain how repeated immunization could produce changes in the quality of the antibody. We shall see below how these and other unexplained phenomena would lead biologists to advance more Darwinian explanations for the origin of antibodies.

The Characteristics of the Immunochemical Research Program. The use of synthetic chemicals (haptens) attached to carrier proteins for the study of antibody specificity helped to clarify the structure of antigen and antibody combining sites, and the thermodynamics of their interaction. From the 1920s to the early 1960s, the forefront of immunological progress was occupied primarily by chemists or chemically oriented biologists, and almost all of the textbooks from which the next generation learned their trade were oriented toward chemical rather than medical/biological topics. This is not to imply that all work in biomedical areas ceased during this period. Research areas which have become well established take a long time to die out altogether. There is always some work that can be done. Thus, clinical allergists continued to study anaphylactic phenomena, purified allergens, and developed desensitization methods; infectious disease personnel prepared better toxoids and better modes of immunization; serologists improved and expanded the application of serodiagnostic procedures; experimental pathologists continued to study the immunology of tuberculosis and other diseases; and from time to time, an effective vaccine would be developed against one or another disease of humans or animals. The results of such studies, however, appeared to lie outside the mainstream of contemporary immunology, were often published in "outside" journals, and seemed to make little impression on those leading the field.

The Immunobiological Revolution

The work of those interested in the more chemical aspects of the immune response significantly advanced our knowledge of the nature of both antigens and antibodies, and of the structural basis and characteristics of immunological specificity. However, biologists were increasingly questioning the apparent inconsistencies in those observations whose explanation was not to be found in the received wisdom of chemical (instructionist) theories of antibody formation. These included the long-term persistence of antibody formation; the enhanced booster response on reexposure to antigen; changes in the quality of the antibody with time; and, most perplexing of all, the observation that immunity to many viral diseases seems to be unrelated to the presence of circulating antiviral antibodies. These and other inconsistencies posed a serious challenge to the reigning immunochemical paradigm. See also **Immunity to Viruses**.

By the mid-1950s, the stage seemed to be set for a major theoretical confrontation between the new immunobiologists and the classical immunochemists. This was stimulated also by several new observations from the fields of biology and medicine. In the years after World War II, Peter Medawar "rediscovered" the laws of transplantation, and demonstrated that the rejection of tissue transplants is based upon some form of immunological mechanism unrelated to humoral antibody. In 1945, it was reported that nonidentical twin calves could not form antibodies to one another's antigens, leading Macfarlane Burnet to postulate the existence of a cell-based immunological mechanism that induces in the fetus an inability to respond to antigens present during development, later named immunological tolerance. Peter Medawar and his coworkers confirmed this hypothesis experimentally, and he and Burnet shared the Nobel Prize in Physiology or Medicine in 1960 for this finding. See also **Medawar, Peter Brian (1915–1987)**.

There were other observations also, for which no ready explanation was available in classical theory. In the early 1950s, several immunological deficiency diseases were described in humans, the explanation of which would go to the very heart of the biological basis of the immune response. In addition, interest in autoimmune diseases was renewed by the discovery of autoimmune hemolytic anemias, autoimmune thyroid disease, and allergic inflammation of the brain, not all of which were mediated by circulating antibodies. A revolution lay waiting in the wings, needing only the support of a new theory to unseat the old regime and its outmoded paradigm. That theory was provided by Macfarlane Burnet. See also **Autoimmune Disease**; and **Immunodeficiency**.

Selection Theories of Antibody Formation. In view of the recent findings, one had now to explain theoretically how the immune response could be aborted by some intrauterine mechanism, in addition to providing an explanation for the characteristics of the active response. The first of the purely biological theories of antibody formation was advanced by Niels Jerne in 1955, in what he called a *natural selection* theory. He proposed, as had Paul Ehrlich before him, that the host forms small quantities of each of the antibodies in the full repertoire. These appear in the blood as "natural" antibodies which then interact specifically with antigen, transport that antigen into appropriate cells which are then stimulated to reproduce more of the same antibody. The phenomenon of immunological tolerance was also dealt with by postulating that any natural antibodies formed against self-antigens would immediately be absorbed by the tissues of the body; they would thenceforth be unavailable to mediate further antibody formation. See also **Jerne, Niels Kaj (1911–1994)**.

Jerne's selection theory was important less for its intrinsic value than for the stimulus it provided Burnet to develop his clonal selection theory of antibody formation. Following Jerne and Ehrlich, Burnet postulated that each antibody is a natural product unique to its host cell, appearing on the cell surface as receptors specific for one or another antigen. When antigen interacts with its specific receptor, it signals a clonal proliferation of that cell. Some cells of the clone become antibody-forming cells, while others remain as immunological memory cells able to participate in later booster responses. Finally, the theory held that immunological tolerance is due to a "clonal abortion," mediated specifically by self-antigens, or by those introduced from without at a critical period in the maturation of the fetus.

Burnet's clonal selection theory of antibody formation soon attained wide acceptance. This was due in part to the application of newer techniques for the study of cells, and in part to developments in the new genetics. Once the DNA control of antibody structure was accepted, and the amino acidsequence of the immunoglobulin chains elucidated by Gerald Edelman and others (for which Edelman shared the Nobel Prize in Physiology or Medicine in 1972 with R. R. Porter), it became possible to find an explanation for the ability to form so many different specific antibodies. The resolution of this repertoire problem is one of the triumphs of twentieth-century molecular genetics. It involves the variable combination of a number of different families of minigene segments, assisted by mutations, to form the large universe of antibody polypeptide chains. See also **Edelman, Gerald Maurice (1929–Present)**; and **Porter, Rodney Robert (1917–1985)**.

The Immunological Synthesis. The history of the discipline of immunology has witnessed three distinct phases, each with its own areas of interest and technological approaches. The first of these lasted from the founding of the field in 1880 to about World War I; its principal concerns were with the new bacteriology and infectious diseases, and it was almost entirely medical in outlook. When interest in this medical approach to immunology declined, due to the lack of meaningful new victories, others took over the reins of immunology. These were individuals interested in chemical approaches to the study of antigens and antibodies. They ushered in an "era of immunochemistry," in which the study of the structural bases of immunological specificity replaced the earlier interest in preventive medicine and therapy. This second period in the life of immunology extended from World War I until the late 1950s and early 1960s.

The 1960s saw an abrupt shift in emphasis in the field of immunology, which might justifiably be termed a scientific revolution. Stimulated by new observations on immunological diseases and on the biological aspects of the immune response, biologists took over command of the discipline from the chemists. In part under the influence of Burnet's clonal selection theory of antibody formation, different questions were now asked, involving the genetic basis of the immune response and the physiological mechanisms involved in its functions. New technologies were developed to study cell function and the molecules that mediate the various activities of antibodies and of those cells which are endowed with immunological functions. However, chemists continued to study the structure and thermodynamics of antibodies while molecular biologists studied the structure and mechanisms of the genes that encode for specific antibodies and cell receptors.

During the immunochemical era, those who dominated the discipline paid little attention to the concerns of the biologists; now that the immunobiologists were in control, they initially had little interest in the continuing activity of the immunochemists. It gradually became evident, however, that the two groups were really working on the same problems, but from different directions and with different techniques. The chemists approached the study of immunology by working back from the final molecular product (the antibody), whereas the biologists were working forward from the initial cellular interactions. As their interests converged, they began increasingly to employ one another's techniques, and even to collaborate. Thus, in recent years, they have clarified the major questions about

the chemical and genetic aspects of antibody structure and formation; the chemistry and biological function of cell receptors and of the molecular messengers with which cells intercommunicate; and the extremely complicated mechanisms whereby the immune response is regulated.

Modern immunology now influences many other disciplines. It has provided evolutionary theory with the model of a complicated mechanism able to anticipate the appearance of new pathogens, rather than merely evolving slowly to meet a new challenge. This peculiarity of evolution occurred not just once but twice! It controls the formation of those specific receptors on the surface of the cells that produce antibodies, and again somewhat differently to guide the formation of receptors on other cell types that protect against viral diseases and which function to reject tissue grafts. It has provided geneticists with the unique example of a super family of genes whose components exercise a broad range of activities that includes the predisposition to various diseases. It has offered to physiologists a variety of examples of how cells may communicate with and influence one another. Finally, immunology has assisted many medical subspecialties in defining the pathogenesis of some of their most important diseases, and has helped to develop preventive measures, diagnostic tools, and therapeutic approaches to combat these diseases.

Additional Reading

Alt, F.W.: *Advances in Immunology*, Elsevier Science & Technology Books, New York, NY, 2007.

Arrhenius, S.: *Immunochemistry*, Macmillan, New York, NY, 1970.

Bibel, D.J.: *Milestones in Immunology*, Springer-Verlag New York, LLC, New York, NY, 1988.

Gallagher, R.B., J. Gilder, G. Salvatore, and J.V.N. Gustav: *Immunology: The Making of a Modern Science: The Making of a Modern Science*, Elsevier Science & Technology Books, New York, NY, 1995.

Goldsby, D.A., J. Kuby, T.J. Kindt, and B.A. Osborne: *Immunology*, 5th Edition, W. H. Freeman Company, New York, NY, 2003.

Johnstone, T.P., and M.W. Turner: *Immunochemistry: A Practical Approach*, Oxford University Press, New York, NY, 1997.

Luttmann, W., D. Myrtek, K. Bratke, and M. Kupper: *Immunology*, Elsevier Science & Technology Books, New York, NY, 2006.

Parrish, H.J.: *A History of Immunization*, Livingstone, Edinburgh, Scotland, 1965.

Silverstein, A.M.: *A History of Immunology*, Elsevier Science & Technology Books, New York, NY, 1989.

Silverstein, A.M.: *Paul Ehrlich's Receptor Immunology: The Magnificent Obsession*, Elsevier Science & Technology Books, 2001.

Tauber, A.I., and L. Chernyak: *Metchnikoff and Origins of Immunology: From Metaphor to Theory*, Oxford University Press, New York, NY, 1991.

ARTHUR M. SILVERSTEIN, Johns Hopkins University School of Medicine, Baltimore, MD

IMPACT. Impact is the action of two bodies in collision, whereby the velocity of one or both bodies is changed. In the case of direct impact, the velocity of the moving bodies is in the direction of the normal (perpendicular) to the bodies at the point of contact. Otherwise the impact is oblique. The impact is central when the centers of gravity of the two bodies lie on the line of impact (normal to the bodies at the point of contact). The momentum of a body is its mass multiplied by its velocity. A law of impact is that the sum of the momentum of the two masses before and after impact is the same, provided the bodies are perfectly elastic, and no energy is absorbed in permanent plastic deformation.

The impact coefficient (coefficient of restitution) is the ratio between the differences of velocities of the two bodies after impact to the same differences before impact. This coefficient would be unity for impact of perfectly elastic bodies, and zero for fully inelastic bodies. To find the energy lost in an imperfect impact or plastic impact (one in which the impact coefficient is some number less than one), the masses of the two bodies, M_1 and M_2 may be substituted into the following formula. This formula has in it also the differences of velocity of the bodies after collision. Let f be the impact coefficient:

$$\text{energy lost} = \frac{M_1 M_2 (1 - f^2)(v_1 - v_2)^2}{2(M_1 + M_2)}$$

Impact Parameter. Consider a situation represented by two molecules as per Fig. 1. In the system of coordinates chosen, molecule 1 is at rest, while molecule 2 moves with the relative velocity g_{12} The axis is parallel

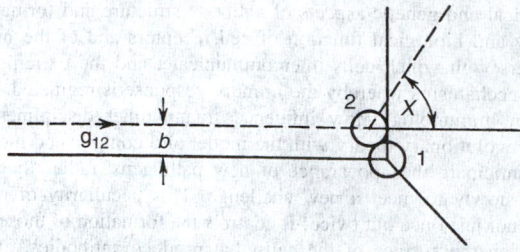

Fig. 1.   Impact parameter. Collision between two hard spheres. The trajectory of sphere 2 in the coordinate system in which 1 is stationary is indicated by the broken line. X is the angle of deflection.

to $\mathbf{g}_{12}$. The *impact parameter* b is the minimum distance at which molecule 2 would pass by molecule 1 if the two molecules did not interact.

For hard spheres, the amount of momentum and energy exchanged during the collision depends on b alone; for other more realistic models, where the interaction potential energy depends on distance, the momentum and energy exchange depends both on the impact parameter and on the initial relative velocity.

IMPACT TESTING.

In the design of metal structures and machines, the static stresses can ordinarily be calculated and a material of suitable strength selected. The impact loading requirements are usually less definitely known, and the translation of these requirements into a material specification is very difficult.

In testing rails, car wheels, and certain structural parts such as railway draft gears, a simple drop test is used. The weight, height of fall, and number of blows to failure give a relative measure of impact resistance.

In determining the impact toughness of metals as materials rather than as finished structures or machine parts, tests are made in a pendulum-type testing machine which breaks a standard beam-type sample with a single blow. The height of swing of the weighted pendulum past the anvil after fracturing the sample is related inversely to the energy absorbed in breaking the sample. For example, a tough steel will absorb a large proportion of the total energy of the pendulum so that the swing past the anvil will be small. The results are expressed in foot-pounds.

While some tests on relatively brittle materials are made on square-sectioned beam samples, in most cases a notch is machined opposite the striking position to localize fracture as in the Izod and Charpy tests. This introduces stress concentration and a state of combined stresses in the vicinity of the notch, which tends to restrict normal plastic deformation and to produce a brittle fracture quite independent of the velocity effect. In fact, some prefer to call this the notched-bar test rather than an impact test because the energy required to break a notched specimen differs little from that required to fracture a similar specimen by slow bending methods. Furthermore, the velocity developed by a free-falling pendulum of reasonable length is relatively small compared to the velocities encountered in moving vehicles, in certain machine parts, or in ballistics; therefore, the so-called impact test is more nearly a static than a dynamic test.

High-velocity impact testers, which break a test specimen in tension, have been developed. The interpretation of test results of this type is still somewhat uncertain.

It is known from experience that impact failures are more likely to occur at low temperatures than at normal room temperatures. Of the ferrous alloys, only austenitic stainless steels and certain other alloy grades containing nickel retain a large proportion of their normal room temperature toughness at low temperatures. The aluminum, copper, and nickel base nonferrous alloys do not develop low-temperature brittleness. The Charpy notched-bar impact test is generally used to determine impact toughness at various temperatures. Other tests such as the tension test fail to detect low-temperature brittleness.

IMPALA.   See **Antelope**.

IMPEDANCE.

The complex ratio of a force-like quantity (force, pressure, voltage, temperature or electric field strength) to a related velocity-like quantity (velocity, volume velocity, current, heat flow, or magnetic field strength). The terms and definitions under the term "impedance" pertain to single-frequency quantities in the steady state, and to systems whose properties are independent of the magnitudes of these quantities. These quantities can be represented mathematically by complex exponential functions of time. Under these conditions, the factors involving time cancel out in the ratios called for, leaving complex numbers independent of time. Solutions based on complex exponential functions under these conditions give the solution for real sinusoidal oscillations. Because of the similarity of electrical, mechanical and acoustical transmission theory, the same terminology is used in the three cases. See also **Alternating Currents**.

IMPEDANCE MATCHING.

Impedance matching is the process of making equal the impedance looking both ways from a junction point of two parts of a circuit. This serves two important functions; it gives a condition for maximum power transfer from one circuit to another for resistive impedances, and also prevents reflection of voltage and current waves. Impedances are usually matched by using the transformer which causes the impedance seen looking into the primary terminals to be equal to the impedance connected to its secondary terminals multiplied by a factor equal to the square of the primary to secondary turns ratio. Other methods include networks of resistances, networks of reactances, tuned circuits, quarter-wave and half-wave transmission lines, and use of open- and short-circuited transmission line segments in parallel with the circuits at proper matching points. Matching is very important in long smooth transmission lines to prevent reflection from the load. The latter must be equal to the characteristic impedance of the line and, if it is not so, must be transformed by one of the methods previously described.

IMPELLER.

1.   A device that imparts motion to a fluid; specifically, in a centrifugal compressor, a rotary disk which, faced on one or both sides with radial vanes, accelerates the incoming fluid outward into a diffuser. Also called *impeller wheel*.

2.   *That part of a centrifugal compressor comprising this disk and its housing.*

IMPERFECTIONS (Solids).

Many properties of solids depend upon the presence of structural imperfections, that is, deviations from a perfect homogeneous crystal lattice. Such properties include luminescence, atomic diffusion, color center absorption, crystal growth, plasticity, semiconduction, and others. The various types of imperfection fall into groups, of which the main ones are vacancies, interstitial atoms, dislocations, and foreign atoms, or impurities. Also involved are various macroscopic features, including mosaic structure, polygonization, growth spirals, and slip lines. See also **Crystal**; and **Semiconductor**.

IMPETIGO.

This is a bacterial infection of the skin and may occur at all ages, but is mainly a disease of preschool children. The most common causative agent is Group A streptococci, although *Staphylococcus aureus* is implicated in bullous impetigo. Commencing as a vesicular infection of the skin, impetigo progresses rapidly to a pustular form, which forms purulent discharge. Golden yellow crusts on lesions after they have dried is a characteristic feature of the disease. The disease is highly contagious because crusts, which are readily removed, contain the infecting organisms. Thus, cleanliness and the use of separate towels and accessories by the patient are mandatory if infection of others is to be avoided. Usually, penicillin is the drug of choice, either in the form of a single injection of long-acting benzathine penicillin or as oral penicillin. Certain strains of streptococci may predispose glomerulonephritis. The threat of this serious complication warrants the use of systemic antibiotics. In persons sensitive to penicillin, erythromycin is usually the substitute used. With effective treatment, the disease usually will complete its course within several days.

IMPLICIT TIME DIFFERENCE.

A finite-difference approximation in which the terms producing time change are specified at the predicted time level.

The approximation

$$(f^{n+1} - f^{n-1})/2\Delta t = g(f^{n+1})$$

(where superscript n denotes a point in time, separated by step Δt from the prior $[n-1]$ and subsequent $[n+1]$ discrete time levels) is an implicit time difference approximation to the differential equation $df/dt = g(f)$. Implicit approximations may be more difficult to implement than explicit time differences. For explicit time differences, $g(f)$ would be specified as $g(f^n)$ or $g(f^{n-1})$. Implicit time differences are relatively more stable and allow larger time steps than explicit time differences.

See also **Leapfrog Differencing**.

AMS

IMPLOSION. The violent shattering of a vessel or container in which the internal pressure is less than the external, e.g., in a highly-evacuated cathode-ray tube when the glass envelope is suddenly broken. Due to the atmospheric pressure against all sides of the tube, the glass moves inward with tremendous force.

The term is also used in connection with fission devices. In the implosion method, fissionable material is assembled into a highly compressed mass by the explosion of a surrounding spherical shell of high explosive. This was one of the methods used in achieving early nuclear explosions.

IMPOTENCE. A disturbance of sexual function in the male, that precludes satisfactory coitus. It varies from premature ejaculation to total loss of erection. Impotence should not be confused with sterility, which, in the male, means an absence of normal spermatozoa and therefore failure of reproduction. An impotent man may be fertile in that his testicles produce spermatozoa; and a sterile individual may be potent. Impotence is considered a major clinical problem of adult men. It is estimated that approximately 10 million adult males suffer from this condition in the United States. Erectile dysfunction results in more than 400,000 outpatient visits and over 30,000 hospital admissions per year.

For many years, this disorder was considered the result of psychic problems or a side effect of certain medications, such as antihypertensive drugs, propanolol, and psychotropic drugs. The disorder also was associated with Cushing's Syndrome, hyperthyroidism, pituitary deficiency, and excessive consumption of alcohol. Association of impotence with systemic sclerosis has been well known for several years. Only recently, however, have the biological pathways that underlie impotence been revealed in some detail. With this knowledge, corrective medications may be developed.

The mechanisms of the disorder have been described. Abnormal vascular responsiveness is the underlying cause of impotence. The failure to retain blood within the sinusoids is the most common cause of vasculogenic impotence. Filling of the sinusoidal spaces compresses the outflow venules against the relatively rigid tunica albuginea, causing engorgement of the corpus cavernosum with blood. Thus, failure of penile erection may result from impaired relaxation of the smooth muscle of the corpus cavernosum. See also **Gonads**.

From a summary of studies (1992) by J. Rajfer (UCLA Medical Center, Los Angeles), "Our findings support the hypothesis that nitric oxide is involved in the nonadrenergic, noncholinergic neurotransmission that leads to the smooth-muscle relaxation in the corpus cavernosum that permits penile erection. Defects in this pathway may cause some forms of impotence."

Nitric oxide was first described in 1979 as a potent relaxant of peripheral vascular smooth muscles. Nitric oxide is synthesized from enogenous larginine by the nitric oxide synthase system located in the vascular endothelium.

For a number of years, surgically adapted prostheses have provided mechanically assisted penile erection.

Additional Reading

Foster, M.: *Impotence*, Blackwell Science, Inc., Malden, MA, 1996.

Hellstrom, W.J.: *Male Infertility and Sexual Dysfunction*, Springer-Verlag, Inc., New York, NY, 1997.

Ledda, A.: *Vascular Andrology: Erectile Dysfunction/Priapism Varicocele*, Springer-Verlag, Inc, New York, NY, 1996.

Palmer, R.M.J., D.S. Ashton and S. Moncada: "Vascular Endothelial Cells Synthesize Nitric Oxide from I-arginine," *Nature*, **333**, 664 (1988).

Rajfer, J., et al.: "Nitric Oxide as a Mediator of Relaxation of the Corpus Cavernosum in Response to Nonadrenergic, Noncholinergic Neurotransmission," *New Eng. J. Med.*, 90 (January 9, 1992).

IMPOUNDING RESERVOIR. An impounding reservoir is a reservoir constructed to store excess flow of a stream so that the water may be available when the flow of the stream is insufficient to supply the demand.

IMPRESSED CURRENT. An electric network may be energized by applying (impressing) either known voltages or known currents on its terminals. For purposes of analysis, it is often convenient to consider given *impressed* currents, with the *resulting* voltages as the unknowns to be determined.

IMPRESSION MATERIALS. See **Polymeric Dental Materials**.

IMPRESSION PLASTERS. See **Dental Materials**.

IMPULSE. A vector quantity defined by the time integral of the force **F** acting on a particle over a finite interval, for example,

$$\int_{t_1}^{t_2} \mathbf{F}\, dt$$

for the interval from t_1 to t_2. The impulse-momentum theorem states that the impulse equals the change in momentum experienced by a particle during the corresponding time interval. See also **Momentum**.

IMPULSE (Nerve). See **Central and Peripheral Nervous Systems**.

INBREEDING. The mating of closely related animals. As practiced in stock breeding, brothers and sisters are often mated, or parents and their offspring. In most human societies the mating of individuals more closely related than first cousins is not approved, and even cousin marriages are regarded as close inbreeding.

Inbreeding is popularly regarded as productive of weakened or abnormal offspring. This is true in many cases because closely related individuals are more likely to possess genes of the same kind than individuals from different hereditary lines. Since most harmful characteristics are brought about by recessive genes, this means that many harmful characteristics which otherwise would be masked by dominant genes from one parent or the other will become homozygous and will be expressed. For instance, the chance of the recessive gene for albinism in man is 34 times greater in the children of first cousins than in the children of nonrelated parents. Congenital ichthyosis occurs 48 times more frequently in the children of first cousins than in nonrelated marriages. Animal breeders have learned that inbreeding among hogs, cattle, and other animals causes fewer and less vigorous offspring.

It is possible, however, for inbreeding to be continued with selection to eliminate the harmful genes, and a strain can be established with normal viability. There are many such inbred strains of mice that show no more abnormalities than outbred strains. The Ptolemies of Egypt practiced brother-sister marriages for many generations and finally succeeded in eliminating many of the harmful genes from their dynasty. Cleopatra was one of the final offspring of this practice.

INCH OF MERCURY. A common unit used in the measurement of atmospheric pressure. 1) One inch of mercury (in Hg) is defined as that pressure exerted by a 1-in. column of mercury at standard gravity and a temperature of $0\,^{\circ}\text{C}$ ($32\,^{\circ}\text{F}$):

$$1 \text{ in Hg} = 25.4 \text{ mm Hg} = 33.864 \text{ mb} = 1.00005 \text{ in Hg}(45^{\circ})$$

This is a unit recommended for meteorological use.

2) One 45° inch of mercury [in Hg (45°)] is defined as that pressure exerted by a 1-in. column of mercury at 45° latitude at sea level and a temperature of $0\,^{\circ}\text{C}$. It is evident that for most purposes these two units are interchangeable. When this is not the case, the unit should be carefully specified. Metric, rather than the English, units of length are used in many branches of science, and in other parts of the world. The early development and continued widespread use of the mercury barometer has fostered this manner of expressing atmospheric pressure. Although it has largely been replaced by the hectopascal (hPa) in most meteorological work, inches of mercury is still used in altimetry, and it remains the most common form of barometer scale calibration.

INCINERATION. Disposal of solid and liquid organic waste materials by burning at temperatures 1200 to 1500 °C. This method is approved by the EPA for use on very toxic organic chemicals and chemical wastes. Use of specially equipped incinerator ships, for burning chemical wastes at sea, has become commonplace.

See also **Wastes and Pollution**.

INCISOR. A Sharp-edged cutting tooth. The incisors are located at the front of the jaws of mammals, between the canines.

INCLINATION. That element of the orbit of a celestial object that indicates the angle between the plane containing the orbit of the object in question and some reference plane. In the case of orbits of members of the solar system, the reference plane is the ecliptic; whereas, in orbits of binary stars, inclination refers to the angle between the plane of the orbit of the stars and the plane perpendicular to the line of sight. See also **Ecliptic**.

INCLINED PLANE. See **Machine (Simple)**.

INCLINOMETER. See **Dip Needle**.

INCLUSION. This term is used in geology to connote a fragment of a foreign rock or mineral in an igneous rock. It may also be used to refer to gas or liquid enclosed in a mineral crystal.

In metallurgy, inclusions are small particles of nonmetallic compounds embedded in iron or steel. The principal types are oxides, sulfides, and silicates, all of which are more or less hard and brittle. They may originate as slag particles or pieces of refractory from the furnace or ladle that become entrapped upon solidification of the molten metal. More commonly they are the result of reactions within the metal itself during the finishing or deoxidation period and during pouring and solidification. The term "sonim," from solid nonmetallic inclusion, is sometimes used.

In high-quality heat treated machine and tool steels subject to impact and repeated stresses, inclusions are considered objectionable. They can be objectionable in any steel if present in large amounts, particularly in segregated areas forming continuous or semicontinuous stringers or plates known as laminations. In normal amounts and when well distributed their effect on strength, ductility, and other properties is negligible.

INCOMPETENT. A term used by many geologists to designate those formations which, when subjected to compressional, deformative processes, tend to be deformed by multiple fracture or crumpling of the relatively weaker formations. Typical incompetent formations are shale and highly carbonaceous sediments, especially when interstratified with such component formations as quartzite, sandstone, limestone or dolomite.

INCOMPRESSIBLE FLUID. A fluid in which the density remains constant for isothermal pressure changes, that is, for which the coefficient of compressibility is zero. Expansion and contraction of an incompressible fluid under diabatic heating or cooling is thus allowed for. In the more usual problem of isothermal processes, the fluid may or may not be stratified (have density differences within it), but motion of a parcel from higher to lower pressure or vice versa will not change the density of that parcel. Stated mathematically, the density gradient ∇ and the local derivative $\partial/\partial t$ may not be zero, but the individual derivative D/Dt vanishes. By the equation of continuity, it follows that the total divergence vanishes:

$$\nabla \cdot \mathbf{u} = \frac{\partial u}{\partial x} + \frac{\partial v}{\partial y} + \frac{\partial w}{\partial z} = 0,$$

where $\mathbf{u}$ is the velocity with components u, v, and w. For many purposes in meteorology, the atmosphere is treated as a heterogeneous fluid in which only vertical motions show compressibility. Together with the assumption of hydrostatic equilibrium, this has the effect of eliminating compression waves (including sound waves).

AMS

INCONTINENCE. See **Kidney and Urinary Tract**.

INCRUSTATION. See **Mineralogy**.

INCUBATION PERIOD. That period which elapses between exposure to an infectious disease and the development of an active infection, with clinical manifestations of the disease.

INCUS (Ear). See **Hearing and the Ear**.

INCUS CLOUD. See **Clouds and Cloud Formation**.

INDEFINITE CEILING. See **Meteorology**.

INDETERMINATE FORM. Limiting processes applied to special combinations of functions sometimes result in meaningless expressions such as $0/0$, ∞/∞, $0 \cdot \infty$, $\infty - \infty$, -1^∞, 0°, ∞°, etc. These are called indeterminate forms. To evaluate them, the L'Hospital rule, or modifications of it may be used, as will now be shown for several special cases.

1. The case 0/0. The limiting value of $f(x)/\phi(x)$ for $x = a$ is $f^{(n)}(a)/\phi^{(v)}(a)$, where the lowest-order derivatives which do not vanish are to be evaluated.
2. The case ∞/∞. Let $f(x) = 1/g(x)$, $\phi(x) = 1/h(x)$ and evaluate $h^{(n)}(a)/g^{(n)}(a)$ as in case (1). The same procedure works when $a = -\infty$.
3. The case $0 \cdot \infty$. If $f(x)\phi(x)$ becomes indeterminate, reduce the expression to case (1) or case (2) by writing it as $f(x)h(x)$ or $g(x)\phi(x)$.
4. The case $\infty - \infty$. The indeterminate form is $f(x) - \phi(x) = 1/g(x) - 1/h(x) = (h - g)/gh$, which is now case (1).
5. The cases 0^0, ∞^0, 1^∞. The function has the form $\phi(x)^{\phi(x)}$. Its logarithm, however, is $\phi(x) \ln f(x)$, which is case (3).

See also **L'Hospital Rule**.

INDETERMINATE STRUCTURE. A statically indeterminate structure is one that cannot be analyzed by means of the equations of static equilibrium alone. If a structure is statically indeterminate a solution can be obtained only if, in addition to the requirements of statics, the requirements of geometry, or continuity, are also considered.

Examples of indeterminate structures are triangular frameworks and trusses containing redundant members, rigid frames, hingeless and two-hinged arches, continuous arches, suspension bridges with stiffening trusses, most building frames, continuous beams, continuous trusses, plates or slabs, shells, and various space frames.

In general, statically indeterminate structures can be analyzed by means of various methods of classical structural mechanics or by modern numerical methods. In the past few years, numerical procedures have been developed for the analysis of frames with girders curved in plan and space frameworks.

See also **Beam (Structural)**.

INDEX NUMBER. An index number, as of prices, is a number intended to exhibit the changes in price that occur over a period of time (or in different areas). In the simplest case of a single commodity, we can express all prices as percentages of the price in a fixed base year; these percentages are called price relatives. When several commodities are involved, all prices can be expressed as relatives to bring them to common units, and an average of these relatives provides an index number. This average may be a simple arithmetic mean, or it may be weighted, for example by the relative amounts spent on the different quantities in the base year. Geometric means can also be used.

This form of index number is intended to reflect changes in price. Another form refers to changes in quantity, apart from changes in price. This is given by the average ratio of the current output to the output in the base period, both being expressed as values at the prices of the base period. This is known as a quantum index.

If the prices of commodities in the basic period are typified by p_{j0} and those in the given period by p_{jn}, a simple index, known as Carli's, is the average of the price relations:

$$I_{0n} = \frac{1}{k} \Sigma \left(\frac{p_n}{p_0} \right)$$

where k is the number of commodities.

If the prices are weighted by quantities in the base period, the index

$$I = \frac{\Sigma (p_r q_0)}{\Sigma (p_0 q_0)}$$

is known as Laspeyres index.

If the weights are chosen at the period n, the index

$$I = \frac{\Sigma (p_n q_n)}{\Sigma (p_0 q_n)}$$

is known as a Pasche index.

Various other index numbers, such as with weights which alter through time, are also in use.

SIR MAURICE KENDALL, International Statistical Institute,
London, UK

INDEX OF ARIDITY. See **Meteorology**.

INDEX REGISTER. The contents of the index register of a computer is generally used to modify the data address of the instruction as the instruction is being read from storage. The modified address is called the *effective data address*. A particular index register is addressed by a specified field in the format of the instruction. Index registers provide an ability to modify the program efficiently as it is being executed. Where an operation to be performed is a repetitive sequence of instructions on a table of data, it is necessary to change only the index register value instead of the data-address portion of each instruction. The data address of the instructions would contain the address of the required data with reference to the start of the table. All instructions that reference table data are indexed by the specified index register, which contains the address of the start of the table. Thus, when the program is to perform these operations on another table of data, the value in the index register is changed to the start address of the new data table. This effectively modifies all the indexed instructions in the sequence.

Index registers may be fixed locations in main storage, or they may be implemented in logic using flip-flops or triggers. In most modern computer designs, the function of index registers is provided through the capabilities of general-purpose registers. In the case of storage-resident index registers, the index-register address contains only the number of bits required to specify the register number uniquely, and a fixed prefix is supplied by the system logic to provide the actual storage address. This technique minimizes the length of the computer instruction inasmuch as most computers have only a few index registers. Thus, only two or three address bits required in the instruction even though a complete storage address may require 10 or more bits.

Index registers also are used as counters by the program. The same register may be used as both a counter and an address-modification value to step through tables. The index register is initialized to contain the number of factors in the table to be operated on, and the instructions to be performed contain the table-start address as a data address which is to be indexed by the same index register. Each time the sequence of instructions is performed, the index register is decremented by one and tested for zero value. Where the result of the test is nonzero, the sequence of instructions is repeated. However, since the index-register value has been reduced by one, the effective data address references the next lower table value. Where the test result is zero, all the factors in the table have been operated on, and the program steps to the next sequential operation.

THOMAS J. HARRISON, IBM Corporation, Boca Raton, FL

INDEX TABLE. A mechanism designed to rotate a workpiece to preset angular positions with rotary motion accomplished manually, or automatically as in the case of numerical control. Position is indicated by table graduations and Vernier scale, optical systems, or digital display. Table types include horizontal, vertical, and tilting. Tops are round, square, or rectangular. Tables differ in size, height, means of rotating, method of clamping, accuracy, and load capacity. Some tables use a mechanical system for table rotation and load support; others use pneumatic or hydraulic means.

Laboratory inspection tables require a greater degree of positioning and repetitive accuracies than shop tables. Accuracies of the former are usually on the order of a few seconds of arc. Shop table accuracies are in the 15–30 seconds of arc range and are designed for heavy-duty operation. Air-bearing tables handle heavier loads than conventional tables and are easier to rotate. The table top rotates on a preloaded film of air. There is no metal-to-metal contact. Friction essentially is eliminated. The total weight of the workpiece and the table top is sustained by concentric air bearing surfaces within the table. Radial thrust is absorbed by a special center bearing.

Automatic indexing tables are designed for use on machine tools for production machining operations. The table moves to the next position at the push of a button; or at the command of a numerical controller. Tables are supplied with automatic positioning locations preset to specifications, with eight equally spaced positions normal. The table is motor driven to its approximate location, an index plunger engages a hardened steel bushing for positive location, and the table clamps in position.

Digital readout tables use the moire fringe principle of measurement to provide digital readout of table position in degrees, minutes, and seconds of arc. Resolution is two seconds of arc. Numerically controlled tables advance to the desired positions, clamping and unclamping under the control of a perforated tape.

INDIAN EQUATORIAL JET. See **Meteorology**.

INDIAN GUM. See **Gums and Mucilages**.

INDIAN OCEAN WATER. Two major oceanic water masses occupy the surface layers of the Indian Ocean, the Indian Equatorial Water to the north, and the Indian Central Water south of it. The latter has a temperature range of 8–15 °C (46.4–59 °F) and a salinity range of 34.3–35.6%. It is thus virtually the same as the Western South Pacific Central Water with which it has so extensive a region of contact. In that region it sinks to an intermediate water level, whence it travels through other oceans. The Indian Equatorial Water is, as might be expected, warmer and more saline—temperature range 10–17 °C (50–62.6 °F), salinity range 34.9–35.3%.

INDIAN SUMMER. A period, in mid- or late autumn, of abnormally warm weather, generally clear skies, sunny but hazy days, and cool nights.

In New England, at least one killing frost and preferably a substantial period of normally cool weather must precede this warm spell in order for it to be considered a true "Indian summer." It does not occur every year, and in some years there may be two or three Indian summers. The term is most often heard in the northeastern United States, but its usage extends throughout English-speaking countries. It dates back at least to 1778, but its origin is not certain; the most probable suggestions relate it to the way that the American Indians availed themselves of this extra opportunity to increase their winter stores. The comparable period in Europe is termed the Old Wives' summer, and, poetically, may be referred to as halcyon days. In England, dependent upon dates of occurrence, such a period may be called St. Martin's summer, St. Luke's summer, and formerly All-hallown summer.

INDICATED AIR SPEED (IAS). See **Meteorology**.

INDICATED ALTITUDE. See **Meteorology**.

INDICATOR. An instrument used to reveal but not necessarily measure the presence of an electrical quantity. Sometimes called *display*. It is used to display the output of a sensing element after suitable amplification and modification. In radar the term is used to refer to the cathode-ray oscilloscopes, or other recording devices, where the echoes returned from targets are presented visually or graphically.

INDICATOR (Chemical). A substance which shows by a color change, or other visible manifestation, some change in, or particular condition of, the chemical nature of a system. Thus acid-base indicators may be used to indicate the end point of a particular neutralization reaction, or they may also be used to indicate the pH value of a system. For example, there are over fifty useful indicators for determining pH, covering the range from 0 to 14. Although indicators still are used in connection with colorimetric pH determinations and are extensively applied, sometimes in the form of dye-impregnated paper tape, for ascertaining the approximate pH of soils, swimming pools, and fish tanks where convenience and cost are predominating factors, indicators for pH and other chemical determinations are not nearly so important as they were before the advent of improved electrometric instrumental analytical techniques. Several pH (hydrogen ion indicators) are listed in Table 1.

TABLE 1. pH RANGES AND COLOR CHANGES OF SELECTED INDICATORS

Indicator	pH Range of Color Change	Color Change with Increasing pH
Alpha naphtholbenzein	0–0.8	Colorless to yellow
Methyl violet	0.2–1.9	Yellow to blue-violet
Para methyl red	1.0–3.0	Red to yellow
Thymolsulfonphthalein (Thymol Blue)	1.2–2.8	Red to yellow
	8.0–9.6	Yellow to blue
Methyl orange	3.3–4.5	Red to yellow
Methyl red	4.2–6.2	Red to yellow
Aurin (rosolic acid)	6.2–7.2	Amber to pink
Phenolsulfonphthalein (Phenol Red)	6.8–8.5	Yellow to red
Phenolphthalein	8.3–10.2	Colorless to purple
Thymolphthalein	9.4–10.7	Colorless to blue
Sodium nitrobenzeneazo-salicylate (Alizarin Yellow R)	10.1–12.0	Yellow to red
Malachite green	11.4–13.0	Blue-green to colorless
1,3,5-Trinitrobenzene	12.0–14.0	Colorless to orange

Indicators also are useful in following oxidation-reduction reactions, precipitation reactions, and, in general, throughout all volumetric analysis, and in many other chemical control operations.

INDICATOR FUNCTION. A signal that is used to decide which subsets of data to include in an analysis. For example, an aircraft flying through a convective boundary layer will fly sometimes within thermal updrafts and sometimes between thermals. Positive vertical velocities that exceed some threshold could be used to indicate when a measurement is being made in a thermal. By averaging only those temperatures that were obtained within thermals as defined by the indicator function, an average temperature for the thermals can be found. This method of using indicator functions to select portions of a larger dataset is called conditional sampling.

INDICATOR, pH. See **pH (Hydrogen Ion Concentration)**.

INDIFFERENT STATES. Let us consider a closed system whose state is determined completely by T, p, the phases $\alpha(\alpha = 1, \ldots, \phi)$ and the mass of each phase, i.e., by the weight fractions w_i^a of each component $(i = 1, \ldots, c)$ in the various variables

$$T, p, w_1^1, \ldots, w_c^\phi, m^1, \ldots, m^\phi \qquad (1)$$

Suppose the system is initially in the state (1).

Let us then consider the set of states accessible to this closed system, i.e., compatible with the conservation of mass (see also **Conservation Laws and Symmetry**). If there exists in this set states which differ from (1) in the mass of at least one of the phases, but in which all the weight fractions are the same, the state (1) is called an indifferent state. The system is then called an *indifferent system*. This terminology is due to Duhem.

If F is the number of degrees of freedom of a system, it can be shown that if F is 2 or more, there are $F - 1$ conditions which have to be satisfied

for the state to be an indifferent state. These $(F - 1)$ relations between the F variables leave only one independent variable. Therefore the indifferent states of the system fall on a line called the *indifferent line*.

A simple example is given by the decomposition of calcium carbonate

$$CaCO_3(s) = CaO(s) + CO_2(g) \qquad (2)$$

where (s) means a solid phase and (g) a gas. $CaCO_3$ and CaO are two distinct solid phases.

If a molecule of a $CaCO_3$ decomposes, then it increases the amount of CaO and CO_2 without altering the composition of any of the phases, each phate being formed by a simple component. All states of the system are indifferent.

The azeotropic systems are also special cases of indifferent systems.

The properties of indifferent systems are very similar to those of azeotropic systems. For example, one has the generalized *Gibbs-Konovalov* theorems: If in any isothermal (isobaric) equilibrium change, the system passes through an indifferent state, then the pressure (temperature) passes through an extreme value, and conversely.

INDIGO. A dye used for coloring cotton or woolen cloth a deep blue color. Prior to the development of synthetic indigo, the dye ingredient was obtained from *Indigofera tinctoria* of the *Leguminoseae* family, a shrub ranging from 4 to 6 feet (1.2 to 1.8 meters) in height and growing wild in southern Asia. The shrub has pinnately compound leaves, which are downy beneath, and bears reddish-yellow flowers. No indication of the dye substance is indicated by the coloration of the plant. The dye is contained in a glucosidal substance, which can be extracted from the shoots with water.

INDIUM. [CAS: 7440-74-6] Chemical element symbol In, at. no. 49, at. wt. 114.82, periodic table group 13, mp 156.6 °C, bp 2,078–2,082 °C, density 7.31 g/cm^3 (20 °C). Elemental indium has a face-centered tetragonal crystal structure. Indium is a silver-white metal, softer than lead, malleable, ductile, and crystalline. It is stable in dry air, but upon heating in air burns with a blue flame to form indium trioxide In_2O_3. Up to a temperature of 100 °C, the element does not decompose H_2O. Indium becomes a superconductor at 3.37 K. The element dissolves in HCl, H_2SO_4, or HNO_3, but not in NaOH. Metallic indium combines readily with chlorine and sulfur. 113In is the only nonradioactive isotope and is in isotopic abundance of 4.23%. 115In, with an extremely long half-life of 6×10^{14} years, accounts for the other 95.77% of naturally-occurring indium. Other radioactive isotopes include ^{107}In through ^{112}In, ^{114}In, ^{116}In, and ^{117}In. The half-lives of these isotopes are relatively short, measured in minutes, hours, or days. First ionization potential 5.785 eV; second, 18.86 eV; third, 28.03 eV. Oxidation potentials In → $In^{3+} + 3e^-$, −0.34 V; In → $In^+ + e^-$, −0.25 V.

Other important physical properties of indium are given under **Chemical Elements**.

Indium occurs in very small amounts in zinc blende, tungsten, tin and iron ores of certain localities. The recovery of indium from zinc flue dust (sometimes, 1 part per thousand) is effected by treating with a slight deficiency of HCl and allowing to stand. The residue is subjected to a series of treatments until finally pure indium sulfate is obtained, a solution of which when electrolyzed yields compact indium metal. A thin surface layer of indium is used on some bearings.

As the result of spectroscopic studies, the element was discovered by Reich and Richter in 1863.

On the scale of nonferrous metals, the production of indium is very limited, annual production probably not exceeding 1.3 million troy ounces (40.4 million grams). The availability of the element is affected by zinc production because it is a minor coproduct in the refining of zinc ores. The metal, in the form of an electroplate over lead and silver, has been used in aircraft bearings, the primary benefit being improved corrosion resistance. Indium also has been used as a dopant for germanium diodes and transistors. Several significant semiconductor compounds have been formulated, including InAs, InSb, and InP. The oxide has been used in electroluminescent panels. Indium alloys readily with several metals and it has been found particularly effective as a low-melting point fusible alloy when alloyed with bismuth, lead, tin, and cadmium. The eutectic alloy of indium-tin is an effective solder for glass-to-glass or glass-to-metal seals. With a melting range of 700–800 °C, copper-gold-indium and copper-silver-indium alloys are used as brazing materials. The eutectic alloy of mercury, thallium, and indium has a solidifying temperature of $-63\,^{\circ}$C, considerably below the mp of mercury, a feature that makes the alloy attractive for seals, switches, and thermometers for low-temperature applications. Control rods for nuclear reactors sometimes are produced from an alloy of silver, indium, and cadmium. Indium is also used in the manufacture of low-pressure sodium lamps. Indium for electroplating generally is furnished as the normal sulfate $In_2(SO_4)_3 \cdot 9H_2O$, the acid salt $In_2(SO_4)_3 \cdot H_2SO_4 \cdot 7H_2O$, or the basic salt $In_2O(SO_4)_2 \cdot 6H_2O$.

Chemistry and Compounds: Since indium has only three electrons in its valence shell, it is an electron acceptor. Indium trihalides include the trifluoride, trichloride, tribromide, and triiodide. They can be prepared by heating the metal or oxide in the halogen acid, or in the case of the trichloride and tribromide, by use of the halogen acid, or in the case of the trichloride and tribromide, by use of the halogen itself. Indium sesquisulfate forms double salts like the alums with alkali metal sulfates. A monohydrogen sulfate, $HIn(SO_4)_2 3\tfrac{1}{2}H_2O$, is known. Other compounds of indium(III) include the oxide (and its gelatinous hydrate), nitride, the nitrate, and the sulfide, selenide, and telluride.

Indium(II) compounds include the oxide, sulfide, fluoride, and chloride. They are prepared either by reduction of the corresponding trivalent compounds or, in the case of the chloride, by heating the metal in hydrogen chloride. They disproportionate, under suitable conditions, to give as end products the metal and the stable trivalent compound. Like gallium, indium(II) chloride is diamagnetic, having the structure $In^+[InCl_4]^-$.

Indium(I) compounds are formed by reduction of the corresponding In(III) compounds with hydrogen (on heating) or with indium metal, as in the case of the chloride. They are reactive compounds, the chloride disproportionating with water to form the metal and $InCl_3$; the oxide being oxidized on heating in air to the sesquioxide; and the sulfide reacting in dilute acids to form the sesquisulfide.

A number of indium trialkyls have been prepared, starting from the trimethyl, and some diaryl compounds are known, such as the diphenyl bromide. The lower trialkyls are tetramers. Like aluminum, indium forms a polymeric hydride, $(InH_3)_n$, from which tetrahydroindates, such as the lithium compound, $LiInH_4$, can be derived.

Like aluminum and gallium, indium forms a number of chelated oxycompounds, almost all of which are of 6-coordinate type. They include the stable crystalline inner complexes of which the β-diketones coordinate in the proportion of 3 molecules of diketone per atom of indium. Trioxalato as well as dioxalato salts are known, and compounds such as 8-quinolinol and substituted 8-quinolinols form trimolecular chelate rings involving nitrogen and donor oxygen.

Additional Reading

Davis, J.R.: *Metals Handbook*, 2nd Edition, ASM International, Materials Park, OH, 1998.

Greenwood, N.N. and A. Earnshaw: *Chemistry of the Elements*, 2nd Edition, Butterworth-Heinemann, Inc., Woburn, MA, 1997.

Krebs, R.E.: *The History and Use of Our Earth's Chemical Elements: A Reference Guide*, Greenwood Publishing Group, Inc., Westport, CT, 1998.

Lide, D.R.: *CRC Handbook of Chemistry and Physics*, 88th Edition, CRC Press, LLC., Boca Raton, FL, 2007.

INDOLE. Indole is a heteroaromatic compound consisting of a fused benzene and pyrrole ring, specifically benzo[*b*]pyrrole. The systematic name, 1*H*-indole distinguishes it from the less stable tautomer 3*H*-indole [CAS: 271-26-1]. 1*H*-Indole [CAS: 120-72-9] is also more stable than the isomeric benzo[*c*] pyrrole, which is called isoindole, (2*H*, and 1*H*). A third isomer benzo[*a*]pyrrole is a stable compound called indolizidine [CAS: 274-40-8]. The indole ring is incorporated into the structure of the amino acid tryptophan [CAS: 6912-86-3] and occurs in proteins and in a wide variety of plant and animal metabolites. Much of the interest in the chemistry of indole is the result of efforts to understand the biological activity of indole derivatives in order to develop pharmaceutical applications.

Indole was first obtained (and its structure elucidated) in 1866 by Adolf von Baeyer. The original preparation was by zinc-dust pyrolysis of oxindole which had been obtained by the reduction of isatin, an oxidant product of the natural dyestuff, indigo.

Indole is a colorless crystalline solid, mp $52-54\,^{\circ}$C, which is easily soluble in most organic solvents but sparingly soluble in water. The heat of combustion at constant volume is 4,268 MJ/mol (1020 kcal/mol). Indole has a musty odor which is very persistent and its derivatives have some applications in the formulation of fragrances.

The industrial source of indole has been isolation from coal-tar distillate. Several patents for the manufacture of indole have been issued with aniline and ethylene glycol, aniline and ethylene oxide, 2-ethylaniline, and *N*-ethylaniline as the starting materials.

Indole is a heterocyclic analogue of naphthalene. The basic reactivity patterns of indole can be understood as resulting from the fusion of an electron-rich protein pyrrole ring with a benzene ring.

Reactions include electrophilic aromatic substitution (e.g., halogenation, nitration, *C*-acylation, and alkylation), *N*-alkylation, arylation, lithiation and subsequent transformations, and oxidation.

Although there are a wide variety of indole ring syntheses, most of the more useful examples fall within a small number of groups. Indole syntheses usually start with an aromatic compound, either monosubstituted or orthodisubstituted.

Processes include the Fischer indole synthesis from arylhydrazones and related sigmatropic syntheses, reductive cyclizations of nitro compounds, the Madelung synthesis from anilides and related base-catalyzed condensations, and transition-metal catalyzed cyclizations.

Biologically Active Indole Derivatives

Thousands of indole derivatives have been prepared and evaluated as potential pharmaceuticals. Of those which have been put into use perhaps the most important are the nonsteroidal antiinflammatory agent, indomethacin [CAS: 53-86-1], and the β-adrenergic blocker, pindolol [CAS: 13823-86-9].

Many derivatives of indole are found in plants and animals where they are derived from the amino acid tryptophan. Several of these have important biological function or activity. Serotonin [CAS: 73-31-4] functions as a neurotransmitter and vasoconstrictor. Melatonin [CAS: 73-31-4] production is controlled by the circadian cycle and its physiological level influences daily and seasonal rhythms in humans and other species. Indole-3-acetic acid [CAS: 87-51-4] is a plant growth stimulant used in several horticultural applications.

The largest single class of naturally occurring indoles are the plant alkaloids. These occur with a wide range of structural diversity and are typically derived from tryptophan and terpenoid structural units. Several of these compounds are pharmacologically significant. Reserpine [CAS: 50-55-5] acts as a tranquilizer and hypotensive agent. The dimeric vinca alkaloids, vincristine [CAS: 57-22-7] and vinblastine [CAS: 865-21-4], are used in the treatment of Hodgkin's disease, leukemia, and other forms of cancer. Derivatives of the ergot alkaloid lysergic acid are used in the treatment of migraine and the diethylamide is lysergic acid diethylamide. See also **Alkaloids**.

There are several documented cases where indole derivatives, both natural and of synthetic origin, have been linked to pathological effects in humans. 3-Methylindole [CAS: 83-34-1], which is produced by bacterial fermentation in cattle, can lead to pulmonary edema. The pyridoindoles Trp-P-1 and Trp-P-2 are genotoxic substances which originate from pyrolysis of tryptophan and have been identified in foods cooked at excessively high temperatures. 4-Chloro-6-methoxyindole, which can be extracted from fava beans, yields a potent mutagen on interaction with nitrate ion.

Additional Reading

Atta-ur-Rahman, and A. Basha: *Indole Alkaloids*, Taylor & Francis, Inc., Philadelphia PA, 1998.

Bird, C.W. and G.W.H. Cheeseman, eds.: *Comprehensive Heterocyclic Chemistry*, Vol. 4, Pergamon Press, Oxford, 1984, Chapts. 3.04, 3.05, and 3.06.

Houlihan, W.J. eds.: *The Chemistry of Heterocyclic Compounds*, Vol. 25, Parts 1, 2, and 3, Wiley-Interscience, New York, NY, 1972.

Saxton, J.E.: *Monoterpenoid Indole Alkaloids: Supplement to Part 4*, John Wiley & Sons, Inc., New York, NY, 1994.

Sundberg, R.J.: *The Chemistry of Indoles*, Academic Press, Inc., New York, NY, 1970.

Sundberg, R.J., O. Meth-Cohn, C.S. Rees, and Alan R. Katritzky: *Indoles*, Elsevier Science & Technology Books, New York, NY, 1996.

INDUCTANCE. The inductance of a circuit (such as a coil) is the rate of increase in magnetic linkage with increase of current. If we have a coil of several turns, carrying a steady current, a certain magnetic flux will, as a result, be linked with the coil, depending upon the size and shape of the coil, the number of turns, and the material occupying the surrounding space. If the current is now slightly increased, the resulting increase in flux may or may not be proportional to the change in current; if not, we shall have to consider a very small increase in each. The "linkage" is the product of the flux through the coil by the number of turns. Since magnetic flux is ordinarily expressed in maxwells (emu) or webers (mks), the linkage may be expressed in maxwell-turns or weber-turns. The inductance unit called the henry corresponds to a rate of linkage increase of 10^8 maxwell-turns or one weber-turn per ampere of current. This is a rather large unit, hence the millihenry and microhenry are commonly used.

The inductance of a coil wound on a ferromagnetic core depends on the magnitude of the current, because of hysteresis effects. By convention, the inductance of such a coil is usually taken as

$$L = X/2\pi f$$

where X is the reactance (see also **Alternating Currents**) and f is the frequency. The impedance used in determining this reactance is taken as the ratio of the effective voltage to the effective current, neglecting harmonics produced by the variability of the inductance.

Critical inductance is the minimum inductance required to prevent the current from going to zero during any part of the cycle in the input choke of a choke input filter for a full-wave rectifier circuit. The value of this inductance (in henries) is equal to the load resistance (ohms) divided by three times the supply frequency (radians per second). The input choke must have a value equal to or greater than the critical value if the best performance in the way of regulation is to be realized.

Incremental inductance is the inductance an iron-cored coil will offer to A.C. when it is superimposed on D.C. through the coil. This condition occurs very frequently in communication and electronic circuits since many of these involve direct currents for establishing an operating point and then superimpose the A.C. signal. The D.C. produces a certain amount of saturation in the core so the flux conditions presented to the A.C. are not the same as if no D.C. were present. When the core is in this partially saturated condition the flux changes produced by the A.C. are not as great as they would be otherwise, and hence the back emf of the coil or its inductive effect is reduced. Since the actual inductance presented depends upon the degree of saturation the rating of a coil which is designed to carry both types of current should include both the D.C. value of the current and the inductance (which is understood in this case to be the incremental inductance). The effect on incremental inductance is due to the change of the permeability of the core, the permeability which is effective in this case being called the incremental permeability. It is given by

$$\mu_\Delta = \frac{\Delta B}{\Delta H}$$

where μ_Δ is the incremental permeability, ΔB the change in flux density produced by the A.C., and ΔH the change in magnetizing force produced by the A.C.

INDUCTION (Electric/Magnetic). The production of an electric charge or magnetic field in a substance by the approach or proximity of an electrified body, a magnet, or any other source of an electric or magnetic field. The term induction implies that there is a relatively nonmagnetized medium between the body in which the electric or magnetic effect is induced and the electrified body, or other source of the electric or magnetic field.

Magnetic Induction is the basic observable property of a magnetic field. It is directly associated with the force on a current element or the electromotive force induced on a moving conductor. The mechanical force on a length $d\mathbf{l}$ of a circuit carrying a current I is given by

$$d\mathbf{F} = I d\mathbf{l} \times \mathbf{B}$$

The electromotive force induced in a conductor of length $d\mathbf{l}$ moving with a velocity $\mathbf{V}$ is given by

$$d\mathbf{E} = \mathbf{B} \times \mathbf{V} d\mathbf{l}$$

Thus, the concept of induction extends to the "induction" of a current in a conducting circuit by variation of the magnetic flux linking the circuit.

Induction Field. This is the magnetic field set up around a conductor by the current in the conductor. If the field changes, it causes a back electromotive force to be induced in the conductor. The magnetic fields present around dc and low-frequency circuits are considered to be of this sort since the error in the assumption is entirely negligible. All the energy stored in this field is returned to the circuit when the current flow is stopped. The induction field is contrasted with the radiation field which has a large value at radio frequencies and whose energy is not returned to the circuit but is radiated outward giving electromagnetic waves in space (the waves by which radio communication is effected). Many short-distance wireless communication systems utilize the induction field since it does not extend very far and hence does not interfere with regular radio reception at any great distance.

Induction Forces. When a charged particle a (for example an ion) interacts with a neutral molecule, the charged particle a *induces* on the neutral molecule b a dipole moment. If the polarizability of molecule b is α_b, the energy of interaction between the charge e_a and this induced moment is

$$\phi(r) = -\frac{e_a \alpha_b}{2r^4} \tag{1}$$

Similarly, there exists a potential energy of interaction between a point dipole μ_a and an induced dipole produced in a neutral molecule of polarizability α_b. When averaged over the angles, the result is

$$\phi(r) = -\frac{\mu_a^2 \alpha_b}{r^6} \tag{2}$$

It is important to note that (1) and (2) correspond always to an *attraction*. This effect was discussed for the first time by Debye.

Nuclear induction is magnetic induction in material samples (which may be solid, liquid or gaseous) that has its origin in the magnetic moments of the constituent nuclei. This effect is due to the unequal population of energy states available when the material is placed in a magnetic field. Nuclear induction is usually weak, but may be readily observed in the Bloch type of experiment depending upon the occurrence of nuclear magnetic resonance.

INDUCTION (Mathematics). This is a general method of proof in which a positive integral variable is involved. It consists of two main parts: (1) direct verification of the theorem for the smallest admissible value of the positive integer involved; (2) the algebraic proof that if the theorem is true for any value of the integer, it is true for the next greater value. In conclusion, the theorem is proved by combining the two parts.

INDUCTION MOTOR. See **Motor (Electric).**

INDUCTIVE INTERFERENCE. When a telephone line is paralleled by a power line or even another telephone line there is almost certain to be induced interference in the telephone line. This interference is due to voltages and corresponding currents induced in the line by voltages or currents in the paralleling line. If A.C. flows in a line it causes an alternating flux to be set up around the wires. This flux extends outward for a considerable distance and may link another line inducing voltages in series with this second line. Because of their extremely small magnitude, the effects produced by telephone lines in power lines are of no consequence, but those produced by the power line in the telephone line are very serious, since, although they may be small, they are comparable to the normal signal voltages. Sixty-cycle currents are below the transmission limits of most telephone equipment but may cause trouble in telegraph circuits. However, harmonics of the power circuit, and especially high-frequency transients induced by switching and lightning, cause objectionable interference. Another type of induced voltage is caused by the electrostatic flux from the high-voltage transmission line. The telephone line in this field will assume a potential corresponding to its capacitance with respect to the ground and the power line. Noise currents in the terminal equipment of the phone circuits may be eliminated or materially reduced by accurate balancing of the various line and equipment impedances with respect to ground in the telephone system and by a coordinated system of line transpositions. By transposing the lines (both telephone and power), very nearly equal voltages will be induced in both wires of the telephone line, and hence there will be no net voltage across them or in series with them to cause the flow of noise currents. Telephone lines are also transposed to avoid crosstalk between them. These transpositions must be carefully worked out for all lines concerned if full benefit is to be realized. Inductive interference from power lines is largely eliminated by the shielding by the sheath of a cable and is completely eliminated by the use of the coaxial cable.

INDURATION. The state of increased resistance or hardness in any tissue or organ. This may be an indication of inflammation, abscess, or tumor.

INDUS. A southern constellation located near Sagittarius.

INDUSTRIAL BIOTECHNOLOGY. In the broad sense, *industrial biotechnology* is the practical application of scientific principles of biology learned over many decades to the development and manufacture of useful products. In modern terms, this includes the use of knowledge gained from studies in cell biology, molecular biology, and gene-transfer techniques. These topics are respectively discussed in articles on **Cell (Biology); Molecular Biology**; and **Genetics and Gene Science**. In a more restricted sense, industrial biotechnology is frequently referred to as *industrial microbiology*—because in a large number of bioprocesses, living substances in the form of yeasts, molds, bacteria, etc. are used as raw materials. For centuries, processes that depend upon living substances have existed, fermentation being a notable example. The scope of biology-related manufacturing, so to speak, was greatly expanded with the appearance of antibiotics, pioneered by Alexander Fleming in 1929 and soon followed by the bioprocessing of microorganisms to produce antibiotics in large quantities. As is covered in the article on **Genetics and Gene Science** and in other articles in this encyclopedia, the science of biology was impacted in the 1950s by a breakthrough in our knowledge of the DNA molecules, which led to gene recombination technology. The impact was felt early and continues today in industrial biological processes and products. As bioprocessing progressed from the traditional fermentation industries through the massive production of antibiotics and other pharmaceuticals to the current and potential application of gene-transfer techniques on an industrial scale, the nomenclature changed. A preferred term for the foregoing activities is now *industrial biotechnology*.

Starting Materials—Microorganisms

Four kinds of microorganisms make up the raw materials for most biochemical processes. They are:

A. *Eukaryotes*—cells or organisms whose DNA is organized into chromosomes with a protein coat and surrounded by a nuclear membrane. Eukaryotes contain organelles, such as mitrochondria. The latter function and furnish the cells with their main energy supply. Yeasts and molds, the eukaryotes in common use, are fungi.

1. *Yeast (Ascomycetes)*, notably the *Saccharomyces* (called *sugar fungus* by Meyen, 1837). Among the most industrially important are *S. cerevisiae* (used in making alcoholic beverages and bread); *S. cerevisiae* var. *ellipsoideus*, *S. bayanus*, and *S. beticus* (used in making wine); *S. uvarium* (used in brewing); and *Kluyveromyces fragilis* (used in whey disposal). *K. fragilis* was formerly called *S. fragilis*.

Yeasts play a major role in biochemical processes, but notably participate in (a) *fermentation*, where simple sugars and other chemicals are transformed into the desired intermediate or end-products of the process; and (b) *respiratory (oxidative) metabolism*. This respiratory activity of oxidative dissimilation is characteristic of many species of yeast. For example, during aerobic growth, sugar is oxidized to carbon dioxide and water, with the release of large amounts of energy. Other biochemical processes in which yeasts participate include amination, condensation, deamination, decarboxylation, esterification, hydrolysis,

lipolysis, pectinolysis, and proteolysis, among others. See also **Yeasts and Molds**.

2. *Molds* are filamentous and of numerous varieties. They grow as a branched system of threadlike hyphae rather than as single cells. In bioprocessing molds play a positive role, as in the instance of *Penicillium roqueforti* used as a culture in making certain cheeses; or *Penicillium chrysogenum* used in the production of penicillin antibiotics. Molds also have a significant negative side in biochemical processing because of the degradative roles they play in causing food spoilage and human disease, not to mention plant and crop damage. When food processing vessels and machines are not kept clean, so-called *dairy* or *machinery mold* may be found in the equipment. The presence of this mold may be indicative of serious microbial contamination.

Molds are widely distributed throughout nature. Critical to the growth of molds is availability of sufficient moisture. Some molds are quite resistant to adverse conditions, including high temperature.

B. *Prokaryotes*—cells or organisms that have only one chromosome. They have no nuclear membrane or mitochondria.

3. *Bacteria* (normally unicellular), of which many are used in bioprocessing (see also **Bacteria**), including: *Lactobacillus bulgaricus* (used in making yogurt); *Gluconobacter suboxidans* (vinegar); *Clostridium acetobutylicum* (acetone and butanol); *Corynebacterium glutamicum* (flavor enhancing nucleotides); *Methylophius methylotrophus* (single-cell proteins); *Propionibacterium* (vitamin B_{12}); *Bacillus* (enzymes—proteases); *Xanthomonas campestris* (polysaccharides—xanthan gum); *Mycobacterium* (pharmaceuticals—steroids); *Streptomyces* (antibiotics—amphotericin, streptomycin, tetracyclines, etc.); *Escherichia coli* (by way of recombinant DNA technology to produce insulin, human growth hormone, somatostatin, interferon, etc.); and *Bacillus thuringiensis* (bioinsecticides).

4. *Actinomycetes*—a group of branching unicellular organisms, which reproduce either by fission or by means of special spores or conidia. They usually form a mycelium which may be of a single kind, designated as substrate or vegetative, or of two kinds, substrate and aerial. The actinomycetes are closely related to the filamentous bacteria and some authorities regard them as prototypes from which both fungi and bacteria were derived. These microorganisms have been prolific sources of several thousand antibiotics, although only a relatively few of these have been produced commercially in very large quantities.

Classes of Bioproducts Made from Microorganisms

The principal commercially important products made from the microorganisms previously mentioned include:

1. *Microorganisms per se*—for use in a wide variety of bioprocesses.
2. *Large molecules*, including enzymes.
3. *Primary metabolic products* (compounds required for their growth).
4. *Secondary metabolic products* (compounds that are not essential to their growth).

The primary and secondary metabolites that are important commercially normally are of relatively low molecular weight, usually 1500 daltons[1] or less. For comparison, an enzyme may range from about 10,000 to millions of daltons.

Commercial Applications of Microbial Cells

There are two major classes of the commercial use of microbial cells:

1. *Protein sources*, of which the most important current product is called *single-cell protein*, used in animal feedstuffs. See also **Protein**.
2. *Bioactive ingredients* for use in bioprocessing. In bioprocessing, chemical reactions are usually referred to as *biological conversions* (even though they proceed at the molecular level) in processes where microorganisms are the major participants. The term *microbial transformation* is also used.

---

[1] Dalton = atomic mass unit.

For large numbers of end-objectives, microorganisms as participants have a number of advantages over nonbiological reactants. For example, the latter involve substantial energy exchange (either requiring heating or cooling). Also, nonbiological processes usually are conducted in a solvent medium and often in the presence of inorganic catalysts. Both solvent and catalyst are possible sources of product pollution. Further, a large percentage of biological conversions do not yield undesirable byproducts which must be removed in separate purification operations. Once separated, profitable uses must be found for the byproducts, or they must be disposed in a costly, nonpolluting way.

In biological conversions, water is usually the solvent and temperatures and pressures are at reasonable levels, as dictated by the properties of most natural living substances. When a specific enzyme is used in a biological conversion, outstanding specificity maintains because a given enzyme usually will catalyze but one specific kind of reaction. As pointed out by Demain and Solomon (see reference), an enzyme can be caused to select one isomer, or molecular form of a compound, in a mixture of forms to produce a single isomer of the product. These characteristics account for the high yields that are typical of biological conversion, sometimes reaching nearly 100 percent.

The production of enzymes is the major target of numerous biological conversions. In years past, the principal sources of enzymes were extraction products from plant and animal sources. The application of DNA technology has made large inroads in the production of "synthetic" enzymes.

Products of commercial importance extend well beyond enzymes and include the production of large molecules, such as polysaccharides.

Xanthan is an example. This is a gum widely used in food processing. See also **Xanthan Gum**.

Fermentation Industry. *Primary metabolites* of importance in the fermentation field include amino acids, purine nucleotides, vitamins, and organic acids. Specific products include citric acid, riboflavin (vitamin B_2), and cobalamin (vitamin B_{12}). Check alphabetical index pertaining to specific vitamins. Of the *secondary metabolites*, antibiotics are the most important. In the past, about three-quarters of all antibiotics have been obtained from the actinomycetes, and a very large percentage of these have stemmed from a single genus, *Streptomyces*. See also **Antibiotic**.

The process of fermentation has been known for at least 4000 years, notably in connection with the arts of making wine, leavened bread, brews (beer), and for decades, for example, in connection with naturally produced vinegar. In most fermentation, the microorganisms effect the desired biological transformations (serving in a way that is somewhat comparable to intermediate chemicals used in conventional organic synthesis and not present in the final product). Usually, at some point in the bioprocess, the earlier invaluable microorganisms must be killed because their presence in the final product would lead to numerous undesirable characteristics, including spoilage.

See also **Fermentation**; and **Grape**; and alphabetical index for related topical matter.

Examples of organic synthesis by way of fermentation are given in Figs. 1 and 2.

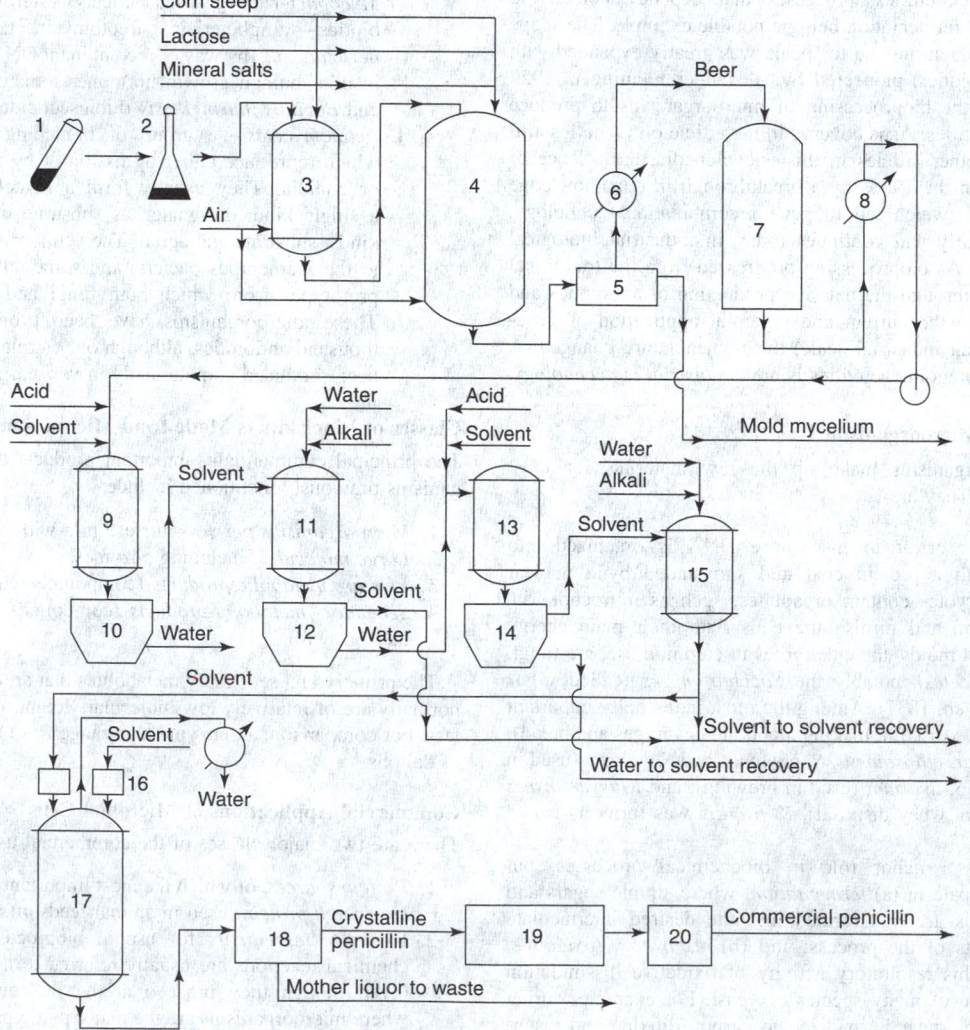

Fig. 1.   Schematic representation of materials flow in penicillin manufacturing: (1) agar slant culture, (2) bran spore culture, (3) seed tank, (4) fermentor, (5) filter, (6) brine cooler, (7) storage tank, (8) brine cooler, (9) mixing tank, (10–15) separation operations, (16) bacteriological filters, (17) crystallizer, (18) filter, (19) dryer, (20) finishing operations.

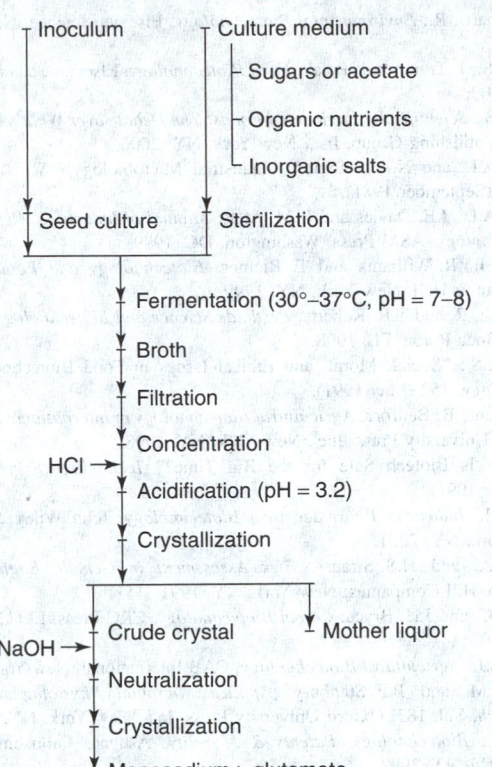

Fig. 2.   Preparation of monosodium l-glutamate by fermentation. Although not involving genetic engineering, this process as pioneered by Japanese microbiologists was very advanced when introduced in 1956. Currently, almost all common amino acids can be produced by amino acid fermentation and in terms of very high tonnage production.

Bioprocessing Methodologies

Bioprocessing equipment must be designed to meet the environmental requirements of whatever microorganisms are used. As well understood for a century or more, enzymes created by the microorganisms catalyze the desired biological conversions in a highly efficient manner, as exemplified by the conversion of sugars into ethanol and carbon dioxide. A major advancement in fermentation occurred a number of years ago when brewers found that instead of relying on microorganisms to create the desired enzymes, they could add the enzymes manufactured separately and available from commercial sources to the fermentation vessel directly. Many bioprocesses are not so simple as producing alcohol from sugar, but require cadres of enzymes to effect the biological transformation of the substrate by initiating a number of reactions (transformations) that involve numerous specific enzymes.

The environmental requirements of microorganisms are quite demanding. Process parameters must be carefully controlled to gain full efficiency, or indeed to permit the bioprocess to continue at all. This contrasts with a considerably greater flexibility permitted when dealing with nonliving reactants and catalysts. Such parameters include temperature control and pH (hydrogen ion concentration) within very narrow limits. Sufficient water of specified purity acts as the processing medium. Although enzymes can be preserved by drying, there is no catalytic activity in the absence of water.

An Additional requirement, not encountered in effecting nonbiological reactions, is the need to provide nutrition for the living microorganisms. They require a source of carbon that normally furnishes energy for metabolism. In some cases, this requirement is met by one of the starting raw materials, such as carbohydrates (sugar in the case of alcohol fermentations). Considerable investigation for certain bioprocesses has targeted other sources of carbon, including hydrocarbons. It has been learned that some industrially important microorganisms can exist on these nutritional carbon sources. Sometimes a period of adaptation is required and because of lack of efficiency, the use of such alternative carbon sources is frequently uneconomic. But with so many products now being produced

by microorganisms, research continues and effective uses are expected to be found. Hydrocarbon sources considered have included petroleum and natural fats (soybean oil, etc.). The important point is that the demonstrated versatility of enzyme mechanisms bodes well for much further exploitation of those characteristics in the future.

In addition to provision of carbon, other nutrients required by microorganisms embrace nitrogen, phosphorus, and oxygen, all elements of which are part of the structural and functional molecules of the cell. Smaller quantities of micronutrients are needed. The requirement for cobalt in the synthesis of cobalamin is one of these obvious requirements.

In connection with oxygen, there is an interesting situation because some microorganisms are *anaerobic* (requiring *absence* of oxygen), while others are *aerobic* (needing an ample supply of oxygen). Normally, oxygen is furnished by pumping large volumes of air through the mixture. An advancement in this procedure is the use of enriched air (over 21% oxygen).

A further important parameter in bioprocessing is that of providing adequate mixing to ensure that the several ingredients will be within immediate vicinity of each other. Depending upon the particular microorganism, they remain continually suspended in the watery medium, desirable from a processing standpoint, but some tend to collect in clusters; others tend to take the form of slimes.

To date, bioprocessing is primarily conducted on a batch basis. This provides flexibility in shifting the products to be made from time to time and, in particular, any failure to provide aseptic conditions may result in the condemnation of only one batch versus what could happen in the case of a continuous process. Notably, in the case of pharmaceutical biologicals, it is practical to keep track of the product by batch number from start to final use. Even with these kinds of problems, however, the many attractions of continuous processes are being thoroughly studied and applied in limited instances.

Genetic Engineering—State of the Art

As development of the concept of recombinant DNA progressed after its discovery in the early 1970s, great promise was given for the "engineering" of new plant species and varieties, new designer drugs, and new processes. In the 1970s, there were continual predictions of an "explosion" in such new product development. As of the early 1990s, however, there have been fewer dramatically improved products developed than had been initially contemplated. Actually, only comparatively few products are now moving from the laboratory to the processing plant. Over 77 small-scale field trials of genetically engineered tomato, potato, alfalfa, cucumber, corn (maize), and cotton have been conducted in several growing regions. The first food processing aid produced by a genetically engineered microorganism (the enzyme rennet) was approved by the U.S. Food and Drug Administration in March 1990. Most scientists in the field, however, maintain their confidence that the applications of genetic engineering in the food system alone are seemingly unlimited. Because of hunger problems throughout so much of the world, genetically engineered agriculture may prove to be the ultimate solution.

The principal steps required to genetically engineer a "new plant" are illustrated very schematically in Fig. 3.

In 1991, an expert in food biotechnology observed the following. (1) The public, in general, is not enamored with food biotechnology. (2) Public concerns continue pertaining to the potential long-term unanticipated effects of modifications of food. (3) Although willing to take modest risks, the public becomes very conservative in terms of food modifications on infants, children, and the chronically ill. (4) The public challenges corporation-sponsored university research in this area of technology. (5) The public has lost confidence in governmental regulatory actions pertaining to genetic engineering in the food field.

Attempts to provide sociological answers to scientific problems are beyond the purview of this encyclopedia. Provided such answers are forthcoming, biotechnology can be applied effectively to the solution of some of the following technical problems, particularly in terms of agriculture and food production:

- Develop temperature-tolerant plants that can survive in warmer or cooler climates. Frost damage causes more than $ 14 billion per year worldwide in crop losses.

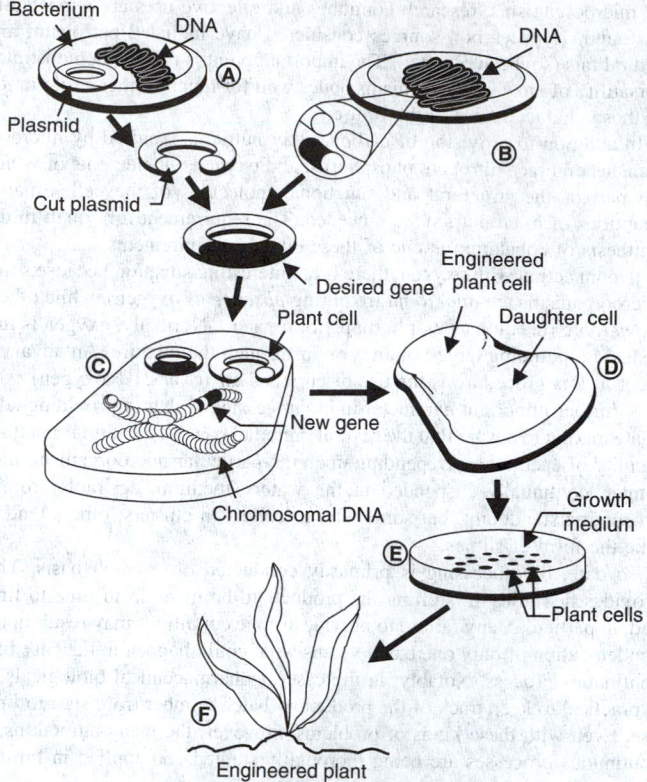

Fig. 3.   Major steps in producing a genetically engineered plant: (**a**) Plasmid (a circular piece of DNA found outside the chromosome in bacteria) is removed from the bacterium and cut open by using restriction enzymes. Plasmids are the main tool for inserting new genetic information into the microorganisms of plants. Restriction enzymes are proteins that recognize specific gene sequences on a chromosome and cut DNA at these sites. (**b**) The gene of interest is cut out of the chromosomal DNA of another organism and "pasted" into the plasmid by using ligase enzymes. A ligase enzyme is one that splices segments of DNA together. (**c**) The plasmid is put back into the bacterium and mixed with plant cells. The bacterium duplicates the plasmid and transfers the new gene into the chromosomal DNA of the plant cell. (**d**) When the plant cell divides, each daughter cell receives the new gene, giving the whole plant a new trait or characteristic. (**e**) Plant cells are placed on special growth media to promote the formation of callus (unorganized tissue). After shoots grow from the callus, the plantlets are transferred to traditional media that stimulates roots to grow. (**f**) The plantlets are transferred to soil and grow to maturity. (*After Volpo and Monsanto.*)

- Engineer plants that can withstand drought conditions. For example, if salt-tolerant varieties can be developed, sea water could be used for irrigation.
- Improve ways for certain crop plants, such as corn (maize) and wheat to fix their own nitrogen from the atmosphere, thus reducing fertilization costs by $ billions per year.
- Engineer insect and pest resistance into plants, thus not only reducing the cost for chemical control applications, but also alleviating the environmental problems related to agricultural chemicals.
- Develop plants that have nutritional values superior to those obtainable from the existing natural varieties. One example, would be that of increasing various amino acids contained in the edible portions of the plant.

The foregoing are only part of a larger genetic research agenda. See Harlander (May 1991) reference listed.

See also **Bioprocess Engineering (Biotechnology)**; and **Genetic Engineering**.

Additional Reading

Acharya, R.: *The Emergence and Growth of Biotechnology: Experiences in Industrialized and Developing Countries*, Edward Elgar Publishing, Inc., Cheltenham, UK, 1999.

Bhamidimarri, R.: *Environmental Biotechnology*, Elsevier Science, New York, NY, 1998.

Bielecki, S., J. Tramper, J. Polak: *Food Biotechnology*, Elsevier Science, New York, NY, 2000.

Cobb, A.B.: *Scientifically Engineered Foods: The Debate over What's on Your Plate*, Rosen Publishing Group, Inc., New York, NY, 2000.

Demain, A.L. and N.A. Solomon: "Industrial Microbiology," *Sci. Amer.*, **245**(3), 66–75 (September 1981).

Demain, A.L., J.E. Davies and R.M. Atlas: *Manual of Industrial Microbiology and Biotechnology*, ASM Press, Washington, DC, 1999.

Goldberg, I., R. Williams and E. Riemer: *Biotechnology and Food Ingredients*, Chapman & Hall, New York, NY, 1999.

Greenhalgh, R. and T.R. Roberts: *Pesticide Science and Biotechnology*, CRC Press, LLC., Boca Raton, FL, 1998.

Harlander, S.: "Social, Moral, and Ethical Issues in Food Biotechnology," *Food Technology*, 152 (May 1991).

Ives, C.L. and B. Bedford: *Agricultural Biotechnology in International Development*, Oxford University Press, Inc., New York, NY, 1998.

Kiely, T.: "Is Biotech Safe for the Big Time?" *Technology Review (MIT)*, 24 (October 1991).

Klefenz, H.: *Industrial Pharmaceutical Biotechnology*, John Wiley & Sons, Inc., New York, NY, 2001.

Levin, M.A. and H.S. Strauss: *Risk Assessment in Genetic Engineering*, The McGraw-Hill Companies, New York, NY, 1991.

Morris, P.C. and J.H. Bryce: *Cereal Biotechnology*, CRC Press, LLC., Boca Raton, FL, 2000.

Persley, G.J.: *Agricultural Biotechnology*, CAB International, New York, NY, 2000.

Rhodes, P.M. and P.F. Stanbury: *Applied Microbial Physiology: A Practical Approach*, Vol. 183, Oxford University Press, Inc., New York, NY, 1997.

Sterckx, S.: *Biotechnology, Patents & Morality*, Ashgate Publishing Company, Brookfield, VT, 1997.

Uhlig, H.: *Industrial Enzyes and Their Applications*, John Wiley & Sons, Inc., New York, NY, 1998.

Welch, R.A.S., S.R. Davis, A.I. Popay, et al.: *Milk Composition, Production and Biotechnology*, CAB International, New York, NY, 1997.

INDUSTRIAL CLIMATOLOGY.   See **Meteorology**.

INEQUALITY.   The notation $a < b$ means: a is less than b, and the notation $a > b$ means: a is greater than b. The notation $a \leq b$ means: a is either less than or equal to b; similarly $a \geq b$ means: a is either greater than or equal to b.

The rules for operating with these relations, which are called inequalities, are: (1) if $a < b$, and $b < c$, then $a < b$, (2) if $a < b$, then $(a + c) < (b + c)$; (3) if $a < b$ and $c > 0$, then $ac < bc$. If the sense of the inequality is the same for all values of the symbols for which its members are defined, the inequality is called an absolute or unconditional inequality. If the sense of an inequality holds only for certain values of the symbols involved, but is reversed or destroyed for other values of the symbols, the inequality is called a conditional inequality. The sense of an inequality is not changed if both members are increased or decreased by the same number nor is it changed if both members are multiplied, or divided, by the same positive number. The sense of an inequality is reversed if both members are multiplied, or divided, by the same negative number.

INERT.   A term used to indicate chemical inactivity in an element or compound. Helium, neon, and argon are practically inert gaseous elements; carbon dioxide is a gaseous compound of low activity. Ingredients added to mixtures chiefly for bulk and weight purposes are said to be inert.

INERT GASES (The).   The elements of group 18 of the periodic classification sometimes are referred to as the Inert Gases, or the Noble Gases. In order of increasing atomic number, they are helium, neon, argon, krypton, xenon, and radon. The elements of this group are characterized by their closed shells or subshells of electrons. They generally are considered as having zero valence. The name of this group derives from the lack of chemical activity which the elements display, forming compounds only under abnormal conditions (high pressures, strong electrical fields, etc.).

INERTIA.   A property manifested by all matter, representing the resistance to any alteration in its state of motion. Mass is the quantitative measure of inertia.

By inertia of an object is meant the property of opposing any change in motion of the object. To change the motion of an object, i.e., to accelerate or decelerate the object, a force, a push or pull, is required. A comparison of the masses of two objects can be made, by placing the two objects on a reasonably frictionless horizontal surface with a compressed spring between them. When the masses are released, they are accelerated in opposite directions and the larger acceleration occurs with the smaller inertia or mass of the object. Newton's second law is given as: The net applied force $\mathbf{F}$ on an object of inertia or mass m gives it an acceleration of $\mathbf{a}$, or $\mathbf{F} = m\mathbf{a}$.

INERTIAL COORDINATE SYSTEM.   A system in which the (vector) momentum of a particle is conserved in the absence of external forces. Thus, only in an inertial system can Newton's laws of motion be appropriately applied. For all purposes in meteorology, a system with origin on the axis of the earth and fixed with respect to the stars (absolute coordinate system) can be considered an inertial system. When relative coordinate systems are used, moving with respect to the inertial system, apparent forces arise in Newton's laws, such as the Coriolis force. See also **Coriolis Force**.

Additional Reading

Pedlosky, J.: *Geophysical Fluid Dynamics*, 2nd Edition, Springer-Verlag New York, LLC, New York, NY, 1998.

AMS

INERTIAL CURRENT.   A current in which the dominant balance is between the inertial and the Coriolis terms in the equation of motion, causing streamlines to be curved to the right (left) in the Northern (Southern) Hemisphere. If one thinks of the streamlines in an inertial current as being locally circular with radius of curvature R, and speed V along the streamlines, the balance of forces in the radial direction is

$$V^2/R = fV,$$

where f is the Coriolis parameter. Thus, the radius of curvature is V/f, which is about 10 km for a 1 m s^{-1} current.

See also **Coriolis Parameter**.

AMS

INERTIAL FLOW.   Flow in the absence of external forces; in meteorology, frictionless flow in a geopotential surface in which there is no pressure gradient. The centrifugal and Coriolis accelerations must therefore be equal and opposite, and the constant inertial wind speed V_i is given by

$$V_i = fR,$$

where f is the Coriolis parameter and R the radius of curvature of the path. The inertial path is anticyclonic in both hemispheres, more strongly curved near the poles than near the equator, resembling a series of similar loops called inertial circles. The inertial frequency with which these loops are described by the parcel is approximately equal to $f/2\pi = 2 \sin\phi$ per sidereal day, where ϕ is the latitude. All the loops are bounded on the north and south by the same parallels of latitude, but the parcel has a net longitudinal movement as it describes them. The inertial period (the reciprocal of inertial frequency) is just one-half pendulum day.

AMS

INERTIAL FORCE (OR INERTIA FORCE).   A force in a given coordinate system arising from the inertia of a parcel moving with respect to another coordinate system. For example, the Coriolis acceleration on a parcel moving with respect to a coordinate system fixed in space becomes an inertial force, the Coriolis force, in a coordinate system rotating with the earth. See also **Apparent Force**; and **Coriolis Acceleration**.

INERTIAL GUIDANCE SYSTEMS.   Early inertial guidance systems, developed in Germany during World War II, simply gyrostabilized the airframe to the desired flight attitude, and used a single accelerometer to measure acceleration along the longitudinal (thrust) axis. When the

integrated acceleration reached the desired injection velocity the engines were cut off.

More sophisticated inertial systems provide a gyro-stabilized platform which is gimbal-mounted to permit unlimited vehicle motion without disturbing the stable element. On the stable element are mounted linear accelerometers to measure the two or three components of the vehicle's acceleration vector. These components of acceleration are inputs to a computer, which solves the navigation equations (see Fig. 1), adding computed gravitation, integrating to find velocity, and integrating again to determine position.

$$\mathbf{R} = \int_0^t \int_0^t (\mathbf{A} + \mathbf{G})\, dt^2 + \mathbf{V}_0 t + \mathbf{R}_0$$

where $\mathbf{R}$ = position vector
$\mathbf{A}$ = nongravitational acceleration vector (sensed acceleration)
$\mathbf{G}$ = gravitational vector (calculated)
$\mathbf{V}_0$ = initial velocity vector (inserted)
t = time
$\mathbf{R}_0$ = initial radius vector (inserted)

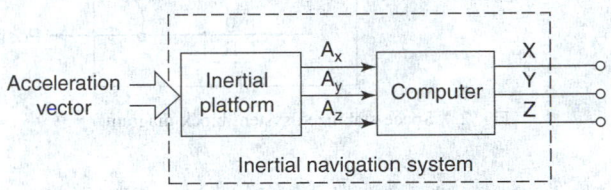

Fig. 1.   Function of the basic inertial system.

The foregoing basic inertial navigation equation points up some of the fundamental characteristics of inertial systems. (1) The inertial system must have initial position and initial velocity information (the two constants of integration). (2) The accelerometer senses all nongravitational forces (including thrust, drag, lift, and structural support). (3) The gravitational field is not sensed; it must be calculated from known field equations.

Distinctive characteristics of inertial systems include: (1) They give continuous, rather than discrete, information on acceleration, velocity, position, and vehicle attitude; (2) they require no signals from outside the system and thus they are jamproof, and can be used in vehicles launched in salvo; (3) they do not radiate signals and hence they are difficult to detect in military applications; (4) they can be launched quickly, but are most accurate when adequate prelaunch time is available for warmup, trim, and alignment; (5) they have errors that are a function of time rather than speed or distance; and (6) they provide byproduct signals, such as stabilization for flight control or radar antennas, and velocity for mapping cameras.

Systematic errors in pure inertial systems based on error in the knowledge of the gravity vector have a characteristic Schuler oscillation corresponding to orbital period (84.5 minutes at the earth's surface) in the horizontal components of the navigation position vector and an unstable exponential error in the local vertical component.

In the basic space-stabilized system (Fig. 2), the three accelerometer input axes are stabilized to any desired orientation in space by the gyroscope stabilization control loops. An often-used orientation for space vehicle launches is to have the Z axis vertical at time $t = 0$, the Z and X axes in the orbital or launch plane, and the Y axis orthogonal to Z and X. The gravitational force which starts out parallel to the Z axis is continuously computed as a function of the vehicle's position.

Another system mechanization is the local vertical system that maintains the Z axis vertical and the X axis north throughout flight for convenience in surface and air navigation. This requires biasing each of the gyros with a turning rate that is a function of the earth's rotation rate plus the vehicle angular velocity around the curved earth's surface:

$$\omega_x = \Omega \cos lt + \frac{V_e}{R}$$

$$\omega_y = \frac{V_n}{R}$$

$$\omega_z = \Omega \sin lt + \frac{V_e}{R} \tan lt$$

where $\omega_{x,y,z}$ = the computed bias signals to the x, y, z (north, east, and vertical) gyros

Ω = earth's sidereal rotation rate (15.041 degrees/hour)

lt = local latitude of the vehicle

$V_{e,n}$ = vehicle east and north velocity

R = local earth's radius plus altitude

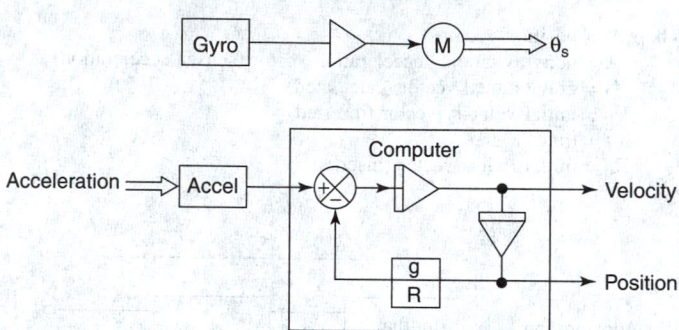

Fig. 2.   Space-stabilized system block diagram.

In the local vertical system, the accelerometers measure acceleration in a north, east, and vertical reference system which rotates in space and, therefore, requires coriolis corrections. The explicit gravitational calculation is avoided in the two level axes, and the vertical axis is often unnecessary in two-dimensional surface navigation. This mechanization is shown in Fig. 3.

During the navigation mode, the gyros hold the stable element to the desired attitude, but this attitude must be assumed before the navigation mode begins. This can be done during a self-alignment mode. The stable element is placed in a local level attitude by using information from the two level accelerometers, whose outputs are null when their input axes are level. Azimuth orientation information is derived from the east gyro, which senses no component of the earth's rate of rotation when oriented east.

In addition to the gimbaled gyro-stabilized systems, there are several gimballess or strapdown configurations. One is the relatively simple accelerometer, vehicle attitude control system first described. Another uses three accelerometers whose vehicle-referenced outputs pass through

a dynamic coordinate transformation matrix in the computer. Three *rate* gyros provide the computer with vehicle attitude rate information necessary to compute the matrix. Although this approach eliminates gimbaling, it requires precision rate gyros and relatively large computer capacity to integrate attitude rate and provide a dynamic matrix.

A third gimballess system concept uses an inertial reference, such as electrostatically suspended gyros to give vehicle *attitude* information to the computer for use in calculating the dynamic coordinate conversion matrix. Attitude rate integration is thereby avoided.

An inertial navigation system primarily provides vehicle velocity or position. Guidance can be provided by the inertial system by supplying target location information to the computer, which then compares vehicle position with target position and calculates steering and (in the case of space flight), engine shutoff commands. See also **Space Vehicle Guidance and Control**.

INERTIAL INSTABILITY.   Generally, instability in which the only form of energy transferred between the steady state and the disturbance is kinetic energy. See also **Barotropic Instability**; and **Helmholtz Instability**.

2. The hydrodynamic instability arising in a rotating fluid mass when the velocity distribution is such that the kinetic energy of a disturbance grows at the expense of kinetic energy of the rotation.

For a small plane-symmetric displacement (wavenumber zero) using the parcel method, this criterion for instability is that the centrifugal force on the displaced parcels is larger than the centrifugal force acting on the environment. On the assumption that absolute angular momentum is conserved, this states that the fluid is unstable if absolute angular momentum decreases outward from the axis;

$$R\frac{\partial \omega_a}{\partial R} + 2\omega_a < 0,$$

where ω_a is the absolute angular speed and R the distance from the axis. If this criterion is applied to rotation of the westerlies about the earth's axis, the angular speed of the earth is so large that the inequality fails and the disturbance is stable. If applied to a system rotating about a local vertical, the criterion might be satisfied in low latitudes where the component of the earth's rotation about the local vertical is small. Inertial instability has been suggested in connection with the genesis of hurricanes.

Additional Reading

Holton, J.R.: *An Introduction to Dynamic Meteorology*, 4th Edition, Elsevier Science & Technology Books, New York, NY, 2004.

AMS

INERTIAL MASS.   See **Gravitation**.

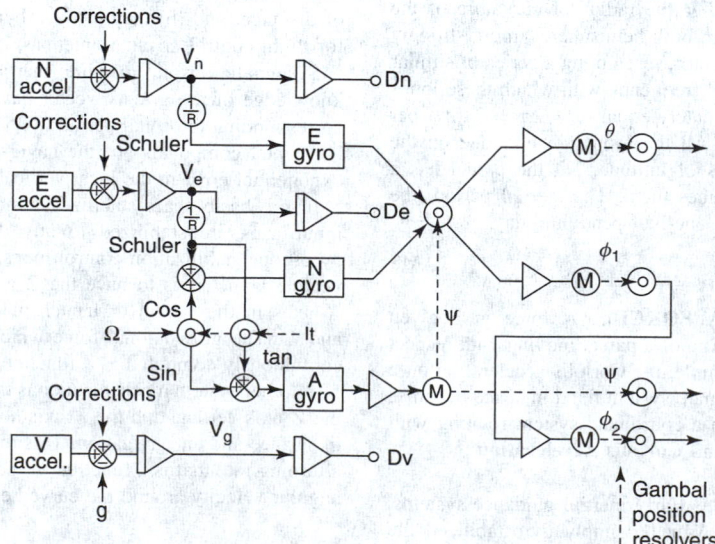

Fig. 3.   Local vertical system block diagram.

INERTIAL REFERENCE FRAME. Within Newtonian mechanics, a reference frame relative to which every point mass not subjected to a net force is unaccelerated.

Within relativistic mechanics, a reference frame is inertial in a (local) region of space and time if every point mass in this region remains in uniform motion. According to the principle of relativity, all the laws of physics have the same form (and contain the same numerical constants) when expressed relative to any inertial reference frame.

Additional Reading

Taylor, E.F., and J.A. Wheeler: *Spacetime Physics*, 2nd Edition, W. H. Freeman Company, New York, NY, 1992.

INERTIAL STELLAR COMPASS. See **Space Technology 6 (ST6) MISSION**.

INERTIA (Moments and Products of). In the general case of the motion of a particle or aggregate of particles with respect to a single fixed point, the angular momentum can be written as having three components with respect to a coordinate system based at the fixed point:

$$H_x = \omega_x \Sigma m_i (y_i^2 + z_i^2) - \omega_y \Sigma m_i x_i y_i - \omega_z \Sigma m_i x_i z_i$$

$$H_y = -\omega_x \Sigma m_i x_i y_i + \omega_y \Sigma m_i (x_i^2 + z_i^2) - \omega_z \Sigma m_i y_i z_i$$

$$H_z = -\omega_x \Sigma m_i x_i z_i - \omega_y \Sigma m_i y_i z_i + \omega_z \Sigma m_i (x_i^2 + y_i^2)$$

where ω_x, ω_y, ω_z = components of angular velocity, m_i = mass of ith particle, x_i, y_i, z_i = coordinates of ith particle.

The terms $\Sigma m_i(y_i^2 + z_i^2)$, $\Sigma m_i(x_i^2 + z_i^2)$, $\Sigma m_i(x_i^2 + y_i^2)$ are called moments of inertia with respect to the x, y, and z axes, respectively, and are symbolized by I_{xx}, I_{yy}, and I_{zz}.

The terms $\Sigma m_i x_i y_i$, $\Sigma m_i x_i z_i$, etc., are called the products of inertia and are symbolized by I_{xy}, I_{xz}, etc.

For a continuous rigid body, the summations are replaced by integrals over the volume of the body. In a rigid body, it is sometimes easier to choose coordinate axes, called moving axes, which are fixed in the body. There always exists one set of such axes, called principal axes, such that the products of inertia vanish and the angular momentum can be expressed in terms of the moments of inertia alone. Following are formulas for the moments of inertia of certain homogeneous solids with respect to the axes specified (M is the total mass of the body in each case):

Particle distant r from axis	Mr^2
Sphere of radius R, with respect to any diameter	$\frac{2}{5}Mr^2$
Cube of edge L, with respect to axis through center parallel to edge	$\frac{1}{6}ML^2$
Rectangular plane, dimensions $A \times B$, with respect to axis perpendicular to it at center	$\frac{M}{12}(A^2 + B^2)$
Cylinder of length L and radius R, with respect to axis perpendicular to its length at center	$M\left(\frac{L^2}{12} + \frac{R^2}{4}\right)$
Cylinder of radiaus R, with respect to its own longitudinal axis	$\frac{1}{2}Mr^2$
Any body with respect to any axis distant r from the center of mass, the value for a parallel axis through that point being I_0	$I_0 + Mr^2$

Experimental methods of obtaining moments of inertia by the use of a torsion pendulum are explained in any laboratory manual of elementary dynamics.

For certain purposes it may be desirable to know at what one distance from the axis all the particles of the body of mass M would have to be placed to give it the same moment of inertia I that it actually has. This distance is the radius of gyration, and is expressed by the formula

$$R\sqrt{\frac{I}{M}}$$

The "principal axes" of a body through a given point are axes of maximum or minimum moment of inertia.

The quantity expressed by

$$I\int r^2 da$$

in reference to any plane figure, is called the areal moment of inertia of the figure with respect to a given straight line in its plane. The figure is divided into elements of area da, each element multiplied by the square of its distance r from the axis, and the products summed as indicated above to get the areal moment of inertia. This quantity is purely geometric and has of course no actual connection with inertia or mass. One of its important applications is in the theory of flexure of elastic rods or beams. If E is the (Young's) elastic modulus of the material and I the areal moment of inertia of the cross section with respect to the neutral axis, the bending moment or flexural torque required to bend the rod to a curvature C is given by

$$T = EIC$$

INERTIA WAVE. 1. Any wave motion in which no form of energy other than kinetic energy is present. In this general sense, Helmholtz waves, barotropic disturbances, Rossby waves, etc., are inertia waves. See also **Rossby Wave**.

2. More restrictedly, a wave motion in which the source of kinetic energy of the disturbance is the rotation of the fluid about some given axis.

In the atmosphere a westerly wind system is such a source, the inertia waves here being, in general, stable. A similar analysis has been applied to smaller vortices, such as the hurricane.

INFANT DEATH. See **Sudden Infant Death Syndrome**.

INFANTILISM. See **Pituitary Gland**.

INFECTION. The invasion of body tissues by pathogenic microorganisms, such as germs, viruses, and fungi, resulting in the process termed *disease*. The term infection is not used in connection with invasion of the body by higher-order organisms, such as worms. The latter condition is termed *infestation*. The infectious disease process includes the successful establishment and multiplication of the microorganisms in body tissue to produce tissue damage, both locally and at distant sites by the toxic products of the metabolism of the microorganisms. The successful establishment of infection is determined as to the invading organism, chiefly by the size of the invading force and the ability of the organisms to produce toxins. In the case of the subject invaded, the general physiological state of the body, with particular reference to age and physical vitality, together with the existence of any local devitalization of the tissues at the site of infection, are the factors most concerned. Once established, the infection tends to run a course characteristic of the organism concerned, modified by the reaction of the tissues of the host and by the influence of such therapeutic measures as may be used against it. See also **Inflammation**.

The term infection also applies to the transfer of disease from one part of the body to another, more specifically termed *autoinfection*. A common example is the spread of infection from diseased tonsils via the bloodstream to joints whereupon inflammatory arthritis may be produced. A large number of infectious diseases are communicable, depending upon the virility of the microorganism and the manner in which transfer may take place — by physical contact, by presence of air, in water, etc.

Additional Reading

Dalhoff, A.: *Bacterial Infections*, S. Karger Publishers, AG, Basel, Switzerland, 1999.

Evans, A. and P. Brachman: *Bacterial Infections of Humans: Epidemiology and Control*, 3rd Edition, Kluwer Academic Publishers, Norwell, MA, 1998.

Powers, D. (Editor), et al.: *Aging, Immunity, and Infection*, Springer Publishing Company, New York, NY, 1994.

INFECTIOUS ARTHRITIS. See **Arthritis (Infectious)**.

INFECTIOUS MONONUCLEOSIS. This disease which affects lymphoid tissues throughout the body is caused by one of the four main herpes viruses, a virus known as the Epstein-Barr virus (EBV). See also **Virus**.

Replication of the virus occurs in lymphocytes and nasopharyngeal epithelial cells. These cells have receptors that are specific for EBV. The primary manifestation of EBV in humans is *infectious mononucleosis*. However, EBV has been detected in biopsies and cells cultured from human cancers, these cancers also involving the cells affected in infectious mononucleosis. It is estimated that one-quarter and possibly a considerably greater fraction of the population in the United States carries antibody to EBV. Statistics indicate that individuals most likely to carry antibody reside in a warm climate and are at the lower socioeconomic level, in contrast with persons who live in cold climates.

According to an earlier common definition, infectious mononucleosis was the "kissing" disease and this has been shown to be reasonably descriptive. However, the disease may well be spread by aerosol and not necessarily by close contact. Attempts to transmit the virus by blood or plasma transfusion have not been successful. Replication of the virus in the throat produces pharyngitis and fever. After replication in the throat, EBV infects B lymphocytes and spreads throughout the body. In the peripheral blood, the presence of proliferating mononuclear cells of varying size, derived from lymphocytes, is characteristic. The cells are probably T lymphocytes and contain T lymphocyte-specific antigens, but lack B-cell markers. EBV infects B cells and causes them to proliferate, but this is controlled by the T lymphocytes. Nervous system complications, such as encephalitis, are rare. Replication of the virus in lymphocytes is usually limited, although EBV in the laboratory has been demonstrated to transform these cells into malignant lymphocytes. The connection between EBV and malignancies is now under intense investigation.

Diagnosis of infectious mononucleosis requires clinical detection of EBV antibodies, some of which are present during the early stages of the disease, but are not found in later stages. It is very difficult to determine when a patient may have stopped shedding the virus from the throat, and it is also difficult to determine a precise incubation period for the disease. Specific treatment of the disease has not been forthcoming. The physician directs attention toward alleviating symptomatic features—fever, sore throat, etc. Where airway obstruction or severe hepatitis develop as complications, corticosteroids have been used with some success.

Rupture of the spleen is an uncommon complication of the disease, but is probably responsible for most of the few deaths reported in infectious mononucleosis.

Frank jaundice has been reported in up to 25% of cases and the commonest form of neurological involvement is meningitis; there are occasional clusters of this in a benign form.

Because of the oncogenic (cancer characteristics) of EBV, the current outlook for an effective vaccine for infectious mononucleosis is guarded.

Additional Reading

Berkow, R. and M.H. Beers: *The Merck Manual*, 17th Edition, Merck & Company, Inc., Whitehouse Station, NJ, 1999.

Schlossberg, D.L.: *Infectious Mononucleosis*, Springer-Verlag, Inc., New York, NY, 1989.

Shader, L. and J. Zonderman: *Mononucleosis and Other Infectious Diseases*, Chelsea House Publishers, Broomall, PA, 2000.

R.C. VICKERY, M.D., D.Sc., Ph.D., Blanton/Dade City, FL

INFERIOR MIRAGE. A mirage in which the image or images are displaced downward from the position of the object. If only a single image of distant objects is seen, then the term *sinking* is often applied: A horizontal surface appears to curve downwards with increasing distance and terminate in a relatively nearby optical horizon. The inferior mirage is most striking when it exhibits two images; the second, lower image is always inverted and of reduced magnification. Sometimes textbooks suggest that there is but a single image: the lower, inverted one. The upper erect image is claimed to be the object. However, both are images, and have positions and magnifications that differ from that of the object. Also, the lower inverted image is sometimes misinterpreted as having resulted from a reflection and when this is seen over land, it leads to the assumption that there must be water in the distance causing the reflection. This is the origin of the long association of the mirage and illusory water, and this leads to the assumption that water is present on a dry surface. The mirage owes its name (from se mirer, to look in a mirror) to this impression of arising from a reflection, having been named by French mariners for

images seen at sea. For vertical objects seen beyond the optical horizon, typically the lower portion of the object cannot be seen, and an upper portion of the object is seen twice: erect, and inverted. The farther away the object, the more of the lower portion of it will have vanished so that, for example, the upper decks of a distant ship might appear erect and inverted and apparently floating above and disconnected from the optical horizon while the lower decks will not be seen at all. Sometimes a scene such as this is misinterpreted as resulting from a *superior mirage* by a person who thinks the ship's images have been lifted up from the horizon. Actually, in this case, everything is displaced, but the horizon has merely been displaced more. Inferior mirages occur over a surface when the temperature decreases with height. The formation of a two-image inferior mirage also requires that the temperature gradient decrease with height. These conditions are met when the surface is relatively warm, resulting in an upward heat flux, such as over sun-warmed ground or a lake at night. See also **Looming**; **Mirage**; **Omega Sun**; **Sinking**; **Stooping**; **Superior Mirage**; and **Towering**.

INFERTILITY. The inability to reproduce; sterility. The cause in any one case may lie in either or both of the partners of the sexual union. Often the term is confused with sexual impotence in the male, which is an entirely different condition associated with the inability to perform the sexual union. However, sexual potency does not necessarily imply that the male is fertile, because sterility in the male usually is the result of some defect in the number or the structure of the sperm cells. Until relatively recent years, male sterility was underestimated. A better understanding of human reproduction has led to more accurate diagnosis of sterility and the male as well as the female should undergo careful examination before a final diagnosis is made.

Infertile unions sometimes result from a lack of understanding of the sexual cycle in females. The average woman is fertile only during a short period of each month, i.e., during the time when the egg cell is in the Fallopian tube and is still viable. The high peak of fertility occurs approximately 12 to 16 days after the beginning of the last menstrual period in a woman with a 28-day menstrual cycle. In any event, the period of ovulation may be difficult to predict with certainty, since the menstrual cycle may vary in length, as well as in time of ovulation, among individuals. Carefully charting rectal temperature can help in determining the time of ovulation because then there is a sharp, but transient rise in body temperature. At this time, the egg has been ejected from the ovary and descends gradually into the Fallopian tube. This represents the time of maximum fertility. The egg is viable for only 1 or 2 days at most—then loses its fertility. Similarly, the spermatozoa are viable in the genital tract for only a short period, perhaps no more than 2 days. Thus, any union that precedes or follows the time of ovulation by more than 2 days will probably be unsuccessful.

Infertility in Women

Infertility in women may be caused by a variety of conditions. Pelvic diseases and infections may be conducive to infertility. These conditions have decreased in a marked fashion, however, since the availability of antibiotic drugs. Inflammation of the genital organs is often responsible for the production of mucus that is considered toxic or poisonous to spermatozoa. The mucus also tends to obstruct passage of the male germ cells into the uterus. The orifice of the cervix may become obstructed with mucus.

Two major causes of infertility in women are: (1) failure to ovulate (about 60% of cases); and (2) obstruction of the Fallopian tubes (about 40% of cases).

Anovulation. Originally developed as an oral contraceptive, a drug known as *clomiphene* was found, in actuality, to encourage fertility rather than to discourage it. This has been described as one of the major turnarounds in modern pharmacology. This drug binds to estrogen receptors on the hypothalamus and prevents estrogen from binding there. In a complex hormonal pathway (described in the entry on **In-Vitro Fertilization**), this drug produces the end result of hormonal stimulation of the ovaries, thus causing ovulation. Some authorities estimate that 30% of infertile patients who take clomiphene (citrate) ovulate and become pregnant. When there is no positive response to clomiphene (Clomid®), a substance known as *human menopausal gonadotrophin* (HMG) may be prescribed. HMG (Pergonal®), extracted from the urine of postmenopausal women, is rich in follicle-stimulating hormone (FSH) and also contains some luteinizing hormone (LH). These hormones stimulate the ovaries to ovulate. A problem sometimes encountered with HMG is overstimulation. In some instances, it has been reported that the ovaries greatly enlarge, precipitating a life-threatening situation.

In an effort to develop a substance with milder ovary stimulation, some researchers have produced a synthetic analog of luteinizing hormone releasing hormone (LHRH). Acting on the pituitary, LHRH stimulates the increased excretion of luteinizing hormone (LH). This therapy is in the testing stage.

In polycystic ovary syndrome (PCO), symptoms reflect an excess of androgen, including increased body hair, but true virilism, with balding and deepening of the voice, is uncommon. Generally, one or both ovaries are enlarged, but in many patients the ovaries are cystic, with thickened capsules, yet not palpably enlarged. Enlargement of the ovaries can be demonstrated by echography, pneumogynogram, and laparotomy or laparoscopy. (Laparoscopy is described in the entry on **In-Vitro Fertilization**.) Polycystic ovary syndrome arises from imbalances in the hormonal systems of reproduction. For a number of years, a procedure known as wedge resection was performed on one or both ovaries. Many patients ovulated and had normal periods within a few months after the procedure, but the results were seldom permanent. Wedge resection was used mainly when prompt pregnancy was desired. In recent years, clomiphene therapy has been used by many physicians in the treatment of this syndrome.

It has been found that failure to ovulate may arise from the oversecretion of the hormone *prolactin*. This condition is called *hyperprolactinemia*. One researchers has suggested that nearly 30% of women with menstrual abnormalities are hyperprolactinemic. The manner in which prolactin inhibits ovulation is poorly understood. A number of years ago, the drug *bromocriptine* was found to inhibit prolactin secretion. This drug has been used in Europe to treat infertility, but was not introduced into the United States until 1978. Several hundred infants born to women who have taken the drug have shown no evidence of any birth defects. The drug also is effective in treating *galactorrhea* (milk in breasts). Bromocriptine also has been used in connection with pituitary tumors which secrete large amounts of prolactin.

Surgical Reconstruction of Blocked Fallopian Tubes. Introduction of microsurgical techniques into procedures for surgically unblocking the tubes has increased the success rate, although not by a wide margin. By magnifying the surgical field some 4 to 25 times, the surgeon can distinguish the three layers of the tubes and sew them up one layer at a time. This contrasts with former procedures where surgeons looped stitches around the outer edges of the tubes and allowed the layers of the tubes to grow together. Microsurgery has been found useful only when there are small obstructions at the narrowest end of the tube (end nearest the uterus). Usually the operation has been most successful to reverse a former sterilization operation, where a section of the tubes had been looped out and tied. Success rate on these cases has been about 75%. Successful reconstruction depends upon how badly damaged the tubes may be. Damage from extensive tubal infections (as caused by gonorrhea) reduces success rate to about 20%, and even in such instances an ectopic pregnancy (where the embryo becomes embedded in tube and ruptures it, with accompanying spontaneous abortion) may occur.

Infertility in Men

In males, the semen may be inadequate to produce conception if there are large numbers of abnormal cells in proportion to normal cells. The average sperm cell is composed of a head and a relatively long thin tail. The movements of the tail are whiplike and serve as a means of propulsion. The head constitutes a major portion of the cell and contains the germ plasm. Examination of semen often reveals abnormal cell forms, such as a double-headed cell, a split tail, or a shortened tail. Motility becomes an important factor when the cells are in the vagina, where they swim toward the uterus. Although the only decisive proof of fertility in the male is the production of offspring, a semen examination can be of great value. When the testes are involved in a case of mumps, the infection is known as *orchitis*. This condition occurs once in every 4 to 5 cases of mumps among males between the ages of 15 and 26. Young boys are seldom affected. When both testes are involved, sterility may result. Normally,

the testes descend from inside the abdomen into the scrotum by the time of birth, but sometimes the descent is interrupted in one of the various stages of development. In a small percentage of cases, the testes remain in the abdomen after birth (*cryptorchism*). In this position, the testes do not function, because the temperature of the body is too high for the production of spermatozoa. Undescended testes may provoke a series of complications in addition to that of sterility. See also **Gonads**; and **Impotence**.

Additional Reading

Burfoot, A.J.: *Encyclopedia of Reproductive Technologies*, Perseus Books Group, Boulder, CO, 1999.

Carcio, H.A.: *Management of the Infertile Woman*, Lippincott Williams & Wilkins, Philadelphia, PA, 1998.

Carr, B.R. and R.E. Blackwell: *Textbook of Reproductive Medicine*, Appleton & Lange, Old Tappan, NJ, 1998.

Cook, R.: *Infertility: A Psychosocial and Medical Problem*, John Wiley & Sons, Inc., New York, NY, 1998.

Glover, T.D. and C.L. Barratt: *Male Fertility and Infertility*, Cambridge University Press, New York, NY, 1999.

Jequier, A.M.: *Male Infertility: A Guide for the Clinician*, Blackwell Science, Inc., Malden, MA, 2000.

Kupesic, S. and D. De Ziegler: *Ultrasound and Infertility*, Parthenon Publishing Group, New York, NY, 2000.

McElreavey: *The Genetic Basis of Male Infertility*, Vol. 28, Springer-Verlag, Inc., New York, NY, 2000.

Mortimer, D. and R. Jansen: *Towards Reproductive Certainty: Fertility and Genetics beyond 1999*, Parthenon Publishing Group, New York, NY, 1999.

Naz, R.K.: *Endocrine Disruptors: Effects on Male and Female Reproductive Systems*, CRC Press, LLC., Boca Raton, FL, 1999.

Sohoham, Z., H.S. Jacobs and C.M. Howles: *Female Infertility Therapy: Current Practice*, Blackwell Science, Inc., Malden, MA, 1998.

INFILTROMETERS. Infiltrometers allow measuring of the rate of infiltration of water into a given medium. Rain simulators, a particular type of infiltrometer, are discussed elsewhere. See also **Rain Simulators**. Unlike rain simulators, other infiltrometers do not simulate raindrop activity but measure the rate of infiltration from a reservoir of water at the ground surface. Their main advantage is their simplicity and ease of use. There are two main types of infiltrometer, a ponded ring infiltrometer and a disk infiltrometer.

Ponded Ring Infiltrometer

A ponded ring infiltrometer is the most common type of infiltrometer used. It is inexpensive to construct or operate and requires relatively little water for measurements. One person can set up and run several tests simultaneously. A cylindrical ring of stainless steel or plastic pipe is used, and the ring is pushed a few centimeters into the ground (Fig. 1). Care is taken to minimize disturbing the soil surface and soil structure during installation. Water is then flooded into the ring. The water inside the ring gradually infiltrates the soil, so measurements of the depth of water in the ring over time can be used to provide an infiltration rate usually in units of length per unit time (e.g., mm hr^{-1}) allowing easy comparison with rainfall intensity data. If the ring empties, it can be easily refilled.

There are two main problems in this method: (1) lateral flow below the ring; (2) changing pressure of water in the ring as the water level decreases in the ring during infiltration. To get around the lateral flow problem, it is advisable to use two concentric rings; the larger ring forms a buffer compartment around the inner ring (Fig. 2). While water is topped up in the outer ring, measurements are taken only of infiltration from the inner ring. It is preferable that a constant head of water is maintained within the rings to avoid changing pressures. The head of water should also be less than 5 cm (2 in) deep. The constant head can be supplied by a Mariotte bottle placed above the rings, and readings of water usage can be made from the Mariotte bottle (Fig. 2). Thus the optimum design for a ring infiltrometer is a constant-head, double-ring device whose rings have large diameters, so that they can represent the soil surface more readily. A plot of water uptake against time should show a leveling off as the infiltration capacity is reached. It often takes several hours before a constant rate of infiltration is achieved from the infiltrometer, depending on soil type, texture, and antecedent soil moisture.

The method of placement is serious limitation to the use of ring infiltrometers. Knocking rings into the ground can result in destruction

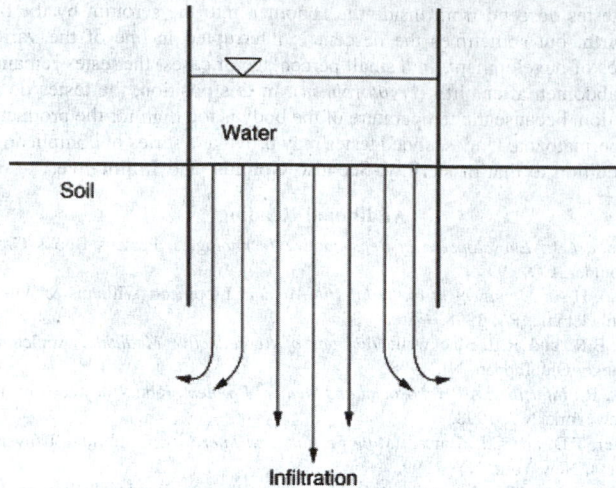

Fig. 1.   Infiltration from a single ring infiltrometer.

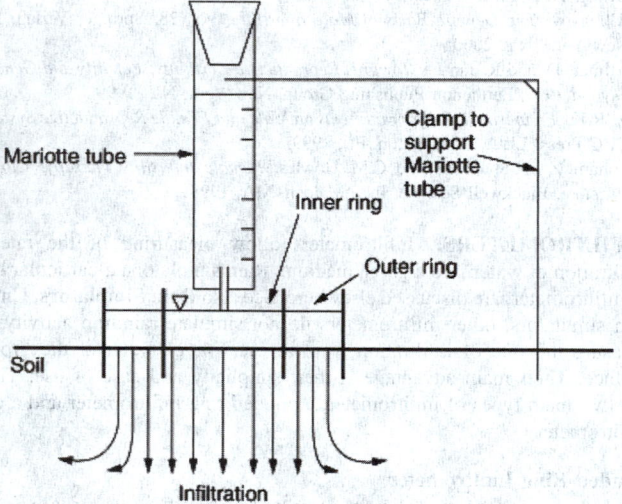

Fig. 2.   Double-ring infiltrometer with constant-head supply.

of the soil structure or compression. If a soil is shattered, it can disturb the contact between the ring edge and the soil, resulting in leakage and high infiltration rates. Research has demonstrated that ring infiltrometers give higher infiltration rates than rain simulators because of the effect of a pond on the soil surface, lateral seepage, and soil cracking.

Disk Infiltrometer

Tension disk infiltrometers are a standard tool for *in situ* determination of saturated and near-saturated soil hydraulic properties. These infiltrometers have a porous membrane disk at their base which rests on the soil surface of interest. To assess the role of matrix and macropore flow, a tension infiltrometer allows infiltration of water into the soil matrix, while preventing flow into larger pores that may otherwise dominate the infiltration process. The infiltrometer provides a source of water at a small negative pore water pressure at the surface. The supply pressure head is controlled with a Mariotte bottle. A schematic diagram of a tension disk infiltrometer is given in Fig. 3. As for a ring infiltrometer, infiltration rates can be measured manually by observing the volume of water lost from the Mariotte bottle over time. The negative pressure prevents the larger pores that fill at greater pore water pressures from wetting and short-circuiting the flow. Hence, by subtraction, the hydrological role of larger pores during the infiltration process can be evaluated. Further details on designs for these infiltrometers are given in Ankeny et al. and Zhang et al. Most studies using tension disk infiltrometers have been conducted at the soil surface, although Azevedo et al. looked at infiltration properties of an Iowa loamy soil at 0.15 m depth, and Logsdon et al., and Messing and Jarvis conducted measurements at different depths under different agricultural tillages.

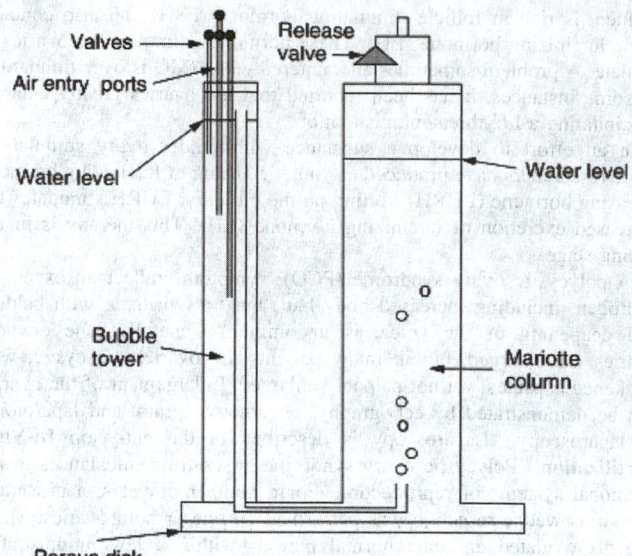

Fig. 3.   Schematic diagram of a tension disk infiltrometer.

Careful preparation of the soil surface is required to use a disk infiltrometer. This is to ensure even and sound contact of the disk with the soil surface. At each location, vegetation must be cut back to the soil surface, and a fine layer of moist fine sand of the same diameter as the circular base of the infiltrometer should be applied. This must be smoothed out to remove any irregularities at the soil surface and improve contact between the disk and soil surface. Moist sand is essential as air-dry sand may readily fall down into surface-vented macropores, forming 'wicks'. The infiltrometer is then placed on the sand. The weight of the infiltrometer must not be too great so as compress the soil surface significantly, as this will restrict the water flux. Therefore, it is usual to make the water supply reservoir narrow so that the total volume of water held in the infiltrometer is low, resulting in reduced weight and also aiding accurate measurements of discharge.

If a range of supply heads is to be used, then infiltration tests are normally conducted with the lowest supply head first. Reversal of this may lead to hysteresis where drainage occurs close to the disk while wetting continues near and at the infiltration front. Infiltration measurements should proceed until a steady state is achieved. Users should be careful to ensure that sunlight does not heat the supply reservoir significantly: this can be reduced by shading.

Methods for analyzing the data from disk infiltrometers (e.g., hydraulic conductivity values and macropore contribution to infiltration) are given in Reynolds and Elrick. Typically, Wooding's solution for infiltration from a shallow pond is combined with Gardner's unsaturated hydraulic conductivity function. A range of assumptions is involved in using of these techniques, including that the hydraulic conductivity before the test is much less than that imposed under the infiltration experiment and that the soil below the tension disk is homogeneous, isotropic, and uniformly saturated. These are unlikely to be satisfied fully in most soils, and so it is necessary to evaluate the potential errors involved.

Water levels in supply bottles for both the ring and disk infiltrometers can be measured automatically by using electronic pressure sensors with data recorders, thus allowing an experiment to continue unattended for several hours if the infiltration rates are very low.

This article originally appeared in the *Water Encyclopedia*, 5 Vols. Lehr, J. H., and J. Keeley, Editors, John Wiley & Sons, Inc, Hoboken, NJ.

Additional Reading

Ankeny, M.D., T.C. Kaspar, and R. Horton: "Design for an Automated Tension Infiltrometer," *Journal of the Soil Science Society of America*, **52**, 893–896 (1988).

Azevedo, A.S., R.S. Kanwar, and R. Horton: "Effect of Cultivation on Hydraulic Properties of an Iowa Soil Using Tension Infiltrometers," *Soil Science*, **163**, 22–29 (1998).

Gardner, W.R.: "Some Steady State Solutions of the Unsaturated Moisture Flow Equation with Application to Evaporation from a Water Table," *Soil Science*, **85**, 228–232 (1958).

Holden, J., T.P. Burt, and N.J. Cox: "Macroporosity and Infiltration in Blanket Peat: The Implications of Tension Disc Infiltrometer Measurements," *Hydrological Processes*, **15**, 289–303 (2001).

Logsdon, S.D., J.L. Jordahl, and D.L. Karlen: "Tillage and Crop Effects on Ponded and Tension Infiltration Rates," *Soil and Tillage Research*, **28**, 179–189 (1993).

Messing, I. and N.J. Jarvis: "Temporal Variation in the Hydraulic Conductivity of a Tilled Clay Soil as Measured by Tension Infiltrometers," *Journal of Soil Science*, **44**, 11–24 (1993).

Reynolds, E.D. and D.E. Elrick: "Determination of Hydraulic Conductivity Using a Tension Infiltrometer," *Journal of the Soil Science Society of America*, **55**, 633–639 (1991).

Wooding, R.A.: "Steady Infiltration from a Shallow Circular Pond," *Water Resources Research*, **4**, 1259–1273 (1968).

Zhang, Y., R.E. Smith, G.L. Butters, and G.E. Cardon: "Analysis and Testing of a Concentric-disk Tension Infiltrometer," *Soil Science Society of America Journal*, **63**, 544–553 (1999).

JOSEPH HOLDEN, University of Leeds, Leeds, UK

INFINITY.   The word infinity is usually defined only as part of a phrase. Thus, a function $f(x)$ is said to approach infinity at a point $x = a$ if after choice of any number N, a number $\delta > 0$ can be found such that $f(x) > N$ for all x if $|x - a| < \delta$.

INFLAMMATION.   The response of the body to infection or irritation. It is characterized by redness, heat, swelling, and pain. The redness and heat are due to the increased blood supply to the involved area. The blood vessels are dilated and engorged, and there is a loss of plasma fluid from them into the tissue spaces. This results in edema or swelling. The swelling distends the tissues, compresses nerve endings, and thus causes pain. The white blood cells or leucocytes play an important part in inflammation. They escape from the capillaries, crowd the tissue spaces, carry on their work as phagocytes, picking up bacteria and cellular debris. They aid in walling off an infection and preventing its spread. As the inflammatory reaction subsides, repair of the damaged tissue takes place. If the tissue is one capable of complete regeneration, new cells of the same type may completely replace the old ones. This phenomenon is seen in minor inflammations of the skin. In other tissues, such as nervous tissue, regeneration may be very limited or absent; the damaged cells will then be replaced by fibrous scar tissue. This latter form of repair occurs with all inflammations of great size which cause marked cellular destruction.

INFLUENCE LINE.   An influence line is a graphical way of representing the effect of a certain variable circumstance upon a given condition. In particular, the influence line as applied in structural engineering represents the variable effect of a single moving unit concentrated load upon the shear, bending moment, reaction, or any other function of a structure such as a beam, truss, or bridge. The influence line is plotted in reference to a base or zero line. Positive or tensile effects are represented above the line and negative or compressive effects below. The ordinate of the influence line is the ratio of the effect to the concentrated load producing it. If the load is in pounds or tons the effect is in pounds or tons. It is very useful for locating the position of the load that will produce maximum effect. For instance, the influence line for bending moment at the center of the beam shown in Fig. 1 indicates that the maximum moment for this point will occur when the moving load that may be taken as unity is directly over the point. Any other ordinate such as ab represents the bending moment at the center due to a load of unity at point A. The maximum moment at the center of this beam due to a uniform load of w pounds per linear foot, covering the entire length, may be computed by multiplying the area of the influence line by w:

$$\text{area} = \frac{1}{4} \times l \times \frac{1}{2} = \frac{l^2}{8}$$

$$\text{maximum moment at center} = \frac{wl^2}{8}$$

Such quantitative results can also be obtained by the usual analytical methods. However, since influence lines are almost invariably drawn to

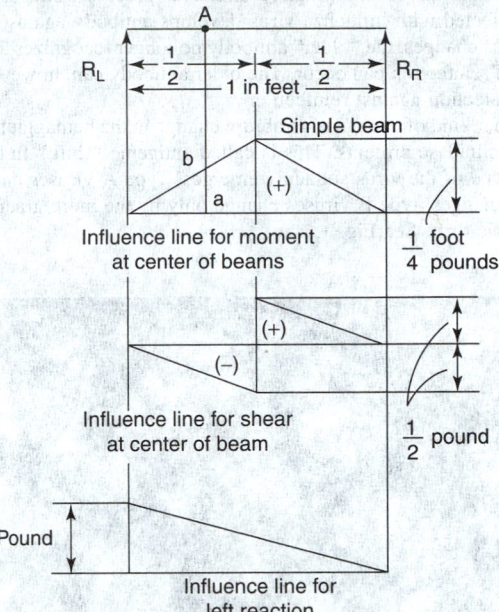

Fig. 1.   Examples of influence lines.

indicate how a structure subjected to moving loads should be loaded, it is usually convenient, and faster, to use the ordinates and areas of the influence line to obtain moments, stresses, etc.

Influence lines for statically determinate structures are composed of straight lines. Those for indeterminate structures are curved, or have straight segments, the intersections of which lie on curves.

See also **Determinate Structure**; and **Indeterminate Structure**.

INFLUENT.   In hydrology, a term designating that portion of a stream which contributes water to, rather than derives water from, the groundwater zone. The term also may be used to describe the charge or feedstock that is introduced into a chemical process.

INFLUENZA.   Hippocrates described an epidemic, now believed to one of influenza, as early as 412 B.C. and the first well-described pandemic of influenza-like disease occurred in 1580. Since that time, 31 such possible influenza pandemics have been documented, with three occurring in the 20th century: in 1918, 1957, and 1968. The disease today still affects large sections of the population each year. Its ability to kill stems from the fact that the causative virus can mutate quickly, often producing new strains against which human beings have no immunity. When this occurs, mortality from influenza can be high.

The Influenza Viruses

The nature of this respiratory viral infection varies with the particular virus responsible for the infection. Influenza viruses are divided into three types, designated A, B, and C. Influenza types A and B are responsible for epidemics of respiratory illness that occur almost every winter and are often associated with increased rates for hospitalization and death. Influenza type C differs from types A and B in some important ways. Type C infection usually causes either a very mild respiratory illness or no symptoms at all; it does not cause epidemics and does not have the severe public health impact that influenza types A and B do. Efforts to control the impact of influenza are aimed at types A and B. Influenza type A viruses are divided into subtypes based on differences in two viral proteins called the hemagglutinin (H) and the neuraminidase (N). The current subtypes of influenza A are designated $A(H_1N_1)$ and $A(H_3N_2)$. Influenza $A(H_1N_1)$, $A(H_3N_2)$, and influenza B strains are included in each year's influenza vaccine.

Influenza type A viruses undergo two kinds of changes. One is a series of mutations that occur over time and cause a gradual change in the virus. This is called antigenic "drift." This constant changing enables the virus to evade the immune system of its host, so that people are susceptible to influenza virus infection throughout life. This process works as follows: a person infected with influenza virus develops antibody against that virus; as the virus changes, the "older" antibody no longer recognizes the "newer" virus, and reinfection can occur. The older antibody can, however, provide partial protection against reinfection.

The other kind of change is an abrupt change in the hemagglutinin and/or the neuraminidase proteins. This is called antigenic "shift." In this case, a new subtype of the virus suddenly emerges. Type A viruses undergo both kinds of changes; type B viruses change only by the more gradual process of antigenic drift. See Fig. 1.

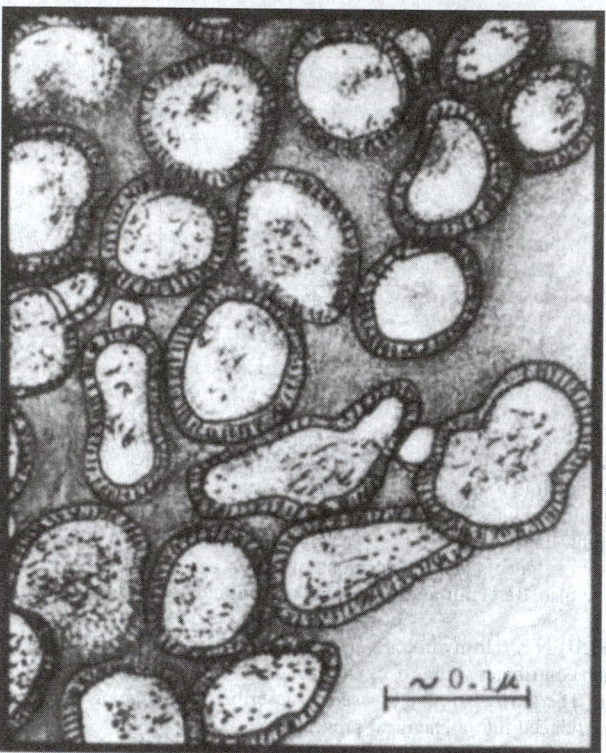

Fig. 1.   Reasonable facsimile of electron micrography of the influenza virus. Hemagglutinin and neuraminidase antigens are borne on projecting surface spikes. These surface proteins are attached to a lipid layer. The latter envelops the membrane protein. The nucleocapsids contain the genome of the virus. This consists of single strands of DNA.

Natural History of Human Influenza

Influenza also occurs as worldwide epidemics called pandemics due to major antigenic changes (antigenic shift) that are independent of season. Such an antigenic shift occurs only occasionally. When it does occur, however, pandemic influenza, which could affect large portions of the population, results. This is because of the absence of immunity against the "new" virus. Before World War I, influenza was not considered a major public health threat. But that changed with the great "Spanish flu" pandemic in 1918–1920.

Mortality associated with pandemics: 1918–1919 "Spanish flu" $A(H_1N_1)$, caused the highest known influenza-related mortality: approximately 500,000 deaths occurred in the United States, 20 million worldwide; 1957–1958 "Asian flu" $A(H_2N_2)$, 70,000 deaths in the United States; and 1968–1969 "Hong-Kong flu" $A(H_3N_2)$, 34,000 deaths in the United States.

The pandemics in 1957 and 1968 together killed more than 1.5 million people and caused an estimated $32 billion in economic damages worldwide due to productivity losses and medical expenses.

There is evidence that the viruses which caused these pandemics originated from animals (1918 = swine, 1957 and 1968 = avian strains). In 1976, a new influenza virus from pigs caused human infections and severe illness, while in 1997–1998, an outbreak of influenza whose origins were avian occurred in Hong Kong, Special Administrative Region of China.

A vaccine against swine influenza was developed and administered in some countries in 1976, although no pandemic in fact occurred.

The emergence of the "Hong Kong flu" in 1968–1969 marked the beginning of the type $A(H_3N_2)$ era. When this virus first emerged, it was associated with fewer deaths than that caused by the two previous pandemic viruses. There are several possible reasons for this. First, only the hemagglutinin changed from the "Asian" strain [type $A(H_2N_2)$]; the neuraminidase (N2) stayed the same, and therefore, people previously infected with H_2N_2 viruses may have had some protection against the H_3N_2 virus. A second possibility is that a virus with a similar hemagglutinin may have circulated from the late 1890s to the early 1900s. If this were the case, people who were in their sixties and older in 1968 may have had some protection from antibody acquired in their youth.

Many things about influenza viruses still are not understood. Although the newly emerged type $A(H_3N_2)$ virus caused many fewer deaths in 1968 compared with other pandemic viruses, it has continued to cause many deaths during annual epidemics. In the years since its emergence, type $A(H_3N_2)$ epidemics have caused more than 400,000 deaths in the United States alone, and more than 90% of these deaths have occurred among people age 65 and older. Of the influenza viruses currently in worldwide circulation, $A(H_3N_2)$ still has the most severe overall impact.

The other influenza A subtype currently in circulation, type $A(H_1N_1)$, also has an interesting history. After the devastating pandemic of 1918–19, this subtype continued to circulate and undergo antigenic drift. It periodically caused large epidemics, but never on the scale of the 1918–1919 pandemic. When the "Asian" strain [$(A(H_2N_2))$] emerged in 1957, the $A(H_1N_1)$ viruses disappeared (as did the $A(H_2N_2)$ viruses when the "Hong Kong" virus emerged in 1968). In 1977, the $A(H_1N_1)$ viruses reappeared and have cocirculated with $A(H_3N_2)$ viruses ever since. However, the impact of $A(H_1N_1)$ has been different during its most recent appearance. The virus that reappeared in 1977 was virtually identical to an $A(H_1N_1)$ virus that circulated in 1950. Therefore, most people born before 1950 had protective antibody, and epidemics caused by $A(H_1N_1)$ viruses since 1977 have primarily affected younger people. The fact that the elderly appear to have natural protection against current $A(H_1N_1)$ viruses probably explains the low mortality associated with recent epidemics in which this subtype was the predominant strain. However, as $A(H_1N_1)$ viruses continue to evolve, they could begin to have a more severe impact on the elderly.

Clinical Features of Influenza

Influenza, commonly called "the flu," is an infection of the respiratory tract caused by the influenza virus. Compared with most other viral respiratory infections, such as the common cold, influenza infection often causes a more severe illness. Typical influenza illness includes fever, usually 100–103 °F in adults and often even higher in children. Respiratory symptoms, such as cough, sore throat, runny or stuffy nose, as well as headache, muscle aches, and often extreme fatigue. Although nausea, vomiting, and diarrhea can sometimes accompany influenza infection, especially in children, these symptoms are rarely the primary symptoms.

The term "stomach flu" is a misnomer that is sometimes used to describe gastrointestinal illnesses caused by organisms other than influenza viruses.

Most people who get the flu recover completely in 1 to 2 weeks, but some people develop serious and potentially life-threatening medical complications, the most common of which is pneumonia. This may be attributed to the primary viral invasion, or to secondary bacterial pathogens. Thus, any person generally considered of high risk will be given penicillinase-resistant penicillin (or other antibiotic) shortly after onset of symptoms. A purely *viral pneumonia* will be evidenced within 1 to 1.5 days of onset of influenza. It is the overwhelming nature of viral pneumonia in many cases that accounts for the high fatality. In an average year, influenza is associated with more than 20,000 deaths nationwide and more than 100,000 hospitalizations. Age and fundamental health are major factors in the course of viral pneumonia. The severity of this disease increases with the age of the patient, and in pregnant women and in persons having a history of cardiorespiratory, renal, or metabolic disease. Less frequently occurring is infection of the heart (myocarditis). In persons under 16 years of age, Reye's syndrome may appear after influenza, but relatively infrequently. See also **Reye's Syndrome**.

In what may appear an otherwise uncomplicated case of influenza, a secondary bacterial superinfection may be manifested at just about the time the patient is feeling better and on the way to recovery. The causative organisms usually involved are *Staphylococcus aureus* or pneumococcus. See also **Pneumonia**.

Other less frequently occurring complications of influenza include sinusitis, otitis media (infection of middle ear), lung abscess, and meningitis.

Influenza Vaccine

Influenza virus vaccines were first developed in the 1940s and consisted of partially purified preparations of influenza viruses grown in embryonated eggs. Because of substantial contamination by egg-derived components, these killed (formaldehyde-treated) vaccines were highly pyrogenic and lacking in efficacy. A major breakthrough came with the development of the zonal ultracentrifuge in the 1960s (invented by Norman G. Anderson). This technology, which originated from uses for military purposes, revolutionized the purification process and industrial production of many viruses for vaccines. To this day, it remains the basis for the manufacturing process of our influenza virus vaccines.

Much of the illness and death caused by influenza can be prevented by annual influenza vaccination. Influenza vaccine is specifically recommended for people who are at high risk for developing serious complications as a result of influenza infection. These high-risk groups include all people aged 50 years or older and people of any age with chronic diseases of the heart, lung or kidneys, diabetes, immunosuppression, or severe forms of anemia. Other groups for whom vaccine is specifically recommended are residents of nursing homes and other chronic-care facilities housing patients of any age with chronic medical conditions, women who will be more than three months pregnant during the influenza season, and children and teenagers who are receiving long-term aspirin therapy and who may therefore be at risk for developing Reye syndrome after an influenza virus infection. Influenza vaccine is also recommended for people who are in close or frequent contact with anyone in the high-risk groups defined above. These people include health-care personnel and volunteers who work with high-risk patients and people who live in a household with a high-risk person.

Although annual influenza vaccination has long been recommended for people in the high-risk groups, many still do not receive the vaccine. Some people do not receive influenza vaccine because they believe it is not very effective. There are several reasons for this belief. People who have received influenza vaccine may subsequently have an illness that is mistaken for influenza, and they believe that the vaccine failed to protect them. In other cases, people who have received vaccine may indeed have an influenza infection. Overall vaccine effectiveness varies from year to year, depending upon the degree of similarity between the influenza virus strains included in the vaccine and the strain or strains that circulate during the influenza season. Because the vaccine strains must be chosen 9 to 10 months before the influenza season, and because influenza viruses mutate over time, sometimes mutations occur in the circulating strains between the time vaccine strains are chosen and the next influenza season is over. These mutations sometimes reduce the ability of the vaccine-induced antibody to inhibit the newly mutated virus, thereby reducing vaccine efficacy.

Vaccine efficacy also varies from one person to another. Studies of healthy young adults have shown influenza vaccine to be 70–90% effective in preventing illness. In the elderly and those with certain chronic medical conditions, the vaccine is often less effective in preventing illness than in reducing the severity of illness and the risk of serious complications and death. Studies have shown the vaccine to reduce hospitalization by about 70% and death by about 85% among the elderly who are not in nursing homes. Among nursing home residents, vaccine can reduce the risk of hospitalization by about 50%, the risk of pneumonia by about 60%, and the risk of death by 75% to 80%. When antigenic drift results in the circulating virus becoming different from the vaccine strain, overall efficacy may be reduced, especially in preventing illness, but the vaccine is still likely to lessen the severity of the illness and to prevent complications and death.

Some people are not vaccinated because of misconceptions about influenza and the vaccine. Many people are not aware of the seriousness of influenza infection and some believe that the vaccine can cause the flu. Influenza vaccine produced in the United States cannot cause influenza. The only type of influenza vaccine that has been licensed in the United States is made from killed influenza viruses, which cannot cause infection. An influenza vaccine that is made with live influenza viruses has been developed and may be marketed in the future. This vaccine is made with viruses that can confer immunity but do not cause classic influenza symptoms.

Some people worry about the side effects of influenza vaccine. While influenza vaccine, like any other vaccine or medicine, is capable of causing serious problems such as severe allergic reactions, the risk of the vaccine causing serious harm, or death, is extremely small. Almost all people who get influenza vaccine have no serious problems from it. The most common side effect from influenza vaccination is soreness at the site of the injection. The soreness can last up to two days but is usually mild and does not affect a person's ability to perform their normal daily activities. Some people, usually children who have not been exposed to influenza virus in the past, may have fever and body aches after vaccination. These symptoms, if they occur, usually start 6–12 hours after vaccination and can continue for one or two days.

Less common side effects that can occur after vaccination include allergic reactions and Guillain-Barré syndrome (GBS), a severe paralytic illness. Life-threatening allergic reactions are very rare, but can happen in people who have severe allergy to any vaccine component, most commonly allergy to eggs. The influenza viruses used in the vaccine are grown in hens' eggs. People who have an allergy to eggs or who have ever had a serious allergic reaction to a previous dose of influenza vaccine should consult with a doctor before getting an influenza vaccination.

In 1976, swine flu vaccine was associated with an increased number of cases of GBS. Influenza vaccines since then have not been clearly linked to GBS. However, if there is a risk of GBS from current influenza vaccines, it is estimated at one or two cases per million persons vaccinated. Much less than the risk of severe influenza, which can be prevented by vaccination.

Although only a few different influenza viruses circulate at any given time, people continue to become ill with the flu throughout their lives. The reason for this continuing susceptibility is that influenza viruses are continually changing, usually as a result of mutations in the viral genes. Currently, there are three different influenza virus strains, and the vaccine contains viruses representing each strain. Each year the vaccine is updated to include the most current influenza virus strains. The fact that influenza viruses continually change is one of the reasons vaccine must be taken every year. Another reason is that a person's production of antibody after being vaccinated declines over time, and antibody levels are often low one year after vaccination.

In the United States, influenza usually occurs from about November until April, with activity peaking between late December and early March. The optimal time for vaccination of persons at high risk for influenza-related medical complications is usually the period from October to mid-November. However, to avoid missed opportunities for vaccination, vaccine should be offered to high-risk persons who are hospitalized or seen at their physician's office starting in September and continuing through the winter. It takes about one to two weeks after vaccination for antibody against influenza to develop and provide protection.

Influenza vaccine is strongly recommended for any person aged 6 months or older who, because of age or underlying medical condition, is at increased risk for complications of influenza. In addition, health-care workers and others (including household members) in close contact with persons in high-risk groups should be vaccinated to decrease the risk of transmitting infection to persons at high risk. Influenza vaccine also can be administered to any person who wishes to reduce the chance of becoming infected with influenza (the vaccine can be administered to children as young as 6 months).

Vaccine for the 2000–2006 Influenza Season

The trivalent influenza vaccine prepared for the 2000–2001 season includes A/Moscow/10/99 (H_3N_2)-like, A/New Caledonia/20/99 (H_1N_1)-like, and B/Beijing/184/93-like antigens. For the A/Moscow/10/99 (H_3N_2)-like antigen, U.S. manufacturers use the antigenically equivalent A/Panama/2007/99 (H_3N_2) virus and for the B/Beijing/184/93-like antigen, they use the antigenically equivalent B/Yamanashi/166/98 virus; these viruses are used because of their growth properties and because they are representative of currently circulating A (H_3N_2) and B viruses.

Current influenza virus vaccines consist of 3 components: an H1 N1 (hemagglutinin [HA] subtype 1; neuraminidase [NA] subtype 1), an H3 N2 influenza A virus, and an influenza B virus. Specifically, the 2005–2006 vaccine formulation is made up of the A/New Caledonia/20/99 (H1 N1), A/California/7/2004 (H3 N2), and B/Shanghai/361/2002 viruses. Changes in the HA of circulating viruses (antigenic drift) require periodic replacement of the vaccine strains during interpandemic periods. The World Health Organization publishes semiannual recommendations for the strains to be included for the Northern and Southern Hemispheres. To allow sufficient time for manufacture, in the United States the US Food and Drug Administration (FDA) determines in February which vaccine strains should be included in the following winter's vaccine. Unfortunately, FDA recommendations are not always optimal. For example, in 2003 FDA rejected the use of the most appropriate H3 N2 strain, A/Fujian/411/2002, and instead again used the same strain as in the 2002 formulation. This decision was made primarily because the A/Fujian/411/2002 strain had first been isolated in Madin Darby canine kidney (MDCK) cells rather than in embryonated eggs. Use of MDCK cells for virus isolation is not allowed by FDA's rules, which do not yet encompass advanced technologies or scientifically sound purification procedures based on limiting dilutions or cloning with DNA. Because of this bureaucratic roadblock, the H3 N2 component of the 2003–2004 influenza virus vaccine was antigenically "off" and showed suboptimal efficacy. One hundred fifty-three pediatric deaths were associated with influenza infections during the 2003–2004 season in 40 states, whereas only 9 such deaths had been reported in the following season. Also, because the cumbersome classical reassortment technique used for preparing the appropriate seed strains makes the yearly process of manufacturing influenza virus vaccines unnecessarily lengthy, new variants first appearing early in the season are rarely considered for the vaccine formulation of the following winter.

Currently Licensed Influenza Virus Vaccines

Most influenza virus vaccines used in the United States and Europe consist of embryonated egg-grown and formaldehyde-inactivated preparations, which, after purification, are chemically disrupted with a nonionic detergent (for example, Triton X-100). The split virus preparations show lower pyrogenicity than whole virus vaccines. In general, 1 dose for adults contains the equivalent of 45 µg HA (15 µg HA for each of the 3 antigenic components). This dose is approximately the amount of purified virus obtained from the allantoic fluid of 1 infected embryonated egg. If 100 million doses of killed influenza virus vaccine are prepared, the manufacturer has to procure 100 million embryonated eggs. Clearly, this manufacturing process is dependent on the timely availability of embryonated eggs and the vaccine seed strains to be used in a particular season. Most of these prototype seed strains are provided to the manufacturers by government agencies, which create high-yielding strains through classical reassortment with a high-yielding laboratory strain, A/PR/8/34, following the procedures designed by Kilbourne. Unfortunately, only (high-yielding) influenza A viruses can be made in this way, and even with the A types, the 6:2 reassortants (HA and NA from recently circulating strains and the remaining 6 genes from A/PR/8/34 virus) are sometimes not easily obtained. This time-consuming process of reassortment is then followed by repeated passaging of the strain in embryonated eggs to allow for egg adaptation and growth enhancement. Influenza B virus prototype strains with good growth characteristics are usually obtained by direct and repeated passaging in embryonated eggs without attempting to generate reassortants. Although the manufacturing process is time-consuming, these killed influenza A and B virus vaccines are the workhorses for vaccination against influenza and have been shown time and again to be highly effective.

The second major class of viral vaccines consists of live viruses. The only FDA-licensed product against influenza is the cold-adapted attenuated vaccine. It is based on work originally done by Maassab's laboratory and later by Murphy and colleagues. Influenza virus was passaged at 25 °C (77 °F) in tissue culture (chicken kidney cells) and in embryonated eggs. This modified Jennerian approach resulted in a cold-adapted, temperature-sensitive, and highly attenuated master strain. The annually updated vaccine strains are generated in the laboratory by reassortment with viruses more closely related to the currently circulating ones. The resulting vaccine strains (both A and B types) are 6:2 reassortants with the 6 nonsurface protein genes derived from the cold-adapted master strains and the HA and NA from circulating A and B viruses, reflecting the changing antigenicity. These cold-adapted influenza virus vaccines are easily administered by nasal spray. They induce local mucosal neutralizing immunity and cell-mediated responses that may be longer lasting and more cross-protective than those elicited by chemically inactivated (killed) vaccine preparations. Vaccine efficacy in vaccine-naive children 6 months to 18 years of age is high (range 73%–96%). In children revaccinated for a second season, vaccine efficacy climbs to 82% to 100%.

Despite the obvious efficacy of both killed and live influenza virus vaccines, there is room for new developments. Among the critical issues in developing new and better vaccines are the following: price per dose, speed of production, ease of production, choice of substrates to grow the virus in or to express viral antigens, cross-protection for variant strains, efficacy in general and in immunologically naive populations, safety, and acceptance by the regulatory agencies and the public.

Genetically Engineered Live and Killed Influenza Virus Vaccines

As indicated, current FDA-licensed influenza vaccines are based on technologies developed in the 1960s and earlier. Through the breakthrough of reverse genetics techniques, infectious influenza viruses from plasmid DNAs transfected into tissue culture cells can now be rescued. This technology permits the construction of high-yield 6:2 seed viruses by mixing the 6 plasmid DNAs from a good-growing laboratory strain with the HA and NA DNAs obtained by cloning relevant genes from currently circulating viruses. Thus, within a 1- to 2-week period, the appropriate seed viruses could be generated for distribution to the manufacturers. The backbones of the 6:2 recombinant viruses could be prepared, tested, and distributed in advance. Similar approaches can be envisioned for the manufacturing of live, cold-adapted influenza virus vaccines. In this case, the backbone would consist of the 6 genes of the cold-adapted master strain. Again, the HA and NA of the currently circulating strains would be cloned and used for rescue in the plasmid-only reverse genetics system. Such an approach would have several advantages over the present manufacturing process. First, it would dramatically accelerate the timeframe for obtaining seed viruses for annual production and thus allow more time to select the appropriate antigenic seed strains. Second, it would standardize the seed viruses to be used. Regulatory agencies do not insist on a sequenced product to be given to humans but instead allow only partially characterized products for annual immunization. Third, DNA cloning may eliminate any adventitious agents present in the throat washings of the original isolate. Finally, in the case of the current highly pathogenic H5 strains, viruses with that HA (containing a multibasic HA1/HA2 cleavage site) kill embryonated eggs, making it difficult to use eggs as growth substrate. Also personnel involved in manufacturing those vaccines might be in danger of becoming infected. Thus, the HA of these virulent strains will need to be modified. Removal of the basic cleavage peptide by reverse genetics results in a virus that is attenuated for embryonated eggs, thus allowing high yields to be attained. Modification by reverse genetics results in a product that is easier to manufacture and safer to handle (this includes safety considerations for all persons working with the virus).

Universal Vaccines

Influenza viruses continue to undergo antigenic drift, which is mostly reflected in accumulating changes in the HA. This fact requires us to change the vaccine formulation or at least to reexamine the seed strains on an annual basis. Unfortunately, predicting the evolutionary change of the viral HA has not been reliable. Thus, short of developing 20/20 foresight, predicting strain variation or the emergence of a particular pandemic strain (avian or otherwise) is unlikely. A more realistic approach is the design of more cross-protective vaccines for use in interpandemic years and during pandemics. Neirynck et al. have designed vaccines based on the conserved extracellular portion of the M2 protein fused to the hepatitis B core protein. Such an immunogen may induce a cross-reactive response in the vaccinated host. Similarly, immunization with the NA antigen is likely to induce responses that are more cross-reactive than those by the more variable HA. In both cases, however, protection will require immune responses that are more vigorous than what is seen after natural infection. Antibodies against NA and M2 proteins in infected humans are generally not protective. Thus vaccines consisting of NA or M antigens would need to be made to induce a dramatically enhanced immune response. Alternatively, genetically engineered viruses could be generated, which would express several variant antigens or epitopes, thereby achieving a more cross-protective immunization. Chimeric HA recombinant viruses that express an additional 140 amino acids have recently been described). Such genetically engineered viruses may present several conserved immunogenic epitopes on the viral surface, which would be a first step toward a more universal influenza vaccine. See also **Vaccines**; and **Vaccine Technology**.

Target Groups for Vaccination

Groups at High Risk for Influenza-Related Complications. Vaccination is recommended for the following groups of persons who are at increased risk for complications from influenza or who have a higher prevalence of chronic medical conditions that place them at risk for influenza-related complications: (1) Persons aged 50 years or older; (2) residents of nursing homes and other chronic-care facilities that house persons of any age who have chronic medical conditions; (3) Adults and children who have chronic disorders of the pulmonary or cardiovascular systems, including asthma; (4) Adults and children who have required regular medical follow-up or hospitalization during the preceding year because of chronic metabolic diseases (including diabetes mellitus), renal dysfunction, hemoglobinopathies, or immunosuppression (including immunosuppression caused by medications); (5) children and teenagers (aged 6 months to 18 years) who are receiving long-term aspirin therapy and therefore might be at risk for developing Reye syndrome after influenza; and (6) women who will be in the second or third trimester of pregnancy during the influenza season.

Persons who can Transmit Influenza to those at High Risk. Infected persons can transmit influenza virus to persons at high risk for complications from influenza. Efforts to protect members of high-risk groups against influenza might be improved by reducing the likelihood of influenza exposure from their care givers. Therefore, the following group should be vaccinated: (1) Physicians, nurses, and other personnel in both hospital and outpatient-care settings; (2) employees of nursing homes and chronic-care facilities who have contact with patients or residents; (3) employees of assisted living and other residences for persons in high-risk groups; (4) persons who provide home care to persons in high-risk groups; and (5) household members (including children) of persons in high-risk groups.

Pregnant Woman. Women who will be beyond the first trimester of pregnancy (greater than or equal to 14 weeks' gestation) during the influenza season, should be vaccinated. Pregnant women who have medical conditions that increase their risk for complications from influenza should be vaccinated before the influenza season, regardless of the stage of pregnancy. Because currently available influenza vaccine is an inactivated vaccine, many experts consider influenza vaccination safe during any stage of pregnancy.

Breast-Feeding Mothers. Influenza vaccine does not affect the safety of mothers who are breast-feeding or their infants. Breast-feeding does not adversely affect immune response and is not a contraindication for vaccination.

Travelers. The risk of exposure to influenza during travel depends on the time of year and destination. In the tropics, influenza can occur throughout the year, whereas most influenza activity occurs from April through September in the temperate regions of the Southern Hemisphere. In temperate climate zones of the Northern and Southern Hemispheres, travelers also can be exposed to influenza during the summer, especially when traveling as part of large organized tourist groups containing persons from areas of the world where influenza viruses are circulating.

Persons at high risk for complications of influenza should consider receiving influenza vaccine before travel if they were not vaccinated with influenza vaccine during the preceding fall or winter and they plan to (1) travel to the tropics; (2) travel with large organized tourist groups at any time of year; or (3) travel to the Southern Hemisphere from April through September. Persons at high risk who received the previous season's vaccine before travel should be revaccinated with the current vaccine in the following fall or winter.

Because influenza vaccine might not be available during the summer in North America, persons aged 50 years or older and others at high risk might wish to consult with their physicians before embarking on travel during the summer to discuss the symptoms and risks of influenza and advisability of carrying antiviral medications for either prophylaxis or treatment for influenza.

General Population. Physicians should administer influenza vaccine to any person who wishes to reduce the likelihood of becoming ill with influenza (the vaccine can be administered to children as young as 6 months). Persons who provide essential community services should be considered for vaccination to minimize disruption of essential activities during influenza outbreaks. Students or other persons in institutional settings (e.g., those who reside in dormitories) should be encouraged to receive vaccine to minimize the disruption of routine activities during epidemics.

Influenza Vaccination Activities

The timing and administration route of vaccination is considered, as well as the simultaneous administration of other vaccines.

Timing. Beginning each September, influenza vaccine should be offered to persons at high risk when they are seen by health-care providers for routine care or as a result of hospitalization. The optimal time to vaccinate persons in high-risk groups is usually from October through mid-November, because influenza activity in the United States generally peaks between late December and early March. Although vaccine generally becomes available in August or September, vaccine availability in any location cannot be assured consistently in the early fall. Therefore, persons planning large organized vaccination campaigns may consider scheduling these events after mid-October to reduce the possibility that the vaccination campaign will need to be canceled because vaccine is unavailable. Administering vaccine too far in advance of the influenza season should be avoided in facilities such as nursing homes, because antibody levels can begin to decline within a few months of vaccination. If regional influenza activity is expected to begin earlier than December, vaccination programs can be undertaken as soon as current vaccine is available. Vaccine should be offered to unvaccinated persons even after influenza virus activity is documented in a community.

Vaccination Administration Route. The intramuscular route is recommended for influenza vaccine. Adults and older children should be vaccinated in the deltoid muscle; a needle length of one inch or longer can be considered for these age groups. Infants and young children should be vaccinated in the anterolateral aspect of the thigh.

Simultaneous Administration of Other Vaccines, Including Childhood Vaccines. The target groups for influenza and pneumococcal vaccination overlap considerably. For persons at high risk who have not previously been vaccinated with pneumococcal vaccine, health-care providers should strongly consider administering pneumococcal and influenza vaccines concurrently. Both vaccines can be administered at the same time at different sites without increasing side effects. However, influenza vaccine is administered each year, whereas pneumococcal vaccine is not. Children at high risk for influenza-related complications can receive influenza vaccine at the same time they receive other routine vaccinations.

Antiviral Drugs for Influenza. Although influenza vaccine is the primary prevention measure for influenza, antiviral drugs for influenza are an important adjunct to influenza vaccine for the control and prevention of influenza. Four currently licensed agents are available in the United States: amantadine, rimantadine, zanamivir, and oseltamivir.

Amantadine and rimantadine are indicated for the prophylaxis and treatment of influenza A infection. Zanamivir and oseltamivir are neuraminidase inhibitors with activity against both influenza A and B viruses. Both zanamivir and oseltamivir were approved in 1999 for treatment of uncomplicated influenza, but neither is approved for prophylaxis.

When administered prophylactically to healthy adults or children, both amantadine and rimantadine are approximately 70%–90% effective in preventing illness from influenza A infection. Chemoprophylaxis is not a substitute for vaccination, but may be considered for the following situations.

- Protection of high-risk persons vaccinated after the influenza season has begun during the time needed to develop protective antibody levels
- Protection of persons who cannot receive influenza vaccine
- Use by unvaccinated close contacts of high-risk persons during the peak of the influenza season in order to reduce the chances of spreading influenza to high-risk persons

- Protection of immunocompromised persons who may not develop protective levels of antibody following vaccination
- Outbreak control in an institution housing or caring for high-risk people

When amantadine or rimantadine is administered as prophylaxis, factors related to cost, compliance, and potential side effects should be considered when determining the period of prophylaxis.

When administered as treatment within 2 days of illness onset in healthy adults, amantadine and rimantadine can reduce the severity and duration of signs and symptoms of influenza A illness, and zanamivir and oseltamivir can reduce the duration of uncomplicated influenza A and illness by approximately one day. To reduce the emergence of antiviral drug-resistant viruses, amantadine or rimantadine treatment should be discontinued as soon as clinically warranted, generally after 3–5 days of treatment or within 24–48 hours after the disappearance of signs and symptoms. The recommended duration of treatment with either zanamivir or oseltamivir is 5 days.

Both amantadine and rimantadine can cause central nervous system (CNS) and gastrointestinal side effects. However, the incidence of CNS side effects (e.g., nervousness, anxiety, difficulty concentrating, and lightheadedness) is higher among persons taking amantadine than among those taking rimantadine. Dosages of both drugs should be lower for older persons and persons with reduced kidney function. Consult the package insert for more information on dosing and side effects.

Zanamivir is given as an inhaled powder. Among persons with chronic lung disease, such as asthma or chronic obstructive pulmonary disease, inhalation of zanamivir may result in difficulty breathing. If, after considering the potential risks and benefits, this drug is prescribed for someone with chronic respiratory disease, the drug should be used with caution. Precautions should include proper monitoring and availability of supportive care, including availability of short-acting brochodilators.

The primary side effects associated with oseltamivir use are nausea and vomiting. These side effects may be diminished if the drug is taken with food. See also **Antiviral Drugs**.

Additional Reading

Crosby, A.W.: *America's Forgotten Pandemic: The Influenza of 1918*, Cambridge University Press, New York, NY, 1990.

Kilbourne, E.D., R.B. Couch, J.A. Kasel, W.A. Keitel, T.R. Cate, J.H. Quarles, et al..: "Purified Influenza A Virus N2 Neuraminidase Vaccine is Immunogenic and Non-toxic in Humans," *Vaccine*, **13**, 1799–803 (1995).

Li, Z.N., S.N. Mueller, L. Ye, Z. Bu, C. Yang, R. Ahmed, et al..: "Chimeric Influenza Virus Hemagglutinin Proteins Containing Large Domains of the Bacillus Anthracis Protective Antigen: Protein Characterization, Incorporation into Infectious Influenza Viruses, and Antigenicity," *J. Virol.* **79**, 10003–10012 (2005).

Nicholson, K.G. and A. Hay: *Textbook of Influenza*, Blackwell Science, Inc., Malden, MA, 1998.

Noble, G.R.: "Epidemiological and clinical aspects of influenza," in: A.S. Beare ed., *Basic and Applied Influenza Research*, CRC Press, LLC., Boca Raton, FL, 1982.

Neirynck, S., T. Deroo, X. Saelens, P. Vanlandschoot, W.M. Jou, and W. Fiers: "A Universal Influenza A Vaccine Based on the Extracellular Domain of the M2 Protein," *Nat Med.*, **5**, 1157–1163 (1999).

Palese P. "Influenza: Old and New Threats'," *Nat Med.* **10**, S82–87 (2004).

Simonsen, L. etal: "The Impact of Influenza on Mortality in the USA," in: L.E. Brown etal, eds., *Options for the Control of Influenza*, Elsevier Science, New York, NY, 1996.

Simonsen L. etal: "The Impact of Influenza Epidemics on Hospitalizations," *J. Infect. Dis.*, **181**, 831–837 (2000).

Staff: Center for Disease Control, "Prevention and Control of Influenza: Recommendations of the Advisory Committee on Immunization Practices," *MMWR*, **49**(PR-3) (2000).

Staff: Center for Disease Control, "Influenza activity—United States, 1999–2000," *MMWR*, **48**, 1039–1042 (1999).

Staff: World Health Organization, *International Statistical Classification of Diseases and Related Public Health Problems*, 10th Rev., World Health Organization, Geneva, Switzerland, 1993.

Web References

Centers for Disease Control and Prevention. http://www.cdc.gov/health/diseases. htm#I http://www.cdc.gov/mmwr/preview/mmwrhtml/mm5408a1.htm

World Health Organization. http://www.who.int/emc/diseases/flu/index.html http://www.who.int/csr/disease/influenza/vaccinerecommendations1/en/

INFORMATION. 1. A collection of facts or other data especially as derived from the processing of data.

2. The word "information" occurs frequently in statistics with its ordinary meaning. In a specialized sense in the theory of estimation, the amount of information about a parameter θ from a sample of n independent observations drawn at random from a population with frequency function $f(x, \theta)$ is defined as

$$nE\left(\frac{\partial \log f}{\partial \theta}\right) \equiv n \int_{-\infty}^{\infty} \left(\frac{\partial \log f(x, \theta)}{\partial \theta}\right)^2 f(x, \theta)\, dx$$

Under some general regularity conditions, the reciprocal of the information gives a lower bound for the variance of unbiased estimators of q, so that the greater the variance, the less the "information."

INFORMATION THEORY. A branch of statistical communication theory that studies the information content of messages or physical observations and its relation to the problem of transmitting this information from one place to another. The term *information*, as used in the context of information theory, is not related to the meaning, usefulness, or correctness of a message or observation, but rather to the uncertainty or randomness of that message or observation. Since uncertainty can be modeled mathematically in terms of probabilities, information theory has also emerged as a branch of probability theory, and many key results have profound mathematical significance entirely apart from their application to communication theory. However, the discussion here stresses the communications aspect of information theory because of its greater impact on modern science.

Although many early writers had grappled with the problems of information transmission, and had recognized its statistical nature, the consolidation and extension of these concepts into a complete and cohesive theory of communication is quite properly attributed to Shannon. His original paper in 1948 is a remarkable document that has survived the tests of time and become a genuine classic whose relevance is increasingly impressive as the years pass. Although information theory has become more precise and more complete in the past 25 years, no fundamental concepts of major importance have been added to or significantly altered from those originally proposed by Shannon.

After surviving numerous ill-conceived and usually fruitless attempts to apply these concepts to the whole range of human existence, information theory has returned in its maturity to the original structure from which it emerged. This structure contains three major parts that are almost independent so far as analytical techniques are concerned, but when taken together form a complete description of the communication problem. These parts pertain to:

1. The information content of messages or observations, the rate at which such information is produced, and the relationship between information rate and the accuracy with which messages can be reproduced at the distant end of a communication system (rate-distortion theory).
2. The rate at which a transmission medium (channel) can convey information without error or with a specified amount of error (channel capacity).
3. The construction and analysis of coding techniques that are used to control errors in the channel (coding theory).

The relationship of the three areas outlined above to the general communication system is clearly indicated in the diagram of Fig. 1. It is apparent that all three aspects are essential to a complete understanding of the communication problem. The discussion here follows the structure indicated and attempts to show the inter-relations of the various parts.

Quantitative Measure of Information. In order to develop a mathematical theory of information it is necessary to have a quantitative measure for the information content of messages. Since there is no known method of quantifying the semantic information content of messages, a reasonable alternative is to relate information to the probability structure of the message. In this sense, the information produced by a message source may be viewed as a measure of the uncertainty about the message that is removed by the occurrence of the message. Similarly, the information conveyed by the channel is the uncertainty about the message that is removed at the

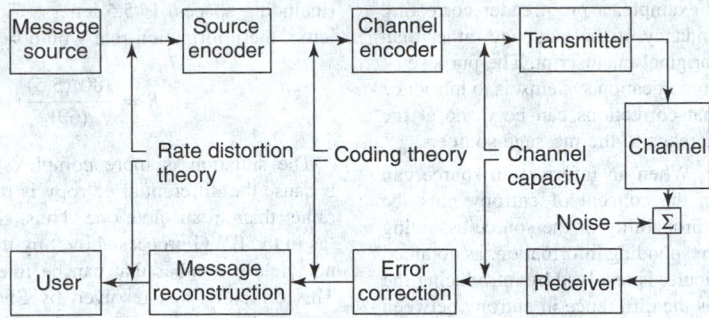

Fig. 1. Illustration of how the three main parts of information theory relate to a general communication system.

destination by the reception of the message. Because of noise or distortion in the channel, the information at the receiver is usually less than that at the transmitter.

The relationship between uncertainty about a message and its probability structure is almost intuitive. Thus, a message that is certain to occur, does so with probability one, and no information is produced. On the other hand, if a particular message is very improbable, the fact that it occurs, rather than some other message, conveys a great deal of information. Hence, the functional relationship between probability and information should reflect this intuitive concept.

Another intuitive property that an information measure should possess is that the information associated with two or more independent messages should be the *sum* of the information contents of the individual messages. Since the joint probability associated with independent events is given by the *product* of their individual probabilities, it follows that information should be a *logarithmic* function of probability. Thus, if a message x_i has a probability $P(x_i)$, its information content is defined to be

$$I(x_i) \triangleq - \log P(x_i) \tag{1}$$

Although any base might be used for the logarithm in (1), it is common to specify that one unit of information should be associated with a probability of one-half. On this basis, the logarithmic base becomes two and the unit of information is the binary digit or *bit*. The base e is also used in analysis, and the corresponding unit of information is the nat. Unless otherwise indicated, the base two is used throughout this discussion.

The Discrete Information Source. A discrete information source is one that produces a sequence of symbols from a finite set of symbols, $x_1, x_2, \ldots, x_n$. If these symbols are produced independently, with probabilities of $P(x_i)$, $i = 1, 2, \ldots, n$, then the *average* information per symbol is

$$H = E[I(x)] = - \sum_{i=1}^{n} P(x_i) \log P(x_i), \quad \text{bits/symbol} \tag{2}$$

where $E[\cdot]$ implies the mathematical expectation. The similarity between this result and certain formulations in statistical thermodynamics led Shannon to use the term *entropy* for the average information, H. Although the heated arguments as to whether information entropy is in fact the same as thermodynamic entropy have long since faded away, with no clear consensus on either side, the term *entropy* has persisted in the literature.

It may be noted that for a discrete source, H is always positive since every $P(x_i)$ is positive and less than unity. Furthermore, it is easy to show that H possesses the following properties:

a. $H = 0$, if and only if $P(x_i) = 1$ for one i and $P(x_i) = 0$ for every $j \neq i$.
b. H is a maximum when $P(x_i) = 1/n$, $i = 1, 2, \ldots, n$, and has a maximum value of

$$H_{\max} = \log n, \quad \text{bits/symbol} \tag{3}$$

An example of an information source that produces independent symbols might be a computer producing a sequence of decimal digits. If the digits $0, 1, \ldots, 9$ are all equally probable, which is a reasonable model, then the entropy is

$$H = H_{\max} = \log 10 = 3.322, \quad \text{bits/digit}$$

For more general discrete sources, such as sequences of English letters, the symbols are not produced independently, but each is strongly influenced by the symbols that preceded it. Except in the simplest cases, such as those that might be modeled by a finite-order Markov process[6], actual calculation of the entropy from the probability structure of the source is almost impossible. Nevertheless, it is often possible to estimate the entropy of printed languages, and many studies of this sort have appeared in the literature. For example, Shannon has shown[7] that the entropy of printed English is on the order of 1 bit per letter. It is of interest to compare this value with the 5 binary digits per letter that are required to encode English on a letter-by-letter basis (e.g., the tele-type code), and with the value of 4.065 bits per letter that would be given by Equation (2) if the letters were assumed to be produced independently with their usual probability of occurrence.

Redundancy. An information source whose entropy, H, is less than the maximum entropy, $H_{\max}$, that could be obtained with the same set of symbols is said to possess redundancy. A common definition of redundancy is:

$$\text{Redundancy} \triangleq 1 - \frac{H}{H_{\max}} \tag{4}$$

and the resulting number is simply the fraction of the available information from each symbol that is being wasted by virtue of its dependence on previous symbols. For the sequence of independent decimal digits mentioned above, the redundancy is zero since $H = H_{\max}$. However, for printed English with 27 symbols (26 letters and a space)

$$H_{\max} = \log 27 = 4.755, \quad \text{bits/letter}$$

and the redundancy (using the entropy estimate of 1 bit per letter) is

$$\text{Redundancy} = 1 - \frac{1}{4.755} = 0.79$$

The redundancy of an information source is an extremely important concept because it indicates the extent to which the efficiency of a communication system can be improved by the use of *source encoding*. For example, in the case of printed English, sufficiently elaborate techniques for encoding long sequences of letters into binary digits would enable a communication system to transmit long messages with only one binary digit per letter (on the average) rather than the five binary digits per letter that are presently required.

The redundancy of black-and-white pictorial information sources (e.g., maps and half-tone pictures) is even greater than that of printed English, with most estimates indicating a redundancy larger than 0.99. This suggests that communication systems transmitting pictorial data can become very much more efficient (by factors of 100 or more) than existing systems if effective and practical source encoding techniques are developed. The search for such techniques is one of the most active areas of research in information theory at the present time, and the rewards for success are certainly greater than can be achieved by any other improvement in communication technology.

Although redundancy in message sources reduces the efficiency of communication systems, it is not an entirely useless aspect of messages. When the communication channel has noise in it, so that some symbols are received in error, the presence of redundancy may make it possible to correct some or all of the errors. For example, a proofreader correcting English text relies mostly on the redundancy of the language rather than a letter-by-letter comparison with the original manuscript. The purpose of error correcting codes, as applied to communication systems, is to introduce redundancy in a known manner so that corrections can be made at the receiving point without any further reference to the message source.

The Continuous Information Source. When an information source can assume a continuous range of values, the concept of entropy must be altered somewhat. This is because the probability of the source assuming any particular value is zero, and the corresponding information, as obtained from Equation (1), is infinite. This difficulty is resolved by introducing the concept of *differential entropy*, which is the difference in entropy between that of the source random variable and a standard random variable having a uniform probability density function of unit width.

If the source random variable is denoted as x and has a probability density function of $p(x)$, then the differential entropy is defined as

$$H(x) \triangleq - \int_{-\infty}^{\infty} p(x) \log p(x)\, dx, \text{ bits} \qquad (5)$$

It is important to note that differential entropy can be either positive or negative since it is the entropy relative to that of a standard random variable. This is in contrast to the entropy of a discrete source, which is always positive.

Although it is possible to determine $H(x)$ for almost any probability density function, the Gaussian density function is the one of greatest interest. The reason for this interest is that the Gaussian density function yields the largest value of $H(x)$ of any random variable whose variance has a specified value. Thus, for the Gaussian case

$$p(x) \triangleq \frac{1}{\sqrt{2\pi}\sigma} e^{-(x-\bar{x})^2/2\sigma^2} \qquad (6)$$

where $\bar{x}$ is the mean value of x and σ^2 is its variance. The differential entropy becomes

$$H(x) = \tfrac{1}{2} \log(2\pi e\sigma^2), \text{ bits} \qquad (7)$$

in which e is the base of the natural logarithms. No other random variable with a variance of σ will have a differential entropy larger than the value given by Equation (7).

The discussion so far has considered only the entropy of a single random variable rather than that of a random process. Since most continuous message sources produce time functions that may be modeled as ensemble members of a random process, it is necessary to extend this simple case to the more realistic situation. An elementary way of doing this is to consider the special situation in which the message comes from a bandlimited source, with a bandwidth of W Hz, and has a spectrum that is constant over the bandwidth. This approach avoids the complexities that arise in more general situations, but it does reveal the essential features that are common to all continuous sources.

The sampling theorem indicates that a bandlimited time function can be uniquely represented by sample values of that time function taken at time instants separated by $\frac{1}{2}W$ seconds. If the time function comes from a Gaussian process, and if its spectrum is constant over the bandwidth [i.e., a bandlimited, white, Gaussian (BWG) process], then the random variables associated with successive samples are statistically independent and the entropy of each sample is given by Equation (7) as

$$H(x) = \tfrac{1}{2} \log(2\pi e\sigma^2), \text{ bits/sample} \qquad (7)$$

Information Rate of Sources. The rate at which a message source produces information is an important property that determines the required capability of any communication channel that is to convey the message. In the case of discrete information sources this rate is determined readily by the average information (entropy) per source symbol and the rate at which source symbols are produced. Thus, if a discrete source produces m symbols per second, the information rate is

$$R = mH, \text{ bits/second} \qquad (8)$$

As an example of the discrete case, suppose that a typist is capable of typing 80 words per minute and that the average length of English words

(including spaces) is 5.5 letters. Using the entropy estimate of 1 bit per letter, the information rate would be

$$R = \frac{(80)(5.5)}{(60)}(1) = 7\tfrac{1}{3} \text{ bits/second}$$

The situation is more complex for a continuous information source because the differential entropy is only a relative measure of information rather than an absolute one. Thus, simply multiplying the bits per sample (as in the BWG process above) by the number of samples per second does not yield a number that can be interpreted properly as information rate. This difficulty was resolved by Shannon by introducing the concept of a *distortion measure*. The philosophy here is that although it requires an infinite amount of information to reproduce a message value exactly, only a finite amount of information is required to reproduce the message value with a prescribed amount of distortion. Furthermore, as more distortion is permitted, the amount of information required is reduced. This concept makes it possible to describe the information rate of a continuous source as a function of distortion.

Although many distortion measures are possible, the most common one is the mean-square distortion. If the message random variable at any instant of time is x, and the reproduced random variable is y, then the mean-square distortion is defined as

$$d \triangleq E[(x-y)^2] \qquad (9)$$

The functional relationship between the information rate of the source and the distortion is denoted $R(d)$ and decreases monotonically as d increases, reaching zero when d equals the mean-square value of x.

The evaluation of rate-distortion functions is difficult to carry out in general, but can be done for continuous messages that are Gaussian. For the special case of the BWG process the result is quite simple and can be expressed as

$$R(d) = W \log \frac{\sigma^2}{d}, \text{ bits/second} \qquad (10)$$

This result is sketched in Fig. 2. It may be noted that the rate distortion function is unbounded as d approaches zero. This will be true for all continuous information sources.

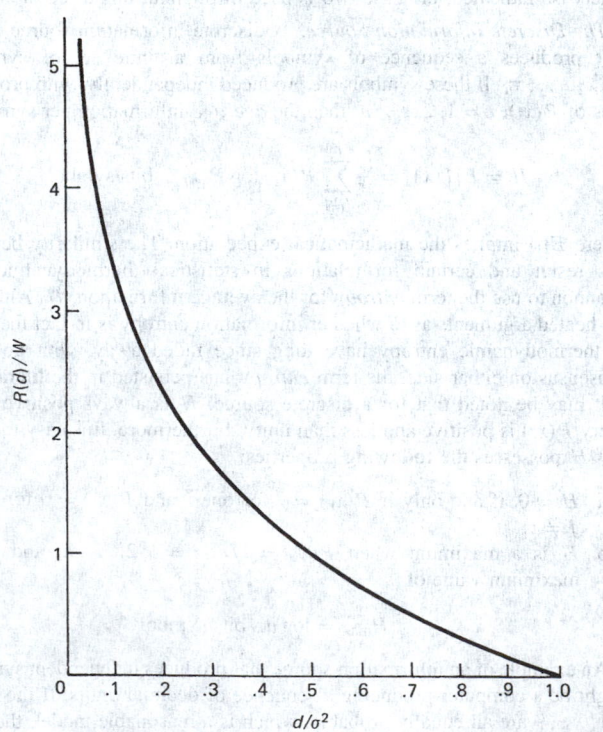

Fig. 2.   Rate-distortion function for a bandlimited, white, Gaussian message source.

It is also possible to define a rate-distortion function for discrete information sources. In this case as d approaches zero,

$$R(0) = R = mH \tag{11}$$

which is the information rate described above for discrete sources.

Channel Capacity. The function of a communication channel is to convey the information of the message source to its destination. In order to do this, the channel must be able to handle information at least as rapidly as it is produced by the source. Hence, it is of interest to determine the maximum rate at which information can be conveyed through the channel. This maximum rate is the *channel capacity*.

The reason that a physical communication channel has a finite capacity is that there is noise in the channel. In the case of discrete channels, the presence of noise may cause the symbol appearing at the receiver to be different from the one produced by the source. When the channel is continuous, the magnitude of the received signal at any instant of time may be different from the one transmitted. In either case, the effect of noise is to reduce the information capacity of the channel.

The significance of channel capacity is apparent from the fundamental theorem of information theory, which may be stated as:

If an information source has a rate $R(d)$ for a specified distortion d, and a channel has a capacity C, then it is possible to encode the output of this source and transmit it over the channel with a distortion arbitrarily close to d if $R(d) \leq C$. This is not possible if $R(d) > C$.

This remarkable theorem reveals that in the case of discrete sources, for which $R(0)$ is finite, it is possible to transmit information over a noisy channel with negligible error. It also reveals that for continuous sources, the information can be transmitted with a distortion that is arbitrarily close to any desired value, regardless of noise in the channel. Prior to Shannon's presentation of this theorem, it had been widely believed that noise in the communication channel inevitably degraded the quality of the message received and that there was no way to circumvent this limitation. However, the fundamental theorem asserts that there is a way of encoding messages into channel signals so that the message is not degraded, and establishes the limits on this operation.

Unfortunately the fundamental theorem does not reveal how one can find practical coding methods that achieve, or even approach, the theoretical limits. Although much insight into this problem has been achieved in the past 25 years, the goal itself remains elusive. All known coding schemes approach zero error with an information rate that either vanishes or is a small fraction of the channel capacity. Nevertheless, in spite of the failure to achieve the theoretical results, the pursuit of this goal has resulted in remarkable improvements in the performance of practical communication channels.

Capacity of the Discrete Channel. A discrete channel may be modeled in terms of a set of input symbols, x_i, $i = 1, 2, \ldots, n$ and a set of output symbols, y_j, $j = 1, 2, \ldots, m$. These symbols are not the same as the message symbols, but are the ones that the message is encoded into. For example, printed English may be encoded into the dots and dashes of the Morse Code, or into the binary digits of the teletype code. A continuous message such as speech may be sampled, quantized into a finite number of amplitude levels, and each sample encoded into a block of binary digits as in pulse-code modulation (PCM). Thus, the concept of a discrete channel involves a finite set of *channel symbols*, and such a channel may be used with either a discrete or continuous message source.

When there is noise in the channel, the output symbol may not be the same as the corresponding input symbol. For example, a dash may be confused as a dot, or a binary 1 may be received as a binary 0. The mathematical representation of errors of this sort is in terms of *conditional or transitional* probabilities. Thus, $P(y_j|x_i)$ is the probability that y_j is received given that x_i is transmitted. This is the probability of correct transmission when $i = j$, and is the probability of a particular erroneous transmission when $i \neq j$.

If input symbols are presented to the channel with probability $P(x_i)$, the joint probability of transmitting x_i and receiving y_j is

$$P(x_i, y_j) = P(x_i)P(y_j|x_i) \tag{12}$$

This can also be expressed in terms of the output probabilities, $P(y_j)$, as

$$P(x_i, y_j) = P(y_j)P(x_i|y_j) \tag{13}$$

where $P(x_i|y_j)$ is a conditional probability that is a measure of the uncertainty about what symbol was transmitted when the received symbol is known.

A convenient way of representing a discrete channel and its associated probabilities is by means of a pair of directed linear graphs. Such a pair of graphs for a hypothetical channel involving two input symbols and three output symbols is shown in Fig. 3. Such a channel is known as a *binary erasure channel* (BEC), and demonstrates that it is not necessary for the channel to have the same number of output symbols as input symbols. The physical significance of the erasure symbol ϕ is that the received binary digit is obscured by noise to the extent that a reliable estimate of the binary state is not possible.

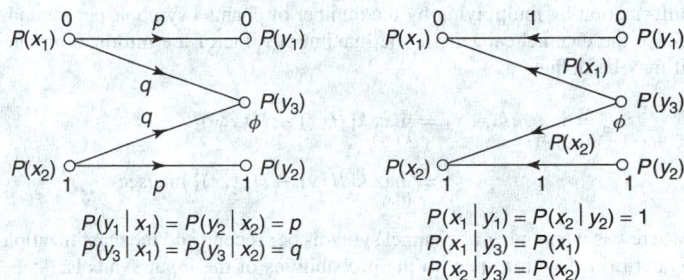

$$P(y_1 | x_1) = P(y_2 | x_2) = p$$
$$P(y_3 | x_1) = P(y_3 | x_2) = q$$

$$P(x_1 | y_1) = P(x_2 | y_2) = 1$$
$$P(x_1 | y_3) = P(x_1)$$
$$P(x_2 | y_3) = P(x_2)$$

Fig. 3. Line diagrams illustrating the conditional probabilities in a binary erasure channel.

Under the assumption that the input channel symbols are produced independently (a condition that any source encoding scheme strives to achieve) it is possible to define a set of four entropies and a mutual information. Thus,

$$H(x) \triangleq -\sum_{i=1}^{n} P(x_i) \log P(x_i) \tag{14}$$

$$H(y|x) \triangleq -\sum_{i=1}^{n}\sum_{j=1}^{m} P(x_i, y_j) \log P(y_j|x_i) \tag{15}$$

$$H(y) \triangleq -\sum_{j=1}^{m} P(y_j) \log P(y_j) \tag{16}$$

$$H(x|y) \triangleq -\sum_{i=1}^{n}\sum_{j=1}^{m} P(x_i, y_j) \log P(x_i|y_j) \tag{17}$$

$$I(x, y) \triangleq -\sum_{i=1}^{n}\sum_{j=1}^{m} P(x_i, y_j) \log \frac{P(x_i|y_j)}{P(x_i)P(y_j)} \tag{18}$$

Consideration of these definitions reveals some Additional relations. First there is a joint entropy that becomes

$$H(x, y) = H(x) + H(y|x) = H(y) + H(x|y) \tag{19}$$

and, secondly, the mutual information can be expressed as

$$I(x, y) + H(x) - H(x|y) = H(y) - H(y|x) \tag{20}$$

The relationship and physical significance of these various quantities is revealed by the Venn diagram of Fig. 4 in which the two circles represent the entropies of the channel input and output. The union of these circles is the joint entropy, while their intersection is the mutual information. The conditional entropy $H(y|x)$ is a measure of the uncertainty about what is received when the transmitted symbol is known, while $H(x|y)$ measures

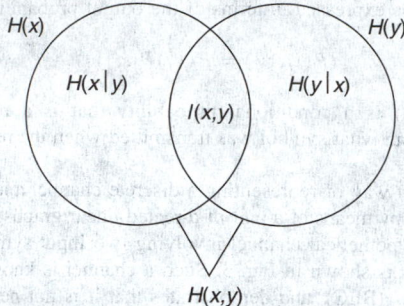

Fig. 4.   Venn diagram illustrating the relationships among the various entropies.

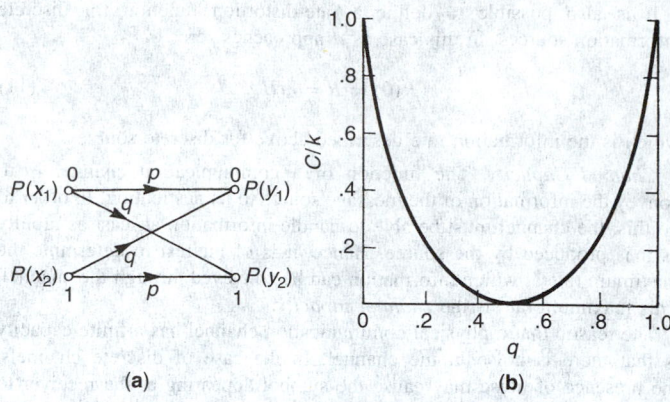

(a)   **(b)**

Fig. 5.   The binary symmetric channel (BSC) and its capacity: **(a)** channel diagram; **(b)** channel capacity.

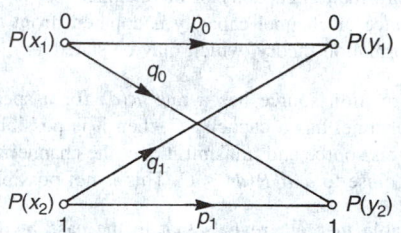

Fig. 6.   The general binary channel.

the uncertainty about what was transmitted when the received symbol is known.

The information rate of the channel is obtained from the mutual information by multiplying by the number of channel symbols per second. Since the channel capacity is the maximum value of the information rate, it may be defined as

$$C \triangleq \max_{P(x_i)} kI(x, y) = \max_{P(x_i)} k[H(x) - H(x|y)]$$

$$= \max_{P(x_i)} k[H(y) - H(y|x)] \text{ bits/sec} \tag{21}$$

where k is the number of channel symbols per second and the maximization is performed with respect to the probabilities of the input symbols. Thus, the objective of any source encoding scheme is to encode the message into a sequence of channel symbols that are not only independent, but also have the probabilities, $P(x_i)$, that maximize Equation (21).

Analytical techniques for finding the channel capacity in the general case are not known, although any specific case could be solved by computer. Some simple channels can be solved analytically, and a commonly considered one is the *binary symmetric channel* (BSC) illustrated in Fig. 5(a). Because of the symmetry, the maximum occurs when the input symbols are equally probable; i.e., when $P(x_1) = P(x_2) = \frac{1}{2}$. The resulting value of channel capability is shown readily to be

$$C = k[1 + p \log p + q \log q], \text{ bits/sec} \tag{22}$$

where $p = P(y_1|x_1) = P(y_2|x_2) q = P(y_2|x_1) = P(y_2|x_1)$ and $p + q = 1$. The quantity q may be interpreted as the probability of error for any binary digit and, because p is uniquely determined by q, the channel capacity can be expressed entirely in terms of q. This is illustrated in Fig. 5(b). It is of interest to note that a binary channel that is always in error can convey as much information as one that is always correct.

The general binary channel (i.e., one that is not symmetric) can also be handled analytically. Such a channel is illustrated in Fig. 6, from which the conditional probabilities are apparent. Perhaps the simplest solution of this more general case is due to Muroga and can be expressed as

$$C = k \log[2^{M_0} + 2^{M_1}], \text{ bits/sec} \tag{23}$$

where

$$M_0 = \frac{q_0 H_1 - p_1 H_0}{p_0 p_1 - q_0 q_1}$$

$$M_1 = \frac{q_1 H_0 - p_0 H_1}{p_0 p_1 - q_0 q_1}$$

and H_0 and H_1 are binary entropy functions defined by

$$H_0 = -[p_0 \log p_0 + q_0 \log q_0]$$

$$H_1 = -[p_1 \log p_1 + q_1 \log q_1]$$

The binary entropy function is widely tabulated in the literature.

The binary erasure channel shown in Fig. 3 is another case that can be solved readily. The channel capacity for this case becomes

$$C = kp, \text{ bits/sec} \tag{24}$$

This result could have been obtained intuitively since all digits not erased are correct and p is simply the probability of correct transmission for each digit. Under some conditions of noise, this method of detecting the channel output may yield a higher capacity than can be obtained from the binary symmetric channel in which a binary decision is made for every received digit.

Capacity of the Continuous Channel. The continuous channel is modeled in terms of an input time function $x(t)$ and a corresponding output time function

$$y(t) = x(t) + n(t) \tag{25}$$

where $n(t)$ is the noise added in the channel. Since these time functions have a continuous distribution of amplitudes, their absolute entropy is infinite even though the differential entropy is not. Fortunately, however, the mutual information between input and output depends upon the difference in entropies. This difference will be the same regardless of what standard density function is used to define the reference. Thus, it is possible to obtain a unique value for mutual information even when the differential entropy is not unique.

For purposes of illustrating the capacity of continuous channels, the input signal $x(t)$ will be assumed to be bandlimited, white, and Gaussian (BWG). If the bandwidth of this signal is B Hz (not to be confused with the message bandwidth of W), then $x(t)$ can be uniquely represented by a set of samples spaced $1/(2B)$ seconds apart. By analogy to the discrete case, as defined in Equation (21), the channel capacity can be defined as

$$C = \max_{P(x)} 2B[H(y) - H(n)], \text{ bits/sec} \tag{26}$$

in which $H(n)$ is the differential entropy of the noise $n(t)$ and is equivalent to the conditional entropy $H(y|x)$ because the noise is additive. The basis for this equivalence is apparent if it is noted from Equation (25) that the randomness of $y(t)$ depends only on $n(t)$ when $x(t)$ is given.

The maximization of Equation (26) must be carried out with respect to the probability density function of the input and requires that certain constraints be imposed. The most common constraint is that the signal has a specified average power, S. If the noise is also Gaussian and has an average power of N, then the output $y(t)$ will have an average power of $S + N$. For a given average power, the maximum entropy is obtained when $y(t)$ is Gaussian, and the value of this maximum entropy is

$$\max H(y) = \tfrac{1}{2}\log[2\pi e(S+N)]$$

Similarly, the noise entropy is

$$H(n) = \tfrac{1}{2}\log(2\pi eN)$$

and the resulting channel capacity is

$$C = B[\log 2\pi e(S+N) - \log 2\pi eN]$$
$$= B\log\left[1 + \frac{S}{N}\right], \text{ bits/sec} \tag{27}$$

This formulation of channel capacity is the one most commonly quoted in the literature, although it is often misused in situations for which it does not apply.

An important implication of Equation (27) is that it is possible to increase the capacity of a channel by increasing either the bandwidth or the signal-to-noise ratio in the channel. Because of the logarithmic dependence on signal-to-noise ratio, it would appear to be more effective to increase bandwidth. This may not be true in actuality because increasing the bandwidth also increases the noise power N. If it is assumed that the noise is white (and not bandlimited) with a one-sided spectral density of n_0 watts/Hz, then the noise power is

$$N = n_0 B$$

and the channel capacity becomes

$$C = B\log\left(1 + \frac{S}{n_0 B}\right) \tag{28}$$

This capacity is shown in Fig. 7 and is seen to asymptotically approach a limit of

$$C_\infty = 1.443\,\frac{S}{n_0}, \text{ bits/sec} \tag{29}$$

as the bandwidth is increased.

Another useful insight that applies to the white noise case, pertains to the signal energy required to transmit each bit of information. This energy is simply

$$E_b = \frac{S}{C}, \text{ watt-sec/bit}$$

and from Equation (28) can be expressed as

$$\frac{E_b}{N_0} = \frac{S/n_0 B}{\log(1 + S/n_0 B)} \tag{30}$$

Figure 8 shows the energy per bit as a function of the channel signal-to-noise ratio and makes it evident that the most efficient operation (smallest E_b) occurs with the smallest signal-to-noise ratio (largest B for a given S).

Error Correction Codes. The study of error correction codes has been by far the most active area of information theory and, in terms of practical results, has also been the most fruitful. However, in spite of the activity

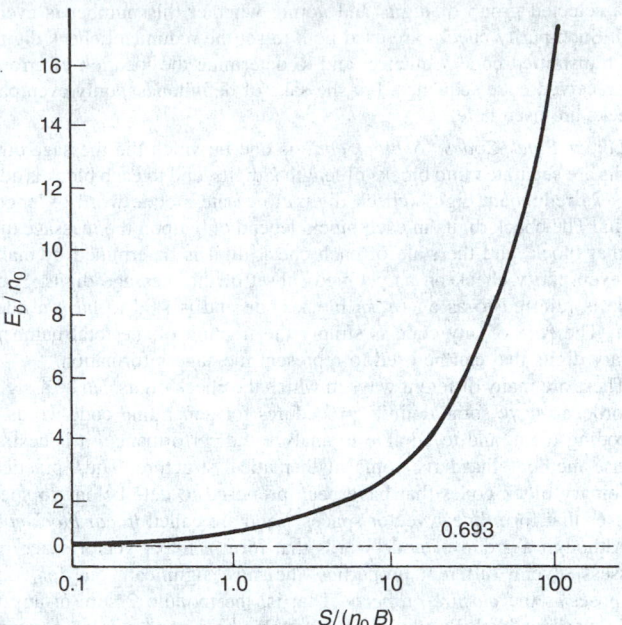

Fig. 8.   Energy required to transmit one bit of information as a function of signal-to-noise ratio.

and a multitude of practical applications, there is no general theory and specific results are limited to a handful of approaches for which a suitable mathematical structure exists. For the most part, coding theorists have divided into two camps; one group favoring linear block codes, while the other group favors convolutional codes. The elementary aspects of both approaches are considered here.

During the first two decades of information theory, coding theory was primarily an intellectual exercise and many of its followers despaired of ever seeing practical applications. However, the revolution in integrated circuits, the availability of inexpensive computers, and the increased emphasis on reliable transmission of digital data have spurred the development of practical coding equipment. The future importance of these techniques and equipment can hardly be overestimated.

As mentioned previously, error correction is achieved by adding redundancy to the message in a known fashion. The easiest place to do this is after the message has been encoded into channel symbols and before the resulting symbols are applied to the channel. Hence, error correction coding takes place in the *channel encoder*, as shown in Fig. 1. There is a corresponding decoder at the receiving end of the communication system. The function of the decoder is to remove the redundancy that was added, and to remove it in such a way that erroneous channel symbols are corrected.

Binary Arithmetic and Parity Checks. Most error correcting codes are designed for binary signals, and their operation and analysis utilize binary arithmetic. For this purpose it is customary to denote one state of the binary signal as 0 and the other state as 1. The operations of addition and multiplication are defined for these symbols as shown in Table 1. All of the calculations needed to encode and decode messages are performed in terms of these two operations.

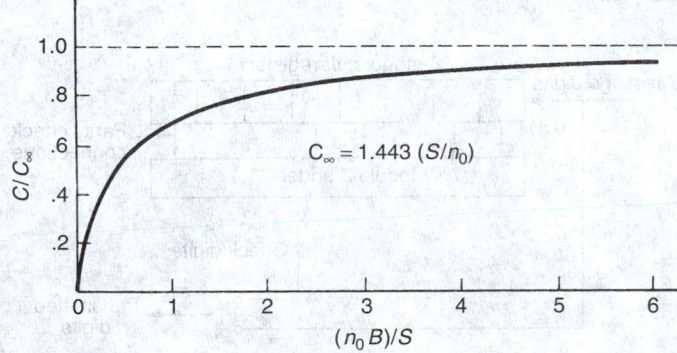

Fig. 7.   Capacity of a white-noise channel as a function of bandwidth.

TABLE 1. BINARY ARITHMETIC OPERATIONS

+ ‖* 0	1	× ‖* 0	1
0 ‖* 0	1	0 ‖* 0	0
1 ‖* 1	0	1 ‖* 0	1
(a) Modulo 2 addition		(b) Modulo 2 multiplication	

Perhaps the most fundamental operation in coding theory is the *parity check*. A parity check is performed by counting the number of 1's, say, in a selected group of digits and noting whether this number is even or odd. Such parity checks are used both to set the redundant check digits in the transmitted code sequence, and to determine the location of errors in the received code sequence. For the sake of definiteness, only even parity checks are used here.

Linear Block Codes. A *block code* is one in which the message binary digits are separated into blocks of length k digits, and to each block is added $(n - k)$ redundant digits (check digits) to create a code word of length n digits. The check digits in each block depend only upon the message digits in that block, and the state of each check digit is determined by making an even parity check on a specified subset of the message digits. Such a code is referred to as an (n, k) block code and is said to have a rate of k/n. The *rate* of any code is simply the fraction of the total number of binary digits that can be used to represent message information.

There are many different ways in which the check digits can be assigned. In order to have some definite procedures for generating code words, for decoding them, and to be able to analyze the performance, it is desirable to use methods that have some mathematical structure. Thus, practically all binary block codes that have been proposed to date belong to special classes that form linear vector spaces[15] and are called *linear block codes*.

Although a group of code words that form a linear vector space must possess several different properties, the most significant one for coding purposes is the *closure property*. That is, the modulo 2 sum of any two code words, added digit by digit, must also be a code word. An example of such a set of code words is the (5,2) code shown below.

$$
\begin{array}{ccccc}
0 & 0 & 0 & 0 & 0 \\
1 & 0 & 1 & 0 & 1 \\
0 & 1 & 1 & 1 & 1 \\
1 & 1 & 0 & 1 & 0 \\
\end{array}
$$

Note that the modulo 2 sum of any combination of these code words (using the addition rules shown in Table 1(a)) results in another code word from this set, and that the first two digits (which are the message digits for $k = 2$) form all combinations of two digits. The last three digits are the check digits and are formed by the *parity check equations*

$$a_3 = a_1 + a_2$$
$$a_4 = a_2$$
$$a_5 = a_1 + a_2$$

where a_i is the binary digit (either 0 or 1) in the ith position (from the left), and the addition is modulo 2. A basic problem in coding theory is that of selecting the parity check equations. When the number of digits in the code word becomes large, there are far too many ways of doing this to make it feasible to select a good set of check equations by simply trying all possibilities. In such cases the algebraic structure is essential to making a selection.

The error correction capability of any block code depends upon how many digits in each block can be received in error and still leave the code word "closer" to the one that was transmitted than it is to any of the other code words. In the (5,2) code shown above, for example, an error in any one digit leaves a word that still differs from any code word in at least two digits. Since an error in one digit is more probable than simultaneous errors in two or more digits, the original code word is the most probable one. Such a code is said to be *single-error correcting*.

Because of the algebraic structure of linear block codes, their error correction capability is determined readily from the *minimum distance* between code words. The distance between any two code words is the number of positions in which the binary digits are different, and, because of the closure property, the minimum distance is simply the smallest number of 1's in any code word that is not all 0's. Thus, in the (5,2) code above, the minimum distance is 3. It can be shown that in order to correct all combinations of t errors, the minimum distance must be

$$d_{\min} \geq 2t + 1$$

Hence, the (5, 2) code shown above cannot correct all combinations of two errors in any one code word, although it may correct some. In general, the minimum distance increases as the number of check digits in each code word increases. Thus, in order to achieve a large amount of error correction

capability with a given value of k, it is necessary to make n large. The code rate, therefore, becomes correspondingly small with the result that the information rate of the system vanishes as the code approaches the condition of correcting all errors.

A special class of linear block codes is the *cyclic codes* for which every cyclic shift of a code word is also a code word. These codes have the desirable property that they can be generated and decoded with shift registers so that even quite large codes become feasible from the equipment standpoint. A subclass of cyclic codes is the Bose-Chaudhuri-Hocquenghem (BCH) codes. Although these codes are limited to a few specific word lengths, they have very desirable error-correction capabilities and are the block codes that have been most widely implemented.

Convolutional Coding. The second major class of error correcting codes does not separate the message digits into blocks. Instead, the parity checks are performed continuously as the message is shifted through a shift register. This operation is illustrated in Fig. 9, which shows a rate $\frac{1}{2}$ convolutional coder. In this case the binary digits applied to the channel are alternately message digits and check digits, and the check digits are the result of an even parity check over the contents of four different stages in the shift register.

Although it is possible to obtain an algebraic representation for convolutional encoding, there is no general theory that describes the error correction capability. Nevertheless, experimental results indicate that such codes are usually as good as the best BCH codes of the same size and require less hardware to implement.

The decoding of convolutional codes can be carried out by either *threshold decoding* or by *sequential decoding*. Threshold decoding is easier to implement but is not an optimum procedure. In this method of decoding, the received message digits and check digits are separated, a new set of check digits is formed from the received message digits (using the same coder as in Fig. 9), and the new check digits compared with the received check digits. Whenever the number of differences between the two sets of check digits exceeds a specified threshold, the corresponding message digit is changed in state. Sequential decoding is optimum (or nearly so) but requires greater computational effort. This procedure utilizes a special purpose computer to examine a sequence of past decisions on the states of binary digits and determine if they are all consistent with the most recent data. Whenever it appears probably that a past decision was in error, that particular digit is changed and the revised sequence is examined. The computational effort for this procedure increases rapidly with the length of sequence being considered. However, a more recent procedure, the *Viterbi* algorithm, is extremely attractive for shorter codes, and many decoders employing this algorithm are now in use.

Source Encoding. Error correction coding takes place in the *channel encoder* block of Fig. 1. Hence, the input to the channel encoder is the *source encoder*, whose function is to convert the message source into a sequence of symbols (usually binary digits) that can be utilized by the channel encoder. In order to achieve efficient communication it is necessary that the source encoder convert the source into binary digits at a rate that is consistent with the rate distortion function $R(d)$. This is particularly important in the case of continuous sources since some distortion must always be accepted in order to transmit the source at a finite rate.

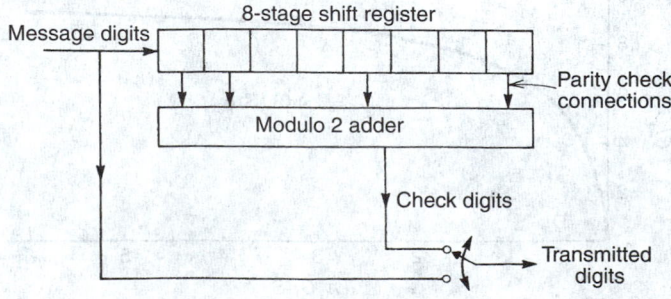

Fig. 9.   Encoder for a one-half rate convolutional code with parity checks that span eight message digits.

Unfortunately, there is no constructive theory of source coding that is analogous to the channel encoding techniques just discussed. There are several reasons why this is so. In the first place, sources arising in practice are extremely difficult to model mathematically and the corresponding measures of distortion are equally difficult. Secondly, the acceptable distortions are ordinarily so small that there is not much to be gained by source encoding techniques. Finally, source encoding is inherently much more difficult than channel encoding.

Transmission of continuous sources through a discrete digital communication system requires that the source be quantized in some fashion. Thus, much of the work that has been done in connection with source encoding pertains to optimum techniques for quantizing continuous sources. It has been shown that it is possible to achieve performance that is within one bit of the theoretical rate distortion curve for any Gaussian source. In addition, Berger provides a detailed summary of quantization techniques, including a discussion of their applicability to sources with memory.

An alternative approach to source encoding is to use channel codes (i.e., error correction codes) in a backward fashion. In this case the channel decoding algorithm becomes the source encoding algorithm. This procedure will work if the decoding algorithm produces the closest code word regardless of the input. Unfortunately, this condition exists only rarely, so that only a few special cases are available.

Additional Reading

Berger, T.: *Rate Distortion Theory*, Prentice-Hall, Inc., Upper Saddle River, NJ, 1971 (A classic reference).

Bouwmeester, D., A. Zeilinger and A.K. Ekert: *Physics of Quantum Information: Quantum Cryptography, Quantum Teleportation, Quantum Computation*, Springer-Verlag, Inc., New York, NY, 2000.

Buchmann, J., J. Stichtenoth and H. Tapia-Recillas: *Information and Coding Theory*, Springer-Verlag, Inc., New York, NY, 2000.

Gravano, S.: *Introduction to Error Control Codes*, Oxford University Press, Inc., New York, NY, 2000.

Green, H.S.: *Information Theory and Quantum Physics: Physical Foundations for Understanding the Conscious Process*, Springer-Verlag, Inc., New York, NY, 2000.

Johannesson, R., K.Sh. Zigangirov and B. Anderson: *Fundamental of Convolutional Coding*, IEEE Press, New York, NY, 1999.

Jones, G.A. and J.M. Jones: *Information and Coding Theory*, Springer-Verlag, Inc., New York, NY, 2000.

Lint, J.H.: *Introduction to Coding Theory*, 3rd Edition, Springer-Verlag, Inc., New York, NY, 1998.

Macchi, C., A. Zeilinger and G.M. Palma: *Quantum Computation and Quantum Information Theory: Collected Papers and Notes*, World Scientific Publishing Company, Inc., Riveredge, NJ, 1999.

Pless, V.S. and W.C. Huffman: *Handbook of Coding Theory*, Vol. 1, Elsevier Science, New York, NY, 1998.

Saito, K. and T. Hida: *Quantum Information: Proceedings of the Second International Conference*, World Scientific Publishing Company, Inc., Riveredge, NJ, 2000.

Shannon, C.E.: "A Mathematical Theory of Communication," *Bell System Tech. J.*, Section 20 (October, 1948). (Continuation of reference 4) (A classic reference).

Staff: *Information Theory: 2000 IEEE International Symposium*, IEEE Standard Office, New York, NY, 2000.

Stanley, W.D. and R.F. Harrington: *Electronic Communication Systems*, Prentice-Hall, Inc., Upper Saddle River, NJ, 2001.

Stepanov, S.A.: *Codes on Algebraic Curves*, Kluwer Academic Publishers, Norwell, MA, 1999.

GEORGE R. COOPER, Purdue University, West Lafayette, IN

INFO-ZIP. See **Data Compression**.

INFRARED ASTRONOMICAL SATELLITE (IRAS). Launched from Vandenberg Air Force Base (California), IRAS assumed a nearly polar orbit at an altitude of 900 kilometers. The satellite pointed roughly 90 degrees away from the sun. The main survey instrument was an array of 62 detectors that covered the major portion of the IR spectrum, from 8 to 120 micrometers, in four bands centered on wavelengths of 12, 25, 60, and 100 micrometers. (By comparison, the visible part of the spectrum extends to about 0.7 micrometer and radio waves begin at about 1000 micrometers.) In addition to the IR detectors, the satellite carried a spectrometer for recording the IR spectrum of bright sources and a photometer for measuring preselected sources with higher spatial resolution. The telescope has a "folded" optical path: radiation struck a 57-centimeter primary mirror, then a small secondary mirror directed it through an opening in the center of the primary. The telescope and associated instrumentation were cooled by some 475 liters of liquid helium. The mirrors were cooled to about 10 degrees Kelvin and the detectors to about 2 degrees K. To conserve coolant by slowing its evaporation the temperature was maintained somewhat below its boiling point. Nevertheless, the helium supply was exhausted approximately ten months later and the project was shut down. During the interim, however, it has been estimated that 95% of the celestial sky had been surveyed. Further it has been estimated that 99.8% of the sources catalogued are indeed real and that no more than 2% of the real sources bright enough to be detected were missed. As observed by Habing and Neugebauer, the IRAS mission resulted in the cataloging of some 250,000 discrete sources, representing about one-third of all sources cataloged during all the history of astronomy. Some of the highlights of infrared astronomy during the past few years include:

Dust Trails in the Orbits of Comets

Sykes and colleagues (University of Arizona) reported in 1986 that analysis of data from the IRAS yielded evidence for narrow trails of dust coincident with the orbits of periodic comets Tempel 2, Encke, and Gunn. Dust was found both ahead of and behind the orbital positions of these comets. The dust was produced by the low-velocity ejection of large particles during perihelion passage. More than a hundred Additional dust trails were suggested by the data, almost all near the detection limits of the IRAS. Some of these trails are suspected to have been derived from previously unobserved comets. Among other possible sources of dust trails are asteroids, which also have generally low inclinations. The researchers observed that since the initial result of an asteroid collision is the distribution of debris along its orbit, some of the dust trails noted by the IRAS may be asteroidal in origin. In commenting further on this phase of the IRAS mission, the investigators observe it is conceivable that in a future comet rendezvous mission, a spacecraft might be able to directly sample and analyze the rocky component of a comet nucleus by approaching the comet along its orbit and sampling its associated dust trail. The relative velocity of such a spacecraft and the trail debris would be quite low (meters per second). Following a comet through perihelion, the spacecraft could monitor trail development and the corresponding processes at the nucleus. Such a mission could determine particle-size distributions, structures, and compositions of this material, which would provide a better understanding of cometary origin and the formation of the solar system. See also **Comet**.

During the mission, IRAS discovered at least five comets and also observed Tempel 2, a comet known from 16 prior appearances in the inner solar system. Earlier, it had been presumed that this comet had no tail and that volatile material has disappeared during earlier passages near the sun. IRAS found, however, that the comet does have a long, narrow tail extending some 30 million km from the cometary nucleus.

New Class of Galaxies

IRAS found that in most galaxies, including our own, the IR and visible luminosities are approximately the same. IRAS did discover, however, what appears to be a different class of galaxies, making up an estimated 5% of all galaxies. They have been likened to a hot frying pan—dim in the visible and bright in the infrared by a factor of 50 to 100. Most dramatic of all was Arp 220, estimated to have a power output equivalent to two trillion suns. There is considerable speculation concerning these unresolved objects. About a third of the IR bright objects appear to be pairs of galaxies in the process of collision or merger. Are shock waves produced in the collision triggering bursts of star formation? It is known that star formation produces lots of IR emission in normal galaxies. However, certain other spectral features are not consistent with normal star formation. Are the objects quasars or Seyfert galaxies which may be shrouded in dust? Is most of the immense energy emanating as heat rather than light? Other spectral features are not consistent with this concept. Are they protogalaxies in a very early stage of star formation?

Solid Matter in Orbit Around a Star

In their 1986 Report on IRAS, Rieke and colleagues rank the discovery of clouds of millimetersized particles around Vega and β Pictoris among the most intriguing. These particle clouds are very different from the planets around our sun. Apparently, they are detectable only because their mass is finely divided and thus they have a large surface area that absorbs energy from the central star. It is estimated that Vega may only be a few hundred million years old as compared with 4.5 billion years for our sun. Thus, it is reasoned that perhaps our solar system at one time went through a similar stage of development. IRAS found some 50 other stars whose excessive IR radiation could be explained by this mechanism, but only in one case is there evidence of millimeter-size particles orbiting a star.

IR Cirrus Clouds

In the 100-micrometer band, IRAS observed fluffy, wispy trails of cold dust distributed over the entire sky. Habing and Neugebauer point out that the nature and location of the infrared cirrus clouds are not yet definitely established. Some of them may be part of the solar system, but early studies cast doubts. The clouds may be as far away as 50,000 or even 100,000 astronomical units, but still be gravitationally bound to the sun. Tentatively, it is believed that most of the infrared cirrus is probably in the interstellar medium, outside the solar system, but within the sun's immediate neighborhood. In studies of IRAS data it was found that some of the cirrus features are coincident with clouds of hydrogen gas observed at radio wavelengths. Should this assumption be true, it follows that the clouds would be made up of gas and dust ejected by dying stars and probably swept up by expanding supernova remnants.

Bulge in the Milky Way

IR observations differed from the visible image of the galaxy by the presence of a bulge near the center of the galaxy. The galactic center cannot be studied at visible wavelengths because it is obscured by large amounts of dust. The knowledge must come from radio and IR observations. The IRAS reconfirmed many of the features of the galaxy, but extended structures, such as wisps of dust that appear above and below the galactic plane near the nucleus were seen clearly for the first time.

It is beyond the scope of this encyclopedia to probe further into the massive amounts of data collected by IRAS (700 million bits of image data in less than one year). Some of these topics are discussed in other articles. In particular, check **Cosmology**; and **Galaxy**.

Improvements in IR Instrumentation

As pointed out by I. Gatley, D.L. DePoy, and A.M. Fowler (National Optical Astronomy Observatories), the majority of infrared detectors used are hybrid devices, resulting from a process in which the detectors and readouts are manufactured separately. "The readout circuit is fabricated on silicon in the same way as integrated circuits, whereas the detectors are made, for example, from mercury cadmium telluride, indium antimonide, platinum silicide, or extrinsic doped silicon. Then the two chips are sandwiched together, with the electrical interconnections being made by indium bumps on each of the chips." Most arrays in use in telescopes have been obtained through the collaboration of industry with NASA and the Department of Defense.

Further strides can be made with array detectors in space, where the thermal background can be eliminated by cooling the telescope. Gains achieved by the previously described IRAS were the result of cooling the instrument. That instrument, however, did not have the benefit of arrays. Plans for all future IR instruments include cooling and the use of arrays.

Infrared astronomy is yielding much information on the formation and survival of galaxies. Gamma radiation also has been an invaluable observing tool. See also **Gamma-Ray Astronomy**; and **Infrared Astronomy**.

Additional Reading

Cowen, R.: "Lifting a Dusty Veil to Clear IRAS' View," *Sci. News*, 182 (September 22, 1990).

Crease, R.P.: "Millimeter Astronomers Push for New Telescope," *Science*, 1504 (September 28, 1990).

Gatley, I., D.L. DePoy and A.M. Fowler: "Astronomical Imaging with Infrared Array Detectors," *Science*, 1264 (December 2, 1988).

Habing, H.J. and G. Neugebauer: "The Infrared Sky," *Sci. Amer.*, 48–57 (November 1984).

Neugebauer, G., et al.: "Early Results from the Infrared Astronomical Satellite," *Science*, **224**, 14–21 (1984).

NOAO: *IRAF—Image Reduction and Analysis Facility*, National Optical Astronomy Observatories, Tucson, Arizona, 1988.

Olson, C.: "Tiny 'Eye' Produces Unique View: Longwave Infared Sensor," *Hughes News*, 1 (August 24, 1990).

Rieke, G.H., et al.: "Infrared Space Observatory," *Science*, 232, 1487 (1986).

Sykes, M.V., et al.: "The Discovery (by IRAS) of Dust Trails in the Orbits of Periodic Comets," *Science*, 232, 115–117 (1986).

Waldrop, M.M.: "A Window Looking Out on Creation," *Science*, 32 (October 5, 1990).

INFRARED ASTRONOMY.   For many years, scientists contemplated the use of an infrared-sensitive telescope to survey the heavens. As early as 1878, Edison used a sensitive infrared (IR) detector to observe a solar eclipse from a site in Wyoming. Edison calibrated his instrument on the bright star Arcturus and suggested that the entire sky could be mapped in search of invisible stars. The emphasis in that period, however, was on the construction of larger and larger optical telescopes. Herschel discovered IR when he measured temperatures in the spectrum of the sun (late 1700s) and it is reported that Lord Rosse detected IR from the moon in 1845. Modern IR astronomy did not commence until the 1950s and 1960s because of the lack of appropriate instrumentation. During that period, the military became very seriously interested in IR detection of earthly objects (example: the "snooperscope," designed for battlefield detection; later the use of IR for reconnaissance). Out of these interests there stemmed an increased sophistication in IR and electrooptical technology, including developments in low-temperature physics and thermometry. Interests temporarily culminated in the launching of the *Infrared Astronomy Satellite* (IRAS) by the United States, the Netherlands, and Great Britain in January 1983.

The hundreds of IR sources in the IRAS catalogue presented a challenge of long duration for study. The IR astronomy community also turned its attention to consideration of the *Space Infrared Telescope Facility* (SIRTF), a helium-cooled, pointed space telescope. Design criteria were established for a thousandfold gain in sensitivity over five octaves of the spectrum.

Infrared astronomy is yielding much information on the formation and survival of galaxies. Gamma radiation also has been an invaluable observing tool. See also **Gamma-Ray Astronomy**.

Uniqueness of Infrared

Infrared observations provide the following unique perspectives on the Universe:

The Cold Universe. There is an inverse relationship between the temperature of an object and the peak wavelength λ of its intrinsic or blackbody radiation: $T(\lambda) = 3700/\lambda$. Here T is measured in degrees kelvin (K) above absolute zero, and λ in microns (μm). Objects with $T < T(\lambda)$ radiate very little at wavelengths less than λ. Observations at infrared wavelengths from $1–1000$ μm are thus uniquely sensitive to astronomical objects whose temperatures are from $\sim$3000 K to $\sim$3 K. These include the coolest stars, planets and interplanetary dust, circumstellar and interstellar matter, and, at the longest wavelengths, the Universe itself.

The Dusty Universe. Interstellar dust—microscopic particles composed of ices, minerals, and common organic and inorganic materials—is a ubiquitous constituent of astrophysical environments. The properties of this material are such that a cloud that is totally opaque in the visible or ultraviolet can be virtually transparent in the infrared; thus infrared wavelengths can probe regions—such as the core of our galaxy—which are inaccessible at shorter wavelengths. Additionally, the dust particles are heated by the shorter wavelength radiation they absorb and reradiate the absorbed power at infrared wavelengths. The majority of the radiant energy from dense, dusty regions such as star-forming clouds—and in some cases from entire galaxies—lies at infrared wavelengths because of this efficient downconversion process.

The Distant Universe . In the expanding Universe, the more distant an object is, the greater the velocity at which it recedes from us. This cosmic expansion shifts the starlight from distant galaxies into the infrared; the more distant the object, the farther out into the infrared. This expansion is characterized by the redshift parameter z: $1 + z =$ (observed wavelength/emitted wavelength). The most distant known objects have $z > 6$, so that radiation from the middle of the visual band is shifted out beyond 3 μm. Because $1 + z$ is also equal to the factor by which the Universe has expanded between the times of emission and absorption of the radiation, objects at $z = 5$ are seen as they were at an epoch when the Universe was only one-sixth of its present size.

The Chemical Universe. The infrared band contains the spectral signatures of a variety of atoms, molecules, ions, and solid substances—some of which will be found in any astrophysical environment. Examples range from cool ices in the interstellar medium to highly excited ions in active galactic nuclei. Infrared spectroscopy can isolate these features, determine their absolute and relative strengths, and provide an important and often unique probe of the chemical and physical conditions in these systems. See also **Interplanetary Medium**.

The Advantages of Space

The space environment presents powerful advantages for conducting infrared astronomical observations, which motivate the technological developments discussed below. First, in space, one is free of the absorption by Earth's atmosphere, which—even from the best mountaintop observatories—is totally opaque at wavelengths from $\sim$30 to $\sim$300 μm. Outside of this region, there are other bands of high and moderate opacity, and atmospheric absorption remains appreciable at aircraft and balloon altitudes. Only from space do we have access to the entire infrared band. A second, equally fundamental benefit is that a space observatory is free of the blackbody radiation of Earth's atmosphere, and the space telescope can be cooled to low temperature to minimize its own blackbody radiation without fear of atmospheric condensation. Infrared observations from Earth are limited by very bright foreground radiation from the atmosphere and the ambient temperature telescope; in space using a sufficiently cold telescope, the limiting background—set by the faint glow of the interplanetary zodiacal dust cloud—is some six orders of magnitude fainter. This is about the same factor by which the night sky at new moon is fainter than the daytime sky at high noon; note that optical astronomy is practiced at night, not during the day. The impact of this million-fold background reduction, in space, is impossible to overestimate because it produces a thousandfold increase in sensitivity, or a millionfold increase in the speed of observations. Thus the first major cryogenic infrared space observatory, the Infrared Astronomical Satellite (IRAS), revolutionized our knowledge of the infrared sky, even though it observed each point for less than about 20 seconds during its 10-month survey of the sky.

Complementary Approaches

Infrared astronomy is pursued very successfully from ground-, aircraft-, and balloon-borne platforms, and these sites present opportunities and capabilities complementary to the very high sensitivity and spectral access of space. The current state of the art for ground-based infrared astronomy is a series of 8- to 10-m diameter telescopes in Hawaii and Chile that provide ongoing scientific opportunities, much higher spatial resolution than achievable from space at present, and a greater variety of focal plane instrumentation—including complex spectroscopic instruments—than typically available in space observatories. Airborne observatories—exemplified by the imminent $\sim$3-m-class Stratospheric Observatory for Infrared Astronomy (SOFIA)—provide capabilities similar to those of large ground-based telescopes in most of the wavelength bands that are inaccessible from the surface of Earth. Balloon-borne instruments have been successful for specialized measurements, most notably survey observations and studies of the CMBR, and will have an important niche in the upcoming era of long-duration balloon flights.

Detectors and Detector Arrays

Modern detectors fall into two classes, bolometers and photoconductors. Bolometers are devices that change resistance when heated by absorbed radiation; they respond to radiation across a wide wavelength band, as long as they effectively absorb across this entire band. Bolometer technology has advanced dramatically in the past few years due to the application of modern semiconductor processing techniques. See also **Bolometer**. Photoconductors are solid-state devices in which incident photons can excite electrons from a bound state—a valence band or an impurity level—into a conduction band to produce a current in response to an applied voltage. Because the valence or impurity levels and the conduction bands are separated by a well-defined energy gap, the energy or wavelength range within which photoconductors respond to radiation is restricted. To cover a broad infrared wavelength range, a combination of different photoconductors generally needs to be employed. The current materials of choice are InSb and HgCdTe photodiodes for wavelengths shorter than $\sim$10 μm, extrinsic (doped) silicon photodetectors (Si:xx) for wavelengths from 5–40 μm, and extrinsic Germanium (Ge:xx) for $\sim$40–200 μm. In this nomenclature "xx" denotes the specific dopant. The preferred dopants for silicon detectors are As, B, Ga, and Sb; Ga, Be, and Sb have been used as dopants for germanium. Bolometers are the detectors of choice for wavelengths longer than 200 μm and for some applications at shorter wavelengths as well.

All detectors are inherently noisy. They register the incidence of arriving photons, generally referred to as "signal," and also any number of other types of events classed as "noise." Notable among the noise sources are thermally excited conduction electrons, or "dark current" in photoconductors, and thermal fluctuations in bolometers. Both sources of noise can be reduced by cooling the detectors to temperatures so low that thermal effects become negligible. In real-life applications, both types of detectors can be degraded by the electronics required to operate them and read them out, although modern circuit design techniques and on-chip integration generally allow minimizing these effects. In space, noise can also be generated by high-energy cosmic rays that traverse the detectors. Effective shielding against such particles becomes a high priority; in addition, special fabrication techniques can be used to reduce the susceptibility of the detectors to this ionizing radiation, as has been done with extrinsic silicon photoconductors of the type used on SIRTF.

A recent major advance in infrared detectors is large arrays of many active elements, or pixels, bonded to a multiplexer that is used to sample and read out the pixels. The impact of this technology on space infrared astronomy—which is similar to the CCDs used in the visible band—will be very dramatic. Used with photoconductors in the low-background space environment, the technology permits on-chip integration, so that the signal can be accumulated on the detector array and read out only when it is large enough to overcome electronic noise. Clever schemes involving multiple, nondestructive readouts of the array have been devised to suppress electronic noise further. Detector arrays are equally applicable to imaging and to spectroscopic instruments—and to photoconductor and bolometer technology, and they will certainly be used very extensively, if not exclusively, for space infrared astronomy in the future. There will continue to be a push for larger format arrays, and the next generation of infrared space experiments should use arrays of at least 1024×1024 pixels, a substantial advance over the 256×256 pixel arrays used on SIRTF.

The ultimate performance goal for detectors for infrared astronomy is that they permit "background-limited" observations, that is, the intrinsic detector and electronic noise should be less than the noise due to the statistical fluctuations in the rate of arrival of photons from ambient and astrophysical backgrounds. Modern infrared detectors achieve this readily in ground-based applications where the warm telescope and emissive atmosphere produce very high backgrounds. For space applications using cryogenic telescopes, the infrared background is that due to the zodiacal dust within the solar system, which is at least a million times fainter than the ground-based foreground sky. Achieving background-limited performance in this environment is quite challenging, even with the benefits of on-chip integration; for example, observations in the 3 to 5 μm window require that dark current and electronic noise contribute less (often much less) than the equivalent of one electron/second/pixel. Detector technologists and astronomers working together have responded to these challenges and improved the performance of infrared detectors by many orders of magnitude in the past two decades. The improvement has come from reducing the noise and also by improving the "quantum efficiency"—the fraction of incident photons that is absorbed by the detector. As a result, the arrays used on SIRTF will achieve background-limited performance for both photometry and low-resolution spectroscopy at all wavelengths.

Cryogenics

Space infrared telescopes invariably require efficient cooling or cryogenic systems; the telescope and the surrounding structure are cooled to reduce their background radiation, and the detectors are cooled to reduce their intrinsic noise and increase their sensitivity. In many applications, these effects together require cooling below 10 K. In many of the rocket and satellite instruments built to date, the entire telescope has been cooled to temperatures as low as 2 K, where no part of the apparatus emits as

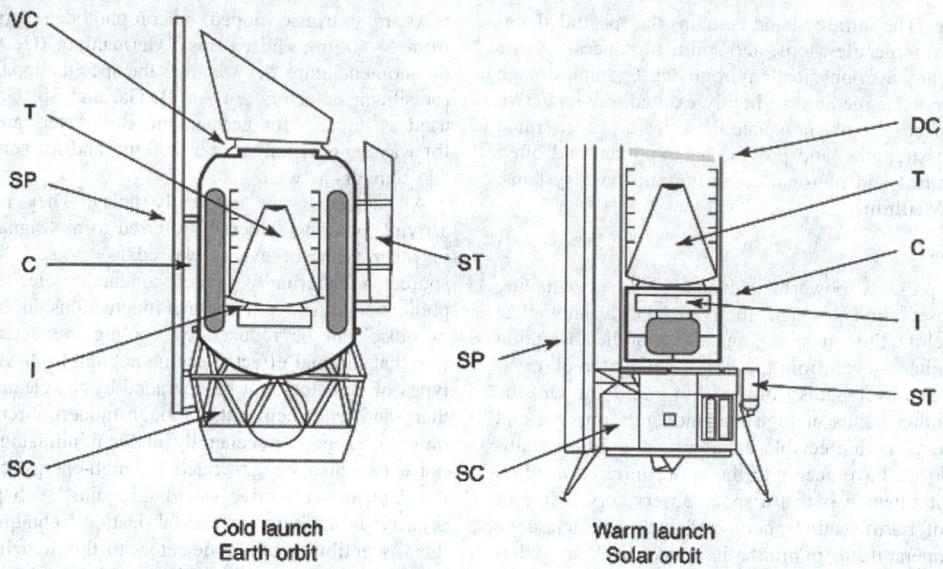

Fig. 1.   This figure compares the cold launch architecture used for the ISO and IRAS observatories (a) with the warm launch architecture used for SIRTF (b). Each is shown in cutaway view. Certain components, such as the spacecraft (SC), the solar panel (SP), and the startracker (ST) are common to both systems. In addition, the telescope (T) and instrument package (I) are identical in size for the two. Each also includes a cryostat ©, containing the liquid helium cryogen in a separate helium tank, which is shown shaded. In the cold launch system, the telescope is located within the cryostat and cooled by direct contact with the cryogen tank. In the warm launch system, the cryostat and cryogen tank can be much smaller, and the telescope is cooled by conduction and by the cold boil-off helium gas. This architecture works in the solar orbit because the cylindrical thermal shields that surround the telescope cool radiatively to 40 K or below, so there is very little parasitic heat diffusing inward toward the telescope. The cryostat must withstand atmospheric pressure, and it is much larger and more massive for the cold launch than for the warm launch system. In the former, it surrounds the entire telescope and supports a heavy vacuum cover (VC). In the warm launch system, the telescope is launched at ambient temperature and pressure, protected only by a lightweight dust cover (DC). The sawed-off conical sunshade at the top of the cryostat is required in an Earth-orbiting system by the Sun–Earth-orbit geometry. A much smaller sunshade is need in the solar orbit system because Earth is not a concern.

much radiation as the 2.73 K CMBR. This has been done by placing the entire telescope structure in direct physical or thermal contact with a pumped-liquid-helium bath, and the vacuum pump is the natural vacuum in space. The IRAS and Infrared Space Observatory (ISO) systems used this architecture, as shown in Fig. 1a. Liquid helium is required to achieve temperatures below ~5 K, but other stored cryogens that provide more cooling power per unit mass are used in applications where higher temperatures are acceptable. In practice, this is equivalent to reducing the long-wavelength limit of the instrument. Other cryogens that have been used—and the approximate temperature that they provide—are solid hydrogen (8 K), liquid neon (30 K), solid nitrogen (50 K), and liquid nitrogen (75 K).

A primary design problem for the cryogenic engineer is to minimize the heat load on cooled surfaces of the apparatus. A first step is to shield these surfaces from the principal heat source, which is solar radiation, and to use suitable combinations of low- and high-emissivity materials to reduce heat transfer within the satellite. A next step is to blanket the container that holds the cryogen with dozens of layers of aluminized mylar loosely packed within a vacuum jacket to isolate the entire system from its ambient-temperature surroundings before launch. Such a vacuum-packed cryogenic system is referred to as a dewar, named for the nineteenth-century Scottish scientist J. Dewar. A well-designed cryogenic system also uses the cold effluent gas generated as the cryogen evaporates or sublimes to cool the surrounding structures and further reduce the heat load on the cryogen.

A second design challenge is the construction of apparatus sufficiently sturdy to survive launch and also to maintain optical alignment between the cooled telescope and the guide telescopes or gyroscopic components that typically operate at ambient temperature within the spacecraft bus. The mechanical rigidity required tends to go hand in hand with high thermal conductivity, which the design must avoid; this requires using low-thermal-conductivity materials that have high strength, such as epoxy-glass composites.

Considerable attention has also focused on minimizing heat loads on the cryogen by passively radiating intercepted heat into cold space; this is referred to as "radiative cooling". The SIRTF telescope, described later, exploits the favorable thermal environment of its heliocentric orbit by using a hybrid cryogenic system in which the instruments and detectors are cryogenically cooled, whereas the telescope is launched warm and is cooled by a combination of radiation, conduction, and effluent cryogen (Fig. 1b). This approach has many advantages over that used in earlier missions such as IRAS and ISO, in which the entire telescope was placed within the cryostat. It leads to a lower mass cryogenic system for a fixed telescope size and decouples the size of the telescope from that of the cryostat. Thus this hybrid approach is certain to be adopted for large infrared telescopes in the future.

In a more extreme application of radiative cooling, it may be possible to cool the entire telescope to a temperature acceptable for many purposes without using cryogens. Just how low a temperature can be reached in practice is still not clear, but many designers now assume that equilibrium temperatures as low as 30 K could be within reach at 1 astronomical unit (the radius of Earth's orbit) from the sun. Lower temperatures might be achieved by a telescope operating in the outer solar system. At such low temperatures, a well-designed telescope that has exceptionally low-emissivity mirrors might radiate at such low levels in the wavelength range shorter than 100 µm that the primary and secondary mirrors require no active cooling at all. Active cooling would be required primarily for the detector arrays and their immediate housings. The reduced cooling requirements of such a system might be satisfactorily met by an acceptable, though still substantial, charge of cryogen, or by closed-cycle refrigerators required only to pump heat at low rates.

A stored cryogenic system always has a limited lifetime: unless replenished (an approach which has not been adopted for any astronomical mission), the cryogen eventually is fully depleted, the system warms up, and the mission comes to an end. To increase mission life spans, a variety of recyclable cryocoolers—both mechanical and electrochemical—have been under intense study. In principle, they could extend lifetimes indefinitely. In practice, a nagging long-term problem has been the limited reliability of closed-cycle, low-temperature refrigerators designed to operate in the vacuum of space. In the laboratory, such systems have often failed catastrophically after only a few months. No refrigerator of this type has ever operated in the laboratory continuously for a 10-year span. Yet this is the expected mission lifetime of many infrared astronomical space facilities now on the drawing boards. In 1998, NASA successfully tested a particular type of mechanical cooler on a Shuttle mission—a reverse Brayton-cycle cryocooler that can cool detectors to temperatures as low as 60–70 K. This test showed that operation in weightless conditions was not a problem for this type of cooler, but long-term reliability is still an open question, though the same coolers have a good record in the laboratory and run reliably for many months to a few years. Reliable closed-cycle cryocoolers are certain to affect critically the design and life spans of future infrared astronomical missions in space. The reverse Brayton-cycle cryocooler described before has been retrofitted to the NICMOS instrument on HST to extend its useful lifetime beyond the almost 2 years achieved with the initial charge of solid nitrogen.

Some types of highly sensitive infrared detectors now being planned for future missions operate effectively only at temperatures in the millikelvin range. Additional cooling beyond that achievable with liquid ^{4}He must be provided for them. In the laboratory, a variety of techniques has already been developed to reach such low temperatures; often, they require a succession of stages that might employ combinations of thermoelectric, liquid ^{3}He, ^{3}He/^{4}He dilution, adiabatic demagnetization, or other refrigerators. For long-duration astronomical space missions, reliable refrigerators will be required to provide these low temperatures continuously or cyclically. Again, these devices are often used in tandem; on the ESA/NASA Planck mission to study the CMBR, a hydrogen sorption refrigeration provides an 18-K heat station for a mechanical cooler which, in turn, provides a 4.5 K stage for a dilution refrigerator that cools the bolometer detectors to ~100 mK.

Light Collectors

With few exceptions, light collectors for the infrared are all-reflecting telescopes whose optical components may be aluminized, or gold-coated, depending on the wavelength range. Conventional telescopes image a portion of the sky onto a focal plane to provide accurate maps. Occasionally, however, the astronomer is interested in diffuse radiation that is not localized but arrives from all over the sky. For such observations, a carefully designed horn, an all-reflecting funnel, is generally employed to gather radiation from a large but well-defined field of view in the sky onto the smallest possible detector. These two types of light collectors were used, respectively, on the ISO/IRAS and COBE spacecraft.

For space applications, there is a premium on lightweight optics because the mass of the entire satellite scales with the mass of the optical system it must support, and, in turn, a more massive satellite requires a larger and more expensive launch vehicle. For an infrared mission, there is the added complication of the increased thermal conductivity of the beefier structure required to support the more massive optical system. Both IRAS and SIRTF, as discussed later, used all-beryllium optical systems because of the favorable strength-to-mass ratio of this material. The 85-cm diameter SIRTF primary, for example, has a mass of 15 kg and an areal density of 26 kg/m^2. By comparison, the Hubble Space Telescope primary mirror has an areal density of 180 kg/m^2. As telescope apertures beyond ~4 m diameter are considered for future missions, another launch vehicle limitation, set by the physical size of the payload shroud, is encountered. Thus planning for the 8-m diameter Next Generation Space Telescope (NGST) is based on ultralightweight panels of glass, beryllium, or composite materials, whose areal density is no greater than 15 kg/m^2. It would deploy after launch to achieve the desired aperture.

Filters

Filters isolate wavelength ranges of particular interest to the astronomer. For imaging and photometry, a well-defined, broad wavelength range needs to be isolated. Carefully designed transmission filters are usually used for this purpose. For spectroscopy, different types of spectrometers that select numerous narrow-wavelength intervals are inserted between the light collector and the detector or detector arrays. The most common types of spectrometers for infrared are prism or grating "dispersive systems" that separate out radiation directionally, according to wavelength, and interferometers. Fabry–Perot interferometers select one narrow-wavelength range at a time; Michelson and other two-beam, multiplex interferometers transmit many wavelengths simultaneously but have to be swept through a range of settings to encode unambiguously and register the flux detected at each wavelength. Later, we describe spectrometers by their spectral resolving power R, defined as $\lambda/\delta\lambda$, where λ is the operating wavelength and $\delta\lambda$ is the finest discernible spectral detail. The unique requirements placed on filters and spectrometers for space applications are largely environmental and have to do with surviving launch or the ionizing radiation in space, or achieving low mass or volume, rather than with the device's functionality or performance.

Early Rocket Instrumentation

In the mid-1960s, a collaborative effort between Cornell University and the Naval Research Laboratory led to the design of liquid-nitrogen-cooled and eventually liquid-helium-cooled, rocket-borne infrared telescopes. Early Cornell designs incorporated a parabolic primary mirror with an 18-cm aperture and focal ratio length ratio $f/0.9$. The entire telescope, except for the entrance aperture, was surrounded by the liquid. Four different types of detectors were flown on many of these flights to sample the spectral range from 5 μm to 1.6 mm. Using this apparatus, the total flux in a field of view roughly 1° in diameter was first successfully measured for the galactic center and four other regions in the central portions of the Milky Way, at 5, 13, 20, and 100 μm. A first spectral measure of the radiation emitted by the solar system's zodiacal dust was also obtained.

Some years later, results from a survey conducted in a series of rocket flights were published by the U.S. Air Force Cambridge Research Laboratories (now the Air Force Geophysical Laboratories). The group initially flew liquid-neon-cooled telescopes that had 10-cm apertures and detectors sensitive to radiation at 12–14 μm. Each of six detectors in a linear array surveyed a $10' \times 10'$ field of view. Later, the group began all-sky surveys using satellite-borne instrumentation and also began observations across wider spectral ranges. Early results of one of these surveys at 4.2, 11.0, 19.8, and 27 μm were cataloged and published by Price and Walker.

Although rockets have not been extensively used for infrared astronomy in recent years, large-format infrared detector arrays may enable significant science in the limited duration of a rocket flight. For example, a 16.5-cm rocket-borne telescope instrumented with a 256×256 InSb array and cooled by supercritical helium has been flown to search for a faint halo of low-mass stars enveloping a nearby edge-on spiral galaxy.

The Infrared Astronomical Satellite (IRAS)

The first true infrared survey of the sky from space was carried out by the Infrared Astronomical Satellite (IRAS), jointly sponsored by the United States, the Netherlands, and Great Britain. See Fig. 2. Approximately two-thirds of the 300-day mission that lasted from January to November 1983 was devoted to an unbiased survey of the sky that succeeded in charting 98% of the celestial sphere in four broad wavelength bands. IRAS was launched into a polar orbit at the day–night terminator which precessed about 1° per day. In this "Sun-synchronous" orbit the Earth/Sun/spacecraft geometry varied only slowly, so that the survey could be executed by a simple scanning strategy. Observations were carried out with an all-beryllium 57-cm aperture, $f/9.6$ Richey–Chrétien telescope whose focal plane was cooled to 3 K and featured a total of ~60 Si:As, Si:Sb, and Ge:Ga discrete photoconductors; each had a separate JFET amplifier readout. The detectors covered, respectively, the 12-, 25-, 60-, and 100-μm bands, using Ge:Ga appropriately filtered to cover the last two. A low-resolution spectrometer covered the wavelength range from 7.5–23 μm.

A measure of the mission's success was the cataloging of some 250,000 celestial sources; the vast majority had never before been detected in the infrared. No area of modern astrophysics was untouched by IRAS. A few of the many scientific highlights include:

1. The discovery of galaxies that emit up to fifty times more energy at far-infrared wavelengths than in the optical domain and also emit from 100 to 1000 times as much total power as our own galaxy, the Milky Way. The existence of such highly luminous infrared galaxies came as a huge surprise.

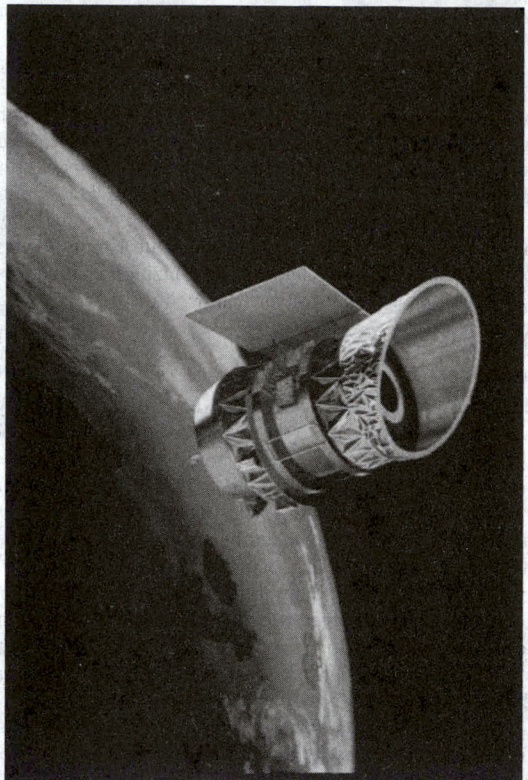

Fig. 2.   The artist's rendering shows the Infrared Astronomical Satellite (IRAS) in its 560-mile-high, near-polar orbit above the Earth. From this vantage point, IRAS searched the sky for stars and other infrared-emitting sources, unhampered by the obscuring effects of Earth's atmosphere. (*Infrared Processing and Analysis Center, Caltech/JPL*).

2. The discovery of disks composed of fine dust grains that orbit around a number of stars that, in many ways, were reminiscent of our own Sun. This dust, it was conjectured, is the remnant of an originally far more massive circumstellar cloud of gas and dust

from which a system of planets had already formed and initiated further astronomical searches for signs of planets around these stars. A sharply defined, though far fainter, set of dust rings was also found orbiting our own Sun, as were enduring trails of dust left by the passage of solar system comets.

3. The successful measurement of spectra for planetary nebulae and a variety of other sources at wavelengths previously inaccessible due to telluric absorption. IRAS also identified patchy infrared emission from the diffuse interstellar medium, referred to as "infrared cirrus" because of its similarity to the thin, streaky clouds in Earth's atmosphere. Infrared cirrus is important as a tracer of matter within our galaxy and as a potential source of interference in observations of distant galaxies. See also **Infrared Astronomical Satellite (IRAS)**.

The Cosmic Background Explorer, COBE

COBE, built by NASA and launched into a polar orbit identical to that of IRAS in 1989, was dedicated to the study of the microwave and infrared background radiation in space. See Fig. 3. It carried three instruments; two of them, the Diffuse Microwave Radiometer (DMR) and the Far Infrared Absolute Spectrophotometer (FIRAS) were used to study the CMBR—the isotropic blackbody radiation whose temperature is $\sim$2.73 K and is believed to be a relic of the Big Bang in which the Universe was born. The third experiment, the Diffuse Infrared Background Experiment (DIRBE), measured the background at infrared wavelengths from 1–200 μm. DIRBE and FIRAS were cooled by liquid helium and mapped the entire sky repeatedly during the $\sim$10-month cryogenic lifetime of COBE. The critical components of DMR were cooled passively to $\sim$140 K; this instrument operated for about 4 years.

The CMBR carries important cosmological information, and DMR and FIRAS were extremely successful, respectively, in measuring the spatial structure in that radiative field and in establishing its blackbody nature at a very high degree of precision. These important cosmological experiments will not be discussed further here. Of course, both FIRAS and DMR also measured the foreground radiation from our own galaxy. FIRAS was a polarizing Michelson interferometer instrumented with helium-cooled bolometers as detectors. It obtained spectra of the galactic emission from $\sim$100 μm to $\sim$3 mm using a 7° field of view. Of particular interest was its detection of emission from oxygen, carbon, nitrogen, and carbon monoxide from gas in the Galaxy.

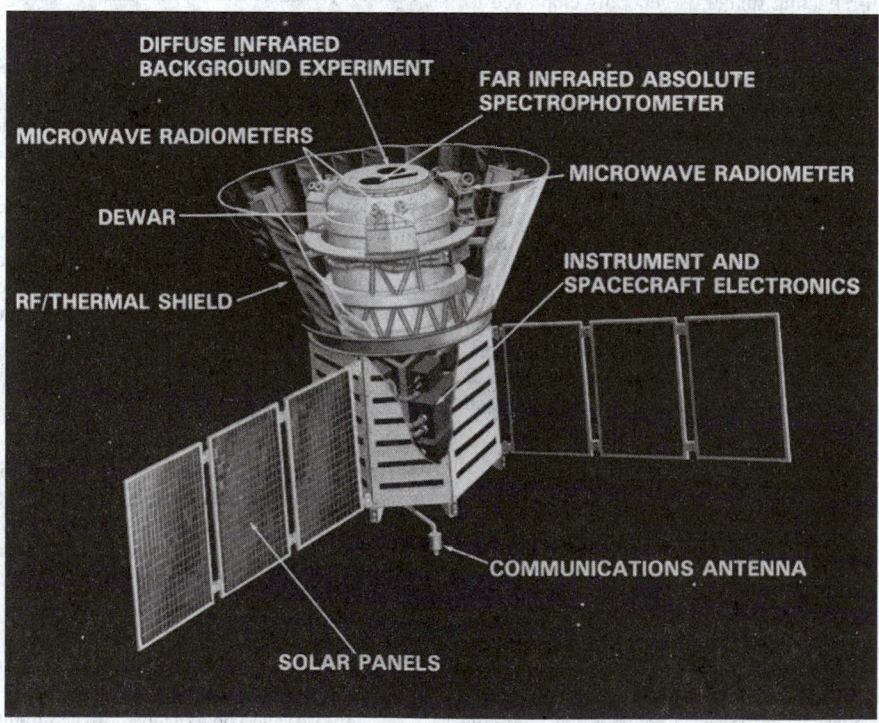

Fig. 3.   Artist Concept of COBE (*image courtesy of NASA*).

DIRBE made measurements in ~15 wavelength bands, using a variety of discrete photodiodes and photoconductors from 1–100 μm and helium-cooled bolometers at 140 and 240 μm. All measurements were referenced to an internal cold, black reference surface so that the absolute sky brightness was determined.

The principal scientific results from DIRBE include:

1. An improved determination of the distribution of the infrared radiation from the zodiacal dust cloud within the solar system, which has led to improved models of the dust cloud and its infrared emission. DIRBE also confirmed IRAS' discovery of a modest enhancement of emission in Earth-trailing direction, which is attributed to temporary gravitational trapping by Earth of zodiacal dust particles that are spiraling inward towards the Sun;

2. Measurements of the large-scale distribution of infrared radiation from the Galaxy, including both the far infrared radiation from 25–200 μm that samples the distribution of heated dust, and the near-infrared radiation from 1–25 μm that is indicative of the large-scale distribution of stars in the Galaxy;

3. Detection of an isotropic background of infrared radiation at 140 and 240 μm that arises from outside the Galaxy and may be attributable to the integrated effects of star-forming galaxies at redshifts $z \sim 1$ to 2.

The Infrared Space Observatory (ISO)

The Infrared Space Observatory, built and launched by the European Space Agency (ESA), was the first comprehensive infrared astronomical space observatory. See Fig. 4. NASA and the Japanese Space Agency (ISAS), provided important technical, operational, and scientific support. ISO mapped celestial sources and analyzed them through spectroscopy, photometry, and linear polarization studies.

Fig. 4.   The ISO Spacecraft. (*ESA, Illustration by Medialab*).

On the night of 16–17 November 1995, an Ariane 4 rocket launched ISO into a highly elliptical, 24-hour, circumterrestrial orbit, where the observatory operated with great success until its helium ran out and instruments began warming up in April 1998. The spacecraft in orbit was 5.3 m (17.4 ft) long, 2.3 m (7.5 ft) wide, and its mass was approximately 2,500 kg (5,512 lbs). At launch, it carried a superfluid helium charge of 2,300 liters (608 gallons), which maintained the Ritchey–Chrétien telescope, the scientific instruments, and the optical baffles at temperatures of 2–8 K (−456.1 to −445.3 °F). The diameter of the telescope's fused silica primary mirror was 60 cm (24 inches). A three-axis-stabilization system provided an absolute pointing accuracy of a few seconds of arc and stability of a fraction of an arc second in both jitter and long-term drift. The telescope was diffraction-limited down to wavelengths of roughly 5 μm. Four instruments formed the core of the scientific payload:

1. A camera containing two 32 × 32 pixel arrays: InSb for the wavelength range 2.5–5.5 μm, and Si:Ga for the range 4–18 μm. Each array could be operated with a selection of filters for broadband spectrophotometry or continuously variable filters (CVF) for low-resolution ($R \sim 40$) imaging spectroscopy and could view sources through three linear polarizers oriented relative to each other at angles of 60°.

2. A photometer covered the entire wavelength range from 2.5–240 μm. It employed Si:Ga detectors that gave peak response at 15 μm, Si:B detectors that gave peak response at 25 μm, unstressed Ge:Ga detectors that gave peak response at 100 μm, and stressed Ge:Ga detectors that gave peak response at 180 μm. Stressed detectors are mounted in a miniature clamp or vise that applies high mechanical pressure to the crystal, thereby extending their wavelength response. At 100 and 200 μm, the instrument housed, respectively, 3 × 3 and 2 × 2 arrays of unstressed and stressed Ge:Ga to facilitate mapping. Multiple apertures, multiple filters, and polarizers were used for photometric and photopolarimetric measurements in each range. Scanning and mapping operations were carried out at all wavelengths. Two grating spectrophotometers, each with a 64-element linear Si:Ga detector array, provided spectra with resolving power $R \sim 100$ at 2.5–5 and 6–12 μm.

3. A short-wavelength spectrometer included both grating and Fabry–Perot (FP) instruments. Grating spectra were available for the entire wavelength range from 2.38 to 45.2 μm, with resolving power $R \sim 1000$–2000. The FP mode covered the 11.4- to 44.5-μm range and gave resolving power of a factor of 20 higher. For the grating mode, the detectors were InSb at 2.38–4.08 μm, Si:Ga at 4.08–29 μm, and Ge:Be at 29–45.2 μm. For the FP mode, Si:Sb was used out to 26 μm, and Ge:Be from 26–44.5 μm.

4. A long-wavelength spectrometer provided coverage from 43–196.9 μm. A grating provided resolving power $R \sim 150$–200. A FP mode permitted observations at $R \sim 6800$–9700. The 10 detectors, arranged in a linear array on a curved surface, were Ge:Be at 43–50 μm, unstressed Ge:Ga at 50–100 μm, and stressed Ge:Ga beyond 110 μm.

Among the scientific highlights of ISO were:

1. The detection of water vapor throughout the interstellar medium of the Galaxy. Before ISO, the infrared emission from interstellar water vapor could not be detected because telluric water vapor absorbs at precisely the emission wavelengths. Water vapor, however, can be one of the primary coolants of interstellar clouds, and the extent of this cooling needed to be understood to assess the extent to which it facilitates protostellar collapse.

2. Detection of polycyclic aromatic hydrocarbons in the spectra of galaxies. These large molecules were well known in our galaxy, but ISO had the sensitivity needed to show that their emission dominates the 5- to 12-μm emission from nearby spiral galaxies as well. The emission from these molecules is due to radiative fluorescence: a molecule is excited by optical or ultraviolet radiation, quickly reradiates the energy of a single absorbed photon, in the infrared and returns to the ground state.

3 Inventories of extragalactic source-counts at wavelengths ranging from 4–175 μm. These are of particular value in understanding the origins of the extragalactic diffuse infrared radiation detected by the COBE mission. Many of the randomly observed galaxies appear to be ultraluminous, indicating that they contain substantial regions of massive star formation or that they harbor an active galactic nucleus that possibly surrounds a massive central black hole.

The Spitzer Space Telescope (formerly SIRTF, the Space Infrared Telescope Facility)

The Spitzer Space Telescope (formerly SIRTF, the Space Infrared Telescope Facility) was launched into space by a Delta rocket from Cape Canaveral, Florida on 25 August 2003. See Fig. 5. During its 2.5-year mission, Spitzer will obtain images and spectra by detecting the infrared energy, or heat, radiated by objects in space between wavelengths of 3 and 180 microns (1 micron is one-millionth of a meter). Most of this infrared radiation is blocked by the Earth's atmosphere and cannot be observed from the ground.

Spitzer is the final mission in NASA's Great Observatories Program—a family of four orbiting observatories, each observing the Universe in a different kind of light (visible, gamma rays, X-rays, and infrared). Other

Fig. 5.   Spitzer 3-D model. (*NASA/JPL-Caltech*).

missions in this program include the Hubble Space Telescope (HST), Compton Gamma-Ray Observatory (CGRO), and the Chandra X-Ray Observatory (CXO). Spitzer is also a part of NASA's Astronomical Search for Origins Program, designed to provide information which will help us understand our cosmic roots, and how galaxies, stars and planets develop and form. See also **Chandra X-Ray Observatory**; **Compton Gama-Ray Observatory (CGRO)**; and **Hubble Space Telescope (HST)**.

The Spitzer Space Telescope (SST) is a NASA mission managed by the Jet Propulsion Laboratory. This website is maintained by the Spitzer Science Center, located on the campus of the California Institute of Technology and part of NASA's Infrared Processing and Analysis Center. http://www.spitzer.caltech.edu/about/index.shtml and http://ssc.spitzer.caltech.edu/

Spitzer culminates the four decades of technology development and scientific progress described above. Spitzer is the first space mission to use exclusively the imaging and spectroscopic power of large format infrared detector arrays. The Spitzer's all-beryllium telescope that incorporates an 85-cm (33.5 in) diameter primary mirror and is diffraction-limited down to 6.5 μm, defines the state of the art for ultralightweight cryogenic optics. The Spitzer telescope and cryogenic system is carried on a fairly standard spacecraft bus that provides pointing control, power, data storage, and communication. The pointing system is built around an external autonomous star tracker that controls and reports the spacecraft orientation at more than 2″ accuracy and has a reaction wheel/gyro control system. Visible light sensors in the cold focal plane can sense stars simultaneously using the external star tracker to track the relative orientation of the telescope and star tracker lines of sight. This pointing system architecture was used on ISO as well. The Spitzer spacecraft also incorporates a nitrogen gas system that is used to unload the reaction wheels if they accumulate too much angular momentum; the magnetic torquer bars used for this function in Earth-orbiting spacecraft would not work on Spritzer because it is far outside Earth's magnetosphere.

Unlike the missions previously described, all of which operated in Earth orbit, Spitzer was placed into an Earth-trailing heliocentric orbit, drifting slowly away to reach a distance of ~0.5 AU from Earth after 5 years.

In this orbit, Spitzer is free of the heat load from Earth and provides good access to the sky for target selection and scheduling. Spitzer was launched with the telescope warm and the instruments at helium temperature. In space, the telescope and its surrounding thermal shields cool radiatively to ~40 K, and the effluent helium from the cryogenic tank cools the telescope down to its operating temperature of ~5.5 K (see Fig. 1 (right). The thermal shields remain at 40 K and below throughout the mission, so that the parasitic heat conducted into the telescope is very small. As a result, the heat load that dissipates the Spitzer cryogen and determines the lifetime of the mission comes largely from the focal plane instruments.

This hybrid, radiative, cryogenic cooling system is facilitated because Spitzer can maintain an attitude in its solar orbit in which the solar panel is always oriented toward the Sun and shades the thermal shields that control the telescope temperature. These structures, in turn, are optimized to minimize the heat transferred from the solar panel and to radiate to space any heat that is transferred. This approach would not work in near-Earth orbit because the heat load from Earth would occasionally be incident on the thermal shields and turn the radiator into an absorber.

This optimized cryogenic system, together with the low-power dissipation of its instruments and the elimination of parasitic heat loads, makes Spitzer a much more efficient system cryogenically than any of its predecessors. Spitzer carries 350 liters (92.5 gallons) of helium at launch, and a lifetime of 5+ years is predicted, based on an average instrument power dissipation of ~5 mW. By comparison, ISO was launched with ~2,300 liters (608 gallons) of helium and achieved a lifetime of ~2.5 years with an average instrument power dissipation of ~10 mW.

Spitzer has three array-based focal plane instruments:

1. A near-infrared camera that provides imaging simultaneously in four bands at 3.6, 4.5, 5.8, and 8 μm. Both of the 3.6- and 5.8-μm channels image the same field of view in the sky; this is made possible by a dichroic filter that transmits 5.8 μm and reflects 3.6 μm. An adjacent field of view is imaged at 4.6 and 8 μm in a similar fashion. Each band uses a 256 × 256 pixel array hybridized to a 256 × 256 MOSFET multiplexer. The detector material is InSb in the 3.6- and 4.5-μm bands and Si:As in the 5.8- and 8-μm bands.

2. A spectrometer that provides low resolving power ($R \sim 60–120$) spectroscopy from 5–40 μm and higher resolution spectroscopy ($R \sim 600$) from 10–38 μm. The spectrometer consists of four physically distinct modules; each contains a 128 × 128 array (Si:Ga for the shorter wavelengths, Si:Sb for the longer wavelengths) illuminated by an optical train of mirrors and gratings. The use of detector arrays allows these modules to be very compact and efficient and obviates the need for moving parts. The higher resolution modules use two diffraction gratings so that, an entire octave of the spectrum can be cross-dispersed across the entire array and measured simultaneously. In the lower resolution modules, a long entrance slit is used to permit obtaining spectra simultaneously at many spatial points. A portion of the array in one of these modules is also used for the precision target acquisition required to place a source on a narrow spectrograph slit, thereby alleviating the absolute pointing requirements placed on the spacecraft.

3. An imager/photometer that provides imaging and low-resolution spectrophotometry at wavelengths between 25 and 160 μm. This instrument uses a 128 × 128 Si:Ga array at 25 μm but incorporates two Ge:Ga arrays for longer wavelength measurements. The 32 × 32 Ge:Ga array used by SIRTF at 70 μm is composed of eight 4 × 32 submodules; each in turn consists of four 1 × 32 linear arrays, coupled to a 1 × 32 amplifier/multiplexer. The 2 × 20 array used at 160 μm is similarly built up of four 2 × 5 pixel modules with the added complication that the module construction allows for the mechanical stress needed to extend the long-wavelength response of Ge:Ga from 120 to beyond 160 μm. These arrays represent substantial advances in the state of the art and point the way toward still larger arrays for future applications. Note that although these arrays have fewer pixels than those described before, they are to be used at longer wavelengths where the diffraction-limited image size is larger compared to the field of view. Thus they provide comparable sampling of the telescope's focal plane; in fact, all three arrays are designed to sample the image fully to allow numerical postprocessing of the data to enhance the spatial resolution.

The scientific return of Spitzer cannot be forecast because its capabilities represent such a great advance beyond what has been possible in the past and also because the bulk of the observing time on Spitzer will be dedicated to programs to be proposed and carried out by the general scientific community. However, based on the science return from the other missions described before, it is anticipate that Spitzer will lead to great advances in our understanding of such problems as:

1. the formation and early evolution of galaxies, stars, and planets;
2. the physical processes that power the objects of highest luminosity in the Universe;
3. the chemical composition of interstellar and circumstellar matter;
4. the nature of the coolest, lowest luminosity stars and star-like objects in the solar neighborhood; and
5. the properties and interrelationships of comets, asteroids, interplanetary dust, and other small bodies in the solar system.

In addition, Spitzer's large arrays, very high sensitivity, and long lifetime give this mission great potential for discovering new phenomena. See also **Spitzer Space Telescope (SST)**.

Other Infrared Missions Already Flown

Other significant infrared space astronomy missions are described briefly here:

Spacelab II Infrared Telescope—1985. A 15-cm (6 in) diameter helium-cooled telescope was flown on Spacelab-2 and made infrared measurements between 2 and 120 μm. It provided data about the structure of the Galaxy and about the infrared background environment on the Space Shuttle.

Midcourse Space Experiment (MSX) a Ballistic Missile Defense Organization satellite, was launched in April 1996. The first ten months of the mission were devoted to mid-infrared observations with a solid hydrogen-cooled telescope. MSX carried a 35-cm (13.8 inches) aperture off-axis telescope and five linear Si:As arrays that mapped the sky in a push-broom fashion in bands from 4.2–26 μm. Although it was primarily designed to scan Earth's limb, it carried out a number of astrophysical experiments and produced excellent images of the entire galactic plane at ~18 arcsec resolution. http://irsa.ipac.caltech.edu/Missions/msx.html.

Infrared Telescope In Space (IRTS)—1995 is the first Japanese orbiting telescope dedicated to infrared astronomy observations. IRTS was an ISAS program with significant NASA participation. IRTS had a 15-cm (6 in) diameter liquid-helium-cooled telescope and four varied focal plane instruments covering wavelengths from 1–1000 μm. It was carried on a Japanese satellite called the Space Flyer Unit and surveyed ~7% of the sky in a 28 day mission. http://www.ir.isas.jaxa.jp/irts/irts_E.html.

Near Infrared Camera and Multi-Object Spectrograph (NICMOS)—1996. NICMOS is a replacement focal plane instrument, which was installed on HST. See also **Hubble Space Telescope (HST)**. It was instrumented with three 256×256 HgCdTe arrays that carried a range of filters that covered the 1- to 2.5-μm spectral band and was optimized for high spatial resolution imaging. NICMOS was cooled by solid nitrogen and achieved a lifetime of slightly less than 2 years. This was somewhat less than expected because of a partial failure of the cryogenic system in orbit.

Submillimeter Wave Astronomy Satellite (SWAS)—1998. SWAS was the first space mission to carry radio-type (heterodyne) receivers for spectroscopic exploration. SWAS has a 55×71 cm (21×28 inches) near-optical quality off-axis primary mirror and two heterodyne radiometers with Schottky barrier diode mixers and a single acousto-optical spectrometer. SWAS is surveying the galactic plane in the emission of atomic carbon, molecular oxygen, water vapor, and carbon monoxide in five transitions between 538 and 615 μm. http://cfa-www.harvard.edu/swas/swas.html.

Future Missions

A number of missions are planned or proposed for the next two decades to go beyond even the great scientific and technical accomplishments described before. These include:

ASTRO-F. (Previously known as IRIS — Infrared Imaging Surveyor) is the second space mission for infrared astronomy in Japan. ASTRO-F is being developed by members of JAXA/ISAS and collaborators. IRAS (Infrared Astronomical Satellite, launched in 1983 by the United Kingdom,

the United States, and the Netherlands) carried out the first all-sky survey at infrared wavelengths and made a huge impact on astronomy. The ASTRO-F mission is an ambitious plan to make an all-sky survey with much better sensitivity, spatial resolution and wider wavelength coverage than IRAS. ASTRO-F has a 68.5 cm telescope cooled down to 6K, and will observe in the wavelength range from 1.7 (near-infrared) to 180 (far-infrared) micron. See Fig. 6. The detectors to be used for this purpose are stressed and unstressed Ge:Ga. ASTRO-F was successfully launched from Uchinoura Space Center at 6 h28 m (JST), February 22nd, 2006 by the M-V-8 rocket. The satellite was given a new name "AKARI" (means "light"). http://www.ir.isas.jaxa.jp/ASTRO-F/index-e.html.

Fig. 6.   ASTRO-F (IRIS; Infrared Imaging Surveyor). (*Image courtesy of JAXA*).

Herschel Space Observatory. Formerly called Far Infrared and Submillimeter Telescope or FIRST is a mission sponsored by the European Space Agency with substantial participation by NASA. Herschel will be the first example of a new generation of space telescopes. It will be the first space observatory covering the full far infrared and sub-millimetre waveband, and its telescope will have the largest mirror ever deployed in space. It will be located 1.5 million kilometers away from Earth at the second Lagrange point of the Earth-Sun system. Herschel's three and a half meter mirror will collect the light from distant and poorly known objects, such as newborn galaxies thousands of millions of light-years away, and will focus it onto three instruments with detectors kept at temperatures close to absolute zero.

The Herschel spacecraft is approximately 7.5 m (24.6 ft) high and 4×4 m (13×13 ft) in overall cross section, with a launch mass of around 3.3 tonnes. The spacecraft comprises a service module, which houses systems for power conditioning, attitude control, data handling and communications, together with the warm parts of the scientific instruments, and a payload module. The payload module consists of the telescope, the optical bench, with the parts of the instruments that need to be cooled, i.e. the sensitive detector units and cooling systems. The payload module is fitted with a sunshield, which protects the telescope and cryostat from solar visible and infrared radiation and also prevents Earth straylight from entering the telescope. The sunshield also carries solar cells for the electric power generation. See Fig. 7.

ESA's Herschel mission has been designed to unveil a face of the early Universe that has remained hidden until now. Thanks to its ability to detect radiation at far infrared and sub-millimetre wavelengths, Herschel will be able to observe dust obscured and cold objects that are invisible to other telescopes. Targets for Herschel will include clouds of gas and dust where new stars are being born, disks out of which planets may form and cometary atmospheres packed with complex organic molecules. However, Herschel's major challenge will be discovering how the first galaxies formed and how they evolved to give rise to present day galaxies like our own.

The Herschel telescope is a Cassegrain design with a primary mirror diameter of 3.5 meters, the largest ever built for use in space. The three scientific instruments are:

Fig. 7.   A computer generated image of the Herschel spacecraft. (*ESA*).

- HIFI (**H**eterodyne **I**nstrument for the **F**ar **I**nfrared), a very high resolution heterodyne spectrometer
- PACS (**P**hotodetector **A**rray **C**amera and **S**pectrometer)— an imaging photometer and medium resolution grating spectrometer
- SPIRE (**S**pectral and **P**hotometric **I**maging **R**eceiver)— an imaging photometer and an imaging Fourier transform spectrometer

The instruments have been designed to take maximum advantage of the characteristics of the Herschel mission. In order to make measurements at infrared and sub-millimetre wavelengths, parts of the instruments have to be cooled to near absolute zero. The optical bench, the common mounting structure of all three instruments, is contained within the cryostat and over 2000 litres of liquid helium will be used during the mission for primary cooling. Individual instrument detectors are equipped with additional, specialized cooling systems to achieve the very lowest temperatures.

Herschel is the only space facility ever developed to cover the far infrared to sub-millimetre parts of the spectrum (from 60 to 670 μm). It will open up an almost unexplored part of the spectrum, which cannot be observed well from the ground.

An Ariane-5 launcher will carry Herschel into space in July 2007. For reasons of cost effectiveness, ESA has decided to launch Herschel together with Planck, a mission to study the cosmic microwave background radiation. The two spacecraft will separate soon after launch and will operate independently. Herschel has a nominal routine operational lifetime of three years, with a possible extension of one year. About 7000 hours of science time will be available per year. Herschel is a multi-user observatory accessible to astronomers from all over the world. http://sci.esa.int/science-e/www/area/index.cfm?fareaid=16

James Webb Space Telescope. Formerly known as the Next Generation Telescope (NGST). The JWST project is an international mission led by NASA. The man whose name NASA has chosen to bestow upon the successor to the Hubble Space Telescope is most commonly linked to the Apollo moon program, not to science. Yet, many believe that James

E. Webb, who ran the fledgling space agency from February 1961 to October 1968, did more for science than perhaps any other government official and that it is only fitting that the Next Generation Space Telescope would be named after him.

The JWST is an orbiting infrared observatory that will take the place of the Hubble Space Telescope (HST) at the end of this decade. It will study the Universe at the important but previously unobserved epoch of galaxy formation. It will peer through dust to witness the birth of stars and planetary systems similar to our own. And using JWST, scientists hope to get a better understanding of the intriguing dark matter problem. The JWST is also a key element in NASA's Origins Program. See also **Origins Program**.

- Proposed Launch Date: no earlier than June 2013
- Proposed Launch Vehicle: Altlas V; Delta IV or Ariane 5
- Mission Duration: 5–10 years
- Total payload mass: Approx 6,200 kg (13,669 lbs), including observatory, on-orbit consumables and launch vehicle adaptor
- Diameter of primary Mirror: ~6.5 m (21.3 ft)
- Clear aperture of primary Mirror: 25 m^2
- Primary mirror material: beryllium
- Mass of primary mirror: about one-third as much as Hubble's
- Focal length: TBD
- Number of primary mirror segments: 18
- Optical resolution: ~0.1 arc-seconds
- Wavelength coverage: 0.6–28 microns
- Size of sun shield: ~ 22 m × 10 m (72 ft × 33 ft)
- Orbit: 1.5 million km from Earth at L2 Point
- Operating Temperature: under 50 K (−370 °F)

The James Webb Space Telescope (JWST) system is made up of three segments: (Fig. 8.) the Observatory (Flight Segment), the Ground Segment, and the Launch Segment. The Observatory is the space-based portion of the JWST system, and is comprised of three elements: the Optical Telescope Element (OTE), the Spacecraft, and the Integrated Science Instrument Module (ISIM).

The JWST Observatory includes a large segmented primary mirror that will unfold, or deploy, to approximately ~6.5 meters (21.3 ft) in diameter and a sunshield that will also deploy to about the size of a tennis court. Visually, the observatory is dominated by the sunshield subsystem, which separates the observatory into a sun-facing side with a temperature around 300 K (80 °F) and a cold anti-sun side. The observatory will be pointed so that the Sun, Earth and Moon are always on one side, and will act like a parasol, keeping the mirrors and the science instruments cool by keeping them in the shade and protecting them from the heat of the sun and warm spacecraft electronics. The sunshield greatly attenuates the incident solar energy allowing the Optical Telescope Element and Integrated Science Instrument Module to passively cool to their cryogenic operating temperatures of around 35 Kelvin (about 400 degrees below zero on the Fahrenheit scale). In addition to providing a cold environment, the sunshield provides a thermally stable environment. This is one essential element to maintaining proper alignment of the primary mirror segments as the telescope changes its orientation to the Sun.

Segmentation of the primary mirror is required to allow the mirror to fold for accommodation in the launch vehicle. Segmentation also accommodates more mirror technologies for a cost-effective architecture. On the down side, the segments must be aligned relative to each other to an accuracy of a few tens of nanometers in order to achieve the full performance of the large 6.5 m (21.3 ft) primary mirror (Fig. 9.). This requires minimizing thermal expansion effects through a combination of good thermal isolation, ultra-low expansion composite structural materials and micro-dynamically stable mechanisms. The other principal disturbing effect is vibration from spacecraft mechanisms, particularly reaction wheels used to point the Observatory. Vibrational effects are controlled through the use of vibration isolation and dampening devices.

The Optical Telescope Element for JWST includes the segmented primary mirror, secondary and tertiary mirrors, wavefront sensing and control algorithms and all related structure, vibration isolation and deployment mechanisms (Fig. 10.). http://www.jwst.nasa.gov/.

Interferometers in Space. All of the missions described before were built around a single telescope. To achieve much higher resolution than

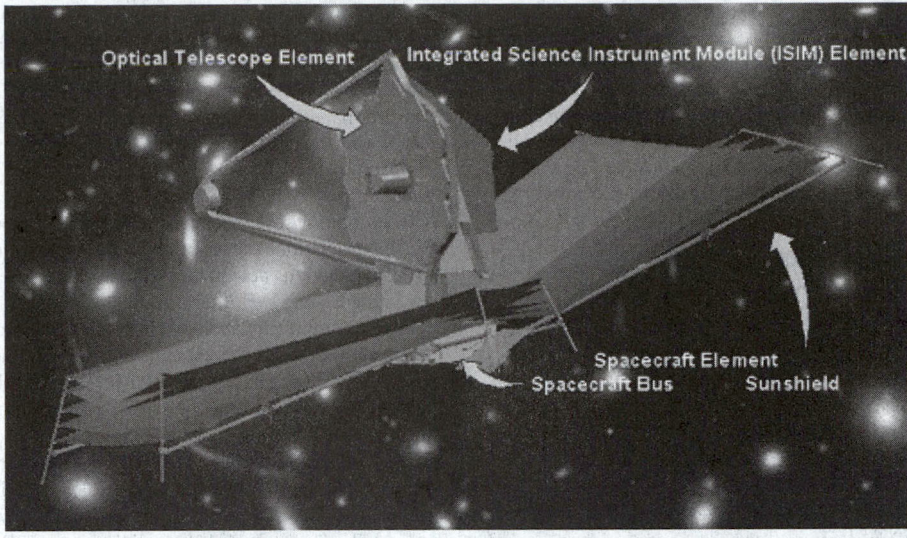

Fig. 8. High-level schematic of JWST. (*image courtesy of Northrop Grumman Space Technology*).

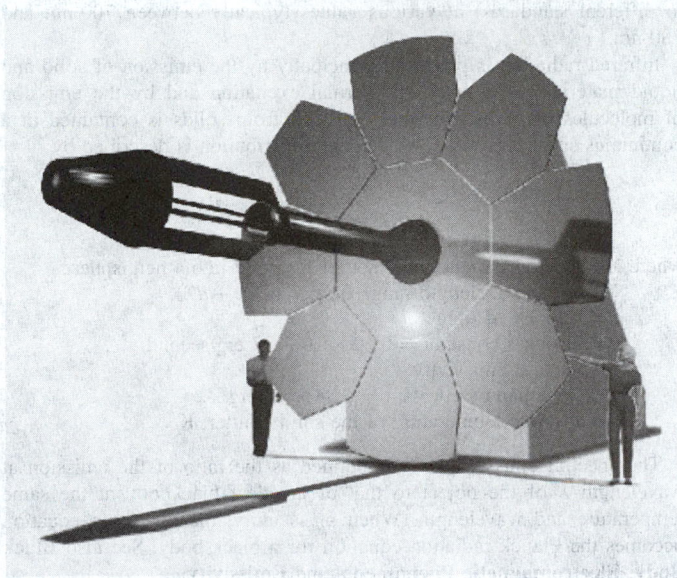

Fig. 9. Scale drawing of 6.5-meter primary mirror (Yardstick design). (*NASA*).

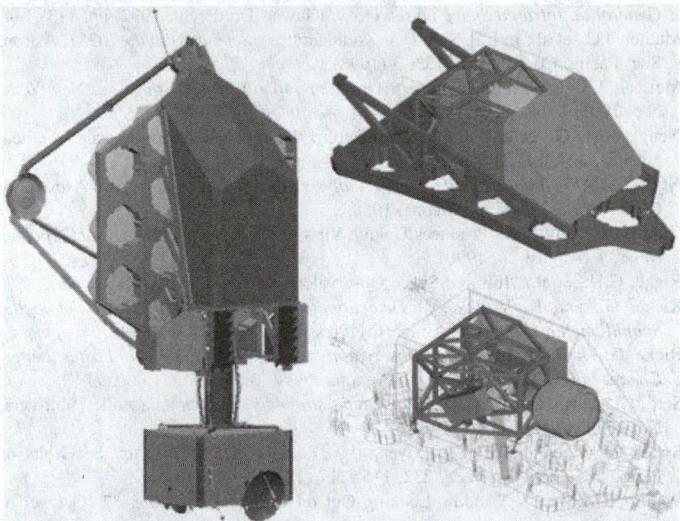

Fig. 10. Left: The ISIM is shown in relation to the mirror structure. Right: The ISIM is shown in more detail including the positions of the science instruments. The ISIM is a distributed system consisting of a cryogenic instrument module that is integrated with the OTE and science processors, software, and other electronics located in the warm SSM. The ISIM provides structure, environment, and data handling for three modular science instruments: Near Infrared Camera (NIRCam), Mid Infrared Instrument (MIRI), Near Infrared Spectrograph (NIRSpec) and the observatory Fine Guidance Sensor (FGS). (*NASA*).

the ∼0.1–1 arcsec achievable with an ∼8-m (26 ft) telescope, it will be necessary to use the techniques of interferometry, in which infrared radiation collected by two widely spaced telescopes can be brought together to achieve angular resolution comparable to that which would be provided by a single telescope whose aperture is equal to the separation of the two telescopes. Ultimately, this technique will be employed in NASA's Terrestrial Planet Finder, scheduled to launch in the 2012–2015 time frame to image Earth-like planets around nearby stars.

The *Terrestrial Planet Finder* (TPF) uses a small collection of high sensitivity telescopes (probably 4 large 3.5-meter (11.5 ft) telescopes) with revolutionary imaging technologies. It will measure the temperature, size, and the orbital parameters of planets as small as our Earth in the habitable zones of distant solar systems. Also, TPF's spectroscopy will allow atmospheric chemists and biologists to use the relative amounts of gases like carbon dioxide, water vapor, methane and ozone to find whether a planet might support life.

One great challenge is how to detect planets against the blinding glare of their parent star. TPF with 4 big telescopes, will reduce the glare of parent stars to see planetary systems up to 50 light-years away.

The primary scientific goal of TPF is the direct detection and characterization of Earth-like planets that orbit nearby stars. Specifically, TPF will seek answers to these questions:

- Are there Earth-like planets in the "habitable zones" around their parent stars where the surface temperature is capable of supporting liquid water over a range of surface pressures?
- What are the compositions of the atmospheres of terrestrial planets orbiting nearby stars? Is water, carbon monoxide, or carbon dioxide present?
- Are there atmospheric components or conditions attributable to primitive life, such as ozone or molecular oxygen, seen in the Earth's atmosphere?
- How do planets form out of disks of solid and gaseous material around young stars?

In a broader scientific context, the goal of TPF is to understand the properties of all planetary constituents. In addition to Earth-like planets, it will study the orbital and physical properties of gas giants and debris disks.

The standard model of solar system formation holds that planets are procreated in a flattened disk of material formed in the collapse of a rotating cloud of dust and gas. While this theory has been strengthened by observations of protostellar disks that span tens to hundred of astronomical units (AU) across, the modern discoveries of extrasolar planets with different orbital characteristics suggest that planetary systems are dynamic and that planets may migrate from the site of their birth.

TPF will deliver fundamental information on the mass and temperature distribution within the disks surrounding young stars, the place where new planets form. This information will hand over considerable clues on physical processes that determine how rocky and gaseous planets form.

See also **Astronomy**.

Additional Reading

Collaudin, B., and T. Passvogel.: *Cryogenics*, **39**, 157 (1999).

Cowen, R.: "Lifting a Dusty Veil to Clear IRAS' View," *Sci. News*, 182 (September 22, 1990).

Crease, R.P.: "Millimeter Astronomers Push for New Telescope," *Science*, 1504 (September 28, 1990).

Gatley, I., D.L. DePoy and A.M. Fowler: "Astronomical Imaging with Infrared Array Detectors," *Science*, 1264 (December 2, 1988).

Glaiser, D.S., et al.: In R.G. Ross, Jr., ed., *Cryocoolers 10*, Kluweer Academic/Plenum, New York, NY, 1999, pp. 1–19.

Glass, I. S., R. Ellis, and J. Huchra: *Handbook of Infrared Astronomy*, Cambridge University Press, New York, NY, 2005.

Habing, H.J. and G. Neugebauer: "The Infrared Sky," *Sci. Amer.*, 48–57 (November 1984).

Hawarden, T.G., et al.: In S.J. Bell Burnell, J.K. Davies, and R.S. Stone, eds., *Next Generation Infrared Space Observatory*, Kluwer, Dordrecht, 1992, pp. 113–144.

Mather, J.C., et al.: In E.P. Smith, A. Koratkar (eds). *Science with the NGST*, Astron. Soc. Pacific Conf. Series, **133**: 3 (1998).

McLean, I.S.: *Electronic Imaging in Astronomy*, John Wiley & Sons, Inc., New York, NY, 1997.

Neugebauer, G., et al.: "Early Results from the Infrared Astronomical Satellite," *Science*, **224**, 14–21 (1984).

NOAO: *IRAF—Image Reduction and Analysis Facility*, National Optical Astronomy Observatories, Tucson, Arizona, 1988.

Olson, C.: "Tiny 'Eye' Produces Unique View: Longwave Infared Sensor," *Hughes News*, 1 (August 24, 1990).

Rieke, G.H., et al.: "Infrared Space Observatory," *Science*, **232**, 1487 (1986).

Rieke, G., and K. Visnovsky.: *Detection of Light: From the Ultraviolet to the Submillimeter*, Cambridge University Press, New York, NY, 1994.

Rieke, G: *The Last of the Great Observatories: Spitzer and the Era of Faster, Better, Cheaper at NASA*, University of Arizona Press, Tucson, AZ, 2006.

Setti, Giancarlo, and G.G. Fazio: *Infrared Astronomy*, Kluwer Academic Publishers, Norwell, MA, 2004.

Sykes, M.V., et al.: "The Discovery (by IRAS) of Dust Trails in the Orbits of Periodic Comets," *Science*, **232**, 115–117 (1986).

Waldrop, M.M.: "A Window Looking Out on Creation," *Science*, 32 (October 5, 1990).

MICHAEL WERNER, Jet Propulsion Laboratory (JPL), Pasadena, CA

MARTIN HARWIT, Cornell University Ithaca, NY

INFRARED INTERFEROMETER SPECTROMETER (IRIS). See **Nimbus Satellite Program**.

INFRARED PHOTOGRAPHY AND IMAGERY. See **Photography and Imagery**.

INFRARED RADIATION. The region of the electromagnetic spectrum between the wavelength limits 0.75 to 1,000 micrometers. The lower wavelength limit is set to coincide with the upper limit of the visible radiation region. Radiation of wavelength greater than 1,000 micrometers is generally considered of the microwave spectrum. Both limits are arbitrary. The infrared region is sometimes broken down into five subregions: (1) the (NIR) *near-infrared region* (0.75–1.5 micrometers); (2) the (SWIR) short wavelength-infrared region (1.5–3 micrometers); (3) the *intermediate-infrared region* (3–8 micrometers); (4) the (LWIR) long wavelength-infrared region (8–15 micrometers) and (5) the (FIR) *far-infrared region* (15–1,000 micrometers). However, these terms are not precise, and are used differently in various studies i.e. near (0.75–5 micrometers (μm) / mid (5–30 μm) / long (30–1,000 μm). Especially at the telecom wavelengths the spectrum is further subdivided into individual bands, due to limitations of detectors, amplifiers and sources.

Infrared radiation is popularly known as "heat" or perhaps "heat radiation," since many physics teachers traditionally attribute all radiant heating to infrared light. This is wrong, and is a very widespread misconception. Light or electromagnetic waves of any frequency will heat surfaces which absorb it. IR light from the sun only accounts for 50% of the heating of the Earth, the rest is caused by visible light. Green lasers can char paper, incandescently hot objects put out visible radiation, and ice cubes emit mostly microwaves. However, it is true that objects at room temperature will emit radiation mostly concentrated in the mid-infrared band.

The common nomenclature is justified by the different human response to this radiation: near infrared is the region closest in wavelength to the radiation detectable by the human eye, mid and far infrared are progressively further from the visible regime. Other definitions follow different physical mechanisms (emission peaks, vs. bands, water absorption) and the newest follow technical reasons (The common silicon detectors are sensitive to about 1,050 nm, while InGaAs sensitivity starts around 950 nm and ends between 1,700 and 2,600 nm, depending on the specific configuration). See also **Sensors**.

The boundary between visible and infrared light is not precisely defined. The human eye is markedly less sensitive to light above 700 nm wavelength, so longer frequencies make insignificant contributions to scenes illuminated by common light sources. But particularly intense light (e.g., from lasers) can be detected up to approximately 780 nm, and will be perceived as red light. The onset of infrared is defined (according to different standards) at various values typically between 700 nm and 780 nm.

Infrared radiation is produced principally by the emission of solid and liquid materials as a result of thermal excitation and by the emission of molecules of gases. Thermal emission from solids is contained in a continuous spectrum, whose wavelength distribution is described by

$$\lambda \, d\lambda = \frac{2\pi c^2 h \varepsilon_\lambda}{\lambda^5} \frac{1}{e^{ch/\lambda kT} - 1} d\lambda$$

where λ = spectral radiant emittance of the solid into a hemisphere in the wavelength range from λ to $(\lambda + d\lambda)$.

c = velocity of light

h = Planck's constant = $6.62 \times 10 - 27$ erg/second

ε_λ = spectral emissivity

k = Boltzmann's constant = 1.38×10-erg/k

T = absolute temperature of the solid emitter, K

The spectral emissivity, ε_λ, is defined as the ratio of the emission at wavelength λ of the object to that of an ideal blackbody at the same temperature and wavelength. When ε_λ is unity, the foregoing equation becomes the Planck radiation equation for a black body. See also **Black Body**; **Electromagnetic Phenomena**; and **Emissivity**.

Gaseous emission of infrared radiation differs in character from solid emission in that the former consists of discrete spectrum lines or bands, with significant discontinuities, while the latter shows a continuous distribution of energy throughout the spectrum. The predominant source of molecular radiation in the infrared is the result of vibration of the molecules in characteristic modes. Energy transitions between various states of molecular rotation also produce infrared radiation. Complex molecular gases radiate intricate spectra, which may be analyzed to give information of the nature of the molecules or of the composition of the gas.

Spectral Emittance

The spectral radiant excitance of a blackbody at various temperatures is shown in Fig. 1. We note that for a given temperature, the emitted radiation has a maximum at a specific wavelength. As the temperature increases, the wavelength at maximum emission shifts progressively to shorter wavelengths, and the relative emission rises. In accordance with Wien's law, the wavelength at maximum emission varies inversely with temperature. For objects near room temperature, for example, the maximum occurs in the long-wave IR region near 10 μm. In general the position of this maximum has a strong influence on the selection of the sensor wavelength band of operation.

Spectral emissivity $\varepsilon(\lambda)$ is a property of macroscopic matter that is a measure of the efficiency of emission of absorption of radiation relative to a perfect blackbody, for which $\varepsilon(\lambda) = 1.0$. For a "graybody," that is an

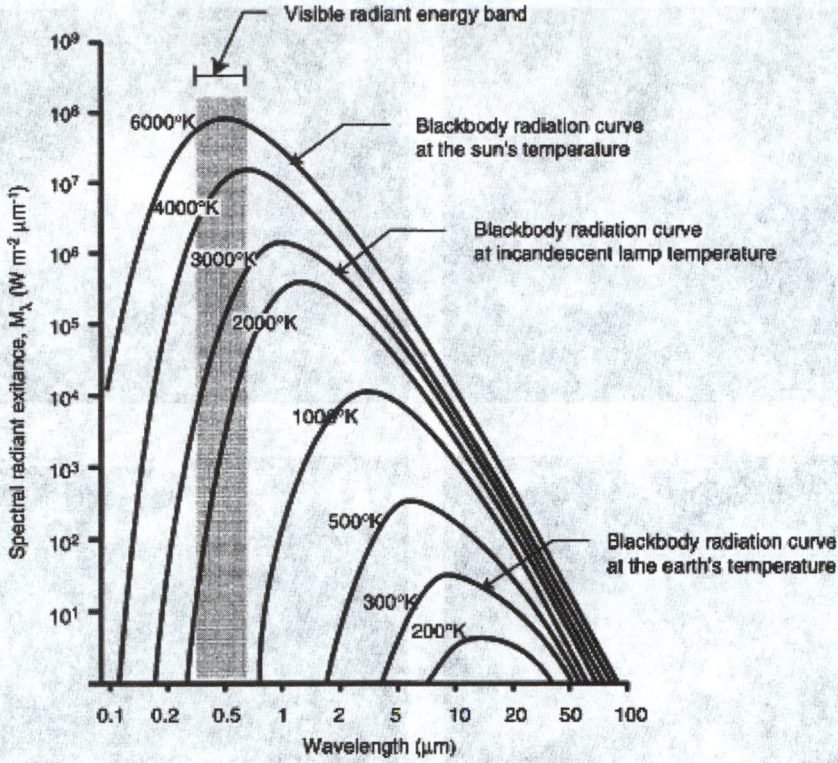

Fig. 1. Spectral distribution of energy emitted per unit wavelength radiated from blackbodies of various. As the temperature rises, the emission both increases and shifts to shorter wavelengths.

object whose emissivity is independent of wavelength, only the magnitude (but not the relative spectral distribution) of the radiation is altered. For nonblackbody radiation where the emissivity is a function of wavelength, the distribution of the radiation may be radically altered from the Planck distribution. In some substances such as gases, vibrational-rotational molecular transitions give rise to characteristic emission and absorption lines in the infrared. These gases can thus be identified by analysis of their corresponding spectra. In general, because of energy-level broadening in the solid state, narrow-line emission or reflectance spectra are not usually observed in solid materials. Relatively broad spectral features are much more common. High-temperature line broadening produces the same effect in the radiation spectra of gases at high temperature. It is the analysis of such spectral features of materials and even living organisms that allows remote discrimination and identification of various materials constituting the observed scene. This technique forms the basis of the science of multispectral remote sensing. Integrating the Planck expression over all wavelengths yields the Stefan–Boltzmann equation, which, under certain assumptions about the object's emissivity, enables an inference of object temperature. Remote infrared temperature-sensing devices use this principle to measure object temperature indirectly. The accuracy of this technique depends largely upon on how well the emissivity is known.

Propagation

Infrared radiation propagates through various media and, in general, is subject to absorption, which varies with the wavelength of the radiation. Molecular vibration and rotation in gases, which are related to the emission of radiation, are also responsible for resonance absorption of energy. The lesser gases in the atmosphere exhibit pronounced absorption throughout the infrared spectrum. However, nitrogen and oxygen do not absorb significantly in the infrared region. Water vapor, carbon dioxide, and ozone are responsible for strong absorption in the infrared. The absorption of radiation is so prevalent that those spectral bands in which relatively little absorption occurs are identified as atmospheric windows.

Solid and liquid materials show, as a rule, strong absorption in the infrared. There are, however, many solids that transmit well in broad regions of the infrared spectrum. Many materials, such as water and silica

glasses, which show little absorption in the visible, are opaque to infrared radiation at wavelengths greater than a few micrometers. Many of the electrically insulating crystals, such as the alkali halides and the alkaline-earth halides, which transmit well in the visible, also are transparent to much of the near- and intermediate-infrared spectrum. Several of the semiconductor materials absorb strongly in the visible, but become transparent in the infrared beyond certain wavelengths characteristic of the semiconductor.

Detection of the presence, distribution and/or quantity of infrared radiation requires techniques which are, in part, unique to this spectral region. The frequency of the radiation is such that essentially optical methods may be used to collect, direct, and filter the radiation. Transmitting optical elements, including lenses and windows, must be made of suitable materials, which may or may not be transparent in the visible spectrum.

The detector for infrared represents the most unique component of the detection system. Photographic techniques can be used for part of the near-infrared region. Photoemissive devices, comparable to the visible- and ultraviolet-sensitive photocells, are available with sensitivity extending to about 1.3 micrometers. The intermediate-infrared region is most effectively detected by photoconductors. These elements, photosensitive semiconductors, are essentially photon detectors, which respond in proportion to the number of infrared photons in the spectral region of wavelength. This wavelength corresponds to the minimum photon energy necessary to overcome the forbidden gap of the semiconductor. All spectral regions from ultraviolet through visible, infrared, and microwaves, can be detected by an appropriately designed thermal element, which responds by being heated by the absorption of the incident radiation. In the infrared region, thermal detectors take the form of thermocouples, bolometers, and pneumatic devices. The thermal devices, in general, are not sensitive or as rapidly responding as photoconductors.

A very practical application of infrared radiation is found in radiant heating. Solid radiators, such as hot tungsten filaments, alloy wires, and silicon carbide rods are used widely as sources of infrared to provide surface heating by radiation. Commercially available infrared lamps are extensively used in specially designed ovens for drying painted and enameled surfaces.

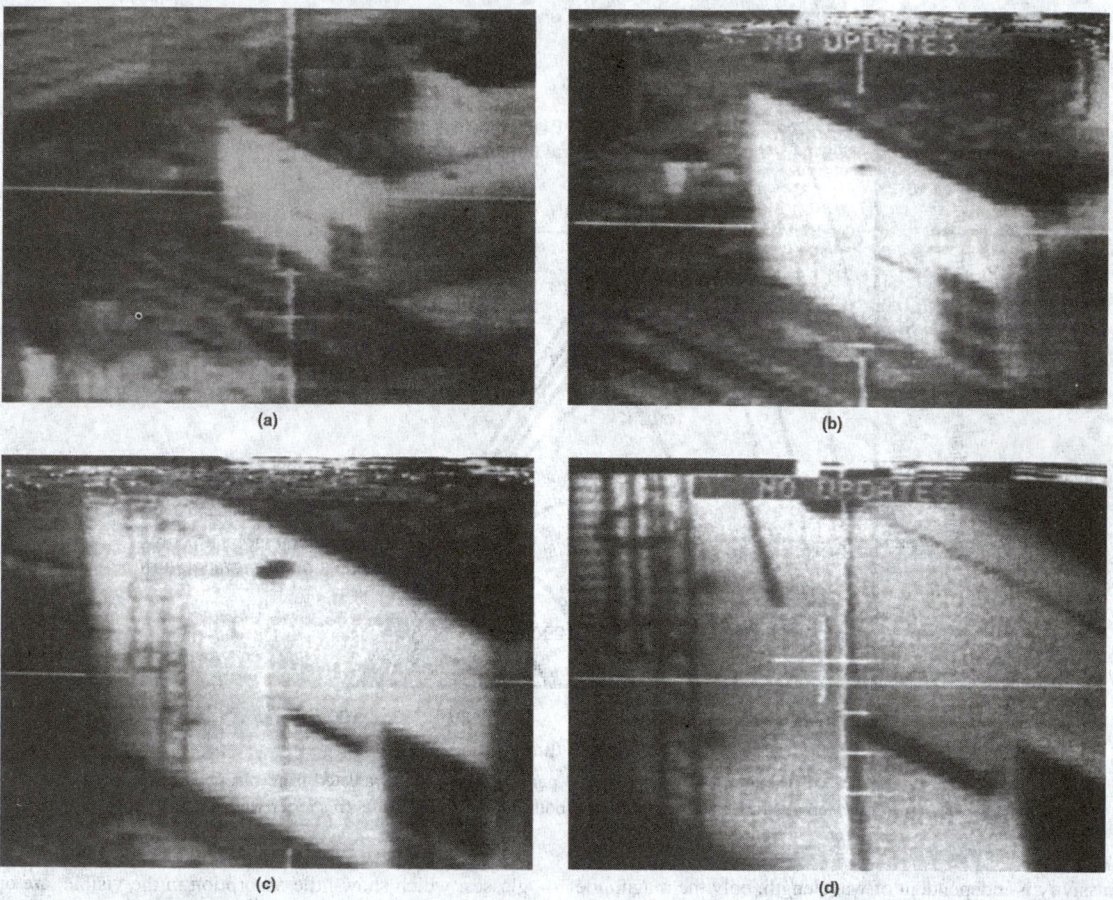

(a)   (b)

(c)   (d)

Fig. 2.   Second-by-second image of target taken by an attack aircraft. (*McDonnell Douglas photo.*)

Infrared Imagery

Infrared technology, sometimes called "night vision," has been used for several years in both military and commercial applications. The effectiveness of IR detectors was dramatically demonstrated during the military operation, Desert Storm, in 1991. These detectors, not previously demonstrated in actual war situations, were tremendous aids in achieving bombing target accuracy, not only from aircraft but from land vehicles as well. See series of images given in Figs. 2 and 3. Infrared imaging of the earth's surface from satellites also has become much more precise over the past decade. These topics are described in more detail in the articles on **Photography and Imagery**; and **Satellites (Scientific and Reconnaissance)**.

Infrared Spectroscopy[1]

Scientists have long used infrared absorption as a means of probing the structure of molecules. Studying the manner in which specific wavelengths of infrared energy excite vibration and rotation in molecules reveals information about the molecule that can be used to determine what and how many molecules are present.

The *infrared spectrophotometer* is the principal instrument used by scientists for these measurements. Most laboratory spectrophotometers are of a dispersive design, i.e., a prism or grating is used to separate the spectral components in the source radiation. Modern infrared spectrophotometers have a wide wavelength range from 2 to 50 micrometers. They find use in research, quality control, and analytical service laboratories.

Ultrafast IR spectroscopy is a comparatively recent development. Chemical reactions can be studied on the picosecond and femtosecond time scale. For example, as described by Stoutland, Dyer, and Woodruff (Los Alamos National Laboratory), the dynamics after CO dissociation

[1] Information on infrared analytical instrumentation furnished by Rodney M. Durham, Instruments Division, Infrared Industries, Inc., Santa Barbara, California.

Fig. 3.   Infrared image of trawler in the North Atlantic.

from CO-ligated hemoglobin and myoglobin have been investigated by monitoring the CO chromophore. These studies have provided not only evidence for the sub-picosecond dissociation of CO, but also structural information. Other researchers have used ultrafast IR spectroscopy to investigate energy transfer dynamics of organometallic molecules both in solution and on surfaces. Such information is central to understanding chemical activation and how thermally activated reactions occur. Recent experiments have provided complementary information on surfaces.

As pointed out by Stoutland, "The general principles of ultrafast laser experiments are well known. All ultrafast experiments are variants on the 'pump-probe' scheme, in which time resolution is obtained by spatial delay of a probe pulse relative to the pump, or excitation, pulse (1 ps = 3.0 mm)."

Short IR pulses can be generated in a number of ways. Typically, these are based on Raman scattering processes or nonlinear mixing schemes or the related optical parametric oscillator. Much more detail is given in the Stoutland reference listed.

Infrared Process Analyzer

This instrument has evolved from the laboratory spectrophotometer to satisfy the specific needs of industrial process control. While dispersive instruments continue to be used in some applications, the workhorse infrared analyzers in process control are predominantly nondispersive infrared (NDIR) analyzers. The NDIR analyzer can be used for either gas or liquid analysis. For simplicity, the following discussion addresses the NDIR gas analyzer, but it should be recognized that the same measurement principle applies to liquids. The use of infrared as a gas analysis technique is certainly aided by the fact that molecules, such as nitrogen (N_2) and oxygen (O_2), which consist of two like elements, do not absorb in the infrared spectrum. Since nitrogen and oxygen are the primary constituents of air, it is frequently possible to use air as a zero gas.

Many different analyzer configurations have been developed to address the diverse needs of the industrial process control industry. The basic constituents of an NDIR analyzer are: (1) a source of infrared radiation; (2) a means of restricting the wavelength range of the source radiation; (3) a means of detecting the infrared radiation; (4) a sample chamber to hold the gas or liquid to be measured; (5) a means of modulating the source radiation; and (6) electronics to process the signal generated by the source energy falling on the detector.

Microphone Detectors

Many IR process analyzers installed over the past several years have utilized *microphone detectors*. These detectors generally are the *Veingerov single-sided microphone* system or the *Luft balanced condenser microphone* system. These detectors are shown schematically in Figs. 4 and 5. See also **Hill-Powell** reference. The microphone detector uses an absorbing gas as its detecting medium. When radiation reaches the detector (that the sensitizing gas will absorb), the gas heats up and expands. This causes a diaphragm to distend. The diaphragm movement varies the condenser microphone capacity, which is part of an electric circuit that generates an electrical output signal. Both analyzers use dual sources, which are chopped to alternately allow energy to pass through a sample cell and a reference cell.

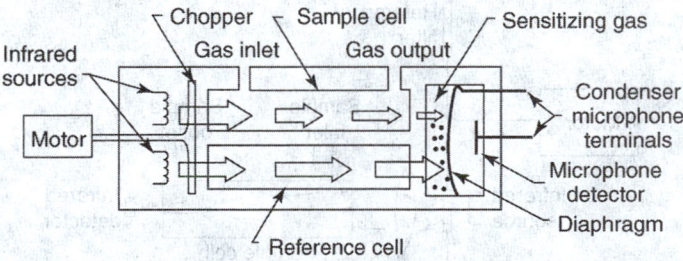

Fig. 4.   Nondispersive infrared analyzer with a Veingerov-type detector

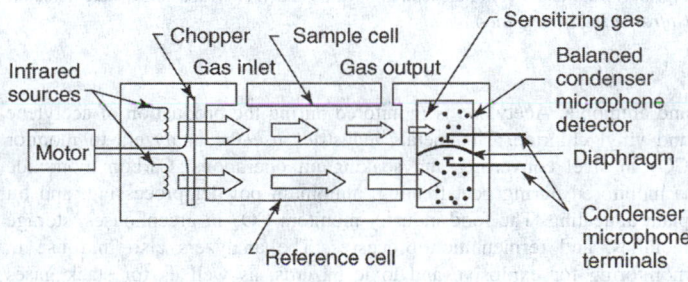

Fig. 5.   Nondispersive infrared analyzer with a Luft-type detector

If the sample cell contains a nonabsorbing zero gas, such as nitrogen, the modulated beams reaching the detector through the two paths are of equal amplitude. In the case of the Veingerov single-sided detector, the chopper is configured so that, at any given time, the sum of the cross-sectional areas of the two beams as seen by the detector equals the total cross-sectional area of a single beam, so that, when no absorbing sample is present, a constant signal is produced, and the output is zero. When a sample is present, the sample and reference path signals become imbalanced and a signal at the chopper frequency is developed. The amplitude of this signal is a function of the concentration of the gas present in the sample cell.

The Luft detector (Fig. 5) operates similarly, but the detector has two chambers separated by a diaphragm. The signal generated by the presence of an absorbing gas in the sample cell is at twice the chopping frequency. This is an advantage over the single-sided microphone detection system since it is less susceptible to vibration caused by imbalance of the chopper motor. Having separate chambers does, however, allow for the possibility of a change occurring in one-half of the detector and not the other and thus resulting in zero drift. More recently, infrared process analyzers have been introduced which use a Luft-type detection system but replace the diaphragm with flow sensors. The flow of gas from one chamber to another is sensed by the flow sensor rather than by using a capacitance detection technique. This is claimed to eliminate one of the major modes of detector failure—failure of the thin diaphragm. For a given path length, process analyzers which use the microphone detector are more effective than those analyzers which use solid state detectors and optical filters at measuring low concentrations of gases which have a lot of structure in their absorption band. This structure results from the molecular rotation spectrum being superimposed on the vibration spectrum and is easily resolved in the simpler molecules such as carbon monoxide, methane, and ammonia.

Solid-State Detectors

The most recent generation of NDIR analyzers have evolved to satisfy the frequently harsh industrial environments encountered. These analyzers utilize solid-state sensors for the detection of infrared radiation. Most frequently used sensors are lead selenide (PbSe), thermopiles, or pyroelectric detectors. The gas analyzers generally are configured as single-path instruments, dual-beam with a reference path, or dual-channel with a reference filter.

Single-beam instruments find use where low cost is important, but where stability requirements are not stringent. Generally, this is true when the measurement period is short and frequent rezeroing is practical. Changes in source intensity due to power variation or changes in detector sensitivity due to temperature fluctuations are reflected directly in the output as zero drift. To avoid this, a reference path is commonly used. The dual-beam configuration is shown in Fig. 6. The source energy is modulated by a chopper blade. This allows the source to alternately pass through the reference and sample paths. The reference path is always free of absorbing gas so the detector is exposed to the source through a path unaffected by the presence of the sample. This signal is monitored by an automatic gain control circuit, which holds the reference signal level constant. If the source intensity or detector sensitivity change, the gain control will correct for it in both the reference and sample channels. Sync pickups monitor the chopper position and alert the electronics when the sample or reference path is irradiated. A narrow bandpass optical filter is located in

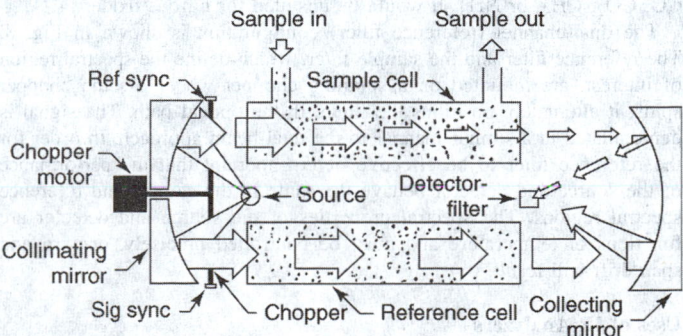

Fig. 6.   Dual-beam infrared analyzer with a solid-state detector. (*Infrared Industries, Inc.*)

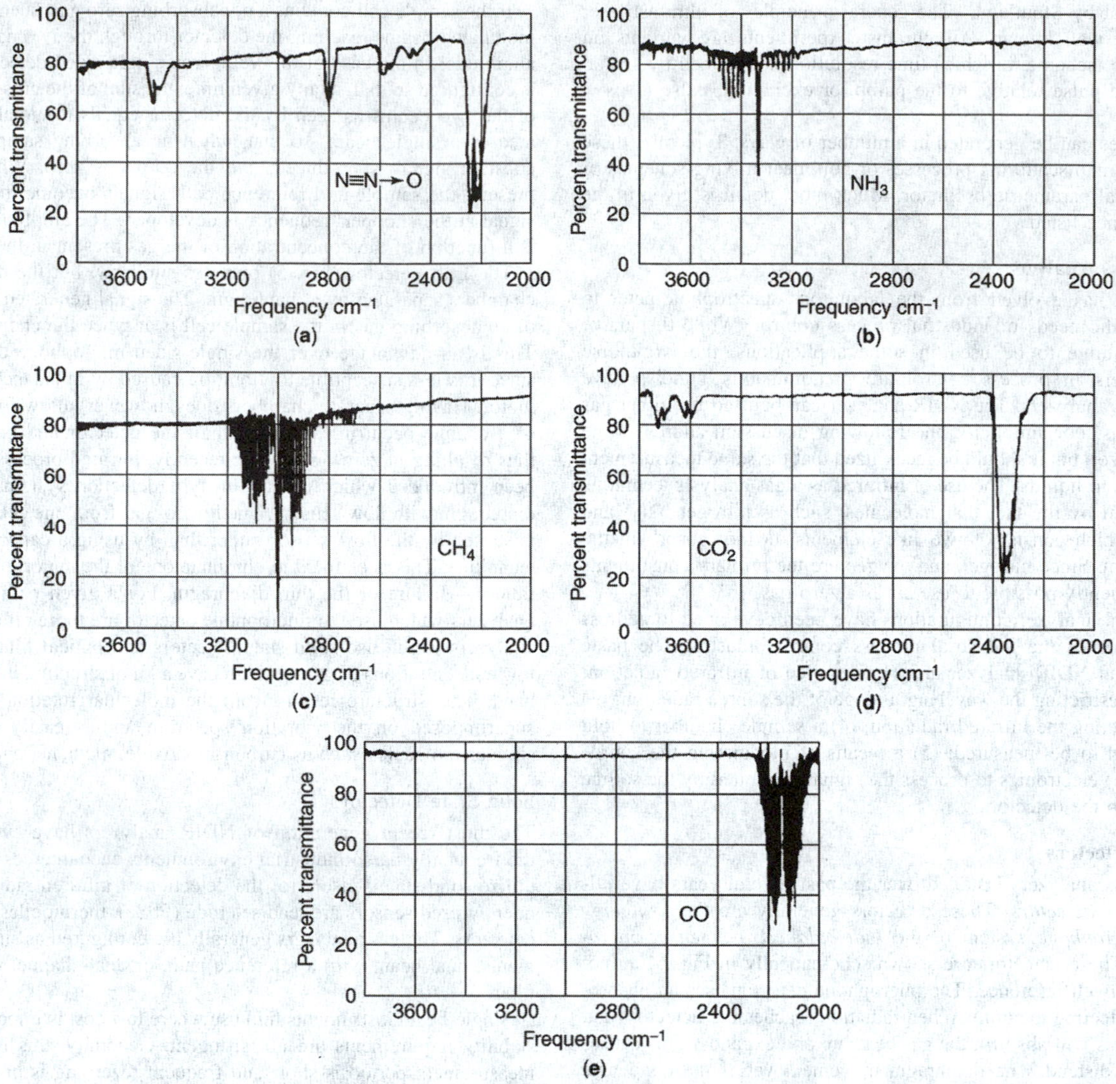

Fig. 7.   Infrared spectra: (**a**) nitrous oxide, (**b**) ammonia, (**c**) methane, (**d**) carbon dioxide, (**e**) carbon monoxide. (*Sadtler Research Laboratories.*)

front of the detector to limit the infrared energy and sensitize the analyzer to a particular gas absorption band. The signals generated by the two optical paths are synchronously demodulated. When an absorbing gas is introducing, the signal reaching the detector through the sample path is attenuated and the magnitude of the detected signal corresponds directly to the concentration of the sample gas present in the sample cell.

A reference optical filter can be used as an alternative to the reference path. This requires that a spectral window exist where is no interference from the sample. The 3.8 to 3.9 micrometer spectral region is frequently used in NDIR gas analyzers for this purpose. The spectral curves of Fig. 7 show that this spectral window will work well as a reference for measuring CO, CO_2, CH_4, or NH_3. It would be unsuited for nitrous oxide.

The dual-channel (reference filter) configuration is shown in Fig. 8. The reference filter and the sample filter, which define the spectral region of interest, are mounted on a spinning chopper wheel. As the chopper spins, it alternately positions the filters in the optical path. The signal is demodulated in a similar manner to the dual-beam approach. In order for the reference filter to be effective, it is important that the performance of the source and detector behave the same in the sample and reference spectral regions. The spectral properties of the source and detector are functions of temperature and must be controlled precisely, or zero and span drift will result.

Uses of IR Analyzers

Many gases can be monitored with IR analyzers. The petrochemical industry, for example, monitors CO_2 in the manufacture of ethylene oxide

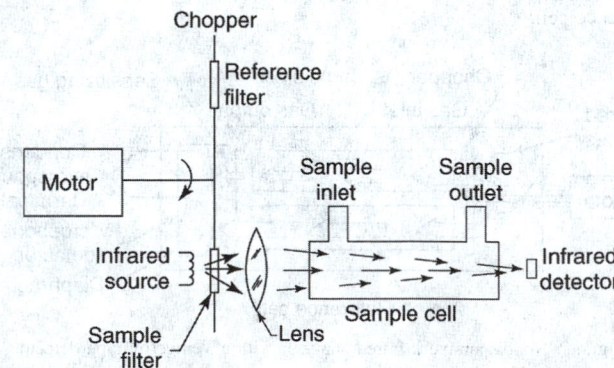

Fig. 8.   Dual-channel nondispersive infrared analyzer with a solid-state detector. (*Infrared Industries, Inc.*)

and ammonia. Acetylene is monitored during the production of acetylene and vinyl chloride. The metals industries use the analyzers to monitor CO_2 in steel converting and soaking pit operations. Carbon monoxide is monitored during heat treating, aluminum powder processing, and tin plant annealing. The food industry monitors CO_2 in greenhouses, storage facilities, and fermentation processes. The analyzers also find use in monitoring for explosive and toxic hazards, as well as for stack gases in pollution control systems.

TABLE 1. REPRESENTATIVE GASES AND CONCENTRATIONS MEASURED BY NDIR ANALYZERS

Gas	Range (Full scale concentration)	
	Minimum (ppm)	Maximum (percent)
Ammonia	1000	100
Butane	300	100
Carbon monoxide	500	100
Carbon dioxide	200	100
Ethylene	1000	100
Ethane	500	100
Ethylene oxide	1500	100
Hexane	400	5
Methane	400	100
Nitrous oxide	150	100
Propane	400	100
Sulfur dioxide	400	100
Water vapor	1500	10

Typical gases and concentrations measured by NDIR analyzers are given in Table 1.

Other Applications

Optical-electronic devices of many kinds have been designed to determine the direction of weakly radiating remote objects by means of detection of their infrared emission. Detailed maps of the earth's surface can be made from aircraft at night by observing the varying infrared emission of the ground. For security, personnel can be detected in total darkness by infrared radiation. Such devices require the detection of low-level radiation in the intermediate-infrared region. Optical lenses or mirrors are used to collect the observed radiation and concentrate it onto the sensitive infrared detector. High-gain, low-noise amplifiers must be used to increase the weak signal from the detector. The wavelength of detection is such that angular resolution capability, as set by diffraction, is much greater with infrared devices than with radars. Infrared photography is described under **Photography and Imagery**.

Astronomy. The potential of using the infrared portion of the electromagnetic spectrum for investigating celestial bodies and interstellar space has been considered by some astronomers for a number of years. This concept was first proposed by William Herschel. It has only been within the last few decades, however, that serious experiments in infrared astronomy have been made. The state of the art is described in entry on **Infrared Astronomy**. The use of various infrared instrumental techniques in connection with space probes is described in specific entries on the planets.

Solar Energy

The sun's total radiation output is approximately equivalent to that of a blackbody at 10,350°R (5750 K). However, its maximum intensity occurs at a wavelength that corresponds to a temperature of 11,070°R (6150 K) as given by Wien's displacement law. A figure plotting solar irradiance versus spectral distribution of solar energy is given in Fig. 9. See also **Solar Energy**.

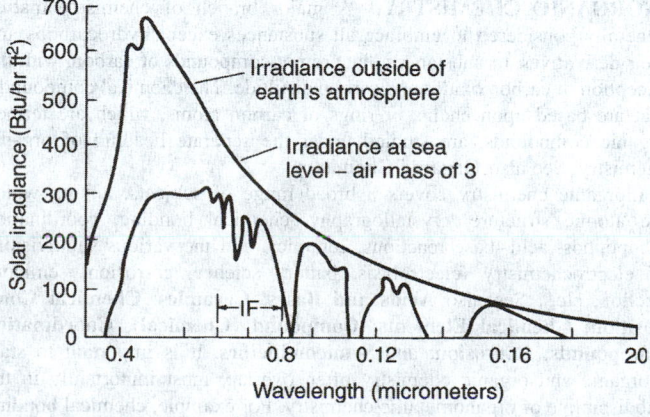

Fig. 9. Spectral distribution of solar energy.

Infrared Laser Chemistry

The application of IF lasers to chemical reactions is described in entry on **Photochemistry and Photolysis**.

Infrared Radiation Sources

Blackbody radiation sources are accurate radiant energy standards of known flux and spectral distribution. They are used for calibrating other infrared sources, detectors, and optical systems. The radiating properties of a blackbody source are described by Planck's law. Energy distribution for blackbody sources at different temperatures is shown in previously mentioned Fig. 1. A blackbody radiation source with its temperature controller is shown in Fig. 10. The specifications of this unit include: temperature range, 100–1,000 °C (212–1,832 °F); cavity diameter, 1 inch (25.4 mm); aperture diameter (in steps) from 0.0125 in (0.3175 mm) to 0.6 in (15.24 mm); emissivity, 0.99 ± .01; field of view, 20 degrees; internal thermocouple, platinum versus platinum-10% rhodium. Radiation sources are also available for use in the infrared region which have an emissivity somewhat less than one. The Nernst glower is a graybody source, which finds frequent use in spectrophotometers. The emissivity of a Nernst glower (Fig. 11) is a function of wave length, averaging approximately 0.6

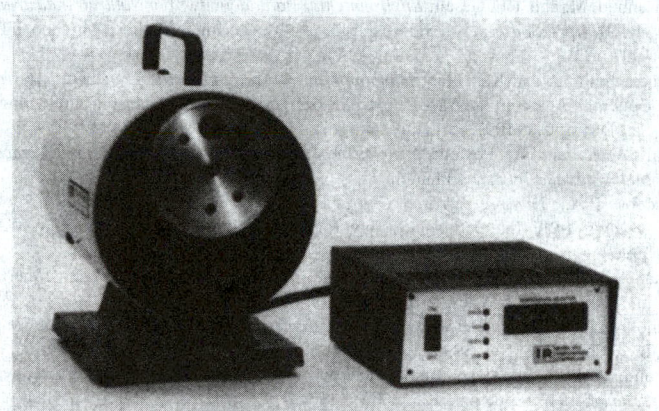

Fig. 10. Infrared radiation source with temperature controller. (*Infrared Industries, Inc.*)

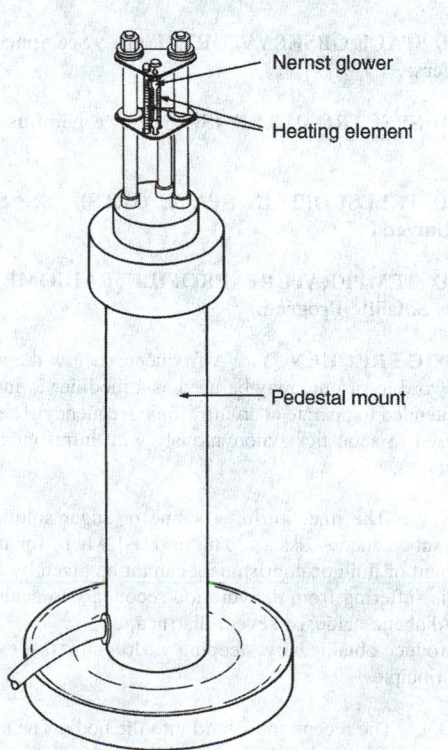

Fig. 11. Nernst glower assembly. (*Infrared Industries, Inc.*)

from 2 to 15 micrometers. The glower element will not conduct electricity when cold, thus the unit must be heated to approximately 400 °C (752 °F) before it will begin to conduct. The unit will operate at a temperature from 1500–1950 K (2,240–3,050 °F) with an expected life of several hundred hours.

See also **Sensors: Infrared**.

Additional Reading

Abid, M.M.: *Spacecraft Sensors*, John Wiley & Sons, Inc., Hoboken NJ. 2005.

Caniou, J.: *Passive Infrared Detection: Theory and Applications*, Kluwer Academic Publishing, Norwell, MA, 1999.

Capper, P. and C.T. Elliott: *Infrared Detectors and Emitters: Materials and Devices*, Kluwer Academic Publishers, Norwell, MA, 2000.

Coleman, P.B.: *Practical Sampling Techniques for Infrared Analysis*, CRC Press, LLC., Boca Raton, FL, 1993.

Colthup, N.B., L.H. Daly and S.E. Wiberley: *Introduction to Infrared and Raman Spectroscopy*, 3rd Edition, Academic Press, Inc., San Diego, CA, 1999.

Drexage, M.G. and C.T. Moynihan: "Infrared Optical Fibers," *Sci. Amer.*, 110 (November 1988).

Durham, R.M.: "Infrared Process Analyzers," in *Process/Industrial Instruments and Controls Handbook*, 5th Edition, (G.K. McMillan, Editor), McGraw-Hill Companies, Inc., New York, NY, 1999.

Eisenbud, M. and T.F. Gesell: *Environmental Radioactivity from Natural, Industrial, and Military Sources*, 4th Edition, Morgan Kaufmann Publishers, Orlando, FL, 1997.

Guelachvili, G. and K.N. Rao: *Handbook of Infrared Standards Two: With Spectral Coverage between 1.4 UM–4 UM and 6.2 UM–7.7 UM*, Academic Press, Inc., San Diego, CA, 1997.

Hill, D.W. and T. Powell: *Non-Dispersive Infrared Gas Analysis in Science, Medicine, and Industry*, Plenum Press, New York, NY, 1968.

Jacobs, P.A.: *Thermal Infrared Characterization of Ground Targets and Backgrounds*, SPIE International Society for Optical Engineering, Bellingham WA, 1996.

Jha, A.R.: *Infrared Technology: Applications to Electro-Optics, Photonic Devices and Sensors*, John Wiley & Sons, Inc., New York, NY, 2000.

Knoll, G.F.: *Radiation Detection and Measurement*, 3rd Edition, John Wiley & Sons, Inc., New York, NY, 1999.

Stoutland, P.O., R.B. Dyer and W.H. Woodruff: "Ultrafast Infrared Spectroscopy," *Science*, 1913 (September 25, 1992).

Wilson, J.S.: *Sensor Technology Handbook*, Elsevier Science & Technology Books, New York, NY, 2004.

Wormhoudt, J.: *Infrared Methods for Gaseous Measurements: Theory and Practice*, Vol. 7, Marcel Dekker, Inc., New York, NY, 2001.

INFRARED SPACE OBSERVATORY (ISO). See **Space Science Missions: Universe**.

INFRARED SPECTROMETER (SIRS). See **Nimbus Satellite Program**.

INFRARED TELESCOPE IN SPACE (IRTS). See **Space Science Missions: Universe**.

INFRARED TEMPERATURE PROFILE RADIOMETER (ITPR). See **Nimbus Satellite Program**.

INFRASONIC FREQUENCY. A frequency below the audiofrequency range. The word infrasonic may be used as a modifier to indicate a device or system intended to operate at an infrasonic frequency. The term subsonic was once used in acoustics synonymously with infrasonic; such usage is now discourages.

INFUSION. 1. The injection of a saline or sugar solution into a vein or into the subcutaneous tissues. This is used when, for any reason, the normal amount of fluid or nourishment cannot be given by mouth or when the patient is suffering from dehydration secondary to acute infection and high fever, diabetic acidosis, severe diarrhea, etc.

2. The product obtained by steeping a drug for the extraction of its medicinal principles.

INGESTION. The reception of food into the body. The simplest type of ingestion is found in some of the one-celled animals, whose protoplasm merely flows around the food particle and engulfs it.

INGOT. A casting designed for reduction by hot working to a semifinished product such as a billet or to a finished product such as a bar, plate, or sheet. Steel ingots are cast in massive cast-iron ingot molds, which extract the heat faster than a sand mold and facilitate both the casting and handling of the ingots. See also **Iron Metals, Alloys, and Steels**.

INGOT IRON. See **Iron Metals, Alloys, and Steels**.

INGUINAL HERNIA. See **Hernia**.

INHIBITOR. (1) A compound (usually organic) that retards or stops an undesired chemical reaction, such as corrosion, oxidation, or polymerization. Examples are acetanilide, which retards decomposition of hydrogen peroxide and salicylic acid, used to prevent prevulcanization of rubber. Such substances are sometimes called negative catalysts. (2) A biological antagonist used to retard growth of pests and insects and in medicine.

INIOMOUS FISHES (*Osteichthyes*). Of the order *Iniomi*, there are several families of iniomous fishes that appear to represent a mid-form between the more primitive and the more advanced isospondylous fishes. Most of the iniomous fishes are found in deep ocean waters, are relatively small (2 feet — 0.6 meter — long or less in most species), possess photophores (light organs) on their sides, with spineless, soft-rayed fins. There are over 300 species, including two suborders (*Myctophoidea* and *Alepisauroidea*), each suborder having about a half-dozen families.

The lizard fishes (family *Synodidae*) are well named because they act and look like reptiles. They prefer the bottom and are inactive much of the time. However, they are capable of very fast movements when a small food fish gets within range. Although the *Saurida undosquamis* (Indo-Pacific species) attains a length of about 20 inches (51 centimeters), most of the lizard fishes do not exceed 12 inches (30 centimeters) in length. The Indo-Pacific species is regarded as an acceptable food fish. Most synodids prefer tropical waters, but some move into temperate waters during summer months. *Synodus foetens* (American Atlantic) ranges from Brazil northward to Cape Cod; *S. lucioceps* (American Pacific) moves north in the summer to the mid-California coast.

Of the lantern fishes (family *Myctophidae*), there are some 150 species. Although a deep-water fish, the lantern fish seeks out light at night and will surface around a boat with a light. They return to the depths during daylight hours. They are from 3 to 6 inches (7.6–15 centimeters) long. The genera of lantern fishes are determined by the positioning and number of the photophores on their sides.

Other members of the order Iniomi include: (1) thread-sail fishes (*Aulopidae*); (2) greeneyes (*Chlorophthalmidae*); (3) grid-eye fishes (*Ipnopidae*); (4) spider fishes (*Bathypteroidae*); (5) Bombay duck (*Harpodontidae*); (6) barracudinas (*Paralepidae*); (7) pearleyes (*Scopelarchidae*); (8) saber-tooth fishes (*Evermannellidae*); (9) lancet fishes (*Alepisauridae*); (10) hammerjaw (*Omosudidae*); and (11) javelin fish (*Anotopteridae*).

INLAND SEA BREEZE. See **Meteorology**.

INORGANIC CHEMISTRY. A major branch of chemistry that is generally considered to embrace all substances except hydrocarbons and their derivatives, or substances that are not compounds of carbon, with the exception of carbon oxides and carbon disulfide. The chemical compounds, that are based upon chains or rings of carbon atoms, which are termed organic compounds, are studied under the separate heading of organic chemistry. See also **Organic Chemistry**.

Inorganic chemistry covers a broad range of subjects, among which are atomic structure, crystallography, chemical bonding, coordination compounds, acid–base reactions, ceramics, and the various subdivisions of electrochemistry (electrolysis, battery science, corrosion, semiconduction, etc.). See also **Acids and Bases**; **Ceramics**; **Chemical Composition**; **Chemical Elements**; **Compound (Chemical)**; **Coordination Compounds**; **Corrosion**; and **Semiconductors**. It is important to state inorganic and organic chemistry often overlap, most importantly in the subdiscipline of organometallic chemistry. For example, chemical bonding applies to both disciplines; electrochemistry and acid–base reaction have their organic counterparts; catalysts and coordination compounds may be either organic or inorganic.

Inorganic chemistry is based upon physical chemistry and forms the basis for mineralogy and materials chemistry. It often overlaps with geochemistry, analytical chemistry, environmental chemistry, and organometallic chemistry. See also **Mineralogy**; and **Physical Chemistry**.

The range of inorganic chemistry includes both molecular compounds, which exist as discrete molecules, and crystals, whose structures are described by infinite lattices of regularly-ordered atoms and which are studied by crystallography and solid-state chemistry. See also **Solid-State Chemistry**.

Regarding the importance of inorganic chemistry, R. T. Sanderson has written: All chemistry is the science of atoms, involving an understanding of why they possess certain characteristic qualities and why these qualities dictate the behavior of atoms when they come together. All properties of material substances are the inevitable result of the kind of atoms and the manner in which they are attached and assembled. All chemical change involves a rearrangement of atoms. Inorganic chemistry [is] the only discipline within chemistry that ... examines specifically the differences among all the different kinds of atoms.

Major branches of inorganic chemistry include:

- Minerals, such as salt, asbestos, silicates, ...
- Metals and their alloys, like iron, copper, aluminum, brass, bronze, ...
- Compounds involving non-metallic elements, like silicon, phosphorus, chlorine, oxygen, for example water
- Metal complexes

Commercially important inorganic substances include silicon chips, transistors, LCD screens, fiber optical cables and many catalysts.

Organometallic chemistry combines aspects of organic chemistry with those of inorganic chemistry, and is formally defined as the study of compounds containing metal-carbon bonds, although many organometallic compounds contain no such bonds. See also **Organometallic Compounds**. Among the simplest organometallic compounds are the metal carbonyls, in which carbon monoxide binds to a metal through the carbon. Vitamin B12, whose active site is similar to that of hemoglobin, is a naturally-occurring, metabolically-important organometallic compound containing large organic components (corrin and protein) and a metal, cobalt, bonded to carbon.

See also **Alloys**; **Iron Metals, Alloys, and Steels**; **Metals (The)**; and **Periodic Table of the Elements**.

Additional Reading

Bard, A.J., E. Gileadi, and M. Urbakh: *Encyclopedia of Electrochemistry, Thermodynamics and Electrified Interfaces*, Vol. 1, John Wiley and Sons, Inc., New York, NY, 2002.

Cotton, F.A., G. Wilkinson and P.L. Gaus: *Basic Inorganic Chemistry*, 3rd Edition, John Wiley and Sons, Inc., New York, NY, 1995.

Cotton, F.A., G. Wilkinson, M. Bochmann, and C.A. Murillo: *Advanced Inorganic Chemistry*, 6th Edition, John Wiley and Sons, Inc., New York, NY, 1999.

Gaus, P.L.: *Student's Solution Manual to Accompany Basic Inorganic Chemistry*, 3rd Edition, John Wiley and Sons, Inc., New York, NY, 1995.

Miessler, G.L., and D.A. Tarr: *Inorganic Chemistry*, 3rd Edition, Prentice-Hall, Inc., Upper Saddle River, NJ, 2003.

INORGANIC LIGHT-EMITTING DIODE DISPLAYS. Light-emitting diodes (LEDs) were developed in the 1960s as an outgrowth of semiconductor technology. The devices emit light when a forward bias voltage is applied to a P–N junction in a single crystal of gallium arsenide or other group III-V compounds. By appropriate doping (adding trace amounts of materials) and/or the use of crystals containing III-V materials, it is possible to produce emissions of red, green, yellow, and (recently) blue light. It is also possible to produce a variety of colors from a single device.

The basic structure of an inorganic LED typically consists of a wafer substrate of either single crystal gallium arsenide (GaAs) or gallium phosphide (GaP), on which a layer of gallium arsenide phosphide (GaAs(1-x)Px) is formed by vapor-phase or liquid-phase epitaxial growth. Doping with zinc, oxygen, or nitrogen creates a P/N junction in the epitaxy layer. Thin films of silicon dioxide and/or silicon nitride are then deposited over the wafer to provide a protection layer for long-term stability.

Gallium phosphide allows better overall efficiency, because it can be rendered transparent to the LED radiation if the proper LED material is used. This, in combination with such factors as the type of encapsulation and the use of lenses and reflectors in the total package, results in significant improvements in the overall luminous efficiency. These improvements are evident when examining the actual luminous output at different levels of input power, as detailed in the data sheet for the specific LED.

Another important characteristic of the LED is that it is indeed a diode and has the usual current–voltage relationship found in diodes. The sharpness of the break at the turn-on point of about 1.5 volts affects the multiplexing and matrix addressing of this type of display device, whereas the linear range, up to the maximum forward current that can be tolerated without destruction, determines the dynamic range that can be achieved. Because the luminous output is directly proportional to the current over the linear range, it is possible to have a gray scale range with LEDs, although the exact output level for a given input may vary by as much as 50% from one LED to another with the same designation.

The mechanism of LED operation is explained with the help of Figure 1. With no voltage or a reversed voltage applied across the P/N junctions, an energy barrier is formed, preventing the flow of electrons and holes. When a forward bias voltage, typically 1.5 to 2 volts, is applied across the junction, the potential barrier height is reduced by allowing electrons to be injected into the P region and holes into the N region. The injected minority carriers recombine with carriers of the opposite sign, resulting in the emission of photons. If the number of photons is sufficiently large, then a useful amount of light is emitted. To obtain a useful LED for visual display applications, there must be (1) efficient radiative recombination, (2) radiation at a useful visible wavelength, and (3) the possibility of manufacturing the LED so as to have a high-minority carrier injection efficiency.

Binary and ternary compounds from GaAs(1-x)Px, with x in the range of 0.4 to 1.0 will meet the three criteria listed previously. GaP doped with zinc and oxygen produces red light; when doped with nitrogen, it produces green light. GaAs(1-x)Px produces red through green light, depending on the dopant and the amount of doping used.

During the past 20 years, many attempts were made to develop bright blue LEDs that could be produced economically. Much of the effort focused on the use of silicon carbide (SiC). But the big breakthrough came in 1993 when Nichia Chemical Company, of Tokushima, Japan, developed a bright blue LED using gallium nitride. Later, indium gallium nitride was also used. Today, Nichia offers a wide range of LEDs, from ultraviolet to yellow to white. This made full-color LED panels possible (these are now in use in large videowall displays in both indoor and outdoor locations).

The voltage drop across an LED is 1.5 to 3 volts, and the current required is a function of the desired light output and the luminous efficiency. Typically the power dissipated ranges from 0.1 to 10 watts/cm^2 of the display area. Because the LED mechanism is electronic, the rise and decay times are very fast, typically 1 to 100 microseconds.

The devices will operate from $-100\,°C$ to $+100\,°C$. At a constant current, the light output from an LED decreases at the rate of 1% for each degree centigrade drop. The forward bias voltage decreases at the rate of 1 millivolt for each degree centigrade drop. These temperature dependencies are unimportant for most applications.

The luminous intensity generated by an LED is proportional to its forward current. When a reverse bias is applied, the current is very small and no light is emitted. It is possible to pulse an LED at a fast rate and obtain high luminosity (brightness). Because the eye integrates the light pulses, this technique is often used in high-brightness displays to circumvent the maximum current limitation of an LED, which is about 25 milliamps for continuous or D.C. operation.

LED devices are now used in a wide range of consumer and industrial products, virtually replacing the incandescent lamp as a status indicator. Numeric and alphanumeric displays now appear in digital clocks, television sets, radios, appliances, test and measuring equipment, and computer terminals.

Currently, more than 20 companies around the world produce LEDs. Some of the largest producers are Toshiba, Hewlett-Packard, Sharp, Siemens, Stanley Electric, Taiwan Liton, and Sanyo Electric. The top manufacturers have focused their research and development on developing a

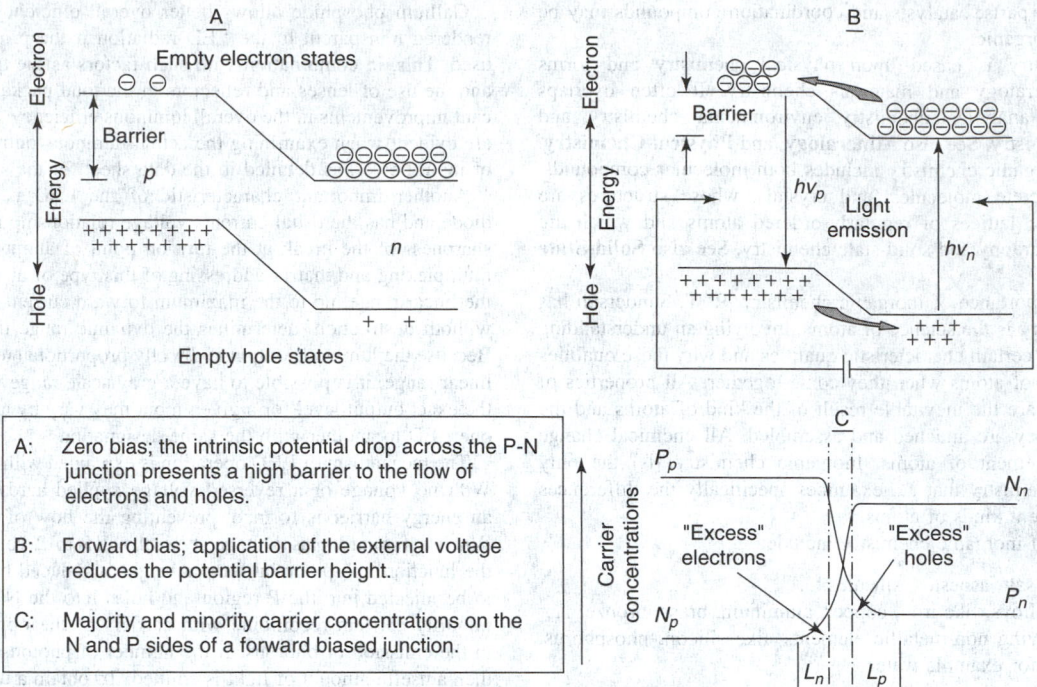

A: Zero bias; the intrinsic potential drop across the P-N junction presents a high barrier to the flow of electrons and holes.

B: Forward bias; application of the external voltage reduces the potential barrier height.

C: Majority and minority carrier concentrations on the N and P sides of a forward biased junction.

Fig. 1.   Mechanism of P-N Junction Operation in LED. (*Stanford Resources, Inc.*)

brighter blue LED device and "intelligent" LED displays (a module consisting of an array of LED devices along with drive electronics and, in some cases, microprocessor capability). There is an ongoing trend to make all the devices brighter. Large-screen displays for public information kiosks and message centers are made by forming an array of LED lamps.

Large-screen LED displays have become quite popular, but the cost rises in direct proportion to the size of the screen. With the recent availability of the blue LED lamp, many full-color, large-area matrix displays have appeared. These displays consume a lot of power, but they fill the gap between the very large cathodoluminescent displays and the smaller color plasma and direct-view CRTs.

Although LED lamps are used in very large matrix displays, they have not been used as high-resolution graphic displays for several decades (they were employed in several military cockpit programs in the 1970s). Two problems are very difficult to overcome in high-resolution LED matrix displays: power consumption and interconnect integrity. LEDs are very inefficient illuminators, requiring huge amounts of power to operate at high luminance levels. Discrete LED chips must be attached to a substrate in a high-density matrix in order to form a large array.

Gur Optics, of Tel Aviv, Israel, has developed an adaptation of LED technology. This approach uses a set of 24 arms, which hold linear arrays of LED lamps. The arms spin, and the lamps are multiplexed to produce a two-dimensional array of $2,000 \times 2,000$ pixels. Displays up to 60 inches are proposed, although only a 7-inch feasibility prototype has been made. Full color is implemented by using red, green, and blue lamps. See also **Diode**; **Organic Light-Emitting Diodes (OLEDs)**; and **Semiconductors**.

Additional Reading

Nakamura, S. and G. Fasol: *The Blue Laser Diode: GaN Based Light Emitters and Lasers*, Springer-Verlag Inc., New York, NY, 1997.

Schubert, F.E.: *Light Emitting Diodes: Research Manufacturing and Applications V*, Vol. 327, SPIE-International Society for Optical Engineering, Bellingham, WA, 1998.

Willardson, R., E. Weber, G.B. Stringfellow, and M.G. Craford: *Semiconductors and Semimetals: High Brightness Light Emitting Diodes*, Morgan Kaufman Publishers, Orlando, FL, 1997.

Web References

Toshiba. http://www.toshiba.com/

Sharp. http://www.sharp-usa.com/

Stanley Electronic Group. http://www.stanley.co.jp/e/product/index.html

Liteon Technology. http://www.liteontc.com.tw/index.html

Sanyo. http://www.sanyo.com/

Stanford Resources, Inc., San Jose, CA

INOSITOL.   A constituent of body tissue. In purified form it is used as a nutrient and dietary supplement in some foods and feed-stuffs. The chemical name of inositol is hexahydroxycyclohexane, $C_6H_6(OH)_6 \cdot 2H_2O$. There are nine isomeric forms of inositol. Myoinositol or meso-inositol (*cis*-1,2,3,5-*trans*-4,6-hexahydroxycyclohexane) is the isomer that possesses essential nutrient activity. The substance, often identified as a vitamin, is found in small amounts in many vegetables, citrus fruits, cereal grains, liver, kidney, heart and other meat. The commercial source is corn (maize) steep liquor. In addition to its use in nutrition, it finds use in medicine and as an intermediate for organic syntheses.

INPUT/OUTPUT DEVICES (Computer System).   An input device is a means by which information can be read into a computer; an output device is a means by which information from a computer can be recorded, transferred, or displayed for further processing or interpretation, including interpretation by the people using the system. These devices are described in a number of articles in this encyclopedia. Check alphabetical index.

INSECT.   A class (*Insecta*) of the phylum *Arthropoda* which is by far the largest taxonomic division of the animal kingdom. See Tables 1 and 2 and Fig. 1 on pp. 1946–1947. Upwards of 700,000 species of insects are known, with at least a few thousand Additional species being identified each year. Thus, some scientists estimate that there may be a million or more different species on earth. There is a 3 to 4 to 1 greater variety among insects than among all other animal species combined. Although one of the earliest of animals to inhabit the earth, the habits of insects continue to be astonishing. In some form, insects inhabit nearly all parts of the earth.

The immense numbers of insects indicate remarkable biological success. They have invaded all possible habitats save the ocean; here only one form, *Halobates*, a genus of marine water striders, is known. On land and in fresh water, insects are found in almost every imaginable habitat. They eat plants, animals, and dead organic matter of all kinds, and many are parasitic. They walk, run, jump, fly, swim, and burrow. Moreover, their small size permits them to live in very limited habitats. Hence, the plant feeders include species which are confined to flowers, leaves, stems, fruits,

TABLE 1. BROAD CLASSIFICATION OF INSECTS

(Kingdom)	ANIMAL
(Phylum)	ARTHROPODA
(Class)	INSECTA
(Subclasses) (Orders)	APTERYGOTA[a] PTERYGOTA[a]

Collembola. The springtails.
Protura. Rare and primitive Insects of small size.
Thysanura. The silverfish or Fish moth and allied species.

Anoplura. True or sucking lice.
Coleoptera. The beetles.
Corrodentia. The book lice and psocids.
Dermaptera. The earwigs.
Diptera. The 2-winged flies, including true flies, mosquitoes, midges, gnats, and others.
Embiidina. Rare insects (no common name).
Ephemeroptera. May flies (shad flies, salmon flies).
Hemiptera. The true bugs.
Homoptera. Cacadas, leaf hoppers, plant lice, scale insects.
Hymenoptera. Sawflies, ants, bees, wasps, and many parasitic forms.
Isoptera. The white ants or termites.
Lepidoptera. Butterflies, skippers, and moths.
Mallophaga. Bird lice or biting lice.
Mecoptera. The scorpion flies.
Neuroptera. Lacewings, dobson flies, hell-grammites, ant lions, alder flies, and others.
Odonata. Dragon flies and damsel flies.
Orthoptera. Grasshoppers, crickets, cockroaches, katydids, mantises, walking-sticks and their allies.
Plecoptera. The stone flies.
Siphonaptera. The fleas.
Thysanoptera. The thrips.
Trichoptera. The caddis flies.
Zoraptera. Rare insects (no common name).

[a]Defined in separate alphabetical entries.

or roots, and parasites are not limited to larger animals but find adequate hosts in other insects: even insect eggs are attacked by parasitic insects. Predacious species are, of course, confined to smaller prey such as other insects and the small members of other groups. This extreme diversity makes it impossible for human beings to avoid experience with insects. The blood-sucking mosquitoes and flies and the scavenger clothes moths and beetles force themselves upon us.

General Profile of Insects

Insects are characterized by the hard exoskeleton and jointed appendages of the phylum. Among other arthropods they are distinguished by several characteristics. (1) The body is divided into head, thorax, and abdomen. (2) Both compound and simple eyes may be present. (3) There is one pair of antennae. (4) The thorax bears a maximum of three pairs of legs. (5) Wings are found in the adults of many species. Two pairs occur in most orders. (6) The abdomen is usually without joined appendages, although a few primitive forms bear modified derivatives of these structures. (7) Respiration is accomplished by means of tracheae, which are air tubes opening from the exterior and branching among the tissues. (8) Metamorphosis is complex in some orders, although lacking in the most primitive forms.

TABLE 2. CLASSIFICATION OF INSECTS AND INSECT-RELATED TERMS DESCRIBED IN THIS VOLUME

Order (and example)	Title of Entry
Anoplura (Sucking Lice)	*Anoplura*
Coleoptera (Beetles)	Asparagus Beetle
	Bark Beetle
	Bean Weevil
	Beetle
	Blister Beetle
	Boll Weevil
	Bombardier Beetle
	Buffalo Carpet Moth
	Cadelle
	Carrion Beetle
	Chafer
	Click Beetle
	Cockchafer
	Coleoptera
	Colorado Potato Beetle
	Corn Rootworm
	Curculio
	Death Watch
	Dermestid
	Firefly
	Flea-beetle
	Glow Worm
	Grub
	Japanese Beetle
	June Beetle
	Khapra Beetle
	Lady Bug
	Pea Weevil
	Plum Curculio
	Raspberry Fruitworm
	Rose Beetle
	Rose Chafer
	Scarab
	Tiger Beetle
	Tumble-Bug
	Water Penny
	Weevil
	White-Fringed Beetle
	Wireworm
Collembola (Springtails)	*Collembola*
Corrodentia (Gnawers)	Book Louse
	Corrodentia
Dermaptera (Earwigs)	*Dermaptera*
Diptera	Bat Tick
	Bee Fly
	Bee-Louse
	Black-Fly
	Blow-Fly
	Bluebottle
	Bot Fly
	Brine-Fly
	Crane Fly
	Deer Fly
	Diptera
	Drosophilia
	Fly
	Fruit Fly
	Fungus Gnat
	Gall Gnat
	Gnat
	Haltere
	Hessian Fly
	Horn Fly
	Housefly
	Leather-Jacket
	Maggot
	Midge
	Mosquito
	Nose Fly
	Petroleum Fly
	Punkie
	Rasberry-Cane Borer

(continued)

TABLE 2. (*Continued*)

Order (and example)	Title of Entry	Order (and example)	Title of Entry
	Robber Fly		Bag-Worm
	Screw-Worm		Bee-Moth
	Sheep Tick		Blue
	Snipe Fly		Bollworm
	Spinach Leaf Miner		Brown-Tail Moth
	Stable Fly		Bud Moth
	Tsetse Fly		Butterfly
	Warble Fly		Cabbage Butterfly
	Wheat Midge		Candle Fly
Embioptera	*Embiidina*		Canker Worm
Ephemeroptera (Mayflies)	*Ephemeroptera*		Carpenter-Moth
Exopterygota	*Exopterygota*		Caterpillar
Hemiptera (Bugs)	Ambush Bug		Clear-Winged Moth
	Apple Redbug		Clothes Moth
	Assassin Bug		Codling Moth
	Black Swimmer		Corn Earworm
	Bed Bug		Cutworm
	Bug		Death's Head Moth
	Chinch Bug		Eastern Tent Caterpillar
	Cotton Stainer		European Corn Borer
	Harlequin Bug		Fritillary
	Hemiptera		Grape-Leaf Folder
	Kissing Bug		Grape-Leaf Skeletonizer
	Negro Bug		Gypsy Moth
	Scale Insects		Hairstreak
	Squash Bug		Hawk Moth
	Stink Bug		Hornworm
	Toad Bug		Leaf Miner
	Water Boatman		Leaf Roller
	Water Measurer		*Lepidoptera*
	Water Scorpion		Meadow-Brown
	Water Strider		Measuring Worm
Hexapoda (Insects)	*Hexapoda*		Metal-Mark
Homoptera (Plant Lice)	Aphid		Moth
	Bark Louse		Mourning Cloak
	Cicada		Oriental Fruit Moth
	Citrus Psylla		Owlet Moth
	Frog Hopper		Peach-Tree Borer
	Ground Pearl		Peach-Twig Borer
	Homoptera		Pink Bollworm
	Lantern Fly		Plume Moth
	Leaf Hopper		Silkworm
	Mealy Bug		Skipper
	Phylloxeran		Squash Borer
	Spittle Bug		Swallowtail
	Tree Hopper		Tobacco Worm
	White Fly		Tomato Worm
Hymenoptera (Bees)	Ant		Tussock Moth
	Bee		Underwing
	Bumblebee		Wax Moth
	Carpenter-Bee		Webworm
	Chalcid Wasp		Woolly Bear
	Cuckoo Wasp		Yucca Borer
	Ensign Fly		Yucca Moth
	Gall Wasp	*Mallophaga* (Chewing Lice)	Bird Louse
	Hornet		Louse
	Horn-Tail		*Mallophaga*
	Hymenoptera	*Mecoptera* (Scorpion Flies)	*Mecoptera*
	Ichneumon		Scorpion Fly
	Jointworm	*Megaloptera* and *Neuroptera*	Alderfly
	Mud Dauber	(Lacewings and Aphislions)	Ant Lion
	Paper Wasp		Aphis-Lion
	Potter Wasp		Corydalis
	Propolis		Fish Fly
	Royal Jelly		Golden-Eye
	Sawfly		Hellgrammite
	Velvet Ant		Mantis Fly
	Wasp		
	Yellow-Jacket		
Isoptera (Termites)	*Isoptera*		
Lepidoptera (Moths)	Angle-Wing		
	Apple Leaf Skeletonizer		
	Army Worm		

TABLE 2. (*Continued*)

Order (and example)	Title of Entry
	Neuroptera
	Sialid
	Snake Fly
Odonata (Dragonflies)	Dragon fly
	Odonata
Orthoptera (Grasshoppers)	Ant Loving Cricket
	Bush Cricket
	Camel Cricket
	Cockroach
	Cricket
	Croton Bug
	Grasshopper
	Katydid
	Leaf Insect
	Locust
	Mantis
	Mole Cricket
	Mormon Cricket
	Orthoptera
	Sand Cricket
	Sword-Bearer
	Walking Stick
Protura (Primitive)	*Protura*
Pterygota	*Pterygota*
Siphonaptera (Fleas)	Flea
Streptsiptera (Stylopids)	*Streptsiptera*
Thysanoptera (Thrips)	*Thysanoptera*
Thysanura (Bristletails)	Fire Brat
	Silverfish
Trichoptera (Caddisflies)	Caddis Fly
	Caddis Worm
Zoraptera (Rare)	*Zoraptera*
Anatomy and Physiology	Anemotaxis
	Bursa Copulatrix
	Caudal Filament
	Chrysalis
	Colleterial Glands
	Corneagen Cell
	Crystalline Cone
	Elytron
	Epipharynx
	Fat Body
	Gnathochilarium
	Gonapophysis
	Honeydew
	Hypopygium
	Hypopus
	Larva
	Metamorphosis
	Naiad
	Nymph
	Paurometabola
	Podical Plate
	Prementum
	Proctodaeum
	Proleg
	Pseudostigmatic Organ
	Pupa
	Scolophore
	Sensillae
	Spiracle
	Stigma
	Trophallaxis
	Winter-Egg
Other Related Terms	Beneficial Insects
	Borer
	Chigger
	Dried-Fruit Insects
	Entomology
	Fruitworm
	Insects
	Grain-Storage Insects

Order (and example)	Title of Entry
	Insect Control and
	Insecticides
	Mite
	Millipede
	Nematodes
	Myrmeocophile
	Termitophile

Superior Features of Insects

In addition to their adaptability to environmental change and tremendous instinctive capabilities (scientists are still not certain in many insects of the true relationship between precise instinctive insect habits and acquired information "intelligence"), most insects possess superior "constructional and design features," which have contributed to their biological success.

Toughness and Ruggedness. Insects display an astonishing application of naturally "engineered" materials of construction. For example, a secretion from the outer skin of many insects hardens to a protective armor. The main ingredient is chitin, which is a lightweight, tough material highly resistant to water and corrosives — a material comparable in many ways to certain synthetic plastics that people have known for only a comparatively few years. Sclerotin or cuticulin is still another covering that some insects use to provide an extra coating of protective material. Cuticulin is much like the substance of fingernails, rendering insect bodies waterproof, not only protecting the insect from severe exposure to external moisture, but also keeping the body moist and from drying out in hot, dry environments. These structural and coating materials, when coupled with streamlined design among many flying insects, contribute to lower frictional resistance during flight.

Exceptional Flying Capability. Flight is a major asset to the numerous winged species, providing a means for quick escape from danger, a much broader range of area for foraging food, and, among many species, a means of migration, for seeking not only food, but also more suitable seasonal environments. Flight is enhanced by the streamlining and light weight of most insects and, when coupled with an amazingly efficient energy utilization system, enables many insects to combine the best features of birds with those of many non-winged creatures. Considering the size of insects, the flying speed of many is outstanding. Among the fastest are the dragonfly, hawkmoth, and horsefly, any of which can reach flying speeds of 25 miles (40 km) per hour (and even greater when necessary). Butterflies, honeybees, hornets, and wasps fly at an average speed of about 12 miles (19 km) per hour; and the housefly at about 5 miles (8 km) per hour.

Endurance. This is characteristic of many insect species, again arising from a very efficient energy system and more-than-ample means for storing the necessities of life, enabling many species to endure long periods of famine and drought. The flight endurance of locusts, for example, when swarms fly thousands of miles is exemplary of great energy and muscular system reserves. Grasshoppers have over 900 muscles, making them very resistant to fatigue.

Superior Sensing Powers. The sensory organs of many insects are highly developed. The antennae are the most noticeable sense organs of most insects, with such "feelers" almost continuously in motion — turning, feeling, twisting, or being groomed. For various species, the antennae have been "customized" over centuries of development and refinement — they are of various lengths and shapes with relationship to the body. Flies have rather short antennae; while moths and butterflies have plumes of a knotted-thread construction that incorporate a sense of smell, position, sound, taste, and temperature, and possibly other variables still not fully understood. The male honeybee's keen sense of smell is enhanced by thousands of odor receptors in its antennae.

Effective Societal Habits. Many insect species feature a specialized work caste system, which provides such insects with all of the advantages of what might be termed "instinctive cooperation." The best qualities of each caste is fully exploited, thus providing the individual member, the colony, and, in turn, the species with a tremendous advantage toward

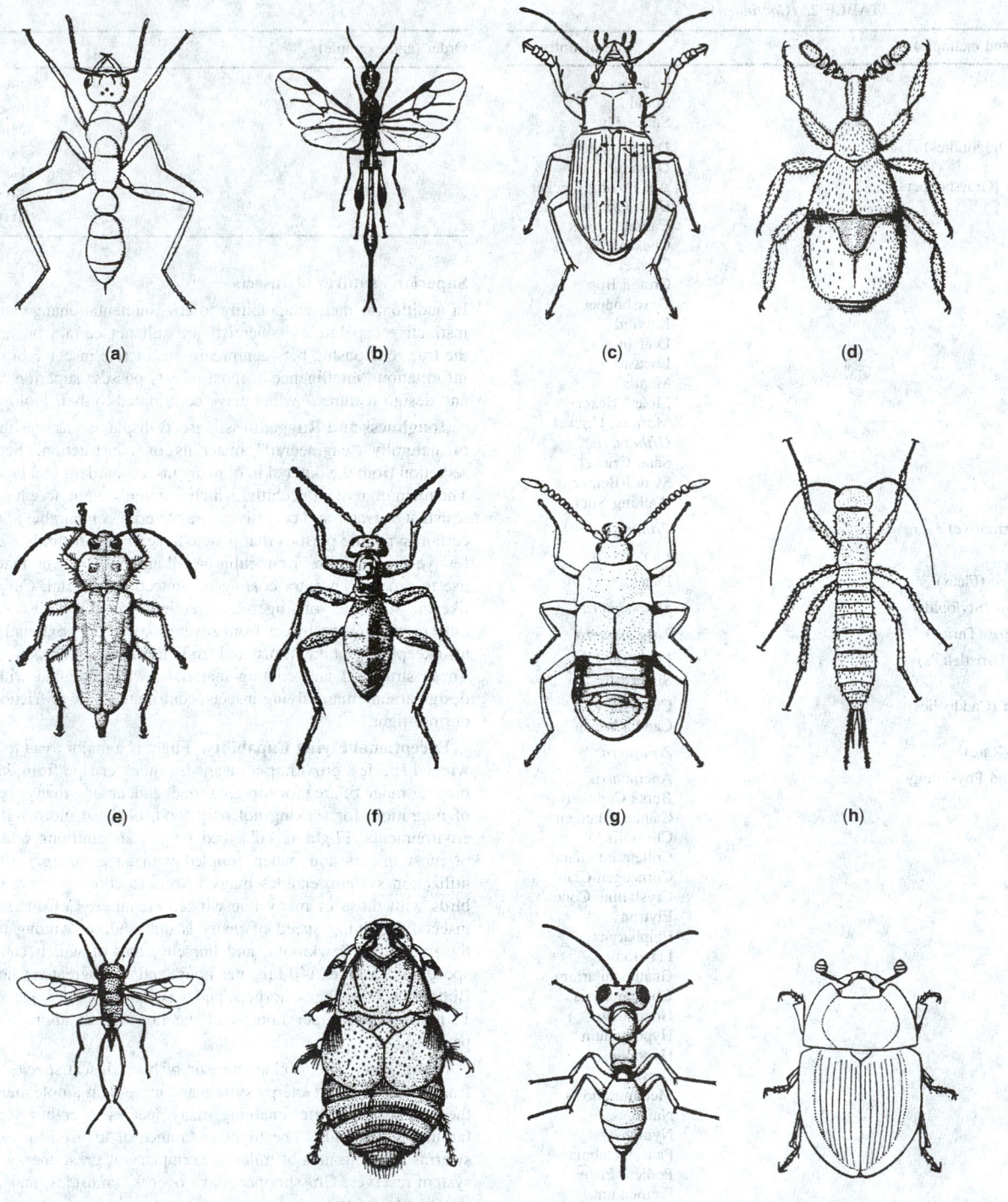

Fig. 1.   The insects are by far the largest class in the entire animal kingdom, with more than three-quarters of a million species. The insects shown in this figure have been selected to indicate the numerous shapes and formats of insects. (**a**) Bulldog ant (*Myrmecia* sp). (**b**) Gasteruptiid wasp, with bulbous swellings of the hind tibiae. (*Gasteruption assectator*) (female). (**c**) Ground beetle (family Carabidae). (**d**) Blind beetle (*Claviger testaceus*); lives in an ant colony. (**e**) Old house borer (*Hylotrupes bajulus*) (female). (**f**) Oak-apple gall wasp (*Biorhiza pallida*) (parthogenetic female (always wingless)). (**g**) Adult beetle (*Lomechusa strumosa*). (**h**) Grouse locust (*Grylloblatta campodeiformis*). (**i**) Ensign wasp (*Evania* sp), distinguished by the short, stalked, laterally compressed abdomen. (**j**) "Beaver Louse" beetle (*Platypsyllus castoris*). The insect does not parasitize beavers, but rather it frees them from parasitic mites. (**k**) Wingless parasitic wasp (*Gelis* sp). (**l**) Sap-sucking beetle (*Amphotis marginata*), which begs from ants. (**m**) *Agriotypus armatus* (female), a parasite of the prepupae and pupae of caddis flies. (**n**) Wingless earwig (*Axixenia jacobsoni*). (**o**) Butterfly (*Hypolimnas mysippus*) (female) (**p**) A symphytan wasp (*Xyela julii*) (female), often found on blooming pine trees in the spring. (**q**) A zoropteran (*Zorotypus guineensis*). (**r**) Web-spinner (*Embia sabulosa*). (**s**) Water measurer (*Hydrometra stagnorum*) walks slowly on six legs. (**t**) Spotted pear psyllid (*Psylla pyricola*). (**u**) Stream-dwelling naiad of the Mayfly (*Prosopistoma foliacea*). (**v**) Diploglossata (*Hemimerus bouvieri*). (**w**) Tree-dwelling tropical springtail (*Campylothorax* sp), with furca extended. (**x**) Rove beetle (*Stenus* spp) catch their prey with a sticky prehensile apparatus that can be shot forward. (**y**) Rhinocerus beetle (*Oryctes nasicornis*) (male). (**z**) Feather-winged beetle (*Acrotrichis sericans*), with wings extending from beneath the elytra.

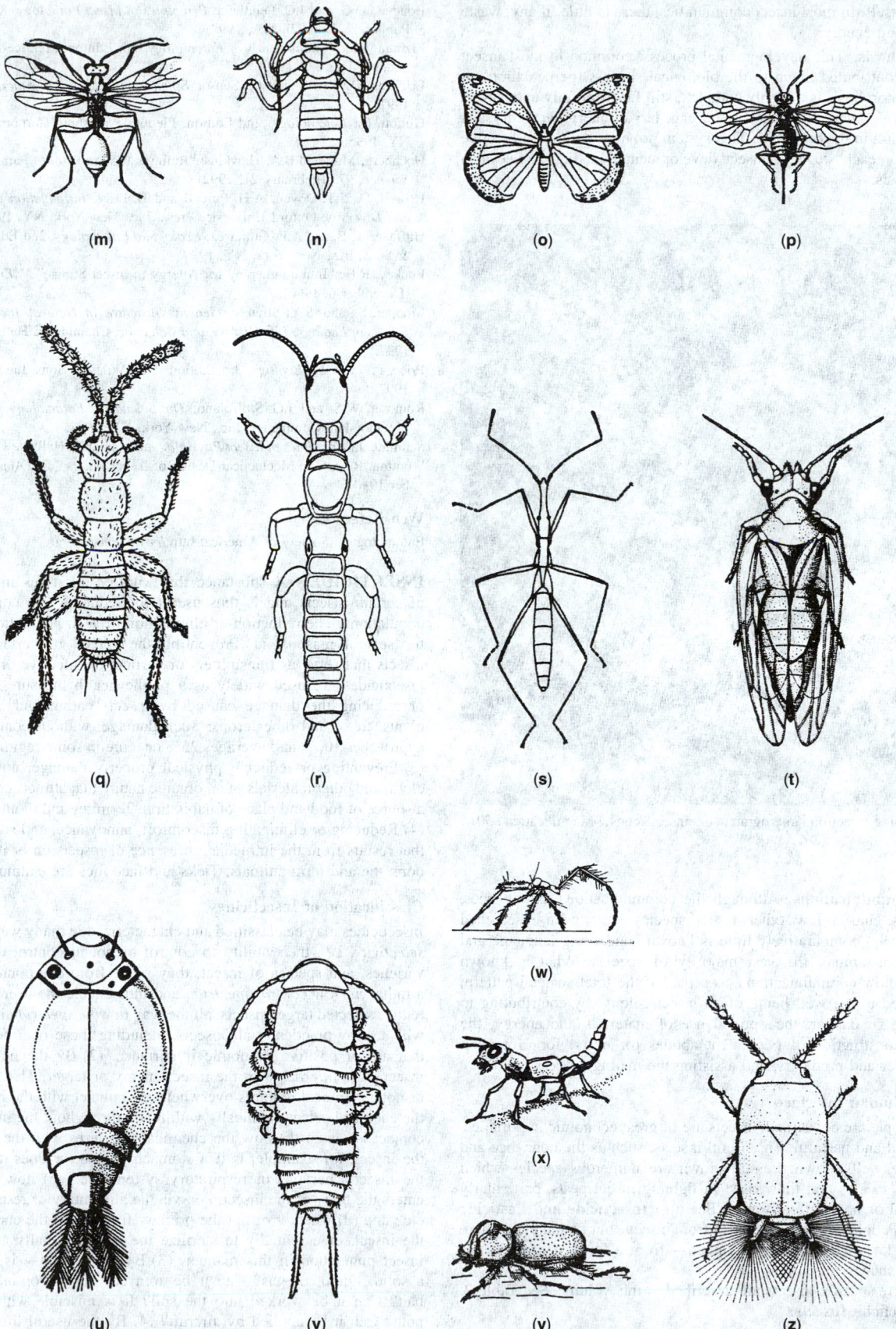

(m) (n) (o) (p)

(q) (r) (s) (t)

(w)

(u) (v) (x) (y) (z)

Fig. 1. (continued)

2719

perpetuating itself. In most insect communities, there is little, if any, waste of materials and energy.

Metamorphosis. This developmental process, common to most insect species, has contributed much to the biological success (perpetuation) of insects. Metamorphosis is a highly complex, still far from fully understood system of developmental changes, (e.g., egg, larva (caterpillar), chrysalis (pupa), and adult in the butterfly). This system provides an ideal protective mechanism for each stage of insect development. See Fig. 2. See also **Metamorphosis**.

Fig. 2. Scanning electron micrograph of insect eggs. Magnification 50×. (*Polaroid.*)

Insect Communications. Although the communications among bees, ants, termites, and a few other insect species have been researched quite extensively, comparatively little is known concerning language and communications among the vast majority of species. What is known demonstrates how communications, as a part of the total societal pattern, has contributed to the well being of an insect colony, by contributing to the avoidance of danger, the optimal use of materials and energy, the cooperative construction of permanent abodes (or nests) for protection against weather and predators, and assisting the mating process.

The Insect/Human Interface

Insofar as people are concerned, insects are of great economic importance, both beneficial and harmful. The useful insects, such as the honeybee and silkworm, are well known. Lesser known are numerous species which destroy other species and thus assist in fighting insect pests, particularly of agricultural or health significance. See also **Insecticide and Pesticide Technology**. People have learned much from insects, but undoubtedly there is far greater knowledge yet to be gained — from a form of life that has been notably successful biologically.

Several score specific-insects are described in this volume. See Table 2. See also **Beneficial Insects**.

Additional Reading

Amato, I.: "Insect Inscriptions," *Science News*, 376 (June 16, 1990).

Borror, D.J., C.A. Triplehorn and N.F. Hohnson: *An Introduction to the Study of Insects*, 6th Edition, Saunders College Publishing, Philadelphia, PA, 1997.

Borror, D.J. and R.E. White: "A Field Guide to Insects," *America North of Mexico*," Houghton Mifflin Company, New York, NY, 1998.

Boucias, D.G. and J.C. Pendland: *Principles of Insect Pathology*, Kluwer Academic Publishers, Norwell, MA, 1998.

Elzinga, R.J.: *Fundamentals of Entomology*, 5th Edition, Prentice-Hall, Inc., Upper Saddle River, NJ, 1999.

Erickson, D.: "An Acoustic Sensor Spies on Insects," *Sci. Amer.*, 131 (February 1991).

Gilliott, C.: *Entomology*, 2nd Edition, Plenum Publishing Corporation, New York, NY, 1995.

Hochberg, M.E. and B.A. Hawkins: "Refuges as a Predictor of Parasitoid Diversity," *Science*, 973 (February 21, 1992).

Howell, D., J.T. Doyen, A.H. Purcell and B. Daly: *Introduction to Insect Biology and Diversity*, Oxford University Press, Inc., New York, NY, 1998.

Huffaker, C.B. and A.P. Gutierrez: *Ecological Entomology*, 2nd Edition, John Wiley & Sons, Inc., New York, NY, 1999.

Lockey, R.F.: "Immunotherapy for Allergy to Insect Stings," *N. Eng. J. Med.*, 1627 (December 6, 1990).

Mopper, S. and S.Y. Strauss: *Genetic Structure in Natural Insect Populations: Effects of Ecology, Life History, and Behavior*, Chapman & Hall, New York, NY, 1997.

Price, P.W.: *Insect Ecology*, 3rd Edition, John Wiley & Sons, Inc., New York, NY, 1997.

Romoser, W.S. and J.G. Stoffolano: *The Science of Entomology*, 4th Edition, The McGraw-Hill Companies, Inc., New York, NY, 1997.

Schmidt, J.: *Principles of Insect Physiology*, Chapman & Hall, New York, NY, 2000.

Wootton, R.J.: "The Mechanical Design of Insect Wings," *Sci. Amer.*, 114 (November 1990).

Web Reference

Entomological Society of America: http://www.entsoc.org/

INSECTICIDE. A substance that kills or interferes in the life cycle of certain insects and is thus useful for reducing and controlling insect populations. The reduction or elimination of such populations is desirable for several reasons. (1) Preventing the spread of certain diseases by insects that serve as transmitters or carriers of infective organisms. Thus, insecticides are used widely as a public health measure. (2) Preventing or reducing the damage caused by insects eating and inhabiting food plants, trees, and other crops. Such damage, without control, sometimes approaches 100% and averages 25% or more in some regions of the world. (3) Preventing or reducing physical property damage, notably of wood, cloth, and other materials of an organic nature that attracts certain insects as a source of food and place of habitation. Termites and moths are examples. (4) Reducing or eliminating discomfort, annoyance, and sometimes injury that results from the immediate presence of insects on or near people and domestic and farm animals. Ticks and face flies are examples.

Classification of Insecticides

Insecticides may be classified and characterized in many ways. (1) By their *selectivity*, i.e., their ability to control or not to control different forms, varieties, and species of insect, they range from compounds which have a rather *narrow control spectrum* and thus enable the user to eradicate or reduce selected target insects, all the way to *wide spectrum* insecticides that will destroy practically all insects, including those of a beneficial nature that are of positive economic importance. (2) By the manner in which insecticides *interact* with the insect enemy or target. That is: (a) whether or not a chemical requires overwhelming contact with the insect, or where the chemical acts systemically within and throughout the insect once local contact is made; (b) how the chemical interferes with the life process of the insect; for example, is it a stomach poison, or does it interfere with the insect's nervous or respiratory system, etc.; (c) how the insecticide enters the body of the insect, i.e., via the alimentary or respiratory system, etc.; and (d) whether or not the primary function of the chemical is to kill the insect, or essentially to sterilize the insect sexually and thus reduce insect population in this manner. (3) By application — is the insecticide a solid, liquid, or gas? Can it be sprayed, dusted, or incorporated into bait? Can it be worked into the soil? It is miscible with oil, water, or both? Can it be applied by aircraft? (4) By the useful life or persistence of the chemical, i.e., a few days, weeks, or months. (5) In terms of safety to humans and livestock and pets. Insecticide dangers range from highly toxic, to moderate, light, and low toxicity. Quite important, is the insecticide harmful to bees, fishes, other wildlife, or to adjacent crops and orchards should a spray or dust of the material drift away in the wind from the immediate point of application? (6) By the chemical structure of the insecticide, i.e., chlorinated hydrocarbon, organic phosphate, etc., and how produced—extracted from natural bacterial and biological materials or synthesized from basic raw materials and chemical intermediates. (7) By cost, a very important and practical consideration for large food producers, health authorities, and other users.

Selectivity and Spectral Range of Insecticides. There is no practical universal insecticide because such a chemical, to successfully eradicate all forms of insect life, would be dangerous to other life forms. There are, however, multipurpose pesticides, such as hydrogen cyanide gas, which are used *most carefully and with the ultimate of safety provisions* to kill not only all insects within a given area or space, but all other animal life forms as well. Some insects are much easier to destroy than others. Thus, numerous compounds are available to control aphid, fly, leafhopper, and trip, whereas only a few chemicals are effective against fire ant and certain beetle and weevil species.

Action of Insecticides. The average user of insecticides is essentially interested in the results of insecticide application rather than the exact manner in which these chemicals act upon the life and life cycle of the insect. But, there are some important distinctions that govern both the timing and selection of a given insecticide. For example, some chemicals are much more effective in destroying larvae or nymphs than when applied to insects in the adult stage, or vice versa. Also, some perennial crop plants and orchards are undamaged by certain chemicals (dormant sprays, for example) during winter inactivity, whereas the same chemicals would cause severe damage if applied during springtime budding. Also, the food producer, by taking advantage of multifunction compounds, can reduce the number of control chemicals required and hence the number of applications.

Control chemical manufacturers frequently offer combined formulations. In addition to blending acaricides with insecticides, molluskicides, etc., it is not uncommon to blend in herbicidal compounds. Usually, however, as the blending becomes more complex, the selection becomes more difficult, because great care must be taken to ensure the following: (1) That the chemicals will mix well without destroying any of their intended effectiveness (many control chemicals are weakened or destroyed, for example, if mixed with strongly alkaline materials, such as lime, or with sulfur-bearing compounds). (2) That the mixture is truly customized to the crop at hand, and that the combination is not used on other crops without specifically checking. And (3) that there will be no unexpected environmental damage. A combination of chemicals may be entirely effective and safe when applied to a specific crop or area, whereas it may be inappropriate if used on other crops and other areas that may be adjacent to pastures and streams or lakes and thus be dangerously polluting. For crops of large economic value and commonly grown, manufacturers offer numerous crop-specific formulations.

Nomenclature of Insecticides. It is estimated that there are well over 100,000 pesticide formulations and perhaps half of these would be basically classified as insecticides. Although the basic chemicals used may number in the hundreds, variations arise from the many thousands of possible formulations, not only combinations of materials, but formats (sprays, wettable powders, emulsifiable concentrates, dusts, baits, etc.). Added to this are the scores of control chemical manufacturers worldwide. Each manufacturer markets products under trade names—names that are essentially coined for their marketing charisma and infrequently connoting the content or purpose of the product. Thus, there are scores of equivalent (or essentially equivalent) products, adding to the difficulty of selecting these chemicals. Unfortunately, the generic chemical names of the majority of insecticide chemicals are long and complex and essentially meaningless to persons who are not well versed in organic and biochemistry. There are also a number of frequently revised directories of control chemicals and considerable information is available from various government agencies and universities. See list of references at end of this entry. This situation of nomenclature is quite similar to that which applies to generic and trade name drugs and pharmaceuticals.

Toxicity of Insecticides. Even though insecticides and other pesticides vary greatly in toxicity, all may be considered hazardous if they are not handled properly and with precautions. The following elementary rules are always worthy of repetition. (1) Observe all directions, restrictions, and precautions on pesticide labels. It is dangerous, wasteful, and, in some regions and countries, illegal to do otherwise. (2) Store all pesticides

behind locked doors in original containers with labels intact. (3) Use pesticides at correct dosage and intervals to avoid excessive residues and injury to plants and animals. (4) Apply pesticides carefully to avoid drifting of the compounds to nearby fields, lakes, and streams. (5) Bury surplus pesticides and destroy used containers so that contamination of water and other hazards will not result. (6) Certain pesticides must not be used during a specified period just prior to harvest because of the danger to workers and pickers who will be handling the product and because of the danger of inadequately washing away or otherwise removing residues prior to releasing the commodity to the consumer. (7) Do not mix two or more compounds without prior knowledge of their compatibility. (8) When in doubt concerning the applicability of a given compound to a given situation, such as a specific crop, seek advice from local sources of expertise, such as extension service representatives, reliable suppliers, neighboring users who have faced similar problems, regional colleges and universities, and agricultural experiment stations. Above all, the final responsibility for safe usage rests with the person who ultimately applied the chemicals.

In the United States, when registering insecticides, pesticides, and other control chemicals, regulatory agencies have been using acute LD_{50} values to determine the toxicity category and the words or symbols that must be placed on labels and containers. For this purpose, the test animals usually are rats, mice, or rabbits, but other mammals are sometimes used. The LD_{50} value is the dosage of the chemical at which 50% of the test animals are killed. It is based on the body weight of the animal and is expressed in milligrams of the chemical per kilogram of animal (mg/kg). One mg/kg = 1 part per million (ppm). Thus, the lower the LD_{50} value, the higher the toxicity of the chemical. The usual way of administering chemicals to test animals is by mouth, application to the skin, and in some cases, by inhalation. Toxicity may be either acute or chronic. Acute refers to rather quick action from a single exposure, whereas chronic refers to the toxic effect of many exposures over a period of time.

Chemistry of Insecticides

In classifying compounds used as insecticides, from the standpoint of chemical characteristics and structure, the most fundamental division separates the inorganic from the organic chemicals. The latter are by far the most widely used. Acaricides, bactericides, fungicides, and nematicides, along with insecticides, are included in the following descriptions.

Inorganic Compounds. These control chemicals are comparatively simple compounds and include calcium and lead arsenates, elementary sulfur and inorganic sulfur compounds, such as calcium polysulfide (lime-sulfur) and sodium thiosulfate. Because of their effectiveness against certain fungus infections, these compounds are more frequently considered as fungicides than as insecticides. This is also true of a number of copper, zinc, and other metal inorganics, such as copper carbonate, copper oxychloride, copper sulfate, and copper-zinc sulfate. It should be pointed out that Paris Green (cupric acetoarsenite), one of the older and once widely used inorganic insecticides, essentially has been phased out in most regions of the world because of the effects of chronic arsenic poisoning of workers and users who come in contact with the substance. This was particularly true in the case of vineyard workers a number of years ago.

A number of other metals, such as iron and tin, enter into insecticide and pesticide compounds, but as part of an organic chemical structure, as exemplified by triphenyltin hydroxide. Such compounds are sometimes referred to as organometallics (or, specifically in the case of tin, as organotins). Mercury compounds are rapidly being phased out because of their long-term toxic residual effects as pollutants, particularly of fresh and saline waters. Regulations vary from one country to another.

A few other inorganic chemicals, more frequently identified as multi-purpose pesticides than as insecticides, do have very strong insecticidal properties. These compounds are used sparingly, often requiring special permits in some places, and with the greatest of safety cautions observed. Such compounds would include calcium cyanide, carbon bisulfide, carbon tetrachloride, hydrogen cyanide, paraformaldehyde, and phosphine.

Organic Compounds. A listing of categories of organic chemicals used reads like the table of contents of an organic chemistry text, with relatively few subfamilies of organic compounds not represented in one way or other.

Alcohols. The open straight or branched chain (alkyl, aliphatic) saturated alcohols, such as methanol, ethanol, up through tetradecanol, etc., are not powerful insecticides, although they are somewhat more effective than the related hydrocarbons (methane, ethane, etc.). As is evidenced in other areas of activity, these alcohols increase in insecticidal effectiveness roughly in proportion to their molecular weight (number of carbon atoms in compound). This is an effect, however, which levels off when from 9 to 12 carbons (molecular weight of 144 to 186) are present in the chain. For example, although not a direct measure of insecticidal activity, nonyl (9 carbons) and decyl (10 carbons) alcohols are most effective in reducing the sprouting of stored potatoes. Beyond this point, a greater number of carbon atoms does not increase the effectiveness.

Unsaturated and Cyclic Alcohols. Although these compounds show somewhat stronger insecticidal effectiveness, as compared with the saturated aliphatic alcohols, this added strength is still not sufficient to warrant their serious considerations as insecticides. Some of these compounds, however, do make effective herbicides.

Aldehydes. The aldehydes possess greater insecticidal effectiveness than the alcohols, formaldehyde, for example, serving as a stomach poison. The compound also exhibits strong bactericidal and fungicidal activity. Formaldehyde tends to polymerize into paraformaldehyde, the pesticidal properties of which are considerably less than those of formaldehyde. The same increasing effectiveness with increasing molecular weight exhibited by the alcohols also applies to the aldehydes.

Metaldehyde, the polymer of acetaldehyde, is a widely used molluskicide and is effective in the control of snails. Some of the unsaturated aldehydes are more potent in their pesticidal and herbicidal effectiveness. However, only a few compounds are of commercial significance, notably acrolein and related compounds, which are used as aquatic herbicides in connection with water reservoirs and systems. Some aldehydes also have growth-regulating properties. See also **Plant Growth Modification and Regulation**.

Amines. This class of organic compounds also exhibits the same relationship between molecular weight and insecticidal activity as previously described. In the case of aliphatic amines, studies of toxic potency against house fly larvae have shown that the most effective compound is di-*n*-octylamine, with the compounds of higher or lower molecular weight in the series proving less toxic. Effectiveness also improves as one proceeds from the aliphatic amines to the aromatic amines, noting, for example, the greater toxicity of aniline as compared with hexylamine. Although *o*-iodoaniline and 2,5-dichloroaniline demonstrate some toxicity for caterpillar and louse, respectively, generally and perhaps surprisingly, inclusion of halogen atoms within the aromatic amine nucleus does not promote greater toxicity. Some toxicity increase is shown, however, when nitro groups are introduced into the nucleus. Diphenylamine, a diarylamine, is effective against lice and at one time enjoyed wide use for troops during wartime.

The insecticidal usefulness of the amines is hindered by their tendency to severely injure plants (phytotoxic effects) to which they may be applied. Advantage of this property, however, is taken by using some amine compounds as herbicides. Examples include benefin, nitralin, and trifluralin.

Carbonic Acid Derivatives. Mixed esters of carbonic acid display toxic potency as acaricides and fungicides, including their inhibiting action against powdery mildew. When sulfur is introduced into the structure, as in the case of the mixed esters of thio- and dithiocarbonic acids, the acaricidal and fungicidal toxicity is further increased. Derivatives of thio- and dithiocarbonic acids used commercially include carbon bisulfide (CS_2) and 6-methylquinoxaline-2,3-dithiocyclocarbonate (*Morestan*), an effective acaricide, fungicide, and insecticide.

Carbamic Acid Derivatives. A rather large number of commercially available food crop control chemicals fall into this category of organic compounds. Carbamic acid (or aminoformic acid) is NH_2COOH, but is best known in the form of its salts and esters. Several insecticides are found among the aryl esters of *N*-methylcarbamic acid, whereas the alkyl esters of *N*-arylcarbamic acids possess strong herbicidal powers, particularly in connection with undesired monocotyledonous plants.

The biological and physiological actions involved in the toxicity of the carbamates to animal and plant life processes are quite complex and not fully understood. It has been established that the esters of *N*-alkylcarbamic acids with insecticidal properties inhibit cholinesterase. Of the *N*-methylcarbamic acid ester series, the most powerful insecticidal

compound is 1-naphthyl-*N*-methylcarbamate, the basis for such commercially produced compounds as carbaryl, naphthyl carbamate, and Sevin. Other important carbamate-type acaricides, fungicides, insecticides, and nematicides include Aldicarb, Allyxycarb, Aminocarb, Bassa, Benomyl, Buffencarb, Carbendazim, Carbofuran, Ethiofencarb, Formetanate, Knockbal, Landrin, Mancozeb, Maneb, Meobal, Metacrate (Tsumacide), Metiram, Mexacarbate, Pirimcarb, Promecarb, and Propoxur (Baygon). Most of these names are proprietary.

Thio- and Dithiocarbamic Acid Derivatives. The derivatives of thiocarbamic acid are essentially herbicidal in nature. See also **Herbicide**. However, excellent nematocidal effectiveness is illustrated by the dithiocarbamic acid derivatives, notably by sodium N-methyldithiocarbamate (Vapam), which serves the multipurpose of not only eradicating nematodes, but many insects and weeds as well. It is frequently used as a soil sterilant.

In the case of the alkali metal salts of alkyldithiocarbamic acids, studies have shown that the fungicidal, nematocidal, and herbicidal effectiveness decreases as the length of the alkyl radical is increased. In terms of nematicidal activity, the esters of alkyl- and dialkylcarbamic acids is greater than the salts. Toxicity is greatest in the methyl and ethyl esters and decreased as the number of carbons in the ester radical increases.

Commonly available (check regulations) bactericides, fungicides, insecticides, and nematicides that fall into this category include: Carbothion, Eptam, Ferban, Nabam, Propineb (zinc bearing), TEC, Thiram, Zineb, and Ziram. Most of the foregoing names are proprietary.

Aliphatic Carboxylic Acids. Studies indicate a very low pesticidal activity for the aliphatic monobasic acids (acetic, propionic, etc.) and the dibasic carboxylic acids (oxalic, fumaric, maleic, etc.). As is usually expected, the pesticidal activity of the acids increases when halogen atoms are introduced to displace hydrogens in the alkyl radicals. Examples of this effect include the monohaloacetic acids, which are sometimes used commercially. The salt *calcium propionate* is used in bread- and cheese making as a preservative for its mild antibactericidal and fungicidal effects. Generally, the fluorine-containing derivatives are more toxic than those compounds with chlorine atoms. As can be expected, the unsaturated compounds exhibit greater toxicity than their saturated counterparts.

Alicyclic Carboxylic Acids. With exception of copper-bearing compounds, such as copper naphthenate, which is strongly fungicidal, the free alicyclic acids are not important as control chemicals. This is not the case, however, for a number of their derivatives. The natural pyrethrins and their synthetic analogs are included in this category. Among them are Allethrin, Barthrin, Bioallethrin, Cinerin I, Cyclethrin, Dimethrin, Furethrin; as well as Neopyanimin, Pyresy, Pyrexcel, and Pyrocide. Most of the foregoing names are proprietary.

The alicyclical carboxylic acid derivatives also include a family of growth-regulating compounds, known as the *gibberellins*. See also **Gibberellic Acid and Gibberellin Plant Growth Hormones**; and **Plant Growth Modification and Regulation**.

The alicyclic carboxylic acid derivatives have been studied extensively and, to date, with the exception of the compounds already mentioned, relatively few have been found to possess commercially important potential as pesticides. An exception is dimethyl carbate (Dimelton), which finds use as a repellent for certain blood-sucking diptera. The compound frequently is mixed with other pesticides.

Aromatic Carboxylic Acids. As in the case of aliphatic and the alicyclic carboxylic acids, the free aromatic acids (benzoic, naphthenic, etc.), their halogen and nitro derivatives, as well as their alkali metal salts, possess a lower insecticidal effectiveness. However, for some species of mite, the benzyl ester of benzoic acid is very effective. As a general rule, the incorporation of chlorine or other halogen atoms in the benzoic acid and benzyl alcohol configurations accentuate the biological power. Introduction of chlorine into the para position of the benzyl radical appears to provide a maximum effect against both egg and adult mite. Further enhancement of acaricidal potency is obtained by the presence of amino, hydroxy, and nitro groups. For codling moth and body louse, the aliphatic esters of anisic, anthranoic, and salicylic acids have toxic effectiveness. Chlorobenzilate is an effective selective acaricide, useful against numerous species of mite, including the tracheal mite which is parasitic to honeybee.

Numerous other aromatic carboxylic acids possess some bactericidal, fungicidal, and herbicidal characteristics, as well as growth-regulating properties. Salicylanilide, the amide of salicylic acid, is effective against leaf mold and tomato brown spot.

Heterocyclic Compounds. Quite a large number of insecticides and other pesticides as well as herbicides are heterocyclic compounds. The variations pertaining to structure and composition are so many that this is strictly a generalized, umbrellalike classification. For the biochemist, organic chemist, entomologist, or other professional concerned in the development and theoretical aspects of control compound chemistry, other more detailed classifications are required. A first step toward this was undertaken by Melnikov. See list of references. Among important fungicides and insecticides in this classification are copper quinolate and Phenazim fungicide.

Aliphatic Hydrocarbons. After extensive research into the biological activity of the aliphatic hydrocarbons, relatively few of the pure (non-derivative) compounds have been found worthy of commercial attention. Popular for use in orchard spraying are the petroleum derivative oil sprays, which possess a good combination of acaricidal and insecticidal activity with low phytotoxicity. These sprays are effective against San Jose scale and mite.

Ethylene. for a number of years, has been used to hasten the ripening of some fruits. Although difficult to apply, ethylene is also an excellent defoliant. The insecticidal and nematocidal effectiveness of the halogenated aliphatic hydrocarbons is in proportion with the general chemical activity of these compounds. A number of these compounds have been used in the form of fumigants, notably in treating stored commodities and storage areas. Some of these include methyl bromide and DD pesticide.

Aromatic Hydrocarbons. Compounds in this category, such as benzene, naphthalene, xylene, etc.) have undergone extensive investigation. Although many of their derivatives are important, the pure compounds find little if any pesticidal use. Some of them, however, and in particular the xylenes, are used as solvents for carrying other control chemicals. For a number of years, naphthalene did enjoy large usage as a control agent against moth species. This has largely been replaced by synthetic compounds.

Halogen derivatives of benzene vary considerably in acaricidal and insecticidal toxicity, depending upon type, location, and number of halogen atoms introduced. Bromine appears to impart maximum effectiveness, followed by chlorine and fluorine. Biological activity increases with halogen atom introduction up to a total of three such atoms. A larger number tends to decrease effectiveness. Dichlorobenzene is more powerful than hexachlorobenzene. Loading a compound with bromine exhibits a greater effect than chlorine in reducing effectiveness.

Although there are eight stereoisomers of benzene hexachloride, only one of these, 1,2,3,4,5,6-hexachlorocyclohexane, is important commercially.

DDT (1,1,1,-trichloro-2,2-*bis* (*p*-chlorophenyl)ethane), now banned in many countries, is a derivative of an asymmetrical diarylethane and an effective insecticide. Paradichlorobenzene is an effective multipurpose pesticide. The compound is useful against sugarbeet weevil and in the control of phylloxera.

Ketones. Because of their rather weak insecticidal effectiveness, the ketones are used mainly as solvents for control chemical formulations.

Mercaptans. The aliphatic mercaptans with four or fewer carbon atoms have rather powerful insecticidal properties and can be used as fumigants against certain insect species. This is not true of those compounds containing over four carbon atoms; and also not true of aromatic mercaptans.

The control chemical interest of the mercaptans essentially is in the derivatives that incorporate chlorine or bromine. Methyl mercaptan rivals hydrogen cyanide as a powerful and useful fumigant. This compound is also an important intermediate in the synthesis of Captan and Folpet fungicides.

Some of the closely related organic sulfides and thioacetals have found commercial pesticidal use. These formulations include Mikazin, Fluoroparacide, and Fluorosulfacide. Names are proprietary in most cases.

Nitro Compounds. Biological activity of organic nitro compounds compares favorably against their pure (nonderivative) hydrocarbon compounds. However, this activity is considerably enhanced in terms of those nitro compounds that contain one or more halogen atoms. Bromine imparts a greater toxic power than chlorine. Many of the halonitro compounds command a wide spectrum of functionality, ranging from acaricidal, bactericidal, fungicidal, herbicidal, insecticidal, and nematicidal attributes. Some of their action is ascribed to their strong oxidizing properties. Examples of effective nitro compounds include chloropicrin and several related compounds, such as dichloronitroethane and chloronitropropane, as well as Binapapacryl acaricide-fungicide, Dicloran fungicide, Dinobuton acaricide, Dinocap acaricide fungicide, and Dinoterb acetate pesticide. Most of these names are proprietary.

Mercury Compounds. The powerful biological activity of mercury in simple compounds, such as the inorganic mercuric chloride, particularly against molds and other bacterial and fungus infections, has been recognized for generations. Also, the toxicity not only to microorganisms, but to higher animal forms also has been known and of major concern for a long time. Because of emphasis on environmental factors and safety during the past few decades, mercury-containing chemicals have been undergoing a phaseout in many countries. In connection with mercury compounds, it is interesting to note that many of these compounds have a good chemotherapeutic rating (or index), i.e., the dosage required to control a plant disease organism is many, many times smaller than the dosage that would be harmful to the plant. Some mercury compounds stimulate plant growth and yield.

Tin Compounds. Several organotin compounds are quite biologically active. Some of the simple inorganic tin salts, such as stannous or stannic chloride, have little if any pesticidal value. The fungicidal effectiveness is achieved by substituting alkyl or aryl groups for the chlorine atoms. A peak of insecticidal activity is achieved with the trialkyl- and triaryltins. It is interesting to note that the tetraalkyl- and tetraaryltins are essentially ineffective. Research has indicated that tributyltin chloride and tributyltin fluoride are the most active of the possible combinations. Most popularly used (check regulations) commercially are triphenyltin hydroxide and triphenyltin acetate.

Copper Compounds. Principally used as fungicides, copper inorganic compounds are widely applied. Of the organic copper compounds, the most commonly used are copper linoleate, copper naphthenate, and copper quinolate.

Zinc Compounds. Zinc is associated in a number of organic pesticides, notably Propineb fungicide, Zineb fungicide, and Ziram fungicide. Names are proprietary.

Phenols. As compared with the aliphatic alcohols, the phenols are more active biologically, but even with this greater activity, most of these compounds are not of practical commercial importance. Introduction of halogen, nitro, thiocyano, and some other groups increase their activity, the nitro group appearing to have the greatest insecticidal power. Included among these compounds are some that date back 50 years or more — dinocap (Karathane) acaricide-fungicide and dinobuton (Dinoseb) acaricide. The phenols tend to severely burn plants, a property that led to their use, commencing in the late 1930s, as contact, selective-type herbicides and desiccants.

Phosphorus Organics. Possibly the most extensive of all categories of organic compounds used as control chemicals, the organophosphorus compounds, are derived from the inorganic acids of phosphorus. It is estimated that the very fundamental compounds in this category number well over one hundred and that the commercial formulations resulting may number into the several hundreds. For study it is sometimes convenient to classify these compounds as derivatives of (1) phosphorous acid, H_3PO_3; (2) phosphoric acid, H_3PO_4; (3) thiophosphoric acid, $PS(OH)_3$; (4) pyrophosphoric acid, $H_4P_2O_7$; and (5) phosphonic acids (phosphine = PH_3).

Phosphorous Acid Derivatives. The principal control chemical potential of these organic phosphite compounds lies with their abilities as herbicides. Acaricidal, fungicidal, insecticidal, and nematicidal powers are comparatively weak. But, as is nearly always the case, the toxic potential increases with many different kinds of derivatives that can be prepared. Commercial products based upon derivatives include: DDVP (Dichlorvos) acaricide-insecticides; *tris*-(2,4-dichlorophenoxyethyl) phosphite (Falone); Gestid (Mevinphos, Phosdrin) acaricide-insecticide; Naled (Dibrom) acaricide-insecticide; and Phosphamidon acricide-insecticide. Most of these names are proprietary.

Phosphoric Acid Derivatives. The biological activity of the phosphates is considerably greater than the phosphites and notably among the mixed esters of phosphoric acid where one of the ester radicals is acidic. The toxicity of the resulting derivative is roughly proportional to the dissociation constant of the parent alcohol, phenol, or acid, the toxicity decreasing as the dissociation constant decreases. Research on the phosphoric acid mixed esters shows that the methyl derivatives are the most toxic. The toxicity of these types of compounds is believed to be the result of (1) high alkylating potential as regards certain biological nitrogen and sulfur constituents; and (2) elevated rates of hydrolysis.

Numerous proprietary examples of the derivatives of phosphoric acid include: Bromophos insecticide; Chlorpyrifos (Dursban) insecticides; Demeton (Mercaptophos) acaricide-insecticide; Diazinon acaricide-insecticide; Fenitrothion insecticide; Fensulfothion insecticide-nematicide; Fenthion acaricide-insecticide; Kitazin fungicide; Methyl Parathion (Metafos) acaricide-insecticide; Oxydemeton-Methyl acaricide- insecticide; Parathion (Thiophos) acaricide-insecticide; Vamidothion acaricide-insecticide; and Zytron insecticide. Most of these names are proprietary. Some of the compounds are banned in some countries.

Thiophosphoric Acid Derivatives. A fortunate combination of characteristics occurs in the phosphoric acid derivatives upon substitution of a sulfur atom for one of the oxygens of the parent compound. Namely, the toxicity of the derivatives to higher forms of life is substantially diminished, while at the same time, the acaricidal and insecticidal powers, with few exceptions, remain strong. Derivatives of thiophosphoric acids may feature a thiono or a thiolo (most toxic) structure. Commercial preparations are usually mixed esters of thiophosphoric acid, as well as of dithio- and trithiophosphoric acids. The trithio compounds usually are markedly less effective than the dithio compounds.

Some commercial formulations based upon derivatives of the dithio-phosphoric acids include: Azinphos-Methyl (Guthion) acaricide-insecticide; Carbophenothion (Trithion) pesticide; Dimethoate acaricide-insecticide; Disulfoton acaricide-insecticide; Malathion (Carophos) acaricide-insecticide; Mecarbam acaricide-insecticide; Menazon acaricide-insecticide; Phorate acaricide-insecticide; Phosalone acaricide-insecticide; and Phosmet (Imidan, Phthalodophos) acaricide-insecticide. Most of these names are proprietary. Some of these compounds are banned in some countries.

Pyrophosphoric Acid Derivatives. A pioneer among the phosphorus-containing organic control chemicals, tetraethyl pyrophosphate (Bladen, TEPP), was developed by Bayer AG in Germany in the early 1940s. The pyrophosphates are powerful contact-type acaricide-insecticides that have little or no tendency to function systemically. In addition to TEPP, some of the commercial formulations in this category include NPD, Pirophos, Schraden, and Sulfotepp. Most of these names are proprietary. Some of these compounds have been banned in some countries.

Phosphonic Acid Derivatives. Very important in this category of proprietary compounds are Trichlorfon (Chlorophos, Dipterex, Dylox) and Trichloronate.

Other Organic Bases. *Quinones.* As compared with the alcohols and aldehydes, the biological activity of the quinones is greater, notably in their actions as fungicides. Although benzoquinones exhibit relatively low activity, the presence of halogen and hydrocarbon radicals in the ring structure, as is true of many other organic compounds, significantly increases the effectiveness. Some of the derivatives of the quinones used commercially include tetrachlorobenzoquinone (*Chloranil*), which is particularly effective for disinfecting seed, and 2,3-dichloronaphthoquinone-1,4 (Dichlone, Phygon).

Sulfonic Acid Derivatives. The sulfonic acids, including their salts, have found value as agents for treating wool fabrics against species of moth for a number of years. One of these products is Eulan, a number of variations of which have been produced. Mitin-FF, a derivative of urea, also has been used in this way. To date, the free sulfonic acids have not played an important role as insecticides for crops.

Impressive activity against mite larva and egg has been shown by some of the aromatic esters of arylsulfonic acids. Some of the commercial formulations in this category include CPCBS (Chlorofenson, Ester Sulfonate, Ovex, Ovotran) acaricide.

Thiocyanates and Isocyanates. The intense biological activity of hydrogen cyanide has been known for scores of years. Similarly, derivative compounds have been known and used for many years. The derivatives of thiocyanic acid are particularly powerful fungicides and pesticides. Research has indicated that the straight-chain thiocyanates are more effective than those with branched chains and, as can be expected, introduction of halogen atoms into the compounds increases their toxicity. Some of the commercial formulations that fall into this overall category include Thiophanate fungicide; and Thiophanate Methyl fungicide. Regulations over usage must be checked.

Urea and Thiourea Derivatives. The value of urea as a fertilizer is well known. See also **Fertilizer**. However, the most elementary derivatives of urea show strong phytotoxic effects. Thus urea derivatives are widely used as herbicides. A number of derivatives of thiourea also show strong bactericidal and fungicidal activity. In this category are found Dodine fungicide and Guazatine fungicide.

Bacterial and Botanical Compounds as Insecticides and Control Chemicals

One of the long-established and better known of the botanical compounds is pyrethrum or pyrethrin insecticide. Known since the early 1800s, the active ingredient of this formulation is obtained from pyrethrum plants found in Africa and South America. Pyrethrin is considered to be one of the safest insecticides, but it is costly. The first synthesis of allethrin, the analog of pyrethrin, was developed in the late 1940s and is now widely used.

A much more recent natural insecticide, developed in the early 1960s, is Bacillus Thuringiensis-Berliner and, as indicated by the name, the compound is developed from living spores of the *Bacillus thuringiensis*, a bacterial strain that causes disease among certain types of insects.

Other naturally derived commercial insecticides include Evisect, Hellebore, nicotine sulfate, and Sabadilla insecticide.

An interesting concept of using specific antibodies as a potential insecticide has been proposed by Nogge, Giannetti, and associates at the Institut für Angewandte Zoologie (Bonn, Germany). See reference listed. It has been learned that many insects are able to absorb orally administered antibodies. The researchers found that when tsetse flies are fed on human blood, the hemolymph of the flies contains human albumin. Then, if the flies ingest antibodies to human albumin, they perish within a short time. It is observed that the albumin fraction in the insect's hemolymph disappears and osmoregulation is severely disturbed. Thus, antibodies may be used as a biological insecticide.

Additional Reading

Ishaaya, I.: *Biochemical Sites of Insecticide Action and Resistance*, Springer-Verlag, Inc., New York, NY, 2000.

Jones, D.G.: *Piperonyl Butoxide: The Insecticide Synergist*, Academic Press, Inc., San Diego, CA, 1998.

McKenzie, J.A.: *Ecological and Evolutionary Aspects of Insecticide Resistance*, R.G. Landes Company, New York, NY, 1996.

Melnikov, N.N. (Gunter, F.A. and R.T. Huber): *Chemistry of Pesticides*, Springer-Verlag Inc., New York, NY, 1971.

Mullin, C.A. and J.G. Scott: *Molecular Mechanisms of Insecticide Resistance: Diversity Among Insects*, American Chemical Society, Washington, DC, 1992.

Narahashi, T. and J.E. Chambers: *Insecticide Action: From Molecule to Organism*, Plenum Publishing Corporation, New York, NY, 1990.

Nogge, G. and M. Giannetti: "Specific Antibodies: A Potential Insecticide," *Science*, **209**, 1028–1029 (1980).

Page, B.G.: *Insecticide, Herbicide, Fungicide Quick Guide, 2000*, Thomson Publications, Washington, DC, 2000.

Staff: *Insecticide and Acaricide Tests: 1993*, Entomological Society of America, Lanham, MD, 1993.

INSECTICIDE AND PESTICIDE TECHNOLOGY. The technology of controlling pests, notably in the area of food production, is undergoing serious examination and reevaluation.

Chemicals are and have been the main weapon for controlling agricultural pests for well over a century and will continue to be important for the foreseeable future. But within the last 20 years and notably the last decade, the total chemical approach to pest control has been subject to questioning and alternative approaches have been sought.

It no longer can be taken for granted that progress in pesticide technology will be confined to the research and development of new and improved chemicals. Progress in pesticide chemistry will continue to be important, but for the long term, actions commenced just a few years ago toward development of a *total systems concept of pest management* may represent the technology of the future. In the long term, this new technology may provide more effective pest control, coupled with a progressively lessened dependence upon chemicals.

A Retrospective View. When an important technology is at a crossroads, it is in order to glance back in an effort to clarify the forces which are bringing about a major change in direction. As early as 1828, it is reported that Persian (Iranian) farmers used pyrethrum (obtained from *Chrysanthemum coccineum*) as an insect control on certain farm crops. Rotenone, another naturally- derived organic chemical, has been used for pest control for over a century. Bordeaux mixture (copper sulfate and hydrated lime) has been used as a fungicide since the early 1880s. Inorganic arsenic compounds were used in German vineyards at the turn of the century and not banned until 1942.

Prior to the period just preceding, but mainly following World War II, pesticide chemicals were either inorganics or naturally derived and extracted organic compounds. Although there was an early awareness of the poisonous nature of most of these compounds, there was not an immediate connection made between poisoning insects and other pests during the growing period of a crop and possible poisoning of persons who might consume the produce after harvest. During this period, most likely it was generally assumed that the chemical insecticide would be confined to the foliage surfaces of the plants, later to be washed off by rain and in preparing produce for market. And, of course, it is true that a number of these compounds are contact-type pesticides, that is, their area of influence is confined to the surfaces to which applied. There was little, if any, understanding or consideration of possible systemic actions of such toxic materials, that is, the absorption of the poisons by the plant. Poisons were transported throughout the plant by its vascular system and thus residuals remained in edible parts that, considering the analysis techniques then available, were extremely difficult to assay in minute quantities, even if suspected.

Numerous important findings of biochemistry, microbiology, and human and plant physiology were still unknown. The concepts of slow, prolonged poisoning processes were unknown and/or unappreciated. Based upon suspicions of arsenic-caused deaths, it was as recent as 1938 when a laboratory in Speyer, Germany studied 336 samples of wine bottled for sale that year and found to contain as much as 14.4 milligrams of arsenic per liter. Earlier vintages were found to contain as much as 24 milligrams per liter. It was later concluded that many vineyard workers, who regularly consumed "house wine" prepared from grape skins, succumbed to cancer of the liver after a latent period ranging from 10 to 35 years. Even in 1972, some 30 years after arsenic use in the vineyards of the Mosel and Kaiserstuhl regions had been banned, oldtime workers of the vineyards were expiring from arsenic-induced liver cancer.

Further, little was known and/or appreciated pertaining to the poisonous nature of the metabolites (compounds resulting from digestive processes) produced from certain pesticide chemicals, particularly those of an organic nature. Only during the last 40 to 50 years (a comparatively short period in terms of the total history of chemical pesticides) has the persistent nature of some chemicals been appreciated. This is also true of awareness of the large capacity of soils to retain chemical residuals for many years. When these two factors are coupled, excessive concentrations of chemicals can be built up over a period of years.

During the pre-World War II period, the world population was much smaller than today and the concentration of agricultural operations much less. Thus, the effect of certain pesticidal chemicals on birds, beneficial insects, livestock, fishes, and other forms of wildlife was less discernible. There certainly were various forms of warning, but these were noted only by comparatively few people. There was no popular awareness and concern as regards environmental problems.

A major alteration in insecticide and pesticide technology occurred as the result of what might be called a tremendous expansion in the field of organic chemicals. Pre-World War II efforts to synthesize dyes from coal tar chemicals were among the first efforts toward commercially expanding organic chemical synthesis. These efforts were shortly followed by the wartime needs for synthetic rubber, improved aircraft fuels, and an improvement of the earlier resins for the manufacture of plastic substitute materials for a host of applications. Knowledge and experience in organic chemistry multiplied several-hundredfold during the postwar period and

spawned the petrochemical industry. Tens of thousands of new, previously unknown organic compounds were produced, for which in many instances uses had to be found.

Further expansion of commercial organic chemicals was greatly aided by the addition of natural gas as an almost ideal raw material. Although used for many years, a booming natural gas industry did not develop until shortly after World War II. The 1930s, 1940s, and 1950s became the age of miracle chemicals — with the introduction of new fibers, new plastics and resins, new coatings, and, importantly, new chemical pesticides. Mass production made it possible to produce many new chemical pesticides at a relatively low cost and their convenience in application was widely accepted by food producers. And the period was essentially without any major worries concerning possible deleterious side effects from their use. This period was also marked by a rapidly expanding worldwide population and greatly expanded food-producing operations. During this period, a great chemical pesticide industry and distribution system was created and, even more importantly, many food producers became highly dependent upon the use of pesticide chemicals. Many earlier farming practices were discarded in favor of wider application of chemicals and, at that time, apparently all for good reasons. Chemicals greatly reduced crop losses to insects and other pests, thus increasing effective yields from a given unit of land and labor. Chemicals still offer these advantages and, as of the late 1980s, this is still the general mode of food production operations in countries with advanced technology. The chemical trend, in fact, was again markedly accelerated a few decades ago by the introduction of herbicides, which, in turn, led to the concept of "no-till" farming.

The Apex of Conventional Chemical Pesticides. The wide acceptance of conventional chemical pesticides, not only by food producers for reasons previously given, but also by society in general, is exemplified by DDT (dichlorodiphenyltrichloroethane). This organic chemical was developed during the early 1930s by Paul Müller, a Swiss research scientist. In the early years of World War II, people were advised to use it without reservation on food and fodder plants, since it was said to be entirely harmless to warm-blooded animals. DDT was welcomed not only as an excellent control chemical for food production, but also as a public health measure against mosquitoes and other annoying and sometimes dangerous insects. Household preparations containing DDT and other related organic compounds were widely sold. Thus DDT in its early days exemplified the best of technology — a scientific breakthrough accompanied by completely positive economic and social benefits, as witnessed by the award of a Nobel Prize in 1948 to Müller.

Knowledge of the true nature of DDT was very slow in arriving. Not until the early 1950s did scientists at the U.S. Department of Agriculture find that although fodder treated with DDT caused no damage to the cows eating it, the health of their calves was severely impaired, sometimes with fatal results. The DDT was being passed along from cow to calf via the milk. These findings were confirmed in 1953 by experiments sponsored jointly by Swiss universities and pesticide manufacturers. The Swiss experimenters found that about one-tenth of the DDT sprayed from aircraft to control the May beetle settled on and was retained on the surface of pasture grass. Again, there was no apparent damage to the cows eating the grass, but their calves suffered the same effects as those previously described in the United States. It was also learned that the damage was principally to the nervous system of the animals.

In 1972, a group of German scientists discovered that a conversion product of DDT, a metabolite known as DDD, has a mutagenic effect. Some 30 or 40 years earlier, instrumental analytical techniques were not available to detect small residuals of highly complex organic chemicals and, further, the knowledge of human biology and biochemistry was but a fraction of that knowledge amassed during the period after World War II. In a way, possibly, the introduction of DDT marked the apex of conventional chemical pesticides, certainly not in terms of tonnages of chemicals produced, but in terms of their apparently trouble-free acceptance by society. Numerous chemical pesticides have been banned since the banning of DDT (in several countries) and the banning and tight governmental regulating trend continues at a rapid pace. In fact, some authorities believe that possibly the present period of questioning and suspicion of chemical pesticides may produce a net deficit for society. That is, society, on the one hand, might be fearful of the long-term effects of pesticidal chemicals, but, on the other hand, having learned to depend upon chemical pesticides for many years, may need all the assistance it can get from modern technology, including chemical pesticides, to feed an ever expanding world population.[1]

Possibly, the root of the fundamental biological problems resulting from widespread application of modern chemical pesticides extends back to the previously mentioned great expansion of organic chemicals that occurred during and just after World War II. With present knowledge of molecular biology and of proven carcinogens, although admittedly this knowledge is still extremely limited, it is almost certain that much more concern would have been expressed pertaining to the public release of many of the chemicals in use today. But once an entire business of chemical pesticide supply is established and once an industry (food production) takes on certain performance patterns, change becomes extremely difficult, particularly when satisfactory substitutes are not in view.

System Concept of Pest Management

The system approach to pest control involves not only a total look at the target pests and the plants to be protected, but an investigation of all elements which make up what might be termed a crop ecosystem. Particular emphasis is placed upon the interactions of all elements; this emphasis, of course, involves a study of all feedback and any feedforward loops that may be present in the system. In essence, the systems approach represents applied ecology, which is basically a system-oriented science, but with the addition of numerous specialist viewpoints — physiology, biochemistry, engineering, entomology, botany, agronomy, economics, meteorology, and climatology, among others.

The long-range of objectives of the system approach are several: (1) to control food crop pests more effectively than is possible with chemical insecticides alone, even when these chemicals are used in excessive dosages; (2) to take full advantage of all natural factors that may act against the pests and in the favor of the plant; (3) to find new methods for combating high pest populations; (4) to improve crop yields as the result of optimizing favorable growth conditions; and (5) one hopes, to reduce the total cost of pest control by lowering the amount of control chemicals required for equivalent or better results.

It is obvious, of course, that the food producer cannot bring all the aforementioned skills together as they may pertain to a given crop ecosystem and make individual decisions as what to do next and when to do it in an effort to control pest populations. How is such very specialized information gathered in the first place? How is such information interpreted and translated into actions for the individual food producer?

The amount of data to be gathered just in the interest of applying the system approach to a limited number of major crops is staggering. A pioneering program developed by the Purdue University Agricultural Experiment Station (West Lafayette, Indiana) may, at some time in the future, point the way toward substituting information for hard chemicals in dealing with insect damage to crop plants. A program of this type requires the interest and intellectual cooperation of crop producers rather than simply depending upon chemical overkill of plant enemies.

In the early Indiana program, approximately one hundred alfalfa growers participated who essentially were dairy farmers who rely on alfalfa as the major source of protein for their herds. One of the initial tasks required was, to assess the level of alfalfa management practiced by program participants.

This was accomplished by developing a 40-item questionnaire, which each cooperator completed and returned. A considerable variation in statewide insecticide and herbicide practices was found.

In developing the early data bank for the system, many actions were required: (1) collection of insect, alfalfa plant, and weed data from other cooperator's fields once each week for a season; (2) collection of weather data from appropriate agricultural weather stations on a daily basis, to be used as input into the alfalfa plant and weevil models to be developed later; (3) comparison of model output with actual grower field data and to make specific pest management recommendations based upon current insect, plant, and weather conditions; (4) input and storage of files of individualized alfalfa pest management advisories on a central computer; and (5) utilization of these data by cooperative extension agents by dissemination of advisories to the cooperating growers, using telephone, local radio station farm broadcasts, etc. Weather, obviously, plays a very important role in any pest management system. Thus, the Purdue program involved an excellent meteorological observing, forecasting, and communicating system.

Crop Ecosystem. What constitutes a crop ecosystem? This will vary considerably from one region within a state to another and probably can be determined practically only through making many observations. But it is possible, for example, that pest control information dispatched to producers in the southern half of one county may be applicable to producers of the same crop in the northern half of the adjoining county. In the long run, the system may take form of daily pest management advisories to food producers, possibly over local radio stations that for years have been featuring farm news of all kinds. Such an advisory might read along the following lines:

"Alfalfa weevil larval populations are on the increase and have reached sufficient levels in many cooperators' fields in southern Indiana to eventually result in economic losses. The alfalfa in these fields is still relatively short, averaging 4.6 inches and should (should not) be treated at this time. For growers in (such and such) areas, insecticide application should be delayed until more larvae have hatched so that a greater number can be controlled. Delay insecticide application for 7 to 10 days from this advisory. For growers in (such and such) areas, there are sufficient larval numbers and spraying should commence in accordance with previously given schedules."

Central to some pest management systems is a computerized simulator which, based upon the analysis of hundreds of past observations and experiments, can accept current weather information, for example, and read out the effects of the weather parameters. It thus provides directions for whatever pest control actions should or should not be taken at any given time. In essence, the simulator takes the place of numerous observers in the field and enables an information center to pass along directives in real time. A number of factors in addition to weather information, of course, can be input into the system. Needless to say, if such a network were established, all manner of other information pertaining to the crop ecosystem could be handled in addition to pest management data.

Such a system should be contrasted with the pest management means available today. Pest control information comes to the food producer from a number of sources: (1) pesticide chemical suppliers, who usually provide fundamental information on the application of their product, including precautionary information, safety measures to be taken, timing directions, etc.; (2) local extension personnel; and (3) special booklets, bulletins, etc., issued by state agriculture departments, giving specific directions as to types of acceptable pesticide chemicals, signs to watch for on the plant to diagnose pest conditions and stages, and general counsel pertaining to the timing of initial and repeat applications of chemicals — together, of course, with safety and precautionary information. But the prime limitation of this kind of information system is that it is not dynamic, it does not function in real time. Inputs of new information may range from one to several years and thus such bulletins do not always reflect the latest in chemical pesticide technology. And possibly the greatest weakness is the fact that this kind of information is based upon traditional pest control methods, as contrasted with the concepts of dynamic pest management.

Biological Pest Control Methods

The role of chemicals generated by insects in affecting the normal metabolism of plants has been known for many years. Less understood has

---

[1] A February 5, 1976 report of the study committee on pest control of the National Academy of Sciences expressed concern that "future agricultural productivity is threatened by a possible breakdown in chemical control of pests." Factors leading to this conclusion included: (1) the appearance of genetic resistance among "target" insect pests; (2) disruption of natural pest control mechanisms when beneficial insects as well as target pests are killed by a chemical compound toxic to a broad spectrum of insect life (the committee gave as an example the use of an organophosphate insecticide by California cotton growers for controlling the lygus bug that also kills certain predators which normally control the bollworm, a late-season pest); and (3) the effect of increasing constraints by laws and regulations on much needed chemical pesticides. Although not decrying the new laws and regulations, the study committee indicated that such regulations make more difficult and expensive the introduction of new pesticides to replace those already banned so that, if pesticide developers and suppliers are sufficiently discouraged, a serious gap in availability of effective control chemicals could result.

been the role of chemicals generated by plants on the metabolic processes of insects. Research is beginning to demonstrate that some plants possess surprising chemical defense against attack by insects and other pests. Such defense chemicals operate in a variety of ways, and it has been mainly during the past 10 to 15 years that operation of these chemical defenses has been explained in a rudimentary way.

Pheromones. Some entomologists believe that insect pests may be controlled economically with a minimum of environmental disruption by exploiting the hormones and pheromones by which an insect regulates its growth, development, and behavior. Pheromones may be defined as chemicals which are secreted by one insect that affect the behavior of other individuals of the same species. Pheromones evoke several behavioral responses, but the sex-attractant pheromones are those most frequently mentioned by entomologists. It is believed that inasmuch as pheromones are natural substances the insects may be less likely to develop a resistance to them than to some synthetic organic insecticides. However, entomologists point out that insects are quite adaptable and that a change in their pheromones is a possibility.

Three approaches in the use of pheromones have appeared in the literature: (1) use of traps baited with sexual attractant material as a means for monitoring the infestation of areas with select insects; (2) similar use of traps, except on a massive scale, to attract males (female sex pheromone used as bait); and (3) "male confusion" technique, in which female sex pheromone is permeated in the air, frustrating the attempts of males to locate females.

The use of pheromones is particularly attractive because of the high selectivity of the method, enabling the destruction of pests by way of large reductions in future populations and doing this without interfering with the normal life and habits of beneficial insects.

Juvenile Hormones. These are organic chemicals that are present in insects during the greater part of the insect's development. It is only during metamorphosis (period when a larva changes into an adult) that these chemicals are absent. When juvenile hormones are applied to insects during metamorphosis, the adults produced are deformed and lack the capacity for further development and soon die. Because juvenile hormones are relatively simple compounds, synthetic analogs are not too difficult to prepare and thus can be used as effective insecticides. However, timing if very critical because effectiveness is limited to the relatively short period of metamorphosis. If applied before or after this period, the compounds are essentially ineffective. Wide use of these chemicals could become practical if tied into a computerized pest management system control center as previously described.

Antiallatotropins. As part of their defense mechanism, some plants contain chemicals with juvenile hormone activity. Plants also have been found that contain chemicals with antijuvenile hormone activity, known as antiallatotropins. Although still not fully understood, it is assumed that the biological control system of the plant distinguishes between the metamorphosis period of an insect (during which time the juvenile hormones would be used as weapons) and the other periods of the insect's life cycle (during which time juvenile hormones are required by the insect, hence use by the plant of the antijuvenile or antiallatotropin compounds). Two antiallatotropins have been isolated from a common bedding plant (*Ageratum houstoniatum*). Chemically, these are 7-methoxy-2,2-dimethylchromene and 6,7-dimethoxy-2,2-dimethylchromene.

Phytoalexins. First reported by K. Müller in Germany in 1940, phytoalexins are lipidlike chemicals that are synthesized by some plants. Research indicates that these compounds are toxic to fungi and bacteria, as well as some other pests. It has been found that the chemicals are produced as the result of an attack upon the plant and an analogy between these compounds and inteferon (an antiviral substance produced in humans in response to a viral infection) has been suggested. To date, about 100 phytoalexins have been isolated. One of their roles is believed to be prevention of germination of fungus spores.

Research by K. Uehara at the University of Osaka, Japan, in 1958 produced the concept of elicitors. Since then, at the University of Colorado (Boulder), the first phytoalexin elicitor was isolated. This was obtained from filtrates of cultures of *Phytophthora megasperma* var. *sojae*, a fungus that attacks soybean. This compound stimulates the accumulation of the phytoalexin glycerollin, previously characterized by research at the University of London.

Viruses. There are epidemics caused by viruses, which occur periodically in the insect populations and thus naturally help to check their spread. Entomologists would like to find ways of infecting pest insects before they can cause serious damage. Most insect viruses are specific for a few closely related hosts. The general structure of the viral particles includes for DNA plus protein, all imbedded in a protein matrix. They are termed nuclear polyhedrosis virus (NPV). The viruses spread when larvae eat contaminated foliage.

Apparently viral infections in the past have helped to control the population of the Douglas-fir tussock moth. This insect undergoes population explosions at intervals of about 10 years. The outbreaks may last up to 3 years, during which time severe damage results. DDT helped to control the tussock moth population before it was banned. Attempts are now being made to combat a current outbreak of tussock moths with virus. Preliminary experiments in spraying virus on trees have been encouraging.

A virus called NPV is commercially produced and is used for the control of the cotton bollworm, a species closely related to the tobacco budworm. With present technology, the viruses can reproduce only in living cells. The insect viruses are usually grown in the appropriate hosts. Investigators are attempting to develop cell culture systems for propagating insect viruses that will eliminate the inconvenience of contamination of large numbers of insects or insect larvae for virus production.

The possible impact of virus insecticides on other forms of life, including humans, in the long term will obviously require some years of actual experience at a high level of usage. Regulatory agencies at this juncture are understandably extremely cautious in approving more than limited use. There appears to be evidence in support of the safety of the viruses and the negative aspects of this juncture appear to be in the category of theoretical possibilities in the absence of any substantive evidence. One scientist at the U.S. Department of Agriculture, with reference to the case of the cabbage looper caterpillar that succumbs to a virus attack, refers to the so-called coleslaw example. The insect body dissolves and sheds onto the leaf of cabbage large quantities of virus, which are not killed in the preparation of coleslaw. In mid-October, when mortality of the loopers is at the highest level, the average bowl of coleslaw will contain about 4 billion live particles of cabbage looper nuclear polyhedrosis virus. The scientist reasons that if the virus were harmful to people, this would have been long evident.

Microbial Agents. Considerable academic and governmental research has been conducted in this area for many years and is also an area to which industry has made significant contributions. The bacterial agents, *Bacillus thuringiensis* and *B. popilliae* are used commercially. Permission also has been granted for the experimental investigation of NPV (nucleopolyhedrosis virus) of *Heliothis zea* (corn earworm). Although it is not generally believed that microbial agents will replace chemical methods, it is felt that such agents will function importantly in integrated pest management programs of the future.

Application of Gene Science. The development and current technology at the gene level (see also **Genetics and Gene Science (Classical)**) and at the molecular level (see also **Molecular Biology**) has great potential for achieving a number of new ways to control insects and pests beyond the more traditional chemical and biological controls. The public concerns with unexpected results of genetic manipulation, particularly at levels of life above microorganisms, may dampen for quite some time serious efforts along these lines. More traditional biogenetics may proceed at a good rate. As early as 1908, Boveri demonstrated that multiple fertilizations of an egg cause chromosomes to be unequally distributed in cells during early cleavages, because the multiple centrioles set up multiple spindle orientation sites and chromosomes proceed to these sites at random. Boveri's study demonstrated that the well-being of the organism depends upon a full complement of chromosomes with corresponding total genic balance. Many years ago, four major areas for insect population control through genetic means were suggested: (1) induced dominant lethality; (2) contrived dominant lethality; (3) induced inherited sterility; and (4) contrived inherited partial sterility.

Insect Adaptability to Climate. Closely related to the foregoing discussion is the concept of genetic suppression of insect populations by their adaptations to climate. Many of the serious insect pests have very broad geographical distributions. For example, the codling moth (*Carpocapsa pomonella* (L.)) is distributed from Canada to Argentina and it occurs in Australia, South Africa, the Mediterranean, and throughout Europe north to the Scandanavian countries. Insects with such broad distributions adapt in various ways to climate.

A number of reports indicate that genetic differences occur between insect populations within a species with regard to: (1) ability to undergo a hibernal diapause; (2) response to diapause-inducing stimuli; (3) duration of diapause; (4) temperature limits of diapause termination; (5) temperature optima for diapause termination; (6) ability to develop cold hardiness; (7) thermal constants and temperature threshold for development; (8) choice of hibernal niches and other behavioral traits associated with surviving inhospitable seasons; and (9) ability for aestival (summer) diapause, its duration and response to conditions that induce or terminate it.

Such genetic differences must exist so that adaptations of insects to climate may be appropriate to their locality. Changes in climate from locality to locality require appropriate changes in adaptations. Insects may synchronize their life cycles with the seasons so that (1) frost-sensitive stages are passed in the frost-free season; (2) feeding stages occur when food is available; and (3) no actively developing stages occur in periods of intense heat or drought. Therefore, insects must be sensitive to stimuli that portend the change in seasons so that they may prepare for adverse periods.

If it is possible to genetically disrupt the seasonal regulations or other climatic adaptations of insects, the insects may not survive. For example, if an insect population at Fargo, North Dakota must respond to a photoperiod of 15 hours in order to enter diapause and be cold-hardy in time for dangerous frosts, and if the population is genetically modified, so that it does not diapause until the day has shortened to 13 hours, then the population may be destroyed by the winter. Further, it may be assumed that this population must remain in diapause until early May, when the danger of killing frost is past and when the host plant has again become available. If the diapause is genetically shortened so that the insects resume development in March, then they would be destroyed either by frost or lack of food.

Inappropriate adaptations to climate are lethal at certain times of the year; they are conditional lethal traits. A conditional lethal trait or combination of conditional lethal traits can be used to suppress or eradicate insect populations. The principle of suppressing insect populations by means of their adaptations to climate has been suggested by a number of investigators.

Related entries include **Herbicide**; and **Insecticide**.

Additional Reading

Alfassi, Z.B. and R.M. Joy: *Pesticides and Neurological Diseases*, CRC Press, LLC., Boca Raton, FL, 1994.

Baringa, M.: "Entomologists in the Medfly Maelstrom," *Science*, 1168 (March 9, 1990).

Best, G.: *Pesticides–Developments, Impacts, and Controls*, CRC Press, LLC., Boca Raton, FL, 1995.

Brooks, G.T. and T. Roberts: *Pesticide Chemistry and Bioscience: The Food Environment Challenge*, The Royal Society of Chemistry, London, UK, 1999.

Cheremisinoff, N.P. and J.A. King: *Toxic Properties of Pesticides*, Marcel Dekker, Inc., New York, NY, 1994.

Dodge, A.D. Editor: *Herbicides and Plant Metabolism*, Cambridge University Press, New York, NY, 1990.

Frehse, H.: *Pesticide Chemistry: Advances in International Research, Development, and Legislation*, John Wiley & Sons, Inc., New York, NY, 1991.

Gibbons, A.: "Moths Take the Field Against Biopesticide (Bacillus thuringiensis)," *Science*, 646 (November 1, 1991).

Greene, C.: "Environmental Concern Sparks Renewed Interest in Integrated Pest Management," *Food Review*, 8 (April–June 1991).

Hayes, W.J., Jr. and E.R. Laws, Jr.: *Handbook of Pesticide Toxicology*, Academic Press, Inc., San Diego, CA, 1991.

Ishaaya, I.: *Biochemical Sites of Insecticide Action and Resistance*, Springer-Verlag, Inc., New York, NY, 2001.

Jacobson, M.: *Glossary of Plant-Derived Insect Feeding Deterrents*, CRC Press, LLC., Boca Raton, FL, 1990.

Kearney, P.C., N.N. Ragsdale and J.R. Plimmer: *Eight International Congress of Pesticide Chemistry*, American Chemical Society, Washington, DC, 1994.

Kirchhoff, J. and Hans-Peter Their: *Manual of Pesticide Residue Analysis*, Vol. 2, John Wiley & Sons, Inc., New York, NY, 1992.

Laird, M., L.A. Lacey and E.W. Davidson: *Safety of Microbial Insecticides*, CRC Press, LLC., Boca Raton, FL, 1990.

Lynch, L.: "Consumers Choose Lower Pesticide Use Over Picture-Perfect Produce," *Food Review*, 9 (January–March 1991).

McKenzie, J.A.: *Ecological and Evolutionary Aspects of Insecticide Resistance*, Academic Press, Inc., San Diego, CA, 1996.

Milne, W.A.: *CRC Handbook of Pesticides*, CRC Press, LLC., Boca Raton, FL, 1994.

Nadasy, M. and GyForgy Matolcsy: *Pesticide Chemistry*, Elsevier Science, New York, NY, 1989.

Reganold, J.P., R.I. Papendick and J.F. Parr: "Sustainable Agriculture," *Sci. Amer.*, 112 (June 1990).

Schaub, J.R.: "Pesticides: How Safe and How Much?" *Food Rev.*, 2 (April–June 1991).

Sherma, J.: "Pesticides (Analysis of)," *Analytical Chemistry*, 118R (June 15, 1991).

Staff: *Handbook of Natural Pesticides, Vol. VI: Microbial Insecticides*, CRC Press, LLC., Boca Raton, FL, 1990.

Staff: "Organically Grown Foods," *Food Technology*, 26 (June 1990).

Strobel, G.A.: "Biological Control of Weeds," *Sci. Amer.*, 72 (July 1991).

Torgersen, T.R.: "Saving Forests the Natural Way," *Amer. Forests*, 31 (January/February 1990).

Van Rie, J., et al.: "Mechanism of Insect Resistance to the Microbial Insecticide Bacillus thuringiensis," *Science*, 72 (January 5, 1990).

Zilberman, D., et al.: "The Economics of Pesticide Use and Regulation," *Science*, 581 (August 2, 1991).

INSECTIVORA.   See **Moles and Shrews**.

INSECTIVOROUS PLANTS.   These are plants able to obtain a part of their nitrogen supply from the bodies of small insects and other animals that are trapped by the plants in various ways. They are also frequently called carnivorous plants. All of them are green and capable of living without this animal nitrogen, but many seem to thrive better if they have it. They are plants that grow in marshy or boggy places, where the supply of available nitrogen may be very slight. Some of them are water plants.

Insectivorous plants may be divided into three distinct groups, distinguished by the manner in which the plant captures the insects. In one group, the insects are attracted to the plant's leaves by a glandular secretion,, which is sticky and holds them fast. In some species in this group, the glandular hairs fold inward over the prey to hold it firmly and to aid in digesting it. The most common plants of this group are the sundews, species of the genus *Drosera*. They are small bog plants of fairly common occurrence. In some species the leaves are linear, in others round and long-petioled. In all, the upper surface of the leaf is covered with long tentacle-like hairs with swollen tips. This tip secretes a copious quantity of a colorless sticky substance, which glistens in the sunlight and attracts many small insects. When the insect alights on the leaf or on one of the hairs, it stimulates the latter to fold inward, gradually carrying the insect toward the center of the leaf. The stimulus is transmitted to other nearby hairs, which fold in likewise, until the insect is carried to the leaf center and pressed firmly against the surface. A digestive enzyme is there secreted, which acts on the proteins of the animal body, changing them to a soluble form that can be absorbed by the leaf. When digestion is completed, the glandular hairs unfold, and the leaf is ready for another victim.

The second group of insectivorous plants comprises all those in which the leaf is variously modified to form a pitcher in which the prey is entrapped. Plants of this group are often very striking objects. They occur in widely scattered regions. In the eastern part of North America are found species of the genus *Sarracenia*, which are commonly called pitcher-plants. They are found in open marshes where plenty of light will reach them. The leaves occur in basal rosettes, and are green, often deeply mottled with red. Each leaf has the form of an open pitcher with a distinct lip or flange at the top, and a green wing down one side of the pitcher. Around the mouth of the pitcher, on the inner side, are numerous glands, which secrete a fluid that attracts insects. Lining the inside of the pitcher are numerous stiff pointed teeth or bristles which project sharply downward, so that it is easy for the insect to crawl down into the pitcher, but practically impossible to crawl up. The lower part of the pitcher is usually full of water, into which the unfortunate insect eventually falls and is drowned. Either due to the action of bacteria or that of secretions from glands in the surface of the leaf, the proteins of the animal body become assimilable by the leaf, which thus obtains nitrogenous matter.

The third group of insectivorous plants is composed of those plants which capture their prey in some sort of movable trap. In this group are found Venus' flytrap, the Bladderworts, and Aldrovanda.

Venus' flytrap, *Dionaea muscipula*, is a small plant found in the Carolinas. The leaves form a basal rosette close to the ground. Each leaf has a broad blade-like petiole, which abruptly narrows at its tip and bears a remarkable blade. The latter is formed of two halves, which are joined by a movable hinge down the center. The edges of each half are fringed by long stiff bristles, while on the upper surface of each are borne three long slender trigger hairs which are sensitive to contact with any solid object. Each trigger hair is jointed at its base. The upper surface of the blade is abundantly supplied with small glands. When any insect alights on the leaf and comes in contact with one of the trigger hairs, a stimulus is given which causes the two halves of the blade to fold together with the bristles around their edges interlocking and so trapping the insect securely. Once caught the insect is slowly digested by the leaf.

INSEMINATION. AL/MAP The introduction of the seminal fluid, bearing the reproductive cells of the male into the genital passages of the female.

INSERTION LOSS. The insertion loss of a piece of apparatus, usually expressed in dB, is the loss introduced in an electrical circuit by the insertion of the apparatus. If P_1 is the power in the load circuit prior to the insertion of the network and P_2 is the power with the network present, the insertion loss is the ratio P_2/P_1. Thus, in many communication circuits, the connecting of essential components of the system may introduce an insertion loss, which must often be compensated for by additional amplifier gain. See also **Bridging Gain**.

INSOLATION. Acronym for "incoming solar radiation." In general, the term means the solar radiation received at the earth's surface. The rate at which direct solar radiation is incident upon a unit horizontal surface at any point on or above the surface of the earth. See also **Heat Balance (Planet)**; and **Solar Energy**.

INSOMNIA. See **Sleep**.

INSTABILITY. 1. The concept of instability is employed in many sciences. It is, in general, a property of the steady state of a system such that certain disturbances or perturbations introduced into the steady state will increase in magnitude, the maximum perturbation amplitude always remaining larger than the initial amplitude. The method of small perturbations, assuming permanent waves, is the usual method of testing for instability; unstable perturbations then usually increase exponentially with time. An unstable nonlinear system may or may not approach another steady state; the method of small perturbations is incapable of making this prediction. The small perturbation may be a wave or a parcel displacement. The parcel method assumes that the environment is unaffected by the displacement of the parcel. The slice method has occasionally been used as a modification of the parcel method to gain a little information about the interaction of parcel and environment. Stability as defined above is an asymptotic concept; other definitions are possible. Precision is required of the user, and caution of the reader.

2. In meteorology the reference is usually to one of the following:

1. Static instability (or hydrostatic instability) of vertical displacements of a parcel in a fluid in hydrostatic equilibrium. (*See* conditional instability, absolute instability, convective instability, buoyant instability.)
2. Hydrodynamic instability (or dynamic instability) of parcel displacements or, more usually, of waves in a moving fluid system governed by the fundamental equations of hydrodynamics, to which the quasi-hydrostatic approximation may or may not apply.

The space scale of unstable waves is important in meteorology: Thus Helmholtz, baroclinic, and barotropic instability give, in general, unstable waves of increasing wavelength. The timescale is also important: A perturbation that grows for two days before dying out is effectively unstable for many meteorological purposes, but this is an initial-value problem and one cannot assume the existence of permanent waves. These meteorological types of hydrodynamic instability must not be confused with the phenomenon often referred to by mathematicians and physicists by the same term. A great deal of study has been devoted to the problem of the onset of turbulence in simple flows under laboratory conditions, and here viscosity is a source of instability. See also **Atmosphere (Earth)**.

INSTABILITY LINE. See **Fronts and Storms**.

INSTRUCTION (Computer System). 1. A set of characters which defines an operation together with one or more addresses, or no address, and which, as a unit, causes the computer to perform the operation on the indicated quantities. The term instruction is preferable to the terms command and order. Command is reserved for a specific portion of the instruction word, i.e., the part which specifies the operation which is to be performed; order is reserved for the ordering of the characters, implying sequence, or the order of the interpolation, or the order of the differential equation.

2. The operation or command to be executed by a computer, together with associated addresses, tags and indices.

Alphanumeric instruction. The name given to instructions that can be used equally well with alphabetic or numeric kinds of fields of data.

Branch instruction. (or *transfer instruction*). An instruction to a computer that enables the programmer to instruct the computer to choose between alternative subprograms depending upon the conditions determined by the computer during the execution of the program.

Breakpoint instruction. 1. An instruction that will cause a computer to stop or to transfer control in some standard fashion to a supervisory routine which can monitor the progress of the interrupted program. 2. An instruction which, if some specified switch is set, will cause the computer to stop or take other special action.

Macro instruction. 1. A pseudoinstruction which causes a sequence of instructions to be inserted into the object routine for performing a specific operation. 2. The more powerful instructions that combine several operations in one instruction.

Micro instruction. A basic or elementary machine instruction.

Multiple-address instruction. An instruction consisting of an operation code and two or more addresses. Usually specified as a two-address, three-address, or four-address instruction.

One-address instruction. An instruction consisting of an operation and exactly one address. The instruction code of a single address computer may include both zero- and multi-address instructions as special cases. Related to *address, one*.

Pseudoinstruction. (or *quasi instruction*). 1. A symbolic representation in a compiler or interpreter. 2. A group of characters having the same general form as a computer instruction, but never executed by the computer as an actual instruction.

Two-, three-, or four-address instruction. An instruction consisting of an operation and 2, 3, or 4 addresses, respectively. The addresses may specify the location of operands, results, or other instructions.

See also **Program (Computer)**.

THOMAS J. HARRISON, IBM Corporation, Boca Raton, FL

INSTRUCTION COUNTER (Computer System). A counter register in a computer that contains the address of the instruction to be accessed in storage. Also known as program counter in some computer designs. Each time an instruction is executed, the register is incremented such that, at the completion of the operation, the instruction counter is able to address the next instruction. When a program is interrupted, the instruction-counter address must be saved so that the program may resume at the point of interruption when the interrupt program is finished. If a "branch or condition" instruction is executed and the branch is taken, the contents of the instruction counter is replaced by the "branch to" address. See also **Counter (Computer System)**.

INSTRUMENT. A tool or device for extending or substituting for manual dexterity and guidance, as a surgical instrument or machine tool.

An apparatus for accomplishing a measurement, such as a gage, meter, indicator, recorder, etc., of any one of numerous variables, including temperature, pressure, flow, specific gravity, etc. Frequently an indicating-recording instrument will incorporate means for automatic controlling of one or all of the variables measured. *Instrumentation* is the science and art of effectively applying measuring and controlling instruments in industrial, laboratory, and other applications.

INSTRUMENTAL VARIABLE. MIS In statistics, a variable which is introduced to resolve some difficulty arising in the model under study; for example, as a substitute for some variable which leads to inconsistent estimates of parameters or to resolve unidentifiable situations.

INSTRUMENTATION.

1. The installation and use of electronic, gyroscopic, and other instruments for the purpose of detecting, measuring, recording, telemetering, processing, or analyzing different values or quantities as encountered in the flight of a rocket or spacecraft.
2. The assemblage of such instruments in a rocket, spacecraft, or the like.
3. A special field of engineering concerned with the design, composition, and arrangement of such instruments.

INSTRUMENTATION (Analytical). See **Analysis (Chemical)**.

INSULATION (Electric). An electrical insulator, when placed between conductors at different potentials, will permit a negligible current (in phase with the applied voltage) to pass through it. In essence, electrical insulators are applied dielectrics. A perfect insulator (dielectric) will pass no current in the foregoing situation. The closest approach to a perfect dielectric is a perfect vacuum. The difference between low-resistance insulators and semiconductors is ill defined. Materials that can be considered as insulators have resistivities greater than 10^{20} ohms down to 10^6 ohms. Because of varying needs for insulators in the construction and use of electrical and electronic equipment and systems, the wide range of materials available, including a considerable span in cost, provides a convenient selection for the designer.

Relative Dielectric Constant. The capacitance between plane electrodes when in a vacuum, neglecting fringing, may be expressed by

$$C = k_0 A/t = 0.0884 \times 10^{-12} A/t \text{ farads}$$

where k_0 is dielectric constant of a vacuum; A is area, cm^2, t is spacing between plates, cm. If the vacuum is replaced by a dielectric material, the capacitance increases for the same applied voltage. With the dielectric between the plates, the capacitance is

$$C = kk_0 A/t$$

where k is the relative dielectric constant of the material. Capacitance relations change, of course, for other commonly occurring situations, such as coaxial conductors, concentric spheres, and parallel cylindrical conductors. The relative dielectric constant, then, is the key criterion in the functioning of an insulator. Values for several representative materials are given in Table 1. Numerous factors, including temperature, frequency, humidity, age, degree of cure (in case of plastics and polymers), and geometry, affect the relative dielectric constant. For this reason, the term dielectric permittivity is often used instead of relative dielectric constant.

Materials Used. With increasing frequency, the permittivity of dielectric decreases. A major factor in the selection of insulation is the ability of the insulation to resist the absorption of moisture. Moisture, of course, can greatly lower resistivity. For wire insulation, synthetic polymers and plastics essentially have replaced the use of natural rubber. Usually, prior to coating a wire with a plastic material, the wire must be treated to assure good contact and adhesion of the insulating material. Copper wire, for example, is treated with hydrogen fluoride, which creates a coating of copper fluoride; in the case of aluminum wire, aluminum fluoride. Thin films of fluoride possess high dielectric strength and resist heat well.

With the possible exception of common line insulators, electrical porcelain is especially formulated and may contain varying percentages of zirconia and beryllia. These ingredients increase both strength (mechanical)

TABLE 1. DIELECTRIC PERMITTIVITY OF REPRESENTATIVE MATERIALS*

Material	k	Material	k
Ceramics and Glasses		*Nonpolar Resins*	
Alumina	8.1–9.5	Polyethylene	2.3
Aluminum silicate	4.8	Polypropylene	2.2
Pyrex (Corning 7740)	5.1	Polystyrene	2.5–2.6
Fused silica	3.8	Polytetrafluoroethylene	2.0
Forsterite	6.2–6.3		
Steatite	5.5–7.0		
		Polar Resins	
High-tension Porcelains		Cellulose cotton (dry)	5.4
Beryl	4.5	Cellophane (dry)	6.6
Magnesia	8.2	Cellulose triacetate	4.7
Mica (glass-bonded)	6.4–9.2	Epoxies (unfilled)	3.0–4.5
Titanates	50–10,000	Nylon	4.0–4.6
Zirconia	8.0–10.5	[a]Phenolics (cellulose)	4–15
		[a]Phenolics (glass)	5–7
Crystals		[a]Phenolics (mica)	4.7–7.5
Aluminum oxide	10.0	Polyvinyl chloride	3.2–3.6
Calcium carbonate	9.2	Polyvinyl acetate	3.2
Boron nitride	4.2	Polyvinyl fluoride	8.5
Barium titanate	4.100	Methylmethacrylate	3.6
Mica, synthetic	6.3	Polycarbonate	2.9–3.0
(fluorophlogopite)		[a]Silicone (glass)	3.1–4.5
Mica (muscovite)	7.0–7.3	Polyethylene	3.25
Magnesium oxide	8.2	terephthalate	
Sodium chloride (dry)	5.5		

[a]Filled with material indicated in parentheses.
*Some insulations incorporating organic materials may be carcinogenic when subjected to high temperatures. Regulations on the use of these materials varies from one country to the next. Thus some of these products are not available throughout the world.

and resistance to high temperatures. Hard porcelains are especially formulated to resist thermal shock as well.

Liquid insulators are required for circuit breakers, transformers, and some cable applications. Natural hydrocarbon mineral oils are commonly used, as well as chlorinated aromatic liquids (desirable because of nonflammability). For high-temperature situations, silicone fluids may be used. Permittivities range between 2 and 7. Insulating liquids function both as electrical insulators and heat-transfer media. See also **Dielectric**; and **Electrical Conductivity**.

INSULATION (Thermal). Thermal insulation is any substance or configuration of materials that resists the flow of heat. Thermal insulation does not stop heat flow, but retards it to rates that suit particular requirements. For example, in the case of buildings or residences in climates or during seasons when the ambient temperature of the atmosphere is uncomfortably hot or cold, thermal insulation will be used to retain heat within a structure during cold weather and to shield or insulate the structure from the penetration of external heat during hot weather. In the one case, by reducing the flow of heat from structure to atmosphere and near outer space during winter, less energy is required to maintain the desired temperature within the structure. In the other case, by reducing the flow of heat from atmosphere and sun to structure during summer, less energy is required to artificially cool the inside temperature. Many parallel instances occur in industry. By thermally insulating processing vessels, piping, etc., where it is desired to maintain warm or hot conditions, energy is not lost to ambient surroundings. The efficient maintenance of temperature is critical to many industrial situations because temperature affects the physical and chemical properties of materials, such as viscosity, and determines the rate at which chemical reactions occur, among numerous other temperature-sensitive properties. In cryo-processing, as in the liquefaction of gases, the freezing of foods, etc., the objective is the maintenance of low temperatures and thermal insulation in such cases restricts the flow of heat from ambient surroundings and thus reduces the amount of energy required to maintain desirable low temperatures.

Although the principles of heat flow have been understood and treated mathematically since the early 19th century (Fourier, LaPlace, Poisson, Peclet, Lord Kelvin, Riemann, and many others), it was not until nearly the beginning of the 20th century that major developments of commercial thermal insulating materials and systems were undertaken.

In addition to increasing thermal efficiency (conservation of energy), thermal insulation is frequently used to protect personnel from injury by burns and to shield adjacent structures from overheating, thus assisting in protection against fire. Thermal insulations are not usually suited to fire protection per se once a fire is in progress, but by restricting heat flow in the first place, insulation plays an important role in preventing some kinds of fires from starting. The behavior of thermal insulation once a fire has started is not necessarily positive in all situations and thus requires the attention of equipment and structure designers. A major concern in fires is thermal diffusivity rather than thermal resistance. Thermal insulation may not be suited for protection against high-velocity radiation.

Although not the primary function, the retardation of moisture migration into insulated spaces, where condensation may occur, is an important engineering consideration in the design of thermal insulation systems.

Insulation and Heat-Flow Principles. Heat flows from places of higher temperature to those of lower temperature by one or more of three modes: (1) Conductance through solids; (2) convection by induced motion of fluids carrying heat; and (3) radiation by heat waves emitted from a surface. The rate of heat flow in solids depends upon temperature difference $T_2 - T_1$ and the resistances encountered. The heat flow, under steady state, is expressed by:

$$Q_{\text{heat flow}} = \frac{T_2 - T_1}{R\text{-value}}$$

R-value is a measure of thermal resistance and varies from one insulating material to the next.[1] Q can be expressed Btu/square foot/hour or as Cal/square meter/hour, depending upon the units used in the equation. Helpful relationships between English and metric units are given in Table 1. See also **Heat Transfer**.

While convection may be a significant factor within processes, in general, convection affects thermal insulation primarily at the surface of the insulation or its jacket, where the air film is a resistance to heat flow from (or to) that surface. Wind reduces the air film resistance. To a lesser extent convection can occur within some low-density fibrous insulations, especially in walls, and to a greater extent within cavities and unfilled spaces within constructions. In walls, it is important that insulation that does not fill the space completely be installed so that remaining spaces are uniform, not skewed. Radiant heat flows through space, either vacuum or gaseous, from a higher-temperature surface toward a lower-temperature substance by the difference in absolute temperatures to the fourth power and the surface characteristic called emittance, *e*, as shown by the Stefan-Boltzman relation:

$$Q_{\text{rad}} = 0.174e \left[\left(\frac{T_2}{100} \right)^4 - \left(\frac{T_1}{100} \right)^4 \right] \text{Btu/ft}^2\text{hr} \quad \text{(on Rankine scale)}$$

$$Q_{\text{rad}} = 5.670e \left[\left(\frac{T_2}{100} \right)^4 - \left(\frac{T_1}{100} \right)^4 \right] \text{W/m}^2 \quad \text{(on Kelvin scale)}$$

A common use of low convection with high-reflectance/low-emittance surfaces is in the food and liquid containers (Dewar flask, Thermos™ bottle, etc.). The double glass-wall space is under high vacuum so convection is virtually eliminated, and the surfaces are coated with silver to reduce heat transfer by the low emittance on the outside of the inner wall and the high reflectance on the inside of the outer wall.

Low emittance can be observed if the hand is held close to a very hot silver teapot without feeling much heat despite the high temperature of the surface of the teapot.

In high-temperature process plants, men have been burned on low-emittance hot metal jackets on insulation because the low emittance did not give them a sense of heat. Yet the thermal resistance from the heat conservation standpoint was excellent.

In some materials, especially foams, the spaces may contain gases other than air and the performance in convective and radiative heat transfer in the spaces affects the overall performance of the material. Also,

[1] Resistance, thermal, *R*-value — the mean temperature difference at equilibrium between two defined surfaces of material, or a construction, that induces unit heat flow rate through unit area. (From ASTM STD. C168-80A)

TABLE 1. CONVERSION FROM U.S. CUSTOMARY UNITS TO METRIC UNITS[1]
(Data are for thermochemical values unless noted. International table values differ slightly.)

W = watt; m = metre; J = joule; kg = kilogram;
C° = temperature difference Celsius

Multiply	By	To Obtain
Btu (mean) British thermal unit	1.055870×10^3	J
Btu ft/h ft² F°	1.729 577	W/m C°
(*k*-factor, thermal conductivity)		
Btu in/h ft² F°	$1.441\,314 \times 10^{-1}$	W/m C°
(*k*-factor, thermal conductivity)		
Btu in/s ft² F°	$5.188\,732 \times 10^2$	W/m C°
Btu/h	2.928751×10^{-1}	W
Btu/ft² h	3.152 481	W/m²
Btu/ft² min	1.891489×10^2	W/m²
Btu/ft² s	$1.134\,893 \times 10^4$	W/m²
Btu/h ft² F°	5.674 466	W/m² C°
(*C*-factor, thermal conductance)		
(*U*-factor, overall thermal conductance)		
Btu/s ft² F°	$2.042\,808 \times 10^4$	W/m² C°
Btu/lb	$2.324\,444 \times 10^3$	J/kg
(Heat capacity)		
Btu/lb F°	$4.184\,000 \times 10^3$	J/kg C°
(Specific heat capacity)		
Btu/ft³	$3.723\,402 \times 10^4$	J/m³
Calorie (mean)	4.190 020	J
Calorie (kilogram)	$4.184\,000 \times 10^3$	J
(Kilocalories)		
Calorie/cm²	$4.184\,000 \times 10^4$	J/m²
Calorie/g	$4.184\,000 \times 10^3$	J/kg
Calorie/g C°	$4.184\,000 \times 10^3$	J/kg C°
Calorie/min	$6.973\,333 \times 10^{-2}$	W
Calorie/s	4.184 000	W
Calorie/cm² min	$6.973\,333 \times 10^2$	W/m²
Calorie/cm² s	$4.184\,000 \times 10^4$	W/m²
Calorie/cm s C°	$4.184\,000 \times 10^2$	W/m² C°
F° h ft²/Btu	$1.762\,280 \times 10^{-1}$	C° m2/W
(*R*-value, thermal resistance)		
F° h ft²/Btu in.	6.928 113	C° m/W
(*ru* -value, thermal resistivity)		
Therm (100,000 Btu)	$1.055\,056 \times 10^8$	J

[1] Most metric units shown are SI (the universally adopted designation for Le Systéme International d'Unités), except SI uses K (kelvin) for both absolute temperature and for temperature differences even though temperatures are determined on the Celsius scale. Since temperatures will usually be measured on the Celsius scale, the symbols used here are °C for temperature Celsius, as in the past, while C° is Celsius degrees difference. Similarly, °F is temperature Fahrenheit and F° is Fahrenheit degrees difference.

the emittance/reflectance performances of the walls of the spaces affect performance overall.

Technically, the performance of materials and systems depends upon all three modes of heat transfer to varying degrees in different materials, so it is the effective or apparent conductance that is to be evaluated. Although such terminology may be correct technically, and is appearing again in the literature, it was discussed many years ago and abandoned because it aroused too many questions by users of insulations with limited technical knowledge. It was felt that those with necessary technical competence would understand that a multimode heat transfer was involved, and the simple thermal conductance would satisfy users so long as the data were correct. R-values are even more readily understood by users, and technical analyses can be made by identifying by subscript that phase of the analysis being evaluated.

Thermal Insulation Systems and Materials

ASTM[2] Committee C-16 on Thermal Insulating Materials defines thermal insulation as a material or assembly of materials used primarily to resist

[2] American Society for Testing and Materials, 1916 Race St., Philadelphia, Pa. 19103.

heat flow. The reference to assembly of materials indicates that the concern is thermal insulating systems, because it is not until materials have been designed into systems that performance can be estimated. Thermal insulating systems include not only the basic materials, but also the auxiliary materials and the methods of application and protection in service.

Time of exposure differentiates the needs for insulation performance when used in relatively continuous exposures, in cyclic increases and decreases of temperature, in processes with wide ranges of temperature in the various phases, especially in pipelines carrying fluids that must not fall below critical temperatures lest they solidify and necessitate dismantling and replacement of the lines.

A special short-time performance of thermal insulation is the ablative protection on the bottom of astronautical capsules returning from outer space when they are heated to sudden high temperatures by impact with the atmosphere. As principles stated below indicate, ablation is the process of resisting heat flow by using absorptance in changes of state from solid to liquid and to vapor of the ablative insulation, which is thereby lost, so that a one-time or at most a few times of exposure is practicable.

Classes of Insulating Materials. See Fig. 1. While glass has a high conductance, if it is fiberized and formed into wool-like masses, the high conductance of the fibers is counteracted by the still air that is held within the mass. Still air (no motion) has high thermal resistance, and at one time it was presumed that still air was the best insulator. A few other materials have been found with somewhat greater thermal resistance than still air, but they are so costly that they are suited only to very special applications. Other fiberized materials perform similarly, and rock, slag, and glass wools are collectively called mineral wools, but each has its own temperature limits.

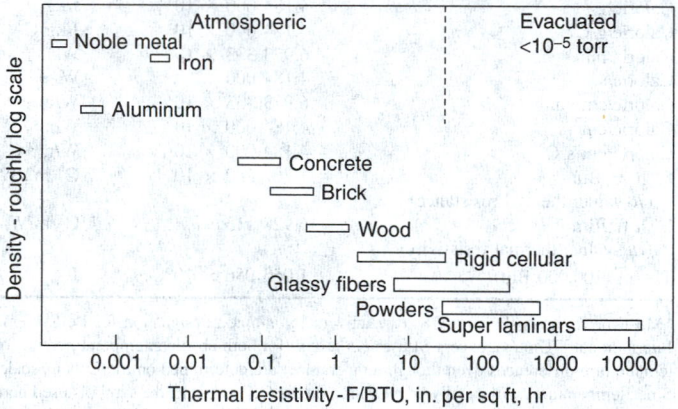

Fig. 1. Thermal resistivity of materials in insulation systems.

All mass-type thermal insulations rely for their thermal resistance upon dispersion of the solid phase with air, or sometimes with gases. Plastics are reduced in density by foaming them into low-density material (0.5 to 2 lb/ft^3; 8 to 32 kg/m^3). Molten glass is foamed so that it performs as thermal insulation; its advantages are high compressive strength and virtually imperviousness to moisture, although thermal resistance is not as high as in some other materials.

Reacted materials, such as hydrous calcium silicate, are made with the solids dispersed to create density on the order of 12 lb/ft^3 (192 kg/m^3) and still provide high compressive strength. While the reflectance and emittance of the solid phase have an effect on heat transfer, it is the still air within the mass that gives it significant thermal resistance.

Still air is the important factor in dispersed solids, such as wood fibers, exfoliated mica, powdered diatomite, and expanded perlite.

The earliest reacted insulation was so-called 85% magnesium that had wide acceptance for many years, but since it was suited to less than 500 °F (260 °C), it has been replaced by other insulations with higher R-values or more desirable physical properties.

Although metals are good conductors of heat, they can also perform as good thermal insulators when their surface properties are used to advantage. While solid metal conducts heat readily, the emittance and

reflectance of some metals are used to provide high overall R-values, especially when they are used in multiple sheets with spacers. Moreover, metals have relatively high temperature tolerances but have no absorption, so that all-metal thermal insulation may be suited to higher-temperature services than some other kinds of materials. While silver and gold are the best metals for high reflectance service, their use is limited to special applications where cost is not governing. For long-time exposure, gold surfaces would maintain high reflectance longest. However, aluminum sheets are used effectively when all-metal insulation is required. The first sheet of aluminum spaced about 12 mm (1/2 in.) from the hot surface reflects a large percentage of heat striking it. The unreflected heat passes through the metal rapidly by conductance, but the low emittance of the reverse surface prevents much of the heat leaving to strike the next sheet of aluminum where, again, the reflectance prevents a large portion of the emitted heat to enter that sheet. The number of reflective sheets is designed for the R-value desired. If the service temperature is above the working temperature of aluminum, about 1000 °F (540 °C), the first one or two sheets can be made of polished stainless steel.

Since aluminum is also used for jackets, it should be noted that although the thermal performance may be acceptable, the low emittance of an aluminum surface on the outside may introduce a personnel hazard, mentioned above.

An opposite effect of emittance occurs when an attempt is made to use aluminum jackets on very cold (cryogenic) piping. In this case, heat reaching the surface from surroundings is reflected away so that the metal surface becomes so cold that it will condense and freeze moisture from the air. To overcome this surface condensation and freezing by use of thicker thermal insulation on the lines would require a very great increase in thickness over that needed if the jacket had been made of a heat-absorbing material that would keep the surface above the dew point.

In the high-temperature case, low emittance is desirable from an operating standpoint, whereas in the cryogenic case low emittance is undesirable.

Two general types of heat flow systems exist: one in which it is desired that heat flow as rapidly as practicable, and another in which heat flow must be resisted as much as practicable. The former is a high thermal conductance type of system, whereas the latter is a high thermal resistance, high R-value, type. Consequently, in heat conserving systems it is simpler to think in terms of thermal resistances, because resistances are additive whereas conductances are not.

However, sometimes both high and low and low R-value materials are needed simultaneously, as in traced lines. See Fig. 2. Although systems are designed with limitation on overall heat loss, high-conductance cements are used for process safety on pipelines carrying hot fluids that would solidify if flow was interrupted or fluids become so viscous that the pumps could not handle the material. In such cases, one or more small pipes called tracers carrying hot fluids are enclosed with the thermal insulation envelope, and high conductance cements are used to improve heat flow from tracers to the main pipe. The size of the pipe insulation must be large enough to enclose the main pipe and the tracers, and also the insulation on fittings.

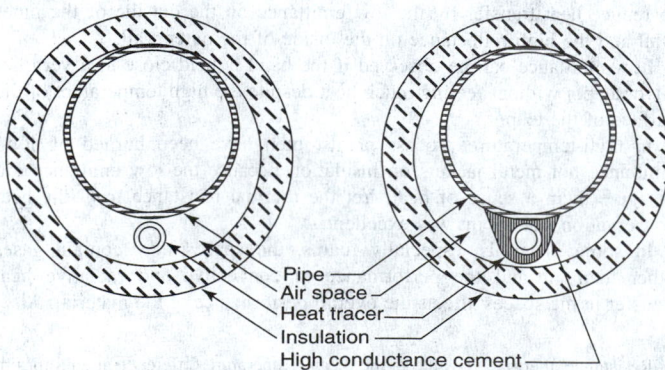

Fig. 2. Pipe insulation over pipe and heat tracer line, with and without high-conductance cement.

Importance of Moisture Migration. Heat conservation has been treated from an energy standpoint as if it were an independent subject, whereas great costs and energy losses have been incurred by premature failures, because it was not realized that in many cases thermal insulations cannot be installed without inducing an effect on the migration of moisture. The problems are usually not great from the standpoint of solutions, but are great in getting people to realize that they have created problems that have been overlooked. In most constructions, heat and moisture performance must be considered jointly, because even high-temperature systems are shut down for alterations, maintenance, or repair. For example, in a house wall, adding thermal insulation makes the indoor wall warmer, as desired, but at the same time it makes the outdoor wall colder, and it is well known that when surfaces are colder than the dewpoint, condensation occurs. The old log cabin with its loose construction had no moisture problems, but it was often only the side of the body that faced the fireplace that felt warm while the rest did its best to accommodate the facts.

Economic Considerations. For many years, cost appraisal of insulation was based on a publication by L.B. McMillan in *ASME Proceedings*, December 1926, in which several factors enter into the analysis, as shown in Fig. 3. At present, the cost of money and the costs of materials and labor are so unstable, that such analyses are of little value except that they do indicate the factors that enter into actual costs to plant management.

To presume that thermal insulation is always necessary is false. For example, when cryogenic fluids are transported from a supply source to a vessel, even in bright sunshine, it is usually most economical not to insulate the line at all. The reason for this is twofold: (1) for the short time of transport the area of the pipe exposed to the sun is much smaller than the area that would be exposed if it were insulated; and (2) the heat that would be in the insulation when the transport starts would have to be removed by the cryogenic fluid. The combination of these two factors often makes the use of insulation undesirable for this type of cold fluid transport system. Moreover, the rapid formation of ice crystals from moisture in the air constitutes a thermal insulation.

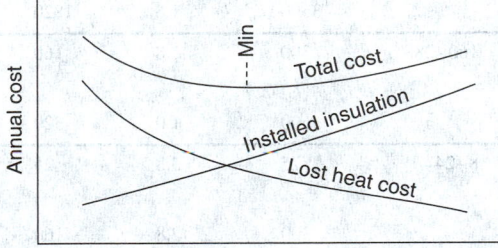

Fig. 3.   Relation of incremental cost of Additional thickness of insulation to the resultant savings and total cost. (After *L.B. McMillan*.)

While thermal insulations are selected first for their resistance to heat flow, their other properties need evaluation for each application. Hence there is no "best" insulation because a material well suited to one service may be poorly suited to another. Economics must be studied in detail because in some services the cost of a highly efficient material per unit of thickness may be overcome by additional thickness of a less efficient material, provided there is room for the greater thickness. All properties of materials must be considered for each exposure, even within the same system. Recognize that when different materials are used on different parts of insulation systems that are close together, the probability of using the wrong material on a particular surface is increased appreciably. Unless there is specific reason for wide use of multiple types of materials, it may be prudent to accept some compromises of properties.

In general, it is desirable to place thermal insulation on or near the outside of constructions, including basement walls, because this location reduces appreciably the temperature stresses in the structure induced by the changes in exposure.

Selection of Thermal Insulation

Test methods, specifications, and some recommended practices are in ASTM Book of Standards, Part 18. Producer's literature gives forms, properties and design data.

1. *Thermal Resistance* — *R*-value — Thermal data in general Tables, as Table 2, are for dry materials, and for one or two mean temperatures, but most materials are not linear with mean temperature, hence, for specific designs, the whole range of resistance should be obtained.

   A factor often overlooked is that space to be occupied by the insulation must be made available; a 3-inch (76-millimeter) iron pipe with 3 inches (76 millimeters) of insulation covered with a protective jacket would have a diameter of about 10 inches (254 millimeters).

2. *Temperature Limit* — Usually the high temperature limit governs because shrinkage then becomes excessive. Generally, high-temperature materials are physically stable at low temperatures but another material may be preferable.

   Note that shrinkage data are usually from tests in soaking heat, whereas the field condition is for heat on only one surface. Hence, such data should not be presumed to mean that the insulation will shrink that amount in service; a 1% shrinkage would imply a change of almost 3/8 inch (9.5 millimeters) in a standard 36-inch (914-millimeter) length and leave a wide crack. However, in service such a large shrinkage does not occur; shrinkage would be 1/16–1/8 inch (1.6–3.2 millimeters). Moreover, hot metals expand, and it is this expansion that must be considered in compensating for openings between adjacent pieces of insulation, usually by use of double-layer insulation with staggered joints.

   For cold-temperature service, moisture ingress is a problem that must be designed against and the material selected for its resistance to moisture and potential freezing.

3. *Corrosion* — While corrosion is usually not a concern on hot surfaces, recognize that all systems have shut down periods, with the probability that moisture will find ingress and condense. Many insulations are alkaline and have little adverse effect on iron and copper, but aluminum is affected adversely. A major concern is stress-corrosion of stainless steel induced by even trace amounts of soluble chlorides; an ASTM Test Method may be used.

4. *Density* — While the density of thermal insulation is low enough not be a problem in most cases, density (or mass-weight) must be considered in airplanes, balloons and other antiterrestrial constructions.

5. *Moisture-Wetting* — If moisture can enter an insulation, the high conductance of water (or ice) reduces the thermal resistance. However, all materials that admit moisture are not affected adversely to the same degree. Some fibrous insulations have a threshold moisture content below which the adverse effect on resistance is not significant; thresholds may be on the order of 8% by weight. The reason is that small droplets adhere to the contact points of one fiber against another so that the volume of relatively still air that provides the thermal resistance is not reduced significantly. Moreover, the temperature difference within the insulation drives the moisture to the cold side, and if it condenses there, the thickness of the wet insulation with decreased resistance is still a small portion of the total resistance.

   In materials that readily absorb water, a decrease in thermal resistance will occur, but consideration must be given to the performance after the material is redried.

   Caution is needed in interpreting absorption tests by immersion because the internal structure may resist displacement of air or gas so that a low absorption is indicated. However, if moisture is induced to flow through the specimen by vapor pressure differential and a condensing temperature is reached, high absorption may occur in service.

   Materials with very high internal surface areas may absorb a measurable amount of moisture; even a monomolecular thickness of moisture on a large area becomes readily measurable. However, exposure to high relative humidity will not "saturate" the material, and as relative humidity changes so will the absorbed moisture. Some materials are tested wet to indicate wet to dry strength ratio.

6. *Handleability* — In the field, materials must have strength properties that enable them to be handled in application by the usual procedures of the industry without excessive breakage. ASTM tests for flexure, impact, and friability are aimed at indicating handling potential. Sometimes vibration tests are indicated if an unusual vibrating condition is to be encountered, although this test is receiving less

TABLE 2. REPRESENTATIVE THERMAL INSULATIONS

Temperature Limit				Density		Thermal Resistivity			
°F	°C	Materials and Usual Forms	Inorganic (I) or Organic (O)	lb/ft^3	kg/m^3	At Mean temp., °F	$\frac{F°\ ft^2\ h}{Btu\ in}$	At Mean temp., °C	$\frac{C°\ m}{W}$
		HIGH-TEMPERATURE SERVICE							
2300	1265	Alumina-silica ceramic fiber, soft mass	I	3–12	48–192	300	3.2	160	22.2
						1000	1.2	540	8.3
2200	1200	Potassium titanate fiber, soft mass	I	15–18	240–288	300	3.0	160	20.8
1900	1040	Diatomaceous silica, bonded, semirigid, preformed block and pipe	I	23–25	368–400	300	1.5	160	10.4
						1000	1.3	540	9.0
1800	1000	Mineral fiber, rock, and slag, loose fill, preformed block and pipe	I	16–24	256–384	400	1.7	204	11.8
						1000	1.3	540	9.0
1600	875	Perlite, expanded, loose granules	I	4–10	64–160	0	3.0	–18	20.9
						1000	0.9	540	6.2
1200	650	Hydrous calcium silicate, may contain unexposed organic reinforcing fibers, preformed rigid block and pipe, compression over 100 psi (0.69 Mpa)	I/O	11–14	176–224	300	2.5	160	17.3
						700	1.7		11.8
1000	540	Glass fiber, no binder, loose mass	I	3–5	48–80	300	2.9	160	20.1
						800	1.4	430	9.7
500	275	Gilsonite, processed pure asphalt powder for underground fill, compacted, impervious	O	35–48	560–640	50	1.9	10	13.2
						3300	1.6	1820	11.1
800	425	Glass, cellular, preformed block, impervious, compression over 100 psi (0.69 Mpa)	I	10–18	160–288	300	1.8	160	12.5
						600	1.0	320	7.0
450	225	Glass fiber, organic binder, loose fill, blankets, batts, preformed block and pipe	I/O	0.5–3	8–24	75	3.8	25	26.4
						300	2.8	160	19.4
200	95	Cellulosic fibers of wood, cane, reused paper, as loose fill	O	0.3–5	4.8–80	75	3.8	25	26.4
		LOW-TEMPERATURE SERVICE							
–225	–140	Plastic Foams Polyurethane (under investigation)	O	1.8–2.2	28.8–35.2	–200	11.0	95	76.3
						100	6.0	40	41.6
–200	–130	Polystyrene	O	1.0–4.0	16–64	40	3.9	4.4	27.0
						75	2.6	25	18.0
–40	–40	Polyvinyl chloride	O	4–25	64–400	75	3.9	25	57.0
–40	–40	Rubber, cellular	O	3–20	48–320	25	4.3	–3.9	29.8
						75	3.3	25	22.9
–400	–245	Glass, cellular, preformed block	I	10–18	160–288	25	2.8	–3.9	19.4
						100	2.4	40	16.7
–450	–270	Mineral Fibers	I	0.5–10	8–160	25	3.7	–3.9	25.7
						100	3.3	40	22.9
–459	–273	Evacuated multilayer foil and fiber mats	I	Various		On the order of 25–100+		On the order of 175–700+	

1. Most high-temperature insulations are usable also at low temperatures, but other materials may be preferable.
2. Maximum temperatures apply to surface service, not soaking heat.
3. Thermal resistivity data are approximate and vary widely with density (for specific designs and temperatures, consult current manufacturers' data).
4. Celsius temperatures are rounded. °C is temperature; C° is temperature difference.

attention than in the past because vibrations are usually designed against.

7. *Reflectances* — When an all-metal system is desired for temperatures to 1000 °F (540 °C) usually to avoid absorptions in case of leaks, multilayer sheets of aluminum are spaced on the order of 1/2 inch (13 millimeters). For temperatures to 1400 °F (760 °C) the first sheets are made of stainless steel.

8. *Fire Behavior* — While thermal insulations are not intended for fire protections (treated elsewhere) their behavior in fire is important, especially from the standpoint of contribution of combustible matter to a fire that has started at the site. Material behavior may be complex, e.g., an absorptive material that would hold a combustible fluid (say, kerosene) would not be a major contribution to fire intensity because the fluid would not flow to the surface to burn as rapidly as it would from a pool of the fluid. Materials that contain organic binders may not be a serious contribution in an open fire, but if they are totally enclosed they may contribute to persistence of fire by smoldering.

While some thermal insulation may not constitute a significant fire hazard, consideration must be given to jackets, coatings, or coverings, and to the methods of attachment, so that even in moderate fires the insulation will not readily fall off the construction they insulate.

Principal Types of Thermal Insulating Materials

Thermal insulations are made from natural or processed materials and combined to provide properties that meet the needs of specific installations. Obviously, all desired properties are not available in any one insulation. Hence, selection of thermal insulation for specific uses involves comparisons for each use, and some high thermal resistance (R-value) may need to be sacrificed in favor of some other property, such as handleability, resistance to compression, thermal diffusivity, avoidance of stress-corrosion, toxicity, or in-fire performance. See Table 2. Based upon ASTM C 168–80a, the principal types of thermal insulation are:

Ablative. Heavy density combination of materials that change state from solid to liquid or vapor in high temperature so that heat absorbed through their change of state reduces substantially the heat transfer rate through the material. Suited to one-time or very few times use.

Calcium Silicate. Composed principally of hydrous calcium silicate, usually containing reinforcing fibers.

Cellular Elastomeric. Composed principally of natural or synthetic elastomers, or both, processed to form flexible, semirigid, or rigid foams which have a predominantly closed-cell structure.

Cellular Glass. Composed of glass processed to form a rigid foam usually having a predominantly closed-cell structure.

Cellular Polystyrene. Composed principally of polymerized styrene resin processed to form rigid foam having a predominantly closed-cell structure.

Cellular Polyurethane. Composed principally of the catalyzed reaction product of polyisocyanate and polyhydroxy compounds, processed usually with fluorocarbon gas to form a rigid foam having a predominantly closed-cell structure. Under investigation by regulators.

Cellulosic Fiber. Composed principally of cellulose fibers usually derived from paper, paperboard stock, or wood, with or without binders.

Diatomaceous Silica. Composed principally of diatomite (diatomaceous earth) with or without binders, usually containing reinforcing fibers.

Gilsonite. A pure form of asphalt processed into powdered form for use as the enclosing insulating mass around pipes or tanks underground.

Mineral Fiber. Composed principally of fibers manufactured from rock, slab, or glass, with or without binders.

Perlite. Composed of natural perlite ore expanded and processed to form particles of various sizes with a cellular structure.

Vermiculite. Composed of natural vermiculite ore expanded and processed to form particles of various sizes with an exfoliated structure.

Wood Fiber. Composed of wood fibers, with or without binders. This is a type of cellulosic fiber insulation.

Principal Forms of Thermal Insulation

Also based upon ASTM C 168–80a, the principal forms of thermal insulation are:

Blanket Insulation. A relatively flat and flexible insulation in coherent form furnished in units of substantial area. Some forms are called batts.

Blanket Insulation, Metal Mesh. Blanket insulation covered by flexible metal-mesh facings attached on one or both sides.

Block Insulation. Rigid insulation preformed into rectangular units.

Board Insulation. Semirigid insulation preformed into rectangular units having a degree of suppleness particularly related to their geometrical dimensions.

Cement, Finishing. A mixture of dry fibrous or powdery materials, or both, that when mixed with water develops a plastic consistency, and when dried in place forms a relatively hard, protective surface.

Cement, Insulating. A mixture of dry granular, flaky, fibrous, or powdery materials that when mixed with water develops a plastic consistency, and when dried in place forms a coherent covering that affords substantial resistance to heat transmission.

Fitting Covers. Manufactured or assembled segments of insulation to form covers for various pipe and vessel fittings such as elbows, tees, crosses, valves, etc. See ASTM Standard C450–76 (or later) for dimensions.

Loose-Fill Insulation. Insulation in granular, nodular, fibrous, powdery, or similar form designed to be installed by pouring, blowing, or hand placement.

Pipe Insulation. Insulation in a form suitable for application to cylindrical surfaces.

Reflective Insulation. Insulation depending for its performance upon reduction of radiant heat transfer across spaces by use of one or more surfaces of high reflectance and low emittance.

Roof Insulation. Rectangular boards or blocks of various thicknesses with properties for use beneath the roofing membrane protected from the weather, or with properties for use above the roofing membrane exposed to the weather.

Underground Systems. Systems that enclose insulated piping in small tunnels that include expansion arrangements and provide drainage. Since accidental general flooding may occur, the insulations should be capable of withstanding "boiling water" effects so that when the system has been dewatered the redried insulation will perform thermally essentially as it did prior to flooding.

Super Insulations. Several insulation systems have been developed that have very higher thermal resistances, such as multi-layer radiation shields, specially selected and matted fibers with small interfiber distances, special powders, ceramic foams, honeycomb composites, often highly evacuated, but they are too costly for usual services with which the general public is familiar.

Insulating Concrete. This should be recognized as relative in performance to usual heavy density concrete, and does not provide thermal resistance in the range of materials understood to be thermal insulations.

Additional Reading

ASHRAE: *Handbook of Fundamentals*, American Society of Heating, Refrigerating, and Air Conditioning Engineers, Inc., New York, NY (Published every 5 years). http://www.ashrae.org/

ASTM: *Annual Book of ASTM Standards: Thermal Insulation: Environmental Acoustics*, Vol. 6, American Society for Testing and Materials, Philadelphia, PA, 1997. http://www.astm.org/ American Society for Testing and Materials.

ASTM: *Book of Standards*, Part 18, American Society for Testing and Materials, Philadelphia, Pennsylvania (Issued annually).

ASTM: *Thermal Insulation Performance*, STP718, American Society for Testing and Materials, Philadelphia, Pennsylvania (1981).

Bynum, R.T. Jr.: *Insulation Handbook*, The McGraw-Hill Companies, Inc., New York, NY, 2000.

Glaser, P.E., et al.: *Thermal Insulation Systems*, National Aeronautics and Space Administration, NASA SP-5027, Washington, DC, 1967.

Staff: *How to Determine Economic Thickness of Insulation*, Thermal Insulation Manufacturers Association, Mt. Kisco, New York, NY, (Revised periodically).

Turner, W.C. and J.F. Malloy: *Thermal Insulation*, Krieger Publishing Company, Melbourne, FL, 1990.

Web Reference

North American Insulation Manufacturers Association. http://www.naima.org/

E.C. SHUMAN, Consulting Engineer, State College, PA

INSULIN. A polypeptide hormone having a molecular weight of 5733. It is formed in the isles of Langerhans located in the pancreas and was so named for this reason. Insulin is composed of 16 amino acids arranged in a coiled chain and cross-linked in several places by the disulfide bonds of cystine residues. The sequence of amino acids has been elucidated. The insulin molecule was synthesized in 1963. In 1977, rat insulin was produced in the bacterium *E. coli* by recombinant DNA techniques. A year later human insulin was generated after chemically synthesized genes were added to *E. coli*. This synthetic insulin is now in commercial production and has been approved by the FDA. Insulin regulates carbohydrate metabolism in the body by decreasing the blood glucose level. A systemic deficiency leads to diabetes.

See also **Carbohydrates**.

History

Aware that in 1889, Joseph von Mering and Oskar Minkowski had demonstrated that removal of the pancreas from a dog produced diabetes, George Zuelzer in Berlin succeeded, in 1908, in controlling sugar levels in a depancreatized dog by injecting a pancreatic extract. He proceeded to administer the extract to eight diabetics whose condition then improved until supplies of the extract ran out. Meanwhile, in Chicago Ernest Scott was preparing his extracts by pre-soaking pancreases in alcohol to inhibit the digestive enzymes that otherwise would destroy the active hormone. These extracts successfully reduced urinary sugar levels in dogs, as did aqueous extracts being produced at the University of Bucharest by Nicolas Paulesco. In 1921, Paulesco published a report in which he claimed to have isolated the antidiabetic hormone, which he called *pancreine*. His extract required further purification before it could be administered in human trials, but Canadian researchers introduced a pure product before this was achieved. See also **Diabetes Mellitus**.

In May 1921, Frederick Banting began working on pancreatic extract in the department of Professor John Macleod at the University of Toronto, assisted by Charles Best who was a final year biochemistry student. After examining a variety of ways of preparing extracts, the Canadian workers eventually took up the suggestion by Macleod that they make use of the method previously developed by Scott. Progress reached a climax at the end of the year, when they subcutaneously injected each other without experiencing untoward effects. The hormone was then given the name insulin. See also **Banting, Sir Frederick Grant (1891–1941); and Macleod, John James Richard (1876–1935)**.

Impressed by their progress, Professor Macleod arranged for biochemist J. B. Collip to assist Banting and Best with the purification. By pouring their aqueous alcoholic extract into several volumes of alcohol, Collip was able to precipitate insulin. After reconstituting this into an aqueous injection for subcutaneous administration, it was administered to Leonard Thompson on 11 January 1922. He was a dangerously ill 12-year-old patient at the Toronto General Hospital and was expected to die within days. His life was saved by the insulin injections and he remained in good health until his accidental death some years later. Since then, regular injections of insulin have saved the lives of millions of diabetics.

After further clinical trials, Best took control of the production of insulin in May 1922, at the Connaught Laboratories of the University of Toronto. Patents on the production of the hormone were administered by the university and all necessary information was provided to enable the American company Eli Lilly to produce it on a commercial scale. Other companies around the world were subsequently licensed to produce insulin, all producing the hormone from cattle or pig pancreases.

Crystals of pure insulin were isolated in 1926 by Professor John Abel at Johns Hopkins University. Its shorter duration of action than the impure product was presumed to be due to the absence of a protein contaminant. This led Hans Christian Hagedorn of the Nordisk Insulin Laboratory in Copenhagen to seek a protein that could safely extend the action of pure insulin, resulting in the introduction of protamine insulin in 1936. D. A. Scott of the Connaught Laboratories then found that the extended duration of this depended on the presence of zinc. Consequently, protamine zinc insulin was found to be more consistent in its duration of action. After World War II, further refinements were made, culminating in the production of ultra-pure insulin from which minute traces of allergy-producing protein contaminants had been removed by chromatographic methods. See also **Hodgkin, Dorothy Mary Crowfoot (1919–1994) and Yalow, Rosalyn Sussman (1921–Present)**.

The Nobel Prize in Physiology or Medicine was awarded to Banting and Macleod in 1923. Both were upset that their colleagues had been ignored, so Banting shared his prize with Best and Macleod with Collip. In 1958, Frederick Sanger of Cambridge received the Nobel Prize in Chemistry for his determination of the chemical structure of insulin, the first protein to have its amino-acid sequence elucidated. His work revealed that there were minor variations in the chemical structures of ox, cow and human insulin. Amid growing concern over the production of antibodies to animal insulin in patients, Eli Lilly employed recombinant DNA technology, involving the insertion of the insulin gene into the *Escherichia coli* bacterial cell, to produce human insulin on a commercial scale. Novo Industries took a different approach, substituting an amino acid in porcine insulin to convert it into human insulin. As a result of these developments, over three million patients have received human insulin and it has gradually displaced porcine and bovine insulin. See also **Sanger, Frederick (1918–Present)**.

INTEGER. See **Number Theory**.

INTEGRAL. In calculus, $\phi(x)$ is an integral of $f(x)$ if $d\phi/dx = f(x)$. The process of finding an integral is integration or the inverse of differentiation. If C is any real number, then $\phi(x) + C$ is also an integral of $f(x)$, the integrand and C is a constant of integration. Thus, if one integral exists, an infinite number of others may be obtained by adding an arbitrary constant. These are called indefinite integrals and indicated symbolically as

$$\int f(x)dx = \phi(x) + C$$

In a precise mathematical sense, one must carefully distinguish between several different kinds of integrals, known by the names of Cauchy, Riemann, Stieltjes, Lebesque, and others, some of which are discussed in entries following. (For others, see James and James, *Mathematics Dictionary*, 5th edition, Van Nostrand Reinhold, 1992.)

Elimination of the constant of integration by appropriate means gives a definite integral. For example, one could take the difference between values of an indefinite integral for two given values of the independent variable. The definite integral of $f(x)$ between the limits a and b is denoted by the symbol $\int_b^a f(x)dx$. Its properties include

$$\int_b^a f(x)dx = -\int_a^b f(x)dx$$

$$\int_a^b f(x)dx = \int_a^c f(x)dx + \int_c^b f(x)dx$$

This integral is the subject of what is known as the fundamental theorem of integral calculus. Let $f(x)$ be a continuous function over an interval form $x = a$ to $x = b$ and let this interval be divided up into n parts of length Δx_i. Choose a point x_i in each subinterval; then

$$\lim_{n \to \infty} \sum_{i=1}^{n} f(x_i)\Delta x_i = \int_a^b f(x)\ dx$$

The definite interval is thus the limiting value of a sum. It is used to evaluate the length of and the area under plane curves, the area of surfaces of revolution, the volume of solids of revolution, and for many other problems in physics and chemistry.

The simple definition of a definite integral as a sum needs some modification when the integrand becomes unbounded within the integration limits or when one of the limits becomes infinite. In the first case, suppose $f(x) \to \infty$ as $x \to a$. Then, provided that the limit exists,

$$\int_a^b f(x)dx = \lim_{\delta \to 0} \int_{a+\delta}^b f(x)dx; \ \delta \longrightarrow 0$$

and it is called an improper integral. Similarly, if $f(x) \to \infty$ as $x \to b$:

$$\int_a^b f(x)dx = \lim_{\delta \to 0} \int_a^{b-\delta} f(x)dx$$

If $f(x) \to \infty$ as $x \to c$, where c is between a and b:

$$\int_a^b f(x)dx = \lim_{\delta \to 0} \int_a^{c-\delta} f(x)dx \lim_{\epsilon \to 0} \int_{c+\epsilon}^b f(x)dx$$

In the second case, if $f(x)$ is continuous for $x1\ a$ and if the definite integral $\int_a^t f(x)dx$ approaches a limit as $t \to \infty$, this limit is denoted by $\int_a^\infty f(x)dx$ and called an infinite integral. The integral with $-\infty$ as a limit is defined in a similar way. Moreover,

$$\int_{-\infty}^{\infty} f(x)dx = \int_{-\infty}^{a} f(x)dx + \int_{a}^{\infty} f(x)dx$$

An integral requiring more than one integration in order to be evaluated is called a multiple integral.

These improper integrals are also called *convergent*.

INTEGRAL DEPTH SCALE.   See **Meteorology**.

INTEGRAL EQUATION.   An integral equation is an equation in which the unknown appears as part of an integral. The general linear integral equation, said to be of the third kind, is

$$g(x)\phi(x) = f(x) + \lambda \int_a^b K(x, z)\phi(z)dz$$

The known functions are $g(x)$, $f(x)$, and $K(x, z)$, the latter being called the kernel or nucleus. The limits of the integral, a and b, are either known functions of x or constants and λ may be either an absolute constant or a parameter. The unknown quantity, found by solving the integral equation, is ϕ as a function of the independent variable x.

Four special cases have been most widely studied. In Fredholm's equation of the first kind, $g(x) = 0$; and in his equation of the second kind, $g(x) = 1$; in both cases, a, b are constants. Volterra's equations of the two kinds are similar except that $a = 0$ and $b = x$. Nonlinear equations also occur. If one or both limits, or the kernel, become infinite, the equation is singular. If $f(x) = 0$, the equation is homogeneous. General methods for solving integral equations include the Liouville-Neumann series, the Fredholm method, the Schmidt-Hilbert method.

A differential equation, together with its boundary conditions, may be formulated as an integral equation. The resulting functions are particularly useful in eigenvalue or Sturn-Liouville equations, which frequently occur in mathematical problems. See also **Abel Equation**.

INTEGRAL LENGTH SCALE.   See **Meteorology**.

INTEGRAL LENGTH SCALES.   Of the three standard turbulence length scales, the ones that are measures of the largest separation distance over which components of the eddy velocities at two distinct points are correlated. They characterize the energy-containing range of eddy length scales. In the most general form, the integral scales (expressed here as a tensor) are functions of position and are defined in terms of the normalized two-point velocity correlations.

See also **Kolmogorov Microscale**; and **Taylor Microscale**.

INTEGRAL (Line).   Given a vector function of position $V(x, y, z)$, which is defined for all points on a curve such as $A - B$ in Fig. 1, one may replace the curve approximately by a series of equal, directed chords $\Delta_1 S, \Delta_2 S, \ldots, \Delta_n S$. The magnitude and direction of the vector V may then be determined at some point in each segment of the curve. The sum of the scalar products:

$$\sum_{j=1}^{n} V_j \cdot |_{g Dj} S$$

can then be obtained. The line integral is defined as

$$\int_A^B V \cdot dS = \lim_{n \to \infty} \sum_{j=1}^{n} V_j \cdot \Delta_j S$$

The usefulness of the line integral will be immediately recognized if a special case is considered. Suppose that V is the force acting on a particle in a field of force. Then the line integral is just the work done on the particle as it moves from A to B under the action of the force.

When the line integral is taken over a closed path, or contour, starting and ending at the same point, it is usually denoted as

$$\int_C V \cdot dS \text{ or } \oint V \cdot dS$$

and is sometimes called a *contour integral* or a *circulatory integral*.

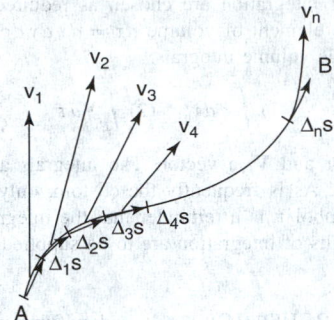

Fig. 1.   Demonstration of line integral.

The choice of the direction in which dS shall be counted as positive is a matter of convention, but cases arise (e.g., in connection with Stokes' theorem) when consistency of convention must be assured. The magnitude of the integral is related by Stokes' theorem to the integral of $\nabla \times V$ (curl V) over any surface bounded by the path.

In the particular case in which V is the gradient of a potential ϕ,

$$\int_A^B V \cdot dS = \phi_B - \phi_A$$

i.e., to the difference of potential between B and A. Over a closed contour, then,

$$V \cdot dS = 0$$

and the field of V is described as *irrotational*.

INTEGRAL (Surface).   To integrate a function $f(x, y)$ over a given surface in the XOY-plane, the methods of calculus show that the result is a definite double integral, usually called the surface integral

$$\int_{a1}^{a2} \int_{b1}^{b2} f(x, y)dx\, dy$$

where the limits of the integration are chosen so that the entire surface S is covered. Vector methods are useful for discussing such integrals, for the surface element dS may be treated as a vector, $dS = dx \times dy$. If ϕ, V are scalar, vector functions, respectively, there are three possible surface integrals:

$$(1) \int_S \phi dS; \quad (2) \int_S V \cdot dS; \quad (3) \int_S V \times dS$$

which give a vector, a scalar, a vector. It is convenient to write only one integral sign, in general, and to understand by the symbol S that the limits of integration are suitably chosen.

In case (2), if V is the product of density and velocity of a fluid (or electric, magnetic, gravitational force; heat, etc.), the integral is the flux of V through the surface. See also **Area**.

INTEGRAL TRANSFORM.   Consider a homogeneous integral equation

$$f(y) = \int K(x, y)F(x)dx$$

with kernel $K(x, y)$. The functions $f(y)$ and $F(x)$ are the integral transforms of each other. Given $F(x)$, presumably $f(y)$ might be found explicitly. Regarding the equality as an integral equation, however, one wishes to solve for $F(x)$, or invert the transform. Thus, if the transform can be inverted, the result will be the solution of the integral equation for a given kernel.

Many special cases have been studied and given special names.

INTEGRAL (Volume).   In elementary calculus, the idea of a surface integral is extended to treat the case of the volume of a solid. If the bounding surface of the solid is given by $f(x, y, z)$, then the volume is the definite triple integral:

$$\iiint dx\, dy\, dz$$

where the limits of integration are chosen as required in each case. In vector notation, the element of volume $d\tau = dx\,dy\,dz$ is a scalar. There are thus two possible volume integrals:

$$(1)\int_\tau \phi d\tau; \quad (2)\int_\tau \mathbf{V} d\tau$$

where ϕ is a scalar and $\mathbf{V}$, a vector. The integrals are, respectively, a scalar and a vector. As is frequently the custom, only one integral sign is used and the symbol τ is a reminder that the integration is triple and that appropriate limits of integration are to be supplied. See also **Volume (Geometry)**.

INTEGRATED CIRCUIT (IC).

A key identifying word in connection with modern electronics is *integrated circuit*. A great majority of articles and books in the literature of electronics refer in some fashion or other to integrated circuitry. The integrated circuit evolved from the printed circuit board (PCB), which shortly after World War II replaced the "spaghetti" wiring formerly found in early radio and television chassis and all manner of electric and electronic gear. The PCB contributed to some size reduction, certainly to neatness and ease of troubleshooting, and to the efficiency (hence lower cost) of automating electronics manufacture. The PCB was a first step toward miniaturization and the concept of "throwaway" electronic subassemblies—it was easier to replace a whole board than to repair or replace a single component. The concept of the integrated circuit, which has moved through a number of key phases, essentially revolutionized electronics, which resulted from microelectronic technology. If the PCB was a major step in advancing the electronics industry, the IC was a super step.

The integrated circuit was first conceived by a radar scientist, Geoffrey W.A. Dummer (born 1909), working for the Royal Radar Establishment of the British Ministry of Defence, and published in Washington, D.C. on May 7, 1952. Dummer unsuccessfully attempted to build such a circuit in 1956. The first integrated circuits were manufactured independently by two scientists: Jack Kilby of Texas Instruments, http://www.ti.com/corp/docs/kilbyctr/jackbuilt.shtml filed a patent for a "Solid Circuit" made of germanium on February 6, 1959. Kilby received patents US3138743, US3138747, US3261081, and US3434015. Robert Noyce of Fairchild Semiconductor was awarded a patent for a more complex "unitary circuit" made of Silicon on April 25, 1961. Fairchild Semiconductor would go on to become one of the major players in the evolution of Silicon Valley in the 1960s.

Integrated circuits made their initial impact on the electronics industry in the early 1960s. In an IC, a number of active and passive circuit elements are inseparably associated on or within a continuous body to perform the function of a circuit. Common forms of integrated circuits include thin-film circuits and monolithic silicon circuits. Hybrid integrated circuits utilize combinations of integrated circuit technologies. A monolithic integrated circuit is one formed in a single-crystal chip of a semiconductor. Transistors, diodes, resistors, and capacitors can be formed in the chip by appropriate diffusion processes. Additional flexibility in the fabrication of resistors and capacitors is obtained by using evaporation techniques to deposit metalized films. Integrated circuits offer a number of advantages in terms of the minute volume needed for an electronic assembly to perform a given function. Because of automated production methods, integrated circuits often improve reliability of operation and a lower unit cost compared with assemblies of discrete components. Much of the electronic equipment described in this volume incorporates integrated circuits. Fig. 1 is a highly schematic illustration of an integrated diode transistor logic circuit. As compared with present-day integrated circuits, the device shown is ultrasimple.

The nomenclature of integrated circuits has changed as the complexity of ICs has increased: The first integrated circuits contained only a few transistors. Called "**Small-Scale Integration**" (**SSI**), they used circuits containing transistors numbering in the tens. SSI circuits were crucial to early aerospace projects, and vice-versa. Both the Minuteman missile and Apollo program needed lightweight digital computers for their inertially-guided flight computers; the Apollo guidance computer led and motivated the integrated-circuit technology, while the Minuteman missile forced it into mass-production. These programs purchased almost all of the available integrated circuits from 1960 through 1963, and almost alone provided the

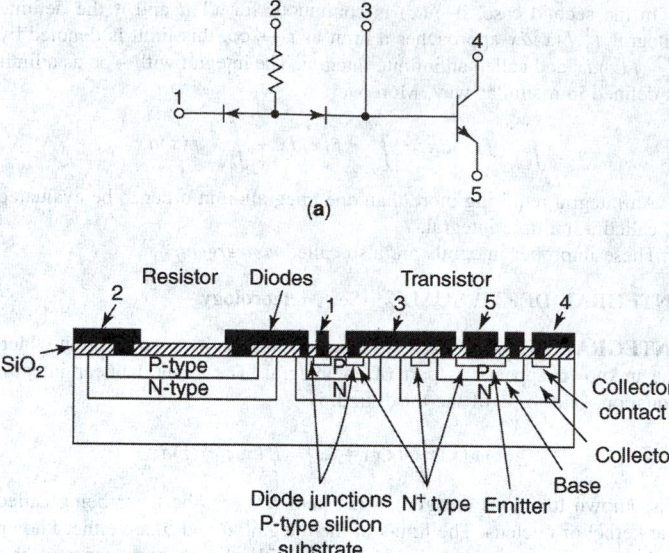

(a)

(b)

Fig. 1.   Diode transistor logic circuit (**a**) shown highly schematically in integrated circuit format. (**b**) Since this early example (shown for its simplicity), integrated circuits have become highly complex.

demand that funded the production improvements to get the production costs from \$1000/circuit (in 1960 dollars) to merely \$25/circuit (in 1963 dollars).

The next step in the development of integrated circuits, taken in the late 1960s, introduced devices which contained hundreds of transistors on each chip, called "**Medium-Scale Integration**" (**MSI**). They were attractive economically because while they cost little more to produce than SSI devices, they allowed more complex systems to be produced using smaller circuit boards, less assembly work (because of fewer separate components), and a number of other advantages.

Further development, driven by the same economic factors, led to "**Large-Scale Integration**" (**LSI**) in the mid 1970s, with tens of thousands of transistors per chip. LSI circuits began to be produced in large quantities around 1970, for computer main memories and pocket calculators.

The final step in the development process, starting in the 1980s and continuing on, was "**Very Large-Scale Integration**" (**VLSI**), with hundreds of thousands of transistors, and beyond (well past several million in the latest stages). For the first time it became possible to fabricate a CPU on a single integrated circuit, to create a microprocessor. In 1986 the first one megabit RAM chips were introduced, which contained more than one million transistors. Microprocessor chips produced in 1994 contained more than three million transistors. This step was largely made possible by the codification of "design rules" for the CMOS technology used in VLSI chips, which made production of working devices much more of a systematic endeavour.

To reflect further growth of the complexity, the term **ULSI** that stands for "**Ultra-Large Scale Integration**" was proposed for chips of complexity more than 1 million of transistors. However there is no qualitative leap between VLSI and ULSI, hence normally in technical texts the "VLSI" term covers ULSI as well, and "ULSI" is reserved only for cases when it is necessary to emphasize the chip complexity, e.g. in marketing.

In the 1980s programmable integrated circuits were developed. These devices contain circuits whose logical function and connectivity can be programmed by the user, rather than being fixed by the integrated circuit manufacturer. This allows a single chip to be programmed to implement different LSI-type functions such as logic gates, adders, and registers. Current devices named FPGAs (Field Programmable Gate Arrays) can now implement tens of thousands of LSI circuits in parallel and operate up to 400 MHz.

The techniques perfected by the integrated circuits industry over the last three decades have been used to create microscopic machines, known as MEMS. These devices are used in a variety of commercial and defense applications, including projectors, ink jet printers, and accelerometers used to deploy the airbag in car accidents.

In the past, radios could not be fabricated in the same low-cost processes as microprocessors. But since 1998, a large number of radio chips have been developed using CMOS processes. Examples include Intel's DECT cordless phone, or Atheros's 802.11 card.

A new generation of IC technology has developed roughly every three years. The Rule of Two holds that approximately for every two generations (six years), the device feature size decreases by two, and other properties such as logic gate speed, chip area, power dissipation, and maximum input/output (I/O) pins increase by two. IC technology has advanced to 0.5-μm circuit geometries for volume production, and is reaching into the deep submicrometer (0.35 μm and smaller) dimensions. Forecasts predict an ambitious scaling of device dimensions of 0.25 μm by 1998, and 0.18 μm by 2001. Moreover, the number of chips per wafer is increasing, as well as the size of the chips and wafers. Silicon wafers are now fabricated in sizes up to 200 mm in diameter, and undergoing development of 300–400 mm sizes. As of 2006, chip areas range from a few square mm to around 250 mm^2, with up to 1 million transistors per mm^2.

Trends in the industry include transferring more of the functions that are found on the supporting printed circuit boards onto the wafers themselves, to reduce the amount of chip packaging and required interconnections. This development is known as wafer-scale integration (WSI) where the goal is to design a complete computer on a wafer. These developments have been supported by concurrent advances in computer-aided design (CAD) tools, both in software and hardware, which have been used to develop integrated CAD design systems that perform IC layout design, simulation, and testing.

Silicon [CAS: 7440-21-3], Si, technology predominates in the semiconductor industry. See also **Semiconductors: Silicon-Based Semiconductors**. Gallium arsenide [CAS: 1303-00-0], GaAs, is considered a possible substitute for silicon substrates, based on its potential for high speed applications where it can operate at high (1.9 GHz) frequencies using low power consumptions and high sensitivity. One reason that GaAs technology has not fulfilled its promise is that silicon technology has dramatically improved in the interim, particularly with improvements in speed, and has reduced the cost-effectiveness of pursuing GaAs development. See also **Semiconductors: Compound Semiconductors**. Expense has limited usage of GaAs to microwave devices, primarily for military use. See also **Microwave Radiation**. However, nonmilitary applications for GaAs devices have been growing, particularly in wireless products such as cellular phones.

Basics of Silicon Technology

VLSI technology is based on the unique attributes of silicon which have allowed the rapid evolution of integrated circuits. As a semiconductor source silicon has an adequate bandgap in its electronic make-up for the movement of electrons, it forms a stable insulating oxide, is abundant, and is inexpensive to make. In constrast, the use of GaAs has been limited to specialty applications because of its costly source and fragile nature.

The property that allows silicon to function in a number of capacities is electronic configuration. Silicon has four electrons in its outer shell. Its crystalline structure allows other elements to reside next to silicon and share electron orbitals with silicon, altering its electrical properties. These other elements are called dopants, and are introduced into the silicon

structure through doping processes. The most common dopant is boron [CAS: 7440-42-8], B, which has three electrons in its outer shell. Silicon doped with the electron-deficient boron is fabricated with an overall positive charge, and is called p-type silicon. A second common dopant is phosphorus [CAS: 7723-14-0], P, having five electrons in its outer shell. Silicon doped with P has an overall negative charge, and is called n-type silicon. Arsenic [CAS: 7740-38-2], as, is another commonly used dopant.

The fabrication of an integrated circuit involves the sequential formation of alternating layers of insulators, semiconductors, and conductors on a silicon wafer. These layers are assembled to form transistor devices that are interconnected to produce particular electrical functions. The layers can be formed by deposition of new material, oxidation of material present on the surface, implantation of additional constituents into surface features, or epitaxial growth of silicon. In order to interconnect the layers, isolate devices from each other, and form integrated circuitry, these layers must be selectively patterned. The patterning is accomplished by photolithography and etching processes.

There are two kinds of integrated circuits (ICs): analogue, or linear ICs, and digital, or logic ICs. Analogue ICs produce, amplify, or respond to various voltages, and are used for any kinds of amplifiers, timers, oscillators, etc. Digital ICs respond to or produce signals that have only two voltage levels. These are used for microprocessors, memories, and microcomputers. It is possible to combine digital and analogue devices on one chip.

Digital IC families are further divided by design and function. The principal IC technologies include p- and n-channel metal-oxide semiconductors (PMOS and NMOS, respectively), complementary metal-oxide semiconductors (CMOS), bipolar, and integrated-injection-logic (I^2L) devices. Of these, CMOS designs are by far the most popular, having an estimated 73% of the worldwide market in 1994. Leading-edge microprocessors, application specific integrated circuits (ASICs), and DRAM ICs larger than 1 megabyte (Mb) are almost entirely fabricated with CMOS technology. Bipolar devices are the choice technology for high speed applications.

There are several reasons for the widespread use and development of CMOS devices, including low power density, relatively good noise immunity and soft error protection, design simplicity, and the capability to include lower power analogue and digital circuitry on the same chip. The most attractive feature has been the ability to scale CMOS technology to smaller dimensions. Processes exist that produce 0.5-μm dimensions under manufacturing conditions. Development is underway to effect 0.18-μm regimes. Another significant development in CMOS technology has been to reduce the power supply voltage from 5.5 to 3.3 V. There is expectation to step down further to 2.5 V as geometries decrease. Developments in equipment and processes are necessary to attain these goals.

Typical CMOS devices use both NMOS and PMOS transistors to form logic devices. A simple NMOS transistor is shown in Figure 2a. Two channels of n-doped silicon are formed in p-silicon to form a source and drain. An NMOS transistor is designed to permit a negative charge to move from the source to the drain in response to a positive charge in the gate. When the charge on the gate is large enough such that the source-to-gate voltage is higher than a threshold voltage V_t, electrons create a conducting path between drain and source causing current to flow. A common CMOS design combines NMOS and PMOS constructions in twin-well (twin-tub) structures, as shown in Figure 2b.

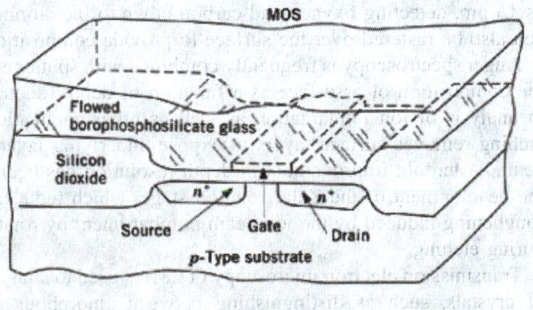
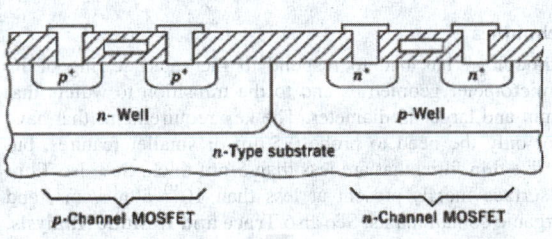

Fig. 2.   Cross sections of electronics devices used in ICs. (**a**) NMOS transistor; (**b**) a twin-tub CMOS device on an n-type substrate. (Courtesy of Custom VLSI Microelectronics.)

Crystal Growth and Wafer Preparation

The single-crystal silicon that is used in IC technology starts with a polycrystalline material called electronic-grade silicon (EGS). Its purity is determined by resistivity measurements made on the test ingot, or by low-ir absorption measurements. EGS is made by starting with a relatively pure form of sand, SiO_2. The sand is melted to form metallurgical-grade silicon (MGS), which reacts with HCl gas to form trichlorosilane [CAS: 10025-78-2], $SiHCl_3$, gas. The $SiHCl_3$ reacts with H_2 in a chemical vapor deposition (CVD) process to form EGS.

$$2\ SiHCl_3(g) + 2\ H_2(g) \rightarrow 2\ Si(s) + 6\ HCl(g)$$

An alternative process involves pyrolysis of silane [CAS: 7803-62-5].

$$SiH_4(g) + heat \rightarrow Si(s) + 2\ H_2(g)$$

This latter process is lower in cost and has fewer harmful by-products.

The polycrystalline EGS is converted to single-crystal silicon via the Czokralski (CZ) crystal growing process, based on the solidification of silicon atoms from the liquid phase at a moving interface. Volume production of 200-mm diameter crystals is standard. Development of crystals having diameters of up to 400 mm has been predicted.

The process of growing a pure crystal is sensitive to a host of process parameters that impact the incorporation of impurities in the crystal, the quality of the crystal structure, and the mechanical properties of the crystal rod. For example, the crystal-pulling mechanism controls the pull rate of the crystallization, which affects the incorporation of impurities in the crystal, and the crystal rotation, which affects the crystal structure.

Two common impurities that must be controlled in silicon crystal growth are oxygen and carbon. Oxygen affects the formation of donor regions, the yield strength of the crystal, and the level of defect generation. Oxygen, as silicon tetraoxide [CAS: 12359-25-0], SiO_4, can act as a donor and change the resistivity of the silicon. Carbon also contributes to the formation of defects, and comes from graphite parts of the melt furnace.

Ingots of EGS are evaluated for resistivity, crystal perfection, and mechanical and physical properties, such as size and mass. The ingots are sliced into wafers using at least 10 machining and polishing procedures. These wafers are sliced sequentially from the ingot, and evaluated for the correct surface orientation, thickness, taper, and bow. As a final procedure, the wafers are chemically cleaned to remove surface contaminants prior to use.

Defects in the silicon crystals affect the electrical, optical, and mechanical properties. Possible defects include point defects, which affect the kinetics of diffusion and oxidation; line defects, which can affect diffusion; planar defects; and volume defects, where impurities have precipitated and can act as sites for dislocation generation. Integrated circuits are typically constructed on wafers having á100ñ crystal orientation. Wafers 200-mm in diameter are being fabricated primarily for high volume, large-area circuits such as 16 Mb DRAMs, although 150 mm and smalled-sized wafers form the majority of product substrates. It is possible to fabricate almost twice as many 16 Mb chips on a 200-mm wafer as on a 150-mm wafer; the higher number of chips offsets the additional costs of the more expensive equipment and reduced throughput found in 200-mm processing. Important factors in the production of 200-mm wafers include a greater sensitivity to flatness, thermal stress, uniformity, and surface microroughness. Particle contamination control has become essential at the wafer fabrication stage to prevent damage in subsequent process steps. IC manufacturers are requiring very low particle counts from wafer suppliers, on the order of 50 or less particles per square centimeter, 0.1–0.15 μm in diameter.

Analytical Techniques

Analytical methodology has had to respond to the rapid scaling of IC designs to submicrometer geometries and to the transition to wafers that are up to 200 mm and larger in diameter. The key requirements that have emerged are not only the need to probe 0.5 μm or smaller features, but also to characterize thin films that are less than 5 nm thick. See also **Thin Films**. Detect surface metals present at less than 10^{10} atoms/cm^3, and identify trace organic contaminants. See also **Trace and Residue Analysis**. These capabilities are essential in process development, process control, and failure mode analysis (FMA), for all stages of fabrication. See also **Process Control**. The purity of gas and liquid starting materials must also be determined.

The physical techniques used in IC analysis all employ some type of primary analytical beam to irradiate a substrate and interact with the substrate's physical or chemical properties, producing a secondary effect that is measured and interpreted. The three most commonly used analytical beams are electron, ion, and photon x-ray beams. Each combination of primary irradiation and secondary effect defines a specific analytical technique. The IC substrate properties that are most frequently analyzed include size, elemental and compositional identification, topology, morphology, lateral and depth resolution of surface features or implantation profiles, and film thickness and conformance. A summary of commonly used analytical techniques for VLSI technology can be found in Table 1.

Electron Beam Techniques

One of the most powerful tools in VLSI technology is the scanning electron microscope (SEM). A sem is typically used in three modes: secondary electron detection, back-scattered electron detection, and x-ray fluorescence (XRF). All three techniques can be used for nondestructive analysis of a VLSI wafer, where the sample does not have to be destroyed for sample preparation or by analysis, if the sem is equipped to accept large wafer-sized samples and the electron beam is used at low (ca 1 keV) energy to preserve the functional integrity of the circuitry. Samples that do not diffuse the charge produced by the electron beam, such as insulators, require special sample preparation.

In the secondary electron mode, a 1–20 keV electron beam is rastered across a surface, causing low energy electrons to be emitted from the surface to produce a high magnification, high resolution image of a surface. The wide depth of field yields three-dimensional images that are focused even with a wide variation of surface feature heights. The back-scattered mode is used to obtain images with a better contrast between elements of differing atomic number. Sems are also typically equipped for xrf analysis, where a primary x-ray beam generates fluorescent x-rays that are analyzed to qualitatively or semiquantitatively identify the elemental composition of the surface, or to map the distribution of elements in the surface.

A state-of-the-art sem uses a field-emission (FE) electron gun (FESEM) to obtain 0.7-nm lateral resolution, equivalent to the resolution obtained by transmission electron microscopy (TEM) but with easier and quicker sample preparation. Dramatic improvements in image quality have resulted from the development of immersion lenses; 1.0-nm resolution is possible using small (2 mm) samples in off-line inspection. High (50–200 keV) energy beams are used to take three-dimensional measurements of high aspect ratio (up to 10:1) contact holes having good depth of field focus. These high energy sources have a sampling depth of up to 20 μm, and can yield images of layers that are covered by other layers, useful for nondestructive detection of voids in underlying films. Lastly, sems are used in nondestructive, on-line, critical dimension (CD) metrology control, where precision as well as speed are required for monitoring dimensional tolerances during processing.

Auger spectroscopy uses a primary electron beam to generate valence-shell electrons that are analyzed for elemental identification and compositional analysis. The two distinguishing features of Auger spectroscopy are that (1) it is very surface sensitive with small spatial resolution. The surface sensitivity stems from the shallow escape depth (0.5–1.0 μm) of the detected electrons, resulting in the characterization of only the uppermost 1–10 monolayers of the surface. (2) In general, Auger spectroscopy can provide quantitative analyses of films as thin as 1.5 nm in areas as small as 15 nm, detecting oxygen and carbon down to 0.1 atomic %. The beam can also be rastered over the surface to provide compositional maps.

Auger spectroscopy is frequently combined with sputter etching to reveal the composition of a surface as a function of depth (depth profile). Used in analysis of ion implantation as well as other applications, the sputter etching removes surface layers to expose underlying layers to the Auger beam. A tenfold improvement in depth resolution has been obtained with the development of the Zalar rotation stage, which reduces crater bottom roughening induced by the ion beam bombardment by rotating the sample during etching.

Transmission electron microscopy (TEM) is used to analyze the structure of crystals, such as distinguishing between amorphous silicon dioxide and crystalline quartz. The technique is based on the phenomenon that crystalline materials are ordered arrays that scatter waves coherently. A crystalline material diffracts a beam in such a way that discrete spots

TABLE 1. ANALYTICAL TECHNIQUES USED IN USED IN VLSI TECHNOLOGY

Energy range, KeV	Secondary signal	Acronym	Technique	Application
Electron beam				
0.020–0.200	Electron	LEED	Low energy diffraction	Surface structure
0.300–30	Electron	SEM	Scanning electron microscope	Surface morphology
1–30	X-ray	EMP	Electron microprobe	Surface region composition
500–10	Electron	AES	Auger spectroscopy	Surface layer composition
100–400	Electron	TEM	Transmission electron microscopy	High resolution structure
100–400	Electron, x-ray	STEM	Scanning TEM	Imaging, x-ray analysis
100–400	Electron	EELS	Electron energy loss spectroscopy	Local small area composition
Ion beam				
0.5–2.0	Ion	ISS	Ion scattering spectrometry	Surface composition
1–15	Ion	SIMS	Secondary ion mass spectrometry	Trace composition vs depth
1–15	Atoms	SNMS	Secondary neutral mass spectrometry	Trace composition vs depth
≥1	X-ray	PIXE	Particle induced x-ray emission	Trace composition
5–20	Electron	SIM	Scanning ion microscope	Surface characterization
>1,000	Ion	RBS	Rutherford back-scattering	Composition vs depth
Photon beam				
>1	X-ray	XRF	X-ray fluorescence	Composition (μm depth)
>1	X-ray	XRD	X-ray diffraction	Crystal structure
>1	Electron	ESCA.XPS	X-ray photoelectron spectroscopy	Surface composition
Laser	Ion		Laser microprobe	Composition of irradiated area
Laser	Light	LEM	Laser emission Microprobe	Trace elements (semiquantitative)
Neutron beam				
Reactor	Gamma	NAA	Neutron activation analysis	Bulk (trace) composition

can be detected on a photographic plate, whereas an amorphous substrate produces diffuse rings. Tem is also used in an imaging mode to produce images of substrate grain structures. Tem requires samples that are very thin (10–50 nm) sections, and is a destructive as well as time-consuming method of analysis.

X-Ray Radiation

After xrf, the most common method of surface analysis utilizing x-rays as the primary beam is x-ray photoelectron spectroscopy (XPS), also known as electron spectroscopy for chemical analysis (ESCA). The x-rays generate photoelectrons emitted with energies that are equal to the difference between the incident photon energy and the binding energy of the electron, and are characteristic for a given element. XPS is very surface-sensitive, similar to Auger analysis, and is particularly useful in analyzing insulators and surface hydrocarbon contaminations. XPS is also used to obtain information about the chemical structure of a film, as the binding energy of an electron is affected by the electronegativity of the surrounding bonded atoms. For example, the carbon $1s$ signal varies in energy maximum and peak shape depending on the electronegativity of the surrounding bonding atoms. A shortcoming of the technology is the large spot size of the x-ray beam, limiting the lateral resolution of the technique; newer instruments have beam sizes of ca 20 nm.

Ion Beam Techniques

Secondary-ion mass spectroscopy (SIMS) uses ion beams having high enough energies to penetrate the surface and break surface bonds, ejecting neutral and ionic species from the surface in a process called sputtering. The primary beam is typically O^+_2. The ejected secondary ions are analyzed and identified according to mass. The sensitivities of ejection or the ratio of ejected ions to atoms present in the substrate varies greatly according to the particular element, the substrate chemistry, or the substrate structure. The principal advantage of SIMS analysis is in its very low detection limits, which are applied to the analysis of doping profiles and the detection and identification of surface contaminants. Time-of-flight secondary mass spectrometry (TOF-SIMS) is a new surface-sensitive technique that analyzes both organic and inorganic contaminants in the top monolayer of a surface at ppm detection limits. SIMS is a destructive analytical process, and requires a large surface area for analysis (5 × 5 mm). See also **Mass Spectrometry**.

Rutherford back-scattering (RBS) is used to determine the composition and distribution of heavy elements in thin films composed of light elements. $^4He^+$ ions having energies are accelerated toward a surface, penetrate the surface, undergo elastic collisions with substrate nuclei, and back-scatter out of the surface for detection and analysis. The energy distributions of these ions are characteristic of the elements in the substrate and the depth within the surface where the collisions occur. RBS analysis is nondestructive, and requires a large sample size of several mm. Using hydrogen forward scattering (HFS), it is possible to quantitatively profile hydrogen concentrations as low as 0.1 atomic % to ±10% accuracy on both conducting and insulating samples.

Newer techniques that are responding to the need for atomic level imaging and chemical analysis include scanning tunneling microscopes (STMs), atomic force microscopes (AFMs), and focused ion beams (FIBs). These are expected to quickly pass from laboratory-scale use to in-line monitoring applications for 200-mm wafers. See also **Scanning Tunneling Microscope**.

Additional Reading

Berlin, L.: *The Man Behind the Microchip: Robert Noyce and the Invention of Silicon Valley*, Oxford University Press, New York, NY, 2005.

Chen, Wai-Kai: *VLSI Technology*, CRC Press, LLC, Boca Raton, FL, 2003.

Cheng, Yi-Kan: *Electrothermal Analysis of VLSI Systems*, Kluwer Academic Publishers, Norwell, MA, 2000.

Dimitrijev, S.: *Principles of Semiconductor Devices*, Oxford University Press, New York, NY, 2005.

Franssila, S.: *Introduction to Microfabrication*, John Wiley & Sons, Inc., New York, NY, 2004.

Grout, I.A.: *Introduction to Integrated Circuit and System Test Engineering*, Springer-Verlag New York, LLC, New York, NY, 2005.

Hurst, P.J., R.G. Meyer, and S.H. Lewis: *Analysis and Design of Analog Integrated Circuits*, 4th Edition, John Wiley & Sons, Inc., New York, NY, 2001.

Jaeger, R.C., R.F. Pierret, and G.W. Neudeck: *Introduction to Microelectronic Frabrication: Vol. 5 of Modular Series on Solid State Devices*, 2nd Edition, Prentice-Hall, Inc., Upper Saddle River, NJ, 2001.

Kingsriter, D.M., and M. Golio: *RF and Microwave Semiconductor Device Handbook*, CRC Press, LLC, Boca Raton, FL, 2002.

Mahapatra, S., and A.M. Ionescu: *Hybrid CMOS Single-Electron-Transistor Device and Circuit Design*, Artech House, Inc., Norwood, MA, 2006.

Noyce, R.N.: "Large-Scale Integration: What is Yet to Come?" *Science*, **195**, 1102–1106 (1977).

Razavi, B.: *Design of Integrated Circuits for Optical Communications*, The McGraw-Hill Companies, Inc., New York, NY. 2002.

Rogers, J., C. Plett, and F. Dai: *Integrated Circuit Design for High-Speed Frequency Synthesis*, Artech House, Inc., Norwood, MA, 2006.

Roy, K. and S. Prasas: *Low Power CMOS VLSI Circuit Design*, John Wiley & Sons, Inc., New York, NY, 1999.

Sait, S.M. and H. Youssef: *VLSI Physical Design Automation: Theory and Practice*, World Scientific Publishing Company, Riveredge, NJ, 2000.

Spanos, C.J.: *Fundamentals of Semiconductor Manufacturing and Process Control*, John Wiley & Sons, Inc., Hoboken, NJ, 2006.

Sudhakar, A.: *The Linear and Digital Integrated Circuits Design Primer*, Thomson Delmar Learning, New York, NY, 2002.

Sze, S.M.: *Physics of Semiconductor Devices*, 3rd Edition, John Wiley & Sons, Inc., Hoboken, NJ, 2006.

Vai, M.M.: *VLSI Design*, CRC Press, LLC, Boca Raton, FL, 2000.

Wang, Laung-Terng, Cheng-Wen Wu, and X. Wen: *VLSI Test Principles and Architectures: Design for Testability*, Morgan Kaufmann, New York, NY, 2006.

Widmann, D., H. Friedrich and H. Mader: *Technology of Intergrated Circuits*, Springer-Verlag, Inc., New York, NY, 2000.

Wong, B.P., G. Starr, A. Mittal, and Y. Cao: *Nano-CMOS Circuit and Physical Design*, John Wiley & Sons, New York, NY, 2004.

Zobrist, G.W.: *VLSI Design Environments*, Gordon & Breach Publishing Group, Newark, NJ, 2000.

Web References

A list of notable manufacturers.

Agere Systems; (formerly part of Lucent, which was formerly part of AT&T): http://www.agere.com/.

Agilent Technologies; (formerly part of Hewlett-Packard, spun-off in 1999): http://www.home.agilent.com/agilent/home.jspx?cc=US&lc=eng&cmpid=4533.

Alcatel: http://www.alcatel.com/.

Altera: http://www.altera.com/.

AMD (Advanced Micro Devices); (founded by ex-Fairchild employees): http://www.amd.com/us-en/.

Analog Devices: http://www.analog.com/.

ATI Technologies; (Array Technologies Inc. acquired parts of Tseng Labs in 1997): http://www.ati.com/.

Atmel; (co-founded by ex-Intel employee): http://www.atmel.com/.

Broadcom: http://www.broadcom.com/.

Fairchild Semiconductor; (founded by ex-Shockley Semiconductor employees: the Traitorous Eight http://traitorous-eight.biography.ms/), and http://www.fairchildsemi.com/.

Freescale Semiconductor; (formerly part of Motorola): http://www.freescale.com/.

Hitachi: http://www.hitachi.com/.

IBM (International Business Machines): http://www.ibm.com/us/.

Infineon Technologies; (formerly part of Siemens):http://www.infineon.com/cgi-bin/ifx/portal/ep/home.do?tabId=0.

Intel; (founded by ex-Fairchild employees):http://www.intel.com/.

Intersil: http://www.intersil.com/cda/home/.

Lattice Semiconductor: http://www.latticesemi.com/.

Linear Technology: http://www.linear.com/index.jsp.

LSI Logic; (founded by ex-Fairchild employees): http://www.lsi.com/index_flash.html.

Maxim IC: http://www.maxim-ic.com/.

National Semiconductor (aka "NatSemi"; founded by ex-Fairchild employees): http://www.national.com/.

Nordic Semiconductor; (formerly known as Nordic VLSI): http://www.nordicsemi.no/.

NEC Corporation; (formerly known as Nippon Electric Company): http://www.nec.com/.

NVIDIA; (acquired IP competitor 3dfx in 2000; 3dfx was co-founded by ex-Intel employee): http://www.nvidia.com/page/home.html.

Parallax Inc: http://www.parallax.com/.

Philips: http://www.usa.philips.com/index.html.

PMC-Sierra; (from the former Pacific Microelectronics Centre and Sierra Semiconductor, the latter co-founded by ex-NatSemi employee): http://www.pmc-sierra.com/.

Realtek Semiconductor Group: http://www.realtek.com.tw/.

Renesas; (joint venture of Hitachi and Mitsubishi Electric): http://www.renesas.com/.

Rohm: http://www.rohm.com/.

Smart Code Corp: http://www.smartcodecorp.com/index.asp.

SMSC: http://www.smsc.com/.

Silicon Optix Inc.: http://www.siliconoptix.com/home.cfm?CFID=5616964&CFTOKEN=58cf803935d947e2-0154D63E-7E90-E2A3-B7F2591DC652108E.

STMicroelectronics; (formerly SGS Thomson): http://www.st.com/stonline/.

Texas Instruments: http://www.ti.com/.

Toshiba: http://www.toshiba.co.jp/worldwide/.

VIA Technologies; (founded by ex-Intel employee) (part of Formosa Plastics Group): http://www.viatech.com/en/index.jsp.

Xilinx; (founded by ex-ZiLOG employee): http://www.xilinx.com/.

ZiLOG; (founded by ex-Intel employees) (part of Exxon 1980-89; now owned by TPG): http://www.zilog.com/.

INTEGRATING-RAMP A/D CONVERTER. This type of analog-to-digital converter converts the unknown input signal into an equivalent pulse-duration signal. The latter is measured by counting pulses that are generated by a precision clock pulse generator. Several design configurations are obtainable, one of the most common being the *dual-ramp* or *dual-slope* integrating A/D converter. The input signal is integrated over a precise time period; then the resultant integral signal is integrated over a variable length of time wherein a reference signal of opposite polarity is used. Inasmuch as the integral of the unknown input signal is proportional to the input signal, it follows that the duration of the second integration also is proportional to the input signal. The second integration period is measured by counting constant-frequency pulses and thus yields a digital representation of the input signal.

A device of this type is shown schematically in the accompanying diagram. When the polarity of the integrator output changes, the comparator at the output of the integrator changes state. The signal to be integrated is determined by the switches at the input of the integrator. The operation sequence of the switches and the required gating of clock signals into the counter is provided by the control logic. The counter also is the output register for the A/D converter.

As shown in Fig. 1, the "start convert" signal causes the counter to clear and closes switch S_1. As shown in Fig. 2, inasmuch as the input signal V_s may be considered essentially constant, the integrator output increases as a linear ramp signal, commencing from an initial integrator offset voltage, $-V_i$. Then, for a fixed time interval (O, t_1), an interval normally determined as the period required to fill the counter one time, the integration is continued. As indicated by an overflow pulse from the counter, when the interval is completed, switch S_2 is opened, while switch S_2 is closed to apply a reference signal $(-V_i)$ to the integrator input. The integrator output decreases as the result of this action. See Fig. 2. It is during the second integration that the interval clock pulses are counted by the counter. Once the integrator output reaches its initial level $(-V_i)$, the comparator changes state and thus pulses are kept from entering the counter. Inasmuch as the integrator output at time t_1 was proportional to the average value of the input signal (during time interval (O, t_1)), the length of the second integration also is proportional to V_s. Thus, the count shown by the counter at time t_2 is a digital representation of the input signal.

Typically, the dual-ramp integrating A/D converter is limited to conversion speeds of from 1,000 to 2,000 samples/second at a resolution of 10 or 12 bits, caused by limitation on the counter counting speed.

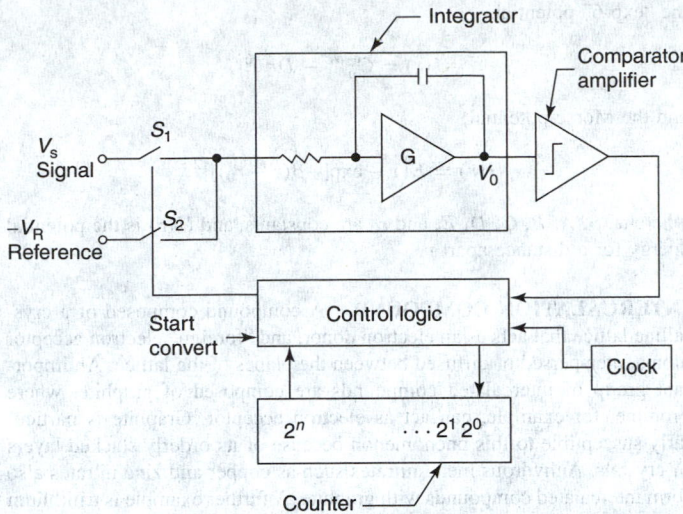

Fig. 1. Dual integrating-ramp analog-to-digital converter.

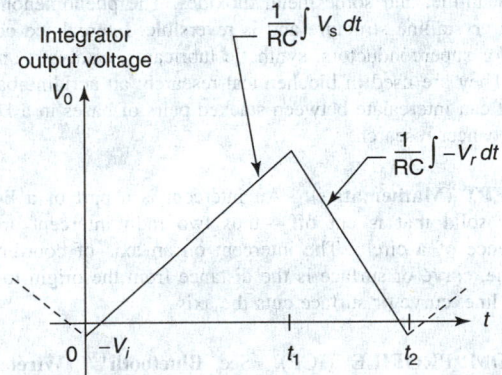

Fig. 2. Integrator-output voltage of dual integrating-ramp analog-to-digital converter.

The conversion rate can be increased to approximately 30,000 samples/second at 14 bits resolution and without markedly increasing the logic requirements by using a two-step integration during the second integration period.

The averaging characteristics and cancellation of errors that usually limit the performance of a ramp-type A/D converter are the principal advantages of the integrating-ramp A/D converter. The integration characteristic provides the *average value* of the input signal during the period of the first integration. Consequently, disturbances such as spurious noise pulses are minimized. The integration compares with a low-pass filter with a 6-dB rolloff and points of infinite attenuation at harmonics of $f = 1/t_1$.

Performance analysis of these converters also indicates that some other types of errors cancel out as well. Long-term drift in the time constant, as may result from temperature changes or aging, do not affect conversion accuracy. Also, long-term alterations in clock frequency have no effect. See also **Analog-to-Digital Converter**.

THOMAS J. HARRISON, IBM Corporation, Boca Raton, FL

INTEGRATION. The process of finding an integral, thus the process inverse to differentiation. If the integrand is simple in form, the integral may be solved by remembering what function would give the integrand by differentiation. It will generally be classifiable as: (I) an algebraic rational function; (II) an algebraic irrational function; (III) a transcendental function. Moreover,

$$\int [Af_1(x) + Bf_2(x) + \cdots]dx = A\int f_1(x)dx + B\int f_2(x)dx + \cdots$$

where A, B, ... are constants or parameters, independent of x. It is then possible to evaluate elementary forms in each of these classes.

The integrand will usually be of a more complicated nature than these elementary types hence it is necessary to reduce it to one of them or some combination of two or more of them. The procedures for this step of the work include: integration by parts; substitution of a new variable; conversion into partial fractions; use of reduction formulas.

In what follows, a, b are fixed constants or parameters; C is a constant of integration; n can be positive or negative, rational or irrational. Moreover, since the general form is

$$\int X(x)dx = I(x) + C$$

we list only X and I in each case, in order to avoid the continued repetition of integral signs and the symbols dx, C, etc.

I. Algebraic Rational Function

	$X(x)$	$I(x)$
1	x^n	$\dfrac{x^{n+1}}{n+1}; n \neq 1$
2	$1/x$	$\ln x$
3	$\dfrac{1}{a^2 + x^2}$	$\dfrac{1}{a}\tan^{-1}\dfrac{x}{a}$
4	$\dfrac{1}{a^2 - x^2}$	$\dfrac{1}{2a}\ln\dfrac{a+x}{a-x}$

II. Algebraic Irrational Function

	$X(x)$	$I(x)$
5	$\dfrac{1}{\sqrt{a^2 - x^2}}$	$\sin^{-1}\dfrac{x}{a}$
6	$\dfrac{1}{\sqrt{x^2 \pm a^2}}$	$\ln(x + \sqrt{x^2 \pm a^2})$

III. Transcendental Function

	$X(x)$	$I(x)$
7	$\sin x$	$-\cos x$
8	$\cos x$	$\sin x$
9	$\tan x$	$\ln \sec x$
10	$\cot x$	$\ln \sin x$
11	$\sec x$	$\ln(\sec x + \tan x)$
12	$\csc x$	$\ln(\csc x - \cot x)$
13	$\sec^2 x$	$\tan x$
14	$\csc^2 x$	$-\cot x$
15	e^x	e^x

The result in (6) is equivalent to $\sinh^{-1} x/a$ for the upper sign and $\cosh^{-1} x/a$ for the lower sign. If $X = ax^2 + bx + c$ and the integrand is of the form $F(x, \sqrt{X})$, the special devices in an earlier paragraph of this article will reduce the integral to one of the elementary forms or combinations of them.

Large collections of integrated expressions are published in tables of integrals, and, in most cases, with sufficient ingenuity, a given integral can be converted to one of these standard forms. It should be noted, however, that many expressions cannot be integrated to give known functions; hence, one must then resort to graphical methods, numerical integration, or series integration.

INTEGRATION BY PARTS. If u and v are functions of a single independent variable, differentiation of their product gives $d(uv) = u\,dv + v\,du$. The inverse formula is that for integration by parts

$$\int u\,dv = uv - \int v\,du$$

It frequently happens that a given function is not integrable directly but a solution may be found by this method. For a definite integral, the formula may be written

$$\int_a^b f(x)\,dg(x) = [f(x)g(x)]_a^b - \int_a^b g(x)\,df(x)$$

INTEGRATION (Numerical). Evaluation of a definite integral from pairs of numerical values of the integrand. Graphical or mechanical methods may be used, but more frequently the integrand is approximated by an interpolation formula, which is then integrated term by term. Special cases are the trapezoidal, Simpson, and Weddle rules; the Gauss, Gregory, Newton-Cotes, and Euler-Maclaurin formulas.

When an integral or the solution to a differential equation can be displayed in the form

$$y = \int f(x)\,dx$$

the mathematician often says it has been reduced to quadrature (or cubature, if there are two integral signs). For this reason, the approximate methods described here are known as quadrature (or cubature) formulas.

INTEGUMENT; INTEGUMENTARY SYSTEM. (For the use of this term in botany, see also **Seed**.) The body of every multicellular animal is covered with a layer of tissue adapted to meet the external conditions that prevail in the normal environment of the species. This covering is the integument, and together with all the specialized structures derived from the cellular layers it constitutes the integumentary system.

The general functions of the integumentary system are protection against mechanical damage and desiccation, the transmission of materials which must pass into or out of the body, the conservation of heat, and the reception of stimuli. Among the invertebrates, such rigid supporting structures as the animal may possess are often developed in the integument, so that it becomes a skeletal system or exoskeleton as well as a covering for the body.

The integument is always at least partly ectodermal in origin. In the simpler animals, such as the coelenterates, it is an epithelial layer containing cells specialized for the reception of stimuli, defensive cells, and simple cells bearing contractile basal processes. It bears cilia in some of the flatworms, in the rotifers and bryozoans, and in some of the mollusks (Mollusca), and so aids in bringing food to the animal and in locomotion. In parasitic flatworms it degenerates into a syncytium and produces a noncellular cuticle, and in roundworms, segmented worms, and arthropods it also secretes an external cuticle, although it remains cellular. It gives rise to the setae of the segmented worms and arthropods and to the external portions of sensory organs, and contains glands of various kinds. In the arthropods, the cuticula is highly developed as an exoskeleton. The integument also lays down the hard deposits of corals and secretes the shells of brachiopods and mollusks.

The integument of vertebrates is a skin composed of two layers, an inner corium or dermis derived from mesoderm and an outer cuticle or epidermis which is ectodermal. It produces various hard structures, including the scales of fishes, scales of reptiles, birds and mammals, feathers, hair, horns, claws, hoofs, and nails. In addition, it contains glands of various kinds, such as the mucus glands of fishes and amphibians and sweat glands of mammals, and forms parts of the sensory organs.

INTENSITY. The concentration of some factor, such as radiation, over a given area, or within a given span of time. Thus, the intensity of illumination, magnetization, or other radiation, or of sound. In photography, the density or opaqueness of an image.

INTERATOMIC DISTANCE. See **Chemical Elements**.

INTERATOMIC POTENTIAL. The potential energy of two atoms. The three most important potentials used are the Lennard-Jones "12–6" potential

$$U(r) = Ar^{-12} - Br^{-6}$$

the "exp-6" potential,

$$U(r) = C_p^{-\alpha r} - Dr^{-6}$$

and the Morse potential,

$$U(r) = E\{1 - \exp[-\beta(r - r_0)]\}^2$$

where α, β, A, B, C, D, E, and r_0 are constants, and $U(r)$ is the potential energy for a distance apart r.

INTERCALATION COMPOUND. A compound composed of a crystalline lattice that acts as an electron donor, and "foreign" electron acceptor atoms interspersed of diffused between the planes of the lattice. An important group of intercalated compounds are composed of graphite, where bromine, for example, can act as electron acceptor. Graphite is particularly susceptible to this phenomenon because of its orderly stacked layers of crystals. Anhydrous metal nitrates such as copper and zinc nitrates also form intercalated compounds with graphite. A further example is trilithium nitride, whose structure consists of a series of layers of dilithium nitride, between which is a layer of lithium atoms. This markedly increases the conductivity, so that the material becomes an effective solid electrolyte in batteries. Other substances having this property are sodium β-alumina, titanium disulfide, and some metal dioxides. The phenomenon does not impair the crystalline structure and is reversible. Intercalated compounds are used for superconductors, synthetic lubricants, catalysts, and storage batteries. They are used in biochemical research; an acridine-based compound that can intercalate between stacked pairs of bases in a DNA helix is used in cancer research.

INTERCEPT (Mathematics). An intercept is a part of a line, plane, surface or solid that is cut off—thus two radii intercept arcs of the circumference of a circle. The intercept on an axis of coordinates of a straight line, curve or surface is the distance from the origin to the point where the line, curve or surface cuts the axis.

INTERCOM PROFILE (ICP). See **Bluetooth^{TM} Wireless Technology**.

INTERCONNECTIONS (Electronics). The continuing push toward microminiaturization has had a heavy impact on connector technology. Many electronic systems now demand a whole new line of connectors with a high-density pitch of 2 mm or less, including 1.5 mm, 1.25 mm, and even 1 mm. Smaller connectors require much tighter production standards, which adds to cost, while, concurrently, more and more connectors are needed for the increased complexity of end-use equipment.

Connectors are designed for a wide range of applications. Currently, printed circuit boards (PCBs) consume approximately one-third of all low-voltage, low-current connectors made. Although fiber optic connectors presently account for only about 2% of the total, this is expected to rise markedly during the next few years.

Modern connector designs, combined with miniaturization, are largely responsible for the neat electronics packaging seen today. This is in extreme contrast with the disheveled array of tangled, hard-soldered wiring once found in the chassis of a radio set.

Some degree of standardization for connector hardware has been in effect for a number of years, as through the efforts of ANSI (American National Standards Institute), but connector sizes and configurations depend largely upon bus configurations. Until fairly recently, original equipment manufacturers (OEMs) have used proprietary bus designs as a competitive advantage. This concept is no longer acceptable to the end user.

As of the early 1990s, the IEEE (Institute of Electrical and Electronics Engineers) is sponsoring a concept known as "Futurebus Plus" to be used as a standard. Some forecasters state that, within just a few years, a wide range of OEMs will support the new bus protocol. Metrification also has plagued connector manufacturers, and it is projected that European OEMs will decree that connector hardware be made to metric specifications within a relatively short time span.

The "Futurebus Plus" standard is based on a 2-mm connector. It is claimed that, with bus widths of 62, 148, and 256 bits, the new standard will provide ten times the speed of present buses. The new bus is asynchronous, meaning that the speed is determined by the semiconductors and interconnects that it supports.

Several state-of-the-art electronic systems and the application of several varieties of connector formats are shown in Fig. 1.

Terms Commonly Used in Connector Terminology. These include:

Contact alignment—Sometimes called *contact float*, the amount of allowable contact movement within the connector body. Permits self-alignment.

Crimp termination—A stripped wire inserted into a barrel or trough, which then is crimped to the conductor, using a special crimping tool.

Contact(s)—The part or parts of a connector that provide circuit continuity between two sections of equipment (i.e., the parts that must be mated electrically). Usually, the back end of a connector is called the *termination*. This is, for example, the point where the wires or PC board is attached to the connector.

DIL—Dual In-Line.

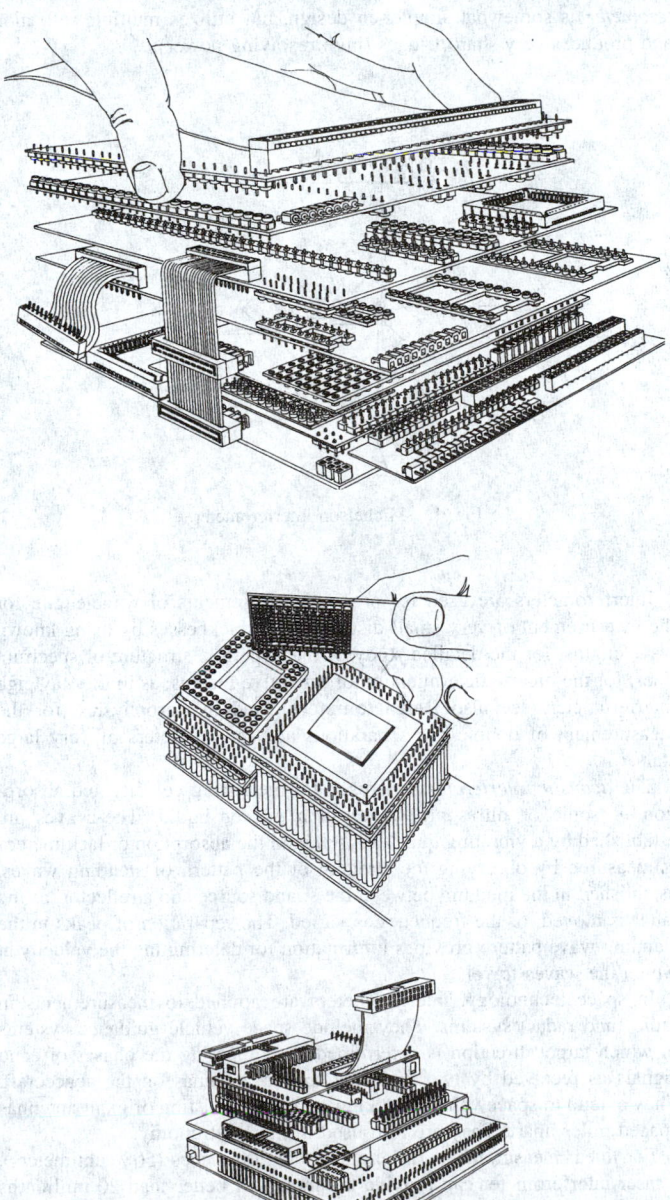

Fig. 1. Examples of precision and intricacy of connectors required in the assembly of modern electronic system hardware. (*Samtec, Inc.*)

DIN—Deutsche Industrie Norm. A German dimensional standard widely accepted throughout Europe.

Printed circuit connector—Used in conjunction with printed circuit boards (PCBs). There are two different styles:

Edgeboard—The printed circuit board edge enters the connector.

Two-piece—One part of the connector is attached physically to the PC board; the other part is attached to some other portion of an assembly, such as a motherboard or cable.

Readout—The arrangement at the terminating edge of a PC board, which may be single or double. A double readout PC board has termination strips on each side that usually are not interconnected.

Ribbon connector—A flat "ribbon" cable that is attached to a round or rectangular connector. In another version, a flat cable may be connected to a flat connector.

Ribbon contact connector—Contacts are in a rectangular connector, which has a self-wiping action. (Not to be confused with ribbon connector.)

Service rating—Maximum rated voltage and/or current for which a connector may be used. Sometimes called *working voltage*. Term also may apply to the number of times that a connector can be mated and separated without faulty performance.

Test voltage—The voltage that a connector must withstand for 1 minute without breakdown, when voltage is applied between connector and shell and any grounding devices of the connector.

Additional Reading

Antelman, L.: "Connectors Shrink Center Spacing to under 2 mm," *Electronic Buyers' News*, 12 (February 25, 1991).

Harman, G.G.: *Microelectronic Interconnections and Assembly*, Kluwer Academic Publishers, Norwell, MA, 1998.

Harper, C.A. and M.B. Miller: *Electronic Packaging, Microelectronics, and Interconnection Dictionary*, The McGraw-Hill Companies, Inc., New York, NY, 1993.

Harper, C.A.: *Electronic Packaging and Interconnection Handbook*, 4th Edition, The McGraw-Hill Companies, Inc., New York, NY, 2000.

Koser, J.R.: "Futurebus + Leads to Connector Systems," *Electronic Buyers' News*, 8 (February 25, 1991).

Kraetzl, M. and F. Hsu: *Parallel System Communications and Interconnections*, CRC Press, LLC., Boca Raton, FL, 1999.

INTERFACE. 1. A common boundary between two parts of a (phase or) system, whether material or nonmaterial.

2. Specifically, in a rocket vehicle or other mechanical assembly, a common boundary between two components.

3. Specifically, in fluid dynamics, a surface separating two fluids across which there is a discontinuity of some fluid property such as density or velocity or of some derivative of these properties in a direction normal to the interface. The equations of motion do not apply at the interface but are replaced by the boundary conditions.

4. Frequently, in industrial processes, it is necessary to measure the interface between two liquids of different densities. Interface measurements also are encountered in the operation of pipelines that are designed to carry different kinds of liquid products.

INTERFACIAL TENSION. The contractile force of an interface between two liquids, resulting from their surface tensions, and the attraction between the molecules of the two liquids. It is commonly determined by measuring the interfacial surface energy.

INTERFERENCE FILTER. 1. A filter used to suppress manmade interference entering a receiver from the power line.

2. A filter which effectively increases the selectivity of a receiver, thus decreasing its sensitivity to strong adjacent-channel, image or intermediate-frequency transmissions.

3. An optical device which transmits only a narrow band of wavelengths, other wavelengths being suppressed by the destructive interference of waves transmitted directly through the filter and those reflected $2n$ times, where n is an integer (from back and front faces of the filter).

INTERFERENCE (Signal).   In a signal transmission system, interference is either extraneous power, which tends to interfere with the reception of the desired signals, or the disturbance of signals which results.

INTERFERENCE (Wave).   The variation of wave amplitude with distance or time, caused by the superposition of two or more waves. As most commonly used, the term refers to the interference of waves of the same or nearly the same frequency. Wave interference is characterized by the phenomenon of the occurrence of local maxima and minima of wave amplitude, which cannot be described by the ray approximation to solutions of the wave equation. In terms of Huygens' principle, interference can occur whenever wave disturbance can be propagated from a source to a region of space by two or more paths of different length. There is (destructive) interference if the phases and amplitudes of the disturbances arriving by the various routes are such as to reduce the square of the resultant amplitude, below the sum of the squares of the amplitudes of the components. Two or more sources may only be used if there is a fixed phase-relation between them. Sound interference results when the waves concerned are sound waves. Optical interference occurs with light waves. Thus, a beam of radiation may be separated into two parts, follow different paths and then brought back to form a single beam. Unless the two paths are of identical optical length, the two beams may not be in phase, and can destructively interfere at some points (dark) and constructively interfere at other points (bright). From the principle of conservation of energy, it is known that there is not a loss in energy due to interference. The energy missing at dark points will be found in the bright points. Interference patterns are commonly light and dark bands, all of equal width. Light beams that can cause interference patterns are called "coherent," while beams that cannot cause interference patterns are "incoherent." See Fig. 1.

INTERFEROMETER.   The term interferometer may be applied to any arrangement whereby a beam of light from a luminous area clearly defined is separated into two or more parts by partial reflections, the parts being subsequently reunited after traversing different optical paths. The two components then produce interference.

The best known instrument is that of Michelson, shown diagrammatically in Fig. 1. The original beam a is separated at the surface AM of a glass plate, part of it $(1, 1')$ going to a mirror M_1 and part $(2, 2')$ going on through a second, exactly similar plate B to the mirror M_2. They reunite at AM and are observed together at E. One of the mirrors, M_1, is mounted on a micrometer screw so that its distance from AM can be varied, the phase difference of the reunited beams thereupon passing through a series of cycles. If M_1 or M_2 is not quite perpendicular to the beam reflected by it, the field at E is crossed by interference fringes, which move across the field as the mirror M_1 is moved. Each complete cycle corresponds to a displacement of M_1 equal to a half-wavelength. The *Fabry* and *Perot interferometer* is somewhat simpler in design, but utilizes multiple reflection and produces very sharp fringes (high resolving power).

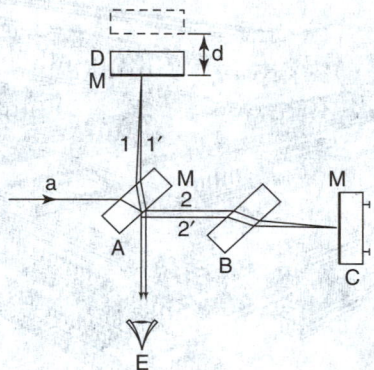

Fig. 1.   Michelson interferometer.

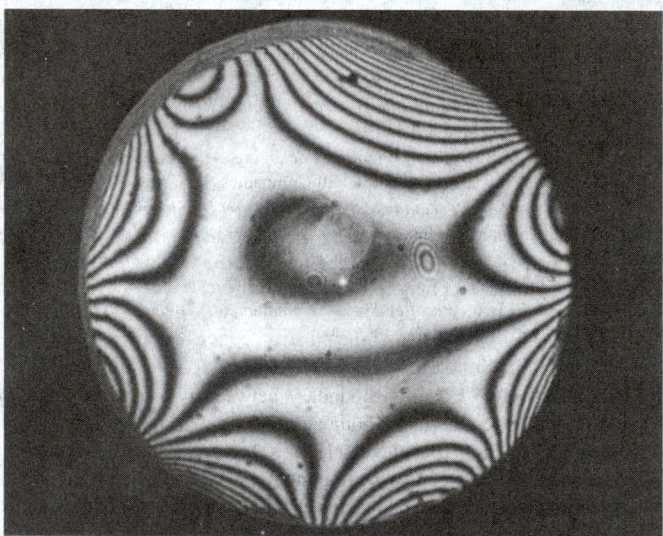

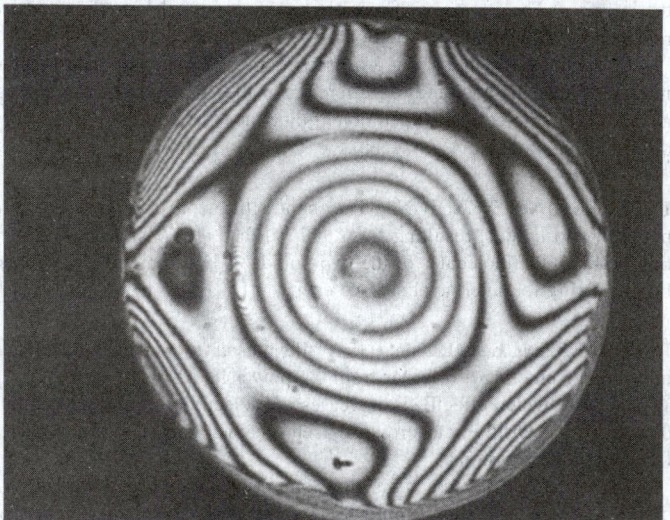

Fig. 1.   Interferograms showing stress patterns induced in a lens that is staked at four points. (*Polaroid*.)

Interferometers are used for precise measurements of wavelength, for the measurement of very small distances and thicknesses by using known wavelengths, for the detailed study of the hyperfine structure of spectrum lines, for the precise determination of refractive indices, as in the Rayleigh interferometer (see also **Refractometers**), and in astrophysics, for the measurement of double-star separations and the diameters of very large stars.

The *acoustic interferometer* is used for measuring velocity and absorption of sonic or ultrasonic waves in a gas or liquid. The waves are established by a vibrating quartz crystal, and the absorption or lack thereof is measured by observing the strength of the pattern of standing waves, established in the medium between the sound source and a reflector, as the latter is moved, or the frequency is varied. The separation of peaks in the standing wave pattern provides information for determining the velocity at which the waves travel.

In space technology, interferometers are applied to measurements in radio (and radar) systems. They include space vehicle guidance systems in which target direction is determined by comparing the phases of echo signals as received by two precisely spaced antennas on the spacecraft. They extend to space vehicle tracking systems consisting of giant antennas spaced miles apart. They are also used in radio astronomy.

For linear measurements of lengths up to 200 inches (508 centimeters), a laser interferometer can provide accuracies of better than 20 millionths of an inch (0.000000127 centimeter). The detection system determines the precise difference in optical path length between a known length and the one to be measured. This technique is used to measure the height of ocean waves or contours of machine parts and earth masses. See Fig. 2.

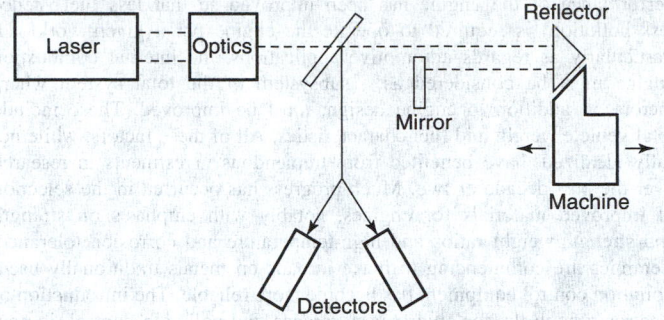

Fig. 2.   Laser interferometer.

As early as 1929, some researchers suggested that atoms can mimic light waves after observing atoms diffracting from the surface of a crystal. The grooves that were responsible for diffracting the atoms were assumed by the early investigators to be the minute rows of atoms in the crystal lattice.

David Pritchard (Massachusetts Institute of Technology) and a research team have developed an interferometer that utilizes the aforementioned principle. As explained by Pritchard, "The device first breaks up the matter wave of each atom into separate components—the interference pattern's *bright spots*. Then, another grating sends two of the wave components toward each other, forcing them to interfere a second time. A third grating channels lines in the pattern to a detector. See Fig. 3. Like conventional interferometers, the device can record very small shifts in the interference pattern. The detector picks up at the *bright lines* in the pattern."

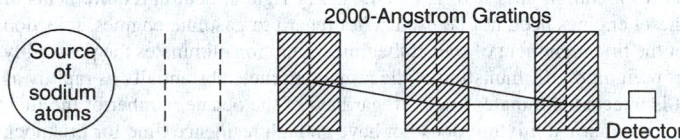

Fig. 3.   Atomic interferometer, as described by David Pritchard.

Other researchers currently are working to construct interferometers based upon atomic interference. These include groups at the Universitäat Konstanz (Germany), at Stanford University, and at Physikalisch Technische Bundesanstalt in Braunschweig. Each group is taking a different approach. The first group mentioned is investigating interference patterns created by helium atoms as they pass through two extremely minute slots in a gold foil. The Stanford group is using laser pulses to split up atom waves instead of foils or gratings. More detail can be found in the excellent summary by F. Flam (*Science*, 921, May 17, 1991).

INTERGLACIAL.   See Climate.

INTERMEDIATE (Chemical). An intermediate generally is considered to be a material (usually a chemical compound) that occurs somewhere in a chemical manufacturing process between the introduction of the basic raw materials and the creation of the final end products. When two or more separate chemical reactions are involved, the intermediate may be the product of one of the *between reactions* and serve as a charge material for a subsequent reaction. For example, in the manufacture of aromatic polyester, several materials and reactions are required. The fundamental raw materials are nitric acid, xylene, methanol, and ethylene glycol. In one reaction, *p*-xylene and nitric acid yield terephthalic acid. The terephthalic acid then is esterified with methanol using sulfuric acid as a catalyst to yield dimethyl terephthalate. The dimethyl terephthalate then undergoes an ester interchange with ethylene glycol which yields *bis*-(β-hydroxyethyl)terephthalate, later condensed to polyethylene terephthalate. This low-molecular-weight polymer then is polymerized to a high-molecular-weight polyethylene terephthalate. In this operation, terephthalic acid and dimethyl terephthalate can be regarded as intermediates. In some instances, a producer will procure intermediate materials

from the outside rather than produce them in-house, particularly in the cases of the pharmaceutical and dye industries. Thus, a number of intermediates are high-tonnage items of commerce. Some intermediates are of low-tonnage requirements and sometimes the economics is in favor of one producer who supplies a number of using firms. A representative list of intermediates would include: *o*-aminophenol-*p*-sulfonic acid; 2,6-dichloro-4-nitroaniline; 4-sulfophthalic acid; *o*-tolidine dihydrochloride; diphenylmethane; diphenylacetaldehyde; methyl cyclopentylphenylglycolate; and 2,3-dichloro-5,6-dicyano-benzoquinone. See also **Synthesis (Chemical)**.

INTERMEDIATE METALS (Law of).   See Thermocouple.

INTERMEDIATE-MOISTURE FOODS. Authorities have various defined intermediate-moisture foods (IMF) as having from 15–40% water at a water activity $a_w = 0.6 - 0.8$. Interest in intermediate-moisture foods for the human diet largely stemmed from the introduction and acceptance of "soft-moist" pet foods, which were first marketed in the 1970s. Research continues at a good pace and considerable progress has been made as, for example, in processed cheese foods with high moisture contents.

More important than moisture content, per se, is the objective of creating a preserved food substance that is stable and can be eaten directly. An idealized IMF system will have (1) a microbial stability at reduced water activity, (2) storage stability without special conditions, (3) reduction of weight and more compactness of product, and (4) can be consumed from the package without rehydration. Some authorities believe that IMF systems represent excellent potential in the development of the snack market.

Two factors have largely delayed expansion of IMF systems: (1) technology and (2) consumer acceptance. Some of the technical problems involved in formulating IMF systems include: (1) rates of lipid oxidation, (2) enzymatic deterioration, and (3) nonenzymatic deterioration. With some products, there are also problems associated with the desired texture.

INTERMEDIATE WATER. As a general term, any water mass found at intermediate depth in the ocean. Antarctic Intermediate Water is the most important of these, followed by Subarctic Intermediate Water and Arctic Intermediate Water. Other water masses identified as intermediate water are Atlantic Intermediate Water in Baffin Bay, also called Polar Atlantic Water, identified by a temperature maximum at a depth of about 500 meters (1,640 feet) resulting from inflow from the West Greenland Current; Arctic Intermediate Water in Baffin Bay, identified by a temperature minimum at a depth between 50 and 200 meters (164 and 656 feet) resulting from inflow of arctic water from the north; and Levantine Intermediate Water in the Eurafrican Mediterranean Sea, identified by a salinity maximum at a depth between 150 and 400 meters (492 and 1,312 feet) and formed when cold winter winds, descending on the region between Rhodes and Cyprus and on the northern and central Adriatic Sea, result in the cooling and sinking of surface water. See also **Ocean**.

INTERMETALLIC COMPOUND. In certain alloy systems, distinct intermediate phases occur where the constituent atoms are in fixed integral ratios, e.g., CuZn (β-brass). Such a compound is held together by metallic bonding and may form a very complicated crystal structure. The constitution of such an alloy is often governed by the Hume-Rothery rules. In some cases, if the electron concentration is such as to just fill a band, the material may even be semiconducting (e.g., InAs). See also **Compound (Chemical)**.

INTERMITTENCY. The property of turbulence within one air mass that occurs at some times and some places and does not occur at intervening times or places. Whereas the classical theory of homogeneous turbulence relies on the assumption that the turbulence energy dissipation rate ε is constant in space, in reality ε is not always constant. Those inhomogeneities may lead to intermittent turbulence. To predict a turbulent or nonturbulent (laminar) behavior within an air mass properly, the complete vertical profiles of either the virtual potential temperature or the buoyancy must be known. Turbulence is often intermittent in the stable boundary layer (e.g., nocturnal) and in the entrainment zone capping the convective mixed layer (e.g., daytime).

INTERNAL COMBUSTION ENGINE.   An internal combustion engine consists primarily of a cylinder, almost always stationary, and a piston, generally single acting, which, together, form a combustion chamber of variable volume. Both of these parts are constructed of metal. As the temperatures attained during combustion are well above the ability of uncooled metals to withstand them, the cylinder of an internal combustion engine must be adequately cooled by transferring through the cylinder wall a certain amount of the heat contained in the gases of combustion. This is accomplished in a practical way by surrounding the cylinder with a jacket of cooling water, or by providing it with an extended outer surface of fins so that air can absorb enough heat to keep the metal cool. The required motion is given to the piston by a crank and connecting rod mechanism; this also serves to take from the piston the power developed by gas pressure.

An engine operating with flaming gas in its cylinder would not last long with simple metal-to-metal contact of the moving parts. Therefore a lubricating system, embodying oil as the lubricant, is an important feature of every internal combustion engine. The most difficult job is that of lubricating the piston in the cylinder. During a portion of the stroke, at least, the lubricated wall is exposed to incandescent gases which tend to burn off the film of lubricating oil. The cooling system must be adequate to maintain the metal surfaces cool enough to save the lubricating film.

The events of the cycle upon which an internal combustion engine works are controlled chiefly by the operation of valves located in ports leading to and from the cylinder. Generally, an admission or inlet valve, and an exhaust valve, are provided in each cylinder. The operation of these valves is derived mechanically from the crankshaft through the valve gear system.

The combustion of fuel in an internal combustion engine is not a continuous affair, but a series of individual explosions, each one requiring a metered amount of fuel to be individually ignited. For this reason, every internal combustion engine must incorporate an ignition system, whose function is to supply in proper time the ignition temperature required for combustion. The internal combustion engine is of a type tending to deliver its power cyclically, and in a fashion which would be very fluctuating unless balanced by the use of heavy flywheel, or by overlapping of power impulses through multicylindered arrangements. It is usual, in fact, to build internal combustion engines with more than one cylinder so that the delivery of power will be more uniform, and flywheel proportions will not be excessive. The supply of fuel to multi-cylindered engines from a common source, and the conduction of exhaust from them, leads to another service feature for the internal combustion engine, namely, the inlet and exhaust manifolds.

The production of power by this type of engine represents a thermodynamic conversion of a portion of the heat energy developed into mechanical energy. The heat energy enters the engine latently in the form of fuel. Mechanical energy appears as power available at the crankshaft. Unavailable or rejected heat is found in exhaust, cooling, and friction. The conversion of the energy of the fuel into useful power takes place about as follows: Air is brought into the cylinder and, either after, before, or during compression, depending on the cycle, fuel is introduced into the air and mixed with it. Upon ignition of this fuel, the heat developed raises the pressure of the products of combustion, or, at least, maintains the pressure during some motion of the piston. The fact that the piston has, against one face of it, a gas pressure greatly exceeding that on the other, inevitably results in the transmission of energy through the train of mechanism consisting of moving piston, wrist pin, connecting rod, and crankshaft. During the motion of the piston, the gases of combustion expand and are cooled somewhat. It has not been found economical to build an engine sufficiently bulky to expand the gases until they reach ordinary atmospheric temperature, and there is always considerable heat loss in the exhaust.

Design Criteria. Although the internal combustion engine in recent years has been identified as a major environmental pollutant, a satisfactory substitute source of power for vehicles, and for many portable tools and machines, had not been found. It is unlikely that the engine will be displaced in a major way for several decades. Meanwhile however, the

performance of the engine has been improved so that less fuel (hence less pollution) is required to operate the engine per a given workload. Particularly as regards automotive applications, the internal combustion engine must be considered as a subsystem of the total system where factors, in addition to engine design, must be improved. These include total vehicle weight and fuel characteristics. All of these factors, while not fully idealized, have benefited from tremendous investments in research over the past decade or two. Much progress has occurred in the selection of improved materials for engines, notably with emphasis on strength and strength/weight ratios and high-temperature and corrosion tolerance. Ceramics are commencing to make inroads on metals traditionally used. Pollution control equipment has become more reliable. The introduction of electronic controls, for engine performance and pollution control, is now widely used. See also **Automotive Electronics**.

Nevertheless, the engine designers continuously refer to a menu of objectives, which include: (1) achieving the best fuel-air ratio; (2) finding the optimal system for introducing fuel and air to the engine and for removing products of combustion; (3) improving heat transfer; and (4) reducing friction, among others. In terms of numbers built per year, the gasoline-using internal combustion engine remains the leader, followed by the diesel engine. See also **Diesel Engine**. In realizing that the principles of internal combustion can be utilized in a number of design configurations, research continues in this very competitive, worldwide field, to seek alternative approaches. The performance comparison of the diesel engine with the spark-ignition gasoline piston engine is a case in point.

In the diesel engine, positive, immediate ignition of the first increment of injected fuel eliminates the need for high compression ratio and cetane number required to achieve compression ignition. It also eliminates the ignition delay characteristics of the compression ignition process and its associated attendant limitations. Combustion loads, noise and shock are correspondingly less severe, and the heavy rigid structural requirements of diesel engines need not be used. With regard to gasoline engines, injection of the first increment of fuel at the time of ignition eliminates the possibility of preignition. Combustion of the remaining fuel substantially as rapidly as it is injected eliminates knock, regardless of the octane number of the fuel, since unburned mixture does not have enough residence time for preknock reactions to occur. Load control does not require throttling of the air charge, but is obtained by regulating the quantity and thereby the duration of fuel injection with full load duration corresponding to about the time for one air swirl. These factors point to the freedom to choose compression ratio and/or supercharge regardless of octane or cetane number of available fuels. Consequently, considerable design latitude exists for optimization of engine parameters. The elimination of octane and cetane numbers as relevant fuel qualities, coupled with the direct cylinder injection of the fuel produces an engine with very broad fuel tolerance.

The rotary engine represents a marked design departure from the conventional gasoline engine. Theoretically, the rotary engine has a number of advantages which, in practice, remain to be achieved. For a given power input, the rotary engine can be up to 50% smaller and 30% lighter than a comparable piston engine. This permits flexibility in the configuration of vehicles. In efficiency and emissions, the engine has offsetting advantages and disadvantages that lead to some penalties. Emission control by rich thermal reactors tends to produce fuel economy penalties and durability problems.

The Stirling cycle engine is considered the most efficient of alternative heat engines. This engine employs the alternate heating and cooling of an enclosed working fluid (hydrogen). The heat source is a continuous-flow external burner. Combustion can be controlled much more accurately than is the case of the intermittent combustion systems. Fundamentally, the engine runs quietly and is comparatively free of vibration, and can be adapted to a rather wide range of liquid fuels, with torque characteristics suitable for application in vehicles. Although invented much earlier, the Stirling engine was manufactured in the Netherlands during the World War II era as a power plant for portable generators. The modern designs of the Stirling engine tend to be complex and do not achieve, by a wide measure, the theoretical efficiency. There are problems in connection with sealing at the point where the piston rods leave the closed system containing the working fluid. There are problems with the design of a low-cost heater head capable of handling the working fluid at 100 atmospheres pressure and a temperature of about 750 °C (400 °F); and with power control. Because the combustion system is external, any alteration of the fuel flow has only an indirect effect on the working fluid.

The gas turbine engine dates back several decades. The development of a turbine power system for a passenger car by Chrysler as early as 1938 is well known. Several inherent advantages of a turbine system include: (1) few moving parts; (2) theoretical engine weight/power ratio is attractive; (3) simpler energy transfer processes (rotary-rotary); (4) good control over emissions because combustion process is isolated, continuous, and relatively easy to control; (5) good cold start characteristics; (6) potentially longer life and less maintenance; and (7) ability to handle a number of fuels. Use of a continuously variable transmission is the solution to the constant turbine speed (where efficiency is the highest) coupling with the variable speed demands of an automobile. The principal problem of the gas turbine and one of great difficulty is the high temperature at which the system must operate if the attractive efficiencies are to be achieved. It is desirable to increase turbine inlet temperatures to at least 1400 °C (2552 °F), which essentially dictates the use of ceramic components.

The strong trend toward smaller cars tends to go counter to the best performance of gas turbine vehicles. As the size of the gas turbine is reduced, the inherent losses of the engine claim an increasing share of its total power, thus adversely affecting fuel economy.

See also **Electric Cars**.

Additional Reading

Ferguson, C.R. and A.T. Kirkpatrick: *Internal Combustion Engines*, John Wiley & Sons, Inc., New York, NY, 2000.

Ganesan, V.: *Internal Combustion Engines*, The McGraw-Hill Companies, Inc., New York, NY, 1995.

Heywood, J.: *Internal Combustion Engine Fundamentals*, The McGraw-Hill Companies, Inc., New York, NY, 1988.

Lumley, J.L. and W.C. Reynolds: *Engines: An Introduction*, Cambridge University Press, New York, NY, 1999.

Mowery, D.C. and N. Rosenberg: *Paths of Innovation: Technological Change in 20th Century America*, Cambridge University Press, New York, NY, 1998.

Pulkrabek, W.: *Engineering Fundamentals of the Internal Combustion Engine*, Simon & Schuster Trade, Kansas City, MO, 1997.

Stone, R.: *Introduction to Internal Combustion Engines*, Society of Automotive Engineers, Washington, DC, 1999.

INTERNAL CONVERSION.   A process in which the energy released in the de-excitation of an excited energy state of an atomic nucleus is transferred through electromagnetic coupling to one of the bound electrons of that atom rather than being released as a photon. The coupling is usually with an electron in the K-, L-, or M-shell of the atom. Kinetic energy, equal in amount to the difference between the transition energy and the emitted electron's binding energy, is transferred to the electron. The internal conversion is followed by emission of Auger electrons or characteristic x-rays in consequence of the necessary rearrangement of the atomic electrons.

The conversion fraction is the ratio of the number of internal conversion electrons to the number of gamma quanta emitted plus the number of conversion electrons emitted in a given time interval by a single nuclidic species during de-excitation of one of its excited energy states. Partial conversion fractions refer to conversion fractions for various electron shells, e.g., K-conversion fractions, etc. Sometimes called *conversion coefficient*.

INTERNAL ENERGY.   A mathematically defined thermodynamic function of state, interpretable through statistical mechanics as a measure of the molecular activity of the system. It appears in the first law of thermodynamics as:

$$du = dq - dw$$

where du is the increment of specific internal energy, dq the increment of heat, and dw the increment of work done by the system per unit mass. The differential du is a perfect differential. Its integral therefore introduces a constant of integration, the *zero-point internal energy*, so that care must be taken when absolute values of the internal energy are employed.

INTERNAL GRAVITY WAVE.   A wave that propagates in density-stratified fluid under the influence of buoyancy forces.

The dispersion relation is given by frequency

$$\omega = \pm (Nk_h)/|\mathbf{k}|,$$

in which N is the buoyancy frequency and k_h is the horizontal component of the wavenumber vector $\mathbf{k}$. For all wavenumbers, internal gravity waves have frequency smaller than N. Their group velocity is perpendicular to the phase velocity such that the vertical component of the group velocity is opposite in sign to the vertical component of the phase velocity. Also called *internal waves*, and *gravity waves*.

AMS

INTERNAL TIDES.   See **Meteorology**.

INTERNAL WATER CIRCULATION.   The conceptual hydrological cycle for a specified continental surface that comprises water evaporating from the surface, condensing in the overlying atmosphere, and falling back to the surface. In reality, some of the evaporated water leaves the region in the wind and is replaced with water vapor brought into the region by wind.

INTERNAL WAVE.   See **Meteorology**.

INTERNATIONAL DATE LINE.   In accordance with the fundamental definition of civil time, the date changes when the mean sun crosses the meridian at lower culmination, i.e., at midnight. This date line is approximately 180° west of Greenwich in longitude, but is adjusted so that, so far as is practical, the Pacific insular possessions of the different countries shall carry the same date as the home nations, and also so that the line shall not cross any land. Ships and aircraft crossing the date line from east to west skip one day. That is, if it is Monday when the craft arrives at the line from the east, it immediately becomes the same hour on Tuesday after crossing the line; and the day is omitted from the log book. On crossing the line from west to east, a day is repeated.

INTERNATIONAL EXTREME ULTRAVIOLET HITCHHIKER (IEH-3).   See **Space Science Missions: Universe**.

INTERNATIONAL GAMMA-RAY ASTROPHYSICS LABORATORY (INTEGRAL) MISSION.   See **Space Science Missions: Universe**.

INTERNATIONAL INDEX NUMBERS.   See **Meteorology**.

INTERNATIONAL SYSTEM OF UNITS (SI).   See **The International System of Units (SI)**.

INTERNATIONAL ULTRAVIOLET EXPLORER (IUE) MISSION.   See **Space Science Missions: Universe**.

INTERNET CONTROL MESSAGE PROTOCOL (ICMP).   See **Transmission Control Protocol (TCP) / Internet Protocol (IP) Suite**.

INTERNET PROTOCOL (IP).   See **Transmission Control Protocol (TCP) / Internet Protocol (IP) Suite**; and **Wireless Communications Applications**.

INTERNET SECURITY STANDARDS.   When the Internet was created, security was left out of its Transmission Control Protocol/Internet Protocol (TCP/IP) standards. At the time, the crude state of security knowledge may have made this omission necessary. Today, however, security expertise is more mature. In addition, the broad presence of security threats on the Internet means that security must be addressed deliberately and aggressively in Internet standards. This article describes three ways to add standards-based security to the Internet. The first is for users to add standards-based security to specific dialogues, that is, to two-way conversations between pairs of communicating processes. Today this can be done largely through the use of virtual private networks (VPNs) using Point-to-Point Tunneling Protocol, IPsec (Internet Protocol security), secure socket layer, and other "add-on" security protocols. This "security overlay" approach is not as desirable as good general Internet security standards, but it solves some specific current needs of users. The second approach is to add more security to central Internet standards. In addition to the "core four" standards (IP, TCP, User Datagram Protocol, and Internet Control Message Protocol), these central Internet standards include supervisory standards (such as the Domain Name System, Dynamic Host Communication Protocol, Simple Network Management Protocol, Lightweight Directory Access Protocol, and routing protocols), and application standards (such as Simple Mail Transfer Protocol, Hypertext Transfer Protocol, and File Transfer Protocol). A strong effort to retrofitting existing Internet standards to add security is underway at the Internet Engineering Task Force (IETF) but is far from complete. The third way to add security is to take a comprehensive look at the Internet to decide how security *should* work on the Internet. This could lead to anything from the creation of *multiple Internets* with different classes of security for different types of users to the creation of international policies to make forensics more effective when there are attacks.

Security Threats and Defenses

Before looking at security in Internet standards, it is necessary to have an understanding of the major security attacks and defenses.

Penetration Attacks. One set of threats against which networks must guard is attackers attempting to hack systems (access them intentionally without authorization or in excess of authorization) or attempting to conduct denial-of-service attacks against them by crashing or slowing them to the point of being useless to their legitimate users.

Attacks on Dialogues. Second, when two parties communicate, they normally engage in a dialogue in which multiple messages are sent in each direction. These dialogues must be secure against attackers who may wish to read, delete, alter, add, or replay messages. In general, dialogue security has been the main focus of IETF efforts to add security to Internet (TCP/IP) standards. Hacking and denial-of-service are sometimes addressed, but this is comparatively uncommon.

Confidentiality. Confidentiality means that eavesdroppers who intercept messages en route over the Internet cannot read them. Encryption is the chief tool against unauthorized reading en route.

Authentication and Integrity. Authentication means that the receiver of messages can verify the identity of the sender to ensure that the sender is not an impostor. There are two forms of authentication. In *initial authentication*, one side proves its identity to the other side at or near the beginning of a dialogue. In *message-by-message authentication*, the sender continues to identify itself with every message, much as parties exchanging letters sign each letter. Message-by-message authentication is crucial to good security.

Integrity means that if a message has been captured and changed en route, then the receiver will be able to detect that a change has occurred. All forms of message-by-message authentication provide message integrity as a by-product. If a message is changed during transmission, either deliberately or through transmission errors, authentication will fail, and the receiver will discard the message.

Message-by-Message Authentication. Impersonation is a serious problem because many Internet standards are supervisory standards that govern how devices on the Internet function. If an attacker can trick a network management program system or other supervisory system into accepting false information, the resultant damage can be widespread. There are two common message-by-message authentication methods: HMACs and digital signatures.

Digital Signatures. In digital signatures, the sender hashes the message to be sent. This creates a message digest. The message digest is then signed (encrypted) with the sender's private key, which only the sender should know. This produces the digital signature, which the sender appends to the message before transmission.

The receiver recomputes the message digest in two ways. First, as the sender did, the receiver first hashes the message to produce the message digest. Second, it decrypts the digital signature with the true party's public key. This should also produce the message digest—if the sender is the true party. Otherwise, the message digests will not match, and the message will be rejected. In addition, digital signatures produce message integrity; if an attacker changes the message en route, the message digests will not match and the message will be rejected.

Digital Certificates. Note that digital signatures require the digital signature to be tested with the *true party's* public key—not the public key of the sender (who may be an impostor). The definitive way to get public keys of a true party is to get a digital certificate from a trusted certificate authority (CA). This digital certificate has the name of the true party and the true party's public key. It also has a digital signature signed by the CA's private key so that the certificate cannot be changed without this change being detected.

One problem with digital certificates is that certificate authorities are not regulated in most countries, leading to questions about their trustworthiness. Although European countries are planning to create regulated CAs, the United States and most other countries are trusting market forces to handle certificate authorities.

Another problem with digital certificates is that a CAs must create complex public key infrastructures to create public key–private key pairs, distribute private keys and digital certificates, and allow users to download certificate revocation lists (CRLs), which are lists of ID numbers for certificates that were revoked before their valid period ends.

Is it possible to distribute public keys without using certificate authorities? Pretty Good Privacy (PGP) offers one way to do so. Each user has a "key ring" of trusted public key–name pairs. If User A trusts User B, User A can trust User B's key ring, which may, of course, contain keys received through further distributed trust. This approach works, but if an impostor can dupe even one user, trust in the impostor can spread widely. This is not a good way to manage large public systems of routers, Domain Name System (DNS) hosts, and other sensitive devices.

It is also possible to distribute public keys manually. This works well in small systems, but it does not scale well to very large systems because of the labor involved.

HMACs. Another tool for message-by-message authentication is the key-hashed message authentication code (HMAC). For HMAC authentication, two parties share a secret key. To send a message, one of the parties appends the key to the message, hashes the combination, and adds this hash (the HMAC) to the outgoing message. To test an arriving message, the receiver also adds the key to the message and hashes the combination. This creates the HMAC. If the received and computed HMACs are the same, the sender knows the secret key, which only the other party should use.

HMAC processing requires little processing power because it only involves hashing, which is a fast process. In contrast, public key encryption and decryption, which are used in digital signatures, are extremely processing intensive.

A secret key needs to be created and distributed for each pair of communicating parties, however. In large systems of routers and hosts, this is highly problematic. One approach to reduce this problem is to use public key authentication for initial authentication, then use Diffie–Hellman key agreement or public key distribution to send the secret keys between pairs of communicating parties. This still requires a public key infrastructure.

Another solution is to use community keys, which are shared by all communicating parties in a community. The problem here, of course, is that if a single member of the community is compromised, attackers reading its community key will be able to authenticate themselves to all other members of the community.

Adding Security to Individual Dialogues. The first approach to creating standards-based security is for users to add security to individual existing dialogues. Effectively, this overlays security on the Internet for limited purposes without requiring the creation of completely new Internet standards.

Cryptographic Systems. Secure dialogues require that confidentiality, authentication, and integrity be protected. This is a somewhat complex process involving several phases.

Cryptographic System Phases. Establishing a secure dialogue typically involves four sequential phases (although the specific operations and the order of operations can vary among cryptographic systems). The first three are handshaking stages at the beginning of the dialogue. After the secure dialogue is established, the two parties engage in ongoing secure conversation.

First, the two parties must select standards options within the range offered by a particular cryptographic system. For instance, in encryption for confidentiality, the system may offer the choice of a half dozen encryption methods. The communicating parties must select one to use in their subsequent exchanges. The chosen methodology may offer several options; these, too, must be negotiated.

Second, the two parties must authenticate themselves to each other. Although this might seem like it should be the first step instead of the second, the two parties need the first phase to negotiate which authentication method they will use to authenticate themselves.

Third, the two parties must exchange one or more secret keys securely. These keys will be used in the ongoing dialogue that will take place after the "handshaking" steps are finished.

Fourth, the security of the communication is now established. The two parties now engage in ongoing dialogue. Typically, nearly all communication takes place during this ongoing dialogue.

The User's Role. Few users have the training to select security options intelligently, much less handle authentication and other tasks. Consequently, cryptographic systems work automatically. The user selects a communication partner, and the systems of the two partners work through the four phases automatically.

At most, users may have to authenticate themselves to their own systems by typing a password, using an identification card, using biometrics, or by using some other approach. This user authentication phase tends to be the weak link in cryptographic systems because of poor security practices on the part of users, such as using weak passwords.

The Policy Role. Different security options have different implications for the strength of a dialogue's security. Users rarely are capable of selecting intelligently among options. Consequently, companies must be able to set policies for which methods and options will be acceptable, and they must be able to enforce these policies by promulgating them and enforcing their use.

Figure 1 shows how policy guides security in IPsec.

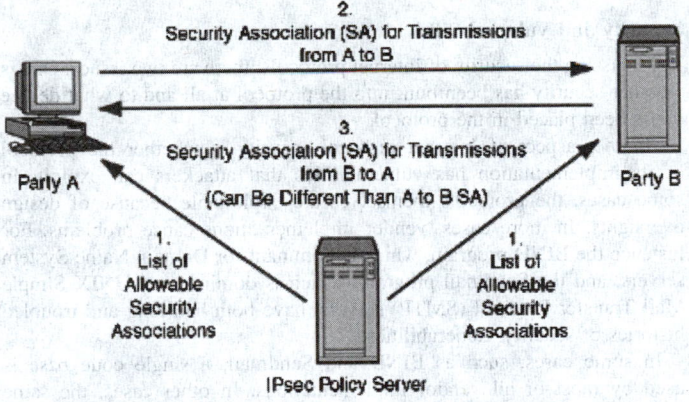

Fig. 1.   Policy-based security associations in IPsec

Adding Security at the Data Link Layer; Dial-Up Security and PPP. Early computer systems used dial-up security. As, the user dialed into a server, generally using the Point-to-Point Protocol (PPP) at the data link layer. Later, companies added security to this approach by creating remote access servers (RAS) for sites. The user dialed into the RAS, authenticated himself or herself (usually with a password), and then received access to all servers at a site or to selected servers.

PPP can be used without security, but PPP offers moderately good encryption and several options for authentication ranging from nothing to strong security options. Again, the dialogue partners must select the security options they wish to employ.

Tunneling. Although PPP is secure, it is limited to a single data link because it operates at the data link layer. When a connection is made over the Internet, however, each connection between a host and a router or between two routers is a separate data link. PPP cannot function over the Internet.

Consequently, Internet-based data link layer security approaches must use tunneling. In tunneling, a data link layer frame is placed within the data field of a packet (the opposite of the usual situation). The packet is sent from user computer to the RAS, across multiple links. The RAS reads the tunneled (encapsulated) frame. Although this effect seems needlessly complex, it allows us to use traditional PPP user-RAS security over the Internet.

Point-to-Point Tunneling Protocol (PPTP). The first tunneling protocol was the Point-to-Point Tunneling Protocol (PPTP). As its name suggests, PPTP is a way to tunnel PPP frames over the Internet. PPTP uses PPP encryption and authentication mechanisms, using them over an entire Internet connection instead of over a single dial-up data link. PPTP has a number of moderate security weaknesses, but it is good for low-threat environments. It offers medium security for moderate threat environments.

Layer 2 Tunneling Protocol (L2TP). Although PPTP works and is attractive, it is limited to transmission over an IP network such as the Internet. The L2TP can work over a number of transmission mechanisms, including IP, frame relay, and ATM, to name just three. L2TP does not offer security by itself, however. It requires users to rely on the IPsec protocol, discussed in the next section, to provide security at the Internet layer during transit. L2TP is a pure tunneling protocol, not a security protocol.

Adding Security at the Internet Layer. The Internet layer (or network layer) is the core layer in TCP/IP internetworking. The Internet Protocol (IP) is the main packet standard at this layer. IP comes in two versions — IP Version 4 (IPv4) and IP Version 6 (IPv6). IPv4 is the dominant version in use on the Internet today, but IPv6 is beginning to grow, especially in Asia, where relatively few IPv4 addresses were allocated when the Internet was first created.

IPsec. The IETF has been working to retrofit IP to be more secure. Their effort has crystallized around a group of standards collectively called IPsec (IP security). Although IPsec was initially planned for IPv6, the IETF has developed it to work with IPv4 as well.

Encryption and Authentication. The dominant way of using IPsec is to employ the encapsulated security protocol (ESP) option, which offers both encryption and authentication.

Tunnel Mode. IPsec can operate in two modes: tunnel mode and transport mode. In tunnel mode, IPsec is handled by IPsec gateways at the two sites of the communicating parties, not by the computers of the communicating parties themselves. The packet to be delivered securely is tunneled by encapsulating it in another packet, encrypting the packet to be delivered securely, adding authentication, and sending the encapsulating packet from one IPsec gateway to the other. Although attackers can read the IP header of the encapsulating packet, they cannot read the secured packet, nor can they change the secured packet without the change being obvious.

Tunnel mode is attractive because it does not require the individual users to have IPsec software on their computers or to know how to use the software. In fact, the users may not even be aware that IPsec is protecting their packets over the Internet. Windows, which dominates client operating systems today, did not get native IPsec support until Windows 2000. Implementing IPsec software on the many older personal computers (PCs) in the firm would require a massive investment in large companies.

The disadvantage of tunnel mode is that it does not provide any protection for the packet as it travels within the two sites. The focus is entirely on security during Internet transmission.

Transport Mode. In contrast, transport mode offers *end-to-end* encryption and authentication between the two computers. This approach provides security not only while packets travel through the Internet but also when the packets are passing through local site networks on their way to and from the Internet. Although the Internet exposes traffic to many attackers, there also are dangers within corporate networks. In fact, some of the worst attacks are made by corporate insiders working within corporate networks.

Although this end-to-end security is attractive, it comes at a price. Larger firms have thousands or even tens of thousands of PCs. Transport mode requires the presence of IPsec software on every PC to be protected. As noted previously, Windows only began shipping with native IPsec software with Windows 2000, and retrofitting older computers would be extremely expensive.

A more modest problem is that transport mode packets must have the IP address of the receiving computer in the destination address field of the packet header. If sniffers can be placed along the route of these packets, attackers will be able to learn the IP addresses of many corporate hosts. This is a first step in most types of Internet-based attacks. In contrast, tunnel mode packets only have the IP addresses of the receiving IPsec gateway in their destination address fields. This only tells attackers about the IP address of a single machine that usually is well hardened against attacks.

Combining Tunnel Mode and Transfer Mode. One solution to the relative weaknesses of the two modes is to use both. The user can employ transport mode end-to-end but also to use IPsec in tunnel mode between sites as a second layer of protection. This will provide end-to-end encryption while still hiding the IP address of communicating hosts. Of course, implementing IPsec protection twice adds to cost.

Adding Security at the Transport Layer: TLS. IPsec is a complex security mechanism, but it has the advantage of transparency. IPsec protects *all* higher layer traffic at the transport and application layers automatically, without requiring higher-layer protocols to do any work or even be aware of this protection.

In some applications, dialogues are limited to World Wide Web or at most the Web and e-mail. Under these conditions, the costly burden of implementing IPsec is not justified, and many firms turn to a simpler protocol, Transport Layer Security (TLS), which was originally created by Netscape as Secure Sockets Layer (SSL) and was then renamed by the IETF when it took over the standard's development. If your URL begins with "https," then you are using TLS.

Although IPsec provides a blanket of protection at the internet layer, TLS creates a secure connection at the transport layer. This secure connection potentially can protect all application layer traffic. Unfortunately, TLS requires that applications be *TLS-compliant* to benefit from this protection. Although all browsers and Web servers can use TLS, only some e-mail systems can use TLS, and few other applications can benefit from TLS protection at the transport layer.

TLS was created for electronic commerce, in which a residential client PC communicates with a merchant server. TLS requires the merchant to authenticate itself to the client using digital certificates and digital signatures. Few residential client PC owners have digital certificates. So although TLS allows client authentication, it makes it optional. This lack of mandatory client authentication is a major security vulnerability that is intolerable in many situations.

In corporate settings, however, the organization can require clients as well as servers to use digital certificates. This eliminates the main security limitation of TLS, although TLS in general offers only moderate security.

Adding Security at the Application Layer. For the application layer, IETF is adding security to a number of application layer standards. In addition, many vendors are adding proprietary security to their applications, such as database applications

Multilayer Security. Security can be applied at several standards layers. A major principle of security is defense in depth. Almost all security protections break down from time to time. Until they are fixed, the attacker has free access — unless the attacker must break through two or more lines of defense, the company still will be protected while it repairs a broken security countermeasure. Consequently, companies would be better protected if they added dialogue security at more than one layer. This is an expensive undertaking, however, so it is uncommon.

Added Dialogue Security and Firewalls. Although adding security to dialogues passing through the Internet is attractive, it creates problems for firewalls. Firewalls are designed to examine all packets coming into or leaving a firm. When firewalls find attack packets created by hackers, they drop them. Almost all cryptographic system packets are encrypted, however. Unless they are decrypted before passing through the firewall, the firewalls cannot scan their packets for attack signatures. Consequently,

companies that buy added security by implementing cryptographic systems tend to lose some scanning security. This places a heavier burden on end stations to do scanning, and many end stations are client computers, the owners of which lack the willingness, much less the knowledge, to do packet scanning.

Adding Security to Individual Standards. Although adding dialogue security to individual dialogues works, it would be better for Internet standards themselves to offer high security. When the Internet was first created, none of its standards offered any security. Today several TCP/IP standards offer security. Unfortunately, the more closely one examines their security features, the less adequate many appear. Fortunately, the IETF has a broad program to add security to a broad range of existing Internet standards individually.

A Broad IETF Commitment: The Danvers Doctrine. In 1995, a meeting of the IETF in Danvers, Massachusetts, reached a consensus that the IETF should develop strong security for all of its protocols, Request for Comments (RFC) 3365. Originally, the consensus was limited to a decision to use strong rather than weak encryption keys that met existing export restrictions. Soon, however, this Danvers Doctrine expanded into a consensus to develop strong security for all TCP/IP protocols. As a first step, all RFCs are now required to include a section on security considerations. Considerable progress has also been made in adding security to individual TCP/IP standards.

Although all TCP/IP standards are to be given strong security, the IETF decided that it should be the option of individual organizations whether to implement security in individual protocols.

An important consequence of the decision to leave security up to organizations is that organizations must decide which security options to use. There are five broad layers of functionality in networking — physical, data link, internet, transport, and applications. Implementing security at all layers would be horrendously expensive and generally unnecessary.

Consequently, many organizations wish to implement security at only one layer. Although this provides security, individual security technologies often are found to have vulnerabilities. To maintain protection during these periods of vulnerability, organizations probably will wish to implement security in the protocols at two layers at least. While one layer's security is being repaired, the other or others will continue to provide protection.

Security and Vulnerabilities

In discussing the security of Internet protocols, there are two issues. One is whether security has been built into the protocol at all and to what degree it has been placed in the protocol.

Another aspect of Internet protocol security is whether the protocol or its implementation has vulnerabilities that attackers can exploit. In some cases, the protocols themselves are exploitable because of design oversights. In more cases, vendor implementations cause problems. For instance the BIND program, which is dominant for Domain Name System servers, and the Sendmail program, which is dominant on UNIX Simple Mail Transfer Protocol (SMTP) servers, have both had long and troubled histories of security vulnerabilities.

In some cases, such as BIND and Sendmail, a single code base is used by most or all vendor implementations. In other cases, the same flaw is discovered in multiple vendor code bases. In the discovery of a vulnerability can suddenly put a significant fraction of the Internet's servers or routers at risk until organizations can download patches.

In addition, many firms fail to install patches quickly or at all. This makes some vulnerabilities exploitable for weeks, months, or even years after they are discovered. All too often, firms only patch vulnerabilities in earnest when a virus or hackers create widespread damage.

Although vulnerabilities are important, they are situation-specific. This chapter must focus on security within Internet standards themselves.

The Core Four. There are four core standards at the heart of the Internet. These are IP, TCP, and UDP, and ICMP.

Internet Protocol (IP). The main job of the Internet Protocol (IP) is to move packets from the source host to the destination host across the Internet, which consists of thousands of networks connected by routers. IP is a hop-by-hop protocol designed to govern how each router handles each IP packet it receives.

A packet may travel over one to two dozen routers as it passes from the source host to the destination host. Many core routers in the Internet backbone handle so much traffic that they can barely keep up with demand. To reduce the work done on each router, IP was designed as a simple, unreliable, and connectionless protocol. Although packet losses on the Internet are modest, IP is a "best effort" protocol that offers no guarantee that packets will arrive at all, much less arrive in order. Given this minimalist vision for IP, it is hardly surprising that security was completely left out of IP's core design. Given continuing router overload, it would be difficult to make IP secure throughout the Internet.

IPsec was created to add security to the Internet Protocol, making IPsec more than a security overlay method. IPsec cannot achieve its promise without a truly worldwide system of certificate authorities, however.

Transmission Control Protocol (TCP). To compensate for IP's unreliability at the internet layer, TCP/IP was given a reliable sibling protocol, TCP at the transport layer. When TCP was created in the early 1980s, however, security technology was far too immature to be implemented. Today there are no plans to add security to TCP given the IETF's focus on IPsec and IPsec's ability to provide security for higher layer protocols.

User Datagram Protocol (UDP). When transport layer error correction is not required or is not practical, applications specify the UDP at the transport layer. Like TCP, UDP was created without security and is also unlikely to receive security extensions because of the IETF's reliance on IPsec.

Internet Control Message Protocol (ICMP). IP merely delivers packets. It does not define any Internet layer supervisory messages. To compensate for this, the IETF created the ICMP to carry supervisory information. ICMP messages are carried in the data fields of IP packets. ICMP messages allow hosts to determine if other hosts are active (by "pinging" them). It also allows hosts to send error messages to other hosts and to tell other hosts or routers to act differently from how they have been acting. ICMP is a powerful tool for network managers. Unfortunately, this power also makes ICMP a popular tool for hackers. Making ICMP less useful to hackers would also tend to make it less useful to network administrators, so the IETF has done nothing to implement ICMP security. For this reason, most corporate firewalls block all ICMP traffic except for outgoing pings and returning pongs.

If ICMP were given good authentication, corporations might be more willing to allow it through firewalls. In addition, authenticated ICMP would protect against attacks generated within the firewalls at individual sites. There are no current IETF activities to add authentication and other security to ICMP, however. The reason may be that ICMP messages are carried in the data fields of IP packets, which can be protected by IPsec, including the authentication of the sending ICMP process.

Administrative Standards. The "core four" Internet standards do most of the work of the Internet, but several other administrative protocols are needed to keep the Internet functioning. The following sections discuss security in the most important of these standards.

Domain Name System (DNS). The DNS allows a source host to find the IP addresses of a destination host if the source host only knows the destination host's host name. In a sense, DNS is like a telephone directory for the Internet.

Most client users only know the host names of the servers they use. Consequently, if the DNS were to fail for some period of time, clients could no longer reach servers. On the positive side, DNS is hierarchical and highly distributed, so that if one DNS server goes down, service continues. The top level of the DNS server hierarchy only has 13 root DNS hosts, however, and one of them periodically feeds changes to the others. Although these root DNS hosts are geographically distributed, most use the same software, leaving them open to common mode attacks if vulnerabilities are found. Also, geographical distribution is not much protection when there are only 13 targets to attack. In later 2002, a brief denial-of-service attack degraded the service given by 9 of the 13 root DNS hosts. Had this attack been more intense, and had it continued over many hours, service disruption across the Internet would result.

The threat of such concentrated resources (rare on the Internet) is real, but the root DNS servers are well protected with both technical and human protections. Below the root servers are a larger number of DNS servers for the various generic and national top-level domains, and there are many corporate DNS servers. At each level, there are more DNS servers to give less concentration, but the loss of nonroot DNS servers near the top could still be widely damaging to the resources within the domain they serve.

To address problems with DNS security, the IETF DNSSEC Working Group is developing an authentication approach based on public key encryption with digital signatures. There are a (relatively) limited number of high-level DNS servers, so giving each a public key and a private key is not too daunting a challenge. As long as DNS servers do not get taken over so that their private keys can be stolen, authentication should be good. The DNSSEC effort is just getting underway, however, so widespread DNS protection will not be realized for some time to come. Nonetheless, methods for the critical issue of authentication have already been created (RFC 3118).

Simple Network Management Protocol (SNMP). The goal of the Simple Network Management Protocol (SNMP) is to allow a central administrative computer (the manager) to manage many individual managed devices. The manager talks to an agent on each device. The manager does this by sending the managed devices Get messages (which ask for information on the device's status) or Set messages (which tell the managed device's configuration).

Obviously, SNMP is a powerful management tool. Unfortunately, it is also a golden opportunity for attackers. It potentially allows them to learn a great deal of information about a network through Get commands. It also allows them to do an endless amount of damage through malicious use of the Set command to create misconfigurations on large numbers of devices. These misconfigurations can make the devices unusable or can even make them interfere with the operation of other devices.

SNMP version 1 had no security at all, making the protocol extremely dangerous, given its power. SNMP Version 2 was supposed to add security, but an inability to settle differences within the IETF prevented full security from being built into it.

Version 2 did receive one authentication advance. Only in version 3 did security get built into SNMP extensively. Version 3 offered confidentiality (optional), authentication, message integrity, and time stamps to guard against replay attacks.

Unfortunately, SNMP version 3 security is based on HMAC authentication and integrity, which requires the network administrator to have a secret bit string and for each managed device to know that secret bit string.

This severe problem could be solved if the standard specified the use of digital certificates for managers. Authentication and integrity only require that managed devices know the manager's public key, and there is no problem with shared public keys. In addition, public key encryption allows the secure exchange of symmetric keys for bulk encryption for confidentiality. Nonetheless, the large number of devices involved and the desire to keep processing loads on managed devices low has made public key authentication and integrity unattractive to standards makers.

In addition, SNMP is connectionless, and it uses connectionless UDP at the transport layer. Connectionless operation reduces the network load created by SNMP. It is difficult to add security to connectionless applications because, as noted earlier, security typically begins with handshaking phases that are difficult (although possible) to implement in a connectionless protocol.

Lightweight Directory Access Protocol (LDAP). Increasingly, companies store security information and other critical corporate information in central repositories called directory servers. To query data in a directory server, devices commonly use the Lightweight Directory Access Protocol (LDAP), which governs search request and responses.

The IETF has long understood the importance of creating good security for LDAP in light of the extreme importance of information contained in the directory. The core security focus of LDAP must be authentication because successful impostors could learn large amounts of damaging information about a firm if they could get wide access to this data by claiming to be a party with broad authorization to retrieve data.

The first version of the standard, LDAP version 1, had no security and generally was rudimentary. LDAP version 2 provided authentication options for the first time. It permitted initial anonymous, simple, and Kerberos 4 authentication.

Anonymous authentication means that no authentication is needed for some users. These users would be given access to certain information on the server, much as anonymous File Transfer Protocol (FTP) users are allowed to download information from selected parts of an FTP server. Generally speaking, anonymous authentication should be turned off on directory servers with sensitive information.

Simple authentication requires the user to transmit a user name and password. This information is sent in the clear, without encryption. Anyone who reads the user name and password will be able to authenticate themselves later as the party.

Kerberos 4 authentication uses a central authentication server separate from the directory server. Unfortunately, Kerberos 4 had known weaknesses. Most notably, Kerberos 4 uses a long-term shared secret key to communicate with each device. If enough traffic uses this key, a cryptanalyst can determine the key and then impersonate the device. (Kerberos version 5 only uses the long-term key for initial authentication and then uses a session key for remaining communication during a time-limited session.)

The latest version of LDAP, version 3, also supports anonymous and simple authentication. Again, anonymous authentication should not be turned on for sensitive servers. In addition, sites that use simple authentication should protect application-level LDAP traffic with lower layer protection, most commonly TLS at the transport layer. This ensures that eavesdroppers will not be able to read the user name and password.

LDAP version 3 also supports the simple authentication security layer (SASL). This method allows the searcher and the directory server to authenticate each other in several ways, including Kerberos 5, HMACs, and "external authentication." External authentication means that the parties can perform authentication any way they wish, essentially taking authentication outside the LDAP process. For instance, if the searcher and directory server create a TLS connection at the transport layer, the authentication that takes place at that layer may be sufficient. SASL is extremely flexible, but companies that use it must ensure that sufficiently strong authentication is selected.

Address Resolution Protocol (ARP). When a router receives a packet, it looks at the packet's destination IP address. Based on this information, the router sends the packet back out another port, to the destination host or to a next-hop router that will handle the packet next. The term "target" used to describe the destination host or next-hop router.

The destination IP address in the packet gives the target's IP address. To deliver the packet, however, the router must encapsulate the packet in a data link layer frame and deliver the frame to the target. This requires the router to know the data link layer address of the target, for instance, the target's Ethernet message authentication code address.

If the router does not know the target's data link layer address, the router must use the Address Resolution Protocol (ARP). The router first broadcasts an ARP request message to all hosts on the subnet connected to the port out which the packet is to be sent. If the subnet is an Ethernet network, the broadcast data link destination address is 48 ones. Switches will deliver frames with this address to all hosts on the subnet.

Within a broadcast frame, the router sends an ARP request message. The ARP request message contains the IP address of the target. All hosts except the target ignore this message. The target, recognizing its IP address and sends an ARP reply message back to the router. This message tells the router the target's data link layer address.

Now that the router knows the target's data link layer address, the router takes the IP packet it has been holding and places it in a frame with the target's data link layer address in the destination address field. The router sends this frame to the target.

The router also places the target's IP address and data link layer address in the router's ARP cache. When the next packet addressed to this IP address arrives, the router will not have to use ARP. It will simply look up the data link layer address in the ARP cache.

ARP assumes that all hosts are trustworthy, and thus an attacker host on the same subnet as the target can use this trust to make an attack. Most simply, the attacker host can flood the router with ARP reply messages associating its data link layer address with every IP address on the subnet. If the router accepts some of these messages, it will suffer ARP cache poisoning.

The next time a packet arrives at the router, the router may rely on the poisoned cache to send the packet to the attacker rather than to the correct target. This allows the target to read the contents of the packet.

The attacker can then simply drop the packet, creating a denial-of-service attack against hosts on the subnet.

Alternatively, the attacker can listen to real ARP response messages to learn the true data link layer addresses of subnet hosts and send the packet on to the correct target. This attacker-in-the-middle attack permits the attacker to continue reading packets without alerting the victim hosts to this fact. This approach may also allow the attacker to hijack the communication session between the source and destination hosts, inserting its own packets into the dialogue. Currently, there are only preliminary efforts within the IETF to add authentication to ARP.

Dynamic Host Communication Protocol (DHCP). It is possible to configure a client PC manually, typing in its IP address and other configuration parameters needed to communicate over the Internet or a corporate internet. Servers typically are configured manually. Client PCs, however, are configured automatically by getting these configuration parameters from a Dynamic Host Configuration Protocol (DHCP) host.

Application Layer Standards. To users, the Internet is attractive because of its application standards for the World Wide Web, e-mail, and other popular services. Although decent security is present for some application standards, many continue to have only weak security, if they have any at all.

Hypertext Transfer Protocol (HTTP). To communicate with a Web server, a browser uses the Hypertext Transfer Protocol (HTTP), a simple protocol that offers no security by itself. All browser and Web server programs support TLS security, however. Although TLS has some theoretical weaknesses, it offers sufficiently strong Web server authentication and confidentiality for consumer e-commerce transactions. For high-volume business-to-business e-commerce, however, TLS security is marginal.

E-mail Security. Security for e-mail is perhaps the great scandal in IETF history. Competing cliques within the IETF have consistently refused to cooperate in selecting security standards for e-mail. Consequently, companies that wish to implement e-mail security have to use a nonstandard method to do so.

Not surprisingly, sending application layer e-mail traffic over a secure TLS transport connection is the most popular way to secure e-mail in organizations today. TLS is well understood, offers consumer-grade security, and has been widely used for HTTP security for several years. Fewer organizations use another approach, S/MIME (Secure Multipurpose Internet Mail Extensions).

A third approach, "Pretty Good Privacy" (PGP), is used primarily by individuals. Organizations have tended to stay away from PGP because it uses user-based transitive trust (if User A trusts User B, and if User B trusts User C, User A may trust User C). As noted earlier, this is not a good security policy. If a single user mistakenly trusts an impostor, others may unwittingly trust the impostor as well. Also, PGP has had a troubled development history.

Remote Access. The first ARPANET application was Telnet, which allows a user to log in to a remote computer and execute commands on it as if the user were local. This allows ordinary users to access remote services. It also allows system administrators to manage servers and routers remotely. This use of remote administration is attractive, but it must be done carefully or hackers will end up "managing" corporate servers and routers.

Unfortunately, Telnet has poor security. For example, Telnet does not encrypt host user names and passwords. Hackers can intercept user names and passwords and then log in as these users with all of their privileges.

Telnet should never be used for remote administration because hackers intercepting root passwords would be able to execute any commands on the supervised machine. Telnet is not alone in this respect. In the UNIX world, rlogin and rsh do not even require passwords to gain access to a computer, although other conditions must apply.

For remote administration, some organizations turn to the Secure Shell (SSH) protocol, which offers good authentication, integrity, and authentication. Unfortunately, SSH version 1 had security flaws, and although these have been fixed in SSH version 2, many version 2 implementations will also allow version 1 connections, thus leaving the system open to attack.

File Transfer Protocol (FTP) and Trivial File Transfer Protocol (TFTP). Another early ARPANET service that continues to be popular on the

Internet is the File Transfer Protocol (FTP), which allows users to download files from a remote computer to their local computers and sometimes to upload files from their local computers to the remote computer. Unfortunately, FTP also sends user names and passwords in the clear, making it dangerous. In addition, although the use of Telnet has declined to the point where quite a few companies simply stop all Telnet traffic at their firewalls, FTP is still widely used. The FTPEXT Working Group is considering security for FTP.

FTP has a simpler sibling, the Trivial File Transfer Protocol (TFTP). TFTP does not require user names or passwords, making it a darling of hackers who often use this protocol after taking over a computer to download their rootkits (collections of hacker programs) to automate their exploitation of the computer they now "own."

Making the Internet Forensic

Forensics is the application of science to criminal prosecution. One of the biggest problems in Internet security today is simply that there is almost no way to stop attackers. Victims must operate almost entirely on the defensive while under constant attack. As any military scientist will attest, this is a poor basis for security.

Few attackers are ever prosecuted. In some cases, the laws of a country are insufficient. In the case of two of the worst viruses—CIH and Love Bug—governments gave up on prosecution because they could not prosecute under their countries' statutes. Although the legal situation has become clearer, prosecutors are reluctant to prosecute security incidents. In many cases, the perpetrator is a minor, and only slap-on-the wrist penalties are possible. In other cases, prosecutors feel that even for adult perpetrators, prosecution is too expensive for the minor penalties that the courts have tended to apply. Post-September 11 laws have toughened penalties somewhat, but prosecutors in general still feel that prosecution for hacking and other attack activities is a low priority. Unless losses are in the tens of thousands of dollars, law enforcement officials probably will not prosecute.

A more promising approach is to sanction perpetrators by cutting off their Internet access and making it difficult or impossible for them to restore Internet access. To take the onus off the ISP, a court could make the decision based on ISP data, and the legal system could even impose ticket-based fines for modest infractions (such as sending probing messages), as is done for automobile misbehavior.

In many cases, attackers do little or nothing to hide their identities. For instance, hackers often send out many probing messages against companies they plan to attack. These messages must have the attacker's IP address in the packet so that the attacker can see replies.

In other cases, attackers hide their identities by working through a string of computers.

To make attackers easier to identify, ISPs could keep connection information far longer than they do today. (Today IP source address information is only stored for a brief period of time for billing purposes if it is stored at all.) Because of the costs involved, ISPs have been unwilling to store connection information for long periods of time. However, several countries are now considering laws to require the long-term storage of connection information plus the ability of law enforcement personnel to get at such information quickly for traceback.

The IETF is now working on a new standard to make traceback easier. This is a new ICMP message type, traceback. Each router will randomly but rarely transmit ICMP traceback messages to the destination host. These traceback messages will list routers along the way.

Even if an IP source address is spoofed, occasional ICMP traceback messages that the victim received in a denial-of-service attack or any other type of sustained attack will still be able to determine the route of the attack. Routers may also send occasional traceback messages back to their sources to aid in the analysis of certain types of reflection attacks. It will take some time for the IETF to ratify the new message type and for many of the Internet core routers to implement the standard, if they implement it at all. (It would cost money for them to do so and would reveal the ISP's internal router architecture, leaving ISPs vulnerable to certain attacks.)

Proactive ISP Efforts to Stop Attacks. More proactively, ISPs could use intrusion detection systems to watch packets coming from their own customers. If many attack packets began to appear, the ISP could inform the attacker and threaten to cut off Internet access. If the attacker failed to desist, the ISP could cut off their service. Generally, ISPs do not wish to do this (although a few already do). First, they do not want to lose customers. Second, intrusion detection and analysis would be expensive. ISPs, in other words, would have little to gain and much to lose. Stopping hackers at their source connection is so attractive, however, that governments should force ISPs to do intrusion detection and to cut off attackers. There is now sufficient danger to the entire Internet system to make such a requirement reasonable. ISPs that fail to do filtering might even be "black holed" by other ISPs, much as some ISPs that fail to stop their users from spamming have been black-holed.

Of course, there would have to be procedural safeguards to protect users who are not really attackers but whose computers have been compromised by attackers. (At the same time, it is reasonable to require users of compromised computers to clean out their computer if they wish to continue service.)

An Internet Network Operations Center?. One way to increase security on the Internet is to create a network operations center (NOC) for the Internet. Data on the status of the Internet would be delivered to the NOC from multiple sensors distributed throughout the Internet. This would include performance data to spot congestion on the Internet. It would also include attack data.

Centralizing attack data and adding trend detection software might allow the early detection of massive attacks so that actions to stop the attack could be undertaken before massive attacks spread too far. Early detection might also improve the probability of finding the sources of attacks.

Unfortunately, this type of NOC could also allow snooping on legitimate traffic for political purposes. Given diverse opinions on hacktivism (hacking for political purposes), separating legitimate from illegitimate traffic might be controversial.

Creating Multiple Internets? Another possibility is to create multiple Internets with different levels of security. During the 1970s, the U.S. military attempted to do this by creating milnet (which became the Defense Communication Network). What we may see in the future is the emergence of a business-class Internet and a consumer Internet (probably today's Internet). The business-class Internet would have high security throughout, and the consumer Internet would either have today's casual security or the type of expanded security we have seen in this article.

Interconnecting Internets with different security levels would be difficult. Although firewalls between high-security and low-security Internets would lessen the danger, air gaps between them would be needed for high security. Even then, users might surreptitiously interconnect multiple Internets by connecting servers or clients to two or more Internets simultaneously.

The State of Internet Security Standards

Today if two communication partners wish to communicate securely, they can do so by adding dialogue security on top of nonsecure Internet transmission. For lightweight needs, they can turn to PPTP or TLS. For industrial-strength security, they can use IPsec.

General Insecurity. More generally, however, the standards that the Internet needs for message delivery (IP, TCP, UDP, and ICMP), Internet supervisory standards (DNS, SNMP, LDAP, etc.), and Internet application standards (the Web, e-mail, etc.) vary widely in security from none to semiadequate. Consequently, the Internet today is rather fragile and open to seriously damaging attacks. Given the pace of security implementation in individual standards, the fragility of the Internet is not likely to change radically in the immediate future. Even after more secure standards are developed, it will take several years for them to be widely adopted.

Additional Reading

Behringer, M.H., and M. Morrow: *MPLS VPN Security*, Cisco Press, Indianapolis, IN, 2005.

Blumenthal, U., and B. Wijen: User-based Security Model (USM) for Version 3 of the Simple Network Management Protocol (SNMPv3), RFC 3413. 1998.

Clark, M.P.: *Data Networks, IP and the Internet: Protocols, Design and Operation*, John Wiley & Sons, Inc., New York, NY, 2003.

Dierks, T., and C. Allen: The TLS Protocol Version 1.0. RFC 2246. 1999.

Doraswamy, N., and D. Harkins: *IPSec*, 2nd Edition, Pearson Education, New York, NY, 2003.

Easttom, C.: *Network Defense and Countermeasures: Principles and Practices*, Prentice Hall, Upper Saddle River, NJ, 2005.

Forouzan, B.A.: *TCP/IP Protocol Suite*, 3rd Edition, McGraw-Hill Science/ Engineering/Math, New York, NY, 2005.

Harrison, R.: LDAP: Authentication Methods and Connection-level Security Methods, Internet draft, 2002.

Johnston, A.B., and D.M. Piscitello: *Understanding Voice over IP Security*, Artech House, Inc., Norwood, MA, 2006.

Lee, H.K.: *Understanding IPv6*, Springer-Verlag New York, LLC, New York, NY, 2005.

Lindskog, H., and S. Lindskog: *Web Site Privacy with P3P*, John Wiley & Sons, Inc., New York, NY, 2003.

Ohrtman, F.D.: *Voice Over 802.11*, Artech House, Inc., Norwood, MA, 2004.

Panko, R.: *Business Computer and Network Security*, Prentice-Hall, Inc., Upper Saddle River, NJ, 2004.

Raab, S., and M. Chandra: *Mobile IP Technology and Applications*, Cisco Press, Indianapolis, IN, 2005.

Rhee, M.Y.: *Internet Security: Cryptographic Principles, Algorithms and Protocols*, John Wiley & Sons, Inc., New York, NY, 2003.

Sennewald, C.A., and J. Sanger: *Effective Security Management*, 4th Edition, Elsevier Science & Technology Books, New York, NY, 2003.

Stallings, W.: *Cryptography and Network Security*, 4th Edition, Pearson Education, New York, NY, 2005.

Stallings, W.: *Network Security Essentials*, 3rd Edition, Pearson Education, New York, NY, 2006.

Thomas, S.A.: *IP Switching and Routing Essentials*, John Wiley & Sons, Inc., New York, NY, 2002.

INTERNET (The History).

The Internet has revolutionized the computer and communications world like nothing before. The invention of the telegraph, telephone, radio, and computer set the stage for this unprecedented integration of capabilities. The Internet is at once a world-wide broadcasting capability, a mechanism for information dissemination, and a medium for collaboration and interaction between individuals and their computers without regard for geographic location.

Today the Internet represents one of the most successful examples of the benefits of sustained investment and commitment to research and development of information infrastructure. Beginning with the early research in packet switching, the government, industry and academia have been partners in evolving and deploying this exciting new technology.

The early history of the Internet revolves around four distinct aspects. There is the technological evolution that began with early research on packet switching and the ARPANET (and related technologies), and where current research continues to expand the horizons of the infrastructure along several dimensions, such as scale, performance, and higher level functionality. There is the operations and management aspect of a global and complex operational infrastructure. There is the social aspect, which resulted in a broad community of Internauts working together to create and evolve the technology. And there is the commercialization aspect, resulting in an extremely effective transition of research results into a broadly deployed and available information infrastructure. Table 1 highlights some of the key events and technologies, which helped shape the Internet, as it is known it today.

The Internet today is a widespread information infrastructure, the initial prototype of what is often called the National (or Global or Galactic) Information Infrastructure. Its history is complex and involves many aspects; technological, organizational, and community. Its influence reaches not only to the technical fields of computer communications, but throughout society as it moves toward increasing use of online tools to accomplish electronic commerce, information acquisition, and community operations.

Origins of the Internet

The first recorded description of the social interactions that could be enabled through networking was a series of memos written by J.C.R. Licklider of MIT in August 1962 discussing his "Galactic Network" concept. He envisioned a globally interconnected set of computers through which everyone could quickly access data and programs from any site. In spirit, the concept was very much like the Internet of today. Licklider was the first head of the computer research program at DARPA,[1] starting in

[1] The Advanced Research Projects Agency (ARPA) changed its name to Defense Advanced Research Projects Agency (DARPA) in 1971, then back to ARPA in 1993, and back to DARPA in 1996. DARPA, the current name is used here.

October 1962. While at DARPA he convinced his successors at DARPA, Ivan Sutherland, Bob Taylor, and MIT researcher Lawrence G. Roberts, of the importance of this networking concept.

Leonard Kleinrock at MIT published the first paper on packet switching theory in July 1961 and the first book on the subject in 1964. Kleinrock convinced Roberts of the theoretical feasibility of communications using packets rather than circuits, which was a major step along the path towards computer networking. The other key step was to make the computers talk together. To explore this, in 1965 working with Thomas Merrill, Roberts connected the TX-2 computer in Mass achusects. to the Q-32 in California with a low speed dial-up telephone line creating the first (however small) wide-area computer network ever built. The result of this experiment was the realization that the time-shared computers could work well together, running programs and retrieving data as necessary on the remote machine, but that the circuit switched telephone system was totally inadequate for the job. Kleinrock's conviction of the need for packet switching was confirmed.

In late 1966 Roberts went to DARPA to develop the computer network concept and quickly put together his plan for the "ARPANET". He published it in 1967. At the conference where he presented the paper, there was also a paper on a packet network concept from the UK by Donald Davies and Roger Scantlebury of NPL. Scantlebury told Roberts about the NPL work as well as that of Paul Baran and others at RAND. The RAND group had written a paper on packet switching networks for secure voice in the military in 1964. It happened that the work at MIT (1961–1967), at RAND (1962–1965), and at NPL (1964–1967) had all proceeded in parallel without any of the researchers knowing about the other work. The word "packet" was adopted from the work at NPL and the proposed line speed to be used in the ARPANET design was upgraded from 2.4 kbps to 50 kbps.[2]

In August 1968, after Roberts and the DARPA funded community had refined the overall structure and specifications for the ARPANET, an RFQ was released by DARPA for the development of one of the key components, the packet switches called Interface Message Processors (IMP's). The RFQ was won in December 1968 by a group headed by Frank Heart at Bolt Beranek and Newman (BBN). As the BBN team worked on the IMPs with Bob Kahn playing a major role in the overall ARPANET architectural design, the network topology and economics were designed and optimized by Roberts working with Howard Frank and his team at Network Analysis Corporation, and the network measurement system was prepared by Kleinrock's team at UCLA.[3]

Due to Kleinrock's early development of packet switching theory and his focus on analysis, design and measurement, his Network Measurement Center at UCLA was selected to be the first node on the ARPANET. All this came together in September 1969 when BBN installed the first IMP at UCLA and the first host computer was connected. Doug Engelbart's project on "Augmentation of Human Intellect" (which included NLS, an early hypertext system) at Stanford Research Institute (SRI) provided a second node. SRI supported the Network Information Center, led by Elizabeth (Jake) Feinler and including functions such as maintaining tables of host name to address mapping as well as a directory of the RFC's. One month later, when SRI was connected to the ARPANET, the first host-to-host message was sent from Kleinrock's laboratory to SRI. Two more nodes were added at the University of California, Santa Barbara and University of Utah. These last two nodes incorporated application visualization projects, with Glen Culler and Burton Fried at UCSB investigating methods for display of mathematical functions using storage displays to deal with the problem of refresh over the net, and Robert Taylor and Ivan Sutherland at Utah investigating methods of 3-D representations over the net. Thus,

[2] It was from the RAND study that the false rumor started claiming that the ARPANET was somehow related to building a network resistant to nuclear war. This was never true of the ARPANET, only the unrelated RAND study on secure voice considered nuclear war. However, the later work on Internetting did emphasize robustness and survivability, including the capability to withstand losses of large portions of the underlying networks.

[3] Including amongst others Vint Cerf, Steve Crocker, and Jon Postel. Joining them later were David Crocker who was to play an important role in documentation of electronic mail protocols, and Robert Braden, who developed the first NCP and then TCP for IBM mainframes and also was to play a long term role in the ICCB and IAB.

TABLE 1. INTERNET TIMELINE

Year	Event
1958	President Dwight D. Eisenhower saw a need for the Advanced Research Projects Agency (APRA) to keep the U.S. at the forefront of technology.
1961	Leonard Kleinrock invents packet-switching technology, which forms the basis for the creation of the Internet
1962	J.C.R. Licklider, head of computer research at ARPA, articulates vision of worldwide network.
1965	Ted Nelson invents the term Hypertext to describe links to other texts embedded in a text. He later designs a long, worldwide hypertext network called Xanadu, but it is not until 25 years later that Tim Berners-Lee makes hypertext available to all network users with his World Wide Web project.
1967	Larry Roberts publishes a paper proposing the ARPAnet network.
1968	The United States Department of Defense Advanced Research Projects Agency (DARPA) launches the ARPANET project to design a distributed computer network.
1969	The ARPANET network is unveiled at UCLA creating the basis for the Internet.
1972	The predecessors of Internet electronic mail are introduced into the ARPANET in 1972. Networks developed for remote access to computers thus become a channel for human communication. The early counterparts to e-mail were generally only discussion systems internal to mainframe systems. E-mail introduced by Ray Tomlinson.
1973–1977	Research on the ARPANET is carried out in the USA. This produces TCP/IP, the basis of the Internet.
1980	The TCP/IP data transmission protocol is adopted as the U.S. Department of Defense's (DoD) offical network standard. The development of Internet applications and protocols receives considerable support from the Federal government.
1981	BITNET ("Because It's Time NETwork") set up to link the mainframes at Yale and the City University of New York using an NJE (Network Job Entry) fixed link developed by IBM. NJE transmitted virtual 80-character punch-card packets, i.e. remote batch jobs, and 132-character-wide output from one machine to another. These virtual punch-card packets could be used by the appropriate programs to transmit batch job files and also to send electronic mail, as well as for simple interactive communication
	BITNET and its European counterpart, EARN, set up in 1982, formed the first network available to the broad academic user community and students. The ARPANET set up previously was only used by the few, rare laboratories that collaborated with the U.S. Department of Defense. BITNET/EARN gave rise to a widespread e-mailing and discussion culture, and to support this software such as LISTSERV and the original prototype for IRC, RELAY, were developed.
	The KERMIT file-transfer protocol was devised at Columbia University in New York. Local Area Networks and fixed Internet links were a speciality of laboratories, and the only shared means of data communication between computers was a slow serial connection and a modem. KERMIT made it possible to transfer files reliably and to establish data-terminal connections with almost any mainframe or microcomputer via a slow serial cable. KERMIT software is still available for most operating systems, and its way of transmitting data is also supported by many other modem-communications programs.
	Hayes brings its Smartmodem 300 modem onto the market. This is intended for consumers to use with their microcomputers at a speed of 300 bit/s. It could be directly controlled from the computer using AT commands, which are still the industry standard for all dial-up modems, even today. Before this, the only modems available were expensive "dumb" modems mainly rented to companies and installed at high prices from telecom companies. This required the acquisition of a special data-telephone connection, and dialing had to be done by hand or using a separate dialing device.
1983	On January 1, the ARPANET goes over from Network Control Protocol (NCP) to TCP/IP. This makes the ARPANET a true network of networks, i.e. the backbone of the Internet.
	The Domain Name System is developed, so that everyone no longer needs a hosts.txt file to serve as an address book.
	Berkeley University releases 4.2BSD Unix, which supports TCP/IP and forms the basis of a great deal of later Internet software.
1986	Eric Thomas in Paris develops the LISTSERV mailing list distribution system, originally for the EARN network. This later also became popular on the Internet, as it is able to handle very large mailing lists
	The Internet Engineering Task Force (IETF) and Internet Research Task Force (IRTF) are set up as collaborative forums for standardizing Internet protocols and for research.
	The National Science Foundation in the USA funds the setting up of the 56 Kbit/s NSFNET network, which links U.S. supercomputer centers together to become the Internet using Fuzzball computers. This makes possible Internet links to the scientific community outside the Department of Defense-funded ARPANET project, who had to put up with other alternatives, such as the BITNET and UUCP networks.
	Cisco Systems, a small company with four workers in Silicon Valley, California, delivers its first Internet routers.
	Diginet, a 64 Kbit/s voice-and-data network is launched as a predecessor to ISDN
1988	Albert Gore, then a Tennessee Senator, proposes the National Research and Education Network, which would provide top computing facilities to research communities and schools.
	On November 2, Robert Morris' Internet Worm program spreads through the Internet, automatically copying itself. The NORDUnet's connection was not yet ready for routing, so the Nordic countries were spared this problem. But, as a consequence, people begin to pay more attention to Internet security.
1989	Tim Berners-Lee dreams about a hypertext system that integrates scattered information resources, improving mutual understanding and the spread of information. Every user could also be an information producer on a global level. He decides to submit a proposal that hypertext be used for information management at his workplace in the European Organization for Nuclear Research (CERN). His proposal is not approved, so he continues to pursue his dream in 1990.
1991	Gopher document retrieval system introduced at University of Minnesota.
1992	Tim Berners-Lee's updated proposal, "WorldWideWeb: Proposal for a HyperText Project" is approved by CERN. Prototype versions of a graphic WWW browser-editor for the NeXT computer and of a line mode interface that works on all terminals are produced, i.e. the Web is born.
	The first audio and video multicasts are broadcast over the Internet.
	The prototype Internet network ARPANET set up in 1969, with network address 10.0.0.0, is officially shut down.
	Over 26 Web servers in use.
	Over a million computers in the Internet.
	The Internet Society (ISOC) and RIPE NCC are founded.
	The first MBONE transmissions: sound in March and video in November.
	Jean Polly writes her article "Surfing the Internet: an Introduction", thus coining the phrase "surfing the Net".

(continued)

TABLE 1. (*Continued*)

1993	Use of the WWW in the USA expands by 341 634%, thanks to the easy-to-use graphics-based NCSA MOSAIC Internet browser, introduced at the University of Illinois by Marc Andreeson.
	The Internet's takeover of the world begins in earnest.
	The United Nations joins the Internet.
1994	Real Audio introduced to Internet, which allows one to hear audio in near real time. Radio HK, first 24-hour Internet-only radio station starts broadcasting.
	One of the Internet's biggest and best-known subject indexes comes on-line at WWW.Yahoo.com.
	Use of the Web spreads and overtakes telnet in popularity, only lagging behind FTP, which is used for file transfer.
	A standard is agreed for 28.8 bit/s V.34 modems.
1995	The Altavista.digital.com Internet search service begins. Anyone can use it free to look for Web pages in various parts of the world.
	Use of the WWW overtakes FTP in the NSFNET statistics, i.e. it is now the world's most frequently used Internet application.
	SUN Microsystems releases the platform-independent Java language and the HotJava browser written using it. Applets written in Java can add animations and various interactive features to Web pages.
	Netscape goes on the stock market.
1996	Various attempts are made to pass laws restricting freedom of speech on the Internet, but without success.
	Telephone companies begin to suspect that Internet calls will take their markets.
	Telecommunications Act of 1996 deregulates data network transmission.
	SUN releases its JavaStation network computer, which runs Java programs straight from the network. The aim is to produce an intelligent Internet terminal that does not require local software maintenance. Enthusiasts for network computers include, for example, the makers of major terminal-operated database environments such as IBM and Oracle, since they could replace millions of ageing mainframe terminals in situations where there is no need for an expensively maintained and complicated microcomputer. There are also plans to use network computers to create an simple Internet user interface for domestic televisions and other appliances.
1997	The World Wide Web Consortium publishes version 4.0 of the HTML language used to create web pages. This includes multimedia features, UNICODE support, for displaying the world's various languages, and features that help people with disabilities use the Net.
	The Internet2 project is announced in the U.S. to develop within two years new Internet services for the research community, such as interactive TV, videoconferencing and remote presence for teaching and research. For this collaboration, the research community began to construct new Internet connections, which initially ran at 620 Mbit/s, increasing to 2.4 Gbit/s at the beginning of 1999.
1998	The World Wide Web Consortium releases the specifications for XML (Extensible Markup Language) version 1.0, which will make it easy to expand future web pages.
1999	150 million users on the Internet. Over 800 million web pages accessible.
2000	Over 407 million users worldwide.

by the end of 1969, four host computers were connected together into the initial ARPANET, and the budding Internet was off the ground. Even at this early stage, it should be noted that the networking research incorporated both work on the underlying network and work on how to utilize the network. This tradition continues to this day.

Computers were added quickly to the ARPANET during the following years, and work proceeded on completing a functionally complete Host-to-Host protocol and other network software. In December 1970 the Network Working Group (NWG) working under S. Crocker finished the initial ARPANET Host-to-Host protocol, called the Network Control Protocol (NCP). As the ARPANET sites completed implementing NCP during the period 1971–1972, the network users finally could begin to develop applications.

In October 1972 Kahn organized a large, very successful demonstration of the ARPANET at the International Computer Communication Conference (ICCC). This was the first public demonstration of this new network technology to the public. It was also in 1972 that the initial "hot" application, electronic mail, was introduced. In March, Ray Tomlinson at BBN wrote the basic email message send and read software, motivated by the need of the ARPANET developers for an easy coordination mechanism. In July, Roberts expanded its utility by writing the first email utility program to list, selectively read, file, forward, and respond to messages. From there email took off as the largest network application for over a decade. This was a harbinger of the kind of activity one sees on the World Wide Web today, namely, the enormous growth of all kinds of "people-to-people" traffic.

The Initial Internetting Concepts

The original ARPANET grew into the Internet. Internet was based on the idea that there would be multiple independent networks of rather arbitrary design, beginning with the ARPANET as the pioneering packet switching network, but soon to include packet satellite networks, ground-based packet radio networks and other networks. The Internet as it is now known, embodies a key underlying technical idea, namely that of open architecture networking. In this approach, the choice of any individual network technology was not dictated by a particular network architecture but rather could be selected freely by a provider and made to interwork with the other networks through a meta-level "Internetworking Architecture". Up until that time there was only one general method for federating networks. This was the traditional circuit switching method where networks would interconnect at the circuit level, passing individual bits on a synchronous basis along a portion of an end-to-end circuit between a pair of end locations. Recall that Kleinrock had shown in 1961 that packet switching was a more efficient switching method. Along with packet switching, special purpose interconnection arrangements between networks were another possibility. While there were other limited ways to interconnect different networks, they required that one be used as a component of the other, rather than acting as a peer of the other in offering end-to-end service.

In an open-architecture network, the individual networks may be separately designed and developed and each may have its own unique interface, which it may offer to users and/or other providers, including other Internet providers. Each network can be designed in accordance with the specific environment and user requirements of that network. There are generally no constraints on the types of network that can be included or on their geographic scope, although certain pragmatic considerations will dictate what makes sense to offer.

The idea of open-architecture networking was first introduced by Kahn shortly after having arrived at DARPA in 1972. This work was originally part of the packet radio program, but subsequently became a separate program in its own right. At the time, the program was called "Internetting". Key to making the packet radio system work was a reliable end-end protocol that could maintain effective communication in the face of jamming and other radio interference, or withstand intermittent blackout such as caused by being in a tunnel or blocked by the local terrain. Kahn first contemplated developing a protocol local only to the packet radio network, since that would avoid having to deal with the multitude of different operating systems, and continuing to use NCP.

However, NCP did not have the ability to address networks (and machines) further downstream than a destination IMP on the ARPANET and thus some change to NCP would also be required. (The assumption was that the ARPANET was not changeable in this regard). NCP relied on ARPANET to provide end-to-end reliability. If any packets were lost, the protocol (and presumably any applications it supported) would come to a grinding halt. In this model NCP had no end-end host error control, since the ARPANET was to be the only network in existence and it would be so reliable that no error control would be required on the part of the hosts.

Thus, Kahn decided to develop a new version of the protocol, which could meet the needs of an open-architecture network environment. This protocol would eventually be called the Transmission Control Protocol/Internet Protocol (TCP/IP). While NCP tended to act like a device driver, the new protocol would be more like a communications protocol.

Four ground rules were critical to Kahn's early thinking:

- Each distinct network would have to stand on its own and no internal changes could be required to any such network to connect it to the Internet.
- Communications would be on a best effort basis. If a packet did not make it to the final destination, it would shortly be retransmitted from the source.
- Black boxes would be used to connect the networks; these would later be called gateways and routers. There would be no information retained by the gateways about the individual flows of packets passing through them, thereby keeping them simple and avoiding complicated adaptation and recovery from various failure modes.
- There would be no global control at the operations level.

Other key issues that needed to be addressed were:

- Algorithms to prevent lost packets from permanently disabling communications and enabling them to be successfully retransmitted from the source.
- Providing for host to host "pipelining" so that multiple packets could be enroute from source to destination at the discretion of the participating hosts, if the intermediate networks allowed it.
- Gateway functions to allow it to forward packets appropriately. This included interpreting IP headers for routing, handling interfaces, breaking packets into smaller pieces if necessary, etc.
- The need for end-end checksums, reassembly of packets from fragments and detection of duplicates, if any.
- The need for global addressing
- Techniques for host to host flow control.
- Interfacing with the various operating systems
- There were also other concerns, such as implementation efficiency, internetwork performance, but these were secondary considerations at first.

Kahn began work on a communications-oriented set of operating system principles while at BBN and documented some of his early thoughts in an internal BBN memorandum entitled "Communications Principles for Operating Systems." At this point he realized it would be necessary to learn the implementation details of each operating system to have a chance to embed any new protocols in an efficient way. Thus, in the spring of 1973, after starting the internetting effort, he asked Vint Cerf (then at Stanford) to work with him on the detailed design of the protocol. Cerf had been intimately involved in the original NCP design and development and already had the knowledge about interfacing to existing operating systems. So armed with Kahn's architectural approach to the communications side and with Cerf's NCP experience, they teamed up to spell out the details of what became TCP/IP.

The give and take was highly productive and the first written version[4] of the resulting approach was distributed at a special meeting of the International Network Working Group (INWG) which had been set up at a conference at Sussex University in September 1973. Cerf had been invited to chair this group and used the occasion to hold a meeting of INWG members who were heavily represented at the Sussex Conference.

[4] This was subsequently published as V.G. Cerf and R.E. Kahn, "A protocol for packet network interconnection" *IEEE Trans. Comm. Tech.*, vol. COM-22, **5**, 627–641, (May 1974).

Some basic approaches emerged from this collaboration between Kahn and Cerf:

- Communication between two processes would logically consist of a very long stream of bytes (they called them octets). The position of any octet in the stream would be used to identify it.
- Flow control would be done by using sliding windows and acknowledgments (acks). The destination could select when to acknowledge and each ack returned would be cumulative for all packets received to that point.
- It was left open as to exactly how the source and destination would agree on the parameters of the windowing to be used. Defaults were used initially.
- Although Ethernet was under development at Xerox PARC at that time, the proliferation of LANs were not envisioned at the time, much less PCs and workstations. The original model was national level networks like ARPANET of which only a relatively small number were expected to exist. Thus a 32 bit IP address was used of which the first 8 bits signified the network and the remaining 24 bits designated the host on that network. This assumption, that 256 networks would be sufficient for the foreseeable future, was clearly in need of reconsideration when LANs began to appear in the late 1970s.

The original Cerf/Kahn paper on the Internet described one protocol, called TCP, which provided all the transport and forwarding services in the Internet. Kahn had intended that the TCP protocol support a range of transport services, from the totally reliable sequenced delivery of data (virtual circuit model) to a datagram service in which the application made direct use of the underlying network service, which might imply occasional lost, corrupted or reordered packets.

However, the initial effort to implement TCP resulted in a version that only allowed for virtual circuits. This model worked fine for file transfer and remote login applications, but some of the early work on advanced network applications, in particular packet voice in the 1970s, made clear that in some cases packet losses should not be corrected by TCP, but should be left to the application to deal with. This led to a reorganization of the original TCP into two protocols, the simple IP which provided only for addressing and forwarding of individual packets, and the separate TCP, which was concerned with service features such as flow control and recovery from lost packets. For those applications that did not want the services of TCP, an alternative called the User Datagram Protocol (UDP) was added in order to provide direct access to the basic service of IP.

A major initial motivation for both the ARPANET and the Internet was resource sharing, for example, allowing users on the packet radio networks to access the time sharing systems attached to the ARPANET. Connecting the two together was far more economical that duplicating these very expensive computers. However, while file transfer and remote login (Telnet) were very important applications, electronic mail has probably had the most significant impact of the innovations from that era. Email provided a new model of how people could communicate with each other, and changed the nature of collaboration, first in the building of the Internet itself (as is discussed below) and later for much of society.

There were other applications proposed in the early days of the Internet, including packet-based voice communication (the precursor of Internet telephony), various models of file and disk sharing, and early "worm" programs that showed the concept of agents (and, of course, viruses). A key concept of the Internet is that it was not designed for just one application, but as a general infrastructure on which new applications could be conceived, as illustrated later by the emergence of the World Wide Web. It is the general purpose nature of the service provided by TCP and IP that makes this possible.

Proving the Ideas

DARPA let three contracts to Stanford (Cerf), BBN (Ray Tomlinson) and UCL (Peter Kirstein) to implement TCP/IP (it was simply called TCP in the Cerf/Kahn paper, but contained both components). The Stanford team, led by Cerf, produced the detailed specification and within about a year there were three independent implementations of TCP that could interoperate.

This was the beginning of long term experimentation and development to evolve and mature the Internet concepts and technology. Beginning with the first three networks (ARPANET, Packet Radio, and Packet Satellite) and their initial research communities, the experimental environment has grown to incorporate essentially every form of network and a very broad-based research and development community. With each expansion has come new challenges.

The early implementations of TCP were done for large time sharing systems such as Tenex and TOPS 20. When desktop computers first appeared, it was thought by some that TCP was too big and complex to run on a personal computer. David Clark and his research group at MIT set out to show that a compact and simple implementation of TCP was possible. They produced an implementation, first for the Xerox Alto (the early personal workstation developed at Xerox PARC) and then for the IBM PC. That implementation was fully interoperable with other TCPs, but was tailored to the application suite and performance objectives of the personal computer, and showed that workstations, as well as large time-sharing systems, could be a part of the Internet. In 1976, Kleinrock published the first book on the ARPANET. It included an emphasis on the complexity of protocols and the pitfalls they often introduce. This book was influential in spreading the lore of packet switching networks to a very wide community.

Widespread development of LANS, PCs, and workstations in the 1980s allowed the nascent Internet to flourish. Ethernet technology, developed by Bob Metcalfe at Xerox PARC in 1973, is now probably the dominant network technology in the Internet and PCs and workstations the dominant computers. This change from having a few networks with a modest number of time-shared hosts (the original ARPANET model) to having many networks has resulted in a number of new concepts and changes to the underlying technology. First, it resulted in the definition of three network classes (A, B, and C) to accommodate the range of networks. Class A represented large national scale networks (small number of networks with large numbers of hosts); Class B represented regional scale networks; and Class C represented local area networks (large number of networks with relatively few hosts).

A major shift occurred as a result of the increase in scale of the Internet and its associated management issues. To make it easy for people to use the network, hosts were assigned names, so that it was not necessary to remember the numeric addresses. Originally, there were a fairly limited number of hosts, so it was feasible to maintain a single table of all the hosts and their associated names and addresses. The shift to having a large number of independently managed networks (e.g., LANs) meant that having a single table of hosts was no longer feasible, and the Domain Name System (DNS) was invented by Paul Mockapetris of USC/ISI. The DNS permitted a scalable distributed mechanism for resolving hierarchical host names (e.g. www.acm.org) into an Internet address.

The increase in the size of the Internet also challenged the capabilities of the routers. Originally, there was a single distributed algorithm for routing that was implemented uniformly by all the routers in the Internet. As the number of networks in the Internet exploded, this initial design could not expand as necessary, so it was replaced by a hierarchical model of routing, with an Interior Gateway Protocol (IGP) used inside each region of the Internet, and an Exterior Gateway Protocol (EGP) used to tie the regions together. This design permitted different regions to use a different IGP, so that different requirements for cost, rapid reconfiguration, robustness and scale could be accommodated. Not only the routing algorithm, but the size of the addressing tables, stressed the capacity of the routers. New approaches for address aggregation, in particular classless inter-domain routing (CIDR), have recently been introduced to control the size of router tables.

As the Internet evolved, one of the major challenges was how to propagate the changes to the software, particularly the host software. DARPA supported UC Berkeley to investigate modifications to the Unix operating system, including incorporating TCP/IP developed at BBN. Although Berkeley later rewrote the BBN code to more efficiently fit into the Unix system and kernel, the incorporation of TCP/IP into the Unix BSD system releases proved to be a critical element in dispersion of the protocols to the research community. Much of the CS research community began to use Unix BSD for their day-to-day computing environment. Looking back, the strategy of incorporating Internet protocols into a supported operating system for the research community was one of the key elements in the successful widespread adoption of the Internet.

One of the more interesting challenges was the transition of the ARPANET host protocol from NCP to TCP/IP as of January 1, 1983. This was a "flag-day" style transition, requiring all hosts to convert simultaneously or be left having to communicate via rather ad-hoc mechanisms. This transition was carefully planned within the community over several years before it actually took place and went surprisingly smoothly (but resulted in a distribution of buttons saying "I survived the TCP/IP transition").

TCP/IP was adopted as a defense standard three years earlier in 1980. This enabled defense to begin sharing in the DARPA Internet technology base and led directly to the eventual partitioning of the military and nonmilitary communities. By 1983, ARPANET was being used by a significant number of defense R&D and operational organizations. The transition of ARPANET from NCP to TCP/IP permitted it to be split into a MILNET supporting operational requirements and an ARPANET supporting research needs.

Thus, by 1985, Internet was already well established as a technology supporting a broad community of researchers and developers, and was beginning to be used by other communities for daily computer communications. Electronic mail was being used broadly across several communities, often with different systems, but interconnection between different mail systems was demonstrating the utility of broad based electronic communications between people.

Transition to Widespread Infrastructure

At the same time that the Internet technology was being experimentally validated and widely used among a subset of computer science researchers, other networks and networking technologies were being pursued. The usefulness of computer networking, especially electronic mail, demonstrated by DARPA and Department of Defense contractors on the ARPANET was not lost on other communities and disciplines, so that by the mid-1970s computer networks had begun to spring up wherever funding could be found for the purpose. The U.S. Department of Energy (DoE) established MFENet for its researchers in Magnetic Fusion Energy, whereupon DoE's High Energy Physicists responded by building HEPNet. NASA Space Physicists followed with SPAN, and Rick Adrion, David Farber, and Larry Landweber established CSNET for the (academic and industrial) Computer Science community with an initial grant from the U.S. National Science Foundation (NSF). AT&T's free-wheeling dissemination of the UNIX computer operating system spawned USENET, based on UNIX built-in UUCP communication protocols, and in 1981 Ira Fuchs and Greydon Freeman devised BITNET, which linked academic mainframe computers in an "email as card images" paradigm.

With the exception of BITNET and USENET, these early networks (including ARPANET) were purpose-built, i.e., they were intended for, and largely restricted to, closed communities of scholars; there was hence little pressure for the individual networks to be compatible, and indeed, they largely were not. In addition, alternative technologies were being pursued in the commercial sector, including XNS from Xerox, DECNet, and IBM's SNA.[5] It remained for the British JANET (1984) and U.S. NSFNET (1985) programs to explicitly announce their intent to serve the entire higher education community, regardless of discipline. Indeed, a condition for a U.S. university to receive NSF funding for an Internet connection was that "... the connection must be made available to ALL qualified users on campus."

In 1985, Dennis Jennings came from Ireland to spend a year at NSF leading the NSFNET program. He worked with the community to help NSF make a critical decision that TCP/IP would be mandatory for the NSFNET program. When Steve Wolff took over the NSFNET program in 1986, he recognized the need for a wide area networking infrastructure to support the general academic and research community, along with the need to develop a strategy for establishing such infrastructure on a basis ultimately independent of direct federal funding. Policies and strategies were adopted (see below) to achieve that end.

---

[5] The desirability of email interchange, however, led to one of the first "Internet books": *!%@:: A Directory of Electronic Mail Addressing and Networks*, by Frey and Adams, on email address translation and forwarding.

NSF also elected to support DARPA's existing Internet organizational infrastructure, hierarchically arranged under the (then) Internet Activities Board (IAB). The public declaration of this choice was the joint authorship by the IAB's Internet Engineering and Architecture Task Forces and by NSF's Network Technical Advisory Group of RFC 985 (Requirements for Internet Gateways), which formally ensured interoperability of DARPA's and NSF's pieces of the Internet.

In addition to the selection of TCP/IP for the NSFNET program, Federal agencies made and implemented several other policy decisions, which shaped the Internet of today.

- Federal agencies shared the cost of common infrastructure, such as transoceanic circuits. They also jointly supported "managed interconnection points" for interagency traffic; the Federal Internet Exchanges (FIX-E and FIX-W) built for this purpose served as models for the Network Access Points and "*IX" facilities that are prominent features of today's Internet architecture.
- To coordinate this sharing, the Federal Networking Council[6] was formed. The FNC also cooperated with other international organizations, such as RARE in Europe, through the Coordinating Committee on Intercontinental Research Networking, CCIRN, to coordinate Internet support of the research community worldwide.
- This sharing and cooperation between agencies on Internet-related issues had a long history. An unprecedented 1981 agreement between Farber, acting for CSNET and the NSF, and DARPA's Kahn, permitted CSNET traffic to share ARPANET infrastructure on a statistical and no-metered-settlements basis.
- Subsequently, in a similar mode, the NSF encouraged its regional (initially academic) networks of the NSFNET to seek commercial, non-academic customers, expand their facilities to serve them, and exploit the resulting economies of scale to lower subscription costs for all.
- On the NSFNET Backbone—the national-scale segment of the NSFNET—NSF enforced an "Acceptable Use Policy" (AUP) which prohibited Backbone usage for purposes "not in support of Research and Education." The predictable (and intended) result of encouraging commercial network traffic at the local and regional level, while denying its access to national-scale transport, was to stimulate the emergence and/or growth of "private", competitive, long-haul networks such as PSI, UUNET, ANS CO+RE, and (later) others. This process of privately-financed augmentation for commercial uses was thrashed out starting in 1988 in a series of NSF-initiated conferences at Harvard's Kennedy School of Government on "The Commercialization and Privatization of the Internet" and on the "com-priv" list on the net itself.
- In 1988, a National Research Council committee, chaired by Kleinrock and with Kahn and Clark as members, produced a report commissioned by NSF titled "Towards a National Research Network". This report was influential on then Senator Al Gore, and ushered in high speed networks that laid the networking foundation for the future information superhighway.
- In 1994, a National Research Council report, again chaired by Kleinrock (and with Kahn and Clark as members again), Entitled "Realizing The Information Future: The Internet and Beyond" was released. This report, commissioned by NSF, was the document in which a blueprint for the evolution of the information superhighway was articulated and which has had a lasting affect on the way to think about its evolution. It anticipated the critical issues of intellectual property rights, ethics, pricing, education, architecture and regulation for the Internet.
- NSF's privatization policy culminated in April, 1995, with the defunding of the NSFNET Backbone. The funds thereby recovered were (competitively) redistributed to regional networks to buy national-scale Internet connectivity from the now numerous, private, long-haul networks.

The backbone had made the transition from a network built from routers out of the research community (the "Fuzzball" routers from David Mills) to commercial equipment. In its $8\frac{1}{2}$ year lifetime, the Backbone had grown from six nodes with 56 kbps links to 21 nodes with multiple 45 Mbps

links. It had seen the Internet grow to over 50,000 networks on all seven continents and outer space, with approximately 29,000 networks in the United States.

Such was the weight of the NSFNET program's ecumenism and funding (\$200 million from 1986 to 1995), and the quality of the protocols themselves, that by 1990 when the ARPANET itself was finally decommissioned[7], TCP/IP had supplanted or marginalized most other wide-area computer network protocols worldwide, and IP was well on its way to becoming THE bearer service for the Global Information Infrastructure.

The Role of Documentation

A key to the rapid growth of the Internet has been the free and open access to the basic documents, especially the specifications of the protocols.

The beginnings of the ARPANET and the Internet in the university research community promoted the academic tradition of open publication of ideas and results. However, the normal cycle of traditional academic publication was too formal and too slow for the dynamic exchange of ideas essential to creating networks.

In 1969 a key step was taken by S. Crocker (then at UCLA) in establishing the Request for Comments (or RFC) series of notes. These memos were intended to be an informal fast distribution way to share ideas with other network researchers. At first the RFCs were printed on paper and distributed via snail mail. As the File Transfer Protocol (FTP) came into use, the RFCs were prepared as online files and accessed via FTP. Now, of course, the RFCs are easily accessed via the World Wide Web at dozens of sites around the world. SRI, in its role as Network Information Center, maintained the online directories. Jon Postel acted as RFC Editor as well as managing the centralized administration of required protocol number assignments, roles that he continues to this day.

The effect of the RFCs was to create a positive feedback loop, with ideas or proposals presented in one RFC triggering another RFC with additional ideas, and so on. When some consensus (or a least a consistent set of ideas) had come together a specification document would be prepared. Such a specification would then be used as the base for implementations by the various research teams.

Over time, the RFCs have become more focused on protocol standards (the "official" specifications), though there are still informational RFCs that describe alternate approaches, or provide background information on protocols and engineering issues. The RFCs are now viewed as the "documents of record" in the Internet engineering and standards community.

The open access to the RFCs (for free, if one has any kind of a connection to the Internet) promotes the growth of the Internet because it allows the actual specifications to be used for examples in college classes and by entrepreneurs developing new systems.

Email has been a significant factor in all areas of the Internet, and that is certainly true in the development of protocol specifications, technical standards, and Internet engineering. The very early RFCs often presented a set of ideas developed by the researchers at one location to the rest of the community. After email came into use, the authorship pattern changed. RFCs were presented by joint authors with common view independent of their locations.

The use of specialized email mailing lists has been long used in the development of protocol specifications, and continues to be an important tool. The IETF now has in excess of 75 working groups, each working on a different aspect of Internet engineering. Each of these working groups has a mailing list to discuss one or more draft documents under development. When consensus is reached on a draft document it may be distributed as an RFC.

As the current rapid expansion of the Internet is fueled by the realization of its capability to promote information sharing, it should be understood that the network's first role in information sharing was sharing the information about its own design and operation through the RFC documents. This unique method for evolving new capabilities in the network will continue to be critical to future evolution of the Internet.

---

[6] Originally named Federal Research Internet Coordinating Committee, FRICC. The FRICC was originally formed to coordinate U.S. research network activities in support of the international coordination provided by the CCIRN.

[7] The decommisioning of the ARPANET was commemorated on its 20th anniversary by a UCLA symposium in 1989.

Formation of the Broad Community

The Internet is as much a collection of communities as a collection of technologies, and its success is largely attributable to both satisfying basic community needs as well as utilizing the community in an effective way to push the infrastructure forward. This community spirit has a long history beginning with the early ARPANET. The early ARPANET researchers worked as a close-knit community to accomplish the initial demonstrations of packet switching technology described earlier. Likewise, the Packet Satellite, Packet Radio and several other DARPA computer science research programs were multi-contractor collaborative activities that heavily used whatever available mechanisms there were to coordinate their efforts, starting with electronic mail and adding file sharing, remote access, and eventually World Wide Web capabilities. Each of these programs formed a working group, starting with the ARPANET Network Working Group. Because of the unique role that ARPANET played as an infrastructure supporting the various research programs, as the Internet started to evolve, the Network Working Group evolved into Internet Working Group.

In the late 1970s, recognizing that the growth of the Internet was accompanied by a growth in the size of the interested research community and therefore an increased need for coordination mechanisms, Vint Cerf, then manager of the Internet Program at DARPA, formed several coordination bodies, an International Cooperation Board (ICB), chaired by Peter Kirstein of UCL, to coordinate activities with some cooperating European countries centered on Packet Satellite research, an Internet Research Group, which was an inclusive group providing an environment for general exchange of information, and an Internet Configuration Control Board (ICCB), chaired by Clark. The ICCB was an invitational body to assist Cerf in managing the burgeoning Internet activity.

In 1983, when Barry Leiner took over management of the Internet research program at DARPA, he and Clark recognized that the continuing growth of the Internet community demanded a restructuring of the coordination mechanisms. The ICCB was disbanded and in its place a structure of Task Forces was formed, each focused on a particular area of the technology (e.g., routers, end-to-end protocols, etc.). The Internet Activities Board (IAB) was formed from the chairs of the Task Forces. It of course was only a coincidence that the chairs of the Task Forces were the same people as the members of the old ICCB, and Dave Clark continued to act as chair.

After some changing membership on the IAB, Phill Gross became chair of a revitalized Internet Engineering Task Force (IETF), at the time merely one of the IAB Task Forces. By 1985 there was a tremendous growth in the more practical/engineering side of the Internet. This growth resulted in an explosion in the attendance at the IETF meetings, and Gross was compelled to create substructure to the IETF in the form of working groups.

This growth was complemented by a major expansion in the community. No longer was DARPA the only major player in the funding of the Internet. In addition to NSFNet and the various U.S. and international government-funded activities, interest in the commercial sector was beginning to grow. Also in 1985, both Kahn and Leiner left DARPA and there was a significant decrease in Internet activity at DARPA. As a result, the IAB was left without a primary sponsor and increasingly assumed the mantle of leadership.

The growth continued, resulting in even further substructure within both the IAB and IETF. The IETF combined Working Groups into Areas, and designated Area Directors. An Internet Engineering Steering Group (IESG) was formed of the Area Directors. The IAB recognized the increasing importance of the IETF, and restructured the standards process to explicitly recognize the IESG as the major review body for standards. The IAB also restructured so that the rest of the Task Forces (other than the IETF) were combined into an Internet Research Task Force (IRTF) chaired by Postel, with the old task forces renamed as research groups.

The growth in the commercial sector brought with it increased concern regarding the standards process itself. Starting in the early 1980s and continuing to this day, the Internet grew beyond its primarily research roots to include both a broad user community and increased commercial activity. Increased attention was paid to making the process open and fair. This coupled with a recognized need for community support of the Internet eventually led to the formation of the Internet Society in 1991, under the auspices of Kahn's Corporation for National Research Initiatives (CNRI) and the leadership of Cerf, then with CNRI.

In 1992, yet another reorganization took place. In 1992, the Internet Activities Board was reorganized and renamed the Internet Architecture Board operating under the auspices of the Internet Society. A more "peer" relationship was defined between the new IAB and IESG, with the IETF and IESG taking a larger responsibility for the approval of standards. Ultimately, a cooperative and mutually supportive relationship was formed between the IAB, IETF, and Internet Society, with the Internet Society taking on as a goal the provision of service and other measures which would facilitate the work of the IETF.

The recent development and widespread deployment of the World Wide Web has brought with it a new community, as many of the people working on the WWW have not thought of themselves as primarily network researchers and developers. A new coordination organization was formed, the World Wide Web Consortium (W3C). Initially led from MIT's Laboratory for Computer Science by Tim Berners-Lee (the inventor of the WWW) and Al Vezza, W3C has taken on the responsibility for evolving the various protocols and standards associated with the Web.

Thus, through the over two decades of Internet activity, we have seen a steady evolution of organizational structures designed to support and facilitate an ever-increasing community working collaboratively on Internet issues.

Commercialization of the Technology

Commercialization of the Internet involved not only the development of competitive, private network services, but also the development of commercial products implementing the Internet technology. In the early 1980s, dozens of vendors were incorporating TCP/IP into their products because they saw buyers for that approach to networking. Unfortunately they lacked both real information about how the technology was supposed to work and how the customers planned on using this approach to networking. Many saw it as a nuisance add-on that had to be glued on to their own proprietary networking solutions: SNA, DECNet, Netware, NetBios. The DoD had mandated the use of TCP/IP in many of its purchases but gave little help to the vendors regarding how to build useful TCP/IP products.

In 1985, recognizing this lack of information availability and appropriate training, Dan Lynch in cooperation with the IAB arranged to hold a three day workshop for ALL vendors to come learn about how TCP/IP worked and what it still could not do well. The speakers came mostly from the DARPA research community who had both developed these protocols and used them in day to day work. About 250 vendor personnel came to listen to 50 inventors and experimenters. The results were surprises on both sides: the vendors were amazed to find that the inventors were so open about the way things worked (and what still did not work) and the inventors were pleased to listen to new problems they had not considered, but were being discovered by the vendors in the field. Thus a two-way discussion was formed that has lasted for over a decade.

After two years of conferences, tutorials, design meetings and work-shops, a special event was organized that invited those vendors whose products ran TCP/IP well enough to come together in one room for three days to show off how well they all worked together and also ran over the Internet. In September of 1988, the first Interop trade show was born. Fifty companies made the cut and 5,000 engineers from potential customer organizations came to see if it all did work as was promised. It did. Why? Because the vendors worked extremely hard to ensure that everyone's products interoperated with all of the other products even with those of their competitors. The Interop trade show has grown immensely since then and today it is held in seven locations around the world each year to an audience of over 250,000 people who come to learn which products work with each other in a seamless manner, learn about the latest products, and discuss the latest technology.

In parallel with the commercialization efforts that were highlighted by the Interop activities, the vendors began to attend the IETF meetings that were held three or four times a year to discuss new ideas for extensions of the TCP/IP protocol suite. Starting with a few hundred attendees mostly from academia and paid for by the government, these meetings now often exceeds a thousand attendees, mostly from the vendor community and paid for by the attendees themselves. This self-selected group evolves the

TCP/IP suite in a mutually cooperative manner. The reason it is so useful is that it is comprised of all stakeholders: researchers, end users, and vendors.

Network management provides an example of the interplay between the research and commercial communities. In the beginning of the Internet, the emphasis was on defining and implementing protocols that achieved interoperation. As the network grew larger, it became clear that the sometime ad hoc procedures used to manage the network would not scale. Manual configuration of tables was replaced by distributed automated algorithms, and better tools were devised to isolate faults. In 1987 it became clear that a protocol was needed that would permit the elements of the network, such as the routers, to be remotely managed in a uniform way. Several protocols for this purpose were proposed, including Simple Network Management Protocol or SNMP (designed, as its name would suggest, for simplicity, and derived from an earlier proposal called SGMP), HEMS (a more complex design from the research community) and CMIP (from the OSI community). A series of meeting led to the decisions that HEMS would be withdrawn as a candidate for standardization, in order to help resolve the contention, but that work on both SNMP and CMIP would go forward, with the idea that the SNMP could be a more near-term solution and CMIP a longer-term approach. The market could choose the one it found more suitable. SNMP is now used almost universally for network based management.

The last few years, have seen a new phase of commercialization. Originally, commercial efforts mainly comprised vendors providing the basic networking products, and service providers offering the connectivity and basic Internet services. The Internet has now become almost a "commodity" service, and much of the latest attention has been on the use of this global information infrastructure for support of other commercial services. This has been tremendously accelerated by the widespread and rapid adoption of browsers and the World Wide Web technology, allowing users easy access to information linked throughout the globe. Products are available to facilitate the provisioning of that information and many of the latest developments in technology have been aimed at providing increasingly sophisticated information services on top of the basic Internet data communications.

History of the Future

On October 24, 1995, the FNC unanimously passed a resolution defining the term Internet. This definition was developed in consultation with members of the internet and intellectual property rights communities. *RESOLUTION: The Federal Networking Council (FNC) agrees that the following language reflects our definition of the term "Internet". "Internet" refers to the global information system that—(i) is logically linked together by a globally unique address space based on the Internet Protocol (IP) or its subsequent extensions/follow-ons; (ii) is able to support communications using the Transmission Control Protocol/Internet Protocol (TCP/IP) suite or its subsequent extensions/follow-ons, and/or other IP-compatible protocols; and (iii) provides, uses or makes accessible, either publicly or privately, high level services layered on the communications and related infrastructure described herein.*

The Internet has changed much in the two decades since it came into existence. It was conceived in the era of time-sharing, but has survived into the era of personal computers, client-server and peer-to-peer computing, and the network computer. It was designed before LANs existed, but has accommodated that new network technology, as well as the more recent ATM and frame switched services. It was envisioned as supporting a range of functions from file sharing and remote login to resource sharing and collaboration, and has spawned electronic mail and more recently the World Wide Web. But most important, it started as the creation of a small band of dedicated researchers, and has grown to be a commercial success with billions of dollars of annual investment.

One should not conclude that the Internet has now finished changing. The Internet, although a network in name and geography, is a creature of the computer, not the traditional network of the telephone or television industry. It will, indeed it must, continue to change and evolve at the speed of the computer industry if it is to remain relevant. It is now changing to provide such new services as real time transport, in order to support, for example, audio and video streams. The availability of pervasive networking (i.e., the Internet) along with powerful affordable computing and communications in portable form (i.e., laptop computers, two-way

pagers, PDAs, cellular phones), is making possible a new paradigm of nomadic computing and communications.

This evolution brings new applications, Internet telephone, and slightly further out, Internet television. It is evolving to permit more sophisticated forms of pricing and cost recovery, a perhaps painful requirement in this commercial world. It is changing to accommodate yet another generation of underlying network technologies with different characteristics and requirements, from broadband residential access to satellites. New modes of access and new forms of service will spawn new applications, which in turn will drive further evolution of the net itself.

The most pressing question for the future of the Internet is not how the technology will change, but how the process of change and evolution itself will be managed. As this article describes, the architecture of the Internet has always been driven by a core group of designers, but the form of that group has changed as the number of interested parties has grown. With the success of the Internet has come a proliferation of stakeholders, stakeholders now have an economic as well as an intellectual investment in the network. In the debates over control of the domain name space and the form of the next generation IP addresses, a struggle to find the next social structure that will guide the Internet in the future will be seen. The form of that structure will be harder to find, given the large number of concerned stake-holders. At the same time, the industry struggles to find the economic rationale for the large investment needed for the future growth, for example, to upgrade residential access to a more suitable technology. If the Internet stumbles, it will not be because of lack of technology, vision, or motivation. It will be because a direction and march collectively into the future cannot be set.

Conclusion

The Internet is still in its infancy. And like an infant, growing in spurts with overabundant energy and wanting to test its capabilities, sometimes it stumbles. Fantastic technological possibilities and human imagination have taken the development of the Internet in many directions. Some of these directions have proven to be impossible to maintain in the business sense. As a result, some technologies and business models have lost favor in the marketplace, have been revised or discontinued. This phenomenon has created opportunities for redirection, convergence, and consolidation in many areas. Areas of strength (value-added telecommunications, software development, Internet infrastructure) have new focus, energy and resources to move forward.

The dominance in growth of the Internet, due to its ability to convey information, has stimulated the convergence of diverse technologies and industries. Telecommunications is converging with mass media to build-out high capacity networks for content delivery. Entertainment is converging with technology to produce interactive on-line devices for the home, office, and mobile uses. Industries are seeking on-line applications, devices and partners that will allow them to reach end users conveniently and expeditiously no matter what their location. These emerging technologies have one common denominator: connectivity to the Internet.

Additional Reading

Baran, P.: "On Distributed Communications Networks," *IEEE Trans. Comm. Systems* (March 1964).

Cailliau, R. and J. Gillies: *How the Web Was Born: The Story of the World Wide Web*, Oxford University Press, Inc., New York, NY, 2000.

Cerf, V.G. and R.E. Kahn: "A Protocol for Packet Network Interconnection", *IEEE Trans. Comm. Tech.*, vol. COM-22, 5, 627–641 (May 1974).

Cole, M.: *Introduction to Telecommunications: Voice, Data, and the Internet*, 2nd Edition, Prentice-Hall, Inc., Upper Saddle River, NJ, 2001.

Crocker, S.: RFC001 Host software, April 7, 1969.

Kahn, R.: *Communications Principles for Operating Systems*, Internal BBN memorandum, Jan. 1972.

Kahn, R. Guest editor: associate guest editors: K. Uncapher and H. van Trees: *Proceedings of the IEEE, Special Issue on Packet Communication Networks*, **66**(11), Nov. 1978.

Kleinrock, L.: *Information Flow in Large Communication Nets*, RLE Quarterly Progress Report, July 1961.

Kleinrock, L.: *Communication Nets: Stochastic Message Flow and Delay*, McGraw-Hill Companies Inc., New York, NY, 1964.

Kleinrock, L.: *Queueing Systems: Vol II, Computer Applications*, John Wiley & Sons Inc., New York, NY, 1976.

Licklider, J.C.R. and W. Clark: *On-Line Man Computer Communication*, August 1962.

Livingston, D.: *Essential Xml for Web Professionals*, Prentice-Hall, Inc., Upper Saddle River, NJ, 2001.

Reid, R.H.: *Architects of the Web*, John Wiley & Sons, Inc., New York, NY, 1997.

Roberts, L. and T. Merrill: *Toward a Cooperative Network of Time-Shared Computers*, Fall AFIPS Conf., Oct. 1966.

Roberts, L.: *Multiple Computer Networks and Intercomputer Communication*, ACM Gatlinburg Conf., October 1967.

Web References

Babbage Institute: http://www.cbi.umn.edu/
Claymath Mathematics Institute Millennium Prize Problems: http://www.claymath.org/prize_problems/index.htm
Cyberface: http://www.globalschoolhouse.org/
Global Internet Project: www.gip.org
Greenwich, CT. http://www.cristina.org/
Institute of Electrical and Electronic Engineers: www.IEEE.org
Internet Architecture Board: www.iab.org
Internet Corporation for the Assignment of Numbers and Names: www.ICANN.org
Internet Engineering Task Force: www.ietf.org
Internet Policy Institute: www.internetpolicy.org
Internet Society: www.ISOC.org
Internet Society's Special Interest Group: www.ipnsig.org
IPv6 Forum: http://www.ipv6forum.com/
Living Internet: http://www.livinginternet.com/
Marcopolo: http://www.worldcom.com/marcopolo/
Marshall Symposium: www.si.umich.edu/marshall
NASA Jet Propulsion Laboratory: www.jpl.nasa.gov
NSF CISE: http://www.cise.nsf.gov/
Planetary Society: www.planetary.org
SpaceRef: www.spaceref.com
Terena Networking Conference 2000 — Pioneering Tomorrow's Internet (RealPlayer Presentation). View the proceedings of the TERENA Networking Conference 2000.
Think Quest: http://www.thinkquest.org/
United States Institute of Peace: www.usip.org
vBNS: http://www.vbns.net/

LEONARD KLEINROCK, ROBERT E. KAHN, DAVID D. CLARK,
VINTON G. CERF, BARRY M. LEINER, DANIEL C. LYNCH,
LAWRENCE G. ROBERTS, and STEPHEN WOLFF

INTERPLANETARY MEDIUM. In the commonness of life on Earth, most would expect that the composition of space beyond our atmosphere might be planets, comets, asteroids, and an occasional meteor. In reality, the space between the planets and other larger objects of our solar system is richly composed of a variety of complex phenomena, which on Earth drive weather, affect communications, and provide beautiful displays with the aurora.

Long considered an empty void, the vast space between the planets and our Sun is actually filled with a tenuous gas comprised of neutral and ionized particles along with small dust grains. The source of the ionized particles comprising this gas is mostly outflows and outbursts of the Sun. Some of this gas is due to outflow of particles from planets, comets, and asteroids. Finally, some of this gas comes from the infall of particles of gas and dust from the surrounding interstellar space. In this chapter, we will explore the boundaries, composition, sources, and dynamics of the particles filling the interplanetary medium.

The Interplanetary Medium — Inner Boundary

The inner boundary of the interplanetary medium (IPM) is derived from specific models of gases in the outer atmosphere of the Sun called the corona. Because the Sun is an extremely dynamic object in space, the inner boundary of the Sun fluctuates with the modes of solar activities. In a simplistic argument, the boundary between the corona and the IPM can be defined as that point where the solar corona becomes less dense than other constituents of the IPM. This definition becomes too limited, though, when we realize that the interactions of solar plasmas are also governed by local magnetic fields, and hence trapped solar plasma can extend into the IPM significantly beyond the boundaries of the solar corona.

A possible alternative boundary point between the corona and the IPM is the point where the subsonic dynamics of the plasma of the corona transit

into the supersonic flow, known as the solar wind. This boundary is best understood by examining the hydrostatic balances of gases that comprise the solar corona. Parker showed in 1959 that the gas comprising the corona must expand due to pressure balances. Because the Sun is in pressure equilibrium, the outward thermal and magnetic pressures balance the gravitational attraction of the mass of the Sun. The solar wind derives from those particles that escape this boundary. The static equilibrium of the solar atmosphere is determined by the balance of gas pressure and solar gravity. Beyond a certain distance from the Sun, gas pressure exceeds gravity, and supersonic outflow ensues — the solar wind. In general, models show that in steady-state conditions, the exterior boundaries of the corona occur at around 1.01 to 10 solar radii depending on the values of the parameters used in solving the equations. See also **Solar Wind**; **Sun (The)**.

The Interplanetary Medium — Outer Boundary: the Heliosphere

The heliosphere is defined as the region that extends from the exterior boundaries of the Sun to the outermost reaches of the influence of the Sun. The heliosphere is a magnetic bubble formed by the effects of the Sun's magnetic fields as it interacts with local interstellar winds. As the solar wind flows outward, it interacts with the flow of local interstellar wind and with infalling neutral particles and dust grains. Because the solar wind is a supersonic flow, the transition from the heliosphere into the local interstellar medium, it is believed, occurs as a shock called the termination shock. Farther out from the termination shock is the heliopause boundary layer. The termination shock is the backup of the pressure wave that develops from the heliopause boundary itself. It occurs due to the initial "collision" of the plasma that composes the interstellar wind with the magnetic forces due to the Sun. The location of the termination shock and the heliopause varies significantly based on the activity of the Sun. During the solar maximum, the solar wind is weaker so that the external pressure on the heliopause forces the heliosphere to shrink. The most recent, high, solar activity levels give a potential opportunity for the far-flung Voyager I and Voyager II spacecraft to encounter the termination shock not just once but several times. (As of January 1, 2002, the Voyager I spacecraft was approximately 85 AU from the Sun, and the Voyager II spacecraft was approximately 67 AU from the Sun.) The shrinkage and expansion of the heliosphere occur much faster than the outward motion of the spacecraft. If the termination shock is currently shrinking past one or both of the spacecraft, as it expands, years later it will again pass across the spacecraft to be encountered again. There is the possibility of many such encounters that will provide the opportunity to understand the nature of the termination shock and the heliopause in great detail. See also **Heliopause**; and **Voyager Missions to Jupiter and Saturn**.

The Interplanetary Medium — Solar Inputs

The primary source of particles in the IPM is from the Sun in the form of the solar wind, coronal mass ejections (CMEs), and solar flares. The general composition of the plasma injected into the IPM is constrained by the composition of the corona. The primary constituents of the solar wind are approximately 95% protons, 4% alpha particles, and 1% minor ions including multiple ionization states of C, O, Si, and Fe. The solar wind also contains electrons in number approximately equal to the ions, and hence the solar wind is considered an electrically neutral plasma. The solar wind contains approximately 1–10 particles per cubic centimeter. The solar wind is a fast stream of particles that leaves the corona at approximately 400 km/s (248.5 mi/s) in the ecliptic plane. The velocity of this stream varies significantly and ranges from 300–1,000 km/s (186–621 mi/s). At high heliolatitudes (above 45°) during the solar minimum, the solar wind leaves the corona at approximately 800 km/s (497 mi/s), again with a very large range of velocities.

Other solar sources of particles include disturbances in the form of coronal mass ejections (CMEs) and solar flares. CMEs are large-scale bubbles of plasma and embedded magnetic fields that are released from the surface of the Sun. CMEs take hours to develop and are released abruptly. The release of a CME occurs across a large portion of the solar surface and can even affect the entire solar disk. On the other hand, solar flares are smaller scale explosions from the surface of the Sun that take just minutes to form. Solar flares tend to be localized to the area surrounding sunspots. Each of these phenomena transports large amounts of solar plasma into the IPM. The plasma released in a CME or solar flare is more energetic than the steady plasma flow of the solar wind. One last particle type emitted by the Sun is the solar energetic particle (SEP). SEPs are very high-energy ions and electrons that are accelerated by processes within the solar corona, including explosive solar flares. SEP energies are typically between 10 and 100 MeV but can exceed 1 GeV. SEPs provide to solar physicists the opportunity to investigate the composition of the Sun and to understand the accelerative processes that energize the particles. See also **Ulysses Mission**.

The Interplanetary Medium — Planetary "Pickup Ions"

Another source of plasma injected into the IPM is due to the interaction of the interplanetary magnetic field with the magnetospheres of magnetized planets. The best examples come from what are called "upstream" particle bursts from the Jovian magnetosphere. Upstream events are characterized by enhancement of ions and electrons, compared to the solar wind. Voyagers I and II, Ulysses, and most recently the Cassini spacecraft have detected short-term enhancements in the overall density of particles as they approached and receded from the Jovian magnetosphere. In general, upstream events seem to occur when the local magnetic field of the IPM is pointed directly toward the planetary magnetosphere. The overall composition of ions during these events is best interpreted as planetary, not solar. The onset of these events is occasionally characterized by the reception of the faster, more energetic ions arriving at the spacecraft before the slower ions. In general, though, the detection of particles during an event occurs at the same time, indicating that the spacecraft is passing through a "flux tube" of magnetically contained plasma that connects directly to the planetary magnetosphere. It is assumed that the magnetic fields associated with these flux tubes have become directly connected to the magnetic fields of the IPM and hence allow for transport of particles away from the planetary magnetosphere. See also **Cassini Mission to Saturn**.

The Interplanetary Medium — Interplanetary Magnetic Field

The other major component of the IPM is the interplanetary magnetic field (IMF) that fills the entire region of the heliosphere. The IMF is mostly due to the transport of solar magnetic structures into the IPM. As the outward moving plasma of the solar wind goes from subsonic in the corona to supersonic just outside the corona, the magnetic field becomes locked within the solar wind plasma. The solar wind then carries the coronal magnetic fields into the IPM. Parker first developed a model of the way the coronal magnetic field is carried into the IPM. The overall model takes into account the shape of the magnetic field lines as the solar wind velocity and the rotation of the Sun determine them. If one assumes a purely radial solar wind and then considers this solar wind in the reference frame of the rotating Sun, then the solar wind has the following components:

$$U_r = u_{sw}; \; U_\varphi = -\Omega_\odot r \sin\theta; \; U_\theta = 0 \tag{1}$$

where u_{sw} is the radial solar wind speed, $\Omega_\odot$ is the solar angular velocity, r is the distance from the sun, and θ is the solar latitude. This solution assumes that the solar rotational rate on the surface of the Sun is constant instead of the observed latitudinal differential rotational rate. The solar magnetic field lines then follow velocity streamlines. The path is defined by the following equation:

$$\frac{1}{r}\frac{dr}{d\varphi} = \frac{U_r}{U\varphi} = \frac{-u_{sw}}{\Omega_\odot r \sin\theta} \tag{2}$$

As discussed in the first section, as the solar wind escapes from the solar surface, the solar wind speed becomes a constant at a critical distance from the solar surface. Using this fact, we can integrate equation 4 and derive the following solution:

$$\Delta R = r - R_\odot = \frac{-u_{sw}}{\Omega_\odot \sin\theta}(\varphi - \varphi_\odot) \tag{3}$$

where $\phi_\odot$ is the initial longitude where the solar wind developed on the surface of the Sun. This equation describes an Archimedean spiral where the magnetic field follows a spiraling path as it moves away from the Sun. For low latitudes (slow moving solar wind), the value of ΔR is approximately 6 AU — 1 AU farther than the orbit of Jupiter. For higher

latitudes (fast moving solar wind), the value of ΔR is approximately 12–2 AU farther than the orbit of Saturn.

The Interplanetary Medium — Corotating Interaction Regions and Shocks

Based on the combined understanding of the latitudinal solar wind dependency coupled with the observed latitudinal differential rotational rate of the surface of the Sun, the overall structure of the IMF becomes complex. In specific regions, on the boundaries between fast moving solar wind and slow moving solar wind, interplanetary spacecraft have observed magnetically complex structures known as corotating interaction regions (CIRs). CIRs most generally occur when the fast moving solar wind overtakes the slower moving solar wind. CIRs are bounded by two shocks at the edges, called forward and reverse shocks. These shocks are characterized by changes in the magnetic field intensity, the particle density, and the overall magnetic pressure. The shocks associated with CIRs provide an explanation for energetic streams of ions propagating from interactive regions.

The effects are best understood by studying how shocks accelerate particles. Magnetized shocks in general are quite efficient in accelerating charged particles and producing an overall increase in the number of energetic ions and electrons in the local plasma environment. Acceleration of charged particles can occur through one of two processes: shock drift acceleration and diffusive shock acceleration. Shock drift acceleration occurs when a charged particle encounters a magnetized structure so that particle acceleration occurs due to induced electric fields parallel to the surface of the shock. The general understanding of shock drift acceleration comes from examining the motion of a charged particle as it encounters a shock. Because the magnetic field strength is larger inside the shock than outside, the gyroradius of a charged particle that impacts a shock decreases. In general, the motion of a particle includes a component of velocity parallel to the shock boundary. In the reference frame of the shock, the particle experiences an electric field parallel to the shock boundary that can either accelerate or decelerate the particle. Acceleration occurs while the particle is outside the shock, and deceleration occurs while the particle is within the shock. Because the gyroradius of the particle is larger outside the shock than inside the shock, the particle spends more time accelerating than decelerating. The factor that determines whether the particle is transmitted through the shock or reflected is based on conservation of the first adiabatic invariant. The condition for reflection is best seen from the following equation that describes the magnetic moment in terms of the pitch angle:

$$\frac{p_1^2 \sin^2 \alpha_1}{B_1} = \frac{p_2^2 \sin^2 \alpha_2}{B_2} \tag{4}$$

where subscript 1 refers to the conditions outside the shock, subscript 2 refers to the conditions inside the shock, p is the momentum of the particle, and p is the pitch angle of the particle. Under the condition

$$\sin^2 \alpha_1 > \frac{B_1}{B_2} \tag{5}$$

conservation of the first adiabatic invariant requires that $\sin \alpha_2 > 1$, which is not possible. This condition indicates that a particle is reflected from the shock boundary. If the condition is not true, then, the particle is transmitted through the shock. In either case, though, the motion of the particle includes acceleration while the particle is within the shock. In general, shock drift acceleration energizes a particle by a factor no larger than 10. However, repeated encounters with a particular shock and/or encounters with multiple shocks can allow the energy of a particle to increase dramatically. Another mitigating factor is the relationship between the direction of the shock normal and the direction of the magnetic field. If these are parallel, then shock drift acceleration is not effective in accelerating particles, but if these are perpendicular, shock drift acceleration is very effective in increasing a particle's energy.

Diffusive shock acceleration occurs when a particle encounters a shock that is approaching. By examining the motion of the particle in the rest frame of the shock, it is seen that the energy of the particle is increased. Denote the velocity of the particle before and after the collision with the shock as v_1 and v_2, respectively. The velocity of the particle in the rest frame of the shock is indicated by primes:

$$v'_1 = v_1 - u_{shock}, -v'_2 = v_2 - u_{shock} \tag{6}$$

In the rest frame, the velocity of the particle is simply reflected:

$$v'_1 = -v_1 \tag{7}$$

The change in energy is then given by combining Eqs. 6 and 7:

$$E_2 - E_1 = \tfrac{1}{2}m(v_2^2 - v_1^2) = 2m(u_{shock}^2 - v_1 u_{shock}) \tag{8}$$

Depending on the value of $v_1 u_{shock}$ the particle either gains or loses energy. If $v_1 u_{shock} < 0$, then the particle gains energy; if $v_1 u_{shock} > u_{shock}^2$, then the particle loses energy.

When a particle encounters the shock multiple times (due to reflection from other magnetic anomalies outside of the shock), then, the particle can experience repeated acceleration. If a distribution of particles encounters a shock, individual collisions with the shock will be stochastic so that some of the particles are accelerated and some are decelerated. This process has a tendency to spread the particle velocity distribution function to include: more slower and faster moving particles.

The Interplanetary Medium — Interstellar Sources of Particles: Pickup Ions

One final source of particles that comprise the constituents of the IPM is the infall of neutral and charged particles from the local interstellar medium. Because the heliopause is a shock boundary, it is difficult for charged particles to penetrate into the IPM. The main component of interstellar particles within the IPM are neutral atoms. Because these particles are neutral, magnetic fields cannot deflect the motion of these particles. As these particles fall farther into the IPM toward the Sun, solar radiation ionizes them. As the particles become ionized, they become bound to the IPM's magnetic field lines, and their overall motion becomes trapped within the outflowing solar wind. Blum and Fahr originally proposed the concept of these particles, but they were not discovered until the Ulysses spacecraft entered into the quiet regions of the high latitude solar wind. The Solar Wind Ion Composition Spectrometer (SWICS) on Ulysses is designed to determine uniquely the elemental and ionic-charge composition, and the temperatures and mean speeds of all major solar-wind ions, from H through Fe, at solar wind speeds ranging from 175 km/s (109 mi/s) (protons) to 1,280 km/s (795 mi/s) (Fe8+). The instrument, which covers an energy per charge range from 0.16 to 59.6 keV/e in ~13 min, combines an electrostatic analyzer with post-acceleration, followed by a time-of-flight and energy measurement. The measurements made by SWICS will have an impact on many areas of solar and heliospheric physics, in particular providing essential and unique information on: (i) conditions and processes in the region of the corona where the solar wind is accelerated; (ii) the location of the source regions of the solar wind in the corona; (iii) coronal heating processes; (iv) the extent and causes of variations in the composition of the solar atmosphere; (v) plasma processes in the solar wind; (vi) the acceleration of energetic particles in the solar wind; (vii) the thermalization and acceleration of interstellar ions in the solar wind, and their composition; and (viii) the composition, charge states and behavior of the plasma in various regions of the Jovian magnetosphere.

Conclusion

Existing NASA satellite programs continue to provide a significant amount of data regarding the nature of the constituents of the interplanetary medium and transport processes. Proposed programs such as the Interstellar Probe trajectory (see **Voyager Missions to Jupiter and Saturn**) represent significant opportunities to clarify further the details of the overall structure of the heliosphere and provide *in situ* measurements of the plasma constituents within the interplanetary medium. To continue developing our understanding of the interplanetary medium and the overall heliosphere, we must maintain a continued presence in space.

Additional Reading

Alurkar, S.K.: *Solar and Interplanetary Disturbances*, World Scientific Publishing Company, Inc., Riveredge, NJ, 1997.

Blum, P.W., and H.J. Fahr: "Interaction Between Interstellar Hydrogen and the Solar Wind," *Astron. Astrophys.*, **4**, 280 (1970).

Chian, A.C.L., I.H. Cairns, and S.B. Gabriel: *Advances in Space Environment Research*, Vol. 1, Kluwer Academic Publishers, Norwell, MA, 2003.

Cravens, T.E.: *Physics of Solar System Plasmas*, Cambridge University Press, New York, NY, 1997.

Gloeckler, G., J. Geiss, E.C. Roelof, L.A. Fisk, F.M. Ipavich, K.W. Ogilvie, L.J. Lanzerotti, R. vonStieger, and B. Wilken: "Acceleration of Interstellar Pickup Ions in the Disturbed Solar Wind Observed on Ulysses," *J. Geophys. Res.*, **99**, 17,637 (1994).

Gombosi, T.I., J.T. Houghton, and A.J. Dressler: *Physics of the Space Environment (Cambridge Atmospheric and Space Science Series)*, Cambridge University Press, New York, NY, 2004.

Krimigis, S.M.: "Observations of Energetic Ions and Electrons at Interplanetary Shocks and Upstream of Planetary Bow Shocks by the Voyager Spacecraft," in K. Szego, ed, *Proc. Int. Symp. Collisionless Shocks*, Hungary, 1987, p. 3.

McComas, D.J., S.J. Bame, B.L. Barraclough, W.C. Feldman, H.O. Funsten, J.T. Gosling, P. Riley, R. Skoug, A. Balogh, R. Forsyth, B.E. Goldstein, and M. Neugebauer: "Ulysses' Return to the Slow Solar Wind," *Geophys. Res. Lett.*, **25** (1), 1 (1998).

Parker, E.N.: "Extension of the Solar Corona into Interplanetary Space," *J. Geophys. Res.*, **64**: 1675 (1959); Bame, S.J., B.E. Goldstein, J.T. Gosling, J.W. Harvey, D.J. McComas, M. Neugebauer, and J.L. Phillips: "Ulysses Observations of a Recurrent High-speed Stream and the Heliomagnetic Streamer Belt," *Geophys. Res. Lett.*, **20**, 2323 (1993).

Sanderson, T.R., R.G. Marsden, K.-P. Wenzel, A. Balogh, R.J. Forsyth, and B.E. Goldstein: "Ulysses High-latitude Observations of Ions Accelerated by Co-rotating Interaction Regions," *Geophys. Res. Lett.*, **21**, 1113 (1994).

Velli, M., R. Bruno, and F. Malara: *Solar Wind Ten*, Springer-Verlag New York, LLC, New York, NY, 2003.

Web References

Interplanetary Medium: http://search.jpl.nasa.gov:8080/cgi-bin/htsearch?words=Interplanetary+Medium&search.x=4&search.y=2

JERRY W. MANWEILER, LLC, Lawrence, Kansas

INTERPOLATION. A process by which an appropriate value is placed between tabulated values of a function. Linear interpolation is based on a principle of proportional parts. If (x_1, y_1) and (x_2, y_2) are neighboring entries in a numerical table, a value of the dependent variable y for a value of the argument x between x_1 and x_2 is given by $y = y_1 + (y_2 - y_1)(x - x_1)/(x_2 - x_1)$. The procedure assumes that y varies linearly with x over the interval considered. It is commonly used for logarithms, trigonometric functions, etc., where the values of the argument are closely spaced. Such tables often contain columns of proportional parts, which facilitate interpolation.

For more accurate work, the relation between the two variables is usually approximated by a polynomial and finite differences are used with formulas of Newton, Lagrange, Bessel, Stirling, etc.

Given $y = f(x)$, in tabulated numerical form, inverse interpolation in a process for finding x at a value of y, intermediate between two tabulated values. Possible procedures include: Lagrange's formula; successive approximations; reversion of series applied to other interpolation formulas.

Extrapolation means estimation of a value outside of the range of tabulated values. This, in general, is a risky procedure since one usually does not know how the function behaves beyond the calculated range. Because of these uncertainties, a graphical method is probably as satisfactory as any other.

INTERPRETER (Computer System). (1) An executive routine which translates a stored program expressed in some pseudocode into machine code and performs the indicated operations, by means of subroutines, as they are translated. Interpreters are used widely for translating some high-level languages, such as BASIC and APL. An interpreter is essentially a closed subroutine that operates successively on an indefinitely long sequence of program parameters, the pseudoinstructions, and operands.

(2) In punched card operations, a device that prints on the card the characters corresponding to hole patterns punched in the cards. See also **Program (Computer)**.

INTERQUARTILE RANGE. MIS The interquartile range is defined as $Q_3 - Q_1$ where Q_3 and Q_1 are the third and first quartiles in a distribution. It is sometimes used as a measure of dispersion.

INTERROGATION. MIS Transmission of a radio signal or combination of signals intended to trigger a transponder or group of transponders.

INTERROGATION RECORDING AND LOCATION SYSTEM (IRLS). See **Nimbus Satellite Program**.

INTERRUPT (Computer System). A signal which causes the central processing unit (CPU) to change state as a result of a specified condition. An interrupt represents a temporary suspension of normal program execution. An interrupt arises from an external condition, an input or output device, or by the program currently being processed in the CPU. Upon recognition of an interrupt, the current program is suspended and replaced by another program. Upon completion of the new program, the control of the CPU is returned to the interrupted program at the exact point where discontinuance occurred. As there is more than one possible interrupt condition in most systems, a priority may be established to determine the sequence for servicing programs. The priority may be established by hardware logic or programming. Where the priority is established by programming, the interrupt condition causes the system to transfer control to an interrupt-service subroutine. This routine determines the specific cause of the interrupt. Based upon the assigned priority of the interrupt condition, the routine schedules the execution of the interrupt-service routine in the correct sequence. In the case of a hardware priority-interrupt structure, the hardware is designed to prevent any interrupt from being recognized should it be of lower priority than the current program.

See also **Program (Computer)**.

INTERSTELLAR BOUNDARY EXPLORER (IBEX) MISSION. See **Space Science Missions: Sun**.

INTERSTELLAR DUST. See **Interplanetary Medium**; and **Meteorology**.

INTERSTELLAR REDDENING. Interstellar space is permeated with gas and dust, and thus the light passing through it is subject to scattering. Since the Rayleigh coefficient for scattering σ_a is proportional to λ^{-4}, we see from

$$I = I_0 e^{-\sigma a x}$$

(where I is the intensity of the scattered light, and I_0 is the intensity of the incident beam), that more light is scattered for small λ than for large λ. Thus, in effect, light is reddened in passing through interstellar space.

INTERSTICE. A small space within a phase or, more commonly, between particles.

INTERSTITIAL. (1) Descriptive of a nonstoichiometric compound of a metal and a nometal whose structure conforms to a simple chemical formula, but exists over a limited range of chemical composition. Interstitial compounds are represented by borides, nitrides, and carbides of the transition metals. (2) Descriptive of an atom of an impurity that causes a defect or dislocation in a crystalline lattice, e.g., an atom of carbon or nitrogen in an iron crystal, or of arsenic in a semiconductor. (3) In a biological sense, the term describes cells located between or within layers of tissue.

INTERTRIGO. See **Dermatitis and Dermatosis**.

INTERTROPICAL CONVERGENCE ZONE. See **Meteorology**.

INTERTROPICAL FRONT. See **Meteorology**.

INTERVALOMETER. Any device that may be set so as to accomplish automatically a series of like actions, such as the taking of photographs, or the closure of electrical circuits, at constant predetermined intervals.

INTESTINAL NEMATODES. These are by far the most common parasites of the human intestine, with probably about half of the world's population being infected by one or the other of the group of roundworms, hookworms, whipworms, and pinworms.

Roundworms. *Ascaris lumbricoides* is the largest nematode parasitizing the human intestine and is the most common human helminthic infection, with worldwide distribution affecting about 25% of the human population. The annual global mortality due to this infection is about 20,000 and morbidity about one million. The highest prevalence can be found in tropical and subtropical regions, and areas with inadequate sanitation. Roundworm also occurs in rural areas of the southeastern United States.

The worm is a geohelminth living, except in humans, in moist, warm soil. The life cycle is simple and characteristic. Adult worms live in the lumen of the small intestine. A female may produce up to 240,000 eggs per day, which are passed with the feces. Fertile eggs embryonate and become infective after 18 days to several weeks, depending on the environmental conditions (optimum: moist, warm, shaded soil). After infective eggs are swallowed, the larvae hatch, invade the intestinal mucosa, and are carried via the portal, then systemic circulation to the lungs. The larvae mature further in the lungs (10 to 14 days), penetrate the alveolar walls, ascend the bronchial tree to the throat, and are swallowed. Upon reaching the small intestine, they develop into adult worms. See Fig. 1. Between 2 and 3 months are required from ingestion of the infective eggs to oviposition by the adult female. Adult worms can live 1 to 2 years. The male worm is about 15–20 cm long by 2.5 mm in diameter. The female is much larger (20–49 cm long by 3–6 mm in diameter) and lays about a quarter of a million eggs per day.

Both sexes retain their position in the intestine by bracing against the walls, feeding with limited selectivity on the intestinal contents. In general, *Ascariasis* is asymptomatic, causing a slight cough as the larvae pass from the lungs to the pharynx. However, there is considerable evidence that the presence of *Ascarii* causes nutritional problems and hinders the development of children. Occasionally patients may develop fever, malaise, urticaria, colic, and diarrhea. The infection may manifest as severe life-threatening disease when a number of worms get entangled to form a bolus and block the intestinal lumen or when ectopic migration occurs resulting in entry of the worm into appendix, bile duct, and pancreatic duct.

Diagnosis is usually made by detecting *Ascaris* eggs in feces. Treatment is effective with levamisole, pyrantel pamoate, or mebendazole.

Hookworm. An intestinal parasite of humans, that usually causes mild diarrhea or cramps. Heavy infection with hookworm can create serious health problems for newborns, children, pregnant women, and persons who are malnourished. Hookworm infections occur mostly in tropical and subtropical climates and are estimated to infect about 1 billion people — about one-fifth of the world's population. Hookworm is the second most common human helminthic infection (after ascariasis). Both *N. americanus* and *A. duodenale* are found in Africa, Asia and the Americas. *Necator americanus* predominates in the Americas and Australia, while only *A. duodenale* is found in the Middle East, North Africa and southern Europe. The human hookworms include two nematode (roundworm) species, *Ancylostoma duodenale* and *Necator americanus*. (Adult females: 10 to 13 mm (*A. duodenale*), 9 to 11 mm (*N. americanus*); adult males: 8 to 11 mm (*A. duodenale*), 7 to 9 mm (*N. americanus*). A smaller group of hookworms infecting animals can invade and parasitize humans (*A. ceylanicum*) or can penetrate the human skin (causing cutaneous larva migrans), but do not develop any further (*A. braziliense, Uncinaria stenocephala*).

One can become infected by direct contact with soil that contains human feces in areas where hookworm is common. Children are the most likely to become infected, because they play in dirt and often go barefoot, or accidentally swallow contaminated soil. Since transmission of hookworm infection requires development of the larvae in soil, hookworm cannot be spread person to person. Contact among children in institutional or child care settings should not increase the risk of infection.

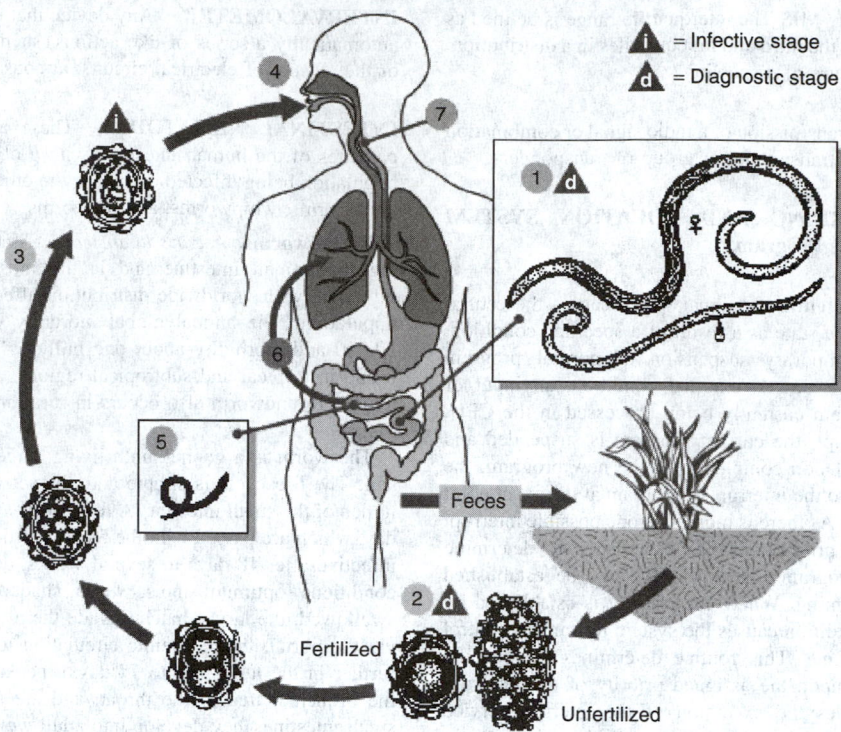

i = Infective stage

d = Diagnostic stage

Feces

Fertilized

Unfertilized

Fig. 1.   Life cycle of the *Ascaris lumbricoides*. (CDC.)

Itching and a rash at the site of where skin has touched soil or sand are usually the first sign of infection. These symptoms occur when the larvae penetrate the skin. While a light infection may cause no symptoms, heavy infection can cause anemia, abdominal pain, diarrhea, loss of appetite, and weight loss. Heavy, chronic infections can cause stunted growth and mental development.

Adult worms live in the lumen of the small intestine, where they attach to the intestinal wall with resultant host blood loss. Eggs are passed in the stool, and under favorable conditions (moisture, warmth, shade), hatch in 1 to 2 days. Larvae are released, grow in the feces and/or the soil, and after 5 to 10 days (and two molts) have become filariform (L-3) larvae that are infective. These infective larvae can survive 3 to 4 weeks in favorable environments. On contact with the human host, the larvae penetrate the skin and are carried through the veins and the heart to the lungs. They penetrate into the pulmonary alveolae, ascend the bronchial tree to the pharynx, and are swallowed. See Fig. 2. Upon reaching the small intestine, they undergo two more molts yielding fourth stage larvae (L4) and then adult worms. Five weeks or more are required from invasion by the L3 to oviposition by the adult female. Most adult worms are eliminated in 1 to 2 years, but longevity records can reach several years. Some *A. duodenale* larvae, following penetration of the host skin, can become dormant (in the intestine or muscle!). In addition, infection by *A. duodenale* may probably also occur by the oral and transmammary route. (*N. americanus*, however, requires a transpulmonary migration phase.)

Accurate diagnosis depends upon identification of hookworm eggs in the feces. In developing countries, tetrachlorethylene is the most widely used therapy, but it requires prior treatment with piperazine, because adult worms may be caused to migrate ectopically. In more advanced countries, thiabendazole and mebendazole are the drugs of choice.

Strongyloides. Worms of this genus *Strongyloides stercoralis* are one of the world's major human intestinal nematode infections, affecting an estimated one hundred million people. Other *Strongyloides* include *S. fülleborni*, which infects chimpanzees and baboons and may produce limited infections in humans. These worms differ from other worms in one important aspect; they undergo a succession of generations within the host so that heavy infection may guild up when conditions are favorable for them without the need for repeated transmission. As a result, the patients may die of fulminating disease many years after leaving an

endemic area, having had few symptoms in the interval. *Strongyloides stercoralis* is widespread in the tropics and subtropics, but cases also occur in temperate areas (including the South of the United States). They are more frequently found in rural areas, institutional settings, and lower socio-economic groups. It has been estimated that 15% of former prisoners of war in the Far East were still infected thirty years after their return home.

The adult female worm is about 3 mm long by 650 micrometers in diameter; the male is 750×45 micrometers. Complex, and unique among helminths in its potential for autoinfection and multiplication within the host. Two types of cycles exist:

Parasitic cycle. Filariform larvae in contaminated soil penetrate the human skin, and are transported to the lungs where they penetrate the alveolar spaces; they are carried through the bronchial tree to the pharynx, are swallowed and then reach the small intestine. In the small intestine they molt twice and become adult female worms. The females live in the lumen of the small intestine and by parthenogenesis produce eggs, which yield rhabditiform larvae. The rhabditiform larvae can either be passed in the stool (see "Free-living cycle", below), or can cause autoinfection. In autoinfection, the rhabditiform larvae become infective filariform larvae, which can penetrate either the intestinal mucosa (internal autoinfection) or the skin of the perianal area (external autoinfection); in both cases the filariform larvae follow the previously described route, being carried successively to the lungs, the bronchial tree, the pharynx, and the small intestine where they mature into adults. (Autoinfection is a feature unique to *Strongyloides* among helminthic infections in humans, and explains the possibility of persistent infections and of hyperinfections in immunodepressed individuals.)

Free-living cycle. The rhabditiform larvae passed in the stool (see "Parasitic cycle" above) can either molt twice and become infective (filariform) larvae (direct development) or molt four times and become free living adult males and females which mate and produce rhabditiform larvae. The latter in turn can either develop into new free-living adults, or into infective (filariform) larvae. The filariform larvae penetrate the human host skin to initiate the parasitic cycle. See Fig. 3.

Clinical Features are typically asymptomatic. Gastrointestinal symptoms include abdominal pain and diarrhea. Pulmonary symptoms (including Loeffler's syndrome) can occur during pulmonary migration of the filariform larvae. Dermatologic manifestations include urticarial rashes in the buttocks and waist areas. Disseminated strongyloidiasis occurs in

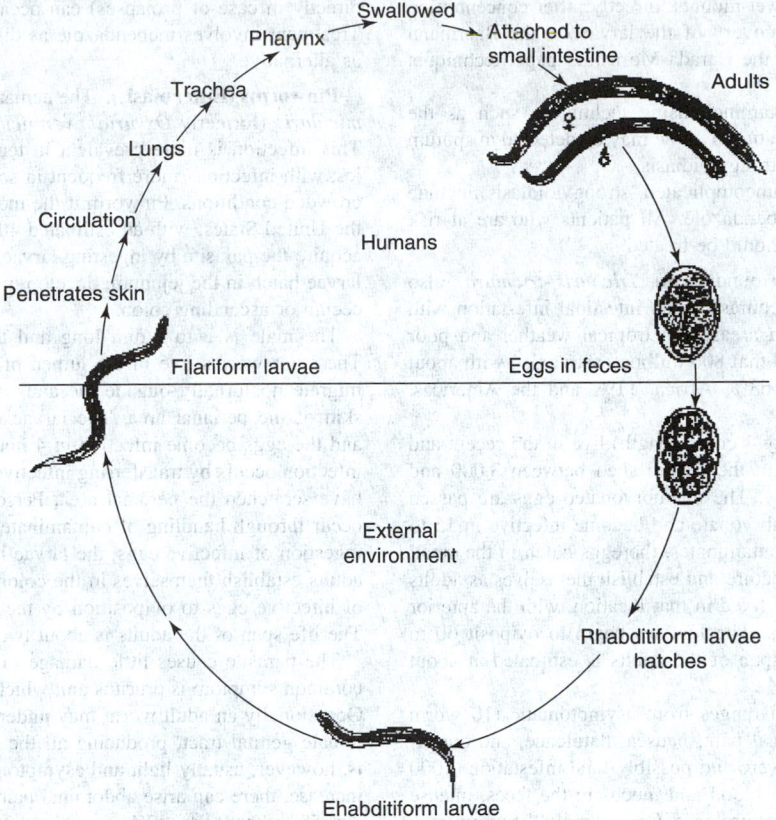

Fig. 2.   Life cycle of the *Ancylostoma duodenale* and *Necator americanus*. (*CDC.*)

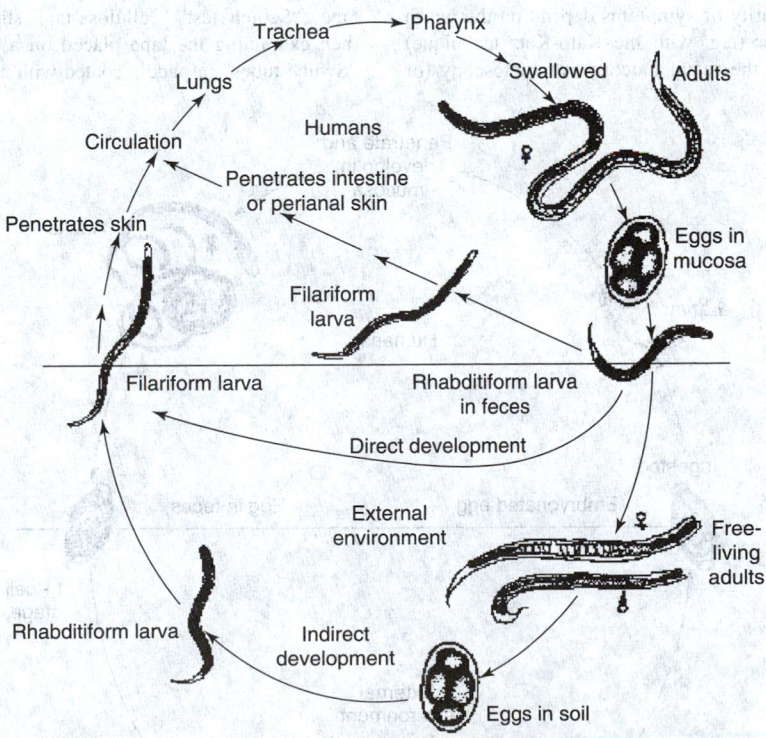

Fig. 3.   Life cycle of the *Strongyloides stercoralis*. (*CDC.*)

immunosuppressed patients, can present with abdominal pain, distension, shock, pulmonary and neurologic complications and septicemia, and is potentially fatal. Blood eosinophilia is generally present during the acute and chronic stages, but may be absent with dissemination.

Diagnosis is made through the microscopic identification of larvae (rhabditiform and occasionally filariform) in the stool or duodenal fluid. Examination of serial samples may be necessary, and not always sufficient, because stool examination is relatively insensitive.

The stool can be examined in wet mounts: directly; after concentration (formalin–ethyl acetate); after recovery of the larvae by the Baermann funnel technique; after culture by the Harada-Mori filter paper technique; or after culture in agar plates.

The duodenal fluid can be examined using techniques such as the Enterotest string or duodenal aspiration. Larvae may be detected in sputum from patients with disseminated strongyloidiasis.

Drugs for the treatment of uncomplicated strongyloidiasis include ivermectin, thiabendazole, and albendazole. All patients who are at risk of disseminated strongyloidiasis should be treated.

Whipworms. The nematode (roundworm) *Trichuris trichiura*, also called the human whipworm. Trichurasis is an intestinal infestation with worldwide distribution, usually in areas with tropical weather and poor sanitation practices. It is estimated that 800 million cases exist, with about the following distribution: Asia, 63%, Africa, 11%, and the Americas, 14%.

The adult worms (approximately 4 cm in length) live in the cecem and ascending colon. Female worms in the cecum shed between 3,000 and 20,000 barrel-shaped eggs per day. The unembryonated eggs are passed with the stool. In the soil they embryonate and become infective in 15 to 30 days. After ingestion of soil contaminates, the eggs hatch in the small intestine, and release larvae that mature and establish themselves as adults in the colon. The adult worms are fixed in that location, with the anterior portions threaded into the mucosa. The females begin to oviposit 60 to 70 days after infection. The life span of the adults is estimated at about 1 year. See Fig. 4.

Clinical expression of infection ranges from asymptomatic (10 worm infection), through lower abdominal pain, nausea, flatulence, and constipation (100 worm infection) to severe and possibly fatal infestation (1000 worms), giving rise to colitis with blood and mucus in the feces, intense abdominal pain, tenesmus and rectal prolapse. Occasionally the worm may lodge in the lumen of the appendix and induce appendicitis.

Diagnosis is based upon microscopic examination of feces. Because eggs may be difficult to find in light infections, a concentration procedure is recommended. Because the severity of symptoms depend on the worm burden, quantification of the latter (e.g. with the Kato-Katz technique) can prove useful. Examination of the rectal mucosa by proctoscopy (or directly in case of prolapses) can occasionally demonstrate adult worms. Treatment involves mebendazole as the drug of choice, with albendazole as alternative.

Pinworms (Enterbiasis). The nematode (roundworm) *Enterobius vermicularis* (formerly *Oxyuria vermicularis*) also called human pinworm. This infection is more prevalent in temperate countries than in the tropics, with infections more frequent in school, or preschool, children and in crowded conditions. Pinworm it the most common helminthic infection in the United States, with an estimated 40 million persons infected. Humans acquire the parasite by ingesting larvae; there is no visceral migration, the larvae hatch in the jejunum, develop in the ileum, and finally locate in the cecum or ascending colon.

The male is 2 to 5 mm long and the females are 8 to 13 mm long. The adult worms live in the lumen of the human colon. Gravid females migrate nocturnally outside the anus and ovipost while crawling on the skin of the perianal area. The larvae contained inside the eggs develop and the eggs become infective in 4 hours under optimal conditions. Self-infection occurs by transferring infective eggs to the mouth with hands that have scratched the perianal area. Person-to-person transmission can also occur through handling of contaminated clothes or bed linens. Following ingestion of infective eggs, the larvae hatch in the small intestine and the adults establish themselves in the colon. The time interval from ingestion of infective eggs to oviposition by the adult females is about one month. The life span of the adults is about two months. See Fig. 5.

The parasite causes little damage to the colonic mucosa and the most common symptom is pruritus ani, which can be very troublesome at night. Occasionally an adult worm may undergo ectopic migration and enter the female genital tract, producing all the symptoms of salpingitis. Infection is, however, usually light and asymptomatic but as the numbers of worms increase, there can arise abdominal pain. The range of symptoms: pruritus ani, 55%; headache, 29%; dysentery, 35%; tenesmus, 11%.

Microscopic identification of eggs collected in the perianal area is the method of choice for diagnosing enterobiasis. This must be done in the morning, before defecation and washing, by pressing transparent adhesive tape ("Scotch test", cellulose-tape slide test) on the perianal skin and then examining the tape placed on a slide. Alternatively, anal swabs or "Swube tubes" (a paddle coated with adhesive material) can also be used.

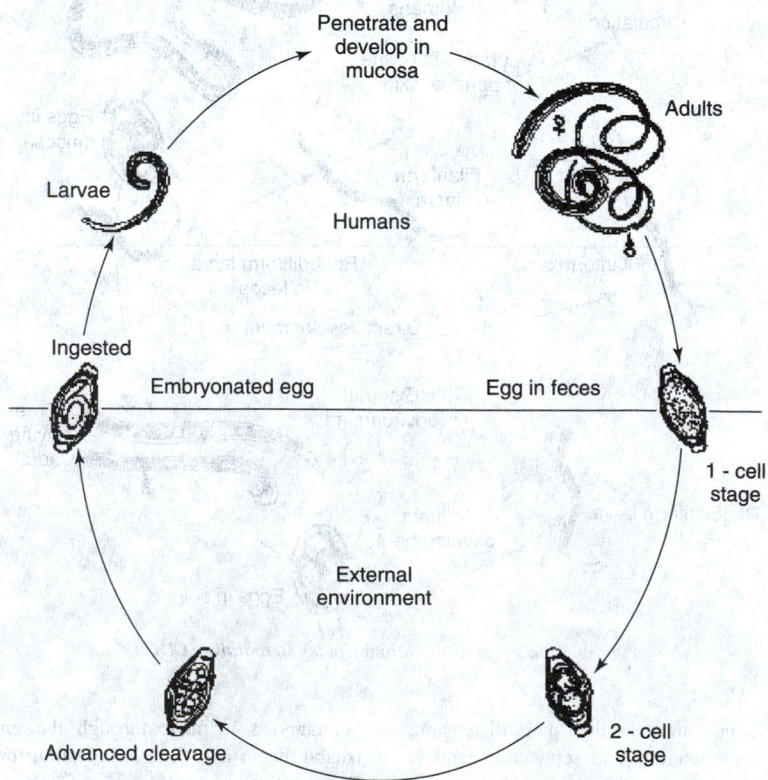

Fig. 4.   Life cycle of the *Trichuris trichiura*. (*CDC.*)

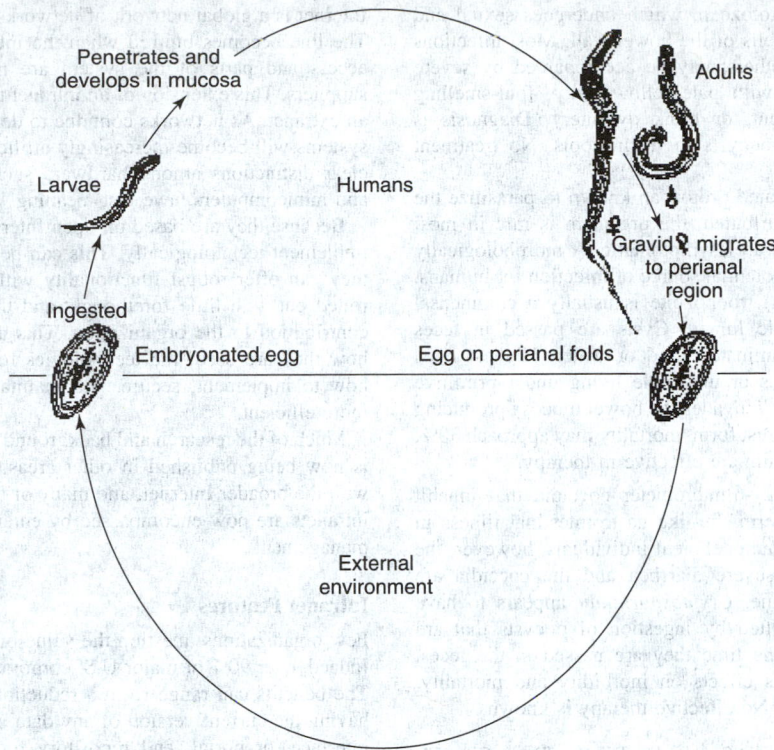

Fig. 5. Life cycle of the *Enterobius vermicularis*. (*CDC.*)

Eggs can also be found, but less frequently, in the stool, and occasionally are encountered in the urine or vaginal smears. Adult worms are also diagnostic, when found in the perianal area, or during ano-rectal or vaginal examinations. The drugs of choice for treatment are pyrantel pamoate, mebendazole, or albendazole.

Additional Reading

Bird, J.F.: *The Structure of Nematodes*, 2nd Edition, Academic Press Inc.,, San Diego, CA, 1997.

Kennedy, M.W.: *Parasitic Nematodes—Antigens, Membranes and Genes*, Taylor & Francis, Inc., Philadelphia, PA, 1991.

Web Reference

Centers for Disease Control and Prevention: http://www.cdc.gov/health/diseases.htm
WHO/OMS: Intestinal nematodes: http://www.who.int/health-topics/intestinal.htm

Ann C. DeBaldo, Ph.D., University of South Florida, Tampa, FL

INTESTINAL PROTOZOA. While most protozoa are free-living, others sometimes invade human tissue, producing disease. See also **Amebiasis (Amebic Dysentery)**; and **Primary Amebic Meningoencephalitis**. A number of flagellates and amoebae may live in the human gastrointestinal tract as commensals, feeding upon bacteria or other materials. These forms are noteworthy because they are widely distributed, must be distinguished from pathogenic forms, and because they are indicators of unsanitary conditions. Typically, the trophozoite is the active feeding stage, which divides by binary fission, giving rise to smaller, resistant cysts that are passed in the feces and infect the new host. Diagnosis is by demonstrating either stage microscopically in appropriately stained fecal smears. Although similar or identical species are common in dogs, cats, and monkeys, most human infections arise from other human cases.

Entameba coli is one of the most common commensal amoebae. Its trophozoite (15–50 micrometers) is distinguished from its pathogenic look-alike, *E. histolytica*, by its sluggish movements, coarse granular cytoplasm, prominent, eccentric nucleolus and coarse chromatin around the margin of the nucleus. The cyst is distinguished by eight nuclei and splinterlike chromatoidal bodies.

Endolimax nana, also a nonpathogen, is a small, sluggish amoeba living in the colon and developing a 4-nucleus cyst. *Indameba buetschlii* is much less common. Its cyst characteristically contains a glycogen vacuole, which is stained brown with iodine. *Dientameba fragilis* is a binucleate amoeba, which may irritate the colon mucosa, producing a mucoid diarrhea with abdominal pain and tenderness. No cyst form is known. *Entameba gingivalis* lives in the odontal tissue crevices and thrives in association with pyorrhea alveolaris. Again, no cyst is known.

The flagellate *Cilomastix mensili* is morphologically distinctive. The trophozoite (6–20 micrometers) is pear-shaped, bears an asymmetrical spiral groove at mid-body, with nucleus and a group of flagellae at the anterior end. Cysts are uninucleate and lemon shaped. No *Trichomonas* species produces a cyst; the trophozoites have four free flagella with a fifth along the outer margin of the undulating membrane. A prominent, rod-like axostyle extends posteriorly beyond the body. Commensal forms include *T. hominis* in the cecum and *T. tenax* in the mouth. *T. vaginalis* is widespread in infection and may cause a sexually transmitted disease with highest prevalence among women between 16 and 35 years. In the vagina there may be inflammation accompanied by a profuse foamy, foul-smelling discharge. Infection of the male sexual organs is usually asymptomatic. Diagnosis is established by examining stained vaginal smears or culture. Metronidazole by mouth is effective for both sexes.

Giardia lamblia is a small intestine flagellate, which usually but not always produces an asymptomatic infestation; it is the most common pathogenic intestinal protozoan seen in the western world. The trophozoite (9–21 micrometers) with its double nuclei remarkably resembles a little face. It is not resistant to an external environment and although found in loose stools does not appear when they are formed. The quadri-nucleate cyst is transmitted by food, by water or by intimate contact, but epidemics are usually waterborne. Asymptomatic cyst passers are not uncommon and are important sources of water contamination. Distribution is worldwide, but most cases are seen in children as well as in institutions and areas of poor sanitation. In symptomatic cases of giardiasis, the incubation period is from 1 to 4 weeks; the walls of the duodenum and jejunum are irritated and inflamed, but not directly invaded. There is a chronic bloodless diarrhea, fat-filled, foul smelling stools (steatorrhea), flatulence, and dull epigastric pain. Finding the parasites in feces or duodenal aspirates does not exclude other causes of duodenitis, such as carcinoma. Metronidazole is the drug of choice, but it is a potential carcinogen and may have a disulfiran-like effect. Quinacrine is also effective. Prevention relates to water supplies, and chlorination may not be effective.

Isospora belli is a coccidian protozoan, which undergoes sexual and asexual multiplication within the cells of the bowel wall. Most infections are probably asymptomatic, but others may be accompanied by severe gastrointestinal distress, diarrhea with pale-yellow, fatty, foul-smelling stools suggesting liver involvement, or frank dysentery. Diagnosis is established by demonstrating the oocysts in fresh stools. No treatment is known.

Balantidium coli is the only ciliated protozoan known to parasitize the human intestinal tract. Widely distributed, the organism is rare in most areas and most cases of infestation are asymptomatic. A morphologically identical form known in hogs is a potential source of infection for humans. The large (up to 200 micrometers) trophozoite is usually a commensal feeding on bacteria in the colonic lumen. Cysts are passed in feces and infect the new host by contaminated food or water. Highest rates of infection are seen in institutions or in people living under primitive conditions. In some instances, *B. coli* invades the bowel mucosa producing ulceration and dysentery. In its severest form, mortality may approach 30%. Oxytetracycline or diiodohydroxyquin are effective in therapy.

Cryptosporidium. protozoa are 2–6 micrometer coccidia that inhabit the microvilli, producing a short-term, flu-like gastrointestinal illness in immunocompetent humans. In immunodeficient individuals, however, the organism produces a prolonged, severe diarrhea and the coccidia are not always confined to the intestine. *Cryptosporidium* appears to have no host specificity and is transmitted by ingestion of oocysts that are fully sporulated and infective at the time they are passed in the feces. *Cryptosporidium* is ominous in its effects on morbidity and mortality, particularly in patients with AIDS. No effective therapy is known.

R.C. VICKERY, M.D., D.Sc., Ph.D., Blanton/Dade City, FL

INTRACAPSULAR CATARACT EXTRACTION (ICCE). A form of cataract surgery developed in the early 1980s, but seldom used today because more advanced techniques are available. In this surgery, the entire natural lens of the eye, including the capsule that holds it in place, is removed. The procedure requires a much larger incision than used in the more recently developed techniques.

To understand how ICCE works, it is important to understand what a cataract is and how it interferes with vision. The eye works like a camera with two lenses. The first lens is the cornea, a clear membrane that covers the front of the eye. The second lens is the eye's natural crystalline lens, which is held in place by a capsule located behind the pupil. The cornea is responsible for about 70% of the eye's focusing power, while the natural lens fine-tunes the image. When the natural lens becomes cloudy, usually because of the aging process, it keeps light rays from passing through or diffuses the light in such a way that vision becomes fuzzy or hazy. This cloudy lens is called a cataract. The object of cataract surgery is to remove this hazy lens and to replace it with a plastic prescription lens that is permanently implanted in the eye.

When performing intracapsular cataract extraction, the surgeon makes a large opening in the eyeball and injects medicine into the eye, causing the zonular fibers that hold the lens in position to dissolve. A special probe is then placed on the lens, and liquid nitrogen is applied to freeze the lens. As the probe is gently withdrawn from the eye, the natural lens is pulled out with it. Once the natural lens is removed, an intraocular lens implant is inserted in front of the iris, the colored part of the eye. (In the newer form of cataract surgery called *extracapsular cataract extractions*, the lens is placed behind the iris.) Several stitches are necessary to close the eye until it heals, which may take as long as six weeks.

In addition to the larger incision and accompanying sutures required with intracapsular cataract extraction, the technique also carries a greater risk for retinal detachment and swelling. It is seldom performed today. See also **Cataract**; **Extracapsular Cataract Extraction (ECCE)**; and **Vision and the Eye**.

Vision Rx, Inc. Elmsford, NY

INTRACLOUD FLASH. See **Clouds and Cloud Formation**.

INTRANETS. An intranet is defined by the use of Internet technologies (HTTP, TCP/IP, FTP, SMTP) within an organization. By contrast, the Internet is a global network of networks connecting myriad organizations. The line becomes blurred when the internal system is opened to remote access and parts of the system are made available to customers and suppliers. This extension of an intranet to selected outsiders is often called an extranet. As networks continue to develop, distinctions among types of systems will become increasingly artificial and contrived, just as the once clear distinctions among hardware such as personal computers, servers, and minicomputers have lost meaning.

Because they are based on open Internet standards, intranets are easy to implement technologically. This can be a blessing and a curse. Although they can offer robust functionality with little investment, they are often rolled out with little forethought and therefore fail to make a significant contribution to the organization. This article outlines what intranets are; how they are used; the technologies for constructing and running them; how to implement, secure, and maintain them; and how to make them more efficient.

Much of the research and background information applicable to intranets is now being published in other areas. The technological issues overlap with the broader Internet, and many of the internal applications offered on intranets are now encompassed by enterprise applications and knowledge management.

Intranet Features

Few organizations question the value of having an intranet in some form. Indeed, over 90% of major U.S. corporations have intranets [Baker, 2000]. The benefits can range from a reduction of paper and headcount, always having the current version of any data or document available, to a central interface, or portal, and repository for all corporate systems and data. Intranets may encompass many different types of content and features. Other benefits include the use of Internet-standard network protocols that facilitate connections to the broader Internet.

Portal. Intranets are often designed to be "portals," serving as the central point of access to all information resources within an organization. The implication of grandiosity is deliberate and helpful, as good design can be a crucial determining factor in the success of an intranet. A portal also needs to be available regularly to a wide community, with access to multiple sources of data, both internal and external, with a useful search mechanism. The term portal often evokes third-party commercial portals such as Excite.com (http://www.excite.com) and vertical portals, also called *vortals*, such as aluminium.com (http://www.aluminium.com). These portals have struggled to establish successful business models, especially given the collapse of Internet advertising rates. Portals inside organizations, in contrast, have thrived; often with the support of the information systems (IS) department and the business users.

One of the greatest frustrations for information technology users inside organizations is the necessity of multiple systems with varied interfaces, including command-line based interfaces. With a portal, users have the ability to go to a single location, use a common, intuitive, and well-established graphical interface, and find their information. Types of the information and services found on intranets are outlined below.

Human Resource Materials. Human resource materials are an obvious and common type of content for intranets. Benefits management is a significant cost for organizations; tax withholding, health care election choices in the United States, and retirement and pension accounting all involve substantial processing costs. These processing costs include printing, distributing, completing, collecting, and entering data from paper forms, providing support to fill out the forms, and managing subsequent changes. With a well-designed intranet, employees can process their own benefits, saving substantial overhead and time. The forms tend to be highly structured, with well-defined fields, and the application can be scaled to all employees, reducing the cost of development per employee. These forms can also change regularly. Changing the online versions is much less expensive, and eliminates the possibility of employees filling out the wrong version of a form.

Human resources are also an excellent choice for intranet applications, particularly early in an intranet rollout, because every employee uses its services. Beginning with a common set of applications with which all or most employees interact may lead users to become familiar with other services located on the intranet as well. Self-service benefits also allow the human resource department to train less computer literate employees on an application that human resources knows intimately and that all employees must use [Meuse, 1999]. This makes it an excellent introduction to the intranet. Unfortunately, most users only need to visit the employee benefits site once or twice a year, so the human resource function cannot be counted on to drive traffic throughout the year. However, newsletters can be published on the intranet, with notifications sent out via e-mail that may bring some traffic regularly.

Purchasing. For many organizations, especially larger ones, purchasing is a major cost that can be difficult to control. Managing the purchasing function and its information requirements is an inconvenience, and much of the cost is incurred in the overhead of processing purchases. The story of Ford's and Mazda's accounts payable departments has been retold many times and is famously cited by Hammer as an example of reengineering. Ford, with 500 people working in accounts payable, decided to compare itself to Mazda as a benchmark for its efficiency. Ford found that Mazda had only five people working in a similar department [Hammer & Champy, 1993]. Although Mazda's efforts predated the Web, information processing was a crucial part of their efforts. This underscores the complexity and overhead often involved in purchasing. Intranets can greatly reduce purchasing overhead by simplifying the process and centralizing the collection of order data. Approval processes can be automated or at least facilitated, reducing paperwork and related clerical work.

Many office and maintenance, repair, and operations suppliers offer customizable versions of their product catalogs to be put onto client company intranets. This customization, which automates the process of billing and shipping, allows employees to order products directly without intervention from corporate purchasing departments. The order data are captured once, at the point of origin, stored in the appropriate database, and reused as needed without reentry, reducing mistakes and clerical staffing needs.

Operations. The intranet can be at the center of all operations for the company, serving in essence as an interface to all enterprise systems. This can include logistics, inventory, project management, and operational systems. With the increasing focus on customer responsiveness, time to market, and value chain integration, such operational transparency is becoming increasingly valued.

Directories. One simple, useful application is an electronic directory of employees. The directory can easily be kept current, and is regularly used by most employees, particularly if paper directories are no longer published.

Menus. Something as simple as cafeteria menus and catering information can be a useful tool to drive traffic to the site.

Calendar Systems. Centralized calendar systems are an excellent intranet application for many reasons. First, it is an extremely valuable service to all users who schedule meetings or other appointments. It is also something most people are likely to use on a daily basis. Benefits include the reduction of time spent in arranging meetings or other events, the elimination of "double booking," and the increased efficiency of all involved staff.

The system may be used to schedule meetings and conference rooms and to coordinate individual schedules. Such technologies can also synchronize personal digital assistants (PDAs), allowing employees to keep their calendars with them and also have them stored centrally. Organizations may also enable the PDAs to access the intranet remotely via wireless technologies including WiFi and Bluetooth. Convergence of standards, advances in security, and decreasing hardware cost make such wireless access relatively inexpensive and viable.

Intranets may also include time clocks. As organizations move closer to activity-based costing, they require additional data on employee productivity. In particular, as more employees become knowledge workers, time tracking becomes an increasingly valuable tool and intranet applications make the input and analysis of these data easier, with widely accessible, familiar, intuitive graphical interfaces.

Group Collaboration. An increasing amount of work in organizations is being done by multiple people working collaboratively in disparate geographic locations, often at different times. Many intranet- and Internet-based applications are available to support such work. These systems include those for group authoring, document management, change control, moderated and threaded message boards, shared whiteboards, and work-flow management software. Benefits of these applications can be realized by employees working on-site and remotely, making the intranet a necessary component of any organization's attempts to expand remote work arrangements such as telecommuting, hoteling, and accommodating those employees who are frequently on the road.

Hewlett–Packard found that such traditional group collaboration tools were not sufficient because they lacked the "causal proximity" necessary for productive group work. They implemented passive cameras to indicate whether someone was at or near their desk, and included an intercom-enabled telephone system and instant messaging, to foster regular, brief conversations [Sieloff, 1999]. Such richer media helped emulate the benefits of working in the same physical location.

Syndicated Data. Many organizations make regular use of purchased external data feeds for strategic and operational work. Stock prices, weather, and news streams may all be purchased and made available through the intranet, customized for users based on their needs. Coordinating these streams requires careful management to balance easy access to current data with security.

Knowledge Management. Knowledge management has received a great deal of attention in the business and popular press. As organizations move away from manufacturing toward services, more knowledge is in the minds of workers than embedded in physical systems, or even documentation. This increases the cost of employee turnover and can impede growth.

Part of the challenge of knowledge management is uncovering, storing, and retrieving tacit knowledge. While many organizations struggle with simply managing explicit knowledge that can be readily articulated and documented, the greatest benefits often come from capturing tacit knowledge. The regular use of an intranet may encourage employees to routinely store information and knowledge and make explicit previously tacit knowledge.

Given the volume of information created in organizations, not only is indexing and retrieval important, but forgetting is also crucial. Archiving functions can move information out of the main databases to improve the relevancy of searches.

Implementing an intranet can fundamentally change the nature of the organization by reorganizing its business processes. By doing so, intranets can be a powerful lever of control [Simons, 1995], serving as a catalyst for change. General Motors credits its intranet, which links GM's 14 engineering centers with computer-aided design software and 3-D simulators, with an increase in creativity and even now recruits from the motion picture industry rather than just fine arts and automotive design programs [Rifkin, 2002] to take advantage of the Web-based skills. Such intranet-driven changes are so common that Pitt et al. have developed I-CAT, an instrument to measure the effectiveness of an intranet as a catalyst for change [Pitt, Murgolog-Poore, & Dix, 2001].

Technology. The technology for intranets is straightforward and similar to any other Web-based system. If the corporate network is TCP/IP, as most are, then starting an intranet site is as simple as placing a machine on the network, or finding one already connected, and installing or enabling a Web server. Current versions of Microsoft Windows desktop and server software come with Web servers installed and can be easily configured by anyone comfortable with personal computers. Once the server is up and running, the server may operate in background on an employee's ordinary PC workstation. Other employees may access the server by entering the IP address of the machine. Obviously, most users would prefer to type in a text domain name rather than a series of numbers. Such a text name would require either registration on the company's domain name server, which maps IP addresses to text names, or the use of Windows Internet Name Service (WINS), which serves the same purpose as DNS, but on Windows networks.

Content may be added to the Intranet in the form of HTML pages, which may be coded by hand, saved as HTML from word processors such as Microsoft Word, or written in HTML editors such as FrontPage and DreamWeaver. Simple programs may also be written to generate content in response to requests on the intranet through scripting in such languages as Cold Fusion, UltraDev, JSP, ASP, and Perl. HTML editors can generate simple scripts automatically for novice users. Such dynamic Web pages allow greater flexibility in the design of intranet Web pages.

Establishing the corporate portal/intranet as the access point for all information services is an appealing architecture for the IS department. Assuming the workstations are Windows-based, then all that needs to be installed on each machine is Microsoft Office with the Internet Explorer Web browser. This simplicity greatly reduces the support necessary for personal computers on the network because every employee in the company receives an identically configured computer with only standard Microsoft applications. This simplicity is one overarching benefit of an intranet over legacy network systems, since the need for individual computing support is significantly reduced by the common hardware and software, and user access is enabled through graphical, largely intuitive, interfaces.

All major software vendors have addressed the intranet market. Microsoft is working aggressively to integrate its desktop productivity software in Office with Web-based intranet tools, including video, Exchange server, and Microsoft Project. Microsoft even ships wizards for the creation of intranets, such as the 60 Minute Intranet Kit. Server software vendors such as Sun, IBM, and open-source Apache all offer intranet configuration suggestions and options for network and systems administrators familiar with those companies' products.

Enterprise system vendors have also focused heavily on Web-based access in the past several years. In that time, SAP, PeopleSoft, Baan, Oracle, and JD Edwards have all launched enterprise portal products to integrate their systems with a Web interface and other corporate intranet applications. Customer relationship management products such as Applix and Siebel have also launched portal interfaces in response to demands from the corporate market. The intranet/Web browser has become a common and accepted interface for corporate applications.

Extensible markup language (XML) is rapidly gaining support as a standard for Web documents. This is good news for intranets for several reasons. First, XML, with its focus on metadata, supports the complex, structured documents that populate most intranets. Second, as XML becomes a standard format for word processors, saving files to the intranet will become even easier. This approach has gained attention since Microsoft announced it intends to base its Office applications on XML. Finally, XML documents can be readily stored in databases and searched and retrieved quickly and easily, improving performance for large intranets.

Dot-coms were a major market for many of the software vendors who also compete in the intranet software market. With the collapse of the dot-coms, prices for Internet, and therefore intranet, software have dropped considerably. In addition, open-source software is becoming more robust, offering much less expensive alternatives to the major software vendors. Many companies use the open-source Apache suite to develop and manage their corporate intranets, and most major software vendors have started to support interface with Apache and other common open-source products.

As intranets grow, and increasing amounts of data are available, there must be a way to find information effectively. Information anxiety [Wurtman, 1989] is an inevitable challenge for intranets, given the volume of data likely to proliferate. Hewlett–Packard, for instance, found that within two years of launching its intranet, there were over two million documents stored on thousands of servers throughout the organization, and fewer than 5% of them were traditional official communications [Sieloff, 1999]. Such volumes make information architecture [Rosenfeld & Morville, 2002] essential. Information architecture is a crucial component of successful intranets, and the hierarchical structure and navigation themes must be well thought out and implemented. An efficient and effective search engine available for the entire intranet is also required. Many organizations license third-party search engines to reduce the problems inherent in searching the wide variety and amount of information stored in an intranet.

Unless entries are key worded with metadata, users' searches are reduced to full-text searches, which are likely to return massive numbers of hits even with the best search engines. Proper metadata coding requires an appropriate taxonomy. Such a taxonomy may be available within the industry, or extant in the organization. If not, a thorough metadata taxonomy must be developed and implemented. Without a pre-hoc taxonomy, it is unlikely the metadata will be very useful. One survey found that only 31% of companies with enterprise portals had implemented a taxonomy, emphasizing the need for corporate information metadata management.

Establishing a taxonomy is only the first step; it must be used correctly and consistently to be effective. This requires all of those entrusted with entering data to use the taxonomy properly. This can be accomplished by simplifying the taxonomy so that there is no ambiguity in the interpretation, and there must be documentation and/or training to ensure consistency of the taxonomy's application at all data entry points. It is also likely that the taxonomy will need to evolve over time, requiring a plan for maintenance and continued training.

It may be possible to post much of the content on the Intranet with limited metadata coding because most users would find the content via browsing, or the search engine would likely find it and rate it as highly relevant.

Personalization. For users, having the most applicable material readily accessible can be extremely powerful and empowering. Customization strategies for intranet access can be individualized or group based. For example, a different interface can be developed for selective categories of employees, offering some customization at low cost, or personalization software can be installed to generate pages based on an individual's expressed or inferred preferences or behaviors. Most Web servers, including Microsoft IIS, iPlanet, and Apache, now come with basic personalization functionality.

However, Web developers have found that personalization is not a silver bullet. Indeed, Yahoo!, one of the largest personalizable sites on the Web, learned that most users, if given the choice, do not customize their interfaces [Manber, 2000]. This makes designing the default pages and customizations more important than extensive customization options. Yahoo! also discovered that any customizations applied in one section of the site should apply to all areas, or users will become frustrated because they expect to need to set up customization only once. Yahoo! did find that a cadre of power users customized their interfaces beyond all expectations of the developers. Lessons learned from these power users can be applied to other users through redesign of the default or group-customized pages, underscoring the power of an information architecture steering group with input from all user groups.

Content Management. The Yankee Group outlines a three-stage content delivery cycle: creation, management, and presentation [Perry, 2001]. Content management systems such as Vignette and Interwoven's TeamSite manage all three of these stages by storing all Web site data in a database. They facilitate content management by accepting external and internal automated data streams through syndicated, operational, and financial data, and receive input from publishers throughout the organization in word processing, spreadsheet, presentation, XML or HTML formats. These systems manage the data with change control, security, and workflow tools. Content management systems facilitate presentation by working with application servers such as BEA's WebLogic and IBM's WebSphere, to create pages based on templates created by designers.

For large intranets, content management can be a worthwhile solution. Much of the data on the intranet changes regularly, and updating a database is more efficient than making changes to all Web pages that contain it. Content management systems facilitate the distribution of intranet content over multiple channels, including multiple Web sites, possibly including externally accessible sites, high- and low-bandwidth versions, wireless/mobile, syndication, and even hard copy. Document management systems such as Documentum and Hummingbird and even document imaging software serve a similar purpose. In general, all of these products are evolving toward enterprise content management.

Ease of Use. The importance of usability is well established for all Web sites. In addition, usability has been linked to effective knowledge management [Begbie & Chudry, 2002]. For intranets to be effective, they need a wide array of users across the organization. Providing extensive training and technical support to all users is not likely to be possible for reasons of cost and logistics. As employee turnover rises, reducing the learning curve for using the intranet becomes more important. Therefore the intranet must be well designed and simple to use.

IS Department Intranet Challenges

While intranets may be developed easily within departments with little support from IS, if the intranet is to scale and to function well, there are several areas the IS department must address. Indeed, intranets are likely to fundamentally change the nature of systems analysis and design, as users become more intimately involved in the process.

Standards vs Flexibility. The question of how much to constrain users in the interest of promoting standards for efficiency has plagued information systems professionals for decades. With intranets, the implications are substantially greater, as many people in the organization become, essentially, systems developers.

Controlling the use of spreadsheets and other personal productivity software was hard enough, but at least those systems were only accessed by one user, or perhaps disseminated over a sneaker net. With intranet data, potentially the entire organization has access to what is developed, and the developers have no formal training in systems analysis or design. Substantial duplication may arise, adding to costs and confusion. Poorly designed applications or content may lead to user rejection of the intranet, despite realizing other benefits.

The power of such distributed systems lies in the empowerment of employees with access to the data, and the ability to enter and manipulate it online. This empowerment is consistent with the principles of reengineering, "just in time," customer relationship management, supply chain management, and other management trends of the past several decades: enter data only once, at the source.

Intranet standards can be set for data storage, color schemes, templates, navigation, file naming, graphics formats, file formats, file sizes, and narrative voice. Content management systems can automate much of this standardization, and also establish approval processes for implementation.

Intranets have the benefit of limited or no access by those outside the organization, and this insulation offers many a false sense of security about the lack of need to protect the brand image. It is likely that much of the data on the intranet will make its way to suppliers and customers, whether indirectly or directly, because a successful intranet will be the primary source of information throughout the organization.

Availability. Once an intranet site is established, IS will be held responsible for ensuring that the site is available whenever needed. This may mean 24/7 in some instances. For a traditional system, with a limited number of professional developers operating in a staged development environment with source control on systems running in a data center, this is a well-solved problem, depending on resource constraints. However, for systems often hosted within departments outside IS control, and developed by untrained personnel, keeping the systems running can be a challenge. One poorly constructed script can bring a system to its knees and make diagnosis difficult. These systems are often not backed up, and nearly all lack any form of source control. However, such carelessness is not inevitable, and traditional IS methodologies may be applied to ensure the reliability of intranets.

Training and Support. As applications and services are moved onto the intranet, and traditional physical channels are closed down, there will be a significant need for training and support. Often the systems are brought live by departments without thought of the need for training personnel, but the support demands are made to the IS department. A training program can obviate many of the support calls, but requires a coordinated effort and a willingness on the part of departments to support in principle, if not financially, such training programs. Given the expense of lost productivity, such programs are often politically unpopular, but necessary for a successful implementation of any internal system.

Support for the systems is also a challenge because these applications may change frequently and the IS department may not have been involved in their construction. It is unrealistic, however, to believe that the call center, or its equivalent in smaller companies, will not be the main source of intranet support, whether de facto or de jure. Successful, sustainable intranets involve the IS department, especially for support.

Access. For the intranet to achieve its potential, it must displace other channels such as paper, phone, in person, and even e-mail. As these other channels are shut down, many employees who did not previously need access to a computer now need it to manage their benefits, enter a request for a repair with physical plant, look up a phone number, etc. This means providing access to all employees, many of whom have never needed computers before. Custodial staff, warehouse employees, and others may not need individual workstations, but do need convenient access to a computer connected to the intranet. This may mean extending the network

to areas of buildings previously left unwired. While modern construction and wireless technologies make this less of an issue than previously, there may still be significant expense and effort to ensure intranet access to all. Access must also be established to legacy software systems that may not be Web enabled.

Networks. Intranets may be more secure and reliable than the external Internet because companies have control over the network and servers on which they run. Intranet access may substantially increase the need for bandwidth on the network. Bandwidth demands can easily grow exponentially when an intranet is introduced, especially as the cost of multimedia hardware continues to drop. Although the cost of bandwidth is also dropping, the complexity of network management increases substantially as more bandwidth is used. Also, as demand for the services increases, the need for network reliability also rises. Good networks are more than bandwidth. As more work is intranet-based and dependent on the corporate network, availability, reliability, service, support, and scalability of the network all become crucial.

Security. Intranets may contain a great deal of sensitive personal and organizational data. These clearly need to be protected. The fact that intranets are for internal use only offers some a false sense of security. This laxity can lead to greater risks if there is any exposure to external access.

Intranet security is similar to security for any Web-based system, but potentially more complex because of the larger number of people with authorized access. When the Navy began implementing its Navy/Marine Corps Intranet (NMCI), it quickly found that the U.S. Department of Defense Information Technology Security Certification and Accreditation Process (DITSCAP) was difficult to apply. The Department of Defense needed to adapt the process to fit the needs of intranet development [Gerstmar, 2002].

Given that most security lapses are from authorized users within the organization, intranets create numerous potential breaches. Access controls for data and process may be set at the operating system, directory, application server, and Web server levels. Authorization can be done through passwords, or even biometrics, depending on the value of the information. Firewalls can be established not only at the border with the Internet to keep those outside the organization out, but also inside the organization to limit access to authorized employees. See also **Firewalls**.

Implementation. It is well established that top management support is a crucial factor in the success of all information systems implementations. However, it is not always sufficient. Many intranets are small, bootstrapped applications implemented with little fanfare. For an intranet to be successful, the goals for the initiative must be articulated. This involves determining the audiences for the system, the services to be offered, and the extent to which they are to be used. Once the goals are known, then metrics can be established for measuring success. With metrics in place, a plan can be created for implementing the system. This is another reason to have a corporate information systems architecture that covers the development, implementation and maintenance of all intranets.

Pilot Study. Because intranet systems are Web-based, prototypes are easy and generally inexpensive to build. This makes pilot testing an effective tool for determining the relevance of this technology in relation to a company's needs. However, it is important to keep in mind the goal of the pilot study. Many people incorrectly view pilot studies as always determining definitively whether a given system will work within the target community. Rather, most initial pilot studies should be tests for proof of concept, and should be repeated several times in different areas of the organization to ensure the generalizability of the concept test.

If the pilot is testing for proof of concept, then the population chosen for the test should be the one likely to adopt it. The initial pilot will be closely watched by many in the organization, and a failure could set the project back significantly. Participation in the pilot study is also an opportunity for potential users of the production system to invest psychologically in it. Potential users may be co-opted into adopting the technology by being brought into the development process as well as testing. IS research has shown that involvement in the design process makes people more likely to support and use the intranet system when implemented, and with intranet technologies, these people can be involved not only in the abstract design but in the actual construction, increasing their actual and perceived investment in the system, and therefore the likelihood of their support.

Successful intranets require momentum to achieve critical mass. Gladwell refers to a tipping point, where phenomena, including technology adoptions, either succeed or fail. Gladwell notes three factors that determine the tipping point of a technology: the law of the few, stickiness, and context.

The law of the few reflects the fact that not all users or decision makers are equally influential. Gladwell identifies three types of people that are particularly influential: connectors, who know and interact with a large number of people; mavens, who are considered experts in a given area; and salespeople, who will advocate for a position in which they believe. When implementing an intranet, identifying and engaging people from each of these groups is crucial to the project's success.

Stickiness reflects the extent to which something is memorable. The term stickiness is often used in reference to Web pages to describe how long visitors stay. Time spent by visitors is an easily gathered measure of user interest in a page or site.

Finally, context refers to the way in which information about the innovation or technology is disseminated. For instance, something advocated by members of the business units is likely to be more successful than something promoted by an e-mail from the IS department.

When the concept has been firmly established as viable, and there is support within the organization, then a representative group that will be more likely to predict success in the organization as a whole may be chosen.

Once the system is implemented, it is essential to eliminate other sources of the same information or service. This can be done drastically or gradually. For instance, calls to the help desk or human resources could ask whether the employee has gone through the Web site. If she/he has not accessed the intranet before calling for support, IS or HR staff could offer to guide the caller through the site for resolution before providing an answer directly.

Maintenance and Management. A major component of a successful intranet is a plan for its control. The first obvious question is who should control it. Four common configurations have emerged: IS, corporate, human resources, and distributed. Any one of the models may work, depending on the goals of the intranet and the structure of the organization.

With the IS department in control of the intranet, the technology is likely to be well aligned with the existing systems and architecture, making it more reliable. However, in terms of prioritization, the IS department is likely to be the worst choice for deciding what needs to be updated first. At a minimum, putting IS in charge adds an additional layer of control beyond the creators of the content. Technologies such as content management can make it more viable for business units to control the system.

In a corporate structure, there is a decision maker within the organization who oversees the business units involved in the intranet. This person makes strategic decisions about the intranet. Tactical decisions may be delegated farther down the organization. The systems themselves may be built by the business units or the IS department.

Although human resources applications are among the most common and heavily used functions of intranets, as more operational information is added, most human resource departments are ill equipped to manage the systems.

The distributed model is appealing because the subject matter experts can be put in control of the data. With content management systems supported by the IS department, this model can be viable as the business users are relieved of responsibility for design, but retain control of content.

A plan for maintenance of the system is crucial. At a minimum, such a plan should address how often the information should be updated, who is responsible for removing out-of-date information, how information will be archived, and how often the design should be revisited. Redesigns should have a formal structure and budget process [Fichter, 2001]. There should also be a disaster recovery plan commensurate with the importance of the intranet. Depending on the organizational value of the intranet, the plan for business continuity in the event of a disaster may range from periodic backups with a restoration plan, to off-site mirrored servers.

An acceptable use policy is an integral part of any intranet plan. Such a plan makes clear what is appropriate for the intranet, and may also include guidelines for Internet access as well. Although filters can be helpful in reducing inappropriate content, they are not a substitute for a well-articulated policy.

Cost/Benefit Calculations. Calculating the value of information technology has been the subject of extensive research over the past decade. The term productivity paradox is often used to describe the argument that despite rapid increases in spending on information technology, there is not a concurrent increase in productivity.

Measuring the return on intranet investments poses the same complexities as other information systems: calculating all of the costs involved is difficult and attributing benefits to them is even more challenging. However, such calculations are feasible, and there is reason to believe that a well-planned intranet can be profitable.

Among the obvious and easily calculated costs of an intranet are the hardware and software required for the servers, including any support contracts, and development software purchased. The costs of development may be relatively easy to measure for IS personnel involved in the system construction if they are contracted, or if the internal IS shop tracks costs on a project basis. However, much of the construction will involve the time of those outside IS, and there may be no extant tools for tracking their time, nor easy access to data on the cost of their time. Network costs for running the intranet may not be easy to track, because both utilization and marginal cost would need to be calculated. A methodological change in how such costs are calculated could make the difference in positive or negative ROI. Training costs must be included, both for the trainers' time in development and delivery and also lost productivity for trainees.

Application service providers can provide all of the infrastructure and functionality of an intranet for an organization with only an Internet connection. Infostreet.com (http://www.infostreet.com), Intranets.com (http://www.intranets.com), and others offer a broad array of cost-effective intranet services to companies of all sizes.

While cost calculation is difficult, benefit calculation is even more difficult. Benefit calculation should include obvious direct cost reductions such as headcount, paper, and software licenses. One commonly used metric for efficiency is to credit the system for the pro-rated salary saved by using the intranet rather than the previous methods. However, this makes the assumption that all of the difference is saved. In many cases, the previous methods were not a complete loss; for instance, workers were doing other work while on the phone, or filled out forms during their commute. Not all of the recovered time will be used productively. Potentially more important than such bottom line cost reductions are top line enhancements in revenue. However, such improvements are harder to measure. Specifically, has the intranet contributed to better service, higher quality products, or improved time to market? If so, how much is that worth to the organization?

One excellent source of metrics for evaluating an intranet is the server access logs. All page accesses are automatically logged, along with the time and date of access, IP address of the user, which can be mapped to the specific intranet user, and the referring page, if any. These data may be analyzed to determine how employees are using the intranet and to identify potential enhancements. Estimates of the most and least popular pages, typical paths through the site, most used search terms, performance data, and other metrics are easily generated with simple analysis tools. This can be supplemented with direct surveys of users, online and through other channels.

Note: *Portions of this article originally appeared in the Internet Encyclopedia, John Wiley & Sons, Inc., Hoboken, NJ.*

Additional Reading

Baker, S.: "Getting the Most from Your Intranet and Extranet Strategies," *Journal of Business Strategy,* **21**(4), 40–43 (2000).

Bleeker, T., B.T. Smith, E. Rosenfeld, and G. Bushey: *Microsoft SharePoint Bible,* 2nd Edition, John Wiley and Sons, Inc., Hoboken, NJ, 2007.

Comer, D. E: *The Internet Book: Everything You Need to Know About Computer Networking and How the Internet Works,* 4th Edition, Pearson Education, Upper Saddle River, NJ, 2006.

Fichter, D.: "The Intranet of Your Dreams and Nightmares: Redesign Issues," *Online,* **25**(5), 74–76 (2001).

Gerstmar, T.: "Legacy Systems, Applications, Challenge Intranet Rollout," *Signal,* **57**(4), 29–31 (2002).

Gladwell, M.: *The Tipping Point: How Little Things Can Make a Big Difference,* Little, Brown, Boston, MA, (2000).

Hammer, M., and J. Champy: *Reengineering the Corporation,* Harper Business, New York, NY, 1993.

Manber, U.: "Experience with Personalization on Yahoo," *Communications of the ACM*, 35–39 (2000).

Meuse, D.: "Making Employee Self-service Work for Employees and Your Company," *Benefits Quarterly*, **15**(3), 18–23 (1999).

Paswan, A.: *Internets, Intranets, and Extranets*, The Haworth Press, Inc., Binghamton, NY, 2003.

Perry, R.: *Managing the Content Explosion into Content-rich Applications (Internet Computing Strategies)*, The Yankee Group, Boston, MA, (2001).

Pitt, L., M. Murgolog-Poore, and S. Dix: "Changing Change Management: The Intranet as Catalyst," *Journal of Change Management*, **2**(2), 106–114 (2001).

Rifkin, G.: "GM's Internet Overhaul," *Technology Review*, **105**(8), 62–67 (2002).

Rosenfeld, L., and P. Morville: *Information Architecture for the World Wide Web*, 2nd Edition, O'Reilly, Cambridge, MA, (2002).

Sieloff, C.G.: "If Only HP Knew What HP Knows: The Roots of Knowledge Management at Hewlett-Packard," *Journal of Knowledge Management*, **3**(1), 47–53 (1999).

Simons, R.: *Levers of Control: How Managers Use Innovative Control Systems to Drive Strategic Renewal*, Harvard Business School Press, Boston, MA, 1995.

Townsend, J.T., D. Schaffer, and D. Riz: *Building Portals with Microsoft.Net*, Addison-Wesley Longman, Reading, MA, 2004.

Wurtman, R.: *Information Anxiety*, Doubleday, New York, NY, 1989.

WILLIAM T. SCHIANO, Bentley College, Waltham, MA

INTRAOCULAR LENSES (IOLs). These are permanent, plastic lenses that are surgically implanted to replace or supplement the eye's natural crystalline lens. They have been used in the United States since the late 1960s to restore vision to cataract patients, and more recently are being used in several types of refractive eye surgery. To understand how an intraocular lens works, it is necessary to understand how the eye's natural lenses function.

The eye works like a camera with two lenses. One lens is the cornea, a clear membrane that covers the front of the eye. The other lens is the eye's natural crystalline lens, which is located behind the pupil. The cornea is responsible for about 70% of the eye's focusing power, while the natural lens "fine-tunes" the image before it is focused on the retina at the back of the eye. The retina works like the film in a camera, receiving light images and sending them through the optic nerve to the brain. If both lenses are working properly, the image is focused precisely on the surface of the retina, and the result is perfect "$\frac{20}{20}$" vision.

A cataract occurs when the eye's natural crystalline lens becomes cloudy, usually because of the natural process of aging. The only way to treat a cataract is to remove the cloudy lens. Because the eye cannot focus properly without this lens, most modern cataract surgery includes the implantation of a permanent prescription intraocular lens. Most intraocular lenses used in cataract surgery can befolded and inserted through the same tiny opening that was used to remove the natural lens. Once in the eye, the lens unfolds to its full size. The opening in the eye is so small that it heals itself quickly without stitches. The intraocular lenses are made of inert materials that do not trigger rejection responses by the body.

IOLs are permanent. They do not get lost nor do they have to be replaced. Careful measurement of the eye prior to surgery can determine the focusing power of the IOL to correct nearsightedness or farsightedness.

Most intraocular lenses implanted during cataract surgery are monofocal lenses, meaning the prescription is set for one distance, usually for distance vision. Therefore, one will probably still need reading glasses after surgery. A new intraocular lens implant, the multifocal lens, functions more like the eye's natural lens by providing clear vision at a distance and good focus for a range of near distances. Although not all patients are good candidates for the multifocal lens, those who can use the lens are generally pleased with the results.

Several new refractive eye surgery procedures now rely on intraocular lenses rather than laser surgery to correct vision deficiencies. Unlike laser surgery, which sculpts the cornea to correct refractive disorders, these new procedures replace or supplement the eye's natural crystalline lens with an intraocular lens. Clear Lens Extraction (CLE) works like cataract surgery in that the eye's natural lens is removed and replaced with a prescription intraocular lens. The only difference is that the natural lens being replaced by the CLE procedure is, as the name implies, clear. Studies indicate that CLE may eventually be the procedure of choice for patients with farsightedness.

Another refractive surgery procedure (currently investigational) that uses an intraocular lens is referred to as Implantable Contact Lens (ICL) surgery. With this procedure, a prescription implantable lens is inserted inside the eye between the iris and the natural lens of the eye. This intraocular lens is prescription-engineered to correct refractive error by complementing the focusing powers of the natural inner lens of the eye.

A different refractive surgery procedure implants a tiny plastic lens similar to a contact lens, within the corneal tissue to correct problems of nearsightedness. These implants, called Intracorneal Lens Implants, are under FDA investigational protocol, as are the ICL lenses.

Also under FDA investigation is an intraocular lens that holds promise for correcting presbyopia, the vision problem that occurs when the natural lens of the eye becomes less flexible. This condition inhibits the reflex ability of the lens to contract or expand in order to focus properly on nearby objects. Because this condition comes about as a person matures, most people over the age of 45 need reading glasses. Known as a Small Diameter Corneal Inlay (SDCI), this prescription lens is inserted into the corneal tissue to create an effect similar to a bifocal contact lens. See also **Cataract**.

Vision Rx, Inc., Elmsford, NY

INTRASTROMAL CORNEAL RING SEGMENTS (ICRS). The implantation of intrastromal corneal ring segments (trade name Intacs) is the first approved nonlaser procedure to correct mild myopia, or nearsightedness. As in all surgical vision correction, the objective of the intrastromal corneal ring segments procedure is to reshape the cornea in order to correct for imperfections. Unlike most other refractive eye surgery procedures, however, the ICRS procedure corrects the vision problems without removing any eye tissue.

The basis of the ICRS procedure is the surgical implantation of two ultrathin arcs in the peripheral area of the cornea. These arcs (or segments) are manufactured from a polymer that has been safely used in cataract surgery for more than 40 years. When the arcs are implanted, they flatten the cornea to the degree required to correct the myopic condition. The procedure does not cut or remove any tissue, making this quite different from other refractive surgery procedures that permanently alter the cornea. They are designed to remain permanently in place, although the segments can later be removed or replaced to correct for possible sight changes as the eye ages.

The best candidates for the corneal ring procedure are usually those patients with mild myopia (−1.00 to −3.00) who have no more than 1.00 diopter of astigmatism. Patients should also be at least 21 years of age with vision that has been stable for at least one year and be free of eye disease.

To understand how the ICRS procedure works, it is first necessary to understand the visual function of the eye. The eye works like a camera with two lenses. The first lens is the cornea, a clear membrane that covers the front of the eye. The second lens is the eye's natural crystalline lens, which is located behind the pupil. The cornea is responsible for about 70 percent of the eye's focusing power, while the natural lens fine-tunes the image before it is focused on the retina at the back of the eye. The retina works like the film in a camera, receiving light images and sending them through the optic nerve to the brain. If both lenses are working properly, the image is focused precisely on the surface of the retina, and the result is perfect 20/20 vision.

Just as people are born with different sizes and shapes of hands, their eyes also vary in form and proportion. A perfect eye has an evenly rounded cornea that allows light to fall exactly on the retina, resulting in perfect vision. If the cornea is too steep or if the eye is too long from front to back, light rays are focused in front of the retina resulting in myopia. The ICRS procedure flattens the cornea, bringing the light rays back in focus on the surface of the retina.

The ICRS surgical procedure takes about 15 minutes per eye and is performed on an outpatient basis. Patients are typically given a mild oral sedative along with topical anesthetic eye drops to numb the eye.

To begin, the surgeon makes a tiny opening (less than 2 mm) near the upper edge of the cornea beneath the eyelid. The arcs are then placed in opposing positions on the outer edges of the cornea away from the central optical zone, the critical area for clear vision. The thickness of the arcs is determined by the degree of myopia that needs correcting. The higher the degree of myopia, the thicker the ring segments. When in place, they form a circle around the edges of the iris and alter the curvature of the eye. In essence, they stretch the cornea into a more flattened shape, thereby correcting the myopic condition. The patient does not feel the rings because they are placed beneath the nerve endings in the cornea.

Although the ring segments are designed to remain permanently in the corneal tissue, they can be replaced with thicker or thinner segments to achieve any adjustments that may be required, or they can be removed permanently. These adjustments are possible because there is no invasion of the visual axis or central optical zone of the cornea, and the procedure does not involve any removal of corneal tissue.

Most patients notice an improvement in vision within a day of surgery and are able to resume normal activities within two or three days.

Vision Rx, Inc., Elmsford, NY

INTRINSIC SAFETY. By definition (National Electrical Code), intrinsically safe equipment and wiring is "incapable of releasing sufficient electrical or thermal energy under conditions of use to cause ignition of a specific hazardous atmosphere mixture." The British first applied intrinsic safety in direct current mine signaling circuits as early as 1913. Worldwide acceptance of the principle in equipment and instrument design did not occur until the late 1960s. Data on this topic can be obtained from the National Fire Protection Association, the Factory Mutual Research Laboratories, the Underwriters Laboratories, and the Instrument Society of America.

INTRINSIC WAVE FREQUENCY. The frequency of oscillation measured by a sensor that moves with the mean wind.

If a wave of wavelength λ has intrinsic frequency f_i and is imbedded in a mean wind speed of U, then the local wave frequency f in Hz measured by a sensor fixed to the ground is

$$f = f_i + U/\lambda.$$

This relationship applies to internal gravity waves in the atmosphere. A special case is for standing mountain waves (in which occur lenticular clouds) where $f = 0$, because the intrinsic frequency exactly counteracts the wind effect.

INTROMISSION. The lateral mixing of ambient environmental air into the edges of mixed-layer convective thermals or cumulus clouds, leaving a large undiluted core in the middle of the thermal. While this term is closely related to lateral entrainment, the phrase "lateral entrainment" is often associated with idealizations where the ambient air is mixed quickly across the whole diameter of the plume, leaving no undiluted core. While the lateral entrainment model has been applied successfully to smokestack plumes of meters to tens of meters in diameter, it is not appropriate for mixed-layer thermals of order 1 km in diameter.

AMS

INTROMISSION ZONE. The outside part of convective thermals that experience mixing with the ambient environment. These thermals are somewhat cylindrically shaped in the convective boundary layer rather than bubble-shaped as was proposed by some classical theories. The inside of a thermal is called a protected core and is usually not contaminated with entrained environmental air.

AMS

INTRUSION DETECTION. Intuitively, *intrusions* in an information system are the activities that violate the security policy of the system, and *intrusion detection* is the process used to identify intrusions. Intrusion detection has been studied for approximately 20 years. It is based on the beliefs that an intruder's behavior will be noticeably different from that of a legitimate user and that many unauthorized actions will be detectable.

Intrusion detection systems (IDSs) are usually deployed along with other preventive security mechanisms, such as access control and authentication, as a second line of defense that protects information systems. There are

several reasons that make intrusion detection a necessary part of the entire defense system. First, many traditional systems and applications were developed without security in mind. In other cases, systems and applications were developed to work in a different environment and may become vulnerable when deployed in the current environment. (For example, a system may be perfectly secure when it is isolated but become vulnerable when it is connected to the Internet.) Intrusion detection provides a way to identify and thus allow responses to attacks against these systems. Second, due to the limitations of information security and software engineering practice, computer systems and applications may have design flaws or bugs that could be used by an intruder to attack the systems or applications. As a result, certain preventive mechanisms (e.g., firewalls) may not be as effective as expected.

Intrusion detection complements these protective mechanisms to improve the system security. Moreover, even if the preventive security mechanisms can protect information systems successfully, it is still desirable to know what intrusions have happened or are happening, so that we can understand the security threats and risks and thus be better prepared for future attacks.

In spite of their importance, IDSs are not replacements for preventive security mechanisms, such as access control and authentication. Indeed, IDSs themselves cannot provide sufficient protection for information systems. As an extreme example, if an attacker erases all the data in an information system, detecting the attacks cannot reduce the damage at all. Thus, IDSs should be deployed along with other preventive security mechanisms as a part of a comprehensive defense system.

Intrusion detection techniques are traditionally categorized into two methodologies: *anomaly detection* and *misuse detection*. Anomaly detection is based on the normal behavior of a subject (e.g., a user or a system); any action that significantly deviates from the normal behavior is considered intrusive. Misuse detection catches intrusions in terms of the characteristics of known attacks or system vulnerabilities; any action that conforms to the pattern of a known attack or vulnerability is considered intrusive.

Alternatively, IDSs may be classified into host-based IDSs, distributed IDSs, and network-based IDSs according to the sources of the audit information used by each IDS. Host-based IDSs get audit data from host audit trails and usually aim at detecting attacks against a single host; distributed IDSs gather audit data from multiple hosts and possibly the network that connects the hosts, aiming at detecting attacks involving multiple hosts. Network-based IDSs use network traffic as the audit data source, relieving the burden on the hosts that usually provide normal computing services.

Conclusion

Intrusion detection continues to be an active research field. Even after 20 years of research, the intrusion detection community still faces several difficult problems. How to detect unknown patterns of attacks without generating too many false alerts remains an unresolved problem, although recently, several results have shown there is a potential resolution to this problem. The evaluating and benchmarking of IDSs is also an important problem, which, once solved, may provide useful guidance for organizational decision makers and end users. Moreover, reconstructing attack scenarios from intrusion alerts and integration of IDSs will improve both the usability and the performance of IDSs. Many researchers and practitioners are actively addressing these problems. Intrusion detection is expected to become a practical and effective solution for protecting information systems.

INTRUSION (Geology). In terms of igneous rocks, intrusion is the process of emplacement of magma in preexisting rock; also the igneous rock mass so formed within the surrounding rock, sometimes referred to as a *pluton*. Intrusions are found in the Bushveld Complex of South Africa and are valuable sources of ores. The Dufek intrusion, located in Antarctica, is the second largest — possibly the largest — intrusion in the world. This intrusion has been under investigation since 1957 by U.S. geologists and geophysicists. Scientists from the U.S.S.R. also have explored the area since 1978. The intrusion extends across the transition from West to East Antarctica along the Transantarctic Mountains. Prior to a 1979–1980 study by members of the U.S. Geological Survey and the Scott Polar Research Institute (Cambridge University, England), the intrusion area was estimated at 34,000 square kilometers, based upon gravity and magnetic surveys. Scientists have observed a remarkable chemical variation and perfection of layering of the Dufek intrusion. In the latest studies, the minimum area of the intrusion has been revised upward to 50,000 square kilometers and thus comparable to the Bushveld Complex. In the study, comparisons were made of the magnetic and subglacial topographic profiles. Aeromagnetic and radio echo ice-sounding measurements were used. Exposed rocks of the intrusion occur in only about 3% of its area, contributing to difficulties of possible later ore exploration and exploitation. Further details can be found in "Aeromagnetic and Radio-Echo Ice-Sounding Measurements Show Much Greater Area of the Dufek Intrusion, Antarctica," *Science*, **209**, 1014–1017 (1980). In terms of sedimentary rocks, an intrusion is a sedimentary injection on a relatively large scale, e.g., the forcing upward of clay, chalk, salt, gypsum, or other plastic sediment, and its emplacement under abnormal pressure in the form of a diapiric plug; a sedimentary structure or rock formed by intrusion.

INTUBATION.   The insertion of a tube into the larynx through the throat to permit breathing when the larynx becomes closed through swelling.

INTUSSUSCEPTION.   The invagination (telescoping) of one part of the intestine into another, usually creating intestinal obstruction.

IN VACUO.   See **Correction to Vacuum**.

INVAR.   See **Nickel**.

INVARIANCE PRINCIPLE.   An invariance principle states that some physical law must be invariant under certain transformations. (1) Some invariance principles are based on symmetry operations. For example, physical laws must be invariant under a pure rotation of spatial coordinate systems.

(2) The equivalence principle of special relativity is an invariance principle, namely that the laws of physics must be variant under a transformation from one inertial system to another (relativistic invariance or Lorentz invariance).

(3) The invariance principle of general relativity states that the laws of motion are invariant for all observers, whether accelerated or not. This leads to the equivalence of gravitational and accelerated frames of reference.

INVARIANCE THEOREM AND CONSERVATION LAWS IN QUANTUM MECHANICS.   In a treatment of most quantum mechanical systems, an exact solution for the problem is not possible and approximation procedures are used. They often involve, as a first step, the neglect of second order terms in the energy of the system. In this approximation, many systems can be considered a combination of two noninteracting systems that can be considered separately. For instance, if the interaction energy between an atom and an electromagnetic field is neglected, the atom-radiation field system can be treated by regarding the atom as one independent system and the radiation field as another. When one of these subsystems is treated, a series of possible stationary states is found, for which wave functions may simultaneously be eigenfunctions of one or more quantum mechanical operators corresponding to various dynamical variables. Since, in this approximation, these variables will be constant in time, they are referred to as *conserved quantities* of the system. When the interaction of this subsystem with the other subsystem is taken into account, it is seen that the "stationary states" of the subsystems are not really stationary, since energy can be exchanged with the rest of the system. During such exchanges of energy, *during the interaction*, it may happen that some of the conserved quantities of the subsystem may change without a compensating change in the rest of the system. In such a case, it is said that these quantities are not conserved during the interaction. The question of whether a given dynamical variable of a system is or is not conserved during an interaction is determined by the invariance of the interaction under various types of spatial transformations. For instance, the interaction between parts of a system will be invariant under space rotation if and only if there is no net torque acting on the system. Thus invariance of an interaction under spatial rotation has as

a consequence the conservation of total angular momentum and vice-versa. Invariance under spatial inversion through origin implies parity conservation. Invariance under spatial translation means linear momentum conservation, etc.

INVARIANT.   1. An adjective used to describe a quantity that remains unchanged, or a relationship that remains unchanged in form, during some operation. E.g., an invariant subgroup; a vector whose magnitude remains unchanged upon rotation of a coordinate system; a physical law that remains unchanged in form with respect to a coordinate transformation, as for example, Newton's laws of motion, which are invariant under a Galilean transformation, but are not invariant under a Lorentz transformation. See also **Covariance**; and **Invariance Principle**.

2. A noun used to denote a quantity that is invariant. Thus the speed of light is an invariant of all inertial frames moving with uniform speed relative to one another. Conserved quantities are invariants of a system.

3. An expression involving the coefficients of an algebraic function which remains constant when a transformation, such as translation or rotation of coordinate axes, is made. See also **Discriminant**; **Quadratic Equation**; and **Vector**.

INVERSE.   If $y = f(x)$, the inverse function is $x = g(y)$. The inverse of an operation is one that undoes what has been done: addition, subtraction; multiplication, division; differentiation, integration. Examples of inverse functions are: the square of a variable and the square-root function; the exponential and the logarithmic functions; the trigonometric or hyperbolic functions and their inverse functions. See also **Inverse Trigonometric Function**.

INVERSE MATRIX.   The inverse of an $n \times n$ matrix $\mathbf{A}$ whose determinant is not zero is the unique $n \times n$ matrix $\mathbf{A}^{-1}$ such that $\mathbf{AA}^{-1} = \mathbf{A}^{-1}\mathbf{A} = \mathbf{I}$, where $\mathbf{I}$ is the unit matrix of order n. Numerically, a rapid method of evaluating the inverse of a given matrix is found under **Cholesky Method of Solving Equations**.

INVERSE TRIGONOMETRIC FUNCTION.   The inverse function to $y = \sin z$ is the angle whose sine is y or symbolically, $z = \arcsin y = \sin^{-1} y$. Other inverse trigonometric functions are indicated in a similar way. If $y^2 < 1$, the following series expansions may be used:

$$\sin^{-1} y = y + \frac{y^3}{6} + \frac{1 \cdot 3}{2 \cdot 4 \cdot 5}y^5 + \frac{1 \cdot 3 \cdot 5}{2 \cdot 4 \cdot 6 \cdot 7}y^7 + \cdots$$

$$= \frac{\pi}{2} - \cos^{-1} y$$

$$\tan^{-1} y = y - \frac{1}{3}y^3 + \frac{1}{5}y^5 - \frac{1}{7}y^7 + \cdots$$

$$= \frac{\pi}{2} - \cot^{-1} y$$

and, for $y^2 > 1$,

$$\tan^{-} y = \frac{\pi}{2} - \frac{1}{y} + \frac{1}{3y^3} - \frac{1}{5y^5} + \cdots$$

$$\sec^{-1} y = \frac{\pi}{2} - \frac{1}{y} - \frac{1}{6y^3} - \frac{1 \cdot 3}{2 \cdot 4 \cdot 5y^5} - \frac{1 \cdot 3 \cdot 5}{2 \cdot 4 \cdot 6 \cdot 7y^7} + \cdots$$

$$= \frac{\pi}{2} - \csc^{-1} y$$

The inverse trigonometric functions are many-valued. For example, if $z = \sin^{-1} y$, the variable y must lie between ± 1 but there are an infinite number of quantities z which satisfy this condition. See Figs. 1, 2, and 3. Therefore, it is customary to select some particular value, known as the principal value of the inverse function, in order to make the solutions unique. This is generally the smallest, or the smallest positive value of the angle. While these conventions are not always followed, a convenient definition of the principal values is $-\pi/2 \leq Y_1 \leq \pi/2$, where Y_1 can be $\sin^{-1} y$, $\tan^{-1} y$, $\cot^{-1} y$, $\csc^{-1} y$ and $0 \leq y_2 \leq \pi$, Y_2 is $\cos^{-1} y$, $\sec^{-1} y$. The general solution can then be written in terms of these principal values

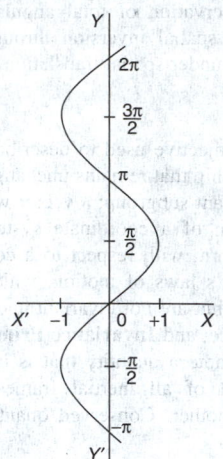

Fig. 1.   Inverse trigonometric function: $y = \sin^{-1} x$.

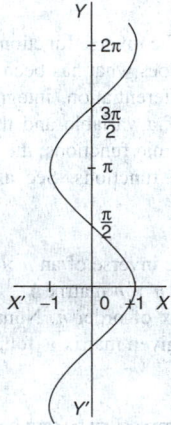

Fig. 2.   Inverse trigonometric function: $y = \cos^{-1} x$.

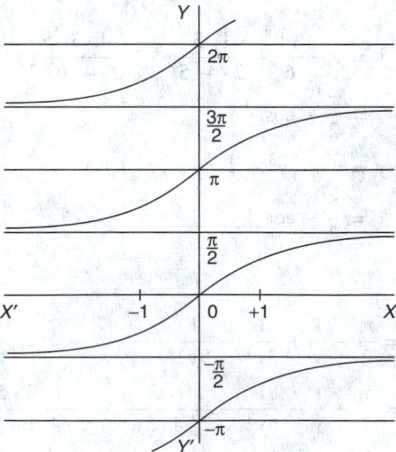

Fig. 3.   Inverse trigonometric function: $y = \tan^{-1} x$.

as follows: $z = (-1)^n \sin^{-1} y + n\pi$; $\pm \cos^{-1} y + 2\pi n$; $\tan^{-1} y + n\pi$, with similar relations for the other functions.

See also **Trigonometry**; and **Trigonometric Curve**.

INVERSION (Fog).   See Fog and Fog Clearing.

INVERSION (Meteorology).   A departure from the usual decrease or increase with altitude of the value of an atmospheric property; also, the layer through which this departure occurs (the "inversion layer"), or the lowest altitude at which the departure is found (the "base of the inversion"). See also **Atmosphere (Earth)**.

INVERTEBRATE PALEONTOLOGY.   The study, description, and geologic use of invertebrate fossils in relation to paleobiological and stratigraphic problems. The science of invertebrate paleontology is primarily founded upon invertebrate zoology. Since thousands of invertebrate fossils, ranging in age from the Cambrian to the Pleistocene, have been figured and described, it is not possible to list them all in a general science encyclopedia. Also there is no single reference work in existence that covers the entire subject. For detailed information the student must consult special bibliographies which list the references in a large number of special papers and monographs. The following condensed classification of the invertebrates serves as an outline of the more significant facts relating to invertebrate paleontology.

I.   *Protozoa*. Single-celled animals. While most of the protozoa are naked, a particular marine group called the *foraminifera* (a term derived from words meaning a hole and to bear) form shells that occur as fossils from the Cambrian to the present. See Fig. 1. *Foraminifera* are an important constituent of chalk. Due to the large number of distinctive and rapidly evolving species, *foraminifera* are useful index fossils, especially in the Mesozoic and Cenozoic periods.

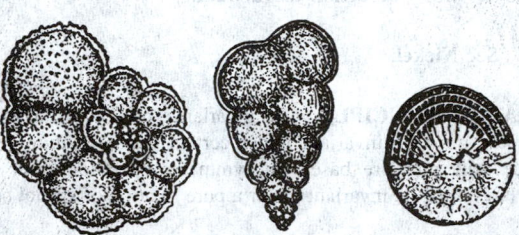

Fig. 1.   Two diagrams at left are of Cretaceous foraminifers, greatly enlarged Diagram at right is of an Eocene foraminifer, *Nummulities*. (*LeConte, "Element of Geology," Appleton-Century Co.*)

II.   *Porifera* or Sponges. The name "porifera" is derived from words meaning a pore and to bear. Fossil sponges occur from the Cambrian to the Pleistocene. Except for a few species they are not particularly valuable index fossils. A particularly interesting genus is *Hydnoceras*, a delicate glass sponge that has left its imprint in the fine-grained muds of Devonian Age.

III.   *Graptolites*. This term is derived from words meaning written and stoned, because the fossils look like pencil marks on slate. Graptolites were colonial marine animals with chitinous skeletons. Their stratigraphic range is from the Ordovician to the Silurian, inclusive. Because the graptolites evolved rapidly and have a worldwide distribution they are excellent index fossils. Figure 2 illustrates two of the principal types of *Graptolites*.

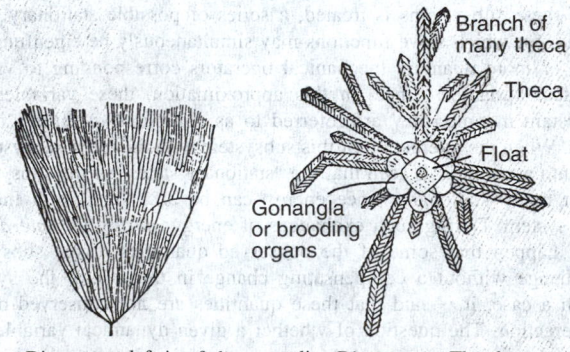

Fig. 2.   Diagram at left is of the graptolite *Dictyonema*. The theca are microscopic and like those shown on *Diplograptus*. Dendroid type of benthonic or anchored graptolite. Diagram on the right is of the graptolite *Diplograptus pristis*. Floating type of graptolite. Only single stipes are usually found fossil. (*Field, "Geology Manual," Part II, Princeton Univ. Press.*)

IV. *Corals*. The geologic record of fossil corals is from the Ordovician to the present day. They are important reef builders in the tropical waters of the present oceans and seas. The earliest forms are both colonial and single. Figure 3 shows a single or rugose coral of Ordovician age. Corals are important and useful index fossils in the strata of certain geologic periods.

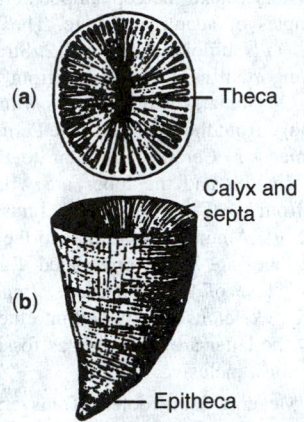

Fig. 3.   Single, rugose or cup coral: (**a**) Top view; (**b**) complete coral skeleton (*Field, "Geology Manual," Part II, Princeton Univ. Press.*)

V. *Vermes*, or *Worms*. Worm trails and worm tubes are found in the sedimentary strata of all ages from the pre-Cambrian to the present. Not particularly useful as index fossils, except in the early Silurian. The jaws and teeth of worms, called conodonts, are useful index fossils.

VI. *Echinoderms*. The term is derived from words meaning hedge hog and skin, because certain types of echinoderms have sharp barbed spurs. Echinoderms are aquatic, marine animals with radial symmetry. The test or skeleton is composed of plates of calcium carbonate, with or without a chitinous covering (when living), and usually in the shape of a cup or "calyx" with or without "arms." Some forms are attached to the sea bottom by means of "stems" and "roots"; others are floating or free swimming. The echinoderms are subdivided into the following characteristic groups. (1) Cystoids (Cystids). Stratigraphic range from the Cambrian to the Mississippian. (2) Blastoids. Stratigraphic range from the Cambrian to the Mississippian. An important and frequently beautifully preserved genus is *Pentramites*. (3) Crinoids, or sea lilies. See Fig. 4. Stratigraphic range Cambrian to present. (4) Asteroids, or starfishes. (5) Echinoids, sea urchins, from which the echinoderms take their name. See Fig. 5. (6) Holothouroids, or sea cucumbers.

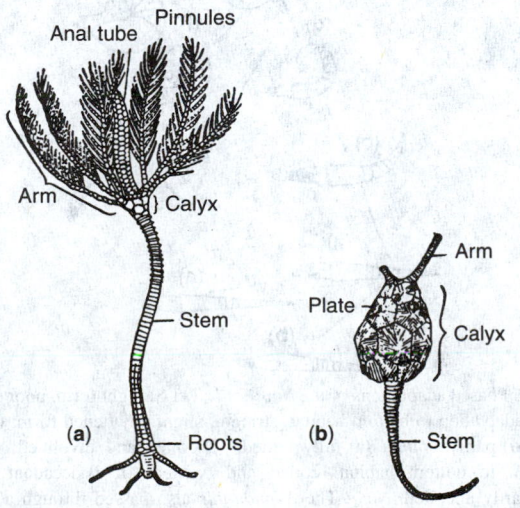

Fig. 4.   Crinoid (**a**); and cystid (**b**). (*Field, "Geology Manual," Part II, Princeton Univ. Press.*)

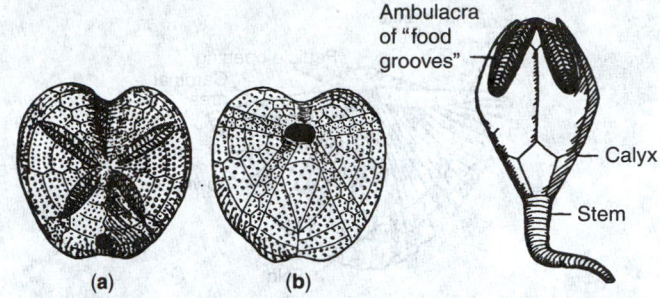

Fig. 5.   Views at left are of the "sea urchin" *Echinoid*: (**a**) dorsal view; (**b**) ventral view. Diagram at right is of the *Blastoid*, an anchored echinoderm with free "arms." (*Field, "Geology Manual," Part II, Princeton Univ. Press.*)

VII. *Molluscoidea*, meaning the form of a mollusk. They are a group of diverse forms of aquatic, and usually marine, animals with chitinous calcareous skeletons, free-swimming only in the young stages. The two principal subdivisions are *bryozoa* (moss-like animals) and the *brachiopoda* (arm-footed), which are bivalves. The bryozoans are aquatic, colonial, and usually marine, animals whose skeletons look something like small colonial corals; important index fossils from the Ordovician to the present; and important reef builders, especially in the Paleozoic. The skeleton of the brachiopod is composed of two parts or valves which are formed of either "horn" or calcium carbonate. The brachiopod skeletons are principally distinguished from the pelecypods (clams, etc.) by a different type of bilateral symmetry. The higher forms have internal skeletons. Brachiopods are excellent index fossils, especially in the Paleozoic, where they share their importance with the graptolites and the trilobites. See Figs. 6, 7, 8, 9, and 10.

VIII. *Mollusca*, meaning soft-bodied animals. This class includes the *Pelecypoda*, meaning axe-footed bivalves whose shells are formed of calcium carbonate with a "horny" covering. No interior skeleton, even in the higher types. The skeleton is distinguished from that of the brachiopods by a different type of bilateral symmetry. See Fig. 11. A few species, such as the oyster, have no symmetry. All pelecypods are aquatic, but may be either freshwater or marine. Most species are attached to the bottom in the adult stage, but few are free-swimming. Pelecypods are not particularly good index fossils except at certain horizons in the Mesozoic and Cenozoic.

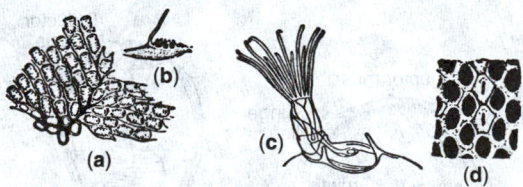

Fig. 6.   Bryozoans: (**a**) portion of modern colony seen from above (× 15); (**b**) an individual, expanded; (**c**) fossil form. (**d**) cross section. (*a–c after Verrill and Smith; d, from Ulrich. Shimer, "Introduction to the Study of Fossils," The Macmillan Co.*)

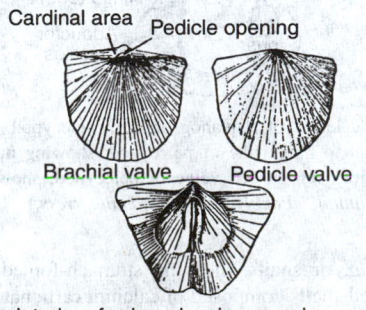

Fig. 7.   Rafinesquina. A calcareous brachiopod having no interior skeleton. (*Field, "Geology Manual," Part II, Princeton Univ. Press.*)

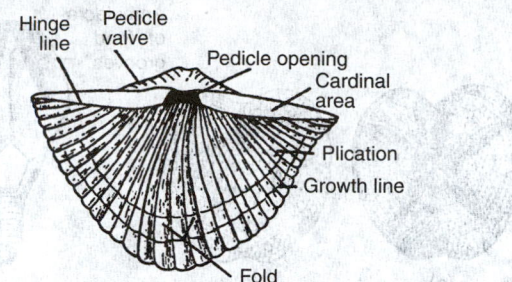

Fig. 8.   Front view of spirifer, showing full view of brachial valve. (*Field, "Geology Manual," Part II, Princeton Univ. Press.*)

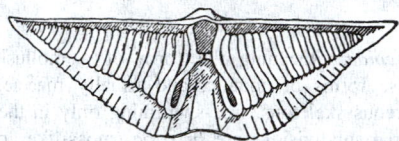

Fig. 9.   Interior of brachial valve of spirifer, showing brachidia, or spires. (*Field, "Geology Manual," Part II, Princeton Univ. Press.*)

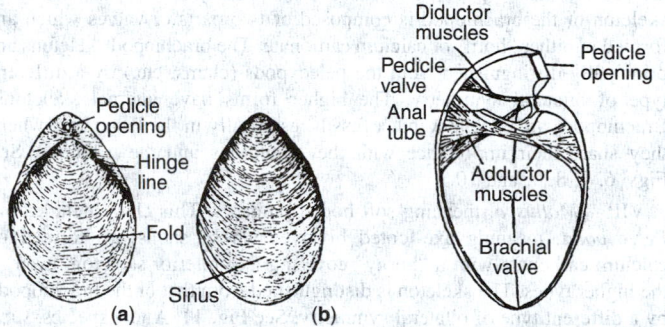

Fig. 10.   Views at left are of the *Brachiopod*, "Lamp shell," showing front view (**a**); rear view (pedicle valve) (**b**). At right is cross section of *Brachiopod* (Lamp shell), showing musculature. (*Field, "Geology Manual," Part II, Princeton Univ. Press.*)

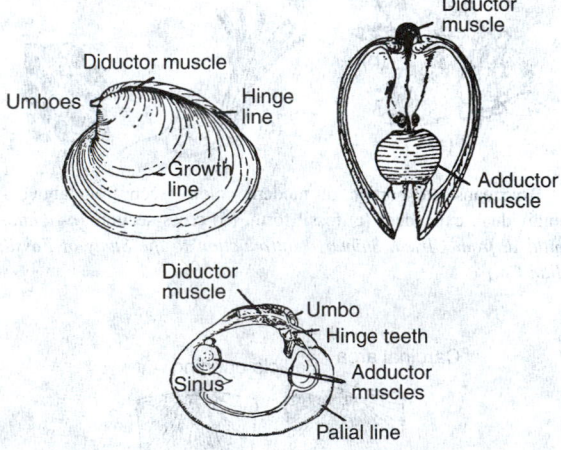

Fig. 11.   Diagram at left is of a quahog, modern pelecypod. At right is shown cross section, normal to hinge line of pelecypod, showing musculature. Lower diagram is an interior view of left valve of pelecypod, showing muscle scars. (*Field, "Geology Manual," Part II, Princeton Univ. Press.*)

IX.   *Gastropoda*, or snails, meaning stomach-footed. *Mollusca* with single unchambered shells composed of calcium carbonate. All shapes and types of ornamentation, frequently well-preserved. Gastropods are only important as index fossils in the Canadian, and at certain horizons in the Mesozoic and Cenozoic.

X.   *Cephalopoda*. Meaning head-footed. Existing forms are nautilus, cuttlefish, octopus, etc. Shells composed either of calcium carbonate or horn, and either external or internal in relation to the living animal. The shell differs from that of the gastropods because the interior is divided into a number of compartments by means of platforms or septa, the animal living only in the outer compartment. The cephalopods are naturally divisible into the following groups, which, because of their anatomical and stratigraphical history, make the cephalopods one of the best-known paleontological examples of adaptive change. The forms with external shells are as follows. (1) Nautiloids. See Fig. 12. Straight to coiled forms with smooth septa. Important as index fossils from the Cambrian to the Devonian, inclusive. (2) Goniates. Coiled forms with septa wrinkled into saddles and lobes. Range from the Silurian to the Permian. Important index fossils in the Devonian. (3) Ceratites. Similar to the goniatites except that the saddles are smooth and the lobes are wrinkled or crenulated. See Fig. 13. Range from the Devonian to the Jurassic. Important index fossils in the Triassic. (4) Ammonites. Similar to the ceratites except that both the saddles and lobes are highly crenulated. Range from the Upper Pennsylvanian to the close of the Cretaceous. Important index fossils. The forms with interior skeletons are divided into the squids, cuttlefishes, and belemnites. Only the latter are important as fossils, ranging from the Triassic to the Cretaceous, inclusive.

XI.   *Arthropoda*. Meaning joint-footed. Transversely segmented animals with mouth and anus at opposite ends of an elongated body that is composed of the following fairly well-defined regions: "head" or cephalon, thorax or pleura, and pygidium. A few or most of the segments bear paired appendages. Arthropods range from the Cambrian to the present, and are both fresh water and marine. The three most important subdivisions of this group, from the paleontological point of view, are the Trilobites, Ostracods, and Eurypterids. The first insects appear as fossils in the Carboniferous and several of the ancestral types of the older order appear in the Permian. See Fig. 14. The fossil insects of the Tertiary are particularly interesting as they prove that social life in the insect world began as long ago as the Oligocene.

From the Paleontological point of view, however, the trilobites are the most interesting. These comprise an extinct group of arthropods, or transversely segmented invertebrates, with mouth and anus at the opposite ends of an elongated body made up of a variable number of segments, each of which bears a pair of appendages. The term trilobite means "three-lobed," referring to the bilateral symmetry. The major anatomical features of the external skeleton are shown in Fig. 15. Although one of the highest orders of the invertebrates, trilobites occur among the oldest known fossils of the Cambrian period, becoming extinct at the close of the Paleozoic

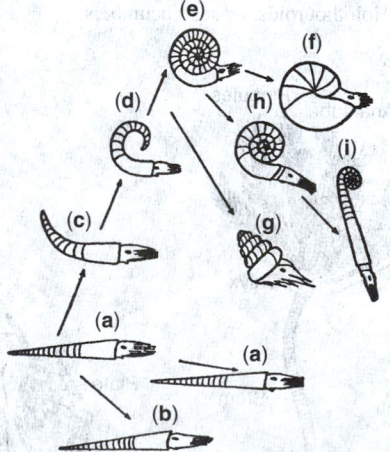

Fig. 12.   Phased adaptations of the *Nautiloids*: (**a**) Straight form, poor swimmers; (**b**) first adaptation to bottom habitat, straight, slightly flattened form; (**c**) slightly coiled; (**d**) partly coiled; (**e**) fully coiled; (**f**) coiled and involuted; (**g**) second adaptation to bottom habitat, coiled and twisted; (**h–i**) decadent (gerontic) stages, partly uncoiled. *Note*: The *Ammonoids* also passed through a somewhat similar series of adaptive phases. (*Dunbar, in Thorpe, "Organic Adaptation to Environment," Yale Univ. Press, 1924.*)

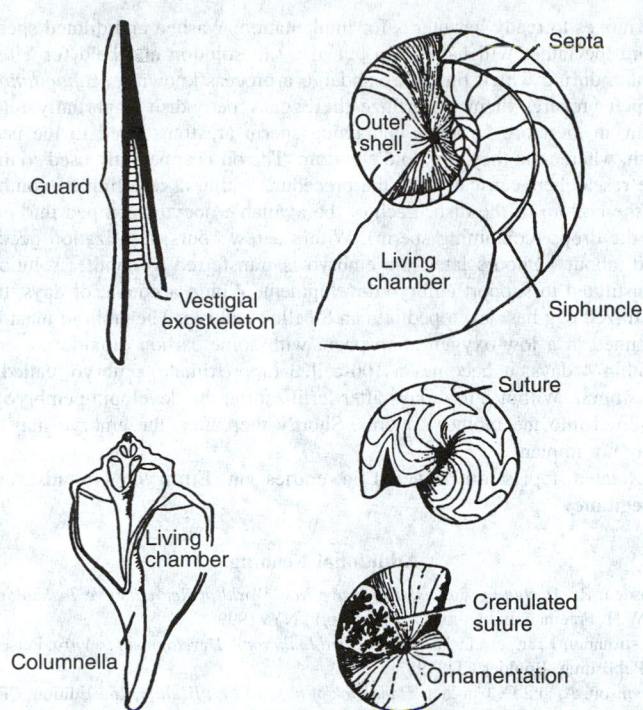

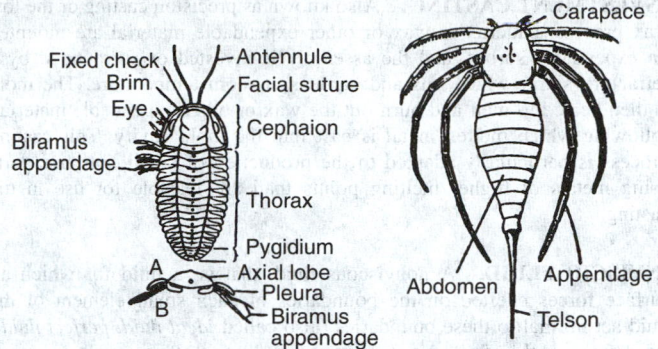

Fig. 13. *Belemnite* shown at top left; *Nautilus*, with outer shell broken away to show septa and siphuncle shown at top right. View at right-center is of *Goniatite*, showing sutures with simple saddles and lobes. *Gastropid*, "conch shell," cut away to show interior whorls is shown in lower-left. *Ammonite*, with exterior shell broken away to show dendritic form of sutures, shown at lower-right. (*Field, "Geology Manual," Part II, Princeton Univ. Press.*)

Fig. 15. Diagram at left is of the *Trilobite*: (A) dorsal view; (B) cross-section of thorax. (*Field, "Outline," Barnes & Noble.*) Diagram at right is of the *Eurypterid.* (*Schubert, "Textbook of Geology," Wiley.*)

Eurypterids, meaning head-winged, are extinct marine or estuarine scorpion-like anthropods, related to the horseshoe crab, with elongated bodies, and appendages attached to the head region only. They differ from the trilobites also in that the former always have a regular number of appendages. The Eurypterids range from the Cambrian to the Permian, and are important index fossils in certain horizons of the Silurian and Devonian.

Additional Reading

Clarkson, E.N.: *Invertebrate Paleontology and Evolution*, Chapman & Hall, New York, NY, 1993.

Doyle, P.: *Understanding Fossils: An Introduction to Invertebrate Palaeontology*, John Wiley & Sons, Inc., New York, NY, 1996.

Enay, R. and T. Reimer: *Paleontology of Invertebrates*, Springer-Verlag New York, Inc., New York, NY, 1993.

Euan, C.: *Invertebrate Paleontology and Evolution*, 4th Edition, Blackwell Science, Inc., Malden, MA, 1998.

Levin, H.L.: *Ancient Invertebrates and Their Living Relatives*, Prentice-Hall, Inc., Upper Saddle River, NJ, 1998.

Kaesler, R.: *Treatise on Invertebrate Paleontology*, Geological Society of America, Inc., Boulder, CO, 1992.

INVERTER. A device for converting direct current into alternating current. The term is commonly used to designate a circuit for performing this function that employs either transistors or gas-filled electron tubes. See Fig. 1. The inverter offers the possibility of generating power as alternating current, then stepping it up to the desired transmission voltage, rectifying it with high-efficiency rectifiers, transmitting as high-voltage direct current with certain advantages, inverting it to alternating current at the receiving end, and stepping down to the normal distribution voltage by using transformers. Inverters using transistors operating from low-voltage batteries are used to power various types of electronic equipment in which high-voltage, low-current supplies are needed.

INVERT SUGAR. See **Sweeteners**.

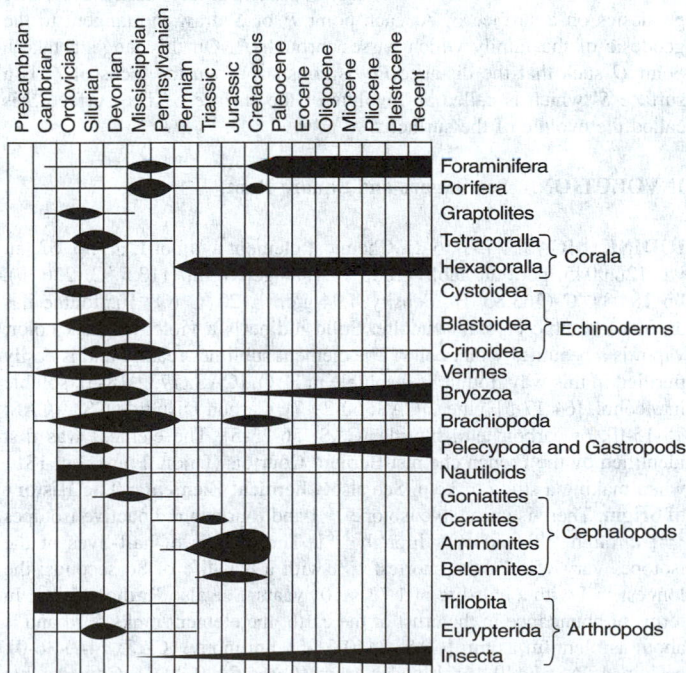

Fig. 14. Geologic range of the invertebrates. Note that the "lifetimes" are swelled during the periods in which there were the greatest number of genera and species, or when the particular class of organism is most important as an "index fossil." (*Field, "Geology Manual," Part II, Princeton Univ. Press.*)

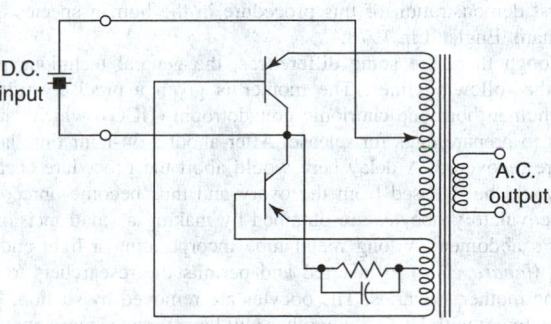

Fig. 1. Transistor inverter.

Era. Trilobites are important as index fossils, especially in the Cambrian, Ordovican, Silurian, and Devonian periods, and are of great aid to the geologist in helping to determine the relative ages of the oldest fossiliferous formations of the geologic time-scale.

INVESTMENT CASTING.   Also known as precision casting or the lost wax process, patterns of wax or other expendable material are mounted on expendable sprues, and the assembly is invested or surrounded by a refractory slurry which sets and hardens at room temperature. The mold is then heated to melt and burn out the wax or other expendable material, following which molten metal is cast into the mold cavity. This casting process is particularly adapted to the production of small, intricate parts using metals of higher melting points than are feasible for use in die casting.

INVISCID FLUID.   A nonviscous fluid, that is, a fluid for which all surface forces exerted on the boundaries of each small element of the fluid act normal to these boundaries (also called *ideal fluid*, *perfect fluid*). By definition, therefore, the stress tensor reduces to the pressure, a point-function scalar in the fluid. Thus, in the dynamics of an inviscid fluid, as opposed to a real viscous fluid, 1) no restraints are placed on the tangential component of the flow at a solid bounding surface, and 2) there is no dissipation of kinetic into thermal energy within the fluid. In the free atmosphere the flow is often treated as inviscid, and the viscous forces may be neglected for many purposes. Where an inviscid fluid flows along a surface, that surface is said to be a free slip surface.

IN VITRO.   An event or process occurring outside a living organism — in an unnatural environment, as in a test tube.

IN-VITRO FERTILIZATION.   As early as 1893, research was conducted on marine animals to determine if it was possible to fertilize the egg of a species with the sperm of that species in an external environment in a laboratory (*in vitro*, literally, "in the glass," i.e., the laboratory vessel). Early research was hampered by a lack of understanding of the complex activities involved in fertilization, particularly of a hormonal nature. See also **Hormones**. Mainly in the latter half of this century, aided materially by new instruments and micro techniques, a much better understanding of the total process of natural fertilization has been achieved — although there remain many more details to be learned. Improved physiological and biochemical understanding made it possible to demonstrate the external fertilization of rabbit eggs in the 1950s. This was rapidly followed by examples in several other mammalian species. Some of the knowledge applicable to experiments with human eggs and sperm stems from research on both fertility and birth control drugs.

It was ultimately established that, in the human female, the reproductive cycle is commenced by a "releasing factor" secreted by the hypothalamus (located at the base of the brain). This substance stimulates the nearby anterior pituitary gland to release two hormones into the circulation — follicle-stimulating (FSH) hormone and leutenizing hormone (LH). The follicles located in the ovary are stimulated to grow by FSH. The follicles, in turn, produce estrogens, notably estradiol. Ovulation and transformation of the follicle to form a *corpus luteum* is triggered by LH. In turn, the corpus luteum secretes progesterone. Circulation of the estrogens and progesterone to the uterus ready the wall of the uterus for implantation of the egg. There is a feedback control system to establish the optimal rate of secretions from the hypothalamus and the pituitary. With a staged system this complex, it is obvious that there is considerable room for error when substituting an artificial environment for the natural environment.

In-vitro fertilization requires an appropriate replica of these conditions in the laboratory. Although simple in concept, it is a complex procedure. The first demonstration of this procedure in the human species occurred in Oldham, England in 1978.

Although there are some differences, the general technique proceeds along the following lines. The mother is given a precisely timed dose of the human hormone chorionic gonadotropin (HCG), which causes her ovaries to prepare eggs for release. After about a 34-hour time lapse, the eggs are recovered. A delay here would abort the procedure because the eggs would be released from the ovary and thus become unrecoverable. The preovulatory oocytes are obtained by making a small incision in the mother's abdomen. A long metal tube incorporating a light and optical system (*laparascope*) in inserted and permits the researchers to directly view the mother's ovaries. The oocytes are removed by suction. Usually, after treatment with HCG, the mother will have three or more oocytes. Just prior to removal of the oocytes, the mother may be treated with Additional

hormones to ready her uterus for implantation. Washed and diluted sperm from the father will have been put in a salt solution and held for a few hours, during which time they undergo a process known as *capacitation*, which prepares them to fertilize the egg. A petri dish is partially filled with an inert oil. Droplets containing sperm are transferred to the petri dish, whereupon they sink to the bottom. The oil is apparently used so that the researchers can carry out the procedure within a very limited volume at the bottom of the dish. Each of the available oocytes is piped into one of the drops (containing sperm). Within a few hours, fertilization occurs and, about 12 hours later, the embryo is transferred to another solution, constituted to support embryo development. Within a couple of days, the fertilized egg has developed into an 8-celled embryo. The embryo must be retained in a low-oxygen atmosphere with some carbon dioxide present. Within 4 days, it becomes a 100-celled (approximate) embryo, called a *blastocyst*. Within 2 to 4 days after fertilization, the developing embryo is inserted into the mother's uterus. Shortly thereafter, the embryo may or may not implant.

Related topics are covered in entries on **Embryo**; **Gonads**; and **Pregnancy**.

Additional Reading

Gosden, R.: *Designing Babies: The Brave New World of Reproductive Technology*, W.H. Freeman and Company, New York, NY, 1999.
Mastroianni, L. Jr., et al.: *Fertilization and Embryonic Development in Vitro*, Perseus Publishing, Boulder CO, 1981.
Trounson, A. and D. Gardner: *Handbook of In Vitro Fertilization*, 2nd Edition, CRC Press, LLC, Boca Raton, FL, 1999.

IN VIVO.   An event or process occurring naturally or spontaneously within a living organism.

INVOLUTE.   With reference to a curve, if the tangents to the curve C are normals to the curve C', then C' is an involute of C and C is an evolute of C'. With reference to a surface, take a singly-infinite system of geodesics on a surface S. At each point P or S draw the tangent to the geodesic of the family which passes through P. On this tangent, take a point Q such that the distance PQ is constant. Then the locus of Q is a surface S' which is called an involute of the surface S. The surface S is called the evolute of the surface S'.

INVOLUTION.   See **Square and Square Root**.

IODINE.   [CAS: 7553-56-2]. Chemical element symbol I, at. no. 53, at. wt. 126.9045, periodic table group 17 (halogens), mp 113.5 °C (236 °F), bp 184.35 °C (363.83 °F), density 4.94 g/cm^3 (20 °C (68 °F). Iodine has an orthorhombic crystal structure. Solid iodine is a violet-to-black color; vapor is a beautiful violet color. The element sublimes readily and is easily purified in this way. Iodine is insoluble in H_2O, [CAS: 7732-18-5], soluble in alcohol, [64-17-5], ether, [CAS: 60-29-7], Carbon Disulfide CS_2, [CAS: 75-15-0], or carbon tetrachloride [CAS: 56-23-5]. The element was first identified by the French chemist Bernard Courtois (Dijon, France) in 1811 when making a study of kelp. See also **Chemical Elements: The History of origin**. There is one stable isotope ^{127}I and fourteen radioactive isotopes ^{122}I through ^{126}I and ^{128}I through ^{136}I. The lengths of half-lives of the isotopes vary widely, the shortest ^{136}I with a half-life of 86 seconds; the longest ^{129}I with a half-life of 1.72×10^7 years. See also **Radioactivity**. In terms of abundance in the crust of the earth, the element ranks 53rd and is about as plentiful as tin, [CAS: 7440-31-5], antimony, [CAS: 7440-36-0], cesium, [CAS: 7440-46-2], and barium [CAS: 7440-39-3]. Considerable quantities of iodine have concentrated in the oceans. The average iodine content of a cubic mile of seawater is 230 tons (50 metric tons per cubic kilometer).

First ionization potential 10.44 eV; second, 19.4 eV. Oxidation potentials $I^- \rightarrow \frac{1}{2}I_2 + e^-$, -0.535 V; $I^- + H_2O \rightarrow HIO + H^+ + 2e^-$, -0.99 V; $I^- + 3H_2O \rightarrow IO_3^- + 6H^+ + 6e^-$, -1.085 V; $\frac{1}{2}I_2 + 3H_2O \rightarrow IO_3^- + 6H^+ + 5e^-$, -1.195 V; $\frac{1}{2}I_2 + H_2O \rightarrow HIO + H^+ + e^-$, -1.45 V; $IO_3^- + 3H_2O \rightarrow H_5IO_6 + H^+ + 2e^-$, ca. -1.7 V; $I^- + 6OH^- \rightarrow IO_3^- + 3H_2O + 6e^-$, -0.26 V; $I^- + 2OH^- \rightarrow IO^- + H_2O + 2e^-$, -0.49 V; $IO + 4OH^- \rightarrow IO_3^- + 2H_2O + 4e^-$, -0.56 V; $IO_3^- +$

$3OH^- \rightarrow H_3IO_6^{2-} + 2e^-$, ca. -0.70 V. Other important physical characteristics of iodine are given under **Chemical Elements**.

Sea plants, particularly kelp found in the waters around California and the Bay of Biscay, have been a source of iodine. Because of pollution, the kelp beds in California are no longer a major source. Iodine also is found in the petroleum oil well brine of California and, in small percentages, in sodium nitrate [CAS: 7631-99-4] of Chile. The latter was once the primary source of the element. Brines now are the major source.

Uses

For many years, iodine tincture (3% to 7% dissolved in ethyl alcohol) has been an important antiseptic. The commercial tinctures also usually contain 5% potassium iodide to provide stability. This form produces a mild burning of the skin and stains both skin and fabrics. A milder preparation is available in which about 2% iodine is contained in an oil-water emulsion that also contains lecithin. The burning effect of the compound is greatly reduced and the mild stains produced usually wash off easily. There are a number of prepared medicines that contain iodine, although some of these have been removed from the market in recent years. At one time, iodoform CHI_3, [CAS: 75-47-8], a yellow, insoluble, crystalline powder with a very penetrating odor, was a very popular antiseptic and used widely in the preparation of gauzes and packings for infected cavities. Because of possible toxic effects with some individuals, the compound largely has been replaced by other less objectionable, less odorous materials.

The medical use of iodine compounds, particularly organo-iodine substances, has been researched with resultant limited applications. The dietary requirement for iodine was established many years ago as needed for the maintenance of cell growth in humans and animals. The largest concentration of iodine occurs in the thyroid gland where the hormone thyroxine $C_{15}H_{11}O_4I_4N$ [CAS: 300-30-1] is present. The waters and soils of some areas, as in the inland areas of the United States, do not contain the minimal trace quantities of iodine required by the normal diet. Thus for many years so-called iodized table salt with a content of about 0.01% potassium iodide has been available. This preventive practice has been accredited with forestalling goiter and associated glandular disturbances in untold thousands of instances. For health reasons, iodine supplements also are added to cattle feeds.

Iodine tablets provide an easy means for sterilizing drinking water in small portions, usually resulting in less odor and taste objections than chlorine compounds for the same purpose. Iodine chemicals are widely used in photography and printing reproduction processes. See also **Photography and Imagery**.

The production of vanadium [CAS: 7440-62-2] metal essentially is the calcium [CAS: 7440-70-2] reduction of vanadium pentoxide [CAS: 1314-62-1] in the presence of iodine and is known as the McKechnie-Seybolt process. The reaction is carried out in a steel bomb at about 700°C (1,292°F). The end products are vanadium metal, lime, [CAS: 1305-78-8] and calcium iodide [CAS: 10102-68-8]. A similar iodide process also is used in the production of high-purity zirconium [CAS: 7440-67-7].

Chemistry and Compounds

Iodine exhibits in common with the other halogens a marked readiness to form singly charged negative ions, as would be expected from the fact that these atoms need only one electron to acquire an inert gas configuration. However, of the four common halogens, iodine has the lowest electron affinity (3.2 eV) due to more effective screening of the nucleus. The iodides range in character from ionic to covalent compounds, many of them, such as hydrogen iodide, [CAS: 10034-85-2], having bonds of intermediate nature. Iodine is the most electropositive of the common halogens, functioning as the positive univalent "ion" I^+, as in the compound iodine perchlorate, which, however, is not a salt, as well as in forming trivalent complex radicals, such as IO^+. Iodine also forms essentially covalent linkages with negative elements, in which it has positive valences 1, 3, 5, or 7.

In binary combination with oxygen, however, only one simple compound has been isolated, iodine pentoxide, I_2O_5 [CAS: 12029-98-0], a white compound. The yellow I_2O_4, prepared from sulfuric and iodic acids, is considered to be made up of 3+ and 5+ iodine, and the structure iodyl iodate, $(IO^+)(IO_3^-)$, is assigned to it, in which the trivalent state is stabilized by the acid radical. Similarly, yellow tetraiodine enneaoxide, I_4O_9, is considered to have the structure $(I^{3+})(IO_3^-)_3$, in which again the trivalent state is stabilized by the acid radicals.

Hydrogen iodide, HI, is the least stable of the four common hydrogen-halides, and correlatively, the best reducing agent, readily reducing vanadic acid, nitrous oxide to ammonium, nitrous acid to nitric oxide, and HNO_3 to nitrous acid. Because it is so readily oxidized, it cannot be prepared by action of H_2SO_4 on an iodide, but can be made by the action of weak acids, e.g., H_2S, upon iodine, or by hydrolysis of certain iodides. It can be prepared by direct combination of hydrogen and iodine vapor on a platinum catalyst. It is also liberated in many organic iodination reactions, such as the reaction of iodine and refluxing tetralin. The H–I bond is considered to be partly covalent. HI is a monoprotic acid, and is stronger than hydrogen chloride [CAS: 7647-01-0] and hydrogen bromide [CAS: 10035010-6].

The oxyacids of iodine are essentially covalent compounds, the known acids being hypoiodous acid, HIO [CAS: 14332-21-9], iodic acid, HIO_3 [CAS: 7782-68-5], and the various periodic acids. Hypoiodous acid is formed, along with iodide ion, by dissolving iodine in dilute alkali, or by action of mercuric oxide upon iodine and water. On standing hypoiodous acid disproportionates to iodic acid and hydriodic acid. Hypoiodous acid is a powerful oxidant. It is also amphiprotic: $H^+ + IO^- \leftrightarrow HIO \leftrightarrow I^+ + OH^-$, $pK_A = 10.4$, $pK_B = 9.49$. The evidence for the formation of the I^+ ion is the existence of a number of compounds of composition $I(r)_nX$, where r is pyridine or some other nitrogen organic base, n is 1 or 2, and X is hydroxyl, nitrate, or chlorate ion. The conductivity of liquid iodine also indicates ionization into solvated I^+ and I^-.

Hydriodic acid (HI) [CAS: 10034-85-2] is a colorless solution formed when hydrogen iodide gas is dissolved in water, commercially of strength 10% HI, frequently colored brown by iodine. There is a maximum constant boiling point 127°C (774 mm) at 57% HI (distillate) for mixtures of hydriodic acid and water. Hydriodic acid is used in the preparation of iodides, and as an important reagent in organic chemistry.

All metallic iodides except silver iodide, [CAS: 7783-96-2], mercurous iodide, [CAS: 7783-30-4], mercuric iodide, [CAS: 7774-29-0], lead iodide, [CAS: 13779-93-6], cuprous iodide, [CAS: 7681-65-4], thallium iodide, [CAS: 7790-30-9], and palladium iodide, [CAS: 7790-38-7], are soluble. The iodides of antimony, bismuth, tin require a little free acid to keep them in solution.

Dilute hydriodic acid reacts with hydroxides, oxides, carbonates, sulfides, metals in a manner chemically analogous to dilute HCl; with solutions of some salts, e.g., silver nitrate, to yield the corresponding iodide, e.g., silver iodide, precipitate. Higher strengths of hydriodic acid react with oxygen of the air upon standing to yield free iodine, which imparts a brown color to the solution, thus indicating the reducing character of the acid.

Hydriodic acid is made by the reaction (1) of iodine and hydrosulfuric acid [CAS: 7783-06-4] (or sulfurous acid), (2) of phosphorus plus iodine plus water, with subsequent distillation in all cases.

Iodic acid, HIO_3, is commonly prepared by oxidizing iodine with HNO_3. It is a strongly oxidizing acid, oxidizing iodide to iodine, sulfite to sulfate, and H_2S to sulfur. It reacts vigorously even with dry carbon, phosphorus, or organic matter.

Iodate ion in aqueous solution appears to be $I(OH)_6^-$, and the iodine atom in crystalline periodates always is coordinated to six oxygen atoms, three nearest neighbors on one side and three next nearest neighbors on the other side. Thus, potassium iodate, KIO_3, has distorted perovkite structure.

In the broad picture, the iodides, like the other halides, range in character from completely ionic structures to covalent ones. The transition is clearly exhibited by the iodides of the first four groups in the (extended) periodic table, potassium iodide, KI [CAS: 7681-11-0], being ionic and titanium (IV) tetraiodide, TiI_4, essentially covalent. On the right-hand side of the periodic table, even the group I elements form bonds with iodine that exhibit a considerable degree of covalence. In general, the ionic iodides are the most soluble of the halides of the given element, e.g., sodium iodide, NaI, is the most soluble sodium halide, while the covalent (or partly covalent) iodides are the least soluble, e.g., silver iodide, AgI. These effects are correlated with the size and polarizing power of the cation and the increasing size and polarizability of the halogen ion, as is increase of reducing character and stability of coordination complexes. A related fact is the readiness of formation of the complex ion, I_3^-, upon dissolving iodine in an aqueous solution of KI. The variation in character of iodine solutions in various organic solvents as well as water, is also attributed to complex formation.

Because iodine is the least electronegative of the four common halogens, it forms a relatively larger number of compounds with the other three. They include iodine trichloride, iodine chloride, iodine bromide, iodine fluoride, iodine trifluoride, iodine pentafluoride, and iodine heptafluoride. These are reactive compounds, especially the lower ones, which enter into reactions with many organic and some inorganic substances. With ICl and organic substances, the end product is usually an iodine or chlorine substitution product, the solvent being important in determining which is formed. With inorganic compounds, the addition product is often the final product, thus ICl and antimony (V) pentaiodide $SbCl_5$ give a product of composition $ISbCl_6$, which ionizes to I^+ and $SbCl_6^-$.

Several oxyfluorides exist: IOF_3 (iodyl hexafluoridate, $[IO_2][IF_6]$), iodyl fluoride, IO_2F, and periodyl fluoride, IO_3F.

Related to the complex ions, as well as the interhalogen molecules, are their association products, the polyhalide complexes. For iodine they include the alkali (and ammonium) triiodides, along with higher compounds such as that with cesium, box Cs_2I_8, ammonium, NH_4I_5, tetraethylammonium iodide, $(C_2H_5)_4NI_7$, tetramethylammonium iodide, $(CH_3)_4NI_9$, and benzene, $KI_9 \cdot 3C_6H_6$. They also include compounds of iodine with alkali metals and one or two other halogens, such as NH_4IBr_2, $KICl_2$, $RbICl_2$, $HICl_4 \cdot 4H_2O$, $CsFIBr$, $RbClIBr$, $CsClIBr$ and $KIF6$. Most of these compounds hydrolyze readily, decompose on heating to give the metal halide of greatest lattice energy, and ionize to give ions such as $ClICl^-$, $BrIBr^-$, and $BrII^-$.

Structural studies show that the complex $HICl_4 \cdot 4H_2O$ contains a positive trivalent iodine atom, and the planar ion ICl_4^- is one of the most stable of the polyhalide anions, as might be expected from consideration of the large size of the iodine atom and the small size of the chlorine atoms.

The difluoroiodate ion. $IO_2F_2^-$, is a trigonal bipyramid with two apical fluorine atoms, two equatorial oxygen atoms and one equatorial nonbonding electron pair surrounding the central iodine atom.

Iodine forms many organic compounds, chiefly by replacement of hydrogen or by addition reactions at double bonds.

Periodic acid H_5IO_6 has been isolated as a colorless solid, mp about $130\,°C$ ($266\,°F$), and at $138\,°C$ ($280\,°F$) begins to decompose, metaperiodic acid HIO_4 being formed and at higher temperatures iodine pentoxide plus oxygen plus water. $H_4I_2O_9$ and H_3IO_5 have been reported as fairly well established in identity; in solution the evidence points to the presence of HIO_4. Prepared by reaction of iodine and perchloric acid.

Paraperiodic acid H_5IO_6, [CAS: 10450-60-9], is obtained from sodium paraperiodate, formed by action of chlorine upon a NaOH solution containing I_2. On vacuum drying, paraperiodic acid yields metaperiodic acid, HIO_4, and dimesoperiodic acid, $H_4I_2O_9$; which form heteropoly acids with a number of oxides and acids. The periodic acids and their salts are strong oxidizing agents both with inorganic and organic compounds. Although, in general, the iodates do not disproportionate to give periodates, barium iodate, $Ba(IO_3)_2$, on ignition gives barium paraperiodate, $Ba_5(IO_6)_2$, iodine and oxygen.

Sodium periodate $Na_2H_3IO_6$ [CAS: 7790-28-5], is formed by reaction of sodium iodate plus sodium hydroxide plus chlorine (sodium chloride also formed), and the periodate separates as crystals from the medium. In solution, it is stated, periodate gradually forms ozone and iodate at the ordinary temperatures.

Metallic periodates are solids, slightly soluble in water. Periodates, when heated, evolve oxygen with simultaneous formation of iodate, which is decomposed at higher temperatures. Periodate in acid solution oxidizes hydrosulfuric acid or sulfurous acid to H_2SO_4, oxalic acid to CO_2, manganous to manganate, and with hydrogen peroxide yields oxygen and iodate.

The biological aspects of iodine are covered in **Iodine (In Biological Systems)**.

Additional Reading

UNIPUB: *Control of Iodine in the Nuclear Industry*, Bernan Associates, Lantham, MD, 1973.

Varvoglis, A.: *The Organic Chemistry of Polycoordinated Iodine*, John Wiley & Sons, Inc., New York, NY, 1992.

Varvoglis, A.: *Hypervalent Iodine in Organic Synthesis*, Morgan Kaufmann Publishers, San Mateo, CA, 1996.

Wirth, T.: *Hypervalent Iodine Chemistry: Modern Developments in Organic Synthesis*, Springer-Verlag New York, LLC, New York, NY, 2003.

IODINE (In Biological Systems). Iodine is not required by plants, but if iodine is present in the soil, it is taken up by most plants and moves on into the diets in forms that are effective in preventing goiter. In areas where the soils are high in iodine, ground water is also high in iodine, but the food supply is still the major source of iodine for people in these areas. Seafoods are good sources of dietary iodine.

Gout is no longer the single target of iodine deficiency concerns. The deficiency is implicated in mental retardation, deaf mutism, short stature, and an increased risk of death during childhood. Iodine deficiency is known to affect development of the central nervous system, particularly during the growing years. Intellectual impairment may range from a mild disorder to one of *cretinism* (significant and abnormal intellectual disturbance). Thus, there is a marked trend today among professionals to group all consequences of iodine deficiency under the category "iodine deficiency disorders."

Uneven Distribution of Iodine. Many of the iodine-deficient regions of the world have been identified. They are generally either mountainous or in the centers of continents, and distant from the oceans in the prevailing wind directions. Studies of the geochemistry of iodine indicate that this element is volatilized from oceans, carried overland by winds, and deposited on the soil by rain. The mountainous areas are low in iodine because little of that volatilized from the seas reaches sufficient altitude to be deposited in high altitudes. In some areas, the younger soils have less iodine than the older ones because of less time for the geochemical processes to build up the iodine level.

Although the amount of iodine in the soil is the primary factor determining iodine levels in food crops from various regions, the level of iodine in plants and the dietary requirements for iodine are modified to some extent by the plants themselves. There are important differences among plant species, and even among varieties of the same species, in their tendency to take up iodine from the soil. Certain plants, especially some of the *Brassica* genus, such as cabbage, contain compounds called *goitrogens*, which interfere with the effect of iodine on the thyroid gland. The amount of iodine required to protect animals, including humans, against goiter and other important iodine-deficiency disorders depends not only upon the kinds of plants in the diet, but also upon the iodine level characteristic of the soils in which the plants are cultivated. Iodine could be increased by adding iodine compounds to the soil, but that is a very inefficient way of insuring adequate dietary levels of the element. Much of the iodine added to the soil would be leached out and returned to the seas before it could be taken up by the crop plants. The use of iodized table salt is such an effective way of supplying this element that there is little need to include iodine in fertilizers.

A more recent method is that of using iodized oil, which has been found to be particularly effective in underdeveloped areas, such as several African countries.

Several comparative regional studies of iodine deficiency have been made. For example, one study shows that 1 of every 12 newborn infants in Freiberg, Germany, had an elevated serum thyrotropin concentration as contrasted with 1 of 1,428 infants born in Stockholm, Sweden, during an equivalent time frame. It has been found that iodine dietary supplementation reduced the incidence of congenital deafness in Switzerland by half. In countries that extend over broad and varied geographical regions, such as the United States, China, and Russia, the iodine content of soil can vary widely from one geographical community to the next. Dietary supplementation of iodine has increased the survival and birth weight of newborns in Zaire, prevented cretinism and decreased the rate of childhood mortality in Papua New Guinea, and advanced educability and economic productivity in China. Inasmuch as the facts leading to past successes have been well established, a meeting of the World Summit for Children in 1990 pledged to eliminate iodine deficiency by the year 2000. The Summit comprised responsible persons from the International Council for the Control of Iodine Deficiency Disorders, the United Nations Children's Fund, and the

World Health Organization. Dunn (see reference listed) makes the interesting observation that, in 1923, David Marine wrote that "simple goiter is the easiest of all known diseases to prevent."

Dunn observes, "One intramuscular injection of iodinated vegetable oil containing 480 milligrams of iodine provides adequate amounts of iodine for up to three years. Oral administration of iodized oil is appealing for its simplicity and safety and millions of doses have been given worldwide. Still, the optimal dose and duration of effect have not been fully defined." Tonglet (see reference) and others have reported that as little as 48 milligrams of iodine administered orally as iodized oil is sufficient supplementation for 6 months. See also **Vegetable Oils (Edible)**.

One must be careful in dose determination because excessive doses of iodine can produce harmful effects.

Biochemistry of Iodine. In historical terms, 3,5-diiodotyrosine (iodogorgoic acid) had been discovered in sponges and corals long before there was any hint concerning thyroid hormone structure even in mammalian forms. Since this time, iodotyrosines have been demonstrated in algae as well as in many animals possessing a horny skeleton. It has been suggested that iodide in the water is activated by peroxidases due to the presence of oxygen in the water and the resulting iodine is accepted by tyrosine in the protein molecule, similarly to the process in vertebrates. In animals possessing an exoskeleton, the presence of benzoquinones is part of the formation of scleroproteins by a quinone tanning process, and the iodination of tyrosine may be related to this in some way. In rotochordates, the endostyle, a structure secreting mucus, has been found capable of iodination of protein present in the mucus, which is then secreted into the alimentary canal.

The next phylogenetic development takes place in the vertebrates, in the most primitive members of which there is a structure located in the hypopharynx similar to a thyroid capable of collecting iodine and of forming iodinated protein, which is broken down by a protease, liberating iodinated amino acids. In some of these forms, small amounts of thyroxine are actually found in addition to the iodinated tyrosines. In other vertebrates, culminating in the amphibia and higher vertebrates, a thyroid gland is present in which the iodinated protein is held in a storage form known as colloid. In some, but not all, of these forms, hormonal material liberated by action of proteolytic enzymes is secreted into the bloodstream and plays an essential role in the development of the young animal as well as in the behavior and metabolic activity of the adult. Because of the process known as ontogeny, it is not surprising that the development of the thyroid gland in the human embryo commences with a structure located near the alimentary canal, which then separates and develops its peculiar follicular structure. Not until this type of development has occurred can a genuine function for the iodinated substances be demonstrated.

In these discrete glands, a series of specific chemical reactions can be demonstrated. See Fig. 1. Some of these reactions take place in the absence of any apparently specific synthesis of iodotyrosines, and it has not been fully explained how many of these steps require enzymes, even in mammalian forms, although these processes are usually considered as enzymatic. It is possible to iodinate tyrosine in soluble proteins in vitro by addition of elemental iodine and under these circumstances thyroxine and triiodothyroxine will also be formed, in company with small amounts of iodohistidine.

In amphibian vertebrates undergoing metamorphosis, thyroxine is known to be essential for this transition from an immature to a mature animal. There is considerable evidence that after metamorphosis has occurred, the thyroid gland is no longer essential, although it may be involved in seasonal changes, such as molting.

There is no such clear cut differentiation as metamorphosis in the mammal, but development is an extremely complex process and has been shown to depend upon the presence of adequate amounts of thyroid hormones. Deficient development, especially of the central nervous system, is marked in children suffering from thyroid deficiency early in life, and this inadequacy cannot be overcome completely by medication commenced after the first few weeks. In the adult, thyroxine is important in the maintenance of energy turnover in most of the tissues of the body, such as the heart, skeletal muscle, liver, and kidney. Other physiological functions, most notably brain activity and reproduction, are also dependent upon thyroxine, although the metabolic rates of the tissues concerned in these functions do not seem to be altered.

A great deal of work has been done on determining the portions of the thyroxine molecule essential to biological activity. The fact that the hormone is an amino acid is almost certainly due to the widespread existence of the excellent iodine acceptor, tyrosine. Thus, the deaminated and decarboxylated metabolic product of thyroxine, tetraiodothyroacetic acid, has been shown to have appreciable biological activity, although quantitatively less than that of thyroxine itself. As for the halogens present on the diphenyl ether portion of the molecule, bromines and chlorines are also active, although diminishing considerably in that order from iodine.

Assuming there are some quite definite structural requirements for thyroid hormone activity, it is important to inquire into the specific actions of this material. It seems clear that thyroxine does not participate directly in any enzyme system, but rather affects the function of many systems, presumably by some far more general process. One of the earliest of such demonstrated actions was the uncoupling of oxidation from formation of high-energy phosphate compounds, such as adenosine triphosphate. However, such uncoupling is also produced by many other substances not showing thyroid hormone-like effects, and it has been shown that thyroxine is actually capable of accelerating coupled reactions under the proper conditions. From this evidence, it is suggested that the principal role of thyroxine may be the acceleration of enzyme processes ordinarily limiting the level of metabolic turnover. Mitochondria isolated from broken

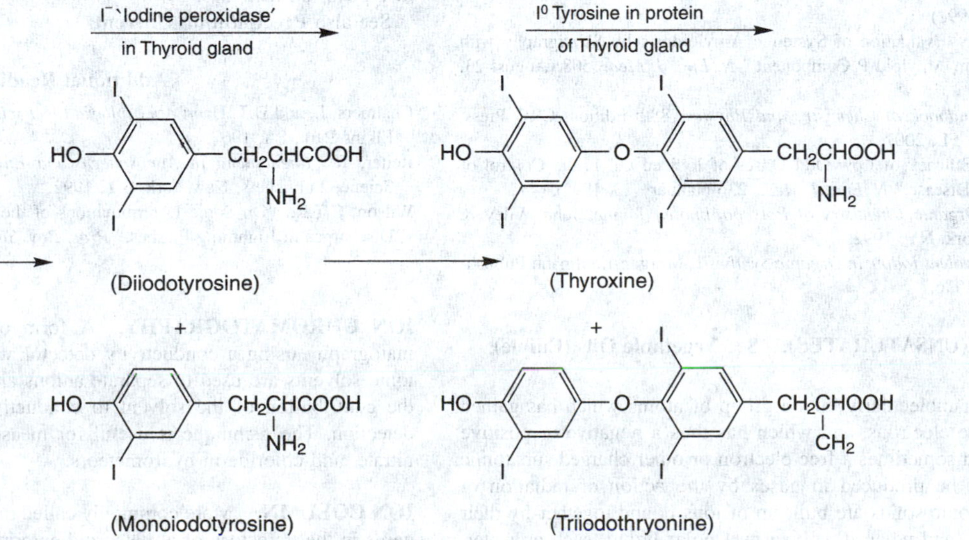

Fig. 1. Series of reactions occurring in thyroid system.

cell preparations by high-speed centrifugation have been shown to swell when placed in contact with thyroxine and similar substances. This may be evidence for a membrane function of the hormone, although it is not fully understood as to how this may alter cellular function in such specific manner as the hormone does in vivo.

An acceleration of protein turnover by thyroxine also has been shown, implying that the hormone may alter various processes by a specific effect on synthesis of certain key proteins involved in enzymatic reactions. Thus, not only does thyroxine increase the rate of formation of new protein material, but it also may be responsible for the transformation of non-enzymatically active protein into protein with enzymatic activity. The hormone has also been shown to be capable of acceleration of the synthesis of urea cycle enzymes and probably is essential for the production of a sodium ion transporting mechanism, both of which are essential in the metamorphic transformation of larval forms into mature amphibia.

Steps in the synthesis of thyroid hormone include the following. (1) Active concentration of inorganic iodides within the thyroid epithelial cells. Concentration achieved is approximately 30 times that of plasma concentration. The so-called trapping of iodide is stimulated by thyroid-stimulating hormone (released by the pituitary gland). When present, this step is competitively opposed by thiocyanate and perchlorate ions. (2) Next, the inorganic iodide is oxidized to an organic form, in which peroxidase participates. The iodine becomes part of tyrosine residues in the thyroglobulin molecule. In this way, monoiodotyrosine (MIT) and diiodotyrosine (DIT) are formed. (3) The MIT and DIT are coupled by way of an ether linkage to form tetraiodothyronine (T4) and triiodothyronine (T3).

It has been observed that non-iodinated thyronine is not found in the thyroid gland.

The storage capacity and slow release mechanism of thyroid hormone appear to be unique among the endocrine glands. Usually a reserve of 100 days' needs (about 80,000 micrograms) are stored in the gland. For diseases related to malfunction of the thyroid gland, see also **Endocrine System**; and **Thyroid Gland**. Sometimes iodides in drugs can cause a condition known as eosinophilia, in which there are reduced counts of eosinophil in the plasma. However, there are many other possible causes of this condition. Eosinophilic granulocytes are components of the blood in which the cytoplasm is filled with coarse acidophilic granules which may be spherical or rod-shaped, and the nucleus is bilobed and stains deeply.

Additional Reading

Braverman, L.E.: *Diseases of the Thyroid*, Vol. 2, Humana Press, Totowa, NJ, 1997.

Delange, F., J.T. Dunn and D. Glinoer: *Iodine Deficiency in Europe: A Continuing Concern*, Kluwer Academic Publishers, Norwell, MA, 1993.

Dunn, J.T. and F. van der Haar: *A Practical Guide to the Correction of Iodine Deficiency*, International Council for Control of Iodine Deficiency, Wageningen, the Netherlands, 1990.

Dunn, J.T.: "Iodine Deficiency — The Next Target for Elimination," *N. Eng. J. Med.*, 267 (January 23, 1992).

Hawkins, P.N., et al.: "Evaluation of Systemic Amyloidosis by Scintigraphy with 123I-Labeled Serum Amyloid P Component," *N. Eng. J. Med.*, 508 (August 23, 1990).

Lide, D.R.: *CRC Handbook of Chemistry and Physics*, 88th Edition, CRC Press, LLC., Boca Raton, FL, 2007.

Tonglet, R., et al.: "Efficacy of Low Oral Doses of Iodized Oil in the Control of Iodine Deficiency Disease," *N. Eng. J. Med.*, 236 (January 23, 1992).

Varvoglis, A.: *The Organic Chemistry of Polycoordinated Iodine*, John Wiley & Sons, Inc., New York, NY, 1992.

Varvoglis, A.: *Hypervalent Iodine in Organic Synthesis*, Morgan Kaufmann Publishers, Orlando, FL, 1996.

IODINE VALUE (UNSATURATES).   See **Vegetable Oils (Edible)**.

ION.   An atom or molecularly bound group of atoms which has gained or lost one or more electrons, and which has thus a negative or positive electric charge, and sometimes a free electron or other charged subatomic particle. Ions may be produced in gases by the action of radiation of sufficient energy; ionic solids are built up of ions bound together by their electrostatic forces, and when dissolved in a polar liquid, such as water, the salt dissociates into its ions, which have an independent existence.

Ions may be characterized in various ways. When they are described by the sign of their electric charge, they are described as *positive*, *negative* or *amphoteric* (or *zwitter*), the latter being an ion which carries both a positive and a negative charge, commonly at opposite ends of a long, or fairly long, chain, as in the case of ions of amino acids. See also **Amino Acids**.

Ions may also be described by their atomic structure, when they consist of more than one atom. Thus, a *complex ion* is a complex electrically charged radical or group of atoms such as $Ag(CN)_2^-$ or $Cu(NH_3)_2^{++}$, which may be formed by the addition to an ion of another ion or ions, or of an electrically neutral radical or molecule. When the particle combined with the ion is a large molecule, and the attachment is essentially by an adsorption process, the complex ion is called a *heteroion*. When the complex particle consists of a simpler ion combined with one or more molecules of water, it is known as an *aquoion* or *hydrated ion*. On the other hand, charged molecules, commonly produced by electrical discharges through gases, are often called molecular ions.

In meteorology, there are two special types of "ions" that enter into atmospheric processes: small "ions" and large "ions."

A *small ion* (also called a "light ion" or "fast ion") is the type that has the greatest mobility; hence, collectively, it is the principal agent of atmospheric conduction. The exact physical nature of the small ion has never been fully clarified, but much evidence indicates that each is a singly charged atmospheric molecule (or, rarely, an atom) about which a few other neutral molecules are held by the electrical attraction of the central ionized molecule. Estimates of the number of satellite molecules range as high as twelve. When freshly formed, by any of several atmospheric ionization processes, small ions are probably singly charged molecules; but after a number of collisions with neutral molecules, they acquire (actually, in a fraction of a second) their cluster of satellites.

A *large ion* (also called a "slow ion" or a "heavy ion") is an ion of relatively large mass and low mobility, produced by the attachment of a small ion to an Aitken nucleus.

ION-CAPTURE THEORY.   A theory of thunderstorm charge separation advanced by C. T. R. Wilson (1916). According to this theory, the lower negative charge of a thundercloud is generated by the accumulation there of raindrops that have captured predominantly negative ions in their descent through the cloud. The preferential capture of negative ions by such drops is said to be due to the polarization of the drops in the normal atmospheric electric field existing between the negatively charged earth and positively charged ionosphere. The lower halves of the falling drops therefore would attract and capture negative charges while their upper halves would be unable to draw in positive charges with comparable efficiency; thus a net negative charge would build up on the drops. This theory is generally regarded today as incapable of accounting for any important portion of thunderstorm charge separation, for it is quantitatively inadequate in view of typical ion densities.

See also **Precipitation Current**.

Additional Reading

Chalmers, J., and D.T. Haar: *Atmospheric Electricity*, Franklin Book Company, Inc., Elkins Park, PA, 1967.

Reiter, R.: *Phenomena in Atmospheric and Environmental Electricity*, Elsevier Science Publishers, New York, NY, 1992.

Wilson, C.T.R.: "On Some Determinations of the Sign and Magnitude of Electric Discharges in Lightning Flashes," *Proc. Roy. Soc. A*, **92**, 555–574.

AMS

ION CHROMATOGRAPHY.   A form of high-pressure liquid chromatography using a conductivity detector where a combination of weak ionic solvents are used to separate anions and cations of a solution, with the contribution of the solvent to conductivity suppressed just prior to detection. The technique is useful for measuring anions such as sulfate, nitrate, and chloride in hydrometeors.

ION COLUMN.   More commonly called meteor trail, the trail of ionized gases in the trajectory of a meteoroid entering the upper atmosphere; part of the composite phenomenon known as a meteor.

ION COUNTER. An apparatus that counts the number of unit charges of electricity that are contained in a sampled volume of the atmosphere.

The general procedure is to pass a sample of the atmosphere through a charged cylindrical condenser. The change in the potential across the condenser is a measure of the ionic charge contained in the sample volume. The change in potential depends upon such factors as the polarizing potential of the condenser, the mobility and charge of the ions, volume and length of the condenser, and sample flow rate. An ion counter is an ionization chamber in which there is no internal amplification by gas multiplication.

ION DENSITY. In atmospheric electricity, the number of ions per unit volume of a given sample of air; more particularly, the number of ions of a given type (positive small ion, negative small ion, positive large ion, etc.) per unit volume of air. Also called *ion concentration*. Measurement of ion density is used in determining efficiency of ionizers in ion engines. See also **Ion**.

ION ENGINE. A reaction engine in which ions, accelerated in an electrostatic field, are used as propellant. Also called *electrostatic engine*.

ION EXCHANGE. Ion exchange is a process in which cations or anions in a liquid are exchanged with cations or anions on a solid sorbent. Cations are interchanged with other cations, anions are exchanged with other anions, and electroneutrality is maintained in both the liquid and solid phases. The process is reversible, which allows extended use of the sorbent resin before replacement is necessary.

Many naturally occurring inorganic and organic materials have ion-exchange properties. Synthetic organic ion-exchange resins became available in the late 1930s with the introduction of phenolic-type products. Styrenic resins appeared in the mid-1940s, and acrylic resins about 20 years later. The ion-exchange market of the early to middle 1990s is dominated by the styrenic resins, but acrylic resins are becoming increasingly important. Phenolic-based resins have almost disappeared. A few other resin types are available commercially but have not made a significant impact. Inorganic materials retain importance in a number of areas where synthetic organic ion-exchange resins are not normally used. Only the latter are discussed here. This article places emphasis on the styrenic and acrylic resins that are made as small beads. Other forms of synthetic ion-exchange materials such as membranes, papers, fibers, foams, and liquid extractants are not included. See also **Extraction (Liquid-Liquid)**; **Fibers**; **Foam**; **Membrane Technology**; and **Paper**.

The primary application for ion exchange is the softening and deionization of water. The remaining applications include waste treatment, catalysis, purification of chemicals, plating, hydrometallurgy, food processing, and pharmaceutical uses. Because ion-exchange resins are insoluble polymeric acids and bases, these resins are also useful in removing acids and bases from gaseous streams via the neutralization of functional groups. See also **Food Processing**; and **Water**.

Weak and strong acid-type resins are for removal of cations and are called cation exchangers. Weak and strong base resins remove anions and are called anion exchangers. In addition to these four resin types, there are specialty resins used in applications where higher specificity for certain ions under challenging conditions is a critical factor.

Continuous columnar operation of ion-exchange systems is preferred over batch operation. Each column must be taken off-stream periodically to remove the adsorbed ions and restore the resin to the ionic form required for the adsorption step. See also **Adsorption (Process)**. In this sense, a columnar ion-exchange operation is not continuous. Installations are usually designed with multiple units to assure a continuous flow of the process stream when one or more of the columns require regeneration. In those installations where the ion-exchange system is not required throughout the day, regeneration is scheduled during idle time and the system requires fewer ion-exchange units. Continuous operation has been approached in a number of designs by moving resin, or vessels containing resin, in a direction opposite to the flow of liquid. Some of these approaches have been abandoned; others are increasing in popularity. See **Ion-Exchange Resins**.

ION-EXCHANGE RESINS. These materials are insoluble solid acids or bases that have the property of exchanging ions from solutions. During the ion-exchange reaction, the ion-exchange resins are converted into insoluble acids, bases, or salts. Cation-exchange resins contain fixed electronegative charges that interact with mobile counterions having the opposite, or positive, charge. Anion-exchange resins have fixed electropositive charges and exchange negatively charged anions. Ion-exchange resins are three-dimensional macromolecules or insoluble polyelectrolytes having fixed charges distributed uniformly throughout the structure.

Early Developments

The first ion-exchange resins were described by Adams and Holmes, a water-treatment expert and polymer chemist respectively, of the British Chemical Research Laboratory (1935). These ion-exchange resins were condensation products of phenol [CAS: 108-95-2] and formaldehyde [CAS: 50-00-0]. The granular-type cation-exchange resin contained sulfonic groups, and the anion exchanger contained aromatic amine groups. They are termed *strong-acid* and *weak-base* ion exchangers. A number of condensation-type ion-exchange resins were manufactured during 1935–1945. The first commercial deionization system was installed in 1939.

The next important step in ion-exchange resin technology was the synthesis of sulfonated styrene divinylbenzene (DVB) cation exchangers. Commercial quantities of strong-base styrene–DVB anion exchangers appeared in 1948. The first anion exchangers, the weak-base type, removed only strong mineral acids from water, such as HCl and H_2SO_4. The strong-base materials remove all acids, thus paving the way for production of water of equal or better quality than distilled water and at a much lower cost. The combination of the styrene strong-acid and strong-base exchange resins in a single tank (the mixed-bed deionizer), commercialized in 1949, produces water containing just a few parts per billion of dissolved salts and at a very low operating cost. The mixed-bed process produces ultrapure water from most freshwater supplies at a fraction of the cost of distillation. This is the basic method used for centralization high-pressure boilers (5,500 psig) in the power industry and for applications in the electronics, chemical, and pharmaceutical industries.

Research has continued apace over the years to develop new organic ion-exchange organic polymers — the details of most are proprietary. Both natural and synthetic zeolites also are used in ion-exchange processes, but their extreme importance as catalysts has tended to overshadow their applications for deionizing purposes. See also **Adsorption (Process)** and **Zeolite Group**.

Classification of Ion-Exchange Resins

Ion-exchange resins are categorized by the nature of functional groups attached to a polymeric matrix, by the chemistry of the particular polymer in the matrix, and by the porosity of the polymeric matrix. There are four primary types of functionality: strong acid, weak acid, strong base, and weak base. Another type consists of less common structures in specialty resins such as those which have chelating characteristics.

Cation-Exchange Resins: Strong Acid. Strong acid cation-exchange resins have sulfonic acid groups, $-SO_3^-H^+$, attached to an insoluble polymeric matrix. When the functional groups are in the hydrogen form and the resin is in contact with a liquid containing other cations, hydrogen ions leave the solid phase and enter the liquid phase as they are replaced by cations from the liquid phase, for example,

$$resin - SO_3^-H^+ + Na^+ + OH^- \rightleftharpoons resin - SO_3^-Na^+ + HOH$$

The liquid phase is free of Na^+ and the functional groups of the resin are converted to a sodium salt. Multivalent cations are removed in a similar manner. Electric charge neutrality must be maintained in both the liquid and solid phases.

It is not always necessary for the resin to be in the hydrogen form for adsorption of cations, especially if a change in the pH of the liquid phase is to be avoided. See also **pH (Hydrogen Ion Concentration)**. For example, softening of water, both in homes and at industrial sites, is practiced by using the resin in the $^+$ form.

$$2\, resin - SO_3^-Na^+ + Ca^{2+} + 2\, Cl^- \rightleftharpoons$$
$$(resin - SO_3^-)_2Ca^{2+} + 2\, Na^+ + 2\, Cl^-$$

Sodium ions are displaced from the resin by calcium ions, for which the resin has a greater selectivity.

In many industrial applications, strong acid cation-exchange resins are used in the hydrogen form to process liquids containing low concentrations of salts.

$$2 \text{ resin} - SO_3^-H^+ + 2 Na^+ + SO_4^{2-} \rightleftharpoons$$

$$2 \text{ resin} - SO_3^-Na^+ + 2 H^+ + SO_4^{2-}$$

This is commonly referred to as a salt splitting reaction. The resin's selectivity for Na^+ is greater than it is for H^+. Anions are removed in a similar manner with an anion-exchange resin.

Ion-exchange reactions are reversible. A regeneration procedure restores the resin to the ionic form it was in prior to the adsorption step. Reversibility of reactions allows resins to be used many times before replacement is considered. Strong acid cation exchangers are returned to the hydrogen, H^+, form with dilute hydrochloric acid [CAS: 7647-01-0] or sulfuric acid [CAS: 8014-95-7]. Other mineral acids are used at times. However, the safety, cost, and methods of disposal must be thoroughly reviewed before using other acids. A 4% acid concentration is common. The use of higher or lower concentrations is dependent upon the process, the design of the system, and the potential for forming insoluble salts of the acid.

Cation-Exchange Resins: Weak Acid. Weak acid cation-exchange resins have carboxylic acid groups, $-COOH$ attached to the polymeric matrix. Although not as versatile in process applications as the strong acid resins, these resins are included in numerous systems where higher operating capacities and greater ease in regeneration can be used advantageously.

Weak acid cation exchangers have essentially no ability to split neutral salts such as sodium chloride [CAS: 7647-14-5]. On the other hand, an exchange is favorable when the electrolyte is a salt of a strong base and a weak acid.

$$\text{resin} - COO^-H^+ + Na^+ + HCO_3^- \rightleftharpoons \text{resin} - COO^-Na^+ + H_2CO_3$$

The sodium form of weakacid resins has exceptionally high selectivity for divalent cations in neutral, basic, and slightly acidic solutions.

$$2 \text{ resin} - COO^-Na^+ + Ca^{2+} + 2 Cl^- \rightleftharpoons$$

$$(\text{resin} - COO^-)_2Ca^{2+} + 2 Na^+ + 2 Cl^-$$

The selectivityis is so great that reversal of the reaction to restore the resin to the Na^+ form is not practical using NaCl solutions at any concentration. Regeneration with dilute acid, followed by conversion of the resulting H^+ form to the Na^+ form with dilute sodium hydroxide [CAS: 1310-73-2], is the preferred alternative.

Anion-Exchange Resins: Strong Base. Strong base anion-exchange resins have quaternary ammonium groups, $-_+NR_3OH^-$, where R is usually CH_3, as the functional exchange sites. These resins are used most frequently in the hydroxide form for acidity reduction. See also **Quaternary Ammonium Compounds**.

$$\text{resin} - \overset{+}{N}(CH_3)_3OH^- + H^+ + Cl^- \rightleftharpoons \text{resin} - \overset{+}{N}(CH_3)_3Cl^- + HOH$$

Hydroxide ions [CAS: 14280-30-9] are released by the resin as anions are adsorbed from the liquid phase. The effect is elimination of acidity in the liquid and conversion of the resin to a salt form. Typically, the resin is restored to the OH^- form with a 4% solution of NaOH.

The hydroxide form is also used in salt splitting applications.

$$\text{resin} - \overset{+}{N}(CH_3)_3OH^- + Na^+ + Cl^- \rightleftharpoons$$

$$\text{resin} - \overset{+}{N}(CH_3)_3Cl^- + Na^+ + OH^-$$

Salt forms of a strong base anion exchangerare used to remove other anions for which the resin has greater selectivity.

$$2 \text{ resin} - \overset{+}{N}(CH_3)_3Cl^- + 2 Na^+SO_4^{2-} \rightleftharpoons$$

$$(\text{resin} - \overset{+}{N}(CH_3)_3)_2SO_4^{2-} + 2 Na^+ + 2 Cl^-$$

Cation-Exchange Resins: Weak Base. Weak base anion-exchange resins may have primary, secondary, or tertiary amines as the functional

group. The tertiary amine $N-(CH_3)_2$ is most common. Weak base resins are frequently preferred over strong base resins for removal of strong acids in order to take advantage of the greater ease in regeneration.

$$\text{resin} - N(CH_3)_2H^+OH^- + H^+ + Cl^- \rightleftharpoons$$

$$\text{resin} - N(CH_3)_2H^+Cl^- + HOH$$

Most weak base anion exchangers adsorb weak organic acids such as formic acid [CAS: 64-18-6] and acetic acid [CAS: 64-19-7], but do not remove weak organic acids such as carbonic acid [CAS: 463-79-6] or silicic acid [CAS: 7669-41-4].

Weak base resins when in the free base (hydroxyl) form are not capable of splitting neutral salts such as sodium chloride. Salt forms of weak base resins release anions to the liquid phase if other ions for which the resin has a greater selectivity are present.

$$2 \text{ resin} - N(CH_3)_2H^+Cl^- + 2 Na^+ + CrO_4^{2-} \rightleftharpoons$$

$$(\text{resin} - N(CH_3)_2H^+)_2CrO_4^{2-} + 2 Na^+ + 2 Cl^-$$

This interchange of ions is similar to that of the strong base resins.

Chromatographic Resins. The ion-exchange reactions illustrated are typical of those which occur in numerous industrial installations when the primary objective is the removal of ions, acids, or bases from a liquid stream. These same resins are useful for chromatographically separating ions having the same valence. Ion-exchange resins have a different selectivity for each ion. Sometimes these differences are too small for a separation to occur. In addition, other factors including flow rate, column design, and resin properties usually govern the success or failure of a chromatographic separation. See also **Chromatography**.

Although not of commercial interest, consider a separation of Ca^{2+} and Mg^{2+}. A solution containing low concentrations of each cation is fed slowly and continuously to a column containing a strong acid-type cation-exchange resin in the Na^+ form. At first the solution leaving the column contains only Na^+ because it is displaced from the resin as Ca^{2+} and Mg^{2+} are adsorbed. As flow continues, Mg^{2+} appears in the effluent, but not Ca^{2+}. Because the selectivity of the resin is greater for Ca^{2+} than for Mg^{2+}, Ca^{2+} displaces the Mg^{2+} previously adsorbed by the functional groups. As flow continues and the column becomes loaded, Ca^{2+} appears in the effluent at increasing concentrations along with Mg^{2+}.

Chromatographic separations are not limited to ionic constituents. For example, glucose [CAS: 50-99-7], $C_6H_{12}O_6$, is separated from fructose [CAS: 57-48-7], $C_6H_{12}O_6$, when using a strong acid cation exchanger in the Ca^{2+} form. The functional groups hinder the forward movement of each sugar at a slightly different rate as a solution containing both sugars flows slowly through a column containing the resin. Water [CAS: 7732-18-5] is fed alternately with the sugar solution to aid in developing the separation. Glucose precedes the appearance of fructose in the effluent, indicating fructose is prevented in its forward motion by the resin to a greater degree than is glucose.

Resins having other types of functional groups are growing in importance. Resins that have metal chelating capabilities include those containing iminodiacetic acid [CAS: 142-73-4] or aminophosphonic acid sites. Resins having thiol functionality are interesting for adsorption of metals which form sulfide precipitates. Resins containing an *N*-methylglucamine [CAS: 6284-40-8] functionality are selective for boron [CAS: 7440-42-8].

Manufacture of Ion-Exchange Resins

The production of ion-exchange resins is a multiple step process. It begins with the polymerization of monomers to form solid intermediate copolymers that are insoluble in both water and solvents. The copolymers are functionalized during additional steps in different reactors from those used for copolymer production. Conversion to another ionic form may be required after functionalization is completed. All resins are thoroughly rinsed with water to cleanse them of residual chemicals. Excess water is removed by vacuum filtration prior to packaging. Complete removal of all water by drying is unusual. Packaging in fiber drums is common. Alternative containers include metal drums, bulk boxes and bags, and smaller plastic, paper, and burlap bags. Polyethylene [CAS: 9002-88-4] liners are used as a barrier between containers and water-containing resin.

Copolymerization. The chemistry of the resin matrix, the type and degree of porosity, the particle size, and the particle size distribution are established in the copolymerization step. Formulations and operating procedures must be strictly followed. Reaction vessels must be well designed. Mistakes made during copolymerization are rarely corrected during functionalization.

The procedure of forming copolymers dates back to the early 1940s when only phenolic resins were available. Copolymers were produced by bulk polymerization of phenol [108-95-2] and formaldehyde [50-00-0]. Because the resulting solid product had the shape of the vessel in which polymerization took place, it had to be reduced to smaller particles by crushing and grinding before being functionalized. A new chemistry, and a new process, appeared in the mid-1940s when it was learned that styrene [CAS: 100-45-2] could be copolymerized with divinylbenzene [CAS: 1321-74-0] (DVB) by a suspension polymerization process, and that the small, hard copolymer beads could be functionalized. These styrenic resins eventually replaced the phenolic resins in practically all applications. Since then, another chemistry has evolved to complement the styrenic resins in process applications. These are the acrylic resins which are functionalized copolymers of an acrylic monomer and DVB produced by the suspension polymerization technique.

Design of copolymer reactors became more complex with the introduction of suspension polymerization. The size and uniformity of the ion-exchange resin to be produced from a copolymer was now dependent on the size and uniformity of a liquid monomer mix dispersed as small droplets in an aqueous medium. The shape, size, and speed of mixers, as well as baffling to control fluid flow within the reactors, became critical factors. These reactors must be closed to the atmosphere, and they must be fitted with appropriate piping and valves to allow rapid transfer of organic and aqueous phases from other tanks to the reactor, and to allow rapid removal of a copolymer slurry when the reaction is over. The reaction kettles are jacketed which provides a means for raising and lowering temperatures during polymerization. Careful selection of materials of construction assures long life, minimizes downtime for maintenance, and guards against contamination of the liquid and aqueous phases with elements which may interfere with the preferred rate of polymerization.

The organic and aqueous phases are prepared in separate tanks before transferring to the reaction kettle. In the manufacture of a styrenic copolymer, predetermined amounts of styrene and divinylbenzene are mixed together in the organic phase tank. Styrene is the principal constituent, and is usually about 90–95 wt % of the formulation. The other 5–10% is DVB. It is required to link chains of linear polystyrene together as polymerization proceeds. DVB is referred to as a cross-linker. Without it, functionalized polystyrene would be much too soluble to perform as an ion-exchange resin. Ethylene–methacrylate [CAS: 97-90-5], and to a lesser degree trivinylbenzene [CAS: 1322-23-2], are occasionally used as substitutes for DVB.

Formulations for acrylic copolymers involve monomers such as acrylic acid [CAS: 79-10-7], methacrylic acid [CAS: 79-14-4], or esters of these acids. Formation of a copolymer from a methylmethacrylate ester, where DVB serves as the cross-linker, can be obtained.

The fraction of DVB used in the monomer mix is governed by the required performance characteristics of the ion-exchange resin. The higher the percentage of DVB, the greater the number of sites at which polymeric chains of styrene, or acrylic monomer, are linked. High DVB levels increase the tightness of the polymer matrix and lessen the ability of ions or other solutes to migrate through the resin phase. Thus the porosity decreases as the cross-linking increases. When the organic phase contains only the monomers that participate in polymerization, the copolymers and resins produced from them are described as having a microporous or gel-type porosity. It is a porosity which cannot be measured by conventional methods. A different type of porosity, macroporous (also called macroreticular), is formed by incorporating a solvent along with the monomers in the organic phase. That solvent should have low aqueous phase solubility and must not participate in the polymerization reaction. When the organic phase is dispersed into the aqueous phase, the solvent is distributed throughout each droplet. The solvent, which remains with the hard copolymer beads during polymerization, is not bound to the resin matrix because it did not react with either monomer. Displacement of the solvent from the copolymer, or removal by vaporization, yields a copolymer with a measurable pore volume and pore size distribution. Both pore volume and distribution are dependent on the solvent and the amount incorporated in the original monomer mix.

The aqueous phase into which the monomer mix is dispersed is also prepared in a separate tank before transferring to the copolymerization kettle. It contains a catalyst, such as benzoyl peroxide [CAS: 94-36-0], to initiate and sustain the polymerization reaction, and chemicals that aid in stabilizing the emulsion after the desired degree of dispersion is achieved. Careful adherence to predetermined reaction time and temperature profiles for each copolymer formulation is necessary to assure good physical durability of the final ion-exchange product.

Continuous processes for copolymer production were developed initially for the microporous resins. The system generally involves injecting the monomer mix into the aqueous phase through orifice plates. Droplet size is controlled by the diameter of the holes in the plate and the rate at which the monomer is injected into the aqueous phase. The continuous process produces copolymer beads which have greater uniformity in size than those produced in batches.

Functionalization. Copolymers do not have the ability to exchange ions. Such properties are imparted by chemically bonding acidic or basic functional groups to the aromatic rings of styrenic copolymers, or by modifying the carboxyl groups of the acrylic copolymers. There does not appear to be a continuous functionalization process on a commercial scale.

A simplified schematic layout of an ion-exchange production facility is presented in Figure 1. Layouts vary from one company to another and are significantly more complex when recycle of streams and environmental controls are incorporated in the schematics.

Separate kettles and backwash towers are frequently used to convert ion-exchange resins from one ionic form to another prior to packaging, and to cleanse the resin of chemicals used in the functionalization reactions. Excess water is removed from the resin prior to packaging by a vacuum drain. Both straight line filters and towers or columns are used for this purpose.

Physical and Chemical Properties of Ion-Exchange Resins

Ion-exchange resins are used repeatedly in a cyclic manner over many years, and deterioration of both physical and chemical properties can be anticipated. Comparison of the properties of used resin with those of new resin is helpful to learning more about the nature and cause of deterioration. Corrective action frequently extends the life of the resin. Comparison of properties must always be made with the resin in the same ionic form.

Particle Shape and Size. With few exceptions, resins are supplied as small, round beads having a diameter between 0.3 and 1.2 mm (0.01 and 0.05 in). Some resins are reduced to a smaller size by grinding to satisfy specific requirements in applications for electric power generation and pharmaceuticals. See also **Pharmaceuticals**.

Resin size is dependent on two factors. One of these is the method by which the copolymer beads are formed, and the other is related to the exchange kinetics within the beads. Copolymer beads are made by batch-wise suspension of a liquid monomer mix in an aqueous medium or by a continuous jetting technique. In either case, formation of beads larger than 1 mm (0.04 in) in diameter is difficult to achieve without agglomeration. Large beads are preferred for pressure drop advantages in columnar operations. However, large beads are subject to a greater rate of breakage than those having a smaller size.

Migration of ions to functional groups within the resin particles is slow compared to the aqueous or organic liquid phase surrounding the particles. The most efficient ion-exchange processes occur when most of the functional groups can be accessed within a short contact time between liquid and resin. The larger the particle size, the greater is the time required to utilize groups deeper in the particles. Thus, the smaller the bead, the better. However, the smaller the size, the greater the pressure drop. A particle size range spanning 0.3–1.2 mm (0.01–0.05 in) in diameter has been a compromise between acceptable kinetics and pressure drop.

Wet screening using a set of U.S. Standard sieves (297, 420, 595, and 841 μm openings; 50, 40, 30, and 20 mesh, respectively) is the standard procedure for determining particle size distribution. The volume percent retained on each is recorded. Wet screening is being replaced in many laboratories by instruments that use a photosensor to record the diameter of each particle in a water suspension as that suspension flows past the sensor.

Density and Specific Gravity. Density generally pertains to the bulk, or pack-out, weight of wet resin per unit volume. The density is characteristic

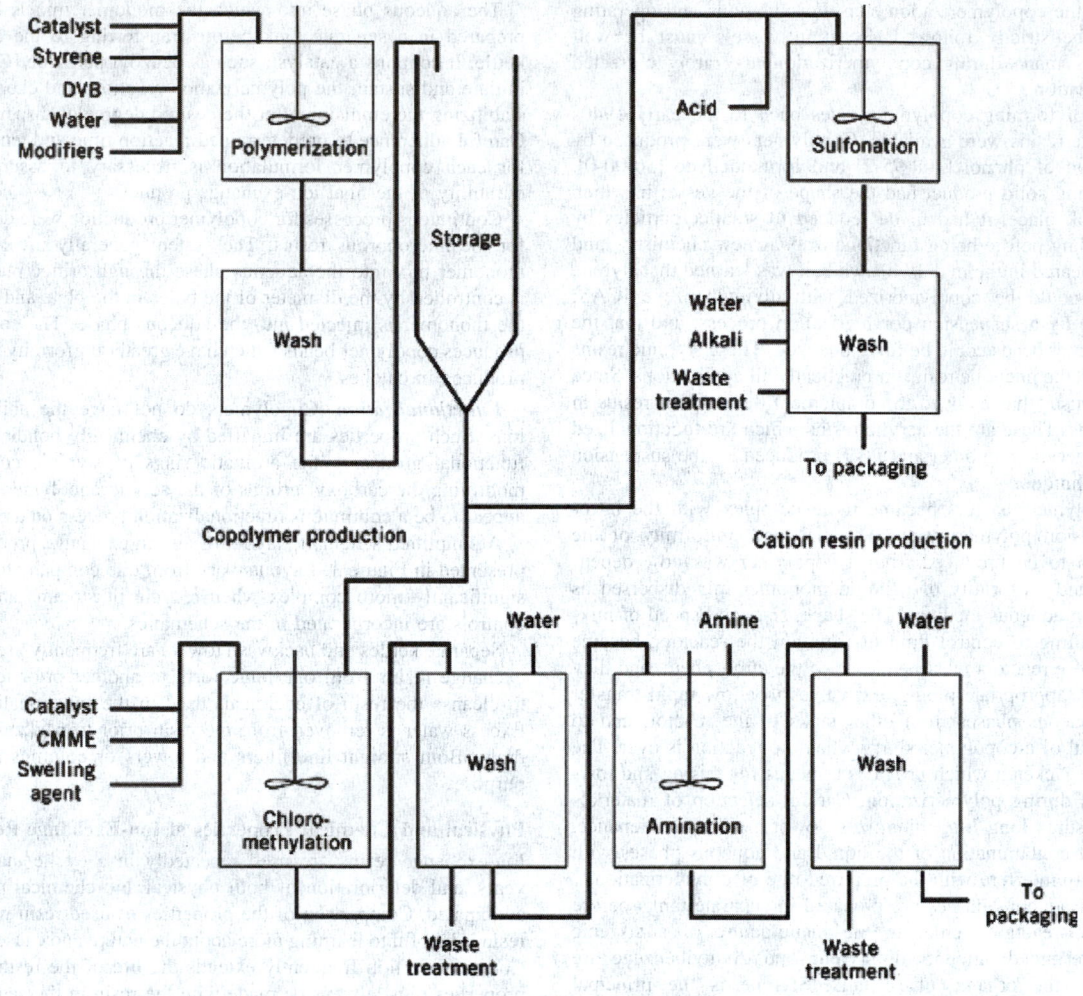

Fig. 1.   Ion-exchange production schematic. CMME is chloromethyl methyl ether.

of the resin and is dependent on the copolymer structure, the degree of cross-linking, the nature of the functional groups, and the ionic form of those groups. A change in density after extended use is a signal that chemical degradation has occurred. The density of most cation exchangers is in 800–900 g/L the range, whereas most anion exchangers are in the 640–740 g/L range.

The specific gravity generally refers to the value determined for wet resin when using a pycnometer. Values range from about 1.04 to about 1.25. Cation exchangers have a greater specific gravity than anion exchangers.

Porosity. The structure of ion-exchange resins is either microporous or macroporous. Microporous resins are more commonly referred to as gel or gelular-type resins. Porosity of this type resin cannot be measured by standard techniques. Gel resins are porous when the particles are swollen with water or another solvent. There is no porosity when the resin is dry. Microporosity is inversely proportional to the degree of cross-linking. Large ions migrate through a low cross-linked resin faster than through the less porous, higher crosslinked resins.

Macroporous resins are also called macroreticular. Macroporous resins have a measurable porosity. It does not disappear when the resin is dry. Porosity is more dependent on the solvent used when manufacturing the copolymer than on the degree of cross-linking.

Capacity. Capacity is a measure of the quantity of ions, acid, or base removed (adsorbed) by an ion-exchange material. The quantity removed is directly correlated with the number of functional groups. Capacity is reported in several different ways, but requires further definition because the word by itself does not cover all situations. Total capacity is a measure of all the functional groups on a resin and is recorded on a weight as well as a volume basis. A manufacturing objective is to place at least

one functional group on each aromatic ring in styrenic-type resins. As a consequence, the degree of cross-linking has little effect on the total dry weight capacity for strong acid cation exchangers in the hydrogen form. Dry resins swell when wet with water. The amount of swelling decreases as the degree of cross-linking increases.

Operating capacity, also called the working capacity or column capacity, is a measure of the quantity of ions, acids, or bases adsorbed, or exchanged, under the conditions existing during batch or columnar operation. Because the adsorption process is terminated in most commercial applications before all functional groups have been utilized, the operating capacity is less than the total capacity. Operating capacities vary from one installation to another, even though the same resin might be used at each location. Such variations are the result of differences in composition of the stream to be treated, flow rate, effluent quality that triggers shutdown for regeneration, and regeneration conditions.

The steeper slope for the weak base resin is an indication that these resins are regenerated with greater efficiency, or ease, than strong base resins. Strong base anion exchangers do not release anions as readily as weak base anion exchangers. For each incremental increase in the amount of chemical used for regeneration there is a greater incremental increase in the operating capacity of the weak base resin than the strong base resin. If two resins have the same total capacity but differ in basic strength and are regenerated under identical conditions, the resin having the lower basicity is restored more completely to a regenerated form and has the higher operating capacity. A similar analogy can be made for weak and strong acid cation exchangers.

Selectivity. A significant exchange of ions does not occur unless the functional group of the resin has a greater selectivity for ions in solution than for ions occupying the functional group, or unless there is a mass action effect, as in regeneration. Selectivity coefficients have been reported in numerous publications for both cations and anions. These coefficients are determined at very low concentrations using the two specific ions in question.

Selectivity for ions having the same charge usually increases as the atomic weight increases (Li < Na < K < Rb < Cs and Mg < Ca < Sr < Ba). Selectivities for divalent ions are greater than for monovalents. Higher selectivity for trivalent and tetravalent ions does not necessarily follow the expected progression. The examples Table 1 have been chosen to illustrate the effect of type of strong base functionality and mole fraction of chloride [CAS: 16887-00-6] in a two-component anion system on selectivity constant.

TABLE 1. SELECTIVITY FOR ANIONS ON ANION EXCHANGERS[a]

Ion	CAS Registry Number	Dowex 1[b]		Dowex 2[c]		Dowex 2[c]	
		X_{Cl}	K_{Cl}	X_{Cl}	K_{Cl}	X_{Cl}	K_{Cl}
Iodide	[20461-54-5]	0.27	8.7	0.27	7.3	0.07	13.2
Nitrate	[14797-55-8]	0.38	3.8	0.36	3.3	0.34	3.3
Nitrite	[14797-65-0]	0.51	1.2	0.52	1.3		
Hydroxide	[14280-30-9]	0.77	0.09	0.56	0.65		
Bicarbonate	[71-52-3]	0.65	0.32	0.63	0.53		
Formate	[71-47-6]	0.70	0.22	0.68	0.22		
Fluoride	[16984-48-8]	0.77	0.10	0.76	0.10		

[a]Selectivity is vs that for Cl^-. Mol fraction of Cl^-, X_{Cl}, is given.
[b]Type I functionality.
[c]Type II functionality.

The need to know selectivity coefficients precisely is rarely necessary in industrial applications. However, knowledge of relative differences is important when deciding if the reaction is favorable or not. Most ion-exchange applications are for the removal of more than one ionic species. In water softening, it is important to know that a cation exchanger has a greater selectivity for divalent cations than for monovalent cations. It is not critical to know the resin prefers Ca^{2+} over Mg^{2+}. In deionization, it is important to know that all cations normally present in a water supply are preferred by a strong acid cation exchanger over a hydrogen ion [CAS: 12408-02-5], H^+, and that all anions are preferred over OH^- by a strong base anion exchanger. Differences in selectivity coefficients take on greater importance in chromatographic separations where flow rates are much slower to allow more time for ionic species to separate as the liquid flows through the resin bed.

Selectivity differences increase as the degree of cross-linking of a resin increases, but these differences are relatively minor. Even greater selectivity for divalent cations is observed for resins having aminophosphonic acid or iminodiacetic acid functionality.

Kinetics. The degree to which an ion-exchange reaction is completed depends on a number of factors which include contact time, ionic concentration, degree of cross-linking, and temperature. Contact time and concentration are interrelated and most important for a successful ion-exchange process. The impact of temperature increases as the viscosity increases. In columnar operations, short contact times associated with high flow rates are acceptable when the concentration of ionic constituents in the influent stream is very low. At higher electrolyte concentrations, the contact time must be increased. If contact time is too short, functional groups on or near the surface are converted to another ionic form sooner than groups deeper in the resin particles. The net result is the appearance of ions that were to be adsorbed in the effluent sooner than would have occurred with a longer contact time, and lower utilization of all the resin's capacity. Flow rates of 8 to 40 bed volumes (BV)/h (1–5 gal/min/ft^3) are common in conventional water treatment systems where ionic concentrations might be as high as several hundred mg/L. Slower flow rates should be considered for ionic concentrations exceeding 500 mg/L. For those systems which have ionic concentrations in the µg/L range, as in condensate polishing and similar recycle applications, flow rates are usually significantly greater than 40 BV/h.

Some industries practice ion exchange in nonaqueous systems. These solvents may cause resin particles to shrink or swell. Shrinkage has a negative effect on the kinetics, whereas swelling opens up the structure and improves migration of those constituents to be adsorbed. Microporous resins usually do not work well in nonaqueous systems because of the disappearance of porosity. Macroporous resins, however, are more satisfactory in these systems since porosity is retained even if the resins are dried completely. More functional groups on outer and inner surfaces are available for exchange as a result of the combined fixed porosity and relatively high surface area. Nevertheless, these systems are operated at slower rates to compensate for slower migration of ions, acids, and bases through organic solvent or process stream surrounding the resin particles.

Several aqueous systems should be considered in a similar manner. For example, in the selective removal of divalent cations from a saturated salt solution, the hydrated resin gives up a portion of its normal water content as it contacts the salt stream. In so doing, the particles shrink, and the inner pathways for ion migration become smaller.

Moisture and Water Content. Resins are thoroughly washed with water upon completion of manufacture and conversion (if necessary) to another ionic form. Excess water is removed by vacuum draining or filtration. Nevertheless, a significant quantity of water associated with the functional groups and adhering to the outer surface of the resin particles remains with the resin as it is discharged into shipping containers. No effort is made to dry the resin, except in a few application areas, since the resins are used in aqueous processes in most installations.

Each resin has a characteristic water content dependent on the resin matrix, the structure of the functional groups, and the ionic form of those groups. Resins are packaged by weight and sold by volume. The dewatering operation prior to packaging is a critical step since removal of too much water is costly to the manufacturer, and removal of too little is costly to the buyer. Analyzing for water content is important to both the seller and user. The quantity of water contained by the resin is recorded on a percentage basis and determined by two methodologies. In each procedure, a small (ca 15 g (0.5 ounce) sample is removed from a larger composite sample collected during pack-out. In one procedure, the sample is accurately weighed before and after placing in a 105 °C (221 °F) oven for at least 8 h. This procedure yields the moisture content typical of resin contained in the shipping containers. In the other procedure, a similar sample is soaked in water, and then filtered under vacuum in a Buchner funnel prior to weighing before and after oven drying. The moisture content reported is a pseudoequilibrium value typical of the specific resin and its ionic form. If the value reported is either greater or lower than the expected range, all manufacturing steps need to be examined for deviation from the standard manufacturing procedure. Values determined by both methods rarely differ by more than 1%. A significant difference is an indication of a procedural change or malfunction of equipment in the plant dewatering step prior to packaging. Water content of strong base resins in the hydroxide form cannot be determined in the same manner since thermal degradation with additional weight loss occurs at 105 °C (221 °F).

Swelling and Shrinking. Ion-exchange resins shrink or swell reversibly as they are converted from one ionic form to another. The degree of change is dependent on the resin matrix, the functional group, and the ions adsorbed by the functional groups. For similar matrixes, the magnitude of volume changes decreases as the level of cross-linking increases. Resins in contact with solutions having a high electrolyte concentration shrink as water is drawn from the resin. These reswell when in contact with more dilute solutions. Wetting the resins with nonaqueous solvents causes shrinking or swelling, depending on the solvent. Swelling is greatest for weak acid cation exchangers (up to 100%). Weak base anion exchangers may swell as much as 50%. Strong base resins swell about 15–25%. Strong acid-type resins swell the least and are usually in the 5–10% range.

The degree of swelling and shrinking is important for design of ion-exchange columns, especially for the location of the distributors used to disperse incoming fluids, and collect outgoing ones, evenly over the cross-sectional area of the resin bed. Once placed, these distributors are not adjustable. The upper distributor should be above (the lower one below) the resin bed, even in the bed's swollen form.

Hydraulic Properties. Both the resistance to liquid flow through a resin bed and the degree to which a resin bed expands during a backwashing step are important design factors for ion-exchange systems.

These characteristics are also critical to those using the resins because movements of resins not only signal the existence of a problem but give indications as to the nature of the problem. Pressure drop and hydraulic information for new resins are available from the resin manufacturer and the supplier of equipment.

Factors which have the greatest impact on pressure drop are the depth of the bed, flow rate, viscosity, temperature, and particle size. Pressure drop is also dependent on the hardness of the resin, or its ability to resist deformation when under compressive forces as in columnar systems with the process stream flowing down. Styrenic resins are more resistant to deformation than acrylics, and macroporous resins are more resistant than microporous. Minor increases in pressure drop as a result of bed packing are to be expected in lengthy downflow runs. Resins which have been used for a large number of cycles may yield a significantly higher pressure drop compared to those when the resin was new. These increases suggest problems with dirt, biological growth, or an increase in resin fragments within the column. Another cause may be oxidative attack of the resin which, in effect, makes it behave as a more compressible lower cross-linked resin.

Backwashing is the upward flow of water through a bed of resin at a flow rate sufficient to fluidize the resin, but not so great that resin is carried out of the column with the exiting water. Resins are backwashed to remove dirt and resin fragments, to classify resin particles by size, and to relieve any packing that may have occurred with previous use. Frequency may be as great as once each cycle. Backwash times are 15 to 30 minutes, unless conditions require more time. Resin columns are designed with adequate space above the resin bed to allow 50–100% expansion during backwash. Each resin has a characteristic bed expansion profile which is dependent on the resin's specific gravity and particle size distribution. Severe accidental losses of resin occur during the colder months when the temperature of water used for backwashing is ignored. Lower temperature increases the water viscosity which increases the buoyancy effect on the resin particles. If the flow rate of the backwash water is not reduced, the bed expansion may be so great that resin particles leave the column with the exiting water. In warmer months, backwash flow rate should be increased.

Chemical Stability. Oxidants, such as dissolved chlorine [CAS: 7782-50-5] in water supplies, react with synthetic ion exchangers to cause a loss of capacity, physical weakening of the resin, and partial solubilization of the resin. Anion-exchange resins are most prone to loss of functionality as the oxidant attacks and severs the linkage between nitrogen and carbon on the polymeric structure. In addition to this form of degradation, the functional groups of strong base anion exchangers partially convert to weak base groups through loss of one or more of the alkyl groups attached to the nitrogen. The net effect is loss of both strong base and total capacity with an increase in weak base capacity. Loss of functional groups from cation-exchange resins by oxidative attack is uncommon.

The point at which two polymeric chains are joined together by a cross-linker such as divinylbenzene, or sites where tertiary hydrogens are located in the structure, are other locations for oxidative attack. In both cation- and anion-exchange resins, oxidative attack results in the removal of cross-linking. The moisture content is higher than when the resin was new. Resins having lesser amounts of cross-linking are more subject to physical deformation when under compressive forces than a resin with a greater amount of cross-linkage. Consequently, a gradual increase in pressure drop across the resin bed is to be expected as oxidative attack continues. In severe cases, the resin breaks into fragments. A resin that has undergone significant oxidative degradation releases small soluble fractions of the polymeric structure to the liquid phase, thus contributing to the biological oxygen demand (BOD). Such groups may or may not interfere with ion-exchange units that follow. An increase in the quantity of water to rinse the column after regeneration is another indicator of oxidative attack, especially for anion-exchange resins. The severance of functional groups leads to the development of carboxylic acid functionality. These groups are converted to the sodium salt when in contact with the NaOH used for regeneration. The water rinse which follows the regeneration step slowly hydrolyzes sodium ions from these groups causing the pH of the effluent to remain high for longer periods of time than realized with new resin.

The rate of oxidative attack is enhanced by the presence of metals such as iron [CAS: 7439-89-6] and copper [CAS: 7440-50-8] which serve as catalysts, by higher temperatures, and by higher concentrations of oxidants. The direct effect of each of these factors is a matter of opinion because all of the factors causing oxidative degradation are present at the same time but to varying degrees over the useful life of the resin. The tolerable limit for residual chlorine in a water supply is an example. At any concentration, the rate of attack is greater in warmer climates than in colder ones, all other factors being equal. A common recommendation is not to exceed 0.3 mg/L residual chlorine; others may recommend a 0.1 mg/L limit, and still others believe 1 mg/L is a safe upper limit. The oxidant concentration can be lowered before the process stream contacts the resin by standard procedures such as the use of carbon [CAS: 7440-44-0] columns and the feeding of sulfite [CAS: 14265-45-3].

Aside from low concentrations of oxidants found in most water supplies, the processing of chemical streams with much higher levels of oxidizing chemicals is practiced occasionally on an industrial basis. The potential dangers are generally recognized. Nevertheless, there is the potential for an uncontrolled reaction that releases heat which converts a liquid to a gas, resulting in the rupture of equipment in an explosive manner. Systems must be designed with appropriate detectors for abnormal performance, and with procedures that reverse or stop the reaction before it gets out of control.

Thermal Stability. Ion-exchange resins should not be used at temperatures above those recommended by the manufacturer. Exceptions are made when frequent replacement of resin is an economic advantage over the operating and capital cost of cooling and reheating the process stream. Functional groups are lost from both cation- and anion-exchange resins when the temperature limit is exceeded. The rate of loss increases exponentially as the temperature rises above the upper limit. Sulfonated cation exchangers can frequently tolerate temperatures up to 125°C (255°F). Strong base anion exchangers having a trimethyl quaternary ammonium structure can be used up to 77°C (170°F) in salt forms and up to 60°C (140°F) in the OH$^-$ form. Those strong base resins with a dimethylethanol quaternary ammonium structure and the acrylic anion exchangers are limited to about 40°C (104°F).

Physical Stability. Excessive pressure drop across the resin bed causes fragmentation of the beads. The point at which this occurs depends on the structure of the resin and in most systems is well above the pressure drop listed in product literature for water systems. Upper limits are about for gel (microreticular) type resins in the 8–12% DVB range. If oxidative attack occurs during use of the ion-exchange resin the maximum pressure drop would be characteristic of a lower cross-linked resin. Resin breakage aggravates the pressure drop problem. Macroporous resins generally can tolerate somewhat higher pressure drop than gel-type resins. Gel-type resins are more resistant than macroporous resins to breakage caused by a shearing motion.

Resins shrink and swell as they are alternately put through adsorption and regeneration cycles. The larger the volume change and the shorter the time involved, the greater the potential for physical damage to the resin particles. In most applications, the greater potential for physical damage occurs during regeneration. However, similar effects occur in a few applications when regenerated resin is contacted with high concentrations of salts, as in the removal of impurities from those salt solutions. The appearance of cracks is the first sign of physical deterioration. Fragmentation into smaller irregularly shaped particles is a sign of further deterioration.

Resins should always be protected from freezing, although that may not always be possible. Generally, a few freeze–thaw cycles do not result in visual damage (cracking or fragmentation). Nevertheless, some weakening of the physical structure occurs because fragmentation is apparent if cycling continues. If operating conditions dictate a lengthy shutdown of the ion-exchange system and the resin columns are in an area that cannot be protected from freezing, the columns may be filled with typical antifreezes without damaging the resin. Neither glycol nor alcohol damage any of the standard cation and anion exchangers. Solutions of NaCl may also be used. When the units are returned to service, they must be thoroughly rinsed with water and, preferably, regenerated before using. The glycol or alcohol must be disposed of in an environmentally approved manner.

Transfer of resin from one vessel to another, as for regeneration, does not physically damage the resin as long as certain practices are avoided. The fluidity of the resin slurry and the linear velocity in transfer lines should be sufficient to keep all particles in suspension. Resin that settles can lead to plugging of the transfer line. Sharp bends in the transfer line should be avoided. If a pump is used for transferring the slurry, it should be a type that will not allow resin particles to be caught in valves when they close. Recessed impeller, peristaltic, and some diaphragm pumps are used in large industrial systems. Recommendations from someone knowledgeable in resin transfer is advised in order to avoid a costly installation that may cause rapid loss of resin through physical damage.

Equipment

Ion-exchange systems in process applications may be batch, semicontinuous, or continuous. Batch operations are not common but, where used, involve a kettle with mechanical agitation. Injecting with air or an inert gas is an alternative. A screened siphon or drain valve is required to prevent resin from leaving with the product stream.

Semicontinuous and continuous systems are, with few exceptions, practiced in columns. Most columnar systems are semicontinuous since flow of the stream being processed must be interrupted for regeneration. Columnar installations almost always involve the process stream flowing down through a resin bed. Those that are upflow use a flow rate that either partially fluidizes the bed, or forms a packed bed against an upper porous barrier or distributor for process streams.

The lower section of a column with downward flow must have a distributor system that not only collects liquid evenly over the cross-sectional area, but also supports the resin bed and prevents resin from leaving the column. The traditional method has been to place a network of pipes with small holes drilled in them (a distributor) in a bed of graded gravel, sand, or anthracite coal, which supports the resin bed. While that practice continues, the trend has been toward other approaches. In one modification, the underbed is eliminated by securely wrapping the pipe elements with small mesh, noncorrosive screening. The size of the screen openings must be sufficiently smaller than the resin particles to avoid plugging. Blockage of the openings increases pressure drop and contributes to uneven or channeled flow. Special pipes formed by spirally winding triangular wire around supports, while carefully controlling the space between the flat side of the wire as it is wound, is another approach that is gaining acceptance. Perforated plates separating the resin from the distributor are used in other installations. Careful design of the distributor is essential, especially for the larger diameter units. If the linear flow rate near the wall of the column is substantially less than the midsection of the column, premature breakthrough, more frequent regeneration, and incomplete utilization of the rated operating capacity for the resin result. See also **Fluid and Fluid Flow**; and **Fluid Mechanics**.

The space immediately above the resin bed may or may not be filled with liquid in downward flow systems, depending on the design. If not filled, water entering the column from the top and impinging on the upper surface of the resin bed forms hills and valleys unless the flow is dispersed over the cross-sectional area. A distributor similar to the one used to collect resin below the bed, or splash plate, is placed a short distance above the resin bed to improve the distribution of the process stream flow.

A distributor is frequently installed at the top of the column for use during backwash. It collects water evenly and prevents resin from escaping the column should unexpected surges develop in the water flow during backwash. Columns lacking an upper distributor or screen to prevent loss of resin should have an external system to prevent resin from being lost to the drain. It is referred to as a resin trap and may consist of a porous bag that fits over the outlet pipe or a tank designed to lower the linear velocity. Resin drops to the bottom of the tank and is returned to the column when convenient.

Mixed-bed columns contain an anion and cation exchanger which must be regenerated independently after separation by backwashing. When regeneration is performed in the same column, a distributor is installed near the expected interface of the resins following the backwash. The distributor is used to feed regenerating solutions, feed water, and to collect spent regenerant solutions. Again, distributor design is critical.

All columns, distributors, and ancillary hardware such as piping, valves, and pumps must be constructed of corrosion-resistant materials, or coated with an appropriate substance. All streams that contact the hardware during each step of the cyclic operation need to be considered in this selection.

Columns are designed to have a larger internal volume than the quantity of resin they will contain. The extra space is to provide the necessary volume for a fluidized bed during backwash. Most units are designed for the space above the resin bed (free-board) to be between 50 and 100% of the packed resin bed. Small columns are, on occasion, designed for one-use applications. Since backwashing is of no importance, there is a tendency to fill the unit with as much resin as possible. That practice can be hazardous, especially if the resin swells as a result of oxidative attack or through conversion from one ionic form to another.

Column dimensions vary considerably from one installation to another, depending on the application, total flow, and overall system design. If a tall narrow column and a short wide column contain the same amount of resin and process a stream at the same flow rate, the wide column will have a more favorable linear velocity and a lower pressure drop. However, bed depths cannot be too low, especially in the larger diameter units. Otherwise achieving uniform flow over the cross-sectional area of the column is impossible. The recommended minimum bed depth is about one meter. Bed depths over three meters are most common in applications involving catalysis and chromatographic separations. Serious consideration should be given to the use of several columns in parallel in place of one very large unit, especially when a system cannot be shut down when regeneration is required.

Feed systems utilizing gravity are rarely used. Line pressure is usually adequate for small systems. Auxiliary pumps are required in larger systems to assure proper flow through all units and to avoid uneven flow should line pressure decrease as other demands for water or the process stream occur elsewhere in the facility.

Regenerating streams require dilution with water before contacting the resin. Eductor systems (which mix two fluids), in-line dilution, and separate storage tanks containing sufficient diluted regenerating solutions are commonly used. Appropriate measuring methods to assure that the correct concentration and volume of regenerant are transferred to the resin column should be incorporated in the system. A low concentration, or insufficient volume, adversely affects the subsequent adsorption step. An unintentional higher concentration increases operating costs and magnifies a waste regenerant disposal problem.

Ion-Exchange Systems

Ion-exchange systems vary from simple one-column units, as used in water softening, to numerous arrays of cation and anion exchangers which are dependent upon the application, quality of effluent required, and design parameters. An illustration of some of these systems, as used in the production of deionized (demineralized) water, is presented in Figure 2.

A single-column installation is satisfactory if the unit can be shut down for regeneration. However, if flow of the stream being processed must be continuous, then two or more columns of the same resin must be installed in parallel. Regeneration of each is staged for different time periods.

Two columns, one containing a cation exchanger and one an anion exchanger, are required for a deionization process. The cation exchange unit must be a strong acid-type resin, except in more complex systems, and it must precede the anion-exchange unit. Placing the anion exchanger first (reverse deionization) generally causes problems with precipitates of metal hydroxides, assuming that cations that form these compounds are present in the entering water or process stream. The resin in the anion-exchange unit may be a weak base resin when removal of anions (silicate [CAS: 12627-13-3], bicarbonate, fluoride, and others) of weak acids is not essential. Otherwise, the column must contain a strong base resin. An alternative is the installation of a third column, which contains the strong base resin, to be placed after the weak base anion-exchanger unit.

A column containing a mixture of cation- and anion-exchange resins is called a mixed bed. Although all types of resin have been considered for these units, the majority consist of strong acid cation and strong base anion exchangers. This system yields a higher quality deionized water or process stream than when the same resins are used in separate columns. Regeneration of a mixed bed is more complicated than of a two-bed system. A mixed bed containing the same total resin volume as in one of the two-bed columns lowers capital costs but must be regenerated at roughly twice the frequency. Mixed-bed units are preferred as final polishers for multiple-bed systems, recycling streams, condensate polishing, and other areas where the electrolyte concentration is in the low mg/L range or less.

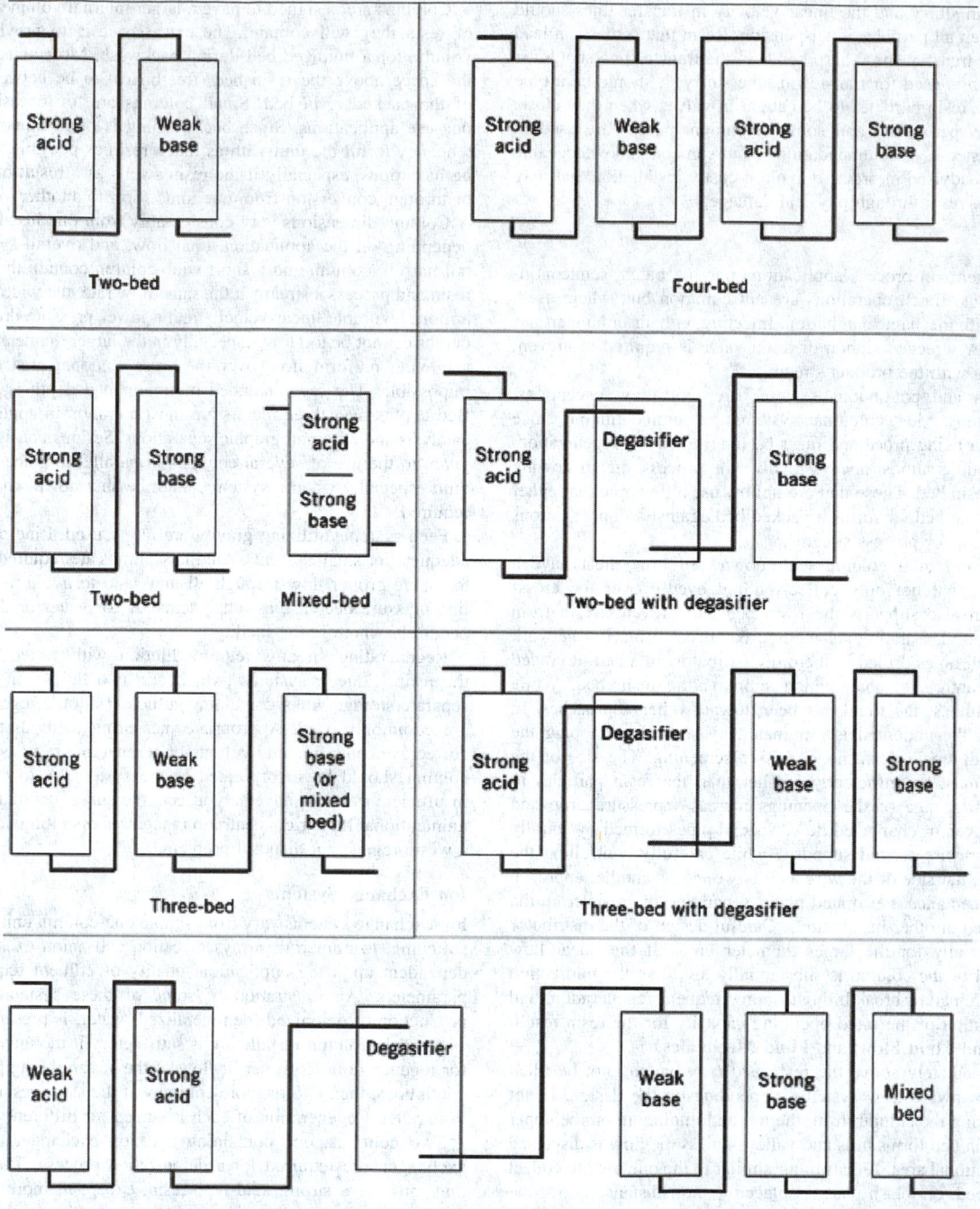

Fig. 2.   Various deionization systems. A degasifier facilitates the removal of dissolved gases.

Most ion-exchange columns are operated concurrently. Both the process stream and the regenerating solution flow through the resin bed in the same upward or downward direction. Downflow is more common. These streams flow in opposite directions in counterflow systems, as with downflow during the adsorption step and upflow during regeneration. Counterflow provides more efficient use of regenerating chemicals, a higher quality effluent, and higher operating capacity than is obtained with the same resins in concurrent systems. Installation costs are somewhat higher. Extra care must be taken during upflow regeneration since the higher specific gravity of the regenerating stream has a more buoyant effect on the resin particles. Fluidization lessens regeneration efficiency.

Weak acid cation exchangers are used in some deionization systems when bicarbonate alkalinity is higher than in normal water supplies. These resins are regenerated with much greater efficiency than strong acid resins, and operating capacity is higher. The column containing the weak acid resin is installed as the first in a chain or series of columns and removes all or part of the divalent cations, depending on the water composition. In an effort to reduce capital costs for each additional column placed in an ion-exchange system, some designs incorporate a layered bed approach. The two-bed weak acid column followed by a strong acid resin column is reduced to one column by layering the weak acid resin on top of the strong acid resin. A similar anion unit consists of a weak base resin layered on top of a strong base resin. The resins must not be mixed; otherwise, the advantages of using the weaker acidic or basic resin before the stronger version disappears. Much more care must be taken during backwash and regeneration to maintain the layers. Precipitation of calcium sulfate in the cation unit, and precipitation of silica in the anion unit, are potential problems during regeneration.

The most demanding requirements with respect to water quality are in the electronics industry and in very high pressure power plants. Although mixed-bed units are recognized for giving practically complete removal of all ionic constituents, the mixed-bed unit will give off trace amounts if systems are not designed to approach 100% separation of the two resins before regeneration. Any cation-exchange resin remaining with the anion exchanger is converted to the sodium form when the anion exchanger is regenerated with NaOH. Likewise, any anion-exchange resin remaining with the cation exchanger is converted to the sulfate form when the cation exchanger is regenerated with sulfuric acid. Resins returned to service, after remixing, gradually release sodium ions or sulfate [CAS: 14808-79-8] from resin exposed to the wrong regenerating solution. Dissociation of water provides the H^+ and OH^- needed for an exchange of ions to occur. Designs to overcome this problem include addition of an inert resin which occupies space between the anion and cation exchangers after backwash is completed. Special backwash towers have been designed for those systems which do not incorporate the inert resin. For example, a narrower diameter where the interface appears after separation minimizes the volume of resin which might contain some of each resin. That portion is saved for the next regeneration while resin above and below that zone is regenerated in other units.

Numerous efforts have been made to develop continuous ion-exchange systems in which resin moves intermittently or continuously in a direction opposite to the flow of all liquids during adsorption, backwash, regeneration, and rinse. These include resin-in-pulp (RIP) systems used in the uranium industry. Resin is placed in banks of baskets constructed from screens having a mesh size that allows ore particles in an acid or alkaline leached slurry to pass through. The baskets are dipped in and out of a trough through which the slurry passes. The banks of resin are advanced through water washes and regenerations by valving changes for solutions flowing through the troughs. In another design, a number of agitated tanks each containing an ion-exchange resin and the leached ore slurry are arranged in series. A vibrating screen is placed between each tank. A mix of resin and ore slurry flowing from a tank was pumped or air lifted to the vibrating screen to separate resin from slurry. Resin moves in one direction to another agitated tank; slurry moves to a different tank in the opposite direction.

Another continuous system consists of columns having numerous perforated plates. Resin enters the top and liquid is pumped into the bottom. The liquid flow rate is adjusted to prevent resin from passing through the openings in either direction. Liquid flow is stopped for a short period to allow resin to drop through openings to the chamber below. Liquid flow resumes before resin can drop through more than one chamber.

A unit referred to as the Higgins Loop has been popular in water treatment, as well as other applications. Resin is pulsed at regular intervals around a rectangularly shaped loop. The diameter of the adsorption section is larger than that of the regeneration section. See also **Water**.

A more recent approach, developed by Advanced Separation Technologies (Lakeland, Florida), involves the placement of a number of columns on a carousel that rotates constantly at an adjustable speed. Instead of having one tall fixed column, this system has one taller column which has been broken down into smaller columns on the rotating table. Each is connected in series which allows each column to be on the adsorption cycle beyond the normal breakthrough point typical of the larger column with no other column in series behind it. The number of columns in series during adsorption, backwash, regeneration, and rinse is variable. Liquids flowing into and out of each column change when the column reaches specific positions on the carousel as it rotates.

Many ion-exchange systems incorporate activated carbon [7440-44-0] columns to adsorb dissolved chlorine or high molecular weight organics. Reverse osmosis units are incorporated at times to lower the electrolyte concentration before ion exchange. See also **Reverse Osmosis**. Ultrafiltration units are installed to lower colloidal solids. Ultraviolet light systems are used to destroy microbiological organisms that tend to propagate in resins and recirculated water systems. Not only do these other water treatment procedures improve the quality of water produced but they also extend resin life. See also **Ultrafiltration**.

Lead-lag or merry-go-round systems are more common in other areas of application than in water treatment. In these systems, two or more columns of the same resin are connected in series and the lead column is retained on the adsorption cycle beyond a typical breakthrough point. The lead column is removed for regeneration when the concentration of the effluent from

that column is about equal to the influent concentration, or when leakage from one of the succeeding columns reaches a specified limit. The column that was second in line becomes the lead column. The column removed from the circuit is regenerated, rinsed, and then either put back into service in the last position or put on standby. These systems are most useful for recovery of valuable products, or for removal of toxic substances from waste streams. In either case, the objective is to use as much of the resin's total capacity as possible. Adsorbing more valuable product in this manner means a smaller amount of contaminating products are coadsorbed. When the system involves toxic substances, the regenerating stream is at higher concentration for precipitation or other means of disposal.

Health and Safety

Ion-exchange resins are not considered hazardous. However, cation exchangers when in the hydrogen form, and anion exchangers when in the hydroxide form, yield acidic and basic solutions, respectively, when in contact with neutral salt solutions. The corrosive potential should not be overlooked, and skin sensitivity has been reported occasionally, especially when gloves are not used when handling resin. Resins which have been used to remove toxic substances may slowly release these materials if the toxic substances are still attached to the resin. Burning resins, if not incinerated properly, release toxic and odorous fumes. Fumes from burning anion exchangers are particularly foul smelling.

A few resins go through unusually large volume changes when converted from one ionic form to another, when changing from one solvent to another, or when wetting dry resin with water or another solvent. Such changes may cause shattering of glass equipment if constrained. Oxidizing chemicals in contact with ion-exchange resins can result in rapid and uncontrolled degradation which may lead to rupture or bursting of the column in an explosive manner. Moreover, resin beads spilled on a floor are especially dangerous and can be the cause of serious accidents resulting from falls.

Applications of Ion-Exchange Resins

Water Treatment. The two primary applications in water treatment are softening and deionization. Other important but less frequently used applications include dealkalization, softening of produced water, desilicizing, and nitrate removal.

Food Processing. The sugar and corn sweetener industries have the largest volume of installed ion-exchange resin in the food processing industry. Lesser quantities are used to process wine, whey, fruit juices, and gelatin. See also **Food Processing**; **Fruit Juices**; **Gelatin** and **Grapes and Wines**.

Sugar, or sucrose [CAS: 57-50-1], is obtained from sugar cane as a juice by pressing cut canes, and from sugar beets by slicing the beets and extracting the sucrose with hot water. Organic and inorganic impurities must be removed from these extracts to obtain a white, crystalline product. See also **Sugar** and **Sugarcane**.

Cane sugar factories produce a raw crystalline sucrose which is shipped to the refinery where the raw sugar is redissolved and impurities are removed by precipitation, bone char or granular carbon, ion exchange, and crystallization. See also **Crystallization**. Resins are used for several purposes. Organic compounds responsible for color in sugar syrups are removed by char or carbon in most refineries. Additional removal is achieved by following the char or carbon system with columns of a strong base anion exchanger in the chloride form. There is a trend toward replacing char and carbon systems with resin for decolorization purposes. Deionization with a weak acid cation exchanger and a strong base anion exchanger is incorporated at those refineries desiring to produce a sucrose syrup with low color and a low ash content. A weak acid resin is selected over a strong acid resin to minimize conversion of sucrose to invert sugar (glucose and fructose). The soft drink industry, on the other hand, prefers invert sugar over sucrose as a syrup for sweetening purposes. Sucrose is inverted, or converted, to an approximate 50–50 mix of fructose and glucose by hydrochloric acid or a strong acid cation exchanger in the hydrogen form. See also **Carbonated Beverages**.

Beet sugar factories do not produce an intermediate raw, crystalline sucrose. Instead, the thin juice obtained from the beets passes directly through purification processes similar to those used by the cane sugar refineries. Ion exchange, as a processing step in beet sugar purification, is more common outside the United States.

Syrup derived from corn starch is called corn sweetener. Starch is converted to glucose enzymatically or by acid hydrolysis. Color, color precursors, and salts are removed by ion exchange. Most systems consist of several pairs of a strong acid cation exchanger followed by a weak base anion exchanger connected in series. The syrup, thus purified, is used in numerous food products. This same purified syrup is enzymatically processed to convert a portion of the glucose to fructose. In a process similar to the recovery of sucrose from molasses, as mentioned above, glucose and fructose are separated chromatographically by using a strong acid resin. See also **Starch**; and **Sweeteners**.

Whey, a by-product in the manufacture of cheese, is deacidified and deionized using a two-bed system consisting of a strong acid cation exchanger followed by a weak base anion exchanger. A mixed bed of a strong acid cation and a strong base anion exchanger is included, at times, after the two-bed system to achieve a higher degree of purification.

Fruit juices can be deacidified with a weak base anion-exchange resin. Removal of compounds which cause a bitter taste is a more popular application. It is accomplished with resins that have no ion-exchange functionality. In essence, they are similar to the copolymer intermediates used by resin manufacturers in the production of macroporous cation and anion exchangers. These products are called polymeric adsorbents. See also **Coffee Tree and Coffee**; and **Tea (Chucamellia Sinensis; Ternstromiaceae)**.

Wines are processed by ion exchange for two purposes: excess acidity, which is responsible for tartness, is removed using a weak base anion exchanger; newly fermented wine is supersaturated with respect to potassium bitartrate [CAS: 868-14-4] and, unless the concentration is reduced, a precipitate eventually forms. Precipitation is hastened in the traditional method of processing wine by storing it at a lower temperature (chill-proofing). The sediment is periodically removed (racking). The chill-proofing process is substituted with an ion-exchange process at numerous wineries. Precipitation is prevented by converting a portion of the potassium bitartrate to the more soluble sodium bitartrate [CAS: 526-94-3] when passing the wine through a strong acid cation exchanger in the sodium form.

Pharmaceutical. Ion-exchange resins are useful in both the production of pharmaceuticals and the oral administration of medicine. Antibiotics, such as streptomycin [CAS: 57-92-1], neomycin [CAS: 1404-04-2], and cephalosporin C [CAS: 61-24-5], which are produced by fermentation, are recovered, concentrated, and purified by adsorption on ion-exchange resins, or polymeric adsorbents. Impurities are removed from other types of pharmaceutical products in a similar manner. Resins serve as catalysts in the manufacture of intermediate chemicals. See also **Antibiotics**; and **Pharmaceuticals**.

Ground ion-exchange resins have been used for many years as carriers for drugs which are ingested. This method of dosing overcomes objectionable odors and tastes. Resins, especially those having strong acid or base functionality, provide slow (or sustained) release over many hours for those medicines adsorbed by them. A low cross-linked modification of a strong base styrene–divinylbenzene resin is dried and ground, then ingested for adsorption of bile acids in the treatment of people having high levels of blood cholesterol [CAS: 57-88-5]. The pharmaceutical generic name for this resin is cholestyramine [CAS: 11041-12-6]. A hemoperfusion system incorporating a polymeric adsorbent was developed to adsorb drug and drug metabolites from the blood of patients who had overdosed. See also **Drug Delivery Systems**; and **Controlled Release Technology, Pharmaceutical**.

Catalysis. Ion-exchange resins, especially the strong acid type, have long been recognized as excellent substitutes for sulfuric acid and other similar catalytic agents. Resins participate in fewer side reactions, and because they are insoluble, remain in the reactor and do not contribute to downstream corrosion problems. Neutralization of the reaction product is not necessary. Regeneration of a resin catalyst is not required unless the incoming reactant streams contain impurities that would be adsorbed by the resin. Separate ion-exchange units are used, at times, to remove the impurities from those streams before they enter the reactor. Numerous large reactor columns have been designed and installed in the petrochemical industry which has no facilities for regeneration. Resin is replaced when performance drops off, which may be well over one to two years of continuing operation.

Chemical Purification. Many organic and inorganic products manufactured in commercial quantities contain objectionable impurities which can be removed by ion exchange. Selection of the appropriate resin is important. Iron (III) is removed from hydrochloric acid using a strong base anion exchanger. See also **Hydrogen Chloride**. Divalent cations, eg, calcium, magnesium, strontium, and barium, are removed from saturated NaCl solutions or other monovalent salt solutions using chelate resins. Formic acid is removed from hot, concentrated formaldehyde [CAS: 50-00-0] using a weak base anion exchanger. Amines are removed from methanol [CAS: 67-56-1] using a strong acid-type cation exchanger. Salts are removed from dimethylformamide [CAS: 68-12-2] with a strong acid cation exchanger followed by a weak base anion exchanger. Gelatin [CAS: 9000-70-8] is purified in a similar manner, although a strong base anion exchanger is generally preferred over a weak base resin. See also **Gelatin**. Oxazole [CAS: 288-42-6] is removed from acrylonitrile [CAS: 107-13-1] using a strong acid resin. See also **Acrylonitrile**.

Metal Processing. Plating, etching, anodizing, pickling, and galvanizing involve chemical solutions or baths that are used repeatedly until the impurity concentration increases to a level where additional use of the bath impairs performance. Destruction of the bath is costly and creates a disposal problem. Ion-exchange units are installed at numerous locations to lower the impurity concentration to an acceptable value. Complete removal is not necessary. Examples include the removal of iron, copper, and trivalent chromium(III) from chromic acid [CAS: 1333-82-0] plating baths, the removal of copper from etching solutions, the removal of aluminum [CAS: 7429-90-5] from anodizing baths, removal of iron from pickling acids, and removal of iron from acidic zinc sulfate [CAS: 7733-02-0] $ZnSO_4$, galvanizing baths. Resin selection is critical for success of the process. See also **Electroplating**.

Hydrometallurgy. Uranium [CAS: 7440-61-1] recovery from sulfuric acid leaching and bicarbonate leaching operations involved the largest use of ion-exchange resins in the hydrometallurgical area. Activity was at its peak in the 1970s.

Uranium ores are leached with dilute sulfuric acid or an alkaline carbonate [CAS: 3812-32-6] solution. Hexavalent uranium forms anionic complexes, such as uranyl sulfate [CAS: 56959-61-6], $UO_2(SO_4)_3^{-2}$, which are more selectively adsorbed by strong base anion exchangers than are other anions in the leach liquors. Sulfate complexes are eluted with an acidified NaCl or ammonium nitrate [CAS: 6484-52-2], NH_4NO_3, solution. Carbonate complexes are eluted with a neutral brine solution. Uranium is precipitated from the eluent and shipped to other locations for enrichment.

Other hydrometallurgical uses for resin have been small in comparison.

Waste Treatment. Environmental concerns have increased the need to treat liquid discharges from all types of industrial processes, as well as runoffs where toxic substances appear as a result of leaks or following solubilization. See also **Wastes and Pollution**. One method of treatment consists of an ion-exchange system to remove the objectionable components only. Another involves complete or partial elimination of liquid discharges by recycling streams within the plant. This method is unacceptable unless a cyclic increase in the impurities is eliminated by removing all constituents prior to recycling.

Numerous toxic metals are lowered below discharge limits through use of the standard cation- and anion-exchange resins, or by using specially formulated resins which have much higher selectivities.

Recycling systems incorporate cation and anion exchangers to remove all electrolytes. Other nontoxic material, common to the process, appear in the process water along with the toxic substances and, unless removed, lower the quality of the water. By using a cation- and anion-exchange resin system, the water produced will be of higher quality than water normally available to the facility. Deionization of recycled water is less costly if the electrolyte concentration is lower than in external water supplies. If the concentration is greater in recycled water, elimination of streams with high electrolyte concentrations and free of toxic substances should be considered. Recycling lowers regeneration chemical costs and minimizes the volume of water that must be purchased from an external source.

Numerous organic compounds cannot be removed efficiently with ion-exchange resins or polymeric adsorbents, but others can.

Toxic substances adsorbed on resins are removed during a regeneration procedure. The resulting spent regeneration solution has a higher concentration of the toxic substance than the stream from which it was removed by the resin. Toxic material in the spent regenerating solution can usually be precipitated, electrodeposited as in an electrolytic cell, or made insoluble by other acceptable procedures.

Gas Adsorption. There are few commercial installations. Ammonia [CAS: 7664-41-7] is adsorbed by a cation exchanger in the hydrogen form and eluted with an acid to give ammonium sulfate or ammonium chloride [CAS: 12125-02-9]. Success has been reported on the removal of sulfur dioxide [CAS: 7446-09-5] on a weak base anion exchanger. See also **Adsorption: (Gas Separation)**. Chemical compounds such as phenol, ethylene dichloride [CAS: 107-06-2], and benzene [CAS: 71-43-2] have been successfully removed on polymeric adsorbents. The concern with systems for removing impurities from air, or other gaseous streams, is the high pressure drop typical of high velocities through beds of small-diameter resin particles. Other concerns are water content of both the resin and the gaseous stream, temperature, and cost effective regeneration procedures, especially for organic substances.

Analytical. Ion-exchange resins have been extremely valuable for a variety of analyses. Total ion electrolyte concentration can be determined by analyzing for total cations or total anions using a cation exchanger in the hydrogen form or an anion exchanger in the hydroxide form. Ions present in solution at very low concentrations are concentrated by adsorption on a resin before eluting and analyzing the effluent by standard procedures. Ions that are interferences for analytical procedures are eliminated by adsorption on an ion exchanger. The progress of a large-scale plant reaction can be monitored by following the disappearance of a reactant. Impurity levels in a finished product are determined by ion exchange if it is adsorbable. Numerous commercial processes have evolved from analytical separations and purifications practiced in the laboratory as an analytical procedure.

Ion chromatography (ic) is a highly valued and growing methodology for analytical analysis of ionic constituents in aqueous streams. In contrast to the chromatographic separations mentioned earlier with conventional resins, ion chromatography uses similar, yet different, resins which yield separations that are measured in minutes, rather than hours and days. See also **Chromatography**. Resins are somewhat smaller in size and have most of the functional groups on or near the outer surface, in contrast to being distributed throughout the resin matrix in conventional resins. The outer surface functionality shortens the path from liquid phase to resin phase and is the factor not only for more rapid separations, but also separations with little overlap in peaks for separate ions. With functionalization limited to the outer shell, the capacity of the resin is significantly reduced. Ion chromatography is generally considered for the more dilute streams where concentrations extend down to the mg/L and µg/L ranges. Ion chromatography resins are placed in a narrow separator column which is followed by a suppressor column and an analytical instrument to pick up signals in the effluent stream. Separate systems are used for cations and anions.

See also **Ion Exchange**; and **Resins**.

Additional Reading

Alper, J.: "Archimedes, Plato Make Millions for Big Oil: Zeolite Structure," *Science*, **1190** (June 8, 1990).

Bauman, W.C.: U.S. Patent 2,684,331 (1954).

Cavender, M.R., H.-L. Chiang and K. Myers: "Optimize Ion Exchange Resins Replacement," *Chem. Eng. Progress*, **56** (September 1992).

Inglezakis, V., and S. Poulopoulos: *Adsorption, Ion Exchange and Catalysis: Design of Operations and Environmental Applications*, Elsevier Science & Technology Books, New York, NY, 2006.

Kerr, G.T.: "Synthetic Zeolites," *Sci. Amer.*, **100** (July 1989).

Korkisch, J.: *Handbook of Ion Exchange Resins*, Vols. I–VI, CRC Press, LLC, Boca Raton, FL, 1989.

Korkisch, J.: *Handbook of Ion Exchange Resins: Their Application to Inorganic Anal*, CRC Press, LLC, Boca Raton, FL, 1999.

Korkisch, J.: *Concise Handbook of Ion Exchange Resins in Analytical Chemistry*, CRC Press, LLC., Boca Raton, FL, 2001.

Kunin, R.: *Ion Exchange Resins*, 2nd Edition, Krieger Publishing Company, Melbourne, FL, 1990.

Lawton, S.L. and W.J. Rohrbaugh: "The Framework Topology of ZSM-18, a Novel Zeolite Containing Rings of Three (Si, Al)-O Species," *Science*, **1319** (March 16, 1990).

Sengupra, A. K.: *Ion Exchange Technology: Advances in Pollution Control*, CRC Press, LLC, Boca Raton, FL, 1995.

Sengupra, A. K., and Y. Marcus: *Ion Exchange and Solvent Extraction: A Series of Advances*, CRC Press, LLC, Boca Raton, FL. 2004.

Wachinski, A. M.: *Ion Exchange Treatment for Drinking Water*, American Water Works Association, Denver, CO, 2006. http://www.awwa.org/.

Weib, J.: *Handbook of Ion Chromatography*, 3rd Edition, John Wiley & Sons, Inc., New York, NY, 2003.

Weitkamp, J. and H.G. Karge: *Zeolite Science, 1994: Recent Progress and Discussions*, Elsevier Science, New York, NY, 1995.

Weitkamp, J. and L. Puppe: *Catalysis and Zeolites: Fundamentals and Applications*, Springer-Verlag, Inc., New York, NY, 1999.

Zoccolante, G.V.: "Produce Ultrapure Process Water," *Chem. Eng. Progress*, **69** (December 1990).

IONIC CHARGE.   Either the total charge carried by an ion or the charge carried by an ion that has unit charge. Since ions owe their charges to gain or loss of electrons, unit charge is the charge on an electron, and all ionic charges are either equal in magnitude to this value or integral multiples of it.

IONIC COMPOUND.   See **Compound (Chemical)**.

IONIC CRYSTAL.   A crystal that consists effectively of ions bound together by their electrostatic attraction. Examples of such crystals are the alkali halides, including potassium fluoride, potassium chloride, potassium bromide, potassium iodide, sodium fluoride, and the other combinations of sodium, cesium, rubidium or lithium ions with fluoride, chloride, bromide or iodide ions. Many other types of ionic crystals are known.

IONIC EQUILIBRIUM.   In a system containing ions, at any particular temperature and pressure, the conditions at which the rate of dissociation of unionized molecules, or other particles to form ions, is equal to the rate of combination of the ions to form the unionized molecules, or other particles so that activities and concentrations remain constant as long as the conditions are unchanged.

IONIC MIGRATION.   The movement of charged particles of an electrolyte toward the electrodes under the influence of the electric current.

IONIC MOBILITY.   1. The ratio of the average drift velocity of an ion in solution to the electric field. It is expressed by the relationship

$$\mu_+ \text{ or } \mu_- = \frac{\lambda_+ \text{ or } \lambda_-}{F}$$

in which μ_+ or μ_- is the mobility of the ion, λ_+ or λ_- is the ion conductance, i.e., the contribution of the particular ion to the equivalent conductance, and F is the Faraday constant.

2. For gaseous ions in an electric field, the quantity k defined by the relationship

$$k = vp/E$$

where v is the drift velocity, p, the gas pressure, and E, the electric field strength.

3. Conduction of electricity in ionic crystals is due to the motion of lattice defects, either of the Schottky or Frenkel type. The mobility is given by

$$\mu = (eD_0/kT)e^{-E/kT}$$

where D_0 is a numerical constant, and E is an activation energy, which depends on the energy required to make a defect and on the height of the energy barrier that must be surmounted in order that the defect may move.

IONIC POTENTIAL. The ratio of the charge on an ion to its radius.

IONIC RADIUS. See **Chemical Elements**.

IONIC STRENGTH. A mathematical quantity used to evaluate the effectiveness of the forces restricting the freedom of ions in an electrolyte, and defined as one-half the sum of the terms obtained by multiplying the total concentration of each ion by the square of its valence, i.e., $\mu = \frac{1}{2}\Sigma c_i z_i^2$, where μ is the ionic strength, c is ionic concentration and z is valence.

ION IMPLANTATION. A process for introducing alloying elements into a host material by accelerating the ions to a high energy (at least tens of kilovolts) and allowing them to strike the surface of the host. The impinging atoms penetrate into the substrate material to a depth of 0.01 to 1 micrometer, depending on the atomic number and energy of the atom, and create a thin alloyed surface layer on the substrate. The process differs from others, such as electroplating, in that it does not produce a discrete coating, but rather it alters the chemical composition near the surface of the base material.

In recent years, the electronics industry has made increasing use of ion implantation as a method of doping semiconductors. Since the number of ions implanted is determined by the charge transferred to the substrate and their depth distribution by the incident energy, ion implantation has improved the controllability and reproducibility of certain semiconductor device processing operations. Also, ion implantation processes do not require the high temperatures needed to introduce impurities by diffusion. Thus the limitations arising from the changes produced in materials by high temperature are eased. Ion implantation also has been used in electronics to change the magnetic properties of substrates used for magnetic bubble devices.

Ion implantation also has promise in other fields involving surface technology; for example, new metallurgical phases with prior unknown properties can be formed. In some cases, such as heavy implantations of tantalum in copper of phosphorus in iron, amorphous or glassy phases can be formed. Or, if the implanted atoms are mobile, inclusions and precipitates can be formed as, for example, implanted argon and helium atoms are insoluble in metals and may form bubbles. The composition of a surface layer can be changed by differential sputtering caused by the implanted ions.

The damage and high concentrations of lattice defects, resulting from atomic displacements produced by the incident atoms, can change the chemical reactivity and mechanical hardness of a treated surface. Implantation can enhance the diffusion of impurities already deposited in a substrate, presumably through the motion of the high concentrations of lattice defects produced by the incident ions.

One of the most promising nonelectronic applications of ion implantation involves surface treatment to improve the hardness and wear resistance, as well as lowered susceptibility to corrosion, of metals. In some experiments, the benefits of ion implantation on wear may persist to a depth 103 times that of the implanted layer thickness. The implanted atoms are apparently transported into the metal as a tool wears. Thus, the technology is of large interest in connection with improving cutting tools and bearings. Some experiments have suggested that nitrogen implantation increases the fatigue life of carbon steel parts. The results are consistent with present understanding of the mechanisms of fatigue failure. It is well known that fatigue cracks start at the surface and that there is a close connection between surface hardness and fatigue life. Compressive stresses due to the presence of additional implanted ions may also play a role in the suppression of crack initiation.

The production of corrosion-resistant materials by alloying is well established, but the mechanisms are not fully understood. It is known, of course, that elements like chromium, nickel, titanium, and aluminum depend for their corrosion resistance upon a tenacious surface oxide layer (passive film). Alloying elements added for the purpose of passivation must be in solid solution. The potential of ion implantation is promising because restrictions deriving from equilibrium phase diagrams frequently do not apply (i.e., concentrations of elements beyond the limits of equilibrium solid solubility might be incorporated). This can lead to heretofore unknown alloyed surfaces which are very corrosion resistant.

Ion plating is another area of surface treatment. Ion plating is carried out in a gaseous electrical discharge in which the substrate to be plated is the cathode. The discharge is created by an applied potential of 500 to 5000 V. The primary component of the gaseous environment usually is an inert gas, most often argon. Atoms of the material to be plated are introduced into the gas by evaporation from a heated source. A fraction of the atoms injected by evaporation are ionized before striking the substrate. In ion plating, atoms arrive at the surface with energies of only a few hundred volts and penetrate no more than a few lattice constants into the substrate. Thus, ion implantation produces an alloyed surface layer whose composition varies continuously with depth because of the rather broad distribution of the ranges of the implanted ions, while ion plating produces a coating, the composition of which is independent of the nature of the substrate.

Semiconductor Applications. In semiconductor manufacture, the area of the workpiece into which ions are implanted is quite small. High homogeneity is sought in semiconductor applications, that is, the concentration of the implanted species should not vary by more than a few percent over the surface of a wafer. The implantation of ions into semiconductors is usually patterned, that is, some areas of the substrate are covered by a mask that stops the incident ions before they enter the substrate. A doped layer in which the implanted atoms are locally in an equilibrium phase is usually desired. Thus, implantation is usually followed by a high-temperature annealing treatment, which removes radiation damage through diffusion of lattice defects to defect sinks and the recrystallization of disturbed regions. Laser annealing has been used successfully. Laser annealing affects only a surface layer approximately equal to the depth of typical implantations, leaving the bulk of the piece unaltered.

See also **Semiconductors**.

IONIZATION. A process that results in the formation of ions. Such processes occur in water, liquid ammonia, and certain other solvents when polar compounds (such as acids, bases, or salts) are dissolved in them. Dissociation of the compounds occurs, with the formation of positively and negatively charged ions, the charges on the individual ions being due to the gain or loss of one or more electrons from the outermost orbits of one or more of their atoms. The ionization of gases is a process by which atoms in gases similarly gain or lose electrons, usually through the agency of an electrical discharge, or passage of radiation, through the gas.

Ionization by collision is an ionization process occurring by removal of an electron or electrons from an atom as the result of the energy gained in a collision with a particle (or quantum of radiation) possessing sufficient energy.

Specific ionization is the number of ion pairs formed per unit distance along the track of an ion passing through matter. This is sometimes called the total specific ionization to distinguish it from the primary specific ionization, which is the number of ion clusters produced per unit track length. The relative specific ionization is the specific ionization for a particle of a given medium relative either to that for (1) the same particle and energy in a standard medium, such as air at 15 °C and 1 atmosphere, or (2) the same particle and medium at a specified energy, such as the energy for which the specific ionization is a maximum.

Total ionization is a term used to denote either the total specific ionization (defined above); or the total electric charge on the ions of one sign when the energetic particle that has produced these ions has lost all of its kinetic energy. For a given gas the total ionization is closely proportional to the initial energy and is nearly independent of the nature of the ionizing particle. It is frequently used as a measure of particle energy.

Minimum ionization is the smallest possible value of the specific ionization that a charged particle can produce in passing through a particular substance. When the specific ionization produced along the path of a charged particle is plotted as a function of the particle energy, minimum ionization appears as a broad dip, bound on one side by a rather sharp rise for decreasing particle energy, and on the other side by a gradual rise for increasing particle energy. For singly charged particles in ordinary air, the minimum ionization is about 50 ion pairs per centimeter of path. In general, it is proportional to the density of the medium and the square of the charge of the particle. It occurs for particles having velocities of 95% of the velocity of light, which corresponds to a kinetic energy of 1 MeV for an electron, 2 BeV for a proton and 8 BeV for an alpha-particle.

Ionization potential is the energy per unit charge, for a particular kind of atom, necessary to remove an electron from the atom to infinite distance. The ionization potential is usually expressed in volts, and is numerically equal to the work done in removing the electron from the atom, expressed in electron-volts. See also **Chemical Elements**.

IONIZATION CHAMBER.   An instrument constructed to measure the number of ions within a gas-filled enclosure between two electrodes, across which a voltage is applied. These electrodes may be in the form of parallel plates or of coaxial cylinders. One of the electrodes may be the wall of the vessel itself. When the gas between the electrodes is ionized by any means, as by x-rays or radioactive emission, the ions move to the electrodes of opposite sign, thus creating an ionization current, which may be measured by a galvanometer or an electrometer.

IONIZED GASES.   Various agencies, such as fast-moving electrons, alpha particles, various forms of radiation, and high temperature, are capable of dislodging electrons from atoms or molecules of a gas and thereby leaving them positively charged. Some of the dislodged electrons may attach themselves to other molecules and render them negatively charged. In some cases, two or more electrons may be removed from the same molecule, or a molecule with a double positive charge may unite with a singly charged negative molecule, forming a singly charged complex, etc. Such charged atoms, molecules or molecular groups are called ions, and their production from neutral molecules is called ionization. The complete separation of an electron from a molecule or an atom requires a definite amount of energy. This may be expressed in ergs, but is more commonly given in electron volts (1.59×10^{12} erg), its value being the ionization potential. A lesser amount of energy may excite the atom or molecule to emit radiation, but will not ionize it.

If an ionized gas is left to itself, the ions soon recombine and become neutral. But if it is subjected to an electric field, as in an ionization chamber, the ions pass to the electrodes, such a migration being an "ionization current." Such currents, commonly called electric discharges, are attended by diverse phenomena and vary widely in character from the silent glow discharge to the lightning stroke.

At ordinary pressures, discharges may be classified into four types: (1) If the voltage between two electrodes in open air is gradually increased, the electrodes become surrounded with a luminosity. This "glow" or "corona" gives way, at the negative electrode first, to (2), a "brush," composed of hair-like branches. (3) Finally, the disruptive spark passes. (4) Under other conditions, an arc may be formed. If, however, the electrodes are enclosed in a tube and the pressure reduced, a point is reached at which the tube becomes filled with a beautiful luminosity. Close examination shows this to have structure. Very close to and surrounding the cathode is a thin, luminous layer *c*, the cathode glow (Fig. 1); and outside this, the Crookes dark space *C*. Next, extending toward the anode, is the short negative glow *n*, then the Faraday dark space *F*. From this to the anode extends the long positive column *p*, with its regular, transverse striations. As the pressure is further reduced, the cathode dark space enlarges and the other features dwindle toward the electrodes until they finally disappear at about 0.001 mm pressure. From this point on, the cathode rays are the predominant feature.

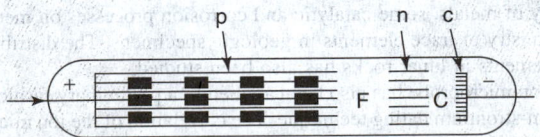

Fig. 1.   Elementary gas-discharge tube.

Upon exploring the discharge in a Crookes tube with suitable probes, it is found that in certain regions the positive and negative ions are so nearly equal in number as to neutralize each other's effect. Such a region is called a "plasma." The plasma may be surrounded by a "sheath" of ionized gas in which ions of one sign greatly predominate, the effect being that of a space charge.

IONIZING ENERGY.   The average energy given up by an ionizing particle in producing an ion pair in a specified gas.

IONIZING PARTICLE.   A particle that produces ion pairs in its passage through a substance. Ionizing particles may be divided into two groups: (1) *directly ionizing particles* — charged particles (electrons, protons, alpha particles, and so on) having sufficient kinetic energy to produce ionization by collision; and (2) *indirectly ionizing particles* — uncharged particles, such as neutrons and photons, which can liberate directly ionizing particles or can initiate a nuclear reaction.

Ionizing radiation is any radiation consisting of directly or indirectly ionizing particles, or a mixture of both. Ionizing radiation, unless controlled, poses a biological and environmental hazard. See also **Cancer Research**.

ION MEAN LIFE.   The average time interval under specified atmospheric conditions between the formation and destruction of an ion of any given type. The mean life of small ions in clean air, for example, over the sea, is four to five minutes, but in polluted air it is generally less than a minute. Large ions have mean lifetimes of as much as 15 to 20 minutes over the oceans, while in very polluted areas, lifetimes may approach an hour.

ION MICROPROBE MASS ANALYZER.   An instrument designed to provide an in situ mass analysis of microvolume of the surface of a solid sample. The analysis is accomplished by bombarding the surface with a high-energy beam of ions which causes the atoms at the surface to be sputtered away. A fraction of the sputtered particles is electrostatically charged and these sputtered ions are collected and analyzed according to their mass-to-charge ratio in a mass spectrometer.

Figure 1 represents a schematic diagram of the instrument. The ions used for sample bombardment are generated in a hollow cathode, dual plasmatron ion source capable of producing ions of a wide variety of gases including those of a highly electronegative character. The ions which can be either positively or negatively charged, are accelerated to energies ranging from 5.0 to 22.5 kilovolts and passed through the primary mass spectrometer. The spectrometer permits the analyst to select and purify, by mass separation, a specific chemical species from those produced in the ion source. The purified ion beam is focused to a small probe in an electrostatic lens column and allowed to impinge on the surface of the sample. The diameter of the ion probe may be varied continuously from about 2 to 500 micrometers. The sample and the point being analyzed can be viewed through an optical microscope while under bombardment.

Sputtered ions are collected and their masses analyzed in a double-focusing mass spectrometer in which the velocity dispersions of the magnetic and electric sectors are matched to permit the acceptance of a wide range of initial energies of the sputtered ions. No entrance slit is used and the bombarded area is stigmatically focused directly onto the resolving slit.

The ion beams are then detected with a high gain device that permits single ion counting. Sputtered ions from the sample eject secondary electrons at the conversion electrode and these are accelerated towards the scintillator of a photomultiplier tube where the light produced by their impact is detected. The resolved ion signals can be read as count rates from scalers which can accommodate rates in the megacycle range within significant dead time losses or as direct-currents on chart recorders.

Analytical Method. The analytical method applied with this instrument is based upon the observation that the yields of sputtered ions are greatly affected by the surface chemistry of the sample. When a metal such as aluminum is bombarded with ions of inert gas such as argon, the yield of positive aluminum ions falls exponentially with time. The ability of the sample to yield positive ions is progressively destroyed by the bombardment. On the basis of the similar behavior of many metals under bombardment by inert gases, it was postulated that the production of sputtered ions is a function of the electronic properties of the surface. The ability to extract positive ions from the sample diminishes as the strongly bonded compounds formed on the surface of the sample through the chemisorption of reactive gases are removed by the eroding action of the bombarding ion beam. It has been shown that the production of positive ions may be maintained at a higher level by controlling the surface

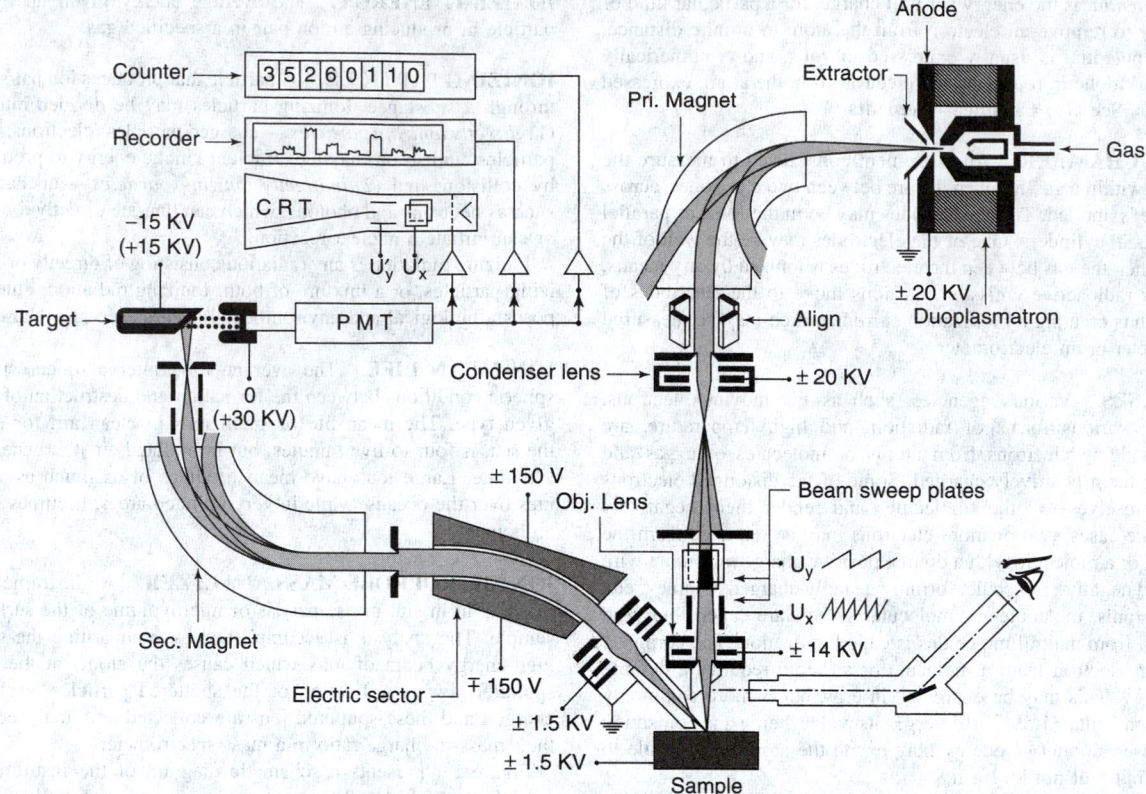

Fig. 1.   Schematic representation of the ion microprobe mass analyzer. (*Bausch & Lomb/ARL.*)

chemistry through a proper selection of the species of bombarding ions. Instead of destroying the necessary chemical compounds with an inert gas, it is possible to reconstitute them by bombarding with a reactive gas. Enhanced stable yields of sputtered positive ions of many pure elements have been produced, by bombarding them with beams of carbon, nitrogen, oxygen, chlorine, and iodine ions.

Figure 2 illustrates the relative sputtered ion intensities of some pure elements subjected to bombardment by oxygen ions $^{16}O^-$. The relative intensity for each isotope has been corrected only for its natural abundance.

Application. Ion sputtering mass spectrometry has been applied to several problems in the analysis of solids with various types of instruments. These include studies of semiconductor devices as shown in Fig. 3, oxygen

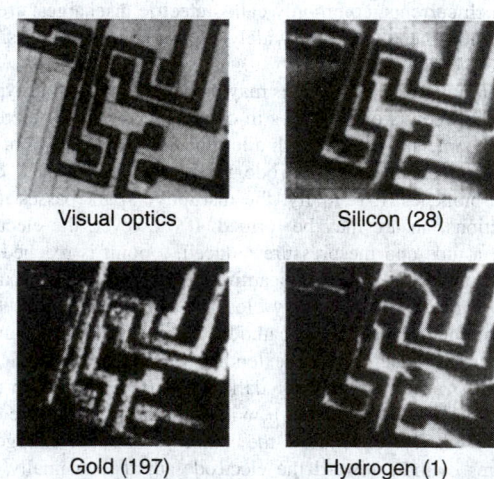

Fig. 3.   Ion images of semiconductor device. (*Bausch & Lomb/ARL.*)

concentrations and concentration gradients and of processes of oxidation in a variety of metals, some catalytic and corrosion processes on metals, and the chemistry of trace elements in geologic specimens. The distribution of trace elements in lunar rocks has also been studied.

The ion microprobe has also been applied in a preliminary fashion to the rubidium-strontium dating technique. The correlation of the ion microprobe results with the independently determined isochron indicates that it may be possible to obtain useful results for samples on a micrometer scale from this dating technique.

The ion microprobe mass analyzer's unique features permit three dimensional microanalysis of all elements in the periodic table and in addition, the determination of their relative isotopic abundances in a given matrix. Both conductors and insulators may be analyzed. The instrument is applicable in many areas of the science of solid materials analysis. Most elements will have optimum yields in the spectrum of positive sputtered ions and will be detected in concentrations of parts per million

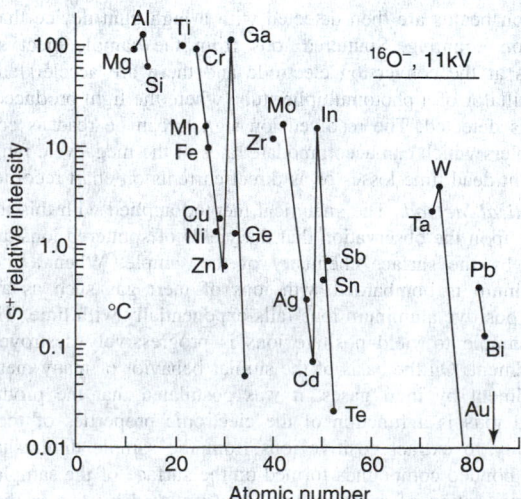

Fig. 2.   Relative sputtered ion intensities of some pure elements subjected to bombardment by oxygen ions $^{16}O^-$.

in micrometer-sized sampling areas. Electronegative elements will be detected with similar sensitivities in the spectrum of negative sputtered ions but inert gases which are ionized with difficulty and have small electron affinities will be detected with considerably poorer sensitivities. In general, it is possible to measure isotope ratios without chemical separation of the constituent elements of the sample. A controlled sputtering process provides mono-layer resolution of depth profiles and also the ability to make precise in-depth analyses of thick and thin films. Detection efficiency of the instrument can yield quantitative accuracy in the parts per billion range in many applications. The precision of an ion microprobe isotope ratio measurement depends basically upon the counting rates involved and its accuracy can approach its precision if auxiliary standards are used.

Additional Reading

Breese, M.B.H., P.J. King and D.N. Jamieson: *Materials Analysis Using Nuclear Microprobe*, John Wiley & Sons, Inc., New York, NY, 1995.

Goldstein, J.I. and H. Yakowitz: *Practical Scanning Electron Microscopy: Electron and Ion Microprobe Analysis*, Perseus Books, Boulder, CO, 1975.

Murr, L.E.: *Electron and Ion Microscopy and Microanalysis: Principles and Applications*, Marcel Dekker, Inc., New York, NY, 1991.

WINSTON G. SHEQUEN, P.E., Bausch & Lomb/ARL, Sunland, CA

IONOMERS. The generic term *ionomer* was introduced by DuPont in 1964 in conjunction with the commercialization of the new Surlyn resins to denote a thermoplastic polymer containing both covalent and ionic bonds, and having properties influenced to substantial effect by the ionic bonding. Since that time, the meaning has been expanded to include many compositions such as the glass ionomers used in dentistry which cannot be melt processed. In the interest of clarity and consistency, it is proposed that the term ionomer be reserved for polymers having melt viscosities suitable for conventional melt processing methods. Descriptions such as ion-containing or ion-linked are appropriate for highly viscous or true thermoset materials.

Despite the broad scope of the field and the unusual property combinations obtainable, commercial exploitation has been confined mainly to the original family based on ethylene copolymers. Within certain industries, such as flexible packaging, the word ionomer is understood to mean a copolymer of ethylene with methacrylic or acrylic acid, partly neutralized with sodium or zinc.

Ethylene-Based Ionomers

The semicrystalline, ethylene-based ionomers of commerce are flexible, transparent polymers notable for high strength and elasticity in both solid and molten states. The ionic bonding is completely reversible and has a strong influence on properties, even at temperatures well above the melting point.

The issue of mechanical property changes over time has been addressed and a structural model has been developed. A correlation was established between stiffness and the size of an endotherm (T_i), normally seen in dsc scans of ionomers at about 50 °C. This endotherm increases in size with increasing neutralization. The T_i endotherm disappears completely when the dsc measurement is repeated immediately, but then gradually reappears during room-temperature storage.

In addition to time-related effects, the solid-state physical properties are also affected by adsorbed water, which functions as a plasticizer. Water pickup is affected by the nature of the cation, with sodium ionomers absorbing about 10 times the level of the zinc equivalent under the same conditions.

Ionomers are much less hazy than the ethylene acid copolymers from which they are derived. Studies with optical and electron microscopes have shown that this is due to suppression of the spherulitic structure by the metal ions. Surprisingly, x-ray diffraction has shown that polyethylene crystallinity is present in the ionomers. A typical level of crystallinity is 30%.

The melt viscosity of an acid copolymer increases dramatically as the fraction of neutralization is increased.

Softening is apparent over a wide range, while the melt is strong and elastic. This gradual melting is beneficial in heat-sealing applications.

Ionic bonding with metal ions decreases solubility in organic solvents. At high neutralization levels with alkali metal ions, many ionomers

spontaneously form colloidal suspensions in water when stirred vigorously at 100–150 °C under pressure. These provide convenient methods for applying thin coatings of ionomers to paper and other substrates.

Due to the comparatively low content of polar groups, most commercial ionomers are very good insulating resins.

Acid copolymers are less permeable to natural oils than conventional homopolymers, and this difference increases greatly when they are neutralized.

In the area of gas permeability, the low crystallinity of a typical ionomer (~30%) results in relatively high permeability to oxygen. For packaging of fresh meat this is advantageous, but in other packaging areas, combination with a barrier layer may be required.

Most commercial processes involve copolymerization of ethylene with the acid comonomer followed by partial neutralization, using appropriate metal compounds.

Many methods for the conversion of acid copolymers to ionomers have been described by DuPont. The chemistry involved is simple when cations such as sodium or potassium are involved, but conditions must be controlled to obtain uniform products. Solutions of sodium hydroxide or methoxide can be fed to the acid copolymer melt, using a high shear device such as a two-roll mill to achieve uniformity. All volatile by-products are easily removed during the conversion, which is run at about 150 °C.

During processing at elevated temperatures, normal precautions are needed to prevent accidental burns. Surlyn ionomers have U.S. Food and Drug Administration clearance for food contact.

Flexible packaging is the largest commercial application area for ethylene ionomers. Uses include monolayer films, coextruded components of multilayer films, and coatings on paperboard or aluminum. The key properties are broad heat-sealing range, ability to seal through oily contamination, high melt strength, oil resistance, clarity, and impact resistance.

The unusual resilience and roughness of ionomers have resulted in sporting goods applications, including golf ball covers and bowling pin coatings. Ionomers are easily foamed due to high melt strength, and the foams are durable, leading to uses in construction, skilifts, and softball cores.

Additional Reading

Eisenberg, A., and Joon-Seop Kim: *Introduction to Ionomers*, John Wiley & Sons, Inc., New York, NY, 1998.

Frisch, K., and H.X. Xiao: *Advances in Urethane Ionomers*, CRC Press, LLC., Boca Raton, FL, 1995.

MacKnight, W.J. and T.R. Earnest: *J. Macromol. Rev.*, **16**, 41 (1981).

Rees, R.W.: in K.C. Frisch, ed., *Polyelectrolytes*, Technomic Publishing Co., Inc., Westport, Conn., 1976, pp. 177–197.

Rees, R.W.: in J.I. Kroschwitz, ed., *Encyclopedia of Polymer Science and Engineering*, 2nd Edition, Vol. 4, Wiley-Interscience, New York, NY, 1986, pp. 395–417.

Schlick, S.: *Ionomers: Characterization, Theory, and Applications*, CRC Press, LLC., Boca Raton, FL, 1996.

Zutty, N.L., J.A. Faucher, and S. Bonotto: in N.M. Bikales, ed., *Encyclopedia of Polymer Science and Technology*, Vol. 6, Interscience Publishers, a Division of John Wiley & Sons, Inc., New York, NY, 1967, p. 420.

IONOSPHERE. The atmospheric region containing significant concentrations of ions and electrons. Its base is at about 70–80 kilometers (44–50 miles) and it extends to an indefinite height. In terms of the standard upper-atmospheric nomenclature, the ionosphere is collocated with the thermosphere and the upper mesosphere, while the outer ionosphere also forms part of the magnetosphere. Most of the early knowledge of the ionosphere came from sounding by ground-based radars known as ionosondes, and the sharply defined echoes produced by these instruments led to the definition of ionospheric layers, implying well-defined regions with clear maxima of ionization. More recent investigations by a number of remote and in situ techniques have shown that such sharply bounded layers do not generally exist (with the exception of the so-called sporadic-E layers), and that the layers are better described as regions. The term layer is still frequently used, however. Ionospheric sounding normally shows an E-layer in the 100- to 120-kilometer height range (62 to 75 miles), an F_1-layer in the 150- to 190-kilometer range (93 to 118 miles), and an F_2-layer above 200 kilometers (124 miles). The D-region, lying in the upper mesosphere below 100 kilometers (62 miles), is responsible for daytime absorption of high-frequency radio waves, but does not usually produce an echo on ionosonde recordings. Early suggestions of a C-layer below the D-region, and a G-layer above the F_2-layer, have not been generally accepted, and these terms are now obsolete. Over most of the earth, the ionosphere is produced by the action of solar radiation of short wavelength (extreme ultraviolet and x-ray radiation) on the atmospheric constituents. At high magnetic latitudes, energetic particles of solar or auroral origin become important, and even dominant, sources of ionization. See also **Atmosphere (Earth)**.

Radio Application

The ionized air high above the earth's surface, the existence of which was surmised independently in 1902 by Oliver Heaviside in England and Arthur E. Kennelly in the United States, and for about 20 years, the idea of a *Kennelly-Heaviside layer* was generally accepted as an electrically conducting layer, at an altitude of about 50 miles, that would reflect radio waves and return them to the earth. But it was not until 1924 that the theory was proven and expanded to include further regions of ionized particles. The importance of the ionosphere in the transmission of radio signals is now well recognized and continues to be the subject of extensive study. In brief, particles are ionized as the result of the absorption of various kinds of radiations from the sun. Waves from a transmitting station, proceeding obliquely upward and encountering the various densities of electrons, are deflected (reflected or refracted) downward, so that a distant receiving station may receive waves from a transmitting station. See Fig. 1. See also **Radio Communication**.

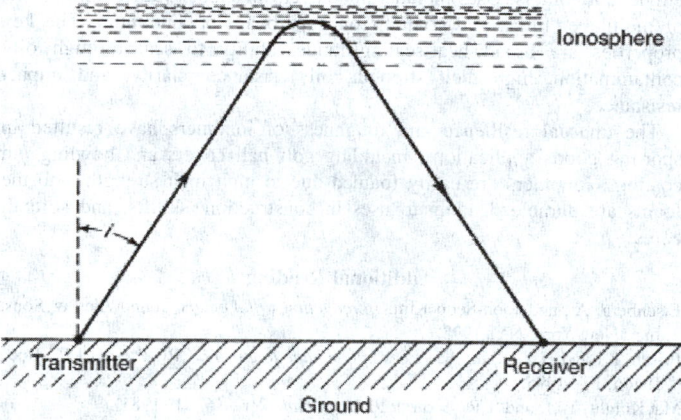

Fig. 1.   Reflection of radio waves by the ionosphere.

IONOSPHERE-THERMOSPHERE STORM PROBES (ITSP) MIS-SION.   See **Living With a Star Program (LWS)**.

IONOSPHERIC SOUNDER.   Also called ionosonde, an instrument used on some satellites to determine electron densities by radio echo techniques. The instrument consists of a radio plus transmitter and an oscillograph.

IONOSPHERIC STORM.   See **Meteorology**.

IONOSPHERIC TIDES.   See **Meteorology**.

IONOSPHERIC TROUGH.   See **Meteorology**.

ION PAIR.   This term is used to denote: 1. A positive ion and a negative ion or electron, having charges of the same magnitude, and formed from a neutral atom or molecule by the action of radiation or by any other agency that supplies energy.

2. As postulated in the Debye-Hückel theory (see also **Electrochemistry**), in concentrated solutions of strong electrolytes (two or more), ions may occasionally approach each other so closely that they may form pairs (or groups) without entering into permanent chemical combination.

ION RETARDATION.   A process based on amphoteric (bifunctional) ion-exchange resins containing both anion and cation adsorption sites.

These sites will associate with mobile anions and cations in solution and thus remove both kinds of ions from solutions. These ions may be eluted by rinsing with water. This process can make clean separations of ionic-nonionic mixtures. It has also been suggested for demineralization of salt solutions.

See also **Ion**.

IP ADDRESS.   See **Transmission Control Protocol (TCP) / Internet Protocol (IP) Suite**.

IPATIEFF, VLADIMIR N. (1890–1952).   Born in Russia, Ipatieff was an army officer as well as a chemist. He was a member of the Academy of Science and carried out organic research at the Institute of Chemistry in Leningrad. He left the former U.S.S.R. under the Stalin regime and at the invitation of Gustav Egloff joined the Universal Oil Products Co. He and his close associate, Herman Pines, did basic development on catalytic alkylation and isomerization of hydrocarbons of the greatest importance for high-octane aviation gasoline.

IPOD.   See **Data Compression**.

IRIDESCENCE.   The exhibition of the colors of the rainbow, commonly by interference of light of the various wavelengths reflected from superficial layers in the surface of a substance.

IRIDESCENT CLOUDS.   See **Clouds and Cloud Formation**.

IRIDIUM.   [CAS: 7439-88-5]. Chemical element symbol Ir, at. no. 77, at. wt. 192.22, periodic table group 9 (transition metals), mp 2,410°C (4,370°F), bp 4,130°C (7,466°F), density 22.42 g/cm^3 (solid at 17°C (63°F), 22.8 g/cm^3 (single crystal at 20°C (68°F). Elemental iridium has a face-centered cubic crystal structure. The two stable isotopes of iridium are ^{191}Ir and ^{193}Ir. The ten unstable isotopes are ^{187}Ir through ^{190}Ir, ^{192}Ir, and ^{194}Ir through ^{198}Ir. In terms of earthly abundance, iridium is one of the scarce elements. Also, in terms of cosmic abundance, the investigation by Harold C. Urey (1952), using a figure of 10,000 for silicon, estimated the figure for iridium at 0.0025. No notable presence of iridium in seawater has been found. The element was identified and named by Tennant (London, England) in 1803.

Electronic configuration

$$1s^2 2s^2 2p^6 3s^2 3p^6 3d^{10} 4s^2 4p^6 4d^{10} 4f^{14} 5s^2 5p^6 5d^9.$$

Metallic Ir is not attacked by any mineral acid unless it is very finely divided. It can be brought into solution by fusion with indium [CAS: 7440-74-6], at 800–1,000°C (1,472–1,832°F) to give a soluble alloy. When fused with Na_2O_2 or an alkaline oxidizing flux, water-soluble iridates(IV) are formed. The finely divided metal is oxidized by air or O_2 at red heat to the dioxide, which decomposes into its elements at higher temperature. The valences of Ir are 1–6, the 3 and 4 valences being most common.

Iridium black is only slightly soluble in aqua regia. When fused with alkalies and alkaline nitrates or Na_2O_2, the metal is converted to an acid-soluble form. The metal at red heat reacts to a small extent with O_2, S, and P. At elevated temperature, the metal is attacked by Cl_2 and F_2. When fused with NaCl and treated with Cl_2, the water-soluble sodium hexachloroiridate(IV), Na_2IrCl_6, is formed.

Iridium(III) hydroxide is a yellow-green or blue-black compound soluble in alkali and insoluble in water. It is made by adding KOH to a solution of potassium hexachloroiridate(III), K_3IrCl_6, in an inert atmosphere. When the trihydroxide is heated, a mixture of iridium(IV) oxide and the metal is formed. Iridium(III) oxide, Ir_2O_3, is made by fusing potassium hexachloroiridate(IV) with Na_2CO_3 and then leaching the mixture with water. At about 1,100°C (2,012°F), both the oxide and the hydroxide decompose into the metal and O_2. When a solution of Ir is heated with $NaBrO_3$ at a pH of approximately 6, the dark-blue precipitate $Ir(OH)_4$ or $IrO_2 \cdot H_2O$ is formed. This water-insoluble compound, when heated to 350°C (662°F) in N_2, loses its H_2O and is converted to the black oxide, IrO_2.

When iridium(III) chloride is heated in Cl_2 at 773–798°C (1,423–1,468°F), iridium(I) chloride is formed. The copper-red crystals are insoluble in acids and alkalies. The compound sublimes in Cl_2 at 790°C (1,454°F) and decomposes into Cl_2 and metallic Ir. Iridium(II) chloride is stable from 763 to 773 °C (1,450 to 1,423 °F). The brown crystals are insoluble in H_2O, acids, and alkalies. Iridium(III) chloride is an insoluble green compound made by reacting the elements at about 600 °C (1,112 °F). The reaction is catalyzed by CO. Iridium(III) chlorides are prepared by reducing the corresponding iridium(IV) chlorides with oxalate or SO_2. Iridium(III) bromide is made by dissolving iridium(III) hydroxide in HBr. The blue solution yields olive-green crystals of $IrBr_3 \cdot 4H_2O$. When heated, the anhydride is formed. The triiodide is formed in analogous fashion as a trihydrate. Iridium(IV) chloride can be made in solution by the action of Cl_2 or aqua regia on ammonium hexachloroiridate(III). The relative insolubility of ammonium hexachloroiridate(IV), $(NH_4)_2IrCl_6$, makes it useful in the purification of Ir. The compound may be reduced in H_2 to the metal. The analogous sodium salt is very soluble, and the potassium salt is relatively insoluble.

Iridium(VI) fluoride is made from the element at 300–400 °C (572–752 °F). The bright yellow solid melts at 44 °C (111 °F) and boils at 53 °C (127 °F). Potassium hexafluoroiridate(V) also has been prepared. Iridium fluoride can be made by heating the hexafluoride with the metal in a sealed tube at 150 °C (302 °F) or heating it with glass above 200 °C (392 °F), at which temperature the glass reduces it to the tetrafluoride. This yellow solid melts at 106–107 °C (223–225 °F) and boils above 300 °C (572 °F). Iridium(III) fluoride is formed by reducing the tetrafluoride with glass for 12–18 hrs. at 430–450 °C (806–842 °F).

Iridium(II) sulfide is formed by burning the metal in sulfur or by heating a higher sulfide at 700 °C (1,292 °F) in N_2. This black solid is insoluble in H_2O, acids, and aqua regia. Iridium(III) sulfide, Ir_2S_3, is formed as a brown-black insoluble precipitate by passing H_2S through a hot acidic solution of an iridium(III) chloride. The precipitation is usually not quantitative. This amorphous black solid is not attacked by HNO_3 but is slowly dissolved by aqua regia or fuming HNO_3. The brown insoluble iridium(IV) sulfide, IrS_2, is partly formed by treating a tetravalent Ir solution with H_2S; when it is prepared in this way, some iridium(III) sulfide also is formed by reduction. Iridium(III) sulfate can be formed by dissolving iridium(III) hydroxide in H_2SO_4 in the absence of air. Trivalent iridium forms numerous cationic and anionic complexes in which it has a coordination number of 6. The amines are extremely stable and, once formed, difficult to destroy. Tetravalent Ir also forms complex ions, but to a lesser extent.

Iridium As a Key to Mass Extinctions. Airborne particles from the January 1983 eruption of Kilauea (Hawaii) volcano indicated exceptionally large concentrations of selenium, arsenic, indium, gold, and sulfur, as expected from a volcanic eruption. Unexpected were exceptionally high levels of iridium. Investigators found that the ratio of Ir to Al was about 17,000 times its value in the associated Hawaiian basalt. Inasmuch as Ir enrichments had not previously been detected in volcanic eruptions, Zoller et al. (1983) suggested that the Kilauea volcano may be part of an unusual volcanic system which may be fed by magma from the mantle. The researchers further suggested that the Ir enrichment may be linked with the high fluorine content of the volcanic gases, indicating that the Ir was released as volatile IrF_6.

In recent years, Zoller and associates (University of Maryland) have studied six active volcanoes (Augustine, Mount St. Helens, El Chichón, Arenal, Poas, and Colima) and have found no evidence of Ir enrichment. The new Kilauea evidence of volcanic action as an Ir source tends to conflict with that of other researchers who have generally attributed the Ir anomaly to an extraterrestrial source, such as resulting from a cataclysmic meteorite or asteroid impact, notably in connection with the Cretaceous-Tertiary (K-T) boundary layer.

Much of the theorizing pertaining to mass extinctions of flora and fauna during the past history of the earth has been based upon finding Ir anomalies. Currently, scientists are attempting to establish the order of mass extinctions. Present opinions seem to suggest that the extinction of terrestrial fauna, including the dinosaurs, was an event that followed rather than occurring concurrently with the catastrophe met by many marine species. See also **Chemical Elements**; **Chemical Elements: The History of the Origin**; **Mass Extinctions**; and **Platinum and Platinum Group**.

Additional Reading

Fox, L.S., et al.: "Gaussian Free-Energy Dependence of Electron-Transfer Rates in Iridium Complexes," *Science*, **1069** (March 2, 1990).

Greenwood, H.N. and A. Earnshaw: *Chemistry of the Elements*, 2nd Edition, Butterworth-Heinemann, Inc., Woburn, MA, 1997.

Krebs, R.E.: *The History and Use of Our Earth's Chemical Elements: A Reference Guide*, Greenwood Publishing Group, Inc., Westport, CT, 1998.

Lide, D.R.: *CRC Handbook of Chemistry and Physics*, 88th Edition, CRC Press, LLC, Boca Raton, FL, 2007.

Stwertka, A. and E. Stwertka: *A Guide to the Elements*, Oxford University Press, Inc., New York, NY, 1998.

Zoller, W.H., J.R. Parrington and J.M. Phelan Kotra: "Iridium Enrichment in Airborne Particles from Kilauea Volcano: January 1983," *Science*, **222**, 1118–1121 (1983).

IRIS. See **Vision and the Eye**.

IRIS DIAPHRAGM (Microwave). In a waveguide, an iris diaphragm is a conducting plate or plates, thin compared to a wavelength, occupying a part of the cross section of the waveguide. When only a single mode can be supported, an iris acts substantially as a shunt admittance.

IRIS DIAPHRAGM (Optics). A diaphragm introduced into an optical system as a stop, and which is so constructed that the diameter of the opening may be changed continuously throughout a considerable range. The iris of the eye has this property so that the intensity of the light falling on the retina will be kept within proper bounds. The effective f-number of a camera is changed by adjusting the iris diaphragm to a desired value to fit the illumination and exposure time to the sensitivity of the film used.

IRISH MOSS. See **Gums and Mucilages**.

IRITIS. Iritis is an inflammation of the iris, the colored part of the eye that controls the size of the pupil opening. The condition is relatively common and responds well to treatment, but it can be sight-threatening if left untreated. It is not uncommon for patients to have multiple episodes of iritis even with prompt, proper treatment. Usually, only one eye is affected at any one time, but the condition can occasionally occur in both eyes.

The iris, which contains two muscles, controls the size of the pupil opening. In dim light or at night, the iris makes the pupil larger to increase the amount of light entering the eye. When too much light is present, the iris causes the pupil to become smaller to reduce excessive light and glare.

Most often, the cause of iritis is unknown, although it can be associated with other inflammatory conditions of the body such as arthritis. Infection of some parts of the body, especially the tonsils, sinuses, kidney, gallbladder, and teeth, can also cause inflammation of the iris. On occasion, iritis can occur following an eye injury or can accompany an ulcer or foreign body on the cornea.

Symptoms of iritis often appear suddenly and develop rapidly over a period of several hours or days. The condition commonly causes pain, tearing, light sensitivity, blurry vision, and, in some cases, red eye.

Some patients also experience floaters, which are small dots or flecks that seem to float in the field of vision. The pupil in the affected eye may also become smaller.

See an eye doctor for an examination when symptoms of iritis occur. The most useful tool in diagnosing iritis is the slit lamp, a table top microscope that illuminates and magnifies the various structures of the eye. A complete medical examination may also be needed because the condition is sometimes associated with another disease.

Iritis usually responds well to treatment. Eyedrops and ointments relieve pain and inflammation, dilate the pupil, and reduce the possibility of scarring. Dilating eyedrops relax the irritated iris, decreasing the discomfort and preventing the inflamed iris from sticking to the lens. In some cases, both steroids and antibiotics are prescribed. Hot compresses may also provide some relief. Severe cases are treated with oral medications and injections. Iritis usually lasts from six to eight weeks. Cataracts, glaucoma, corneal changes, and secondary inflammation of the retina can result from iritis and the medications used to treat this disorder, so careful monitoring is important.

Iritis can be sight-threatening, therefore proper diagnosis and prompt treatment are essential. As soon as symptoms occur, a complete eye examination is necessary so that the condition can be controlled before any vision loss occurs. The sooner treatment is begun, the easier it is to treat. See also **Floaters**.

Vision Rx, Inc., Elmsford, NY

IRMINGER CURRENT.   An ocean current that is one of the terminal branches of the Gulf Stream system (part of the northern branch of the North Atlantic current); it flows toward the west off the south coast of Iceland.

IRON.   [CAS: 7439-89-6] Chemical element symbol Fe, at. no. 26, at. wt. 55.847, periodic table group 8 (transition metals), mp 1,535°C, bp approximately 2,750°C, density 7,874 g/cm3 for the pure solid (20°C); 7.92 for a single crystal of a-iron. Iron has a body-centered cubic crystal structure (a-iron).

Iron is a silver-white metal, capable of taking a high polish; ductile; malleable; can be welded when hot. Pure iron is attracted by a magnet, but does not retain the magnetism. Silicon steel is preferred for electromagnets because it retains magnetism even less than pure iron. See also **Magnetism**. The discovery of iron was prehistoric. There are nine isotopes of iron, 52Fe through 60Fe Isotopes 54, 56, 57, and 58 are fully stable, whereas four others have fairly short half-lives, ranging from 8.9 minutes (53) to 2.94 years (55). The half-life of 60Fe is approximately 3×10^5 years. Iron has valence numbers of 2+ (ferrous) and 3+ (ferric). The hardest of the ductile metals, iron is surpassed only by cobalt and nickel in tenacity. Iron is an extremely versatile construction and engineering material and serves both in relatively pure forms, such as malleable and wrought iron, and in many hundreds of iron-base alloys of major importance, including the numerous types of steel.

Electronic configuration is 1s 22s 22p 63s 23p 63d 64s 2. First ionization potential is 7.896 eV; second 16.5 eV. Oxidation potentials: $Fe \rightarrow Fe^{2+} + 2e^-$, 0.441 V; $Fe \rightarrow Fe^{3+} + 3e^-$, 0.036 V; $Fe^{2+} \rightarrow Fe^{3+} + e^-$, −0.771 V; $Fe + 20H^- \rightarrow Fe(OH)_2 + 2e^-$, 0.877 V; $Fe(OH)_2 + OH^- \rightarrow Fe(OH)_3 + e^-$, 0.56 V; $Fe(OH)_4^- + 4OH^- \rightarrow FeO_4^{2-} + 4H_2O + 3e^-$; E = 0.55 V (80°C; 40% NaOH). Metallic radius, 1.2412 Å; ionic radius Fe 2+, 0.80 Å, Fe3+, 0.67 Å. Other important physical properties of iron are given under **Chemical Elements**. See also Table 1.

Three allotropic forms of iron are known: (1) *alpha iron*, which is present below 769°C; (2) *gamma iron*, which exists between 906° and 1,404°C, and (3) *delta iron*, which occurs between 1,404° and 1,536°C. On slow cooling, the reverse changes occur, but may be slowed or partly or entirely prevented in the presence of alloying elements.

In terms of abundance in the earth's crust, iron ranks fourth, being estimated as comprising about 5% of the weight of igneous rocks. Of course, only a very small portion of this large amount of the element in the earth's crust is obtainable as iron ore. In terms of cosmic abundance, the estimate of Harold C. Urey made in 1952 put iron as number 9 among the elements, having a figure of 7,250 related to a base for silicon of 10,000. Iron is ranked number 23 among the elements in terms of its presence in seawater, an estimated 47–48 tons per cubic mile (10.2–10.4 metric tons per cubic kilometer) of seawater. In this regard, it is approximately equal with aluminum, molybdenum, and zinc.

Iron Ores

Iron occurs abundantly in several materials, mainly in the form of oxides, carbonate, silicates, and sulfides. These are shown in Table 2. Most of the ores shown in the table are described under separate alphabetical listings in this volume. Briefly, the major iron-bearing materials are:

Magnetite, Fe_3O_4, corresponding to 72.4% Fe and 27.6% O_2, dark gray to black, sp gr 5.16–5.18, strongly magnetic, permitting magnetic exploration methods. Some ores contain small amounts of titanium whereupon they are referred to as titaniferous magnetite.

Hematite, Fe_2O_3, corresponding to 69.94% Fe and 30.06% O_2, steel gray to dull red or bright red, earthy to compact or crystalline, sp gr 5.26, most important of iron ores, occurs widely in many types of rocks of varying origin.

Ilmenite, $FeTiO_3$, corresponding to 36.8% Fe, 31.6% Ti, and 31.6% O_2, iron-black, opaque, generally mined for titanium with iron as a byproduct, also called iron titanate.

Limonite, mineralogically composed of various mixtures of the minerals *goethite* and *lepidocrocite*, $HFeO_2$ and $FeO(OH)$, respectively.

TABLE 1. SELECTED PHYSICAL PROPERTIES OF IRON

Electrical Properties

Electrical conductivity, volume, % of annealed copper at 20°C	17.75
Electrical resistivity, microohm-cm,	
at 0°C	8.9
at 20°C	9.7
Temperature coefficient of electrical resistance (0–100°C)	0.65×10^{-4}
Electrode potential (standard hydrogen scale), at 25°C, volts	−0.44
Electrochemical equivalent, milligrams/second-absolute amperes	
$Fe^{2+} - Fe$	0.1929
$Fe^{3+} - Fe$	0.2893
Magnetic susceptibility, at 820°C	$1,000 \times 106$

Thermal Properties

Emissivity, at 0.65 micrometers, %	40
Heat of combustion, cal/gram atom	88.355
Latent heat of fusion, cal/gram	65.5
Latent heat of vaporization, cal/gram	1.598
Specific heat, at 25°C, cal/(°C)(gram atom)	6.55
Thermal conductivity, at 0°C, (cal)(cm)/(second)(cm²)(°C)	0.18
Thermal expansion, linear coefficient,	
cm/(cm)(°C)	11.76×10^{-6}
in/(in)(°F)	6.53×10^{-6}
Average coefficient of linear expansion, at 77°F,	
cm/(cm)(°C)	12.3×10^{-6}
in/(in)(°F)	6.83×10^{-6}

Mechanical Properties

Brinell hardness, at 25°C, 99.9% Fe	82–100
Percent elongation, at 25°C, 99.9% Fe	30–40
Yield strength, psi, at 25°C, 99.9% Fe	10,000–20,000
Tensile strength, psi, at 25°C, 99.9% Fe	30,000–40,000
Modulus of elasticity, psi	28.5×10^6
Modulus of rigidity, psi	11.64×10^6
Poisson's ratio	0.28

Other Properties

Density, liquid, at 1564°C,	
g/cm³	7.00
lb/in³	0.253
Density, solid, at 20°C,	
g/cm³	7.874
lb/in³	0.284
Reflectivity (light from tungsten filament),	
2,500 Å	38
10,000 Å	65
Surface tension, at 1,550°C, dynes/cm	1,835–1,865
Thermal-neutron-absorption cross section, barns	2.53
Viscosity, cP,	
at 1,743°C	4.45
at 1,390°C	7.85

Goethite contains 62.9% Fe, 27% O_2, and 10.1% H_2O, sp gr 3.6–4.0, commonly yellow or brown to nearly black, compact to earthy and ocherous. Limonites are important sources of iron throughout the world.

Siderite, $FeCO_3$, corresponding to 48.2% Fe, 51.8% CO_2, sp gr 3.83–3.88, white to greenish-gray and brown, contains variable amounts of calcium, magnesium, and manganese, varies from dense, fine-grained and compact to crystalline, sometimes referred to as spathic iron ore, or black-band ore. The carbonate ores are calcined before they are charged into the blast furnace; frequently contain sufficient lime and magnesite to be self-fluxing.

Silicate Group Ores. There are comparatively few silicates with iron as the principal base. Often they have a rather complex chemical formula, with sp gr higher than 2.8, occurring in various shades of green or black tones. Important iron-silicate minerals are *chamosite*, *stilpnomelane*, *greenalite*, *minnesotaite*, and *grunerite*. Presently of minor importance as a source of iron.

Sulfide Group Ores. The principal materials in this group are *pyrite*, *pyrrhotite*, and *marcasite*. Pyrite, FeS_2, corresponding to 46.6% Fe,

TABLE 2. PRINCIPAL IRON-BEARING MINERALS

Class and Mineralogical Name	Chemical Composition of Pure Mineral	Common Designation
Oxides		
Magnetite	Fe_3O_4	Ferrous-ferric oxide
Hematite	Fe_2O_3	Ferric oxide
Ilmenite	$FeTiO_3$	Iron-titanium oxide
(Goethite)	$HFeO_2$	
Limonite		Hydrous iron oxides
(Lepidocrocite)	$FeO(OH)$	
Carbonate		
Siderite	$FeCO_3$	Iron carbonate
Silicates		
Chamosite		
Stilpnomelane	Various; often complex	
Greenalite		Iron silicates
Minnesotaite		
Grunerite		
Sulfides		
Pyrite	FeS_2	
(iron pyrites)		
Marcasite	FeS_2	Iron sulfides
(white iron pyrites)		
Pyrrhotite	FeS	
(magnetic iron pyrites)		

53.4% S, sp gr 4.95–5.10, is pale brass yellow, and the most widespread of the iron sulfides. Pyrrhotite (magnetic pyrite), varies in composition from FeS to $FeS + S$, typically contains 59.4% Fe, 40.6% S, bronze yellow to copper red, frequently tarnished, and is often considered an indicator of nickel deposits because of common association with pentlandite. Marcasite (white iron pyrites), FeS_2, corresponding to 46.6% Fe, 53.4% S, is pale brass yellow, commonly associated with limestones, clays, and lignite deposits. It differs from pyrite only in its crystal structure and greater chemical instability. Iron sulfides sometimes are mined for their sulfur content; more commonly because of their association with other valuable metallic elements, such as copper, nickel, zinc, gold, and silver. Iron sometimes is recovered as a byproduct.

Geology and Genesis of Iron Deposits. Iron ores have a wide range of formation in geologic time as well as a wide geographic distribution. They are found in the oldest known rocks of the earth's crust, with an age in excess of 2.5 billion years, as well as in rock units formed in various subsequent ages. Iron ores are forming presently where iron oxides are being precipitated in marshy areas, and where magnetite placers are being formed on certain beaches. Thousands of iron deposits are known throughout the world. The deposits range in size from a few tons to many hundreds of millions of tons. Many of the world's largest deposits of iron ore are located in the oldest geologic series, the Pre-Cambrian.

Iron ores occur in igneous, metamorphic or sedimentary rocks, or as weathering products of various primary iron-bearing materials. For convenience of analysis, iron ores are grouped into (1) igneous ores, (2) contact ores, (3) hydrothermal ores, (4) sedimentary ores, with several subclassifications of the latter ores. Brief definitions follow:

Igneous Ores. These were formed by crystallization from liquid rock materials, either as layered-type deposits that possibly are the result of crystals of heavy iron-bearing minerals settling as they crystallize to form iron-rich concentrations, or as bodies which show intrusive relationship with their wall rocks. These ore bodies may be tabular or irregular and are composed largely of magnetite with varying amounts of hematite.

Contact Ores. Iron-ore deposits formed at or near the contact between igneous rocks and sedimentary rocks, the latter usually limestones, are commonly composed of magnetite and hematite with associated carbonates and pyrite. The ore deposits are commonly in the sedimentary rocks as irregular or tabular replacement bodies.

Hydrothermal Ores. Iron-ore deposits formed by hot solutions which transported iron and replaced rocks of favorable chemical composition with iron minerals to form irregular ore bodies, commonly in limestones,

are termed hydrothermal deposits. The iron often occurs as siderite, or sometimes as oxides.

Sedimentary Ores. There are six subclassifications:

Bedded Ores. These often are composed of oölites of hematite, siderite, iron silicate, or less commonly, limonite in a matrix of siderite, calcite, or silicate. They have a wide geographic distribution associated with other sedimentary rocks. They sometimes contain fossils and fine grains of sand. They often have a fairly high phosphorus content and may be self-fluxing.

Siderite Ores. These ores consist of beds of siderite or siderite nodules associated with shales. They are common in coal-associated beds and commonly contain associated sulfides, with a fairly high sulfur and phosphorus content.

Placer Ores. Iron oxides, when compact, are rather resistant to weathering and erosion, and under favorable conditions may form placer deposits, which, in relatively few instances, constitute iron ores. Generally they are of rather minor importance as sources of iron.

Bog Iron Ores. Bog ores occur in many swampy areas, particularly in glaciated areas in Europe, Asia, and North America. They occur commonly as dark-brown, cellular masses, or granular or fine particles of limonite. Once important when iron furnaces were local and small, they have ceased to be of commercial importance.

Metamorphic Ores. These ores include sedimentary iron-ore deposits, which have been metamorphosed, as well as ores associated with metamorphic rocks, in which the origin of the ore is obscured by recrystallization. Essentially all of the Pre-Cambrian sedimentary iron formations are of this type.

Residual Ores. These ores are commonly products of the surficial weathering of rocks, but may include ores formed by hydrothermal oxidation and leaching. Ores of this kind were formed extensively in Pre-Cambrian iron formations by leaching of silica, which commonly constituted in excess of 50% of the rock. Oxidation changes iron carbonate, silicate minerals, and magnetite to hematite or limonite.

Iron Deposits. The principal world iron deposits include:

Name and/or Type of Ore	Location
Kerch oölitic limonite	Crimea, Russia
Salzgitter limonite and hematite	Germany
Minette limonite and hematite	France, Germany, Luxembourg
Blackband ironstones	British Isles
Siegerland siderite	Germany
Clinton hematites	Alabama (U.S.A.)
Wabana oölitic hematites	Newfoundland
Minas Gerais hematite	Brazil
Krivoi Rog hematites	Ukraine, Russia
Bihar, Orissa, and Bastar hermatites	India
Labrador hematite	Quebec, Labrador (Canada)
Lake Superior taconites,	Michigan, Wisconsin, Minnesota (U.S.A.)
jaspilites, hematites, and magnetites	Ontario (Canada)
Cerro Bolivar and El Pao hematites	Venezuela
Kirunavaara magnetite	Sweden
Hematites	Australia

Iron Ore Processing. Two major developments have occurred during recent years because of the increasing needs for iron: (1) increased search for new supplies of high-grade iron ores, and (2) expansion of iron-ore pellet-plant production, particularly in those industrial nations where supplies of high-grade ores have diminished. More recently, iron ore has been shipped in slurry form in ocean-going tankers. Once the slurry, containing about 75% solids, settles, the excess water is pumped off, leaving a nonshifting cargo in the hold of about 92% solids. Upon arrival at the receiving port, the cargo is reslurried with high-pressure water jets. Considerable savings in dockside loading and unloading expense are thus effected.

Beneficiation. This is a term that describes all processes used to improve the chemical and physical characteristics of ore (not limited to iron ore) for later use. In the case of iron ore, beneficiation makes the ore better for handling by the blast furnace. The principal methods include crushing, screening, blending, grinding, concentrating, classifying, and agglomerating. Concentration operations include jigging, flotation, and magnetic separation. A blast furnace operates best with a permeable burden, which permits not only a high rate of gas flow, but also a uniform gas flow with minimum channeling of the gas. Agglomeration improves burden permeability and thus the gas-solid contact in the furnace. This reduces blast-furnace coke rates and increases the rate of reduction. Agglomeration also decreases the amount of fines blown out of the blast furnace, thus reducing the load on the gas-recovery system.

Practice has shown that the best agglomerate for a blast furnace will contain about 60% or more of iron, a very minimum of undesirable constituents, and a minimum of material larger than 1 inch. However, the agglomerate must have sufficient strength to withstand degradation while in stockpiles and during transportation and handling. The target is to have the material arrive at the blast furnace, after prior handling, with about 85–90% of the material over 1/4 inch. The agglomerate must be able to withstand the high temperature and the degradation forces within the furnace without slumping or decrepitating. Further, the agglomerate should be reasonably reducible so that a satisfactory reduction rate can be maintained in the furnace.

There are four major types of agglomerating processes: (1) sintering, (2) pelletizing, (3) briquetting, and (4) nodulizing. The first two processes have been most popular.

Sinter consists of small particles of iron-bearing materials, which are fused or fritted together at high temperature. The latter is achieved by burning carbon in the form of coke breeze in a sintering-machine feed mix. Optionally, fluxing material may be added to eliminate later additions to the blast furnace. A number of materials can be converted by sintering, such as flue dust, naturally fine ores, ore fines from screening operations, and other iron-bearing material of small particle size. A continuous sintering process is shown in Fig. 1.

A traveling grate conveys a bed of ore fines or other finely divided iron-bearing material intimately mixed with approximately 5% of a finely divided fuel, such as coke breeze or anthracite. Near the head or feed end

of the grate, the bed is ignited on the surface by gas burners and, as the mixture moves along the traveling grate, air is pulled down through the mixture to burn the fuel by downdraft combustion. As the grates move continuously over the windboxes toward the discharge end of the strand, the combustion front in the bed moves progressively downward. This creates sufficient heat and temperature (1,310–1,480 °C) to sinter the fine ore particles together into porous, coherent lumps. Sinter plants with a suction area of up to 5,200 square feet (483 square meters) a strand width of 17 feet (5.2 meters) and a production capacity of about 20,000 tons per day have been built.

In the pelletizing process, the agglomeration of material is effected prior to heat treatment. A green, unbaked pellet or ball (glomerule) is formed and hardened by heating. The iron ores to be pelletized are ground finely to present an adequate surface area for the formation of green balls and mixed with water and a binding agent, such as bentonite clay. A small amount of fine solid fuel may be added to the pellet mix or coated on the pellets to furnish part of the heat required. Oxidation of a pelletized magnetite concentrate to hematite during the firing step may also furnish a substantial portion of the heat requirements. The optimum moisture content for pellets is a function of fineness and use of additives, but usually ranges from 9–12%. To improve pellet strength, soda ash, limestone, or dolomite may be added. Hardening of pellets is effected in several ways, including a traveling-grate, as shown in Fig. 2, a combined grate and rotary kiln, or a shaft furnace. As illustrated by Fig. 3, ore beneficiation operations usually require large amounts of water.

Chemistry of Iron

Reduction of iron ores in preparation of iron and steel making is described under **Iron Metals, Alloys, and Steels**.

With its $3d^6 4s^2$ electron configuration, iron forms Fe^{2+} and Fe^{3+} ions, the latter, involving the removal of one 3d electron. The ferrate ion, FeO_4^{2-}, containing hexavalent iron, is unstable in acidic solution, being a very strong oxidizing agent.

Three oxides of iron are known, FeO, Fe_3O_4, and Fe_2O_3 although pure FeO does not exist. The actual composition of iron(II) oxide may be approximated by replacement of a small proportion of the Fe(II) atoms by two-thirds their number of Fe(III) atoms. If the operation is continued until three-quarters of the Fe(II) atoms have been replaced, then the

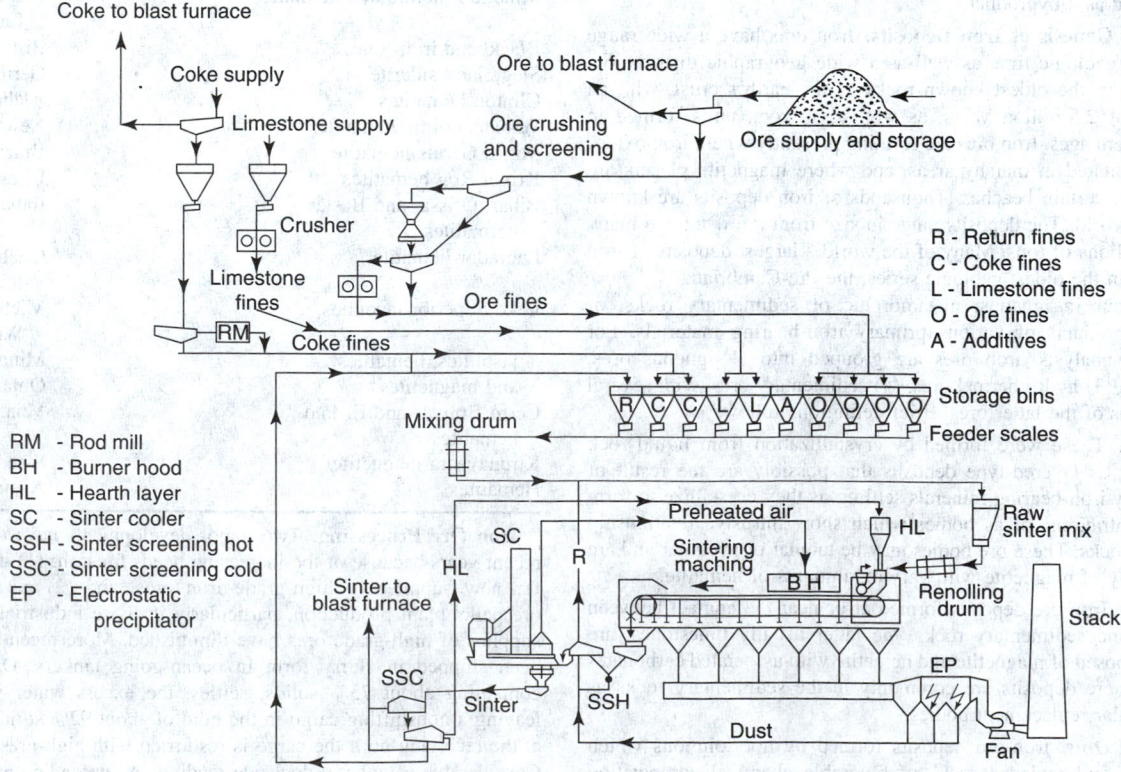

RM - Rod mill
BH - Burner hood
HL - Hearth layer
SC - Sinter cooler
SSH - Sinter screening hot
SSC - Sinter screening cold
EP - Electrostatic precipitator

R - Return fines
C - Coke fines
L - Limestone fines
O - Ore fines
A - Additives

Fig. 1. Schematic flow diagram of continuous iron-ore sintering process. (*Metallgesellschaft A.G.*)

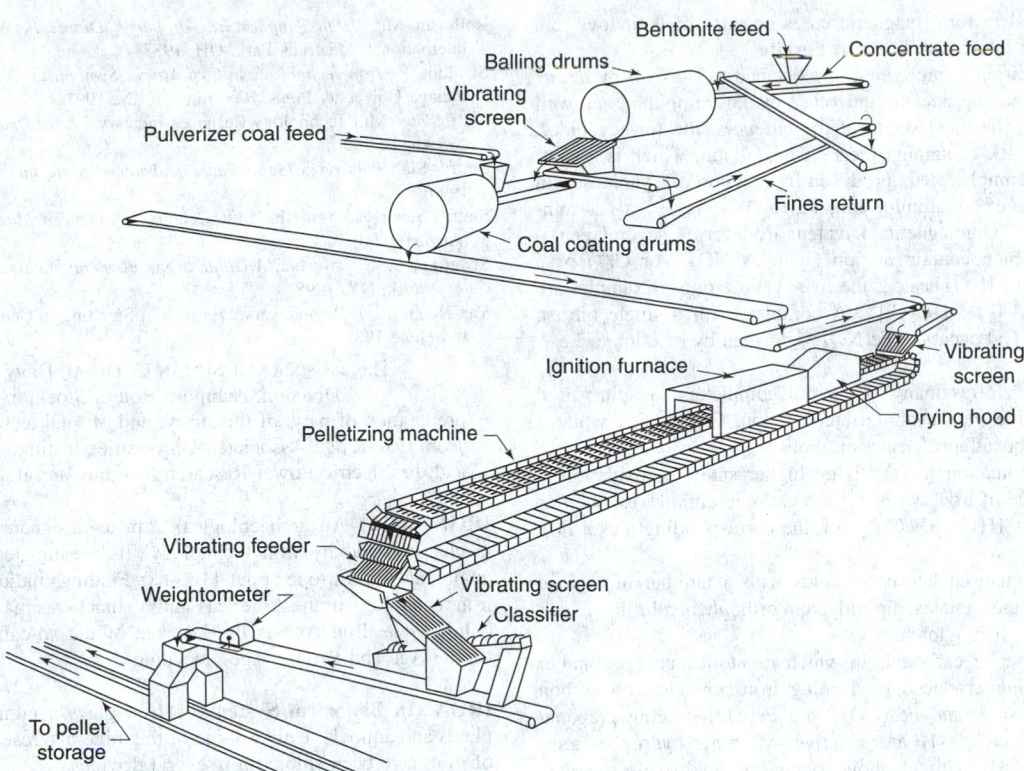

Fig. 2.   Pelletizing process in which a traveling grate is used.

Fig. 3.   The beneficiation of taconite ore on the iron range requires large volumes of water in concentrating by magnetic separation. To eliminate massive waste-disposal problems, huge thickeners, such as the 300-foot (91.5-meter) diameter caisson unit shown here, are used. This system will handle over 70 million gallons (265 million liters) per day or 50,000 gallons (189,250 liters) per minute of liquid and 250 tons per day of suspended solids. Clarifying the waste tailing stream permits reclamation of water on a large scale for plant reuse.

composition Fe_3O_4 is reached, which may thus be described by the formula $Fe^{2+}(Fe^{3+}O^{2-}O^{2-})_2$. Continuation of the replacement until all the Fe(II) atoms have been replaced yields α-Fe_2O_3. The γ-allotrope of Fe_2O_3 and the compound Fe_3O_4 are both ferromagnetic.

Iron(II) sulfide, FeS, also may show considerable departure from stoichiometric proportions, exhibiting an electrical conductivity when in large crystals that resembles an alloy rather than a sulfide. Iron(III) sulfide, Fe_2S_3, cannot be prepared in solution in pure form, because of the oxidizing action of Fe^{3+} upon H_2S, and even upon S^{2-} ions in alkaline solution, and when Fe_2S_3 is prepared by reaction of dry H_2S and the hydrated Fe_2O_3, it breaks up into FeS and FeS_2. The latter is made up of Fe^{2+} and S_2^{2-} ions, and the mineral, pyrites, it has a cubical structure composed of these ions.

Iron forms dihalides with all four of the common halogens, and trihalides with all but iodine. FeF_2 and $FeCl_2$ are readily formed in anhydrous state by action of the hydrogen halide upon the heated metal, and the others can be made directly from the elements. The iron(III) halides, like iron(III) salts generally, are more readily hydrolyzed than the corresponding ferrous compounds, due to the smaller size and greater change of the Fe^{3+} than the Fe^{2+} ion.

Other elements with which iron forms binary compounds, especially at higher temperatures, are boron, carbon, nitrogen, silicon, and phosphorus. Like FeO, these compounds often depart slightly or even considerably from daltonide composition, frequently being interstitial compounds, and in higher elements of groups VB and VIB, merging into the interstitial compound-solid solution picture which iron exhibits with the transition metals.

The oxyacid salts of iron(III) are more numerous than those of iron(II). Among the former, the sulfates are of interest because of the readiness with which iron(III) sulfate replaces aluminum sulfate in the alums, which are hydrated double sulfates formed by certain trivalent and alkali metal (and other monovalent) sulfates. Iron(III) sulfate, $Fe_2(SO_4)_3$, is isomorphous with aluminum sulfate, $Al_2(SO_4)_3$, because the radius of the Fe^{3+} ion is so close to that of the Al^{3+} ion (0.57 Å). For that reason, the isomorphous relationship extends to other salts, i.e., the fluorides and some of the nitrates.

Like the neighboring elements of group 8, iron forms large numbers of complexes. This is due to the availability of two $3d$ orbitals in Fe^{2+} and Fe^{3+} to form hybrid orbitals with the $4s$ and $4p$ orbitals to yield spin-paired complexes (the so-called "covalent" or "inner" complexes).

Many ions contain iron combined with oxygen atoms or hydroxyl groups and are known, respectively, as ferrates or hydroxyferrates. Those of iron(II) include FeO_2^{2-}, $Fe(OH)_4^{2-}$, and $Fe(OH)_6^{4-}$, while those of iron(III) include FeO_2^-, $Fe_2O_4^{4-}$, and $Fe(OH)_8^{5-}$. Their compounds are also called ferrates, except that the name ferrite is used for such compounds of iron(III) as MFe_2O_4, in which M is a divalent metal, this last type of compound being used to form magnetic cores because of their low core losses when properly fabricated. See also **Ferrite**.

The complexes of iron(II) are commonly octahedral (formed by d^2sp^3 hybridization), and include chelate and other cyclic compounds as well as monocyclic ones. The most stable of the latter are the ferrocyanides, or hexacyanoferrates(II), containing the $Fe(CN)_6^{4-}$ ion, which is a spin-paired, diamagnetic complex. It is produced by reaction of cyanides with Fe^{2+} solutions. The Fe^{2+} diammine ion, $Fe(H_2O)_4(NH_3)_2^{2+}$ is a spin-free complex, and is paramagnetic. Divalent iron forms several penta-cyano complexes, which contain an ion (such as NO_2^- or Cl^-) or a molecule (NH_3, CO, or H_2O) besides the five cyano groups. Examples are $[Fe(CN)_5NO_2]^{4-}$ and $[Fe(NH_3)_5Cl]^+$. A complex with a single nitroso group in addition to H_2O occurs in $Fe(NO)^{2+}$, formed by reaction of Fe^{2+} and NO.

The ferric ion also forms many octahedral complexes. A spin-paired type is the ferricyanide (hexacyanoferrate(III)) ion, $Fe(CN_6)^{3-}$, while a spin-free type is the hexafluoroferrate ion, FeF_6^{3-}.

The effect of coordination in stabilizing higher states of oxidation is seen in the occurrence of iron(IV) in certain cationic complexes, such as $[Fe(Cl)_2 \cdot 2C_6H_4(As(CH_3)_2)_2](FeCl_4)_2$ (cf. the corresponding nickel(IV), Ni(IV), complex).

Iron forms polydentate chelate compounds with a number of organic substances, including the oxalates, dipyridyl and orthophenanthroline. Such a structure is found in hemoglobin.

Iron forms a number of carbonyls in which its atomic charge number is zero, $Fe(CO)_5$ being produced by heating iron powder with carbon monoxide under pressure, and $Fe_2(CO)_9$ and $Fe(CO)_{12}$ being prepared from it. These iron carbonyls are reactive, yielding hydrogen compounds, e.g., $H_2Fe(CO)_4$ with alcoholic potassium hydroxide; halogen compounds, $Fe(CO)_4X_2$, with halogens; amino or substituted-amino compounds $Fe(CO_2)_n(Am)_{5-n}$, where Am is an amino group, with ammonia, pyridine or ethylenediamine. Iron carbonyls form compounds with R_3P, R_3As, R_3Sb, or diphosphines, etc., such as o-phenylenediarsine. They also form nitroso derivatives; such as $Fe(CO)_2(NO)_2$, and mercaptides, such as $Fe_2(C)_6(SC_2H_5)_2$. In strong acids (e.g., liquid hydrofluoric acid), $Fe(CO)_5$ gives $Fe(CO)_5H^+$ with the proton attached directly to the iron atom.

Iron also forms a series of similar nitrosyls, such as $Fe(NO)_4$, which is probably $[NO^+][Fe(NO)_3^-]$, $Fe_2(NO)_4I_3$, $Fe(NO)_2I$, $Fe(NO)I$, $Fe(NO)_3Cl$, $Fe_2(NO)_4(SA)_2$, where $A = H$, a metal, a sulfonate group, an alkyl or aryl group, $M^1[Fe(NO)_2S]$ (Roussin's red salt), $M^1(Fe_4(NO)_7S_3]$ (Roussin's black salt), obtained by treating the red salts with alkali, $Fe(NO)_2SR$, where $R = C_2H_5$ or C_6H_5, $M^1(Fe(NO)SSO_3]$ and others described under hexacyanoferrates.

By heating powdered iron with cyclopentadiene

$$CH{=}CH{-}CH{=}CH{-}CH_2$$

the compound ferrocene, insoluble in water, soluble in organic solvents, consisting of two cyclopentadienyl radicals connected to an iron atom, $C_5H_5 - Fe - C_5H_5$ is prepared. It is of great interest because the iron atom is sandwiched between the two parallel and symmetrical rings with delocalized bonding. Alkali metals produce salts of the type $M^1Fe(C_6H_5)_2]$, powerful reductants. Halogens produce water-soluble ferrocenium salts, such as $[Fe(C_5H_5)_2]Cl$. From both of these, sigma-bonded alkyl and aryl derivatives, such as $Fe(C_5H_5)_2CH_3$, can be obtained in which the alkyl group is attached directly to the iron atom.

See also **Iron Metals, Alloys, and Steels**; and **Iron (In Biological Systems)**.

Additional Reading

Greenwood, N.N., A. Earnshaw: *Chemistry of the Elements*, 2nd Edition, Butter-worth-Heinemann, Inc., Woburn, MA, 1997.

Hill, H., etal: *Metal Sites in Proteins and Models: Iron Centres*, Springer-Verlag New York, Inc., New York, NY, 1999.

Krebs, R.E.: *The History and Use of Our Earth's Chemical Elements: A Reference Guide*, Greenwood Publishing Group, Inc., Westport, CT, 1998.

Lide, D.R.: *CRC Handbook of Chemistry and Physics*, 88th Edition, CRC Press, LLC., Boca Raton, FL, 2007.

Rawers, J.C.: "High-Pressure-Nitrogen Alloying of Steels," *Advanced Materials & Processes*, 50 (August 1990).

Rothman, M.: *High-Temperature Property Data: Ferrous Alloys*, A S M International, Materials Park, OH, 1988.

Schmidt, P.: *Iron Technology in East Africa: Symbolism, Science, and Archaeology*, Indiana University Press, Bloomington, IN, 1997.

Staff: "New Mill Technology Reshapes Industry," *Advanced Materials & Processes*, 26 (August 1990).

Staff: "Slag Puts on a Good Face," *Advanced Materials & Processes*, 6 (March 1990).

Staff: "Steelmaking in the 21st Century," *Advanced Materials & Processes*, 31 (August 1990).

Stwertka, A., E. Stwertka: *A Guide to the Elements*, Oxford University Press, Inc., New York, NY, 1998.

Van Noten, F., J. Raymaekers: "Early Iron Smelting in Central Africa," *Sci. Amer.*, 104 (June 1988).

The assistance of SUMAN C. DESAI, Davy McKee Iron & Steel Division, Ashmore House, Stockton-on-Tees, England, in preparation of parts of this entry, and of vital technical inputs received from Oak Ridge Associated Universities Institute for Energy Analysis, and the Electric Power Research Institute, are gratefully acknowledged

IRON AGE. An archaeological term to designate a cultural level that is characterized by iron technology. It is estimated that the Iron Age commenced in Europe about 1100 B.C. Findings indicate that there was no iron technology in the Americas until contact was made with the European culture. The Iron Age is the last age of the so-called three-age system (Stone Age and Bronze Age preceding).

IRON (In Biological Systems). Iron plays a number of vital roles in plants and animals. Unlike some of the minerals, research on the functions of iron have been studied for several decades.

Iron and Crops

Iron deficiency is a serious problem in crop production in certain regions of the world and some nutritionists consider iron deficiency anemia to be one of the most frequently observed mineral element deficiency conditions in humans. But iron fertilization of soils is not likely to be effective in decreasing the incidence of this deficiency. The reasons for this apparent contradiction are based upon the behavior of iron at several stages in the food chain.

In the United States, severe iron deficiency in crop plants occurs most frequently on the alkaline soils of the western states and on very sandy soils, although some plants, especially broad-leaved evergreens, are sometimes iron deficient on many other kinds of soils. Iron deficiency is rarely due to a total lack of iron in the soil. It is nearly always due to the low solubility of the iron that is present. For example, some soils that are red from iron compounds may contain too little available iron for normal plant growth. The relative susceptibilities of cultivated crops to iron deficiency are listed in Table 1.

To correct iron deficiency in plants, it is usually necessary to add a soluble form of iron to the soil or to spray the foliage. Since soluble iron added generally will revert to insoluble forms, these procedures for plants are only temporarily effective. Soil treatments that make alkaline soils more acid, such as incorporating large amounts of sulfur, may offer a more lasting correction. Incorporating large amounts of manure into soil makes the iron more soluble and may be effective in correcting iron deficiency, particularly in fine-textured alkaline soils.

Iron-deficient plants are generally stunted and chlorotic, i.e., normally green leaves are yellow or streaked with yellow. When the iron deficiency is treated by adding soluble iron to the soil, the plants turn green, grow larger and yield more, but sometimes the concentration of iron unit weight of plant material may be no higher than in the stunted, iron-deficient plants. Thus, in terms of forage plants, correction of iron deficiency in the plant does not necessarily improve the plant as a source of dietary iron. The iron-treated plants, however, may contain a higher concentration of carotene or provitamin A than the yellow, stunted, deficient plants. Thus, iron fertilization can be more useful in improving the vitamin A than the iron level in diets.

In livestock, iron deficiency is most common in young pigs raised in confinement on concrete floors. Injecting iron compounds and painting the sow's udder with iron compounds are measures used to prevent this deficiency. Grazing animals seldom suffer from iron deficiency unless they are heavily parasitized.

TABLE 1. RELATIVE SUSCEPTIBILITY OF CULTIVATED CROPS TO IRON DEFICIENCY

Crop	Highly Susceptible	Moderately Susceptible	Relatively Tolerant
Alfalfa		x	x
Barley		x	x
Berries	x		
Citrus	x		
Corn		x	x
Cotton		x	x
Field beans	x	x	
Flax	x	x	x
Forage sorghum	x		
Grain sorghum	x		
Grapes	x		
Grasses		x	
Groundnuts (Peanuts)	x		
Millet			x
Mint	x		
Oats		x	x
Potatoes			x
Rice		x	x
Soybeans	x	x	x
Sudangrass	x		
Sugar beets			x
Tree fruits	x	x	
Vegetables	x	x	x
Walnuts	x		
Wheat		x	x

Note: Crops listed in more than one category have a wide range of tolerance, depending upon variations in soils, crop varieties, and growing conditions.

Source: Martvedt/Wallace/Curley (1977).

Synthetic chelates and some natural organic complexes tend to resist the adverse effects of soil reactions, climate, and management practices that change available iron into unavailable forms. Natural organic complexes, including some lignin sulfonates and polyflavonoids, or synthetic chelates of iron tend to remain in an available form during most of the growing season. Synthetic chelates are mobile, however, and may be lost from the root zone if subjected to excessive leaching due to overirrigation or high rainfall. Not all iron chelates behave the same in soils, since some are differentially fixed onto clay particles. Fixed chelates do not function in delivering iron to plants. Stability or resistance to decomposition of metal chelates is related to soil pH. Some chelates tend to release their iron more readily than others. For application to calcareous soils of pH 7.5 or higher, the principal iron chelate that will correct iron deficiency for most dicotyledonous plants is FeEDDHA or other similar compounds which have high metal chelate stability. Commercial sources of iron for agricultural application include:

Inorganic Sources: Ferrous sulfate; ferric sulfate; ferrous carbonate; ferrous ammonium sulfate; iron frits.

Chelates: FeEDTA (ethylenediaminetetraacetic acid); FeHEDTA (hydroxyethylethylenediaminetriacetic acid)

Organic Complexes: Lignin sulfonate; methoxy phenylpropane complex; polyflavonoid

During recent years, more emphasis has been placed on testing soils for iron deficiency. A commonly used test, developed at Colorado State University, employs a solution of the chelating agent DTPA (diethylenetriaminepentaacetate) and calcium chloride buffered at pH 7.3 for extracting the soil under test. For testing larger areas, a new method of detection and visual assessment of iron chlorosis or deficiency symptoms in growing crops has been developed. This involves aerial infrared photography which records distinctive color differences between chlorotic and normal green plants. Assessment of a spotty chlorotic field by proportion (percent) and three-dimensional projection makes possible economic evaluation of the iron deficiency problem existing in any given field on a large-scale basis. See also **Fertilizer**.

Iron in Human Physiology

The bioavailability of iron has been researched extensively. Investigators have found this to be quite variable because numerous factors influence absorption of iron, including the consumer's needs and the composition of the diet. Chemical factors that affect iron availability include the valence, solubility, and degree of chelation or complex formation of the iron. Researchers have shown that the ferrous valence is considerably more available than the ferric valence. Others have shown that prior to absorption in the gut, iron must be in solution. Further, it has been demonstrated that chelation may augment iron absorption by maintaining the iron in solution under conditions where it would otherwise be insoluble. Because of more ready availability, ferrous sulfate is used in bread and flour enrichment even though it has a greater reactivity with foods. Ascorbic acid has been shown to increase iron availability when in the diet. See also **Diet**.

Iron Absorption. The iron content of the normal adult human is dependent on the size of the individual and the hemoglobin concentration. The distribution of iron in a male weighing 155 pounds ($\sim$70 kilograms) has been estimated at just under 3.5 grams. About 64% of this iron is in hemoglobin as part of the peripheral blood; and 2.5% as hemoglobin in the bone marrow. Another 4% is present as myoglobin which also participates in oxygen transport and storage. Another 13% is present as ferritin and 16% as hemosiderin, which are storage forms. Extremely small amounts are found in cellular cytochrome and in the enzyme catalase.

It is generally agreed that iron is absorbed from the most part in its ferrous form directly into the bloodstream. Radioactive iron has been shown to be absorbed from any portion of the intestinal tract, but its uptake appears greatest in the duodenum. On the basis of experiments done on the absorption of iron from the intestinal tract of guinea pigs, it was earlier postulated that iron is taken into the mucosa cells and ferritin is formed by a combination of a protein, *apoferritin*, with iron. After the cell is saturated with ferritin, absorption no longer takes place until the iron of ferritin is transferred to plasma. For a number of years, this concept of a mucosal block was the accepted explanation for iron absorption. However, later research showed that there is no absolute block to iron absorption. It is found that the absorption of iron in patients and in experimental animals is greater than normal in iron deficiency and in cases where erythropoiesis is accelerated, even when the body iron reserves are elevated. Later evidence indicated that the ferritin concentration in the intestinal mucosa neither controls nor blocks absorption. An active transport mechanism requiring energy is concerned with iron transfer across the intestinal mucosa.

The factors involved in the absorption of iron in food are more complex than those involving inorganic iron. To obtain Fe^{59}-marked foods, radioactive iron has been injected into hens to obtain labeled eggs and meat; plants have been grown in media containing Fe^{59}, and Fe^{59}-enriched bread has been prepared. It has been shown that iron-deficient subjects absorb more food iron than normal subjects. Absorption from liver, hemoglobin, muscle, and "enriched" bread is greater than from eggs or plants. Most probably, the low absorption from egg yolk derives from the presence of a ferric iron-phosphate complex. In such research, large variations in results have been obtained.

In the presence of a large amount of ascorbic acid (vitamin C), the absorption of iron is appreciably enhanced, because of the reduction of Fe^{3+} to the Fe^{2+} form. In the presence of phosphates, carbonates, and phytates, insoluble iron compounds are formed, thus reducing absorption.

It has been estimated that normal subjects ingesting a mixed diet containing 12–15 milligrams of iron retain 5–10% (0.6–1.5 milligram), whereas iron-deficient patients retain 10–20% (1.2–3 milligrams) iron.

Iron Transportation. After iron enters the bloodstream, it is immediately bound by a specific plasma protein that is a β_1-globulin. This protein, *transferrin* (siderophilin), has a molecular weight of about 90,000 and binds two atoms of ferric iron. About 0.25 gram of transferrin in 100 milliliters of plasma is capable of binding about 300 micrograms Fe^{3+}, but normally it is only one-third saturated while the remaining two-thirds are unbound reserve. If a small amount of ionized iron is injected intravenously, it is bound by the transferrin, which may be completely saturated. If the binding limit is exceeded, ionized iron exhibits toxic effects. The transferrin concentration is increased in iron deficiency and during the latter half of pregnancy; it is decreased during infection and a variety of other disorders.

Electrophoretic studies show there are several genetically controlled variants of human transferrins. They all deliver iron in an equivalent manner for utilization and storage. Evidence indicates that iron may be transferred directly to the developing erythroblast. It has been demonstrated that transferrin-bound iron is utilized by reticulocytes for hemoglobin formation. The transfer of iron is not maximal until 25% of the transferrin is saturated.

Excretion. The total loss of iron from an adult is about 1 milligram daily and is distributed in sweat, feces, hair, and urine. Since approximately 1 milligram of iron is normally absorbed daily, the organism is in iron balance. The loss of red cells from the body in normal menstruation would account for 16–32 milligrams of iron, which would amount to an average daily loss of from 0.5–1.0 milligram during the 28-day menstrual cycle. Pregnancy would also represent a loss of iron from the body, but this is compensated by the absence of menstruation. During normal hemoglobin catabolism, about 20–25 milligrams of iron are released per day. The excretion of minute amounts of iron allows the body to conserve and reutilize the iron for the synthesis of hemoglobin. This tenacious conservation has been demonstrated repeatedly by radioactive techniques.

Enzymes. Heme serves as the prosthetic group for catalase, peroxidase, cytochrome oxidase, and the related cytochromes. Catalase and peroxidase iron are presumably present in the ferric form while the iron of the cytochromes may exist in the reduced or oxidized form. A number of flavoproteins, including succinic dehydrogenase, contain iron in the molecule. Iron appears to act as coenzyme for aconitase. A number of other enzymes require the presence of iron for their activities.

Storage Iron. Ferritin and hemosiderin represent practically all the iron present in the reticulo-endothelial cells of the liver, spleen, and bone marrow and in the parenchymal cells of the liver. Ferritin is an iron protein complex containing up to 23% iron. It is composed of a protein, which has a molecular weight of 450,000 and a colloidal ferric-hydroxide-phosphate complex. Preparations of hemosiderin granules contain up to 40% iron and are insoluble in water. It appears to be an iron-loaded organelle, such as mitochondrion. The granule contains a small amount of ferritin, but the remaining material is composed of heterogenous proteins.

Hemoglobin. The approximate formula (molecular weight >52,000) is $(C_{738}H_{1166}FeN_{208}S_2)_4$. Hemoglobin is the respiratory protein of the red blood cells. It transfers oxygen from the lungs to the tissues and carbon dioxide from the tissues to the lungs. Its affinity for carbon monoxide is over 200 times that for oxygen. Hemoglobin is a conjugated protein consisting of approximately 94% globin (protein portion) and 6% heme. Each molecule can combine with one molecule of oxygen to form oxyhemoglobin. The iron (in the heme portion) must be in the reduced (ferrous) state to enable the hemoglobin to combine with oxygen. Heme $(C_{34}H_{32}FeN_4O_4)$ is the nonprotein portion of hemoglobin and myoglobin, consisting of reduced (ferrous) iron bound to protoporphyrin. See also **Hemoglobin**.

Iron Deficiency Disorders

Iron deficiency may occur from several causes, including loss of blood through hemorrhage, an iron-poor diet, or an inability to metabolize iron in a normal fashion. There are a few more complex disorders, some of which have hereditary vectors.

Nutritional Anemias. These disorders may result from nutritional deficiencies or decreased bone marrow function, both of which cause defective blood formation. See also **Anemias**. The least severe but most common of these anemias results from an inadequate amount of iron required for red cell formation. The result is *microcytic hypochromic anemia*. About 100 milligrams of iron per day are needed for hemoglobin manufacture. About 85% of this iron may be obtained from the iron released by breakdown of older red cells. However, some iron is always lost in the excretions and thus must be made up by the diet. Where there is chronic blood loss, as in cases of ulcers or hemorrhoids, or where the iron may not be properly absorbed from foods, the need for iron may be greater. Milk, cereals, and many refined foods, unless artificially supplemented, do not contain much iron. Better sources of iron include meat and leafy vegetables. Iron deficiency is not uncommon.

A common form of iron-deficiency anemia frequently seen in young women during the last century was sometimes called chlorosis, or "green sickness" because of the peculiar hue of the skin. With the discovery that iron salts can effect a cure, the disease almost completely disappeared. Idiopathic hypochromic anemia is another iron-deficiency anemia associated with a lack of proper stomach acidity. When hydrochloric acid in the stomach is lacking, iron cannot be liberated from foods and converted into a form that can be absorbed. Administration of iron in proper form also alleviates this condition.

Iron in Pregnancy. During pregnancy, the mother must furnish greater volumes of blood to support herself and the developing baby. Blood volume is increased and causes dilution unless sufficient iron is available. Vomiting in early pregnancy may increase the danger of an iron deficiency. Usually, babies are born with adequate supplies of iron in their tissues to last several months. However, infants born of a mother with an iron deficiency have low reserve stores of iron and will require a diet that is supplemented with the proper amounts of iron. Milk is a poor source of iron, and infants strictly on a diet of milk almost invariably develop hypochromic anemias. Anemic babies are much more subject to infections, which may in turn further increase the anemia. Thus, such children should be treated early.

Infants with Iron Deficiency. B. Lozoff (Case Western Reserve University) and associates reported that "Several consistent results have emerged from five studies of the behavior and development of infant with iron-deficiency anemia, a condition that affects at least 20 to 25 percent of the world's babies." All five studies used careful definitions of iron status and included comparison groups without anemia. All showed that infants with anemia scored lower on tests of mental development administered before treatment than infants without anemia did.

Iron Overload. An inborn error of metabolism leads to the absorption of excess iron from a normal diet. Hereditary *hemachromatosis* is found mainly within the white population. Black people who live in sub-Saharan Africa, however, show a high incidence of this disorder. At one time, this was attributed to the high content of iron in a home-brewed beer. Recent studies show that it is a combination of a high-iron diet and hereditary disposition.

Thalassemia Major. Transfusion-dependent thalassemia major patients have abnormal growth and sexual maturation at puberty, presumably as a result of pituitary iron overload. Still poorly understood, this disorder is reported to respond well to deferoxamine iron chelation therapy, particularly if administered before the age of maturity.

Mitochondrial Myopathy. A general deficiency of iron may be implicated in mitochondrial myopathy, which is a complex disorder that affects muscular activity. It has been suspected for a number of years that the disorder is caused by a defect of mitochondrial-protein transport. H.H.V. Scharpa and a team of researchers (Royal Free Hospital, London) postulate that a deficiency of an iron-sulfur protein in muscle dehydrogenase may be the specific cause.

Iron Chelation Therapy in Cerebral Malaria. It is estimated that over 1 million children die from severe forms of malaria annually, notably in sub-Saharan Africa. Cerebral malaria is one of the most severe complications of malaria infection (*Plasmodium falciparum*). It is known that iron is an essential nutrient for promoting the growth of the infectious agent. Victor Gordeuk and a team of investigators report that the iron-chelating agent deferoxamine enhances the clearance of the *P. falciparum* parasitemia. Iron chelation inhibits peroxidant damage to the central nervous system, as previously tested in animals. Iron also serves as a redox agent in the generation of free radicals that mediate ischemic and hemorrhagic tissue energy. A report issued in November 1992 concludes: "Iron chelation therapy may hasten the clearance of parasitemia and enhance recovery from deep coma in cerebral malaria."

Additional Reading

Bacon, B.R.: "Causes of Iron Overload," *N. Eng. J. Med.*, 126 (January 9, 1992).

Bronspiegel-Weintrob, N., et al.: "Effect of Age at the Start of Iron Chelation Therapy on Gonadal Function in Thalassemia Major," *N. Eng. J. Med.*, 713 (September 13, 1990).

Bullen, J.J. and E. Griffiths: *Iron and Infection: Molecular, Physiological and Clinical Aspects*, John Wiley & Sons, Inc., New York, NY, 1999.

Gordeuk, V., et al.: "Effect of Iron Chelation Therapy on Recovery from Deep Coma in Children with Cerebral Malaria," *N. Eng. J. Med.*, 1473 (November 19, 1992).

Gordeuk, V., et al.: "Iron Overload in Africa—Interaction between a Gene and Dietary Iron Content," *N. Eng. J. Med.*, 95 (January 9, 1992).

Loretta, J. and P.W. Atkins: *Chemistry: Molecules, Matter and Change*, W.H. Freeman and Company, New York, NY, 1999.

Lozoff, B., Jimenez, E. and A.W. Wolf: "Long-Term Developmental Outcome of Infants with Iron Deficiency," *N. Eng. J. Med.*, 687 (September 5, 1991).

Mielczarek, E.V. and S.B. McGrayne: *Iron, Nature's Universal Element: Why People Need Iron and Animals Make Magnets*, Rutgers University Press, Piscataway, NJ, 2000.

Mortvedt, J.J., A. Wallace and R.D. Curley: "Iron-The Elusive Micronutrient," *Fertilizer Solutions*, 21, 1, 26–36 (1977).

Scharpa, A.H.V.: "Mitochondrial Myopathy with a Defect of Mitochondrial-Protein Transport," *N. Eng. J. Med.*, 37 (July 5, 1990).

Schwertmann, U.: *Iron Oxides in the Laboratory*, 2nd Edition, John Wiley & Sons, Inc., New York, NY, 2000.

Staff: Institute of Medicine, *Iron Deficiency Anemia*, National Academy Press, Washington, DC, 1994.

Sullivan, J.L.: "Retinopathy of Prematurity," *N. Eng. J. Med.*, 648 (Aug. 27, 1992).

Sykes, A.G. and R. Cammack: *Advances in Inorganic Chemistry: Iron-Sulfur Proteins*, Academic Press, Inc., San Diego, CA, 1999.

Symons, M.C.: *Free Radicals and Iron: Chemistry, Biology, and Medicine*, Oxford University Press, Inc., New York, NY, 1998.

Wyler, D.J.: "Bark, Weeds, and Iron Chelators—Drugs for Malaria," *N. Eng. J. Med.*, 1519 (November 19, 1992).

IRON METALS, ALLOYS, AND STEELS. Chemically pure iron is used essentially in powder metallurgy and for chemical applications where the element serves as a catalyst, or as a base ingredient for ferrous and ferric chemicals. Iron principally is used as the dominant ingredient of cast irons and steels. Iron-base alloys notably are known for their physical strength and toughness and, when compared with most other metals for similar applications, reasonable cost. The following properties depend upon the nature and extent of the ingredients, such as carbon, present in the alloys and also upon the mechanical and heat treatments given to the formed metals: (1) impact strength or brittleness, (2) cohesive strength, (3) compressive strength, (4) creep, (5) fatigue, (6) ductility, (7) hardness, (8) malleability, (9) shear strength, (10) yield strength, (11) torsional strength, (12) electrical conductivity, (13) thermal conductivity, (14) thermal stability, (15) thermal expansion, (16) corrosion resistance, (17) magnetic properties, and (18) heat treatability.

The iron metals family of products may be classified into: (1) *the pure irons*, such as ingot iron and wrought iron, which have only traces of carbon (see Table 1) and other elements, and are very ductile; (2) *cast irons*, which are alloys of iron and carbon, with or without other elements, and normally containing from 2.4 to 4.5% carbon; (3) *steels*, which are alloys of iron and carbon, with or without other elements, in which the carbon content seldom exceeds 1.7%; (4) *alloy steels* whose properties mainly are attributed to the presence of one or more elements other than carbon. There are other groups and numerous subgroups in the total iron metals family.

TABLE 1. TYPICAL ANALYSES OF PURE IRONS

Ingredient	Ingot Iron	Electrolytic Iron	Carbonyl Iron	Hydrogen-Purified Iron
	Percent of Total			
Carbon	<0.020	0.006	0.0004	0.005
Manganese	<0.020	—	—	0.028
Phosphorus	0.005±	0.005	—	0.004
Sulfur	0.020±	0.004	—	0.003
Silicon	Trace	0.005	—	0.001
Copper	0.04±	—	—	—
Oxygen	Some	Some	<0.01	0.003
Nitrogen	0.004±	—	—	0.0001

Ironmaking

The starting ingredients for the numerous iron metals are obtained in three major ways: (1) by smelting run-of-mine or beneficiated iron ore in a blast furnace, low-shaft furnace, or electric smelter to yield a liquid, molten product; (2) by reducing run-of-mine or beneficiated iron ore via direct reduction processes to produce sponge iron; and (3) by melting ferrous scrap in a cupola, electric furnace, or fuel-fired furnace. Inasmuch as the majority of iron ores are in the form of oxides of iron, the iron is obtained by employing suitable reducing agents to reduce the oxides. Reducing agents most often used are carbon, carbon monoxide, hydrogen, and hydrocarbons, such as methane. With carbon monoxide, the reduction is exothermic; with the other reductants, it is endothermic.

Blast Furnace. For several decades, the blast furnace was the unchallenged producer of iron (pig iron) from iron ore. With the growing availability of steel scrap over the last several years, scrap as a source of iron for steelmaking became increasingly important, a fact which made electric furnace steel production, to be described later, very attractive. Blast furnaces are large, bulky structures that have been part of the integrated mill concept, a concept that has been threatened economically for several years. Blast furnaces are high-capacity producers and, consequently, essentially unsuited to the current philosophy of spreading steel production geographically (nearer the consumers). It is much easier to commence with scrap than to have to make pig iron. A number of the older style blast furnaces have been dismantled in recent years because of the growing dependability of scrap as a starting material. But, as pointed out later, all steel products do not represent potential usable scrap and, consequently, the iron ore reduction processes, including the blast furnace, will continue to be used in a number of locations.

A typical, traditional blast furnace is shown in Fig. 1. The furnace is a tall, refractory-lined vessel. Raw materials including iron ore (sinter or pellets), coke (the reducing and thermal agent), and limestone (for fluxing the gangue material) are charged into the top of the furnace. A blast of hot air is introduced at the bottom of the furnace to burn the coke and thus to heat, reduce, and melt the charge as it descends toward the bottom of the furnace. Liquid iron and slag collect in the furnace hearth. These materials are tapped at regular intervals. Although the furnace can be damped down for short periods, the process essentially is continuous. The waste gas contains about 28% carbon monoxide with a calorific value of about 90 Btu/cubic foot (800 kcal/cubic meter). After collection at the top of the furnace, dust is removed, and the gas is used as a fuel for heating the hot-blast stoves. Blast furnaces range from 100 to 10,000 tons/day in capacity; the hearth diameter may range from 9 to 46 feet (2.7 to 13.8 meters); the height from 50 to 150 feet (15 to 45 meters).

Air required for combustion is furnished by turboblowers and preheated in hot-blast stoves lined with refractory-brick checker-work. Commonly, three stoves are used per furnace. The stoves are operated alternatively on a regenerative principle, one stove normally providing the hot blast while the other two stoves are heating up to temperature. The checkerwork provides the means for temporarily storing the rather large quantities of heat required. Fuel for the stoves includes the blast-furnace off-gas previously mentioned, augmented by coke-oven gas from a nearby coke-producing facility. The exact fuel-supply arrangement varies with local conditions. The hot blast is supplied by a gas main to a bustle pipe, which encircles the bosh and distributes it to water-cooled tuyeres located below the bosh for injection into the furnace. The metal is tapped into refractory-lined, open-top torpedo or kling-type ladles and thence transported to the pig-casting machine; or directly to the steel plant. A comparatively few blast furnaces are operated with charcoal instead of coke, but these are limited to a capacity of about 300 tons/day because of the low crushing strength of charcoal. Improvements in blast furnace operations over the last several years have resulted from better preparation of the charge, fuel injection through tuyeres, reducing-gas injection in bosh, and oxygen enrichment of blast.

Low-Shaft Furnace. These furnaces are circular or oval in cross section. The oval shape permits greater hearth area without increasing the required depth of penetration of the blast supplied through the tuyeres. Such furnaces are designed for finer raw materials and low-grade coke or lignite. Once considered ideal for small-scale iron production, only a limited number have been installed. The operating principles are essentially similar to those of a blast furnace.

Electric Smelter. Electric energy provides the heat in these designs. Low-grade coke can be used as the reducing agent. Electric smelters are generally limited to areas of low-cost electric power. The most commonly used furnaces of this type employ a submerged arc, using the Söderberg continuous self-baking electrodes. Developed by Tysland and Hole in Norway, the furnace is circular or rectangular in cross section

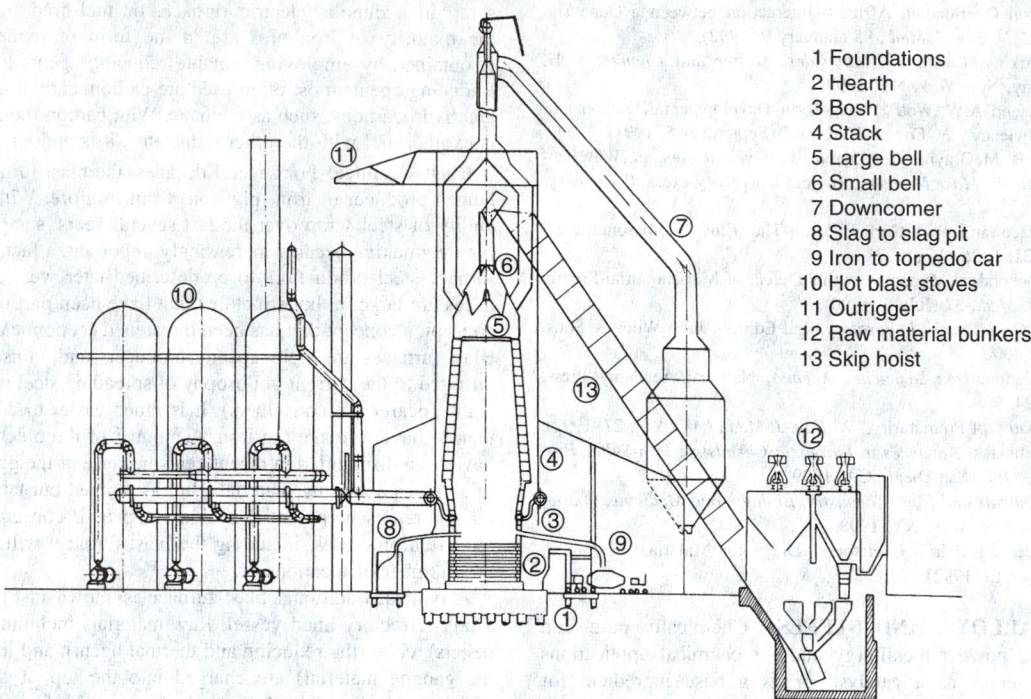

1 Foundations
2 Hearth
3 Bosh
4 Stack
5 Large bell
6 Small bell
7 Downcomer
8 Slag to slag pit
9 Iron to torpedo car
10 Hot blast stoves
11 Outrigger
12 Raw material bunkers
13 Skip hoist

Fig. 1.   Cross section of a traditional blast furnace plant.

with transformer ratings of up to 60,000 kVA. Production capacity can be increased and power and coke requirements lowered by preheating and prereducing the iron-ore charge.

Direct Reduction Process. Numerous schemes over the years have been attempted as an alternative to the blast furnace. These include rotary and stationary kilns and furnaces, reverberatory furnaces, retorts, fluid-bed reactors, pot furnaces, and jet smelting. Similarly a variety of reductants have been used, including lignite, coal, char, fuel oil, tar, and various gases. The direct reduction approach is essentially useful for producing a highly reduced product containing mostly metallic iron and little gangue material, thus providing a substitute for ferrous scrap for steelmaking operations.

SL/RN Process. In this process, using a rotary kiln and solid reductant, high-grade pellets or lump ores and anthracite are used. The ore or pellets, anthracite, dolomite or limestone, and return coal are fed to the rotary kiln. The temperature is controlled by means of shell burners which are furnished with air and gas or oil. A uniform temperature of about 1,100 °C is maintained over about 60% of the kiln length. After leaving the reduction kiln, the charge is passed through a gastight seal into a water-cooled drum, whereupon it is cooled to a temperature below 100 °C to prevent reoxidation of the sponge iron. Most of the sponge iron can be separated by screening, which is augmented by magnetic separators.

HyL Process. This batch-cyclic process reduces rich lump-iron ores by flowing a reducing gas in a fixed-bed reactor. The reducing gas may be prepared by steam reforming of natural gas or other hydrocarbons. A typical reducing gas may contain 74% hydrogen, 13% carbon monoxide, 8% carbon dioxide, and 5% methane (all volume percent). The process requires four reactors, each reactor following four steps in a 12-hour cycle: (1) removal of cold sponge iron and loading with fresh iron ore or pellets; (2) preheating and secondary reduction with partially spent reducing gas from another reactor; (3) primary reduction to sponge iron; and (4) cooling the sponge iron with fresh, cool reducing gas and controlled deposition of carbon where required.

Purofer, Midrex, and Armco Processes. These are continuous processes in which shaft furnaces are used. Iron ore or pellets are charged from the top and the reduced product is withdrawn from the bottom. The reducing gas may be generated by reforming natural gas, or methane rich gas from naphtha may be used. The hot reducing gas flows counter-current to the descending charge.

Fior and U.S. Steel Processes. Similar reducing gases are used in connection with a fluid-bed reactor. The iron ore may require further

grinding and drying before introduction into the fluid bed. The reduced ore may be briquetted or used as fines.

Cupola. This type of furnace is widely used for making iron for casting in foundries and sometimes may be used to augment the iron required by steelmaking operations. A cupola may be operated for just a few hours/day, or up to 2 to 3 months continuously. A cupola is a vertical cylindrical shaft furnace that uses the countercurrent-flow principle to heat and melt the charge as it descends. Cupola capacity may range from 2 to 75 tons/hour. Unlike a blast furnace, a cupola is not a reducing unit, but is essentially used to melt ferrous scrap and cold pig iron or previously reduced sponge iron. The heat required is supplied by nearly complete combustion of coke. Air is injected through tuyeres near the hearth zone. Some cupolas are equipped for hot-blast supply through recuperators. The iron raw material is charged in alternate layers with coke. Limestone is added to flux the ash from the coke and form slag. The molten iron collects in the hearth and may be removed continuously or intermittently through a taphole.

Steelmaking

Raw materials for steelmaking include liquid iron, steel scrap, preproduced sponge iron, or mixtures of the foregoing ingredients. During the past quarter century, steelmaking has undergone numerous changes, essentially motivated by marked increases in energy costs, by the need for better production flexibility, and by the impact of worldwide competition, particularly by nations that could essentially commence serious production with modern technology rather than engaging in tremendously costly modernization of decades-old, large, and sprawling integrated mills. Essentially since the turn of this century, the traditional steelmaking countries such as the United States, United Kingdom, and Germany depended upon the integrated mill concept to produce vast tonnages of steel, largely a system that served them well during two world wars. The basic economics of the steel industry today are strikingly different, and altered economics, even more than the availability of technology, have brought about steel production changes. The competition of other metals and materials also has adversely affected the once-exclusive market for steels. The greater availability of steel scrap has also been a factor.

Historical Open-Hearth Process. For well over a century prior to the 1960s, the open-hearth process was the principal means for making steel. As of the late 1980s, open-hearth steels had dropped from well over 90% of total steel production to less than 10% of steel production. The attraction of the basic open-hearth system over many decades was versatility in handling a variety of raw materials required for most grades of steel. Raw material charges could be 100% scrap, 100% hot metal, or scrap and hot metal in all intermediate ratios.

In the open-hearth process, the iron is kept molten with gas burners for as long a period as needed for reaction of carbon in the metal with oxygen. Oxygen is added along with the gas fuel above the molten material. Oxygen can be added to the melt in the form of iron oxides, the latter sometimes derived from iron ore or rusty steel scrap. In the open-hearths still operating, the percentage of scrap has generally increased. In its many decades of use, the open-hearth process became a mature technology, having benefited from numerous technical improvements. But, these improvements were not sufficient to permit the open-hearth process to compete with other basic steelmaking processes.

Steel was also made for many years in Bessemer converters, wherein liquid iron was refined in a bottom-air-blown converter—a refractory-lined, pear-shaped, cylindrical vessel open at the top to permit charging of materials and to allow the escape of gases. In this process, a considerable amount of heat was wasted in heating the nitrogen in the air. The Bessemer process could melt only 5 to 10% scrap. The nitrogen content of Bessemer steels was high and oxygen-steam or oxygen-carbon dioxide mixtures were used instead of air to produce low-nitrogen steels.

Basic Oxygen Process (BOP). In this process, almost pure oxygen is blown at high velocity onto the surface of the molten iron. Conversion to steel occurs roughly ten times faster than with the open hearth. The BOP produces heat as the oxygen combines with the carbon in the molten metal, which occurs at a very fast reaction rate. Thus, supplemental fuel is not required to keep the melt from solidifying during its conversion to steel. In 1960, only 3% of steels produced in the United States were by BOP. A basic oxygen steel plant is shown in Fig. 2.

In this process, molten iron is refined to steel by top-blowing oxygen at high pressure onto the surface of the metal through a water-cooled lance contained in a tilting furnace. The oxidation of carbon, silicon, manganese, and phosphorus provides sufficient heat for converting molten iron into steel. Because of the excess heat generated, up to about 30% scrap can be charged. A conventional basic oxygen process can refine iron containing up to 0.3% phosphorus into most grades of steel. Where the phosphorus content is higher, a modified process uses injection of powdered lime with the oxygen stream, or double slagging is required. A basic oxygen furnace with 300-ton capacity is shown in Fig. 3.

Numerous modifications of the BOP have appeared during recent years. For example, the OBM/Q-BOP, LWS, and SIP processes have been developed which use bottom blowing of oxygen and a shielding hydrocarbon through tuyeres in the bottom of the converter vessel. The endothermic dissociation of hydrocarbon by its cooling effect prevents excessive refractory wear in the tuyere area. The result is a substantial increase in the life of the bottom refractory plug. The OBM process was initially developed in Germany. The designation Q-BOP, introduced in the United States after further development work, is intended to emphasize the advantages of the new process compared with the Basic Oxygen Process (BOP). The letter, Q, stands for "quiet, quick, quality." Natural gas, propane, or liquefied petroleum gas are used as the gaseous hydrocarbon shield injected through an outer concentric gap around the central oxygen tuyere. A similar development, preferably using fuel oil as a hydrocarbon shield, is termed the LWS process. The OBM tuyere is successfully inserted in the bottom of an open hearth furnace for injecting oxygen into the metal bath for refining. This relatively new steelmaking technique is known as SIP (submerged injection process).

In the Kaldo process developed in Sweden, the refining of molten iron to steel is carried out in a titled pear-shaped basic-lined converter. Oxygen is blown at an oblique angle to the metal bath through a water-cooled lance. Much of the carbon monoxide produced by the carbon/oxygen reaction is burned inside the converter. The heat generated is absorbed by the rotating vessel and transferred to the bath, thus providing a high thermal efficiency and allowing up to 40% scrap in the charge. Both low- and high-phosphorus irons can be handled. Refractory consumption is high, tending to reduce the availability of the furnace.

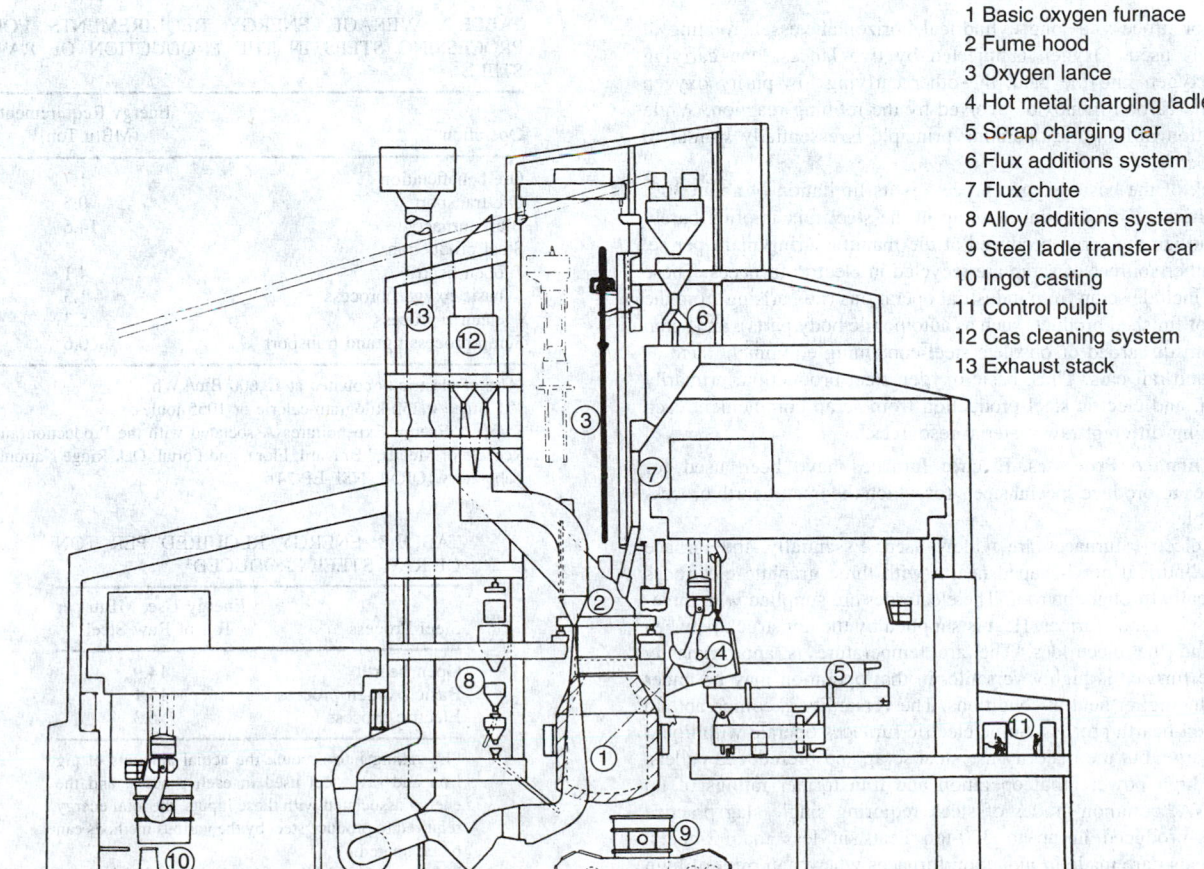

1 Basic oxygen furnace
2 Fume hood
3 Oxygen lance
4 Hot metal charging ladle
5 Scrap charging car
6 Flux additions system
7 Flux chute
8 Alloy additions system
9 Steel ladle transfer car
10 Ingot casting
11 Control pulpit
12 Cas cleaning system
13 Exhaust stack

Fig. 2.   Section through basic oxygen steelplant.

Fig. 3.   A basic oxygen furnace of 300-ton capacity. (*Davy McKee Iron & Steel Division, Stockton-on-Tees, England.*)

In the Rotor process, a long cylindrical horizontal vessel, rotating at 1 to 5 rpm, is used. Oxygen is injected by two lances, one carrying high-purity oxygen into the bath, the other carrying low-purity oxygen for burning the carbon monoxide evolved by the refining reaction. While the configuration differs, the operating principle is essentially similar to the Kaldo process.

A drawback of the basic oxygen process is its limitation of about 30% scrap in its charge. This amount of scrap in the steel mix is often barely adequate to utilize the scrap produced at the manufacturing plant per se. As a result, other sources of scrap are recycled in electric furnaces. These other sources include scrap from industrial operations (i.e., arising from the manufacture of finished products such as automobile body parts), and scrap reclaimed from discarded or obsolete steel-containing equipment such as automobiles and rail cars. Thus, basic oxygen steel production, primarily from pig iron, and electric steel production from scrap complement each other in utilizing different raw material resources.

Electric Furnace Processes. Electric furnaces have been used for several decades to produce special steels for which the open-hearth process was not suitable.

Direct-arc electric furnaces are widely used. Essentially, the furnace is a tilting cylindrical bowl-shaped hearth with three graphite electrodes inserted vertically through the roof. The electrodes are supplied with three-phase current via a transformer. Heat is supplied by the arc struck between the charge and the electrodes. The arc temperature is approximately 3,400 °C. The furnace is highly versatile, in that operation may be under oxidizing, reducing, or neutral conditions. The versatility is comparable to that of the open-hearth process. Some electric furnaces operate with liquid iron in the charge, but the majority use steel scrap and prereduced pellets.

With very high power input operation and transformer ratings of up to 100,000 kVA, common grades of steel, requiring single slag practice only, can be produced in up to 300-ton heats in less than 3 hours. Special steels also are made in induction furnaces where a current of high or medium frequency is passed through a coil surrounding a refractory crucible containing the charge.

Spray Steelmaking Process. In this process, liquid iron is poured through a tundish and refined continuously by injecting powdered lime and oxygen tangentially from a ring onto the surface of the metal stream.

FOS Process. In the fuel-oxygen-scrap process, a vessel similar in shape to an electric-arc furnace is used, but having a greater height-to-diameter ratio. There is a removable roof to permit rapid charging of scrap. Heat is supplied by an oxyfuel burner inserted through a central opening in the roof.

Additional processes include the Wocra and the Irsid processes which are based upon continuous melting and/or refining techniques. The cyclosteel and jet-smelting processes make liquid iron by flash smelting of iron ore.

Comparison of Process Energy Requirements. Energy requirements for the various steps in raw steel production are shown in Table 2. When the energy requirement for each step is weighted according to the different proportions of raw materials used in the three basic processes just described, the total amount of energy per ton of raw steel is determined to be about 15 MBtu per ton for the open hearth and basic oxygen processes, compared to about 6.3 MBtu per ton for electric melting. See Table 3. Note that the value given for electric steel includes the energy required to generate the electricity. Part of the reason for the large difference in energy requirements is that electric steel is made from scrap and, therefore, little energy is required to reduce iron oxide to elemental iron. In this sense, stockpiles of iron and steel scrap represent a significant source of stored energy.

It is, of course, important to observe that currently about 30% of the output of steel products made is not recoverable as scrap. Examples include reinforcing bars incorporated within concrete structures, wire products, such as nails and fencing, and buried piping, such as oil well casings. Other products, such as "tin" cans, may someday be recovered on a large scale from municipal wastes. Presently, much of this steel is wasted. It is this unrecoverable quantity of material that in any long-term equilibrium sense ultimately must be derived from the mining and reduction of iron ore. Consequently, the reduction of ore to iron, as in a blast furnace, and the need for the basic oxygen process or equivalent is self-evident.

TABLE 2. AVERAGE ENERGY REQUIREMENTS FOR PROCESSING STEPS IN THE PRODUCTION OF RAW STEEL

Operation	Energy Requirement* (MBtu Ton)**
Ore beneficiation	1.7
Ore transport	0.5
Blast furnace	14.6
Steel production	
open hearth	4.1
basic oxygen process	1.3
electric process	5.3
Scrap processing and transport	0.6

*Use of electricity counted at 10,600 Btu/kWh.
**1 Btu = 0.252 kilogram-calorie or 1055 joules.
Source: "Energy Expenditures Associated with the Production and Recycle of Metals," Bravard, Flora, and Portal, Oak Ridge National Laboratory (ORNL-NSF-EP-24).

TABLE 3. ENERGY REQUIRED PER TON OF RAW STEEL PRODUCED*

Steel Process	Energy Use, MBtu per Ton of Raw Steel**
Open hearth	14.9
Basic oxygen process	15.1
Electric process	6.3

*By taking into account the actual amounts of pig iron and raw steel used in each process, and the energy associated with those inputs, the total energy required to produce steel by the various methods can be computed.
**1 Btu = 0.252 kilogram-calorie or 1055 joules.
Source: Electric Power Research Institute, Palo Alto, California, August 1986.

Minimill Concept

The switchover from open-hearth to basic oxygen processes for steel production created an opportunity for new producers of electric steel to enter the competition. Small "minimills" using electric furnaces are not tied to the logistical problems of coal, ore, and limestone supply or the economies of scale associated with blast furnace and coking operations. Small minimills can be built with a relatively modest investment. These mills are well suited to take advantage of the greater availability of local scrap and, because of their relatively small size, can be located virtually anywhere, thus avoiding the costs associated with transportation.

Casting Steel

Although steelmaking processes vary considerably, the liquid steel resulting is tapped from the furnace into a ladle. This is a refractory-lined cylindrical container with trunnion attachment for crane lifting to transport the steel. Generally, the bottom of the ladle is fitted with a stopper-rod nozzle or a sliding-gate nozzle for pouring. Lip-poured ladles are occasionally used. Additions of deoxidants, recarburizers, and alloying materials may be made to the ladle during tapping from the furnace so that final composition of the steel may be adjusted. Sometimes, vacuum or gas-stirring treatment is used prior to casting steel.

Treatment of liquid steel under vacuum makes it possible to reduce the amounts of hydrogen, nitrogen, and oxygen and some harmful non-metallic inclusions, thus improving the properties and qualities of the steel. Lengthy heat treatments of up to 2 and 3 weeks can be eliminated by the effective removal of hydrogen from certain forging steels. Vacuum degassing of steel prior to casting includes: (1) ladle degassing in a chamber; (2) stream degassing by pouring from ladle to ladle or ladle to ingot; (3) vacuum-lifter or circulation degassing; (4) mold degassing; (5) a combination of arc heating and degassing; (6) vacuum-furnace degassing; and (7) employing a consumable electrode under vacuum. To equalize temperature and improve steel quality, sometimes inert gases, such as argon or nitrogen, are bubbled through liquid steel.

Once tapped from the furnace, vacuum degassing, or stirring treatment, the liquid steel is teemed into molds as ingots, continuously cast or pressure-poured into semifinished shapes, or poured into molds as steel casting. In conventional casting-pit practice, steel is teemed from the ladle into iron ingot molds of square, rectangular, polygonal, or round cross section where it solidifies as blocks. The ingots can be top-poured directly into individual molds, or bottom-poured simultaneously into a cluster through a trumpet-and-runner arrangement. To counteract shrinkage during solidification of killed steels, hot tops are used on molds. Once teemed, the molds are stripped and the ingots charged into soaking pits for heating for subsequent processing; or allowed to go cold for placement in the stockyard.

Continuous Casting. This process, wherein liquid steel is poured directly into semifinished shapes, such as slabs, blooms, blanks, or billets, is growing in use, mainly because it eliminates the need for heavy rolling-mill equipment.

In continuous casting, 90% or more of the molten steel ends up as finished product. This represents an enormous productivity gain compared with traditional practice. This gain is the result of converting steel finishing from a batch operation (with the ingot removed for reheating at several stages) to a continuous operation (with reheating applied as needed as the ingot moves along).

In principle, all steel could be delivered to finishing operations via continuous casting regardless of the process used to produce the steel. In practice, however, continuous casting operations are most easily and economically introduced in conjunction with new steelmaking facilities where the capacity and design of the steelmaking equipment can be matched with the capacity and layout of the finishing section and with the line of products planned for the operation. This is precisely the situation for the regional minimill specializing in the production of a small variety of high-volume simple shapes. In Japan, where more than 70% of the steelmaking capacity came on line after 1963, more than 80% of steel production was continuously cast by 1982. In that year, the United States by comparison continuously cast only about 25% of its steel production.

A continuous casting line is shown in Fig. 4. Steel is poured via a tundish into a water-cooled copper mold. As casting commences, the bottom of the mold is sealed with a dummy bar onto which the steel solidifies.

The solidified cast product is removed continuously by way of a direct spray-cooling withdrawal roll and cutoff system, maintaining a desired

1 Ladle
2 Ladle car
3 Tundish
4 Tundish car
5 One of the tundish preheating stations
6 Control room
7 Mould
8 Mould reciprocation drive
9 Secondary cooling zone
10 Cooling plates
11 Roller segments
12 Extended roller segment zone
13 Auxiliary hoist for roller segment maintenance
14 Tiltable dummy bar head
15 Dummy bar storage
16 Dummy bar storage cradle elevating mechanism
17 Cast strand
18 Cutting station

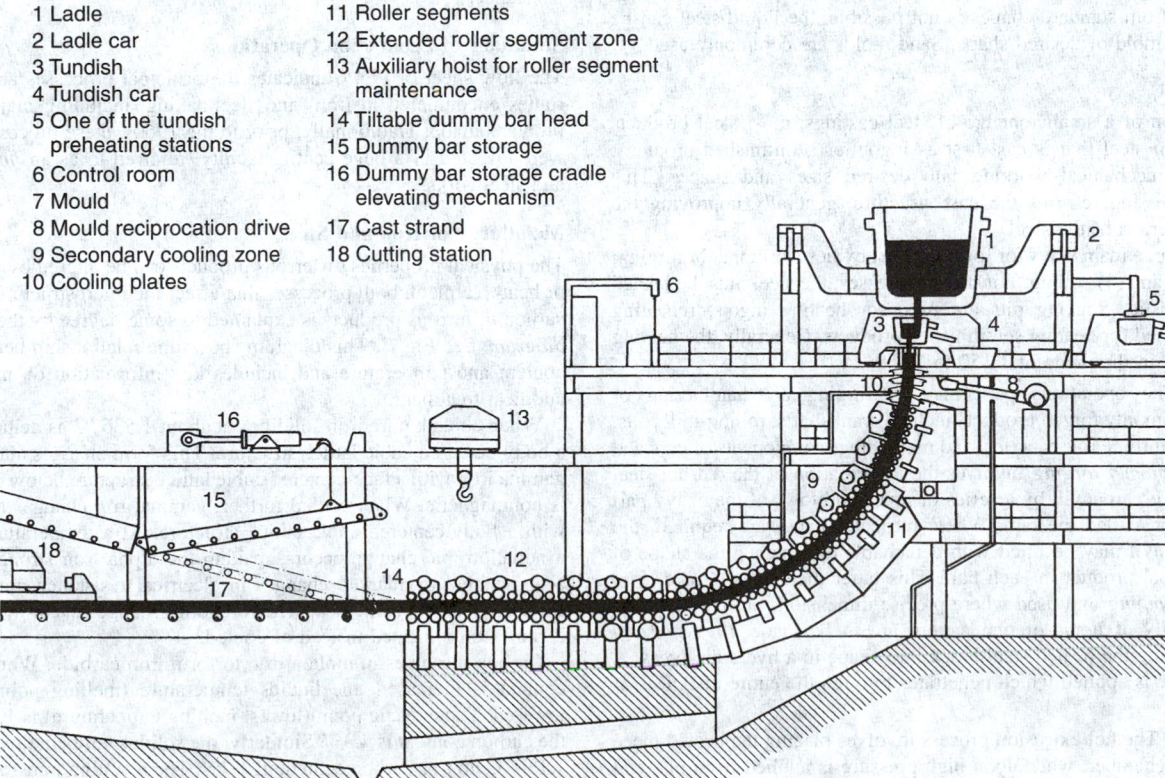

Fig. 4.   Continuous slab casting machine. (*Concast A.G. Zurich.*)

molten metal level in the copper mold. Once cut off, the cast product is discharged onto a cooling bank. Principal types of single or multistrand continuous-casting machines available include: (1) vertical mold with vertical cutoff; (2) vertical mold with blending rolls and horizontal cutoff; (3) curved mold with bending and horizontal cutoff; and (4) machines with direct strand-reduction units to reduce cross section by rolling before cutoff.

Pressure-Pouring Process. In the system shown in Fig. 5, the molten metal is forced up through a refractory tube into a mold by means of compressed air. To cast a number of molds in succession, two systems are used. The ladle may be placed in a stationary airtight pressure chamber and the molds moved over the chamber; or the molds may be stationary and the pressure chamber incorporating the ladle may be transported underneath the molds. The rate of pouring is determined by the rate of air-pressure increase; the height to which the liquid steel is raised is a function of the pressure applied.

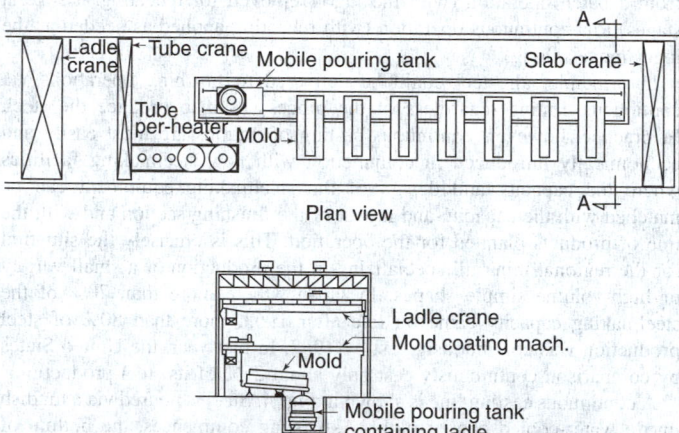

Fig. 5.   Pressure pouring plant showing fixed mold and moving pouring tank layout. (*Davy McKee Iron & Steel Division, Stockton-on-Tees, England.*)

Steel Castings. Where intricate shapes are involved or where mechanical working from standard shapes is not possible, the liquid steel can be poured into a mold of desired shape. Sand molds are commonly used.

Shaping Steel

With exception of a small tonnage of steel castings, most steel products are made from steel that is cast first as ingots or semifinished products, followed by mechanical working into desired sizes and shapes. This working processing reforms the cast structure, generally improving the physical properties of the steel.

There are three main ways for forming steel by hot working: (1) forging; (2) extrusion; and (3) rolling. For these processes, the steel must be heated until it is plastic. Soaking pits are used for heating ingots; reheating furnaces are used for heating semifinished products. Generally, the heating temperature is in the range of 1,150 to 1,350 °C.

Forging. The operations performed in forging are hammering or pressing. Commonly forged products include crankshafts, rolling-mill rolls, boiler drums, turbine rotors, axles, and many other components of cars and machinery. *Hammer forging* involves the deformation of the red-hot steel block, resting on an anvil, by a series of repeated blows of the heavy part of the hammer, called the ram. Where intricate shapes are required, the ram and the anvil may be fitted with detachable dies having the shape of the desired final product in each half. This latter method is termed *drop forging* or *stamping* and used where precise dimensions are required and a large quantity of items of one pattern are to be made. *Press forging* involves forming the heated steel block into shape in a hydraulic press. A steady squeeze is applied which penetrates through the entire thickness of the forging.

Extrusion. The hot-extrusion process involves placing the heated piece of metal in a chamber, whereupon high pressure is applied from one end by means of a hydraulically operated ram, thus causing the metal to flow through a restricted orifice at the other end. Desired shapes (cross sections) may include rounds, squares, and hexagons. For producing tubes, a die and

a mandrel are required. The process normally is limited to stainless and high-alloy steel products because of cost.

Rolling. In excess of 90% of steel production is processed by rolling. A rolling mill is used to produce a variety of semifinished and finished products, notably blooms, slabs, billets, rails, beams, angles, channels, rounds, squares, sheets, plates, and strip. Essentially, the process consists of passing the metal through (between) two rolls which are revolving at the same peripheral speed, but in opposite directions. The rolls may have smooth or grooved surfaces. The tap between the rolls is less than the height (thickness) of the material being rolled. While gripping the material during its passage through them, the rolls effect a reduction in cross-sectional area, with a corresponding increase in length. A final shape can be produced from a large block by making multiple passes between the rolls in a reversing action.

Rolling mills for processing large steel ingots are termed *blooming* or *slabbing* mills and the resulting forms are termed blooms or slabs. Blooms range from 5×5 inches (12.5×12.5 centimeters) upward. Products which commence with blooms are numerous, including: rails, structural shapes for bridges and building construction, window framing, steel partitions, bars of a variety of cross sections (ultimately made into nuts, bolts, shafts, and machinery parts, etc.), rod that ultimately may become wire, and narrow strip, which may become razor blades, tubes, pipes, wheel rims, etc.

Billet mills produce a product of 2 to 5 inches (5 to 12.5 centimeters) square in cross section. Heavy plates are produced in *plate mills.*

Steel sheets, widely used in thousands of products, are produced from slabs from a slabbing mill, continuous casting machine, or pressure-poured molds. These pieces are hot-rolled in a hot strip mill, descaled by pickling in an acid solution, and then cold-rolled and tempered in a cold-reduction and temper mill. The cold-rolled strip may be marketed in coil form, or carried to a side-trimming and sheet-shearing line, where it is customized for specific user needs.

Tin cans used as containers are made from tin plate, that is, a sheet steel thinly coated on each side by hot-dipping or the electrolytic process.

The hot strip is pickled and passed through cold and temper mills, after which it is passed through annealing and tinning lines. Further, zinc, terne, aluminum, or plastic coating can be applied for additional corrosion resistance.

Summary of Production Operations

The flow sheet of Fig. 6 indicates the principal processes and processing routes encountered in iron- and steelmaking, including major classes of products made. Traditionally, prior to the 1960s, these processes generally were effected in a huge central facility, referred to as an *integrated* iron and steelworks.

Metallurgy of Iron and Steel

The physical properties of ferrous products can be altered by cold working or heat treatment, both processes that affect the microstructure. The role of carbon in ferrous products is explained to some degree by the *iron-carbon diagram.* See Fig. 7. This diagram shows the relationship between carbon content and temperature and includes key information on microstructure and heat treatment.

When cooled, pure iron solidifies at about 1,536 °C as delta iron, having a body-centered cubic lattice structure. This form changes allotropically to gamma iron with a face-centered cubic lattice structure below 1,404 °C, and is nonmagnetic. When cooled further, gamma iron changes to alpha iron, with a body-centered cubic lattice structure. At a temperature of 768 °C, a nonallotropic change occurs, making the alpha iron strongly magnetic, accompanied by marked changes in electrical resistance, rate of thermal expansion, and specific heat. The foregoing changes occur in reverse order if pure iron is heated instead of cooled.

Carbon dissolves in molten iron to form iron carbide. When the carbon content is increased, the liquids temperature (melting point of iron) is lowered. The eutectic point (lowest melting temperature) is 1,130 °C when the carbon content is 4.3%. Similarly, the solidus temperature is lowered to 1,130 °C up to a carbon content of 1.7%. The resulting transformations and the products formed after cooling the iron-carbon alloy below the solidus temperature depend on the carbon content, the temperature, and the rate of cooling. Some of these substances are defined briefly below:

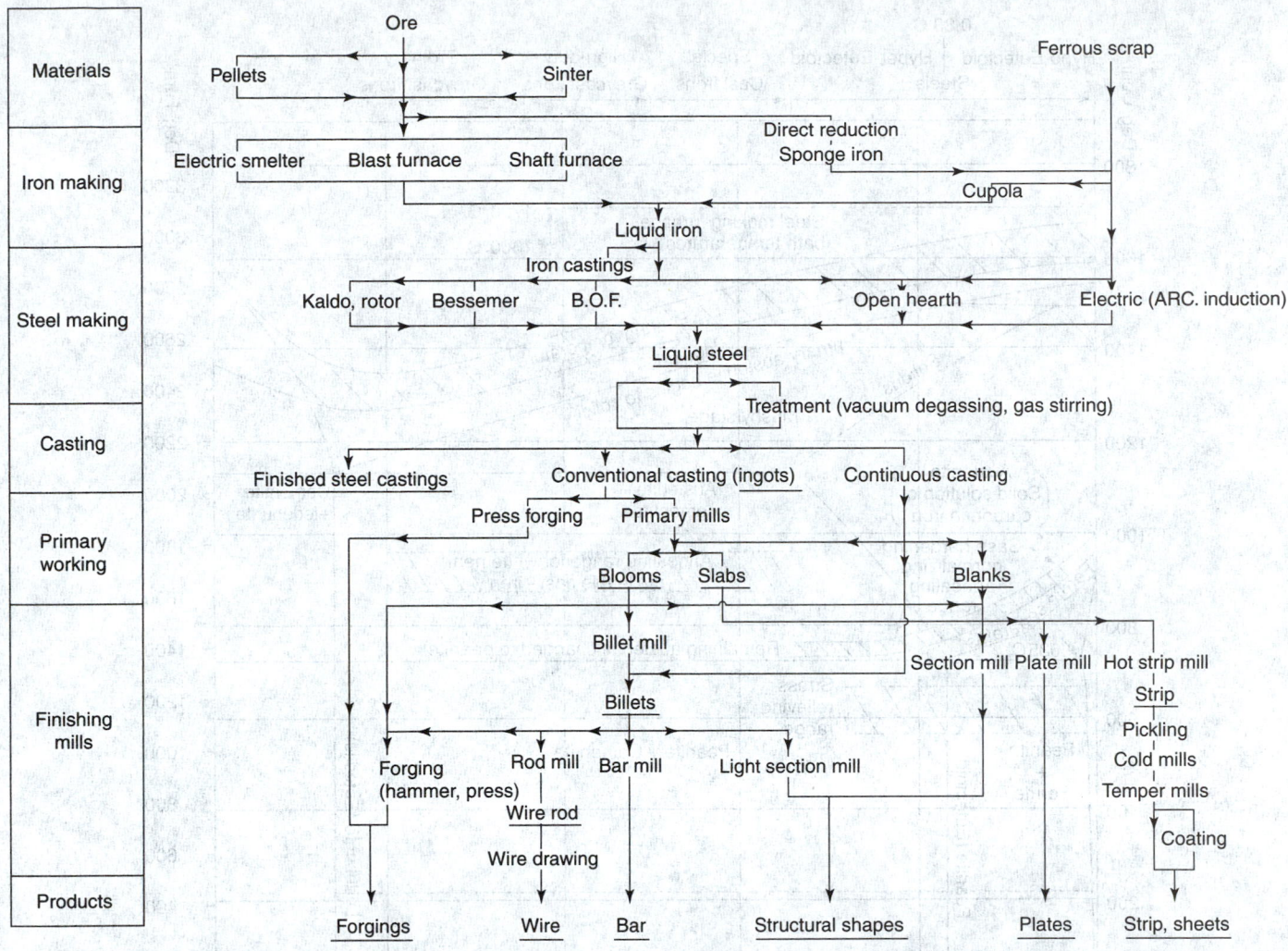

Fig. 6.   Principal processes and processing routes in iron- and steelmaking, indicating major classes of products made.

Austenite. This is an allotropic form of gamma iron with carbon in solid solution. Austenite transforms to other products on cooling below 723 °C. The products depend on the rate of cooling. At ordinary temperatures, austenite containing only carbide is not stable and thus cannot be completely retained by quenching. The stability can be increased by adding certain alloying elements.

Ferrite. This is practically pure iron and can exist in magnetic alpha-iron form in iron, with up to 0.83% carbon. Ferrite exists at room temperature and up to about 910 °C in the absence of carbon. Its upper limit of existence is lowered progressively to about 723 °C as the carbon content increases up to 0.83%. Ferrite cannot dissolve carbon, is soft and ductile, and has poor abrasive resistance.

Cementite. This is iron carbide, Fe_3C, containing 6.67% carbon. The substance is hard, brittle, and crystalline. Cementite is precipitated when austenite cools.

Pearlite. This is a eutectoid comprised of a laminated structure of ferrite and cementite. Pearlite is formed by transformation of austenite upon cooling. The fineness or coarseness of the laminated structure is determined by the rate of cooling. The lamellar arrangement of ferrite and cementite produces a very tough structure. It is responsible for the mechanical properties of steels.

Graphite. This is the free or uncombined carbon usually found in cast irons. Because graphite occurs as flakes, cast irons are easily machinable even though they have a high resistance to abrasion.

Mystery of Damascus Steels Solved. The legendary steel used in Damascus swords, probably first used as early as 320 B.C., has puzzled historians and metallurgists for many centuries. Probably the first serious attempt to explain the superiority of Damascus steel was made by Anosoff

in 1841. A two-volume monograph (*On the Bulat*) was published by Anosoff, who proclaimed, "Our warriors will soon be armed with bulat blades, our laborers will till the soil with bulat plowshares, our artisans will use tools fashioned of bulat, and bulat will supersede all steel employed for the manufacture of articles of special sharpness and endurance." This forecast was not realized. Bréant and Faraday also had investigated the secret of making Damascus steel. The most recent and very serious study of the topic was made by Sherby and Wadsworth (Stanford University), who reported in 1985 their concept as just how the Damascus steel was processed. In outlining a typical manufacturing procedure, the investigators claim that a Damascus sword commenced with the casting of an ultrahigh-carbon steel, called *wootz*, in Indian foundries. (Damascus steels contained more carbon than most modern steels.) Iron ore and charcoal were mixed and heated to about 1200 °C in a shallow stone hearth. The iron was reduced, that is, stripped of oxygen by virtue of reactions with carbon present in the charcoal. At this point, the metal had a spongy consistency. To remove impurities, the sponge iron was hammered, to produce bits of wrought iron with a low carbon content. To increase the carbon content, the investigators envision that pieces of the wrought iron were heated in a clay crucible along with charcoal. To prevent oxidation of the iron, the crucible was sealed. When there was indication of melting within the crucible, the latter was cooled slowly within the furnace. The product was wootz, which incidentally during the period was traded in the form of cakes. A Damascus blade was forged from an individual cake of wootz, which is estimated to have been heated to a temperature of 650–850 °C. (Modern ultrahigh-carbon steels are ductile within that temperature range.) The finished blades were hardened, by reheating them and then quenching them in water, brine, or some other liquid. The investigators attribute their postulations to a study of the iron phase

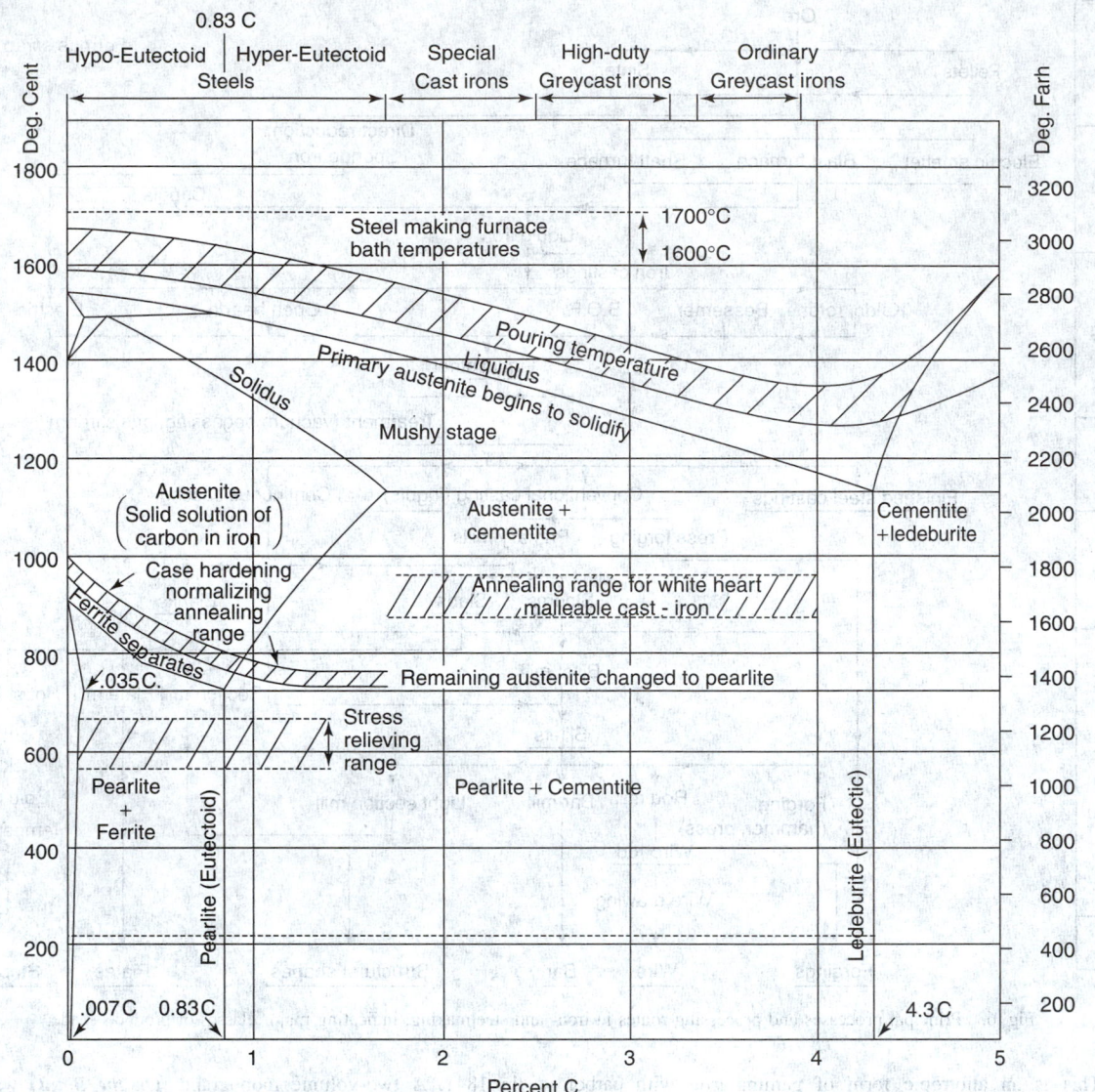

Fig. 7.   Iron-carbon equilibrium diagram.

diagram (previously shown in Fig. 7.). Quoting from their report. "When wrought iron and charcoal were heated to 1200 °C in a crucible, the iron converted into face-centered *austenite*. Carbon from the charcoal could then dissolve in the iron, decreasing its melting temperature. Molten cast iron formed at the surface of the iron particles when the carbon content of the surface layer exceeded 2%. Slow cooling allowed the carbon to diffuse through the metal, producing a steel with an average carbon content between 1.5 and 2%. Slow cooling also allowed the austenite grains to grow to a coarse size. When the temperature fell below about 1000 °C, carbon precipitated out of solution as cementite at the grain boundaries. The coarse *cementite* network was the source of the whitish damask markings. As the temperature fell below 727 °C, face-centered austenite converted into alternating layers of cementite and carbon-poor, body-centered *ferrite*. The blades were hardened by being reheated above 727 °C and then quenched, which converted austenite into martensite." It is assumed that medieval smiths estimated the metal's temperature from its color.

It is interesting to note that a description of the hardening procedure for Damascus steel (*bulat*) was located in the Balgala Temple in Asia Minor. "The bulat must be heated until it does not shine, just like the sun rising in the desert, after which it must be cooled down to the color of the king's purple then dropped into the body of a muscular slave.... the strength of the slave was transferred to the blade and is the one that gave the metal its strength."

Types of Steel

The carbon content of steels usually does not exceed 1.7%. In addition to carbon, *plain carbon steels* contain small amounts of silicon, manganese, phosphorus, and sulfur—derived from the raw materials and fuel used in the steelmaking process. Within limitations, silicon and manganese are beneficial and often are purposely added, mainly because they are deoxidants. Except in free-cutting steels, where sulfur is purposely added, sulfur and phosphorus are deleterious and their content is kept as low as possible. Some of these terms are rooted in steelmaking technology prior to the growing adoption of continuous casting processes.

Killed Steel. Deoxidizing elements are used to remove oxygen by forming solid oxides. Thus the reaction to form carbon and oxygen gas is suppressed and the killed steel lies quiet in mold when poured, shrinking upon solidification, usually with formation of a conical cavity known as a pipe. The most commonly used deoxidizers are silicon and aluminum. Killed steels are used for forging, carburizing, heat treating, and other applications because of their superior uniformity and soundness.

Rimming Steel. This is produced by leaving sufficient oxygen to react with carbon to evolve bubbles of carbon monoxide, conditions best achieved in steels with low carbon and manganese contents through the controlled addition of deoxidants. The effervescing action of carbon monoxide evolution causes a pure outer skin of ferrite and a tougher inner core containing carbon, impurities, and inclusions on solidification of the poured block. The sandwich like macrostructure is retained during subsequent shaping operations. Thin sheets intended for deep drawing or deep pressing, as used in the manufacture of auto bodies and domestic appliances, are made from rimming steel because they provide a smooth surface with adequate strength. Rimming steels also are used for forgings requiring smooth surfaces.

When the top end of the ingot is sealed after pouring, the term *capped steel* is applied. Big-end-down bottle-top molds are used and, after a small addition of aluminum, are sealed by using a heavy metal cap. Steels of this type are used for sheet, strip, skelp, tin plate, wire, and bars.

Semikilled or Balanced Steel. This is produced by adjustment of the silicon and aluminum added to low-carbon steel before teeming. The character is intermediate between the killed and rimming steels. The steel is deoxidized less than the killed steel, leaving sufficient oxygen to react with carbon and form blowholes to compensate in all or in part for the shrinkage that accompanies solidification. Semikilled steels are used for less severe drawing and pressing than rimming steels and for structural shapes, plates, and merchant bar.

Effect of Ingredients on Steels

The iron-carbon diagram (Fig. 7) shows the effects of carbon in steel. Steels containing less than 0.83% carbon are known as *hypoeutectoid steels*. When cooled slowly, the microstructure of these steels consists of pearlite and ferrite. *Eutectoid steel* containing 0.83% carbon consists entirely of pearlite. Steels containing more than 0.83% carbon are known as *hypereutectoid steels*. When cooled slowly, their microstructure is comprised of pearlite and cementite. Each increase in the carbon content of the steel increases the hardness and tensile strength of the steel in the *as-rolled* or *normalized* condition up to 0.83% carbon. The effect is less pronounced above this figure. The maximum hardness attainable after quenching also increases with the carbon content up to about 0.60% carbon. The strength of quenched and tempered steels depends upon the tempering temperature. Ductility decreases as the carbon content increases, and weldability is impaired above certain levels.

Manganese. This element is generally added to bring the amount to between 0.5 and 1.0%. Normally some manganese is present since it occurs in so many iron ores. Manganese contributes to strength and hardness, but the effect is less than like additions of carbon. Manganese lowers difficulty, but again to an extent less than carbon. Surface quality improves with manganese in all carbon ranges, notably in resulfurized steels. Manganese also increases the rate of carbon penetration during carburizing.

A steel qualifies as an alloy steel when the manganese specified is within the limits of 1.65–2.10%. Manganese is of major importance in increasing hardenability — the depth of hardness penetration after quenching. Thirteen percent manganese steel is widely used as a wear-resistant steel.

Phosphorus. A high phosphorus content in some types of steel is undesirable because it decreases ductility and impact toughness. Because of large loss of ductility, phosphorus is notably undesirable in the higher-carbon steels. In lower-carbon steels, phosphorus promotes machinability and, with copper, improves resistance to atmospheric corrosion.

Sulfur. This element is detrimental to surface quality, but beneficial to machinability, particularly in low-carbon and low-manganese steels. Sulfur decreases transverse ductility and impact resistance, but has only a small effect on longitudinal properties. As sulfur content increases, weldability decreases. Sulfur is added to the extent of 0.2–0.4% in free-cutting steels to improve machinability.

Silicon. Rimmed and capped steels contain no significant amounts of silicon. When specified within the limits of 0.60–5.00%, silicon qualifies a steel as an alloy steel. The resiliency of steel for spring applications is increased with silicon content. The element also raises the critical temperature for heat treatment. Silicon promotes the susceptibility of steel to decarburization. Because they have a low hysteresis loss and a high electrical resistance, very low carbon steels with 0.6–5.00% silicon are used as transformer steels. Silicon promotes the adherence of zinc coating on hot-dipped galvanized wire. Silicon is less effective than manganese in increasing strength and hardness.

Aluminum. The main use for this element is to deoxidize steels and to obtain a fine grain size. Aluminum also is used to obtain nonaging characteristics and to prevent the recurrence of stretcher strains in sheets and strip. When added in amounts of about 1%, aluminum promotes

nitriding properties, that is, surface hardening by means of nitrogen-bearing gases at high temperatures.

Copper. This element is beneficial to atmospheric corrosion resistance if present in amounts in excess of 0.20%. Appreciable amounts of copper are detrimental to hot-working operations. Copper also adversely affects forge welding and is detrimental to surface quality. However, copper does not seriously affect arc or acetylene welding. Copper is not removed in the conventional steelmaking processes and hence, because of increasing accumulation in scrap, it is becoming increasingly difficult to control copper within low limits.

Nickel. Aside from manganese, nickel is the most common alloying element for steel. Nickel is used in amounts up to 5% to increase strength and improve shock resistance. The element counteracts the brittleness that develops in most pearlite steels at subnormal temperatures, lowers the critical temperature of steel, widens the temperature range for successful heat treatment, and promotes corrosion resistance. When nickel is used in quantities greater than 5%, the steels fall into the stainless and heat-resistant steel categories. These are described shortly.

Niobium. By addition of amounts of up to 1%, niobium stabilizes chromium and stainless steels. Additions of only about 0.02% increase the yield point of medium-carbon steels by about 50% without any loss of weldability.

Tungsten. When added to steels in amounts up to 20%, tungsten greatly improves the hardness of a steel, a hardness that is maintained at high temperature and very important to high-speed tool steels. Smaller amounts of tungsten are added to hot-working steels.

Zirconium. Small amounts of this element, when added to high-chromium steels, improve their machinability.

Cobalt. This element provides cutting efficiency to high-speed steels and also is a constituent of heat-resisting steels because it conveys a resistance to creep and scaling.

Chromium. This element increases hardness, improves hardenability, and promotes the formation of carbides and for these reasons is used in constructional steels. Chrome steels are relatively stable at elevated temperatures and have outstanding wear resistance. Chromium is an important constituent of stainless and heat-resistant steels to be described shortly.

Molybdenum. Steels with molybdenum are usually less susceptible to temper brittleness. Molybdenum has a major effect on increasing hardenability and a notable effect on increasing the high-temperature tensile and creep strengths of alloy steels, that is, the steels have less tendency toward deformation under stress at elevated temperatures.

Vanadium. A strong deoxidizing agent, vanadium promotes a fine austenitic grain size. Constructional steels contain about 0.03–0.25% vanadium. Larger quantities are used in tool steels. Vanadium additions of about 0.04–0.05% increase the hardenability of medium-carbon steels with a minimum effect on grain size. Further additions, however, decrease the hardenability with normal quenching temperatures. Where the austenizing temperatures are increased, however, the hardenability can be increased with higher vanadium contents.

Titanium. This element acts as a deoxidizer in pearlitic steels. The yield point of plain-carbon steels is increased with titanium in amounts of 0.02–0.05%. Titanium promotes weldability without the need for normalizing.

Boron. This element is added to increase hardenability, but is effective only when added to fully killed steels. Since only a few thousandths of 1% of boron usually remains in the steel, evaluation of boron steels is by increased hardenability rather than chemical content. The hardenability characteristics of elements already present in the steel are intensified by boron, making possible alloy ingredient conservation. Although effective with low-carbon steels, the effectiveness of the element decreases as the carbon content increases.

Industrial Classification of Steels

Numerous systems are used for classifying steels — some based on composition, others on physical properties, special properties, and so on. For convenience, there are two broad categories: (1) plain carbon steels; and (2) alloy steels. Plain carbon steels account for about 95% of all steel production. As described earlier, plain carbon steels are classed as hypoeutectoid or hypereutectoid steels, depending on whether the carbon content is above or below 0.83% (the eutectoid composition). *Low-carbon* steels have a carbon content below 0.20%. *Medium-carbon* steels have a carbon content in the range between 0.20 and 0.50%. *High-carbon* steels have a carbon content in excess of 0.50%.

Alloy Steels. An alloy steel as defined by The American Iron and Steel Institute is: "By common custom alloy steel is considered to be alloy steel when the maximum of the range given for the content of alloying elements exceeds one or more of the following limits: manganese, 1.65%; silicon, 0.60%; copper, 0.60%; or in which a definite range or a definite minimum quantity of any one of the following elements is specified or required within the limits of recognized field of constructional alloy steels: aluminum, boron, chromium up to 3.99%, cobalt, columbium (niobium), molybdenum, nickel, titanium, tungsten, vanadium, zirconium, or any other alloying element added to obtain a desired alloying effect."

High-strength, low-alloy steels have a twofold objective: (1) higher mechanical properties, and (2) greater resistance to atmospheric corrosion than achievable with structural-grade carbon- or copper-bearing steels. Often, these are proprietary steels with specific trade names. A representative steel in this class will have a tensile strength of about 70,000 psi (483 MPa) for a 1/2-inch (~1.3-centimeter) thick section and have a yield point of about 50,000 psi (345 MPa).

Constructional alloy steels are a major part of the tonnage of alloy steels and are used mostly in the automotive and aircraft industries. These steels usually are quench-hardened and tempered with or without carburizing.

Stainless steels have a large degree of resistance to chemical attack. This property sometimes is referred to as *passivity*. This property results when iron is alloyed with at least 11% chromium. The corrosion resistance is further enhanced by higher chromium additions and by the addition of nickel. A steel with 12% chromium will stain, but will not exhibit progressive rusting in normal atmospheres. Under normal circumstances, a steel with 18% chromium will not stain, but may discolor, particularly in heavy industrial areas. When 8% nickel is added to an 18% chromium steel, the metal will be stain-resistant in all but the very worst of atmospheres. Even further enhancement of corrosion and heat resistance results with the addition of molybdenum.

Iron-chromium alloys and their general corrosion-resistant properties were known in England and France nearly 150 years ago, but the phenomenon of passivity was not formally recognized until 1910 (Borchers and Monnartz in Germany). This discovery led to rapid development of a series of commercial stainless steels. Stainless steels fall into three broad categories: (1) *martensitic types* — chromium-iron alloys with chromium in the lower range (12 to 17%) and with a wide range of carbon. A main characteristic is an ability to harden by heat treatment in a manner similar to carbon steels. Tensile strengths range from 70,000 to 105,000 psi (483 to 725 MPa) for annealed steels and 125,000 to 200,000 psi (863 to 1380 MPa) for hardened steels. They are particularly well suited for hot working and forging; the lower-carbon types can be cold-worked. (2) *ferritic types* — chromium-iron alloys with higher chromium in a range of 18 to 30% and with a lower carbon content. They have a microstructure that is predominantly ferritic. The steels are not hardenable by heat treatment. They are ferromagnetic. They have a relatively low coefficient of thermal expansion. These steels exhibit good resistance to oxidation and corrosion; they are frequently selected for high-temperature service, notably for applications involving intermittent heating and cooling, because of their ability to retain the oxide scale that has formed. (3) *austenitic types* — iron-chromium-nickel alloys, with a chromium content ranging from 8 to 30% and a nickel content ranging from 6 to 20%. They retain austenite at room temperature. They are characterized by high ductility of the austenite, work-hardening ability, good corrosion resistance, and superior high-temperature properties. Austenitic stainless steels are inherently tough; well adapted for fabrication by deep drawing. They are easily welded and soldered. Their tensile strength (annealed) approximates 90,000 psi (621 MPa) with a yield strength of about 35,000 psi (242 MPa).

Heat-Resisting Steels, as may be required for steam-generating boilers, pressure vessels, furnaces, distillation equipment, and internal-combustion engines must retain their specified physical properties at elevated temperatures. Where temperatures exceed 540 °C, molybdenum is used along with

chromium as an alloying ingredient. Only 2% chromium provides oxidation protection up to about 620 °C. A chromium content of 10 to 14% is required for temperatures up to about 760 °C. For higher temperatures, stainless steels are used. For service in the temperature range of about 815° to 1,095 °C, steels containing 25% chromium and from 20% to 27% nickel are frequently used.

Electrical Steels. The properties required of a good electrical steel include high electrical resistance, high permeability, and low hysteresis loss. These properties are provided by the addition of 0.6–5.0% silicon to a relatively carbon-free steel. Such steels are used in power transformers, motor and generator rotors and stators, and communications equipment.

Cold Working and Heat Treating Steels

When steel is cold-drawn or cold-rolled, it is said to be *cold-worked*. This process significantly improves mechanical properties, such as increasing the tensile strength, yield strength, torsional strength, hardness, and wear resistance. By suitably combining chemical composition, cross section, method of steel production, and thermal treatment with cold-working, distinctive properties in steels can be achieved. Cold working can impart properties to some steels comparable to those of heat-treated bars. In low-carbon steels, cold-worked steel bars show greatly improved machinability. The ratio of yield strength to tensile strength influences machinability. A high yield-strength ratio, resulting from cold drawing, minimizes plastic flow during machining, thus permitting better utilization of machine tool energy.

Heat Treating Steels. Heat treatment enables the modification of mechanical properties of steels. Three fundamental operations are involved: (1) heating the steel above the critical range, to approach a uniform solid solution of austenite; (2) hardening by quenching in oil, water, or air, to induce the formation of martensite (the hardest micro-constituent of steel); and (3) tempering by reheating to a temperature below the critical range, to secure the desired combination of strength and ductility. Three types of steel generally do not respond to this form of heat treatment: (1) steels which contain very low amounts of carbon; (2) austenitic steels for which the critical ranges are below room temperature; and (3) ferritic stainless steels. Products which normally can be furnished in the quenched and tempered condition include carbon and alloy steel plates; carbon, alloy, and martensitic stainless steel bars; hot-rolled alloy steel sheets; alloy steel tubular products; and carbon steel wire.

Normalizing. This process consists of heating to an appropriate temperature above the critical range, followed by cooling to below that range in still air. This process promotes uniformity of structure. Products which can be normalized include: (1) carbon, alloy, and high-strength low alloy steel hot-rolled bars; (2) carbon, alloy, and high-strength low alloy steel hot-rolled plates; (3) carbon and alloy semifinished steel; (4) carbon alloy, and high-strength low alloy steel hot-rolled sheets; (5) carbon, alloy, and high-strength low alloy steel hot-rolled and cold-rolled strip; (6) carbon and alloy steel tubular products; and (7) carbon and alloy steel wire.

Annealing. For carbon steels, alloy steels, and martensitic and ferritic stainless steels, regular annealing consists in maintaining the steels at a temperature in or near the critical range, followed by cooling at a predetermined rate or cycle. In the case of austenitic stainless steels, these are generally annealed by holding at appropriate temperatures and rapidly cooling to minimize the precipitation of carbides. Annealing provides softness, improves machining, forming, or shearing, reduces stress, improves or restores ductility, and may modify other properties. Usually annealing is used on stainless and heat-resisting steels and may be performed on the same kinds of products as listed under normalizing. See also **Annealing**.

Box Annealing. A process of annealing steel in an appropriate metal container to shield the steel from objectionable oxidation. Sometimes, a reducing atmosphere is used.

Spheroidize Annealing. A process of prolonged heating at a suitable temperature, followed by slow or cyclic cooling to produce a globular condition of the carbide. The structure produced may be attractive for machining or cold-forming, cold-drawing operations, or it may be desirable for subsequent heat treatments.

Stress Relieving. In this process, internal stresses are reduced by heating to a temperature below the critical range and holding it for a sufficient time for equalization of the temperature throughout the piece.

Patenting. This is a continuous heating of individual strands to above the critical range, followed by relatively rapid cooling. The process applies to wire and wire rods and increases toughness for withstanding severe distortion or drawing without breakage.

Isothermal Annealing. This is a process of heating to the correct temperature above the critical for proper austenizing, followed by rapid cooling to a suitable temperature and holding sufficiently long for completion of the transformation.

Additional Reading

Babu, P.B., et al.: "Bar Steel: User Concerns," *Advanced Materials & Processes*, 35 (August 1990).

Baxter, D.F., Jr.: "Users Like Steel's New Look," *Advanced Materials & Processes*, 17 (August 1990).

Colling, D. and T. Vasilos: *Industrial Materials: Metals and Alloys*, Vol. 1, Prentice Hall, Inc., Upper Saddle River, NJ, 1994.

Davis, J.R.: *Metals Handbook*, 2nd Edition, ASM International, Materials Park, OH, 1998.

Decker, R.F.: "Maraging Steels," *Advanced Materials & Processes*, 45 (June 1988).

Dulski, T.R.: "Steel and Related Materials (Analysis of)," *Analytical Chemistry*, 65R (June 15, 1991).

Feinman, J. and D.R. Rae: *Direct Reduced Iron: Technology and Economics of Production & Use*, Iron & Steel Society, Warrendale, PA, 1999.

Fischer, J.J. and J.H. Weber: "Mechanical Alloying," *Advanced Materials & Processes*, 43 (October 1990).

Fromont, R.I.: "NODS Alloy Makes Better Heat-Exchanger Tube," *Advanced Materials & Processes*, 68 (October 1990).

German, R.M.: *Powder Metallurgy of Iron and Steel*, John Wiley & Sons, Inc., New York, NY, 1998.

Gordon, R.B.: *American Iron, 1607-1900*, Johns Hopkins University Press, Baltimore, MD, 1996.

Gupta, V.K.: "New Treatments Toughen Maraging Steels," *Advanced Materials & Processes*, 90 (September 1990).

Ho, C.: *Properties of Selected Ferrous Alloying Elements*, Vol. 3, Taylor and Francis, Inc., Philadelphia, PA, 1989.

Linstroth, R.L.: "Check for Atmospheric Corrosion When Using Stainless Steels," *Chem. Eng. Progress*, 49 (July 1991).

Molloy, W.J.: "Investment-Cast Superalloys a Good Investment," *Advanced Materials & Processes*, 23 (October 1990).

Polmear, I.: *Light Alloys: Metallurgy of the Light Metals*, Hodder Headline PLC, London, UK, 1995.

Pope, G.T.: "One Step Steel," *Sci. Amer.*, 79 (March 1990).

Staff: "Potpourri of New Steel Products," *Advanced Materials & Processes*, 19 (August 1990).

Staff: "Quick-Quenching Steels," *Pop. Mechanics*, 18 (October 1990).

Staff: "Steel Forecasts," *Advanced Materials & Processes*, 31 (January 1991).

Staff: *Annual Book of ASTM Standards 2000: Section 1, Iron and Steel Products: Ferrous Castings: Ferroalloys (Annual Book of ASTM Standards)*," Vol. 01.0, American Society for Testing & Materials, West Conshohocken, PA, 2000.

Wright, P.H.: "Microalloyed Forging Steels," *Advanced Materials and Processes*, 29 (December 1988).

Web References

American Iron and Steel Institute (AISI): http://www.steel.org/
ASM International: http://www.asm-intl.org/
Iron & Steel Society: http://www.idis.com/aime/iss.htm
Steel Links: http://www.steel.org/hotlinks/

IRON MICA. See **Biotite**.

IRON OXIDE. See **Hematite**; **Limonite**; and **Magnetite**.

IRON WINDS. See **Winds and Air Movement**.

IRONWOOD TREE. See **Hornbeam Trees**.

IRRADIANCE. A radiometric term for the rate at which radiant energy in a radiation field is transferred across a unit area of a surface (real or imaginary) in a hemisphere of directions. In general, irradiance depends on the orientation of the surface. The radiant energy may be confined to a narrow range of frequencies (spectral or monochromatic irradiance) or integrated over a broad range of frequencies. Irradiance follows from radiance but not, in general, vice versa. The photometric equivalent of irradiance is illuminance, obtained by integrating spectral irradiance times luminous efficiency over the visible spectrum.

IRRADIATED FOODS. The concept of using ionizing radiation to destroy microorganisms that cause spoilage of food products was proposed in the early 1940s, during that period when "peaceful uses of the atom" received much coverage in the public press. The Atomic Energy Commission, the nuclear regulatory agency in the United States at that time, recommended that the Department of the Army assume the task of correlating and supporting research in food irradiation. This seemed to be a fitting assignment because it was the U.S. Army that, in the early 1800s, was responsible for bringing to practicality the concepts of Apert and preservation of foodstuffs in metal "tin" cans. During the 1950s and 1960s, considerable research in the field was also carried out in a number of other countries, including Japan, the United Kingdom, the Netherlands, Canada, and India, among others. The International Project in the Field of Food Irradiation was established and headquartered in Karlsruhe, Germany. The interests during that time frame were largely concentrated in making foods available to impoverished and underdeveloped countries, prompting the interest of the World Health Organization, the Food and Agriculture Organization of the United Nations, and other international groups.

Basic research continued, but at a relatively slow pace. Public fears and concerns over nuclear radiation cast doubts concerning irradiated product safety, and, even after full safety assurance could be given, it was not sure that the products would sell in the marketplace.

Governmental approval in some countries, including the United States, was given to irradiating certain food products.

A breakthrough occurred on January 25, 1992, when the first irradiated fresh strawberries were sold to consumers in a North Miami Beach, Florida, produce and grocery store. During an introductory period, over 1000 pints were purchased. The irradiated product, of course, was priced higher than nonradiated berries. In most countries, at least for a while longer, regulatory approval will have to be obtained in terms of specific products. A key determination that must be made is radiation dosage.

Absorption of Ionizing Radiation. The most critical factor in designing a food irradiation system is to make certain that the radiation source selected and the equipment configuration provided allows the proper amount of ionizing radiation to reach and be absorbed by the food substance. Unless a minimal required amount is absorbed, the objectives of destroying microorganisms; of reducing their numbers; of inactivating specific pathogenic microorganisms; of destroying insects that infest foods; or of altering physiological processes, such as preventing the sprouting of tubers or the inhibiting of mushroom growth, cannot be achieved. As will be noted from Table 1, the amount of energy that must be absorbed is determined by application as well as of characteristic food substances.

TABLE 1. DOSAGE RANGE FOR VARIOUS FOOD IRRADIATION OBJECTIVES

Application	Examples	Range of Dose (kJ/kg)
PHYSIOLOGICAL ALTERATION		
Inhibition of sprouting	Potatoes; onions	0.05–0.15
RADICIDATION APPLICATIONS		
Destruction of parasites	Meats	0.1–1
Disinfestation of insects	Cereals	0.1–1
Reduction of molds and yeasts	Fruits; vegetables	1–5
RADURIZATION APPLICATIONS		
Extension of refrigerated storage (0°–4°C (32°–39.2°F)	Meats; fish	0.5–10
Elimination of specific pathogens	Salmonellae (meat, poultry, egg, animal feeds)	5–10
RADAPPERTIZATION APPLICATIONS		
Sterilization of certain ingredients	Spices	10–20
Sterilization of animal diets	Feedstuffs	20–50
Sterilization for long-term unrefrigerated storage	Meats; meat products	40–60

The matter of energy absorption can be explained by use of a cobalt-60 energy source as an example. The gamma rays from ^{60}C penetrate thick materials, but at a cost of energy lost. A fraction of the rays is absorbed and their intensity decreases with thickness. In the first 40 centimeters of water, the ^{60}Co gamma ray intensity is reduced by approximately 1.64 percent. The energy absorption process consists of a primary event in which the electromagnetic field in the gamma ray removes an electron from an atom. The atom, so ionized, is raised, thereby to a highly excited state. The deexcitation of this highly excited atom will often result in ionization and excitation of several surrounding atoms. The electron kicked out in the primary process is usually very energetic and will cause most of the overall ionizations and excitations. Close to the ^{60}Co source, each primary ionization leads to approximately 17,000 secondary ionizations. Further away (passing through 63 centimeters of water), the gamma rays become softer (less energetic; longer wavelengths) and, under these conditions, there are only about 1400 secondary ionizations for each primary ionization.

Upon energy absorption, the ionized and activated molecules form unstable intermediate products. This is accompanied by a relatively small rise in temperature, and a small total chemical change, but is not accompanied by any induced radioactivity. On the other hand, where high-energy electrons are used, they generate lower-energy secondary and tertiary electrons. These also lose their energy to the irradiated substance and ultimately are no longer energetic.

Units of Radiation Absorption. Absorbed energy is measured in the unit *rad* (rd), which corresponds to 100 ergs of radiation energy absorbed per gram of substance. Therefore:

$$1 \text{ rad} = 100 \text{ ergs/gram}$$
$$= 6.24196 \times 10^{13} \text{ eV/gram}$$
$$= 10^{-5} \text{ joule/gram}$$
$$= 2.389 \times 10^{-6} \text{ cal/gram}$$

It is also common to use *kilojoules (kJ) per kilogram* as the fundamental unit of absorbed ionizing radiation:

$$1 \text{ rad} = 100 \text{ ergs/gram}$$
$$= 10 \times 10^3 \text{ joules/kilogram (J/kg)}$$
$$1 \text{ kJ/kg} = 1 \times 10^3 \text{ joules/kilogram}$$
$$10 \text{ kJ/kg} = 10 \times 10^3 \text{ joules/kilogram}$$
$$= 1 \text{ Mrad (formerly used)}$$

Dose Values. Some researchers use the concept of dose level (D), the D values expressed in kJ/kg for the quantity of radiation required to inactivate 90% of the target microorganism.

System Design Complexities. From the foregoing, it is evident that in the practical application of engineering physics, radiation chemistry, and microbiology, system design tends to become complex. The amount of energy absorbed by the product being irradiated (sufficient to effect the sterilization, inactivation, etc. goal); the geometry of the irradiating equipment; the thickness, geometry, and other characteristics of the food being irradiated; the capacity desired of the equipment (throughput); and the manner in which the food is packaged or containerized—these are, among other factors, critical to equipment design and operation. Computer and statistical techniques are commonly employed. There are also the problems of providing appropriate instrumentation (dosimeters, etc.). Energy requirements and general economics of the irradiation process also present numerous tradeoffs.

Radappertization. This is a process applicable to precooked (enzyme inactivated) foods that are hermetically sealed-in metal cans, flexible pouches, or aluminum or plastic trays. Radiation energy sources may be any of those previously discussed. A comparatively high dosage of irradiation is used and sometimes the process is called radiation sterilization. To date, it has been found that the process is particularly applicable to precooked red meat, poultry, fin fish, and shellfish, as well as to dry foods, animal feeds, and spices. The resulting radappertized products are free of food spoilage microorganisms and organisms of public health significance, including the pathogens such as *Clostridium botulinum*,

Salmonellae, trichinae, among others. The radappertized products can be stored without refrigeration for long periods (years), the limiting factor being the integrity of the primary packaging material to avoid postprocessing contamination. Although radappertized products are ready to eat, they can also be warmed prior to table serving and additional culinary preparation, using a variety of recipes, can be applied to these foods.

Energy Requirements of Radappertization. The energy used for irradiating (including the energy used in capital investment, such as equipment and buildings) is much less than any other preservation process. For example, about 75 Btu/pound (160 kJ/kilogram) for radappertizing doses, compared with about 2200 Btu/pound for retorting and about 17,000 Btu/pound for freeze-drying. Since many foods, such as meats, must be enzyme-inactivated for extended storage stability at room temperature, such foods are "ready to eat." Irradiated foods are first heated to 73 °C (163.4 °F) and then irradiated while frozen to maintain high quality. The total energy used in all these processes (including the energy for inactivating enzymes and for irradiating in the frozen states) still results in less use of energy than items preserved by other means, such as thermal canning, retorting, freezing, or freeze-drying. The total amount of energy used in the food system for entire processing is 7200 Btu/pound (15,000 kJ/kilogram) for enzyme-inactivated radappertized meats. The numbers for the energy cost of heat sterilization are based upon small cans (4 to 30 ounces; 113 to 851 grams). The time required for retorting is about 280 minutes for 6-pound (2.7-kilogram) cans versus 110 minutes for 30-ounce (851-gram) cans.

Radappertized products also can result in reduction of from 15 to 40% of packaging costs and storage space requirements by eliminating, in many cases, the water or brine additions required in thermal processing for heat penetration. Some authorities believe the products are more nutritious because fewer chemical changes occur in irradiated food than in thermally processed products-with consequently less destruction of amino acids and greater vitamin retention. Because no refrigeration is required in the transport and storage of radappertized products, there are significant Additional energy savings.

Radurization. This term applies to the use of *low-dose* irradiation of foods to extend the shelf life of the products under refrigeration. The intention of this process is to reduce the number of spoilage organisms; to alter their growth patterns, i.e., by extension of lag phase, and thus enhance keeping quality.

Radicidation. This term applies to the use of *low-dose* irradiation of foods with the objective of destroying disease-causing organisms (pathogens) of public health significance.

Irradiation of Specific Foods

Potato Sprout Inhibition. This was one of the earliest irradiation applications to be widely accepted and approved by a number of countries. Canada was the first country to approve this application on foods cleared for human consumption. Approval was given as early as 1960 for use of a 60Co source with a maximum dose of 10 krad. In 1963, the dose was increased to 15 krad maximum.

Onion Sprout Inhibition. Canada was the first country to approve this application (1963), followed later by Spain (1971) and Israel, Thailand, and Russia (1973). Various degrees of approval have been given in other countries.

There has also been considerable research on onion powder during recent years. Galetto observed that, as a result of concern about the microbial population in onion powders and the problems associated with the current method of sterilizing these powders with ethylene oxide, a study was in order to determine the usefulness of gamma-irradiation as a sterilization process. Since irradiation is known to affect some of the chemical components of foods, tests were established to determine these effects in onion powder at the dose necessary to reduce the microbial population to an acceptable level. Onion attributes chosen for testing were volatiles, amino acids, hot water insolubles, color, and gross nutrients. The researchers found and as reported in listed reference that onion powders appear to be very resistant to chemical change when treated with gamma-irradiation.

Chicken. The purpose of irradiating chicken is (a) to prolong storage life and/or (b) to eliminate pathogenic microorganisms from eviscerated chicken stored below 10 °C (50 °F). Within the dosage ranges for these applications: (a) from 200 to 700 krad; (b) from 500 to 700 krad, authorities granted unconditional acceptance of irradiation for chicken.

Strawberry. The purpose of irradiating fresh strawberries is to prolong the storage life by partial elimination of spoilage organisms. Within the dosage range of 100 to 300 krad, authorities granted unconditional acceptance of irradiation for strawberries.

Wheat and Ground Wheat Products. The purpose of irradiating these products is to control insect infestation in the stored product. Within the dosage range of 100 to 220 krad and with the stipulation that whether the products are prepackaged or handled in bulk, they shall be stored under such conditions as will prevent reinfestation, authorities granted unconditional acceptance of irradiation for use on these products.

Rice. The purpose of irradiating rice is to control insect infestation in storage. Within the dosage range of 10 to 100 krad and with the same storage stipulation as applies to wheat products, the authorities granted provisional acceptance of irradiation for use on rice.

Papaya. The purpose of irradiating papaya is to control insect infestation and to improve its keeping quality by delaying ripening. The authorities granted unconditional acceptance for irradiation of this product provided the source of radiation is either ^{60}Co or ^{137}Cs (to provide adequate penetration) and that dosage range will be from 50 to 100 krad.

Additional Reading

Diehl, J.F.: *Safety of Irradiated Foods*, 2nd Edition Marcel Dekker, Inc., New York, NY, 1995.

Marcotte, M.: "Irradiated Strawberries Enter the U.S. Market," *Food Technology*, 80 (May 1992).

Murray, D.R.: *Biology of Food Irradiation*, John Wiley & Sons, Inc., New York, NY, 1990.

Pszcola, D.: "Food Irradiation: Countering the Tactics and Claims of Opponents," *Food Technology*, 94 (June 1992).

Raffi, J.J. and J. Belliardo: *Potential New Methods of Detection of Irradiated Food*, Vol. 133, Bernan Associates, Lanham, MD, 1991.

Staff: *Safety and Nutritional Adequacy of Irradiated Food*, World Health Organization, Geneva, Switzerland, 1994.

Terry, D.E. and R.L. Tabor: "Consumer Acceptance of Irradiated Food Products: An Apple Marketing Study," *J. of Food Distribution*, **XXI** (2) 63–73.

IRRATIONAL NUMBER. A number that cannot be obtained from the set of positive integers using a finite number of the four operations: addition, subtraction, multiplication, and division. This has the consequence that the number, when expressed to any base, never becomes repetitive after the radix point. Irrational numbers can result when the operations of evolution (raising the power) and involution (extraction of roots) are applied to the set of integers. These are the algebraic irrational numbers. There are others that are termed transcendental numbers. See also **Transcendental**.

IRREVERSIBLE PROCESS. A process occurring in a system such that, in order to reverse the direction of the process, a finite change in the parameters of the system must be made, e.g., the compression or expansion of a gas in a cylinder by means of a piston, when friction is present between piston and cylinder. See also **Reversible and Irreversible Processes**.

IRRIGATION. The artificial distribution of water over the ground surface, as by canals, channels, pipes, or by flooding, or by overhead sprinkling and dripping in simulation of rain in order to promote plant growth. Approximately 16% of the arable land of the earth is serviced by some form of irrigation during the growing season. This figure is probably a percentage point or more lower than the actual because statistical reports do not include the use of special-purpose irrigation systems applied to orchards and vineyards.

Asia leads all other continents in terms of the area of land that is irrigated, accounting for nearly 75% of all irrigation. Irrigation is practiced most extensively in China, India, the United States, the former U.S.S.R., and Pakistan. However, in terms of intensity of irrigation, i.e., percentage of total arable land that is irrigated, Egypt leads with nearly 100% of all useful land subjected to some form of irrigation. Although the United States ranks third in terms of installed irrigation systems, the country ranks nearly fortieth in terms of percentage of total land irrigated. In the United

States, the western states, including Hawaii, are heavy users of irrigation. Florida in the southeast and Nebraska in the Plains region also rank high in terms of installed irrigation. Fifteen of the states irrigate less than 1% of their cropland. Only 5.6% of the arable land in Europe is irrigated.

Background. The historical records pertaining to the use of irrigation are rather sparse. Some authorities claim that irrigation was known and practiced in Egypt as early as 2000 B.C. These authorities also postulate that ultrasimplistic methods of watering plants, using portable water containers, date back into antiquity. Simple as though it may seem, recognition by primitive peoples that there is a direct relationship between plants and a water supply was, on any scale, a major discovery.

Irrigation at any significant level did not occur until quite late in the history of the United States, logically developing out of the later maturity of the western states and territories. Growers in the eastern and midwestern states were generally favored with periodic and adequate rainfall for the types of crops grown. By the 1880s and 1890s, recognition of large problems in terms of both water availability and water needs for artificially supplying vast lands in the western states took form. By 1902, the U.S. Congress had passed the Reclamation Act (also known as the Newlands Act) which provided a mechanism for returning much of the funds received by the government for sale of lands back into the improvement of government-held lands. By 1910, there were about 4 million acres (1.6 million hectares) of western land irrigated. As of the early 1980s, about 43 million acres (17 million hectares) of land in the United States are irrigated.

Types of Irrigation

Irrigation systems may be classified by: (1) source of water, and (2) manner in which water is distributed to the land. Principal water sources are surface streams—which may be diverted, held for periods in reservoirs, and transferred over long distances by canals and specially constructed channels; and wells which draw upon water from the water table. In terms of distributing irrigation water, there are four basic methods: (1) flood irrigation, (2) furrow irrigation, (3) subirrigation systems, and (4) overhead (above-ground) systems, which include sprinkler systems of several designs and drip irrigation.

Flooding. Probably the earliest irrigation scheme used, flooding involves long, large irrigation ditches that run between fields, constructed so that when openings are made in the ditch, water will flow from the ditch to adjacent fields. To flood large areas (wild flooding) without further provision for controlling the water flow requires extremely level fields, a relatively infrequent situation in most areas. Flooding, of course is also used where fields are not perfectly level, but in any case, the tolerances from perfect levelness are not large. Sometimes much effort is required to make the land reasonably level so that the irrigation can be effective and efficient and not damaging to the soil. Flooding irrigation has been practiced for centuries in Pakistan, India, Bangladesh, and Egypt.

To afford more control over water flow, a *check* or *levee* system is commonly used.

Leveling is a key practice in on-farm irrigation improvement projects. Equipment with laser beam control is used to finish the leveling job, enabling farmers to make their fields nearly dead level. Laser-guided earthmovers are used. Multiple exposures catch laser light beamed from a sending unit on a tripod to a receiver on a tractor-pulled scraper. The beam is turned at a rate of 5 to 10 revolutions per second. A receiver on the hydraulic scraper locks into the beam and adjusts the blade level automatically. It has been possible with this equipment to level fields to within 0.5 inch (13 millimeters) of zero grade. Where fields are not level, the slope can be made extremely uniform.

Furrow Irrigation. This is a common form of irrigation in many parts of the world and is most applicable to irrigating row crops. In this system, in contrast with flooding, the entire surface of the field is not wetted. In essence, the furrow irrigation scheme utilizes a network of *furrow streams* to distribute water that is supplied by a channel. Furrows prepared for irrigation are also sometimes called corrugations. See Fig. 1. Furrow irrigation is applicable to many situations where flood irrigation is quite impractical, as for uneven topography and where supply streams are relatively small. Furrow irrigation can be adapted to considerable variations in slope. Some slope, of course, is required, optimally falling between drops of 10 and 30 feet per 1000 feet (a range of 1–3%). However, the system

Fig. 1.   Furrow irrigation of young orange trees with lettuce in between. (*USDA photo.*)

will work where slopes are greater, such as a drop between 30 and 60 feet per 1000 feet (3–6%).

The length of the furrows varies considerably—from 200–300 feet (60–90 meters) to as long as 1300 feet (390 meters) and more. A midrange length is preferred by many growers because longer furrows require much greater exposure of water in the area near the supply channel, causing undue percolation of soil and erosion in many cases. See Figs. 1 and 2.

Disadvantages of furrow irrigation include accumulation of mineral salts along furrow walls and, during hot weather, high rates of evaporative losses of water to the atmosphere. Without the use of improvements for accomplishing better control and regulation of water going to the furrows, it is extremely difficult to accomplish maximum efficiency, particularly in times of drought when highly selective use of water between crops must be practiced.

Subirrigation. This form of irrigation occurs when water is supplied naturally or artificially from under the crop root zones. In relatively few areas there is a combination of conditions that favor the lateral flow of water through the soil, including porous-loam or sandy-loam soil, uniform topography with gentle slope, and an impervious subsoil that is at a minimum depth of 6 feet (1.8 meters). Depending upon the balance of these factors, advantage can be taken by providing a series of ditches that are 200 feet (60 meters) or more apart. When these ditches contain water, the water works its way down a gentle slope through the soil laterally without any further direction from above. Occurring quite infrequently are situations where the ditches may be as much as 1000 feet (300 meters) to 0.25 mile (400+ meters) apart.

Much experimentation has gone forth toward the development of an artificial underground irrigation system, but thus far such systems have been applied to crops of very high value and on relatively small tracts of land. The costs of constructing the required underground network of pipes are very high. Further, plant roots tend to seek out openings in the pipes, dictating annual crops that have relatively shallow roots to avoid this complication. The system is impractical, for example, as a method to be considered for tree irrigation. In orchards, however, a vertical approach to piping is sometimes used, locating the pipes a sufficient distance from the trees to make certain that roots do not interfere and depending upon lateral sloping of the soil to carry water to the root zones of the trees.

Sprinkler Irrigation. Sprinkler systems mimic natural rain conditions by distributing water by spraying it over fields. Water is piped under pressure (by pumping assistance or by gravity) through sprinklers or through perforations or nozzles in pipelines and thus forms a spray. Nearly all irrigatable soils can be sprinkler-irrigated. It is difficult, however, to sprinkler-irrigate if the water intake of the soil is less than 0.1 inch (2.5 millimeters) per hour. Sprinkler irrigation is often the most effective method on soils that have high intake rates, on fields that have steep slopes or irregular topography, and on soils that are too shallow to level. Most crops can be sprinkler-irrigated, but there can be difficulty in moving portable lateral lines in tall crops such as corn (maize). Some fruits must be protected from water sprays when they are ripening. Wind distorts spray patterns and usually reduces the efficiency of the system.

Sprinklers may rotate or remain fixed. Perforated lateral pipelines require less pressure than rotating sprinklers and release more water per unit of time than rotating sprinklers. However, their use should generally be restricted to soils that have high intake rates.

Sprinkler systems have been generally classified as portable, semi-permanent, or permanent. The classification depends upon whether the lateral pipeline (including sprinklers), main pipeline, and pumping plant are movable or fixed. Sprinkler systems may be more specifically classified

Fig. 2.   Furrow irrigation of carrots near Lompoc, California. Syphoning arrangement for transferring water from head ditch to furrows is shown. (*USDA photo*.)

according to special mechanical features that are used to move the lateral pipelines. See Figs. 3, 4, and 5.

Permanent-type systems have fixed main and lateral pipelines and pumping plant. Equipment and installation cost per area is higher than that of any other system and can range from several hundred to well over $1000 per acre (over $2000/hectare) in 1980 dollars. However, labor requirements are lower than those for other configurations. These systems are best adapted to long-lived crops requiring full-season irrigation, such as permanent pastures, orchards, citrus groves, vineyards, and nurseries.

Semipermanent systems consist of portable lateral pipelines and permanent main pipelines and pumping plant. Cost of these installations is usually less than half that of a fully permanent system. Labor requirements are moderate. These systems are especially well suited to areas requiring full-season irrigation.

Portable sprinkler systems have movable pipelines from the pumping plant to the last sprinkler lateral. The pumping plant may be fixed or movable, although in a completely portable system, everything will be movable. Costs for these systems can be as low as 20% of the cost for a fully permanent system. Labor requirements are higher. These systems are particularly well adapted for occasional or supplemental irrigation. With most systems of the portable and semiportable types, portable lateral pipelines are moved over the field manually. The need to reduce labor requirements has led in recent years to mechanization. Side-roll, side-move, pull-type wheel, drag, self-propelled continuously moving, giant sprinkler, and solid-set systems have been developed. While labor costs are less, initial costs are high and the highly mechanized systems require level or relatively uniform sloping fields.

In the *side-roll system*, the lateral pipe is used as an axle. Wheels, which are from 4 to 5 feet (1.2 to 1.5 meters) in diameter are mounted on it. See Fig. 4. Hand or power-driven devices are then utilized to move the lateral into a new position. In the *side-move system*, the lateral pipe is

Fig. 3.   Sprinkler irrigation system for contour-planted potatoes. (*USDA photo*.)

supported on carriages spaced from 40 to 60 feet (12 to 18 meters) apart along the sprinkler lateral pipe. Small 1-to-3-inch (2.5-to-7.5-centimeter) diameter trailing pipelines having from 1 to 5 or more sprinklers spaced on each pipeline may be connected to the lateral pipe. The trailing pipelines

Fig. 4.   Side-roll lateral irrigation system located on field of potatoes. (*USDA photo.*)

Fig. 5.   Electrically driven sprinkler system with capability of delivering varying amounts of water. (*USDA photo.*)

are towed by the lateral when the system is moved. In the *pull-type wheel system*, a fixed or swiveling 2-wheel carriage supports the lateral 1 foot (0.3 meter) or more above the ground. The lateral is then towed endways by a tractor or truck to the new setting. The design is best adapted to close-growing forage crops. However, it can be adapted for use on most crops. Some growers sow grass strips in row crops and use the strips when moving the laterals of a pull-type wheel system. The *drag-type system* has laterals similar to pull-type systems. A skid pan or outrigger attachment is substituted for the wheeled carriage. The skid pan or outrigger helps to stabilize the laterals, but frequent moves in abrasive soils may cause excessive wear. This system is best adapted to well-sodded forage crops.

In the *self-propelled continuously moving system*, two configurations are used: (1) a circular center-pivot system, and (2) a straight lateral system. The lateral pipe in both systems is mounted on wheeled supports, with each wheel being driven by hydraulic power or electric motors. Valves on the hydraulic systems and switches on the electric systems are controlled by safety devices to keep the various sections of the lateral in alignment as it moves continuously around the field in the circular center-pivot system; or across the field in the straight-moving lateral system.

In the *giant sprinkler systems*, individual sprinklers are usually mounted on a stand or trailer and moved by a tractor truck. Each sprinkler has from one to eight or more nozzles. Pressures vary from 60 to 120 pounds per square inch (8.2 atmospheres). Sprinkler discharges range from 150 to 2000 gallons (5.7 to 45.4 hectoliters) per minute. Areas covered by the spray range from one to many acres (0.4 to many hectares) per set. Minimum application rates range from 0.3 inch (7.6 millimeters) to over 0.6 inch (15.2 millimeters) per hour. Giant-sprinkler systems have been converted to a continuously moving sprinkler traveler system by mounting the sprinkler on a trailer and connecting the sprinkler to the main pipeline with a length of high-pressure hose. Giant-sprinkler systems are not well adapted to areas exposed to high winds. Some of the systems can be used only on high-infiltration-rate soils. Others are limited to fairly uniform fields.

Solid-set systems have portable or buried laterals. There are enough laterals and sprinklers to irrigate the field without repositioning any lateral. Laterals may be operated individually or in blocks of laterals, depending upon the water supply. Automatic sequencing valves sometimes are used. These are placed on each riser, allowing only one sprinkler on each lateral to operate, thus enabling the use of smaller lateral pipe. Electric, hydraulical, and pneumatic valves are used in sequencing solid-set systems. Solid-set systems may be used for irrigation, fertilization, temperature, and humidity control, and for the application of insecticides, herbicides, and other control chemicals, providing that corrosion is not a problem.

Drip Irrigation

Sometimes called *trickle irrigation*, a drip irrigation system delivers water slowly to growing plants and limits the delivered water to the immediate vicinity of the roots, creating a wetted pattern at the root zone. The system is well named because of the small quantities of water applied at any given time. The water is supplied frequently, often with small amounts of soluble fertilizer. The objective is to replace daily loss of plant water through evapotranspiration and, at the same time, retain an optimal balance of air and moisture in the soil.

Developed in Israel, drip irrigation is considered a major breakthrough in agricultural technology of recent years. Drip systems consist of emitters, lateral lines, main lines, and a control station. Emitters control the flow of water from the lateral line to the soil at the base of a plant, generally at a rate of from 1 to 4 gallons (3.8 to 15.1 liters) per hour. They are available in a hundred or more styles. Some are simply perforations in lateral tubing; others, which look like pillboxes or small garden-hose nozzles, are more complex and adjust water pressure. The number of emitters used ranges from less than one per plant to eight or more for large trees. See Fig. 6. At the control station or "head," water is measured, filtered or screened, and treated with fertilizer. Chemicals are injected to prevent clogging of emitters. Water pressure is regulated as well as timing of applications, generally for one to 16 hours every one to three days, depending upon crop and environmental conditions.

Fig. 6. Drip irrigation by multiple $\frac{1}{8}$-inch (3-mm) diameter tubes. Rate of flow is controlled by screw on top of emitter. (*USDA Soil Conservation Service*.)

Soil scientists have reported that corn (maize) requires 14-acre-inches of water under drip irrigation, whereas 31 acre-inches are required with furrow irrigation.

Additional Reading

Borelli, J., et al.: *Advances in Irrigation and Drainage: Surviving External Pressures*, American Society of Civil Engineers, Reston, VA, 1983.

Burt, C.M.: *Selection of Irrigation Methods for Agriculture*, American Society of Civil Engineers, Reston, VA, 1999.

Dasberg, S.: *Drip Irrigation*, Springer-Verlag, Inc., New York, NY, 1999.

Mabry, J.: *Canals and Communities: Small-Scale Irrigation Systems*, University of Arizona Press, Phoenix, AZ, 1996.

Pereira, L.S.: *Water and the Environment: Innovation Issues in Irrigation and Drainage*, John Wiley & Sons, Inc., New York, NY, 1998.

Smith, S.W.: *Landscape Irrigation: Design and Management*, John Wiley & Sons, Inc., New York, NY, 1996.

Southorn, N.: *Irrigation and Drainage*, Butterworth-Heinemann, Inc., Woburn, MA, 1998.

Van Bentum, R. and I. Smout: *Buried Pipelines for Surface Irrigation*, Intermediate Technology Publications, London, UK, 1994.

William, R.J.: *Dams and Disease: Ecological Design and Health Impacts of Large Dams, Canals, and Irrigation Systems*, Routledge, New York, NY, 1999.

IRROTATIONAL. Applied to a vector field having zero vorticity or curl throughout the field. Two equivalent properties of an irrotational field are that there is no circulation about any reducible curve within the fluid, and that a potential exists. An autobarotropic fluid is irrotational for all time if it is irrotational at any time. Meteorological motions of the smaller scales, for example, gravity waves, may be treated as irrotational, but when the scale is large enough to take the rotation of the earth into account, only rotational motions are of interest.

See also **Helmholtz's Theorem;** and **Solenoidal**.

ISALLOBAR. See **Atmosphere (Earth)**.

ISALLOBARIC. See **Meteorology**.

ISALLOBARIC CHART. See **Meteorology**.

ISALLOBARIC WIND. See **Winds and Air Movement**.

ISCHEMIC HEART DISEASE. Ischemia is a temporary or permanent deficiency of blood flow to an organ. Two major cardiac diseases arise from ischemia—angina pectoris and acute myocardial infarction (heart attack). Ischemia involving the heart muscle may result from coronary arterial disorders—atherosclerosis (damaged arterial walls) or by spasms. Two conditions generally lead to a heart attack—atherosclerosis (a slow process) and clotting of blood to form a thrombus (a fast process). See also **Arteries and Veins (Vascular System)**; and **Heart and Circulatory System (Physiology)**.

Angina Pectoris

Because the heart is an extremely active muscle, it requires a continuous and adequate supply of oxygen from the blood. Any impediment in the arteries supplying the heart muscle may impair the cardiac blood supply. Lack of oxygen in the blood (*anoxemia*) also may cause an inadequate supply of oxygen to the heart muscle. Under such circumstances, persons who exert themselves to only a limited extent may suffer from pain in the chest or the area below the collar bones. Such pain, frequently excruciating, is referred to as *angina pectoris*. It usually occurs in persons over 40 years of age, and its alleviation depends upon the cause of the heart condition; in other words, angina pectoris is not, properly speaking, a disease, but rather it is symptomatic of *myocardial anoxia*. The heart's oxygen balance (supply and demand) is, of course, dependent upon the individual's particular activity at any given time. This activity affects heart rate and metabolic rate, but there are also other major determinants of myocardial oxygen consumption, including wall tension (systolic intraventricular pressure, ventricular size, and ventricular wall thickness) and contractile state (inotropic state). Minor determinants include electrical depolarization, electromechanical coupling, maintenance of active tension, and muscle shortening per se. When increased demand from any of these factors cannot be met, i.e., when the coronary artery flow cannot be increased accordingly, then ischemia results. The clinical features are manifested by *angina pectoris* (the milder consequence), by *myocardial infarction* (death of parts of the heart muscle), and by *electrical instability*. Many authorities attribute the latter factor as the major cause of sudden death in ischemic heart disease. See also **Arrhythmias (Cardiac)**.

In addition to pain, some patients with angina pectoris complain of a "pressure" or "heavy sensation," sometimes using a tightened fist to demonstrate the kind of sensation felt. Frequently such discomfort will also be felt in the neck and lower jaw and on the left side, including the shoulder and arm. Sometimes the right arm also will be affected by the pain. Seldom, however, do the pains of angina pectoris penetrate below the diaphragm. Symptomatic of angina pectoris is aggravation of pain when exercising and prompt relief when resting. Some authorities believe that precipitating factors also may include cold weather, a large meal, or emotional tension, but these observations are not well supported by scientific evidence. The diagnosis of angina pectoris in some patients is complicated by atypical chest pain. The sensations encountered in esophageal spasm may mimic angina pectoris. Other symptomatic conflicts may arise from pericarditis and pleurisy, particularly if the latter is due to a pulmonary embolus.

In North America, the most frequently used therapy for angina pectoris is a combination of beta-adrenergic blocking agents, such as propranolol, and the nitrates (nitroglycerin). Propranolol blocks beta receptors in both the cardiac and pulmonary systems, a serious disadvantage in the cardiac patient who has bronchospastic disease. In such cases, metoprolol, which is more cardio-selective, may be used because it frequently does not produce adverse side effects in bronchospastic patients. This latter drug has been much more readily available in Europe than in the United States. Propranolol must be used with caution in patients with borderline congestive heart failure. There are numerous other, less serious side effects to the beta-adrenergic blocking agents.

Nitroglycerin functions to cause venodilation of the peripheral vasculature and to some extent, arteriolar dilation. In essence, this decreases the return of blood to the heart, thus reducing left ventricular filling pressure and left ventricular volume. Arteriolar dilation reduces peripheral resistance, reducing the impedance to left ventricular outflow. These effects, in combination, lower the myocardial oxygen consumption. In some patients, particularly the elderly, the hypotension and headache induced by nitrate therapy requires careful initial administration and monitoring of the drug's actions. Frequently these problems can be overcome by a gradual buildup of dosage levels.

Some patients may require additional medication, including the antihypertensive agents (thiazides or methyldopa) and antiarrhythmic agents for the control of tachyarrhythmias and bradyarrhythmias. See also **Arrhythmias (Cardiac)**. The use of anticoagulation therapy in angina pectoris remains controversial. As pointed out in the entry on **Arteries and Veins**, some authorities have suggested that platelet aggregation may play a role in reducing coronary blood flow and this has motivated some physicians to administer antiplatelet agents. These agents are described in the aforementioned entry.

Unstable Angina Pectoris — Coronary Artery Spasms. Some cases of apparent arterial constriction, where the cause was believed to be atherosclerotic plaques, have been found to be the result of coronary artery spasms. Although such spasms have been known for a long time, it is now suggested by some investigators that such spasms may be a more prevalent cause of angina pectoris and heart attacks than previously believed. A new line of drugs is being intensively researched for the treatment of these spasms.

In the early 1900s, William Osler suggested that coronary vasospasms could be an important cause of certain heart disorders, but especially since the 1940s the predominant view among specialists has been that angina pectoris and heart attacks result from the buildup of atherosclerotic plaques in the coronary arteries. In retrospect, certain past observations tend to support the role of spasms, namely, the finding at autopsy that a heart attack victim had no prominent atherosclerosis. Also, over the years, some angina pectoris patients have complained of chest pains, not during exercise as is usually the case, but rather during rest—a fact that does not square with traditional explanations. Some investigators found that in certain angina pectoris patients the experience of chest pains followed a rhythmic pattern (frequently early in the morning) rather than showing any correlation with exercise and rest. To describe such situations, some investigators (Cedars of Lebanon Hospital, Los Angeles; University of Colorado; University of Pisa, Italy; and London University Medical School, among others) proposed the term *variant angina* to distinguish this condition from angina pectoris caused by mechanical obstruction. In the literature, this has sometimes been referred to as *Prinzmetal's angina* because of Myron Prinzmetal's observations in 1959. The condition may affect only about 5% of patients identified with angina pectoris.

The role of spasms in initiating heart attack also is being investigated. In one study, persons with heart attack had detectable spasms within 6 hours after the attack. These spasms may have resulted from the attack, or they may have been the cause. Nevertheless, there is suspicion that a transient constriction in a partially occluded vessel may cause a heart attack.

Acute Myocardial Infarction

As contrasted with angina pectoris, where there is a deficiency of blood supply to the heart muscle—a deficiency that ranges rather widely among various patients and during various stages of rest and exercise, as previously described—in acute myocardial infarction there is an absolute insufficiency of blood supply. Myocardial infarction (necrosis of the heart muscle) accounts for about 700,000 deaths per year in the United States alone. It is estimated that about half of these deaths occur before a patient can be transported to a hospital. The exact mechanism that brings about a very serious or fatal heart attack still is rather poorly understood. At one time, it was believed that a sudden obstruction of a coronary artery by thrombosis (*coronary thrombosis*) was the principal cause of acute myocardial infarction. Research in recent years has shown this not to be so. Some researchers have suggested that this cause may be present in only about 10% of cases. Intramural hemorrhage has been suggested as a major cause, but one researcher found this to be the cause in only 25% of cases (autopsies in fatal coronary disease). What is known, of course, are the numerous predisposing conditions, such as valvular disease, infective endocarditis, cardiomyopathy, and atrial fibrillation. In recent years, there has been greatly accelerated interest in the role of spasms, both in cases of angina pectoris and acute myocardial infarction. As previously mentioned, this topic is being intensively researched.

The symptoms of acute myocardial infarction are well known, but in sudden, severe, often fatal cases, where death may occur in a matter of minutes, such symptoms may not have been previously indicated to a physician. Early symptoms include chest pain (similar to angina pectoris but more severe). Such pain will not be relieved by nitroglycerin.

The pain radiates as in angina, but there usually is also pain near the epigastrium, which mimics the discomfort of indigestion.

Frequently premonitory signs are divulged only after the patient has received a checkup during which electrocardiograms are made. A fact not widely reported is that myocardial infarctions occur with some frequency in hospitals after surgery. Some researchers report that the risk of such incidents is considerably higher when elective surgery is performed within six months after a previous infarction. The incidence of "within hospital" infarctions is also higher after thoracic and upper abdominal operations, after surgical procedures that may exceed one hour, and after periods of hypotension (low blood pressure) during anesthesia.

Cases of acute myocardial infarction require hospitalization, preferably in a specially equipped coronary care unit (CCU). Continuous electrocardiographic monitoring is the principal means for keeping track of the patient's condition on a second-by-second basis. Because of the availability of instruments and skills, even where acute myocardial infarction is suspected but not fully diagnosed, the physician will prefer to place the patient in a CCU. The severe pain of acute myocardial infarction will be relieved by several small doses (intravenous) at about 10-minute intervals of morphine sulfate. Later sedation is usually achieved with a drug such as diazepam. Oxygen therapy is routinely used. During the first 24 hours after attack, the patient will be restricted to clear liquid diets with careful control over sodium. Once common and still continued by some practitioners, anticoagulation therapy can be used to reduce the incidence of venous thrombosis in the legs in cases of acute myocardial infarction. There has been a lessening trend to routinely use anti-coagulation therapy.

Additional Reading

Beller, G.A. and E. Braunwald: *Chronic Ischemic Heart Disease*, Mosby-Year Book, Inc., St. Louis, MO, 1995.

Buxton, B.: *Ischemic Heart Disease*, Harcourt Health, San Diego, CA, 1999.

Herregods, M.: *Echocardiography in Ischemic Heart Disease*, Leuven University Press, Leuven, Belgium, 1994.

Maseri, A.: *Ischemic Heart Disease: A Rational Basis for Clinical Practice and Clinical Research*, W.B. Saunders Company, Philadelphia, PA, 1995.

Picano, T.: *Echocardiography in Ischemic Heart Disease*, Gordon & Breach Science Publishers, Newark, NJ, 1996.

Van Der Wall, E.: *Advances in Imaging Techniques in Ischemic Heart Disease*, Kluwer Academic Publishers, Norwell, MA, 1995.

Willerson, J.T.: *Atlas of Ischemic Heart Disease: Clinical and Pathologic Aspects*, W.B. Saunders Company, Philadelphia, PA, 1997.

ISENTROPIC. Of equal or constant entropy (or, in meteorology, potential temperature), with respect to either space or time. Thus, an *isentrope* is a line of equal or constant entropy; an *isentropic surface* is a surface in space in which entropy, or potential temperature, is everywhere equal; *isentropic mixing* refers to any atmospheric mixing process that occurs within an isentropic surface; and an *isentropic change* is a change accomplished without any increase or decrease of entropy.

ISENTROPIC ANALYSIS.   See **Meteorology**.

ISENTROPIC CHART.   See **Meteorology**.

ISLAND ARC.   See **Earth Tectonics and Earthquakes**; **Ocean**; and **Volcano**.

ISLETS OF LANGERHANS.   See **Diabetes Mellitus**.

ISOBAR.   1. A line connecting points at equal pressure, such as that which appears on a meteorological chart. The pressures on such a chart are not the observed pressures, but are corrected for elevation, i.e., to sea level.

2. One of two or more nuclides, which have the same mass number, but which differ in atomic number.

ISOBARIC HEATING AND COOLING. See **Precipitation and Hydrometeors**.

ISOBATH.   1. (Sometimes called Fathom Curve.) A contour of equal depth in a body of water, represented on a bathymetric chart.

2. In hydrology, a line on a map connecting all points at which there exists an equal vertical distance between the earth's surface and the water table, or equal depths to the upper or lower surface of an aquifer.

See also **Isopach**.

ISOBRONT.   1.   (Sometimes called homobront.) A line drawn through geographical points at which a given phase of thunderstorm activity occurred simultaneously.

2.   In climatology, a line drawn through geographical points that have the same average number of days with thunder in a given period; a type of isoceraunic line.

ISOCHORE.   Also called isometric, a graph representing the state of a system as a function of two variables (e.g., pressure, temperature), the volume remaining constant. Hence any process that occurs without a change of volume.

ISOCHRONE.   A line connecting points having the same time values, as points of the same gelation time for colloidal solutions.

ISOCHRONISM.   See **Pendulum Clock**.

ISOCLINE (Geodesy).   Also called isoclinal or isoclinic line, a line drawn through all points of the earth's surface having the same magnetic inclination. The particular isoclinic line drawn through points of zero inclination is given the special name, *aclinic line*.

ISOCLINE (Geology).   Vertical duplication of geological formations by close folding.

ISODESMIC STRUCTURE.   An ionic crystal structure in which there are no distinct groups formed within the structure, i.e., where no bond is stronger than all the others.

ISODIAPHERE.   One of two or more nuclides having the same difference between the number of neutrons and protons in their nuclei. In alpha-particle decay, for example, the parent and daughter nuclides are isodiapheres.

ISODIMORPHISM.   The condition, double isomorphism, in which both crystalline forms of a dimorphous substance which is isomorphous with a second dimorphous compound are isomorphous with both forms of the second compound. Example: arsenious oxide and antimonious oxide, which crystallize in rhombs and also in regular octahedra.

ISOELECTRIC POINT.   See **Amino Acids**.

ISOELECTRONIC.   Pertaining to similar electronic arrangements. This term is applied, for example, to two or more atoms or atomic groups having an analogous arrangement of the same number of valency electrons, and similar physical properties.

ISOGAMY.   A type of sexual reproduction in which the male and female germ cells are similar in form and size.

ISOGEOTHERM.   Depths of equal temperature in the earth.

ISOGONIC LINE.   In the study of terrestrial magnetism, a line drawn through all points on the earth's surface having the same magnetic declination. This is not to be confused with magnetic meridian, which is the horizontal line oriented, at any specified point on the earth's surface, along the direction of the horizontal component of the earth's magnetic field at that point.

A particular case of an isogonic line is the *agonic line*. This is the line through all points on the earth's surface at which the magnetic declination is zero; that is, the locus of all points at which magnetic north and true north coincide. The position of this line exhibits variations in time, but is now so located that it emanates from the north magnetic pole, trends southward and slightly eastward through the Great Lakes region, leaves the American mainland near eastern Florida, cuts across South America to near Buenos Aires, thence through the south magnetic pole, and up in an irregular path on the other side of the earth to return to the north magnetic pole. At the present time, the North American segment of the agonic line is drifting very slowly westward.

ISOGRAM.   Also called an isoline, a line on a given reference surface, drawn through all points where a given quantity has the same numerical value. Sometimes used in meteorology for drawing lines through geographical points that experience the same frequency of a selected meteorological event.

ISOHALINE.   1. Of equal or constant salinity.

2. A line on a chart connecting all points of equal salinity; an isopleth of salinity.

ISOKINETIC SAMPLING.   Any technique for collecting airborne particulate matter in which the collector is so designed that the airstream entering it has a velocity equal to that of the air passing around and outside the collector. The advantage of isokinetic sampling consists in its freedom from the uncertainties due to selective collection of only the larger, less easily deflected particulates. In principle, an isokinetic sampling device has a collection efficiency of unity for all sizes of particulates in the sampled air.

ISOLEUCINE. See **Amino Acids**.

ISOMAGNETIC.
Lines of equal magnetic force, but not necessarily isogonic. Isomagnetic lines may represent local magnetic anomalies such as are caused by magnetic ore bodies, magnetic minerals in sediments, or the vertical rather than the horizontal deviation of the compass or magnetic needle.

ISOMALT. See **Sugar Alcohols**; and **Sweeteners**.

ISOMERASES. See **Enzyme**.

ISOMERISM.
If two chemical compounds incorporate the same elements in exactly the same numbers, the compounds are referred to as *isomers* or *isomerides*. An excellent example of relatively simple isomers is the case of normal butane [CAS: 1-6-97-8], $CH_3CH_2CH_2CH_3$, which is an open, straight chain of four carbon atoms, and of isobutane [CAS: 75-28-5], $(CH_3)CHCH_3$, wherein one of the carbons lies in a short branch from the main chain of three carbon atoms. Obviously, as the number of carbon atoms in a compound increases, the possibility of branches and sub-branches increases. Normally, then, compounds with high carbon counts, at least theoretically, are capable of numerous isomers.

In *geometric isomerism*, the isomeric relationship can be explained in terms of two dimensions—as shown by the relationship of the two isomers, maleic acid [CAS: 110-16-7] and fumaric acid [CAS: 110-17-8]:

H — C — COOH   H — C — COOH
‖   ‖
H — C — COOH   HOOC — C — H
maleic acid   fumaric acid

Where the identical atoms or groups are in juxtaposition, as in maleic acid, the compound is designated as the *cis* form. (*Cis* = "on this side" in Latin.) Where the identical atoms or groups are on the opposite sides, as in fumaric acid, the compound is designated as the *trans* form.

In *stereoisomerism*, three dimensions must be considered. In stereoisomerism (also termed optical isomerism), there is no plane of symmetry in the molecule, so that the two forms are mirror-images, and thus cannot be turned into a position of coincidence. Thus, compounds containing a carbon atom (or other tetravalent atom) to which four different atoms or radicals are bonded are optical isomers. They receive this name from the fact that one isomer rotates the plane of polarized light to the right (*dextro form*); the other rotates it to the left (*levo form*). Lactic acid is an example. See also **Hydroxycarboxylic Acids**, and formulas below:

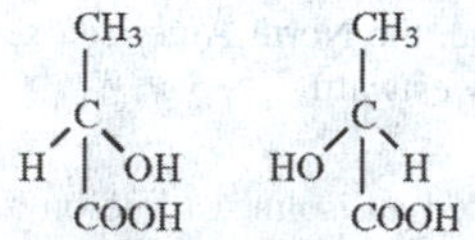

CH₃   CH₃
| |
C   C
H | OH   HO | H
COOH   COOH
d-lactic acid   *l*-lactic acid

The carbon atom, to which are attached the four different groups to produce stereoisomerism, is known as *asymmetric*, and when written (not shown structurally), that carbon may be printed more prominently than the nonasymmetric carbon atoms.

The projection formulas of the four forms of tartaric acids [CAS: 87-69-4] are shown below. Note that the arrows indicate the direction of rotation of light by the asymmetric carbon atoms.

The two optically active tartaric acids when crystallized differ in the arrangement of the faces—one is the mirror image of the other. Pasteur (1848), observing this difference, was able to separate the two optically active forms of ammonium sodium tartrate crystals made from racemic tartaric acid.

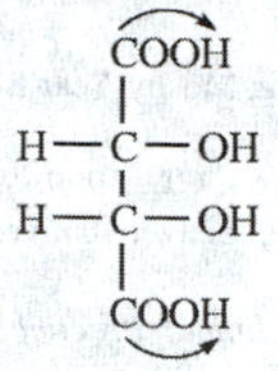

COOH
|
H — C — OH
|
H — C — OH
|
COOH

inactive or *meso*-tartaric acid (internally compensated; possesses a plane of symmetry; optically inactive).

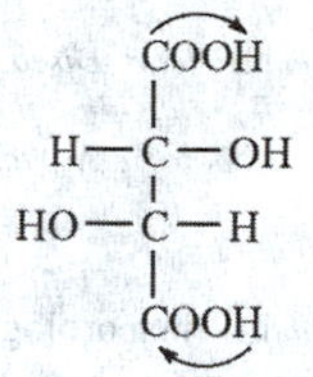

COOH
|
H — C — OH
|
HO — C — H
|
COOH

dextro-tartaric acid (arrangement of groups around each asymmetric carbon atom is cumulative; optically active; dextrorotatory).

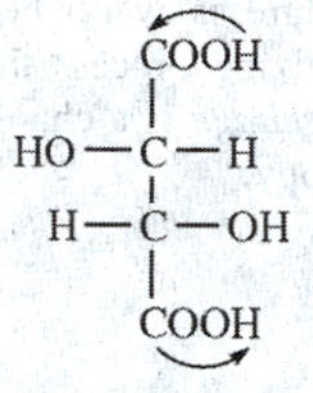

COOH
|
HO — C — H
|
H — C — OH
|
COOH

levo-tartaric acid (arrangement of groups around each asymmetric carbon atom is cumulative; optically active; levorotatory).

{ *d*-tartaric acid
  l-tartaric acid }

dextrolevo-tartaric acid, racemic tartaric acid (externally compensated; optically inactive; can be resolved into *d* and *l* components).

Tautomerism is a form of isomerism in which a substance exists in two forms which are in equilibrium and exhibit characteristic reactions; either one may predominate, depending upon the conditions. Thus acetoacetic ester may react as a ketone or an enol (a compound containing a carbon atom having both an alcoholic hydroxyl group and a double bond) depending upon the conditions.

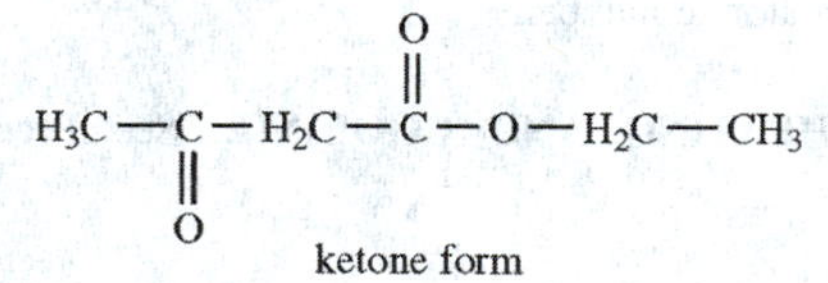

O
‖
H₃C — C — H₂C — C — O — H₂C — CH₃
‖
O
ketone form

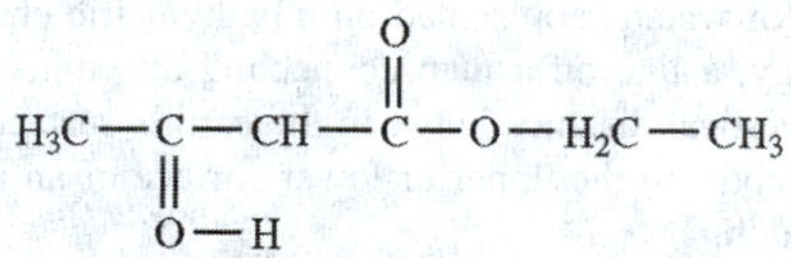

O
‖
H₃C — C — CH — C — O — H₂C — CH₃
‖
O — H
enol form
acetoacetic ester

ISOMERIZATION.
The rearrangement of the structural configuration of a molecule without changing its molecular weight. Although structural changes of this type occur in other processes, e.g., catalytic reforming and cracking, isomerization can be the principal reaction desired in some processes. In petroleum refining, isomerization processes are used to change the structural configuration of C_4 paraffins (alkanes), such as normal butane [CAS: 106-97-8], into isobutane [CAS: 75-28-5] in order to supplement other sources to provide enough butane for alkylation with olefins (alkenes) in the production of motor fuel. C_5 and C_6 paraffins are isomerized to the more highly branched structures to improve their antiknock ratings. Isomerization is also applied to a lesser extent in C_8 aromatic hydrocarbons. See also **Petroleum Refining**.

One isomerization process (UOP) is shown in the accompanying diagram. This unit is arranged to process a C_5/C_6 mixture with fractionating facilities to provide for the recycling of both *n*-pentane and *n*-hexane. A desulfurized C_5/C_6 blend first is fractionated to remove the native isopentane [CAS: 78-78-4] as a net product. The de-isopentanizer bottoms are desiccant-dried before being joined by n-hexane recycle abd brought to reaction temperature by heat exchange and suitable preheating. Before entering the reactor, the combined feed stream is joined by hydrogen recycle gas, which functions to suppress catalyst-deposit formation.

The fixed-bed reactor effluent is cooled and passed to a high-pressure separator. Gas from the separator, along with a small quantity of dried make-up hydrogen, is recycled to the reactor. The separator liquid is stabilized as a next step to remove any C_4 and lighter hydrocarbons that may be introduced with the make-up hydrogen, plus a very minor amount of light hydrocarbons formed by hydrocracking in the reactor. Hydrogen dissolved in the separator liquid is also removed by the stabilizer.

The next fractionator in series receives the stabilized liquid, from which it separates an equilibrium isopentane-n-pentane mixture that is routed back to the deisopentanizer for separating the isopentane as a net product. Thus, the n-pentane content of the feed is converted entirely to isopentane in the arrangement shown in Fig. 1.

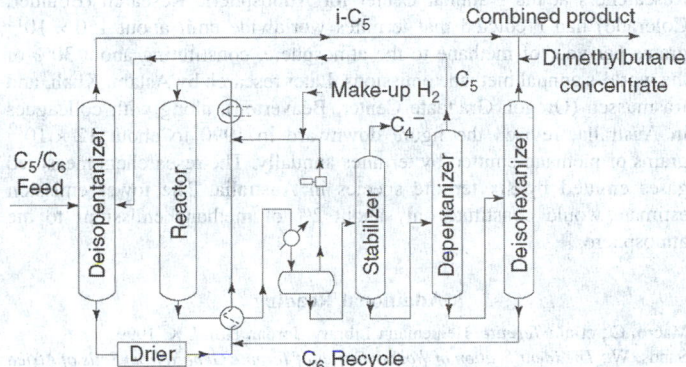

Fig. 1. C_5H_6 Isomerization unit. (*UOP Process Div.*)

As a final step in the fractionation sequence, the hexane fraction is separated into a dimethylbutanes [CAS: 38719-68-5] concentrate as a net overhead product and an n-hexane-rich bottoms stream to be recycled for the further isomerization of the n-hexane [CAS: 110-54-3] and methylpentanes. With economically practical fractionation, the methylpentanes [CAS: 43133-95-5] split between the overhead and bottoms of the deisohexanizer column. For the C_5 fraction, the boiling points of the two isomers are far enough apart to make a relatively clean split economically feasible. For the C_6 fraction, the greater number of isomers and the bunching of some of their boiling points preclude precision separation in columns having a reasonable number of plates.

Once-through processing of a typical C_5/C_6 (68–70 octane number) straight-run fraction results in a product having a Research Method octane number (clear) of about 83. By recycling the unconverted n-pentane, n-hexane, and most of the methylpentanes, a product having an octane number of about 93 would result. Obviously, any octane number between 83 and 93 could be produced, depending on the amount and quality of the equipment installed to separate the reactor effluent into net product and recycle streams.

ISOMER (Nuclear). One of two or more nuclides that are both isotopes (same atomic number) and isobars (same mass number) of each other, but which have some measurably different physical property, such as half life. Of any two isomeric states, one must be an excited metastable state of the other. Ultimately the nuclide in the excited state decays with a measurable lifetime to a lower energy state, usually its ground state. At present about 200 nuclear isomeric states with half-lives longer than 10^{-6} seconds are known. Metastable isomeric states are denoted by adding the letter m to the mass number where it appears in the nuclidic symbolism, as for example ^{80m}Br. In this particular case, ^{80m}Br and ^{80}Br are nuclear isomers. Ordinary excited states with lifetimes too short to be measured are generally *not* considered to be isomeric states, but this is only a matter of convention. On rare occasions, such as for ^{124}Sb, more than two isomeric states may exist for a single atomic number and a single mass number (isotopic isobar).

ISOMETRY. 1. An isothermic map.

2. A length-preserving map (also called Isometric Map). In the map given by $x = x(u, v)$, $y = y(u, v)$, $z = z(u, v)$, lengths are preserved if, and only if, the fundamental coefficients of the first order satisfy $E = G = 1$, $F = 0$. The coordinates u, v are called *isometric parameters*. The above functions and the functions $x = \bar{x}(u, v)$, $y = \bar{y}(u, v)$, $z = \bar{z}(u, v)$ give *length-preserving maps* between corresponding surfaces S and $\bar{S}$ if, and only if, the corresponding fundamental coefficients of the first order satisfy $E = \bar{E}$, $F = \bar{F}$, $G = \bar{G}$. Then the surfaces S and $\bar{S}$ are said to be applicable.

3. A one-to-one correspondence of a metric space A with a metric space B such that, if x corresponds to x^* and y to y^*, then the "distances" $d(x, y)$ and $d(x^*, y^*)$ are equal. It is then said that A and B are isometric.

See also **Metric Space**; and **Topological Space**.

ISOMORPHISM. An isomorphism is a one-to-one correspondence of a set A with a set B (the sets A and B are then said to be *isomorphic*). If operations such as multiplication, addition, or multiplication by scalars are defined for A and B, it is required that these correspond between A and B in the ways described in the following. If A and B are groups (or semigroups) with the operation denoted by $\cdot$, and x corresponds to x^* and y to y^*, then $x \cdot y$ must correspond to $x^* \cdot y^*$. An isomorphism of a set with itself is an automorphism. If A and B are rings (or integral domains or fields) and x corresponds to x^* and y to y^*, then xy must correspond to x^*y^*, and $x + y$ to $x^* + y^*$. If A and B are vector spaces, multiplication and addition must correspond as for rings and scalar multiplication must also correspond in the sense that if a is a scalar and x corresponds to x^*, then ax corresponds to ax^*. If the vector space is normed (e.g., if it is a Banach or Hilbert space), then the correspondence must be continuous in both directions. Objects that are isomorphic are mathematically indistinguishable. Thus properties known for one immediately carry over to the other. See also **Mineralogy**.

ISOPACH. A line, on a geological (stratigraphical or structural) map, all points on which show equal thickness of a formation.

ISOPLETH. 1. In common meteorological usage, a line of equal or constant value of a given quantity, with respect to either space or time; same as isogram.

2. (Also called Isarithm.) More specifically, a line drawn through points on a graph at which a given quantity has the same numerical value (or occurs with the same frequency) as a function of the two coordinate variables.

3. A straight line along which lie corresponding values of a dependent and independent variable.

ISOPROPYL ALCOHOL. [CAS: 67-63-0] Also called dimethylcarbinol or secondary propyl alcohol, formula $(CH_3)_2$ CHOH, *isopropyl alcohol* is a colorless liquid at room temperature. Pleasant odor, bp 82.4 °C, specific gravity 0.7863 (20/20 °C), autoignition temperature, 400 °C. The compound is soluble in water, ethyl alcohol, and ether.

Two basic methods of production are in commercial use: (1) absorption of propylene in sulfuric acid to form alkyl hydrogen sulfate, followed by the hydrolysis of the ester; and (2) by direct hydration with water, using a catalyst. An inherent disadvantage in the first process is the need to handle sulfuric acid. Further, the first process yields little more than 70% isopropanol as compared with the second process, in which liquid propylene is used as the charge stock. All direct-hydration processes can be represented by: $C_3H_6 + H_2O \longrightarrow C_3H_7OH + heat$.

Isopropyl alcohol is a widely used chemical, finding use as a starting material for making acetone and its derivatives; in the manufacture of glycerol and isopropyl acetate; as a solvent for essential and other oils, alkaloids, gums, and resins; as a latent solvent for cellulose derivatives; as a deicing agent for liquid fuels; in pharmaceuticals, perfumes, and lacquers; in extraction processes; as a dehydrating agent; and as a preservative.

See also **Alcohol**; and **Organic Chemistry**.

ISOPTERA (Termites). The white ants or termites (see accompanying illustration). An order of insects of great economic importance and biological interest, made up of social species which eat wood and other

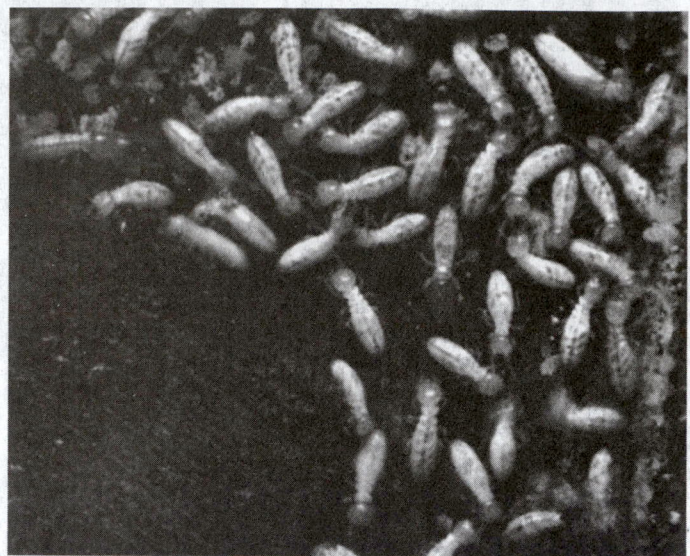

Fig. 1.   Isoptera (termites). (*Hugh Spencer from National Audubon Society.*)

vegetable matter. Most termite species are tropical or subtropical but a few live in temperate regions. See Fig. 1.

Termites have biting mouth parts and are moderate to small soft-bodied insects. They live in dark nests and tunnels except when the winged sexual individuals emerge to leave the parent colony. The bodies of these flying individuals are dark, but the termites that remain in the nest are whitish with dark heads. They do not resemble ants in form; hence their similar habits are probably responsible for the name white ant. The temporary wings of termites are long and slender and the two pairs are similar in form. In most species the veins near the anterior margin are strong and the rest are faintly marked. The wings are shed after the swarming termites find a new nesting place.

The termite colony contains workers, soldiers, and reproductive individuals of both sexes. The workers are developed in subordinate castes in several species. Soldiers have large heads and strong jaws. The queen in some colonies becomes relatively enormous through the expansion of the abdomen as the eggs develop, and is quite helpless. The workers feed and groom her and carry away her eggs.

It has been shown conclusively that termites depend on protozoans in the intestine for the digestion of the wood that they eat. Very few animals can digest cellulose, which is the chief compound in wood, but the protozoans do so and the termites utilize the products developed by the protozoans. This relationship is one of the finest examples of symbiosis among animals.

Because of their wood-eating habits, termites sometimes do great damage to buildings. Their habit of building tunnels wherever they go and of remaining concealed in the wood where they work often results in their presence being unsuspected until the honeycombed timbers give way. When they once enter buildings they are not restricted to wood but damage papers, books, clothing, carpets and many other things. In regions where termites are plentiful, no timber in construction should be left in contact with the ground. Even a small contact may be a point of entry. Where timber must be exposed to attack it can be protected by impregnation with creosote, but the most effective type of construction demands masonry wherever contact with the ground must be made. Even in such structures, termites may traverse several feet of masonry, building tunnels as they go, and may work through small cracks into the wooden parts of the building. Where termites have already entered, blocking their entrance and destroying the colony both inside the building and out with creosote or fumigants are usually effective methods of control. Special equipment and methods are available commercially for this work.

An entire colony of termites may number three million. Although destructive to structures, termites provide a useful service in wet forests where they constantly chew debris from fallen trees and convert the material to nourishing materials for new trees.

Mounds constructed by termites in Africa resemble large mushrooms and may be as much as 20 feet (6 meters) in height. In Australia, some termites construct mounds approximately 10 feet (3 meters) in length, 12 feet (3.6 meters) high, and 4 feet (1.2 meters) across. Such mounds are constructed of soil or wood or both. Small bits of construction materials are bound together by saliva, intestinal fluid, or partly-digested wood to form a very tough cementing material. In desert areas, termites have been known to tunnel down as much as 130 feet (39 meters) to reach water, which they then carry up their nest. The evaporation of this water maintains the humidity in the nest close to the saturation point, which is the comfort zone for the termites.

Research has indicated that nitrogen fixation, measured by the reduction of acetylene to ethylene, was found in workers of the dry-wood termite *Kalotermes minor*. Nitrogen fixation can be a significant source of nitrogen for some termites.

Earlier research had indicated the threat that termites may be to global warming, as the result of their generating copious amount of methane. Researchers at the National Center for Atmospheric Research (Boulder, Colorado) had predicted that termites worldwide emit about 150×10^{12} grams per year of methane to the atmosphere, constituting about 30% of the world's annual methane emissions. Later research by Aslam, Khali, and Rasmussen (Oregon Graduate Center, Beaverton), along with colleagues in Australia, revised the figure downward in 1990 to about 12×10^{12} grams of methane emitted by termites annually. The researchers measured gases emitted by six termite species in Australia. The lower emission estimate would constitute only about 2% of methane emissions to the atmosphere.

Additional Reading

Macro, C., et al.: *Termite*, Heinemann Library, Jordan Hill, UK, 1999.
Sands, W.: *The Identification of Worker Castes of Termite Genera from Soils of Africa and the Middle East*, CAB International, New York, NY, 1998.
Vander Meer, R., et al.: *Pheromone Communication in Social Insects: Ants, WASPs, Bees, and Termites*, Westview Press, Boulder, CO, 1998.

ISOSPIN.   See **Quantum Number (Isospin).**

ISOTHERMAL.   A term used to denote the following: 1. Of constant temperature, with respect to either space or time. Isothermal processes are those conducted without temperature change.

2. A line or curve expressing a relationship between variables such as pressure and volume, for all values of which the temperature remains constant.

3. A line joining points at the same temperature.

ISOTHERMAL ATMOSPHERE.   An idealized atmosphere in hydrostatic equilibrium in which the temperature is constant with height and in which, therefore, the pressure decreases exponentially upward. Also called *exponential atmosphere.*

In such an atmosphere the thickness between any two heights is given by

$$Z_B - Z_A = \frac{R_d T_v}{g} \ln \frac{P_A}{P_B},$$

where R_d is the gas constant for dry air, T_v the virtual temperature (°K), g the acceleration of gravity, and P_A and P_B the pressures at the heights z_A and z_B, respectively. In the isothermal atmosphere there is no finite height at which the pressure vanishes.

See also **Atmosphere (Earth).**

ISOTHERMAL EQUILIBRIUM.   The state of a hypothetical atmosphere, at rest and uninfluenced by radiative heating or cooling, in which the conduction of heat from one part to another has, after a sufficient length of time, produced a uniform temperature throughout its entire mass. Also called *conductive equilibrium.*

If such an atmosphere consisted of more than one gas, the pressure of each gas would be distributed exponentially according to Dalton's law, so that

$$P_n = P_{no}^{-(m_n g/kT)^h},$$

where p_{no} is the surface pressure and the nth constituent gas, m_n its mean molecular mass, g the acceleration of gravity, h the geometric height, k Boltzmann's constant (1.3804×10^{-23} J K^{-1}), and T the absolute temperature. At a sufficiently great height the lighter gases will predominate. The time necessary for the establishment of isothermal equilibrium in a mixed atmosphere by diffusion of one gas through another has been estimated to decrease rapidly with height from about one year near 100 kilometers (62 miles) to a matter of seconds near 200 kilometers (124 miles).

See also **Diffusive Equilibrium**; and **Isothermal Atmosphere**.

ISOTONE. One of two or more nuclides having the same number of neutrons in their nuclei.

ISOTOPE. An isotope is one of two or more nuclides that have the same number of protons in their nuclei. Any two isotopes have the same atomic number, Z. However, their mass numbers, A, differ. Isotope is a term that stems from the Greek words, *isos* (same) and *topos* (place), to designate substances having different atomic weights and yet having chemical properties so much alike that in the early days of research it was not possible to perform a chemical separation of the isotopes of a given element.

Sometimes, the term *nuclide* is confused with isotope. A nuclide may be defined as a species of atoms, with a specified atomic number and mass number. Different nuclides having the same atomic number should be described as isotopes. This is evident from the accompanying table. Different nuclides having the same mass number are termed *isobars*.

The existence of isotopes first became evident in the early years of this century, from the investigation of natural radioactivity. Then it was found that the natural radio elements underwent successive nuclear disintegrations, that they could be arranged in radioactive series according to these changes, and that in these series there were several instances in which atoms of the same atomic number (or as then stated, atoms occupying the same place in the periodic table) differed widely in their radioactive behavior. For example, it was found that radium C, radium E, thorium C and actinium C were all identical in their chemical properties with bismuth (atomic number 83) but differed in their radioactive properties and origins.

In the long course of research that led to the conclusion that more than one stable isotope of an element may exist, an important milestone was the method of positive-ray analysis. As applied by J.J. Thomson, an electric discharge was passed through a vessel containing a gas at low pressure. The effect of the discharge was to produce ions in the gas, and these ions,

because of their electric charge, could be formed into beams, deflected and otherwise directed by applied electric and magnetic fields. An experimental apparatus was designed so that the amount of this deflection would depend upon the masses of the particles of the gas. By using neon gas (atomic weight 20.183) in the tube, Thomson obtained photographs showing two beams of particles, one of them in a position calculated for particles having a mass of about 20, and the other for a mass of about 22. Although the conclusion was not immediately reached, it was later concluded (from the work of Aston) that neon (and other elements) consisted of atoms of more than one mass. Aston expressed this conclusion in the whole number rule, according to which, all atomic weights of individual atomic species are close to whole numbers, and the whole number plus decimal values calculated chemically for the atomic weights of elements are due to the presence of two or more isotopes, each of which has an atomic weight that is approximately a whole number. The fact that the chemically determined values for the elements of naturally occurring materials from different sources are the same is because the isotopic composition of naturally occurring materials (except those of radioactive origin) is essentially the same.

Aston's work was founded upon accurate measurements of the deflections of charged particles. These measurements were made in an instrument he devised, the mass spectrograph. Many later instruments were developed following Aston's work, or following the Dempster instrument, which was built before Aston's. The direction-focusing mass-spectrographs and the later velocity-focusing instruments and composite instruments facilitated the determination, not only of the masses (and hence mass numbers) of the isotopes of an element, but their quantities as well. As a result of the immense amount of research in this field, the isotopic composition of the stable elements has been closely determined. And this data can now be said to be subject to only slight revision, as more refined methods, and the possible discovery of stable isotopes present in very small quantities, are found.

It will be seen from Table 1 that naturally occurring elements differ widely in the number of isotopes they contain. Some (usually of odd atomic number) are composed entirely of atoms of one mass number. Others have many stable isotopes. For example, the element of atomic number 50, which is tin, has at least ten stable isotopes and many radioactive ones.

The great importance of isotopes is due to two facts. (1) Since the atomic number of an atom determines its chemical properties, all the isotopes of a given element exhibit essentially the same chemical behavior; that is, all atoms of the same atomic number undergo essentially the same reactions with atoms of other atomic numbers. Thus, all three of the isotopes of hydrogen (protium [CAS: 1333-74-0], deuterium [CAS: 7782-39-0] and tritium [CAS: 10028-17-8]) undergo essentially the same reactions with

TABLE 1. RELATIONSHIP OF ATOMIC AND MASS NUMBERS IN DESIGNATING ISOTOPES

Element	Mass Number A	Atomic Number Z	Atomic Number Z	Atomic Number Z	Atomic Number Z	
Hydrogen	1	1				
Hydrogen	2	1				
Hydrogen	3	1				
Helium	3		2			
Helium	4		2			
Helium	5		2			
Lithium	5			3		
Helium	6		2			
Lithium	6			3		All nuclides
Lithium	7			3		
Beryllium	7				4	
Lithium	8			3		
Beryllium	8				4	
Beryllium	9				4	
Beryllium	10				4	
	Isobars - - - - - - - - - - - - - - - - - Isotopes - - - - - - - - - -					

Table indicates that hydrogen has three isotopes, each with same atomic number, but with differing mass numbers, and designated as ^{1}H, ^{2}H, and ^{3}H (thus using the mass number to designate a particular isotope. Similarly, the four lithium isotopes (all with atomic number 4) are designated by their mass number 4) are designated by their mass numbers, ^{5}Li, ^{6}Li, ^{7}Li, and ^{8}Li. Although redundant, because the element symbol implies the atomic number, symbols sometimes are sometimes are written to indicate both mass and atomic numbers, as $^7_3$Li.

oxygen, carbon, and all the other elements. Therefore, if we add to the ordinary form of an element (which has a known isotope composition) a measured amount of an isotope of that element, we can follow the course of our sample through chemical reactions, especially those in which the same element enters at other points. This instance cited is representative of the many applications of isotopes which will be discussed in this article. (2) The other fact that accounts for the great present-day importance of the knowledge of isotopes is that, while the isotopes of an element exhibit similar chemical behavior, the nuclear characteristics of these isotopes often differ greatly. A most important example of this difference is that between ^{235}U and ^{238}U See also **Uranium**, where the separation of these and other isotopes is described.

The isotopes of all elements are tabulated under **Chemical Elements**. Also, there are descriptions of isotopes under the alphabetical entries for each element. The nature and importance of radioisotopes are described under **Radioactivity**.

Additional Reading

Clark, I. and P. Fritz: *Environmental Isotopes in Hydrogeology*, CRC Press, LLC, Boca Raton, FL, 1997.

Clayton, D.D.: *Handbook of Isotopes in the Cosmos: Hydrogen to Gallium*, Cambridge University Press, New York, NY, 2003.

Cook, P.: *Enzyme Mechanism from Isotope Effects*, CRC Press, LLC, Boca Raton, FL, 1991.

Faure, G., and T.M. Mensing: *Isotopes: Principles and Applications*, 3rd Edition, John Wiley & Sons, Inc., Hoboken, NJ, 2004.

Fritz, P. and J. Fontes: *Handbook of Environmental Isotope Geochemistry*, Elsevier Science, New York, NY, 1989.

Jackson, M. and N. Lowe: *Advances in Isotope Methods for the Analysis of Trace Elements in Man*, CRC Press, LLC, Boca Raton, FL, 2000.

Stevenson, N.R.: *Isotope Production and Applications in the 21st Century: Proceedings of the 3rd International Conference*, World Scientific Publishing Company, Inc., Riveredge, NJ, 2000.

ISOTOPE RATIO DETERMINATION. See **Mass Spectrometry**.

ISOTOPIC ABUNDANCE. See **Chemical Elements**.

ISOTOPIC MEDIUM. A medium whose properties are the same, in whatever direction they are measured. Such a medium has only two independent elastic moduli or constant, and only one refractive index, dielectric constant, magnetic susceptibility, etc.

ISOTROPIC TARGET. In radar or lidar, a target that scatters the same intensity of radiation in all directions. Specifically, for plane-wave incident radiation, an isotropic target scatters in all directions the same power per unit solid angle. Real targets are generally not isotropic, but scatter different intensities in different directions. The concept of an isotropic target is an artifice that allows the scattering properties of a real target to be described in terms of those of an equivalent isotropic target. Thus, the backscattering cross section of a target is the cross-sectional area of an isotropic target that scatters the same intensity in the direction of the radar receiver as the real target.

AMS

ISOTROPIC TURBULENCE. Turbulence in which the products and squares of the velocity components and their derivatives are independent of direction, or, more precisely, invariant with respect to rotation and reflection of the coordinate axes in a coordinate system moving with the mean motion of the fluid. Then all the normal stresses are equal and the tangential stresses are zero. Atmospheric turbulence is generally nonisotropic, although isotropic turbulence is that most easily produced in wind tunnel experiments and forms the basis of much of the theoretical analysis of turbulent flow. A related but less restricted type of turbulence is known as homologous turbulence, in which the fluctuations differ only in scale at every point in the flow. See also **Aerodynamics and Aerostatics**; and **Turbulent Flow**.

ITCHING. See **Pruritus**.

ITCHY EYES. Most cases of itchy eyes are related in some way to allergic reactions. The eye is one of the most sensitive organs of the body. Because it is moist and exposed, it is easy for airborne allergens, such as pollens and pet dander, to adhere to the ocular surface causing allergic reactions. The eye is also more sensitive to systemic allergies, such as those to food or cosmetics. Allergic reactions in the eye are also called allergic conjunctivitis because the source of the problem is in the conjunctiva, the mucous membrane that lines the inner surface of the eyelid and the front of the eyeball.

Any condition that causes the eye to be irritated can also cause itching. Other common eye disorders that may be associated with itching are blepharitis and dry eye syndrome. For this reason, one should see an eye doctor if one experiences to itchy eyes.

Like all allergies, the allergies that cause eyes to itch are overreactions to foreign substances by the body's immune system. In response to these foreign substances, the body releases chemical mediators that are designed to protect the body, the most common being histamine. In people without allergies, the release of these substances are controlled and produces few, if any symptoms. In an allergic person, the mediators are released in excessive amounts causing itchy eyes, swelling of the eye area, inflammation, irritation, and excessive tearing. In essence substances produced by the body itself are responsible for itchy eyes and other allergic symptoms.

Allergic reactions in the eyes, or ocular allergies, can be caused by something in the air such as pollen, animal dander, dust, or molds. They can be seasonal, or they may be perennial, affecting sufferers unpredictably throughout the year. Bacteria, food sensitivities, cosmetics, fabrics, soaps, and other substances may cause year-round allergies. Plant pollens and molds are the most common causes for chronic seasonal allergies. Ocular allergies in contact lens patients can be either environmental or because of contact lens solutions

Itchy eyes often look red and are accompanied by symptoms of irritation, burning, and a full feeling. People with this condition have a tendency to rub their eyes and might also have blurry vision and/or some discharge. Allergic conjunctivitis may or may not be accompanied by other familiar allergy symptoms such as congestion and sneezing.

The best way to protect your eye from itching due to allergies is to avoid allergens completely. Some of the most common are cigarette smoke, cat dander, smog, some houseplants, and petroleum solvents. It is important to determine the cause of allergies where possible. An allergist can provide a comprehensive battery of tests.

Rubbing your eye aggravates the itching condition. Instead, apply cool compresses to the eyelids. Artificial tears can also help reduce the itching, and they wash away some of the allergens and the mediators produced by the immune system.

Antihistamines are medicines designed to block the release of histamine, which causes many of the symptoms associated with eye allergies. Oral antihistamines are convenient to take, but often take more than an hour to work. Some can also produce side effects such as drowsiness, irritability, and dryness. These work best when taken before exposure to the allergen, a situation that is not always possible.

Many prescription and over-the-counter eye drops containing both antihistamines and decongestants are available to treat eye allergies. Some of these ease the symptoms of allergies. While others work to prevent the release of histamines. In severe cases, steroid eye drops may be used to relieve itching.

Because itchy, burning eyes may be an indication of conditions other than allergies, consult an eye doctor to determine the exact cause. See also **Blepharitis**; **Conjunctiva**; **Conjunctivitis**; and **Dry Eye**.

Vision Rx, Inc., Elmsford, NY

ITERATED LOGARITHM (Law of). A law in probability due to Khintchin which states that if S_n is the number of "successes" in binomial trials with probability p of success,

$$\lim_{n \to \infty} s_{np} \frac{S_n - np}{(\ldots, npq \log \log n)^{1/2}} = 1.$$

where $q = 1 - p$.

ITERATIVE METHODS (for Solving Equations). These methods are, in fact, methods of successive approximation in which, having given one or more approximations to a solution, it is used in computing an improved one. Only the case of a single equation in a single variable will be considered here (see also **Matrix Inversion**).

If the equation to be solved is

$$f(x) = 0$$

let

$$\phi(x) = x - g(x)f(x)$$

where throughout some region containing α, the root to be determined, $g(x)$ nowhere vanishes or becomes infinite. Then

$$\alpha = \phi(\alpha)$$

and if, for some x_0 in this region, every

$$x_{i+1} = \phi(x_i)$$

is again in the region and the sequence of x_i converges, it necessarily converges to a root. A sufficient condition for this is that

$$|\phi'(x)| \leq k < 1$$

at every point of the region. Moreover, if ϕ is analytic in some circle about x_0 and if it can be shown that for some positive $k < 1$,

$$|\phi(x_1) - \phi(x_2)| < k|x_1 - x_2|$$

whenever both x_1 and x_2 are in the circle, then it can be concluded that every x_i will, in fact, fall within the circle and that the equation has a root α to which the sequence converges.

Newton's method is obtained with

$$g(x) = 1/f'(x)$$

and one is assured of convergence if, for real α,

$$f(x_0)f''(x_0) > 0$$

and neither f' nor f'' changes sign between α and x_0.

The singly and doubly primed terms are the first and second derivatives with respect to the independent variable.

IVORY.   See **Elephant**.

IZOD TEST.   See **Impact Testing**.

J

JACAMAR. See **Woodpeckers and Toucans**.

JACANA *(Aves, Charadriiformes)*. Plover-like birds found in central and northern South America, in much of Africa south of the Sahara, and in parts of Southeast Asia. With their toes spread, the jacanas are able to walk across the floating leaves of water plants that cover inland waters in their tropical homelands. Since their feet are not as well adapted to walking on the ground, jacanas are shy on land. They can run quickly on graveled beaches. In flight, the long legs are especially conspicuous because they extend far beyond the end of the tail.

In body shape and movements, jacanas resemble graceful long-footed common moorhens; some species have colored plates on the forehead. With seven species distributed over six genera, jacanas occur in the tropic regions of the whole world. They are inconspicuous inhabitants of fresh waters rich in aquatic plants.

The sexes in all jacanas have the same coloration; the females are usually larger. Jacanas feed on insects and small mollusks, probably also on small fishes and Seeds of aquatic plants. The nest is built from loose plant debris; it rests or floats on the water.

A pheasant-tailed jacana *(Hydrophasianus chirurgus)* is shown in Fig. 1. The naturalist, Alfred Hoffman, made a thorough study of the polyandry in the pheasant-tailed jacana of the palace lakes in Peking (Beijing) in 1946. He found that the female is able to "present" her males with a full clutch of four eggs each, at intervals of from 9 to 12 days. "With great care," Hoffman reports, "the cock approaches the clutch; he carefully preens himself, especially the breast, so that the feathers are quite dry and airy for the brooding. Then he spreads his pale green legs with those disproportionately long toes and lowers his breast slowly and cautiously onto the precious clutch, while he supports himself on his wings as one would on one's hands. He then pushes the eggs together under his belly with his wings, and swaying slightly from side to side, he scoops them up from the wet ground with his white wings so that they lie warm and protected between his breast and the underside of his wings."

Fig. 1. Pheasant-tailed jacana *(Hydrophasianus chirurgus)*.

After torrents of rain and rising water, the pheasant-tailed jacana moves the nesting site, according to Hoffman's repeated observations. The bird also moves the nesting site when disturbed. Moving does not occur regularly, but it is very common. The move usually does not exceed a few meters. During the incubation period, the jacana may move three or more times with the same clutch. See also **Waders, Shorebirds, and Gulls**.

JACCARANDA TREES. See **Catalpa Trees**.

JACKAL. See **Canines**.

JACK RABBIT. See **Rabbits and Hares**.

JACOB, FRANÇOIS (1920–Present). Jacob was a French bacteriologist and co-discoverer of messenger RNA, who is also renowned for his research on bacterial sexuality and genetics.

François Jacob was born in Nancy, France on 17 June 1920. He entered the University of Paris (the Sorbonne) with the goal of becoming a surgeon. However, World War II intervened and when Germany invaded France in 1940, Jacob fled to London where he joined the French Free Forces. His service in the war earned him France's highest military honor, the Croix de la Liberation. Although he sustained severe damage to his hands in combat, which dashed his hopes of becoming a surgeon, his injuries did not prevent him from completing his medical doctorate in 1947.

In 1950, Jacob became an assistant at the Pasteur Institute. Working under André Lwoff, he studied the *Pseudomonas* species of lysogenic bacteria, which are destroyed through the invasion of bacteriophage viruses. Lwoff had discovered that bacteriophages can exist in the bacterial cell in a noninfectious stage, which he called the prophage. However, the prophage could be stimulated (e.g. by ultraviolet light) to produce the infective virus. Jacob earned his PhD in 1954 for his research with Lwoff on lysogenic bacteria and their prophage. See also **Bacteriophage**; and **Lwoff, André Michel (1902–1994)**

From 1954 to 1958, Jacob and Ellie Wollman researched bacterial sexuality. They found that bacteria exchange genetic material between cells prior to division and that the bacterial chromosome has a circular shape. Jacob and Wollman also used an electron microscope to capture various stages of bacterial conjugation, and thus were able to determine the time needed to transmit various hereditary characteristics. In their research, they discovered that bacteria adapt not only to chemical and physical environmental changes, but also to the presence of drugs and bacteriophages.

Working with Jacques Monod, Jacob discovered messenger RNA, one of the three types of RNA. Messenger RNA conveys genetic information from DNA in the nucleus of the cell to ribosome's in the cytoplasm, where protein is synthesized. Monod and Jacob also discovered that DNA contains two different types of genes—structural and regulatory—and that bacteriophage DNA contains both of these types. Structural genes transmit genetic information between generations, while regulatory genes communicate with the structural genes and regulate the biochemical processes of the cell, enabling it to adapt to changing environments. See also **Monod, Jacques Lucien (1910–1976)**.

Jacob shared the 1965 Nobel Prize in Physiology or Medicine with Lwoff and Monod, for their research on the genetic control of enzyme and virus synthesis, which furthered knowledge of the transmission of genetic information in biochemical processes. Jacob is also recognized for his research on the bacterium *Escherichia coli*; in the 1950s, he determined its nutritional requirements, and between 1968 and 1971, he and other researchers described its process of cell division. He has also written a number of historical and philosophical essays on biology—such as *The Logic of Life: A History of Heredity* (1973) and *The Possible and the Actual* (1982), a discussion of evolutionary theory and its limits—and an autobiography, *The Statue Within* (1988).

Additional Reading

Fox, D, M., M. Meldrum, and I. Rezak: *Nobel Laureates in Medicine or Physiology*, Garland, New York, NY, 1990.

Jacob, F.: *The Logic of Life: A History of Heredity*, Pantheon, New York, NY, 1973.

Jacob, F.: *The Possible and the Actual*, University of Washington Press, Seattle, WA, 1982.

Jacob, F.: *The Statue Within*, Basic Books, New York, NY, 1988.

C. CHRISTOPHER SMITH, Indiana University, Bloomington, IN

JACOBIAN. If $y_1, y_2, \ldots, y_n$ are functions of $x_1, x_2, \ldots, x_n$ then the functional determinant or Jacobian is

$$J = \frac{\partial(y_1, y_2, \ldots, y_n)}{\partial(x_1, x_2, \ldots, x_n)} = \begin{vmatrix} \dfrac{\partial y_1}{\partial x_1} & \dfrac{\partial y_1}{\partial x_2} & \cdots & \dfrac{\partial y_1}{\partial x_n} \\ \dfrac{\partial y_2}{\partial x_1} & \dfrac{\partial y_2}{\partial x_1} & \cdots & \dfrac{\partial y_2}{\partial x_n} \\ \cdots & \cdots & \cdots & \cdots \\ \dfrac{\partial y_n}{\partial x_1} & \dfrac{\partial y_n}{\partial x_2} & \cdots & \dfrac{\partial y_n}{\partial x_n} \end{vmatrix}$$

The second form is a symbolic abbreviation for the Jacobian, which is often used. Even more briefly one could write $J(y_i/x_i)$ or, if the independent variables were understood, even $J(y_i)$. Some properties of the Jacobian are: (a) $J(y_i/x_i)J(x_i/y_i) \equiv 1$; (b) $J(y_i/z_i)J(z_i/x_i) \equiv J(y_i/x_i)$. If a relation $F(y_1, y_2, \ldots, y_n) = 0$ exists between the dependent variables, then $J(y_i/x_i) = 0$, and, conversely, if the Jacobian vanishes, the y_i's are related by at least one equation. See also **Hessian**.

JACOBI ELLIPTIC FUNCTION. If an incomplete elliptic integral of the first kind is indicated by z then z is defined as a function of k and 1, where k is the modulus of z, and 1 is the amplitude of the integral. Calling the integral u and designating the inverse function by $1 = am\,u$, then $x = \sin(am\,u)$. The latter quantity, generally written $x = sn\,u$, is a Jacobi elliptic function. Related ones are $cn\,u = \cos(am\,u) = \pm\sqrt{1 - x^2}$; $dn\,u = \pm\sqrt{1 - k^2 x^2}$. They have simple poles, are doubly periodic, and are generalizations of the trigonometric and hyperbolic functions, which are singly periodic. See also **Theta Function**.

JACOBI POLYNOMIAL. A solution of the differential equation

$$x(1 - x)y'' + [c - (a+1)x]y' + n(n+a)y = 0$$

with a, c real; $c > 0$; $a > (c - 1)$; n, an integer. The nth polynomial is given by the series

$$J_n(a, c, x)$$
$$= 1 + \sum_{k=1}^{n}(-1)^k \binom{n}{k} \frac{(a+n)(a+n+1)\cdots(a+n+k-1)}{c(x+1)\cdots(c+k-1)} x^k$$

where c is not a negative integer. The polynomials are a special case of the Gauss hypergeometric function

$$J_n(a, c, x) = F(-n, a+n, c; \, x)$$

while the Legendre and Chebyshev polynomials are special cases of the Jacobi polynomials

$$P_n(x) = J_n(1, 1, z); \quad T_n(x) = J_n\left(0, \tfrac{1}{2}, z\right)$$

where $2z = (1 - x)$.

See also **Polynomial**.

JADE. Jade is a general term for a compact green mineral substance much prized for ornamental purposes. Jade is either a compact actinolite called nephrite, a variety of amphibole, or jadeite, a monoclinic pyroxene. It is easily worked, and many prehistoric implements have been found of this material in Mexico, Switzerland, France, Greece, and Egypt. The word jade is derived from the Spanish *pietra di hijada*, kidney stone, because it was supposed to be beneficial to diseases of the kidneys. Nephrite is derived from the Greek word for kidney, the allusion being the same as in the case of jade.

See also **Amphibole**; **Jadeite**; and **Pyroxene**.

JADEITE. The mineral jadeite, essentially sodium-aluminum silicate, [CAS: 73987-94-7] $Na(Al, Fe^{2+})Si_2O_6$, is a monoclinic pyroxene usually appearing in crystalline masses, or may be granular, fibrous, or compact. It has a prismatic cleavage; splintery fracture; hardness, 6 in crystals, 6.5–7 massive variety; specific gravity, 3.24–3.43; luster, vitreous to pearly;

color, various shades of green, bluish-green, greenish-white or almost white; translucent to opaque. The processes that have acted to form this mineral are little understood both because of the confusion that exists between jadeite and nephrite, and the fact that the localities are not well known. Jadeite is found in Myanmar and China and has been reported from Mexico. It has probably resulted from the metamorphism, at great depths, of rocks rich in soda and aluminum, such as nephelite syenites. Its association in Myanmar with serpentine suggests its origin in more basic igneous rocks. See also **Pyroxene**.

Jadeite is a tough and yet rather easily worked substance, and has long been used for ornamental purposes. Evidence has been found in Europe, Mexico, Egypt, and elsewhere that it was used in prehistoric times for both ornaments and implements. The word jadeite is the general term used for all green-colored tough compact stones that have been used as indicated above.

JAGUAR. See **Cats**.

JAMESONITE. This mineral is commonly called brittle *feather-ore* from its brittle character and usual habit in acicular mats of crystals. It crystallizes in the monoclinic system and is of metallic luster, opaque with gray-black color and streak. It has a hardness of 2.5 and a specific gravity of 5.63–5.67. It is a sulfide of lead and iron antimony, $Pb_4FeSb_6S_{14}$.

Jamesonite occurs in low- to moderate-temperature hydrothermal veins with other lead sulfosalt minerals. Exceptional specimens are found in Bolivia, at Potosi and Oruro, and as choice felted masses on pyrite at Zacatecas in Mexico. Jamesonite is also found in Arkansas, Idaho, and Utah in the United States and in Ontario and British Columbia, Canada. It is named after the English mineralogist, Robert Jameson, from specimens obtained from Cornwall, England. It is a minor ore of lead.

JAMES WEBB SPACE TELESCOPE (JWST). The James Webb Space Telescope (JWST) was formerly known as the "Next Generation Space Telescope" (NGST). JWST was renamed in Sept. 2002 after a former NASA administrator, James Webb.

The James Webb Space Telescope (JWST) will be a large infrared telescope with a 6.5 -meter (21.3 foot) primary mirror. Launch is planned for 2013. JWST will be the premier observatory of the next decade, serving thousands of astronomers worldwide. It will study every phase in the history of our Universe, ranging from the first luminous glows after the Big Bang, to the formation of solar systems capable of supporting life on planets like Earth, to the evolution of our own Solar System.

JWST is an international collaboration between NASA, the European Space Agency (ESA), and the Canadian Space Agency (CSA). The NASA Goddard Space Flight Center is managing the development effort. The prime contractor is Northrop Grumman Space Technologies; the Space Telescope Science Institute will operate JWST after launch.

Several innovative technologies have been developed for JWST. These include a folding, segmented primary mirror, adjusted to shape after launch; ultra-lightweight beryllium optics; detectors able to record extremely weak signals, microshutters that enable programmable object selection for the spectrograph; and a cryocooler for cooling the mid-IR detectors to 7K. The long-lead items, such as the beryllium mirror segments and science instruments, are under construction. All mission enabling technologies will be demonstrated by January 2007.

There will be four science instruments on JWST: a near-infrared (IR) camera, a near-IR multi-object spectrograph, a mid-IR instrument, and a tunable filter imager. JWST's instruments will be designed to work primarily in the infrared range of the electromagnetic spectrum, with some capability in the visible range. It will be sensitive to light from 0.6 to 27 micrometers in wavelength.

JWST has four main science themes: The End of the Dark Ages: First Light and Reionization, The Assembly of Galaxies, The Birth of Stars and Protoplanetary Systems, and Planetary Systems and the Origins of Life.

JWST in Contrast With HST. The James Webb Space Telescope (JWST) has been called the successor to the Hubble Space Telescope (HST). But what does this really mean? How will JWST be different than HST? There are some similarities—both telescopes are (or will be) in space. They both seek to improve our understanding of processes like star birth and the evolution of galaxies. However, there are many differences between HST and JWST. See also **Hubble Space Telescope (HST)**.

For starters, JWST will primarily look at the Universe in the infrared, while HST studies it at optical and ultra-violet wavelengths. JWST also has a much bigger mirror than HST. This larger light collecting area means that JWST can peer farther back into time than HST is capable of doing. HST is in a very close orbit around the earth, while JWST will be 1.5 million miles away at the second Lagrange (L2) point.

JWST will observe primarily in the infrared and will have four science instruments that can take images and spectra of objects. These instruments will provide wavelength coverage from 0.6 to 28 micrometers (or "microns"; 1 micron is 0.6×10^{-6} meters). The infrared part of the electromagnetic spectrum is generally 0.75–30 microns, so JWST's instruments can cover nearly the entire infrared part of the spectrum.

The instruments on HST can observe a small portion of the infrared spectrum from 0.8 to 2.5 microns, but its primary capabilities are in the ultra-violet and visible parts of the spectrum from 0.1 to 0.8 microns.

HST is 13.2 meters (43.5 ft.) long and its maximum diameter is 4.2 meters (14 ft.) It is about the size of a large tractor-trailer truck. By contrast, JWST's sunshield is about 22 meters by 12 meters (72 ft × 39 ft). A Boeing 737-200 is 100 feet long!

JWST will have a 6.5 meter (21.3 foot) diameter primary mirror, which would give it a significant larger collecting area than the mirrors available on the current generation of space telescopes. See Fig. 1. HST's mirror is a much smaller 2.4 meters (7.8 feet) in diameter and its corresponding collecting area is 4.5 m², giving JWST around 7 times more collecting area! JWST will have significantly larger field of view than the NICMOS camera on HST (covering more than ~15 times the area) and significantly better spatial resolution than is available with the infrared Spitzer Space Telescope. See also **Spitzer Space Telescope (SST)**.

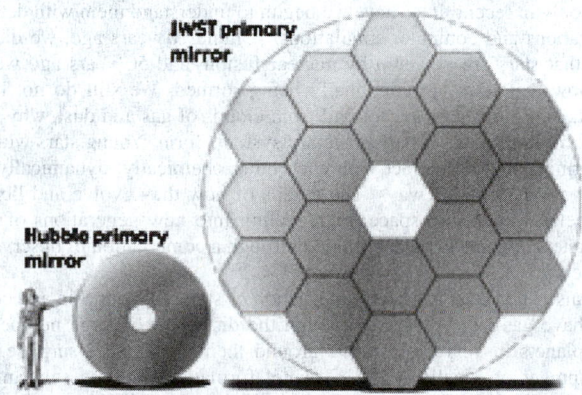

Fig. 1. Comparison between JWST and HST's primary mirror. (*NASA*).

The Earth is 150 million km from the Sun and the moon orbits the earth at a distance of approximately 384,500 km. The Hubble Space Telescope orbits around the Earth at an altitude of ~570 km above it.

JWST will not actually orbit the Earth—instead it will sit at the L2 Lagrange point, 1.5 million miles away! Because HST is in earth orbit, it was able to be launched into space by the space shuttle. JWST will be launched on an Ariane 5 rocket and because it won't be in earth orbit, it is not designed to be serviced by the space shuttle.

JWST's Orbit. JWST will observe primarily the infrared light from faint and very distant objects. But all objects, including telescopes, also emit infrared light. To avoid swamping the very faint astronomical signals with radiation from the telescope, the telescope and its instruments must be very cold. Therefore, JWST has a large shield that blocks the light from the Sun, Earth, and Moon, which otherwise would heat up the telescope, and interfere with the observations.

To have this work, JWST must be in an orbit where all three of these objects are in about the same direction. The most convenient point is the second Lagrange point (L2) of the Sun-Earth system, (Fig. 2) a semi-stable point in the gravitational potential around the Sun and Earth.

The L2 orbit is an elliptical orbit about the semi-stable second Lagrange point. It is one of the five solutions by the mathematician Joseph-Louis Lagrange in the 18th century to the three-body problem. Lagrange was searching for a stable configuration in which three bodies could mutually orbit each other and stay in the same position relative to each other. He

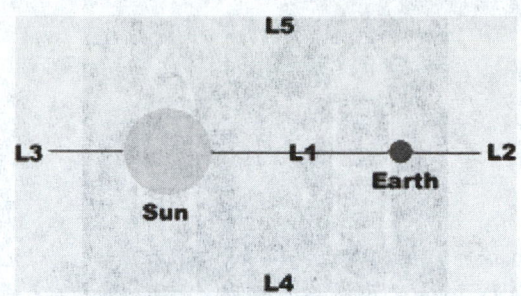

Fig. 2. The five Lagrangian points of the Earth-Sun system. (*NASA*).

found five such solutions, and they are called the five Lagrange points in honor of their discoverer.

In three of the solutions found by Lagrange, the bodies are in line; in the other two the bodies are at the points of equilateral triangles. The five Lagrangian points for the Sun-Earth system are shown in the figure (not drawn to scale) below. An object placed at any one of these 5 points will stay in place relative to the other two.

In this case, the 3 bodies involved are the Sun, the Earth and the JWST. (The gravitational pull of other bodies in the Solar System on JWST will be very small—small enough that we don't have to worry about it in this discussion.) Normally, an object circling the Sun further out than the Earth would take more than one year to complete its orbit. However, the balance of gravitational pull at the L2 point means that JWST will keep up with the Earth as it goes around the Sun.

The JWST will be launched from Arianespace's ELA-3 launch complex at European Spaceport located near Kourou, French Guiana. See also http://www.arianespace.com/site/spaceport/ariane5_sub_index.html.

The Launch Segment has 3 primary components:

1. Launch Vehicle: an Ariane 5 ECA with the cryogenic upper stage. It will be provided in the single launch configuration, with a long payload fairing providing a maximum 4.57 meter (15 foot) static diameter and useable length of 16.19 meters (53 feet). See Fig. 3.
2. Payload Adapter, comprising the Cone 3936 plus ACU 2624 lower cylinder and clamp-band, which provides the separating mechanical and electrical interface between the JWST Observatory and the Launch Vehicle.
3. Launch campaign preparation and launch campaign.

The European Space Agency (ESA) will provide the launch vehicle and the payload adapter to the JWST Mission. The launch campaign preparation and launch campaign is the mutual responsibility of NASA, ESA, and Arianespace.

JWST Science

The James Webb Space Telescope (JWST) will be a giant leap forward in our quest to understand the Universe and our origins. The JWST will examine every phase of cosmic history: from the first luminous glows after the Big Bang to the formation of galaxies, stars, and planets to the evolution of our own solar system. The science goals for the JWST can be grouped into four themes:

The End of the Dark Ages. First Light and Reionization seeks to identify the first bright objects that formed in the early Universe, and follow the ionization history.

Theory and observation have given us a simple picture of the early universe. The Big Bang produced dark energy (the cosmic acceleration force), dark matter, hydrogen, helium, cosmic microwave and neutrino background radiation, and trace quantities of lithium, beryllium, and boron. As the universe expanded and cooled, hydrogen molecules formed, enabling the formation of the first individual stars, at about 180 million years after the Big Bang. According to theory, the first stars were 30 to 300 times as massive as the Sun and millions of times as bright, burning for only a few million years before meeting a violent end. Each one exploded as a supernova that might have collapsed further to form a black hole. The supernovae enriched the surrounding gas with the chemical elements produced in their interiors, and future generations of stars contained these heavier elements. The black holes started to swallow gas and other stars to become mini-quasars, which grew and merged to become the huge black holes now found at the centers of nearly all massive galaxies.

Ariane 5 Generic   Ariane 5 ECA

Fig. 3. This performance ensures that Ariane 5 will be able to loft the heaviest telecom satellites in production or on the drawing boards, and enables Arianespace to match up most spacecraft for highly efficient dual launches — a capability of that has been proven by the company in Ariane missions since the 1980s.

Ariane 5 ECA is a follow-on to the highly mature Ariane 5 Generic launcher version, which carries 6,700 kilogram (14,771 pound) payloads to GTO and has demonstrated its reliability in 16 successful missions.

With the Ariane 5 ECA, Arianespace is able to offer customers even greater performance, flexibility and competitiveness to meet the satellite telecommunications industry's evolving launch services demands. (*Image courtesy of Arianespace and ESA*)

The supernovae and the mini-quasars could be individually observable by JWST.

The emergence of the first sources of light in the universe, shown at left in the famous Hubble Ultra Deep Field, marks the end of the "Dark Ages" in cosmic history, a period characterized by the absence of discrete sources of light. Understanding these first sources is critical, since they greatly influenced subsequent structures. The current leading models for structure formation predict a hierarchical assembly of galaxies and clusters. The first sources of light act as seeds for the successive formation of larger objects, and by studying these objects we will learn the processes that formed the nuclei of present day giant galaxies.

Some time after the appearance of the first sources of light, hydrogen in the intergalactic medium was split apart into electrons and protons (a process called "reionization"). Results from the Wilkinson Microwave Anisotropy Probe combined with data on quasars from about a billion years after the Big Bang show that this reionization had a complex history. See also **Explorers Program**. Although there are indications that galaxies produced the majority of the ultraviolet radiation which caused the reionization, the contribution of quasars could be significant.

JWST will address several key questions in this theme: What are the first galaxies? When and how did reionization occur? What sources caused reionization?

To find the first galaxies, JWST will make ultra-deep near-infrared surveys of the universe, and follow up with low-resolution spectroscopy and mid-infrared photometry. To study reionization, high resolution near-infrared spectroscopy will be needed.

Assembly of Galaxies will determine how galaxies and dark matter, including gas, stars, metals, physical structures (like spiral arms) and active nuclei evolved to the present day.

Theory predicts that galaxies are assembled through a process of the hierarchical merging of dark matter concentrations. Small objects formed first, and were drawn together to form larger ones. This dynamical build-up of massive systems is accompanied by chemical evolution, as the gas (and dust) within the galaxies are processed through successive generations of stars. The interaction of these luminous components with the invisible dark matter produces the diverse properties of present-day galaxies, organized into the Hubble Sequence of galaxies.

This galaxy assembly process is still occurring today, as the Magellanic Clouds fall into the Milky Way, and as the Andromeda Nebula heads toward the Milky Way for a possible future collision. To date, galaxies have been observed back to times about one billion years after the Big Bang. While most of these early galaxies are smaller and more irregular than present-day galaxies, some early galaxies are very similar to those seen nearby today.

Despite all the work done to date, many questions are still open. We do not really know how galaxies are formed, what controls their shapes, and what makes them form stars. We do not know how the chemical elements are generated and redistributed through the galaxies, and whether the central black holes exert great influence over the galaxies. We do not know the global effects of violent events as small and large parts join together in collisions.

JWST will address several key questions in this theme: When and how did the various kinds of galaxies form? How did the heavy elements form? What physical processes determine galaxy properties? What are the roles of high-mass star formation and black holes in galaxy evolution?

To answer these questions, JWST will observe galaxies back to their earliest precursors, so that we can understand their growth and evolution. JWST will conduct wide-area spectroscopic surveys of thousands of galaxies, and observational studies of individual ultra-luminous infrared galaxies (ULIRGs) and active galactic nuclei (AGN).

The Birth of Stars and Protoplanetary Systems focuses on the birth and early development of stars and the formation of planets.

While stars have been the main topic of astronomy for thousands of years, only in recent times have we begun to understand them with detailed observations and computer simulations. A hundred years ago, we did not know that stars are powered by nuclear fusion, and 50 years ago we did not know that stars are continually being formed. We still do not know the details of how stars are formed from clouds of gas and dust, why most stars form in groups, or how planetary systems form. Young stars within a star-forming region interact with each other chemically, dynamically and radiatively in complex ways. The details of how they evolve and liberate the "metals" back into space for recycling into new generations of stars and planets remains to be determined through a combination of observation and theory.

We also know that a substantial fraction of stars, like our Sun and cooler stars, have gas-giant planets, although the discovery of large numbers of these planets in very close orbits around their stars was a surprise. The development of a full theory of planet formation requires substantially more observational input, including observations of disks around young stars and older debris disks in which the presence of planets can be traced.

JWST will address several key questions in this theme:

- How do clouds collapse to form stars?
- How does environment affect star formation and vice versa?
- Do objects that are too low in mass to become stars follow the same law for formation as stars?
- How do proto-planetary systems form? How do gas and dust coalesce to form planetary systems?

To unravel the birth and early evolution of stars, from infall onto dust-enshrouded protostars, through the formation of planetary systems, JWST will provide near- and mid-IR imaging and spectroscopy to observe these objects.

Planetary Systems and the Origins of Life studies the physical and chemical properties of solar systems (including our own) and where the building blocks of life may be present.

Understanding the origin of the Earth and its ability to support life is an important objective for astronomy. Key parts of the story include understanding how the building blocks of planets are assembled. We do not know how planets reach their present orbits, and how the large planets affect the smaller ones in solar systems like our own. We want to learn about the chemical and physical history of the small and large objects that formed the Earth and delivered the necessary chemical precursors for life. The cool objects and dust in the outer Solar System are evidence of conditions in the early Solar System, and are directly comparable to cool objects and dust observed around other stars.

WST will address several key questions in this theme: How do planets form? How are disks around stars like our Solar System? How are habitable zones established?

JWST will determine the physical and chemical properties of planetary systems including our own, and investigate the potential for the origins of life in those systems. JWST will provide near- and mid-IR imaging and spectroscopy to observe these objects.

The observations needed to accomplish these goals require a telescope that can study the Universe in infrared light. For this, it needs to be cooled till it is just a few tens of degrees above Absolute Zero. To accomplish this goal, the JWST will reside far from the Earth at the L2 region.

See also http://www.jwst.nasa.gov/resources/JWST_SSR_JPG.pdf

JWST Observatory

The term JWST Observatory refers to the space-based portion of the JWST system and is comprised of three elements: the Integrated Science Instrument Module (ISIM), the Optical Telescope Element (OTE), and the Spacecraft Element (Spacecraft Bus and Sunshield). See Figs. 4 and 5.

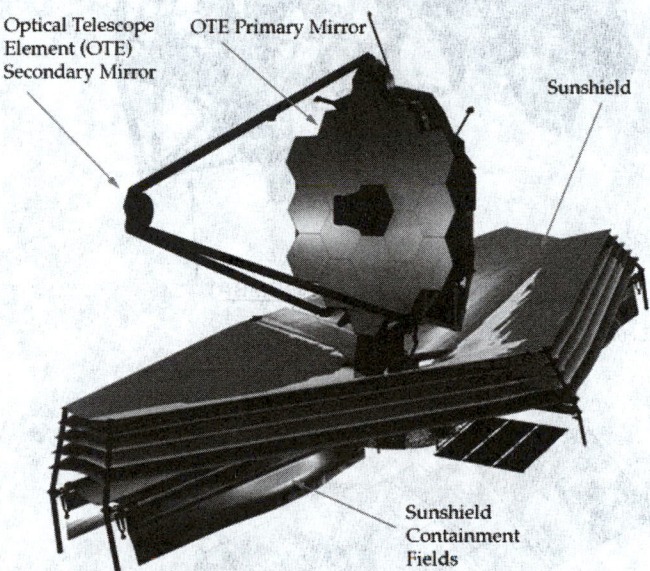

Fig. 4.   Artist's conception of the front view of JWST. (*NASA*)

The ISIM includes the following science instruments:

- Mid-Infrared Instrument (MIRI)
- Near-Infrared Spectrograph (NIRSpec)
- Near-Infrared Camera (NIRCam)
- Fine Guidance Sensor (FGS)

The OTE is the eye of the JWST Observatory. The OTE gathers the light coming from space and provides it to the science instruments located in the ISIM.

The sunshield subsystem separates the observatory into a warm sun-facing side (spacecraft bus) and a cold anti-sun side (OTE and ISIM). The sunshield keeps the heat of the Sun, Earth, and spacecraft bus electronics away from the OTE and ISIM so that these pieces of the Observatory can be kept very cold (The operating temperature has to be kept under 50 K or −370°F).

The spacecraft bus provides the support functions for the operation of the JWST Observatory. The bus houses the six major subsystems needed to operate the spacecraft: the Electrical Power Subsystem, the Attitude Control Subsystem, the Communication Subsystem, the Command and Data Handling Subsystem, the Propulsion Subsystem, and the Thermal Control Subsystem.

The Optical Telescope Element (OTE). The Optical Telescope Element (OTE) is the eye of the JWST Observatory. The OTE gathers the light coming from space and provides it to the science instruments. JWST needs a large mirror to collect as much light as possible to see galaxies from the beginning of the Universe. The JWST scientists and engineers have determined that a primary mirror 6.5 meters (21.3 feet) across is needed to measure the light from these galaxies. See Fig. 6.

The OTE has 2 other components: the Fine Steering Mirror (FSM) and the structural pieces to hold everything together [the secondary mirror support structure (SMSS) and the primary mirror backplane assembly (PMBA)]. The OTE tertiary mirror and the fine steering mirror are both contained within an OTE subsystem known as the Aft Optics Subsystem. The PMBA, in addition to holding the OTE together will be where the science instrument, in the Integrated Science Instrument Module (ISIM), is installed in the Observatory.

No launch vehicle (rocket) is large enough to hold a 6.5-meter mirror if it was all one piece. The JWST team decided to build the mirror with 18 hexagonal primary mirror assembly segments, which can be folded up (shown at left) to fit into the launch vehicle and then unfold after launch.

So that the segments work together as a single large mirror, the 18 segments have been divided into 3 groups of six mirrors, each group having a slightly different shape (prescription). See Fig. 7. A system known as

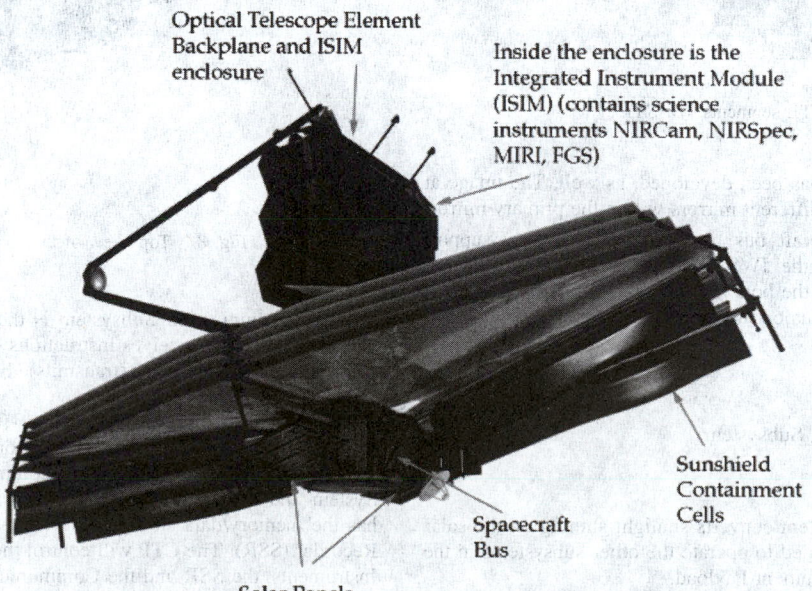

Fig. 5.   Artist's conception of the back view of JWST (*NASA*).

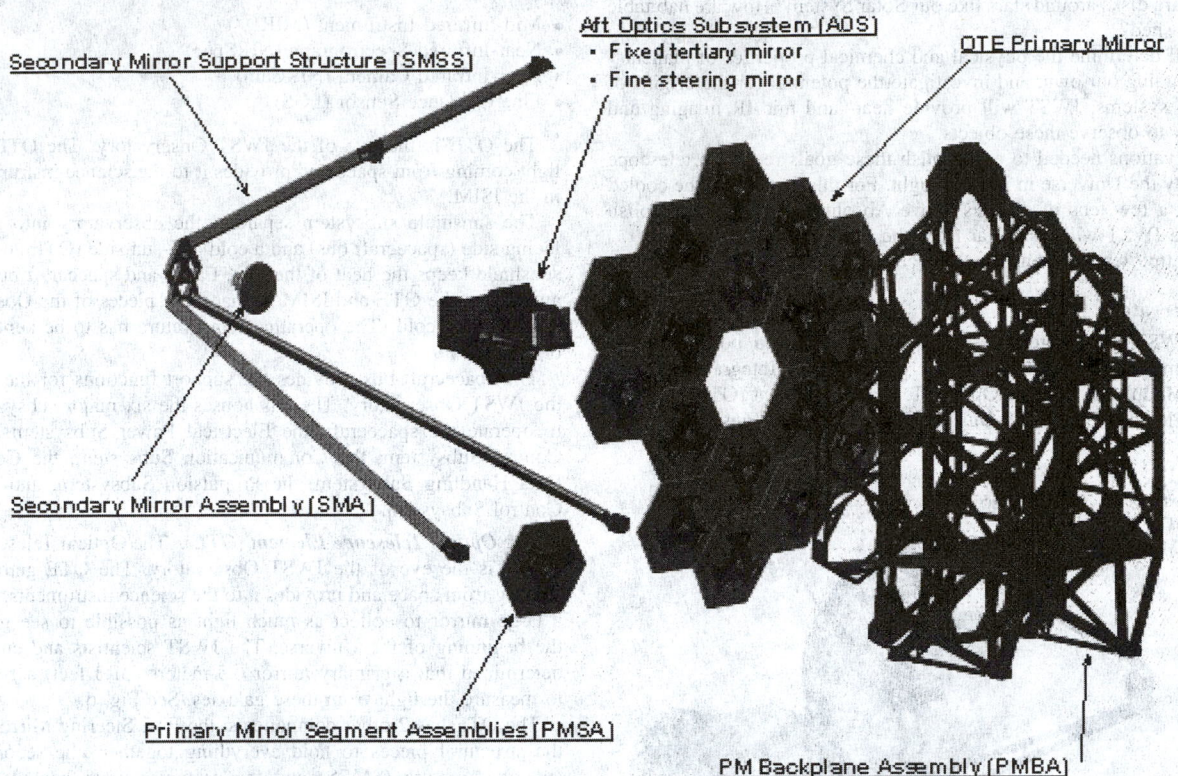

Fig. 6.   The Optical Telescope Element (OTE) (*NASA*).

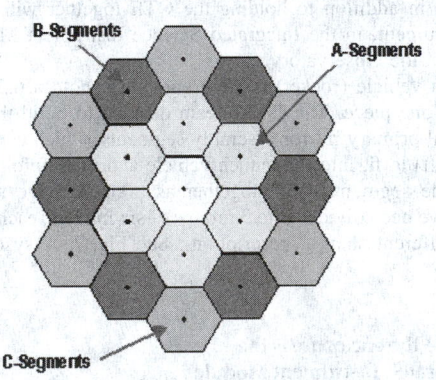

Fig. 7.   OTE segments. (*NASA*)

Wavefront Sensing & Control has been developed, as well. The image at right shows the location of the different mirrors within the primary mirror.

Spacecraft Bus. The spacecraft bus provides the necessary support functions for the operation of the JWST Observatory. A Top View of the bus is shown in Fig. 8, and the bottom view is illustrated in Fig. 9.

The bus is the home for six major subsystems:

- Electrical Power Subsystem
- Attitude Control Subsystem
- Communication Subsystem
- Command and Data Handling Subsystem
- Propulsion Subsystem
- Thermal Control Subsystem

The Electrical Power Subsystem converts sunlight shining on the solar array panels into the power needed to operate the other subsystems in the bus as well as the Science Instrument Payload.

The Attitude Control Subsystem senses the orientation of the Observatory, maintains the Observatory in a stable orbit, and provides the coarse pointing of the Observatory to the area on the sky that the Science Instruments want to observe.

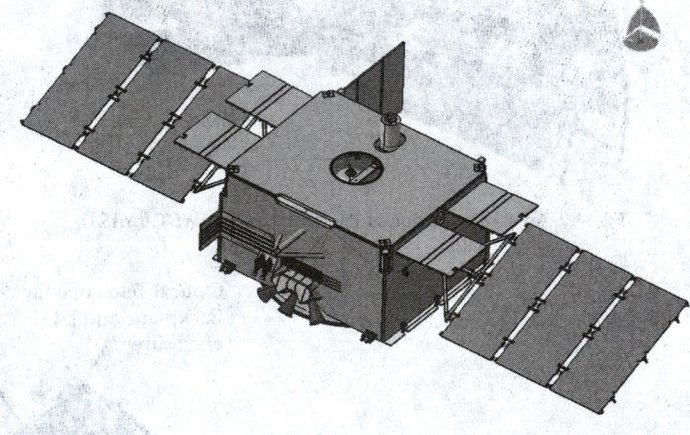

Fig. 8.   Top View of the spacecraft bus. (*NASA*)

The Communication Subsystem is the ears and mouth for the Observatory. The system receives instructions (commands) from the Operations Control Center and sends (transmits) the science and status data to the OCC.

The Command and Data Handling (C&DH) System is the brain of the spacecraft bus. The system has a computer, the Command Telemetry Processor (CTP) that takes in the commands from the Communications System and directs them to the appropriate recipient. The C&DH also has the memory/data storage device for the Observatory, the Solid State Recorder (SSR). The CTP will control the interaction between the Science Instruments, the SSR and the Communications System.

The Propulsion System contains the fuel tanks and the rockets that, when directed by the Attitude Control System, are fired to maintain the orbit.

The Thermal Control Subsystem maintains the operating temperature of the spacecraft bus.

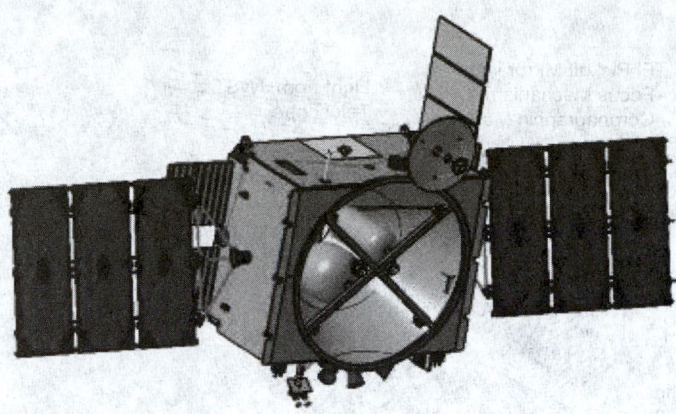

Fig. 9. Bottom View of spacecraft bus (*NASA*).

Sunshield. The observatory is dominated visually by the sunshield subsystem, which separates the observatory into a warm sun-facing side and a cold anti-sun side.

JWST will observe primarily the infrared light from faint and very distant objects. But all objects, including telescopes, also emit infrared light in the form of heat energy. To avoid swamping the very faint astronomical signals with radiation from the telescope, the telescope and its instruments must be very cold, at an operating temperature of under 50 K (−370 °F).

The observatory will be pointed so that the Sun, Earth and Moon are always on one side, and the sunshield will act like a parasol, keeping the Optical Telescope Element and the Integrated Science Instrument Module cool by keeping them in the shade and protecting them from the heat of the sun and warm spacecraft bus electronics.

In addition to providing a cold environment, the sunshield provides a thermally stable environment. See Fig. 4. This is essential to maintaining proper alignment of the primary mirror segments as the telescope changes its orientation to the Sun. When fully deployed, the sunshield will be about the size of a regulation tennis court.

JWST Instruments

The Integrated Science Instrument Module, or ISIM, is the heart of the JWST, what engineers call the main payload. This is the unit that will house the four main JWST instruments. The ISIM is like a chassis in a car providing support for the engine and other components.

The four instruments that attach to the ISIM (Fig. 10) are:

- Mid-Infrared Instrument, or MIRI — provided by the European Consortium with the European Space Agency (ESA), and by the NASA Jet Propulsion Laboratory (JPL)
- Near-Infrared Camera, or NIRCam — provided by the University of Arizona
- Near-Infrared Spectrograph, or NIRSpec — provided by ESA, with components provided by NASA/GSFC.
- Fine Guidance Sensor, or FGS — provided by the Canadian Space Agency. The FGS contains a dedicated Guider and a Tunable Filter Camera.

Fig. 10. The Integrated Science Instrument Module (*NASA*).

Goddard Space Flight Center provides the ISIM infrastructure subsystems required for operation of these instruments. These include the ISIM Structure Subsystem; ISIM Thermal Control Subsystem; ISIM Control and Data Handling Subsystem; ISIM Flight Software; and ISIM Harness Assemblies.

As you can imagine, integrating four major instruments and numerous subsystems into one payload, the ISIM, is a daunting endeavor. To simplify integration, engineers have divided the ISIM into three regions.

The "region 1" component is the cryogenic instrument module. This chills the detectors down to 39 K, a necessary first-stage cooling effort so that the spacecraft's own heat doesn't interfere with the infrared light (a form of heat) detected from distant cosmic sources. The ISIM/OTE Thermal Management Subsystem provides passive cooling. Other devices will get the detectors even colder.

The "region 2" component is the ISIM Electronics Compartment, which provides the mounting surfaces and ambient thermally controlled environment for the instrument control electronics.

The "region 3" component, located within the Spacecraft Bus, is the ISIM Command and Data Handling subsystem, with integral ISIM flight Software, and the MIRI cryocooler compressor and control electronics.

Mid-Infrared Instrument (MIRI). The Mid-Infrared Instrument (MIRI) is an imager/spectrograph that covers the wavelength range of 5 to 27 micrometers, with a possible spectrographic coverage up to 29 micrometers. See Fig. 11. The MIRI has three Arsenic-doped Silicon (Si:As) detector arrays. The camera module provides wide-field broadband imagery, and the spectrograph module provides medium-resolution spectroscopy over a smaller field of view compared to the imager. The nominal operating temperature for the MIRI is 7K. This level of cooling cannot be attained using the passive cooling provided by the Thermal Management Subsystem. Instead, there is a two-step process: A Pulse Tube precooler gets the instrument down to 18K; and a Jules-Thomson Loop heat exchanger knocks it down to 7K. The cryocooler compressor assembly and control electronics are housed in the Spacecraft Bus. See also http://www.stsci.edu/jwst/instruments/miri/.

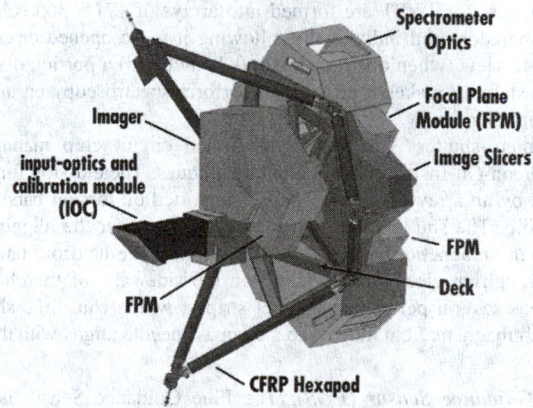

Fig. 11. MIRI Instrument (*STSI*).

Near Infrared Camera (NIRCam). The Near Infrared Camera (NIRCam) is an imager with a large field of view and high angular resolution. The NIRCam covers a wavelength range of 0.6 to 5 micrometers and has ten mercury-cadmium-telluride (HgCdTe) detector arrays. These are analogous to CCDs found in ordinary digital cameras. The NIRCam is a science instrument but also an Optical Telescope Element wavefront sensor, which provides something similar to instant LASIK vision correction.

The NIRCam will be built by a team led by the University of Arizona, Tucson, with Principal Investigator Dr. Marcia Rieke. The industrial partner is Lockheed-Martin Advanced Technology Center, Palo Alto, California. See also http://ircamera.as.arizona.edu/nircam/.

Near Infrared Spectrograph (NIRSpec). The Near Infrared Spectrograph (NIRSpec) enables scientists to obtain simultaneous spectra of more than 100 objects in a 9-square-arcminute field of view. This instrument provides medium-resolution spectroscopy over a wavelength range of 1 to 5

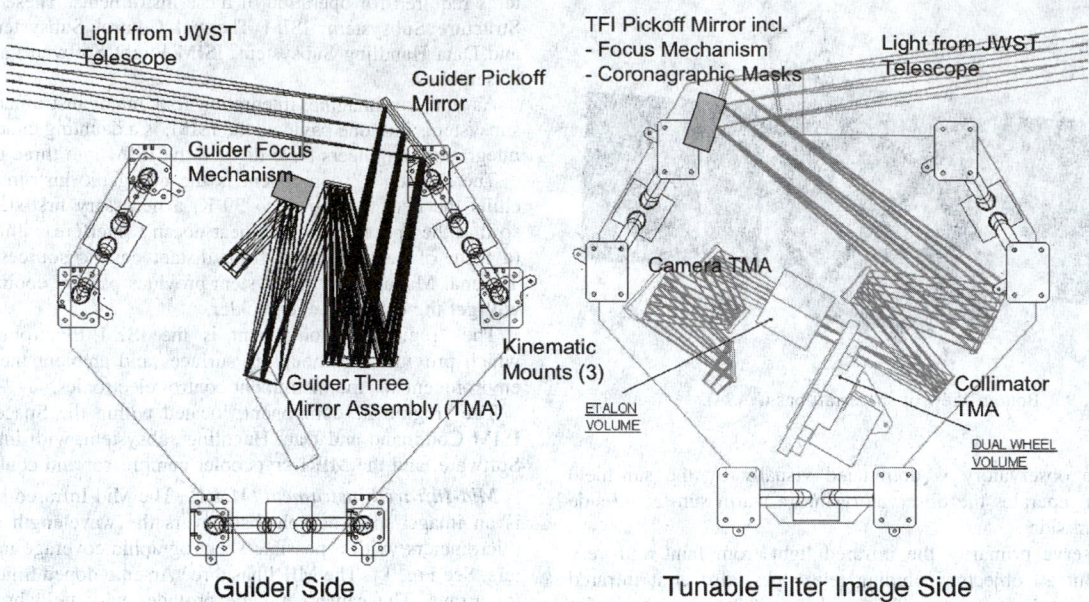

Fig. 12.   Optical-mechanical packaging of the FGS (*STSI*).

micrometers and lower-resolution spectroscopy from 0.6 to 5 micrometers. The NIRSpec employs a micro-electromechanical system "micro-shutter array" for aperture control, and it has two HgCdTe detector arrays.

The European Space Agency (ESA) will be providing the NIRSpec instrument (NASA will provide the detectors and the MEMS aperture masks) as part of their contribution to JWST. See also http://www.stsci.edu/jwst/instruments/nirspec/.

Microshutters. Micro-shutters are tiny, 200-micron-wide cells with lids that open and close in response to the application of a magnetic field. The micro-shutters for JWST are formed into arrays of 171×365 cells. Each cell can be addressed individually, allowing it to be opened or closed as required to view (when open) or block (when closed) a portion of the sky. This adjustability makes it possible to perform spectroscopy on up to 100 targets simultaneously.

The micro-shutter is a custom integrated circuit chip manufactured using Silicon-On-Insulator fabrication techniques. Theield-stop function is provided by an array of shutter "flaps" supported on torsion bars residing on the chip. The shutters are opened and closed under the aligning force of a moving magnetic flux field, and the shutters are held or "latched" in place through an electrostatic charge on the sidewalls of the chip. Each shutter has several permanent magnet stripes, which cause the shutter to align with magnetic field much like a compass needle aligns with the North Pole.

Fine Guidance Sensor (FGS). The Fine Guidance Sensor is a very broadband guide camera that is incorporated into the cryogenic instrument payload in order to meet the image motion requirements of the JWST. See Fig. 12. This sensor is used for both "guide star" acquisition and fine pointing. The sensor operates over a wavelength range of 1 to 5 micrometers and has two HgCdTe detector arrays. Its field of view is sufficient to provide a 95 % probability of acquiring a guide star for any valid pointing direction.

The Fine Guidance Sensor Tunable Filter Camera is a wide-field, narrow-band camera that provides imagery over a wavelength range of 1.6 to 4.9 micrometers, with a gap between 2.6 and 3.1 micrometers, via tunable Fabry-Perot etalons that are configured to illuminate the detector array with a single order of interference at a user-selected wavelength. The camera has a single HgCdTe detector array.

The Canadian Space Agency will provide the Fine Guidance Sensor to the the JWST Project. The prime contractor is ComDev. The Principal Investigator is John Hutchings of the Herzberg Institute of Astrophysics, National Research Council of Canada. See also http://www.stsci.edu/jwst/instruments/guider/.

Web References

Canadian Space Agency (CSA): http://www.space.gc.ca/asc/eng/default.asp
European Space Agency (ESA): http://www.esa.int/esaCP/index.html
Goddard Space Flight Center: http://www.nasa.gov/centers/goddard/home/index.html
NASA: http://www.nasa.gov/
Northrop Grumman Space Technologies: http://www.st.northrop-grumman.com/
Space Telescope Science Institute: http://www.stsci.edu/resources/
The Lagrange Points: http://map.gsfc.nasa.gov/m_mm/ob_techorbit1.html

JONATHAN P. GARDNER, National Aeronautics and Space Administration

JAPANESE BEETLE (*Insecta, Coleoptera*).   A small, bronze-green beetle, *Popillia japonica*, about $\frac{3}{8}$ inch (6–7 millimeters) long, related to the May beetles. It is native to Japan and was first noticed as a pest in North America in 1916, when it was found near Riverton, New Jersey. The pest has continued to spread and is now present from southern Maine southward into South Carolina and Georgia and westward into Kentucky, Illinois, Michigan, and Missouri. See Fig. 1.

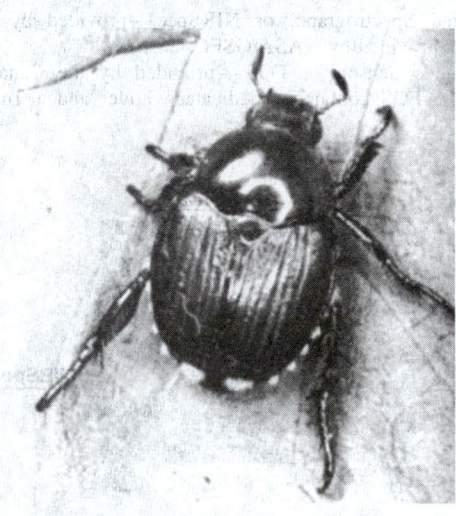

Fig. 1.   Adult Japanese beetle. (*USDA.*)

The Japanese beetle attacks fruits, shrubs, forest and shade trees, and several crops. The larva damages roots of grasses. Adult beetles will feed on more than 275 plants. In some cases, damage is slight. In other cases, insecticide and other control methods are required to prevent serious plant injury. The beetles often congregate and feed on flowers, foliage, and fruit of plants and trees exposed to bright sunlight. Beetles feeding on leaves usually chew out the tissue between the veins, leaving a lacy skeleton. They may eat many large, irregular areas on some leaves. A badly attacked tree or shrub may lose most of its leaves in a short time. The beetles often mass on ripening fruits and feed until nothing edible is left. They seldom touch unripe fruit. They seriously injure corn by eating the silk as fast as it grows, thus keeping kernels from forming.

There are a number of natural controls on both grubs and adults. Weather, disease, parasites, and other natural enemies can play a part in determining whether or not a beetle infestation will occur, and how serious it may be. Cooperative national and state government efforts to retard Japanese beetle spread into new areas are centered at airports and elsewhere in order to obtain information on distribution of the insect. Mass trapping in large areas has reduced injury as much as 30%, but this is not considered a major control method. Studies are in progress to determine further the value of mass trapping and to improve available beetle attractants.

Diseased and poorly nourished trees and plants are especially susceptible to attack by the beetles. Prematurely ripened or diseased fruit is also attractive. The odor of diseased fruit definitely attracts the beetles. Certain weeds should be removed to help control infestations. The beetles are particularly fond of bracken, elder, evening primrose, Indian mallow, sassafras, poison ivy, smartweed, wild fox grape, and wild summer grape.

Japanese beetle adults seldom become a problem where regular spray schedules for the control of other insects are followed. However, when there is a specific need for beetle control, foliage and fruit of most plants can be protected by sprays of carbaryl and of malathion. These chemicals should not be applied during the blooming season because the chemicals are highly toxic to honeybees. Small infestations sometimes can be controlled simply by collecting the beetles by hand, particularly in the morning when it is cool and the beetles are quiet.

A biological type of control has been used quite successfully. This consists of a mixture of the spores of a milky-disease organism, *Bacillus popilliae*, and powdered talc. By spreading this material at 10-foot (3 meter) intervals in infested areas, the disease, which kills the larvae of the beetles, is established much more promptly and effectively than through its natural spread.

JAPANESE ENCEPHALITIS VIRUS (JEV). See **Arthropod-borne Viruses**.

JAPP-KLINGEMANN REACTION. Formation of hydrazones by coupling of aryldiazonium salts with active methylene compounds in which at least one of the activating groups is acyl or carboxyl. This group usually cleaves during the process.

JAROSITE. The mineral jarosite is a basic hydrous sulfate of potassium and iron corresponding to the formula $KFe_3(SO_4)_2(OH)_6$. It is formed in the outcrops of ore deposits during oxidation of iron sulfides. It is a hexagonal mineral with basal cleavage; is brittle; hardness 2.5–3.5; specific gravity 2.9–3.26; luster vitreous to dull; color, dark yellow to yellowish-brown; pale yellow streak; translucent to opaque. Jarosite was originally reported from and named for Barranco Jaroso in the Sierra Almagrera, Spain. It has been found in Bohemia, France, the Island of Elba, Siberia, and Bolivia. In the United States it is found in Arizona, Colorado, Texas, New Mexico, Utah, Nevada, and South Dakota.

JASON 1. See **Earth Observing System (EOS)**.

JASPER. A variety of chert, always quartz, that is associated with iron ores and thus contains iron-oxide impurities which give the rock a variety of colorations, often red, but ranging through yellow, green, grayish-blue, brown, and even black. The term also has been used with reference to any red chert or chalcedony. The material is dense, cryptocrystalline, and usually opaque, although it may be slightly translucent.

JASPILITE. A rock made up of alternating layers of siliceous material such as quartz or chalcedony, and red jasper or hematite. Jaspilites are often contorted and brecciated to a considerable degree, and are sometimes polished for ornamental use.

JATO (From jet-assisted take-off). 1. A take-off utilizing an auxiliary jet-producing unit or units, usually rockets, for additional thrust. Hence JATO bottle, Jato unit, etc.; a rocket or unit so used. Where rockets are the auxiliary units, RATO (which see) is the more specific term.

2. A JATO bottle or unit; the complete auxiliary power system used for assisted take-off.

JATROPHONE. Jatrophone [CAS: 29444-03-9] $C_{20}H_{24}O_3$, is diterpenoid growth inhibitor isolated from an alcohol extract of the plant *Jatropha gossypiifolia*. Its unique structure includes a 12-membered ring and is readily attacked by nucleophiles. *Jatrophone* is useful in the study of tumor growth inhibition and other biochemical research.

JAUNDICE. When an excess of bile pigment is released into the blood, a yellowish staining of the skin and mucous membranes results, which is referred to as jaundice. It is a symptom rather than a disease. Jaundice may be caused by the production of bile pigment in excess of the amount the liver can excrete, or it may result from liver damage, the liver being unable to excrete a normal amount of bile pigment. The pigment is derived from hemoglobin of broken-down red blood cells. Normally, most bile pigment is excreted by the liver into the intestines and then is absorbed from the intestines back to the liver, so that little pigment is present in the blood.

Jaundice may arise from a number of causative factors. These include hepatitis, bacteremia, chemical poisons, chronic pancreatitis, clostridial uterine infections, murine typhus, obstructions in the biliary tract, Q fever, sickle cell anemia, Weil's disease, and Wilson's disease. See also **Bile**; **Gallbladder and Biliary Tract Diseases**; **Hepatitis**; and **Wilson's Disease**.

JAWFISH. See **Bass**.

JAY (*Aves, Passeriformes*). Birds of several species related to the magpies and crows. Most jays live in the Northern hemisphere, but a few species occur in the Oriental region and northern Africa. They are noisy birds, often brightly colored, and sometimes interesting in habits, but some are too destructive of the eggs and young of other species and too quarrelsome to be desirable.

The blue jay, *Cyanocitta cristata*, is the most widely known of the North American species. In appearance, it is blue above and white below, with white bands on the wings, a black band around the neck, narrow black bars on the wings and tail, and a crest on the head. More than a dozen other species occur in the western states, where the Oregon jay, *Perisoreus obscurus*, and the Canada jay or whiskey jack, *P. canadensis*, are friends of campers, visiting camps and human habitations freely in search of food.

Generally jay birds are omnivorous; they are intelligent, bold, and curious; they nest in trees or old buildings; they have strong, coarse feet; the eggs are pale blue with brown markings, and there usually are five.

JBIG2. See **Data Compression**.

J-DISPLAY. In radar, a modified A-display in which the time base is a circle. The target signal appears as a radial deflection from the time base. Also called *J-scan, J-scope,* or *J-indicator*.

JEJUNUM. That portion of the small intestine that extends from the end of the duodenum to the beginning of the ileum. The jejunum is about 8 feet (2.4 meters) long and continues into the ileum without any line of demarcation.

See also **Digestive System (Human)**.

JELLY. See **Colloid System**.

JELLYFISH. See **Hydratuba (Jellyfish)**; **Medusa**; and **Scyphozoa**.

JENNER, EDWARD (1749–1823). Jenner was an English physician, surgeon and naturalist who introduced smallpox vaccination in 1798. A native of Berkeley, Gloucestershire, Jenner is often depicted as a 'country doctor' because he preferred rural life. However, he studied in London with John Hunter and so received excellent training.

As a rural practitioner, Jenner investigated the idea that farm workers recovered from the mild disease cowpox were immune to smallpox. After collecting circumstantial supporting evidence he performed his first cowpox inoculation (later called vaccination, from *vacca* =cow), in 1796. A small series of vaccinations followed in 1798 in which cowpox was transferred arm-to-arm, and those tested successfully resisted smallpox inoculation (variolation). Later that year Jenner published his *Inquiry*, proposing cowpox inoculation as a safe alternative to variolation.

Jenner's conclusions were based on slender evidence and needed confirmation. His most serious error was to claim that vaccination conferred lifelong immunity. This started a controversy which intensified as the need for re-vaccination became apparent. A particularly astute observation by Jenner, long before the development of laboratory microbiology, was the clinical distinction between 'true cowpox', which protected against smallpox, and varieties of 'spurious cowpox', which did not. See also **Smallpox**; and **Virology (The History)**.

Jenner has been regarded as genius, charlatan and simple country doctor. His powers of observation and insight were considerable, although he did make mistakes and theorize on slender evidence, and his basic observation that an animal virus could safely protect against smallpox proved true.

Jenner's work as a naturalist should not be forgotten. He studied animal behavior and was the first to record the murderous habits of the newly hatched cuckoo, for which he was elected Fellow of the Royal Society in 1789.

DERRICK BAXBY, University of Liverpool, UK

JERNE, NIELS KAJ (1911–1994). Niels Jerne was a Danish immunologist who proposed a selection theory of antibody formation, the functional network of antibodies and lymphocytes and distinction of self from nonself by T lymphocytes.

Niels Kaj Jerne was born in London to Danish parents. His ancestors were from western Jutland. He studied at Leiden, Holland, where he completed his baccalaureate when he was 16 but spent the next 12 years deciding on a profession. At the age of 28, he began the study of medicine in Copenhagen with the plan to become a village doctor. After taking a part-time job in a scientific laboratory, he became interested in science, especially immunology. His doctoral dissertation was on the theoretical foundation for the study of antibody avidity. He earned the MD in 1951.

Jerne followed in his father's footsteps by moving frequently. He worked at the State Serum Institute in Copenhagen between 1943 and 1954. Thereafter, he left for America, working at the California Institute of Technology in Pasadena where he wrote a paper that was the death knell for the instructionist theories of antibody synthesis and a forerunner of the clonal selection theory. He next accepted a staff position with the World Health Organization in Geneva, Switzerland. Here he coined new immunological terms such as "epitope" and "idiotype." He also served as Professor of Biophysics at the University of Geneva between 1960 and 1962. Thereafter, he returned to the United States to serve for 4 years as Professor and Chairman of the Department of Microbiology at the University of Pittsburgh, where he developed a plaque assay for antibody-forming cells. He was then appointed Director of the Paul Ehrlich Institute at the University of Frankfurt, Germany, and began formulating a theory on the role of self antigens in the generation of antibody diversity. In 1969, Jerne became Director of the Basel Institute for Immunology, where he developed a network theory of the immune system. He was elected a Fellow of the Royal Society in 1980 and after retiring from the Basel Institute, served for a year at the Institut Pasteur in Paris. Taking up residence in France, he shared the Nobel Prize in Medicine or Physiology in 1984 with Georges Köhler and Cesar Milstein. See also **Epitopes**; and **Milstein, Cesar (1927–2002)**.

Additional Reading

Carson, D.A., P.P. Chen, and T.J. Kipps: *Idiotypes in Biology and Medicine*, S. Karger Publishers, Inc., Farmington, CT, 1989.

Greene, M.I., and A. Nisonoff: *The Biology of Idiotypes*, Plenum Press, New York, NY, 1984.

Jerne, N.K.: "The Natural Selection Theory of Antibody Formation," *Proceedings of the National Academy of Sciences of the USA*, **41**, 849 (1955).

Jerne, N.K.: "Immunological Speculations," *Annual Review of Microbiology*, **14**, 341–358 (1960).

Jerne, N.K.: "Towards a Network Theory of the Immune System," *Annals of Immunology*, **125 C**, 373–389 (1974).

Jerne, N.K., C. Henry, A.A. Nordin, et al.: "Plaque Forming Cells: Methodology and Theory," *Transplantation Reviews*, **18**, 130–191 (1974).

Jerne, N.K.: "The Generative Grammar of the Immune System," *Science*, **229**, 1057–1059 (1985).

Silverstein, A.M.: *A History of Immunology*, Elsevier Science & Technology Books, New York, NY, 1989.

J.M. CRUSE
R.E. LEWIS,
Universtiy of Mississippi Medical Center, Jackson, MS

JET.   1.   A strong well-defined stream of fluid either issuing from an orifice or moving in a contracted duct, such as the jet of combustion gases issuing from a reaction engine, or the jet in the test section of a wind tunnel.

2.   A tube, nozzle, or the like through which fluid passes, or from which it issues, in a jet, such as a jet in a carburetor.

3.   A jet engine, as, an airplane with jets slung in pods.

JET AIRCRAFT.   See **Airplane**; **Helicopters and V/STOL Craft**; and **Supersonic Aerodynamics**.

JETAVATOR.   A control surface that may be moved into or against a rocket's jetstream, used to change the direction of the jet flow for thrust vector control.

JET ENGINE.   1.   Broadly, and engine that ejects a jet or stream of gas or fluid, obtaining all or most of its thrust by reaction to the ejection.

2.   Specifically, an aircraft engine that derives all or most of its thrust by reaction to its ejection of combustion products (or heated air) in a jet and that obtains oxygen from the atmosphere for the combustion of its fuel (or outside air for heating, as in the case of the nuclear jet engine), distinguished in this sense from a rocket engine. A jet engine of this kind may have a compressor, commonly turbine-driven, to take in and compress air (turbojet), or it may be compressorless, taking in and compressing air by other means (pulsejet, ramjet).

JET FUEL.   A fuel for jet (turbine) engines, usually a petroleum distillate similar to kerosine. A number of types with somewhat different compositions and properties have been used. See also **Petroleum**.

JET-LAG.   "Jet-lag" is a name for the physiological asynchrony and subjective sensations that come with abrupt east-west change to a different time zone, excluding those that also come with north-south travel. Fatigue, sleeplessness, dehydration, and minor culture-shock are common in any direction, but the "lag" in "jet-lag" connotes its specific dependence on changing the phase of entrainment of the **circadian clock**. See also **Circadian Clock**.

Jet-lag is believed to have two components:

- *first*, the inconvenience of being out of synchrony with people at destination, e.g., having low body temperature and being sleepy during business hours, and
- *second*, the discomfort of internal asynchrony during several days while physiological processes catch up to the environment at destination, but at different rates or even in opposite directions, for example, one advancing 9 hours while another delays 15 hours.

Both components are presumably minimized by promptly resetting the master clock in the suprachiasmatic nuclei of the brain, a pair of sub-millimeter specializations containing about 100,000 cells. This is best done by light affecting ganglion cells in the retina of the eye, from which fibers enter these nuclei just above the crossing of the optic nerves. Other factors (the time of day or departure or arrival, the duration of flight, the quality of sleep, food, and drink during travel, the extent of latitude change, melatonin, extra-ocular illumination) seem to play at most relatively minor roles in phase resetting. Something similar to jet-lag may afflict alternating shift workers, with this difference, that their pattern of light exposure generally differs from that of travelers.

In humans, as in all other organisms, light is the dominant phase-setter, so the *timing* of light exposure strongly affects the process of phase-shifting to the new time zone (or shift). As with any other oscillating mechanism, the circadian clock advances or delays in reaction to a given stimulus to an extent that depends on its phase when the stimulus arrives

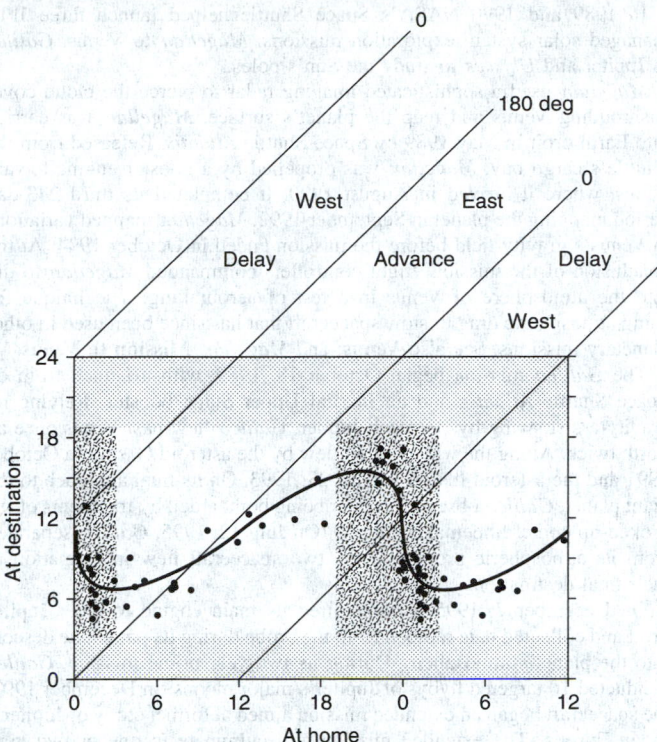

Fig. 1. The human phase resetting curve in response to a rather complicated dim light exposure is here supposed similar to that measured in response to a single, short, more intense exposure. Horizontally old phase, and vertically new phase, are indicated numerically in hours after the minimum of body temperature late in the night. The shading is personal night. Loci of fixed phase *shift* (new phase minus old phase) are diagonal lines. For any needed shift, equal to the longitude change, follow its diagonal to the curve and read off new phase on the vertical axis to the left and old phase on the horizontal axis below. Notice the time at home relative to the shaded block on the old phase axis and on the new phase axis notice time at destination relative to the shaded block when and where the exposure is taken that resets the clock from that old phase to that local new phase. Notice that entrainment to the prior light/dark cycle (shading) so phased the clock that its hours of negligible phase shift (along the 0° diagonals occur in mid-day, with modest advances occurring around dawn and balancing delays occurring around dusk (in the prior environment, i.e., personal habits of daylight exposure at home). The hour of &plusmin; 12-hour resetting is slightly after the middle of darkness or "personal midnight" at home, near the minimum of body temperature. Notice that in terms of time at destination, i.e., the time to which the clock is thus reset, these are all in daylight, viz., early morning for modest advances, late afternoon for modest delays, and mid-day of the personal light/dark cycle for the largest advances or delays, both of which reset to the opposite side of the cycle. (*Adapted from Czeisler* et al., *1989 Figure 5, through Winfree, 2000 Figure 19-0.2, with permission.*)

[see also Figure 1 or **circadian clock**]. For the first such stimulus (the first sunlight exposure), that is the unperturbed phase of a clock left in the home time zone, unaltered by travel without sunlight exposure. The amount of phase shift needed to match clocks at destination is the shift of longitude between home and destination. Because the human clock's reaction to strong sunlight exposure is topologically "Type 0", there is one and only one phase in its cycle when this particular phase-shift will result from such exposure. We look up this phase on the "phase resetting curve" of the circadian clock, see Fig. 1: this is like the back plane of Figure 1 in **Circadian clock**, representing resetting at fixed large stimulus magnitude. It is an update of Daan and Lewy (1984) but using more recent human data.

To more simply look up the best timing for sunlight exposure by entering your home and destination airport names, or by rotating a world map, go to *http://cochise.biosci.arizona.edu/~art* (Find "jet-lag"). It should be noted that, although carefully timed light is universally the most effective way to immediately phase-shift the circadian clock in any species, and the data points in Figure 1 do come from human trials in a laboratory setting, the

needed double-blind clinical trial of its efficacy for jet-lag in travelers remains for the future.

Additional Reading

Czeisler, C.A. et al.: "Bright Light Induction of Strong (Type 0) Resetting of the Human Circadian Pacemaker," *Science*, **244**, 1328–1333 (1989).
Czeisler, C.A.: "The Effect of Light on the Human Circadian Pacemaker," *Ciba Foundation Symposia*, **183**, 254–320 (1995).
Daan, S. and A.J. Lewy: "Scheduled Exposure to Daylight: A Potential Strategy to Reduce Jet-Lag following Transmeridian Flight," *Psychopharm. Bull.*, **20**, 566–568 (1984). http://www.geocities.com/chuutoro/JetLagDeleter.html
Rusak, B.: "Open Forum: Human Phase-resetting Sensitivity to Light," *J. Biol. Rhy.*, **8**, 339–361 (1993).
Winfree, A.T.: *The Timing of Biological Clocks*, Scientific American Books, New York, NY, 1986.
Winfree, A.T.: *The Geometry of Biological Time*, 2nd Ed., Springer-Verlag, New York, NY, 2000.

A.T. WINFREE, University of Arizona, Tucson, AZ

JET NOZZLE.   A nozzle, usually specially shaped, for producing a jet, such as the exhaust nozzle on a jet or rocket engine.

JET PROPULSION LABORATORY (JPL).   JPL is a federally funded research and development facility managed by the California Institute of Technology for the National Aeronautics and Space Administration.

A new generation of space missions to explore the solar system and the universe beyond is unfolding at the Jet Propulsion Laboratory.

The American space age began January 31, 1958, with the launch of the first U.S. satellite, *Explorer 1*, built and controlled by JPL. In the more than four decades since then, JPL has led the world in exploring all of the solar system's known planets, except Pluto, with robotic spacecraft. The tools developed at JPL for its spacecraft expeditions to other planets have also proved invaluable in providing new insights and discoveries in studies of Earth, its atmosphere, climate, oceans, geology, and the biosphere.

Approaching the new millennium as the 21st century begins, JPL continues as a world leader in science and technology, breaking new ground in the miniaturization and efficiency of spacecraft components. At the same time, the Laboratory is pushing the sensitivity of space sensors and broadening their applications for a myriad of scientific, medical, industrial, and commercial uses on Earth.

JPL's Beginnings

JPL's history dates to the 1930s, when of Cal-Tech Professor Theodore von Kármán conducted pioneering work in rocket propulsion. Von Kármán, head of Caltech's Guggenheim Aeronautical Laboratory, gathered with several graduate students to test a primitive rocket engine in a dry riverbed wilderness area in the Arroyo Seco, a dry canyon wash north of the Rose Bowl in Pasadena, California. Their first rocket firing took place there on October 31, 1936.

After the Cal Tech group's successful rocket experiments, von Kármán, who also served as a scientific adviser to the U.S. Army Air Corps, persuaded the Army to fund development of strap-on rockets (called JATO, for "jet-assisted take-off") to help overloaded Army airplanes to take off from short runways. The Army helped Cal Tech acquire land in the Arroyo Seco for test pits and temporary workshops. Airplane tests at nearby air bases proved the concept and tested the designs. By this time, World War II had begun and the rockets were in demand.

As the group wound up the JATO work, the Army Air Corps asked von Kármán for a technical analysis of the German V-2 program just discovered by Allied intelligence. He and his research team then proposed an U.S. research project to understand, duplicate, and reach beyond the guided missiles beginning to bombard England. In the proposal, the Cal Tech team referred to their organization for the first time as "the Jet Propulsion Laboratory."

Funded by Army Ordnance, the Jet Propulsion Laboratory's early efforts would eventually involve technologies beyond those of aerodynamics and propellant chemistry, technologies that would evolve into tools for space flight, secure communications, spacecraft navigation and control and planetary exploration.

The team of about 100 rocket engineers began to expand, and the team began testing in the California desert of small unguided missiles (named *Private*) that reached a range of nearly 18 kilometers (about 11 miles). They experimented with radio telemetry from missiles, and began planning for ground radar and radio sets. By 1945, with a staff approaching 300, the group had begun to launch test vehicles from White Sands, New Mexico, to an altitude of 60 kilometers (200,000 feet), monitoring performance by radio.

Control of the guided missile was the next step, requiring two-way radio as well as radar and a primitive computer (using radio tubes) at the ground station. The result was JPL's answer to the German V-2 missile, named *Corporal*, first launched in May 1947, about two years after the end of war with Germany.

Developing a missile that would fly and survive in the field involved testing its aerodynamic design and durability under vibration and other stresses. The team developed a supersonic wind tunnel and an array of environmental test technologies, all of which had wider use and came to support outside customers. Developing so complex a device as a missile to fly unaided and beyond reach of repair meant a new degree of quality, new test techniques, and a new discipline called *system engineering*.

Subsequent Army work further sharpened the technologies of communications and control, of design and test and performance analysis. This made it possible for JPL to develop the flight and ground systems and finally to fly the first successful U.S. space mission, *Explorer 1*. The entire three-month effort began in November 1957 and culminated with the successful launch on January 31, 1958.

On December 3, 1958, two months after NASA was created by Congress, JPL was transferred from Army jurisdiction to that of the new civilian space agency. It brought to the new agency experience in building and flying spacecraft, an extensive background in solid and liquid rocket propulsion systems, guidance, control, systems integration, broad testing capability and expertise in telecommunications using low-power spacecraft transmitters and very sensitive Earth-based antennas and receivers.

The Laboratory now covers some 72 hectares (177 acres) adjacent to the site of von Kármán's early rocket experiments. Jet propulsion is no longer the focus of JPL's work, but the world-renowned name remains the same.

Planetary Exploration

In the 1960s, JPL began to conceive and execute robotic spacecraft to explore other worlds. This effort began with the *Ranger* and *Surveyor* missions to the Moon, paving the way for NASA's Apollo astronaut lunar landings. See also **Moon (Earth's)**. During that same period and through the early 1970s, JPL carried out *Mariner* missions to Mercury, Venus and Mars.

Mariner 2 became the first spacecraft to fly by another planet when it was launched August 27, 1962, to Venus (*Mariner 1* was lost because of a launch vehicle error). Other successful Mariners included: *Mariner 4*, launched in 1964 to Mars; *Mariner 5*, launched in 1967 to Venus; *Mariner 6*, launched in 1969 to Mars; *Mariner 7*, launched in 1969 to Mars; and *Mariner 9*, launched in 1971 to orbit Mars. See also **Mars**; and **Venus**.

Mariner 10 became the first spacecraft to use a "gravity-assist" boost from one planet to send it on to another; a key innovation in space flight that would later enable the exploration of the outer planets that would have otherwise been impossible. *Mariner 10's* launch in November 1973 delivered the spacecraft to Venus in February 1974, where a gravity-assist swing-by allowed it to fly by Mercury in March and September that year. See also **Mercury**.

The first search for life on Mars was conducted in 1975, when NASA launched the *Viking* mission's two orbiter spacecraft and two Martian landers. The elaborate mission was divided between several NASA centers and private U.S. aerospace firms, with JPL building the *Viking* orbiters, conducting mission communications and eventually assuming responsibility for management of the mission. See also **Mars**; and **Viking Mission to Mars**.

Credit for the single mission that has visited the most planets goes to JPL's *Voyager* project. Launched in 1977, the twin *Voyager 1* and *Voyager 2* spacecraft flew by the planets Jupiter (1979) and Saturn (1980–1981). *Voyager 2* then went on to an encounter with the planet Uranus in 1986 and a flyby of Neptune in 1989. Early in 1990, *Voyager 1* turned its camera around to capture a series of images assembled into a "family portrait" of the solar system. Still communicating their findings as they speed out toward interstellar space, the Voyagers are expected to communicate information about the Sun's energy field until perhaps the second decade of the 21st century. In February 1998, *Voyager 1* passed NASA's Pioneer 10 to become the most distant human-made object in space. See also **Voyager Missions to Jupiter and Saturn**.

In 1989 and 1990 NASA's Space Shuttle helped launch three JPL-managed solar system exploration missions: *Magellan* to Venus, *Galileo* to Jupiter and *Ulysses* to study the Sun's poles.

Magellan used a sophisticated imaging radar to pierce the cloud cover enshrouding Venus and map the planet's surface. *Magellan* was carried into Earth orbit in May 1989 by Space Shuttle *Atlantis*. Released from the Shuttle's cargo bay, *Magellan* was propelled by a booster engine toward Venus, where it arrived in August 1990. It completed its third 243-day period mapping the planet in September 1992. *Magellan* mapped variations in Venus's gravity field before the mission ended in October 1994. At the conclusion of the mission, flight controllers commanded *Magellan* to dip into the atmosphere of Venus in a test of aerobraking; a technique for using atmospheric drag to slow spacecraft that has since been used in other planetary missions. See also **Venus**; and ***Magellan* Mission to Venus**.

The *Galileo* mission began October 18, 1989, with a launch from on Space Shuttle *Atlantis* and an Inertial Upper Stage booster. Relying on gravity-assist swing-bys to reach Jupiter, *Galileo* flew past Venus once an Earth twice. Along the way *Galileo* flew by the asteroid Gaspra in October 1991 and the asteroid Ida on August 28, 1993. On its final approach to the giant planet, *Galileo* observed Jupiter being bombarded by fragments of the broken-up comet Shoemaker-Levy 9. On July 12, 1995, *Galileo* separated from its atmospheric probe and the two spacecraft flew in formation to their final destination.

On December 7, 1995, *Galileo* fired its main engine to enter Jupiter orbit and collected data radioed from the probe during its parachute descent into the planet's atmosphere. During its two-year prime mission, *Galileo* conducted 10 targeted flybys of Jupiter's major moons. In December 1997, the spacecraft began an extended mission aimed at further study of Jupiter's moon Europa. The extended mission will culminate in one or two very close flybys of Jupiter's volcanic moon Io in October and November 1999. See also ***Galileo* Mission to Jupiter**; **Jupiter**; and **Venus**.

NASA's shuttle fleet again launched a probe bound for other parts of the solar system when the Space Shuttle *Discovery* carried aloft *Ulysses* in October 1990. A joint mission between NASA and the European Space Agency, this project for the first time sent a spacecraft out of the ecliptic, the plane in which Earth and other planets orbit the Sun, to study the Sun's north and south poles. *Ulysses* first flew by Jupiter in February 1992, where the giant planet's gravity flung it into an unusual solar orbit nearly perpendicular to the ecliptic plane. The prime mission concluded in September 1995. In April 1998, *Ulysses* began its second orbit of the Sun. See also **Jupiter**; and ***Ulysses* Mission**.

The mission of *Mars Observer*, launched aboard a *Titan* III rocket September 25, 1992, ended with disappointment in August 1993 when contact was lost with the spacecraft shortly before it was to enter orbit around Mars. Some science instruments from *Mars Observer* are currently being re-flown on *Mars Global Surveyor*.

The next JPL planetary launches were those of *Mars Global Surveyor* and *Mars Pathfinder*, launched in November and December 1996, respectively. *Mars Pathfinder* put a lander and rover on the surface of the Red Planet in a highly successful landing July 4, 1997; the project fulfilled all the objectives of its prime mission and lasted considerably longer than originally designed before the lander fell silent in September 1997. *Mars Global Surveyor* went into orbit around the red planet on September 12, 1997 (September 11 EDT/PDT), and spent a year and a half lowering its orbit using the technique of aerobraking. The spacecraft began its prime mission in spring 1999 and is currently making highly detailed maps of the Martian surface. See also **Mars**; and ***Pathfinder* Mission to Mars**.

Another disappointment at Mars occurred in late 1999, however, with the loss of the orbiter and lander developed and launched under the Mars' 98 project, named *Mars Climate Orbiter* and *Mars Polar Lander*, respectively. Climate Orbiter entered the planet's atmosphere too low and did not survive orbit insertion on September 23, 1999. *Polar Lander* and two *Deep Space 2* microprobes piggybacking on it to Mars were lost during arrival at the planet December 3, 1999. JPL is working with NASA to reevaluate other missions planned under the Mars program, designed to take advantage of launch opportunities every two years.

JPL designed and built the *Cassini* mission to Saturn, launched on October 15, 1997. *Cassini* is carrying a probe, *Huygens*, provided by the European Space Agency, which will descend to the surface of Titan, Saturn's largest moon, upon arrival at the ringed planet in 2004. Titan appears to boast organic chemistry, possibly like that which led to the existence of life on Earth. *Cassini* flew by Venus in April 1998 and June

1999, followed by an Earth flyby in August 1999. The spacecraft will fly by Jupiter in December 2000 before arriving at Saturn. See also **Saturn**; and *Cassini* **Mission to Saturn**.

In 1995, NASA selected a proposal by a team affiliated with JPL to develop and fly a mission called *Stardust* under the space agency's Discovery program of low-cost missions. Launched in February 1999, *Stardust* will fly within about 100 kilometers (60 miles) of the comet Wild-2 in the year 2004 and collect dust and volatile materials. Those materials will be returned to Earth in a return capsule that will parachute to a landing on a dry lake bed in Utah in 2006. See also *Stardust* **Mission**.

JPL will also provide project management for another Discovery mission, *Genesis*. Launched in 2001, Genesis will collect samples of charged particles in the solar wind and return them to Earth laboratories in 2003 for detailed analysis.

Another major initiative for a new breed of NASA spacecraft is New Millennium, designed to flight-test new technologies so that they can be reliably used in science missions of the 21st century.

The first New Millennium spacecraft, *Deep Space 1*, was launched in October 1998 to test an ion engine and 11 other new technologies. *Deep Space 1* tested autonomous navigation and two advanced science instruments when it flew by the asteroid 1992 KD on July 29, 1999 (July 28 PDT). Under *Deep Space 2*, two microprobes to test the Martian soil for water vapor piggybacked on *Mars Polar Lander*, but were lost at arrival in December 1999. The concept for the third mission, called Space Technology 3, calls for launching three spacecraft to fly in formation to create a space-based interferometer, an optical instrument designed to look for planets around other stars. The New Millennium program also includes *Terra*, a mission managed by NASA's Goddard Space Flight Center that will launch an advanced imager designed to study Earth in December 1999.

JPL is responsible for the Outer Planets/Solar Probe project, which is studying three potential missions to study the outer solar system and the Sun. In early 1998 NASA selected one of these missions, *Europa Orbiter*, for development; the spacecraft will be launched in 2003 to orbit Jupiter's moon Europa, which scientists believe may harbor a vast ocean under its frozen crust. The project also includes the proposed Pluto-Kuiper Express, which would send a spacecraft out to the distant planet Pluto and the Kuiper Belt, believed to be the birthplace of many comets; and Solar Probe, a mission to the Sun.

JPL is also developing a robot rover that will be launched on Japan's Mu Space Engineering Spacecraft (*MUSES-C*) in January 2002 to land on the surface of an asteroid.

In addition to directing spacecraft that visit planets, asteroids and comets, JPL scientists are active in many programs of observations from the ground. The Laboratory created the Near-Earth Asteroid Tracking (NEAT) system, an automated system used at an Air Force observatory in Hawaii to scan the skies for asteroids or comets that could threaten Earth. In June 1999, the project announced that it will expand its operations to also use the 1.2-meter-diameter (48-inch) Oschin telescope on Palomar Mountain, California. In 1998, NASA designated JPL as home of the agency's Near-Earth Objects Office to coordinate observations of Earth-crossing asteroids and comets by various NASA scientists.

Earth Sciences

In the late 1970s, JPL engineers and scientists realized that the sensors they were developing for interplanetary missions could be turned upon Earth itself to better understand of the planet. This has led to a series of highly successful Earth-orbiting missions that have evolved into a major segment of the Laboratory's activities, now sponsored by NASA's Office of Earth Sciences.

In 1978, JPL built an experimental satellite called *Seasat* to test a variety of oceanographic sensors including imaging radar, altimeters, radiometers and scatterometers. Many of the later Earth-orbiting instruments developed at JPL owe their legacy to the *Seasat* mission.

The imaging radar flown on *Seasat* led to a pair of missions flown on the Space Shuttle, 1981's Shuttle Imaging Radar-A (SIR-A) and 1984's Shuttle Imaging Radar-B (SIR-B). These were followed by Spaceborne Imaging Radar-C (SIR-C), an experiment teamed with the German/Italian X-Band Synthetic Aperture Radar and flown on the Space Shuttle twice in 1994. SIR-C/X-SAR's goal was to study a variety of scientific disciplines - geology, hydrology, ecology and oceanography — by comparing the radar images to data collected by teams of people on the ground.

Imaging radar will be re-flown on the Space Shuttle under the *Shuttle Radar Topography Mission* (SRTM) scheduled in January 2000. See also **Space Shuttle**.

Seasat also tested an altimeter that measured sea level heights from space. This concept led to a full-scale satellite mission developed jointly by JPL and the French space agency, *Topex/Poseidon*. The oceanographic satellite, launched August 10, 1992, on an *Ariane 4* rocket from Kourou, French Guiana, has provided scientists with unprecedented insight into global climate and ocean interactions, currents, eddies, and new details about the global ocean seafloor. U.S. and French teams are currently working on *Jason-1*, a follow-on satellite planned for launch in late 2000. See also **Climate**.

Another mission with heritage in Seasat is the JPL-built NASA Scatterometer (NSCAT), an instrument that measures near-surface ocean winds from space. NSCAT was launched in August 1996 on the Advanced Earth Observing Satellite (ADEOS) prepared by Japan's National Space Development Agency (NASDA), and continued operating until the ADEOS satellite failed in early 1997. JPL prepared a rapid replacement, *QuikScat*, which was launched in June 1999 from Vandenberg Air Force Base, California. Also being readied is a next-generation scatterometer, *Seawinds*, was to be launched by Japan in 2000. See also **Scatterometry**.

JPL also designed and built an instrument called the Microwave Limb Sounder that studies the chemistry of Earth's upper atmosphere, relaying important data on topics such as ozone depletion. Early versions flew as payloads on the Space Shuttle, followed by an instrument onboard NASA's Upper Atmosphere Research Satellite (UARS) launched in September 1991. Currently, a new-generation version of the instrument is being developed to fly on a satellite for launch in 2002 under NASA's Earth Observing System (EOS) program. See also **Earth Observing System (EOS)**.

JPL is responsible for several other instruments being flown under the EOS program. They include the Multi-angle Imaging Spectro Radiometer (MISR), launched on NASA's *Terra* satellite in December 1999 to study the role of clouds in global climate; the *Atmospheric Infrared Sounder* (AIRS), due for launch in 2000, which will relay data on temperature and humidity in the atmosphere helping to understand how heat is exchanged between land, air, sea and the atmosphere; and the Tropospheric Emission Spectrometer (TES), planned for launch in 2002, which will help scientists understand the causes of acid rain and track trends in atmospheric chemistry on a global scale.

The Active Cavity Radiometer Irradiance Monitor (ACRIM) is an instrument that measures the Sun's total output of optical energy from ultraviolet to infrared wavelengths, called the total solar irradiance, an important factor in the study of Earth's climate. ACRIM was flown on several shuttle missions and satellites in the 1980s and 1990s. A dedicated satellite called *AcrimSat* was launched in December 1999.

A JPL-teamed mission called the Gravity Recovery and Climate Experiment (GRACE) will launch twin satellites to conduct global high-resolution studies of Earth's gravity field. GRACE is planned for launch in June 2001.

Astrophysics

In addition to studying Earth itself and other bodies within the solar system, JPL has produced missions that have peered deeper into the universe (see also **Universe (The)**).

JPL designed and built the Wide Field/Planetary Camera (WFPC), the main observing instrument on NASA's *Hubble Space Telescope*. See also **Hubble Space Telescope (HST)**. After a flaw was discovered in the space telescope's main mirror, JPL created a second-generation camera, WFPC-2, that compensated for the optical problem - essentially like fitting Hubble with a set of corrective eyeglasses. WFPC-2 was installed by spacewalking astronauts during a shuttle mission in December 1993, allowing *Hubble* to fulfill its promise in producing unprecedented views of the cosmos.

JPL was U.S. manager of the *Infrared Astronomical Satellite* (IRAS), a joint project with the Netherlands and the United Kingdom. Launched in 1983, IRAS was an Earth-orbiting telescope which mapped the sky in infrared wavelengths invisible to the eye. IRAS data have led to a wealth of discoveries about the formation of galaxies, stars and planets, including the first-ever direct evidence of an emerging planetary system around a star besides the Sun; material orbiting Vega, 26 light-years away. Previously unseen phenomena found by IRAS has led to gains in other areas of astronomy and astrophysics ranging from studies of comets to cosmology. See also **Comets**; and **Cosmology**.

In 1996, NASA assigned JPL programmatic responsibility for the space agency's Origins program. The program ties together a variety of proposed instruments and spacecraft missions that will study the formation of galaxies, stars and planets, and search for Earth-like planets around nearby stars. The *Space Interferometry* Mission is being developed for launch in 2005 to search for planets around other stars. Among other missions under study is the *Terrestrial Planet Finder*, being considered for launch in 2011.

Starburst galaxies, vast clouds of molecular gas cradling the sites of newborn stars, were to be the target of the *Wide-field Infrared Explorer* (WIRE). WIRE is a small, cryogenically cooled infrared telescope launched on a *Pegasus XL* vehicle in 1999 as part of NASA's Small Explorer program. The telescope's coolant was lost shortly after launch, effectively ending the mission; some science objectives will be picked up by the SIRTF mission when it launches in 2001. In October 1997, NASA selected another JPL-teamed mission for development under the Small Explorer program, the Galaxy Evolution Explorer, due for launch in 2001.

To support the Origins program from the ground, JPL is involved in planning and designing a system that will link two telescopes at the Keck Observatory in Hawaii. The combined telescopes will function as an interferometer to detect large planets and dust clouds around nearby stars. See also **Origins Program**.

Telecommunications

To provide tracking and communications for planetary spacecraft, JPL designed, built and operates NASA's Deep Space Network (DSN) of antenna stations. DSN communications complexes are located in California's Mojave Desert, in Spain and in Australia. In addition to NASA missions, the DSN regularly performs tracking for international missions sending spacecraft to deep space. DSN stations also conduct experiments using radar to image planets and asteroids, as well as experiments using the technique of very long baseline interferometry (VLBI) to study extremely distant celestial objects. See also **Antenna (Communications)**; and **Deep Space Network (DSN)**.

The DSN is playing a major role in Space Very Long Baseline Interferometry (Space VLBI), a radio astronomy project combining orbiting spacecraft with ground antennas to examine extremely distant objects. An international team is arraying ground antennas with a Japanese spacecraft launched in 1997 to make science observations.

Technologies

In the three decades it has led the nation's planetary exploration program, JPL has honed several skills and areas of innovation, including deep space navigation and communication, digital image processing, imaging systems, intelligent automated systems, instrument technology, microelectronics and more. Many of these disciplines found applications outside the planetary spacecraft field, from solar energy to medical imagery.

In the mid-1970s, in response to a world energy crisis, JPL worked to develop and apply both alternative sources of electricity, such as solar energy, for the Department of Energy, and electric vehicles and other alternative transport systems, for the Department of Transportation.

The Laboratory has also applied space-based operational, communication, and information processing techniques to the needs of the Department of Defense, Federal Aviation Administration and other federal agencies. Its active technology transfer program with the industrial community dates back to the early days of the missile program.

JPL's Technology and Applications Programs Directorate oversees technology development projects both for NASA and for sponsors other than NASA. Non-NASA projects at JPL have included *Firefly*, an aircraft-borne infrared fire-mapping system for the U.S. Forest Service; a document monitoring system to help the National Archives safeguard the U.S. Constitution, Declaration of Independence and Bill of Rights; medical projects such as robot-assisted microsurgery and medical imaging systems;- and Internet-based telemedical systems; and varied projects in such fields as advanced spacecraft and sensor technology, microelectronics, supercomputing and environmental protection.

JPL work for the Department of Defense has included the *Miniature Seeker Technology Integration* (MSTI), a satellite built and launched in November 1992 to demonstrate miniature sensor technology and a rapid development system. JPL also managed the U.S. Army's All Source Analysis System (ASAS) project, a battlefield information management system.

Research and development activities at JPL include an active program of automation and robotics supporting planetary rover missions and NASA's Space Station program. In supercomputing JPL has pioneered work with new types of massively parallel computers to support processing of enormous quantities of data to be returned by space missions in years to come.

Installations

In addition to the Laboratory's main Pasadena site and the three DSN complexes around the world, JPL installations include an astronomical observatory at Table Mountain, California, and a launch operations site at Cape Canaveral, Florida.

In 1999, JPL had a workforce of about 4,900 employees and 710 on-site contractors, and an annual budget of approximately $1.15 billion.

Web References

Jet Propulsion Laboratory. http://www.jpl.nasa.gov/
JPL Fact Sheets. http://www.jpl.nasa.gov/facts/
JPL Picture Archive. http://www.jpl.nasa.gov/pictures/archive.html

JETSET. A fast-setting cement developed by the Portland Cement Association. Reported to harden in 20 minutes after pouring. Accelerating agent has not been disclosed.

Web Reference

Portland Cement Association: http://www.portcement.org/index.asp

JET STREAM. A jet issuing from an orifice into a medium with much lower velocity, such as the stream of combustion products ejected from a reaction engine.

JET STREAMS. A jet stream in the atmosphere is any relatively narrow, flat tube of air moving more rapidly than environing air. The wind speed in the jet stream must be at least 50 knots as a threshold value. A jet stream is similar to a narrow current in a river — moving faster than adjacent waters. Jet streams have been measured up to 400 knots in very narrow (50 miles; 80 kilometers) and flat (2 miles; 3.2 kilometers) streaks.

Jet streams in the atmosphere were postulated prior to the 1940's, but were first encountered by high-flying bombers in World War II when they met headwinds greater than the true air speed of the planes. The ground speed in some encountered jet streams became negative and the aircraft moved rearwards over the earth.

Two polar jet streams are recognized, one each in the northern and the Southern hemisphere. They are at an altitude between 25,000 and 35,000 feet (7,620 and 10,668 meters); they meander between 30° and 70° latitude; they have maximum speeds in excess of 200 knots.

Two subtropical jet streams are recognized, one in each hemisphere. They are at an altitude between 30,000 and 45,000 feet (9,144 and 13,715 meters) and most frequently occur from 35,000 to 40,000 feet (10,668 to 12,192 meters). They meander between 20° and 50° latitudes; they have maximum speeds in excess of 300 knots. See Figs. 1 and 2.

Other, shorter and temporary jet streams have been observed. Some of these flow from east to west. Their maximum speeds are less than those in the polar or subtropical jet streams.

Jet streams usually are located just below stable layers of air and particularly just below the tropopause. In addition, there is very frequently an abnormally steep slope downward in the stable layer on the left side of the jet stream looking downwind. In the Southern hemisphere, the slope is to the right of the jet stream looking downwind. The two principal jet streams are associated with the tropopause as shown in Fig. 3. See also **Winds and Air Movement**.

The major jet streams in both hemispheres are integral parts of the global general circulation. They make up the main stream of the circumpolar whirl about both the North and South Poles. Middle-lattitude-traveling storms are related to the fluctuations of wind speeds in the jet stream core and with the north-south meandering of the jet streams. Much of the clear air turbulence encountered by high-flying aircraft is caused by, or associated with the jet streams.

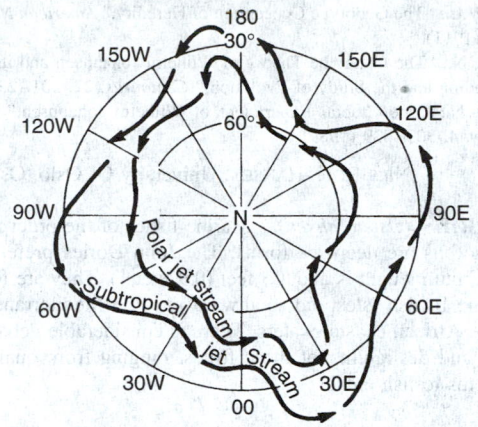

Fig. 1. Principal jet streams in the Northern hemisphere.

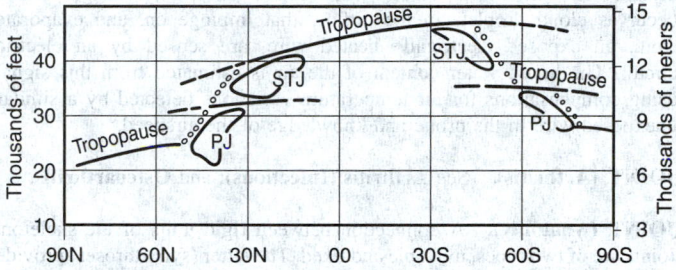

Fig. 2. Polar jet stream (PJ) and subtropical jet streams (STJ). Cross section looking eastward in northern hemisphere (winter).

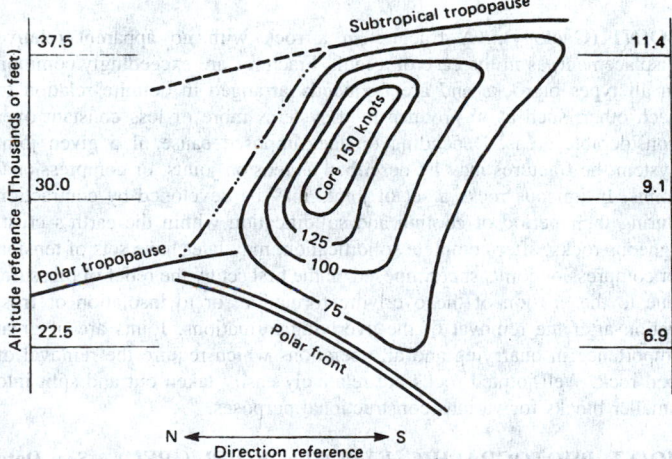

Fig. 3. Cross section of polar jet stream looking east. Subtropical tropopause bounds it on the top, the tropopause leaf on the left, and the polar front on the bottom. Note the large velocity gradient on the left side.

Both major jet streams weaken in summer in their respective hemispheres. They also travel poleward in summer about half as much latitudinal distance as the sun travels.

Additional Reading

Aguado, E. and J.E. Burt: *Understanding Weather and Climate*, 2nd Edition, Prentice-Hall, Inc., Upper Saddle River, NJ, 2000.

Dunlop, S.: *A Dictionary of Weather*, Oxford University Press, Inc., New York, NY, 2001.

Kjelgaard, M.J.: *Engineering Weather Data*, The McGraw-Hill Companies, Inc., New York, NY, 2001.

Reynolds, R.: *The Cambridge Guide to the Weather*, Cambridge University Press, New York, NY, 2000.

PETER E. KRAGHT, Certified Consulting Meteorologist, Mabank, TX

JET THRUST. The thrust of a fluid, especially as distinguished from the thrust of a propeller. The thrust of a rocket engine is calculated in the same manner as gross thrust of a jet engine.

JET VANE. A vane, either fixed or movable, used in a jetstream, especially in the jetstream of a rocket, for purposes of stability or control under conditions where external aerodynamic controls are ineffective. Also called *blast vane*.

JEVONS EFFECT. The effect of the presence of the rain gauge on the rainfall measurement. In 1861, W. S. Jevons pointed out that the rain gauge causes a disturbance in airflow past it, which carries past the gauge part of the rain that would normally be captured. The effect is a function of the wind speed and the height of the gauge from the ground. Rain-gauge shields have been devised to minimize this loss.

JH. (methyl-*cis*-10,11-epoxy-7-ethyl-3,11-dimethyl-*trans*,*trans*-2,6-tri decadienoate). A synthetic hormone containing a 13-carbon chain; said to have possibilities as an insecticide. It acts by preventing insects from maturing. Its future depends on the possibility of large-scale production.
  See also **Juvenile Hormones**.

JIMSPHERE. A spherical balloon made of metallized polyester film. It uses a valve to produce a fixed overpressure during ascent. The balloon contains surface protuberances to maintain turbulent flow throughout the ascent. It is tracked by precision radar to obtain accurate wind profile data.

JOHANNSEN, WILHELM LUDWIG (1857–1927). Wilhelm Johannsen was a Danish biologist, theoretician and experimentalist, who clarified the basis of classical genetics and its relation to evolution by introducing the concepts of gene, genotype and phenotype.

By birth Wilhelm Johannsen belonged to the class of civil servants, the cultured higher bourgeoisie that ran the state and governed society in Protestant continental Europe in the nineteenth century. His father was a military officer whose limited means did not permit a regular academic education. After an early general education at the best schools in Denmark Johannsen trained as a pharmacist apprentice in Germany and Denmark, steadily cultivating an interest in nature and natural science that had first been stimulated by his mother. This was a period of educational reforms throwing off the yoke of traditional academic formalism. Natural science was still in the early phase of specialization and professionalization, and the strengths of enthusiastic autodidacts with practical experience were appreciated.

A final year of study at the University of Copenhagen concluded Johannsen's preparations for the pharmacist examination. During this period he developed close ties with the professor of Botany, Eugenius Warming, famous as a founder of ecology. In 1881 he became an assistant at the Carlsberg Laboratory, the research establishment of the Carlsberg Brewery. Among his senior colleagues was Johan Kjeldahl, internationally famous for his contributions to analytic chemistry, and Emil Chr. Hansen known for growing "pure lines" of yeast. Supported by Danish stipends, Johannsen visited several German universities to study with leading plant physiologists of the day. In 1892 he was appointed lecturer and later professor of plant physiology at the Royal Agricultural College (Den Kongelige Veterinær- og Landbohøjskole), and in 1905 he was called to the University of Copenhagen as professor of plant physiology.

An early discovery by Johannsen was that dormancy in plants can be broken with ether or chloroform. This discovery was used extensively in flower business in Denmark and other countries. At Carlsberg, Johannsen had started studying variability and selection in barley. In a small 1896 book on *Arvelighed* (heredity) he deplored that the popularity of Darwinism had led to neglect of pedigree selection and other pre-Darwinian methods. Heredity had to be studied independently of evolution, he argued. While an understanding of the evolution of species was essentially dependent upon theories of heredity, the converse was not the case. Johannsen favored August Weismann's idea of a stable germplasm transmitted from parents to offspring, and Francis Galton's parallel idea of a continuous stirp[1] running through the generations. But he rejected Weismann's cytological ideas about the nature of the hereditary material

[1] A root or roostock running through the generations and form which each individual organism grows

as loose speculations, and he was critical of natural selection as the all-powerful force molding evolution. Johannsen insisted that natural selection does not create hereditary change but only selects among variations originating from other mechanisms.

In 1901, Johannsen started his classic selection experiment on beans. He deliberately chose a self-fertilizing plant knowing that the descendants of one individual, which he defined as a pure line, will in general be highly homogeneous in heredity. Johannsen followed up the work of Hugo de Vries who had found that often in a stock of plants there is a characteristic limit to change for each property, and no amount of selection can force the property beyond this limit. This contradicted the orthodox Darwinian idea of ubiquitous variation as the basis for evolutionary change. More specifically the aim of Johannsen's bean experiment was to investigate to what extent the so-called "individual" (or "fluctuating") variations were inherited. After only two growing seasons he concluded, somewhat to his own surprise, that there was no effect of selection on weight or form of the seeds within a pure line of beans. When the Biometricians apparently found such effects it was because they worked with populations that contained from the start a mixture of hereditary types. The observed effect was due to selection between types and not due to hereditary change within types.

Johannsen published his bold generalization in a small monograph, *Ueber Erblichkeit in Populationen und in Reinen Linien* (On heredity in populations and pure lines). The leading biometricians W.F.R. Weldon and Karl Pearson reacted with ridicule. But within a few years Johannsen's theory of "hard" heredity for quantitative as well as qualitative traits was confirmed in a wide variety of plants and animals. In the classical textbook, *Elemente der exakten Erblichkeitslehre* (elements of an exact theory of heredity), of 1909 Johannsen introduced the new terms "gene," "genotype" and "phenotype" to describe his theory of heredity.

Early twentieth century plant and animal breeding was largely shaped by Johannsen's theory of pure lines. He created the paradigm experiment as well as the concepts that quickly made hard heredity an accepted fact. His *Elemente* became a highly influential textbook on the European continent, but like most of his publications it was never translated into English. Nevertheless, Johannsen had a deep influence on American genetics in the formative period around 1910. This was clearly demonstrated during the winter of 1911–1912 when he traveled the Eastern and central United States for about 5 months giving a series of public lectures, scientific talks and participating in scientific meetings. Johannsen's genotype theory was a main input to the discussions that convinced American geneticists of the hardness of biological heredity. In animals as in plants the hereditary factors (genes) are generally stable and change only infrequently and stepwise (mutations). On this foundation T.H. Morgan and his students Herman Muller, Calvin Bridges and Alfred Sturtevant a few years later erected the epoch-making chromosome theory of heredity in their classic monograph *The Mechanism of Mendelian Heredity*. See also **Morgan, Thomas Hunt (1866–1945)**; and **Muller, Hermann Joseph (1890–1967)**.

The small scientific milieu in Copenhagen around 1900 was centered on two national and interdisciplinary institutions, the university and the academy of sciences. This promoted a broad interdisciplinary scientific and scholarly culture. Johannsen's work on heredity drew heavily on ecology, systematics and taxonomy. He also found inspiration in philosophy and history of science. In explaining his concept of genotype Johannsen refers to the Aristotelian concept of form, and his revealing critique of *False Analogies with Respect to Similarity, Kinship, Heredity, Tradition and Evolution* (Falske Analogier med Henblik paa Likhed, Arv, Tradition og Udvikling, 1914) was inspired by Francis Bacon's classical analysis of human illusions.

Johannsen was a typical representative of the Nordic enlightenment of the nineteenth and early twentieth century spanning science and scholarship as well as literature and art. Darwinism developed on the background of a progressive modern, naturalistic and biological view of man evident in the plays of Henrik Ibsen and August Strindberg as well as the paintings of Edward Munch. Biological heredity was a popular theme and the geneticist Johannsen was a prominent public figure taking active part in cultural and political debates, for instance by striving to inject some clear thinking and political restraint into the debates over eugenics.

Additional Reading

Churchill, F.B.: "William Johannsen and the Genotype Concept," *Journal of the History of Biology*, 7, 5–30 (1974).

Dunn, L.C.: "Johannsen, Wilhelm Ludwig," *Dictionary of Scientific Biographies*, 7, 113–115 (1973).

Johannsen, W.L.: "The Genotype Conception of Heredity," *American Naturalist*, 45, 129–159 (1911).

Roll-Hansen, N.: "The Genotype Theory of Wilhelm Johannsen and its Relation to Plant Breeding and the Study of Evolution," *Centaurus*, 22, 201–235 (1978).

Roll-Hansen, N.: "The Crucial Experiment of Wilhelm Johannsen," *Biology and Philosophy*, 4, 303–329 (1989).

NILS ROLL-HANSEN, University Of Oslo, Oslo, Norway

JOHN DORIES (*Osteichthyes*). Marine fishes of the order *Zeomorphi*. Most zeomorphs are deep-sea forms. The John Dories prefer mid-depth water. Maximum length is about 3 feet (0.9 meter). They are found in the waters of the British Isles and southward into the Mediterranean as well as along the African coastal waters. There is considerable debate as to the food value and desirability of these fishes, ranging from qualification as gourmet items to fish meal.

JOHNSON- WILLIAMS LIQUID WATER PROBE. A hot-wire type instrument for measuring the liquid water content of clouds *in situ*. The probe is most often used on research aircraft, but also occasionally at mountain-top installations and in wind tunnels. Resistivity changes, which occur as cloud droplets in the airflow that impinge on, and evaporate from, an exposed electrically heated wire, are sensed by an electric circuit. The liquid water content of the air is estimated from this signal using compensations for air temperature variations detected by a similar unexposed wire in the probe and knowledge of the airspeed.

JOINT (Arthritis). See **Arthritis (Infectious)**; and **Osteoarthritis**.

JOINT (Anatomy). A connection between rigid units of the skeleton. Joints are of two types, movable and fixed. The latter (synarthroses) provide for the formation of firm skeletal structures where more than one rigid part is involved, and movable joints (diarthroses) enable the separate divisions of the body and jointed appendages to act together as systems of levers, moved by muscles. See also **Bone**; and **Skeletal System**.

JOINT (Geology). A fracture in a rock with no apparent relative displacement, as in the case of a fault. Fractures are exceedingly common in all types of rocks and are frequently arranged in definite relation to each other, such as to produce joint systems more or less constant over considerable areas. Depending on the supposed cause of a given joint system the fractures may be described as tension joints, or compressional joints. In igneous rocks a set of joints may be developed by contraction during their period of cooling and solidification within the earth's crust. Igneous rocks, after complete solidification, may later have sets of tension or compression joints superimposed on the first set by the relief of pressure, due to the erosion of the overlying formation, or to insolation or frost action after the removal of the overlying formations. Joints are of great importance in quarrying and all operations which require the removal of bed rock. Well-jointed rocks are relatively easily taken out and split into smaller blocks for various constructional purposes.

JOINT PHOTOGRAPHIC EXPERT GROUP (JPEG). See **Data Compression**.

JOINTWORM (*Insecta, Hymenoptera, Harmolita tritici Fitch*). Small insects whose larvae work in the stems of grains and grasses, sometimes causing swelling at the joints or the formation of growths. Their work may weaken the stem and cause it to break. On the positive side, however, these insects help to rid plants, especially wheat, of noxious caterpillars. The insect is generally distributed in the United States east of the Mississippi River. Most damage from the insect has occurred in the state of Missouri. The adult, jet-black insects are from $\frac{1}{10}$ to $\frac{1}{8}$ inch (2.5 to 3.5 millimeters) in length. In some wheat-producing areas, this pest is regarded second to the Hessian fly in severity.

Burning of stubble will kill most of the overwintering jointworms. Plowing the stubble under shortly after harvest is considered a good control measure. Infestations usually occur because of failure to heed these control measures. Rotation of crops is also effective in keeping the jointworm population within control. The jointworms are usually found inside the straw at a height of 8 to 10 inches (20 to 25 centimeters) above ground level.

JOLIOT-CURIE, FREDERICK (1900–1958). A French physicist who, along with his wife Irene Joliot-Curie, won the Nobel Prize in chemistry in 1935. His important discoveries included artificial radioactivity. He did much work on atom structure, dematerialization of electrons, and inverse transformation. Work on hormone synthesis and thyroid substances containing radioactively labeled elements was significant. Sc.D. from the University of Paris was followed by a distinguished career filled with honors and appointments.

JOLIOT-CURIE, IRENE (1897–1956). A French nuclear scientist who won the Nobel Prize for chemistry with her husband Frederick Joliet-Curie. Their joint work involved production of artificial radioactive elements by using α-rays to bombard boron. They discovered that hydrogen-containing material when exposed to what they considered γ-rays would emit protons. They were involved in many firsts: they gave the first chemical proof of artificial transmutation and of capture of alpha particles, and were the first to prepare positron emitter. Her career started with a Sc.D. at the University of Paris, and included scores of honors and awards.

JONES OXIDATION. The oxidation of primary and secondary alcohols to acids and ketones by the addition of the calculate amount of chromic anhydride in dilute sulfuric acid to a solution of the alcohol in acetone. This procedure does not attack triple bonds or shift double bonds into conjugation with the ketone formed in the oxidation.

JORDAN RIFT. See **Ocean**.

JORDAN SUNSHINE RECORDER. A sunshine recorder of the type in which the timescale is supplied by the motion of the sun. It consists of two opaque metal semicylinders mounted with their curved surfaces facing each other. Each of the semicylinders has a short narrow slit in its flat side. Sunlight entering one of the slits falls on light-sensitive paper (blueprint paper) that lines the curved side of the semicylinder. One semicylinder covers morning hours, the other afternoon hours. The sensitivity of the recording paper is variable, and this introduces an uncertainty in the evaluation of the record.

JOSEPHSON, BRIAN DAVID (1940–). Josephson was born in Wales and was recognized even in his undergraduate studies at Cambridge University as having great natural insight on scientific ideas. In 1973, he won the Nobel Prize in Physics jointly with Leo Esaki and Ivar Giaever for his work in experimental superconductivity. He is remembered for the "Josephson effect" also explained as the tunneling phenomena, which is an important idea for modern physics. His discoveries helped advance technology, especially in the development of high-speed switching circuits used in instrumentation and computers.

Josephson was greatly influenced by transcentdental meditation techniques of the 1960s and 1970s and believed that mystical experience should be incorporated into science. His latter research also included work in the theory of intelligence.

See also **Josephson Junctions**; **Superconductivity**; and **Thin Films**.

J. M. I.

JOSEPHSON TUNNEL-JUNCTION. As early as 1962, B. Josephson recognized the implications of the complex order parameter for the dynamics of the superconductor, in particular when one considers a system consisting of two bulk superconductors connected by a "weak link." The basic requirement for the weak link is that the amplitude of the order parameter at the link should be substantially smaller than in the bulk regions. In early experimentation, such a situation was realized in a variety of ways—two evaporated films separated by a thin (less than 20 angstroms) oxide layer; a light point contact between two bulk superconductors; a single hourglass-shaped evaporated film, with the constriction of dimensions small compared to the coherence length; or even a bare niobium wire with a pendant frozen blob of soft solder, where the weak links, indeterminate in number, are formed by solder bridges through pinholes in the surface oxide. Collectively, all such weak link junctions are referred to as *Josephson junctions*.

Both the dc and ac Josephson effects have found interesting and novel applications. The high sensitivity to magnetic field of the dc Josephson current in certain circuit configurations has been used to develop a family of devices called squids (super conducting $q$uantum interferometric $d$evices) which can be used to measure extremely small currents, voltages, and magnetic fields. The ac effect has been useful in making precise measurements in connection with research toward improving the maintenance of the U.S. legal volt.

Much interest is presently exhibited in the use of Josephson superconducting devices of the tunnel-junction type (operated at near absolute zero) in connection with building superfast computers. In such devices, the electrical signals have only a millimeter or two to travel. Switching time is about 10^{-11} second. The power dissipation of Josephson devices permits high circuit density. It is estimated that the power dissipation of transistors is about 100 times that of Josephson devices. Different materials are being studied to improve the original lead alloy thin-film materials. These include lead–indium–gold alloys and niobium-tin alloys.

Additional Reading

Kircher, C.J. and J. Murakami: "Josephson Tunnel-Junction Electrode Materials," *Science*, **208**, 944–950 (1980).
Likharev, K.K.: *Dynamics of Josephson Junctions and Circuits*, Taylor & Francis, Inc., Philadelphia, PA, 1986.

JOSHUA TREE. See **Palm Trees**.

JOULE CYCLE. See **Gas and Expansion Turbines**.

JOULE, JAMES PRESCOTT (1818–1889). Joule was an English physicist. He was born in a family of brewers and as he grew up he became knowledgeable about steam engines used in the brewery. Also, the town he grew up in, Manchester, England had factories with machinery that depended on the steam engine and so it was almost natural that with his curiosity of science, Joule became interested in mechanical energies. By 1837, Joule published his first scientific paper on electric motors.

He is remembered for Joule's Law that describes the rate at which heat is produced by an electric current. Joule's work showed there were different kinds of energy, which can be changed into each other. He established the mechanical equivalence of heat. His work led to the law of conservation of energy. Also, he collaborated with William Thomson (Lord Kelvin) and verified experimentally the Joule-Thomson refrigeration effect.

See also **Gas and Expansion Turbines**; **Joule Law**; **Joule-Thomson Effect**; and **Oxygen**.

JOULE LAW. The quantity of heat generated by a steady electric current is proportional to the resistance of the conductor in which the heat is generated, to the square of the current, and to the time of its duration: $H = KRI^2 t$. If the resistance is in ohms, the current in amperes, the time in seconds, and the heat in calories, the constant K has the value 0.2390 calories/joule.

JOULE-THOMSON EFFECT. In passing a gas at high pressure through a porous plug or small aperture, a difference of temperature between the compressed and released gas usually occurs. This phenomenon is called the Joule-Thomson effect. The equation for this effect contains two partial derivatives and is

$$\left(\frac{\partial T}{\partial P}\right)_H = \frac{T\left(\frac{\partial V}{\partial T}\right)_P - V}{C_P}$$

where the expression on the left is the rate of change of temperature with pressure at constant enthalpy (heat content) (since no heat is supplied to, or removed from the system). The expression on the right has in its numerator the difference between the product of the temperature and the rate of change of volume with temperature at constant pressure, from which the volume is subtracted; the denominator contains the molar specific heat at constant pressure. The term on the left of the equality sign is called the *Joule-Thomson Coefficient*. It varies with the temperature and pressure of the gas, passing from positive values through zero to negative values. The temperature at which it is zero is called the *Joule-Thomson Inversion Temperature* and varies with the particular gas. It is to be noted, however, that for hydrogen, and also for helium, the temperature is low, far below $0°C$. For other gases, however, much higher values are found, the maximum value for oxygen being 1,058 K (785 °C). At temperatures above their inversion temperature, gases are warmed, while at temperatures below it, they are cooled by the effect. For that reason, this type of expansion is often used in industrial processes for cooling gases. An interesting application is the application is the process for producing solid carbon dioxide by expanding carbon dioxide through an aperture.

For an ideal gas, $PV = RT$; thus, $(\partial V/\partial T)$ is equal to R/P, that is, to V/T, so that the numerator of the right-hand term in the equation above becomes 0; thus an ideal gas shows no Joule-Thomson effect.

JOVIAN.   An adjective applying to Jupiter—as Martian to Mars, lunar to the earth's moon, etc. See also **Jupiter**.

J PARTICLE.   See **Psi Particle**.

JPEG (JOINT PHOTOGRAPHIC EXPERT GROUP)..   See **Data Compression**.

JUMP.   In computer programming, to cause the next instruction to be selected conditionally or unconditionally from a specified storage location.

JUNCO *(Aves, Passeriformes; Junco)*.   Small North American birds of several species, mostly colored in shades of gray and white. The *Junco hiemalis* is slate gray above, on the throat and breast, with belly and outer tail feathers white. Juncos breed chiefly in colder regions and in higher altitudes, migrating to the valleys and to more southern latitudes, although they thrive even in cold winters.

JUNE BEETLE *(Insecta, Coleoptera)*.   There are two kinds of beetle, both members of the family *Scarabaeidae*, which are called June beetle: (1) Several brownish-black types (*Phyllophaga* or *Lachonesterna* spp.), which are also called May beetle, "dae bug," or simply June bug in some areas; and (2) a green beetle which is correctly designated as the green June beetle (*Cotinis nitida*, Linne).

Brownish-Black June Beetle. The larvae of this beetle are severely destructive soil insects and they are commonly referred to as *white grubs*. They are particularly injurious to young corn (maize). Poorly developing plants will be found to be practically rootless as the result of the eating done by the white grubs. Inspection has shown that as many as 200 such grubs may feed from the root structure of one corn plant. The grubs range from $\frac{1}{2}$ to 1 inch (13 to 25 millimeters) in length. They are white and have six legs and a brown head. Although the white grubs are notably damaging to corn plants, they also attack bean, potato, strawberry, nursery stock, and almost any other cultivated crop if their food favorites are not nearby. These insects winter both as larvae and adults. The adult brownish-black June beetle flies mostly at night and appears from mid-to-late spring, depending upon locale. The adult is a very noisy, buzzing, clumsy insect, with a habit of banging against doors, screens, and windows. When the adult is noticed, it can be assumed that numerous larvae are nearby consuming the roots of whatever crops are available.

The complete life cycle of this insect is comparatively long, ranging from 3 to over 5 years, depending upon species. Consequently, severe infestations tend to occur at 3-year intervals. Control is essentially preventive. Any field in which white grubs are found should not be planted to the favorite targets of the insect, notably corn and potato. Because the adult female prefers to lay her eggs in grassy and weedy plots, such areas, particularly near prospective corn and potato plantings, should be cleaned up whenever practical. Turning of hogs into field during the summer also rids the soil of a high percentage of the white grubs. Plowing of fields during mid-to-late summer also helps to destroy the pupating insects. This action also brings the insects to the surface where they are prey for birds, notably crows and blackbirds, as well as for certain kinds of wasps.

In planning all control actions, the food producer should be aware of the 3-year cycle of events in this insect's life. Major infestations have been reliably noted. Soil fumigants and insecticides useful against the grubs include chlordane and lead arsenate.

Green June Beetle. The adults of this insect can be quite damaging to vegetable crops, as well as lawns and ripening fruit. Unlike the brownish-black June beetle, which occurs throughout North America, the green June beetle is found in the southern portions of the United States, to as far north as the Hudson River in the east and the Ohio River in the Midwest. These beetles do not follow the complex cycle of the brownish-black June beetle, but have one generation per year. An effective measure against the grub of this beetle includes heptachlor or toxaphene, where such chemicals can be used safely. The adults are much more difficult to control. Trapping of the adults in peach orchards has been practiced in some areas. This involves baiting trapping pails with a mixture of fermented malt extract and molasses and geraniol. Caproic acid also can be used.

The foregoing beetles are not to be confused with a June bug, also known as the wingless May beetle, which attacks cotton.

JUNE BERRY.   See **Rose Family**.

JUNGLE FOWL *(Aves, Galliformes)*.   Wild game birds of India and other parts of the Oriental region. One species, the red jungle fowl, is ancestral to the domestic gamecock and resembles it in color. The jungle fowls are believed to be remotely ancestral to all domestic fowls. See also **Galliformes**.

A special position among the pheasants is held by the jungle fowl. Their faces are almost bare, possessing fleshy excrescences, such as combs, wattles, and ear lobes, all of which are strongly formed. The tail has fourteen or sixteen feathers and a roof-like slope.

The red jungle fowl have been domesticated in their homeland since antiquity. As early as the fourteenth and fifteenth centuries B.C., domestic fowl were exported from India to China. The ancient Egyptians, too, kept domestic fowl and bred them, as did the Chinese. Soon thereafter, the domestic fowl reached southern Europe from Egypt. The Old Testament does not mention domestic fowl, yet they were present in Greece by the fifth and fourth centuries B.C. Sacred fowl enjoyed special honor among the ancient Romans. They also reached the Germanic and Celtic tribes centuries before the Christian era. They were brought to America, about 470 years ago, with the European conquests.

When domestic fowl were first bred in Europe, there already must have been three groups:

1. The common chickens, widespread in Europe, resembling the red jungle fowl and having white ear lobes and white-shelled eggs.
2. The Asiatic cochins, having soft, thick plumage, red ear lobes, and yellowish-brown–to–brown eggs.
3. The gamecocks, which have erect, muscular bodies, red ear lobes, and brownish-yellowish eggs.

The aforementioned three types constitute the majority of present European domestic fowl. Domestic fowl in North America are described in article on **Poultry**.

One jungle fowl, the cheer pheasant (*Catreus wallichi*), is shown in Fig. 1.

Fig. 1.   Cheer pheasant (*Catreus wallichi*).

TABLE 1. RECORD JUNIPER TREES IN THE UNITED STATES[1]

Specimen	Circumference[2]		Height		Spread		Location
	Inches	Centimeters	Feet	Meters	Feet	Meters	
Alligator juniper (1995) (*Juniperus deppeana*)	328	833	46	14	49	14.9	Arizona
Alligator juniper (1998) (*Juniperus deppeana*)	308	782	57	17.4	76	23.2	Arkansas
Ashe juniper (1999) (*Juniperus ashei*)	138	351	57	17.4	48	14.6	Texas
Common juniper (1993) (*Juniperus communis*)	37	94	46	14	28	8.53	Michigan
Rocky Mountain Juniper (1989) (*Juniperus scopulorum*)	247	627	40	12.2	21	6.4	Utah
Utah juniper (1996) (*Juniperus osteosperma*)	268	681	38	11.6	46	14	California
Western juniper (1983) (*Juniperus acidentalis*)	480	1220	86	26.2	58	17.7	California

[1] From the "National Register of Big Trees," American Forests (by permission).
[2] At 4.5 feet (1.4 meters)

JUNIPER TREES. Of the family *Cupressaceae*, genus *Juniperus*, there are numerous species, subspecies, and varieties of these evergreen trees or shrubs. In many ways, the junipers run counter to other conifers. Junipers grow slowly, preferring lots of sunshine and little shelter. They do well in dry, mineral soil. Instead of woody cones, the junipers have berries. Junipers do not have both male and female flowers on the same tree as do most conifers. These qualities tend to detract from the attraction of junipers for forest projects, but they do offer advantages for a number of garden situations. The common juniper occurs naturally in North America, Europe, and Asia. The berries of this species are used to flavor gin. A common juniper is difficult to find in a nursery, yet numerous cultivated varieties are widely available. Only some of the many species and varieties of junipers can be listed. Important species not included in Table 1 include:

Cherrystone juniper	*J. monosperma*
Chinese juniper	*J. chinensis*
Himalayan juniper	*J. squamata*
Hollywood juniper	See *Chinese juniper*
Pencil juniper	*J. virginiana*
Phoenician juniper	*J. phoenicea*
Prickly juniper	*J. oxyedrus*
Syrian juniper	*J. drupacea*
Temple juniper	*J. rigida*

The common juniper prefers dry slopes and rocky pastures and in North America ranges from Newfoundland south and west into Pennsylvania and in the mountains of North Carolina and New Mexico. The alligator juniper is a small conical tree with waxy blue leaves. The tree flourishes in Mexico and is normally from 10 to 35 feet (10.5 meters) in height. The cherrystone juniper is found in the southern United States and can attain a height of from 30 to 50 feet (9 to 15 meters). The leaves are of a gray-blue coloration.

The Chinese juniper is native to Himalaya, China, and Japan. A normal height is 75 feet (22.5 meters) when fully grown. It can adapt to almost any soil. The foliage is dark green. There are numerous variants suitable for garden planting, including the "Hetzii," "Hollywood," "Kaizuka," "Keteleeri," "Pfitzeriana," "Pyramidalis," "San Jose," "Variegata," and so on.

The *J. virginiana* is commonly referred to as the eastern red cedar and is described in the entry on **Cedar Trees.** It also has numerous variants for garden planting, including the "Canaertii," "Glauca," "Pendula," "Pseudocupressus," and "Skyrocket." The *J. virginiana* is also sometimes referred to as the pencil juniper.

JUPITER. Fifth planet from the Sun, known primarily for the banded appearance of its upper atmosphere and it centuries-old Great Red Spot, a massive, hurricane-like storm as big as three Earths. Jupiter is by far the most massive planet in the solar system, being nearly 318 times as massive as earth, and well over three times as massive as the second most massive planet, Saturn. The volume of Jupiter is over 1400 times that of Earth, but only $\frac{1}{4}$ as dense, since it is composed primarily of hydrogen (89 percent) and helium (10 percent). Jupiter generates the biggest and most powerful planetary magnetic field, and it radiates more heat from internal sources than it receives for the Sun. The equatorial diameter is 142.740 ± 508 kilometers (88.500 ± 315 miles). Other statistics are given in the entry on **Planets and the Solar System.**

Missions to Jupiter

The environs of Jupiter have been explored, by five United States spacecraft and one European-American sponsored spacecraft, with one more spacecraft scheduled to reach its environs in 2007:

1. *Pioneer 10*, in 1973, passed the equator of the planet within a distance of about 80,000 mi (126,720 km).
2. *Pioneer 11*, in 1974, came within about 30,000 mi (48,270 km) of the planet.
3. *Voyager 1*, in 1979, viewed the planet, at its closest approach, from a distance of about 216,790 mi (348,890 km).
4. *Voyager 2*, several months later in 1979, visited the planet at a somewhat greater distance.
5. *Ulysses* flyby of Jupiter, passing through the planet's magnetosphere during the time frame 2 to 16 February 1992.
6. *Galileo*, designed specifically to explore Jupiter, entered into orbit around Jupiter on December 7, 1995 and successfully completed its two-year primary mission. That was followed by a two-year extended mission called *Galileo Europa*, and successfully completed its two-year mission on December 1999. *Galileo* is now continuing its studies under yet another extension, called the *Galileo Millennium* mission.
7. *Europa Orbiter*, designed specifically to study Jupiter's fourth largest moon, Europa. Following launch in 2003 the *Europa Orbiter* is scheduled to arrive at Jupiter's system in 2007 and enter orbit around Europa in 2009.

With the exception of a few land-based studies and observations from the Hubble space telescope, the majority of the early scientific information pertaining to Jupiter is the large body of data assembled by the early *Pioneer* probes and by the *Voyager I* and *Voyager II* missions. Some important details were added by the *Ulysses* flyby in 1992. The knowledge base of Jupiter expanded when *Galileo* began orbiting Jupiter and its moons. See also *Voyager* **Missions to Jupiter and Saturn.**

Galileo **Mission.** Named for the Italian astronomer Galileo Galilei, who discovered four of Jupiter's moons in 1610, the *Galileo* spacecraft was conceived, designed, and built during the 1980s. The mission was designed to increase, by a wide margin, the understanding of Jupiter.

Reaching its ultimate destination after an already eventful space journey of more than six years and 2.3 billion miles, NASA's Galileo mission arrived at the giant planet Jupiter on Dec. 7, 1995. The Galileo mission should uncover new clues about how the Sun and the planets formed, and about how they continue to interact and evolve.

Galileo's scientific instruments represent the most capable payload of experiments ever sent to another planet. The data returned promises to revolutionize our understanding of the Jovian system and reveal important clues about the formation and evolution of the solar system.

The 2,223-kilogram (2-1/2-ton) *Galileo orbiter* spacecraft carries 10 scientific instruments; the 339-kilogram (746-pound) probe carries six more instruments. The spacecraft radio link to Earth and the probe-to-orbiter radio link serve as instruments for additional scientific investigations.

NASA's Jet Propulsion Laboratory, Pasadena, CA, built the *Galileo orbiter* spacecraft and manages the overall mission. *Galileo's* atmospheric probe is managed by NASA's Ames Research Center, Mountain View, CA. See also **Deep Space Network**; and *Galileo* **Mission to Jupiter**.

Ulysses Flyby of Jupiter. This spacecraft was launched from the space shuttle *Discovery* on October 6, 1990. The mission is sponsored by a joint venture of the National Aeronautics and Space Administration (NASA) and the European Space Agency (ESA). The primary mission of *Ulysses* is that of investigating the sun and solar wind, the heliosphere, interstellar matter, and signals from the galaxy. The mission designers required the gravity assistance of another planet to achieve *Ulysses's* ultimate trajectory. Jupiter is the nearest celestial body capable of meeting these requirements. The use of the gravity of Jupiter to carry out the primary mission made it possible for the spacecraft to fly through the Jovian magnetosphere. The closest approach to the planet occurred on February 8, 1992.

A summary of *Ulysses's* findings is given at the end of this article. Unfortunately, no dramatic photo images of the planet were obtained.

Hubble Space Telescope Imaging of Jupiter. As reported by Caldwell, Turgeon, and Hua (Institute for Space and Terrestrial Science, Canada), "The first direct images of the Jovian aurora at ultraviolet wavelengths were obtained by the Hubble Space Telescope Faint Object Camera near the time of the Ulysses spacecraft encounter with Jupiter. The auroral oval is not uniformly luminous. It exhibits a brightness maximum in the vicinity of longitude 180°. In the few images available, the brightest part of the oval occurs in late afternoon, Jovian time. The observed oval is not concentric with calculated ovals in the O6 model of Connerney. The size of the oval is consistent with auroral particles on field lines with magnetic L parameter greater than 8, indicating significant migration from the satellite Io, its torus, or both, if these are their origins." See also **Hubble Space Telescope (HST)**.

Highlights of *Voyagers'* Scientific Findings. After digesting the mass of data returned by the two *voyager spacecraft*, a study task that is far from complete, scientists at NASA (National Aeronautics and Space Administration), and Jet Propulsion Laboratory (JPL), selecting the following highlights:

1. Features of broadly different sizes in Jupiter's atmosphere move with uniform velocities, indicating that mass motion and not wave motion was observed.
2. Material associated with the Great Red Spot (GRS) moves in a counterclockwise direction. The rotation period at the outer edge is about 6 days; little motion was observed at the center.
3. Cloud-top lightning bolts, similar to superbolts on Earth were observed.
4. Auroras similar to Earth's northern lights were observed in both ultraviolet (UV) and visible light.
5. Whistler emissions were detected in the Jovian atmosphere. They were interpreted as lightning whistlers. (A whistler is a radio wave (rf) generated by lightning and traveling long distances along a magnetic field.)
6. Jupiter has a ring system. Its outer edge is 80,000 mi (128,000 km) from the center of the planet. The ring is no more than 20 mi (30 km) thick.
7. Photos showed at least 8 volcanoes erupting on the Galilean satellite Io. (This was the first volcanic activity, except on Earth, previously noted in the planetary system.)
8. A torus of sulfur, oxygen, and sodium surrounds Jupiter at the distance of Io. The satellite appears to be the source of the material.
9. An electric current of about 3 million amperes was detected in the flux tube connecting Io and Jupiter. It is three times stronger than predicted.
10. Europa, a Jovian satellite, displayed a large number of intersecting linear markings on its surface. They exhibited no topographic relief.
11. Ganymede, another satellite, showed two distinctly different kinds of terrain — cratered and grooved. Its ice-rich crust appears to have been under tension from global tectonic processes. Ganymede is the largest satellite in the solar system, with a radius of 1640 mi (2638 km).
12. Callisto, a satellite, has an ancient, heavily cratered crust with remnant rings of enormous impact basins.

More detail on spacecraft and trajectory are given in article on *Voyager* **Missions to Jupiter and Saturn**.

Geophysics and Chemistry of Jupiter

Jovian Weather System. In contrast with the Earth, which has only one zone of weather (troposphere) with its continuing cycle of evaporation, condensation, and precipitation of water. Jupiter appears to have at least three weather-producing zones, consisting of water, ammonium hydrosulfide, and ammonia. Below the atmosphere, it is envisioned that the planet is principally liquid hydrogen and helium. Even with the sparse pre-*Voyager* data, scientists have been attempting to create models of the planet for many years. Some models have included a small iron-silicate core only a few thousand miles (km) in diameter. The core is inferred because cosmic abundances of the elements include small amounts of iron and silicates. The temperature in the core has been estimated to be about 30,000 K (53.500 °F; 29.704 °C). Surrounding the speculative core is a thick layer in which hydrogen is the most abundant element. Further, the hydrogen is separated into two layers. In both layers, it is believed to be liquid, but in different states. An inner layer, to about a 28,700 mi (46.000 km) radius is proposed to be liquid metallic hydrogen (electrically conductive), a form of hydrogen not found on Earth. On Jupiter, it is thought to exist at temperatures of about 11,000 K (19,300 °F; 10,704 °C) and at pressures about 2 million Earth atmospheres (at sea level). The next layer is proposed to be liquid hydrogen in its molecular form and extending to a radius of about 44,000 mi (70,000 km). Above this second hydrogen layer and reaching to the cloud tops for another 600 + mi (1000 km) is the planet's "atmosphere." See Fig. 1. If this concept is correct. Jupiter has no solid surface, but exists as a rapidly spinning ball of gas and liquid.

One of the puzzles of Jupiter is that it radiates about 1.5 to 2 times the amount of heat that it absorbs (from sunlight). Early models postulated nuclear reactions inside the planet, or heat from gravitational contraction as one may expect from a star. These concepts are no longer believed valid.

Great Red Spot as a Clue to Jovian Atmospheric Circulation. One of the most prominent features on Jupiter is the Great Red Spot, which has been observed almost continuously since its discovery some 300 years ago by Cassini (who also discovered a gap in the rings of Saturn). The GRS is about 8700 mi (14,000 km) wide, but its length varies between 18,720 and 24,960 mi (30,000 and 40,000 km). The GRS, simply from an appearance standpoint, is similar to an immense hurricane on Earth, although it is much larger and persists, as contrasted with short-lived storms on Earth. At one time, scientists believed it might be a phenomenon known as a Taylor column, caused by a mountain or depression on the surface. This postulation has been abandoned because no current model of Jupiter presumes a solid surface, and besides it appears that the same GRS has wandered in longitude several times around the planet. The GRS and the constant changes occurring on the planet are documented by Fig. 2.

The significant changes in Jovian cloud patterns had been noted from ground observations during the five-year period prior to the *Voyager* encounters and thus came as no surprise. However, the earlier *Pioneer*

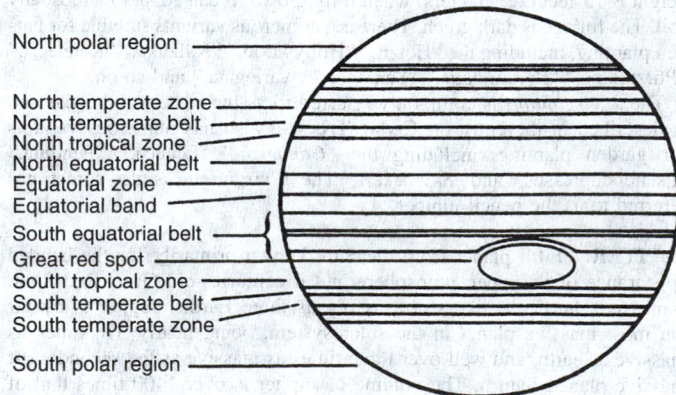

Fig. 1.   Principal visible features of Jovian atmosphere.

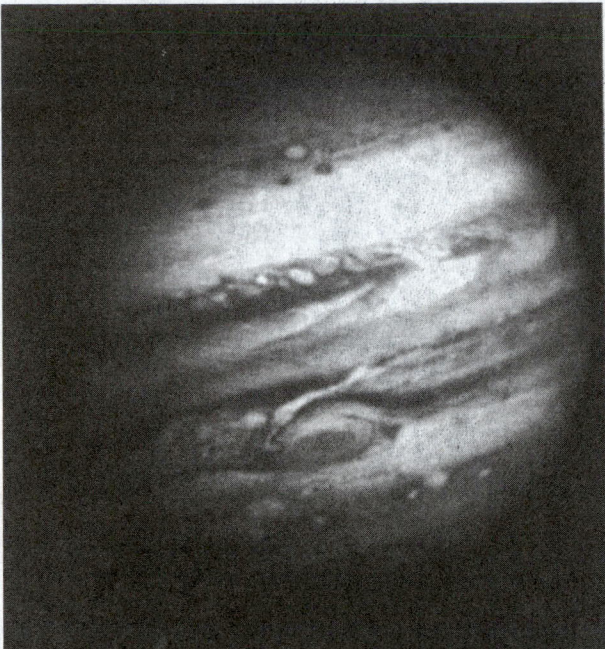
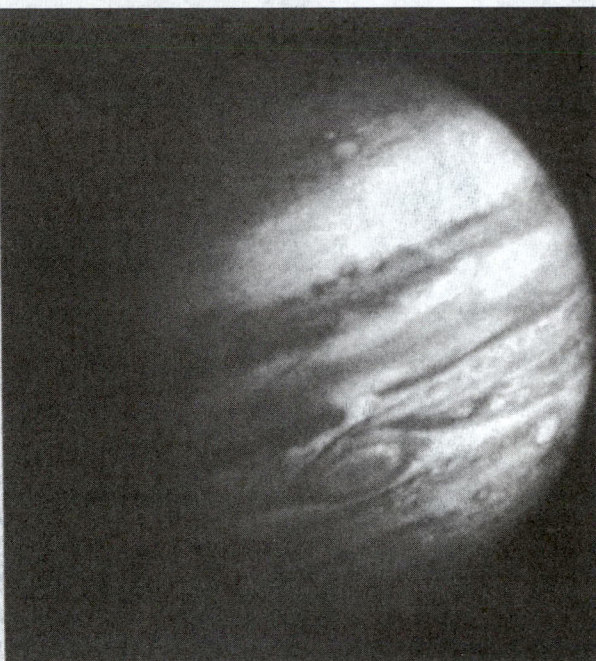

Fig. 2.   Photos taken of Jupiter almost 4 months apart by *Voyagers 1* and *2* show that the planet's atmosphere undergoes constant change, presenting an ever-shifting face to observers. The photos show that, although individual clouds in the Jovian atmosphere are long-lived, winds blow at greatly different speeds at different latitudes, causing the clouds to move independently and pass each other. Photo at left was taken on January 24, 1979 by *Voyager 1* while 40 million kilometers (25 million miles) from Jupiter. Photo at right was taken on May 9, 1979 when *Voyager 2* was 46.3 million kilometers (28.7 million miles) distant from the planet. One of the white ovals, located southwest (below and left) of the Great Red Spot, had drifted 60 degrees eastward (toward right) around the planet since the earlier, late-January observation allowing another feature, just to the west (left) of the oval in the *Voyager 1* photo, to move directly beneath the Great Red Spot. The bright tongue extending upward from the red spot is interacting with a thin, bright cloud above it that has traveled twice around Jupiter in 4 months. The only feature in the *Voyager 2* photo not observed by *Voyager 1* is a dark spot that developed along the north edge of the dark equatorial region. A similar feature was observed by Pioneer 10 in 1973. Ganymede is visible at bottom of *Voyager 1* photo. (*NASA: Jet Propulsion Laboratory, Pasadena, California.*)

missions, which were about one year apart, did not indicate the same kinds of major short-term changes. Some investigators have suggested that perhaps relatively recently the planet has commenced a transition in appearance from a prior decades-long period when the GRS Seemed to reside in apparently featureless surroundings. It is also possible that the differences largely resulted from changes in observational techniques.

Even with the much-increased information from the *Voyager* encounters, Jovian weather remains somewhat mysterious. Some scientists are fascinated by the so-called "one storm of a kind" that appears to be represented by the GRS. One explanation of the GRS likens the spot to a standing wave, of a type rarely observed on Earth. Close-ups of the GRS are given in Figs. 3 and 4. Other oval spots (while rather than reddish-orange as the GRS) also tend to be relatively long-lived, but move longitudinally in their zones about the planet. They appear to behave like solitary waves (simple wave crests with no other crest ahead or behind), persisting for several decades. One observer has drawn the parallel of a stone dropped in a pond that hypothetically creates but one single ripple instead of the succession of expanding crests and troughs that actually occur. Although it is now believed by some authorities that solitary waves may be more durable than previously thought, the postulate is regarded by some investigators as tenuous.

Over the years, some researchers have regarded the Jovian atmosphere as earthlike. This school has compared the regular bands on Jupiter with the jet streams on Earth. On Jupiter, it has been postulated, "weather" that is turbulent on a comparatively small scale is systematically transferred to the better-organized jet streams of the planet. The transfer process appears to be much more effective on Jupiter than on Earth. The visible bands and zones, if in fact they have not altered much as indicated by Earthbound observations for nearly a century, would indicate a stability not to be compared with the atmosphere of the Earth. From whence does this stability on Jupiter arise? Perhaps the immense and deep interior of Jupiter steadily intersects with the visible circulation, acting as a reliable engine to steady the circulation. Jupiter and Earth theoretically could have fundamentally similar weather systems, provided accommodations are made for the large differences between the planets. On Jupiter, upward flow of heat from a massive depth of liquid hydrogen to the outermost clouds

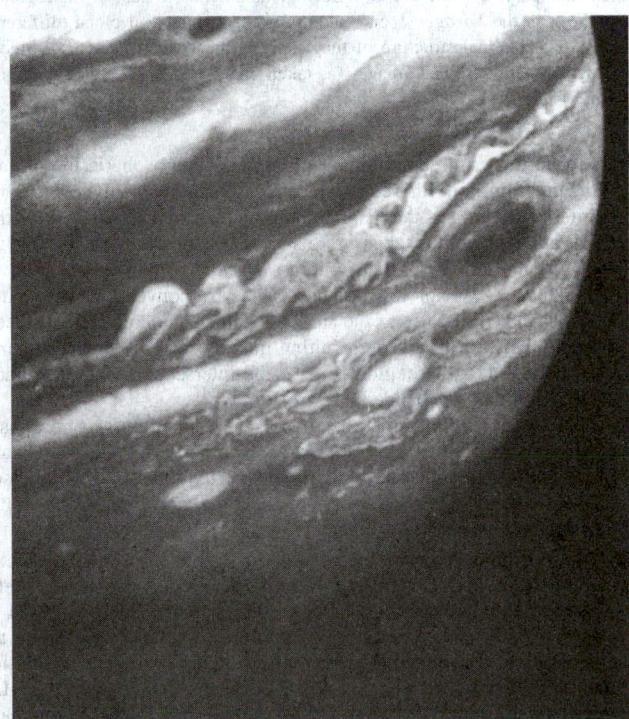

Fig. 3.   Photo taken on June 9, 1979 of the long-lived region west of the Great Red Spot. The white oval to the lower left of the GRS has a similar chaotic region of clouds to its west. This white oval, which is not the same one as that Seen below the GRS earlier by *Voyager 1* (March 1979), is moving to the right relative to the red spot. At the time this composite was taken, the spacecraft was over 24 million kilometers (15 million miles) from the planet. The smallest features which can be Seen are roughly 450 kilometers (280 miles) across. (*NASA: Jet Propulsion Laboratory, Pasadena, California.*)

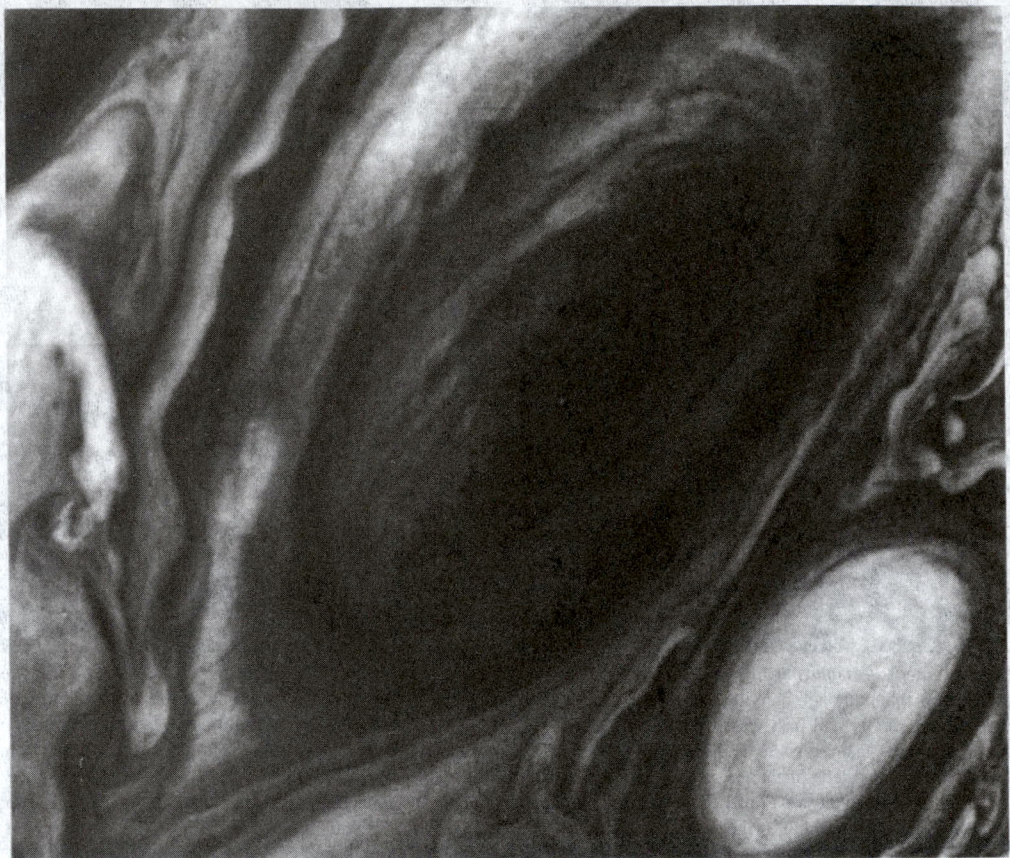

Fig. 4.   This mosaic of the Great Red Spot shows that the region had changed significantly since the *Voyager 1* encounter some three months before. Around the northern boundary, a white cloud is Seen. This extends to east of the region. The presence of this cloud prevents small cloud vertices from circling the spot in the manner Seen in the *Voyager 1* encounter. Another white oval cloud (different from the one present in this position three months earlier) is Seen south of the Great Red Spot. The internal structure of these spots appears to be identical. Since both spots rotate in an anticyclonic manner, these observations indicate that they are meterologically similar. This image was taken on July 6, 1979 from a range of 2,633,003 kilometers (1,632,462 miles). (*NASA; Jet Propulsion Laboratory, Pasadena, California.*)

appears to be the prime determinant of Jupiter's atmosphere. In contrast, on Earth, the sun alone drives the jets. Some scientists suggest that the Jovian interior dominates the visible circulation, which might suggest that the Jovian atmosphere has taken more starlike than earthlike patterns. It is further suggested, however, that, within a star, fluid motion must convey heat from the hot interior toward the surface by convection, on Jupiter, heat is not carried vertically to the surface. Rather, the convection process is twisted by the Coriolis effect, resulting in convective cylinders centered about the body's axis of rotation. It is at the intersection of the cylinders with the surface that the banded jets are formed.

An elucidation of the numerous models of Jupiter thus far proposed would require many pages. A post-*voyager* summary of such models and other aspects of both Jupiter and Saturn is expertly delineated by Ingersoll (California Institute of Technology), in the reference listed. One of the models explained is shown in Fig. 5.

Atmospheric Chemistry. Studies of the spectrum of sunlight reflected from Jupiter date back a half-century or more. Scientists sometimes refer to the apparent sources of a rather large variety of substances detected by the *Voyager* missions as "chemical disequilibrium," further indicating the dynamic nature of the planet. Infrared studies of Jupiter reveal, in addition to water vapor, the existence of phosphine (PH_3), germane (GeH_4), hydrogen cyanide (HCN), and carbon monoxide (CO). Heavy methane (methane molecules incorporating the heavy isotope of either hydrogen-2, i.e., deuterium, or carbon-13), ethane (C_2H_6), and acetylene (C_2H_2) have also been detected on the disk of Jupiter (and Saturn as well). It is observed that these chemicals would not be present in a hydrogen-rich atmosphere if it were in chemical equilibrium. Normally, carbon compounds would revert to methane and nitrogen would form ammonia. What causes this disequilibrium? Ultraviolet radiation from the sun may create free radicals (CH_2 and CH_3) by breaking down methane (CH_4) Lightning as noted on

Jupiter, may be a similar cause. Scientists find, however, the presence of carbon monoxide somewhat more puzzling. What is the source of oxygen? Perhaps in the planet's clouds: perhaps water may be present below the clouds. It is observed that thermal and pressure conditions may be such as to favor a reaction of methane and oxygen to form carbon monoxide. It has also been suggested that the oxygen existent above Jupiter's clouds may not emanate from Jupiter itself, but may be picked up with ejecta from the planet's volcanic satellite, Io. Oxygen could be derived from SO_2 in the Io vapors.

As pointed out by Ingersoll, the solid and liquid particles that constitute the clouds of Jupiter (and Saturn) give further evidence of chemical disequilibrium. The most abundant condensable vapors in a mixture of solar composition are water, ammonia, and hydrogen sulfide (H_2S). At chemical equilibrium, they form crystals of water ice, of ammonia, and of ammonium hydrosulfide (NH_4HS). Liquid drops of water and of ammonia-water solutions also are conceivable, as suggested by Lewis and colleagues (Massachusetts Institute of Technology). There is a knotty problem, however, because these condensates are white, whereas some of the clouds of Jupiter (and Saturn) are colored. One suggestion put forth is that molecular sulfur (S_n) can exist in several forms, i.e., n has several values. See also **Sulfur**. Thus, the colors for varying n values range from the familiar yellow to shades of brown and red. This suggestion appears plausible until one considers the fact that, to date, H_2S, the most likely parent of S, has not been detected. Phosphorus, which exists in yellow and red forms, could also be a source of coloration. See also **Phosphorus**. By combining observations in the visible and in the infrared (IR), the color and temperature of the clouds can be compared. Thus, the highest clouds on the planet noted (cool to IR) are visibly red; at the next lower level, they are visibly white; and at the lowest level they are visibly brown. This could lead to a conclusion that the compounds present in the clouds at

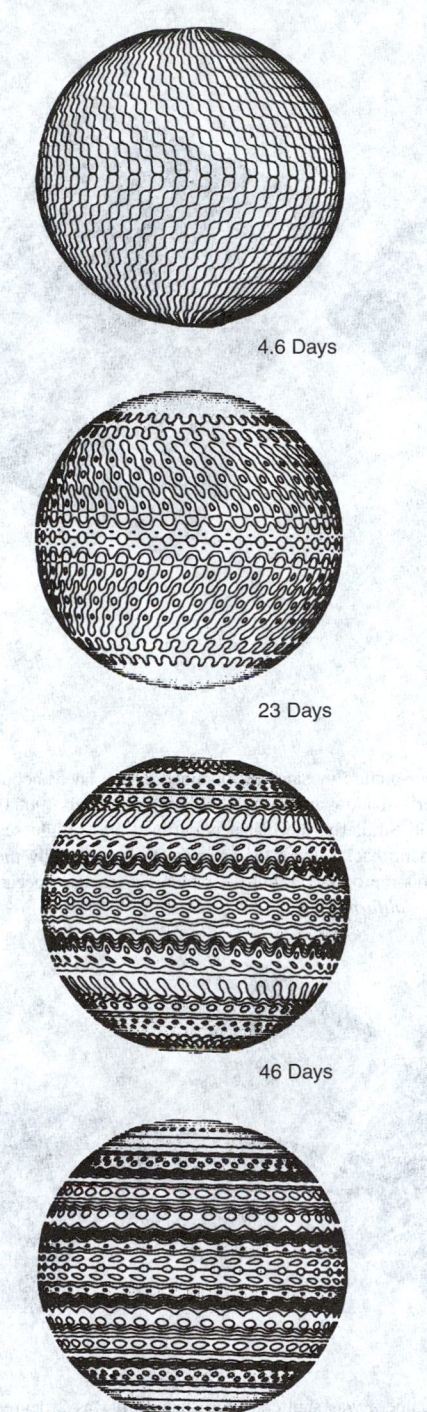

4.6 Days

23 Days

46 Days

73.3 Days

Fig. 5.   Reasonable facsimile of computer simulated planetary weather model generated in the early 1980s (G.P. Williams, Geophysical Fluid Dynamics Laboratory, National Oceanic and Atmospheric Administration). Eddy current patterns are markedly reminiscent of visible patterns Seen on Jupiter (and Saturn). The model assumes that the process shaping the weather on the planets is similar to the Earth's weather-producing machine. The model assumes an uneven distribution of solar radiation across the disk of the planet, creating eddy currents in a thin sunlit layer of the atmosphere. Like the visible patterns of Jupiter, trains of spots arise in the model, but isolated ovals as directly viewed on the planet do not. The state of the computer simulation is indicated by number of days from start.

different levels are affected by the different temperatures and amounts of solar radiation received by the respective locations.

There is consensus among the experts that much remains to be learned from future explorations, such as the proposed *Galileo* mission to Jupiter,

of the processes that take place below the cloud tops before conjecture can be converted to sound concepts of the Jovian atmosphere. One advantage of compiling the unanswered questions is the influence of such a list on the design of instrumentation for future explorations.

Jupiter's Ring

Observations from *Voyager 2* indicated distinct inner and outer boundaries of the brightest part of the ring system. See Figs. 6, 7, and 8. Although viewed in backscattered sunlight under somewhat adverse conditions, no obvious gaps in the rings were noted. Observations indicated that the ring system is optically very thin with a well-defined outer edge. The outer part of the ring system is composed of a relatively bright segment about 800 kilometers (~500 miles) wide surrounded by a dimmer and broader segment about 5200 kilometers (~3200 miles) across. Less distinct divisions were also visible. Some scientists believe that the ring system represents a steady state between loss and supply, being neither a leftover from the original accretion and condensation events that formed the planet (as may have been the case with Saturn), nor fragments of a disrupted satellite. Material sources on the supply side may be cometary and meteoritic debris, impact ejecta from the inner satellites, and volcanic ejecta remove from the Io (perhaps by magnetospheric forces). It also has been suggested that the recently discovered inner satellite may explain the location of the outer boundary of the ring system.

As explained by Pilcher (1980), Jupiter's innermost Galilean satellite, Io, appears to be the source of a host of atomic and perhaps molecular nebulae that are distributed around the inner Jovian magnetosphere. Neutral sodium and potassium, singly ionized sulfur, and oxygen have been detected by ground-based instruments. Optical instruments on spacecraft have also detected emissions of doubly and triply ionized sulfur, doubly ionized oxygen, and possibly neutral hydrogen. Observations on two successive nights in April 1979 showed that the ring characteristics may change markedly during a 24-hour period. As observed on the first night, the ring was narrow and confined to the magnetic equator inside Io's orbit, whereas on the following night it was confined symmetrically about the centrifugal symmetry surface and showed considerable radial structure, including a "fan" extending to Io's orbit. It has been suggested that these differences in the ring can be explained in terms of differences in sulfur plasma temperature.

Jupiter's Satellites

One of the most rewarding aspects of the two *Voyager* encounters with Jupiter pertained to the large amount of information returned from observations of the satellites of the planet. As will be noted from the following descriptions, the moons of Jupiter exhibit considerable contrast, ranging from the long-time geologically "dead" periods of Ganymede and Callisto to the currently geologically active moon, Io.

Callisto. This is the outermost of the four Galilean moons, darker than the others, with a diameter of about 4890 kilometers (3038 miles). It is darker probably because more rock material is exposed. The entire surface of Callisto has a high density of craters which are several tens of kilometers or larger in diameter.

Callisto appears to be the least geologically active major moon or planet observed thus far in the solar system. The meteorite craters are quite ancient. One half of the satellite was viewed by *Voyager 1*; the other half by *Voyager 2*. The cratered terrains on Callisto appear to date back to the late torrential bombardment period some 4 billion years ago, as do the cratered highlands of the earth's moon.

Observations indicate that the surface of Callisto is largely water (frozen or liquid). A first impression of Callisto is that it has proportionately far fewer very large craters (hundreds of kilometers across) than do the earth's moon or the planets Mars and Mercury. This would suggest different meteorite size distributions in the outer and inner solar system. It is further suggested, however, that the weak, ice-logged surface of Callisto may have allowed crater walls to slump and the floors to rise after formation. These events would cause the craters to appear nearly two-dimensional. Where planetary bodies are made of rock, the large craters persist as detectable, large, regional depressions, but on ice-covered bodies, these depressions may disappear. Also, because small meteorites are much more prevalent than large ones, it is possible that younger, smaller meteorites may have obliterated the large depressions over a long time span. See Fig. 9.

Considering the foregoing observations, it is interesting to note the presence of three relatively large craters on Callisto, which appear as large,

Fig. 6. A brilliant halo around Jupiter, the thin ring of particles discovered by *Voyager 1*. This four-picture mosaic was obtained by the later encounter by *Voyager 2* on July 10, 1979, with the spacecraft's wide-angle camera while the spacecraft was deep in Jupiter's shadow and some 1,448,000 kilometers (900,000 miles) beyond the planet. The ring is unusually bright because of forward scattering from small particles within it. Similarly, the planet is outlined by sunlight scattered toward the spacecraft from a haze layer high in Jupiter's atmosphere. On each side, the arms of the ring curving back toward the spacecraft are cut off by the planet's shadow as they approach the brightly outlined disk. The night side of Jupiter appears completely dark in this reproduction, but the pictures were also specially reprocessed to search for evidence of lightning storms and auroras. (*NASA; Jet Propulsion Laboratory, Pasadena, California.*)

Fig. 7. High-resolution pictures of Jupiter's ring taken by *Voyager 2* on July 10, 1979. As the camera was shuttered, the spacecraft was 2 degrees below the ring plane and at a range from Jupiter 1 of 1,550,000 kilometers (961,000 miles). The forward scattering of sunlight reveals a radial distribution and density gradient of very small particles extending inward from the ring toward Jupiter. There is an indication of structure within the ring, but the spacecraft motion during these long exposures blurred out the highest resolution detail, particularly in the frame at the right. Note: The black dots are part of the imaging process. (*NASA; Jet Propulsion Laboratory, Pasadena, California.*)

bright patches surrounded by multiple concentric bright rings. The largest of these is about 2896 kilometers (~1800 miles) across. By observing the number of younger, small craters superimposed on these large objects, scientists infer that these are the youngest large features to be observed in the solar system to date. It has been suggested by some observers that the rings are fractures in the ground caused by the shock of meteoric impact. But, others point out that they are more closely spaced, more numerous, and extend much farther from the central crater than do similar structures on rock bodies. Thus, the central bright patch found in each of these fractures may be the original crater, buried under its own ejected debris. Such phenomena may simply arise because of the lower strength of the icy surface (Loudon, 1979).

Ganymede. This is one of the largest satellites in the solar system. It has a diameter of 5216 kilometers (3241 miles). The surface of Ganymede is marked by highly complex tectonic patterns and consists of four basic terrain types[1]. (1) Dark, densely cratered terrain; (2) grooved terrain of diverse patterns and ages; (3) a young, rugged impact basin and ejecta

---

[1] From joint report prepared by scientists at the California Institute of Technology. Cornell University, Jet Propulsion Laboratory. NASA Headquarters, New Mexico State University, Rand Corporation, Smithsonian Astrophysical Observatory, State University of New York, University College (London), University of Arizona, University of Hawaii, University of Wisconsin, and the U.S. Geological Survey.

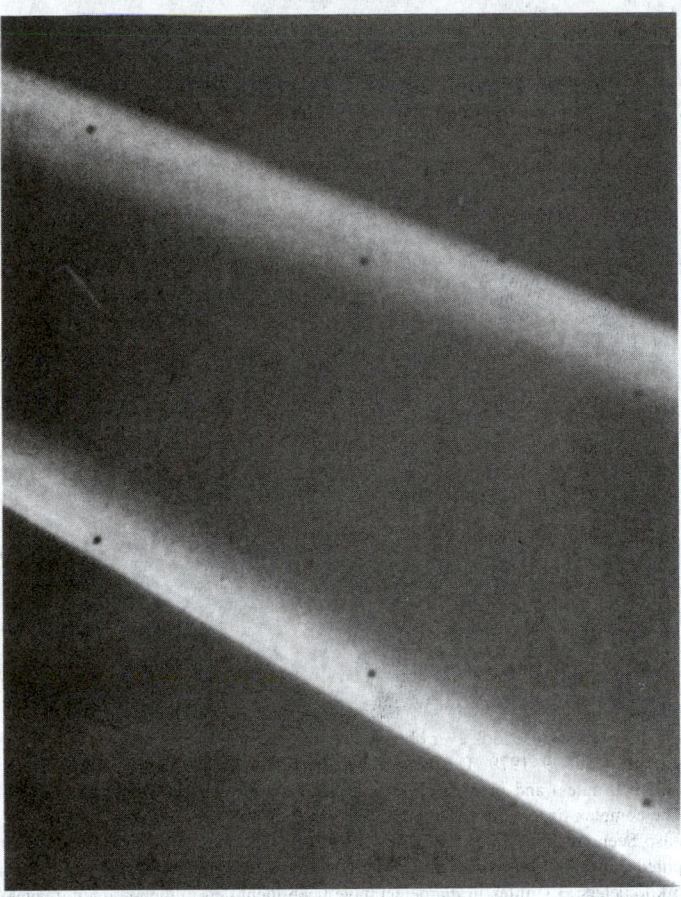

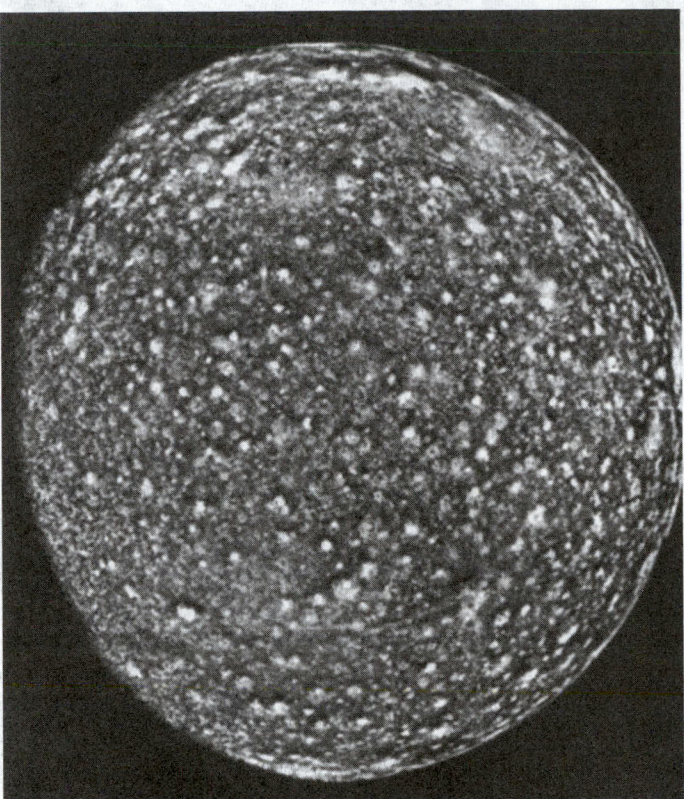

Fig. 8. This high-resolution view of Jupiter's ring further suggests that the ring may be divided into several components (as are the rings of Saturn). The V-shaped figure to the left is caused by a star image which was trailed out as the camera moved slightly during the long exposure. Distance of spacecraft from Jupiter at this time was 1.5 million kilometers (930,000 miles). The ring was unexpectedly bright, apparently due to forward scattering of sunlight by small ring particles. Note: The black dots are part of the imaging process. (*NASA; Jet Propulsion Laboratory, Pasadena, California.*)

Fig. 9. Photomosaic of Callisto taken on July 8, 1979 at a distance of 390,000 kilometers (240,000 miles) from the planet. The mosaic is composed of nine frames. The impact crater distribution is quite uniform across the disk. Notable are the very bright rayed craters that probably are very young. Near the limb is a giant (probable) impact structure. Several large structures on the other side of the moon were discovered by *Voyager 1*. There are about fifteen concentric rings surrounding the bright central spot. Many hundreds of moderate-sized impacts are also visible, a few with bright ray patterns. The limb is very smooth, confirming that no high topography has been noted, an observation which is consistent with its icy composition. (*NASA; Jet Propulsion Laboratory, Pasadena, California.*)

blanket; and (4) smooth terrain, which occurs in patches that locally obscure older terrains. The largest area of old cratered terrain, about 4000 kilometers (2480 miles) across, contains part of a giant concentric ring structure that is similar in some respects to the giant ring structures on Callisto. See Fig. 10.

Voyager 1 obtained good coverage of the side of the moon facing Jupiter; *Voyager 2* observed the anti-Jupiter hemisphere. Global patterns are similar. The equatorial region consists of a mosaic of dark polygons separated by brighter bands (grooved terrain), across which are scattered numerous bright-rayed craters. A general brightening in the polar regions is evident. However, the cap boundaries are ill-defined and fuzzy. The bluish-white polar caps are clearly evident.

Based upon the most recent observations, it is suggested that the ancient craters flattened soon after they formed. The crust was weak at that time, but became more rigid with the passage of time. The comparatively young, mountainous regions of Ganymede are believed to be considerably older than the large lava sheets (maria) found on earth's moon. From these observations, one can conclude that Ganymede has been geologically inactive for perhaps 3.5 billion years.

Radar observations of Ganymede, at X-band, show that the surface is unusually bright and has unusual polarization properties. Some scientists have proposed a model of the surface based upon large numbers of random ice facets (hence vacuum-ice interfaces) to account for these characteristics.

Europa. This satellite has a diameter of 3066 kilometers (1905 miles). Authorities estimate that Europa's surface must be at least 108 years old if the flux of impacting bodies near Jupiter is similar to that near earth and the moon, even allowing for the enhancement of the flux by Jupiter's

gravity. If the flux on Europa is far less than that on earth's moon, as is now suspected, Europa's surface may be several billion years old. Europa may have frozen once, after the torrential bombardment was over, and simply been subjected to continual erosion by sputtering since that time. In fact, the crater density on Europa gives a lower limit for the flux after the torrential bombardment. However, Europa's surface probably is not more than 4.5 billion years old, or it would show scars of ancient craters formed during that bombardment. This gives a lower limit (3 craters in 4.5 billion years), or about one-tenth the cratering rate of earth's moon.

Europa has a smooth surface crisscrossed with highly complex fracture patterns that show little or no relief except for thin, bright ridges perhaps a hundred meters high. See Fig. 11.

Voyager 1 density observations indicated that Europa consists mainly of rock, but with about 20% of the outer volume made up of water — frozen at the surface, but probably liquid underneath as the result of tidal heating. This would provide Europa with a surface, since it is partially frozen, that is much weaker structurally than the surfaces of Callisto and Ganymede. Observers generally conclude that Europa is the flattest body yet observed in the universe, with mountains (ice ridges) only a few hundred feet high. No valleys were noted.

The thousands of brown lines that seam the surface of Europa have not been fully explained. One school of thought suggests that the global icecap at one time expanded, possibly caused by freezing of underlying water. Cracks were formed and these were filled with a water-borne darker material from below. When the icecap retreated at a later time, the cracks could not close because they were filled with material, thus forming the presently observed ice ridges and lines. The presence of underlying dark material has been confirmed by the observation of at least one crater on

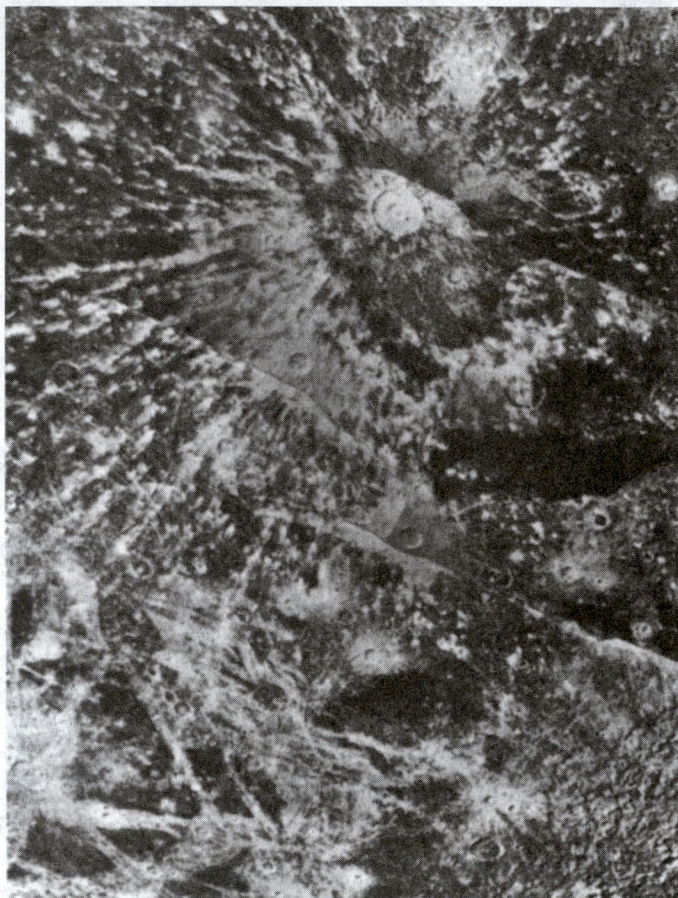

Fig. 10.   Photomosaic of Ganymede taken on July 9, 1979 by *Voyager 2* at a range of about 100,000 kilometers (62,000 miles). View shows numerous impact craters, many with bright ray systems. The rough, mountainous terrain at lower right is the outer portion of a large, fresh impact basin which postdates most of the other terrain. At bottom of view, portions of grooved terrain transect other portions, indicating they are younger. This may be the result of the intrusion of new icy material which comprises the crust of Ganymede. The dark patches of heavily cratered terrain (right center) are probably ancient icy material formed prior to the grooved terrain. The large rayed crater at upper center is about 150 kilometers (93 miles) in diameter. A variety of ray patterns is Seen around the craters. Some craters have dark halos; others have diffuse bright rays. The variation in albedo patterns around the craters may indicate layering in the surface materials. The intensity of craters suggests that the dark areas are extremely old. (*NASA; Jet Propulsion Laboratory, Pasadena, California.*)

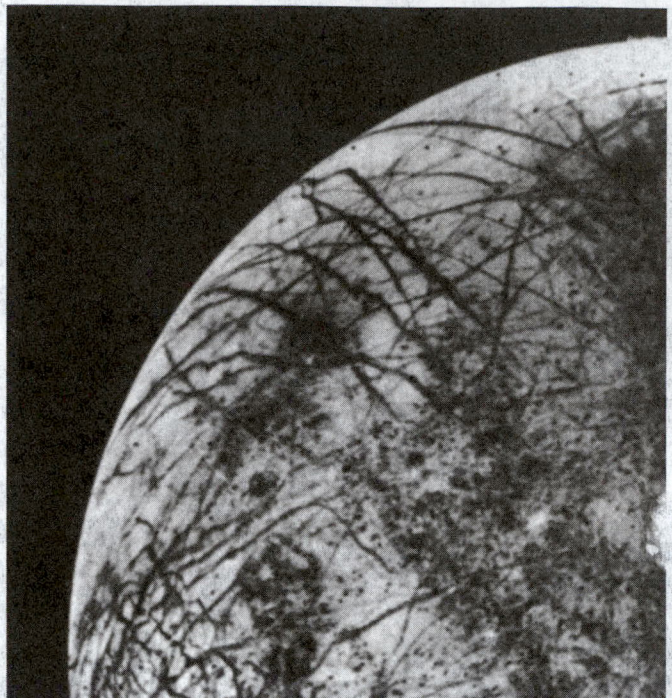

Fig. 11.   First close look ever obtained of Jupiter's satellite Europa, taken by *Voyager 2* on July 9, 1979. This picture was made at a range of 246,000 kilometers (152,000 miles) and provides a resolution of about four kilometers (2.5 miles). The complex linear features appear like cracks or huge fractures in these images. Also Seen are somewhat darker mottled regions which appear to have a slightly pitted appearance, perhaps due to small craters. No larger craters (more than 5 kilometers; 3.1 miles in diameter) have been identified in the *Voyager* images, possibly suggesting that this satellite has a young surface relative to those of Ganymede and Callisto, although probably not as young as that of Io. Various models for Europa's structure will be tested during analysis of these images over a period of research during the 1980s, and will include the possibility that the surface is a thin ice crust overlying water or softer ice and that the fracture systems Seen are breaks in that crust. Resurfacing mechanisms, such as the production of fresh ice or snow along the cracks and cold glacierlike flows are being considered as possibilities for removing evidence of impact events. Present information indicates that Europa has properties intermediate between Ganymede and Io. (*NASA; Jet Propulsion Laboratory, Pasadena, California.*)

Europa that apparently ejected dark material to form streaks. There is another suggestion that the observed ridges were more resistant to erosion. Such erosion could have been caused by subatomic particles that comprise the intense radiation belts of Jupiter. It has been suggested that these particles could have removed a depth (0.5 mile; 0.8 kilometer) of ice over a 4.5 billion-year period. Observers suggest that a process of this kind would affect only Europa of the Galilean satellites because Io is currently active and Ganymede and Callisto are situated where the intensity of the radiation belt is too low.

Io. This Jovian satellite is about 3636 kilometers (2260 miles) in diameter and displays some of the most unusual phenomena observed in the solar system. Io has red polar caps. For about 15 minutes after it emerges from behind Jupiter it is several per cent brighter than usual. The brightening occurs only about half the time. Perhaps the color change arises from an alteration of the colored material on the surface of Io when it is behind Jupiter. One of the most exciting discoveries of the *Voyager* missions to Jupiter was that of volcanic activity on Io. Volcanic activity discovered by *Voyager 1* was continuing during the *Voyager 2* encounter. Of seven volcanic plumes that were observed by *Voyager 1* six were still erupting a few months later when *Voyager 2* encountered the satellite. However, the largest plume observed earlier had ceased activity. But an

additional plume evidently became active during the time between the two *Voyager* encounters. See Fig. 12.

Several large-scale changes in the appearance of Io's surface occurred between the two *Voyager* encounters. Two of the more prominent changes were associated with the regions near two of the volcanic plumes. The cleft in the southern area of a heart-shaped dark ring associated with one of the plumes appeared to be filled in by darker material. Another important change was the increased albedo of the northern part of the black annulus located just south of the eruption sites. The annulus was observed by the infrared interferometer on *Voyager 1* to be a prominent "hot spot"—possibly this albedo change was the result of pyroclastic deposition from the plume over part of the annulus.

Io was observed by the *Voyager 2* television system intermittently for a 5-day period at resolutions varying between 90 and 21 km/lp (lp = line pair). The satellite was monitored with a nearly continuous 8-hour sequence of television images at high phase angles (115–155°) and a resolution of about 20 km/lp. Plumes having heights of a few tens of kilometers were identified with confidence only when they were within 10° of the bright limb. *Voyager 2* surveyed 80% of the surface in this manner. Thus, all of the plumes 100 kilometers (62 miles) or higher that were active during both encounters probably were detected.

Enhancement by Jupiter's gravity makes the Io crater production rate close to or greater than the present cratering rate on the moon, using a plausible range of crater production rates at Jupiter's heliocentric distance. Hence, the absence of craters on Io strongly indicates that it must be the youngest and most dynamic surface in the solar system.

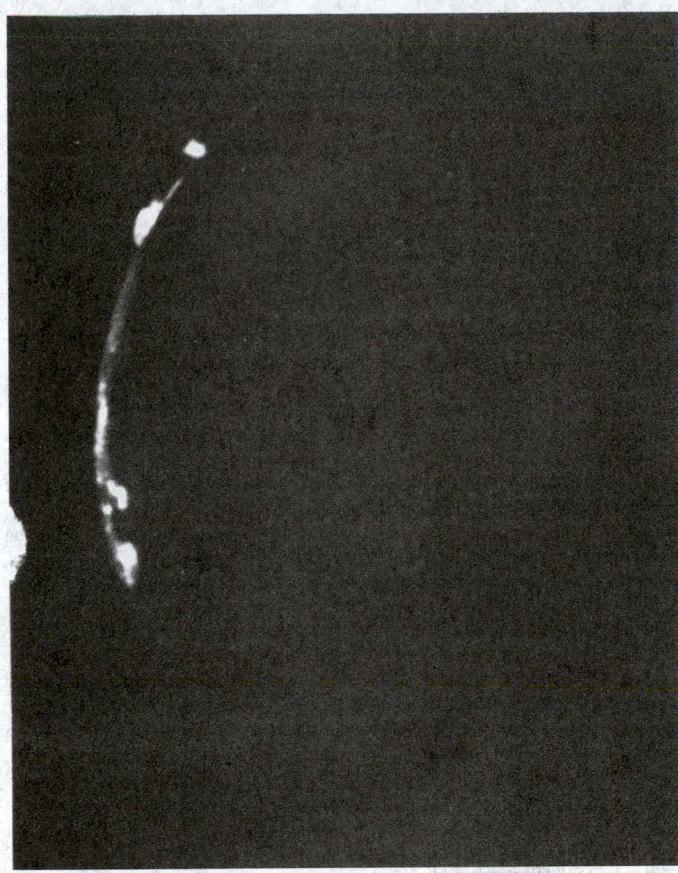

Fig. 12.   Picture of Io taken by *Voyager 2* on Jul 9, 1979. On the limb of Io are two volcanic eruption plumes about 100 kilometers (62 miles) high. These two plumes were first Seen by *Voyager 1* in March 1979. Thus, they had been erupting for four months or longer when this view was taken. (*NASA; Jet Propulsion Laboratory, Pasadena, California.*)

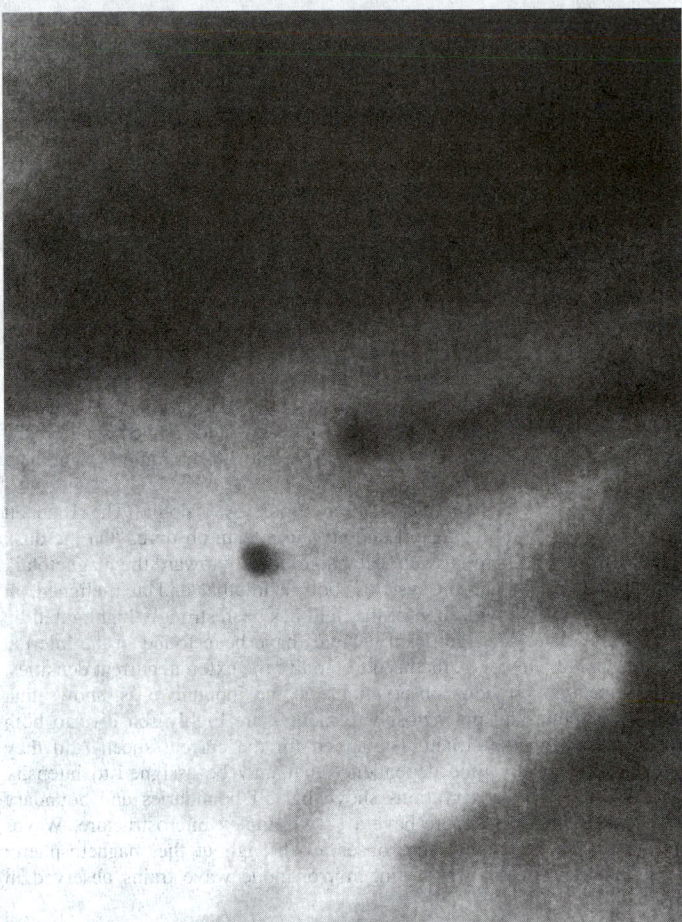

Fig. 13.   The newly discovered 15th moon of Jupiter (black spot at bottom Seen against the face of the planet. This view was taken on March 4, 1979, when the *Voyager 1* spacecraft was 880,681 miles (1,417,320 km) from Jupiter and 21 hours before the craft's closest approach to the planet. The satellite, discovered from images by Dr. Stephen P. Synnott, of JPL, is about 43 to 50 miles (70 to 80 km) in diameter. The satellite orbits Jupiter every 16 hours, 16 minutes, at a distance of 93,000 miles (151,000 km) above Jupiter's cloud tops. The new satellite, identified temporarily as 1979 J2, is the second discovered from Voyager data. (*NASA; Jet Propulsion Laboratory, Pasadena, California.*)

Amalthea. This is the closest satellite to Jupiter and is much smaller than the Galilean moons. One observer has suggested that it is so small that it would fit into one of the Great Lakes (Huron). *Voyager 1* observed that Amalthea emits about 30% more heat than it receives from the sun. Thus, instead of having a temperature of −235 °F (−148 °C), the temperature of the moon is −125 °F (−87 °C). Because of its size, scientists have ruled out the tidal mechanism as a source of heat, as has been suggested for Io and Europa, as well as heat from internal radioactivity. Many scientists at this time suggest that the heat may derive from collisions with radiation-belt particles. These are denser around Amalthea than around any of the other satellites of Jupiter.

Newly Discovered Satellites. A detailed examination of imaging data taken by *Voyager 1* when it was close to Jupiter showed a shadowlike image. Analysis of the motion of the image showed that it was not an atmospheric feature, nor could it have been a shadow of any known satellite. Later observations from *Voyager 2* confirmed the satellite, which has been identified as 1979-J1. See Fig. 13. The moon orbits at the edge of the Jupiter ring at a distance of 57,000 kilometers (36,000 miles) from the top of the Jovian clouds. It is estimated to be 30–40 kilometers (18–25 miles) in diameter and orbits Jupiter faster than any satellite in the solar system—at 30 kilometers per second (67,000 miles per hour), It circles Jupiter each 7 hours, 8 minutes.

Analysis of Voyager 1 data also revealed a second previously unknown Jovian satellite, now identified as 1979-J2, thus making a total of 15 Jovian moons. It is estimated that this satellite is some 70–80 kilometers (43–50 miles) in diameter. The moon orbits Jupiter every 16 hours, 16 minutes, at a distance of 151,000 kilometers (93,900 miles) above the Jovian cloud tops.

Magnetosphere

The magnetosphere, which differs markedly from that of the earth, appears to be powered by the rapid rotation of the planet. Particles are added by the satellite Io. Occasional jets of neutral sodium that can be Seen emanating from Io Seem to support the concept that sputtering is a likely way to rip atoms from Io for later ionization in the torus.

Jupiter's magnetic field, like that of the earth, is tied to a magnetic dipole in the planet. Ideally, the magnetic field spins around at the same rate Jupiter rotates. In the earth's magnetosphere, the motion of plasma is controlled by the earth's rotation only very close to the atmosphere and ionosphere; elsewhere the plasma flow is governed by the solar wind. *Voyager* observations indicate that inward from the orbit of Io, the plasma rotates with Jupiter just as if it were rigidly attached, but beyond Io's orbit the plasma slows down and requires some 20 hours to circle the planet. Other instrumentation showed, however, that essentially the plasma is rotating with the planet regardless of how deep it is located in the magnetosphere. This conflict is not likely to be resolved until the two experimental teams reduce data taken at the same time and place. Resolution may have to await the *Galileo* spacecraft.

See also *Voyager* **Missions to Jupiter and Saturn.**

Findings from the *Ulysses* Flyby (February 1992)

In an overview of the Ulysses encounter with Jupiter, E.J. Smith and D.E. Page (JPL) and K.-P. Wenzel (European Space Research and Technology Center) observe, "The *Ulysses* scientific investigations were well suited to observations of the Jovian magnetosphere, and the encounter has resulted in a major contribution to our understanding of this complex and dynamic plasma environment. Among the more exciting results are: (1) possible entry into the polar cap; (2) the identification of magnetospheric ions originating from Jupiter's ionosphere, Io, and the solar wind; (3) observation of longitudinal asymmetries in density and discrete wave-emitting regions of the Io plasma torus; (4) the presence of counterstreaming ions and electrons, field-aligned currents, and energetic electron and radio bursts in the dusk sector on high-altitude magnetic field lines; and (5) the identification of the direction of the magnetic field in the dusk sector, which is indicative of tailward convection.

Regarding the volcanic activity of Io, J.R. Spencer and a team of investigators summarize, "The population of heavy ions on Io's torus is ultimately derived from Io volcanism. Ground-based infrared observations of Io between October 1991 and March 1992, contemporaneous with the February 8, 1992 *Ulysses* observations of the Io torus, show that volcanic thermal emission was at the low end of the normal range at all Io longitudes during this period. In particular, the dominant hot spot Loki was quiescent. Resolved images show that there were at least four hot spots on Io's Jupiter-facing hemisphere, including Loki and a long-lived spot on the leading hemisphere (Kanehekili), of comparable 3.5 micrometer brightness but higher temperature."

Pertaining to magnetic field observations, A. Balogh (The Blackett Laboratory, Imperial College, London) and a team observe, "On the dusk side, the magnetic field is swept back significantly toward the magnetotail. The importance of current systems, both azimuthal and field-aligned, in determining the configuration of the field has been strongly highlighted by the *Ulysses* data. No significant changes have been found in the internal planetary field. However, the need to modify the external current densities, with respect to previous observations on the inbound pass, shows that Jovian magnetic and magnetosphere models are highly sensitive to both the intensity and the structure assumed for the current sheet. And they are sensitive to any time dependence that may be assigned to intensity and structure. The observations show that all boundaries and boundary layers in the magnetosphere have a very complex microstructure. Waves and wave-like structures were observed throughout the magnetosphere; these included the longest lasting mirror-mode wave trains observed in space."

Observations of the hot plasma environment at Jupiter are summarized by L.J. Lanzerotti (AT&T Bell Laboratories) and a team of researchers from various institutions: "The Jovian magnetosphere was found by *Ulysses* to be very extended, with the day-side magnetopause located ∼105 Jupiter radii. The heavy ion (sulfur, oxygen, and sodium) population in the day-side magnetosphere increased sharply at −76 Jupiter radii. This identified the *Voyager* hot plasma measurements."

R.G. Stone (NASA) and other researchers from NASA and other institutions reported on the *Ulysses* radio and plasma wave observations: "The URAP (Unitred Radio and Plasma Wave) experiment produced new observations of the Jovian environment, owing to the unique capabilities of the instruments and the transversal of high Jovian latitudes. Broadband continuum radio emission from Jupiter, and in-situ plasma waves, have proved valuable in delineating the magnetospheric boundaries. Simultaneous measurements of electric and magnetic waves have yielded new evidence of whistler-mode radiation within the magnetosphere. Thermal noise measurements of the Io torus densities yielded values in the densest portion that are similar to models suggested on the basis of *Voyager* observations in 1979. The URAP measurements also suggest complex beaming and polarization characteristics of Jovian radio components. In addition, a new class of kilometer-wavelength striated Jovian bursts has been observed."

Other reports on the *Ulysses* observations of Jupiter are listed in the references. See also *Voyager* **Missions to Jupiter and Saturn**.

Additional Reading

Balogh, A. et al.: "Magnetic Field Observations During the Ulysses Flyby of Jupiter," *Science*, 1515 (September 11, 1992).

Bame, S.J. et al.: "Jupiter's Magnetosphere: Plasma Description from the Ulysses Flyby," *Science*, 1539 (September 11, 1992).

Beebe, R.: *Jupiter: The Giant Planet*, Smithsonian Institution Press, Washington, DC, 1996.

Belcher, J.W.: "The Jupiter-Io Connection," *Science*, **238**, 170–175 (1987).

Bird, M.K. et al.: "Ulysses Radio Occultation Observations of the Io Plasma Torus During the Jupiter Encounter," *Science*, 1531 (September 11, 1992).

Brown, R.A. and W.H. Ip: "Atomic Clouds as Distributed Sources for the Io Plasma Torus," *Science*, **213**, 1493–1495 (1981).

Caldwell, J., B. Turgeon, and X.-M. Hua: "Hubble Space Telescope Imaging of the North Polar Aurora on Jupiter," *Science*, 1512 (September 11, 1992).

Geiss, J. et al.: "Plasma Composition in Jupiter's Magnetosphere: Initial Results from the Solar Wind Ion Composition Spectrometer," *Science*, 1535 (September 11, 1992).

Goguen, J.D. and W.M. Sinton: "Characterization of IO's Volcanic Activity by Infrared Polarimetry," *Science*, **230**, 65–69 (1985).

Gold, T.: "Electrical Origin of the Outbursts on Io," *Science*, **206**, 1071–1073 (1979).

Goldberg, B.A., G.W. Garneau, and S.K. LaVoie: "IO's Sodium Cloud," *Science*, **226**, 512–516 (1984).

Gore, R.: "What Voyager Saw: Jupiter's Dazzling Realm," *National Geographic*, **157**, 1 (1980).

Grüm, E. et al.: "Ulysses Dust Measurements Near Jupiter," *Science*, 1550 (September 11, 1992).

Hartline, B.K.: "Voyager Beguiled by Jovian Carrousel," *Science*, **208**, 384–386 (1980).

Hill, T.W.: "Corotation Lag in Jupiter's Magnetosphere: Comparison of Observation and Theory," *Science*, **207**, 301–302 (1980).

Hockey, T.: *Galileo's Planet: Observing Jupiter before Photography*, Institute of Physics Publications, London, UK, 1999.

Howell, R.R. and M.T. McGinn: "Infrared Speckle Observations of Io: An Eruption in the Loki Region," *Science*, **230**, 63–65 (1985).

Ingersoll, A.P.: "Jupiter and Saturn," *Sci. Amer.*, **245**(6), 90–108 (1981).

Johnson, R.E. et al.: "Erosion of Galilean Satellite Surfaces by Jovian Magnetosphere Particles," *Science*, **212**, 1027–1029 (1981).

Johnson, T.V. and L.A. Soderblom: "Io," *Sci. Amer.*, **249**(6), 56–67 (1983).

Johnson, T.V. et al.: "Volcanic Hotspots on Io: Stability and Longitudinal Distribution," *Science*, **226**, 134–137 (1984).

Keppler, E. et al.: "An Overview of Energetic Particle Measurements in the Jovian Magnetosphere with the EPAC Sensor on Ulysses," *Science*, 1553 (September 11, 1992).

Kivelson, M.G., J.A. Slavin, and D.J. Southwood: "Magnetospheres of the Galilean Satellites," *Science*, **205**, 491–493 (1979).

Lanzerotti, L.J. et al.: "The Hot Plasma Environment at Jupiter: Ulysses Results," *Science*, 1518 (September 11, 1992).

Loudon, J.: "Voyager 2: The Jovian Moons Revisited," *Technology Review (MIT)*, **82**, 2, 70–73 (1979).

Mandillo, M., B. Flynn, and J. Baumgardner: "Imaging Observations of Jupiter's Sodium Magneto-Nebula During the Ulysses Encounter," *Science*, 1510 (September 11, 1992).

Nelson, R.M. et al.: "Io: Longitudinal Distribution of Sulfur Dioxide Frost," *Science*, **210**, 784–786 (1980).

Pilcher, C.B. and J.S. Morgan: "Detection of Singly Ionized Oxygen around Jupiter," *Science*, **205**, 297–298 (1979).

Pilcher, C.B.: "Images of Jupiter's Sulfur Ring," *Science*, **208**, 181–183 (1980).

Rogers, J.: *The Giant Planet Jupiter*, Vol. 6, Press Syndicate of University of Cambridge, Middlesex, UK, 1995.

Schneider, N.M. et al.: "Eclipse Measurements of IO's Sodium Atmosphere," *Science*, **238**, 55–58 (1987).

Simpson, J.A. et al.: "Energetic Charged-Particle Phenomena in the Jovian Magnetosphere: First Results from the Ulysses COSPIN Collaboration," *Science*, 1543 (September 11, 1992).

Smith, E.J., K.-P. Wenzel, and D.E. Page: "Ulysses at Jupiter: An Overview of the Encounter," *Science*, 1503 (September 11, 1992).

Spencer, J.R. et al.: "Volcanic Activity on Io at the Time of the Ulysses Encounter," *Science*, 1507 (September 11, 1992).

Stone, E.C. and A.L. Lane: "Voyager 1 Encounter with the Jovian System (Contains 15 papers relating to Voyager 1 encounter with Jupiter)," *Science*, **204**, 945–1008 (1979).

Stone, E.C. and A.L. Lane: "Voyager 2 Encounter with the Jovian System (Contains 13 papers relating to Voyager 2 encounter with Jupiter)," *Science*, **206**, 925–997 (1979).

Stone, R.G. et al.: "Ulysses Radio and Plasma Wave Observation in the Jupiter Environment," *Science*, 1524 (September 11, 1992).

Trauger, J.T.: "The Jovian Nebula: A Post-Voyager Perspective," *Science*, **226**, 336–341 (1984).

Ulysses Flyby of Jupiter.

Web References

Site Directory. http://www.jpl.nasa.gov/directory/
The Nine Planets. http://www.seds.org/billa/tnp/
Views of the Solar System. http://www.hawastsoc.org/solar/eng/homepage.htm

JURARA (*Reptilia, Testudinata*). The giant tortoise, Podocnemis expansa, of the Amazon River basin.

JURASSIC PERIOD.   A major subdivision of the Mesozoic Era of the geologic time-scale. Type locality, the Jura Mountains, Switzerland. The formations of this system were first studied in the south of England by William Smith, the father of stratigraphy, and the period was named by A. Brongniart in 1829. The Jurassic period began approximately 150,000,000 years ago and lasted for 40,000,000 years. In North America the formations of this system are best exposed on the Pacific Coast, where they occur both in the Rocky Mountain and Pacific geosynclines. No upper Jurassic strata are known to occur in eastern North America. Aeolian "Red Beds" of Early and Middle Jurassic age occur in the western interior and are especially well exposed in the Colorado Plateau. Marine Jurassic formations occur in the Arctic, also Africa, South America, Australia, New Zealand, Asia, the Himalayas, and Japan. The principal economic products of the Jurassic age are: gold (Sierra Nevada), coal, and lithographic limestone from Solenhofen, Bavaria. Plant life during this period was essentially like that of the Triassic. Among the marine invertebrates the pelecypods and cephalopods are the most important fossils. Sharks and the modern type of fishes were abundant. The complete adaptive or radial evolution of the reptiles (Saurians) during this period is proved by the fossil skeletons of Ichthyosaurs (fish lizards), Plesiosaurs (marine lizards), Teleosaurus (ancestral crocodile), Pterosaurs (flying reptiles) (See **Fossil Reptiles**), and a number of terrestrial herbivorous and carnivorous dinosaurs, such as Diplodocus, Stegosaurus, Ceratosaurus, and Allosaurus. Perhaps the most famous fossil of this or any other geologic period is Archaeopteryx, the "missing link" between the reptiles and the birds. Severe crustal deformations occurred near the close of the Jurassic, especially in the Cordilleran region, with the birth of the Sierra Nevada and the Cascade Mountains. These mountain-building movements, which typify the close of the Jurassic Period in North America, are referred to as the Sierra Nevada Revolution.

JURY PROBLEM.   A differential equation, solved numerically by a method of successive approximations, which fits the solution to given boundary conditions.

Elliptic equations, such as the Poisson equation, lead to jury problems. A partial differential equation, may combine a jury problem and a marching problem.

JUVENILE HORMONES.   One of several hormones, that retard the development of insects in the larval stage. So called because they prevent the insect from maturing by maintaining its juvenile characteristics. Obtained naturally from silk moths; various syntheses indicate possible use as insecticides, especially for fire ants. Composition of one type is $C_{18}H_{30}O_3$.

See also **JH**; and **Insecticide**.

K

KACHCHAN.   See **Winds and Air Movement**.

KAFFIR.   See **Sorghum**.

KAHAMSIN.   See **Winds and Air Movement**.

KAMCHATKA CURRENT.   See **Ocean Currents**.

KAME.   Irregularly shaped mounds and depressions associated with terminal moraines. A kame topography is usually the result of a rather mixed set of glacial conditions, including both stratified and unstratified drift, and frequent kettle holes. One peculiarity of the term is that it is never used for a single mound or depression but rather to designate the character and origin of the general kame type of topography found only in glaciated regions.

KAMMERER, PAUL (1880–1926).   At the start of the twentieth century Paul Kammerer was considered the brightest star of the biology institute, the Vivarium, of the University of Vienna. He was a convinced Lamarckian and was intent on demonstrating the inheritance of acquired characteristics. His most celebrated experiments, which led eventually to his downfall, were performed on *Alytes obstericans*, the midwife toad. This, normally terrestrial, animal can be made to breed in water, when the male develops pigmented swellings, or "nuptial pads" on its hands, with which to grip the slippery back of the female while mating. Kammerer reported that males from eggs selected for the capacity to survive under water emerged with ready-made nuptial pads. He interpreted this to mean that the progeny had inherited the environmentally imposed characteristic of their father, but he probably deceived himself in his characterization of vestigial swellings. See also **Lamarck, Jean-Baptiste Pierre Antoine de Monet de(1744–1829)**.

World War I interrupted Kammerer's work, and it was not until 1923 that he gave a demonstration of his results in Cambridge, UK. They were met with skepticism. Kammerer defended himself vigorously and finally in 1926 he sent the last surviving toad to an American zoologist for dissection. Kingsley Noble found that, when pierced, the dark swellings exuded a black pigment with the appearance of Indian ink. Kammerer rejected the implication of fraud and suggested sabotage by persons unknown. Some months later, on a hillside outside Vienna, he shot himself, leaving a note in which he reaffirmed his innocence. Kammerer, who was a socialist, was shortly to have taken up an appointment in Moscow, where his Lamarckian views had found great favor. The story is told in a rather biased, lightly fictionalized form, in Arthur Koestler's book, *The Case of the Midwife Toad*.

Additional Reading

Gratzer, W.: *The Undergrowth of Science*. Oxford University Press, Oxford, UK, 2000.
Hamilton, D.: *The Monkey Gland Affair*, Chatto and Windus, London, UK, 1986.
Koestler, A.: *The Case of the Midwife Toad*, Random House, New York, NY, 1972.

WALTER GRATZER, King's College, London, UK

KANGAROO.   See **Marsupialia**.

KANGAROO RAT.   See **Rodentia**.

KAOLINITE.   Kaolinite [CAS: 1332-58-7] is the most common mineral of a group of hydrous aluminum silicates, which result from the breaking down of aluminum-rich silicate rocks, such as the feldspars and nepheline syenites, either through weathering or hydrothermal activity. Kaolinite, when pure, corresponds to the formula $Al_2Si_2O_5(OH)_4$, and occurs in white, clay-like masses. It has a perfect basal cleavage; is flexible, but not elastic; hardness, 2–2.5; specific gravity, 2.6–2.63; luster, pearly to dull; color, white when pure, as described above, but may be yellow, red, blue, or brown; translucent to opaque.

Kaolinite is a mineral of widespread occurrence, well distributed throughout the world. The finest kaolinite locality in Europe is said to be in France, from whence the clay is obtained for porcelain ware. Cornwall and Devonshire in England supply large quantities of this mineral. In the United States, Pennsylvania, Virginia, Colorado, Georgia, and South Carolina contain deposits of kaolinite. The word kaolin or kaolinite is said to be a corruption of a Chinese word *kauling*, the name of a locality where this mineral is found. Kaolinite is very important commercially in the manufacture of china and pottery.

KAON.   See **Muon**.

KAPOK.   See **Silk Cotton Trees**.

KARLE, JEROME.   (1918–).   An American physical chemist who won the Nobel Prize for chemistry along with Herbert A. Hauptman in 1985. He developed a series of mathematical equations that allow determination of phase information from X-ray crystallography intensity patterns. The advent of computers allowed the use of the equations to determine the conformation of thousands of chemicals. The work was done at the Naval Research Laboratory in Washington, DC., where Karle headed the Laboratory for the Structure of Matter.

KÁRMÁN VORTICES.   Sometimes referred to as the Kármán "Street of Eddies," this is the name given to the regular array of two-dimensional vortices that form behind a circular cylinder when the velocity of flow is such that the Reynolds number is between 45 and 150. They are equally spaced on both sides of the wake, alternating from one side to the other. Benard first observed them, and T. von Kármán showed that only one configuration could be stable against slight displacement of their positions. See also **Fluid and Fluid Flow**.

KARRER, PAUL (1889–1971).   A recipient of the Nobel Prize for chemistry in 1937 with Walter N. Haworth. Although born in Moscow, he attended European universities and received his doctorate in Zurich. He Initiated work on flavins, carotenoids, and vitamins A and B, and accomplished work on structure and synthesis of vitamin B_2 as well as vitamins A and E.

KARST.   See **Hydrology**.

KASHMIR CYPRESS.   See **Cypress Trees**.

KATABATIC WIND.   1. Most widely used in mountain meteorology to denote a downslope flow driven by cooling at the slope surface during periods of light larger-scale winds; the nocturnal component of the along-slope wind systems. The surface cools a vertical column of the atmosphere starting at the slope surface and reaching perhaps 10–100 m deep. This column is colder than the column at equivalent levels over the valley or plain, resulting in a hydrostatic pressure excess over the slope relative to over the valley or plain. The horizontal pressure gradient, maximized at the slope surface, drives an acceleration directed away from the slope, or downslope. Although the pressure-gradient forcing is at its maximum at the slope, surface friction causes the peak in the katabatic wind speeds to occur above the surface, usually by a few meters to a few tens of meters. The depth of the downslope flow layer on simple slopes has been found to be 0.05 times the vertical drop from the top of the slope. Surface-wind speeds in mountain–valley katabatic flows are often 3–4 ms^{-1}, but on long slopes, they have been found to exceed 8 m s^{-1}. Slopes occur on many scales, and consequently katabatic flows also occur on many scales. At local scales katabatic winds are a component of mountain-valley wind systems. At scales ranging from the slopes of individual hills and mountains to the slopes of mountain ranges and massifs, katabatic flows represent the nocturnal component of mountain-plains wind systems mountain-plains wind systems. Besides diurnal-cycle effects, surface cooling can also result from cold surfaces such as ice and snow cover. Katabatic flows over such surfaces have been studied as glacier winds in valleys and as large-scale slope flows in Antarctica and Greenland. The large-scale katabatic wind blowing down the ice dome of the Antarctic continent has sometimes reached 50 m s^{-1} on the periphery of the continent. The persistence of the surface forcing and the great extent of the slopes on these great landmasses means that the flows are subject to Coriolis deflection, and thus they are not pure katabatic flows. 2. Occasionally used in a more general sense to describe cold air flowing down a slope or incline on any of a variety of scales, including phenomena such as the bora, in addition to thermally forced flows as described above.

From its etymology, the term means simply "going down" or "descending," and thus could refer to any descending flow; some authors have further generalized it to include downslope flows such as the foehn or chinook even though they do not represent a flow of cold air. This concept has given rise to the expression katafront, which indicates flow down a sloped cold-frontal surface.

See also **Chinook**; and **Winds and Air Movement**.

KATABOLISM.   See **Basal Metabolism**.

KATATHERMOMETER.   A type of cooling-power anemometer based upon the principle that the time constant of a thermometer is a function of its ventilation. The form developed in the early nineteenth century consisted of a liquid-in-glass thermometer having two calibration markers on the stem corresponding to 38.5° and 35 °C (101° and 95 °F). The thermometer was heated to 40 °C (104 °F), and the time required for the column to fall from 38° to 35 °C (100° to 95 °F) was measured by a stopwatch and used to compute the wind speed. It was especially useful for very low wind speeds. The katathermometer was used also, in human bioclimatology, to determine cooling power.

KATOPTRIC SYSTEM.   If an optical system is convergent, it is called katoptric if the object space focal length is positive and the image space focal length is negative. If a lens system is divergent, it is called katoptric if the object space focal length is negative and the image space focal length is positive.

KATYDID (*Insecta, Orthoptera*).   Large winged insects with long hind legs formed for jumping and very long slender antennae. They belong to the family of long-horned grasshoppers. The true katydid is found throughout the United States east of the Rockies and sings normally in three syllables, which have been interpreted as ka-ty-did. Other insects of a different subfamily are commonly called katydids because of their similar appearance and habits. All of these insects are bright green and have leaf-like wings. See Fig. 1. A pink form of the katydid occasionally appears.

The katydid "sings" only at night. The male has a shrill call produced by highly developed organs at the base of the wing covers. These are drum-like and the action of the wings on these organs causes a loud sound. The call of the male is answered by a female chirrup and can be heard for about one-quarter mile. Katydids are harmless to people. See also **Grasshopper**.

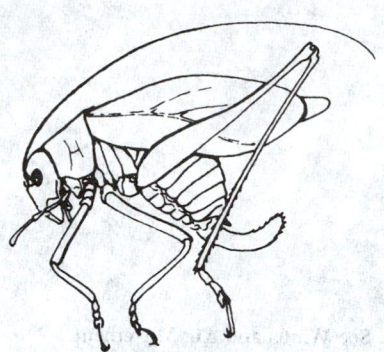

Fig. 1.   Katydid. (*USDA*.)

KATZ, BERNARD (1911–2003).   Sir Bernard Katz was a German-born British biophysicist, best known for his work on the mechanisms of neural transmission. Katz was born and educated in Leipzig, Germany. After qualifying in medicine he left Nazi Germany in 1935, and began physiological research in the laboratory of A. V. Hill at University College, London. He started working on problems associated with the electrical stimulation of nerves and the processes of neuromuscular transmission, and gained his PhD in 1938. Just before the outbreak of World War II he joined (Sir) John Eccles at the Kanematsu Institute in Sydney, Australia, where he collaborated in further neurophysiological experiments with Eccles, before enlisting as a radar officer in the Australian Royal Air Force. In 1946 he returned to University College, London. Here he remained for most of his career although he spent a substantial part of the late 1940s in Cambridge and Plymouth, working with Sir Alan Hodgkin and Sir Andrew Huxley on the mechanisms by which the nerve impulse is transmitted, using the giant axon of the squid. For the next three decades Katz's work focused on the neuromuscular junction; he showed that the chemical neurotransmitter acetylcholine is stored in nerve terminals, in small packets called vesicles that can be seen in the electron microscope, and is released in specific quantities called quanta when stimulated by the arrival of the neural impulse. This work won him the Nobel Prize in Physiology or Medicine in 1970, which he shared with Ulf von Euler and Julius Axelrod. See also **Axelrod, Julius (1912–2004)**; **Eccles, John Carew (1903–1997)**; **Hill, Archibald Vivian (1886–1977)**; **Hodgkin, Alan Lloyd (1914–1998)**; and **Huxley, Andrew Fielding (1917–)**.

Additional Reading

Frank, R.: Sir Bernard Katz, In: Fox, D.M., M. Meldrum, and I. Rezak, eds., *Nobel Laureates in Medicine or Physiology: A Biographical Dictionary*, Garland Publishing, New York, NY. 1990, pp. 298–303.

E.M. TANSEY, Wellcome Institute for the History of Medicine, London, UK

KAURI-BUTANOL VALUE.   A measure of the aromatic content and hence the solvent power of a hydrocarbon liquid. Kauri gum is readily soluble in butanol but insoluble in hydrocarbons. The KB value is the measure of the volume of solvent required to produce turbidity in a standard solution containing kauri gum dissolved in butanol. Naphtha fractions have a KB value of about 30, and toluene about 105.

K-DISPLAY.   In radar, a modified A-display in which a target appears as a pair of vertical deflections or blips instead of a single deflection. When the radar antenna is correctly pointed at the target in azimuth, the blips are of equal height. When not correctly pointed, the difference in blip height is an indication of direction and magnitude of azimuth pointing error. Also called *K-scan*, *K-scope*, or *K-indicator*.

KEKULE, AUGUST. (1829–1896).   Born in Darmstadt, Germany, Kekule laid the basis for the ensuing development of aromatic chemistry. His idea of a hexagonal structure for benzene in 1865 was a monumental contribution to theoretical organic chemistry. This had been preceded in 1858 by the remarkable notion that carbon was tetravalent and the carbon atoms could be joined to each other in molecules. The theory of the benzene ring has been called the "most brilliant piece of scientific prediction to be found in the whole field of organic chemistry, for besides promulgating the idea, he had predicted the number and types of isomers which might be expected in various substitutions on the ring" (L. B. Clapp).

KELOID. See **Burn**.

KELP. See **Ocean Resources (Energy)**.

KELVIN BRIDGE. See **Bridge Circuits (Electrical/Electronic)**.

KELVIN-HELMHOLTZ INSTABILITY. An instability of the basic flow of an incompressible inviscid fluid in two parallel infinite streams of different velocities and densities.

If the overlying fluid has velocity U_2 and density $_2$, and the underlying fluid has velocity U_1, and density $_1$, disturbances of the form e^{ikx} (where k is the wavenumber) are unstable if

$$g(\rho_1^2 - \rho_2^2) < k\rho_1\rho_2(U_1 - U_2)^2,$$

where g is the acceleration of gravity. Thus, the flow is always unstable to short waves (high wavenumber) if $U_1 \neq U_2$.

Additional Reading

Drazin, P.G., and W.H. Reid: *Hydrodynamic Stability*, 2nd Edition, Cambridge University Press, New York, NY, 2004.

AMS

KELVIN LAW. When a system of rigid circuits does mechanical work under constant current conditions, the energy of the circuits increases at the same rate as work is done.

KELVIN TEMPERATURE SCALE. See **Absolute Zero**; and **Temperature**.

KELVIN (UNIT OF THERMODYNAMIC TEMPERATURE). The definition of the unit of thermodynamic temperature was given in substance by the 10th CGPM (1954) which selected the triple point of water as the fundamental fixed point and assigned to it the temperature 273.16 K, so defining the unit. The 13th CGPM (1967) adopted the name *kelvin* (symbol K) instead of "degree Kelvin" (symbol °K) and defined the unit of thermodynamic temperature as follows:

The kelvin, unit of thermodynamic temperature, is the fraction 1/273.16 of the thermodynamic temperature of the triple point of water.

Because of the way temperature scales used to be defined, it remains common practice to express thermodynamic temperature, symbol T, in terms of its difference from the reference temperature $T_0 = 273.15$K, the ice point. This temperature difference is called a Celsius temperature, symbol t, and is defined by the quantity equation

$$t = T - T_0.$$

The unit of Celsius temperature is the degree Celsius, symbol °C, which is by definition equal in magnitude to the kelvin. A difference or interval of temperature may be expressed in kelvins or in degrees Celsius (13th CGPM, 1967). The numerical value of a Celsius temperature t expressed in degrees Celsius is given by

$$t/\,^\circ\mathrm{C} = T/\mathrm{K} - 273.15.$$

The kelvin and the degree Celsius are also the units of the International Temperature Scale of 1990 (ITS-90) adopted by the CIPM in 1989.

See also **The International System Of Units (SI)**; and **Units and Standards**.

KELVIN WAVE. A type of low-frequency gravity wave trapped to a vertical boundary, or the equator, which propagates anticlockwise (in the Northern Hemisphere) around a basin. The flow is parallel to the boundary and in geostrophic balance with the pressure gradient perpendicular to the boundary. The velocity normal to the boundary is identically zero. For a homogeneous ocean, the wave is called a barotropic or external Kelvin wave, and for a stratified ocean, the wave is called a baroclinic or internal Kelvin wave. Near a boundary in a rotating system, a Kelvin wave propagates with wave crests perpendicular to the side wall and wave height greatest at the side wall to the right of an observer looking in the direction of wave propagation. The wave height decreases exponentially from the side wall with an *e*-folding length scale equal to the Rossby radius of deformation c/f, in which f is the Coriolis parameter and c is the phase speed of the wave in the along-boundary direction. In the shallow water approximation the waves are nondispersive with frequency

$$w = \pm ck$$

in which k is the along-boundary wavenumber and the phase speed

$$c = (gH)^{1/2}$$

with g the acceleration of gravity and H the mean fluid depth. Related to Kelvin waves in a channel are Poincare waves.

KELVIN, WILLIAM THOMSON, 1ST BARON (1824–1907). An Irish mathematician and physicist, Kelvin designed several kinds of electrometer. He did fundamental research into thermodynamics, especially wave-motion and vortex-motion, and helped develop the absolute temperature scale now referred to as Kelvin. He was made Baron Kelvin of Largs for his work on undersea telegraph cables. His work was instrumental in enabling signals to be sent across the Atlantic Ocean.

See also **Absolute Zero**; **Bridge Circuits (Electrical/Electronic)**; **Electrical Instruments**; **Kelvin Law**; **Kinetic Theory**; and **Temperature**.

J. M. I.

KENDALL, EDWARD CALVIN (1886–1972). Kendall graduated from Columbia University in 1908 and remained there for his postgraduate education: MS (1909) and PhD in chemistry (1910). After unsuccessful short periods at Parke Davis and Co. and St Luke's Hospital, New York, he joined the Mayo Clinic in charge of a new biochemistry department. Here, in December 1914, he isolated crystalline thyroxine.

Over the next few years Kendall attempted to determine the chemical structure of thyroxine, but was anticipated in 1925 by C. R. Harington in London. Kendall turned to understanding the effects of thyroxine on the basal metabolism and elucidated the chemical structure of glutathione in 1929. In the 1930s he switched to the hormones of the adrenal cortex, attempting to find the active substances in adrenal extract useful in treating adrenal insufficiency. See also **Adrenaline and Noradrenaline**.

Initially researchers thought in terms of a single adrenal hormone, but by the mid-1930s it was apparent that there were several active compounds and these were steroid derivatives. Success in identifying these fell first to a rival group of organic chemists led by Thadeus Reichstein. Kendall continued his research using a colony of adrenalectomized dogs at the Mayo Clinic and, despite setbacks, his team and researchers at the pharmaceutical firm Merck synthesized cortisone in 1948. However, Kendall needed to prove the utility of this steroid beyond relatively rare cases of adrenal insufficiency. In collaboration with Philip Hench, arthritic patients were successfully treated with cortisone. Rather wild claims were made for the general efficacy of this drug and further work was necessary before its side effects and proper uses were determined. However, it has proven immensely useful in suppressing inflammation. See also **Hench, Philip Showalter (1896–1965)**; and **Reichstein, Tadeus (1897–1996)**.

In 1950, Kendall was awarded the Nobel Prize in Physiology or Medicine, with Reichstein and Hench, for work on the hormones of the adrenal cortex.

Additional Reading

Fox, D.M., M. Meldrum, and I. Rezak: *Nobel Laureates in Medicine or Physiology*, Garland, New York, NY, 1990.

Ingle, D.: Edward C. Kendall, *Biographical Memoirs, National Academy of Science*, **47**, 249–292 (1975).

HELEN J. POWER, Wellcome Trust Centre, London, UK

KENDREW, JOHN COWDERY (1917–1997).   John Kendrew was an English chemist and molecular biologist who builtthe first atomic model of a protein molecule.

John Kendrew was born in Oxford to academic parents. His father taught geography at the University and his mother, an art historian, was an expert on the Italian Renaissance. Kendrew received his schooling at Clifton College in Bristol. Inspired by an outstanding chemistry teacher, he studied natural sciences at Cambridge University and had begun to work for a PhD in physical chemistry when war broke out and he was drafted into research on radar. He later switched to operational research and ended the war with the honorary rank of Wing Commander in the Royal Air Force and as scientific adviser to Lord Mounbatten's South Eastern Command in Ceylon. There he met the X-ray crystallographer John Desmond Bernal who, in 1934, had discovered the rich X-ray diffraction pattern of a crystalline protein, the enzymepepsin. Bernal convinced Kendrew that the structure of proteins was the most fundamental unsolved problem in biochemistry and suggested that he join Max Perutz at the Cavendish Laboratory in Cambridge. Perutz had begun someyears earlier to work on the structure of haemoglobin, the protein of the red blood cell. See also **Biochemistry (The History)**.

Kendrew joined Perutz in October 1945 and embarked on a comparative study of adult and fetal sheep haemoglobin. Nowadays, protein structures are being solved every day, which makes it hard to imagine how courageous Kendrew's decision was then to take up protein crystallography. Most crystallographers and physicists regarded the problem of protein structure as insoluble, because the X-ray diffraction pattern from a crystal contains only half the information needed to determine its structure — the amplitudesof the diffracted rays, but not their phases — and there seemed to beno practical method of determining them. Not unexpectedly, Kendrew's study of sheep hemoglobin told him essentially nothing, which made him decide to take up an X-ray study of the simpler protein myoglobin, which has only a quarter of the molecular weight of haemoglobin and acts as an oxygen store in muscle. See also **Perutz, Max Ferdinand (1914–2002)**.

Even so, there seemed no way of solving the structure of such a largemolecule when the structures of most amino acids were still unknown, but in1953 Perutz discovered that the phase problem could be solved by attaching a heavy atom, such as mercury, to hemoglobin and comparing the diffraction patterns of hemoglobin crystals with and without the mercury atoms. Soon several adventurous young collaborators came to join Kendrew and managed to diffuse a variety of heavy atoms into his myoglobin crystals. By 1957, their joint efforts resulted in the first rough molecular model which revealed the course of its polypeptide chain and the position of its haem. The chain followed a convoluted course, devoid of any of the simplifying regularities which Kendrew and Perutz had hoped to be present in earlier attempts to interpret their X-ray patterns. It was the first glimpse of a protein molecule, but the resolution was still too low to give any chemical information. Two years later, in 1959, Kendrew and his collaborators raised the resolution sufficiently to build an atomic model of myoglobin, the first of any protein molecule ever. It showed that for most of its course, the polypeptide chain was coiled into α-helices of precisely the structure that Linus Pauling had predicted in 1951. The haem was embedded in a pocket between the α-helices, its iron atom bound on one side to the amino acid histidine, and on the other side to a water molecule which had replaced the oxygen molecule originally present there.

At the same time, Perutz and his collaborators solved the structure of hemoglobin at low resolution, but sufficient to show the fold of its four polypeptide chains to be the same as that of the single chain of myoglobin and to reveal that fold as a fundamental pattern of nature, evolved to allow haem proteins to act as oxygen carriers. The two structures were solved independently by physical methods without any assumptions about the chemical nature of the proteins. Since no combination of errors could have led to two such similar structures, the results proved themselves and were never seriously doubted. They led to the award of the 1962 Nobel Prize for chemistry jointly to Kendrew and Perutz and they gradually opened up the subject of protein crystallography worldwide. 1962 also marked the end of Kendrew's scientific research and the shift of his interests to international scientific collaboration. He was deeply concerned that European universities and research institutes were slow in grasping the promise of molecular biology and that Europe was falling behind the United States in training young people in the subject. In 1963, he became one of the founders of the European Molecular Biology Organization (EMBO) http://www.embo.org/, which started a program of traveling fellowships and summer schools that has been anoutstanding success. He also founded and remained for many years editor-in-chief of the *Journal of Molecular Biology*. He took great pride in it, and it became the journal where nearly all important papers in the subject appeared. In 1974, after four years of skillful diplomacy, he persuaded governments to build a European Laboratory of Molecular Biology in Heidelberg, and he became its first director. This great laboratory will stand as his monument.

After his retirement from the Heidelberg laboratory Kendrew became President of St John's College Oxford. He was a connoisseur of art and architecture, and used to be one of the trustees of the British Museum in London. For many years he was President of the International Council of Scientific Unions, not as a figurehead, but as its driving force. Almost to his death he remained an indefatigable traveler in the service of international scientific collaboration.

Additional Reading

Kendrew, J.C.: The Three-dimensional Structure of a Protein Molecule, *Scientific American*, 96–104. (December 1961).

MAX PERUTZ, Cambridge University, Cambridge, UK

KENNEDY SPACE CENTER.   See **Spaceports U.S.**

KENNELLY-HEAVISIDE LAYER.   See **Ionosphere**.

KENTUCKY BLUEGRASS.   See **Grasses**.

KENTUCKY COFFEE TREE.   Sometimes simply called *coffeetree*, this tree is of the genus *Gymnocladus* and is not related to the coffee (beverage) tree of the *Rubiaceae* (madder) family. However, the Kentucky coffeetree was so named because in early pioneering days, the seeds of the tree were roasted, ground and used as coffee. The tree is allied to the locusts and redbuds of the family *Caesalpinaceae*. The Kentucky coffeetree is rather sparsely distributed throughout the middle region of the Mississippi Valley. When young, the bipinnate leaves of the tree are relatively large. The wood can be used in cabinet making.

The champion Kentucky coffeetree *G. dioicus* in the United States selected by American Forests is located in Maryland. The circumference at 4.5 feet (1.4 meters) above the base is 187 inches (475 centimeters); the height is 84 feet (25.6 meters); and the spread is 80 feet (24.4 meters).

A closely related species is *G. canadensis*. which is found in Canada and over much of the United States. This tree is called Chicot and sometimes *stump tree* because during winter it appears to be dead.

KEPLER, JOHANNES (1571–1630).   Kepler was a German mathematician and astronomer. He worked for a time with Tycho Brahe and used many of his accurate positional determinations of the planets. After Tycho Brahe died, Kepler began to intensely study the orbits of the planets. He dismissed the ancient idea that planets moved in perfect circles and concentrated on the planet, Mars. His astronomical observations helped Kepler to formulate his three laws of planetary motion. The first law established that planets travel in elliptical orbits. His second law describes planetary velocity. Kepler loved the beauty and structure in the universe. He was guided by a theory of harmony and his third law is the relationship between orbital periods of the planets and their distances from the Sun.

Kepler is also remembered for being instrumental in the development of telescopes. He also invented the convex eyepiece. Kepler's Rudolphine Tables (1627) was a compilation of accurate tables of planetary motion. Kepler was an important scientist in the 17th-century scientific revolution.

See also **Kepler's Laws of Planetary Motion**; and **Planets and the Solar System**.

J. M. I.

KEPLER MISSION. Planned for launch in June of 2008, Kepler will monitor 100,000 stars similar to our sun for four years. The results will be extremely important either way. If Kepler detects many habitable, Earth-size planets, it could mean the universe is full of life. Kepler would then be a stepping stone to the next extensive search for habitable planets and life, the Terrestrial Planet Finder. If nothing is found, it may mean we're alone in the galaxy. See also **Terrestrial Planet Finder (TPF)**.

The centuries-old quest for other worlds like our Earth has been rejuvenated by the intense excitement and popular interest surrounding the discovery of giant planets like Jupiter orbiting stars beyond our solar system. With the exception of the pulsar planets, all of the extrasolar planets detected so far are gas giants, approximately 150 as of 2005. The challenge now is to find terrestrial planets (habitable planets like Earth), which are 30 to 600 times less massive than Jupiter.

The *Kepler Mission* a NASA Discovery mission, is specifically designed to survey our region of the Milky Way galaxy to detect and characterize hundreds of Earth-size and smaller planets in or near the habitable zone. The habitable zone encompasses the distances from a star where liquid water can exist on a planet's surface. See also **Discovery Program**.

Results from this mission will allow us to place our solar system within the continuum of planetary systems in the Galaxy.

The Kepler mission is managed for NASA by the Jet Propulsion Laboratory, Pasadena, CA. The Principal Investigator is William Borucki of NASA Ames Research Center. The mission team is composed of twenty-five scientists from sixteen institutions. Ball Aerospace and Technologies Corp., Boulder, CO, will build the spacecraft.

Scientific Objective

The scientific objective of the *Kepler Mission* is to explore the structure and diversity of planetary systems. This is achieved by surveying a large sample of stars to:

Goal 1: Determine the percentage of terrestrial and larger planets there are in or near the habitable zone of a wide variety of stars. The frequency of planets is derived from the number and size of planets found and from the number and spectral type of stars monitored. Even a null result would be highly meaningful because of the large number of stars searched and the low false alarm rate.

Goal 2: Determine the distribution of sizes and shapes of the orbits of these planets. The planet's area is found from the fractional brightness decrease and the stellar area. For a detection with a statistical significance of >8 sigma, the uncertainty of the planetary area is about 14% and the planetary radius of 7%.

The planet's semi-major axis is derived from the measured period and stellar mass, using Kepler's Third Law. An uncertainty in the semi-major axis of about 1% results from a 3% uncertainty in the mass of the central star, derived from ground-based spectroscopic observations and stellar modeling.

Goal 3: Estimate the frequency of planets and orbital distribution of planets in multiple-stellar systems.This goal is achieved by comparing the number of planetary systems found in single versus multi-stellar systems. Multiple-stellar systems are identified from ground-based spectroscopic measurements if they are tightly bound or from high angular resolution observations if they are widely spaced systems.

Goal 4: Determine the distributions of semi-major axis, albedo, size, mass and density of short-period giant planets. Short-period giant planets are also detected from variations in their reflected light. As above, the semi-major axis is derived from the orbital period and the stellar mass.

Transits should also be seen in about 10% of the cases and the size of the planet determined. These planets are found in the first few months of the mission. From the planet size, semi-major axis and the amplitude of reflected light modulation, the albedo is determined. The density is calculated when the planet is seen both in transit (to yield its size) and when Doppler spectroscopy is used (to determine the planet's mass for stars with $m_v < 13$ and cooler than F5) as was done for the case of HD209458b.

Goal 5: Identify additional members of each photometrically discovered planetary system using complementary techniques. Observations using both the Space Interferometry Mission (SIM) and ground-based Doppler spectroscopy are used to search for additional massive companions, which do not transit, thereby providing greater details of each planetary system discovered.

Goal 6: Determine the properties of those stars that harbor planetary systems.The spectral type, luminosity class, and metalicity for each star showing transits are obtained from ground-based observations. Additionally, rotation rates, surface brightness inhomogeneities and stellar activity are obtained directly from the photometricdata. Stellar age and mass is determined from *Kepler* p-mode measurements (asteroseismology).

The *Kepler Mission* also supports the objectives of future NASA Origins theme missions Space Interferometry Mission (SIM) and Terrestrial Planet Finder (TPF), By identifying the common stellar characteristics of host stars for future planet searches, By defining the volume of space needed for the search, and By allowing SIM to target systems already known to have terrestrial planets.

Transit Method of Detecting Extrasolar Planets. When a planet crosses in front of its star as viewed by an observer, the event is call a transit. Transits by terrestrial planets produce a small change in a star's brightness of about 1/10,000 (100 parts per million, ppm), lasting for 2 to 16 hours. This change must be absolutely periodic if it is caused by a planet. In addition, all transits produced by the same planet must be of the same change in brightness and last the same amount of time, thus providing a highly repeatable signal and robust detection method.

Once detected, the planet's orbital size can be calculated from the period (how long it takes the planet to orbit once around the star) and the mass of the star using Kepler's Third Law of planetary motion. The size of the planet is found from the depth of the transit (how much the brightness of the star drops) and the size of the star. From the orbital size and the temperature of the star, the planet's characteristic temperature can be calculated. From this the question of whether or not the planet is habitable (not necessarily inhabited) can be answered.

Kepler Mission Design. For a planet to transit, as seen from our solar system, the orbit must be lined up edgewise to us. The probability for an orbit to be properly aligned is equal to the diameter of the star divided by the diameter of the orbit. This is 0.5% for a planet in an Earth-like orbit about a solar-like star. (For the giant planets discovered in four-day orbits, the alignment probability is more like 10%.) In order to detect many planets one can not just look at a few stars for transits or even a few hundred. One must look at thousands of stars, even if Earth-like planets are common. If they are rare, then one needs to look at many thousands to find even a few. *Kepler* looks at 100,000 stars so that if Earths are rare, a null or near null result would still be significant. If Earth-size planets are common then *Kepler* should detect hundreds of them.

Considering that we want to find planets in the habitable zone, the time between transits is about one year. To reliably detect a sequence one needs four transits. Hence, the mission duration needs to be at least four years.

The *Kepler* instrument is a specially designed 0.95-meter (3 foot) diameter telescope called a photometer or light meter. It has a very large field of view for an astronomical telescope — 105 square degrees — or about the area of both your hands held at arm's length, in order to observe the necessary large number of stars. It stares at the same star field for the entire mission and continuously and simultaneously monitors the brightnesses of more than 100,000 stars for the life of the mission — 4 years.

The diameter of the telescope needs to be large enough to reduce the noise from photon counting statistics, so that it can measure the small change in brightness of an Earth-like transit. The design of the entire system is such that the combine differential photometric precision over a 6.5 hour integration is less than 20 ppm (one-sigma) for a 12th magnitude solar-like star including an assumed stellar variability of 10 ppm. This is a conservative, worse-case assumption of a grazing transit. A central transit of the Earth crossing the Sun lasts 13 hours. And about 75% of the stars older than 1 Gyr are less variable then the Sun on the time scale of a transit.

The photometer must be spacebased to obtain the photometric precision needed to reliably see an Earth-like transit and to avoid interruptions caused by day-night cycles, seasonal cycles and atmospheric perturbations, such as, extinction associated with ground-based observing.

Extending the mission beyond four years provides for:

1. Improving the signal to noise by combining more transits to permit detection of smaller planets.
2. Finding planets in orbits with larger periods.
3. Finding planets around stars that are noisier either due to being fainter or having more variability.

Mission Design

Photometer and Spacecraft. The *Kepler* photometer is a simple single purpose instrument. It is basically a Schmidt telescope design with a 0.95-meter (3 foot) aperture and a 105 deg^2 (about 12 degree diameter) field-of-view (FOV). It is pointed at and records data from just a single group of stars for the four year duration of the mission. See Fig. 1.

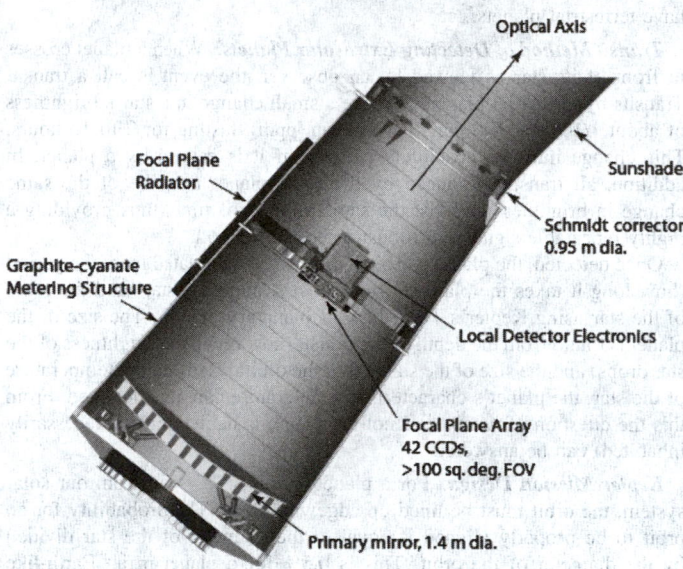

Fig. 1.   The *Kepler* Photometer.

The photometer is composed of just one "instrument," which is, an array of 42 CCDs (charge coupled devices). Each 50×25 mm CCD has 2200×1024 pixels. The CCDs are read out every three seconds to prevent saturation. Only the information from the CCD pixels where there are stars brighter than $m_v = 14$ is recorded. (The CCDs are not used to take pictures. The images are intentionally defocused to 10 arc seconds to improve the photometric precision.) The data are integrated for 15 minutes.

The instrument has the sensitivity to detect an Earth-size transit of an $m_v = 12$ G2V (solar-like) star at 4 sigma in 6.5 hours of integration. The instrument has a spectral bandpass from 400 nm to 850 nm. Data from the individual pixels that make up each star of the 100,000 main-sequence stars brighter than $m_v = 14$ are recorded continuously and simultaneously. The data are stored on the spacecraft and transmitted to the ground about once a week.

The spacecraft provides the power, pointing and telemetry for the photometer. See Fig. 2. Pointing at a single group of stars for the entire mission greatly increases the photometric stability and simplifies the spacecraft design. Other than the gimbaled antenna for communications, the small gyroscopes used to maintain the pointing and an ejectable cover, there are no other moving or deployable parts and there are no liquids to slosh. This enhances the pointing stability and the overall reliability of the spacecraft.

Launch Vehicle and Orbit. The continuous viewing needed for a high detection efficiency for planetary transits requires that the field-of-view (FOV) of the photometer be out of the ecliptic plane so as not to be blocked periodically by the Sun or the Moon. A star field near the galactic plane that meets these viewing constraints and has a sufficiently high star density has been selected, with galactic coordinates centered on l = 70° and b = 5° (RA = 19h 45 m, dec = +35°).

An Earth-trailing heliocentric orbit with a period of 372.5 days provides the optimum approach to meeting of the combined Sun-Earth-Moon avoidance criteria within the Boeing D2925-10 (Delta-II) launch vehicle capability. In this orbit the spacecraft slowly drifts away from the Earth and is at a distance of 0.5 AU (worst case) at the end of four years. Telecommunications and navigation for the mission are provided by NASA's Deep Space Network (DSN).

Another advantage of this orbit is that it has a very-low disturbing torque on the spacecraft, which leads to a very stable pointing attitude. Not being in Earth orbit means that there are no torques due to gravity gradients, magnetic moments or atmospheric drag. The "largest" external torque then is that caused by solar pressure. This orbit also avoids the high radiation dosage associated with an Earth orbit, but from time to time is subject to solar flares.

The mission must last long enough to detect and confirm the periodic nature of the transits of planets in or near the HZ. A four year mission is proposed which enables a four-transit detection of all orbits up to one year in length and a three-transit detection of periods up to 1.33 years. This mission duration also provides three-transit detections for 50% of 1.6 year orbits and 10% of 1.9 year orbits.

We have also proposed a two year mission extension which greatly enhances the ability to detect planets smaller than Earth and reliably detect Earth-size planets in orbits corresponding to that of Mars (2 year periods).

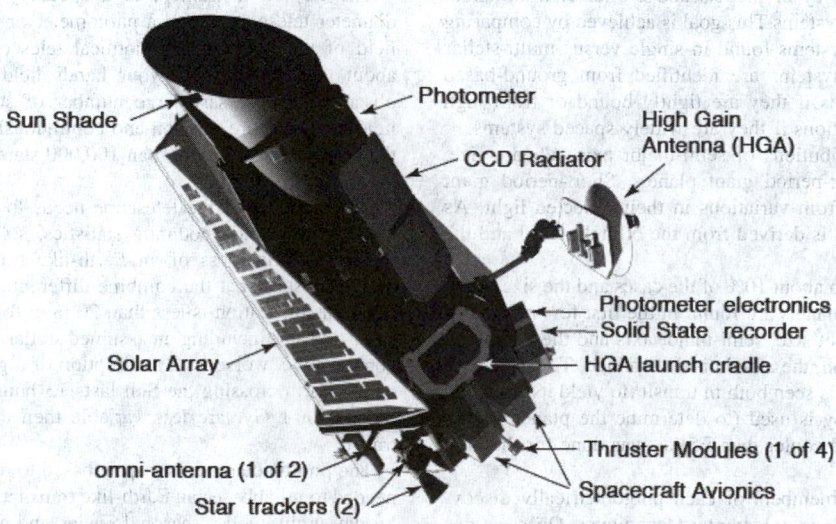

Fig. 2.   Photometer ounted on the pacecraft.

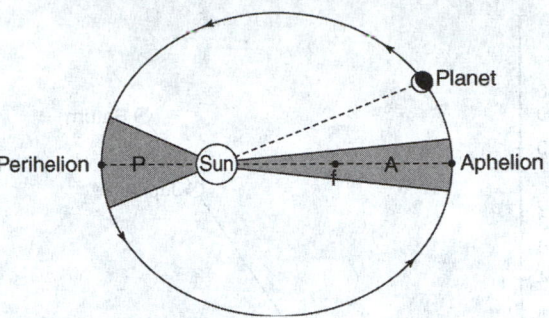

Fig. 1.   Elliptic orbit of planet with sun at one focus. Areas *A* and *P* are described in equal length of time.

When first published, the Keplerian laws were without theoretical foundation, being empirically derived from observational data. They created a tremendous stir, and were declared heretical, since they abandoned the circle as the only possible path for a planet. About 50 years later, Newton was able to show that the Keplerian laws are a direct consequence of motion under the action of the law of universal gravitation. It has since been shown that all types of orbital motion are characterized by these Keplerian laws.

The harmonic law, as first stated by Kepler, was independent of the masses of the bodies involved; but Newton was able to show that the masses are actually involved, and he obtained a more rigorous expression for it. Assume M_1 to be the mass of one body, M_2 to be the mass of the other, R to be the mean distance between the two bodies, P the sidereal period of revolution of one about the other, and K a numerical constant involving the constant of universal gravitation, and we have

$$K(M_1 + M_2) = \frac{R^3}{P^2}$$

The masses of the planets are all inappreciable in comparison with the mass of the sun; accordingly, the expression $K(M_1 + M_2)$ is virtually a constant when considering the motions of the planets about the sun. In cases such as binary star orbits, and motions of satellites about primaries, the rigorous expression of the harmonic law is of immense value in the determination of masses of celestial objects.

Although the development of the theory of relativity has necessitated slight modifications in the original laws of planetary motion, nevertheless, together with the law of universal gravitation, those laws form the basis upon which rests the entire structure of celestial mechanics and related fields.

It is interesting to see how easily the third law of Kepler's may be "discovered" by the use of techniques not available to Kepler. Kepler knew from Tycho's observations the relative distances from the sun to any of the planets (out as far as Saturn); such relative distances can be calculated without knowing the actual distances which were not found for another century. If one plots these relative distances against the planetary periods, using log-log graph paper, there results a fine straight-line graph of points, one point for each planet, for which the slope is $\frac{3}{2}$. See Fig. 2.

This shows at once that there is a constant ratio of R^3/T^2; or the ratio of the cube of R, mean distance from the sun, to the square of the sidereal period T, is the same ratio for all planets. Thus, in a few minutes, one may discover what it took Kepler ten years to find.

Kepler's laws are, of course, applicable to the motion of artificial or launched satellites circling the earth. One may draw a graph similar to that drawn for the planets, except that now the earth rather than the sun is the controlling body. Just as no satellite could possibly continue to orbit the sun if its mean distance were less than the radius of the sun, no satellite can continue to orbit the earth if its orbit intersects the earth. (In fact, nearby satellites are captured by the atmosphere and are likely to be burned up in the upper atmosphere as their kinetic energy is rapidly turned into heat.) It is interesting, however, to draw a straight line on our log-log graph from the point representing Sputnik I, the first artificial satellite to be launched in 1957 with its period of 96 minutes, to the moon with its period of 27.32 days. It will be found (Fig. 3) that other satellites lie as points on this line. The semimajor axis (or mean distance) of each satellite's orbit may

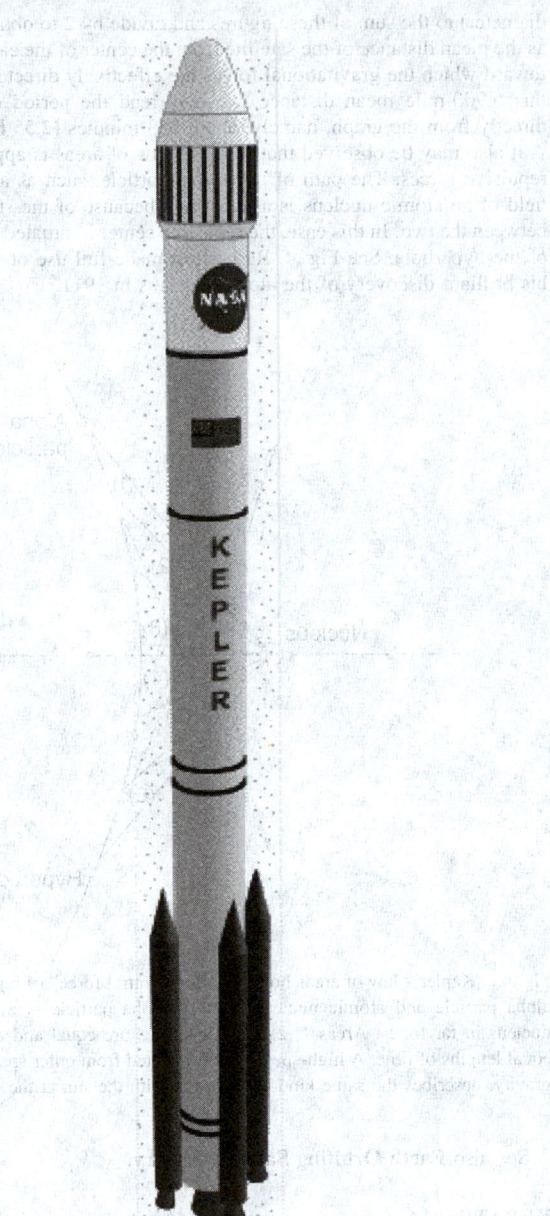

Fig. 3.   Delta 2925-10 (Delta-II) launch vehicle.

Web References

Kepler Mission: Detection of Giant Planets. http://kepler.nasa.gov/sci/basis/giants.html
Kepler Mission: Expected Results. http://kepler.nasa.gov/sci/basis/results.html
Kepler Mission: In Depth Science. http://kepler.nasa.gov/sci/basis/index.html

KEPLER'S LAWS OF PLANETARY MOTION.   After years of labor in attempting to develop a theory of planetary motion that would satisfy the accumulation of planetary positions determined by Tycho Brahe, Johann Kepler decided to abandon the idea that the circle is the only perfect curve and, hence, must be followed by the planets. In the early part of the seventeenth century, Kepler announced three fundamental laws of planetary motion that satisfied Tycho Brahe's observations. These laws may be stated as (see Fig. 1):

1. Each planet moves in an ellipse with the sun at one focus.
2. The radius vector of each planet passes over equal areas in equal intervals of time. (The law of areas.)
3. The square of the period of revolution of a planet about the sun is proportional to the cube of the mean distance of the planet from the sun.

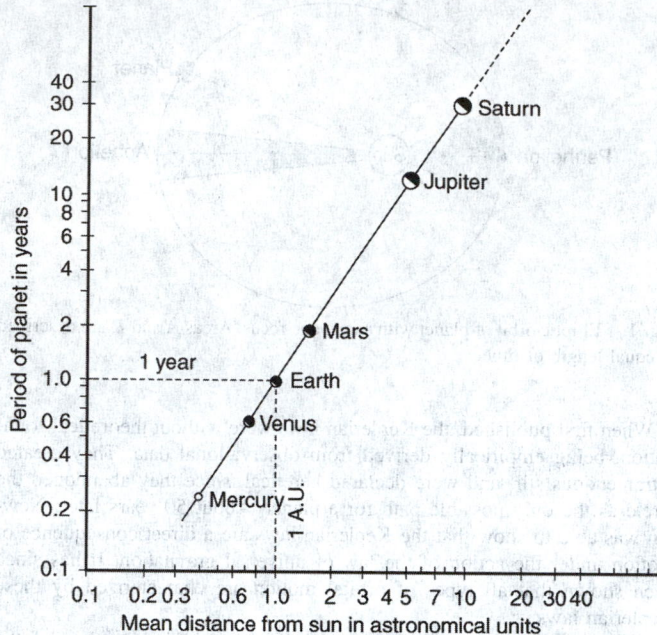

Fig. 2. Log-log graph of relative mean distances of planets from sun, in Astronomical Units (where sun-earth distance is 1 A.U.) versus period of planets about the sun in years. Straight-line graph with slope $\frac{3}{2}$ means constant ratio of R^3 to T^2, which Kepler found in 1619. Planets beyond Saturn, unknown to Kepler, can be added.

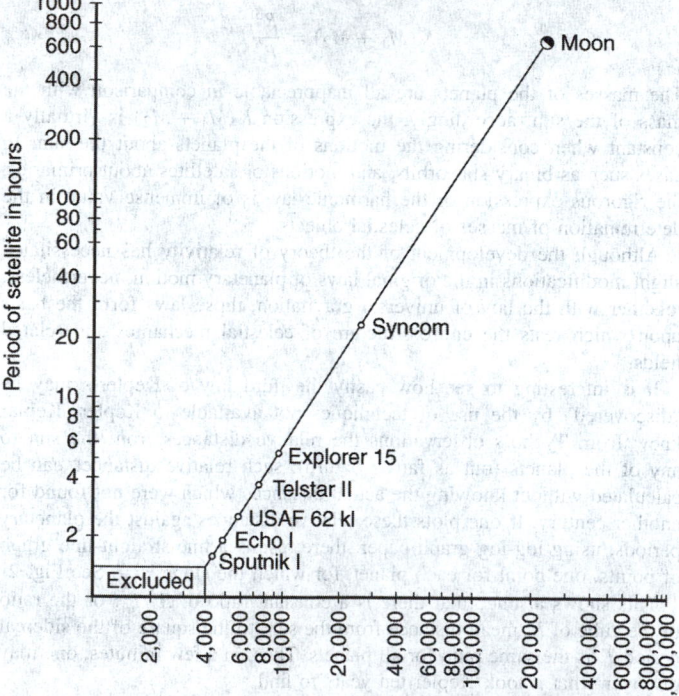

Fig. 3. Satellites of the earth—log-log graph of mean distance from center of earth versus period of satellite, from Sputnik I (1957) to moon. This figure is similar to Fig. 2 except that the earth rather than the sun controls the satellite periods. Again, the R^3/T^2 constant ratio shows up in the $\frac{3}{2}$ slope of the straight-line graph. Several representative satellites are listed.

be found from the addition of the earth's diameter to the usually published perigee and apogee distances, most commonly offered as minimum and maximum distances of the satellite above the earth's surface. Thus, for a satellite with a minimum distance 1,730 miles and maximum distance 2,120 miles from the earth's surface, we add 7,950 miles (for the earth's

diameter) to the sum of these figures and divide by 2 to obtain 5,900 miles as the mean distance of the satellite from the center of the earth — the point toward which the gravitational forces are effectively directed. Armed with this 5,900-mile mean distance, we may read the period of the satellite directly from the graph, namely, about 153 minutes (2.55 hours).

It also may be observed that Kepler's law of areas is applicable also to repulsive forces. The path of an atomic particle, such as a proton, in the field of an atomic nucleus is a hyperbola because of the strong repulsion between the two. In this case, the repulsive center is situated at the far focus of the hyperbola. See Fig. 4. Rutherford made full use of this concept in his brilliant discovery of the atomic nucleus in 1911.

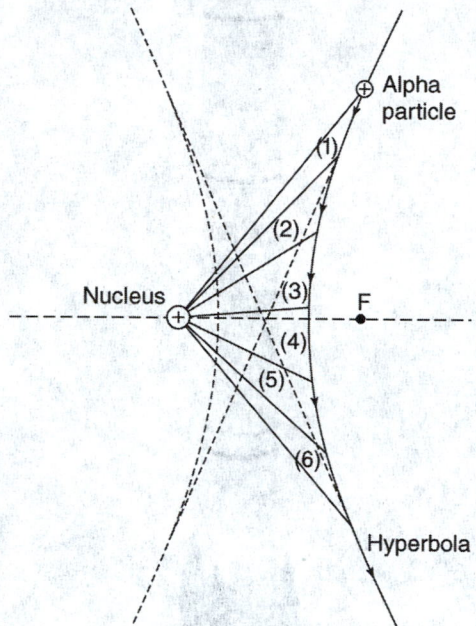

Fig. 4. Kepler's law of areas holds also for "central force" of repulsion between alpha particle and atomic nucleus; path of alpha particle is a hyperbola with nucleus in far focus. Areas 1, 2, 3, 4, 5, 6, etc., are equal and are described in equal lengths of time. A high-speed comet injected from outer space might, under gravity, describes the same kind of path, but with the sun at the near focus F.

See also **Earth Orbiting Satellite Theory.**

KERATIN. A class of natural fibrous proteins occurring in vertebrate animals and humans, they are characterized by their high content of several amino acids, especially cystine, arginine, and serine. They are generally harder than the fibrous collagen group of proteins. The softer keratins are components of the external layers of skin, wool, hair, and feathers, while the harder types predominate in such structures as nails, claws, and hoofs. The hardness is largely due to the extent of cross-linking by the disulfide bonds of cystine by the mechanism shown below:

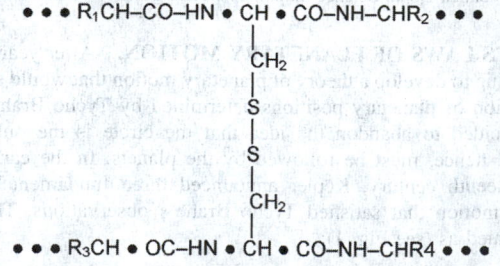

Kertains are insoluble in organic solvents but do absorb and hold water. The molecules contain both acidic and basic groups and are, thus, amphoteric.

Uses for kertains include: tablet coatings which dissolve only in the intestines, foam extinguishers and protein hydrolyzates.

KERATINASE.   A water-soluble, proteolytic enzyme having the ability to digest the keratin in wool and other forms of hair, converting a portion of it to a water-soluble form. It, thus, acts as a depilatory and is used in removing hair from pelts and hides, as well as from human skin. It is inactivated by heating to 100 °C.

KERATITIS.   Keratitis is an infection or inflammation of the cornea, the transparent front covering of the eye that covers the iris and pupil. Depending on the cause and with early treatment, most cases are curable.

The symptoms of keratitis may include eye pain, light sensitivity, the feeling that something irritating is in the eye, excessive tearing and blurred vision. A minor decrease in visual acuity may also occur.

The most frequent causes of keratitis include dry eye resulting from an eyelid disorder or insufficient tear formation, Vitamin A deficiency, foreign object in the eye, intense light (such as welding), allergies, and misuse or overuse of contact lenses. Viral, bacterial, and fungal infections are also causes of keratitis. The most common infection is the herpes simplex virus infection. Herpes simplex keratitis occurs as a result of the herpes simplex infection. It commonly leads to recurring inflammation of the cornea, scarring, and loss of vision. Anyone suspecting herpes simplex keratitis should seek immediate medical attention for prompt treatment. In severe, rare cases where the cornea is damaged, an eye care professional may recommend surgery. If left untreated, keratitis can cause ulcers on the cornea, glaucoma, permanent scarring of the eye, and permanent loss of vision.

The biggest risk factors for developing keratitis include poor nutrition, crowded or unsanitary living conditions, lowered resistance as a result of another illness, wearing contact lenses, or the presence of viral infections elsewhere in the body such as cold sores or genital herpes.

To prevent keratitis from recurring, wear glasses if work involves eye hazards. Visit their physician if you suspect any viral, bacterial, or fungal infection. In addition, modify your diet to ensure adequate Vitamin A.

After it is diagnosed, keratitis can be treated with eye medications such as antibiotics or antiviral eye drops and ointments. Nonprescription eye drops containing topical corticosteroids should not be used. Eye patches aid in the protection and healing of the infected eye. The doctor may also recommend discontinued use of contact lenses until the infection clears. See also **Herpes Simplex Virus Diseases**; and **Vision and the Eye**.

Vision Rx, Inc., Elmsford, NY

KERATOCONUS.   Keratoconus, also called conical cornea, results in the cornea changing from dome-shaped to cone-shaped through the progressive thinning of the cornea. Because the cornea is the part of the eye that refracts most of the light entering, the change of its shape prevents incoming light from being focused properly, ultimately distorting vision (astigmatism). In addition, vision may be impaired as a result of swelling and scarring of tissue. Keratoconus is a slow-developing disorder that almost always affects both eyes, though each eye may be affected differently.

Most cases have no known or traceable cause, though some cases have traced causes to heredity, others to injury, and yet others to diseases or medical disorders such as Down's Syndrome, Ehlers-Danlos Syndrome, Marfan's Syndrome, and Addison's Disease. In some cases, keratoconus has been linked to excessive eye rubbing and long-term use of hard contact lenses.

Keratoconus usually appears in puberty (late teens or early 20s), and progresses over a 10- to 20-year span, although it sometimes halts progression during that time. Not all cases are the same. Some progress faster and with more severity.

The earliest signs of keratoconus include blurred vision where things far away appear out of focus, objects appear tilted, and glare and light sensitivity increases. As keratoconus progresses and if scarring has occurred, patients may have moderate to severe vision distortion including double vision, cloudy lenses, seeing halos around lights, and sometimes, pain or discomfort.

Patients in the early stages of keratoconus can have their vision corrected with eyeglasses or contact lenses. As the astigmatism progresses and the cornea continues to thin and change shape, patients must rely on specially fitted, rigid gas permeable lenses to improve vision. Because changes continue, frequent fittings and follow-up care is essential for optimum visual acuity. In most cases, the cornea stabilizes over time.

Though rare, corneal transplantation is advised when patients are unable to tolerate contact lens wear or when adequate vision can no longer be maintained with the use of contact lenses. One of the most successful corneal transplants, keratoconus corneal transplantation generally helps 9 out of 10 patients. Despite the success of corneal transplants, however, patients often still require the uses of glasses or contact lenses to correct nearsightedness or farsightedness. See also **Corneal Topography (Videokeratography)**.

Vision Rx, Inc., Elmsford, NY

KERATOID.   See **Singular Point of a Curve**.

KERATOPATHY.   See **Vision and the Eye**.

KERATOSIS.   See **Dermatitis and Dermatosis**.

KERMA.   Of ionizing particles, the kerma is $\Delta E_K / \Delta m$, where ΔE_K is the sum of the initial kinetic energies of all the charged particles liberated by indirectly ionizing particles in a volume element of the specified material, and Δm is the mass of the matter in that volume element. In these definitions, the symbol Δ precedes the letters E_K and m to denote that these letters represent quantities that can be deduced only from multiple measurements that involve extrapolation or averaging procedures. Since ΔE_K is the sum of the initial kinetic energies of the charged particles liberated by the indirectly ionizing particles, it includes not only the kinetic energy these charged particles expend in collisions, but also the energy they radiate in bremsstrahlung. The energy of any charged particles is also included when these are produced in secondary processes occurring within the volume element. Thus the energy of Auger electrons is part of ΔE_K. In actual measurements, Δm should be so small that its introduction does not appreciably disturb the radiation field. This is particularly necessary if the medium for which kerma is determined is different from the ambient medium; if the disturbance is appreciable, an appropriate correction must be applied.

KERNEL.   A known function, also called a nucleus $K(x, z)$ which occurs in an integral equation. It is symmetric if $K(x, z) = K(z, x)$; Hermitian, if $K(x, z) = K^*(z, x)$; skew Hermitian, if $K(x, z) = -K^*(z, x)$, where the asterisk indicates the complex conjugate. A polar kernel has the form $K(x, z) = u(z)G(x, z)$, where $G(x, z)$ is symmetric. It may be transformed into a symmetric kernel by a change of the dependent variable. A definite kernel, positive or negative, satisfies the requirement

$$\int f(x)\, dx \int K(x, z) f(z)\, dz \lessgtr 0$$

where $f(x)$ is any finite function defined over the range for which the kernel is defined. A singular kernel has a discontinuity, one or more singular points within its integration limits, or an infinite integration limit. A solving kernel or resolvent is an infinite series of iterated kernels which appears in the Liouville-Neumann series. See also **Hermite Equation**; and **Integral Transform**.

KERNICTERUS.   A fetal jaundice resulting from Rh incompatibility between the mother and the fetus. The disorder can cause hearing loss, cerebral palsy, and mental retardation.

KERNITE.   This hydrated sodium borate mineral, $Na_2B_4O_7 \cdot 4H_2O$, occurs in a single known world locality at Kramer, Kern County, California. It is found here as veins and interbedded masses in clay, hundreds of feet thick, thought to be a product of heating and dehydration of a large buried body of borax by intrusive igneous rocks. Crystals are monoclinic and attain individual size to 3 by 8 feet. Cleavage is very perfect to both the macro and basal pinacoids, with hardness, 2.5–3, and specific gravity of 1.908; possesses vitreous luster grading to satiny on cleavage surfaces; colorless to white; transparent. Surface alters readily to tincalconite, a white powder. It is a major source of borax and boron compounds. See also **Borax**.

KEROGEN. The organic component of oil shale, it is a bitumen-like solid whose approximate composition is 75–80% carbon, 19% hydrogen, 2.5% nitrogen, 1% sulfur, and the balance oxygen. It is a mixture of aliphatic and aromatic compounds of humic and algal origin and comprises a substantial proportion of the shale; after fractionating and refining, the oil is reported to yield 18% gasoline, 30% kerosene, 27% gas oil, 15% light lube oil, and 10% heavy lube oil.

See also **Oil Shale**.

KEROSINE. See **Petroleum**.

KERR EFFECT. The occurrence of double refraction in a substance, when it is placed in an electric field. A more exact name is the *Kerr electro-optic effect*. It was first discovered for glass, in 1875, but it exists for liquids and gases also.

An ordinary dielectric medium has a single value for the index of refraction at any given wavelength of electromagnetic radiation, independent of polarization or direction of propagation of the radiation. In a uniaxial crystalline substance, there are two indices of refraction: the ordinary index which refers to radiation polarized with its electric vector perpendicular to the optic axis (a direction in the crystal), and the extraordinary index which refers to radiation with its electric vector parallel to the optic axis.

A substance that exhibits the Kerr effect is ordinarily singly refracting, but it becomes doubly refracting when an external electric field is applied. The substance behaves like a uniaxial crystal, with the optic axis parallel to the electric field. Liquids generally show larger effects than either solids or gases.

The difference in indices of refraction is proportional to the square of the applied electric field strength, and hence the Kerr effect is sometimes called the quadratic electro-optic effect. The expression for the optical path difference is given by

$$\Delta = (n_E - n_O)l = K E^2/\lambda$$

where n_E = ordinary index
l = path length in the medium
E = electric field strength
λ = wavelength in vacuum
K = Kerr constant.

The constant is usually positive, decreasing with an increase in wavelength and with an increase in temperature. The effect is caused by either natural or induced optical anisotropy of the individual molecules of the medium, and a lining up of the molecules by the applied electric field. The effect is particularly large for liquids made up of polar molecules with large anisotropies. Nitrobenzene, for example, gives an optical path difference of approximately $\frac{1}{40}$ wavelength in the visible region of the spectrum, for a field strength of 10,000 volts/cm and a 1-cm path length. The Kerr effect can be useful in helping to determine the structure of complex molecules.

Later investigations have shown the existence of an enormous quadratic electro-optic effect in certain crystals, notably perovskites in the para-electric phase. The effect can be several thousands times larger than that observed for nitrobenzene and is independent of temperature. The theoretical explanation requires knowledge about both electronic and ionic states of the crystalline lattice constituents.

Kerr cells are light shutters constructed from substances showing large Kerr effects. One type of widely used cell contains nitrobenzene placed between capacitor plates or electrodes. The cell itself is situated between a crossed polarizer and analyzer combination. In the absence of an electric field, no light is transmitted. When the electric field is applied, its strength is chosen so that the cell becomes a half-wave plate (the optical path difference is one-half wavelength). The plane of polarization of the light is therefore rotated by 90°, and the light is transmitted. An important property of the cell is its extinction ratio, or ratio of the light transmitted when the cell is closed to that transmitted when it is open. A cell filled with

extremely pure nitrobenzene can give a ratio as low as 10^{-5} to 1. Cells of the perovskites are closer to 10^{-2} to 1. Another important property of the Kerr cell is the frequency with which it can be turned on and off. Frequencies as high as 10^{10}/second are attainable in some instances. The inductance and capacitance of the cell, as well as relaxation effects in the medium, must be given considerable attention in the design of a cell. Also, the transit time of the light across the cell must be much less than the period of the applied field, in order for the cell to be effective.

A second effect is known as the *Kerr Magneto-optic effect*. It is the change in polarization of light reflected from a polished metal pole of a magnet. Linearly polarized light incident normally, is reflected as elliptically polarized light.

KETENES. Members of a class of compounds which contain the functional group =C=C=O. Examples are ketene [CAS: 463-51-4] itself (CH_2CO), methylketene (1-Propen-1-one) [CAS: 6004-44-0] $CH_3CH=C=O$, dimethylketene (2-Methyl-1-propen-1-one) [CAS: 598-26-5] $(CH_3)_2C=C=O$ diphenylketene (Ethenone, 2,2-diphenyl-) [CAS: 525-06-4] $(C_6H_5)_2C=C=O$, and carbon suboxide [CAS: 504-64-3] $O=C=C=C=O$. Of these, ketene is by far the most important. It is a reactive, colorless gas of considerable industrial importance. Physiologically, it is extremely poisonous and care must be taken to avoid breathing it.

The availability of ketone by pyrolysis of acetone (or acetic acid) is the reason for the attention it has received, contrasted to other ketenes which are relatively unavailable. Ketene may be prepared also by pyrolysis of acetic anhydride or phenyl acetate or diketene. Other sources are quite unsatisfactory from a standpoint of yield. Small quantities may be made conveniently by heating acetone in a "ketene lamp." This is a glass apparatus containing a Nichrome filament, heated electrically to red heat. Larger amounts are made by passing acetone or acetic acid through a tube at 700 °C. A very brief contact time is required, so that much of the acetone is undecomposed and has to be condensed and recycled. Also, it is imperative that the reaction tube be of inert material such as porcelain, glass, quartz, copper or stainless steel. A copper tube, if used, should be protected from oxidation by an iron sheath. Inert packing may be used (glass, vanadium pentoxide, porcelain), but just as good yields are obtained with empty tubes. No catalyst is known which accelerates this decomposition at significantly lower temperatures.

Methyl ethyl ketone is totally unsatisfactory as a source of methylketene by pyrolysis, but pyrolysis of propionic anhydride in a quartz tube at 400–600 °C and low pressures does produce it in a stated yield of about 90%. Another synthetic approach is to prepare methylketene dimer by allowing a mixture of propionyl chloride and triethylamine to stand at 25° for 24 hours and then to pyrolyze the dimer.

The known disubstituted ketenes include dialkyl-ketenes, diarylketenes, and the ester analogs. Dimethylketene may be made from α-bromoiso-butyryl bromide by reaction with zinc in boiling ether. Diphenylketene may be made similarly, but the usual way to prepare it is to oxidize benzil hydrazone with yellow mercuric oxide to benzoylphenyldiazomethane which, on heating in benzene solution, decomposes into the ketene.

The best way to make carbon suboxide is a pyrolytic method, starting with tartaric acid. The latter is converted into diacetyltartaric anhydride and then pyrolyzed at 625–650 °C (either in an empty tube or in a ketene lamp) into acetic acid and carbon suboxide, the latter in 35–50% yields.

Some recent synthesis of ketenes involve interesting chemistry. A butadienylketene is obtainable at −100 to −150 °C by photolysis (mercury lamp) of the appropriate cyclohexadienone:

The reaction is reversed on warming, but the ketene may be captured by an amine to form an amide.

1-Ethoxy-1-alkyne, $R-C\equiv C-OC_2H_5$, pyrolyzes at 120° into ethylene and alkylketene, but the latter is consumed by the original alkynyl ether to form 2,4-dialkyl-3-ethoxy-2-cyclobutenone,

$$R-CH-C=O$$
$$C_2H_5O-C=C-R'$$

in high yield.

Ketene and diimide are formed during the alkaline decomposition of chloroacetic hydrazide. The diimide, however, spontaneously changes into hydrazine plus nitrogen, and the hydrazine consumes the ketene to form acetohydrazide, $CH_3CONHNH_2$. If an olefin is present in the reaction mixture it is reduced to a paraffin by the diimide, thus preventing hydrazine formation.

The formulas of acetic acid, acetic anhydride and ketene show that the anhydride differs from the acid by 0.5 mole of water, and that ketene differs by 1.0 mole. Ketenes, therefore, may be regarded as super acid anhydrides, and this viewpoint leads to a good appreciation of their reactions. Because of an original lack of understanding of this relationship, the monosubstituted ketenes were once classed as "aldoketenes," and the disubstituted ketenes as "ketoketenes." Such terms are quite misleading, however, and should be abandoned.

Ketene is absorbed in sodium hydroxide solution, yielding sodium acetate. Aniline adds to ketene to form acetanilide. Both of these reactions are quantitative and are used to assay the ketene in a gas stream.

Primary alcohols react readily with ketene to form acetic esters but tertiary alcohols require the catalytic help of sulfuric acid. Even with primary alcohols, as 1-butanol, it has been established that addition of ketene ceases at about the 75% conversion point unless a little sulfuric acid is present as catalyst. Phenol, which is inert toward ketene at ordinary temperature, may be converted into phenyl acetate by reaction at the boiling point of phenol or by reaction at room temperature if a trace of sulfuric acid is present.

An important industrial synthesis of acetic anhydride is via acetic acid and ketone. Mixed acetic anhydrides are made similarly by passing ketene into the acid in question: $RCOOH + CH_2CO \rightarrow RCO-O-COCH_3$. This is the basis of a good method of synthesizing symmetrical anhydrides in view of their formation from the mixed anhydrides on heating:

$$2 \ RCOOCOCH_3 \rightarrow (RCO)_2O + (CH_3CO)_2O.$$

Comparable reactions of ketene are those with mercaptans to form thio esters (CH_3COSR), with amino acids (in water) to obtain N-acetyl derivatives ($CH_3CONHCHRCOOH$), with hydroxylamine to yield aceto-hydroxamic acid [CAS: 564-88-3] ($CH_3CONHOH$), dimethylchloroamine to form chloroacetic dimethylamide ($ClCH_2CON(CH_3)_2$), and Grignard reagents to form ketones ($RCOCH_3$).

Ketene adds to pyridine in a 4:1 ratio to form a yellow, crystalline compound,

Quinoline [CAS: 91-22-5], isoquinoline [CAS: 119-65-3], and phenanthridine [CAS: 229-87-8] resemble pyridine in reacting comparably, and diketene may be substituted for ketene.

Aromatic aldehydes take up ketene in the presence of potassium acetate [CAS: 127-08-2] in the manner of a Perkin reaction. The product is a cinnamic acid [CAS: 62182-9]:

$$ArCHO + CH_2CO \xrightarrow{AcOK} ArCH=CHCOOH.$$

Friedel-Crafts catalysts are effective in converting formaldehyde and ketene into β-propiolactone [CAS: 57-57-8]. This is a process of industrial importance. Similarly, furfural [CAS: 98-01-1] and ketene give rise to 3-(2-furyl)-propionolactone.

When aluminum chloride [CAS: 7446-70-0] is used as catalyst, ketene reacts with benzene [CAS: 71-43-2] to form acetophenone [CAS: 98-86-2]: $C_6H_6 + CH_2CO \rightarrow C_6H_5COCH_3$. Also, methyl chloromethyl ether [CAS: 107-30-2] under such conditions reacts to yield 3-methoxypropionyl

chloride:

$$CH_3OCH_2Cl + CH_2CO \xrightarrow{AlCl_3} CH_3OCH_2CH_2COCl.$$

Ketones such as acetone [CAS: 67-64-1], ethyl acetoacetate [CAS: 141-97-9], ethyl levulinate [CAS: 539-88-8] or acetylacetone [CAS: 123-54-6] react smoothly with ketene if a trace of sulfuric acid is present. The products are enol acetates of the ketones, acetone giving rise to isopropenyl acetate [CAS: 108-22-5],

$$CH_2=C(CH_3)-OCOCH_3,$$

and ethyl acetoacetate changing into ethyl 3-acetoxycrotonate. Cyclobutanone derivatives are made quite easily by addition of ketenes to styrene or to vinyl ethers or to enamines:

$$R_2C=CH-NR_2 + O=C=CH_2 \xrightarrow{0°} \begin{array}{c} R_2C-CH-NR_2 \\ OC-CH_2 \end{array}$$

One of the most characteristic reactions of ketene or dialkyl-ketenes is that of polymerization into dimers. Diarylketenes do not display this tendency. The dimer from ketene, or "diketene," is a liquid that boils without decomposition at 43° (28 mm), but the compound tends to decompose (into dehydroacetic acid and resinous substances) on distillation (b.p. 127°C) at atmospheric pressure. The structure of diketene was in doubt for many years, but recent critical chemical and physical evidence indicates the structure as 3-buteno-β-lactone. Diketene is an acetoacetylating agent. Thus, with aniline it yields acetoacetanilide, and with methanol (catalyzed by sulfuric acid) it produces methyl acetoacetate. These reactions are useful industrially.

Additional Reading

Kroschwitz, J.I., and M. Howe-Grant: *Kirk-Othmer Encyclopedia of Chemical Technology*, John Wiley & Sons, Inc., New York, NY, 2001.

Lide, D.R.: *CRC Handbook of Chemistry and Physics*, 88th Edition, CRC Press, LLC., Boca Raton, FL, 2007.

Patai, S. *Chemistry of Ketenes, Allenes and Related Compounds*, John Wiley & Sons, Inc., New York, NY, 1980.

Tidwell, T.T.: *Ketenes*, John Wiley & Sons, Inc., New York, NY, 1994.

CHARLES D. HURD, Northwestern University, Evanston, IL

KETIMINE. A type or class of curing agent for epoxy resins that makes it possible to use very-high-solids content coatings in spray equipment. Reacts with epoxies very slowly and thus delays curing time, which prevents setting up of the resin during spraying operation. In presence of water or water vapor, ketimine breaks down to a polyamine and a ketone. Epoxy coatings cured with ketimine should not exceed a thickness of 10 mils.

KETONES. The homologous series of ketones (like aldehydes) has the formula $C_nH_{2n}O$. Structurally, the ketones consist of a carbonyl group (C:O) linkage between two radicals (R–CO–R'). R and R' may be the same as in acetone [CAS: 67-64-1] (dimethyl ketone) CH_3–CO–CH_3; or they may differ as in methylethyl ketone [CAS: 78-93-3] CH_3–CO–C_2H_5. The latter may be referred to as a *mixed ketone*. A ketone is isomeric with the aldehyde that contains the same number of carbon atoms. Thus, acetone C_3H_6O is isomeric with propaldehyde [CAS: 123-38-6] C_3H_6O. Where R and R' are alkyls, the ketone may be called an *alphyl ketone* and may be considered to be derived from the secondary alcohols. Ketones also may be formed from aromatic alcohols as in the case of benzophenone [CAS: 119-61-9] C_6H_5–CO–C_6H_5. The latter compound is a fully aromatic or *diaryl ketone*. Further, there are mixed *aryl-alphyl ketones* as in the case of acetophenone [CAS: 98-86-2] (phenylmethyl ketone) C_6H_5–CO–CH_3. The foregoing examples illustrate the use of trivial names for the ketones. In another system, the ketone may take its name from the alcohol from which it may be derived — thus, propione (from propyl alcohol); or the ketone may be named for the acid to which it may be oxidized — thus, acetone (acetic acid the oxidation product).

Essentially ketones exhibit the following properties: (1) all ketones up to C_{11} are neutral, mobile, volatile liquids. Ketones above C_{11} are solids under usual ambient conditions; (2) all ketones have a reasonably agreeable odor; (3) all ketones except those with a very high carbon count are soluble in H_2O, the solubility decreasing with a rise in formula weight; (4) most ketones are soluble in alcohol or ether, and (5) the specific gravity of ketones rises uniformly to about 0.83 with a rise in formula weight. The physical properties of some common ketones are listed in Table 1.

The presence of the double bond (carbonyl group C:O) markedly determines the chemical behavior of ketones. The hydrogen atom connected directly to the carbonyl group is not easily displaced. The chemical properties of the ketones may be summarized as follows: (1) they are readily reduced to form *secondary alcohols*, particularly in the presence of a catalyst. This property is used to advantage in the production of numerous organic compounds where ketones serve as a starting material or as an intermediate; (2) unlike the aldehydes, ketones are considerably more stable and do not combine readily with the alcohols, nor generally with NH_3 at ordinary temperatures and pressures; (3) they do not reduce alkaline solutions of metals and also, unlike aldehydes, ketones do not undergo polymerization; (4) they combine with hydroxylamine to yield *ketoximes*; (5) they react with hydrazine to form *hydrazones*; (6) they combine with semicarbazine to form *semicarbazones*; (7) with H_2SO_4 or HCl present, ketones can be induced to undergo a cyclic trimerization with the loss of H_2O; (8) when oxidized, ketones decompose to form two acids. Each acid will contain fewer carbon atoms than the originating ketone; and (9) ketones react with HCN to form *cyanohydrins*.

When ketones are reduced to secondary alcohols, varying amounts of ditertiary alcohols (pinacols) are produced. In a reaction known as the pinacol-pinacoline rearrangement, effective in synthesis of difficult compounds, ketones when treated with magnesium amalgam, after hydrolysis, yield a 1, 2-glycol. Important to the purification of methyl and some cyclic ketones is their ability to form crystalline additive compounds with sodium bisulfite solutions. Ketones also react with phosphorus pentachloride and pentabromide to form dihalogen derivatives of the alkyls in which instances the oxygen atom of the carbonyl group is replaced by two hydrogen atoms.

Ketones are widely used as starting and intermediate ingredients in the production of numerous synthetics, such as resins, and they find wide application as solvents. Other important ketones produced on a tonnage basis include methylethyl ketone (MEK) and methylisobutyl ketone (MIBK). Commercially, MEK may be manufactured by the direct oxidation of butylene in which air is used, along with a catalyst solution comprising copper chloride and palladium chloride. The overall reaction is $C_4H_8 + \frac{1}{2}O_2 \rightarrow CH_3COC_2H_5$. During the reaction, the palladium chloride is reduced to elemental palladium and HCl. Cupric chloride causes reoxidation. The resulting cuprous chloride is reoxidized to cupric chloride during the catalyst regeneration cycle. The process proceeds under moderate pressures at a temperature of about $100\,°C$. The MEK product must be treated with sodium bisulfite and caustic soda, followed by distillation, to yield pure MEK.

In the production of MIBK, acetone may be the chargestock. In a first step, acetone is converted to diacetone alcohol (DAA) by condensation under pressure with an alkaline catalyst. The latter may be calcium or barium hydroxide, both of which are slightly soluble in H_2O. After cooling, in a second step, mesityl oxide (MSO) is produced by dehydrating the DAA. For this step, an acid catalyst is used and the step proceeds in the temperature range of 100 to $120\,°C$. When the MSO is formed from the DAA, the latter partially decomposes into acetone. Thus, a distilling phase is required to separate and recover the acetone. In a third step, the MSO is hydrogenated and a mixture of MIBK and methylisobutyl carbinol (MIBC) is produced. These must be separated by a further fractionating phase. The hydrogenation reactions are: $(CH_3)_2C{=}CHCOCH_3 + H_2 \rightarrow (CH_3)_2CHCH_2COCH_3$ (MIBK); and $(CH_3)_2 CHCH_2COCH_3 + H_2 \rightarrow (CH_3)_2CHCH_2CHOHCH_3$ (MIBC). The temperature of the process and the hydrogen mole ratio used determines the ratio of MIBK/MIBC produced and thus the manufacturer has effective control over the amounts of end-products (MIBK and MIBC) which can be manipulated in accordance with market requirements.

Related Compounds

A polyhydric ketone is refereed to as a *ketose*. Monosaccharoses are examples of open-chain polyhydroxyaldehydes or ketones. The aldehyde sugars are termed *aldoses*; the ketone sugars are called ketoses. Fructose (Arabino-hexulose; beta-D-fructopyranose; Fructopyranose) [CAS: 7660-25-5] is a ketose. Dihydroxyacetone [CAS: 96-26-4] $HO \cdot CH_2 \cdot CO \cdot CH_2 \cdot OH$ is a very simple ketose. Compounds that contain both a carbonyl and a carboxylic group are termed *ketonic acids*. They display the reactive properties of both an acid and a ketone. Pyroracemic acid $CH_3 \cdot CO \cdot CO_2H$, acetoacetic acid $CH_3 \cdot CO \cdot CH_2 \cdot CO_2H$, and laevulic acid $CH_3 CO \cdot CH_2 \cdot CH_2 \cdot CO_2H$ are examples of monobasic ketonic acids. *Ketonic hydrolysis* is a term used to describe such actions as the hydrolysis of the ethyl ester of acetoacetic acid into acetone, CO_2, and ethyl alcohol. Compounds such as ketene $CH_2{:}CO$, methyl ketene $CH_3CH{:}CO$, dimethyl ketene $(CH_3)_2C{:}CO$, and diphenyl ketene $(C_6H_5)_2C{:}CO$ are of the family known as *ketenes*.

Health and Safety Factors

Ketones are flammable substances that do not exhibit a known high degree of chronic toxicity. Low molecular weight (C_3-C_{12}) saturated aliphatic ketones, which represent the bulk of industrially important ketones, may be classified among the solvents of comparatively low toxicity hazard. The eight-hour threshold limit value is generally above 100 ppm, although the odor threshold is in the range 5–25 ppm. High vapor concentrations of these volatile ketones induce anesthesia, however, the vapors are so irritating to the eyes and mucous membranes of the respiratory system that the atmosphere generally becomes intolerable before toxic concentrations are achieved. Many ketones are also powerful drying and degreasing agents, and prolonged skin contact can cause dermatitis.

The C_3-C_{12} are all highly flammable liquids with flash points varying from $-18\,°C$ for acetone to $85\,°C$ for isobutyl heptyl ketone. Ketones float on water, and become only partially soluble in water with increasing molecular weight. Thus, ketones typically require copious quantities of water to extinguish pool fires. Saturated ketones are in general stable at ambient conditions, and do not undergo hazardous polymerization in normal environment. Most ketones are incompatible with strong oxidizing and reducing agents; some ketones are also incompatible with bases and/or acids.

Additional Reading

Bingham, E. et al.: *Patty's Toxicology: Ketones, Alcohols and Ester Compounds*, Vol. 6, John Wiley & Sons, Inc., New York, NY, 2000.

Gmehling, J. et al.: *Vapor-Liquid Equilibrium Data Collection: Ketones*, Supplement 1, Scholium International, Inc., Port Washington, NY, 1993.

Kroschwitz, J.I., and M. Howe-Grant: *Kirk-Othmer Encyclopedia of Chemical Technology*, John Wiley & Sons, Inc., New York, NY, 2001.

Lide, D.R.: *CRC Handbook of Chemistry and Physics*, 88th Edition, CRC Press, LLC., Boca Raton, FL, 2007.

McKetta, J.: *Encyclopedia of Chemical Processing and Design: Hydrogen Cyanide to Ketones Dimethyl (Acetone)*, Vol. 27, Marcel Dekker, Inc., New York, NY.

KETOSES. See **Carbohydrates**.

KETOXIMES. See **Hydroxylamine**.

KETTLE HOLE. See **Glacial Deposits (or Drift)**.

KEW BAROMETER. A type of cistern barometer. No adjustment is made for the variation of the level of the mercury in the cistern as pressure changes occur; rather, a uniformly contracting scale is used to determine the effective height of the mercury column.

KEYES EQUATION. An equation of state for a gas, deduced from the concept of the nuclear atom. This equation is designed to correct the van der Waals equation for the effect upon the term b of the surrounding molecules. The equation is written as

$$P = \frac{RT}{V - Be^{-\alpha/V}} - \frac{A}{(V + l)^2}$$

in which P is pressure, T is absolute temperature, V is volume, R is the gas constant, e is the base of natural logarithms, $2.718\ldots$, and A, α, and B, and l are constants for each gas.

TABLE 1. PHYSICAL PROPERTIES OF KETONES

Systematic name (common name)	Mol wt	Fp, °C	Bp at 101.3 kPa,[a] °C	Refractive index, n^{20}D	Sp gr 20/20, °C	Viscosity at 20°C, mPa·s (=cP)	Surface tension, mN/m (=dyn/cm) at 20°C	H_{vap} at 101.3 kPa,[a] kJ/mol[b]	Liquid specific heat capacity at (T)°C, J/(kg·K)[b]	Flash point, open cup, °C (closed)	Soly at 20°C, wt % In water	Soly at 20°C, wt % water in
METHYL ALKYL KETONES												
2-propanone (acetone)	58.08	−94.7	56.1	1.3590	0.7905	0.33	24.0	29.53	2224 (30)	−16 (−18)	complete	complete
2-butanone (methyl ethyl ketone)	72.10	−85.9	79.57	1.3780	0.8062	0.41	24.6	31.64	2203 (20)	−6 (−6)	26.8	11.8
2-pentanone (methyl propyl ketone)	86.13	−77.8	102.4	1.3902	0.8076	0.51 (28.3°C)	23.2	33.39		(7)	4.3	3.3
3-methyl-2-butanone (methyl isopropyl ketone)	86.13	−92	94.2	1.3882	0.8044	0.43 (25°C)	24.6 (25°C)	30.63		(6)	6.53	
4-methyl-2-pentanone (methyl isobutyl ketone)	100.16	−84.0	116.2	1.3957	0.8020	0.61	23.6	35.60	1920 (20)	23 (16)	1.6	1.9
2-hexanone (methyl n-butyl ketone)	100.16	−55.8	127.5	1.4007	0.8125	0.62	25.4	36.05	2228 (25)	(35)	1.75	3.7
3-methyl-2-pentanone (methyl sec-butyl ketone)	100.16	−83	117.4	1.4001	0.8142			35.12			2.26	
3,3-dimethyl-2-butanone (pinacolone)	100.16	−50	106.4	1.3986	0.8070			33.5			2.0	1.8
2-heptanone (methyl amyl ketone)	114.18	−35	151.5	1.4087	0.8166	0.77	26.1	39.25		47 (49)	0.43	1.45
5-methyl-2-hexanone (methyl isoamyl ketone)	114.18	−73.9	144.9	1.4069	0.8127	0.77	25.3 (25°C)			41 (35)	0.54	1.28
2-octanone (methyl hexyl ketone)	128.22	−20.5	173.3	1.4153	0.8197	0.95 (25°C)	26.6 (25.5°C)	40.88		(62)		
4-hydroxy-4-methyl-2-pentanone (diacetone alcohol)	116.16	−44.2	169.2	1.4226	0.9406	3.2	31	41.6	1883	61 (47)	complete	complete
DIALKYL KETONES												
3-pentanone (diethyl ketone)	86.3	−39.4	101.8	1.3923	0.8155	0.47	24.7	33.69	2215 (25)	(13)	3.4	2.6
2,4-dimethyl-3-pentanone (diisopropyl ketone)	114.19	−69	125.0	1.399	0.8076	1.02		39.31			0.05	0.75
2,6-dimethyl-4-heptanone (diisobutyl ketone)	142.24	−46	169.4	1.4172	0.8076		22.2	35.66		49 (49)	1.57	
3-hexanone (ethyl propyl ketone)	100.16		123.2	1.4003	0.8174		25.04			35	0.43	
3-heptanone (butyl ethyl ketone)	114.19	−39	147.3	1.4088	0.8197	0.84	25.7	36.59		41 (46)		0.78
3-octanone (ethyl amyl ketone)	128.22	−46	167–168	1.4150	0.8220							
2,6,8-trimethyl-4-nonanone (isobutyl heptyl ketone)	184.32	−75	218.2	1.4257	0.8180	1.9		44.56		90 (88)	0.01	0.2

(continued)

TABLE 1. (Continued)

Systematic name (common name)	Mol wt	Fp, °C	Bp at 101.3 kPa,[a] °C	Refractive index, n^{20}D	Sp gr 20/20, °C	Viscosity at 20°C, mPa·s(=cP)	Surface tension, mN/m (=dyn/cm) at 20°C	H_{vap} at 101.3 kPa,[a] kJ/mol[b]	Liquid specific heat capacity at (T)°C, J/(kg·K)[b]	Flash point, open cup, °C (closed)	Soly at 20°C, wt% In water	water in
UNSATURATED KETONES												
3-buten-2-one (methyl vinyl ketone)	70.09	−6	81.4	1.4130								
3-methyl-2-buten-2-one (methyl isopropenyl ketone)	84.12	−54	98	1.4236	0.855							
4-methyl-3-penten-2-one (mesityl oxide)	98.15	−53	129.5	1.4414	0.8521	0.6	28.4	43.1	2176 (20)	29 (31)	3.1	3.4
4-methyl-4-penten-2-one (isomesityl oxide)	98.15		121.5	1.4458	0.8548							
3,5,5-trimethyl-2-cyclohexen-1-one (isophorone)	138.21	−8.1	215.3	1.4775	0.9229	2.6	32	43.4	1799 (20)	104 (85)	4.3 (25 °C)	1.2 (25 °C)
3,5,5 trimethyl-3-cyclohexen-1-one (β-isophorone)	138.21		181–191		0.89						0.03	
DIKETONES												
2,3-butanedione (diacetyl)	86.09	−2.5	90.2	1.3938	0.9843			34.3				
2,3-pentanedione	100.12	−52	111				31	35.4	1983(20)			
2,4-pentanedione (acetylacetone)	100.11	−23.5	140.4	1.4510	0.9753	0.58		36.55	1956.2			
2,5-hexandione	114.15	−5.4	192.3	1.4256	0.9734	1.6			(15)			
CYCLIC KETONES												
cyclopentanone (adipic ketone)	84.12	−50.6	130.8	1.4359	0.9512	1.2	33.35	36.53			29	14
cyclohexanone (pimelic ketone)	98.15	−31.1	155.7	1.4510	0.9482	2.21	35.2	37.62	2039.8 (30.8 °C)	46 (43)	2.5	8.0
cycloheptanone	112.17	−21	179	1.4611	0.888		26.4			72 (62.5)	0.3	1.4
3,3,5-trimethylcyclohexanone	140.22	−10	188.8	1.4455		2.54						
AROMATIC KETONES												
acetophenone (methyl phenyl ketone)	120.15	19–20	201.7	1.5342	1.0296	0.93		45.69	93 (82)		0.55	1.65
benzophenone (diphenyl ketone)	182.22	48–49.5	305									
1-phenyl-2-propanone (phenylacetone)	134.18	−15		1.5158								
propiophenone (phenyl ethyl ketone)	134.17	18.2	218	1.5265	1.012		37.4	45.44		96 (85)	0.01(25 °C)	

[a]To convert kPa to mm Hg, multiply by 7.5.

[b]To convert J to cal, divide by 4.184.

2882

KEYES PROCESS. A distillation process involving the addition of benzene to a constant-boiling 95% alcohol–water solution to obtain absolute (100%) alcohol. On distillation, a ternary azeotropic mixture containing all three components leaves the top of the column while anhydrous alcohol leaves the bottom. The azeotrope (which separates into two layers) is redistilled separately for recovery and reuse of the benzene and alcohol.

KEY (Machine). An element for preventing relative rotation in machinery. The sunk key is the commonest form, and must be carefully fitted to prevent rotation in the keyway, and consequent crushing or deformation. The Pratt and Whitney, or dropseat, key is applicable to conditions where the hub must move on the shaft and key, but the keyway must be end-milled. The Woodruff key fits in a semicircular recess in the shaft, and automatically adjusts itself in instances where both the hub and shaft are tapered. The gibhead taper key will prevent both axial and rotative motion, since it is driven into place against a tapered keyway in the hub. It is usually equipped with a head, as shown, to permit withdrawal when required. The feather key is employed to prevent relative rotation but permits axial motion. It may be demonstrated that the axial force necessary to move a hub along a keyed shaft is twice as great when one key is used as when two or more keys are used. For this reason, and because of strength considerations, multisplined shafts are employed where shafts or hubs must move axially under load. Modern production methods have made it possible to machine both shaft and hole with comparative ease. See Fig. 1.

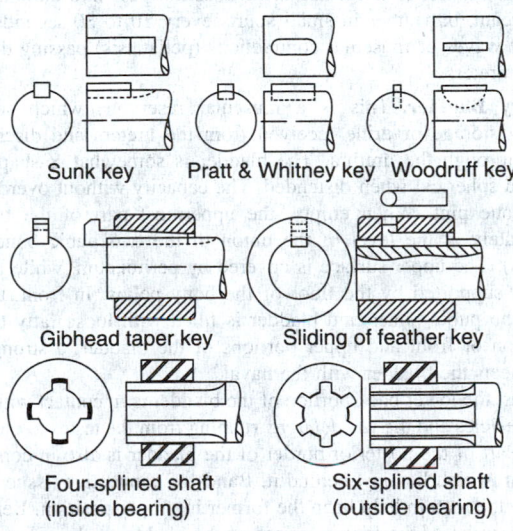

Sunk key   Pratt & Whitney key   Woodruff key

Gibhead taper key   Sliding of feather key

Four-splined shaft (inside bearing)   Six-splined shaft (outside bearing)

Fig. 1.   Types of keys used on machines.

KEYPUNCH. A device, usually operated by a keyboard, for recording information on cards or tape by punching holes in the card or tape to represent letters, digits, and special characters.

KHAPRA BEETLE *(Insecta, Coleoptera).* A major pest of stored grains and cereal products as well as fried vegetable products. This beetle *(Trogoderma granarium,* Everts), first described in 1898, is a native of India, Sri Lanka, and Malaysia. The insects occur in the United Kingdom, Japan, and the Philippines in large numbers; and is also present in several countries of Europe, Asia, and Africa. Although it is believed that the pest may have been in the United States as early as 1939, it was not until 1953 that it was discovered in California. Subsequently, the beetle was found in Arizona, New Mexico, Texas, and Mexico. Infested properties were immediately scheduled for fumigation when they were found. The khapra beetle, as of the late 1970s, has been successfully eradicated in the United States. Control procedures are in continuous effect, including quarantines and careful inspections at ports of entry. The beetle attacks stored cereal products and feeds on whole kernels of wheat, corn, barley, oats, rye, and rice. It also attacks stored seed, cottonseed meal, nut meats, dried fruits, and many other products of plant and animal origin. Severe

losses of infested products can be expected. Total loss usually occurs when infested grain is stored undisturbed for long periods. Many times each year, the khapra beetle is intercepted in foreign cargo shipped into the United States. In control measures, the governments of the United States and Mexico cooperate.

Development of the khapra beetle has four stages—egg, larva, pupa, and adult. The life cycle in total varies from 26 days (at 93–95°F; 34–35°C) to 220 days at 70°F (21°C). The eggs are about $\frac{1}{64}$ inch (0.4 millimeter) long and white in coloration. The larvae are yellowish-white upon hatching; they change to reddish-brown as they shed their skins 2 to 11 times. They are about $\frac{1}{4}$ inch (6 millimeters) long when fully grown. The pupae are about the same length and are enclosed in the last larval skin.

The adult males are small, brownish-black beetles, about $\frac{1}{8}$ inch (3 millimeters) long. Adult females are a bit larger and lighter in color. Young larvae are unable to feed on whole kernels of grain and depend upon damaged kernels or grain products for food. Older larvae can feed on whole grains. The amount of food present is a major factor in the speed of their development, but larvae can survive for long periods without food. Adults feed very little, if at all.

The khapra beetle feeds only on stored products; it has not been found on growing crops. The beetle cannot fly, thus its movement is limited. Activity of the insect is generally confined to the top 12 inches (0.3 meter) of grain, although it has been found as deep as 12 feet (3.6 meters), particularly in corners and along walls of bins. The insects crowd into cracks in masonry, woodwork, and cartons, into creases of sacks, in debris, and in other protected places where they are difficult to find.

The khapra beetle cannot be eradicated by using conventional insecticide sprays or the customary dosages of grain-treatment fumigants. Special measures must be employed. Methyl bromide and hydrogen cyanide gases are effective fumigants when used at concentrations higher than those required to kill other grain pests. These high concentrations should be applied only under the supervision of trained federal or state pest control specialists.

KIANG. See **Horses, Asses, and Zebras.**

KIDNEY AND URINARY TRACT. The kidneys (two in number) are the principal organs of the human urinary system. (The excretory systems of various kinds of animals are described in entry on **Excretory System.**) The word *renal* is commonly used in referring to the urinary system and is essentially interchangeable with the word *kidney,*[1] used as an adjective. The medical science concerned with this system is called *nephrology*: the medical practice dealing with the treatment of diseases and abnormalities of the urinogenital system is sometimes called *urology.*

Similar to an industrial process that converts raw materials into energy and products and wastes, the body must process and remove its wastes. The blood, which is continuously recycled through the body, is purified by the action of the kidneys, which remove the undesired waste products as constituents of the urine. Functioning as ultrafilters, the glomeruli of the kidneys filter 170–180 liters (45–47.5 gallons) of blood plasma in a 24-hour period. Most of this protein-free filtrate (about 99%) is reabsorbed as it passes through the tubules—as evidenced by the comparatively small amount of urine produced during the same period (1.5 liters or 1.6 quarts). The kidneys selectively remove water and soluble salts (urea, uric acid, creatinine) that result from the metabolism of proteins as well as the neutralization of acids. In the complex reduction process, certain intermediate substances, such as hippuric acid and ammonia (from amino acids mainly glutamine) are formed. Coincidentally, the kidneys also remove some carbon dioxide and heat.[2] The composition and characteristics of urine are described under **Urine.**

[1] Related to the word *reins* (rāns), found in traditional Bibles (*Pss.* 7:9, 16:2; *Jer.* 17:10; *Job* 19:27) where it is used to designate the *inward parts,* particularly the kidneys, which were once thought by the Israelites to be the seat of the emotions.

[2] Looking at the human excretory system as a whole, the kidneys are assisted by (1) the lungs, which principally eliminate carbon dioxide and incidentally some water and heat; (2) the alimentary canal, which eliminates solids and secretions and incidentally some water, carbon dioxide, salts, and heat; and (3) the skin, which principally eliminates heat and incidentally some water, carbon dioxide, salts, hair, nails, and dead skin.

In addition to their primary function of eliminating waste products, toxic materials, and basic and nonvolatile acid radicals, the kidneys serve other important purposes. These include the maintenance of a constant volume of circulating blood and the regulation of the fluid content of the body as a whole (homeostasis), the regulation of osmotic pressure relationships of the blood and tissues, and the maintenance of optimum concentrations of various constituents of the plasma. As pointed out by Marriott many years ago, blood becoming as acid as distilled water or as alkaline as ordinary tap water would be incompatible with life. Renal function is key to the electrolyte (water, sodium, potassium) system of the body. This will be apparent later in this entry. For the background chemistry of this topic, see **Water**; also see the entry on **Potassium and Sodium (In Biological Systems)**. See also **Blood**; and **Homeostasis**.

As shown by Fig. 1, the principal elements of the urinary system are two *kidneys* (right and left); two *ureters* (right and left), which are ducts for conveying urine from the kidneys to the urinary bladder; one *urinary bladder*, a reservoir for the reception and storage of urine; and one *urethra*, a tube for conveying the urine from the bladder.

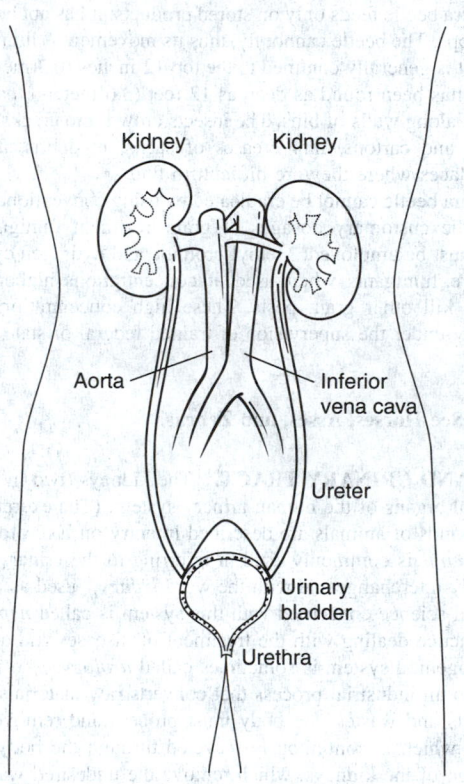

Fig. 1.   Principal elements of the urinary system.

The kidneys are two bean-shaped organs located on each side of the spine, approximately in the middle of the back. Normally, the upper parts of the kidneys lie beneath the last two ribs. Each kidney is about 4.5 inches (11.4 centimeters) long, 2 inches (5 centimeters) wide, and 1 inch (2.5 centimeters) thick. Because of the proximity of the liver, the right kidney is positioned a bit lower than the left kidney. The top portion of each kidney is attached by strong fibers to the diaphragm, so that the kidneys rise and fall about 0.5 inch (12 millimeters) during breathing. The lower ends of the kidneys are generally from 1 to 2 inches (2.5 to 5 centimeters) above the crest of the hipbone. Lying in front of the right kidney are portions of the liver and of the gastrointestinal tract; in front of the left kidney are parts of the stomach, spleen, pancreas, loops of the small intestine, and parts of the colon. The *adrenal glands* (cap-like bodies) are located on the upper ends of the kidneys. Inasmuch as the kidneys are located behind and outside of the lining membrane (*peritoneum*) of the abdominal cavity, it is possible to operate on the kidneys without opening the peritoneum. Each kidney is imbedded in a mass of fatty tissue, which is surrounded by fibrous tissue (*renal fascia*). A thin, tough envelope of fibrous tissue covers each kidney.

Each kidney is composed of a *cortex* or outer portion, made up of cortical substance, surrounding a *medulla* or middle layer, which consists of 20 or fewer radially striated cones (renal pyramids), the bases of which are toward the circumference of the kidney and the apexes of which converge into projections (*papillae*), which are best described as cup-like cavities (calyxes) in the *pelvis* of the kidney. (In Fig. 1, the ureters extend from the pelvis of the organ.) Most of kidney substance (cortex and medulla) is made up of minute tubes (tubules) which are very closely packed together. The functional unit of the kidney is called a *nephron*; it incorporates substance from both the cortex and medulla of the organ. At one extreme end of the nephron unit is the *glomerulus* (a pinpoint filtering unit) and its surrounding envelope (capsule), and at the other end is the collecting tube, from which urine is transmitted into the renal pelvis. See Fig. 2. There are, at a minimum, one million nephrons in each kidney. The collecting tubules themselves finally unite with other collecting tubules to form larger channels known as *papillary ducts*. A detailed description of the operation of the kidney is far beyond the scope of this encyclopedia. A scanning electron micrograph of a rat kidney (Fig. 3) is illustrative of the complex structure.

The two ureters which connect the kidneys to the urinary bladder are each about 12 to 14 inches (30.5 to 35.5 centimeters) long. The ureters are thick-walled, muscular tubes and have an average caliber of about 0.2 inch (5 millimeters). In males, the ureters lie close to the seminal vesicle and are crossed by the vas deferens. In females, they pass close to the mouth of the uterus and the upper part of the vagina. The ureters enter the urinary bladder at its lower and back portion. Muscle fibers of the ureter exert a clamp-like (sphincter) effect and urine does not pass into the bladder in a steady stream, but rather in small spurts every 10 to 30 seconds. This is caused by waves of muscular contractions (peristalsis) passing downward along the ureters.

Urinary Bladder. This is a muscular reservoir, which serves for temporary storage of urine received from the ureters and discharged at intervals through the urethra. The bladder is somewhat Y-shaped when empty and spherical when distended. The capacity without overdistension is about one pint. When empty, the uppermost part of the bladder is approximately at the level of the union of the two pubic bones (*pubic symphysis*). The upper surface is covered by peritoneum, while the lower portion is supported by the floor of the bony pelvis. In front, the space between the pubic bones and bladder is filled with loose fatty tissue. At the junction of front and upper portions of the bladder, a strong fibrous cord connects the bladder with the naval.

In males, the lower back portion of the bladder is in contact with the two seminal vesicles and the *vas deferens* running from the testes to the seminal vesicles. Part of the posterior border of the bladder is also in contact with the rectum, which lies just behind it. Bands of connective tissue from the bladder to the rectum help keep the former in proper position. Beneath the posterior portion of the lower surface of the bladder is the prostate gland, through which the urethra passes. Ligaments from the prostate to the walls of the pelvis also help support the bladder. In females, the posterior lower part of the bladder is in contact with the neck (*cervix*) of the womb (*uterus*) and the upper part of the front wall of the vagina. The bladder is attached to these structures by connective tissues. The triangular area between this opening and the opening of the two ureters is known as the *trigone*.

Since the bladder occupies a position between the kidneys and the urethra, it is often called the middle portion of the urinary tract. Most abnormal conditions of the bladder are associated with, or the result of, conditions affecting the upper or lower portions of the tract.

Rare abnormalities of the bladder may be listed as complete absence, double bladder, or the more or less incomplete division of the bladder by *septa*, such as the hourglass bladder. Outpocketings (*diverticula*) of the bladder are relatively common, and may be present at birth or develop as the result of increased pressure caused by obstruction to the normal flow of urine. In the condition known as *exstrophy*, the front wall of the bladder and the abdominal wall covering the organ are absent. The back (*posterior*) wall of the bladder protrudes and the ureters excrete urine to the outside of the body continuously. Exstrophy has been corrected surgically with some success.

The tissues that hold the bladder in place may become weakened and allow the bladder to bulge into various abnormal positions. If the bladder protrudes into the vagina, the condition is known as *cystocele*. It is often

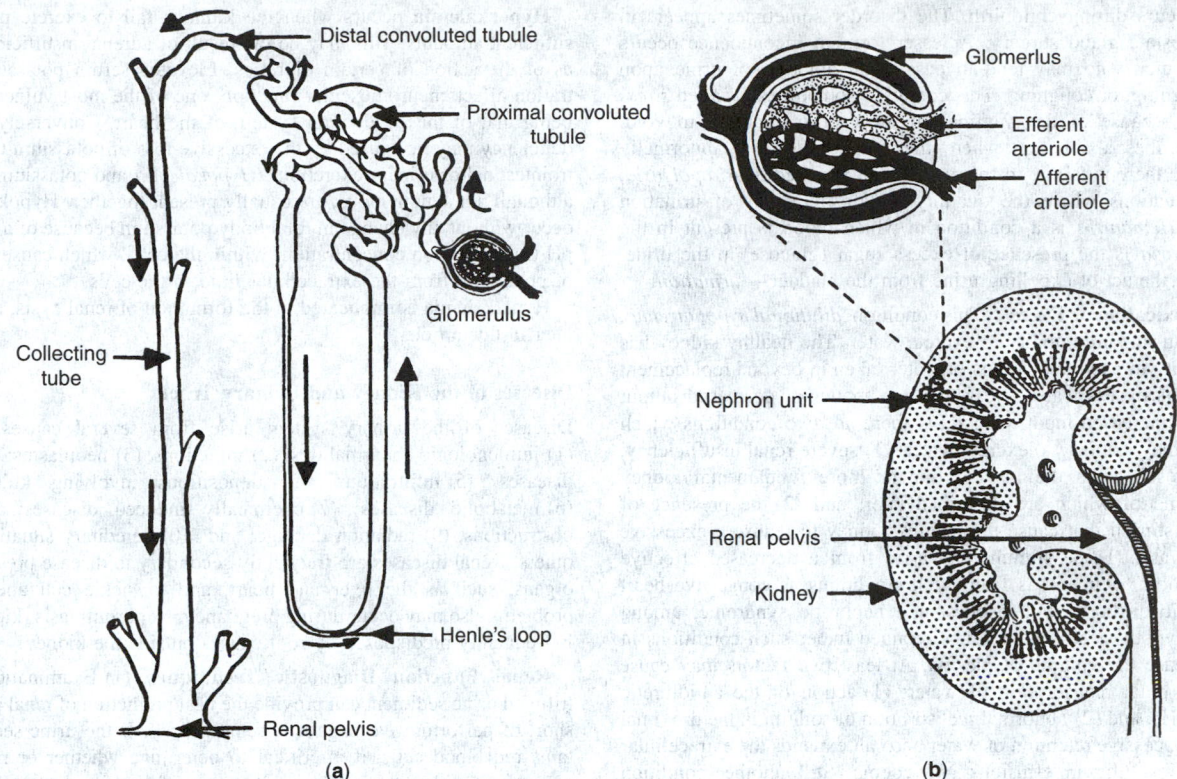

Fig. 2. (a) Very enlarged and simplified artist's concept of a nephron unit of the human kidney. Over one million of these units filter fluids from the blood and convert filtrate fluids into urine in the tubules. The collecting tubes then transfer the urine to the renal pelvis. (b) Very schematic cross section of a kidney, showing position of glomerulus and nephron unit in the outer layers. Renal pelvis and center of kidney serve as a collecting system for urine formed in the nephron units. Arterioles carry blood to and from the glomerulus.

Fig. 3. Scanning electron micrography of rat kidney. Magnification about 400×.

secondary to childbirth, and is due to relaxation of perineal support. It can be repaired surgically.

Abnormal channels allowing the passage of urine from the bladder into other hollow organs, or directly to the outside of the body through the skin, are known as *fistulae*. Fistulae are of many types. In *vesico-intestinal fistula*, a communication exists between the bladder and some part of the intestine. In the more common *vesico-vaginal fistula*, the channel lies between the bladder and vagina. Although other causes exist, vesico-vaginal fistula may result from childbirth. In such women, urine leaks almost constantly from the bladder into the vagina. Treatment for patients with fistulae consists usually of closing the abnormal opening surgically.

The urethra, a narrow passageway through which the urine flows from the urinary bladder to the outside, is, in the female, about 1.5 inches (3.8 centimeters) long and the external opening (meatus) lies above the vagina, between it and the clitoris. In the male, it measures about 8 inches (20 centimeters) in length. The prostatic portion is about 1 inch (2.5 centimeters) long and extends downward from the bladder through the prostate gland. In its central part, the prostatic urethra is enlarged to form a bulge; into this bulge open the ducts of the prostate, the ejaculatory ducts, and the prostatic utricle. The middle or membranous part of the urethra is approximately 0.5 inch (12 millimeters) in length and passes through the muscular urogenital diaphragm. The remainder of the urethra is called the cavernous or spongy portion because it is surrounded by the cavernous portion of the underside of the penis. The urethra can be the site of acute and chronic infection.

Voiding is a complex function, generally controlled at will. Contraction of the bladder muscle (detrusor) forcibly widens the bladder outlet. In addition, the voluntary relaxation of the muscles in the pelvic floor aids in the dropping and funneling of the bladder outlet and thus helps to widen the bladder outlet.

Incontinence is the involuntary intermittent or continuous flow of urine from the bladder as the result of an ineffective sphincter muscle, which normally closes the urethra. The muscle may be weak, as is relatively common in aged persons (particularly bed patients), or it may have been torn or otherwise damaged. The most common cause of damage in women occurs during childbirth. The disorder sometimes appears in men after prostate gland surgery. A lesser form of incontinence occurs mainly in women who may tend to pass small amounts of urine upon sneezing, laughing, or coughing. This condition sometimes is called *stress incontinence* because of the minimal (if any) stress required to void. *Oliguria* describes a situation where the flow of urine is abnormally reduced; when there is failure to secrete urine, it is called *anuria polyuria*, applies to situations where the quantity and/or frequency of urination is excessive. *Hematuria* is a condition in which blood is present in the urine. *Glycosuria* is the presence of excess sugar (glucose) in the urine. *Micturition* is the act of expelling urine from the bladder — *urination*.

Water Intoxication or Excess. This condition, *dilutional hyponatremia*, results from an impaired ability to excrete water. The healthy kidney has a marked ability to excrete amounts of water taken in beyond replacement needs, and thus excess water is essentially a matter of decreased output rather than of increased input. Generally, there are two conditions which impair the kidney's ability to excrete water: (1) severe renal insufficiency, which may result from the loss of one or more fundamental kidney functions, some of which are described later; and (2) the presence of physiological stimuli that cause the healthy kidney to reabsorb excessive amounts of water. These stimuli may arise from a decreased effective circulating blood volume — as from such conditions as congestive heart failure, hepatic cirrhosis with ascites, and nephrotic syndrome, among others. It is hypothesized that water is retained under such conditions in order to increase the volume of plasma. At least two factors may cause an inordinate renal reabsorption of water: (1) action of the antidiuretic hormone (ADH); and (2) abnormal reabsorption of sodium in the proximal tubules. The excessive retention of water and salt expands the extracellular fluid (ECF), resulting in effusions and edema. Still another condition may be present — *inappropriate antidiuretic hormone secretion (SIADH) syndrome*. This has the same characteristics of an overdose of ADH. There are numerous causes of SIADH, such as certain lung diseases, various endocrine insufficiencies, tumors, surgical stress, positive-pressure breathing, and a relatively large number of drugs.

Water intoxication can be quite serious because of the intolerance of the brain, encased as it is in a rigid skull, to increased volume. This is manifested by the central nervous system (sleepiness, mental dullness and, in the extreme, convulsions and coma). The therapy for water intoxication depends upon the extent and seriousness of the condition and may range from the use of osmotic diuretics (such as mannitol), hemodialysis, peritoneal dialysis, saline diuretics (when renal function is normal), and the administration of hypertonic sodium chloride.

Water Deficiency — Dehydration. This may result simply from insufficient water intake to make up for the water normally lost by the lungs, skin, and kidneys, or the causes may be more complex. This condition often appears in persons who have a deficiency of ADH, as in diabetes insipidus, or whose kidneys do not show a normal response to ADH, as in nephrogenic diabetes insipidus. The condition may be primary, or it may be secondary to other conditions, such as hypercalcemia, hypokalemia, sickle cell anemia, chronic pyelonephritis, and partial urinary tract obstruction, among other causes. See also "Water Deficiency" in the entry on **Water**.

Sodium and Potassium Abnormalities. As previously mentioned, the kidneys are the keys to the management of the body's electrolyte system, and this includes the concentrations of potassium and sodium. See also **Potassium and Sodium (In Biological Systems)**. Sodium is present in all extracellular body fluids — hence when there is a loss of these fluids, there is an accompanying loss of sodium. The most common causes of sodium loss are through vomiting, diarrhea (extreme in some diseases, such as cholera), and by excessive perspiration. Diuretics, metabolic acidosis, the inappropriate ADH secretion syndrome, and renal disease which affects those mechanisms which control renal reabsorption of sodium may cause a net loss of sodium in the urine. Because sodium deficiency decreases plasma volume, there is a reduction of cardiac output, manifested by a drop in blood pressure and increased pulse rate. Conversely, an excess of total body sodium can result in edema, ascites, and effusions. The healthy kidney has an exceptional ability to adapt to the body's needs for sodium, but frequently excesses of sodium must be reduced through dietary controls and by administration of diuretics (where the kidneys respond) to increase the output of sodium in the urine. See also **Diuretics**.

Hyperkalemia occurs when the kidneys fail to excrete potassium in sufficient amounts. This may be the result of adrenal insufficiency as well as of the action of certain diuretics. Elevated serum potassium concentration affects neuromuscular function, one of the most vulnerable tissues being that of the conduction system of the heart. Conversely, potassium deficiency may be caused by the excessive loss of potassium through gastrointestinal or urinary excretions. *Hypokalemia* and potassium deficiency, although not synonymous, are usually present together. Hypokalemia may occur without any change in total body potassium because of a lowering of pH (hydrogen ion concentration) within the cells, which causes movement of potassium from the extracellular fluid to the cells.

Hyperkalemia is implicated in the formation of renal cysts, as described later in this article.

Diseases of the Kidney and Urinary Tract

Diseases of the urinary system arise from several causes, including: (1) immunologic abnormalities; (2) infections; (3) neoplasms; (4) vascular diseases; (5) infiltrations and depositions involving kidney tissue; (6) metabolic diseases; (7) chemically induced diseases; (8) physical obstructions; (9) radiation damage; and (10) hereditary situations, among others. Renal diseases are frequently secondary to disease present in other organs, such as the liver and heart, and to sickle cell anemia. Renal problems also may occur during pregnancy. Nephrolithiasis (kidney stones) is frequently predisposed by root causes outside the kidney.

Renal Function Diagnostic Techniques. (1) Examination of centrifuged urine sediment can provide the best prediction of renal morphology short of performing a diagnostic renal biopsy. If the urine sediment contains red blood cells, it is critical to determine whether or not the cells are present within casts. If the cells are not found in casts, they may have originated anywhere in the genitourinary tract, including the kidney. If red cells are present in casts, a proliferative glomerulitis is almost always present. (2) Intravenous urography (sometimes called intravenous pyelography, or IVP) is the basic method for evaluating gross renal structure. Generally regarded as a safe procedure, incidents have been reported of patient's experiencing episodes of acute decline in renal function following intravenous urography. At particular risk are persons who have reduced renal function, diabetes mellitus, or multiple myeloma, and persons over 60 years of age. It has been found that liberal administration of normal saline solution during major angiographic procedures will reduce the incidence of renal insufficiency. Thus the procedure is approached with considerably more caution than was used in earlier years. (3) Ultrasound is a well accepted renal diagnostic tool Ultrasound can determine kidney size in lieu of x-ray examination. The technique also is valuable in determining the presence or absence of hydronephrosis, quantity and quality of renal masses (cyst versus tumor), early detection of cysts in suspected polycystic kidney diseases, and medullary cystic disease. In many medical centers, ultrasound has become the preferred initial renal imaging technique. (3) Computed tomography (CT) provides information similar to that obtained from ultrasound, particularly regarding mass lesions, but at greater cost and effort. CT is a next step in renal examination if the physician is not satisfied with the information from ultrasound. (4) Radionuclide imaging is suggested in a variety of clinical situations, including renal failure, suspected urinary tract obstruction, and special situations involving hypertension. It is also considered a next step in evaluation where intravenous urography has been inadequate, particularly in identifying glomerular filtration rate and renal blood flow. A principal hurdle with radionuclide imaging is that the impaired renal function of the patient interferes with the uptake and accumulation of the radioisotopes used in the procedure. (5) Some authorities believe that nuclear magnetic resonance (NMR) holds great promise as an imaging technique with advantages over both ultrasound and tomography. (6) Renal biopsy, which entails a risk (mortality rate from percutaneous renal biopsy is about 0.1%), is usually approached with caution. Further, in a 1982 study, the results of renal biopsy were shown to have influenced therapeutic decisions in only 20% of patients.

Bright's Early Findings and Major Syndromes. Diseases of the kidney were coarsely described in the time of Hippocrates, but very little was learned concerning kidney function until the findings of Richard Bright, a distinguished English physician of Guy's Hospital, London. In 1827, Bright demonstrated that the presence of albumin in the urine indicated kidney disease. Bright first connected albuminuria with dropsy (edema) and a "disease" (actually a series of diseases) was named for him, *Bright's disease*. It was determined later that at least three facets or separate diseases are characterized by a combination of protein in the urine and abnormally large amounts of fluid in the intercellular spaces of the body. (In recent years, with the availability of modern instrumentation, the combinations and permutations of kidney disease sometimes appear limitless.) In these abnormalities the kidney is affected, but not infected. The first of Bright's diseases is a *degenerative nephrosis*, where there is necrosis (death) of the outer layer of cells in the tubules of the kidney. Albumin and dead cells in the urine are the principal signs of this condition. The second of Bright's diseases is a *hemorrhagic disease*, believed to be caused by substances created by infections found elsewhere in the body. Kidney function is reduced and there is edema present. In this disease, there is inflammation of the capillary blood vessels in the filtering units (glomeruli) of the kidney. Later this type of condition was named *glomerulonephritis*. The third Bright's disease is characterized by hardening (sclerosis) of the kidney arteries (arterioles), and was later called *arteriolar nephrosclerosis*. As the walls of the arterioles thicken, the blood supply to the kidney lessens and kidney substance shrinks, the kidney ultimately taking on a finely granular appearance as function becomes less and less satisfactory. Particularly in the absence of any effective therapy, the Bright's diseases are degenerative and progress toward kidney failure.

Glomerulonephritis. Only about 50 years ago (Addis, 1950), glomerulonephritis was considered to be a single disease, but with several facets. It is now known that the condition is much more complex than initially envisioned, based upon knowledge gained in recent years through the use of percutaneous renal biopsy and the examination of the tissue specimen with electron microscopy and immunofluorescence microscopy. The complexities and variations involved are evidenced by the names given to specific situations, such as *immune complex nephritis* and *antiglomerular basement membrane antibody nephritis*.

With some reorganization of approach to these diseases, the major syndromes with which the modern clinician must cope are: (1) the *acute nephritic syndrome*; (2) the *nephrotic syndrome*; (3) *asymptomatic urinary abnormalities*; and (4) end stage of chronic nephrotic processes—*chronic renal failure*.

Acute Nephritic Syndrome. Characteristics include sudden onset, hematuria (with red cell casts), proteinuria and usually accompanied by edema and circulatory congestion, oliguria, and decreased glomerular filtration rate (GFR). The condition is found more frequently in children than in adults. The syndrome can arise from a large variety of causes, including bacterial, viral, or parasitic infections, post-streptococcal glomerulonephritis, and such less understood conditions as allergic vasculitis. Goodpasture's syndrome, cryoglobulinemia, Wegener's granulomatosis, etc. Diagnosis is complex and therapy is varied.

Nephrotic Syndrome. This is a protein-wasting disease and is believed to result from increased glomerular permeability by plasma proteins. This, of course, increases the filtration rate of proteins and they find their way to the urine, although some are reabsorbed, but are changed to amino acids during the process. This renal condition may be primary or secondary to a systemic disease. The possible combinations of cause and effect are large, adding to difficulties of diagnosis and assignment of effective therapy.

One of the clinical tests used is that of determining the protein content of the urine (*proteinuria*). A distinct nephropathy sometimes is associated with acquired immunodeficiency syndrome and is characterized by a sudden onset of proteinuria. This nephropathy may be one of the very first manifestations of HIV infection.

Primary Kidney Infections. Where only the kidney pelvis is involved, an infection is called *pyelitis*, but if both pelvis and kidney are involved, the condition is known as *pyelonephritis*, which is the most common type of kidney infection. Infection and inflammation of the bladder only is referred to as *cystitis*. *Nephritis* is a rather general term used to describe inflammation of the kidney. Admittedly, the nomenclature of kidney infections and other diseases is somewhat inconsistent.

In the normal kidney, bacteria do not pass through the filtering apparatus (glomeruli), but they may in the case of kidney damage caused by obstruction (stones, etc.). Bacteria may invade the urinary tract by way of the bloodstream, the lymphatic system, and directly from the urethra to the

bladder, the latter route by far being the most common. Over 90% of these infections are caused by gram-negative bacilli or enterococci. Urinary tract infections are among the most prevalent bacterial infections in humans, and they can occur any time during the life span. It has been estimated that between 10 and 20% of women develop a urinary tract infection at some age, the incidence of bacteriuria increasing up to the sixth decade of life. Urinary tract infections among male adults are much less frequent except as they may occur in the presence of prostatic disease or anatomic abnormalities. Males are protected by their anatomical differences and also by the presence of antimicrobial substances in prostatic secretions. Chronic prostatitis is the major cause of recurrent urinary tract infection in males. However, particularly in male infants, bacteriuria and pyelonephritis occur with some frequency and, in general, urinary tract infections are frequently found in children.

In women, colonization of the vaginal vestibule by *Escherichia coli* (present in normal stool flora) is a frequent cause of bladder infection. In 30–50% of bladder infections, unless promptly and effectively controlled, the condition will spread via the ureter to the kidney. Symptoms associated with lower urinary tract infection include frequency of urination, burning sensation when voiding, and suprapubic pain. Signs associated with upper urinary tract infection include loin pain, fever, rigor, nausea and vomiting, and macroscopic hematuria. A less frequent kidney infection is *xanthogranulomatous pyelonephritis*, a chronic condition that can occur at any age and, in a majority of cases, it is associated with a *Proteus* infection.

Except in occasional situations in which infection is localized in the outer layer (cortex) of the kidney, pus and bacteria (bacteriuria) are found in the urine of most persons with pyelonephritis. The white blood cell count is usually high and fever may run up to 106 °F (41.1 °C).

Acute, uncomplicated urinary tract infections are usually successfully treated with antibiotics, although the rate of recurrence is moderately high. It is sometimes difficult to initially distinguish between upper and lower urinary tract infection, although the most appropriate therapy may differ for the two situations. In the case of renal infection, deep tissue is involved, whereas in bladder infection, the affected tissue is superficial mucosa.

In 3–4% of persons with tuberculosis, renal infection occurs. This is the most common form of extrathoracic tuberculosis, with the highest incidence among young and middle-aged adults. When present, the tuberculin skin test is almost always positive. Several other clinical manifestations are present, but the disease, sometimes described as insidious, can cause significant kidney damage if left undetected. Urine cultures are the most certain key to diagnosis. Surgery no longer is indicated to remove infected tissue, but may be required for the reconstruction of damaged tissues. Through application of chemotherapy, the prognosis is excellent.

Urethritus is an infection of the urethra and commonly is associated with sexually transmitted diseases.

Neoplasms. Conditions such as Hodgkin's disease, multiple myeloma, lymphomas, and leukemia, among other neoplastic diseases, can impact on renal functions in varying degrees. Nephrotic syndrome may arise from tumor antigen-antibody complexes deposited on the glomerular capillary membrane, but there are numerous other causes of the syndrome which first should be evaluated. Hypercalcemia, which occurs frequently in cancer patients, may involve the kidneys and may be indicated by increased polyuria and resultant polydipsia and, in advanced cases, oliguric renal failure may occur.

Blockage of a ureter by a neoplasm can occur and the symptoms may mimic those of ureteral stones (calculi). Pelvic tumors also can block both ureters (bilateral obstruction). In situations of this type, attention is concentrated on control of the neoplasm through radiation or chemotherapy, while at the same time actions such as percutaneous nephrostomy are taken to alleviate the condition during the period of antitumor therapy.

There is an exceptionally high incidence of bladder cancer in Egypt. Authorities relate this to a combination of bladder infections resulting from bilharziasis and bacteria. Bilharziasis, also known as schistosomiasis, is an invasion by blood flukes of the genus *Schistosoma*.

Wilm's Tumor. This tumor accounts for about 85% of all kidney cancers in children. It may be defined as an adenosarcoma with mixed carcinomatous elements that occurs fetally and may lie dormant for years. The tumor is second only to neuroblastoma in frequency of solid tumors of childhood. With early diagnosis and prompt treatment, a 2-year survival rate of up to 80% has been experienced, with a 5-year cure rate in excess of 50%. Correction usually requires surgical intervention, followed by chemotherapy.

In 1989, a team of researchers at Massachusetts Institute of Technology and the University of Colorado found the candidate gene for Wilm's tumor on chromosome 11. This gene, among a few others found, has tumor-suppressing qualities. As reported, however, the question of how Wilm's tumor begins remains poorly understood. The tumor usually is diagnosed between ages 2 and 5 because of severe abdominal pain evinced.

Carcinoma of the Kidney. Of adult cancers, kidney cancer accounts for only 1–2% of cancer cases. Two-thirds of these occur in males. The most common indication is hematuria (blood in urine). Treatment usually consists of radical nephrectomy with regional node dissection. The overall 5-year survival rate is about 45%. As of the mid-1990s, therapy with progestational agents and either cortisone derivatives or testosterone had not proved successful.

Vascular Diseases. The kidney has a generous blood supply and complex vascular system and, like other organs, is subject to malfunction and accompanying renal insufficiency as the result of renal arterial disease, renal vein thrombosis, and related conditions. Obstruction of the major renal arteries, which produce renal insufficiency, may be surgically reversible. The usual therapy for renal vein thrombosis is anticoagulation therapy.

Infiltrations and Depositions. Diffuse *amyloid deposition* in glomeruli may occur and is generally followed by progressive renal failure. Amyloidosis is characterized by protein deposits in various organs, including the kidney. Some success has been reported with the use of melphalan therapy and, in some instances of end-stage renal failure, hemodialysis is employed. *Gout patients* are prone to renal failure for a number of reasons, including renal vascular disease, renal calculi with pyelonephritis, and less frequently the deposition of urate crystals. When uric acid exceeds its saturation point in acid urine, uric acid crystals may precipitate in the collecting tubules, renal pelvis, or ureters. Acute obstruction to the flow of urine may occur, leading to oliguric renal failure. Therapy includes use of diuretics, and hemodialysis, which is effective in removing uric acid and restoring renal function. *Calcium deposition* (nephraocalcinosis) within the kidney can produce a gradual, but substantial decrease in glomerular filtration rate (GFR). This deposition is not to be confused with renal stones (nephrolithiasis). Calcium deposition is most frequently associated with hypercalcemia (faulty calcium control by parathyroid hormone, PTH), as well as with sarcoidosis, hyperparathyroidism, milk-alkali syndrome, vitamin D excess, metastases to bone, and multiple myeloma.

Nephrolithiasis (Renal Stones). Stones of varying composition and arising from a variety of causes may occur in the urinary tract. About 80% of renal stones seen in cases in the United States are calcium oxalate; 10% are magnesium ammonium phosphate (struvite), usually mixed with some calcium phosphate (apatite); 5–10% are uric acid; 1% are calcium phosphate (apatite or apatite and brushite); 1% are cystine; and, more rarely, xanthine stones. All are opaque to x-ray, except uric acid and xanthine which are lucent. See Fig. 4.

Predisposing factors for calcium oxalate stones are *hypercalciuria* (excessive amounts of urinary calcium, possibly caused by an increased gastrointestinal absorption of calcium); *hyperoxaluria*; and possibly *hyperuricosuria*. Underlying causes of calcium oxalate stone formation include primary hyperparathyroidism, vitamin D intoxication, certain malignancies, and sarcoidosis (granuloma formation). The predisposing factor for magnesium ammonium phosphate stones is a persistently alkaline urine which frequently is due to infection with urea-splitting organisms. Similarly, persistently alkaline urine is the predisposing factor for calcium phosphate stones, usually as the result of renal tubular acidosis, the generous use of oral antacids, and carbonic anhydrase inhibitors. In the case of uric acid stones, the urine is persistently acid, which may be due to an inability to excrete ammonia buffer. Gout is also a predisposing factor for uric acid stones. Cystine present in the urine is a predisposing factor for cystine stones and may be caused by defective renal tubular absorption. Xanthine stones form as the result of xanthine oxidase deficiency.

Very important to developing a rational therapy is identification of the structural and crystalline composition of the stone. Passage and collection

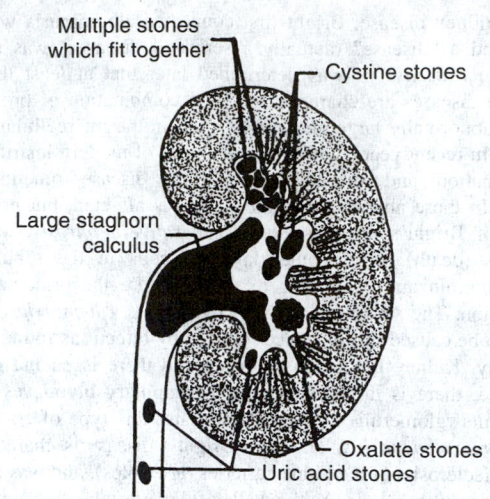

Fig. 4.   Cross section of human kidney illustrating the various types of kidney stones which may occur.

of a stone is highly advantageous for diagnosis. Laboratory procedures, sometimes extending over a period of days, have been developed for evaluating the patient who has passed a stone. Where a stone is not available for evaluation, X-ray appearance and urinary findings are used as criteria.

Several means are available, in addition to surgery, for treating nephrolithiasis. In addition to treating the underlying cause, the patient's stone-forming tendency can be treated physiologically—mainly (1) by lowering the urinary concentration of the stone-causing constituents; and (2) by increasing the solubility of the stone constituents—frequently with both procedures used. Allopurinol has been used to reduce uric acid excretion; low-calcium diets are sometimes followed; in some countries, cellulose phosphate is used to lower intestinal calcium absorption. By maintaining the urine pH above 6.0, the precipitation of uric acid can be minimized; sodium bicarbonate can be used to adjust the pH. However, if the pH is elevated too much, this may favor the formation of calcium stones. In the case of calcium phosphate stones, ammonium chloride is sometimes used to lower the urine pH to below 6.0. For stones that do not respond to adjustment of urine pH, certain chemicals have been used with some success. Increasing urine volume is successful in some cases, particularly in the cases of uric acid and cystine stones.

Struvite calculi form within the urinary tract as a result of the action of the bacterial enzyme urease, which is produced by bacteria, such as *Proteus* species, that may infect the urinary tract. Much has been learned in recent years concerning the etiology of this type of stone, and accordingly some revisions in therapy have resulted. Often this type of stone (sometimes referred to as the *infection stone*) branches as it grows, involving two or more calyceal systems and incorporating within its intricacies the offending bacteria, which thus are protected from antibiotics in the urine. This type of stone formation may be secondary, occurring in patients who have an underlying metabolic disorder.

In addition to traditional surgical removal procedures, *percutaneous lithotripsy* with ultrasound and electrohydraulic sources of energy can be used to fragment and remove the stones from the urinary tract. The procedure is *invasive*, involving a small skin incision (about 2.5 cm) in the flank. In patients with struvite stones, in whom it is critical to remove all of the stone material, the procedure can be conducted under direct vision and also allows irrigation of the collecting system.

Another, but *non-invasive* procedure is also used. This is *extracorporeal shock-wave lithotripsy*, a method that induces shock waves in water and focuses them on the stone within the urinary tract. The stone is fragmented to the size of sand so that it subsequently passes down the patient's ureter and out of the urinary tract. Obviously, the patient's urinary tract must be free of obstruction so that the stone fragments can pass. Period of hospitalization and recovery time is usually quite short.

Still another technique has been introduced—namely, *laser surgery*. Stones are subjected to energy from a laser beam. The beam ultimately creates an ionized plasma, which in turn produces shock waves to break apart the stones.

Renal Cysts. As pointed out by V.E. Torres and researchers, "Renal cysts arise from diverticula or segmental dilatations of the renal tubules that develop as part of the normal aging process. Many poorly understood genetic, environmental, and local factors can enhance the development of cysts and thus result in various renal cystic diseases."

As previously mentioned, hypokalemia may be a primary cause of renal cyst development. In concluding a study involving 55 patients, the aforementioned research team concluded: "Chronic hypokalemia is accompanied by enhanced renal cystogenesis and may lead to interstitial scarring and renal insufficiency. Renal cysts are thus dynamic structures whose growth can be influenced by hormonal or pharmacologic interventions."

Metabolic Diseases. The principal metabolic disease that may predispose renal disease is *diabetes mellitus* (see also **Diabetes Mellitus**), although other metabolic diseases, such as gout, previously mentioned, oxalosis, cystinosis. Wilson's disease, and Fabry's disease may also impact on the renal system. Among persons with juvenile-onset diabetes, renal failure is the most common cause of death. Diabetic glomerulopathy is characterized by progressive thickening of the glomerular basement membranes. Steroids or cytotoxic agents have been used in treatment, but they are not considered effective. Patients who reach end-stage renal failure must turn to hemodialysis or transplantation.

Chemically Induced Renal Disease. There are numerous chemicals (nephrotoxic agents) which vary in their intensity in causing renal damage. Among relatively common drugs, chemical agents and other substances reported to produce acute or chronic renal failure are: certain analgesics (heroin, phenacetin, phenylbutazone, and salicylates); anesthetics (methoxyflurane); anticonvulsants (paramethadione and trimethadione); antimicrobials (aminoglycosides, amphotericin B, bacitracin, cephalosporins, colistin, semisynthetic penicillins, polymyxin, rifampin, sulfonamides, viomycin); chelating agents (EDTA and penicillamine); radio-contrast agents; miscellaneous drugs (allopurinol, clofibrate, lithium, methysergide, pyridium); metals (arsenic, bismuth, cadmium, copper, lead, mercury, uranium); organic solvents (carbon tetrachloride, diesel fuel, ethylene glycol); biologic substances (hemoglobin, myoglobin, mushroom poisons, venoms); and other substances, such as lysol and phenol.

It also has been reported that exposure to x-rays exceeding 1200 rads/kidney can lead to renal damage. Consisting of various degenerative changes, this condition is sometimes referred to as radiation nephritis.

In an excellent report by M.R. Cullen, M.G. Cherniack, and L. Rosenstock in connection with the Occupational Medicine Program (University of Washington School of Medicine, Seattle), the researchers observe, "Because of the considerable amount of intrinsic renal disease whose origin is unexplained and because of the role of the kidney in detoxification and filtration, new methods for evaluating industrial nephrotoxicity continue to be explored. Although lead poisoning sufficient to cause direct renal injury occurs only rarely in the United States, moderate exposure remains commonplace. Organic solvents, such as trichloroethylene, carbon tetrachloride, and toluene have long been recognized as causes of renal tubular injury, but recent data and the reinterpretation of previous findings implicate this wide class of agents as potentially important causes of acute and chronic glomerular disease."

Polycystic Disease of the Kidney. Several hereditary renal diseases have been established. These include polycystic and medullary cystic renal disease, as well as Alport's syndrome, which is a hereditary glomerulonephritis frequently associated with nerve deafness as well as ocular lens abnormalities and abnormal platelets. Other hereditary diseases seen with hematuria, proteinuria, and renal insufficiency include Fabry's disease, the nail-patella syndrome, sickle cell disease, and progressive lipodystrophy.

Polycystic disease is a condition in which the kidney is filled with numerous cysts. Both kidneys are usually involved. In this condition, a decreased urine concentrating ability develops early. Later the kidneys enlarge, usually confirmed by intravenous pyelography or ultrasonic examination. Because the disease is localized to the diseased kidneys, polycystic patients are favorable candidates for hemodialysis or transplantation. In medullary cystic disease, the cysts appear in the medullar portion of both kidneys. The kidneys do not enlarge. Diagnosis is very difficult, but

because the disease is most frequently familial, a family history is very helpful.

Solitary cysts also appear in the kidney, but usually have no functional significance.

In a recent study by P.S. Parfrey and a team of researchers (see reference listed), it is reported that autosomal dominant polycystic kidney disease is responsible for 6–9% of cases of end-stage renal disease in North America and Europe. Seventeen families with this trait were studied. The researchers concluded: "At present, in most persons with a 50 percent risk of autosomal dominant polycystic kidney disease, imaging techniques are the only mode of reaching a diagnosis before symptoms appear. In such persons a negative ultrasonographic study during early adult life indicates that the likelihood of inheriting a PKD1 mutation is small. In the few who inherit a non-PKD1 mutation for polycystic kidney disease, renal failure is likely to occur relatively late in life."

Renal Disease and Pregnancy. Pregnancy in women with chronic renal disease is accompanied by an increase in such complications as hypertension and proteinuria and by increases in prematurity and fetal loss. Although fertility is decreased in women with advanced renal insufficiency, pregnancy sometimes occurs. However, transplantation restores fertility to many women with end-stage renal disease. Hypertension poses the greatest danger to the mother's life, with its risk of central nervous system bleeding. Some women with the nephrotic syndrome have massive edema during pregnancy. Salt restriction may limit the development of edema, but is rarely helpful once edema has occurred. The use of diuretics in pregnancy, particularly for the treatment of hypertension, is controversial. Women who have renal diseases that tend to progress should be encouraged to complete their childbearing while their renal function is preserved. When moderate renal insufficiency is present, they must be advised that about a third of women will have more than the expected decline in renal function. There is insufficient information present today, however, to accurately predict which women may have this problem. Women with transplants need to be advised about the best timing of pregnancy and the potential hazards of immunosuppressive drugs and possible viral infections for the fetus.

Uremia and End-Stage Renal Disease (ESRD). With malfunctioning kidneys, certain poisonous substances, instead of being excreted with the urine, tend to collect in the blood. Such a condition is called *uremia*. This is a condition, of course, that must be dealt with immediately. Sometimes referred to as the *uremic syndrome*, the related infection accounts for approximately 20% of the deaths of patients with chronic renal failure.

For patients with kidney conditions that do not respond to medication or surgery, this condition is referred to as *end-stage renal disease*. Currently, this condition allows for only two remedial procedures: (1) *hemodialysis* and (2) *kidney transplantation*.

Hemodialysis sometimes is referred to as an *artificial kidney*. It should be pointed out that this principle also finds application in other medical situations—as a direct or supportive measure during the treatment of certain heart conditions and scleroderma (deposition of fibrous connective tissue in the skin and other organs), and in the removal of metabolites from methyl alcohol poisoning, among other special situations.

In a noteworthy study, P. Rut, F. Gomez, and A.D. Schreiber of the University of Cadiz, Spain, and the University of Pennsylvania School of Medicine found that macrophage Fcγ receptor function is impaired in patients with end-stage renal disease who are undergoing hemodialysis. This impairment probably contributes to the observed immunodepression and high prevalence of infection among such patients. The details of this finding are beyond the scope of this article, but reference to the Rutz et al. reference listed is suggested.

Hemodialysis. This is a mechanical-chemical approach for accomplishing the important kidney functions of impurity and fluid removal. The artificial kidney can perform these functions by utilizing the processes of diffusion and ultrafiltration. In dialysis, the blood is circulated from the patient's body into the artificial kidney, where it flows across a semipermeable membrane bathed in a special "wash" or dialysate solution. The impurities pass through the membrane into the dialysate solution (by diffusion). Because there is a difference in pressure between the blood and dialysate solution, excess fluid in the body is removed (by ultrafiltration). The total process is known as *hemodialysis*. The dialysis patient must undergo this treatment for 4 to 6 hours, 2 or 3 times per week. Portability and ease of use of dialysis machines have been stressed in recent years.

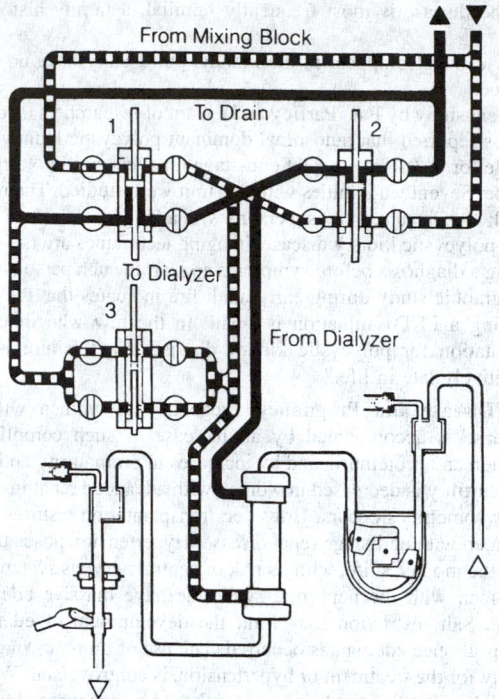

Fig. 6.   One-time-use dialyzer unit.

Fig. 5.   Schematic diagram of artificial kidney system. A known and equal quantity of fluid (approximately 500 milliliters/minute) is moved into and out of the dialyzer by two positive displacement pumps. These pumps (1 and 2) alternate the functions of bringing dialysis fluid from the mixing block and sending spent dialysate to the drain. A third pump (3) extracts the programmed amount of fluid from the dialysis fluid circuit. This circuit uses seals and valves to create a watertight section. Since a known and equal quantity of dialysis fluid is sent to the kidney and demanded from the kidney, and the circuit is also watertight, the programmed ultrafiltrate must come across the dialyzer membrane. Within established safe limits, the trans-membrane pressure is allowed to change freely to any value that results from the combination of dialyzer ultrafiltration rate chosen. (*Cordis Dow Corp.*)

Hemodialysis service is available to patients in essentially three ways: (1) *center dialysis*, in which the treatment is carried out within a hospital environment (the most costly approach); (2) *limited-care dialysis*, where treatments are given in clinic-type settings and where professional technologists supervise the treatments among a number of patients; and (3) *home dialysis*, where someone within the home is able to assist the patient, if necessary, to operate the machine. In 1972, the U.S. Congress extended Medicare coverage to patients with kidney failure. This provision also covered kidney transplants. By far, the funding for hemodialysis exceeds that of kidney transplants, of which there are only a few thousand per year.

The dialysis machine involves a number of sensors (blood flow, blood leaks, blood pressure), three pumps, several valves, flow pistons and flow diverters, and one-time dialyzer units. One system is shown schematically in Fig. 5. The flashlight-like dialyzer, filled with thousands of hollow fibers, specially designed to remove excess fluid and impurities from the blood stream, is shown in Fig. 6.

Unquestionably, dialysis is extending the lives of many persons and is a great scientific contribution to society. It is not, however, without its own sociological problems.

In addition to persons who are subjected to chronic hemodialysis, some patients require the machines only during periods of other therapy, which may repair reversible kidney abnormalities. Patients awaiting kidney transplantation may also require hemodialysis.

Kidney Transplants. Traditionally, a key to successful renal transplantation is the closeness of genetic relationship between recipient and donor. Research has shown, in terms of organ and patient survival over a 5-year period, that success rates range downward from 61% (organ) and 71% (patient) for identical twins (donor and recipient), to 42% (organ), 62% (recipient) for parent donors, and a somewhat lower range for other relatives. The lowest survival rate is 35% (organ), 51% (recipient) with cadaver

transplants, where organs are removed from donors at or immediately after death. However, other researchers have reported significantly higher rates, particularly with the comparatively recent stress on blood transfusion prior to transplantation. This procedure appears to improve graft survival. With this technique, some institutions have reported an 80–90% success rate for grafts of kidneys donated by living relatives and about 70% for all transplantations where kidneys taken from individuals who have died have been used.

The role of tissue matching in the transplantation of cadaver kidneys has been a subject of controversy for several years and is continuously being reviewed.

The importance of histocompatibility antigens (HLA) to the survival rate of kidney transplants from cadaveric donors has been studied by S. Takemotto and a team of researchers at the University of California at Los Angeles [under sponsorship of the United Network for Organ Sharing (UNOS)]. The study concludes: "The collaborative renal-transplantation program for HLA matching of donors and recipients yielded an increased rate of one-year graft survival and an estimated half-life for matched grafts twice that for mis-matched grafts. An increased role for HLA matching in kidney allocations is therefore indicated."

W.E. Braun of the Cleveland Clinic has observed, "It is time to expand procedures that will improve the success rates for cadaveric renal allo-grafts, not just for 1 to 2 years, but for 10 to 30 years. The number of human kidneys available for transplantation is extremely limited and there is no ethically acceptable repository that can suddenly be tapped to improve the terrible imbalance between supply and demand. Every kidney counts."

Many of the opportunities for improving graft survival are already available or can be made so (for example, providing immunosuppressive drugs, refining our understanding of atypical sensitization, and widening the use of better recipient matching).

In concluding this article, it should be noted that Joseph E. Murray, a pioneer in kidney transplantation, and E. Donnall Thomas shared the Nobel Prize in Physiology or Medicine in 1990 for their work in life-saving organ and cell transplant techniques. Thomas is known for his work in bone marrow transplantation.

Additional Reading

Addis, T.: *Glomerular Nephritis, Diagnosis and Treatment*, MacMillian, New York, NY, 1950.

Berkow, R. and M.H. Beers: *The Merck Manual*, 17th Edition, Merck & Company, Inc., Whitehouse Station, NJ, 1999.

Berns, M.W.: "Laser Surgery," *Sci. Amer.*, 84 (June 1991).

Better, O.S. and J.H. Stein: "Early Management of Shock and Prophylaxis of Acute Renal Failure in Traumatic Rhabdomyolysis," *N. Eng. J. Med.*, 825 (March 22, 1990).

Bostwich, D.G. and J.N. Eble: *Urologic Surgical Pathology*, Mosby-Year Book, Inc., St. Louis, MO, 1996.

Bourgoignie, J.J. and V. Pardo: "HIV-Associated Nephropathies," *N. Eng. J. Med.*, 729 (September 3, 1992).

Brace, R.A., M.A. Hanson, and C.H. Rodeck: *Body Fluids and Kidney Function*, Vol. 4, Cambridge University Press, New York, NY, 1998.

Brady, L.W., Z. Petrovich, and L. Baert: *Carcinoma of the Kidney*, Springer-Verlag, Inc., New York, NY, 1998.

Braun, W.F.: "Every Kidney Counts," *N. Eng. J. Med.*, 883 (September 17, 1992).

Brenner, B.M. and F.C. Rector, Jr: *Brenner and Rector's the Kidney*, 5th Edition, W.B. Saunders Company, Philadelphia, PA, 1995.

Cameron, J.S.: "Membranous Nephropathy — Still a Treatment Dilemma," *N. Eng. J. Med.*, 638 (August 27, 1992).

Cameron, J.S.: *Kidney Failure: The Facts*, Oxford University Press, Inc., New York, NY, 1996.

Cattell, W.R.: *Infections of the Kidney and Urinary Tract*, Oxford University Press, Inc., New York, NY, 1996.

Converse, R.I. et al.: "Sympathetic Overactivity in Patients with Chronic Renal Failure," *N. Eng. J. Med.*, 1912 (December 31, 1992).

Cullen, M.R. et al.: "Occupational Medicine (Renal Disorders)," *N. Eng. J. Med.*, 598 (March 1, 1990).

Cundiff, G.W. and A.E. Bent: *Endoscopic Diagnosis of the Female Lower Urinary Tract*, W.B. Saunders Company, Philadelphia, PA, 1999.

Danovitch, G.M.: *Handbook of Kidney Transplantation*, 3rd Edition, Lippincott Williams & Wilkins, Philadelphia, PA, 2000.

Donnelly, P.K. and R. Henderson: "Matching for Age in Renal Transplantation," *N. Eng. J. Med.*, 851 (March 22, 1990).

Greenberg, A., A. Cheung, T.M. Coffman, et al.: *Primer on Kidney Diseases*, Morgan Kaufmann Publishers, Orlando, FL, 1997.

Greenberg, A.: *Primer on Kidney Diseases*, National Kidney Foundation, New York/New Jersey, 2001.

Hoffman, M.: "One Wilms' Tumor Gene is Cloned; Are There More?" *Science*, 1387 (December 15, 1989).

Hook, J.B. and R.S. Goldstein: *Toxicology of the Kidney*, 2nd Edition, Lippincott Williams & Wilkins, Philadelphia, PA, 1996.

Kunin, C.M.: *Urinary Tract Infections: Detection, Prevention, and Management*, 5th Edition, Lippincott Williams & Wilkins, Philadelphia, PA, 1997.

Leslie, S.W.: *The Kidney Stones Handbook: A Patient's Guide to Hope, Cure and Prevention*, 2nd Edition, Four Geez Press, Roseville, CA, 1999.

Levine, D.Z. and R. Kersey: *Care of the Renal Patient*, 3rd Edition, W.B. Saunders Company, Philadelphia, PA, 1997.

Levinsky, N.G. and R.A. Rettig: "The Medicare End-Stage Renal Disease Program," *N. Eng. J. Med.*, 1143 (April 18, 1991).

Martin, D.H. et al.: "A Controlled Trial of a Single Dose of Azithromycin for the Treatment of Chlamydial Urethritis and Cervicitis," *N. Eng. J. Med.*, 921 (September 24, 1992).

Mitch, W.E. and S. Klahr: *Handbook of Nutrition and the Kidney*, 3rd Edition, Lippincott Williams & Wilkins, Philadelphia, PA, 1998.

Mobley, H.L. and J.W. Warren: *Urinary Tract Infections: Pathogenesis and Clinical Management*, ASM Press, Washington, DC, 1997.

Morgan, S.H. and Jean-Pierre Grunfeld: *Inherited Disorders of the Kidney: Investigation and Management*, Oxford University Press, Inc., New York, NY, 1998.

Murphy, W.M. and L. Day: *Urological Pathology*, 2nd Edition, W.B. Saunders, Philadelphia, PA, 1997.

O'Callaghan, C.A. and B.M. Brenner: *The Kidney at a Glance*, Blackwell Science, Inc., Malden, MA, 2000.

Olsen, S.: *"Tumors of the Kidney and Urinary Tract,"* W.B. Saunders Company, Philadelphia, PA, 1998.

Parfrey, P.S. et al.: "The Diagnosis and Prognosis of Autosomal Dominant Polycystic Kidney Disease," *N. Eng. J. Med.*, 1085 (October 18, 1990).

Ruiz, P. et al.: "Impaired Function of Macrophage Fcg Receptors in End-Stage Renal Disease," *N. Eng. J. Med.*, 717 (March 15, 1990).

Schrier, R.W., W.M. Bennett, and W.L. Henrich: *Atlas of Diseases of the Kidney*, Vol. 5, Blackwell Science, Inc., Malden, MA, 1994.

Sidransky, D. et al.: "Identification of p53 Gene Mutations in Bladder Cancers and Urine Samples," *N. Eng. J. Med.*, 706 (May 3, 1991).

Symmers, W.S., R.C. Pugh, and K.A. Porter: *Kidney and Urinary Tract*, 3rd Edition, Churchill Livingstone, Inc., Philadelphia, PA, 1992.

Takemoto, S. et al.: "Survival of Nationally Shared, HLA-Matched Kidney Transplants from Cadaveric Donors," *N. Eng. J. Med.*, 834 (September 17, 1992).

Tolkoff-Rubin, N.E. and R.H. Rubin: "Uremia and Host Defenses," *N. Eng. J. Med.*, 770 (March 15, 1990).

Torres, V.E. et al.: "Association of Hypokalemia, Aldosteronism, and Renal Cysts," *N. Eng. J. Med.*, 345 (February 8, 1990).

Wiebers, D.O. and V.E. Torres: "Screening for Unruptured Intracranial Aneurysms in Autosomal Dominant Polycystic Kidney Disease," *N. Eng. J. Med.*, 953 (September 24, 1992).

Web Reference

National Kidney Foundation: http://www.kidneynynj.org/ and http://www.kidney.org/

KIDSAT. NASA's pilot education program that used an electronic still camera aboard the Space Shuttle to bring the frontiers of space exploration to a growing number of U.S. middle school classrooms via the Internet.

KidSat is a research and development project that links middle school, high school and university students to Space Shuttle missions. The mission of KidSat was to understand and demonstrate how middle school students can actively make observations of the Earth by using mounted cameras onboard the Space Shuttle to conduct scientific inquiry in support of their middle school curricula. Students engaged in a process to select and analyze images of the Earth during Shuttle flights and use the tools of modern science (computers, data analysis tools and the Internet) to widely disseminate the images and results. A team environment, modeling scientific research and space operations and promoting student growth, discovery and achievement, while helping students participate in solving real-world problems, is implemented.

These students remotely operated a Kodak electronic still camera, mounted in the right overhead window on the flight-deck of the Space Shuttle, to take digital photographs of the Earth. Middle school students are responsible for planning the photo requests, which involves calculating the longitude and latitude of a region, as well as the exact time the Shuttle flies over it. High school and university students then compiled the requests into a single control file which was forwarded, by KidSat representatives at the Johnson Space Center (JSC) in Houston, to the IBM Thinkpad connected to the camera. Using special flight software, the Thinkpad automatically commands the camera to snap the pictures requested by the middle schools. These pictures then retrace their path back down to Earth where they reach their final destination — a computer archive. Students then can access their pictures from this archive, using the Internet.

KidSat has flown on three Shuttle missions: the first was in March 1996 (STS-76) and the second in January 1997 (STS-81). The third and final mission of this pilot program was flown on the Space Shuttle *Atlantis* (STS-86) in September 1997. Three U.S. middle schools participated in the first flight. Since then, KidSat has been growing; there were fifteen schools participating in STS-81, and 52 schools participated in STS-86.

Over 300 photos were taken during STS-76, another 500 were taken during STS-81, and over 700 photos were taken during STS-86. These photos can be accessed at the following URL: *http://kidsat.jpl.nasa.gov/*

The three-year pilot program was a partnership between NASA's Jet Propulsion Laboratory (JPL), the University of California at San Diego (UCSD) and the Johns Hopkins University Institute for the Academic Advancement of Youth (JHU-IAAY). The curriculum used by the middle school students and teachers was developed by the JHU-IAAY and UCSD.

During the Shuttle mission, the KidSat mission operations center at UCSD, was staffed by undergraduate and high school students. The center is modeled after Mission Control at JSC (Johnson Space Center). The students received telemetry from the Shuttle on their computer monitors and could listen to and receive instructions from NASA's flight controllers over direct channels to JSC.

The KidSat mission operations team monitored the Shuttle's progress around the clock and continually provided up-to-date information to the middle schools, who were using the Internet to send instructions to photograph specific regions of the Earth. Since any change in the Shuttle's orbit could affect the students selections, UCSD constantly updated this information so that the middle schools could re-plan their photographic requests if necessary. This was done through a sophisticated World Wide Web site that allowed students access to interactive maps of orbit ground tracks to aid in photo selection.

When the image requests had been verified by KidSat mission operations, they were compiled into a single camera control file and forwarded electronically to the KidSat representatives at JSC. They passed this file on to flight controllers who uplinked it to an IBM Thinkpad connected to the KidSat camera. Software on the Thinkpad, developed by students working at JPL, used these commands to control the camera. These same students trained the astronauts on the use of the software and the installation of the KidSat camera in the Shuttle's overhead window.

Some of the topics the students explored during the KidSat missions were weather, biomes, the relationship between history and geography and the patterns of rivers on the landscape. Additional interests for these missions included searching for impact craters and studying the relationships of center pivot irrigation fields to available water supply.

The KidSat pilot program was sponsored by NASA's Office of Human Resources and Education, with support from the Offices of Space Flight, Mission to Planet Earth and Space Science.

KILLDEER.   See **Waders, Shorebirds, and Gulls**.

KILOGRAM.   At the end of the 18th century, a kilogram was the mass of a cubic decimeter of water. In 1889, the 1st CGPM sanctioned the international prototype of the kilogram, made of platinum-iridium, and declared: *This prototype shall henceforth be considered to be the unit of mass.*

The 3rd CGPM (1901), in a declaration intended to end the ambiguity in popular usage concerning the use of the word "weight", confirmed that:

The kilogram is the unit of mass; it is equal to the mass of the international prototype of the kilogram.

The complete declaration appears below.

It follows that the mass of the international prototype of the kilogram is always 1 kilogram exactly, $m(K) = 1$ kg. However, due to the inevitable accumulation of contaminants on surfaces, the international prototype is subject to reversible surface contamination that approaches 1 µg per year in mass. For this reason, the CIPM declared that, pending further research, the reference mass of the international prototype is that immediately after cleaning and washing by a specified method (PV, 1989, 57, 104–105 and PV, 1990, 58, 95–97). The reference mass thus defined is used to calibrate national standards of platinum-iridium alloy (Metrologia, 1994, **31**, 317–336).

The kilogram is the last remaining base unit of the SI that is still defined by a material artefact. The base unit of mass in the International System, equal to 1,000 grams (2.2046 pounds).

See also **Units and Standards**.

KILOHERTZ.   Formerly called *kilocycle*; abbreviated kHz. A unit of frequency equal to one thousand (10^3) cycles per second. It is most commonly used in connection with radio wave frequencies.

KILOJOULE (KJ).   A unit of energy in the International System of Units (SI). One kilojoule is equal to 10^3 joules, or to 10^{10} ergs, or to 238.9 calories.

KIMBERLITE.   The name applied to a mica peridotite which occurs at Kimberley and other places in the Republic of South Africa, the source of rich deposits of diamonds. These valuable gem stones were originally found in the decomposed kimberlite which, being colored yellow by limonite, was termed "yellow ground." Deeper workings disclosed the less altered rock, kimberlite, which the miners call "blue ground."

KINASES (Protein).   See **Central and Peripheral Nervous Systems**.

K INDEX.   A stability index that is a measure of thunderstorm potential based on temperature lapse rate, moisture content of the lower troposphere, and the vertical extent of the moist layer. Also called *George's index*.

The K index is determined by the following equation:

$$K = (850 \text{ mb } T - 500 \text{ mb } T) - 850 \text{ mb } Td$$
$$- (700 \text{ mb } T - 700 \text{ mb } Td),$$

where T is the temperature and Td is the dewpoint in degrees Celsius at the pressure levels indicated. The higher (positive) the K index, the greater the likelihood of thunderstorm development.

KINEMATIC BOUNDARY CONDITION.   The condition that the fluid velocity directed perpendicular to a solid boundary must vanish on the boundary itself. This may be stated mathematically by the expression

$$\mathbf{n} \cdot \mathbf{u} = 0,$$

where $\mathbf{n}$ is a unit vector normal to a solid surface and $\mathbf{u}$ is the fluid velocity vector. In meteorology, this boundary condition is often employed in considering flow near the earth's surface. When the boundary is a fluid surface or interface, this condition applies to the vector difference of velocities across the interface and requires that the interface, although in motion, will at all times consist of the same fluid parcels. In meteorology, such a condition must be applied at fronts and other surfaces of discontinuity.

See also **Dynamic Boundary Condition**.

AMS

KINEMATIC CHAIN.   A kinematic chain consists of a number of links connected to one another in such a way as to allow motion to take place in combination. A kinematic chain is not necessarily a mechanism; it becomes one when so constructed as to allow constrained relative motion between its links.

KINEMATIC FLUX.   Transport of a variable per unit area per unit time (i.e., a dynamic flux), but divided by the average air density (or in the case of heat flux, divided by the product of air density times specific heat of air at constant pressure). The resulting kinematic flux has the same units as velocity times the variable being transported. For example, a vertical turbulent heat flux is $\overline{w'\theta'}$, while a vertical heat flux associated with mean wind is $\overline{w\theta}$, where w is vertical velocity, θ is potential temperature, an overbar denotes an average, and a prime denotes a deviation from the average. Kinematic fluxes are more closely related to meteorological variables that can be easily measured (such as temperature and wind) than are the associated dynamic fluxes (such as J m^{-2} s^{-1}). Statistically, kinematic fluxes are covariances.

AMS

KINEMATICS.   The physics of abstract motion without regard to forces or bodies of matter. Two aspects need special consideration: (1) the motion of points, and (2) the motion of rigid figures. Whatever system of space coordinates is found simplest may be used, and may be transformed from one system to another as desired.

The motion of a point is completely specified by giving each of its three rectangular coordinates (for example) as a function of the time; that is, by writing its "equations of motion":

$$\left.\begin{array}{l} x = f_1(t), \\ y = f_2(t), \\ z = f_3(t) \end{array}\right\} \tag{1}$$

In many cases, the information represented by these equations is supplied only indirectly in the statement of a problem. For example, an expression may be given for the component of the linear velocity or the linear acceleration, as a function of either the time or the distance. In such a case, a differential equation is first obtained. Thus, if it is specified that the X-component of the acceleration is constant and equal to a, the differential equation is

$$\frac{d^2x}{dt^2} = a \tag{2}$$

and the corresponding equation of (uniformly accelerated) motion is its solution,

$$x = \tfrac{1}{2}at^2 + v_0t + x_0 \tag{3}$$

in which the integration constants x_0, v_0, stand, respectively, for the initial distance (value of x when $t = 0$ and the initial velocity. By eliminating t from the equations of motion (1), three geometric equations may be arrived at, one in x, y, one in x, z, and one in y, z, any two of which determine the path of the moving point in space.

Again, if the Y-component of acceleration is proportional to the coordinate y and of opposite sign,

$$\frac{d^2y}{dt^2} = -k^2y \tag{4}$$

may be written, in which k^2 is a positive constant. The solution of this is

$$y = A \sin kt + B \cos kt \tag{5}$$

which is a general equation of simple harmonic motion.

A rigid 3-dimensional figure has six degrees of freedom. These require six equations of motion, which may, for example, express the three components of linear motion of the centroid of the figure with respect to the three rectangular axes, and the three components of angular motion or rotation about the same three axes. In such a case, it is often convenient to use the notation of vector analysis. It may be shown that any motion of a rigid figure is at any given instant equivalent to a linear motion of its centroid in some definite direction, plus a rotation of the figure about some definite axis through the centroid.

The laws and equations of kinematics are of constant service in problems of kinetics.

KINEMATIC VISCOSITY.

A coefficient defined as the ratio of the dynamic viscosity of a fluid to its density. The kinematic viscosity of most gases increases with increasing temperature and decreasing pressure. For dry air at $0\,°C$ ($32\,°F$), the kinematic viscosity is about $1.46 \times 10^{-5}\ m^2\ s^{-1}$. Common symbols for these variables are μ for dynamic viscosity and v for kinematic viscosity.

Additional Reading

List, R.J.: *Smithsonian Meteorological Tables*, 6th Edition, Smithsonian Institution Press, Washington, DC, 2000.

AMS

KINETIC ENERGY.

That part of the energy of a body that the body possesses as a result of its motion. A particle of mass m, moving with a speed v, has a kinetic energy $E = \frac{1}{2}mv^2$ according to classical mechanics, i.e., if $v \ll c$, the speed of light. Similarly, a rigid body of mass m, whose center of mass is moving with a speed v and which is simultaneously rotating about an axis through its center of mass with an angular velocity ω, possesses a kinetic energy

$$E = \tfrac{1}{2}mv^2 + \tfrac{1}{2}I\omega^2$$

where I is the moment of inertia of the body about the axis of rotation.

When the speed of a particle is sufficiently high so that relativity theory must be used, the kinetic energy is

$$E = (m - m_0)c^2 = m_0c^2 \left[\frac{1}{(1 - v^2/c^2)^{1/2}} - 1 \right]$$

where m is the observed mass of the body and m_0 is the rest mass. This expression always gives a higher kinetic energy than does the classical one; it is approximated by the classical form when $v \ll c$ and increases without limit as v approaches c. The kinetic energy of a body, in either classical or relativistic mechanics, is different for two observers who are in relative motion.

The separation of the total energy of a mass into kinetic and other forms of energy depends on the scale of observation. Thus, the speed of the center of mass of a large-scale body is the quantity that enters into the computation of its kinetic energy, whereas the individual molecules that make up the body may be moving at high speeds relative to the center of mass. So, the sum of their kinetic energies may be much greater than the kinetic energy of the body as a whole. The kinetic energies associated with the random motion of the molecules contribute to the internal energy of the body.

Negative Kinetic Energy. In quantum mechanics there may be a finite probability of finding a particle in the classically forbidden region where the total energy of the particle is less than its potential energy. In such a region, the kinetic energy, given as the difference between the total energy and the potential energy, would be negative. If an attempt is made to observe a particle in this classically forbidden region, one gives to the particle an indeterminate amount of energy, sufficient to lead to a positive measured value for its energy (Heisenberg uncertainty principle), with the result that negative kinetic energy is not observable. Particles pass through such classically forbidden regions of negative kinetic energy in the tunnel effect.

The term is also used to describe the energy states of particles in those energy levels that were first predicted by the Dirac electron theory.

KINETIC MEASUREMENTS.

Kinetic measurements are studies of the rates at which chemical reactions occur. Generally, these studies involve preparing a chemical system using reagent concentrations different from the equilibrium values and then monitoring the concentration changes as the system approaches equilibrium, although other, less direct strategies are sometimes exploited. Chemical kinetic data are used in materials science, biochemistry and molecular biology, earth and atmospheric science, and many branches of engineering. Related concepts appear in nuclear physics, but presuppositions and methods are different there.

Kinetic information is acquired for two different purposes. First, data are needed for specific modeling applications that extend beyond chemical theory. These are essential in the design of practical industrial processes and are also used to interpret natural phenomena such as the observed depletion of stratospheric ozone. Compilations of measured rate constants are published in the United States by the National Institute of Standards and Technology (NIST). Second, kinetic measurements are undertaken to elucidate basic mechanisms of chemical change, simply to understand the physical world. The ultimate goal is control of reactions, but the immediate significance lies in the patterns of kinetic behavior and the interpretation in terms of microscopic models.

Explaining chemical change by postulating mechanisms in terms of macroscopic concentrations is expected to continue for the foreseeable future. For a fundamental understanding of very simple reactions, however, traditional kinetics is being challenged by theoretical and experimental methods that focus directly on the behavior of individual atoms.

Macroscopic Behavior and the Rate Law

Chemical changes are discussed with the aid of the equations used to treat equilibrium, i.e., the reaction of reactants A, B, C, and so on, to produce products P, Q, and so forth.

The essential information implied by the chemical equation is the stoichiometry at the macroscopic level, i.e., if a moles of A react, then b moles of B do also; p moles of P formed, etc. No inference can be made about behavior at the atomic level, eg, there is no implication that p molecules of P appear simultaneously. There may also be intermediates that appear and disappear in the course of the reaction but are not shown in the overall equation.

A kinetic study typically prepares some set of initial concentrations not at equilibrium and describes the subsequent evolution of each. A basic assumption is that each component evolves according to some differential equation where t represents time.

$$d[A]/dt = f([A], [B], \ldots, [P], \ldots, \text{ other conditions}) \tag{1}$$

In general, the differential equation could be very complicated, e.g., the concentrations may be functions of spatial coordinates as well as time. Experimental measurements are arranged to ensure that simplified equations apply.

A key assumption of most laboratory kinetic measurements is that of a well-mixed solution of reactants. Then any component can be characterized by a single time-dependent concentration, applicable to the entire system.

In particularly simple cases, which occur frequently, one may assume a dependence on powers of reactants and ignore products

$$d[A]/dt = -h[A]^x[B]^y[C]^z \tag{2}$$

Verification of a Rate Law

It is possible to prepare a system having an initial concentration for each component, and then measure a finite, but small, change in the concentration of one component, $\Delta[A]$ for example, over a known interval of time, Δt. The experimental velocity $\Delta[A]/\Delta t$ and the concentrations can be substituted into a proposed rate law, like equation 2, along with postulated values for the exponents x, y, $\ldots$ to determine an observed constant k_{obs}. If this process is repeated for a reasonable range of concentrations, and the postulated rate law having the same exponents always yields the same k_{obs}, then it is asserted that the rate law has been verified for those concentration ranges and the rate constant determined. This approach is a reasonable strategy for an initial survey of a totally unknown system; but it is wasteful, in that it extracts very little data from each set of initial conditions. More often, the integrated form of the rate law is fit to multiple concentration measurements recorded at different times for each set of initial conditions.

One of two very different strategies are used in kinetic measurements to produce the initial, nonequilibrium concentrations of reactants. Either the separate reagents are mixed or a system previously at equilibrium is perturbed.

Mixing known quantities of reagents to produce desired initial concentrations can be carried out with either a continuous flow or a stopped flow apparatus. Engineering details become important for fast reactions, especially those occurring in less than one millisecond.

Perturbation methods can be divided further into two categories. One uses a flash of light (flash photolysis) or other radiation (radiolysis) to create a homogeneous distribution of a desired reagent from some precursor molecule. The other perturbation method does not affect reagents but instead changes some intensive thermodynamic property, such as temperature, pressure, or an electric field.

Indirect and Novel Methods. A direct measurement is a record of changing concentrations as a function of time. In principle, the same information is available as a Fourier transform in the frequency domain. Since the latter part of the nineteenth century, it has been possible to measure the absorption of electromagnetic radiation as a function of frequency (or wavelength) and interpret lineshapes to yield kinetic information on the picosecond time scale. In gases, line widths increase at high pressure owing to collision broadening. Dissociation or ionization may also determine a linewidth. More subtle effects occur in liquids. More recently, chemical reactivity on slower time scales has been measured by lineshape analysis in Mossbauer spectroscopy and magnetic resonance.

The experimentally measured dependence of the rates of chemical reactions on thermodynamic conditions is accounted for by assigning temperature and pressure dependence to rate constants.

Microscopic Models in Kinetics

Mechanism is a technical term, referring to a relatively detailed, microscopic description of a chemical transformation, which, nevertheless, still falls far short of a complete dynamical description at the atomic level. A mechanism for a reaction is sufficient to predict the macroscopic rate law of the reaction. This deductive process is valid only in one direction, i.e., an unlimited number of mechanisms are consistent with any measured rate law. A successful kinetic study postulates a mechanism, derives the rate law, and demonstrates that the rate law is sufficient to explain experimental data over some range of conditions. New data may be discovered later that prove inconsistent with the assumed rate law and require that a new mechanism be postulated. Mechanisms state, in particular, what molecules actually react in an elementary step and what products these produce. An overall chemical equation may involve a variety of intermediates, and the mechanism specifies those intermediates.

Additional Reading

Claesson, S.: *Fast Reactions and Primary Processes in Chemical Kinetics*, 5th Edition, Wiley-Interscience, New York, NY, 1967.

Espenson, J.H.: *Chemical Kinetics and Reaction Mechanisms*, 2nd Edition, The McGraw-Hill Companies, Inc., New York, NY, 2002.

Kustin, K. *Fast Reactions, Methods in Enzymology*, Vol. 16, Academic Press, Inc., New York, NY, 1969.

Moore, J.W. and R.G. Pearson: *Kinetics and Mechanism*, 3rd Edition, Wiley-Interscience, New York, NY, 1981.

Strehlow, H.: *Rapid Reactions in Solution*, VCH, Weinheim, Germany, 1992.

KINETICS. The branch of mechanics that deals with the effects of forces or of torques upon the motions of material bodies. The basis of this subject in terms of classical theory consists of Newton's laws of dynamics, which may be extended to include d'Alembert's principle of kinetic equilibrium.

The motion of a particle or of a rigid body may be either "free" or "constrained." That is, it may be at liberty to move in any manner in obedience to the applied forces or torques, or there may be present material barriers that limit its linear motion to a certain path or surface, or its rotation to a certain axis. Thus, a projectile or a planet is free to travel and to rotate as it will, but a railway train must follow a fixed track; a grindstone must revolve, and a pendulum must swing about a fixed axis. These latter cases may be dealt with by considering that the barriers offer opposing forces or torques in equilibrium with those components tending to cause motion

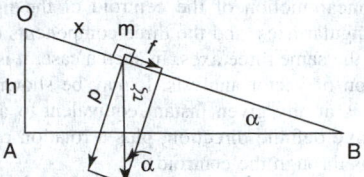

Fig. 1.   Block and inclined plane.

against them, and by treating as "effective" only those components acting in directions in which motion is permissible.

A simple problem of this type is that of a block of mass m sliding without friction down an inclined plane at an angle α. See Fig. 1. Its weight mg (in which g is the acceleration due to gravity) is resolved into components f and p, along and perpendicular to the plane, only the former of which is effective. This force $f = mg \sin \alpha$, acting on the mass m, gives it the constant acceleration $a = f/m = g \sin \alpha$. Therefore, the motion of the block along OB is expressed by

$$x = \tfrac{1}{2} g \sin \alpha \cdot t^2 + v_0 t + x_0 \qquad (1)$$

in which $x = Om$, x_0 is the initial value of x, and v_0 is the initial velocity (see also **Kinematics**). An interesting problem of constrained motion is to determine in what path from O to B a block would slide down in the least time; it turns out to be not a straight line, but a cycloid.

Many kinetic problems are greatly simplified by applying the conservation of energy principle. Thus for any path, the final speed of the block starting at O and sliding to B is $\sqrt{2gh + v_0^2}$, where h is the vertical height of O above B. This may be shown by using Equation (1) as an equation of pure kinematics, with $x_0 = 0$ and $x = OB = h/\sin \alpha$. It is necessary first to solve the equation, which now takes the form

$$\frac{h}{\sin \alpha} = \tfrac{1}{2} g \sin \alpha \cdot t^2 + v_0 t$$

for t, obtaining as the (positive) root

$$t = \frac{\sqrt{v_0^2 + 2gh} - v_0}{g \sin \alpha} \qquad (2)$$

Then the final speed is $V = at + v_0 = g \sin \alpha \cdot t + v_0$, which, using the value of t from Equation (2), gives

$$V = \sqrt{2gh + v_0^2} \qquad (3)$$

But let us now use instead the fact that the kinetic energy at B is equal to that at O plus the loss of potential energy in descent through height h, which is mgh. The kinetic energy at O and at B are, respectively,

$$E_0 = \tfrac{1}{2} m v_0^2, \quad E_B = \tfrac{1}{2} m V^2$$

so that

$$\tfrac{1}{2} m V^2 - \tfrac{1}{2} m v_0^2 = mgh \qquad (4)$$

Solution of this gives at once

$$V = \sqrt{2gh + v_0^2}$$

which is the same as Equation (3). The most significant feature of this is that the solution is here reached without considering the nature of the motion at all. Indeed, if a simple pendulum is pulled aside until the ball is at an elevation h above its lowest level and then given a push with initial speed v_0, it will attain the speed given by Equation (3) as it descends to that lowest level, although no algebraic equation corresponding to Equation (1) can be derived to describe its motion on the way down. The problem has thus become one of energetics rather than of kinematics.

Problems involving impacts or collisions are best handled by means of the principles of conservation of momentum and conservation of energy. In this way, for example, it may be shown that if an elastic sphere A strikes centrally a stationary elastic sphere B of greater mass, B will take part of the energy and momentum and A will rebound. If A and B are of equal mass, B will take all the energy and momentum and A will stop; while if A is more massive than B, B will take part of the energy and momentum and A will continue more slowly in its original direction.

By the use of d'Alembert's principle, problems of kinetics become equivalent to problems of statics. In case a body or a system is in frictionless, uniform rectilinear motion, in which there are no accelerations, classical dynamics introduces no element of the problem due to the motion; the situation is exactly as if the system were at rest. But with accelerations there develop what may be called "inertia forces," namely, the reactions opposing the forces causing the accelerations. If there is appreciable friction, that factor enters also as another opposing force.

Suppose a heavy boat is at rest in the water, in static equilibrium. Its mass is $m = 5,000$ kilograms. A tow line begins pulling on the boat. When, at the end of 30 seconds, the boat has attained a speed of 0.6 meters per second, the force of water friction on the boat is 3 kilograms. It is desired to find the tension in the tow line at that moment.

We take the direction of motion as positive and let the required tension be f. The reaction force of inertia is negative and in gravitational units is equal to ma/g, where a is the acceleration. We then have the equation of equilibrium,

$$f + \text{inertia force} + \text{friction force} = 0$$

or

$$f - \frac{5{,}000 \text{ kilograms} \times 0.02 \text{ meters/second}^2}{9.8 \text{ meters/second}^2} - 3 \text{ kilograms} = 0$$

from which $f = 13.2$ kilograms. The last two force components enter just as if the boat were at rest and two other ropes were pulling backward on it, one with a force of 10.2, the other with 3 kilograms. The same principle applies, using vector addition, when the forces concerned are not collinear.

Additional Reading

Billing, G. and K. Mikkelsen: *Advanced Molecular Dynamics and Chemical Kinetics*, John Wiley & Sons, Inc., New York, NY, 1997.

Do, D.: *Adsorption Analysis: Equilibria and Kinetics*, World Scientific Publishing Company, Inc., Riveredge, NJ, 1998.

Perthame, B.: *Advances in Kinetic Theory and Computing: Selected Papers*, World Scientific Publishing Company, Inc., Riveredge, NJ, 1994.

KINETIC TEMPERATURE (Atmosphere). See **Atmosphere (Earth)**.

KINETIC THEORY. A theory (proved by experiment) that explains the phenomena of heat and pressure as due to the kinetic motion and elastic collisions of atoms and molecules. The phenomena include gas and vapor pressure, evaporation, and diffusion of fluids.

Gases expand indefinitely when released, not because of repulsion between the molecules as formerly supposed (though the Joule-Thomson effect under certain conditions may involve this), but because the molecules are in rapid motion and do not stop unless they collide with something. Air is not "forced" out through a tire puncture; only those air molecules pass out which, in their aimless wanderings, happen to encounter the opening. Molecules also pass in from the outside; but since there are several times as many per unit volume inside as outside, many more pass out than in. This continues until, a statistical equilibrium being reached, the air inside is no more dense than that outside, and the tire is "flat." The rapidity with which this takes place emphasizes the speed of the molecular motion and the relative insignificance of the "internal friction" opposing it.

What appears to be a steady pressure is due to the incessant impacts of the gas molecules on any surface exposed to them. If n molecules of equal mass m are released in an enclosure of volume v, and if their speeds are $u_1, u_2, \ldots u_n$, it is easy to show that the average pressure set up by these impacts, neglecting the effects of collisions and gravity, is

$$p = \frac{m}{3v}(u_1^2 + u_2^2 + \cdots + \cdots u_n^2) \qquad (1)$$

This may be written

$$P = \frac{1}{3}\frac{nm}{v}\frac{\sum (u^2)}{n}$$

or, since nm/v is the gas density ρ, and $\rho\,(u^2)/n$ is the mean square molecular speed $\overline{u^2}$,

$$p = \frac{1}{3}\rho\overline{u^2} \qquad (2)$$

This relation gives the mean square speed as $\overline{u^2} = 3p/\rho$; which is the square of the effective speed, corresponding to average kinetic energy. From this it may be shown that the average speed is

$$\overline{u} = \sqrt{\frac{8p}{\pi\rho}} \qquad (3)$$

easily determined since p and ρ are measurable.

Again, Equation (1) may be written

$$pv = \frac{2}{3}\left(\frac{1}{2}mu_1^2 + \frac{1}{2}mu_2^2 + \cdots + \frac{1}{2}mu_n^2\right) = \frac{2}{3}E \qquad (4)$$

in which E is the total kinetic energy of linear motion of the molecules. From this it follows that the absolute temperature T of the gas bears a constant ratio to this total kinetic energy, and hence to the average translational kinetic energy of the molecules. See also **Ideal Gas Law**.

Further analysis shows that when gravity is considered, the pressure in an undisturbed pure gas of uniform temperature, at an elevation h, is given by

$$p = p_0 e^{-3gh/\overline{u^2}} \qquad (5)$$

in which p_0 is the pressure at the zero of elevation. This is a form of Laplace's "law of atmospheres," useful in barometric altitude determinations.

A quantity much used in kinetic theory is the "mean free path," which is the average distance traversed by a molecule between collisions. There are ways of calculating this and also the effective diameters of molecules, and these data lead to many conclusions as to frequency of collisions and rate of diffusion.

Irreversibility. The kinetic theory assumes that the velocity of a molecule may depend on the conditions in the region where it has just suffered a collision, but is otherwise random—in other words, independent of its previous history. This assumption permits one to use the methods of probability theory even though, in classical mechanics, the actual motions of the molecules are regarded as completely determined by their initial configurations. As long as one uses the theory only to calculate properties of a gas that can actually be measured during a relatively short time, the assumption of randomness leads to no serious errors. However, it introduces an element of irreversibility that is inconsistent with the reversibility of the laws of classical mechanics. A reversible process is one that can go equally well forwards or backwards, in contrast to an irreversible process. Lord Kelvin pointed out the importance of irreversible processes in 1852. The irreversible aspect of the kinetic theory is shown most clearly by Boltzmann's "H-theorem," which has led to a considerable controversy over the foundations of kinetic theory. Boltzmann showed in 1872 that a certain quantity, later called H, which depends on the velocity distribution, must always decrease with time, unless the velocity distribution is Maxwell's distribution, in which case H remains constant. In the latter case, which corresponds to the equilibrium state, H is proportional to the negative of the entropy. Thus, the H-theorem provides a molecular interpretation of the second law of thermodynamics or, in particular, the principle that the entropy of an isolated system must always increase or remain constant.

Irreversible processes are those in which entropy increases. The entropy itself can be regarded as a measure of the degree of randomness or disorder of the gas, although it must be recognized that disorder really means a lack of knowledge about the details of molecular configurations. The equilibrium state represents the maximum possible disorder; the H-theorem implies that a gas which is initially in a nonequilibrium (partly ordered) state will eventually reach equilibrium and then stay there forever if it is not disturbed.

If the long-term consequences of the H-theorem were applicable to all matter in the universe, one might expect that the universe would eventually "run down"—although the total energy might always remain the same, no useful work could be done with this energy, because all matter would be at the same temperature. This final state has been called the "heat death" of the universe.

The contradiction between the *H*-theorem and the laws of classical mechanics is shown by two well-known criticisms of the kinetic theory: (1) the reversibility paradox and (2) the recurrence paradox. The first paradox is based on the fact that Newton's laws of motion are unchanged if one reverses the time direction, so that it would seem to be impossible to deduce from these equations a theorem that predicts irreversible behavior. Kelvin discussed this paradox in 1874, and concluded that while any single sequence of molecular motions could be reversed, leading to an ordered state, the number of disordered states is so much greater than the number of ordered states that it is virtually impossible to stay in an ordered state for any period of time. Thus, irreversibility is a statistical, but not an absolute consequence of kinetic theory. Boltzmann gave a similar answer when the problem was pointed out to him by Loschmidt a few years later.

The second paradox is based on a theorem of Henri Poincaré—if a mechanical system is enclosed in a finite volume, then after a sufficiently long time it will return as closely as one likes to its initial state. Hence *H* must return to its original value; if it has decreased during some period of time, it must increase during some other period. The time between successive recurrences of the same state for the molecules in one cubic centimeter of air is much longer than the present age of the universe, so one does not have to be concerned about recurrences in any actual experiment. In his attempt to resolve the recurrence paradox, Boltzmann was finally led to a remarkable psycho-cosmological speculation; he suggested that the direction of time as perceived by an animate being is determined by the direction of irreversible processes in his environment and in his body. Thus, when the time comes for a recurrence, entropy will decrease, but subjective time will flow in the opposite direction. The concept of alternating time-directions in cosmic history was further explored by Reichenbach in 1956 and has been proposed again in recent theories of the expanding (and contracting) universe.

Updating of Kinetic Theory. Since the late 1940s, there has been a revival of interest in the classical kinetic theory of gases, based on the assumptions of Clausius, Maxwell, and Boltzmann, and ignoring quantum effects except insofar as these may determine the intermolecular force law. In part this interest grew out of applications involving high speed aerodynamics and plasma physics and, in part, from renewed attempts to construct reliable theories of liquids as well as dense gases. Methods for obtaining accurate solutions of the Boltzmann equation were developed by Grad, Pekeris, Ikenberry, and Truesdell, among others. These solutions were used to describe the behavior of gases in many circumstances more complex than those treated in the 1800s (including the interactions of charged particles and magnetic fields). Problems such as the propagation and dispersion of sound waves were treated by Uhlenbeck and his collaborators.

In 1946, three general formulations of kinetic theory were published (Born and Green; Kirkwood; and Bogoliubov). In each case, the goal was to derive a generalized Boltzmann equation in a form that would be valid when simultaneous interactions among more than two molecules have to be taken into account, and thence to obtain solutions of the equation from which transport properties of dense gases and liquids could be calculated. In each formulation, certain approximations had to be made in order to obtain practical results. Because of the difficulty in estimating the error involved in these approximations, and the great complexity of the equations involved, there was no clear evidence that the results for properties such as the viscosity coefficient would be significantly more accurate than those obtained by Enskog from his modified kinetic theory for dense gases, published in 1922. Eventually, in the early 1960s, attention was centered on the systematic derivation of series expansions for the transport coefficients in ascending powers of the density, together with attempts to calculate the first few terms in such series for special molecular models, such as elastic spheres.

In the meantime, an alternative and apparently more rigorous method for deriving theoretical expressions for transport coefficients, based on the fluctuation-dissipation theory introduced in 1928 by H. Nyquist in electrical engineering problems, was developed by Green, Mori, and Kubo. This method had the heuristic advantage of bringing out clearly the connection between transport theory and the description of fluctuations in equilibrium statistical mechanics. Later, it was proved that the Green-Mori-Kubo method gives results precisely equivalent to those that could be obtained from the Born-Green, Kirkwood, and Bogoliubov methods. Thus, just as in the case of quantum mechanics, several alternative approaches are equally valid in modern kinetic theory.

After intensive efforts to calculate terms in the density expansion of transport coefficients, it was finally discovered in 1965 that such a density expansion does not actually exist, for mathematical reasons associated with the persistence of weak correlations between colliding particles over very long times. The divergence of the expansion (and thus the inadequacy of the approximations on which most earlier theories had been based) was established by a number of investigators, including Dorfman and Cohen, Weinstock, and Goldman and Friedman. The result was an increased overall interest in kinetic theory.

Additional Reading

Brush, S.G., and N.S. Hall: *Kinetic Theory of Gases: An Anthology of Classic Papers with Historical Commentary*, World Scientific Publishing Company, Inc., Riveredge, NJ, 2003.

Liboff, R.L.: *Kinetic Theory*, 3rd Edition, Springer-Verlag New York, LLC., New York, NY, 2003.

Pauli, W., and C.P. Enz: *Thermodynamics and the Kinetic Theory of Gases*, Vol. 3, Dover Publications, Inc., Mineola, NY, 2000.

Perry, R.H., and D. Green: *Perry's Chemical Engineers' Handbook*, 7th Edition, The McGraw-Hill Companies, Inc., New York, NY, 1997.

KINGBIRD *(Aves, Passeriformes).* A large North American flycatcher. The most common species, *Tyrannus tyrannus*, is called the bee martin and tyrant flycatcher as well as the kingbird. In color, it is dark slate gray above, with a concealed orange spot on the top of the head, white below, and has a white band across the tip of its black tail. This species and the several others found in the West and South are noisy and quarrelsome birds.

See also **Passeriformes**.

KING COBRA. See **Snakes**.

KING CRAB. See **Xiphosura**.

KINGDOM. See **Taxonomy**.

KINGFISHERS AND OTHER CORACIIFORMES. There are many species of kingfisher, mostly with strong sharp beaks used for catching fish. Some species eat insects and reptiles more than fish and have broader beaks, hooked at the tip. One of the species, *Dacelo gigas*, with a broad beak, lives in Australia and the Papuan islands and has been named from its peculiar call, the laughing jackass. Kingfishers are found on every continent. Other *Coraciiformes* include the following:

Bee-Eater (Merops apiaster), believed to be African in origin. There are 25 species in Europe and Asia. The bird is relatively small. The male and female both work in construction of their domicile, a hole in the ground ranging from 2 to 10 feet (0.6–3 meters) in depth and cut into the earth at an angle. These birds eat bees. After a 13-day incubation period, the young bird is hatched naked from a glassy white egg. The young are covered with a waxy material derived from the honey consumed by the parents. When the young are about ready to depart from the nest, the waxy sheath splits away, exposing the natural plumage. The young are fed by their parents for many days after leaving the nest. The bee-eater consumes numerous insects in addition to bees. The numerous species are much alike, but range in color—from green, blue, orange, and russet. The common bee-eater is found in Denmark, Finland, and Scotland, and to a lesser extent throughout other parts of the Old World. There are about eight species of square-tailed bee-eaters in Africa, smaller in size and of more primitive characteristics.

Hoopoe is of the Old World with a very high crest and a long sharp beak. One species (*Upupa epops*) lives in Europe; the others are found in the Oriental region, Africa, and Madagascar. The wood hoopoe is an African bird of a family closely related to the true hoopoes. The wood hoopoes differ in the metallic gloss of the plumage and the long wedge-shaped tail and in being crestless with a more curved bill. They also are called chatterers.

Hornbill is a large bird of the African and Oriental regions, characterized by a very large beak, and, in many species, a large prominence above the base of the upper mandible extending back onto the head. See Fig. 1.

Fig. 1.   Great Indian hornbill. (*Sketch by Glenn D. Considine.*)

Kiroombo is a rather peculiar bird of Madagascar and several similar species of other Pacific islands. It is related to the rollers.

Motmot is a bird with a long tail and serrated beak, found in Mexico, Central and South America. Several species are reported to nest in burrows in the banks of streams. The motmot (*Eumomota superciliosa*) is of a green coloration and is about 12 inches (30 centimeters) in length. The tail is about one-third the length of the bird and is beautifully tapered. There is black and blue plumage about the throat. Larger and more common is the *Momotus momata*, which attains a length of about 18 inches (46 centimeters). It has a cobalt-blue colored crown, with green body and black face.

Roller is a bird found in Africa, the Oriental region, Europe and Asia. There are several species, mostly of bright colors. The bird is named for its habit of turning over during flight.

Tody is a small green-and-red insect-eating bird (*Todus viridis*), found in the West Indies. The legs are relatively small and the beak is long and flattened. The bird nests in tunnels along the banks of streams, much as the kingfishers. See also **Coraciiformes**.

KING SNAKE.   See **Snakes**.

KININS.   Chemical substances that stimulate differentiation of plant cells that otherwise may have lost permanently the power of differentiation. The kinins contain adenine, which seems to give these substances their biological activity. Small bits of a carrot taken from a region containing no cambium can be stimulated to growth and differentiation to produce entire new carrot plants through the addition of kinins and a small amount of indoleacetic acid.

See also **Adenine**.

KINKAJOU.   See **Raccoons**.

KIRCHHOFF, GUSTAV ROBERT (1824–1887).   Kirchhoff was a German physicist born in Prussia. He studied at the University of Konigsberg and made his first contribution to physics in the field of electromagnetism. He derived equations now known as Kirchhoff's laws as a student.

He is best remembered for his general theory of motion of electricity in conductors, his law of absorption and emission of radiation by material bodies, his helping to establish the science of spectrum analysis and the invention of the spectroscopy, his discovery in 1860 of the elements cesium and rubidium.

See also **Cesium**; **Electric Circuits**; **Heat Transfer**; **Kirchhoff Laws of Networks**; and **Rubidium**.

J. M. I.

KIRCHHOFF LAWS OF NETWORKS.   Two laws relating to electric networks. The general case is that of *n* points or junctions, each one of which is connected with each of the $n - 1$ remaining points by a conductor containing a source of electromotive force or electric current in series with elements of resistance, capacitance, and inductance. Kirchhoff's two statements are as follows:

1. If conductors forming part of a network carrying currents meet at one point, or node, the sum of the instantaneous currents flowing toward the node is equal to the sum of those flowing away from it. Expressed alternatively, the algebraic sum of the instantaneous currents entering the node is zero. As applied to the node *P* in Fig. 1, application of the law above results in the following equation for current equilibrium at any instant of time:

$$i_1 + i_2 - i_3 - i_4 - i_5 = 0.$$

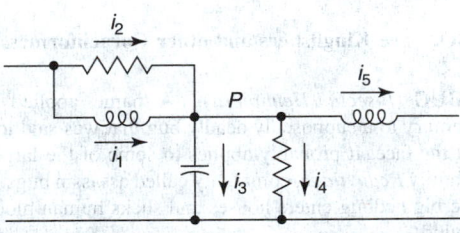

Fig. 1.   Application of Kirchhoff's first statement to analysis of node *P*.

2. Associate a current having an arbitrary direction with each connection between two junctions of the network. Associated with the assumed current direction there will be a unique change of electric potential from plus to minus (in the direction of the current) in passing through each resistive, inductive, and capacitive element. Furthermore, there will be appropriate changes of potential in passing through the sources of electromotive force, the polarity of each potential change being a definite property of the applied sources of potential. Each element in the network now has a change of potential associated with it that is unique for the assumed current directions and polarity of the sources. The second law states that, starting at any one of the junctions of such a network and following any succession of the conductors and elements that form a closed path, around either way to the starting point, the algebraic sum of the changes in potential encountered in the traversal is equal to zero, where a change in potential from plus to minus (voltage drop) in passing through an element is assigned a positive sign in the sum and a change in potential from minus to plus (voltage rise) is assigned a negative sign. The law is often expressed in the two alternative forms "the sum of the voltage rises equals the sum of the voltage drops" or "the sum of the voltage rises is equal to zero." Voltage rises and voltage drops as used in the last two expressions have the same meaning as indicated above, and a voltage drop may be considered to be a negative voltage rise of the same magnitude. An example of the application of the law is the formulation of the voltage equilibrium equation for the simple series circuit shown in Fig. 2. The sources of electromotive force in the mesh shown are designated $e_1(t)$ and $e_2(t)$. The current in each portion of

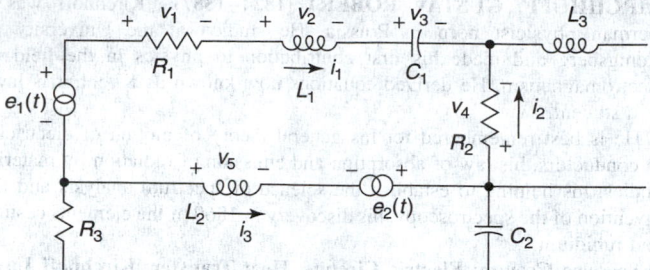

Fig. 2.   Formulation of voltage equilibrium equation for simple series circuit.

the circuit is shown in Fig. 2. In terms of the indicated potential changes the equilibrium equation is

$$-e_1(t) + v_1 + v_2 + v_3 - v_4 + e_2(t) - v_5 = 0$$

If the parameters of the system do not depend on the instantaneous currents and voltages, the foregoing equation can be expressed in terms of the indicated currents and element values as

$$e_1(t) - e_2(t) = R_1 i_1 + L_1 \frac{di}{dt} + \frac{1}{C_1} \int i_1\, dt - R_2 i_2 - L_2 \frac{di_3}{dt}$$

Maxwell has set forth a general method of calculating the currents and the relative potentials of the junctions when the elements and electromotive forces in the several branches of a network are given. For a network of n points, this involves the solution of $n - 1$ simultaneous equations. The work is often simplified, however, by the circumstance that some of the conductors or some of the electromotive forces are absent. See also **Electric Circuits.**

KIROOMBO.   See **Kingfishers and other Coraciiformes.**

KISSING BUG (*Insecta, Hemiptera*).   A name applied late in the nineteenth century to a supposedly deadly bug that was said to bite human beings about the face. It probably applies to some of the large predacious bugs of the family *Reduviidae*, commonly called assassin bugs. One species known as the big bedbug enters houses and sucks human blood, inflicting a painful wound.

See also **Chaga's Disease (South American Trypanosomiasis).**

KISTIAKOWSKY, GEORGE B. (1900–1982).   Born in Kiev, Russia, where he fought in the White Russian Army, he studied in Germany under Maxwell Bodenheim, where he obtained his doctorate in chemistry. In 1926, he came to the U.S. and became an American citizen in 1933. For 41 years, he was Professor of Chemistry at Harvard; in addition to many chemical awards, he was the recipient of the Priestley Medal as well as distinguished Medals of Honor from three Presidents of the U.S. President Eisenhower appointed him as his assistant for science and technology, and he was chairman of the Science Advisory Committee from 1957 to 1963. Among his many achievements in both chemistry and physics, he was a world-famous authority on explosives. A key member of the Manhattan Project, he devised the detonating mechanism for the first experimental atomic bomb in New Mexico, at which time he was head of the Los Alamos Laboratory. Though he ranks high among those who developed the bomb, he perceived the awesome destructive potential of nuclear weapons and became an ardent opponent of their future use. Resigning from the Pentagon in 1967, he returned to teaching at Harvard. Among other distinguished organizations, he was a member of the Royal Society of London, the AAAS, and the ACS.

KITASATO, SHIBASABURO (1852–1931).   Kitasato was a Japanese bacteriologist who demonstrated that tetanus antitoxin could prevent tetanus.

Kitasato studied medicine at the University of Tokyo, graduating in 1883. Initially he worked at the new Public Health Bureau, becoming research assistant to Masanori Ogata.

In 1886 he was sent to Germany by the Japanese Government to study bacteriology under Robert Koch. His initial research was on tetanus, and he was the first to obtain a pure culture of *Clostridium tetani*. Working alongside Emil von Behring (1854–1917) in 1890, he proved that animals injected with tetanus toxin produced an antitoxin which could then be used to prevent the disease. The same phenomenon was shown to be true of diphtheria. This research initiated the era of serum therapy. See also **Koch, Heinrich Hermann Robert(1843–1910).**

He returned to Japan in 1892, where he became Director of the Institute for Infectious Diseases. In 1894 he was sent to Hong Kong to investigate the outbreak of bubonic plague. Kitasato has been credited with the identification of the causative bacteria of the disease, *Pasteurella pestis*. Although he identified a bacillus ten days before his rival, Alexandre Yersin, a fervent supporter of Pasteur, there is still controversy over who isolated the true causative factor of the disease. Today the bacillus is more commonly known as *Yersinia pestis*.

In 1914 Kitasato founded his own Institute and inspired many promising students to continue his own bacteriological researches. In recognition of his services he was created a Baron by the Emperor of Japan in 1924.

CLAIRE E.J. HERRICK, London, UK

KIWI (*Aves, Apterygiformes, Apterygidae*).   The most peculiar of the ratites, and also differs most from the other families. See also **Ratites.** Among the many birds found only in New Zealand, they are the most characteristic of these islands.

There is only one genus (*Apteryx*). The length is 48 to 84 centimeters (19 to 33 inches), the standing height reaches 35 centimeters (14 inches), and the weight is 1.25 to 4 kilograms (3 to 9 pounds). The beak is long, pliable, and sensitive to touch, and the nostrils are lateral at the tip of the beak. The feathers have no aftershaft; rather, the shafts project much like coarse hair from the vanes, which lack barbules and are therefore loose. There is no tail, only a small pygostyle. The legs are strong but short, and the claws are sharp. Kiwis inhabit wooded parts of New Zealand up into the high mountains; today they also inhabit partly cleared bush and deserted farmland.

There are two readily distinguishable species: (a) Common Kiwi (*Apteryx australis*); (b) Owen's Kiwi (*Apteryx owenii*).

Kiwis have little in common with other ratites aside from the lack of a sternal keel and the structure of the bony palate. Superficially they do not appear to be related to any other group of birds. The missing "breast" gives them a strange outline. Their senses of smell and hearing are well developed; the eyes are small. Kiwis are truly nocturnal and spend the day hidden in caves that are generally surrounded by dense vegetation. They search for their food at night in the same habitat. They feed on insects, particularly their larvae, and also on worms and fallen berries. While probing for hidden worms and larvae in the soft forest floor, they use their long beaks in the same manner as snipes.

The northern common kiwi generally lays two eggs in late winter at an interval of several days; often it may be a week or longer until the second egg is in the nest. The kiwis of the south island, on the other hand, apparently lay only one egg. The shiny, white, thin-shelled, elongated egg weighs about 450 grams ($15\frac{1}{2}$ ounces), 14% of the female's weight. The smaller male incubates for 57 to 80 days in a hidden nest made of fallen leaves and leaf mould. When they hatch, the young are entirely covered by a soft juvenile plumage. It appears that they stay in the nest for the first five days and do not eat during this time, but use up the remaining yolk in their body cavity. Then they leave the nest, led by their father, and independently seek food. The father sometimes scratches the ground for them. The young grow slowly; they are said to be sexually mature in 5 or 6 years.

KIWIFRUIT.   The *Chinese gooseberry* or *yang-tao* plant (*Actinidis chinensis*) was long regarded by horticulturists essentially as an ornamental plant. However, since the early 1960s, the effort to commercialize the fruit because of its attractive features as an edible fruit has expanded. Over the years, the fruit, shown in Fig. 1, has been variously called *Ichang gooseberry*, *monkey peach*, *kiwi*, and *kiwi berry*, in addition to the names already mentioned. The name now commonly accepted in the North American market for the fruit is *kiwifruit*. The growing seasons for

the kiwifruit in New Zealand and California are complementary and thus extend the market availability over much of the year. As recently as 1979, the United Fresh Fruit and Vegetable Association, with reference to the kiwifruit, stated: "Those who have not yet met this fascinating fruit can look forward to a new and exciting taste experience. We predict a bright future for this exotic commodity." Advocates of the fruit proclaim it as a dessert and snack fruit, as well as a source of juice, as an ingredient of shortcakes, pies, and cakes, as a beverage additive, and as a cooked fruit for adding to ice creams, cream pies, and milkshakes. The fruit also contains enzymes, similar to papain in papaya, that are useful in tenderizing meats.

Fig. 1.   Kiwifruit or Chinese gooseberry (*Actinidis chinensis*). (*USDA photo.*)

Although kiwifruit is relatively new to the Western Hemisphere, the fruit has been established in China for hundreds of years, originating in central China along the Yangtze River valley. The fruit was introduced into New Zealand in 1906, where it was called Chinese gooseberry. During the first several decades in New Zealand, nurserymen selected large-fruited seedlings, out of which commercial varieties were developed. The New Zealand kiwifruit industry developed after World War II and is, today, primarily centered in the Bay of Plenty area near Tauranga on New Zealand's north island. In 1935, the U.S. Plant Introduction Station at Chico, California received plants of a large-fruited variety. These plants are the ancestors of most of the *Hayward* variety growing in California today.

KJELDAHL TEST.   An analytical method for determination of nitrogen in certain organic compounds. It involves addition of a small amount of anhydrous potassium sulfate to the test compound, followed by heating the mixture with concentrated sulfuric acid, often with a catalyst such as copper sulfate. As a result ammonia is formed. After alkalyzing the mixture with sodium hydroxide, the ammonia is separated by distillation, collected in standard acid, and the nitrogen determined by back-titration.

KLIPSPRINGER.   See Antelope.

KLUG, AARON S. (1926–).   A South African-born chemist who won the Nobel Prize for chemistry in 1982 for his work with the electron microscope and research into the structure of nucleic and protein complexes. His use of crystallographic electron microscopy to analyze the structures of biologically important complex chemicals was noteworthy. He was cited in particular for his establishment of Fourier microscopy.

KLYDONOGRAPH.   A device attached to electric power lines for estimating certain electrical characteristics of lightning by means of the figures produced on photographic film by the lightning-produced surge carried over the lines. The size of the figure is a function of the potential and polarity of the lightning discharge.

KLYSTRON.   A power amplifier tube used to amplify weak microwave energy (provided by a radio- frequency exciter) to a high power level for a radar transmitter. A klystron is characterized by high power, large size, high stability, high gain, and high operating voltages. Electrons are formed into a beam that is velocity modulated by the input waveform to produce microwave energy. A klystron is sometimes referred to as a linear beam tube because the direction of the electric field that accelerates the electron beam coincides with the axis of the magnetic field, in contrast to a crossed-field tube such as a magnetron. Klystrons provide a coherent transmitted signal appropriate for Doppler radar and pulse-compression applications. They are used in many operational radars, for example, NEXRAD (Next Generation Weather Radar) and TDWR (Terminal Doppler Weather Radar).

See also **Microwave Tubes**; and **Magnetron**.

AMS

KNIFEFISH.   See Gymnotid Eels.

KNOPP HARDNESS TEST.   See Hardness.

KNUDSEN DIFFUSION.   See Diffusion.

KNUDSEN FLOW.   This is free molecular diffusion — the flow of a gas through a long tube at pressures such that the mean free path is much greater than the tube radius.

KNUDSEN GAGE.   A gage which measures pressure in terms of the net rate of transfer of momentum by molecules between two surface maintained at different temperatures and separated by a distance smaller than the mean free path of the gas molecules. Also called *radiometer vacuum gage*. Various types of Knudsen gage are distinguished by the names of the inventors and differ mainly in the shape and method of suspension of the movable element.

KOALA.   See Marsupialia.

KOCH, HEINRICH HERMANN ROBERT (1843–1910).   Koch was a German medical bacteriologist, who supplied the principal tools, methods and theories of bacteriological hygiene and established the bacterial aetiology of various infectious diseases.

Heinrich Hermann Robert Koch was born in the small north German mining town of Clausthal on 11 December 1843. The family of Hermann and Mathilde Koch was lower middle class, with Koch's father working as a senior official in a local mine. While at school in Clausthal, Koch — apart from observing and collecting plants and animals in a general way — displayed little interest in the micro-cosmos whose researcher he would become later on. Of rather more relevance is that Koch who was to spend good parts of his work on the micro-cosmos in the macro-cosmos, i.e. traveling around the world, showed a considerable passion for traveling, had fantasies of emigration (to America) and related issues in his early years. Such plans and desires were not uncommon in the Koch family: six out of eight brothers emigrated from Germany and Koch's own such plans were a continual source of worry to his parents — only to be settled by Koch's first marriage in1867 to Emmy Fraatz, who seems to have had a similar dislike for such endeavors.

From 1862 Koch was a student in nearby Göttingen, starting in botany, physics and mathematics and then turning to medicine in 1863. Koch was a talented student, finishing first in a prize essay competition on the issue of the existence and distribution of uterine nerves in the ganglia. His doctoral thesis was on urinary excretion. Research included a heroic self-experiment of eating several kilograms of butter to test the presence of succinic acid in urine. Results were published in Jacob Henle's *Zeitschrift für rationelle Medizin* in 1865. The anatomist Henle, who was amongst Koch's teachers, had been a pioneer of medical bacteriology somewhat earlier, by giving a theoretical treatment of the possible existence and proof of pathogenic germs in his *Von den Miasmen und Kontagien...* in 1840. However, Koch himself later in his life denied having received any particular training for his career in bacteriology at university. It seems that what he took with him upon graduating from Göttingen in 1866 was a sound training in microscopical anatomy and physiology which he was given, apart from Henle, by the physiologist Georg Meissner and the pathologist Karl Hasse. See also **Henle, Friedrich Gustav Jakob (1809–1885)**.

Due in large part to the recent work of French researchers such as Louis Pasteur, who had just about refuted the existence of spontaneous generation, microbiology was becoming a fashionable issue in the 1860s and it is likely that a cholera epidemic which hit Germany in 1866 gave Koch his own first experience in working on pathogenic germs. While working as an assistant physician at a hospital in Hamburg, Koch seized the opportunity to investigate the microscopical pathology of cholera and observed the structures that came to be known as the comma bacillus almost two decades later. See also **Microbiology**.

However, there are no indications that Koch was heading for a scientific career at that time. Upon leaving university, he instead got married and decided to go into private practice to make himself a living. That proved difficult to achieve and included several failed attempts, renewed plans for emigration, and practice as a military doctor in the Franco-Prussian war of 1870–1871, where he gained experience in wound infections and typhoid. In 1872 Koch finally became district physician (Kreisphysikus) in the town of Wollstein (today Wolsztyn, Poland), where he continued to be a successful and increasingly wealthy practitioner for the next seven years.

The Development of Medical Bacteriology

In Wollstein, Koch finally started his work in medical bacteriology and the case he chose for a starting point was a veterinary disease, anthrax, which frequently occurred among local sheep populations. Starting from Casimir Joseph Davaine's earlier research, which had indicated that the disease was caused by small rod-like structures in the blood, Koch succeeded in describing the full life cycle of what became known as *Bacillus anthracis* and in experimentally reproducing the disease. A crucial point in Koch's proof of the bacillus's stability as a species and its pathogenic effects was the demonstration of a spore stage: it provided evidence of the constant presence of the microorganism in infected animals, offered a new explanation concerning phenomena that had seemed to indicate pleomorphism or spontaneous generation and explained the epidemiology of the disease, most notably the duration of the bacteria under unfavorable conditions. Koch's achievement lay less in discovering a microorganism than in fully describing its life cycle and in establishing a causal link between this bacterial species and a particular disease. On the conceptional level this meant that he applied knowledge about bacteria that had been developed by botanists such as Ferdinand Julius Cohn to the medical question of disease causation and disease definition. Finally his work included proposals for hygienic measures to prevent further propagation of anthrax epidemics, e.g. by properly disposing of cadavers. See also **Davaine, Casimir-Joseph (1812–1882)**.

For publication of his results Koch in 1876 turned to F. J. Cohn. Cohn, on whose bacteriological work Koch's approach relied, was a professor of botany at the University of Breslau. Cohn not only enthusiastically accepted Koch's results, but brought him into contact with other local researchers in the medical faculty who crucially influenced Koch's further career. Carl Weigert introduced Koch to bacterial staining and to his nephew, Paul Ehrlich, who would later be one of Koch's most important collaborators and a master of staining technique himself. Equally important was the pathologist Julius Cohnheim, because he was an expert in experimental pathology and animal experimentation, methods that were to be crucial for Koch's work. See also **Cohn, Ferdinand Julius (1828–1898)**; and **Ehrlich, Paul (1854–1915)**.

Still living in Wollstein, Koch spent the next three years expanding his experimental technology and methodology. In 1877 he published his "*Verfahren zur Untersuchung, zum Konservieren und Photographieren der Bakterien*" in which, in addition to other inventions such as cultivation techniques, he introduced microphotography. Not only was this seen as a means of providing an objective image, it also facilitated rapid dissemination of the newly acquired knowledge about bacteria. Later in 1877, Koch published on the aetiology of wound infections and for the

first time made systematic statements about criteria to be fulfilled in experimentally establishing disease causation. These came to be called *Koch's postulates* by his colleague Friedrich Löffler somewhat later. In their classical form these postulates include first the proof of a specific microorganism in infected tissues, second its isolation and cultivation into pure cultures and third the ability to bring about the same disease by using these cultures in animal experiments. The term *postulates* is, however, somewhat misleading, since the most constant trait of the various statements Koch gave on the issue was not the chain of identification, cultivation, inoculation, but the flexibility with which he modified his methods in any given case. Even though Koch mostly emphasized the crucial and indispensable character of animal experiments, he showed no hesitation in denying their necessity in other cases. See also **Bacteriology**.

Koch's undeniable talent for polemics became obvious in 1877 when he delivered a ferocious attack on Carl von Naegeli's theory of infectious diseases. Since Naegeli was the most prominent advocate of bacterial pleomorphism, which is of the concept that among bacteria no distinct species could be found, the object of his polemic was well chosen. Koch's medical bacteriology was entirely reliant on the specificity of bacterial species to which he linked constant pathological effects as one more property of such organisms.

Eventually Koch tried to leave his practice in Wollstein and to embark on a career in scientific research. Attempts by the Breslau faculty of medicine in 1879 to create an extraordinary professorship in hygiene for him failed. An attempt to move his medical practice to Breslau, where he was to be city physician (Stadtarzt), proved to be economically disastrous. One year later, in 1880, aided by Julius Cohnheim, Koch became an employee in the recently founded Imperial Health Office in Berlin, where he set up a small hygiene laboratory. His first collaborators where Ferdinand Loeffer and Georg Gaffky. Later the team started to expand. The following five years, which Koch spent at the Imperial Health Office, were the most productive phase of his career and laid the foundations for his fame as the "Bacillenvater" (father of bacilli). First the team concentrated on methodology, which resulted in Koch's treatise "Zur Untersuchung von pathogenen Mikroorganismen," published in the first volume of the *Mitteilungen aus dem Kaiserlichen Gesundheitsamte*. The team also did groundwork on disinfective measures, particularly by the application of hot air, which they showed to be superior to Joseph Lister's carbolic acid. A demonstration of Koch's new techniques at the International Medical Congress in London 1881 brought him favorable recognition and the personal acquaintance of Lister and Pasteur. See also **Lister, Joseph (1827–1912)**.

In 1882 Koch demonstrated the powers of his methods on the single most deadly disease of the age, tuberculosis. He managed to establish a bacterial aetiology, thus verifying older claims for tuberculosis being an infectious disease that had most notably been put forward by Jean Antoine Villemin in 1865 and leaving in ruins the complex theories that other German physicians, most notably Rudolf Virchow, had developed on tuberculous processes. In the context of his work on tuberculosis, Koch came to develop a distinctive concept of infectious diseases as bacterial invasions that were defined by the presence, distribution and behavior of bacteria in an infected organism. This included a profound disregard for clinical appearances as a means of identification of diseases and an almost exclusive reliance on aetiological arguments with regard to the definition of diseases. In the particular case of tuberculosis this resulted in a redefinition of the disease's essence and boundaries by identifying phthisis, caseous pneumonia, lupus etc., as a single disease. See also **Tuberculosis**; **Villemin, Jean-Antoine (1827–1892)**; and **Virchow, Rudolf Carl (1821–1902)**.

Interest in bacteria was growing amongst professionals and the wider public and it culminated in 1883–1884 when Koch went on an expedition to Egypt and India to identify *Vibrio cholerae*, the *comma bacillus* as the bacterial agent causing cholera. Koch's cholera expedition was first of all a public relations success. The imperial government on whose commission he was traveling actively secured publicity, which was all too easy, since the German team found itself in competition with a group of French researchers headed by Louis Pasteur. Even though the expedition was given a triumphal reception upon returning — with Koch and his fellow travelers being given financial donations and medals, the scientific results were somewhat inconclusive. Even though pure cultures of the *comma bacillus* could be attained and knowledge about the pathological anatomy of the disease was expanded, no satisfactory results were reached in animal experimentation and Koch had thus failed to meet his own standards. His bacterial aetiology for cholera was instead based on epidemiological observations which, however, might just as well have supported Max von Pettenkofer's miasmic theory of the disease. Nevertheless Koch's bacterial aetiology proved to be robust enough to be accepted by most of members of the cholera commission of the German empire, where Koch's and Pettenkofer's opposing theories clashed from late 1884 on.

National rivalry and chauvinism was an important feature in the relations between French microbiologists and German bacteriologists. This became obvious in the Koch–Pasteur controversy that started at the International Congress for Hygiene in Geneva in 1882. Even though the factual focus of the controversy was on questions of proper bacteriological technique, vaccines and priority-issues related to anthrax, it quickly became a highly politicized event in which the German public learnt to consider bacteriological hygiene as a matter of national pride. Apart from this, the controversy served to further develop Koch's theoretical concepts and methodology: the emphasis he put on bacterial specificity, his skepticism regarding virulence and Pasteur's vaccines and finally the conviction that "proper" bacteriological work required the methods and tools of his school were sharpened in the conflict. Added to this there was an undeniable element of personal jealousy in Koch's ferocious critique of Pasteur's vaccines, which had their spectacular appearance at the latter's famous demonstrations of anthrax vaccines at Pouilly-le-Fort. The results of Pasteur's trials were published almost simultaneously with the outbreak of the controversy.

During the early 1880s Koch's institutional position had been constantly improving and following the cholera expedition he became deputy director to the Imperial Health Office. Likewise his group of collaborators and pupils was growing fast, as was the reputation of his methods and discoveries both in expert circles and among the public. Bacteria became widely regarded as ever so plausible embodiments of diseases and Koch's public image acquired heroic dimensions.

Koch himself felt a need to improve his own institutional position in the period. However, plans for his own independent, government-funded imperial institute for bacteriological research failed and instead in 1885 Koch found himself with a university chair without ever having had the formal qualification of a Habilitation and became director of a newly created institute for hygiene at Berlin University. With regard to the institutional and disciplinary growth of bacteriological hygiene this was a fortunate move. The institute set up training courses in hygiene for physicians from all over the world and for civil servants and thus made Koch's medical bacteriology a standard method for many hygienists. The discipline itself underwent changes in this process. In France at the Pasteur Institute similar programs were being set up, so that bacteriology, which had previously been almost a secret science, practiced in only a few places, now became more and more well known. The rapid increase in the number of those who were familiar with its methods meant that the discipline was getting more pluralistic and discursive from the late 1880s on.

Aside from this institutional growth, however, Koch's own career ran into a crisis. The chair for hygiene had been created against stiff resistance by the medical faculty, and his position remained precarious. Extended everyday duties such as teaching and conducting examinations annoyed him. His deteriorating health imposed interruptions on his work. Private problems arose, which in 1890, led to Koch's separation from his first wife.

Koch was also facing a conceptual problem. His successes had so far relied on the introduction of methods and the establishment of aetiologies and it was obvious that such a strategy was unlikely to continue to produce spectacular results in the long term. Since everybody, including Koch himself, had taken the new knowledge about bacteria to be almost a promise of their control the next obvious step was therefore into therapy. However, the tools to fight diseases that German medical bacteriology had so far deployed remained restricted to preventive medicine and unspecific hygienic measures such as disinfection or filtration of drinking water. Therapies for infected patients were nowhere in sight. Koch thus seems to have felt an enormous pressure to continue in this direction, a pressure that was certainly increased by Pasteur's spectacular rabies vaccine presented in 1885, which—even if it was not therapeutic—was a highly specific tool based on microbiological knowledge. See also **Rabies**; and **Virology (The History)**.

Epidemiology and Tropical Diseases

The Hamburg cholera epidemic of summer/autumn 1892 offered an opportunity for the institute that was still under criticism, to demonstrate the skills of its director and staff. Furthermore, it was while working on cholera in Hamburg that Koch's research took a turn that would be a central strand of his future work. Koch did not, like many of his younger collaborators such as Paul Ehrlich, Emil von Behring or August Wassermann, enter the rising fields of immunology or serology. Instead he supplemented his bacteriological work with an epidemiological interest, which had far-reaching implications for his further work. See also **Immunology (The History)**; and **Von Behring, Emil Adolf (1854–1917)**.

Certain epidemiological observations made while working on the Hamburg epidemic caused him to modify his original model of infectious diseases as bacterial invasion. Phenomena such as subclinical infections, atypical infections and—most of all—healthy carriers forced him to rearrange his ideas about infectious diseases. The result was a growing awareness of the importance of epidemiological concepts in such diseases and a stress on the behavior of diseases in populations rather than the infection of individuals, which had been his previous focus. Koch had always been a dedicated traveler and much of his work to come was into epidemic parasitic diseases which Koch researched on extensive travels to tropical countries.

There were, however, continuities. In the case of tuberculin, Koch continued to keep faith in its curative effect and never realized the erosion of his work that followed from the disaster. In fact, his 1902 proclamation of the non-identity of human and bovine tuberculosis still showed strains of the older conceptions and was soon vigorously criticized. Koch's disregard of clinical evidence gave his critics a particularly good angle from which to attack his assumptions.

Koch's work on tropical diseases was diverse and apart from the fact that it was almost exclusively concerned with parasitic diseases and epidemiological questions it is not easy to see the common motive in it. One important strain was a connection to eradication campaigns which Koch conducted or developed proposals for. In 1896 Koch traveled to South Africa on commission of the British Colonial Office to study and control rinderpest and in 1902—again on British commission—to Rhodesia to research East-coast fever, which was at that time a threat to local cattle farming. This and other research into parasitic diseases also reflected Koch's increasing interest in epidemiology and the new measures for control of epidemics in populations that were made possible by the definition of the carrier state. In 1901 Koch was offered the chance of a large-scale demonstration of such methods on a human population at home. A few, isolated cases of typhoid in the Trier area provided the opportunity for a large-scale eradication campaign, carried out in close collaboration with military officials, which included testing of the local population for healthy carriers of the disease. The area in question was, of course, to become one of the most important deployment areas of the Schlieffen plan, Germany's strategic plan for World War I. Eradication campaigns were thus needed to create sufficiently hygienic spaces for the deployment of big, modern armies. See also **Parasitology (The History)**.

Another important focus of Koch's work was on malaria, which he researched on various expeditions to Italy, Africa and New Guinea. In this case his most important contributions were that he added new information about the life cycle of the parasite to Ronald Ross's aetiological argument and did epidemiological work on acquired immunity amongst local populations in New Guinea. In this case—as in some others—voices were heard that accused Koch of a self-interested style of scientific work, not always giving due credit to fellow workers and colleagues.

Later Years

Koch's directorship of his Berlin institute had never been a particularly active one. After the mid-1890s, when Koch spent much of his time on his extensive travels, most of the day-to-day work was left to others, most notably to Richard Pfeiffer. He also suffered from poor health, so it was a logical move to retire at the age of 60 in 1904. Koch did not have to leave the institute upon retirement, but continued to have his own laboratory there. His successor as director was Georg Gaffky, who had been his assistant in the early days of the Imperial Health Office.

Following his retirement, Koch went on his longest expedition ever. He had done work on sleeping sickness earlier and in 1906/1907 traveled to

East Africa for the purpose of therapeutic research on the disease. His trust in atoxyl as a means of therapy turned out to be a fatal choice. Instead of curing the disease, it led only to temporary disappearance of the parasites of sleeping sickness, the trypanosomes, and at the same time about 25% of the patients treated lost their eyesight. See also **African Trypanosomiasis**.

The twentieth century saw many decorations and also travels of a more private sort, such as to the United States and Japan in 1908, where many honors were bestowed on him. He received the Pour le mérite (1906), Germany's most prestigious medal, was elected external member of the French Academy of Science in 1902, which was in fact two years before he became a member of the Prussian Academy of Science. The rather late call to the Prussian Academy was due to Virchow's delaying resistance and it is worth noting that Koch was elected to the French Academy as the latter's successor. In 1905 Koch received the Nobel Prize, notably four years later than his ex-pupil Emil von Behring, with whom he had been on bad terms for scientific and personal reasons since the late 1890s. Shortly after, a foundation for fighting tuberculosis and a medal for outstanding medical research were given his name.

In particular with regard to his early work it seems fair to label Koch a founder of medical bacteriology. He contributed essential methods of investigation, technologies and key concepts to the field. Equally the application of such knowledge was a central issue for him. His career was closely linked to the growth of public health institutions and large-scale research in late nineteenth century Germany. That Koch failed to live up to the successes of his early days after 1890 is obvious. Nevertheless, his work indicates an epochal shift in medical science. All in all it was with Koch's medical bacteriology that laboratory science finally entered everyday medical practice. See also **Bacteriology (The History)**.

Additional Reading

Brock, T.D.: *Robert Koch: A Life in Medicine and Bacteriology*, Science Tech Publishers, Madison, WI, 1988.
Gradmann, C.: Robert Koch and the Pressures of Scientific Research: Tuberculosis and Tuberculin, *Medical History*, **45**, 1–32. (2001).
Haddad, G.E.: Medicine and the Culture of Commemoration, Representing Robert Koch's Discovery of the Tubercle Bacillus, *Osiris*, **14**, 118–137 (2000).

CHRISTOPH GRADMANN, University of Heidelberg, Heidelberg, Germany

KODIAK BEAR. See **Bears**.

KOEL. See **Turacos**; and **Cuckoos and Coucals**.

KÖHLER, GEORGES JEAN FRANZ (1946–1995). Köhler was born in Munich and educated at the University of Freiburg, Germany, where he received a diploma in biology in 1971 and a PhD in 1974. His research for the doctorate was conducted at the Institute for Immunology in nearby Basel. Visiting the Basel Institute in 1973 to present a seminar, Milstein invited Köhler to join him at Cambridge as a postdoctoral fellow in 1974. Köhler accepted and later returned to Basel in 1976 as a Member of the Institute.

The pivotal advance made by Köhler and Milstein was neither planned nor was it an accident. Michael Potter, in 1962, had devised technologyto produce myelomas in mice and described their maintenance. Milstein and Cotton had already used rat and mouse myelomas to investigate whether or not different chromosomes encode variable and constant regions of immunoglobulins. Their results revealed that both variable and constant regions were always of one species even though heavy and light chain hybrids did occur.

Köhler's dissertation research had shown that one thousand different murine immunoglobulins were able to react with a single epitope. When Köhler began his postdoctoral fellowship in Milstein's laboratory at Cambridge University in 1974, he embarked upon a project with Milstein on antibody gene mutations. Encountering difficulties with the research, Köhler sought ways to synthesize myeloma antibodies of known specificity. He conceived the idea of producing hybridomas at the end of 1974 while lying in bed halfway between wake fulness and sleep. On learning his concept the next morning, Milstein encouraged him to proceed. Initial failures were traced to the use of toxic reagents but eventually the experiments succeeded. For this work Milstein and Köhler were awarded the Nobel Prize in Medicine or Physiology for 1984, together with the Danish immunologist, N. K. Jerne. See also **Jerne, Niels Kaj (1911–1994)**; and **Milstein, Cesar (1927–2002)**; and **Monoclonal Antibodies**.

Monoclonal antibodies have broad applications in both basic science and clinical research. They have been developed against a multitude of antigens, including tissue markers that are of critical importance in surgical pathological diagnosis of tumors, to cite but one example. The thousands of articles and dozens of monographs on monoclonal antibodies published in the last quarter century reflect their widespread use and significance.

Additional Reading

Köhler, G.F., and C. Milstein: Continuous Cultures of Fused Cells Secreting Antibody of Predefined Specificity, *Nature*, **256**, 495–497 (1975).
Godling, J.W.: *Monoclonal Antibodies: Principles and Practice. Production and Application of Monoclonal Antibodies in Cell Biology, Biochemistry and Immunology*, 2nd Edition, Academic Press, London, UK. 1986.
Hurrell, J.G.R.: *Monoclonal Hybridoma Antibodies: Techniques and Applications*, CRC Press, LLC, Boca Raton, FL, 1982.
McMichael, A.J., and J.W. Fabre: *Monoclonal Antibodies in Clinical Medicine*. Academic Press, London, UK, 1982.
Sherby, L.S., and W. Odelberg: *The Who's Who of Nobel Prize Winners 1901–2000*, 4th Edition, Greenwood Publishing Group, Inc., Westport, CT, 2001.
Wade, N.: Hybridomas: The Making of a Revolution, *Science*, **215**, 1073–1075. (1982).

J.M. CRUSE
R.E. LEWIS
University of Mississippi Medical Center, Jackson, MS

KOHLRABI. See **Brassica**.

KOHLRAUSCH LAW. The conductivity of a neutral salt in dilute solution is the sum of two values, one of which depends upon the cation, the other upon the anion. In other words, each ion contributes a definite amount to the total conductance of the electrolyte, independent of the nature of the other ion.

See also **Electrochemistry**.

KOHOUTEK (Comet). Named after its discoverer, the Czech astronomer, Kohoutek, this comet made its appearance to earthly observers toward year-end 1973. Predictions in mid-1973 indicated that it would be one of the spectacular comets, possibly rivaling Halley's comet last seen in 1910. It had been predicted that Kohoutek would shine almost as brightly as Mars and would produce a band of light stretching as far as 30° across the evening sky on December 30, 1973. Kohoutek did not live up to these expectations insofar as the lay observer was concerned and its weaker-than-expected performance was compounded by rather poor weather conditions. Nevertheless Kohoutek did furnish some scientific information that will be helpful toward solving some of the unanswered questions pertaining to comets. See also **Comet**.

On January 5, 1974, an Aerobee 200 rocket was launched from the White Sands Missile Range, New Mexico to study ultraviolet emissions from Kohoutek. At the time of the launch, the comet-sun-earth angle was $20.3°$; the sun was 5° below the rocket horizon at apogee (232 kilometers). Scattered sunlight near the horizon made it difficult for the star tracker of the attitude control system to lock on the comet, whose magnitude was estimated visually at $m_v \approx +2.5$, approximately three magnitudes weaker than predicted. However, several seconds of data on the coma were obtained, and observations of the tail were obtained during rocket descent after the star tracker locked on when the solar depression angle exceeded 7°. In addition to the Lyman α line of atomic hydrogen (HI) at 1,216 Å and the (0, 0) and (1, 1) bands of OH at 3,090 and 3,142 Å, which were previously observed in Comet Bennett (1970 II) by the Orbiting Astronomical Observatory OAO-2, both atomic oxygen (OI) (1,304 Å) and atomic carbon (CI) (1,657 and 1,561 Å) were detected; the derived luminosities were commensurate with those observed by Opal three days later. No other spectral features were clearly identified, from either the coma or the tail. Details on these observations are reported by Feldman, Takacs, and Fastie (Johns Hopkins

University) and Donn (NASA) in *Science*, **185** (4152), 705–707 (August 23, 1974).

On January 8, 1974, ultraviolet cameras and photometers were flown on another Aerobee rocket at White Sands by scientists of the U.S. Naval Research Laboratory to observe the comet. At that time, the comet was 0.43 A.U. (astronomical units) from the sun and twice that far from the earth. The sun-comet-earth angle was nearly 90°. Emissions of atomic oxygen (1,304 Å), atomic carbon (1,657 Å), and atomic hydrogen (1,216 Å) from the comet were observed. Analysis of the Lyman alpha halo at 1,216 Å gave an atomic hydrogen production rate of 4.5×10^{29} atoms per second. For comparison, a production rate derived from the prior data on Comet Bennett was about 6×10^{28} atom second^{-1} ster^{-1} when the comet was 0.8 A.U. from the sun. If one assumes an inverse square law dependence of production rate on sun-comet distance, then the production rate for Bennett at the sun-comet distance of these observations was some six times greater than the production rate for Kohoutek. Details on the ultraviolet spectrophotometry observations of Kohoutek are reported by Opal, Carruthers, Prinz, and Meier (U.S. Naval Research Laboratory) in *Science*, **185** (4152), 702–705 (August 23, 1974).

KOLMOGOROFF-SMIRNOFF TEST.

In statistics a test whether a sample can have come from a specified population, or alternatively to set confidence limits to the population from the samples. The test depends on the maximum absolute value of the difference between the observed and theoretical distribution functions.

KOLMOGOROV CONSTANT.

The proportionality constant α in Kolmogorov theory, which states that the spectral energy S in the inertial subrange is $S = \alpha \varepsilon^{2/3} k^{-5/3}$, for ε representing the viscous dissipation rate of turbulence kinetic energy, and k the wavenumber (inversely proportional to the wavelength or eddy size).

Measurements of the 1D longitudinal spectrum of the wind in the planetary boundary layer show that this constant is equal to about 0.5.

AMS

KOLMOGOROV MICROSCALE.

Of the three standard turbulence length scales, the one that characterizes the smallest dissipation-scale eddies. See also **Turbulence Length Scales**. As the turbulence kinetic energy cascades from the largest scales down to the smallest scales, the dynamics of the small eddies become independent of the large-scale eddies. At the smallest scales, the rate at which energy is supplied must equal the rate at which it is dissipated by viscosity. Thus, parameters available to form length and velocity scales are the dissipation rate, ε, and the kinematic viscosity, v. The Kolmogorov length, η, and velocity, u, scales are:

$$\eta = \left(\frac{v^3}{\varepsilon}\right)^{1/4}$$

$$v = (v\varepsilon)^{1/4}.$$

Note that the Reynolds number formed from the Kolmogorov microscale is equal to one. Based on the observation that the large eddies lose their energy in about one large eddy turnover time, the dissipation rate may be scaled as $\varepsilon = u^3/L$ where L is the appropriate length scale. Thus, the ratio of the largest to the smallest length scales is

$$\left(\frac{uL}{v}\right)^{3/4} = R^{3/4},$$

and the number of grid points necessary to resolve a turbulent flow in a numerical model is therefore proportional to $R^{9/4}$.

See also **Integral Length Scales**; **Isotropic Turbulence**; and **Taylor Microscale**.

Additional Reading

Bohr, T., G. Paladin, and A. Vulpiani: *Dynamical Systems Approach to Turbulence*, Cambridge University Press, New York, NY, 2005.

Tennekes, H., and J.L. Lumley: *A First Course in Turbulence*, MIT Press, Cambridge, MA, 1972.

AMS

KONA.   See **Winds and Air Movement**.

KOPPEN CLASSIFICATION.   See **Meteorology**.

KÖPPEN, WLADIMIR PETER (1846–1940).

Wladimir Köppen was a German meteorologist, climatologist and botanist. He elaborated the Köppen climate classification system, which is still commonly used today to group climates into similar types (albeit with modifications). See also **Climate**.

Although Köppen's parents were Germans in the service of the Russian emperor (Wladimir's father Peter Köppen, was Russia's first ethnographer, and grandfather royal physician), he himself was born in Russia and attended a school in Crimea. While being at the school, it was the first time that Köppen was attracted by the environment and especially by the relationship between plants and the climate they grow in. Later, he studied at the universities of Heidelberg and Leipzig in Germany where he graduated in 1870. His student dissertation dealt with the effects of temperature on plant growth.

Between 1872 and 1873 Köppen was employed in the Russian meteorological service. In 1875, he moved back to Germany and became the chief of the new the Division of marine meteorology at the German naval observatory (Deutsche Seewarte) based in Hamburg. There he was responsible for establishing a weather forecasting service for the northwestern part of Germany and the adjacent sea areas.

Köppen began a systematic study of the climate and also experimented with balloons to obtain data from upper air. In 1884, he published the first version of his map of climatic zones in which the seasonal temperature ranges were plotted. This work led to the development of the Köppen climate classification system around 1900, which he kept improving for the rest of his life. The full version of his system appeared first in 1918 and, after several modifications, the final version was published in 1936.

Apart from the description of various climate types, he was acquainted with paleoclimatology as well. In 1924 he and his son-in-law Alfred Wegener published a paper called *Die Klimate der Geologischen Vorzeit* (The climates of the geological past) giving a crucial support to the Milankovic theory on ice ages. See also **Wegener, Alfred Lothar (1880–1930)**.

Köppen Climate Classification System

The Köppen Climate Classification System is the most widely used for classifying the world's climates. Köppen divided the Earth's surface into climatic regions that generally coincided with world patterns of vegetation and soils. The Köppen system recognizes five major climate types based on the annual and monthly averages of temperature and precipitation. Each type is designated by a capital letter.

A — Moist Tropical Climates are known for their high temperatures year round and for their large amount of year round rain.

B — Dry Climates are characterized by little rain and a huge daily temperature range. Two subgroups, **S** — semiarid or steppe, and **W** — arid or desert, are used with the **B** climates.

C — In Humid Middle Latitude Climates land/water differences play a large part. These climates have warm, dry summers and cool, wet winters.

D — Continental Climates can be found in the interior regions of large land masses. Total precipitation is not very high and seasonal temperatures vary widely.

E — Cold Climates describe this climate type perfectly. These climates are part of areas where permanent ice and tundra are always present. Only about four months of the year have above freezing temperatures.

Further subgroups are designated by a second, lower case letter which distinguish specific seasonal characteristics of temperature and precipitation.

f — Moist with adequate precipitation in all months and no dry season. This letter usually accompanies the **A**, **C**, and **D** climates.

m — Rainforest climate in spite of short, dry season in monsoon type cycle. This letter only applies to **A** climates.

s — There is a dry season in the summer of the respective hemisphere (high-sun season).

w — There is a dry season in the winter of the respective hemisphere (low-sun season).

To further denote variations in climate, a third letter was added to the code.

a — Hot summers where the warmest month is over 22 °C (72 °F). These can be found in **C** and **D** climates.

b — Warm summer with the warmest month below 22 °C (72 °F). These can also be found in **C** and **D** climates.

c — Cool, short summers with less than four months over 10 °C (50 °F) in the **C** and **D** climates.

d — Very cold winters with the coldest month below −38 °C(−36 °F) in the **D** climate only.

h — Dry-hot with a mean annual temperature over 18 °C (64 °F) in **B** climates only.

k — Dry-cold with a mean annual temperature under 18 °C (64 °F) in **B** climates only.

See also **Climatic Classification**.

Additional Reading

Allaby, M.: *Encyclopedia of Weather and Climate*, Facts On File, Inc., New York, NY, 2002.

Köppen, W.P., and R. Geiger: *Handbuch der Klimatologie*, Berlin: Gebruder Borntraeger, **6** vols, 1930–1939.

McKnight, T.L., and D. Hess: *Physical Geography: A Landscape Appreciation*, Prentice-Hall, Inc., Upper Saddle River, NJ, 2004.

Thornthwaite, C.W.: The Climates of North America According to a New Classification. *Geogr. Rev.*, **21**, 633–655 (1931).

Web Reference

Köppen's Climate Classification Map: http://www.blueplanetbiomes.org/images/climate_map.gif

World Map of the Köppen-Geiger Climate Classification: http://koeppen-geiger.vu-wien.ac.at/

KOPP'S LAW. The molecular heat of a solid compound is an additive function of the atomic heat capacities of its individual atoms. The molecular volume of a liquid is equal to the sum of the atomic volumes of its constituent atoms.

KOREAN HEMORRHAGIC FEVER. The causative agent of this disease has not yet been isolated, but it is believed to be transmitted by some local arthropod or from infected rodents to humans by means of virus-contaminated rodent excreta. Although sporadic cases may occur at any time throughout the year, the majority of cases are seen from May to June and October to November. The clinical course has been divided into five phases. In the febrile phase, the onset is abrupt, often with a chill accompanied by fever, headache, backache, and abdominal pain. Anorexia and thirst are general, and sometimes nausea and vomiting are observed. Temperatures range from 38 to 41 °C (100.4–105.8 °F), and peak on the third or fourth day, falling by lysis. A typical early observation is diffuse reddening of the skin, most marked over the face and neck. Petachiae appear by the third to fifth day at pressure areas, such as the axillary folds, chest, belt line, hips, and thighs. Mild lymphadenopathy is found and abdominal and costovertebral tenderness are constant. About the fifth day of the illness, during the last day or two of the febrile phase, the hypotensive phase begins. Headache often diminishes, but thirst continues. Most patients have warm, dry skin, but as the systolic blood pressure decreases and pulse pressure narrows, the skin becomes cool and moist and tachycardia replaces the earlier bradycardia. The hemorrhagic phase starts on or about the eighth day of illness.

Blood pressure returns to normal and oliguria becomes prominent. Backache and thirst become more severe, hematemesis, melena, and hematuria appear, and the enlarged lymph nodes become tender. Signs of renal failure appear and, if this progresses, the patient goes into an acute uremia and dies from renal failure, the kidneys at autopsy being necrosed and degenerative. Where, however, the oliguric phase passes into the diuretic one, improvement is the rule, but evidence of damage to the renal epithelium is shown in the convalescent phase by passage of large quantities of dilute urine. This period lasts some 3 to 6 weeks, and weight is regained slowly, as is the general stamina of the patient. Mortality ranges from 10 to 15% and even in survivors, residual renal tubular dysfunction appears up to ten years after the disease.

The clinical hallmark of severe Korean hemorrhagic fever is extreme vascular instability with shock and bleeding. There is no therapy at present for the disease. Attention must be paid to maintenance of blood pressure, fluid, and electrolyte balance. Isolation of patients other than the use of gloves and gowns for hospital personnel is not necessary because no instances of person-to-person spread have been observed.

The disease is endemic in Korea, where several hundred persons must be hospitalized each year. The disease also occurs in eastern regions of Russian, the Balkans, and Japan.

KORIGUM. See **Antelope**.

KORNBERG, ARTHUR (1918–1997). Arthur Kornberg was an American biochemist who devised the artificial enzymatic synthesis of DNA.

Arthur Kornberg was born in Brooklyn, New York, on 3 March 1918. He graduated in chemistry and biology from the City College, New York in 1937 and gained his MD at Rochester University in 1941. From 1942 until 1953 he worked on enzymes and intermediary metabolism at the National Institutes of Health, Bethesda, Maryland. Here he contributed to the discovery of chemical reactions in the cell leading to the formation of flavin–adenine dinucleotide (FAD) and diphosphopyridine nucleotide (DPN), coenzymes important as hydrogen-carrying intermediates in biological oxidations and reductions. During these years he also served as an officer in the US coastguard.

In 1953 he was appointed Professor and Director of the Microbiology Department at Washington University, St Louis, Missouri. He continued to study the production of nucleotides in living organisms; compounds consisting of an organic base containing nitrogen linked to a pentose such as ribose or deoxyribose and a phosphate group. Kornberg believed that the solution of the problem of nucleic acid synthesis lay in understanding the biogenesis of such nucleotides and the coenzymes. It was known that DNA contains equal numbers of purine and pyrimidine bases, a discovery used by J. D. Watson and F. H. C. Crick in proposing the helical structure of DNA in 1953. Fundamental to this structure was the pairing of the organic bases, adenine with thymine and cystosine with guanine. In 1956 Kornberg set out to discovera way of synthesizing DNA using ^{14}C labeled nucleoside triphosphates with adenosine triphosphate (ATP) and an extract of *Escherichia coli* containing an enzyme, which he called *DNA polymerase*. He found that precise replicas of short DNA molecules were formed in the absence of living cells, but only when the reaction mixture included some natural DNA to act as a primer. This showed that DNA is an essential template for the synthesis of new DNA molecules and that it controls the formation of enzymes and tissue proteins through the mediation of RNA. Kornberg's work clarified the way in which DNA polymers, the components of chromosomes that carry and transmit genetic information, are formed. In 1959 he shared the Nobel Prize in Physiology or Medicine with Severo Ochoa, who discovered a bacterial enzyme that synthesized RNA. These discoveries of enzymes that catalyzethe synthesis of nucleic acids *in vitro* provided an entirely new approach to the biology and chemistry of RNA and DNA. They revealed the means by which DNA molecules are duplicated in the bacterial cell, and provided a method of reconstructing the duplication process in the test tube. They were of fundamental importance in genetics, in the study of viruses, in understanding cell and tissue differentiation and in the biosynthesis of specific proteins. See also **Ochoa, Severo (1905–1993)**.

In 1959 Kornberg moved to the Biochemistry Department at Stanford University, Palo Alto, California. He was Chairman of the Department until 1969 and remained Professor of Biochemistry at Stanford until 1988 when he was made Emeritus Professor. His research during this period was concerned with the synthesis of phospholipids and many of the reactions of the Krebs (tricarboxylic acid) cycle.

Additional Reading

Kornberg, A.: *For the Love of Enzymes: the Odyssey of a Biochemist*, Harvard University Press, Cambridge, MA, 1989.

Kornberg, A., and T.A. Baker: *DNA Replication*, 2nd Edition, W. H. Freeman, New York, NY, 2005.

NOEL G. COLEY, The Open University, Milton Keynes, UK

KOSSAVA. See **Winds and Air Movement**.

KOSSEL, ALBRECHT (1853–1927). Kossel, a German scientist, was a pioneer in the chemistry of nucleic acids. Early in his teenage years he developed a passion for botany and became an expert on the flora of the Rostock region of Germany. He entered the University of Strasbourg as medical student but never became a practicing doctor, but instead did research in biochemistry. In 1883, he became the chemical director of the Institute of Physiology in Berlin.

Kossel spent his life doing research on the substances in the cell nucleus. By 1884, he identified histone proteins in the nucleus. By 1885, he had discovered adenine and guanine in nucleic acids. In 1891, he had identified the presence of a carbohydrate in nucleic acids, in 1894, he discovered cytosine and thymine in nucleic acids, and by 1896 he discovered the amino acid histidine.

Kossel was awarded the Nobel Prize in Physiology or Medicine in 1910 for his important work on the chemistry of cell constituents. His findings are important in today's new molecular genetics.

J. M. I.

KRAFT PULP PROCESS:. See **Pulp (Wood) Production and Processing**.

KRAKATOA WINDS. See **Winds and Air Movement**.

KREBS CITRIC ACID CYCLE. See **Carbohydrates**.

KREBS, HANS ADOLF (1900–1981). Hans Adolf Krebs was born on 25 August 1900 at Hildesheim near Hamburg, Germany. His family was Jewish; his father was a physician. He entered the University of Göttingen in 1918 but transferred to the University of Freiburg in the following year. Krebs admired Franz Knoop's work on the oxidation of fatty acids in rabbits. Graduating in medicine in December 1923, Krebs began research on the permeation of charged dyes across cell membranes, in the hope of improving the design of drugs.

In December 1925 Krebs became research assistant to Otto Warburg at the Kaiser-Wilhelm Institute for Biology in Berlin-Dahlem. Warburg's research concerned the enzymes and coenzymes affecting cell respiration, and metabolism in normal and tumor tissues. The uptake of oxygen was measured using special manometers designed by Warburg. In 1929 Warburg persuaded the German Physiological Society to support Krebs's attendance at the 13th International Physiological Congress in Boston. Krebs declined the offer of a temporary post in Philadelphia, preferring to return to Berlin, but Warburg had decided that Krebs should move on and in 1930, given notice to leave, he was forced to seek a new post. With difficulty he secured a post in the Municipal Hospital at Altona, near Hamburg on condition that he worked on a topic proposed by Warburg, but in April 1931 Krebs returned to the University of Freiburg where he could pursue research of his own choice. See also **Warburg, Otto Heinrich (1883–1970)**.

Warburg's manometer technique was applied to the degradation of carbohydrates in thin slices of tissue and Krebs used the same methods to investigate urea metabolism in the liver. He found that ammonia and certain amino acids, especially ornithine, increased the quantities of urea formed in liver tissue. Urea was known to be formed in the liver by the action of arginase on arginine. Krebs found that arginine was transformed into ornithine by citrulline, a degradation product of arginine. In the presence of ammonia citrulline catalyses the production of urea and in this way the metabolic cycle is completed. Krebs went on to study the oxidative degradation of amino acids by various other tissues. He found that the kidney is a major site of amino acid metabolism and that both the liver and the kidney contain an enzyme (D-amino acid oxidase). Krebs's research was brought to a sudden halt in June 1933 when he was summarily dismissed in the Nazi purge of Jewish scholars. See also **Urea**.

In anticipation of some such action Sir Frederick Gowland Hopkins had already approached Krebs with an offer of help and in June 1933 he arrived in England with his experimental equipment. By early July he was installed in a laboratory at Cambridge where he resumed the work he had begun in Freiburg. In May 1935 he accepted a lectureship in pharmacology at Sheffield, later changed to biochemistry. Here he engaged a technical assistant, Leonard Eggleston, took on a graduate student, W. A. Johnson, and began work on the oxidation of di- and tricarboxylic acids in pigeon breast muscle. This laid the foundations for the tricarboxylic acid cycle, providing explanations of intermediary metabolism, the stepwise release of energy during the complete oxidation of foods, and the carbon skeletons of many cell constituents. Krebs also continued his work on amino acid metabolism and suggested avenues of research to link it with the metabolism of carboxylic acids. See also **Hopkins, Frederick Gowland (1861–1947)**.

During the war Krebs compared the digestibility of 75% extraction flour with that of 85% extraction "national wheatmeal" flour. He found that the latter reduced the uptake of calcium. In experiments on the optimum requirement of vitamins A and C, Krebs found that the body could withstand prolonged vitamin A deficiency, but on a diet low in vitamin C, scurvy appeared within 6 months. To avoid this a daily intake of 30 mg of vitamin C was recommended. In 1945 the Medical Research Council established a Unit for Research in Cell Metabolism with Krebs as Honorary Director and the University of Sheffield established a Chair of Biochemistry for him. His first task was to press the University to give him more space and staff to establish a department. The offer of a Chair at Harvard gave him bargaining power at Sheffield, as did his FRS in 1947 and the Nobel Prize for Physiology or Medicine in 1953, but in 1954 Krebs was appointed to the Whitley Chair of Biochemistry at Oxford in succession to Sir Rudolf Peters.

In accepting this appointment one condition he made was that the Unit he had founded at Sheffield should go with him. This caused administrative difficulties within the Oxford collegiate system and his fight to gain due recognition for his colleagues made him unpopular. Nevertheless his research flourished as he set out to discover the energy exchanges within living cells. He showed that food materials released energy in ways that could be grouped simply and by 1957 was able to announce the third metabolic cycle, the glyoxylate cycle, a variant of the tricarboxylic acid cycle. Krebs went on to study the regulation of metabolic pathways in animal cells, opening up a new field, which he was to pursue for many years.

After his retirement in 1967, Krebs elected to remain in Oxford, accepting a place in the Nuffield Department of Clinical Medicine at the Radcliffe Infirmary. The MRC supplied financial support for 3 years and with this "The Metabolic Research Laboratory" was established. It continued in existence for 14 years, during which Krebs himself published over a hundred papers. At the same time, over a hundred papers were published independently by his colleagues. Krebs never sought to claim credit that was rightly theirs and this, in part, explains the respect and affection of his co-workers. This great volume of research formed an important contribution to understanding the regulation of biochemical processes. Although Krebs did not write a textbook of biochemistry, his articles, with their clear, concise style, became necessary reading for students and research workers. He wrote a biography of Otto Warburg, and an autobiography, both of which were completed in 1981, the year in which he died at Oxford on 22 November. See also **Biochemistry (The History)**.

Additional Reading

Sherby, L.S., and W. Odelberg: *The Who's Who of Nobel Prize Winners 1901-2000*, 4th Edition, Greenwood Publishing Group, Inc., Westport, CT, 2001.

NOEL G. COLEY, The Open University, Milton Keynes, UK

KRILL. See **Crustaceans (Edible)**.

KROLL PROCESS. A widely used process for obtaining titanium metal. Titanium tetrachloride is reduced with magnesium metal at red heat and atmospheric pressure, in the presence of an inert gas blanket of helium or argon. Magnesium chloride and titanium metal are produced. The reaction is $TiCl_4 + 2\,Mg \rightarrow Ti + 2\,MgCl_2$. Essentially the same process is also used for obtaining zirconium.

KROTO, SIR HAROLD W. (1939–). A British chemist who won the Nobel Prize for chemistry along with Robert F. Curl, Jr., and Richard E. Smalley in 1996, the 100th anniversary of Alfred Nobel's death. The trio won for the discovery of the C_{60} compound called buckminsterfullerene. He received a Ph.D. from the University of Sheffield.

See also **Buckminsterfullerene (Buckyballs)**; **Curl, Robert F., Jr. (1933–)**; and **Smalley, Richard E.**

KRYPTON. [CAS: 7439-90-9]. Chemical element, symbol Kr, at. no. 36, at. wt. 83.80, periodic table group 18 (inert or noble gases), mp $-156.6\,°C$ ($-250\,°F$), bp $-152.3\,°C$ ($-242\,°F$), density 3.4 g/cm^3 (solid at $-273\,°C$ ($-523\,°F$). Solid krypton has a face-centered cubic crystal structure. At standard conditions, krypton is a colorless, odorless gas and does not form stable compounds with any other element. Due to its low valence forces, krypton does not form diatomic molecules, except in discharge tubes. It does form compounds under highly favorable conditions, as excitation in discharge tubes, or pressure in the presence of a powerful dipole. Krypton forms a hydrate at 14.5 atmospheres pressure and $0\,°C$ ($32\,°F$). The element also forms addition compounds with a number of organic substances, such as $Kr \cdot 2C_6H_5OH$ with phenol, which has a dissociation pressure of 6 to 10 atmospheres at $0\,°C$ ($32\,°F$). Krypton also forms compounds, possibly clathrates, with certain substances in nonstoichiometric proportions. Examples are its crystalline compounds with benzene [CAS: 71-43-2] and the compounds formed in aqueous solutions of hydroquinone [CAS: 123-31-9], under 40 atmospheres pressure of krypton, which contains 15.8% krypton by weight. First ionization potential, 13.996 eV; second, 26.4 eV; third, 36.8 eV.

Krypton occurs in the atmosphere to the extent of approximately 0.000114% and thus is the second least abundant of the rare gases in ordinary air. In terms of abundance, krypton does not appear on lists of elements in the earth's crust because it does not exist in stable compounds. However, because of its limited solubility in H_2O, krypton is found in seawater to the extent of approximately 1.4 tons per cubic mile. Commercial krypton is derived from air by liquefaction and fractional distillation. With exception of very special applications, krypton usually is not prepared in pure form, but supplied along with other rare gases, such as argon and neon, for filling fluorescent and incandescent light. As a filter in lamps, the gas assists in reducing filament evaporation and enables higher operating temperatures for lamps. Very high-candlepower aircraft-approach lamps contain krypton. When contained in an electric-discharge tube, krypton, when pure, emits a characteristic pale-violet light; when impure, it emits a brilliant red color characteristic of so-called neon tubes. Krypton also has been used in lasers. There are five natural isotopes ^{78}Kr, ^{80}Kr, and ^{82}Kr through ^{84}Kr, and five radioactive isotopes ^{76}Kr, ^{77}Kr, ^{79}Kr, ^{81}Kr, and ^{85}Kr. The latter isotope is generated in atomic reactors, and is a beta emitter with a half-life of approximately 10.6 years. This isotope, in solid form combined with a hydroquinone, has been used for activating phosphors and in luminous paints. ^{81}Kr, has a particularly long half-life of 2.1×10^5 years. While investigating the properties of liquid air in 1898, Ramsay and Travers found krypton in the residue remaining after nearly all of the liquid air had boiled away. The element then was identified spectroscopically. The element emits a characteristic brilliant green and yellow line in its spectrum.

In 1989, physicists at the Los Alamos National Laboratory's High-Energy Physics facility developed a krypton-fluoride laser at a cost of about $1 million. The laser's energy beam is capable of striking a very small target. Key to the success of the small laser is the ability to compress modest energy into pulses of less than 1 picosecond duration. The laser is being used in research on new materials. With the laser, tiny targets can be heated to temperatures approaching those of the sun's core.

By international agreement in 1960, the *fundamental unit of length*, the *meter*, is defined in terms of the orange-red spectral line of ^{86}Kr. This corresponds to the transition $5p\,[O_{1/2}]_1 - 6d\,[O_{1/2}]_1$. One meter = 1,650,763.73 wavelengths (in vacuo) of the orange-red line of ^{86}Kr.

Meteorological Effects of Krypton-85 in the Atmosphere. As pointed out by Boeck (1976), projections indicate that ^{85}Kr, a radioactive, chemically inert gas, may be produced and released in such quantities that it will create atmospheric ions at rates comparable to the present ion production rate near the tropical ocean surface. Krypton-85 is a by product in nuclear fission reactors and explosions. The ^{85}Kr is sealed in fuel elements of reactors, but during reprocessing of the fuel or plutonium separation, ^{85}Kr is released in a controlled manner to the atmosphere. The radioactive half-life of ^{85}Kr is 10.76 years. With no apparent natural mechanism to remove it, the gas will accumulate in the atmosphere until a balance between the rate of release and decay is reached. Ultimately this could produce a unique form of atmospheric radiation as contrasted with such natural radioactivity as emanates from uranium, thorium, etc., which are limited to producing ions near ground level. In the atmosphere, of course, are several radioactive isotopes (half-lives are given in parentheses): ^{22}Na (2.6 years); ^{32}P (14.22 days); ^{7}Be (53.6 days); ^{33}P (24.4 days); ^{35}S (87.1 days), among others. Not much concern has been given to these isotopes because of their comparatively short half-lives. Boeck 1976 makes a case for the need for more fundamental research into the possible effects of ^{85}Kr in the atmosphere, including possible climatic alterations.

Krypton in Meteorites. Along with other rare gases, krypton has been found in meteorites, notably the Murchison carbonaceous chondrite. Some scientists have been predicting for years that meteorites may entrap materials that date back to the beginning of the universe. For many years, the field was dominated by the dogma of an isotopically and chemically uniform early solar system (Srinivassan and Anders, 1978). However, careful analyses of the Murchison meteorite have produced anomalies in the form of unexpected differences in the ratios of the various rare element isotopes present.

See also **Chemical Elements**; **Chemical Elements: The History of the Origin** and **Meteorite**.

Additional Reading

Alexander, E.C., Jr. and M. Ozima: *Terrestrial Rare Gases*, Japan Scientific Societies Press, Tokyo, 1978.

Baker, H.: Getting to the Heart of the Matter (Krypton Laser), *Advanced Materials and Processes*, **8** (September 1989).

Birgeneau, R.J. and P.M. Horn: Two-Dimensional Rare Gas Solids, *Science*, **232**, 329–336 (1986).

Boeck, W.L.: Meteorological Consequences of Atmospheric Krypton-85, *Science*, **193**, 195–198 (1976).

Lide, D.R.: *CRC Handbook of Chemistry and Physics*, 88th Edition, CRC Press, LLC, Boca Raton, FL, 2007.

Newton, D.E. and L.W. Baker: *Chemical Elements: From Carbon to Krypton*, Vol. 1–3, Gale Group, Inc., Farmington Hills, MI, 1999.

Srinivassan, B. and E. Anders: Noble Gases in Murchison Meteorite, *Science*, **201**, 51–55 (1978).

Styra, B. and D. Butkus: *Geophysical Problems of Krypton-85 in the Atmosphere*, Hemisphere Publishing Corporation, New York, NY, 1991.

k-STATISTICS. The pth k-statistic of a sample is defined to be the unique polynomial, symmetric in the observations, whose expectation is k_p, the pth cumulant. The k-statistics are readily computed from the moments or from the sums of powers of the observations. Thus, in a sample of n,

$$k_1 = s_1/n = m_1$$

$$k_2 = (ns_2 - s_1^2)/n(n-1) = nm_2/(n-1)$$

$$k_3 = (n^2 s_3 - 3ns_2 s_1 + 2s_1^3)/n(n-1)(n-2)$$

$$= n^2 m_3/(n-1)(n-2)$$

$$k_4 = ((n^3 + n^2)s_4 - 4(n^2 + n)s_3 s_1 - 3(n^2 - n)s_2^2$$

$$+ 12ns_2 s_1^2 - 6s_1^4)/n(n-1)(n-2)(n-3)$$

$$= n^2((n+1)m_4 - 3(n-1)m_2^2)/(n-1)(n-2)(n-3)$$

where $s_p = \Sigma x^p$, m_1 is the mean and m_2, m_3, m_4 the moments about the mean. The sampling properties of the cumulants are much simpler than those of the moments.

KUDZU *(Pueraria lobata)*.   This is a long-lived, coarse-growing vine, introduced into the United States from Japan in the mid-1870s. It was not considered seriously as a forage crop until the early 1900s. With a nutritive value only slightly lower than that of alfalfa, the plant has found value for pasture and hay in the humid southern region of the United States. The plant is subject to winterkill and is seldom found north of Kentucky and Tennessee. The plant develops very long runners, which, if not frost damaged, become woody. The long runners root at each node, giving rise to new plants. Kudzu will produce for many years after it is well established, a fact that is an annoyance to many persons who dislike the extremely heavy growth of kudzu found along the roads and highways of the South during summer. To others, the kudzu provides an item of interest. Kudzu cannot withstand continual close grazing or overly frequent cuttings for hay. Kudzu is effective in reducing erosion on rough land and is particularly good for deep gully control. An interesting evaluation of kudzu, pros and cons, is given by B. Weekes (*American Forests*, 36–39, August 1982).

KUHN, RICHARD   (1900–1967).   Kuhn was born in a suburb of Vienna and educated at the local grammar school; his class produced two Nobel laureates. Following brief service in the First World War he studied chemistry at the universities of Vienna and Munich, taking a PhD in 1922 with a thesis on enzyme specificity in carbohydrate metabolism. He taught at Munich and in Zürich before his 1929 appointment as head of the chemistry department of the Kaiser Wilhelm Institute for Medical Research at Heidelberg, where he remained for the rest of his life.

Kuhn is best known for his work on the chemistry of plant pigments and vitamins. Having shown that diphenyl polyenes $[C_6H_6(CH{=}CH)_nC_6H_6]$ are colored when n is greater than 2, he began a study of carotenoids, which have a similar structure. During the early 1930s he discovered three isomers of carotene, isolated carotenoids from a wide range of natural sources, and determined the structure of vitamin A, of which carotene is a precursor (*Journal of the Society of Chemical Industry*, 1933). Turning next to the B vitamins, he isolated vitamin B_2 from milk, synthesized it and showed that it is important to respiratory enzyme action. By 1939 he had isolated vitamin B_6, having won (but been forced by Nazi law to refuse) the Nobel Prize in Chemistry the previous year.

Kuhn was a fine pianist, and enjoyed tennis, billiards and chess. A disciplined and dedicated scientist, he became increasingly interested in the practical applications of his work to medicine and agriculture.

Additional Reading

Sherby, L.S., and W. Odelberg: *The Who's Who of Nobel Prize Winners 1901-2000*, 4th Edition, Greenwood Publishing Group, Inc., Westport, CT, 2001.

Katherine D. Watson, University of Oxford, Oxford, UK

KUIPER AIRBORNE OBSERVATORY (KAO).   See **Space Science Missions: Universe**.

KUMQUAT.   See **Citrus Trees**.

KUROSHIO.   See **Ocean Currents**.

KURTOSIS.   Kurtosis is a property of a distribution or probability function depending on its fourth moment. It was originally thought that this moment adequately expressed the degree of "peakedness" of the distribution, but this is now known not to be necessarily so.

KURU.   See **Virus**.

KWASHIORKOR.   See **Nutritional Science (The History)**; and **Protein**.

KYANITE.   The mineral kyanite [CAS: 1302-76-7] is an aluminum silicate, corresponding to the formula Al_2SiO_5. It is triclinic, and has a good cleavage parallel to the macropinacoid. Its hardness varies considerably, depending on the crystallographic direction from 5 to 7.5; specific gravity, 3.56–3.67; luster, vitreous to pearly; color, commonly blue to white, but sometimes gray to green or nearly black; transparent to translucent; usually found in long-bladed crystals or columnar to fibrous structures. Kyanite is found in some metamorphic rocks as gneisses or mica schists. Of the many European localities for fine specimens might be mentioned the Ural Mountains of the former U.S.S.R.; the Czech Republic and Slovakia; Austria; Trentino, Italy; the St. Gotthard region of Switzerland; and France. In the United States, Chesterfield, Massachusetts; Litchfield, Connecticut; and Gaston County, North Carolina, have furnished fine specimens. Kyanite derives its name from the Greek word meaning *blue*, in reference to the delicate blue of the inner portions of the bladed crystals.

KYPHOSIS.   See **Bone**.

L

LABARIA *(Reptilia, Sauria).* A poisonous South American snake belonging to the pit vipers. It ranges from eastern Brazil north into the Guianas. Related to the jararaca.

LABRADOR CURRENT. An ocean current that flows southward from Baffin Bay, through the Davis Strait, thence southeastward past Labrador and Newfoundland. East of the Grand Banks, the Labrador current meets the Gulf Stream, and the two flow east separated by the cold wall.

LABYRINTH FISHES *(Osteichthyes).* Of the suborder *Anabantoidea*, and family *Anabantidae*, this is a group of tropical freshwater fishes, quite small in size, and usually found in Africa and southeast Asia. The pelvic fins bear a long slender filament, apparently sensory. They are named because of their highly specialized breathing apparatus. In using this labyrinth-type anatomical mechanism, the fish draws in a bubble of air from which it extracts the oxygen, and when the fish next surfaces, it expels old air out of the gill covers. This is an accessory apparatus that enables these fishes to tolerate water deficient in oxygen. One member of this family is the walking fish (sometimes called climbing perch) (*Anabas testudineus*) which attains a length of about 10 inches (25 centimeters). It is found in Malaya, the Philippines, and India. Its name stems from the fact that it apparently can walk for considerable distances on land, possibly seeking another body of water. In walking, the fish uses its gill plates, which fortunately are equipped with spiny edges, serving as "feet." The fish can cover about 10 feet (3 meters) in a minute by this method of locomotion.

Another labyrinth fish is the *Betta splendens* (Siamese fighting fish), found in Thailand. The males of this species are renowned for their ability to fight other males of their species. They have a rather dull color and achieve a length of about 2 inches (5 centimeters). The so-called "gourami" (*Osphronemus goramy*) is the largest of the labyrinths, attaining a length of 2 feet (0.6 meter). The fish is used as food in the Orient. Because of its unusual habits, the kissing gourami (*Helostoma temmincki*) is popular among tropical-fish hobbyists. There are several varieties of gourami. Favorites among fanciers are the genus *Colisia*, paradise fishes (genus *Macropodus*), and the croaking gourami (*Trichopsis vittatus*), the males of which make an odd croaking noise, particularly at night, when they come to surface for air. See Fig. 1.

LACCOLITH. An intrusive type of igneous rock. Studies of the forms of igneous rock masses have shown that the openings followed by volcanic lavas in their upward journey toward the surface are of two dominant types, *tubular* and *tabular*. Tubular openings are approximately circular in outline. These openings may vary from a few centimeters to many meters. Dikes are thin, tabular, parallel-walled masses of igneous rocks. Essentially, they have a vertical position and are formed by the injection of lava into fissures and joints in rocks. Sills are another form, but they are flat and essentially horizontal. Laccoliths, although somewhat similar to sills, differ from them in that the overlying beds are arched. The horizontal area occupied by a laccolith usually is smaller than that occupied by a sill. It is believed that the lava forming a laccolith was too viscous to flow far between the beds of the rock and thus pushed them up to form a dome. A laccolith generally is fed through a conduit. Laccoliths may merge into sills, and they are also commonly associated with dikes. Several laccoliths may occur in the same area. Laccoliths form numerous buttes and mountains in the western United States. See Figs. 1 and 2.

Giant gourami

Belontia signata

Trichopsis vittatus

Paradise fish

Climbing fish

Fig. 1.   Species of labyrinth fishes.

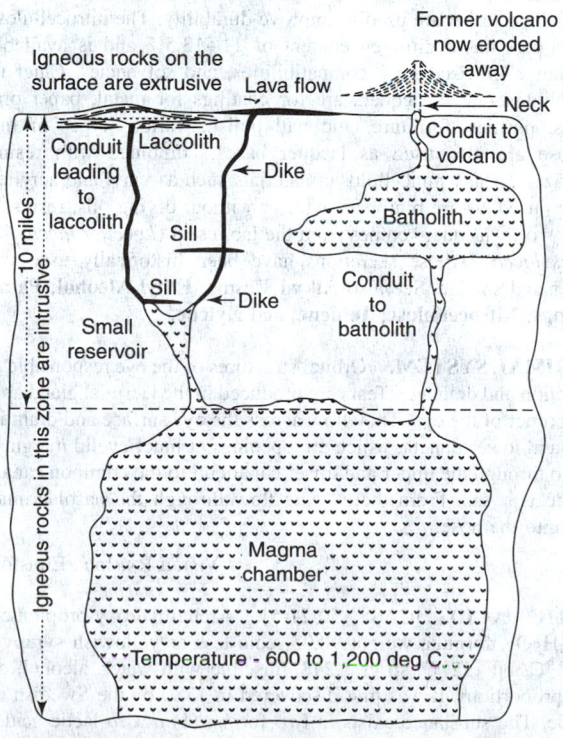

Fig. 1.   An idealized magma, showing the various forms of igneous rocks, such as dikes, sills, batholiths, and laccoliths.

Fig. 2.   An idealized laccolith, with associated dikes and sills.

The formal definition (American Geological Institute) is "A concordant igneous intrusion with a known or assumed flat floor and a postulated dike-like feeder somewhere beneath its thickest point. It is generally lens-like in form and roughly circular in plan, less than 5 miles in diameter and from a few meters to several hundred meters in thickness."

LACHRYMATOR.   A chemical substance that causes tears to form in the eyes. See also **Chlorinated Organics**.

LACQUER.   A protective or decorative coating that dries primarily by evaporation or solvent, rather than by oxidation or polymerization. Lacquers were originally comprised of high-viscosity nitrocellulose, a plasticizer (dibutyl phthalate or blown castor oil), and a solvent. Later, low-viscosity nitrocellulose became available; this was frequently modified with resins, such as ester gum or rosin. The solvents used are ethanol [CAS: 64-17-5], toluene [CAS: 108-88-3], xylene [CAS: 1330-20-7], and butyl acetate [CAS: 123-86-4]. Together with nitrocellulose [CAS: 9004-70-0], alkyd resins are used to improve durability. The nitrocellulose used for lacquers has a nitrogen content of 11–13.5% and is available in a wide range of viscosities, compatibilities, and solvencies. Chief uses of nitrocellulose-alkyd lacquers are for coatings for metal, paper products, textiles, plastics, furniture, and nail polish. Various types of modified cellulose are also used as lacquer bases, combined with resins, and plasticizers. Many noncellulosic materials such as vinyl and acrylic resins are also used, as are bitumens, with or without drying oils, resins, etc.

The word "lacquer" comes from the lac insect (*Laccifer lacca, formerly Coccus lacca*), whose secretions have been historically used to make lacquer and shellac. See also **Alkyd Resins**; **Ethyl Alcohol**; **Paints and Coatings**; **Nitrocellulose**; **Toluene**; and **Xylene**.

LACRIMAL SYSTEM.   Orbital structures of the eye responsible for tear production and drainage. Tears are produced in the lacrimal gland above the outer corner of the eye. They flow across the eye surface and drain into the upper and lower puncta, which are openings at inner eyelid margins. They then go through the upper and lower canaliculi to the common canaliculus, into the tear sac. From there, tears flow through the nasolacrimal duct, down into the nose.

Vision Rx, Inc., Elmsford, NY

LACTIC ACID.   [CAS: 50-21-5]. Alpha-hydroxy-propionic acid, $H-C_3H_5O_3$, formula weight 90.05, colorless or yellowish syrupy liquid, mp 18 °C, bp 122 °C, sp gr 1.248, miscible with water, alcohol, or ether in all proportions. It was first discovered in 1780 by the Swedish chemist Scheele. The substance exists in two forms: (1) *dextro* lactic acid, which rotates the plane of polarized light to the right; and (2) *levo* lactic acid which rotates the plane of polarized light to the left. A mixture of these two forms is ordinary lactic acid, which does not rotate the plane of polarized light. Ordinary lactic acid is termed dextrolevo lactic acid. Lactic acid is a product of corn refining.

Lactic acid was one of the first biological substances to be investigated from the standpoint of the existence of the two optically active forms.

Lactic Acidosis. Lactic acid is the cause of one of many possible disorders in human acid-base metabolism. Lactic acidosis represents an accumulation of lactic acid in the blood and tissues. This condition gradually depletes the natural buffers in the body and there is a consequent lowering of pH. As described in the entry on **Glycolysis**, lactic acid is the end product of that process. Lactic acid blood levels are determined by at least four factors. The rate of generation of lactic acid; the rate of transport from tissues to plasma and from plasma to the liver (point of utilization of lactic acid); the rate of utilization; and excretion of lactic acid by the kidneys. Normally, all of these functions are maintained in balance to give a normal blood lactate concentration of about 1 mEq/1.

On the generation side, three factors are involved. (1) The availability of oxygen is a major controlling determinant of lactic acid generation because, as adenosine triphosphate (ATP) generation from oxidative phosphorylation diminishes, the cells naturally respond with a greater rate of glycolysis. This increases tissue lactate levels and ultimately lactate blood levels. See also **Phosphorylation (Oxidative)**. (2) If, as may be caused by various factors, there is an increase in pH, the activity of phosphofructokinase will increase (this is the rate-limiting enzyme of glycolysis). With increases in pH, the enzyme is more active and more lactate is formed. (3) Factors that affect the biological oxidation-reduction potentials also influence the rate at which glucose is metabolized to lactate.

Fundamental predisposing conditions causing an increased generation of lactate include: decreased tissue perfusion associated with shock, which may occur in cardiac arrest; increased skeletal muscle activity (the rate of glycolysis increases with exercise; this also may be associated with convulsive states that may follow severe exercise—brought about by increased blood lactate concentrations); large tumors, since tumors (leukemias, lymphomas, etc.) may have an increased rate of glycolysis even in the presence of a sufficient supply of oxygen; and both cyanide and carbon monoxide poisoning, which can increase lactate levels because of insufficient oxygen supply.

On the utilization side, there are a number of influencing factors. The liver is the principal lactic acid utilization center. In liver failure, a surplus of lactate builds up, a condition which may be associated with reduced hepatic perfusion, hepatocyte failure, and hepatocytes replaced by tumor. Blood lactate concentrations are elevated in persons with diabetic ketoacidosis. The observed elevation of lactic acid levels in cases of alcoholism is not fully understood: the condition may increase generation or by decreasing utilization. The latter effect is now favored by many authorities, the theory being that ethanol completes for electrons in the liver, thus decreasing utilization of lactic acid in that organ.

Specifications, Quality Control, and Analytical Methods

Lactic acid is generally sold under four general product categories: synthetic: a highly purified product from a chemical synthesis process. It is water-white, has excellent heat stability, and can be used in both food and industrial applications; fermentation: a food-grade product from carbohydrate fermentation refined by ion exchange and activated carbon. The product contains residual carbohydrate or protein impurities and is not heat stable; heat-stable fermentation: a highly refined, heat-stable product from esterification of fermentation-derived lactic acid, followed by hydrolysis of the recovered ester to produce the acid; and technical: a crude product from either a synthetic or fermentation process, used in industrial applications where high purity is not required.

The fermentation-derived food-grade product is sold in 50, 80, and 88% concentrations; the other grades are available in 50 and 88% concentrations. The food-grade product meets the Food Chemicals Codex III and the pharmaceutical grade meets the FCC and the *United States Pharmacopoeia XX* specifications. Other lactic acid derivatives such as salts and esters are also available in well-established product specifications. Standard analytical methods such as titration and liquid chromatography can be used to determine lactic acid, and other gravimetric and specific tests are used to detect impurities for the product specifications. A standard titration method neutralizes the acid with sodium hydroxide and then back-titrates the acid. An older standard quantitative method for determination of lactic acid was based on oxidation by potassium permanganate to acetaldehyde, which is absorbed in sodium bisulfite and titrated iodometrically.

Lactic acid is generally recognized as safe (GRAS) for multipurpose food use. Lactate salts such as calcium and sodium lactates and esters such as ethyl lactate used in pharmaceutical preparations are also considered safe and nontoxic. The U.S. Food and Drug Administration lists lactic acid (all isomers) as GRAS and sets no limitations on its use in food other than current good manufacturing practice.

Uses

Currently, the principal use of lactic acid is in food and food-related applications, which in the United States accounts for approximately 85% of the demand. The rest (~15%) of the uses are for nonfood industrial applications. The expected advent of the production of low cost lactic acid in high volume can open new applications for lactic acid and its derivatives, because it is a versatile molecule that can be converted to a wide range of industrial chemicals or polymer feedstock's.

As a food acidulant, lactic acid has a mild acidic taste, in contrast to other food acids. It is nonvolatile, odorless, and is classified GRAS for general-purpose food additives by the FDA in the United States and by other regulatory agencies elsewhere. It is a good preservative and pickling agent for sauerkraut, olives, and pickled vegetables. It is used as acidulant, flavoring, pH buffering agent, or inhibitor of bacterial spoilage in a wide variety of processed foods such as candy, breads and bakery products, soft drinks, soups, sherbets, dairy products, beer, jams and jellies, mayonnaise, processed eggs, and many other processed foods, often in conjunction with other acidulants. An emerging new use for lactic acid or its salts is in the disinfection and packaging of carcasses, particularly those of poultry and fish, where the addition of aqueous solutions of lactic acid and its salts during processing increase shelf life and reduce the growth of anaerobic spoilage organisms such as *Clostridium botulinum*.

Polymers of lactic acids are biodegradable thermoplastics that can be made from a variety of renewable carbohydrate resources. A fairly wide range of properties is obtainable by copolymerization with other functional monomers such as glycolide, caprolactone, polyether polyols, etc. The polymers are transparent, which is important for packaging applications. They offer good shelf life because they degrade slowly by hydrolysis which can be controlled by adjusting the composition and molecular weight. The properties of lactic copolymers which approach that of large-volume petroleum-derived polymers such as poly-styrene, flexible poly(vinyl chloride) (PVC), vinylidene chloride, etc, have been summarized by Lipinsky and Sinclair. There are numerous patents and articles on lactic acid polymers and copolymers, their properties, potential uses, and processes, dating back to the early work by Carothers at Du Pont.

See also **Hydroxydicarboxylic Acids**.

Additional Reading

Bozoglu, T., Faruk, and B. Ray (Editors): *Lactic Acid Bacteria: Current Advances in Genetics, Metabolism, and Application of Lactic Acid Bacteria*, Springer-Verlag New York, Inc., New York, NY, 1996.

Dave, P. and co-workers: *Polymer Prep.*, **31**(1), 442–443, 1990.

Goldberg, I., and R. Williams: *Biotechnology and Food Ingredients*, Chapman & Hall, New York, NY, 1999.

Lipinsky, E.S. and R.G. Sinclair: *Chem. Eng. Progr.*, **82**(8), 26–32, 1986.

Nakamura, T. and co-workers: Adv. Biomater. *Biomater. Clin. Appl.*, **7**, 759–764, 1987.

Salminen, S.: *Lactic Acid Bacteria: Microbiological and Functional Aspects*, 3rd Edition, Marcel Dekker, Inc., New York, NY, 2004.

Staff: *Code of Federal Regulations*, U.S. Food and Drug Administration, Washington, DC., April 1, 1992.

U.S. Pats. 4,045,418,4,057,537 (1977), R.G. Sinclair (to Gulf Oil Corp.).

Wood, B., and P.J. Warner: *Genetics of Lactic Acid Bacteria*, Kluwer Academic Publishers, Norwell, MA, 2003.

Wood, B.: *The Lactic Acid Bacteria in Health and Disease*, Aspen Publishers, Inc., Gaithersburg, MD, 1999.

Wood, B. and W. Holzapfel: *The Genera of Lactic Acid Bacteria*, Blackie Academic & Professional, UK, 1999.

Zhu, J. and co-workers: *Proc. of C-MRS Intl. Symp.*, **3**, 387–390. 1990.

LACTIC ACIDOSIS.   See **Lactic Acid**.

LACTITOL.   See **Sugar Alcohols**.

LACTONE.   An inner ester of a carboxylic acid formed by intramolecular reaction of hydroxylated or halogenerated carboxylic acids with elimination of water. They occur in nature as odor-bearing components of various plant products, also made synthetically. See also **Carboxylic Acids**.

LACTOSE.   Lactose [CAS: 63-42-3] (milk sugar) is the only commercially available sugar that is derived from animal rather than plant sources. It is a disaccharide consisting of one galactose and one glucose moiety, 4-*O*-β-D-galactopyranosyl-D-glucose (1). The concentration of lactose in milk products ranges from 4.8 wt% in whole milk to 73.5 wt% in sweet dried whey. There have been reports of the presence of lactose in plant materials, e.g., sapote and acacia, but this has not been confirmed.

(1)

Lactose is isolated commercially as the crystalline α-monohydrate from the whey by-products of cheese or caseinate production. It is available in varying degrees of purity. Fermentation grade is 98 wt% pure, whereas USP lactose is refined to 99.8 wt% purity. Although the α-monohydrate is the commercially available form of lactose, the sugar can be crystallized at high temperature to give the β-anhydride [CAS: 56907-28-9]. The sugar is not very soluble in water (ca 22 g/100 g water at 25 °C (77 °F), nor is it very sweet (ca one-fifth the sweetening power of sucrose). Lactose is a reducing sugar that reacts with amines and amino compounds with resultant browning.

Uses of lactose production by application include baby and infant formulations, human food, pharmaceuticals, and fermentation and animal feed. It is used as a diluent in tablets and capsules to correct the balance between carbohydrate and proteins in cow-milk-based breast milk replacers, and to increase osmotic property or viscosity without adding excessive sweetness. It has also been used as a carrier for flavorings, volatile aromas, and synthetic sweeteners. Physiologically, lactose promotes the absorption of calcium, phosphorus, and essential trace minerals; has low cariogenicity; and is more slowly and gradually absorbed than sucrose [CAS: 57-50-1], therefore of potential benefit to diabetes mellitus patients.

Lactose, and the lactose in substances such as milk and whey, has been hydrolyzed commercially by enzymes to yield products that can be tolerated physiologically much more easily by people who have a lactose intolerance.

See also **Carbohydrates**; **Milk And Milk Products**; **Sucrose**; and **Sweeteners**.

LACTULOSE.   See **Sweeteners**.

LACUNOSUS.   See **Clouds and Cloud Formation**.

LADDER POLYMER.   An ordered molecular network of double-stranded chains connected by hydrogen or chemical bonds located at regular intervals along the chains. Many complex proteins, including DNA, are of this nature.

LADY BEETLE *(Insecta, Coleoptera).*   Small oval beetles, strongly convex and with relatively small legs. The common name and also the name lady bug apply chiefly to the more common red species, marked with black and white, but the family *Coccinellidae* to which they belong contains many others.

Lady beetles are found on plants and trees and deposit their eggs on the underside of leaves. They are practically round in shape and quite small, approximately $\frac{1}{8}$ inch (3 millimeters) across. The color varies with the species. They are harmless and often carried around by children as "pets."

The worm-like maggots eat plant lice. Aphids are a favorite food and thus the presence of lady beetles helps with gardening and growing flowers. See also **Beneficial Insects**.

LAG (Angle of).   When two related quantities, such as an alternating voltage and an alternating current, vary sinusoidally with time and have the same frequency, they may be expressed as

$$Q_1 = A \begin{Bmatrix} \sin \\ \cos \end{Bmatrix} (\omega t + \phi)$$

$$Q_2 = B \begin{Bmatrix} \sin \\ \cos \end{Bmatrix} \omega t$$

where A, B, and ω are constants. It is then said that Q_2 lags (behind) Q_1 and ϕ is known as the angle of lag if it is positive. If ϕ is negative its magnitude is the angle of lead and Q_2 is said to lead Q_1.

LAGARITH LOSSLESS VIDEO CODEC.   See **Data Compression**.

LAGOON.   See **Estuary**.

LAGRANGE FORMULA FOR INTERPOLATION.   Used when $(n + 1)$-pairs of values are given for $y = f(x)$, but not necessarily at equally spaced increments of x or y. Let the given number pairs be (x_0, y_0), $(x_1, y_1), \ldots, (x_n, y_n)$. Then for any desired value of x within this interval,

$$y = y_0 L_0^{(n)}(x) + y_1 L_1^{(n)}(x) + \cdots + y_n L_n^{(n)}(x)$$

where

$$L_i^{(n)}(x) = \frac{(x - x_0) \cdots (x - x_{i-1})(x - x_{i+1}) \cdots (x - x_n)}{(x_i - x_0) \cdots (x_i - x_{i-1})(x_i - x_{i+1}) \cdots (x_i - x_n)}.$$

The quantities L_i, known as Lagrange coefficients, are independent of y; hence they may be calculated once for a given set of x values and used unchanged to obtain results for varying y. Moreover, it will be found that they remain unchanged with a change of variable to $u = (x - a)/h$, where h and a are constants.

Because of the symmetry in the equation, x and y may also be interchanged so that inverse interpolation may be effected.

See also **Interpolation**.

LAGRANGIAN COORDINATES.   Sometimes called material coordinates. A system of coordinates by which fluid parcels are identified for all time by assigning them coordinates which do not vary with time. Examples of such coordinates are (a) the values of any properties of the fluid conserved in the motion; or (b) more generally, the positions in space of the parcels at some arbitrarily selected moment. Subsequent positions in space of the parcels are then the dependent variables, functions of time and of the Lagrangian coordinates.

Few observations in meteorology are Lagrangian: this would require successive observations in time of the same air parcel. Exceptions are the constant-pressure balloon observation, which attempts to follow a parcel under the assumption that its pressure is conserved, and certain small-scale observations of diffusion particles. See also **Eulerian Coordinates**.

LAGRANGIAN FUNCTION.   Also called kinetic potential, the difference between the kinetic energy and the potential energy of a dynamic system. It is generally symbolized by L.

LAGRANGIAN POINT.   One of the five solutions by Lagrange to the three-body problem in which three bodies will move as a stable configuration. In three of the solutions the bodies are in line; in the other two the bodies are at the vortices of equilateral triangles.

Lagrange predicted in 1772 that if the three bodies form an equilateral triangle revolving about one of the bodies, the system would be stable. This prediction was fulfilled in 1908 with the discovery of the asteroid Achilles approximately 60° ahead of Jupiter in Jupiter's orbit. Since then other asteroids have been discovered 60° ahead and 60° behind Jupiter.

LAGUERRE DIFFERENTIAL EQUATION.   The linear equation $xy'' + (1 - x)y' + ny = 0$, having a simple pole at the origin. Its solutions are the *Laguerre polynomials*. Differentiation of the equation k times and replacement of the kth derivative by y gives

$$xy'' + (k + 1 - x)y' + (n - k)y = 0$$

which is the associated Laguerre equation with solutions as associated Laguerre polynomials. These functions occur in the quantum mechanical problem of the hydrogen atom.

The associated polynomials may be defined by the equivalent expressions

$$L_n^{(k)}(x) = \frac{e^x x^{-k}}{n!} \frac{d^n}{dx^n} (e^{-x} x^{n+k})$$

$$= \sum_{i=0}^{n} \binom{n + k}{n - i} \frac{(-x)^i}{i!}$$

The special case of $k = 0$ gives the Laguerre polynomials

$$L_n(x) = 1 - \binom{n}{1} x + \binom{n}{2} \frac{x^2}{2!} - \binom{n}{3} \frac{x^3}{3!} + \cdots$$

Both kinds may also be expressed in terms of the Gauss hypergeometric series and by generating functions. They are also related to the Hermite polynomials and the Bessel functions.

LAKE.   See **Earth**; and **Limnology**.

LAKE BREEZE.   See **Winds and Air Movement**.

LAKE EFFECT.   See **Meteorology**.

LAKE-EFFECT SNOW.   See **Meteorology**.

LAKES.   *Lakes* are waterbodies of considerable size that are larger than ponds. *Shallow lakes* are mixed by the wind from the surface to the ground, and the daylight reaches the bottom. *Stratified lakes* are deep enough to be vertically divided into a lighted surface-near *euphotic* zone, where photosynthesis of green plants is possible, and a deep-water *aphotic* zone, which is too dark for photosynthetic activities. In many lakes, thermal vertical stratification is observed as a warm layer near the surface, the epilimnion, and a colder layer below, the hypolimnion. The epilimnion is wind-exposed and well mixed, whereas the deep water layer of the hypolimnion is not included in the mixis process.

Lake basins containing standing waterbodies were formed by different processes; tectonic forces; volcanic, glacial, fluviatile activities; or by solution of easily dissolvable rock such as limestone. Hutchinson ordered the lakes by their mode of origin into about 80 categories. Typical examples are listed by Meybeck, who also gives numbers of the types of origin, of size, and of age.

Tectonic lakes are found in geological rift valleys; the deepest are Lake Baikal and Tanganyika. The Caspian Sea and the Aral Sea are rest areas of the ancient Thetis Sea which was closed by tectonic movements, and they are the largest inland waters in area (374,000 km^2, 64,500 km^2). The oldest inland waters are found among tectonic lakes, some of these, it is estimated, are older than 20 to 30 million years.

Volcanic processes led to the origin of different lake types. *Maar lakes* emerged after singular volcanic events (steam explosions in contact zones of magma and groundwater). Lava dams or emerging volcanoes resulted in reservoir-like lakes, such as L. Kivu and Laguna del Laja (Chile). Large lakes were formed on lava plateaus (L. Myvatn, Iceland) and in *caldera* basins, whereas *crater lakes* are usually small. Lakes at active volcanoes are often heavily acidified by acidic brines rinsing into the lake water.

The main types of glacially formed lakes are (1) small *kar lakes* at the slopes of mountains at different altitudes, (2) *piedmont lakes* where glaciers from valleys reached the foreland at the foot of the mountains, and (3) extended *lake districts from low altitude glaciations* as on the Canadian Shield and the circum-Baltic area. Postglacial lakes of temperate climatic zones are only 8 to 12 thousand years old.

In the lowland area of large rivers, the flow of the running rivers often changes, leaving standing waters, oxbows or side waters, as *fluvial lakes*, which were parts of the rivers in the past or are permanently in open connection with the river system.

Lakes are more numerous as size decreases. Meybeck gives the estimated worldwide number of lakes, ordered by size classes and by the origin of lake basins. A list of the deepest and largest lakes, by area and by water volume, is given by Herdendorf, including many uncertainties on the basic data of even the 50 most important inland waters of the world.

Water Budget and Salinity of Lakes

The oceans contain 83.5% of the world's water, the inland waters only 0.015%. Lakes contain a total water volume of 205,000 km^3; about 50% of this total volume is freshwater, on the one side, and saline water on the other side. The deepest and oldest lake, Baikal, has a volume of about 20% of the total freshwater volume of lakes and is the worldwide largest singular freshwater resource. The annual fluxes through the pools of the atmosphere, rivers, and groundwater are 496, 38, and 12 thousand km^3; exchange times are 10 days, 16 days, and 700 years, respectively. The volume of freshwater lakes is 100 thousand km^3; exchange times are 1 month to 500 years.

Lakes are standing waters that are usually connected with and included in the flow system of river catchments. The lakes within the catchment systems can be ordered by their degree of connection to the overall water flows. Lakes with intensive through-flow have high rates of flushing and water exchange and short retention times. The lake volume compared with the volume of all inflows per year gives the virtual *filling time*. The relation of the outflow and the lake volume gives the net exchange rate of the lake water. Including groundwater exchange, precipitation on the lake surface, and evaporation, the relation of all inputs to all losses of water gives the gross exchange rate; its reciprocal value is the *retention time* or average age of the water.

Depending on the given regional water balance, lake basins are permanently flushed, have an outflow, are in a steady state of the flow of water and of dissolved salts, and, thereby, contain freshwater. In areas with a negative water balance, the river catchments are *endorheic* systems, and terminal lake basins occur without outflows; here, the water finally evaporates, and the inflowing salts accumulate, leaving *saline lakes*. Saline lakes are typical of (semi)arid areas and contain nearly half of the world's lake water volume. Saline lakes (with salinities >3%) are found in all climatic zones, warm, temperate, and cold regions, where negative water balances are given by large-scale climatic zones or by regional orographic conditions (rain-lea east of the Rocky Mountains in North America, east of the Andes in South America, or in the Iranian basin). Such endorheic river catchments, in which the water flow ends in terminal lakes, are the Jordan River system with the Dead Sea, the system of Lake Titicaca with Lake Poopó as the terminal saline lake, and the system of Lake Tahoe and Pyramid Lake in North America. In Central Australia, Lake Eyre is a large, endorheic saline basin. The highest possible salt concentrations are found in the Dead Sea (~300 g/L) and in Deep Lake (Antarctica) which never freezes; the water reaches $-14.5\,^\circ$C during the winter.

Standard Composition of Freshwaters

Freshwaters, soft waters as well as hard waters, are chemically dominated by the carbonate system, CO_3^{2-} and HCO_3^- as anions, and Ca^{2+} and Mg^{2+} as cations; together they are the main constituents and the buffering system of neutral freshwaters. This is caused by rainwater that is weakly acidic from CO_2 and, thereby, dissolves the carbonatic minerals of soils and rock. In areas poor in carbonates, the water contains low concentrations of dissolved salts, and in other areas of dominating limestone, the water is rich in dissolved carbonates. The qualitatively similar composition of freshwaters is described as a worldwide "standard water composition". See also **Freshwater**.

The gases in air, 78.09% N_2, 20.95% O_2, and 0.03% CO_2, are in physical equilibrium between air and water. At low temperatures, higher concentrations of gases dissolve in the water than at high temperature. The maximum content of dissolved oxygen and CO_2 in water is 14.5 (8.9) mg O_2/L and 1.005 (0.51) mg CO_2/L at 0 $^\circ$C (20 $^\circ$C).

Stratification and Mixing

Lakes are usually stratified into an upper layer, the epilimnion, which is well mixed by the wind, and a deep waterbody, the hypolimnion, which is not included in the mixing process. Stratification occurs seasonally in the temperate climatic zones, or in a diurnal–nocturnal rhythm in the tropical zones (Fig. 1). Ice covers block energy inputs by wind into the water column during the winter. The regime of stratification and mixis of the water column, as given by the climate, defines the types of lakes as (1) *monomictic*, (2) *dimictic*, or (3) *polymictic*, according to the number of periods of total mixing (*holomixis*) per year. Tropical lakes at low altitude are polymictic; they mix during the night and stratify during the day. Dimictic lakes have a stable stratification with a warm epilimnion during the summer and an ice cover during the winter; they show holomixis during the two seasonal periods of homothermy. One holomixis occurs in spring after ice melting when the surface water has the same temperature as the hypolimnion, usually 4 $^\circ$C (39 $^\circ$F), and a second during late autumn before freezing. Monomictic lakes mix only once per year. The warm-monomictic lakes of oceanic-temperate regions never freeze and show holomixis during the winter. Cold-monomictic lakes are found in polar climatic regions; they are covered by ice during large parts of the year and melt only in summer when the water column reaches 4 $^\circ$C (39 $^\circ$F) and mixes from the surface to the ground.

The stability of stratification of a warm epilimnion and a cold hypolimnion determines whether wind forces mix only the uppermost layers, or the entire water column from the surface to the bottom. Stratification stability depends on the density gradient between the epilimnetic and the hypolimnetic partial waterbodies. Indicative numbers are the *Wedderburn Number* or the *Lake Number*, which, at given threshold values, predict optional holomixis with the next wind-forcing.

Climatic Zones and Lake Temperatures

Lakes are integrators of the regional climate and weather development. The surface temperatures follow the actual air temperature with a time delay, and lakes of the same region show time courses of similar surface temperatures. The deep water of lakes in polar and temperate continental climatic zones is at 4 $^\circ$C (39 $^\circ$F), the temperature of water at its maximum density. The deep water of warm-monomictic lakes of temperate oceanic climatic zones in the Southern Hemisphere, in New Zealand and South America and also in western and southern Europe, has a temperature above 4 $^\circ$C (39 $^\circ$F). The hypolimnetic temperatures remain from the last holomixis at the end of winter, and changes reflect long-term climatic conditions.

Waves and Oscillations

Driven by wind, waves emerge on the surfaces of lakes; their heights can be approximated by $h = 0.105$ (fetch)$^{0.5}$ (wave height and fetch given in cm; the *fetch* is the maximum length of the wind blowing across an open lake surface). Less peculiar are oscillations along the lake axis, which have amplitudes of few centimeters at the lake surface (*seiches*). Their frequencies depend on the length of the lake axis, between minutes in small

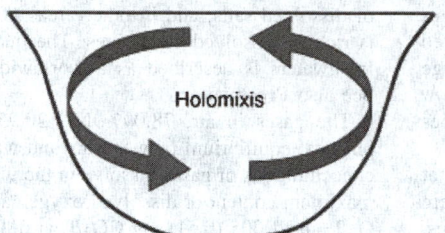

Wind-induced partial mixis of the warmer epilimnion in stratified lakes, as occurring in temperate zones during summer, and during the day in tropical climatic zones. The colder water of the hypolimnion is not included in the mixing process.

Wind-induced holomixis of the entire water column from surface to bottom in unstratified, homothermic lakes, occurs in spring, fall, or winter in temperate zones and during the night in tropical climatic zones.

Fig. 1. Seasonal or diurnal succession of stratification into warm epilimnion and cold hypolimnion and unstratified water column when the water is homothermic with complete overturn of the water from surface to ground (holomixis).

lakes and one day in large lakes. At the boundary layer between the epi- and hypolimnion, however, internal oscillations can reach wave amplitudes of 10 m or more. The periods of these internal oscillations are slower and reach 1 month in the largest lakes, such as Baikal or Tanganyika.

Ecosystems of Lakes

The ecosystems of lakes belong to the best described and understood ecosystems because the composition of the community is limited in number of species and the populations are limited by the lake size. The seasonal development of the populations, their change in species composition, and the growth and decrease of the populations are well investigated and analyzed in terms of numbers of individuals as well as in terms of energetics and species composition.

Lakes have characteristic compartments as partial habitats of their ecosystems. The ecosystem of the pelagic zone of lakes is dominated by planktonic microorganisms, larger zooplankton, and fish. The phytoplankton is autotrophic and limited to the euphotic layers near the surface where light intensities are sufficient for photosynthesis and oxygen production. The limit between the *euphotic* and the *aphotic zone* is usually at depths where the light intensity decreases to <1% of that at the lake surface.

The *pelagic zone* is the body of free water of the lake, the *littoral zone* is the euphotic part at the lakeshore, and the *profundal* is the aphotic dark layer of the *benthic zone* below the littoral. The benthic community outside the littoral zone, the profundal, consists of the organisms living in and on the sediments, microorganisms, macrozoobenthos, and benthic fish species. Below the light limit for photosynthesis, the species of the benthic community, macro- and microorganisms, are heterotrophic and consumers of oxygen. All particles in the water eventually sink to the lake bottom, form annual layers of growing sediments (varves), and leave an archive of residuals. Therefore, the history and the development of the lake can be reconstructed from the microlayers of a sediment core, with respect to eutrophication and nutrient levels and impacts of climatic changes.

The littoral zone is the euphotic part of the benthic zone. The living community consists of autotrophic macrophytes and algae and of heterotrophic animals and microorganisms. The belt of macrophytes appears as an interface between land and water; wetland plants are on the land side of the belt, followed by plants rooting in the water and emerging in the air, swimming plants, and lastly submerged plants that cover the light bottom of the deeper littoral zone. The partial ecosystem of the littoral zone connects the land and water, and simultaneously, as a biologically reactive zone, retains and changes the flow of nutrients and other allochthonous inputs of matter into the lake's pelagic zone.

Trophic State and Productivity

Trophy is a measure of the level of nutrients and of the biological productivity resulting from nutrients supplies. The organismic productivity

of aquatic ecosystems, measurable as the formation of biomass, is limited by the availability of the rarest nutrient element (C,N,P,S,Fe, . . .). In temperate zones, phosphorus is usually the least available element because there is no gaseous transport form through the air and P is tightly sorbed on soil particles and retained on the land. The production of organic substance is limited by the availability of phosphorus, so the trophy (productivity) of lakes can be related to the P-content of the water at the start of the growing season. The process of growth needs a proportion of 106 C:16 N:1P as the atomic ratio of biomass. The primary production of phytoplankton and macrophytes—depending on P—reaches annual rates of 10 to 50 g C/m^2 of organic carbon in nutrient-poor lakes and more than 500 g C/m^2 in very nutrient-rich systems.

The biomass produced sediments to the lake bottom, where this dead organic substance is decayed by oxygen-consuming bacteria. Depending on the trophic state, the oxygen conditions in the hypolimnic water body reach different end points in late summer and fall. In oligotrophic lakes, the nutrients and productivity are low, as well as the following decomposition and oxygen losses in deep water. In eutrophic lakes, however, the nutrient supply and the resulting productivity are high. After decomposition in deep water layers, the oxygen content in the hypolimnion may reach critically low limits in late summer. The consequences of high nutrient supplies are mitigated in deep lakes with high hypolimnic volumes but are more severe in shallower lakes with small hypolimnic volumes and small oxygen reserves therein (Fig. 2).

Food Chains, Filtrators, and Size Spectra in Plankton

Within the pelagic zone of lakes, the species of the plankton community can be structured by the food chain from the primary producers, the green algae, to herbivores, zooplankton species that feed on these algae, and carnivores that feed on herbivorous zooplankton. The efficiency of energy transfer from the lower to the higher trophic levels is about 10% from plants to animals and up to 30% from animal prey to their predators. Because there are losses of unused particles and dissolved organic matter (DOM) on all levels, recycling by bacteria as decomposers of DOM is important. Based on the bacteria as food, leading via bacteria-feeding protozoans to larger predators, the "lost" DOM is recycled again into the "normal" food chain by the "microbial loop." This secondary food chain is important for the overall efficiency of the ecosystem as a processor of energy and its ability to recycle nutrient elements.

In oligotrophic systems, the organisms, prey and predators, are present in low concentrations, and a predator has to find and to grasp each single prey individually. The prey in such food chains is usually smaller than its predator by a factor of 10 to 20 by body length. Therefore, in oligotrophic planktonic systems, the food chain from green algae, via zooplankton to fish needs four to five subsequent trophic levels. The

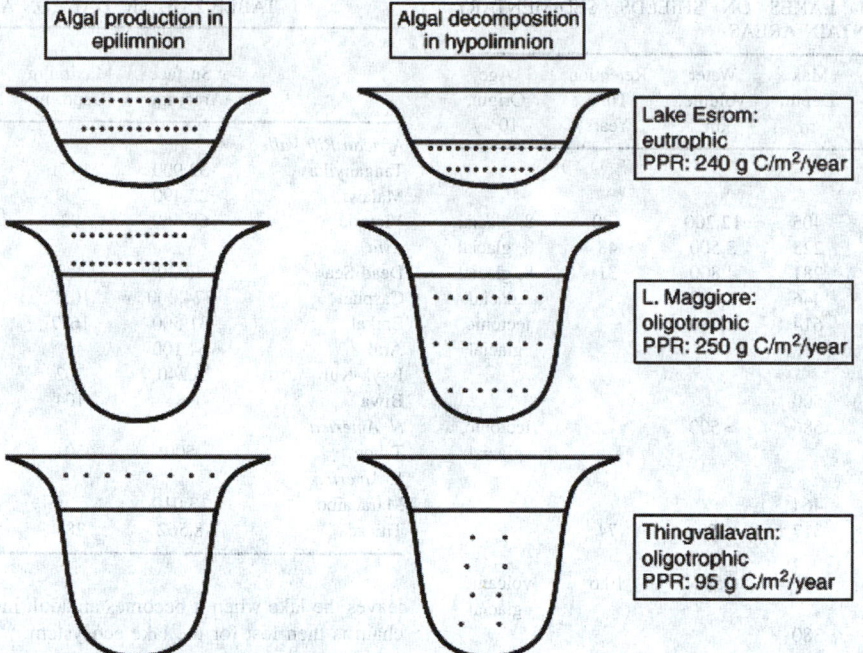

Fig. 2. The relationship between the supply and demand of oxygen in deep water is determined by the intensity of the respective decomposition rate and the volume of the hypolimnetic waterbody of a given lake. The oxygen demand with algal decomposition consumes a large percentage of the supply in the small hypolimnic waterbody of L. Esrom, it is medium in L. Maggiore with its big hypolimnion, and it is unimportant in Thingvallavatn because the mass of algae is small and the big volume of the hypolimnion has a big oxygen reserve. L. Esrom is eutrophic, L. Maggiore has the same annual primary production (PPR) of algae, but is oligo-/mesotrophic because of its morphology. Thingvallavatn is a typical oligotrophic lake because of low primary production and its depth. (*redrawn from Jónasson, P. Verh. Internat. Verein. Limnol.* **26**: *1996*).

overall transfer efficiency with its losses at any transfer is small. In the more productive systems of meso- to eutrophic lakes, however, the concentrations of organisms from algae and bacteria to zooplankton and fish are high, and additional, new types of filtering predators appear. Filtrators use a filtering system, comb-like sieves in cladocerans (*Daphnia*), fine gill structures in filter-feeding fish (Coregonids), to collect large numbers of very small prey particles. The linear size relation between prey and filter-feeding predators is $100\times$ to $1000\times$, and, thereby, the length of the food chain becomes short, three levels from algae to fish. In temperate lakes, we find *Daphnia* as the keystone filtrator on the level of herbivores all over the world. The filtering, zooplankton-feeding fish of the genus *Coregonus* are found only in the Northern Hemisphere. In the ancient Lake Tanganyika, an evolution of a "new" filtrator occurred, *Limnotrissa*, where a zooplankton-feeding sardine, led to very high fish production. In other ancient tropical lakes, herbivorous *Tilapia* species appeared that filter-feed on algae.

The food web of any ecosystem might be described on the level of species composition and populations, according to its taxonomic structure. The system, however, might be alternatively viewed with an ataxonomic approach and explained functionally as a system of trophic levels linked by energy transfers: the "trophic-dynamic system".

Another way to describe the planktonic food web and to explain some of its functions and properties is to view its size structure and the continuity of sizes as a spectrum. Investigations of the entire planktonic system revealed the plankton as a spectrum of size classes, in which the biomass of all size classes is equal. Many small organisms are present with the same sum of biomass, as are the fewer organisms of larger sizes. The regular pattern of body sizes over all organisms of plankton corresponds to growth and physiological rates in a reciprocal way, fast overturn in the pools of small organisms and slow overturn in the big ones.

Regional Lake Limnology

Glacial Lakes. The glaciated areas of the Canadian Shield, the Scandinavian mountains, and the adjacent sedimentary platforms of the circum-Baltic countries have been eroded by glaciation that left ten thousands of large and small lake basins. The large basins of the Hudson Bay and the Baltic Sea are corresponding central basins. The Baltic Sea was intermittently an inland lake and then became a brackish shelf sea. The surrounding areas from Finland to northern Germany have extended lake districts that cover large parts of the land. The "limnic ratio" is a measure of the percentage of the recent lake area of the deglaciated land area: world total deglaciated area: 6.8%; Scandinavian Shield: 12.2%; Finland: 9.4%; Norway: 13.9%; Sweden: 8.55%; South Baltic Platform: 2.2%; Canadian Shield: 10.3%; and Patagonia: 3.0%. The lakes of this origin are the largest in Europe, Lakes Ladoga and Onega in western Russia and the large southern Swedish lakes, and in North America, the Laurentian Great Lakes and the large lakes of Canada (Table 1).

Formerly glaciated mountain districts are found on all continents (Table 1). The fjord lakes and foothill lakes in the deglaciated mountain areas are found in the Northern and the Southern Hemispheres of the temperate climatic regions in Scandinavia, the loghs/lochs of Ireland and Scotland, the English Lake District, the Alps, New Zealand, the Patagonian mountain areas of Chile and Argentina, and the Central and Northern Asian mountains. *Fjord lakes* are the deepest glacial lakes. Their mountain valley erosion basins have peculiarly steep borders and flat bottoms with depths of 500 m (1,640 ft) or more. Where the glacial erosions from valleys extend into the foothill areas of the foreland, the characteristic *piedmont lakes* cover large areas, but they are shallow compared with fjord lakes or fjord-like parts of the same lake (e.g. L. Garda, L. Argentino, L. Buenos Aires, show both, deep fjord-like parts and large, shallow foothill areas).

Lakes in the (sub-) polar regions are covered by ice permanently or during the largest part of the year. Lake Vanda (Antarctica) has a permanent, clear, ice cover. The deepest water layers are stabilized by saline inputs and trap the down-welling light energy, leading to a temperature of 25°C (77°F) above the bottom (63 m (207 ft)). Some large lakes in the Antarctic were found by satellite radar imaging under several thousand meters of glacial ice cover. The largest of these lakes, L. Vostok, is the size of Lake Ontario.

TABLE 1. LARGE GLACIAL LAKES ON SHIELDS, SEDIMENTARY PLATFORMS, AND IN MOUNTAIN AREAS

Glacial Lakes	Surface Area, km^2	Max. Depth, m	Water Volume, km^3	Retention Time Years	Age Origin, 10^3 y
Canadian					
Shield lakes					
Superior	82,100	405	12,200	180	8, glacial
Huron	59,500	223	3,500	48	8, glacial
Michigan	57,750	281	4,800	31	8, glacial
Great Bear	31,326	446	2,200		glacial
Great Slave	28,568	614	2,088		tectonic, glacial
Andean lakes					
Argentino	1,410	500			
Gral. Carrera, Chile/Buenos Aires,Arg.	1,892	586	>500		tectonic, glacial
Nahuel Huapi, Arg.		464			
Llanquihue, Chile	871	317	159	74	
New Zealand					
Taupo	616	165	59	10.6	volcanic, glacial
Waikitipu	289	380			
Wanaka	180	311			
Baltic lakes					
Ladoga, Russia	18,130		908		>8, baltic
Onega, Russia	9,700		292		>8, baltic
Vänern, Sweden	5,648		152	8.8	6.5, baltic
Vättern, Sweden	1,912	128	74	58	6.5, baltic
Mälaren, Sweden	1,096		14	2.3	6.5, baltic
Alpine piedmont lakes					
Geneva, W.-Alps	580	310	89.9		piedmont
Constance, N.-Alps	472	253	47.7	4.15	piedmont
Garda, S.-Alps	370	346	49	26.6	piedmont
Maggiore, S.-Alps	213	370	37.5	4.1	piedmont
Como, S.-Alps	146	410	22.5	4.5	piedmont

TABLE 2. BASIC DATA ON ANCIENT LAKES

	Surface Area, km^2	Maximum Depth m	Water Volume, km^3	Age & Origin, Years
African Rift Valley				
Tanganyika	32,900	1471	18,900	20 mio, tectonic
Malawi	22,490	709	6,140	2 mio, tectonic
Victoria	68,460	92	2,700	20,000, tectonic
Asia				
Dead Sea	1,020	433	188	tectonic
Caspian	374,000	1025	78,200	>5 mio, tectonic
Baikal	31,500	1637	23,000	20 mio, tectonic
Aral	64,100		1,020	>5 mio, tectonic
Issyk-Kul	6,240	702	1,730	25 mio, tectonic
Biwa	681	104	27.6	2 mio, tectonic
N. America				
Tahoe	500	501	156	2 mio, tectonic
S. America				
Maracaibo	13,010		280	36 mio, tectonic
Titicaca	8,562	284	903	2.8 mio, tectonic

leaves the lake when it becomes an adult insect, so the branch of the food chain is then lost for the lake ecosystem.

Man-Made Lakes: Reservoirs and Pit Lakes

Man-made lakes numbers are built in great numbers, ten thousands of reservoirs and pit lakes in the voids of surface mining. They show limnological peculiarities that come primarily from the management regime for reservoirs and from geogenic impacts on water chemistry for pit lakes in coal and ore mining districts. The state of reservoir building and the problems are presented by the World Commission on Dams http://www.dams.org.

Regional Programs for Protecting Lakes and Collecting Data

The protection of lakes and other surface waters is different regionally. The Environmental Protection Agency (EPA) http://www.epa.gov/, has set standards in North America, and the European Union in the EU countries http://www.eurunion.org/. The regulations of the EU-Water Framework Directive have initiated many activities to register, describe, and improve surface waters and their state in Europe (see documents of the European Environmental Agency (EEA) http://www.eea.eu.int/main_html, and EU-Commission http://europa.eu.int/comm/index_en.htm).

See also **Earth**; and **Limnology**.

This article originally appeared in the *Water Encyclopedia*, 5 Vols. Lehr, J.H., and J. Keeley, Editors, John Wiley & Sons, Inc, Hoboken, NJ.

Tropical and Ancient Lakes

The lakes of the (sub-)tropical climatic zone are described in textbooks by Beadle and Serruya. The most important difference from temperate lakes is the diurnal variability of temperatures that are as large as the seasonal variability between summer and winter in temperate lakes. The restricted time of nocturnal holomixis presumably limits the maximum depth of mixis in deep tropical lakes. Therefore, the very deep lakes Tanganyika and Malawi are meromictic, and, consequently, have anoxic, deep waterbodies. The deep African Rift Valley Lakes are ancient lakes and, therefore, have peculiarities beyond their tropical sites.

The largest and deepest lakes of the world are also the oldest lakes: Baikal, Tanganyika, Malawi, and Issik-Kul. The most important ancient lakes are listed in Table 2. The old lakes originate from the Tertiary and have a long development with changing water levels that could be reconstructed by paleolimnological investigations. The best investigated ancient lake is L. Biwa, whose history is based on sediment cores that are 200 and 1,000 m (656 and 3,281 ft) long, covering 0.5 to 1 million years. The oldest and deepest of the ancient lakes gave time to their biota for evolutionary processes which led to adaptive radiations of — mostly endemic — species, found 10 to 100× more numerously than in young lakes. There are >400 species of Cichlid fish species in Tanganyika and Malawi and >200 species of Gammarid crustacean species in Baikal. The ecosystem structure is severely changed by new keystone species in the food web and new top carnivores: in Lake Baikal the pelagic Gammarid *Macrohectopus* and Cottoidei fish and the freshwater seal *Phoca sibirica*. In Lake Tanganyika, the species *Limnotrissa* directs the food chain to high production of fish, whereas in the other large African lakes the insect larva *Chaoborus* plays the role of the main zooplankton feeder. This species

Additional Reading

Allan, R.J,, M. Dickman, C.B. Gray, and V. Cromie: *The Book of Canadian Lakes*, Monograph Series No. 3. The Canadian Association on Water Quality, Gloucester, Ontario. 1994.

Awange, J.L., and O. Ong'ang'a: *Lake Victoria: Ecology, Resources, Environment*, Springer-Verlag New York, LLC, New York, NY. 2006.

Beadle, L.C.: *The Inland Waters of Tropical Africa: Introduction to Tropical Limnology*, Longman, London, UK. 1981.

Day, T.: *Lakes and Rivers*, Facts on File, Inc., New York, NY. 2006.

Hammer, U.T.: *Saline Lake Ecosystems of the World*, Junk, Dordrecht, The Netherlands. 1986.

Herdendorf, C.E.: In: *Large Lakes, Ecological Structure and Function*, M.M. Tilzer, and C. Serruya (Eds.), Springer, Berlin, pp. 3–38. 1990.

Hutchinson, G.E.: *A Treatise on Limnology 1: Geography, Physics, and Chemistry*, John Wiley & Sons, Inc., New York, NY. 1957.

Hutchinson, G.E.: *A Treatise on Limnology 2: Introduction to Lake Biology and the Limnoplankton*, John Wiley & Sons, Inc., New York, NY. 1967.

Hutchinson, G.E.: *A Treatise on Limnology 3: Limnological Botany*, John Wiley & Sons, Inc., New York, NY, 1975.

Hutchinson, G.E.: *A Treatise on Limnology 4: The Zoobenthos*, John Wiley & Sons, Inc., New York, NY, 1993.

Johnson, T.C. and E.O. Odada: *The Limnology, Climatology and Paleoclimatology of the East African Lakes*, OPA, Amsterdam. 1996.

Kozhova, O.M. and L.R. Izmest'eva: *Lake Baikal*, 2nd Edition, Biology of Inland Waters, Backhuys, Leiden, German. 1998.

Le Cren, E.D. and R.H. Lowe-McConnell: *The Functioning of Freshwater Ecosystems*, International Biological Programme, 22, Cambridge University Press, Cambridge, UK, 1980.

Maitland, P.S., P.J. Boon, and D.S. McLusky: *The Freshwaters of Scotland*, John Wiley & Sons, Inc., New York, NY, 2002.

Meybeck, M.: *Les Lacs et Leur Bassins*. In: R. Pourriot and M. Meybeck, Eds. Limnologie Générale, Masson, Paris, 1995, pp. 6–59.

Meybeck, M.: In: *Physics and Chemistry of Lakes*, A. Lerman, D.M. Imboden, and J.R. Gat, Springer, Berlin, 1995, pp. 1–36.

Moriarty, Ch.: *Studies of Irish Rivers and Lakes*, Marine Institute, Dublin, Ireland, 2002.

Munawar, M. and R.E. Hecky: *The Great Lakes of the World (GLOW) Food-Web, Health and Integrity*, Backhuys, Leiden, Germany, 2001.

Odum, E.P., and G.W. Barrett: *Fundamentals of Ecology*, 5th Edition, Brooks/Cole, New York, NY, 2004.

Parsons, T.R., M. Takahasi, and B. Hargrave: *Biological Oceanographic Processes*, 3rd Edition, Butterworth-Heinemann, Woburn, MA, 1984.

Robertson, A. and D. Scavia: In: *Lakes and Reservoirs: Ecosystems of the World*, F.B. Taub, and D.W. Goodall (Eds.). Elsevier Science, New York, NY, 1984, pp. 135–176.

Serruya, C. and U. Pollinger: *Lakes of the Warm Belt*, Cambridge University Press, New York, NY, 1983.

Stumm, W. and J.L. Schnoor: in A. Lerman, D.M. Imboden, and J.R. Gat, Eds., *Physics and Chemistry of Lakes*, Springer, Berlin. pp. 185–215. 1995.

Stumm, W. and J.J. Morgan: *Aquatic Chemistry*, 3rd Edition, John Wiley & Sons, Inc., New York, NY, 1995.

Taub, F.B.: *Lakes and Reservoirs: Ecosystems of the World*, Elsevier Science, New York, NY, 1984.

Tilzer, M.M. and C. Serruya: *Large Lakes—Ecological Structure and Function*, Springer, Berlin, 1990.

Wetzel, R.G.: *Limnology—Lake and River Ecosystems*, 3rd Edition, Academic Press, San Diego, CA, 2001.

WALTER GELLER, UFZ-Centre for Environmental Research, Magdeburg, Germany

LAKES (Colors).   See **Colorants (Foods)**; and **Dyes (Textile)**.

LAMARCK, JEAN-BAPTISTE PIERRE ANTOINE DE MONET DE (1744–1829).

Jean-Baptiste Lamarck was a French botanist, zoologist and natural historian; early exponent of transformism, of biology as the science of living bodies, and a holistic view of nature.

Lamarck came from a minor noble family in Picardy and was educated by Jesuits at Amiens. In his youth he essayed soldiering, but, injured in a prank, he moved to Paris in 1769 where he became a disciple and protégé of Comte Buffon, Director of the Jardin des Plantes. Ten years later, Lamarck published *Flore Françoise* and with Buffon's patronage was elected to the Académie des Sciences. After the French Revolution his interests moved more towards animals and in 1793 he obtained a chair at the Muséum d'Histoire Naturelle, with responsibility for 'insects and worms' (*viz.* invertebrates). See also **Buffon, Georges Louis (1707–1788)**.

As the title of his major work—*Philosophie Zoologique* (1809)—suggests, Lamarck was concerned to develop a philosophy of the natural world, as well as studying the specifics of anatomy or taxonomy. His thinking was wide-ranging, embracing chemistry, mineralogy, geology andmeteorology, as well as botany and zoology. He was concerned with the nature of life, and coined the term "biology" to refer to the study of livingbodies generally. See also **Taxonomy (The History)**.

Like Buffon, Lamarck was skeptical of artificial taxonomic systems such as Linnaeus's, with their discrete "pigeon-holes." And, unlike Linnaeus, he saw plants and animals as having significant analogies, so that, for example, in his contribution to the *Encyclopédie Méthodique* (Vol. 2, 1786) both were given six main subdivisions, represented in two parallel columns. About 1798–1799, Lamarck was faced with the problem of fossils and their differences from known living forms. Rejecting the idea of extinction, he proposed (1) that, due to the Earth's rotation, the eastern sides of land masses were continually being eroded while sediments were deposited on their western margins; (2) the hypothesis of "transformism" (equivalent to what is today termed evolution); and (3) a great agefor the Earth, allowing the slow geological and evolutionary changes. The ideas were published in *Hydrogéologie* (1802) and *Recherches sur l'Organisation des Corps Vivans* (1802). See also **Linnaeus, Carl Linné) (1707–1778)**; and **Paleontology (The History)**.

Lamarck's transformism depended on the activity of living matter in relation to changing environmental circumstances. He supposed that simple life forms were generated spontaneously and through time more complex ones evolved. In his *Histoire Naturelle des Animaux sans Vertèbres* (1815–1822), Lamarck summarized the essential features of the processes involved in the form of four laws. First, life, by its internal forces, tends to increase the size of organisms, up to a limit (as, for example, when a tree or a baby grows.) Second, new animal organs are produced in response to new *besoins* or *needs* (and new *habitudes* or *habits* are adopted in response to changing circumstances). Third, the development of organs and their power of action are proportional to their use. Fourth, changes acquired during organisms' lifetimes are transmitted to their offspring. (This was the notorious principle of 'inheritance of acquired characteristics'.) Given these assumptions, transformism followed. The theory could satisfactorily explain adaptations such as the long legs of wading birds. See also **Evolution (History)**.

By internal forces, Lamarck meant those of life itself, which supposedly operated through "subtle fluids" such as caloric, electrical fluid, or nervous fluid. These were thought to be active in used organs and caused them to expand—as muscles grow when exercised. Conversely, disuse over generations led to atrophy of organs (such as snakes' legs). Lamarck also proposed the notion of *sentiment intérieur* (*inner feeling* or *sensation*), found in more advanced animals; and humans had the power of thought. For simple animals, evolutionary changes occurred inresponse to the interactions of living bodies with their environments. Forhigher forms, *inner feelings* were involved, and humans could slowly change through will, in accordance with their changing needs.

Lamarck's scheme gave a basis for classification, in that later evolved types were naturally separable from the earlier ones. He also had a kind of psychological theory. When attention is focused on something—the *sentiment intérieur* being stimulated by some need or interest—nervous fluid flows to the part of the brain to which the sensation of the object is related; and then the animal has an idea of the object.

Lamarck was a radical and original thinker, his politics being influenced by Rousseau, and his psychology by the so-called *idéologues*. He thought of *nature* as having (or being) a quasi-autonomous power, producing bodies according to natural laws. God supposedly created matter, the Earth, and *nature* by will; but God was entirely separate from its creation—a deistic conception.

Lamarck's late career was dogged by poverty; and he became blind. His views were unwelcome to his *Muséum* colleague, Georges Cuvier, who in his Academy *éloge* misrepresented Lamarck's theory, suggesting that he supposed animals evolved by their own acts of will. (The ambiguity in Lamarck's term *besoins*—meaning either *needs* or *desires*—made this possible.) Cuvier thought Lamarck's theory was unscientific and speculative: it transcended his specialized knowledge and was based on outmoded physical theory, including the old four-element theory of earth, water, air and fire. Further, it treated topics like the nature of the human mind that were deemed beyond the scope of science. More specifically, Lamarck and Cuvier differed on the question of the age of the Earth, the empirical evidence thought to pertain to transformism (such as Egyptian mummies), and the continuity or discontinuity of animal groups. See also **Cuvier, Georges Léopold Chrétien Frédéric Dagobert Baron de (1769–1832)**.

So Lamarck's reputation was in eclipse at the time of his death. However, there have been many latter-day Lamarckians, and Darwin himself utilized the inheritance of acquired characteristics, the truth or falsity of which principle has been actively debated and subjected to tests in numerous experiments. Though most biologists reject the notion today, a few pursue investigations claimed to exhibit "Lamarckian" phenomena.

Lamarck is also remembered for ideas akin to the modern notion of the biosphere, with his recognition of the interchanges of matter between living and non-living things. Indeed, his ideas are sometimes regarded as forerunners of the Gaia hypothesis, which envisages the Earth and the biosphere as a quasi-living entity. Certainly human culture is "Lamarckian" in that acquired characteristics (artifacts and knowledge) can be passed from onegeneration to the next—which makes human evolution quite different from that of all other species.

Additional Reading

Burkhardt, R.W.: *The Spirit of System: Lamarck and Evolutionary Biology*, Harvard University Press, Cambridge, MA, 1995.

Corsi, P.: *The Age of Lamarck: Evolutionary Theories in France 1790–1830*, University of California Press, Berkeley, CA, 1989.

DAVID OLDROYD, The University of New South Wales,
Sydney, New South Wales, Australia

LAMBDA PARTICLE. A hyperon with a rest-mass energy of 1115.6 MeV, an isospin quantum number zero, an angular momentum spin quantum number $\frac{1}{2}$, and a strangeness quantum number 1. Symbol, λ.

LAMBERT. See **Units and Standards**.

LAMBERT PROJECTION. The Lambert modified conformal conic projection (commonly known as the Lambert, and sometimes as the Gauss conformal) has been used for many years in the construction of maps. During the twentieth century, this projection gained favor rapidly for use in constructing charts for air navigators.

The projection is actually a mathematical type, but may be quite accurately described as a conical projection. It differs from the simple conical in that the cone is not tangent to the earth's surface, but cuts it on two latitude parallels known as standard parallels. This type of projection is particularly valuable for portraying large longitudinal areas, e.g., the entire United States. The graticule of the Lambert chart shows parallels of latitude as concentric circles centered at the nearer pole, and the meridians of longitude as straight lines converging on this pole. From simple geometric considerations, it is obvious that the meridians and parallels must be perpendicular. The angle of convergence of the meridians, and therefore the radii of the parallels, depends upon the distance of the nearer pole from the center of the area.

The great advantage of the Lambert projection is that the scale of distance is uniform, for all practical purposes, all over the chart. To indicate the accuracy of this statement, consider the Lambert projection on which the series of aeronautical charts of the United States are constructed by the Coast and Geodetic Survey. The standard parallels for these charts are N 45° and N 33°. The scale of distance, if considered as unity on the standard parallels, is 0.994 at the central parallel, expands to 1.010 at the extreme north boundary of the United States, and to 1.023 at the tip of Florida. Comparing the Lambert with the mercator distance scales, we find that if we consider the mercator scale as unity at N 39°, the scale at the northern limits of the United States is 1.154, whereas, at the tip of Florida, it is 0.846.

The nonorthogonal graticule of the Lambert chart is a distinct disadvantage for general navigational problems, since neither the rhumb line nor the great circle is straight. However, the uniformity of scale is of such great advantage that air navigators have begun to use this type of chart even in preference to the mercator, particularly when navigating in good visibility over land, or where good radio aids are available.

A straight line between two points on a Lambert chart is referred to as a Lambert line. Since the meridians on the graticule are convergent, this will not be a rhumb line. However, the distance measured along this line will be less than that along the rhumb line, and only slightly greater than the great-circle distance between the two points. This line is frequently used by aviators during conditions of good visibility. The standard procedure in using the Lambert line is to draw a straight line on the chart and pick out conspicuous landmarks separated by about 25 miles. The rhumb line indicated by measurement of the angle between the Lambert line and the meridian nearest the starting point is then followed. This will lead to a point at some distance from the first landmark but within visibility. From this first landmark, a new heading is adopted for the second, and so on to destination. If a rhumb line is desired for the entire route, the Lambert line is drawn as before, and the course is measured from the meridian halfway between the point just left behind and the point to be arrived at. This rhumb line will appear as a curve on the Lambert chart, but can be laid down by any one of a number of standard methods with sufficient accuracy for the selection of landmarks.

The convergence of the meridians prevents the use of the Lambert chart for graphical solution of dead-reckoning problems, and introduces difficulties in plotting lines of position obtained either by radio bearings or by celestial observation. The ease with which such problems can be solved on the mercator chart seems to cast doubt on any statement to the effect that "the Lambert Chart will completely supersede the mercator for all navigational purposes."

See also **Course**; **Great-Circle Course**; **Line of Position**; **Mercator Sailing**; **Navigation**; and **Rhumb Line**.

LAMBERT'S COSINE LAW. The intensity from a surface element of a perfectly diffuse radiator is proportional to the cosine of the angle between the direction of emission and the normal to the surface. An element of a surface that obeys this law will appear equally bright when observed from any direction.

LAMB SHIFT. The displacement between the $2S_{1/2}$ and $2P_{1/2}$ levels of hydrogen, which in the absence of radiative corrections would be zero due to the Coulomb degeneracy. The experimental value obtained by Lamb and Rutherford

$$E_{2S_{1/2}} - E_{2P_{1/2}} = 1057.8 \pm 0.1 \text{ megahertz}$$

is in agreement with the theoretical value. Of this, 27 MHz arises from vacuum polarization, the rest from self-energy corrections. The term is now used to indicate the displacement of any bound state level due to radiative corrections. See also **Field Theory**.

LAMÉ EQUATION (Generalized). The most general second-order (linear) differential equation with five regular singularities, one of them being the point at infinity, with preassigned exponents differing from each other by $\frac{1}{2}$ at each singularity, all other points of the complex plane being ordinary points. This equation is remarkable because of the large number of important equations (Legendre, Bessel, etc.) obtainable from it by confluence. Letting a_1, a_2, a_3, a_4 and ∞ be the singular points, with exponents α_1, $\alpha_1 + \frac{1}{2}$, $\cdots \alpha_4$, $\alpha_4 + \frac{1}{2}$, μ_1, $\mu_1 + \frac{1}{2}$, the equation has the form

$$\frac{d^2w}{dz^2} + P\frac{dw}{dz} + Qw = 0$$

where

$$P = \sum_{i=1}^{4} \frac{\frac{1}{2} - 2\alpha_i}{z - a_i},$$

$$Q = \sum \frac{\alpha_i\left(\alpha_i + \frac{1}{2}\right)}{(z - a_i)^2} + \frac{Az^2 + 2Bz + C}{(z - a_1)\cdots(z - a_4)}.$$

A is expressible in terms of the α_i, and B, C are arbitrary constants. For example if the confluence takes the form $a_1 = a_2 = 0$, $a_3 = a_4 = \infty$, then choosing all $\alpha_i = 0$ and setting $z = \zeta^2$, with proper choice of B and C, we get

$$\zeta^2\frac{d^2w}{d\zeta^2} + \zeta\frac{dw}{d\zeta} + (\zeta^2 - n^2)w = 0,$$

which is Bessel's equation.

LAMELLA (Botany). The middle lamella is the compound layer composed of the primary walls and the cement-like intercellular substance that occurs between the primary walls of two cells. This composite layer is usually made up of pectic materials (colloids which have a great affinity for water), one of which is calcium pectate. The function of the middle lamella is to hold adjoining cells together. Sometimes, particularly in mature fruits, the middle lamella substance breaks down. As a result, the cells of the fruit separate easily, giving to the fruit a meal-like character.

The middle lamella is the common source of pectin, which is added to concentrated fruit juices in the preparation of jellies. The term lamella is also applied to each of the concentric growth layers in large starch grains.

LAMELLAE. See **Bone**.

LAMELLA (Zoology). 1. A thin leaf or plate, such as a lamella of bone. See also **Bone**.

2. A flat plate formed by the fusion of ctenidial filaments in the bivalve mollusks. Two lamellae united by bridges of tissue form a gill through which water circulates under the influence of ciliary action in the persisting open spaces. This form of gill is the source of the name *Lamellibranchiata* applied to the class containing these animals.

LAMELLIBRANCHIATA.   The bivalve mollusks, a class of the phylum *Mollusca* including the clams, mussels, oysters, scallops and related species. Many of these animals are valuable for food, and they produce pearls and mother-of-pearl. The class is also named *Pelecypoda*.

Bivalve mollusks differ from other members of the phylum in the following characters: (1) The body is bilaterally symmetrical and transversely compressed. (2) The mantle forms two lobes extending down along the sides of the body. In most species these lobes unite at the posterior end to form two passages, an upper excurrent and a lower incurrent siphon. Currents of water carry food and oxygen into the mantle cavity through the lower opening and a current bearing wastes passes out of the upper. (3) Each mantle fold secretes a valve of the shell formed of calcareous matter covered outside by a horny periostracum and inside by nacre, commonly called mother-of-pearl. The two valves of the shell are joined by a hinge ligament and the articulation is strengthened in some species by interlocking teeth. They are closed ventrally by the contraction of one or two adductor muscles. (4) The gills are thin plates on each side of the body in most species. They are formed of united ctenidial filaments. (5) The foot is a muscular wedge-shaped protuberance at the anterior end of the body. The head is rudimentary.

All bivalves are aquatic. Most species creep slowly by thrusting the foot into the muddy or sandy bottom but some propel themselves by jets of water squirted from the siphons or forced from the mantle cavity by rapidly closing the valves. The species vary from freshwater forms about $\frac{1}{8}$ inch long to giant marine shells more than a yard long.

The class is divided into four orders:

Order *Protobranchiata*. Gills in the form of small leaflets, two rows on each side of the body. Marine species.

Order *Filibranchiata*. Marine mussels, scallops, etc. Gills composed of filaments united only by ciliary junctions.

Order *Eulamellibranchiata*. Gill filaments united to form continuous plates. Freshwater clams or mussels, marine clams, oysters, shipworms, etc.

Order *Septibranchiata*. Gills replaced by a horizontal partition between the upper and lower divisions of the mantel chamber. A few marine species.

See also **Clam**; **Mollusca**; **Mussel**; and **Pearl**.

LA METTRIE, JULIEN OFFRAY DE (1709–1751).   La Mettrie, the son of a textile merchant, studied medicine at the University of Paris before gaining his medical degree from Rheims (1733). He spent a further year under the tutelage of the famous medical teacher Hermann Boerhaave at Leiden (1734). La Mettrie practiced medicine in the Saint-Malo district until 1742; he then moved to Paris, but failed to settle. From 1743 to 1746, he served with the army during the War of Austrian Succession, rising to the post of medical inspector of armies in the field. His scathing attacks on the medical profession and his atheistic medical philosophy prompted a self-imposed exile in Holland from 1747. After publication of his best-known work *L'homme machine* in 1748, he moved to Prussia. At the court of Frederick II he was appointed a physician to the King and made a member of the Royal Academy of Sciences. See also **Boerhaave, Herman (1668–1738)**.

Boerhaave's mechanistic medicine inspired La Mettrie, who published a series of annotated translations of his mentor's work from 1735 onwards. The translations are rather more significant than his own medical writings, which tended to be abstract and theoretical. La Mettrie's medical philosophy — man as a machine — in part a combination of Boerhaavian mechanism and Descartes' automatism, was based upon the assumption that only knowledge learned through the medical sciences was reliable. La Mettrie marshaled various kinds of evidence in support of his position including Albrecht von Haller's demonstration of irritability in muscles, reflex action, the reliance of psychological states on internal and external physical factors such as hunger and climate. His overtly anti-religious stance compelled him to argue that mental processes were not derived from the soul but from the functioning of the cerebral and neural part of the body, making him an early exponent of physiological psychology. See also **Haller, Albrecht von (1708–1777)**.

Additional Reading

Gillispie, C.C.: *Dictionary of Scientific Biography*, Charles Scribner's Sons, New York, NY, 1972.

Wellman, K.: *La Mettrie: Medicine, Philosophy, and Enlightenment*, Duke University Press, Durham, NC, 1992.

H.J. POWER, Wellcome Trust Centre for the History of Medicine at UCL, London, UK

LAMINAR BOUNDARY LAYER.   An interfacial region in which flow is smooth and nonturbulent. Above a surface, a laminar layer will develop and fluid velocity will increase with distance from the surface, but not indefinitely. At some point, flow will become turbulent, with the laminar sublayer separating the turbulent layer from the surface. In the real world, most laminar boundary layers are extremely thin (order of 1 mm 0.04 in), but can be of biological importance, for example, next to plant leaves or as invertebrate refuges in streams.

LAMINAR FLOW.   A condition of fluid flow in a closed conduit in which the fluid particles or "streams" tend to move parallel to the flow axis and not mix. This behavior is characteristic of low flow rates and high viscosity fluid flows. As the flow rate increases (or viscosity significantly decreases), the streams continue to flow parallel until a velocity is reached where the streams waver and suddenly break into a diffused pattern. This point is called the *critical velocity*.

Laminar flow is characterized by a parabolic flow profile where the maximum velocity at or near the center of the conduit is approximately twice the average velocity in the profile. Laminar flow often is referred to as *viscous flow*, streamline flows, and *low-Reynolds number flow*. Special attention must be paid to the constancy of coefficient of most flowmeters in the region of laminar flow.

Laminar Sublayer. When a fluid is in turbulent flow past a rigid surface, fluctuations of velocity in the direction normal to the surface are inhibited, and very close to the surface they may be negligible. Then the Reynolds shear stress is small compared with the viscous stresses, and it has been common to describe the region as a laminar sublayer. In fact, turbulent fluctuations of velocity in planes parallel to the wall are considerable in comparison with the mean velocity.

Viscosity and Ostwald's Experiment

Viscosity is a measure of how easily the fluid deforms or flows under an applied stress. Alternatively, viscosity is a measure of the fluid's internal resistance to deformation or internal friction between the fluid layers. Experimentally, under conditions of laminar flow, the force required to move a plate at constant speed against the resistance of a fluid is proportional to the area of the plate and to the velocity gradient perpendicular to the plate. The constant of proportionality is called the viscosity. Consider a simple fluid flow between the two plates shown in Fig. 1.

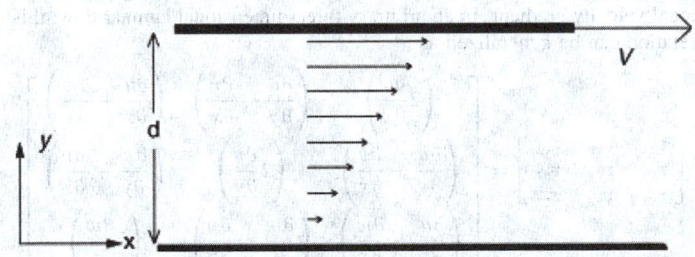

Fig. 1.   Viscous drag between two plates; the bottom plate is held fixed.

If the force per unit area of the top plate is measured as F, we can write that as

$$\frac{F}{A} = \mu \frac{v}{d} \tag{1}$$

with d as the separation between the plates, which essentially defines the absolute or dynamic viscosity, μ, which has the units of kg/ms. The ratio of dynamic viscosity to density of the fluid is known as kinematic viscosity and has the units of m^2/s. Experimentally, the viscosity can be measured using specially designed viscometers, whose operation is based on simple flow between one plate fixed and another moving with constant velocity v. If the gap d between the plates is small compared with the length of the plates, the flow with reasonable accuracy can be assumed to be one-dimensional. For such flows, the velocity is constant in the direction of flow but varies in the normal direction. For example, in Fig. 1, the velocity in x-direction is a function of y coordinate, i.e., $v = v(y)$. Equation 1 also represents that the viscous stresses are directly proportional to the velocity gradient, i.e.,

$$\textit{Viscous stresses} = \mu(\textit{Rate of deformation})$$

These stresses developing in the viscous transfer can be divided into normal and shear stresses. For example, τ_{xx} is the normal stress, whereas τ_{yx} is the shear stress resulting from the viscous forces. At each point in a continuous medium, whether it is solid or fluid, we need six components, each of them representing a component of force per unit area, to define the local stress completely. The local stresses acting in the x-y plane have been shown in Fig. 2.

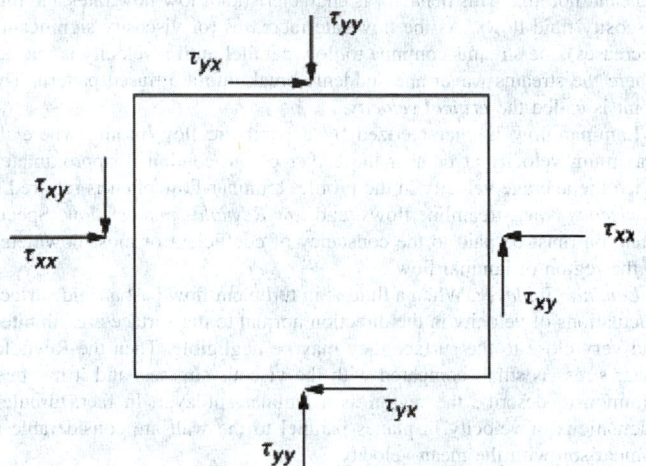

Fig. 2.   Local stresses acting in a x-y plane on a cubic element.

Equation (1) can be cast in a more generalized form as given by Equation (2). The shear stress exerted in the x-direction on a fluid surface of constant y by the fluid in the region of lesser y is designated as τ_{yx}, and the x-component of the fluid velocity vector is designated as v_x

$$\tau_{yx} = -\mu \frac{du}{dy} \tag{2}$$

This law is commonly known as Newton's law of viscosity. It states that the shear force per unit area is proportional to the negative of the local velocity gradient. In an arbitrary three-dimensional laminar flow, this relation can be generalized as

$$
\begin{bmatrix}
\tau_{xx} & \tau_{xy} & \tau_{xx} \\
\tau_{xy} & \tau_{yy} & \tau_{yz} \\
\tau_{xz} & \tau_{yz} & \tau_{zz}
\end{bmatrix}
=
\begin{bmatrix}
\left(2\dfrac{\partial u}{\partial x}\right) & \left(\dfrac{\partial u}{\partial y}+\dfrac{\partial v}{\partial x}\right) & \left(\dfrac{\partial u}{\partial z}+\dfrac{\partial w}{\partial x}\right) \\
\left(\dfrac{\partial u}{\partial y}+\dfrac{\partial v}{\partial x}\right) & \left(2\dfrac{\partial v}{\partial y}\right) & \left(\dfrac{\partial w}{\partial y}+\dfrac{\partial v}{\partial z}\right) \\
\left(\dfrac{\partial u}{\partial z}+\dfrac{\partial w}{\partial x}\right) & \left(\dfrac{\partial w}{\partial y}+\dfrac{\partial v}{\partial z}\right) & \left(2\dfrac{\partial w}{\partial z}\right)
\end{bmatrix}
\tag{3}
$$

Flows in which fluid viscosity is important can be of two types, namely laminar and turbulent. The basic difference between these flows was demonstrated by Osborne Reynolds in 1883. A thin stream of dye was introduced into the flow of water through the tube. At low flow rates, the dye stream was observed to follow a well-defined straight path, indicating that the fluid moves in parallel layers (*laminas*) with no macroscopic mixing across the layers, which is known as *laminar flow*. As the flow rate was increased beyond certain critical values, the dye streak broke up into an irregular motion and spread throughout the cross section of the tube, indicating the macroscopic mixing motions perpendicular to the direction

of flow. Such a chaotic fluid motion is known as *turbulent flow*. See also **Turbulent Flow**.

Reynolds Number

The Reynolds Number (Re) is a quantity that engineers use to estimate if a fluid flow is laminar or turbulent. Consider the flow of an incompressible fluid through a circular pipe as shown in Fig. 3.

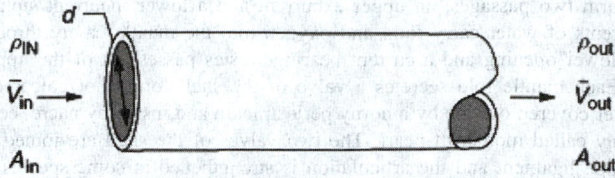

Fig. 3.   Flow through a pipe.

The variables associated with this pipe-flow configuration are the dimensions of the pipe, the rate of flow, and the physical properties of the fluid.

d = Internal diameter of the pipe

ρ = Density of the fluid and in an incompressible flow

$\rho_{\text{in}} = \rho_{\text{out}}$ because it is constant

$\bar{v}$ = Average velocity

Q = Volumetric flow rate

A = Cross-sectional area and is constant, i.e., $(A_{\text{in}} = A_{\text{out}})$

We can then represent the Reynolds Number as a dimensionless entity as

$$\text{Re} = \frac{\rho \bar{v} d}{\mu} = \frac{\rho \left(\dfrac{Q}{A}\right) d}{\mu} = \text{Unitless Number}$$

The Reynolds Number also indicates the relative significance of the inertial forces compared with the viscous forces. Laminar flow occurs at low Reynolds Numbers, where viscous forces are dominant, and is characterized by smooth, constant fluid motion, whereas turbulent flow occurs at high Reynolds Numbers and is dominated by inertial forces, producing random eddies, vortices, and other flow fluctuations. Different ranges of Reynolds Numbers determine the nature of different flow regimes, as shown in Fig. 4.

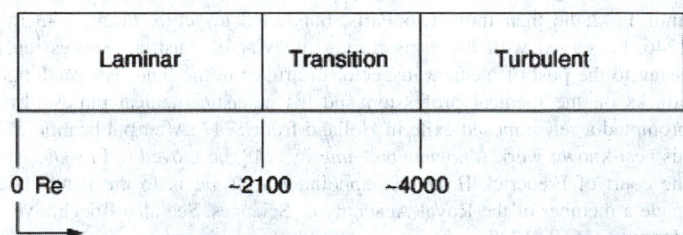

Fig. 4.   Different flow regimes and Reynolds Number in pipe flow.

In laminar flow, the particles in the fluid follow streamlines (see Fig. 4), and the motion of particles in the fluid is predictable. If the flow rate is very large, or if objects obstruct the flow, the fluid starts to swirl in an erratic motion. No longer can one predict the exact path a particle on the fluid will follow. This region of constantly changing flow lines (see Fig. 4) is said to consist of turbulent flow.

Range 1:laminar flow (see Fig. 4). Generally, a fluid flow is laminar from Re = 0 to some critical value Re = 2100, at which transition flow begins.

Range 2:transition flow (see Fig. 4). Flows in this range may fluctuate between laminar and turbulent flow. The fluid flow is on the verge of becoming turbulent. In this zone, 2100 < Re < 4000.

*Range 3:*turbulent flow (see Fig. 4). The fluid flow has become unstable. In turbulent flow, increased mixing exists that results in viscous losses, which are generally much higher than in those in laminar flow. In this zone, the values of Reynolds Number is greater than 4000.

It should be noted that the critical Reynolds Number depends on the flow type and the definition of the Reynolds Number, which varies for different geometries. See also **Reynolds Number**.

Conservation Laws of Fluid Flow Systems

The behavior or dynamics of any fluid flow system can be described by laws of conservation of mass, momentum, and energy. In order to yield the detailed descriptions of variables of interest within the flow system, the control volume must be of infinitesimal dimensions that can shrink to zero, yielding a point volume. This approach reduces the quantities to point variables. The application of the conservation principles to this infinitesimal system produces microscopic or differential conservation equations. The governing momentum, energy, and mass balance differential equations for a given flow system can be derived using the shell balance concepts or directly from equations of change, which have been discussed in detail in Bird, et. al., and Hanspal, et. al. References. Shell balance concepts are generally valid for simple flow systems, whereas equations of change can describe the flow behavior in complex flow real-life situations.

The governing differential equations for any flow system can be expressed in Lagrangian and Eulerian coordinate systems. In the Lagrangian approach, the fluid is assumed to be composed of particles, and the motion of individual particles is tracked. However, in the Eulerian approach, the attention is instead focused on a fixed coordinate system that defines space. We consider a prime variable of interest in fluid dynamics, that is, velocity. In the Lagrangian approach, this parameter is a function of time, whereas in Eulerian systems, it is a function of both time and spatial coordinates. Therefore, the acceleration of a particle that at time t occupies a fixed location can be calculated either by using the material or substantial derivatives as

$$a = \frac{DV}{Dt} \tag{4}$$

or by chain rule of differentiation as

$$a = \frac{\partial v}{\partial t} + \frac{\partial v}{\partial x}\frac{\partial x}{\partial t} + \frac{\partial v}{\partial y}\frac{\partial y}{\partial t} + \frac{\partial v}{\partial z}\frac{\partial z}{\partial t}$$

Equations (4) and (5) are the same; Equation (4) represents the Lagrangian approach, whereas Equation (5) is representative of an Eulerian system.

$$\frac{DV}{Dt} = \frac{\partial v}{\partial t} + \frac{\partial v}{\partial x}\frac{\partial x}{\partial t} + \frac{\partial v}{\partial y}\frac{\partial y}{\partial t} + \frac{\partial v}{\partial z}\frac{\partial z}{\partial t} \tag{6}$$

See also **Fluid and Fluid Flow**.

Equations of Change in Laminar Flow Systems

The governing differential equations corresponding to mass and momentum conservation are solved to obtain field variables of interest in fluid dynamics (i.e., velocities and pressures), which are also known as the continuity equation and equations of change, which describe the laminar flow of a pure isothermal fluid.

Equation of Continuity

Equation of continuity is derived by applying the law of conservation of mass to a small 3-D volume element (see Fig. 5) within a flowing fluid. In vector notation, this equation can be represented as

$$\frac{\partial \rho}{\partial t} = -(\nabla . \rho v) \tag{7}$$

Here, (Ñ. ρv) is called the "divergence" of ρ**v**. This vector ρ**v** is the mass flux and is the net rate of mass efflux per unit volume. Equation (7) states that the rate of increase of the density within a small volume element fixed in space is equal to the net rate of mass influx to the element divided by its volume. A very special form of the equation of continuity is Equation (8), which is applicable for fluids of constant density, that is, incompressible fluids.

$$(\nabla . v) = 0 \tag{8}$$

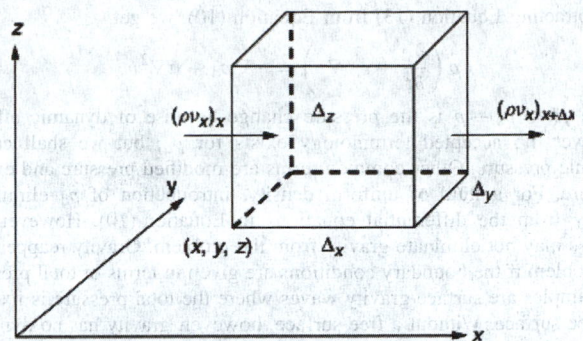

Fig. 5.   Region of volume in $\Delta x \Delta y \Delta z$ through which the fluid is flowing.

Equation of Motion

The equation of motion is developed by implementing the shell momentum balance concept on a 3-D volume element, as shown in Fig. 5.

It should be noted that momentum is a vector quantity, and hence the equation of motion will have components in each of the three coordinate directions x, y, and z, respectively. Adding the three components of equation of motion, the equation of motion can be represented in vector notation as

$$\frac{\partial}{\partial t}\rho \mathbf{v} = \qquad [\nabla . \rho \mathbf{vv}] \qquad - \qquad [\nabla p]$$

Rate of increase of momentum per unit volume	Rate of momentum gain by convection per unit volume	Pressure force on element per unit volume

$$\qquad\qquad -[\nabla \cdot \tau] \qquad + \qquad \rho \mathbf{g}$$

Rate of momentum gain by viscous transfer per unit volume	Gravitational force on element per unit volume

$$\tag{9}$$

τ is known as "stress tensor" and has nine components, [Ñ. ρvv] represents the rate of loss of momentum per unit volume by fluid flow, and (Ñ. ρ v) represents the rate of loss of mass per unit volume. In order to use Equation (9) to determine velocity distribution, appropriate expressions for various stresses in terms of velocity gradients and fluid properties are substituted, which yields the equations of motion for a Newtonian fluid with varying density and viscosity.

For constant ρ and μ, the equation of motion can be written as

$$\rho \left(\frac{\partial v}{\partial t} + v.\nabla v \right) = -\nabla p + \mu \nabla^2 v + \rho g \tag{10}$$

Equation (10) is the well-known Navier–Stokes equation, first developed by Navier in France in 1822.

The equations of motion together with equation of continuity, equation of state $p = p(\rho)$, viscosity models for different fluids, and the initial and the boundary conditions, complete the problem specification for any laminar flow problem in hand. These equations, when solved mathematically, determine completely the density, pressure, and velocity components in an isothermal laminar flow situation.

Equation (10) can be written in nondimensional form as

$$\left(\frac{\partial v'}{\partial t} + v'.\nabla v' \right) = -\nabla p' + \frac{1}{Re}\nabla^2 v' + g \tag{11}$$

where the primed quantities are dimensionless. The ratio of second term to the fourth term in Equation (11) is of order Re, which indicates the physical representation of Reynolds Number as

$$Re = \frac{\text{inertia forces}}{\text{viscous forces}} \tag{12}$$

Concept of Dynamic Pressure

If the body of fluid is at rest, the pressure is hydrostatic and

$$\rho g - \nabla p = 0 \tag{13}$$

Subtracting Equation (13) from Equation (10), we get

$$\rho\left(\frac{\partial v}{\partial t} + v \cdot \nabla v\right) = -\nabla p_d + \mu \nabla^2 v \tag{14}$$

where $p_d = p_s - p$ is the pressure change because of dynamic effects. However, no accepted terminology exists for p_d, but we shall call it dynamic pressure. Other common terms are modified pressure and excess pressure. For a fluid of uniform density, introduction of p_d eliminates gravity from the differential equations in Equation (10). However, the process may not eliminate gravity from the problem. Gravity reappears in the problem if the boundary conditions are given in terms of total pressure p. Examples are surface gravity waves where the total pressure is fixed at the free surface. Without a free surface, however, gravity has no dynamic role, and its only effect is to add a hydrostatic contribution to the pressure field.

Equations (7)–(14) have been presented in rectangular coordinate systems. However, in real complex flow situations, depending on the geometrical configurations of the flow system, other coordinate systems such as spherical and cylindrical can be used. Hence, in these cases, Equations (7)–(14) can be cast in alternative coordinate systems using coordinate transforms.

Steady-State Flow between Parallel Plates

A laminar flow within two parallel plates is considered in Figure 6.

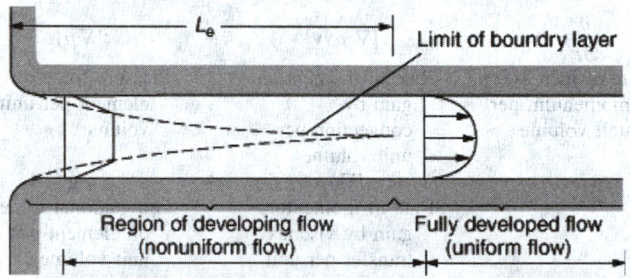

Fig. 6.   Developing and fully developed flows in a channel. Flow is fully developed after the boundary layers merge.

The term fully developed in the figure signifies that we are considering regions beyond the developing stage near the entrance, where the velocity profile changes in the direction of flow because of the development of boundary layers from the two walls. The discussion of boundary layer theory is beyond the scope of the present text, and we shall only consider the flows under fully developed situations, i.e., far away from the entrance.

Consider the fully developed stage of a two-dimensional flow between two infinite parallel plates, as shown in Figure 7.

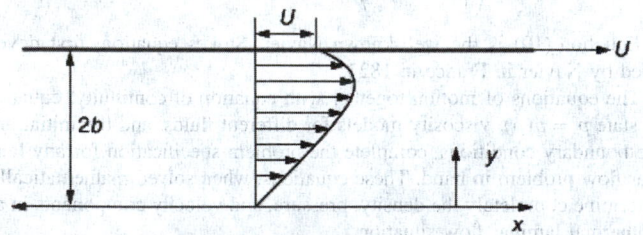

Fig. 7.   Steady-state flow between two parallel plates.

The flow is driven by an externally imposed pressure gradient, and the motion of the upper plate is maintained at velocity U. The x coordinate axis is parallel to the direction of flow. Two-dimensionality of the flow requires that $\frac{\partial}{\partial z} = 0$. Flow characteristics are also invariant in the x-direction, so that continuity requires $\frac{\partial v}{\partial z} = 0$. As $v = 0$ at $y = 0$, it follows that $v = 0$ everywhere, which implies that the flow is parallel to the walls. The momentum balance equations in x and y directions can be written as

$$0 = -\frac{1}{\rho}\frac{\partial p}{\partial x} + \mu\frac{d^2 u}{dy^2} \tag{15}$$

$$0 = -\frac{1}{\rho}\frac{\partial p}{\partial y} \tag{16}$$

The y-momentum in Equation (16) shows that p is not a function of y. In the x-momentum in Equation (15), the first term can only be a function of x, whereas the second term can only be a function of y. The only way this can be established is that both the terms are constants. The pressure gradient is hence a constant, implying that the pressure varies linearly across the channel. Integrating Equation (15) twice, we get

$$0 = \frac{y^2}{2}\frac{dp}{dx} + \mu u + Ay + B \tag{17}$$

where p is a function of x only. The constants of integration are determined by using the conditions imposed on lower and upper plates. For the lower boundary, $u = 0$ at $y = 0$ requires that $B = 0$. The upper boundary condition $u = u$ at $y = 2b$, requires that $A = b\frac{dp}{dx} - \mu\frac{u}{2b}$.

The velocity profile from Equation (17) then becomes

$$u = \frac{yU}{2b} - \frac{y}{\mu}\frac{dp}{dx}\left(b - \frac{y}{2}\right) \tag{18}$$

The volumetric flow rate per unit width of the channel can be calculated using

$$Q = \int_0^{2b} u\,dy = Ub\left[1 - \frac{2b^2}{3\mu U}\frac{dp}{dx}\right] \tag{19}$$

The average velocity can further be written as

$$\overline{V} = \frac{Q}{2b} = \frac{u}{2}\left[1 - \frac{2b^2}{3\mu U}\frac{dp}{dx}\right] \tag{20}$$

Two special cases of interest have been discussed in the subsequent section,

1. **Plane Couette Flow** The flow driven by the motion of the upper plate, without the imposition of any external pressure gradient, is called a plane Couette flow.

   Equation (18) then reduces to a linear profile represented by Equation (21) and can be seen in Figure 8.

$$u = \frac{yU}{2b} \tag{21}$$

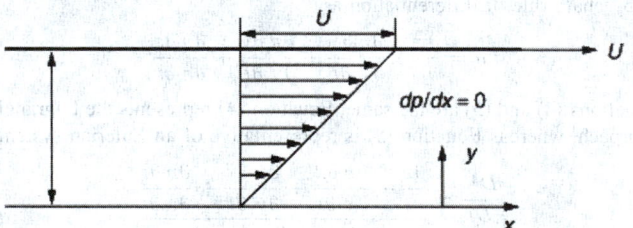

Fig. 8.   Plane Couette flow between two parallel plates.

The magnitude of shear stress is given by

$$\tau_{yx} = \mu\frac{dU}{dy} = \frac{\mu u}{2b} \tag{22}$$

which is uniform across the channel.

2. **Plane Poiseuille Flow** The flow driven by an externally imposed pressure gradient through two stationary flat walls is called a plane Poiseuille flow. In this case, Equation (18) reduces to Equation (23), which represents a parabolic profile, as seen in Figure 9.

$$u = \frac{y}{\mu}\frac{dp}{dx}\left(b - \frac{y}{2}\right) \tag{23}$$

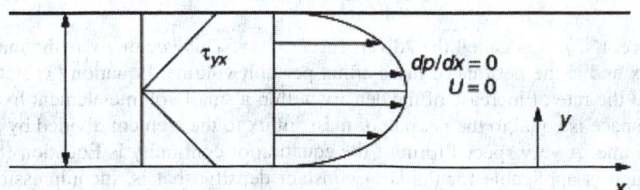

Fig. 9.   Plane Poiseuille flow between two parallel plates.

The magnitude of shear stress is given by

$$\tau_{yx} = \mu \frac{du}{dy} = (b - y)\frac{dp}{dx} \tag{24}$$

which shows that the stress distribution is linear with a magnitude of $b(dp/dx)$ at the walls (as seen in Fig. 9).

It should be noted that the constant pressure gradient and the linearity of the shear stress distribution are general results for a fully developed flow and even hold in a turbulent flow situation.

Steady-State Flow in a Pipe

Consider the fully developed laminar motion of a fluid through a circular pipe of radius a as shown in Figure 10.

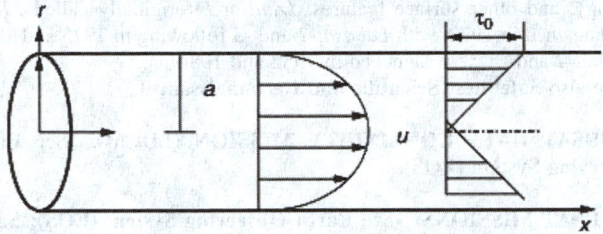

Fig. 10.   Laminar flow through a circular pipe.

This type of flow is commonly known as circular Poiseuille flow. In this case, cylindrical coordinates (r, θ, x) have been used, with the x-axis coinciding with the axis of the pipe. The only nonzero component of velocity is the axial velocity u(r), and none of the variables depend on θ. The equations of motion are written in cylindrical coordinates. The radial equation of motion gives,

$$0 = \frac{\partial p}{\partial r} \tag{25}$$

showing that p is a function of x only. The x-momentum equations gives

$$0 = \frac{dp}{dx} + \frac{\mu}{r}\frac{d}{dr}\left(r\frac{du}{dr}\right) \tag{26}$$

As the first term can only be a function of x and the second term can only be a function of r, it follows that both terms should be constants. The pressure, therefore, falls linearly along the length of the pipe. Integrating twice, we get

$$0 = \frac{r^2}{4\mu}\frac{dp}{dx} + A\ln r + B \tag{27}$$

As u must be bounded at $r = 0$, we must have $A = 0$. The wall condition, $u = 0$ at $r = a$, gives $B = -\left(\frac{a^2}{4\mu}\right)\left(\frac{dp}{dx}\right)$. The velocity profile is hence parabolic and can be represented as in Equation (28),

$$u = \frac{r^3 - a^2}{4\mu}\frac{dp}{dx} \tag{28}$$

The shear stress at any point can be expressed using Equation (29).

$$\tau_{xr} = \mu \frac{du}{dr} = \frac{r}{2}\frac{dp}{dx} \tag{29}$$

The shear stress distribution is linear with a maximum value of $\frac{a}{2}\frac{dp}{dx}$ at the walls.

The volumetric flow rate is given by

$$Q = \int_O^a u2\pi r\,dr = \frac{\pi a^4}{8\mu}\frac{dp}{dx} \tag{30}$$

Equation (30) is also commonly known as the Hagen–Poiseuille law. It gives the relationship between the volumetric flow rate and the forces causing the flow, that is, the forces associated with the pressure gradient and the viscosity.

The average velocity over the circular cross section can be written as

$$\bar{v} = \frac{Q}{\pi a^2} = \frac{a^2}{8\mu}\frac{dp}{dx} \tag{31}$$

Creeping Flow Past a Sphere

When the motion of fluid is "very slow" (Re ≪ 1), the flow is said to be creeping or Stokes flow. Examples of such flows are the motion of settling of particles near the ocean bottom and the fall of moisture drops in the atmosphere. In creeping motions, viscous forces are dominant and, hence, the nonlinear inertia term present in Equation (10) is neglected. We then have

$$\rho \frac{\partial v}{\partial t} = -\nabla p + \mu \nabla^2 v + \rho g \tag{32}$$

Equation (32) is the well-known **Stokes** equation, which is used to model creeping flows. The solution of creeping flow around a sphere was first given by Stokes in 1851. The detailed solution of velocity and pressure distributions around a sphere is quite advanced for the present text. The positive pressure difference in front and negative in the rear of the sphere creates pressure drag on the sphere. Consider a fluid flowing with velocity U past a sphere of radius r. The total drag force on a sphere is obtained by calculating the total viscous stress and the normal stress and then multiplying it with the surface area of the sphere, which is found out to be

$$D = 6\pi \mu r U \tag{33}$$

One-third of the total drag comprises pressure drag, whereas two-thirds is the viscous or skin drag. It can be seen from Equation (33) that the resistance in a creeping flow is proportional to the velocity, and is commonly known as Stokes Law.

Drag coefficient can be defined as the drag force nondimensionalized by $\rho u^2/2$ and the projected area of the sphere, which is πr^2, and stated as follows:

$$C_D = \frac{d}{\frac{1}{2}\rho u^2 \pi r^2} = \frac{24}{Re} \tag{34}$$

where Re = $2rU\rho/\mu$

Most of the problems in fluid dynamics fall under two broad categories, namely flow in channels and flow around submerged objects. Relatively simple geometrical configurations were presented in prior sections. However, real-life examples of flows in channels include pumping oil in pipes, flow of water in channels, extrusion of plastics through dyes, and fluid flows in filters. Examples of flow around submerged objects are the flow of air around an airfoil, an airplane wing, motion of fluid around particles undergoing sedimentation, etc. Flow in channels are mainly concerned with establishing a relationship between the volumetric flow rate and the pressure gradient. However, in flow past submerged objects, one is more interested in determining the relationship between the velocity of the fluid approaching the object and the drag force. As in many flow systems velocity and pressure profiles cannot be determined, the concept of friction factors/drag coefficient is found to be very useful. In other words, standard correlations are used to estimate the flow in geometrically complex systems. See also **Aerodynamics and Aerostatics**.

Portions of this article originally appeared in the 5 Vols. Lehr, J. H., and J. Keeley, Editors, John Wiley & Sons, Inc, Hoboken, NJ.2005.

Additional Reading

Alexandraou, A.: *Principles of Fluid Mechanics*, Prentice-Hall, Inc., Upper Saddle River, NJ. 2001.

Aris, R.: *Vectors, Tensors and the Basic Equations of Fluid Mechanics*, Dover Publications, New York, NY. 1997.

Batchelor, G.K.: *Introduction to Fluid Dynamics*, Cambridge University Press, New York, NY, 2000.

Batchelor, G.K., H.K. Moffatt, and M.G. Worster: *Perspectives in Fluid Dynamics: A Collective Introduction to Current Research*, Cambridge University Press, New York, NY, 2003.

Bird, R.B., R.C., Armstrong, and O. Hassager: *Dynamics of Polymeric Liquids: Fluid Mechanics*, John Wiley & Sons, Inc., New York, NY. 1997.

Bird, R.B., W.E. Stewart, and E.N. Lightfoot: *Transport Phenomena*, 2nd Edition, John Wiley & Sons, Inc., New York, NY. 2001.

Hanspal, N.S., D.B., Das, and V. Nasschi: "Viscous flow," *Encyclopedia of Water*, John Wiley & Sons, Inc., New York, NY. 2004.

Nassehi, V.: *Practical Aspects of Finite Element Modelling of Polymer Processing*, John Wiley & Sons, Inc., New York, NY. 2002.

Lighthill, M.J.: *Informal Introduction to Theoretical Fluid Mechanics*, Oxford University Press, New York, NY. 1988.

Papanastasiou, T.C., G.C. Georgiou, and A.N. Alexandrou: *Viscous Fluid Flow*, CRC Press, LLC, Boca Raton, FL, 2000.

Petrila, T., and D. Trif: *Basics of Fluid Mechanics and Introduction to Computational Fluid Dynamics*, Springer-Verlag New York, LLC, New York, NY, 2005.

Tritton, D.J.: *Physical Fluid Dynamics*, 2nd Edition, Oxford University Press, New York, NY. 1988.

White, F.M.: *Viscous Fluid Flow*, 3rd Edition, The Mc-Graw-Hill Companies, Inc., New York, NY, 2005.

NAVRAJ S. HANSPAL, Loughborough University, Loughborough, UK

RAJ SHARMA, University of KwaZulu-Natal, Durban, South Africa

LAMINATE. Laminates are materials made up of plies or laminae stacked up like a deck of cards and bonded together. Laminates are a form of composite material made of any one of several types of thermosetting plastic (phenolic, polyester, epoxy or silicone) bonded to paper, cloth, asbestos, wood, or glass fiber. High tensile and dielectric strength and low moisture absorption are characteristic of these products. Available as sheet, rod, or tubing in mechanical, electrical, and general-purpose grades (National Electrical Manufacturers Association). Plywood is composed of a veneer with grain oriented at a 90-degree angle on successive layers and bonded with a thermosetting adhesive of the urea or phenol—formaldehyde type to give a high strength, dimensionally stable, weather-resistant construction material. It can be made nonflammable by treatment with salt solution. Polyvinyl butyral sheet is used in safety glass.

See also **Composite Materials**; and **Reinforced Plastics**.

LAMP. See **Illumination**.

LAMPBLACK. A black or gray pigment made by burning low-grade heavy oils or similar carbonaceous materials with insufficient air and in a closed system such that the soot can be collected in settling chambers. Lampblack is markedly different from carbon black, strongly hydrophobic and nonflammable.

Uses for lampblack include: Black pigment for cements, ceramic ware, mortar, inks, linoleum, surface coating, crayons, polishes, carbon paper, soap, etc.; ingredient of insulating compositions, liquid-air explosives, matches, fertilizer, furnace lutes, lubricating compositions, carbon brushes; reagent in cementation of steel.

See also **Carbon Black**.

LAMPREYS (*Agnatha*).   A jawless fish of the family *Petromyzontidae*, the lamprey appears much like an eel. However, it is not an eel. Normally, various species of lampreys occur on both sides of the Atlantic. Like its close relative, the hagfish, the lamprey is characterized by the primitive features of jawless fishes—no scales, no sympathetic nervous system, a cartilage skeleton, and single nostril. Since about the mid-1800s, lampreys have not been considered of commercial value. However, they were eaten during the middle ages. Probably the parasitic landlocked lamprey *Pretomyzon marinus* is best known because of the very extensive damage it has done to many freshwater fish species in the Great Lakes. The lamprey attaches itself to a host and literally sucks life-giving juices from it. In the saliva of the lamprey, there is an anticoagulating material that continually dilutes the blood of its victim. Once the host is drained of vital juices, the lamprey moves on to another victim.

As an example of this damage, just a few decades ago, the Great Lakes yielded an annual catch of nearly 12 million pounds (5.4 million kilograms) of lake trout. Within a period of about 30 years, lake trout practically disappeared from the lakes. Initially, the lampreys migrated from their normal marine-water habitat to fresh water for spawning. Inasmuch as the Great Lakes are interconnected to the sea through various waterways, including the Welland Canal and the New York State Barge Canal, the lampreys ultimately invaded the lakes. During the latter 1970s, considerable progress was made toward specifically combating the lampreys without harming other species. Population of trout in the lakes is again rising.

LAND BREEZE.   See **Meteorology**.

LANDING GEAR.   The apparatus comprising those components of an aircraft or spacecraft that support and provide mobility for the craft on land, water, or other surface. The landing gear consists of wheels, floats, skis, bogies, and treads, or other devices, together will all associated struts, bracing shock absorbers, etc.

LANDING SKID.   A skid or runner used in the main landing gear of an aerodynamic vehicle, upon which the vehicle slides over the ground.

LANDSAT.   A series of U.S. satellites designed for remote sensing and mapping of land areas. These satellites are launched into sun-synchronous orbits that return to image the same swath, two degrees wide, every 16 days. The primary imaging instrument on the first generation of Landsat satellites was a multispectral scanner, a four-channel radiometer with a ground resolution of 80 meters (262 feet). *Landsat-4* and *-5* added a second imaging instrument, a thematic mapper, a seven- channel instrument with a ground resolution of 30 meters (98 feet), used to study vegetation, geology, and other surface features. *Landsat-1* (originally called *ERTS*) was launched in 1972, with *Landsat-2* and *-3* following in 1975 and 1978. *Landsat-4* and *-5* were launched in 1982 and 1984.

See also **Satellites (Scientific and Reconnaissance)**.

LANDSAT DATA CONTINUITY MISSION (LDCM). See **Earth Observing System (EOS)**.

LANDSAT MISSIONS.   See **Earth Observing System (EOS)**; **Satellites (Scientific and Reconnaissance)**; and **Space Science Missions: Earth**.

LANGLEY.   A unit of energy per unit area once commonly employed in radiation theory; equal to one gram-calorie per square centimeter. The langley is almost always used in conjunction with some time unit to express a flux density; but the time unit has been purposely separated in order that it may be chosen in a manner convenient to each particular problem. The unit was named in honor of the American scientist Samuel P. Langley, 1834–1906, who made many contributions to the knowledge of solar radiation. Modern meteorologists tend to use the mks unit W m^{-2}.

LANGLEY, SAMUEL PIERPONT (1834–1906).   Born in the Boston suburb of Roxbury, Massachusetts. Samuel Langley was one of America's most accomplished scientists. His work as an astronomy, physics, and aeronautics pioneer was highly regarded by the international science community. Ironically though, Langley's formal education ended at the high school level, but he managed to continue his scientific education in Boston's numerous libraries.

Langley began his career as a civil engineer in Chicago, continuing later in St. Louis, before returning to Boston to accept an assistantship at the Harvard Observatory. Heading south once again, Langley later taught mathematics at the U.S. Naval Academy in Annapolis, Maryland. Then, from 1867–1887, he served as professor of physics and astronomy as well as director of the Allegheny Observatory at the Western University of Pennsylvania (now known as the University of Pittsburgh). After 1887, Langley was appointed Secretary of the Smithsonian Institution in Washington DC.

Langley's chief scientific interest was the sun and its effect on the weather, and believed that all life and activity on the Earth were made possible by the sun's radiation. In 1878, he invented the bolometer, a radiant-heat detector that is sensitive to differences in temperature of one hundred-thousandth of a degree Celsius (0.00001 °C). Composed of two thin strips of metal, a Wheatstone bridge, a battery, and a galvanometer (an electrical current measuring device), this instrument enabled him to study solar irradiance (light rays from the sun) far into its infrared region and to measure the intensity of solar radiation at various wavelengths. See Fig. 1.

Bolometers have been flown on numerous NASA missions including the Earth's Radiation Budget Experiment (ERBE) and the Clouds and Earth's Radiant Energy System (CERES), which provided accurate regional and global measurements of the components of the Earth's radiation budget. Langley's highly original and innovative research earned him honorary doctorates, awards, and medals from universities and scientific societies around the world.

In addition to his solar interests, Langley was the only professional scientist of his day who believed that the human race was destined to fly.

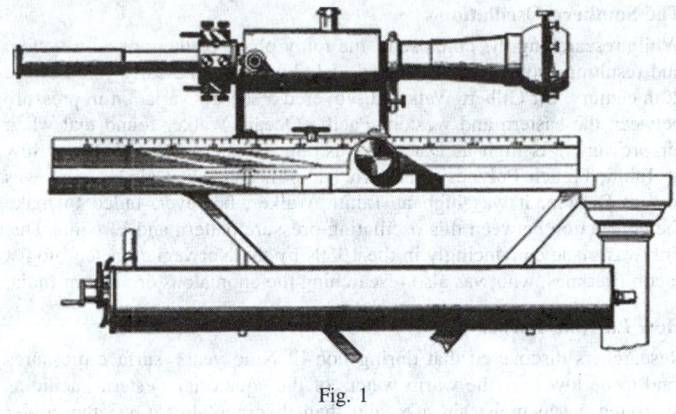

Fig. 1

While at the Allegheny Observatory, he made important experiments on the lift and drag of an aircraft moving through the air at a measured speed. Backed by these experiments, he was the first to offer a clear explanation of the way birds soar and glide without appreciable wing movement.

In 1886, he undertook a series of experiments on a rotating rig to measure the power needed to propel objects through the air. Encouraged by his findings, Langley set out to build a series of large working models of steam-powered flying machines he called "aerodromes," and, in 1896, became the first to build heavier-than-air machines capable of sustained (although uncontrolled) flight. Langley built two unmanned craft, each of which had two sets of 14-foot (4.3-meter) wings, weighed 26 pounds (11.8 kilograms), and were powered by steam engines.

Langley's first manned aircraft, powered by a five-cylinder air-cooled gasoline engine designed by Charles M. Manly did not fair as well as his unmanned craft. Piloted by Manly, the aircraft snagged upon launching from a catapult, and crashed into the Potomac River for the second and last time on Dec. 8, 1903, just nine days before the successful flights of the Wright brothers near Kitty Hawk, NC. This aircraft had a wingspan of 48 feet (14.6 meters) and a total weight (with pilot) of 850 pounds (386 kilograms). Some authorities believe that if his catapult had not failed, Langley would have been the first to achieve sustained flight in a manned heavier-than-air machine.

Langley's memory lives on in the names of the NASA Langley Research Center, the adjacent Air Force base, and several place names across the country. Our nation's first aircraft carrier, CV-1, built at the Norfolk Navy Yard in the early 1920's, was also named after Langley.

See also **Bolometer**.

Web References

Icarus on the Mall: http://www.150.si.edu/chap6/six.htm
Langley's Feat—and Folly (from Smithsonian Magazine): http://smithsonian-mag.com/smithsonian/issues97/nov97/object_nov97.html

LANGMUIR CIRCULATION.   Roll circulations approximately aligned with the surface stress vector that frequently occur in the upper boundary layer of oceans or lakes. Although similar in form to atmospheric longitudinal roll vortices, Langmuir circulations are thought to be driven by nonlinear interactions between the surface gravity wave field and the larger-scale turbulent motions within the mixed layer. They are sometimes called *windrows* because they form lines of surface debris or bubbles in their surface convergence zones. Their spatial scale is related to the depth of the mixed layer and their characteristic velocity is on the order of $8u_*$, where u_* is the friction velocity in water. As a result of this scaling, Langmuir circulations generally require surface winds of at least 8 m s^{-1} in order to form.

See also **Coherent Structures** and **Longitudinal Rolls**.

Additional Reading

Etling, D., and R.A. Brown: "Roll Vortices in the Planetary Boundary Layer: A Review," *Bound.-Layer Meteor.*, **65**, 215–248 (1993).

LANGMUIR, IRVING (1881–1957).   Langmuir was an American scientist whose fields of contribution include chemistry, physics, and technology. He graduated as a metallurgical engineer from the School of Mines at Columbia University in 1903. Postgraduate work in Physical Chemistry under Nernst in Göttingen earned him the degrees of M.A. and Ph.D. in 1906.

Returning to America, Dr. Langmuir became Instructor in Chemistry at Stevens Institute of Technology, Hoboken, New Jersey, where he taught until July 1909. In 1909, Langmuir began working for the General Electric Company in Schenectady, New York where he eventually became Associate Director.

His work on filaments in gases led directly to the invention of the gas filled incandescent lamp and to the discovery of atomic hydrogen. He later used the latter in the development of the atomic hydrogen welding process.

He was the first to observe the very stable adsorbed monatomic films on tungsten and platinum filaments, and was able, after experiments with oil films on water, to formulate a general theory of adsorbed films. He also studied the catalytic properties of such films.

Dr. Langmuir received twenty-three scientific medals and prizes, including the Nobel Prize in Chemistry in 1932 for his work on surface chemistry.

See also **Molecular and Supermolecular Electronics**.

J. M. I.

LANGMUIR LAYER.   A trace organic impurity in water with a polar or similar chemical group that disperses the material as a surface near monomolecular layer and influences surface properties as water vapor deposition coefficient, vapor pressure, and surface tension.

LANGMUIR NUMBER.   A nondimensional number representing the ratio of viscous to inertial forces in the nondimensionalized governing equations for Langmuir circulation:

$$\text{La} = [(v_T\beta/u_*)^{3/2}][(S_0/u_*^{-1/2}],$$

where v_T is the eddy viscosity, $2S_0$ and $1/2\beta$ are the surface value and e-folding depth of the Stokes's drift current, respectively, and u_* is the friction velocity.

This parameter can also be interpreted as expressing the balance between the rate of diffusion of streamwise vorticity by eddy viscosity and the rate of production of streamwise vorticity by vortex stretching accomplished by the Stokes's drift. The Langmuir number is inversely related to the Reynolds number.

LANGMUIR PROBE.   A probe used to measure the electron temperature of ionized plasmas. Langmuir probes mounted on spacecraft are often used to measure the electron temperature of the earth's ionosphere.

LANGUAGE (Computer).   A communications means for transmitting information between human operators and computers. The human programmer describes how the problem is to be solved using the computer language. A computer language consists of a well-defined set of characters and words, coupled with a series of rules (termed syntax) for combining them into computer instructions or statements. There is a wide variety of computer languages, particularly in terms of flexibility and ease of use. There are three levels in the hierarchy of computer languages: (1) machine languages; (2) procedure-oriented languages; and (3) problem-oriented languages.

Machine Language (1) A language designed for interpretation and use by a machine without translation. (2) A system for expressing information which is intelligible to a specific machine; e.g., a computer or class of computers. Such a language may include instructions that define and direct machine operations, and information to be recorded by or acted upon by these machine operations. (3) The set of instructions expressed in the number system basic to a computer, together with symbolic operation codes with absolute addresses, relative addresses, or symbolic addresses. In this case, it is known as an Assembler Language. See also **Assembler (Computer System)**.

Procedure-oriented Language. A machine-independent language which describes how the process of solving the problem is to be carried out. For example, FORTRAN, ALGOL, PL/I, and COBOL.

Problem-oriented Language. A language designed for convenience of program specification in a general problem area. The components of such a language may bear little resemblance to machine instructions and often incorporate terminology and functions unique to an application. Also known as Applications Language.

Other computer languages include:

Algorithmic language. An arithmetic language by which numerical procedures may be precisely presented to a computer in a standard form. The language is intended not only as a means of directly presenting any numerical procedure to any appropriate computer for which a compiler exists, but also as a means of communicating numerical procedures among individuals.

Artificial language. A language specifically designed for ease of communication in a particular area of endeavor, but one that is not yet "natural" to that area. This is contrasted with a natural language that has evolved through long usage.

Common machine language. A machine-sensible information representation, common to a related group of data processing machines.

Common business-oriented language. A specific language by which business data processing procedures may be precisely described in a standard form. The language is intended not only as a means for directly presenting any business program to any appropriate computer for which a compiler exists, but also as a means of communicating such procedures among individuals.

Object language. A language which is the output of an automatic coding routine. Usually, object language and machine language are the same; however, a series of steps in an automatic coding system may involve the object language of one step serving as a source language for the next step and so forth.

THOMAS J. HARRISON, IBM Corporation, Boca Raton, FL

LANGUR.   See Monkeys and Baboons.

LA NIÑA. The coupled atmosphere–ocean phenomenon known as El Niño is frequently followed by a period of normal conditions in the equatorial Pacific Ocean. Sometimes, but not always, El Niño conditions give way to the other extreme of the El Niño-Southern Oscillation (ENSO) cycle. This cold counterpart to El Niño is known as La Niña, Spanish for "the girl child." See Fig. 1.

The Southern Oscillation

While researching the collapse of the rainy phase of the monsoon system and resulting drought that occurred in India during the early years of the 20th century, Sir Gilbert Walker discovered a seesaw variation in pressure between the eastern and western Pacific Ocean. Walker found that when air pressure was high at Darwin, Australia (western Pacific) it was low at Tahiti, French Polynesia (eastern Pacific), and when air pressure was low at Darwin, it was high at Tahiti. Walker, however, failed to make the connection between this oscillating pressure pattern and El Niño. This link was made convincingly in the 1960s by the Norwegian meteorologist Jacob Bjerknes, who was also researching the anomalous drought in India.

How La Niña Forms

Researchers discovered that during non-El Niño years, surface pressures tend to be low over the warm waters of the equatorial western Pacific as overlying warm moist air rises and then diverges aloft. Over the colder waters of the eastern equatorial Pacific, surface pressures tend to be higher as converging winds aloft contribute to the sinking of cool air. In much the same way as a ball rolls down a hill, air flows from high pressure in the east to low pressure in the west along this equatorial pressure gradient. This contrast in pressure is what drives the trade winds, the prevailing large-scale surface winds that blow from east to west. As these winds blow along the surface of the equatorial waters, there is a net transport of ocean water in a westward direction. As this occurs, cold, nutrient-rich water rises up (or upwells) along the coast of South America to replace the westward-moving surface water. This upwelling brings nutrients to the surface waters off the coast allowing the fish population living in these upper waters to thrive.

During La Niña years, the trade winds are unusually strong due to an enhanced pressure gradient between the eastern and western Pacific. As a result, upwelling is enhanced. The cycle of La Niña is completed as the strengthened trade winds allow enhanced westward currents to flow, carrying cool water from the eastern Pacific along the coast of South America, contributing to colder than normal surface waters over the eastern tropical Pacific and warmer than normal surface waters in the western tropical Pacific. See Figs. 2 and 3.

The Effects of La Niña

Changes in global atmospheric circulation patterns accompany La Niña and are responsible for weather extremes in various parts of the world that are typically opposite to those associated with El Niño. These patterns result from colder than normal ocean temperatures inhibiting the formation of rain-producing clouds over the eastern equatorial Pacific region while at the same time enhancing rainfall over the western equatorial Pacific region (Indonesia, Malaysia, and northern Australia.) These patterns affect the position and intensity (weakening) of jet streams and the behavior of storms outside of the tropics in both the Northern and Southern hemispheres.

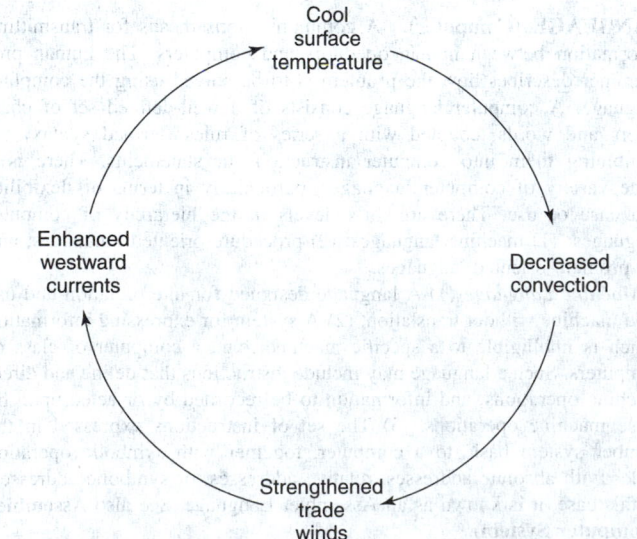

Fig. 1. Processes that affect La Niña. The cycle of La Niña is completed as the strengthened trade winds allow enhanced westward currents to flow, carrying cool water from the eastern Pacific.

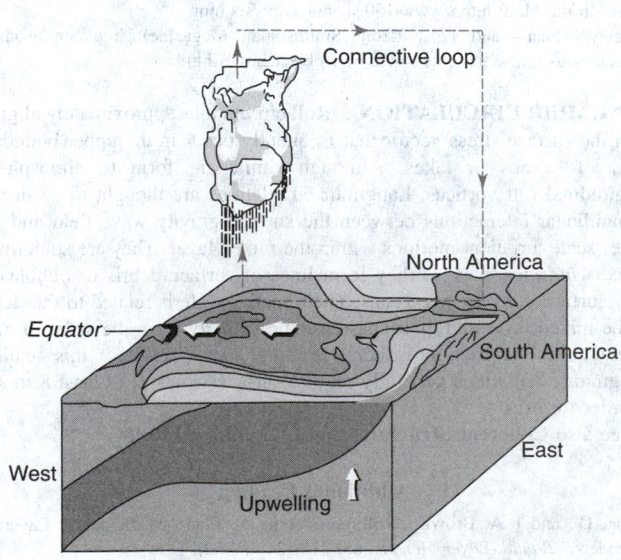

Fig. 2. Normal conditions over the Pacific basin.

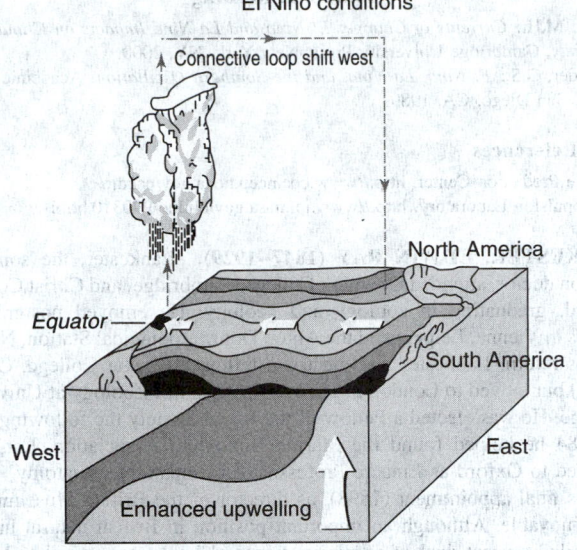

El Niño conditions

Fig. 3. Disturbed conditions over the Pacific basin during a La Niña.

Inside the tropics, ENSO strongly affects tropical cyclone activity around the world. During La Niña, weakened jet streams contribute to an increase in the number of Atlantic tropical storms and hurricanes. During El Niño, strengthened jet streams contribute to a decrease in tropical cyclone activity in the Atlantic and Australian basins.

U.S. La Niña Impacts

The first three months of the year during a La Niña typically feature below normal precipitation in the Southwest, the central and southern sections of the Rockies and Great Plains, and Florida. Meanwhile, the odds of surplus precipitation increase across the Pacific Northwest, in the northern Intermountain West, and over scattered sections of the north-central states,

Ohio Valley, and upper Southeast. La Niña features unusually cold weather in the Northwest and (to a lesser extent) northern California, the northern Intermountain West, and the north-central states. Farther south, higher than normal temperatures are slightly favored in a broad area covering the southern Rockies and Great Plains, the Ohio Valley, the Southeast, and the mid-Atlantic states.

Global La Niña Impacts

Globally, La Niña is characterized by wetter than normal conditions west of the equatorial central Pacific over northern Australia and Indonesia during the northern hemisphere winter, and over the Philippines during the northern hemisphere summer. Wetter than normal conditions are also observed over southeastern Africa and northern Brazil, during the northern hemisphere winter season. During the northern hemisphere summer season, the Indian monsoon rainfall tends to be greater than normal, especially in north-west India. Drier than normal conditions are observed along the west coast of tropical South America, and at subtropical latitudes of North America (Gulf Coast) and South America (southern Brazil to central Argentina) during their respective winter seasons. See Fig. 4.[1]

NASA and NOAA Missions to Study La Niña

Over the years, several NASA missions have studied the effects associated with La Niña and El Niño, such as changes in sea-surface temperature (SST) and cloud cover. These studies are augmented by data from operational satellites of the National Oceanic and Atmospheric Administration (NOAA).

Initial efforts at mapping SST and cloud cover were conducted using data from NASA's Nimbus series of satellites. The four-channel Advanced Very High Resolution Radiometer (AVHRR), flown on NOAA's TIROS-N weather satellite in 1978 and on the NOAA-6 satellite in 1979, greatly increased the accurate measurements of El Niño effects. ("Four channel" means that the instrument views in four different parts of the electromagnetic visible and infrared spectrum.) Still further increases in

[1] National Centers for Environmental Prediction-Climate Prediction Center (NOAA).

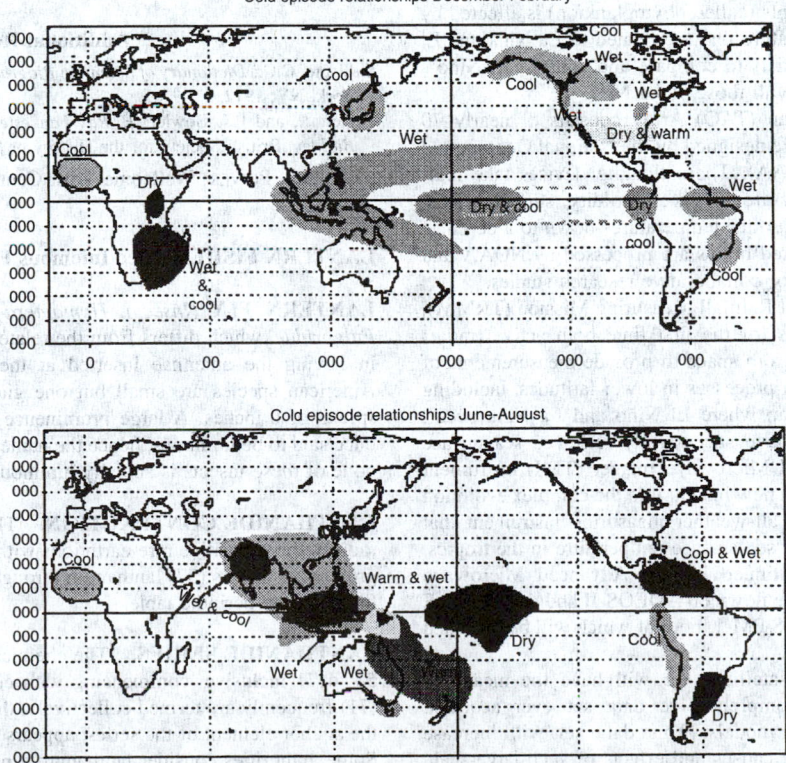

Fig. 4. Anomalous precipitation and temperature patterns associated with La Niña. (*National Oceanic and Atmospheric Administration.*)

accuracy resulted when a fifth channel was added to the AVHRR instrument flown on NOAA-7 in 1981, and on subsequent NOAA satellites. The fifth channel improved the measurement of SST by providing corrections for atmospheric water vapor that otherwise would have interfered with the temperature measurements.

The joint U.S.-French TOPEX/Poseidon mission was launched in 1992 and is providing global determinations of changes in ocean surface currents that are related to the La Niña and El Niño phenomena. Data retrieved from TOPEX/Poseidon are important because they provide measurements of the depth to which the cold or warm anomaly extends.

A NASA scatterometer called NSCAT flew on the Japanese Advanced Earth Observing System (ADEOS) spacecraft, which was launched in August 1996. NSCAT provided very high quality data on the speed and direction of ocean-surface winds worldwide. Unfortunately, after nine months in orbit, a spacecraft failure brought to an end the stream of NSCAT data. Recognizing the important contributions to Earth science made by NSCAT, NASA launched the QuikSCAT satellite in June 1999 to bridge the gap remaining before launch of the Japanese spacecraft designated ADEOS II (planned for 2000). The SeaWinds instrument onboard QuikSCAT and ADEOS II will provide detailed measurements of the winds above the oceans.

In addition to the scatterometer measurements, which use active microwave radar systems to determine surface wind speeds and directions over the ocean, surface wind speeds are also being obtained from the Special Sensor Microwave Imager (SSM/I), a passive microwave sensor onboard a Department of Defense spacecraft.

Key sources of data related to El Niño have been retrieved from the five-channel AVHRRs flown on NOAA-7, 9, and 11. These historic data sets cover the period 1981 through 1992 and beyond and will permit more accurate SST determinations than were previously available. These data are important to the development and testing of a new generation of computer models in which the interacting processes of the land, the atmosphere, and the oceans are coupled. These coupled models will lead the way to an increased understanding of phenomenon such as La Niña and the teleconnections that link La Niña with changes in weather patterns throughout the world.

NASA's SeaWiFS (Sea-viewing Wide Field of View Sensor) was launched on the OrbView-2 satellite in August 1997. SeaWiFS is designed to detect ocean color, which is an indicator of microscopic plant life in the ocean. The growth of such plants (called phytoplankton) is affected by the changes in sea surface temperature that are related to La Niña and El Niño. SeaWiFS data enable scientists to compare and contrast El Niño's impacts on the marine biosphere with those of La Niña.

The Tropical Atmosphere Ocean (TAO) Array consists of nearly 70 moored buoys in the tropical Pacific designed by the National Oceanic and Atmospheric Administration (NOAA). These floating devices take real-time measurements of air temperature, relative humidity, surface winds, sea surface temperatures, and subsurface temperatures down to a depth of 500 meters. Data from these moored buoys are processed by NOAA and then made available to scientists for collaborative research studies.

The joint U.S.-Japanese Tropical Rainfall Measuring Mission (TRMM), launched in November 1997, uses for the first time both active (radar) and passive microwave detectors from space to provide measurements of precipitation, clouds, and radiation processes in lower latitudes, including those portions of the Pacific Ocean where El Niño and La Niña occur. TRMM research team members have successfully retrieved sea-surface temperature data from the TRMM Microwave Imager (TMI) instrument onboard the spacecraft, giving them new insight into the complex evolution of the La Niña event. TMI is an all-weather measuring instrument that can see through clouds to measure sea-surface temperature in the tropics. Similar observations will be continued by the Advanced Microwave Scanning Radiometer (AMSR) to be flown on ADEOS-II and the AMSR-E instrument to be flown onboard EOS PM-1, both of which will be launched in the year 2000.

With the launch of the EOS satellites, we will have the means to collect and analyze the most comprehensive data set ever acquired for the development of coupled models. This data set will increase markedly our understanding of the causes and effects of such large-scale ocean–atmosphere phenomena as La Niña and El Niño.

See also **El Niño**.

Additional Reading

Glantz, M.H.: *Currents of Change: El Nino and La Nina Impacts on Climate and Society*, Cambridge University Press, New York, NY, 2000.
Philander, G.S.: *El Nino, La Nina, and the Southern Oscillation*, Academic Press, Inc., San Diego, CA, 1989.

Web References

Climate Prediction Center. http://www.cpc.ncep.noaa.gov/pacdir/
Jet Propulsion Laboratory. http://www.jpl.nasa.gov/elnino/990310.html

LANKESTER, EDWIN RAY (1847–1929). Lankester, the son of a London doctor, attended Downing College, Cambridge, and Christ College, Oxford, graduating in zoology and geology. He enjoyed postgraduate studies in Vienna, Leipzig and the Anton Dohrn Zoological Station, Naples. On his return, Lankester was elected a fellow at Exeter College, Oxford (1872) but moved to London in 1874 as Professor of Zoology at University College. He was elected a Fellow of the Royal Society the following year. In 1884 he helped found the Marine Biological Association. Lankester returned to Oxford as Linacre Professor of Comparative Anatomy (1891) but his final appointment (1898), as director of the British Museum, was less enjoyable. Although an important position in British natural history, it entailed a great deal of administration and Lankester retired following a knighthood in 1907. In retirement he produced several books for the nonscientist. With their engaging titles — *Science From an EasyChair*; *Diversions of a Naturalist*; *Great and Small Things* — these volumes provided an easy entry point to the science of natural history.

Lankester was active during a period when an energetic and gifted naturalist could work on the major animal groups, living and fossilized, ranging in size from large mammals to microscopic protozoa. Indeed under his editorship, the *Quarterly Journal of Microscopical Science* became internationally renowned for its high standards of scholarship. He analyzed thestructure and determined the affinities between spiders, scorpions and horseshoe crabs. He introduced among others the terms *blastopore* and *invagination* as part of his systematization of the field of embryology.

Lankester was a friend of Darwin, accepting and promoting through hiswork Darwin's theories of evolution, speciation and natural selection. He clearly enunciated the distinction between the concepts of homology and homoplasy and provided a logical basis for rejecting Lamarckian inheritance ofacquired characteristics.

Additional Reading

Gillispie, C.C.: *Dictionary of Scientific Biography*, Charles Scribner's & Sons, New York, NY, 1972.
Lester, J., and P.J. Bowler: *E. Ray Lankester and the Making of Modern British Biology*, British Society for the History of Science, Farringdon, UK, 1995.

HELEN J. POWER, Wellcome Trust Centre for the History of Medicine at UCL, London, UK

LANTERN FISHES. See **Iniomous Fishes**.

LANTERN FLY (*Insecta, Homoptera*). Any member of the family *Fulgoridae*, which differs from the related leaf hoppers and other families in having the antennae inserted at the sides of the head. The North American species are small but one giant Brazilian species has a wing spread of 6 inches. A large prominence on the head of this species was once said to be luminous, hence the name lantern fly has persisted although none of these insects is actually luminous.

LANTHANIDE CONTRACTION. The decreasing sequence of crystal radii of the tripositive rare-earth ions with increasing atomic number in the group of elements (57) lanthanum through (71) lutetium of the Lanthanide Series in the periodic table.

LANTHANIDE SERIES. The chemical elements with atomic numbers 58 to 71 inclusive, commencing with cerium (58) and through lutetium (71) frequently are termed collectively, the Lanthanide Series. Lanthanum, the anchor element of the series, appears in group 3b of the periodic table. Some authorities consider lanthanum a part of the series. Members of the series, along with lanthanum and yttrium, are described under **Rare-Earth Elements and Metals**. See also **Actinide Series**.

LANTHANUM. [CAS: 7439-91-0]. Chemical element, symbol La, at. no. 57, at. wt. 138.91, periodic table group 3, homolog of the Lanthanide Series of elements, mp 918 °C (1,684 °F), bp 3,464 °C (6,267 °F), density 6.146 g/cm^3 (20°C (68°F). Elemental lanthanum has a double close-packed hexagonal crystal structure at 25°C (77°F). The pure metallic lanthanum is silver-gray in color, but with a luster that remains only briefly upon exposure to air, rapidly oxidizing to a white powder. The oxide is hygroscopic and tends to spall, thus exposing fresh surfaces of the metal for oxidation. Thus, the metal must be handled in an inert atmosphere. Chips and powdered lanthanum are quite pyrophoric. Under required inert atmospheric conditions, the metal is easy to work with normal tools, paralleling tin in its workability. There are two natural isotopes ^{139}La and ^{138}La. The latter is mildly radioactive with a half-life of $10^{10}-10^{15}$ years. The element becomes a superconductor below 6 K. There are 19 known artificial isotopes, all radioactive. Of the light (or cerium-group) rare-earth metals, lanthanum is the second most plentiful and ranks 57th in abundance of elements in the earth's crust, exceeding gold [CAS: 7440-57-5], tantalum [CAS: 7440-25-7], platinum [CAS: 7440-06-4], mercury [CAS: 7439-97-6], bismuth [CAS: 7440-69-9], and several other commonly-used elements. The element was first identified by C.G. Mosander in 1839. See also **Chemical Elements: History of the Origin.** Electronic configuration

$$1s^2 2s^2 2p^6 3s^2 3p^6 3d^{10} 4s^2 4p^6 4d^{10} 5s^2 5p^6 5d^1 6s^2.$$

Ionic radius La^{3+} 1.045 Å. Metallic radius 1.879 Å. First ionization potential 5.577 eV; second, 11.06 eV. Oxidation potentials La $\rightarrow$ La^{3+} + $3e^-$, 237 V; La + 3OH$^-$ $\rightarrow$ La(OH)$_3$ + $3e^-$, 2.76 V.

Other important physical properties of lanthanum are given under **Rare-Earth Elements and Metals.**

Much of the commercial lanthanum production uses bastnasite, a rare-earth fluorocarbonate found in Inner Mongolia and Southern California, as the source. See also **Bastnasite.** The element is separated from other rare-earth elements by a liquid solvent extraction or an ion-exchange process after acid leaching of bastnasite (or monazite) minerals. Pure lanthanum is obtained by (1) electrowinning from the oxide La$_2$O$_3$ in a molten fluoride electrolyte, (2) electrolysis of fused anhydrous LaCl$_3$, or (3) metallothermic reduction of LaF$_3$ by calcium in a reactor under an inert atmosphere.

Lanthanum metal dissolves readily in dilute mineral acids. The oxide is dissolved by concentrated mineral acids and acetic and formic acids. The metal is a component of mischmetal used for lighter "flints" and in the cores of carbon electrodes for high-intensity lighting. See also **Cerium.** Several of the best grades of optical glass require pure lanthanum oxide as an ingredient for lowering the dispersion of light and for improving the index of refraction. The oxide melts at 2,310 °C (4,190°F). The oxide ranks eleventh among the most refractory metal oxides, but finds limited use because of its highly hygroscopic nature. The oxide also is used as a host matrix for fluorescent phosphors and in thermistors and capacitors and other elements of electronic circuitry. Lanthanum oxide combined with transition metal (e. g. Mn, Co, Cr) oxides are being used in solid oxide fuel cells (SOFC) as electrodes. By far, the largest use of lanthanum (mixed with other rare-earths) is for molecular-sieve catalysts for cracking crude petroleum.

As an alloying metal, lanthanum finds broad use. Although lacking mechanical strength, lanthanum has a high affinity for oxygen [CAS: 7782-44-7], sulfur [CAS: 63705-05-5], nitrogen [CAS: 7727-37-9], and hydrogen [CAS: 1333-74-0] and thus makes an effective component for scavenging gases from molten metals. Cobalt-base alloys containing lanthanum have shown increased resistance to oxidation and hot corrosion. One of the intermetallic compounds of lanthanum LaCo$_5$ possesses excellent magnetic properties — well in excess of those of alnico and platinum cobalt permanent magnets. The intermetallic compound LaNI.C. "eye"$_5$ shows exceptional properties for absorbing and desorbing large amounts of hydrogen at room temperature. It is the major component of rechargeable metal hydride batteries, which is a rapidly expanding major application.

See references listed at ends of entries on **Chemical Elements;** and **Rare-Earth Elements and Metals.**

K.A. GSCHNEIDNER Jr., B. EVANS, Iowa State University, Ames, IA

LAPIS LAZULI. See **Lazurite.**

LAPLACE EQUATION. A second-order partial differential equation of elliptic type which, in vector form, is $\nabla^2 \phi = 0$. It is the homogeneous case of Poisson's equation. Its solutions, the scalar quantity ϕ, occur in problems involving steady-state temperatures, gravitational and electric potentials, hydrodynamics of ideal fluids, and many other physical phenomena. The equation is usually solved by the method of separation of variables in a suitable curvilinear coordinate system and with boundary conditions imposed by physical requirements. Such solutions are called harmonic functions. In two dimensions, an analytic function of the complex variable must satisfy Laplace's equation. Thus, $w = u + iv$ is an analytic function of $z = x + iy$ for $u_{xx} + u_{yy} = 0$ and $v_{xx} + v_{yy} = 0$.

We shall here cite two familiar physical examples to which Laplace's equation applies; there are many others.

1. Consider a region of space in which there is an electric field (due to electric charges in the vicinity) but no free electricity, that is, no such space charge as exists in a vacuum tube in operation. At any point in this space there is an electric potential, which varies with the position of the point and is therefore a function of its coordinates. It is shown in electrostatic theory that if the potential satisfies Laplace's equation, it is possible to trace the lines of force and equipotential surfaces in the region. This is done by means of special solutions of Laplace's equation, called harmonic functions, which satisfy the "boundary conditions" imposed by the arrangement of the neighboring charges. As to which form of the equation is to be used, this depends upon which system of coordinates is most conveniently adapted to the shape and arrangement of the charged bodies.

2. If specified parts of the surface of a solid thermal conductor (such as a block of copper) are kept at different specified constant temperatures, heat will flow from the warmer toward the colder boundaries within the conductor. When this flow has become steady, the temperature of the conductor takes on a definite constant value at each point, dependent upon the location of the point. If the temperature is designated by u, it now satisfies Laplace's equation, and the lines of flow and isothermal surfaces may be determined accordingly. A similar procedure is applied to potential in the steady flow of electricity through a metallic conductor.

LAPLACE THEOREM. This limit theorem, of which the Bernoulli theorem is a corollary, states that if there are n independent trials, in each of which the probability of an event is p, and if this event occurs k times, then

$$P\left\{z_1 \leq \frac{k-np}{\sqrt{npq}} \leq z_2\right\} \longrightarrow \frac{1}{\sqrt{2\pi}} \int_{z_1}^{z_2} e^{-1/2z^2} \, dz$$

as $n \rightarrow \infty$, whatever the numbers z_1 and z_2. Roughly speaking, the theorem states that the number of successes k in n trials is normally distributed for large n.

LAPLACE TRANSFORM. This class of integral transform has so many applications in modern technology that it justifies the more extended discussion in this special entry.

The behavior of ac electric circuits is readily expressed in terms of integro-differential equations, as is understood by recalling that the voltage drop across an inductance is a differential function of the current, whereas the voltage across a capacitance is an integral function of the current. Moreover, many types of nonelectrical systems — mechanical, acoustical, hydraulic — lend themselves conveniently to analysis by analogy to electrical systems, so providing a convenient means of generalizing and unifying the analysis of complex systems involving subsystems of various kinds. This fact, together with the adaptability of the Laplace transform method to the solution of differential and integral equations, and the analysis of discontinuous functions, has brought about its wide application to control problems.

The Laplace transform of a function $f(t)$ is defined as the function given by

$$\mathcal{L}[f(t)] = \int_0^\infty f(t)e^{-st} \, dt = F(s) \tag{1}$$

Here the double equation means that the Laplace transform may be written as $\mathcal{L}[f(t)]$ or as $F(s)$, since the transformation process results in a new function $F(s)$ of a variable s instead of the original function f of time, t. The variable s may be real but is usually complex. The transformation is effected by integrating the function $f(t)$ by the use of the factor e^{-st}, where e is the natural logarithmic base (2.71828 ...). The integral is, as shown, in the improper form, its upper value being considered to increase indefinitely, approaching ∞ as a limit.

Evaluating Laplace Transforms. One of the simplest discontinuous functions is the *unit function*, which has only two values, 1 for positive or zero values of t, and 0 for negative values of t; it is often expressed as u (t). Thus for values of $t \geq 0$,

$$\mathcal{L}[u(t)]\,\mathcal{L}(1) = \int_0^\infty 1e^{-st}\,dt = -\frac{1}{s}e^{-st}\Big]_0^\infty$$

$$= -\frac{1}{s}0 - \left(-\frac{1}{s}1\right) = \frac{1}{s} \qquad (2)$$

The *unit ramp function* is a linear and continuous function of t, thus expressed as $f(t)$ for positive or zero values of t, so that

$$\mathcal{L}[f(t)] = \int_0^\infty te^{-st}\,dt$$

$$= -\frac{te^{-st}}{s}\Big]_0^\infty + \frac{1}{s}\int_0^\infty e^{-st}\,dt$$

$$= 0 + \left[\frac{1}{s}\cdot\frac{-1}{s}e^{-st}\right]_0^\infty = \frac{1}{s^2} \qquad (3)$$

The Laplace transform of a *power of t* is found by a change of the variable of integration, as from t to $w = st$, where s is real and positive:

$$\mathcal{L}(t^p) = \int_0^\infty w^p e^{-st}\,dt = \frac{1}{s^{p+1}}\int_0^\infty w^p e^{-w}\,dw \qquad (4)$$

The last integral at the right is essentially the gamma function, which is given in tables, and which is a generalization of the factorial $n!$, that is defined only for integers and zero. The gamma function may be written as

$$\Gamma(p) = \int_0^\infty w^{p-1}e^{-w}$$

and when $p = n$, the value of the function reduces to $n!$. Therefore, we can write for Equation (5),

$$\mathcal{L}(t^p) = \frac{1}{s^{p+1}}\Gamma(p+1) \qquad (5)$$

To find the Laplace transform of the *exponential function* e^{-ct} (for positive values of t) where c is a real number:

$$\mathcal{L}(e^{-ct}) = \int_0^\infty e^{-ct}e^{-st}\,dt = \int_0^\infty e^{-(s+c)t}\,dt$$

$$= -\frac{1}{s+c}e^{-st}\Big]_0^\infty = \frac{1}{s+c} \qquad (6)$$

To find the Laplace transform of the *trigonometric function* $\sin at$, where a is a real, positive number:

$$\mathcal{L}(\sin at) = \int_0^\infty \sin at\; e^{-st}\,dt$$

Now in the preceding examples, it was found that the evaluation of the Laplace transform of an exponential was quite simple because e^{-st} is also an exponential. So in the present example we replace $\sin at$ by its exponential expression $(e^{iat} - e^{-iat})/2i$, where $i = \sqrt{-1}$, giving

$$\mathcal{L}(\sin at) = \frac{1}{2i}\int_0^\infty (e^{iat} - e^{-iat})e^{-st}\,dt$$

$$= \frac{1}{2i}\left(\frac{1}{s-ia} - \frac{1}{s+ia}\right)$$

$$= \frac{a}{s^2+a^2} \qquad (7)$$

A similar substitution is used to find the Laplace transform of $\cos at$. Since $\cos at = (e^{iat} + e^{-iat})/2i$, we have

$$\mathcal{L}(\cos at) = \frac{1}{2i}\int_0^\infty (e^{iat} + e^{-iat})e^{-st}\,dt$$

$$= \frac{s}{s^2+a^2} \qquad (8)$$

Properties of the Laplace Transform. An important property of the Laplace transform is its linearity, so that the Laplace transform of the sum of two functions is equal to the sum of the individual Laplace transforms

$$\mathcal{L}[f_1(t) \pm f_2(t)] = F_1(s) \pm F_2(s) \qquad (9)$$

Here the original functions are denoted by $f_1(t)$ and $f_2(t)$, and their Laplace transforms by $F_1(s)$ and $F_2(s)$.

(I) The full statement of the *linearity* (also called *superposition*) *theorem* introduces two constants (a and b below):

$$\mathcal{L}[af_1(t) \pm bf_2(t)] = aF_1(s) + bF_2(s) \qquad (10)$$

conveying the further information that constant factors of functions appear unchanged in their Laplace transforms.

Since by Equation (2) the Laplace transform of $t = 1$ was found to be $1/s$, then it follows directly that the Laplace transform of any constant is expressed by

$$\mathcal{L}(c) = c\mathcal{L}(1) = \frac{c}{s} \qquad (11)$$

It also follows from the linearity theorem that the Laplace transform of any polynomial is equal to the sum of the Laplace transforms of its terms:

$$\mathcal{L}(a_0 + a_1t + a_2t^2 + \cdots + a_nt^n) = a_0\mathcal{L}(1) + a_1\mathcal{L}(t)$$

$$+ a_2\mathcal{L}(t^2) + \cdots + a_n\mathcal{L}(t^n) \qquad (12)$$

The general expression for this series of terms is, therefore,

$$\sum_{}^n a_i\frac{\Gamma(i+1)}{s^{i+1}} = \frac{a_0}{s} + \frac{a_1}{s^2} + \frac{2!a_2}{s^3} + \frac{3!a_3}{s^4}\cdots\frac{n!a_n}{s^{n+1}} \qquad (12a)$$

Note that, as derived, Equation (12a) is valid only for positive and real values of s. In fact, in all operations with Laplace transforms constant attention must be paid to (1) whether the transform of a function exists at all, and, if it does, (2) for what values of the variables it has corresponding real values and for what values it becomes zero or indefinitely great.

(II) The *real differentiation theorem* asserts that if a function and its first derivative have Laplace transforms and if $\mathcal{L}[f(t)] = F(s)$, then

$$\mathcal{L}\frac{df(t)}{dt} = sF(s) - f(t \to 0+) \qquad (13)$$

which states that the Laplace transform of the first time derivative of a function is equal to the product by s of the Laplace transform of the function itself minus the value of the function as t approaches 0 from the positive side.

(III) The *real integration theorem* asserts that if a function of t has a Laplace transform and if $\mathcal{L}[f(t)] = F(s)$, then

$$\mathcal{L}\left[\int f(t)\,dt\right] = \frac{1}{s}\left[F(s) + \int_{0+} f(t)\,dt\right] \qquad (14)$$

which states that the Laplace transform of the integral of the function is equal to the product by $1/s$ of the Laplace transform of the function plus the value of the integral of the function as t approaches zero from the positive side.

The usefulness of these two theorems is due to the fact, as stated earlier, that so many of the circuit and control equations contain differentials or integrals or both.

(IV) There are two translation theorems used with Laplace transforms. The first of these is the *time-delay theorem*, which is written

$$\mathcal{L}[f(t - t_0)] = e^{-t_0 s}F(s) \qquad (15)$$

where t_0 is real and ≥ 0, provided $f(r) = 0$ for $t < 0$ and, in the case of the $0+$ transform, $f(t)$ in addition contains no impulse functions at $t = 0$.

A direct corollary of the time-delay theorem is the *unit step function*. In Equation (2) it was found that for values of *t* of unity, the Laplace transform was $1/s$. The unit step function is $u(t - t_0)$, where *t* has a value of 1 for all positive and zero values of the function, i.e., for all values of *t* greater than or equal to t_0, while *t* is zero for all its other values. This function expresses the commencement of a steady-state operation or process of unit value at time t_0.

Equations (2) and (15) give the unit step function directly, for by (2) the Laplace transform of unity is $1/s$, which corresponds to $F(s)$ in (15), giving

$$\mathcal{L}u(t - t_0) = e^{-t_0 s} \cdot \frac{1}{s} = \frac{e^{-t_0 s}}{s}$$

(V) The *time advance theorem* is written

$$\mathcal{L}[f(t + t_0)] = e^{et_0 s} F(s) \tag{16}$$

Note the change of sign of t_0 on both sides of the equation from those in the time delay theorem of Equation (15). The same conditions apply: that t_0 is real and ≥ 0, provided $f(t) = 0$ for $t < t_0$ and, in the case of the $0+$ transform, $f(t)$ in addition contains no impulse functions at $t = t_0$.

(VI) The *frequency shift theorem* is written

$$\mathcal{L}[e^{at} f(t)] = F(s - a) \tag{17}$$

where *a* is any finite constant, real or complex.

(VII) The *final value theorem* asserts that if the function $f(t)$ and its first derivative have Laplace transforms and if $\mathcal{L}[f(t)] = F(s)$, then

$$\lim_{s \to 0} s F(s) = \lim_{t \to \infty} f(t) \tag{18}$$

provided that $s F(s)$ is analytic on the imaginary axis and on the right half-plane.

(VIII) The corresponding *initial value theorem* asserts that if a function and its first derivative have Laplace transforms, and if $\mathcal{L}\phi f(t)] = F(s)$, then

$$\lim_{s \to \infty} s F(s) = \lim_{t \to \infty} f(t) \tag{19}$$

provided that the $\lim s F(s)$ as *s* approaches infinity exists.

(IX) The last of these Laplace theorems that is widely used in elementary system calculations is the *convolution theorem*. The convolution of two functions $f_1(t)$ and $f_2(t)$, denoted by $f_1^* f_2 = f_2^* f_1$ is defined by the integral

$$f_1^* f_2 = \int_0^t f_1(t - \tau) f_2(\tau) \, d\tau$$

If $f_1 = t^n$ and $f_2 = t^m$ (*n* and *m*, two positive integers), it can be verified by direct calculation that

$$\mathcal{L}\{f_1^* f_2\} = F_1(s) F_2(s)$$

where $F_1(s)$ is the Laplace transform of $f_1(t)$ and $F_2(s)$ is the Laplace transform of $f_2(t)$. Thus, if $f_1 = t^2$ and $f_2 = t^3$, then

$$f_1^* f_2 = \int_0^t (t - \tau)^2 \cdot \tau^3 \, dt = \frac{t^6}{60}$$

On the other hand, if

$$\mathcal{L}(t^2) = 2/s^3 \quad \text{and} \quad \mathcal{L}(t^3) = 6/s^4$$

then

$$\mathcal{L}(t^2) \cdot \mathcal{L}(t^3) = \frac{12}{s^7} = \mathcal{L}\left(\frac{t^6}{60}\right)$$

From this result it follows immediately, if $f_1(t)$ and $f_2(t)$ can be expanded in two convergent power series, that

$$(f_1^* f_2) = F_1(s) F_2(s) \quad \text{(convolution theorem)} \tag{20}$$

In fact, the theorem is true in more general cases. The convolution theorem can be used to derive additional Laplace transforms and also to solve so-called "integral equations," i.e., functional equations where the unknown function appears behind an integral sign.

The Inverse Laplace Transform. In solving problems by means of the Laplace transform, the inverse operation of finding functions which correspond to transforms is, of course, as important as the direct one of finding the transforms. For this purpose, it is convenient to have values of transform pairs, which are given in Table 1. The justification for the inverse operation rests, of course, upon the assumption of uniqueness, that for each transform there is only one function. This is true, although its proof is not given here.

TABLE 1. LAPLACE FUNCTION-TRANSFORM PAIRS

	$f(t)$	$F(s)$
1.	1	$\dfrac{1}{s}$
2.	c	$\dfrac{c}{s}$
3.	t	$\dfrac{1}{s^2}$
4.	t^p	$\dfrac{\Gamma(p + 1)}{s^{p+1}}$
5.	$\sin at$	$\dfrac{a}{s^2 + a^2}$
6.	$\cos at$	$\dfrac{s}{s^2 + a^2}$
7.	e^{-et}	$\dfrac{i}{s + c}$
8.	$e^{at} t^n$	$\dfrac{n!}{(s - a)^{n+1}}$
9.	$t \sin at$	$\dfrac{2as}{(s^2 + a^2)^2}$
10.	$t \cos at$	$\dfrac{s^2 - a^2}{(s^2 + a^2)^2}$

The symbol for the operation of inverse (Laplace) transformation is $\mathcal{L}^{-1}$.

There is one kind of inverse transform, however, that is not readily tabulated. It is the inverse transform of a fraction. Since fractions occur quite frequently in the problems solved by these methods, the following discussion of one case of finding their inverse is important.

Given a Laplace transform in the form of a rational algebraic fraction,

$$F(s) = \frac{A(s)}{B(s)} = \frac{a_m s^m + a_{m-1} s^{m-1} + \cdots + a_1 s + a_0}{b_n s^n + b_{n-1} s^{n-1} + \cdots + b_1 s + b_0} \tag{21}$$

where the a_i and b_i are all real constants and *n* and *m* are positive integers. Where the roots of the denominator are all real or zero and different, a direct method is available to find $\mathcal{L}^{-1}$.

Let the roots of the denominator be $s_1, s_2, \ldots, s_n$. The $F(s)$ takes the form

$$F(s) = \frac{a_m s^m + a_{m-1} s^{m-1} + \cdots + a_1 s + a_0}{(s - s_1)(s - s_2) \cdots (s - s_n)}$$

This equation may be expressed as the sum of partial fractions

$$F(s) = \frac{k_1}{s - s_1} + \frac{k_2}{s - s_2} + \cdots + \frac{k_n}{s - s_n}$$

where the k_i are coefficients to be determined. The procedure is to multiply both sides of the equation by $(s - s_i)$, which is then equated to 0. Thus expressions for each coefficient are obtained of the general form

$$k_i = \left[(s - s_i) \frac{A(s)}{B(s)}\right]_{s=s_i}$$

where the term $A(s)/B(s)$ represents the transforms whose inverse is to be found.

Then using for each coefficient transform pair number 8 in the table of transforms, we have

$$\mathcal{L}^{-1}\left[\frac{k_i}{s - s_i}\right] = k_i e^{s_i t} \tag{22}$$

which can be found for each fraction, so that the inverse transform of $F(s)$ that is sought is the sum of all *i* values

$$\mathcal{L}^{-1}[F(s)] = \sum^i k_i e^{s_i t}, \quad 0 \leq t \tag{23}$$

Applications of the Laplace Transform. (a) *Simple resistance-capacitance circuit.* A simple example of the application of the Laplace transform in solving equations is provided by the RC-circuit. Let us find, for example, the behavior of the instantaneous current $i(t)$ from the time that switch S in Fig. 1 is closed. The current produced by a given electromotive forces (E) varies inversely as the resistance $(i(t) = E/R)$ or $E = i(t)R$, and the integrated current (with respect to time) varies directly as the capacitance $(\int i(t) \, dt = CE)$ or $(E = (1/C) \int i(t) \, dt)$ so that the equation for a circuit having both capacitance and resistance is

$$Ri(t)\frac{1}{C} \int i(t) \, dt = E \tag{24}$$

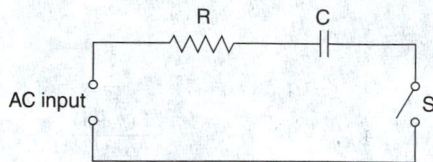

Fig. 1.   Resistance-capacitance (RC) circuit.

Taking the Laplace transform of both sides, we have

$$\mathcal{L}\left[Ri(t) + \mathcal{L}\left[\frac{1}{C}\int i(t) \, dt\right]\right] = \mathcal{L}(E) \tag{25}$$

By theorem I, and the definition of the Laplace transform, $\mathcal{L}[Ri(t)] = RI(s)$, where $I(s)$ is the effective value of the current, by theorem II,

$$\mathcal{L}\left[\frac{1}{C}\int i(t) \, dt\right] = \frac{1}{sC}\left[I(s) + \int i \, dt(0+)\right]$$

Therefore the equation becomes

$$RI(s) + \frac{1}{sC}\left[I(s) + \int i \, dt(0+)\right] = \frac{E}{s} \tag{26}$$

by using transform pair number 2 from the table to show that $\mathcal{L}(E) = E/s$. If it is assumed that the initial condition of the circuit is with the switch open, then

$$\frac{1}{sC}\int i \, dt(0+) = 0$$

so that

$$RI(s) + \frac{1}{sC}I(s) = \frac{E}{s}$$

Transposing, we have

$$I(s) = \frac{E/s}{R + \dfrac{1}{sC}} = \frac{E}{s\left(R + \dfrac{1}{sC}\right)}$$

$$= \frac{E/R}{s + \dfrac{1}{RC}}$$

Substitute P for RC, so that

$$I(s) = \frac{E}{R} \cdot \frac{1}{s + 1/P} \tag{27}$$

Taking the inverse transform, we have

$$\mathcal{L}^{-1}[I(s)] = i(t) = \frac{E}{R}\mathcal{L}^{-1}\frac{1}{s + 1/P}$$

Then by inverse transform number 7 in the table,

$$i(t) = \frac{E}{R}e^{-t/P} = \frac{E}{R}e^{-t/RC} \tag{28}$$

which is the instantaneous value of the current for given values of E, R, and C, at time t.

(b) *Generalized resistance-capacitance* circuit. In investigating the behavior of control and servo systems, an effective approach is from the point of view of the ac electrical circuit. Thus, voltage or electromotive force in the electrical system has as its analogs force in a mechanical system, and pressure in an acoustical or hydraulic system. Current in an electrical system is analogous to velocity in a mechanical system or an acoustical system, and to rate of flow in a hydraulic system. Electrical

resistance has as its analogs friction in a mechanical system or a hydraulic system, and the real component of acoustic impedance in an acoustical system. These analogies can be extended to capacitance, inductance, impedance, etc.

Because of these analogies between systems, the use of generalized concepts in control problems is particularly valuable. Let us take the first step in this direction by using again the simple resistance-capacitance circuit of the previous application, but denoting the input voltage by $\theta_i(t)$ to differentiate it from the voltage available to an external load by connecting the latter across the condenser, the latter being called the output, $\theta_0(t)$. See Fig. 2. Then the electrical equations for these two voltages are

$$\theta_i(t)Ri(t) + \frac{1}{C}\int i(t) \, dt \tag{29}$$

$$\theta_0(t) = \frac{1}{C}\int i(t) \, dt \tag{30}$$

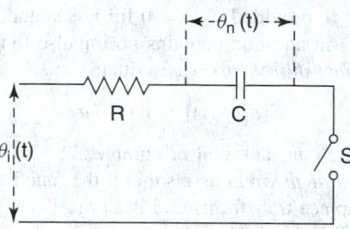

Fig. 2.   Resistance-capacitance (RC) ac circuit showing input and output connections.

Assuming, as in the previous application, that there is no initial charge on C (switch open), then the Laplace transforms of $\theta_i(t)$ and $\theta_0(t)$ are

$$\mathcal{L}[\theta_i(t)] = \theta_i(s) = RI(s) + \frac{1}{sC}I(s) \tag{31}$$

$$\mathcal{L}[\theta_0(t)] = \theta_0(s) = \frac{1}{sC}I(s) \tag{32}$$

We now generalize these terms; $\theta_i(t)$ is called the driving function of the system; $\theta_0(t)$ is called the response function of the system; $\theta_i(s)$ is called the driving transform of the system; and $\theta_0(s)$ is its response transform. The ratio of the response transform to the driving transform, $\theta_0(s)/\theta_i(s)$ is called the transfer function, and is a characteristic of the system independent of its particular input or output: it is often denoted by $G(s)$.

Then we have, from the foregoing equations,

$$G(s) = \frac{\theta_0(s)}{\theta_i(s)} = \frac{\dfrac{1}{sC}I(s)}{RI(s) + \dfrac{1}{sC}I(s)} = \frac{1}{RCs + 1} \tag{33}$$

(c) *Unit step driving function.* The first Laplace transform evaluated in this section (Equation (2)) was that of the unit function $u(t)$ for the two values 1 and $t \geq 0$, and 0 for $t < 0$. Its Laplace transform was found to be $1/s$.

Therefore, if the driving function, $\theta_i(t)$, of a system is of the unit kind, as shown in Fig. 3, its driving transform is

$$\theta_i(s) = 1/s \tag{34}$$

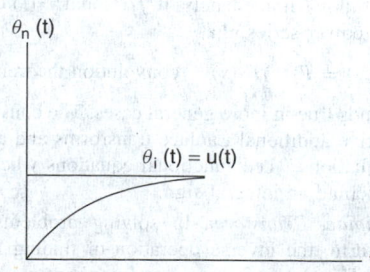

Fig. 3.   Response of RC circuit to unit input.

and its response transform is, from Equation (33),

$$\theta_0(s) = G(s)\theta_i(s) = \frac{1}{RCs+1} \cdot \frac{1}{s} + \frac{1}{s(RCs+1)} \quad (35)$$

Then applying the partial fraction method expressed in Equations (22) and (23), we have

$$\theta_0(s) = \frac{k_1}{s} + \frac{k_2}{RCs+1} = \frac{1}{s} + \frac{-RC}{RCs+1} \quad (36)$$

Therefore,

$$\theta_0(s) = \frac{1}{s} - \frac{1}{s - 1/RC}$$

Then using transform paris numbers 1 and 7 in the table,

$$\mathcal{L}^{-1}[\theta_0(s)] = \theta_0(t) = 1 - e^{-t/RC} \quad (37)$$

Figure 3 shows how the value of $\theta_0(t)$ approaches the unit value asymptotically as t increases, as is evident from the equation.

(d) *Unit ramp driving function.* The second Laplace transform evaluated in this treatment, Equation (3), was that of the unit ramp function $f(t)$ for values of $t \geq 0$. Its Laplace transform was found to be $1/s^2$.

Therefore, if a driving function $\theta_i(t)$ is of the Heaviside unit ramp kind, as shown in Fig. 4, its driving transform is

$$\theta_i(s) = 1/s^2$$

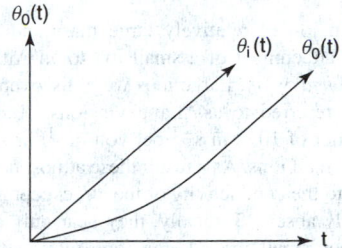

Fig. 4.   Response of *RC* circuit to Heaviside unit ramp input.

and its response transform is, from Equation (35),

$$\theta_0(s) = \frac{1}{s^2(RCs+1)} \quad (38)$$

Then applying the partial fraction method expressed in Equations (22) and (23), we have

$$\theta_0(s) = \frac{k_1}{s^2} + \frac{k_2}{s} + \frac{k_3}{RC+1}$$

$$= \frac{1}{s^2} - \frac{RC}{s} + \frac{(RC)^2}{RCs+1}$$

$$= \frac{1}{s^2} - \frac{RC}{s} + \frac{RC}{s+1/RC}$$

Then by use of transform pairs numbers 1, 2, and 7 in the table, we have so

$$\mathcal{L}^{-1}\theta_0(s) = \mathcal{L}^{-1}\left[\frac{1}{s^2} - \frac{RC}{s} + \frac{RC}{s+1/RC}\right]$$

$$\mathcal{L}^{-1}\theta_0(s) = \theta_0(t) = t - RC(1 - e^{-t/RC}) \quad (39)$$

Figure 4 shows how the value of $\theta_0(t)$ lags the unit ramp function as t increases. See also **Time Constant.**

LAPLACIAN.   The vector operator

$$div \text{ grad} = \nabla \cdot \nabla = \nabla^2$$

In rectangular coordinates its components are $\partial^2/\partial x^2$, $\partial^2/\partial y^2$, $\partial^2/\partial z^2$; in spherical polar coordinates

$$\frac{1}{r^2}\frac{\partial}{\partial r}\left(r^2\frac{\partial}{\partial r}\right), \frac{1}{r^2\sin\theta}\frac{\partial}{\partial\theta}\left(\sin\theta\frac{\partial}{\partial\theta}\right), \frac{1}{r^2\sin^2\theta}\frac{\partial^2}{\partial\phi^2}$$

and in cylindrical coordinates

$$\frac{1}{\rho}\frac{\partial}{\partial\rho}\left(\rho\frac{\partial}{\partial\rho}\right), \frac{1}{\rho^2}\frac{\partial^2}{\partial\phi^2}, \frac{\partial^2}{\partial z^2}$$

See also **Del; Divergence (Mathematics);** and **Gradient (Mathematics).**

LAPPING.   The process of producing an extremely accurate, highly finished surface by means of a lap, which is a block charged with abrasive. Lapping reduces the possibility of wear on close-fitting running parts or on the surfaces of measuring equipment, by reducing the minute ridges and serrations left by machining and grinding operations to a more uniform bearing surface. Lapping may be done by hand or by machine. If a part is to be hand lapped to a final accurate dimension, a lap or mating form is made from a metal somewhat softer than the part to be finished. The surface of the lap is charged with a fine abrasive, or a small amount of abrasive mixed with grease, oil, or alcohol. A flat lap has a carefully trued surface with a series of grooves in it. The lapping compound is smeared on the face of the lap and the work is rubbed over the face along an ever-changing path. The grooves in the face of the lap act as channels for any excess abrasive and oil. Very little pressure is used in order to eliminate the danger of scoring the work or stripping the lap. Hand lapping requires skill and time. The amount of material removed by lapping should not exceed 0.0002 to 0.0005 inch (0.005 to 0.013 millimeter).

LAPSE.   See **Atmosphere (Earth).**

LAPSE RATE.   The rate of change of an atmospheric variable, usually temperature, with height. A steep lapse rate implies a rapid decrease in temperature with height (a sign of instability) and a steepening lapse rate implies that destabilization is occurring.

LAPWING.   See **Waders, Shorebirds, and Gulls.**

LARCH TREES.   Members of the family *Pinaceae* (pine family), these trees are of the genus *Larix*. The larches are different among the conifers in that they are deciduous, turning a beautiful yellow color in the fall. They prefer a lot of sunlight and do well in most soils except dry, yellow, chalky soils. The principal species, in addition to those listed on accompanying table, include:

Dahurian larch	*Larix gmelini*
Dunkeld hybrid larch	*L. eurolepsis*
Japanese larch	*L. kaempferi*
Red larch	See Japanese larch
Siberian larch	See Dahurian larch
Siberian larch (red)	*L. olgensis*
Sikkim larch	*L. griffithiana*

The term *tamarack* frequently refers to the eastern larch, but is also used in connection with certain other larches, including the western larch.

The eastern larch is found through Alaska, Canada, south to Minnesota and through the northern parts of the midwestern United States, and south and eastward through West Virginia, Pennsylvania, and New York. In Canada, the tree follows the Mackenzie River northward into Yukon-Northwest Territory. The tree bark is thin, bright red-brown. The wood weighs about 38 pounds per cubic foot (609 kilograms per cubic meter). It is strong, durable, hard, and close-grained. In the spring, the flowers are of bright yellow. The needles are bright green from 1/2 to 1 inch (1.3 to 2.5 centimeters) long. The cone is winged, chestnut brown, and from 1/2 to 3/4 inch (1.3 to 1.9 centimeters) long. The eastern larch or tamarack of the far north does not grow so large as those found in more southern climes, but it grows very straight and is ideal for telephone poles and railroad ties. However, it is not as extensively used for timber as the western larch. Other names sometimes used for the tree are American larch, black larch, and eastern hackmatack. See Table 1.

The western larch is found principally from British Columbia to Oregon, Montana, and into Idaho. It grows at elevations up to about 7000 feet (2100 meters). The bark is thick (3 to 6 inches) (7.6 to 15.2 centimeters) and is reddish-brown with deep wide ridges. The twig is stout and brittle. The leaf is pale green, 1e1/2 to 2 inches (3.8 to 5 centimeters) in length and grows in

TABLE 1. RECORD LARCH TREES IN THE UNITED STATES[1]

Specimen	Circumference[2]		Height		Spread		Location
	Inches	Centimeters	Feet	Meters	Feet	Meters	
European larch (1996) (*Larix decidua*)	183	465	92	28	72	21.9	Vermont
Subalpine larch (1993) (*Larix lyallii*)	236	599	94	28.7	56	17.1	Washington
Western larch (1993) (*Larix accidentalis*)	230	584	189	57.6	35	10.7	Washington
Western larch (1995) (*Larix accidentalis*)	264	671	153	46.6	34	10.4	Montana

[1]From the "National Register of Big Trees," American Forests (by permission).
[2]At 4.5 feet (1.4 meters).

clusters of 20 to 30. The wood is heavy, durable, strong, and close-grained. The heart wood is red-brown in color; the sap wood is thin and nearly white. In the green condition, the moisture content of western larch is 58%, with a weight of 48 pounds per cubic foot (769 kilograms per cubic meter). When air-dried to 12% moisture content, the weight is 36 pounds per cubic foot (577 kilograms per cubic meter), and 1,000 board-feet (2.36 cubic meters) weigh about 3,000 pounds (1360 kilograms). The compressive or crushing strength parallel to the grain of the green wood is 3,990 pounds per square inch (27.5 MPa); of the air-dried wood, 8,110 pounds per square inch (56 MPa). The tensile strength perpendicular to the green of green wood is 330 pounds per square inch (2.3 MPa); of the air-dried wood, 430 pounds per square inch (3 MPa). The wood is valuable for many uses, including interior finished products, furniture, and cabinetry, but its major value lies in its use for heavy-duty construction, rough timbers for mining, railroad ties, and telephone poles. The tree usually grows very straight. In some areas, the tree has been extensively cut and is now under government protection. The tree frequently is planted for ornamental purposes where the size is appropriate. Douglas fir timber frequently is shipped along with the western larch timbers. Other commercial names for the wood include mountain larch, Montana larch, and hackmatack.

As noted from the list of species, larches also occur widely in parts of Europe and Asia. The European larch is an important source of wood in Russia, and other eastern European countries.

See also **Conifers**.

LARGE BINOCULAR TELESCOPE INTERFEROMETER (LBTI).
See **Origins Program**; and **Space Science Missions: Universe**.

LARGE EDDY MODEL.
A numerical model with an averaging volume sufficiently small that the eddies contained entirely within it (i.e., the unresolved eddies) are of inertial-range scales and smaller.

The larger energy-containing eddies are resolved explicitly away from the boundaries. In the boundary layer the energy-containing eddies are not resolved near the surface and near the inversion at the top. In these regions a subgrid-scale parameterization is needed.

Additional Reading
Nieuwstadt, F., and H. Van Dop: *Atmospheric Turbulence and Air Pollution Modelling*, Springer-Verlag New York, LLC, New York, NY, 2002.

AMS

LARGE EDDY SIMULATION (LES).
1. A three-dimensional numerical simulation of turbulent flow in which large eddies (with scales smaller than the overall dimension of the problem in question) are resolved and the effects of subgrid-scale eddies, which are more universal in nature, are parameterized.

Large eddies are important in characterizing one turbulent flow from another. The difference between a large eddy simulation and the traditional phenomenological modeling of turbulence is that in the latter case all scales of turbulent motion are parameterized.

2. A modeling technique in which spatial resolution extends into the inertial subrange, but does not resolve the smallest scales of motion. The effects of the latter are approximated using subgrid-scale models, which usually draw heavily on the Kolmogorov theory of the inertial subrange.

AMS

LARGE EDDY SIMULATION MODELS.
Computer codes that numerically integrate in three dimensions the time-dependent Navier—Stokes equations filtered over a grid volume much smaller than the size of the energy-containing eddies or wavelengths of turbulence. See also **Navier-Stokes Equations**.

The solutions of these models consist of energy-containing large eddies that carry most of the turbulent fluxes. The net effect of small subgrid-scale eddies is treated as locally diffusive and dissipative, typically modeled based on inertial-subrange theory.

AMS

LARGE ION.
An ion of relatively large mass and low mobility that is produced by the attachment of a small ion to an Aitken nucleus. Also called *slow ion* and *heavy ion*. Large ions were discovered by P. Langevin and are sometimes referred to as "Langevin ions." Large ions have ion mobilities of the order of 10^{-8} m s^{-1} per volt m^{-1}, or some 10 000 times lower than those of small ions. As a result these atmospheric ions contribute practically nothing to the conductivity of the air, except in rare cases where small ions are nearly absent. Typically, they bear only a single electronic charge, as is true of small ions. Large ions move so slowly that they are not destroyed by being neutralized by still other large ions of paired signs, for such collisions are too infrequent. Instead, they are neutralized by union with a small ion of opposite sign. Their mean lifetimes are of the order of 15–20 minutes over the oceans, but may approach 1 h in very polluted air. The ion density of large ions varies widely depending upon the degree of atmospheric pollution. Representative low-altitude values might be 10^9 m^{-3} in clean country air, 10^{10} m^{-3} in an industrial area, and 10^8 m^{-3} over the oceans.

Additional Reading
Wait, G.R., and W.D. Parkinson:in *Compendium of Meteorology*, American Meteorological Society, Boston, MA, 1951.

AMS

LARGEMOUTH BLACK BASS. See **Sunfishes**.

LARGE NUMBERS (Law of). See **Law of Large Numbers**.

LARGE SCALE. See **Meteorology**.

LARK (*Aves, Passeriformes*).
Song birds of many species, confined to the northern hemisphere. The skylarks of Europe and Asia are the most famous members of the group because of the quality of their song. The common European species, *Alauda arvensis*, is established in Oregon and North America also has a native species, the horned lark, *Otocoris alpestris*. The meadowlark is more closely related to the blackbirds and orioles. See also **Meadowlark**.

LARMOR'S THEOREM.
This theorem concerns the motion which a charged particle or system of charged particles subject to a central force directed toward a common point experiences when under the influence of a small uniform magnetic field. If a coordinate system is chosen which

rotates about the direction of the magnetic field with an angular velocity,

$$\omega = -\frac{q}{2mc}H$$

where ω is the angular velocity, q is electric charge in electrostatic units, m is mass, c is the velocity of light, H is magnetic field strength in electromagnetic units, then the motion in this system of coordinates is the same as the motion referred to a coordinate system fixed in space without a magnetic field. Application of this principle arises almost exclusively in the description of the motion of atoms and electrons in magnetic fields.

LARVA. An immature form of animals, that undergoes metamorphosis between emergence from the egg and the attainment of adult life. The larva is often very different from the adult.

The larvae of many invertebrates of sessile habit, such as the sponges and some coelenterates, are ciliated (cilia) organisms, which swim about for a time before attaching themselves to the permanent support where they are to develop. In the two phyla mentioned the larva is little more than a ciliated gastrula. It is filled with solid endoderm in the coelenterates and is called a planula.

The flukes also begin life as a ciliated larva known as a miracidium. This form gives rise to more complex larvae called rediae and these in turn produce tailed larvae called cercariae. The cercaria is transformed into the adult. In the same phylum the tapeworms hatch as 6-spined hexacanth larvae and pass through a bladder-worm stage before becoming adults.

Roundworms of many species pass through one or more larval stages in which they are worm-like but differ in habits and in some structural details from the adults.

Some of the segmented worms, mollusks, echinoderms, and chordates, hatch as complex larvae with localized zones of cilia for locomotion. The trochophore or trochosphere larva of annelids and mollusks has a ciliated alimentary tract with mouth and anus 90 degrees apart and a belt of cilia around the middle of the body. Larvae of echinoderms also have a bent alimentary tract but the cilia are arranged in one or more bands, sometimes of intricate form. These larvae bear various names: bipinnaria, auricularia, pluteus, of the starfishes, sea cucumbers and sea urchins and brittle stars, respectively, or collectively the dipleurula. The bipinnaria resembles the tornaria larva of Balanoglossus.

Among the arthropods the high development of metamorphosis is accompanied by great diversity of larval forms. In the class Crustacea these forms seem to represent previous evolutionary stages and are named after groups of the class whose adults they resemble. Among them are the nauplius, cypris, and cyclops larvae and many others. A single individual may pass through several of these stages in the course of its development. Insects present an entirely different type of metamorphosis in which the larval characteristics appear to have been acquired later in the course of evolution than those of the adult, as an adaptation to special conditions (Cenogenesis). In species with complete metamorphosis, only the first immature stage is called the larva. In this stage, butterflies and moths are called caterpillars; some of the beetles, grubs; and many flies, maggots. Species with less complex metamorphosis are called nymphs or naiads during development (in gradual and incomplete metamorphosis respectively).

See also **Metamorphosis**; and **Pupa**.

LARYNX. Sometimes called voice box, the larynx is a cartilaginous organ of the throat that contains the vocal cords and produces most of the sound in phonation. The larynx is shaped fundamentally like a tube, wide at the top and narrow at its lower portion. Generally, it appears somewhat like a three-cornered tube with a prominent ridge on the front side. It is made up of several pieces of firm elastic tissue (cartilage), which are held together by muscles and ligaments. The thyroid cartilage is the largest cartilage of the larynx and consists of two plates standing on end, which meet in the front of the neck and form a ridge (Adam's apple).

The interior of the larynx extends from the pharynx above to the windpipe (trachea) below. The inner tube of the larynx is divided horizontally into two parts by the projection of the muscular *vocal folds*, which contain the two *vocal cords*. The cords are really folds in the lining of the larynx rather than true cords. These cords produce the sound that is converted into speech by the movements of the mouth and tongue. When the vocal cords are tightened, the air being exhaled causes the cords

to vibrate, and sounds are produced. The tighter the cords, the higher the tone. These sounds are made into words by the tongue, teeth, and lips. The degree to which a person can tighten and relax his vocal cords determines his tone range in singing and speaking. When the vocal cords are fully relaxed, no sound is made.

Laryngitis is an inflammation of the mucous membrane of the voice box. Common microbial flora found in the larynx include *Haemophilus influenzae*, influenza virus, parainfluenza virus, adenovirus, coxsackievirus, rhinovirus, and respiratory syncytial virus. Uncommonly present in the larynx are pneumococci, Group A streptococci, *Mycobacterium tuberculosis*, and *Mycoplasma pneumoniae*. Viral laryngitis in children is commonly called *croup*.

Cancer of the larynx makes up only about 2% of all human cancers. It occurs most often in men between ages 40 and 60 years, although all age groups and both sexes may be affected. The main symptom is hoarseness. Any voice change that persists more than two weeks should be investigated by a physician. There may be a tickling sensation in the throat, or simply a discomfort of the throat. There may be difficulty in swallowing and pain on speaking. Other ensuing developments may be a cough, shortness of breath, wheezing, and halitosis. Lymph nodes in the neck may become enlarged. Occasionally the patient may expectorate blood.

In the United States, new cases of laryngeal cancer number about 12,000 per year, of which 3,700 are ultimately fatal. Epidemiologic studies from many parts of the world have demonstrated an increase in the relative risk of laryngeal cancer in smokers as compared with nonsmokers, ranging from 2.0 to 27.5, with a strong dose-response relation. Laryngeal cancer, in accordance with recent statistics on smoking, may be increasing among women and somewhat decreasing among men. As with lung cancer, the time necessary after smoking cessation for the risk levels for laryngeal cancer to approach those in nonsmokers is about 10 to 15 years.

Radiation and/or surgical therapy may be used in connection with laryngeal cancers. Surgery is usually indicated if the disease involves a large portion of the larynx. When the larynx is removed, the patient's windpipe is attached to the skin of the neck. Thus, the patient no longer breathes through the nose, but rather through a hole in the neck. The patient must take precaution to avoid breathing in foreign matter because of the lack of protection normally afforded by the nose. As soon as healing permits, the patient will normally commence speech lessons to learn to speak by application of one of two natural procedures, the esophageal method or the pharyngeal method.

Where sustained effort to learn either of these methods does not lead to success, the patient may use an artificial larynx, a device that is held against the side of the throat. When activated, the device transmits vibrations to the pharynx and mouth, providing intelligible but unnatural sounds.

Alternate Treatment. As pointed out in a 1991 study by the Department of Veterans Affairs, a Laryngeal Cancer Study Group reports, "Because laryngectomy results in substantial functional morbidity, including the loss of the natural voice, alterations in deglutition (swallowing), and the creation of a permanent tracheostoma (hole) in the neck, alternative forms of treatment have been developed. In selected patients with moderately advanced cancers, partial laryngeal resections that spare vocal function or primary radiation therapy achieve survival rates comparable to those obtained with total laryngectomy, permitting preservation of the larynx in 40 to 70 percent of patients. For patients with more advanced disease, however, treatment by radiation alone, with salvage by surgery, if necessary, results in lower survival rates."

The Veterans Affairs Group made a study of 352 patients, who were randomly assigned to receive either three cycles of chemotherapy (cisplatin and fluorouracil) and radiation therapy or surgery and radiation therapy. After two cycles of chemotherapy, the clinical tumor response was complete in 31% of patients and partial in 54% of patients. The report concludes: "These preliminary results suggest a new role for chemotherapy in patients with advanced laryngeal cancer and indicate that a treatment strategy involving induction chemotherapy and definitive radiation therapy can be effective in preserving the larynx in a high percentage of patients, without compromising overall survival."

Injuries. Freak, uncommon accidents can affect the functioning of the larynx and related organs. In 1991, T.C. James, T.C. Li (Mayo Clinic), and D. Gunderson (Park Nicollet Medical Center, Minneapolis) report that a young man suffering from asthma awoke in the middle of the night and reached for a metered-dose inhaler that he had been using. Unfortunately, still groggy from sleep, the man did not remove the cap of the inhaler and aspirated it into the larynx. He suffered some vocal cord damage, but recovered within a period of about 2 months. In another instance, a middle-aged man had acute respiratory distress after using an inhaler. A laryngoscopy revealed that he had inhaled a loose coin that he kept in his pocket along with the inhaler. Radiology revealed a coin in the right mainstem bronchus. This was removed, and the patient recovered without sequelae. This latter case was reported by Carol Shultz (Henry Ford Hospital) and associates in 1991.

Obviously, persons who are given inhalers should be instructed to check the mouthpiece before each use and to cap the device after each use.

Myasthenic laryngitis signifies a weakness and exhaustion of the muscles in the larynx resulting from overuse of the voice box. Resting the voice part of each day is important, as well as refraining from shouting, talking in a loud voice, and all unnecessary uses of the voice. Usually the voice returns to normal when these precautions are observed.

Additional Reading

Bailey, B.J. and H.F. Biller: *Surgery of the Larynx*, W.B. Saunders Company, Philadelphia, PA, 1998.

Ferlito, A.: *Diseases of the Larynx*, Arnold Publishing, London, UK, 2000.

Ferlito, A. and W. Arnold: *Cancer of the Larynx: Current Concepts in the Treatment of the Neck*, S. Karger Publishers, Inc., Farmington, CT, 2000.

Fried, M.P. and R. Hurley: *The Larynx: A Multidisciplinary Approach*, Mosby-Year Book, Inc., St. Louis, MO, 1995.

Hawkshaw, M., R.T. Sataloff, and J.R. Speigel: *Atlas of Laryngology*, Singular Publishing Group, Inc., San Diego, CA, 1999.

Hirano, M. and K. Sato: *Histological Color Atlas of the Human Larynx*, Singular Publishing Group, Inc., San Diego, CA, 1993.

Kavuru, M.S. and A.C. Merta: "Treatment of Recurrent Respiratory Papillomatosis (Juvenile largyngotracheobronchial papillomatosis)," *N. Eng. J. Med.*, 204 (January 16, 1992).

Lang, J.: *Clinical Anatomy of the Oral Cavity, Pharynx, and Larynx*, Thieme Medical Publishers, Inc., New York, NY, 1998.

Li, T.C., T.C. James, and D. Gunderson: "Inhalation of the Cap of a Metered-Dose Inhaler," *N. Eng. J. Med.*, 431 (August 8, 1991).

Schultz, C.H., S.W. Hargarten, and J. Babbitt: "Inhalation of a Coin and a Capsule from Metered-Dose Inhalers," *N. Eng. J. Med.*, 431 (August 8, 1991).

Veterans Affairs Laryngeal Cancer Study Group: "Induction Chemotherapy Plus Radiation Compared with Surgery Plus Radiation in Patients with Advanced Laryngeal Cancer," *N. Eng. J. Med.*, 1685 (June 13, 1991).

Stafford, N.: *Cancer of the Pharynx and Larynx*, Blackwell Science, Inc., Malden, MA, 1999.

Tucker, H.M.: *The Larynx*, Thieme Medical Publishers, Inc., New York, NY, 1992.

LASER ALTIMETER. See **Moon (Earth's)**.

LASER DOPPLER FLOWMETER. See **Flow Measurement (Liquids and Gases)**.

LASER GLASS. See **Glass**.

LASER IN-SITU KERATOMILEUSIS (LASIK). LASIK is now the most popular of all laser vision correction procedures. It was estimated that $1\frac{1}{2}$ to 2 million procedures would be completed in the United States in the year 2000. This highly successful procedure combines the minimal postoperative discomfort and rapid visual recovery of the Automated Lamellar Keratoplasty (ALK) procedure with the computer-controlled precision of the Photorefractive Keratectomy (PRK) procedure.

The first step in the LASIK procedure is the creation of a flap of tissue from the outer layer of the central zone of the cornea using the microkeratome. The flap is then folded back out of the way, but it is held ready for replacement upon completion of the procedure. The Excimer laser is then used to sculpt the remaining central zone in accordance with predetermined data that have been entered into the laser system's computer. Under this precise control, the laser reshapes the curvature of the cornea to correct for nearsightedness, farsightedness or astigmatism. This part of the procedure takes only 30 to 60 seconds, after which the corneal flap is replaced. No sutures are used, and the surface of the eye will normally heal itself. Most patients can see quite well within 24 hours or less. Complete healing of the cornea takes about one month.

The LASIK procedure is performed on an outpatient basis. Although the actual laser procedure takes only a few seconds per eye, the procedure requires a couple of hours at the surgery center. Some of this time is spent preparing the patient for the procedure. A few minutes are required afterwards for postoperative instructions and departure preparation.

Surgical Procedure

Step 1: Eye preparation. Before the procedure begins, a nurse or technician talks to the patient about any immediate health problems that may affect readiness for the procedure. Antibiotic and anesthetic eye drops are then placed in the eye to numb it and prevent infection. The eye is swabbed with a sterile solution. The eyelid is then propped open with a lid retainer, and a paper or plastic "mask" is placed over the eye to keep eyelashes out of the way. Then the cornea is marked with a blue "dye ring," which serves as a reference point for the surgeon throughout the procedure. Because the cornea is numb, most patients experience little if any discomfort during these pre-operative preparations.

Step 2: Creating the flap. Next, the doctor creates a flap from the central zone of the cornea using the microkeratome. This precision instrument works much like a miniature carpenter's plane. It contains a disposable cutting blade that is preset according to the thickness of the cornea, usually about 160 to180 microns or 1/3 the depth of the cornea. The microkeratome operates in conjunction with a suction ring that holds the eye perfectly still, and when activated by a vacuum tube, it raises and flattens the cornea so it can be reached easily for cutting the flap.

Step 3: The Excimer laser. After the flap has been folded back from the center of the eye, the doctor dries the underlying cornea with a sponge-tipped swab and aligns the Excimer laser's microscope with the central corneal area in order to monitor the laser's sculpting pulsations. The patient is asked to focus on a fuzzy red light inside the laser. As the doctor activates the laser, there is a "popping" or "tacking" sound, and there is a slight odor similar to that of hair burning, but no discomfort for the patient. The number of laser pulsations will depend on the nature of the refractive vision problem that is being corrected. This phase of the procedure takes only a minute or so. The doctor then carefully folds the flap back in place and irrigates the eye with a sterile saline solution. The corneal area may be dried with a gentle blower, which helps seal the flap. In addition, a contact lens may be placed in the eye.

Step 4: Post-operative measures. When the procedure is complete, additional antibiotic drops are placed in the eye, and it may be covered with a plastic shield. For a short while after the procedure, the eye is numb from the anesthetic drops. As the numbness wears off, the patient may experience some light sensitivity and a scratchy or dry sensation as though something is in the eye. This feeling usually goes away within a few hours. Patients must not drive themselves home following the procedure.

The patient returns to the doctor's office the next day for a post-operative examination. The doctor checks the flap to see if it is healing properly. If a contact lens was placed after surgery, it will be removed at this time. Vision is checked and, for most patients will range from 20/20 to 20/40 depending on the number of laser pulsations received. For some patients, vision may continue to improve for several weeks before totally stabilizing.

The LASIK procedure has several advantages over both the ALK and the PRK procedures on which it is based. Although it employs the Excimer laser precision control and accuracy of the PRK procedure, the LASIK procedure does not remove any part of the epithelium, the thin, filmlike protective outer layer of the cornea, as does PRK, and there is less chance of scarring. Thus, the primary healing process is the resealing of the corneal flap, which usually happens within 24 hours and with little postoperative discomfort. The LASIK procedure can also handle successfully higher degrees of myopia than PRK and can be used to treat cases of farsightedness and astigmatism.

See also **Automated Lamellar Keratoplasty (ALK)**; **Lasers (Eye Surgery)**; **Photorefractive Keratectomy (PRK)**; and **Refractive Eye Surgery**.

Vision Rx, Inc., Elmsford, NY

LASER INTERFEROMETER SPACE ANTENNA (LISA). How did the Universe begin? Does time have a beginning and an end? Does space have edges? These are the questions we've struggled to answer for centuries. Science and technology have now reached the point where answers to these questions are finally within our grasp. The Laser Interferometer Space Antenna (LISA) may supply some of these answers as the mission studies the mergers of supermassive black holes, tests Einstein's Theory of General Relativity, probes the early Universe, and searches for gravitational waves—the primary objective. See Fig. 1.

Fig. 1.   Artist's concept of the Lisa Mission. (*Image courtesy of NASA/JPL.*)

As the first dedicated space-based gravitational wave observatory, LISA will detect waves generated by binaries within our Galaxy, the Milky Way, and by massive black holes in distant galaxies. Although gravitational wave searches have previously been made, they were conducted for short periods by planetary missions that had other primary science objectives. Some current missions are using microwave Doppler tracking to search for gravitational waves. However, LISA will use an advanced system of laser interferometry for detecting and measuring them. And, LISA will *directly* detect the existence of gravitational waves, rather than inferring it from the motion of celestial bodies, as has been done previously. Additionally, LISA will make its observations in a low-frequency band that ground-based detectors can't achieve. Note that this difference in frequency bands makes LISA and ground detectors complementary rather than competitive. This range of frequencies is similar to the various types of wavelengths applied in astronomy, such as ultraviolet and infrared. Each provides different information.

In space, LISA will not be affected by the environmental noise that affects ground detectors on Earth's surface. Due to earthquakes and other vibrations, ground detectors can only make observations at frequencies above 1 hertz. However, other environmental factors *will* impact LISA. Such factors include the drift of the spacecraft, charging of the test masses, and buffeting by the solar wind. Making these small disturbances negligible is a technological challenge of the mission. Meeting this challenge will help to ensure the detection of gravitational waves.

LISA is jointly sponsored by the European Space Agency (ESA) http://sci.esa.int/science-e/www/area/index.cfm?fareaid=1, as a Cornerstone mission in ESA's Cosmic Vision Programme, and NASA's Astronomy and Astrophysics Division, http://astrophysics.gsfc.nasa.gov/, as part of the Structure and Evolution of the Universe 2003 roadmap, "Beyond Einstein: From the Big Bang to Black Holes" (http://universe.nasa.gov/be/roadmap.html). This program studies the building blocks of our own existence at the most basic level: the matter, energy, space, and time that make up the Universe. LISA is one of the program's Great Observatories, http://universe.nasa.gov/program/observatories.html. See also **Beyond Einstein Program**.

Mission Description

The Laser Interferometer Space Antenna (LISA) consists of three spacecraft, shaped like hockey pucks or pillboxes, (Fig. 2) freely flying (not connected to each other) five million kilometers (a little more than three

million miles) apart, in an equilateral triangle. See Fig. 3. The spacecraft will carry delicate instruments to track each other and, in concert, measure passing gravitational waves. These waves, as predicted by Einstein, are space-time distortions generated from massive celestial bodies that are accelerated or disturbed. Rippling outward, gravitational waves affect any type of matter they encounter—a solid body should vibrate if a distortion of about the same size hits it. Widely separated bodies will move in and out, with respect to one another, as the distortion passes them. Although the resulting motion would be very small, it would still be measurable with modern techniques such as laser interferometry. This is the goal of the LISA mission. LISA will make measurements with 100 times greater precision than has been achieved ever before, using identical instruments in each spacecraft. Each of these instruments is made up of two optical assemblies, which contain the main optics, lasers, and a free-falling gravitational reference sensor.

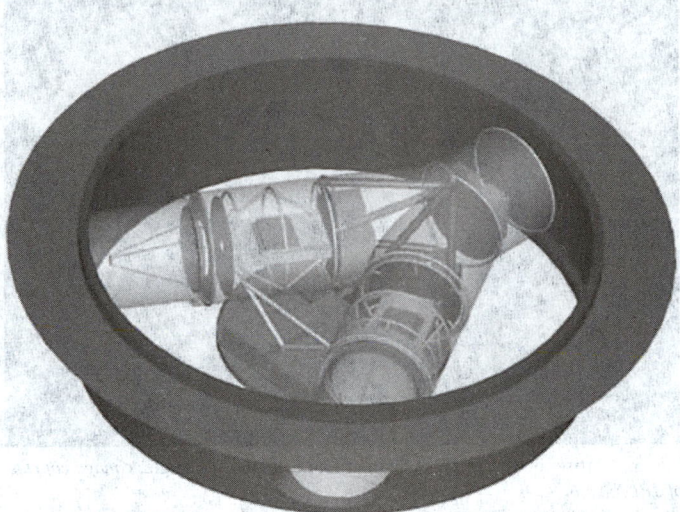

Fig. 2.   Cutaway of LISA's instrument. The actual spacecraft will be protected with a cover.

The sensor, used to control the motion of the spacecraft, is considered the "heart" of the instrument. It contains the "test masses"—two cubes allowed to float freely within the spacecraft. These cubes, so small they can both fit inside a Chinese-food take-out carton with room to spare, are shielded from external and internal disturbances so that they detect *only* the force of gravity. See Fig. 4.

The cubes are highly polished to enable them to reflect laser light. In this way, they act as mirrors in an interferometer. The relative motion of these cubes on different spacecraft are what will detect passing gravitational waves.

LISA will operate 50 million kilometers (about 30 million miles) above Earth. The center of LISA's triangle will follow Earth's orbit around the Sun, trailing 20 degrees behind. It will maintain a distance of 1 AU (astronomical unit) from the Sun, the average distance between the Earth and the Sun. LISA's operational position was chosen as a compromise between the need to minimize the effects on the spacecraft of changes in the Earth's gravitational field and the need to be close enough to the Earth for easy communication. See Fig. 5.

Individually, each spacecraft will travel in its own specially selected orbit, chosen to minimize the changes in the distances between the spacecraft. This distance, LISA's interferometer arm length, will allow observation in the frequency band of most of the interesting sources.

LISA will search for gravitational waves at low frequencies that ground-based detectors can't achieve. See Fig. 6. However, ground detectors, such as the Laser Interferometry Gravitational-Wave Observatory (LIGO) or VIRGO, and LISA will complement each other. In space, LISA will "hear" the long, low rumble of space-time swells. On Earth, LIGO and others will "hear" higher-frequency space-time ripples. LISA will observe binaries thousands of years before they collide. Ground detectors will observe other binaries just before they collide, when their orbiting speeds are much

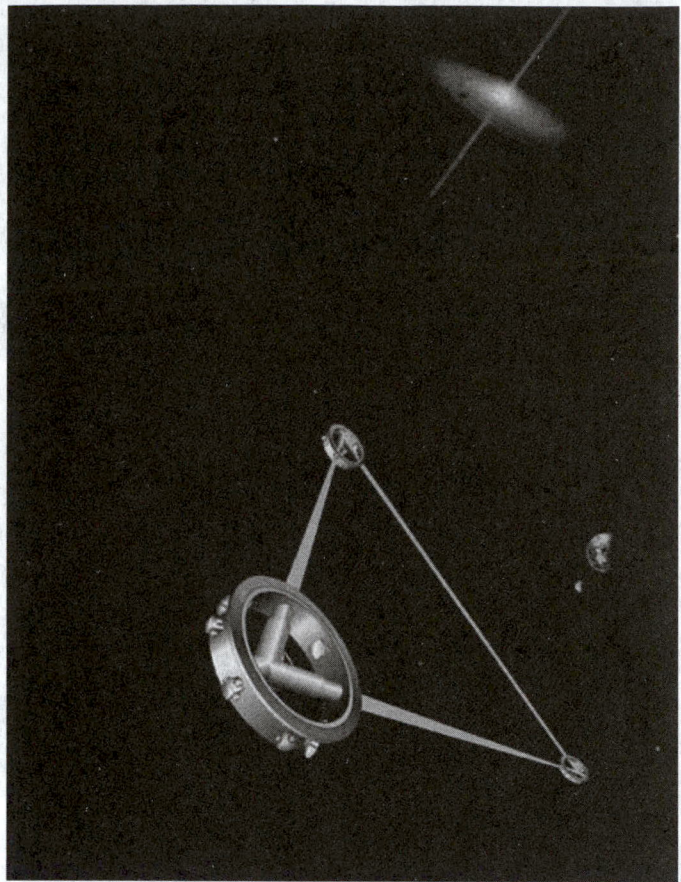

Fig. 3.   Artist concept of LISA flying in an equilateral triangle. (*Image courtesy of JPL.*)

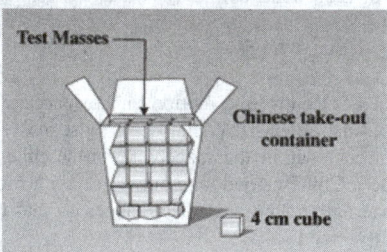

Fig. 4.   Twenty seven Test mass cubes can fit inside the average quart-sized take-out container!

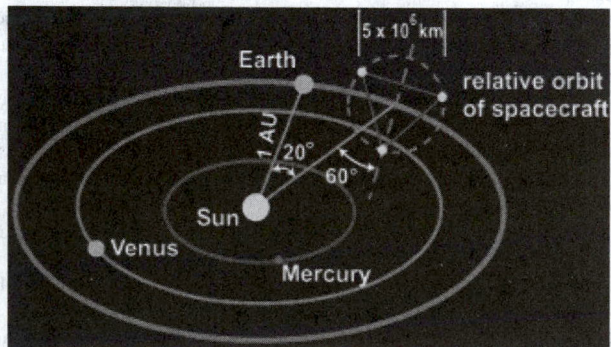

Fig. 5.   LISA's orbit

higher. Both types of observatories are needed to hear the broad spectrum of ripples in space-time.

LISA will detect gravitational wave sources from all directions in the sky. These sources will include all the thousands of compact

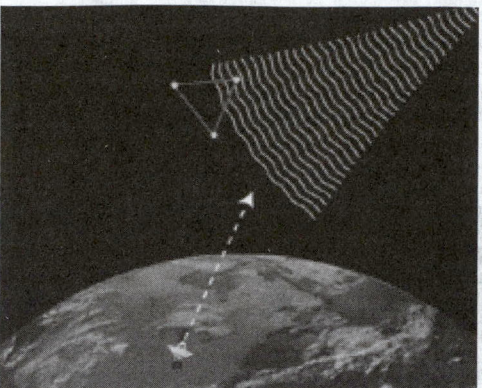

Fig. 6.   LISA will be complemented by ground detectors.

binary systems—many containing neutron stars, black holes, or white dwarfs—in our own Galaxy, plus merging supermassive black holes in distant galaxies. During the five-year lifetime of the mission, LISA is expected to yield 163 gigabytes of significant data for analysis.

Ground Detectors. Current and planned laser ground-based gravitational wave observatories include:

> **AIGO,** the Australian International Gravitational Observatory, is the first laser interferometer observatory in the southern hemisphere. Its location makes AIGO a major component of a global array of detectors. See also http://www.gravity.uwa.edu.au/AIGOinterior.html.
> **GEO600,** located in Germany, is a British-German collaboration. GEO600 has short interferometer arms (only 600 meters) but makes up for this by using advanced optical techniques to improve its sensitivity. See also http://www.geo600.uni-hannover.de/.
> **TAMA,** located in Japan, has even shorter arms (300 meters), which limits its use as a true observatory. However, TAMA is an important testbed center for interferometry development. See also http://tamago.mtk.nao.ac.jp/.
> **VIRGO,** located in Italy, is a collaboration between Italy and France. Appropriately, this detector is based near PISA where Galileo carried out some early experiments on gravity. VIRGO has arms that are 3 kilometers long and is designed so that multiple reflections from the mirrors at each end of the arms can give an effective armlength of 120 kilometers. It uses an elaborate seismic isolation system to reduce interference from external sources of noise. See also http://www.virgo.infn.it/central.html.
> **LIGO,** The Laser Interferometer Gravitational-Wave Observatory, is funded by the National Science Foundation in the United States. LIGO consists of three interferometers in two widely separate facilities. Two interferometers are located in the state of Washington and the third is located Louisiana. All three operate in unison as a single observatory. See also http://www.ligo.caltech.edu/.

Gravitational Waves

Gravitational waves offer a new way to observe the Universe. Through them we will learn more about the mergers of giant black holes and the death spirals of stars that black holes capture and swallow.

Gravitational waves in some ways resemble optical (visible light) waves. Both are traveling waves, spreading outward from their sources—like waves on an ocean. They both tell us something about the matter that generated them. For example, light carries information about the structure of the matter that generated it—what it is made of or its temperature. Gravitational waves tell us something different about the matter that generated them. They tell us about the its mass and motion.

Gravitational waves, generated by the motion of massive bodies, are a varying strain or distortion of space-time. This means that there are changes in the distance between points of space-time, and the size of the change is proportional to the distance between the points. This is much like changes in the distance between points of a vibrating rubber sheet, one illustration of space-time. On this rubbery sheet, stars, planets, and galaxies ride. So, when a celestial object embedded in this sheet moves, the movement generates ripples of strain and distortion that spreads. This spreading is caused in the same way someone bouncing on a trampoline causes the canvas to jiggle.

Waves of different frequencies are caused by different motions of mass. Differences in the phases of the waves (the timing of when the wave crests pass) allow us to figure out the direction of the source and the motions, or dynamics, of the matter that generated them. Even though they may have been generated billions of years ago, and their weak signals make them hard to detect, they will be able to tell us about their sources. This is because they travel across space essentially unaffected by intervening matter such as planets, stars, gases, or dust in space.

How Do We Know Gravitational Waves Exist? Although there has been no direct detection of gravitational waves, powerful indirect evidence does exist. This evidence was gleaned from the observations of a binary neutron star system by astrophysicists[1] Joseph Taylor and Russell Hulse.

Sources of Gravitational Waves. Almost all motions of bodies in space produce gravitational waves. However, because gravitational waves become weaker the further they travel, only strong sources of gravitational waves can be detected with our current instruments.

LISA will be able to detect the motions of binary systems, made up of neutron stars, white dwarfs, ordinary stars, or black holes in our Galaxy, and the motion of the mergers of black holes where distant galaxies have interacted. These massive black holes are thought to exist in the centers of most galaxies and may have masses that are more than a million times the mass of our Sun!

Mergers of these massive binaries, spinning around one another in timeframes ranging from thousands of seconds to years before they finally collide, will generate the most powerful gravitational waves. And, LISA will be able to detect them. See also **Binary Stars**.

We also hope to be able to detect gravitational waves generated by the Big Bang. This cataclysmic initial event left behind an echo in the electromagnetic spectrum. Called the *cosmic microwave background* (CMB), this echo has already been observed. The Big Bang most likely also left gravitational waves that we can observe. Other violent astrophysical events, such as the explosion of a star in our own Galaxy, should also produce measurable gravitational waves, although at frequencies where ground-based instruments are sensitive.

The detection of gravitational waves may yield experimental tests of fundamental physical laws that cannot be made any other way. For example, comparison of arrival times of light and gravitational waves from the same source would test Einstein's prediction that light and gravitational waves travel at the same speed. Measurements could verify or disprove Einstein's prediction that gravitational waves only change or disturb distances perpendicular (at a right angle) to the direction of propagation (where they are spreading).

Detection of gravitational waves will open up a new window on the Universe for observational astronomy. When gravitational waves from supernovae or from galactic nuclei and quasars are detected, we will have the first observations of the interiors of strong-gravity regions of space.

---

[1] Einstein's Theory of General Relativity predicts that as binaries (pairs of objects) orbit each other, they lose energy by radiating gravitational waves. This causes them to spiral towards each other.

Astrophysicists Joseph Taylor and Russell Hulse provided the first indirect evidence of the existence of gravitational waves with their observations of a now famous binary system. Known as PSR 1913 + 16 (PSR means pulsar, the numbers give the location in the sky), this binary system consists of two neutron stars, separated by a distance only a few times larger than the distance from the Earth to the moon.

The neutron stars in this system orbit each other once every eight hours. One of these stars is a pulsar. It generates a very regular radio signal (pulses), which enables the orbital motion (the way it orbits) of PSR 1913 + 16 to be well determined. During their observations, Hulse and Taylor found that the rate of in-spiral matched the value predicted by General Relativity to better than 1%. Their observations earned them the 1993 Nobel prize.

Ground-based detectors such as the Laser Interferometer Ground Observatory will not be able to detect PSR1913 + 16 directly because the frequencies are too low. The frequency of gravitational waves emitted by this source *will* be within the range of the Laser Interferometer Space Antenna. However, the amplitude (size) of the waves will be well below its instrumental noise, and therefore will not be directly detectable.

Benefits

The Laser Interferometer Space Antenna (LISA) mission will use advanced technologies to achieve its science goals: the direct detection of gravitational waves, the observation of signals from compact (small and dense) stars as they spiral into black holes, the study of the role of massive black holes in galaxy evolution, the search for gravitational wave emission from the early Universe, and testing Einstein's theories.

The technology used by the LISA mission will benefit future missions involving multiple spacecraft, flying in formation, with instruments that use waves of light to determine distances or speeds. For example, LISA technology may be used to map the gravity fields of the Earth or other planets. Mapping will allow scientists to monitor water tables, study ocean currents driving climate, and identify locations of oil and other resources. LISA technology may also be used to study properties of planetary crust and ocean currents. Such technology will reduce the cost and risks of these types of future missions.

NASA's Beyond Einstein Program addresses mankind's drive to understand the Universe and our place in it. LISA's findings are expected to confirm, or perhaps modify, Einstein's theories. Direct detection will open a completely new window on the Universe, using neither light nor its invisible counterparts (X-rays, radio etc.), but the tremors of space itself.

Sources of gravitational waves detectable by LISA should include newly forming black holes, colliding black holes, and pairs of stars orbiting close together. There may even be gravitational waves from as far back as the origin of the Universe.

Spacecraft Design

Each of the three Laser Interferometer Space Antenna (LISA) spacecraft is designed as a short structural cylinder, 1.8 meters (5.9 feet) in diameter and 0.48 (a little over 1.57 feet) meters high. They look like pillboxes or hockey pucks. However, with this design the spacecraft size and shape are optimized to contain the instrument's optical assemblies.

Solar panels, mounted on a sun shield, help to power the spacecraft. Spacecraft equipment, such as the telecommunications (not shown) and micronewton thrusters, are mounted on the wall of the cylinder.

LISA's telecommunications equipment includes two 30 centimeter (a little over 11.8 inches) diameter X-band radio antennas, which communicate with ground stations: the NASA Deep Space Network at Goldstone in the Mojave Desert, near Madrid in Spain, and near Canberra in Australia.

LISA's spacecraft positions must be precisely controlled to follow the motion of the test masses. LISA's micronewton thrusters provide the very small thrust needed to keep each spacecraft centered about its test mass. They perform attitude and position control of the spacecraft.

Each spacecraft supports an inner, Y-shaped tube that houses an identical science instrument. See Fig. 7. In this way, each of LISA's three spacecraft can act as independent interferometers at the same time, with each one acting as the beamsplitter of three separate interferometers.

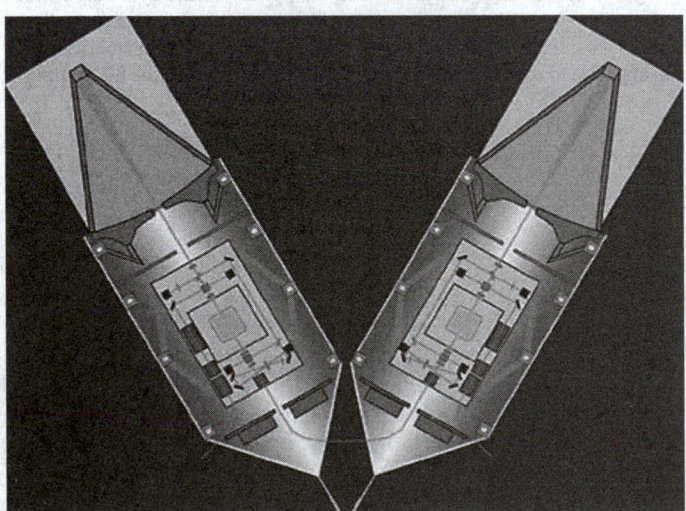

Fig. 7. Y-Payload.

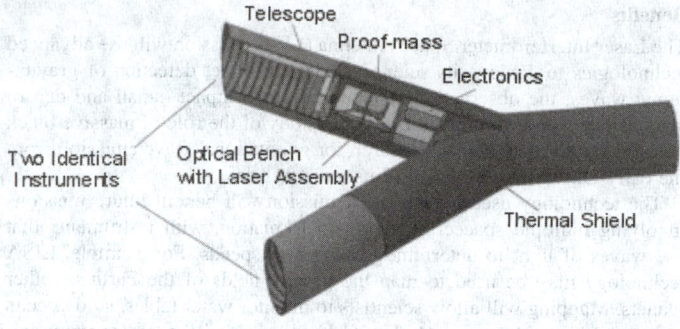

Fig. 8.   LISA instrument.

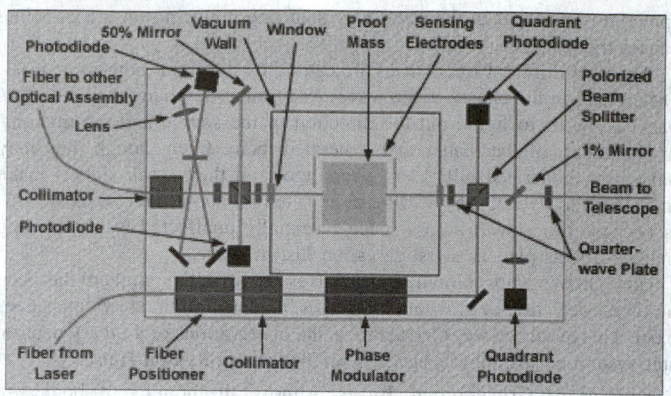

Fig. 10.   Cross section of one optical bench.

Instrument Package. The Laser Interferometer Space Antenna (LISA) carries identical science instruments on each of the three spacecraft. Each spacecraft's instrument is made up of two optical assemblies and lasers, mounted on a disk-shaped radiator, housed inside a Y-shaped tubular structure that is supported by the spacecraft's cylindrical walls. See Fig. 8.

The instrument is protected at all times by thermal shielding. This thermal shield, gold-coated and suspended by stressed fiberglass bands, protects the instruments from solar brightness. A cover across the top of the spacecraft cylinder prevents sunlight from striking the thermal shield, and thus the instrument, during science operations. Inside the Y-shaped tube, the instrument's optical assemblies are surrounded by additional thermal shielding.

Each optical assembly contains a 30 centimeter (about 12 inches) diameter telescope (f/1 Cassegrain) for transmission and reception of laser signals to another spacecraft. The primary mirror is a double-arch light-weight ultra-low expansion design. The secondary mirror is supported by a three-leg "spider" made of a graphite-epoxy material with low thermal expansion. See Fig. 9.

The optical bench within each assembly contains the main optics: the laser beam injection, detection, and beam shaping optics, as well as the gravitational reference sensor. This bench consists of a solid glass block to which all components are rigidly attached. See Fig. 10.

Behind the bench is a support where electronics for the gravitational reference sensor preamplifiers and the laser readout are mounted. Most components on the optical bench are passive, meaning they don't move. Exceptions are a motorized positioner for fiber selection and focusing, photodiodes for signal detection, and a phase modulator that allows transfer of information between spacecraft.

The lasers on the assembly transmit a signal between each pair of LISA's spacecraft to measure the distance between them. This signal is a continuous 1 watt of output power in the infrared (1 micron wavelength). See Fig. 11.

Light from the laser is delivered to the optical bench by a single-mode optical fiber. The fiber positioner can be moved to optimize the delivery of light from the fiber into the optical bench and to the telescope for transmission to the other spacecraft. This, combined with the 30 centimeter

Fig. 11.   LISA's lasers.

telescopes, provides the capability to measure distances to the desired accuracy.

But, the heart of the LISA instrument is the freely falling gravitational reference sensor. See Fig. 12. This sensor houses LISA's test masses—two freely floating cubes, so small they can both fit inside a Chinese-food take-out carton, with room to spare. These cubes are highly polished to enable them to reflect laser light, allowing them to to act as mirrors at the end of the interferometer arms. See Fig. 4

The test masses are carefully shielded from external and internal disturbances so that they move only under the influence of gravitational forces, not other forces. The relative motion of these cubes on different spacecraft are what will detect passing gravitational waves.

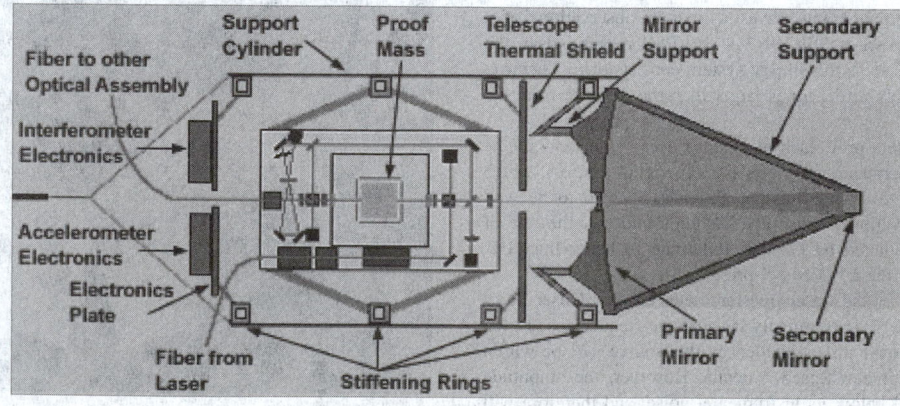

Fig. 9.   Cross section of one optical assembly.

Fig. 12. Free-falling gravitational reference sensor.

LISA Interferometry. LISA uses a technique known as interferometry to detect gravitational waves. The spacecraft are arranged in an approximately equilateral triangle to form a giant Michelson interferometer. Light from the central spacecraft is sent out to the other two spacecraft. Each spacecraft contains freely floating test masses that act as mirrors and reflect the light back to the source spacecraft where it hits a detector causing an interference pattern of alternating bright and dark lines. If the light from the two arms is in phase then there will be a bright spot at the center of the detector. When two beams of light with the same wavelength are in phase then the peaks of the waves will arrive at the same time. If they are out of phase the peaks arrive at different times, causing the bright spot to move away from the center of the detector. If the arms are the same length then the light waves will be in phase. The passage of a gravitational wave will lengthen one arm relative to the other one, and so the light waves will be out of phase.

Each LISA spacecraft is identical. They can each act both as the light source and as the mirrors at the end of the arms. LISA therefore consists of three independent Michelson interferometers, with each spacecraft being the light source of one interferometer. See Fig. 13. The additional information that can be gained by using the three interferometers simultaneously will allow polarization of the gravitational waves to be determined. Gravitational waves can have two different polarizations known as 'cross' and 'plus'. They can be distinguished by the different ways in which they stretch or compress spacetime.

See also **Space Science Missions: Universe**.

Web Reference

Laser Interferometer Space Antenna: http://lisa.nasa.gov/index.html.

LASER IONIZATION. See **Mass Spectrometry**.

LASERS. An acronym for *light amplification by stimulated emission of radiation*. The device is identical in theory of operation to the maser except that it operates at frequencies in the optical region of the electromagnetic spectrum, rather than in the microwave. See also **Maser**. By common usage, these devices are all called lasers, although more precise terminology would aptly use such terms as ultraviolent maser, optical maser, infrared maser, etc. Although the original microwave maser offers an extremely stable frequency source, its main use has been as an amplifier with very low noise output. In contrast, one of the principal attributes of the laser is its ability to produce a single frequency at high intensity in the optical region. Not only may the output be a single monochromative wave, but the wave may be coherent, or in phase, over the whole face of the radiator. In this mode of operation, the laser is an oscillator whose output depends upon the selective amplification of one of the single-frequency modes of the resonant cavity containing the active laser medium.

The decade of the 1980s represented the period when laser technology received profound acceptance, to the point where lasers, like semiconductors and digital computers before them, are now considered important components (hardware) of larger systems. The development of laser technology continued apace during the early 1990s, with numerous new and practical applications occurring at an accelerated rate. Lasers now range from microlasers that measure but a few millionths of a meter, for use in optical communications and information processing, to the building-sized x-ray research laser that can deliver 100,000 joules of energy in less than one-billionth of a second—that is, a 10^{14}-watt pulse. One other feat, accomplished in the mid 90's, is a veritable tour de force. A group at MIT managed to coax single excited barium atoms crossing a miniature resonator to serve as a laser. The crucial part is the extreme precision of the mirror arrangement that keeps the photons emitted by the atoms resonating. So far the main goal of this work has been pure physics: to test the theory of quantum electrodynamics (QED).

Historical Perspective

The fundamental theoretical concepts of the maser and laser date back many years, to the early workers in quantum mechanics who appreciated that an incident electromagnetic beam of an appropriate resonant frequency, passing through a medium, might stimulate molecules in an upper quantum energy state to return to a lower quantum energy state, and thus reinforce the primary beam by negative absorption. As early as 1940, Fabrikant (Russia) suggested that experiments be made to prove negative absorption. Fabrikant was the first scientist to introduce the term "collisions of the second kind" (later to prove of importance in laser patent litigation). By this he meant a collision in which some of the kinetic energy of motion of colliding particles is converted to internal energy (or a change in energy state of at least one of the colliding particles).

It is also interesting to note that, as early as 1950, researchers Lamb and Retherford (Columbia University), observed that, if an upper quantum energy state could be caused to be more highly populated (as compared with a lower quantum energy state), the result would be a net induced emission to an incident beam. They further indicated that such a population inversion would probably occur between the $2p$ and $2s$ levels in hydrogen. There shortly followed experiments by Purcell and Pound (Harvard University) who used magnetic techniques to invert the population of a pair of nuclear spin states in lithium fluoride and thus were the first researchers known to directly observe negative absorption of an applied pulse, a phenomenon which they called negative temperature.

Townes (Columbia University) is generally credited with first recognizing that stimulated emission could be utilized in the making of practical

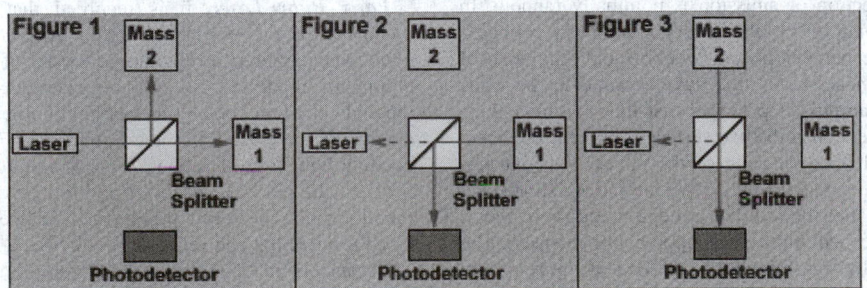

Fig. 13. Michelson interferometers.

hardware. In 1951, Townes described an approach wherein an ammonia beam would be divided into two portions along the lines of experimentation carried out in Germany, wherein a quadrupolar focusing technique was used for separating a beam of molecules into two portions, one of which contained molecules predominantly in the upper of two energy states. Townes proposed to pass the high-energy portion through a cavity resonant at the frequency corresponding to the energy separation of the two states and so reported this proposal. Instead of a millimeter wave generator, a microwave oscillator was used. The latter was given the name maser. This led to the award of U.S. Patent No. 2,879,439, which covered the use of stimulated emission for the amplification and/ or generation of oscillatory electromagnetic energy. This patent subsequently was licensed to laser manufacturers.

Following the development of the microwave maser, Schawlow (AT&T Bell Laboratories) and Townes in 1958 proposed that optical maser action could be obtained by placing an active medium in an optical cavity. The medium would be a gas or a solid excited electrically or by light in such a manner that any optical wave present would be amplified as it moved through the material. This work led to the ultimate awarding of the basic laser patent in March 1960 (U.S. Patent No. 2,929,922). During the 1960s, laser research was carried on at a rapid pace at AT&T Bell Laboratories, among others. Considering the work at AT&T Bell Laboratories, in 1960 a laser capable of emitting a continuous beam of coherent light (using helium-neon gas) was developed; in 1961, the continuous-wave solid-state laser (neodymium-doped calcium tungstate) was developed. As a refinement of the helium-neon laser, in 1962 the basic visible light helium-neon laser was developed, of which several hundred thousand are in use as of the early 1980s. In 1964, the carbon dioxide laser (highest continuous-wave power output system known to date) was developed. Other developments during 1964 included the neodymium-doped yttrium aluminum garnet laser, the continuously operating argon ion laser, the tunable optical parametric Oscillator, and the synchronous mode-locking technique, a basic means for generating short and ultrashort pulses. In 1967, the continuous wave helium-cadmium laser (utilizing the Penning ionization effect for high efficiency) was developed. These lasers find use in high-speed graphics and biological and medical applications. In 1969, the magnetically tunable spin-flip Raman infrared laser, used in high-resolution spectroscopy, as well as in pollution detection in both the atmosphere and the stratosphere, was developed. Laser developments continued and, in 1970, semiconductor heterostructure lasers capable of continuous operation at room temperature were introduced. The distributed feedback laser, a mirror-free laser structure compatible with integrated optics, was introduced in 1971. These were followed by the tunable, continuous-wave color-center laser (1973), techniques for creating optical pulses of less than one-trillionth second duration (1974), and, in the late 1970s, long-life semiconductor lasers for lightwave communications.

Laser Principles

The basic concept of the laser may be described in general terms as follows. Optical maser action can be obtained by placing an active medium in an optical cavity. The medium may be a gas, solid, or liquid which can be excited electrically, by light, chemically, or thermally in such a manner that any optical wave present will be amplified as it moves through the material. One of the first cavities proposed, for example, is a Fabry-Perot resonator — two plane, parallel reflecting plates with a small transmission through which the radiation can escape. Upon excitation of the material, light is emitted with a band of frequencies determined by the particular material. In addition, the direction of emission is nominally random. In the presence of the cavity, some of the waves escape after several back-and-forth reflections from the parallel plates, "walking off" the edge of the reflectors, so to speak. Those waves that travel normal to the walls remain in the cavity and are amplified, provided that they reinforce each other after each round-trip reflection at the two surfaces. this reinforcement or resonance is only satisfied if the spacing of the plates is an integral multiple of one-half the wavelength in the medium. Thus, after a short time, only that frequency which satisfies the resonant condition and those waves traveling normal to the reflector will build up an appreciable intensity. The resultant light which is partially transmitted through one of the reflectors will thus be a single frequency, or several discrete frequencies if there is more than one cavity resonance within the band of frequencies emitted by the laser material. In addition, the wave front will be in phase across the surface of the reflector since waves striking the surface at normal incidence are amplified most strongly. The resultant beam will then be diffraction limited, i.e., the beam will spread by an angle in radians given approximately by the ratio of the wavelength to the diameter of the beam. In actual practice, single-mode operation is obtained only under special conditions. Several frequency modes may be present because of the multiple resonances of the cavity and numerous "off-axis" modes may be found that correspond to resonant waves, which travel at small angles from the normal to reflectors. These waves "walk off" so slowly that they still are amplified appreciably. Refinements of the simple cavity may consist of concave reflectors that decrease the diffraction losses, or several parallel reflectors, which limit the oscillation to a frequency common to each pair in the set, among other approaches.

Basic Requirements of Lasing Media. The key to successful laser operation is the active medium that amplifies the wave. Qualitatively, a material that fluoresces or exhibits luminescence is an obvious candidate. In fluorescence, electrons are excited to an upper-energy state by short-wavelength light, such as ultra-violet, while luminescence is produced by passing an electron current through the medium, such as in a gaseous discharge. In either process, stimulated emission can occur only if more electrons are produced in the upper-energy state than in the lower or terminal state for the radiating transition. In this case, an incident photon will stimulate further transitions and amplification will result. If the final state were more heavily populated, then the photon would cause more upward or absorbing transitions and the net effect would be absorption.

Laser source requirements vary widely among specific applications. Practical systems require a large range of wavelengths, output powers, spatial and temporal beam characteristics, among other features. In many applications, the following factors apply: (1) an optimum wavelength exists; (2) a specific minimum power level is required; (3) capital and operating costs of the laser must be minimized; (4) size and weight constraints must be met; (5) the laser should be capable of operating for extended periods with little maintenance; (6) the laser output should have a specific temporal and spatial characteristics; and (7) operator safety must be considered in design and use of the overall laser system.

Wavelength characteristics of available laser types are given in Table 1.

TABLE 1. WAVELENGTH RANGE OF SOME LASERS

Type of Laser	Wavelength (Micrometers)
Solid state	1.06
Ion	0.514, 0.488
Carbon dioxide	10.6
Diode	0.65–1.8
Dye	Visible (tunable)
Helium-neon	0.63
Rare gas halide	0.35, 0.25, 0.19
Helium-cadmium	0.422

Types of Lasers

From the above list of media requirements, it is obvious that a majority of materials are not candidates for making an effective laser. When the laser was first conceived, the possibility of finding numerous materials, as has turned out to be the case, initially was not envisioned by most researchers in the field.

Early Ruby Laser. It is recorded that the first optical laser was demonstrated by Maiman (Hughes Aircraft Research Laboratories) in 1960. Maiman used a ruby single crystal of aluminum oxide doped with chromium impurities. By applying semitransparent reflective coatings on the ends of a rod about 2 inches (5 centimeters) long, Maiman made the cavity and the crystal an integral unit. See Fig. 1. Exposure to an intense exciting light from a xenon flashtube was found to invert the population between the red-emitting level and the ground or lowest-energy state of the electrons. The result was a burst of intense red light emanating in a beam through the end reflectors. This was a powerful laser.

Because of off-axis modes and multiple resonances, the output is not a single frequency, single plane-wave mode, but generally consists of the order of 100 separate modes. The beam is still quite narrow, being

Fig. 1.   Schematic diagram of early giant pulsed ruby laser.

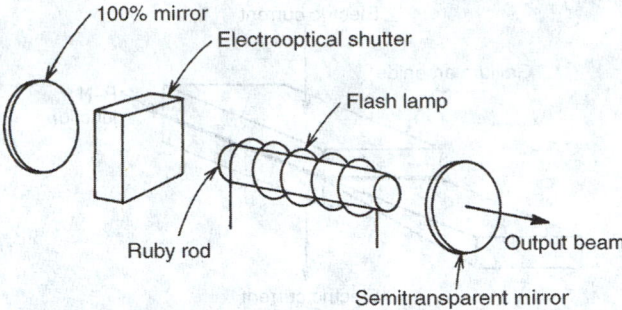

Fig. 2.   Schematic diagram of early helium-neon laser with external spherical reflectors. Curvature of reflectors is exaggerated.

of the order of 1 milliradian or 0.05 degree. As a comparison with conventional light sources, the energy radiated from 1 square centimeter of the brightest flash lamp is less than 10 kW and is distributed over the entire visible spectrum. In addition, the radiation is incoherent and is spread out uniformly in all angles from the source. Thus, the directivity and spectral purity of the laser source are many orders of magnitude superior to that of an incandescent source. The ruby laser suffers from low efficiency, about 1%, and except with elaborate cooling systems, only operates on a pulsed basis. The ruby laser was first reported in the literature (Maiman, 1960) and, in 1967, Maiman was awarded U.S. Patent No. 3,353,115 for the optically pumped ruby laser.

Other crystalline or glass systems with impurity ions have been developed, which yield wavelengths from the ultraviolet to approximately 3 micrometers wavelength in the infrared. Some, such as neodymium-doped yttrium aluminum garnet (YAG), operate in a continuous mode at the one-watt level, while peak powers have reached values as high as 10^{14} W, or greater, in pulses of the order of 10^{-12} second. These ultrahigh powers are obtained in neodymium-doped glass systems, using several stages of amplification and novel pulse-forming techniques.

Gas Lasers.   In 1961, Javian, Bennet, and Harriott demonstrated laser action in a gaseous discharge of helium and neon. The parallel-plate reflector cavity was used, but with much greater spacing. Later, concave mirrors were used to decrease the loss of energy out the sides of the cavity. See Fig. 2. The gas laser operated continuously and delivered power up to about one watt. Pulsing the gas discharge yielded peak power as high as 100 watts. The first laser radiated at 1.15 micrometers in the infrared, while further development with different gases yielded output from the ultraviolet to 330 micrometers or 0.33 millimeters in the far infrared. In contrast to the ruby laser, the gaseous laser beam may be diffraction limited and the frequency is pure, i.e., oscillation may be limited to one mode. By careful design, the frequency may be stabilized to within a few thousand cycles per second, or approximately one part in 10^{13} or better. Although the original gas laser utilized electrical excitation of electronic transitions, later versions use vibrational transitions in molecules, such as carbon dioxide, and the excitation may be electrical, chemical, or thermal. The helium-neon laser is commonly used in supermarket bar-code scanners.

In the chemical laser, atomic species such as hydrogen and fluorine can be reacted to produce molecules in an excited vibrational state, which, in turn, yields amplification or oscillation. Electrically excited lasers, particularly those using carbon dioxide at 10 micrometers, can operate at atmospheric pressure, using spark discharges or pre-ionization voltages in the 100-kV range. The high pressure and the powerful electrical excitation result in peak powers in the 10 to 100 MW region. For continuous laser operation, the gas may be circulated rapidly to avoid excessive heating.

In a gas dynamic laser, an appropriate fuel is burned to produce carbon dioxide and nitrogen at high temperature and pressure. When released through a nozzle into the optical resonator region, the gas cools rapidly in terms of its kinetic or translational energy, but the population of the vibrational energy levels of the carbon dioxide molecules becomes inverted

since the lower level of the laser transition relaxes much more rapidly. In addition, the vibrationally excited nitrogen molecules are in near resonance with the upper laser state of the carbon dioxide and transfer energy with high efficiency to maintain the inversion. Lasers of this type are capable of producing continuous power at a relatively high level.

Semiconductor Laser.   In the semiconductor laser, a solid (semiconductor) material is used. The electron current flowing across a junction between p- and n-type material produces extra electrons in the conduction band. These radiate upon making a transition back to the valence-band or lower-energy states. If the junction current is large enough, there will be more electrons near the edge of the conduction band than there are at the edge of the valence band and a population inversion may occur. To utilize this effect, the semiconductor crystal is polished with two parallel faces perpendicular to the junction plane. The amplified waves may then propagate along the plane of the junction and are reflected back and forth at the surfaces. See Fig. 3.

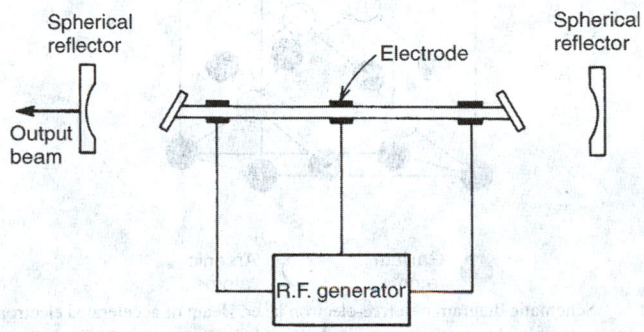

Fig. 3.   Schematic diagram of early gallium arsenide laser.

A major advantage of the gallium arsenide (GaAs) laser is that it has the electron distribution of a semiconductor. The main difference between electrons in semiconductors and electrons in other laser media is that in semiconductors all of the electrons occupy and thus share the entire crystal volume. Although all semiconductors possess this property, not all of them can be used as lasers. See Fig. 4.

The gain in the material is high enough so that the reflection at the semiconductor-air interface is sufficient to produce oscillation without special reflective coatings. The first such device used gallium arsenide and radiated at 8400 angstroms, or just beyond the visible region in the infrared. The efficiency of this laser, first demonstrated in 1962, was high (about 40%) and the power source was low-voltage direct current.

The compactness and efficiency of the semiconductor laser make it particularly attractive for systems use. Other substances used have included indium arsenide, indium phosphide, indium antimonide, and alloys such as gallium-arsenide-phosphide. These lasers may be tuned over several percent of their normal frequency of operation by varying the current flow through the device. The tuning results from the variations in temperature with current, which, in turn, changes the index of refraction and the resultant resonant frequency of the cavity.

Free Electron Laser.   Lasers of this type depart markedly from conventional lasers. Free electron lasers employ an electron beam and a magnetic

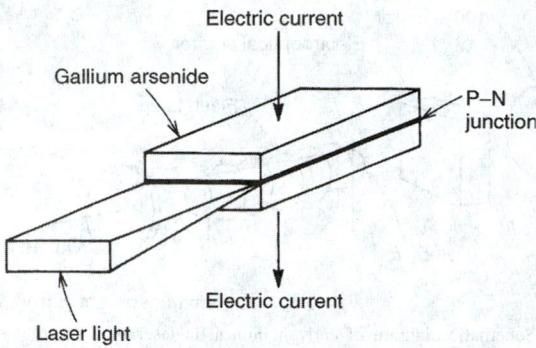

Fig. 4. Schematic diagram of a gallium arsenide (GaAs) crystal, which is commonly referred to as a zinc-blende structure. The structure consists of a face-centered cubic lattice of gallium atoms with arsenic atoms positioned on the body diagonals. The arsenic atoms also lie on a face-centered cubic lattice displaced relative to the gallium lattice by one-fourth the body diagonal of the cube.

field. A "free" electron may be defined as an electron that is not bound into atoms or molecules. Traditional lasers use bound electrons. Thus, the conventional laser is limited to producing light (radiation) that is consistent with those frequencies that are specific to the vibration of a given atom or molecule. Carbon dioxide and helium lasers, for example, are so inhibited.

Free electrons are caused to vibrate by passing them through an alternating magnetic field. By changing (tuning) the apparatus, a broader range of frequencies is obtainable. Coherent radiation can range from the far-infrared to the far-ultraviolet regions of the spectrum. See Fig. 5.

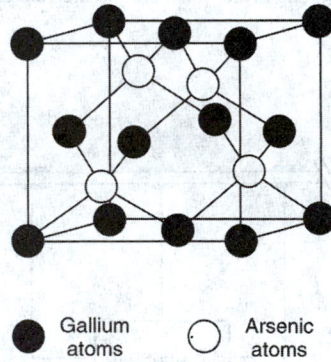

Gallium atoms   Arsenic atoms

Fig. 5. Schematic diagram of a free-electron laser. Beam of accelerated electrons passes through a field of alternating magnetism (wiggler magnets). The coherent light is generated and contained in an optical cavity defined by mirrors. (*Kim and Sessler.*)

Currently, free electron lasers are large and costly, but are ideal for certain kinds of research. Developers believe that ultimately the free electron laser will find numerous applications beyond research. Kim and Sessler suggest that applications may include surgery, fixing polymers, pharmaceutical manufacture, and lithography. It is predicted that the free electron laser will continue to expand in research usage, including condensed matter studies, nonlinear plasma studies, nonlinear quantum electrodynamics, nonlinear optics, and nonlinear microwaves, as well as microscopy, DNA studies, and cell response research in biology. One study is in progress to determine the feasibility of adapting the free electron laser to perform precision radar measurements in space. Researchers will study the potential of the laser as a compact space radar transmitter for discriminating objects in space, as would be required, for example, in connection with the Strategic Defense Initiative (SDI) program. The program will take advantage of the electron laser's inherent tunability, high power and efficiency, and ability to operate in frequency bands of 100 GHz and higher. The program's ultimate goal is a space-based, multiband, adaptive laser capable of operating efficiently at randomly chosen, stable frequencies.

Researchers have found that electrostatic accelerators are well suited for the far-infrared spectrum. An early device of this type was built at

the University of California at Santa Barbara for the main objective of free-electron research and studies in solid-state physics and biophysics. The accelerator has an operating range of 2–6 MV, corresponding to wavelengths in the range from 100–800 micrometers. Pulse duration is from 3–30 microseconds.

Solid-State Laser Development. Over the years, the progress of solid-state lasers has depended heavily on the improvement of old and the establishment of new pump sources. The helical lamp used by the early ruby laser was replaced by the linear flash lamp and arc discharge lamps. The next step was that of using a diode laser to pump another solid-state laser. This latter approach is advantageous because the diode laser emits optical radiation into a narrow spectral band. If the emission of the wavelength of the diode laser lies within the absorption band of the ion-doped solid-state laser medium, diode laser optical pumping can be very efficient and accompanied by little excess heat generation. By contrast, flash lamp pumping is limited by the broad emission spectrum and by excess heat production.

As pointed out by R.L. Byer (Stanford University), "The diode laser is essentially a continuous wave device with low energy storage capability, whereas the solid-state laser can store energy in the long-lived metastable ion levels." Stored energy can be extracted by Q-switching (rapid switching) to provide peak power levels that are orders of magnitude greater than obtainable from the diode laser per se. Important, too, is the fact that a solid-state laser can collect output from several diode lasers and thus furnish greater average power than is obtainable from a single diode laser. Furthermore, the line width of the diode laser-pumped solid-state laser is many times less than that of the diode laser source. Finally, the solid-state laser source emits optical radiation in a diffraction-limited spiral beam that is easily focused into a fiber or small space.

As early as 1982, a diode laser-pumped miniature Nd:YAG laser with a linewidth of less than 10 kHz was demonstrated. The research in this area continued apace at Stanford University and by a number of commercial electronics firms, with emphasis placed on the development of three-level lasers, Q-switched and mode-locked operation, single-frequency operation (monolithic nonplanar ring oscillator), visible radiation by harmonic generation, and array-pumped solid-state lasers. See Fig. 6.

A new class of lasers, so called vibronic solid state lasers, can emit — in contrast to other solid state lasers — a comparatively broad range of wavelengths. The lower level in these lasers is a band of energy levels that is caused by interaction between the electron motion and lattice vibrations. With the help of the usual tools (filters, etalons etc.) a narrow and tunable frequency bandwidth is selected for the laser output. The most popular materials are Ti-sapphire, i.e., titanium doped Al_2O_3 with output range from 600 to 1180 nm and Alexandrite, i.e., chromium doped $BeAl_2O_4$ lasing between about 700 and 825 nm.

Laser Microminiaturization — Target Optical Computer. For many years, scientists have accepted the fact that optical computers would not become a reality until optical components of micro size and exceptional performance equivalent to the already existing electronic switches and circuits could be developed. Thus, the optical computer became a major driving force toward the development of optical components. The problem was extraordinarily complex because size reductions of several orders of magnitude were mandatory.

As early as 1989, researchers at AT&T Bell Laboratories fabricated more than a million micron-size lasers (microlasers) on a single semiconductor chip, about 7 mm wide by 8 mm long. Individual devices ranged in size from 1 to 5 microns. Thus, these devices were two orders of magnitude smaller than conventional diode lasers. Researchers feel that it may be feasible to manufacture such devices that measure only between 1/2 and 1/4 micron. Traditional devices that measure a few microns wide by several hundred microns long have been well established for use in compact-disk players and fiber-optic communications. Thus, the *much, much smaller* devices coming out of research comprise a major step toward achieving the needs of optical computing. Much competitive research continues during the early 1990s because the market demand for microminiature lasers can be immense.

As pointed out by Jewell and associates (Bellcore), "The principles of operation underlying a diode laser are the same as those for any laser. Atoms in a part of the laser called the *amplifying medium* — typically a solid, liquid, or gas- are pumped, or energized, either electrically or with

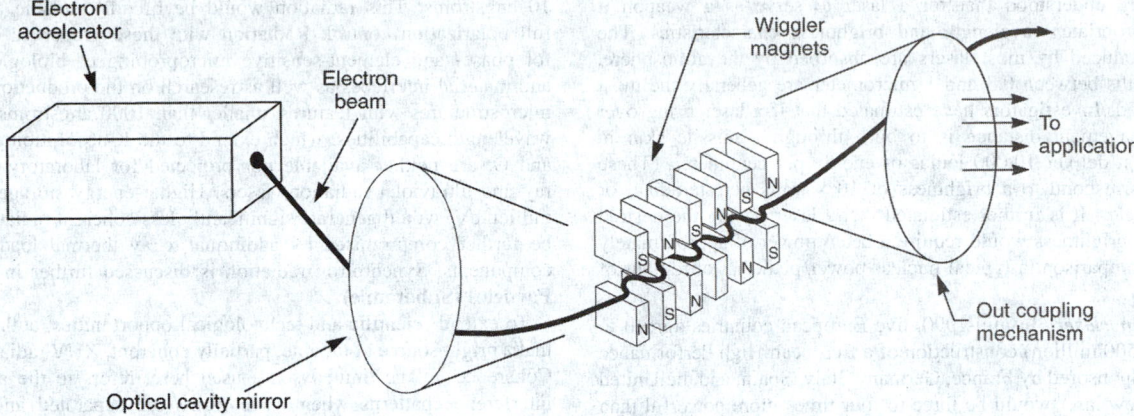

Fig. 6. Diagram of high-average power slab laser oscillator pumped by an array of diode lasers. Such an arrangement offers lower cost, ease of power scaling, and long-term reliability. (*After Byer.*)

a source of electromagnetic radiation. When a light wave of a specific wavelength traveling through the amplifying medium encounters a pumped atom, it can induce the atom to release its energy in the form of a light wave at the same wavelength. The process is coherent, which is to say that the crests and troughs of the waves match up, and the intensity of the light increases. Mirrors on each end of the amplifying medium form a cavity, and they force the light to bounce back and forth many times through the medium, maximizing the increase in intensity.

The differences in construction of a microlaser from a conventional gas laser or conventional diode laser are shown in Fig. 7.

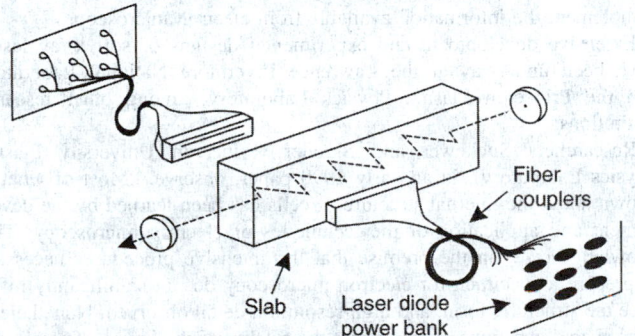

Fig. 7. Comparison of constructional configurations of (**a**) conventional helium-neon laser, (**b**) traditional diode laser, and (**c**) microlaser. Several orders of magnitude in size difference cannot be depicted here. The helium-neon laser ranges from 100 to 1,000 times longer than a traditional diode laser. The latter is some 100 times longer than a microlaser. (*After Jewell, Harbison, and Scherer.*)

To date, manufacture of the very tiny microlasers depend upon critical production techniques. For example, molecular beam epitaxy allows the basic material of each laser to be built up from layers of semiconducting materials. A typical microlaser may comprise 500 or more individual layers.

It was reported in late 1991 that a semiconductor laser that emits a blue-green light had been developed (3 M Company). Manufacturers of optical disks and other consumer electronic devices have been seeking a laser that would produce light in this range of the spectrum. The detailed development procedures required are beyond the scope of this article. However, it has been reported that the layered device is comprised of a gold electrode, a *p*-type zinc selenide, a quantum well (cadmium zinc selenide), an *n*-type zinc selenide, a substrate (gallium arsenide), and an indium electrode. By the end of the decade, progress in semiconductor research has led to reliable diode lasers with output in the blue and green part of the spectrum. In particular, SiC and GaN based systems promise to yield successful applications (See further reading by Fasol and by Fasol, Nakamura and Davies).

Liquid Lasers. Lasers operating in liquid media utilize rare earth ions in such organic hosts as chelates. Laser action is obtained in liquids using a flash tube or another laser as the pump. Early versions used rare earths in an organic liquid, while organic dyes, introduced somewhat later, were found to be more efficient, but required a separate laser for the exciting radiation. The dye laser has the special attraction that one laser may be tuned over a significant fraction of the visible spectrum by using a reflection grating as one of the cavity mirrors. One of the major areas of applications for dye lasers is the field of spectroscopy—for both basic research and applications such as combustion diagnostics or trace element analysis (See section on **Laser Spectroscopy**).

Another type of liquid laser utilizes a different principle and depends upon stimulated Raman scattering. Raman laser action was discovered by Woodbury in 1962, using a ruby laser and nitrobenzene. Here the laser excites the nitrobenzene, which in turn shows amplification at a frequency displaced from the ruby line by the vibrational frequency of the molecule. There is not true inverted population in this case. The incident photon is scattered by the molecule, which absorbs an amount of energy determined by its vibrational energy. The molecule is left in an excited state and the scattered photon is frequency-shifted by the energy loss. The process may be stimulated inasmuch as the rate at which the scattered photons are produced is proportional to the number of photons already present in the cavity at the scattering wavelength. As in the normal stimulated emission case, the frequency and phase of the output wave are identical with the wave that stimulates the scattering.

The Raman laser normally operates using the Stokes line, or the wavelength corresponding to the loss of one vibrational quantum. Other modes of operation utilize the second or third Stokes lines corresponding to double or triple vibrational absorptions. Similarly, higher-order effects in the medium may produce a series of anti-Stokes lines, which correspond to vibrational energy being added to the initial energy of the photons from the driving laser. The wavelength range of Raman lasers using different liquids is from the visible to the near-infrared.

X-Ray and Other Very High-Power Lasers. Almost since the inception of laser technology, the laser has been of interest to the military in connection with a variety of applications—weapons, radars, illuminators, rangers, etc. Among the highly sophisticated laser applications long envisioned by the military is the use of a laser to propel a spacecraft. In early concepts, a laser beam produced at ground level would vaporize an appropriate fuel, which could be water, and the supersonic jet caused by vaporization would be sufficient to place the vehicle into orbit without any chemical energy being expended by the craft itself. In the late nineties, experiments with tethered vehicles have shown initial albeit limited success in this quest. A small test craft without any on-board fuel was propelled 75 feet upward by 450 joule pulses from a powerful carbon dioxide (CO_2) laser. The small craft operates by expelling air that is rapidly heated by the absorbed laser energy. No longer motivated by SDI, the stated long-term goal is to loft small "picosatellites" into orbit (See further reading by Appell). Within the last few years, a major interest of the military has been directed toward the development of an x-ray laser for possible application in connection with the SDI (Strategic Defense Initiative) program. There are, of course, also strictly scientific interests in tunable coherent x-rays.

It is generally understood that for a laser to serve as a weapon it must have appropriate wavelength and brightness characteristics. The wavelengths produced by most lasers are absorbed by the atmosphere. Laser wavelengths between 0.3 and 1 micrometer are generally the most easily transmitted. Investigators have estimated that if a laser firing over a 3000 km engagement distance is to burn through a missile skin in 1 second, it must deliver 10,000 joules of energy per centimeter. (These requirements correspond to a brightness of 10^{21} watts per steradian, or unit of solid angle.) It is further estimated that a laser having the desired wavelength and brightness would require a beam power of approximately 100 MW. By comparison, a typical nuclear power plant has an output of about 1000 MW.

European Superlaser. In mid-1990, five European countries agreed to fund ($200 to $500 million) construction of a European High Performance Laser Facility. Sponsored by France, Germany, Italy, Spain, and the United Kingdom, the new laser would be three to four times more powerful than the Lawrence Livermore National Laboratory's NOVA, currently the world leader. The program was to progress in two principal stages, commencing with two intermediate power lasers—one neodymium glass and one KrF laser, built side by side. The major goal was to free European laser scientists from the dependence on high-powered machines in the United States and Japan. A superlaser of this type can be used to investigate some fundamental problems of physics. The intense pulse of the proposed laser would create conditions even hotter than the core of a burning star.

As of the late 1980s and early 1990s, four kinds of laser were under development. (1) *Chemical lasers* utilize chemical reactions between two gases to generate radiation. This technology is probably the most mature of the four kinds of lasers. It has been reported that the brightest of the chemical lasers is the MIRACL (mid-infrared advanced chemical laser), which in a demonstration at the White Sands Missile Test Range (New Mexico) destroyed a mock-up of a missile standing about a half-mile distant from the laser. The MIRACL was estimated to have a brightness of about 10^{17} watts per steradian, short of the SDI goal by a factor of about 10,000. (2) The *Excimer* (meaning excited dimer) consists of an unstable compound made up of two molecules. An electric discharge excites the molecules to form the dimer, and in breaking down, the dimer emits radiation. In some way, this radiation triggers a cascade of reactions that produce a laser beam. Radiation in the beam is estimated between 0.2 and 0.4 micrometer. A krypton-fluoride laser, tested at the Los Alamos National Laboratory, is estimated to operate at a wavelength of about 0.25 micrometer, delivering 10,000 joules of energy in a 380 nanosecond pulse. (One nanosecond = One billionth of a second.) It is reported that while the energy produced meets SDI goals, the pulse duration is off by a factor of about 3 million. (3) In the *free-electron laser*, a beam of electrons passes by a series of so-called wiggler magnets, which cause the electrons to vibrate and emit radiation. The wavelength can be tuned, theoretically, to any value between about 0.1 and 20 micrometers. The smaller the wavelength, the greater the energy. It is reported that a free-electron laser operated at wavelengths down to 10 micrometers in tests at Los Alamos. More recent research targets a 1-micrometer radiation of 100 microsecond pulses, containing 30 kW of power. As of the mid-1990s, both excimer and free-electron lasers were relatively poor converters of electric energy into beam energy, requiring massive power supplies—hence difficulties in locating the needed equipment in space. See prior description of free-electron laser in this article. (4) The *X-ray laser* also appears to be plagued with heavy and costly support equipment. Essentially, the x-ray laser consists of a nuclear explosive surrounded by an array (cylindrical configuration) of metal fibers. The emission of x-rays during the nuclear explosion stimulates the emission of a beam of x-rays from the fibers. This occurs within a microsecond prior to immolation of the device per se. For obvious reasons, further details remain sparse. It has been reported that to make an effective weapon, a particle beam would require energy of 250 MeV. If one assumes an acceleration gradient of 10 MeV per meter, it follows that the structure must be 25 meters long. When accounting is made of the mass of the power supply and its fuel, the weight of the weapon is found to be 50 to 100 tons. (Current typical payloads weigh a comparatively few tons.) Much additional work of a guarded nature continues.

Soft X-Ray Laser. A modern 1- to 2-billion-eV synchrotron radiation facility, based on high-brightness-electron beams and magnetic undulators, would generate coherent, laserlike soft x-rays of wavelengths as short as 10 angstroms. This radiation would be broadly tunable and subject to full polarization control. Radiation with these properties could be used for phase- and element-sensitive microprobing of biological assemblies and material interfaces as well as research on the production of electronic microstructures with features smaller than 1000 angstroms. These short-wavelength capabilities, which extend to the K-absorption edges of C, N, and O, are neither available nor projected for laboratory XUV (soft x-ray and ultraviolet radiation) lasers. Higher-energy storage rings (5 to 6 million eV) would generate significantly less coherent radiation and would be further compromised by additional x-ray thermal loading of optical components. Synchrotron radiation is discussed further in the article on **Particles (Subatomic)**.

To extend scientific and technological opportunities, authorities suggest that a bright source of tunable, partially coherent, XUV radiation is needed. Coherence, in the limited sense used here, refers to the ability to form interference patterns when wave fronts are separated and recombined. The availability of a tunable source of coherent soft x-rays, combined with other developments in x-ray optical techniques, would make it possible to construct an x-ray microprobe of sufficient intensity to permit fundamentally new, phase-sensitive experimentation in a number of scientific and technological fields. Various imaging and scattering techniques would be enhanced by the greatly increased photon flux available to study small samples, as well as providing the capability of tuning the radiation to the wavelength of interest. For example, with soft X-rays well matched to the absorption edges of elements, such as carbon (284), nitrogen (400), and oxygen (532), as well as other elements of relatively low atomic numbers (Na, P, S, K, and Ca), it should be possible to study elemental distributions and motion within biological specimens without the need for dehydration, fixing, or staining. Three-dimensional imaging, made possible by combining partially coherent undulator radiation and x-ray microholographic techniques, would complement the information available from electron microscopes.

Extensive development and experimental designs of soft x-ray lasers have been underway at the Lawrence Livermore National Laboratory and the Princeton Plasma Physics Laboratory, among other research institutions.

Researchers Suckewer and Skinner (Princeton University Plasma Physics Laboratory), in an early 1990 paper, observe, "Most of what is known about the internal structure of cells has been learned by the development and application of the techniques of electron microscopy. This knowledge rests on the premise that the intensive procedures necessary to prepare a specimen for electron microscopy do not significantly influence the structure, form, and high-resolution detail observed. Nonetheless, unanswered questions remain about the fidelity of the image of a cell that has been fixed, stained with heavy metals, and sectioned to the original living cell. X-ray microscopy offers a new way to look at unaltered cells in their natural state." X-ray laser microscopy can offer numerous advantages in this regard.

In summarizing the current status of soft x-ray laser research, the aforementioned researchers comment, "The general impact soft X-ray lasers will have in science and technology will depend on improvements in their performance and cost. It is necessary for their successful commercialization that these devices operate routinely at high gain-lengths ($GL > 4$), with the use of a low-cost driver laser, and this needs more system development and engineering. Most applications of visible-wavelength lasers are based on the fact that the brightness of these lasers is several orders of magnitude greater than that of conventional spontaneous emission sources, and this is achieved principally by the laser cavity mirrors. This technology is significantly more difficult in the x-ray region because of intrinsic limitations of x-ray absorption in materials and present limits in the soft x-ray laser pulse lengths. Nevertheless, a 'revolution' in x-ray optics is under way and the precedent of visible-wavelength lasers illustrates the potential benefits awaiting the creative inventor of applications of this technology to novel fields."

Laser Applications

During the early phases of laser development (late 1950s to early 1970s), there was a high tempo of research activity and confidence in the ultimate potential for practical applications. Some of the early suggestions for laser use included instrumental applications in metrology and spectroscopy, as well as working tools for industry, such as cutting, welding, and annealing. But during that period it was also observed by some researchers that the laser was an invention for nonexistent needs. The decade of the 1980s removed all such doubts, and as science entered the 1990s, the laser had become established as an essential component in numerous laboratory research programs and industrial and medical applications. Just a cursory inspection of the literature during the late 1980s and early 1990s is indicative of the wide scientific interest in the laser.

Atomic Cooling and Trapping. During the last few years, the ability to control the position and velocity of isolated atoms and microscopic particles has progressed markedly. By the end of 1998 molecules have also been laser cooled and several groups have obtained sufficiently low temperatures and high densities to achieve Bose-Einstein condensation of trapped atoms. In such a state of matter—predicted by Bose and Einstein—the quantum nature of the atoms causes them to lose their individual existence and to coalesce into one collective system. The tremendous success of these efforts has been recognized by the award of the 1997 Nobel Prize in physics to Steven Chu, Claude Cohen-Tanoudji and William Phillips for development of methods to cool and trap atoms with laser light. As pointed out by S. Chu (Stanford University) in a late 1991 paper, "Light can exert forces on an atom because photons carry momentum. The exchange of photon momentum with an atom can occur *incoherently*, as in the absorption and reemission of photons, or *coherently*, as in the redistribution (or lensing) of the incident field by the atom."

Coherent interaction is called the *dipole force*. The incoherent interaction that alters the momentum of an atom is called the *scattering force*.

Successful atom manipulation, however, often depends more upon cooling the atoms than upon exciting the aforementioned forces. Dramatic cooling of atoms to extremely low temperatures is accomplished by employing counterpropagating laser beams, arranged along *x, y,* and *z* axes—in essence, creating three-dimensional cooling. As pointed out by Chu, "Because the cooling force is viscous (linearly proportional to the velocities of the atom for low velocities), we named the laser beams that generate the drag force, 'optical molasses'."

In 1991, a research team (Ecole Normale Superieure, Paris) reported the cooling of a sample of cesium atoms to 2.5 μK. At about the same time, a research group (Joint Institute for Laboratory Astrophysics, Boulder, Colorado) reported the achievement of 5 μK. The aforementioned "optical molasses" technique was used in both cases.

Laser cooling, trapping, and related techniques are finding numerous research and practical applications. For example, practical laser-cooled atomic clocks are now possible, constituting a major improvement in accuracy over present atomic clocks. As mentioned by Chu, "A cesium time standard based on a sealed design for which the cooling, manipulation, and detection of the atoms are all done with diode lasers should exceed the stability of the best present-day time standards."

In an excellent paper, Chu (See reference) observes, "Perhaps the most exciting applications in the field of laser cooling and trapping will come out of the ability to study problems in polymer physics and biology on a single molecular basis. Normally one examines the behavior of a large number of molecules, and the fundamental chemistry of the molecules must be inferred from the average behavior of the entire ensemble. On the other hand, the processes that govern the behavior of a single molecule are important: for example, the nucleus of a cell has a single molecular copy of its genetic blueprint, and its chemistry depends in part on the chemistry of single molecules."

In a 1990 paper, Zewail (See reference) describes how atoms can collide, interact, and give birth to molecules in less than a trillionth of a second. As an example of how high-speed imagery has improved over the years, he compares photos of a galloping horse (10 meters per second) taken in 1887 with quantitative observations (made in 5 trillionths of a second) of hydrogen iodide colliding with carbon dioxide to form carbon monoxide, hydroxide, and iodine.

Lasers as Mini-Manipulators. As scientists continue to probe the very minute aspects of natural organs and substances (nanotechnology), small lasers have been found to possess "manipulative" abilities of a kind not envisioned in the early years of laser technology. Scientists (Massachusetts Institute of Technology) in 1990 reported of how lasers can be used effectively as manipulators at the microscopic level. In a study of "mechanoenzymes," which are responsible for the rotary motions

of flagella, laser light was used to lift up, move, and position microscopic objects with the "pressure" of the laser light itself, a phenomenon that has been described by Amato as "akin to a blast of air levitating a plastic ball."

Laser mini-tweezers also have been used to clip off regions of chromosomes, moving organelles around inside cells, pushing molecules tiny distances within crystals, and, when used as tiny scalpels or scissors, to catch, trap, puncture, and splice subcellular structures. Recently, measurements have been made of the elastic properties of DNA. Also, it has been found that bacteria can be moved around in a water solution without apparent damage to the organism. Medical applications are described later.

Laser Spectroscopy. In the early years of laser technology, spectroscopy was one of its major uses, an application that has expanded markedly during the past few decades. The review by Gupta provides a good starting point for further reading since it includes an–at the time–up-to date resource letter featuring a large number of annotated references. The techniques of laser spectroscopy parallel those of microwave or radio-frequency spectroscopy, but because lasers are imbued with high spectral purity, they permit vastly improved resolution of fine detail. Early lasers were limited to molecular lines that were coincident with the laser wavelengths. Then lasers using fluorescent organic dyes appeared. These instruments had relatively wide emission bands, offering a tuning capability. Both continuous-wave and pulsing dye lasers have been widely used in most of the visible and near-visible ranges of the spectrum. During the interim, much progress has been made, particularly in providing tunability to lasers. For example, a methyl fluoride molecular gas laser is continuously tunable over broad portions of the far infrared, a region that previously had been difficult. A highly schematic diagram of the operating principle used in early laser-probe emission spectrography is given in Fig. 8.

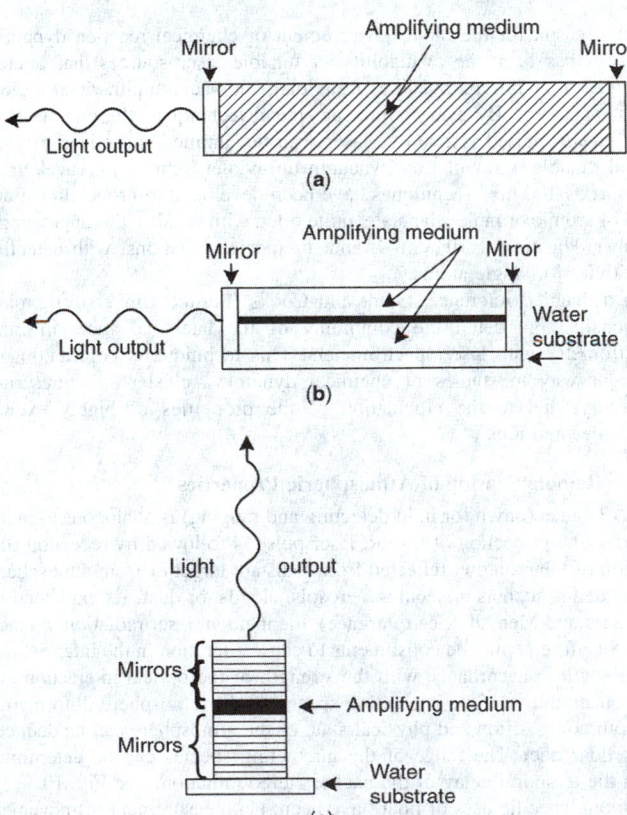

Fig. 8. Operating principle of laser-probe emission spectrography.

A particularly interesting development is that of the so-called "atomic fountain." As noted by Chu in 1991, "The precision of a spectroscopic measurement depends on both the high $Q(Q =$ quality factor of the resonance defined by $Q = V/\Delta V)$ and the signal-to-noise ratio of the signal. Thus, it is important to create a high-flux source of cold atoms.

Also, many applications would benefit from a continuous beam of atoms instead of the pulsed sources." An extreme limit of a slow beam is an "atomic fountain," which first was envisioned by Zacharias in the early 1950s. A group of Stanford University scientists has constructed an atomic fountain by first trapping atoms from a thermal beam in a magneto-optic trap and then pushing the atoms upward with a pulse of light from a continuous-wave laser. See Fig. 9.

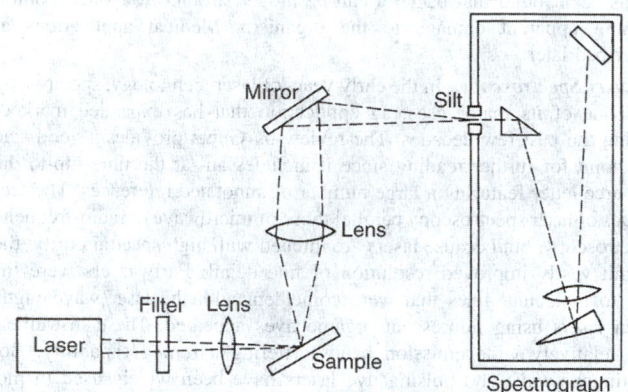

Fig. 9.   The atomic fountain makes it possible to determine precisely the energy states of atoms. Upon injection, the atoms in question are slowed down by a laser beam. Then, the atoms are captured and cooled by means of a magnetic field and several light beams. The cooled atoms follow a ballistic trajectory through a radio frequency (rf) waveguide and a resonant photoionization detection region. (*After Kasevich, Riis, Chu, and DeVoe.*)

One reason for the rapid advancement of chemical reaction dynamics research has been the availability of tunable laser sources that operate throughout the infrared, visible, ultraviolet, and vacuum ultraviolet regions of the spectrum. By using nonlinear optical techniques, the outputs from high-power, pulsed visible dye lasers can be summed and mixed to yield useful tunable ultraviolet and vacuum ultraviolet light, with wavelengths as short as 100 nm. Techniques have been developed to probe almost any kind of atomic or molecular state, quite often with sensitivities approaching number densities of 10^5 cm^{-3}, and, in special situations, with detection sensitivity for single atoms.

In a detailed reference, Grant and Cooks (Purdue University) explain in considerable detail the combining of the latest advances in mass spectrometry with laser spectrometers. This technique is contributing in a major way to studies of chemical dynamics, cluster structures, and reactivity, and to the elucidation of the properties of highly excited molecules and ions.

Laser Remote Sensing of Atmospheric Properties

LIDAR (an acronym for light detecting and ranging) is analogous to radar. In lidar, the projection of a short laser pulse is followed by reception of a portion of the radiation reflected from a distant target or from atmospheric constituents, such as molecules, aerosols, clouds, or dust. As explained by Killinger and Menyuk (See reference), the incident laser radiation interacts with the aforementioned constituents to cause alteration in the intensity and wavelength in accordance with the strength of the optical interaction and the concentration of the interacting species in the atmosphere. Information on both composition and physical state of the atmosphere can be deduced from lidar data. The range of the interacting species can be determined from the temporal delay of the backscattered radiation. See Fig. 10.

Among specific uses of lidar have been: (1) measurement of movement and concentration of urban air pollution; (2) determinations of chemical emissions from and in the vicinity of industrial plants; (3) determination of atmospheric trace chemicals in the atmosphere; (4) measurement of the velocity and direction of winds near storms and airports, including windshear and gust fronts; and (5) determination of the global circulation of volcanic ash emitted into the atmosphere, relatively recent examples including Mount Pinatubo and Kilauea; among several other applications.

Shortly after the discovery of lasers (early 1960s), Fiocco and Smullin bounced a laser beam off the moon (1962). These researchers also

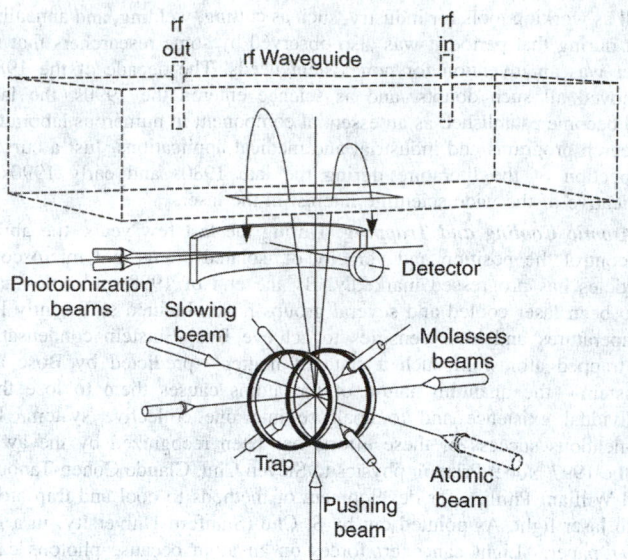

Fig. 10.   Basic components of lidar system used for remote sensing of the atmosphere. Backscattered information sometimes will contain spectral information useful for determining composition and physical characteristics of the cloud or of the intervening atmosphere.

investigated the turbid layers in the upper atmosphere. As early as 1963, Ligda used a ruby laser to obtain the first lidar measurements of cloud heights and tropospheric aerosols. In 1964, Scotland used a temperature-tuned ruby laser to detect water vapor in the atmosphere. Lidar, in recent years, has been greatly improved because of the availability of several kinds of laser sources and improvements made in optical instrumentation and data processing.

As summarized by Killinger and Kenyuk, the future of laser remote sensing is promising and will depend upon several factors, including: (1) development of practical, eye-safe laser sources that cover certain spectral gaps where lidar is currently weak; (2) a further simplification of lidar systems, including lowering size and cost of equipment needed; and (3) more experience to be gained from promising new applications. Among these new applications are: (a) detection of methane gas leaks in coal mines, using a diode laser lidar system; (b) detection of methane and natural gas leaks in industrial plants, using a laser coupled to a low-loss optical fiber network; (c) measurement of global wind fields through the use of Doppler lidar systems mounted in a satellite as a means for improving weather forecasting; and (d) the planned use of lidar on the NASA space-borne Earth Observing System for measurements of global temperature, water vapor, and pressure.

Classification of Lidar System. Lidar systems can be classified on the basis of particular optical interactions which they utilize. Classes of lidar include:

1. *Atmospheric backscatter lidar,* wherein the lidar system transmits one laser wavelength and detects changes in the backscatter due to the aerosols or dust in the atmosphere. This is the most common type of lidar and consists of a nontunable, high-power, pulsed laser. Atmospheric constituents having comparatively large optical scattering cross sections are relatively easy to detect. These systems are used in tracking turbid effluent and gas plumes from factories as well as for mapping rain, snow, ice crystals, and dense clouds in the atmosphere. This type of system was used for checking volcanic ash in the atmosphere.

2. *Differential-absorption lidar* (DIAL), a system which measures the concentration of a molecular species in the atmosphere. This is accomplished by transmitting two wavelengths, only one of which is absorbed. The difference in the intensity of the returns at the two wavelengths is measured. Backscatter in DIAL may come from a hard target or aerosols and dust. One wavelength will be absorbed by the target molecules; the other wavelength will not be absorbed. Many DIAL studies have been carried out in the infrared (IR) range, where almost all molecules of interest

have extensive absorption bands. Molecules so far studied include SO_2, NH_3, O_3, CO, CO_2, HCl, NO, N_2H_4, N_2O, and SF_6.

3. *Fluorescence lidar* uses two wavelengths (as in DIAL) plus spectrometric techniques for separating the wavelength-shifted fluorescence signal from the strong Rayleigh backscatter in the atmosphere. The laser is tuned to an absorption line of the species to be measured. Reradiated fluorescence is detected by selective spectral filtering of the returned radiation. The fluorescence radiation may be at the same wavelength as the excitation wavelength, or it may have a longer wavelength because of the red-shift. The backscatter coefficient for fluorescence is greater in the ultraviolet (UV) than in the IR — this due to combined effects of absorption cross section, which is greater in the UV than in the IR. For some applications, fluorescence lidar is limited for remote sensing because of detector sensitivity coupled with solar background radiation. The latter tends to confine fluorescence measurements to nighttime studies and to wavelengths shorter than 1 micrometer, where photomultiplier detection can be used. Nevertheless, some investigators have been quite successful in using the method, particularly in the study of alkali metal (Na, K, Li, and Ca) profiles at altitudes of 80 to 100 km. The method also has been useful for studying the hydroxyl free radical (OH). This radical is of principal interest because of the catalytic role which it exerts in atmospheric chemistry. The OH radical, along with chlorine and nitrogen oxides, is involved in the ozone destruction cycle.

4. *Raman lidar*, a method that is limited by the small optical interaction strength for Raman scattering. High-energy pulsed lasers are employed in this method. The method is limited to the UV or visible regions to permit the use of sensitive photomultiplier tubes for detection. Raman lidar has been used effectively for species that either are at close range or in high concentration, such as N_2, O_2, and H_2O. As mentioned by Killinger and Menyuk, this method does have some attractive features, the most noteworthy of which is that the laser wavelength need not be tuned across an absorption line because the spectral information is given by the frequency shift of the emission (independent of laser wavelength).

5. *Doppler lidar*, a method that detects only a very narrow spectral range ($\sim 10^{-5}$ nm) that encompasses the Doppler-shifted backscatter lidar return. Doppler shifts in the return lidar signals have been used to measure wind velocities and to differentiate between molecular and aerosol returns in the atmosphere. Optical heterodyne techniques are used to detect the shifts which are very small — for example, a fractional change in frequency of about 10^{-8} for a velocity of 1 m/sec at a wavelength of 10 micrometers. Carbon dioxide lasers which provide high power and stable single-frequency operations, are commonly used in Doppler lidar systems. These are in the 10-micrometer range. The system has provided information on boundary layer flow near storm gust fronts and wind shears near airports. They also have been used to measure aircraft vortices and clear air turbulence. See Fig. 11.

Doppler broadening effects also have been used to separate backscattered lidar signals into molecular and aerosol components. Characteristics of some lidar systems are summarized in Table 2.

Laser Techniques in High-Pressure Geophysics

Laser techniques used in conjunction with the diamond cell make it possible to study the high-pressure properties of material that heretofore had to be inferred from samples found on the surface. Spontaneous

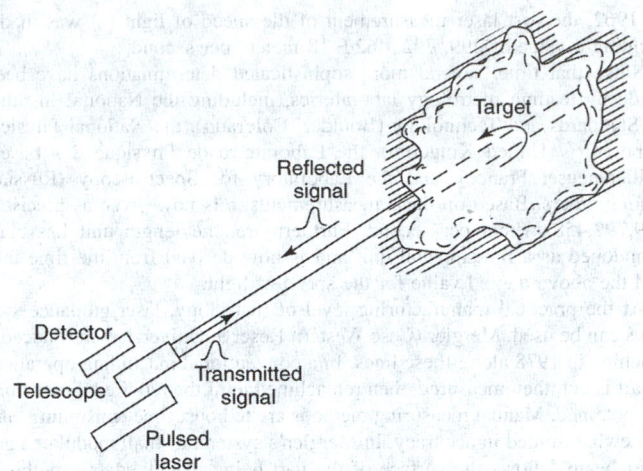

Fig. 11. Doppler lidar measurement of wind direction and velocity near an airport during a storm. White arrow points to presence of a strong, localized downburst-gustfront.

Raman scattering of crystalline and amorphous solids at high pressure demonstrates that dramatic changes in structure and bonding occur on compression. High-pressure Brillouin scattering is sensitive to the pressure variations of single-crystal elastic moduli and acoustic velocities. Laser heating techniques with the diamond anvil cell can be used to study phase transitions, including melting, under deep-earth conditions. Laser-induced ruby fluorescence has been essential for the development of techniques for generating maximum pressures now possible with the diamond anvil cell, and currently provides a calibrated in situ measure of pressure well above 100 gigapascals. Hemley, Bell, and Mao (Carnegie Institution of Washington) point out that applications of new spectroscopic techniques, such as double resonance, ultrafast kinetics, Fourier-transform, Raman, and nonlinear optical methods, are likely prospects in future work on geophysical problems with the diamond anvil cell. Recent high-pressure studies involving the use of picosecond spectroscopy and hyper-Raman scattering of perovskites may be representative of this trend. Time-resolved studies may permit the detailed investigation of the kinetics of high-pressure phase transitions and the rheology of minerals under in situ deep-earth conditions. The combination of spectroscopic and x-ray diffraction probes with laser-heating techniques may yield detailed structural information on earth materials at high temperatures and pressures, thus advancing an understanding of the connection between atomic-scale properties and global deep-earth processes.

Laser Metrology

Lasers are widely used for making precision measurements of geometric variables. In the early 1960s, laser pioneers demonstrated precise measurements with the device. Lasers introduced the concept of frequency metrology as contrasted with wavelength metrology. In an early experiment with a super-stabilized laser, scientists at the Massachusetts Institute of Technology during the early 1960s worked out a laser version of the famous Michelson-Morley experiment at Case Institute of Technology in Cleveland. The MIT scientists concluded that an advance in measurement sensitivity by a factor of 1000 over the Michelson-Morley data was potentially available through the use of frequency rather than length metrology.

TABLE 2. SUMMARY OF LIDAR SYSTEM CHARACTERISTICS

Type of Lidar	Type of Laser Used	Nominal Accuracy	Range (km)	Atmospheric Targets
Atmospheric backscatter	Ruby, Nd:YAG	1–10%	10–50	Dust, clouds, volcanic ash, smoke plumes
DIAL, Raman	Dye, CO_2, optical parametric amplifier, CO:MgF_2	1 ppb–100 ppm	1–5	H_2O, O_3, SO_2, NO, NO_2, N_2O_3, C_2H_4, CH_4, HCl, CO_2, CO, Hg, SF_6, NH_3
Fluorescence	Dye	10^2–10^7 atoms/cm	1–90	OH, Na, K, Li, Ca, Ca^+
DIAL, Raman	Dye, Nd:YAG	1 K, 5 mbar	1–30	Temperature, pressure
Doppler	CO_2	0.5 m/sec	15	Wind speed

After Leone (1987).

In 1962, the first laser measurement of the speed of light (c) was made, yielding a value of 299, 792, 462± 18 meters per second.

Since that time, several more sophisticated determinations have been made by leading metrology laboratories, including the National Institute of Standards and Technology (Boulder, Colorado), the National Physical Laboratory (United Kingdom), the Laboratoire de Physique des Lasers (Villetaneuse, France), and the Laboratory for Spectroscopy (Russia), among others. Based on these measurements, c is now given as precisely 299,792,458 meters per second. Furthermore, the length unit has been abandoned as a fundamental unit and is now derived from the time unit and the above quoted value for the speed of light.

At the practical manufacturing level of metrology, laser guidance systems can be used. Mergler (Case Western Reserve University) introduced a machine in 1978 along these lines. In a conventional machining operation, a part is cut, then measured, then remachined until the required dimensions are obtained. Manual measuring methods are tedious, time consuming, and somewhat limited in accuracy. In Mergler's system, a small modulated gas laser beam follows the surface of the part being machined and, within a precision of 1/5000 inch (0.005 millimeter) measures the piece as it is being cut. In later systems, the gas laser was replaced by a solid-state laser which occupies less space.

Laser Doppler Flowmeter

As shown in Fig. 12 fluid flow can be determined by measuring the doppler shift in laser radiation scattered from particles in the moving fluid stream. No sensor is required in the moving stream. The laser radiation focal point can be moved across the flow tube to measure velocity profiles. Fluid linear flows from 0.01 to 5000 inches (0.03 centimeter to 127 meters) per second have been measured. Contaminants, such as smoke, may have to be added to gases to provide scattering centers for the laser beam.

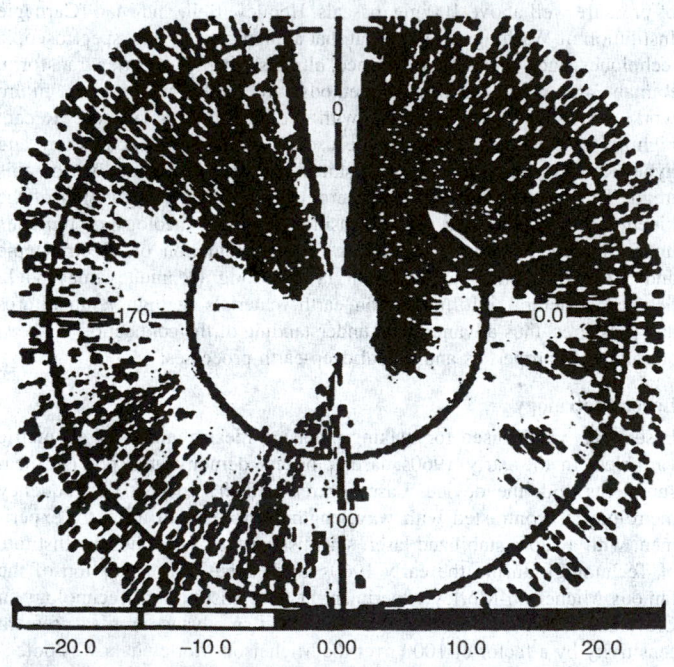

Fig. 12.   Operating principle of laser doppler flowmeter.

Laser Gyroscope

As early as the beginning of the 20th century, some investigators suggested that light will exhibit gyroscopic behavior, that is, the time required by light to traverse a circular pathway depends on whether the pathway is stationary or rotating. Thus the time difference can be used as a measure of the amount of rotation. The practical application of this observation, however, had to await vast improvements in optical systems, including the discovery of the laser, advances in fiber-optics, and better reflective mirrors. Within recent years, this principle has been applied in two configurations — fiber gyroscopes and ring-laser gyroscopes. The latter is described briefly here. As of the late 1980s, several aircraft depend upon ring-laser gyroscopes

instead of their mechanical counterparts. The ring-laser gyroscope is more sensitive, has virtually no moving parts, and is as accurate as the best mechanical instruments. The rotation-induced difference in length of light path traversed is called the Sagnac effect, after the researcher who first demonstrated the phenomenon in 1913.

As previously mentioned in this article, a laser usually incorporates a resonant cavity. C.V. Heer (Ohio State University) in 1958 proposed that a resonant cavity could be used to measure rotation rates. In such an instrument, light circulates many times around a given path, not just back and forth between two mirrors. The first gyroscopes of this kind were constructed on a large scale, consisting of four glass tubes, each a meter long and arranged in a square. Light was made to travel around the device by placing a mirror in each corner. Over the last several years, the device has been markedly reduced in size (fits in the palm of the hand). Contemporary gyroscopes of this type are made from a single block of glass, into which a square channel is drilled. The channel is filled with a mixture of helium and neon. The laser is completed by attaching a small number of electrodes and four mirrors. As explained by Anderson, some ring-laser gyroscopes have a triangular channel and three mirrors; others have a hexagonal channel and six mirrors.

Beyond the scope of this article, Anderson explains the operation of the gyroscope in intimate detail and describes two problems that have proved most vexing to manufacturers of the ring-laser gyroscope, namely, frequency locking at low rotation rates and the bias effect. Improvements in this instrument are expected in the relatively near future because of what scientists have recently learned pertaining to the phenomenon of optical phase conjugation.

Lasers in Manufacturing Operations. A principal advantage of the laser in manufacturing operations is its ability to apply an extremely high flux of energy to the surface of a workpiece, as compared with traditional heat sources, such as flames, torches, electric arcs, and plasma jets. For manufacturing operations, lasers are usually placed in two categories. (1) *Light-duty lasers* range from a few tens of watts to a few hundred watts. Typical applications include cutting and drilling ceramic substrates in the electronics industry, drilling gems (for example, rubies in watchmaking), and cutting light-gauge metals as well as cloth, plastics, wood, and a variety of materials. Light-duty lasers that have been used include ruby lasers, neodymium-doped glass lasers, and neodymium-doped yttrium aluminum garnet lasers, among others. Depending upon the particular laser selected, the laser may operate in a pulsed or continuous mode. Argon and CO_2 lasers usually are operated in the continuous-wave mode. (2) *Heavy-duty lasers* range from a few kilowatts to a few tens of kilowatts. Typical applications include pipeline welding, automobile part welding, surface heat-treating of engine and other parts, with the applications expanding as experience is gained.

The high flux of electromagnetic energy applied to the surface of a workpiece by a laser is absorbed in an outer layer only about 10 nanometers thick. Thus the heat source is confined essentially to a thin film. Through careful design of equipment, the heat energy required is maintained in a comparatively small region, thus preventing or reducing thermal damage to the rest of a given part, and achieving a very high energy efficiency, estimated to range from 10 to 1000 times greater than can be achieved with conventional energy sources.

The electronics industry utilizes laser welders for joining dissimilar materials, fixing electrodes to batteries and connectors to a host of devices. A whole new area of laser technology, sometimes called laser microchemistry, has been exploited in the microstructure engineering of semiconductors. Lasers are used to initiate chemical reactions that result in deposition of material at a surface, for removing materials, and for alloying or diffusively mixing two or more solids on microscopic spatial scales. Lasers thus have played a major role in establishing new dimensions in microfabrication technology. It is possible to use a single laser to produce both gas-phase photolysis and surface heating. As described by Christensen (See reference), solar cells have been fabricated by using a UV laser to photo dissociate trimethylboron, $B(CH_3)_3$, over a silicon surface in the manufacture of solar cells. The laser also heats the surface so that the boron atoms absorbed on the surface after the photolytic step rapidly diffuse into the bulk of the material. After irradiation, the silicon is heavily doped with boron near the surface, and the p-n junction thus formed functions as a photovoltaic cell. In some other applications, it has proved advantageous to use two lasers of different wavelengths to separately achieve photolysis and heating.

Perspective. The industrial applications for lasers developed comparatively slowly. As previously mentioned, lasers depend upon raising active molecules of the lasing medium to what might be called an *upper laser level* of energy, after which they relax to a *lower laser level*. Energy is given up during this process. Part of this energy is represented by photons of which the laser beam is composed. The other part is waste heat, which raises the temperature of the lasing medium. Thus, an excess of waste heat be removed so that the upper-level population can be maintained, a significant problem in the case of a continuously emitting high-power laser. Higher packets of energy in pulses can be attained, but waste heat must be removed by conduction between pulses. The end result is a pulsed high-power laser, but one that has a comparatively low average energy level simply because of the pauses in between.

In early laser designs, the quantity of waste heat generated was a limiting factor and consequently the average power output was low. Such lasers were excited by diffuse longitudinal electric discharges in long tubes with relatively large diameters. Heat generated at the center of the tube diffused to the side walls essentially by conduction, and the rate of heat transfer varied inversely with the tube radius and essentially directly with the length of the tube. Thus, the length of laser tubes increases as greater output power was sought.

Various component cooling schemes were proposed and used, but a major improvement was made when the concept of cooling a flowing laser medium was proposed, thereby taking advantage of the far more effective cooling by convection than by conduction. Gas lasers were considered the most apt for application of this concept and this led to the gas dynamic laser.[1]

With this concept, gas dynamic lasers increased in power outputs from less than 10 kilowatts by a factor of 13 to 14 within less than a decade (by the late 1960s). Success with the early gas lasers in this respect catalyzed a number of other refinements and improvements. However, the problem of maintaining a high-pressure glow discharge remained. Population inversion can be produced when electrons in an ionized gas are at a temperature relatively high as compared with the kinetic temperature of ions or molecules. This is a condition referred to as *glow discharge*. But an arc may form when the discharge is destabilized as the result of greatly increased gas pressure. Overheating of the molecules and ions destroys the population inversion. This problem was overcome by the concept of the ionizer/sustainer.

With the availability of high-power continuous electric discharge lasers capable of operating at up to 20 kilowatts output, a number of the previously predicted applications for lasers became practical. One of the first uses of a laser beam strictly for its power in cutting (exploding) a material was the fabric cutting system developed by Hughes Aircraft Company (circa 1966-1967) for which U.S. Patent No. 3,761,675 was awarded to W.J. Mason, D.W. Wilson, D.M. Considine, F.J. Viosca, and J.P. Wade on September 25, 1973. See Fig. 13. In this system cloth is carried in a single layer into a cutting area where a laser beam focused on the cloth is directed by computer commands to travel within the cutting area so as to cut many patterns in the cloth rapidly and accurately. The cut produced by the focused laser beam is sharp and narrow, leaving the fabric unfrayed. With synthetic materials, such as nylons and Dacrons®, the laser beam also serves to seal the cut edges by melting them during the cutting process. Unlike a mechanical blade, the laser beam does not dull; its cut remains uniform and is effective in cutting a wide range of materials, even those having metallized threads.

In the early 1980s, a helium-neon laser was used in a scanner system for inspecting textiles. The system uses laser output split into three beams,

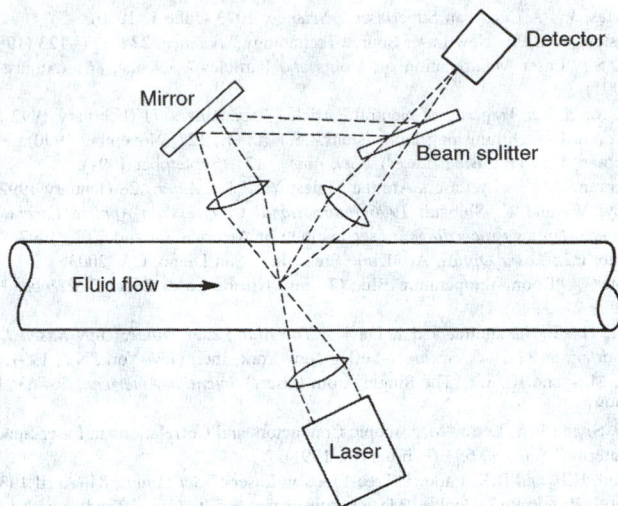

Fig. 13. A cloth cutting system wherein cloth is carried in a single layer into a cutting area, where a laser beam is focused on the cloth and is directed by computer commands to travel within the cutting area so as to cut a plurality of patterns through the cloth rapidly and accurately. Invented by W.J. Mason, D.W. Wilson, D.M. Considine, and J.P. Wade (*Hughes Aircraft Company*) in 1973, this was one of the very early and successful industrial applications of the laser. Diagram is part of U.S. Patent 3,761,675.

each of which scans the fabric independently in a pattern covering the entire surface. The system, moving at a rate of four meters per second, detects flaws through changes in reflected light and flags these areas for elimination or repair. In terms of economics, one laser system working one shift performs the same function as human inspectors at two plants working two shifts.

High-power lasers can perform many metalworking operations, including welding, cutting, surface hardening, and surface alloying. For small devices, the laser can perform much as a conventional electron beam, but without requiring the need for operation under a vacuum. High power densities can be achieved—up to 10^6 watts per square centimeter. It has been shown that a 16-kilowatt laser can make a 0.75-inch (1.9-centimeter) penetration weld in stainless steel at a rate of about 30 inches (76 centimeters) per minute. The laser beam can be directed by mirrors, thus making it effective for welding pipe from the inside. It also has been shown that a continuous-wave carbon dioxide device (15 kilowatts) can be used for welding half-inch (1.2-centimeter) thick steel plates at the rate of about 50 inches (127 centimeters) per minute. If the laser is focused to a spot size of about 0.03 inches (0.08 centimeter) in diameter, power densities of some 2200 kilowatts per square centimeter are produced.

Laser Recording. For many years, it has been known that lasers can be used to encode information on materials that respond in an irreversible manner to exposure to high-intensity light. However, it is only comparatively recently that the concept has been reduced to commercial practice—with the almost sudden appearance of optical disk recording (compact disk) in the entertainment field. It is because of the coherence and relatively short wavelength of laser radiation that such large volumes of information can be written onto a very small space of the recording medium. The potential for microlasers in this field is discussed earlier in this article.

Additional Reading

Adams, C.S. and E. Riis: "Laser Cooling And Trapping Of Neutral Atoms," *Progress in Quantum Electronics*, 1–79 (1997).

Amato, J.: "Moving Tiny Things by Optical Tweezers," *Science News*, 148 (March 10, 1990).

Anderson, D.Z.: "Optical Gyroscopes," *Sci. Amer.*, 94–99 (April 1986).

Appell, D.: "High-Power Laser Beam Launches Fuel-Less Craft," *Laser Focus*, 90 (March 1998).

Attwood, D., K. Halbach, and K. Kwang-Je, Kim: "Tunable Coherent X-rays," *Science*, **228**, 1265–1272 (1985).

Byer, R.L.: "Diode-Laser-Pumped Solid-State Lasers," *Science*, 742 (February 12, 1988).

---

[1] Invented by Kantrowitz in the late 1960s. In essence, the device had two compartments separated by a nozzle. In the first compartment, gas was held at a temperature of about 1400 K and pressure of 17 atmospheres. This high-pressure compartment held about 10% of the active CO_2 molecules in the total system. Expansion of this gas through an orifice caused cooling. Because of the cooling, the lower-level population essentially vanished a few centimeters downstream from the nozzle. This occurred before the upper-level population had an opportunity to decline significantly. The population "inversion" resulting was adequate for effecting a laser beam of considerable power.

Cherfas, J.: "A European Superlaser?" *Science*, 1073 (June 1, 1990).

Christensen, C.P.: "New Laser Source Technology," *Science*, **224**, 117–123 (1984).

Chu, S.: "Laser Manipulation of Atoms and Particles," *Science*, 861 (August 23, 1991).

Chu, S.: "Laser Trapping of Neutral Particles," *Sci. Amer.*, 71 (February 1992).

Corcoran, E.: "Diminishing Dimensions," *Sci. Amer.*, 122 (November 1990).

Corcoran, E.: "True Blue (Laser)," *Sci. Amer.*, 171 (September 1991).

Corcoran, E.: "Tacky Lasers Are the Tiniest Yet," *Sci. Amer.*, 28 (January 1992).

Duley, W. and K. Shibata: *1996 International Congress on Applied Lasers and Electro-Optics Proceedings*, Laser Institute of America, Orlando, FL, 1997.

Duarte, F.J.: *Laser Optics*, Academic Press, Inc., San Diego, CA, 2003.

Fasol, G.: "Room-Temperature Blue Gallium Nitride Laser Diode," *Science*, 1751 (June 21, 1996).

Fasol, G., S. Nakamura, and I. Davies: *The Blue Laser Diode: GaN Based Light Emitters and Lasers*, Springer-Verlag New York, Inc., New York, NY, 1997.

Feld, M.S. and K. An: "The Single Atom Laser," *Scientific American*, 56–63 (July 1998).

Feng, S. and P.A. Lees: "Mesoscopic Conductors and Correlations in Laser Spackle Patterns," *Science*, 633 (February 8, 1991).

Freund, H.P. and R.K. Parker: "Free-Electron Lasers," *Sci. Amer.*, 84 (April 1989).

Grant, E.R. and R.G. Cooks: "Mass Spectrometry and Its Use in Tandem with Laser Spectroscopy," *Science*, 61 (October 5, 1990).

Hemley, R.J., P.M. Bell, and H.K. Mao: "Laser techniques in High-Pressure Geophysics," *Science*, **237**, 605–612 (1987).

Hannaford, P., H. Bachor, and A. Sidorov: *Laser Spectroscopy*, World Scientific Publishing Company, Inc., Riveredge, NJ, 2004.

Hirschfelder, J.O., R.E. Wyatt, and R.D. Coalson: *Lasers, Molecules, and Methods*, John Wiley & Sons, Inc., New York, NY, 1989.

Jewell, J.L., J.P. Harbison, and A. Scherer: "Microlasers," *Sci. Amer.*, 86 (November 1991).

Killinger, D.K. and N. Menyuk: "Laser Remote Sensing of the Atmosphere," *Science*, **235**, 37–45 (1987).

Kinoshita, J.: "Atomic Fountain: Laser Light Slows Atom Beam to a Trickle," *Sci. Amer.*, 26 (June 1990).

Kim, K. Kwang-Je, and A. Sessler: "Free-Electron Lasers: Present Status and Future Prospects," *Science*, 88 (October 5, 1990).

Lamb, W.E., Jr. and R.C. Retherford: *Phys. Rev.*, **79**, 549 (1950).

Langreth, R.N.: "Laser Cooling Made Simpler, Cheaper," *Science News*, 216 (October 6, 1990).

Maddox, John: "The Wonders Of The Microlaser," *Nature*, 101 (January 12, 1995).

Matthews, D.L. and M.D. Rosen: "Soft X-Ray Lasers," *Sci. Amer.*, 86 (December 1988).

Maiman, T.H.: *Br. Commun. Electron.*, **7**, 674 (1960).

Meyers, R. (Editor): *Encyclopedia of Lasers and Optical Technology*, Academic Press, Inc., San Diego, CA, 1990.

Misaelides, P.: *Application of Particle and Laser Beams in Materials Technology*, Kluwer Academic Publishers, New York, NY, 1995.

Morrison, D.C.: "An Unsung Legacy of the First Lunar Landing (Laser)," *Science*, 447 (October 27, 1989).

Murname, M.M. et al.: "Ultrafast X-ray Pulses from Laser-Produced Plasmas," *Science*, 531 (February 1, 1991).

Narayan, J.: "Surfaces, Interfaces, and Films: New Tools (Lasers) Aid Engineering," *Adv. Materials & Processes*, 51 (January 1988).

Numai, T.: *Fundamentals of Semiconductor Lasers*, Springer-Verlag New York, LLC., New York, NY, 2004.

Pepper, D.M., J. Feinberg, and N.V. Kukhtarev: "The Photorefractive Effect," *Sci. Amer.*, 62 (October 1990).

Phillips, W.D., P.L. Gould, and P.D. Lett: "Cooling, Stopping, and Trapping Atoms," *Science*, 877 (February 19, 1988).

Pool, R.: "Making Atoms Jump Through Hoops," *Science*, 1076 (June 1, 1990).

Pool, R.: "Laser Cooling Hits New Low," *Science*, 1077 (June 1, 1990).

Purcell, E.M. and V. Pound: *Phys. Rev.*, **81**, 279 (1951).

Ruthen, R.: "Surfing Photons," *Sci. Amer.*, 12D (August 1989).

Silfvast, W.T.: *Laser Fundamentals*, Cambridge University Press, New York, NY, 2004.

Svelto, O. and D.C. Hanna: *Principles of Lasers*, 4th Edition, Perseus Publishing, Boulder, CO, 1998.

Staff: "Miniature Lasers Reach Mass Production," *Chem. Eng. Progress*, 15 (August 1991).

Taylor, N.: *Laser: The Inventor, the Nobel Laureate, and the Thirty-Year Patent War*, Simon & Schuster Trade, New York, NY, 2000.

Suckewer, S. and C.H. Skinner: "Soft X-Ray Lasers and Their Applications," *Science*, 1553 (March 30, 1990).

Townes, C.H.: "Harnessing Light," *Science*, **84**, 153–155 (November 1984).

Vander Been, M.R.: "Gallium Arsenide Sandwich Lasers," *Adv. Materials & Processes*, 39 (May 1988).

Waterbury, R.C.: "Catalysts Enable Sealed Carbon Dioxide Laser," *Instrumentation Technology*, 80 (April 1990).

Yamamoto, Y., M. Susumu, and W.H. Richardson: "Photon Number Squeezed States in Semiconductor Lasers," *Science*, 1219 (March 6, 1992).

Zewail, A.H.: "The Birth of Molecules," *Sci. Amer.*, 76 (December 1990).

LASERS (Eye Surgery). The word "laser" is an acronym for light amplification by stimulated emission of radiation. In most lasers used in ophthalmology, an electric current is passed through a tube that contains an amplifying medium, usually a gas or solid material, which serves to intensify the energy. This energy is emitted as a narrow light beam, which when focused through a microscope, will either cut, burn, or dissolve various tissues.

Different types of lasers emit specific colors of light and are used to treat various eye problems. The lasers are usually named for the amplification materials used. For instance, the carbon dioxide laser is called a CO_2 laser, while the YAG laser contains a solid material made up of yttrium, aluminum, and garnet.

Ophthalmic lasers allow precise treatment of a variety of eye problems without risk of infection. Most laser procedures are also relatively painless and can be done on an outpatient basis. This combination of safety, precision, convenience, and reduced cost make lasers one of the most successful medical tools available to physicians.

Types of Lasers and Their Uses

Excimer laser. The Excimer laser is perhaps the best known of all lasers because of its use in laser vision correction surgery such as laser *in-situ* keratomileusis (LASIK) and photorefractive keratectomy (PRK). The Excimer or pulsed gas laser, emits an ultraviolet light beam, vaporizing tissue by breaking down molecular tissue bonds in a minute targeted area. It is called a *cold laser* because it does not produce heat that could have harmful effects to the surrounding tissue.

The most important feature of the Excimer laser for surgical applications is its ability to focus powerful energy on a microscopic target without affecting the surrounding area. Each pulse of the laser removes about 1/500 of the thickness of a human hair, which is about 125 microns in diameter. These two factors of precise depth and area control are of particular significance in surgical applications such as refractive vision correction.

YAG laser. An acronym for yttrium–aluminum–garnet, the YAG laser produces short-pulsed, high-energy light beams to cut, perforate, or fragment tissue. This laser may also be called a neodymium–YAG or ND–YAG laser.

Cataract patients often have the misconception that a YAG laser is used to remove their cataracts, but no lasers are used in cataract surgery. This misconception occurs because up to 75% of cataract patients develop a condition known as posterior capsular opacification, a clouding of the residual lens capsule left in place after cataract surgery. This gradual loss of vision resembles the symptoms of cataract development, making some people believe that their cataracts have grown back.

The YAG laser is commonly used to vaporize a portion of the capsule, allowing light to pass through to the retina. The procedure is completely painless, takes only a few minutes in the office, and is effective in eliminating the cloudy condition.

Holmium laser. Also known as the infrared holmium YAG laser, this laser is used in a refractive surgery procedure called laser thermal keratoplasty (LTK) to correct mild to moderate cases of farsightedness and some cases of astigmatism. Unlike the Excimer laser, which reshapes the cornea by removing or ablating tissue, the Holmium laser produces infrared light that reshapes the cornea by causing tissue to constrict. The pulsations from the Holmium laser are computer-controlled to produce a pattern of 8 to 16 tiny beams in concentric rings around the periphery of the cornea. The heated fluid in the spots where these beams hit the cornea creates a series of tiny craters. The subsequent shrinkage pulls in the periphery of the cornea, causing the center to bulge, much like tightening a belt, and thus correcting farsightedness.

CO_2 laser. The CO_2 laser is a specialized laser that is filled with carbon dioxide gas and uses an infrared emission for cutting tissue through heat absorption. It is one of the most common lasers used in surgery and is good for precise cutting and vaporization of tissue, such as that needed in the treatment of superficial lesions or removing small volumes of tissue.

The CO_2 laser is used by ophthalmic plastic surgeons to remove fine wrinkles from around the eyes. This laser precisely removes the outermost layer of skin and the underlying dermis, allowing the regrowth of wrinkle-free new skin.

Erbium laser. The Erbium laser, or erbium–YAG laser, is also used in skin resurfacing and is considered to be more precise and accurate than the CO_2 laser. It is able to remove finer wrinkles with less damage to the skin. The depth of penetration is about 5 microns compared with the 20 microns typical of the CO_2 laser. The Erbium laser also causes less irregular skin pigmentation in darker skinned individuals, because it produces a thinner laser area and less heat. Because the Erbium laser produces minimal thermal scatter, the healing time is less than the healing time with the CO_2 laser.

The Erbium laser is also being used in a promising new clinical procedure to emulsify the eye's natural lens during cataract surgery. Most cataract surgeons currently use a piece of equipment called a phacoemulsifier to break up and remove the cloudy lens. The Erbium laser was chosen for the new technique because of its high absorption rate in water, a primary component of the eye's natural crystalline lens.

Argon laser. The argon laser is filled with argon gas that produces blue/green wavelengths. These particular wavelengths are absorbed by the cells that lie under the retina and by the red hemoglobin in blood, but the blue-green wavelengths can pass through the fluid inside the eye without damage. For this reason, the argon laser is used extensively in the treatment of diabetic retinopathy, a severe disorder of the retina that causes blood vessels to leak. The argon laser can burn and seal these blood vessels.

Retinal detachment is another serious eye problem that can be treated by the argon laser. The laser is used to weld the detached retina to the underlying choroid layer of the eye.

Several forms of glaucoma, which is a leading cause of blindness, are also treated with argon lasers. The very serious angle closure glaucoma, for instance, is sometimes treated by using the laser to create a tiny opening in the iris, allowing excess fluid inside the eye to drain to reduce pressure.

Macular degeneration, a severe condition that affects central vision in older adults, is sometimes treated with an argon or krypton laser. In this treatment, the laser is used to destroy abnormal blood vessels so that hemorrhage or scarring will not damage central vision. See also **Laser In-Situ Keratomileusis (LASIK); Laser Thermal Keratoplasty (LTK); Photorefractive Keratectomy (PRK);** and **Refractive Eye Surgery**.

Vision Rx, Inc., Elmsford, NY

LASER (Telephony). See **Telephony (Telecommunications)**.

LASER THERMAL KERATOPLASTY (LTK). LTK is a refractive surgery procedure that uses a Holmium laser to reshape the cornea for correction of low ranges of hyperopia (farsightedness). The Holmium laser is an infrared (thermal) laser that uses heat to shrink corneal tissue.

On the other hand, the Excimer laser uses a cool beam to vaporize corneal tissue. The Excimer laser is used in the laser *in-situ* keratomileusis (LASIK) and photorefractive keratectomy (PRK) procedures.

In LTK, the Holmium laser is used to gently heat stromal collagen in a ring around the outside of the pupil. The heat causes the tissue to shrink, thereby creating an effect like tightening a belt. The periphery of the cornea is pulled, causing the center to bulge. Because the cornea of a farsighted eye is too flat, this bulging effect, when carefully controlled, corrects the problem.

People with mild hyperopia, +0.75 to +2.75, are prime candidates for LTK. It is also being used to treat presbyopia (age-related loss of focus) and overcorrection from radial keratotomy (RK), PRK, and LASIK procedures. LTK is currently under investigation by the FDA for approval in the United States, and patients can only be treated as part of the investigational protocol.

To understand how LTK works, it is first necessary to understand the visual function of the eye. See also **Visual Function (Eye)**.

If the cornea is too flat or the eye is too short from front to back, light rays are theoretically focused behind the retina, resulting in hyperopia (farsightedness). The objective of LTK is to make the cornea steeper in order to correct for farsightedness.

The LTK procedure is performed on an outpatient basis with topical anesthetic eye drops to numb the eye. Based on the patient's prescription, the laser's computer is calculated to deliver the number of pulses and the diameters of the circles needed to provide the proper amount of correction. After aligning the pupil with the use of a slit-lamp microscope, the surgeon activates the laser, and it transmits tiny beams of infrared light in two concentric rings around the periphery of the cornea. Because moisture in the cornea absorbs the energy in the laser pulses, the tissue shrinks slightly creating tiny craters, which tighten the cornea and result in a steeper surface. The laser never touches the eye, and the entire process takes just a few seconds per eye.

The LTK procedure is painless, although the patient may have blurry vision and a mild scratchy sensation for a couple days. Antibiotic eye drops are normally used for about a week, and, if needed, Tylenol and ice packs can be used to relieve discomfort. Most eyes are fully healed in three days, and, although many patients report almost instant vision improvement, vision stabilization usually occurs within two weeks. See also **Hyperopia (Farsightedness); Laser In-Situ Keratomileusis (LASIK); Photorefractive Keratectomy (PRK); Presbyopia;** and **Refractive Eye Surgery**.

Vision Rx, Inc., Elmsford, NY

LASSA FEVER (African Hemorrhagic Fever). Lassa fever is an acute viral illness that occurs in West Africa. The illness was discovered in 1969 when two missionary nurses died in Nigeria, West Africa. The cause of the illness was found to be Lassa virus, named after the town in Nigeria where the first cases originated. The virus is a member of the virus family Arenaviridae. It is a single-stranded RNA virus and is zoonotic or animal-borne.

In areas of Africa where the disease is endemic, Lassa fever is a significant cause of morbidity and mortality. Although the disease is mild or has no observable symptoms in about 80% of people infected with the virus, the remaining 20% have a severe multisystem disease. Lassa fever is also associated with occasional epidemics, during which the case-fatality rate can reach 50%.

Lassa fever is an endemic disease in portions of West Africa. It is recognized in Guinea, Liberia, Sierra Leone, as well as Nigeria. However, because the rodent species that carry the virus are found in other regions outside of West Africa, the actual geographic range of the disease may extend to other portions of Africa. The number of Lassa virus infections per year in West Africa is estimated at 1,00,000 to 300,000, with approximately 5,000 deaths. Unfortunately, such estimates are crude because surveillance for cases of the disease are not uniformly performed. In some areas of Sierra Leone and Liberia, it is known that 10–16% of people admitted to hospitals have Lassa fever, which indicates the serious impact of the disease on the population of this region.

The reservoir, or host, of Lassa virus is a rodent known as the "multimammate rat" of the genus *Mastomys natalensis*. It is not certain which species of Mastomys are associated with Lassa. At least two species carry the virus in Sierra Leone: *M. huberti* and *M. erythroleucus*. Mastomys rodents breed very frequently, produce large numbers of offspring, and are numerous in the savannas and forests of West, Central, and East Africa. In addition, some species, like *M. huberti*, prefer to live in human homes. All these factors together contribute to the relatively efficient spread of Lassa virus from infected rodents to humans.

There are a number of ways in which the virus may be transmitted or spread to humans. The Mastomys rodents shed the virus in urine and droppings. Therefore, the virus can be transmitted through direct contact with these materials, through touching objects or eating food contaminated with these materials, or through cuts or sores. Because Mastomys rodents often live in and around homes and scavenge on human food remains or poorly stored food, transmission of this sort is common. Contact with the virus also occurs when a person inhales tiny particles in the air contaminated with rodent excretions. This is called aerosol or airborne transmission. Finally, because Mastomys rodents are sometimes used as a food source, infection may occur via direct contact when they are caught and prepared for food.

Lassa fever may also spread through person-to-person contact. This type of transmission occurs when a person comes into contact with virus in the blood, tissue, secretions, or excretions of an individual infected with the Lassa virus. A person may also become infected by breathing in small airborne particles which an already infected person may produce by actions like coughing. The virus cannot be spread through casual contact (including skin-to-skin contact without exchange of body fluids). Person-to-person transmission is common in both village settings and in health care settings, where, along with the above-mentioned modes of transmission, the virus also may be spread in contaminated medical equipment, such as reused needles (this is called nosocomial transmission).

Symptoms of Lassa fever typically occur 1–3 weeks after the patient comes into contact with the virus. These include fever, retrosternal pain (pain behind the chest wall), sore throat, back pain, cough, abdominal pain, vomiting, diarrhea, conjunctivitis, facial swelling, proteinuria (protein in the urine), and mucosal bleeding. Neurological symptoms have also been described, including hearing loss, tremors, and encephalitis. Because the symptoms of Lassa fever are so varied and nonspecific, clinical diagnosis is often difficult.

Lassa fever is most often diagnosed by using enzyme-linked immunosorbent serologic assays (ELISA), which detect IgM and IgG antibodies as well as Lassa antigen. The virus itself may be cultured in 7 to 10 days. Immunohistochemistry performed on tissue specimens can be used to make a postmortem diagnosis. The virus can also be detected by reverse transcription-polymerase chain reaction (RT-PCR); however, this method is primarily a research tool.

The most common complication of Lassa fever is deafness. Various degrees of deafness occur in approximately one-third of cases, and in many cases hearing loss is permanent. As far as it is known, severity of the disease does not affect this complication: deafness may develop in mild as well as in severe cases. Spontaneous abortion is another serious complication.

Approximately 15–20% of patients hospitalized for Lassa fever die from the illness. However, overall only about 1% of infection with the Lassa virus result in death. The death rates are particularly high for women in the third trimester of pregnancy, and for fetuses, about 95% of which die in the uterus of infected pregnant mothers.

Ribavirin, an antiviral drug, has been used with success in Lassa fever patients. It has been shown to be most effective when given early in the course of the illness. Patients should also receive supportive care consisting of maintenance of appropriate fluid and electrolyte balance, oxygenation and blood pressure, as well as treatment of any other complicating infections.

Individuals at risk are those who live or visit areas with a high population of Mastomys rodents infected with Lassa virus or are exposed to infected humans. Hospital staff are not at great risk for infection as long as protective measures are taken.

Primary transmission of the Lassa virus from its host to humans can be prevented by avoiding contact with Mastomys rodents, especially in the geographic regions where outbreaks occur. Putting food away in rodent-proof containers and keeping the home clean help to discourage rodents from entering homes. Using these rodents as a food source is not recommended. Trapping in and around homes can help reduce rodent populations. However, the wide distribution of Mastomys in Africa makes complete control of this rodent reservoir impractical.

When caring for patients with Lassa fever, further transmission of the disease through person-to-person contact or nosocomial routes can be avoided by taking preventive precautions against contact with patient secretions (together called VHF isolation precautions or barrier nursing methods). Such precautions include wearing protective clothing, such as masks, gloves, gowns, and goggles; using infection control measures, such as complete equipment sterilization; and isolating infected patients from contact with unprotected persons until the disease has run its course.

Further educating people in high-risk areas about ways to decrease rodent populations in their homes will aid in the control and prevention of Lassa fever. Other challenges include developing more rapid diagnostic tests and increasing the availability of the only known drug treatment, ribavirin. Research is presently under way to develop a vaccine for Lassa fever.

See also **Viral Hemorrhagic Fevers**.

Additional Reading

Buckley, S.M. and J. Casals: "Pathobiology of Lassa Fever," *Int. Rev. Exp. Path.*, **18**, 97 (1978).
Holmes, G.P. et al.: "Lassa Fever in the United States," *N. Eng. J. Med.*, 1120 (October 18, 1990).
Jarling, P.B. et al.: "Lassa Virus Infection," *J. Inf. Dis.*, **141**, 580 (1980).
Johnson, K.M. and T.P. Monath: "Imported Lassa Fever — Reexamining the Algorithms," *N. Eng. J. Med.*, 1139 (October 18, 1990).
Walker, D.H.: "Lassa Fever in Man," *Am. J. Path.*, **107**, 349 (1982).

Centers for Disease Control and Prevention, (CDC), Atlanta, GA

LAST INTERGLACIAL. The most recent time (115,000 to 125,000 years ago) during which global temperatures were as high as or higher than in the postglacial, when continental glaciers were limited to the Arctic and Antarctic, and sea levels were near current positions.

LASTUS RECTUM. See **Parabola**.

LATENT HEAT. Heat gained by a substance or system without an accompanying rise in temperature during a change of state. As examples, the latent heat of fusion is the amount of heat necessary to convert a unit mass of a substance from the solid state to the liquid state at the same temperature, the pressure being that to allow coexistence of the two phases. A considerable part of the latent heat arises from the entropy increase consequent on the greater disorder of the liquid state. The latent heat of sublimation is the amount of heat necessary to convert a unit mass of a substance from the solid state to the gaseous state at the same temperature, the pressure being that to allow coexistence of the two phases.

LATERAL. A force that acts on a structure or a structural member in a transverse direction is sometimes called a lateral load. The wind blowing upon the exposed surface of a bridge or building at right angles to its length or upon the stationary or moving traffic using the bridge constitutes one type of lateral load. The sway of a moving train on a bridge or the centrifugal force transmitted if the bridge is on a curve is a type of lateral loading. A moving crane supported on girders exerts a side thrust on the girders that may also be included in this classification.

Trusses and girders, which constitute the main load-carrying members of bridges, are not ordinarily designed to carry side loads of this nature, and consequently have very little strength in that direction. For this reason, the trusses of girders of a bridge are joined together in a horizontal plane by a system of lateral bracing composed of struts and diagonals. These members are often referred to as the laterals or the lateral system. This lateral bracing stiffens the whole bridge and opposes any sidewise deflection or vibration.

The term lateral is also used in connection with sewerage systems. Any sewer which serves the abutting property owners and in which each owner has an equal right is a common sewer. A lateral sewer is one that has no other common sewer flowing into it.

LATERAL LINE. See **Fishes**.

LATERAL MIRAGE. A mirage in which the image (or images) is displaced laterally from the position of the object. This is not a difficult mirage to find, especially along the sun-warmed walls of buildings. In many cases, it appears as nothing but an *inferior mirage* turned on its side. However, there are often interesting subtleties. Easiest to find, perhaps, are the high-order multiple images that result from inhomogeneities along the wall. These can arise both from the wall having a slightly wavy surface and from the periodic variations in the internal structure of the wall that alter the thermal conductivity and so produce periodic temperature variations. Curiously, unlike the inferior mirage, the lateral mirage seems to be capable of producing three images even in the absence of inhomogeneities. The temperature profiles normal to horizontal and vertical surfaces are slightly different. In the case of the inferior mirage, gravity acts normal to the surface, while in the case of the lateral mirage, gravity is parallel to the surface. This produces a flow up the wall that results in a temperature profile capable of giving the three-image mirage. Lateral temperature gradients in the free atmosphere, away from vertical surfaces, are not sufficient to produce lateral mirages; the rare reports of such sightings undoubtedly arose from misinterpretations of observations. See also **Inferior Mirage**; and **Mirage**.

LATERITE. The sub-aerial decay of rocks in tropical regions, having a distinctly moist or rainy climate, results in the development of a residual, reddish, and usually sticky soil frequently containing concretions. The principal products of laterization are the hydrated oxides of aluminum and iron either in the crystalline or amorphous form. If the concentration of iron oxide is sufficiently high the laterite may be valuable as an iron ore. If, on the other hand, the concentration of alumina is high the laterite may be valuable as an ore of that metal.

LATEX. Latex is a milky liquid drawn from any of 200 plants. It is a complex emulsion in which such substances as proteins, alkaloids, starches, sugars, oils, tannins, resins, and gums are found. In most plants the latex is white; but in some it is yellow; in others, orange or scarlet.

The cells or vessels in which latex is found make up the laticiferous system. There are two very different ways in which this system may be formed. In many plants the laticiferous system is formed from cells laid down in the meristematic region of the stem or root. Rows of these cells are formed. The cell walls separating them are dissolved, so that continuous tubes, called latex vessels, are formed. This method of formation is found in the poppy family; in the rubber plant, *Hevea brasiliensis*; and in the *Cichorieae*, a section of the composite family distinguished by the presence of latex in its members. Dandelion, lettuce, hawkweed, and salsify are members of the *Cichorieae*.

Latex Technology

This methodology encompasses colloidal and polymer chemistry in the preparation, processing, and conversion of natural and synthetic latices into useful products.

Latex technology vernacular is not always consistent. By definition a *latex* is a colloidal suspension of polymer particles stabilized by dispersing agents in an aqueous medium. The dispersing agents are conventional ionic or nonionic surfactants or polymeric surfactants made from block or graft copolymers derived from monomers with different hydrophobicities. An *emulsion* is a dispersion of two or more immiscible liquid phases (one being water) stabilized by amphiphilic materials. A latex is a specific type of emulsion; one where the organic phase is a polymer particle. The terms latex and emulsion are often used interchangably partly because emulsion polymerization is the principal synthetic route to latices. Emulsion polymerization is one type of heterophase polymerization involving organic and aqueous phases. Others include suspension, dispersion, and precipitation polymerization which generate water-borne particles different from latices. Heterophase polymerization can also involve two immiscible organic phases as in nonaqueous polymer dispersions and polymer microgels. Finally, polymers prepared via homogeneous polymerization, whether in solution or neat, can be inverted into a polymer dispersion in water using surfactants. Similar to latices, these materials can have very different molecular weight and functional group distributions and thus form a separate class of materials. See also **Emulsions**.

Latices have been in use for a very long time and the history of latices and polymer development are closely linked. The Mayas, around 1600 BC, used the sap of trees like the sparse *Hevea brasiliensis* of South America to make rubber products and waterproof clothing. The Mayas called the sap "caa o-chu," literally translated as "weeping tree." *Caoutchouc* is now the French word for rubber. The natural rubber derived from this sap was shown to be 93–95% *cis*-1-4-isoprene by Faraday in the early 19th century. Goodyear's invention of vulcanization, and later the automobile, increased natural rubber demand through the beginning of the 20th century. Large rubber plantations in Malaya, Ceylon, Indonesia, and Indochina increased the world's natural rubber production to 200,000 t by 1920. Enhanced supply led to rapid growth of natural rubber products and improvements in latex processing. The Allied blockade of Germany during World War I led to the first process for making synthetic latex. Gottlob and others were early developers of emulsion polymerization (at first using methods to duplicate how natural rubber is produced in nature). Soon thereafter, U.S. companies began producing commercial synthetic latices: Buna S (butadiene–styrene copolymer), also known as Government Rubber–Styrene or GR–S rubber; Neoprene (polychloroprene); and Thiokol (polysulfides). See also **Rubber (Natural)** and **Styrene-Butadiene Rubber**. Japan's seizure of the Southeast Asia rubber plantations during World War II led to intensive research in synthetic rubber production. Today synthetic latex production accounts for 60% of the 18×10^6-t total rubber market which has been growing at 2–5% over the last decade. Synthetic latices account for ca 2.5% of the world polymer market. Allergic reactions to proteins in natural latex and other natural latex market drivers have recently created more opportunities for synthetic latices. Over 70% of the natural latex market is converted to solid polymer for use in tires.

Many synthetic latices exist. They contain butadiene and styrene copolymers (elastomeric), styrene–butadiene copolymers (resinous), butadiene and acrylonitrile copolymers, butadiene with styrene and acrylonitrile, chloroprene-copolymers, methacrylate and acrylate ester copolymers, vinyl acetate copolymers, vinyl and vinylidene chloride copolymers, ethylene copolymers, fluorinated copolymers, acrylamide copolymers, styrene–acrolein copolymers, and pyrrole and pyrrole copolymers. Many of these latices also have carboxylated versions.

Traditional applications for latices are adhesives, binders for fibers and particulate matter, protective and decorative coatings, dipped goods (especially without allergens), foam, paper coatings including waterproofing paper, backings for carpet and upholstery, modifiers for bitumens and concrete, and thread and textile modifiers to improve feel or properties such as flame retardence. More recently latices have found use in biomedical applications as protein immobilizers; as visual detectors in immunoassays; as release agents in drug delivery, wound treatment, and synthetic blood, in electronic applications as photoresists for circuit boards; in batteries, conductive paint, copy machines; as key components in molecular electronic devices; in specialty coatings for seeds, in artificial turf plastics; and as an important component of oil recovery techniques.

Latex Properties

The observable properties of a latex, ie, stability, rheology, film properties, interfacial reactivity, and substrate adhesion, are determined by the colloidal and polymeric properties of the latex particles. Important colloidal properties include ionic charge, stability, particle size and morphology distribution, viscosity, solids, and pH. Important polymer properties include molecular weight distribution, monomer sequence distribution, glass-transition temperature, crystallinity, degree of cross-linking, and free monomer. Methods for analyzing each of these properties exist, depending on the end use of the product. See also **Colloid System**.

Stability. For a latex to be a useful product, control of polymer isolation is crucial. The individual polymer particles must be stable enough to avoid coagulation resulting from perturbances like high temperature, freeze–thaw cycles, high shear in handling, electrolyte addition, and organic solvent addition during processing, but not so stabilized that polymer isolation is impossible. Stability is related to the surface properties of the latex particles, and these are usually determined during latex manufacture. Visual detection of coagulation is easy; more sophisticated optical techniques are possible. The types of initiator, emulsifier, and monomers used are the key determinants.

Rheology. Flow properties of latices are important during processing and in many latex applications such as dipped goods, paint, inks, and fabric coatings. For dilute, nonionic latices, the relative latex viscosity is a power—law expansion of the particle volume fraction. The terms in the expansion account for flow around the particles and particle–particle interactions. For ionic latices, electrostatic contributions to the flow around the diffuse double layer and enhanced particle–particle interactions must be considered. A relative viscosity relationship for concentrated latices was first presented in 1972. In practice, latex viscosity measurements are carried out with rotational viscometers.

It is possible to increase the viscosity of a latex after manufacture using thickeners. Thickening occurs through increases in medium viscosity or polymer particle aggregation. If considerable aggregation occurs without a corresponding increase in medium viscosity, undesirable separation or creaming occurs. Methylcellulose, caseinates, and polyacrylate salts are typical thickeners. Ease of adding the thickener, ability to maintain viscosity, and undesirable side effects must be considered when selecting a thickener. Some thickeners slowly hydrolyze in the latex and lose their effectiveness over time. The full range of the effects of adding thickener develops over time, some of them much faster than others. To avoid exceeding the desired viscosity, it is advisable to add thickener in small increments, waiting after each for the viscosity to reach equilibrium before adding the next one.

The viscosity of the latex can also be dependent on pH. In the case of some latices, lowering the pH with a weak acid such as glycine is an effective method for raising the viscosity without destabilizing the system.

Latices made with poly(vinyl alcohol) as the primary emulsifier can be thickened by increasing the pH with a strong alkali.

Improving Properties Through Compounding. The potential value of most polymers can be realized only after proper compounding. Materials used to enhance polymer properties or reduce polymer cost include antioxidants, cross-linking agents, accelerators, fillers, plasticizers, adhesion promoters, pigments, etc.

Latex Applications

Adhesives. Latices are used as additives in the construction market, in tires and belt fabrication, in furniture manufacture, in packaging, and in tapes, labels, envelopes, and bookbinding. The adhesives are used in wet or dry laminations.

Binders. Latices are used as fiber binders in the paper and textile industries. The two principal methods of application are (1) wet-end addition, wherein the ionic latex is added to a fiber slurry and then coagulated in the slurry prior to sheet formation; and (2) saturation of the latex into a formed fiber web, wherein the latex is coagulated by dehydration. Latices are also used as binders for particulate matter such as rubber scrap.

Coatings. Latices are used in residential and industrial paints, coated paper and paperboard, seeds, fabric coatings, backing for carpet, upholstery, and drapery; as basecoats for wallpaper and flooring; and in insulation coatings. See also **Conversion Coatings.** Application methods include brushing, squeegee, spraying, dipping, and frothing.

Dipped Goods. Latices are used in various dipping processes to produce balloons, bladders, gloves, extruded thread, and tubing. Manufacturing techniques include multiple dip and dry, and coagulant dipping employing a colloidal destabilizer.

Foam Products. Latices are made into foams for use in cushioning applications. The latices are frothed with air and then chemically coagulated for thick applications, or heated to induce coagulation for thinner applications. The latter method allows for infinite pot life during production.

Modifiers. Latices are added to bitumens, mortars, and concrete to improve impact resistance and reduce stress cracking. Key to the use of latices in these technologies is compatibility between the latex and the construction materials.

Synthetic Latex Manufacture

The history of emulsion polymerization has been well documented. Early efforts to produce synthetic rubber coupled bulk polymerization with subsequent emulsification. The first attempts at emulsion polymerization arose from problems controlling the heat generated during bulk polymerization. In emulsion polymerization, hydrophobic monomers are added to water, emulsified by a surfactant into small particles, and polymerized using a water-soluble initiator. The result is a colloidal suspension of fine particles, 50–1000 nm in diameter, usually comprising 30–50 wt% of the latex product. By 1935 emulsion polymerization became the method of choice in making synthetic rubber because of its many advantages: (1) the reaction mass viscosity remains low throughout the polymerization, providing for improved heat transfer, agitation, and product handling; (2) the sensible heat of the water in the emulsion balances the heat of reaction generated by free-radical polymerization; and (3) the rate of reaction is rapid, while producing very high molecular weight.

Kinetics and Mechanisms. Early researchers misunderstood the fast reaction rates and high molecular weights of emulsion polymerization. In 1945 the first recognized qualitative theory of emulsion polymerization was presented. This mechanism for classic emulsion preparation was quantified and the polymerization separated into three stages—stage I: particle nucleation; stage II: growth in polymer particles saturated with monomer; and stage III: growth in polymer particles with a decreasing monomer concentration.

Basic Components. The principal components in emulsion polymerization are deionized water, monomer, initiator, emulsifier, buffer, and chain-transfer agent. A typical formula consists of 20–60% monomer, 2–10 wt % emulsifier on monomer, 0.1–1.0 wt % initiator on monomer, 0.1–1.0 wt % chain-transfer agent on monomer, small amounts of various buffers and bacteria control agents, and the balance deionized water.

Process. Commercial processes manufacturing latex can be divided into batch, semibatch, and continuous methods.

Additional Reading

Beswick, R.H.D.: Latex 2002 International Conference on Latex and Latex Based Products, Rapra Technology Ltd., Shrewsbury, U.K., Dec. 2002, pp. 45–59.

Morton, M.: *Introduction to Rubber Technology,* Reinhold Publishing Corp., New York, NY, 1959.

Seidel, A.: *Kirk-Othmer Encyclopedia of Chemical Technology,* John Wiley & Sons, Inc., Hoboken, NJ, 2004.

Warson, H.: *Applications of Synthetic Resin Latices, Latices in Diverse Applications,* Vol. 3, John Wiley and Sons, Inc., New York, NY, 2001.

Warson, H., and C.A. Finch: *Applications of Synthetic Resin Latices, Fundamental Chemistry of Latices & Applications in Adhesives,* Vol. 1, John Wiley & Sons, Inc., New York, NY, 2001.

Winspear, G.G.: *The Vanderbilt Latex Handbook,* R. T. Vanderbilt Co., New York, NY, 1954.

LATIN SQUARE. An experimental design based on a $p \times p$ array of p letters such that each letter occurs once and only once in each row and column; e.g., for $p = 4$ and the letters, A, B, C, D:

$$
\begin{array}{cccc}
A & B & C & D \\
B & D & A & C \\
C & A & D & B \\
D & C & B & A
\end{array}
$$

A layout of this type, for example, may correspond to 16 plots and four treatments represented by the letters.

The design may be generalized to allow for a further treatment represented by Greek letters and it is then known as a Graeco-Latin square, e.g.,

$$
\begin{array}{cccc}
A\alpha & B\beta & C\gamma & D\delta \\
B\gamma & A\delta & D\alpha & C\beta \\
C\delta & D\gamma & A\beta & B\alpha \\
D\beta & C\alpha & B\delta & A\gamma
\end{array}
$$

In this case, no combination of Roman and Greek letters occurs more than once. More general designs are sometimes known as Hyper-Graeco-Latin squares. The purpose of all these designs is to provide independent comparisons of row, column, and treatment effects. See also **Orthogonal Squares.**

LATITUDE. The celestial latitude of a point on the celestial sphere is the spherical coordinate measured from the plane of the ecliptic along a great circle passing through the object and the poles of the ecliptic.

Because of the fact that the earth is not a perfect sphere, there are several different sorts of terrestrial latitude in use. In Fig. 1, we have an ellipse $PEP'E'$ representing a section of the earth in the plane of a meridian. C is the geometric center of the earth, and the line COZ' is the line to the geocentric zenith of the point. O. The angle $ECO(\phi')$ is the geocentric latitude of the point O.

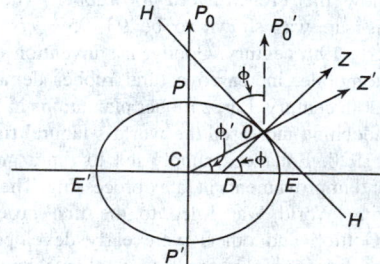

Fig. 1.   Ellipse representing a section of the earth in the plane of a meridian.

The line DOZ represents the direction of gravity at the point O and extends to the astronomic zenith of O. The angle $EDZ(\phi)$ is the astronomic latitude of the point O. The difference between the astronomic and geocentric latitude of a point, the angle $COD = \phi - \phi'$, is defined as the reduction of latitude for the point O.

Because of local influences, such as massive mountains in the vicinity, the direction of the plumb line may not be strictly perpendicular to the surface of the earth. The geographical latitude of a point is the angle, measured in the plane of the local meridian, between the equator and a line drawn perpendicular to the theoretical geoid (surface of the earth) through the point in question. The difference between astronomic and geographic latitude is always relatively small, but by no means an inappreciable angle, and is known as station error. Station error is commonly between 4 and 6 seconds of arc, but occasionally amounts to 30 or 40 seconds.

In Fig. 1, CP represents the axis of rotation of the earth, which, if extended, will pierce the celestial sphere in its pole of rotation. The parallel line OP_0' is the line from the observer at O to the pole of rotation of the celestial sphere, and the line HOH represents the plane of the astronomic horizon at O. HOP_0' is the altitude of the pole of rotation at O. Inspection of the figure will indicate that this is equivalent to the angle EDZ. This gives rise to the common definition of the astronomic latitude of a point as the altitude of the pole of rotation of the celestial sphere at the point.

Astronomic latitude may be determined in a variety of ways by observation of the celestial objects. The most direct method is to observe the altitude of some object on the meridian whose declination is known. In Fig. 2, we have a representation of the celestial sphere drawn in the plane of the local meridian of the point O. In the figure, HOH' represents the plane of the horizon; $HPZQH'$ represents the local meridian; OP the direction of the pole of rotation; OQ the direction of the equator. $HOP = \phi$ (the astronomic latitude of O), and $H'OQ = 90 - \phi$. Since $H'S$ represents the altitude of a celestial object, S, which is on the meridian, and QS represents the declination, d, of the object, we have at once the relation: $\phi = \delta + 90 - \text{altitude}$. This is the method of determination of latitude most commonly used at sea, and it presents two fundamental difficulties to the navigator. The instant that the object is on the meridian must be accurately known and, also, the declination of the object must be observed. If both the Greenwich time and the longitude are known, the instant that the object should reach the meridian may be calculated in advance from the right ascension of the object; and the observation of altitude is taken at the predetermined instant. (Before chronometers came into use, it was necessary to watch the object very carefully and to record the maximum altitude attained by the object. If the object was the sun, the time that the maximum altitude was obtained became the local apparent noon, and was used by the navigating officer for setting the watch time for the ship.) If the observed object is a star, the declination may be immediately obtained from star catalogues; but if the sun, whose declination is changing rapidly, is the observed object, the Greenwich time of observation must be used to obtain the declination from the ephemeris.

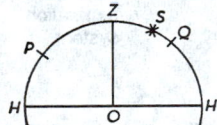

Fig. 2.   Representation of the celestial sphere drawn in the plane of the local meridian of the point O.

Should the meridian observation be missed, because of cloud cover or for any other reason, the astronomical triangle may be solved to obtain the latitude if the local time of observation and the declination of the object are both known. If the object is observed very close to the meridian, and if the approximate latitude as well as the local time is known, the observation may be "reduced to the meridian" by tables. See also **Celestial Sphere and Astronomical Triangle**.

Modern navigational methods for determining latitude are discussed elsewhere. See also **Navigation**.

A meridian altitude of an object is always effected by the correction for astronomical refraction, which is always subject to error unless the object observed is close to the zenith. For accurate determination of latitude for purposes of geodetic surveying, the zenith telescope is used. See also **Earth**.

LATITUDE (Geomagnetic). See **Geomagnetic Latitude**.

LATTICE COMPOUNDS. Chemical compounds formed between definite stoichiometric amounts of two molecular species that owe their stability to packing in the crystal lattice, and not to ordinary valence forces.

LATTICE CONSTANT. A length representing the size of the unit cell in a crystal lattice. In a cubic crystal, this is just the length of the side of the unit cell, but such a simple definition is not in general possible, and the lattice constant must be chosen according to the geometry of the structure in each case.

LATTICE DESIGNS. Lattice designs form a class of experimental designs enabling a large number of unrelated treatments to be compared in randomized blocks of a reasonable size. If there are n treatments where $n = p \times q$, the treatments are thought of as generated by the combinations of two pseudofactors A and B, one at p, the other at q levels (see also **Factorial Experiment**). Two types of replicates are then laid down, confounding the main effect of A in one type and the main effect of B in the other. The most useful case is that in which $p = q$, which gives rise to blocks of equal size.

LATTICE DIMENSIONS. According to the Bragg formula the spacing of the atomic planes can be deduced from the X-ray diffraction pattern and a knowledge of the X-ray wavelength, which can itself be measured by diffraction from a ruled grating.

LATTICE ENERGY OF CRYSTAL. The decrease in energy accompanying the process of bringing the ions, when separated from each other by an infinite distance, to the positions they occupy in the stable lattice. It is made up of contributions from the electrostatic forces between the ions, from the repulsive forces associated with the overlap of electron shells, from the van der Waals forces, and from the zero-point energy.

LATTICE (Mathematics). A set S of elements $a, b, \ldots,$ is *partially ordered* if a binary relation often denoted by the symbol $\leq$, which is reflexive, antisymmetric and transitive, is defined for certain of its elements. For example, let $a, b, \ldots$ denote the subsets of S, and let a stand in the given relation to b if the subset a is included in the subset b. A partially ordered set is a lattice if for any two elements a, b there exists an element c which is a least upper bound for a, b; that is, such that $a \leq c$, $b \leq c$ and if $a \leq e$, $b \leq e$, then $c \leq e$, and also an element d which is a greatest lower bound for a, b; that is, such that $d \leq a$, $d \leq b$ and if $f \leq a$, $f \leq b$ then $f \leq d$. These elements c and d are called the *join* (or *union*) and the meet (or *intersection*), respectively, of a and b, and are denoted by $c = a \cup b$ and $d = a \cap b$. The terms cup and cap are also used, and it is common to write $a + b$ for $a \cup b$ and ab or $a \times b$ for $a \cap b$. See also **Boolean Algebra**.

LATTICE WATER. See **Hydrate**.

LAUAN TREE. See **Mahogany Trees**.

LAUNCH VEHICLES: AIR AND SHIP-BASED. In 1957, the Soviet Union placed the first man-made object in orbit around the earth. Since then, numerous launch vehicles have been developed to improve the performance, reliability, and cost of placing objects in orbit. By one estimate, roughly 75 active space launch vehicles either have established flight records or are planning an inaugural launch within the year. This does not include the numerous launch vehicles from around the world that are no longer operational such as the Jupiter, Redstone, Juno, Saturn, Scout, Thor, Vanguard, and Conestoga family of rockets from the United States or the N-1 from the former Soviet Union, to name just a few. Despite the many differences among all of these launch vehicles from both past and present, one common element can be found in all but four of them: they are ground-launched. Of the four exceptions, two are air-launched (NOTSNIK and Pegasus), one is ship-launched (Sea Launch), and one is submarine-launched (Shtil). It is important to keep in mind that numerous air-launched and ship-launched suborbital launch systems are in use by militaries, commercial entities, and educational institutions. However, the four mentioned are the only mobile launch systems that can place objects into a sustainable Earth orbit.

Mobile Space-Launched Vehicles

Project Pilot (NOTSNIK) NOTSNIK is the oldest and, until recently, the least well known of the four mobile space-launched systems. Following the launch of Sputnik by the Soviet Union, President Eisenhower's administration elicited proposals to launch a satellite into orbit. The Naval Ordinance Test Station (NOTS) located at China Lake in California proposed launching a rocket from a jet fighter. The idea is the same as that of the current Pegasus vehicle: reduce the amount of energy needed to place a payload into orbit by launching it above the denser portion of the atmosphere. In this fashion, the engineers at NOTS designed a vehicle from existing rocket motors that could place a 2-pound satellite in a 1500-mile-high orbit. The engineers recognized the energy savings from such a launch concept and also the utility of such a flexible platform. Launching from a jet fighter could, theoretically, place a satellite into any orbit from anywhere in the world at any time.

The U.S. Navy accepted the proposal from NOTS in 1958, by some accounts as a safety net in the event that the ongoing Vanguard project was unsuccessful. The program was officially called Project Pilot, but the engineers at NOTS preferred the name NOTSNIK in direct reference to the Soviet satellite that was currently orbiting above them and the rest of the world. A Douglas Aircraft F4D-1 Skyray was the carrier aircraft for the rocket and consequently was considered the first stage. The second and third stages were modified antisubmarine missiles. The final stage was taken from a Vanguard rocket. The entire launch vehicle measured a mere 14 feet in length and had four fins at the aft end that provided a span of 5 feet.

The NOTSNIK was launched six times from an altitude of about 41,000 ft. Four of those launches ended in known failures. However, the results of two have never been verified. Some in the program insist that they achieved their goal of placing the small payload of diagnostic instruments in orbit. At least one ground station in New Zealand picked up a signal in the right place at the right time. However, confirmation that the signal was from the NOTSNIK payload was never established. Even the possibility of a success was veiled in secrecy for more than 40 years for, by all accounts, two critical reasons. The first was that in the days following the early embarrassments of Vanguard, the Eisenhower administration did not want to claim success unless it was absolutely certain. The second reason was that a mobile air-launched system that could reach orbit had extremely appealing military applications. However, the tactical advantages of such a system were far outweighed by the strategic consequences, as stated in the Antiballistic Missile (ABM) Treaty between the United States and the former Soviet Union that was concluded in 1972: http://www.state.gov/www/global/arms/treaties/abm/abm2.html.

Further, to decrease the pressures of technological change and its unsettling impact on the strategic balance, both sides agree to prohibit development, testing, or deployment of sea-based, air-based, or space-based ABM systems and their components, along with mobile land-based ABM systems. Should future technology bring forth new ABM systems 'based on other physical principles' than those employed in current systems, it was agreed that limiting such systems would be discussed, in accordance with the Treaty's provisions for consultation and amendment.

Pegasus. Roughly 30 years later, while NOTSNIK remained an official government secret, the idea of launching payloads into space from an airborne platform was revisited in the form of the Pegasus launch vehicle. The driving forces behind NOTSNIK and Pegasus were essentially the same. An air-launched space vehicle provides several advantages compared with ground-based counterparts. As an example, Pegasus is launched at an altitude of 39,000 ft, which is above a significant portion of the atmosphere. As mentioned, with NOTSNIK, this eliminates the need for extra performance that would otherwise be needed to overcome atmospheric forces. This also implies that the structural components of the vehicle can be lighter, which improves the efficiency of the rocket as a whole. The energy required from the launch vehicle is also reduced by the speed already achieved by the carrier aircraft. An air-launched system also allows applying more of the impulse of the first stage along the velocity vector. This is a more efficient use of the vehicle's energy than that of ground-launched vehicles that must first apply the thrust almost perpendicular to the velocity vector already imparted by Earth's rotation. These factors combine to produce a requirement for a velocity increment that is on the order of 10% less than a comparable ground-launched rocket.

The Pegasus vehicle is a winged, three-stage, solid rocket booster (Fig. 1). It is the first space-launched vehicle developed solely with commercial funding. Three versions have been developed and flown over the years: Standard, Hybrid, and XL. The XL is the only vehicle within the Pegasus family currently in production. The XL is roughly 10,000 lbm heavier than the Standard or Hybrid models and is roughly 6 ft longer. Because the XL extends farther aft beneath the L-1011 carrier aircraft, the port and starboard fins become an obstacle to the landing gear doors. To correct this problem, the port and starboard fins were modified to include an anhedral of 23°. To maintain commonality between the various members of the Pegasus family of vehicles, the same anhedral was introduced into the Standard vehicle, which was then given the designation Pegasus Hybrid.

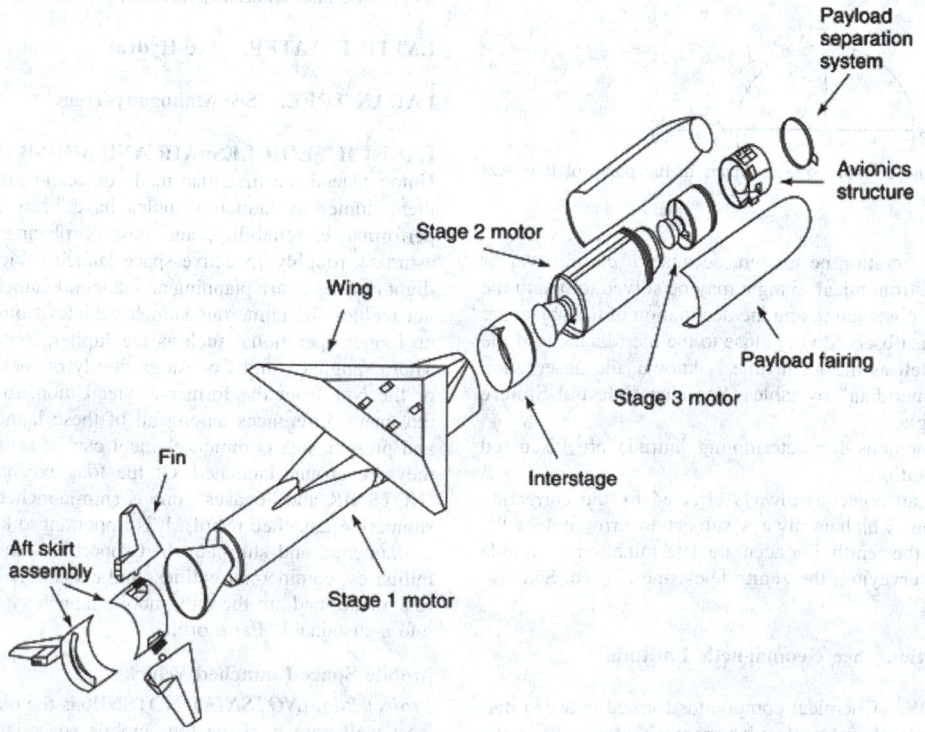

Fig. 1.   Disassembled version of standard Pegasus launch vehicle.

Other than the anhedral of the fins, the Standard and Hybrid vehicles are exactly the same. The Standard, the first Pegasus vehicle built, was flown on six missions. The Hybrid vehicle has flown four times. The XL vehicle has flown 21 times. Of 31 Pegasus launches, only three missions failed to reach orbit.

The Pegasus XL was designed and developed to provide increased performance above and beyond that provided by the Standard and Hybrid vehicles. A typical Pegasus XL vehicle weighs roughly 51,000 lbm at launch, is 55.4 ft long and 50 inches in diameter, and the wingspan is 22 ft. At launch, the Pegasus XL is carried aloft by the company's carrier aircraft, a modified L-1011, which originally saw commercial service with Air Canada. The vehicle is dropped from an altitude of 39,000 ft at Mach 0.8. Five seconds after release from the L-1011, the first stage ignites and the vehicle's on-board flight computer continues the sequence of events that eventually lead to orbital insertion. The brief coast period between drop and stage one ignition is designed to provide a safe distance between the L-1011 and the launch vehicle.

The Pegasus Standard vehicle was originally dropped from a NASA-owned and operated B-52. The Pegasus vehicle was attached to one of the pylons underneath the starboard wing much in the same manner as the early supersonic and hypersonic test vehicles such as the X-15. For a variety of reasons, Orbital purchased and modified the L-1011 to facilitate all future launches.

Unlike the B-52 that supported initial Pegasus launches, the L-1011 carries the Pegasus vehicle underneath the fuselage rather than underneath the wing. Once Pegasus is ready to be mated to the carrier aircraft, it is towed from Orbital's integration facility at VAFB to the plane on the Assembly and Integration Trailer (AIT). Regardless of where the launch is to take place, the Pegasus is always integrated and mated to the L-1011 at VAFB. From there, the launch system can travel to any location in the world for launch. There is enough ground clearance for the L-1011 to take off and land with Pegasus attached underneath. However, the added height of the AIT underneath Pegasus requires raising the L-1011 off the ground slightly by hydraulic jacks to mate Pegasus to the carrier aircraft (Fig. 2). While mated to the L-1011, the vertical rudder actually protrudes into the plane's fuselage in a compartment specifically designed for this purpose. When mating the Pegasus to the L-1011, the rudder is usually detached from the Pegasus vehicle and placed inside the housing first. Then the Pegasus is rolled underneath the L-1011 and attached to the rudder and then to the plane. Removing the rudder first minimizes the height to which the L-1011 needs to be raised for the mating process. The entire mating process from rollout to mating takes about 6 hours. Pegasus is attached to the L-1011 using four hooks on the center box of the wing and a fifth hook on the forward portion of the vehicle. The inside of the airplane has been stripped of all unnecessary equipment and hardware. Up front in what would normally be the first class cabin are eight seats for personnel during ferry flights from VAFB to the launch site of interest and two computer stations from which personnel can monitor the health of the vehicle and the payload. The rest of the interior of the cabin has been completely gutted. Access to the rear portion of the aircraft cabin is obtained through a galley door.

Unlike most other launch vehicles in the U.S. fleet, the Pegasus launch vehicle is integrated horizontally on the AIT (Fig. 3). Horizontal integration facilitates easy access to the vehicle and eliminates the need for high bays and large cranes. Components are received as needed either from groups within Orbital Sciences or from outside vendors. To ensure that all of the major flight hardware and software is thoroughly tested before flight, Pegasus, like many other vehicles, is subjected to a series of "fly to orbit" simulations at various stages of the integration process. Four flight tests are normally performed. The first tests the three stages individually. The second test is conducted after the three stages are electrically mated together. The third test is performed after the three stages are electrically and mechanically mated and the stack is electrically mated to the payload. The fourth and final flight test is performed once the payload has been mechanically mated to the rest of the vehicle and the half of the fairing that includes the pyro devices necessary for jettisoning the shroud is electrically mated. These tests are intended to verify that various systems function and also respond as expected to known disturbances. If the inertial measurement unit (IMU) onboard receives data to indicate that an unexpected attitude change has occurred, will the fins or thrust vector control systems respond accordingly? Are all the commands to the various

Fig. 2.   Fully assembled Pegasus launch vehicle being mated to the L-1011 aircraft.

Fig. 3.   Horizontal integration of Pegasus launch vehicle.

subsystems appropriate, and do those subsystems respond appropriately? Once the Pegasus vehicle has been mated to the L-1011 carrier aircraft, one last test is performed, called the Combined Systems Test (CST). This test verifies that the launch vehicle and the carrier aircraft are communicating as expected. This is particularly important since the vehicle's health can be monitored both from telemetry that is broadcast from the vehicle to the ground via antennas on Pegasus and also by the computer stations inside the L-1011 via hardwired electrical connections. More importantly, some data and commands are sent to the Pegasus vehicle before launch. The only method currently available for accomplishing this transfer of data is through the electrical connections between the Pegasus vehicle and the carrier aircraft.

To be fully mobile, the Pegasus launch system must also be fully self-contained. Except for those services provided by the range (such as radar coverage), the L-1011 can transport all of the equipment required to support a launch of Pegasus, including, of course, Pegasus itself (Fig. 4). Some launches take place off the coast of California where the Western Range (based at VAFB) is the lead range. In these instances, no ferry flight is required. The L-1011 simply takes off from VAFB and flies to the designated drop point roughly 100 nmi out to sea. The checklist that is processed in the control room on the day of launch requires about 4 to 5 hours to complete. The L-1011 usually takes off an hour before the scheduled launch time. If all systems are "go," as determined by the mission team members in the control room, the launch conductor on the ground commands the pilot of the L-1011 to drop the Pegasus from the carrier aircraft.

Shtil. In a classic example of turning swords into plowshares, the Russian Navy developed a satellite delivery system for nonmilitary applications that uses a submarine-launched. The SS-N-23 (NATO's designation) is a three-stage liquid-fueled vehicle that can deliver small

Fig. 4.   L-1011 aircraft taking off with Pegasus.

satellites to low Earth orbit. Very little is known about this launch vehicle service including performance to various altitudes and inclinations. What is known is that two satellites belonging to the Technical University of Berlin were successfully launched in 1998 from a Russian submarine for the stunningly low price of $150,000. Some sources indicate that the typical commercial price for a Shtil launch is actually in the neighborhood of $500,000. There are two possible reasons for the low cost of a Shtil launch. The first is that more than 200 missiles have already been produced by the Russian military. There is also speculation that offering commercial launch services provides a way to maintain proficiency in launching missiles without using precious military funding. One disadvantage of this system is that the Shitl vehicle likely does not have enough performance to achieve circular orbits in the medium to high low earth orbit (LEO) altitudes. This is a direct result of the Shtil's heritage as a ballistic missile first and foremost.

Sea Launch. The most recent mobile launch system is the Sea Launch vehicle which is launched from a converted oil-drilling platform along the equator (Fig. 5). Sea Launch is both the name of the launch vehicle and the name of the international joint venture that provides the launch services. The partnership is comprised of Boeing, KB Yuzhnoye of Ukraine, which provides the two Zenit stages, and RSC Energia of Russia, which provides the Block DM-SL upper stage. The launch vehicle and payload integration takes place at the vehicle's home port of Long Beach, California. Once integration is complete, the launch vehicle is loaded onto the converted oil-drilling platform and towed to a predetermined launch location at the equator, specifically 154° West. Once on site, the Zenit 3SL is raised into its launch attitude (vertical) and launched. A second ship that houses mission personnel and the control room monitors the launch from nearby. The vehicle itself is a little less than 200 ft long and roughly 13 ft in diameter. The performance to geosynchronous transfer orbit (GTO) is approximately 5,250 kg (11,574 lbs). "In terms of spacecraft mass in final orbit, this would be equivalent to approximately 6,000 kg (13,228 kbs) of payload capability if launched from Cape Canaveral, because the spacecraft does not need to perform a plane change maneuver during the geosynchronous earth orbit (GEO) circularization burn".

There are three key phases in the integration of a Sea Launch vehicle. Phase I takes place in the Payload Processing Facility (PPF). This phase includes receipt of the spacecraft, processing of the spacecraft, testing, and enclosure within the payload fairing. Phase II takes place on the Assembly and Command Ship (ACS). This entails mating the encapsulated spacecraft to the launch vehicle and testing the integrated stack. Phase III takes place on the Launch Platform (LP) once the vehicle has been transferred from the ACS. While still in port, the integrated launch vehicle is raised to its vertical launch attitude so that a series of tests can be conducted. The launch vehicle is then lowered back into a horizontal position, stored in an environmentally controlled room, and transported to the equator while on board the launch platform. At the launch site, the launch vehicle is rolled out to the launch pad, raised to a vertical attitude again, and fueled. The launch is performed by an automated system and monitored by the Assembly and Command Ship which is moved for launch to a distance 6.5 km away (Fig. 6).

Fig. 5.   Computer simulation of Sea Launch.

Fig. 6.   Sea Launch successfully lifts DIRECTV 1-R satellite into orbit.

The Assembly and Command Ship for Sea Launch serves as the launch vehicle integration and testing facility. In addition to acting as the temporary home for launch crews, the ship also houses the Launch Control Center (LCC) and the equipment necessary to track the initial ascent of the rocket. Unlike the Pegasus carrier aircraft that was modified after serving in a different capacity, the ACS was designed and constructed specifically to suit the unique requirements of Sea Launch. The ship is roughly 660 ft long and 110 ft in beam and has an overall displacement of approximately 30,830 tonnes.

The rocket assembly facility is on the main deck of the ACS where the launch vehicle integration takes place. This activity is conducted before setting sail for the equator and simultaneously with spacecraft processing. After the spacecraft has been satisfactorily processed, it is encapsulated and transferred to the rocket assembly compartment, where it is mated to the launch vehicle. Following integration and preliminary testing, the integrated launch vehicle is transferred to the launch platform.

Then both ships begin the journey to the equator, which takes roughly 12 days.

The launch platform has all of the necessary systems for positioning and fueling the launch vehicle, as well as conducting the launch operations. Once the launch vehicle has been erected and all tests are complete, personnel are evacuated from the launch platform to the ACS using a link bridge between the vessels or a helicopter. Redundant radio-frequency links between the vessels permit personnel on the ACS to control all aspects of the launch, even when the command ship has retreated to a safe distance before launch. The launch platform, which was converted from an oil drilling platform, is very stable. It is supported by a pair of large pontoons and is propelled by a four-screw propulsion system (two in each aft lower hull). Once at the launch location, the pontoons are submerged to a depth of 70.5 ft to achieve a more stable attitude for launch, level to within approximately 1°.

Advantages of Mobile Space-Launched Systems

NOTSNIK, Pegasus, Sea Launch, and Shtil were never intended to replace the existing fleet of ground-launched rockets. Rather, they effectively supplement the existing worldwide capability by providing additional services to a targeted market of payloads that benefit greatly from the mobility and flexibility of these unique space-launch systems. These vehicles can provide services similar to ground-launched vehicles for payloads within their weightclass. In fact, all four vehicles have fixed launch locations for standard services. For example, Pegasus uses the launch location of 36°N, 237°E for all high-inclination missions that originate from VAFB. In this regard, the mobile launch systems are no different from ground-launched vehicles in that they repeatedly launch from a fixed location, albeit a location that is not on land. However, they can also offer services and performance that avoid many of the restrictions inherent in being constrained to a particular launch site. Few of those restrictions are trivial. They include inclination restrictions, large plane changes required to achieve low-inclination orbits from high-latitude launch sites, large plane changes required to transfer from GTO to GEO when launching from certain ranges, and low-frequency launch opportunities for missions that require phasing such as those involving a rendezvous with another spacecraft already in orbit.

Inclination Restrictions. Inclination restrictions stem from range safety considerations. To understand these restrictions fully, it is first necessary to understand two concepts: (*1*) transfer orbits and (2) instantaneous impact-point tracks.

Transfer Orbits. Transfer orbits are intermediate orbits established by the various stages of a launch vehicle that provide a path to the final desired orbit. The transfer orbits for early stages are mostly suborbital, meaning that some portion of the orbit intersects Earth's surface. The most efficient way to transfer between two orbits is to apply thrust at opposite apses. An application of thrust in the right direction at the perigee of the initial orbit will raise the apogee. Coasting to the new apogee and applying thrust (again in the appropriate direction) at this apsis will then raise the perigee. This provides a stair-step approach to raising the altitude of a vehicle's orbit. The ascent of a launch vehicle from launch to orbit follows a similar trend with one critical caveat. The impulse of initial stages is usually not sufficient, individually, to raise the perigee above Earth's surface. This means that using the optimal Hohmann transfer approach would bring the launch vehicle back to Earth before another transfer burn could be made. As a result, initial launch vehicle stages usually apply their thrust at places within a transfer orbit other than the apses and usually always on the ascending side of the orbit.

Consider a modest three stage, ground-launched rocket launching into a circular low Earth orbit as an example. Before launch, the vehicle is effectively sitting at the apogee of an orbit (Fig. 7). If the surface of Earth were not present to support the rocket, it would be drawn downward along a path that would take it closer and closer to Earth's center before swinging back to an apogee altitude equal to the radius of Earth. This is essentially the first of several transfer orbits and the rocket has not even been launched. When the rocket lifts off, it applies its thrust at an apsis, but in a direction that is perpendicular to the initial velocity vector of the rocket, which itself is in the direction of Earth's rotation. During the first burn, the vehicle slowly tilts over so that the thrust is applied in a direction that is increasingly parallel to Earth (Fig. 8). This has the effect

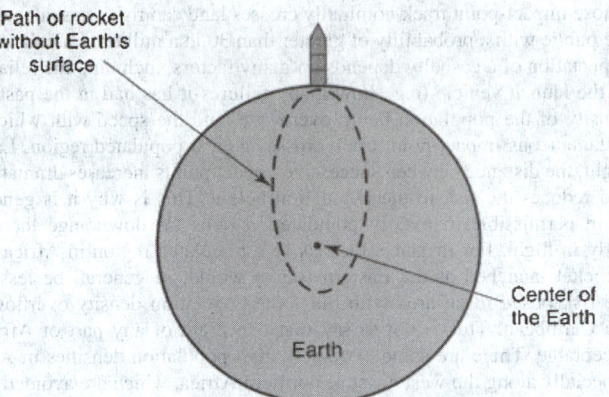

Fig. 7.   Path of rocket without earth's surface.

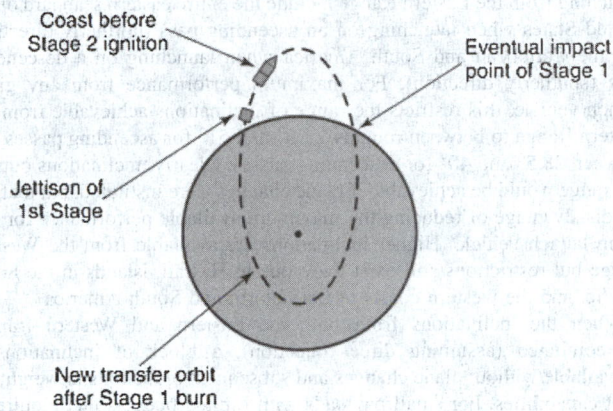

Fig. 8.   Path of rocket after launch.

of increasing both the apogee and perigee. The perigee will most likely still be suborbital at the end of the burn. The apogee will be increased sufficiently that the launch vehicle can coast up to a location near the new apogee, following the first stage burnout, and ignite the second stage. The key consideration here is that the second stage will be ignited near but not at the apogee. Again, this is not the most energy-efficient way to transfer orbits, but it is necessary because the opposite apsis is still below Earth's surface, and the second stage may not have sufficient impulse to raise it above the atmosphere. Igniting the second stage at a location other than the apogee again has the effect of raising both the perigee and the apogee. In this case, because only one stage is left, the burn is designed to raise the apogee to the desired altitude of the final orbit. After the second stage burns out, the vehicle coasts up to the new apogee and ignites the third stage. This will raise the perigee up to the final orbit altitude without changing the altitude of the apogee.

Impact-Point Tracks. By always burning on the ascending side of the trajectory and iteratively raising the apogee while the transfer orbit remains suborbital, anything jettisoned before the final burn will reenter the atmosphere and either burn up or impact Earth's surface. As the burn of each stage progresses, the point at which the transfer orbit intersects the Earth extends farther and further downrange until, at some point late in the final burn, there is no longer a point of intersection. These points of intersection comprise the instantaneous impact-point track. Clearly, as the vehicle is coasting, the instantaneous impact point does not change. Conversely, during a motor burn, it is constantly changing and each point represents the location of impact on Earth if, in fact, the thrust were to be instantly terminated either by design or due to some sort of failure. It is this impact-point track and the need for it to avoid populated areas that is a primary source of inclination restrictions from various ranges.

For any rocket launch, whether it be space-based, suborbital, ground-launched, ship-launched, or air-launched, the public-safety considerations that must be satisfied are very stringent. Those stages of a rocket that are jettisoned before reaching orbit should avoid land. And no launch vehicle whose impact-point track nominally crosses land can risk a casualty among the public with a probability of greater than 30 in a million. Calculating the expectation of a casualty depends on many factors, including the reliability of the launch vehicle (e.g., how many failures it has had in the past), the density of the population being overflown, and the speed with which the instantaneous impact-point track crosses over a populated region. Late in flight, the distance between successive impact points increases dramatically and reduces the risk to the population below. This is why it is generally more permissible to overfly populated regions far downrange than it is early in flight. For instance, the risk to a populated region in Africa from a rocket launched at the Eastern Range would, in general, be less than the risk posed to an area with the same population density overflown in the Caribbean. This is not to say that overflight of any part of Africa is acceptable. There are some extremely high population densities in Africa, especially along the west coast of northern Africa, which are avoided at all costs. And it is this very consideration that constrains the paths of many launch vehicles from the existing ranges.

The key land masses that must be avoided early in flight for vehicles launching from the Eastern Range include the entire eastern seaboard of the United States when launching on an ascending pass (northerly direction) and the Caribbean and South America when launching on a descending pass (southerly direction). For maximum performance from any given launch vehicle, this restricts the range of inclinations achievable from the Eastern Range to between roughly 28.5° and 51° for ascending passes and between 28.5° and 40° for descending passes. Clearly, inclinations outside this range would be achievable if plane changes were instituted, but that has the disadvantage of reducing the maximum available performance for any given launch vehicle. Higher inclinations are available from the Western Range but restrictions still exist there due to Hawaii, islands in the South Pacific, and the western coasts of both North and South America.

When the inclinations from both the Eastern and Western Ranges are combined (assuming direct injection), a block of inclinations is unavailable without plane changes and subsequent reductions in weight-to-orbit capabilities. For small payloads with limited budgets that require an inclination outside what is directly available from the existing ranges, the cost of launching on the heavy-lift launchers that can execute the necessary plane changes can be prohibitive. And reducing launch costs by flying as a secondary or even tertiary payload is advantageous only in the rare event that a primary payload can be found that requires the same final orbit. For these customers, Pegasus and Shtil provide an alternative due to their relatively low cost, mobility, and self-contained launch infrastructure. Sea Launch provides a similar alternative for the heaviest satellites that are intended for either GEO or low Earth orbits.

Plane Changes Required to Achieve Low Inclinations. The inclination of an orbit represents the angle between the equatorial plane and the orbital plane around Earth. This also happens to be similar to the definition of lines of latitude. It is no coincidence then that the maximum latitude of the ground track for any object in space is roughly equivalent to the inclination of the object's orbit. The only reason that the maximum latitude is not exactly equal to the inclination is because Earth is not a perfect sphere. Conversely, this implies that the minimum inclination attainable by a launch vehicle is roughly equivalent to the latitude of the location from which it is launched. The maximum is 180° minus the latitude of the launch point. This leads to the important conclusion that the only latitude from which all inclinations are directly accessible is 0° (the equator). The Eastern Range is at a latitude of roughly 28.5°. Therefore, the minimum inclination attainable without plane changes is roughly 28.5°. Lower inclinations can be achieved by launching into any available inclination, achieving a preliminary orbit, and then making an inclination correction burn when the satellite is over the equator or at any latitude that is numerically less than the desired inclination. The significant disadvantage of this process is that inclination changes while in orbit require a great deal of energy. The larger the change in inclination required, the more energy must be expended. Depending on the final orbit desired, this usually requires an additional stage to correct the inclination and achieve the final orbit. The most common recipient of this type of orbit maneuver is a satellite headed to geosynchronous orbit. However, there are low Earth orbit payloads that require low inclinations as well. The ability of Pegasus and Shtil to move the drop point to a latitude from which such energy-intensive plane changes would not be required permits smaller launch vehicles to achieve the same

orbit from lower latitudes that larger vehicles can achieve from higher latitudes. The difference in cost, complexity, and performance can often mean the difference for some customers between launching or not.

Some launch locations maintained by other countries are at significantly lower latitudes than those in the United States. For some customers, such ranges can provide the necessary services. However, many satellites in the United States, especially government sponsored, are required to contract with a U.S. launch service provider and use a U.S. controlled range.

Phasing. An object's orbit is essentially a locus of points that defines the path of the satellite. Those points define a plane that goes through the center of Earth. To define an object's precise position within an orbit, that plane and every position in it is defined with respect to both Earth and a coordinate system, one of whose axes always points toward the vernal equinox. Every position of a satellite as it orbits Earth is defined in terms of an epoch (time), the semimajor axis, and eccentricity, measured from Earth's center, inclination and argument of perigee, which are both referenced to Earth's equator, and the right ascension of the ascending node, which is referenced to the vernal equinox frame.

A rendezvous between two objects in space involves a series of maneuvers designed to make the orbital elements of both objects the same, hence confirming the fact that they have, in fact, become a single object orbiting Earth. Just as motor burns can raise or lower the perigee or apogee of an orbit or change the inclination, so too can motor burns be used to change every orbital element that defines a satellite's motion. However, changing some of those elements, especially those that require plane changes, requires large amounts of energy, and they are considered "expensive" in the parlance of orbital mechanics. One way to avoid paying the high price of actively changing the orbit of a satellite with a motor burn is to do it passively through the aid of various external forces. Several naturally occurring forces cause every orbital element to change over time. These include atmospheric drag, solar radiative pressure, the gravitational attraction of the Moon, Sun, and planets, and the nonuniform gravitational forces due to Earth's oblateness. These forces can be used to one's advantage when planning a rendezvous mission. However, some changes resulting from these forces can take a very long time to reach significant levels. This means that the initial differences between the rendezvousing satellite and the target must be initially small to avoid spending too much unproductive time in orbit. This can be accomplished by simply timing the launch appropriately so that at the time of orbital insertion, the satellite that has newly arrived in orbit is very close to the orbital plane of the target satellite.

To accomplish this maneuver, the launch must occur when the target satellite passes almost directly overhead. It also must be passing in the same direction as the intended launch. In other words, if the satellite being launched is to head off in a southerly direction (along the descending pass), the target satellite must be overhead and also on its descending pass as well. Otherwise, the two satellites will end up with right ascensions that are 180° apart which would be excessively expensive (either in terms of time or energy) to correct once in orbit.

For ground-launched vehicles, the wait between successive passes of the target satellite could be as much as several days, depending on the target orbit because the distance between ground tracks on successive passes depends on the period of the orbit, which depends on the orbit's semimajor axis. Clearly, the ground track of an object that requires only 90 minutes to orbit Earth will be more closely spaced than the ground track of an object whose period is several hours. These ground tracks will pass to the east and west of the given launch site on a daily basis, but the distance between the ground track and the launch site will only be minimized by a periodicity of the order of days.

Mobile assets, however, can eliminate the wait by essentially choosing a launch point that is ideally suited for a rendezvous. Instead of waiting for the ground track to come to the launch point, the launch point is moved to the ground track. In this way, the launch opportunities can be reduced from one every two to three days to at least once a day if not twice a day, if the launch vehicles have the flexibility to launch on both ascending and descending passes.

Consider an example of a satellite being launched by a Pegasus XL to rendezvous with a satellite currently in orbit at an altitude of 400 km circular. A normal ground-launched vehicle would require a wait of about 2 days between successive launch attempts. However, the mobility of

Pegasus permits two launch opportunities every day, which is graphically represented in Figs. 9 and 10. Two key assumptions need to be kept in mind when viewing these figures. The first is that the maximum range of the Pegasus carrier aircraft is roughly 1000 nmi. This includes a captive carry to the launch site, an aborted launch, and a return to base with Pegasus still attached. The second assumption is that for launches that do not require the full advantage of Pegasus' mobility, the standard launch point for Pegasus out of the Eastern Range is 28°N, 281.5°E. The vertical axes in Figs. 9 and 10 represent the difference in argument of latitude between the two satellites (the angular separation within the same orbital plane). The horizontal axis represents the launch point as the difference in degrees from the nominal point listed before. The diagonal lines represent the difference in argument of latitude for each day in the first week of October, which was chosen simply as an example. Figure 9 represents the difference in argument of latitude for northerly launches (launch along the ascending pass). Figure 10 represents the difference in argument of latitude for southerly launches (launch along the descending pass). The horizontal lines simply demarcate zero angular separation between the two satellites.

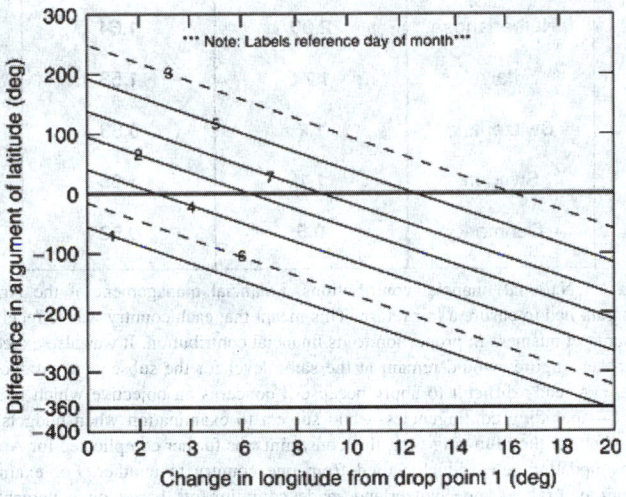

Fig. 9.   Graph of difference in argument of latitude for northerly launches.

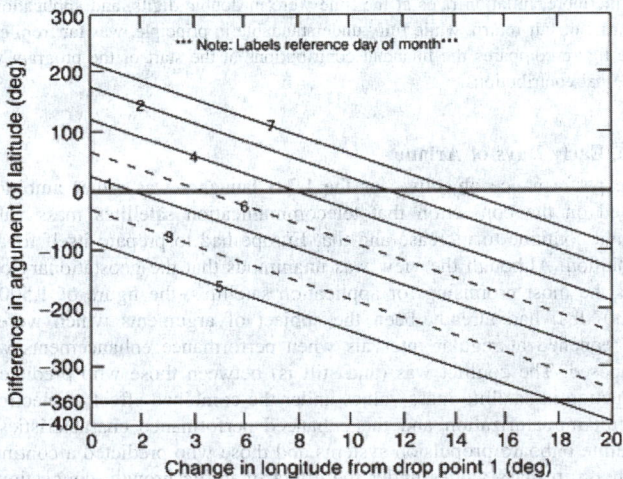

Fig. 10.   Graph of difference in argument of latitude for southerly launches.

The intersection of a diagonal line with a horizontal line defines a drop point within the range of the Pegasus carrier aircraft from which Pegasus can be launched and effectively deliver its satellite to the front door of the target satellite at the time of orbital insertion. Realistically, this is not how a rendezvous would normally be achieved. Ideally, the satellite being launched would be placed in a temporary parking orbit slightly

below and behind the target satellite. Over the course of several orbits the distance separating the two objects would be slowly decreased using several controlled burns of the satellite just placed in orbit. This would imply that a drop point is needed not to achieve 0° difference in argument of latitude but some finite value. The example is still valid. Simply shift the horizontal lines up or down until the desired difference in argument of latitude is matched. Again, an intersection between a diagonal line and the horizontal line defines a launch point within the range of the Pegasus carrier aircraft that would result in the desired difference in argument of latitude.

As can be seen from Figs. 9 and 10, every day except two in the first week of October provides two launch opportunities. A southerly launch on the 3rd does not provide a drop point within the range of the carrier aircraft that will achieve the desired result. However, a drop point can be found on that day if the launch is along the ascending pass instead. Likewise, a suitable drop point cannot be found within the range of the carrier aircraft on the 6th of October when launching along the ascending pass, but one can be found if launching in a southerly direction. The same qualitative results would be obtained for any other time frame. The quantitative results might be slightly different. For instance, instead of having only one launch opportunity on the 3rd and 6th it may be the 4th and the 7th. But the end result is the same. The mobility of Pegasus and, by definition, Sea Launch, and Shtil provides ideal rendezvous launch opportunities at least once a day and in most cases twice a day.

Clearly there are disadvantages with all of these mobile assets. Pegasus is limited in its size due to the restrictions of the L-1011 and, more importantly, the mechanical limitations of the hooks that hold the vehicle to the plane. Sea Launch has somewhat of a temporal disadvantage in that it requires almost 2 weeks to travel to the launch site. Those problems are exacerbated for Shtil because its home port is farther north. Nonetheless, for some specific missions, the mobility and flexibility that are provided by these unique space-launched assets provide valuable supplemental services to the fleet of existing ground-launched vehicles.

http://www.orbital.com/

Additional Reading

Isakowitz, S.J.: *International Reference Guide to Space Launch Systems*, 3rd edition AIAA, Washington, D.C., 1999.

Powell, J.: "The China Lake Launches," *Air and Space*, 367–378, (Feb/Mar 1997).

Staff: *Pegasus Users Guide, Release 5.0*, Orbital Sciences Corporation, August 2000.

Staff: *Sea Launch User's Guide, rev. B*, Boeing Commercial Space Company, July 2000.

Dale Fenn, Orbital Sciences Corporation, Dulles, VA

LAUNCH VEHICLES: EUROPEAN. In 1972, the European space community was in a state of crisis. It is interesting to look back at the situation which preceded this crisis to understand the reasons that led to the decision to proceed with the Ariane program and have subsequently guided the program up to the present time.

The first European initiative in the launcher field was taken by the United Kingdom in 1961, which put forward proposals to France, and later Germany, for manufacturing a launcher based on the Blue Streak, roughly similar to the American Atlas, designed and developed up to that time as a first missile stage. The Blue Streak stage was available, following a change of British policy in this domain. This was the basis for the Europa 1 project that was designed to place payloads of 1 metric ton into low orbit.

Unfortunately, the organization set up already contained the seeds of failure because the United Kingdom was supplying a first stage, whose characteristics were frozen, whereas France took on responsibility for the second stage, and Germany for the third stage. Each adopted its own technology and retained an autonomous stage, creating a total absence of anything resembling a system study or an attempt at general optimization. Furthermore, the program was created by cooperation among sovereign states, coordinated by a secretariat under the direction of a diplomat.

By 1966, there was no lack of technical and financial problems. The British threatened to withdraw from the project, which they did two years later. Therefore, the program was obliged to buy the first stage directly from United Kingdom industrial firms. The program was redirected toward geostationary orbit launches (a satellite placed into geostationary orbit appears to be in a fixed position with respect to Earth), and a solid propellant fourth stage was added. At the same time, the Australian launch range was abandoned in favor of a new launch center to be constructed in French Guiana. The payload specification was then 270 kg (595 lb) in geostationary orbit, and the new program was dubbed Europa 2.

At the same time, geostationary orbit application prospects became clearer, and in 1970, Europe decided to commence studies and predevelopment work on a more powerful launcher (Europa 3) that could place 1,500 kg (3,307 lb) payloads into geostationary transfer orbit (GTO). Unfortunately, the lessons to be learned from the Europa 1 failures were ignored insofar as program organization was concerned. Furthermore, although the first-stage technical configuration remained prudent, based on experience acquired by France from its Diamant program (first unit acquired in November 1965), the second stage was much too ambitious for Europe and involved a high-pressure liquid hydrogen/liquid oxygen topping cycle engine delivering 200 kN of thrust.

The period from 1969 to July 1973 was difficult for the European space community for the following reasons:

- In the technical context, Europa 1 launches F7, F8, and F9, made from the Australian base, all failed. The launcher on flight F11, the first launch Europa 2 made from the new range in French Guiana, exploded 150 seconds after lift-off on 5 November 1971.
- In the political sphere, Germany decided to withdraw from the Europa 2 program in December 1972 (which led to stopping that program in April 1973) and temporarily suspend its participation in the Europa 3 program. The Germans were also considerably attracted by collaborative proposals made by a NASA glowing from the success of the Apollo program. Germany considered that European efforts in the space transportation sector could be limited essentially to technological development, in parallel with major participation in the post-Apollo program. Nevertheless, negotiations between NASA and European representatives on that subject were difficult and engendered frustrations in Europe. At first invited to be involved in developing specific Shuttle hardware, the European contribution progressively narrowed to a science module that would fit into the Shuttle cargo bay.

France adopted a different approach that expressed a triple objective:

- acquisition of absolute control of space applications;
- founding of this control on an autonomous launch capability, with particular reference to geostationary satellites; and
- adoption of these first two objectives by its European partners to assemble sufficient financial capacity and sufficient volume to make production feasible.

However, these French ambitions were momentarily weakened by two launch failures in the Diamant national program in 1971 and 1973, despite the fact that they followed a run of six successful launches.

Finally, the Ariane program may well owe its very existence to the difficult negotiations undertaken with NASA in connection with the launch of the two "Symphonie" experimental telecommunications satellites. The extremely harsh conditions imposed on the German and French negotiators, including an embargo on using these two satellites for any commercial purpose in particular strengthened the determination to achieve the autonomy proposed by France. European agreement was finally achieved in July 1973, following a period of intensive negotiation.

It was decided to embark on three simultaneous programs:

- the "L3S" heavy launch program proposed by France at the end of 1972 (this program was renamed "Ariane" shortly afterward);
- the Spacelab program in cooperation with NASA, backed by Germany; and
- the "Marecs" maritime telecommunications program backed by the United Kingdom.

The principles of setting up the European Space Agency (ESA) were also defined, and the Agency was officially formed in 1975.

The commitment made by France in connection with this agreement was very considerable and corresponded to more than 60% funding for the program, plus an undertaking to fund excess cost above 120% of the initial figure of MFF 2,060 (about $ 447 M 1973) up to a maximum of 15% of this figure. In exchange, France obtained agreement that the European

Space Agency (ESA) would delegate management of the program to the French Space Agency (CNES). The obligation to ensure a workload return for each participating country in proportion to its contribution was and is indeed a particular aspect of the ESA programs to be emphasized (Fig. 1).

Country	Contribution %	
	Initial	Final
France	63.87	73.55
Germany	20.12	11.47
Belgium	5.00	4.71
United Kingdom	2.47	2.27
Spain	2.00	2.38
Netherlands	2.00	1.64
Italy	1.74	1.58
Switzerland	1.2	0.83
Sweden	1.1	1.05
Denmark	0.5	0.52

Fig. 1.   National financial contributions. Financial management of the Ariane program had to ensure a fair return. This meant that each country was to receive a volume of business in proportion to its financial contribution. It was also expected that these figures would remain at the same level for the subsequent production phase. Already difficult to apply because it concerns an objective which is both final—including contingencies—and subject to examination when budgets are voted on for the following year, this constraint was further complicated for Ariane 1 by updating rules which varied from one country to another. For example, Belgium, France, and Switzerland made commitments based on a percentage of initial development cost, whereas Germany undertook to make an annual contribution, expressed in its national currency, which would be revised only once in midprogram, according to monetary parities and observed inflation rates. The United Kingdom was involved only through a specific agreement with France. Furthermore, inflation rates at the time were in double digits, and application of the rule of fair return, while fully understandable in principle, was far from easy. The figure compares the financial contributions at the start of the program with the final contributions.

The Early Days of Ariane

The performance objective for the L3 S launcher was rather ambitious, based on the conviction that telecommunication satellites mass values would continue to increase and that Europe had to prepare itself for this evolution. Although the view was unanimous that the geostationary orbit was the most promising for application satellites, the figure of 1,500 kg (3,307 lbs) had already been the subject of arguments which were to be reopened at regular intervals when performance enhancements were proposed. The conflict was (and still is) between those who predicted a reduction in satellite mass values under the combined effect of electronic circuit miniaturization and the enhanced performance characteristics of satellite onboard propulsion systems and those who predicted a continued increase in mass values under the effect of traffic growth, congestion of the geostationary orbit, and the resultant reduction in in-orbit transponder cost. Finally, the payload mass objective adopted for the Europa 3 launcher was confirmed. It was set at 800 kg (1,764 lbs) for geostationary orbits, corresponding to about 1,500 kg (3,307 lbs) in geostationary transfer orbit (about 200 km by 36,000 km (124 miles by 22,369 miles). This represented almost twice the performance of the American Delta launcher, which had previously launched most application satellites for the Western world. This performance was comparable with that of the Atlas-Centaur, which was used at the time only for few heavy satellites. This decision provided a substantial growth potential for European satellites and demonstrated a determination to design the Ariane launcher for an extended lifetime.

Past experience, as mentioned at the beginning of this article, demonstrated the absolute need for imposing three major principles at the very start of the program, namely, a genuine system approach, strong management, and a simple design.

The System Approach. This made it necessary to regard the Ariane launcher (and its ground support facilities) as a single entity, not merely a stack of independently designed stages; to base the development of these stages and other subsystems on the results of studies conducted at the highest level (trajectory, flight mechanics, general loads, thermal, guidance and control, etc.); to design electrical systems in terms of the complete launcher, installing only actuator devices and equipment specific to its flight phase in each stage; and to impose design rules common to each discipline. For example, it was in this way that the need for POGO control systems was demonstrated at the beginning of the development phase. This meant that corrector devices could be designed and integrated into the propulsion systems at the outset. This was of even greater interest because Ariane was the first launcher designed entirely for geostationary orbit missions, in contrast to the other launchers existing at the time that derived to a greater or lesser degree from ballistic missiles. CNES entrusted a specific contractor (Aérospatiale) with this task and also associated Aérospatiale with the reviews conducted during the development of the various launcher systems and subsystems.

Strong Management. After obtaining delegated management authority, CNES spent the first year of the program establishing the basis for a strong management structure founded on a number of specific principles:

- unique management link between CNES, main contractors and other contractors;
- clearly defined industrial organization, with precise definition of the tasks allocated to each party. For the Ariane 1 program, CNES used the same French level 1 contractors as for its Diamant program, thus ensuring design unity in the main disciplines. These were Aérospatiale for the launcher stages, SEP (Société Européenne de Propulsion) for the propulsion systems, Matra for the vehicle equipment bay, and Air Liquide for the cryogenic third-stage tanks. Other contractors subsequently achieved level 1 status, including Contraves (Switzerland) for the fairing, DASA (Germany) for the second-stage and then the Ariane 4 liquid propellant boosters, Fiat-BPD (Italy) for the Ariane 3 and 4 boosters, etc.;
- a common understanding of the content of work to be done by each contractor, including achievements expected and reports to be submitted.

These principles led CNES to issue a set of management specifications applicable to each launcher system, subsystem and component, and also its ground facilities. These specifications covered overall planning and milestones, work breakdown structure, industrial organization, technical work coordination, schedule and cost reporting, monitoring of the development of critical elements, quality, and reliability. Application of such specifications in the context of a program involving 10 countries that differ in language and culture was a new departure for Europe and generated initial difficulties with companies each of which had their own particular methods of working. Nevertheless, this proved essential to maintain both visibility and coherence. These basic principles were also applied to the Ariane complementary development programs. They proved their effectiveness and made it possible to achieve effective control of technical development, costs, and time schedules.

Simple Design. This final point meant that the Ariane design should be based only on technologies which were either already available or involved only low development risks. The Europa 3 first stage had been designed on this principle, applying technologies and experience acquired with the first stage of the French Diamant launcher. This stage was powered by four rugged "Viking" storable propellant engines, for which the prototype, designed in France, had been tested with very encouraging results before this program was terminated. Therefore, the Europa 3 stage was selected as the basis for the first stage of the Ariane launcher.

This could not be the case for the Europa 3 second stage that was considered much too ambitious for Europe in 1973. However, although a number of difficulties were to be anticipated if liquid hydrogen and liquid oxygen were used, the efficiency of these propellants would make it possible to reduce the lift-off mass of the launcher by a factor of almost 2. Furthermore, a certain amount of experience had been acquired on a low-thrust gas generator cycle prototype engine developed within the French national program. Work conducted in France and Germany during the Europa 3 program meant that this approach could be adopted with a satisfactory level of confidence, provided that there was no major deviation from the experience acquired from the engine and propellant and fluid system. Thus, the tanks were designed for 8 tons of propellant and a maximum possible diameter using available tooling not to exceed 2.6 m (8.5 ft). Having made this choice, it was found necessary to add a second, intermediate stage to achieve the performance target. This stage was designed on the basis of elements and technologies developed for the first or third stage, including the Viking engine used in the first stage (with a suitably adapted nozzle) and the light alloy tanks of the third stage.

Ariane 1 Launcher

Ariane 1 was a three-stage launcher that had a total height of 47.8 m (157 ft) and a lift-off mass of 210 tons. The first stage had a dry mass of 13.3 tons, a height of 18.4 m (60.4 ft) and a diameter of 3.8 m (12.5 ft). It had four Viking 5 engines (Fig. 2) that developed 2500 kN thrust on lift-off. Burn-time in flight was 146 s. The 148 tons of propellant (UDMH and N_2O_4) were contained in two identical steel tanks protected again internal corrosion by an aluminum layer and connected via a cylindrical skirt. The four turbopump Viking engines were mounted symmetrically on a thrust frame and articulated in pairs on two orthogonal axes to provide for three-axis attitude control. An annular water tank, located inside the propulsion bay, provided for cooling the gas obtained from the engine gas generators that was used to pressurize the propellant tanks and supply the hydraulic motors of the attitude control actuator systems. Four fins with a surface area of 2 m^2 provided aerodynamic stability. This first stage was designed to destruct approximately 30 seconds after stage one to two separation.

The second stage had a dry mass of 3.13 tons (excluding the interstage conical skirt and the jettisonable acceleration rockets), its height was 11.6 m (38 ft), and its diameter was 2.6 m (8.5 ft). It had a single Viking 4 engine that developed 740 kN thrust in vacuum for a burn-time of 136 s. The motor was linked to the conical thrust frame via a gimbal with two degrees of freedom, that provided pitch and yaw control. Auxiliary nozzles supplied with hot gas from the engine gas generator provided the roll control function. The two aluminum alloy tanks that had a common intermediate bulkhead were pressurized with helium gas (3.5 bars) and contained 34.1 tons of propellant (UDMH and N_2O_4). The second stage was also designed to destruct about 30 seconds after stage two to three separation. During the prelaunch waiting period on the pad, a thermal shroud, ventilated with cold air, which restricted heat exchange between the propellants and the environment, protected the second-stage tanks. This shroud was jettisoned on launcher lift-off.

The third stage weighed 1.164 tons dry, was 9.08 m (29.8 ft) high, and had a diameter of 2.6 m (8.5 ft). This was the first cryogenic stage produced in Europe. It was equipped with a type HM7A engine that developed 62 kN thrust in vacuum for a burn-time of 545 s. This engine was designed by Société Européenne de Propulsion (SEP), based on experience acquired with an earlier cryogenic engine, the HM 4, which delivered 40 kN thrust, and tested over the period 1962–1969. The HM 7a engine uses the conventional gas generator cycle technology, and achieves a specific impulse of 443 s with a mixture ratio of 4.43 at pump inlet. The combustion chamber is supplied with propellants pressurized by a turbopump, via a set of injection valves. The pump turbine is driven by gas supplied by a generator, the latter receiving a small proportion of propellant tapped off at the pump outlet. The liquid hydrogen and liquid oxygen tanks that contained a total of 8.23 tons of propellant were made of an aluminum alloy and had a common intermediate bulkhead (double skin under vacuum). The tanks were covered with external thermal protection to avoid warming the propellant. Both tanks were pressurized in flight, using hydrogen gas tapped at the outlet from the regenerative chamber and helium. The motor was linked to the conical thrust frame via a gimbal that provided pitch and yaw control. Auxiliary nozzles ejecting hydrogen gas were used for roll control.

The stages were separated by pyrotechnic cutter devices fitted on the rear skirts of the second and third stages. The separating stages were

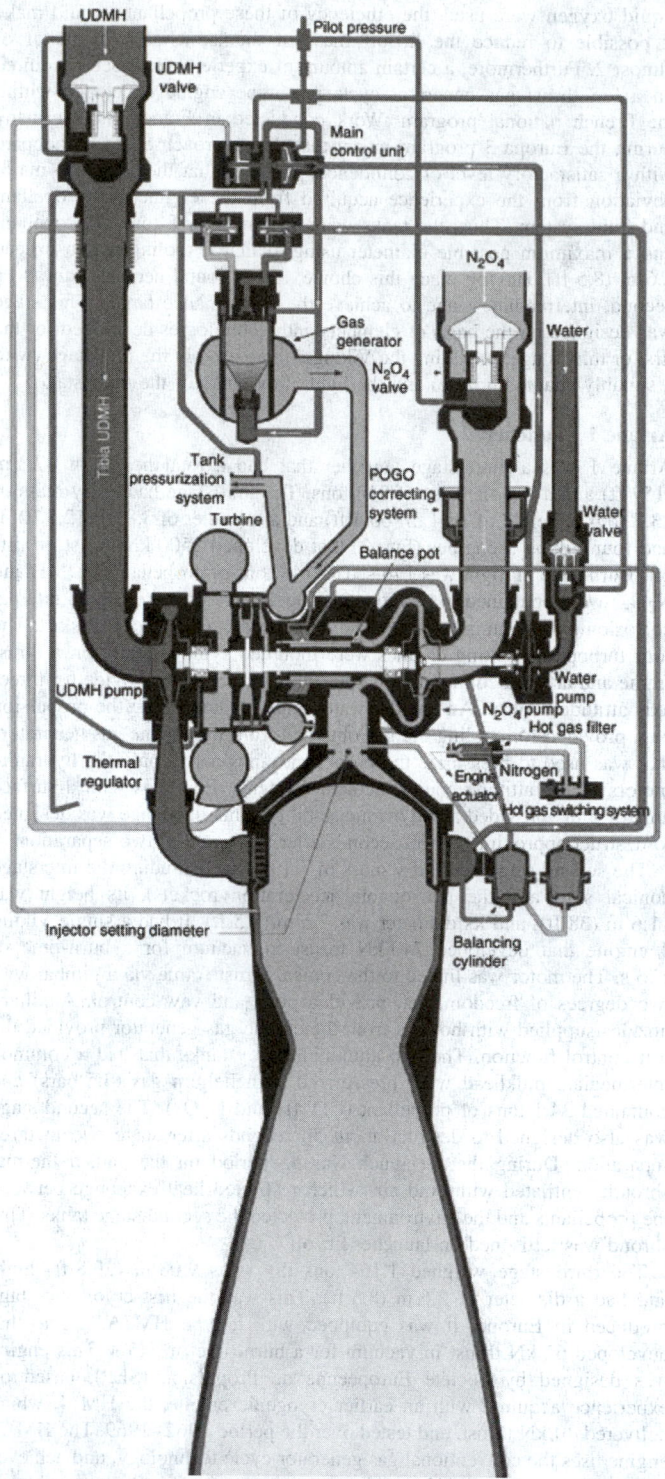

Fig. 2. Viking engine flow diagram. The Viking engine uses the gas generator cycle, and the gas generator itself operates at a stoichiometric ratio. The gas produced is cooled by injecting water to reduce gas temperature to values compatible with the turbopump turbine, pressurization of the main propellant tanks, and operation of the power pack for the hydraulic servoactuators. The three pumps (UDMH, N_2O_4, and water) are mounted on a single shaft that rotates at 10,000 rpm. (For Ariane 1, the chamber pressure was limited to 53.5 bars.) The hydropneumatic regulation system slaves the combustion pressure to a reference pressure value by adjusting the gas generator feed and, thus, the rotational speed of the turbopump. The mixture ratio is maintained at a constant level by a regulator that equalizes propellant pressures before injection into the combustion chamber. Engine ignition is induced by the pressure in the propellant tanks. When the valves open, the propellants, which are hypergolic, are delivered to the combustion chamber and gas generator and ignite spontaneously. The turbopump speed then builds up to the value set by the regulating system in 1.3 s.

distanced from each other by retrorockets incorporated in the lower stage and acceleration rockets mounted on the upper stage. Stage 1 to stage 2 separation was controlled by the onboard computer, on detection of first-stage thrust decay (propellant exhaustion). Stage 2 to stage 3 separation was controlled by the onboard computer when the second-stage speed increase reached 1500 m/s.

The vehicle equipment bay (VEB) weighed 316 kg (697 lbs), had a diameter of 2.6 m (8.5 ft), and was of 1.15 m (3.8 ft) high. Mounted on the third stage, the VEB contained the electronic equipment of the launcher and also served as a base for the payload and fairing. In addition to the onboard computer, the VEB housed all of the electrical equipment required for executing the launcher mission, namely, the sequencing unit, guidance and navigation control, and the location, safety, and telemetry systems. Only the power systems and actuator devices were located elsewhere in the launcher stages.

The two half-fairings were ejected parallel to the main axis of the launcher under the control of the onboard computer, as soon as the calculated thermal flux dropped below the specified level. The fairing was jettisoned by two pyrotechnic systems, a horizontal system at the interface with the VEB and a vertical system which also served to impart horizontal velocity to each half-fairing. The fairing was 3 m in diameter and was compatible with Atlas-Centaur class satellites.

The launch facilities in French Guiana (ELA 1:Ariane launch site No1) were designed to make use of the earlier investment for the Europa 2 launcher. The stages were erected and assembled directly on the launch pad. A mobile gantry sheltered the launcher and provided for assembly of the payload with the launcher and closure of the fairing under clean room conditions. The gantry withdrawal commenced 6 hours before lift-off. The main fueling of the third stage with liquid hydrogen and liquid oxygen and final topping off were performed using a cryogenic arm system carrying a set of umbilical valve plates. Disconnection and retraction of the arms commenced at time T−3 s. Control of the launcher on the pad was fully automatic from time T−6 minutes. The launcher was then controlled by two ground computers, one for the electrical systems and the other for the propellant and fluid systems. The two computers also cross-checked each other. The first stage engine was ignited at time T, and the lift-off command signal was sent at T + 3 s, following satisfactory verification of the Viking engines. The first flight took place on Christmas eve 1979 (Fig. 3).

When the Ariane program was first initiated in 1973, few imagined a commercial career for the launcher. Things began to move in 1976 and led to a series of actions relating to launcher performance, production, and marketing. As regards performance, the idea was to propose Ariane for launching the Intelsat satellites. The success of such a venture would represent both an exceptional reference for the program and strong motivation for the players involved. However, the first task was to augment performance objectives up to 1600 kg minimum in GTO. Fortunately the prudent approach adopted for the basic definition of the launcher, combined with the results of the first flight, demonstrated propulsion performance in excess of specifications and made it possible to exceed the initial objective and achieve a figure of 1,850 kg (4,079 lbs).

As far as production was concerned, it was obvious that there was no hope of selling the launcher if one waited to receive orders before commencing manufacture. It was on these lines that discussions were opened at the European Space Agency, and a promotional phase was duly adopted. Apart from the production of six launchers, the objective of this phase was to achieve full operational qualification, develop and validate the dual launch capability, and adapt the launcher comprehensively to meet user needs by providing for the construction of payload preparation facilities in French Guiana. Analysis of marketing aspects led to two actions: initiation of the Ariane 3 program and formation of Arianespace.

Ariane 3 Program

At the end of the 1970s, the application satellite market was organized around two launcher groups. These were the Delta launchers, for which GTO performance in 1978 was around 1,200 kg (2,646 lbs) and the Atlas-Centaur launchers whose performance figure was close to 1,800 kg (3,968 lbs). Apart from the Intelsat satellites, all other payloads corresponded to the Delta class, for which Ariane could not pretend to be competitive in its actual state. The idea then emerged to adapt the launcher

Fig. 3. Launch L0 1 (24 December 1979). The decision to proceed directly to flight tests, using three active stages in the final flight configuration, was made following very lengthy discussions at the start of the Ariane program. This was in total opposition to the highly progressive, but also extremely costly approach adopted for the Europa 2 program. Initial ignition occurred on 15 December 1979, following a faultless countdown. Unfortunately, the ground computers did not authorize liftoff, and the engines shut down automatically at time T + 10 s and aborted the launch. Subsequent analysis showed that an explosion in a small measurement pipe had damaged the sensors, whose signals were used for operational diagnosis of the engines. The case of an aborted launch, for which the probability was infinitely small, had, nevertheless, been taken into account during the development phase. Procedures for return to flight configuration had been written and validated by tests conducted in Europe. This allowed restarting the count on 23 December, following 8 days of round-the-clock work. Technical problems, combined with adverse meteorological conditions, prevented a launch on 23 December. On the following day, Ariane made a practically faultless launch. Only two anomalies were identified, and these were corrected before the following flight. These concerned minor pollution of the payload, caused by the second-stage retrorockets and low amplitude vibrations (POGO effect) at the end of the second-stage flight. The first Ariane launch thus took place only 6 months after the initial target date set in 1973.

The second launch in May 1980 resulted in a failure. Destruction of the launcher occurred at T + 63.75 s, as a result of high frequency combustion instability in one of the first-stage Viking engines, 2.75 seconds after liftoff. Corrective measures essentially comprised modifying the propellant injection orifices in the combustion chamber. A total of 95 tests that represented 4300 seconds of burn-time were conducted to qualify the new injector under conditions substantially more severe than those encountered in actual flight. The Ariane 1 development phase terminated in December 1981, following three successes out of the four launches made. The total cost of the program was within 120% of the initial estimate.

to enable it to launch two Delta class satellites simultaneously, thus raising the performance objective to twice 1200 kg (2,646 lbs), plus the mass of the dual launch structure designed to isolate the two satellites from each other. This corresponded to a total GTO mass of 2,500 kg (5,511.56 lbs). The launcher also had to be a competitive with the Atlas-Centaur. See also **Launch Vehicles: U.S. Expendable**.

This program was proposed by France to the European Space Agency and was approved in July 1980, despite the failure of the second Ariane flight (L02) in May of that year. CNES was charged with managing the Ariane 3 program, under conditions very similar to those for Ariane 1.

Analysis conducted from 1976 and aimed at increasing the performance of Ariane 1, had in fact identified modifications whose feasibility was checked out during the final development test phase. These modifications were adopted in part or in full for the Ariane 3 program, in accordance with the policy of minimizing development risks. The addition of two solid propellant boosters achieved the objective of competitiveness, with Ariane 3 for two satellites of 1,200 kg (2,646 lbs) each, and Ariane 2, with no boosters, for satellites of 2,000 kg (4,409 lbs).

These performance objectives were achieved on the basis of the Ariane 1 launcher by introducing the following modifications:

- increased thrust for the first- and second-stage Viking engines by augmenting combustion pressure by 10% (53.5 bar to 58.5 bar). This was obtained mainly by adding hydrazine hydrate to UDMH;
- adding two solid propellant boosters with a unit thrust of 600 kN and a burn-time of about 40 seconds. The low level of performance sensitivity to the structural mass of the boosters made it possible to adopt extremely prudent technical solutions;
- increase in the third-stage propellant load from 8 to 10 tons;
- enhancement of Stage 3 performance by increasing combustion chamber pressure by 5 bar and stretching the nozzle by 200 mm (HM7 B);
- adaptation of the SYLDA dual-launch system and the fairing to the volume required for 1200 kg-class-satellites.

Figure 4 shows a SYLDA dual-launch structure and half a fairing. The lower satellite is placed inside an egg-shaped compartment. Protected by the fairing, this carbon fiber structure is subject to reduced loads by comparison with an external structure. Furthermore, the operational constraints induced by the need to carry two satellites remain within acceptable limits.

The SYLDA structure makes it possible to achieve complete separation of the two payloads. The long orientation sequence, spin-up phase and separation of the satellites and upper part of the SYLDA, are achieved by the SCAR attitude and roll control system, using third stage pressurization hydrogen gas to operate this system.

Tested successfully on the sixth launch, this concept was unquestionably one of the keys to the success of Ariane. The cost of a launcher is not proportional to its size, and a number of functions must exist whether the launcher is small or large. A dual-launch capability, taking advantage of this scale effect, represents a major competitive plus factor.

The industrial organization was the same as that for the Ariane 1 program; the only main modification was an increase in the Italian contribution, which led to entrusting the development of the solid propellant boosters to FIAT-BPD. The first Ariane 3 was launched successfully in mid-1984.

Arianespace

The lengthy discussions which took place at the European Space Agency and led to the decision to go ahead with production of six Ariane 1 launchers, also demonstrated that this multinational organization, whose basic purpose was research and development, was not suitable for engaging in commercial and production activities. At the same time, the promoters of Ariane reached another paramount conclusion. For European autonomy to be effective in access to the geostationary orbit, the Ariane launcher had to be credible. To this end, it had to be both reliable and available, in other words produced in sufficient quantities and in advance of actual launch needs. This called for world-scale marketing, and at the same time made it possible to spread the fixed production costs over a larger production volume, and thus reduce the cost of the autonomy-related strategy (in contrast to its American competitors, Ariane did not have a major captive market for launching governmental satellites, to which all or

Fig. 4.   SYLDA, Ariane dual-launch system.

part of the fixed costs could be allocated). Only a private commercial entity, responsible for both Ariane production, marketing, and launch operations, could take up this challenge.

In December 1977, Frédéric d'Allest, then head of the CNES launcher division and future Chairman and CEO of Arianespace, proposed forming a company to market the Ariane launcher. *Arianespace* was incorporated on 26 March 1980, well before qualification of the Ariane 1 program. *Arianespace* is a French business corporation. Its capital is held by CNES, the leading European manufacturers that participate in Ariane production, and a number of banks. *Arianespace* is responsible for producing, launching, and marketing the Ariane launcher or launchers, whose development and qualification have been, are, or will be conducted by the European Space Agency.

The first Arianespace commercial launch was made successfully on 22 May 1984, inaugurating the very first commercial space transportation service.

Quality. A number of technical difficulties were encountered during the early days of the Ariane program, resulting from residual design problems or omissions in the production files. The European launcher industry was in its infancy was in the process of discovering the problems inherent in launcher batch production, and had to acquire expertise in this demanding discipline. Major efforts were made during the 1980s to amplify and correct the production files, identify and analyze all defects irrespective of their importance, and carry out regular tests on equipment sampled from the production line. The severity of these tests frequently exceeded the levels specified for the qualification tests. The flight data measurement plan that covered more than 700 parameters for which results were transmitted to the ground during the qualification flights — this figure was reduced to 400 for operational flights — made it possible to analyze each flight in the finest detail and thus acquire in-depth knowledge of the launcher. This unprecedented quality construction program, conducted by a European team highly motivated at all levels, meant that the 90% reliability objective assigned to the Ariane in 1973 was quickly exceeded. By the end of

October 1999, a total of 118 Ariane launchers (versions 1 to 4 inclusive) had flown, and the reliability figure for Ariane 4 exceeded 97% after a string of 48 consecutive successful flights.

Marketing. Ariane marketing operations achieved rapid success, despite a number of technical difficulties encountered at the start of the program. The success of the development plan, combined with the appraisals conducted by potential customers, bred a high degree of confidence in the European launcher and the French Guiana ground facilities. The first non-European customer to place an order for Ariane launch services in 1978 was Intelsat (International Telecommunications Satellite organization), well known in the satellite world for its technical expertise. See also **Satellites (Communications and Navigation)**. This choice was extremely important for Ariane and induced confidence on the part of other customers, who then decided to regard the Ariane launch system objectively. Apart from its own inherent merits, Ariane had the advantage of a favorable market situation. Commissioning of the American Space Shuttle led NASA to stop producing its conventional Delta and Atlas-Centaur launchers. Delays with the Shuttle, due to technical difficulties, discouraged satellite operators, obliging them to seek other launch possibilities. Furthermore, the high U.S. dollar exchange rate during the early 1980s made Ariane prices extremely attractive compared with the American launch systems. This marketing success grew further with the passage of time. The first Ariane 3 flight in August 1984 qualified the dual-launch capability of two Delta class spacecraft and significantly increased both the launch capacity and competitiveness of the Ariane system.

In 1985, Ariane launched the same number of commercial satellites as the Space Shuttle, and in the following year, *Arianespace* signed no fewer than 16 launch contracts. By 1987, Arianespace had 57 contracts (satellites to be launched or already launched) from the European Space Agency, the member countries of the European space community, international organizations, national organizations in non-European countries and private companies, representing a total sales value exceeding FF 16 billion (about $2.25 billion). *Arianespace* had acquired a share of more than 50% of the open satellite market, a position that the company has succeeded in maintaining despite increasingly severe competition.

Ariane Launch Site No. 2 (ELA 2)

France decided to construct a launch range in French Guiana in April 1964, after the deciding to discontinue launch operations from the Hammaguir range in southern Algeria. The French Guiana site was chosen as a result of a comparative study of a number of possible locations. Paradoxically, if we compare the 1964 situation with that of today, equatorial launches, if they were even mentioned, did not initially constitute a priority criterion. However, this view was quick to change, and in July 1966, the European launcher organization accepted the French proposal to contribute to constructing an equatorial launch base in French Guiana, initially intended for Europa 2. It should be noted that at that time, 33 years ahead of the Sea Launch project, the French Guiana site was in competition with a floating marine platform project proposed by Italy, similar to the San Marco platform which it was then operating offshore from Kenya.

A decision was quickly made in favor of the Kourou site (the population of the village of Kourou was 660 in 1964) on the Atlantic seaboard 70 km northwest of Cayenne in a sparsely populated region. The exceptional geographical characteristics of this site combined a wide launch arc over the ocean and favorable climatic conditions (infrequent storms, no cyclonic or seismic activity, and temperatures varying only slightly round a mean figure of 25 °C). All types of mission (polar and equatorial orbits in the range −10.5° to +93.5°) could be planned in complete safety, an advantage that no other operational base possessed then. The closeness of Kourou to the equator (5° 2¢N) is ideal for placing telecommunications satellites into geostationary orbit. At this latitude, the slingshot effect induced by the rotation of Earth is near its maximum, and propellant consumption for adjusting the plane of the geostationary orbit is minimum. In global terms, the mass gain achieved with a launch from Kourou is approximately 17% compared with a launch from the Cape Canaveral, using an identical launcher.

The Guiana Space Centre (CSG) facilities were first operationally tested in April 1968 with the launch of a Véronique sounding rocket, and the Centre was inaugurated officially in 1969. The first satellite was launched on 10 March 1970 with the French Diamant B launcher. The first Europa

2 flight was in November 1971, using the new CSG launch pad. The failure of this launch had a series of consequences already mentioned at the beginning of this article. For both economic and political reasons, it was essential for the Ariane 1 launcher to make the best possible use of the heavy investments made in Europa 2. Nevertheless, certain facilities were unsuitable for launch operations with Ariane because of insufficient capacity and lack of operational flexibility and due to an absence of any growth potential. This made these facilities largely incompatible with the Ariane performance enhancements already under study. After proposals from the French Space Agency (CNES), the European Space Agency (ESA) decided to construct Ariane launch site No. 2 (ELA 2) in July 1980. This new facility was required to meet the following specifications:

- provision for 10 launches per year (Ariane 2, 3, or 4) and execution of two launches within an interval of less than one month;
- provision for replacing a launcher already on the pad, in case of need, by the launcher scheduled for the next flight;
- provision for preparing larger payloads with improved facilities for payload/launcher integration.

The design of the new launch site differed from that of ELA 1. The time spent by the launcher on the pad had to be reduced to a minimum to allow for executing launch operations and postlaunch rehabilitation of the pad within a maximum 1 month. This assumed locating a launcher preparation zone on the edge of the safety perimeter.

Following their arrival from Europe, the stages are erected, assembled, and tested in the rear preparation zone. Then, the launcher on its mobile launch table is transferred to the pad via a dual rail track (Fig. 5). There it is connected to the propellant and fuel circuits and undergoes a launch rehearsal procedure, including filling of the third stage with liquid hydrogen and liquid oxygen to check the absence of ground and onboard leakage. The success of this operation authorizes assembly of the upper part (payloads, adaptors, and fairing were integrated beforehand in a dedicated building) with the launcher. The principles of the launch sequence qualified on the ELA 1 site have been retained. The ELA 2 pad uses many of the support facilities previously used for ELA 1.

Fig. 5.   ELA 1 and ELA 2 launch sites. ELA 1 site (foreground): Ariane 3 launcher during third-stage fueling tests on day D−9. At rear: Ariane 4 launcher recently arrived on the ELA 2 pad. Note the double rail track between the pad and the preparation zone (background). The white circle midway down the rail track is a turntable, used to allow two launchers to pass each other if necessary.

Ariane 4 Program

Ariane 3 represented a short-term response that enabled the Ariane launcher to position rapidly vis à vis its two American competitors, Thor Delta and Atlas-Centaur. The arrival of the Space Shuttle at the end of the 70s and the policy adopted by NASA of marketing Shuttle flights at extremely attractive prices required a more comprehensive response from the Ariane program.

It was obvious that the NASA price policy would lead to a profound change in satellite design, broadly on the following lines:

- increase in Delta class satellite size and mass, up to the point of occupying, in a vertical position, the maximum volume available in the Shuttle cargo bay. This led to satellites of between 1,400 and 1,500 kg (3,087 and 3,307 lbs) for injection into geostationary transfer orbit;
- appearance of a family of satellites installed in a longitudinal position in the Shuttle cargo bay, designed to take advantage of the large diameter available and reduce launch cost (dependent on the height of the satellite). Corresponding changes, that involved problems of both mass and diameter were more difficult to predict.

As regards satellite mass, there were some projects for TV satellites of 2,500 kg (5,512 lbs). However, these were rare and still only moderately credible at the end of 1980, in particular among those who predicted a reduction in satellite mass values from miniaturization of satellite electronics. However, things moved in 1981, and the objective of 2,500 kg (5,512 lbs) was finally adopted.

The diameter problem was even more delicate. The Ariane program team foresaw major aerodynamic problems if the fairing diameter exceeded by too great an extent the 2.6 m (8.5 ft) diameter of the third stage on which it was mounted. However, determination of this value required aerodynamic tests that were too lengthy to conduct before making the decision. The diameter of 3.65 m (12 ft) finally available for payloads was in fact the result of a compromise between what the program team considered possible at the least risk and the sacrifice which satellite customers were prepared to accept for the advantage of a second launch service source.

Competitiveness requirements dictated the pursuit of the dual-launch policy implemented for Ariane 3. The GTO performance objective assigned to Ariane 4 was consequently an ability to execute simultaneous orbit injection of two payloads, one of 1,400 kg (3,089 lbs) and the other of 2,500 kg (5,512 lbs), giving a total of 4,300 kg (9,480 lbs), including the mass of the dual-launch structure. (Note that this objective had increased from 3,400 to 4,300 kg (7,496 to 9,480 lbs) in 1981 and that the maximum performance demonstrated in 1999 exceeded 4,900 kg (10,803 lbs).

This program was to be regarded as a continuation of the Ariane family, in other words as a complementary performance enhancement phase, taking full advantage of work carried out for Ariane 2 and 3, and innovating as little as possible in propulsion. This was decided to take the fullest advantage of experience already acquired and to improve the reliability target from 0.90 to 0.92. Given the ambitious performance objective that had been set, the configuration adopted was required to embody a certain degree of flexibility with regard to lower mass payloads to apply an attractive pricing structure for a wide range of satellites. Single launches of heavier satellites like Intelsat 6 — 3600 kg — had to be possible, of course.

These performance objectives were achieved on the basis of Ariane 3 by introducing the following modifications:

- First stage propellant load increased from 147 to 227 tons, while retaining the same operating point for the Viking engines as qualified for the Ariane 3 program. Stage 1 burn-time was increased to 205 s.
- Development of a liquid propellant booster, corresponding to a reduced scale copy of the first stage as it uses the same engine with an appropriately adapted pressure ratio, and the same tank pressurization system. Identical stainless steel tanks carry 39 tons of propellant.
- Adaptation of the Ariane 3 solid propellant boosters. Carrying 9.5 tons of propellant, these boosters burn for 36 s and deliver 650 kN thrust each.
- Development of a new water tank, located on top of the UDMH tank. This tank is constructed in composite materials, and supplies the first stage and liquid propellant boosters in blow-down mode.
- Modification of the vehicle equipment bay structure, to achieve better payload integration flexibility and easier transition to the new fairing diameter.
- Development of an external Ariane dual-launch carrier structure (SPELDA), providing for dual launches of large-diameter satellites.
- Development of a new fairing with a useful diameter of 3.65 m.

This program was proposed by France to the European Space Agency and formally accepted in January 1982. The same program management principles were adopted as those for Ariane 1 and 3.

Launcher Development. The development phase did not introduce any major innovations in propulsion. First, the first-stage propulsion bay had performed extremely satisfactorily on the test bed in the Ariane 3 development program for burn-times very close to those required for Ariane 4. Furthermore, the booster propulsion system was based on a configuration already used to develop the Viking engine. Consequently, insofar as propulsion was concerned, the task was one of optimization or of demonstrating new operating margins because the burn-time for the Viking engine had been increased substantially. On the other hand, work relating to the launcher system was substantially more complicated than first thought. It was found necessary to adopt the complete system approach for each lower composite configuration. Furthermore, the increase in the height of the launcher and the size of the upper composite were reflected in a considerable increase in general loads that required substantial modification of a number of structures, including connecting flanges, in particular. Furthermore, first-stage in-flight stability was a major source of concern during the first few years of the program. Digital control was adopted to introduce a large degree of flexibility in the development time schedule and also with actual operation of the launcher by making it possible to finalize the configuration of each launcher only 2 months before lift-off. Today, Ariane 4 can place from 2.1 to 4.9 tons in a typical GTO orbit, depending on the number of liquid (L) or solid (P) propellant boosters fitted on the first stage (AR 40,42P,42L,44P,44LP,44L). Nine different configurations of SYLDA, SPELDA, and fairings were available for payload accommodation. The Ariane 4 maiden flight, initially planned for late 1985, took place successfully on 15 June 1988.

The Birth of Ariane 5

Initial reflection at CNES on Ariane 5 dates back to 1978. At that time, the Ariane 5 launcher was regarded more as a means to access low-orbit, manned-flight missions. Nevertheless, the launching of application satellites into geostationary transfer orbit continued as an objective, in particular in triple-launch mode, to pursue the Ariane scale effect competitiveness approach. The two missions rapidly acquired identical importance.

In contrast to the approach adopted for the American Space Shuttle, the primary principle adopted was that priority should be given to competitiveness for commercial flights. Manned spaceflight induces substantial costs that it is absurd to impose on operations that can be conducted by an unmanned vehicle. The system proposed had to be capable of manned flight and unmanned flight missions without increasing the cost of the latter. Consequently, the new launcher had to be regarded and optimized as a single entity at the design stage to ensure optimum use of available resources, and the manned flight aspect had to benefit from lessons learnt from unmanned missions, while also providing for substantially dissociated use when the operational phase was reached. The manned module was consequently designed as a "special payload" mounted on top of the launcher. The capsule and spaceplane concepts were then analyzed in parallel. The latter presented the advantage of a considerable reentry cross-range and a high degree of orbit return flexibility that provided for a soft landing and consequent reuse. The onboard intelligence provided launcher "brain" functions during the ascent phase and thus eliminated the need for a vehicle equipment bay required for unmanned flight. This, then, was the Ariane 5—Hermes concept at the end of 1979.

At that time, Ariane 5 comprised the Ariane 4 first stage, and a new second stage burning liquid hydrogen and liquid oxygen (H 55), carrying Hermes for manned missions or the Ariane 4 third stage (H 10), the vehicle equipment bay, and fairing for unmanned flight. This study, which emphasized the need for a large cryogenic engine for future Ariane improvements, made it possible to propose and obtain funding for a 3-year French national program in 1980 and to commence work on the basis of design for an engine to deliver 600 kN thrust. Germany, Sweden, and Belgium joined this project under the terms of bilateral agreements.

Partially Reusable Concept Studies. In 1980, the impact of the Space Shuttle and its price policy led to strong criticism of the conservative characteristics of the Ariane 5 configuration. As a matter of principle, reusable concepts were regarded as more economical, the more so because the maintenance costs for such systems were ignored and operating costs were substantially underestimated. Considerable attention was then paid during that year to improving the cost levels and analyzing the operating costs announced for the Space Shuttle and those observed for the early stages of the Ariane operational phase. In parallel, the problems induced by the rehabilitation of equipment used during the Ariane 1 development

phase demonstrated the importance of the corresponding work and the complementary qualification cost that would be required.

A number of concepts were considered:

- recovery of the first stage by parachutes. This was tried unsuccessfully with the first stage of the Ariane launcher used for flight 14. The liquid propellant boosters for Ariane 4 were also initially designed to facilitate this type of recovery.
- design of a first stage comprising a stack of several H 55 stages, recovered individually using a delta wing;
- consideration of a winged first stage, etc.

This line of approach failed to produce any attractive, realistic solutions, and was abandoned early in 1982.

Comparative Configuration Studies. A systematic review of all possible concepts that matched the performance objectives was then undertaken. More than 24 different configurations were drawn and assessed for performance and cost. A configuration involving a large cryogenic core stage flanked by two large solid propellant boosters was examined and then abandoned because it was impossible to identify a derivative solution, with reduced performance, which would have made it possible to adapt a launcher configuration to the mission model in the same way as with Ariane 4. This configuration was looked at again, when the desired degree of flexibility had been introduced by offering the choice of an Ariane 4 third stage (H 10), or alternatively a highly simplified storable propellant stage, and after further studies had made it possible to set an acceptable price objective for manufacturing the solid propellant.

Finally, these three configurations were selected for more detailed comparative analysis in mid-1983:

- a solution that corresponded to direct continuity with Ariane 4 but introduced a cryogenic second stage. This solution was derived directly from experience acquired from Ariane 4 and made it possible to evaluate cost and performance extremely precisely.
- a solution based on the above, replacing the Ariane 4 first stage with a cryogenic stage equipped with four or five engines identical to that used for the second stage. This solution had the advantage of requiring only development of a single propulsion engine, namely a high-thrust LOX/LH$_2$ engine.
- a solution involving a large cryogenic core stage, flanked by two large solid propellant boosters. This basic composite carried a simple storable propellant upper stage for low-performance missions or a cryogenic stage that carried 10 tons of LOX/LH$_2$ (Ariane 4 third stage) for high-performance missions.

All three configurations required developing of a new LOX/LH$_2$ engine. This conclusion led France to propose the development of an engine of this type, based on the project commenced in 1980, to the European Space Agency. A further year was then devoted to a detailed comparative study of the three configurations.

The configuration that had large solid propellant boosters was finally selected on the basis of intrinsic reliability, recurrent cost, greater potential, and development time schedule control criteria. Only the safety criterion remained questionable. Compared with the abrupt failure of solid propellant boosters, their liquid propellant counterparts are always presented as easy to control. Nevertheless, in-depth analysis of the complete system demonstrates that this is not always the case, and that even in this favorable situation, there are still flight phases where the spaceplane (in contrast to a capsule) simply cannot accomplish its true rescue mission. (Long after this choice had been made, this frequently impassioned discussion was stimulated once more following the Challenger failure.)

Therefore, the configuration that incorporated two solid propellant boosters was finally presented in 1985 and adopted as a European Space Agency program in 1986 and 1987. Further optimization led to incorporating the following aspects:

- abandonment of the triple-launch concept in a basic mission context due to the excessive operational constraints relating to the satellites (availability);
- increase in payload mass values: the nominal performance objective was then dual launches into geostationary transfer orbit for satellites of 2950 kg each;

- mass problems encountered with the Hermes program. This led to an increase in the performance of the lower composite and abandonment of the cryogenic upper stage.

Ariane 5 Specifications. Final specifications for the Ariane 5 program were as follows:

- simultaneous launch of two satellites of 2,950 kg (6,504 lbs) and a diameter of 4.53 m (15 ft) into geostationary transfer orbit under environmental conditions and with a degree of precision, etc., comparable with Ariane 4 launches, representing a single-launch performance equivalent of 6,800 kg (14,991 lbs);
- launch of an 18-ton payload into a circular orbit of 550 km (342 miles), inclined at 28°5′;
- Hermes launch: this mission was not to introduce constraints liable to penalize unmanned flight. Consequently, performance requirements for this mission were deduced from those specified for unmanned flight, taking due account of safety constraints;
- Reliability of unmanned flight was set at 0.98, almost 10 times higher than the initial specification for Ariane 1. This ambitious target was justified by the high cost of insurance for both launcher and satellites (over 20% in 1988) and the major consequences of a flight failure. The safety of the crew was consequently provided forby ejection of the Hermes spaceplane on detection of an operating anomaly. The safety objective for the crew was set at a figure of $1 - 10^{-3}$.
- Cost objective per launch, on the basis of eight launches per year, was set at 90% of the figure for an Ariane 44 L. Given the respective performance of the two launchers, this corresponded to a reduction of 45% in the cost per kilogram in orbit for an equivalent fill factor.
- The target for a maiden flight was initially set for April 1995. It was subsequently put back by one year due to a number of economic constraints.

As for earlier Ariane programs, management was delegated to the French Space Agency (CNES) by the European Space Agency. The conditions for this delegation of authority were more restrictive in this case, however, due to the need for close coordination with the Hermes program and the now powerful image of the Ariane program.

The three main management principles stated at the beginning of this article with regard to Ariane 1 were retained and in fact strengthened in two areas:

- Management specifications incorporated a design-to-cost objective.
- The safety/reliability approach was further emphasized through systematic integration of past experience, both from incidents and accidents that occurred during testing, in the course of previous Ariane flights, or in the context of other programs (Challenger, etc.).

Furthermore, faced with an identified failure mode, the principle of a dual approach, involving simultaneous reduction of the probability of that failure and improvement of system tolerance was imposed.

Ariane 5 Generic Launcher

The Ariane 5 launcher stands approximately 51. 5 m (169 ft) high; the actual figure depends on the upper composite configuration, and the lift-off mass is 746 tons. Lift-off thrust is 11,660 kN. On the ground, the central core vehicle that has an outside diameter of 5.46 m (18 ft) is suspended at the level of the first-stage forward skirt between two solid propellant boosters. This connection, through which the thrust of each booster is introduced in the core stage, is made of alternate elastomer and metallic shims, designed to provide a damping effect on vibrations induced by booster combustion (Fig. 6).

The cryogenic main stage, developed under Aérospatiale as the prime contractor, is 30.5 m (100 ft), and its dry mass is 12.2 tons. This stage contains 158 tons of liquid hydrogen and liquid oxygen. The Vulcain engine (Fig. 7) is mounted on a thrust cone that distributes thrust evenly at the base of the liquid hydrogen tank. The engine can be oriented along two orthogonal axes by hydraulic actuators operating on the lost fluid principle. This fluid is stored in tanks operating in the blowdown mode. Connecting struts between the main stage and the solid propellant boosters provide rigidity for the rear part. The light alloy propellant tanks have a common bulkhead, insulated with expanded polyurethane. The hydrogen tank has a volume of 390 m^3 and is pressurized at values that vary according to

Fig. 6. This is a cutaway drawing of Ariane 5. The core is the Vulcain Engine with a payload of two satellites shown above the fuel tanks. The two solid strap-on motors are shown on either side of the core.

the flight phase (between 2.15 and 2.35 bar), using hydrogen tapped at the outlet from the engine regenerative circuit. The oxygen tank (120 m^3) is in the upper position and is pressurized with helium gas (3.5 dropping to 2.85 bar), obtained by heating liquid helium in a heat exchanger located in the oxygen turbine exhaust line. This helium (1.15 m^3) is stored in a superinsulated tank mounted on the thrust frame. The engine control system is supplied with helium gas stored under pressure in separate tanks. A thrust frame, secured to the upper part of the oxygen tank, receives the thrust of the solid propellant boosters, which is then distributed evenly toward the upper composite.

Europropulsion (joint subsidiary of SNECMA/SEP and FIAT/BPD) was the prime contractor for development of the Ariane 5 solid propellant boosters.

Each booster is 31.16 m (102 ft) high, with a diameter of 3.05 m (10 ft) and a post-combustion mass of 39.3 tons. It contains 238 tons of solid propellant grain, a composite with an ammonium perchlorate and polybutadiene base charged with aluminum. The booster casing is made of a high-strength low-alloy carbon steel and comprises seven cylindrical sections and two bulkheads. The sections are flowturned to a thickness of 8 mm from forged preforms and then assembled using a tang and clevis connection. The sections and bulkheads are assembled to form three segments, each of which is loaded independently with propellant. Internal thermal insulation, made of rubber-based, silica or fiber-filled material, protects the structure from hot combustion gasses. The forward segment is loaded with a 20 tons, star-shaped solid propellant block in Italy. In view of the mass and size of the boosters, a dedicated plant has been constructed in French Guyana for fuelling the central and rear segments (approximately 110 tons of propellant each).

The nozzle, with a flexible bearing made of alternate elastomer and metallic shims, can be steered up to 6° to control the thrust vector. The hydraulic actuators are driven by fluid stored within high pressurized, carbon-fiber vessels operating in blow-down. This fluid is ejected at the nozzle exit. The nozzle, which is highly integrated with the motor, represents a prudent extrapolation of the nozzles developed and qualified for defense applications. The throat in carbon/carbon material ensures minimum erosion during flight. The exit cone is composed of a light alloy housing, with phenolic carbon and silica insulation.

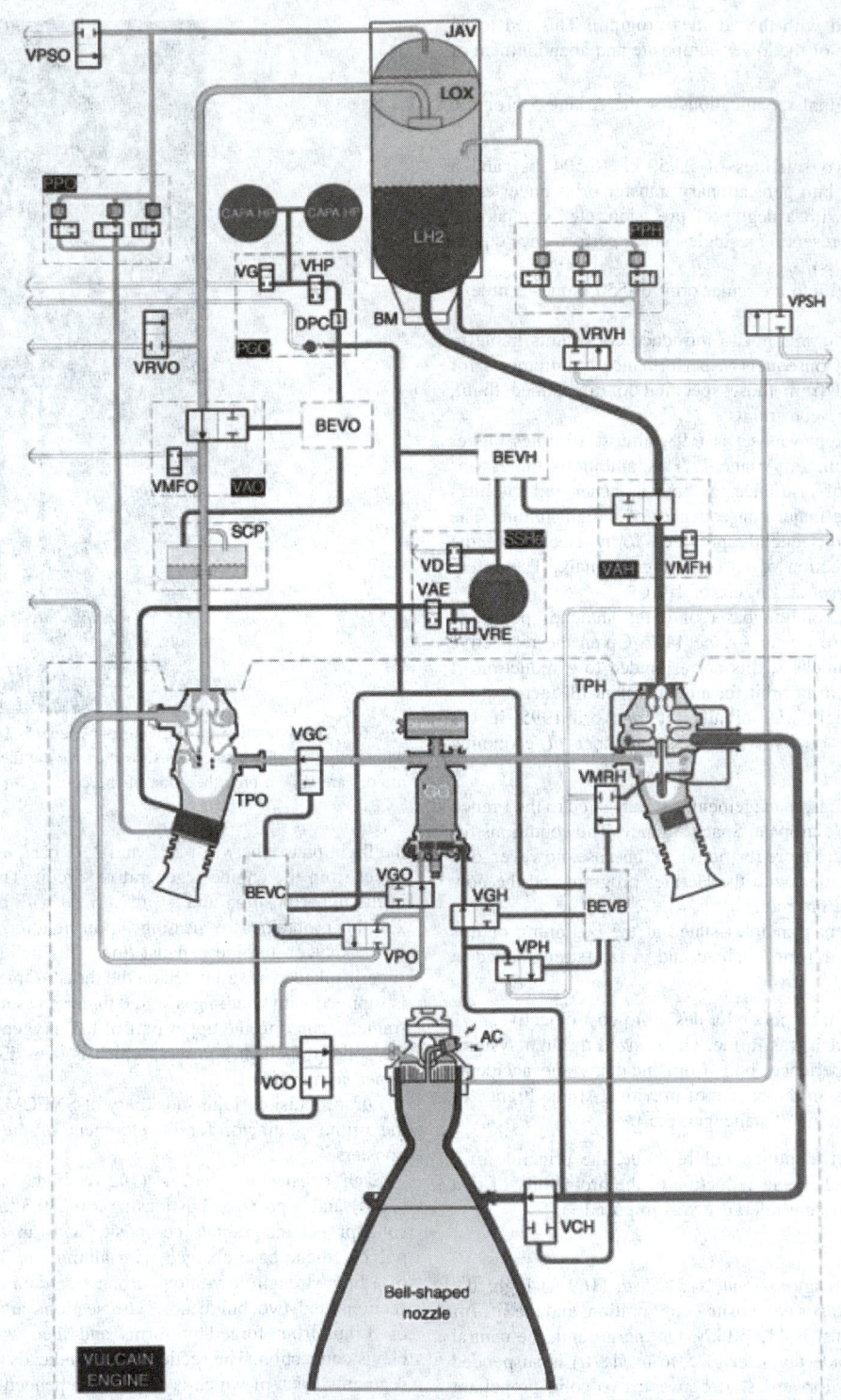

Fig. 7. ARIANE 5—Vulcain flow diagram. The Vulcain engine delivers 1140 kN thrust in vacuum and has a specific impulse of 432 sec. It uses conventional gas generator cycle technology. The propellants are delivered by two independent turbopumps. The liquid hydrogen unit operates at 34,000 rpm and comprises a two-stage centrifugal pump, preceded by an inducer that ensures favorable intake characteristics. This pump is driven by a 12-MW, two-stage turbine. The pump delivers up to 560 L/s of liquid hydrogen at a pressure of 17 MPa. The single-stage liquid oxygen turbopump operates at 13,400 rpm and delivers 177 L/s of propellant at 13 MPa (3.7 MW). The two turbines are driven in parallel by a single radial injection gas generator, that operates at a combustion pressure of 8 MPa. The generator is supplied with propellant tapped off at the pump outlet. The combustion chamber pressure is 110 bar. The liquid hydrogen enters the propulsion chamber via an annular distributor. Most of this flow is routed through channels integrated in the double-walled structure of the combustion chamber and throat assembly. The nozzle is cooled by a simple process known as dump cooling: the remaining hydrogen flow is routed through 460 spirally welded iconel tubes, whose diameters increase to give a continuous, bell-shaped surface, then escapes through micronozzles set along the bottom rim of the main nozzle. Although these gases do not undergo combustion, they are heated during the trip and contribute to overall thrust. The turbopumps, gas generator, and combustion chamber ignitions are started by pyrotechnic cartridges. The mixture ratio (mean value 5.25) is adjusted by a two-way valve, used to modify this ratio to terminate combustion on quasi-simultaneous depletion of the two propellants.

The boosters are jettisoned from the core stage by means of pyrotechnic cords. Solid fuel thrusters located in the rear part and in the forward cone are used to distance the boosters from the launcher in a radial plane.

The booster thrust evolves in flight in order to limit general loads, with a maximum of 6,700 kN and an average value of 4,900 kN in vacuum. It must remain within very tight tolerances to limit any thrust differential between the two boosters at any moment. The maximum combustion pressure is 60 bar, and the specific impulse in vacuum is 270 s.

The vehicle equipment bay is 1.56 m (5 ft) high and constitutes a linkage structure between the first stage, second stage, and fairing.

Developed under the prime contractorship of DASA, the second stage is fitted inside the vehicle equipment bay. This is an internal and relatively compact stage, with a diameter of only 3.94 m (13 ft) and a height of 3.36 m (11 ft). The dry mass is about 1,250 kg (2, 756 lbs). This second stage carries the payload adaptor (Fig. 8). The propulsion system comprises an "Aestus" engine, burning storable hypergolic propellants (MMH and N204), loaded in helium-pressurized tanks and consequently requiring no turbopump. This technological solution was adopted for its reliability, and its simple operation and re-ignition. A total of 6,550 kg (14,440 lbs) of N204 and 3,200 kg (7,055 lbs) of MMH (maximum load mass) are contained in two pairs of identical cylindrico-spherical tanks, arranged axisymmetrically two by two. The lift-off mass is 11 tons.

Fig. 8.   Ariane 5 second stage.

The engine combustion chamber, extended by a nozzle in refractory steel, is cooled by an MMH circuit. The mixture ratio is adjusted to a value of 2.05 by calibration on acceptance testing. The engine can be oriented through an arc of ±4.8° on two axes, using electrical servo actuators. The Aestus engine develops 29.1 kN thrust and has a specific impulse of 324 s. The burn-time is about 19 min.

Other details of Aestus engine are as follows:

Propellant feed rate: 8.77
Nozzle expansion ratio: 84
Mass: 111 kg (245 lbs).

The bay carries all functional electrical equipment used throughout the mission and a hydrazine attitude control system operating in the "blowdown" mode. This system is used to control roll throughout the propulsive flight, except during the solid booster burn phase, and during the attitude control phase after shutdown of the second-stage engine. By comparison with the Ariane 1 to 4 programs, the functional electrical systems have been fully duplicated and operate in the full active redundancy mode. This might allow for progressive incorporation of standard electrical components.

The vehicle equipment bay is separated from the main stage by dilation of a sealed pyrotechnic tube, which causes a fragile flange to rupture. The upper composite is then distanced by pyrotechnic actuators.

The assembly sequence and facilities in French Guiana (Launch site N° 3 : ELA3) are based totally on experience with Ariane 4, taking due account of the new mass and dimensional values involved, and carrying the policy of stripping the pad down to the bare minimum a step further.

As for the previous Ariane launchers, the synchronized sequence commences at time T-6 min, 30 s. It is subdivided into two separate automatic procedures that involve the fluid systems, on the one hand, and the electrical systems, on the other. At T-4.5 s, an execution command signal is sent to the onboard computer, which then activates the inertial guidance systems and authorizes ignition of the Vulcain engine at time T. Following verification of correct engine combustion at T + 6.2 s, the solid propellant boosters are ignited at T + 7 s. The boosters burn for about 130 s. They are jettisoned after burnout, when an acceleration limit of about 5.2 ms^2 is detected. For a typical GTO launch, booster separation occurs at an altitude of approximately 60 km (37 miles) at about T + 140 s. By this time the launcher relative speed has reached approximately 2000 m/s. The fairing is jettisoned on the basis of a thermal flux calculation, at about T + 3 min 10 s. Shutdown of the Vulcain engine is initiated by measuring propellant depletion at about T + 9 min 46 s. Launcher speed is then approximately 7600 m/s. After separation, fallback of the main stage is controlled so that the stage hits the ocean approximately 1,800 km (1,118.5 miles) from the Colombian coast. The upper stage is ignited at about T + 10 minutes. During the launch, onboard telemetry is transmitted to Kourou via ground stations located at Natal (Brazil), Ascension Island, Libreville (Gabon), Hartebeesthoek (South Africa) and Malindi (Kenya). The upper stage is fully passivated after separation of the payloads to minimize orbital pollution.

The first Ariane 5 flight was a failure due to a software fault. The second flight (502) took place on 3 October 1997. During the second flight, the only anomaly observed was an excessive roll torque at the limit of acceptability, following jettisoning of the solid propellant boosters and during the Vulcain engine burn phase. This fault was corrected very simply by a minor modification to the orientation of the turbine exhaust nozzles.

The third flight demonstrated the exceptional flexibility of the launcher, based on the simultaneous execution of two largely different missions. The first involved placing a reentry capsule into a suborbital trajectory. The precision of the point of impact in the Pacific was excellent. The other mission involved injecting a payload into a geostationary transfer orbit, again achieved with excellent precision and followed by a second-stage maneuver. This mission involved several successive reignitions and shutdowns of the second stage.

Comparison Between Ariane 44 L And Ariane 5

Unquestionably, Ariane 5 marks a technical break with earlier members of the Ariane family. However, continuity in human resources, management principles, and industrial organization ensured that the Ariane 5 program benefited from all experience previously acquired and made it possible to improve the intrinsic reliability by one order of magnitude. The Ariane 5 launcher has four active stages, whereas Ariane 44 L has seven. The number of separation sequences has thus been reduced. Ariane 5 does not have a launch table that is active in positive time, as is the case with Ariane 4. Nor does Ariane 5 have cryogenic fuelling arms that retract at the last moment before ignition and lift-off. Ariane 5 fluid connections are passive and are pulled free on lift-off. The aborted launch case, which requires extremely complex revalidation operations for Ariane 4, is thus simplified to the extreme. Vulcain engine ignition on the ground was deliberately chosen (following ignition failures observed on Ariane 3 flights 15 and 18), as was also the decision to wait to establish a stationary regime for the complete stage (following the flight 36 failure) before checking the engine and authorizing ignition of the solid propellant boosters. The price of this option is a loss of performance (Vulcain engine thrust was set to ensure minimum acceleration of the launcher after boosters were jettisoned). On the other hand, the Vulcain nozzle could have been adapted more efficiently in the case of in-flight ignition). The production of hydraulic power required to operate the booster and first-stage attitude control actuators is obtained from simple pressurized tanks on Ariane 5, compared with the hot-gas-fed engines used for Ariane 4. This list, which covers only simplifications visible at launcher system level, can be extended for each subsystem and component.

Ariane 5 and Hermes

It has frequently been said or written that the Hermes project led to the design of a launcher which was too large, or nonoptimum, in stage configuration. The need to improve performance yet again, to face up to the competition and remain competitive, clearly demonstrates that the first criticism was foundless.

No choice was made to the detriment of optimization for GTO launches. On the other hand, where a number of possibilities were equivalent for this optimization, choices were based on a Hermes criterion. Speaking in 1999, one can be thankful that this approach was adopted for at least three reasons:

- The Ariane 5 launcher is fully adapted for all types of missions, from constellations to heavy GEO spacecraft and servicing of the Space Station. The 503 flight demonstrated a high degree of flexibility during a single mission.
- The growth potential of Ariane 5 is very substantial, including replacement of the existing upper stage by a cryogenic stage.
- Manned spaceflight represented an excellent stimulus for the construction of launcher reliability. In this area again, the approach adopted was in no way unfavorable for unmanned missions. Abandonment of the Hermes program changed none of the choices made before this decision.

Ariane 5 Evolution Program

The increase in individual mass values of satellites in the *Arianespace* order book continued steadily throughout the 1980s. Extrapolation of this growth demonstrated that Ariane 5 performance was likely to be insufficient by about one ton shortly after qualification of the new launcher. However, this demonstration failed to convince at the beginning of the 1990s. The Hermes spaceplane program was encountering difficulties, including mass-related problems, and the proposal to increase the performance of the Ariane 5 lower composite was consequently regarded as a subterfuge for increasing the size of a launcher, still regarded as too big for launching commercial satellites. Furthermore, the multiplication of constellation projects involving small satellites in LEO and the increased credibility of plasma propulsion techniques for satellite orbit control provided ammunition for those who supported stopping further increases in the mass of geostationary satellites. When the decision to go ahead with the proposed enhancement program was finally made in 1995, this was based more on the desire to preserve cryogenic propulsion expertise than on a genuine need to enhance Ariane 5 performance, even with no change in production costs. The new enhancement program, designated "Ariane 5 Evolution" is aimed at achieving capacity for simultaneous launch of two 3,300-kg (7,275 lb)-class satellites into geostationary transfer orbit.

The following modifications were adopted to meet this specification:

- Increase in Vulcain engine thrust from 1140 to 1350 kN, while changing the mixture ratio from 5.3 to 6, but without modifying the external dimensions of the first-stage tank (although the position of the intermediate bulkhead changed). First-stage mass is increased up to 170 tons of liquid hydrogen and liquid oxygen.
- To recover the specific impulse lost through this modification of the mixture ratio, the turbine exhaust gases are injected into the nozzle, whose expansion ratio area is increased. This technical solution, not adopted for the basic Vulcain program to avoid excessive integration of elements supplied by different contractors, was validated by a technological program in parallel with the main development program.
- Replacement of connections within the three main segments of the solid propellant boosters by welds. In addition to a substantial booster mass gain and lower costs, this simplifies application of internal thermal protection.
- Adaptation of the dual-launch system (SYLDA) used with Ariane 2, 3, and 4 to the dimensions of Ariane 5.

Between 1996 and 2003 the Ariane 5 Generic launcher has made 16 launches from Europe's Spaceport. Depending on the payload, the configuration of the upper part of the launcher has been adapted for each

launch to ensure the optimum accommodation of single, dual or, for Flight 162 on 27 September 2003, a triple payload.

Apart from the initial qualification flight which ended in failure 39 s after liftoff, and the 10th flight where a problem with the storable propellant stage (EPS) placed one satellite in an unrecoverable orbit, the Ariane 5 Generic launcher has demonstrated its robustness and reliability.

To improve the performance of the initial Ariane 5 version and to respond to evolution in market demand, a set of small but important changes have been introduced on a limited number of launchers. Launchers with these modifications are referred to as the Ariane 5 Generic Plus.

These changes are:

- introduction of the lighter P2001 nozzle on the EAP boosters
- modifications to the EPS upper stage: increased capacity of the two MMH tanks (+300 kg of ergols); modified mixture ratio of the Aestus engine
- modifications to the vehicle equipment bay (VEB): replacing the aluminium structure with a lighter composite version; a new separation system for the VEB and the main stage to reduce the shock created during separation; new electrical equipment and components

All these modifications have led to an increase in performance for a standard geostationary transfer orbit of 150 kg (331 lbs) net payload.

The Ariane 5G+ operated flawlessly on the three launches made in 2004.

In the mid 1990s an analysis of the launch market was carried out. This showed that the majority of customers require launches into geosynchronous transfer orbit and identified the need for further upgrades of the Ariane launch capacity.

Decisions to modify and improve Ariane 5 launcher versions were taken leading to the development of the:

- Ariane 5 ECA
- Ariane 5 GS
- Ariane 5 ES ATV

Ariane 5 ECA. The latest version of the Ariane-5 launcher, Ariane 5 ECA, is designed to place payloads weighing up to 9.6 tons into geostationary transfer orbit. With its increased capacity Ariane 5 ECA can handle dual launches of very large satellites.

This new version of the Ariane 5 will help to maintain Europe's competitiveness in the commercial space transport sector by offering customers the opportunity to launch a wider range of heavier satellites while reducing launcher production costs.

Ariane 5 ECA is a higher capacity Ariane 5 Generic launcher. Although it has the same general architecture, a number of major changes have been made to the basic structure of the Ariane 5 Generic version to increase thrust and enable it to carry heavier payloads into orbit.

The EAP boosters' upper segment of the Ariane 5 ECA (also known as segment S1) carries 10% — approximately 2.5 tons — more propellant. This extra propellant gives the Ariane 5 ECA an additional 50 tons of thrust in the first 20 seconds following liftoff. This means that together the twin boosters deliver a thrust of 1,300 tons at liftoff, nearly 10 times the level delivered by the new engine of the central stage.

The boosters are also equipped with a new nozzle that has fewer parts and is easier and cheaper to produce.

An improved version of the Vulcain engine has been developed and implemented on the Ariane 5 ECA version. The Vulcain cryogenic engine has been modified to increase its thrust by 20% up to 137 tons. This new version, called Vulcain 2, operates under slightly higher pressure with a mixture ratio that has 20% more liquid oxygen than the Vulcain 1. Because of this change in the mixture, a new oxygen turbopump has been developed.

In addition, the Vulcain 2 turbopumps to be re-injected into the main system, thus improving engine performance at high altitude. The Vulcain 2 engine is a key contributor to the additional lift capability of the new Ariane 5 ECA version into geostationary transfer orbit.

To hold the extra liquid oxygen needed for the operation of the central core stage, the capacity of the liquid oxygen tank has been increased by 16 tons. This was achieved by re-locating the common tank bulkhead between the liquid oxygen tank and the liquid hydrogen tank of the Ariane 5 Generic EPC stage and reinforcing the structure elements.

The new fuel load capacity of the EPC is:

- Liquid oxygen: 150 tons
- Liquid hydrogen: 25 tons

The EPC stage operates for nearly 540 s. It performs also the roll control function during all its propulsion phase. At the extinction, at an altitude between 160 and 210 km (99 and 130.5 miles) depending on the mission's trajectory, the EPC stage separates from the upper composite and re-enters the atmosphere above the Atlantic Ocean.

Ariane 5 GS. The Ariane 5 GS is the latest evolution of the Ariane-5 Generic launcher, produced after the end of the Ariane 5 G+ series.

This version integrates more recent elements produced in the frame of the Ariane 5 ECA and Ariane 5 ES ATV developments. It incorporates:

- EAP boosters with more propellant in the S1 segments
- a composite VEB with the electrical equipments identical to those produced for the Ariane 5 ECA version
- an EPS stage loaded with 300 kg more propellant

The central core stage (EPC) uses the Vulcain 1B engine which delivers a thrust level of 110 tons.

A limited number of Ariane 5 GS are being produced and these will be launched in the coming two years. The first flight of the Ariane 5 GS version took place successfully on 11 August 2005.

Ariane 5 ES ATV. The ES ATV version of the Ariane 5 has been designed to place ESA's Automated Transfer Vehicle**(ATV) into a 260 km (162 miles) circular low Earth orbit* inclined to 51.6°.

From this orbit the ATV will use its own propulsion system to automatically reach and dock with the International Space Station (ISS) and provide logistic support to the on-orbit infrastructure. The maiden flight of the Ariane 5 ES ATV is planned for 2007 with the first ATV (Jules Verne) mission to the ISS.

The Ariane 5 ES ATV is derived from the Ariane 5 G+ and Ariane 5 ECA versions. It is composed of the same lower composite as Ariane 5 ECA and will use the EAP boosters identical to those of the Ariane 5 ECA version and the same cryogenic main stage equipped with the Vulcain 2 improved engine. The upper composite is composed of a new reinforced vehicle equipment bay (VEB) to withstand flight loads with the ATV and a re-ignitable storable propulsion stage (EPS). EPS re-ignition capability has been incorporated to maximise the launcher's performance into the target orbit and to fulfil the mission requirements.

A first EPS ignition takes place immediately following the separation of the cryogenic main stage (EPC). The EPS engine is then cut-off and the composite VEB-EPS-ATV commences a ballistic phase of about 45 minutes, at the end of which a second EPS ignition occurs for a short duration before the ATV is separated and injected into the target Low Earth Orbit (LEO).

A third and last ignition is then carried out to enable the nearly empty VEB-EPS composite to safely re-enter the Earth's atmosphere. There it safely destructs in the upper layers of the atmosphere.

Vega Program

Although there is a growing tendency for satellites to become larger, there is still a need for a small launcher to place 300 to 2,000 kg (661 to 4,409 lb) satellites, economically, into the polar and low-Earth orbits used for many scientific and Earth observation missions. See Fig. 9.

Europe's answer to these needs is Vega, named after the second brightest star in the northern hemisphere. Vega will make access to space easier, quicker and cheaper.

Costs are being kept to a minimum by using advanced low-cost technologies and by introducing an optimised synergy with existing production facilities used for Ariane launchers.

Vega has been designed as a single body launcher with three solid propulsion stages and an additional liquid propulsion upper module used

---

** The ATV will supply the ISS with pressurized cargo, water, air, nitrogen and oxygen, and attitude control propellant. It will also be used to remove waste from the station and to re-boost the ISS to a higher altitude to compensate for atmospheric drag.

* Low Earth orbit (260 × 260 km, 51.6° inclination)

Fig. 9. Artist's view of Vega, Europe's new small launcher. (*Image courtesy of ESA-J. Huart*).

for attitude and orbit control, and satellite release. Unlike most small launchers, Vega will be able to place multiple payloads into orbit.

Development of the Vega launcher started in 1998. The first launch is planned for 2007 from Europe's Spaceport in French Guiana where the Ariane-1 launch facilities are being adapted for its use.

Soyuz Program

In the second half of 2008, a Soyuz launcher will take off from Europe's Spaceport in French Guiana. See Fig. 10. This will be an historic event as it will be the first time that a Soyuz launcher lifts off from a spaceport other than Baikonur or Plesetsk. It will also be a milestone in the strategic cooperation that exists between Europe and Russia in the launcher's sector. See also **Launch Vehicles: Russian**.

Fig. 10.   Artist's impression of a Soyuz liftoff in French Guiana. (*ESA*).

The decision to develop the launch infrastructure to enable Soyuz to be launched from French Guiana is of mutual interest to both Europe and Russia and benefits from funding from the European Community.

Soyuz is a medium-class launcher. Its performance will complement perfectly that of the ESA launchers Ariane and Vega, and enhance the competitiveness and flexibility of the exploitation of Ariane launchers on the commercial market.

The Soyuz launch vehicle that will be used at Europe's Spaceport is the Soyuz-2 version called Soyuz-ST. This includes the Fregat upper stage and the ST fairing. Soyuz-2 is the last version of the renowned family of Russian launchers that began the space race more than 40 years ago by launching Sputnik, the first satellite to be placed into orbit, and then sending the first man into space — Yuri Gagarin.

Soyuz-2 will have improved performance and be able to place up to 3 tons into geostationary transfer orbit, compared to the 1.7 tons that can be launched from Baikonur in Kazakhstan.

Soyuz Launch Site. Geological and topographic surveys began at the site selected for Soyuz at Europe's Spaceport in 2003. The site, called ELS, is located 13 km (8 miles) northwest of the Ariane launch site and will consist of three main zones: the launch platform where the launcher will be erected for liftoff; the preparation area (MIK) where the three stages will be horizontally assembled and controlled; and the launch control centre.

Earthworks for the Soyuz infrastructure have now been completed and building will soon commence. The MIK is connected to the launch platform by a 700 m (2,297 ft) railway which will be used to transport the launcher in a horizontal position. The launch control centre is located 1 km (1,094 yards) from the launch pad.

Human Spaceflight. The Soyuz rocket is the workhorse of the Russian human spaceflight missions and has been used for that purpose longer than any other spacecraft. In the 1950s it began transporting cosmonauts into space and then to the Soviet stations Salyout and Mir. Along with the

Space Shuttle it is the only other spacecraft available for the transport of crews to and from the International Space Station.

To ensure that Soyuz will still be able to carry out these missions from Europe's Spaceport, the launch infrastructure has been designed so that it can be smoothly adapted for human spaceflight, should this be decided upon.

See also **Space Shuttle**; and **Space Stations**.

Next Generation Launchers

The European Space Agency (ESA) is justifiably proud of its past achievements such as the very successful Ariane-4 launcher, but knows that if Europe is to continue to be at the forefront of new developments in space then it must look towards the future.

A programme dedicated to the preparation of this future, the Future Launchers Preparatory Programme (FLPP), began in February 2004 and aims to have a Next Generation Launcher (NGL) operational around 2020. This programme is divided into periods: the first covering 2004 to 2006 and the second from 2006 to 2009.

The FLPP will conduct system studies and technology activities, including ground and in-flight tests, to foster new technology capabilities within Europe to enhance the reliability and competitiveness of European launchers.

A cornerstone of Europe's strategy for launchers is to make optimum use of available resources. The FLPP will harmonise European launcher technological activities and encourage the progressive restructuring of the European launcher industrial sector.

Developing an NGL will challenge the best minds in European universities and industry. Research and technology will be needed in many areas including:

- launcher system studies to define possible architecture for an NGL
- in-flight experimentation
- technologies including rocket propulsion, materials and structures, aerothermodynamics, launcher-health management systems and avionics

See also: http://www.esa.int/SPECIALS/Launchers_Home/SEMNCI1 PGQD_0.html and www.esa.int/SPECIALS/Launchers_Access_to_Space/index.html

ROGER VIGNELLES, Corbeil-Essonnes, France

LAUNCH VEHICLES: RUSSIAN FEDERATION. The history of cosmonautics and rocket building in Russia may be considered to have begun at the start of the twentieth century. The founder of these disciplines was the great Russian scientist K.E. Tsiolkovsky; major theoretical contributions were made by F.A. Tsander and Yu. V. Kondratyuk. Even in the prewar years, experimental work was being conducted to develop rocket technology for a number of purposes. In 1921, the Gas Dynamics Laboratory (GDL) was founded in Moscow, and in the early 1930s, the Group for the Study of Reaction Propulsion (GIRD) was formed. In 1933, the Reaction Scientific Research Institute (RNII) was created from the GDL, and Moscow GIRD, and in the late 1930s, Design Bureau (KB-7) was founded. All of this resulted in the development of the world's best volley fire (Katyusha) and aircraft missile systems. At the same time, experimental, jet aircraft and liquid-fueled guided missiles were being designed and developed. The production facilities and supporting fundamental and applied science for these areas developed rapidly, serving as the basis for the start of the powerful space and rocket industry in the 1950s. See also **Tsiolkovsky, Konstantin E (1857–1935)**.

S.P. Korolev, the director of Special Design Bureau No. 1 (OKB-1) is considered to be the founder of practical cosmonautics and rocket building. He was the Chief Designer of the first Russian rockets, including the first intercontinental ballistic missile, R-7, which was the base rocket for the development of the earliest space launch vehicles (LV). The country's success in developing rocket technology, in discoveries of the space era and the conquest of space are also associated with the names of the following outstanding designers: V.P. Glushko, N.A. Pilyugin, M.S. Ryazanskiy, V.N. Chelomey, V.I. Kuznetsov, V.P. Barmin, M.K. Yangel, V.F. Utkin, G.N. Babakin, V.S. Budnik, A.M. Isayev, S.A. Kosberg, V.P. Makeyev, M.F. Reshetnev, V.P. Mishin, V.N. Solovyev, A.D. Nadiradze, and V.M. Kovtunenko; the scientists: M.V. Keldysh and B.N. Petrov; the military specialists: V.I. Voznyyk, A.G. Karas, A.I. Sokolov, and A.G. Mrykin;

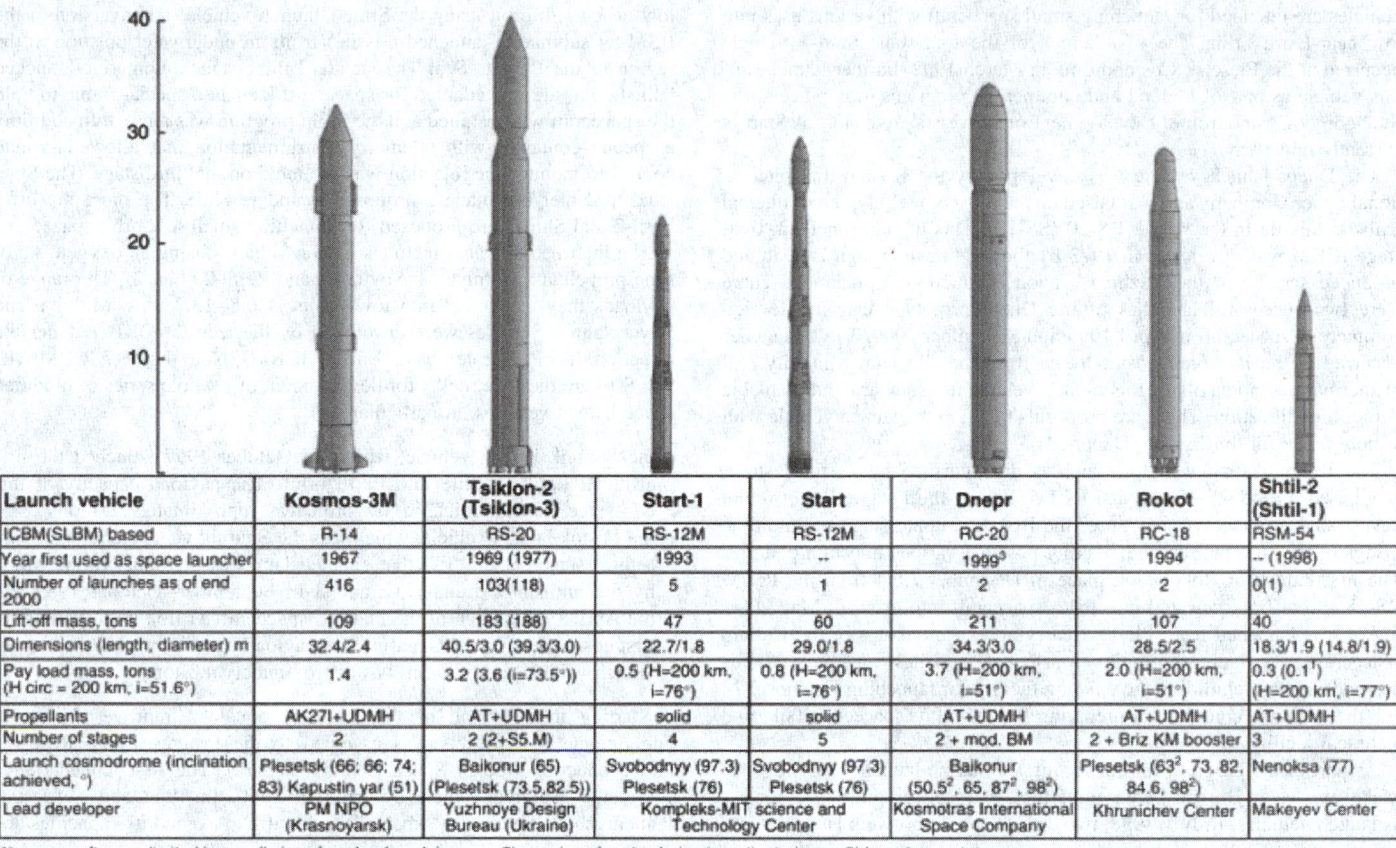

Launch vehicle	Kosmos-3M	Tsiklon-2 (Tsiklon-3)	Start-1	Start	Dnepr	Rokot	Shtil-2 (Shtil-1)
ICBM(SLBM) based	R-14	RS-20	RS-12M	RS-12M	RC-20	RC-18	RSM-54
Year first used as space launcher	1967	1969 (1977)	1993	--	1999³	1994	-- (1998)
Number of launches as of end 2000	416	103(118)	5	1	2	2	0(1)
Lift-off mass, tons	109	183 (188)	47	60	211	107	40
Dimensions (length, diameter) m	32.4/2.4	40.5/3.0 (39.3/3.0)	22.7/1.8	29.0/1.8	34.3/3.0	28.5/2.5	18.3/1.9 (14.8/1.9)
Pay load mass, tons (H circ = 200 km, i=51.6°)	1.4	3.2 (3.6 (i=73.5°))	0.5 (H=200 km, i=76°)	0.8 (H=200 km, i=76°)	3.7 (H=200 km, i=51°)	2.0 (H=200 km, i=51°)	0.3 (0.1¹) (H=200 km, i=77°)
Propellants	AK27I+UDMH	AT+UDMH	solid	solid	AT+UDMH	AT+UDMH	AT+UDMH
Number of stages	2	2 (2+S5.M)	4	5	2 + mod. BM	2 + Briz KM booster	3
Launch cosmodrome (inclination achieved, °)	Plesetsk (66; 66; 74; 83) Kapustin yar (51)	Baikonur (65) (Plesetsk (73.5,82.5))	Svobodnyy (97.3) Plesetsk (76)	Svobodnyy (97.3) Plesetsk (76)	Baikonur (50.5², 65, 87², 98²)	Plesetsk (63², 73, 82, 84.6, 98²)	Nenoksa (77)
Lead developer	PM NPO (Krasnoyarsk)	Yuzhnoye Design Bureau (Ukraine)	Kompleks-MIT science and Technology Center		Kosmotras International Space Company	Khrunichev Center	Makeyev Center

1) spacecraft mass limited by small size of payload module;   2) opening of routes being investigated   3) launcher prototype

Fig. 1.   Light-class launch vehicles.

directors and managers: K.N. Rudnev, D.F. Ustinov, S.A. Afanasyev, L.V. Smirnov, G.A. Tyulina, Yu.A. Mozzhorin and many others.

In addition to OKB-1 (which is currently Korolev Rocket and Space Corporation Energia (Korolev, Moscow region) such well-known enterprises as Khrunichev State Research and Production Space Center (Moscow), Progress State Research and Production Rocket Space Center (Samara), Makeyev Design Bureau (State Rocket Center) (Miass), Polyot Production Association (Omsk), Research and Production Association, Yangel State Design Bureau Yuzhnoe (Dnepropetrovsk, Ukraine) and many others were engaged in activity related to rocket and space technology.

At present time, the development of space rocket building in Russia is associated with directors and chief designers such as Yu. N. Koptev, Yu. P. Semenov, D.I. Kozlov, A.A. Medvedev, S.D. Kulikov, G.A. Yefremov, Yu. S. Solomonov, V.G. Degtyar and others.

The current Russian system of launch vehicles evolved during the process of formulating and addressing a whole set of tasks in the interests of various users, employing the experience accrued through design and use of previously developed military rockets and space complexes. The available capacities of existing launch vehicles, in general, satisfies the requirements of the spacecraft that are in current use and planned for the near future for injection into circular and elliptical orbits of various altitudes and inclinations and interplanetary flight trajectories. Spacecraft are generally launched from the Baikonur and Plesetsk cosmodromes, whose infrastructures were developed based on progressive evolution of space activity. A few light launch vehicles are launched from the Svobodnyy cosmodrome and Kapustin Yar range. Current space rocket systems have a high level of reliability and cost efficiency.

The current launch vehicles includes the expendable launch vehicles Kosmos-3 M, Tsiklon-2,3, Molniya-M, Soyuz-U, Zenit-2, Proton-K and a number of launch vehicles developed as part of the military rocket conversion program. The launch vehicles used can be divided into light, middle, and heavy classes depending on the payloads launched.

The light class includes the Kosmos-3 M, Tsiklon-2, 3 and also the converted Dnepr, Rokot, Strela, Start, Start-1, and Shtil launch vehicles

(Fig. 1); the mass of payloads inserted by these rockets into low orbits is 0.3–3.7 tons. All launch vehicles, aside from the solid-fuel rockets of the Start type, use AT (nitrogen tetroxide), AK-27I (nitric acid + 27% N_2O_4 + iodine) + UDMH (unsymmetrical dimethyl hydrazine) propellants.

The prototype launch vehicle Kosmos-3 M (RN 65 S3) was developed by the Yuzhnoye Design Bureau (Dnepropetrovsk, Ukraine) on the basis of the intermediate ballistic missile R-14 (SS-5). Kosmos-3 M is the result of a number of updates of the prototype vehicle and has been used since 1967. Since 1970, it has been serially produced by the Polyot Production Association (Omsk).

The two-stage Kosmos-3 M launch vehicle can inject spacecraft into orbits up to 1700 km high by twice firing the second-stage liquid fuel sustainer engine and using a low thrust system. This launch vehicle was used to launch light spacecraft as part of Federal programs, and starting in the late 1990s, was also used to launch commercial spacecraft. At present, Kosmos-3 M is no longer produced.

The two- and three-stage Tsiklon-2 and Tsiklon-3 launch vehicles were developed by the Yuzhnoye Design Bureau in 1969 and 1977, respectively. Both modifications were based on the R-36 (SS-9) military missile. The first two stages of these vehicles are virtually standardized. As its third stage, Tsiklon-3 uses the S5.M unit and repeatedly fires the sustainer engine, making it possible to expand the range of injection orbits. Production of the main components of the Tsiklon was handled by the Yuzhmashzavod Production Association (Dnepropetrovsk, Ukraine). At the present time the production of these rockets has been discontinued.

The Tsiklon System is highly automated and does not personnel at the launch facility between the time the rocket is delivered to it and the moment of launch.

In the early 1990s, development work started on launch vehicles as part of the program to convert intercontinental ballistic missiles and submarine-launched ballistic missiles that were to be eliminated in accordance with the arms reduction treaty. Solid-fuel launch vehicles of the Start class were developed strategic by the Kompleks-MIT Science and Technology Center based on the RS-12 M (SS-25) intercontinental ballistic missile. These vehicles are intended for launching small spacecraft with various uses into low near-Earth orbit. The first launch of the four-stage Start-1 vehicle occurred at the Plesetsk Cosmodrome in March 1992. Further launches of this vehicle as part of Federal and commercial programs took place at the Svobodnyy Cosmodrome. Development of a five-stage version of Start is currently underway.

The Dnepr launch vehicle was developed by the Kosmostras International Space Company and was based on the most powerful intercontinental ballistic missile in the world, RS-20 (SS-18). This liquid-propellant, two-stage ICBM was developed in 1973 by the Yuzhnoye Design Bureau and produced serially at the Yuzhmashzavod Production Association. There were two successful launches of the Dnepr prototype carrying foreign commercial spacecraft in April 1999 and September 2000. These launches occurred at the Baikonur cosmodrome from the silo launch facility. All of the main components of this launch vehicle are standard and available without modification. There are proposals to equip the launch vehicle with a more powerful third stage (Dnepr-M).

The Rokot launch vehicle, which is distinguished by its third stage, was based on the RS-18 (SS-19) ICBM. For its third stage, Rokot (chief developer Khrunichev Center) uses the Briz-KM upper stage and multiple firings of the sustainer engine, making possible various injection patterns. The first launch of Rokot took place in December 1994 from the ICBM RS-18 silo at Baikonur and injected a spacecraft into orbit. In May 2000, Rokot was launched from Plesetsk carrying two mock-ups of the Iridium spacecraft. The launch complex for Rokot launches was built at this cosmodrome by rebuilding the existing facility for launching Kosmos-3 M and the technical complex for preparing Rokot and its spacecraft Tsiklon-3 technical facility.

As part of the program for converting submarine-launched missiles, the Makeyev Design Bureau Center undertook to adapt certain rockets for use as launch vehicles. In July 1998, two Tubsat spacecraft were launched into low near-Earth orbit using the Shtil-1 launch vehicle (a conversion of the RSM-54 submarine-launched missile) from an underwater position in the region of the Barents Sea. The standard three-stage, submarine-launched ballistic missile was adapted for spacecraft launch: a special frame to hold the spacecraft was installed and the flight program was altered. In addition, a special container with telemetry instrumentation that allowed ground control to monitor the injection was mounted on the third stage. The State Rocket Center has made a proposal to update Shtil. The more powerful Shtil-2 and Shtil-3 are proposed for launching small low-orbit spacecraft.

The intermediate class includes launch vehicles using an oxygen–kerosene propellant: Molniya-M, Soyuz-U, and Zenit-2 (Fig. 2). The range of payloads they can inject into low orbits is 6.8–13.7 tons. Molniya and Soyuz launch vehicles were developed by the team at OKB-1 under the direction of S.P. Korolev based on R-7 ICBMs. Note that R-7 ICBMs (R-7A, S-6) are the base rocket for development of a whole series of modified space launch vehicles, in particular,

- the Sputnik launch vehicle, which, in October 1957 launched the first artificial Earth satellite into Earth orbit. The payload capacity of this two-stage launch vehicle in low orbit was approximately 2.0 tons.
- the Vostok launch vehicle, which was the Sputnik vehicle equipped with a third stage (block "E") that enabled launches of flights to the Moon by the unmanned Luna-1, -2 and 3 in September–October 1959 and the April 1961 launch of the manned spacecraft Vostok with the Earth's first cosmonaut, Yu.A. Gagarin. The payload capacity of this three-stage launch vehicle in low orbit was approximately 4.8 tons.

Starting in 1961, all of the work to develop, improve, flight-test, and operate R-7 rockets was assigned to the Progress State Research and Production Rocket Space Center (Samara). The first launch of the four-stage Molniya vehicle occurred in 1960. In 1964 and 1985, this launch vehicle underwent substantial updating to expand its capacities and

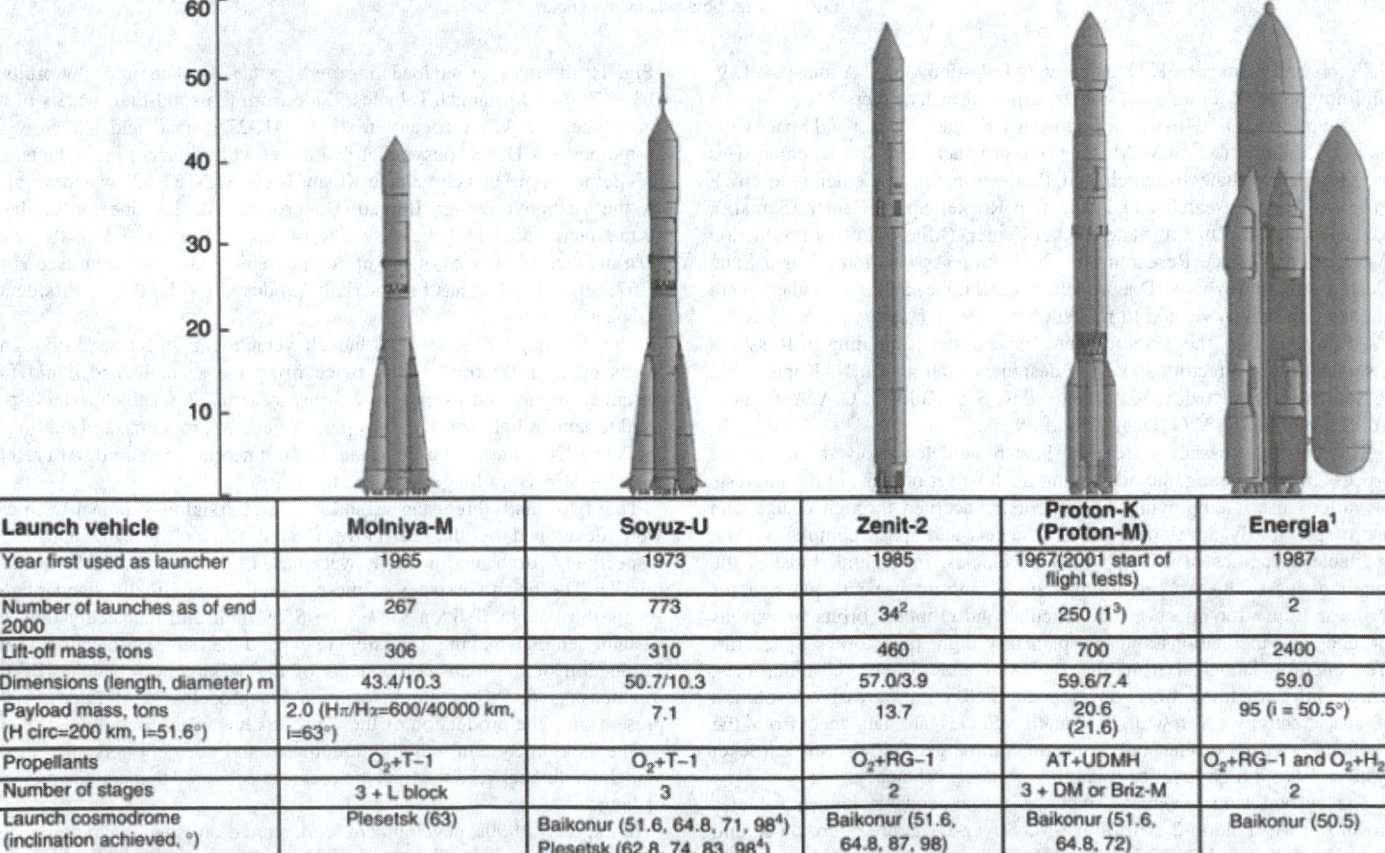

Launch vehicle	Molniya-M	Soyuz-U	Zenit-2	Proton-K (Proton-M)	Energia[1]
Year first used as launcher	1965	1973	1985	1967(2001 start of flight tests)	1987
Number of launches as of end 2000	267	773	34[2]	250 (1[3])	2
Lift-off mass, tons	306	310	460	700	2400
Dimensions (length, diameter) m	43.4/10.3	50.7/10.3	57.0/3.9	59.6/7.4	59.0
Payload mass, tons (H circ=200 km, i=51.6°)	2.0 (Hπ/Hα=600/40000 km, i=63°)	7.1	13.7	20.6 (21.6)	95 (i = 50.5°)
Propellants	O₂+T-1	O₂+T-1	O₂+RG-1	AT+UDMH	O₂+RG-1 and O₂+H₂
Number of stages	3 + L block	3	2	3 + DM or Briz-M	2
Launch cosmodrome (inclination achieved, °)	Plesetsk (63)	Baikonur (51.6, 64.8, 71, 98[4]) Plesetsk (62.8, 74, 83, 98[4])	Baikonur (51.6, 64.8, 87, 98)	Baikonur (51.6, 64.8, 72)	Baikonur (50.5)
Lead developer	Progress KB Center	Progress KB Center	Yuzhnoe KB (Ukraine)	Khrunichev Center	Energia

1) production of Energia has been terminated
3) first and currently only launch of Proton-M occured in April 2002
2) number of launches as of May 2001
4) opening of routes being investigated

Fig. 2.   Intermediate- and heavy-class launch vehicles.

maintanance safety. At present, the Molniya-M is being used. It has an "L" upper stage with a liquid propellant engine fired in weightlessness. This module contains the control system, which controls the flight of both modules L and I. The Molniya-M is designed to launch spacecraft for interplanetary flights and into highly elliptical orbits. It has been used for launches of spacecraft of the Luna, Venera, Mars, Molniya, and Prognoz type spacecraft.

The three-stage Soyuz-U is an updated version of the Soyuz launch vehicle, used from 1966–1973. It can launch various kinds of spacecraft, including manned spacecraft from the Baikonur Cosmodrome and unmanned craft from Plesetsk. Between 1982 and 1995, a modification of Soyuz-U, called Soyuz-U2, was used for the manned space programs. Unlike Soyuz-U, Soyuz-U2 used the "sintin" propellant instead of kerosene T-1 in its core module. At present, this launch vehicle is no longer manufactured.

At present, the Progress State Research and Production Rocket Space Center (Samara), is conducting work on a phased modernization of Soyuz launch vehicles. The tasks of modernizing Soyuz involve the phased development of a standardized three-stage launch vehicle, Soyuz-2 (Fig. 3).

During the initial phase, to allow launch of Soyuz-TMA spacecraft and the cargo spacecraft Progress-MA, work is being conducted to provide a minor update of the Soyuz launch vehicle. The essence of this update is the use of updated stage I and II engines on the basic Soyuz vehicle, and minimal changes of the existing control system, the tank drainage and synchronization, and indicating speed regulation systems. This launch vehicle, designated Soyuz-FG, using a nose cone 3.0 m in diameter, can inject a payload weighing up to 7.4 tons into low Earth orbit. Flight tests of Soyuz-FG began in 2001.

The next phase, along with the use of launch vehicle engines on stages I and II, will involve

- use of a control system based on a highly efficient digital computer that has modern components and advanced software;
- use of a new digital radiotelemetry system.

The new control system will provide very precise injection into orbit and also will increase the diameter the nose fairing from 3.3 to 4.1 m. The Soyuz-2 launch vehicle will have a payload capacity for injection into low Earth orbits of up to 6.9 tons.

During the last phase, the Soyuz-2 launch vehicle will be equipped with a more powerful stage III and a new liquid propellant engine developed by the Design Bureau for Chemical Automation, which will increase the payload mass for low near-Earth orbit to 8.1 tons.

To deploy Globalstar telecommunications satellites, the Starsem Company (a Russian-French joint venture), used the Soyuz-Ikar system. The Ikar upper stage was developed at the Progress Center using as a base the equipment module of the Kometa spacecraft. Since 1999, there have been six successful launches of Soyuz-Ikar.

To expand the altitude range for spacecraft orbital injection (right up to geostationary orbit), the Lavochkin Research and Production Association developed the upper stage Fregat, based on the propulsion unit of the Fobos spacecraft, for use with Soyuz launch vehicles (Soyuz-Fregat launch vehicle). At present, the certification of the system has been completed, so that it is ready for use. In 2000–2001, four successful launches of this vehicle were completed as part of the Cluster program. The Soyuz-Fregat can insert a payload with a mass of 3640 kg into highly elliptical orbits (for $H_\pi/H_\alpha = 236/18,000$, $i = 64.9°$).

The Starsem Company has commissioned development of a commercial variant of this launch vehicle, under the name Soyuz-ST. The major difference between this vehicle and Soyuz-2 is a larger nose cone based on the nose fairing from the Ariane-4 launch vehicle (diameter 4.1-m) to increase the space available for the payload. This launch vehicle can be launched from the Baikonur and Kourou cosmodromes.

The two-stage launch vehicle Zenit-2, developed by the Yuzhnoye Design Bureau and manufactured in Ukraine, is intended for injecting midclass unpiloted spacecraft into low and intermediate altitude circular orbits (including sun-synchronous orbits). This system requires almost no manual labor for service, when it is being prepared for launch and when it is removed from the launch facility if the launch has been aborted. When

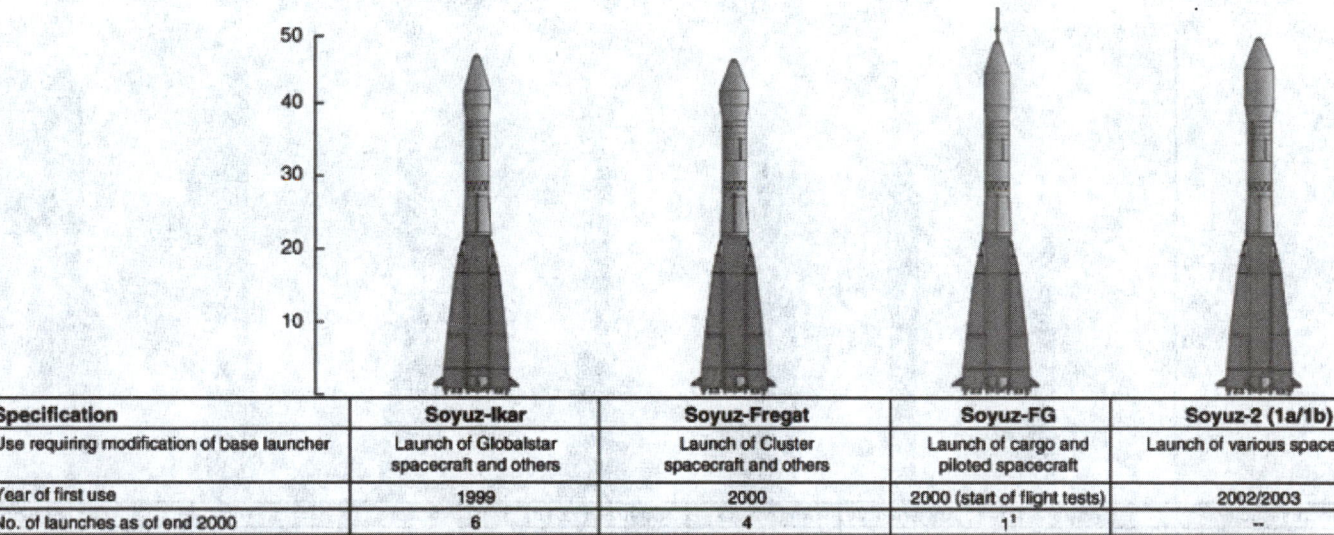

Specification	Soyuz-Ikar	Soyuz-Fregat	Soyuz-FG	Soyuz-2 (1a/1b)
Use requiring modification of base launcher	Launch of Globalstar spacecraft and others	Launch of Cluster spacecraft and others	Launch of cargo and piloted spacecraft	Launch of various spacecraft
Year of first use	1999	2000	2000 (start of flight tests)	2002/2003
No. of launches as of end 2000	6	4	1[1]	–
Lift-off mass, tons	307.1	307.1	307.4	307.4/308.6
Dimensions (length, diameter) m	45.0/10.3	44.4/10.3	50.0/10.3	50.0/10.3
Payload mass, tons (Hcir=200 km, i=51.6°) (Hcir=200 km, i=62.8°)	2100 (H_π/H_α=910/952 km, i=52°)	3640 (H_π/H_α=236/18000 km, i=64.9°)	7460 6900	6900/8080 6600/7450
Nose cone, diameter/length m	3.3/8.34	3.7/7.7	3.0/10.14	3.7/10.7
Propellants	O_2+T–1 AT+UDMH	O_2+T–1 AT+UDMH	O_2+T–1	O_2+RG–1 AT+UDMH
Number of stages	3 + Ikar	3 + Fregat	3	3 + Fregat
Launch cosmodrome (inclination achieved, °)	Baikonur (51.6, 64.8, 71)	Baikonur (51.6, 64.8, 71, 98[2]) Plesetsk (62.8, 74, 83, 98[2])	Baikonur (51.6, 64.8, 71)	Baikonur (51.6, 64.8, 71, 98[2]) Plesetsk (62.8, 74, 83, 98[2])
Lead designer		Progress KB Center		

1) number of launches as of july 2001   2) opening of routes being investigated

Fig. 3. Specification of Soyuz type launch vehicle modifications.

launched from Baikonur, Zenit-2 injects payloads of up to 13.7 tons into low orbits.

The Zenit and the DM upper stage (developed by RSC Energia), advanced based on innovative technology, formed the basis for the international commercial Sea Launch project. When launched from the equatorial zone, Zenit-3SL with the DM-SL upper stage can inject a payload weighing 2.9 tons into geostationary orbit. In 1999, the first demonstration launch of this rocket took place from the Odyssey floating platform in the vicinity of Christmas Island. After that, Sea Launch began commercial operations. In mid-2001, there were seven launches of Zenit-3 SL; one of which was unsuccessful.

Work has started to develop the next generation of intermediate-class launch vehicles. RSC Energia, is developing the Avrora midclass launch vehicle. The Aurora is planned for launches of spacecraft for Federal programs and commercial customers. For commercial launches, the Asia Pacific Space Center (APSC) has commissioned a cosmodrome on Christmas Island (Australia). According to the draft project published by RSC Energia in 2001, the launcher will deliver payloads with a mass of 12.0 tons to low circular orbit and 2.1 tons to geostationary orbit, when the launch occurs from the cosmodrome on Christmas Island, and payloads of 11.0 and 0.9 tons, respectively, for launches from Baikonur.

The heavy class of launch vehicles includes Proton-K, developed by the Khrunichev Center using AT + UDMH propellants (Fig. 2). Two versions of this launcher are used: a three-stage for delivering spacecraft into low orbits (Mir modules and International Space Station modules, and heavy spacecraft) and the four-stage version with the DM transfer stage to inject spacecraft into high-energy orbits (including geostationary transfer orbits, geostationary orbits, and interplanetary trajectories).

The Proton-K has earned a reputation as the most reliable (flight reliability = 0.97), well-tested (230 launches have been performed), and cost efficient launch vehicle of the heavy class. Progressive design changes incorporated as it was built have allowed it to keep pace with growing demands during almost three decades. At present, its payload capacity for orbits 200 km high and inclination of 51.6° is 20.6 tons, and its payload capacity in geostationary orbit has been increased to 2.6 tons.

The Khrunichev Center is modernizing the Proton-K. The development of the Proton-M launcher is being conducted to replace the outmoded analog control system by a modern system with a onboard digital computer and a flexible injection program; to reduce environmental pollution; to provide a smaller impact zone for the separated elements; and also to improve energy performance (in low orbit, up to 21.6 tons and in geostationary orbit using a DM or Briz-M upper stage up to 3.0 tons). In April 2001, the first launch of Proton-M took place, starting its flight-test phase. In the future, plans call for using an oxygen–hydrogen upper stage, which will enable an increase in payload mass to geostationary orbit to 4.0 tons.

In the 1970s, Special Design Bureau-1 (in 1966, renamed the Central Design Bureau for Experimental Machine Building) developed the super-heavy N-1 launch vehicle for the manned Lunar program. The launch mass of the rocket was approximately 2800 tons, and the length of the rocket with payload was approximately 100 m. Propellants were supercooled oxygen and kerosene. The N-1 launcher was supposed to inject a payload of 95 tons into a parking orbit of 220 km. The first test launch of the rocket occurred in 1969. There was a total of four launches. Unfortunately, all launches failed and in 1974, it was decided to discontinue work. Note that these N-1 developments were subsequently used to develop the Energia launcher. The Energia, which was flight-tested in the late 1980s, used oxygen and kerosene and oxygen and hydrogen as propellants and had a payload capacity for low orbit of 95.0 tons, belongs to the superheavy class (Fig. 2). After two launches (both were successful) in May 1987 and October 1988 as part of the program to create the reusable Energia-Buran space shuttle system, the production of Energia was terminated.

In the near future, the development of the Russian system of launch vehicles (Fig. 4) stipulates the use of Soyuz-2 and Proton-M launchers to implement Federal and commercial programs. The Zenit launcher may also be used to perform particular launches in the Federal program.

Creation of the new generation of launch vehicles is linked, first and foremost, with development of the Angara launch vehicle family (lead development at the Khrunichev Center). The main goal of the project is to develop the Angara-A5 heavy-class launcher. This launcher is being

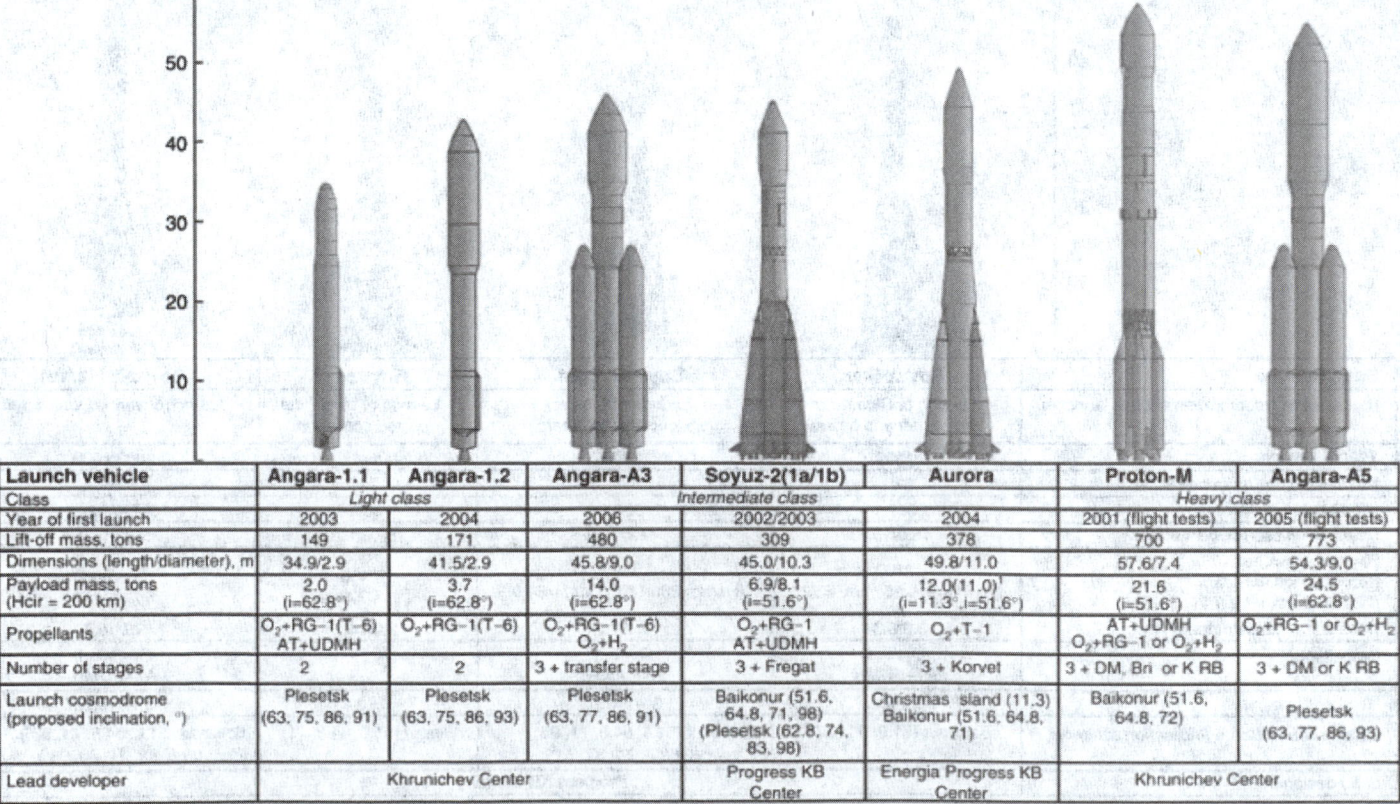

Launch vehicle	Angara-1.1	Angara-1.2	Angara-A3	Soyuz-2(1a/1b)	Aurora	Proton-M	Angara-A5
Class	Light class		Intermediate class			Heavy class	
Year of first launch	2003	2004	2006	2002/2003	2004	2001 (flight tests)	2005 (flight tests)
Lift-off mass, tons	149	171	480	309	378	700	773
Dimensions (length/diameter), m	34.9/2.9	41.5/2.9	45.8/9.0	45.0/10.3	49.8/11.0	57.6/7.4	54.3/9.0
Payload mass, tons (Hcir = 200 km)	2.0 (i=62.8°)	3.7 (i=62.8°)	14.0 (i=62.8°)	6.9/8.1 (i=51.6°)	12.0(11.0)[1] (i=11.3°,i=51.6°)	21.6 (i=51.6°)	24.5 (i=62.8°)
Propellants	O₂+RG–1(T–6) AT+UDMH	O₂+RG–1(T–6)	O₂+RG–1(T–6) O₂+H₂	O₂+RG–1 AT+UDMH	O₂+T–1	AT+UDMH O₂+RG–1 or O₂+H₂	O₂+RG–1 or O₂+H₂
Number of stages	2	2	3 + transfer stage	3 + Fregat	3 + Korvet	3 + DM, Bri or K RB	3 + DM or K RB
Launch cosmodrome (proposed inclination, °)	Plesetsk (63, 75, 86, 91)	Plesetsk (63, 75, 86, 93)	Plesetsk (63, 77, 86, 91)	Baikonur (51.6, 64.8, 71, 98) (Plesetsk (62.8, 74, 83, 98)	Christmas sland (11.3) Baikonur (51.6, 64.8, 71)	Baikonur (51.6, 64.8, 72)	Plesetsk (63, 77, 86, 93)
Lead developer	Khrunichev Center			Progress KB Center	Energia Progress KB Center	Khrunichev Center	

1) for launch from Australia s Christmas sland, in parentheses Baikonur launch

Fig. 4. Future launch vehicles.

developed to provide Russia with guaranteed access to space from the Plesetsk cosmodrome (including launches into geostationary orbit) for implementing the Russian National Space Policy and future advancement of Russian launch vehicles in the global market. A modular technique for building this family of launchers of different classes has been proposed (Fig. 4). As the core module, designers are considering the new universal rocket module (URM-1), a development with an oxygen–kerosene engine, developed by NPO Energomash. This module will serve as the basis for constructing launchers of various classes, starting with light-class launchers Angara-1.1 and Angara 1.2, which will use one URM as the first stage and different second stages. Successful development of URM as a component of relatively inexpensive light-class launchers, and also the possibility of obtaining additional funds from potential commercial clients for their launches, shall become the basis for developing more powerful launchers in the Angara family. The heavy class Angara-A5 will have a cluster of five URMs and the midclass Angara A3 will use a cluster of three URMs. Launches of Angara family launchers are planned for the Plesetsk cosmodrome with maximal use of the existing launch and technical facilities at that site.

To expand the spectrum of orbits with regard to altitude and inclination, upper stage modules are used for Russian intermediate- and heavy-class launch vehicles. Figs. 5, and 6 show the designations and characteristics of current and modernized (future), upper stage modules for intermediate- and heavy-class launchers. Depending propellant used, these, modules can be divided into three groups:

- Upper stage modules using storable toxic propellants (Ikar, Fregat, Briz-M);
- Upper stage modules using oxygen–kerosene propellants (block DM type)
- Upper stage modules using oxygen and hydrogen propellants.

Further improvement of the lift capacities of heavy- and intermediate-class launchers is associated with the development and use of transport modules with low thrust engines, particularly, electric rocket engines (ERE). The current technical level of development and cumulative flight experience in the application of ERE and solar cells has created the preconditions for developing sustainer engine systems for energy-intensive transport operations in near-Earth space. The use of low thrust engines with substantially better efficiency than traditional liquid propellant engines will require implementing new patterns of injection into high-energy orbits, including geostationary orbits. The time required for orbital injection would be increased from several hours to several months, which demands the development of a spacecraft adapted to such prolonged injection.

The main line of development of space launch, system in the twenty-first century, first and foremost, is based on the gradual shift to reusable space transportation systems, including space-rocket and air-space systems. As research has shown, the shift to reusable launch vehicles of a new generation can:

1. decrease the cost of launching a unit of payload mass compared with the traditional expendable launch vehicles:

   - by a factor of 1.5–2 if reusable space transport systems are built with the optimum percentage of reusable components using a level of technology that will be available in the near future;
   - by a factor of 5 to 7 if a completely reusable vehicle were built using high level advanced technology.

2. increase the mission performance probability and crew safety by a factor of at least 5 compared to the level associated with current launchers.

3. virtual elimination of environmental pollution by maximum decrease, up to complete elimination, of the hazard zones along the launch trajectories, as a result of use of nontoxic propellants, and also injection of spacecraft into orbit without accompanying debris of spent launcher parts.

At present, more than 15 Russian scientific, research, and design organizations are working on a program to develop the scientific and technical requirements for a reusable transport system. Results of a comparative analysis of various vertical and horizontal launch reusable transport vehicle designs allowed to select four reusable space transportation system versions. Two of them are based on technology available in the near future:

- A system based on a two-stage reusable all-azimuth rocket with a vertical launch and multiple return (to the launch site) rocket booster in the first stage and an expendable injection module as the second stage. The payload in this design could be a spacecraft under the nose fairing, if injection is to take place according to the traditional pattern, or an orbital transport spacecraft, consisting of a reusable orbital spacecraft and an expendable cargo module, for performing transport and technical maintenance operations. At present, the Khrunichev Center in cooperation with the NPO Molniya, is developing a reusable module

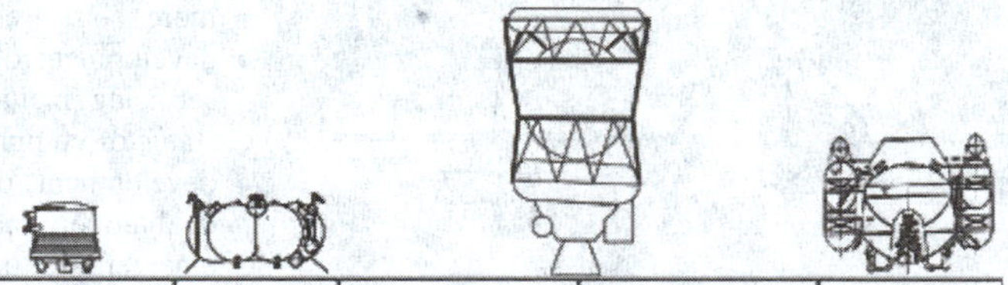

Upper stage					
Launch vehicle					
Year first used	1999	2000 (fl test)	1999	1967	1999 (fl.test)
No of launches, as of end 2000	6	4	7[1]	232	2[2]
Propellants	AT+UDMH	AT+UDMH	O_2+RG–1	O_2+RG–1/ sintin	AT+UDMH
Payload mass in orbit, tons Geost. orbit from Baikonur	2.2[3] **	0.3	2.9 (sea launch)	2.4/2.6	2.1/2.3[4]
Mass of fueled stage, tons	3.3	6.5	18.5	18.2	22.4
Stage dimensions (DxL)m	2.7 × 2.6	3.3 × 2.6	3.7 × 6.3	3.7 × 6.3	4.1 × 3.1
Mass of expendables, tons	0.9	5.4	15.0	15.0	5.2/14.7[5]
Total trust of cruise engine, tons sec	0.3	2.0	8.5	8.5	2.0
Specific impulse, kg x s/kg	324	327	353	352/361	325.5
Developer	Progress KB Center	Lavochkin NPO	Energia	Energia	Khrunichev Center

1) number of launches as of May 2001
2) number of launches counting the first launch of Proton-M in April 2001
3) on launch into mid-near Earth orbit (Hπ/Hα=906/948km,i=52°)
4) numerator-direct injection into geostationary orbit, denominator- 10 hour mode (with intermediate orbital phasing)
5) numerator-central module, denominator-auxiliary propellant tank ejected in flight

Fig. 5.   Upper stages used for intermediate- and heavy-class launch vehicles.

Block	Korvet	DM (11C861-03)	KVRB
Launch vehicle	Aurora	Proton-M Angara-A5	Proton-M Angara-A5
Propellants	O_2+RG–1	O_2+RG–1	O_2+H_2
Payload mass in orbit, t Geostationary from Plesetsk from Bayonur	-- 0.9(2.1)[1]	3.0 (Angara-A5) 3.0 (Proton-M)	4.0 (Angara-A5) 4.0 (Proton-M)
Mass of fueled stage, tons	9.8	22.3	22.0
Stage dimensions (D×L),m	3.4 × 4.5	3.7 × 6.3	4.1 × 8.0
Mass of expendables, tons	8.5	19.0	18.5
Total trust of sustainer engine, tons sec	8.0	8.5	7.4
Specific impulse, kg × s/kg	355	353	470
Developer		Energia	Khrunichev Center

1) in parentheses are data for launch from Christmas Island cosmodrome

Fig. 6.   Modernized and future upper stages for intermediate- and heavy-class launch vehicles.

of the first stage Baykal, which could be used in an all-azimuth rocket launch vehicle in the near future (Fig. 7). The reusable Baykal module has a landing mass of about 18.5 tons, a length of the order of 29 m, and a diameter of 2.9 m. It has the configuration of a high-wing monoplane with a rotating wing attachment above the body of the fuselage. An RD-191 M liquid propellant engine is installed in the Baykal tail section. In the nose, there is an RD-33 jet engine unit for cruise flight on return to the launch area and landing at an airport. A full size technological mock up of the Baykal was displayed in June 2001 at an aerospace exhibit at Le Bourget and in August 2001 at the MAKS-2001 airshow.

- a two-stage reusable aerospace system with horizontal launch, consisting of a hypersonic aircraft booster with a combined propulsion system and a rocket orbital boost stage.

The development of a reusable space transportation system is linked to the solution of a whole series of technical problems; the key ones are

- development of reusable sustainer liquid propellant engines;
- development of reusable returnable booster rockets for the first stage;
- development of materials, designs, technologies, and components for decreasing the mass of the assemblies and systems by 30–60% and more, compared to the current level;
- development of methods for minimizing work entailed in turnaround servicing, including methods of monitoring and diagnosing the postflight status of equipment, including the large cryogenic containers;
- development of technical principles and methods for providing a qualitatively new level of safety and preservation of equipment in emergency situations.

Thus, proposals call for creating a new generation of space transportation systems by 2010–2015, which, because of the use of advanced engineering designs and technologies (reusability, advanced materials, designs, components propulsion units, control systems, ground servicing technologies, etc.) will make it possible (by a factor of 2–4) to decrease annual expenditures significantly on transportation support, expand the consumer potential of launch vehicles, improve their performance characteristics (reliability, safety, readiness for launch, and others) and make Russian transportation systems competitive on the launch vehicle world market.

See also **Space Stations**.

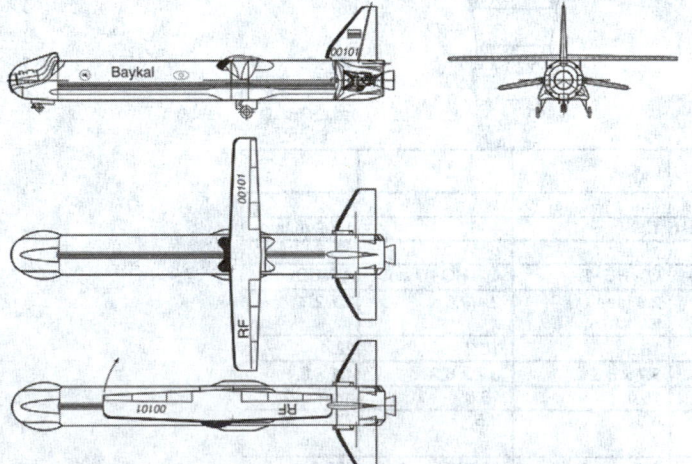

Fig. 7.   First stage of the reusable Baykal module.

- A two-stage reusable aerospace vehicle system with horizontal launch, consisting of an An-225 subsonic carrier aircraft and a reusable orbital aircraft with an expendable external propellant tank.

Two versions were selected based on more advanced future technology:

- a single stage fully reusable space rocket plane with vertical launch and horizontal landing;

Additional Reading

Burleson, D.: *Space Programs Outside the United States: All Exploration and Research Efforts Outside the United States*, McFarland & Company, Incorporated Publishers, Jefferson, NC, 2005.

Godwin, R.: *Launch Vehicles*, Collector's Guide Publishing, Wheaton, IL, 2006

Karpenko, A.V., A.F. Utkin, and A.D. Popov. *Russian Strategic Rocket Systems*. Nevskiy Bastion, St. Petersburg, 1999.

Kiselev, A.I., A.A. Medvedev, and V.A. Menshikov: *Cosmonautics on the Threshold of the Millennium. Conclusions and Prospects*. Mashinostroyeniye-Polet, Moscow, 2001.

Novosti kosmonautiki (News of Cosmonautics), No. 1–12, 2000, No. 1–7 (2001).

Umanskiy, S.P.: *Launch Vehicles and Cosmodromes*, edited by Yu.N. Koptev Director of the Russian Aerospace Agency. Redstart+, Moscow, 2001.

Web Reference

Launch Vehicles Russian: http://www.russianspaceweb.com/rockets_launchers.html

ALEXANDER N. KUZNETSOV, Russian Aviation and Space Agency,
Russian Federation

LAUNCH VEHICLES: U.S. EXPENDABLE. Robert H. Goddard was a Professor of Physics at Clark University in Worcester, Massachusetts, a post he held for 26 years from 1919 until his death in 1945. During these years, Goddard performed a classic series of experiments in which he, almost single-handedly, worked out the basic elements of liquid-fueled rocket technology. Goddard was interested in using rockets for space flight from the very beginning. He felt that for this purpose, liquid fuels, specifically kerosene or gasoline and liquid oxygen would be best. This fuel would be more efficient, that is, it has a higher specific impulse, and it has the advantage that by closing and opening valves, the rocket engine could be stopped and restarted. This latter point was especially important because it gave liquid–fueled rockets a truly decisive advantage over those that operated with solid fuel. In solid-fueled rockets, once the motor is started, the rocket is committed, and the engine cannot be turned off.

In 1919, Goddard published a paper titled "A Method for Reaching Extreme Altitudes" in the *Journal of the Smithsonian Institution* in which he outlined his plans. These came to fruition on 16 March 1926, when a liquid-fueled rocket built by Goddard performed the first successful liftoff and flight and reached an altitude of about 80 feet. This experiment was carried out in an open field on a farm near Auburn, Massachusetts. Goddard realized that this was not the best place for these experiments. He took advantage of an offer from Daniel Guggenheim, the heir to a huge mining fortune and an enthusiastic supporter of American aviation, to support the continuation of his experiments and to set him up on a large ranch near Roswell, New Mexico. It was here that Goddard built and flew the first gyroscopically stabilized liquid-fueled rocket on 31 May 1935; it reached an altitude of more that 7000 feet.

Goddard's remarkable success was unfortunately not pursued. Because of the depression, Guggenheim could no longer support Goddard's work, and eventually, the work at the Roswell ranch was abandoned. The U.S. military showed no interest in Goddard's work. When the Second World War started, the military did initiate a vigorous effort to develop solid-fueled rockets for use as airborne and ground-based weapons, in which Goddard participated. Goddard's untimely death in 1945 precluded his participation in the rapid development of large liquid-fueled rockets in the United States following World War II.

While all of this was happening in the United States, a group of Germans was also interested in developing liquid-fueled rockets for the same reasons that motivated Goddard. The intellectual leader of this group was Professor Hermann Oberth, a Romanian-German, who wrote the first book on planetary exploration that was published in 1923, "Die Rakete zu den Planetenraumen". He advocated using rocket propulsion. In this book, he worked out the necessary mathematics to accomplish this objective and also elaborated on the infrastructure that would have to be built. (Both Goddard and Oberth were unaware of the work of Konstantin E. Tsiolkovski, a Russian mathematics professor who had worked out the rocket equation and had speculated on the use of rockets for space travel 20 years before Goddard and Oberth wrote about their work.) Oberth's book attracted the attention of a small group of German scientists and engineers who had a great interest in space travel and rockets. Along with Oberth, these included Rudolf Nebel, Klaus Riedel, Willy Ley, and Max Valier. On 5 July 1927, these people and a few others founded the "Verein fûr Raumschiffarht," the "Society for Space Travel." By 1929, the Society had almost 900 members, one was a 17-year-old high school senior named Wernher von Braun. Von Braun was a member of a prominent and affluent Prussian family. His father served as Minister of Agriculture and also as Minister of Education under the democratic German Weimar Republic.

The Society acquired a small tract of land a few miles north of Berlin that was a German army ammunition storage depot and could be used for rocket flight experiments. The area was named the "Rocket Flight Area" (Raketenflugplatz, in German), and the Society began to conduct flight experiments. The technical leaders of the Society were convinced—as was

Goddard—that only liquid-fueled rockets were practical for space travel. They knew about Goddard's work, and they began to build and test liquid-fueled rockets. The first test models were called "Mirak" which stood for the German "Minimum Rakete." They were small test rockets designed only for static tests. They also built flight models called "Repulsors" which looked very much like Goddard's first rocket. One of these flew for the first time on 10 May 1931 and reached an altitude of 18 meters. The rockets following this event were somewhat more sophisticated, but none reached the technical complexity of Goddard's gyrostabilized rocket. See also **Goddard, Robert H.**

Both Oberth and Nebel had talked and written about military applications of rockets. There was interest in rockets in Germany because the Versailles Treaty that ended Word War I contained provisions that seriously limited German artillery development. In 1929, the Army Ordnance Department established a small group to study the possible military applications of rockets. A year later, a young German artillery captain, Walter Dornberger, was named to head this group. Dornberger was aware of the experiments conducted by Oberth, Nebel, and the others at the Raketenflugplatz. He decided to help the group by providing modest financial support. He also quickly came to the conclusion that the Raketenflugplatz site was not adequate, and he offered the facilities of the German artillery range at Kummersdorf for the rocket experiments. These moves accelerated technical progress, but they also weakened the Society because several of the people who worked on the Society's rockets joined Dornberger's group. Wernher von Braun, one of them, started to work at Kummersdorf on 1 October, 1932.

When Adolf Hitler and the Nazi party came to power in January 1933, things changed drastically. The civilian organizations interested in rocketry disappeared, and the work on military rockets was substantially accelerated. During the mid–1930s, the Germans developed a well-organized, systematic research program on liquid-fueled rockets. The first of these rockets, called the A-2, was successfully launched in December 1934 and reached an altitude of 2000 meters, about 6000 feet. It was not as sophisticated as Goddard's gyrostabilized rocket that was launched a few months later, but the A-2 did work. Wernher von Braun had acquired a Ph.D. and was the technical director of the Kummersdorf enterprise reporting directly to Dornberger. In any event, the Germans had now roughly caught up to where Goddard was when he was forced to quit.

By the mid-1930s, the Germans had made enough progress that they decided to develop a test site which would make it possible for them to test large rocket vehicles. They chose an isolated region on the north German Baltic coast where the River Peene reached the sea. The place was called Peenemunde and by 1939, test operations were started there. The A-2 was followed by the A-3 and the A-4, the latter was successfully flown on 3 October, 1942, almost a decade to the day that von Braun joined Dornberger's unit. The A-4 was the prototype of the V-2, the first long-range weapon based on rocket technology. It could throw a payload of 1 metric ton about 250 miles. The rocket engine of the A-4 developed a thrust of 60,000 pounds. The rocket burned ethanol as a fuel and used liquid oxygen as an oxidizer. Due to the success of the A-4 launch and the loss of the Battle of Britain in 1941, Hitler decided, on 7 July 1943, to put the development and production of the V-2 missiles at the highest level of priority. See also **Rocketry**.

Because of good air reconnaissance and brilliant photo interpretation, the British were aware that new and dangerous weapons were being developed and tested at Peenemunde. They mounted a massive air raid on the test site, which was partially successful. The Germans realized that they would have to disperse their test and production facilities because more allied air raids attacking the Peenemunde complex were inevitable. The Germans built a massive underground facility in central Germany called the "Mittlewerk" for producing V-2 missiles. They also built a test site in Poland which was out of range of long range bombers based in England. A little more than 6,250 V-2s were produced. About 3700 were fired at targets in Great Britain and on the continent, of which about 700 failed! About 2000 were in storage at the end of the war, 300 were expended in tests, and 250 were taken to the United States at the end of the war. Although the outcome of the war was not influenced by the use of large numbers of V-2 rockets, the creation of the V-2 missile and its successful operation was still a technical achievement of the first rank. It has been said that if the Manhattan Project to produce nuclear weapons was the technical "tour de force" of the United

States during World War II, then the development of the V-2 played the same role in Germany.

The End of World War II

The V-2 rocket team that Wernher von Braun organized at Peenemunde was dispersed at the end of World War II. Wernher von Braun and most of the 100 or so senior technical people were housed in the small mountain village of Oberjoch on the Austrian-German border. They wanted to surrender to the U.S. Army, which had some units in the neighborhood, but they did not know how to make contact. At the same time, in 1945, the U.S. Army realized that it would be important to capture the Germans who had the knowledge to develop and build long-range missiles. An operation dubbed "Paperclip" was initiated to locate and detain the people who had this expertise. It was headed by U.S. Army Colonel Holger N. Toftoy. Eventually, the von Braun group surrendered to a U.S. Army unit and was taken into custody. Other veterans of the V-2 program wound up as prisoners of the Soviets. There were a few who actually volunteered to go to work for the Soviets.

Wernher von Braun, Walter Dornberger, and other leaders of the V-2 program arrived at Newcastle Air Base in Wilmington, Delaware on 19 September 1945. They were now under contract to the U.S. Army and were going to work on missile development. They were taken to Fort Bliss near El Paso, Texas, the U.S. Army's Center for Anti-Aircraft Artillery. Fort Bliss is located at the southern end of the White Sands Missile Test Range in New Mexico. It would be the job of von Braun and his colleagues to launch some of the V-2s brought over from Germany, to perform research, and to train Americans in the art and science of rocketry. In doing so, a large number of V-2 rockets was used; some were modified to measure the upper atmosphere and to photograph Earth, the Sun, and other astronomical objects. All of these activities were very important to the future of U.S. ballistic missile and space flight programs. See also **Von Braun, Werhner (1912–1977)**.

At the same time that the Germans were working for the U.S. Army, a group at Douglas Aircraft Company in Santa Monica, California, published a far-reaching document in 1946. The group was led by some outstanding scientists and engineers including Francis Clauser, David Griggs, and Louis Ridenour among others; they later left Douglas to form the RAND Corporation. The title of their work was "Preliminary Design of an Earth-Orbiting Spaceship". The group anticipated that large rockets, based on extensions of the V-2 designs, would be developed both by the United States and the Soviet Union to carry nuclear weapons at ranges from 8,000 to 10,000 miles. Such rockets could be modified to place significant payloads into Earth orbit and to other places in the solar system. Thus, the means to realize the early ideas of Tsiolkovsky, Goddard, and Oberth would soon be available, and therefore, it was time to make some detailed calculations about spaceflight. The remarkable thing about "Preliminary Design of an Earth-Orbiting Spaceship" is that it was a serious engineering study — about 250 pages long — that predicted nearly all of what has happened in spaceflight since 1946 [Clauser, Griggs, Ridenour, et al. Ref.]. Weather observations, intelligence gathering, communications, and other things now being done by Earth orbiting vehicles were treated. In addition, there were estimates of what it would take to put people in space. Thus, the stage was set for humanity's first steps into space. See also **Tsiolkovsky, Konstantine E. (1857–1935)**.

The Development of U.S. Space Launch Vehicles

As the authors of the 1946 report anticipated, the United States embarked on a vigorous program to develop and then build large liquid-fueled rockets for military reasons. Two military services, the U.S. Army and the U.S. Air Force, were charged with the responsibility for creating the missiles that would carry nuclear weapons. In 1950, the Army reached the decision that the Germans had finished their work at White Sands and that they would be moved to the Redstone Arsenal near Huntsville, Alabama, to initiate the development of new ballistic missiles.

The U.S. Army Rocket Development Program. Wernher von Braun, the leader of the German rocket engineers, arrived in Huntsville on 15 April 1950. It is safe to say that the town has not been the same since. The Army's attitude toward long-range rockets was that they were an extension of artillery. Thus, in the beginning, the objective was to develop a second-generation V-2 missile. The military requirements, written in

early 1951, called for a rocket that had a range of about 200 miles but had a substantially increased payload capability to carry the nuclear weapons then being developed. The rocket was named the Redstone and there were some important differences between the Redstone and the V-2. One was that the payload was designed to separate from the rocket, so that the entire vehicle would not have to reenter the atmosphere. This made it possible to build the rocket from aluminum rather than steel, as was the case for the V-2. The Redstone's engine had a thrust of 80,000 lb, rather than the 60,000 lb of the V-2. The first Redstone was launched on 20 August 1953, from the new launch site on the east coast of central Florida at Cape Canaveral. The flight was a failure, but subsequent flights proved that the Redstone design was sound. The Chrysler Corporation was given the contract to produce the Redstone in quantity.

Wernher von Braun and his German colleagues never lost their interest in spaceflight. In 1952, von Braun, along with Fred Whipple, the Harvard astronomer, and Joseph Kaplan, a distinguished atmospheric scientist at UCLA, were the leading authors of a series of articles published between 1952 and 1954 in *Collier's* magazine entitled "Man Will Conquer Space Soon"(5). There were articles about what became the space shuttle, space stations, and journeys to the Moon and the planets. The articles attracted considerable attention and encouraged von Braun and his colleagues to press for an upgrade of the Redstone so that a man-made satellite could be put into Earth orbit. On 15 September 1954, von Braun submitted a proposal to Colonel Toftoy — who by now was chief of the rocket branch at the Redstone Arsenal — to upgrade a Redstone rocket with suitable upper stages so that a small satellite could be put into Earth orbit. This was technically feasible, but the proposal was rejected for political reasons.

The Eisenhower Administration was anxious to keep the effort to create an Earth orbiting satellite as a civilian project. The scientific community had designated 1957 as the first "International Geophysical Year," and orbiting a satellite would be part of the program of research. The satellite would be launched by a Vanguard rocket that would be developed by the Martin Company for this purpose. Even though people from the U.S. Naval Research Laboratory were involved in developing the Vanguard, the program would be managed by a civilian organization.

In spite of this edict, the Redstone was upgraded. The first of these modifications was called the Jupiter A, which was used as a rocket test vehicle. The second, Jupiter C, which had three solid-fueled upper stages, was used in developing heat-resistant materials for ballistic reentry vehicles that would carry nuclear warheads. The first Jupiter C was launched on 20 September 1956 carrying a one-third scale warhead. It reached an altitude of 682 miles and flew a distance of 3,335 miles, a record that stood until the first intercontinental missiles were tested. It was only a short step from the Jupiter C to a rocket that could orbit a satellite. Von Braun secured permission to continue studies to achieve this objective in spite of the fact that his Army superiors were under strict orders not to develop orbital vehicles. Finally, an extended version of the Redstone was used to develop the Jupiter MRBM (medium range ballistic missile) that had a total takeoff thrust of 150,000 lbs. See also **Rocketry**.

The Soviets were not bound by the rules imposed on the U.S. Army's group at Huntsville. On 4 October 1957, using a military rocket, the Soviets placed the first man-made object, Sputnik I, into an orbit around Earth. (How this was done is described in an article elsewhere in this *Encyclopedia* .) The Soviet spacecraft created a sensation and caused great consternation in the United States. To add insult to injury, the Soviets launched a much larger satellite, Sputnik II that carried a dog named "Laika" and weighed 1100 lbs, a month later on 3 November. The final embarrassment came on 6 December 1957 when the first Vanguard was launched and then fell back to Earth two seconds later. At this point, the decision was finally made to ask von Braun's group to use the Jupiter C to orbit an American satellite. The Huntsville group recruited William Pickering, the Director of the Jet Propulsion Laboratory at the California Institute of Technology, then funded by the Army, to build a fourth stage for the Jupiter C. Professor James A. Van Allen was recruited to build a payload for Pickering's fourth stage so that if it went into Earth orbit, some scientific results would be obtained. Von Braun's team estimated that they could put a satellite into Earth orbit within sixty days.

The modified Jupiter C, now called the Juno I, blasted off from Cape Canaveral on 31 January 1958 and put Explorer I, the first American satellite into Earth orbit. It was launched 56 days after the Huntsville group was given the job to go ahead. Professor Van Allen's scientific payload was the first to measure the radiation fields above the Earth's atmosphere. This led to the discovery of permanent "belts" of radiation surrounding the Earth, now called the "Van Allen Belts." Juno I was used to launch two more satellites, Explorers 3 and 4, and was then replaced by further upgrades of the Redstone, Juno II, and the Mercury Redstone. Juno II, was used to launch a series of "Pioneer" satellites that explored the newly discovered radiation belts. Finally, the Mercury-Redstone was used to launch Alan Shepard on a suborbital flight on 5 May 1961 followed by an identical flight by Gus Grissom on 21 July. Both of these flights were carried out after Yuri Gagarin achieved the first orbital flight by a human being on 12 April 1961. Thus, the United States was still substantially behind the Soviet Union in spaceflight technology.

The flights of Shepard and Grissom were the last to use the Redstone as the core of the launch vehicle. Thus, the period of playing catchup with the Soviet Union would not be over until launch vehicles more powerful than the Redstone were available.

The U.S. Air Force Rocket Development Program. The U.S. Air Force, unlike the Army, was given a broader mission in rocket development. In addition to short- and intermediate-range missiles, the Air Force would also develop rockets with intercontinental ranges. Another difference between the Army and the Air Force rocket programs was that the leaders of the Air Force program were Americans rather than people from Germany who had been captured at the end of the Second World War. Although a number of members of the German rocket team went to work for contractors who built the rockets, none of them went to work in the Air Force management organization, the Western Development Division in Los Angeles. Walter Dornberger, who had risen to the rank of Major General in the German Army before the end of World War II, went to work at Bell Aircraft, and he was soon joined by Krafft Ehricke. Ehricke later joined Convair to work on the Air Force Atlas rocket development. Dr. Hans Friedrich also joined Convair to work on the Atlas. Dr. Adolf K. Thiel joined Ramo/Woolridge, later TRW, and Dr. Martin Schilling became a vice president of Raytheon Corporation.

Wernher von Braun and the rocket team that remained with him in Huntsville eventually were transferred to NASA when part of the Redstone Arsenal was turned over to the new civilian space agency in 1960. The new organization was called the George C. Marshall Space Flight Center. There they developed the Saturn V launch vehicle that eventually put humans on the Moon in 1969. See also **Spaceflight: U.S. Manned (Mercury, Gemini, Apollo, Skylab, Apollo-Soyuz Project and Space Shuttle Program)**.

The rocket development program that the Air Force eventually adopted grew out of a careful study conducted by the Air Force Strategic Missiles Evaluation Committee. This group was headed by Professor John von Neumann of Princeton University who played a major role in the development of nuclear weapons during World War II. Von Neumann first broached the idea of putting nuclear warheads on top of large rockets, which he called "intercontinental artillery." In 1954, the Missile Evaluation Committee urged the development of relatively small intercontinental ballistic missiles (ICBMs) that were able to carry the advanced "second-generation" nuclear and thermonuclear weapons then being developed. Von Neumann's detailed familiarity with nuclear weapons development influenced the Evaluation Committee to make this judgment. He knew that nuclear explosives much smaller and more efficient than those used at Hiroshima and Nagasaki were being developed, and the rockets, therefore, would be tailored to carry the new weapons. Ironically, this was the reason, among others, that the Soviets gained a significant advantage over the United States during the first years of orbital space operations. Their nuclear weapon technology was well behind that of the U.S. Accordingly, they had to develop larger and more powerful rockets to carry their heavier nuclear weapons.

To develop the new ICBMs, the Air Force established the Western Development Division at Los Angeles. This move was in accord with the recommendation of the von Neumann Committee to create an organization that would have full authority and responsibility for ICBM development. The Western Development Division would have the technical support of a new organization, the Aerospace Corporation, which was a nonprofit organization created by the transfer to the Air Force of the Space Technology Laboratory of the Ramo-Woolridge Corporation. The first commander of the Western Development Division was a brilliant young Air Force Brigadier General, Bernard A. Schriever. He later achieved four-star rank as the leader of the U.S. Air Force Systems Command. The basic organization described here is still in existence, although the Western Development Division has undergone periodic name changes. Today, it is the Air Force Space and Missiles Division. This organization still receives technical support from the Aerospace Corporation just as it did half a century ago.

The Western Development Division was given the job of developing three missile systems, two ICBMs, the Atlas and the Titan, and one intermediate range ballistic missile (IRBM), the Thor.

The Thor Missile and the Delta Space Launcher. The Thor first stage is a missile that has roughly the same performance characteristics as the Army's Redstone missile. The decision to go ahead with the Thor was controversial because of the apparent duplication of effort. Eventually, the Joint Chiefs of Staff approved the proposal to go ahead with both missiles. In December 1955, the Douglas Aircraft Company was given the contract to develop the missile. The first successful flight of the Thor was carried out on 20 September 1957, and the first operational missiles carrying warheads were deployed in the United Kingdom in 1958, 3 years after the contract was given to Douglas. This remarkably rapid development cycle was due to General Schriever's push for "concurrency." This means essentially that calculated risks are taken in the development process to speed the schedule and, in this case, the adoption of "concurrency" succeeded.

Both the Thor and the Jupiter MRBM had pressurized, regeneratively cooled rocket motors that developed thrusts of about 150,000 lb. Unlike the Redstone, which burned ethanol and liquid oxygen, the Thor burned a kerosene liquid oxygen mixture. It was also designed to be very rugged, and this is the feature that has led to the large number of versions of the Thor, so that it has been called, justifiably, the workhorse of space launchers. It is truly remarkable that space launch vehicles based on the Thor started in the late 1950s with the capability of placing a few hundred pounds in near Earth orbit and now can place a payload of more than 8000 lb. into a geostationary transfer orbit.

The Thor has carried a number of upper stage vehicles. One is the Agena, built originally by the Bell Aerospace Company and later by the Lockheed Corporation. It will be described later. The most important booster stage was the "Delta" solid-fueled rocket built by Aerojet General Corporation that had a thrust level of 8,000 to 10,000 lb depending on the version used. The Delta stage was used so frequently with the Thor that the combination is now called the "Delta."

The second important thrust augmentation for the Thor is the Castor rocket built by Thiokol Corporation. These are powerful solid-fueled rockets that are strapped to the bottom skirt of the Thor rocket. Each unit has a thrust level of 50,000 lbs, and they can be strapped on in numbers between three and nine. The Castor rocket capability adds both power and flexibility to the Delta space launch vehicle system. More than 200 successful Delta rocket launches have been conducted since the first Thor was launched in 1957.

The Delta launch family originated in 1959 when the NASA Goddard Space Flight Center awarded a contract to Douglas Aircraft Company to produce and integrate 12 launch vehicles. Using components from the U.S. Air Force Thor intermediate range ballistic missile (IRBM) program and the U.S. Navy Vanguard launch program, the Delta rocket was available 18 months after the award. On 13 May 1960, the first Delta was launched from Cape Canaveral Air Force Station, Florida, carrying a 178-pound Echo I passive communication satellite. Although the first flight was a failure, the ensuing series of successful launches established Delta as one of the most reliable of all U.S. boosters.

The Delta II Space Launcher. For more than 40 years, the Delta system has consistently demonstrated its robust design, launch flexibility, and value to launch service customers. A second generation, the Delta II launch system, was developed to include multiple configurations to suit the needs of the U.S. Air Force, the National Aeronautics and Space Administration (NASA), and to accommodate the emerging commercial satellite market.

From 1985 through 1987, the space industry was impacted by an unprecedented string of failures of various launch systems, which seriously impeded U.S. space launch capability. In one of several steps to revitalize assured access to space, the U.S. Air Force held a competition for a medium launch vehicle that primarily would launch Global Positioning Satellites (GPS). The contract was awarded to McDonnell Douglas (now

Boeing) for its Delta II series vehicles. With a 98.1% mission success rate since its inception in 1989 the Delta II has become the industry workhorse for deploying remote sensing satellites for U.S. government and commercial applications, GPS, commercial satellite systems/constellations, and numerous planetary missions for NASA.

In addition to the demonstrated reliability of the Delta II launch system, Delta vehicles provide incremental performance capability with three, four or nine Castor solid rocket motor (SRM) configurations. These configurations provide a broad range of performance, from 2000 to 4000 pounds to geosynchronous transfer orbit (GTO), using the highly reliable Rocketdyne RS-27A main engine, and two- and three-stage configurations. The latest version, which is designated Delta II Heavy, integrates the solid rocket motors (SRMs) used on Delta III with the Delta II standard upper stage, resulting in approximately a 12% increase in payload lift performance from the standard nine SRM configuration to more than 4700 pounds to GTO (Fig. 1).

The Delta II launch system payload accommodations provide various mechanical interfaces, separation systems, and deployment systems that are designed for compatibility with launch industry standard interfaces. Payload accommodations enable deploying single, dual, or multiple satellites in a single Delta II launch. The multiple manifest, spacecraft dispensers have successfully deployed 55 Iridium spacecraft, five per launch on eleven launches, and 24 Globalstar spacecraft, four per launch on six launches.

The Delta II is launched in the United States from three launch pads, two on the East Coast at Cape Canaveral Air Force Station that has a capacity of 12 launches and one on the West Coast at Vandenberg Air Force Base, California, that has a potential of nine launches per year. Launch sites on both coasts enable the Delta II launch system to launch to virtually any orbit and provides customers with launch schedule assurance.

The demand for Delta II launches increased significantly in the mid- and late 1990s and has continued to maintain a steady backlog of government and commercial customers. In addition to continuing GPS deployment missions, Delta II has been selected for numerous NASA payloads because of NASA's focus on smaller, less expensive spacecraft and its demand for proven reliable launch systems. NASA remains a critical customer for Delta launches, Delta II has assigned and planned missions through 2009.

Building on the success of the Delta II launch system, McDonell-Douglas developed the two-stage Delta III in the mid- to late 1990s to address the trend toward increasing mass of commercial satellites. The Delta III nearly doubled the payload lift capability of the Delta

II launch system by deploying an 8400-pound payload to GTO. Delta III evolved from the highly reliable Delta II by maximizing the use of common components and infrastructure. Both launch systems use the same Rocketdyne RS-27 main engine, first stage LO$_2$ tank, flight avionics, and launch operations infrastructure. Most notable of Delta III's evolved features are a 13-foot diameter cryogenic second stage, 13-foot diameter composite biconic payload fairing and nine, larger, more powerful SRMs (46-inch dia.). The second stage uses a single RL10B-2 engine fueled by LH$_2$ and LO$_2$ that incorporates an extendable nozzle.

The Delta III established itself as an operational launch system in August 2000 by launching a 9460-pound demonstration payload to a planned subsynchronous GTO. The mission was successful; all of the systems and subsystems performed as planned. Intended to be a transition vehicle to the Delta IV, the Delta III has enabled demonstration and flight qualification of several critical components that would be used on the Delta IV.

The Atlas Intercontinental Ballistic Missile (ICBM) and Space Launcher. Concurrently with the development of the Thor, the U.S. Air Force also undertook to develop the first large truly intercontinental missile — the Atlas. This missile included a number of innovative design features. The first and most important of these was that the body of the rocket vehicle was also the wall of the pressurized fuel and oxidizer tanks, and unlike the Thor and Redstone, both of which had conventional braced aluminum air frames, the Atlas skin was manufactured from a thin sheet of stainless steel. The Atlas could only stand on a launch pad if the tanks were pressurized; if they were not, the vehicle would collapse. The pressurized steel tank thus assumes some of the structural burden. This design saves enough weight, so that the Atlas dry weight fraction was the smallest of any large rocket yet built.

The propulsion system of the Atlas was also unique. It consisted of three engines, one mounted on the centerline of the vehicle and the other two fitted on a skirt and placed on either side of the center engine. At launch, all three Rocketdyne MA5 engines would be running, generating a total thrust of 370,000 lb. The skirt, called the "booster section," could be jettisoned at an appropriate time, and the center engine, called the "sustainer," would continue to run, generating 60,000 pounds of thrust until the propellant is consumed. This concept is called "stage and a half" because no fuel is carried in the booster section, so that no fuel tank is dropped. The engines burned RP-1 (kerosene) using liquid oxygen as the oxidizer. The engines were gimballed for thrust vector control, as was the Vanguard rocket mentioned earlier.

The Atlas program was initiated in 1951, and the contract to develop the rocket was given to Convair/General Dynamics Corp. The first successful

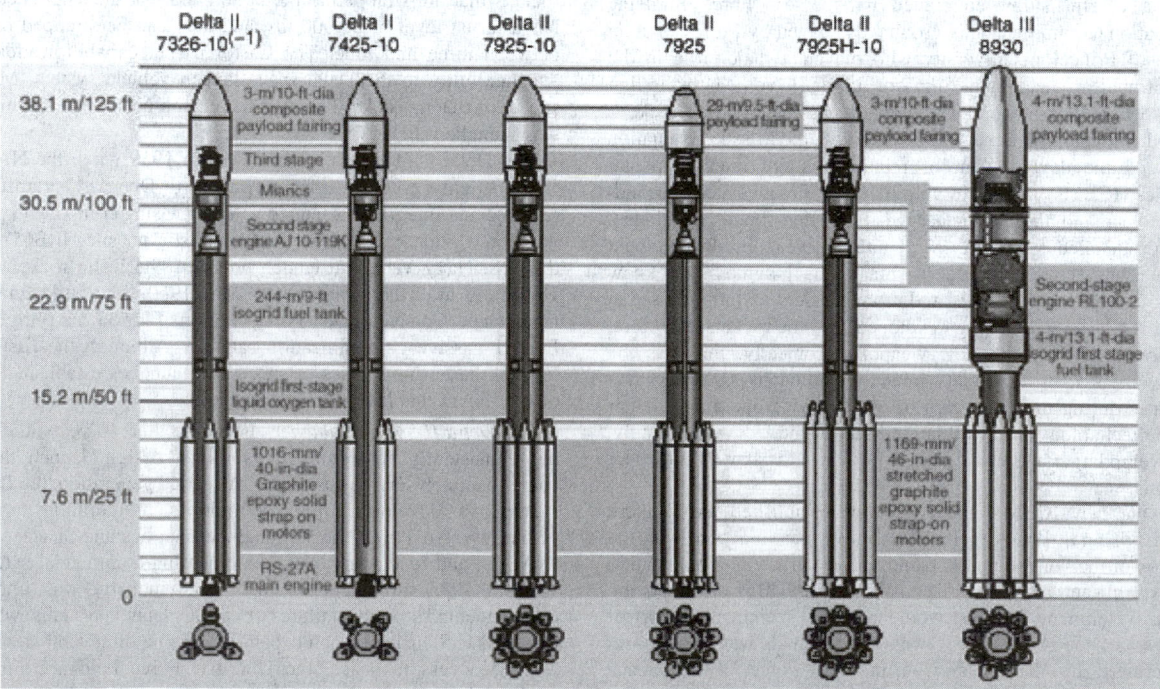

Fig. 1.   Delta II and III space launch vehicles.

flight of an Atlas ICBM occurred in December 1957, and the first employment as a space vehicle launcher occurred in December 1958 when it was used to put an active communications satellite that weighed about 200 pounds into Earth orbit.

Since these first flights, the Atlas, in combination with a number of different upper stages, has been a very successful space launch vehicle. Perhaps the most important launch of an Atlas booster was John Glenn's first flight in the Mercury program. The Mercury spacecraft, "Freedom Seven," was placed into Earth orbit by a modified Atlas D ICBM on 20 February 1962 (Fig. 2). The Atlas has also been used very effectively with Agena, Centaur, Delta, and Burner upper stages. (The technical details of these upper stages will be discussed more thoroughly later.) The Atlas/Agena combination was used for some of the first Mariner Missions (Mariner 4) that returned the first pictures from a Mars flyby in July 1965. The Atlas/Agena was also used to send a Ranger spacecraft to the Moon for a hard landing, and it placed the Lunar Orbiter around the Moon as well. The latter spacecraft was particularly important because the high resolution pictures it obtained were used to select the landing sites for the Apollo landings. The Atlas/Centaur combination is a more capable launch vehicle that was used to put the first soft landing spacecraft, the Surveyors, on the Moon. These spacecraft were also very important precursors for the Apollo missions because they determined the mechanical properties of the lunar surface.

Fig. 2.　Launch of John Glenn's Mercury "Freedom Seven" spacecraft by an Atlas space launch vehicle. (*photo courtesy NASA*.)

The most interesting application of the Atlas/Centaur was launching the Pioneer 10 spacecraft on 3 March 1972, as well as its sister ship, Pioneer 11, 13 months later. Pioneer 10 was the first spacecraft to pass beyond the orbit of Mars, to fly through the asteroid belt, to fly past Jupiter, and finally, it became the first human artifact to leave the solar system. Pioneer 11 repeated this performance a year later, and, in addition, it became the first spacecraft to fly by Saturn. See also **Mars.**

The Atlas II family was developed in the mid-1980s to address the growing demand for large commercial geosynchronous satellites. The Atlas II family has a 100% success rate and 56 consecutive launches, a reliability

record unmatched in the industry. During the past decade, the Atlas II vehicle has been continually stretched and upgraded to improve payload performance. The Atlas IIAS is the most powerful and has the highest lift capability of the Atlas II family. Other configurations include the Atlas IIA and the Atlas II, which was retired in March 1998. Currently, Atlas II vehicles are being flown from both U.S. ranges.

Four solid fuel rockets are used to augment payload performance of the Atlas IIAS. All three MA-5 engines are ignited prior to liftoff. Approximately 180 seconds into first-stage ascent, the two larger MA-5 booster engines are shut down and jettisoned, reducing weight and improving payload performance. The center MA-5 sustainer engine burns for an additional 100 seconds up to main engine cutoff and staging. The Centaur second stage uses two RL-10A-4 engines to place up to 6700 lb (Atlas IIA) and 7950 lbs (Atlas IIAS) into a GTO orbit.

In the early 1990s, General Dynamics (now a part of Lockheed Martin) decided to upgrade the Atlas first-stage propulsion system. The prime modification was replacing the two MA-5 first-stage and single MA-5 sustainer engines with a single Russian NPO Energomash RD-180. Furthermore, Lockheed Martin simplified vehicle construction by drastically reducing the total part count. The reengined vehicle, known as the Atlas IIAR, has since been renamed the Atlas III. The maiden flight of the Atlas III launch vehicle, the replacement for the Atlas II, occurred on 24 May 2000.

Two versions of the Atlas III are currently available. The Atlas IIIA has a single RL-10A-4-2 engine powering the Centaur upper stage, whereas the Atlas IIIB has two RL-10A-4-2 engines and a stretched Centaur upper stage to increase GTO performance to just under 10,000 lb (Fig. 3)

The Titan ICBM and Space Launch Vehicle. One of the limitations of the Redstone, the Thor, and the Atlas as military missiles was that they could not be kept on "instant alert." This meant that they could not be launched on very short notice because the liquid oxidizer (liquid oxygen) is a liquid only at 210 °C below room temperature so that it cannot be stored on the missile itself. Special cryogenic storage facilities had to be build at each of the military launch sites, and upon the order to launch the missile, the liquid oxygen had to be transferred from the storage tank to the missile. Such operations take, at best, something of the order of half an hour, which means that an instant "launch on alert" is not possible.

The Titan missile actually started as a replacement for the Atlas because the fragility of the Atlas was deemed undesirable for deployed military rockets. The Titan was thus designed using the more rugged monocoque technique in which an appropriately braced aluminum "fuselage" took the stresses of the launch. Originally, the first version of this missile, the Titan I, was designed for conventional fuels and would be placed in hardened underground shelters. A successful test was conducted in February 1959, and the Titan I system became operational in 1962. In spite of this, the system was awkward to operate, and the Air Force decided to convert the Titan I to a missile that could use storable fuels so that it could be maintained on instant alert. The implementation of this idea led to the Titan II ICBM.

The Titan I ICBM was somewhat more capable in range and payload than the Atlas. The Titan II had to be substantially more capable. The storable fuel chosen for the Titan II was a mixture of nitrous oxide (N_2O_4) and "Aerozine," a hydrazine-based fuel. These two substances are liquids at room temperature, and they ignite when they come into contact, that is, they are hypergolic. Thus, with appropriate care, they can be stored on the missile itself. However, the hypergolic fuel mixture does not have the same high specific impulse of the liquid oxygen/hydrocarbon fuels, so that the rocket is not as effective as a launch vehicle. Thus, the Titan II ICBM was designed from the very beginning as a two-stage system, which gave it substantially better performance than the Atlas. The Titan II second-stage motor was also fueled by a hypergolic mixture, so that the entire vehicle could be stored at room temperature.

The first successful test of the Titan II was carried out on 16 March 1962, and the first operational missile was installed in its hardened underground bunker in December 1962. Ultimately, 51 Titan II missiles were deployed at three sites near Wichita, Kansas; Tucson, Arizona; and Little Rock, Arkansas. The Titan II was a formidable part of the U.S. nuclear deterrent force; it carried a multimegaton warhead that could reach targets 8000 to 10,000 miles away. In September 1980, a technician at one of the Titan II missile sites in Arkansas dropped a wrench down the silo and

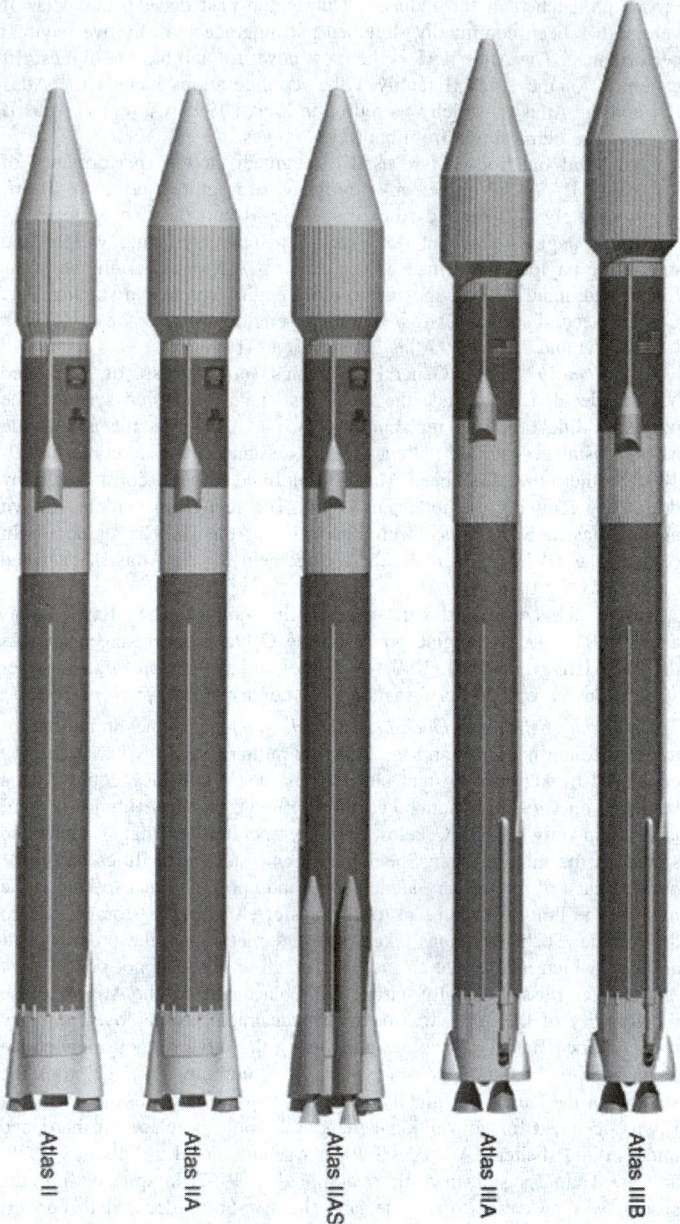

Fig. 3.   The Atlas space launch vehicle family.

Atlas II

Atlas IIA

Atlas IIAS

Atlas IIIA

Atlas IIIB

Force Base in Tucson, Arizona. Lockheed Martin was awarded a contract in January 1986 to refurbish, integrate, and launch 14 Titan II ICBMs for government space launch requirements.

Tasks involved in converting the Titan II ICBMs into space launch vehicles included modifying the forward structure of the second stage to accommodate a payload; manufacturing a new 10-ft diameter payload fairing with variable lengths plus payload adapters; refurbishing the Titan's liquid rocket engines; upgrading the inertial guidance system; developing command, destruct, and telemetry systems; modifying Vandenberg Air Force Base Space Launch Complex 4 West to conduct the launches; and performing payload integration. Six Titan II Space Launch Vehicles have been launched from Vandenberg Air Force Base, California, since 5 September 1988.

The major modification of the Titan III could be made because of the sturdy monocoque construction of the original Titan II missile. This was the addition of two very large solid-fueled strap-on rockets on opposite sides of the Titan II core vehicle. These rockets have a thrust of just over 1,000,000 pounds each, which raises the takeoff thrust of the entire vehicle in the launch configuration to about 3,000,000 pounds. This configuration of the launch vehicle is called Titan IIIC, and with a Transtage, it can place about 30,000 pounds in low Earth orbit. It can place a payload of about 3600 pounds in a geosynchronous orbit by using the appropriate upper stages. Another important variation is the Titan IIID, which has no upper stages but can put a 13,000-pound payload into near Earth polar orbits. This is the launch vehicle that, when launched from the West Coast at Vandenberg Air Force Base, puts most of the U.S. highly capable reconnaissance satellites in orbit. The Titan IIID was declared operational in 1971.

In 1977, the Titan IIIE/Centaur was fielded. This is the most capable expendable space launch vehicle in the current American inventory. The addition of the Centaur upper stage makes the difference (Note: The upper stage will be described in more detail later.) The Titan IIIE/Centaur has launched a whole galaxy of spacecraft to explore everything from the outer planets using exquisitely designed cameras to putting very sophisticated payloads into the atmosphere of Jupiter. This would include Helios, Galileo Jupiter with the probe, the two Voyagers, the two Vikings, and a substantial number of others. A related version of the Titan IIIE is the Titan 34D which, instead of carrying the Centaur, carries a solid-fueled upper stage called the "inertial upper stage" or IUS. This stage was developed by the Air Force to put large military and intelligence gathering satellites into polar orbits.

Titan IV consists of two solid-propellant, stage-zero motors, a liquid propellant two-stage core and a 16.7-ft diameter payload fairing. Upgraded three-segment solid rocket motors increase the vehicle's payload capability by approximately 25% (Fig. 4). In 1985, the U.S. Air Force selected the Martin Marietta Astronautics division (now Lockheed Martin) in Denver to build and launch 10 Titan IVs. In 1986, the contract was increased to 23 vehicles, and, in November 1989, the contract was extended to 41. Titan IV has been used exclusively to launch U.S. government satellite missions. It provides primary access to space for critical national security and civil payloads.

The operating success rate of Titan launch systems is better than 95%. It can place 47,800 lb into low Earth orbit or more than 12,700 lb into geosynchronous orbit.

Titan IV is launched from Launch Complex 40 at Cape Canaveral Air Force Station, Florida, and from Space Launch Complex 4E at Vandenberg Air Force Base, California. The first Titan IV B was successfully flown from Cape Canaveral Air Force Station on 23 February 1997. This configuration improves reliability and operability and increases lift capability by 25%. Advancements also include improved electronics and guidance. The Titan IV B has standardized vehicle interfaces that increase the efficiency of vehicle processing. Additionally, the more efficient programmable aerospace ground equipment is used to monitor and control vehicle countdown and launch.

Upper Stages. In describing the evolution of military missiles to become space launchers, we mentioned a number of upper stages used in combination with the military rockets to place spacecraft into Earth orbit and beyond. In this section, we will provide brief descriptions of the most important.

The Centaur, developed and built by Convair/General Dynamics, is the most capable of these vehicles. The Centaur uses liquid oxygen to burn

it punctured the fuel and oxidizer tanks as it fell. Eventually, the two liquids mixed, ignited, and the resulting conflagration destroyed the missile and silo. Fortunately, the nuclear warhead, which had a hardened design, survived the accident unharmed. Partly as a result, all of the deployed Titan II ICBMs were decommissioned by 1987.

Like the Thor and the Atlas, the Titan II ICBM was also converted into a very flexible and capable space launch system. The first of these, fielded in 1964, was called the Titan IIIA; it consisted of the Titan II two-stage ICBM plus a third stage called "Transtage" that could put payloads of more than 3000 pounds into low Earth orbit. This was followed quickly by the Titan IIIB/Agena in 1966 that could put 8500-pound payloads into near Earth orbit. This was accomplished by increasing the thrust of the first stage from 430,000 to 463,000 pounds and using the somewhat more capable Agena rather than the Transtage. All of these modifications were carried out by the Martin Company.

The Titan II space launch vehicle is a two-stage liquid-fueled booster designed to provide small-to-medium weight class capability. It can lift approximately 4200 pounds into a polar, low Earth circular orbit. Titan IIs were also flown in NASA's Gemini manned space program in the mid-1960s. Deactivated Titan II missiles are in storage at Davis-Monthan Air

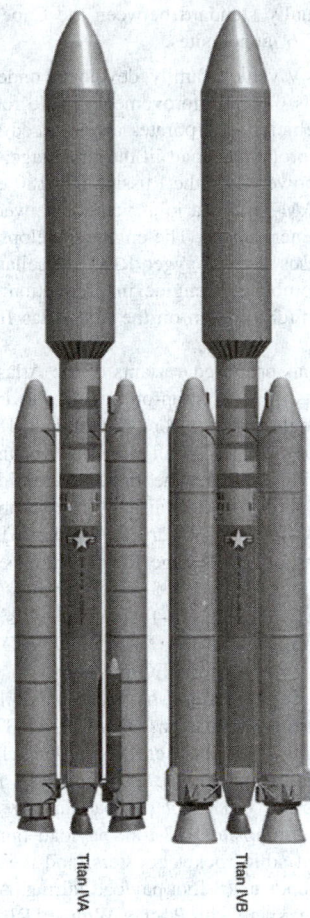

Fig. 4.   The Titan IV space launcher.

hydrogen fuel. This mixture is the most potent of all chemical rocket fuel combinations because it provides the highest specific impulse to the rocket. Thus, rockets using this combination have the highest propulsive efficiency. The Centaur can be shut down and restarted in space, which is most important because it permits executing complex space maneuvers. However, because the fuels are cryogenic, these maneuvers have to be carried out shortly after launch, so that the fuel and oxidizer are not lost by evaporation. The Centaur has two Pratt and Whitney RL-10 engines that develop a total thrust of 30,000 pounds. The first Centaur was flown in 1962, and since then, Centaur upper stages have been used on top of Atlas and Titan boosters.

The Agena is the second important liquid-fueled upper stage. The Agena also uses a hypergolic fuel/oxidizer mixture of (N_2O_4) and unsymmetrical dimethyl hydrazine (UDMH) to provide a thrust of about 15,000 pounds. The Agena also has shutdown and restart capability like the Centaur. The difference is that the fuel/oxidizer mixture can be stored at ambient temperature, so that the Agena vehicle can stay in orbit in a dormant condition much longer than the Centaur. The Agena has been used on Thor, Atlas, and Titan rockets and has proven to be the most ubiquitous of the high-performance liquid upper stage vehicles. The Agena was originally designed and built by the Bell Aerospace Company for the Atlas and was later taken over by the Lockheed Missiles and Space Company (now called Lockheed Martin Co.). It was flown for the first time in 1960 on an Atlas booster.

Solid-fueled rockets have been used extensively as upper stages for space launch vehicles ever since the first orbital flights. The Juno I launcher, which placed Explorer I in Earth orbit in January 1958, had three solid-fueled upper stages, including a small solid booster on the Explorer I satellite itself. Solid-fueled rockets have the advantage that they are safely storable and also extremely versatile. They have ranged from the small rockets already mentioned that might develop tens of pounds of thrust to the huge strap-on rockets of the Titan IIIC that develop thrust levels of more than a million pounds. The primary disadvantage of solid-fueled rocket

boosters is that, except in certain cases, such as the two-grain systems used in antimissile applications, they cannot be turned off once they are lit. Another disadvantage is that failure of solid-fueled rocket is usually catastrophic — which was unfortunately graphically illustrated in 1986 by the failures of a solid rocket booster in January that caused the loss of "Challenger" and then again in April that caused the loss of a Titan 34D flight.

There are too many solid-fueled upper stages to list here. Probably the most important and capable solid-fueled upper stage is the inertial upper stage (IUS) that was built by the Air Force for use with the Space Shuttle and the Titan 34D system.

The Evolved Expendable Launch Vehicle Program (EELV)[1]

The EELV Concept. The Evolved Expendable Launch Vehicle program, born of studies conducted in the late 1980s and 1990s, represents a commitment to reducing significantly the cost of access to space. An industry/government partnership has developed two competing EELV systems to meet space transportation needs during the next 20 years.

Conceived as a "system of systems" to improve operability and reduce recurring and infrastructural costs, EELV is using streamlined manufacturing and improved mission assurance processes. Its facilities and operations are designed to lower costs.

Requirements call for a 25–50% reduction in recurring operational cost compared to current systems and for improving system reliability and availability. The U.S. Air Force interest is that EELV will replace the Titan, Atlas, and Delta vehicles and their launch infrastructures supported by DOD. The program implements DOD acquisition excellence goals by streamlining the government's role and replacing its oversight of contractors with less intrusive "insight." The objective is to enhance U.S. launch industry competitiveness in the international market by reducing costs across the entire system.

In October 1998, the government awarded $500 million contracts each to Boeing and Lockheed Martin. Development costs are shared between the contractors and the government, resulting in a national, dual-use launch service. The government program office has virtually unlimited access to all but some highly sensitive and proprietary cost and pricing data. The Air Force simultaneously awarded initial launch service contracts to both firms.

The strategy enabled two further benefits: competition and assured access to space. Having two competitors throughout the life cycle of the program is key to achieving price competitive procurement. Two providers using a standard payload interface maintain payload interchangeability between Delta IV and Atlas V and enhance assured access to space.

Each delivery order for a launch service has a standard 24-month period for performance. Individual launch service plans, however, are highly flexible and can be tailored to spacecraft customer needs.

EELV will support U.S. military intelligence, civil, and commercial mission requirements using contractor-provided commercial launch services. The two are the Boeing Delta IV and Lockheed Martin Atlas V, both designed to meet the full range of government launch requirements.

The EELV program has three key performance parameters: specific payload mass-to-orbit requirements; vehicle design reliability of 0.98 (threshold) at a 50% confidence level; and standardization, including standard payload interface for each class of vehicle and standard launch pads that can accommodate all configurations in an EELV family.

Delta IV. Delta IV was developed under a U.S. Air Force EELV contract. The Delta IV launch vehicle uses a new liquid oxygen/liquid hydrogen 16.7-foot diameter common booster core (CBC) powered by the new Boeing-Rocketdyne RS-68 main engine. This is the first large liquid-fueled engine to be developed in the United States since the SSME (Space Shuttle main engine). The RS-68 is a gas generator liquid oxygen/hydrogen booster engine. The bell-nozzle RS-68 develops 650,000 pounds of sea level thrust and uses a simple design approach that has drastically reduced the total part count compared to engines of equivalent size or performance. The vehicle's cryogenic upper stage, which uses the Pratt & Whitney RL10B-2 engine, is substantially similar to that flown on the Delta III.

There are several variants of the Delta IV launch vehicle. The Delta IV-M (medium) is a single-core variant that combines the CBC with

---

[1] Excerpts from input by James Simpson, The Boeing Company.

a version of the Delta III liquid oxygen/hydrogen second stage and a stretched 4-meter fairing that provides 9200 pounds to GTO. The Delta IV-M+ variants augment the single core with two or four solid rocket strap-on Alliant Techsystems graphite epoxy motors (GEMs) and provide two variations of upper stages and payload accommodations, the 13-foot Delta III derivative and the 16.7-foot version that has greater fuel capacity and greater payload volume. The M+ variants enable payload deployment of 14,700 pounds to GTO. The Delta IV vehicle family will have a GTO lift capability of up to 29,500 pounds and is available in three major variants. The largest variant, the Delta IV-Heavy, combines three CBCs with the 16.7-foot upper stage. The payload accommodations include either a 16.7-foot isogrid aluminum fairing based on the existing Titan IV or a newly developed composite fairing based on the Delta II and Delta III designs. The 13-foot fairing is the existing Delta III composite fairing lengthened by 3 ft (Fig. 5).

Improvements include the new CBC, the newly developed and simplified main cryogenic engine, the focused factory facility, and simplified launch-processing operations.

Parts for the medium-plus and heavy CBCs are, respectively, 88% and 93% common relative to the medium CBC. All are manufactured using a common factory production list. CBC innovations include friction stir-welded tanks, spun-formed domes, and the use of composite structures. The RS-68 has reduced operating pressure, 80% lower part count, 95% less labor, uses cast versus welded parts, and has no special coatings. However, more than 85% of the upper stage part count is a Delta III heritage, and much of the avionics are from Delta II and III.

Full integration, assembly, and checkout testing take place before each vehicle leaves the factory. Delta IV's horizontal booster processing flow and vehicle stage mating in the horizontal integration facility allow parallel integration, reduced hazardous lifting operations, and decreased pad time. Total vehicle time at the launch base is less than 1 month and only 8–11 days on the pad. Delta IV features launch sites on both the East Coast (Cape Canaveral Air Force Station, Florida) and West Coast (Vandenberg Air Force Base, California). Each pad can launch all configurations, and

launch pads are virtually standard between the Cape Canaveral SLC-37 and Vandenberg SLC-6 launch sites.

Atlas V. The Atlas V vehicle family, developed under the U.S. Air Force EELV contract, builds on the improvements made for the Atlas III. The Atlas V family of vehicles incorporates a reinforced first-stage structure, as well as increased propellant load in the first stage, called the common core booster (CCB) powered by the Russian RD-180 engine. The RD-180 is produced by RD AM-ROSS, a joint venture between Pratt & Whitney and Russian's NPO Energomash. The engine develops 860,000 pounds of thrust at sea level, uses liquid oxygen/RP-1 propellants, and is the only high-thrust, staged-combustion engine in production. It has been tested extensively and was flight proven on the first Atlas IIIA mission in May 2000 (Fig. 6).

Lockheed Martin has proposed variants of the Atlas V that incorporate different arrangements of solid strap-on boosters to increase the payload performance of the single common core variant up to 18,000 lb to GTO. To differentiate the various Atlas V configurations, Lockheed Martin devised a secondary numbering system. The first number identifies the fairing diameter in meters (3,4, or 5-meter fairing). The second number identifies the number of solid strap-on boosters (0 through 5). The final number identifies the number of second stage RL-10 engines (either 1 or 2). As an example, an Atlas 5 532 has a 5-meter fairing, three solid strap-on boosters, and two second-stage RL-10 engines. A single engine RL-10 second stage is used for high altitude (MEO and GTO) missions, whereas the two-engine variant is used for LEO missions.

Atlas V's several configurations have the flexibility to meet varied performance requirements for missions from LEO to GTO. Options include the addition of one to five Gencorp Aerojet strap-on solid rocket motors for intermediate lift capability or the use of three CCBs for heavy payloads. The Atlas 400 series has a 13-foot payload fairing and a single CCB; the 500 series has a composite 16.7-foot payload fairing, a single CCB, and up to five Aerojet solid rocket boosters; and the heavy launcher has three CCBs and a composite 18-foot payload fairing. All three series use a common Centaur upper stage with Pratt & Whitney RL10A-4-2 engines(s).

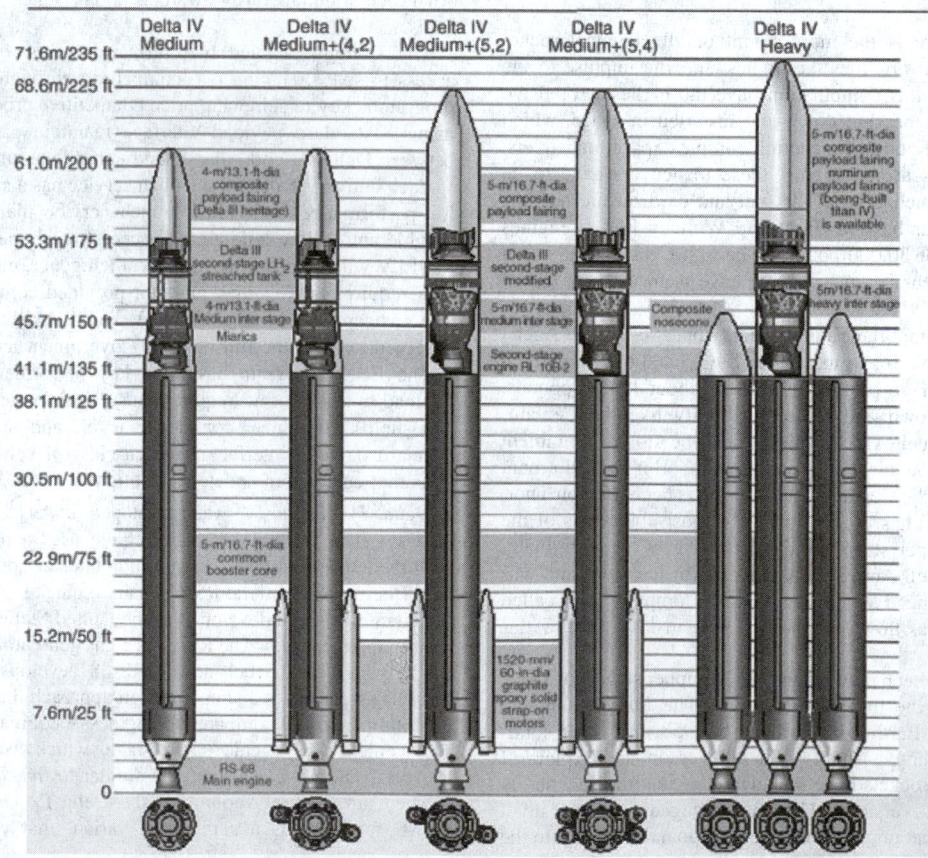

Fig. 5.   Evolved Delta.

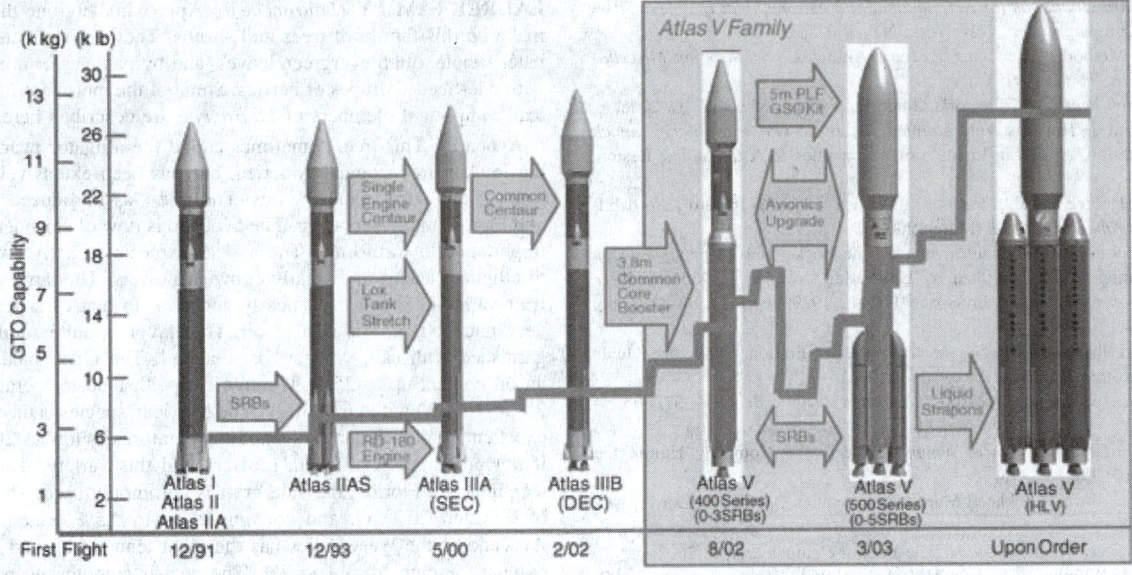

Fig. 6.   Evolved Atlas.

These modifications, combined with the stretched Atlas IIIB Centaur upper stage, allow the Atlas V to place more than 10,000 lb in a geosynchronous transfer orbit with the single CCB. The heaviest variant of the family currently planned for production will incorporate an arrangement of five solid strap-on boosters to increase the payload performance of the single common core variant up to 18,000 pounds to GTO. Lockheed Martin has designed a three common booster variant capable of placing more than 13,000 lb directly into a geostationary orbit that is currently not being marketed commercially.

Atlas V, which uses the same Centaur upper stage as the Atlas IIIB, can be configured with either one or two RL10A-4-2 engines. A hydrazine attitude control system provides precise in-orbit maneuvering. The 18-foot payload fairing is a new design derived from the Ariane V fairing manufactured by Contraves Space of Switzerland. It will be offered in two lengths, one optimized for communications satellites and the other for accommodating large-volume spacecraft missions. The 13-foot payload fairing is the same one used on Atlas II and III. Among Atlas V's innovations is the RD-180 engine capability for continuous throttle between 47 and 100% of nominal thrust, which allows substantial control over launch vehicle and payload environments. Others include reduced manufacturing cycle time and simplified launch processing. Atlas V also includes the Air Force EELV standard payload interface that allows payload interchangeability with Delta IV.

Atlas V incorporates efficient launch site processing, including use of an off-pad vertical integration facility (VIF) for the vehicle and parallel processing of the encapsulated payload in separate installations. Launch site processing has been reduced from 28–38 days for Atlas II to just 18–26 days. The encapsulated payload will be transported to the VIF and mated to the launch vehicle. After combined systems testing, the fully integrated Atlas V/encapsulated payload will be transported to the nearby "clean launch pad." All vehicle configurations use common processing procedures and can be launched from the same clean pad. On-pad time has been reduced to less than 1 day.

Current Status. The Lockheed Martin-built Atlas 5 401 (single CCB, no solids, and single engine Centaur) booster launched flawlessly on 21 August 2002 deploying the Hot Bird 6, a communications satellite for Eutelsat.

The 12.5-foot diameter rocket was the largest to launch from Cape Canaveral since the Saturn 5 sent Apollo astronauts to the Moon. This flight gives the Atlas family a string of 61 consecutive successful launches during 9 years using the Atlas II, Atlas III, and Atlas V vehicle configurations.

On 20 November 2002 the Delta IV-M+4,2 carried the W5 communications satellite for Eutelsat to a precise geostationary transfer orbit.

The Department of Defense and related agencies (e.g., NASA), as well as the commercial sector, will be major customers for both vehicles. The Pentagon has scheduled 29 launches aboard the rockets to date.

Conclusions

It is truly remarkable that from Goddard's first liquid-fueled rocket flight to the landing on the Moon was a mere 43 years — well within the lifetimes of many people living today. The spurs for this achievement were clearly the Second World War (1939–1945) and later, the Cold War (1948–1991). In both cases, the technology of rocket propulsion was considered critical by all concerned to prevailing in the conflicts. However, this is only half the story. The other half is that a group of unusually talented and motivated people from many nations contributed to the successes that we have described. The principal technical conclusion that can be drawn from what has been said here is that the technology of chemically fueled rockets is mature. For the past 40 plus years, since the development of the ICBMs in the early 1960s, no new propulsion technology has been developed and applied. What has happened is that the proliferation of liquid- and solid-fueled chemical rockets has made it possible to develop a very large number of launch vehicle combinations tailored to meet a great many different requirements for expendable space launchers. For the foreseeable future, this evolution will continue and will be made possible by advances in guidance and control systems, more accurate timing and navigation, and other auxiliary technologies. It is a tribute to the designers of the early ICBM rockets that their products are still in our front line expendable space launchers.

What of the future? We believe that there are now signs on the horizon that new technologies for space launch systems will be required. We are on the threshold of initiating human exploration of the solar system. The International Space Station will be the staging base for this new phase of space exploration. We believe that both electric propulsion systems and nuclear rockets will be assembled at the space station and will eventually take people and equipment on journeys around the solar system. Hopefully, the designers and builders of these new propulsion systems will display the same skill and virtuosity as the people who created the space launchers described in this article.

See also **Ion Propulsion (Rockets)**; **Rocket Propellants**; **Rockets: Liquid-Fueled**; and **Rockets: Solid Fuel**; **Space Shuttle**; and **Space Stations**.

Additional Reading

Clauser, F. and D. Griggs, L. Ridenour, et al.: "Preliminary Design of an Experimental Earth-Orbiting Spaceship," Report No. 5M-11827 (Contract W33-038 ac 14105, Douglas Aircraft Company, Inc. May 2, 1946.

Davies, M: *Standard Handbook for Aeronautical and Astronautical Engineers,* The McGraw-Hill Companies, Inc., New York, NY, 2002.

Goddard, R.: "A Method of Reaching Extreme Altitudes," *Smithsonian Miscellaneous Collections,* **71** (2) (1919).

Godwin, R.: *Launch Vehicles,* Collector's Guide Publishing, Wheaton, IL, 2006.

Isakowitz, S.J., and J. Hopkins: *International Reference Guide to Space Launch Systems,* 4th Edition, American Institute of Aeronautics & Astronautics, Reston, VA, 2004.

Knauf, J.M., L.R. Drakee and P.L. Portonova: "EELV Evolving Toward Affordability," *Aerospace America,* 38–42 (March 2002).

Oberth, H.: *Die Rakete zu den Planetenraumen* (The Rocket Into Planetary Space), Oldenburg, Munich, 1923, reprinted by Uni-Verlag, Nürnberg, 1960 and *Wege Zur Raumschiffahrt* (Ways Toward Space Travel), 1929, reprinted by Kriterion, Bucharest, 1974.

Ordway, F. III and Sharpe, M.R.: *The Rocket Team,* 3rd Edition, Collector's Guide Publishing, Wheaton, IL, 2003.

Prandini, E: "Arianespace Under Pressure," *Interavia Business Technol.,* **57**, 665, **54** (3) (2002).

Staff: *Delta and SeaLaunch Technical Summary,* The Boeing Company, Huntington Beach, CA, April 2002.

Staff: *Atlas and Titan Data,* Lockheed-Martin, Astronautics Operations, Denver CO, September, 2002.

Staff: *Delta IV Payload Planners Guide,* The Boeing Company, April 2002.

von Braun, W., F.L. Whipple, J. Kaplan, H. Haber, W. Leyy and C. Ryan: Man Will Conquer Space Soon, Collier's Magazine, a series of articles starting on March 22, 1952 and ending April 30, 1954.

Web References

Atlas 5-Summary. Space and Tech Database Expendable LVs: www.spaceandtech .com/spacedata/elvs/delta4

Delta II Rockets: http://www.boeing.com/defense-space/space/delta/delta2/delta2 .htm

Launch Vehicles: http://spacelist.org/launch.html

Launch Vehicle History (1963–1998): http://www.nasa.gov/centers/glenn/about/ history/lvpo.html

NASA Kennedy Space Center, Launching Rockets: http://www.nasa.gov/centers/ kennedy/launchingrockets/index.html

Taurus Rocket: http://www.orbital.com/SpaceLaunch/Taurus/index.html

U.S. Expendable Launch Vehicle Data for Planetary Missions: http://www1.jsc.nasa. gov/bu2/ELV_US.html

Michael I. Yarymovych, (Retired) Boeing Space and Communications
Seal Beach, CA
Hans Mark, Austin, TX

LAUNCH WINDOW.   The postulated opening in the continuum of time or space, through which a spacecraft or missile must be launched in order to achieve a desired encounter, rendezvous, impact, or the like.

LAUREL FAMILY (*Lauraceae*).   Approximately one thousand species make up this family of trees and shrubs. They are characterized by alternate, simple, often evergreen leaves, and by panicles or umbels of flowers with one-Seeded drupes or berries. Some of the more familiar and economically important members of *Lauraceae* are described here. See Table 1.

Avocado. This tree, sometimes called the alligator pear, is a native of the lowlands of tropical America, but has been extensively cultivated in tropical and subtropical regions. The tree was introduced into California and Florida many years ago. The avocado is now of considerable economic importance in California. The avocado tree is attractive, with large oval to elliptical leaves and small yellowish flowers. The large green-to-brown fruit varies in shape from nearly spherical to that of a pear. The fruit is very nutritious and is rich in oil. The flavor is quite subtle and often is garnished with salt, vinegar, or salad oil. The Guatemalan avocado has an oil content up to 25%. The tree can withstand temperatures as low as $25\,°F$ ($-4\,°C$.) without damage. The Mexican species is the hardiest and of excellent quality. It can withstand temperatures as low as $20\,°F$ ($-6.7\,°C$.) if not prolonged. California growers bud this variety, using patch-a-bud technique. In Florida, the side graft is commonly used. The pulp is about 69% water, 20% oil, and contains close to 2.5% protein. The Western Avocado of the West Indies has the most tender fruit and is of a low oil content, ranging from 4 to 7%. The peel is smooth and purple. The tree cannot withstand temperatures below $28\,°F$ ($-2.2\,°C$.).

J.M. Haller (American Forests magazine, p. 29, May 1982) observes some of the unusual characteristics of the avocado. "Flowers, which in other species are certifiably and consistently male, female, or hermaphroditic, may on the avocado be male in the morning and female in the afternoon! Other species are rigorously grafted to prevent reversion to a primitive type bearing inferior fruit. But the avocado, though it may be and often is similarly grafted, will produce an astonishing variety of viable types from seed, most of them equal to any given grafted line and many of them superior. Other trees are either deciduous (leaf-shedding) or evergreen. The avocado manages both at the same time, shedding its leaves regularly each spring but not until the new season's crop is ready as a replacement (hence always green)."

Bay. This tree is native to the West Indies, but is found in France, Germany, and the coastal areas of the Americas. It is a small-to-medium-size tree, attaining a height of from 35 to 40 feet (10.5 to 12 meters), with a trunk of about 5 to 12 inches (12.7 to 30.5 centimeters) in diameter. The tree is related to the allspice and sassafras trees. The fruit is a berry. The oil from the fruit is yellow and aromatic and is the basis of bay rum. The bay tree sometimes is called the wax myrtle tree.

Cinnamon. The spice is obtained from a small tree native to Sri Lanka and India, where it is cultivated extensively. The tree grows to a height of

TABLE 1. RECORD LAUREL TREES IN THE UNITED STATES[1]

Specimen	Circumference[2]		Height		Spread		Location
	Inches	Centimeters	Feet	Meters	Feet	Meters	
Avocado laurel (1999) (*Persea americana*)	172	437	75	22.9	60	18.3	California
Californialaurel (1997) (*Umbellularia californica*)	546	1387	108	32.9	118	36	California
Loblolly bay laurel (1993) (*Gordonia lasianthus*)	164	417	95	29	60	18.3	Florida
Mountain laurel (1999) (*Kalmia latifolia*)	56	142	20	6.1	19	5.8	Georgia
Redbay laurel (typ.) (1993) (*Persea borbonia var. borbonia*)	152	386	77	23.5	52	15.8	Florida
Sassafras laurel (1995) (*Sassafras albidum*)	262	665	78	23.8	69	21	Kentucky
Swampbay laurel (1999) (*Persea borbonia var. pubescens*)	63	160	36	11	47	14.3	North Carolina
PAWPAWS							
Common pawpaw (1986) (*Asimina triloba*)	26	66	63	19.2	29	8.8	Virginia
Smallflower pawpaw (1993) (*Asimina parviflora*)	21	53	24	7.3	17	5.2	Florida

[1]From the "National Register of Big Trees," American Forests (by permission).
[2]At 4.5 feet (1.4 meters).

from 25 to 40 feet (7.5 to 12 meters) and has shiny dark green, leathery leaves, small whitish flowers which have a rather disagreeable odor, and dark purple fruits. The bark of young twigs is smooth and somewhat mottled; in the older branches and the main stem, the bark becomes thick, rough and of little value. To insure the desideratum of many young branches, the limbs are severed so that many slender branches will form, a practice known as coppicing. From these slender stems, the bark is removed by lengthwise splitting and partial loosening from the stem. As it dries, it rolls back. It is then removed from the stem, the dry useless periderm scraped off, and the inner bark remaining allowed to dry completely. During drying, its color changes from pale yellow to deep brown. The tight rolls of dried bark are packed together in bundles, called pipes, and are ready for marketing as cinnamon. The bark contains considerable amounts of a powerful drug, which in large doses is a dangerous poison. The principal use of cinnamon is as a spice for pastries. By distillation of cinnamon stems and leaves, oil of cinnamon is obtained, used in flavoring candy and in scenting soaps.

Laurel. The laurel tree grows along the coastal mountains and in the Sierra Nevada mountains of California — at an altitude of about 4,000 feet (1220 meters). The California laurel and Oregon myrtle are essentially the same tree. The tree attains a height of from 50 to 80 feet (15 to 24 meters), with a trunk of 2 to 3 feet (0.6 to 0.9 meter) in diameter in mature trees. The branches are erect, long, and thick. The bark is thin, scaly, and dark brown. The leaves are about 2 to 5 inches (5 to 12.7 centimeters) in length and one-half to one or more inches (1.3 to 2.5 centimeters) wide. The underside is light green; the top side is leathery, glossy, and thick. The flowers are in clusters, pale yellow and small. The fruit of approximately 1 inch in length hangs in clusters of two and three. It is about the size of an olive and is a yellow-green color, containing one Seed. Laurel wood, used for cabinet work, veneers, and garden tool handles, is fine-grained and hard and weighs approximately 40.5 pounds per cubic foot when dry (649 kilograms per cubic meter).

Pawpaw. This tree is found in the southern and midwestern states of the United States. It is related to the banana plant. The tree is small, usually grows wild, and is found in woodland areas. The fruit is exotic in appearance with a rich golden color. The fruit ranges from 3 to 5 inches (7.6 to 12.7 centimeters) in length and hangs from the tree in clusters. Pawpaw wood is spongy and weak and of no commercial value. The flower is purple and fragrant.

P. Stevenson presents an interesting portrait of the pawpaw in the March/April 1990 issue of Amer. Forests, page 46.

Sassafras. Frequently more of a shrub than a tree, the sassafras plant is found in the New England states, west to Wisconsin, and south to the Gulf coast. In many areas, the sassafras grows into a large tree, ranging up to 30 or 40 feet (9 to 12 meters) in height, with a trunk measuring from 8 inches to 2 feet (20 centimeters to 0.6 meter) in diameter. The record sassafras tree in the United States, as reported by American Forests Association, is located in Owensboro, Kentucky. See table.

All parts of the sassafras tree have a characteristic fragrance. The branch grows horizontal. The leaf is bright green and glossy and well known for its "mitten" shape. In the autumn, the tree turns a golden red and is quite showy.

The flower is small, yellow, and occurs in clusters. The flower appears before the leaf and is staminate with a six-lobed calyx, orange stalked glands, and nine stamens. The fruit is a dark blue, thin and fleshy berry of oblong shape. Although eaten readily by birds, the fruit is not enjoyed by humans. The bark is thick, scaly, and gray with longitudinal ridges. The wood is brittle and coarse grained and, although it resists moisture decay well, it is seldom considered of commercial value. Sassafras oil is sometimes used in soaps and toiletries.

H. Clepper elucidates further details of the sassafras tree in the March/April 1989 issue of Amer. Forests, page 33.

LAURIC ACID. [CAS: 143-07-7]. Also called dodecanoic acid, formula $CH_3(CH_2)_{10}COOH$. A fatty acid that occurs in many vegetable oils and fats as the glyceride, especially in coconut oil and laurel oil. See also **Vegetable Oils (Edible).** Combustible. It takes the form of colorless needles at room temperature. Specific gravity 0.833; mp 44°C; bp 225°C (100 millimeters pressure). Insoluble in water; soluble in alcohol and ether. It is derived by the fractional distillation of coconut oil.

Lauric acid is used in alkyd resins; rubber & latex; plastics; lubricants; pharmaceuticals; wetting agents; soaps; detergents; cosmetics; insecticides; and food additives.

LAVA. Molten material that has poured out on the surface of the earth and, due to relief of pressure, may have lost much of its original gas and water content during its relatively rapid consolidation. The term lava is used for both the liquid and the consolidated state of the igneous material. Lava may be erupted either by volcanoes or from fissures. Flowing lava is shown in Fig. 1. The most extensive lava flows are fissure eruptions, such as the Columbia Plateau basalts in Oregon or the plateau basalts of the Deccan, India, which are derived from basic magma. Had this magma, either basic or acid, cooled slowly beneath the surface of the earth under great pressure and with all its original gases, the resulting rock would have had a coarser texture and somewhat different mineral content.

See also **Earth**; **Ocean**; and **Volcano**.

Fig. 1.   Lava flow.

LAVAGE.   See **Empyema**.

LAVOISIER, ANTOINE LAURENT (1743–1794). Antoine Laurent Lavoisier was born in Paris on 26 August 1743. His father was a lawyer; his mother died when he was 5 years old. He was educated at the Collège Mazarin from 1754, where he received a sound literary and classical training and the best scientific education then available in Paris. In 1761 he transferred to the Faculty of Law, gaining the baccalaureate in 1763 and his license in 1764. In 1762–1763 he attended the chemical lectures of Guillaume François Rouelle at the Jardin du Roi. In the summer of 1763 and 1764 he accompanied the botanist Bernard de Jussieu on his *promenades philosophiques* (philosophical walks) in the Paris region. He also worked with the geologist Jean-Étienne Guettard with whom he made many expeditions collecting specimens and assembling data for a geological and mineralogical atlas of France. Lavoisier added a quantitative character to the observations, using the barometer systematically to measure the heights of mountains, observing the inclination of strata and collecting samples of rocks, minerals and spring waters for chemical analysis.

These varied experiences provided the basis for his later development in science, for they encouraged him to use his observations to construct a theory of the earth, while his quantitative skills were also tested and improved. His first chemical investigation, in 1764, was a study of gypsum, which he found to be a neutral salt of sulfuric acid with a chalky base. He observed that natural gypsum contained "water of crystallization," driven off when it was heated to make plaster of Paris.

In 1768 Lavoisier became assistant to the Fermier Général (a financier in the *Ancien Régime* who purchased the right from the Crown to collect certain taxes and whose reward was the remainder of the taxes over the agreed sum to be paid to the Crown), François Baudon, and in 1780 he became a Fermier Général himself, a fateful step which ultimately led to his execution. In 1768 he was a candidate for the Académie des Sciences. His rival Gabriel Jars, an older man, was elected, but Lavoisier was awarded a provisional place and in 1769, when Jars died, he was duly elected. He became a salaried Academician in 1778, was appointed director of the Académie des Sciences in 1785 and became its treasurer in 1791. In 1771 he married Marie-Anne Paulze, daughter of Jacques Alexis Paulze, a Fermier Général. She later assisted him by translating English and other foreign language publications into French, by recording his experiments and making the illustrations for the *Traité de Chimie* (1789).

In his early chemical work, begun about 1770 with observations and speculations about the nature of the four ancient elements, Lavoisier examined the supposed conversion of water into earth, the analysis of mineral waters and the destruction of a diamond by fire. He was influenced by Stephen Hales's observations of "airs" evolved on heating various different substances. Lavoisier now noticed that effervescence often produced a cooling effect, suggesting that the release of "air" required the uptake of "fire." His analysis of gypsum had already shown that water could be fixed in solid substances. About 1772 Lavoisier began the investigations of combustion which culminated in his discovery of the role of oxygen, the overthrow of the phlogiston theory and the "chemical revolution." The fundamental importance for the development of chemistry of Lavoisier's progress towards its reform has been investigated and reassessed many times. The main steps involved include:

1. His experiments on the air, showing that it is not an element but a mixture of gases of which oxygen is the most important.
2. His experimental demonstration that water is a compound of oxygen with hydrogen.
3. His pragmatic definition of the chemical element as any substance that cannot be analyzed further, a definition that enabled the identification of elements and the experimental discovery of new ones.
4. The new system of chemical nomenclature introduced by Lavoisier, Guyton de Morveau and Fourcroy in the *Méthode de nomenclature chimique* (1787) (*Method of Nomenclature Chemical*).
5. The new scientific journal, *Annales de chimie* (*Annals of Chemistry*), established by Lavoisier in 1789 to allow the publication of researches using the language and ideas of the new chemistry.
6. Lavoisier's own treatise, *Traité élémentaire de chimie* (*Elementary Treaty of Chemistry*), also published in 1789, containing a list of elements and a succinct account of the antiphlogistic theory of chemistry.

Lavoisier's new system also enabled him to recognize the qualitative composition of simple organic compounds in terms of their main elementary constituents, carbon, hydrogen, oxygen and nitrogen. He introduced the method of combustion analysis in which the carbon dioxide and water formed are collected, and weighed. From the results the proportions of carbon, hydrogen and oxygen in the organic substance may be determined. Lavoisier showed that sugar contains carbon, combined with hydrogen and oxygen in the same proportions as water. He also recognized that organic compounds contain "radicals," groups of elements acting as one, an observation with far-reaching consequences for theoretical organic chemistry.

Like Karl Scheele and Joseph Priestley, Lavoisier recognized the similarity between respiration and combustion. In 1777 he suggested that during respiration oxygen combines with the blood in the lungs at the same time displacing fixed air. His experiments on respiration led him to the conclusion that animal heat is released in the lungs by the combination of compounds of carbon and hydrogen in the blood with oxygen. The distribution of heat throughout the body could be explained by the rapidity of the circulation.

Lavoisier also investigated vinous fermentation, concluding that during this process sugar is converted into carbonic acid gas and spirit of wine (alcohol). Putrefaction, he thought, was a form of fermentation, which occurs more readily if nitrogen and certain other elements are present. The combination of nitrogen with hydrogen then forms ammonia, while sulfur and phosphorus yield hydrogen sulfide and phosphoretted hydrogen, respectively. In acetous fermentation alcohol is oxidized to form acetic acid in the presence of air. Thus Lavoisier began to lay the foundations of modern organic chemistry by showing that it could be studied using methods similar to those used in mineral chemistry. See also **Biochemistry (The History)**.

Lavoisier always aimed to apply the social and economic benefits of science, but in 1794 his life was cut short when he was arrested, tried and executed as an enemy of the people on a charge arising from his connections with the tax farm.

Additional Reading

Donovan, A.: "Lavoisier and the Origins of Modern chemistry," In: Donovan, A. *The Chemical Revolution: Essays in Reinterpretation, Osiris*, 2nd Series, Vol. 4. University of Pennsylvania Press, Philadelphia, PA, 1988.

Donovan, A.: *Antoine Lavoisier: Science, Administration and Revolution*, Blackwell Publishing, Oxford, UK.

Guerlac, H.: "Lavoisier, Antoine Laurent," In: Gillispie, C.C.: *Dictionary of Scientific Biography*, 18 Vols, 1970–1980, Vol. 8, Charles Scribner's Sons, New York, NY, 1972, pp. 66–91.

Holmes, F.L.: *Lavoisier and the Chemistry of Life*, University of Wisconsin Press, Madison, WI.

Poirier, J-P.: *Lavoisier: Chemist, Biologist, Economist*, University of Pennsylvania Press, Philadelphia, PA. 1996.

NOEL G. COLEY, The Open University, Milton Keynes, UK

LAW OF AREAS.   See **Kepler's Laws of Planetary Motion**.

LAW OF COSINES.   See **Direction Cosine**; and **Pythagorean Theorem**.

LAW OF LARGE NUMBERS.   There are various laws of large numbers but the essential idea is exactly the same in each case. If the size of a sample is increased indefinitely or becomes very large, good sample estimates of population parameters will tend to concentrate more and more closely about the true value. Bernoulli's theorem is perhaps the simplest illustration of a law of large numbers.

Put another way, such laws state conditions under which random variables converge in probability to constants as some parameter n(usually a sample number) tends to infinity. *Strong laws* are concerned with showing that, for example, a variable $x$converges to a value μ with probability unity. Weak laws consider conditions under which the probability that $|x - \mu|$ is greater than some given ϵ, tends to zero.

LAWRENCE, ERNEST O. (1901–1958).   An American physicist who invented the cyclotron in 1929. Both the element lawrencium and the Lawrence Livermore Research Laboratory at the University of California were named after him.

See also **Cyclotron**.

LAWRENCIUM.   [CAS: 22537-19-5]. Chemical element, symbol Lr, at. no. 103, at. wt. 257 (mass number of known isotope), radioactive metal of the Actinide series, also one of the Transuranium elements. ^{103}Lr was identified in 1961 by A. Ghiorso, T. Sikkeland, A. Larsh, and R. Latimer at the University of California at Berkeley.

This method used to produce and identify lawrencium was similar to that used in the later, direct-counting experiments performed in connection with the production of nobelium at Berkeley. About 3 micrograms of a mixture of californium isotopes were bombarded with boron ions accelerated in the heavy-ion linear accelerator. The atoms of lawrencium recoiled from the target into an atmosphere of helium, where they were electrostatically collected on a copper conveyor tape. This tape was then periodically pulled into place before radiation detectors to measure the emission rate and the energy of the alpha particles being emitted. By this means, it was possible to identify the lawrencium isotope ^{257}Lr, with a half-life of 8 seconds. At present, because of the short half-life and the lack of a suitable daughter isotope, available in the case of nobelium, it has not been possible to perform a chemical identification.

Another isotope, ^{256}Lr, half-life about 45 seconds, was reported by the Soviet Union in 1965. It was produced by impact of oxygen atoms (^{18}O) on americium (^{243}Am). It decayed by alpha-particle emission and electron capture to form ^{252}Fm. See also **Chemical Elements**.

Lawrencium has been found to behave quite differently from dipositive nobelium and, in fact, it is comparable to the tripositive elements that appear earlier in the Actinide series.

Additional Reading

Eskola, K., Eskola, P., Nurmia, M., and A. Ghiorso: "Studies of Lawrencium Isotopes with Mass Numbers 255 through 260," *Phys. Rev.*, **4**, 2, 632–642 (1971). (A classic reference.)

Fuger, J. and L.R. Morss: *Transuranium Elements: A Half Century*, American Chemical Society, Washington, DC, 1992.

Ghiorso, A., T. Sikkeland, Larsh, A.E., and R.M. Latimer: "New Element, Lawrencium, Atomic Number 103," *Phys. Rev., Lett.*, **6**, 9, 473–475 (1961). (A classic reference.)

Greenwood, N.N. and A. Earnshaw: *Chemistry of the Elements*, 2nd Edition, Butterworth-Heinemann, Inc., Woburn, MA, 1997.

Lide, D.R.: *CRC Handbook of Chemistry and Physics*, 84th Edition, CRC Press, LLC., Boca Raton, FL, 2003.

Seaborg, G.T. and W.D. Loveland: *The Elements beyond Uranium*, John Wiley & Sons, Inc., New York, NY, 1990.

LAWSON CRITERION.   See **Nuclear Reactor**.

LAWSONITE.   Named for Andrew Cowper Lawson (1861–1952), a Scottish-American geologist. This calcium aluminum silicate mineral, $CaAl_2(Si_2O_7)$ $(OH)_2 \cdot H_2O$, is found as grains and veins within the metamorphic rocks, gneisses, and schists. It was found originally on the Tiburon Peninsula, San Francisco Bay, California, but also occurs in schistose rocks in France and New Caledonia. The mineral has a hardness of 7; specific gravity of 3.09. It is colorless, pale blue to bluish gray, translucent, with vitreous to greasy luster. The mineral crystallizes in the orthorhombic system.

LAXATIVE.   See **Constipation**.

LAZULITE.   Lazulite is named from an Arabic word for *heaven* in allusion to its sky blue color. This mineral crystallizes within the monoclinic system, a basic phosphate of magnesium and aluminum, $MgAl_2(OH)_2$ $(PO_4)_2$. Ferrous iron can substitute for the magnesium and the isomorphous mineral scorzalite is the product. Usually occurs massive but acute pyramidal crystals are not uncommon. Color is azure-blue to bluish-green, usually translucent (rarely transparent), with vitreous luster. It has a hardness of 5.5–6, with specific gravity of 3–3.1.

Lazulite is a rare mineral found principally within high-grade metamorphic rocks. Notable world crystal occurrences are Salzburg, Austria; Syria; Hörnsjöberg, Sweden; Madagascar; Brazil; and Graves Mountain, Georgia. When transparent, the mineral can be cut into gem stones.

LAZY EYE.   See **Vision and the Eye**.

L-BAND.   A frequency band used in radar extending approximately from 0.390 gigacycles per second to 1.55 gigacycles per second.

L-DISPLAY.   In radar, a display in which a target appears as two horizontal blips, one extending to the right and one to the left, from a central vertical time base. When the radar antenna is aligned in azimuth at the target both blips are of equal amplitude. When not correctly pointed the relative blip amplitude indicates the pointing error. The position of the signal along the baseline indicates target distance. The display may be rotated 90 degrees when used for elevation instead of azimuth aiming. Also called *L-scan, L-scope,* or *L-indicator*.

L-DOPA.   See **Parkinson's Disease**.

LEACHING.   Leaching, sometimes referred to as *eluviation*, is the phenomenon whereby a liquid, usually water, moves through a matrix from one location to another, dissolving or suspending materials along the way. The matrix must be porous enough to allow fluid movement and is generally soil, but fractured bedrock or unconsolidated mineral or organic (humus) material can also leach liquids. Normally, this occurs under gravity and movement is downward. Occasionally, this movement is lateral when forces such as capillary action, soil matric potential, or confinement pressure draw the liquid to an area of high water tension or push the liquid from an area of high pressure. The liquid that migrates, generally referred to as leachate, contains a mix of dissolved and/or suspended ions or compounds. These materials, either present naturally in the environment or applied to the ground surface, include dissolved salts, fertilizers, soluble minerals, organic material such as humic or fulvic acids, plant nutrients, and natural or synthetic chemicals. Because water is a polar liquid, it can dissolve or transport a wide variety of substances from one location to another. Even nonpolar molecules can form micelles and be moved great distances by leaching.

Leaching is a function of several factors that affect the amount, rate, direction, and quality of the leachate: the hydraulic conductivity of the matrix, impeding layers (aquitards or aquicludes) (see also **Aquifers**), partitioning coefficients of any chemicals present, antecedent soil water content (saturated vs. unsaturated flow), osmotic or matric potentials, the cation exchange capacity of the soil, the ionic makeup of the dissolved compounds in the percolating fluid, the water-holding capacity of the soil (largely attributable to soil texture), and the macropore flow (a function of soil porosity, structure, disturbance, burrowing animals, and root channels).

If water is applied at a rate slower than the ability of the soil surface to accept it (infiltration rate), water percolates downward once the water holding capacity of soil is exceeded. If water is applied at a faster rate, a portion of the applied water runs off the soil surface (potential erosion). See Table 1.

TABLE 1. WATER-HOLDING CAPACITY OF VARIOUS SOILS (INCHES OF WATER PER INCH OF SOIL DEPTH)

Textural Class	Capacity
Sand	0.070
Loamy sand	0.065
Sandy loam	0.095
Sandy clay loam	0.150
Loam	0.165
Clay loam	0.185
Silt loam	0.20
Silt	0.23
Silty clay loam	0.25
Silty clay	0.22
Clay	0.20

Most nonindustrial leaching occurs on agricultural land or under antiquated or modern landfills. Leaching rates are highest in humid regions or on land under irrigation and lowest in arid or permanently frozen regions. Leaching is governed by a few basic processes.

Mass Flow

This is the dominant leaching process in most systems and accounts for the greatest movement of water and materials, often over large distances. When water infiltrates a dry, permeable soil, a discrete wetting front is formed that has a higher water content (lower tension) above than below. The depth of this wetting front depends on the amount of water added, the permeability of the surface (pore size, volume, and connectivity), and its initial wetness. To a lesser extent, wetting front migration is a response to the rate at which the water is applied. After water is no longer applied, the wetting front moves but at a slower rate due to reduced pressure from above.

Water infiltrating a dry soil can dissolve and transport salts or chemicals that are not strongly adsorbed on soil solids. Water can also dislodge and transport materials in suspension such as organic colloids or fine clays that are not readily soluble. This dissolved and/or suspended load is transported to depth via the percolating water, but only as far as the wetting front progresses. It can be moved more deeply by subsequent rainfall or irrigation, provided it is of sufficient volume to move percolating water past the previously established wetting front.

If a substance is loosely adsorbed onto solid surfaces in the soil, the depth to which the adsorbed chemical moves is reduced or the rate of movement slowed. This retardation factor, R, is related to the soil-water partition coefficient, K_d, of the particular substance being moved (Fig. 1.)

$$R = 1 + \frac{\rho K_d}{\theta} \qquad (1)$$

where ρ is the dry bulk density of the soil and θ is its volumetric water content. The retardation factor varies from a low of 1 for nonadsorbed chemicals to values in excess of 100 for some highly adsorbed ones. The depth to which a substance is leached largely depends on the

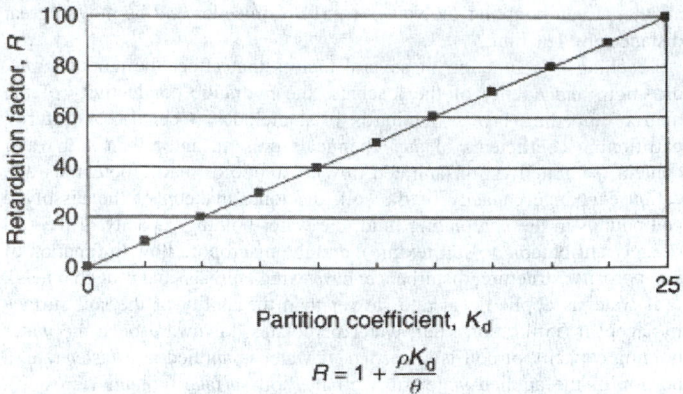

$$R = 1 + \frac{\rho K_d}{\theta}$$

Fig. 1.   Retardation factor as partition coefficient.

adsorbent relationship between a substance and the matrix through which it percolates.

The wetting front in an initially dry soil corresponds to the leading edge of the infiltrating liquid. Surfaces that are not initially dry behave similarly except in some soils, where water entering a moist soil can displace water already present and push that water ahead of the front. In such cases, a dissolved or suspended substance would move only to the depth of the infiltrating water, not all the way to the wetting front. Alternatively, because macropore flow (preferential flow paths along cracks in structured soils or fractured rock) is usually present, some water at the wetting front may represent new water. Most likely the situation is a combination of these two mechanisms of water dispersal, which leads to uncertainty in the real location or concentration of a dissolved substance and also suggests that a dissolved chemical is not likely to be concentrated in a sharp pulse but will be more diluted or dispersed.

Dispersion

Water moving through soil pores travels at different rates due to porosity, tortuosity, and connectivity differences. A pulse of chemical traveling with the water has its leading and trailing edges spread out, a phenomenon called dispersion. Under natural conditions in the field, water rarely moves through a matrix at a constant rate at all times, even if its physical properties (texture, structure, ionic exchange capacity, etc.) are extremely uniform and water is applied at a constant rate.

Diffusion

Dissolved or suspended substances also move from regions of higher to lower concentration due to osmotic potential referred to as diffusion. Diffusion is usually much less significant than mass flow; it merely spreads the chemical out in an already wetted area. Its influence is generally small and its rate slow due to adsorption (on matrices high in clay and organic matter content).

Sorption

A chemical added to a soil partitions itself between a solution phase and an adsorbed solid phase. When the cation exchange capacity (CEC) of a soil is high (e.g., soils high in smectite clays or organic matter), this bonding is strong because cations are adsorbed to the negatively charged surfaces (Table 2). In other cases (soils with low-CEC clays or lacking clays and organic matter), bonding is quite weak.

TABLE 2. NORMAL RANGE OF VALUES FOR COMMON SOIL GROUPS

Soil Groups	CEC in cmol (+)/kg
Light colored sands	3–5
Dark colored sands	10–20
Light colored loams and silt loams	10–20
Dark colored loams and silt loams	15–25
Dark colored silty clay loams and silty clays	30–40
Organic soils	50–100

Sorption or binding of dissolved chemicals on surfaces is significant because they are not free to move with percolating water. Although sorption retards the rate of chemical movement, it does not necessarily alter the rate of movement of the percolating solution itself.

Substances in a liquid–soil environment tend to establish a balance between the amount on solid surfaces and the amount in solution. Some chemicals exist primarily in the liquid phase; others are strongly adsorbed and exist primarily on solid surfaces. Molecules tend to move from one phase to another to maintain this balance. The manner in which the molecules are partitioned into solid and liquid phases depends on both matrix and chemical properties. These relationships, called sorption isotherms, can be determined experimentally. The center of the chemical pulse on a particular day is greater for substances that have smaller partition coefficients than for those that have larger ones because substances that have smaller partition coefficients are adsorbed less strongly to solid surfaces and can be dislodged and moved more readily by the percolating solution.

Degradation

Chemicals in the environment can undergo biological or chemical transformations, be absorbed by plants, volatilize into the atmosphere, be lost to surface water by surface water flow and erosion, or remain in the surface matrix and be available for leaching to groundwater. The importance of these different processes depends on the unique properties of the substance.

Many natural and synthetic chemicals break down or degrade over time from microbiological and chemical reactions. Substances with low persistence degrade quickly to other products, and highly persistent ones linger. Persistence is usually specified in terms of the degradation half-life, a measure of the time it takes for one-half of the original amount of a substance to be degraded (first-order degradation). Degradation rates are generally temperature and moisture dependent, and most biological breakdown occurs in the root zone of plants because more microbes exist there than elsewhere. Therefore, as compounds are leached to lower depths (below the root zone), they become more persistent and can leach to groundwater before surface organisms have had a chance to degrade them. Typically, root zones can vary from 1 to 6 feet below the surface, depending on the plant and its stage of development.

The chemical makeup of leachate can be altered or its volume reduced through several processes. Roots of actively photosynthesizing plants can scour out essential nutrients or absorb percolating water, denitrification and mineralization can remove or immobilize dissolved nitrogen, chemical alterations such as oxidation to less motile forms can occur (e.g., iron transformed in an oxidizing environment from a mobile, reduced ferrous Fe^{2+} state to an immobile, oxidized ferric Fe^{3+} state), cations (e.g., Na^+, NH_4^+, Ca^{2+}, Mg^{2+}, etc.) can be adsorbed onto negatively charged surfaces, and shallow subsurface agricultural tile drains can siphon off leachate before it percolates more deeply (into groundwater).

The concern with leaching is that it accounts for the bulk of nonpoint source pollution of groundwater and subsequently to any surface water to which it is hydrologically connected. Any chemicals or suspended solids that are transported to groundwater may impair water quality for aquatic life or render it nonpotable or unfit for human contact (drinking, bathing, swimming). Lawn fertilizers, agricultural chemicals (pesticides, herbicides, fungicides, miticides, etc.), road salts, petrochemical spills, atmospheric particulate fallout (Hg, Pb, oxides of sulfur or nitrogen, etc.) can all end up in groundwater via leaching.

High-risk environments that are of primary concern include:

- sandy soils
- shallow-rooted plants
- high rainfall areas or excess irrigation
- shallow unprotected groundwater
- wellhead protection zones
- homesteads with open wells

Leaching affects the receiving waterbody (e.g., a change in the trophic state of a lake receiving discharge from contaminated groundwater), and also in agricultural systems, represents a two-pronged economic loss: fertilizer purchases go wasted if nutrients end up in places other than the root zone, and plant nutrients made unavailable by leaching result in lower harvests. Essential plant nutrients, naturally occurring or supplemented by fertilizers, that are lost by leaching include:

- **Calcium (Ca)** ion generally in the largest concentration in leachate
- **Magnesium (Mg), Sulfur (S), Potassium (K)** next highest concentrations, depending on soil composition and acidity of leachate
- **Nitrogen (N)** ranges from high (recent fertilizer application on porous soil prior to intense rainfall) to low (natural systems with no disturbance or low precipitation), depending on plant growing conditions and natural soil fertility
- **Phosphorus (P)** limited concentration in leachate due to low solubility and high affinity for sorption onto soil particles

The U.S. Environmental Protection Agency (EPA) regulates several pollutants that routinely make their way into drinking water supplies and pose health risks for consumers in the United States. For example, the EPA currently stipulates that drinking water cannot exceed 10 mg per liter (ppm) nitrate (NO_3^-); this poses significant problems for drinking water suppliers in rural areas that rely on aquifers under agricultural fields for their water source. The EPA limits lead (Pb) in drinking water to 15 µg per liter (ppb), a challenge in older urban areas that still have lead service pipes as part of their underground water systems. Recently, the EPA issued guidelines for arsenic (Ar) in drinking water; the standard was lowered to a maximum of 10 ppb beginning January 23, 2006. In addition to these specific thresholds for drinking water, the EPA also regulates waterbodies (surface, groundwater, drinking water aquifers) for dissolved salts, fecal coliform bacteria, harmful chemicals, and more, some of which are attributable to leaching processes.

Acknowledgment

This article originally appeared in the *Water Encyclopedia* , 5 Vols. J. H., Lehr, and J. Keeley, Editors, John Wiley & Sons, Inc, Hoboken, NJ, 2005.

Additional Reading

Brady, N.C. and R.R. Weil: *The Nature and Properties of Soils*, 13th Edition, Pearson Education, Upper Saddle River, NJ, 2002.

Hillel, D.: *Environmental Soil Physics: Fundamentals, Applications, and Environmental Considerations*, Academic Press, San Diego, CA, 1998.

Jury, W.A. and R. Horton: *Soil Physics*, 6th Edition John Wiley & Sons, Inc., New York, NY, 2004.

McBride, M.B.: *Environmental Chemistry of Soils*, Oxford University Press, New York, NY, 1994.

Nolan, B.T., B.C. Ruddy, K.J. Hitt, and D.R. Helsel: "Risk of Nitrate in Groundwaters of the United States — A National Perspective," *Environ. Sci. Technol.*, **31**, 2229–2236 (1997).

Sparks, D.L.: *Environmental Soil Chemistry*, 2nd Edition, Elsevier Science & Technology Books, New York, NY, 2002.

Sawhney, B.L. and K. Brown: *Reactions and Movement of Organic Chemicals in Soils*, Soil Science Society of America, Spec. Publ. No. 22, Madison, WI. 1989, http://www.soils.org/

DAVID W. KELLEY, University of St. Thomas, Saint Paul, MN

LEAD. Lead [CAS: 7439-92-1]. Chemical element, symbol Pb. at. no. 82, at. wt. 207.2, periodic table group 14, mp 327.5 °C. bp 1740 °C, density 11.35 g/cm³. (20 °C). Elemental lead has a face-centered cubic structure with an edge length of 4.950 Å.

Lead is a white to bluish-gray metal, soft, malleable, and slightly ductile; tarnishes in air, forming a film of oxide, forms oxide scum upon heating the molten metal in air; soluble in dilute HNO_3; HCP or H_2SO_4 attack lead only slightly, the extent depending markedly upon the concentration and the temperature; slowly dissolves in H_2O and consequently the use of lead constitutes a health hazard due to its toxic effect; attacked by solutions of organic acids or sodium hydroxide. Lead is one of the four most largely produced and utilized metals, and considerable scrap metal is recovered. Used (1) in construction and apparatus where workability is demanded, and definite resistance to corrosion is supplied by the metal, (2) as a constituent of various alloys, especially solder, type metal, pewter, and fusible alloys, (3) for storage battery plates, (4) for shot and bullets, and (5) as a protective coating for iron and steel.

Lead has four naturally occurring isotopes. In order of abundance, these are ^{208}Pb, ^{206}Pb, ^{207}Pb, and ^{204}Pb. There are ten unstable isotopes, 200–203, 205, and 209–214. See also **Radioactivity**. In terms of abundance, lead is scarcely represented in the earth's crust, the average composition of igneous rocks containing only 0.002% Pb by weight. In terms of cosmic abundance, an estimate made by Harold C. Urey in 1952,

using silicon as a basis with the figure of 10,000, lead had an abundance figure of less than 0.02. In terms of presence in seawater, lead is 27th among the elements, with an estimated 14 tons per cubic mile (3 metric tons per cubic kilometer) of seawater. In this regard, it is comparable to tin, copper, arsenic, protactinium, and selenium.

The atomic weight varies because of natural variations in the isotopic composition of the element, caused by the various isotopes having different origins: ^{208}Pb is the end product of the thorium decay series, while ^{207}Pb and ^{206}Pb arise from uranium as end products of the actinium and radium series respectively. Lead-204 has no existing natural radioactive precursors. Electronic configuration $1s^2 2s$; $&2 2p^6 3s^2 3p^6 3d^{10} 4s^2 4p^6 4d^{10} 4f^4 5s^2 5p^6 5d^{10} 6s^2 6p^2$. Ionic radius Ph^{2+} 1.18 Å. Pb^{44} 0.70 Å. Metallic radius 1.7502 Å. Covalent radius (sp^3) 1.44 Å. First ionization potential 7.415 eV; second, 14.97 eV. Oxidation potentials Pb → Pb^{2+} + 2e$^-$, 0.126 V; Pb^{2+} + 2H$_2$O → PbO$_2$+ 4H$^+$ + 2e$^-$, −1.456 V; Pb + 2OH$^-$ → PbO + H$_2$O + 2e$^-$, 0.576 V; Pb+ 3OH$^-$ → HPbO$_2^-$ + H$_2$O + 2e$^-$, 0.54 V. Other physical properties are given under **Chemical Elements**.

Lead is of interest as being the terminal product of radioactive decay. Thus, while ordinary lead has the atomic weight 207.19 (being composed of 1.37% ^{204}Pb, 26.26% ^{206}Pb, 20.8% ^{207}Pb and 51.55% ^{208}Pb), the isotopic composition, and hence the atomic weight, varies somewhat in lead from meteorites, from deep-seated rocks and from uranium ores (the last being somewhat less dense, as would be expected from the fact that ^{206}Pb is the end-product of the uranium series). These variations in isotopic composition of lead permit of calculations of the age of the earth (and the meteorites).

Lead Melting Point as a Standard

Melting, defined as the equilibrium transition between crystalline and liquid states, is of large concern in the development of the physical and materials sciences. To date, some of the purest crystals of silicon, diamond, and other technologically important materials have been produced from melts. Studies of melts also are of significance in understanding the interiors of terrestrial plants and, in fact, of Earth. In research at the University of California (Berkeley), studies of the effects of high pressure on the fusion temperature of lead have been underway. The advantages of studying lead are outlined by the investigators as: (1) the melting temperature of lead at ambient pressures is low and well determined, (2) lead is highly compressible and therefore should show the effects of pressure, (3) the behavior of lead under pressure is relatively simple, involving only one known polymorphic transition (from face-centered cube to hexagonal close-packed crystal structure), and (4) shock-wave experiments have been carried out previously to document the compression of both crystalline and molten lead at simultaneously high pressures and temperatures.

Occurrence and Processing

Galena [CAS: 1314-87-0], PbS, lead sulfide, is the source of over 95% of the lead currently produced. Bodies containing galena range from 3% to 30% lead. One of the most widely distributed sulfide minerals, galena frequently occurs along with sphalerite [CAS: 1314-98-3], ZnS. The lead-zinc ores processed usually contain recoverable quantities of copper [CAS: 7440-50-8], silver [CAS: 7440-22-4], antimony [CAS: 7440-36-0], and bismuth [CAS: 7440-69-9]. Principal sources being worked are in Australia's Broken Hill area in New South Wales, the western United States, Canada, Mexico, Peru, former Yugoslav Republics, and the former Soviet Union. When groundwater reacts with galena, cerussite, PbCO$_3$, is formed; when galena is in contact with sulfate solutions generated by the oxidation of sulfide minerals, anglesite, PbSO$_4$, may be formed. See also **Anglesite**; **Cerussite**; and **Galena**.

In processing, the ore first is crushed, wet-ground, and classified to a point where it is at least 90% less than 200 mesh. Separation of the sulfide ore from the gangue is aided by flotation agents. The resulting concentrates contain from 45% to 60% lead, from zero to 15% zinc, and often a few ounces (∼50 grams) of gold and up to 50 ounces (1.4 kilograms) of silver per ton. Copper content may be as much as 3%, arsenic, 0.4%, and antimony, 2%. The sulfur content (10 to 30%) is reduced by roasting in a Dwight-Lloyd sintering machine. This sulfur reduction is necessary because PbS is not reduced by carbon or carbon monoxide at blast-furnace temperatures. Once formed, the sinter, together with limestone and coke, is fed into a blast furnace. Further oxidation and electrolytic methods may be used to refine the lead. Lead is commercially produced to standards of very high purity. The minimum lead content permitted by specifications for Pig Lead (7 classifications) is 99.73%. Fully refined lead averaging 99.99% lead is obtainable. Large quantities are used for production of chemicals. At one time, primary uses for lead chemicals were in the production of paint pigments and lead tetraethyl gasoline additive.

Lead Metals and Alloys

Lead is soft and ductile and is readily worked by common methods, predominantly by rolling and extruding. Lead is easily formed and readily joined by welding (burning), or by soldering and can be bonded to steel, or used as a liner for steel, wood, concrete, and other materials. Lead is widely used in this manner because of its excellent resistant to atmospheric and soil corrosion, and attack by sulfuric and phosphoric acids. Lead generally does not resist the action of the organic acids, nor the oxidizing mineral acids, such as HNO_3. Lead is attacked by alkalies.

Due to its low melting point, pure lead will very gradually flow or creep at room temperature. Thus, lead sheeting used as a roofing material on old buildings will usually be thicker at the lower edge than at the upper edge. Other examples of creep occur under low sustained stresses due to the oil pressure in lead-covered power conducting cable, for example, or due to the weight in the case of a deep tank lined with sheet lead. To counter the effects of creep, lead containing 0.06% copper (*chemical lead* or *acid lead*) is preferred.

The addition of antimony in amounts up to 12% greatly improves the casting properties and increases the hardness very materially. These properties make possible the casting of intricately shaped antimonial lead storage-battery grids which, including the weight of the lead oxide paste applied to them, constitute the largest single use for the metal.

Tin and lead in various proportions form a highly useful series of alloys generally known as the soft solders which are used for joining copper, iron, nickel, lead, zinc and even glass. The solders can be applied by means of a soldering tool, by wiping, by hot-dipping, or by special machines as in the tin-can industry. Numerous compositions are used, the most popular of which are listed in the accompanying table.

Further additions of bismuth, cadmium, and antimony to the tin-lead alloys result in the low melting or "fusible" alloys widely used as safety devices, the melting points of which can be varied to suit a wide range of requirements. The type metals of the printing industry are lead-tin-antimony alloys having the requisite hardness and good casting properties needed for high-fidelity reproduction.

Babbitt metals (white-metal bearing alloys) are generally classified as either tin-base or lead-base. The true tin-base Babbitts contain only tin, antimony and copper, and have been used for many years. The practice of adding up to 25% lead to the tin Babbitts to reduce their cost is to be avoided since the net result is an expensive series of alloys with inferior properties to the inexpensive lead-base Babbitts. The lead-base bearing alloys of the older type usually contain lead, antimony and tin, and while not considered the equal of the tin-base alloys for severe service have been widely employed due to their low cost. The lead-base alloy containing arsenic has found extensive use and has come to the fore of this group since it has successfully met many automotive and other severe service requirements. All of these alloys render their most efficient service when used in the form of a thin lining bonded to a bronze or steel shell. See Table 1.

Lead Eliminated from Free-Cutting Alloys

Among the numerous efforts being made to eliminate lead from the environment, including the potable water plumbing systems, free-cutting copper alloys that contain no lead have been developed. As reported in late 1991, bismuth, as a replacement, has significant potential as a nontoxic alternative to lead to enhance the machinability of copper. When bismuth is used alone, however, the element embrittles copper because of its tendency to "set" grain boundaries. J.T. Plewes (see reference) ascribes this characteristic to the large difference in surface tension between copper and bismuth. It has been found that adding a third element in modest amounts removes this limitation of bismuth. Such elements include phosphorus, indium, and tin.

Chemistry of Lead

A number of oxides of lead are known, but not all are daltonide compounds. Thus, lead(I) oxide [CAS: 1309-60-0], Pb_2O, made by heating lead(II)

TABLE 1. REPRESENTATIVE LEAD AND TIN ALLOYS

Name	Pb	Sn	Sb	Cu	Bi	Ag	Cd	Typical Application
Lead Alloys								
Chemical or acid lead	99.9			.06				Tank linings, coils, etc., power cable sheath.
Cable sheath	98.9		1.0					Telephone cable sheath.
Hard lead	96–92		4–8					Cast shapes, wrought sheet and pipe.
Battery grid metal	92–88	.25	8–12					Cast battery grids.
Solders								
Soft solder	50	50						General purposes, most popular solder.
Wiping solder	60	40						For wiping joints in cables, lead pipes, etc.
	60	37.5	2.5					
"Fine solder"	40	60						For making joints at low temperature.
Solder	95–97.5					5–2.5		High temperature solder.
Fusible Alloys								
Wood's metal	25	12.5			50		12.5	Melts in hot water at 154 °F. Wets glass. Wide range of melting points possible with changes in composition for automatic sprinkler systems and other safety devices.
Matrix metal	28.5	14.5	9		48			For anchoring punches, etc., in jigs and fixtures. Expands on freezing.
Bending alloy	26.5	13.5			50		10	Filler for tubes, etc., during bending. Melts out in hot water.
Type Metals								
Electrotype	93	3	4					
Linotype	84	4	12					
Stereotype	80.5	5.75	13.75					
Monotype	76	8	16					Single type.
Tin Base Babbitts								
	89	7.5	3.5					General usage.
	83.3	8.3	8.3					Hard Babbitt.
Lead Base Babbitts								
	82.5	1.0	15	.5	1.0 As			General usage.
	80	5	15					General usage.
	75	10	15					General usage.

Notes: Figures given in percent. Wood's metal melts at ~68 °C in water.

oxalate, has been shown by x-ray analysis to be a mixture of the metal and lead(II) oxide [CAS: 1317-36-8], PbO. The latter is obtained by heating lead in air, which yields a yellow, rhombic material, which has a peculiar layer structure having each lead atom attached to four oxygen atoms all lying on the same side of it, forming a square pyramid with the lead at the apex. Each oxygen atom is surrounded tetrahedrally by four lead atoms. Another form of PbO, somewhat more stable and soluble in water, red in color, and tetragonal in structure, may be obtained along with the yellow form by alkaline dehydration of $Pb(OH)_2$. PbO is amphiprotic, but only weakly acidic. Lead(IV) oxide [CAS: 1314-41-6], Pb_2O, is obtained by action of chlorine on alkaline solutions of lead(II) oxide or acetate. The reaction is $Pb(OH)_3^- + ClO^- \rightarrow PbO_2 + Cl^- + OH^- + H_2O$. PbO_2 can also be produced on a lead or platinum anode by electrolysis in acidic solution. Like the lower elements of main group 4, lead(IV) forms tetrahedral bonds exhibiting sp^3 hybridization. In its relatively more stable salts, however, the $6s^2$ electrons are unused, and Pb^{2+} ions are formed by loss of the $6p^2$ electrons. These facts explain the marked difference between the essentially covalent character of many of the tetravalent compounds and the essentially electrovalent character of the divalent compounds, as well as the peculiar structure of PbO and many other Pb(II) compounds.

The dioxide, Pb_2O, has rutile structure, and the compound is a strong oxidizing agent. It is also amphiprotic, giving unstable lead(IV) salts with acids, and orthoplumbates, $M_4^1PbO_4$, or metaplumbates, $M_2^1PbO_3$, upon fusion with alkalies. Lead dioxide dissolves in aqueous alkali with formation of the ion $Pb(OH)_6^{2-}$, the alkali salts of which are isomorphous with the corresponding stannates and platinates. Lead sesquioxide, Pb_2O_4, has been shown not to exist as a stable phase.

Lead orthoplumbate, Pb_2PbO_4, red lead, is similarly described as a salt, in this case an orthoplumbate of divalent lead, Pb_2PbO_4, because on treatment with nitric acid, two-thirds of the lead dissolves and one-third remains as PbO_2. It is prepared in the red form by atmospheric heating of PbO, and in a black form by reaction of PbO with pure oxygen. Red lead is formed of PbO_6 octahedra (with one common edge) linked by lead atoms covalently bonded to three oxygen atoms.

The lead dihalides are known for all four of the common halogens. They are not strictly ionic in the anhydrous state, but they dissolve in (hot) water to give Pb^{2+} ions, more or less hydrated. They are much less soluble in cold water. They also form complex compounds such as M_2PbCl_4, MPb_2Cl_5, M_4PbF_6, and $MPbF_3$, where M is an alkali metal. The compound formed, especially of the fluoroplumbates(II) depends somewhat on the alkali metal, some of which form/nondaltonide (berthollide) compounds. Of the lead tetrahalides, only PbF_4 and $PbCl_4$ are known, the fluoride being prepared by fluorination of PbF_2. The chloride, which easily loses chlorine, is made by careful acidification of a hexachloroplumbate(IV). $PbCl_4$ forms the complex compound ammonium hexachloroplumbate, $(NH_4)_2PbCl_6$, upon addition to its solution of solid ammonium chloride.

Lead(II) inorganic compounds and salts of organic acids are far more numerous than those of lead(IV), as is to be expected from the essentially covalent character of the latter. In addition to the oxides and halides already discussed, there are lead(II) compounds of essentially all of the common anions, including many basic compounds. Thus lead(II) chloride forms such basic compounds as $PbCl_2 \cdot Pb(OH)_2$, $PbCl_2 \cdot PbCl_2 \cdot 2PbO$, $PbCl_2 \cdot 3PbO$, and $PbCl_2 \cdot 7PbO$. In fact, a whole series of lead salts are derived from the hydroxide, some of which are double compounds, such as $PbX_2 \cdot 2Pb(OH)_2$ and some of which, of composition Pb(OH)X, have been shown to be dimeric of the general formula

$$\left[Pb \begin{pmatrix} HO \\ \\ HO \end{pmatrix} Pb \right] X_2$$

Other Lead Compounds

Lead Acetates. Anhydrous lead acetate [CAS: 301-04-2] (plumbous acetate), $Pb(C_2H_3O_2)_2$, is a white, crystalline solid that decomposes on heating above its melting point; some physical properties are given in Table 2. Because of its high solubility in water, lead acetate is often used for the preparation of other lead salts by the wet method. Lead acetate is made by dissolving lead monoxide (litharge) or lead carbonate in strong acetic acid. Several types of basic salts are formed when lead acetates are prepared from lead monoxide in dilute acetic acid or at high pH. The basic

TABLE 2. PHYSICAL PROPERTIES OF LEAD ACETATES

Property	Anhydrous	Basic	Trihydrate	Tetraacetate
mol wt	325.28	807.69	379.33	443.77
mp, °C	280	75 (200 dec)	75 (200 dec)	175
d, g/cm³	3.25		2.55	2.228
refractive index, n_D			1.567[a]	
soly, g/100 mL H_2O				
at 15 °C	44.3[b]	6.25	45.61	
100 °C	221[c]	25	200	

[a] Along the $\beta\beta$-axis.
[b] At 20 °C.
[c] At 50 °C.

salts of lead acetate are white crystalline compounds, which are highly soluble in water and dissolve in ethyl alcohol.

Basic Lead Acetate. Basic lead acetate [CAS:1335-32-6] (lead subacetate), $2Pb(OH)_2Pb(C_2H_3O_2)_2$, is a heavy white powder which is used for sugar analyses. Some physical properties are given in Table 2. Reagent grade is available in 11.3-kg cartons and in 45- and 147-kg fiber drums.

Lead Acetate Trihydrate. Lead acetate trihydrate [CAS: 6080-56-4] (plumbous acetate trihydrate), $Pb(C_2H_3O_2)_2 \cdot 3H_2O$, is a white, monoclinic crystalline solid; some physical properties are given in Table 2. Upon heating it loses some of its water of crystallization, and after melting, decomposes at 200 °C. The trihydrate is highly soluble in water but insoluble in ethyl alcohol. It has an intensely sweet taste, hence it is sometimes called sugar of lead, but it is poisonous. The trihydrate is made by dissolving lead monoxide in hot dilute acetic acid solution; on cooling, large crystals separate, sometimes up to 60-cm long.

Lead acetate trihydrate, the usual commercial form, is used in the preparation of basic lead carbonate and lead chromate, as a mordant in cotton dyes, as a reagent for the manufacture of lead salts of higher fatty acids, as a water repellant, as a component in combined toning and fixing baths for daylight printing papers, and as a means of treating awnings and outdoor furniture to prevent removal of mildew- and rot-proofing agents by rain or laundering. Other uses include preparation of rubber antioxidants; processing agent in the cosmetic, perfume, and toiletry industries; component of coloring agents for adhesives; and preparation of organic lead soaps as driers of paints and inks.

Lead Tetraacetate. Lead tetraacetate [CAS: 546-67-8] (plumbic acetate), $Pb(C_2H_3O_2)_4$, is a colorless, monoclinic crystalline solid that is soluble in chloroform and in hot acetic acid, but decomposes in cold water and in ethyl alcohol. Some physical properties are given in Table 2. Lead tetraacetate can be prepared by adding warm, water-free, glacial acetate acid to red lead, Pb_3O_4, and subsequent cooling. The salt decomposes with the addition of water to give PbO_2, but the yield can be improved by passing in chlorine gas. Lead tetraacetate is available in laboratory quantities as colorless to faintly pink crystals stored in glacial acetic acid.

Oxidation with lead tetraacetate is often used in organic syntheses, because the lead salt is highly selective in the splitting of vicinal glycols. The rate of oxidation of cis glycols is more rapid than of the trans isomers, a property widely used in the structural determination of sugars and other polyols. Lead tetraacetate readily cleaves α-hydroxy acids as oxalic acid at room temperature. Another use is the introduction of acetoxy groups in organic molecules, as in the preparation of cyclohexyl acetate and the acetoxylation of cyclohexanol. At high temperature, methylation takes place. In these reactions, the organic molecule must contain double bonds or activating substituents.

Lead Antimonate. Lead antimonate [CAS: 13510-89-9] (Naples yellow), $Pb_3(SbO_4)_2$, mol wt 993.07, is an orange-yellow powder that is insoluble in water and dilute acids, but very slightly soluble in hydrochloric acid. Lead antimonates are modifiers for ferroelectric lead titanates, pigments in oil-base paints, and colorants for glasses and glazes. They are made by the reaction of lead nitrate and potassium antimonate solutions, followed by concentration and crystallization.

Lead Arsenate. Lead arsenate [CAS: 10031-13-7], arsenate of lead $Pb_3(AsO_4)_2$, white precipitate, formed by reaction of soluble lead salt solution and sodium arsenate solution. Used as an insecticide. Banned or tightly controlled in some countries.

Lead Azide. Lead azide [CAS: 13424-46-9], $Pb(N_3)_2$, mol wt 291.23, crystallizes as colorless needles. It is a sensitive detonating agent, exploding at 350°C. Lead azide is commonly prepared by the reaction between dilute solutions of lead nitrate and sodium azide. For safety, it is stirred vigorously to prevent formation of large crystals, which may detonate. Lead azide is usually precipitated with a protective material, such as gelatin, and then granulated. Lead azide is also used to prepare electrophotographic layers and for information storage on styrene–butadiene resins.

Lead Benzoate. Lead benzoate monohydrate [CAS: 6080-57-5], $Pb(C_6H_5CO_2)_2 \cdot H_2O$, mol wt 467.43, is a white crystalline powder that loses its water of hydration when heated to 100°C. It is slightly soluble in cold water (0.16 g/ 100 mL at 20°C) and somewhat more soluble in warm water (0.31 g/100 mL at 49.5°C). The salt may be prepared by adding benzoic acid to a slurry of litharge, PbO, or by the reaction between solutions of sodium benzoate and a soluble lead compound. Lead benzoate is used as an antioxidant in organolead engine lubricants, as a catalyst in a blowing agent for polyethylene foams, and in fluorescence quenching of organic phosphors.

Lead Borate. Lead borate monohydrate [CAS: 14720-53-7] (lead metaborate), $Pb(BO_2)_2 \cdot H_2O$, mol wt 310.82, $d = 5.6$ g/cm^3 (anhydrous) is a white crystalline powder. The metaborate loses water of crystallization at 160°C and melts at 500°C. It is insoluble in water and alkalies, but readily soluble in nitric and hot acetic acid. Lead metaborate may be produced by a fusion of boric acid with lead carbonate or litharge. It also may be formed as a precipitate when a concentrated solution of lead nitrate is mixed with an excess of borax. The oxides of lead and boron are miscible and form clear lead-borate glasses in the range of 21 to 73 mol% PbO.

The main use of lead metaborate is in glazes on pottery, porcelain, and chinaware, as well as in enamels for cast iron. Other applications include as radiation-shielding plastics, as a gelatinous thermal insulator containing asbestos fibers for neutron shielding, and as an additive to improve the properties of semiconducting materials used in thermistors.

Lead Carbonates. Lead carbonate [CAS: 598-63-0], $PbCO_3$, mol wt 267.22, $d = 6.6$ g/cm^3, forms colorless orthorhombic crystals; it decomposes at about 315°C. It is nearly insoluble in cold water (0.00011 g/100 mL at 20°C), but is transformed in hot water to the basic carbonate, $2PbCO_3 \cdot Pb(OH)_2$. Lead carbonate is soluble in acids and alkalies, but insoluble in alcohol and ammonia. It is prepared by passing CO_2 into a cold dilute solution of lead acetate, or by shaking a suspension of a lead salt less soluble than the carbonate with ammonium carbonate at a low temperature to avoid formation of basic lead carbonate.

Lead carbonate has a wide range of applications. It catalyzes the polymerization of formaldehyde to high molecular weight crystalline poly(oxymethylene) products. It is used in poly(vinyl chloride) friction liners for pulleys on drive cables of hoisting engines. To improve the bond of polychloroprene to metals in wire-reinforced hoses, 10–25 parts of lead carbonate are used in the elastomer. Lead carbonate is used as a component of high pressure lubricating greases; as a catalyst in the curing of moldable thermosetting silicone resins; as a coating on vinyl chloride polymers to improve their dielectric properties; as a component of corrosion-resistant, dispersion-strengthened grids in lead–acid storage batteries; as a photoconductor for electrophotography; as a coating on heat-sensitive sheets for thermographic copying; as a component of a lubricant-stabilizer for poly(vinyl chloride); as a component in the manufacture of thermistors, and as a component in slip-preventing waxes for steel cables to provide higher wear resistance.

Basic Lead Carbonate. Basic lead carbonate [CAS: 1319-46-6] (white lead), $2PbCO_3 \cdot Pb(OH)_2$, mol wt 775.67, $d = 6.14$ g/cm^3, forms white hexagonal crystals; it decomposes when heated to 400°C. Basic lead carbonate is insoluble in water and alcohol, slightly soluble in carbonated water, and soluble in nitric acid. It is produced by several methods, in which soluble lead acetate is treated with carbon dioxide. For example, in the Thompson-Stewart process, an aqueous slurry of finely divided lead metal or monoxide, or a mixture of both, is treated with acetic acid in the presence of air and carbon dioxide. High quality, very fine particle-size basic lead carbonate is produced, ranging in carbonate content from 62 to 65% (vs 68.9% $PbCO_3$, theoretical).

Although white lead was the oldest white hiding pigment in paints, it has been totally replaced by titanium dioxide, which has better covering power and is nontoxic. Nevertheless, basic lead carbonate has many other uses, including as a catalyst for the preparation of polyesters from terephthalic acid and diols; a ceramic glaze component; a curing agent with peroxides to form improved polyethylene wire insulation; a pearlescent pigment; a color-changing component of temperature-sensitive inks; a red-reflecting pigment in iridescent plastic sheets; a smudge-resistant film on electrically sensitive recording sheets; a lubricating grease component; a component of ultraviolet light reflective paints to increase solar reflectivity; an improved cool gun-propellant stabilizer which decomposes and forms a lubricating lead deposit; a heat stabilizer for poly(vinyl chloride) polymers; and as a component of weighted nylon-reinforced fish nets made of poly(vinyl chloride) fibers.

Lead Chromates. Lead chromate, [CAS: 7758-97-6], "chrome yellow" $PbCrO_4$, yellow precipitate, by reaction of soluble lead salt solution and sodium dichromate or chromate solution, melting point of lead chromate 844°C. Used as a pigment; basic lead chromate, red solid, insoluble, formed by heating lead chromate and sodium hydroxide solution.

Lead Halides

Lead Fluoride. Lead difluoride, PbF_2, is a white orthorhombic salt to about 220°C where it is transformed into the cubic form; some physical properties are given in Table 3. Lead fluoride is soluble in nitric acid and insoluble in acetone and ammonia. It is formed by the action of hydrofluoric acid on lead hydroxide or carbonate, or by the reaction between potassium fluoride and lead nitrate.

Lead fluoride has been used in low power fuses; as a catalyst for the manufacture of picoline. See also **Pyridine and Derivatives**. In glass coatings for infrared reflection; in low melting glasses; in phosphors for television-tube screens; in activators for electroless plating of nickel on glass; in electrooptical coatings; and in zinc oxide varistors.

Lead Tetrafluoride. Like all the lead tetrahalides, lead tetrafluoride [CAS: 7783-59-7], PbF_4, is very reactive. It is relatively the most stable halide, however. PbF_4 is a white crystalline powder, which is highly moisture sensitive, turning yellowish brown in moist air owing to hydrolysis. It should be handled in a dry box or under an atmosphere of dry nitrogen. Properties for PbF_4 are in Table 3.

Lead Chloride. Lead dichloride, $PbCl_2$, forms white, orthorhombic needles; some physical properties are given in Table 3. Lead chloride is slightly soluble in dilute hydrochloric acid and ammonia and insoluble in alcohol. It is prepared by the reaction of lead monoxide or basic lead

TABLE 3. PHYSICAL PROPERTIES OF LEAD HALIDES

Property	PbBr$_2$	PbCl$_2$	PbF$_2$	PbF$_4$	PbI$_2$
CAS Registry Number	[10031-22-8]	[7758-95-4]	[7783-46-2]	[7783-46-2]	[10101-63-0]
mol wt	376.04	278.1	245.21	283.2	461.05
mp, °C	373	501	855	600	402
bp, °C	916	950	1290	decomposes	954
d, g/cm^3	6.66	5.85	8.24	6.7	6.16
soly, g in 100 mL H$_2$O					
at 0°C	0.455	0.673		^{a}	0.044
20°C		0.99	0.064		0.063
100°C	4.71	3.34			0.41

^{a}Material hydrolyzes to PbO_2 and HF.

carbonate with hydrochloric acid, or by treating a solution of lead acetate with hydrochloric acid and allowing the precipitate to settle. It easily forms basic chlorides, such as $PbCl_2 \cdot Pb(OH)_2$ [CAS: 15887-88-4], which is known as Pattinson's lead white, an artist's pigment.

Lead dichloride is the starting material for a number of organolead compounds. It has been used in asbestos clutch or brake linings; as a co-catalyst for acrylonitrile production; as a catalyst for polymerization of olefins to highly crystalline, stereoregular polymers; as a cathode for magnesium-lead dichloride seawater batteries; to make rectifying junctions on gallium arsenide; as a flame retardant in polycarbonates and nylon-6,6 wire coatings; as a flux for the galvanizing of steel; as a solid-phase chemical scrubber for ozone and hydrogen sulfide removal from gas; as a photochemical-sensitizing agent for metal patterns on printed circuit boards; and as a sterilization indicator on tapes that darken with zinc sulfide at 121 °C in moist air.

Lead Bromide. Lead dibromide, $PbBr_2$, forms white orthorhombic crystals; some physical properties are given in Table 3. Lead(II) bromide is slightly soluble in ammonia and highly soluble in potassium bromide solutions owing to complex formation, soluble in acetic acid, but insoluble in alcohol. On exposure to light, lead dibromide decomposes slowly and darkens because of release of lead. It is prepared from lead monoxide or carbonate and hydrobromic acid, or lead diiodide and lead(IV) bromide, or by treating an aqueous solution of lead nitrate with hydrobromic acid or a soluble metal bromide and allowing the precipitate to settle.

Lead bromide is a photopolymerization catalyst for acrylamide monomer and is used in photoduplication at exposures of 365-nm radiation. Black-gray positive images are obtained on a white background by applying a methyl alcohol solution of lead bromide in a poly(vinyl butyral) binder to a suitable substrate and then exposing the coating to light through a negative film. In another photographic process, the latent image is developed by reduction of $PbBr_2$ with a sulfur-containing reducing agent, such as mercaptoacetic acid. Lead dibromide used as an inorganic filler in fire-retardant polypropylene, polystyrene, and acrylonitrile–butadiene–styrene (ABS) plastics reduces the requirements of chlorinated hydrocarbon flame-resistant additives. For welding aluminum or aluminum-base alloys to other metals, such as iron, nickel, copper, zinc, or their alloys, an aqueous paste containing $PbBr_2$ serves as an excellent general-purpose welding flux.

Lead Iodide. Lead diiodide, PbI_2, forms a powder of yellow hexagonal crystals; some physical properties are given in Table 3. Lead diiodide is soluble in alkalies and potassium iodide, and insoluble in alcohol. It is made by treating a water-soluble lead compound with hydroiodic acid or a soluble metal iodide. It is readily purified by recrystallization in water.

Lead iodide decomposes when exposed to green light at about 180 °C, thereby making it possible to record optical images on thin lead iodide films. other applications of lead iodide include photographic emulsions with thiols; aerosols for cloud seeding to produce rain artificially; asbestos brake linings; primary thermal batteries with iodine; mercury-vapor arc lamps; thermoelectric materials; lubricating greases; electrosensitive recording papers; and filters for far-infrared astronomy.

Lead Nitrate. Lead nitrate, [CAS: 10099-74-8], $Pb(NO_3)_2$, mol wt 331.23, sp gr 4.53, forms cubic or monoclinic colorless crystals. Above 205 °C, oxygen and nitrogen dioxide are driven off, and basic lead nitrates are formed. Above 470 °C, lead nitrate is decomposed to lead monoxide and Pb_3O_4. Lead nitrate is highly soluble in water (56.5 g/100 mL at 20 °C; 127 g/100 mL at 100 °C), soluble in alkalies and ammonia, and fairly soluble in alcohol (8.77 g/ 100 mL of 43% aqueous ethanol at 22 °C). Lead nitrate is readily obtained by dissolving metallic lead, lead monoxide, or lead carbonate in dilute nitric acid. Excess acid prevents the formation of basic nitrates, and the desired lead nitrate can be crystallized by evaporation.

Lead nitrate is used in many industrial processes, ranging from ore processing to pyrotechnics to photothermography. Thus lead nitrate is used as a flotation agent in titanium removal from clays; in electrolytic refining of lead; in rayon delustering; in red lead manufacture; in matches, pyrotechnics, and explosives; as a heat stabilizer in nylon; as a coating on paper for photothermography; as an esterification catalyst for polyesters; as a rodenticide; as an electroluminescent mixture with zinc sulfide; as a means of electrodepositing lead dioxide coatings on nickel anodes; and as a means of recovering precious metals from cyanide solutions.

Lead Oxalate. Lead oxalate, PbC_2O_4, white precipitate, formed by reaction of soluble lead salt solution and ammonium oxalate solution, yields lead suboxide on heating at 300 °C out of contact with air.

Lead Phthalates. Two commercial forms of lead phthalates, both dibasic, are widely used as heat stabilizers in poly(vinyl chloride) (PVC) polymers and copolymers. During processing, usually extrusion, and in actual service, thermal degradation of PVC occurs principally by a dehydrochlorination mechanism. Thus one of the primary functions of dibasic lead phthalate stabilizers is to neutralize and inactivate the resulting hydrogen chloride. Such stabilizers are ideally suited for high temperature applications of PVC because of their low reactivity with plasticizers, particularly of the polyester type. Moreover, dibasic lead phthalates provide the long-term stability and retention of elongation required in 90 and 105 °C Underwriters' Laboratories classes of wire insulation.

Dibasic Lead Phthalate. [CAS: 17976-43-1], $2PbO \cdot Pb(O_2C)_2C_6H_4 \cdot \frac{1}{2}H_2O$, is a white, crystalline powder, mol wt 826.87, sp gr 4.6, lead oxide content 79.8% PbO, moisture loss (2 h at 105 °C), 0.3%; sieve analysis, 99.9% through 44 μm (325 mesh); water solubility, nil. In PVC, it provides excellent heat stability; excellent processibility, allowing high extrusion rates; excellent electrical properties over a wide range of temperatures and insulation classifications; good compatibility and low reactivity with a broad range of plasticizers; and for vinyl foams, it is an effective activator for azodicarbonamide-type blowing agents. Other applications include flexible extruded and molded PVC compounds, where it provides good resistance to early color development during processing. In vinyl plastisols, it provides low viscosity build-up on aging.

Lead Phosphate. Lead phosphate [CAS: 7446-27-7], $Pb_3(PO_4)_2$, white precipitate, by reaction of soluble lead salt solution and sodium phosphate solution.

Lead Phosphite. In commercial applications of poly(vinyl chloride) polymers where weathering resistance, thermal stability, and electrical insulating properties are required, a stabilizer system base on dibasic lead phosphite provides a unique balance of properties. Its plasticizer reactivity is in the same range as dibasic lead phthalate, its electrical properties are superior, and it is the only stabilizer known that can provide the required electrical properties and weathering resistance in the absence of carbon black pigmentation. A properly formulated PVC electrical insulation compound containing a dibasic lead phosphite stabilizer, in combination with rutile-type titanium dioxide, remains in serviceable condition for up to 20 years. This superior performance results from the high absorption of the ultraviolet portion of sunlight, as well as the antioxidant activity of the phosphite anion. See also **Antioxidant**. The high PbO content of this dibasic lead salt makes it a very effective acid acceptor for HCl during PVC processing.

Dibasic Lead Phosphite. Dibasic lead phosphite [CAS: 12141-20-7], $2PbO \cdot PbHPO_3 \cdot \frac{1}{2}H_2O$, is a white crystalline powder, mol wt 742.63, sp gr 6.9, refractive index 2.25, lead oxide content 90.2%, sieve analysis, 99.8% through 44 μm (325 mesh) (wet), water solubility, nil. Fields of application for dibasic lead phosphite stabilizers in PVC include garden hose, flexible and rigid vinyl foams (as high temperature activator for azodicarbonamide-type blowing agents), coated fabrics, plastisols, electrical insulation, and extruded profiles for outdoor use. In general, at five parts per hundred of resin (5 phr), dibasic lead phosphite provides good heat stability and superior outdoor weathering properties in PVC. This stabilizer should be stored in closed containers, away from open flames, and at temperatures below 200 °C. Exposure to sparks or static electricity should be avoided by grounding all electrical equipment and using wooden scoops.

Lead Silicates. Lead monosilicate [CAS: 10099-76-0] (lead pyrosilicate), $1.5PbO \cdot SiO_2$, is a light yellow trigonal crystalline powder, insoluble in water. Its composition, by weight, is 85% PbO and 15% SiO_2. Lead monosilicate is commercially available as granular, <1.68 mm (10 mesh), and ground, 97% through 44 μm (325 mesh). It provides the most economical method of introducing lead into a ceramic glaze. It is also used as a source of PbO in the glass industry.

Lead Bisilicate. Lead bisilicate [CAS: 11120-22-2], $PbO \cdot 0.03Al_2O_3 \cdot 1.95SiO_2$, is a pale yellow powder, insoluble in water. Its composition, by weight, is 65% PbO, 1% Al_2O_3, and 34% SiO_2. Lead bisilicate is available as granular, <1.68 mm (10 mesh), and ground, 88% through

44 µm (325 mesh). It was developed as a low solubility source of lead in glazes, where its high viscosity and low volatility are equally important.

Tribasic Lead Silicate. Tribasic lead silicate [CAS: 12397-06-7], 3PbO · SiO$_2$, is a reddish yellow powder, sparingly soluble in water. Its composition by weight is 92% PbO and 8% SiO$_2$. Tribasic lead silicate is available as granular, <1.68 mm (10 mesh), and ground, 99% through 44 µm (325 mesh). It is used primarily by glass and frit manufacturers and has the lowest viscosity of the three commercial lead silicates. Commercial lead silicates are generally prepared by melting lead monoxide and silica in the desired ratio.

Lead Sulfates. Lead forms a normal and an acid sulfate and several basic sulfates. Basic and normal lead sulfates are fundamental components in the operation of lead-sulfuric acid storage batteries. Basic lead sulfates also are used as pigments and heat stabilizers in vinyl and certain other plastics.

Lead Sulfate. Lead sulfate [CAS: 7446-14-2], PbSO$_4$, is soluble in concentrated acids and alkalies, forming hydroxyplumbites; some physical properties are given in Table 4. It is prepared by treating lead oxide, hydroxide, or carbonate with warm sulfuric acid, or by treating a soluble lead salt with sulfuric acid. Lead sulfate forms in lead storage batteries during discharge cycles. It has been used in photography in combination with silver bromide and is used in the stabilization of clay soil for adobe structures, earth-fill dams, and roads.

TABLE 4. PHYSICAL PROPERTIES OF LEAD SULFATES

Property	PbSO$_4$	PbOPbSO$_4$
CAS Registry Number	[7446-14-2]	[12765-51-4]
mol wt	303.25	526.44
mp, °C	1170[a]	977
d, g/cm^3	6.2	6.92
soly, g/100 mL H$_2$O		
at 25 °C	4.25×10^{-3}	4.4×10^{-3} [b]
40 °C	5.6×10^{-3}	
crystal structure	orthorhombic, monoclinic	monoclinic

[a] Decomposes above 900 °C.
[b] At 0 °C.

Monobasic Lead Sulfate. Monobasic lead sulfate PbOPbSO$_4$, is very slightly soluble in hot water and slightly soluble in sulfuric acid; some physical properties are given in Table 4. Basic lead sulfate can be prepared by fusing PbO and PbSO$_4$ or by boiling aqueous suspensions of these two components. The resultant white solid is filtered and dried. Basic lead sulfate is used in paints as a white pigment, in poly(vinyl chloride) (PVC) plastics as a heat stabilizer, in rubbers as an inert filler, and as additives in textile dyeing and printing

Dibasic Lead Sulfate. Dibasic lead sulfate [CAS: 12036-76-9], 2PbOPbSO$_4$, is a white powder, mol wt 749.70, mp 961 °C. The dibasic compound can be prepared by fusion of the two components. It has been sold as a PVC stabilizer in Japan and is sold in Europe in combination with dibasic lead phosphite.

Tribasic Lead Sulfate. Tribasic lead sulfate [CAS: 12202-17-4], 3PbO · PbSO$_4$ · H$_2$O is a fine white powder, mol wt 890.93, sp gr 6.9, refractive index 2.1, lead oxide content 90.1%, sieve analysis, 99.8% through 44 µm (325 mesh) (wet), water solubility at 18 °C. Tribasic lead sulfate is by far the most widely used basic lead sulfate for the stabilization of PVC polymers. It may be prepared by boiling aqueous suspensions of lead oxide and lead sulfate. The anhydrous compound decomposes at 895 °C. Tribasic lead sulfate provides efficient, long-term heat stability in both flexible and rigid PVC compounds, it is easily dispersible and has excellent electrical insulation properties, and it is an effective activator for azodicarbonamide-type blowing agents for vinyl foams. Applications for tribasic lead sulfate stabilizers include thermal stabilization of flexible PVC wire insulation compounds containing phthalate-type plasticizers, wire insulation designed to meet Underwriters' Laboratories specifications through 80 °C, rigid and flexible PVC foams, rigid vinyl profiles, and PVC plastisols. The usual range of tribasic lead sulfate required in PVC is between two and seven parts per hundred of resin (2–7 phr), depending on the intended application of the vinyl product.

Lead Sulfide. Lead sulfide, plumbous sulfide, [CAS: 1314-87-0], PbS, brownish-black precipitate, formed by reaction of soluble lead salt solution and hydrogen sulfide or sodium or ammonium sulfide, soluble in dilute nitric acid.

In the great majority of organometallic compounds of lead, the metal is tetravalent and covalently bonded, although the organolead group includes many compounds with both organic radicals and halogen atoms attached to Pb which are not to be described merely as covalent compounds. More than five hundred organometallic compounds of lead have been reported, many of which are named as substituted plumbanes, although PbH$_4$ is not a starting point in their production. Tetraethyl lead, Pb(C$_2$H$_5$)$_4$, is made from a sodium–lead alloy and ethyl chloride.

Like carbon and silicon, and to a lesser extent, germanium and tin, lead forms binary compounds with metals, such as Na$_4$ Pb$_7$ and Na$_4$Pb$_9$. These materials are essentially salt-like, and contain polyplumbide anions. They are of theoretical interest, because they are intermediate in character between stoichiometric compounds (daltonide compounds) and intermediate phases. The two compounds cited dissolve in liquid ammonia, electrolyze in such solutions to give the metals, and apparently form ions such as [Pb$_7$]$^{4-}$ and [Pb$_9$]$^{4-}$ which readily form amine complexes.

Lead Telluride. Lead telluride [CAS: 1314-91-6], PbTe, forms white cubic crystals, mol wt 334.79, sp gr 8.16, and has a hardness of 3 on the Mohs' scale. It is very slightly soluble in water, melts at 917 °C, and is prepared by melting lead and tellurium together. Lead telluride has semiconductive and photoconductive properties. It is used in pyrometry, in heat-sensing instruments such as bolometers and infrared spectroscopes, and in thermoelectric elements to convert heat directly to electricity. Lead telluride is also used in catalysts for oxygen reduction in fuel cells, as cathodes in primary batteries with lithium anodes, in electrical contacts for vacuum switches, in lead-ion selective electrodes, in tunable lasers, and in thermistors. See also **Fuel Cells**; and **Lasers**.

Lead Titanate. Lead titanate [CAS: 12060-00-3] (lead metatitanate), PbTiO$_3$, mol wt 302.09, d = 7.52 g/cm^3, forms yellow tetragonal crystals below 490 °C and cubic crystals above 490 °C. It is insoluble in water. In hydrochloric acid, lead titanate decomposes into PbCl$_2$ and TiO$_2$. It can be formed by calcining an equimolecular mixture of lead monoxide and titanium dioxide and has been used in surface coatings as a pigment in outdoor paints, as a component of ceramic electrical insulators, in ceramic ferroelectric-piezoelectric compositions, in ceramic electrical capacitors, in ceramic glazes, in transducers, in low melting glass sealants, and in oxidation catalysts for manufacturing acrylonitrile from propylene and nitrous oxide.

Lead Zirconate. Lead zirconate [CAS: 12060-01-4], PbZrO$_3$, mol wt 346.41, has two colorless crystal structures: a cubic perovskite form above 230 °C (Curie point) and a pseudotetragonal or orthorhombic form below 230 °C. It is insoluble in water and aqueous alkalies, but soluble in strong mineral acids. Lead zirconate is usually prepared by heating together the oxides of lead and zirconium in the proper proportion. It readily forms solid solutions with other compounds with the ABO$_3$ structure, such as barium zirconate or lead titanate. Mixed lead titanate-zirconates have particularly high piezoelectric properties. They are used in high power acoustic-radiating transducers, hydrophones, and specialty instruments

Lead in Biological Systems — Toxicity

Lead has been identified as a biological system deterrent for decades. It is only recently, however, that studies pertaining to low-dosage exposures of lead have been published despite the fact that probably millions of words have appeared in various publications on the overall topic of lead poisoning.

From a qualitative standpoint, exposure to lead results in a clinical picture of hypertensive encephalopathy, neuropathy, and hemolytic anemia characterized by coarse basophilic stippling in red blood cells. The mechanism of lead's action on human tissue is complex. For one thing, lead blocks heme synthesis. This leads to a build-up of red blood cell protoporphyrin. Lead interferes with cell metabolism by causing a deficiency of pyrimidine 5'-nucleotidase. Lead attacks erythrocyte membrane phospholipids with resultant loss of potassium and interference with the sodium-potassium balance. Diagnosing lead poisoning may involve a determination of the free erythrocyte protoporphyrin level as well as determination of blood and urine levels. Once confirmed, further exposure to lead must be stopped immediately. Chelating compounds, such as $CaNa_2EDTA$, may be administered intravenously over an 8-hour period for several days. This may be followed by treatment with oral penicillamine for several days.

Lead poisoning can lead to chronic renal failure. In its effect on kidney function, lead acts much like cadmium. Chronic exposure to or ingestion of practically any heavy metal, such as lead, is the most common path to polyneuropathy. Where effects of heavy metals on the peripheral nervous system are suspected, many physicians will require testing for metal in hair, fingernails, serum, and urine of the patient. Habitual sniffing of leaded gasolines can lead to lead poisoning. Scientists have compared the lead concentration in the diets of present Americans (0.2 part per million) with the diets of prehistoric peoples (estimated to be less than 0.002 part per million). Some investigators believe that the presence of "natural" lead contamination has been grossly overestimated and that what has appeared to be natural has been the result mainly of a gradual build-up of lead pollution in the air derived from anthropogenic sources. The principal sources of atmospheric lead contamination include (1) natural sources, such as wind-blow volcanic dust, sea spray, forest foliage, and volcanic sulfur compounds; and (2) anthropogenic sources, such as lead alkyls (present in fuels), iron smelting, lead smelting, zinc and copper smelting, and the burning of coal. Much remains by way of research into the sources of lead contamination, including the contributions of atmospheric pollution, of food containers, and of food processing equipment.

In 1990, H.I. Needleman and co-researchers (University of Pittsburgh, Boston University, and Harvard University) reported their findings on the long-term effects of exposure to low doses of lead in children. An abstract of the report is as follows:

To determine whether the effects of low-level lead exposure persist, we reexamined 132 of 270 young adults who had initially been studied as primary school-children in 1975 through 1978. In the earlier study, neurobehavioral functioning was found to be inversely related to dentin lead levels. As compared with those we restudied, the other 138 subjects had somewhat higher lead levels on earlier analysis, as well as significantly lower IQ scores and poorer teachers' ratings of classroom behavior.

When the 132 subjects were reexamined in 1988, impairment in neurobehavioral function was still found to be related to the lead content of teeth shed at the ages of six and seven. The young people with dentin lead levels >20 ppm had a markedly higher risk of dropping out of high school (adjusted odds ratio, 7.4; 95 percent confidence interval, 1.4 to 40.7) and of having a reading disability (odds ratio, 5.8; 95 percent confidence interval, 1.7 to 19.7) as compared with those with dentin lead levels <10 ppm. Higher lead levels in childhood were also significantly associated with lower class standing in high school, increased absenteeism, lower vocabulary and grammatical-reasoning scores, poorer hand—eye coordination, longer reaction times, and slower finger tapping. No significant associations were found with the results of 10 other tests of neurobehavioral functioning. Lead levels were inversely related to self-reports of minor delinquent activity.

We conclude that exposure to lead in childhood is associated with deficits in central nervous system functioning that persist into young adulthood.

An interesting professional critique of the Needleman report is summarized by J. Palca (reference listed).

The lead elimination and clean-up problem has numerous parallels with the asbestos pollution problem. The main problem is not one of finding substitutes for these substances, because lead-free paints, for example, have been available for several years, just as substitute insulating materials for asbestos have been found. As pointed out in an excellent article by Pollack (reference listed), the problem (or dilemma) lies with the cleanup of old structures that have such materials installed; and to what limits must one go, within the limitations of financial resources, to remove such materials; and, once removed, how to dispose of them safely. While removing all lead-painted surfaces from schools, for example, ultimately can provide assurance that children will not be exposed to the long-term effects of lead, a great deal of new exposure to workmen and the immediate neighborhoods of such removal projects can occur. Modern technology has designed equipment to protect the safety of restoration or demolition crews, but

there remains the enforcement of their using such equipment. One crux of the problem is that posed by the apparent effects of lead in very low dosages. Such pollutants of low concentration, as may be typified by the creation of dust and windborne aerosols, definitely exacerbates the problem. Obviously, the problem becomes one more hampered by socioeconomic measures than by technology.

Additional Reading

Bodwal, B.K. et al.: "Ultralight-Pressure Melting of Lead: A Multidisciplinary Study," *Science*, **462** (April 27, 1990).

Davis, J.R.: *Metals Handbook*, 2nd Edition, ASM International, Materials Park, OH, 1998.

Greenwood, N.N. and A. Earnshaw: *Chemistry of the Elements*, 2nd Edition, Butterworth-Heinemann, Inc., Woburn, MA, 1997.

Holden, C.: "Resurrected Lead," *Science*, **192** (October 11, 1991).

Holden, C.: "Toxic Waste Program Lacks Science Base," *Science*, **797** (November 8, 1991).

Krebs, R.E.: *The History and Use of Our Earth's Chemical Elements: A Reference Guide*, Greenwood Publishing Group, Inc., Westport, CT, 1998.

Lide, D.R.: *CRC Handbook of Chemistry and Physics*, 88th Edition, CRC Press, LLC., Boca Raton, FL, 2007 .

Needleman, H.L. et al.: "The Long-Term Effects of Exposure to Low Doses of Lead in Childhood," *N. Eng. J. Med.*, **83** (January 11, 1990).

Palca, J.: "Get-the-Lead-Out Guru Challenged," *Science*, **842** (August 23, 1991).

Plewes, J.T. and D.N. Loiacono: "Free-Cutting Copper Alloys Contain No Lead," *Advanced Materials & Processes*, **23** (October 1991).

Pollack, S.: "Solving the Lead Dilemma," *Technology Review (MIT)*, **22** (October 1989).

Raloff, J.: "Beverages Intoxicated by Lead in Crystal," *Science News*, **54** (January 26, 1991).

Stwertka, A. and E. Stwertka: *A Guide to the Elements*, Oxford University Press, Inc., New York, NY, 1998.

LEAD-ACID BATTERY.   See **Battery**; and **Electric Car**.

LEADER (or Leader Streamer).   The electric discharge that initiates each return stroke in a cloud-to-ground lightning discharge. It is a channel of high ionization that propagates through the air by virtue of the electric breakdown at its front produced by the charge it lowers. The stepped leader initiates the first stroke in a cloud-to-ground flash and establishes the channel for most subsequent strokes of a lightning discharge. The dart leader initiates most subsequent strokes. Dart-stepped leaders begin as dart leaders and end as stepped leaders. The initiating processes in cloud discharges are sometimes also called leaders but their properties are not well measured.

LEAD SCREW.   The screw that controls the longitudinal motion of a tool on a lathe or other machine tool. Also, the screw that drives the cutting head across a recording disk in the initial recording process.

LEAD SULFIDE.   See **Galena**.

LEAF.   The food-manufacturing organ of a plant. Typically, leaves consist of a broad thin blade borne on a slender stalk. The important chemistry taking place within a leaf is described under **Photosynthesis**.

The leaf originates as a small protuberance from the surface of the growing tip of the stem. Numerous divisions of the cells of this protuberance produce a structure from five to eight cells thick. Many of these leaf primordia are borne together on the stem tip, and, together with any protecting scales that may cover them, form the buds of the stem. At first, all the cells of these leaf primordia are alike. Very early in their existence, however, certain cells become distinct by their somewhat elongated shape. These cells are the beginnings of the vascular elements. Cells divisions continue in these small bodies until there are present in the bud recognizable but very small leaves, which are folded in various ways. In woody plants, this development takes place in the year previous to that in which the leaf will unfold. With the advent of the new growing season, growth of the many minute cells of these tiny leaves is very rapid, so that within a few days' time the leaf has unfolded and grown to its mature size. During this enlargement, many changes have taken place in the cells of the leaf.

The mature leaf is commonly composed of two distinct parts, the broadly expanded, thin green blade, and the petiole or stalk which supports it and

Fig. 1.   Leaf of apple, illustrating all parts — blade, petiole, and stipules.

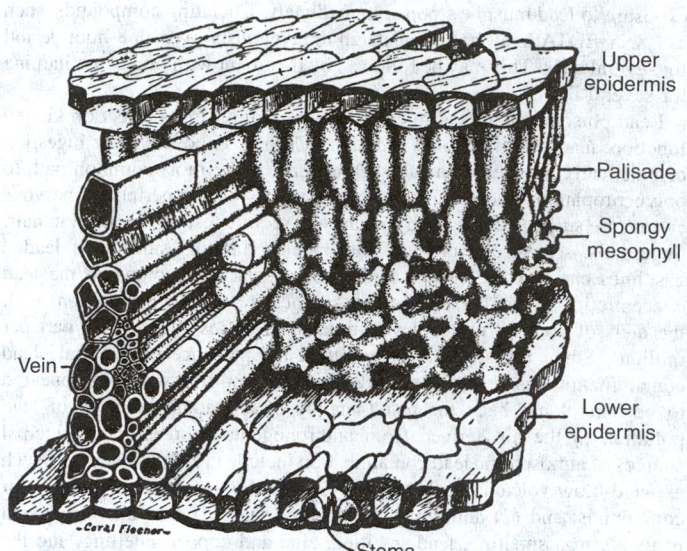

Fig. 2.   A portion of the blade of a leaf cut so as to show the internal structure. The cell contents are not shown.

connects it with the stem. In many plants there is formed at the base of the petioles a pair of outgrowths called stipules, which in some plants may take the form of a complete sheath. See Fig. 1. This sheath is well developed in members of the Umbelliferae. Sometimes the petiole is completely lacking, the blade being attached directly to the stem; leaves of this kind are called sessile leaves. Less frequently the blade of the leaf is lacking, the petiole being expanded into a flattened object looking much like a blade. Certain Australian trees, species of *Acacia* and *Eucalyptus*, exhibit this peculiarity. Leaves of such plants often show progressive changes from those having well-developed blades to those in which the blade is completely lacking, showing clearly that the flattened portion present is a modified petiole. Such flattened petioles are not uncommon, but usually the blade is present, as is the case in the lemon tree. Leaves may be deciduous, falling off at the end of a single growing season, or evergreen and persistent through several seasons. In nearly all cases the leaf fall is brought about by the development of a definite abscission layer. In many plants, such a layer is formed not only at the base of the petiole, but also at the point where the petiole joins the blade.

The shape of the blade is extremely varied, ranging from very slender linear leaves to those that are broader than they are long. The margin of the leaf may be entire, that is, without indentations of any sort, or toothed or lobed in various ways, until some are incised nearly to the midrib. If the leaf is completely divided into separate segments, it is said to be a compound leaf, in contrast with the undivided leaves, which are simple leaves, no matter how deeply they may be lobed. If the sections of a compound leaf all come from a common point, the leaf is said to be palmately compound; if they are borne along a central axis, the leaf is pinnately compound. While such infinite variations do exist, the leaves of any single species of plant are recognizably constant in shape.

The blade of the leaf is supported by a framework of veins, which are also very characteristically arranged. In many leaves, especially in dicotyledons, one vein, usually extending through the center of the blade, is more prominent than the others. This is called the main vein or midrib. The others are lateral veins. In most dicotyledons, the veins branch abundantly to form an intricately anastomosing network, which reaches all parts of the leaf. In most monocotyledons, the midrib and lateral veins extend in parallel lines from base to apex of the leaf. Between these, many minute veinlets exist, too small to be readily seen, reaching all parts of the leaf.

The cellular structure in leaves is very constant. See Fig. 2. Covering the entire surface of the leaf is the epidermis, a layer of tabular cells. On the upper surface of the leaf, the epidermal cells are frequently covered with a layer of cutin, a waxy substance that is impervious to water and so greatly reduces the loss of water by evaporation from the leaf surface. Epidermal cells contain a scant peripheral cytoplasm, and a large central vacuole full of cellsap. Usually, there are no chloroplasts present in the epidermal cells. The cells of the epidermis of the lower surface are similar to those of the upper, but with a less evident cuticle. In the epidermis of the leaf, particularly that of the lower surface, there are many minute openings, called stomata, which permit a ready exchange of gases between the interior of the leaf and the external air. Each stoma is surrounded by a pair of guard cells containing chloroplasts. These cells close the stoma by collapsing and open it by expanding. All cells occurring between the upper and lower epidermal layers are called mesophyll cells. Beneath the upper epidermis the mesophyll cells form a very distinct layer, called the palisade mesophyll. These are elongated cells with their long axis perpendicular to the surface of the leaf. They contain large numbers of chloroplasts. In them,

furthermore, active photosynthesis takes place. Occupying all the rest of the leaf is a loose tissue composed of irregularly arranged rounded cells known as the spongy mesophyll. Numerous intercellular spaces separate these cells from one another. Ramifying through the leaf just below the palisade cells are the veins. Each vein is composed of three types of cells. Some of them are thick-walled xylem cells, which carry water and dissolved mineral matter to all parts of the leaf. Others are phloem cells, which carry food away from the green cells of the leaf where they are elaborated. The xylem cells are toward the top of the leaf, the phloem cells toward the bottom. Outside these and often forming a conspicuous tissue are the masses of fibers, or collenchyma, thick-walled cells that give support to the leaf.

Leaves are often greatly modified. See Fig. 3. In many plants, they become greatly enlarged and fleshy, and serve as organs of storage of water and food. Many rock garden plants, such as species of *Sedum*, have leaves of this type. Of similar nature are the scale-like leaves that form the greater part of many bulbs, such as those of many lilies. The common onion is composed of the closely enwrapped bases of leaves, swollen with food material. In other plants, modification of the leaves becomes extreme, as, for example, in the pitcher plants and bladderworts. See also **Insectivorous Plants**.

In other plants, such as the common barberry, the leaf is reduced to sharp-pointed branched spines; in many cases all gradations between these spines and typical leaves may be found on a single branch. In some plants, as the Locust, *Robinia pseudoacacia*, only the stipules are modified to short sharp spines. Many plants have leaves modified into tendrils, slender thread-like objects, which twine tightly around any suitable object with which they may come in contact. Sometimes only the tip of the blade functions in this way, and sometimes only the stipules are thus modified, as in the Carrion flower, *Smilax herbacea*. Many plants of the legume family have pinnately compound leaves, some of the segments of which are changed into tendrils. Weirdest of all are the leaves of species of *Nepenthes*, one of the pitcher plants. (See **Insectivorous Plants**, where this leaf is described.)

In a few plants, the leaf becomes a vegetative reproductive body, having in the notches of its margin, at its tip, or less commonly on its surface, groups of meristematic cells, which, when the leaf is mature, give rise to tiny plants that remain attached to the parent leaf for some time. Among the plants in which reproduction of this type occurs are species of *Bryophyllum* and *Kalanchoë*.

The principal function of the leaf is to carry on photosynthesis. To do this, the leaf must receive adequate light. Leaves are not distributed haphazardly on the stem, but in a very definite way which assures them the maximum of light. In many plants, the leaves are in pairs on opposite sides of the stem. Each successive pair usually grows out at right angles to the pair beneath it, thus preventing overshadowing. Leaves may occur

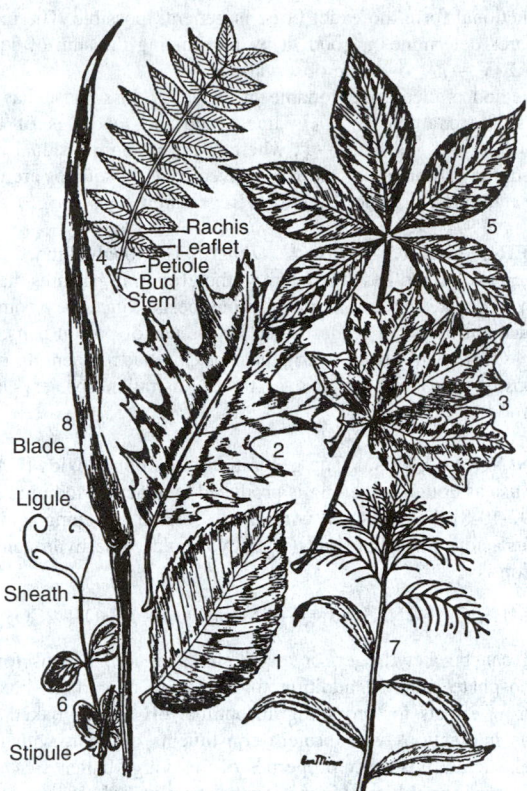

Fig. 3. Types of leaves: (1,2) elm leaf and oak leaf, both pinnately netted veined; (3) maple leaf, palmately netted veined; (4) black walnut leaf, pinnately compound; (5) buckeye leaf, palmately compound; (6) a pea leaf, with stipules, tendrils, and two unmodified leaflets; (7) portion of a plant of the water mermaid, *Proserpinaca*, with upper leaves modified by immersion in water; (8) grass leaf.

to gardens. It inserts a virus with its tiny mouth parts, causing leaves to curl. The species *Nephotettix* is a pest to rice fields and has caused much damage in India. The insect stunts the growth of the plant by robbing juices and causing wilting.

Other leaf hopper species include the *apple leaf hopper* (*Empoasca mali*). The adults are of several colors, ranging from green to brown and yellow; or they may be striped. The insects are wedge-shaped and attain a length of about $\frac{1}{8}$ inch (3 millimeters). The nymphs are similar to the adults, but smaller, and crawl sideways like crabs. Action of the insects causes leaves to curl and turn yellow or reddish brown. The apple leaf hopper is found throughout the United States.

Young trees are the most seriously infested by leaf hoppers. The insects usually attack the underside of the leaves. Control should be directed toward the young nymphs; adult leaf hoppers often escape by flying away when disturbed. To control young trees infested by leaf hoppers, the tips of affected branches can be dipped into a container of soap solution, using about 1 pound of soap per 8 gallons of water (about 250 grams of soap per 15 liters of water). Dipping, which kills some of the young leaf hoppers, should be done in the latter part of June and again one month later. This is the period when the maximum number of nymphs will be found on the trees. Many adult leaf hoppers can be captured as they fly away by placing a shield covered with a sticky substance close to the tree.

Four generations of leaf hoppers are produced each year. The eggs, laid in blisters under the bark of the tree, winter over. Eggs laid in the summer are placed in leaf veins and petioles.

The *grape-vine leaf hopper* (*Typhlocyba comes*) is a small yellow-colored insect, sometimes mistakenly called thrips. This pest is most prevalent in the western United States. It sucks sap from the underside of leaves, causing the leaves to become brown and brittle. In treating for this hopper, care must be taken to thoroughly reach the underside of the foliage.

LEAF INSECT (*Insecta, Orthoptera*). Large insects of the Old World tropics related to the walking-stick insects. They have leaf-like wings and in some species the body and legs are extended in flat processes which also resemble leaves.

LEAF MINER (*Insecta, Lepidoptera*). Larval insects that work in the soft tissue of leaves between the upper and lower epidermis. They are necessarily small and are sometimes able to complete their development on a very small part of the food available in a single leaf. The burrow or mine shows as a brownish or transparent patch in the leaf and its form is characteristic of the insect making it. The larvae of many of the smallest moths and of some sawflies are leaf miners.

LEAF ROLLER (*Insecta, Lepidoptera*). Also sometimes referred to as the *leaf tyer* or leaf sewer, these small moths are usually members of the family *Tortricidae*. Their descriptive name derives from their practice (in caterpillar stage) of rolling all or part of a leaf into a cylindrically shaped case, tying the case with natural gum threads, and then lining the case with silk and thus forming a cocoon wherein the insect transforms into the pupa stage. Researchers have observed that several larvae may work cooperatively to form a common "nest." There are several species, each of which builds a characteristic nest.

The adults of the *apple leaf roller* (*Archips argyrospila*) are brown moths with light markings on the wings and having a wingspan of about $\frac{3}{4}$ inch (18–19 millimeters). The larvae are from pale yellow to a dirty green in color, with brown or black heads, and ranging up to $\frac{3}{4}$ inch (18–19 millimeters) in length. Light yellow, green, or grayish eggs are laid on branches in masses of 10–19. The red-banded leaf roller has a broad, reddish-brown band across the wings. The larvae feed on buds, fruit, and leaves. The leaves are webbed together to form a tent or cocoon as previously described. The larvae eat irregular holes in leaves and fruit.

Distribution is throughout the United States. The red-banded leaf roller is confined to the eastern United States and ranges as far west as the Mississippi Valley.

A number of parasites and predators attack the leaf rollers. Toads eat many caterpillars that drop from the trees; birds also prey upon the caterpillars. Since the insect overwinters in the egg stage and deposits its eggs on the twigs and bark of the tree, it is possible to control the first

in whorls, in which case there will be three or more leaves growing from each node of the stem. In many plants the leaves are alternate, each node bearing a single leaf. In every case, alternate leaves arise from the stem in such a way that a line passing around the stem and through the junction of the petiole with the stem forms a regular spiral. Examination of this spiral shows that the leaves are distributed on it in a very exact mathematical arrangement. In the simplest case, the leaves are in two longitudinal rows along the stem, every third leaf being directly above the first; in the next arrangement there are three longitudinal rows, the fourth leaf of the spiral being above the first. Other more complicated arrangements are found. The arrangement of leaves on a stem is called phyllotaxy.

Sometimes the exact arrangement is more or less obscured by twisting of the stem during growth. The leaves themselves turn considerably during their growth, petioles twisting to one side or the other, or elongating unequally, in such a way as to bring the blade into a position to receive the most favorable light.

LEAF HOPPER (*Insecta, Homoptera*). Any insect of the large family *Cicadellidae* or *Jassidae*. They are small to moderate jumping insects, and they often come freely to light.

Many species are of economic importance and, since they have sucking mouths, they must be attacked with contact poisons such as nicotine sulfate or kerosene emulsion. These sprays are effective against the tender immature insects.

The body of the insect is slender and long, with a round head. The four legs are stout and strong. A hair-like antenna is below each eye. Sometimes they are called "dodgers" because of their habit of dodging around various objects to escape attention.

The leaf hopper exudes a sweet liquid from its abdomen in a fashion similar to the aphid.

In Australia, the *Eurymela* group lives on eucalyptus leaves. These insects are attended by ants, much as the aphids. The potato and apple leaf hopper is the *Empoasca fabae*. The species *Eutettix tenellus* is a pest brood by spraying with a dormant fruit tree oil spray. Also, the folded leaves can be pinched by hand to destroy the caterpillars inside. Debris should be burned after picking.

The *avocado leaf roller* (*Amorbia emipratella*) is a yellowish-green caterpillar with a pinkish-brown stripe approximately an inch (2.5 centimeters) long when fully mature. The insect rolls the leaves and eats small holes into the fruit, making it unmarketable.

The *strawberry leaf roller* (*Ancylis comptana*) has similar habits, with the larva, usually less than $\frac{1}{2}$ inch (12 millimeters) long, feeding and folding the leaves. The insect produces two broods per year.

LEAKAGE CURRENT. This is the current that flows or "leaks" along the surface or through the body of an insulator. Except under abnormal conditions such as dirty or moist surfaces or in electronic circuits having very minute currents the leakage is usually negligible.

LEAKAGE REACTANCE. This is the inductive reactance caused by the flux that links only one coil of a transformer. The useful flux, of course, links both windings and is the medium of transfer of energy between them. Leakage reactance is one of the major internal impedance components of the transformer.

LEAK DETECTION. See **Mass Spectrometry**.

LEAPFROG DIFFERENCING. A finite-difference approximation to a time evolution equation in which the time derivative is approximated with values one time step before and one time step ahead of the values that specify other terms of the equation.

The scheme $(f^{n+1} - f^{n-1})/2\Delta t = g(f^n)$ (where superscript n denotes a point in time, separated by step Δt from the prior $[n-1]$ and subsequent $[n+1]$ discrete time levels) is a leapfrog approximation to the differential equation $df/dt = g(f)$.

See also **Implicit Time Difference**.

AMS

LEAPFROGGING. The process of phasing, or delaying the ranging pulse of a tracking radar in order to move, or shift (on the radarscope presentation) the target blip past the target blip from another radar.

LEAPFROG TEST. In computer operation, a check routine which eventually occupies every possible position in the memory.

LEAPING MAMMALS. See **Rabbits and Hares**.

LEARNING DIFFICULTY. See **Dyslexia**.

LEASAT COMMUNICATION SATELLITES. See **Satellites (Scientific and Reconnaissance); Syncom Communication Satellites; and Weather Technology**.

LEAST ENERGY PRINCIPLE. A principle relating to stable equilibrium, and having very wide application. If a system is in stable equilibrium, any slight change in its condition or configuration, requiring the performance of work, will put it out of equilibrium, so that, if the system is now left to itself, it will return to its former state, and in so doing it will give up the energy imparted when it was disturbed. Consider, for example, a block of wood floating in a pail of water. If the block is lifted slightly, work is done and the center of mass of the wood-water system as a whole is raised, so that it now has more potential energy. The same would be true if the block were pushed a little farther into the water. In either case, when the block is released, it resumes its former level and the potential energy of the system diminishes to its former minimum value. This illustrates the general principle, which is that a system is in stable equilibrium only under those conditions for which its potential energy is at a minimum.

The principle of least energy is one aspect of the principle of virtual work.

LEAST SQUARES. Suppose that it is required to fit an equation of functional form $y = f(x_1, x_2, \ldots, x_p)$ to a series of observations on y and the x's. If the number n of observations exceeds the number of constants

in the functional form, no exact fit is, in general, possible. The method of least squares determines a good fit by minimizing the sum of squares of residuals $\Sigma(y - f)^2$ over the observations.

The method is clearly reasonable in all cases. In some it has optimal properties; for example, if f is linear and the model is of the type $y = \beta_0 + \beta_1 x_1 + \cdots + \beta_p x_p + \varepsilon$, where ε is a random residual normally distributed, the estimators of the β's derived by least squares are unbiased and have minimum variance. See also **Regression**.

LEATHER-JACKET. 1. *Insecta, Diptera.* The tough-skinned larvae of some species of crane flies. They live in the ground in pastures, hay fields, and grain fields and are sometimes serious pests. Since they come to the surface at night they can be destroyed by the use of poison baits.

2. *Pisces.* File fish related to trigger fish. Coastal; frequenting reefs and rocks, mostly poisonous but two or three Australian species said to be good as food.

LEAVENING AGENTS. The generation of carbon dioxide [CAS: 124-38-9] for use as dough leavening is produced by reacting sodium carbonate [CAS: 497-19-8] (baking soda) with one of several leavening acids. In the case of an acidic phosphate salt (with two replaceable hydrogen atoms), the reaction is:

$$MH_2PO_4 + 2NaHCO_3 \longrightarrow MNa_2PO_4 + 2H_2O + 2CO_2$$

where M can be a hydrogen or an alkali metal ion. Claims for use of acidic phosphate salts, in addition to formation of carbon dioxide, are the buffering effects for providing an optimal pH for the baked product, as well as interactions with protein constituents of flour, with resulting optimal elastic and viscosity properties of the dough batter.

Other leavening acids used in modern bakeries include sodium aluminum sulfate, $Na_2SO_4 \cdot Al_2(SO_4)_3$; sodium aluminum phosphate hydrate (and anhydrous) [CAS: 7785-88-8]; potassium bitartrate [CAS: 868-14-4], $C_4H_5KO_6$ (cream of tartar); and glucono-delta-Iactone [CAS: 90-80-2] $C_6H_{10}O_6$. The baker is concerned with (1) *dough rate of reaction* (DRR), a measure of the rate at which the leavening acid reacts with the baking soda during both the mixing stage and the holding period after mixing (bench action); and (2) *neutralizing value* or neutralizing strength, i.e., the weight of leavening acid required to neutralize a given weight of sodium bicarbonate. This value is used to compute the amount of leavening acid required to yield the needed amounts of leavening gas as well as its effect upon the pH of the baked goods.

Properties of the principal leavening acids are given in Table 1, which shows the most appropriate baking applications for each.

Baking powders, as prepared for the home baker and for use in premixes, usually incorporate, along with sodium bicarbonate, one of the following leavening acids: (1) potassium hydrogen tartrate (2 parts for 1 part sodium bicarbonate); (2) tartaric acid (infrequent), 1 part; (3) calcium hydrogen phosphate (crystallized), 1.5 parts; or (4) sodium aluminum sulfate or ammonium aluminum sulfate, 1.8 parts. With 7 parts by weight of this finely powdered mixture, there is usually mixed about 3 parts by weight of starch to diminish the effects of moisture in storage. In some cases, dry powdered egg albumin is added to decrease the loss of carbon dioxide upon wetting the flour and baking powder mixture when used. For some purposes, ammonium carbonate can be used alone, since upon heating this material furnishes both ammonia and carbon dioxide gases to make the product light. These gases escape from the product during the baking process. In selecting a baking powder, one must keep in mind the speed with which the components react at room temperature: alum-containing baking powders act slowly; phosphate baking powders have a medium speed; and tartrate baking powders act quickly to produce carbon dioxide. Hence, when using the latter type, it is necessary to bake quickly after mixing to eliminate the loss of too much gas.

Within the last several years, advantage has been taken of mixing different leavening acids in premixes and household baking powders. Because the use of emulsifiers in most cake mixes reduces the need for early leavening action, it is common practice to use combinations of slow-acting leavening acids that retain much of their leavening reaction for the baking stage. In mixes, the leavening process must be regarded as a system because, in addition to gas generation, the leavening system controls the pH of the finished product and thus affects crumb and crust

TABLE 1. PROPERTIES OF LEAVENING ACIDS

Chemical Name and Formula	Abbreviation	Relative Speed at Room Temperature	Neutralizing Value[1]
Sodium aluminum phosphate (anhydrous)	SALP	Medium	110
Sodium aluminum sulfate, $Na_2SO_4 \cdot Al_2(SO_4)_3$	SAS	Slow	100
Monocalcium phosphate (anhydrous), $CaH_4(PO_4)$	MCP	Slow	83
Monocalcium phosphate (monohydrate), $CaH_4(PO_4) \cdot H_2O$	$MCP \cdot H_2O$	Quite fast	80
Sodium acid pyrophosphate, $Na_2H_2P_2O_7$	SAPP	Medium	72
Glucono-delta-lactone, $C_6H_{10}O_6$	GDL	Slow	55
Potassiumbitartarate, $C_4H_5KO_6$	—	Medium to fast	50
Dicalcium phosphate dihydrate, $CaHPO_4 \cdot 2H_2O$	DCP	Very slow	33
Sodium aluminum phosphate hydrate	$SALP \cdot H_2O$	Slow	100

[1]Values in this column indicate the parts of sodium bicarbonate that will be neutralized by 100 parts of the leavening acid under nominal conditions. Values vary with composition of dough.

color, the intensity of flavor, as well as other properties. For various cakes, the optimum pH values are: white cakes, 6.9–7.2; yellow cakes, 7.2–7.5; chocolate or devil's food cakes, 7.1–8.0. Monocalcium phosphate (anhydrous) and sodium aluminum phosphate are frequently used together in white and yellow cake mixes; monocalcium phosphate and sodium acid pyrophosphate or dicalcium phosphate dihydrate are used in chocolate cake mixes. Generally, the combination will be comprised of 10–20% fast-acting leavening acid and 80–90% slow-acting leavening acid.

For pancake and waffle mixes, a common blend of leavening acids is 20–30% monocalcium phosphate monohydrate or monocalcium phosphate (anhydrous), combined with 70–80% sodium aluminum phosphate. A batter of this type can be prepared several hours in advance if retained under refrigeration. It has been observed that such a batter will sour before a serious loss of leavening power occurs.

Prepared biscuit mixes made of flour, shortening, and salt usually contain 30–50% monocalcium phosphate (anhydrous) and 50–70% sodium aluminum phosphate or sodium acid pyrophosphate. Self-rising flours and corn meals usually contain flour or corn meal, salt, soda, and leavening acid. Usually used in these products are combinations of sodium aluminum phosphate and monocalcium phosphate (anhydrous).

Refrigerated doughs available for preparation of biscuits, dinner rolls, and various sweet rolls, usually contain flour, water, shortening, nonfat milk solids (or dried whey solids), sugar (or corn sugar), salt, soda, and a leavening acid. Long-term refrigerated storage requires that only slow-acting leavening acids be used, frequently the sodium acid pyrophosphates. The latter have the disadvantage of possibly producing orthophosphates under certain conditions. The orthophosphates have a rather disagreeable, astringent flavor.

Unleavened Products. The principal unleavened bakery product is pie crust, which is low in moisture and high in fat content. The ingredients and method of preparation prevent the formation of a continuous gluten network through the dough mass. The porosity associated with leavened products is not desirable because the crust literally acts as a container and requires some strength.

Additional Reading

Amendola, J., and N. Rees: *Understanding Baking*, 3rd Edition, John Wiley & Sons, Inc., New York, NY, 2002.
Figoni, P.I.: *How Baking Works: Exploring the Fundamentals of Baking Science*, John Wiley & Sons, Inc., New York, NY, 2003.

LE CHÂTELIER'S PRINCIPLE. Let us perturb a system that is initially in stable equilibrium to a neighboring nonequilibrium state. Since the initial equilibrium is supposed to be stable, the system will return to an equilibrium state.

Theorems governing the behavior of perturbed systems are often known as *theorems of constraint or theorems of moderation*. The best known thermodynamic theorem of moderation is that of Le Châtelier-Braun, which in the form stated by Le Châtelier is:

"Any system in chemical equilibrium undergoes, as a result of a variation in one of the factors governing the equilibrium, a *compensating* change in a direction such that, had this change occurred alone it would have produced a variation of the factor considered in the *opposite direction*."

However, this principle suffers from a number of important exceptions. It is therefore preferable to study the "moderation" starting from the usual thermodynamic formalism without invoking a special principle.

LECITHIN. Lecithin and other phospholipids are of universal occurrence in living organisms. They are constituents of biological membranes and are involved in permeability, oxidative phosphorylation, phagocytosis, and chemical and electrical excitation.

Lecithin is not only used in the strict scientific sense to describe pure phosphatidylcholine, but also to describe crude phospholipid mixtures containing phosphatidylcholine (PC), phosphatidylethanolamine (PE), phosphatidylinositol (PI), other phospholipids, and a variety of other compounds such as fatty acids, triglycerides, sterols, carbohydrates, and glycolipids. Commercial lecithin is currently available in more than 40 different formulations varying from crude oily extracts from natural sources to purified and synthetic phospholipids. Many of these products are defined according to the stage of the purification process from which they are obtained and fall into three broad categories (Table 1) varying in their constituents both qualitatively and quantitatively.

TABLE 1. CATEGORIES OF COMMERCIAL LECITHIN

Natural	Refined	Modified
Plastic	*Deoiled*	*Physically*
unbleached		custom-blended
bleached		natural and refined
doubled-bleached		
Fluid	*Fractionated*	*Chemically*
unbleached	alcohol-soluble	
bleached	alcohol-insoluble	
double-bleached		*Enzymatically*

Industrial lecithins from a variety of sources are utilized. The main sources include vegetable oils (eg, soybean, cottonseed, corn, sunflower, rapeseed) and animal tissues (egg and bovine brain). However, egg lecithin and in particular soy lecithin are by far the most important in terms of quantities produced, so much so that the term soy lecithin and commercial lecithin are often used synonymously.

Properties

Commercial crude lecithin is a brown to light yellow fatty substance with a liquid to plastic consistency. Its density is 0.97 g/mL (liquid) and 0.5 g/mL (granule). The color is dependent on its origin, process conditions, and whether it is unbleached, bleached, or filtered. Its consistency is determined chiefly by its oil, free fatty acid, and moisture content. Properly refined lecithin has practically no odor and has a bland taste. It is soluble in aliphatic and aromatic hydrocarbons, including the halogenated hydrocarbons; however, it is only partially soluble in aliphatic alcohols. Pure phosphatidylcholine is soluble in ethanol.

Commercial lecithin is soluble in mineral oils and fatty acids but is practically insoluble in cold vegetable and animal oils. It is insoluble but infinitely dispersible in water. Commercial lecithin is a wetting and emulsifying agent. Lecithin is one of the very few natural and edible surface-active agents of this type that is soluble or dispersible in oil.

In general, the presence of fatty acid groups in the phospholipid molecule permits reactions such as saponification, hydrolysis, hydrogenation, halogenation, sulfonation, phosphorylation, elaidinization, and ozonization.

Manufacture and Processing

Crude soy lecithin is obtained as a by-product during the degumming process of soy oil. Only a minor proportion of the total lecithin that is potentially available in the vegetable processing industry is produced.

Separation of neutral and polar lipids, so-called deoiling, is the most important fractionation process in lecithin technology. A classic solvent for the deoiling is acetone.

Due to the possible environmental problems with acetone, new technologies are being developed for the production of deoiled lecithins like an ethanol-based extraction and fractionation or a process involving treatment of lipid mixtures with supercritical gases or supercritical gas mixtures.

Commercial Grades

There are six common grades of lecithin available including (1) clarified lecithins, (2) fluidized lecithins; (3) compounded lecithins; (4) hydroxylated lecithins; (5) deoiled lecithins and (6) fractionated lecithins. Fractions with different phosphatidylcholine content are commercially available. Besides these common commercial grades, more special products are available, eg, enzymatically modified lecithin and phospholipids, semisynthetic phospholipids, and acetylated lecithins.

Health and Safety Factors

The phospholipids are biodegradable, but their presence in streams and water resources, especially in the form of soap stock, is undesirable. Fatty acid recovery from phospholipids is less than with neutral oils because of the lower fatty acid content. There are no known health hazards involved in the production of commercial lecithin from crude vegetable oils because the phospholipids are nonvolatile and are a non-irritating food material.

Uses

The worldwide uses of lecithin break down as follows: margarine, 25–30%; baking/chocolate and ice cream, 25–30%; technical products, 10–20%; cosmetics, 3–5%; and pharmaceuticals, 3%.

One to five percent lecithin moisturizes, emulsifies, stabilizes, conditions, and softens when used in products such as skin creams and lotions, shampoos and hair treatment, and liquid and bar soaps. Since the introduction of Capture in 1986, liposomes produced from phospholipids are commercially available worldwide.

Lecithin and especially purified phosphatidylcholine can act as excipients in pharmaceutical (drug) formulation to enhance and control the bioavailability of the active component. Moreover, phosphalidylcholine can be utilized as a diedelic source, as it involved in the cholesterol metabolism and the metabolism of fats in the liver; also, it can be utilized as a precursor of brain acetylcholine, as neurotransmitter.

Additional Reading

Hanin, I. and G.B. Ansell, eds.: "Lecithin: Technological, Biological and Therapeutic Aspects," *Advances in Behavioral Biology*, Vol. 33, Plenum Press, New York, NY 1987.

Hanin, I. and G. Pepeu, eds.: *Phospholipids: Biochemical, Pharmaceutical, and Analytical Considerations*, Plenum Press, New York, NY 1990.

Szhaj, B.F. and G.R. List, eds.: *Lecithins*, American Oil Chemist's Society, Champaign, IL 1985.

Szhaj, B.F. ed.: *Lecithins: Sources, Manufacture and Uses*, American Oil Chemist's Society, Champaign, IL, 1989.

LEDERBERG JOSHUA (1925–PRESENT).

Joshua Lederberg, who was born in Montclair, New Jersey, USA, carriedout the first experiments demonstrating the existence of genetic recombination in bacteria, similar to that occurring during sexual reproduction in higher organisms. In 1946, while he was still a student at Columbia University, the twenty-one-year-old Lederberg carried out these experiments in collaboration with Edward Tatum during a 6-week stay at Yale. Some years previously, with George Beadle, Tatum had obtained different mutants of the mould *Neurospora crassa* that were affected in their metabolic pathways and unable to grow on minimal media. He then extended the same approach to the bacterium *Escherichia coli*. These metabolic mutants were the essential tool used by Joshua Lederberg to demonstrate the existence of a genetic exchange process in *E. coli*: recombination between the different mutants could be easily observed by using appropriate culture media.

Lederberg subsequently described his discovery as "postmature," in the sense that there was delayed recognition of this mechanism: the absenceof any form of genetic recombination or sexual reproduction in bacteria appeared more and more to be a puzzling exception in the living world in the 1940s. The mechanism of genetic recombination in bacteria was further characterized in the following years by Lederberg, William Hayes, Elie Wollman and François Jacob. The amplitude of genetic exchange is limited, and in hindsight Lederberg's success can be considered to be the result of chance choices in the design of the experimental system.

In 1950, in collaboration with Norton Zinder, Lederberg described a second mechanism of genetic exchange between bacteria–transduction. Transduction is the process by which bacterial viruses (called *bacteriophages*) carry bacterial genes, which are tightly associated with their genomes, from one bacterium to another. Both phenomena discovered by Lederberg–bacterial mating (called *conjugation*) and phage transduction—were essential tools for the development of bacterial genetics and therefore of molecular biology: they led to Lederberg being awarded the 1958 Nobel Prize in Physiology or Medicine.

Lederberg's third major contribution to molecular biology provided a molecular interpretation of the clonal theory of antibody production proposed by the Australian immunologist Sir Frank MacFarlane Burnet. Lederberg confirmed that each antibody-producing cell synthesizes one unique form of antibody. He suggested that each different antibody corresponds to a specific gene sequence. As the genome is not large enough to harbor as many genesas there are different forms of antibodies, he hypothesized that these different genic forms result from a process of somatic mutation. This was confirmed experimentally some years later.

Lederberg was Professor at the University of Wisconsin, and then at Stanford University School of Medicine before going back to Rockefeller University in 1978. In addition to his research endeavors in bacterial genetics and microbiology, he is also known as the editor of the four-volume *Encyclopedia of Microbiology*. Lederberg also had many responsibilities inscience policy: in particular, he was President of Rockefeller University.

Lederberg's most remarkable trait is probably his capacity to anticipate future developments in science, as well as emerging problems at the intersection between science, medicine and social interests. He foresaw the development of genetic engineering technology, and participated in its first steps. As early as 1960, he encouraged the development of exobiological studies, a proposal that was only accomplished 30 years later with the rise of astrobiology.

His main concerns at the end of his career were public health, biomedical science and developing countries, the problems of toxicology and environment, and the future of infectious diseases considered from an evolutionary point of view with the threat of new diseases. In many cases, such as the problem of bioterrorism, his analyses preceded the social concerns. This prospective work was supported by his deep interest in the past developments of science, a historical scope extending far beyond his own scientific contribution.

Additional Reading

Brock, T.D.: *The Emergence of Bacterial Genetics*, Cold Spring HarborLaboratory Press, Cold Spring Harbor, NY, 1990.

Lederberg, J., and E.L. Tatum: "Gene Recombination in *Escherichia coli*," *Nature*, **158**, 558 (1946).

Lederberg, J.: "Genes and Antibodies," *Science*, **129**, 1649–1653 (1959).

Lederberg, J.: "Exobiology: Approaches to Life Beyond the Earth," *Science*, **132**, 393–400, (1960).

Lederberg, J.: "A Fortieth Anniversary Reminiscence," *Nature*, **324**, 627–628 (1986).

Lederberg, J.: "Genetic Recombination in Bacteria: A Discovery Account," *Annual Review of Genetics*, **21**, 23–46 (1987).

Lederberg, J.: *Encyclopedia of Microbiology*, Vols. 1–4, 2nd Edition, Elsevier Science & Technology Books, New York, NY, 2000.

Lederberg, J., and W.S. Cohen: *Biological Weapons: Limiting the Threat*, BCSIA Studies in International Security, Cambridge, MA, 1997.

Morange, M.: *A History of Molecular Biology*, Harvard University Press, Cambridge, MA, 1998.

Zinder, N., and J. Lederberg: "Genetic Exchange in *Salmonella*," *Journal of Bacteriology*, **64**, 679–699 (1952).

Zuckerman, H., and J. Lederberg: "Postmature Scientific Discovery?" *Nature*, **324**, 629–631 (1986).

MICHEL MORANGE, Ecole Normale Supérieure, Paris, France

LEDS.   See **Automotive Electronics**; Luminescence; **Semiconductors**.

LEDUC'S RULE.   States that the volume occupied by a gas mixture is equal to the sum of the volumes occupied separately by each constituent at the same temperature and pressure as the mixture.

LEE CYCLOGENESIS.   See **Winds and Air Movement**.

LEEK.   Of the family *Amaryllidaceae* (amarylis family), the leek (*Allium porrum*) is related to a great number of other species of the genus and of similar odor and taste. Closely related species are chive, garlic, onion, shallot, and Welsh or Japanese onion. The leek resembles the onion in its adaptability and cultural requirements. Instead of forming a bulb, the leek produces a thick, fleshy cylinder that has the characteristics of a large, green onion. See Fig. 1.

Fig. 1.   Leek (*Allium porrum*), closely related to the onion. (*USDA photo.*)

Usually, the seeds are sown in a shallow trench so that the plants can be more easily hilled up as growth proceeds. Leeks are ready for use any time they reach a proper size. Under favorable conditions, they will grow to 1.5 inches (4 cm) in diameter or more, with white parts from 6 to 8 inches (15 to 20 cm) in length. They may be lifted in the fall and stored like celery in a dry, cool place.

LEE TROUGH.   See **Winds and Air Movement**.

LEE, TSUNG-DAO (1926).   Lee was born in Shanghai, China. He began his science studies at the National Southwest Associated University but in 1946 came to the University of Chicago. Here, he became friends with another student named Chen Ning (Frank) Yang. Here, Lee also completed his PhD. on white dwarf stars under Enrico Fermi. At this point, Lee starting researching with Chandrasekhar for about a year, and then went to UC, Berkeley and researched with Yang.

The rest of his scientific career, Yang spent at Columbia University. There he taught and did research in statistical mechanics. He is known for the "Lee model." Later, he again began working with Yang again on the nonconservation of parity. For this research, Lee and Chen Ning Yang shared the Nobel Prize in Physics in 1957.

J. M. I.

LEEUWENHOEK, ANTONI VAN (1632–1723).   Leeuwenhoek, the son of a basket-maker, did not continue his formal education beyond grammar school. He was apprenticed to a cloth merchant and established his own draper's business in his home town of Delft; he did notbegin his scientific pursuits until he was 39. Leeuwenhoek worked in relative isolation, and as he had not been educated in classical or foreign languages relied upon translations and friends such as Regnier de Graaf and Constantijn Huygens. In addition to a large private correspondence, he wrote to the Royal Society in London. Translations in the *Philosophical Transactions* provided the major vehicle for dispersing his findings to the scientific world before he began to publish himself in 1684. In 1699 the Paris Académie des Sciences appointed him a correspondent. He entertained a large number of visitors. Among the learned were also the curious and the famous: Peter the Great, Frederick the Great and the Grand Duke Cosimo III of Tuscany.

In 1671 Leeuwenhoek began making the first of an estimated 550 lenses. The ground globules of glass–a unique method of making a microscope lens, inspired by the glasses drapers used to inspect cloth and not bettered until the nineteenth century–were mounted between perforated metal plates and placed in front of a specimen holder capable of revolving in three planes. The samples of animal, vegetable and inorganic matter were permanently fixed and each specimen therefore required a new apparatus. Although this was a drawback, over the remainder of his life Leeuwenhoek viewed a remarkable range of materials through his microscopes. See also **Light Microscopy**.

Leeuwenhoek's observations were underpinned by his belief in two basic concepts: similarity in the organic and inorganic parts of nature and similarity in the form and function of all living things. Some of his early discoveries were among the most significant. In 1674, for example, he beganto describe in his diary observations on microorganisms, tiny moving objects which he termed 'little animals'. His letters to the Royal Society (1676) created a furor among the members and Leeuwenhoek was obliged to provide testimony from reliable witnesses in support of his claims. Once satisfied, the Royal Society received around 30 further submissions on this subject describing bacteria, protozoa, rotifers and ciliate reproduction. In 1680 he was elected a fellow of the Royal Society.

Leeuwenhoek used his microscope to investigate his two major preoccupations in natural history: the nature of sexual reproduction, including refuting theories of spontaneous generation, and the nature of nutrient transport in plants and animals. Leeuwenhoek argued that sperm were not generated spontaneously by putrefaction but were a normal part of semen found not only in human seminal fluid but throughout the animal kingdom, and he described those belonging to other mammals, arthropods, mollusks, fish, amphibians and birds. He argued, but did not observe, that the sperm penetrated the egg in the process of fertilization, but ascribed the formation of new lifeto the sperm while the egg passively provided nutrient: an animalculist theory of reproduction. Ascribing the flower only an aesthetic purpose misguided his theories of plant reproduction. The study of nutrient transport relied upon his belief in a system of pipes and vessels in plants and he argued from analogy to explain what he regarded as similar structures in animals. In the process of these studies, however, he made important contributions to knowledge of the structure of blood vessels and cells and the three-dimensional nature of roots, stems and leaves.

Additional Reading

Ford, B.J.: *The Leeuwenhoek Legacy*, Farrand Press, London, UK, 1991.
Ruestow, E.G.: *The Microscope in the Dutch Republic: The Shaping of Discovery*, Cambridge University Press, New York, NY, 1996.

HELEN J. POWER, Wellcome Trust Centre for the History of Medicine at UCL, London, UK

LEEUWIN CURRENT.   See **Ocean Currents**.

LEE-WAVE SEPARATION.   See **Meteorology**.

LEEWAY.   The difference between the actual direction in which a ship is moving relative to the surface of the water and the direction in which the keel of the ship is pointing. Leeway is usually produced by the pressure of the wind against the side of the vessel and is much more pronounced in the case of sailing vessels than in internally powered ships. The amount of leeway can best be determined by observing the angle between the wake of the vessel and the line of the keel.

In determining the true course of the vessel, the leeway is treated in the same manner as a compass correction. If the wind is blowing against the left side of the vessel, the vessel is said to be on the port tack, and the true course will be to the right of the course indicated by the keel. Hence, for a ship on the port tack, leeway has the same effect as an east or positive compass correction; on the starboard tack, leeway is applied as a west or negative correction.

See also **Course**; and **Navigation**.

LEE, YUAN T. (1936–). Awarded the Nobel Prize in chemistry in 1986 jointly with John C. Polanyi and Dudley R. Herschbach for their contributions concerning the dynamics of chemical elementary processes. A former student of Herschbach, Lee refined molecular-beam and laser techniques, combining them with theory to perform definitive studies of reactions of individual complex molecules. Lee received his Doctorate from the University of California at Berkeley in 1965.

LEFT-HANDEDNESS (System).   See **Handedness (Right- and Left-)**.

LEGENDRE DIFFERENTIAL EQUATION.   A second-order equation

$$(1 - x^2)y'' - 2xy' + n(n + 1)y = 0$$

It is a special case of the associated Legendre equation

$$(1 - x^2)y^n - 2xy' + \left[n(n + 1) - \frac{m^2}{1 - x^2}\right]y = 0$$

which in turn is a special case of the Gauss hypergeometric equation. Both Legendre equations have singular points at $x = \pm 1$, ∞ and if m, n are integers the solutions are Legendre polynomials. For nonintegral values of these parameters, the solutions are Legendre functions. These differential equations occur in the quantum mechanical problems of the rigid rotator and the hydrogen atom.

The associated polynomials may be defined by the expression

$$P_n(x) = (-1)^m (1 - x^2)^{m/2} \frac{d^m P_n(x)}{dx^m}$$

and the special case $m = 0$ is the Legendre polynomial of degree n

$$P_n(x) = \frac{1}{2^n n!} \frac{d^n}{dx^n}(x^2 - 1)^n$$

$$= \frac{(2n)!}{2^n (n!)^2}\left[x^n - \frac{n(n - 1)}{2(2n - 1)}x^{n-2}\right.$$

$$\left. + \frac{n(n - 1)(n - 2)(n - 3)}{2 \cdot 4(2n - 1)(2n - 3)}x^{n-4} \pm \cdots\right]$$

The first definition, in terms of the nth derivative, is called the *Rodrigues formula*. A second set of polynomials, linearly independent of P_n, is composed of polynomials of the second kind, Q_n. The general solution of the Legendre equation is then $y = AP_n(x) + BQ_n(x)$, where A, B are arbitrary constants. Many other definitions and relations for these polynomials are known.

One integral representation is the *Schläfli formula*

$$P_n(z) = \frac{1}{2\pi i} \int_c \frac{(t^2 - 1)^n}{2^n (t - z)^{n+1}} dt$$

where the contour encircles the point z in the counterclockwise direction in the complex plane. Another such representation is the *Heine formula*, which for the associated polynomial becomes

$$P_n^m(x) = (n + 1)(n + 2) \cdots (n + m)(-1)^{m/2}$$

$$\times \frac{1}{\pi} \int_0^\pi [x + \sqrt{x^2 - 1} \cos\phi]^n \cos m\phi \, d\phi$$

See also **Generating Function**.

LEGENDRE SYMBOLS.   See Number Theory

LEGIONELLOSIS.   Legionellosis is an infection caused by the bacterium *Legionella pneumophila*. This disease has two distinct forms:

- Legionnaires' disease, (LD), is the more severe form of legionellosis and is characterized by pneumonia.
- Pontiac fever, is an acute-onset, flu-like, nonpneumonic illness.

Legionnaires' disease acquired its name in 1976 when an outbreak of pneumonia occurred among persons attending a convention of the American Legion in Philadelphia. Later, the bacterium causing the illness was named *Legionella*.

An estimated 8,000 to 18,000 cases of Legionnaires' disease occur each year in the United States; 23% are nosocomial. Most LD cases are sporadic with 10–20% being linked to outbreaks. Pontiac fever has been recognized only during outbreaks. Some people can be infected with the *Legionella* bacterium and have mild symptoms or no illness at all.

Outbreaks of Legionnaires' disease have received significant media attention. However, this disease usually occurs, as a single, isolated case not associated with any recognized outbreak. When outbreaks do occur, they are usually recognized in the summer and early fall, but cases may occur year-round. About 5 to 15% of people who have Legionnaires' disease die. A substantially, higher proportion of fatal cases occur during noscomial outbreaks. Pontiac fever is a self-limited disease that requires no treatment.

Patients with Legionnaires' disease usually have fever, chills, and a cough, which may be dry or may produce sputum. Some patients also have muscle aches, headache, tiredness, loss of appetite, and occasionally, diarrhea. Laboratory tests may show that these patients' kidneys are not functioning properly. Chest X-rays often show pneumonia. It is difficult to distinguish Legionnaires' disease from other types of pneumonia by symptoms alone; other tests are required for diagnosis. Persons with Pontiac fever experience fever and muscle aches and do not have pneumonia. They generally recover in 2 to 5 days without treatment. The time between the patient's exposure to the bacterium and the onset of illness for Legionnaires' disease is 2 to 10 days; for Pontiac fever, it is shorter, generally a few hours to 2 days.

The diagnosis of legionellosis requires special tests not routinely performed on persons with fever or pneumonia. Therefore, a physician must consider the possibility of legionellosis in order to obtain the right tests. Several types of tests are available. The most useful tests detect the bacteria in sputum, find *Legionella* antigens in urine samples, or compare antibody levels to *Legionella* in two blood samples obtained 3 to 6 weeks apart.

People of any age may get Legionnaires' disease, but the illness most often affects middle-aged and older persons, particularly those who smoke cigarettes or have chronic lung disease. Also at increased risk are persons whose immune system is suppressed by diseases such as cancer, kidney failure requiring dialysis, diabetes, or AIDS. Those that take drugs that suppress the immune system are also at higher risk. Pontiac fever most commonly occurs in persons who are otherwise healthy.

Erythromycin is the antibiotic currently recommended for treating persons with Legionnaires' disease. In severe cases, a second drug, rifampin, may be used in addition. Other drugs are available for patients unable to tolerate erythromycin. Pontiac fever requires no specific treatment.

Outbreaks of legionellosis have occurred after persons have breathed mists that come from a water source (e.g., air conditioning cooling towers, whirlpool spas, showers) contaminated with *Legionella* bacteria. Persons may be exposed to these mists in homes, workplaces, hospitals, or public places. Legionellosis is not passed from person to person and there is no evidence of persons becoming infected from automobile air conditioners or household window air-conditioning units.

Legionella organisms can be found in many types of water systems. However, the bacteria reproduces to high numbers in warm, stagnant water (90–105 °F), such as that found in certain plumbing systems and hot water tanks, cooling towers and evaporative condensers of large air-conditioning systems, and whirlpool spas. Cases of legionellosis have been identified throughout the United States and in several foreign countries. It is believed to occur worldwide.

During outbreaks, the CDC and health department investigators seek to identify the source of disease transmission and recommend appropriate prevention and control measures, such as decontamination of the water source. The improved design and maintenance of cooling towers and plumbing systems will also limit the growth and spread of *Legionella* organisms. Current research will likely identify additional prevention strategies. See also **Bacterial Diseases**.

Additional Reading

Addis, D.G. and J.P. Davis: "Sporadic Cases of Legionnaires' Disease," *N. Eng. J. Med.*, 1699 (June 18, 1992).

Bhopal, R.S., R.J., Fallon, and M.R.C. Path: "Sporadic Cases of Legionnaires' Disease," *N. Eng. J. Med.*, 1699 (June 18, 1992).

Breiman, R.F. and J.C. Butler: "Legionnaires' Disease: Clinical, Epidemiological, and Public Health Perspectives," *Semin in Respir Infects*, **13**, 84–89 (1998).

Fiore, A.E. and J.C. Butler: "Detecting Nosocomial Legionnaires' Disease," *Infections in Medicine*, **15**, 625–635 (1998).

Fiore, A.E., J.P. Nuorti, O.S. Levine, et al.: "Epidemic Legionnaires' Disease two Decades Later: Old Sources, New Diagnostic Methods," *Clin. Infect. Dis.*, **26**, 426–433 (1998).

Hoge, C.W.: "Sporadic Cases of Legionnaires' Disease," *N. Eng. J. Med.*, 1700 (June 18, 1992).

Kool, J.L., J.C. Carpenter, and B.S. Fields: "Effect of Monochloramine Disinfection of Municipal Drinking Water on Risk of Nosocomial Legionnaires' Disease," *Lancet*, **353**(9149), 272–277 (1999).

Kool, J.L., A.E. Fiore et al.: "More than Ten Years of Unrecognized Nosocomial Transmission of Legionnaires' Disease Among Transplant Patients: Difficulties of Legionella Control in a Complex Water System," *Infect. Contr. Hosp. Epidemiol.*, **19**, 898–904, 7 (1998).

Lowry, P.W. et al.: "A Cluster of Legionella Sternal-Wound Infections Due to Postoperative Topical Exposure to Contaminated Tap Water," *N. Eng. J. Med.*, 109 (January 10, 1991).

Marston, B.J., H.B. Lipman, and R.F. Breiman: "Surveillance for Legionnaires' Disease: Risk Factors for Mortality and Morbidity," *Arch. Intern. Med.*, **154**, 2417–2422 (1994).

Morse, D.L., Birkhead, G.S., and S. Kondracki: "Sporadic Cases of Legionnaires' Disease," *N. Eng. J. Med.*, 1700 (June 18, 1992).

Straus, W.L., J.F. Plouffe, T.M. File Jr. et al.: "Risk Factors for Domestic Acquisition of Legionnaires' Disease: Ohio Legionnaires' Diseases Group," *Arch. Intern. Med.*, **156**(15), 1685–1692 (1996).

Yu, V.L. and J.E. Stout: "Sporadic Cases of Legionnaires' Disease," *N. Eng. J. Med.*, 1701 (June 18, 1992).

LEGUMINOSAE. The Pea Family is second only to the Composite Family among the dicotyledons, with respect to the number of species it includes. Of its more than 10,000 species in early 500 genera, many are trees or shrubs, especially those in tropical regions. Herbaceous species are numerous in temperate regions. Many climbing plants, also, are found in the family. Leguminous plants are found in all sorts of environments and climates.

Nearly all the plants in this family have pinnately compound leaves. The stipules present in the leaves are sometimes modified to persistent spines. The flowers are either regular or irregular. When regular, the flowers have five sepals, commonly more or less united, five petals, a varying number of stamens, and a single pistil. Irregular flowers are of the type known as papilionaceous, a name given because of the fancied resemblance of the flower to a butterfly. In flowers of this type, the calyx has five unequal, more or less united sepals, which frequently persist during development of the fruit, five separate petals, showing very constant difference in form. The upper one, called the standard, is large and showy; the two lateral to this, called wings, are smaller in size; and the two lower ones are more or less united into one unit, called the keel or carina. Within this keel are the ten stamens, which may be separate but in many genera are united in groups, the nine lower ones having their filaments more or less completely joined, while the tenth stamen remains free. The pistil has a somewhat flattened ovary, a long style, and a terminal stigma. The ovary contains several ovules. The mature fruit is called a pod or legume, which when mature often splits open with sufficient force to eject the seeds to considerable distances. The seeds in most cases have large food reserves stored in the thick cotyledons.

There are three subfamilies in the *Leguminoseae*. The *Mimosoideae* have regular flowers and valvate corolla. The *Caesalpinioideae* have irregular (zygomorphic) flowers. These two subfamilies are essentially tropical. The *Papilionoideae* have irregular (papilionaceous) flowers, and include most of the important cultivated forms.

Many members of this family supply man with important foodstuffs, such as beans, peas, and peanuts, while others are important forage crops for domestic animals. The high protein content of the plant is the principal reason for its importance as a food source. Clovers and alfalfa are not only valuable as forage plants, but furnish an excellent hay. Legumes are also of immense value because of their association with nodule-forming bacteria, resulting in a considerable accumulation of nitrogenous substance, which is later liberated into the soil, greatly enriching it. Other members of the family yield valuable dyes, gums and resins, and oils; many are sources of timber.

The leaves of many genera of legumes are interesting because of their ability to move. In many of them, the leaflets fold together at night, so that the blade of the leaflet is vertical. Of particular interest in this connection is the Sensitive plant, *Mimosa pudica*, the leaves of which respond very quickly to external stimuli. A light blow will cause the many leaflets to fold together, and the whole section of the compound leaf to bend down. A sudden breeze or change of temperature will produce the same result. Recovery from the shock is gradual. When stimulated by a series of successive shocks, the plant recovers more and more slowly each time.

See also **Locust Trees**.

LEHN, JEAN-MARIE PIERRE (1939–). Awarded the Nobel Prize for chemistry, together with Donald J. Cram, in 1987 for work in elucidating mechanisms of molecular recognition, which are fundamental to the enzymic catalysis, regulation, and transport. He also studied three-dimensional cyclic compounds that maintained a rigid structure, accepting substrates in a structurally preorganized cavity. Lehn named these compounds cryptands, while Cram called them cavitands. Awarded Doctorate by the University of Strasbourg, France in 1963.

LEIBNIZ'S THEOREM OF CALCULUS. A relationship between the derivative of an integral and the integral of a derivative, that is,

$$\frac{d}{dt}\left[\int_{S_1(i)}^{S_2(i)} A(t,s)\,ds\right] = \int_{S_1(i)}^{S_2(i)}\left[\frac{\partial A(t,s)}{\partial t}\right]ds + A(t,S_2)\frac{dS_2}{dt}$$
$$- A(t,S_1)\frac{dS_1}{dt},$$

where S_1 and S_2 are limits of integration, s is a dummy distance or space variable such as height z, t is time, and A is some meteorological quantity, such as potential temperature or humidity, that is a function of both space and time. If the limits of integration are constant with time, then the last two terms are zero, and the derivative of the integral equals the integral of the derivative. However, there are many atmospheric situations, such as a growing atmospheric boundary layer, where the limits of integration can change with time. Namely, if one wishes to integrate over the depth of the boundary layer (between limits $z = 0$ to $z = z_i$) to find a boundary layer average, for example, but the top of the boundary layer at height $z_i(t)$ is rising with time, then one must use the full form of Leibniz's theorem to account for this effect.

AMS

LEISHMANIASIS. Leishmaniasis affects some 12 million humans annually in an area where 350 million are at risk. It is a complex of at least two protozoan diseases, consisting primarily of cutaneous and visceral forms. A mucocutaneous form is considered by some to be another distinct variety. Clinical manifestations of the disease range from an asymptomatic infection to an infection in which there is considerable destruction of cutaneous tissue and mucous membranes. Leishmaniasis can often be fatal, especially in the visceral form. The seriousness of the disease depends on the state of the immunological system of the host and the species of parasite that inflicts the damage. The different species of *Leishmania* that are responsible for the disease can be difficult to distinguish from one another and methods to identify them have occupied the efforts of many researchers. The various species may be localized geographically or may overlap, and each is responsible for somewhat different symptoms. Although the human clinical manifestations may be categorized into different types, infections by more than a single species may be present at one time.

The disease is initiated by a bite from an infected female sand fly of the genus *Phlebotomus*. The injected parasites usually multiply near the site of the original infection, causing a lesion. The sand fly vector becomes infected when it acquires a blood meal from the skin or peripheral blood of a parasitized host and ingests the protozoans in the amastigote stage. The amastigotes develop into promastigotes in the intestine of the sand fly and eventually migrate to its proboscis. The saliva of the sand fly appears to enhance the survival of the injected promastigotes which, after phagocytosis, are transformed into the amastigote phase. The numerous reservoirs for the protozoans include humans, dogs, foxes, cats, rodents, and horses.

Cutaneous leishmaniasis is characterized by one or more slowly healing superficial ulcers that may be painful. These lesions are liable to further infection and may remain as open sores or become hard, wartlike nodules. The form of cutaneous leishmaniasis referred to as New World disease is caused by species of the *Leishmania (Viannia) braziliensis* complex (*L. braziliensis*, *L. guyanensis*, *L. panamanesis*, and *L. peruviana*) and species of the *Leishmania (Leishmania) mexicana* complex (*L. mexicana*, *L. amazonensis*, and *L. venezuelensis*) in the western hemisphere. Old World disease (oriental sore, Delhi or Baghdad boil) is caused by *Leishmania tropica* and *Leishmania major* in Asia, Africa, and southern Europe, and *Leishmania aethiopica*, restricted to eastern Africa. The incubation period for cutaneous leishmaniasis ranges from a few weeks to several months. Spontaneous cures can take place in a period ranging from one month to several years.

Visceral leishmaniasis is also known as kala azar, meaning black fever or black sickness in Hindi, because of the characteristic pigmentary changes that occur in the skin. The disease is endemic in much of the tropical and subtropical regions of southern Europe and the Mediterranean, the Middle East, India, Africa, and South and Central America. It is caused by species of the *Leishmania (Leishmania) donovani* complex [*L. donovani* (Old World), *L. infantum* (Old World), and *L. chagasi* (New World)]. In this systemic form of leishmaniasis, the parasites invade internal organs. It is characterized in patients by an enlarged liver and spleen, fever, weight loss, hemorrhage, leukopenia, and anemia. There is abdominal discomfort due to the spleen and liver involvement. Death due to severe diarrhea, pneumonia, and gastrointestinal bleeding may ensue if the disease is not treated. Following treatment, patients may relapse with a leishmanial form that solely affects the skin (post-kala azar dermal leishmaniasis). Rarer forms of leishmaniasis include chronic relapsing or recidivans forms of the cutaneous disease and diffuse cutaneous leishmaniasis.

Certain species of the *L. braziliensis* complex may cause mucocutaneous leishmaniasis, a form seen most frequently in northern and central South America. The patient first develops skin infections which later metastasize to produce highly disfiguring mucocutaneous lesions of the oronasopharynx.

Antimony compounds have been used to treat leishmaniasis ever since tartar emetic (antimony potassium tartrate) was discovered early in the 20th century to have efficacy against the mucocutaneous form of the disease. The cutaneous form has been treated with tartar emetic formulated in an ointment. Many side effects have been seen with this trivalent antimonial, some of which can be ascribed to the difficulty of obtaining pure antimony for its manufacture. These side effects include toxicity to the heart, liver, and kidneys. Other promising trivalent antimonials have been abandoned in favor of pentavalent antimonials with lower toxicity.

Pentavalent antimony preparations constitute the primary treatments for all forms of leishmaniasis, the most important of which are sodium stibogluconate (**1**, Pentostam® [16037-91-5]) and glucantime (*N*-methylglucamine antimonate, $C_7H_{18}NO_8Sb$, meglumine antimonate) (Table 1). The actual chemical structure of the antimonials is unknown. Studies indicate that they are mixtures of compounds with widely differing molecular weights. Use of glucantime is favored in Latin America and French-speaking countries. Cardio- and hepatotoxicity have been observed when the drug is administered over the typical 3–4 weeks of treatment. Liposome preparations of Pentostam® and glucantime substantially increased the activity of these antimonials under experimental conditions. In 1990, an estimated 10,000 out of 200,000 cases of kala azar reported in India were unresponsive to treatment with antimonials; in Kenya, the strains were even more resistant to antimony. Although antimony-resistant strains of leishmania can be treated with diamidines, the latter compounds are more toxic. Pentamidine [CAS: 100-33-4] (See also **Babesiosis**) as the isethionate, is the most widely used diamidine and has a high cure rate. Another diamidine, stilbamidine (**2**, 4,4′-diamidinostilbene), is also effective.

Amphotericin B (Funizone® [CAS: 1397-89-3] is an antifungal macrolide antibiotic produced by *Streptomyces nodosus* that has been used as an alternative, albeit more toxic, drug to the antimonials. See also **Amebiasis (Amebic Dysentery)** . It acts as a leishmanicide against the visceral and mucocutaneous forms of the disease. To overcome its

TABLE 1. LEISHMANIASIS ANTIPROTOZOAL AGENTS

Structure number	Compound name	CAS Registry Number	Molecular formula	Structure
(1)	Sodium stibogluconate	[16037-91-5]	$C_{12}H_{20}O_{17}Sb_2 \cdot 9H_2O \cdot 3Na$	
(2)	Stilbamidine	[122-06-5]	$C_{16}H_{16}N_4$	
(3)	Allopurinol	[315-30-0]	$C_5H_4N_4O$	
(4)	Ketoconazole	[65277-42-1]	$C_{26}H_{28}Cl_2N_4O_4$	

potentially severe nephrotoxicity, the drug must be administered over an extended period of time.

Because of the outbreak of antimony-resistant leishmaniasis and the need to develop an orally-administered therapy, the use of many other compounds has been considered. Those that appear to have clinical utility are allopurinol 3, ketoconazole 4, and both systemically and topically applied paromomycin.

See also **Antiparasitic Agents, Antimycotics**; and **Antiparasitic Agents, Antiprotozoals**.

Additional Reading

Berman, J.D.: "Human Leishmaniasis: Clinical Diagnostic and Chemotherapeutic Developments in the Last 10 Years," *Clin. Infect. Dis.,* **24**, 684–703 (1997).

Desjeux, R.: "Leishmaniasis: Public Health Aspects and Control," *Clin. Dermatol,* **14**, 417–423 (1996).

Farrell, J.P.: *Leishmania (World Class Parasites, Vol 4)*, Springer-Verlag New York, LLC, New York, NY, 2002.

Hart, D.: *Leishmaniasis: The Current States and New Strategies for Control*, Perseus Books, Boulder, CO, 1989.

Herwaldt, B.L.: "Leishmaniasis," *Lancet,* **354**, 1191–1199 (1999).

Herwaldt, B.L., S.L. Stokes, and D.D. Juranek: "American Cutaneous Leishmaniasis in U.S. Travelers," *Ann. Inter. Med.,* **118**, 779–784 (1993).

Hide, G., et al.: *Trypanosomiasis and Leishmaniasis*, CAB International, New York, NY, 1997.

Parker, J.N., and P.M. Parker: *The Official Patient's SourceBook on Leishmaniasis (The Official Patient's Guide Series)*, Icon Press, Inc., San Diego, CA, 2002.

Staff: *Leishmaniasis—a Medical Dictionary, Bibliography, and Annotated Research Guide to Internet References*, Icon Health Publications, San Diego, CA, 2004.

Tapia, F., et al.: *Molecular and Immune Mechanisms of Pathogenesis of Cutaneous Leishmaniasis*, Landes Bioscience, Georgetown, TX, 1999.

Web References

Centers for Disease Control and Prevention (CDC): http://www.cdc.gov/Ncidod/dpd/parasites/leishmania/default.htm.

World Health Organization (WHO): http://www.who.int/tdr/diseases/leish/default.htm.

DANIEL L. KLAYMAN, Walter Reed Army Institute of Research

LEISHMAN, WILLIAM BOOG (1865-1926).

Leishman was a Scottish bacteriologist, pathologist and protozoologist who shared the discovery of the parasite of kala-azar with Charles Donovan.

Leishman graduated from the University of Glasgow in 1886 and joined the Army Medical Service (AMS) a year later. Stationed at home and then inIndia, he returned to the AMS school at Netley in 1899 in charge of the medical wards and became assistant professor of pathology to Almroth Wright in1900. In 1903 he succeeded to full professor when the AMS school moved to Millbank, London and held this post until 1913. See also **Wright, Almroth Edward (1861–1947)**.

The years at Netley and Millbank afforded Leishman the opportunities for research that he desired. He modified Romanowsky's stain and used thisto find the parasitic protozoa responsible for kala-azar or visceral leishmaniasis in blood samples from patients at Netley. Known also as Dumdum fever, this was a significant cause of death among troops in India and among laborers on the tea-gardens of Assam. Leishman failed to publish the results of his work in 1900, waiting until 1903 when Lieutenant-Colonel Charles Donovan confirmed the role of the parasites (which were given the name *Leishmania donovani*). Leishman also experimentally elucidated the life cycle of *Spironema duttoni* , the cause of relapsing fever. See also **Leishmaniasis**.

Leishman continued to develop Wright's anti-typhoid vaccine, maintaining a large store, which was immediately available on the outbreak of WorldWar I. In January 1914, retiring from the professorship, Leishman became War Office expert on tropical diseases for the army medical advisory board. However, in October he served as pathology adviser with the British Expeditionary Force in France, where his bacteriological and administrative skillswere successfully combined with chairing committees on trench fever and trench nephritis. He was also an original member of the Medical Research Committee (later Council).

See also **Parasitology (The History)**.

HELEN J. POWER, Wellcome Trust Centre, London, UK

LELOIR, LUIS F. (1906–1987).

A French-born biochemist who won the Nobel Prize for chemistry in 1970 for work in biosynthesis of carbohydrates. He discovered chemical compounds that affect the storage of chemical energy in humans and animals. He headed the Department of Biochemistry at the University of Buenos Aires for many years.

LEMMA. A proposition which is stated for the purpose of later use in the proof of another proposition. Strictly, the proposition is assumed. In current use, a short proof is often given.

LEMMING. See **Rodentia**.

LEMNISCATE OF BERNOULLI. A special case of a general type of higher plane curves known as lemniscates (*L. lemniscus*, loop) and also of the oval of Cassini, when $k = a$. Its equation in rectangular coordinates is

$$(x^2 + y^2)^2 + 2a^2(y^2 - x^2) = 0$$

or, in polar coordinates,

$$r^2 = 2a^2 \cos 2\theta$$

The curve is symmetrical about both X- and Y-axes and the node at the origin has tangents given by $(x^2 - y^2) = 0$. See Fig. 1. If $\cos 2\theta$ is replaced by $\sin 2\theta$ in the polar equation for the curve it is rotated about the origin by $45°$ and is then called a two-leaved rose lemniscate. (See also **Rose Curve**). A more general type of curve, known as Booth's lemniscate, has the equation

$$(x^2 + y^2)^2 = a^2 x^2 \pm b^2 y^2$$

The curve $x^4 = a^2(x^2 - y^2)$ is known as the lemniscate of Gerono.

See also **Curve (Higher Plane)**.

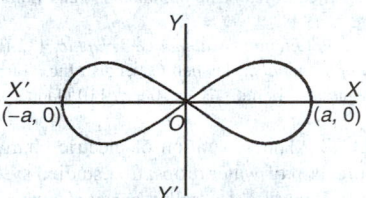

Fig. 1.   Lemniscate.

LEMON TREE. See **Citrus Trees**.

LEMUR (Flying). See **Dermaptera**.

LEMUR (*Mammalia*). Lemurs are the most primitive animals of the order of Primates. They have a well-developed thumb and great toe, like other primates, but the second toe bears a sharp claw instead of a nail. The body is generally more like that of a squirrel than like the apes and monkeys and the face in many species is peculiarly expressionless, with large staring eyes. Lemurs constitute a family *Lemuroids*, containing, in addition to the species named as lemurs, the indri, the sifakas or propitheques, the galagos, the awantibo, the pottos, the lorises or slow lemurs, and the avahi. They center in Madagascar, but a few species occur in eastern Africa, southern India, and on the islands of the Oriental region. The tarsiers and the aye-aye are closely related to the true lemurs. A ring-tailed lemur is illustrated in Fig. 1.

The African slow lemur is named the *potto* and resembles the lorises of Asia in many ways. The potto is remarkable for the rudimentary index finger, which is a stub without a nail or joints, and for the very short tail. The two species are Bosman's potto and the awantibo, *Perodicticus calabarensis*. The sifaka is the genus *Propithecus*, with several species found in Madagascar. They are related to the indri lemur, but have long tails and shorter muzzles.

The galago is an African lemur, also sometimes called bush-baby. The several species have long bushy tails, large ears, large staring eyes, their thumbs and big toes opposed, and thick woolly soft fur. Their nearest relatives are the mouse lemurs of Madagascar.

Fig. 1.   Ring-tailed lemur. (*New York Zoological Society.*)

Additional Reading

Gould, E. and G. McKay: *Encyclopedia of Mammals*, 2nd Edition, Academic Press, Inc., San Diego, CA, 1998.
Jolly, A.: "Madagascar's Lemurs," *National Geographic*, 132 (August, 1988).
Staff: *Encyclopedia of Mammals*, Marshall Cavendish Inc., Tarrytown, NY, 1997.
Tattersall, I.: "Madagascar's Lemurs," *Sci. Amer.*, 110 (January, 1993).

LENARD EFFECT.   The separation of electric charges accompanying the aerodynamic breakup of water drops, first studied systematically by the German physicist P. Lenard. Also called *spray electrification*, or *waterfall effect*. Experiments have shown that the degree of charge separation in spray processes depends upon the drop temperature, presence of dissolved impurities, speed of the impinging airblast, and contact with foreign surfaces. The largest fragments of the broken drops are observed to carry positive charges and the fine spray of drops carried off in the impinging air current carries a net negative charge.

LENGTH OF A CURVE.   The length of a straight line is interpreted experimentally to mean the number of times another straight line of unit length can be superimposed on the given line. Since this operation cannot be conveniently applied to a curve the concept is generalized to the limit of the sum of chords to the curve. As the number of chords increases without limit each chord separately approaches zero as a limit.

In rectangular coordinates, if the curve is described by $f(x, y) = 0$, its length from the point (a, c) to the point (b, d) is given by either one of the definite integrals

$$s = \int_a^b \sqrt{1 + y'^2}\,dx = \int_c^d \sqrt{1 + x'^2}\,dy$$

where $y' = dy/dx$ and $x' = dx/dy$. If t is a parameter, $x = f_1(t)$, $y = f_2(t)$, the arc length between $t = a$ and $t = b$ is

$$s = \int_a^b \sqrt{x_t^2 + y_t^2}\,dt$$

where $x_t = dx/dt$ and $y_t = dy/dt$. In polar coordinates, if $r = f(\theta)$, $r' = dr/d\theta$ and $\theta' = d'/dr$,

$$s = \int_{\theta_1}^{\theta_2} \sqrt{r^2 + r'^2}\,d\theta = \int_{r_1}^{r_2} \sqrt{r^2\theta'^2 + 1}\,dr$$

See also **Circular Curves**, and **Coordinate System**.

LENGTH OF RECORD.   See **Meteorology**.

LENSES.   See **Mirrors and Lenses**.

LENS (Eye).   See **Vision and the Eye**.

LENTICELS.   The young stems of plants are covered with a single layer of cells known as the epidermis. As the stem grows older, this epidermis is lost and replaced by a thicker protective tissue known as cork. The cork cells have walls that are suberized and impervious to gases. The living cells within the stem require an exchange of gases with the outside atmosphere. This exchange occurs through lenticels. A lenticel is a mass of thin-walled parenchyma cells loosely arranged so that air spaces are numerous. Through the lenticel gases pass readily. Lenticels appear on the surface of the stem as rough masses, usually protruding somewhat, and either circular or somewhat elongate in shape. They are very irregularly distributed.

LENTICULARIS.   See **Clouds and Cloud Formation**.

LENZ' LAW.   A general law of electromagnetic induction, stated by H.F.E. Lenz in 1833. It points out that the electromotive force induced by the variation of magnetic flux (with reference to a conductor, in the manner discovered by Faraday) is always in such direction that, if it produces a current, the magnetic effect of that current opposes the flux variation responsible for both electromotive force and current. An outstanding illustration is the drag on a generator armature; if the armature circuit is closed, the rotation is opposed by a torque arising from the reaction between the field and the current in the armature conductors. Power must therefore be applied to drive the machine; and the greater the armature current, the more power is required. The effect known as magnetic damping also depends upon Lenz' Law. A copper disk, when spun between the poles of a strong magnet, quickly comes to rest because of the opposing torque. This arrangement serves as a speed regulator in watt-hour meters.

LEONIDS.   A name applied to a meteor shower that has probably attracted more attention than any other. Each year, about the 12th of November, a number of meteors are observed coming from a radiant point in the constellation of Leo. Records of the appearance of this shower are found back as far as 585 A.D. The Leonid shower is one in which meteors are distributed all along the orbit, so that a radiant point may be determined practically every year. There is, also, a very strong condensation of the meteors into a swarm through which the earth used to pass every 33 years. Probably the greatest display was in November 1833. In Silliman's Journal of that year, we find: "To form some idea of the phenomenon, the reader may imagine a constant succession of fireballs, resembling rockets, radiating in all directions from a point in the heavens." One observer counted 650 during 15 minutes. During the interval between 1833 and 1866, a great deal of computing was done on the Leonid shower, and November 13 was predicted as the date of passage of the earth through the main swarm. The prediction was fulfilled, and at Greenwich, England, eight observers actually counted 8000 meteors, 4860 of them being counted between one and two o'clock in the morning. However, brilliant as the shower was at that time, it apparently was not as striking as the display in 1833. In 1899, there was a moderately good display of the Leonids, but nothing comparable to the showers of 1866 and 1833. The newspapers had promised so much to the general public that the failure of the shower to come up to the expectations proved a rather serious blow to astronomy. The explanation for the failure of the shower to live up to its prediction is to be found in the fact that Jupiter passed very close to the swarm during 1899 and deflected it from the earth's orbit. Further perturbations have so deflected the orbit that no real shower was observed in 1932, 1933, and 1934, although enough meteors were seen during each November to permit a determination of the radiant point. In November 1966, however, perturbations again brought the main swarm into such a position that the earth passed through it, giving rise to showers comparable to the one of 1833. See Fig. 1.

Fig. 1.   Leonid shower as seen from Kitt Peak near Tucson, Arizona.

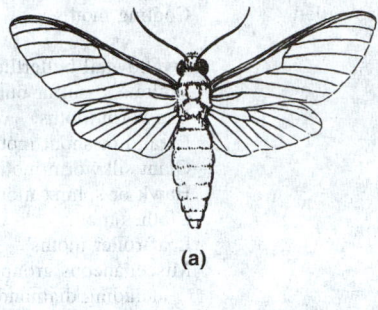

(a)

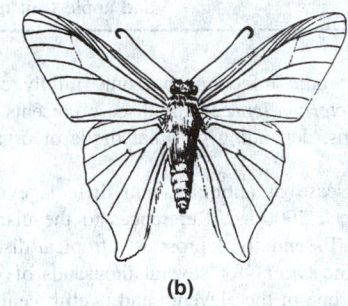

(b)

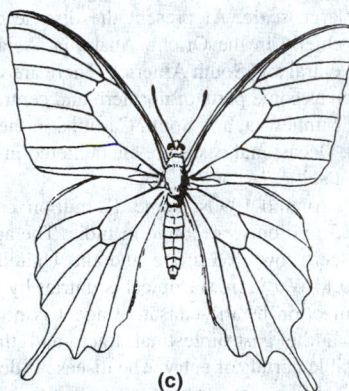

(c)

Fig. 1.   Typical members of Lepidoptera: (a) moth; (b) skipper; (c) butterfly. (*USDA*.)

LEOPARD-CATS.   See **Cats**.

LEO (the lion).   One of the most easily distinguished of all the zodiacal constellations. The "sickle" of Leo, the fifth sign of the zodiac, is known to all watchers of the spring and early summer skies. The brightest star in the group, Regulus, is a double star, but cannot be resolved with telescopes having smaller than a 3-inch aperture because of the fact that, in small instruments, the bright star masks the fainter one. Gamma Leonis is one of the finest of the double stars, having two components of approximately the same magnitude, one yellow and the other orange in color.

This constellation is also noted for the location of the radiant point of the Leonids, one of the best known meteor showers. (See map accompanying entry on **Constellations**.)

LEPIDOLITE.   This member of the mica group of minerals is a silicate of potassium, lithium and aluminum, sometimes with sodium, fluorine, or rarely rubidium. A general formula is $K(Li, Al)_3(SiAl)_4O_{10}(F, OH)_2$. Crystals of lepidolite are monoclinic but often pseudo-hexagonal; cleavage, basal and perfect, being susceptible of splitting into thin laminae; hardness, 2.5–4; specific gravity, 2.8–3.3; luster, pearly; color, reddish to violet, grayish-blue, gray to white. A variety carrying rubidium is yellowish-gray; translucent. It usually is found as granular to scaly masses, in short stocky prisms or less often in easily cleavable sheets. Lepidolite is characteristic of pegmatite veins, frequently being associated with other lithium-bearing minerals such as tourmaline, spodumene, amblygonite, and others. It occurs in the Ural Mountains, the Czech Republic and Slovakia, the Island of Elba, and Madagascar, where it is often found in large sheets. In the United States, it is found in the pegmatites of New England, California, South Dakota, and New Mexico. The name lepidolite is derived from the Greek, meaning scale. See also **Lithium**.

LEPIDOPTERA.   The butterflies, skippers, and moths. An order of insects characterized by sucking mouths in the adult stage with two pairs of wings covered at least in part with a vestiture of flattened scales, complete metamorphosis and a larva with biting mouth-parts. The second order of insects in size, with about 90,000 known species.

Butterflies and moths are widely distributed and because of their bright colors are among the animals known to everyone. The butterflies are diurnal and so are readily observed, but most moths are nocturnal, hence many beautiful species are rarely seen unless they are sought. Skippers are an intermediate group more nearly like the butterflies. The most magnificent species of all three forms are tropical, but representatives are found even in the Arctic regions. See also **Butterfly**.

The adults visit flowers for nectar or take no food. In the larval stage most species are plant feeders, but a few carnivorous forms are known and some are scavengers. The order includes many economic species, among them the clothes moths, the bee moth, and the cut worms. See Fig. 1.

The main families of *Lepidoptera* include:

Aegeriidae (also called *Sesiidae*)	Clear-winged moths
Arctiidae	Tiger moths
Citheroniidae (also called *Ceratocampidae*)	Royal moths
Coleophoridae (also called *Haploptiliidae*)	Casebearers
Cossidae	Carpenter moths
Eucleidae (also called *Limacodidae* or *Cochidiidae*)	Slug-caterpillar moths
Gelechiidae	Gelechid moths
Geometridae	Measuring-worm moths
Gracilariidae	Leaf blotch miners
Hepialidae	Swifts (swift-flying)
Hesperiidae	Common skippers
Incurvariidae	Yucca moths
Lasiocampidae	Lappet moths, tent caterpillars
Lycaenidae	Blues, hairstreaks, gossamer wings
Micropterygidae	Mandibulate moths
Nepticulidae	Nepticulid moths
Noctuidae (also called *Phalaenidae*)	Owlet or cutworm moths
Notodontidae	The prominents
Oecophoridae	Parsnip webworm

Olethreutidae (also called Eucosmidae)	Codling moths
Papilionidae	Swallowtail butterflies
Pieridae	White and sulfur butterflies
Psychidae	Bagworm moths
Pyralididae	Pyralid or snout moths
Saturnidae	Giant silkworm moths
Sphingidae	Hawk or sphinx moths
Tineidae	Clothes moths
Tortricidae	Leaf-roller moths
Yponomeutidae	Miscellaneous grouping, including diamondback moth and apple fruit moth.

LEPROSY. A chronic infectious but only mildly contagious disease caused by *Mycobacterium leprae*. The disease presents a great variety of signs and symptoms, depending on what tissue or organ of the body is involved.

Leprosy is a disease of antiquity, and there is evidence that it has existed at least since 2000 B.C. References to the disease are found in the Old Testament. Essentially, leprosy is a tropical disease. While it has been common in the Orient for several thousands of years, it appeared as a scourge in Europe in the eleventh and twelfth centuries, and did not subside until the sixteenth century when segregation of the victims was carried out on a large scale. At present the disease occurs endemically and sporadically, chiefly in the Orient, Australia, Asia, in Mediterranean countries, and in Central and South America. There are various other foci of sporadic cases such as some parts of northern and central Europe, the West Indies, Louisiana, Minnesota, and South Carolina in the United States, and several in Canada. Occasional cases are encountered in the larger seaports, both Atlantic and Pacific.

As of the early to mid-1990s, some 12 million cases existed in the world, of which 3.5 million were found in India. The approximate number of 25,000 cases seen now in Europe and the United States have been imported since World War II from tropical countries by way of immigrants.

The mode of infection by the causative agent is not definitely known. The nasal mucosa, the gastrointestinal tract, and the skin have been considered as possible portals of entry. The disease is definitely contagious, but years of exposure and contact seem to be necessary for its transmission.

The organism causing leprosy is similar to the tubercle bacillus in appearance and staining characteristics. It is found in great numbers in the nodules occurring under the skin, in discharges from the nose and throat, and in discharges from ulcers. It was first discovered in 1873 by Hansen.

Much difficulty has been encountered in culturing, or rather maintaining, the organism, since no artificial medium has been found which will sustain it. Instead, mouse foot-pad inoculations have been the major source of organisms for study. More recently use of the immunodeficient mouse and the armadillo have provided almost adequate supplies of *M. Leprae*. To a great extent, leprosy may be considered an immunological disease. The leprosy bacillus is virtually nontoxic and most symptoms of the disease are due to immune reactions against antigenic constituents of *M. leprae*.

The period of incubation is variable. It may be from a few months to 20 to 30 years. There are two general types of the disease: (1) nodular leprosy in which the skin is primarily involved; and (2) maculoanaesthetic leprosy, in which there is an involvement of nervous tissue. A mixed form also occurs showing symptoms of both forms.

In the first stages of nodular leprosy, brownish-red spots appear on the skin, usually on the limbs and face, covering large and small areas of the skin. Later nodular thickenings appear at these sites. The face may show the so-called leonine appearance, due to the thickening of the skin in the region of the forehead, eyes, lobes of ears and around the nose and mouth. The entire skin assumes an unhealthy, dusky appearance. Some of the thickened areas ulcerate and fingers and toes may atrophy. Ulceration also appears in the nose and throat and the voice becomes hoarse. The eyes are affected similarly, and blindness may result. This form of the disease may last 10, 20 years, or longer without treatment. Many of the patients die of complicating disorders such as pneumonia, nephritis, tuberculosis and malnutrition.

Maculo-anaesthetic leprosy is characterized by flat, red to brown lesions on the skin, distributed symmetrically. These lesions gradually become insensitive to pain (anaesthetic); trauma, even burns, may occur without the patients feeling pain. Ulceration of the area and contractions will produce deformities.

At one time, chaulmoogra oil was widely used in the treatment of leprosy. Cortisone drugs have been used successfully in the treatment of certain manifestations and for certain phases of the disease, notably for relief of acute lesions of the eye. Derivatives of diaminodiphenylsulfone (Dapsone), administered orally or by injection, have brought about marked improvements, particularly in advanced cases, and for ridding the body of *Mycobacterium leprae* in from two to four years. Para-aminosalicylate and streptomycin also have been used successfully.

Most recently, Dapsone resistance has been seen in the disease organisms so that emphasis is being placed upon the development of a vaccine. This is proving to be a difficult problem, but meanwhile, what appears to be a successful intermediate mode of treatment, has been devised in Venezuela. Purified *M. leprae* bacillus is combined with *Bacillus Calmette-Guerin* (BCG). When this is used to vaccinate leprosy victims, the bacilli were cleared from the patients' bodies in 18 months and progress of the disease was halted. Further work is being done to confirm the value of this form of prophylaxis.

Additional Reading

Antia, N. and V. Shetty: *The Peripheral Nerve in Leprosy and Other Neuropathies*, Oxford University Press, Inc., New York, NY, 1998.
Courtright, P. and S. Lewallen: *Guide to Ocular Leprosy for Health Workers*, World Scientific Publishing, Inc., Riveredge, NJ, 1993.
Hastings, R.C.: *Leprosy: Medicine in the Tropics*, 2nd Edition, Churchill Livingstone, Inc., Philadelphia, PA, 1994.
Staff: *Chemotherapy of Leprosy: Report of a WHO Study Group*, World Health Organization, Geneva, Switzerland, 1994.

Web References

American Leprosy Foundation: http://users.erols.com/lwm-alf/
Centers for Disease Control and Prevention: http://www.cdc.gov/ncidod/dbmd/diseaseinfo/hansens_t.htm
Global Alliance for Leprosy Elimination: http://www.foundation.novartis.com/leprosy/global_alliance/index.htm
Introduction to Leprosy: http://web.raex.com/~bbeechy/introduction.html World Health Organization: http://www.who.int/lep/

R.C. VICKERY, M.D., D.Sci.; PhD, Blanton/Dade City, FL

LEPTONS. The electron, muon, and two kinds of neutrino are collectively called *leptons*. The leptons are considered to be point-like particles without structure and thus truly elementary. Leptons can interact with other particles through the weak interactions. Electrons and muons also can interact through electromagnetic and gravitational forces, but they appear to be without capability of interaction through the strong (nuclear) forces. The neutral members, the electron neutrino, the muon neutrino, and their antiparticles have extremely weak interaction with matter and do not participate in electromagnetic interactions. Leptons make excellent probes in particle physics experiments. The other major family of subatomic particles is referred to as *hadrons*. See also **Electron; Muon; Neutrino;** and **Particles (Subatomic)**.

The name *lepton* from its derivation means "light," referring to the fact that the masses of the leptons are all lighter than that of the lightest meson. The properties of the electron are discussed in that entry; here it will merely be noted that the term *electron* is used to denote the negative electron (often called the "negatron" when ambiguity might arise). Its antiparticle is the positron (also called positive electron).

LEPTOSPIROSIS. Caused by pathogenic spirochetes (*Leptospira interrogans*), leptospirosis may take a number of forms. The serotype *icterohemorrhagiae* causes the syndrome known as Weil's disease; the *autumnalis* serotype causes pretibial fever (Fort Bragg fever); serotypes *canicola*, *icterohemorrhagiae*, and *pomona* may cause aseptic meningitis (leptospiral meningitis), which differs from viral meningitis. In recent years, infections with the *canicola* serotype have occurred most frequently in the United States.

Humans may be infected through contact with the urine or affected tissues of an infected animal and the entry portal may be the skin (hands, etc.), the conjuctiva, or oral mucous membrane. Contaminated soil and water also can be infective. The 1964 outbreaks (approximately 150 cases) in the United States largely reflected water-related infections. Leptospiras can survive for several weeks in a warm, neutral or alkaline medium. Transmission of the disease between humans is rare. At one time, it was believed that leptospirosis was mainly associated with farming, husbandry, veterinary work, meat packing, and sewage facility work, but in recent years outbreaks have occurred among persons engaged in recreational activities, particularly involving contaminated water. The disease is usually seen during the summer months and shows no geographical patterns. Since 1950, the average number of cases per year has been rising slowly, but in a cyclic manner. A peak of 76 cases was reported in 1964. Over the last few years, average cases per year have numbered about ninety.

The incubation period ranges from 1 to 3 weeks. Frequently, the disease is self-limiting. The more serious cases involve secondary developments during the course of the disease. Symptoms usually include moderate to high fever, chills, headache, and prostration. Onset is usually quite abrupt. Fever and chills may persist for about one week. In some patients, cough and chest pain are also present. Muscle tenderness (calves, thighs) and stiff neck are common complaints. Conjunctivitis is present in some patients. This initial phase of the disease abates in about one week, followed by a period of a few days of apparent recovery, during which period the leptospira disappear from the blood. Then the second phase commences, usually with somewhat milder symptoms. In very serious cases, this period of apparent remission may not occur. Generalizations are difficult because of the numerous courses the disease can take.

In the Weil's disease syndrome, the second phase features high fever; the liver enlarges and becomes tender. There may be purpura and gastrointestinal bleeding. Renal damage may occur, with accompanying jaundice. Oliguria (reduction in urine flow) may be serious. Weil's disease is the most serious of the leptospirosis infections, with a mortality of 15–40 percent.

The use of antimicrobials is usually not effective unless their administration is commenced within the first few days after onset. Both penicillin G and tetracycline antibiotics have been used. Further treatment depends upon the course of the disease, with each feature addressed specifically. In cases of renal damage, hemodialysis or peritoneal dialysis may be indicated.

The vaccination of dogs, sometimes implicated in leptospirosis, is not fully protective because even healthy animals can shed the leptospiras in their urine.

Additional Reading

Staff: "Leptospirosis Annual Summary," in *Center for Disease Control*, Atlanta, Georgia (Issued annually).

Tappero, J.W., D.A. Ashford, and B.A. Perkins: "Leptospirosis," in: G.L. Mandell, J.E. Bennet, and R. Dolin, Editors: *Principles and Practice of Infectious Diseases*, 5th Edition, Churchill Livingstone, Inc., Philadelphia, PA, 1999.

Weyant, R.S., S.L. Bragg, and A.F. Kaufmann: "Leptospira and Leptonema," in: P.R. Murray et al.: *Manual of Clinical Microbiology*, 7th Edition, ASM Press, Washington, DC, 1999.

Web References

Centers for Disease Control and Prevention: http://www.cdc.gov/ncidod/dbmd/diseaseinfo/leptospirosis_t.htm

University of Leicester, Department of Biology: http://www.leicester.ac.uk/biology/gat/virtualfc/weil.html

New York State Department of Health: http://www.medhelp.org/

R.C. VICKERY, M.D., D.Sci.; PhD, Blanton/Dade City, FL

LEPUS (the hare). A southern constellation located near Canis Major and Orion.

LESCH-NYHAN SYNDROME. A sex-linked genetic disorder characterized by random, uncontrollable movements, mental retardation, and a self-destructive psychotic behavior. An enzyme deficiency results from the absence of a gene (or presence of a defective gene) on the X chromosome. The disorder affects only males because males have only one X chromosome. Because females have two X chromosomes, the presence of a normal gene on one of these suffices and there is no resulting disorder.

LESION. An injured or otherwise impaired tissue of the body.

LESSER CATS. See **Cats**.

LETTUCE. See **Rose Family**.

LEUCITE. The mineral leucite is a metasilicate of potassium and aluminum corresponding to the formula $KAlSi_2O_6$. It is isometric at a temperature of about 600 °C (1112 °F) and pseudoisometric at lower temperatures, at which leucite is tetragonal but retains an external isometric crystal form, usually trapezohedral. It has a conchoidal fracture; is brittle; hardness, 5.5–6; specific gravity, 2.47–2.50; luster, vitreous; color, white or some shade of gray; translucent to opaque. It is commonly found in the more recent lavas of high alkali content. Leucite is seldom reported from plutonic rock types. It is a relatively rare mineral. It is found plentifully at Vesuvius and Monte Somma and elsewhere in Italy, and Germany in the Tertiary volcanic district of the Eifel. In the United States, leucite has been found in the Leucite Hills of Wyoming, the Highwood Mountains of Montana, and as pseudomorphs (pseudoleucites) representing a mixture of nepheline, orthoclase, analcime, and aegerine from New Jersey, Arkansas, and Montana. Its name is derived from the Greek word leukos, referring to its white color.

LEUCKART, KARL GEORG FRIEDRICH RUDOLF (1822–1898). Karl Leuckart was a German zoologist who made major contributions to comparative anatomy and classification of invertebrates, and helped establish the formal study of parasitology.

Leuckart studied medicine at Göttingen (1842–1845) before qualifying to teach zoology in 1847. After field research studying marine invertebrates, he became associate professor at the University of Giessen in 1850.

After the field work Leuckart created a new phyla, Coelenterata, and reorganized the metazoa into six phyla: Coelenterata, Echinodermata, Annelida, Arthropoda, Mollusca and Vertebrata. His subtle morphological studies, part of an increasingly scientific zoology, enjoyed a mixed reception. In the 1850s, his approach became physiologically orientated and he worked on the sexual organs and development of invertebrates. He investigated the processes of parthenogenesis, the reproduction by a single gamete without fertilization, and polymorphism, the latter being a term he originated, and in this context referring to successive forms passed through during an organism's development. Increasingly he worked with parasitic worms of humans and animals, part of the emerging discipline of parasitology. See also **Parasitology (The History)**.

Leuchakrt studied liver flukes, *Strongyloides stercoralis*, *Onchocerca volvulus* and tapeworms (*Taenia*). Through close attention to morphology and developmental cycles he determined that *T. saginata* are found only in cattle and *T. solium* only in pigs. During the 1860s, he was one of several zoologists working on *Trichina (Trichinella) spiralis*. All these worms were significant threats to public health and his work contributed to meat inspection laws in Germany. In 1869, Leuckart moved to the University of Leipzig, where he continued his research and attracted gifted students from an international field.

Additional Reading

Grove, D.I.: *A History of Human Helminthology*, CAB International, Wallingford, UK, 1990.

H.J. POWER, Wellcome Trust Centre for the History of Medicine at University College, London, UK

LEUCOCYTES. See **Blood**.

LEUCOPLASTS. See **Plastics**.

LEUKEMIAS. The leukemias, like the anemias, represent a group of disorders of the blood cells rather than a single disease and they rank as a significant cause of death in children.

Normally, precursor (hematopoietic) blood cells follow a fixed sequence in their development and differentiation into erythrocytes (red blood cells) and leukocytes (white blood cells) and produce the complement of blood cells required by the human body. In the leukemias, however, the development and differentiation do not cease and the group of diseases is characterized by the undifferentiated proliferation of malignant cells derived from the hematopoietic precursors with resultant replacement of the normal bone marrow and often infiltration of other organs. The cell clone has (1) poor responsiveness to normal regulatory mechanisms, (2) a tendency to have a diminished capacity for normal cell differentiation, (3) an ability to expand at the expense of normal myeloid or lymphatic tissue, and (4) an ability to suppress or impair normal myeloid cell growth.

The cause of this erratic behavior of hematopoietic cells is unclear. Many factors have been suggested in the etiology; these include RNA viruses, ionizing radiation, and a number of drugs and chemicals which have been shown to have adverse effects upon the body. Genetic effects also appear to be involved, especially those conditions associated with chromosomal damage.

About $1-2$ kg of leukemic cells in the body ($1-2 \times 10^{12}$ cells) appear to be sufficient to cause death from leukemia. This number would be about the total marrow volume of an average adult. Because the diagnosis of leukemia cannot usually be made unless the cell burden is 10^9 cells, it is apparent that only ten tumor cell doublings separate the smallest detectable number from a potentially lethal cell number. When a patient is said to be in remission, it means only that the leukemic cell number is not clinically detectable and is, therefore, less than 10^9.

From a standpoint of terminology, the traditional basic classification of leukemia has been the acute and chronic forms of the disease. In acute leukemia, the predominant hematopoietic cell is a primitive precursor, which may be undifferentiated or have the features of lymphoblast, myeloblast, monocyte, or erythroblast. In the chronic disease there is more obvious differentiation into myeloid or lymphoid cells. Apart from this general classification, the leukemias are classed according to the kind of cell primarily involved.

The several different kinds of leukemia differ in symptoms and in life expectancy of the patient. The acute form occurs most frequently in young children; the chronic form usually in persons over 35 years old. *Lymphocytic leukemia* principally involves the white blood cells, which arise from the lymph nodes and spleen. In *granulocytic leukemia*, one or more of the three types of granulocytes, originating in the bone marrow, are involved. *Monocytic leukemia* is typified by the appearance of excessive numbers of monocytes derived from connective tissue. In lymphocytic leukemia, an early symptom is enlargement of the lymph nodes and acute leukemia is often first detected by prolonged hemorrhage following a minor surgical procedure, such as tooth extraction. Fever, arthralgia and anemia are also early symptoms. In some acute leukemias the total white cell count may be normal or even below normal, although the particular type of cell is found in excessive numbers in other tissues.

Three conditions characterize chronic leukemias: (1) enlargement of the spleen; (2) increased numbers of white cells, and (3) anemia. Because of the numerous symptoms and variation in degree, the diagnosis of leukemia is confirmed only by microscopic examination of blood and bone marrow which exemplifies severe anemia and reduced blood platelet counts.

Some of the clinical manifestations of acute leukemias can be related to bone marrow failure, leading to infection, anemia, and bleeding, together with the metabolic effects of a large tumor mass and infiltration of various organs by malignancy.

Treatment of leukemia dates back several decades. In 1948, Farber initiated the folic acid antagonistic route involving daily injections of drugs, such as prednisone, l-aspariginase, vincristine, methotrexate, cytabarine, or cyclophosphamide. Often, within a few weeks, the patient was free of measurable signs of the disease, but when therapy was discontinued, relapse occurred and symptoms recurred within three months. Modern drug treatment follows along similar lines. It is a long-term maintenance procedure, requiring from 3 to 5 years and continued maintenance as required if death is to be averted. Probably less than 50 percent of patients survive after stopping all therapy beyond the initial period of 5 years or longer. Ironically, the chemotherapy used to arrest other cancers has become an important causation of leukemia.

Treatment-Related Leukemias. The treatment methodologies used for primary cancers and the resulting effect upon the incidence of leukemia have been established, but require further elucidation. Much remains to be researched pertaining to the biologic nature of chemotherapy and radiation treatment effects that bring about leukemia. Past studies have indicated cytogenetic abnormalities, including the loss of entire chromosomes 5 and 7, produced by treatment of cancers as well as from a variety of environmental toxins, such as benzene. C.A. Coltman, Jr. and S. Dahlberg

observe, "This consistent picture of resistant leukemia associated with defects of chromosomes 5 and 7, regardless of the underlying disease process and the type of carcinogen, implies a cause-and-effect relation. The deletion of these genes, which are critical to the proliferation and differentiation of hematopoietic (blood forming) tissues, must have a central role in the basic biology of treatment-related leukemia. The exact relation remains to be elucidated, but it is clearly a common pathway of carcinogenesis for ionizing radiation, chemotherapy, and environmental toxins in a wide variety of benign and malignant conditions."

It has been demonstrated that patients who have received chemotherapy with alkylating agents (cyclophosphamide, chlorambucil, melphalan, and other leukemogenic substances) for ovarian cancer have an increased risk of acute leukemia, particularly of the myeloid (resembling bone marrow) type. J.M. Kaldor (International Agency for Research on Cancer, Lyon, France) reported on an extensive study in 1990 and concluded, "The extent to which the relative risks of leukemia are offset by differences in chemotherapeutic effectiveness is not known." A further observation of the Kaldor report: "The trend in treatment for ovarian cancer is toward more use of chemotherapy and in particular an increasing reliance on combinations of drugs. Despite some promising recent findings (Tropé reference), there is limited evidence so far that the effectiveness of such chemotherapy improves with the intensity or number of agents, and our study clearly shows dramatic increases in the long-term risk of leukemia at high dosages."

The Kaldor group also investigated the effect of different treatments for Hodgkin's disease on the risk of leukemia. The conclusion: "We conclude that chemotherapy for Hodgkin's disease greatly increases the risk of leukemia and unaffected by concomitant radio therapy. In addition, the risk is greater for patients with more advanced stages of Hodgkin's disease and for those who undergo splenectomy." However, in another observation, "In the case of Hodgkin's disease, a substantial risk of leukemia following combination chemotherapy is clearly offset by enormous gains in survival. However, after ovarian cancer, the extent to which the increased risk of leukemia is offset is still unclear."

Other Approaches to Leukemia Treatment. Distinct from chemotherapy, red cell transfusion has been used for severe anemia at the outset of treatment, and although granulocyte transfer is not clearly understood, bone marrow transplantation is recognized as a definitive modality for aplastic anemia, acute leukemia, and chronic myeloid leukemia. The effect of such treatment becomes apparent after 2 to 4 weeks, and it is marked by a rise in circulating granulocytes and later by an increase in the platelet count. However, because allogenic grafts of marrow contain immunocompetent cells as well as marrow stem cells, the engrafted cells may mount an attack against the host. This becomes apparent in approximately half of the cases so treated.

A major goal in the treatment of acute leukemia is to reduce the amount of therapy and to minimize the adverse side effects without cutting the effectiveness of established treatments. For related topics, See also **Blood**; **Bone**; and **Cancer**.

Additional Reading

Abraham, N.G., R. Haas, G. Brittinger et al.: *Molecular Biology of Hematopoiesis and Treatment of Leukemias and Lymphomas*, S. Karger Publishers, Inc., Farmington, CT, 1998.

Berkow, R. and M.H. Beers: *The Merck Manual*, 17th Edition, Merck & Company, Inc., Whitehouse Station, NJ, 1999.

Coltman, C.A., Jr. and S. Dahlberg: "Treatment-Related Leukemia," *N. Eng. J. Med.*, 52 (January 4, 1990).

Erickson, D.: "Molecular Trickster: Antisense RNA Pulls A Fast One on the Leukemia Virus," *Sci. Amer.*, 26 (July, 1991).

Freireich, E. and H. Kantarjian: *Leukemia: Advances in Research and Treatment*, Kluwer Academic Publishers, Norwell, MA, 1993.

Goldman, J.M.: *Understanding Leukemia and Related Cancers*, Blackwell Science, Inc., Malden, MA, 1999.

Henderson, E.S. and T.A. Lister: *Leukemia*, 6th Edition, W.B. Saunders Company, Philadelphia, PA, 1996.

Juliusson, G. et al.: "Response to 2-Chlorodeoxyadenosine in Patients with B-Cell Chronic Lymphocytic Leukemia Resistant to Fludarabine," *N. Eng. J. Med.*, 1056 (October 8, 1992).

Kaldor, J.M. et al.: "Leukemia Following Chemotherapy for Ovarian Cancer," *N. Eng. J. Med.*, 1 (January 4, 1990).

Kaldor, J.M. et al.: "Leukemia Following Hodgkin's Disease," *N. Eng. J. Med.*, 7 (January 4, 1990).

Kantarjian, H. and M. Talpaz: *Medical Management of Chronic Myelogenous Leukemia*, Vol. 16, Marcel Dekker, Inc., New York, NY, 1998.

Polliack, A.: *Leukemia and Lymphoma Reviews*, Vol. 6, Gordon & Breach Publishing Group, Newark, NJ, 1999.

Pui, C.: *Childhood Leukemias*, Cambridge University Press, New York, NY, 1999.

Schumacher, H.R.: *Acute Leukemia*, Lippincott Williams & Wilkins, Philadelphia, PA, 1997.

Stass, S.: *Acute Leukemias: Biologic, Diagnostic and Therapeutic Determinants*, Marcel Dekker, Inc., New York, NY, 1987.

Tropé, C.: "Melphalan With and Without Doxorubicin in Advanced Ovarian Cancer," *Obset. Gynecol.*, **582**, 70 (1987).

Weeks, J.C. et al.: "Cost Effectiveness of Prophylactic Intravenous Immune Globulin in Chronic Lymphocytic Leukemia," *N. Eng. J. Med.*, 81 (July 11, 1991).

Weirsma, S.R. et al.: "Clinical Importance of Myeloid-Antigen Expression in Acute Lymphoblastic Leukemia of Childhood," *N. Eng. J. Med.*, 800 (March 21, 1991).

Web References

Leukemia Information Library: http://www.meds.com/leukemia/leukemia.html
Leukemia Resource Center and Links: http://rodneyporter.com/leukaemia/
The Leukemia & Lymphoma Society, Formerly Leukemia Society of America: http://www.leukemia.org/
The Myeloproliferative Disorders: http://www.acor.org/diseases/hematology/MPD/

R.C. VICKERY, M.D.; D.Sc.; Ph.D., Blanton/Dade City, FL

LEUKOCYTOSIS. An abnormally high level of leukocytes (white cells) in the blood, a common manifestation of infection.

LEUKOPENIA. A decrease in the normal number of leucocytes in the bloodstream. This is a usual accompaniment of certain stages of some infectious diseases, while in other infections it is of grave prognostic significance, and indicates a failure of one of the lines of defense of the body. In agranulocytosis, leukopenia is the chief manifestation of the disease.

Drugs and chemicals sometimes implicated in leukopenia include arsenic, chloramphenicol, cimetidine, 5-fluorocytosine, phenytoin, sulfopyridine, thiazides, and trimethoprim. Leukopenia is sometimes associated with Colorado tick fever, meningococcemia, pneumococcal pneumonia, and salmonelolosis. Leukopenia is also a result of starvation.

LEVANTE. See **Winds and Air Movement**.

LEVEL OF ESCAPE. Sometimes called *critical level of escape*, that level, in the atmosphere, at which a particle moving rapidly upwards will have a probability of $1/e$ of colliding with another particle on its way out of the atmosphere (where e is the number **e**, the base of natural logarithms). It is also the level at which the horizontal mean free path of an atmospheric particle equals the scale height of the atmosphere. The critical level of escape is the base of the exosphere, which is the outermost, or topmost, portion of the atmosphere. See also **Atmosphere (Earth)**.

LEVEL (Surveyor's). An instrument for determining differences of elevation. It is of use in obtaining comparative elevations of two points, or in defining the profile of a certain path, such as a roadway, drainage ditch, etc. A level of this type has an accurately made bubble level attached to and made exactly parallel with a telescope. In the wye level this telescope rests in Y-shaped supports. These supports are held in turn by the instrument base, which may be adjusted by hand, and which is attached, usually by screwing, to the top of a tripod, upon which the instrument rests when in use. The bubble tube being parallel to the center of the telescope, the latter will automatically be leveled, ready for a horizontal sight, when the bubble tube is level. The dumpy level has the telescope and supports cast in one piece or rigidly connected. Lasers are also used in leveling operations, particularly for grading irrigated areas. See also **Irrigation**.

LEVEL WIDTH (Excitation Energy). A measure of the spread in excitation energy of an unstable state of a quantized system. In emission or absorption spectra, variations in the intrinsic line widths show that some energy levels in atomic or nuclear systems are broad and others are narrow. In nuclear physics, level widths have been observed chiefly in connection with neutron and charged-particle resonances, which are found to have nonuniform breadths in energy. The level width is related to the mean life of the level by the expression

$$\Gamma = \hbar/\tau$$

where Γ is the level width, $\hbar$ is $\hbar/2\pi$ (h is Planck's constant), and τ is the mean life. Level widths usually show themselves as the widths of resonance peaks observed when the cross section for the particular reaction is plotted as a function of the energy of the incident particle. The quantitative value of the level width is usually taken as the full width at half maximum of the resonance peak.

If a system at a given level has several alternate modes of disintegration, there is associated with each a partial level width proportional to the probability of disintegration by the particular mode. The total level width is the sum of the partial level widths. See also **Broadening of Spectral Lines**.

LEVENE-HUDSON PHENYLHYDRAZIDE RULE. The direction of rotation of the phenylhydrazides of the sugar acids indicates the configuration of the hydroxyl on the α-carbon atom. If the phenylhydradzide rotates to the right, the hydroxyl on the α-carbon is to the right, and vice-versa. The rule was shown to be valid for salts, amides, and corresponding acylated nitriles. In connection with the rule, Hudson mentioned that, "the sugar benzylphenylhydrazones rotate to the left when the asymmetric α-carbon atom of the configuration has its hydroxyl to the right, and vice-versa."

LEVERS. See **Machine (Simple)**.

LEVI-MONTALCINI, RITA (1909–Present). Born in Turin, Italy, to an engineer father and a painter mother, Rita Levi-Montalcini received an MD from the Medical School at the University of Turin in 1936. She stayed on as a research assistant to study the development of nerve cells with one of her teachers, Guiseppe Levi. Her work was interrupted in 1938 by the anti-Semitic policies of Mussolini's Fascist government. She was not, as a Jew, allowed to hold an academic position at a university or to practice medicine. After a brief period in 1939 at the Neurological Institute in Brussels, Belgium, she returned to Turin and spent the war years in hiding. Stimulated by the work of Victor Hamburger, an American embryologist, Levi-Montalcini still managed from 1940 to 1943 to study the nerve development in chicken embryos by setting up a small research laboratory in her "bedroom with a few indispensable pieces of equipment, such as an incubator, a light, a stereo-microscope, and a microtome" alongwith "sewing needles transformed with the help of sharpening stones into micro-instruments." With the German occupation, she had to escape to Florence. The difficulties she experienced are described in her autobiographical account (1975). When the Germans were defeated in southern Italy in 1944, she volunteered as a physician to treat war refugees.

Her research on nerve growth in chicken embryos was resumed first with Levi, then with an invitation in 1946 to work with Victor Hamburger at Washington University, St Louis, Missouri. In 1952, she went to work with Hertha Meyer at the Institute of Biophysics in Rio de Janeiro in Brazil on tissues grown under glass, which enabled faster results to be obtained than with live animals. In exploring the effect of mouse tumors *in vitro* on the nerve tissue of chick embryos, Levi-Montalcini noted that within hours the chick nerve tissue was surrounded by a new growth of nerve fibers. Following up with a series of critical experiments, she discovered that a substance in the mouse tumor was responsible for nerve growth.

Returning to Washington University in 1953, Levi-Montalcini began herlong and productive association with the biochemist Stanley Cohen. They identified further sources of the nerve growth factor (NGF) including the salivary glands of rodent embryos and snake venom glands (in attempting to stop nerve growth). Their research established that the NGF was most effectiveon cells in the early stages of differentiation. They also prepared antibodies to NGF by injecting NGF into rabbits and then withdrawing the serum manufactured by the rabbits' own immune system. When the serum was injected into newborn mice, antibodies blocked the growth of nerve cells. From 1961 to 1969 Levi-Montalcini divided her time between Washington University, where she became a full professor, and the Higher Institute of Health in Rome. In 1968 she was elected to the National Academy of Sciences in the USA, and also began to direct the newly established Laboratory of Cell Biology in Rome, where she continued her research as professor emeritus after her retirement in 1977. Following her Nobel Prize in Physiology or Medicine (shared with Stanley Cohen) in 1986, Levi-Montalcini was awarded the United States National Medal of Science in 1987. See also **Cohen,Stanley (1922–Present)**.

Levi-Montalcini's work created a new field of research which led to the discovery of other growth factors, contributed to the understanding of the development of the nervous system, the mechanisms of tissue repair and regeneration, as well as the conversion of normal cells to cancerous cells.

Additional Reading

Levi-Montalcini, R.: "Growth Control of Nerve Cells by a Protein Factor and its Antiserum," *Science*, **143**, 105–110 (1964).

Levi-Montalcini, R.: NGF: an Uncharted Route, In: Worden, F.G., J.P. Swazey, and G. Adelman: *The Neurosciences: Paths of Discovery*, MIT Press, Cambridge, MA, 1975.

Levi-Montalcini, R, and P. Calissani: "The Nerve-growth Factor," *Scientific American*, **240**, 68–77 (1979).

Levi-Montalcini, R., and P. Calissani: *Molecular Aspects of Neurobiology*, Springer-Verlag, Berlin, Germany, 1986.

Levi-Montalcini, R.: "The Nerve Growth Factor 35 years Later," *Science*, **237**, 1154–1162 (1987).

Levi-Montalcini, R.: *In Praise of Imperfection. My Life and Work*, Attardi, L. (trans.), Basic Books, [Originally published as *Elogio dellimperfezione*, 1987.] New York, NY, 1988.

Levi-Montalcini, R.: "The Saga of Nerve Growth Factor, Preliminary Studies, Discovery, Further Development," *World Scientific Series in 20th Century Biology*, Vol. 3. River Edge, Singapore, 1997.

Magill, F.N.: *The Nobel Prize Winners: Physiology or Medicine*, SalemPress, Pasadena, CA, 1991.

McGrayne, S.B.: *Nobel Prize Women in Science*, Carol, New York, NY, 1993.

Olson, R., and R. Smith: *Biographical Encyclopedia of Scientists*, Marshall Cavendish, New York, NY, 1998.

G.A. RUSSELL , Texas A&M University System Health Science Center, College Station, TX

LEVOROTATORY COMPOUNDS. Having the property when in solution of rotating the plane of polarized light to the left or counterclockwise. Levorotatory compounds may have the prefix *l-*to distinguish them from their dextrorotatory or *d-*isomers, but the minus sign (−) is preferred.

See also **Asymmetry (Chemical)**; **Isomerism**.

LEWIS ACID. Any molecule or ion (called and electrophile) that can combine with another molecule or ion by forming a covalent bond with two electrons from the second molecule or ion. An acid is thus an electron acceptor. Hydrogen ion (proton) is the simplest substance that will do this, but many compounds such as boron trifluoride, BF_3, and aluminum chloride, $AlCl_3$, exhibit the same behavior and are therefore properly called acids. Such substances show acid effects on indicator colors and when dissolved in the proper solvents.

LEWIS BASE. A substance that forms a covalent bond by donating a pair of electrons, neutralization resulting from reaction between the base and the acid with formation of a coordinate covalent bond. It is also called a nucleophile.

See also **Lewis Electron Theory**.

LEWIS ELECTRON THEORY. A theory involving acid and base formation, neutralization, and related phenomena on the basis of exchange of electrons between substances and the formation of coordinate bonds. It represented an important advance in chemical theory, largely replacing earlier concepts. Advanced in 1923 by Gilbert N. Lewis, it contributed much to the development of coordination chemistry in which the base is represented by the ligand and the acid by the metal ion.

LEWIS, GILBERT N. (1875–1946). An American chemist, native of Massachusetts, professor of chemistry at MIT from 1905 to 1912 after which he became dean of chemistry at the University of California at Berkeley. His most creative contribution was the electron-pair theory of acids and bases, which laid the groundwork for coordination chemistry. He was also a leading authority on thermodynamics.

LEWIS, THOMAS (1881–1945). Thomas Lewis was a British doctor who did pioneer work in electrocardiography, cardiology, pain, and blood vessels of the skin and who founded the Medical Research Society.

Lewis was born in Cardiff, Wales of Welsh parents, his father being amining engineer. Whilst still a student in Cardiff, he wrote at the age of only twenty a 54-page paper illustrated by himself on the hemolymph glands, which became a standard work. This was a fine beginning to what was to be an illustrious career in medical science spanning three major disciplines of cardiology, blood vessels of the skin and pain. He went to University College Hospital London for his clinical training and was attracted to cardiac physiology by Ernest Henry Starling, in whose laboratory he worked for a while. After qualifying in 1905, he wrote seven papers on the pulse and blood pressure whilst still a house physician. In 1908 Lewis met Dr James Mackenzie, an authority on cardiac arrhythmias, and he encouraged Lewis to study that subject. Fortunately, a new instrument for recording the heartbeat called a string galvanometer had been invented a few years earlier by Willem Einthoven in Holland, and it enabled electrocardiograms to be recordedwith great precision. Lewis obtained an electrocardiograph machine and after only two years he wrote in 1911 a monograph, *The Mechanism of the Heart Beat*, which was hailed as the bible of electrocardiography. Many young physicians from the USA, Canada, and Australia were attracted to his laboratory at University College Hospital, which remained his medical home forthe rest of his life. His research work into cardiac electrophysiology ledto his election as a Fellow of the Royal Society. During World War I he was in charge of the Military Heart Hospital, where he proved that what had been known as soldier's heart was a non-cardiac disorder, which he called the effort syndrome. For this nationally important work he was knighted in 1921. See also **Arrhythmias (Cardiac)**; and **Starling, Ernest Henry (1866–1927)**.

But although Lewis did renowned work in cardiology, his real mission in life was that of applying the scientific method to the study of clinicalproblems, a discipline that he called *clinical science*. He founded the Medical Research Society (http://www.mrc.ac.uk/index.htm) in 1930 to encourage young doctors to take up what was then an unusual and difficult choice of selecting clinical research as a full-time career, and he created posts to enable them to do this. Lewis gave up cardiac research in 1925and then made an intensive study of the reaction of the skin to injury. This led to his hypothesis that release of a histamine-like substance was thecause of the skin responses. Next he turned his attention to the difficultsubject of pain, and here he formulated another causative hypothesis, namely that muscular pain was due to the release of a chemical compound which he labeled the "P substance." His further work on pain, which was of a high intellectual order, defined the mechanism by which disease of an internal organ leads to pain being referred to various segments of the body. Lewis's contributions to medical research were recognized by the Royal Society'shighest award, the Copley Medal. This was a rare distinction for a clinician, for Lewis was no backroom scientist. He gave careful and personal attention to his patients, both in the wards and in the outpatient clinic, and he was a good bedside doctor.

Lewis had other important attributes. In 1909, with Mackenzie, he founded the journal *Heart*, which he changed into *Clinical Science* in 1933. He was editor of these journals for a record 33 years. He was anexcellent clinical teacher, insisting on the accurate observation of patients, and he distilled his teaching into a highly acclaimed book *Diseases of the Heart*, which was notable for the clarity of his writing.

At work Lewis was a hard taskmaster, and his assistants found it difficult to keep up with his intense and unremitting activity and insistence on very high standards. But they all came to have great personal regard for him and for his quietly done acts of kindness.

At the age of 45 Lewis had a myocardial infarction but it was not until 17 years later that his coronary heart disease led to heart failure fromwhich he died a few months after his 63rd birthday.

Additional Reading

Burchell, H.B.: "Sir Thomas Lewis: His Impact on American Cardiology," *British Heart Journal*, **46**, 1–4 (1981).

Hollman, A.: *Pioneer Cardiologist and Clinical Scientist*, Springer-Verlag London, London, UK, 1997.

ARTHUR HOLLMAN, Pett, East Sussex, UK

LEWIS, WARREN P. (1882–1974). Born in Laurel, Maryland, graduated from MIT in 1905, and received is Ph.D. from the University of Breslau, Germany in 1908. He became professor of chemical engineering at MIT in 1910. He is often regarded as the "father" of chemical engineering in the U.S., as his outstanding books and other publications did much to establish the fundamental principles of this field.

LEYDEN TEMPERATURE SCALE. A low-temperature thermometer scale based on a boiling point of hydrogen equal to $-252.74\,°C$ and of oxygen equal to $-182.95\,°C$.

L'HOSPITAL RULE. If two functions $f(x)$ and $g(x)$ together with their derivatives up to order $(n-1)$ vanish at $x = a$, and if their derivatives

$$\lim_{x \to a} \frac{f(x)}{g(x)} = \frac{f^{(n)}(a)}{g^{(n)}(a)}$$

of nth order do not both vanish there or both become infinite, then
See also **Indeterminate Form**.

LIBBY, WILLARD F. (1908–1980). An American chemist, famous for his role in the development of radiocarbon dating, a process which revolutionized archaeology.

After the start of World War II he worked on the Manhattan Project at Columbia University with Nobel laureate Harold Urey. Libby was responsible for the gaseous diffusion separation and enrichment of Uranium-235 which was used in the atomic bomb on Hiroshima.

In 1945 he became a professor at the University of Chicago. In 1954, he was appointed to the U.S. Atomic Energy Commission, and in 1959, he became Professor of Chemistry at University of California, Berkeley, a position he held until his retirement in 1976. He also started the first Environmental Engineering program at UCLA in 1972.

Libby won the Nobel Prize for chemistry in 1960, for his method to use carbon-14 for age determination in archaeology, geology, geophysics, and other branches of science.

LIBECCIO. See **Winds and Air Movement**.

LIBRA. (the scales or the balances). A small constellation, best known as the seventh sign of the zodiac; its symbol is taken for the autumnal equinox (i.e., the point where the sun apparently crosses the celestial equator from north to south). The brightest star in Libra is a wide double star, which can easily be resolved with a field glass. This star carries the Arabic name Zuben el Genubi, the southern scale. (See map accompanying entry on **Constellations**.)

LIBRATIONS. A term applied in astronomy to many periodic oscillations; in particular, to slight apparent oscillations of the moon, whereby observes on the earth are enabled to observe somewhat more than 50% of the moon's surface. There are three principal librations of the moon: a libration in lunar latitude, a libration in lunar longitude, and a diurnal (or daily) libration.

The libration in lunar latitude arises from the fact that the orbit plane of the moon is inclined at about 6.5° to the plane of the moon's equator. This produces an effect relative to the earth similar to the terrestrial effects relative to the sun, which produce the seasons. During one-half of the month, the north lunar pole is directed slightly toward the earth, and during the remainder of the month the south lunar pole is toward the earth.

The libration in lunar longitude is due to the fact that, although the rotation of the moon is quite uniform, with a period equal to the period of revolution about the earth, the orbital motion is not uniform, but is in accordance with the second Keplerian law of planetary motion. Suppose that a certain point on the surface of the moon is directly toward the earth at a time when the moon is in perigee. Since the moon is moving more rapidly in its ellipse at perigee than at any other part of its orbit, by the time the elongation has increased to 90%, the selected point has not completed one-quarter of a rotation and is not directly toward the earth. The result is that the observer will see a slightly different hemisphere of the moon than he did at perigee. By the time the moon has reached apogee, the selected point will again be toward the observer, and he will have the same view of the moon as at perigee. At this point, the moon's orbital motion is

a minimum; the selected point will complete a quarter rotation before a quarter revolution is completed, and another slightly different hemisphere of the moon will be visible.

The diurnal libration is due to the fact that the observer sees slightly "over the top" of the moon when the moon is rising, and slightly under the bottom when the moon is setting. This is really a libration of the observer rather than a libration of the moon, but is classed with the latter.

The combined result of the librations is that about 41% of the moon is always visible from the earth (or would be if the sun were shining upon it), 41% is never visible, and the remaining 18% is either visible or invisible, depending upon the particular position of the moon relative to the earth.

The term libration is also applied to certain periodic perturbations in the orbits of members of the solar system.

See also **Perturbation (Astronomy)**.

LICHEN. Perennial plants, which are a combination of two plants growing together in an association so intimate that they appear as one. Either of the component plants alone possesses none of the characteristics shown by the two in combination. Lichens often are cited as examples of symbiosis, i.e., two organisms living together of benefit to each other. The components of a lichen are always an alga and a fungus. See also **Algae**; and **Fungus**. The algal constituent is usually one of the simple green algae, or, more rarely, a blue green one. The alga can live perfectly well by itself, and is often found growing free on rocks or tree trunks in regions where the lichen would exist. The fungal component is usually a member of the ascomycetes. In lichens growing in cooler regions it is always one of this group. There are certain lichens found in the tropics, however, in which the fungal component is a basidiomycete. In the lichen, the fungus alone is capable of fruiting, although the algal cells do divide and so increase in number. It is very difficult to see how such an association came about, and to determine whether it is really a case of symbiosis or whether it is not parasitism, one of the plants living on the other. Unquestionably the fungus benefits from the presence of the alga, since the latter carries on photosynthesis, making food materials which are used by the fungus. The latter, lacking chlorophyll, cannot manufacture its own food. It is possible that the alga benefits by the added moisture gathered by the fungus, that the latter protects the alga against desiccation. Certainly the association is well established, and seemingly has been so for a long period of time. Lichens have been "made" artificially, that is, the two components have been grown separately in pure cultures, and when brought together have produced a lichen. So lichens can be formed anew, always with a very constant appearance characterizing the particular form considered.

Lichens are found nearly everywhere where civilization has not killed them. They are found on the surface of rocks and soil; they occur on the bark of trees; in the tropics they may be found on the surface of thick evergreen leaves of trees; a few species even grow on rocks submerged by the tides.

The shape of the lichen body or thallus is very diverse. Some species are flat crusts growing on or even in the surface of the substratum, whether the latter be trunk of tree or barren rock. Lichens of this type are called crustose. In others the thallus is split up into many radiating divisions, and is called a foliose lichen. Many others have an erect, often much-branched thallus and are called fruticose lichens. The color of the thallus may be yellow, orange, brown, gray, or black.

The greater part of the lichen thallus is composed of fungus hyphae, which form a compactly tangled mass. The algal cells occur in an irregular loosely arranged layer near the outer surface of the thallus. Short irregular branches from the fungus hyphae grow tightly around each algal cell, sending into it short, absorbing structures called haustoria. The surface of the lichen is composed of enlarged thick-walled fungus cells, which form a compact layer over the more loosely arranged central portions. In many of the crustose and foliose lichens there are many rhizoids which anchor the plant firmly.

One of the ways in which a lichen reproduces is by means of soredia. These are minute bits of lichen, formed on the surface, and composed of one or more of the algal cells together with a small mass of closely associated hyphae. Often these soredia are so numerous as to give to the lichen a powdery appearance. Either through disintegration of the lichen body, or because the continuity of the hyphae breaks down, soredia become free from the thallus. They are then easily spread by wind or by water.

They may even be carried about unintentionally by the many small insects and other animals that feed on lichens. Lichens also reproduce by means of spores; that is, the fungus component forms special reproductive structures very similar to those formed by similar fungi not forming lichen thalli. In the lichens composed of ascomycetes these reproductive structures are open cups or mounds, called apothecia. Commonly these apothecia are of a different color from that of the thallus.

Reindeer moss, *Cladonia rangiferina*, is the principal food of the reindeer, and may also be used as fodder for other animals. Reindeer moss, not a moss at all, is an erect much-branded lichen which grows abundantly over wide stretches of barren soil. The dense grayish-green tufts grow continuously at the tops, becoming 6–10 inches (15 to 25 centimeters) tall and attaining great age. Another lichen, *Cetraria islandica*, or Iceland moss, may be used as stock food. In habit it resembles reindeer moss but is coarser and less branched. See Fig. 1.

Fig. 1.   Lichen (reindeer moss). (*A.M. Winchester.*)

Litmus paper is prepared from *Lecanora tartarea*, a common lichen found in the Netherlands.

LICORICE. This herbaceous native plant of southern Europe (*Glycyrrhiza glabra* L.) of the family *Leguminosae* is the source of popular flavorings used in the food industry. Stolons or roots (at least 2 years old) are extracted with hot water. The taste is sweet and rich; the essence is sometimes described as slightly spicy. The principal constituent of licorice is the potassium-calcium-magnesium salt of glycyrrhizic acid, which upon hydrolysis produces glycyrrhetic acid and 2 moles of glucuronic acid. Licorice roots also contain triterpene, flavonoids, and B vitamins. The powerful sweetening power of licorice (estimated at 50 times that of sucrose) is derived from the glycyrrhizin. In food and drug applications, licorice is used both to enhance and also to mask or subdue flavors, particularly of a bitter nature.

The licorice plant ranges from 4 to 5 feet (1.2 to 1.5 meters) tall and has many pinnately compound, pale-green leaves and purplish flowers resembling those of the perennial pea. Cultivated in southern Europe, the plant also is found wild in eastern Europe. Licorice root and extract are used in nonalcoholic beverages (up to 130 parts per million); in ice cream (200 ppm); in candies (as high as 2500 ppm); in baked goods (200 ppm); in gelatins and puddings (5 ppm); in syrups (50 ppm); and in some chewing gums (as high as 22,000 ppm). Licorice is also frequently used in tobacco products. A synthetic licorice flavoring is made and is the ammonium salt of glycyrrhizic acid.

LIDAR. Acronym for Light Detection and Ranging. Like RADAR, except lidar used light (laser) instead of radio waves.

An instrument combining a pulsed laser transmitter and optical receiver (usually a telescope) with an electronic signal processing unit used for the detection and ranging of various distant targets in the atmosphere, analogous to the principles of operation of microwave radar. Normally,

the transmitter and receiver are coaligned and placed closely together to measure the laser energy backscattered by the target into the direction of the receiver. The use of lasers allows the light pulses to be exceptionally short, highly focused, and monochromatic, but laser light suffers from strong, range-limiting attenuation in many types of clouds. Depending on the spectral characteristics of the laser and detector and the number of receiver channels, a large variety of lidar applications for atmospheric research are in use. Simple one-channel laser ceilometers measure cloud-base heights and internal cloud structures; polarization lidars measure cloud phase and hydrometeor type; differential absorption (DIAL) and Raman lidars measure the concentrations of selected molecular species; high spectral resolution lidars (HSRL) measure the separation of molecular and aerosol or cloud constituents; and Doppler lidars measure the radial velocity of aerosol or cloud targets. Lidar wavelengths range from the near-ultraviolet to the midinfrared ($\approx 0.3 - 12$ μm).

LIDAR EQUATION. An equation, which may appear in different forms depending on the particular system or application, that describes the relation between the received power p measured in a lidar receiver channel from range r, and the characteristics of the lidar system and the transmission medium (usually the atmosphere) through which the laser pulse propagates.

The most common form of the equation is for plane-polarized radiation and single scattering, for which

$$p(r) = \frac{C\beta(r)t^2(r)}{r^2},$$

where β is the volume backscattering coefficient at range r, t^2 is the two-way transmittance to range r, and C is the lidar constant, which depends on such system parameters as the transmitted power, pulse duration, and receiver characteristics. The transmittance is related to the volume extinction coefficient γ by

$$t^2(r) = \exp\left[-2\int_0^r \gamma(s)ds\right].$$

Normally scattering and extinction of the lidar beam are caused by the combined effects of molecules, aerosols, and hydrometeors, so that β and γ represent the sum of their separate contributions.

AMS

LIEBIG, JUSTUS VON. (1803–1873). A German chemist who founded the *Annalen*, a world-famous chemical journal. He was a great teacher of chemistry, training such men as Hofmann, August Whilhelm, who did basic work on organic dyes. Liebig contributed original research in the fields of human physiology, plant life, soil chemistry, and was first to discover chloroform, chloral, and cyanogen compounds. He was the first to recommend the addition of nutrients to soils and thus may be considered the originator of the fertilizer industry.

LIE GROUP. A group that is also an analytic manifold in which the group operations, multiplication and formation of inverse, are analytic. Historically, the first Lie groups were continuous transformations of the points of a manifold. Thus, consider the set of transformations of the points of a Euclidean plane,

$$x_1 = \phi(x, y, a), \quad y_1 = \psi(x, y, a)$$

for some range of values of the parameter a, given by functions ϕ and Ψ such that, if two transformations of the set are carried out in succession, the result is again a transformation of the set. An easily visualized example is the group of rotations

$$x_1 = x \cos a - y \sin a, \quad y_1 = x \sin a + y \cos a$$

Lie groups of transformations owe their practical importance partly to their usefulness in systematizing the solutions of differential equations, both ordinary and partial, and partly to the fact that many of the standard groups of linear transformations are Lie groups. For example.

The *full linear group* is the group of all nonsingular matrices with complex numbers as elements.

The *real linear group* is the group of all nonsingular n × n matrices with real numbers as elements.

The *unimodular group* is the group of all complex matrices with determinant equal to unity.

The *real unimodular* group is the group of all real matrices with determinant equal to unity.

The *unitary group* is the group of all unitary matrices.

The *unitary modular group* is the group of all unitary matrices with determinant equal to unity.

The *real orthogonal group* is the group of all real orthogonal matrices.

The *rotation group* is the group of all real orthogonal matrices with determinant equal to plus one.

See also **Group**.

LIFE, ORIGIN (*Biogenesis*).   The succession of chemical events that led up to the appearance of living organisms on earth about 3.3 billion years ago. According to one theory, substantiated by experimental evidence, this occurred as follows. The inorganic compounds originally present were carbides, water, ammonia, and carbon dioxide. The carbides reacted with water to form methane, which in turn reacted with ammonia and water vapor as a result of an electric impulse to form amino acids, porphyrins, and nucleotides (or their precursors). All these compounds have been created artificially in the laboratory. It has further been shown that amino acids and nucleotides can be concentrated into proteins (and probably nucleic acids) by the action of zinc-bearing clays, which were present along the shores of the primeval oceans. Little or no free oxygen existed in the primordial atmosphere, which consisted chiefly of reducing gases. The complex chemical reactions that eventually resulted in the formation of DNA took place in an anaerobic aqueous environment, and the earliest living organisms developed in a nutrient solution in which free oxygen finally appeared as the result of photosynthesis by blue-green bacteria. Another theory advances the idea that essential life chemicals such as purines and amino acids were formed under primitive conditions from aqueous solutions of hydrogen cyanide. Both of these theories are based on research carried out by highly competent biochemists.

LIFE SCIENCES.   The field of scientific disciplines encompassing biology, physiology, psychology, medicine, sociology, and other related areas.

LIFETIMES.   The time required for the concentration of a chemical species to decrease to $1/e$ of its original value. The definition arises from assuming a pseudo-first-order decay of the species concentration, $[A] = [A]_0 \exp(-kt)$. When the time t is equal to the inverse of the first-order rate coefficient k, the concentration $[A]$ will have decreased to $1/e$ of the initial concentration $[A]_0$. The atmospheric lifetime of a species in steady state can also be defined as the total atmospheric burden divided by either the total loss rate or the total production rate.

LIGAMENT.   See **Tendon**.

LIGAND.   Any atom, radical, ion, or molecule in a complex (poly-atomic group) which is bound to the central atom. Thus, the ammonia molecules in $[Co(NH_3)_6]^{3+}$, and the chlorine atoms in $[PtCl_6]^{2-}$ are ligands. Ligands are also complexing agents, as for example, EDTA, ammonia, etc. See also **Chelates and Chelation**.

Ligand field theory incorporates elements from the valence bond theory of Pauling and the molecular orbital method of Hund, Mullikan, and others. As pointed out by Mortimer, the chemists of the late nineteenth century had difficulty in understanding how "molecular compounds" or "compounds of higher order" are bonded. The formation of a compound such as $CoCl_3 \bullet 6NH_3$, was baffling, particularly in this case since simple $CoCl_3$ does not exist. In 1893, Alfred Werner proposed a theory to account for compounds of this type. Werner wrote the formula of the cobalt compound as $[Co(NH_3)_6]Cl_3$. Werner assumed that the six ammonia molecules are symmetrically coordinated to the central cobalt atom by "subsidiary valencies" of cobalt, while the "principal valencies" of cobalt are satisfied by the chloride ions. Werner devoted over 20 years preparing and studying coordination compounds and perfecting and proving his theory. Although modern work has amplified his theory, it has required relatively little modification.

In ligand field theory, one is concerned with the origin and the consequences of splitting the inner orbitals of the central metal by the surrounding ligands. The most satisfactory correlations have been demonstrated with the first transition series, in which the $3d$-orbitals are split into different energy levels. To appreciate the effect of a ligand field, imagine that a symmetrical group of ligands is brought up to a charged ion from a distance. First, the electrostatic repulsions between the ligand electrons and those in the d-orbitals of the metal will raise the energy of all five d-orbitals equally. Then, as the ligands approach to within bonding distances, the repulsion interactions will take on a directional character that will vary with the particular d-orbitals under consideration. This arises because of the different shapes and orientations of the five d-orbitals in space along a Cartesian coordinate system. The splitting of the orbitals for a given central metal ion is dependent on the set of ligands.

Applications of ligand field theory to many transition metal complexes have played an important role in the interpretation of visible absorption spectra, magnetism, luminescence, and paramagnetic resonance spectra.

Additional Reading

Bohm, Hans-Joachim, G. Schneider, and R. Mannhold: *Protein-Ligand Interactions: From Molecular Recognition to Drug Design*, John Wiley & Sons, Inc., New York, NY, 2003.

Figgis, B.N.: *Ligand Field Theory and Its Applications*, John Wiley & Sons, Inc., New York, NY, 1999.

Harding, S.E., and B.Z. Chowdhry: *Protein-Ligand Interactions: A Practical Approach*, Oxford University Press, New York, NY, 2001.

Russo, N., M. Witko, and D.R. Salahub: *Metal-Ligand Interactions: Molecular-, Nano-, Micro-, and Macro-Systems in Complex Environments*, Kluwer Academic Publishers, Norwell, MA, 2003.

LIGASES.   See **Enzyme**.

LIGATURE.   A threadlike material or wire used for tying off blood vessels or other structures of the body during surgical operations. The material may be absorbable (catgut) or nonabsorbable (silk, nylon or linen) or metal wire. Ligatures are made in various grades of thickness and tensile strength.

LIGHT.   Although the use of the word *light* has been broadened over the decades, general usage still refers to the visible portion of the electromagnetic spectrum. The wavelengths of visible light extend approximately from 4000 to 7000 Å (1 Å $= 10^{-8}$ centimeter). The speed of light in vacuum (symbol c) is generally published as 299,792,500 meters per second (~186,292 miles per second). The direct determination of the velocity of light, usually performed in air, conventionally has been based upon the measurement of the time for a light pulse to cover a known distance. (See reference to Michelson-Morley experiment in entry on **Laser**.) Such a pulse means an increase, followed by a decrease, of the amplitude of the light vibrations. What is observed is energy exchange associated with the amplitude changes and, as a result, the propagation velocity of light is obtained. A change of amplitude is, however, equivalent to an interference among a series of adjacent wavelengths, inasmuch as that change is created by just such an interference. The light pulse, therefore, consists of a whole group of adjacent wavelengths interfering with each other. Interference is a sum-product. If the participating waves have different velocities, one can find by simple addition of two sine oscillations that the group formed has a velocity different from those of the waves creating the group. An interesting experiment in the use of a laser to measure the speed of light was made in 1962. This experiment was later followed by a number of other, even more sophisticated laser measurements.

A review of the numerous experiments to measure the velocity of light makes fascinating reading for the student of science. An excellent starting review of the experiments of Galileo (1676), Bradley (1725), Foucault (1850), Kohlrausch (1856), Blondlot (1891), Michelson (1879 and 1926), Karolus and Mittelstaedt (1928), Anderson (1940), Jones and Conford (1947), Alaskon (1949), Essen and Gordon-Smith (1947–1950), Bergstrand (1950), Froome (1950–1958), Rank, Bennet and Bennet (1955), Mackenzie (1953), Kolibayev (1958–1963), Mockler (1961), and Cohen (1972) is given in "*The Encyclopedia of Physics*," (R.M. Besançon, editor), Van Nostrand Reinhold, New York.

Light has a physical character similar to that of radio waves. However, the frequency of light waves is almost a billion times higher and the wavelengths a billion times shorter, than the waves of standard radio broadcast bands. The perception of color depends upon the distribution of the electromagnetic energy over the visible wavelengths. White light is a superposition of waves at many frequencies. It can be decomposed into its monochromatic spectral components by a prism or other spectral apparatus. The violet end of the spectrum is near 4000 Å. The red end is near 7000 Å. Whereas light, in its narrow definition, should be confined to this relatively narrow portion of the electromagnetic spectrum, in recent years it has become customary to extend the definition to the ultraviolet and infrared portions of the spectrum. One sometimes speaks loosely of ultraviolet and infrared light, although electromagnetic waves at these frequencies are not detectable by the human eye. Instruments and photographic films, however, can be made sensitive to both the shorter and much longer wavelengths. See also **Infrared Radiation**; **Photography and Imagery**; **Spectrochemical Analysis (Visible)**; **Spectro Instruments**; **Spectroscope**; and **Ultraviolet Radiation**.

The study of the human eye as a detector of light is the task of *physiological optics*. The impression of light is not necessarily always connected with the simultaneous presence of electromagnetic energy at the retina. The eye is capable of creating false images, as when one "sees stars" from a heavy mechanical blow in the dark. The impression of light is retained for about 0.1 second after the light source is shut off. This fact is made use of in motion pictures to create the impression of motion through use of a series of still images. The eye is a detector with a relatively long response time. Photoelectric cells can react more than a million times faster. Color vision is also subject to physiological peculiarities that are quite complex. See also **Vision and the Eye**.

The quantum theory of radiation applies to light, the energy quanta of which are called photons. Some of the optical phenomena so readily interpreted on the wave theory, such as reflection, refraction, interference, diffraction, and polarized light, offer difficulties when studied in terms of quanta. The laws of photoelectric phenomena, photoconductivity, and the spectrum, on the other hand, appear more readily understandable when studied in terms of quanta.

The property of light most immediately accessible to observation is its propagation along straight lines. If light rays pass from one medium to another, their direction is changed according to the law of refraction. See also **Mirrors and Lenses**; **Refraction**; and **Refractive Index**. If light in medium I propagates with velocity v_1 and makes an angle v with the normal to the boundary between media I and II, the direction v in medium II, with a velocity of propagation v_2 is given by Snell's law, $\sin v_1 / \sin v_2 = v_1/v_2 = n$. The constant n is called the relative index of refraction of medium II with respect to medium I. These laws are the basis of *geometrical optics*. This branch of the science of light describes the paths of light rays, the formation of images by mirrors and lenses, the action of telescopes, microscopes, prisms, and numerous other optical instruments.

The wave character of light becomes apparent by more refined observations. The phenomena of diffraction, interference and polarization are the subjects of *physical optics*. Diffraction describes how waves are bent around obstacles. They represent corrections to the deviations from the laws of geometrical optics. These effects become pronounced only when the material has a characteristic dimension comparable to the wavelength of the wave. When light waves reach the same point along different paths, the resulting intensity may be smaller than that produced by each individual wave separately. The relative phases of the waves may be such that they interfere destructively, when the arrival of one wave with maximum positive deflection coincides with that of another wave with maximum negative deflection. Observations of light in crystals of calcite (Iceland spar) first showed that there are two different modes of vibrations for each direction of propagation. These are called the two transverse modes of polarization.

All phenomena of geometrical and physical optics are described consistently by Maxwell's equations of electromagnetic theory. Optical phenomena are, therefore, closely related to other electric and magnetic phenomena. Very early in the present century, the prevailing opinion was that the wave character of light was unambiguously established and the nature of light well understood.

There was, however, a mathematical difficulty with the intensity of radiation of ultraviolet and higher frequencies. The photoelectric effect could also be interpreted only by considering light to have a quality of particles. The number of electrons emitted from a photosensitive surface is proportional to the intensity of the light. The energy of the individual electrons is, however, determined by the light frequency. This led to the postulate of *light quanta* with energy $h\nu$, where h is Planck's constant. This duality in nature, in which wavelike and particle-like properties are combined, is described without internal contradiction by quantum mechanics. The combined particle-and-wave character of light is revealed by the combination of properties of the light sources, the electromagnetic field describing the light waves, and the detectors.

The combination of the laws of quantum mechanics and electromagnetic theory gives a consistent description of the generation, propagation and detection of light. Since these same laws also describe many other properties of matter, such as electronic structure, chemical binding, electricity, and magnetism, it may be said that the nature of light is well understood. In this context, it is not necessary and not even desirable to pose the question, "What is it, precisely, that vibrates in a light wave in vacuum?" The electromagnetic fields acquire meaning only through their relationships with detectors and sources. Human knowledge or understanding is here used in the operational sense that a relatively simple framework of physical concepts and mathematical relationships exist, which gives an accurate description of the wide variety of optical phenomena at present accessible to observation or verification in experimental situations.

The study of the interaction of light waves with matter in the sources and detectors is the subject of *spectroscopy*. This is a wide field, which encompasses atomic and molecular spectroscopy, parts of solid-state physics, and photochemistry. The quantum theory was largely developed on the basis of spectroscopic data. A light quantum is emitted by an excited atom, molecule, or other material system when an electron in such a particle makes a transition or "quantum jump" from a state with higher energy to a state with lower energy. The energy difference between these states is equal to the quantum energy $h\nu$. Similarly, the absorption of light quanta is accompanied by an electronic transition from a state with a lower energy to a state with an energy higher by an amount $h\nu$. In this manner, the frequencies of spectral lines are characteristic for the electronic energy levels in each material. The frequency of the light may be said to correspond to the frequencies of the vibrating charges or oscillators, which are represented by electrons.

Light sources are thus bodies with a sizeable population of electrons in excited states. This may be accomplished by raising the temperature of the material. The most important source of light is the sun. The moon and other planets are visible only because they reflect sunlight, just as all objects on earth which we can see by daylight, but not at night.

Human-engineered light sources range from primitive fire, candles, and oil and kerosene lamps to electric light bulbs, fluorescent-gas discharge tubes, arcs, and many others. See also **Illumination**. In early sources, the material particles of smoke or wick were heated by the chemical reaction of oxidation or burning; in incandescent lamps, a wire is heated to a very high temperature by an electric current. There are so many energy levels, in these luminous solid materials or gases at high pressures, that the emitted light is essentially white and contains all frequencies. The higher the temperature, the more radiation is emitted and the higher the average frequency of radiation. It should be realized that most of the energy is emitted as invisible (infrared) radiation, even in the better incandescent lamps. Hot gases in flames may also emit sharp spectral lines characteristic of the atoms occurring in the flame. The yellow color, which arises when sodium chloride is sprinkled in a flame, is due to the characteristic yellow spectral line of sodium atoms.

In gas discharge tubes, atoms or molecules are excited by collisions with electrons in ionized gas. The energy is provided by the generator, which provides the voltage necessary to maintain the discharge current. An arc is a discharge in air or in a high-pressure vapor. Mercury and sodium discharges are used for street lighting. Fluorescent tubes use a gas discharge with a substantial ultraviolet component. This ultraviolet light excites electrons in fluorescent centers on the walls of the tube. The electrons drop immediately from the highly excited state to an intermediate state with a lower energy. From this state, they finally drop down to the original ground energy level, with emission of visible light. Gas discharges at relatively low pressure may serve as spectroscopic sources to study the emission spectra of atoms, ions, and molecules. From the relationship between the energy levels and the frequency of radiation, it follows that a material, when heated, can emit precisely those frequencies that it absorbs when it is in the lower energy level at low temperature.

All of these light sources are incoherent in the sense that there is no phase relationship between the light waves emitted by the different atoms in the source. This is quite different from the property of the usual sources of electromagnetic radiation at lower frequencies. In electronic oscillators used in radio and microwave transmitters, all electrons move and vibrate in step with each other. Unlike the light sources previously mentioned, lasers emit a coherent beam of light. In lasers, the original, spontaneously emitted light forces the other excited atoms to emit their radiation in step, or coherently. If stimulated emission thus dominates the spontaneous emission, a laser results. This requires a high concentration of excited atoms and a sufficient feedback mechanism of light by mirrors. In its simplest form, a laser consists of a gas discharge in a tube of suitably chosen dimensions and gas pressure between a set of parallel mirrors. Because the atoms in the laser source all act constructively in step, these sources provide a more efficient means to transmit light energy.

The high light intensities available in focused laser beams have led to the development of the branch of *nonlinear optics*. The optical properties of materials are different at high intensities, because the electronic oscillators are driven so hard that enharmonic properties become evident. A typical effect is the harmonic generation of light in which red laser light is converted into ultraviolet light at exactly twice the frequency when the high-intensity beam traverses a suitable crystal, such as quartz. It thus becomes feasible to duplicate at light frequencies all nonlinear effects known from the field of radio communications, such as modulation, demodulation, frequencing mixing, among others. It is no longer correct to say that the propagation of a light wave is independent of the presence of other light waves. At high intensities, there is a noticeable interaction between light waves of different frequencies.

Optical Phase Conjugation

Even though the laser enjoys very wide application (see also **Laser**), its full potential has not been realized because of distortion that occurs in light waves when they pass through optical systems where inhomogeneities are present. There is a real need for means to reduce or compensate for static and dynamic distortions (noise) which frequently occur. High-power lasers, tracking systems, atmospheric communication networks, and photolithographic systems, among others, are degraded by such noise. A relatively new area of optical system research, known as *optical phase conjugation*, promises to provide at least a partial solution to this problem.

That light beams possess the property of reversible propagation has been known for many years. For example, assuming a *perfect* optical system, a coherent light beam introduced at point A in a system and traveling through a complex of components will exit at point B undistorted. If a light beam of exactly the same characteristics were introduced at point B, it would travel the same exact path and exit, undistorted, at point A. In other words, under perfect conditions, the propagation is reversible. What was not known concerning reverse propagation until the early 1970s was that a distorted (noisy) light beam, if propagated in reverse through a given optical system, will during the course of its backward track remove all the distortions introduced into it during its prior forward track, and thus exit at the point of origin as a "clean" beam, free of distortion. Thus, the concept of reversible propagation holds not just for a theoretically perfect system, but for a practical imperfect system as well.[1] The backward-traveling

wave, in essence, is a "time-reversed" replica of the original incident wave.[2]

Producing a Phase-Conjugated Wave. Two general approaches have been investigated thus far for implementing the concept of optical phase conjugation in a practical way for the purposes previously mentioned. These two approaches employ different physical principles, but both rely on the laser light itself to interact with the nonlinear optical properties of a specific medium to initiate phase conjugation or a turnabout of the distorted light waves: (1) *stimulated Brillouin scattering* (SBS) and (2) *degenerate four-wave mixing* (DFWM). When any material — gas, liquid, or solid — is penetrated by light of intensity great enough to compete with the atomic forces that bind the material together, the material is modified, as is also the light penetrating it. This nonlinear interaction generates the SBS or DFWM time-reversed waves. The success in the reversal of wave motion is due to an extreme simplification of the problem: the quantum-mechanical and thermal motions of atoms and electrons that radiate light do not need to be reversed. It is sufficient for practical purposes, as observed by Shkunov and Zel'dovich, to reverse the temporal behavior of macroscopic parameters describing the averaged motion of a large number of particles.

In SBS, the modified material generates sound waves that serve as an appropriate reflective surface to produce the time-reversed waves.[3] In DFWM, the interaction uses a holographic process in a nonlinear material to generate the conjugate, or reversed light waves.

Examples of nonlinear mediums include semiconductors, crystals, liquids, plasmas, liquid crystals, aerosols (as in the atmosphere), and atomic vapors. The term *nonlinear*, as used here, pertains to media that are altered or affected by light. Linear materials, in contrast, are not so affected.

Prospective Uses of Optical Phase Conjugation. Currently, a number of scientific laboratories, in addition to those in Russia, have been conducting intensive research in this area in attempts to refine the technology and to find practical applications. Among the active laboratories are the California Institute of Technology, the AT&T Bell Laboratories, Philips Research Laboratories (the Netherlands), the University of Southern California, the University of Waikato (New Zealand), the University of Arizona, the National Institute of Standards and Technology, IBM, and Hughes Aircraft Research Laboratories. A number of scientists envision several important applications within the next decade. Admittedly, the field remains in an investigative state. Some potential applications as reviewed by Pepper (see reference) and others are reviewed briefly here.

Brillouin scattering) and, acting as a mirror, the gas reflected the beam backward. The investigators were surprised to find that once the reflected wave passed back through the same piece of frosted glass, a nearly perfect, undistorted optical beam emerged. Thus, they found that the distortions introduced by the glass during the forward passage were, in essence, canceled out during the backward passage. (The phase difference between any two points of the reversed beam has a sign opposite to that of the phase difference between the same points of the original beam. As described by Shkunov and Zel'dovich, the mathematical operation of changing a phase sign is known as conjugation, and thus the coining of the term optical phase conjugation.)

[2] During the early phases of explaining this unexpected phenomenon, a number of homely analogies were developed, the most common being that of comparing the retracing of light waves back through the distorting media as "making a film run backward." As described by Shkunov and Zel'dovich, "The relation between the wave fronts of two mutually reversed waves is analogous to the relation between the positions of two opposing armies on a military map. The front line of each army coincides with that of the other, and the directions of desirable movement are opposite. One can say that the front lines are mutually reversed: a convex part of one army's front corresponds to a concave part of the other."

[3] Traditionally, the frequency of scattered light has been regarded as identical to that of the incident light. In actual practice, as first predicted by Brillouin in 1914, a slight line broadening occurs due to motion of the scatterers (Doppler effect) and also due to variations in the directions or magnitudes of their polarizability tensors (due to chemical reactions). The Brillouin effect, simply stated, is as follows: upon the scattering of monochromatic radiation, a doublet is produced, in which the frequency of each of the two lines differs from the frequency of the original line by the same amount, one line having a higher frequency, and the other having a lower frequency.

---

[1] In an experiment at the P.N. Labedev Physical Institute (Moscow) in 1972. Boeia Ya Zel'dovich and coworkers observed a curious phenomenon while doing an experiment. The researchers intentionally distorted an intense beam of red light from a pulsed ruby laser by directing it through a frosted glass plate. The degraded beam was directed down a long tube filled with methane gas under high pressure. Interactions occurred between the beam and the molecules of the gas (stimulated

Some scientists have suggested that phase-conjugated systems may be used for image transfer in photolithography. Researchers at IBM have demonstrated image transfer. Light from a laser passes through the mask pattern, a semitransparent mirror, and then an amplifier. The intensity of the beam increases, but at the cost of introducing distortions into the beam. When the image is returned through an amplifier by a phase-conjugating mirror, the "time-reversed" beam is both powerful and free of distortions. Conventional methods, such as compensating for optical aberrations, are no longer required. Another way in which lensless-imaging schemes may be used include fiber-optic communications and associated memory as used in pattern-recognition devices. As stressed by Pepper (see reference), the emergence of optical phase conjugation has unified many areas of applied and fundamental optical physics. Spectroscopy, the study of the interaction of matter and radiation, has particularly benefited. The concepts, techniques, and basic applications of optical phase conjugation can, in principle, be applied to most other areas of the electromagnetic spectrum. Microwave phase conjugation would have major applications in radar, millimeter-wave imaging systems, and high-frequency temporal signal processing, as well as microwave spectroscopy. Researchers are planning experiments in the acoustic-wave area. Acoustic signal-processing devices and sonar may benefit from such research.

Squeezed Light. Although research has been proceeding for over a decade, the study of "squeezed" light is still essentially in an experimental stage. For a number of years, scientists have recognized that a beam of light is not free from random fluctuations, but in electronic terms it is noisy. Light is used frequently to make measurements and to observe numerous instrumental phenomena. Thus, faulty light contributes to errors, if ever so small. This situation is explained by the quantum theory, which implies that light must be accompanied by a certain minimal amount of light fluctuation.

A beam of light has been defined as an oscillating electromagnetic field, which can be viewed as a smooth wave. The shape of the wave can be foretold with absolute certainty, but its slope must fit within a particular "envelope" of uncertainty. Even in darkness, quantum physicists would allow that the wave exists but is flat, with some degree of uncertainty.

Researchers Slusher and Yurke (AT&T Bell Laboratories) have observed, "In practical terms, this means that even in a vacuum, with no external light sources, there must still be small fluctuations in the electromagnetic field. Noise limits the precision of spectroscopy, in which the frequency and intensity of the radiation emitted by atoms or molecules yield information about their properties." It also is envisioned that quantum noise will also limit those technologies concerned with optical computing and communications.

In early investigations, researchers at several laboratories had experimented with so-called *squeezed light*. This field of study involves quantum fluctuations, and largely stemmed from studies of the coherent light generated by a laser. A number of theoretical physicists more than 30 years ago studied the statistical properties of coherent light. Even the presumed perfectly coherent light of an ideal laser was found to have a Poisson distribution of photons rather than a single, well-defined number. Thus, although not so noisy as a conventional incoherent light beam, the laser is also noisy and, as lasers are employed more frequently in very sophisticated research and practical applications, optical devices could ultimately be limited by such noise.

As related by Robinson (see reference), for squeezed states, the statistical variance of the photon number is not so important as the variances of the amplitude of the electric field, which is related to the photon number, and of its phase. Researchers at AT&T Bell Laboratories have successfully "squeezed" noise or unwanted fluctuations in *phase* at the cost of increasing it in amplitude and, conversely, they have also squeezed noise in *amplitude* at the cost of increasing it in phase. In essence, these researchers generated the squeezed light by using a technique known as *four-wave mixing*, previously mentioned under optical phase conjugation. The researchers directed two laser beams at each other from opposite directions. The laser beams met in a material (nonlinear medium). In their experiment, the nonlinear medium was a beam of sodium atoms oriented at right angles to the laser beams. The research team found that the laser beams interacted with the Na atoms in such a way that two output beams emerged in opposite directions along an axis tilted at a slight angle with respect to the axis of the two input laser beams. Attributed to the properties of the nonlinear medium,

when the output beams were combined by mirrors, the result was a single beam of squeezed light. By varying the position of the mirrors that direct the beams emerging from the nonlinear medium, it was found possible to reduce the noise in either the phase or the amplitude of the squeezed light. The degree of success of the experiment was relatively small—about a 7% drop in noise. More recently, other researchers (University of Texas) have reported a 42% reduction in noise.

Practical applications of squeezed light must await further research findings.

Subwavelength Illumination. Superresolution light microscopy permits researchers to optically study specimens without being limited by the diffraction properties of visible light. This technique, however, requires the efficient emission of light from *subwavelength* light sources. In an experimental setup, an electromagnetic wave is generated to emerge from an aperture. First, the wave is highly collimated to the aperture dimension. As pointed out by K. Lieberman and coworkers (Hebrew University, Jerusalem), "It is only after the wave has propagated a finite distance from the aperture that the diffraction that limits classical optical imaging takes effect. Thus, in the near-field region, a beam of light is present that is largely independent of the wavelength and is determined solely by the size and shape of the aperture." In a 1990 paper (see reference listed), an approach for producing sources of light with subwavelength dimensions is described.

The methodology is based on the packaging of photons as molecular excitons, effectively reducing the volume of the light beam by 109 and making possible propagation through dimensions of 1 nanometer. The researchers further observe, "Molecular microcrystals are grown in the tips of micropipettes that have inner diameters of 100 nanometers or less. Measurements are presented that demonstrate this improvement in transmission for pipettes of various diameters. The ultrasmall dimensions of these light sources, the wavelength range (UV to IR) of their emission, their ease of production, and their expected unique abilities for high efficiency excitation-imaging of surfaces portend significant applications for this methodology."

Light Reflections in Computer Graphics. The technology of computer graphics has enabled the production of objects in realistic three dimensions from two dimensions, a technique that frequently has replaced the need for clay models in a number of fields, notably in the automotive and appliance fields and for creating realistic backgrounds in simulators as, for example, in aircraft flight training. To achieve realism of painted, metal, glass, and other reflective surfaces, "reflection" algorithms can be created.

Reflection patterns vary markedly from one type of product surface to the next. These patterns are unique for given materials or material families, including, for example, steel, chromium, rubber, and plastics, because of the wide range of diffuse and specular reflections for given materials. Through careful experimentation, appropriate algorithms can be developed. Considerable detail pertaining to these methodologies is given in the D.P. Greenberg reference listed. See also **Hurlbert Poggio reference listed**.

Extending the Light-Sensing Range. Since so-called "night vision" sensors were developed several decades ago, the practical use of the infrared (IR) portion of the light spectrum has continued in its importance, particularly in military operations such as tank and helicopter maneuvering. Traditionally, these sensors have depended upon mercury cadmium telluride (*mer-cal*) semiconductors. These sensors pick up wavelengths between 3 and 5 and 8 and 12 microns. A greater sensitivity over the whole IR range has been sought. The materials for these sensors also have been difficult to process. After some years of research, a group at AT&T Bell Laboratories has developed ways to produce gallium arsenide optical crystals that are claimed to approach atomic perfection. The new detectors, which have greatly increased sensitivity, feature what is known as a quantum well, which precisely controls the flow of electrons. Experimental arrays with more than 16,000 pixels have been built and tested. Such detectors can sense temperature differences as small as a ten-thousandth of a degree. Cool objects appear dark, with warmer objects ranging over a gray scale. The quantum-well infrared photodetectors (QUIPs) now are being studied in terms of mass producing them.

Sky Colors. As early as 1899, Rayleigh claimed that the sky is blue on a clear day because of scattering and color separation by the molecules contained in the air of the atmosphere. Rayleigh viewed the phenomenon as a composite of all the colors in the visible spectrum. See also **Scattering**. He proposed that short-wavelength light (blue) is scattered more than red light, which has a longer wavelength. Based upon the ratio of the blue and red wavelengths ($\sim$1.68), Rayleigh reasoned that the blue-scattered light is approximately eight times that of the red-scattered light.

Sunsets are reddened because the light reaching the observer has passed through a much longer path, allowing domination of the red end of the spectrum in contrast with the situation when the sun is viewed high in the sky.

Other investigators have attributed sky color to layers of ozone in the upper atmosphere. It has been postulated that, in these layers, molecules have absorption bands that favor the absorption of light at the red end of the spectrum. A number of postulations have emerged pertaining to the development of several colors, including the purple sky coloration. These are explored in more detail in the Meinel reference listed.

Some correlation of sunset color with volcanic eruptions has been made. As examples, the explosion of Krakatau (near Java) in 1883 produced brilliant sunsets over a span of about 5 years; the explosion of Agung on Bali in 1963 affected sunset coloration for about 3 years.

Optical Computer

The pace of research to develop an optical computer based upon beams of light rather than electric currents have increased markedly during the past decade. An optical analogue of the transistor was demonstrated in the early 1980s. An optical computer that might operate a thousand times faster than contemporary electronic computers is indeed a powerful incentive. Theoretically, the operations carried out by a computer (logical and arithmetic) could be effected by numerous means and, in fact, this already has been demonstrated when one considers that early computers depended exclusively on vacuum tube switches. Although the transistor that replaced the tube was a great step upward in electronics, the transistor was a significant hardware departure — thus indicating that computer designers will quickly adapt to new hardware when there are significant demonstrable advantages. Thus, among some scientists, there is an attitude that tends to regard the optical computer as an inevitable development for the future. Some forecasts indicate that an optical computer may be capable of a trillion operations per second. Even though the current electronic computers are undergoing significant changes in the way they are organized to process information, such reorganization today probably would be easier to achieve if the system were basically optical rather than electronic — as considered by some contemporary scientists. See also **Digital Computer Systems**.

Additional Reading

Andreev, A.V. et al.: *Quantum Optics*, SPIE International Society for Optical Engineering, Bellingham, WA, 1999.

Born, M. and E. Wolf: *Principles of Optics: Electromagnetic Theory of Propagation, Interference and Diffraction of Light*, 7th Edition, Cambridge University Press, New York, NY, 1999.

Buchwald, J.D.: "The Rise of the Wave Theory of Light," University of Chicago Press, Chicago, IL, 1990.

Corcoran, E.: "Body Heat — QWIPs Offer a New Way to See in the Dark," *Sci. Amer.*, 123 (October, 1991).

Feynman, R.P.: *QED: The Strange Theory of Light and Matter*, Princeton University Press, Princeton, NJ, 1988.

Flam, F.: "Through a Glass — Darkly (Light Scattering)," *Science*, 29 (April 5, 1991).

Flam, F.: "Scopes with a Light," *Science*, 30 (April 5, 1991).

Gower, M. and D. Proch: *Optical Phase Conjugation*, Springer-Verlag, Inc., New York, NY, 1994.

Greenberg, D.P.: "Light Reflection Models for Computer Graphics," *Science*, 166 (April 14, 1989).

Hakfoort, C.: *Optics in the Age of Euler: Conceptions of the Nature of Light, 1700–1795*, Cambridge University Press, New York, NY, 1994.

Harris, N.: "Lighting and Its Uses," *Science*, 543 (August 4, 1989).

Hirota, O.: *Squeezed Light*, Elsevier Science, New York, NY, 1992.

Hurlbert, A.C. and T.A. Poggio: "Synthesizing a Color Algorithm from Examples," *Science*, 482 (January 29, 1988).

Leonhardt, U.: *Measuring the Quantum State of Light*, Cambridge University Press, New York, NY, 1997.

Lieberman, K. et al.: "A Light Source Smaller than the Optical Wavelength," *Science*, 59 (January 5, 1990).

Loudon, R.: *The Quantum Theory of Light*, 3rd Edition, Oxford University Press, Inc., New York, NY, 2000.

Marshall, E.: "Science Beyond the Pale," *Science*, 14 (July 6, 1990).

Maxwell, J.C. and P.M. Harman: *Scientific Letters and Papers, 1846–1862*, Vol. 1, Cambridge University Press, New York, NY, 1990.

Maxwell, J.C. and P.M. Harman: *The Scientific Letters and Papers of James Clerk Maxwell: 1862–1873*, Vol. 2, Cambridge University Press, New York, NY, 1995.

Meinel, A. and M. Meinel: "Sunsets, Twilights, and Evening Skies," Cambridge University Press, Cambridge, Massachusetts, 1983.

Pepper, D.M.: "Application of Optical Phase Conjugation," *Sci. Amer.*, 74–83 (January, 1986).

Peterson, I.: "Putting a Far Finer Point on Visible Light," *Science News*, 7 (January 6, 1990).

Robinson, A.L.: "Bell Labs Generates Squeezed Light," *Science*, **230**, 927–929 (1985).

Ruthen, R.: "Surfing Photons," *Sci. Amer.*, 12D (August, 1989).

Saki, Jun-Ichi: *Phase Conjugate Optics*, The McGraw-Hill Companies, Inc., New York, NY, 1992.

Schivelbusch, W.: "Disenchanted Night: The Industrialization of Light in the Nineteenth Century," University of California Press, Berkeley, CA, 1988.

Shkunov, V.V. and B.Y. Zel'dovich: "Optical Phase Conjugation," *Sci. Amer.*, 54–59 (December, 1985).

Siegel, D.M.: *Innovation in Maxwell's Electromagnetic Theory: Molecular Vortices, Displacement Current, and Light*, Cambridge University Press, New York, NY, 1991.

Slusher, R.E. and B. Yurke: "Squeezed Light," *Sci. Amer.*, 50 (May, 1988).

Staff: "Optics Source Book," McGraw-Hill, New York, NY, 1988.

Walker, J.: "The Colors Seen in the Sky Offer Lessons in Optical Scattering," *Sci. Amer.*, 102 (January, 1989).

Weber, M.J., Ed.: "CRC Handbook of Laser Science and Technology," CRC Press, Boca Raton, FL, 1987.

Weiss, P.L.: "Reflections on Refraction," *Science News*, 236 (October 13, 1990).

LIGHT-ACTUATED CELL. See **Photoelectric Effect**.

LIGHT-COUPLED SWITCH. A switch in which the switching signal is transmitted to the device in the form of light energy. The switching element may be a phototransistor, a photodiode, or field-effect transistor (FET). The receipt of light energy by such a device changes the transmission characteristics of the device, permitting conduction between two terminals. When no light is present, the resistance of the device is high. When excited by photon energy, the resistance drops to a much lower value. Various light sources are used in connection with light-coupled switches. Although common incandescent sources may be used, gas-discharge sources, such as neon lamps and solid-state devices, such as gallium arsenide light-emitting diodes, are more commonly used in instrumentation systems. As compared with incandescent sources, the other sources produce less heat and possess higher speed and reliability.

A major advantage of the light-coupled switch is isolation that can be obtained between signal and drive source. Signals can be controlled without introducing errors due to the drive source. Further, isolation helps in maintaining a high common-mode rejection ratio in analog signal-handling equipment.

See also **Analog Switch**.

LIGHT CURVE (Astronomical). In the study of variable stars, and in kindred problems in astronomical research, it is desirable to represent graphically the variation of radiation intensity with time. A diagram in which light intensity, on any convenient scale, is plotted as ordinate against time as abscissa is known as a light curve. As the number of observations increases, it frequently becomes possible to detect a periodic variation in the light intensity. After a provisional period has been determined, some convenient epoch is selected, and all the observations are reduced to the cycle of variation embracing the selected epoch by the use of the provisional period. In order that the resulting points may fall on a regular curve, it is frequently necessary to apply a number of corrections to the provisional period. The curve drawn through the plotted points, all reduced to the selected epoch by means of the repeatedly corrected period, is known as the mean light curve. Examples of light curves will be found in the articles on **Cepheids; Eclipsing Binary**; and **Variable Star**.

LIGHT (Effects on Plants). See **Photoperiodism**.

LIGHT-EMITTING DIODES (LED). A light-emitting diode (LED) is an electronic device that emits ultraviolet, visible, or infrared electromagnetic radiation when an electric current is passed through it. Light-emitting diodes are compact, mechanically rugged, energy efficient, are of low voltage, and have long operating lifetimes. They are found in a wide variety of applications, and tens of billions of LEDs are manufactured and sold worldwide each year.

The basic device consists of a single $p-n$ junction. Most commonly, the junction is formed in a compound semiconductor composed of elements from columns III and V of the periodic table of the elements, such as GaAs, GaP, AlGaAs, GaAsP, GaAsInP, AlInGaP, or AlInGaN. Some LEDs are formed of "semiconductors" that are amorphous or polycrystalline organic molecules or polymers. The specific material used for the LED is chosen on the basis of the color of light emitted by the material, the efficiency at which the material converts electric current into photons, and the cost of manufacture.

A cross section of a common LED lamp is shown in Fig. 1. The semiconductor chip is usually $250 \times 250 \times 250$ μm^3 and is mounted on one of the electrical leads. A bond wire electrically connects the top of the chip to the other lead. The epoxy dome serves as a lens to focus the light and as a structural member to hold the device together.

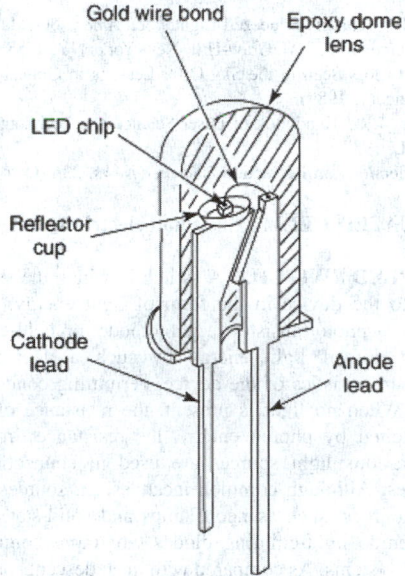

Fig. 1. Cross-section view of a typical LED lamp. The LED chip, typically $250 \times 250 \times 250$ μm^3, is mounted in a reflective cup formed in a lead frame. Clear epoxy acts as a lens, as an index-of-refraction matching medium to help extract light from the chip, and a structural member to hold the package together.

To operate an LED, a source of electric power is connected between the two leads. The LED is reverse-biased if the cathode lead is raised in electrostatic potential (voltage) relative to the anode lead. In this case, essentially no electric current passes through the LED and no light is emitted. If the LED is forward-biased (the anode is at higher voltage than the cathode), an electric current will flow through the LED and photons will be emitted. The amount of emitted light is proportional to the current flow. The energies of the emitted photons are nearly identical and roughly equal to the energy gap of the semiconductor material in the active region of the LED.

Typical operating currents for the LED shown in Fig. 1 are 1 to 50 mA, and typical forward voltages are 2 to 4 V. The fraction of electrical input power that is converted to optical output power ranges from less than 0.1 to 30%. LEDs can be operated continuously for more than 10,000 h in a variety of ambient temperature and humidity conditions, and they are immune to mechanical shock or vibration.

Because of the low operating voltage and current of LEDs, they can be easily integrated with Si transistor electronics. Consequently, visible LEDs are widely used for indicators and displays on radios, cellular telephones, computers, and other electronic devices. LEDs that emit infrared light have high efficiencies and output powers, and the magnitude of the light output can be modulated rapidly; therefore, these LEDs are used to transmit data between electronic devices through the air or through fiber-optic cables. The efficiencies of the highest-performance visible LEDs exceed the efficiencies of incandescent bulbs by several times. Thus, in colored light applications such as traffic signals, variable-message displays, full-color large-area displays, and automotive exterior lighting, LEDs have significantly displaced bulb-type light sources.

The first LEDs were simple devices. Over time, more refined designs have been used in order to improve efficiency, to obtain different colors (and addressing problems and opportunities that have often attended the use of new materials), or to otherwise optimize operating characteristics. In most cases, more sophisticated LEDs are more costly. For this reason, and because higher-performance LEDs tend to open new applications and markets as opposed to replacing existing LED technologies, several general types of LEDs are commercially available. From the user's perspective, the distinguishing characteristic of LEDs is the color (wavelength) of light that is emitted. From a design standpoint, the greatest distinction is the semiconductor material from which the device is fabricated. Several fundamental principles of LED behavior and design exist. See also **Automotive Electronics**; **Luminescence**; and **Telephony (Telecommunications)**.

Basic Characteristics and Designs

The most basic element of an LED is a $p-n$ junction, which determines the current–voltage characteristics of an LED. The color (wavelength) of light emitted by the LED depends upon the semiconductor material in which the $p-n$ junction is formed. The efficiency at which the LED converts electrical input into optical output can be measured by several different figures of merit. Regardless of the measurement, the LED's efficiency depends upon the semiconductor material (internal efficiency) and the configuration or shape of the LED (extraction efficiency). In order to ensure reasonable internal efficiency, only a few semiconductor materials are considered suitable for LEDs.

p-n Junction and Current-Voltage Characteristics. An LED is among the simplest of semiconductor devices since it consists of a single $p-n$ junction, where p-type (principally containing positive charge carriers called *holes*) and n-type (principally containing negative charge carriers called electrons) semiconductor materials are in single-crystal contact with each other.

The p-type region of the LED is formed by including atoms in the semiconductor crystal, which are deficient of one electron relative to the other atoms in the crystal (for example, Zn, which has 2 valence electrons, replacing Ga in a GaAs crystal). The n-type region is formed by including atoms with an excess of valence electrons (for example, Si replacing Ga in a GaAs crystal). The inclusion of impurity atoms in the semiconductor crystal in order to affect the type of charge carrier is known as doping.

When the junction is biased in the forward direction, electrons are injected into the p-type region, and holes are injected into the n-type region. These injected minority carriers return to thermal equilibrium by recombining with the opposite type of carrier, in the process giving up energy, hopefully as a photon. LED junctions begin to conduct significantly at voltages approximately equal to the semiconductor energy gap of the material involved. This can range from 0.8 to 1.5 V for infrared emitters and from 1.5 to more than 3.0 V for visible emitters. When biased in the reverse direction, LED junctions conduct very little current until relatively large voltages (10 to 50 V) are applied. The current–voltage characteristic of an LED is "rectifying" in that conduction is relatively easy in the forward-biased direction and difficult in the reverse-biased direction.

The slope of the current–voltage characteristic after turn-on in the forward direction is known as the dynamic resistance. The area of the $p-n$ junction, the metals used for the ohmic electric contact to the semiconductor, and the conductivity of the semiconductor materials involved determine the dynamic resistance. The dynamic resistance can range from about 1 Ω for an infrared device designed for high-current operation to 10 Ω or more for a visible LED that is optimized for high light-output efficiency and designed to operate at 50 mA or less.

Emission Wavelength. The wavelength of the light emitted by an LED is inversely proportional to the energy of the emitted photon, E, according to the relationship $\lambda = hc/E$, where λ is the wavelength of light, h is Planck's constant, and c is the velocity of light. The energy of the photon is equal to the energy gap, E_g, of the semiconductor material in which the

$p-n$ junction is formed if free electrons and holes are recombining, or to the energy gap minus the energy of the trap, impurity level, or phonon if these processes are involved. This can be represented as $\lambda = hc/(E - E_x)$ where E_x is the energy of the trap, and so on. Typical n-type dopants have level depths of a few thousandths of an electron volt and p-type dopants have depths of a few tens of thousandths of an electron volt, both of which are small compared to the energy gaps of LEDs. Therefore, for most types of LEDs, the peak emission wavelength $\lambda \approx hc/E_g$.

In order to achieve different colors of emission, semiconductor materials with different energy gaps must be utilized. One of the best ways to achieve different energy gaps is to utilize an alloy semiconductor. Ternary (three-element) and quaternary (four-element) alloys are used for many types of LEDs. Examples are $GaAs_{1-x}P_x$, $Al_xGa_{1-x}As$, $(Al_xGa_{1-x})_{0.5}In_{0.5}P$, and $Al_x In_y Ga_{1-x-y}N$ (where $0 \leq x \leq 1$ and $0 \leq x + y \leq 1$). See also **Semiconductors: Compound Semiconductors**.

Efficiency. The process by which a photon is finally emitted from an LED can be divided into two steps: internal conversion of an electron into a photon, and extraction of the photon from the interior of the chip and into free space. The efficiency at which electrons are converted into photons is strongly dependent upon the properties of the semiconductor material from which LED is made. This "internal quantum efficiency" can vary from less than 0.1 (1 photon for 1000 electrons) to 100%. Because of the difficulty in extracting photons from the chip, however, the external quantum efficiency of LEDs is usually much lower than the internal quantum efficiency. This extraction efficiency is determined by the shape and geometry of the chip, as well as absorption of the photons by parts of the chip, including the contacts.

Carrier Confinement. The radiative recombination rate increases with injected carrier concentration. One way to increase the carrier concentration is to inject a larger number of carriers (increase the drive current). Another way is to reduce the volume into which the carriers are injected.

Homojunctions. The simplest and generally least expensive LEDs are homojunctions in which the entire structure, and sometimes the substrate, consists of the same compound or alloy composition. The $p-n$ junction can be formed by p-type diffusion into an n-type epitaxial film, or can be grown by changing the dopant type during the epitaxial growth process. The optimum junction depth is a compromise between two conflicting requirements.

The junction must be deep enough so that only a small fraction of the injected carriers reaches the surface (where they generally recombine nonradiatively), and so that the sheet conductivity of the p-type region is high enough to avoid current crowding around the contact. On the other hand, the junction should be as shallow as possible to minimize absorption. This latter requirement is particularly important for direct band-gap devices because the absorption coefficient for near band-gap radiation is $\sim 10^4$ cm^{-1}. Consequently, junctions in direct band-gap homojunctions, which utilize near-band-edge recombination processes, are generally less than 3-μm deep.

In the case of indirect semiconductors, the absorption coefficient near the band edge is only $\sim 10^3$ cm^{-1}, and the junction can be deeper. In addition, most indirect band-gap LEDs utilize isoelectronic impurities where the transition energy is below the band edge and the absorption is substantially reduced. Consequently, the junction depth is often more than 10 μm in indirect band-gap LEDs, particularly for devices with grown-in junctions. The doping levels on the p and n sides of the junction are optimized both to maximize the injection of minority carriers into the preferred active region of the devices and to have the radiative lifetime in the active region be as short as possible to minimize nonradiative recombination.

Single Heterostructures. A single-heterostructure device has two different materials or alloy compositions on opposite sides of the junction. Devices of this type are generally not much more difficult to produce than homojunction LEDs, and they have several advantages. One limitation of this structure is that some of the injected electrons can penetrate quite deeply into the p-type region before they recombine, and much of the radiation that is then generated will be absorbed before it can escape. Also, any radiation that is emitted downward will be absorbed. These limitations can be minimized with the growth of double heterostructures.

Double Heterostructures. A double heterostructure is an active region with a lower energy gap that is sandwiched between two layers with higher energy gaps: upper and lower "confining" layers. This structure has the advantages of a single heterostructure and, in addition, reduces the absorption of the generated radiation. The active region can be quite thin because the wider band gap of the lower confining layer keeps the injected electrons from penetrating beyond the heterostructure interface. A typical active region thickness is 1 to 3 μm for AlGaAs LEDs, and may be as thin as 500 Å for high-performance AlGaInP LEDs. The confinement of the carriers to a narrow region results in fast and efficient recombination. The issue of fast recombination is important especially for applications related to electronic switching or communication. Infrared LEDs for these applications also often have small junction areas to minimize capacitance.

The thin active region minimizes absorption, and if the confining layer is relatively thick, some of the light that is generated escapes from the edges of the confining layer, thereby increasing the external efficiency of the device.

Quantum Wells. A quantum well is a double heterostructure wherein the thickness of the lower-energy-gap active region is comparable to the "size" of an electron or hole. Depending upon the material, this thickness is 25 to 100 Å. LEDs with complex active regions containing one or more quantum wells are common in the high-brightness LEDs fabricated from AlGaInN materials. Quantum wells confine electrons and holes into a small region and thereby increase their probability of recombining with each other and emitting a photon. On the other hand, quantum wells are so small that they are unable to contain many electrons and holes, so usually more than one quantum well in the active region is required. One disadvantage of the quantum well is that the wavelength of the emitted light depends upon the well's thickness and band gap. Thus an additional variable that affects the device performance is introduced, and this variable must be controlled in a high-volume manufacturing environment. Quantum-well heterostructures are usually grown by the MOVPE technique because it is relatively simple to grow atomically abrupt interfaces and very thin layers.

Extraction Ratio. The external quantum efficiency of LEDs (the number of photons collected at a detector outside of the semiconductor chip, divided by the number of electrons passed through the device) is much lower than the internal quantum efficiency because of the difficulty of extracting light from the semiconductor chip. The ratio of external quantum efficiency to internal quantum efficiency can be defined as an extraction ratio C_{ex}.

Figures of Merit for Efficiency. The internal quantum efficiency and extraction ratio cannot be measured independently of each other. Instead, they must be inferred from measurements of external efficiency, and some type of calculation or estimate. The external efficiency of an LED, which is ultimately of most importance, can be characterized in several ways.

The external quantum efficiency is the internal quantum efficiency multiplied by the extraction ratio, $\eta_{ext} = C_{ex}\eta_{int}$. The power efficiency, or wall-plug efficiency, (P_E) is the optical power output divided by the electrical power input:

$$P_E = \frac{(\text{photons/second}) \times (\text{emission energy/photon})}{(\text{electrons/second}) \times \text{applied voltage}}$$

$$= \eta_{ext} \frac{E_R}{v}$$

where E_R is the energy of the emitted radiation.

The applied voltage is the voltage applied directly to the junction (approximately the energy gap) plus the voltage across the series (dynamic) resistance of the LED diode: $V \gg E_g + IR$. E_R is generally approximately equal to the semiconductor energy gap for near-band-edge recombination processes. The turn-on voltage for the LED is also approximately equal to the energy gap. Therefore, for low currents, where the voltage drop due to series resistance is relatively small, we have $P_E \gg \eta_{ext}$.

This approximation is accurate to within 10% for most LEDs since the series resistance is typically less than 10 Ω, resulting in less than a 0.2-V increase in voltage at a typical operating current of 20 mA, while the energy-gap voltage is typically 2 V.

Luminous Efficiency. For visible LEDs, the human eye is the detector, and the luminous performance, expressed in either lumens/electrical watt or lumens/ampere, is the key performance parameter. In order to facilitate comparison to bulb-type lighting technologies, it is preferred to express performance in lm W^{-1}.

The luminous efficiency, or the sensitivity of the eye to radiometric energy of different wavelengths, has been established by international agreement by the Commission Internationale de l'Éclairage (the so-called "CIE curve". See also **Colorimetry**. The relative eye sensitivity peaks at 1.0 at a wavelength of 555 nm. At this wavelength, 1 W of radiometric power yields 680 lm. The relative eye sensitivity $V(\lambda)$ drops sharply on either side of 555 nm and is about 0.5 at 510 and 610 nm.

The luminous performance of a visible LED in units of lumens per watt is determined by multiplying the radiometric performance by the relative eye sensitivity curve: luminous performance $(\text{lm } W^{-1}) = 680 \cdot V(\lambda) \cdot P_E$. In some cases, it is desirable to have the luminous performance expressed in lumens/ampere, which can be obtained as follows: luminous performance $(\text{lm } A^{-1}) = 680 \cdot V(\lambda) \cdot \eta_{\text{ext}} \cdot E_R$.

Effect of Temperature. The temperature of the $p-n$ junction affects the color and efficiency of most LEDs. The energy gap of most semiconductors shrinks with increasing temperature. For example, the energy gap of a typical AlGaInP LED emitting at 615 nm decreases by 0.6 meV per degree Centigrade of junction temperature. For some applications, such as full-color displays, and general white lighting with phosphor-converted LEDs, the shift in color over a reasonable range of ambient operating temperatures (-20 to $100\,^{\circ}\text{C}$ (-4 to $212\,^{\circ}\text{F}$) can be aesthetically problematic. In other applications, such as traffic signals and automotive exterior lighting, the LED's emission could end up outside the range of colors considered permissible by regulatory agencies.

For most LEDs, the internal quantum efficiency drops at the rate of between 0.5 and 2% per degree centigrade measured at the $p-n$ junction. There are several factors contribute to this decrease.

The extraction ratio generally does not vary with temperature, therefore the external quantum efficiency and power efficiency temperature dependencies are similar to the internal quantum efficiency dependency. For visible LEDs, the luminous efficiency can decrease at a faster rate than the quantum efficiency because of the tendency of the emission wavelength to become longer. For LEDs in the yellow-to-red portion of the visible spectrum, the eye sensitivity decreases with increasing temperature, so as AlGaInP LEDs are heated, for example, the wavelength gets longer, the internal efficiency decreases, and the luminous efficiency decreases even more.

Effect of Current. LED performance will change with current (or current density), depending upon a variety of factors such as the concentration of nonradiative centers in the vicinity of the junction and heating due to both nonradiative recombination and Joule heating associated with the dynamic resistance. Typically, LED efficiency will go through three stages—superlinear, linear, and sublinear—as a function of increasing current.

At low currents, the light output is superlinear, since nonradiative centers provide a shunt path that saturates as the current density increases. GaAsP devices with high-defect densities typically show more superlinearity than AlGaInP devices, which are lattice-matched and have low-defect densities.

At moderate current densities, the light output remains linear with current (and the efficiency is constant). It is generally desirable to operate the LED in this current region, since it represents the point of highest efficiency.

At higher current densities, heating causes the efficiency to drop and the light output becomes sublinear with current and will eventually decrease with increasing current as the heating becomes severe. The effect of heating can be ameliorated by reducing the thermal resistance of the LED's package.

Reliability. Degradation in the range of 5 to 20% is usually observed after 1000 h of operation at normal operating currents at $55\,^{\circ}\text{C}$ ($131\,^{\circ}\text{F}$). The light output generally decreases roughly linearly on a semilog plot as shown, and so the device should not reach half-brightness for more than 10^6 h. In many applications, a change in light output by less than a factor of 2 will not be noticed, and so LED reliability is generally more than adequate, particularly for indoor applications. High-performance LEDs used for outdoor applications, requiring low- and high-temperature operation, are more susceptible to package-related degradation, and must be carefully engineered to avoid nonuniform and occasionally severe degradation.

Radiation- and Current-enhanced Degradation. LEDs are inherently highly reliable devices since the recombination is a normal solid-state electrical process that does not necessarily give rise to any damage to the semiconductor. However, there is a substantial amount of energy released locally (1–2 eV) with each electron—hole recombination event, and this can cause the formation or migration of point defects or dislocations, resulting in "radiation-enhanced" light-output degradation. The degradation is generally accelerated by high-current densities and high temperatures. Under extreme drive conditions (at current densities greater than ~ 300 A cm^{-2}), networks of dislocations can form and grow rapidly around crystalline defects that may be present. Carriers tend to recombine nonradiatively when they are near these dislocation networks, and "dark lines" appear in the light-emitting area. Such dark-line defects (DLDs) often occur in small-emission-area AlGaAs LEDs that are used as light sources for fiber-optic communications. Because the emission areas are small, extremely large current densities are common. Such devices must typically pass a "burn in" test that verifies that they are free of the crystalline defects that lead to DLD formation before they can be incorporated into final products.

The light output of AlGaInP LEDs degrades more rapidly with higher drive-current density, but DLDs are not typically observed. It is suspected that point defects or defect complexes form under the influence of current and radiation and that these defects act as nonradiative recombination centers in the active region.

AlGaInN LEDs can easily be damaged by electrostatic discharge (ESD), probably because of the highly defected crystal structure that results from growing the LED on a substrate with a large lattice mismatch. No other LED exhibits any substantial susceptibility to ESD damage. The end user must take special handling precautions with these LEDs unless ESD protection devices have been integrated into the package. A typical ESD protection device is a diode that is electrically in parallel with the LED but in the opposite polarity. When the visible LED is reverse-biased, the junction in the protection diode is forward-biased and will readily conduct current. Circuits that supply power to the LEDs must be designed to account for this type of arrangement where the typical "rectifying" characteristic of the LED is lost.

Package-related Degradation. Degradation can also be caused by the packaging methods. The encapsulating epoxy is particularly troublesome, since it has a much higher thermal coefficient of expansion than the semiconductor chip. When the LED is cooled to low temperatures, typically $-20\,^{\circ}\text{C}$ ($-4\,^{\circ}\text{F}$) or below, the epoxy contracts and substantial compressive stress is applied to the chip. When the LED is operated at low temperatures, this stress can provide a driving force for defect formation and/or migration, which results in accelerated degradation. In some cases, particularly with AlGaAs devices, DLDs can form and grow. When the LED is cooled, or temperature-cycled, but not forward biased, this degradation does not occur.

Another type of package-related degradation, which occurs in outdoor applications and in AlGaInN LEDs, is the discoloration of the epoxy by exposure to ultraviolet light. The discolored epoxy absorbs some of the light emitted from the LED chip and reduces the light output of the package. White LEDs, which contain a short-wavelength-emitting AlGaInN LED coated by a phosphor, are particularly susceptible to this type of degradation. The energies of the photons emitted by the LED are ultraviolet or nearly so, and the intensity of the radiation is high. Special epoxies and silicones are utilized in the packaging of these LEDs. Ultraviolet LEDs may be mounted in a package that contains a small quartz window and a brass body, with no encapsulating material, in order to eliminate this degradation mechanism, although the penalty in initial efficiency may be substantial.

Semiconductor Material Issues. In order to fabricate high-performance LEDs, it is desirable to choose compounds or alloys from the set that have the following characteristics:

1. A direct energy gap corresponding to the emission wavelength desired.
2. Higher-energy-gap materials that can be lattice-matched to the active layer for the growth of window and confining layers for heterostructures.
3. A low defect-density substrate that is lattice-matched to the epitaxial structure and that is transparent to the emitted radiation.
4. The ability to add impurities to the structure to obtain a low defect-density $p-n$ junction and a low series resistance.

5. A materials growth technology that can reproducibly and cost-effectively produce the structure as designed.

One would also like to have the lowest possible cost. Unfortunately, it is not often possible to have all of these desires met simultaneously.

Visible Emitters

Visible LED Technologies. The AlGaInP and AlGaInN devices have the highest levels of performance and quantum efficiency. AlGaInP devices satisfy most of the selection criteria discussed in the section (**Semiconductor Material Issues**) previously discussed, such as direct band gap, lattice-matched epitaxial structures and substrates, and double-heterostructure device structures. The internal quantum efficiency of 650-nm AlGaInP LEDs is nearly 100%, so further advances in external quantum efficiency will be the result of extraction ratio improvements.

AlGaInN devices have a direct band gap and a quantum-well device structure, but the substrate is grossly lattice-mismatched. The efficiency of AlGaInN LEDs is expected to improve significantly over the next decade, as the physics of the device and material are better understood.

Lower-cost homojunction GaAsP and GaP devices dominate high-volume LED applications even though their performance is an order of magnitude lower. As new applications for high-performance LEDs emerge, and as LEDs penetrate applications that light bulbs traditionally dominate, it is expected that the volume of AlGaInP and AlGaInN LEDs will ultimately exceed the GaAsP and GaP volumes. AlGaInP LEDs have largely displaced AlGaAs LEDs for applications requiring red visible light because the reliability characteristics of AlGaInP LEDs are superior to AlGaAs, the luminous efficiency is better for AlGaInP, and the cost is similar.

Applications. Traditionally, visible LEDs have been used for indicators on instruments, computers, and consumer electrical equipment. LEDs are used in the form of discrete LED lamps, numeric displays, or alphanumeric displays. These applications are particularly cost sensitive and are dominated by lower-cost and lower-performance LED technologies. LED markets such as outdoor large-area displays, traffic signaling, and exterior lighting for automobiles often require LEDs with the maximum possible performance and, in addition, require devices that are robust when operated in outdoor environments.

Visible LED markets will grow and new applications will emerge as LED performance, especially the total luminous flux output, improves. The standard LED technologies will continue to be used for applications where their light output is adequate. The higher-performance technologies will decrease in cost and will be used where high light output or high efficiency is important. Organic LEDs will be used in applications, such as portable consumer electronics, where moderate-performance full-color displays are needed.

The established markets for AlGaInP and AlGaInN LEDs, large-area displays and vehicular lighting, will continue to grow. Emerging markets for high-performance colored LEDs include architectural decorative lighting (such as the use of strips of LEDs to illuminate the edge of a building or roof), backlighting of full-color liquid crystal displays (LCDs) displays and televisions, and turn signaling in automobiles. Adoption of LEDs in the architectural decorative lighting market is expected to be rapid because LEDs offer styling possibilities that are expensive or awkward to create with conventional bulb lighting.

White LEDs are expected to make major advances into lower-optical-flux markets such as forward lighting in automobiles and commercial product display lighting. The pace of market adoption will follow improvements in the flux, efficiency, color-rendering ability, and cost. The barrier to penetration in general illumination is the relatively low light output of LEDs, compared to even small incandescent bulbs. Although the energy efficiency of the LED white light is approaching the performance of fluorescent tube lamps, the flux output is more than 10 times lower. The cost of producing a white LED is presently at least 10 times higher than the cost of producing a bulb-type lamp with equivalent luminous flux. For the highest power applications, conventional lighting will continue to have a substantial and fundamental cost advantage over LEDs because the incremental cost for making a higher flux conventional light bulb is very small, whereas the cost of an LED varies nearly linearly with chip area. To the extent that the energy efficiency, reliability, and styling advantages of LEDs are valued, at least some penetration into all lighting markets is expected.

Organic LEDs are expected to grow in the low-power color alphanumeric and video display market. It is expected that the power efficiency and reliability of these devices will continue to improve, and breakthroughs in these areas could make organic LEDs suitable for general illumination, considering the ability to process these materials in large batches (for example, with roll printing) and therefore at low cost.

Infrared Emitters

The detectors used for infrared LEDs are photodiodes or phototransistors instead of the human eye. The radiometric power output of the LED is therefore a key performance parameter. The rate at which data can be transmitted is determined in part by the LED switching speed, and this is also an important performance parameter for IR LEDs (one that is totally unimportant for most visible applications). A summary of performance characteristics for several types of IR emitters is shown in Table 1.

Current-voltage Characteristics. As with the visible LEDs, the turn-on voltage of infrared LEDs varies with the band gap of the material. InGaAsP emitters have turn-on voltages of ~0.8 V while GaAs:Si and AlGaAs emitters have turn-on voltages of 1.2 V. The operating current varies dramatically, depending upon the application. Some small-emission-area AlGaAs LEDs used in data communication applications can be driven at currents as low as 10 mA, while GaAs:Si emitters used in remote control applications are pulsed up to 1.5 A. Large chips (400×400 μm^2) are used in the high-current devices to reduce both the thermal and the dynamic

TABLE 1. CHARACTERISTICS OF INFRARED LEDs

LED type	Wavelength (nm)	External quantum efficiency (%)	Output power[a] (mW at 50 mA)	90 to 10% fall time (ns)	Device structure[b]	Substrate	Epitaxial growth method (p–n junction) formation)
GaAs:Si	940	9–14	6–9	1000	Homo.	GaAs	LPE (grown)
AlGaAs:Si	880	11–18	8–13	500	Homo	GaAs	LPE (grown)
AlGaAs	780–880	4–6	3–5	20–40	SH	GaAs	LPE (grown)
AlGaAs	780–880	10–14	7–10	20–40	DH	GaAs	LPE (grown)
AlGaAs	780–880	20–27	15–20	20–40	DH-TS	GaAs	LPE (grown)
SEA[c] AlGaAs	820–880	–	–	4–15	DH	GaAs	LPE/MOVPE (grown)
SEA[b] InGaAsP	1300–1500	–	–	~2.5	DH	InP	LPE/MOVPE (grown)

[a] Encapsulated lamp form.
[b] Homo = homojunction, SH = single heterostructure, DH = double heterostructure, DH-TS = $double heterostructure much transparent, epitaxially grown "substrate".
[c] SEA stands for small-emission area.

resistances of the device. To keep the total forward voltage below 3.0 V (two 1.5-V batteries in series) at a current of 1.5 A, the dynamic resistance of the LED must be kept below ~1 Ω.

Again, as with the visible LEDs, the reverse breakdown voltage is limited by avalanche breakdown and is determined by the doping level on the most lightly doped side of the $p-n$ junction. As the infrared emitters are usually more heavily doped than visible emitters, their breakdown voltages are typically lower, in the 7- to 60-V range.

Applications for Infrared Emitters

Remote Controls. One of the highest volume applications for IR LEDs is for remote controls that are used in televisions, audio equipment, and VCRs. Here, the information is sent from the remote control through the air in a sequence of pulses and a high-sensitivity Si p-i-n detector is used as the receiver. The higher the LED output power, the further one can be from the receiver and still have it work. The amount of data to be transmitted, however, is quite small, and slow GaAs:Si or AlGaAs:Si emitters are sufficient. These LEDs are driven at currents up to 1.5 A in order to maximize the transmitter–receiver distance.

Additional Reading

Bulovic, V., and S.R. Forrest: *Semiconductors and Semimetals*, Vol. 65, Academic Press, San Diego, CA, 2000.

Casey, Jr, H.C., and M.B. Panish: *Heterostructure Lasers*, Academic Press, Orlando, FL, 1978.

Gillessen, K., and W. Schairer, W.: *Light Emitting Diodes—An Introduction*, Prentice Hall, Inc., Upper Saddle River, NJ, 1987.

Holonyak, Jr, N., and S.F. Bevacqua: *Appl. Phys. Lett.*, **1**, 82–83 (1962).

Kalinowski, J.: *Organic Light-Emitting Diodes: Principles, Characteristics, and Processes (Optical Engineering, Vol. 91)*, CRC Press, LLC, Boca Raton, FL, 2004.

Kressel, H., and J.K. Butler: *Semiconductor Lasers and Heterojunction LEDs*. Academic Press, San Diego, CA, 1997.

Mueller, G.: *Electroluminescence I and Electroluminescence II, in: Semiconductors and Semimetals*, Vols. 64 and 65, Academic Press, San Diego, CA, 2000.

Nakamura, S., and G. Fasol: *Blue Laser Diode: GaN Based Light Emitters and Lasers*, Springer-Verlag New York, LLC, New York, NY, 1997.

Nalwa, H.S., and L.S. Rohwer: *Handbook of Luminescence, Display Materials and Devices*, 2nd Edition, American Scientific Publishers, Stevenson Ranch, CA, 2003.

Rosencher, E., and B. Vinter: *Optoelectronics*, Cambridge University Press, New York, NY, 2002.

Schubert, E.F.: *Light-Emitting Diodes*, 2nd Edition, Cambridge University Press, New York, NY, 2006.

Shinar, J.: *Organic Light-Emitting Devices*, Springer-Verlag New York, LLC, New York, NY, 2003.

Staff: Joint Industry and Traffic Engineering Council Comm., *Vehicle Traffic Central Signal Heads: Light Emitting Diode (Led) Vehicle Traffic Signal Module*, Institute of Transportation Engineers, Washington, DC, 1998.

Staff: Joint Industry and Traffic Engineering Council Comm., *Vehicle Traffic Central Signal Heads: Light Emitting Diode (Led) Circular Signal Supplement*, Institute of Transportation Engineers, Washington, DC, 2005.

Streetman, B.G.: *Solid State Electronic Devices*, 5th Edition, Prentice Hall, Inc., Upper Saddle River, NJ, 1999.

Stringfellow, G.B., and M.G. Craford: *High Brightness Light Emitting Diodes, in: Semiconductors and Semimetals*, Vol. 48, Academic Press, San Diego, CA, 1997.

Wyszecki, G., and W.S. Stiles: *Color Science: Concepts and Methods, Quantitative Data and Formulae*, 2nd Edition, John Wiley and Sons, Inc., New York, NY, 2000.

Yosyhikawa, A., M. Kobayashi, and K. Kishino: *Blue Laser and Light Emitting Diodes*, I O S Press, Inc., Burke, VA, 1996.

Zhigang, Li: *Organic Light-Emitting Diodes*, CRC Press, LLC, Boca Raton, FL, 2006.

Zukauskas, A., M.S. Shur, and R. Gaska: *Introduction to Solid-State Lighting*, John Wiley & Sons, Inc., New York, NY, 2002.

LIGHT FREEZE.   The occurrence of air temperature below 0 °C (32 °F) that kills some, but not all, annual vegetation. This often occurs in the 0° to −1 °C (32°–30 °F) range. See also **Climate**.

LIGHTING.   See **Illumination**.

LIGHT MICROSCOPY.   The images produced by microscopes have fascinated observers for several centuries. Today, the instrument is often used as the definitive symbol of scientific investigation. It is unfortunate,

then, that the understanding of what the microscope really does is poorly taught, and poorly understood. To start with, a microscope will only produce an image of the specimen (crystal, bacterium or whatever else) that is being observed; we see an image, not the specimen itself. The importance of this distinction is that different instruments could well produce different images of the same specimen. An understanding of microscope performance is therefore a requirement to help distinguish a good image from an inferior one.

The most recognizable judgment of microscope performance is its ability to faithfully reproduce in the image the fine detail that is in the specimen. The term resolving power is used for the smallest distance—a linear measurement—that a lens system can separate or resolve. The smaller this distance, the closer together are the linear elements that a lens is capable of resolving, and the 'better' we think of the lens. There is an unfortunate confusion in language that defines high resolution performance with a small numerical value for resolving power. The term 'resolving power' is not a particularly good one; more recently the phrase minimum resolved distance (MRD) has been introduced, which is an accurate description of the concept.

The lens MRD is easy to understand and is reasonably easy to measure. But possibly of more practical importance in the study of microscope performance is the ability of a lens system to reproduce in an image the weak intensity variations that are present in the specimen, especially living biological specimens. The emphasis that is placed in introductory texts (and this account) on MRD should not override the very great significance of contrast transfer; but the subject is more difficult to explain in a simple way and is therefore confined to more theoretical dissertations.

The Basic Microscope

The specimen is normally placed on a 75 mm × 25 mm (3 in × 1 in) glass slide, that is 1 mm (1/32 in) or less in thickness. The specimen should be very thin, less than 10 mm (0.4 in) in thickness, in a very small quantity of mountant; the mountant is sometimes water but the specimen is more often permanently embedded in a polymerized plastic. A coverslip protects the specimen and prevents it from drying out. The slide preparation is supported on a large flat stage, usually rectangular in shape. It is helpful for higher-magnification studies to have a mechanical stage to facilitate holding and maneuvering of the specimen in the horizontal x and y directions. One disadvantage of a mechanical stage is a lazy tendency of the observer to avoid examining the whole specimen. Specimen movement in the vertical or z direction is with the focusing system, with a coarse focus and a fine focus, although these may be on the one axis (coaxial). The focusing movement may operate on the microscope tube that holds the objective and eyepiece, or on the microscope stage (a feature of later instrumentation).

Microscopes for the observation of thin specimens normally use a transmitted light illumination system in which light is directed through the transparent specimen. Better microscopes will have a transformer or rheostat to permit variation of the intensity of illumination. This is the microscope illustrated in Figure 1. Opaque specimens are much less frequently encountered in the biological laboratory. For such specimens, an epi-illumination system is necessary ('epi' here meaning 'above'), in which the light source is placed on the same side of the specimen as the observation optics. Although epi-illumination is essential to fluorescence microscopy, a very important technique in the life sciences, this account is confined to transmitted light systems.

Beneath the specimen is another lens system, the condenser. Its function and purpose, often misunderstood, are discussed below.

Microscope Theory

Unless there is an understanding of very basic lens theory, it is not possible to appreciate the complexities of image control provided by the condenser. Figure 2 shows a lens bringing parallel rays to a focus, but the arrows in the diagram have been reversed as a legitimate optical design practice. To look at the performance of an objective we should consider light coming from a specimen at the point of focus of the lens. According to Abbe (a German physicist, 1840–1905) the resolving power (MRD) of the lens is given by eq. 1.

$$d = \frac{\lambda}{2(n \sin \alpha)} = \frac{\lambda}{2NA} \tag{1}$$

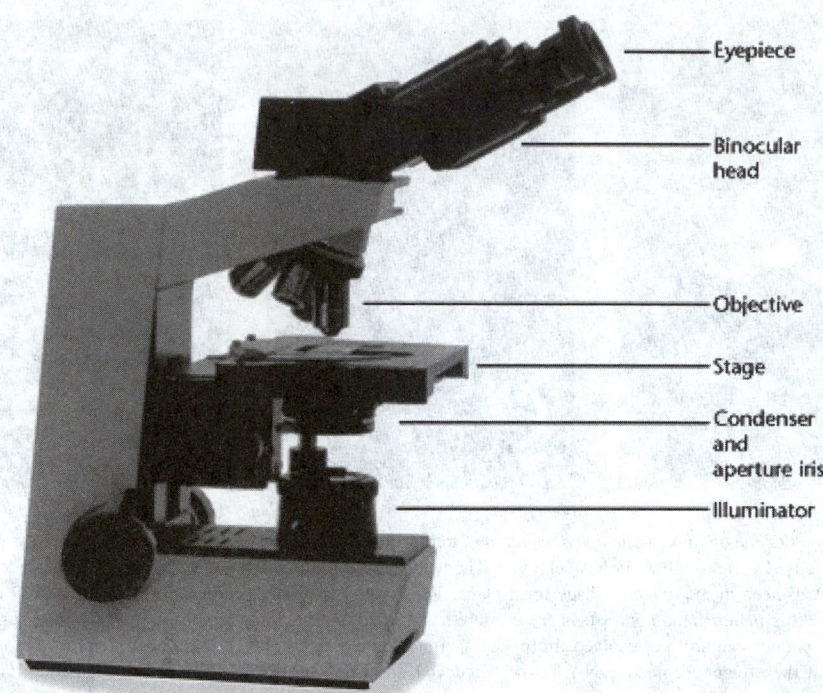

Fig. 1.   A modern classroom microscope with mechanical stage and binocular head.

Where d = MRD, λ = wavelength of illumination, n = refractive index of the medium between the lens and the specimen (usually air, $n = 1.0$), α = half the angular intake of the lens, and $(n \sin \alpha)$ is defined as the numerical aperture (NA) of the lens.

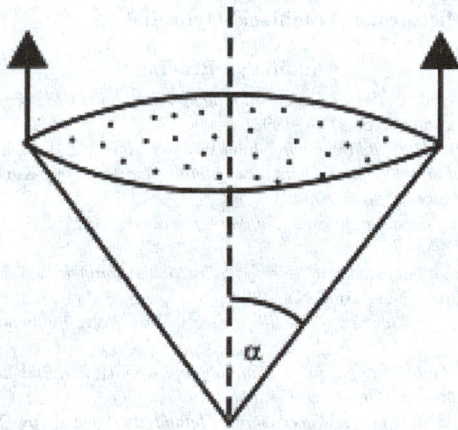

Fig. 2.   Defining numerical aperture (NA).

For improved resolution, the MRD value should be small, and therefore the wavelength should be as small as possible. In a biological laboratory, the wavelength cannot be varied much but, on theoretical grounds, blue light would give better resolution than red light. However, the equation predicted the MRD capability of very low-wavelength sources (ultraviolet light, X-rays, electrons, ions), and microscopes — the instrumental name persisted — were ultimately designed to utilize these sources.

The Abbe equation suggests that the lens NA should be maximized to keep the MRD as small as possible. The NA is a physical property of the lens, a function of the lens intake angle. This is determined by the lens designer and would seem to be beyond the control of the user, but incorrect use of the microscope invariably leads to a decrease in the effective NA of the objective.

Numerical Aperture Control with the Condenser

To completely fill the objective NA with the required cone of light, another lens system, the condenser, must be used to provide that cone. The reason

for having a condenser at all is to provide that cone of light. Two identical NAs are required, the condenser NA exactly matching the objective NA. If the condenser NA exceeds that of the objective, contrast will be lost by uncontrolled reflections in the objective. If the condenser NA is less than that of the objective, the objective NA will not be filled, and the objective will have its resolution performance compromised with an effective NA lower than its true NA.

This implies the need for a separate condenser for each objective; when the objective is changed, the condenser should also be changed. No-one does this. Instead, a condenser of high NA is chosen that has its NA reduced by an iris diaphragm–the aperture iris diaphragm–whenever a lower NA objective is used. The correct position of the aperture diaphragm is recognized by looking at the back (focal plane) of the objective. Remove an eyepiece and look down the tube of the microscope. Inside the objective there is an image of the aperture iris diaphragm, which should be adjusted so that some 7/8 of the diameter of the objective is filled with light. It is an adjustment to be made with each change of objective. Theoretically, exactly matching NAs of the condenser/objective would have the objective 8/8 filled with light by the aperture iris, but the imperfections of objectives require some small closure (1/8) of the aperture iris to maintain contrast. The aperture iris controls the effective NA of the condenser; it is used to match the NA of the condenser to the NA of the specific objective that is in use. See Figure 3.

The effects of closing the aperture iris diaphragm are to

- Reduce the effective NA of the condenser (appropriate only for low-NA objectives)
- Increase contrast (not required with well-stained material)
- Increase depth of field (not required with very thin preparations)
- Reduce light intensity (not the best way of doing this — use the transformer)

The normal position for the condenser is fully raised, to within 1 mm (1/32 in) of the microscope slide. Some would say that it is a sure sign of lack of microscope training if ever the condenser is lowered.

The Objective

Lens Correction. The MRD of the microscope is established by the front lens of the objective. Other lens elements that are a part of the objective are there to correct for deficiencies in image formation resulting from aberrations of the system. Objectives are categorized by the efficiency with which they minimize chromatic aberration (mainly) and curvature of

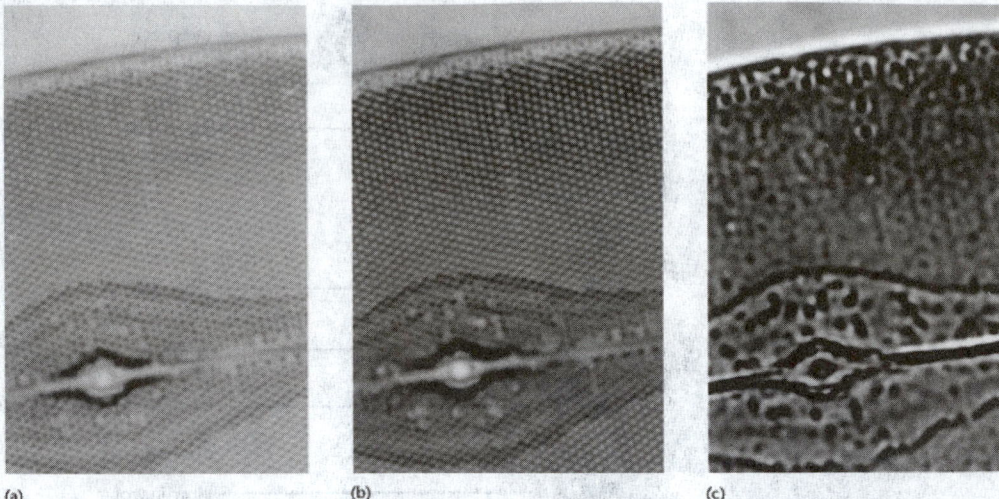

(a) (b) (c)

Fig. 3. Images of the diatom *Pleurosigma angulatum*, to illustrate the function of the condenser aperture iris diaphragm. The objective used was a × 40, 0.65 NA achromat. The field size is 60 μm × 80 μm. In (b) aperture iris set to 7/8 of the objective NA as seen in the objective back focal plane. The image has optimal compromise between MRD (the ability to resolve the line pattern) and contrast. In (a) aperture iris opened too far. Flare has resulted in loss of image contrast, although the pattern can still be resolved. In (c) the aperture iris closed too far. Contrast and depth of field have been increased, but the effective objective NA has been reduced, the MRD adversely affected and the pattern is no longer resolved.

the image plane. The very large majority of objectives are achromatic, in which blue and red rays, but not the green, are brought to the same focus. The images of sharp boundaries are characterized by green/magenta fringes as the focus is slightly changed (magenta = red plus blue). Apochromatic objectives have the blue, green and red rays focused together. A separate correction made by lens designers is for field curvature. The very best and most expensive objectives are called *planapochromats*.

Magnification. Since it is the NA of the objective that determines its MRD, it is unfortunate that we identify objectives by their magnification. Magnification serves only to enlarge the image resolved by the objective to a size in which all the image detail is resolved by the eye. The limit of useful magnification is about 1000 times the objective NA. Further magnification introduces artefact, not specimen detail.

Immersion Objectives. The highest NAs are achieved by increasing the refractive index of the medium between the specimen and the objective; this is achieved in immersion objectives. They are always marked in some way to identify their special function, and to specify the proper immersion fluid, usually immersion oil but occasionally glycerin or water. These objectives have their lens mounts especially sealed to prevent seepage of the immersion fluids into the system. 'Dry' (nonimmersion) objectives should be carefully protected from contact with oil or similar fluids.

The Coverslip. Objectives of NA higher than 0.4, which is those with magnifications in excess of about × 20, are sensitive to variation in the thickness of the coverslip (coverglass); the higher the NA, the more critical is the problem. By 'coverslip thickness' is meant the distance between the actual point of focus in a specimen and the front surface of the objective. When entire cells are being examined, this poses a major difficulty, for the cell itself may have considerable thickness. Most objectives for the life sciences are designed to have a coverslip thickness of 0.17 mm, but objectives are also made for use without a coverslip, a requirement of metallurgical specimens. One important advantage of oil immersion objectives is that the oil is chosen to have optical properties homogeneous with those of the coverslip, so that variations in relative thickness of oil/coverslip, even the absence of a coverslip, are consistent with high-quality imaging. A No. 1 coverslip, about 0.15 mm thick, is the normal compromise for coverslip choice in a biological laboratory.

Advances in Light Microscopy

Living biological material suffers from a lack of contrast, for there is little variation in light absorption by the cellular components that we wish to distinguish. One of the functions of specimen staining is to provide intensity and color contrasts, but chemical (histochemical) information about the specimen is also a significant result. The emphasis with more advanced microscopical techniques is to introduce contrast optically, as in darkfield, phase contrast and differential interference contrast microscopy. Digital image capture techniques, image enhancement and computer reconstruction of images, especially as applied to the confocal microscope are transforming the expectations of the light microscope as a modern research tool.

See also **Microscope (Traditional-Optical)**.

Additional Reading

Bradbury, S., and B. Bracegirdle: *Introduction to Light Microscopy*, Taylor & Francis, Inc., Philadelphia, PA, 1998.

Heath, J.P.: *Dictionary of Microscopy*, John Wiley & Sons, Inc., Hoboken, NJ, 2005.

James, J.: *Light Microscopic Techniques in Biology and Medicine*, Kluwer Academic Publishers, Norwell, MA, 2004.

Kapitza, H.G.: *Microscopy from the Very Beginning*, Carl Zeiss, Oberkochen, Germany, 1994.

Lacey, A.J.: *Light Microscopy in Biology: A Practical Approach*, 2nd Edition, Oxford University Press, New York, NY, 1999.

Oldfield, R.: *Light Microscopy: An Illustrated Guide*, Wolfe Publications, London, UK, 1994.

Schade, K.H.: *Light Microscopy, Technology and Application*, 2nd Edition, Verlag Moderne Industrie, Landsberg, Germany, 1995.

Spector, D.L., and R.D. Goldman: *Basic Methods in Microscopy: Protocols and Concepts from Cells*, 2nd Edition, Cold Spring Harbor Laboratory Press, Cold Spring Harbor, NY, 2005.

RONALD J. OLDFIELD, Macquarie University, Sydney, Australia

LIGHTNING. Under favorable circumstances, large electrical potential differences that are generated in the earth's lower atmosphere may be neutralized within an extremely short span of time in the form of lightning. The lightning may occur fully within the atmosphere, as by intra- or intercloud lightning, or between the atmosphere and the earth's surface, particularly the higher surfaces (mountains) or objects such as buildings and trees which extend upward from the ground for short distances. Favorable paths of electrical conductance predominantly determine the geometry of lightning. It has been observed that conductivities and air-earth current densities may be as much as ten times greater in the vicinities of high mountains than at sea level. This correlates well with other observations which indicate that lightning at sea occurs with a frequency only about 10% of the frequency over land. Conductivity decreases when humidity increases. Considering the earth as a whole, lightning is very common, occurring with a frequency of about 100 lightning events per second. Scores, hundreds, and even thousands of lightning events may occur in connection with a given thunderstorm. The frequency and variation of lightning characteristics ranges widely—with location, season of the year, and local time of day. Lightning over land occurs with the greatest frequency during late afternoon and early evening; and over the seas and oceans during late evening to an hour or two after midnight local time. There is, however, no time of day or night when lightning may not occur. Lightning can strike the same object or location not just once, but many times even during a given storm. This is particularly true of high structures. Trees do not provide protection against lightning, but represent an unsafe location for persons to be during a lightning-productive storm. Some authorities have suggested that a metal-bodied automobile may be one of the safest places to be during a storm of this type if other hazards (falling trees, flying debris, etc.) are minimal. For persons caught in open fields, golf courses, etc., lying prone on the ground is considered best in the interest of protection from lightning.

Although lightning is an extremely common phenomenon of the lower atmosphere, it remains rather poorly understood, particularly in quantitative terms. In recent years, lightning has been found to be much more complex and variable than previously believed. High-speed camera techniques, notably with improvements that permit the analysis of lightning during daytime hours, have produced much new evidence both of a qualitative and quantitative nature. The unaided eye or still camera, incapable of breaking down the individual events which occur during what appears to be a single stroke of lightning, at best can provide very rough data—estimates of patterns, brilliance, and distances and heights. In recent years, concerted programs of research in this field have contributed much toward expanding the knowledge of lightning. Such efforts include work done in connection with the Thunderstorm Research International Program (TRIP), in which several organizations such as the Kennedy Space Center (NASA), the University of Florida, the University of Arizona, the New Mexico Institute of Mining and Technology, Rice University, and the State University of New York at Albany, among others, have participated. Uman et al. (1978) reported on the physics and meteorology of a lightning flash at Kennedy Space Center; Orville and Lala 1980 reported on the development and use of a streaking camera for producing daylight time-resolved photographs of lightning.

Characteristics of Lightning

From the standpoint of geometry, the fundamental types of lightning are: (1) cloud-to-cloud (intercloud); (2) between two portions of the same cloud (intracloud); (3) cloud-to-earth; and (4) earth-to-cloud. The last two types often occur as part of what visually appears to be a single event. Lightning also may be classified as (a) long time span and low current; and (b) short time span and high current. The first of these is generally the more damaging to objects, such as igniting forest and structural fires.

Rarely do the long-time-span lightning events persist up to or over one second, although it was reported by Godionton in 1896 that a single lightning flash lasted up to 15 or 20 seconds. If this observer was accurate, it is indeed an exceedingly rare case. Observations since then have not included events of this time magnitude. Most lightning events persist for fractions of a second, ranging from 10^{-2} second through an average of about 0.2 second, up to 1 second (unusual).

In the usual cloud-to-ground lightning event, the phenomenon commences with what is called a *stepped leader*. High-speed camera techniques have been invaluable in confirming this characteristic. See the very schematic sketch in Fig. 1. The stepped leader may be of relatively low luminosity and possibly one to two meters in diameter. The leader sets up the conditions for the much more dramatic *return stroke* from earth to cloud. The time interval between steps of the stepped leader may range from a minimum of 30 microseconds to 50 microseconds (typical) to a maximum of 125 microseconds. The minimum length of a step may be as low as about 3 meters, ranging up to 50 meters (typical) to a maximum of about 200 meters. The average velocity of propagation of the stepped leader, at a minimum, may be 1.0×10^5 meters/second, ranging up to 1.5×10^5 meters/second (typical) to a maximum of about 2.6×10^6 meters/second. The charge deposited on a stepped-leader channel may be as low as 3 coulombs, ranging up to 5 coulombs (typical) to a maximum of about 20 coulombs. The return stroke may have a diameter measurable in terms of centimeters rather than meters, with a peak current ranging

from 10 to 20 (typical), but up to a maximum of 110 kiloamperes. The channel length may range from 2 kilometers up to 5 kilometers (typical) to a maximum of about 14 kilometers. The velocity of propagation may range from 2.0×10^7 meters/second to 5.0×10^7 meters/second (typical) up to a maximum of about 1.4×10^8 meters/second. Temperatures within the return stroke range up to 30,000 K at pressures up to one MN/m^2 stroke.

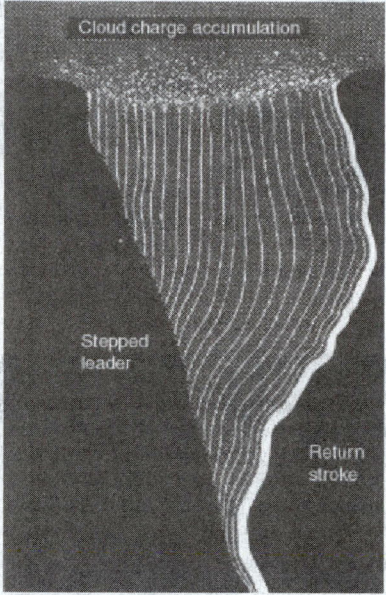

Fig. 1. Stepped leader and return stroke in cloud-to-earth lightning event.

Upon completion of the return stroke, a dart leader may again move downward because of residual potential differences, causing a second return stroke. As many as 40 secondary discharges (return strokes) may occur in what is called a multiple stroke flash. See various lightning configurations in Figs. 2, 3, 4 and 5.

Fig. 2. Dramatic display of several lightning strokes in vicinity of Kitt Peak, Arizona. Buildings of the National Optical Astronomical Observatories are shown in the light of the lightning. *Nikkormat* FTn with 28 mm lens with tripod and cable release; *Kodachrome II* film; *f* 3.5; exposure, about 1 minute. (*Copyright Gary Ladd, 1972.*)

Types of Lightning

Some lightning strikes take on particular characteristics, and scientists and the public have given names to these various types of lightning.

Ball Lightning. Also called globe lightning. A rare and randomly occurring bright ball of light observed floating or moving through

Fig. 3.   Massive display of lightning over city in western United States. (*Electric Power Research Institute.*)

Fig. 4.   Lightning.

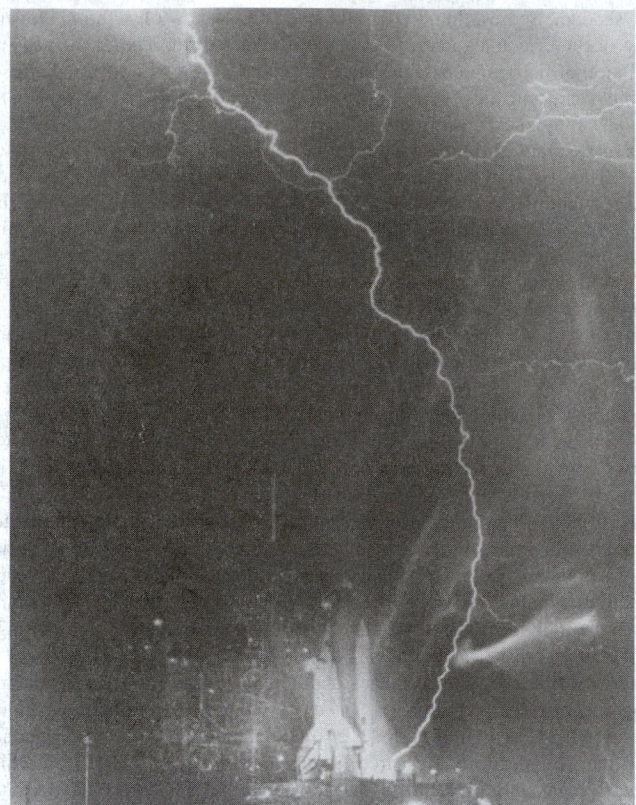

Fig. 5.   Lightning strike to Space Shuttle launch pad. (*NASA*).

the atmosphere close to the ground. Observations have widely varying identifying characteristics for ball lightning, but the most common description is that of a sphere having a radius of 15–50 centimeters (6–20 inches), orange or reddish in color, and lasting for only a few seconds before disappearing, sometimes with a loud noise. Most often ball lightning is seen in the vicinity of thunderstorms or a recent lightning strike, which may suggest that ball lightning, is electrical in composition or origin. Considered controversial due to the lack of unambiguous physical evidence for its existence, ball lightning is becoming more accepted due to recent laboratory recreations resembling ball lightning. Despite the observations and models of these fire balls, the exact mechanism(s) for naturally occurring ball lightning is unknown.

Bead Lightning, Ribbon Lightning, Staccato Lightning. Another special type of cloud-to-ground lightning is bead lightning. This is a regular cloud-to-ground stroke that contains a higher intensity of luminosity. When the discharge fades it leaves behind a *string of beads* effect for a brief moment in the leader channel. A third special type of cloud-to-ground lightning is ribbon lightning. These occur in thunderstorms where there are high cross winds and multiple return strokes. The winds will blow each successive return stroke slightly to one side of the previous return stoke, causing a ribbon effect. The last special type of cloud-to-ground lightning is staccato lightning, which is nothing more than a leader stroke with only one return stroke.

Cloud-to-Ground Lightning, Anvil-to-Ground Lightning. Cloud-to-ground lightning is a great lightning discharge between a cumulonimbus cloud and the ground initiated by the downward-moving leader stroke. This is the second most common type of lightning. One special type of cloud-to-ground lightning is anvil-to-ground lightning, a form of positive

lightning, since it emanates from the anvil top of a cumulonimbus cloud where the ice crystals are positively charged. In anvil-to-ground lightning, the leader stroke issues forth in a nearly horizontal direction till it veers toward the ground. These usually occur miles ahead of the main storm and will strike without warning on a sunny day. They are signs of an approaching storm and are known colloquially as "bolts out of the blue".

Cloud-to-Cloud Lightning. Cloud-to-cloud or intercloud lightning is a somewhat rare type of discharge lightning between two or more completely separate cumulonimbus clouds.

Cloud-to-Ground Lightning. Cloud-to-ground lightning is a lightning discharge between the ground and a cumulonimbus cloud from an upward-moving leader stroke. These thunderstorm clouds are formed wherever there is enough upward motion, instability in the vertical, and moisture to produce a deep cloud that reaches up to levels somewhat colder than freezing. These conditions are most often met in summer. Lightning occurs less frequently in the winter because there is not as much instability and moisture in the atmosphere as there is in the summer. These two ingredients work together to make convective storms that can produce lightning. Without instability and moisture, strong thunderstorms are unlikely. Lightning originates around 15,000 to 25,000 feet (4,572 to 7,620 meters) above sea level when raindrops are carried upward until some of them convert to ice. For reasons that are not widely agreed upon, a cloud-to-ground lightning flash originates in this mixed water and ice region. The charge then moves downward in 50 yard (46 meter) sections called step leaders. It keeps moving toward the ground in these steps and produces a channel along which charge is deposited. Eventually it encounters something on the ground that is a good connection. The circuit is complete at that time, and the charge is lowered from cloud-to-ground. The return stroke is a flow of charge (current) which produces luminosity much brighter than the part that came down. This entire event usually takes less than half a second.

However, it has been proven by movies taken of typical lightning strikes and then, using single-frame examination (looking at each frame of a sequence), that a typical lightning strike is made up of anywhere from 8 to 12 or more individual discharges, with each successive discharge being less intense and farther apart in time. This is easily explained by a process known in the electronics industry as *damped oscillation*, which is sustained by the magnetic field that is built up in the surrounding air during current flow in each discharge, and then that magnetic field starts collapsing when current flow starts decreasing at the end of the current flow. This causes induced current that continues in the same direction, sustaining current flow beyond the point where the original charge voltage would have been depleted, and possibly reversing the charge voltage polarity, bringing on the next successive discharge, as long as sufficient charge is available to sustain another discharge. (This is almost exactly the type of current-flow used in alternating-current circuits to drive motors, lamps, etc.).

Heat Lightning or Summer Lightning. Heat lightning (or, in the UK, "summer lightning") is nothing more than the faint flashes of lightning on the horizon or other clouds from distant thunderstorms. Heat lightning was named because it often occurs on hot summer nights. Heat lightning can be an early warning sign that thunderstorms are approaching. In Florida, heat lightning is often seen out over the water at night, the remnants of storms that formed during the day along a sea breeze front coming in from the opposite coast.

Some cases of "heat lightning" can be explained by the refraction of light or sound by bodies of air with different densities. An observer may see nearby lightning, but the sound from the discharge is refracted over his head by a change in the temperature, and therefore the density, of the air around him. As a result, the lightning discharge appears to be silent.

Intracloud Lightning, Sheet Lightning, Anvil Crawlers. Intracloud lightning is the most common type of lightning which occurs completely inside one cumulonimbus cloud, and is commonly called an anvil crawler. Discharges of electricity in anvil crawlers travel up the sides of the cumulonimbus cloud branching out at the anvil top.

Sprites, Elves, Jets, and Other Upper Atmospheric Lightning. Reports by scientists of strange lightning phenomena above storms date back to at least 1886. However, it is only in recent years that fuller investigations have been made. This has sometimes been called *megalightning*.

Sprites are now well-documented electrical discharges that occur high above the cumulonimbus cloud of an active thunderstorm. They appear as luminous reddish-orange, neon-like flashes, last longer than normal lower stratospheric discharges (typically around 17 milliseconds), and are usually spawned by discharges of positive lightning between the cloud and the ground. Sprites can occur up to 50 km from the location of the lightning strike, and with a time delay of up to 100 milliseconds. Sprites usually occur in clusters of two or more simultaneous vertical discharges, typically extending from 65 to 75 km (40 to 47 miles) above the earth, with or without less luminous filaments reaching above and below. Sprites are preceded by a sprite halo that forms because of heating and ionisation less than 1 millisecond before the sprite. Sprites were first photographed on July 6, 1989, by scientists from the University of Minnesota and named after the mischievous sprites in the plays of Shakespeare. These Sprites may be the result of the neutralization of accumulated charge from the Earth sweeping up particles from the Solar Wind.

Recent research carried out at the University of Houston in 2002 indicates that some normal (negative) lightning discharges produce a *sprite halo*, the precursor of a sprite, and that *every* lightning bolt between cloud and ground attempts to produce a sprite or a sprite halo. Research in 2004 by scientists from Tohoku University found that very low frequency emissions occur at the same time as the sprite, indicating that a discharge within the cloud may generate the sprites. More probably, as said before, they may be generated from interaction with the upper atmosphere's *neutralizing a charge derived from the Earth's movement through the* Solar Wind.

Blue jets differ from sprites in that they project from the top of the cumulonimbus above a thunderstorm, typically in a narrow cone, to the lowest levels of the ionosphere 40 to 50 km (25 to 30 miles) above the earth. They are also brighter than sprites and, as implied by their name, are blue in colour. They were first recorded on October 21, 1989, on a video taken from the space shuttle as it passed over Australia. Again, this could be currents being generated from potential differences in the upper atmosphere caused by the same derivation of charge from the *Solar Wind*.

Elves often appear as a dim, flattened, expanding glow around 400 km (250 miles) in diameter that lasts for, typically, just one millisecond. They occur in the ionosphere 100 km (60 miles) above the ground over thunderstorms. Their colour was a puzzle for some time, but is now believed to be a red hue. Elves were first recorded on another shuttle mission, this time recorded off French Guiana on October 7, 1990. Elves is a frivolous acronym for *Emissions of Light and Very Low Frequency Perturbations From Electromagnetic Pulse Sources*. This refers to the process by which the light is generated; the excitation of nitrogen molecules due to electron collisions (the electrons possibly having been energised by the electromagnetic pulse caused by a discharge from the Ionosphere).

On September 14, 2001, scientists at the Arecibo Observatory photographed a huge jet double the height of those previously observed, reaching around 80 km (50 miles) into the atmosphere. The jet was located above a thunderstorm over the ocean, and lasted under a second. Lightning was initially observed travelling up at around 50,000 m/s in a similar way to a typical *blue jet*, but then divided in two and sped at 250,000 m/s to the ionosphere, where they spread out in a bright burst of light.

On July 22, 2002, five gigantic jets between 60 and 70 km (35 to 45 miles) in length were observed over the South China Sea from Taiwan, reported in Nature http://sprite.phys.ncku.edu.tw/new/news/0626_presss/nature01759_r.pdf. The jets lasted under a second, with shapes likened by the researchers to giant trees and carrots.

Researchers have speculated that such forms of upper atmospheric lightning may play a role in the formation of the ozone layer. Rather, they may be due to differences in potential that result in current *from the* ozone layer.

Streak Lightning. Most of all lightning is streak lightning. This is nothing more than the return stroke, the visible part of the lightning stroke. Because most of these strokes occur inside a cloud, we do not see many of the individual return strokes in a thunderstorm.

Triggered Lightning. Lightning has been triggered directly by human activity in several instances. Lightning struck the Apollo 12 soon after takeoff, and has struck soon after thermonuclear explosions. It has also been triggered by launching rockets carrying spools of wire into thunderstorms. The wire unwinds as the rocket climbs, making a convenient path for the lightning to use. These bolts are typically very straight.

Energy Sources of Lightning

Numerous theories have been advanced pertaining to the accumulation of electric charge in the lower atmosphere. Advanced as early as 1885, the influence theory suggested that the earth's field is usually negative with relation to positive cloud charges, thus setting up the right conditions for lightning. This concept was probably in keeping with the general scientific knowledge of that time. It was also generally proposed that particles (hydrometeors, such as rain, snow, hail, fog, etc., associated with storms) developed a charge as the result of frictional forces with the atmosphere. In the free ionization theory, it was suggested that droplets of different sizes selectively collect available ions, thus establishing large potential differences. It is now generally believed that electric conduction current in the atmosphere is carried almost exclusively by fast ions. Positive ions include $(H_3O)^+(H_2O)_n$ and negative ions include $O_2^-(H_2O)_n$ or $CO_4^-(H_2O)_n$, where the probable values of n for positive ions in the troposphere are 6 or 7; in the stratosphere, 4; and in the mesosphere, 3 or 2. The value of n is dependent upon the water vapor pressure and temperature of the atmosphere, as well as the lifetime of the ion.

A few highlights pertaining to the energy in lightning would include:

1 Typical voltage drop in ground or other conductors after lightning impact in the neighborhood of 10 kV (dangerous!).

2 Energy delivered to an average stroke of lightning, about 100 kJ/m. Intracloud lightnings have been observed up to 100 km in length. Height from which a lightning stroke points at a target estimated several decameters, depending on conductivity distribution in ground.

3 Long-lasting, low-current (hundreds of amperes) flashes are more dangerous to people and objects (for example, forest fires) than short, high-current flashes.

Damage Wrought by Lightning

The natural high-voltage phenomena occurring during lightning are not only of general scientific interest and value. A greater knowledge of these phenomena is of practical value as well, in particular because of the danger from lightning to life and to susceptible structures, notably electric power systems and communications equipment. For many years, the direct and induced effects of lightning discharges have been simulated in industrial laboratories by means of high-voltage impulse generators. These frequently use the Marx method of first slowly charging a number of condensers in parallel and then suddenly connecting them in series by spark-gap switches which, at the same instant, impress the multiplied voltage upon the test circuit. A typical voltage wave produced by such impulse generators rises to its peak value of several million volts in one microsecond and then diminishes exponentially, reaching half voltage in about ten microseconds.

The Electric Power Research Institute (Palo Alto, California) has for a number of years conducted research on the characteristics of lightning, means of predicting lightning, and means of protecting electric utility equipment from damage, particularly overhead transmission lines. Although improvements in calculating lightning performance of transmission lines have been achieved, little information has been available for accurately predicting the lightning flashover performance of multiple-circuit transmission lines. See Fig. 6. In most double-circuit lines, when one circuit flashes over, the second circuit will simultaneously flash over 40–60% of the time. EPRI has generated a computer program, known as MULTI-FLASH and available to electric utility operators, that enables the transmission line designer to accurately predict the lightning performance of lines containing up to 12 ac phases, 12 dc poles, or any combination thereof on the same transmission tower, with a variety of tower shapes and insulator strings. All transmission voltage and significant corona effects are included, as well as the statistical distribution of footing resistance. The output includes an analysis of expected shielding failure performance, followed by a detailed tabulation of the expected flashover frequency of each of the phases for dc poles that are involved. Using this computer analysis, the designer can explore new and innovative shapes of materials and accurately predict the degree of improvement that may be achieved.

Fig. 6.   Much research in recent years has been conducted to understand and to reduce the effects of lightning flashover in multiple-circuit transmission lines. (*Electric Power Research Institute.*)

Research also has been conducted pertaining to arresters. These devices play a vital role in protecting substation equipment from high-voltage surges that originate from either substation switching equipment or lightning strikes. Through the proper selection of arresters, undesirable overvoltages can be limited. New materials having an extremely nonlinear voltage/current characteristic have been used in the construction of gapless surge arresters. Materials include zinc oxide (ZnO). The new designs have been under evaluation, including some at Tennessee Valley Authority substations. See Fig. 7.

A few years ago, some utility system designers concluded that surge protection devices could not be further optimized without an improved database of lightning stroke characteristics. The EPRI thus established a project along these lines. In this study, lightning stroke current and energy magnitude were estimated by comparing the gaps of 2800 surge arresters that accumulated 32,000 arrester years of service on utility lines with gaps that had discharged known currents. The stroke energy statistics

Fig. 7.   New materials, including metal oxides (ZnO), have been used in innovative surge arresters to protect electric power substation equipment from high-voltage surges that originate either from substation switching equipment or lightning strokes. The new materials have extremely nonlinear voltage/current characteristics that lend themselves well to gapless surge arrester designs. (*Electric Power Research Institute.*)

obtained are of particular interest to surge protection engineers because little information along these lines has been available. Under present standards, arrester durability is determined in part by tests using 65-kA and 8 × 20-microsecond current discharges that have a charge of 1.25 coulomb. The study showed that arresters in service are more frequently exposed to longer-duration, lower-current waves having charges of up to 4.2 coulombs, than to higher-current, lower-energy waves.

In another, associated study, the geographic density of lightning flashes was probed. As a starter, EPRI has mapped the location of lightning flashes in the eastern United States. Such data will provide surge protection engineers with information on the number of lightning flashes that can be expected to strike distribution and transmission lines directly or nearby. Data on charge polarity, number of strokes per flash, and peak field strength radiated from the first stroke are being gathered.

Lightning Safety

Thunderstorms are the primary source of lightning. Because people have been struck many kilometers (miles) away from a storm, seeking immediate and effective shelter when thunderstorms approach is an important part of lightning safety. Contrary to popular notion, there is no 'safe' location outdoors. People have been struck in sheds, makeshift shelters, etc. A better location would be inside a vehicle (a crude type of Faraday cage). It is advisable to keep appendages away from any attached metallic components once inside (keys in ignition, etc.).

Several different types of devices, including lightning rods, lightning arresters, and electrical charge dissipaters, are used to prevent lightning damage and safely redirect lightning strikes.

Nearly 2000 persons per year in the world are injured by lightning strikes, and between 25 to 33% of those struck die. Lightning injuries result from three factors: electrical damage, intense heat, and the mechanical energy which these generate. While sudden death is common because of the huge voltage of a lightning strike, survivors often fare better than victims of other electrical injuries caused by a more prolonged application of lesser voltage.

Lightning can incapacitate humans in 4 different ways:

- Direct strike
- 'Splash' from nearby objects struck
- Ground strike near victim causing a difference of potential in the ground itself (due to resistance to current in the Earth), amounting to several thousand volts-per-foot, depending upon the composition of the earth that makes up the ground at that location. (Sand being a fair insulator versus wet, salty and spongy earth being more conductive).
- EMP or electromagnetic pulse from close strikes—especially during positive lightning discharges

In a *direct hit* the electrical charge strikes the victim first. Counterintuitively, if the victim's skin resistance is high enough, much of the current will *flash* around the skin or clothing to the ground, resulting in a surprisingly benign outcome. *Splash* hits occur when lightning prefers a victim (with lower resistance) over a nearby object that has more resistance, and strikes the victim en route to ground. *Ground* strikes, in which the bolt lands near the victim and is conducted through the victim via his or her connection to the ground (such as through the feet, due to the voltage *gradient* in the earth, as discussed above), which can cause great damage.

The most critical injuries are to the circulatory system, the lungs, and the central nervous system. Many victims suffer immediate cardiac arrest and will not survive without prompt emergency care, which is safe to administer because the victim will not retain any electrical charge after the lightning has struck (of course, the helper could be struck by a separate bolt of lightning in the vicinity). Others incur myocardial infarction and various cardiac arrhythmias, either of which can be rapidly fatal as well. The intense heat generated by a lightning strike can burn tissue, and cause lung damage, and the chest can be damaged by the mechanical force of rapidly expanding heated air. Either the electrical or the mechanical force can result in loss of consciousness, which is very common immediately after a strike. Amnesia and confusion of varying duration often result as well. A complete physical examination by paramedics or physicians may reveal ruptured eardrums, and ocular cataracts may develop, sometimes more than a year after an otherwise uneventful recovery.

The lightning often leaves skin burns in characteristic Lichtenberg figures, sometimes called *lightning flowers*; they may persist for hours or days, and are a useful indicator for medical examiners when trying to determine the cause of death. They are thought to be caused by the rupture of small capillaries under the skin, either from the current or from the shock wave. It is also speculated that the EMP created by a nearby lightning strike can cause cardiac arrest.

There is sometimes spectacular and unconventional lightning damage. *Hot lightning* (High-current Lightning) which lasts for more than a second can deposit immense energy, melting or carbonizing large objects. One such example is the destruction of the basement insulator of the 250-meter-high (820 feet) central mast of longwave transmitter Orlunda, which led to its collapse. It also causes sand to fuse into glass tubes and branches like a tree with an empty center (know as *fulgarites*), at the site of a bolt hitting the ground. These *fulgarites* are very collectable.

Lightning Terminology

An abridged glossary of lightning terms would include:

Atmospheric Electricity—

1. Electrical phenomena, regarded collectively, that occur in the earth's atmosphere. These phenomena include not only such striking manifestations as lightning and St. Elmo's fire, but also less noticeable but more ubiquitous effects such as atmospheric ionization, the air–earth current, and other quiescent electrical processes. The existence of separated electric charges in the atmosphere is a consequence of many minor processes (spray electrification, dust electrification, etc.) and a few major processes (cosmic-ray ionization, radioactive-particle ionization, and thunderstorm electrification). The details of thunderstorm charge separation are poorly understood. The maintenance of the prevailing atmospheric electric field is now widely believed to be due to thunderstorm effects.

2. The study of electrical processes occurring within the atmosphere.

Ball Lightning—(Also called globe lightning.) A rare and randomly occurring bright ball of light observed floating or moving through the atmosphere close to the ground. Observations have widely varying identifying characteristics for ball lightning, but the most common description is that of a sphere having a radius of 15–50 cm, orange or reddish in color, and lasting for only a few seconds before disappearing, sometimes with a loud noise. Most often ball lightning is seen in the vicinity of thunderstorms or a recent lightning strike, which may suggest that ball lightning is electrical in composition or origin. Considered controversial due to the lack of unambiguous physical evidence for its existence, ball lightning is becoming more accepted due to recent laboratory recreations resembling ball lightning. Despite the observations and models of these fire balls, the exact mechanism(s) for naturally occurring ball lightning is unknown.

Forked Lightning—The common form of cloud-to-ground discharge always visually present to a greater or lesser degree that exhibits downward-directed branches from the main lightning channel. In general, of the many branches of the stepped leader, only one is connected to the ground, defining the primary, bright return stroke path; the other incomplete channels decay after the ascent of the first return stroke. See **Streak Lightning**; and **Zigzag Lightning**.

Heat Lightning—Nontechnically, the luminosity observed from ordinary lightning too far away for its thunder to be heard. Since such observations have often been made with clear skies overhead, and since hot summer evenings particularly favor this type of observation, there has arisen a popular misconception that the presence of diffuse flashes in the apparent absence of thunderclouds implies that lightning is somehow occurring in the atmosphere merely as a result of excessive heat.

Lightning Arrester—Any device designed to carry to the ground (to "ground") the short-duration surge currents that appear on power lines and telephone lines during severe thunderstorms. Early forms of arresters consisted merely of spark gaps to ground, but in cases where the circuit power maintained the spark after termination of the surge, service could only be restored by momentarily shutting off circuit power. Hence, more elaborate arresters have been developed, especially for high-voltage lines, with features to ensure that the circuit power will not maintain the ground circuit after the surge dies out. See **Lightning Rod**.

Lightning Channel—The irregular path through the air along which a lightning discharge occurs. The lightning channel is established at the start of a discharge by the growth of a leader, which seeks out a path of least resistance between a charge source and the ground or between two charge centers of opposite sign in the thundercloud or between a cloud charge center and the surrounding air or between charge centers in adjacent clouds.

Lightning Discharge—The series of electrical processes taking place within 1 s by which charge is transferred along a discharge channel between electric charge centers of opposite sign in a thundercloud (intracloud flash), between a cloud charge center and the earth's surface (cloud-to-ground flash or ground-to-cloud discharge), between two different clouds (intercloud or cloud-to-cloud discharge), or between a cloud charge and the air (air discharge).

It is a very large-scale form of the common spark discharge. A single lightning discharge is called a lightning flash.

Lightning Counter—A device for measuring the number of lightning events within a specified time interval.

Lightning Flash—The total observed lightning discharge, generally having a duration of less than 1 s. A single flash is usually composed of many distinct luminous events that often occur in such rapid succession that the human eye cannot resolve them.

Lightning Rod—A grounded metallic conductor with its upper extremity extending above the structure that is to be protected from damage due to lightning. The upper extremity, called the air terminal, should be raised well above the top of the structure, to yield an adequate radius of protection. The path to ground must consist of a conductor of low total resistance and must contain no points of high local resistance. Lower ends should be buried deeply enough in the earth that they will always be in good contact with soil moisture. Connection to water pipes usually affords a good ground.

Lightning Stroke—In a cloud-to-ground discharge, a leader plus its subsequent return stroke. In a typical case, a cloud-to-ground discharge is made up of three or four successive lightning strokes, most following the same lightning channel.

Lightning Suppression—Procedures to prevent the occurrence of lightning. Seeding below cloud base with 10-cm fiber chaff in a Colorado study resulted in corona discharges that caused a discharging current to flow within developing or active thunderstorms. Electric fields below thunderstorms seeded with chaff decayed much faster than electric fields below nonseeded storms, and chaff seeding of existing thunderstorms greatly reduced cloud-to-ground flashes compared to nonseeded storms. Recent evidence suggests that chaff releases may result in a significant decrease in downwind cloud-to-ground lightning. Another experimental approach is to use lasers to discharge lightning in an overhead cloud in order to divert the flash from striking people or highly sensitive equipment on the ground; more research is needed to make this a realistic method of lightning suppression. In the 1960s, seeding with silver iodide was considered in order to produce an excess of ice crystals to cause numerous coronal discharges within the thunderstorm and reduce the need for the flash to reach the ground, but the test results were complex and difficult to identify. Finally, electric space charge was released into the atmosphere from a network of high-voltage wires on the ground to produce corona discharge, but a field test showed minimal effects on suppressing lightning.

M Component—In a lightning discharge, an increase in channel luminosity accompanied by a rapid electric field variation, itself called an M electric field change. The M components occur when the channel is already faintly luminous. Downward-moving leaders have not been observed to precede M components. The M components may be confused with branch components, the increases in channel luminosity that occur between each branch and the ground when the upward-propagating return stroke reaches that branch, since the higher branches are obscured by the cloud. The M component is a minor surge of current that reilluminates the channel in a negative cloud-to-ground flash. It may occur within microseconds or up to a few milliseconds of a return stroke. Evidence exists that M components are a result of the sequence of fast in-cloud negative leaders (K changes) contacting a conducting ground channel and renewed ground potential wave that reilluminate the channel.

Negative Cloud-to-Ground Lightning—A lightning flash or stroke between a cloud and the ground that lowers negative charge to the ground.

Pilot Streamer—A relatively slow-moving, nonluminous lightning streamer, the existence of which has been postulated but not verified, to provide a physical explanation for the observed intermittent mode of advance of a stepped leader as it initiates a cloud-to-ground lightning discharge. Whereas the stepped leader descends at an average speed of the order of 10^5 m s^{-1} during its downward motion, it advances only about 50 m at a time with a speed greater than average and then pauses for 50–100 μs before resuming its downward movement. The average downward speed has been associated with an invisible streamer, the pilot streamer, which is postulated to descend at a uniform speed only slightly in excess of the ionizing speed of electrons in air and lay down a trail of weak residual ionization along which the stepped leader moves very rapidly in a pulsating manner. The idea of a pilot leader has been supplanted by more modern theory based on studies of sparks over long distances in the laboratory.

Positive Cloud-to-Ground Lightning—A lightning flash or stroke between a cloud and the ground that lowers positive charge from the cloud to the ground.

Return Stroke—The intense luminosity that propagates upward from earth to cloud base in the last phase of each lightning stroke of a cloud-to-ground discharge. In a typical flash, the first return stroke ascends as soon as the descending stepped leader makes electrical contact with the earth, often aided by short ascending ground streamers. The second and all subsequent return strokes differ only in that they are initiated by a dart leader and not a stepped leader. It is the return stroke that produces almost all of the luminosity and charge transfer in most cloud-to-ground strokes. Its great speed of ascent (about 1 ×

10^8 m s^{-1}) is made possible by residual ionization of the lightning channel remaining from passage of the immediately preceding leader, and this speed is enhanced by the convergent nature of the electric field in which channel electrons are drawn down toward the ascending tip in the region of the streamer's electron avalanche. Current peaks as high as 3×10^5 A have been reported, and values of 3×10^4 A are fairly typical. The entire process of the return stroke is completed in a few tens of microseconds, and even most of this is spent in a long decay period following an early rapid rise to full current value in only a few microseconds. Both the current and propagation speed decrease with height. In negative cloud-to-ground flashes the return stroke deposits the positive charge of several coulombs on the preceding negative leader channel, thus charging earth negatively. In positive cloud-to-ground flashes, the return stroke deposits the negative charge of several tens of coulombs on the preceding positive leader channel, thus increasing positive charge on the ground. In negative cloud-to-ground flashes, multiple return strokes are common. Positive cloud-to-ground flashes, in contrast, typically have only one return stroke. The return streamer of cloud-to-ground discharges is so intense because of the high electrical conductivity of the ground, and hence this type of streamer is not to be found in air discharges, cloud discharges, or cloud-to-cloud discharges.

Ribbon Lightning—Ordinary cloud-to-ground lightning that appears to be spread horizontally into a ribbon of parallel luminous streaks when a very strong wind is blowing at right angles to the observer's line of sight. Successive strokes of the lightning flash are then displaced by small angular amounts and may appear to the eye or camera as distinct paths. The same effect is readily created artificially by rapid transverse movement of a camera during film exposure.

Rocket Lightning—A form of cloud discharge, generally horizontal and at cloud base, with a luminous channel appearing to advance through the air with visually resolvable speed, often intermittently.

Rocket-Triggered Lightning—A form of artificial lightning discharge initiated with a rocket trailing wire that may or may not be connected to the ground. The first phase of the discharge is a unidirectional leader starting from the tip of the wire. When the low end of the wire is not connected to ground, bidirectional leader development occurs from both ends of the wire, similar to lightning initiation from aircraft. In the case of negative space charge overhead (usual summer thunderstorm condition), a triggered lightning may only be a positive leader or may become a sequence of dart leader—return stroke processes following the initial positive leader. The latter is analogous to the subsequent return stroke process in a negative cloud-to-ground flash with the initial positive leader being analogous to the first return stroke. In the case of positive space charge overhead (usual winter storm condition), the triggered lightning is a single negative leader.

Sferics Receiver (Also called lightning direction finder.) An instrument that measures, electronically, the direction of arrival, intensity, and rate of occurrence of atmospherics; a type of radio direction finder, it is most commonly used to detect and locate cloud-to-ground lightning discharges from distant thunderstorms. In its simplest form the instrument consists of two orthogonally crossed antennas that measure the electromagnetic field changes produced by a lightning discharge and determine the direction from which the changes arrived. Negative and positive polarity cloud-to-ground discharges can be distinguished. Cloud-to-cloud discharges can be distinguished based on characteristics of the received signal, and the geometry of nearby discharge channels may be determined.

Sferics Source That portion of a lightning discharge that radiates strongly in the frequency interval 10–30 kHz. The physical source is generally identified with the return stroke in flashes to ground and the K change in the case of intracloud flashes.

Sheet Lightning—(Also called luminous cloud.) A diffuse, but sometimes fairly bright, illumination of those parts of a thundercloud that surround the path of a lightning flash, particularly a cloud discharge or cloud-to-cloud discharge. Thus, sheet lightning is no unique form of lightning but only one manifestation of ordinary lightning types in the presence of obscuring clouds.

Sprite — Weak luminous emissions that appear directly above an active thunderstorm and are coincident with cloud-to-ground or intracloud lightning flashes. Their spatial structures range from small single or multiple vertically elongated spots, to spots with faint extrusions above and below, to bright groupings that extend from the cloud tops to altitudes up to about 95 kilometers (59 miles). Sprites are predominantly red. The brightest region lies in the altitude range 65–75 kilometers (40–47 miles), above which there is often a faint red glow or wispy structure that extends to about 90 kilometers (56 miles). Below the bright red region, blue tendril-like filamentary structures often extend downward to as low as 40 kilometers (25 miles). High-speed photometer measurements show that the duration of sprites is only a few milliseconds. Current evidence strongly suggests that sprites preferentially occur in decaying portions of thunderstorms and are correlated with large positive cloud-to-ground flashes. The optical intensity of sprite clusters, estimated by comparison with tabulated stellar intensities, is comparable to a moderately bright auroral arc. The optical energy is roughly 10–50 kJ per event, with a corresponding optical power of 5–25 MW. Assuming that optical energy constitutes 10^{-3} of the total for the event, the energy and power are on the order of 10–100 MJ and 5–50 GW, respectively. Early research reports for these events referred to them by a variety of names, including upward lightning, upward discharges, cloud-to-stratosphere discharges, and cloud-to-ionosphere discharges. Now they are simply referred to as sprites, a whimsical term that evokes a sense of their fleeting nature, while at the same time remaining nonjudgmental about physical processes that have yet to be determined.

Streak Lightning — Ordinary lightning, of a cloud-to-ground discharge, that appears to be entirely concentrated in a single, relatively straight lightning channel.

Thunderstorm — (Sometimes called electrical storm.) In general, a local storm, invariably produced by a cumulonimbus cloud and always accompanied by lightning and thunder, usually with strong gusts of wind, heavy rain, and sometimes with hail. It is usually of short duration, seldom over two hours for any one storm. A thunderstorm is a consequence of atmospheric instability and constitutes, loosely, an overturning of air layers in order to achieve a more stable density stratification. A strong convective updraft is a distinguishing feature of this storm in its early phases. A strong downdraft in a column of precipitation marks its dissipating stages. Thunderstorms often build to altitudes of 40 000–50 000 ft in midlatitudes and to even greater heights in the Tropics; only the great stability of the lower stratosphere limits their upward growth. A unique quality of thunderstorms is their striking electrical activity. The study of thunderstorm electricity includes not only lightning phenomena per se but all of the complexities of thunderstorm charge separation and all charge distribution within the realm of thunderstorm influence. In U.S. weather observing procedure, a thunderstorm is reported whenever thunder is heard at the station; it is reported on regularly scheduled observations if thunder is heard within 15 minutes preceding the observation. Thunderstorms are reported as light, medium, or heavy according to 1) the nature of the lightning and thunder; 2) the type and intensity of the precipitation, if any; 3) the speed and gustiness of the wind; 4) the appearance of the clouds; and 5) the effect upon surface temperature. From the viewpoint of the synoptic meteorologist, thunderstorms may be classified by the nature of the overall weather situation, such as airmass thunderstorm, frontal thunderstorm, and squall-line thunderstorm. See also **Fronts and Storms**.

Streak Lightning — Ordinary lightning, of a cloud-to-ground discharge, that appears to be entirely concentrated in a single, relatively straight lightning channel.

Thunder The sound emitted by rapidly expanding gases along the channel of a lightning discharge. Some three-fourths of the electrical energy of a lightning discharge is expended, via ion–molecule collisions, in heating the atmospheric gases in and immediately around the luminous channel. In a few tens of microseconds, the channel rises to a local temperature of the order of 10,000 °C (18,032 °F), with the result that a violent quasi-cylindrical pressure wave is sent out,

followed by a succession of rarefactions and compressions induced by the inherent elasticity of the air. These compressions are heard as thunder. Most of the sonic energy results from the return streamers of each individual lightning stroke, but an initial tearing sound is produced by the stepped leader; and the sharp click or crack heard at very close range, just prior to the main crash of thunder, is caused by the ground streamer ascending to meet the stepped leader of the first stroke.

Thunder is seldom heard at points farther than 15 miles (24 kilometers) from the lightning discharge, with 25 miles (40 kilometers) an approximate upper limit, and 10 miles (16 miles) a fairly typical value of the range of audibility. At such distances, thunder has the characteristic rumbling sound of very low pitch. The pitch is low when heard at large distances only because of the strong attenuation of the high-frequency components of the original sound. The rumbling results chiefly from the varying arrival times of the sound waves emitted by the portions of the sinuous lightning channel that are located at varying distances from the observer, and secondarily from echoing and from the multiplicity of the strokes of a composite flash.

Zigzag Lightning — Ordinary lightning of a cloud-to-ground discharge that appears to have a single, but very irregular, lightning channel.

Facts and Trivia

A bolt of lightning can reach temperatures approaching 28,000 Kelvin's (50,000 degrees Fahrenheit) in a split second. This is about five times hotter than the surface of the sun. The heat of lightning which strikes loose soil or sandy regions of the ground may fuse the soil or sand into glass channels called fulgurites. These are sometimes found under the sandy surfaces of beaches and golf courses, or in desert regions. Fulgurites are evidence that lightning spreads out into branching channels when it strikes the ground.

Trees are frequent conductors of lightning to the ground. Since sap is a poor conductor, its electrical resistance causes it to be heated explosively into steam, which blows off the bark outside the lightning's path. In following seasons trees overgrow the damaged area and may cover it completely, leaving only a vertical scar. If the damage is severe, the tree may not be able to recover, and decay sets in, eventually killing the tree. Occasionally, a tree may explode completely, as in this Giant Sequoia struck in Geneva http://www.pinetum.org/lightning.htm. It is commonly thought that a tree standing alone is more frequently struck, though in some forested areas, lightning scars can be seen on almost every tree.

Of all common trees the most frequently struck is the oak. It has a deep central root that goes beneath the tree, as well as hollow water-filled cells that run up and down the wood of the oak's trunk. These two qualities make oak trees better grounded and more conductive than trees with shallow roots and closed cells.

- The odds of an average person living in the USA being struck by lightning once in his lifetime has been estimated to be 1:3000 (NOAA, National Weather Service).
- The city of Teresina in northern Brazil has the third-highest rate of occurrences of lightning strikes in the world. The surrounding region is referred to as the *Chapada do Corisco* ("Flash Lightning Flatlands").
- The United States is home to "Lightning Alley", a group of states in the American Southeast that collectively see more lightning strikes per year than any other place in the US. The most notable state in Lightning Alley is Florida.
- The saying "lightning never strikes twice in the same place" is false. The Empire State Building is struck by lightning on average 100 times each year, and was once struck 15 times in 15 minutes.
- Ukrainian President Viktor Yushchenko is probably the highest-ranked modern statesman to be struck by lightning (which happened in 2005 with no reported health consequences)
- Jim Caviezel, the actor who played Jesus in the film *The Passion of the Christ*, is reported to have been struck twice by lightning during shooting. The assistant director Jan Michelini was struck twice http://news.bbc.co.uk/2/hi/entertainment/3209223.stm.
- Golfers Retief Goosen and Lee Trevino have both been struck by lightning while playing golf. http://www.golfeurope.com/almanac/players/trevino.htm.
- Although commonly associated with thunderstorms, lightning strikes can occur on any day, even in the absence of clouds.

• Lightning interferes with AM (amplitude modulation) radio signals much more than FM (frequency modulation) signals, providing an easy way to gauge local lightning strike intensity.

Estimating distance of a lightning strike: The flash of a lightning strike and resulting thunder occur at roughly the same time. But light travels at 186,000 miles in a second, almost a million times the speed of sound. Sound travels at the slower speed of one-fifth of a mile in the same time. So the flash of lightning is seen before thunder is heard. By counting the seconds between the flash and the thunder and dividing by 5, you can estimate your distance from the strike (in miles). Similarly, by dividing by three, you can estimate the distance in kilometers.

Additional Reading

Bazelkilan, E.M.: *Lightning Physics and Lightning Protection*, Iop Publishing, Philadelphia, PA, 2000.

Franz, R.C., R.J. Nemzek, and J.R. Winckler: "Television Image of a Large Upward Electrical Discharge Above a Thunderstorm System," *Science*, **48** (July 6, 1990).

Horváth, T.: *Understanding Lightning and Lightning Protection: A Multimedia Teaching Guide*, John Wiley & Sons, Inc., Hoboken, NJ, 2006.

Marks, J.A., Ed.: *Electrical Systems*, Electric Power Research Institute, Palo Alto, CA, 1985. http://www.epri.com/

Ohkubo, A., H. Fukunishi, Y. Takahashi, and T. Adachi: "VLF/ELF Sferic Evidence for in-cloud Discharge Activity Producing Sprites," *Geophys. Res. Lett.*, 32, 2005.

Orville, R.E. and G.G. Lala: "Daylight Time-Resolved Photographs of Lightning," *Science*, **201**, 59–61 (1978).

Rycroft, M.J., M. Fallekrug, and E.A. Mareev: *Sprites, Elves and Intense Lightning Discharges*, Springer-Verlag New York, LLC, New York, NY, 2006.

Tahiliani, V.: *Metal Oxide Surge Arresters for Gas-Insulated Systems*, Electric Power Research Institute, Palo Alto, CA, 1983.

Uman, M.A.: *The Lightning Discharge*, Academic Press, San Diego, CA, 1987.

Uman, M.A. and E.P. Krider: "Natural and Artificially Initiated Lightning," *Science*, 457 (October 27, 1989).

Waterbury, R.C.: "Safir Forecasts Lightning Strikes," *Instrumentation Technology*, 72 (July, 1990).

Web References

Ball Lightning: http://www.eskimo.com/billb/tesla/ballgtn.html

Colorado Lightning Resource Center: http://www.crh.noaa.gov/pub/ltg.php

Darwin Sprites: http://www.physics.otago.ac.nz/space/darwin97/darwin97.html

European Cooperation for Lightning Detection: http://www.euclid.org/

Lightning & Atmospheric Electricity Research at the GHCC: http://thunder.msfc.nasa.gov/

Lightning Protection Institute: http://www.lightning.org/?page=home

Lightning Safety: http://www.lightningsafety.noaa.gov/overview.htm

NASA Fact Sheet Lightning: http://www-pao.ksc.nasa.gov/kscpao/nasafact/lightnin.htm

National Lightning Safety Institute: http://www.lightningsafety.com/

Questions and Answers about Lightning: http://www.nssl.noaa.gov/edu/ltg/

Resources: Lightning Science and Safety: http://www.usatoday.com/weather/resources/basics/wlightning.htm

Sprites and Elves: http://www-star.stanford.edu/vlf/optical/press/elves97sciam/

United States Precision Lightning Network: http://www.uspln.com/index2.html

LIGHTNING ARRESTER.   See **Lightning**.

LIGHTNING BUG.   See **Firefly**.

LIGHT (Polarized).   See **Polarized Light**.

LIGHT POLLUTION.   Today, people who live near large cities have lost much of their view of the universe. The spectacular view of the night sky that their ancestors had above them on clear dark nights no longer exists. The great increase in urban population has caused an ensuing rapid increase in urban sky glow due to outdoor lighting. This has brightened the heavens to such an extent that the only view most people have of the Milky Way or most stars is when they are well away from cities. This excess light in the sky has had an adverse impact on the environment and seriously threatens to remove forever one of mankind's natural wonders—the dark sky.

The extent of vast illuminated cosmopolitan areas of the United States is dramatically illustrated by the "map" of Fig. 1.

Effects on Professional Astronomy. While this increased urban sky glow brightens the night sky for the general public and for amateur astronomers, it is a special threat to professional astronomy. Advances in frontier astronomy require observations of very faint objects that can be studied only with large telescopes located on prime observing sites, well away from sources of air pollution and from urban nighttime sky glow. For example, most observations of cosmological interest deal with extremely remote sources—galaxies or quasars at such distances that the light has traveled several billion years, sometimes twice the age of our solar system, before reaching us. This light is then often lost in the glare of anthropogenic sky glow.

Observations of such extremely faint extragalactic sources, and even of many objects of interest in our own galaxy, can be done only during the dark moon period. The sky background is much too bright when moonlight is present. Artificial lighting of the sky due to cities has, unfortunately, the same adverse effect on nighttime sky brightness in limiting observations.

Fig. 1.   Satellite view of the United States at night, on a night when most of the country was clear. The upward light produced by the urban areas is evident in the photograph. Some of the light is direct-up light; the rest is light reflected from the ground. (*United States Air Force*.)

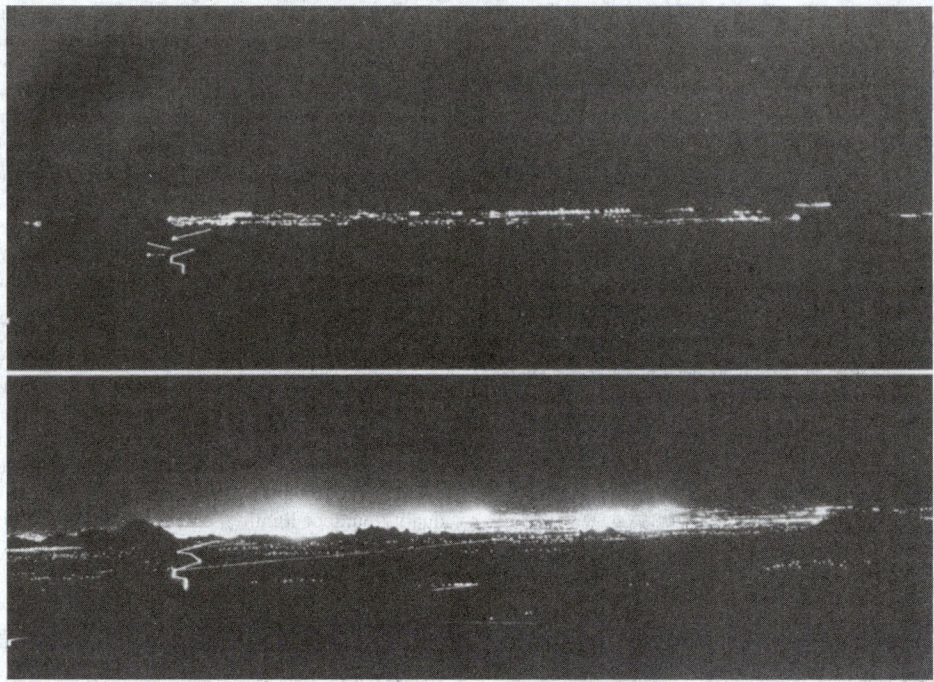

Fig. 2.   Two photographs of Tucson, Arizona as Seen from Kitt Peak, (about 35 miles 56 kilometers) west of the city. The growth of the city over 21 years (top view, 1959; bottom view, 1980) is evident. It is this type of growth, with its associated growth in outdoor lighting, that is comprising the research done at most of the major observatories located near cosmopolitan areas. (*National Optical Astronomy Observatories.*)

An interesting contrast in the situation at Tucson, Arizona, which is located near the Kitt Peak National Observatory, is given in Fig. 2. The worsening of the problem over a 20-year period is demonstrated.

This increased sky glow which adversely affects the environment and compromises astronomical research is called *light pollution*, for it is wasted light that does nothing to increase nighttime safety, utility, or security. It only serves to waste energy and money. An example of an improperly designed billboard that pollutes the night sky is given in Fig. 3.

Ground-Based Astronomical Observations Still Much Needed. The argument that all astronomy can be done from space is not correct; the largest telescopes will continue to be ground-based for a long time because of cost factors. It does not make sense to do in space, at much higher cost, what can be done from the ground. There are many things that can only be done in space and the demand for that type of research is severe. The experience of more than two decades of space astronomy, however, has greatly increased the demand for ground-based facilities. Planning for several ground-based telescopes much larger than anything now in existence is already underway. There are exciting times ahead for astronomers, using present and future ground-based telescopes, which *complement* the telescopes in space. See also **Telescope (Astronomical-Optical)**.

Solutions to the Light Pollution Problem. Control programs are underway now in a number of communities to reduce the effects of light pollution of the night sky. Programs like these are critical to the long-term success of astronomical research, and to preserve people's view of the universe. Unlike dealing with the problems of water and air pollution, vast sums of money are not required to alleviate light pollution.

At present, the lack of awareness rather than resistance is generally the biggest problem in controlling light pollution. Educating the public, government officials and staff, and lighting professionals has been the major thrust of the current programs. These efforts have helped. The increase of light pollution near major observing sites is moderating. More can and must be done. Astronomers, amateurs and professionals, and many others are urging such cooperation.

Astronomers are *not* against night lighting. They have the same needs for quality lighting as everyone else. They advocate the best possible lighting for the task, with lighting designs that allow for all the relevant factors, such as glare control, efficiency, and the need for dark skies. An important added advantage is that everything that is done to minimize light pollution also saves energy by improving the efficiency and utility of the nighttime lighting.

There are other adverse effects of poor quality lighting. Light that comes out of a fixture essentially horizontally does little or nothing to light the ground, but it does cause glare. Such glare is annoying to the eye of the beholder, and it even can cause discomfort or disability. Its blinding effects have often even caused accidents. Glare never helps; it is always a sign of poor-quality lighting.

Confusion or clutter is another adverse effect of poor lighting. Nighttime lighting should provide guidance, providing help and safety, not confusion. Some installations mislead a driver, for example, leading to accidents. In addition, the clutter of outdoor lights that we see today in most cities is often just as trashy a sign of poorly controlled urban growth as is the litter of garbage we See.

Light trespass is another adverse effect of poor outdoor lighting. Light from a fixture that falls in someone else's yard or in through one's window is usually unwanted. It is indeed "trespass." This results in black paint on one side of the fixture, irate phone calls to the owner of the light or to the police or to the sports park owners. Such trespass is the sign of a poorly shielded light fixture, not of a quality lighting design or installation. There is no need for such an adverse effect.

It is a sad state that many people are not aware of quality lighting. Many even think that lighting does not exist if it does not exhibit the adverse effects, for they are so used to the associated glare and light trespass of the all too common poor lighting. Quality lighting does exist. It should always be used. There are many examples of it being used, and it should be used for *all* installations. Professional lighting designers are well aware of quality lighting, and use it whenever possible. Unfortunately, so much of today's outdoor lighting is not done by lighting professionals, or by people aware of or sensitive to quality lighting. That is what must be changed.

Specific Corrective Measures. What aspects of quality lighting can be used to help solve the light pollution problem? Following are some solutions that will greatly minimize light pollution without in any way compromising nighttime safety, security, or utility:

1.   Use night lighting only when necessary. Turn off lights when they are not needed. Timers can be very effective. Use the correct amount of light for the need, not overkill. Until recently, when energy conservation became an important issue, it often Seemed that the only "design" criterion used for outdoor lighting was: "If a certain amount of light is good, double

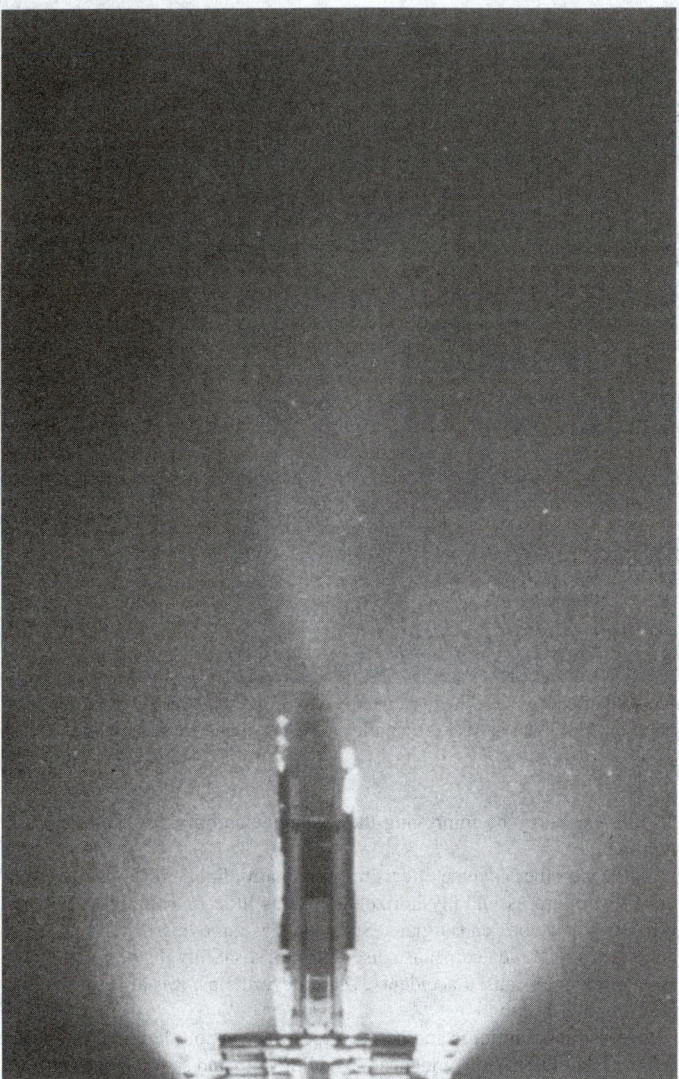

Fig. 3. Photograph of a billboard, showing a typical example of poor lighting. Most of the light output by the fixture Seems to miss the billboard. Better quality fixtures would put most of the light onto the board, thus less waste and less light pollution, all at less cost. To minimize light pollution, the quality fixtures should be mounted at the top of the board, thus minimizing the direct and reflected upward light. Also, the lights should be time-clocked to go off at 10 or 11 PM. There are few potential viewers after then, and the sky would be darker after those hours. Energy and money would be saved. (*National Optical Astronomy Observatories*.)

it and things will be better." That is not good design, as any professional lighting designer will agree.

2. Direct the light downward, where it is needed. The use and effective placement of well-designed lighting fixtures will achieve excellent control. Shielding the light source to avoid the upward light helps. When possible, retrofit present poor fixtures, ones that spray light everywhere, especially directly up into the sky. In all cases, the goal is to use fixtures that control the light well, minimizing glare, light trespass, and light pollution. All of this also minimizes the energy waste. Light is used when and where needed, and not wasted.

3. Use low pressure sodium (LPS) light sources whenever possible. This is the best possible light source to minimize adverse sky glow effects on astronomy. LPS lamps are the most energy efficient light sources that exist. Areas where LPS is especially good are street lighting, parking lot lighting, security lighting. There are some applications where LPS should not be the sole lighting source, for applications where color rendering is critical. But, for most applications, LPS should be considered. It is an excellent, low-cost light source and helps greatly to minimize the adverse sky glow.

4. Avoid growth near the observatories, and apply rigid controls on nighttime lighting when such growth is unavoidable. Such controls do not compromise safety, security, or utility. Lighting ordinances have been enacted by many communities to enforce quality, effective lighting. These communities have found that the quality of lighting has improved, usually at lower cost.

All of these solutions to the problem of adverse nighttime lighting say, really: "Do the best possible professional lighting design for the task. Include all relevant factors, such as glare, light trespass, and light pollution." All the solutions needed for protecting astronomy have positive side benefits of maximizing the quality of the lighting, and of saving energy. See also **Illumination**.

The American Astronomical Society has a Committee on Light Pollution. The Illuminating Engineering Society of North America has a Committee on Light Trespass and another one on the Environmental Impact of Outdoor Lighting. Other groups are responding in a similar matter. Fortunately, lighting technology is advancing and there is an increasing interest in quality lighting. Such advances in technology, and increasing awareness of the problems of light pollution, will help greatly in promoting quality outdoor lighting, and thus minimizing light pollution.

DAVID L. CRAWFORD, Ph.D., Kitt Peak National Observatory,
Tucson, AZ

LIGHT SECOND. See **Units and Standards**.

LIGHT TIME. The elapsed time taken by electromagnetic radiation to travel from a celestial body to the observer at the time of observation. The *American Ephemeris and Nautical Almanac* uses a light time of 498.8 seconds for 1 astronomical unit.

LIGHT WATER REACTOR. See **Nuclear Power Technology**.

LIGHT-YEAR. A popular method of expressing large distances; specifically, the distance that light will travel in the course of one year. The velocity of light is established (International Astronomical Union (Hamburg 1964)) at 299,792.5 kilometers per second, or about 186,282 statute miles per second. There are approximately 31,560,000 seconds in a mean solar year. Accordingly, a light-year represents a distance of approximately 9.454×10^{12} kilometers; 5.875×10^{12} miles. See also **Astronomical Unit**; and **Parsec**.

LIGNIN. Approximately 25% of the content of most woods is lignin. Lignin concentration in wood substance is greatest in the middle lamella (the zone around each individual fiber cell), decreasing in concentration through the cross section of the fiber, reaching a concentration of about 12% at the inner layer of the fiber adjacent to the fiber cavity, or lumen. Lignin and hemicellulose cement the fiber cells together, providing rigidity to the fibrous wood structure. In the destructive distillation of wood, the methanol produced is derived from the lignin. In the manufacture of paper pulp, it is necessary to remove the lignin, usually accomplished by treatment of the wood fibers with such agents as sulfur dioxide, calcium bisulfite, and sodium sulfate/sodium sulfide solutions. Sodium hydroxide is sometimes used. An important byproduct of the paper pulp industry is dimethyl sulfoxide, $(CH_3)_2SO$ which is produced from the lignin released during wood pulping by the Kraft process. Dimethyl sulfoxide has a number of industrial uses — as an intermediate in organic syntheses, as a solvent in spinning synthetic fibers, and in some pharmaceuticals.

The wall material of plant cells is one of their distinguishing characteristics. As a result, lignin, cellulose, and other wall constituents have been studied in many plant tissue cultures. Phenylpropanoids, for example, have been shown to be precursors of lignin formation in white pine, *Sequoia*, lilac, rose, carrot, and geranium tissue cultures. Moreover, the biosynthesis of lignin has been shown to be affected by kinetin, boron, and major elements, such as calcium.

Lignin is a major source of vanillin.

LIGNITE AND BROWN COAL. Lignite and brown coal are common names for coals having properties intermediate between peat and bituminous coal as a result of limited coalification. In general, brown coal designates a geologically younger, i.e., less coalified, material than the firmer, fibrous lignite. In many English-speaking countries, the consolidated coals are termed lignite, and unconsolidated coals are termed brown coal. In Australia, Germany, and a number of other European countries, the generic term brown coal is used for the whole class, including some coals that are included in the ASTM classification as subbituminous. Lignite signifies the firmer, fibrous, woody variety. Herein lignite is used as the comprehensive term.

Selection of coal for a particular use requires a knowledge of composition greater than that supplied from the ASTM classification. Progress is being made toward classifying all kinds of coal, including lignite, by correlating properties with composition and other qualities.

Lignite is less valuable than coals of higher rank, primarily because its much higher (30–70% as mined) water content and high chemically combined oxygen content result in a relatively low heating value (LHV). In the past, the expense of shipping limited the market largely to the vicinity of the mine. However, in the United States the low sulfur content of lignite has made long distance shipments economically feasible, in order to limit sulfur oxide emissions at electric power generation plants. The increasing worldwide demand for energy together with desire for national self-sufficiency has increased the importance of low heating value coals.

Geology

Lignite was deposited relatively recently (ca $2.5–60 \times 10^6$ yr ago), mainly during the Tertiary era. U.S. deposits include those in the Dakotas, Alaska, Montana, and Wyoming. The Miocene period provided the brown coal deposits that are up to 300 m thick in the Latrobe Valley of Victory in Australia.

Composition, Properties, and Analysis

Macroscopic Appearance. Lignitic coals vary from brown to dull black when moist, although the color may appear considerably lighter when the coal is dried. Breakage is easiest for the unconsolidated coals. Strength and toughness increase as coalification increases.

Physicochemical Structure. Water-filled pores and capillaries of differing diameters permeate the organic gel material that makes up asmined lignite.

Properties. The apparent density of lignite is 0.8–1.35 g/cm^3, which is lower than values given for higher ranking coals. Therefore, greater volume is required for storage, transportation, and lignite reactors than is needed for an equivalent weight of more mature coals. Lignite generally has lower elasticity and greater plasticity than more mature coals. The tar yield is usually higher for lignite than for more mature coals. Tar yields are important in determining selection for carbonization and for liquid fuel production by pyrolysis.

Oxidation. The high reactivity of lignites with oxygen requires special care during mining, transportation, and storage to avoid spontaneous combustion from heat generation.

Resources and Production

The importance of a coal deposit depends on the amount that is economically recoverable by conventional mining techniques. The world total recoverable reserves of lignitic coals were 3.28×10^{11} metric tons at the end of 1990, of which ca 47% was economically recoverable as of 1992. These estimates of reserves change as geological survey data improve and as the resources are developed.

The extent of lignite production is generally not proportional either to total resources or to known economic reserves. Lack of energy alternatives is a strong motive to developments in lignite production.

Main Deposits and Production Areas. The eastern European reserves of lignitic coals provide the primary solid fuel for the eastern part of Germany, the former Czechoslovakia, Hungary, the former Yugoslavia, and Bulgaria. The importance of lignite as an energy source is great enough in Germany to permit long-range planning that includes removal and relocation of towns or villages situated on deposits in order to permit more complete recovery of the lignite resource. Hard coal is more important in most of the western European countries, with the exception of Austria and Italy.

In the U.S., lignite deposits are located in the northern Great Plains and in the Gulf states. Subbituminous coal is found along the Rocky Mountains. The lignite deposits of North Dakota and Montana extend into Canada as far as Saskatchewan. Canadian deposits are also located in Alberta, Yukon, the Northwest Territories, Ontario, and Manitoba.

Production. The mining or winning of lignitic coal typically involves deposits near the surface. The open-cast, open-cut, or stripimining techniques employed involve mobile equipment built to provide a range of capacities to over 200,000 m^3/d. The rate of production can be increased rapidly, and the amount of labor per ton of coal mined is less than for underground mining. The quality of the coal, ratio of overburden thickness to seam thickness, stratigraphy, and distance to location of consumption are important in determining the cost to the consumer.

Concern about spontaneous ignition has led some operators to try to match the mining and consumption rates, so that there is little if any reserve, as in minemouth power generation stations. When the coal must be stockpiled, careful stacking minimizes oxygen reaction and overheating. To limit drying, spraying with cold water is useful.

For short distances from the mine, transportation is by truck or conveyer belt. Rail transportation is generally used for greater distances. Slurry pipelines are being considered as an alternative. Drying can be accomplished by evaporative, hydrothermal, or other thermal processes.

Health and Safety Factors

Because lignite mining is carried out by surface methods, the hazards associated with underground mining typically do not exist. See also **Coal**. The principal hazards involve the tendency of the coal toward spontaneous combustion as the coal dries, especially at the exposed seam.

Uses

Most of the world's coal supply is used for combustion to generate steam for electric power production. This is especially true of lignitic coals. Other uses for lignite, such as briquetting, for domestic and industrial fuels; carbonization, to provide coke and liquid by-products; gasification, to provide gaseous fuels; chemical feedstocks, for making fertilizers and other liquid fuels; and direct liquefaction are being developed.

Additional Reading

Berkowitz, N.: *An Introduction to Coal Technology*, 2nd Edition, Elsevier Science & Technology Books, New York, NY, 1993.

Durie, R.A. ed.: *The Science of Victorian Brown Coal: Structure, Properties and Consequences for Utilisation*, Butterworth Heinemann, Oxford, 1991. An excellent reference not only for Victorian Brown Coal, but for lignitic coals of the world.

Miller, B.: *Coal Energy Systems*, Elsevier Science & Technology Books, New York, NY, 2004.

Staff: *Clean Combustion of Brown Coal and Lignite*, United Nations Publications, New York, NY, 1990.

Symposia on the Technology and Use of Lignite have been held in conjunction with the University of North Dakota Energy and Environmental Research Center and the preceding organizations.

The World Energy Council issues Conference reports on reserves, resources and production at six-year intervals. More limited reports are issued at two-year intervals.

LIGNOCELLULOSE. See **Cellulose**.

LIGNUM VITAE (*Guaiacum*). The heartwood of a tree native to the West Indies. The tree also is found in other regions of moderate climate. It is a valuable, tough, resinous wood and very heavy, being the heaviest of all commercial woods; a cubic foot weighs 76 pounds (34.5 kilograms). Lignum vitae has been used in the making of bowling balls, pulley sheaves, and mallet heads.

The tree attains a height of about 40 feet (12 meters); the trunk may range from 2 to 4 feet (0.6 to 1.2 meters) in diameter and normally grows quite crooked. The branches are knotty.

The record *G. sanctum* (roughbark) tree growing in the United States is located in Biscayne National Park, Florida. See Table 1.

The record *G. angustifolium* (Texas lignum vitae) tree growing in the United States is located at Alamo, Texas. See Table 1.

LIKELIHOOD. If $P(x, \theta)$ denotes a probability function depending on one or more parameters collectively denoted by θ, the likelihood of a sample $x_1, x_2 \cdots x_n$ is defined as $L = P(x_1, \theta) \cdot P(x_2, \theta) \cdots P(x_n, \theta)$.

The method of maximum likelihood consists in estimating the parameters θ by choosing those values which maximize L (or log L).

TABLE 1. RECORD LIGNUMVITAE TREES IN THE UNITED STATES[1]

Specimen	Circumference[2]		Height		Spread		Location
	Inches	Centimeters	Feet	Meters	Feet	Meters	
Roughbark lignumvitae (1995) (*Guaiacum sanctum*)	37	94	31	9.4	39	11.9	Florida
Texas lignumvitae (1974) (*Guaiacum angustifolium*)	32	81	26	7.9	22	6.7	Texas

[1]From the "National Register of Big Trees," American Forests (by permission).
[2]At 4.5 feet (1.4 meters).

Under general conditions, maximum likelihood estimates are consistent and efficient and tend to be normally distributed in large samples; further, that they are sufficient if sufficient statistics exist. The large sample variances or covariances of the maximum likelihood estimates are given by the elements of the inverse matrix to

$$\left[nE\left(\frac{\partial^2 \log P}{\partial \theta_i \, \partial \theta_j} \right) \right]$$

where E denotes the expected value.

LILIACEAE. The Lily Family has representatives in all parts of the world, and more especially in the drier regions of the temperate zone. Several members of the family are important vegetables, notably asparagus and onions, while a great many more are cultivated for ornament. Among the latter are the true lilies (the genus *Lilium*), tulips, and hyacinths.

Most members of the Lily Family are herbaceous plants with a shallow fibrous root system. A few species of *Aloe* and *Dracaena* are shrubby or even small trees. Characteristic of the family are underground rhizomes or bulbs, storage organs which enable the plant to survive in regions where protracted dry seasons occur. As a rule, these plants have linear undivided leaves which do not show division into petiole and blade. The inflorescences of the family are very diverse. In some genera the flowers are solitary, in others they occur in racemes, while umbels occur in still others. The perianth of the flower has six separate members in two whorls of three, which are very much alike in size, shape, and color. The stamens have conspicuous anthers. The ovary is superior, 3-celled, and bears a single style with a 3-lobed stigma. The fruit is a capsule or a berry.

LIMACON. A higher plane curve, also known as Pascal's snail (named for Stefan Pascal, the father of Blaise Pascal, 1623–62, the famous French philosopher, mathematician, and physicist). Its equation in rectangular coordinates is $(x^2 + y^2 - 2ax)^2 = k^2(x^2 + y^2)$ and in polar coordinates, $r = 2a \cos\theta \pm k$. The curve is closed and symmetric to the X-axis. When $k < 2a$, there is an internal node at the origin and the limaçon cedilla is called hyperbolic. The loop disappears when $k > 2a$, so that a conjugate point exists and the limaçon is now elliptic. The case of $k = 2a$ is the cardioid, with a cusp at the origin. See Fig. 1.

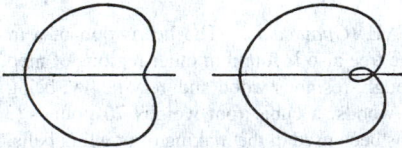

Fig. 1.   Limaçon. In case of diagram at left, $k > 2a$; right, $k < 2a$.

The limaçon (its name comes from Latin, *limax*, snail) can be generated as follows. Let ODM be a circle with diameter $OD = 2a$ on the X-axis. A radius vector OM meets the circle at M and to it is added and subtracted a fixed length $MP = -MP' = k$. The locus of P and P' is the curve.

The various types of limaçons are special cases of the Cartesian oval. See also **Curve (Higher Plane).**

LIMAN CURRENT.   See **Ocean Currents.**

LIMB DARKENING.   The darkening of the limb of the sun or stars is due to the line of sight passing through greater thicknesses of cooler gases at the edge. Limb darkening follows the simple relation

$$I = I_0(1 - u + u \cos\theta)$$

where I is the observed intensity at a point, making the angle θ between the observer, the center of the star, and the point in question; I_0 is the intensity at the center of the disk; and u is the coefficient of limb darkening. See also **Sun (The).**

LIMB INFRARED MONITOR OF THE STRATOSPHERE (LIMS). See **Nimbus Satellite Program.**

LIMB RADIANCE INVERSION RADIOMETER (LRIR).   See **Nimbus Satellite Program.**

LIME AND LIMESTONE. The term *lime* includes a variety of chemicals manufactured from limestone or derived from chemical processes that utilize calcium compounds. According to the composition of the parent limestone, lime may be designated as *high calcium lime* or *dolomitic lime*. Both *quicklimes*, CaO and CaO · MgO, and *hydrated limes*, CaO · H$_2$O, Ca(OH)$_2$ · MgO, and CaO · MgO · 2H$_2$O, are conventionally called lime. Precise terminology requires complex wording, e.g., *dolomitic quicklime* to denote CaO · MgO. The various lime oxides and hydroxides are among the lowest-cost and most widely used sources of alkali for the chemical and metallurgical industries. About 80% of the lime used in the United States is used by the chemical and related industries, mostly as quicklime. About 10% is dead-burned dolomite, and less than 10% goes into construction uses, mostly as hydrate. Very little lime is imported into or exported by the United States.

Limestone

[CAS: 1317-65-3] is a naturally occurring mineral that consists principally of calcium carbonate [CAS: 471-34-1] and variable quantities of magnesium carbonate [CAS: 546-93-0] as a secondary component. Limestone is found in many forms and is classified in terms of its origin, chemical composition, structure, and geological formation as previously mentioned. Deposits are distributed widely throughout the world. Limestone is an essential raw material for many industries, including lime production. Its suitability for a particular application depends profoundly on its physical and chemical properties. *High-calcium limestone* contains 5% or less of MgCO$_3$ and occurs in two mineral forms, calcite and aragonite. See also **Aragonite**; and **Calcite.** *Dolomitic limestone* usually contains over 35% MgCO$_3$, with the remainder CaCO$_3$. See also Dolomite. *Magnesian limestone* is predominantly CaCO$_3$, but contains from 5 to 35% MgCO$_3$. All limestones evolve carbon dioxide and bubble in dilute hydrochloric acid. Dolomite reacts only with dilute HCl, while calcite will decompose in cold dilute HCl.

Limestones vary greatly in color and texture, the latter ranging from dense and hard limestone, e.g., marble or travertine, which can be sawed and polished, to soft, friable forms, e.g., chalk and marl. Chalk is a very fine-grained white limestone, while marl is an impure deposition product that contains clay and sand. Texture, hardness, and porosity appear to be functions of the degree of cementation and consolidation during the formation of these materials. Color variations arise from the presence of impurities. Some impurities, such as sulfur and phosphorus, make limestone unattractive for metallurgical uses.

A high percentage of all limestone is quarried; the balance is mined underground. Although limestone occurs widely, good chemical- and metallurgical-grade limestone is less plentiful. Along the seacoasts, oyster or clam shells are dredged as a source of $CaCO_3$. Limestone is normally processed through a series of crushing, screening, and grinding operations. Because of transportation costs, the proximity of limestone sources to points of use is highly desirable. The major uses of limestone are in construction (asphalt filler, road stone, riprap, and bituminous aggregate); in Portland cement; in agriculture; and in metallurgy.

Precipitated $CaCO_3$ is produced in a number of chemical processes. Sometimes it is economical to dry and calcine the byproduct to regenerate CaO or $Ca(OH)_2$. Some precipitated $CaCO_3$ is made to specific particle size and shape, whiteness, and purity for use as functional filler for paper coatings, paint, and polymers. These products command a premium price as compared with pulverized limestone fillers.

Manufacture of Lime

The basic processes are calcination and hydration. Commencing with high-calcium limestone, the reactions are:

$$CaCO_3 + heat \rightleftharpoons CaO + CO_2 \tag{1}$$

$$CaO + H_2O \rightleftharpoons Ca(OH)_2 + Heat \tag{2}$$

If dolomitic limestone is used, the reactions are:

$$CaCO_3 \cdot MgCO_3 + heat \rightleftharpoons CaO \cdot MgO + 2\,CO_2 \tag{3}$$

$$CaO \cdot MgO + H_2O_{(liq)} \rightleftharpoons Ca(OH)_2 \cdot MgO + heat \tag{4a}$$

or

$$CaO \cdot MgO + 2H_2O_{(gas)} + pres \rightleftharpoons Ca(OH)_2 \cdot Mg(OH)_2 + heat \tag{4b}$$

High-calcium limestone dissociates at 900 °C (1650 °F) in 100% carbon dioxide atmosphere at 1 atm pressure. Under similar conditions, dolomitic limestone dissociates over 727–900 °C (1340–1650 °F). The heat of reaction required to convert $CaCO_3$ to CaO is about 2.8 million Btu per ton (0.64 million kg-Cal/metric ton) of CaO. In practice, heat input may vary from 4 to 10 million Btu/ton (0.9 to 2.3 million kg-Cal/metric ton) of lime. Calcination of limestone particles proceeds by a receding-surface mechanism. To attain reasonable rates of heat transfer into the center of the rock or pebble-sized stone, operating temperatures in lime kilns are 980–1260 °C (1800–2300 °F). Reaction rate is increased and opportunity for recarbonation of the oxide is decreased by rapid removal of CO_2 from the kiln.

Except for very old mixed-feed vertical kilns, lime kilns operate with countercurrent flow of raw material and heat. Modern lime kilns utilize coolers to preheat air by recuperating heat from the hot quicklime. Lime kilns may be fired directly with coal, oil, or gas.

Two major types of lime calciners are the rotary (Fig. 1) and the vertical kiln. In North America, rotary kilns are widely used for lime calcination, whereas in Europe vertical kilns are most popular. Rotary kilns typically have higher output [up to 600 tons (545 metric tons)/day] and lower labor cost. Vertical kilns can be designed for higher fuel efficiency and lower capital investment. They handle down to about $\frac{3}{4}$-in. (19-cm) stone, but normally the rock feed is at least 3×6 in. (7.5 × 15 cm). Rotary kilns can handle down to $\frac{1}{4}$-in. (0.6 cm) stone.

The long-established use of lime is as a structural material in masonry mortars, wall plasters, sand-lime brick, and for soil stabilization. Double-hydrated dolomitic lime or specially processed high-calcium lime mixed with gypsum plaster is troweled on interior walls or ceilings to provide a hard, white, finished surface. It is mixed with cement and sand to make exterior plaster or stucco. Mason's mortar used to lay up bricks or blocks usually contains lime. Lime provides plasticity, water retention, and easy troweling. Sand-lime bricks are more popular in Europe than in the United States. About 10% hydrated lime is mixed with graded sand and water, pressed into shape, and put into autoclaves for 4–8 hours at 150–205 °C (300–400 °F). The reaction product, calcium silicate, results in a strong, white brick. Dead-burned dolomite, formed by calcining dolomite at about 1650 °C (300 °F) to convert MgO to periclase, is used as a *refractory*. Lime is used in the sulfate process for making paper. In the seawater process for producing magnesium metal, lime reacts with $MgCl_2$ to precipitate $Mg(OH)_2$. Calcium metal is made from lime by reducing CaO with coke. In water treatment, hydrated lime can be added to remove temporary hardness, or in the lime-soda process to remove permanent hardness. Dolomitic lime removes silica from boiler feedwater due to silica absorption by $Mg(OH)_2$. For acid neutralization of industrial wastes, lime is widely used.

Additional Reading

Boynton, R.: *Chemistry and Technology of Lime and Limestone*, 2nd Edition, John Wiley & Sons, Inc., New York, NY, 1980.

Gerhartz, W.: *Ullmann's Encyclopedia of Industrial Chemistry*, John Wiley & Sons, Inc., New York, NY, 1987.

Oates, J.: *Lime and Limestone: Chemistry and Technology, Production and Uses*, John Wiley & Sons, Inc., New York, NY, 1999.

Web Reference

National Lime Association: http://www.lime.org/

LIME (Citrus). See **Citrus Trees.**

LIMIT. A finite number s approached by a sequence $\{s_n\}$ if, for every positive number $\in < 0$, there exists a number N for all $n > N$, where $|s_n - s| < \in$. The sequence is then said to converge, or symbolically,

$$s_n \rightarrow s; \quad n \rightarrow \infty$$

or

$$\lim_{n \rightarrow \infty} s_n = s$$

If the limit does not exist, the sequence diverges, but see also **Series.**

The limit of a variable or a function can be defined in a similar way. If the limit is zero, the function is called an infinitesimal. If the function increases without limit, it remains greater than any assigned number however large and it is said to be come infinite or approach infinity. A variable or function that decreases without limit approaches minus infinity.

The following definition is sometimes needed. If $\in$ is a positive number, no matter how small, δ is defined by $0 < (x - a) < \delta$, and $|f(x) - A| < \in$, then A, called the right-hand limit at $x = a$, is symbolized by

$$\lim_{x \longrightarrow a+} f(x)$$

A left-hand limit, $\lim_{x \rightarrow a+} f(x)$ can be defined in a similar way.

See also **Sequence.**

LIMIT SWITCH. An enclosed electromechanical device which makes or breaks electrical circuits when actuated by an external force, such as by a machine member (lever, cam, or dog), or other object when a preselected position of travel or limit is reached. Circuits switched may be for safety stop, position indicating, or be part of a total sequence of automatic or semiautomatic operations. The limit switch is important in many automated systems, such as machine tools, conveyors, and other materials handling equipment. Unlike photoelectric and proximity switches, the limit switch requires physical contact for actuation. These switches are designed to operate reliably for tens of millions of actuations in industrial

Fig. 1. Large rotary lime kiln designed to operate 24 hours per day year around. Rugged construction is required for handling abrasive limestone rock at very high temperature.

environments. The switches are obtainable in numerous current and voltage ratings, ranging from the switching needs of solid-state devices to motor-controlling relays and large solenoids. Voltage requirements may be as high as 600 volts.

Limit switches are actuated in several ways, but the two principal methods are (1) the roller lever switch, and (2) the plunger switch. Low-force actuators, termed *cat-whisker* or *wobblestick*, also are available. See Fig. 1.

See also **Proximity and Object Detectors**.

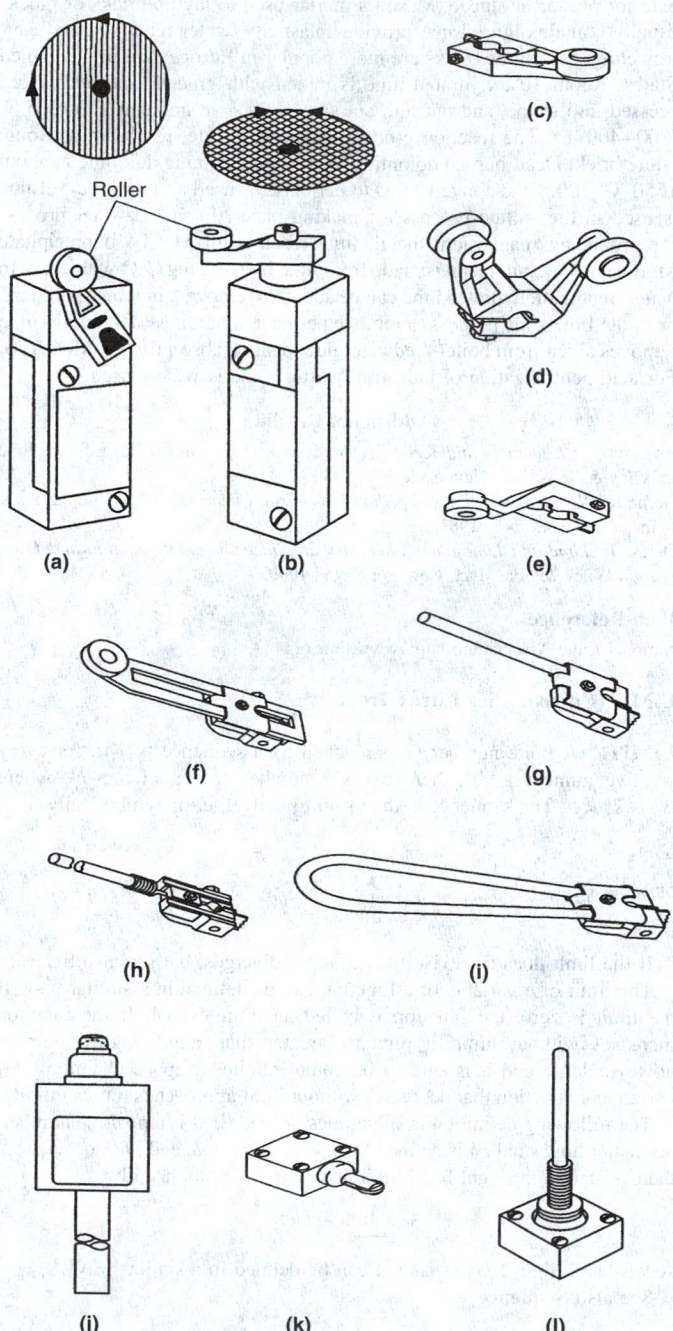

Fig. 1. Various configurations of electromechanical limit switches. (**a**) Side-mounted roller. (**b**) Top-mounted roller. (**c**) Standard roller lever. (**d**) Yoke roller lever. (**e**) Offset roller lever. (**f**) Adjustable-radius roller lever. (**g**) Rod lever. (**h**) Spring-rod lever. (**i**) Flexible loop lever. (**j**) Top plunger. (**k**) Side roller plunger. (**l**) Wobbler switch. (*MICRO SWITCH*.)

LIMNOLOGY. The science of lakes, a synthesis of many disciplines, drawing its specialists from various scientific fields. The field includes the study of inland waters, although it is principally concerned with the physicochemical nature of lakes, their flora and fauna. Stream study has lagged behind lake investigation, although the ecological approach to rivers falls within the realm of limnology.

Lake basins owe their origins to diverse causes, many of which are geologic accidents, or catastrophic. Geologically, lakes are temporary phenomena in geomorphic evolution. Tectonic events have created some of the oldest and deepest lakes of the world: the African rift lakes and Lake Baikal of Siberia, with an ancient, largely endemic fauna. Vulcanism, glacial activities, solution of calcareous substrates, aeolian forces, and even meteoritic impact have created lake basins. In 1957, Hutchinson summarized 76 major categories and 8 subdivisions of these events, which have resulted in lake genesis.

No matter what its origin, a lake is doomed to eventual extinction because of its concave nature and accumulation of autochthonous and allochthonous materials,[1] which gradually obliterate the depression. Thus, a lake passes from a youthful stage to maturity, senility, and extinction. The rate of succession depends upon various factors. For example, introduction of domestic sewage enriches the lake and accelerates the aging process. The youthful lake may be described as *oligotrophic*, the mature lake as *eutrophic*. Many intermediate stages between extreme oligotrophy and extreme eutrophy occur, and the term *mesotrophy* can be applied to them. Senility is characterized by much shallow water and by the conspicuous encroachment of large aquatic plants upon the open water. Extinction often involves a marshy meadow, which is later colonized by plants typical to terrestrial situations. If drainage is poor and the lake is protected from wind, a floating bog mat may close over and eventually obliterate the open water. Bog lakes are acid, or at least circumneutral, and are typified by characteristic marginal vegetation contributing to the floating mat. When calcium content is low in bog-lake waters, decay of organic matter is reduced greatly. Plant fragments from the bog mat accumulate in flocculent layers, and the water may become tea-colored from humic matter. Flocculated humic colloids contribute to bog sediments to form a characteristic deposit termed *dy* by Scandinavian researchers. Under such conditions, nutrients are not recycled by decay, and the lake approaches extinction as a *dystrophic lake*.

In recent years, pollution of lakes has been of major concern. The effect of acid rain on certain lakes, usually at relatively high elevation (alpine lakes) is described in the entry on **Water Pollution**. Accelerated eutrophication, resulting primarily from phosphorus additions due to anthropogenic activities, is generally regarded as one of the major causes of the deterioration of the Great Lakes water quality. A mathematical model of the Great Lakes total phosphorus budgets indicates that a milligram per liter effluent restriction for point sources would result in significant improvement in the trophic status of most of the system. However, because large areas of their drainage basins are devoted to agriculture or are urbanized, western Lake Erie, lower Green Bay, and Saginaw Bay may require non-point source controls to effect significant improvements in their trophic status.

Volcanic lakes are discussed under **Volcano**.

Additional Reading

Bronmark, C. and Lars-Anders Hansson: *The Biology of Lakes and Ponds*, Oxford University Press, Inc., New York, NY, 1998.

Brown, A.C. and A. McLachlan: "Ecology of Sandy Shores," Elsevier, Amsterdam, The Netherlands, 1990.

Ellis, W.S. and D.C. Turnley: "The Aral Sea Lies Dying," *National Geographic*, 73 (February, 1990).

George, D.G.: *Management of Lakes and Reservoirs during Global Climate Change*, Kluwer Academic Publishers, Norwell, MA, 1998.

Imberger, J.: *Physical Limnology*, Kluwer Academic Publishers, Norwell, MA, 2001.

Lampert, W.: *Limnoecology: The Ecology of Lakes and Streams*, Oxford University Press, Inc., New York, NY, 1996.

Lerman, A., D.M. Imboden, and J.R. Gat: *Physics and Chemistry of Lakes*, Springer-Verlag, Inc., New York, NY, 1995.

Niemi, T.M., J.R. Gat, and Z. Ben-Avraham: *The Dead Sea: The Lake and Its Setting*, Oxford University Press, Inc., New York, NY, 1996.

Talling, J.F. and J. Lemoalle: *Ecological Dynamics of Tropical Inland Waters*, Cambridge University Press, New York, NY, 1998.

Wetzel, R.G.: *Limnology*, 3rd Edition, Academic Press, Inc., San Diego, CA, 2001.

[1] Defined in separate entries in this encyclopedia.

Classic References

Cole, G.A.: "Limnology," in *The Encyclopedia of Geochemistry and Environmental Sciences* (R.W. Fairbridge, Ed.), Van Nostrand Reinhold, New York, NY, 1972.

Hutchinson, G.E., published by Wiley, New York: "A Treatise on Limnology." 1957. "Chemistry of Lakes," 1957. "Introduction to Lake Biology and the Limnoplankton," 1966. "Limnological Botany," 1975.

LIMONITE. The mineral limonite, hydrated oxide of iron, corresponds to the formula $Fe_2O_3 \cdot nH_2O$, but is often very impure due to the admixture of sand and clay. It is not found crystallized but grades from loose porous material to compact masses. Its hardness is variable but pure material is 5–5.5; specific gravity is variable at 2.9 to 4.3 (average to above average); usual luster, dull to earthy but may be silky to submetallic; color, various shades of yellowish-brown, sometimes nearly black; streak, yellowish-brown; opaque. Limonite is a secondary mineral from the alteration of various other iron-bearing ores or minerals; it is of widespread occurrence and used both as an ore of iron and as a pigment. Limonite has been formed in marshy and boggy areas and is frequently called bog iron ore. Limonite is an important ore of iron in Lorraine, Luxemburg, Bavaria and Sweden. It is found in Saxony, Austria and England. In the United States, limonite is found particularly in Connecticut, Massachusetts, Pennsylvania, New York, Virginia, Tennessee, Georgia and Alabama, but these deposits are of little economic importance at the present time.

LIMPET *(Mollusca, Gasteropoda).* Marine and fresh-water animals related to the snails, with a low conical shell, not spirally twisted. In the common limpets the shell is solid and in the keyhole limpets it is either notched in front or perforated between that point and the apex. Mollusks of the family *Capilidae*, more closely related to some of the species with coiled shells than to the true limpets, also have shells which are not spiral and are called limpets. One form, *Crucibulum*, is called the cup and saucer limpet and another, *Crepidula*, the boat limpet or slipper shell.

LIND, JAMES (1716–1794). James Lind was a Scottish naval physiian who strongly advocated the curative role of citrus fruits in cases of scurvy, which he established through an innovative clinical trial.

Born in Edinburgh, Lind attended a local grammar school and then was apprenticed to an Edinburgh physician. In 1739 he entered the British navy as a surgeon's mate, becoming a full surgeon in 1747. He left the navy the following year, when he returned to Edinburgh to take his MD and establish himself as a physician. In 1758 he was appointed Physician to the Royal Naval Hospital at Haslar, in the south of England. He retired in 1783.

Lind's active naval service was spent on several ships during the War of Austrian Succession, where cases of scurvy were frequent. Although the protective and curative value of citrus fruits (as well as some other fresh fruits and vegetables) was part of common knowledge among seamen, in 1747 Lind devised a clinical trial to compare citrus fruits with five other possible remedies, including cider and vinegar. Dividing twelve sailors suffering from scurvy into six groups of two, he showed that the two patients receiving citrus fruit — two oranges and a lemon daily — recovered almost immediately. He published his results in his *Treatise of the Scurvy* (1753). He continued to see many cases of scurvy at Haslar, and a later edition of his *Treatise* contains additional material relating to these cases.

Lind also wrote an important general monograph on disease prevention on ships, *An Essay on the Most Effectual Means of Preserving the Health of Seamen in the Royal Navy* (1757), and a pioneering work on tropical medicine, *An Essay on Diseases Incidental to Europeans in Hot Countries* (1768). All of his work is characterized by an Enlightenment concern with common sense and openness.

The Royal Navy did not put Lind's recommendations about citrus fruit into practice until after his death. See also **Ascorbic Acid (Vitamin C)**; and **Nutritional Science (The History)**.

Additional Reading

Carpenter, K.J.: *The History of Scurvy and Vitamin C*, Cambridge University Press, Cambridge, UK, 1986.

Gillispie, C.C.: (1970–1980) *Dictionary of Scientific Biography*, Charles Scribner's Sons, New York, NY, 1981.

W.F. BYNUM, Wellcome Trust Centre for the History of Medicine at UCL, London, UK

LINEAR. An adjective often used in mathematics to describe certain properties. Given the quantities $x_1, x_2, x_3, \ldots, x_n$, a linear combination of them is $a_1x_1 + a_2x_2 + \cdots + a_nx_n = 0$. The quantities x_i are linearly dependent provided all a_i are not zero; otherwise they are linearly independent. The test for such dependence may be made by means of the Gram determinant or the Wronskian.

A linear function is a polynomial of the first degree in its variables and it usually means a polynomial in one variable. Thus, with the linear function, $y = mx + b$, a plot of it would be a straight line of slope m and intercept b on the Y-axis. The general case of a linear algebraic equation would be $a_1x_1 + a_2x_2 + \cdots + a_nx_n = a_0$, where the x_i are variables and the a_i are constants.

A set of simultaneous linear equations is

$$\sum_{j=1}^{n} a_{ij}x_j = b_i$$

$i = 1, 2, \ldots, n$ and a_{ij}, b_i are constants.

A linear differential equation is $A_0(x)y + A_1(x)y' + A_2(x)y'' + \cdots + A_n(x)y^{(n)} = f(x)$ where the $A_i(x)$ are functions of the independent variable only and y', y'', $\ldots$ are the first, second, etc., derivatives. These equations are also inhomogeneous. If the right-hand side is zero in any case, the equation is still linear but homogeneous. Similarly, if all b_i are zero in the case of simultaneous algebraic equations, they are also homogeneous.

See also **Transformation (Mathematics)**.

LINEAR ACCELERATOR. See **Particles (Subatomic)**.

LINEAR ENERGY TRANSFER. The linear energy transfer of charged particles passing through a medium was defined by the ICRU in 1962 as dE_L/dl, where dE_L is the average energy locally imparted to the medium by a charged particle of specified energy in traversing a distance dl. The term locally imparted may refer either to a maximum distance from the track or to a maximum value of discrete energy loss by the particle beyond which losses are no longer considered as local. In either case the limits chosen should be specified. The concept of linear energy transfer is different from that of stopping power. The former refers to energy imparted within a limited volume, the latter to loss of energy regardless of where this energy is absorbed.

LINEAR GRAPH. See **Graph (Mathematics)**.

LINEAR HYPOTHESIS. In statistics, generally, any hypothesis which is linear in the parameters entering into it. More specifically, the term has been used to refer to hypotheses that are linear in the means of a set of normal distributions, many of the simpler statistical hypotheses being capable of being thrown into such a form.

LINEAR INEQUALITIES. A system of relations among variables x_i, possibly including linear equations among them, but also including at least one inequality of the form

$$\sum a_i x_i \geq b$$

(In practice a strict inequality is seldom required.) Such a system may be incompatible (e.g., $x_1 \geq 0$, $x_2 \geq 0$, $-x_1 - x_2 - 1 \geq 0$), may define a unique point (e.g., $x_1 \geq 0$, $x_2 \geq 0$, $-x_1 - x_2 \geq 0$), or else will define a region in space, not necessarily bounded (e.g., $x_1 \geq 0$, $x_2 \geq 0$, $-x_1 - x_2 + 1 \geq 0$ define a bounded region, $x_1 \geq 0$, $x_2 \geq 0$, $x_1 + x_2 - 1 \geq 0$ an unbounded region).

The inequality written above can be replaced by the equivalent pair

$$x_0 + \sum a_i x_i = b_i, \quad x_0 \geq 0$$

and, in general, it is possible to replace a system of inequalities by a system in the special form

$$\mathbf{Ax} = \mathbf{b}, \quad \mathbf{x} \geq 0$$

where $\mathbf{A}$ is a rectangular matrix, $\mathbf{x}$ and $\mathbf{b}$ are vectors.

LINEARITY. With reference to industrial and scientific instruments, the Scientific Apparatus Makers Association defines linearity as the closeness to which a curve approximates a straight line. Linearity is usually measured as a *nonlinearity* and expressed as *linearity*; e.g., a maximum deviation between an average curve and a straight line. The average curve is determined after making two or more full range traverses in each direction. The value of linearity is referred to the output unless otherwise stated. As a performance specification, linearity should be expressed as *independent linearity*, *terminal-based linearity*, or *zero-based linearity*. When expressed simply as linearity, it is assumed to be independent linearity. See also **Conformity**.

Independent Linearity. The maximum deviation of the actual characteristic (average of upscale and downscale readings) from a straight line so positioned as to minimize the maximum deviation. See Fig. 1(a).

Terminal-Based Linearity. The maximum deviation of the actual characteristic (average of upscale and downscale readings) from a straight line coinciding with the actual characteristics at upper and lower range-values. See Fig. 1(b).

Zero-Based Linearity. The maximum deviation of the actual characteristic (average of upscale and downscale readings) from a straight line so positioned as to coincide with the actual characteristic at the lower range-value and to minimize the maximum deviation. See Fig. 1(c).

LINEARITY CONTROL.   1. In cathode-ray tube equipment, an adjustment that tends to correct any distortion in the sawtooth current or voltage waves used for deflection.

2. In television, a control which varies the distribution of scanning speed throughout the trace interval.

LINEARITY THEOREM.   See **Laplace Transform**.

LINEARLY INDEPENDENT VECTORS.   The vectors $A_1, A_2, \ldots, A_m$ are linearly independent if the equation $c_1 A_1 + c_2 A_2 + c_2 A_2 + \cdots + c_m A_m = 0$ implies that $c_1 = c_2 = \cdots = c_m = 0$. Any n-dimensional space contains n linearly independent vectors. They are a base of the vector space. Any other vector can be written as a linear combination of these base vectors. It is convenient to choose the base vectors mutually perpendicular or orthogonal and of unit length. In ordinary three-dimensional space, these vectors are denoted by i, j, k.

LINEAR MAGNIFICATION.   For each optical surface in a system, the linear size of the image, I_i, is to the linear size of the object, O_i, as is the distance of the image, Q_i, to the distance of the object, P_i. Then

$$I_i/O_i = Q_i/P_i = m_i$$

The total linear magnification of the system is then the sum of the products of the magnifications of the parts.

LINEAR PREDICTIVE CODING (LPC).   See **Data Compression**.

LINEAR PROGRAMMING.   A technique of mathematics and operations research for solving certain kinds of problems involving many variables where a best value or set of best values is to be found. This technique is not to be confused with computer programming, although problems using the technique may be programmed on a computer. Linear programming is most likely to be feasible when the quantity to be optimized, sometimes called the objective function, can be stated as a mathematical expression in terms of the various activities within the system, and when this expression is simply proportional to the measure of the activities; i.e., is linear, and when all the restrictions are also linear.

In 1947, mathematician George B. Dantzig devised a linear programming algorithm (the *simplex algorithm*). Long before the advent of modern computers, the method enabled machines to handle complex problems with hundreds of constraints. Over the years, the method has been improved—to the point where it can process problems with 15,000 to 20,000 constraints. Beyond this point, the simplex algorithm may become prohibitively slow and cumbersome. Many of today's problems, especially those of the telecommunications industry, are larger and may reach many

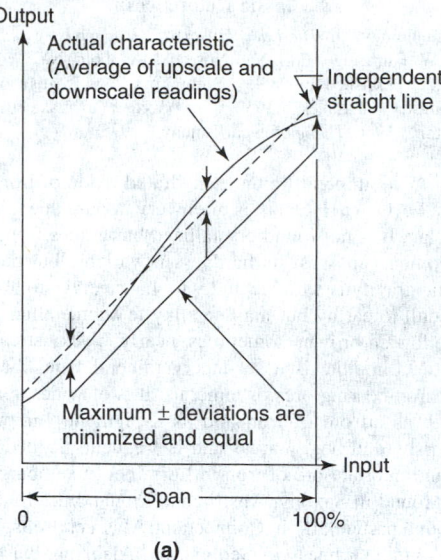

(a)

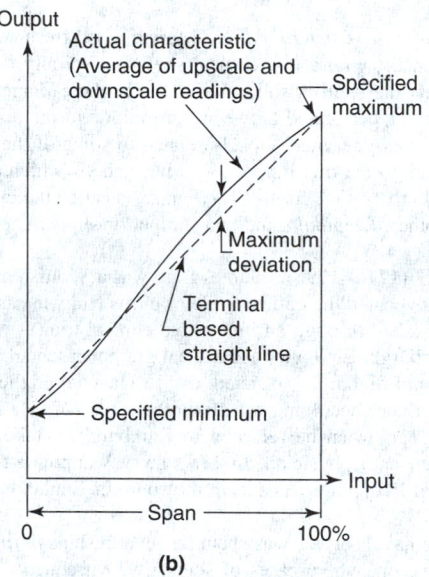

(b)

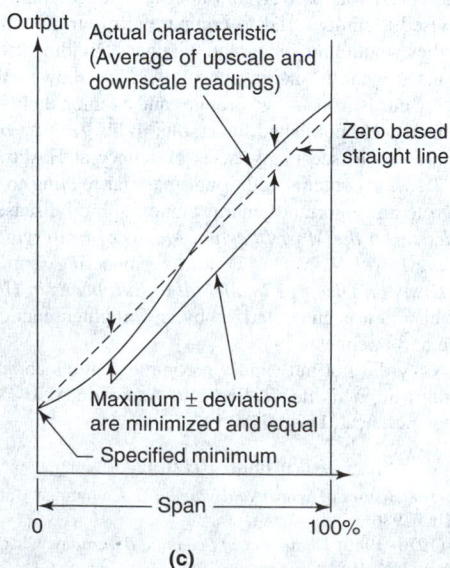

(c)

Fig. 1.   Fundamental relationships pertaining to linearity: (**a**) independent linearity; (**b**) terminal-based linearity; (**c**) zero-based linearity.

tens of thousands or more constraints. A large problem may require 12 to 24 hours or longer of computer running time.

In 1984, Narendra Karmarkar (AT&T Bell Laboratories) demonstrated a new algorithm (called the *Karmarkar* or *AT&T algorithm*) that greatly reduces computation times for many practical problems and it is still being explored to determine its ultimate capabilities and limitations. In an early test of the algorithm, it was applied to the solution of a communication network planning problem involving approximately 42,500 variables and 15,000 constraints. For sake of comparison, the same problem was run with a widely used conventional simplex linear programming package. The results showed the Karmarkar method to be faster than the simplex method by more than an order of magnitude, i.e., every 10 hours of time required by the older method was cut to 1 hour for the new method.

Essence of the New Algorithm. A linear programming problem, such as an overseas communication facilities planning project, can be assembled as a matrix equation that represents variables, constraints, and an objective function (cost) to be optimized. The problem's possible solutions can be envisioned as a geometric shape (a polyhedron, referred to by mathematicians as a *polytope*). The boundary of the polytope is formed from multisided, irregular, flat planes called polygons. Each polygon corresponds to an equation describing a constraint. The optimal solution always lies on one of the vertices (corner points) of the polytope. In the simplex algorithm, the optimal solution is arrived at by starting at one of the vertices of the polytope and "hopping" to the most appropriate adjacent vertex. This vertex must be selected from many available, and must lie in the direction of optimal cost. The process is repeated many times as the computer searches from vertex to vertex for the optimal solution, using the simplex algorithm as its guide ("steering" the computer). The zig-zag path of iterations crosses the surface of the polytope until it ends at the final vertex. Large problems are difficult to solve using the simplex method because a lot of vertices need to be checked, and because movement in the direction of optimal cost is limited to only one vertex at a time. Many thousands of such movements (iterations) are required to solve the problem.

With the Karmarkar algorithm, the lengthy process is cut short, so to speak, by moving through the polytope's interior, finding a much more direct route to the solution. For this kind of problem, the solution does not consist of a single, short answer. The final result defines the optimal value for each of the thousands of variables. It also ensures that the many thousands of constraints have been satisfied as well. This massive solution is typically read onto a disk, from which the facility planner can extract all or part of the information in printed form.

Details and examples of the Karmarkar algorithm can be found in the "Record," 4–13 (March 1986), published by AT&T Bell Laboratories, 150 John F. Kennedy Parkway, Short Hills, New Jersey 07078.

LINEAR SPACE (Vector Space). A generalization of certain algebraic aspects of addition in Euclidean space.

A linear space is a set V with a binary relation called addition making it a group and a multiplication of the elements of V by scalars. The elements of V are called *vectors*. The scalars are usually the real or complex numbers. The subset $x_1, x_2, \ldots, x_n$ of vectors in V is said to be *linearly independent* if the only scalars $\lambda_1, \lambda_2, \ldots, \lambda_n$ for which $\lambda_1 x_1 + \lambda_2 x_2 + \ldots + \lambda_n x_n = 0$ are $\lambda_1 = \lambda_2 = \cdots = \lambda_n = 0$. A subset $x_2, x_2, \ldots x_n$ is said to be a *basis* for V if every vector x in V can be expressed $x = \lambda_1 x_1 + \lambda_2 x_2 + \cdots + \lambda_n x_n$ for scalars $\lambda_1, \lambda_2, \ldots, \lambda_n$. The number of vectors in a linearly independent basis for V is said to be the (linear) dimension of V. The linear dimension of three-dimensional space is three.

If $e_1, e_2 \ldots, e_n$ is a linearly independent basis for V, then each vector in V corresponds to the n-tuple $(\lambda_1, \lambda_2, \ldots, \lambda_n)$ of scalars for which $x = \lambda_1 e_1 + \lambda_2 e_2 + \cdots + \lambda_n e_n$. Conversely, the set of n-tuples of scalars form a linear space.

A mapping L from V to the scalars is said to be a *linear functional* on V if it satisfies: $L(\lambda_1 x_1 + \lambda_2 x_2) = \lambda_1 L(x_1) + \lambda_2 L(x_2)$. The set of linear functionals on V can be made into a linear space in a natural way and this linear space is called the *adjoint* or *conjugate space* of V.

Since many systems describing physical situations (or a first-order approximation) are linear, that is, satisfy the principle of superposition, the space of possible states of the system is a linear space. Also linear

spaces are important in discussing linear partial differential equations such as the wave equation or the heat equation.

See also **Euclidean Space**.

LINEAR SYSTEMS. Systems such that the interrelated quantities comprising the system are related by linear differential or differentio-integral equations. Such equations and therefore such systems obey the principle of superposition, namely, the combined effect of a number of causes acting together is the sum of the effects of the several causes acting separately.

LINEAR TOPOLOGICAL SPACE. A generalization of the algebraic and topological aspects of Euclidean space to infinite dimensional linear spaces.

A linear topological space is a linear space X with scalars either the real or complex numbers along with a topology $\mathscr{T}$ so that addition of vectors and scalar multiplication are continuous. Elementary examples of linear topological spaces are the Euclidean spaces with the usual topology. More representative examples are Hilbert space and the space of infinitely differentiable functions on some Euclidean space with the topology of uniform convergence of each derivative on closed bounded subsets.

A function $x \rightarrow \|x\|$ from a linear space V to the nonnegative reals is said to be a *norm* if (1) $\|x\| = 0$ implies $x = 0$, (2) $\|\lambda x\| = |\lambda| \|x\|$, and (3) $\|x + y\| \leq \|x\| + \|y\|$ and V is said to be a *normed linear space*. The "metric topology" defined on V by the metric $\rho(x, y) = \|x - y\|$ makes V into a linear topological space. V is said to be a *Banach space* if it is complete relative to the metric $\rho(x, y)$.

The *conjugate space* V^* of a linear topological space V is the set of all continuous linear functionals on V.

A continuous linear operator ϕ is a mapping between linear topological spaces V and W that is continuous and satisfies $\phi(\lambda_1 x_1 + \lambda_2 x_2) = \lambda_1 \phi(x_1) + \lambda_2 \phi(x_2)$.

The study of linear topological spaces provides the theoretical underpinning for the modern theory of distributions or generalized functions. Although used by engineers and physicists during most of this century, generalized functions were first defined in a mathematically rigorous way by L. Schwartz in the late forties to be the linear functionals on the space of infinitely differentiable functions.

R.G. Douglas, State University of New York at Stony Brook

LINE (Mathematics). The path described by a moving point. If, as is generally meant, the line is straight its equation in a plane and in rectangular coordinates is $Ax + By + C = 0$, which is a degenerate conic section. Other forms of its equation are: (a) $y = mx + b$, where m is the slope and b the y-intercept; (b) $x/a + y/b = 1$, where a, b are the $x-$, y-intercepts, respectively; (c) $y - y_1 = m(x - x_1)$, where (x_1, y_1) is a point on the line; (d) $(y - y_1)/(y_2 - y_1) = (x - x_1)/(x_2 - x_1)$, where (x_2, y_2) is another point on the line; (e) $x \cos\theta + y \sin\theta = p$, where the perpendicular from the origin to the line has length p and makes an angle θ with the X-axis; (f) $rr_1 \sin(\theta - \theta_1) + r_1 r_2 \sin(\theta_1 - \theta_2) + rr_2 \sin(\theta_2 - \theta) = 0$ where (r, θ) are polar coordinates of any point on a straight line passing through two points (r_1, θ_1) and (r_2, θ_2); (g) $x = x_0 + at$, $y = y_0 + bt$, where t is a parameter for the line of slope b/a which passes through the point (x_0, y_0).

If the straight line is located in a three-dimensional rectangular coordinate system, its equation may be taken as

$$(x - x_1)/\lambda = (y - y_1)/\mu = (z - z_1)/v$$

or

$$(x - x_2)/(x_2 - x_1) = (y - y_1)/(y_2 - y_1) = (z - z_1)/(z_2 - z_1)$$

where (x_1, y_1, z_1) and (x_2, y_2, z_2) are two points on the line and λ, μ, v are its direction cosines or numbers proportional to them. A straight line in space is also determined by the equations of two planes, intersecting to form the given line. Thus, such a line can be defined by the simultaneous equations for two planes, $Ax + By + Cz + D = 0$; $A'x + B'y + C'z + D' = 0$.

See also **Coordinate System**; and **Direction Cosine**.

LINE OF APSIDES.   A line that contains the major axis of an ellipse is known as the line of apsides of the ellipse. In astronomy, the term is used to indicate the line joining perihelion and aphelion points in an orbit and extending to infinity to cut the celestial sphere. See also **Orbit (Astronomy)**.

LINE OF NODES.   The astronomical term applied to a line of intersection of any two fundamental planes. The line of nodes for the moon is the line of intersection of the plane containing the moon's orbit with the plane of the ecliptic. The line of nodes for any member of the solar system, other than satellites, is the line of intersection of the plane of the orbit of the object with the plane of the ecliptic. The line of nodes for the earth is the line of intersection of the plane of the earth's equator with the plane of the ecliptic. For binary stars, the line of nodes is the line caused by the intersection of the plane of the orbit with the plane perpendicular to the line of sight and containing the center of gravity of the system.

LINE OF POSITION.   In navigation any line on the surface of the earth upon which a ship is known to be located. If two or more lines of position are known, the ship must be at their point of intersection. A position determined by the intersection of lines of position is known as a fix.

Lines of position are usually circles, either great or small. In most cases, the distance of the center is so great in comparison with the length of the plotted line that the curvature is not apparent in the plot. In some cases, however, where both points of intersection of two circles of position appear on the plot, a dead-reckoning position will indicate which one is the desired fix. Lines of position are obtained by several methods. See also **Course**; **Navigation**; and **Pilotage (or Piloting)**.

LINE PRINTER.   A printer, often used in conjunction with a computer, which is capable of printing an entire line of characters at one time.

LINE-REVERSAL PYROMETER.   A thermometer for high-temperature gases in which the temperature of a calibrated radiator is adjusted until the spectral areal radiant intensity of its continuum radiation is equal to the intensity of radiation from some suitable characteristic spectral line emitted by the gas. The comparison is made at the wavelength of the spectral line. Seeding is often used to create such a line.

LINE WIDTH.   A measure of the spread in wavelength (or energy) of radiation that is normally characterized by a single wavelength (or energy) value. In practice, the width is usually measured at one-half the maximum intensity of the line. The three phenomena that contribute to line broadening are Doppler broadening, pressure broadening, and the intrinsic level width. Due to the finite resolution of measuring apparatus, a broadening due to the characteristics of the instrument also must be considered.

LINKE SCALE.   A type of cyanometer; an instrument used to measure the blueness of the sky. The Linke scale is simply a set of eight cards are numbered 2 to 16, the odd numbers to be used by the observer if he judges the sky color to lie between any of the given shades. Also called *blue-sky scale*. Sky-blueness study, or cyanometry, is a means of studying atmospheric turbidity.

LINNAEUS, CARL (LINNÉ) (1707–1778).   Linnaeus was a Swedish naturalist whose classification scheme became the basis of modern taxonomy. Carl Linnaeus had a lasting influence on the biological sciences through the adoption of his classification scheme and binomial system of nomenclature for living beings. His classification scheme has now more or less been abandoned but the modern system of biological names, in which the first word indicates the genus and the second word the species, was first brought into common use by him. He intended the double name to serve as an identifying label: 'A plant is completely named if it is provided with a generic name and a specific name' (*Philosophia Botanica*, 1751, p. 219). Nowadays, by international agreement all names date back to the first edition of his Species *Plantarum* (1753), linked with the fifth edition of his *Genera Plantarum* (1754). Many of Linnaeus' original plant and animal specimens are therefore 'type' specimens: i.e. specimens that determine the usage of the name.

Much of Linnaeus' life is embellished with fictional anecdotes, some apparently originating with Linnaeus himself. He was always conscious of his position and wrote several autobiographical commemorations at different points in his life. As a boy, natural history was his favorite study. His father laid out a flower garden and early on introduced him to an appreciation of plants. In 1716, aged 9, he entered school in the nearby cathedral city of Växjö where the noted botanist Johann Rothman encouraged him. In 1727, he entered the University of Lund to study medicine. At that time, medicine included the study of botany especially in relation to plant-derived remedies. Linnaeus became acquainted with Kilian Stobaeus, a medical doctor and polymath, who put his library in Lund at Linnaeus' disposal. In 1728, Linnaeus moved to the medical school in Uppsala, which had a greater reputation than Lund. Here, he found a better-stocked botanic garden and influential patrons, including Olaf Rudbeck the younger, and Olaf Celsius with whom Linnaeus studied the local flora. He earned his living by giving botanical lectures and demonstrations for Rudbeck, soon widening these to include lectures on minerals and assaying (chemical testing for ores). He wrote a number of useful guides for local botany students which were set out in the form of easily memorized aphorisms, rather like a religious catechism. In these early days in Uppsala, Linnaeus also introduced a system of guided field trips for students and other interested people.

In 1735, Linnaeus left Sweden to obtain his MD in Holland, the usual route for Swedes to enter medical practice, but for him also a way of opening doors to the European community of natural philosophers. He spent only 8 days at the University of Harderwijk defending his thesis on the causes of ague. After receiving his degree he went to Leiden, by far the bigger intellectual centre. He stayed in the Netherlands for 3 years, a period of uninterrupted success. He mostly lived and worked in the household of the wealthy merchant George Clifford, the owner of an incomparable botanic garden at the Hartekamp near Haarlem, serving as the superintendent of the garden until 1736. During this time he produced a beautiful folio, *Hortus Cliffordianus* (1736), celebrating the rare plants in Clifford's garden introduced from the Dutch East Indies and North America. He also met influential patrons including Hermann Boerhaave. During a short visit to England in 1736 he met Sir Hans Sloane and JJ Dillenius.

Linnaeus' most significant publication was published here in the Netherlands, the *Systema naturae* (1735). It was very brief, only seven pages. In this tract, he presented a schematic arrangement of a new classification for animals, plants and minerals. The plants were arranged according to what became known as the "sexual system," the first widely distributed account of this innovative scheme. Ambitious and dictatorial, Linnaeus knew he was proposing radical changes in botanical practice. He produced a flurry of other botanical works that supplied rules for identification and arrangement. In *Fundamenta botanica* (1736) he spelled out principles for reformulating botanical theory. In *Critica botanica* (1737) he redesigned plant nomenclature. In *Classes plantarum* (1738) he reviewed the classification schemes of earlier botanists from Cesalpino's time. In *Genera plantarum* (1737) he redefined all the 935 plant genera known at the time. On the crest of an intellectual wave, Linnaeus set out for Paris in 1738. There he met the famous botanical brothers Antoine, Joseph and Bernard de Jussieu.

Although his fame was spreading in Europe, there was no immediate recognition for Linnaeus in Sweden. Between 1738 and 1742 he worked as a doctor in Stockholm, specializing in syphilis and gaining a lucrative aristocratic clientele as well as an appointment to the Swedish navy. He played an active role in Stockholm's scientific circles and in 1739 he was one of the six founders of the Swedish Academy of Science, becoming its first president. Modern party politics were then emerging in Sweden and the new academy emphasized utilitarian sciences that could be harnessed to unite the various parts of the nation. Linnaeus attended to the improvement of the Swedish agrarian state, interesting himself in the acclimatization of introduced species and the development of fisheries, freshwater pearl and silk industries. In economic terms, he was a cameralist, believing in protectionist trade, domestic monopolies and state investment that would modernize productivity by using local or acclimatized natural resources. In his eyes, a polity should aim for economic and political independence. This interest in rational, centralized power structures can also be seen in his biological thought.

After an unseemly fight for the appointment, Linnaeus moved to Uppsala in 1741 to become professor of practical medicine. The following year he exchanged this position for the chair of botany, dietetics and materia medica. He remained in this post for the rest of his life, living in the professorial residence in the botanic garden during the academic term and purchasing a small estate for the summer months at Hammarby. One of his first tasks was to renovate the garden. By the end of his life it was full of innumerable species from all over the world. He also made his home into a museum, crammed with objects, specimens, scientific instruments, stuffed animals, portrait engravings of botanists, shells, ethnographic curiosities from the travels of his students pressed plants, and so forth. The household included parrots and a monkey. Anecdote has it that he furnished some of his rooms with branches in which some 30 species of songbirds nested.

Over succeeding decades Linnaeus built up a centre for botany at Uppsala with himself as its presiding deity. He took many students into his home and heart. Others flocked to his lectures and field excursions. He persuaded a large number of these to obtain the medical doctorate. Ever eager to publish his own ideas, he often wrote their dissertations himself, eventually collected under the title *Amoenitates academicae* (1749–1769). It was then customary for students to defend their professor's ideas in this fashion and even to pay for publication. A shortened English translation was published by two devotees, Benjamin Stillingfleet and FJ Brand, in 1744. On the whole he had a good sense of profitable connections, kept up a useful system of patronage with royalty and corresponded with nearly every botanist of the world. His hope was to receive seeds or specimens for his garden and herbarium. But he also became court physician and in 1762 was elevated to the Swedish nobility as von Linné. He was a member of many foreign scientific societies. One award that pleased him was the Order of the Polar Star, awarded by King Adolf Fredrik in 1753.

Linnaeus' ambition in Uppsala was to complete his plans for the reform of botany. In 1751 he published *Philosophia botanica*, his most influential work but actually an expanded version of *Fundamenta botanica*. He oversaw the publication of numerous regional floras starting with *Flora suecica* in 1745. He did not restrict himself to plants and produced *Fauna suecica* the following year. The remaining inventory of the world's plants was arduous. In 1753 he completed a huge *Species plantarum* which gave the diagnostic characteristics of about 8000 known species. Linnaeus or one of his students had personally examined a large proportion of these. He also expanded and continually updated the *Systema naturae*. It became the bible of natural history knowledge for the period. The tenth edition (1758–1759), which included animal and plant species, ran to more than a thousand pages.

Linnaeus' life closed with several years of clouded mental health. He suffered a fit of apoplexy in 1774 that forced him to retire. Another stroke followed in 1776. He died early in 1778.

Additional Reading

Blunt, W. *Linnaeus: The Compleat Naturalist*, Revised Edition, Princeton University Press, Princeton, NJ, 2001.

Broberg, G.: *Linnaeus: Progress and Prospects in Linnaean Research*, Hunt Botanical Library, Pittsburgh, PA, 1980.

Cain, A.J.: "Logic and Memory in Linnaeus' System of Taxonomy," *Proceedings of the Linnean Society of London*, **169**, 144–163 (1958).

Frangsmyr, T.: *Linnaeus: The Man and His Work*, University of California Press, Berkeley, CA, 1983.

Freer, S.: *Linnaeus' Philosophia Botanica*, (trans), Oxford University Press, Oxford, UK, 2003.

Koerner, L.: *Linnaeus: Nature and Nation*, Harvard University Press, Cambridge, MA, 1999.

Schiebinger, L.: "Why Mammals are Called Mammals: Gender Politics in Eighteenth-century Natural History," *American Historical Review*, **98**, 382–411 (1993).

Smith, J.E.: *A Selection of the Correspondence of Linnaeus and Other Naturalists*, Longman, Hurst, Rees, Orme and Brown, London, UK, 1821.

Stafleu, F.A.: *Linnaeus and the Linnaeans: The Spreading of their Ideas in Systematic Botany, 1735–1789,*: Oosthoek, Utrecht, Netherlands, 1971.

Winsor, M.P.: "Cain on Linnaeus: The Scientist-historian as Unanalysed Entity," *Studies in History and Philosophy of Biological and Biomedical Sciences*, **32C**(2), 239–254 (2001).

JANET BROWNE, Wellcome Trust Centre for the History of Medicine at UCL, London, UK

LINOLEIC ACID. [CAS: 60-33-3]. Also called linolic acid, formula $CH_3(CH_2)_4HC{:}CHCH_2CH{:}CH(CH_2)_7COOH$. This is a polyunsaturated fatty acid (two double bonds) existing in both conjugated and unconjugated forms. It is a plant glyceride essential to the human diet. It is found in linseed oil, safflower oil, and tall oil. See also **Vegetable Oils (Edible)**. At room temperature linoleic acid is a colorless to straw-colored liquid. Specific gravity 0.905 ($15/4\,°C$); mp $-5\,°C$; bp $228\,°C$ (at 14 millimeters pressure). Insoluble in water; soluble in most organic solvents. Combustible. Sources are the oils previously mentioned. Linoleic acid is used in soaps; special driers for protective coatings; emulsifying agents; pharmaceuticals; livestock feeds; and margarine.

LINOLENIC ACID. [CAS: 463-40-1]. Also called 9,12,15-octadecatrienoic acid, formula $CH_3CH_2CH{:}CHCH_2CH{:}CHCH_2CH{:}CH(CH_2)_7COOH$. This is a polyunsaturated fatty acid (three double bonds). It occurs as the glyceride in many seed fats. It is an essential fatty acid in the diet. See also **Vegetable Oils (Edible)**. At room temperature, linolenic acid is a colorless liquid; soluble in most organic solvents; insoluble in water. Specific gravity 0.916 ($20/4\,°C$); mp $-11\,°C$; bp $230\,°C$. Combustible. Linolenic acid finds use in various pharmaceuticals and drying oils.

LION. See **Cats**.

LIOUVILLE EQUATION. In the statistical mechanics of an ensemble of systems, each containing N particles of mass m, it is useful to introduce a density or probability function $P_N(q_N, p_N)$, representing the probability that a system of the ensemble will have its point in phase space fall within the volume bounded by $q_1, q_2, \ldots, q_N, p_1, p_2 \ldots, p_N$, and $q_1 + \delta q_1, q_2 + \delta q_2, \ldots, q_N + \delta q_N, p_1 + \delta p_1, p_2 + \delta p_2, \ldots, p_N + \delta p_N$. If no new systems are created or destroyed, P_N will satisfy the Liouville equation:

$$\frac{\delta P_N}{\delta t} + \sum_{j=1}^{N} \left(\nabla_j \frac{P_N p_j}{N} + \delta_j P_N \dot{p}_j \right) = 0$$

where

$$\nabla_j = \partial/\partial q_j \quad \text{and} \quad \delta_j = \partial/\partial p_j$$

The p-terms are values in generalized coordinates for the positions of the systems, the δ-terms are their momenta, the δ-terms in p and q denote small changes, $\partial P_N/\partial t$ is the partial derivative, and ∇_j and δ_j are vector differential operators. The dot over p_j denotes its first derivative with respect to time.

LIOUVILLE-NEUMANN SERIES. An infinite series

$$\phi(x) = \sum_{n=0}^{\infty} \lambda^n \phi_n(x)$$

which is a unique, continuous solution of a Fredholm integral equation of the second kind. If the nth iterated kernel is defined as

$$K_n(x, z) = \int \int \cdots \int K(x, y_1) K(y_1, y_2) \cdots K(y_{n-1}, z) dy_1 dy_2 \cdots dy_{n-1}$$

then

$$\phi_n(x) = \int K_n(x, z) f(z) \, dz$$

The resolvent or solving kernel is given by

$$K(x, z; \lambda) = \sum_{n=0}^{\infty} \lambda^n K_{n+1}(x, z)$$

hence the solution of the integral equation becomes

$$\phi(x) = f(x) + \lambda \int K(x, z; \lambda) f(z) \, dz$$

Similar methods may be used to solve the Volterra equations.

LIPIDOSES. Disturbances of lipid metabolism in which abnormal deposits of lipids are found in various groups of cells in the body. Primary lipidoses, i.e., those due to an inborn error of metabolism include: (1) *Gaucher's disease*, in which the product accumulated is ceramide glucoside, caused by a beta-glucosidase enzyme defect. In the infantile form, there is mental retardation. In the adult form, there are hepatosplenomegaly and bone changes. (2) *Niemann-Pick disease*, in which the product accumulated is sphingomyelin, caused by a sphingomyelinase enzyme defect. In this disease, there is mental retardation and hepatosplenomegaly. (3) *Tay-Sachs disease*, in which the product accumulated is ganglioside G_{M2}, caused by a hexosaminidase A enzyme defect. In this disease, there is mental retardation and sometimes blindness. (4) *Generalized gangliosidosis*, in which the product accumulated is ganglioside G_{M1}, caused by a beta-galactosidase enzyme defect. In this disease, there is mental retardation and hepatosplenomegaly (enlargement of liver and spleen). In secondary lipidoses there may be abnormal overgrowth of reticulum cells, in which lipids are present. These include the Letterer-Siwe syndrome (destructive lesions chiefly in bone, fatal in infancy), Hand-Schäuller-Christian syndrome (bone lesions with fibrosis of the lungs, diabetes and stunted growth) and the eosinophilic granuloma of bone, a solitary, benign and often self-limiting tumor of children and young adults. See also **Gaucher's Disease.**

Additional Reading

Berkow, R. and M.H. Beers: *The Merck Manual*, 17th Edition, Merck & Company, Inc., Whitehouse Station, NJ, 1999.

Bruyn, G.W. and H.W. Moser: *Neurodystrophies and Neurolipidoses*, Elsevier Science, New York, NY, 1996.

Figueroa, M.L. et al.: "A Less Costly Regimen of Alglucerase to Treat Gaucher's Disease," *N. Eng. J. Med.*, 1632 (December 5, 1992).

Garber, A.M.: "No Price Too High?" *N. Eng. J. Med.*, 1676 (December 3, 1992).

Gatt, S., L. Douste-Blazy, and R. Salvayre: *Lipid Storage Disorders: Biological and Medical Aspects*, Perseus Books, Boulder, CO, 1988.

Schwandt, P.: *Risk Prevention of Arterial Lipidoses*, Warren H. Green, Inc., St. Louis, MO, 1990.

LIPIDS. A heterogenous group of substances which occur ubiquitously in biological materials. They may be categorized as a group by their extractability in nonpolar organic solvents, such as chloroform [CAS: 67-66-3], carbon tetrachloride [CAS: 56-23-5], benzene [CAS: 71-43-2], ether [CAS: 60-29-7], carbon disulfide [CAS: 75-15-0], and petroleum ether. Structural types within the group range from simple, straight-chain hydrocarbon molecules to complex ring structures with varying side chains. A useful classification of the lipids is: (1) fatty acids; (2) neutral fats; (3) phosphatides; (4) glycolipids; (5) aliphatic alcohols and waxes; (6) terpenes; and (7) steroids. See also **Steroids.**

Many lipids, especially the phospholipids, have a strong tendency to form complexes with each other and with various substances. Complex formation is due to the electrostatic attraction of polar groups and to the mutual solubility of the long hydrocarbon chains. Thus, the lipoproteins and proteolipids are complexes of proteins and a variety of lipids, such as cholesterol, phospholipids, glycerides, and glycolipids. The lipids are linked to the proteins by several types of forces. Electrostatic forces, Van der Waals forces, hydrogen bonding, and hydrophobic bonding hold these complexes together. Because of their attraction for water, the polar groups of the protein and phospholipid arrange themselves on the outside of the complex, while the hydrocarbon groups of the lipids are folded into the center. Thus, there is presented to the aqueous phase those groups which have an affinity for water. This arrangement accounts for the solubility of the complexes in water. The phospholipids, owing to their polar groups, act as water solubilizers for the nonpolar lipids. The arrangement may be different in the proteolipids of the brain and nerves, since they are not soluble in water. In these complexes, the lipids may completely envelop the protein.

Knowledge of lipid metabolism has increased at an accelerated rate during the past few decades. The detailed biochemical reactions whereby the fatty acids are synthesized and oxidized; how phospholipids, glycolipids, and cholesterol are synthesized; and how lipids are absorbed and transported have been elucidated. Fatty acids are synthesized from acetyl coenzyme A and malonyl coenzyme A thiol esters. The vitamin biotin plays a vital part in the fixing of carbon dioxide to form malonyl coenzyme A, an important intermediate in fatty acid synthesis. The hormone insulin also favors fatty acid synthesis. The oxidation of fatty acids occurs as their coenzyme A esters in the Krebs cycle of the mitochondria. Cholesterol is biosynthesized from acetyl coenzyme A. Cholesterol in humans is converted to bile acids, fecal sterols, and to steroid hormones. The synthesis of lecithin is mediated via phosphatidic acid and diglyceride precursors. Cytidine nucleotides play a role in the transfer of choline (as phosphorylcholine) to a diglyceride to form lecithin. Uridine nucleotides act to transfer sugar residues in the synthesis of glycolipids.

The transport of lipids in the blood plasma is effected by complex formation with proteins to yield lipoproteins. The liver is the major organ for the synthesis of the lipoproteins. Analysis of serum lipoprotein patterns is important in the understanding of vascular disease (atherosclerosis). The clearing of lipemic blood, such as may occur after a heavy fat meal, is brought about by enzyme known as lipoprotein lipase. This enzyme yields free fatty acids which combine immediately with the plasma albumin to form complexes known as NEFA (nonesterified fatty acids). NEFA act as important transport vehicles for transport of triglycerides and the levels of blood NEFA are very sensitive to hormonal control and neural control. Certain hormones, such as epinephrine, stimulate the membrane-bound adenyl cyclase which converts ATP (adenosine triphosphate) to cyclic AMP (adenosine monophosphate). The latter stimulates adipose tissue lipase and mobilizes depot fat.

The excess utilization of lipids and excess oxidation of fatty acids causes an increase in acetoacetic acid in the body. This condition is known as *ketosis* and can lead to *acidosis*. This situation is common in severe diabetes and can occur whenever carbohydrate utilization is severely decreased.

Research continues in an effort to gain a more thorough understanding of how lipoproteins are synthesized; how lipids are arranged and combined with proteins to form cell membranes; what specific role lipids play in transport across cell membranes; how hormones act to regulate lipid metabolism; the biochemical basis of such abnormal lipid metabolic states as Gaucher's disease, Niemann-Pick's disease, etc.; how lipids per se permeate cell membranes; and how many phenotypic lipoproteins occur in serum.

See also **Cholesterol**.

Additional Reading

Akoh, C.C.: *Food Lipids: Chemistry, Nutrition, and Biotechnology*, Marcel Dekker, Inc., New York, NY, 1998.

Akoh, C.C.: *Handbook of Functional Lipids*, Taylor & Francis, Inc., Philadelphia, PA, 2005.

Bornscheuer, U.T.: *Enzymes in Lipid Modification*, John Wiley & Sons, Inc., New York, NY, 2000.

Chang, T.Y. and D.A. Freeman: *Intracellular Cholesterol Trafficking*, Kluwer Academic Publishers, Norwell, MA, 1999.

Daum, G.: *Lipid Metabolism and Membrane Biogenesis*, Springer-Verlag New York, LLC, New York, NY, 2004.

Feher, M.D. and W. Richmond: *Lipids and Lipid Disorders*, Mosby-Year Book, Inc., St. Louis, MO, 1997.

Fox, P., and P. McSweeney: *Advanced Dairy Chemistry: Lipids*, Vol. 2, 3rd Edition, Plenum Publishing Corporation, New York, NY, 2006.

Gatt, S., L. Douste-Blazy, and R. Salvayre: *Lipid Storage Disorders: Biological and Medical Aspects*, Perseus Books, Boulder, CO, 1988.

Gotto, A.M. and H.J. Pownall: *Manual of Lipid Disorders: Reducing the Risk for Coronary Heart Disease*, 2nd Edition, Lippincott Williams & Wilkins, Philadelphia, PA, 1999.

Gurr, J., J. Harwood, K.N. Frayn: *Lipid Chemistry*, 5th Edition, Blackwell Science, Inc., Malden, MA, 2001.

Hainik, T. and V. Passechnik: *Bilayer Lipid Membranes: Structures and Mechanical Properties*, Kluwer Academic Publishers, Norwell, MA, 1995.

Katsaros, J. and T. Gutberlet: *Lipid Bilayers: Structure and Interactions*, Springer-Verlag, Inc., New York, NY, 2000.

Keane, W.F. and B.L. Kasiske: *Lipids and the Kidney*, S. Karger Publishers, Inc., Farmington, CT, 1997.

Tamm, L.K.: *Protein-Lipid Interactions: From Membrane Domains to Cellular Networks*, John Wiley & Sons, Inc., Hoboken, NJ, 2005.

Vance, D. and J. Vance: *Biochemistry of Lipids Lipoproteins and Membranes*, 3rd Edition, Elsevier Science, New York, NY, 1997.

Wrolstad, R.E., S.J. Schwartz, Decker, E.A., and P. Sporns: *Handbook of Food Analytical Chemistry, Water, Proteins, Enzymes, Lipids, and Carbohydrates*, Vol. 1, John Wiley & Sons, Inc., Hoboken, NJ, 2004.

LIPMANN, FRITZ (1899–1986). A German-born biochemist who won the Nobel Prize in 1953 for the discovery of coenzyme A (CoA). He earned doctorates at the University of Berlin in both chemistry and medicine. He worked at Cornell and Harvard Universities. He founded the biochemistry department of Brandeis University and later joined the faculty of Rockefeller University.

LIPOMA. A benign tumor made up of fat cells. Lipomas occur commonly in the subcutaneous tissues about the head and neck. They cause no symptoms.

LIPOSOMES. Liposomes (or synthetic lipid vesicles) form when amphiphiles (surface active molecules with a hydrophilic group and a hydrophobic chain at opposing ends) are exposed to water. Under appropriate conditions of amphiphile to water mass ratio and temperature, amphiphiles are arranged as one or more concentric bilayers (lamellae) alternating with aqueous compartments. In the process of their formation, liposomes entrap solute molecules (e.g. drugs) present in the aqueous medium. Alternatively, lipid-soluble drugs mixed with the amphiphile can be accommodated in the lipid phase to become components of the bilayers. The bilayer structures vary in size (diameter) from about 30 nm to many micrometers and are usually known as small unilamellar vesicles (SUVs), multilamellar vesicles (MLVs) and large unilamellar vesicles (LUVs). Depending on the gel to liquid crystalline phase transition temperature (T_c) of the amphiphile used (i.e. the temperature at which the hydrophobic chains melt), amphiphiles in liposomal bilayers can be in a gel or fluid state at ambient temperature. Moreover, the composition of liposomes can be tailored to include a sterol (e.g., cholesterol), which contributes to bilayer stability and fluidity, and charged amphiphiles, which render bilayers negatively or positively charged. Physical characteristics of liposomes, such as vesicle size and surface charge, lipid composition and bilayer fluidity, play important roles in determining vesicle behavior within the biological milieu and pharmacological activity of the drugs they carry. The structural versatility of liposomes, their innocuous nature and ability to incorporate a plethora of different pharmacologically active molecules have all contributed to their use in the delivery and targeting of drugs in therapeutics. See also **Lipids**.

Methodology

The numerous methods developed so far to meet widely different requirements in liposome-based therapeutics are broadly divided into two categories: those involving physical modification of existing bilayers; and those that depend on the generation of new bilayers by the removal of the lipid-solubilizing agents. Examples of the first category include: the preparation of drug-containing SUVs from drug-containing MLVs by ultrasonic irradiation, where, however, drug entrapment yield is very low; and the preparation of drug-containing liposomes by the dehydration–rehydration method, where preformed "empty" (water-containing) SUVs are freeze-dried in the presence of drug destined for entrapment. Vesicles produced by this method (dehydration–rehydration vesicles; DRVs) are multilamellar and exhibit a high yield of drug entrapment. Another approach to high-yield drug entrapment is based on the remote loading of preformed liposomes. According to the method, neutral molecules shuttle by diffusion through the bilayers into the liposomes, where they not only become charged as a result of the local pH but also become entrapped because the diffusion rates of charged molecules are very low. Alternatively, drugs remain entrapped because of changes in their solubility affected by ions already present within the preformed liposomes.

Methods of the second category include hydration of a dry amphiphile film to produce MLVs with low drug entrapment yield, and the more efficient reverse-phase evaporation method, in which an aqueous phase containing the drug is emulsified together with a solution of amphiphile in ether, followed by evaporation of the solvent to produce LUVs. Most of such methods will produce large liposomes of heterogeneous size, which can be a disadvantage. Size and heterogeneity, however, can both be reduced by a variety of techniques, including sonication, microfluidization, high-pressure homogenization and extrusion. Many of the procedures available have been scaled up successfully for industrial production.

Interactions with the Biological Milieu and Applications

Liposomes were originally described in 1965 by Bangham, et al.: Since then, owing to their similarity with cell membranes, liposomes have served successfully as models for the study of membrane biophysics. Following their introduction in the early 1970s by Gregoriadis and Ryman as a drug delivery system, liposomes were investigated extensively, both in terms of their fate *in vivo* and their application in therapeutics [Gregoriadis and Ryman, 1972]. It was established that intravenously injected liposomes interact with blood opsonins, which cause their removal from the blood circulation at rates that depend on vesicle size, lipid composition and surface charge, to end up (via opsonin recognition by appropriate cell receptors) in the tissues of the reticuloendothelial system, mostly fixed macrophages in the liver and spleen. On the other hand, intramuscularly or subcutaneously injected liposomes end up in the lymphatic system including the lymph nodes draining the injected site. The stability of liposomes in blood or interstitial fluids, vesicle clearance rates and tissue distribution can be controlled to satisfy particular needs by tailoring the structural characteristics of the vesicles. This includes coating of the vesicle surface with ligands (e.g. antibodies, glycoproteins) that exhibit selective affinity for receptors on the surface of target cells, or with polymers, such as polysialic acids and polyethyleneglycol, which render the vesicle surface highly hydrophilic. A hydrophilic surface helps to repel opsonins and thus contributes to longer vesicle circulation times and, therefore, to opportunities of interaction with target cells other than those of the reticuloendothelial system. Accumulated knowledge of the fate of liposomes *in vivo* and its control have resulted in the successful use of the system in a wide range of therapeutic applications where the pharmacokinetics of drugs is optimized by their liposomal carrier, leading, in turn, to improved pharmacological activity. Liposome-based products approved for clinical use include those employed in the treatment of systemic fungal infections (e.g. AmBisome), the treatment of certain cancers, for instance Kaposi sarcoma (DOXIL, DaunoXome), and vaccination against hepatitis A (Epaxol-Berna) and influenza (Infexal Berna V). Most of these applications have relied either on the ability of liposomes to transport drugs into macrophages where target pathogens reside, and to certain tumors with increased blood vessel permeability, or to their ability to bypass tissues (e.g. heart or kidneys) that are sensitive to certain drugs.

Future Developments

Future challenges that could potentially be met include manipulation of the immune system by the use of tailored liposomes, leading to the production of vaccines (both conventional and genetic) with much higher potency and selectivity, *in vivo* targeting to tumors and gene therapy. In all such cases, designer liposomes carrying drugs, vaccines or genes should be able to recognize their cell target, interact with it specifically and deliver their therapeutic content intracellularly, directly into the appropriate target organelle (e.g., the nucleus). See also **Human Gene Therapy**.

Additional Reading

Bangham, A.D., M.M. Standish, and J.C. Watkins: "Diffusion of Univalent Ions Across the Lamellae of Swollen Phospholipids," *Journal of Molecular Biology*, **13**, 238–252 (1965).

Gregoriadis, G., and B.E. Ryman: "Fate of Protein-containing Liposomes Injected into Rats. An Approach to the Treatment of Storage Diseases," *European Journal of Biochemistry*, **24**, 485–491 (1972).

Gregoriadis, G.: *Liposome Technology*, Vol. 1, 3rd Edition, Informa Healthcare, London, UK, 2006.

Gregoriadis, G.: *Liposome Technology: Entrapment of Drugs and Other Materials into Liposomes*, Vol. 2, 3rd Edition, CRC Press, LLC, Boca Raton, FL, 2006.

Gregoriadis, G.: *Liposome Technology: Interactions of Liposomes with the Biological Milieu*, Vol. 3, 3rd Edition, CRC Press, LLC, Boca Raton, FL, 2006.

Lasic, D.D., and D. Papahadjopoulos: *Medical Applications of Liposomes*, Elsevier Science & Technology Books, New York, NY, 1998.

Nastruzzi, C.: *Lipospheres in Drug Targets and Delivery*, CRC Press, LLC, Boca Raton, FL, 2004.

Williams, A.P., and V. Weissig: *Liposomes: A Practical Approach*, 2nd Edition, Oxford University Press, New York, NY, 2003.

GREGORY GREGORIADIS University of London, London, UK

LIPSCOMB, WILLIAM N. (1919–). An American chemist who won the Nobel Prize for chemistry in 1976 for his studies on the structure and bonding mechanisms of boranes. Much of the research concerned structure and function of enzymes and natural products in organic and theoretical chemistry. He studied at the Universities of Kentucky, California, and Minnesota.

LIQUATION. 1. A process of magmatic differentiation believed to take place as a result of the separation of two immiscible liquids from the parent magma.

2. The separation of a more readily fusible substance from a less fusible one by controlled heating.

LIQUEFACTION. See **Natural Gas**.

LIQUID. Matter in a fluid state, but relatively incompressible. An ideal liquid offers no permanent resistance to a shear stress, but is incompressible. It has then a constant volume and incompletely fills any container of less than this volume. A real liquid is appreciably compressible, and the liquid state of a substance might be defined as the denser, and less compressible, phase of the two-phase fluid system which can exist in equilibrium at temperatures below the critical temperature. X-ray diffraction experiments show that, near the melting-point, the molecules of a liquid show a considerable degree of short-range order and that, in small volumes, they are arranged much as in a solid crystal. This crystalline structure persists over volumes comparable with the intermolecular distances, but cannot be traced beyond. This local or short-range order means that the average molecule is at any moment surrounded by a number of molecules occupying nearly the same relative positions as they would in the solid state. The degree of short-range order is described by the radial distribution function.

This concept of a liquid as an imperfect crystal requires that the molecules in a liquid are packed sufficiently loosely for comparatively free movement, i.e., the energy required to move a molecule from a lattice site to a vacant space is not large compared with thermal energies. Under these conditions, shear flow of the liquid resembles closely the high temperature creep of crystalline solids. A number of theories of the liquid state have this concept as their starting point.

With a few exceptions, including helium, the universal phase diagram applies for all pure compounds (see Fig. 1). The triple point is the single point at which all three phases (crystal, liquid, and gas) are in equilibrium. The triple point pressure is normally below atmospheric. Those substances, such as carbon dioxide, where $P_t = 3,885$ millimeters, $T_t = -56.6\,°C$, sublime without melting at atmospheric pressure. From the triple point, the melting curve defines the equilibrium between crystal and liquid, usually rising with small but positive dT/dP, and presumably always with positive dT/dP at sufficiently high P values. The line is believed to extend infinitely without a critical point (it has been followed to $T \cong 16T_c$ for helium, and calculations indicate that hard spheres would show a gas-crystal phase change). The gas–liquid equilibrium line, the vapor pressure curve, has dT/dP always positive and greater than the melting curve. The

vapor pressure curve always ends at a critical point, $P = P_c$, $T = T_c$, above which the liquid and gas phase are no longer distinguishable. Since the liquid can be continuously converted into the gas phase without discontinuous change of properties by any path in the $P - T$ diagram passing above the critical point, there is no definite boundary between liquid and gas. Two liquids of similar molecules are usually soluble in all proportions, but very low solubility is sufficiently common to permit the demonstration of as many as seven separate liquid phases in equilibrium at one temperature and pressure (mercury, gallium, phosphorus, perfluoro-kerosene, water, aniline, and heptane at $50\,°C$, 1 atmosphere).

See also **Liquid State**.

Additional Reading

Fourkas, J.T.: *Liquid Dynamics: Experiment, Simulation, and Theory*, American Chemical Society, Washington, DC, 2002.
Hansen, J.P., and I.R. McDonald: *Theory of Simple Liquids*, 2nd Edition, Elsevier Science & Technology Books, New York, NY, 1990.
March, N.H.: *Chemical Physics of Liquids*, Taylor & Francis, Inc., Philadelphia, PA, 1990.
March, N.H., and M.P. Tosi: *Atomic Dynamics in Liquids*, Dover Publications, Inc., Mineola, NY, 1991.
March, N.H., and M.P. Tosi: *Introduction to Liquid State Physics*, World Scientific Publishing Company, Inc., Riveredge, NJ, 2002.
Stephan, K., and K. Lucas: *Viscosity of Dense Fluids*, Plenum Publishing Corp., New York, NY, 1979.

LIQUID CHROMATOGRAPHY. An analytical method based on separation of the components of a mixture in solution by selective adsorption. All systems include a moving solvent, a means of producing solvent motion (such as gravity or a pump), a means of sample introduction, a fractionating column, and a detector. Innovations in functional systems provide the analytical capability for operating in three separation modes: (1) liquid-liquid: partition in which separations depend on relative solubilities of sample components in two immiscible solvents (one of which is usually water); (2) liquid-solid: adsorption where the differences in polarities of sample components and their relative adsorption on an active surface determine the degree of separation: (3) molecular size separations which depend on the effective molecular size of sample components in solution.

Solvents, often referred to as carriers, include isooctane, methyl ethyl ketone, acetone/chloroform, tetrahydrofuran, hexane, and toluene.

Packing materials in columns of various lengths include silica gel, alumina, glass beads, polystyrene gel, and ion exchange resins.

High-performance liquid chromatography (HPLC) is the term applied to new and more effective instrumental techniques developed in recent years that have greatly increased the scope of this analytical method. It can now be applied to biological as well as chemical research. Among the separations possible are peptides (by reverse phase chromatography), proteins and enzymes (hydrophobic and size exclusion modes of chromatography), amino acids, and inorganic and organometallic compounds (neutral species, including, clusters, by adsorption and size exclusion, and ionic species including coordination compounds). A comparatively recent development is the use of supercritical fluids as solvents, e.g., carbon dioxide and sulfur hexafluoride.

See also **Gas Chromatography**; and **Thin-Layer Chromatography**.

Additional Reading

Kromidas, S.: *Practical Problem Solving in HPLC*, John Wiley & Sons, Inc., New York, NY, 2000.
Patel, D.: *Liquid Chromatography: Essential Data*, John Wiley & Sons, Inc., New York, NY, 1999.
Sadek, P.C.: *HPLC Solvent Guide*, 2nd Edition, John Wiley & Sons, Inc., New York, NY, 2002.
Swadesh, J.K.: *HPLC: Practical and Industrial Applications*, 2nd Edition, CRC Press LLC., Boca Raton, FL, 2000.

LIQUID CRYSTAL DISPLAY TECHNOLOGY. Liquid crystal materials are organic compounds that have dual properties over some temperature range: They flow like viscous fluids, but at the same time, they have many of the optical characteristics of solid crystals. Liquid crystals were first discovered in 1888 by Reinitzer, who noted their dual liquid and crystalline nature.

The liquid crystalline state is called a *mesophase*. It exists between the normal liquid state (also known as the *isotropic phase*), which has no

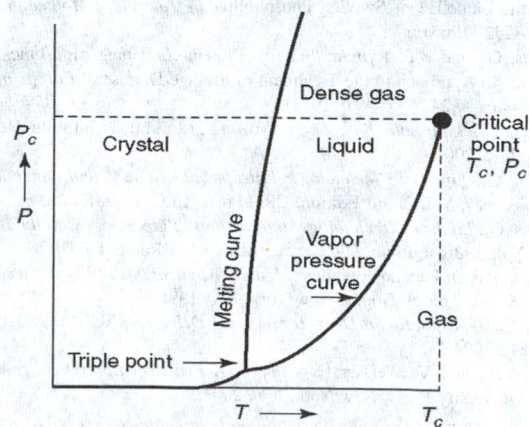

Universal phase diagram

Fig. 1.   Universal phase diagram

positional or orientational order, and the highly ordered solid crystalline state. That is, the liquid crystal phase exists above T_m, the melting point of the solid crystal, and below T_c, the "clearing point" at which the liquid becomes isotropic. Compounds that exhibit liquid crystallinity are called *mesomorphic materials*, and the liquid crystalline state is often referred to by the general term *mesomorphism*. In the mesophase, the molecules spontaneously adopt an ordered arrangement (a process called *self-alignment*), but some movement of the molecules is allowed, which is essential for the functioning of the liquid crystal display (LCD).

These thermal qualities make the operation of LCDs highly temperature dependent. Extremes of temperature render the display inoperative. Even within the temperature range that defines the mesophase, the liquid crystal material becomes more viscous when cooled. This greatly reduces the response time at $0\,°C$ or below, making for a sluggish display, because it is more difficult for the liquid crystal molecules to rotate under the influence of an electric field physically. Furthermore, temperature fluctuations change the dielectric and optical anisotropy of the liquid crystal material. Generally, careful combinations of different liquid crystal compounds produce samples with more-stable temperature characteristics.

Liquid crystal molecules tend to be rod- or disc-shaped. The most important ones for display technologies are the rod-shaped molecules in mesophases called *nematic, smectic*, and *cholesteric*, distinguished by their molecular order. Nearly all of the LCDs made today use materials that exhibit the nematic mesophase, which is covered here in the greatest detail. A bulk sample of a nematic liquid crystal material in its mesophase is a white, opaque liquid that flows easily; it appears cloudy, because the ordering of the molecules occurs only in domains on the order of tens of microns.

The nematic liquid crystal is typically a mixture of various individual compounds, which consist of elongated molecules held together at their ends by polar forces (similar to those that hold the two helical strands of molecules together in DNA (deoxyribonucleic acid, the basic building block of all biological structures). Under a microscope, the liquid crystal material appears to consist of long strands, or threads, that move together like a series of flexible molecular chains. See Fig. 1. This gives liquid crystals the unique property of *cooperative alignment*, which means that the direction of alignment of one molecule influences the alignment of the others in its vicinity. Such preferential alignment gives liquid crystals their *anisotropic* properties; that is, various physical properties, such as the dielectric constant, refractive index, conductivity, viscosity, and magnetic susceptibility, have different values in the directions parallel and perpendicular to the molecular axis. The long axis direction that the molecular chains adopt is primarily determined by the structural nature of the surface on which they lie. However, the alignment of the molecular chains is also influenced by applied electric (or magnetic) fields.

These are the features that enable liquid crystals to be useful in display devices. Liquid crystal materials are highly tailored to have a wide operating temperature range, a fast response time, stability, and an appropriate response to an applied electric field. See also **Liquid Crystals**.

Liquid Crystal Displays

Table 1 summarizes the main types of LCD technology. The categorization is not designed to be comprehensive, but to highlight the key applications

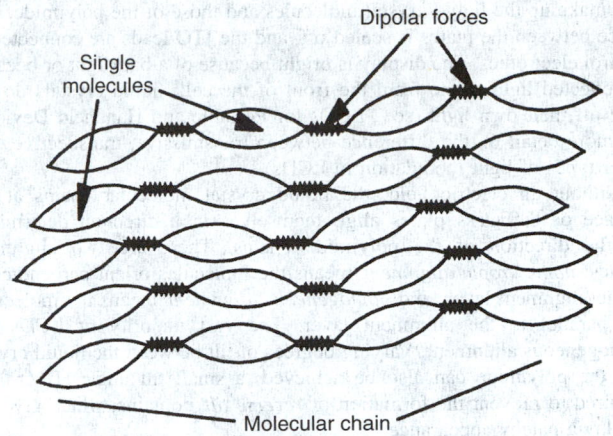

Fig. 1.   Microscopic structure of liquid crystals.

and indicate where they fit into the overall structure. Some categorizations are made on the basis of liquid crystal type, others on addressing technique, depending on which is more relevant to highlight the key applications.

The linear presentation of all of the types of LCDs is complicated by the fact that they can be grouped by addressing technique or by liquid crystal type, and the two groups are partially, but not completely, overlapping. This discussion adheres to a structure based primarily on addressing scheme. All liquid crystal types are covered, but not within one linear section. The liquid crystal types not presented in "LCD Addressing Schemes" are discussed in "Other Types of LCDs."

Basis of LCD Operation

This section outlines the fundamental properties of LCD operation that are common to all types of LCDs. The descriptions are very general because the particular type of liquid crystal or addressing scheme has a major effect on the overall functioning of the display. The most common type of liquid crystal used in displays is the *twisted-nematic* (TN) type, whose operation is described in detail under "LCD Addressing Schemes."

In its simplest form, an LCD is made by creating a sandwich of liquid crystal material several microns thick between two glass plates; this sandwich is sometimes called a *cell*.

A thin, transparent conductive coating, typically indium–tin oxide (ITO), is patterned to form an array of electrodes on each plate. A surface aligning agent, made from a solid polymer material (usually polyimide), is then deposited over the ITO electrodes and buffed so that when the liquid crystal is flowed between the plates, the molecules align in the proper way to react to an applied field. Although still the subject of research, one theory holds that the buffing action produces enough localized heating to realign the polyimide chains with their long axes in the direction of the buffing wheel. When the liquid crystal is applied to such a buffed polymer surface, a single layer of liquid crystal molecular chains becomes anchored to that surface with the chain axis parallel to the direction of alignment in the polymer film due to the strong polar forces between the benzene rings

TABLE 1. MAIN TYPES OF LCD TECHNOLOGY

Liquid Crystal Displays					
Active Matrix				Passive Matrix	
Thin-Film Transistor (TFT)			Liquid-Crystal-on-Silicon (LCOS) (projector)	Supertwisted Nematic (STN)		Twisted Nematic (TN) (watch, calculator)
Amorphous Silicon (notebook computer, desktop monitor)	Polysilicon		CdSe, continuous grain silicon, single crystal silicon	Multiplexed (notebook computer)	Other addressing schemes and alignments (handheld)	
	HTPS (projector, viewfinder)	LTPS (hand-held)				

that make up the liquid crystal molecules and those of the polyimide. The space between the plates is sealed off, and the ITO leads are connected to control electronics. The display is bright because of a backlight or because of reflected light shone from the front of the cell; liquid crystals do not give off their own light (see "LCDs for Portable and Handheld Devices" for more detail on the difference between transmissive, transflective, and other types of light modulation in LCDs).

Without an electric field, the liquid crystal molecular chains at the surface of the glass plates align themselves in a direction determined by the direction of the polyimide chains. There are two alignment modes: *homeotropic alignment* means the molecules orient perpendicular to the alignment layer, and *homogeneous alignment* means the molecules are parallel to the alignment layer. The vast majority of LCDs use homogeneous alignment. Varying degrees of tilt between the liquid crystal and the polyimide can also be achieved; a small tilt angle ($1° \angle 5°$) is required to prevent the formation of *reverse tilt* domains, which give the display a patchy appearance.

The top and bottom plates may have different alignment directions. For example, in a twisted-nematic cell the surface orientation of the polyimide on the bottom plate is rotated 90° from that of the top plate, so through the width of cell, the liquid crystal molecules twist continuously through 90°.

When an external electric field of sufficient strength is applied to the cell, it overrides the surface alignment, and the molecules align uniformly over the entire sample in the direction of the field. The "turn-on" time is 10–15 milliseconds for TN liquid crystals and as short as microseconds for ferroelectric liquid crystals. This alignment occurs because liquid crystal molecules have different dielectric constants, *e*, along the long and short axes. This property is called *dielectric anisotropy* ($\Delta \varepsilon$) and is represented by

$$\Delta \varepsilon = \varepsilon_{\text{parallel}} - \varepsilon_{\text{perpendicular}}$$

Under this electric field excitation, the liquid crystal molecules orient themselves with the axis of greatest dielectric constant parallel to the field in order to be in the lowest (most stable) energy state. Thus, samples with positive dielectric anisotropy, where the dielectric constant is greatest along the long axis, will align parallel to the field (perpendicular to the plates), and samples with negative dielectric anisotropy align perpendicular to the field. When the field is removed, the sample returns to its previous semiordered state over the turn-off time, which is generally slightly longer than the turn-on time.

When the liquid crystal molecules are oriented either by a surface alignment layer or an electric field, the bulk sample has all the optical properties of an anisotropic crystal. Anisotropic crystals have two distinct optical states, which can be viewed using polarized light. Thus, LCDs have a polarizer associated with each glass plate; the second polarizer is often termed the *analyzer*. Light passes through the first polarizer, is modulated by the optical properties of the liquid crystal molecules, then encounters the analyzer. By changing the liquid crystal molecules' optical properties through an applied field, the light can be passed or blocked in a controllable way.

In the (more common) *normally white* (NW) TN cell with positive dielectric anisotropy, the polarizers have their polarization direction parallel to the polyimide alignment direction (recall that the top and bottom plates differ in this direction by 90°). When there is no applied field, light from the backlight rotates its own polarization direction through 90° as it encounters the continuously twisting liquid crystal molecules. In this way, the light passes through the crossed polarizer-analyzer pair when there is no field, so the display looks bright. Under the action of the field, when the molecules are all aligned in a single direction, the light passes straight through the cell and is blocked by the crossed polarizers; the display looks dark. The *normally black* (NB) cell blocks light when the field is off and passes light when the field is on.

This light-modulating effect is the basis for a display because it is a controllable way of producing visible *contrast*. The contrast of a display is an important quality and is defined as the difference in brightness between a fully *on* (bright) state and a fully *off* (dark) state. The process of electronically modulating the cell between the two distinct optical states described above is called optical rotation. There are other optical effects in anisotropic media; for instance, double refraction (birefringence) and optical scattering. They are useful for displays. Optical scattering works by modulating the cell between a transparent state and a scattered

or cloudy state (instead of a totally dark state). Birefringence uses the refractive anisotropy of liquid crystals and is the basis for many enhanced LCDs.

It is important to note that the liquid crystal cell rotates the light because of its anisotropic properties, not because it acts as a polarizer. Optically anisotropic materials have the property of creating elliptically polarized light from incident linearly polarized light (from the front polarizer) that is tilted with respect to one of the refractive axes. The different indices of refraction create a time delay (phase shift) in light propagating along the two orthogonal axes, resulting in elliptically polarized light. This light has a component perpendicular to the original incident polarized light, and so the light appears to have rotated. The amount of rotation is determined by the optical path difference; effectively, light polarized parallel to the incident direction travels a different distance through the material than light polarized perpendicularly. The optical path difference is written $\Delta n \times d$, where Δn, the optical anisotropy, is the difference between the index of refraction parallel to the axis of incoming light (n_{parallel}) and the index of refraction perpendicular to the axis of incoming light ($n_{\text{perpendicular}}$), and where d is the cell thickness. Typical TN cell spacing, *d*, is 7 to 10 microns; typical $\Delta n \times d$ values are 0.5 to 1.5 microns.

The liquid crystal layer in a TN cell cannot be made extremely thin because the refractive index of the liquid crystal material is wavelength dependent; i.e., red light is modulated differently than blue light. This can produce undesirable colors on the display. Cells with a thick liquid crystal layer show less of this effect, but it is noticeable if the cell spacing is thinner than 3 to 4 microns.

Figure 2 shows how liquid crystals respond to the applied electric field. Figure 2 illustrates the voltage/brightness (brightness of light transmitted or reflected) relationship for a typical TN-LCD cell.

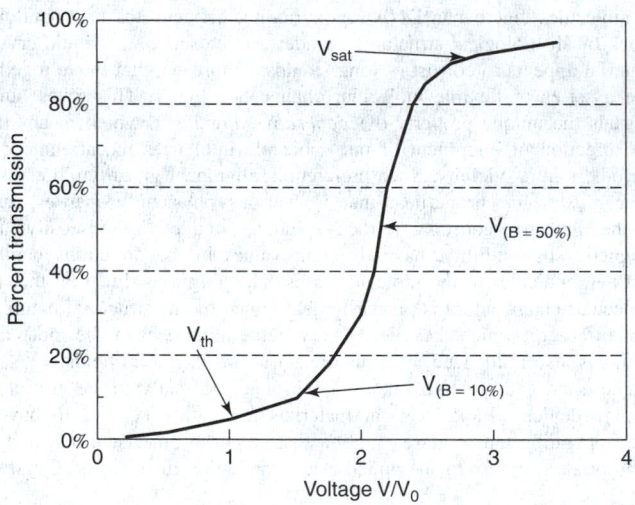

Fig. 2.   Typical liquid crystal response curve.

As shown in Fig. 2, LCDs do not have a linear response; instead, they exhibit a threshold behavior. Significant optical changes do not occur until the voltage exceeds V_{th}, the threshold voltage. Voltages above the saturation voltage, V_{sat}, produce no further darkening. A typical sample has V_{th} of 1 volt and V_{sat} of 2.5 volts. An important evaluation criterion is the performance coefficient, P, which is a measure of the nonlinearity of the display. P is usually defined as

$$P = \frac{V_{(B=50\%)} - V_{(B=10\%)}}{V_{(B=10\%)}},$$

where

$V_{(B=50\%)}$ = the driving voltage required to produce 50% of the maximum display brightness

$V_{(B=10\%)}$ = the driving voltage required to produce 10% of the maximum display brightness

The more nonlinear the response, the smaller the P value will be; a perfect step function will have a P value of zero. P is not simply the slope of the active region, however, but a relationship involving both the slope and the closeness of the transition to the origin. Although increased nonlinearity means better contrast, there is a tradeoff with how much gray scale can be achieved. A display with a step-function response curve can only be black or white; in fact, limited grayscale is a problem for the highly nonlinear ferroelectric liquid crystal displays.

During the 1920s and 1930s, work on liquid crystal materials and the electro-optic effects they produced was conducted in France, Germany, the USSR, and Great Britain. Perhaps the first patent on a light-value device (that is, one controlling the bright and dark states by passing or blocking a backlight) that used liquid crystals was awarded to the Marconi Wireless Telegraph Company (now part of GEC) in 1936. In the mid-1950s researchers at the Westinghouse Research Laboratories discovered that cholesteric liquid crystals could be used as temperature sensors. It was not until the 1960s, however, that serious studies on the materials and the effects of electric fields on them were carried out.

The idea of using liquid crystal materials for display applications was probably first conceived by Richard Williams and George Heilmeier at the David Sarnoff Research Center (then the central research arm of RCA Corporation) in Princeton, New Jersey, in 1963. Later, a larger group, headed by George Heilmeier and including Louis Zanoni, Joel Goldmacher, Lucian Barton, and Joseph Castellano, spearheaded the work to develop LCDs for the fabled "television on a wall" concept, a dream of the late television pioneer David Sarnoff. Between 1964 and 1968, this group discovered many effects, including dynamic scattering, dichroic dye LCDs, and phase change displays.

One of the major breakthroughs was the discovery that by mixing pure nematic liquid crystalline compounds together, it was possible to produce stable, homogeneous liquid crystal solutions that could operate over a broad temperature range. These early materials were composed of Schiff base and ester compounds. Although the Schiff bases worked well in all-glass sealed packages, they were susceptible to breakdown by hydrolysis (from moisture penetration) when less-expensive, nonhermetic plastic seals were used. To eliminate the problem, cyanobiphenyl materials were developed by Gray, Harrison, and Nash in 1973. These materials had improved properties and even broader temperature ranges. Later, in 1977, cyanophenylcyclohexanes, which showed improved operating properties in displays, were developed by researchers at E. Merck & Company in Darmstadt, Germany.

Once the principle of passing and blocking light through the action of liquid crystals was established, the next step was to fabricate a display with many areas whose brightness and darkness could be individually controlled so that words and pictures could be displayed. These areas are of course the *pixels* of modern displays, named for "picture elements."

Twisted-nematic field-effect displays were the first LCDs to undergo widespread commercial development, due to their inherent simplicity and ease of construction. Figure 3 illustrates the construction of a single pixel in a TN-LCD with positive dielectric anisotropy. The alignment direction on the upper glass plate is rotated at an angle of 90° to that on the lower plate; it is a standard normally white pixel as described above (most displays are normally white; the normally black mode is more difficult to control and tends to have lower contrast).

LCD Addressing Schemes

The challenge in making a display with many pixels lies in finding a way to address all of the pixels. The addressing technique affects important image qualities such as grayscale (and hence color), visual contrast, and speed.

The simplest way to address an LCD is by direct addressing, in which each "pixel" (which could actually be a large area) is addressed by sending a direct command to its particular electrodes. Examples of this type of display are low information content segmented displays, such as the 7-electrode numeric displays common in watches and calculators.

To achieve acceptable contrast over a broad viewing angle in applications requiring high information content, one must go beyond direct addressing. There are two general approaches for accomplishing this in LCDs: passive matrix and active matrix. *Passive matrix* refers to improvements in LCD geometry and the use of different physical/optical LCD

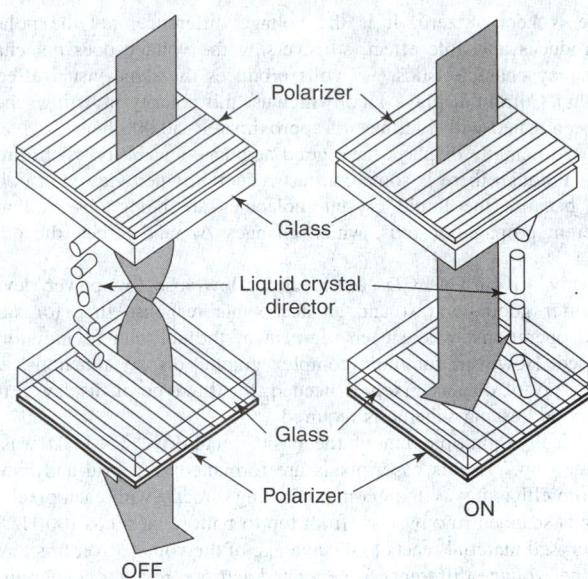

Fig. 3. Single-pixel construction in a TN-LCD.

effects to produce a cell with improved contrast characteristics. Examples include multiplexed displays, as well as liquid crystal devices with memory, such as ferroelectric-smectic and cholesteric devices and cells that incorporate a greater-than-90°-twist (supertwist). Virtually all enhancements use the optical anisotropy of liquid crystal molecules to produce visible change. *Active matrix* refers to the addition of some type of nonlinear switching device, such as a diode or transistor, at the pixel site to improve the nonlinear response of the liquid crystal material. With active matrix devices, the pixel structure itself is relatively unchanged, but the devices can be complex.

The quest for more-efficient passive and active matrix displays has come about because of the limitations inherent in the performance of multiplexed displays. The fundamental property required for a large-area display with a large number of characters or pixels is that it can be written very quickly and produce a high-contrast image over a wide viewing angle. Cathode ray tubes (CRTs) excel in this regard, as the electron beam can address and activate any pixel on the faceplate within milliseconds. The main drawbacks to CRTs are their considerable weight, bulky size, and high power consumption. Passive matrix and, to a greater extent, active matrix LCDs (AMLCDs) are rapidly achieving performance comparable to CRTs, without these undesirable features.

Twisted-Nematic (TN) LCD Operation

There are three main methods to obtain grayscale in TN-LCDs. One is voltage modulation, whereby the cell receives a partial *on* voltage that produces only partial rotation of the liquid crystal molecules, resulting in partial light transmission. Another is spatial modulation, also known as *halftoning*, whereby the density of *on* to *off* pixels in a certain area is changed to create the illusion of a gray level over that area. The third, most common, method is time modulation, or pulse-width modulation. Because of the liquid crystal's finite response time (milliseconds to tens of milliseconds), it responds not to instantaneous voltage but to the time-averaged voltage over approximately its response period. Thus, applying a 10-volt, 1 kHz frequency with a duty cycle of 70% (70% at 10 volts, 30% at 0 volts) means the effective voltage is 7 volts. Applying such pulse-width modulation to individual pixels is an efficient way to produce grayscale.

For a very different reason, time modulation over the entire display is important. The time-averaged voltage across the liquid crystal cell must be zero in order to prevent deterioration by electrolytic effects over time. If the liquid crystal material and aligning layer were perfectly pure, this voltage inversion would not be necessary, but inevitable amounts of trace impurities move toward the plates when a net DC field is applied, ultimately degrading the performance. In all LCD cells, the write voltage is positive one cycle and negative the next, so the time-averaged control voltage is kept at zero. It is the voltage difference, not the polarity, that produces a visible effect, so reversing the voltage does not change any display characteristics (−7 volts produces the same visual effect as +7 volts). Current displays, all of which use this polarity-inverting scheme, have been tested with lifetimes of approximately 50,000 hours.

Twisted-nematic displays have good angular viewability when directly driven, although there is some contrast reduction when viewed at oblique angles because the liquid crystal molecules are being observed along a different propagation axis, which changes Δn and affects the optical rotation.

Directly driven TN-LCDs are ideal for low-cost, low-power devices. They offer good contrast and an acceptable response time for simple readout operations, where each segment of the cell can be individually addressed. However, for more complex graphic display functions, a dot matrix array of display pixels is needed; for this type of display, a more complex addressing scheme is required.

Historically, multiplexing of the basic direct-drive TN-LCD was the "next step up." In this case, pixels are formatted as a grid and handled in a more efficient way than communicating directly with each pixel. The display is scanned row by row, from top to bottom, at 60 to 100 Hz. The liquid crystal material reacts to the average of the voltage over time. When the proper voltage difference is generated across a row and a column, the liquid crystal material at the intersection of these electrodes, which is known as the *selected pixel*, is activated. A result of this arrangement is that the nonselected pixels immediately adjacent to the selected pixels also receive some fraction of the voltage. Thus, the liquid crystal molecules in nonselected pixels are partially aligned by the electric field, reducing the contrast between the *off* and *on* elements of the display. Highly multiplexed TN-LCDs suffer from contrast inversion at certain oblique angles, where the on state pixels become dimmer than the *off* state pixels. This is especially troublesome with grayscale displays, because the different gray levels fade together.

Multiplexed TN-LCD Operation

The most common multiplexing scheme is shown schematically in Figure 4. The horizontal row lines are the *address lines*, and the vertical column lines are the *data lines*. The row and column bus lines are on opposite glass plates to eliminate the possibility of shorts due to manufacturing defects. The driver circuitry sequentially selects the address lines by giving them a nonzero voltage, while the data lines simultaneously send a nonzero select voltage to all pixels in the selected row.

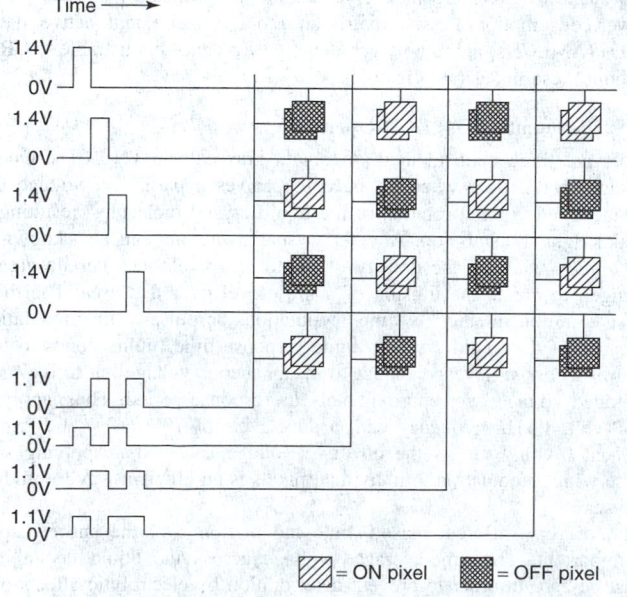

Fig. 4.   Standard Multiplexing Scheme ($V_{th} = 1.5$ volts).

The select voltages from the row and column lines are individually below the threshold voltage, but the sum of the two is greater than the

saturation (on) voltage. Ideally, for a pixel to be on, both its address and data lines must have a select voltage. If either the data line or the address line is in its nonselect state, the pixel voltage is insufficient to make a change in contrast.

With multiplexing, the turn-*on* and turn-*off* times become even more important than in direct addressing. In the TN-LCD shown in Figure 4, each *on* pixel has a select voltage applied to it for only one-quarter of the frame write cycle, meaning it spends the majority of its time having no applied voltage. There is only one-quarter of a cycle for an *on* pixel to distinguish itself from one in the *off* state, not an easy task when the turn-on time is tens of milliseconds. Furthermore, a low temperature will slow the response time even more.

This type of response, in which the molecules respond to the time-averaged voltage (as opposed to the instantaneous or maximum voltage) is known as a *root-mean-square* (RMS) response. A selected pixel, which spends most of its time with its driving voltage below V_{th}, has only a slightly higher RMS voltage than a nonselected pixel. All contrast in the display needs to be produced with only this slight difference in voltage. A highly nonlinear liquid crystal material is required to produce viewable images with adequate contrast; a small difference in $V_{off} - V_{on}$ at the pixel needs to produce a large difference in contrast. The more lines a display has, the smaller $V_{on} - V_{off}$ at the pixel will be, thus the more nonlinear the liquid crystal sample needs to be.

As mentioned earlier, nonselected pixels in the same row or column of a selected pixel will receive partial voltages not intended for them. Simply raising the driving voltages to the rows and columns to the pixel response time will only turn the entire display on. The row and column select voltages must both be under V_{th}, but when added together, they must be above V_{th}. For example, if V_{th} of a liquid crystal material is 1.5 volts, the selected row may be placed at 1.4 volts, while all nonselected rows are kept at zero volts. The selected data lines are supplied with 1.1 volts, while nonselected data lines are kept at zero volts. Thus, every pixel in the selected row, regardless of whether it should be *on* or *off*, has a voltage of at least 1.4 volts. Likewise, every pixel in a selected column has a voltage of at least 1.1 volts. Because these voltages are below threshold, they produce no noticeable change in contrast. Only pixels where both the row and column are selected to produce a voltage difference of 2.5 volts exhibit a change in contrast. But a selected pixel spends only a fraction of the frame time at 2.5 volts; its RMS value will not be significantly higher than that of the nonselect state.

The values for V_{on} and V_{off} are chosen using the results of Alt and Plesko. They showed that the efficiency of multiplexed (and passive matrix) addressing is optimized by choosing

$$V_{on}/V_{off} = [(N^{1/2} + 1)/(N^{1/2} - 1)]^{1/2}$$

where N is the number of rows. Thus, in a multiplexed display with 100 lines, the ratio of V_{on}/V_{off} is only 1.11. This means that if the liquid crystal sample being used has $V_{(B=10\%)} = 2$ volts, then for a contrast ratio of 12:1, the sample must have $V_{(B=50\%)} = 2.22$ volts.

As N increases, the ratio of voltages approaches 1, meaning the display has zero contrast and is not viewable. Hence, this addressing scheme is inherently limited as displays are developed with more and more rows. Liquid crystal cells with low P values (high nonlinearity) are better suited for highly multiplexed displays because they are better able to produce acceptable contrast with only a small voltage change, but even these have increasingly worse performance as N increases.

To achieve a video rate with this addressing scheme, LCDs with 525 lines operating at 30 frames per second must produce adequate contrast with a V_{on} to V_{off} ratio of 1.045. A high definition television, with a line count of 1,100, must produce its contrast with a V_{on} to V_{off} ratio of 1.031 and a line scan time of $1/(1,100 \times 30)$, or 30 microseconds. (Note that the frame update frequency is a function of the IC driver speed and is a separate issue from the liquid crystal response time. The slower of these two factors usually determines overall display speed). Even with advanced liquid crystal materials, however, TN-LCDs have multiplexing capabilities limited to approximately 100 lines.

There are some ways to improve upon the above equation's limitations through display design. For instance, the number of pixels in a multiplexed display can be doubled without increasing the liquid crystal material's nonlinearity. By using a split-screen bus driving scheme, the display is essentially driven as two separate displays, arranged with one half on the top and one half on the bottom. The disadvantage of this approach is that a defect in manufacturing will be much more damaging to final display performance. Also, this approach requires twice the number of column drivers as the regular approach does. A break in a column line will blank out all of the pixels between the break and the end of the column line, while in a conventional multiplexed display (provided that both ends of the columns are attached to the drivers), a break will not affect the display unless two breaks occurred in the same bus line.

Passive Matrix LCD Operation

Supertwisted-nematic (STN) LCDs use available liquid crystal materials, fabrication processes, and drive electronics to achieve wider viewing angles and increased contrast ratios than TN-LCDs, but they require more precise fabrication. Particularly when combined with film compensation, STN-LCDs can produce superior contrast with a neutral color scheme. The drawbacks of relatively slow response time and low color quality have limited the use of STN-LCDs in video applications, although research is continuing to improve this technology.

An STN-LCD cell is drawn schematically in Figure 5. *Supertwist* refers to the fact that the aligning layers are rotated more than 90° (typically between 180 and 270°) relative to each other. The increased twist angle is facilitated by a combination of a high pre-tilt, as much as 35°, and doping of the liquid crystal mixture with a chiral additive of the correct pitch and handedness such that the larger rotation is possible. Care must be taken in the fabrication; if the pre-tilt is insufficient, or if the wrong chiral additive is used, the molecules will rotate in the opposite direction and the display will not function. Increasing the twist angle has the effect of steepening the voltage-contrast response curve shown in Figure 2.

STN-LCDs operate much like TN-LCDs, modulating light through liquid crystal anisotropy such that the effects of the modulation can be observed with two polarizers. Increasing the voltage across the cell rotates the liquid crystal molecules so they are more perpendicular to the plates. By changing the twist angle or the anisotropy of the liquid crystal, and/or by placing the front and rear polarizer at angles other than 90° to each other, different *on* and *off* colors can be attained. By placing the analyzer parallel to the Y-axis in the figure, each pixel will have a faint bluish appearance in the select state and a brighter greenish appearance in the nonselect state. Other common color modes for STN-LCDs include yellow/green,

purple/green, and yellow-green/purple-pink. STN-LCDs operating in these modes have attained contrast ratios of 50:1 multiplexed at 100 lines, a result of the highly nonlinear response curve.

The type of STN-LCD described so far suffers from poor viewing angle dependence. An observer viewing the display at oblique angles sees both a different Δn (optical path difference) and a different d (cell spacing), as with a TN-LCD, but the effect is more dramatic because of the greater nonlinearity. Minimizing oblique-angle distortions requires manufacturing a very uniform gap that is slightly narrower than for a TN-LCD, approximately 4 to 5 microns. Such narrow cell gaps can be difficult to manufacture reliably. Other ideal technical specifications for STN-LCDs include the following:

- Small optical path difference and narrow cell spacing
- Large *bend/splay* ratio (terms related to the material)
- Small twist/splay ratio
- Small $D\varepsilon/\varepsilon_{perpendicular}$
- Large pre-tilt angle (20–35 degrees)
- Uniform cell thickness of 4 to 5 microns

Early STN-LCDs were superior to TN-LCDs in terms of multiplexability and contrast, but they were inferior in that they were not truly black and white. In addition to being aesthetically unpleasant, STN-LCDs increased the likelihood of eyestrain. All the current designs for large-area STN-LCDs employ some form of optical compensation in order to obtain color neutralization. The first acceptable black-and-white displays used an optical compensator, which consisted of an identical liquid crystal cell, except with a twist in the opposite direction of the active cell. However, the liquid crystal compensator cells increased the expense and weight of the display and increased the cell thickness and susceptibility to damage. As a result, it has been replaced by the film compensation type.

Film-Compensated STN (FSTN) LCD. Optical film compensation uses anisotropic organic retardation films (RF) or one or more polymer compensation layers laminated to the display. See Fig. 6. The polymer retardation film is typically a monoaxially stretched polycarbonate or poly(vinyl alcohol) film, approximately 80 microns thick. The film imitates the function of the liquid crystal compensator cell, but without the cost, bulk, weight, and fragility. One difficulty with film compensation is matching the thermal response and oblique viewing characteristics to the active cell. The compensating film needs to be formulated carefully to

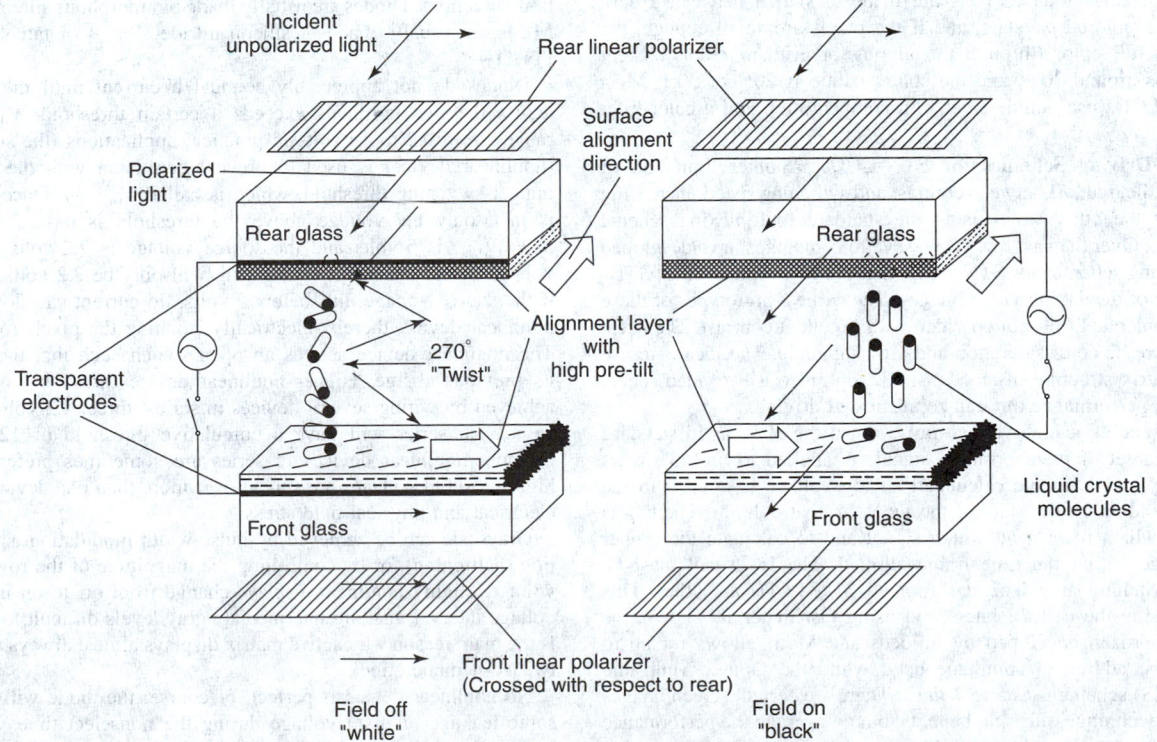

Fig. 5.   Supertwisted-nematic field effect LCD operation.

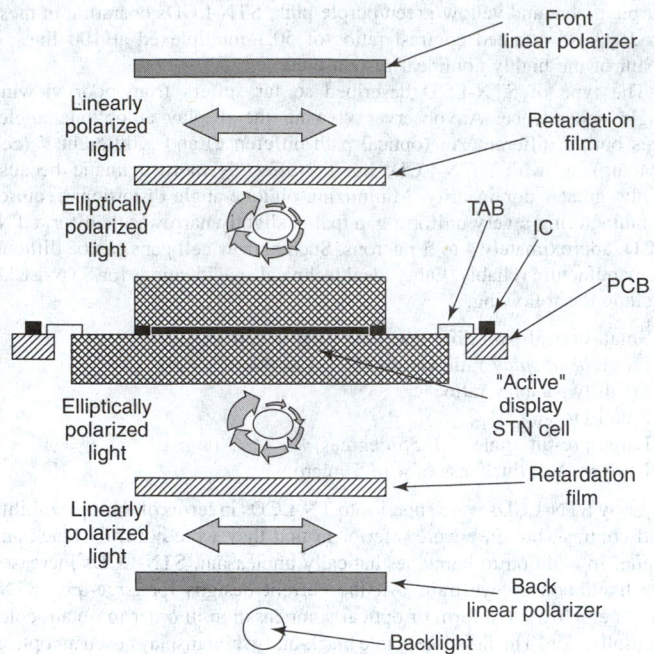

Fig. 6.   Operation and structure of a film-compensated STN-LCD.

compensate for the active STN cell at as many temperatures and viewing angles as possible.

With film-compensated STN-LCDs, issues of alignment of optical axes between the polarizers and film become very complex. Optimizing the relation of a film's optical axis relative to the crossed polarizers is the subject of continuing research. The film is usually sandwiched between the front polarizer and the liquid crystal cell; alternate layerings are possible and will change display performance. These placement issues are even more unwieldy in the double-layer, film-compensated STN.

Color STN-LCDs. The sophisticated compensation techniques for STN-LCDs have resulted in displays with excellent black-and-white quality. This has allowed for the development of color STN-LCDs through the addition of color filters. Pixels are made to switch between a dark *off* state and a colored *on* state, and if the pixels are small enough, the eye perceives full color (through the mixing of primary colors). Light transmission is around 20%, and the contrast ratio is 20:1 or 30:1. Most current STN-LCDs use double-layer film compensation and a color-filter system.

Enhanced Driving Schemes for STN-LCDs. Another limitation of STN-LCDs is the tradeoff between contrast and switching speed: increasing one decreases the other when using the standard multiplexing scheme outlined above. Over the past few years, several companies have developed enhanced driving schemes for STN-LCDs that promise high-contrast STN-LCDs capable of displaying video images. The earliest prototypes of these displays demonstrated full-motion video with excellent contrast. The latest focus is to increase color saturation and viewing angle. The idea is to use the simple matrix structure of an STN-LCD, optimize it for video speed, and adjust the performance through sophisticated driving.

Instead of sending a high-voltage pulse down the rows as in standard multiplexing, a set of predetermined signals is applied to multiple rows simultaneously and a unique calculated analog voltage is applied to the columns, depending on the state of the pixels in that column. The key is to determine which rows need which signal and to calculate the proper column voltage within the time frame allotted. The LCD cell must be capable of switching at video rates (not all STN-LCDs are able). This means decreasing the cell thickness and using fast material. The Active Addressing approach developed by InFocus and Motif allows for up to 255 lines to be addressed simultaneously, while the Optrex Multi-line Selection (MLS) scheme uses 5 to 7 lines. There are certain screen images for which this technique offers no benefits, but on average the performance over standard multiplexing is noticeable. Two basic requirements are that the signals should be practical to generate and that the technique should

produce no artifacts. Also, the entire calculation cycle must fit within the frame cycle.

Active Matrix Addressed LCDs

Active matrix technology is the most broadly pursued technology for increasing the number of addressable pixels in LCDs. Active matrix addressing leaves the basic mode of optical transformation unchanged, but uses some type of nonlinear device external to the liquid crystal cell to enhance the performance of high-information-content displays. The active matrix elements act as switches, electrically isolating a pixel until its input voltage exceeds a certain specified threshold.

Active matrix addressed devices do not respond to RMS time-averaged voltage, but rather to instantaneous peak voltage. Active matrix devices are driven at duty cycles close to 100%. This means the voltage at the pixel remains more or less constant during each frame, as if the pixel were directly driven. During nonaddress times, the nonlinear element has very high impedance, which electrically isolates the pixel from the drive circuitry. The pixel itself acts as a capacitor, storing charge and maintaining voltage throughout the nonaddress period. An additional storage capacitor can be added at the pixel site, in parallel to the pixel, to help maintain charge during nonselect periods.

Active matrix devices consist of two-terminal or three-terminal gain-producing devices. The currently researched two-terminal devices are primarily thin-film diode (TFD) networks (a single diode is unsuitable because of its rectification behavior) and metal-insulator-metal (MIM) devices. Three-terminal devices are almost exclusively thinfilm transistors (TFTs).

Active matrix devices perform much better than almost all passive technologies, but at the cost of increased manufacturing complexity and lower production yields.

Three-terminal devices are generally higher quality than two-terminal types, but they are usually more expensive to manufacture. In a price/performance comparison between passive matrix, two-terminal, and three-terminal types, there is no clear-cut winner. Lower-quality, text-only applications usually can use passive matrix STN-LCDs. High-quality, full-color video applications generally require AMLCDs.

Two-Terminal, Nonlinear Devices. The terms for the two primary types of two-terminal nonlinear devices used in AMLCDs diodes and MIM devices are sometimes used interchangeably, although they are different devices. In this article, a *diode* is a device that passes current in one direction only (rectification), and a *MIM* is a device that passes current in both directions. Diodes are usually made of amorphous silicon (a-Si), while MIMs are usually made of silicon nitride (SiN_x) or tantalum pentoxide (Ta_2O_5).

Diodes do not appreciably conduct a current until the voltage drop across the two terminals exceeds a certain threshold, $V_{th(NLD)}$, which ranges from 0.2 to 15 volts in practical applications (the subscript, NLD [nonlinear device] is used to prevent confusion with the liquid crystal optical switching threshold, which is called $V_{th(LC)}$). Once the threshold is met, only the voltage above the threshold is passed. For example, if $V_{th(NLD)}$ is 5 volts and the source voltage is 7.2 volts, then V_{lc}, the voltage across the liquid crystal cell, will only be 2.2 volts. Furthermore, if the source voltage dips below 5 volts, no current can flow through the nonlinear device, thereby electrically isolating the pixel from the source. The nonlinear device acts as an open switch such that the pixel retains its previous charge. Higher nonlinear device threshold voltages can be achieved by wiring several devices in series; three, 4.2-volt V_{th} nonlinear devices in series will have a cumulative threshold of 12.6 volts. Two or more nonlinear devices in series are sometimes preferable, because high-threshold voltages are spread over more than one device, minimizing electrical and physical field stress.

Grayscale can be achieved by pulse-width modulation, spatial modulation (halftoning), or by controlling the magnitude of the row (data) select voltage. Highly nonlinear displays change from off to on in a very short voltage interval, making intermediate gray levels difficult to achieve. This is the main reason why active matrix displays almost always use the simple twisted-nematic effect.

No nonlinear device is perfect, of course; the diode will always allow some leakage of pixel voltage during the nonselect time. An additional storage capacitor is often placed parallel to the liquid crystal pixel in order to help offset the effect of nonlinear device *off* state leakage current. Adding a capacitor, though, has several undesirable side effects. A capacitor requires extra fabrication steps, and pinhole defects between the capacitor plates can short out a nonlinear device. Furthermore, pixels with capacitors require higher *on* state charging currents, placing a greater load on both the driver circuitry and the nonlinear device *on* state conductance.

Another very important consideration for the active matrix display is conductance symmetry. As described previously, LCDs need to be driven in such a way that there is no net D.C. voltage. Therefore, either the nonlinear device needs to conduct voltages in both directions equally well, or else the driver circuitry needs to compensate for any asymmetrical behavior. Otherwise, the display will exhibit *image sticking* (the same as *ghosting* from the early black-and-white television era), flicker, and rapid aging from electrolytic effects. Much of the current two-terminal device research is aimed at refining the nonlinear device and/or driver circuitry such that it has more balanced response characteristics. The most common means of assuring driving symmetry is to wire together two or more nonlinear devices such that the asymmetries negate each other. In one of the few examples of TFT-LCDs being simpler than two-terminal devices, this symmetry balance is easier to achieve in a TFT-LCD.

Diodes. Two types of thin-film diodes (TFDs) are used in LCDs: Schottky and PIN (positive-doped/insulator/negative-doped). All current TFD networks use a-Si. Much of what is known about a-Si diodes is the result of research into thin-film photovoltaic solar cells, which were first heavily investigated in the 1970s. Although the physics and fabrication techniques were studied and understood during the brief push to develop lowcost solar cells to replace fossil fuel plants, interest dwindled after 1981. It was around this time when research into flat panel displays with thin-film diodes began.

Roughly speaking, diodes act by passing voltage signals of one sign and blocking those of the opposite sign. Because a single diode passes voltage only of one sign, it will convert an A.C. signal into a series of pulses of only one sign, a process called *rectification*. In practice, however, diodes do not abruptly turn on with an applied forward bias or turn off under reverse bias. Rather, they exhibit well-defined transition regions, which do not have sharp corners and are not located precisely at zero volts. It is this threshold behavior, not rectification, that is applied in thin-film devices for LCDs. In addition to forward rectification, many diodes also exhibit a reverse breakdown voltage under higher reverse biases.

A single diode is unsuitable for use in a pixel element because rectification would quickly lead to a net D.C. buildup, damaging the cell. Therefore, diodes are always found in geometries of two or more, such that they cancel out each other's asymmetrical response. The diode networks for each pixel are generally confined to a small physical space to minimize the effects of gradual manufacturing variation over the entire display surface. Diodes that are located far from each other can have different performance characteristics due to variations in doping levels or deposition thickness. State-of-the-art diode networks not only offer superior voltage symmetry but also optimize threshold voltage, *off* pixel current leakage, and *on* state resistance.

A major advantage of using a-Si is that thin-film materials can be deposited at a maximum substrate temperature of less than 300 °C. The economic implications of low-temperature fabrication are substantial. Instead of using polished fused quartz substrates or high-temperature glass, common soda-lime glass or relatively inexpensive alkali-free glass, such as electronics-grade Corning 7059, can be used.

Schottky diodes are formed as a simple junction between an intrinsic (pure) semiconductor and a metal. The capacitance of the diodes must be made much smaller than the capacitance of the liquid crystal cell in order to maximize both the number of addressable lines and the contrast ratio; the leakage of current through the capacitor formed by the diodes creates the limitation on addressability. Schottky diodes also have a reverse breakdown voltage, which means at large reverse voltages (approximately 15 volts) they show a threshold behavior and quickly become conductors. This reverse breakdown voltage is the principle of operation in the back-to-back diode configuration discussed below.

The PIN diodes consist of layers of conductive positive-doped semiconductor, an insulating layer, and a layer of conductive negative-doped semiconductor. The PIN diode is the basis of the a-Si solar cell. However, in a solar cell, the object is to maximize the photon-generated current,

while in a display, the photocurrent must be reduced and the reverse current must remain very low. PIN diodes are formed by plasma-enhanced chemical vapor deposition (PECVD), usually with a Cr electrode above and below. The insulating layer is typically 5,000 angstroms thick, while the P and N layers are 500 angstroms thick. The on/off ratio in such a device can exceed 100:1.

As mentioned previously, a single diode is unsuitable for display applications because of its strongly asymmetrical response. Therefore, diodes must be arranged in a balancing network. The simplest networks consist of two diodes. There are a number of diode geometry variations:

- Back-to-back (BTB) diode structures
- Ring diodes
- Stacked-ring diodes
- Stacked-ring diode equivalent circuits
- Two-diode switches
- Double diode plus reset (D2R)

While some of these have faded into obscurity after some initial effort, Phillips Research Laboratories developed a D^2R circuit that eliminates the extra row electrodes while maintaining the benefits of the two-diode addressing scheme. The number of external connections is the number of rows plus the number of columns plus one. Current D^2R displays show no flicker or crosstalk and a contrast ratio higher than 100:1. The essential feature is that the precision of the two-diode switch is accomplished with a similar level of fabrication complexity (four photomask steps), but without the extra bus lines. However, this bus line reduction comes at the price of a more complicated addressing scheme, and drive voltages must be precisely regulated to avoid net DC current. This technology has been used in production by Philips Flat Display Systems. See also **Organic Light-Emitting Diodes (OLEDs)**; and **Inorganic Light-Emitting Diodes Displays**.

Metal-Insulator-Metal (MIM) Devices. The MIM device as applied to LCDs has a longer history of development than other types of two-terminal devices and also has a simple fabrication process. The MIM device was first investigated as a method of providing threshold control for displays in the late 1970s at Bell Northern Research in Ottawa, Canada. Most this technology is used in production of 2- and 3-inch pocket television displays.

A MIM-addressed display relies on the extremely nonlinear current-voltage characteristics of a very thin layer of semiconducting material placed between two metal plates. Like a diode, the MIM allows current to flow only when a threshold voltage is exceeded. Conduction in this material occurs by field-assisted ionization of localized impurity states in the band gap. It does not conduct at low voltages because the impurity states are filled, neutral, and far from the conduction band. The application of a high field distorts the potential well around the impurity and reduces the barrier to electron flow. The structure, doping, and base material chosen influence the conduction properties significantly. Presently, the two most common MIM materials are anodized tantalum pentoxide (Ta_2O_5) and nonstoichiometric silicon nitride (SiN_x).

One of the most pressing issues in MIM device development is the reduction of parasitic capacitance (unwanted capacitance that occurs as a by product of device geometry). Because MIMs are constructed by sandwiching a nonconductor between two metal electrodes, they produce a very efficient capacitor. This is exacerbated if the insulator has a high dielectric constant. Large MIM capacitance slows response time, because charge that is intended for the liquid crystal pixel capacitor is shared by the MIM capacitor, reducing the amount of charge that finally reaches the liquid crystal pixel.

Tantalum pentoxide (Ta_2O_5) is an especially good material for MIM devices because it exhibits a self-healing process during fabrication and when heated. Due to surface diffusion, pinholes in thin films of tantalum pentoxide are found to diminish after annealing. This property is expected to ease the fabrication of high-quality products at high-volume levels and is probably what allowed MIM displays to be first demonstrated years before any other type of thin-film device addressed display was possible.

There are two types of tantalum pentoxide MIM devices currently under investigation, identified by their structure. One is called a *lateral MIM*, because the active area is grown on the side of a thin-film layer of tantalum. The other type is called *a cross-patterned MIM*; the semiconductor is grown on the surface of a thin film of tantalum. A cross-patterned MIM display is shown in Figure 7.

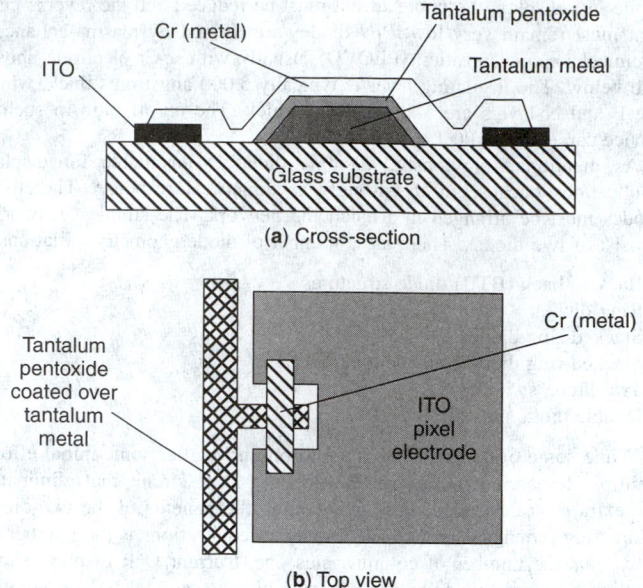

Fig. 7. Metal-insulator-metal (MIM) device (**a**) cross section, (**b**) top view.

Although the cross-patterned MIM is very simple in construction, parasitic capacitance at the intersection of the Cr and Ta conductors through the tantalum pentoxide layer is a serious problem. As noted previously, the larger the capacitance, the more difficult it is to charge the pixel capacitor and, thus, to address each pixel. There are only two straightforward ways to reduce capacitance: increase the thickness of the dielectric layer or decrease the area of the electrodes. The thickness, however, is fairly well defined by the desired voltage characteristics needed to address the display, so the only remaining choice is to reduce the area of the electrodes. This means reducing the line widths of either the Cr or the Ta spur. Both choices increase the likelihood of problems with open circuits and reproducibility of very small relative geometries.

The lateral MIM uses a modified fabrication technique to reduce the area of the capacitor/junction electrodes, resulting in a much lower device capacitance. This is accomplished by anodizing the tapered edge of a thin film of Ta, allowing submicron geometries to be achieved without actually using submicron photolithography. The resulting lateral width of the insulator is less than a micron, so the capacitance of the crossed conductors is small, yet no photolithography below 30-micron accuracy is required (TFTs require 5-micron linewidths).

The lack of any photolithographic patterning below 30 microns is the most significant advantage of the lateral MIM. The only process that needs to be consistently controlled for device performance is the anodization of the Ta electrode to form the insulator. This is a significant advantage over TFTs and even over diodes that require annealing and hydrogenation. The very high nonlinearity of the I-V characteristics and the low leakage through the insulator allow for highly multiplexed displays with performance similar to directly driven displays. The driving voltages for MIMs are necessarily high (15 to 20 volts), due to the conduction mechanism and properties of the insulator film. However, this level is still within the range of highly integrated CMOS drivers.

The University of Stuttgart, Germany, has developed a new geometry for tantalum pentoxide MIMs. This geometry has the low parasitic capacitance of lateral MIMs, can be made with only two photolithography steps, and has independent adjustment of both the gain (nonlinearity) and threshold of the device. The key to this efficient procedure is to use the ITO layer that is already present on the LCD as one of the MIM electrodes. In a later refinement, a second dielectric layer (SiN_x) on top of the Ta_2O_5 was used to further reduce the parasitic capacitance of the MIM without affecting the gain or threshold. The same two-step process was used, except a layer of SiN_x was deposited by RF plasma chemical vapor deposition (CVD) by mixing silane gas (SiH_4) and ammonia (NH_3). The SiN_x and Ta_2O_5 were etched simultaneously. This was used to make a 4.8-inch display with 96×128 pixels, a maximum contrast ratio of 35:1, and a response time

of 20 milliseconds, which is fast enough for television rates. The display had eight gray levels, with excellent uniformity.

An entirely new MIM material is being developed at the Tokyo University of Agriculture and Technology, Tokyo, Japan. It uses a 20-nanometer-thick layer of Langmuir-Blodgett (LB) deposited polyimide film for the insulator. LB deposition is a liquid process that is far less expensive than vacuum deposition and can more easily cover large areas. One of the advantages of LB polyimide is its small dielectric constant (3.2), meaning the MIM parasitic capacitance is kept to a minimum. The LB polyimide MIMs are less nonlinear than other types of MIMs, but early samples had acceptable nonlinearity. A sample pixel has been made with a contrast ratio of 20:1 when driven with a frame frequency of 30 Hz and a duty ratio of 1/100.

Thin-film transistors (TFTs) devices are the most popular and widely investigated three-terminal devices for display applications; more than 30 companies around the world have made significant investments in research and production of this technology. Active matrix LCDs using TFTs provide a combination of performance, form factor, and affordability unmatched by any other display technology, and they dominate the flat-panel display market.

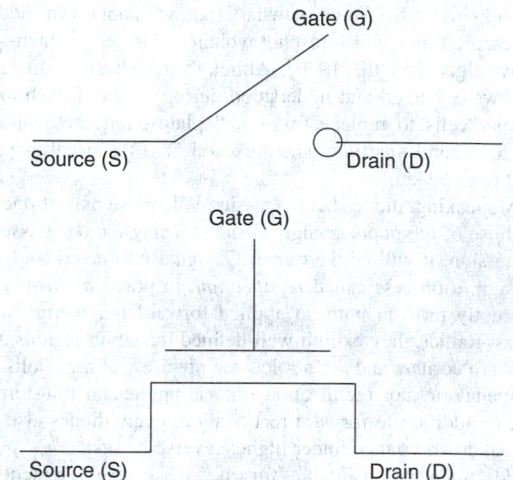

Fig. 8. Schematic of a Thin-Film Transistor. (*Operation as a Switch.*)

A schematic of a TFT is shown in Figure 8; it operates as an electronically controlled switch. By modulating the gate (G) voltage (with respect to ground), the maximum current flow from the source (S) to the drain (D) can be controlled. Larger gate voltage leads to larger maximum source-drain currents. The TFT is a very good conductor between source and drain if the gate voltage is large, and it is a very poor conductor if the gate voltage is small. No matter how large the drain voltage, no current will flow if the gate voltage is near zero. This behavior contrasts with diodes, which have a set threshold voltage.

A schematic of standard TFT layout and addressing is shown in Figure 9. Typically, the rows are connected to the gates, the columns to the source, and the ITO pixel electrode to the drain. However, assigning the source to the columns and the drain to the pixel is somewhat arbitrary; in the case of a TFT, there is symmetry between the source and the drain so that they can be swapped without any change in electrical behavior. (This symmetry does not apply to regular junction and field-effect transistors).

The upper pixel electrodes are all connected to ground. Thus, in contrast to most two-terminal displays, the bottom electrode contains both row and column lines, necessitating crossovers. The liquid crystal cell is usually the standard TN type. A row address line is selected by raising it above the gate saturation voltage; this electrically connects the source and the drain. A pixel can now be written to by raising or lowering the source voltage beyond the switching threshold. To maintain voltage symmetry, the column lines are alternately raised and lowered with successive frame cycles. Because the TFT conducts symmetrically in both directions, this addressing scheme ensures that there is no net-D.C. current. One feature of TFT addressing is that only the column (source) voltages need to be

Time ——→

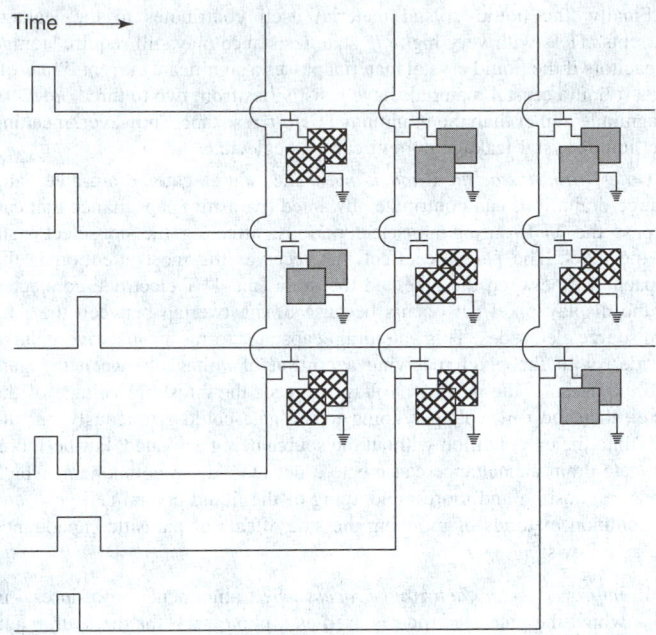

Fig. 9.   Standard TFT layout and addressing.

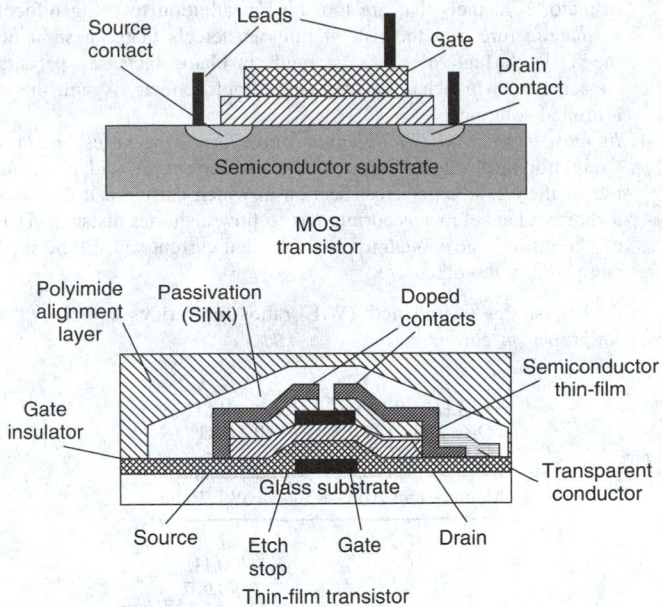

Fig. 10.   Transistor cross sections.

inverted. The row (gate) voltages can have a D.C. bias because they are only control voltages and are not directly attached to the pixel.

Contrast is retained between frame cycles because a nonselected pixel is electrically isolated from the driving circuitry and only loses its charge from small leakage currents. Like two-terminal displays, a small storage capacitor is sometimes added at the pixel site, parallel to the pixel, to help retain charge. Thus, the voltage at the pixel is essentially the same as if it were directly driven.

Operation of TFTs. TFT behavior closely approximates the performance of metal-oxide semiconductor field-effect transistors (MOSFETs), the main difference being the source-drain symmetry noted previously.

However, there is a limitation to the analogy with bulk silicon FET devices, due to the thin-film nature of TFTs. Bulk silicon devices use high-purity, single-crystal materials that exhibit very high mobilities. Thin films, on the other hand, are highly imperfect materials, usually having the same purity level, but lower degrees of crystallinity, especially in the case of a-Si (see "TFT Materials"). Furthermore, the mere thinness of the film acts as a defect, strongly influencing its properties. This limitation should be kept in mind for the remainder of this section.

A typical TFT cross section is shown in Fig. 10, next to a conventional metal oxide semiconductor (MOS) transistor for comparison. Conductive doped-semiconductor contacts are used to ensure ohmic (linear) voltage-current flow. If the metal electrodes were placed directly against the undoped semiconductor, the current flow would be similar to the current flow in a Schottky device, due to the poor matching of the electron energy levels between the very different materials.

The TFT can be thought of as a simple capacitor with the added feature of being controllable. When the gate-source voltage is zero, there is no field between the gate and the channel area. The purified semiconductor has so few free electrons that all current is effectively blocked. The semiconductor electrons are all firmly held in place. The capacitance is effectively infinite.

If, however, the gate voltage is increased such that an upward field is established between the gate and the channel, extra electrons will move from the gate to the channel, facilitating conduction. It is as if the capacitance has decreased to a finite value. The drain current is proportional to the source voltage (that is, ohmic) until the device reaches the saturation voltage and current, at which point all available conduction electrons are being used. Source voltages above saturation will not appreciably increase drain current. The greater the gate voltage, however, the more free electrons will be available and the higher the saturation current will be. The gate controls the supply of current-enabling electrons.

A small reverse-biasing of the gate pulls electrons to the gate plate, which even further reduces conduction from the low value at zero voltage.

Any electrons from the semiconductor lattice with enough thermal energy to break free will be sucked up by the gate before they reach the drain. Further reverse-biasing of the gate voltage, though, increases conductivity by creating holes. In other words, conduction switches from majority carriers (electrons) to minority carriers (holes). See also **Transistor (Invention and Development)**.

TFT Size Limitations. In theory, TFT-LCDs can be built as large as desired; in practice, though, there are limitations that become progressively more severe as the display size increases.

Finite On State Conductance. One of the most important performance criteria of TFT-LCDs is the maximum on current. This current is determined by the source-drain resistance, which must be low enough (that is, the on conductance high enough) that the pixel capacitor can be charged during the row-select time. A higher on current assures this, with the caveat that the gate voltage must also be high (15 to 20 volts) so that the TFT will remain in the linear region and will not be limited by a saturation current. Larger displays have shorter select times for each row of pixels, meaning more current must flow in a shorter amount of time.

The larger the TFT, the larger the on current. However, this simple solution is not a good option because it will greatly increase the parasitic capacitance (see below).

There are four main strategies for ensuring that on state resistance and current requirements have been met. In practice, parasitic effects discussed later will further degrade performance. In large displays these parasitic effects can become highly significant. The methods for increasing source-drain current flow are as follows:

1. *Increase source driving voltage.* A given driver chip has an upper limit to the amount of voltage and current it can practically provide. However, this solution has a limited ability to improve the on current. Even though TFTs have been made to withstand up to 500 volts, the typical maximum high-speed addressing voltage that standard CMOS chips handle is approximately 40 volts.

2. *Increase semiconductor mobility (m).* This second option is also of limited use. Mobility is primarily a function of the semiconductor used in the channel and can only be increased by employing a different semiconductor or by altering its phase. In general, higher mobility semiconductors have higher *off* leakage currents (which is undesirable), offsetting the benefit of using higher-mobility semiconductors. High-mobility semiconductors also usually require higher substrate deposition temperatures.

3. *Increase channel capacitance per unit area (C).* The only ways to increase the capacitance per unit area are to decrease the thickness of the insulating layer or to use a higher dielectric constant

insulator. Channels that are too thin, in addition to being difficult to manufacture, run the risk of pinhole defects that can short out the TFT. A high dielectric strength insulator increases parasitic capacitance, which has its own undesirable effects. Again, this is a limited solution.

4. *Increase aspect ratio (width/length ratio).* This is the easiest and most straightforward way to improve the *on* current. A wider channel means there is a larger cross section in which current can flow, and a shorter channel means current has to flow a shorter distance. TFTs can be made large enough to pass sufficient current and still be small compared to the pixel size.

Table 2 illustrates width/length (W/L) ratios for various mobilities that result in proper *on* current.

TABLE 2. WIDTH/LENGTH RATIOS FOR MOBILITIES RESULTING IN PROPER ON CURRENT

Mobility (cm²/volt-sec)	W/L Ratio
0.2	11
20	0.11
30	0.007

As the table shows, low mobilities require very wide channels relative to the length. The capabilities of the photolithographic processes and the expected yields limit the minimum feature size that can be practically fabricated. Device geometries smaller than a few microns are not economical for any type of large-area devices, including flat panel displays. In designing TFTs, the lower limit for L is typically fixed at 5 microns for practicality, and then the width is adjusted to achieve the correct W/L ratio. A TFT made with a material with a mobility of 0.2 (amorphous Si:H) and with a channel length of 5 microns would have to be 55 microns wide in order to provide the minimum current to drive the cell described previously. The length of the channel runs alongside the pixel. For exceptionally long channels, a serpentine pattern is sometimes used.

TFTs made with high-mobility materials, which could operate with W/L ratios of 0.01 or less, are still subject to the minimum device geometries imposed by fabrication and yield requirements. Using the guideline of a 5-micron-minimum feature size, the channel dimensions for a high-mobility device would be minimized at a 5 × 5 micron square, even though a much shorter channel width would be acceptable with this length. Such a design has problems, however, with *off* state leakage current.

Off State Leakage Current. The property that gives a TFT a high *on* state conductance also gives it a low *off* state resistance, which is undesirable because it allows the pixel charge to leak off the pixel before the next refresh period. There are two sources of leakage: the TFT and the liquid crystal material itself. The visual result of such leakage is reduced contrast.

A general rule is that the charge should not decrease by more than 10% during the interval between frame updates. The most common solution is to place a capacitor in parallel with the liquid crystal cell in order to help maintain the voltage while the display is not being addressed. However, storage capacitors have several drawbacks. First, the capacitor decreases the aperture ratio of the display by including additional inactive area. Second, the addition of another component makes fabrication more complex and increases the potential for yield losses. Large capacitors are also difficult to charge and discharge because of the large currents involved. This places an upper limit on the storage capacitor size.

Integrated drivers fabricated directly on the display substrate can help with leakage current by shortening the link from pixel to driver. High-mobility semiconductor materials are attractive because they can be used for integrated driving circuits mounted directly on the display substrate. However, geometry considerations complicate the matter. High-mobility materials tend to result in channel dimensions that are hundreds of times larger than required to drive the cell, aiding current flow. Low-mobility materials, however, require such large W/L ratios in the driving circuitry that the device dimensions can become too large to be practical. Significant advances in surface-mounted drivers, though, are offsetting some of the need for integrated drivers.

Finally, the liquid crystal material itself contributes to the leakage currents. TFTs with very high *off* state resistance may still require storage capacitors if the liquid crystal material passes a significant current. State-of-the-art liquid crystal materials have resistivities from two to three orders of magnitude higher than the minimum TFT *off* resistance, however, meaning the liquid crystal leakage current can be neglected.

Gate-Drain Parasitic Capacitance. The source-gate, drain-gate, and source-drain links can contribute unwanted (parasitic) capacitance that can bypass the TFT during nonselect periods, affecting the nonselect state of the pixel. The parasitic effect that receives the most attention is the capacitance between the gate and the drain (the TFT electrode connected to the display pixel). It occurs because of the overlap between the gate and source electrodes. This gate-drain capacitor forms a capacitive voltage divider, with the pixel receiving a portion of voltage between the gate and the drain. If the parasitic voltage exceeds the threshold voltage of the pixel, then the row voltages going to the gate could erroneously put the TFT into the on condition without the source being activated. Furthermore, the gate-drain capacitance can create a net D.C. drive component, which leads to ghosting and more rapid aging of the liquid crystal cell.

Common methods of reducing the side effects of parasitic capacitance are as follows:

1. *Improved manufacturing process.* Self-alignment procedures, in which the gate electrode is used as a photomask for the source and drain electrodes, can be accomplished by using more precise steppers that minimize the overlap area.

2. *Lower drive voltage.* Because the drain voltage is proportional to the gate driving voltage, reducing the driving voltage also reduces the voltage shift caused by the unwanted voltage divider. The lower limit for the gate driving voltage is set by the minimum on state current requirement. Greater TFT uniformity allows a lower drive voltage without affecting performance.

3. *New addressing schemes.* Techniques in which nonselected rows adjacent to selected rows are given different voltages than the more remote nonselected rows have been shown to be effective in compensating for the voltage shift from the parasitic capacitance. If the nth row is being selected, then the $(n - 1)_{th}$ row and the $(n + 1)_{th}$ row receive a different voltage than the other nonselect rows. In many of these new addressing schemes, the top-substrate counterelectrode is connected to a variable or nonzero voltage.

4. *Multiple TFTs.* In addition to offering redundancy against defects and lower off state leakage currents, multiple TFTs per pixel can be configured to offset the drain voltage shift. Usually these multiple TFT schemes require a modified driving scheme, like the one described previously. Multiple TFT driving schemes often need additional row or column lines to address the extra TFT terminals, increasing the general complexity of the device.

Uniformity and Other Parasitic Effects. Other parasitic effects are also important, even though they do not receive as much attention as those mentioned previously. One example is poor contacts between the semiconductor channel and the metal electrode. Ideally, the contact should be ohmic, meaning that the current flowing through the interface is linearly proportional to the voltage across it. Furthermore, the interface resistance should be as low as possible.

However, in practice the interface resistance can be high, and it can obey a nonlinear current-voltage law, such as Schottky or Poole-Frenkel conduction. Nonideal contacts, generally caused by such manufacturing imperfections as low-purity materials or contamination between deposition layers, limit the maximum on state current. TFT architectures that use only two or three mask steps have a lower probability of contamination because several thin-film layers are deposited without breaking vacuum. Other solutions include varying the doping amounts of the semiconductor-electrode interface to ensure better matching between the electronic energy levels.

TFTs, like any nonlinear element, must be made with a maximum degree of uniformity. Small variations in the voltage-current characteristics reduce display quality, especially when the display incorporates a large number of gray levels. The above discussion about *on* and *off* state currents makes the tacit assumption that all of the TFTs were equally on or off. In practice, though, small variations in threshold voltage produce above-minimum off state leakage currents and below-maximum on state charge currents. TFTs are known to vary the most when the gate voltage is near the threshold voltage, making threshold-voltage stability especially important.

Increased TFT uniformity also allows lower driving voltages, because there is no longer the need to overcompensate for uncertain threshold or linear-saturation transition voltages. Currently, the threshold is not precisely known, so gate drivers deliver ample voltage to guarantee that the TFT remains in the linear region. However, if this transition were accurately controlled, the driver circuitry could deliver exactly the voltage needed, without requiring extra just to be certain. Lower drive voltages put less strain on the driving circuitry and reduce the crosstalk from gate-drain parasitic capacitance.

The only way to improve device uniformity is through more careful fabrication, although the effects of nonuniformity can be decreased by alternate drive schemes (which will be discussed later). For video rates, uniformity in threshold voltage should be kept below 0.025 volt for small distances and below 0.05 volt for the entire display. Greater nonuniformity leads to visible flicker (at half the drive frequency) because each pixel charges and discharges slightly differently during the positive and negative bias write cycles. There is one positive side to nonuniformity, however: Unless the drive frequency is above 64 Hz, which is rare for current LCDs, the flicker component from nonuniformity will be below the visible threshold of 32 Hz.

Other Performance Issues

Grayscale. For low levels of grayscale (3 bits), the D.C. column data voltage can be adjusted by a digital-to-analog (D/A) converter, with the data voltage height proportional to the gray level. Therefore, the grayscale of the pixel is determined by the D.C. voltage at the pixel source during the address time. But, integrating a 4-bit (16-level) D/A converter into the drive circuitry for each column greatly increases driver chip cost, complexity, and size. One technique for obtaining a 16-level (4-bit) grayscale without using costly D/A converters is to use a ramp generator. When used in a color display, this provides up to 16 shades per color, or 4,096 different colors. The arrangement is shown in Figure 11.

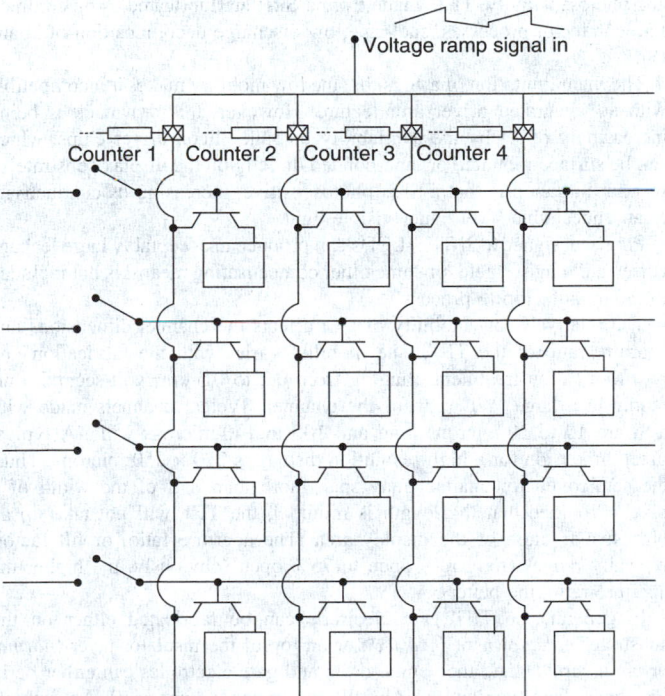

Fig. 11.   Grayscale control using ramp signal voltage.

Each column is connected to a common ramp waveform, which has a period that is equal to the length of time to address one line. The voltage to which the pixel charges is determined by the length of time it is connected to the ramp waveform. Because the display has been set

up so the pixel voltage very closely follows the ramp voltage, the ramp cutoff time determines the gray level of the display. The ramp polarity is reversed each frame to ensure that there is no net D.C. at the pixel site.

Viewing Angle. No matter how advanced active matrix addressing schemes become, the final display quality is still limited by the viewing-angle dependence of the twisted-nematic material used in almost all active matrix pixels. As the viewing angle becomes more oblique, contrast is reduced, making it difficult to distinguish the *on* state from the *off* state. This is particularly troubling with grayscale because all the gray levels wash together. As LCDs move into the desktop monitor market, there have been serious efforts to make them more than simply detached notebook computer displays. In particular, a number of improved methods have emerged to increase the viewing angle of LCDs.

One method for improving the viewing angle and reducing the reverse-image problem at severe angles (where light pixels appear dark, and vice versa) is the *multidomain process*. This involves dividing each pixel into two parts, with one part using a high pre-tilt of the surface layer of liquid crystal molecules and the other using a low pre-tilt. Other methods use two different rub directions, but still maintain a low pre-tilt. Still others use four domains; a development by LG Electronics uses four domains combined with ultraviolet-sensitive alignment film and lighting from four directions. A process has been developed to form protrusions on the alignment layer to align the liquid crystal molecules vertically, resulting in a claimed viewing angle of 160 degrees (horizontal and vertical). The company also claimed, in 1999, to have improved transmittance by 4.5% over previous MVA displays. All of these approaches work both by counteracting the viewing-angle problems at high angles and by equalizing the viewing-angle problems over many viewing angles so that the problem is less dramatic at oblique angles. All are effective in improving the viewing angle, but at the expense of increased manufacturing costs due to additional processing and masking steps.

To date, perhaps the most effective method used to improve the viewing angle is *inplane switching* (IPS). This mode uses a structure with both electrodes on the bottom substrate in an interdigitated configuration. The top substrate (color-filter plate) has no electrodes, in contrast to the conventional LCD. The use of interdigitated electrodes in LCDs was first demonstrated in 1970.

The structure of a Super TFT-LCD is shown in Figure 12. Unlike a conventional TFT-LCD, the device does not use transparent ITO electrodes. Instead, the source electrode is used as the pixel electrode and is formed with the same metal layer as the drain electrodes. The counterelectrode, also a metal layer, is formed at the same time as the gate electrode. There are three main advantages to this arrangement. First, the distance between the pixel and the counterelectrode can be precisely fixed; it is not determined by the cell gap (spacing between the two substrates). Second, fewer steps are involved in the fabrication process. Finally, an auxiliary capacitor is automatically formed between the pixel electrode in the source-drain layer and the counter electrode in the gate layer, so an additional capacitor is not needed.

The operation of a Super TFT-LCD is shown in Figure 13, revealing the reason for the name *in-plane switching*. The nematic liquid crystal molecules are aligned homogeneously between the substrates, with the optical axis of the molecules parallel to the polarization axis of the polarizer, but perpendicular to that of the analyzer. This is accomplished by rubbing the alignment layer on both substrates in the same direction, but at a 45-degree angle relative to the long axis of the pixel and counterelectrode stripes. In the *off* state, light entering through the polarizer passes through the cell and is blocked by the crossed analyzer, so the *off* state is black (this is the opposite of what occurs in a conventional TN-LCD).

When an electric field is applied between the pixel electrode and the counterelectrode, the field lines extend horizontally through the liquid crystal material, rather than vertically between the top and bottom electrodes. The liquid crystal molecules are rotated in-plane in the direction of the field to a maximum of 45° with respect to those firmly anchored at the surface. This sets up a twisted configuration of the molecules in the on state, so that the polarized light passing through the cell is rotated and can thus pass through the analyzer. Consequently, the *on* state in the Super TFT-LCD is white (again, this is the opposite of what occurs in a conventional TN-LCD).

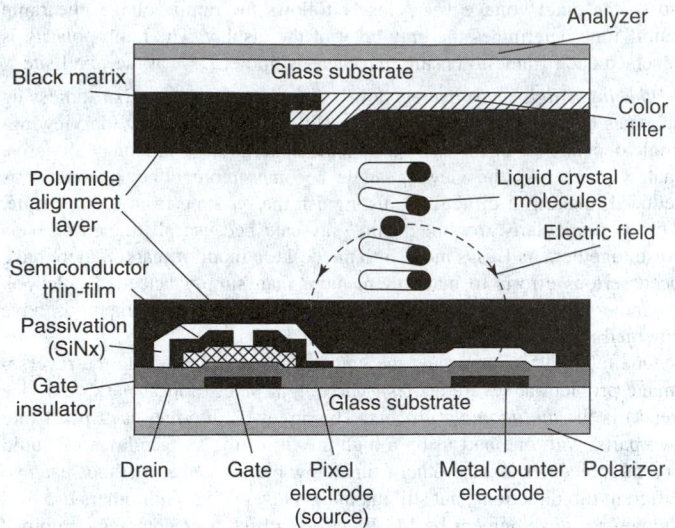

Fig. 12.   Structure of super TFT-LCD using in-plane switching.

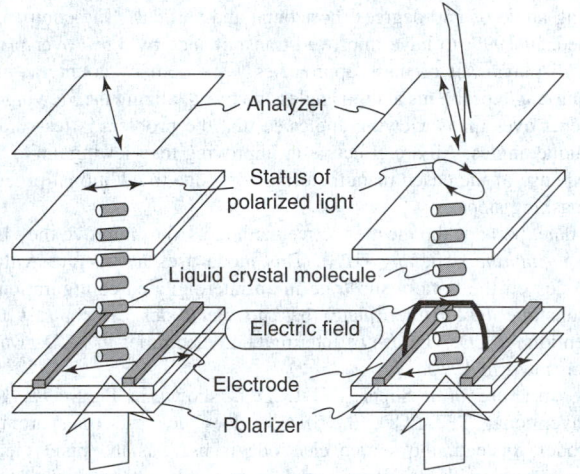

Fig. 13.   Operation of super TFT-LCD using in-plane switching.

The equation for light transmission through the IPS configuration differs from that for the standard TN-LCD. To obtain maximum light transmission and a high contrast ratio, the cell gap d should be adjusted such that

$$(\Delta n) \times d = (\lambda)/2$$

where l is the wavelength of the incident light and n is the optical anisotropy. Super TFT-LCDs provide a much wider viewing angle than conventional TFT-LCDs because of the greater transmission at oblique angles attained through the IPS mode.

There are three disadvantages of Super TFT-LCDs. They have a slower pixel response time, somewhat higher current consumption, and smaller aperture ratio compared to TN-LCDs. However, many design and performance improvements have been made in the past year. Hitachi's IPS display, for example, has been demonstrated to have a 25-millisecond response time, which is comparable to that of TN-LCDs and adequate for full-motion video.

Super TFT-LCDs are still quite expensive, and although they feature some impressive specifications, they have not fully penetrated the market.

Sharp takes a different approach to increasing viewing angle, combining a horizontal liquid crystal alignment with a perpendicular driving field. The *Advanced Super-V* mode offers improved performance through the use of retardation films made with new materials. The ASV LCD has the fastest response time of just 15 milliseconds (8 ms rising, 7 ms falling). Sharp has demonstrated an 18-inch-diagonal SXGA panel using this technology.

Several variations on the above techniques also exist. Hyundai developed a variation of IPS called *fringe field switching* (FFS). The liquid crystal molecules are deliberately placed in the highly spatially varying fringe field of the electrodes. An early demonstration model of this type of display showed a transmission comparable to that of a TN-LCD, but with a far superior viewing angle. An additional type is Samsung Electronics' *patterned vertical alignment* (PVA). The development of advanced modes for wide viewing angles is continuing vigorously.

Amorphous Silicon (a-Si) Thin-Film Transistors

Amorphous silicon (a-Si) is currently the most widely used semiconductor material for AMLCD applications. However, there is growing interest in polycrystalline silicon (poly-Si), which originally required a very high temperature process over 900 °C (over 1,650 °F) but is moving toward a more manageable lower-temperature process under 500 °C (under 932 °F). There are also a number of less-common TFT materials. One reason for the popularity of a-Si is the large base of relevant knowledge resulting from intensive research performed in the 1970s on low-cost, high-efficiency a-Si solar cells. For both solar cells and TFTs, the field-effect mobility of the semiconductor material is essential in determining efficiency; in addition, the following properties are crucial:

- The ability to deposit material uniformly over a large area
- The stability of electrical properties over time and under various conditions
- Low manufacturing costs

The main advantage of a-Si is that it is deposited at low temperatures around 300 °C (572 °F), meaning that relatively inexpensive glass substrates can be used. Its low mobility makes for TFTs with very small *off* state currents, meaning that additional storage capacitors are rarely necessary, but the mobility is nonetheless high enough to provide an adequate *on* current.

The conveniently small threshold voltage (around 3 volts), combined with the high gain of the channel current, is another factor that makes a-Si a good choice for a TFT material. The result of having a low threshold is that CMOS drivers can be used to supply gate voltages to the TFT array. Typically, hydrogenated a-Si (a-Si:H) is used to reduce dangling bond defects, which are thought to increase *off* leakage current and photosensitivity. TFTs made from a-Si:H are fabricated using common semiconductor processes, such as glow discharge decomposition of silane (SiH_4).

The main limitation of a-Si is that the low mobility makes it incompatible with an integrated driver architecture. However, this drawback is being increasingly offset by the availability of bulk-silicon driver chips, which can be surface-mounted or tape-bonded directly on the display substrate. A second limitation is that a-Si is photosensitive, increasing its conductivity to an undesirable level under visible light.

Poorly designed a-Si TFT-LCDs can produce unacceptably large leakage currents if a light shield, or some other compensating means, is not included in the manufacturing process.

The relatively low mobility of a-Si affects the channel dimensions and aperture ratio of the TFT. The mobility varies with the fabrication and post-fabrication treatment, ranging from 0.2 to 0.5 cm²/volt-second, thus requiring a large W/L ratio of the channel. Typical channels made with a-Si are 10 to 20 microns long and 100 to 140 microns wide. A typical pixel dimension for a high-resolution display is 220 × 250 microns. Thus, the semiconductor channel may span more than half of the width of a pixel. Provided that the length is minimal, the TFT will not take up an obtrusive fraction of the display area. The aperture ratio, or fill factor, typically ranges from 50% open to 75% open (obviously, the higher the aperture ratio, the better).

In general, the TFT gate electrode can be fabricated either on the substrate (*gate down* or *inverted*) or on top of the insulator layer (*normal* or *noninverted*), and the source-drain and gate electrodes can either be in the same plane (*coplanar*) or in different planes (*staggered*). Amorphous silicon TFTs are made inverted or normal and are usually staggered. Amorphous silicon offers a variety of structural possibilities because it can be deposited at low temperatures, making it suitable for use with a wide variety of materials.

The inverted-staggered structure provides good control and uniformity over TFT switching parameters, with low parasitic effects. Normal staggered fabrication is similar in terms of photolithographic steps and basic complexity. Fabrication begins with deposition and patterning of the source and drain contacts, followed by the interconnect bus lines. The rest of the TFT is fabricated in the same manner as the inverted-staggered structure.

A particular type of inverted structure is the *self-aligned* structure, in which the gate electrode is used as a photomask during subsequent fabrication steps. Self-alignment has the advantage of providing a very good alignment between the gate electrode and the source-drain electrodes, greatly reducing parasitic capacitance. Large, high-yield aspect ratios have been achieved using a self-aligned serpentine geometry. This geometry would be very difficult with standard alignment techniques because of inevitable amounts of overlap. The self-alignment can be used to fabricate TFTs in a two-photomask-step process that is both high yield and low cost. It was first developed by LETI in France in 1986 and continued by Matsushita Electric of Japan, which showed a video display with 372 × 240 pixels fabricated with this potentially economical procedure. (Because the TFT operation depends on the color-filter/black mask deposition, it really needs three photolithographic steps for proper operation. However, Matsushita does not consider this a step, because it is not electrically part of the TFT and therefore does not appreciably affect yields). A further advantage of the two-step process is that by ensuring that the black mask between the color-filter stripes covers the TFT, no additional light shield is required to guard against the a-Si's photosensitivity.

The two-step manufacturing process is far more economical than earlier methods. Each photolithographic step requires removal from the vacuum for photoresist coating, patterning, etching, and cleaning, which increases the chance of impurities becoming trapped between the thin-film layers. Because the vacuum is not broken during the deposition of the insulator and semiconducting materials in the two-step process, the interfaces between the layers are very clean and uncontaminated.

However, two-step-process displays lack some of the performance of their more complicated predecessors. They must have aluminum, tantalum, or some other metal placed over the source bus lines to lower their resistivity. Furthermore, they lack uniformity across the substrate, which can affect display quality. The threshold voltage ranges from 2 to 3 volts, a wide variation that causes problems when using large numbers of gray levels, especially on larger displays. Two-step-process displays are not yet a major factor in the market.

Photo-induced current, the property that led to the widespread investigation of a-Si for photovoltaic cells, is an undesirable quality when a-Si is used as a flat panel display control device. Functionally, a photovoltaic cell is a diode that converts light into electric current by raising the energy level of the charge carriers in the semiconductor layer. The same semiconductor layer exists in a-Si TFTs, although it usually has a lower concentration of charge carriers than do photovoltaic cells. When the a-Si TFT is exposed to light, the *off* resistance of the channel is decreased, thereby increasing undesirable leakage current. This leads to a washed-out display in bright ambient light conditions.

One remedy for the light sensitivity of a-Si is the incorporation of a light shield over the channel area of each TFT. A common shield material is a metal layer made of the same material as the gate. This type of light shield requires an additional photolithographic step, but because it is electrically passive, the chance of yield decreases is minimal. Alternatively, the black mask between the color-filter stripes can be aligned over the TFTs to act as a light shield with no additional photolithography.

There is some disagreement about whether light shields are truly necessary for a-Si TFTs. Some insist that it is a requirement to build light shields, while others claim the problem is minimal. When a-Si TFTs are used in backlit displays, particularly those with color filters that require very bright backlights, a light shield probably offers improved performance. The most cautious reports claim that the *off* resistance in light and dark conditions can differ by up to a factor of 1,000; with a light shield, this can be reduced to below a factor of 10.

There are two other approaches besides light shields. Very thin films of silicon show substantial immunity from light sensitivity. For a-Si layers less than 30 nm thick, Matsushita demonstrated a contrast ratio of greater than 100:1 in a projection television system with a very high brightness. Another method of reducing light sensitivity is to introduce atomic defects in the middle of the a-Si energy gap. The defects presumably act to stop electrons generated by incident photons in the layer before they can generate an external current.

There is a perception that the a-Si process for fabricating TFTs is nearing the lower limit in feature size. Most a-Si TFT-LCDs have a resolution of 70 to 100 dpi. This perception of diminishing returns is pushing the conversion to poly-Si (described in the next section), which saves space by allowing the TFT to be physically separated from the pixel it controls.

However, new advances in a-Si processing made in 1999 show that this material still has much to offer in terms of improved resolution. IBM announced a new LCD monitor called the Roentgen, which offers nearly 200 dpi with an a-Si process. It is only the latest improvement that began with a more modest 157 dpi (10.5-inch-diagonal SXGA) display. IBM's success is not due to any single innovation; it has come from incremental changes and a systemic improvement of the a-Si process. Also, NEC demonstrated a 9.4-inch diagonal a-Si UXGA display (1,600 × 1,280 pixels), which thus has 212 dpi. Such high resolution was achieved by fabricating the color filter on the TFT array substrate (a process called *CFonTFT*).

Polycrystalline Silicon (Poly-Si) Thin-Film Transistors

Polycrystalline silicon (also called poly-Si, or polysilicon) offers different qualities than a-Si, most of which bring both advantages and disadvantages. Poly-Si has electron mobilities of more than 100 cm^2/V-sec (much higher than a-Si, which is typically less than 1 cm^2/V-sec), resulting in increased switching speeds and reduced device size. The high mobility also leads to higher *on* current levels, reducing the liquid crystal switching time. But higher mobility also means higher *off* state leakage current. With poly-Si, the TFTs and driving circuitry can be fabricated during the same photolithographic steps, producing a self-contained unit at lower fabrication costs than if they were made separately. Connectors, driver chips, and their installation equipment are unnecessary; instead, totally integrated, onboard drivers are deposited with the TFTs. Onboard drivers offer smaller device packages; the device size of poly-Si TFTs is currently limited only by fabrication machinery, which can produce geometries as small as a few microns.

However, recent techniques for attaching drivers to a-Si devices cast doubt on the need for onboard drivers at all. Because poly-Si TFTs take up less space on the display substrate, the aperture ratio is larger, resulting in higher brightness. This is especially useful in light valves, in which a large number of pixels are packed into a small area. The high *off* state leakage current requires the use of an auxiliary capacitor, which partly offsets the aperture ratio advantage by taking up space; nonetheless, poly-Si TFTs are smaller overall than a-Si TFTs. Finally, because poly-Si TFTs are not light sensitive, no special precautions need to be taken when used in high lighting situations, such as projector displays.

Because of these properties, poly-Si is used in two major applications: video/digital camera viewfinders, which benefit from very small pixels, and projection LCDs (particularly for front projection, where poly-Si devices will dominate for the next five years).

To date, nearly all poly-Si TFT-LCDs have been fabricated using high-temperature techniques on quartz substrates, but there has been a great deal of effort applied to low-temperature manufacturing on glass, because of reduced costs.

High-Temperature Poly-Si

Poly-Si fabrication is more complex and expensive than a-Si fabrication. Most poly-Si TFT-LCDs produced to date have been made at 600 °C (1,100 °F) and higher, requiring the use of quartz substrates. High-temperature poly-Si manufacturing is similar to the manufacturing processes for silicon-on-sapphire or silicon-on-insulator integrated circuits. All poly-Si TFT fabrication geometries used to date have been noninverted coplanar. Self-aligned fabrication requires six photolithography steps. As with a-Si, self-alignment has the advantage of very little parasitic capacitance between the gate and the drain, but unlike a-Si, it requires doping of the poly-Si gate periphery to assure ohmic contact. Such doping means employing the expensive machinery used for ion implantation, which also reduces yield. An alternative fabrication procedure uses thermally grown SiO$_2$, which eliminates a photolithographic step and offers better alignment, but requires even higher temperatures because the oxide must be grown at 1,000 °C (1,830 °F). As is the case with a-Si, hydrogenation is shown to have several beneficial effects. Hydrogenation reduces the leaky *off* current of the material, as well as the deleterious effects of dangling bonds, which lower the response time by trapping charges along the doping interfaces.

The full benefits of poly-Si TFTs are not realized unless they are made with fully integrated drivers on the display substrate. The circuitry needed to drive an array of TFTs consists of a combination of shift registers and buffer amplifiers. These devices can be built mostly with poly-Si TFTs with channel dimensions of approximately 8×8 microns; one or two transistors will be required to carry larger currents (that is, dimensions up to 50×8 microns.) The entire amplifier and delay network can be fabricated in an area of $1,400 \times 200$ microns, allowing the drivers to be built in parallel to the pixel rows. The pixel rows for a high-resolution display for a video camera viewfinder can have a pitch of up to 100 microns. Driver circuits with a 200-micron pitch can be alternated from side to side in order to fit on a single display panel.

However, poly-Si integrated drivers are receiving increasing competition from driver chips mounted directly on the display substrate. These drivers can be used with a-Si TFTs, combining the advantages of onboard drivers and simple a-Si fabrication. The superiority of integrated drivers versus surface-mounted drivers promises to be one of the most debated active matrix issues in the coming years.

The effect of poly-Si's high mobility propagates through the TFT design. The high mobility has the undesirable side effect of increasing the *off* state leakage current through the TFT. Because of fewer grain boundary defects, poly-Si has a higher intrinsic conduction. Traditionally, this is compensated for by simply including a pixel storage capacitor, but newer techniques include adding a doping interface or a double-gate structure. Seiko Epson has experimented with a double-gate structure in which all current must pass through two TFTs, halving the *on* and *off* state currents, but maintaining the *on/off* current ratio. Due to the small channel dimensions of each TFT, two poly-Si TFTs are still smaller than a single a-Si device. The yield loss from having to manufacture two TFTs per pixel is less than the yield loss from including a storage capacitor, making this procedure cost effective. This technique has been applied to displays for pocket-size televisions with 240×240 pixels.

The higher *off* leakage current of poly-Si is somewhat offset by the higher on current, making the *on/off* current ratio comparable to a-Si devices. Therefore, it is not troublesome for TFTs to handle the additional current needed to charge and discharge the larger pixel capacitance. This, however, is not a total solution. The capacitors can be difficult to manufacture and can reduce yields by shorting out the pixels. Furthermore, the capacitors reduce the display aperture ratio, partly offsetting the advantage of smaller channels.

As noted previously, poly-Si TFTs are currently used in viewfinder and projection applications. Their prospect for penetrating other markets is extremely limited, however. The requirement for quartz substrates imposed by high-temperature processing limits display sizes to 2-inches in diameter and results in high manufacturing costs. This limitation is the driving force behind research to produce poly-Si TFTs at temperatures below $500\,°C$, which would allow for the use of standard glass types that are less expensive and are available in large-area substrates. Furthermore, a low-temperature poly-Si (LTPS) process could use fabrication equipment that could be integrated into existing a-Si manufacturing lines, saving vast sums that would otherwise go toward building new fabrication lines.

Low-Temperature Poly-Si

The most direct method for fabricating poly-Si TFTs at low temperatures is the deposition of silicon film on glass in the polycrystalline phase, typically using chemical vapor deposition (CVD). But this approach, called *as-deposited poly-Si*, often results in poor film quality, both in terms of defects and surface roughness. Research in this area has included modifying the substrate layer to eliminate surface roughness and reducing background contaminants to decrease defects, but acceptable films cannot yet be mass produced.

Given the problems associated with the as-deposited approach, most efforts have focused on phase transformation from an a-Si film (deposited on glass) to a poly-Si film. This method, called *crystallized poly-Si*, involves imparting energy, typically in the form of laser irradiation or thermal energy, to crystallize the silicon on the glass substrate. The appeal of this approach is that these films are higher quality than as-deposited LTPS and can be produced on larger areas and at lower temperatures

than high-temperature poly-Si. The two types of crystallization for low-temperature poly-Si are referred to as *thermal annealing* and *laser annealing*.

Thermal annealing involves heating the substrate to recrystallize the silicon. The main challenge is, of course, to keep the temperature low. One approach is to use a silicon–germanium alloy instead of silicon for recrystallization. Another approach, developed by Intervac, is called *rapid thermal annealing* (RTA); this approach uses an arc lamp and a reflector to produce a controlled exposure of the film as it passes by in an in-line configuration. This approach requires temperatures above $600\,°C$, however, and furthermore tends to produce low mobility.

Laser annealing, which has come to be the preferred method, typically uses an excimer laser, such as XeCl (308 nm wavelength), to crystallize the silicon film on the glass substrate, heated to $300–400\,°C$ ($572–752\,°F$). The laser beam is rectangular, and it is scanned across the substrate. The long axis of the beam defines the area of recrystallized poly-Si; equipment is available with beam widths of 200 mm, which corresponds to a 12.1-inch display diagonal. Increasing the beam size incurs difficulties with laser power and beam uniformity. Films resulting from this process have shown mobilities in the range of $100–300\ cm^2/V\text{-sec}$, on par with high-temperature poly-Si. However, there are some drawbacks: The laser equipment is expensive, the processing speed is slow, laser scanning must be carefully controlled, and there can be variability between laser pulses. Many companies are still working on improving the laser annealing process.

LTPS processes typically involve more mask steps than their high-temperature counterparts. Ion doping is required to suppress leakage currents at the pixel TFTs and to produce CMOS-type driver circuits. The only means of avoiding time-consuming ion implantation is to use a non-self-aligned architecture in which doped source-drain of contacts are deposited separately from the poly-Si channel such a non-self-aligned poly-Si TFT structure; in the addition of storage capacitors usually requires an extra step, as do double-layer, low-resistance bus lines. Integrated drivers can usually be made simultaneously with the TFT manufacturing process, but they may also require additional steps.

Choosing inverted (gate-down) versus normal structure for the laser-annealed LTPS TFT is a matter of efficiency versus performance. Inverted TFTs allow the laser to simultaneously recrystallize the channel and activate the doped ions, so the fabrication is faster and cheaper than for normal TFTs. However, normal orientation, with its separate crystallization and activation steps, results in higher mobility and less gate/source parasitic capacitance, which means superior performance.

Despite the great advances in LTPS processing in 1999 and early 2000, much work remains before the technology can be considered robust. Areas of active research include identifying optimal doping formulas and temperatures, achieving the appropriate recrystallized grain size, making interfaces cleaner, and further reducing the process temperature.

With drastic price reductions for large a-Si LCDs and declines in driver chip costs, it will be hard for LTPS LCDs to compete in notebook computer and desktop monitor applications. For small and medium LCDs, it is nominally cheaper to make LTPS than a-Si because of the cost savings resulting from integrating driver functions, but low yield and throughput remain problems for poly-Si. With production improvements, poly-Si will be used increasingly in applications such as digital cameras and automobile navigation applications. TFT-LCD panels have gained acceptance in the navigation market, especially in Japan, and there are opportunities for future navigation and passenger entertainment systems.

Single-Crystal Silicon Transistors for Displays

Although single-crystal silicon has the highest mobility of all types of silicon, it is difficult to use in displays because it cannot easily be deposited on glass, which provides a poor growth surface.

An innovative approach to using single-crystal silicon transistors has been developed for a high-resolution display. Instead of fabricating standard a-Si devices, the array is built on a single-crystal silicon wafer using MOS technology, Then the array is physically removes and redeposits it on a glass substrate. The single-crystal silicon properties are preserved. Because the process is based on silicon wafer technology that uses standard semiconductor processing techniques to form TFTs on a silicon wafer, the process of building the transistors can be accomplished in any existing semiconductor fabrication facility. High-resolution devices for near-eye applications such as viewfinders and head-mounted displays have been developed.

A granular material consists of many tiny crystalline regions. Usually, these microcrystalline grains are jumbled together randomly, as is the case for poly-Si, for instance. However, in continuous-grain silicon (CGS) the silicon grains align so that the crystal axes of one grain are parallel to the crystal axes of its neighbor. The only reason CGS is not a single crystal is that the grains start growing from multiple nucleation points so that the atoms do not quite match up when the grains become large enough to touch. CGS has mobility of about 300 cm^2/volt-sec for n-type and 140 cm^2/volt-sec for p-type, which is three times higher than that of poly-Si and almost equal to that of single-crystal Si. The high mobility of CGS allows TFTs to be physically smaller, increasing the aperture ratio and resulting in a brighter display. CGS could also enable system-on-panel (SOP) technologies (with the interface and all other peripheral circuits incorporated into a single substrate) that allow display systems to be extremely thin. Faster driving results in a higher contrast ratio.

The successful completion of a product masks the challenges involved in using CGS to make the TFT-LCD. Sharp claims that CGS manufacturing employs "similar" processes to those involved with poly-Si. This has the advantage of possibly allowing CGS to be made on poly-Si manufacturing lines. However, considering that the time-intensive and costly nature of poly-Si manufacturing is its main limitation, it is clear that adopting CGS is no simplification of the current situation.

Further, realizing the full potential of the CGS-based display required construction of a new optical engine for the projector product. Because the device employs two lamps for high brightness, a fly eye lens system is introduced to collect and direct the light from both sources. The high heat generated from two lamps requires significant cooling to prevent degradation of the polarizers. However, standard fans introduce too much room dust into the system, resulting in an unacceptable image. Hence, the projector includes a closed cooling system with three internal fans and multiple radiator fins. The elements are arranged carefully to provide the most cooling to the blue LCD panel, since it produces the most radiative heat. The projection lens system consists of a nine-group spherical lens that is tailored to handle light with a two-lobed intensity distribution (arising from the two light sources).

Although its price means CGS is not yet competitive with present RPTVs, costs will be reduced over time.

Cadmium Selenide (CdSe) Thin-Film Transistors

Cadmium selenide is of mainly historical interest in LCD development; it has been used for TFT development since the early 1970s, when it replaced tellurium as the material of choice at Westinghouse. Research continues at several companies and universities. CdSe offers excellent performance capabilities rivaling those of poly-Si, but its disadvantages bar it from being commercially viable.

The most important feature of CdSe is its high mobility. CdSe films can be made with mobilities from 40 to 450 cm^2/V-sec (hundreds of times higher than a-Si), at temperatures below 400 °C (752 °F), making the process compatible with inexpensive soda lime glass. The high mobility allows CdSe to be used to make TFTs with channel dimensions much smaller than those of a-Si. Typical *on* current for TFTs with a W/L channel ratio of 7 is 1 microamp with 20 volts on the gate and a 10-volt source-drain voltage. The *off* current for these TFTs is approximately 0.1 nanoamps, for an *on/off* ratio of .00001. The TFT's frequency and current-carrying capability are proportional to its channel mobility, so CdSe devices can operate close to 1 MHz (compared to the order of kHz in a-Si) and carry currents closer to milliamps (rather than microamps). This allows the construction of integrated row and column drivers, and also the system-on-panel possibility noted previously for CGS.

In addition, one of the most problematic steps of silicon processes, forming doped semiconductor source-drain contacts, is not necessary with CdSe. If the CdSe is deposited, annealed, and covered with the aluminum electrodes all during the same pumpdown cycle, the CdSe-electrode contacts will be ohmic.

Despite the impressive capabilities of CdSe, three disadvantages severely limit its application. It is a very toxic material, its performance degrades quickly over time, and manufacturing has proved to be complex and costly.

Ferroelectric Liquid Crystal Display (FLCD)

Ferroelectricity is analogous to the more widely known phenomenon of *ferromagnetism* — the hysteresis, or memory effect, originally observed in iron (ferrite) magnets. The key property of hysteresis is *bistability* (two stable, distinct states). In the case of a display, bistability leads to a property called *write-and-forget*: When the driving voltage is removed from a ferroelectric liquid crystal display (FLCD), it remains in its *on* or *off* state. This is in contrast to TN- or STN-LCDs, which always revert to a unique, unpowered state. Because the liquid crystal material used is of the smectic type, the effect is referred to as the *ferroelectric-smectic effect*. Another term used is *surface stabilized ferroelectric liquid crystal (SSFLC)*, which refers to the structure of an FLCD: These displays are made such that the top and bottom substrates are spaced much more closely than the ferroelectric helix pitch, and the liquid crystal molecules are trapped in a nonhelical geometry. An applied voltage allows the material to be rapidly switched between two optically distinct configurations.

The key feature of FLCs is their fast response time, which is 10–100 microseconds, nearly 1,000 times faster than that of nematic liquid crystals. This shifts the limitation of display speed away from the liquid crystal material to the driving circuitry. For large-area displays, this speed advantage is blunted by comparable speeds achievable in plasma display panels (PDPs) and electroluminescent (EL) displays, but it is important for small displays.

The write-and-forget property of FLC is especially useful in highly multiplexed displays because the pixel remains in its state during nonselect periods; there is no problem with off state leakage. In an FLCD, the pixels respond not to the average RMS voltage, but to the most recent voltage above or below V_{on} or V_{off}. All the driving circuitry needs to do is latch the display to its *on* or *off* state.

Researchers, encouraged by the rapid development of TN and STN-LCDs, hoped that FLCDs would follow the same course. But as FLCD technology progressed, new problems and limitations emerged that were not calculated into the original development timetables. The main trouble arises from the fact that the FLCD substrates must be very close to each other (only a few microns), and there is little tolerance for variation in the spacing. This means the glass and deposited films must be very smooth, pressure sensitivity is a concern, and cleanroom conditions are more stringent because particle contamination is so damaging. All of this adds up to a difficult fabrication process that is costly in both time and money.

Despite considerable effort, working and reliable ferroelectric displays have been much more difficult to produce than originally anticipated.

Polymer-Dispersed Liquid Crystal Display (PDLCD)

Optical scattering is the principle used in the polymer-dispersed liquid crystal display (PDLCD). The device uses nematic liquid crystals encapsulated in micron-sized polymer droplets. The droplets are suspended in an emulsified film several microns thick that is sandwiched between the substrate plates. In the unpowered *off* state, the droplets strongly disperse light, because of the lack of parallel molecular alignment both within each droplet and from droplet to droplet. The display appears cloudy or milky, rather than truly dark. Applying an electric field causes the liquid crystal within each droplet to become parallel and the droplet as a whole to align with the field, making for a transparent display. By mounting a mirror behind the cell, the mode of operation can be changed from transmissive to reflective. This system does not involve any polarizers, which are a main source of absorption on LCDs. Hence, the transmission coefficient is limited only by the small absorption and internal reflection from the glass or plastic substrate.

A major advantage of PDLCDs is their fast (submillisecond) response time. Because they have only weak anchoring bonds, they can easily realign to an electric field. Increasing the driving voltage further increases the response time. This fast response time is offset by the lack of a sharp optical threshold and a high turn-on voltage. In a typical sample, $V(B = 10\%)$ is approximately 5 volts, while $V(B = 75\%)$ is close to 12 volts. A 7-volt difference is far too high for any large, directly multiplexed display to produce visible contrast, especially when compared to STN-LCDs that accomplish their optical switching with less than a 0.05 V difference. Improved manufacturing processes, which maintain uniformity of droplet size (thus increasing nonlinear threshold behavior), may make PDLCDs suitable for small to medium multiplexed displays. It is not expected that PDLCDs will be produced in large, direct-drive multiplexed displays for some time.

Other important features of PDLCDs are their relative ease of fabrication (there is no need to seal the cell) and insensitivity to cell spacing. The cell plates simply form a container for the polymer film matrix; their spacing does not affect display performance. Thus, low-quality unpolished glass (and even plastic) can be used, creating the possibility for flexible displays. Unfortunately, there have not been as many applications for flexible displays as PDLCD developers hoped. Because spacing is not critical, very large panels can be constructed. Raychem Corporation, Menlo Park, California, used to make 1×3-meter (3.28×9.84 feet) glass panes that electronically switched between an opaque state and a transparent state, intended for use in meeting rooms as an alternative to Venetian blinds. However, production has recently ceased. Again, the number of applications has proved disappointing.

For color performance, the polymer droplets can be doped with a dichroic dye, as in *guest-host* displays. The display will modulate between a colored *off* state of strong absorption and high scattering and an on state of weak absorption and high transmittance. In general, there is a tradeoff between brightness and contrast. Dyes that produce high contrast tend to be dark in the *on* state.

In the reflective mode, an alternate technique for improving contrast without sacrificing brightness is to place a color filter between the mirror and the PDLCD cell. In the *off* state, the polymer droplets scatter enough light so that most of it does not reach the mirror, giving the cell a dark, neutral color. In the *on* state, the cell becomes transparent, allowing the bright, saturated filter color to show through. In addition to increasing the actual (measured) contrast, this type of color change greatly improves the perceived contrast. The use of fluorescent reflectors will further improve perceived contrast.

These displays offer many potential advantages such as simple manufacturing and high durability, which has sparked much interest in PCLCDs. However, such benefits have not been proven, and amidst all the research activity, there are only a few prototype PDLCDs. Research groups that have recently published papers on the subject include GEC-Marconi Ltd, Britain; Hughes Research Laboratories, California; Kent State University and Kent Display Systems, Ohio; OIS Optical Imaging Systems, Michigan; Polytronix Inc., Texas; Raychem Corp., California; and the University of Hawaii, Hawaii. See also **Colloid Systems**.

Electronically Controlled Birefringence (ECB) Liquid Crystal Displays

An electronically controlled birefringent (ECB) display is similar to an STN-LCD, except it does not incorporate a twist. There are two basic structures for an ECB display. The first uses liquid crystal material with positive dielectric anisotropy in a cell with homogeneous (parallel) alignment. The second, more popular, configuration uses liquid crystal material with negative dielectric anisotropy in cells with homeotropic (perpendicular) alignment. This geometry is also known as Vertically Aligned Nematic (VAN). The application of an electric field causes the liquid crystal molecules to rotate perpendicular to their original resting position, effecting a change in birefringence.

When the molecules are aligned perpendicular to the cell, whether by electric field or boundary conditions, they exhibit no birefringence, because they are aligned symmetrically about an axis parallel to the normal. Birefringence is only observed as the molecules move parallel to the plates, because they lose their symmetrical alignment. Just as in STN cells, well-placed polarizers will produce contrast between the *on* state and the *off* state. Usually the polarizers are rotated 90° with respect to each other and 45° with respect to the alignment direction of the liquid crystal molecules (homogeneous director vector).

In the more common homeotropically aligned cell (VAN), it is important for the axes of rotation to be parallel to each other (that is, for the rotation to be uniform). To accomplish this, the cell boundaries are also treated with a slight homogenous rubbing so that the rotated molecules have a preferred *on* state direction. Both types of ECB usually incorporate pre-tilt to prevent degenerate rotation. ECB has a much faster response time than STN does, because the molecules' rotation is not retarded by the strong aligning forces that exist in twisted liquid crystal structures.

Unless some type of compensation is used, ECB cells have relatively poor viewing-angle dependence. Oblique angle viewing increases d and changes Δn, affecting the display contrast. Keeping the optical path length small helps minimize viewing angle dependence because of smaller absolute changes at off-angles. But if $\Delta n \times d$ is decreased too much, contrast is reduced because it is the birefringent effects that make the display work. Usually these factors are balanced in a fairly large $\Delta n \times d$, in the range of 0.5 to 1 micron. Advances in film compensation techniques are reducing these contrast and viewing-angle drawbacks, however.

Instability is also a problem with VAN devices. VAN cells are most prone to instability when they are thin cells (<5 microns) operating at a high temperature (>50 °C) and/or a low frequency (>50 Hz). Changing these parameters to reduce instability, however, negatively affects other performance criteria: Increasing cell thickness reduces turn-on and turn-off time, high temperatures cannot always be avoided, and practical limitations of circuit technology limit driving frequency. Adding dopants reduces instability, but at the cost of reducing response time. VAN ECB displays may have high quality and good color, but when left running for too long, infectious color fringes spread over the display as a result of instability. This effect is not permanent; turning the display off, and then on, will restore its original quality.

VAN displays that do not suffer from color fringes have been built and research continues.

Dynamic Scattering Mode (DSM) Liquid Crystal Display

The dynamic scattering mode (DSM) for LCDs has been commercialized but never widely adopted. It works by switching between a scattered, opaque *off* state and a transparent, clear *on* state. The addition of filters could create a colored display. DSM-LCDs were the very first LCDs developed and commercialized, occurring in the mid-1960s and early 1970s. They use liquid crystal material with negative dielectric anisotropy that is doped with a slightly conductive agent. The displays work without polarizers, but they require more than 15 volts for good operation and use more current than TN-LCDs or STN-LCDs.

Guest-Host (GH) Liquid Crystal Display

The *guest-host* (GH) display is structured like the TN-LCD, except that it uses only one polarizer and the surface alignment on both glass surfaces is in the same direction (no twist). It relies on the effects of having a dichroic dye (the guest) added to a liquid crystal material (the host) with positive dielectric anisotropy. The device is based on the fact that the dichroic dye either absorbs light or transmits light of a specific color, depending on its orientation, which is switched by the applied voltage. The effect was first reported by researchers at RCA in the late 1960s. A modification of the original design uses a chiral liquid crystal that imparts twist and eliminates the polarizer altogether. The latter types of GH displays are made in very small quantities for low information content instrument displays used in aircraft and for some clock displays.

There has been some renewed interest in these GH effects to make reflective monochrome or color LCDs. Three-layer GH displays were first developed in the late 1960s under NASA-sponsored contracts, but they were used only in the transmissive mode. A color reflective display was developed using a three-layer structure with magenta, cyan, and yellow GH cells; it showed practically no parallax and had good color purity, as well as high reflectance. More recently, technology had been developed for reflective displays (see "LCDs for Portable and Handheld Devices" for more information on reflective displays).

Cholesteric Liquid Crystal Display

Cholesteric liquid crystals (CLCs) arrange themselves in stacked planes. All molecules in a given plane are aligned, but the alignment axis of adjacent planes is rotated by a fixed amount. The result is a helical stack, with the molecules in planes like steps of a spiral staircase. This structure has two important optical properties: First, it passes light only of one handedness (a left-handed film transmits right-handed light) and in one wavelength range (a small band around the value np, were n is the index of refraction and p is the helical pitch); and second, it reflects all light not transmitted (there is no absorption). Under the application of a voltage in the direction "up the staircase," the molecules all rotate 90° to align with the field, making the film transparent. In this standard configuration, a CLC film makes an excellent narrow-band polarizing filter.

CLCs can also be made into broadband polarizers, allowing the fabrication of an LCD. By adding a chiral material to the CLC, the self-aligned helix will have a varying pitch over the length of the helix and will pass light throughout the entire wavelength range implied by the pitches and index of refraction. If this range spans the visible spectrum, the CLC cell acts as a normally black LCD.

Because the CLC reflects light that is not transmitted, the display can be made very efficient by placing a mirror behind the device and its backlight. The left-handed light reflected from a left-handed CLC film travels backward, bounces off the mirror, and becomes right-handed. It can then pass through the film instead of being wasted. By employing CLC films as the polarizers and color filters that are also needed in an LCD, the entire device can be made 10 times brighter than a standard TN-LCD (that is, it passes about 50% of the backlight's brightness, compared to the usual 5%). Such a bright display is obviously valuable for today's handheld devices because of its lower power requirement and reduced battery draw.

CLC displays are very difficult to fabricate, however. The chemistry of CLCs is delicate and not yet compatible with the processing required to make TFTs. Direct-drive prototypes of CLC displays have been made.

Liquid-Crystal-on-Silicon (LCOS) Devices

Of all the types of AMLCDs, *liquid-crystal-on-silicon* (LCOS) devices most directly use semiconductor manufacturing techniques, as the active matrix array and associated driver circuits are designed and built on a silicon wafer, using complementary metaloxide semiconductor (CMOS) processing techniques. The LCOS displays are the most heavily researched types of *microdisplays*, which are devices built on silicon (or quartz) in a miniature form and designed to be viewed indirectly, either by imaging into the eye or in projection mode.

Instead of thin-film transistors on glass, LCOS displays use bulk silicon devices to control the pixels from outside the optical path. The silicon-based CMOS chip serves as both the active matrix and the reflective layer, on top of which a thin layer of liquid crystal, glass plate, and polarizer are deposited. Instead of the a-Si or poly-Si TFTs used in transmissive LCD cells, LCOS devices use the high-speed switching capability provided by single-crystal silicon. LCOS devices also have a simplified LCD structure because the device is reflective, not transmissive, like most LCDs. For example, only one polarizer is required, and the thickness of the layer of liquid crystal can be reduced, allowing for faster switching.

Two important design choices that are made when developing an LCOS device are the type of liquid crystal material and the type of control circuit to be used. The nematic and ferroelectric liquid crystal modes are the most popular types for such a display. The nematic types provide a good contrast ratio and are typically coupled with a DRAM switching array. Ferroelectric liquid crystals provide very fast switching speeds, but they do not allow for grayscale; the pixel is either black or white. Grayscale can be created temporally by taking advantage of the fast switching speed to dither the binary value within a frame period. Ferroelectric liquid crystals are typically coupled with SRAM devices, which provide the fast frame transfer rates needed to provide temporally dithered grayscales and which are also binary devices.

Liquid crystal microdisplays can also be configured to modulate unpolarized light, an approach that has the advantage of much greater light transmission by eliminating the significant absorption of polarizers. There are two ways to use unpolarized light. In one approach, light is scattered when the liquid crystal molecules are arranged in one fashion and reflected when the molecules are rearranged, a process that is controlled by the application of a voltage across the cell. Scattering displays typically use polymerdispersed liquid crystal materials. The second approach uses the principle of diffraction, in which a periodic arrangement of *on* or *off* pixels (forming a grating) causes light to constructively or destructively interfere. This interference pattern can then be filtered by a Schlieren stop to pass light when in certain grating modes; by altering the grating structure, amplitude and color may be controlled. Twisted-nematic liquid crystal material is the most common type used in diffractive systems.

Color can be achieved in LCOS devices in several ways. The most popular approach for projection applications is to use three LCOS chips and dichroic mirrors to separate red, green, and blue components. Color-filter wheels have also been used to present sequential color to a single chip. For personal viewers, red, green, and blue light sources (typically light-emitting diodes) are used to illuminate sequential frames of data at three times normal video rates.

A number of engineering issues remain with LCOS. First is the light source, which should be as close to a point source as possible because the display is so tiny. Also, improvements in the optics to handle the fine pixel pitch would increase image quality—in the rear-projection screen and the polarizing beamsplitter, in particular.

LCOS displays require a thin cell gap (1–3 µm, with a tolerance of 100–200 nm) as well as specialized filling and sealing procedures. Pixel pitches in microdisplays are as much as five times smaller than in direct-view displays. Current LCOS manufacturing is a tradeoff between circuit and assembly yield.

IBM has reported what is perhaps the most impressive device demonstrated to date. It is a CMOS device with $2,048 \times 2,048$ pixels and a TN-LCD layer. The device can show 65,536 colors at a frame rate of 74 Hz and has been implemented in the form of a 28-inch rear-projection monitor. One problem with the device is that it requires a rather large piece of silicon; because the pixels are 17 µm on a side, the active area is nearly 2 inches in diameter.

LCOS devices are just beginning to take hold in the display market. While there is potential for high growth in the coming years, market acceptance will be contingent on improvements in price, performance, and ergonomics of these products.

The manufacturing of LCOS microdisplays is also not yet a mature practice; manufacturing lines for common LCDs and TFTs are generally not appropriate for LCOS displays.

Reflective and Transflective Liquid Crystal Displays

The solution to making LCDs viewable in bright ambient light is the *reflective* LCD. Instead of using a backlight, this configuration relies on ambient light, incorporating a mirror-like surface behind the device to reflect the light up through the liquid crystal material.

Reflective LCDs can be made in two configurations: outer-surface reflective and inner-surface reflective. Outer-surface reflective devices are constructed similarly to transmissive devices, with the backlight replaced by a reflective surface behind the rear polarizer. While simple, this configuration has the disadvantage of poor contrast and color at a wide viewing angle, because the incident and reflected beams pass through different color filters (leading to both greater absorption of the light and color mixing). Also, there can be parallax problems associated with having the reflector behind both glass substrates. Outer-surface reflection is widely used in watch and calculator displays (and has been for more than 20 years) and may be acceptable for handheld devices, but the quality is too poor for sophisticated displays such as notebook computers.

A superior construction (although more difficult to fabricate) is the inner-surface reflective device, which incorporates the reflective layer into the bottom electrode. Because the reflection occurs much closer to the color-filter layer, the viewing-angle problems are greatly diminished.

The reflective LCD is a tremendous boon for some mobile devices because it completely eliminates the need for a backlight. However, reflective LCDs cannot, of course, be used in darkness, which is a requirement for some devices, such as cellular telephones and PDAs. Hence, a hybrid device has evolved, called the *transflective* LCD. In this case, both transmissive and reflective modes are employed; the device has a backlight, but it also makes use of ambient light through reflection.

Seiko Epson's transflective configuration, which it calls *semi-transparent*, simply replaces the reflective layer in an outer-surface reflective device with a semitransparent plate. The backlight partially shines through, and ambient light is partially reflected. A 6.5-inch thin-film diode LCD has been demonstrated. With half-VGA pixel format, the display acts like a transmissive product when the backlight is on and like a reflective product when the backlight is off. A reflective film for the display that has a 10:1 contrast ratio and a 512-color capability has been developed; power consumption is expected to be 0.12 watt.

Sharp's transflective configuration takes advantage of inner-surface reflection; each bottom electrode consists of a central transparent area surrounded by a reflective region. Sharp calls this the Advanced TFT and has two prototype displays based on this design. One is a 2-inch panel with 560×220 pixels, and the other is a 7-inch, wide-screen model with

$1,440 \times 234$ pixels. Both models have a 100:1 contrast ratio in darkness with the backlight on, but only 5:1 in reflective mode.

A hybrid backlight had been developed, which collects and reflects ambient light under bright conditions and emits light under dark conditions.

In addition to changing the method of light modulation, changes can be made in other components of the LCD to increase the overall light transmission, making the display more suitable for portable and handheld applications. The division between these two approaches is less simple that it appears, however, because changing the method of light modulation often results in the need for additional or different auxiliary components.

In particular, reflecting displays require a diffuser, which spreads the reflected light for the viewer's eyes. The diffuser is a controlled reflector that distributes the intensity of the display's light over an appropriate angle so the display looks natural. The diffuser affects both the brightness and the contrast ratio of the display, and its design must be coupled to the reflection technique and type of liquid crystal used. One such diffuser is the Lumisty™ film made by Sumitomo Chemical Corp.

The two largest sources of transmission reduction are the polarizer sheets (typically, two are used) and the color-filter layer. Therefore, many manufacturers have experimented with improving or even eliminating one or both of these components. Displays have been made with various combinations of two, one, or zero polarizers and one or zero color filters, as shown in Table 3, first compiled by Matsushita.

TABLE 3. CLASSIFICATION OF REFLECTIVE/TRANSFLECTIVE COLOR LCDS

Number of Polarizers	2	1	0
Color filter	(A) "Standard" STN, TN	(B) STN, TN, and others; inner-surface reflection	(C) polymer-dispersed; guest-host
No color filter	(D) field-sequential color, electronically controlled birefringence	(E) none	(F) cholesteric, polymer-dispersed; guest-host

Group A is the "standard" configuration, with two polarizers and one color filter. Matsushita has demonstrated a color STN-LCD that uses a single polarizer instead of the usual two polarizers (Group B). This configuration takes advantage of the fact that light changes its polarization state upon reflection, so a single polarizer is all that is necessary. Using inner-surface reflection, the 7.8-inch-diagonal VGA prototype has 15% reflectance, a 14:1 contrast ratio, and a power consumption 89 mW. It is targeted for mobile business device applications. Another single-polarizer system is made by Sony using LTPS TFT technology; this prototype color display has 34% reflectance and a 19:1 contrast ratio.

Completely eliminating the polarizer sheets (Groups C and F) requires using liquid crystal materials that operate on a principle other than polarization rotation, such as the scattering mechanism of polymer-dispersed liquid crystals, guest-host effects, or electrophoresis. Scattering-type reflective displays have the advantage of simple, lowcost manufacturing, but their limited color and contrast, which have hindered development as transmissive displays, will also hinder development as reflective displays. A *double-layer guest-host* (DGH) LCD, offers a superior combination of contrast and reflectivity. Complications remain in achieving optimal color with this display (using color filters reduces the reflectivity, requiring the addition of special reflectors), so overall it is still in an early stage of development. A prototype monochrome DGH LCD had a 15:1 contrast ratio and a reflectivity of 60%. A related pursuit has been finding ways to produce a color LCD without a color-filter layer (Groups D and F). Several approaches are being investigated: field-sequential color, electronically controlled birefringence (described previously), and stacked LCD layers.

Field-sequential color is accomplished by sequentially illuminating the panel with red, green, and blue backlights at a rate of 180 Hz, or three times the frame rate. The eye integrates the three colors within the time span of a single frame of data. The backlights can be made of monochromatic fluorescent tubes, or with LEDs. One challenge has been identifying liquid crystal materials that have a sufficiently fast response rate while simultaneously allowing the generation of grayscales. Researchers at Tohoku University have developed an optically compensated bend liquid crystal material that has a response time as fast as 2 milliseconds. Another challenge has been to develop fast-reacting phosphors for fluorescent backlights. Ichiko Industries has developed red, green, and blue fluorescent lamps with 0.1 millisecond *on* and *off* times.

Stacked LCD layers use three layers of a dyed liquid crystal material (such as guest-host) in subtractive color (cyan, magenta, yellow) arrangements, which absorb half as much light as additive (red, green, blue) types do. This arrangement provides for higher throughput of the illumination, but requires great care in manufacturing to avoid a misalignment of the layers. Also, TFTs must be constructed on each of the three substrates.

There are additional benefits that arise from eliminating color filters, beyond making the LCD more suitable for handheld devices. First, color-filter fabrication is a significant fraction of the manufacturing cost of LCDs, so there is a direct cost benefit. Also, color filters require making three subpixels for each addressable pixel; eliminating this redundancy lowers manufacturing costs or allows for increased resolution at the same cost, and reduces the number of drivers required.

Overall, there is great motivation to improve the light transmission of LCDs through modifying the backlight and eliminating the color filters, which is spurring intense research and development efforts. Some companies have even launched systems-level efforts to improve many components of the LCD synergistically. An example is Casio's Hyper Amorphous Silicon TFT-LCD (HAST), which combines improvements in aperture ratio with higher-transmittance color filters and a reflective design.

Liquid Crystal Materials

Over the past 20 years, there has been an enormous effort to improve the operating parameters of liquid crystal materials. Many thousands of compounds and mixtures have been prepared to tailor the materials to particular display types or applications. Thus, the types of materials used in TN-LCDs for auto dashboards are quite different from those used in STN-LCDs for computers. Similarly, other LCD technologies (for example, AMLCDs, ferroelectric-smectic LCDs, polymer-dispersed LCDs, and electronically controlled birefringent displays) use unique materials specifically designed to optimize the performance of the display. The major classes of materials for low information content displays are the cyanobiphenyls; larger, higher information content displays use the phenylcyclohexane compounds and derivatives, as well as esters, for high-level multiplexing. The most popular type of liquid crystal material is currently the "F-system" line of products from Merck.

The explosion of new types of display applications in recent years (for notebook computers, handheld devices, cars, and specialty applications) has led to a mini-renaissance of liquid crystal development. The creation of liquid crystal materials has progressed to the point at which rational design of compounds and mixtures is routinely done. In designing materials for STN-LCDs, for example, certain structures are used to optimize the elastic constants and the viscosity in order to improve the steepness of the voltage-versus-contrast ratio curve and to increase the switching speed. Current materials are designed for use with duty ratios of 1:240 and 1:480 and response times of tens to hundreds of milliseconds, with the goal of achieving a contrast ratio greater than 20:1 over a broad viewing angle.

For TN-LCDs used in outdoor applications or in severe environments (for example, gasoline pumps, auto dashboards, aircraft cockpits, and marine instruments), a wide operating temperature range is required, in addition to other performance factors that are specific to the application. Some materials are now available that enable LCDs to operate over a temperature range of $-40\,°C$ to $+110\,°C$. Even at $-20\,°C$, the average response time is 280 milliseconds.

Within the last few years, the development of improved materials for AMLCDs has also progressed dramatically. Materials with low threshold voltages (less than 1.5 volts) have been developed by increasing dielectric anisotrophy; these materials are needed particularly in notebook computers, which have low driving voltages to conserve power. TFT-LCDs with a broader viewing angle have been made possible by such techniques as lowering the optical anisotropy, reducing the incidence of inverse contrast, multidomain switching, in-plane switching, and retardation compensation. And, to improve the contrast ratio, materials with very high resistivity (1,012 ohm-centimeters) have been developed. The reason for this is that in an AMLCD, the RC (resistance times capacitance) time constant must be large compared to the frame time, because the charge on any one pixel must be maintained until the next addressing pulse, typically 20 milliseconds. The search for stable materials with these high resistivities led to the development of fluorinated compounds, which provide high thermal stability, low viscosity, and low birefringence, making the materials ideally suited for use in AMLCDs.

Surface Alignment Materials

An alignment layer (typically a polymer) is required for producing both TN- and STN-LCDs, in order to properly orient the liquid crystal molecules at the substrate surface and enable them to tilt slightly in a preferred direction. The process seems relatively simple to achieve, yet it is perhaps the most complex and least understood of all the processes used for LCD manufacturing.

To produce the tilt angle bias, the films are rubbed or buffed with short-nap polyester or cellulose acetate materials, either in a static or a rotating configuration. The latter may be a simple buffing machine with a paint roller attachment for small operations or it may be a more sophisticated machine with controls for buffing-wheel speed, roller pressure, and substrate travel. Buffing or rubbing of the film with materials that have higher melting points than the film is believed to produce enough localized heating to cause the long-chain polymer molecules to become oriented with their chains parallel to the buffing (or rubbing) direction. The liquid crystal molecules that come into contact with such an oriented film are then aligned in the same direction as the polymer chains. A preferred tilting of the molecules occurs as a result of the interaction of the liquid crystal molecules with the chemical structure of the film. Therefore, different liquid crystal molecules will tilt at different angles, depending on the structure of the two substances.

Early displays used poly(vinyl alcohol), polyesters, polysiloxanes, and low- to medium-molecular-weight methyl cellulose as the alignment material, but today polyimides are the materials of choice. New liquid formulations (solutions) are available in viscosities suitable for spinning, dipping, brushing, or even offset printing. The formulations for LCDs have high optical clarity and excellent adhesion to glass, silicon dioxide, and ITO. Polyimides are also used as high-temperature structural adhesives in aircraft and other applications. Care must be taken when using polyimide materials for LCDs so that only very pure starting materials with low metal ion content are used. Stability at high temperatures is the main advantage of using polyimides. The disadvantage is that expensive, potentially hazardous solvents must be used in its application.

The science of polyimide design and formation can be complex; the molecular weight and molecular structure can be altered to meet specific requirements. The most important properties of polyimides are (1) thermal stability, (2) solubility in suitable solvents, (3) resistance to solubility in the liquid crystal material, (4) adhesion to the substrate, and (5) ability to withstand the high local temperature and mechanical effects of the rubbing or buffing process. Thermal stability is important because one does not want the alignment material to decompose at the temperatures used for sealing the LCD substrates together. Also, the polyamic acid solution must be stable. Because the polyamic acid solution is to be printed on the substrates, solubility in suitable solvents is critical. Presently, *N*-methyl-2-pyrrolidinone and gamma-butyrolactone are the solvents commonly used in conjunction with cellosolves and carbitol additives to make the resulting solution printable.

To prevent the polyimide film from dissolving in the liquid crystal material or in the solvents used to clean the displays, various functional groups are incorporated in the diamine or dianhydride used. This is also done to improve the adhesion on the substrates and ensure the resistance to mechanical and thermal wear during the rubbing process. Because of the need to meet all of these requirements, polyimide material suppliers offer different formulations for the various types of LCDs. Thus, there are specific formulations designed for the fabrication of TN-LCDs, STN-LCDs, AMLCDs, and ferroelectric LCDs.

Color Filters for Displays

The most common way to achieve full-color displays is to extract the primary colors from a white light source and combine them as needed at each pixel. By selectively filtering and combining the three primary color wavelengths — red (700 nanometers), green (546.1 nanometers), and blue (435.8 nanometers) — full-color images can be reproduced. Displays best suited for the addition of color filters are those that switch between neutral black and white states. The materials and array production process are described in greater detail in the following subsections.

The addition of multicolor capability introduces several problems that add costs to the display module. The most obvious is that the number of pixels and driver circuits must be increased by a factor of three or four in order to maintain the same resolution as a monochrome display. This increases driver costs and complicates manufacturing by decreasing line widths or adding crossovers. A second problem is that the color filters cut the brightness of the display, requiring a more powerful backlight.

Performance standards for color filters are still elusive. The National Television Standards Committee (NTSC) has defined the primary color vertices on the standard chromaticity diagram for color television CRT phosphors. LCD manufacturers consequently strive for their displays to be as close to these vertices as possible. To provide a good representation or mixing of colors, the filter system must be definable in high resolutions with accuracy, have good transparency and chromaticity, and be resistant to light and heat, as well as the chemicals used in making the flat panel display.

The color filter is one of the most complex, expensive, and troublesome components in an LCD.

The production of the color-filter array is a technique-sensitive process that is not widely practiced. Consequently, color-filter arrays are available from only a few sources, and filter-plate prices are quite high.

The major suppliers of finished color-filter plates for LCDs are all located in Japan. One of the few non-Japanese companies known to be producing color filters for LCDs is located on Rolla, Missouri. In addition, several LCD manufacturers have onsite color-filter manufacturing facilities, which were set up by their suppliers.

The color-filter matrix manufacturing process employs many steps and consumes many intermediate materials. Besides the basic color-filter material, there is a need for photoresist, etchants, dyes, or other chemicals, depending on the process. An overcoat or planarization layer is usually put over the finished filter matrix to prevent damage to the filters, smooth the surface, increase transmission, and prevent contamination of the liquid crystal. An acrylic resin is the most commonly used overcoat material. The color-filter material that endows each subpixel with its R, G, or B color is usually a pigment dispersed in a viscous medium that is then deposited on one of the substrates. There are also dye and inorganic colorants, although they tend to have inferior resistance to heat, light, water, and chemicals; the pigment-dispersion materials became superior around 1997. The pigment is typically applied by a photolithographic process, as noted below.

The color-filter array and the black mask, a light-absorbing material that defines the colored subpixels, are usually located on the upper substrate on the inside of the liquid crystal cell. The filter fabrication process is constrained by the temperature sensitivity and the requirements for a smooth surface. The insulating properties of color-filter materials mandate that the conductive ITO layer be deposited over the filters, rather than under them (directly on the glass plate). This means that the filter materials must be capable of enduring the elevated temperature processing of the ITO deposition, usually a bit less then 200 °C. The filter material must be chemically resistant to the liquid crystal material because they are in proximity. The filter will also be illuminated by a fairly intense light source, so it must be resistant to changes due to exposure to the light, including ultraviolet light. Further, the colors must conform as closely as possible to the NTSC standards, a property known as **color purity**; state-of-the-art materials from Toppan Printing have color purity near 47%. In summary, the key requirements of color-filter materials for LCDs are optical performance (light transmittivity); dimensional precision; flatness; thermal resistance; light stability; color purity and chemical resistance.

The reproducibility and control of color-filter properties is an ongoing concern. It should be made clear that none of the color-filter manufacturers is now satisfied that the optimum process has been developed. Color-filter arrays are very expensive, due in large part to the stringent requirements for the finished product. Although many processes have been developed, only two are in widespread production: the photolithographic dye system using photosensitive gelatin and the photolithographic pigment system using pigmented photoresist.

A relatively recent lithographic/laminating process, described by T.P Brody, uses a sandwich of red, green, blue, and black layers. Pigmented, photosensitive acrylic films are successively laminated onto a plastic carrier sheet. The films are exposed through a photomask, one step each for the red, green, and blue layers. A black mask layer is also deposited. The process is said to be capable of up to 200 lines per inch and produces filters up to 20×24 inches. Eliminating the stepper and the resist is expected to result in a substantially lower filter manufacturing cost.

In 1999 there were many development efforts to produce color-filter plates in larger sizes than those used for notebook computers. As LCDs move into the desktop monitor market, forward-thinking color-filter-plate manufacturers are stepping up efforts to fabricate large-area plates. The newest manufacturing lines in development in 2000 will be capable of up to 42-inch-diagonal color-filter panels.

Polarizers and Film Compensators for Liquid Crystal Displays

To view the electro-optic effects of nearly all types of LCDs, a set of two polarizers is necessary (however, as noted previously, there are reflective and transflective LCDs with one or even no polarizers). These polymer films are bonded to the front and rear surfaces of the display. In advanced supertwisted-nematic types, a third optical film is added in order to neutralize or compensate for the wavelength-distorting effects of the liquid crystal cell.

Polarizers typically consist of a poly(vinyl alcohol)–iodine polarizing element sandwiched between layers of a protective film that prevents moisture from attacking the polarizing element. The protective layer typically consists of a film of cellulose triacetate or cellulose acetate butyrate. Films to be used in a transmissive LCD also have a pressure-sensitive adhesive layer, release liner, and surface protective film. For transflective or reflective LCDs, the back polarizer, when present, is combined with a reflective film.

The most prominent property of the polarizer is its light absorption; the transparency of the polarizer is one of the prime factors determining the final brightness of the LCD. Absorptive polarizers, such as the standard PVA–iodine type, convert unpolarized light to linearly polarized light by absorbing the orthogonal polarization—that is, half of the light. Further inefficiencies in the material reduce this value further; state-of-the-art polarizers transmit 44% of the incident brightness. Nitto Denko has developed a "wet drawing" process, in which the PVA is stretched (to create the dichroism) under water, in order to better align the acrylic and iodine molecules for maximum transmittance.

A newer method for increasing transparency is to use a nonabsorptive polarizer, which reflects or refracts incident light instead of absorbing it. Such polarizers have traditionally been constructed from three-dimensional optics such as beamsplitters, but these are not feasible for flat panel displays. Materials that can be made into thin-film non-absorptive polarizers are being aggressively developed as demands for increased LCD brightness and power efficiency mount. One possibility is cholesteric liquid crystal; others are stacked films with varying refractive indices.

There are several approaches to laminating the polarizing (or polarizer-retardation composite) film to the LCD glass. One method uses die-cut pieces of polarizer, while another involves the use of strips or reels of the material. Some facilities use sheets of the material to laminate an array of displays; the lamination process uses pressure, but in some cases it uses heat as well. The array is then cut into individual displays using a laser, hot knife, rotary saw, or razor. In the past, problems occurred during this process in which air bubbles or dust particles were trapped, resulting in the formation of visually unacceptable blemishes. Currently, however, the quality of the polarizer materials and the methods of application have improved to the point at which these problems are minimal.

In applications for which antireflection and antiglare properties are desired, the polarizer has a clear, hard antiglare coating, as well as an antireflection coating. Nitto Denko offers such a product, which was developed in cooperation with OCLI. Other products add a coating of ITO to the top layer to provide electromagnetic shielding.

For STN-LCDs in which optical retardation is required, a stretched polymer film is laminated to the polarizing film. While these composite films are more expensive than conventional polarizers, the cost of applying them to the LCD is less than it would be if two separate films were to be applied.

There are also films designed to broaden the viewing angle of TN TFT-LCDs, and these are sometimes integrated with polarizers to reduce the total number of films applied. While such films are not as effective at increasing viewing angles as novel LC modes, this method has the advantage of allowing improvement without modifying the LCD manufacturing process, which can be a major undertaking. Fuji Photo Film, Ltd. offers a TAC (triacetyl cellulose) film that incorporates both the polarizer and their previous product, WV film (for "wide view"), to make a thinner and more transparent optical coating for the LCD cell.

See also **Cathode-Ray Tube**; **Television (TV)**; **Computer Graphics**; and the family of articles catalogued under **Flat Panel Display Technology**.

For additional reading, refer to Flat Panel Display Technology entry.

Stanford Resources, Inc., San Jose, CA

LIQUID CRYSTALLINE MATERIALS.

Liquid crystals represent a state of matter with physical properties normally associated with both solids and liquids. Liquid crystals are fluid in that the molecules are free to diffuse about, endowing the substance with the flow properties of a fluid. As the molecules diffuse, however, a small degree of long-range orientational and sometimes positional order is maintained, causing the substance to be anisotropic as is typical of solids. Therefore, liquid crystals are anisotropic fluids, and thus, a fourth phase of matter. There are many liquid crystal phases, each exhibiting different forms of orientational and positional order, but in most cases these phases are thermodynamically stable for temperature ranges between the solid and isotropic liquid phases. Liquid crystallinity is also referred to as mesomorphism.

Many thousands of organic substances, some rigid-rod polymers, and other macromolecules exhibit liquid crystallinity. The general common molecular feature is either an elongated or flattened, somewhat inflexible molecular framework, which is usually depicted as either a cigar- or disk-shaped entity. The orientational and positional order in a liquid-crystal phase is only partial, with the intermolecular forces striking a very delicate balance involving both attractive and repulsive interactions. As a result, liquid crystals are extraordinarily sensitive to external perturbations, e.g., temperature, pressure, electric and magnetic fields, shearing stress, or foreign vapors. For this reason, liquid crystals are used to design practical devices to either monitor ambient changes of various kinds or to transduce an environmental fluctuation into a useful electrical or optical output.

Besides being used in the scientific study of cooperative phenomena and complex fluid phases, liquid crystalline phenomena have received a good deal of attention due to the possibility of practical applications. Liquid crystals are widely used in electrooptic displays. Other applications include radiation and pressure sensors, optical switches and shutters, and thermography. The liquid crystalline structures formed by amphiphilic molecules form the basis for emulsions and are studied thoroughly by researchers in the food, drug, and oil industries. Polymers that form an anisotropic fluid phase are important in the fabrication of lightweight, ultrahigh strength, and temperature-resistant fibers, and are beginning to be used in electrooptic displays. Liquid crystals also appear to play an important role in the structure and biochemical function of living tissue, where the characteristic combination of order and flow mobility is particularly suited to life processes. Certain disease states, e.g., atherosclerosis, sickle cell anemia, or cancer, may be associated with physical changes in the liquid-crystalline order within biological structures.

Orientational and Positional Order in Fluids

Solids of mesogenic (liquid-crystal forming) molecules melt to form fluids in which some of the long-range molecular order is retained. At the simplest level, the elongation or flattening of the mesogenic molecules prevents the immediate dissolution of the parent, solid-state order. The loss of positional order of the centers of mass of the molecules in the parent solid may be either partial or complete upon melting, but some degree of orientational order is always retained. The fluid retains many solid-like properties, which are finally eliminated when the substance passes into the normal, isotropic liquid phase at a higher temperature (a second melting point). Solid-like features return if the substance is cooled from the isotropic state; this intermediate state is usually thermodynamically reversible, but in some cases it only forms upon cooling. Partial dissolution of solid-state order also may occur in certain substances by the use of solvents. In this case the molecules are either orientationally ordered in the solvent (some macromolecules), or form aggregates, in which the molecules exhibit long-range positional and/or orientational order. Liquid crystals that are established solely by the adjustment of temperature are referred to as thermotropics, whereas those that form through the addition of a solvent are called lyotropics.

In a liquid crystal a snapshot of the molecules at any one time reveals that they are not randomly oriented. There is a preferred direction for alignment of the long molecular axes. This preferred direction is called the director, and it can be used to define an orientational distribution function, $f(\theta)$, where $f(\theta)\sin\theta\,d\theta$ is proportional to the fraction of molecules with their long axes within the solid angle $\sin\theta\,d\theta$.

It is useful to describe the amount of orientational order with a single quantity. X-ray, UV, optical, IR, and magnetic resonance techniques are used to measure the order parameter in liquid crystals.

In addition to orientational order, some liquid crystals possess positional order in that a snapshot at any time reveals that there are parallel planes which possess a higher density of molecular centers than the spaces between these planes. If the normal to these planes is defined as the z-axis, then a positional distribution function, $g(z)$, can be defined, where $g(z)dz$ is proportional to the fraction of molecular centers between z and $z+dz$. Since $g(z)$ is periodic, it can be represented as a Fourier series (a sum of a sinusoidal function with a periodicity equal to the distance between the planes and its harmonics). To represent the amount of positional order, the coefficient in front of the fundamental term is used as the order parameter. The more the molecules tend to form layers, the greater the coefficient in front of the fundamental sinusoidal term and the greater the order parameter for positional order.

In some cases, although the lattice of points of high density of molecular centers parallel to the planes are not correlated from layer to layer, the two principal directions of the lattice are the same for all layers. In these materials, the interactions between the planes do not prevent the planes from translating relative to each other, but do prevent them from rotating relative to each other.

Thermotropic Liquid Crystals

Thermotropic liquid crystals result from the melting of mesogenic solids due to an increase in temperature. Both pure substances and mixtures form thermotropic liquid crystals. In order for a mixture to be a thermotropic liquid crystal, the different components must be completely miscible. Examples include nematic liquid crystals (*p*-methoxybenzylidene-*p'*-*n*-butylaniline (MBBA); *p*-azoxyanisole (PAA); *p*-*n*-hexyl-*p'*-cyanobiphenyl; di-4-methoxyphenyl-*trans*-1,4-cyclohexane-dicarboxylate; and *p*-quinquephenyl), cholesteric liquid crystals ((−)-2-methylbutyl 4-(4′-methoxybenzylideneamino)cinnamate), and smectic liquid crystals (ethyl 4-(4′-ethoxybenzylideneamino)benzoate; ethyl 4-(4′-ethoxybenzylideneamino)cinnamate; *p*-*n*-octyloxybenzoic acid; 4-(4′-*n*-octadecyloxy-3′-nitrophenyl) benzoic acid; diethyl *p*-terphenyl-*p*, *p''*-carboxylate; 2-(*p*-pentylphenyl-5-(*p*-pentyloxyphenyl)-pyrimidine; and 4-ethyl-4′-butyloxybenzylodeneaniline). Much more is known about calamitic (rod-like) liquid crystals then discotic (disk-like) liquid crystals, since the latter were discovered only recently.

Nematic. In a nematic liquid crystal, the long axes of the molecules remain substantially parallel, but the positions of the centers of mass are randomly distributed. Therefore, there is orientational order and a nonzero orientational order parameter, but there is no positional order.

If the molecules of a liquid crystal are optically active (chiral), then the nematic phase is not formed. Instead of the director being locally constant as is the case for nematics, the director rotates in helical fashion throughout the sample. Within any plane perpendicular to the helical axis the order is nematic-like. In other words, as in a nematic there is only orientational order in chiral nematic liquid crystals, and no positional order.

Smectic. Smectic liquid crystals are distinguished from nematics by the presence of some positional order (a tendency to form layers) in addition to orientational order. The direction of preferred orientational order is perpendicular to the layers in a smectic A liquid crystal and at an angle with the layer normal in a smectic C liquid crystal.

In much the same way as a chiral compound forms the chiral nematic phase instead of the nematic phase, a compound with a chiral center forms a chiral smectic C phase rather than a smectic C phase. In a chiral smectic C liquid crystal, the angle the director is tilted away from the normal to the layers is constant, but the direction of the tilt rotates around the layer normal in going from one layer to the next.

Frustrated Phases. Chiral molecules normally form chiral phases, but in some cases this is done in an interesting way. For example, it is not unusual for a chiral molecule to form a smectic A phase, which is not chiral. If the molecule is highly chiral, however, twist is sometimes introduced into the smectic A phase by an array of grain boundaries which are perpendicular to the smectic A layers and parallel to the director. In one compound the normal to the layers is rotated by roughly 17° on either side of a grain boundary and the grain boundaries are separated by about 24 nm, giving this twist grain boundary (TGB) phase a pitch of a little more than 500 nm. In a sense the frustration of an achiral phase of chiral molecules has been relieved by the introduction of these twist grain boundaries.

Discotic Phases. Molecules which are disk-shaped rather than elongated also form hermotropic liquid-crystal phases. Usually these molecules have aromatic cores and six lateral substituents, although the predominance of six lateral substituents is solely historical; molecules with four lateral substituents also can form liquid-crystal phases. Although the flatness of these molecules creates a steric effect promoting alignment of the normal to the disks, the fact that disordered side chains are also necessary for the formation of these phases (as is often the case for liquid crystallinity in elongated molecules) should not be ignored. The most simple discotic phase is the nematic phase, in which the normal to the disks are preferentially aligned along a single direction (director). If the molecules are chiral or if a chiral dopant is added to a discotic liquid crystal, a chiral nematic discotic phase can form.

Metallomesogens. It is also possible to synthesize compounds based on metal atoms which possess liquid-crystal phases. The series based on dithiolene complexes where M = Ni, Pd, or Pt, contains a number of compounds which show the liquid-crystal phases typical of rod-like molecules.

Disk-shaped molecules based on a metal atom possess discotic liquid-crystal phases.

Lyotropic Liquid Crystals. Some molecules in a solvent form phases with orientational and/or positional order. In these systems, the transition from one phase to another can occur due to a change of concentration, so they are given the name lyotropic liquid crystals. Of course, temperature can also cause phase transitions in these systems, so this aspect of thermotropic liquid crystals is shared by lyotropics. The real distinctiveness of lyotropic liquid crystals is the fact that at least two very different species of molecules must be present for these structures to form.

Amphiphilic Molecules. In just about all cases of lyotropic liquid crystals, the important component of the system is a molecule with two very different parts, one that is hydrophobic and one that is hydrophilic. These molecules are called amphiphilic because when possible they migrate to the interface between a polar and nonpolar liquid.

Even more interesting phenomena occur when amphiphilic compounds are put into water−oil mixtures. If the oil concentration is low, the amphiphilic molecules form micelles and the oil collects inside the micelles. As the oil concentration is increased, the micelles continue to swell with oil until it is safe to say that the system is really composed of volumes of water and volumes of oil separated by a single amphiphilic layer. This type of system is called an emulsion, and thus amphiphiles can serve as emulsifiers.

When a highly polar liquid, a slightly polar liquid, and an amphiphile are mixed together at the right temperature and in the right concentrations, the micelles which form are not spherical. Within this vary narrow concentration range, the micelles are rod-shaped for one part of this range and disk-shaped for another part. In either case the micelles themselves orient their symmetry axes (the long axis for the rod-shaped micelles and the short axis for the disk-shaped micelles) just like a thermotropic liquid crystal.

Polymorphism

A liquid crystal compound in more cases than not takes on more than one type of mesomorphic structure as the conditions of temperature or solvent are changed. In thermotropic liquid crystals, transitions between various phases occur at definite temperatures and are usually accompanied by a latent heat.

An exception to the rule that lowering the temperature causes transitions to phases with increased order sometimes occurs for polar compounds which form the smectic A_d phase (a layered structure formed by molecular dimers). Decreasing the temperature causes a transition from nematic to smectic A_d, but a further lowering of the temperature produces a transition back to the nematic phase (called the reentrant nematic phase). Electric or magnetic fields also may induce mesomorphic phase transitions.

Synthesis

Just because a molecule is long, narrow, and meets the requirement of geometric anisotropy does not ensure that it will have a liquid crystal phase. The particular phase structure that occurs in a compound, i.e., smectic, nematic, or chiral nematic, not only depends on the molecular shape but is intimately connected with the strength and position of the polar or polarizable groups within the molecule, the overall polarizability of the molecule, and the presence of chiral centers.

Molecular interactions that lead to attraction include dipole–dipole interactions, dipole-induced dipole interactions, dispersion forces, and hydrogen bonding.

In order for dipole–dipole and dipole-induced dipole interactions to be effective, the molecule must contain polar groups and/or be highly polarizable. Ease of electronic distortion is favored by the presence of aromatic groups and double or triple bonds. These groups frequently are found in the molecular structure of liquid crystal compounds. The most common nematogenic and smectogenic molecules are of the type shown in Table 1. In general, if the X link is rigid, a liquid crystal phase is favored.

TABLE 1. SOME CENTRAL LINKAGES FOUND IN LIQUID CRYSTALLINE COMPOUNDS

$$R_1 - \bigcirc - X - \bigcirc - R_2$$

X	Series Name
$-CH=N-$	Schiff bases
$-N=N-$	Diazo compounds
$-N=N- \\ \quad \downarrow \\ \quad O$	Azoxy compounds
$-CH=N- \\ \quad \downarrow \\ \quad O$	Nitrones
$-CH=CH-$	Stilbenes
$-C\equiv C-$	Tolans
$-OC- \\ \quad \| \\ \quad O$	Esters
$-$(nothing)	Biphenyls

The importance of unsaturation is illustrated by the fact that 2,4-nonadienoic acid forms a liquid-crystal phase, whereas the *n*-aliphatic carboxylic acids do not. The two double bonds enhance the polarizability of the molecule and bring intermolecular attractions to a level that is suitable for mesophase formation. The overall linearity of the molecule must not be sacrificed in potential liquid-crystal candidates. Bulky, even if highly polarizable, functional groups or atoms that are attached anywhere but on the end of a rod-shaped molecule are usually less favorable for liquid-crystal formation.

In the case of carboxylic acids, hydrogen bonding can induce liquid-crystal phases by lengthening the molecular unit through dimerization:

$$R-C\begin{smallmatrix}O--H-O\\\\O-H--O\end{smallmatrix}C-R$$

On the other hand, hydrogen bonding may lead to nonlinear molecular associations that disrupt the parallelism. Hydrogen bonding associations may also be so strong that by the time the solid reaches its melting point the thermal energy is too intense to permit substantial order to remain within the fluid.

Although it is difficult to predict exactly which type of liquid-crystal phase will be formed by a molecule meeting the general requirements, rough trends can be recognized. The presence of functional groups that lead to strong lateral interactions, e.g., dipoles operating across the long molecular axis, favor the layered smectic structure. When these structural elements are not present but the molecule is otherwise suitable for mesomorphism, i.e., is long and narrow, the nematic phase is likely. Longer terminal groups favor the smectic phase over the nematic phase. An asymmetric center on the molecule causes the chiral nematic and chiral smectic C phases in place of the nematic and smectic C phases.

Goals in liquid crystal synthesis include the design of room temperature thermotropics which are stable, colorless liquid crystalline over a wide range of temperature, and operate at low voltage and power levels.

A good deal of synthesis effort has been devoted to chiral liquid crystals, especially those with chiral smectic C phases. The chiral smectic C phase is ferroelectric, which gives it properties quite useful for applications. Perhaps the most important property of these phases is that a lateral dipole can produce a spontaneous polarization.

Polymer Liquid Crystals

Both polymer melts and polymer solutions sometimes form phases with orientational and positional order. Thermotropic polymer liquid crystals possess at least one liquid crystal phase between the glass-transition temperature and the transition temperature to the isotropic liquid. Lyotropic polymer liquid crystals possess at least one liquid crystal phase for certain ranges of concentration and temperature.

Examples of polymers which form anisotropic polymer melts include petroleum pitches, polyesters, polyethers, polyphosphazines, α-poly-*p*-xylylene, and polysiloxanes. Synthesis goals include the incorporation of a liquid crystal-like entity into the main chain of the polymer to increase the strength and thermal stability of the materials that are formed from the liquid-crystal precursor, the locking in of liquid crystalline properties of the fluid into the solid phase, and the production of extended chain polymers that are soluble in organic solvents rather than sulfuric acid.

Polymer Solutions. Perhaps the most extensively studied macromolecular liquid crystals are the synthetic polypeptides, such as poly(γ-benzyl L-glutamate) (PBLG). PBLG is a homopolymer of the L-enantiomorph of a single amino acid with the following repeat unit.

$$\begin{array}{c} \quad\quad O \\ \quad\quad \| \\ \text{-(NHCHC)-}_n \\ \quad | \\ (CH_2)_2 \\ \quad | \\ COOCH_2 - \bigcirc \end{array}$$

The polyamides are soluble in high strength sulfuric acid or in mixtures of hexamethylphosphoramide, N,N-dimethylacetamide, and LiCl. The liquid-crystal phase is optically anisotropic and the texture is nematic. The nematic texture can be transformed to a chiral nematic texture by adding chiral species as a dopant or incorporating a chiral unit in the main chain as a copolymer.

Applications. The polyamides have important applications. The very high degree of polymer orientation that is achieved when liquid crystalline solutions are extruded imparts exceptionally high strengths and moduli to polyamide fibers and films. DuPont markets such polymers, e.g., Kevlar, and Monsanto has a similar product, e.g., X-500, which consists of polyamide and hydrazide-type polymers. Liquid-crystal polymers are also used in electrooptic displays.

Liquid Crystals in Biological Systems

Many biological systems exhibit the properties of liquid crystals. Considerable concentrations of liquid crystalline compounds have been found in many parts of the body, often as sterol or lipid derivatives. A liquid crystal phase has been implicated in at least two degenerative diseases, atherosclerosis and sickle cell anemia. Living tissue, such as muscle, tendon, ovary, adrenal cortex, and nerve, show the optical birefringence properties that are characteristic of liquid crystals. The liquid crystal state has been identified in many pathological tissues, particularly in areas of large lipid deposits. Massive deposits of liquid crystalline cholesterol derivatives have been found in the kidneys, liver, brain, spleen, marrow, and aorta walls. Certain living sperms possess a liquid crystalline state.

The fluid mosaic model of the cell membrane is one in which the phospholipids provide the basic order and integrity of the cell through amphiphilic interaction with the aqueous environment.

There are two important classes of fibers found in the cytoplasm of many plant and animal cells that are characterized by nematic-like organization. These are the micro-filaments and microtubules which play a central role in the determination of cell shape, either as the dynamic element in the contractile mechanism or as the basic cytoskeleton.

In certain cellular organelles, deoxyribonucleic acid (DNA) occurs in a concentrated form. Striking similarities between the optical properties derived from the underlying supramolecular organization of the concentrated DNA phases and those observed in chiral nematic textures have been described. Concentrated aqueous solutions of nucleic acids exhibit a chiral nematic texture *in vitro*.

Additional Reading

Bahadur, B.: *Liquid Crystals: Applications and Uses*, Vols. 1–3, World Scientific, Singapore, 1990–1992.

Collings, P.J.: *Liquid Crystals: Nature's Delicate Phase of Matter*, Princeton University Press, Princeton, NJ, 1990.

Collings, P.J., and M. Hird: *Introduction to Liquid Crystals: Chemistry and Physics*, Taylor & Francis, Inc., Philadelphia, PA, 1997.

de Gennes, P.G. and J. Prost: *The Physics of Liquid Crystals*, Clarendon Press, Oxford, UK, 1993.

Dierking, I.: *Textures of Liquid Crystals*, John Wiley & Sons, Inc., New York, NY, 2003.

Lynch, M.L., and P.T. Spicer: *Bicontinuous Liquid Crystals*, Marcel Dekker, Inc., New York, NY, 2005.

Sonin, A.A.: *Freely Suspended Liquid Crystalline Films*, John Wiley & Sons, Inc., New York, NY, 1999.

White, A.M. and A.H. Windle: *Liquid Crystalline Polymers*, Cambridge University Press, Cambridge, UK, 1992.

LIQUID CRYSTAL POLYMERS. These materials (LCPs) exhibit a highly ordered structure in the melt, solution, and solid states. A tightly packed and highly ordered morphology particularly susceptible to orientation during processing is characteristic. Commercial applications for LCP resins include chemical pumps, tower packings, coil bobbins, connectors, sockets, etc. for electronic components, and various automotive parts. LCPs have an excellent combination of chemical and flame resistance, dimensional stability, and ease of processing. Their thermal stability makes them suitable for dual ovenable cookware and where thermal resistance for both conventional and microwave oven service is important.

Compared with other polymeric materials, LCPs have very high unidirectional properties. *Vectra*™ (Celanese Corp.) resins are primarily aromatic polyesters based on *p*-hydroxybenzoic acid and hydroxynaphthoic acid monomers. *Xydar*™ (Celanese Corp.) injection molding resins are polyesters based on terephthalic acid, *p,p'*-dihydroxybiphenyl and *p*-hydroxybenzoic acid. Differences in monomers are primarily responsible for the differences in specific properties and end uses. The fibrous nature of the polymers imparts good impact strengths.

LIQUID-IN-GLASS THERMOMETER. This instrument consists of a glass envelope, a responsive liquid, and an indicating scale. The envelope is in two parts fused together: a bulb completely filled with the liquid, and a capillary scale section containing the liquid in excess of that required to fill the bulb. The position of the end of the liquid capillary column or index serves, by prior calibration, to indicate the temperature of the bulb. The scale may be marked directly on the capillary tube, as in the laboratory or

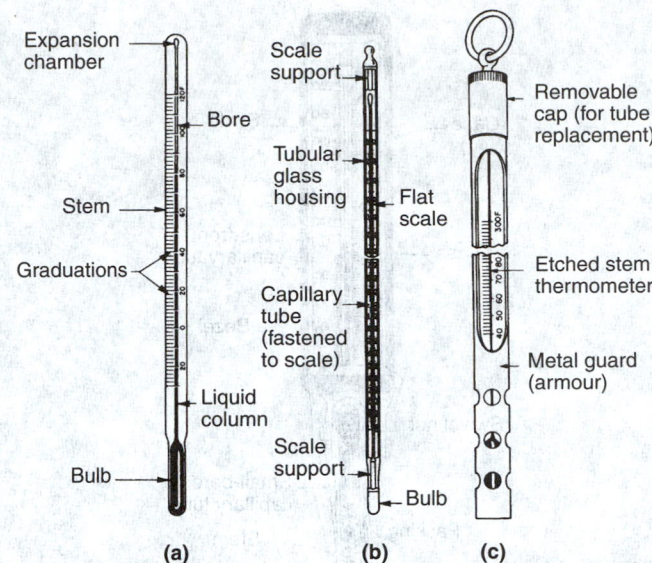

Fig. 1.   Laboratory-type liquid-in-glass thermometers: (**a**) traditional; (**b**) Einschluss; (**c**) armored.

clinical versions, or may be on a separate member mounted alongside the capillary tube, as in the domestic and industrial forms.

Invented over three centuries ago, the liquid-in-glass thermometer reached its zenith as a temperature-measuring device in the early 1800s. It was used as a standard for the dissemination of the scale (Normal Thermometric Scale, adopted internationally) from the International Bureau of Weights and Measures to standardizing laboratories throughout the world, until the later adoption, in 1927, of the International Temperature Scale. Over the years, many practical applications were found for the liquid-in-glass thermometer in addition to its earlier use as a primary standard of temperature. Although still used widely in meteorology, medicine, and industry, and for domestic purposes, the glass thermometer has been replaced by various electrical and electronic temperature measurement methodologies for many other applications.

Various liquids are used in liquid-in-glass thermometers. Mercury is the choice for higher temperatures or where accuracy is critical. Its advantages are a broad temperature span between its freezing and boiling points, a nearly linear coefficient of expansion, the relative ease of obtaining mercury in a very pure state, and its nonwetting-of-glass characteristic. For measurements below the freezing point of mercury, various organic liquids, such as toluene, other hydrocarbons, and organic phosphates, have been used. Representative versions of the glass thermometer are shown in Fig. 1. An industrial version is shown in Fig. 2.

LIQUID JUNCTION. To avoid the unknown liquid junction potential in measuring the potential of a half-cell against a reference electrode, the two half-cells are frequently connected via a salt bridge, usually a concentrated solution of potassium chloride. Since its anion and cation have almost the same velocity, a negligible diffusion potential is set up across the liquid junctions at the ends of the bridge.

LIQUID-PROPELLANT ROCKET ENGINE. A rocket engine using a propellant or propellants in liquid form. Also called *liquid-propellant rocket*. Rocket engines of this kind vary somewhat in complexity, but they consist essentially of one or more combustion chambers together with the necessary pipes, valves, pumps, injectors, etc.

LIQUID STATE. Because of the theoretical and practical importance to the era of electronics, which commenced nearly a half-century ago, the solid state of matter has become better known and understood than the physics of fluids (liquids and gases). Much practical engineering knowledge has been amassed pertaining to substances in the fluid state, but much research of a fundamental nature on fluids remains to be finished. Particularly, the transition of liquids to solids (and vice versa) at the theoretical level has not been fully explored and explained.

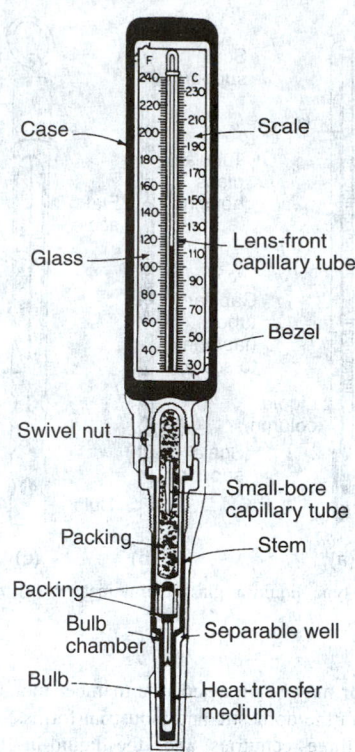

Fig. 2.   Industrial-type liquid-in-glass thermometer.

Prestigious scientists have commented on the mysteries that confront them. Russell J. Donnelly (University of Oregon) has observed, "Most flows of fluids, in nature and in technology, are turbulent. Since much of the energy expended by machines and devices that involve fluid flows is spent in overcoming the drag caused by turbulence, there is a strong practical motivation to understand the phenomenon. The study of turbulent flows, however, is one of the most formidably difficult subjects in physics and engineering. At present (1988), there is no substantial aspect of turbulent flow that can be understood fully from first principles."

Steve Granick (University of Illinois) has commented (1991), "Apart from structure, what are the dynamics of liquids in intimate contact with a solid boundary? This question has proven to be one of the most baffling aspects of liquids, in spite of long-standing interest."

Sir Samuel F. Edwards (Cavendish Laboratory, University of Cambridge) noted (1987), "Liquids are everywhere in our lives, in scientific studies and in our everyday existence. The study of their properties, in terms of the molecules of which they are made, has been the graveyard of many theories put forward by physicists and chemists. The modern student of liquids places his faith in the computer, and simulates molecular motion with notable success, but this still leaves a void where simple equations should exist, as are available for gases and solids. There is a powerful reason for the failure of analytical studies of liquids, i.e., the difficulty experienced in finding simple equations for simple liquids. We can explain the origin of the trouble and show that it does not apply to what at first might seem a much more complex system, that of polymer liquids where, instead of molecules like H_2O or C_6H_6, one has systems of molecules like $H_2(CH_2)_{10,000}$ or $H_2(CHC_6H_6)_{2,000}$ which behave like sticky jellies and yet have complex properties that can be predicted successfully."

Jacob N. Israelachvili and Patricia M. McGuiggan (University of California, Santa Barbara) observe, "The subtleties that can occur in the last few nanometers as two surfaces, particles, or solute molecules approach each other in a medium can be quite remarkable. Sometimes the forces are well described by 'continuum' or 'mean-field' theories, such as the DLVO (Derjaguin, Landau, Verwey, and Overbeck) theory, but more often they are not. Important fundamental questions remain concerning the origin of long-range attractive and repulsive hydration forces in water, the spontaneous nucleation of a bulk liquid or vapor phase between two surfaces close together, and the nature of entropic-fluctuation forces between two

fluid-like interfaces. The elucidation of these interactions both at the fundamental level and when applied to specific systems (where a number of different interactions may be occurring simultaneously) present a challenge to experimentalists and theoreticians. On the purely experimental side, new techniques are constantly being introduced for extending the range and scope of surface force measurements. For example, one may anticipate that the atomic force microscope will soon provide the first direct measurements of the forces between molecules, as opposed to between surfaces."

General Properties of Liquids

A liquid is matter in a fluid state that is relatively incompressible. An *ideal liquid* offers no permanent resistance to a shear stress, but is incompressible. A liquid has a constant volume and incompletely fills any container of less than this volume. A *real liquid* is appreciably compressible, and the liquid state of a substance might be defined as the denser and less compressible phase of the two-phase fluid system that can exist in equilibrium at temperatures below the critical temperature. X-ray diffraction experiments show that, near the melting-point, the molecules of a liquid show a considerable degree of short-range order and that, in small volumes, they are arranged much as in a solid crystal. This crystalline structure persists over volumes comparable with the intermolecular distances, but cannot be traced beyond. This local or short-range order means that the average molecule is at any moment surrounded by a number of molecules occupying nearly the same relative positions as they would in the solid state. The degree of short-range order is described by the radial distribution function.

This concept of a liquid as an imperfect crystal requires that the molecules in a liquid are packed sufficiently loosely for comparatively free movement, i.e., the energy required to move a molecule from a lattice site to a vacant space is not large compared with thermal energies. Under these conditions, shear flow of the liquid resembles closely the high temperature creep of crystalline solids. A number of theories of the liquid state have this concept as their starting point.

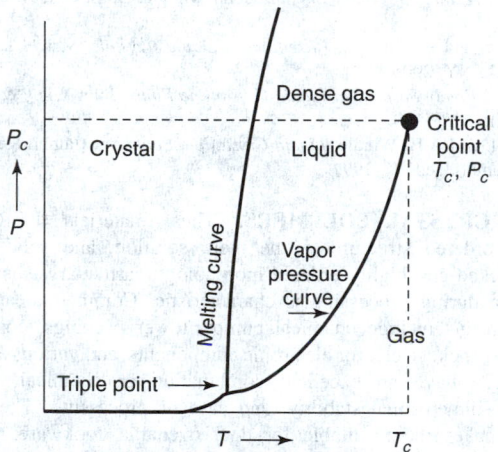

Fig. 1.   Universal phase diagram.

With a few exceptions, including helium, the universal phase diagram shown in Fig. 1 applies for all pure compounds. The triple point is the single point at which all three phases (crystal, liquid, and gas) are in equilibrium. The triple point pressure is normally below atmospheric. Those substances, such as carbon dioxide, where $P_t = 3,885$ millimeters, $T_t = -56.6\,°C$, sublime without melting at atmospheric pressure. From the triple point, the melting curve defines the equilibrium between crystal and liquid, usually rising with small but positive dT/dP, and presumably always with positive dT/dP at sufficiently high P values. The line is believed to extend infinitely without a critical point (it has been followed to $T \cong 16T_c$ for helium, and calculations indicate that hard spheres would show a gas-crystal phase change). The gas-liquid equilibrium line, the vapor pressure curve, has dT/dP always positive and greater than the melting curve. The vapor pressure curve always ends at a critical point,

$P = P_c$, $T = T_c$, above which the liquid and gas phase are no longer distinguishable. Since the liquid can be continuously converted into the gas phase without discontinuous change of properties by any path in the $P - T$ diagram passing above the critical point, there is no definite boundary between liquid and gas. Two liquids of similar molecules are usually soluble in all proportions, but very low solubility is sufficiently common to permit the demonstration of as many as seven separate liquid phases in equilibrium at one temperature and pressure (mercury, gallium, phosphorus, perfluoro-kerosene, water, aniline, and heptane at $50\,°C$, 1 atmosphere).

Stability Limits.[1] With the exception of helium and certain apparent exceptions discussed below, Fig. 1 gives a universal phase diagram for all pure compounds. The triple point of one P and one T is the single point at which all three phases, crystal, liquid, and gas, are in equilibrium. The triple point pressure is normally below atmospheric. Those substances, e.g., CO_2, $P_t = 3885$ mm, $T_t = -56.6\,°C$, for which it lies above, sublime without melting at atmospheric pressure.

From the triple point, the melting curve defines the equilibrium between crystal and liquid, usually rising with small but positive dT/dP, and presumably always with positive dT/dP at sufficiently high P values. The line is believed to extend infinitely without a critical point (it has been followed to $T \cong 16T_c$ for He, and calculations indicate that hard spheres would show a gas-crystal phase change). The gas-liquid equilibrium line, the vapor pressure curve, has dT/dP always positive and greater than the melting curve. The vapor pressure curve always ends at a critical point. $P = P_c$, $T = T_c$ above which the liquid and gas phase are no longer distinguishable. Since the liquid can be continuously converted into the gas phase without discontinuous change of properties by any path in the $P - T$ diagram passing above the critical point, there is no definite boundary between liquid and gas.

The term *liquid* is commonly reserved for $T < T_c$, and "dense gas" is used for $T > T_c$. However, certain properties, such as the ability to dissolve solids, change rather abruptly at the critical density. In many respects, the dense gas resembles the low-temperature liquid of the same density more closely than it does the dilute gas.

The slope, dT/dP, of all phase equilibrium lines obeys the thermodynamic Clapeyron equation:

$$dT/dP = \Delta V/\Delta S = T\Delta V/\Delta H \qquad (1)$$

with ΔV, ΔS, and ΔH the differences, for the two phases, of volume, entropy, and heat content or enthalpy, respectively. The quantity ΔH is the heat absorbed in the phase change at constant P. Since always $S_{cr} < S_{liq} < S_{gas}$, and usually $V_{cr} < V_{liq} < V_{gas}$, one usually has $dT/dP > 0$; the relatively rare cases, including water, for which $V_{liq} < V_{cr}$ at low pressures leads to $dT/dP < 0$ for the melting curve near the triple point.

Figure 1 gives the P-T boundaries of the stable liquid phase. Clean liquids can readily be superheated or supercooled and, in vessels having walls to which the liquid adheres, they can be made to support negative pressures of several tens of atmospheres. Thus the properties of the metastable liquid can be investigated outside the limits shown in the diagram.

Two apparent exceptions to the universality of the phase diagram of Fig. 1 deserve mention. First, many of the more complicated molecules decompose at temperatures below melting or boiling, and the diagram is unobservable. Secondly, some liquids, notably glycerine and SiO_2 and many multicomponent solutions, supercool so readily that crystallization is difficult to observe. In these cases, there is a continuous transition on cooling to a glass, which has the elastic properties of an isotropic solid. The structure of the glass is qualitatively that of the high-temperature liquid, lacking long-range order. Since glass and liquid are not sharply differentiated, the term *liquid* is sometimes used to include glasses, although common parlance reserves *liquid* for the state in which flow is relatively rapid.

Quantum Liquids. The one real exception to the phase diagram of Fig. 1 is that of helium, Fig. 2. Both isotopes, ^{4}He and ^{3}He, have no triple point, the liquid is stable to 0 K below about 20 atm for ^{4}He and below about

30 atm for ^{3}He. The liquids have zero entropy at 0 K in both cases. This is also the only case in which isotopic mixtures form two liquid phases at equilibrium, the isotopic solution separating below 1 K. The isotope ^{4}He has itself two phases, He I above the dotted λ-line of the diagram, and He II with remarkable properties of superfluidity, second sound, etc., below the λ-line. The phase transition along the λ-line is second order; that is, whereas S and V are continuous, heat capacity and compressibility change discontinuously across the λ-line.

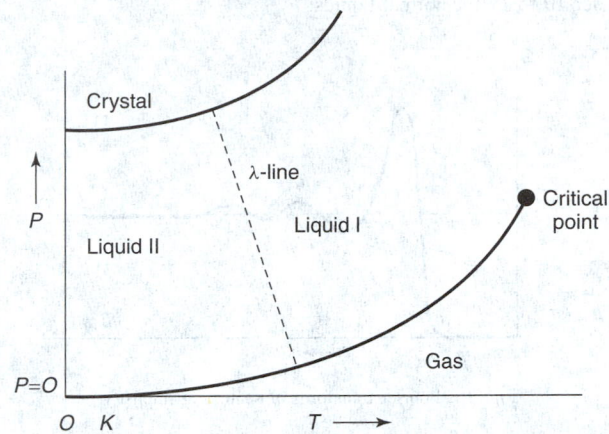

Fig. 2.   Quantum liquid exception to phase diagram of Fig. 1.

Although no completely satisfactory single theory of liquid helium has yet been formulated, one can say that most of the remarkable properties are qualitatively understood and are due to the predominance of quantum effects, including the difference in the statistics of the even and odd isotopes. Thus helium is the one example in nature of a quantum liquid, all other liquids showing only minor deviations from classical behavior.

Structure. Considerable confusion in the description of liquid structure exists, due primarily to difficulties of precise formulation of verbal concepts. The geometric arrangement of any small number (say 10 to 12) of close-lying molecules resembles the arrangement in the crystal, but the order rapidly disappears as larger groups are considered. Long-range order is lacking. The fact that numerical theories based on a lattice or cell structure have some success is evidence only that most properties depend on the configuration of near neighbors alone. Insofar as the arrangement of nearest neighbors is describable in terms of that of the crystal, the structure of the normal liquid is probably characterized best by a somewhat closer spacing than the crystal of the same molecules. The reduced density arises from a considerable number of vacancies in the lattice; the coordination number, or number of nearest neighbors, is lower than in the crystal. The exception is water, in which the low coordination number, 4, of the crystal, is increased by interstitial molecules in the liquid, leading to a higher density of the liquid.

Structural descriptions of this nature usually lack the possibility of precise formulation. It is, however, possible to define for any disordered array of molecules in three-dimensional space an arrangement of contiguous cells, each containing one and only one molecule, the faces of the cells being the loci of the midpoints of neighboring molecules. The statistics of the fraction of cells with n faces and of the distances of the faces from the molecules would give the fraction of molecules having a given number of nearest neighbors and the distance distribution of these in a precisely defined manner. Neither present experimental information nor present theories lend themselves to analysis in such terms.

The only clearly defined manner of describing liquid structure in use at present involves the concept of a set of probability density functions, ρ_n, for ascending numbers, n, of molecules. The function ρ_n depends on the vector coordinates $\mathbf{r}_1, \mathbf{r}_2, \cdots \mathbf{r}_n$ of n molecules, and

$$\rho_n \mathbf{r}_1, \mathbf{r}_2, \cdots, \mathbf{r}_n, d\mathbf{r}_1, \cdots, d\mathbf{r}_n$$

is defined as being the probability that in the liquid of definite P and T, there will be, at any instant of time, one molecule at each position, $\mathbf{r}_i$, within the volume element, $d\mathbf{r}_j$. For a fluid, unlike a perfect single crystal,

---

[1] The following several paragraphs by Joseph E. Mayer are part of a large article that appears in "The Encyclopedia of Physics" (Robert M. Besancon, Editor), Van Nostrand Reinhold, New York, 1984.

$\rho_i(\mathbf{r})$ is a constant independent of $\mathbf{r}$ and equal to the number density: the number, ρ, of molecules per unit volume. The first significant member of the set is then the pair density function, $\rho_2(\mathbf{r}_1, \mathbf{r}_2)$, which depends only on the distance, $\mathbf{r} = |\mathbf{r}_1 - \mathbf{r}_2|$, between the two molecules. At large distances $\rho_2(\mathbf{r} \to \infty) = \rho^2$. This function can be found experimentally from the x-ray scattering intensities of the liquid (it is the three-dimensional Fourier transform of the scattering intensity at angle θ vs $(4\pi/\lambda)/\sin(\theta/2)$). A typical plot is shown in Fig. 3. The area under the ill-defined first peak integrated over $4\pi \mathbf{r}^2 d\mathbf{r}$ is the average number of nearest neighbors, and is of order 10 to 11 for normal liquids.

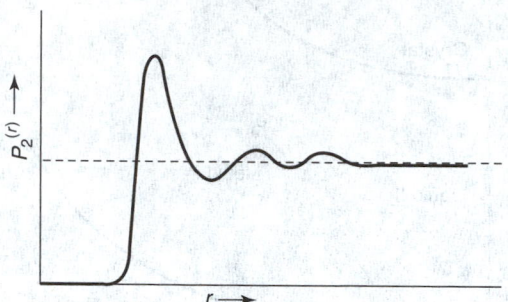

Fig. 3.   Fourier transform of scattering intensity.

The quantity of dimensions of energy,

$$W_n(\mathbf{r}_1, \cdots, \mathbf{r}_n) = -kT \ln[\rho^{-n} \rho_n(\mathbf{r}_1, \cdots, \mathbf{r}_n)]$$

can be shown to be the potential of average force of n molecules located at the positions $\mathbf{r}_1, \cdots, \mathbf{r}_n$. That is, if there are n molecules at these positions, there will be some average force, f_{xi}, along the x-coordinate of molecule i. This average is the sum of the direct force due to the other $n - 1$ plus the average of a fluctuating force due to the others, whose average position is affected by that of the n specified ones. This average force is

$$f_{xi} = -(\partial/\partial x_i) W_n(\mathbf{r}_i, \cdots, \mathbf{r}_n)$$

One frequently assumes that W_n is a sum of pair forces only,

$$W_n(\mathbf{r}_i, \cdots, \mathbf{r}_n) = \sum_{n \geq i >} \sum_{j \geq l}' W_2(\mathbf{r}_{ij})$$

although this assumption is known to be only approximate. With this assumption, the pair average force potential, $W_2(\mathbf{r}_{ij})$, can be computed as the solution of an integral equation, and the solutions agree quite well with the experimental curves.

The knowledge of the complete set of functions ρ_n plus that of the intermolecular forces would permit the computation of all equilibrium properties of the liquid, and indeed if the intermolecular forces are the sum of pair forces, only a knowledge of ρ_2 at all P, T values is necessary. An adequate, although numerically difficult, theory of the transport properties also exists, using the equilibrium functions, ρ_n. At present, only qualitative success is obtained in the completely *a priori* use of the equations.

Associated Liquids. The description given above is adequate only for liquids composed of spherically symmetric molecules or molecules that are nearly so. These constitute the so-called normal liquids, which obey reasonably well the law of corresponding states, for which the entropy of vaporization at the boiling point has the Trouton's rule value of approximately 21 cal/deg. For molecules containing large dipole moments, or those forming mutual hydrogen bonds, the concept of the probability density functions must be extended to include angles or other internal degrees of freedom in the coordinates. Such inclusion is conceptually easy, but incredibly complicates the already difficult numerical evaluation of any equations. However, certain qualitative statements may be made.

Liquids composed of molecules with large dipole moments are frequently referred to as associated. Although in some instances relatively stable dimer or definite polymer units of relatively fixed orientation may exist, in many cases, notably water, it is extremely doubtful if an exact knowledge of the structure would reveal any distinguishable entities of associated molecules other than that of the whole liquid. In such cases, one

would, however, expect that certain mutual angular orientations between neighboring molecules will be highly preferred, whereas in the dilute gas this will not be the case. The effect of this restriction on the internal coordinates will be to decrease the entropy of the liquid markedly compared to the gas. This effect is qualitatively the same as in association, and the properties of these liquids, particularly the high entropy of vaporization, will simulate those of a liquid composed of definite associated complexes.

Traditional Views of Forces Between Surfaces in Liquids

For many years, four kinds of forces have been recognized to operate between surfaces or particles in liquids:

1. *Van Der Waals forces* — Normally, these are monotonically attractive and occur between all molecules. See also **Van Der Waals Forces**.
2. *Repulsive electrostatic (double-layer) forces* — These forces are apparent when ionizable surfaces have a net electric charge, the common case in water. See also **Electrostatics**.
3. *Structural, hydration, or solvation forces* — These forces may be attractive, repulsive, or oscillatory and depend upon the structuring or ordering of liquid molecules. (Solvation may be defined as the adsorption of a microlayer of film of water or other solvent on individual dispersed particles of a solution or dispersion.)
4. *Repulsive entropic (steric or fluctuation) forces* — As defined by Israelachvili and McGuiggan, these are "forces which arise from the thermal motions of protruding surface groups (such as polymers or lipid head groups) or from the thermal fluctuations of flexible fluid-like interfaces (or surfactant or lipid bilayers)."

Although, in a vacuum, only the Van Der Waals forces are important; in liquids, all forces may operate simultaneously. In liquids, it is extremely difficult to separate the effects of each of the aforementioned forces.

In the 1950s, the DLVO theory was based largely upon forces (1) and (2) defined above. The DLVO theory became the basis for studying the properties of colloidal and biocolloidal systems.

Electrorheological Fluids

Complications continue in the theoretical exploration of liquid behavior, but, in attempting to learn about the complexities, leads toward a more fundamental understanding of liquids may emerge. One of these complexities is a class of fluids referred to as *electrorheological*.

In a normal setting, these liquids are liquid in the conventional sense, but, when they are subjected to a strong electric field, they become solids. A common example is a mixture of cornstarch and vegetable oil. The viscous, sticky starting mixture of these two components can be converted into a hardened solid material with the application of an appropriate electric field. In recent years, through random researching, investigators have found numerous combinations of materials that qualify as electrorheological fluids, but to date no satisfactory explanation of the effect from a theoretical standpoint has been developed. The effect of the ER effect can be observed readily by microscopic examination. In a normal liquid or in an electrorheological mixture not in an electric field, examination shows particles moving in random fashion throughout a container, as may be expected. But, when a field is applied, long strands of particles appear to be "solidly" linked together, thus providing rigidity to the mix. No retentivity is involved, however, because liquid normalcy is returned immediately upon cessation of the electric field. This action intrigues a number of designers of equipment, as in the electronics and valve fields, where such materials may be used in future circuits, valves, and any number of other "on-off" devices.

Artificial Magnetic Fluids. The concept of magnetizable liquids dates back over several decades. The first breakthrough occurred in the mid-1960s, with the production of stable colloids of subdomain solid ferromagnets, which variously were called magnetic fluids, magnetizable fluids, or simply ferrofluids. These may be prepared by size reduction or precipitation. It has been a rather remarkable feat to grind bulk material down to a size of 100 micrometers. Grinding is done in the presence of a dispersing agent and a solvent. In chemical precipitation, iron(II) and iron(III) ions in aqueous solution are coprecipitated in the molar ratio of about 2:1 using ammonium hydroxide. To maintain the magnetic product in a small colloidal size range, a peptization step is included in which the particles are transferred to a heated organic phase containing the dispersing agent.

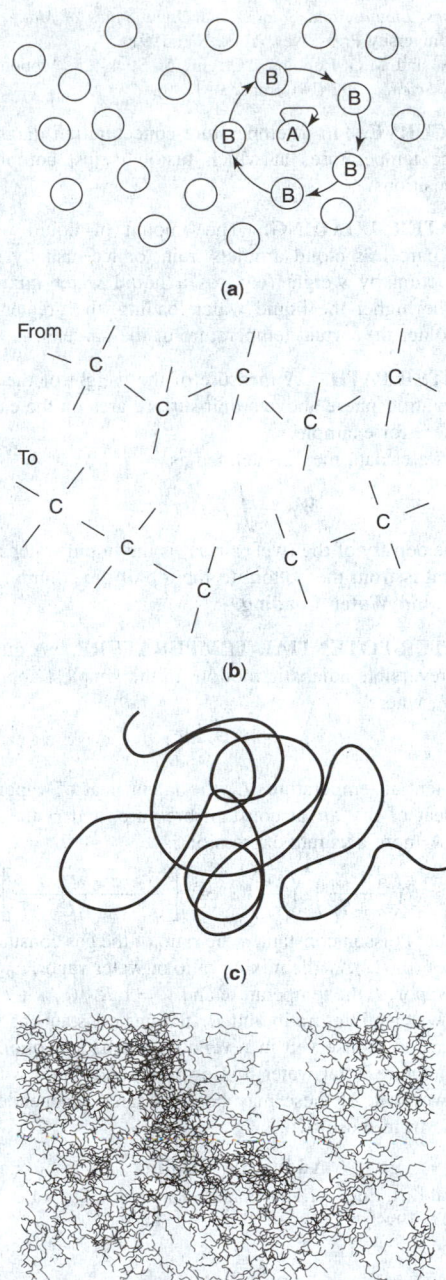

Fig. 4. Sketches by Edwards for schematically illustrating the behavior of polymer liquids. (**a**) Central molecule A moves around a (temporary) average position, occasionally escaping the barrier of the molecules B when some fluctuation of their positions permits it to do so. But, in addition to this single molecule motion, the molecules B can move around cooperatively, one of many cooperative motions that the observer may "invent." The numerous possibilities immediately derail the formulation of a simple theory. As explained by Edwards, "If the motion only involved one molecule at a time, quite a reasonable theory can be put together, but it will always be inadequate to describe the whole motion, and possibly, this always will be the case. The very apparatus of mathematical equations cannot handle this level of complexity, even though the human mind has no difficulty in seeing where the trouble is. The best current way to deal with this problem is to put all the molecules onto the computer and study their molecular dynamics." Continual changes are possible, and the long-chain molecule can be regarded as a flexible string in continual motion. A single coil may be drawn, as shown in (**c**). A computer model could be generated to show many polymers projected onto a plane, as indicated in (**d**). Edwards observes, "So the liquid looks like a seething pit of wriggling motion, a kind of living spaghetti, but with less smooth shapes than real spaghetti. How can one describe such a system? How do the molecules move, and how does the material flow?" (*Sketches after Sir Samuel F. Edwards.*)

The behavior of these artificial fluids offers techniques for achieving efficient heat and mass transfer, drag reduction, wetting, fluidization, sealing, damping, and other process and product potential uses.

Polymer Liquids

Sir Samuel F. Edwards of the Cavendish Laboratory, previously mentioned, observes, "Polymer liquids are liquid because the temperature is high enough to change the configuration of the molecules easily." See Fig. 4.

Scientists and engineers experienced in the production and application of polymeric liquids have learned from thousands of examples of how polymeric liquids behave, and can even classify some of them into behavioral families. But the complexity of these substances to date has eluded the achievement of precise designs and predictive behavior.

Mixing of Liquids

In the chemical process industries and in food manufacturing, the mixing of liquids is an important and frequently used operation. Over the years, the mixing operation has been poorly understood and essentially remains so. Progress in equipment design has been achieved mainly through the development of extensive empirical information rather than upon the creation of precise mathematical and theoretical relationships. J.M. Ottino (University of Massachusetts) observes, "There are many fundamental questions regarding mixing in slow three-dimensional flows, and unfortunately some of the intuition we have obtained from our study of two-dimensional flows does not necessarily carry over to flows in three dimensions." The Ottino reference listed describes studies of chaotic and nonchaotic flows in laboratory setups.

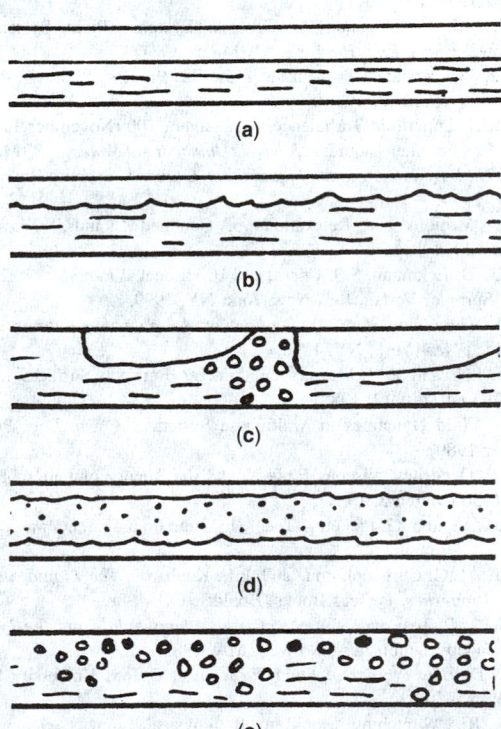

Fig. 5. Types of two-phase flow in a horizontal pipeline: (**a**) Stratified smooth flow where gas velocity is low. Liquid flows along bottom portion of pipelines with essentially a smooth surface. (**b**) Stratified flow with a wavy surface, the waviness caused by increased gas flow velocity. (**c**) Liquid bridges the pipeline cross section, thus causing slugs or plugs of liquid, which move at a velocity approximately that of the flowing gas. (**d**) Annular flow, in which the liquid essentially flows as an annular film on the pipe wall while gas flows as in a central core of the pipe. (**e**) Dispersed bubble flow usually results when liquid flow rates are high and gas rates are low. Because of comparative density differences, most bubbles are found above the pipe center line. Conditions vary somewhat when the pipeline is in a vertical orientation. (*After Cindric, Gandhi, and Williams.*)

Multiphase Fluid Flow

It is quite common in industrial and cross-country liquid transport problems to encounter mixtures of liquids, vapors, and gases — that is, the presence of two phases of matter. As pointed out by D.I. Koch (Cornell University), "Research programs in this area at Cornell involve studies that fall outside of the traditional realm of chemical engineering, including blood flow in capillaries, the transport of contaminants in groundwater, the dynamics of geothermal reservoirs, enhanced oil recovery, the processing of fibrous composites, melt-spinning processes, and the growth of silicon crystals."

As previously mentioned in this article, liquid behavior presents an immense variety of puzzling problems that are difficult to comprehend and hence difficult to forecast precisely. As just one example, in the study of large drops of liquid at high flow rates, the inertia of the drops and the surrounding fluids play an important role. When such drops are propelled toward one another, they may coalesce into a single drop or may rebound like a pair of elastic balls, a phenomenon that is partially (but not fully) dependent upon the comparative velocities of the two drops.

In traditional industrial two-phase flow situations, as occur in process piping, engineers classify flow patterns, as indicated in Fig. 5. This type of classification and the development of empirical data from past experimentation and practice assist much in simplifying the problems from a practical, if not from a theoretical, standpoint. Multiphase flow behavior considerations are essential in calculating pipe diameters, pumping capacities, and energy consumption. Multiphase flows also exhibit different characteristics when being pumped uphill or flowing downhill. See also **Fluid and Fluid Flow**.

Additional Reading

Amato, I.: "Liquids That Tiptoe on the Edge of Solidity," *Science News*, 342 (December 1, 1990).

Cindric, D.T., S.L. Gandhi, and R.A. Williams: "Designing Piping Systems for Two-Phase Flow," *Chem. Eng. Progress*, 51 (March 1987).

Coker, A.K.: "Understand Two-Phase Flow in Process Piping," *Chem. Eng. Progress*, 60 (November 1990).

Donnelly, R.J.: "Superfluid Turbulence," *Sci. Amer.*, 100 (November 1988).

Edwards, S.F.: "Polymer Liquids," *Review (University of Wales)*, 58 (March 1987).

Egelstaff, P.A.: *Introduction to the Liquid State*, Oxford University Press, Inc., New York, NY, 1994.

Granick, S.: "Motions and Relaxations of Confined Liquids," *Science*, 1§374 (September 20, 1991).

Grimmett, G., B. Eckmann, S.S. Chern, and H. Hironaka: *Percolation*, 2nd Edition, Vol. 321, Springer-Verlag, Inc., New York, NY, 1999.

Heyes, D.M.: *The Liquid State: Applications of Molecular Simulations*, John Wiley & Sons, Inc., New York, NY, 1998.

Israelachvili, J.N. and P.M. McGuiggan: "Forces Between Surfaces in Liquids," *Science*, 795 (August 12, 1988).

Koch, D.L.: "Fluid Dynamics in Multiphase Systems," *Chem. Eng. Progress*, 74 (November 1989).

Langer, J.S.: "Dendrites, Viscous Fingers, and the Theory of Pattern Formation," *Science*, 1150 (March 3, 1989).

Lounasmaa, O.V. and G. Pickett: "The ³He Superfluids," *Sci. Amer.*, 104 (June 1990).

Luessen, L.H., L.G. Christophorou, and E.E. Kunhardt: *The Liquid State and Its Electrical Properties*, Perseus Books, Boulder, CO, 1988.

March, N.H., M.P. Tosi and R.A. Street: *Amorphous Solids and the Liquid State*, Kluwer Academic Publishers, Norwell, MA, 1985.

McComb, W.D.: *The Physics of Fluid Turbulence*, Oxford University Press, Inc., New York, NY, 1992.

Monastersky, R.: "Stretching Liquid to Its Physical Limit," *Science News*, 87 (August 11, 1990).

Ottino, J.M.: "The Mixing of Fluids," *Sci. Amer.*, 56 (January 1989).

Ottino, J.M. et al.: "Morphological Structures Produced by Mixing in Chaotic Flows," *Nature*, 419 (June 2, 1988).

Pool, R.: "The Fluids with a Case of Split Personality," *Science*, 1180 (March 9, 1990).

Schmidt, W.F.: *Liquid State Electronics of Insulating Liquids*, CRC Press, LLC., Boca Raton, FL, 1997.

Snedden, R.: *States of Matter: Solids, Liquids and Gases*, Heinemann Library, Oxford, UK, 2001.

Stixrude, L. and M.S.T. Bukowinski: "A Novel Topological Compression Mechanism in a Covalent Liquid," *Science*, 541 (October 26, 1990).

Tabor, D.: *Gases, Liquids, and Solids: And Other States of Matter*, 3rd Edition, Cambridge University Press, New York, NY, 1991.

Thompson, P.A. and M.O. Robbins: "Origins of Stick-Slip Motion in Boundary Lubrication," *Science*, 792 (November 9, 1990).

LIQUIDUS CURVE. In a temperature-concentration diagram, the line connecting the temperatures at which fusion is just completed for the various compositions.

LIQUID WATER LOADING. The amount of liquid water present within an air parcel as cloud droplets, rain, or ice, usually expressed in percent or fraction by weight (e.g., as a liquid water mixing ratio r_L) or volume. The higher the liquid water loading, the greater the average density and colder the virtual temperature of the parcel.

LIQUID WATER PATH. A measure of the weight of the liquid water droplets in the atmosphere above a unit surface area on the earth, given in units of kg m^{-2}, for example.

The liquid water path may be defined as

$$W_p = \int \rho_{air} r_L dz,$$

where$_{air}$ is the density of the (wet) air, r_L is the liquid water mixing ratio, and the integral is from the bottom to the top of the column.

See also **Liquid Water Loading**.

LIQUID WATER POTENTIAL TEMPERATURE. A quantity that is conserved in reversible adiabatic motion. In the simplest approximation it is given by θ_L, where

$$\theta_L \approx \theta - \frac{L_v}{C_{pd}} r_L$$

with θ the potential temperature, L_v the latent heat of vaporization, c_{pd} the specific heat of dry air at constant pressure, and r_l the liquid water mixing ratio. A more accurate expression is

$$\theta_L = \theta \left(\frac{\varepsilon + r_v}{\varepsilon + r_i} \right)^x \left(\frac{r_v}{r_l} \right)^{-\gamma} \exp \left[\frac{-L_v r_l}{(C_{pd} + r_t C_{pv})T} \right],$$

in which X is the Poisson constant, ε the ratio of the gas constants of dry air and water vapor (0.622), r_v the mixing ratio of water vapor, c_{pv} the specific heat of water vapor, T the temperature and $\gamma = r_t R_v / (c_{pd} + r_t c_{pv})$, where r_t is the total water mixing ratio and R_v the gas constant for water vapor. Three quantities are conserved in reversible adiabatic motion: equivalent potential temperature, total water mixing ratio, and liquid water potential temperature. Any two of these may be considered independent, with the third deducible from those two.

Additional Reading

Deardorff, J., and K.A. Emanuel: *Atmospheric Convection*, Oxford University Press, New York, NY, 1994.

AMS

LISSAJOUS, JULES ANTOINE (1822–1880). Lissajous was a French mathematician. He is remembered for frequency patterns (Lissajous figures). The generation of Lisajous figures on a cathode-ray tube is a common method of frequency comparison.

See also **Frequency (Electric) Measurement**; and **Harmonic Motion**.

J. M. I.

LISTERIOSIS. A disease of animals including humans caused by *Listeria monocytogenes*, a thin Gram-positive bacillus having several serotypes. The organism is a soil saprophyte which is present in the intestines of many animals and birds. These reservoirs are potential sources of exposure to humans, but despite the ubiquity of *Listeria sp.*, human disease caused by the organism is uncommon. Most infections occur in the first month of life where, in early onset, the mortality may be as high as 40–60%; or beyond the age of 55, where there is usually an underlying predisposing illness.

The most common form of listeriosis is meningitis, with bacteremia occurring in 5–30% of cases, and endocarditis, osteomyelitis, and cholecystitis also sometimes evidenced.

Ampicillin or penicillin are the antibiotics of choice, with treatment being continued for at least ten days after the patient becomes afebrile.

R. C. V.

LISTER, JOSEPH (1827–1912).   Joseph Lister was a British surgeon whose introduction of antiseptic methods in surgery helped create modern operative technique.

Lister was born into a large and prosperous Quaker family. His father, Joseph Jackson Lister (1786–1869), was a wine merchant who, with Thomas Hodgkin, developed the achromatic microscope in the 1820s. This technological improvement greatly enhanced the reliability of the instrument and helped make it a central tool of biological and medical research. Joseph Lister thus grew up in a scientific household and was taught microscopy as a child. He attended Quaker schools in London and then University College, where he obtained a BA (1847) and stayed on to study medicine (MB, 1852). He was influenced there by the physiologist William Sharpey and the ophthalmic surgeon Joseph Wharton Jones. He engaged in microscopic research as an undergraduate, but determining on a career in surgery, Lister went to Edinburgh to study under James Syme, whose daughter, Agnes, Lister married in 1856. Although continuing his microscopic studies, especially on inflammation and on blood clotting, he also obtained a post as assistant surgeon at the Royal Infirmary in Edinburgh, which involved teaching a course on the principles and practice of surgery. Elected FRS in 1860, he was appointed Regius Professor of Surgery in Glasgow in the same year. Initially hampered by lack of access to patients, he obtained formal clinical responsibilities the following year. See also **Sharpey, William (1802–1880)**.

Lister developed his antiseptic techniques in Glasgow. Aware of Pasteur's researches on the role of microorganisms in such processes as fermentation and putrefaction, and also of the use of phenol (carbolic acid) in the disinfection of sewage, he reasoned that surgical sepsis might be controlled if surgical sites were protected from "germs" by covering the wounds in dressings soaked in carbolic acid. He also soaked his surgical instruments in the fluid, and as his antiseptic system developed, he introduced a carbolic acid spray, to destroy germs carried in dust particles in the air. Between August 1865 and April 1867, Lister treated eleven patients with compound fractures of their legs with his new technique; nine survived. He published his results in *The Lancet* in 1867 and was also able soon to report that his surgical wards were free of "hospital sepsis," a frequent cause of postoperative mortality. Lister's ideas of what actually occurs in surgical putrefaction and sepsis were complicated and changed over time. His antiseptic results were sufficiently favorable to raise a good deal of interest, and he continued to work on his system during his second Edinburgh phase, from 1869 to 1877, where he served as Syme's successor as professor of clinical surgery. He returned to London in 1877, as professor of surgery at King's College Hospital. By then, he had an international reputation, his antiseptic system having been accepted (and often modified) by a number of surgeons in Britain, North America and on the Continent.

Although Lister was clearly inspired by Pasteur's researches in his initial work, the actual role of bacteria in surgical wound sepsis was only slowly elaborated, Lister himself doing important bacteriological investigations in the 1870s and 1880s. He also continued to search for antiseptics less toxic than carbolic acid. The "Listerian revolution" was further developed by aseptic techniques, i.e. procedures that prevented bacteria from contaminating the surgical area in the first place, healthy tissues being free of bacteria. These procedures were made easier by Robert Koch's work on sterilization techniques and the introduction of gowns, masks and gloves into surgical theatres. Lister followed and approved these innovations but did not contribute much to them. See also **Koch, Heinrich Hermann Robert (1843–1910)**.

Lister received many honors in his old age. He served as President of the Royal Society, was the first medical man elevated to the British peerage (created Baron Lister of Lyme Regis in 1897), and appointed to the Order of Merit. He was an ardent supporter of medical research.

Additional Reading

Worboys, M.: *Spreading Germs: Disease Theories and Medical Practice in Britain, 1865–1900*, Cambridge University Press, Cambridge, UK, 2000.

W.F. BYNUM, Wellcome Trust Centre for the History of Medicine at UCL, London, UK

LITCHI TREE.   Of the family *Sapindaceae* (soapberry family), the *Litchi chinensis* is probably best known for its fruit, which when dried is called the litchi "nut." A native of southern China, the tree has spread extensively in cultivation through many southern Asiatic countries. The tree has pinnately compound leaves, the leaflets of which are lanceolate and leathery. The small flowers are borne in panicles, and have no petals. The fruit is roughly spherical, 1 to $1\frac{1}{2}$ inches in diameter, with a hard, brittle rind. Within this rind is a fleshy, translucent pulp, the aril, the part which is eaten. When fresh, it is delectable; when dried into "nuts," it is much shrunken. The fruit contains a single seed.

LITHIFICATION.   To lithify is, literally, to turn to stone. Lithification is a term commonly applied to the consolidation and hardening of sediments so as to form a sedimentary rock.

LITHIUM.   [CAS: 7439-93-2]. Chemical element, symbol Li, at. no. 3, at. wt. 6.941, periodic table group 1, mp 180.54 °C, bp 1342 °C (at 760 torr), density 0.534 g/cm^3 (20 °C). Lithium is lightest in weight of all the chemical elements that are solid at standard conditions. Elemental lithium in the solid phase has a body-centered cubic crystal structure. In comparison with other members of the alkali metal series, lithium has the smallest ionic radius, the highest ionization potential, the highest electronegativity, and the greatest heat capacity. Generally, lithium is the least reactive of the alkali metals. Lithium is a silver-white metal, harder than sodium, but softer than lead. It is tough and may be drawn into wire or rolled into sheets. The element tarnishes rapidly in air and often is preserved under naphtha. The reaction with H_2O is vigorous, producing LiOH (lithium hydroxide) and hydrogen. There are two naturally occurring isotopes, ^6Li and ^7Li. They are not radioactive. Two radioactive isotopes have been identified, ^5Li and ^8Li, both with very short half-lives, measured in fractions of seconds. Among elements occurring naturally in the earth's crust, lithium ranks 28th with an estimated average content of about 10–20 ppm. In terms of content in seawater, lithium ranks 17th with an estimated content of approximately 950 tons of lithium per cubic mile of seawater. The element was first identified by Johann August Arfvedson in 1817 in the laboratory of Berzelius. The name of the element is accredited to Berzelius.

First ionization potential 5.39 eV. Oxidation potential $Li \rightarrow Li^+ + e^-$, 3.02 V. Other physical properties of lithium are given under **Chemical Elements**.

The main sources of lithium are pegmatites and brines. The most important pegmatite mineral is *spodumene*, $LiAlSi_2O_6$, which contains a theoretical content of 8.03% Li_2O. *Petalite*, $LiAlSi_4O_{10}$, contains between 4 and 4.5% Li_2O. *Lepidolite*, a complex mica, contains between 3 and 4% Li_2O. See also entries on **Lepidolite**; **Petalite**; and **Spodumene**. Brines contain normally a few hundred to a few thousand parts per million (ppm) of lithium. The only commercial source of spodumene in North America is located in North Carolina. Abundant resources of lithium pegmatites occur in Canada, the African continent, and unconfirmed sources in Russia and China. Significant quantities of lithium (as carbonate) are produced from the brines of Clayton Valley, Nevada. A recently discovered, lithium-rich brine deposit has been located in the Atacama desert of Chile. More detail on lithium resources is given in the next entry.

There are three major processes for extracting lithium from pegmatite ores. (1) An acid process, wherein the spodumene concentrate, after calcining at about 1095 °C, is reacted with sulfuric acid, followed by water leaching of the resulting lithium sulfate, Li_2SO_4. The sulfate is then converted to the carbonate with soda ash. (2) An alkaline process, wherein the ore is reacted with lime or limestone at high temperatures followed by water leaching of the resulting lithium hydroxide, (3) A base exchange method, whereby the ore is reacted with an alkaline chloride or sulfate at a high temperature in an aqueous phase to yield a soluble lithium salt. The sulfuric acid leaching method is the only commercial process for extraction of lithium from spodumene in practice today.

Lithium metal was first prepared by Sir Humphry Davy in 1818 by electrolyzing lithium oxide. At about that same time, Brande also isolated the metal. In 1855, R. Bunsen and A. Matthiessen prepared gram amounts by electrolyzing fused lithium chloride. Modest commercial quantities were first made in Germany during World War I when the metal was considered as a potential alloying material. Limited production did not commence in the United States until the early 1930s. Present commercial methods were pioneered by Guntz in 1893 and involve electrolyzing a low-melting mixture of LiCl and KCl. Graphite anodes and mild steel cathodes are used. Lithium is formed at the cathode and rises to the surface, from which it is skimmed periodically. Pure lithium chloride is added to the bath as required. Chlorine gas is liberated at the anodes. The process yields a lithium metal of about 99.8% purity. The metal normally is cast into ingots of different sizes, but is also available as extruded rod, ribbon, or wire. The metal also is available as "sand" — fine dispersions in the 10–30 μm range.

Lithium in Metallurgy and Alloys

In metallurgy, lithium metal is used as a deoxidizer, desulfurizer, and degasifier in the production of a number of molten metals, notably copper and copper alloys. Lithium also is an ingredient of an increasing number of alloys, particularly with aluminum. Early alloys included aluminum alloy X2020 (1% Li), which is a structural alloy with improved high-temperature strength. In another early Li alloy, about 14% Li is alloyed with magnesium in the LA 141 alloy, designed for very light-weight structural applications, notably in aerospace applications.

In late 1989, a new proprietary (Martin Marietta Corp.) family of weldable, high-strength Al-Li alloys was introduced. With a 690-MPa (100×10^3 psi) yield strength, the material is claimed to be twice as strong as the previous leading Al-Li alloys. This alloy was developed specifically for space launch systems. The alloy is claimed to maintain a high strength under thermal conditions ranging from cryogenic to elevated temperatures. A primary use is for fuel and oxidizer tanks, where its weldability is a marked advantage. Sheet, plate, extrusion, and ingot products of the new alloy also are available.

The addition of lithium to aluminum castings has been found to be particularly advantageous. Lithium produces a lower density and higher stiffness over conventional aluminum alloys used in aerospace applications. Lithium has one of the highest solubilities of any aluminum alloying element. About 4.2% Li can be dissolved in Al at the 602°C (1116°F) eutectic temperature. The hardness of Al-Li alloys improves with aging temperature. Al_3Li precipitates are formed, producing higher hardness. Yield strength also increases with higher aging temperature and higher Li content.

Lithium alloyed with silver has been used for fluxless brazing.

Lithium Batteries. For many years, lithium has been considered for use in batteries, particularly with the growing emphasis on the electric car. See also **Battery**.

Chemistry and Compounds

Lithium has the highest ionization potential (i.e., Li → Li$^+$ in the vapor) of the alkali metals. However, the measured value of its oxidation potential against a normal aqueous solution of its ion is 3.02 V, which does not differ from those of the other main group I metals by as much as the difference in ionization potentials. That difference, attributed to the high heat of hydration of Li$^+$, explains why lithium is a vigorous reductant in aqueous systems, but reacts slowly with H$_2$O, and not at all with dry oxygen except above 100°C.

The single 2s electron in the outer shell of lithium is easily removed to form the positive ion, and stability of the remaining 1s^2 electron pair requires too high a potential (75.62 eV) for any further ionization (by chemical means) so that lithium is exclusively monovalent in its compounds.

Because of the reactivity of lithium with water to form its hydroxide, LiOH, and hydrogen, its properties when dissolved in other solvents have been studied extensively. It does not decompose liquid NH$_3$, but does form a blue solution, which decomposes to yield its amide, LiNH$_2$, and hydrogen, when catalyzed by metallic salts. With the elements of main groups 2 to 7, lithium in liquid NH$_3$ reacts to form binary compounds, which may vary from simple halides, as with the halogens, to intermetallic phases, as with cadmium and mercury. Lithium amide in liquid NH$_3$ is regarded in the same class as a hydroxide in aqueous solution.

Many other lithium compounds not obtainable in aqueous solution can be produced from the solution of lithium in liquid ammonia. Thus the acetylide is obtained by action of acetylene.

$$C_2H_2 + LiNH_2 \rightarrow LiC_2H + NH_3$$

$$LiC_2H \rightarrow Li_2C_2 + C_2H_2$$

The amide, as stated above, is produced by catalyzed decomposition of the liquid NH$_3$ solution, and the nitride, Li$_3$N, by heating the amide or by direct combination of the elements.

Lithium salts exhibit general high solubility and a high degree of dissociation in other nonaqueous solvents than liquid ammonia, such as liquid sulfur dioxide and acetic acid.

Like the other alkali metals lithium forms compounds with virtually all of the anions, inorganic as well as organic. The lithium salts are in many instances different in their solubility properties from the corresponding salts of the other alkali metals. Thus lithium fluoride, phosphate, and carbonate are the least soluble alkali metal fluoride, phosphate, and carbonate, the solubilities for the other alkali metals increasing with increasing ionic radius. Lithium chlorate and dichromate are, on the other hand, the most soluble alkali chlorate and dichromate, the solubilities for the other alkali metals decreasing with increasing ionic radius. These differences are partly explained, as was that in the oxidation potential, by the considerable hydration of the lithium ion, which also explains the fact that lithium salts generally crystallize as hydrates. Lithium salts, probably because of the small size of the lithium ion, do not form mixed crystals with the other alkali salts, but they do form double salts, notably the two series of lithium-sodium and lithium-potassium sulfates.

Lithium forms several organic compounds. Most of them are lithium salts or lithium acid salts of organic acids or other oxygen-connected lithium compounds. The number of lithium-carbon bonded compounds that have been reported is very small including, in addition to the carbide, methyllithium, CH$_3$Li, ethyllithium, C$_2$H$_5$, *n*-propyllithium, C$_3$H$_7$Li, *n*-butyllithium, C$_4$H$_9$Li, benzyllithium, C$_6$H$_5$·CH$_2$Li, and methylene-dilithium LiCH$_2$Li.

The alkyllithium compounds are usually colorless, soluble in organic solvents, and capable of distillation or sublimation. They are nonelectrolytes and are widely used in synthetic organic chemistry, since, like other lithium compounds, they resemble in their properties the corresponding magnesium compounds.

Lithium carbonate. [CAS: 554-13-2.] Li$_2$CO$_3$, mp 72.6°C, slightly soluble in H$_2$O. Used in glass, enamel, and ceramic formulations, in the electrowinning of aluminum, and in the manufacture of other lithium compounds. The compound also has been used in the treatment of manic-depressive psychoses.

Lithium hydride. [CAS: 7580-67-8.] LiH, mp 686.4°C, reacts vigorously with H$_2$O. With NH$_3$, it forms the amide. The compound is used to produce LiAlH$_4$ and other double hydrides. Lithium hydride is an excellent light-weight source of hydrogen. One pound yields 45 cubic feet of hydrogen (one kilogram yields 2.8 cubic meters of hydrogen) at standard conditions. The compound also can serve as a light-weight shield for thermal neutrons.

Lithium hydroxide monohydrate. [CAS: 1310-65-2.] LiOH · H$_2$O loses water at 101°C. LiOH melts at 450°C. The compound is soluble in water. The compound is used in the formulation of lithium soaps used in multipurpose greases; also in the manufacture of various lithium salts; and as an additive to the electrolyte of alkaline storage batteries. LiOH also is an efficient, light-weight absorbent for carbon dioxide.

Lithium bromide. [CAS: 7550-35-8.] LiBr, mp 550°C, soluble in H$_2$O or alcohols. The compound is very hygroscopic and forms four hydrates. Major use has been in absorption-refrigeration air-conditioning systems in which H$_2$O is the refrigerant — strong LiBr is used to absorb H$_2$O vapor.

Lithium chloride. [CAS: 7447-41-8.] LiCl, mp 608°C, soluble in H$_2$O or alcohols. Very hygroscopic and forms four hydrates like the bromide. The compound is a component of brazing fluxes for aluminum and magnesium. It is used in dehumidification systems, as an additive to the electrolyte of dry cells for low-temperature applications; and it is used in low-freezing fire-extinguishing systems; as an ingredient of fused-salt baths to lower fusing temperature; and, as a coating, in humidity-sensing instruments.

Lithium fluoride. [CAS: 7789-24-4.] LiF, mp 848°C, soluble in H$_2$O (slight). Used in enamel and glass formulations; as a component of welding and brazing fluxes; in the electrowinning of aluminum; and as an ingredient of molten salts.

Lithium in Biological Systems

Although much remains to be learned, there is considerable evidence that lithium can play an active role (positive and negative) in biological systems. Possibly most widely known is the use of lithium salts, notably lithium carbonate, in the therapy for mania (a condition where the patient is mentally and physically hyperactive, associated with an elevated mood and disorganized behavior). Frequently associated with mania is the broad swinging of the patient's mood (*bipolar disorder*). Studies have shown that persons with mania have a defect in the transmissions of impulses between and along nerve cells in the brain, which depends upon the regulated movement of ions across the membranes of those cells. Lithium antagonizes synaptic transmission of catecholamines in the brain by inhibiting norepinephrine and dopamine release. This is the result of weakly increasing their re-uptake by the presynaptic neuron and by decreasing storage. Lithium also interferes with the ability of several hormones to stimulate adenylate cyclase, a property that is believed to decrease the action of catecholamines at the postsynaptic receptor sites.

Additional Reading

Birch, N.J., V.S. Gallicchio, and R.W. Becker: *Lithium: 50 Years of Psychopharmacology: New Perspectives in Biomedical and Clinical Research*, Weidner Publishing Group, Riverton, NJ, 1999.

Carr, S. et al.: "Increase in Glomerular Filtration Rate in Patients with Insulin-Dependent Diabetes and Elevated Erythrocyte Sodium-Lithium Countertransport," *N. Eng. J. Med.*, **500** (February 22, 1990).

Davis, J.R.: *Metals Handbook*, 2nd Edition, ASM International, Materials Park, OH, 1998.

Greenwood, N.N. and A. Earnshaw: *Chemistry of the Elements*, 2nd Edition, Butterworth-Heinemann, Inc., Woburn, MA, 1997.

Julien, C. and Z. Stoinov: *Materials for Lithium-Ion Batteries*, Kluwer Academic Publishers, Norwell, MA, 2000.

Krebs, R.E.: *The History and Use of Our Earth's Chemical Elements: A Reference Guide*, Greenwood Publishing Group, Inc., Westport, CT, 1998.

Kubel, E.J., Jr.: "New Al-Li Alloy," *Advanced Materials 7 Processes*, **10** (October 1989).

Lewis, R.J. and N.I. Sax: *Sax'x Dangerous Properties of Industrial Materials*, 10th Edition, John Wiley & Sons, Inc., New York, NY, 1999.

Lide, D.R.: *CRC Handbook of Chemistry and Physics*, 88th Edition, CRC Press, LLC., Boca Raton, FL, 2007.

Sapse, Anne-Marie and P. von Rague Schleyer: *Lithium Chemistry: A Theoretical and Experimental Overview*, John Wiley & Sons, Inc., New York, NY, 1994.

Schou, M.: *Lithium Treatment of Manic Depressive Illness: A Practical Guide*, S. Karger Publishers, Inc., Farmington, CT, 1993.

Swartz, C.M.: "Serum Lithium During Treatment of Bipolar Disorder," *N. Eng. J. Med.*, **1159** (April 19, 1990).

Taketani, H.: "Properties of Al-Li Alloy 2091-T3 Sheet," *Advanced Materials and Processes*, **113** (April 1990).

Wakihara, M. and O. Yamamoto: *Lithium Ion Batteries: Fundamentals and Performance*, John Wiley & Sons, Inc., New York, NY, 1998.

LITHOLOGY.   Literally, the graphic study of rocks, hence a synonym for petrography, but not petrology. This term is usually restricted, however, to the purely descriptive macroscopic study of rocks, without the aid of the petrographic microscope.

LITHOMETEOR.   Solid matter suspended in the atmosphere, as smoke, dust, dry haze, etc., as contrasted with hydrometeor.

LITHOPRISM.   See **Prism (Optics)**.

LITHOSPHERE.   The solid part of the Earth or other spatial body. Distinguished from the atmosphere and the hydrosphere. See also **Earth**.

LITTLE, ARTHUR D. (1863–1935).   Born in Boston, Little was a pioneer in the field of industrial research and chemical consulting. Originally an authority on paper technology, he established a consulting industrial chemical laboratory in 1886, which has since become a large institution of worldwide reputation, located in Cambridge, MA. It has served as a prototype of many industrially oriented consulting firms that have become a significant factor in the growth of research in the last half century. It has made significant contributions in such fields as flavors, food chemistry and acceptability, paper chemistry, and rubber chemistry, as well as in corporate management.

LITTORAL.   Inhabiting the shoreline of the ocean in shallow waters and in the tidal zone, which is periodically exposed to the air.

LITUUS.   A transcendental plane curve, a special kind of spiral. Its equation in polar coordinates is $r^2\theta = a^2$. It begins at infinity, constantly approaches the pole but never reaches it, and has the polar axis as an asymptote. See Fig. 1.

See also **Curve (Plane)**.

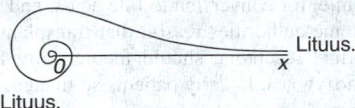

Fig. 1.   Lituus.

LIVER.   The largest and one of the most complex organs in the human body, consisting of four lobes and located in the upper abdominal cavity. The liver performs multiple functions, including: (1) secretion and excretion; (2) blood-related activities, including regulation of blood volume and the formation and disposal of various blood components; (3) storage for certain vitamins and minerals; (4) metabolic functions, including fat and protein processing; and (5) detoxification. The approximate size of the adult human liver is 8–9 inches (20–22.5 centimeters) side to side, 4–5 inches (10–12.5 centimeters) front to back, and 6–7 inches (15–17.5 centimeters) top to bottom along the thickest portion. The organ weighs between 42 and 56 ounces (1.2 and 1.6 kilograms). Five ligaments attach the liver to the anterior walls of the abdomen and undersurface of the diaphragm. The lobes are usually identified as the *right* (largest); the *left* (somewhat smaller and wedge shaped; the *quadrate* (roughly square-shaped); and the *caudate* (tail-like configuration). Principal diseases and disorders with liver involvement are cirrhosis, hepatitis, and jaundice. Primary hepatic carcinomas are also seen, but less frequently. (The word *hepatic* indicates a condition of, or affecting the liver.) The liver is subject to adverse effects caused by a number of substances, including certain antimicrobials, such as isoniazid, rifampin, and pyrazinamide.

The liver secretes bile, which is discharged into the intestine: absorbs from the blood the products of carbohydrate digestion and stores them as glycogen; acts on nitrogenous wastes and returns them to the blood in the form of urea and related compounds; and destroys "worn-out" red corpuscles. The liver also produces fibrinogen and prothrombin. The bile discharged through the intestine plays an important role in the digestion of fat and carries with it some of the more complex waste products of the body. See also **Bile**. In structure, the liver is very complex. It develops as a hollow outgrowth of the embryonic gut just behind the stomach, which forks to produce the gallbladder and the liver. The connection with the gut persists as the common bile duct. In the adult, the liver cells are arranged in cords, separated by blood channels with incomplete lining known as sinusoids. Within the cords, minute bile capillaries between the cells converge to larger and larger ducts, which ultimately form the main hepatic duct. The gland receives blood from an arterial supply and also from the portal vein. The latter drains blood from the capillaries of the intestine and breaks up into sinusoids in the liver. These small passages are drained by the hepatic vein. The formation of stones in the biliary tract is described in **Gallbladder and Biliary Tract Diseases**.

A more detailed description of the physiology and biochemistry of the liver is given toward the end of this entry.

Familial Hypercholesterolemia.   This disease has been treated by way of liver transplantation. It is a genetic disease caused by mutations in the gene encoding the low-density lipoprotein (LDL) receptor. This cell-surface receptor normally removes cholesterol-carrying LDL from the circulation. Persons with two mutant LDL-receptor genes produce few or no LDL receptors and, therefore, remove LDL from plasma at a reduced rate. As a result, LDL accumulates in plasma to levels up to 8 times normal. Patients almost always have severe atherosclerosis in childhood, with death from myocardial infarction often occurring before age 20 years. A single mutant LDL-receptor gene is inherited and occurs at a frequency of 1 in 500 in the general population: affected persons accumulate twice the normal level of LDL and symptomatic atherosclerosis usually occurs in the fourth and fifth decades of life. Typically, each heterozygote for familial hypercholesterolemia will transmit the mutant gene to half of the children, who then become heterozygotes. When two heterozygotes marry (estimated to be 1 in 250,000 marriages), one-fourth of the offspring will inherit a copy of the mutant LDL-receptor gene from both parents, and these offspring will be homozygotes.

Until recently, traditional approaches to the disease have not been effective. Inasmuch as the liver manufactures large numbers of LDL receptors (because the organ requires a large amount of cholesterol for secretion into the bile, for conversion to bile acids, and for the production of lipoproteins), some authorities reason that transplantation of a normal liver, with its normal receptors, should theoretically lower LDL levels profoundly in homozygotes. In early patients so treated, this has proved to be a correct assumption.

In mid-1990, Reihner and associated researchers (Karolinska Institute, Stockholm, Sweden) reported on the use of an inhibitor of cholesterol biosynthesis in the treatment of hepatic metabolism disorders. The cholesterol production rate-limiting enzyme (pravastatin) is 3-hydroxy-3-methyl glutaryl coenzyme A (HMG-CoA) reductase. In a study of ten patients over a period of three weeks, the group found that pravastatin therapy reduced total plasma cholesterol by 26 percent and LDL cholesterol by 39 percent. The report concludes: "Inhibition of hepatic HMG-CoA reductase by pravastatin results in an increased expression of hepatic LDL receptors, which explains the lowered plasma levels of LDL cholesterol."

Diseases of the Liver

Cirrhosis. In the Western Hemisphere, *chronic* diseases of the liver are comparatively infrequent—with exception of cirrhosis of the liver, which occurs frequently in Europe and the United States. In cirrhosis, the hepatic parenchyma (functioning tissue, as contrasted with connective tissue) is progressively destroyed and replaced by collagen (gelatinous substance found in connective tissue and bone). During surgery or autopsy, the organ will exhibit bands of collagen extending between the lobes and connecting portal areas. This process, if left untreated, ultimately grossly disorganizes the liver and leads to cessation of the organ's metabolic functions.

Alcoholic cirrhosis is the most common type of cirrhosis seen in the United States. Infrequently, the liver will shrink in this disease, but more frequently the organ will enlarge and may weigh two kilograms or more. Ingestion of large quantities of alcohol over a period of years[1] is the primary cause of alcoholic cirrhosis. It is no longer generally believed that poor nutrition, which often accompanies heavy alcohol consumption, is a primary cause of the disease, although it may be a secondary contributing factor to degeneration of the health of the individual.

In treatment, the logical first step for the patient is to stop drinking alcohol. Statistics indicate that the 5-year survival for patients who continue to drink is less than 50 percent. This may reach 80% in cases where the patient maintains abstinence. Treatment is directed toward maintaining good nutrition and preventing serious complications. Total fluid intake is supervised to effect an optimum fluid and electrolyte balance. Diuretics may be used, particularly where massive ascites (accumulation of serous fluid in abdominal cavity) are present. Vitamin K therapy is sometimes used. Protein intake may be restricted where hepatic encephalopathy may be suspected.

Cirrhosis increases the risk of gallstones as the result of elevated bilirubin. See also **Bile**. The risk of peptic ulcer is also increased twofold in cirrhosis. Also, in well-established cirrhosis, the *hepatorenal syndrome* may be seen. Simply defined, this is functional renal (kidney) failure. Mortality can range from 60 to 90 percent. Treatment is directed toward eliminating exogenous sources of ammonia, usually accomplished by restricting dietary protein. Means are also taken to control gastrointestinal bleeding. Where alcoholic liver disease is well established, some 10% of patients may develop *spontaneous bacterial peritonitis*. The exact mechanism underlying this condition is not fully understood. *Bleeding esophageal varices* are a serious complication of alcoholic cirrhosis.

The veins of the portal venous system transport all blood from the abdominal gastrointestinal tract, spleen, pancreas, and gallbladder, returning it to the heart by way of the liver. Portal hypertension in patients with cirrhosis causes gastrointestinal bleeding and esophageal varices. T. Poynard and a team of researchers (Franco-Italian Multicenter

Study Group) conducted a study to determine the effectiveness of beta-andrenergic-antagonist drugs in the prevention of gastrointestinal bleeding in patients with cirrhosis and esophageal varices. Prior studies had not been conclusive. Generally, it had been reported that the continuous administration of beta-adrenergic-antagonistic drugs had induced a sustained decrease in portal pressure in patients with cirrhosis. Conclusion of the Poynard team findings (March 1991): "Propranolol and nadolol are effective in preventing first bleeding and reducing the mortality rate associated with gastrointestinal bleeding in patients with cirrhosis regardless of severity."

As reported by Massimo Colombo, et al. "Hepatocellular carcinoma is a highly malignant tumor with an extremely poor prognosis and an estimated incidence of about 1 million cases per year worldwide. Patients with cirrhosis of the liver have been identified as being at risk for this carcinoma." The causation of this particular type of tumor in association with cirrhosis is poorly understood. A study group at the University of Milan has concluded: "In the West, as in Asia, patients with cirrhosis of the liver are at substantial risk for hepatocellular carcinoma, with a yearly incidence rate of 3 percent. Our screening program did not appreciably increase the rate of detection of potentially curable tumors."

Effective treatment is difficult and bleeding from this source may be fatal in 70% of cases.

Primary biliary cirrhosis, relatively uncommon, is typically seen in women during the fourth to sixth decade of life. Symptoms in the early phase include a generalized itching of the skin (pruritus) and minor, continuing fatigue. Often the condition persists for a long period before a physician is consulted. In this disease, there is a significant drop in biliary secretion, causing a marked rise in serum cholesterol level. Xanthomas (small, yellow neoplastic growths) may occur about the eyes and tuberous xanthomas may be seen over the extremities. Bone pain, resulting from chronic malabsorption of fat-soluble vitamins A, D, E, and K, may be apparent. Diagnosis can be difficult because of similarities of primary biliary cirrhosis with subclinical cholangitis or other biliary tract diseases. See also **Gallbladder and Biliary Tract Diseases**.

Although the disease progresses slowly, the long-term prognosis is usually not good (10–20 years). Since there is no effective and specific therapy for the condition, treatment is directed toward alleviating symptoms.

Hemochromatosis. This is an infrequent liver disease in which inordinately large quantities of iron are deposited in the parenchymal cells of the organ. Eventually, these deposits destroy and scar the liver as in cirrhosis. Males have this disease at ten times the rate of females. Although onset may be earlier, symptoms usually develop during the fourth and fifth decades of life. Because of malfunctions in processes which govern iron absorption, the iron deposits not only in the liver, but also in the skin, pancreas, and heart muscle. The symptoms include a brown cutaneous pigmentation (caused by melanin) and grayish appearance (due to iron). The liver may be enlarged. There may be weight loss, decrease in body hair, and weakness. Other symptoms parallel those of diabetes mellitus, congestive heart failure or arrhythmias, and stiffness in the joints. Therapy at one time was directed toward ameliorating the aforementioned symptoms (related disorders), but comparatively recently it has been found that removal of a point (0.47 liter) of blood at regular intervals (*phlebotomy*), depending upon the patient's specific condition, is effective in depleting the iron stores. General improvement occurs if arthropathy or pituitary insufficiency are not present.

Wilson's Disease. Related to excessive deposits of copper, this is a liver-related disease and discussed in the entry on **Wilson's Disease**.

Hepatitis. An inflammatory and necrotic disease of liver cells. Commonly, the disease will be virus-induced, although hundreds of drugs are also known to cause hepatitis. It is difficult to determine the source. Among the known viruses that produce acute hepatitis are: (1) Hepatitis type A, once called infectious or short-incubation hepatitis virus; (2) hepatitis type B, serum or long-incubation hepatitis virus; (3) the non-A, non-B hepatitis viruses; (4) Epstein-Barr virus, which also causes infectious mononucleosis; and (5) cytomegalovirus. See also **Virus**. Hepatitis continues to appear after blood transfusions—in about 30,000 cases per year in the United States even though research has centered on preventing these occurrences. It should be pointed out that, although excellent tests are available for hepatitis B, unfortunately most post-transfusion cases (up to 90%) develop as the result of the presence of non-A, non-B viruses.

---

[1] Ten percent of heavy drinkers (1 pint of whiskey daily for a number of years) run a high risk of developing cirrhosis.

There is some evidence that type A and type B hepatitis may be decreasing in the United States. The fatality rate of type A hepatitis is low, probably not exceeding 0.2 percent. The rate is higher in type B hepatitis, ranging from 0.3 to 15 percent. It is estimated that nearly 45% of the population has an immunity to type A infections; 5–10% of the population for type B infections. Immunity for non-A, non-B infections is unknown. Type A virus is transmitted by the fecal-oral route. Sewage-contaminated shellfish are sometimes implicated. The transmission of type B virus may be percutaneous (penetration through skin), oral-oral, or venereal. With non-A, non-B viruses, the route is percutaneous. The incubation period varies with each type of virus: Type A hepatitis, 20–37 days, but in extremes may range from 15 to 49 days; type B hepatitis, 60–110 days, but in extremes, from 25 to 160 days; non-A, non-B viruses, 37–70 days, but in extremes, from 21 to 84 days.

The course of hepatitis infections also varies with the causative agent. Type A hepatitis does not progress to chronic liver disease, whereas about 10% of cases of Type B infections will lead to chronic liver disease. The risk of chronic liver disease with non-A, non-B virus infections is higher, ranging from 10–40% of cases. In situations where exposure to the virus is known, but disease has not developed, the administration of pooled gamma globulin is effective in the cases involving type A and non-A, non-B viruses, but is not effective in type B cases. Where there has been exposure to type B virus, the use of specific hepatitis B immune globulin is effective.

It is not surprising that the onset and course of acute viral hepatitis vary considerably because of the several possible causative factors. Onset may be sudden or gradual. In general, all or some of the early symptoms will include combinations of fatigue, lassitude, drowsiness, loss of appetite, nausea and, most specific to the disease, dark urine. Headache and very mild fever, in the absence of chills, may be present. There is usually mild and generalized abdominal discomfort. Movement tends to aggravate this discomfort. Itching of the skin may occur. Mild arthritis may develop, although this symptom is usually limited to type B virus infections. As the disease progresses, jaundice will be evident. See also **Jaundice**.

Particularly in older people with type B infections, recovery may be quite long—several months to a year—with recurring intermittent symptoms (relapse). A relapse is milder and of shorter duration than the initial attack. Rarely, the course of the disease will be rampant, leading to coma and even death. Fatal complications of hepatitis may include aplastic anemia, hemolytic anemia, hypoglycemia, and polyarteritis. Some people do not recover completely from type A and non-A, non-B viruses and develop *chronic hepatitis*.

Many authorities agree that treatment seldom alters the course of acute viral hepatitis. Sensible suggestions are made to the patient—bed rest when there is excessive fatigue and serious discomfort, which may be present in the initial phase. Controlled studies have shown that vigorous physical exercise during the recovery phases does not increase the risk of relapse or chronic disease. It has been reported that a high-calorie diet (3000 + calories/day) for a few days at the appropriate time in the recovery stage may shorten the duration of the disease by a few days. Low-fat diets have not been shown to alter the course of the disease. During periods of nausea and vomiting, hospitalization may be required so that intravenous fluid and electrolyte replacement can be effected. Although alcohol has not been shown to aggravate the disease, abstinence is usually suggested in the interest of limiting any additional load on the hepatic and related systems.

In instances of severe acute viral hepatitis where the patient becomes encephalopathic, corticosteroids (not proven effective), hyperimmune gamma globulin, keto analogues of essential amino acids, and exchange transfusions, among other drugs and procedures, have been used. However, some authorities currently feel that acute encephalopathy responds little, if any, to treatment.

Chronic hepatitis may be *chronic active* or *chronic persistent*. Diagnosis is important because the therapy differs for the two conditions. Frequently, a percutaneous liver biopsy will be required. In chronic active hepatitis, the disease is variably progressive and eventually causes cirrhosis and hepatic failure. On the other hand, chronic persistent hepatitis does not progress. Persons with untreated chronic active hepatitis have a 5-year survival expectancy of less than 50%—possibly up to 90–95% where corticosteroid therapy is effective. This therapy in responsive patients brings about improvement of liver function within several months. In contrast, chronic persistent hepatitis is benign and usually patients lead a normal, active life, even though serum aminotransferase abnormalities may continue for many years.

Drug-induced hepatitis is often difficult to differentiate from the disease caused by a virus. Although uncommon, in one case in 9,000–10,000 patients the administration of halothane (see also **Anesthesia**) will produce hepatic necrosis. This occurrence is most common in overweight females or after a second exposure to the drug, and is frequently fatal. Drugs that also are hepatitis related include isoniazid, methyldopa, phenytoin, and the sulfonamides. Although such side effects are uncommon, the physician will be on guard for signs of hepatitis and liver damage in deciding on starting or continuing therapy with these drugs. There is some evidence that cysteamine or acetylcysteine may reverse the actions of these drugs if noted promptly. Poison derived from the wild mushroom *Amanita phalloides* is capable of producing overwhelming hepatic necrosis. See also **Foodborne Diseases**. The adverse effects of certain antimicrobial agents were mentioned earlier in this entry.

Reye's Syndrome. This is an often-fatal systemic disorder that follows viral infection in children. A number of cases of what appears to be Reye's syndrome have also been described in recent years in young and middle-aged adults. Present in the syndrome are encephalopathy and fatty liver. The syndrome develops suddenly, with onset of intractable vomiting occurring a few days after the viral illness. Sensorial impairment appears and soon afterward may progress to coma. Seizures also may occur. The liver is usually enlarged. Specific therapy is not available. Supportive measures include lactulose to control hyper-ammonemia; fresh frozen plasma to replenish clotting factors; mannitol or dexamethasone to lower increased intercranial pressure; and mechanical ventilation. A case fatality rate of 23% has been reported. Epidemiologic evidence strongly links Reye's syndrome with outbreaks of viral disease, especially influenza. Although the mechanism is unknown most physicians recommend that salicylates (aspirin and aspirin-containing mendicants) not be given to children with chicken pox or influenza.

Physiology and Biochemistry of the Liver

The blood returning from the intestine to the heart is shunted through a capillary system, the hepatic sinusoids, which are surrounded by epithelial cells arranged in plate forms. These plates cross each other in space at different angles, to permit the greatest possible contact between the blood and these polygonal epithelial cells. The resulting sponge-like organ located under the diaphragm and covered by the connective tissue capsule of Glisson is the largest organ of the body. Under normal circumstances, the major part of its blood, between 66 and 75%, comes from the portal vein, which drains the splanchnic capillaries, particularly those of intestine, pancreas and spleen. Approximately one quarter to one-third of the hepatic blood comes from the hepatic artery originating from the aorta at the celiac axis. Both hepatic artery and portal vein enter at the hilus of the liver and divide in a dichotomic fashion into parallel running branches. They are surrounded by ramified extensions of Glisson's capsule. The hepatic artery sends branches to the capillary plexus of the portal tracts, whereas the bulk of its blood is released into the sinusoids parenchyma, as does the portal vein, which forms by confluence of superior mesenteric, inferior mesenteric and splenic vein, and receives additional internal radicles from the portal capillary plexus. The sinusoids are blood spaces, normally without the basement membrane otherwise seen in capillaries; they are, therefore, characterized by great permeability for serum protein. Moreover, some of their lining endothelial cells, which are star-shaped and called Kupffer cells, are part of the reticuloendothelial system. The lining cells form the sinusoidal wall and leave small stomata open through which macromolecular substances pass into a tissue space between liver cell plates and sinusoids and extending between neighboring hepatocytes almost to the bile canaliculus. Tissue fluid is drained toward the lymphatics in either the central canal or, in the human, mainly the portal tract. Arterial and venous blood, mixed to a varying degree, flows toward the tributaries of the hepatic veins which combine to larger ones into which frequently small branches enter at almost right angles. The largest branches enter into the vena cava inferior behind the liver. Vascular sphincter mechanisms in various locations regulate hepatic blood flow and thus function. The portal tracts and the central canals around the tributaries of the hepatic veins cross each other in space and are throughout the liver about 0.3 mm apart. The direction of the blood flow from the portal tracts to the central canals produces the concentric arrangement of the liver cell plates characterizing the liver lobule, which conventionally is considered the structural unit of the liver.

The liver forms bile, which is released into slits between the liver cells, the bile canaliculi, which are arranged in a chicken-wire-like fashion; the wall of the canaliculi is formed by part of the hepatocellular plasma membrane. The bile canaliculi are drained by small tubes with an independent cuboidal epithelial lining, the ductules or cholangioles. Under normal circumstances hardly any are found within the lobule, the majority being in the periportal zone or in the portal tract. Under abnormal circumstances, they increase in number and are then found deep within the lobule. The ductules continue into the bile ducts located in the portal tracts, which unite in dichotomic fashion to finally form the common hepatic duct; this duct leaves the liver where hepatic artery and portal vein enter it. It combines with the cystic duct draining the gallbladder, which concentrates bile by water reabsorption to form the common duct running toward the duodenum. This entrance is controlled by the choledochoduodenal sphincter of Oddi. Bile is produced at an almost constant rate but released from the biliary system in human beings and many animals only if food appears in the duodenum. As a result of this or other mechanisms, the sphincter of Oddi relaxes and the gallbladder contracts. This leads to proper utilization of bile which, while being partly an excretory product, is a secretion essential in intestinal digestion and absorption.

In the liver, several fluid currents exist. Blood and some tissue fluid flow toward the central canal, while bile and most of the tissue fluid (at least in the human) are flowing toward the portal canal. The normal liver consists of approximately 60% hexagonal epithelial cells (hepatocytes), 30% littoral endothelial or Kupffer cells, and about 2% each of bile duct cells, connective tissue and blood vessels. The hepatocytes have three types of borders. Where they are in contact with each other, the border is straight indicating limited, if any, exchange of substance between individual cells. The border toward the tissue space is elongated by narrow extensions of the space between neighboring hepatocytes, and particularly by the formation of irregularly shaped finger-like projections in the form of microvilli. This tremendous elongation of the border of the hepatocytes and the preferential location of enzymes in this location reflects structurally the extensive exchange of substances between hepatocytes, tissue space and blood. Much shorter is the border toward the bile canaliculus, also thrown into microvilli, which are far more regular and disappear upon impairment of biliary secretion. Preferential accumulation of ATPase in the villi indicates the intensity of the metabolic processes in bile secretion.

The nucleus is normally vesicular and has conspicuous nucleoli. It varies considerably under normal and pathologic conditions, the majority being tetraploid in adult rodents. Binucleated cells increase in regeneration. The cytoplasm normally contains many and relatively large mitochondria in the matrix, of which the citric acid cycle enzymes and, in the double membrane, the electron transfer enzymes can be demonstrated. Ribosomes as ribonuclear protein are arranged around messenger RNA usually in helix form as polysomes. These polysomes as the site of protein biosynthesis may be either free in the cytoplasm or attached to the extensive endoplasmic reticulum, which thus becomes granular, and the site of secretion of protein such as serum proteins. The endoplasmic reticulum is also the site of steroid synthesis, and the smooth endoplasmic reticulum is the site of detoxification and of glucose-6-phosphatase. In addition, one notes the Golgi apparatus responsible for secretion and the perinuclear dark bodies, the lysosomes, which are the site of various hydrolytic enzymes mainly with peak activity in acid medium. They serve to segregate intracellular material after pinocytosis or for storage, secretion and separation of organelles undergoing destruction in the form of autophagic vacuoles. The soluble fraction of the cytoplasm, the hyaloplasm, corresponding to the supernatant fluid in cytochemical analysis, contains proteins and enzymes and cofactors related to carbohydrate metabolism and activation of amino acids and nucleic acids. In addition, in the normal liver, glycogen and few fat droplets are found as well as some ferritin crystals which, under abnormal circumstances, become hemosiderin deposits giving histochemical iron reaction.

The main functions of the liver cells are: (1) secretion of substances into the bloodstream of which the serum proteins particularly albumins,

alpha-globulin, the proteins concerned with blood clotting, haptoglobin and transferrin, as well as some blood enzymes (e.g., esterase), serum cholesterol, and blood glucose are the most important; in contrast to all other tissues which utilize but do not form blood glucose, the liver cells are the main source of the blood glucose because of a specific phosphatase system; (2) storage of various metabolites particularly glycogen, proteins, fat and vitamins; (3) transformation of various compounds into each other, e.g., fats into carbohydrates and vice versa; (4) detoxification mainly by oxidation or conjugation, the latter mainly for better solubility and urinary excretion; (5) formation of the bile into which bile pigment is transmitted by conjugation and bile acids and cholesterol by transformation.

A variety of sinusoidal cells are seen. Some are flat endothelial cells, others are Kupffer cells with a cytoplasm of varying and irregular outlines and ameboid extensions. They contain few mitochondria but varying inclusions. They are engaged in phagocytosis of circulating exogenous and endogenous macromolecular or corpuscular elements, including bacteria, as well as of hepatocellular breakdown products, they are active in transformation of blood pigment to bile pigment. Other sinusoidal lining cells, rare under normal circumstances, form serum gamma-globulin and correspond to plasma cells. Also fibroblasts can be seen around the sinusoids.

The liver, as a whole, because of its strategic situation near the right heart and because of its sheer bulk, influences circulating blood volume, as well as electrolyte and water metabolism.

Liver Transplantations

It was just about two decades ago when a liver transplantation was covered during the prime time news. By the early 1990s, liver transplantation had become an established therapy, with such transplantations numbering in the thousands. Because of a very serious shortage of livers for donation, the waiting list numbers in the thousands. Liver transplantation ethics has become a major topic of discussion among medical and health professionals. A simple question typifies the current dilemma: "Who should receive a liver for transplantation? A young mother, whose prospects for surviving with a transplant is only 20 percent, or a 65-year-old alcoholic, whose chance of surviving is 80 percent if the patient stops drinking?" Some medical professionals have observed that, if the criteria for selecting patients become too narrow, this could have a dampening effect on liver transplantation technology.

A few experts believe that alternate non-transplantation therapies ultimately will develop, thus making liver transplantation procedures obsolete in the long run. However, for the immediate time, simple optimism for the future does not suffice.

When it was established that a human can regenerate a partly removed liver, some researchers proposed that pieces of liver tissue may be regenerated in the laboratory, thus maximizing the effectiveness of available whole organs from donors, who have been in short supply since the beginnings of liver transplant technology. Research along these lines commenced in the chemical and biochemical engineering laboratories at Massachusetts Institute of Technology. In the early phase of the project, liver cells were mounted on polymer mesh and treated with enzymes. The aim of the program is to grow the cells to about 10% of the mass of a whole liver. This tissue then would be transplanted into the patient and would, within several weeks, become fully-grown and replace the original liver. Research is continuing.

In a large series of childhood liver transplantations, nearly one-third have been performed because of metabolic disorders. Liver transplantation now is a well-accepted treatment for inherited metabolic disorders. Glycogen storage (Type IV) disease can be reversed by successfully transplanting a liver with normal amounts of branching enzyme. Inherited abnormalities of glycogen metabolism have been recognized for many years, mainly as the result of pathologists identifying gross accumulations of glycogen in tissues during postmortem examination. As pointed out by R.R. Howell (University of Miami), "The specific patterns of tissue involvement permitted recognition of a number of clinical types of glycogen storage disease long before the biochemical pathways had been identified and the specific causes of the inborn errors of metabolism understood." Some researchers had postulated that, after liver transplantation, progressive and probably fatal myopathy, cardiomyopathy, or encephalopathy would develop, since the enzyme defect is present in all the affected tissues, and that the abnormal glycogen would continue to accumulate in them. Surprising and fortuitous are findings that the predicted and ultimately fatal conditions have not occurred. A satisfactory explanation remains to be developed.

R.W. Strong and associates (Royal Children's Hospital, Brisbane, Australia) have reported that orthotopic liver transplantation is an effective therapy for end-stage liver disease and has proved to be a major advance in treating liver disease in children. The foremost obstacle faced by transplantation units worldwide has been the relative paucity of infant and child donors. The principle of transplanting a portion of the liver from an adult into a child has been accepted in many centers. Ethical issues must be considered when contemplating liver transplantation from parent to child. These issues are similar to those associated with the transplantation of renal grafts from living, related donors. Experience in many centers has shown that the risk to the donor is considered minimal. The risk to the recipient is considered no greater than that with the transplantation of a reduced-size graft from a cadaver donor. In the opinion of many specialists, living-donor liver transplantations are not justified when there are sufficient numbers of cadaver donors. In special instances, however, the procedure can be justified, as, for example, with a patient with fulminant hepatic failure when no cadaver donor is available and when the recipient has a reasonable chance of a successful outcome.

Cyclosporine is the basis of the immunosuppressive regimen in most patients undergoing orthotopic liver transplantation. The drug is a cyclic polypeptide that is produced by two species of fungi and is essentially insoluble in water. Its oral form consists of a solution of 100 mg of cyclosporine per milliliter of olive oil containing 12.5% ethanol by volume. Cyclosporine is absorbed much as fat and other fat-soluble substances are. Children, particularly infants, require much higher doses of orally administered cyclosporine than adults after liver transplantation. Bowel length appears to be the main determinant of the large difference between the adult and infant populations. These factors are reported by Whitington and associates (University of Chicago).

Additional Reading

Bacon, B.R. and A.M. Di Bisceglie: *Diseases of the Liver*, Churchill Livingstone, Inc., Philadelphia, PA, 1999.

Blumberg, B.S.: *Hepatitis B and the Prevention of Cancer of the Liver: Selected Publications of Baruch S. Blumberg*, 3rd Edition, World Scientific Publishing Company, Inc., Riveredge, NJ, 2000.

Blumgart, L.H. and Y. Fong: *Surgery of the Liver and Billary Tract*, Vol. 1, 3rd Edition, W.B. Saunders Company, Philadelphia, PA, 2000.

Clavien, Pierre-Alain and K. Lyerly: *Malignant Liver Tumors: Current and Emerging Therapies*, Blackwell Science, Inc., Malden, MA, 1999.

Colombo, M. et al.: "Hepatocellular Carcinoma in Italian Patients with Cirrhosis," *N. Eng. J. Med.*, 675 (September 5, 1991).

Geller, S.A. and L. Petrovic: *Biopsy Interpretation of the Liver*, Lippincott Williams & Wilkins, Philadelphia, PA, 2001.

Hellerstein, M.: "Influence of Pravastatin on Hepatic Metabolism of Cholesterol," *N. Eng. J. Med.*, 128 (January 10, 1991).

Howell, R.R.: "Continuing Lessons from Glycogen Storage Diseases," *N. Eng. J. Med.*, 55 (January 3, 1991).

Killenberg, P.G. and Pierre-Alain Clavien: *Medical Care of the Transplant Liver Patient*, 2nd Edition, Blackwell Science, Inc., Malden MA, 2001.

Krawitt, E.L.: *Medical Management of Liver Disease*, Marcel Dekker, Inc., New York, NY, 1999.

Krawitt, E.L., R.H. Wiesner, and M. Nishioka: *Autoimmune Liver Disease*, 2nd Edition, Elsevier Science, New York, NY, 1999.

Maddrey, W.C. and M. Feldman: *Atlas of the Liver*, 2nd Edition, Current Medicine, New York, NY, 2000.

Paumgartner, G. and U. Leuschner: *Immunology and Liver*, Kluwer Academic Publishers, Norwell, MA, 2000.

Poynard, T. et al.: "Beta-Adrenergic-Antagonist Drugs in the Prevention of Gastrointestinal Bleeding in Patients with Cirrhosis and Esophageal Varices," *N. Eng. J. Med.*, 1532 (May 30, 1991).

Reihner, E. et al.: "Influence of Pravastatin, A Specific Inhibitor of HMG-CoA Reductase, on Hepatic Metabolism of Cholesterol," *N. Eng. J. Med.*, **323**(4), 224 (July 26, 1990).

Sauerbruch, T., W.H. Caselmann, and U. Spengler: *Digestion: Complications of Liver Cirrhosis*, S. Karger Publishers, Inc., Farmington, CT, 1999.

Sherlock, S. and J. Dooley: *Diseases of the Liver and Billary System*, 11th Edition, Blackwell Science, Inc., Malden, MA, 2001.

Sorrell, M.F. and W.C. Maddrey: *Schiff's Diseases of the Liver*, 8th Edition, Lippincott Williams & Wilkins, Philadelphia, PA, 1999.

Spital, A. and M. Spital: "The Ethics of Liver Transplantation from a Living Donor," *N. Eng. J. Med.*, 549 (February 22, 1990).

Spital, A.: "The Shortage of Organs for Transplantation," *N. Eng. J. Med.*, 1243 (October 24, 1991).

Steigmann, G.V. et al.: "Endoscopic Sclerotherapy as Compared with Endoscopic Ligation for Bleeding Esophageal Varices," *N. Eng. J. Med.*, 1527 (June 4, 1992).

Strong, R.W. et al.: "Successful Liver Transplantation from A Living Donor to Her Son," *N. Eng. J. Med.*, 1505 (May 24, 1990).

Suchy, F.J.: *Liver Diseases in Children*, 2nd Edition, Lippincott Williams & Wilkins, Philadelphia, PA, 2000.

Veatch, R.M.: "An Alternative to Presumed Consent," *N. Eng. J. Med.*, 1246 (October 24, 1991).

Whitington, P.F. et al.: "Small-Bowel Length and the Dose of Cyclosporine in Children After Liver Transplantation," *N. Eng. J. Med.*, 733 (March 15, 1990).

Wright, R., G.H. Millward-Sadler, and M.J. Arthur: *Wright's Liver and Billary Disease: Pathophysiology, Diagnosis and Management*, 3rd Edition, W.B. Saunders Company, Philadelphia, PA, 1999.

LIVESTOCK SAFETY INDEX (LSI).   An index of animal heat stress. Categories defined from the temperature—humidity index (THI) are based upon an increasing death rate of livestock as the THI value becomes larger. LSI has been related to other heat-related responses of mammals such as weight gain, milk production, and blood chemistry.

THI	LSI	Effect
<75 °F (23 °C)	No stress	No heat-related problems
75°–78 °F (23°–25 °C)	Livestock alert	Reduced weight gain, lower milk production, increased respiration rate
79°–83 °F (26°–28 °C)	Livestock danger	Reduced weight gain, lower milk production, potential mortality if animals further stressed
>84 °F (28 °C)	Livestock emergency	Mortality possible if animals further stressed by activity, lack of water, lack of external cooling

LIVING WITH A STAR PROGRAM (LWS).   We live in the extended atmosphere of an active star. While sunlight enables and sustains life, the Sun's variability produces streams of high-energy particles and radiation that can affect life.

Under the protective shield of its the magnetic field and atmosphere, the Earth is an island in the solar system where life has developed and flourished. The origins and fate of life on Earth are intimately connected to the way the Earth responds to the Sun's variations. Understanding the changing Sun and its effects on the Solar System, life, and society is the goal of the Sun-Earth Connection Theme.

There are two groups of mission spacecraft in the LWS Space Weather Research Network:

- Solar dynamics elements (Solar Dynamics Observatory/Sentinels) that observe the Sun, track the disturbances originating there through the heliosphere, and
- Geospace dynamics elements (Geospace Missions Network) consisting of spacecraft located in the magnetosphere and ionosphere to define the Geospace response to varying solar and solar wind input.

Space Weather

Everyone is familiar with changes in the weather on Earth. But "weather" also occurs in space. Just as it drives weather on Earth, the Sun is responsible for disturbances in our space environment.

Besides emitting a continuous stream of plasma called the solar wind, the Sun periodically releases billions of tons of matter in what are called *coronal mass ejections*. These immense clouds of material, when directed towards Earth, can cause large magnetic storms in the magnetosphere and the upper atmosphere.

The term space weather generally refers to conditions on the Sun and in the solar wind, magnetosphere, ionosphere, and thermosphere that can influence the performance and reliability of space-borne and ground-based technological systems and can endanger human life or health. See also **Space Weather**.

The consequences of space weather include: **Aeronautics**; high altitude aircraft exposure to radiation, especially near the Poles and navigational and GPS interference.

Astronautics; threat of sporadic radiation to astronauts, potential damage to spacecraft electronics.

Science; understand how the Sun works, understand how Geospace responds to solar and heliospheric variations.

Technology Infrastructure; disruption/failure of communication satellites, power grid disruption problems/equipment failure.

Climate Change; global temperature variations, predictive capability for climatic changes.

Magnetic storms produce many noticeable effects on and near Earth: Aurora borealis, the northern lights, and aurora australis, the southern lights; communication disruptions; radiation hazards to orbiting astronauts and spacecraft; current surges in power lines; orbital degradation; and corrosion in oil pipelines.

Objectives

Living With a Star is a cross-cutting program whose goals and objectives have the following links to each of the four NASA Strategic Enterprises:

1. **Space Science** LWS quantifies the physics, dynamics, and behavior of the Sun-Earth system over the 11-year solar cycle.
2. **Earth Science** LWS improves understanding of the effects of solar variability and disturbances on terrestrial climate change.
3. **Human Exploration and Development** LWS provides data and scientific understanding required for advanced warning of energetic particle events that affect the safety of humans.
4. **Aeronautics and Space Transportation** LWS provides detailed characterization of radiation environments useful in the design of more reliable electronic components for air and space transportation systems.

Missions

Solar Dynamics Observatory Mission (SDO). With the exception of the slow evolutionary changes in solar structure over the last 4.5 billion years, all solar variability is magnetic in origin. The solar cycle is a magnetic cycle in which the Sun's magnetic poles reverse with a periodicity of approximately 11 years, and intense magnetic fields erupt through the surface in sunspots whose numbers wax and wane with the cycle. Solar flares and coronal mass ejections occur when magnetic fields are stressed beyond their limits. The very structure of the corona and the solar wind is determined by the structure of the magnetic fields inside the Sun. The heating of the Sun's corona and the acceleration of the solar wind are thought to be due to interaction between small-scale magnetic elements. SDO is designed to help us understand the Sun's influence on Earth and Near-Earth space by studying the solar atmosphere on small scales of space and time and in many wavelengths simultaneously. The NASA SDO, with its launch planned for August 2008 on as Atlas V launch vehicle, is the first mission for the NASA Living With a Star (LWS) program. The observatory will be delivered into a geosynchronous transfer orbit (GTO). SDO's propulsion system will then perform a circularization maneuver to boost the spacecraft into geosynchronous orbit (GEO). The main engine is a bi-prop system using monomethyl hydrazine (MMH) fuel and c (NTO) oxidizer. Once the final orbit is reached, thrusters will be used to prevent SDO from drifting to another orbit.

SDO Science. SDO will help us to understand the how and why of the Sun's magnetic changes. It will determine how the magnetic field is generated and structured, and how the stored magnetic energy is released into the heliosphere and geospace. SDO data and analysis will also help us develop the ability to predict the solar variations that influence life on Earth and humanity's technological systems.

SDO will measure the properties of the Sun and solar activity. There are few types of measurements but many of them will be taken. For example, the surface velocity is measured by HMI. This data can be used for many different studies. One is the surface rotation rate, which must be removed to study the others. After subtracting the rotation, you have the oscillation and convective velocities. The latter look like billows of storm clouds covering the Sun. Hot gas moves outward at the center of the billows and downward at the edges—just like boiling water. By looking at these velocities you can see how sunspots affect the convection zone. By looking at a long sequence of data (more than 30 days), you see the oscillations of the Sun (like the picture). These patterns can be used to look into and through the Sun.

Mission Science Objectives. The scientific goals of the SDO Project are to improve our understanding of seven science questions:

- What mechanisms drive the quasi-periodic 11-year cycle of solar activity?
- How is active region magnetic flux synthesized, concentrated, and dispersed across the solar surface?
- How does magnetic reconnection on small scales reorganize the large-scale field topology and current systems and how significant is it in heating the corona and accelerating the solar wind?
- Where do the observed variations in the Sun's EUV spectral irradiance arise, and how do they relate to the magnetic activity cycles?
- What magnetic field configurations lead to the CMEs, filament eruptions, and flares that produce energetic particles and radiation?
- Can the structure and dynamics of the solar wind near Earth be determined from the magnetic field configuration and atmospheric structure near the solar surface?
- When will activity occur, and is it possible to make accurate and reliable forecasts of space weather and climate?

The research of the SDO teams is organized into four main areas: (1) Explain how the Sun's magnetic field is created and heats the corona; (2) explain solar EUV and UV spectral irradiance; (3) predict what the Sun will do tomorrow; and (4) understand the Solar Cycle.

SDO Instruments. SDO contains a suite of instruments that will provide observations leading to a more complete understanding of the solar dynamics that drive variability in the Earth's environment. This set of instruments will:

- Measure the extreme ultraviolet spectral irradiance of the Sun at a rapid cadence;
- Measure the Doppler shifts due to oscillation velocities over the entire visible disk;
- Make high-resolution measurements of the longitudinal and vector magnetic field over the entire visible disk;
- Make images of the chromosphere and inner corona at several temperatures at a rapid cadence; and,
- Make those measurements over a significant portion of a solar cycle to capture the solar variations that may exist in different time periods of a solar cycle.

The Science Teams will receive the data from SDO. They will process, analyze, archive, and serve the data. The three instruments aboard SDO are.

HMI (Helioseismic and Magnetic Imager). The Helioseismic and Magnetic Imager will extend the capabilities of the SOHO/MDI http://soi.stanford.edu/, instrument with continuous full-disk coverage at higher spatial resolution. See also http://hmi.stanford.edu/.

AIA (Atmospheric Imaging Assembly). The Atmospheric Imaging Assembly (AIA) for the Solar Dynamics Observatory (SDO) is designed to provide an unprecedented view of the solar corona, taking images that span at least 1.3 solar diameters in mutiple wavelengths nearly simultaneously, at a resolution of about 1 arcsec and at a cadence of 10 seconds or better. The primary goal of the AIA Science Investigation is to use these data, together with data from other SDO instruments and from other observatories, to significantly improve our understanding of the physics behind the activity displayed by the Sun's atmosphere, which drives space weather in the heliosphere and in planetary environments. See also http://aia.lmsal.com/.

EVE (Extreme Ultraviolet Variability Experiment). The Extreme Ultraviolet Variability Experiment will measure the solar extreme-ultraviolet (EUV) irradiance with unprecedented spectral resolution, temporal cadence, and precision; Measures the solar extreme ultraviolet (EUV) spectral irradiance to understand variations on the timescales which influence Earth's climate and near-Earth space.

The EVE instrument consists of three subsystems: MEGS, ESP, and EEB.

- MEGS—Multiple Euv Grating Spectrograph: set of 2 Rowland-circle grating spectrographs that measure the 5–105 nm spectral irradiance with 0.1 nm spectral resolution and with 10-second cadence. Part of the MEGS-A CCD is directly illuminated to measure the individual X-ray photons in the 0–7 nm range with 1 nm or better spectral resolution. Si photodiodes are used as zeroth order traps that also provide calibrations for the MEGS CCD sensitivity changes.
- ESP—Euv Spectrophotometer: transmission grating spectrograph that measures 0.1–7 nm, 17–34 nm, and 58–63 nm to provide solar X-ray measurement shortward of 5 nm, calibrations for MEGS sensitivity changes and higher time cadence (0.25-second). The ESP is very similar to the SOHO SEM instrument.
- EEB—EVE Electronics Box: electronics that control the MEGS and ESP instruments and provides an interface to/from the SDO spacecraft.

Geospace Missions. Geospace–the region of space that stretches from the Earth's upper atmosphere to the outermost reaches of the Earth's magnetic field—has a large impact on human technologies. Due to their potential effects on human technological systems, space weather phenomena within the radiation belts and the ionosphere/thermosphere are of particular concern. Space weather phenomena within the former region energize particles that can endanger both astronauts and electronic systems, while space weather phenomena within the ionosphere/thermosphere disrupt radio communication, aircraft navigation, and spacecraft operations, and can also have a deleterious impact on power line transmission and oil pipeline operations on the ground. All of these effects become more pronounced during severe geomagnetic storms, which energize radiation belt particles, create complex changes in the upper atmosphere densities, and produce strong ionospheric density gradients that spawn ionospheric irregularities and scintillations.

The LWS Geospace missions have been designed to address space physics research problems that directly impact society. Given their importance, the first two Geospace missions will focus on the radiation belts and the ionosphere/thermosphere at mid-to-low latitudes (where most people live) in order to provide the level of scientific understanding needed to predict potentially hazardous effects. The knowledge gained from these missions will be used to (1) understand the fundamental physical processes governing the radiation belts and ionosphere/thermosphere, (2) improve space weather forecasts and (3) better the design and operations of new technology on Earth and in space.

Geospace/RBSP: Radiation Belt Storm Probes (RBSP). Energetic ions and electrons within the Earth's radiation belts pose a hazard to both astronauts and spacecraft. The LWS Geospace program will launch two spacecraft, the Radiation Belt Storm Probes, to quantify the source, loss, and transport processes that generate the radiation belts and cause them to decay. Observations from the two spacecraft will enable the development of empirical and physics-based models for the radiation belts. The empirical models will be used by engineers to design radiation-hardened spacecraft, while the physics-based models will be used by forecasters to predict geomagnetic storms and alert both astronauts and spacecraft operators to potential hazards. Science investigations will be selected in 2006; the RBSP spacecraft will be launched in 2012. See also http://www.lws.nasa.gov/geospace/HTML/RBSP_HTML/Mission/gsRBSPabout.html.

Geospace/ITSP: Ionosphere-Thermosphere Storm Probes (ITSP). In addition to an imager to be placed on a non-LWS high-altitude spacecraft, the LWS Geospace program will launch two Ionosphere-Thermosphere Storm Probes to investigate the middle and low latitude distributions of ionospheric and thermospheric densities, ionospheric irregularities, and geomagnetic disturbances as a function of varying solar and geospace conditions. The spacecraft will observe the composition, chemistry, density, and dynamics of the Earth's ionosphere and thermosphere along with radio scintillation properties. Research on these observations will enable the development of models that not only predict intervals during which communication and GPS navigation will be disrupted but also the optimal frequencies for radio communications. An improved knowledge of neutral densities within the thermosphere will help mission planners evaluate drag effects on spacecraft orbits, and facilitate the tracking of the increasing number of small space objects in low Earth orbit. See also http://www.lws.nasa.gov/geospace/HTML/Mission/LWSgeospaceITSP.html.

Solar Probe. Solar Probe (SP) will be a historic mission, flying into one of the last unexplored regions of the solar system, the Sun's atmosphere or corona, for the first time. Approaching as close as 3 RS above the Sun's surface, Solar Probe will employ a combination of in-situ measurements and imaging to achieve the mission's primary scientific goal: to understand how the Sun's corona is heated and how the solar wind is accelerated. Solar Probe will revolutionize our knowledge of the physics of the origin and evolution of the solar wind. Moreover, by making the only direct, in-situ measurements of the region where some of the deadliest solar energetic particles are energized, Solar Probe will make unique and fundamental contributions to our ability to characterize and forecast the radiation environment in which future space explorers will work and live. Solar Probe is currently under study as part of the Sun-Solar System Connection http://sec.gsfc.nasa.gov/, within NASA's Science Mission Directorate http://science.hq.nasa.gov/.

Solar Probe is scheduled for launch in 2007. It will arrive at the Sun along a polar trajectory perpendicular to the Sun-Earth line with a perihelion of 4 solar radii (R_S) from the Sun's center. Two perihelion passages will occur, the first in 2010 (near solar sunspot maximum) and the second in 2015 (near solar minimum), ensuring measurement of both coronal hole and streamer-related solar wind properties. To reach the Sun, the probe must first fly to Jupiter and use a gravity assist to lose its angular momentum about the Sun. The Jupiter flyby also rotates the probe's orbital plane 90° away from the ecliptic.

The baseline Solar Probe is a 3-axis stabilized spacecraft designed to survive and operate successfully in the intense thermal environment that it will encounter during its voyage around the Sun. The spacecraft's most prominent feature is the Thermal Protection System (TPS), comprising a large 2.7 meter (9 foot) diameter carbon–carbon conical primary shield with a low-conductivity, low-density secondary shield attached to its base. The TPS protects the spacecraft bus and instruments within its umbra during the solar encounter. The bus consists of a hexagonal equipment module and a cylindrical adapter. It provides an efficient mechanical structure that accommodates the instruments and spacecraft subsystems and handles the loads from the TPS and the launch loads.

Solar Probe will be powered by three multi-mission radioisotope thermoelectric generators (MMRTGs). Simple monopropellant will be used for ÆV maneuvers and attitude control. The Guidance and Control System consists of two redundant star trackers, an inertial measurement unit, digital Sun sensors, 4 reaction wheels, and 12 thrusters. The spacecraft is equipped with one highgain antenna for data downlink during the first solar encounter; a medium-gain antenna, the primary antenna during the cruise phase of the mission; and two low-gain antennas for emergencies or periods when the pointing of the medium and high-gain antennas is precluded. The X band will be used for both data downlink and command uplink; the Ka band will be used only for data downlink.

The imagers, CD, EPI, NGS, and one FEA are mounted on the Solar Probe bus. The FIA, the second FEA, and the ICA are mounted on a movable ram-looking arm, which will be gradually retracted as the spacecraft approaches the Sun. This arrangement provides viewing to near (2° inside of) the edge of the TPS umbra. To enable imaging of the solar wind source regions, a retractable, thermally robust periscope will be used to extend the PSRI optics beyond the TPS umbra. Both the side-looking arm and the periscope are designed to be failsafe. The MAG is mounted to the 2 meter (6.5 foot) axial boom that extends from the bottom deck of the spacecraft and that also accommodates a solar horizon sensor used for attitude safing during the solar encounter. The PWI consists of three actuator-controlled 1.75 meter (5.7 foot) antennas mounted to the bottom deck. The design of the Solar Probe spacecraft is based on rigorous engineering studies that demonstrate the technical feasibility and affordability of the mission.

The Solar Probe Mission science objectives include:

Determine the Structure and Dynamics of the Magnetic Fields At the Sources of the Solar Wind (a) How does the magnetic field in the solar wind source regions connect to the photosphere and the heliosphere? (b) How do the observed structures in the corona evolve into the solar wind? (c) Is the source of the solar wind steady or intermittent?

Trace the Flow of the Energy That Heats the Solar Corona and Accelerates the Solar Wind (a) How is energy from the lower solar atmosphere transferred to and dissipated in the corona? (b) What coronal processes shape the non-equilibrium velocity distributions observed throughout the heliosphere? (c) How do the processes in the corona affect the properties of the solar wind in the heliosphere?

Determine What Mechanisms Accelerate and Transport Energetic Particles (a) What are the roles of shocks, reconnection, waves, and turbulence in the acceleration of energetic particles? (b) What are the seed populations and physical conditions necessary for energetic particle acceleration? (c) How are energetic particles transported radially and across latitudes from the corona to the heliosphere?

Explore Dusty Plasma Phenomena and Their Influence on the Solar Wind and Energetic Particle Formation (a) What is the dust environment of the inner heliosphere? (b) What is the origin and composition of dust in the inner heliosphere? (c) What is the nature of dust — plasma interactions and how does dust modify the spacecraft environment close to the Sun? (d) What are the physical and chemical properties of dust-generated?

Web References

Geospace/Ionosphere-Thermosphere Storm Probes (ITSP): http://www.lws.nasa.gov/geospace/HTML/Mission/LWSgeospaceITSP.html

Geospace/ Radiation Belt Storm Probes (RBSP): http://www.lws.nasa.gov/geospace/HTML/RBSP HTML/RBSP index.html

Solar Dynamics Observatory: http://sdo.gsfc.nasa.gov/

Solar Probe: http://solarprobe.gsfc.nasa.gov/

Solar Probe Images: http://solarprobe.gsfc.nasa.gov/solarprobe_gallery.htm

LIZARDS (Reptilia, Sauria). Animals closely related to the snakes but having eyelids and the ventral surface of the body as well as the upper covered with small scales. They are usually elongate, with short legs and a long tail. Most lizards are small, though some species attain a length of several feet.

The classification and nomenclature of these animals is confused. The lizards are sometimes grouped with the snakes but some authorities regard them as a separate order of *Reptilia*.

There are many species of lizards but with the exception of the poisonous Gila monster and the edible iguanas they are of no economic importance. A few of the smaller species are eaten to a limited extent.

Table 1 gives the classification of lizards. Containing over 3000 species, the order *Squamata* (scaly reptiles) embraces about one-half of all modern reptiles. As shown by the table, there are several infraorders, families, and subfamilies, some of which are described in separate articles in this encyclopedia. Thus, please refer to **Agamids** (the agamas and chameleons); **Geckos** (geckos, snake lizards, and diabinids); **Iguanids** (iguanas, basalisks, and anoles); and **Skinks**.

This immediate article describes the remaining principal lizards, including girdled lizards, whiptails, so-called true lizards, shovel-snouted legless lizards, beaded lizards, monitors, and earless and ringed lizards, among others.

In many families of lizards, there are special "break points" in the bodies of the tail vertebrae; the tail beyond these can be discarded in various emergencies. Although the tail may be regenerated later, the new tail is usually shorter than the old and is supported not by vertebrae, but by a central rod of cartilage. The pattern of scales and coloration of the new tail frequently differ from the discarded tail.

Girdle-Tailed Lizards. Of the family *Cordylidae*, these lizards are essentially limited to sub-Saharan Africa. These lizards are well armored. The legs are reduced. The body is covered with longitudinal and transverse rows of rectangular scales, each having a bony element beneath it. The tongue is simple, with only a slight notch and covered with papillae. Because these animals have adapted to a dry habitat, they are not sensitive to reasonable extremes of temperature. Thus, most of these species are well adapted to terrariums. They can also adapt to a wide variety of easily available foods. Principal species include the club-tailed lizard (*Cordylus*); the yellow-brown sungazer (*C. giganteus*), which is about 38 centimeters (15 inches) long and lives in South Africa (Fig. 1); the armadillo lizard (*C. cataphractus*), a rather slow, heavily armored animal, the nostrils of

which are conspicuously elongated; and the common cape girdled lizard, a rather flat animal about 20 centimeters (8 inches) long, with a somewhat spiny tail and black through dark-brown coloration, and that lives in the Cape Province of South Africa; and the blue-spotted girdled lizard (*C. caeruleopunctatus*), which is about 18 centimeters (7 inches) long and found in eastern Africa. The aforementioned lizards feed on insects and small animals.

Fig. 1.   Yellow-brown sungazer (*Cordylus giganteus*). Length is approximately 38 centimeters.

Of the genus *Pseudocordylus* (false club tails), the leathery crag lizard is representative. This animal (*P. microlepidotus*) is about 32 centimeters (12.5 inches) long, with a rather broad head. Dorsal scales are small and rounded. The upper side of the body is dark, usually with a dark upper side and frequently with pale bands or transverse patterns. The flanks are yellow-orange, and the belly is light. The animal inhabits the coastal mountains of Cape Province, South Africa. The habitat is characterized by shale, in deep crevices of which the lizard can hide. The diet is one of insects, other small animals, and lichens and other plants.

Of the genus *Platysaurus* (flat lizards), there are several species, including Wilhelm's red-tailed lizard and the Transvaal red-tailed lizard. The latter is the smallest of the genus and has a transparent window in the lower eyelid. In two subspecies, the males are a glowing green, with a red tail. Other species are red-brown. They live in the Soutpansberg and Drakensberg Mountains of South Africa. The females lay two elongated eggs in a crack in the rock during mid-winter. When the sun has warmed the rock surface in early morning, the *Platysauarus* emerges from an overnight shelter. Most of the daylight hours are spent under the sun. However, the mid-day heat drives the animals into shade. They eat primarily locusts and beetles. Males define their territories and display their colors to competitors by raising up. When sought, these lizards rush to narrow crevices and expand their bodies, thus making it virtually impossible to retrieve one alive.

Of the genus *Chamaesaura* (snake lizards), these animals are quite snakelike in appearance. They have pointed heads, slender bodies, and gradually tapering tails, which can be three times longer than the head and trunk together. The Cape snake lizard ranges up to 63 centimeters (25 inches) in length. Snake lizards prefer the grassy mountain slopes and high plateaus in southern and central Africa, where they feed mostly on grasshoppers. They move in grass with essentially the same ability as snakes. Their tiny limbs are seldom used. Generally, they are of a brown coloration.

The subfamily Gerrhosaurinae (plated or rock lizards) is distributed across Africa and is also found in Madagascar. They have well-developed bony plates under large horny scales. When well gorged with food, a fold makes it possible to expand the width of the body. This provision is also utilized when the female is carrying eggs. These animals are oviparous. The tail is long and can be autotomized. The better understood members are the African plated lizard (*G. major*), which is cylindrical in shape and grows to approximately 56 centimeters (22 inches) in length. The animals frequent southeastern Africa. Smith's plated lizard (*G. validus*) is the largest species in the genus. Its diet consists of insects, spiders, millipedes, or scorpions and smaller lizards. The female usually lays four eggs with leathery shells, most frequently locating them in the cracks of rock.

Night Lizards (Xantusiidae). These lizards range in length from 12 to 15 centimeters (4.5 to 6 inches). The animals have four normally developed limbs, each with five toes. The scales are small. The ventral shields are large and rectangular. The pupil is vertical. In place of movable eyelids,

TABLE 1. CLASSIFICATION OF LIZARDS

CLASS: Reptilia (Reptiles)
ORDER: Squamata (Scaly Reptiles)
SUBORDER: Sauria (Lizards)
INFRAORDER: Geckos (Gekkota)

FAMILY: Geckos (Gekkonidae)
Examples: House Geckos (*Hemidactylus*); Common Gecko (*Tarentola mauritanica*), Asian Tokay (*Gekko gekko*); Banded Gecko (*Coleonyx variegatus*); Banded Leaf-toed Gecko (*Hemidactylus fasciatus*); Smooth Gecko (*Thecadactylus rapicauda*); Madagascar Geckos (genus *Phelsuma*); Japanese Gecko (*Gekko japonicus*); Least Geckos (genus *Sphaerodactylus*); African Tropical Gecko (*Hemidactylus mabouia*); Turkish Gecko (*H. turcicus*); European Leaf-fingered Gecko (*Phyllodactylus europaeus*); Sand Gecko (*Chondrodactylus angulifer*); Bibron's Gecko (*Pachydactylus bibronii*); Spotted Gecko (*Pachydactylus maculatus*); Web-footed Gecko (*Palmatogecko rangei*); Emerald Gecko (*Gekko smaragdinus*); Leaf-tailed Gecko (*Uroplatus fimbriatus*); Kuhl's Gecko (*Ptychozoon kuhli*); Panther Gecko (*Eublepharis macularius*); Padless Gecko (*Gonatodes albogularis*); Fat-tailed Gecko (*Oedura marmorata*).

FAMILY: Snake Lizards (Pygopodidae)
Examples: Western Scaly-foot (*Pygopus nigriceps*); Sharp-snouted Snake Lizard (*Lialis burtonis*); New Guinean Snake Lizard (*L. jicari*); Common Scaly-foot (*Pygopus lepidopodus*); Bailey's Scaly-foot (*Pygopus baileyi*).

FAMILY: Dibamids (Dibamidae)
Includes only one genus with three species. Some zoologists consider the dibamids as offshoots of the skinks.

FAMILY: Iguanids (Iguaninae)
SUBFAMILY: Sceloporinae
Examples: Southern Fence Lizard (*Sceloporus undulatus*); Western or Pacific Fence Lizard (*S. occidentalis*); Desert Spiny Lizard (*S. magister*); Tree Lizard (*Urosaurus ornatus*); Side-blotched Lizard or Ground Uta (*Uta stanburiana*); Banded Rock Lizard (*Petrosaurus mearnsi*); Fringe-toed Lizard (*Uma notata*); Greater Earless Lizard (*Callisaurus draconoides*); Short-horned Lizard or Pigmy Horned Lizard (*Phrynosomas douglasii*); Texas Horned Lizard (*P. cornutum*); Collared Lizard (*Crotaphytus collaris*); Leopard Lizard (*Gambelia wislizenii*).
SUBFAMILY: Tropidurinae
Examples: Spiny-tailed Iguanid (*Uracentron azureum*); Smooth-throated Lizards (genus *Liolaemus*); Crested Keeled Lizards (genus *Leiocephalus*); Narrow-tailed Lizards (genus *Stenocercus*); Madagascan Iguanid (*Oplurus sebae*); Weapon-tailed (*Hoplocercus spinosus*).
SUBFAMILY: Iguanidae
Examples: Common Iguana (*Iguana iguana*); West Indian Iguana (*I. delicatissima*); Rhinoceros Iguana (*Cyclura cornuta*); Marine Iguana (*Amblyrhynchus*); Spiny-tail Iguana (*Ctenosaura pectinata*); Desert Iguana (*Dipsosaurus dorsalis*); Chuckwalla (*Sauromalus ater*);
SUBFAMILY: Basiliscinae
Examples: Common Basilisk (*Basiliscus basiliscus*); Double-crested Basilisk (*B. plumifrons*); Banded Basilisk (*B. vittatus*); Helmeted Lizard (*Corytophanes*); Casque-headed Lizard (*Laemanctus serratus*).
SUBFAMILY: Anolinae
Examples: Long-legged Lizard (*Polychrus marmoratus*); Brazilian Tree Lizard (*Enyalius catenatus*); Patagonian Lizard (*Diplolaemus darwinii*); Sword-tailed Lizard (genus *Xiphocercus*); Cuban Water Anoles (genus *Deiroptyx*); False Chameleon (*Chamaeliolis chamaeleontides*); Carolina Anole (*Anolis carolinensis*); Knight Anole (*A. equestrus*).

FAMILY: Agamids (Agamidae)
Examples: Common Agama (*Agama agama*); Black-necked Agama (*A. atricollis*); Kirk's Rock Agama (*A. kirkii*); Bibron's Agama (*A. bibroni*); Desert Agama (*mutabilis*); Hardun (*A. stellio*); Caucasian Agama (*A. caucasica*); African Spiny-tailed Lizard (*Uromastyx acanthinurus*); Egyptian Spiny-tailed Lizard (*U. aegyptius*); Indian Spiny-tailed Lizard (*U. hardwickii*); Toad-headed Agamids (genus *Phrynocephalus*); Bearded Lizard (*Amphibolurus barbatus*); Spotted Agama (*A. maculatus*); Australian Bloodsucker (*A. muricatus*); Lake Eyre Agama (*A. maculosus*); Frilled or King's Lizard (*Chlamydosaurus kingii*); Lesuer's Water Dragon (*Physignathus lesueurii*); Oriental Water Dragon (*P. cocincinus*); Soa-soa (*Hydrosaurus amboinensis*); Philippine Water Lizard (*H. pustulatus*); Weber's Sailing Lizard (*H. weberi*); Bornean Angle-headed Lizard (*Gonocephalus liogaster*); Lyre-headed Lizard (*Lyriocephalus scutatus*); Indian Bloodsucker (*Calotes versicolor*); Bornean Bloodsucker (*C. cristatellus*); Ceylon Deaf Agamid (*Cophotis ceylanica*); Butterfly Lizard (*Leiolepis belliana*); Sita's Lizard (*Sitana ponticeriana*); Flying Dragon (*Draco volans*); Black-bearded Dragon (*D. melanopogon*); Five-lined Dragon (*D. quinquefasciatus*); Indian Dragon (*D. dussumieri*).

FAMILY: Chameleons (Chamaeleontidae)
Examples: European Chameleon (*Chamaeleo chamaeleon*); African Chameleon (*C. africanus*); Common Chameleon (*C. dilepis*); Two-lined Chameleon (*C. bitaeniatus*); Dwarf Chameleon (*C. pumilus*); Oustalet's Chameleon (*C. oustaleti*); Panther Chameleon (*C. pardalis*); Short-horned Chameleon (*C. brevicornis*); Mountain Chameleon (*C. montium*); Armored Chameleon (*Leandria permeata*).

INFRAORDER: Skinks and Allies (Scincomorpha)

FAMILY: Skinks (Scincidae)
SUBFAMILY: Tiliquinae
Examples: Giant Skink (*Corucia zebrata*); Cape Verde Skink (*Macroscincus cocteaui*); Stump-tailed Skink (*Tiliqua rugosa*); Blue-tongued Skink (*T. scincoides*); Spiny-tailed Skink (Genus *Egernia*).

SUBFAMILY: (Scincinae)
Examples: Common Skink (*Scincus scincus*); Eastern Skink (*S. mitranus*); Hemprich's Skink (*S. hemprichi*); Arabian Skink (*S. philbyi*); Persian Sand Skink (*Ophiomorus persicus*); Speckled Sand Skink (*O. punctatissimus*); Algerian Skink (*Eumeces algeriensis*); Schneider's Skink (*E. schneideri*); Five-lined Skink (*E. fasciatus*); Broad-headed Skink or Greater Five-lined Skink (*E. laticeps*); Great Plains Skink (*E. obsoletus*); Cylindrical Skinks (genus *Chalcides*).
SUBFAMILY: (Lygosominae)
Examples: East Indian Brown-sided Skink (*Mabuya multifasciata*); Keeled Indian Mabuya (*M. carinata*); Müller's Tree Skink (*Sphenomorphus muelleri*); Emerald Skink (*Dasia smaragdina*); Spotted Skink (*D. vittata*) Schmidt's Helmeted Skink (*Tribolonotus schmidti*); Florida Sand Skink (*Neoseps reynoldsi*).

FAMILY: Feylinidae

FAMILY: Anelytropsidae
Example: Mexican Blind Lizard (*Anelytropsis papillosus*).

FAMILY: Girdle-tailed Lizards (Cordylidae)
SUBFAMILY: Cordylinae
Examples: Sungazer Giant Girdled Lizard (*Cordylus giganteus*); Armadillo Lizard or Armadillo Girdled Lizard (*C. cataphractus*); Common Cape Girdled Lizard (*C. cordylus*); Blue-spotted Girdled Lizard (*C. caeruleopunctatus*); Leathery Crag Lizard (*Pseudocordylus microlepidotus*); Transvaal Red-tailed Rock Lizard (*Platysaurus guttatus*); Transvaal Snake Lizard (*Chamaesaura aenea*); Cape Snake Lizard (*C. anguina*).
SUBFAMILY: Gerrhosaurinae
Examples: Smith's Plated Rock Lizard (*Gerrhosaurus validus*); Whip Lizards (genus *Tetradactylus*); Girdled Lizards (genus *Zonosaurus*).

FAMILY: Night Lizards (Xantusiidae)
Examples: Cuban Night Lizard (*Cricosaura typica*); Granite Night Lizard (*Xantusia henshawi*).
FAMILY: Whiptails (Teiidae)
Examples: Chilean Spotted Lizard (*Callopistes maculatus*); Six-lined Racerunner (*Cnemidophorus sexlineatus*); Spotted Whiptail or Blue-bellied Racerunner (*C. sackii*).
FAMILY: True Lizards (Lacertidae)
Examples: Sand Lizard (*Lacerta agilis*); Dwarf Lizard (*L. parva*); Jewelled Lizard (*Timon lepida*); Common Lizard (*Zootoca vivipara*); Plated Lacertids (genus *Psammodromus*); Snake-eyed Lacertids (genus *Ophisops*); Fringe-toed Lacertids (genus *Acanthodactylus*).

INFRAORDER: Anguimorpha

FAMILY: Lateral Foldl Lizards (Anguidae)
SUBFAMILY: Diploglossine Lizards (Diploglossinae)
SUBFAMILY: Alligator Lizards (Gerrhonotinae)
Examples: Glass Lizards (genus *Ophisaurus*); Sheltopusik (*O. apodus*); Eastern Glass Lizard (*O. ventralis*).
SUBFAMILY: Anguine Lizards (Anguinae)
Example: Slow-worm (*Anguis fragilis*).

FAMILY: Shovel-snouted Legless Lizards (genus *Anniella*).

FAMILY: Xenosaurids (Xenosauridae)
Example: Crocodile Lizards (genus *Shinisaurus*); Chinese Crocodile Lizard (*S. crocodilurus*).

INFRAORDER: Varanomorphs

FAMILY: Aigalosauridae

FAMILY: Dolichosauridae

FAMILY: Mosasauridae

FAMILY: Bearded Lizards (Helodermatidae)
Examples: Gila Monster (*Heloderma suspectum*); Mexican Bearded Lizard (*H. horridum*).

FAMILY: Monitors (Varanidae)
Examples: Desert Monitor (*Psammosaurus griseus*); Nile Monitor (*Polydaedalus niloticus*); Cape Monitor (*Empagusia albigularis*); Yellow Monitor (*E. flavescens*); Two-banded Monitor (*Varanus salvator*); Giant Monitor (*V. giganteus*); Dwarf Monitor (*Odatria storri*).

FAMILY: Earless Monitors (Lanthanotidae)
Example: Borneo Earless Monitor (*Lanthanotus borneensis*).

INFRAORDER: Worm Lizards (Amphisbaenia)

FAMILY: Bipedidae
Example: Common Two-Legged Worm Lizard (*Bipes biporus*).

FAMILY: Ringed Lizards (Amphisbaenidae)
Examples: White-Bellied Worm Lizard (*Amphisbaena alba*); Darwin's Ringed Lizard (*A. darwini*); King's Worm Lizard (*A. kingi*); Florida Worm Lizard (*Rhineura floridana*).

FAMILY: Trogonophids (Trogonophidae)
Example: Wiegmann's Worm Lizard (*Trogonophis wiegmanni*).

Note: Some of the better known as well as other species selected at random are included in the various examples given. The examples represent only some of the species of lizards. The process of classifying the lizards continues at a relatively slow pace toward refinements in classification as well as with the nomenclature used. A few classifications remain controversial among authorities.

there is a transparent "spectacle" something like that found in the geckos. Also, the vertebrae, skull, tongue, and eyes resemble those of the geckos. These animals feed on insects and spiders. The common night lizard also consumes vegetation. Their range extends from the southwestern United States and the offshore islands as far as Panama. One species is found in Cuba. Their hiding places are cracks in rocks, under roots, and in the bark of trees. All are viviparous, bearing from one to nine young at a time. There are four genera, but only a total of twelve species.

The Cuban night lizard (*Cricosaura typica*) was not discovered until 1863 (by German zoologist Gundlach). The large-headed or common night lizard occurs on the rocky islands of San Clemente, San Nicholas, and Santa Barbara off the coast of California. They consume both seeds and flowers. This particular animal is not strictly nocturnal.

Of the three species of *Xantusia*, the granite night lizard (*X. henshawi*) is native to California. Total length ranges up to about 17 centimeters (6.5 inches). It is found not only in California, but also in the southwestern United States. The animal is quite flat, allowing it to creep into very small crevices, often to avoid sunshine. The desert night lizard (*X. vigilis*) prefers soft yucca plants for its diet. This species has been studied in detail on the southwestern edge of the Mojave Desert in California. They are frequently found at the foot of yuccas. In 1950, it was discovered that the formation of a placenta, rare among lizards, occurred in the Xantusiidae. After a gestation period of 90 days, from one to three young are born.

Whiptails (Teiidae). These animals vary widely in length, from 7.5 to 140 centimeters (3 to 55 inches). They are sometimes referred to as the New World counterpart of the true lizards (Lacertidae). Some species, such as *Ameiva*, have well-developed limbs and scales and are formed so as to appear very much like European lizards. In the Teiidae family, there are some 45 genera and about 200 species. With the exception of northern and northeastern states, the teiids range widely throughout the United States and in Central America, the West Indies, and parts of Argentina and Chile. Their wide distribution accounts for so many species that have evolved through adaptation to numerous living and survival conditions, including arboreal, aquatic, and ground forms in rain forests, plains, deserts, seacoasts and inland regions, lowlands, and high mountains, such as the Andes, to the tropical forests along the Amazon river.

Generalizing, the teiids have lizard-like tails, ranging from cylindrical in form to a laterally flattened form found in aquatic environments. Their scales range widely — rounded, elliptical, elongated, hexagonal, smooth, or keeled. Depending upon particular species, a variety of bands may run longitudinally, transversely, or diagonally. Often, the teiid body is covered with large scales. Often, the tongue can be extended considerably and possesses a deep notch at the end. Depending upon location, the teiid ranges from herbivorous to insectivorous. Mode of reproduction has not been observed in detail for many of the species.

Racerunner (genus *Cnemidophorus*) are whip-tail lizards, the most commonly encountered teiid in the United States, and are also found in southern regions extending to northern Argentina. The term *racerunner* stems from the fact that these animals run fast, stop suddenly to scan for their enemy, and continue immediately in a series of rapid dashes, frequently in a different direction. It appears difficult for these animals to resist almost constant motion. Members of *Cnemidophorus* include the six-lined racerunner (*C. sexlineatus*), which is usually close to 8 centimeters (3.2 inches); the checkered whiptail with a pale blue body; the seven-lined racerunner (*C. deppei*), which can achieve a length of 24 centimeters (9.5 inches), is found from Mexico to Costa Rica, and has a light blue throat, a turquoise belly, with blue spots on the sides and a brick red lateral stripe; the spotted whiptail (*C. sackii*), which is about 25 centimeters (9.8 inches) long. Those animals with striking colorations use these features when reacting to threats from pursuers.

A striking member of the teiid family is the common tegu (*Tupinambis teguixin*), which can achieve a length of 120 to 140 centimeters (47 to 55 inches). The animal is essentially black, with numerous crossbands that incorporate round yellow spots. See Fig. 2. Other animals of this type include the northern tegu or jacura (*T. nigropunctatus*), with a length of up to 120 centimeters (47 inches). With their rather squat bodies, these teiids live in wooded areas, with dense undergrowth and sunny clearings and an obviously abundant food supply. Their meat is prized by local natives, who fish for them with meat-baited hooks. The yellow fat of the lizard's legs is particularly prized. Medicinal properties are ascribed by local people to

several parts of the common tegu. Locally terms ascribed to this lizard testify to its reputation among the local populace — "chicken wolf" and "egg thief," among others.

Fig. 2.   Common tegu (*Tupinambis teguixin*), which achieves a length of about 120 to 140 centimeters (47 to 55 inches) and which is sometimes referred to by native peoples in Central America and northern South America as a "chicken wolf" or "egg thief."

Another teiid of the genus *Crocodilurus*, the dragon lizardet (*C. lacertinus*), achieves a length of about 50 centimeters (19.5 inches) and is found in Central America and northern South America. The animal prefers a swampy habitat. The tail is somewhat flattened and features a double keel. The animal prefers fish and frogs in its diet. The animal seeks prey by hiding underwater in holes of a stream bank or under the roots of trees. Once seizing its prey, the animal returns to its hiding place to consume its catch. See Fig. 3.

Fig. 3.   Dragon lizardet (*Crocodilurus lacertinus*), achieves a length of about 50 centimeters (19.5 inches). The animal prefers the swampy habitats of Central America and northern South America.

True Lizards (Lacertidae). These lizards are the Old World counterparts (analogues) of the Teiidae of the New World. Although these animals have not extended to Madagascar, New Guinea, and Australia, they are found elsewhere on the European and Asian continents and in most of South America as well. They range from the tiniest of Mediterranean islands to the far north of the Arctic Circle. All species of lacterids have well-developed legs and a long tail. They range in length from about 12 to 90 centimeters (4.5 to 35 inches). The lacertids lack many of the characteristics found in other lizard families. They do not have dorsal crests or dewlaps. There are no other movable or expandable skin appendages. They have little or no ability to change their color. However, most feature an autonomous tail, part of which can be sacrificed to an enemy and later regenerated.

Many lacertids are found in dry regions. They adapt well to alternative habitats. Some species live in loose ground and elongated scales under the toes form combs at one or both sides, facilitating rapid locomotion on shifting sand. Some have a modified snout for digging. In some species there is a movable lower eyelid, often with a transparent window. Nearly all lacertids are egg layers. The species of the genus *Lacerta* are familiar to Europeans. Although subject to considerable research by noted herpetologists, the classification and relationships of species remains unclear. General characterizations include an unspecialized body structure, the band of enlarged scales around the neck, and the round or slightly compressed fingers and toes, which lack fringes or scales. There are numerous examples of the lacertids, including the sand lizard (*L. agilis*), an animal that is about 30 centimeters (12 inches) long; the emerald lizard (*L. viridis*), with a length up to 45 centimeters (18 inches); the dwarf lizard (*L. parva*); and the largest lacertid, the jeweled or eyed lizard (*L. lepida*), which reaches a length up to 80 centimeters (31.5 inches); among many others.

Lateral Fold Lizards (Genus Anguidae). These animals historically are considered a rather young group that arose in the Cretaceous period. They are closely related to the monitors, which are described later. These animals generally lack limbs, but American species have four well-developed limbs. As with many other lizard families, limb degeneration in the lateral fold lizards usually can be correlated with adaptation to a wide variety of native habitats. Lateral fold lizards may be shaped like a snake or like a lizard. The limbless species can achieve a length of from 50 to 100 centimeters (20 to 39 inches), whereas the four-limbed species are smaller, some 20 to 40 centimeters (8 to 16 inches) in length. The most familiar lateral fold lizard is the slow-worm (*Anguis fragilis*), which ranges from 35 to 50 centimeters (14 to 20 inches) in length. This animal is not to be confused with the worm lizards (*Amphisbaenia*), to be described later.

The eastern glass lizard (genus *Ophisaurus ventralis*) is the longest lizard encountered in the United States. See Fig. 4. It can achieve a length of about 1 meter (3.3 feet). Closely related species found in the Mississippi basin include the slender glass lizard (*O. attenautus*) and the island glass lizard (*O. compressus*), all generally confined to the southeastern United States, including Florida. Also closely related to the lateral fold lizards are the xenosaurids (*Xenosauridae*). There are two genera. One genus (*Shinisaurus*) embraces only one species, the Chinese crocodile lizard, which was first captured as recently as 1928 and first described in 1930 by the German herpetologist, Ernst Ahl. This lizard inhabits Kwangsi Province in southwestern China. The horny scales on the tail form a double row, as found in crocodiles. The teeth are fang-like. The animal inhabits areas near water. Chinese term the species the "lizard with great sleepiness." However, when threatened it will bite very quickly. When attacked, the lizard usually attempts to escape to water because of its diving and swimming agility.

Fig. 4. Eastern glass lizard (*Ophisaurus ventralis*), the largest lizard occurring in the United States, achieves a length of 1 meter (3.3 feet).

Beaded or Venomous Lizards (Helodermatidae). These animals are the only poisonous lizards. They are closely related to the ancestors of snakes. They are characterized by massive, un-snakelike bodies, and by the presence of four fully developed limbs. The head is broad and somewhat flattened. It is joined with a short neck and an elongated, cylindrical body, ending with a thick, rounded tail. The back is covered with large, bony scales. The legs are short and powerful. Each foot has five clawed toes. The back is covered with large, bony scales. The belly bares flat, regularly arranged scales that are not fully ossified. The lower jaw teeth have single grooves on the front and back that permit the venom to flow into the wound. The two species of the beaded lizard are (1) the gila monster (*Heloderma suspectum*), which achieves a length up to 60 centimeters (24 inches); and (2) the Mexican beaded lizard (*H. horridum*), which may be up to 80 centimeters (31.5 inches) long.

The gila monster (Fig. 5) has a distinctive coloration of pink and black spots. Of two species, these animals are encountered from southern Nevada and southeastern Utah to Sonora in northwestern Mexico. Generally, these beaded snakes are active at night, but during cold weather they may appear by day. The diet consists of nestling rodents and birds and bird and reptile eggs. Beaded lizards first move very slowly and clumsily, but as the night hour's progress, they become quite agile. During long fast periods, demanded by their habitat, the animals survive on fat reserves stored up during more favorable times of the year. Fat is stored in the tail, which

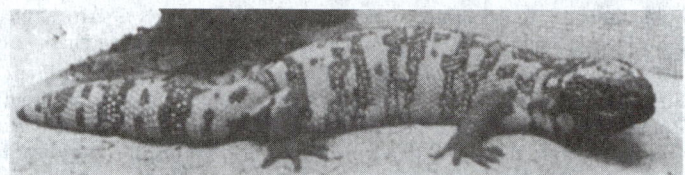

Fig. 5. Gila monster. (*A.M. Winchester.*)

swells noticeably, but after a long fasting period, the tail becomes quite thin. Although seldom near water, they are good swimmers. Observations have shown that the gila monster can survive years of drought. In captivity, gila monsters can achieve an age of 20 years. They may be fed raw chicken eggs mixed with lean meat and supplemented with lime and vitamins.

Herpetologists have reported that snakebite antidote has no neutralizing action against gila monster venom. Typical localized consequences of a gila monster bite may include severe swelling of the victim's arm or part bitten and may be extremely painful for nearly 2 weeks. Evidence indicates that there is no damage to the victim's nervous system or vital organs, such as kidneys and liver. Experts insist that these animals belong only in the hands of very experienced keepers.

Monitors (Varanidae). As indicated by Fig. 6, the members of this genus have somewhat the appearance of legendary dragons. The Nile monitor has been known since antiquity. Herodotus (who died around 424 B.C.) described the desert monitor as a "terrestrial crocodile," while the ancient Egyptians, who often depicted monitors on their monuments, knew the Nile monitor so well that they never confused it with crocodiles. Monitors occur in numerous sizes. However, they all are classified in one genus (*Varanus*). They range from 30 centimeters (12 inches) up to 3 meters (10 feet) in length. The body is usually quite massive, with four powerful legs, each bearing five clawed toes. The tail is thick and may be used as a prehensile organ or as a potent weapon. Monitors are diurnal and reach their full activity level when the sun is up and their habitat has warmed. They are good runners and climbers. Some species are arboreal; others prefer proximity to water. Such species dive and swim well. Their best weapons are their sharp teeth and dagger-sharp claws, which can inflict dangerous wounds. All monitors feed on other animals, such as insects, small lizards, and the nestlings of small mammals. The larger species seek prey, such as crustaceans, fishes, frogs, birds, rats, and snakes. It is known that the giant Komodo dragon can even take small deer and wild pigs, and it has been reported that two-banded monitors have fed on human corpses that have been interred in trees. Monitors especially like eggs in their diet and are known to eat eggs of their conspecifics. Most monitors, in their native habitat suffer at the hands of humans. Their meat and eggs are eaten, and the animals are often used to produce various "medications" and amulets. The fat and oil from the paired fatty organis is used by Chinese druggists, whose buying agents travel as far as Australia. The skin of larger monitors is processed into leather.

Fig. 6. The Nile monitor (*Varanus niloticus*) is found south of the Sahara, both in deserts and on the plains and especially along rivers. The animal feeds heavily on eggs.

Monitors are found in the tropical and subtropical parts of Africa, the Near East, southern Asia, the Indo-Australian islands, and Australia.

Earless Monitors (Lanthanotids). In 1878, Viennese zoologist Franz Steindachner described the single animal before him as a *new* lizard from Borneo. Since it lacked external ears, Steindachner named it *Lanthonotus borneensis*, from the Greek (hidden ear). He realized that this new lizard species belonged to a unique family, which he classified as relatives of gila monsters. During the following 80 years, only six additional earless monitors were identified. Even experienced herpetologists failed to find like specimens of the animal in Borneo. It was not until 1961 that archeologists on the island of Borneo dug out and captured alive an unusual lizard. This was the first earless lizard to be captured within a 45-year period. The local Dayak tribesmen were promised a bounty for capturing live lanthanotids. By 1960, over 60 were found. These were sent to zoo collections in Europe and the United States. Zoologists are much interested in earless monitors because they may be survivors of the animal group that gave rise to snakes. Many experts believe that snakes progressed from subterranean lizards in which the limbs and eyes regressed, the body became longer, and the number of vertebrae increased. the earless monitors possess these same characteristics.

Worm Lizards (Amphisbaenia). Questions persist concerning the accurate classification of worm lizards, which combine several characteristics of snakes and only some of the characteristics that typify lizards. Some herpetologists have even questioned whether these animals are reptiles, let alone lizards. It is notable that the characteristics are not intermediate between snakes and lizards, but rather represent evolvement over some special path. Worm lizards have been observed and their full identity determined only since the early 1800s. Identification was difficult because they are subterranean and come out of their underground locations only after sunset and return just before sunrise. Their size and coloration are such that they can be easily mistaken for earthworms.

As described by Carl Gans (Department of Biology, State University of New York at Buffalo), the typical amphisbaenian body is cylindrical, bearing a loose skin with fairly defined rings. Their color may be reddish or brownish and many incorporate a pattern of dark-brown or black spots on a lighter background. Length ranges from 8 to 80 centimeters (3 to 31.5 inches); diameter is between 1.5 and 30 millimeters (0.06 and 1.2 inch). Their shovel-like head structure is well adapted for digging. The eyes and ears are buried beneath the skin. Typically, worm lizards lay eggs, although a few species bear fully developed young. Worm lizards occur in several tropical areas of both the Old and New World.

In lizard folklore, one finds references to unusual association between ants and amphisbaenians. For example, in Brazil, some species are referred to as "ant mothers" or "ant kings," where it is alleged that worm lizards are raised and fed by ants. "Since worm lizards are often found in ant and termite colonies, the lizards probably have no trouble catching their prey. However, subterranean ant and termite colonies are not just food resources, but also serve as incubation chambers for egg-laying lizards. Most, if not all, worm lizard eggs discovered have been found in ant and termite colonies."

Some protection for the worm lizard is the enemy's difficulty in distinguishing the head from the tail. Both head and tail wag through the air in some species. Where Portuguese is spoken, the amphisbaenians may be called *cobras de dois cabeccas*, or two-headed snakes. Upon being grabbed by a carnivore, the outer tip of the tail may break off. The broken part twists about extensively and may momentarily divert the attention of the enemy. Unlike many other kinds of lizards, however, the broken tail does not regenerate. In the days of the early explorers, any worm lizard found was mistakenly regarded as poisonous.

By way of a series of underground tunnels, many worm lizards can hear the sounds of oncoming prey; their very long tongues can detect surrounding odors. In terrariums, amphisbaenians have been observed to crawl forward and backward with ease in their tunnels when they seek food. Also, their long tongue also assists in gathering insects for their diet. Their shovel-like head allows them to enter tunnels with surprising speed. The shape of their head can be a disadvantage, however, because it requires short jaws and reduces the number of teeth. Thus, in the overall, it has been suggested that species with less-effective digging mechanisms enjoy wider distribution and numbers, which are definite advantages in the long term.

The teeth of worm lizards interlock with an odd number of teeth in the upper jaw and an even number in the lower jaw. Thus, a biting action leaves a cut like that made by jagged scissors. The jaws tear or rip triangular pieces of flesh from the prey. The cheeks have strong muscles to assist the biting. Unlike the case of snakes, the amphisbaenians can rip out comparatively large pieces of flesh from small mammals.

Habits and Lifestyles

Lizards show a great range of shapes and sizes and their habits range from diving in the sea to gliding between trees, and from burrowing in the soil to stalking insects in the tree tops. The major groups of lizards are listed in Table 1, together with an outline of the habits and lifestyles of the major groups.

The iguanas are often brightly colored forms, with ornamental frills and throat fans. They are herbivores, mainly arboreal, and are found mainly in tropical South America. Large ground-dwelling iguanas live on islands in the Caribbean and on the Galapagos Islands, where their most striking representative is the marine iguana. This iguana, *Amblyrhynchus*, feeds on seaweed, which it gathers from the seabed at depths of 10 meters (33 feet) or more. The agamids are another large group of modest-sized lizards, found in the Old World, and they live on the ground and in trees. Close relatives are the chamaeleons, tree-living lizards found mainly in Africa that feed on insects by flicking out an elongate tongue with lightning speed, and seizing the prey on its sticky end. Chamaeleons sit still in trees, gripping the branches with their unusual feet (two fingers opposing three in front, three opposing two behind) and their tails. Their eyes can swivel from side to side, and they can change colour to match the background; both these adaptations allow them to remain concealed until they are ready to strike. See also **Agamids** and **Iguanids**.

Geckos are small agile forms that are mainly nocturnal. Many of them have adhesive pads on their feet, made from ridged scales, and they can run about on the walls and ceilings of houses. They are found on all continents except Antarctica. Pygopodids, close relatives, lack their forelimbs and have reduced hindlimbs. They are found in Australia and New Guinea. See also **Geckos**.

The amphisbaenians are sometimes classified as a separate Suborder from remaining lizards. However, cladistic analyses suggest they fall within the clade Lacertilia. Amphisbaenians, or "worm lizards," are highly specialized burrowers. Their bodies are cylindrical, superficially like a large worm, and most have entirely lost their limbs. The eyes and ears are hidden under the skin, and the skull is modified into a short blunt object that is used to batter through the earth and form tunnels. Amphisbaenians live in tropical South America and Africa, and there is one species in Florida, USA.

The scincomorphs include skinks, often long-bodied and short-tailed, with short limbs. There are nearly 2,000 species and they are found on all continents except Antarctica. See also **Skinks**. The European and Asian lacertids, close relatives, are also small, ground-living lizards that feed on insects and worms. The teiids, herbivorous lizards, are mainly small forms that inhabit the New World. The cordylids are small to medium-sized terrestrial lizards with heavy bony scutes beneath their scales. They are found in subSaharan Africa and on Madagascar. The xantusiids are tiny terrestrial lizards found in the New World.

The anguimorphs are a diverse group of perhaps six families. They include the varanids, or monitor lizards, a widespread Old World tropical group of ground-dwelling predatory lizards. Varanids include the largest living lizard, the Komodo monitor, *V. komodoensis*, as well as other giant fossil forms including a range of extinct marine forms such as the Late Cretaceous mosasaurs, 3 to 10 meters (10 to 33 feet) in length. The living anguids are small elongate forms, often limbless, and found in the Americas, Europe and Asia. They include the glass lizard, *Ophisaurus* and the slow worm, *Anguis*. The helodermatids include the Gila monster, a bulky lizard of desert regions of south-west North America, which is one of only two venomous lizards. Unlike snakes, it stores venom in its lower jaw. Other anguimorphs include the anniellids, a family of two species of small burrowers from North America; the xenosaurids, represented by two species in Mexico and one in China; and the lanthanotids, a single species of semiaquatic lizard from Borneo.

Lizards are all ectotherms, in other words, animals that do not have direct internal, physiological, control of their body temperatures. Lizards use behavioral means to control their body temperatures. When the sun rises, they crawl out of their hiding places and bask on a rock. They absorb heat by direct radiation from the sun, but also by conduction through the rock, and by reflection from surrounding rocks. When it has absorbed enough heat to become active, a lizard goes about its business, until the air temperature becomes too high in the middle of the day. Then, it may shelter in a dark spot until air temperatures cool down. At night, as body heat is lost, the lizard becomes inactive.

Lizards can operate efficiently over a range of body temperatures as much as 4–10 °C (39–50 °F), unlike birds and mammals, which must maintain a constant body temperature within 1–2 °C (34–35.5 °F). The rate of heat absorption and loss can be controlled by varying the orientation of the body to the sun, the body shape, the body color, and the rate of blood flow in peripheral regions of the body. These mechanisms allow lizards to divorce themselves from changes in the temperature of the air to a great extent: active lizards living above the timberline in temperate regions may be 30 °C (86 °F) warmer than air temperature.

Lizards are well adapted to conserving water, and this allows them to live in arid regions, both hot and cold. Indeed, lizards are the most successful desert-dwelling vertebrates. Their water conservation strategies depend on three attributes.

1. They have low metabolic rates, as a result of their ectothermy; a lizard generally eats 1/10th or 1/15th of the amount of food required by a mammal of the same size.
2. They excrete nitrogenous waste as semisolid uric acid and salts, instead of as urea as in mammals, which must be dissolved in water to be passed as urine.
3. Many lizards have additional pathways of salt secretion via salt glands in the nasal region; salt is excreted and is then expelled by sneezing or shaking the head.

Life Histories

Many lizards engage in premating rituals. Species-specific signals are given by specialized crests, throat pouches and other structures that may be brightly colored, and, which are expanded and waved about during display activity. The display activities may involve specific movements–species of *Anolis* lizards on Costa Rica raise their bodies on their forelimbs, bob their heads, and expand and contract the throat pouch in particular ways unique to each species. The site chosen for display is also indicative of species.

Most lizards lay eggs (ovipary), often elongate and with leathery shells, instead of hard mineralized shells. Lizards lay from 1–2 to 60 eggs in a clutch. The nest is usually abandoned, and when the young hatch they must fend for themselves immediately. This leads to a high mortality rate among juveniles, with only 15–40% reaching sexual maturity, depending on the species and environmental conditions.

Some have suppressed egg-laying, and produce live young (vivipary), a habit seen also in some snakes. However, this form of live-bearing is different from that of placental mammals in that the mother lizard or snake retains the eggs inside until hatching time. This habit evolved many times, perhaps on 45 occasions, and it is seen most in lizards and snakes that live in colder areas, so the mother can bask in the sun to keep the eggs warm.

Lizards grow to adult size rapidly, and sexual maturity is reached within 1–3 years, but this figure has been determined in only a tiny number of lizard species in the wild. Lizards typically have rather short lives. Survivorship to first breeding is often in the range 10–50% of juveniles, and survivorship to second breeding is similarly 10–50%. In most species, individual lizards are unlikely to survive for more than 1–2 years in all. Larger lizards grow more slowly. The Komodo dragon, *V. komodoensis*, probably reaches sexual maturity at an age of 4–6 years when about 1.5 meters (5 feet) long. It is estimated that the largest Komodo dragons, at 3.5 meters (11.5 feet) long, are about 20 years old, perhaps the maximum age attained by any lizard. Problems in determining these figures are that wild lizards must be marked and recaught year after year to follow their progress, and figures from zoos are hard to compare, since captive animals may be protected and fed in ways that allow them to live much longer than any animal in the wild.

Fossil History

The first lepidosaurs, members of the broader group that includes lizards, are the sphenodontids of the Late Triassic. These small- to medium-sized herbivorous and insectivorous forms are similar to the living tuatara, *Sphenodon* and probably indicate what the ancestor of lizards was like.

The oldest fossils of unequivocal lizards date from the Mid Jurassic of England and Scotland, and they include fragmentary remains of a possible gekkotan, an anguimorph, and several scincomorphs. These three groups radiated further in the Late Jurassic, and more complete fossils are known from North America and Europe from a number of locations.

The other lizard groups, Iguanid and Amphisbaenia, as well as the snakes, must have originated in the Late Jurassic, because of the nature of the phylogenetic tree, but fossils are not known until later. The oldest iguanian is an agamid from the Late Cretaceous of Mongolia. The fossil record of amphisbaenians is limited, consisting only of isolated vertebrae from the Paleocene and Eocene of North America and France. The first snake fossils date from the Early Cretaceous. Several lizard groups radiated in the Late Cretaceous, not least the Late Cretaceous dolichosaurs and mosasaurs, both relatives of varanids, and both highly successful marine predators, but which became extinct at the end of the Cretaceous.

Phylogeny

Snakes and lizards, as well as the tuatara, the sole surviving sphenodontid, are grouped together in the Order Squamata. The phylogeny of the lizards is debated, mainly on the basis of morphological data, which offer a number of possible resolutions of the phylogeny. Particularly problematic are the placements of amphisbaenians and of snakes. New character information, new fossils and new information from molecular sequencing may resolve the patterns. See also **Snakes**.

One view (Fig. 7) is that Iguania are the basal lizards. The remaining groups, termed collectively Scleroglossa, all share a keratinized (horn-covered) tongue and other features of the skull. Within Scleroglossa, the Gekkota may be the outgroup of the remaining four infraorders, but this is not clear, and Amphisbaenia and Serpentes may fit somewhere near this point in the cladogram. Serpentes was traditionally placed here, but new evidence from fossils suggests that snakes may be related more closely to the mosasaurs and other fossil aquatic groups, hence Varanoidea. The Anguimorpha and Scincomorpha seem to pair off as the clade Autarchoglossa, based on the shared possession of no contact between the jugal and squamosal bones, and other features.

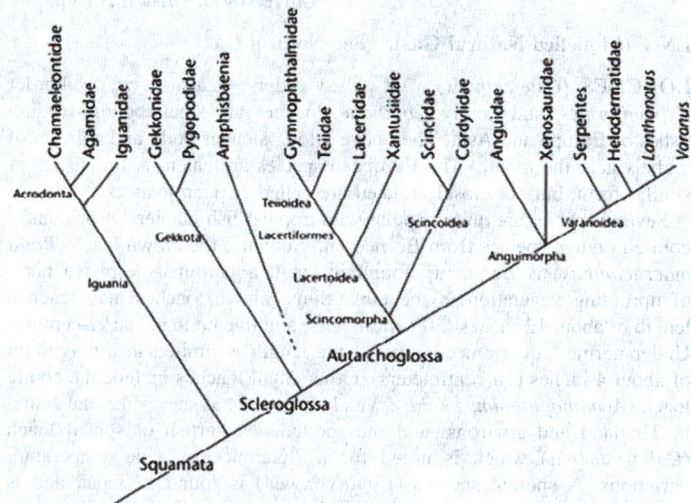

Fig. 7. Cladogram of lizard relationships, based on data in Estes *et al.*[1988]. Lizard relationships are highly contentious, and this cladogram is a 'conservative' estimate.

Conclusion

Lizards are a highly successful vertebrate group, slightly more diverse than mammals. Most lizards are 50–300 mm long, long-tailed, agile and adapted to arid conditions. However, there are many exceptions, and the group is more diverse in habits and lifestyles than is often assumed. Lizards occupy all parts of the world except the polar regions, they live on the ground, in trees, and some live a partially marine existence. Lizards feed on plants, insects and larger prey. Lizards range in size from 20 mm to 3 m. Several lizard groups independently have lost their limbs (mainly to permit burrowing), others have evolved a kind of vivipary (live birth), some can glide, and two species are poisonous.

Additional Reading

Affenberg, W.: *The Behavioral Ecology of the Komodo Monitor*, University of Florida Press, Gainesville, FL, 1981.

Benton, M.J.: "Classification and Phylogeny of the Diapsid Reptiles," *Zoological Journal of the Linnean Society*, **84**, 97–164 (1985).

Benton, M.J.: *Vertebrate Palaeontology*, 3rd Edition, Blackwell Publishers, Oxford, UK, 2005.

Estes, R., K. de Queiroz, and J. Gauthier: "Phylogenetic Relationships within Squamata," In: Estes, R., and G. Pregill: *Phylogenetic Relationships of the Lizard Families*, Stanford University Press, Stanford, CA, 1988, pp. 119–281.

Evans, S.E.: "At the Feet of the Dinosaurs: the Early History and Radiation of Lizards," *Biological Reviews*, **78**, 513–551 (2003).

Gans, C.: *Biology of the Reptilia*, **19** Vols. Academic Press, New York, NY, 1969.

Grzimek, B., M. Hutchins: *Grzimek's Animal Life Encyclopedia*, 2nd Edition, Thomson Gale, Farmington Hills, MI, 2003.

Halliday, T.R., and K. Adler: *The Encyclopedia of Reptiles and Amphibians*, Facts on File, New York, NY, 1986.

Manthey, U. and N. Schuster: *Agamid Lizards*, TFH Publications, Neptune, NJ, 1997.

Mattison, C.: *Encyclopedia of North American Reptiles and Amphibians: An Essential Guide to Reptiles and Amphibians of North America*, New York, NY, 2005.

Pough, F.H., R. Andrews, J. Cadle, et al.: *Herpetology*, 2nd Edition, Prentice-Hall, Inc., Upper Saddle River, NJ, 2003.

Pough, F.H., C.M. Janis, and J.B. Heiser: *Vertebrate Life*, 6th Edition, Prentice-Hall, Inc., Upper Saddle River, NJ, 2002.

Roughgarden, J.: *Anolis Lizards of the Caribbean: Ecology, Evolution and Plate Tectonics*, Oxford University Press, Inc., New York, NY, 1994.

Sprackland, R.: *Giant Lizards*, TFH Publications, Neptune, NJ, 1992.

Vitt, L.J., and E. Pianka: *Lizard Ecology: Historical and Experimental Perspectives*, Princeton University Press, Princeton, NJ, 1994.

Young, J.Z.: *The Life of Vertebrates*, Clarendon Press, Oxford, UK. 1981.

Zug, G., L.J. Vitt, and J.P. Caldwell: *Herpetology: an Introductory Biology of Amphibians and Reptiles*, 2nd Edition, Academic Press, New York, NY, 2001.

DOUGLAS M. CONSIDINE
MICHAEL J. BENTON
University of Bristol, Bristol, UK

LNG (Liquefied Natural Gas). See **Natural Gas**.

LOACHES (*Osteichthyes*). Of the order *Ostariophysi*, suborder *Cypriniformes*, and family *Cobitidae*, loaches are small bottom-feeding fishes of Europe and Asia. They have a long slender body and a group of barbels near the mouth. The European species are eaten. A few fishes of similar form, but not closely related are called African loaches.

Several species are quite popular with tropical-fish fanciers. A brilliantly colored orange species from Borneo and Sumatra, the clown loach (*Botia macracanthus*) is frequently found in small aquariums, despite a habit of uprooting vegetation in the tank. Some clown loaches may reach a length of about 12 inches (30 centimeters) and live up to a quarter-century. Under normal aquarium conditions, the length is limited to a maximum of about 4 inches (10 centimeters). Other small loaches include the coolie loach (*Acanthophthalmus kuhli*), which is somewhat snake-like and found in Thailand and environs; and the spotted weatherfish or spined loach (*Cobitis taenia*) which is noted for its tolerance to wide temperature variations. A spotted species (*Cobitis biwae*) is found in Japan and is sometimes considered a food item. *Misgurnus fossilis* is the European weatherfish, known for its sensitivity to barometric pressure—hence the name, weatherfish. It is quite a large loach, sometimes attaining a length of about 20 inches (51 centimeters), but usually not exceeding 10 inches (25 centimeters).

LOAD FACTOR (Electric). The ideal electric load, from the standpoint of equipment needed and operating routine, is one of constant magnitude and steady duration. A load of this type is shown in the upper part of Fig. 1(a). The cost to produce an elementary area of this load curve could

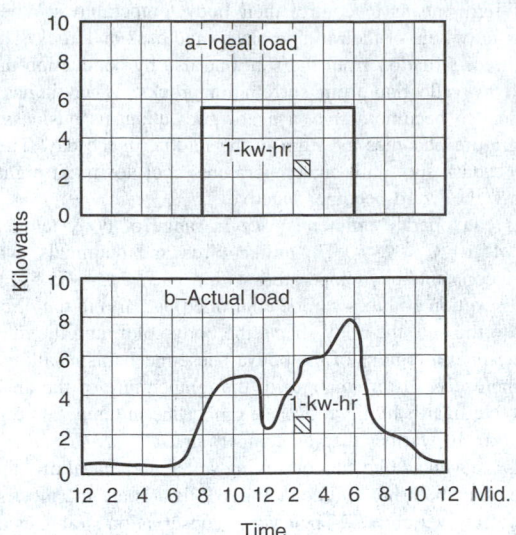

Fig. 1.   Comparison of (**a**) ideal and (**b**) actual electric power loads.

be from one-half to three-fourths of that to produce the same unit under the more frequently realized condition shown in the lower curve (b). The problem of variable load is of vital importance to utilities whose chief concern is to put each kilowatt-hour on the transmission line at the lowest possible production cost. Interconnections between utilities and regional power generating facilities help in matching load with generation. See also **Electric Power Production and Distribution**.

LOADING COIL.   1. An inductance inserted at regular intervals along a long transmission line or cable to increase the line's characteristic impedance and reduce its attenuation constant.

2. An inductance inserted in series with an antenna to increase its electrical length.

LOAD MATCHING.   1. Maximum power is delivered to a load when the impedance of the generator is the image impedance of the load. The adjustment of a circuit to provide this condition is called load matching.

2. In induction heating and dielectric heating usage, the process of adjustment of the load circuit impedance to produce the desired energy transfer from the power source to the load.

LOBE.   In an antenna pattern, a region of local maximum in the emitted intensity. The strongest lobe is in the pointing direction of a directional antenna and is called the main lobe. The configuration of lobes is determined by three factors: 1) wavelength; 2) geometrical properties of the antenna and feed system; and 3) mutual interference between the direct and reflected rays for an antenna situated above a reflecting surface. The sidelobes or minor lobes are an unavoidable consequence of the finite size of the antenna. Though undesirable, they ordinarily contain much less power than the main lobe.

See also **Antenna Pattern**.

LOBECTOMY.   Surgical removal of a lobe of a gland or organ, such as the lung.

LOCAL APPARENT TIME.   The arc of the celestial equator, or the angle at the celestial pole, between the lower branch of the local celestial meridian and the hour circle of the apparent or true sun, measured westward from the lower branch of the local celestial meridian through 24 hours; local hour angle of the apparent or true sun, expressed in time units, plus 12 hours.

LOCAL AREA NETWORKS.   As automation in the factory increases, the need for communication between computers, controllers, and other "intelligent" machines has become critical. In the past, when factory automation was limited to the use of programmable controllers, numerical

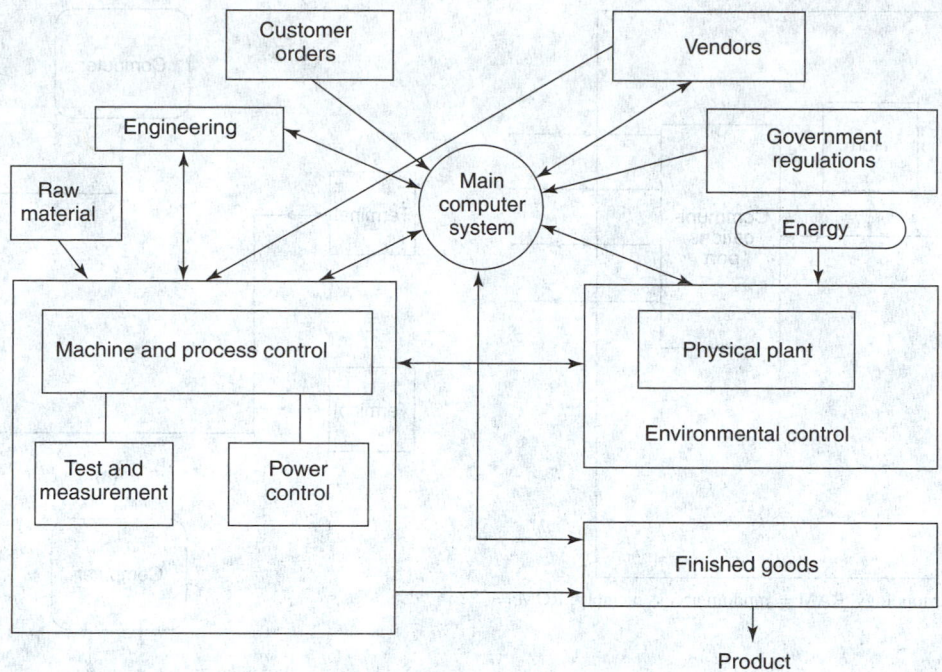

Fig. 1.   Intercommunications in a generalized automated factory situation.

control of machines, and similar traditional approaches, communication was not a major limiting factor. Each tool was essentially self-contained and the communication requirement was mainly a user interface for controlling and updating machine operation. With the accelerated growth of automated tools and processes, communication between these entities is required to control not only their operation, but also their interrelationships. To this is added the desire to overlay environmental control, energy management, and materials requirement planning (MRP) to the factory operation. The end result is that intercomputer/controller communication has become the largest single problem to be addressed for factory automation. These interrelationships are shown in a generalized fashion in Fig. 1.

Early communication needs were served with point-to-point data links, as simply indicated in Fig. 2. Communication was relatively simple. The *star topology* for the communication system (Fig. 3) was developed so that multiple computers could communicate. The central or "master" node uses a communications port with multiple drops as shown in Fig. 4. The master is required to handle traffic from all the nodes attached, poll the other nodes for status, and, if necessary, accept data from one node to be routed to another. This heavy software burden on the master is also shared to a lesser degree among all the attached nodes. In addition, star topology requires routing a separate wire for every piece of equipment attached. This makes it difficult to wire and even more difficult to change. Also, the star topologies are inflexible regarding the number of nodes that can be attached. Either the user must invest in unused connections for further expansion, or have a system that cannot grow.

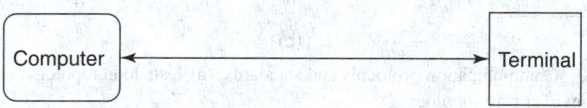

Fig. 2.   Point-to-point communication.

To overcome some of these disadvantages, multidrop protocols were established and standardized. Data loops, such as SDLC (Synchronous Data Link Control), were developed as well as other topologies, including buses and rings. Some of the early standards are shown in Fig. 5. The topology of these standards makes it easy to add (or subtract) nodes on the network. The wiring is also easier because a single wire is routed to all nodes. In the case of the ring and loop, the wire is also returned to the master. Inasmuch as wiring and maintenance are major costs of data communication, these

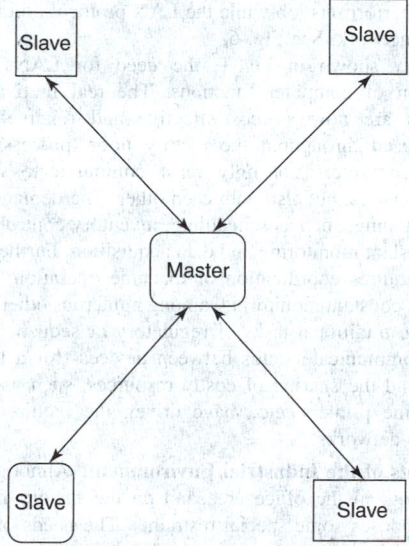

Fig. 3.   Star topology.

topologies have virtually replaced star networks. These systems do have a common weakness, i.e., one mode is the "master," with the task of determining which station may transmit at any given time. As the number of nodes increases, throughput becomes a problem because: (1) a great deal of "overhead" activity may be required to determine who may transmit, and (2) entire messages may have to be repeated, because some protocols only allow master-slave communications. That is, a slave-to-slave message must be sent first to the master and then repeated by the master to the intended slave receiver. Reliability is another problem. If the master fails, communication comes to a halt. The need for multinode networks without these kinds of problems and restraints led to the development of *Local Area Networks* (LANs), which use peer-to-peer communication. In this system, no one node is in charge and all nodes have an equal opportunity to transmit.

A LAN is a distributed communication network with the following characteristics: (1) peer-to-peer oriented (no master); (2) from 2 to 200 data devices (nodes) may be incorporated; (3) distance is limited to less than 2 km (1.2 mi); and (4) from 1 to 20 M bits per second data rates.

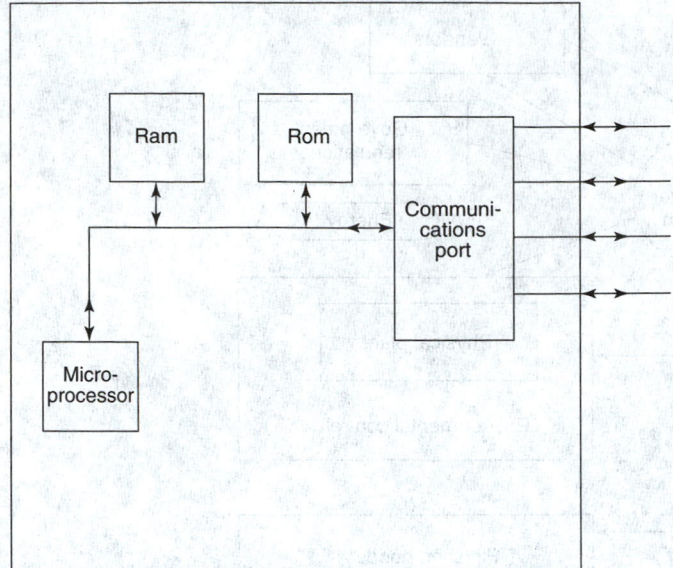

Fig. 4.   Master node for star topology. RAM = random access memory; ROM = read only memory.

In a local area network, each node is an independent computer system. Since there is no master to control traffic, each node must determine when it has the right to transmit. In a typical system, the host computer of the node is free to perform its job while the LAN protocol unit is moving data on and off the network. See Fig. 6.

As previously shown in Fig. 1, the need for LANs is driven by the proliferation of computer functions. The real motivating factor is that computers are now so cost-effective and relatively inexpensive that they are used throughout the factory floor, processing plant, and office. These computers not only must communicate with the large mainframe computers, but also with each other. The demands of materials requirement planning, such as scheduling, inventory control, management, etc. require constant monitoring and data acquisition. Further, factory floor management requires coordination of machine operation; environmental control requires constant monitoring, among numerous other factors that go well beyond the traditional tasks of regulatory or sequencing controllers. The need to communicate status between devices (for a total integrated environment) and the sharing of costly resources, such as large-capacity disk storage, line printers, etc., have driven the requirement for LAN communication networks.

Requirements of the Industrial Environment. Although the need for LANs exists both in the office area and on the factory floor, the latter environment imposes some special restraints. The needs of the industrial environment require:

1. *Noise Tolerance.* Since a LAN will have long cables running throughout the factory, the amount of noise picked up can be large. The LAN must be capable of performing reliably in an electrically noisy area. The physical interface should be defined to provide a significant degree of noise rejection. The protocol must allow for easy recovery from data errors.

2. *Fast Response.* The LAN in an industrial situation should have a guaranteed maximum response time, i.e., the network must be able to transmit an urgent message within a specified time limit. The real-time characteristic of industrial control demands communication within a known time frame.

3. **Ability to Handle Priority Messages**. On the factory floor, both control and status data will be carried over the same network. A control message should have a higher priority and be transmitted before other messages.

Common Standards. A local area network standard that serves the harsh factory environment well can also be used for the less demanding office and administrative areas. Unless the requirements for the factory add too much cost, the factory floor standard can be the choice for the entire network. A common standard is advantageous inasmuch as system

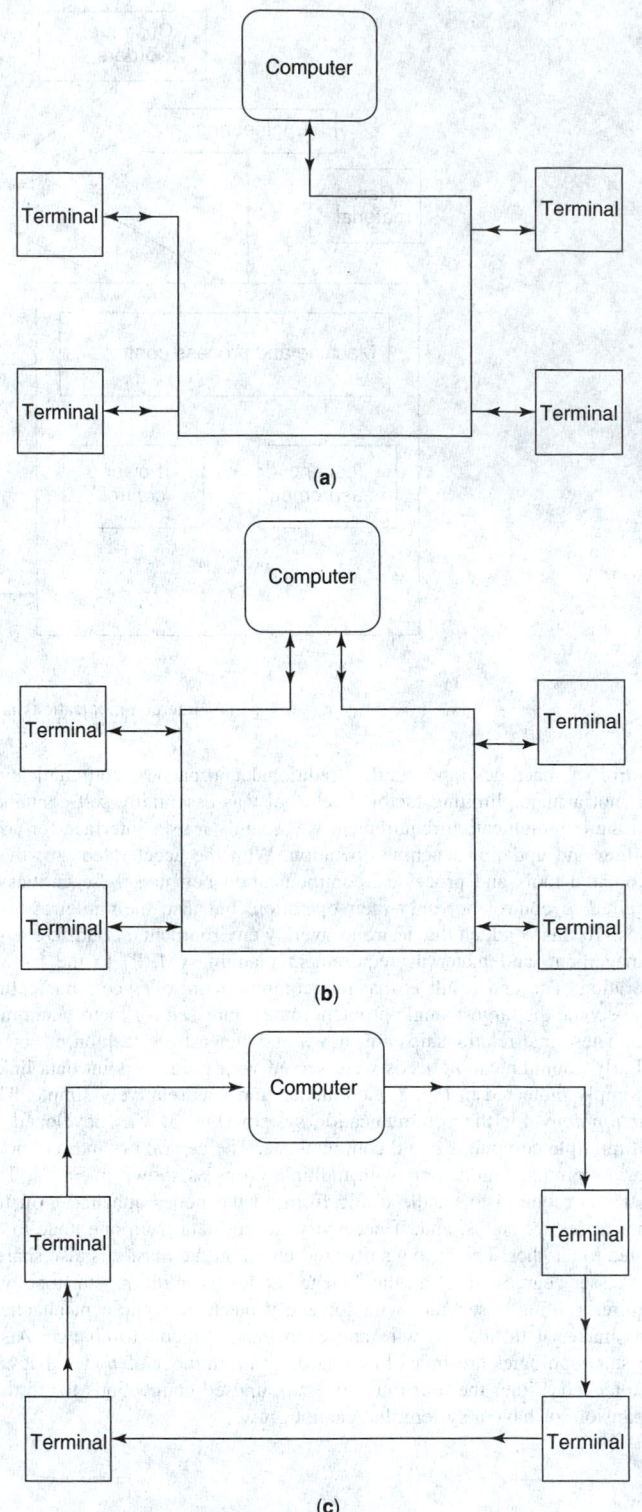

Fig. 5.   Communication protocols and standards: (**a**) Data loop topology; (**b**) bus topology; (**c**) ring topology.

and information handling elements located in both office and factory environments must communicate with each other.

To meet current communication needs, different types of LANs are possible. Many of these already have been developed. All LANs provide the same basic service—to allow computers to pass data. However, since the major function of a network is to connect many different computers, standards are needed. The standard not only should describe how nodes are connected, but the protocol followed in transferring data as well. Ideally, the standard should be sufficiently comprehensive to permit any computer

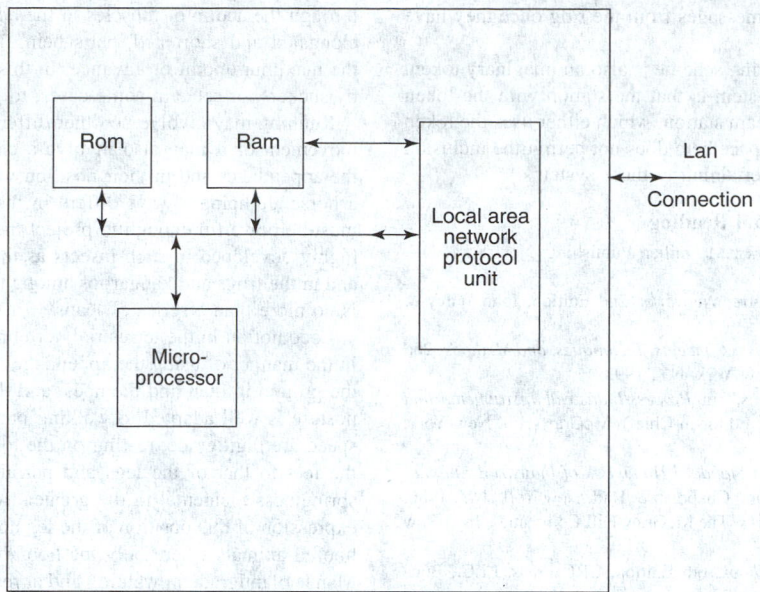

Fig. 6.   A local area network node. RAM = random access memory; ROM = read only memory.

following it to pass data to any other computer that follows the same standard.

Several organizations have been working on the standards problem for a number of years, including the International Standards Organization (ISO), the Institute of Electrical and Electronics Engineers (IEEE), as well as some major users of LANs, such as General Motors Corporation, which has developed MAP (Manufacturing Automation Protocol).

Examples of Protocols

Carrier Sense Multiple Access with Collision (CSMA/CD) This is a baseband system with a bus architecture. See Fig. 5(b). Baseband is a term used to describe a system where the information being sent over the wire is not modulated, i.e., the "ones" are represented by one voltage level and the "zeros" by another voltage level. Normally, only one station transmits at any one time. All other stations hear and record the message. The receiving stations then compare the designated address of the message with their address. The one station, which matches, will pass the message to its upper layers, while the others will throw it away. Obviously, if the message is affected by noise (detected by the frame check sequence), all stations will throw the message away.

The CSMA/CD protocol requires that a station listen before it can transmit data. If the station hears another station already transmitting (*carrier sense*), the stations wanting to transmit must wait. When a station does not hear any other station transmitting (*no carrier sense*), it can start to transmit. Since more than one station can be waiting, it is possible for multiple stations to start transmitting at the same time. This causes the messages from both stations to become garbled (called a "collision"). A collision is not a freak accident, but is a normal way of operation for networks using CSMA/CD. The chances of collision are increased by the fact that signals take a finite amount of time to travel from one end of the cable to the other. If a station on one end of the cable starts transmitting, a station on the other end will "think" that no other station is transmitting during this travel-time interval and that transmission can be resumed. After a station has started transmitting, it must detect when another station is also transmitting. If this happens (*collision detection*), the station must stop transmitting. Before quitting the transmission, however, the station must make sure that every other station is aware that the last frame is in error and must be ignored. To do this, the station sends out a "jam" which is simply an invalid signal. This jam guarantees that the other colliding station also detects the collision and quits transmitting. Each station that was transmitting must then wait before trying again. To make certain that the two (or more) stations that just collided do not collide again, each station selects a random time to wait. The first station to time out will then look for silence on the cable and retransmit its message.

Token Bus. This standard has been selected for use in the previously mentioned MAP protocol. Token bus is also a bus topology, but differs in two ways from CSMA/CD: (1) The right to talk is controlled by passing a "token," and (2) the data on the bus are always carrier modulated. In the token bus system, one station is said to have an imaginary token. This station is the only one on the network that is allowed to transmit data. When this station has no more data to transmit (or it has held the token beyond the specific maximum time limit), it passes the token to another station. This token pass is accomplished by sending a special message to the next station. After this second station has used the token, it passes it to the next station, and so on. After all other stations have used the token, the original station is passed the token again. A station (example, station A) will normally receive the token from one station (B) and pass the token to the third station (C). The token ends up being passed around in a logical token ring (A to C to B to A to C to B . . .). The exception to this is when a station wakes up or dies. For example, if a fourth station (D) gets in the logical token ring between A and C, station A would then pass the token to D so that the token would go: (A to D to C to B to A to D . . .). Only the station with the token can transmit so that every station gets its turn to talk without interfering with anyone else. Obviously, the protocol also has provisions that allow stations to enter and to leave the logical token ring.

The second difference between token bus and CSMA/CD is that with the token bus, data are always modulated before being sent. The data are not sent out as a level, but as a frequency. There are three different modulation schemes allowed—two single-channel and one broadband. Single-channel modulation permits only the token bus data on the cable. The broadband method is similar to CATV (community antenna television) and allows many different signals to exist on the same cable, including video and voice, in addition to the token bus data. The single-channel techniques are simpler, less costly, and easier to implement than broadband. Broadband, however, is of higher performance, permitting much longer distances and, very important, satisfies both present and future communication needs by allowing as many channels as needed (within the bandwidth of the cable).

Token Ring. Originally, token ring and token bus used the same protocol with different topologies. The two systems still remain rather similar. As shown by Fig. 5, any one node will only receive data from the "upstream" node and will only send data to the "downstream" node. All communication is done on a baseband point-to-point basis. This would seem to imply that one node can talk only to its downstream node. This is not the case, inasmuch as each station repeats what it hears from the upstream station to the downstream station. Since the last station is connected to the first (forming a ring), any station can send data to any other station. Precaution must be taken to prevent a short message from being retransmitted around the ring forever. This is prevented, by having the transmitting station remove its messages from the ring once they have gone around the ring one time.

The "right to talk" for the token ring scheme is also an imaginary token. The simplicity of the token ring system is that the station with the token simply sends it to the next downstream station, which either uses the token or passes it on to the next station. Space here does not permit the inclusion of numerous other pros and cons pertaining to these systems.

Additional Reading

Felt, S.: *Local Area High Speed Networks*, Macmillan Publishing, New York, NY, 2000.

Goldman, J.E. and P.T. Rawles: *Local Area Networks*, 2nd Edition, John Wiley & Sons, Inc., New York, NY, 2000.

Held, G.: *Internetworking LANs and WANs: Concepts, Techniques, and Methods*, 2nd Edition, John Wiley & Sons, Inc., New York, NY, 1998.

King, J.P.: "Distributed Control Systems," in *Process/Industrial Instruments and Controls Handbook* (D.M. Considine, Editor-in-Chief), McGraw-Hill, New York, NY, 1993.

Loyer, B.A.: "Local Area Networks," in *Standard Handbook of Industrial Automation* (D.M. and G.D. Considine, Editors) Chapman & Hall, New York, NY, 1986.

Parnell, T.: *Building High-Speed Networks*, The McGraw Hill Companies, Inc., New York, NY, 1999.

Slone, J.P.: *Local Area Networks Handbook*, 6th Edition, CRC Press, LLC., Boca Raton, FL, 1999.

Stallings, W.: *Local and Metropolitan Area Networks*, 6th Edition, Prentice-Hall, Inc., Upper Saddle River, NJ, 2000.

Stamper, D.A., J.C. Van Horne, and J.M. Wachowicz: *Local Area Networks*, 3rd Edition, Prentice-Hall, Inc., Upper Saddle River, NJ, 2000.

Tanenbaum, A.S.: *Computer Networks*, 3rd Edition, Prentice-Hall, Inc., Upper Saddle River, NJ, 1996.

Thompson, A.: *Understanding Local Area Networks: A Practical Approach*, Prentice-Hall, Inc., Upper Saddle River, NJ, 1999.

LOCAL ASTRONOMICAL TIME.   Mean time reckoned from the upper branch of the local meridian.

LOCAL LUNAR TIME.   The arc of the celestial equator, or the angle at the celestial pole, between the lower branch of the local celestial meridian and the hour circle of the moon, measured westward from the lower branch of the local celestial meridian through 24 hours; local hour angle of the moon, expressed in time units, plus 12 hours.

LOCAL SIDEREAL TIME (LST).   Local hour angle of the vernal equinox, expressed in time units; the arc of the celestial equator, or the angle at the celestial pole, between the upper branch of the local celestial meridian and the hour circle of the vernal equinox, measured westward form the upper branch of the local celestial meridian through 24 hours.

LOCOMOTION.   The process of moving from place to place, a characteristic power of most animals and a lesser distinction between them and the majority of plants.

Locomotion is necessary to animals because their food is organic and in most environments does not reach the animal through external forces. Even the sessile animals, which may or may not be capable of some locomotion, often accomplish the same end by bringing food within reach through their own activities.

In the water the weight of the surrounding medium is so great that the animal may float, and the resistance offered to its body is sufficient to be utilized for propulsion. The body is so shaped that resistance is little in the direction of locomotion but great where propulsive effort is expended. Projections from the surface that beat against the water like oars or push or pull by undulating are common organs of locomotion here. They include cilia and flagella in one-celled and small multicellular forms, specialized jointed appendages of arthropods, and fins and flippers of vertebrates. Undulation of the body itself is a sufficient means of propulsion in some animals.

Some aquatic forms rest on the bottom and the terrestrial animals are forced to rest on some solid support at least intermittently because the air is too light to float them. In many of these forms the friction of contact with a solid is utilized by the development of movable supporting appendages which are shifted alternately to change the animal's position. This means of locomotion is known as walking. Other animals, notably the worms, creep through the action of muscles in the body wall. The body is progressively elongated and shortened, parts being thrust ahead and then drawn up to the maximum point of advance. In this type of locomotion they are aided by suckers, or setae in some cases, to grip the supporting surface.

Running may involve no other difference from walking than more rapid movement or it may also involve a change in the order of movement of the appendages and in their position when used, as in the various gaits of a horse. Jumping always differs in that the appendages set farthest back must be powerful enough to project the entire animal through the air. It is highly developed in such insects as the flea beetles and the grasshoppers and in the frogs and kangaroos among the vertebrates. In this class a gallop is no more than a series of leaps.

Locomotion in the terrestrial vertebrates also shows progressive change in the manner of using the appendages. The entire sole of the foot rests on the ground in man and the apes, and they are said to be plantigrade. This posture is well adapted to walking but not to running. Animals that need speed are digitigrade, resting on the toes. This position adds the length of the feet to that of the legs and permits a longer stride. It also adds the springiness incidental to the greater freedom of the ankle joint. The final expression of this position of the leg appears in the unguligrade (Ungulata) hoofed animals where only the hoof comes into contact with the ground. Man is plantigrade in walking and at rest but rises to his toes when he runs.

The locomotion of snakes is a highly specialized creeping process in which the ribs serve as the movable appendages and the grip of the body on the ground is provided by the broad scales of the ventral surface which project backward.

Climbing animals may merely run along branches, aided by sharp claws to provide a secure grip. The sloths, however, have the claws developed as great hooks, which suspend them in an inverted position. They walk as well as hang upside down. The primates show the most extreme specialization for a form of locomotion in the trees called brachiation. Their pectoral appendages are arms, adapted for grasping and suspension, and the pelvic appendages are supporting legs. They move by swinging from branch to branch or by shifting from one hold to another, as human beings climb.

Locomotion in the air is the highly specialized process of flight.

Additional Reading

Blake, R.: *Fish Locomotion*, Cambridge University Press, New York, NY, 1983.

Kimura, T. et al.: *Development and Control in Primate Locomotion*, S. Karger AG, Bassel, Switzerland, 1996.

Leach, D. and H. Schamhardt: *Animal Locomotion*, S. Karger AG, Bassel, Switzerland, 1993.

Patla, A.: *Adaptability of Human Gait: Implications for the Control of Locomotion*, Elsevier Science, New York, NY, 1991.

LOCUST BEAN GUM.   See **Food Additives**.

LOCUST *(Grasshopper; Insecta, Orthoptera).*   The term "locust" is more properly applied to the "short-horned" grasshoppers that include the migratory locusts appearing in Africa, Asia and the plains of North America as serious crop pests. These animals have always been an important food of aboriginal peoples, and are still consumed in many parts of the world. The term grasshopper is now generally restricted to the common non-migratory forms. The name locust is also commonly but wrongly applied to the Cicada or tree cricket.

The adults measure about 2 inches (5 centimeters) in length; some are much larger. The body is thick, strong, tapering to the end of its folded wings. Large eyes are high on the head and located just above the two short antennae. A stiff, thick jacket joining the head covers and protects the back of the neck. The wings, shaped much like some aircraft wings, are held close to the body when at rest. The wings are strong, large, heavily veined and twice as wide at the center as at the ends, and about $\frac{1}{2}$ inch (12 millimeters) longer than the body.

The six legs of the locust are comprised of several segments, connected to the body; the two shorter front legs are located under the shoulders. The two legs located near the center of the body are long and powerful and used for making long hops. The thigh of the leg is oar-shaped and heavily veined. It is connected to a strong joint at the body, which aids greatly in the thrust of the kick when jumping.

The "song" of the locust is created by spurs or spines on the inner area of the hind legs which are rubbed against the wings which have a raised, spurred area on the outside surface. This gives a rasping sound.

The desert locust (*Schistocera gregaria*) of India swarms during the summer monsoons. In autumn, these insects migrate to Iran, Arabia, Soviet Asia, Syria, and Egypt. In early winter, they return to India and East Africa to breed.

Possibly the most impressive of all insect flights will be that of a swarm of billions of locusts. In 1889, a flight crossed the Red Sea estimated to be 2000 square miles (5180 square kilometers) in extent. Desert locusts have appeared in England, apparently flying from southern Algeria, possibly assisted by a tail wind. Swarms of locusts have been reported since ancient times. The Book of Joel describes the army that blackened out the sun—behind them a desolate wilderness and nothing escaping them.

There are about 2,000 species of locusts, of which about 20 important species are capable of causing crop desolation and accompanying famine in some areas of the world. The huge migrations are caused by hunting for food, and where locusts find food they will eat their weight daily if it is plentiful. Fortunately, the mortality rate of the insect is high when a swarm encounters stormy weather. See also **Cicada**.

LOCUST (*Seventeen year; Insecta, Homoptera, Cicadidae*). The 17-year locust (*Magicicada septendecim, Linne*), not a true locust, is named for its life cycle of 17 years. There are, however, many broods that overlap, with adults appearing in different years. There is also the 13-year locust (*M. septendecim tridecim*, Riley). These insects are sometimes confused with the *dog-day cicada* (*Tibicen linnei*) (Smith and Grossbeck), also commonly termed a locust, which has a 9-year cycle. The latter insect is not nearly so damaging as the other two species just mentioned. See also **Cicada**.

The 17-year and 13-year locusts (properly known as *periodical cicadas*) make rough punctures in twigs and small branches of apple and numerous other fruit trees. Damage does not result from the feeding of the insects, but from the puncturing of the twigs when the female deposits her eggs. In the United States, the 13-year locust ranges in a line from Virginia to southern Iowa and thence southward to the Gulf of Mexico and eastward to the Atlantic shore. The 17-year locust is found from the New England states westward to Wisconsin, southwestward into Kansas and Missouri, and south as far as Alabama and northern Georgia. It is most abundant east of the Mississippi River and greatly infests much of Wisconsin, Iowa, Tennessee, and South Carolina, and nearly all of Illinois, Indiana, Ohio, Virginia, Kentucky, West Virginia, Pennsylvania, Connecticut, and New Jersey.

The 17-year locust has the longest period of development of any insect known. As previously described, the eggs are laid in twig punctures in late spring and mid-summer. Each female deposits 400–600 eggs, with about 20 eggs per puncture. Within 6–7 weeks the eggs hatch. The young are ant-like in appearance and, when hatched, drop to the ground where they enter into cracks in the soil. They feed on sap from tree rootlets. Feeding is very slow and their presence cannot be detected by any apparent deterioration of the tree, even though there may be many thousands of these creatures at the base of an affected tree. Depending upon the species, these nymphs require from 13 to 17 years to achieve maturity. At that time, the insects are about an inch (2.5 centimeters) long and appear something like a crayfish. The insects burrow to the surface, sometimes forming mud cones or chimneys that may protrude 2–3 inches (5–8 centimeters) above ground level. Massive numbers of these nymphs emerge within a very short period, usually climbing the tree after sunset. They temporarily take hold of the bark until the adult insect removes itself from the nymphal shell. Leaving the empty skins on the tree, they take flight and are ready to mate during a period of 30–40 days. It is estimated that over 40,000 adults may emerge from a single large tree within a period of a few days.

LOCUST TREES.   Several leguminous trees are commonly called locust trees. Most notable is the black locust, *Robinia pseudoacacia*. It is a medium-sized tree native to the Appalachian and Ozark mountains. The twigs have a pair of spines about $\frac{1}{2}$ inch (1.3 centimeters) long at the base of each leaf, although spineless varieties have been developed. The leaves are compound, 8–14 inches (20.3–35.5 centimeters) long. In the spring the tree produces its flowers, which are white or pink and very fragrant, hanging in clusters 4–8 inches (10–20.3 centimeters) long. The wood is hard and tough, and is used for fence posts, mine timbers, and rough construction. The tree is commonly planted for ornament and shade, or for erosion control.

The honey locust, *Gleditsia*, is also frequently known simply as locust. A variety of this genus, the Moraine locust, has become popular as a shade tree.

Record locust trees, as reported by American Forests, are listed in the Table 1.

LOEB, JACQUES (1859-1924).   Jacques Loeb was born in Mayen, Germany to a prosperous Jewish importer, and grew up in a well-educated home reading the classics of eighteenth-century European philosophical thought. When he first entered university in 1880 in Berlin, it was with the intention of studying the philosophy of the will, but he was so disillusioned by his experience that he abandoned these plans and moved to the University of Strasbourg to study science. He obtained an MD in 1884 and in 1886 moved to the University of Wurzburg as an assistant to the physiologist Adolf Fick. Toward the end of his tenure there he spent a winter at the marine biological laboratories in Naples. In 1890, Loeb married an American student named Anne Leonard and, disturbed by the German political scene at the time, emigrated to the USA the following year. There he taught at a number of universities including Bryn Mawr College (1891–1892), the University of Chicago (1892–1902), and the University of California at Berkeley (1902–1910); he finally worked at the newly established Rockefeller Institute of Medical Research in New York City where he remained until his death. Throughout this period, Loeb spent his summers in the marine biological laboratories at either Pacific Grove in California or Woods Hole in Massachusetts, where he did most of his experimental work.

Perhaps because of the popular interest in the subject, Loeb is best known for his rather controversial work on artificial parthenogenesis in

TABLE 1. RECORD LOCUST TREES IN THE UNITED STATES[1]

Specimen	Circumference[2]		Height		Spread		Location
	Inches	Centimeters	Feet	Meters	Feet	Meters	
Black locust (1974) (*robinia pseudoacacia*)	280	711	96	29.3	92	28	New York
Clammy locust (1996) (*Rabinia viscosa*)	19	48	35	10.7	21	6.4	North Carolina
Honeylocust (1999) (*Gleditsia triacanthos*)	226	574	100	30.5	88	26.8	Maryland
Honeylocust (1993) (*Gleditsia triacanthos*)	233	592	90	27.4	88	26.8	Pennsylvania
New Mexico locust (1997) (*Robinia neomexicana*)	90	229	71	21.6	28	8.5	Arizona
Waterlocust (1993) (*Gleditsia aquatica*)	110	279	74	22.6	73	22.3	Pennsylvania

[1]From the "National Register of Big Trees," American Forests (by permission).
[2]At 4.5 feet (1.4 meters).

animals, but in fact he made several contributions to biology long before becoming active in that field. He was one of the pioneers in the area of objective analysis of animal behavior and is often credited with introducing experimental rigor into this area of biology. He operated under what he termed a "mechanistic" conception of life (which he used as the title of what emerged as the most widely read of his books), namely that all of life's processes, including behavior, could be explained in terms of physical or chemical reactions or mechanisms. This mechanistic theory appears as a common thread in most of the science that Loeb undertook–beginning with his medical school thesis, which focused on the localization of brain function, specifically the effects of injuries to the cerebral cortex on blindness.

In 1888 he began a line of research that demonstrated animal tropisms, namely the instinctual movement of certain animals towards external stimuli such as light, electric currents or chemicals. Specifically, he devised experiments to show that species of caterpillars invariably preferred to move toward light sources, typically the sun. During his winter in Naples, Loeb's interests were turned to problems in development and embryology, and he began to apply his mechanistic philosophies to problems in these fields as well. By adjusting salt concentrations in the environment of sea urchin eggs, he was able to induce artificial parthenogenesis, namely the division of eggs to give rise to larvae without prior fertilization with sperm from the animal. In 1918 he embarked on a third line of research through which he attempted to explain the action of proteins in living systems in terms of their properties in solutions, and made important contributions to understanding the colloidal behavior of these molecules. Loeb was one of the few scientists of his generation who was as well known to the general public as within the field. Perhaps the best evidence for this is in Sinclair Lewis' 1925 novel *Arrowsmith*, where the character modeled on Loeb is the only figure within the fictitious scientific institution (an obvious satire of the Rockefeller Institute) to emerge in a positive light.

NEERAJA SANKARAN, Yale University, New Haven, CT

LOESS. Loess is a buff-colored, wind-blown deposit of fine silt or marl, usually unstratified, which is often exposed in bluffs with steep to vertical faces. Loess is found in the United States in the Mississippi valley from Louisiana to Iowa, and along the course of the Missouri. The average thickness of the loess here is about 20 feet (6 meters), but may range to 50–100 feet (15–30 meters). Loess also occurs in central Europe, Mongolia and China where it is said to attain a thickness as great as 300 feet (90 meters). The loess of the United States and Europe is believed to be the finer materials first transported and deposited by the waters of the melting ice sheets of the glacial period, and later blown to considerable distances and sometimes deposited in lakes. The Asiatic loess seems to be wholly wind transported, the source of the dust being, perhaps, the great deserts of central Asia. In the latter case the accumulation of such thick deposits is attributed to the binding power of successive generations of grasses whose former existence is suggested by a network of narrow tubes.

See also **Erosion.**

LOEWI, OTTO (1873–1961). Otto Loewi was a German physiologist and pharmacologist who discovered the chemical transmission of nerve impulses; he became an American citizen in 1946. He became interested in clinical medicine and pharmacology as a student at the universities of Strasbourg and Munich, graduating in 1896. His first research project concerned the effect of drugs on isolated frog hearts. Work in a tuberculosis ward convinced him that he preferred laboratory research to clinical medicine. He spent the next 12 years working in Marburg with H. H. Meyer. Loewi and Meyer were primarily interested in metabolic problems, such as the induction of experimental diabetes. To learn more about new physiological research techniques, Loewi visited England in 1902 and met Sir Frederick Gowland Hopkins, Ernest Starling, Henry H. Dale, and also T. R. Elliott who was studying autonomic nerves and their response to chemical stimulation. After returning to Marburg, Loewi began an investigation of the chemistry of the autonomic nervous system. In 1909 he was appointed Professor of Pharmacology at the University of Graz in Austria. His research projects included carbohydrate metabolism, experimentally induced diabetes, the metabolic characteristics of different organs, the physiological control mechanisms of heart action, and the

effects of various drugs on the heart. See also **Dale, Henry Hallett (1875–1968)**; **Hopkins, Frederick Gowland (1861–1947)**; and **Starling, Ernest Henry (1866–1927)**.

When Loewi began his studies of neurochemical transmission in the 1920s, the electrical aspect of the nerve impulse was well known, and the work of Elliott, Dale and others suggested that a chemical transmitter might be involved in the autonomic nervous system. Working on an inspiration, which occurred to him in a dream, Loewi carried out experiments that showed the existence of a chemical neurotransmitter. He placed two isolated frog hearts in separate tissue baths and then stimulated the vagus nerves to the first heart; this caused the heartbeat to slow down. When the fluid bathing the heart was collected and transferred to the second heart, the heartbeat responded in the same manner. Similarly, stimulating the sympathetic nerves of the first heart increased the heartbeat. When the fluid bathing the first heart was added to the second heart, it too beat faster. Therefore, it was clear that a chemical released by stimulating the nerves of the first heart could affect the second heart. Further refinements of the technique and the identification of the specific chemicals involved confirmed the concept of the chemical mediation of nerve impulses. Loewi called the vagal transmitter Vagusstoff and in 1926 he tentatively identified it as acetylcholine. Subsequent work by Dale and others confirmed this hypothesis. Loewi and Dale shared the Nobel Prize in Medicine or Physiology in 1936 for their discoveries relating to the chemical transmission of nerve impulses. See also **Central and Peripheral Nervous Systems**; **Neurochemistry (The History)**; and **Neurotransmitters**.

In 1938, during the Nazi occupation of Austria, Loewi and two of his sons were arrested. Members of the international scientific community were able to secure their release and Loewi fled to England. His wife, however, was unable to leave Austria until the Nazis had confiscated all of their assets. In 1940 Loewi was offered a research professorship in the Department of Pharmacology at the New York University School of Medicine. In 1946, he became an American citizen.

Additional Reading

Bacq, Z.M.: *Chemical Transmission of Nerve Impulses*, Pergamon, Oxford. UK, 1975.

Dale, H.H.: "Otto Loewi," In: *Biographical Memoirs of Fellows of the Royal Society*, **8**, 67–89 (1962).

Geison, G.L.: "Otto Loewi," in: Gillispie, C.C., ed., *Dictionary of Scientific Biography*, vol. **8**, Charles Scribner's Sons, New York, NY, 1973, pp. 451–457.

Loewi, O.: "Problems Connected with the Principle of Humoral Transmission of Nervous Impulses, Ferrier Lecture," *Proceedings of the Royal Society of London*, *Series B 118*, 299–316 (1935).

Loewi, O.: "The Chemical Transmission of Nerve Action. Nobel Lecture," Reprinted in *Nobel Lectures, Physiology or Medicine*, vol. **2** (1922–1941), pp. 416–432, Elsevier, Amsterdam, 1965.

Loewi, O.: "Reflections on the Study of Physiology," *Annual Review of Physiology*, **16**, 1–10 (1954).

Loewi, O.: "An Autobiographical Sketch," *Perspectives in Biology and Medicine*, **4**, 1–25 (1960).

LOIS N. MAGNER, Purdue University, West Lafayette, IN

LÖFFLER, FRIEDRICH AUGUST JOHANN (1852–1915). Löffler was a German Bacteriologist and Virologist who identified the organism that causes diphtheria, and described the causative agent of foot-and-mouth disease.

Löffler studied medicine in Würzburg and at the Academy of Military Medicine (Pepinere) in Berlin, where he graduated in 1874. Working as a military physician from 1876, he was commanded to the newly founded Imperial Health Office in 1879 and subsequently became one of the most important members of its bacteriological laboratory. Following his Habilitation (postdoctoral qualification) in 1886, he became Professor of Hygiene in Greifswald, where he also taught history of medicine. In 1913, as a successor to Georg Gaffky, he became director of the Robert Koch Institute in Berlin.

Löffler's achievements lay chiefly in the field of bacteriology and veterinary medicine. In 1882 and 1885, he identified the pathogens of two important veterinary diseases in pigs, glanders and erysipelas. In 1884, he demonstrated the aetiological relation of *Corynebacterium diphtheriae* to diphtheria and thereby facilitated a clear distinction of this disease from a host of other infections with similar clinical manifestations. After receiving his call to Greifswald, Löffler continued to work on the microbiology of veterinary diseases. In 1887, he co-founded an influential journal, the *Centralblatt für Bakteriologie und Parasitenkunde*. In 1891, he identified the pathogen of murine typhoid and tried to employ that knowledge for the control of mouse populations that tormented agriculture. Together with Paul Frosch he pioneered virus research and in 1898 described what is today known as the virus of foot-and-mouth disease as an ultra visible, filterable bacterium. He seems to have coined the phrase of 'Koch's postulates' to describe the essential steps necessary to establish a bacterial aetiology and also became the first historian of medical bacteriology. His book '*Vorlesungen über die geschichtliche Entwicklung der Lehre von den Bakterien*' of 1887 is still an invaluable source of information on the history of nineteenth century bacteriology.

See also **Foot-and-Mouth Disease**.

CHRISTOPH GRADMANN, University of Heidelberg, Heidelberg, Germany

LOFTING. The phenomenon where the upper part of a smoke plume diffuses more rapidly upward than the bottom part diffuses downward. This generally occurs when the boundary layer near the ground is more stable than it is aloft.

LOGARITHM. If B is an arbitrarily chosen number greater than unity, then the logarithm L of any other number N is defined by $N = B^L$; $L = \log_B N$. The chosen number B is the base of the system of logarithms. For any base, $\log_B 1 = 0$; $\log_B B = 1$. The fundamental properties of logarithms are $\log ab = \log a + \log b$; $\log a/b = \log a - \log b$; $a^n = n \log a$, where a, b are positive numbers and n may be greater than unity or less than unity (thus a power or a root of the number, rational or irrational). Two systems of logarithms are generally used: common and natural.

The former, also called *Briggs logarithms*, uses the base 10 and is particularly useful for numerical calculations. The common logarithm of a number N could be indicated by $\log_{10} N$ but the notation $\log N$ is more usual. Since any number may be written in the form $N = 10^n \times M$, where n is an integer, positive or negative, its common logarithm is $\log N = n + \log M$. The first part of this sum is the *characteristic* of the logarithm, and it may be obtained by inspection of the given number. The second part, called the *mantissa*, is an irrational number, less than unity and usually given in decimal form. Tables of the mantissas of common logarithms are available with four, five, six, etc., significant Figures so that any required accuracy can be obtained in numerical calculations.

If a logarithm is given, the number that corresponds to it is the *antilogarithm*. The *cologarithm* is the logarithm of the reciprocal of a number. Thus $\operatorname{colog} N = \log(1/N) = -\log N$. In principle, it could simplify calculations involving quotients, for negative mantissas would not occur. This is not a serious handicap, however, and although cologarithm tables are sometimes given in handbooks, they are seldom used.

The *natural logarithm* has the irrational base $e = 2.71828\ldots$ and is also called the Napierian or hyperbolic system. Such logarithms occur as the result of differentiation, integration, etc., and they often appear in equations representing physical phenomena. Instead of the more exact symbol $\log_e N$, the abbreviated notation $\ln N$ is customary, especially in scientific work. The modulus of the common system relative to the natural system of logarithms is $\log e = 0.434294\ldots$ and, inversely, $\ln 10 = 2.302585\ldots$ is the modulus of the natural system. Conversion of one system of logarithms to another is made from the resulting relations: $\log N = 0.434294\ldots \ln N$; $\ln N = 2.302585\ldots \log N$. Although tables of natural logarithms are available, it is normally simpler to compute with common logarithms and then convert to the natural base.

A logarithmic scale is one in which the distance from the origin to any scale mark is proportional to the logarithm of the number attached to that mark.

Thus, Fig. 1 shows a (common) logarithmic scale with the numbers $1, 2, 3, \ldots, 10, \ldots$ attached to the division marks; if we take OI as the unit of length, then the distances OA, OB OC, etc., are represented by $\log 2$, $\log 3$, $\log 4$, etc., so that $OI = \log 10 = 1$. In going from left to right, the scale marks will become closer and closer together.

The logarithmic scale is applied in the slide rule and in logarithmic paper. In the latter case, if both abscissa and ordinate are marked

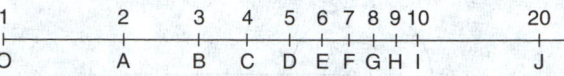
Fig. 1. Logarithmic scale.

logarithmically, the paper is called log-log paper; if the abscissa is equally spaced and only the ordinate is logarithmic, it is semilog paper. See also **Curve Fitting**; and **Logarithmic Function**.

LOGARITHMIC CHART. A graph where one or both axis are scaled in terms of logarithms of the variables. The chart may be called a semi- or double-logarithmic chart according to whether only the ordinate or both the ordinate and abscissa are on a logarithmic scale. In general, the logarithmic method of plotting is used when relative changes are important, since equal linear displacement on a logarithmic scale indicates equal proportional changes in the variable itself.

LOGARITHMIC DECREMENT. For a system having an oscillatory response that decays with increasing time, the logarithmic decrement is defined as the negative natural logarithm of the ratio of two consecutive excursions of the response about the bias or steady-state value. The ratio is taken as shown in Fig. 1. This quantity is used as a measure of the internal friction in an oscillating body. The logarithmic decrement equals one half the specific damping capacity.

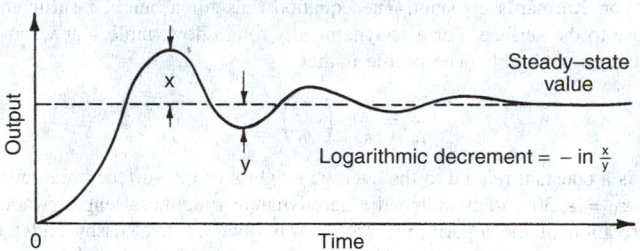
Fig. 1. Logarithmic decrement of oscillatory response.

LOGARITHMIC DIFFERENTIATION. Finding derivatives by taking the logarithm of both sides of an equation and then differentiating.

In meteorology this process is applied, for example, to the equation of state for a perfect gas, giving

$$\frac{dP}{P} = \frac{d\rho}{\rho} + \frac{dT}{T},$$

or to the Poisson equation for potential temperature, giving

$$\frac{d\theta}{\theta} = \frac{dT}{T} - K\frac{dP}{P},$$

where P is pressure, density, T temperature, θ potential temperature, and K Poisson's constant.

LOGARITHMIC FUNCTION. Its equation in rectangular coordinates is $y = a \ln x$ and the exponential function, $x = e^{y/a}$ is its inverse. When plotted on ordinary graph paper, the ordinate increases in arithmetic progression and the abscissa in geometric progression. Semi-logarithmic paper is more convenient for such a graph (see also **Logarithm**). A plot of $x + \log y = $ constant or $y + \log x = $ constant is then a straight line. See also **Curve Fitting**.

LOGARITHMIC SCALE. See **Scale**.

LOGARITHMIC SPIRAL. A transcendental plane curve, with equation in polar coordinates, $\ln r = a\theta$. It is also known as an equiangular spiral since it cuts the radius vectors at a constant angle. See Fig. 1. As θ increases, the curve winds around the pole at ever-increasing distances; when θ is negative, the curve continually approaches, but never reaches, the pole. Its evolute is a congruent logarithmic spiral. It is said that the tombstone of James Bernoulli (1654–1705), the celebrated Swiss

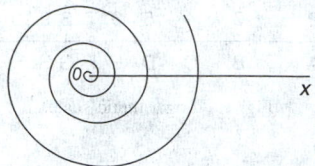

Fig. 1.   Logarithmic spiral.

mathematician, is inscribed with this curve and the words "Eaden mutato resurgo." See also **Bernoulli Number and Polynomial**.

LOGARITHMIC VELOCITY PROFILE.

The variation of the mean wind speed with height in the surface boundary layer derived with the following assumptions: 1) the mean motion is one-dimensional; 2) the Coriolis force can be neglected; 3) the shearing stress and pressure gradient are independent of height; 4) the pressure force can be neglected with respect to the viscous force; and 5) the mixing length l depends only on the fluid and the distance from the boundary, $l = kz$.

Near aerodynamically smooth surfaces, the result is

$$\frac{\overline{u}}{u_*} = \frac{l}{k}\ln\left(\frac{u_* z}{v}\right) + 5.5,$$

that is, the logarithmic velocity profile, where u_* is the friction velocity and v the kinematic viscosity. $k \cong 0.4$ and has been called the Kármán constant or von Kármán's constant. The equation fails for a height z sufficiently close to the surface. For aerodynamically rough flow, molecular viscosity becomes negligible. The profile is then

$$\frac{\overline{u}}{u_*} = \frac{l}{k}\ln\left(\frac{z}{z_0}\right), z \geq z_0.$$

z_0 is a constant related to the average height ε of the surface irregularities by $z_0 = \varepsilon/30$ and is called the aerodynamic roughness length. Another derivation of the logarithmic profile was obtained by Rossby under the assumption that for fully rough flow the roughness affects the mixing length only in the region where z and z_0 are comparable. Then $l = k(z + z_0)$ and

$$\frac{\overline{u}}{u_*} = \frac{l}{k}\ln\left(\frac{z + z_0}{z_0}\right).$$

For statically nonneutral conditions, a stability correction factor can be included.

AMS

LOGICAL DESIGN.

1.   The planning of a computer or data processing system prior to its detailed engineering design.

2.   The synthesizing of a network of logical elements to perform a specified function.

3.   The result of 1 and 2 above, frequently called the *logic of the system, machine,* or *network.*

LOGICAL ELEMENT.

In a computer or data processing system, the smallest building blocks which can be represented by operators in an appropriate system of symbolic logic. Typical logical elements are the AND gate and the flip-flop, which can be represented as operators in a suitable symbolic logic.

LOGICAL OPERATION (Computer System).

1. A logical or Boolean operation on N-state variables, which yields a single N-state variable; e.g., a comparison on the 3-state variables A and B, each represented by $-$, 0, or $+$, which yields: $-$ when A is less than B, 0 when A equals B, and $+$ when A is greater than B. Specifically, operations such as AND, OR, and NOT on two-state variables which occur in the algebra of logic; i.e., Boolean algebra.

2. The operations of logical shifting, masking, and other nonarithmetical operations of a computer.

See also **AND (Circuit)**; **NAND (Circuit)**; **NOR (Circuit)**; **NOT (Circuit)**; and **OR (Circuit)**.

LOGIC (Computer System).

In hardware, a term referring to the circuits that perform the arithmetic and control operations in a computer. In designing digital computers, the principles of Boolean algebra are employed. The logical elements of AND, OR, INVERTER, EXCLUSIVE OR, NOR, NAND, NOT, etc. are combined to perform a specified function. Each of the logical elements is implemented as an electronic circuit which, in turn, is connected to other circuits to achieve the desired result. The word logic is also used in computer programming to refer to the procedure or algorithm necessary to achieve a result.

LOGIC DIAGRAM (Computer System).

A drawing that indicates the interconnection of the individual logic elements in a computer. A logic diagram incorporates all of the information needed for wiring a computer. The logic diagram given in Fig. 1 shows the logic blocks and electrical interconnections. The logic block in the diagram contains the name of the function performed by it, such as AND, OR, and FLIP-FLOP; and the physical location of the circuit within the computer. Input and output signal connections for the function are given by the logic block. The logic diagram also shows the connection of a given logic block to other logic blocks. The computer supplier often provides wiring lists for the machine and a printed logic diagram. The logic diagram is used as an aid in troubleshooting the system. When utilizing large-scale integrated (LSI) circuits, each block may contain a complex logic function, such as an adder or register, rather than single elemental logic functions. Although not needed for physical wiring or troubleshooting, these complex logic functions can, in turn, be expressed in terms of elemental logic similar to that shown in Fig. 1.

THOMAS J. HARRISON, IBM Corporation, Boca Raton, FL

LOGISTIC CURVE.

The logistic curve is a growth curve used to describe functions which continually increase, gradually at first, more rapidly in the middle growth period, and slowly again, reaching a maximum at the end of the growth. We write its equation

$$y = \frac{k'}{1 + e^{a+bt}}, \quad b < 0$$

the symmetrical logistic, where t represents time and y is the population size. A more general form is

$$y = k_1 + \frac{k_2}{1 + e^{a+bt-ct2}}, \quad c < 0$$

the asymmetrical or skew logistic. This usage of the word "logistic" has nothing to do with the military connotation or with the European meaning of "formal logic."

LOGIT.

A transformation of a variable used particularly in the analysis of dose-response relationships. If p is the probability of a certain response on dose x, the logit is defined as

$$y = \log_e\{p/(a - p)\}$$

and analysis proceeds by considering the relation between y and x.

LOG MEAN TEMPERATURE DIFFERENCE (LMTD).

An average temperature difference between the hot and cold side of a piece of equipment, e.g., a heat exchanger, for use where the temperature difference may vary along the equipment. Its form is

$$\frac{\Delta T_1 - \Delta T_2}{2.3 \log \dfrac{\Delta T_1}{\Delta T_2}}$$

where ΔT_1 is the temperature difference at one end of the equipment and ΔT_2 is the temperature difference at the other.

LOG (Navigation).

A term used with two different meanings: a speed-measuring device, and a record book. Prior to the middle of the nineteenth century, the speed of a ship relative to the surface of the water was measured by the log chip and line. "Heaving the log" was a duty performed every time the ship's bell was struck, i.e., every half-hour, and the speeds determined were entered in a book, which came to be known as the log book. Since this log book was always available to the watch officer, it

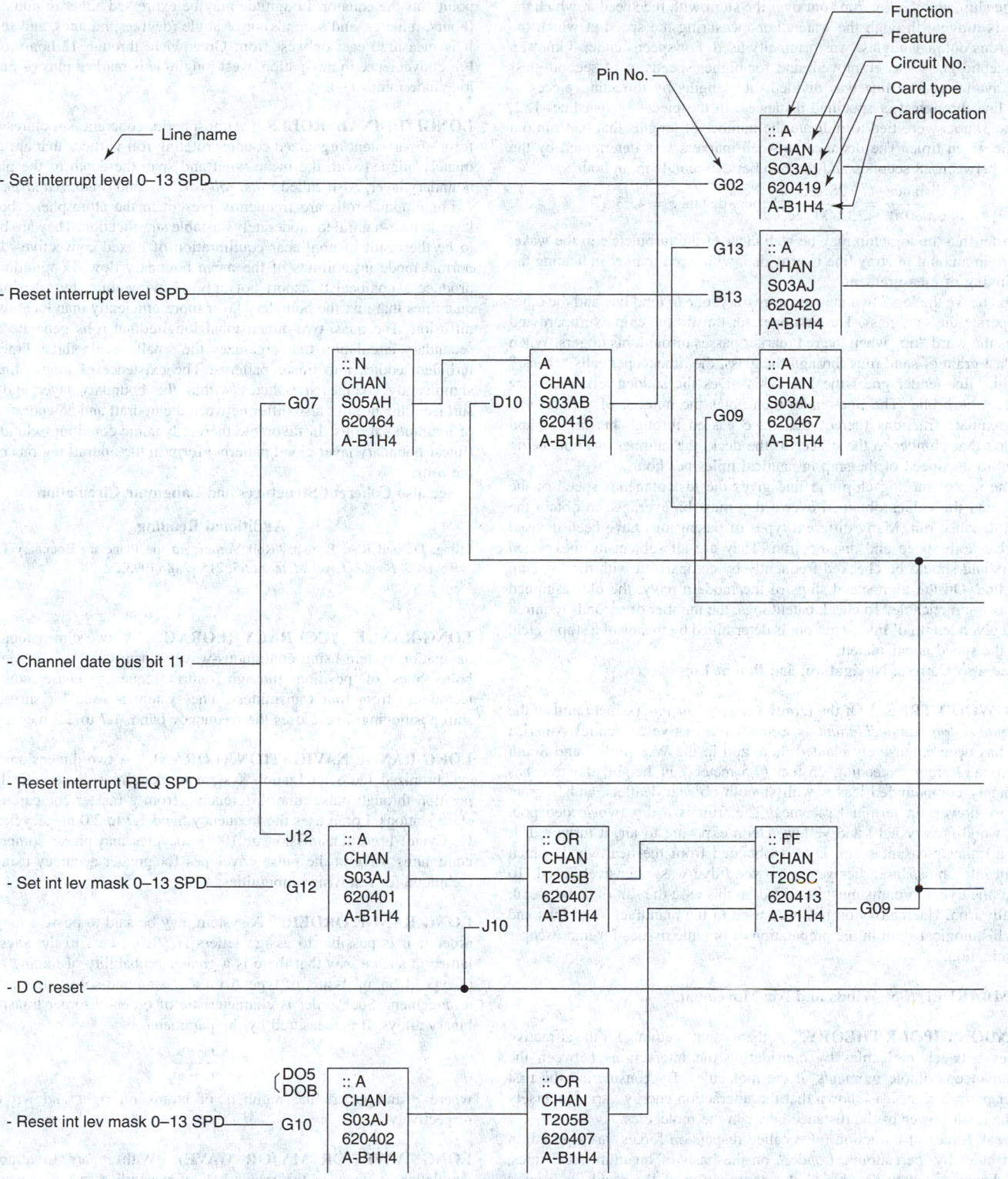

Fig. 1.   Computer logic diagram.

became customary to enter all important incidents relating to the operations of the ship, behavior of members of the crew, conditions of the weather and sea, and, in fact, anything the watch officer might think worthy of recording. This "deck log" was turned over to the captain, who abstracted all important material and made up the ship's log. At present, practically every department of the ship, deck, engine room, ordnance, steward, etc., keeps its individual rough log, and the ship's log is made up from these under the supervision of the captain.

The oldest and perhaps even now, the most accurate and reliable method for determining the speed of a ship relative to the water is by use of the log chip and line. The log chip is a wooden quadrant, loaded along its circular edge so that it will float upright in the water. It is attached to the log line by a 3-legged bridle, the upper leg of which is attached to the apex of the log chip in such a manner that a sharp jerk will free it. Then the log chip may be easily hauled back to the ship. When the log chip is thrown overboard, it floats, nearly submerged, with the flat surface perpendicular to the motion of the ship, and the line runs out over the stern with the speed at which the ship is moving through the water. For measuring the speed at which the line runs out, a sandglass was originally used. For speeds under 4 knots, a 28-second-glass was employed, and for higher speeds, a 14-second-glass was available. The line was divided into lengths by threading pieces of fish line through it at specified distances. In the pieces of fish line, 1, 2, 3, etc., knots were tied to indicate the number of lengths that had run out in the given time. The distance between markers was determined by the ratio between 28 seconds and the number of seconds in an hour:

$$\frac{\text{distance}}{6,080 \text{ ft.}} = \frac{28 \text{ sec.}}{3,600 \text{ sec}} ; \text{hence, distance} = 47'3.5''$$

In order that the log chip may be well clear of the turbulence in the wake, a certain amount of stray line is provided, with a red marker indicating the beginning of measurement.

To "heave the log," two men are necessary, one to tend line and the other to operate the sandglass. The first operator throws the chip overboard and gives the word "tip" when the red marker passes through his fingers. When the last grain of sand runs through the glass, the timekeeper calls "check," and the line-tender grabs the line. This gives the sudden jerk necessary to free the bridle. The line-tender then notes the number of "knots" and approximate fractions thereof that have passed through his fingers and reports that number to the officer of the deck, the number of knots being equal to the speed of the ship in nautical miles per hour.

The use of the log chip and line gives the instantaneous speed of the ship, and the values obtained over a day must be averaged to obtain the total distance run. Many different types of patent logs have been devised to give both speed and distance run. They are all subject to unexpected errors and should be checked frequently by comparison with the log chip and line. On the high-speed ships of the modern navy, the old-fashioned line is impracticable. To check patent logs, the number of seconds required for a given length of line to run out is determined by means of a stop watch, and the speed is calculated.

See also **Course**; **Navigation**; and **Patent Log**.

LOGWOOD TREE.   Of the family *Caesalpiniaceae* (senna family), the *Haematoxylon campechianum* is a small tree native to central America and has been extensively planted there and in the West Indies and South America. Rarely exceeding 25 feet (7.5 meters) in height, the tree has pinnately compounded leaves with smooth obovate leaflets, and fragrant yellow flowers in terminal racemes. The fruit is a dry two-seeded pod. The wood is very hard and yellow. Upon exposure to air, it turns red. It has a rather pleasant scent. Dye is obtained from the heartwood, which is cut into chips. In earlier years, logwood dye was extensively used. To make the dye, mordants must be added, in this case the salt of some metal, usually iron. Haematoxylon has been used in the manufacture of inks and as a histological stain in the preparation of organic tissues for microscopic examination.

LOMBARDE.   See **Winds and Air Movement**.

LONDON DIPOLE THEORY.   A theory that accounts for the attractive forces between molecules by considering the interactions between the instantaneous dipole moments of the molecules. By considering the first order perturbation, it is shown that the interaction energy varies inversely as the sixth power of the distance between the molecules.

Weak forces of attraction (also called dispersion forces) are exerted on each other by inert atoms. London, on the basis of quantum mechanics, has shown that they are due to the perturbation of the repulsive ground state by the higher electronic states of the system consisting of the two atoms. This perturbation at large internuclear distances r, gives a potential energy decreasing as $-(1/r^6)$ toward smaller r values. At smaller distances r the strong repulsion of the zero-valent atoms sets in, so that only a very shallow minimum at a relatively large internuclear distance results. Analogous forces also add to the mutual attraction of atoms with free valences and of molecules with or without permanent dipole moments. See also **Van der Waals Forces**.

LONGITUDE.   The longitude of a point on the surface of the earth is the angular distance measured along the earth's equator from the meridian, through Greenwich, England, to the point where the local meridian of the

point cuts the equator. Longitude may be expressed either in units of time (hours, minutes, and seconds) or of angle (degrees, minutes, and seconds). It is measured east or west from Greenwich, through 12 hours or 180°. For convenience in navigation, west longitude is marked plus (+) and east longitude minus (−).

LONGITUDINAL ROLLS.   Atmospheric coherent structures in the form of persistent organized counter rotating roll vortices that are approximately aligned with the mean wind and span the depth of the planetary boundary layer. Also called *rolls*, *roll vortices*, and *organized large eddies*.

Longitudinal rolls are frequently present in the atmospheric boundary layer in near-neutral to moderately unstable stratification. They are believed to be the result of nonlinear equilibration of mixed convective–dynamic normal mode instabilities of the mean boundary flow. Longitudinal rolls produce a nonlocal transport not only of momentum, but also of scalar quantities that mix the boundary layer more efficiently than local turbulent diffusion. The quasi-two-dimensional longitudinal rolls generate a mean secondary circulation that organizes the smaller-scale three-dimensional turbulent eddies into linear patterns. The existence of longitudinal rolls significantly changes the fluxes within the boundary layer and at the surface. Flux profiles also differ between the updraft and downdraft regions of longitudinal rolls. In favorable thermodynamic conditions cloud streets (linear boundary layer cloud patterns) form in the updraft regions between the rolls.

See also **Coherent Structures** and **Langmuir Circulation**.

Additional Reading

Etling, D., and R.A. Brown: "Roll Vortices in the Planetary Boundary Layer: A Review," *Bound.-Layer Meteor.*, **65**, 215–248 (1993).

AMS

LONG-RANGE ACCURACY (LORAC).   A two-dimensional radio navigation system using continuous-wave transmission to provide hyperbolic lines of position through radio frequency phase comparison techniques from four transmitters. The system is used for surveying or ship-positioning. Lorac uses the frequency band, 1.7 to 2.5 megacycles.

LONG-RANGE NAVIGATION (LORAN).   A two-dimensional pulse-synchronized radio navigation system to determine hyperbolic lines of position through pulse-time differencing from a master compared to two slave stations. Loran uses the frequency band 1.7 to 2.0 megacycles; loran C (Cytac) uses transmission at 100 kilocycles and phase compares the continuous wave in the pulse envelopes for greater accuracy using pulse technique for resolving ambiguities.

LONG RANGE ORDER.   A system may be said to possess long range order if it is possible to assign letters A, B, C, etc., to the sites of the lattice in such a way that there is a greater probability of finding an atom of type A on an A-site, of type B on a B-site, and so on, than any other arrangement. Such order is characteristic of order–disorder transitions in binary alloys. It is measured by the parameter

$$S = \frac{r - w}{r + w}$$

where r and w are the numbers of atoms on right and wrong sites respectively.

LONG WAVE (OR MAJOR WAVE).   With regard to atmospheric circulation, a wave in the major belt of westerlies that is characterized by large length and significant amplitude. The wave length is typically longer than that of the rapidly moving individual cyclonic and anticyclonic disturbances of the lower troposphere. The angular wavenumber of long waves is generally taken to be from 1 to 5. Also called *planetary wave*. See also **Rossby Wave** and **Short wave**.

2. (Also called *shallow water wave*.) A wave with a relatively long wave length and period. For ocean waves, this is typically a wave of period greater than about 10 s and wave length greater than about 150 meters (492 feet). See also **Ocean waves**.

LOOMING.   A mirage in which the image of distant objects is displaced upward. Because the displacement increases with distance, a horizontal surface, such as that of a body of water, appears to bend upward and one's perception is that of being inside a broad shallow bowl. Indeed, the upward bending surface results in an (optical) horizon that can be much farther from the observer than in the absence of a mirage. Looming is an example of a *superior mirage*. The opposite of looming is *sinking*. Looming occurs when the concave side of light rays from a distant object is down, and this in turn occurs when the refractive index of the atmosphere decreases with height. This is very common, but the displacement is usually sufficiently small as to be unremarkable. However, when there is a temperature inversion over the surface, the looming can be striking. See also **Mirage**; **Sinking**; and **Superior Mirage**.

LOON *(Aves, Gaviiformes, Gaviidae)*. Large birds sometimes called divers which range in length from 58 to 90 centimeters (23 to 35 inches) and in weight from 1 to 6.4 kilograms (2.2 to 14.1 pounds). See also **Gaviiformes**. There are extensive webs between the three front toes, and they have a short tail with 16 to 20 tail feathers and 11 primaries. In flight the head and neck are somewhat lowered.

There is only 1 genus, *Gavia*, with four species found in the northern tundra and forest zones of the Old and New Worlds. Three of the species have a black and white checkered pattern on the wings of the breeding plumage: (1) Arctic Loon (Black-Throated Diver, *Gavia arctica*; Fig. 1); the length is 70 centimeters (28 inches) and the weight is 2–3.5 kilograms (4–8 pounds); the nape is gray. (2) The Common Loon or Great Northern Diver *(Gavia immer)*. The length is 75 centimeters (30 inches) and the weight is 4 kilograms (9 pounds); the nape and beak are black. (3) The Yellow-Billed Loon or White-Billed Northern Diver *(Gavia adamsii)*. The length is 87 centimeters (34 inches) and the weight is 4.5–6.4 kilograms (10–14 pounds). The beak is ivory-colored in adults. (4) The fourth species, with small white stripes on the nonbreeding plumage, is the Red-Throated Loon or Diver *(Gavia stellata)*. The length is 58 centimeters (23 inches) and the weight is 1–2.4 kilograms (2–5 pounds).

Fig. 1.   Arctic loon. *(Sketch by Glenn D. Considine.)*

The loon or diver is found in Canada, the northern parts of the United States, particularly the mountain lakes of New York and Pennsylvania and the lakes of Michigan as well as in California and south to Mexico. Some species of loon are found in Europe and essentially in most parts of the world.

LOOP ANTENNA.   An antenna consisting of a conducting coil, of any convenient cross section (generally circular), which emits or receives radio energy. The principal lobe of the radiation pattern is wide and is in the direction perpendicular to the plane of the coil. Also called *loop*.

LOOP (Computer System).   A sequence of instructions that may be executed repeatedly while a certain condition prevails. The productive instructions in the loop generally manipulate the operands, while bookkeeping instructions may modify the productive instructions, and keep count of the number of repetitions. A loop may contain any number of conditions for termination, such as the number of repetitions or the requirement that an operand be nonnegative. The equivalent of a loop could be achieved by the technique of straight line coding, whereby the repetition of productive and bookkeeping operations is accomplished by explicitly writing the instructions for each repetition.

See also **Program (Computer)**.

LOOP CURRENT.   See **Ocean Currents**.

LOOP GAIN.   In feedback terminology, the gain around the feedback loop, numerically equal to the product of the forward gain by the gain of the feedback network when the circuit configuration permits meaningful identification of these two separate transmissions. The feedback network is also called the *beta-network*.

See also **Gain (Magnitude Ratio)**.

LOOP (Mathematics).   A closed path. An immediate consequence of this definition is that a finite graph contains only a finite number of loops, a conclusion which is critical to the practical application of Kirchoff's law for voltage.

LORENTZ FRAME.   Any of the set of coordinate systems in Minkowski space for which the square of the interval between two events is $c^2 dt^2 - (dr)^2$. Any such coordinate system may be obtained from another by means of a Lorentz transformation (together perhaps, with an orthogonal transformation of the space axes). With each Lorentz frame may be associated a point observer, each of whom moves with constant velocity relative to the others.

LORENTZ, HENDRIK ANTOON (1853–1928).   Lorentz was born in the Netherlands. He received a Ph.D. from Leiden in the area of electromagnetism. He became the first chair of theoretical physics at the University of Leiden in 1877.

From Lorentz stems the conception of the electron; his view that his minute, electrically charged particle plays a role during electromagnetic phenomena in ponderable matter made it possible to apply the molecular theory to the theory of electricity, and to explain the behaviour of light waves passing through moving, transparent bodies.

The so-called Lorentz transformation (1904) was based on the fact that electromagnetic forces between charges are subject to slight alterations due to their motion, resulting in a minute contraction in the size of moving bodies. It not only adequately explains the apparent absence of the relative motion of the Earth with respect to the ether, as indicated by the experiments of Michelson and Morley, but also paved the way for Einstein's special theory of relativity.

It may well be said that Lorentz was regarded by all theoretical physicists as the world's leading spirit, who completed what was left unfinished by his predecessors and prepared the ground for the fruitful reception of the new ideas based on the quantum theory.

Lorentz's most important contributions include his proposal that charged particles in matter oscillate when struck by light waves, and his applying his electron theory to explain the Zeeman effect. In 1902, he shared the Nobel Prize in Physics with Pieter Zeeman.

See also **Field Theory**; **Lorentz Transformation**; and **Relativity and Relativity Theory**.

J. M. I.

LORENTZ INVARIANCE.   The equivalence principle of special relativity, which states that physical principles must be invariant under a transformation from one coordinate system to another. Since this is a Lorentz transformation, the invariance itself is often called by the name of Lorentz.

LORENTZ TRANSFORMATION.   Relations connecting the space and time coordinates of an event as observed from two Lorentz frames. If S' moves relative to S with velocity v in the x-direction, x, y, z, t denote position and time coordinates of two events as measured by S, and x', y', z', t' the corresponding quantities for S', then

$$x' = \gamma(x - vt)$$
$$y' = y, z' = z$$
$$t' = \gamma\left(t - \frac{x}{c^2}\right)$$

where $\gamma = (1 - v^2/c^2)^{-1/2}$. These relations were shown by Einstein (1905) to be a consequence of the special relativity theory. See also **Relativity and Relativity Theory**; and **Vector**.

LORISOIDS *(Mammalia).* Primitive animals of the order *Primates* and similar to lemurs. See also **Lemur**. Lorises are found in the warmer parts of southeastern Asia and the East Indian islands. They have large staring eyes and from their very slow movements have sometimes been referred to as slow lemurs. The lorises are forest animals of nocturnal habits. The slender loris is known as *Loris gracilis*; the slow loris as *Nycticebus tardigradus*.

LOSSER. A dielectric material that dissipates energy. A dissipative element placed in a circuit to prevent oscillation.

LOSS FACTOR. The rate at which heat is generated in a dielectric is proportional to its loss factor, which is equal to the product of its dielectric constant by its power factor. Both the dielectric constant and power factor are usually functions of frequency; therefore, the loss factor changes with changing frequency.

LOSS FUNCTION. In decision theory, a function of the decision and the true underlying distributions which expresses the loss incurred in taking that decision. If there are a number of possible situations, the array of losses according to situation and decision is called the *loss matrix*. It is analogous to the payoff matrix of games theory.

LOSSLESS DATA COMPRESSION. See **Data Compression**.

LOSS (Transmission). A general term used to denote a decrease in signal power in transmission from one point to another. Usually expressed in decibels.

LOSSY DATA COMPRESSION. See **Data Compression**.

LOUDNESS LEVEL. The loudness level of a sound is the sound-pressure level of a standard tone (usually 1,000 Hz) which sounds equally as loud as the sound under measurement. In 1933, Fletcher and Munson published their *loudness level contours* for pure tones. See Fig. 1. The curves commonly are referred to as Fletcher-Munson curves or equal-loudness contours. The numbers on the contours are the loudness levels

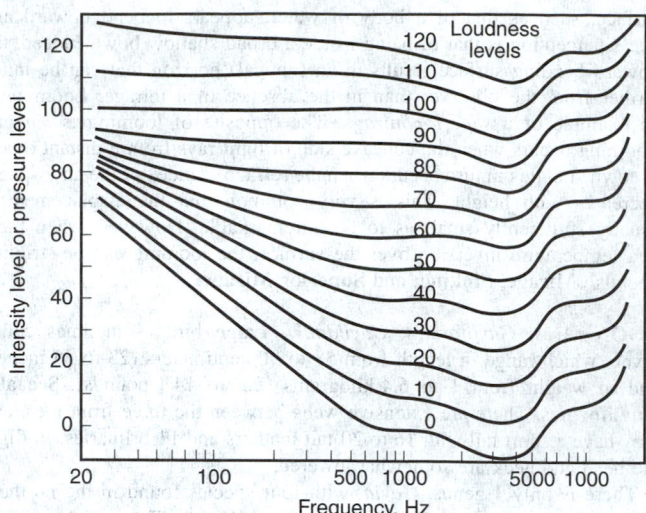

Fig. 1.   Fletcher-Munson curves indicating equal-loudness contours.

of the sound in phons (the sound-pressure level of a 1,000 Hz tone that is equally loud). Thus, if a certain complex sound wave sounds equally as loud as a 1,000 Hz tone having a sound-pressure level of 60 db re 0.0002 dyne/cm², the complex wave is said to have a loudness level of 60 phons regardless of its sound-pressure level. Other sets of equal-loudness contours which deviate in some respects from the Fletcher-Munson curves have been developed since by other investigators.

See also **Acoustics**.

LOUDSPEAKER. A transduction device, usually based on the dynamic (moving-coil) principle, that converts electrical energy into mechanical energy or sound. A coil of wire located in the magnetic field of a permanent magnet is attached to a paper cone. See Fig. 1. The cone, at its outer edges, is flexibly attached to a support ring. When an electric current is passed through the coil, a force is created that acts upon the cone. Cone movement generates sound waves that are proportional to the frequency of the exciting current. Loudspeakers usually are low in efficiency, about 5%. By using horns of a gradually increasing cross sectional area, efficiencies of 30 to 50% can be achieved.

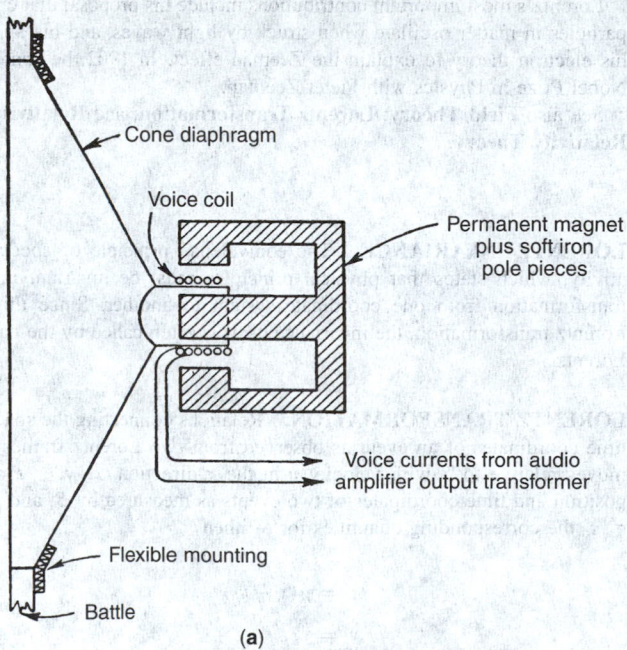

(a)

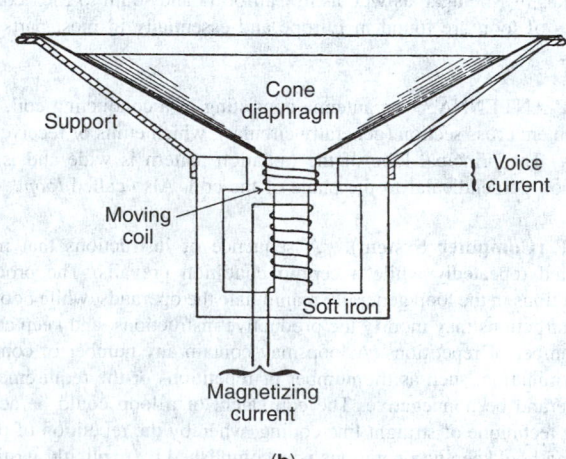

(b)

Fig. 1.   Two common types of loudspeakers: **(a)** permanent magnet type; **(b)** electromagnetic type.

Liquid-Phase Projectors. Piezoelectric, magnetostrictive, and electromagnetic are the principal types of transduction used in transmitters to excite acoustical waves in liquids. Designed to be resonant, these devices usually operate at their fundamental frequency. Piezoelectric liquid-phase projectors are effective from 20 kHz to above 100 MHz; while magnetostrictive transducers will handle the range from 10 to 100 kHz. With bandwidth ratios of 5 to 20, both types have efficiencies on the order of 0.5 to 4 watts per square centimeter. Increasing the pressure of the liquid mass and providing special cooling will raise radiation intensities up to 50 watts per square centimeter. Many liquid-phase projectors are made up of arrays of individual transducers to control directivity and to increase power-handling capacity.

A large number of projectors cannot be used as receivers. The latter are designed to excite acoustical vibrations in air and are not reversible. Nonreversible projectors include modulated air-flow speakers, whistles, and sirens. In a modulated air-flow speaker, a valve controlled by an electrical signal modulates the flow of the airstream. Used for public address systems, these speakers have high efficiency and power output, but also have high distortion and poor frequency response. Whistles have high efficiency, but suffer from frequency and amplitude instability unless driven by an auxiliary device, such as the resonant cavity of an organ pipe. A siren, in which a stream of compressed air is chopped by a series of rotating blades, combines high efficiency and high intensities, with stable, easily controlled frequencies.

LOUSE *(Insecta).* An external parasitic insect found on warm-blooded animals, both mammals and birds. Two kinds of lice occur: (1) The true or sucking lice (order *Anoplura*); and (2) the bird or biting lice (order *Mallophaga*). As the titles suggest, the sucking lice have piercing and sucking mouth parts; the bird lice have biting mouth parts. See Fig. 1.

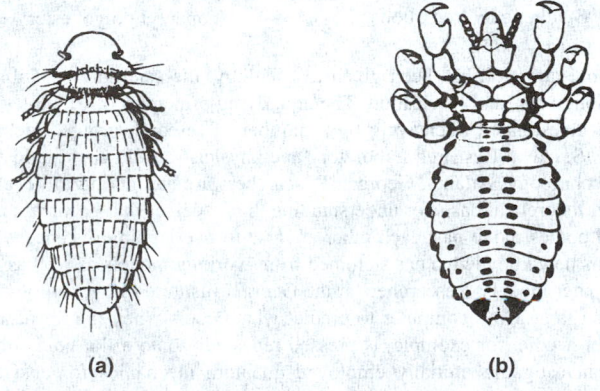

(a)   **(b)**

Fig. 1. (**a**) Biting louse (*Mallophaga*); (**b**) sucking louse (*Anoplura.*) (*USDA.*)

Livestock Lice

Hog louse (Haematopinus suis or *adventicius,* Linne). Of the order *Anoplura.* This is a wingless, rather large, flat, gray-colored louse, about $\frac{1}{4}$-inch (6 millimeters) long that infests hogs. It does not affect other livestock. It is the largest of the blood-sucking lice. The head and legs are comparatively long. The insect is equipped with a hook-like member for clasping hairs of the host. A favorite habitat is in between folds of the hog's skin. The entire life cycle of the insect occurs on the host. The hog has only two commonly encountered insect parasites—the hog louse and the mange mite. Control is by dipping or spraying with Co-Ral, lindane, malathion, methoxychlor, ronnel, or toxaphene, where available and permitted. Medicated hog wallows, which incorporate a surface film of petroleum or pine oil, can be effective. The louse dies within a few days if not on a host.

Sheep and Goat Lice. There are several species:

Bloodsucking body louse (Haematopinus orvillus, Neumann). Of the order *Anoplura.*

Bloodsucking foot louse (Linognathus pedalis, Osborn). Of the order *Anoplura.* A pale, rather slender louse, about $\frac{1}{12}$ inch (2 millimeters) long.

Sheep-biting louse (Boricola or *Trichodectes oris,* Linne). Of the order *Mallophaga.* A pale-brown insect with a red head, about $\frac{1}{20}$ inch (1 +

millimeter) in length. This louse eats the wool fibers, causing fibers to become soiled and tangled. Its full life cycle is spent on the host. Treatment for this and other sheep and goat lice mentioned may include dipping or spraying lindane, methoxychlor, rotenone, or toxaphene. All of these lice can be quite damaging to the wool of the animals. The clip of mohair from an Angora goat, for example, may be reduced severely if there is an infestation.

Cattle and Horse Lice. These also affect mules and donkeys.

Long-nosed cattle louse (Linognathus or *Haemotopinus vituli,* Linne). Of the order *Anoplura.* This is a red louse, about $\frac{1}{8}$ inch (3 millimeters) long. The insect pierces the skin and sucks blood. In heavy infestations over long periods, animals become emaciated. Patches of skin become bare and sore. The insect seems to prefer unhealthy, poorly fed animals. Control is by spraying or dipping with lindane, malathion, methoxychlor, toxaphene, or ronnel, when available and approved. Special directions must be followed for spraying dairy cows to avoid any contamination of milk.

Horse-sucking louse (Haemotopinus asini, Linne or *H. macrocephalus,* Burmeister). Of the order *Anoplura.* This is a medium-size louse, about $\frac{1}{8}$ inch (3 millimeters) long. The bite is very painful. There are several generations per year. Treatment is about the same as for cattle louse.

Poultry Lice. A number of species attack poultry and wild fowl and birds, including pigeons. These lice are of the order *Mallophaga* and they do not suck blood, but rather they feed on bits of skin, scabs, feathers, and other organic debris found on the bird's body. Irritation is caused by the sharp mouth parts and sharp claws. Nibbling of the lice prevents rest and sleep, and causes loss of appetite, diarrhea, droopy wings, leading to progressive emaciation, reduction in egg production, and death of the birds when left unattended. Poultry lice are distributed worldwide. They are wingless, with six legs, flat bodies, and round heads. Treatment is by spraying or dusting with malathion or rotenone, where approved. Nests and litter also should be sprayed. Painting the roosts with 40% nicotine sulfate or 3% malathion can be effective. Some of the species include:

Chicken-head louse (Cuclotogaster heterographus, Nitzsch). An insect about $\frac{1}{10}$ inch (2.5 millimeters) long that severely irritates the birds around the neck and head area. They are particularly irritating to young turkeys and chicks.

Chicken-body louse (Menacanethus stramineus). An insect about $\frac{1}{8}$ inch (3 millimeters) long that irritates areas under the wings and about the vent. It attacks both young and old birds. Records indicate that over 35,000 of these lice were found on one chicken.

Common body louse (Menopan gallinae, Linne). Also called *shaft louse* or *small body louse,* about $\frac{1}{16}$ inch (1.5 millimeters) long. Lives mostly on the feathers and one of the most commonly encountered on fowl.

Fluff louse (Goniocotes gallinae, De Geer). Prefers operating in the fluff and under the vent; about $\frac{1}{16}$ inch (1.5 millimeters) long.

Brown chicken louse (Goniodes dissimilis, Denny). A medium-size louse, about $\frac{1}{10}$ inch (2.5 millimeters) long.

Large chicken louse (Goniodes gigas, Taschenberg). A comparatively large louse, about $\frac{5}{32}$ inch (2.5 millimeters) long.

Wing louse (Liperus caponis, Linne). Prefers the barbules of the wing feathers; about $\frac{1}{10}$ inch (2.5 millimeters) long.

Small pigeon louse (Campanulotes or *Goniocotes bidentalus,* Burmeister). Inhabits feathers of both young and old pigeons.

Slender pigeon louse (Columbicola columbae, Linne). Much like the small pigeon louse.

NOTE: Plant lice are not true lice, but members of the order *Homoptera* and commonly called aphids.

LOW. (Sometimes called *depression*). In meteorology, an "area of low pressure," referring to a minimum of atmospheric pressure in two dimensions (closed isobars) on a constant-height chart or a minimum of height (closed contours) on a constant-pressure chart. Since a low is, on a synoptic chart, always associated with cyclonic circulation, the term is used interchangeably with cyclone.

See also **High**.

LOW EARTH ORBIT (LEO). A low Earth orbit is normally at an altitude of less than 1,000 km (621 miles) and could be as low as 160 km (99 miles) above the Earth. Satellites in this circular orbit travel at a speed of around 7.8 km (4.9 miles) per second. At this speed, a satellite takes approximately 90 minutes to circle the Earth.

In general, these orbits are used for remote sensing, military purposes and for human spaceflight as they offer close proximity to the Earth's surface for imaging and the short orbital periods allow for rapid revisits. The International Space Station is in low Earth orbit.

Most manned spaceflights have been in LEO, including all Space Shuttle and various space station missions; the only exceptions have been suborbital test flights such as the early Project Mercury missions and the flights of the X-15 rocket plane (which was not intended to reach LEO), and the Project Apollo missions to the Moon (which went beyond LEO).

Most artificial satellites are placed in LEO, where they travel at about 27,400 km/h (8 km/s), making one revolution in about 90 minutes. The primary exception are communication satellites that require geostationary orbit. However, it requires less energy to place a satellite into a LEO and the satellite needs less powerful transmitters for data transfer, so LEO is still used for many communication applications. Because these orbits are not geostationary, a network of satellites is required to provide continuous coverage. Lower orbits also aid remote sensing satellites because of the added detail that can be gained. Remote sensing satellites can also take advantage of sun synchronous LEO orbits at an altitude of about 800 km and near polar inclination. ENVISAT is one example of an earth observation satellite that makes use of this special type of LEO.

The LEO environment is becoming congested, not least with space debris. The United States Space Command tracks more than 8,000 objects larger than 10 cm (4 inches) in LEO.

See also **Earth-Orbiting Satellites (Data Receiving And Handling Facilities); Orbit (Astronomy); Satellites (Communications and Navigation);** and **Satellites (Scientific and Reconnaissance).**

LOWEST ASTRONOMICAL TIDE. See **Meteorology**.

LOW-TEMPERATURE HYGROMETRY. The study that deals with the measurement of water vapor at low temperatures. The techniques used differ from those of conventional hygrometry because of the extremely small amounts of moisture present at low temperatures and the difficulties imposed by the increase of the time constants of the standard instruments when operated at these temperatures.

See also **Hygrometry and Psychrometry**.

LOW VACUUM. The condition in a gas-filled space at pressures less than 760 torr and greater than some lower limit. It is recommended that this lower limit be chosen as 25 torr corresponding approximately to the vapor pressure of water at 25 degrees C and to 1 inch of mercury.

L-SECTION. This refers to an elementary section of a network such as a filter where the components are connected in the form of an L, i.e., one component in series with one side and the other in shunt across the two sides of the circuit. See Fig. 1.

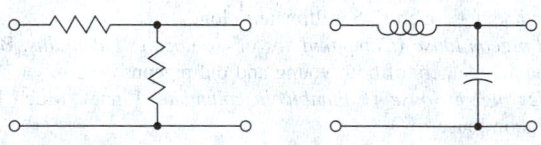

Fig. 1. L-section.

LUBRICANT. A material used to diminish friction between the moving surfaces of machine parts; also to decrease friction between a cutting tool and the material being cut. A wide variety of materials is used for manufacturing lubricants. Animal lubricants are obtained from the fat of common animals and can be classified as hard fats (stearin) and soft fats (lard) or naturally occurring combinations. Vegetable lubricants include rape seed oil, cottonseed oil, soybean oil, castor oil, and linseed oil. They range in properties from solid to liquid. Petroleum and mineral oil lubricants, because of their greater stability, are usually preferred for machine applications. Lubricants range from light oils to very heavy solid greases. Graphite, a solid, is also used as a lubricant.

Because of increased requirements for lubricants, including higher temperature and pressure applications, greater durability, and tolerance to wide changes in ambient temperature conditions, numerous synthetic lubricants have been developed. These include synthetic hydrocarbons, carboxylic acid esters, silicones, polyethers (polyalkylene glycols), phosphate esters, silicate esters, highly fluorinated compounds, and polyaromatics (polyphenyls and polyphenyl ethers). In selecting a lubricant, the following characteristics are considered: (1) lubricity and antiwear properties; (2) fluid range; (3) viscosity index; (4) additive response of base oil; (5) oxidation stability; (6) thermal stability; (7) hydrolytic stability; (8) fire resistance; (9) compatibility with petroleum products; (10) compatibility with paints, plastics, and elastomers; and (11) cost.

Over a number of years, the early polyol esters, the formulas of some of which are shown below, appeared to be adequate for coping with the increasing rigorous properties required for increasingly difficult lubrication problems. They continue to be used, but some professionals in the field have developed a number of proprietary formulations that are claimed to be superior to the polyol esters.

$$\left[\text{Polyaromatics (polyphenyls)} \right]_n \qquad R-O-\overset{\overset{\displaystyle O}{\|}}{C}-(CH_2)_n-\overset{\overset{\displaystyle O}{\|}}{C}-O-R$$
Polyaromatics (polyphenyls) — Diester

$$\left[-CH_2-\overset{\overset{\displaystyle R}{|}}{C}-O- \right]_n \qquad Si-(O-R)_4$$
Polyether (polyalkylene glycols) — Silicate ester

$$[-CH_2-CH_2-CH_2-]_2 \qquad C-(CH_2-O-\overset{}{C}-R)_4$$
Synthetic hydrocarbon — Neopentyl polyol ester

More attention has been given to tribology, the scientific discipline of friction, wear, and lubrication. The underlying principles of tribology have been investigated intensively by a number of research groups, including the U.S. Naval Research Laboratory and physicists at the Georgia Institute of Technology, Atlanta, Georgia. Researchers are attempting to develop a better theoretical basis for understanding the processes that occur when two solid bodies move past each other at close to overlapping distances. Most of this new knowledge has stemmed from working models as well as from computer models. Researchers at the Georgia Institute of Technology have used a large Cray computer to predict what occurs when the tip of a thin nickel needle, for example, is pressed repeatedly onto a flat gold surface. The investigators initially employed quantum mechanics for answering such questions as adhesion, cohesion, and the making and breaking of chemical bonds. Researchers at the Naval Research Laboratory have used an atomic force microscope as a tool to check their experiments. The present goal is to model more complex systems. As pointed out by one researcher, "To make progress in the molecular engineering of lubricants, you need to know the molecular details of the process."

Additional Reading

Klaman, D.: *Lubricants and Related Products: Synthesis, Properties, Applications, International Standards*, John Wiley & Sons, Inc., New York, NY, 1984.

Mang, T., and W. Dresel: *Lubricants and Lubrication*, John Wiley & Sons, Inc., New York, NY, 2000.

Pirro, D.M., A.A. Wessol, and J.G. Wills: *Lubrication Fundamentals*, 2nd Edition, Marcel Dekker, Inc., New York, NY, 2001.

Rudnick, L.R.: *Lubricant Additives: Chemistry and Applications*, Marcel Dekker, Inc., New York, NY, 2003.

Rudnick, L.R. and R.L. Shubkin: *Synthetic Lubricants and High-Performance Functional Fluids*, 2nd Edition, Marcel Dekker, Inc., New York, NY, 1999.

Sequiera, A.: *Lubricant Base Oil and Wax Processing*, Marcel Dekker, Inc., New York, NY, 1994.

Staff: Society of Automotive Engineers: *Heavy Duty Diesel Engine Lubricants*, Society of Automotive Engineers, Warrendale, PA, 1996.

LUBRICANTS AND RELEASE AGENTS (FOODS). See **Food Additives**.

LUBRICATING AGENTS. There are two fundamental classes of lubricating agents—natural and synthetic.

Lubricating oils are fluids whose function is the reduction of friction and wear between solid surfaces (generally metals), in relative motion. This function is accomplished in either of two ways: (1) by formation of adsorbed films on the two opposed surfaces, which can be more easily sheared than the solid substrate, or (2) by interposition of a fluid film between the two opposed surfaces. In the former case the shear strength of the film, and in the latter, the viscosity of the fluid determine the magnitude of the work which must be done to maintain the opposed surfaces in relative motion. In most cases, bearings are designed to operate under fluid film conditions and thus the viscosity of an oil is its most important property in classifying it for lubricating purposes. The range of viscosities required is approximately 1 centistoke at $100\,°F$ $(38\,°C)$ for high-speed, lightly loaded spindle bearings to approximately 4000 centistokes at $100\,°F$ $(38\,°C)$ for some gear units. The great majority of oils used fall within a much narrower viscosity range of about 20–200 centistokes.

Aside from the primary function of friction and wear control, lubricating oils are often called on to serve other purposes, such as corrosion prevention, electrical insulation, power transmission and cooling. This last is particularly important in metal cutting and grinding.

As petroleum consists of a highly complex mixture of hydrocarbons and other organic compounds, the raw material must be carefully processed to separate the fractions which are desired for lubricating purposes. This is generally accomplished by a combination of fractional distillation, solvent extraction and adsorption treatments to obtain various cuts which are blended if necessary, to produce the desired viscosity. A typical sequence of operations would be:

1. Preliminary distillation of the crude petroleum to strip off volatile matter, i.e., dissolved gases and light hydrocarbons through the fuel oil range.
2. Secondary vacuum distillation to separate the various lubricating oil fractions, which in some cases, may include the *residuum*. Since hydrocarbons tend to undergo thermal decomposition above $600\,°F$ $(316\,°C)$, the distillation is generally carried out under vacuum to enable the higher-boiling fractions to be distilled.
3. Solvent extraction to remove waxes, unsaturates, aromatics, asphalt, and nonhydrocarbon material.
4. Contact with an adsorbent, e.g., activated clay, for final removal of polar impurities.
5. Incorporation of special-purpose additives and blending to obtain the desired viscosity.

The lubricating oil fractions thus obtained consist of some of the largest and most complex hydrocarbon molecules to be found in crude petroleum. Their molecular weight ranges from about 250–1000 or more, based on structures containing 20–70 carbon atoms. They may be classified broadly as (1) straight-chain paraffins, (2) branched chain paraffins, (3) naphthenes (one or more saturated 5 or 6 membered rings with paraffin side chains), (4) aromatics (benzene ring structures with paraffinic side chains), and (5) mixed aromatic-naphthene–paraffin.

Of the above classes, the most desirable are the branched chain paraffins and the naphthenes. The refining processes are directed to increasing the concentration of these and removal of the others, as well as of organic compounds containing sulfur, oxygen, and nitrogen which are found in varying proportions in all crude petroleum.

Most hydrocarbons in the lubricating oil range are subject to low-temperature oxidation ($180\,°F$; $82\,°C$ and above) by atmospheric oxygen, particularly under the influence of catalysts such as metals (copper is particularly active), moisture, and occasionally nonhydrocarbon impurities. The paraffinic types show the greatest resistance to oxidation, followed by the naphthenes and the aromatics. An empirical rule states that the rate of oxidation approximately doubles for each $20\,°F$ ($11\,°C$) rise in temperature. Thus, in practice where long life of the oil is desired, bearing temperatures are held to around $150\,°F$ ($66\,°C$) as a maximum. In an internal combustion engine where temperatures at the sliding surfaces of the order of 400–$500\,°F$ (204–$260\,°C$) are encountered, shorter life will result, e.g., 100 hrs. in an engine, compared with several years in a water-wheel generator.

Besides viscosity, the viscosity-temperature variation of lubricating oils is a most important property. It has been found that this behavior can be accurately expressed by the relation: $\log_{10}\log_{10}(v+0.8) = b\log_{10}(t\,°F+460)+c$ where v is the viscosity in centistokes, t the temperature in $°F$, and b and c are constants. In most cases an oil with the smallest viscosity–temperature variation is considered the more desirable. As a generalization, it may be said that the oils with the highest paraffin-to-naphthene ratios have the smallest change, while low paraffin-to-naphthene ratios correspond to a higher viscosity–temperature change. However, it is possible to dissolve materials in oil (e.g., olefinic polymers) which will effect a decrease in this variation.

While most bearings depend on the viscosity of the oil for the formation of a lubricating film (0.0001 to 0.002 in, 0.0025 to 0.051 mm thick), there are many cases where this is not feasible and successful lubrication depends on the formation of very thin films (10 to 10^3 Å thick). This was generally thought to occur through preferential adsorption by the solid surface of polar compounds (e.g., napthhenic acids, sulfur compounds) present in small amounts in the oil.

Another factor possibly contributing to the lubricating ability of very thin films is the variation of viscosity with pressure. The viscosity–pressure relationship is approximately $\log_{10} v = kP + c$ where P is the pressure and k and c are constants. Thus, at very high unit pressures of the order of 10^5 psi (690 MPa), viscosity increases of the order of 1000–10,000 times are encountered. Ths can account for a significant increase in load-carrying capacity. More recent work on rolling contact bearings has shown that these extremely high viscosities may be partially responsible for the observed formation of true hydrodynamic lubricating films in the 10^{-6} to 10^{-5} inch (0.00254–0.000254 mm) range.

Though the major portion of industrial lubricants is derived from petroleum, a small but significant portion is being obtained from other sources.

1. Natural liquid fatty esters, such as lard oil, palm oil, sperm oil, etc. These are good lubricants, but have poor chemical stability.
2. Synthetic hydrocarbons, prepared by polymerization of olefinic hydrocarbons; these have good stability when saturated and good viscosity–temperature coefficients.
3. Polyalkylene glycol oils, made by reaction of alcohols with polymerized ethylene and propylene glycols. They are either water-soluble or -insoluble. Fair stability (improved with additives), good viscosity–temperature coefficient, and good lubricating qualities.
4. Synthetic esters, (a) primarily esters of dibasic acids such as adipic and sebacic, though in Europe some monobasic acid esters have been prepared and used; (b) organic esters of phosphoric and silicic acid, which have some advantage of being more fire-resistant than the other organic compounds but which are subject to hydrolysis on exposure to water.
5. Silicone oils, which are linear and cyclic siloxane polymers of the formula $(-SiR_2O)_n$. They generally possess good thermal and oxidative stability, and good viscosity–temperature coefficients. When $-R$ is a hydrocarbon substituent, their lubricating ability is poor. This can be somewhat improved by the introduction of aromatic halogen in $-R$.
6. Halogenated hydrocarbons (chlorinated or fluorinated). These have good lubricating properties, but very poor viscosity–temperature relations. The fluorinated materials are extremely stable.
7. Perfluorinated polyalkylene glycols which have better viscosity–temperature behavior and temperature behavior and temperature resistance.
8. Polyphenyl ethers are very stable organic fluids which can be used in the 500–$700\,°F$ (260–$371\,°C$) temperature range although, like the silicones, they do not have good boundary lubrication properties. These reportedly, however, can be improved with suitable additives.

Generally speaking, these synthetic materials are at present used primarily as specialty lubricants where the need for a particular property outweighs high cost, e.g., diesters as military aircraft engine lubricants. As synthetic processes are improved more extensive application may be expected.

Synthetic Lubricants

Chemistry has been the guiding science in creating the various classes of chemical lubricants. Each class is named in terms of the chemical structure involved and each has at least one property that is unobtainable with naturally occurring materials. In varying degrees and combinations, synthetic lubricants provide excellent lubricity over extremely wide temperature ranges, high thermal and oxidative stability, fire-resistance, and outstanding resistance to nuclear radiation. Other noteworthy advantages of synthetic lubricants are the ease with which physical or performance properties can be altered by chemical modification or the incorporation of additives. The facility with which modifications can be achieved makes it possible for the chemists to tailor individual members of the family to meet exactly the requirements of a particular application. Compromises in properties are likewise made to obtain a satisfactory lubricant at lowest cost.

Polyglycols. $RO(CH_2CHR'O)_x R''$, where R's can be hydrogen and/or organic groups. Terminal groups determine the type of polyglycol such as diol, monoether, diether, ether–ester, etc. Various alkylene oxides including mixtures are the basic raw materials. The major attributes of the polyglycols are viscosity–temperature characteristics, volatility of products of decomposition, relatively low cost, and the wide variety of properties (water-soluble to organic-soluble) obtainable by structural variations. Major uses are as lubricants in automotive hydraulic brake fluids and water-based hydraulic fluids, as compressor lubricants, textile lubricants and rubber lubricants, and in greases where the polyglycol is the carrier for solid lubricants.

Phosphate Esters. $R'OP(O)(OR'')(OR')$ where at least one R represents an organic group while the remaining represent organic groups or hydrogen. These products are prepared from phosphorus oxychloride or phosphoryl chlorides plus phenols, alcohols or their sodium salts. The tertiary phosphate esters are often classified as triaryl, trialkyl, and alkyl aryl phosphates. The primary and secondary phosphates are used extensively as lubricant additives in various chemical forms. Phosphate esters are best known for their fire-resistant characteristics, and as a result have found extensive application as industrial and aircraft lubricants and hydraulic fluids.

Dibasic Acid Esters. These can include both simple and complex materials. The simple dibasic acid esters, $ROOCR'COOR''$ are made by reacting a dibasic acid, such as sebacic acid, with a primary branched alcohol, such as ethyl hexanol. Complex esters are prepared by reacting a dibasic acid with a polyglycol, such as polyethylene glycol, and capping the chain with a branched primary alcohol or a monobasic acid. The outstanding characteristics of the dibasic acid esters are favorable viscosity–temperature characteristics, excellent lubricating ability, and high stability. Because of this combination of properties, these products are now used as lubricants in almost all aircraft turbine engines.

Chlorofluorocarbons. The polymerization or telomerization of chlorotrifluoroethylene is the route to these relatively low-molecular-weight synthetic lubricants. Manufacture involves elaborate polymerization techniques and stabilization methods to eliminate all of the hydrogen and terminal chlorine introduced into the polymer chain by peroxide fragments or the chain transfer agent. The major characteristics of the chlorofluorocarbons are chemical inertness and thermal stability. Industrial and aerospace applications involving exposure to corrosive or oxidizing atmospheres are the largest uses for these lubricants.

Silicones.

$$\begin{array}{c} R \\ | \\ SiO \\ | \\ R \end{array} \left(\begin{array}{c} R \\ | \\ SiO \\ | \\ R \end{array} \right) \begin{array}{c} R \\ | \\ SiR \\ | \\ R \end{array}$$

where the R's may be the same or different organic groups. The properties are varied by the use of different types of organic substituents; the most popular are methyl, phenyl, and chlorophenyl groups. Recent advances involve fluorine-containing substituents. Manufacture entails the preparation of organochlorosilane intermediates, hydrolysis and condensation of these intermediates, and polymer finishing. In addition to good stability and low volatility, silicones have the best viscosity–temperature characteristics of any lubricant. Although they perform well under many conditions of lubrication, silicone lubricants are generally unsatisfactory for situations involving sliding contact of steel-on-steel. The many and varied uses

include lubricating electric motors, precision equipment, plastic and rubber surfaces, and as greases for antifriction bearings.

Silicate Esters. $ROSi(OR')(OR'')(OR''')$, where the R's may be similar or dissimilar groups. The best-known types are the tetraalkyl, tetraaryl, and mixed alkylaryl orthosilicates. The classic means of preparation is through the reaction of phenol or alcohol with silicon tetrachloride. A closely related group of products, the hexaalkoxy- and hexa-aryloxydisiloxanes, is also generally included in the silicate esters classification. These products, the so-called "dimer silicates," are conveniently made by the reaction of an alcohol or phenol with hexachlorodisiloxane. Notable characteristics of the silicate esters are low volatility, low-temperature fluidity, and thermal stability. The hydrolytic stability varies from poor to good, depending upon chemical structure. The products are used as high-temperature heat-transfer fluids, wide-temperature range hydraulic fluids, electronic coolants, and automatic weapon lubricants.

Neopentyl Polyol Esters. These polyesters are prepared by the esterification of 5-carbon polyfunctional alcohols with mono-functional acids. Because the beta carbon of the starting alcohol does not contain hydrogen, these esters are superior in thermal stability to the diesters. Most of the other characteristics are similar to those of the diesters. As a result of their superior stability, the neopentyl esters are finding increasing use as the lubricant for aircraft turbine engines.

Polyphenyl Ethers. Both alkyl-substituted and unsubstituted polyphenyl ethers are included in this class of synthetic lubricants. General preparation involves the Ullman ether synthesis. The unsubstituted polyphenyl ethers have outstanding thermal, oxidative, and radiation resistance, however, poor low-temperature characteristics are a major drawback. Alkyl substitution improves low-temperature viscosity, but detracts from stability. Most lubricant uses are developmental in nature and involve aircraft and aerospace applications.

Lubrication

Lubrication is the process of separating two rubbing solid surfaces by means of a layer which is effectively "softer" than either surface. Depending upon circumstances, the "soft" layer (the lubricant) may be a gas, a liquid, a solid, or a combination of various phases. The many ways of accomplishing lubrication can be conveniently grouped into two categories: hydrodynamic and solid lubrication.

Hydrodynamic Lubrication. The separation of moving surfaces by a "fluid" (gas, liquid, or gel) is accomplished according to the laws of hydrodynamics. One of the surfaces "swims" on the lubricant, i.e., it is lifted through the simultaneous possession of velocity relative to the lubricating fluid and of an acute angle of attack. This so-called glider bearing is the basis of design of nearly all fluid-lubricated bearings, including the journal bearing, which is just a circular glider bearing.

The basic relation which expresses the load-carrying ability of a glider bearing can be given in the form:

$$W/\eta U = a(r/h) \tag{1}$$

where W is the load on the glider, U the velocity of the glider relative to the stator, h is the minimum distance between the rubbing surfaces, r is a dimension characterizing the angle of attack, a is a numerical coefficient (somewhere between 1 and 10) and η is the viscosity of the lubricating fluid under the temperature and pressure conditions of the application. The energy expenditure required for the service performed by the lubricant film is usually expressed in terms of the friction coefficient, f, defined as the ratio of the frictional force required to move the glider (or the journal) to the load carried by it. For the case under discussion:

$$f = K\sqrt{\eta U/W} \tag{2}$$

where K is a numerical coefficient which depends upon the geometry of the system and varies between about 2 and 6.

While Equations 1 and 2 provide only a qualitative guide, the detailed calculations for specific bearings are quite complicated, they clearly indicate that the only variable at the disposal of the chemist, the viscosity, will be chosen such as to give the maximum load-carrying ability of the bearing consistent with a reasonable amount of frictional energy lost in the lubricant.

The low friction losses (f is usually of the order 10^{-3}) and the naturally wear-free operation generally make the maintenance of hydrodynamic lubrication the primary aim of bearing design. This goal is attained perfectly only with journal and with glider (pad) bearings, and to some extent, in roller bearings. It can be achieved by a special mechanism (thermal expansion of the flowing fluid) in parallel thrust bearings. Its attainment is uncertain (and not readily subject to calculation) in ball bearings and in gear lubrication.

Since the viscosity of readily available fluid varies over a 10^{10}-fold range between gases and the thickest liquids, and the temperature and pressure coefficients of viscosity vary similarly over an about 10^4-fold range, the choice of a suitable lubricant is usually determined by the ancillary conditions of temperature, volatility, chemical stability, etc. The use of sulfuric acid as lubricant for oxygen compressors is a typical illustration of a choice dictated by chemical conditions. The evaporation and deterioration of liquid lubricants in high energy particle fluxes and/or in hard vacua ($<10^{-9}$ torr) has led to increased use and rapid development of gas lubricated bearings. These call generally for extremely high and therefore very costly standards of workmanship. Hence the use of gas lubrication is at present largely restricted to precision instruments and to military and space applications.

The most important source of lubricants is petroleum. There is hardly a chemical species (esters, ethers, sulfides, metal-organic compounds, etc.) which has not contributed to the array of synthetic lubricants now available. Some chemicals, such as the perfluorocarbon compounds and the siloxane polymers were originally synthesized for just this service.

Elastohydrodynamic Lubrication. Well designed and equally well built machine elements with very smooth bearing surfaces permit the imposition of very high bearing loads. The resulting elastic deformation of the bearing may then change the geometry of the load-bearing surfaces substantially. Hence the elastic properties of the load-bearing materials enter the bearing calculations. The change of lubricant properties with pressure, temperature, and with the duration of exposure to the pressure and shear regime also enter the bearing calculations. The required viscoelastic properties of lubricants at high-frequency deformations are only beginning to be determined.

Hydrostatic Lubrication. Slow-moving or even static "gliders" can be separated from the bearing surface by a fluid film if the hydrostatic pressure required to carry the load is provided by an external source, as for instance by a pump. The very large journals of turbo generators, or heavy thrust bearing pads are generally lifted by these means before the onset of rotation in order to avoid scoring damage to the valuable bearings. Exceedingly low friction coefficients are obtained in these hydrostatic bearings, the most spectacular being that of the Mt. Palomar 200-inch (508 cm) telescope mirror pad bearings, where $f = 0.000004$, such that the 500-ton (450 metric ton) structure is easily moved by a $\frac{1}{2}$ HP clock motor.

Solid Lubrication. While fluid film separation of rubbing surfaces is the most desirable objective of lubrication, it is often unattainable, especially when bearings are too small or unsuitable for liquid lubricants. Even bearings built for full fluid lubrication during most of their operating periods experience solid-to-solid contact when starting and stopping.

Solid surfaces in rubbing contact are characterized by friction coefficients varying between 0.04 (Teflon on steel) and >100 (pure metals *in vacuo*). Solid lubrication, in contrast to fluid lubrication, is generally accompanied by a certain amount of wear of the rubbing parts. Optical inspection of the surfaces after rubbing reveals macroscopic (i.e., bulk) damage of the metal both when unlubricated and when lubricated. Quantitative differences between the two cases are easily measured radiographically when a radioactive glider has been used. In this manner it is found that effective solid lubrication can reduce the amount of wear, compared to the dry case, by a factor as high as 10^5.

Typical solid lubricants are the soft metals lead, indium, and tin, the layer lattice crystals graphite and molybdenum disulfide, many soft organic solids, such as metallic soaps, and waxes as well as the crystalline polymers Teflon (polytetrafluorethylene), polythene (polyethylene), and nylon. The integral bonding of these solids to the surface of the hard solid to be lubricated is essential for good performance. The bonding is accomplished either by alloying (copper–lead bearing metals), by flash coating, by introduction of the lubricating solid into the interstices of the sintered metal bearing (Teflon emulsion into sintered bronze), by chemical coating (phosphate coatings), or by anchorage to a phosphate or bonded plastic coating.

Special cases of solid lubrication are boundary and EP (extreme pressure) lubrication. In both cases the solid lubricant is formed by chemical reaction of special compounds, usually applied as oil solutions, with the metallic rubbing surfaces. Typical boundary lubricants are the fatty acids which react with the metal surface to form metallic soaps which then carry the load. Strongly adsorbed but nonreacting substances of linear structure, such as long chain fatty alcohols, can also act as boundary lubricants but only under very mild conditions.

Under the very severe conditions, sometimes encountered in automobile transmissions and especially in hypoid gear differentials as well as in machining operations, only those substances act as lubricants which contain chemically active chlorine, sulfur, or phosphorus to form the corresponding iron chloride, sulfide or phosphide by instantaneous attack on the surface hot spots resulting from the collisions of surface asperities. The chemical stability of these so-called E.P. agents is designed to permit activity at the temperature near the rubbing surface, say $200\,°C$ and above, but not be corrosive under normal conditions.

Mixed Film Lubrication. Mixed film lubrication is almost invariably the true state of affairs when boundary and EP lubrication are encountered, i.e., an appreciable fraction of the load is carried by the fluid film in the "valleys" of the surface while the asperities in contact are permitted to carry the balance of the load without seizure through the beneficent intervention of the boundary or EP lubricant. The very important breakin process of rubbing surfaces consists in the controlled reduction of the number and the size of the surface asperities so that fluid lubrication will prevail for most of the time.

"Real" Lubrication. In "real" lubrication of machinery, such as automotive engines, turbines, etc., the lubricating oil has to perform many functions besides lubrication. The most important of these is the cooling of the bearings. It also has to keep internal combustion engines clean by dispersing the partial combustion products of the fuel and its own degradation products, it has to carry chemicals to counteract wear, and, in common with many other lubrication applications, it must prevent corrosion of the equipment and be inhibited against its own deterioration in service. A relatively large amount of synthetic organic chemicals is therefore carried by many oils to perform the additional functions which one must expect from a modern lubricant.

Ideally one should always have full fluid separation of rubbing surfaces. But in inaccessible locations, or reactive environments, or under conditions of very slow motion or of intermittent operation, recourse must be had to solid lubrication.

LUDWIG, CARL WILHELM FRIEDRICH (1816–1895). Carl Ludwig was a German physiologist, who introduced scientific measurement and instruments into physiological research; important teacher in physiology, and author of numerous investigations of the cardiovascular and secretory systems and metabolism.

The son of a military officer, Ludwig was born in Witzenhausen, Germany and studied medicine at Marburg and Erlangen from 1834 until 1839. After his MD degree in 1840 he started his scientific career in the Department of Chemistry with Wilhelm Bunsen and one year later as prosector (first assistant) with Ludwig Fick in the Department of Anatomy. In 1842 he published his "habilitation" on the blood vessels in the kidney. From that time Ludwig was engaged in the problems of physiology of circulation and its measurement. In 1846 he constructed his first instrument, the kymograph, which registered blood pressure over time. From 1847 Ludwig had a close friendship with Hermann von Helmholtz, Emil Du Bois-Reymond and Ernst Brücke, the young generation of German physiologists who aimed to elucidate physical and chemical explanations of vital functions. In 1849 he was appointed Professor of Anatomy and Physiology at Zürich, Switzerland. In 1855 he moved to Vienna, Austria as Professor of Physiology and Zoology. In his 40th year, Ludwig wrote an influential compendium of physiology based on a mechanistic view of the organism. Finally, in 1865 he was appointed to the chair of the new Institute of Physiology in Leipzig, Germany, where he worked until his death in 1895. See also **Du Bois-Reymond, Emil Heinrich (1818–1896)**; and **Physiology (The History)**.

In Leipzig, Ludwig exerted an enormous influence on the development of physiology in Europe and the USA. Within the next 25 years more than 200 students from Germany and abroad studied and worked in his laboratory (I. P. Pavlov, H. Quincke, J. Bernstein, E. Mach, H. Öhrvall, W. H. Welch, O. Frank). Together with Elias Cyon he studied successfully heart functions from the isolated perfused heart (1868); with his disciple Henry Pickering Bowditch Ludwig discovered both the fundamental laws of neurophysiology (the all-or-none law of the heart (1871) and the staircase phenomenon), and with Luigi Luciani he investigated the nontetanizability of the heart (1872). Besides his numerous investigations of the anatomy and the functions of the cardiovascular system (the first tone of the heart, cardiac innervation, nervous regulation of the blood pressure), Ludwig described the filtration theory of urine function (glomeruli), gland secretion and the secretory nerves, the absorption and metabolism of fat, protein and sugar, the first vacuum blood gas pump (1859), the "Stromuhr" (flow meter, 1867) and mercury blood gas pump (1868). Furthermore Ludwig's school served as a model for scientific teamwork. See also **Pavlov, Ivan Petrovitch (1849–1936)**.

B. LOHFF, Medizinischen Hochschule, Hannover, Germany

LUMEN. A unit of photometric power. The lumen is equal to the amount of photometric power radiated into a unit solid angle (steradian) from a small source having a luminous intensity of one candela. Tungsten-filament light bulbs produces approximately 15 lumens per watt.

See also **Illumination** and **Units and Standards**.

LUMINESCENCE. A characteristic nonthermal emission of electromagnetic radiation by a material upon some form of excitation. Some luminescent materials are called phosphors. E. Wiedemann defined the term in 1888 as "all those phenomena of light not solely conditioned by the rise in temperature."

Whereas the output from blackbody radiators consists of broad-band emissions which follow the Stefan-Boltzmann temperature relationships, luminescence emission from phosphors consists of relatively narrow bands, which do not follow the blackbody laws. Thus, light emission due solely to the temperature of a source is referred to as *incandescence*, while *luminescence*, unlike incandescence, is a function of the specific material involved. Although *fluorescence* and *phosphorescence* are sometimes used synonymously with luminescence, a more rigid definition of *fluorescence* would be luminescence having a persistence (afterglow) shorter than about 10^{-8} second, with *phosphorescence* being longer than 10^{-8} second.

The luminescence process itself involves (1) absorption of energy; (2) excitation; and (3) emission of energy, usually in the form of radiation in the visible portion of the spectrum. The *type* of luminescence is usually defined by the excitation means, i.e., *cathodo*luminescence where excitation is by cathode rays, as in a television kinescope. The most commonly encountered types of luminescence are listed in Table 1.

TABLE 1. TYPES OF LUMINESCENCE

Luminescence Type	Excitation Source	Example
Photoluminescence	Photons	ZnS Ag
Cathodoluminescence	Cathode Rays	$Zn_2SiO_4 \cdot Mn$
Electroluminescence	Electric Fields	$Zn (S \cdot Se) \cdot Cu$
Chemiluminescence	Chemical Reactions	Oxidation of Luminol
Bioluminescence	Biochemical Reactions	Luciferin
Triboluminescence	Mechanical Disruption	$ZnS \cdot Mn$

The luminescent material may be considered as a transformer of energy, i.e., from ultraviolet photons to photons of lower energy; from cathode rays to photons; from electric fields to photons, etc. An inorganic luminescent material, or phosphor, usually consists of a crystalline host material to which is added a trace of an impurity (activator and coactivator).

Chemiluminescence

Numerous chemical reactions produce heat, but relatively few release their energy as light. This latter phenomenon is termed *chemiluminescence*. As pointed out by researchers at the University of Wales College of Medicine, Cardiff, "Absorption of energy by an atom or molecule raises an electron to a higher energy level. This is known as an 'excited state' and is inherently unstable. When the electron drops back to its ground state the energy must either be transferred to another atom or molecule, be released as heat or be emitted as light. The decay of the electron to ground state is very fast, occurring within 1–10 nanoseconds (10^{-9}–10^{-8} second)."

In chemiluminescence, the chemical reaction raises an electron to a higher level, which then decays back to the ground state, releasing a photon of light, the energy of which is predictable by Einstein's equation. When the energy drop is large, the light is blue; if small, the light is red. Inasmuch as the electronic excitation-decay process is extremely fast, the intensity of light in chemiluminescence is determined by the kinetics of the chemical reaction.

The distinction of chemiluminescence from fluorescence results from two factors:

1. When an atom or molecule fluoresces, it remains chemically unchanged and can be immediately be reexcited once light emission has occurred.
2. In chemiluminescence, each molecule only reacts once to form an excited state, while the excited product (actual emitter) has a different chemical structure from the initial substrate.

See also **Bioluminescence**.

Light-Emitting Diode (LED)

Recombination of injection electroluminescence was first observed in 1923 by Lossew, who found that when point electrodes were placed on certain silicon carbide crystals and current passed through them, light was often emitted. Explanation of this emission has been possible only with the development of semiconductor theory. If minority charge carriers are injected into a semiconductor, i.e., electrons are injected into *p*-type material or "positive holes" into *n*-type material, they recombine spontaneously with the majority carriers existing in the material. If some of these recombinations result in the emission of radiation, electroluminescence results. Minority-carrier injection may occur not only at point contacts, but also at broad area rectifying junctions; in this case, the junction must be biased in the forward or "easy flow" direction, and the electric field in the junction is lower when the voltage is applied than in its absence. This type of emission has been observed in several materials, including SiC, diamond, Si, Ge, CdS, ZnS, ZnSe, ZnO, and some of the so-called III-V compounds, such as AlN, GaSb, GaAs, GaP, InP, and InSb. The emission of many of these materials lies in the infrared region of the spectrum. For radiation in the visible region (instead of the infrared), the energy difference between the holes and electrons (band gap of the semiconductor) must be more than 1.8 eV. Numerous materials satisfy this requirement, notably those used for cathode-ray tube phosphors, but the materials present difficulties in fabricating *p-n* junctions and thus are not candidates for light-emitting diodes.

The list of materials for LEDs includes GaP, GaAsP, GaAlAs, GaN, and SiC. The two materials of choice to date have been GaP and GaAsP. Early commercial LEDs were made from $GaAs_{0.6}P_{0.4}$ deposited epitaxially as a thin layer on a GaAs crystal substrate. With these, *p-n* junctions were made, using diffusion techniques similar to those used in making silicon diodes. The band gap is 1.92 eV. There is an emission band of red light with a peak at about 650 nm, resulting from direct recombination of electrons and holes.

GaAsP has a high index of refraction, and consequently only light emitted toward the surface (4%) is usable—the remainder is reflected back. A diode can be encapsulated in epoxy material to take on the shape and form of a lens. These diodes are particularly effective where a number are fabricated in close proximity on a single-crystal chip.

Diodes that emit light in shorter wavelengths (green, yellow, etc.) can be made by increasing the phosphorus content, but only up to about 40% because of rapid decrease in efficiency. Efficiency can be increased by incorporating nitrogen atoms into crystals. The N atoms act as isoelectronic centers, trapping electron-hole pairs in an excited state. Three types of nitrogen-doped diodes have gained some importance: $GaAs_{0.65}P_{0.35}$ (orange light); $GaAs_{0.85}P_{0.15}$ (yellow light); and GaP (green light). Zinc and oxygen doping are also used. Diodes operate more efficiently if driven with periodic pulses of high current rather than with constant current. The short response time of junction diodes to current pulses (a fraction of a microsecond) and their rectifying property (they block current flow in the reverse, nonemitting direction) combine to make the diodes a good choice for X-Y addressing arrangements. See also **Light Emitting Diodes (LED)**.

Additional Reading

Aitken, M.J.: *An Introduction to Optical Dating: The Dating of Quaternary Sediments by the Use of Photon-Stimulated Luminescence*, Oxford University Press, Inc., New York, NY, 1998.

Burgess, C. and D. Jones: *Spectrophotometry, Luminescence and Colour: Science and Compliance: Papers Presented at the 2nd Joint Meeting of the UV Spectrometry Group of the U.K. and the Council for Optical Radiation Measurements of the U.S.A., Rindge, NH, U.S.A., 20–23*, Elsevier Science, New York, NY, 1995.

Mueller, G.: *Electroluminescence*, Vol. 64, Elsevier Science & Technology Books, New York, NY, 1999.

Ropp, R.: *Luminescence and the Solid State*, Elsevier Science, New York, NY, 1991.

Schubert, E.F.: *Light-Emitting Diodes*, Cambridge University Press, New York, NY, 2003.

Schulman, S.G.: *Molecular Luminescence Spectroscopy: Methods and Applications*, Vol. 3, John Wiley & Sons, Inc., New York, NY, 1993.

Stanley, P.E. and L.J. Kricka: *Bioluminescence and Chemiluminescence: Fundamental of Applied Aspects*, John Wiley & Sons, Inc., New York, NY, 1996.

Vij, D.R. and N. Singh: *Luminescence and Related Properties of II-VI Semiconductors*, Nova Science Publishers, Inc., Huntington, NY, 1997.

Vij, D.R.: *Luminescence of Solids*, Kluwer Academic Publishers, Norwell, MA, 1998.

Ziegler, M.M. and T.O. Baldwin: *Bioluminescence and Chemiluminescence, Part C*, Vol. 305, Academic Press, Inc., San Diego, CA, 2000.

LUMINOSITY FUNCTION.
Because of the variable sensitivity of the human eye to radiation of different wavelengths, a standard function has been established. For the standard conditions chosen in establishing this standard luminosity function (photopic vision), the luminously effective radiant intensity in lumens of radiation of spectral energy distribution J_Λ watts/unit wavelength is given by

$$680 \int_{\lambda=0}^{\lambda=\infty} y_\lambda J_\lambda d\lambda$$

where y_Λ is the standard luminosity function normalized to a value of unity at 555 nanometers. The numerical values for y_λ are commonly given as a luminosity curve.

For very low levels of intensity (scotopic vision) the peak of the luminosity function curve shifts toward the violet for young eyes (507 nanometers) with an absolute value of 1,746 lumens/watt.

LUMINOUS.
1. In general, pertaining to the emission of visible radiation. 2. In photometry, a modifier used to denote that a given physical quantity, such as luminous emittance, is weighted according to the manner in which the response of the human eye varies with the wavelength of the light.

LUMINOUS COEFFICIENT.
A coefficient that measures the integrated fraction of the radiant power that contributes to its luminous properties as evaluated by means of the standard luminosity function.

Luminous coefficient

$$= \int_{\lambda=0}^{\lambda=\infty} y_\lambda J_\lambda d\lambda / \int_{\lambda=0}^{\lambda=\infty} J_\lambda d\lambda$$

where y_λ is the standard luminosity function and J_λ is the spectral energy distribution of the radiant intensity. The luminous coefficient is unity for a narrow band of wavelengths at 555 nanometers.

LUMINOUS EFFICIENCY.
The ratio of the radiant energy sensed by the average human eye at a particular wavelength to that received. This ratio reaches a maximum inside the visible portion of the spectrum and falls to zero outside it. Luminous efficiency is dimensionless but is often given the units of lumens per watt. Photometric quantities are obtained by multiplying the corresponding radiometric quantities by the luminous efficiency and so often bear the adjective luminous, for example, luminous flux. However, when the radiometric quantities, radiance and irradiance, are transformed into photometric ones, they are given the special names luminance and illuminance. The luminous efficiency of cones differs from that of rods. See also **Illumination**.

LUNA LUNAR MISSIONS. See **Lunar Exploration**.

LUNAR ATMOSPHERIC TIDE. An atmospheric tide due to the gravitational attraction of the moon. The only detectable components are the 12-lunar-hour or semidiurnal, as in the oceanic tides, and two others of very nearly the same period. The amplitude of this atmospheric tide is so small that it is detected only by careful statistical analysis of a long record, being about 0.06 millibar in the tropics and 0.02 millibar in the middle latitudes.

LUNAR DAY. 1. The time required for the earth to rotate once with respect to the moon, that is, the time between two successive upper transits of the moon. The mean lunar day is approximately 1.035 times as great as the mean solar day, or 24 hours 50 minutes. Also called *tidal day*.

2. In astronomy, the time required for the moon to revolve once, relative to a fixed star, about its own axis.

LUNAR ECLIPSE. See **Eclipse**.

LUNAR EXPLORATION. The first leap in Lunar observation was caused by the invention of the telescope. Especially Galileo Galilei made good use of this new instrument and observed mountains and craters on the Moon's surface. See also **Telescope**.

The Cold War-inspired a space race between the Soviet Union and the United States of America. What was the next big step depends on the political viewpoint: In the US (and the West in general) the landing of the first humans on the moon in 1969 is seen as a culmination, indeed of the space race in general. On the other hand, many scientifically important steps, such as the first photographs of the until then unseen far side of the moon in 1959, were first achieved by the Soviet Union.

The first man-made object to reach the Moon was the unmanned Soviet probe Luna 2, which made a hard landing on September 14, 1959, at 21:02:24 Z. The far side of the Moon was first photographed on October 7, 1959 by the Soviet probe Luna 3. Luna 9 was the first probe to soft land on the Moon and transmit pictures from the Lunar surface on February 3, 1966. It was proven that a lunar lander would not sink into a thick layer of dust, as had been feared. The first artificial satellite of the Moon was the Soviet probe Luna 10 (launched March 31, 1966). The first robot lunar rover to land on the Moon was the Soviet vessel Lunokhod 1 on November 17, 1970 as part of the Lunokhod program. Moon samples were brought back to Earth by three Luna missions (nrs. 16, 20, and 24).

On December 24, 1968 the crew of Apollo 8, Frank Borman, James Lovell, and William Ander became the first human beings to see the far side of the Moon with their own eyes (as opposed to seeing it on a photograph). On July 20, 1969, Neil Armstrong became the first human being to set foot on the Moon. The first step onto the Lunar surface from the Apollo 11 Lunar Module, the Eagle, fulfilled the promise of President John F. Kennedy that the U.S. would land a man on the Moon before the end of the decade. It was the highlight of an extended U.S. program to study and map the Moon, beginning with Ranger 7 impacting the Moon on July 31, 1964 and culminating with Apollo 17, which left the Moon on December 14, 1972. The last man to stand on the Moon was Eugene Cernan. The scientific return from these missions was immensely important and included nearly complete high-resolution imaging of the lunar surface, lunar samples, topographic, seismic, and gravity data, and information on the lunar environment.

On January 14, 2004, President George W. Bush called for a plan to return manned missions to the Moon by 2020. NASA's plan to accomplish that goal was announced on March 19, 2005, and was promptly dubbed Apollo 2.0 by critics.

The European Space Agency has plans to launch probes to explore the Moon in the near future, too. European spacecraft Smart 1 was launched September 27, 2003 and entered lunar orbit on November 15, 2004. It will survey the lunar environment and create an X-ray map of the Moon.

The People's Republic of China has expressed ambitious plans for exploring the Moon and is investigating the prospect of lunar mining, specifically looking for the isotope Helium-3 for use as an energy source on Earth. As of December 2005, China has been making substantial progress in reaching this goal, with the unmanned orbiter and rocket entering production and testing. The craft is expected to launch in 2007, with the eventual plan to land astronauts (called "taikonauts" by the Chinese government) on the moon before 2020. Japan has two planned lunar missions, Lunar-A and Selene; even a manned lunar base is planned by the Japanese Space Agency (JAXA). India will also try an unmanned orbiting satellite, called Chandrayaan-1.

From the mid-1960's to the mid-1970's there were 65 moon landings (with 10 in 1971 alone), but after Luna 24 in 1976 they suddenly stopped. The Soviet Union started focusing on Venus and Space Stations and the US on Mars and beyond. In 1990 Japan visited the moon with the Hiten spacecraft, becoming the third country to orbit the moon. The spacecraft released the Hagormo probe into lunar orbit, but the transmitter failed rendering the mission scientifically useless.

In 1994 the United States launched Clementine a joint project between the Strategic Defense Initiative Organization and NASA. The mission mapped most of the lunar surface at a number of resolutions and wavelengths from UV to IR. In 1998 the US launched the Lunar Prospector mission. The Lunar Prospector was designed for a low polar orbit investigation of the Moon, including mapping of surface composition and possible deposits of polar ice, measurements of magnetic and gravity fields, and study of lunar outgassing events. In 2008 the US plans on launching the Lunar Reconnaissance Orbiter (LRO). This mission will emphasize the overall objective of obtaining data that will facilitate returning men safely to the Moon where testing and preparations for an eventual manned mission to Mars will be undertaken.

Lunar Missions 1959-1967

Pioneer 1. Launch Date/Time: October 11, 1958 at 08:42:00 UT from Cape Canaveral, United States. Pioneer 1, the second and most successful of three project Able space probes and the first spacecraft launched by the newly formed NASA, was intended to study the ionizing radiation, cosmic rays, magnetic fields, and micrometeorites in the vicinity of the Earth and in lunar orbit. Due to a launch vehicle malfunction, the spacecraft attained only a ballistic trajectory and never reached the Moon. It did return data on the near-Earth space environment.

Pioneer 1 consisted of a thin cylindrical midsection with a squat truncated cone frustrum on each side. The cylinder was 74 cm (2.43 ft) in diameter and the height from the top of one cone to the top of the opposite cone was 76 cm (2.49 ft). Along the axis of the spacecraft and protruding from the end of the lower cone was an 11 kg (24.25 lb) solid propellant injection rocket and rocket case, which formed the main structural member of the spacecraft. Eight small low-thrust solid propellant velocity adjustment rockets were mounted on the end of the upper cone in a ring assembly which could be jettisoned after use. A magnetic dipole antenna also protruded from the top of the upper cone. The shell was composed of laminated plastic. The total mass of the spacecraft after vernier separation was 34.2 kg (75.4 lb), after injection rocket firing it would have been 23.2 kg (51 lb).

The scientific instrument package had a mass of 17.8 kg (39.24 lb) and consisted of an image scanning infrared television system to study the Moon's surface to a resolution of 1 milliradian, an ionization chamber to measure radiation in space, a diaphragm/microphone assembly to detect micrometeorites, a spin-coil magnetometer to measure magnetic fields to 5 microgauss, and temperature-variable resistors to record spacecraft internal conditions. The spacecraft was powered by nickel-cadmium batteries for ignition of the rockets, silver cell batteries for the television system, and mercury batteries for the remaining circuits. Radio transmission was at on 108.06 MHz through an electric dipole antenna for telemetry and doppler information at 300 mW and a magnetic dipole antenna for the television system at 50 W. Ground commands were received through the electric dipole antenna at 115 MHz. The spacecraft was spin stabilized at 1.8 rps, the spin direction was approximately perpendicular to the geomagnetic meridian planes of the trajectory.

The spacecraft did not reach the Moon as planned due to an incorrectly set valve in the upper stage which caused an accelerometer to give faulty information leading to a slight error in burnout velocity (the Thor second stage shut down 10 seconds early) and angle (3.5 degrees). This resulted in a ballistic trajectory with a peak altitude of 113,800 km (70, 712 miles) around 1300 local time. The real-time transmission was obtained for about 75% of the flight, but the percentage of data recorded for each experiment was variable. Except for the first hour of flight, the signal to noise ratio was good. The spacecraft ended transmission when it reentered the Earth's atmosphere after 43 hours of flight on October 13, 1958 at 03:46 UT over the South Pacific Ocean. A small quantity of useful scientific information was returned, showing the radiation surrounding Earth was in the form of bands and measuring the extent of the bands, mapping the total ionizing flux, making the first observations of hydromagnetic oscillations of the magnetic field, and taking the first measurements of the density of micrometeorites and the interplanetary magnetic field.

Pioneer 3. Launch Date/Time: December 6, 1958 at 05:45:12 UT from Cape Canaveral, United States. Pioneer 3 was a spin stabilized spacecraft launched by the U.S. Army Ballistic Missile agency in conjunction with NASA. The spacecraft failed to go past the Moon and into a heliocentric orbit as planned, but did reach a maximum altitude of over 102,000 km before falling back to Earth. The revised spacecraft objectives were to measure radiation in the outer Van Allen belt area using Geiger-Mueller tubes and to test the trigger mechanism for a lunar photographic experiment.

Pioneer 3 was a cone-shaped probe 58 cm high (1.9 ft) and 25 cm (0.82 ft) diameter at its base. See Fig. 1. The cone was composed of a thin fiberglass shell coated with a gold wash to make it electrically conducting and painted with white stripes to maintain the temperature between 10 and 50 degrees C. At the tip of the cone was a small probe which combined with the cone itself to act as an antenna. At the base of the cone a ring of mercury batteries provided power. A photoelectric sensor protruded from the center of the ring. The sensor was designed with two photocells which would be triggered by the light of the Moon when the probe was within about 30,000 km (18,641 miles) of the Moon. At the center of the cone was a voltage supply tube and two Geiger-Mueller tubes. A transmitter with a mass of 0.5 kg (1.10 lb)delivered a phase-modulated signal of 0.1 W at a frequency of 960.05 MHz. The modulated carrier power was 0.08 W and the total effective radiated power 0.18 W. A despin mechanism consisted of two 7 gram (0.25 ounce) weights which could be spooled out to the end of two 150 cm (4.92 ft) wires when triggered by a hydraulic timer 10 hours after launch. The weights would slow the spacecraft spin from 400 rpm to 6 rpm and then weights and wires would be released.

Fig. 1.   Pioneer 3 probe. (*National Space Science Data Center* (*NSSDC*).

The flight plan called for the Pioneer 3 probe to pass close to the Moon after 33.75 hours and then go into solar orbit. However, depletion of propellant caused the first stage engine to shut down 3.7 seconds early preventing the spacecraft from reaching escape velocity. The injection angle was also about 71 degrees instead of the planned 68 degrees. The spacecraft reached an altitude of 102,360 km (63,604 miles) (109,740 km (68,189 miles) from the center of the Earth) before falling back to Earth. It re-entered Earth's atmosphere and burned up over Africa on 7 December at approximately 19:51 UT (2:51 p.m. EST) at an estimated location of 16.4 N, 18.6 E. The probe returned telemetry for about 25 hours of its 38 hour 6 minute journey. The other 13 hours were blackout periods due to the location of the two tracking stations. The returned information showed that the internal temperature remained at about 43 degrees C over most of the period. The data obtained were of particular value since they indicated the existence of two distinct radiation belts.

Luna 1. Launch Date/Time: January 2, 1959 at 16:41:21 UT from Tyuratam (Baikonur Cosmodrome), U.S.S.R. Luna 1 was the first spacecraft to reach the Moon, and the first of a series of Soviet automatic interplanetary stations successfully launched in the direction of the Moon. The spacecraft was sphere-shaped. Five antennae extended from one hemisphere. See Fig. 2. Instrument ports also protruded from the surface of the sphere. There were no propulsion systems on the Luna 1 spacecraft. Because of its high velocity and its announced package of various metallic emblems with the Soviet coat of arms, it was concluded that Luna 1 was intended to impact the Moon.

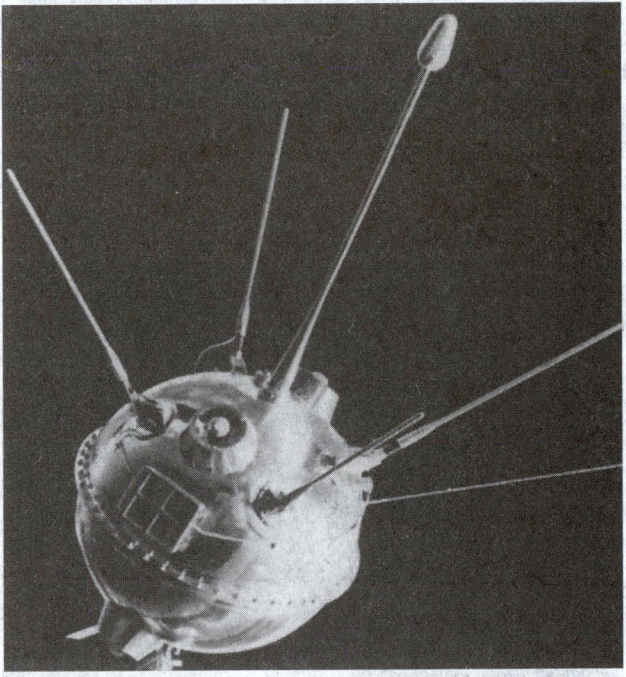

Fig. 2.   Luna 1 Spacecraft. (*National Space Science Data Center (NSSDC).*

On 2 January 1959, after reaching escape velocity, Luna 1 separated from its 1,472 kg (3,245 lb) third stage. The third stage, 5.2 m (17 ft) long and 2.4 m (7.9 ft) in diameter, travelled along with Luna 1. On 3 January, at a distance of 113,000 km (70,215 miles) from Earth, a large (1 kg (2.2 lb) cloud of sodium gas was released by the spacecraft. This glowing orange trail of gas, visible over the Indian Ocean with the brightness of a sixth-magnitude star, allowed astronomers to track the spacecraft. It also served as an experiment on the behavior of gas in outer space. Luna 1 passed within 5,995 km (3,725 miles) of the Moon's surface on 4 January after 34 hours of flight. It went into orbit around the Sun, between the orbits of Earth and Mars.

The spacecraft contained radio equipment, a tracking transmitter, and telemetering system, five different sets of scientific devices for studying interplanetary space, including a magnetometer, geiger counter, scintillation counter, and micrometeorite detector, and other equipment.

The measurements obtained during this mission provided new data on the Earth's radiation belt and outer space, including the discovery that the Moon had no magnetic field and that a solar wind, a strong flow of ionized plasma emmanating from the Sun, streamed through interplanetary space.

Pioneer 4. Launch Date/Time: March 3, 1959 at 17:11:00 UT from Cape Canaveral, United States. Pioneer 4 was a spin stabilized spacecraft launched on a lunar flyby trajectory and into a heliocentric orbit making it the first US probe to escape from the Earth's gravity. It carried a payload similar to Pioneer 3: a lunar radiation environment experiment using a Geiger-Mueller tube detector and a lunar photography experiment. It passed within 60,000 km (37,282 miles) of the Moon's surface. However, Pioneer 4 did not come close enough to trigger the photoelectric sensor. No lunar radiation was detected. See Pioneer 3 for spacecraft design and subsystems; and Fig. 1.

After a successful launch Pioneer 4 achieved its primary objective (an Earth-Moon trajectory), returned radiation data and provided a valuable tracking exercise. The probe passed within 60,000 km (37,282 miles) of the Moon's surface (7.2 E, 5.7 S) on 4 March 1959 at 22:25 UT (5:25 p.m. EST) at a speed of 7,230 km/hr (4,493 miles/hr). The distance was not close enough to trigger the photoelectric sensor. The probe was tracked for 82 hours to a distance of 655,000 km (406,998 miles) and reached perihelion on 18 March 1959 at 01:00 UT. The cylindrical fourth stage casing (173 cm (5.7 ft) long, 15 cm (0.5 ft) diameter, 4.65 kg (10.25 lb) went into orbit with the probe.

Luna 2. Launch Date/Time: September 12, 1959 at 22:02:24 UT from Tyuratam (Baikonur Cosmodrome), U.S.S.R. Luna 2 was the second of a series of Soviet spacecraft launched in the direction of the Moon. The first spacecraft to land on the Moon, it impacted the lunar surface east of Mare Serenitatis near the Aristides, Archimedes, and Autolycus craters. Luna 2 was similar in design to Luna 1, a spherical spacecraft with protruding antennae and instrument parts. The instrumentation was also similar, including scintillation- and geiger-counters, a magnetometer, and micrometeorite detectors. The spacecraft also carried Soviet pennants. There were no propulsion systems on Luna 2 itself.

After launch and attainment of escape velocity on 12 September 1959 (13 September Moscow time), Luna 2 separated from its third stage, which travelled along with it towards the Moon. On 13 September the spacecraft released a bright orange cloud of sodium gas which aided in spacecraft tracking and acted as an experiment on the behavior of gas in space. On 14 September, after 33.5 hours of flight, radio signals from Luna 2 abruptly ceased, indicating it had impacted on the Moon. The impact point, in the Palus Putredinus region, is roughly estimated to have occurred at 0 degrees longitude, 29.1 degrees N latitude. Some 30 minutes after Luna 2, the third stage of its rocket also impacted the Moon. The mission confirmed that the Moon had no appreciable magnetic field, and found no evidence of radiation belts at the Moon.

Luna 3. Launch Date/Time: October 4, 1959 at 02:24:00 UT from Tyuratam (Baikonur Cosmodrome), U.S.S.R. Luna 3, an automatic interplanetary station, was the third spacecraft successfully launched to the Moon and the first to return images of the lunar far side. The spacecraft returned very indistinct pictures, but, through computer enhancement, a tentative atlas of the lunar farside was produced. These first views of the lunar far side showed mountainous terrain, very different from the near side, and only two dark regions which were named Mare Moscovrae (Sea of Moscow) and Mare Desiderii (Sea of Dreams). (Mare Desiderii was later found to be composed of a smaller mare, Mare Ingenii (Sea of Ingenuity) and other dark craters.)

The spacecraft was a cylindrically shaped cannister with hemispherical ends and a wide flange near the top end. The probe was 130 cm long (4.27 ft) and 120 cm (3.94 ft) at its maximum diameter at the flange. See Fig. 3. Most of the cylindrical section was roughly 95 cm (3.12 ft) in diameter. The cannister was hermetically sealed and pressurized at 0.23 atmospheres. Solar cells were mounted along the outside of the cylinder and provided power to the chemical batteries stored inside the spacecraft. Jalousies for thermal control were also positioned along the cylinder and would open to expose a radiating surface when the interior temperature exceeded 25 °C (77 °F). The upper hemisphere of the probe held the covered opening for the cameras. Four antennae protruded from the top of the probe and two from the bottom. Other scientific apparatus (micrometeoroid and cosmic ray detectors) was mounted on the outside of

Fig. 3.   Luna 3 Spacecraft. (National Space Science Data Center (*NSSDC*).

the probe. Gas jets for attitude control were mounted on the outside of the lower end of the spacecraft. Photoelectric cells were used to maintain orientation with respect to the Sun and Moon. The spacecraft had no rockets for course adjustment. The interior of the spacecraft held the cameras and film processing system, radio equipment, propulsion systems, batteries, gyroscopic units for attitude control, and circulating fans for temperature control. The spacecraft was spin stabilized and was directly radio-controlled from Earth.

The imaging system on Luna 3 was designated Yenisey-2 and consisted of a dual lens camera, an automatic film processing unit, and a scanner. The lenses on the camera were a 200 mm (7.87 inches) focal length, f/5.6 aperture objective and a 500 mm (19.69 inches), f/9.5 objective. The camera carried 40 frames of temperature- and radiation resistant 35-mm isochrome film. The 200 mm objective could image the full disk of the Moon and the 500 mm could take an image of a region on the surface. The camera was fixed in the spacecraft and pointing was achieved by rotating the craft itself. A photocell was used to detect the Moon and orient the upper end of the spacecraft and cameras towards it. Detection of the Moon signaled the camera cover to open and the photography sequence to start automatically. After photography was complete, the film was moved to an on-board processor where it was developed, fixed, and dried. Commands from Earth were then given and the film was moved to a scanner where a bright spot produced by a cathode ray tube was projected through the film onto a photelectric multiplier. The spot was scanned across the film and the photomultiplier converted the intensity of the light passing through the film into an electric signal which was transmitted to Earth. A frame could be scanned with a resolution of 1000 lines, the transmission could be done at a slow rate for large distances from Earth and a faster rate at closer range.

After launch on an 8K72 (number I1-8) on a course over the Earth's north pole the Blok-E escape stage was shut down by radio control from Earth at the proper velocity to put the Luna 3 on a trajectory to the Moon. Initial radio contact showed the signal from the probe was only about half as strong as expected and the interior temperature was increasing. The spacecraft spin axis was reoriented and some equipment shut down resulting in a drop in temperature from 40 C to about 30 C. At a distance

of 60,000 to 70,000 km from the Moon, the orientation system was turned on and the spacecraft rotation was stopped. The lower end of the station was oriented towards the Sun, which was shining on the far side of the Moon. The spacecraft passed within 6,200 km of the Moon near the south pole at its closest approach at 14:16 UT on 6 October 1959 and continued on to the far side. On 7 October the photocell on the upper end of the spacecraft detected the sunlit far side of the Moon and the photography sequence started. The first image was taken at 03:30 UT at a distance of 63,500 km from the Moon's surface and the last 40 minutes later from 66,700 km. A total of 29 photographs were taken, covering 70% of the far side. After the photography was complete the spacecraft resumed spinning, passed over the north pole of the Moon and returned towards the Earth. Attempts to transmit the photographs to Earth began on 8 October but were believed to be unsuccessful due to the low signal strength. As Luna 3 got closer to Earth a total of 17 resolvable but noisy photographs were transmitted by 18 October. Contact with the probe was lost on 22 October. The probe was believed to have burned up in the Earth's atmosphere in March or April of 1960, but may have survived in orbit until after 1962.

Ranger 1. Launch Date/Time: August 23, 1961 at 10:02:00 UT from Cape Canaveral, United States. Ranger 1 was a spacecraft whose primary mission was to test the performance of those functions and parts necessary for carrying out subsequent lunar and planetary missions using essentially the same spacecraft design. A secondary objective was to study the nature of particles and fields in interplanetary space.

The spacecraft was of the Ranger Block 1 design and consisted of a hexagonal base 1.5 m (4.92 ft) across upon which was mounted a cone-shaped 4 m (13 ft) high tower of aluminum struts and braces. Two solar panel wings measuring 5.2 m (17 ft) from tip to tip extended from the base. See Fig. 4. A high-gain directional dish antenna was attached to the bottom of the base. Spacecraft experiments and other equipment were mounted on the base and tower. Instruments aboard the spacecraft included a Lyman-alpha telescope, a rubidium-vapor magnetometer, electrostatic analyzers, medium-energy range particle detectors, two triple coincidence telescopes, a cosmic-ray integrating ionization chamber, cosmic dust detectors, and solar X-ray scintillation counters.

Fig. 4.   Ranger Block 1 Spacecraft. (National Space Science Data Center (*NSSDC*).

The communications system included the high gain antenna and an omni-directional medium gain antenna and two transmitters, one at 960.1-mhz with 0.25 W power output and the other at 960.05-mhz with 3 W power output. Power was to be furnished by 8680 solar cells on the two panels, a 57 kg (126 lb) silver-zinc battery, and smaller batteries on some of the experiments. Attitude control was provided by a solid-state timing controller, Sun and Earth sensors, and pitch and roll jets. The temperature was controlled passively by gold plating, white paint, and polished aluminum surfaces.

The Ranger 1 spacecraft was designed to go into an Earth parking orbit and then into a 60,000 × 1,100,000 km (37,282 × 683,508 miles) Earth orbit to test systems and strategies for future lunar missions. Ranger 1 was launched into the Earth parking orbit as planned, but the Agena B failed to restart to put it into the higher trajectory, so when Ranger 1 separated from the Agena stage it went into a low Earth orbit and began tumbling. The satellite re-entered Earth's atmosphere on 30 August 1961. Ranger 1 was partially successful, much of the primary objective of flight testing the equipment was accomplished but little scientific data was returned.

Total research, development, launch, and support costs for the Ranger series of spacecraft (Rangers 1 through 9) was approximately $170 million.

Ranger 2. Launch Date/Time: November 18, 1961 at 08:09:00 UT from Cape Canaveral, FL, United States. Ranger 2 was of the Ranger Block 1 design and was almost identical to Ranger 1. The spacecraft consisted of a hexagonal base 1.5 m (4.9 ft) across upon which was mounted a cone-shaped 4 m (13 ft) high tower of aluminum struts and braces. Two solar panel wings measuring 5.2 m (17 ft) from tip to tip extended from the base. A high-gain directional dish antenna was attached to the bottom of the base. Spacecraft experiments and other equipment were mounted on the base and tower. Instruments aboard the spacecraft included a Lyman-alpha telescope, a rubidium-vapor magnetometer, electrostatic analyzers, medium-energy-range particle detectors, two triple coincidence telescopes, a cosmic-ray integrating ionization chamber, cosmic dust detectors, and scintillation counters.

The communications system included the high gain antenna and an omni-directional medium gain antenna and two transmitters at approximately 960-mhz, one with 0.25 W power output and the other with 3 W power output. Power was to be furnished by 8680 solar cells on the two panels, a 53.5 kg (118 lbs) silver-zinc battery, and smaller batteries on some of the experiments. Attitude control was provided by a solid-state timing controller, Sun and Earth sensors, gyroscopes, and pitch and roll jets. The temperature was controlled passively by gold plating, white paint, and polished aluminum surfaces.

The spacecraft was launched into a low earth parking orbit, but an inoperative roll gyro prevented Agena restart. The spacecraft could not be put into its planned deep-space trajectory, resulting in Ranger 2 being stranded in low earth orbit upon separation from the Agena stage. The orbit decayed and the spacecraft reentered Earth's atmosphere on 20 November 1961.

Ranger 3. Launch Date/Time: January 26, 1962 at 20:30:00 UT from Cape Canaveral, FL, United States. Ranger 3 was designed to transmit pictures of the lunar surface to Earth stations during a period of 10 minutes of flight prior to impacting on the Moon, to rough-land a seismometer capsule on the Moon, to collect gamma-ray data in flight, to study radar reflectivity of the lunar surface, and to continue testing of the Ranger program for development of lunar and interplanetary spacecraft. Due to a series of malfunctions the spacecraft missed the Moon.

Ranger 3 was the first of the so-called Block II Ranger designs. The basic vehicle was 3.1 m (10 ft) high and consisted of a lunar capsule covered with a balsawood impact-limiter, 65 cm (2.13 ft) in diameter, a mono-propellant mid-course motor, a 5080-pound thrust retrorocket, and a gold- and chrome-plated hexagonal base 1.5 m (4.9 ft) in diameter. A large high-gain dish antenna was attached to the base. Two wing-like solar panels (5.2 m (17 ft) across) were attached to the base and deployed early in the flight. See Fig. 5. Power was generated by 8680 solar cells contained in the solar panels which charged a 11.5 kg (25.35 lbs)1000 W-hour capacity AgZn launching and backup battery. Spacecraft control was provided by a solid-state computer and sequencer and an earth-controlled command system. Attitude control was provided by Sun and Earth sensors, gyroscopes, and pitch and roll jets. The telemetry system aboard the spacecraft consisted of two 960 MHz transmitters, one at 3 W power output and the other at 50 mW power output, the high-gain antenna, and an omni-directional antenna. White paint, gold and chrome plating, and a silvered plastic sheet encasing the retrorocket furnished thermal control.

The experimental apparatus included: (1) a vidicon television camera, which employed a scan mechanism that yielded one complete frame in

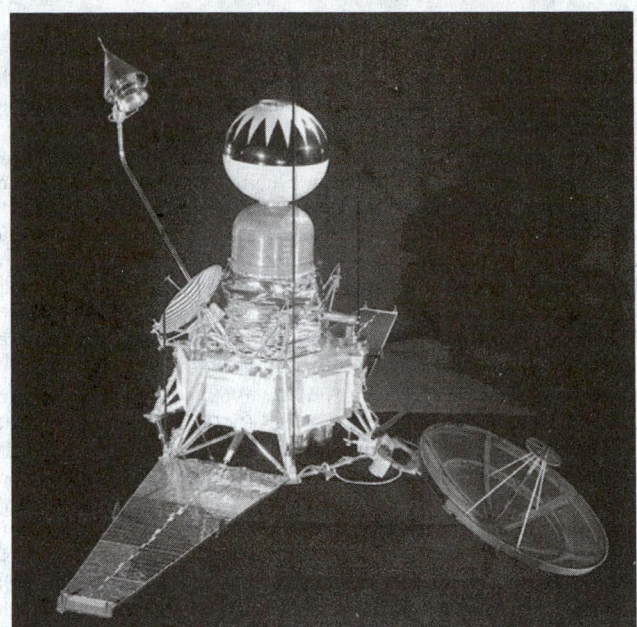

Fig. 5. Ranger 3 Block II Spacecraft. (National Space Science Data Center (*NSSDC*).

10 s; (2) a gamma-ray spectrometer mounted on a 1.8 m boom; (3) a radar altimeter; and (4) a seismometer to be rough-landed on the lunar surface. The seismometer was encased in the lunar capsule along with an amplifier, a 50-milliwatt transmitter, voltage control, a turnstile antenna, and 6 silver-cadmium batteries capable of operating the lunar capsule transmitter for 30 days, all designed to land on the Moon at 130 to 160 km/hr (80–100 mph). The radar altimeter would be used for reflectivity studies, but was also designed to initiate capsule separation and ignite the retro-rocket.

The mission was designed to boosted towards the Moon by an Atlas/Agena, undergo one mid-course correction, and impact the lunar surface. At the appropriate altitude the capsule was to separate and the retrorockets ignite to cushion the landing. A malfunction in the booster guidance system resulted in excessive spacecraft speed. Reversed command signals caused the spacecraft to pitch in the wrong direction and the TM antenna to lose earth acquisition, and mid-course correction was not possible. Finally a spurious signal during the terminal maneuver prevented transmission of useful TV pictures. Ranger 3 missed the Moon by approximately 36,800 km (22,865 miles) on 28 January and is now in a heliocentric orbit. Some useful engineering data were obtained from the flight.

Ranger 4. Launch Date/Time: April 23, 1962 at 20:50:00 UT from Cape Canaveral, FL, United States. Ranger 4 was designed to transmit pictures of the lunar surface to Earth stations during a period of 10 minutes of flight prior to impacting on the Moon, to rough-land a seismometer capsule on the Moon, to collect gamma-ray data in flight, to study radar reflectivity of the lunar surface, and to continue testing of the Ranger program for development of lunar and interplanetary spacecraft. An onboard computer failure caused failure of the deployment of the solar panels and navigation systems, the spacecraft impacted on the far side of the Moon without returning any scientific data. Ranger 4 was a Block II Ranger spacecraft virtually identical to Ranger 3.

The mission was designed to boosted towards the Moon by an Atlas/Agena B, undergo one mid-course correction, and impact the lunar surface. At the appropriate altitude the capsule was to separate and the retrorockets ignite to cushion the landing. Due to an apparent failure of a timer in the spacecraft's central computer and sequencer following launch the command signals for the extension of the solar panels and the operation of the sun and earth acquisition system were never given. The instrumentation ceased operation after about 10 hours of flight. The spacecraft was tracked by the battery-powered 50 milliwatt transmitter in the lunar landing capsule. Ranger 4 impacted the far side of the Moon

(229.3 degrees E, 15.5 degrees S) at 9600 km/hr (5,965 miles/hr) at 12:49:53 UT on April 26, 1962 after 64 hours of flight.

Ranger 5. Launch Date/Time: October, 18, 1962 at 16:59:00 UT from Cape Canaveral, FL, United States. Ranger 5 was a Block II Ranger spacecraft similar to Rangers 3 and 4.

The mission was designed to boosted towards the Moon by an Atlas/Agena B, undergo one mid-course correction, and impact the lunar surface. At the appropriate altitude the capsule was to separate and the retrorockets ignite to cushion the landing. Due to an unknown malfunction after injection into lunar trajectory from Earth parking orbit, the spacecraft failed to receive power. The batteries ran down after 8 hours, 44 minutes, rendering the spacecraft inoperable. Ranger 5 missed the Moon by 725 km (450.5 miles). It is now in a heliocentric orbit. Gamma-ray data were collected for 4 hours prior to the loss of power.

Luna 4. Launch Date/Time: April 2, 1963 at 08:04:00 UT from Tyuratam (Baikonur Cosmodrome), U.S.S.R. Luna 4 was the USSR's first successful spacecraft of their "second generation" lunar program. The spacecraft, rather than being sent on a straight trajectory toward the Moon, was placed first in an Earth orbit and then an automatic interplanetary station was rocketed in a curving path towards the Moon. Luna 4, the second attempt of this program, achieved the desired trajectory but missed the Moon by 8,336.2 km (5,179.9 miles) (at 13:25 UT on 5 April 1963) and entered a barycentric $90,000 \times 700,000$ km ($55,923 \times 434,960$ miles) Earth orbit. The intended mission of the probe is not known, it was speculated the probe was designed to land on the Moon with an instrument package based on the trajectory and on the later attempted landings of the Luna 5 and 6 spacecraft. (And the fact that a lecture program entitled "Hitting the Moon" was scheduled to be broadcast on Radio Moscow at 7:45 p.m. the evening of April 5 but was cancelled.) The spacecraft transmitted at 183.6 MHz at least until 6 April.

Ranger 6. Launch Date/Time: January 30, 1964 at 15:49:00 UT from Cape Canaveral, FL, United States. Ranger 6 was designed to achieve a lunar impact trajectory and to transmit high-resolution photographs of the lunar surface during the final minutes of flight up to impact. The spacecraft carried six television vidicon cameras, 2 wide angle (channel F, cameras A and B) and 4 narrow angle (channel P) to accomplish these objectives. The cameras were arranged in two separate chains, or channels, each self-contained with separate power supplies, timers, and transmitters so as to afford the greatest reliability and probability of obtaining high-quality video pictures. No other experiments were carried on the spacecraft. Due to a failure of the camera system no images were returned.

Rangers 6, 7, 8, and 9 were the so-called Block 3 versions of the Ranger spacecraft. The spacecraft consisted of a hexagonal aluminum frame base 1.5 m (4.9 ft) across on which was mounted the propulsion and power units, topped by a truncated conical tower which held the TV cameras. Two solar panel wings, each 73.9 cm (29 inches) wide by 153.7 cm (60.5 inches) long, extended from opposite edges of the base with a full span of 4.6 m 15.09 ft), and a pointable high gain dish antenna was hinge mounted at one of the corners of the base away from the solar panels. A cylindrical quasiomni directional antenna was seated on top of the conical tower. The overall height of the spacecraft was 3.6 m (11.81 ft).

Propulsion for the mid-course trajectory correction was provided by a 224-N thrust monopropellant hydrazine engine with 4 jet-vane vector control. Orientation and attitude control about 3 axes was enabled by 12 nitrogen gas jets coupled to a system of 3 gyros, 4 primary Sun sensors, 2 secondary Sun sensors, and an Earth sensor. Power was supplied by 9792 Si solar cells contained in the two solar panels, giving a total array area of 2.3 square meters (7.55 square feet) and producing 200 W. Two 1200 Watt-hr AgZnO batteries rated at 26.5 V with a capacity for 9 hours of operation provided power to each of the separate communication/TV camera chains. Two 1000 Watt-hr AgZnO batteries stored power for spacecraft operations.

Communications were through the quasiomni directional low-gain antenna and the parabolic high-gain antenna. Transmitters aboard the spacecraft included a 60 W TV channel F at 959.52 MHz, a 60 W TV channel P at 960.05 MHz, and a 3 W transponder channel 8 at 960.58 MHz. The telecommunications equipment converted the composite video signal from the camera transmitters into an RF signal for subsequent transmission through the spacecraft high-gain antenna. Sufficient video bandwidth was provided to allow for rapid framing sequences of both narrow- and wide-angle television pictures.

Ranger 6 was launched into an Earth parking orbit and injected on a lunar trajectory by a second Atlas-Agena B burn. The midcourse trajectory correction was accomplished early in the flight by ground control. On February 2, 1964, 65.5 hours after launch, Ranger 6 impacted the Moon on the eastern edge of Mare Tranquillitatis (Sea of Tranquility). The orientation of the spacecraft to the surface during descent was correct, but no video signal was received and no camera data obtained. A review board determined the most likely cause of failure was due to an arc-over in the TV power system when it inadvertently turned on for 67 seconds approximately 2 minutes after launch during the period of booster-engine separation.

Ranger 7. Launch Date/Time: July 28, 1964 at 16:50:00 UT Cape Canaveral, FL, United States. Ranger 7 was designed to achieve a lunar impact trajectory and to transmit high-resolution photographs of the lunar surface during the final minutes of flight up to impact. See **Ranger 6** for spacecraft design and subsystems.

The Atlas 250D and Agena B 6009 boosters performed nominally at launch inserting the Agena and Ranger into a 192 km (119.30 mi) altitude Earth parking orbit. Half an hour after launch the second burn of the Agena engine injected the spacecraft into a lunar intercept trajectory. After separation from the Agena, the solar panels were deployed, attitude control activated, and spacecraft transmissions switched from the omniantenna to the high-gain antenna. The next day, 29 July, the planned mid-course maneuver was initiated at 10:27 UT, involving a short rocket burn. The only anomaly during flight was a brief loss of two-way lock on the spacecraft by the DSIF tracking station at Cape Kennedy following launch.

Ranger 7 reached the Moon on 31 July. The F-channel began its one minute warm up 18 minutes before impact. The first image was taken at 13:08:45 UT at an altitude of 2110 km. Transmission of 4,308 photographs of excellent quality occurred over the final 17 minutes of flight. The final image taken before impact has a resolution of 0.5 meters (1.64 ft). The spacecraft encountered the lunar surface in direct motion along a hyperbolic trajectory, with an incoming asymptotic direction at an angle of −5.57 degrees from the lunar equator. The orbit plane was inclined 26.84 degrees to the lunar equator. After 68.6 hours of flight, Ranger 7 impacted in an area between Mare Nubium and Oceanus Procellarum (subsequently named Mare Cognitum) at 10.70 S latitude, 339.33 E longitude. (selenocentric—Apollo 16 Preliminary Science Report, 29-40, 1972.) Impact occurred at 13:25:48.82 UT at a velocity of 2.62 km/s. The spacecraft performance was excellent.

Ranger 8. Launch Date/Time: February 17, 1965 at 17:05:00 UT from Cape Canaveral, FL, United States. Ranger 8 was designed to achieve a lunar impact trajectory and to transmit high-resolution photographs of the lunar surface during the final minutes of flight up to impact. See **Ranger 6** for spacecraft design and subsystems.

The Atlas 196D and Agena B 6006 boosters performed nominally, injecting the Agena and Ranger 8 into an Earth parking orbit at 185 km (114.95 miles) altitude 7 minutes after launch. Fourteen minutes later a 90 second burn of the Agena put the spacecraft into lunar transfer trajectory, and several minutes later the Ranger and Agena separated. The Ranger solar panels were deployed, attitude control activated, and spacecraft transmissions switched from the omniantenna to the high-gain antenna by 21:30 UT. On 18 February at a distance of 160,000 km (99,419 miles) from Earth the planned mid-course maneuver took place, involving reorientation and a 59 second rocket burn. During the 27 minute maneuver, spacecraft transmitter power dropped severely, so that lock was lost on all telemetry channels. This continued intermittently until the rocket burn, at which time power returned to normal. The telemetry dropout had no serious effects on the mission. A planned terminal sequence to point the cameras more in the direction of flight just before reaching the Moon was cancelled to allow the cameras to cover a greater area of the Moon's surface. Ranger 8 reached the Moon on 20 February 1965. The first image was taken at 9:34:32 UT at an altitude of 2510 km 1,559.6 miles). Transmission of 7,137 photographs of good quality occurred over the final 23 minutes of flight. See Figs. 6 and 7. The final image taken before impact has a resolution of 1.5 meters (4.9 ft). The spacecraft encountered the lunar surface in a direct hyperbolic trajectory, with incoming asymptotic direction at an angle of −13.6 degrees from the lunar equator. The orbit plane was inclined 16.5 degrees to the lunar equator. After 64.9 hours of flight, impact occurred at 09:57:36.756 UT on 20 February 1965 in Mare Tranquillitatis

Fig. 6.   Ranger 8 image of the Moon from 302 km (187.65 miles). The image was taken on 20 February 1965 and 9:55 UT, two and a half minutes before the spacecraft impacted on the lunar surface. The two large craters at upper center are Ritter (above left) and Sabine, each about 30 km (18.6 miles) in diameter. The Apollo 11 landing site is just off the right edge of the image at about 4:00. The image is about 130 km (80.78 miles) across and north is up (*Image courtesy of NSSDC*).

lunar surface during the final minutes of flight up to impact. See Ranger 6 for spacecraft design and subsystems.

The Atlas 204D and Agena B 6007 boosters performed nominally, injecting the Agena and Ranger 9 into an Earth parking orbit at 185 km (114.95 miles) altitude. A 90 second Agena 2nd burn put the spacecraft into lunar transfer trajectory. This was followed by the separation of the Agena and Ranger. 70 minutes after launch the command was given to deploy solar panels, activate attitude control, and switch from the omniantenna to the high-gain antenna. The accuracy of the initial trajectory enabled delay of the planned mid-course correction from 22 March to 23 March when the maneuver was initiated at 12:03 UT. After orientation, a 31 second rocket burn at 12:30 UT, and reorientation, the maneuver was completed at 13:30 UT.

Ranger 9 reached the Moon on 24 March 1965. At 13:31 UT a terminal maneuver was executed to orient the spacecraft so the cameras were more in line with the flight direction to improve the resolution of the pictures. Twenty minutes before impact the one-minute camera system warm-up began. The first image was taken at 13:49:41 at an altitude of 2,363 km (1,468 miles). Transmission of 5,814 good contrast photographs was made during the final 19 minutes of flight. The final image taken before impact has a resolution of 0.3 meters. The spacecraft encountered the lunar surface with an incoming asymptotic direction at an angle of -5.6 degrees from the lunar equator. The orbit plane was inclined 15.6 degrees to the lunar equator. After 64.5 hours of flight, impact occurred at 14:08:19.994 UT at 12.91 S latitude, 357.62 E longitude (selenocentric—Apollo 16 Preliminary Science Report, 29-40, 1972.) in the crater Alphonsus. See Figs. 8. and 9. Impact velocity was 2.67 km/s. The spacecraft performance was excellent. Real time television coverage with live network broadcasts of many of the F-channel images (primarily camera B but also some camera A pictures) were provided for this flight.

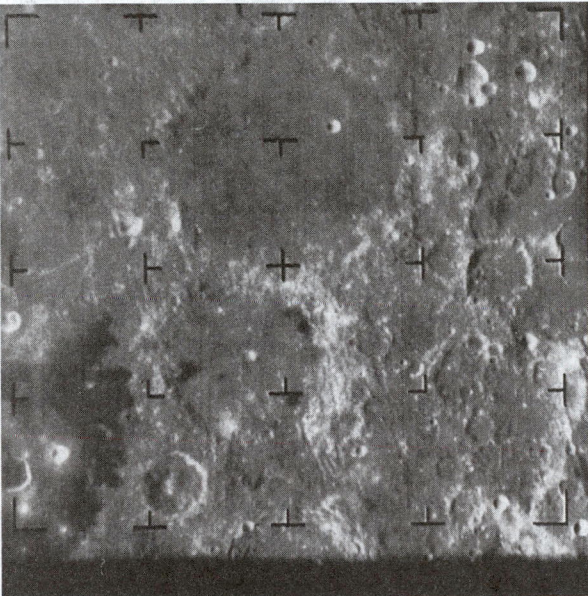

Fig. 7.   The first image taken by the Ranger 8 camera B on 20 February 1965 shows the 164 km (101.90 miles) diameter Ptolemaeus crater, top center, and below it Alphonsus crater (diameter 108 km (67.11 miles)). The Davy crater chain can be seen immediately to the right of the left-middle reticle mark as a dotted white line extending SWW to NEE. The picture was taken 23 minutes before impact from a range of 2,545 km (1,581 mieles). The image is about 360 km (223.69 miles) across and north is up. Ranger 8 was the second successful lunar impact mission in the series. (*Image courtesy of NSSDC*).

Fig. 8.   Ranger 9 image of Alphonsus crater (diameter 108 km (67.11 miles) from a distance of 442 km (274.65 miles), taken about 3 minutes before impact in the upper right portion of the crater. At left is the northeastern edge of Mare Nubium. The crater adjacent to Alphonsus at the bottom is the 39 km (24.23 miles) diameter Alpetragius. Davy crater is at upper left. North is at 12:30. Ranger 9 impacted the Moon on 24 March 1965 at 14:08:20 UT. (*Image courtesy of NSSDC*).

at approximately 2.71 degrees N, 24.81 degrees E. (selenocentric—Apollo 16 Preliminary Science Report, 29-40, 1972.) Impact velocity was slightly less than 2.68 km/s. The spacecraft performance was excellent.

Ranger 9. Launch Date/Time: March 21, 1965 at 21:37:00 UT from Cape Canaveral, FL, United States. Ranger 9 was designed to achieve a lunar impact trajectory and to transmit high-resolution photographs of the

Luna 5. Launch Date/Time: April 9, 1965 at 07:55:00 UT from Tyuratam (Baikonur Cosmodrome), U.S.S.R. The Luna 5 automatic interplanetary station was designed to continue investigations of a lunar soft landing. The retrorocket system failed, and the spacecraft impacted the lunar surface at the Sea of Clouds.

Fig. 9. Ranger 9 B-camera image from 2500 km (1,553 miles) showing Ptolemaeus, Alphonsus, and Albategnius craters. Ptolemaeus is the large (164 diameter 102 miles) flat-floored crater at the top. Alphonsus, diameter 108 km (67.11 miles), is at lower left and the 114 km (70.8 miles) Albategnius crater is at lower right. The terminator runs through the lower corner. Ranger 9 impacted in Alphonsus crater 18.5 minutes after this image was taken. North is at 12:30. (*Image courtesy of NSSDC*).

Fig. 10. Zond 3 Spacecraft. The spacecraft design was similar to Zond 2, in addition to the imaging equipment it carried a magnetometer, ultraviolet (0.25–0.35 micron and 0.19–0.27 micron) and infrared (3–4 micron) spectrographs, radiation sensors (gas-discharge and scintillation counters), a radiotelescope and a micrometeoroid instrument. It also had an experimental ion engine. (*Image courtesy of NSSDC*).

Luna 6. Launch Date/Time: June 8, 1965 at 07:41:00 UT from Tyuratam (Baikonur Cosmodrome), U.S.S.R. Luna 6 was intended to travel to the Moon, but, because a midcourse correction failed, it missed the Moon by 159,612.8 km (99,178.80 miles).

Zond 3. Launch Date/Time: July 18, 1965 at 14:38:00 UT from Tyuratam (Baikonur Cosmodrome), U.S.S.R. Zond 3 was launched from a Tyazheliy Sputnik (65-056B) earth orbiting platform towards the moon and interplanetary space. The spacecraft was equipped with an f106 mm camera and TV system that provided automatic inflight film processing. See Fig. 10. On July 20 lunar flyby occurred approximately 33 hours after launch at a closest approach of 9,200 km (5,717 miles). 25 pictures of very good quality were taken of the lunar farside from distances of 11,570 to 9960 km over a period of 68 minutes. The photos covered 19,000,000 square km (11,806, 052 square miles) of the lunar surface. Photo transmissions by facsimile were returned to earth from a distance of 2,200,000 km (1,367,017 miles) and were retransmitted from a distance of 31,500,000 km (19,573,193 miles) (some signals still being transmitted from the distance of the orbit of Mars), thus proving the ability of the communications system. After the lunar flyby, Zond 3 continued space exploration in a heliocentric orbit.

It is believed that Zond 3 was initially designed as a companion spacecraft to Zond 2 to be launched to Mars during the 1964 launch window. The opportunity to launch was missed, and the spacecraft was launched on a Mars trajectory, although Mars was no longer attainable, as a spacecraft test. For more information on Zond 2, see: http://nssdc.gsfc.nasa.gov/database/MasterCatalog?sc=1964-078 C

Luna 7. Launch Date/Time: October 4, 1965 at 07:55:00 UT from Tyuratam (Baikonur Cosmodrome), U.S.S.R. The Luna 7 spacecraft was intended to achieve a soft landing on the Moon. However, due to premature retrofire and cutoff of the retrorockets, the spacecraft impacted the lunar surface in the Sea of Storms.

Luna 8. Launch Date/Time: December 3, 1965 at 10:48:00 UT from Tyuratam (Baikonur Cosmodrome), U.S.S.R. Luna 8 was launched with the probable mission of achieving a soft landing on the Moon. However, the retrofire was late, and the spacecraft impacted the lunar surface in the Sea of Storms. The mission did complete the experimental development of the star-orientation system and ground control of radio equipment, flight trajectory, and other instrumentation.

Luna 9. Launch Date/Time: January 31, 1966 at 11:45:00 UT from Tyuratam (Baikonur Cosmodrome), U.S.S.R. The Luna 9 spacecraft was the first spacecraft to achieve a lunar soft landing and to transmit photographic data to Earth. See Fig. 11. The automatic lunar station that achieved the soft landing weighed 99 kg (218.3 lbs). It was a hermetically sealed container with radio equipment, a program timing device, heat control systems, scientific apparatus, power sources, and a television system. See Fig. 12. The Luna 9 payload was carried to Earth orbit by an A-2-E vehicle and then conveyed toward the Moon by a fourth stage rocket that separated itself from the payload. Flight apparatus separated from the payload shortly before Luna 9 landed. After landing in the Ocean of Storms on February 3, 1966, the four petals, which formed the spacecraft, opened outward and stabilized the spacecraft on the lunar surface. Spring-controlled antennas assumed operating positions, and the television camera rotatable mirror system, which operated by revolving and tilting, began a photographic survey of the lunar environment. Seven radio sessions, totaling 8 hours and 5 minutes, were transmitted as were three series of TV pictures. When assembled, the photographs provided a panoramic view of the nearby lunar surface. The pictures included views of nearby rocks and of the horizon 1.4 km (1,531 yards) away from the spacecraft.

Luna 10. Launch Date/Time: March 31, 1966 at 10:48:00 UT from Tyuratam (Baikonur Cosmodrome), U.S.S.R. The Luna 10 spacecraft was launched towards the Moon from an Earth orbiting platform. The spacecraft entered lunar orbit on April 3, 1966 and completed its first orbit 3 hours later (on April 4, Moscow time). Scientific instruments included a gamma-ray spectrometer for energies between 0.3 — 3 MeV, a triaxial magnetometer, a meteorite detector, instruments for solar-plasma studies, and devices for measuring infrared emissions from the Moon and radiation conditions of the lunar environment. See Fig. 13. Gravitational studies were also conducted. The spacecraft played back to Earth the 'Internationale' during the Twenty-third Congress of the Communist Party of the Soviet Union. Luna 10 was battery powered and operated for 460

Fig. 11.   Luna 9 spacecraft and flight apparatus. (*Courtesy of Alexander Chernov and the Virtual Space Museum*).

Fig. 13.   Luna 10 Spacecraft. (*Image courtesy of NSSDC*).

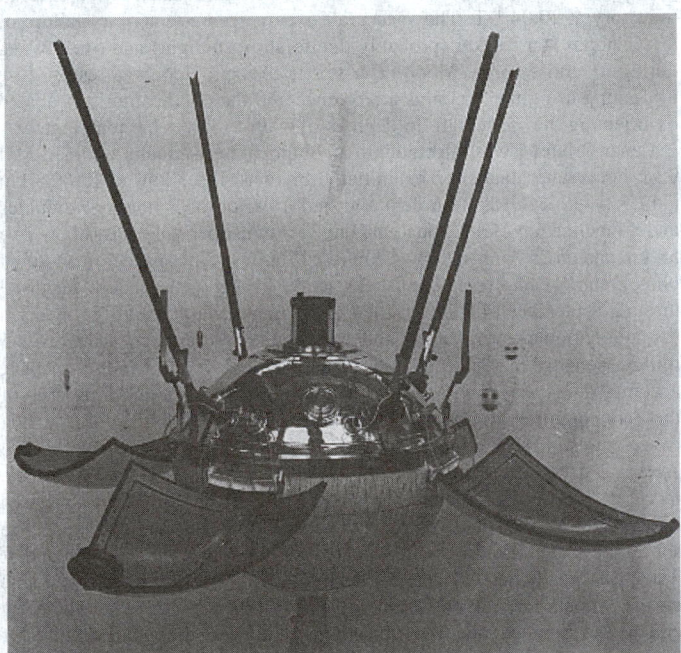

Fig. 12.   Luna 9 Lander. (*Image courtesy of NSSDC*).

lunar orbits and 219 active data transmissions before radio signals were discontinued on May 30, 1966.

Surveyor 1. Launch Date/Time: May 30, 1966 at 14:41:00 UT from Cape Canaveral, FL, United States. Surveyor 1 was the first spacecraft launched in the Surveyor program and the first soft landing on the Moon by the United States. The mission was considered a complete success and demonstrated the technology necessary to achieve landing and operations on the lunar surface. The primary objectives of the Surveyor program, a series of seven robotic lunar softlanding flights, were to support the

coming crewed Apollo landings by: (1) developing and validating the technology for landing softly on the Moon; (2) providing data on the compatibility of the Apollo design with conditions encountered on the lunar surface; and (3) adding to the scientific knowledge of the Moon. The specific primary objectives for this mission were to: (1) demonstrate the capability of the Surveyor spacecraft to perform successful midcourse and terminal maneuvers, and to achieve a soft landing on the Moon; (2) demonstrate the capability of the Surveyor communications system and Deep Space Network to maintain communications with the spacecraft during its flight and after a soft landing; and (3) demonstrate the capability of the Atlas/Centaur launch vehicle to inject the Surveyor spacecraft on a lunar intercept trajectory. Secondary objectives were to obtain engineering data on spacecraft subsystems used during cruise, descent and after landing. Tertiary objectives were to obtain postlanding TV pictures of a spacecraft footpad, the surface material immediately surrounding it and the lunar topography, and to obtain data on radar reflectivity and bearing strength of the lunar surface and on spacecraft temperatures.

The basic Surveyor spacecraft structure consisted of a tripod of thin-walled aluminum tubing and interconnecting braces providing mounting surfaces and attachments for the power, communications, propulsion, flight control, and payload systems. A central mast extended about one meter above the apex of the tripod. Three hinged landing legs were attached to the lower corners of the structure. The legs held shock absorbers, crushable, honeycomb aluminum blocks, and the deployment locking mechanism and terminated in footpads with crushable bottoms. The three footpads extended out 4.3 meters (14.11 ft) from the center of the Surveyor. The spacecraft was about 3 meters (9.84 ft) tall. The legs folded to fit into a nose shroud for launch. See Fig. 14.

A 0.855 square meter (2.81 square feet) array of 792 solar cells was mounted on a positioner on top of the mast and generated up to 85 Watts of power which was stored in rechargeable silver-zinc batteries.? Communications were achieved via a movable large planar array high gain antenna mounted near the top of the central mast to transmit television images, two omnidirectional conical antennas mounted on the ends of folding booms for uplink and downlink, two receivers and two transmitters. Thermal control was achieved by a combination of white paint, high IR-emittance thermal finish, polished aluminum underside.

Fig. 14.   Surveyor probe.

Two thermally controlled compartments, equipped with superinsulating blankets, conductive heat paths, thermal switches and small electric heaters, were mounted on the spacecraft structure. One compartment, held at 5–50 °C (41–122 °F), housed communications and power supply electronics. The other, held between −20 and 50 °C (−4 and 122 °F), housed the command and signal processing components. The TV survey camera was mounted near the top of the tripod and strain gauges, temperature sensors, and other engineering instruments are incorporated throughout the spacecraft. One photometric targets was mounted near the end of a landing leg and one on a short boom extending from the bottom of the structure. Other payload packages, which differed from mission to mission, were mounted on various parts of the structure depending on their function.

A Sun sensor, Canopus tracker and rate gyros on three axes provided attitude knowledge. Propulsion and attitude control were provided by cold-gas (nitrogen) attitude control jets during cruise phases, three throttlable vernier rocket engines during powered phases, including the landing, and the solid-propellant retrorocket engine during terminal descent. The retrorocket was a spherical steel case mounted in the bottom center of the spacecraft. The vernier engines used monomethyl hydrazine hydrate fuel and MON-10 (90% N_2O_2, 10% NO) oxidizer. Each thrust chamber could produce 130 N to 460 N of thrust on command, one engine could swivel for roll control. The fuel was stored in spherical tanks mounted to the tripod structure. For the landing sequence, an altitude marking radar initiated the firing of the main retrorocket for primary braking. After firing was complete, the retrorocket and radar were jettisoned and the doppler and altimeter radars were activated. These provided information to the autopilot which controlled the vernier propulsion system to touchdown.

No instrumentation was carried specifically for scientific experiments, but considerable scientific information was obtained. Surveyor 1 carried two television cameras—one mounted on the bottom of the frame for approach photography, which was not used, and the survey television camera. Over 100 engineering sensors were on board. Surveyor 1 had a mass of 995.2 kg (2,194 lbs) at launch and 294.3 kg (648.82 lbs) at landing.

Surveyor 1 was launched on an Atlas/Centaur from Complex 36-A of the Eastern Test Range directly into a lunar impact trajectory. After a midcourse correction at 06:45 UT on 31 May the spacecraft reached the Moon about 63 hours after launch. At an altitude of 75.3 km (46.8 miles)

and a velocity of 2612 m/s the main retrorocket, signaled by the altitude marking radar, ignited for a 40 second burn and was jettisoned at an altitude of roughly 11 km (6.8 miles) having slowed the spacecraft to 110 m/s. Descent continued with the vernier engines under control of the altimeter and doppler radars. Engines were turned off at a height of 3.4 m (11.15 ft) above the lunar surface and the spacecraft fell freely from this height. Surveyor 1 landed on the lunar surface on 2 June 1966 at 6:17:36 UT (1:17:36 a.m. EST) at about 3 m/s. The landing site was at 2.45 S, 316.79 E (selenographic) on a flat area inside a 100 km crater north of Flamsteed Crater in southwest Oceanus Procellarum.

Surveyor 1's first hour on the Moon was spent performing engineering tests. Photography sessions were then initiated throughout the remainder of the lunar day. The television system transmitted pictures of the spacecraft footpad and surrounding lunar terrain and surface materials. Some 10,338 photos were returned prior to nightfall on June 14. The spacecraft also acquired data on the radar reflectivity of the lunar surface, bearing strength of the lunar surface, and spacecraft temperatures for use in the analysis of the lunar surface temperatures. Surveyor 1 was able to withstand the first lunar night and near high noon on its second lunar day, July 7, photos again were returned. On 13 July at 7:30 UT (2:30 a.m. EST), after a total of 11,240 pictures had been transmitted, Surveyor 1's mission was terminated due to a dramatic drop in battery voltage just after sunset. Engineering interrogations continued until January 7, 1967. All mission objectives were accomplished. The Surveyor program involved building and launching 7 Surveyor spacecraft to the Moon at a total cost of $469 million.

Lunar Orbiter 1. Launch Date/Time: August 10, 1966 at 19:26:00 UT from Cape Canaveral, FL, United States. The Lunar Orbiter 1 spacecraft was designed primarily to photograph smooth areas of the lunar surface for selection and verification of safe landing sites for the Surveyor and Apollo missions. It was also equipped to collect selenodetic, radiation intensity, and micrometeoroid impact data. The spacecraft was placed in an Earth parking orbit on 10 August 1966 at 19:31 UT and injected into a cislunar trajectory at 20:04 UT. The spacecraft experienced a temporary failure of the Canopus star tracker (probably due to stray sunlight) and overheating during its cruise to the Moon. The star tracker problem was resolved by navigating using the Moon as a reference and the overheating was abated by orienting the spacecraft 36 degrees off-Sun to lower the temperature.

Lunar Orbiter 1 was injected into an elliptical near-equatorial lunar orbit 92.1 hours after launch. The initial orbit was 189.1 km × 1,866.8 km (117.5 miles × 1,159.98 miles) and had a period of 3 hours 37 minutes and an inclination of 12.2 degrees. On 21 August perilune was dropped to 58 km and on 25 August to 40.5 km (25.17 miles). The spacecraft acquired photographic data from August 18 to 29, 1966, and readout occurred through September 14, 1966. A total of 42 high resolution and 187 medium resolution frames were taken and transmitted to Earth covering over 5 million square km (3,106,855,961 square miles) of the Moon's surface, accomplishing about 75% of the intended mission, although a number of the early high-res photos showed severe smearing. It also took the first two pictures of the Earth ever from the distance of the Moon. See Fig. 15. Accurate data were acquired from all other experiments throughout the mission. Orbit tracking showed a slight "pear-shape" to the Moon based on the gravity field and no micrometeorite impacts were detected. The spacecraft was tracked until it impacted the lunar surface on command at 7 degrees N latitude, 161 degrees E longitude (selenographic coordinates) on the Moon's far side on October 29, 1966 on its 577th orbit. The early end to the nominal one year mission was due to the small amount of remaining attitude control gas and other deteriorating conditions and was planned to avoid transmission interference with Lunar Orbiter 2.

The main bus of the Lunar Orbiter had the general shape of a truncated cone, 1.65 meters (5.41 ft) tall and 1.5 meters (4.92 ft) in diameter at the base. The spacecraft was comprised of three decks supported by trusses and an arch. The equipment deck at the base of the craft held the battery, transponder, flight progammer, inertial reference unit (IRU), Canopus star tracker, command decoder, multiplex encoder, traveling wave tube amplifier (TWTA), and the photographic system. Four solar panels were mounted to extend out from this deck with a total span across of 3.72 meters (12.20 ft). Also extending out from the base of the spacecraft were a high gain antenna on a 1.32 meter (4.33 ft) boom and a low gain antenna on a 2.08 meter (6.82 ft) boom. Above the equipment deck, the

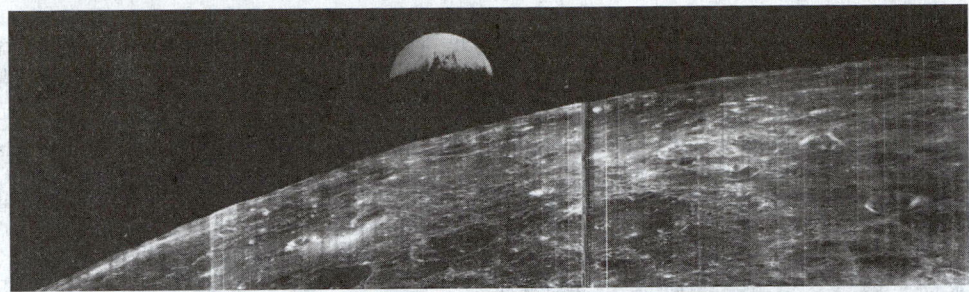

Fig. 15.   Lunar Orbiter 1 view of the Moon and crescent Earth. This is the first good image of the Earth taken from the vicinity of the Moon, 380,000 km (236,121 miles) away. The Earth sunset terminator runs through Odessa, Istanbul, and slightly west of Capetown. The center of the lunar surface corresponds to the location of the crater Pasteur, just on the eastern farside at 10 S,105 E, but the high sun angle makes it hard to see the craters. The horizon covers about 550 km (341.75 miles), and north is to the right in this west facing image. (*Image courtesy of NSSDC*).

middle deck held the velocity control engine, propellant, oxidizer and pressurization tanks, Sun sensors, and micrometeoroid detectors. The third deck consisted of a heat shield to protect the spacecraft from the firing of the velocity control engine. The nozzle of the engine protruded through the center of the shield. Mounted on the perimeter of the top deck were four attitude control thrusters. See Fig. 16.

Fig. 16.   Lunar Orbiter 1 or (Lunar Orbiter-A). (*Image courtesy of NSSDC*).

Power of 375 W was provided by the four solar arrays containing 10,856 n/p solar cells which would directly run the spacecraft and also charge the 12 amp-hr nickel-cadmium battery. The batteries were used during brief periods of occultation when no solar power was available. Propulsion for major maneuvers was provided by the gimballed velocity control engine, a hypergolic 100-pound-thrust Marquardt rocket motor. Three-axis stabilization and attitude control were provided by four one-lb nitrogen gas jets. Navigational knowledge was provided by five Sun sensors, Canopus star sensor, and the IRU equipped with internal gyros. Communications were via a 10 W transmitter and the directional 1 meter diameter high gain antenna for transmission of photographs and a 0.5 W transmitter and omnidirectional low gain antenna for other communications. Both antennas operated in S-band at 2295 MHz. Thermal control was maintained by a multilayer aluminized mylar and dacron thermal blanket which enshrouded the main bus, special paint, insulation, and small heaters.

The Lunar Orbiter program consisted of 5 Lunar Orbiters which returned photography of 99% of the surface of the Moon (near and far side) with resolution down to 1 meter (3.28 ft). Altogether the Orbiters returned 2180 high resolution and 882 medium resolution frames. The micrometeoroid experiments recorded 22 impacts showing the average micrometeoroid flux near the Moon was about two orders of magnitude greater than in interplanetary space but slightly less than the near Earth environment. The radiation experiments confirmed that the design of Apollo hardware would protect the astronauts from average and greater-than-average short term exposure to solar particle events. The use of Lunar Orbiters for tracking to evaluate the Manned Space Flight Network tracking stations and Apollo Orbit Determination Program was successful, with three Lunar Orbiters (2,

3, and 5) being tracked simultaneously from August to October 1967. The Lunar Orbiters were all eventually commanded to crash on the Moon before their attitude control gas ran out so they would not present navigational or communications hazards to later Apollo flights. The Lunar Orbiter program was managed by NASA Langley Research Center and involved building and launching 5 spacecraft to the Moon at a total cost of $163 million.

Luna 11. Launch Date/Time: August 24, 1966 at 08:09:00 UT from Tyuratam (Baikonur Cosmodrome), U.S.S.R. Luna 11 was launched towards the Moon from an earth-orbiting platform and entered lunar orbit on August 28, 1966. The objectives of the mission included the study of: (1) lunar gamma- and X-ray emissions in order to determine the Moon's chemical composition; (2) lunar gravitational anomalies; (3) the concentration of meteorite streams near the Moon; and, (4) the intensity of hard corpuscular radiation near the Moon. A total of 137 radio transmissions and 277 orbits of the Moon were completed before the batteries failed on October 1, 1966.

Surveyor 2. Launch Date/Time: September 9, 1966 at 12:32:00 UT from Cape Canaveral, FL, United States, aboard an Atlas-Centaur launch vehicle. This spacecraft was the second of a series designed to achieve a soft landing on the moon and to return lunar surface photography for determining characteristics of the lunar terrain for Apollo lunar landing missions. It was also equipped to return data on radar reflectivity of the lunar surface, bearing strength of the lunar surface, and spacecraft temperatures for use in the analysis of lunar surface temperatures. The target area proposed was within Sinus Medii. During the midcourse maneuver, one vernier engine failed to ignite, resulting in an unbalanced thrust that caused the spacecraft to tumble. Attempts to salvage the mission failed.

Luna 12. Launch Date/Time: October 22, 1966 at 08:38:00 UT from Tyuratam (Baikonur Cosmodrome), U.S.S.R. Luna 12 was launched towards the Moon from an earth-orbiting platform and achieved lunar orbit on October 25, 1966. The spacecraft was equipped with a television system that obtained and transmitted photographs of the lunar surface. The photographs contained 1100 scan lines with a maximum resolution of 14.9 — 19.8 m. Pictures of the lunar surface were returned on October 27, 1966. The number of photographs is not known. Radio transmissions from Luna 12 ceased on January 19, 1967, after 602 lunar orbits and 302 radio transmissions.

Lunar Orbiter 2. Launch Date/Time: November 6, 1966 at 23:21:00 UT from Cape Canaveral, FL, United States, aboard an Atlas-Agena D launch vehicle. The spacecraft was placed in a cislunar trajectory and injected into an elliptical near-equatorial lunar orbit for data acquisition after 92.5 hours flight time. The initial orbit was 196 km × 1,850 km (121.79 miles × 1,149.54 miles) at an inclination of 11.8 degrees. The perilune was lowered to 49.7 km (30.88 miles) five days later after 33 orbits. A failure of the amplifier on the final day of readout, 7 December, resulted in the loss of six photographs. On 8 December 1966 the inclination was altered to 17.5 degrees to provide new data on lunar gravity.

The spacecraft acquired photographic data from November 18 to 25, 1966, and readout occurred through December 7, 1966. A total of 609 high resolution and 208 medium resolution frames were returned, most of excellent quality with resolutions down to 1 meter. These included a spectacular oblique picture of Copernicus crater which was dubbed by the news media as one of the great pictures of the century. See Fig. 17. Accurate data were acquired from all other experiments throughout the mission. Three micrometeorite impacts were recorded. The spacecraft was used for tracking purposes until it impacted the lunar surface on command at 3.0 degrees N latitude, 119.1 degrees E longitude (selenographic coordinates) on October 11, 1967. See **Lunar Orbiter 1** for spacecraft and subsystem design.

Fig. 17. Lunar Orbiter 2 oblique northward view of the interior of the 100 km (62.14 miles) diameter Copernicus crater on the Moon. The central peaks are in the middle of the image, rising about 400 m (1,312 ft) above the crater floor, and stretching for about 15 km (9.32 miles). The northern wall of the crater is in the background. (*Image courtesy of NSSDC*).

Luna 13. Launch Date/Time: December 21, 1966 at 10:19:00 UT from Tyuratam (Baikonur Cosmodrome), U.S.S.R. The Luna 13 spacecraft was launched toward the Moon from an earth-orbiting platform and accomplished a soft landing on December 24, 1966, in the region of Oceanus Procellarum. The petal encasement of the spacecraft was opened, antennas were erected, and radio transmissions to Earth began four minutes after the landing. On December 25 and 26, 1966, the spacecraft television system transmitted panoramas of the nearby lunar landscape at different sun angles. Each panorama required approximately 100 minutes to transmit. The spacecraft was equipped with a mechanical soil-measuring penetrometer, a dynamograph, and a radiation densitometer for obtaining data on the mechanical and physical properties and the cosmic-ray reflectivity of the lunar surface. It is believed that transmissions from the spacecraft ceased before the end of December 1966.

Lunar Orbiter 3. Launch Date/Time: February 5, 1967 at 01:17:00 UT from Cape Canaveral, FL, United States, aboard an Atlas-Agena D launch vehicle. The spacecraft was placed in a cislunar trajectory and injected into an elliptical near-equatorial lunar orbit on 8 February at 21:54 UT.

The orbit was 210.2 km × 1,801.9 km (130.61 miles × 1, 119.65) with an inclination of 20.9 degrees and a period of 3 hours 25 minutes. After four days (25 orbits) of tracking the orbit was changed to 55 km × 1,847 km (34.18 miles × 1,147.67 miles). The spacecraft acquired photographic data from February 15 to 23, 1967, and readout occurred through March 2, 1967. The film advance mechanism showed erratic behavior during this period resulting in a decision to begin readout of the frames earlier than planned. The frames were read out successfully until 4 March when the film advance motor burned out, leaving about 25% of the frames on the takeup reel, unable to be read.

A total of 149 medium resolution and 477 high resolution frames were returned. The frames were of excellent quality with resolution down to 1 meter (3.28 ft). Included was a frame of the Surveyor 1 landing site, permitting identification of the location of the spacecraft on the surface. See Fig. 18. Accurate data were acquired from all other experiments throughout the mission. The spacecraft was used for tracking purposes until it impacted the lunar surface on command at 14.3 degrees N latitude, 97.7 degrees W longitude (selenographic coordinates) on October 9, 1967.

Fig. 18. Lunar Orbiter 3 high resolution image of the Surveyor 1 spacecraft and landing site. This image shows the area within the Flamsteed ring in Oceanus Procellarum on the moon. The spacecraft is the bright spot at the center of the red circle. The shadow of the 1 meter (3.28 ft) wide solar panels can also be seen. The width of the framelets (the spacing between the horizontal lines on the image) is about 220 meters (722 ft). (*Image courtesy of NSSDC*).

Surveyor 3

Launch Date/Time: April 17, 1967 at 07:05:01 UT. Surveyor 3 was similar in design to Surveyors 1 and 2 but had several changes in the payload. It carried a survey television camera, soil mechanics experiments, and devices to measure temperature and radar reflectivity as on the earlier missions, but the TV camera had an extended glare hood. A surface sampler, consisting of a 12 cm (4.72 inches) long by 5 cm (1.97 inch) wide scoop mounted on a 1.5 meter (4.92 ft) pantograph arm, replaced the approach television camera. Two flat auxilliary mirrors were attached to the frame to provide the camera with a view of the ground beneath the engines and one of the footpads. Surveyor 3 had a mass of 1,026 kg (2,262 lbs) at launch and 296 kg (653 lbs) at landing.

Surveyor 3 was launched on an Atlas-Centaur from complex 36B of the Eastern Test Range at Kennedy Space Center. After separation from the Atlas, the Centaur burned for approximately 5 minutes, putting the spacecraft into a 167 km (104 miles) (circular Earth parking orbit. The Centaur was restarted 22 minutes, 9 seconds later, injecting the spacecraft into a selenographic trajectory. A midcourse maneuver 21.9 hrs after liftoff aimed the Surveyor towards the selected landing point. On 20 April at

00:01:06 UT, at 76 km (47.2 miles) altitude traveling at 2626 m/s, the vernier and main retrorocket were ignited by a signal from the altitude marking radar, slowing the spacecraft to 137 m/s at time of retro burnout and ejection. Descent continued under control of the vernier engines and the doppler and altimeter radars.

A few seconds before touchdown the radars lost lock, apparently due to high scintillating reflections from the landing site. The guidance system automatically switched to an inertially controlled mode which prevented vernier engine cutoff. Touchdown on the lunar surface occurred three times because the vernier engines continued to fire during the first two touchdowns causing the spacecraft to lift off the surface. The distance between the first and second touchdown sites was about 20 meters (66 ft) and between the second and third 11 meters (36 ft). Engines were shut off 34 seconds after initial touchdown by an engine cutoff command transmitted from Earth. Initial touchdown occurred at 00:04:17 UT and final touchdown at 00:04:53 UT on April 20, 1967 (19:04:53 April 19, EST) at 3.01 S, 336.66 E (selenocentric). The spacecraft slid about 30 cm (11.8 inches) following final touchdown. Surveyor 3 came to rest on a 14 degree slope inside a subdued 200 meter (656 ft) crater in southeast Oceanus Procellarum roughly 370 km (230 miles) south of Copernicus crater.

Initial photos were received within an hour of landing and the surface sampler was used two days later. Surveyor operated throughout the lunar day until after local sunset on 3 May. The lunar sampler was operated for a total of 18 hours, 22 minutes, digging trenches as deep as 18 cm (7 inches), and the television camera returned 6326 pictures. A large volume of new data on the strength, texture, and structure of lunar material was transmitted by the spacecraft. Images of an eclipse of the Sun by the Earth and related thermal measurements were recorded. The last data were returned on 4 May 1967 at 00:04 UT and Surveyor 3 failed to come back to life following the two week lunar night. Excessive glare in some of the images has been attributed to dust or erosion effects on the mirror due to the extended operation of the engines during touchdown.

On 19 November 1969 the Apollo 12 Lunar Module (LM) landed within about 180 m (591 ft) of the Surveyor 3 spacecraft. Astronauts and Alan Bean visited the spacecraft on their second moonwalk on 20 November, examining Surveyor 3 and its surroundings, taking photographs, and removing about 10 kg (22 lbs) of parts from the spacecraft, including the TV camera, for later examination back on Earth. See Fig. 19.

Fig. 19.   Astronaut Pete Conrad removing parts from Surveyor 3 with Apollo Lunar Lander in background. (*Image courtesy of NSSDC*).

The Surveyor 3 camera is now on display in the Smithsonian National Air and Space Museum in Washington, D.C.

Lunar Orbiter 4. Launch Date/Time: May 4, 1967 at 22:25:00 UT from Cape Canaveral, FL, United States aboard an Atlas-Agena D launch vehicle. Lunar Orbiter 4 was designed to take advantage of the fact that the three previous Lunar Orbiters had completed the required needs for Apollo mapping and site selection. It was given a more general objective, to "perform a broad systematic photographic survey of lunar surface features in order to increase the scientific knowledge of their nature, origin, and processes, and to serve as a basis for selecting sites for more detailed scientific study by subsequent orbital and landing missions. It was also equipped to collect selenodetic, radiation intensity, and micrometeoroid impact data. The spacecraft was placed in a cislunar trajectory and injected into an elliptical near polar high lunar orbit for data acquisition. The orbit was 2,706 km × 6,111 km (1,681 miles × 3,797 miles) with an inclination of 85.5 degrees and a period of 12 hours.

After initial photography on 11 May 1967 problems started occurring with the camera's thermal door, which was not responding well to commands to open and close. Fear that the door could become stuck in the closed position covering the camera lenses led to a decision to leave the door open. This required extra attitude control manuevers on each orbit to prevent light leakage into the camera which would ruin the film. On 13 May it was discovered that light leakage was damaging some of the film, and the door was tested and partially closed. Some fogging of the lens was then suspected due to condensation resulting from the lower temperatures. Changes in the attitude raised the temperature of the camera and generally eliminated the fogging. Continuing problems with the readout drive mechanism starting and stopping beginning on 20 May resulted in a decision to terminate the photographic portion of the mission on 26 May. Despite problems with the readout drive the entire film was read and transmitted. The spacecraft acquired photographic data from May 11 to 26, 1967, and readout occurred through June 1, 1967. The orbit was then lowered to gather orbital data for the upcoming Lunar Orbiter 5 mission.

A total of 419 high resolution and 127 medium resolution frames were acquired covering 99% of the Moon's near side at resolutions from 58 meters (190 ft) to 134 meters (440 ft). See Fig. 20. Accurate data were acquired from all other experiments throughout the mission. Radiation data showed increased dosages due to solar particle events producing low energy protons. The spacecraft was used for tracking purposes until it impacted the lunar surface due to the natural decay of the orbit no later than October 31, 1967, between 22 — 30 degrees W longitude. See **Lunar Orbiter 1** for spacecraft design and subsystems.

Surveyor 4. Launch Date/Time: July 14, 1967 at 11:53:00 UT from Cape Canaveral, FL, United States aboard an Atlas-Centaur launch vehicle. This spacecraft was the fourth in a series designed to achieve a soft landing on the moon and to return photography of the lunar surface for determining characteristics of the lunar terrain for Apollo lunar landing missions. After a flawless flight to the moon, radio signals from the spacecraft ceased during the terminal-descent phase, approximately 2.5 min before touchdown. Contact with the spacecraft was never reestablished, and the mission was unsuccessful.

Lunar Orbiter 5. Launch Date/Time: August 1, 1967 at 22:32:00 UT from Cape Canaveral, FL, United States, aboard an Atlas-Agena D launch vehicle. Lunar Orbiter 5, the last of the Lunar Orbiter series, was designed to take additional Apollo and Surveyor landing site photography and to take broad survey images of unphotographed parts of the Moon's far side. It was also equipped to collect selenodetic, radiation intensity, and micrometeoroid impact data and was used to evaluate the Manned Space Flight Network tracking stations and Apollo Orbit Determination Program. The spacecraft was placed in a cislunar trajectory and on 5 August 1967 was injected into an elliptical near polar lunar orbit 194.5 km × 6,023 km (120.86 miles × 3,742.52 miles) with an inclination of 85 degrees and a period of 8 hours 30 minutes. On 7 August the perilune was lowered to 100 km (62.14 miles) and on 9 August the orbit was lowered to a 99 km × 1,499 km (61.52 miles × 932.44 miles), 3 hour 11 minute period. The photographic portion of the mission ended on 18 August. The spacecraft acquired photographic data from August 6 to 18, 1967, and readout occurred until August 27, 1967. A total of 633 high resolution and 211 medium resolution frames at resolution down to 2 meters were acquired, bringing the cumulative photographic coverage by the 5 Lunar Orbiters to 99% of the Moon's surface. See Fig. 21. Accurate data were acquired from all other experiments throughout the mission. The spacecraft was tracked until it impacted the lunar surface on command at 2.79 degrees S latitude, 83 degrees W longitude (selenographic coordinates) on January 31, 1968. See **Lunar Orbiter 1** for spacecraft design and subsystems.

Fig. 20.   Lunar Orbiter 4 global image showing the Moon's southern hemisphere near the easternlimb. The dark area at top center is Mare Smythii. Below this and to the left is the 200 km (124.3 miles) diameter Humboldt crater. The dark crater at the middle right edge is Tsiolkovsky at 20 S, 130 E. Schrodinger crater at 75 S, 135 E, is the large crater at the bottom, with the Vallis Schrodinger extending upward. The south pole is just to the left of Schrodinger. (*Image courtesy of NSSDC*).

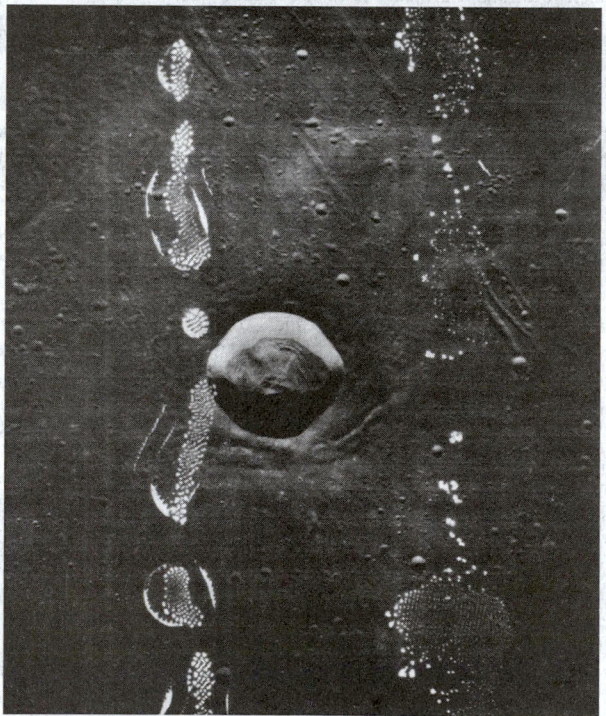

Fig. 21.   Lunar Orbiter 5 image of Dawes crater on the Moon. Dawes crater is 18 km in diameter, and appears to be a relatively young lunar crater. The crater is about 600 m deep. North is up. (*Image courtesy of NSSDC*).

Surveyor 5.   Launch Date/Time: September 8, 1967 at 07:57:01 UT from Cape Canaveral, FL, United States. Surveyor 5 was the third spacecraft in the Surveyor series to achieve a successful lunar soft landing and the first mission to obtain in-situ compositional data on the Moon. The specific

objectives for this mission were to perform a soft landing on the Moon in Mare Tranquillitatis and obtain postlanding television pictures of the lunar surface. The secondary objectives were to conduct a vernier engine erosion experiment, determine the relative abundances of the chemical elements in the lunar soil by operation of the alpha-scattering instrument, obtain touchdown dynamics data, and obtain thermal and radar reflectivity data. See Surveyor 1 for spacecraft design and subsystems.

Surveyor 5 was launched from the Eastern Test range launch complex 36B at Cape Kennedy on an Atlas-Centaur rocket. The Centaur placed the spacecraft into an Earth parking orbit and then restarted 6.7 minutes later and injected Surveyor 5 into a lunar transfer trajectory. A midcourse trajectory correction involving a 14.29 second firing of the verier engines was performed at 1:45 UT on 9 September. Immediately following the maneuver the spacecraft began losing helium pressure. It was concluded that the helium pressure valve had not reseated tightly and the helium was leaking into the propellant tanks, causing an overpressure which opened the relief valves, discharging the helium. A new emergency landing plan was adopted. Early vernier engine firings were made while there was still helium to slow the spacecraft, reduce its mass, and leave more free volume in the propellant tanks for the helium.

The new descent profile worked flawlessly and Surveyor 5 touched down on the lunar surface on 11 September 1967 at 00:46:44 UT (8:46:44 p.m. EDT 10 September) at 1.41 N, 23.18 E (selenographic coordinates) on a 20 degree slope of a 9 × 12 meter (29.5 × 39.4 feet) rimless crater in southwest Mare Tranquillitatis. See Fig. 22. Touchdown was 29 km (18 miles) from the original target. All experiments were performed successfully. Surveyor 5 returned 18,006 television pictures during its first lunar day. The alpha-scattering instrument was deployed and performed the first in-situ analysis of an extraterrestrial body, returning 83 hours of data on lunar soil composition during the first lunar day, A vernier engine erosion experiment was conducted on 13 September, about 53 hours after landing, consisting of a firing of the vernier engines for 0.55 seconds while the spacecraft sat on the ground to examine the effects of the engines on the surface. The spacecraft shut down from September 24 to October 15, 1967 over the first lunar night. An additional 1048 pictures and 22 hours of alpha-scattering data were received during the second lunar day. On 18 October Surveyor 5 acquired thermal data during a total eclipse of the Sun. Transmissions for the second day were received until November 1, 1967, when shutdown for the second lunar night occurred about 200 hours after sunset. Transmissions were resumed on the third and fourth lunar days, with the final transmission occurring at 04:30 UT on December 17, 1967. Pictures were transmitted during the first, second, and fourth lunar days. A total of 19,118 pictures were transmitted.

Fig. 22.   Surveyor 5 image of the footpad resting in the lunar soil. The trench at right was formed by the footpad sliding during landing. Surveyor 5 landed on the Moon on 11 September 1967 at 1.41 N, 23.18 E in Mare Tranquillitatis. The spacecraft landed on the inside edge of a small rimless crater at an angle of about 20 degrees, explaining the sliding. The footpad is about half a meter in diameter. (*Image courtesy of NSSDC*).

Alpha-scattering results indicated soil composition, resembling Earth basaltic rock, of 53% to 63% oxygen, 15.5% to 21.5% silicon, 10% to 16% sulfur, iron, cobalt, and nickel; 4.5% to 8.5% aluminum, and small quantities of magnesium, carbon, and sodium. The quantity of material adhering to the magnet was consistent with a mixture of pulverized basalt and 10% to 12% magnetite with no more than 1% metallic iron. The vernier engine experiment produced minor but observable erosion of the surface.

Surveyor 6. Launch Date/Time: November 7, 1967 at 07:39:00 UT. Surveyor 6 was the fourth of the Surveyor series to successfully achieve a soft landing on the Moon. The specific primary objectives for this mission were to perform a soft landing on the Moon in the Sinus Medii region and obtain postlanding television pictures of the lunar surface. The secondary objectives were to determine the relative abundance of the chemical elements in the lunar soil by operation of the alpha-scattering instrument, obtain touchdown dynamics data, obtain thermal and radar reflectivity data, and conduct a vernier-engine erosion experiment. See Surveyor 1 for spacecraft design and subsystems.

Surveyor 6 launched from complex 36B of the Eastern Test range at Cape Kennedy. The Atlas-Centaur booster put the spacecraft into an initial Earth parking orbit from which it was injected into a lunar-transfer trajectory at 8:03:30 UT. A midcourse correction manuever was performed at 2:20:00 UT on 8 November 1967. Surveyor 6 touched down on the lunar surface on 10 November 1967 at 01:01:06 UT (8:01:06 EST 9 November) in Sinus Medii, a flat, heavily cratered mare region, at 0.49 N, 358.60 E (selenographic), the center of the Moon's visible hemisphere.

On 17 November at 10:32 UT the vernier engines were fired for 2.5 seconds, causing Surveyor to lift off the lunar surface 3 to 4 meters (10 to 13 feet) and land about 2.4 meters (7.87 ft) west of its original position. This lunar "hop" represented the first powered takeoff from the lunar surface and furnished new information on the effects of firing rocket engines on the Moon, allowed viewing of the original landing site, and provided a baseline for stereoscopic viewing and photogrammetric mapping of the surrounding terrain. The mission transmitted images until a few hours after sunset on 24 November, returning a total of 29,952 images. The alpha-scattering experiment acquired 30 hours of data on the surface material.

The spacecraft was placed into hibernation for the lunar night on 26 November. Contact with the spacecraft was resumed on 14 December for a short period, but no useful data were returned and the last transmission was received at 19:14 UT on 14 December 1967. The results of the experiments showed that the surface had a basaltic composition, similar to that found at the Surveyor 5 landing site. Engineering and soil mechanics data indicated the bearing strength of the surface was more than adequate to support human landings. This spacecraft accomplished all planned objectives.

Lunar Missions 1968–1989

Surveyor 7. Launch Date/Time: January 7, 1968 at 06:30:00 UT. Surveyor 7 was the fifth and final spacecraft of the Surveyor series to achieve a lunar soft landing. The specific objectives for this mission were to: (1) perform a lunar soft landing (in a highland area well removed from the maria to provide a type of terrain photography and lunar sample significantly different from those of other Surveyor missions); (2) obtain postlanding TV pictures; (3) determine the relative abundances of chemical elements; (4) manipulate the lunar material; (5) obtain touchdown dynamics data; and, (6) obtain thermal and radar reflectivity data. Surveyor 7 was the only Surveyor craft to land in the lunar highland region. See Surveyor 1 for spacecraft design and subsystems.

Surveyor 7 was similar in design to Surveyor 6, but the payload was the most extensive flown during the Surveyor program. It carried a television camera with polarizing filters, an alpha-scattering instrument, a surface sampler similar to that flown on Surveyor 3, bar magnets on two footpads, two horseshoe magnets on the surface scoop, and auxiliary mirrors. Of the auxiliary mirrors, three were used to observe areas below the spacecraft, one to provide stereoscopic views of the surface sampler area, and seven to show lunar material deposited on the spacecraft. It also carried over 100 items to monitor engineering aspects of spacecraft performance. Surveyor 7 had a mass of 1,039 kg (2,291 lbs) at launch and 306 kg (674.61 lbs) at landing.

Surveyor 7 was launched on an Atlas-Centaur from launch complex 36A of the Eastern Test Range at Cape Kennedy. The spacecraft was put into an Earth parking orbit and then transferred to a lunar trajectory by a second burn of the Centaur upper stage. Surveyor 7 separated from the Centaur at 07:05:16 UT. A midcourse maneuver was performed at 23:30:10 UT on 7 January 1968. Touchdown occurred at 01:05:36.3 UT on 10 January 1968 (8:05:36 p.m. EST 9 January) at 40.86 S, 348.53 E (selenographic) on an ejecta blanket about 29 miles north of the rim of Tycho crater in the lunar highlands. See Fig. 23.

Fig. 23.   Surveyor 7 mosaic of the rim area of Tycho from the highland region north of the crater. Surveyor 7 landed 10 January 1969 at 40.88 S, 11.45 W and took about 21,000 photos over a month, some of which were used to make up this mosaic. The block in the foreground is about half a meter across and the crater is about 1.5 meters (4.9 ft) in diameter. The hills on the horizon are about 13 km (8 miles) away. (*Image courtesy of NSSDC*).

Science operations commenced shortly after landing. The TV camera returned 20,993 pictures on the first lunar day. The alpha-scattering instrument failed to deploy fully, but the surface sampler was used to force it to the ground. The sampler was later used to set the alpha-scattering instrument on a rock and then into a trench it had dug. Approximately 66 hours of alpha-scattering data were obtained during the first lunar day on the three sites. Operations were continued after sunset and included pictures of the Earth, stars, and the solar corona. Operation was terminated at 14:12 UT on 26 January, 80 hours after sunset. Second lunar day operations began at 19:01 UT on 12 February 1968 and included an additional 45 pictures for a total of 21,038 and 34 hours of alpha-scattering data from inside the trench. Operations were terminated on 21 February at 12:24 UT (7:24 a.m. EST). The lunar surface sampler operated flawlessly for a total of 36 hours, 21 minutes, digging trenches and moving and manipulating four rocks.

Results were generally consistent with earlier missions except that the chemical analysis of the highland crust showed it to be poorer in iron group elements than the previous samples, all from the lunar maria. The magnet experiments showed the presence of magnetic constituents in amounts comparable to those at the Surveyor 5 and 6 sites. The lander also successfully detected laser beams transmitted from Earth. The mission objectives were fully satisfied by the spacecraft operations.

Luna 14. Launch Date/Time: April 7, 1968 at 10:09:32 UT from Tyuratam (Baikonur Cosmodrome), U.S.S.R. The Luna 14 spacecraft entered a 160 × 870 km (99.4 × 540.6 miles) lunar orbit with an inclination of 42 degrees at 19:25 UT on April 10, 1968. The spacecraft is believed to have been similar to Luna 12 and the instrumentation was similar to that carried by Luna 10. It provided data for studies of the interaction of the earth and lunar masses, the lunar gravitational field, the propagation and stability of radio communications to the spacecraft at different orbital positions, solar charged particles and cosmic rays, and the motion of the Moon. This flight was the final flight of the second generation of the Luna series.

Zond 5. Launch Date/Time: September 14, 1968 at 21:42:11 UT. Zond 5 was launched from a Tyazheliy Sputnik (68-076B) in earth parking orbit to make scientific studies during a lunar flyby and to return to Earth. En route to the Moon the main stellar attitude control optical surface became contaminated and was rendered unusable. Backup sensors were used to guide the spacecraft. On September 18, 1968, the spacecraft flew around the Moon. The closest distance was 1,950 km (1,211.7

miles). High quality photographs of the Earth were taken at a distance of 90,000 km (55,923 miles). A biological payload of turtles, wine flies, meal worms, plants, seeds, bacteria, and other living matter was included in the flight. Additionally, according to the Russian Academy of Sciences, in the pilot's seat was a 175 cm (5 ft 9in) tall, 70 kg (154 lb) mannequin containing radiation detectors. Returning to Earth another attitude control sensor failed, making the planned guided entry impossible and forcing the spacecraft controllers to use a direct ballistic entry. On September 21, 1968, the reentry capsule entered the Earth's atmosphere, braked aerodynamically, and deployed parachutes at 7 km (4.35 miles). The capsule splashed down in the backup area in the Indian Ocean at 32.63 degrees S, 65.55 degrees E and was successfully recovered, safely returning the biological payload. It was announced that the turtles (steppe tortoises) had lost about 10% of their body weight but remained active and showed no loss of appetite. The spacecraft was planned as a precursor to crewed lunar spacecraft. It represented the first successful Soviet circumlunar mission.

Zond 6. Launch Date/Time: November 10, 1968 at 19:11:31 UT from Tyuratam (Baikonur Cosmodrome), U.S.S.R. Zond 6 was launched on a lunar flyby mission from a parent satellite (68-101B) in earth parking orbit. The spacecraft, which carried scientific probes including cosmic-ray and micrometeoroid detectors, photography equipment, and a biological payload, was a precursor to manned spaceflight. Zond 6 flew around the moon on November 14, 1968, at a minimum distance of 2,420 km (1,504 miles). Photographs of the lunar near and farside were obtained with panchromatic film. Each photo was 12.70 by 17.78 cm (5 by 7 inches). Some of the views allowed for stereo pictures. The photos were taken from distances of approximately 11,000 km (6,835 miles) and 3,300 km (2,051 miles). Controlled reentry of the spacecraft occurred on November 17, 1968, and Zond 6 landed in a predetermined region of the Soviet Union.

Apollo 8. This spacecraft was the first of the Apollo series to successfully orbit the moon, and the first manned spacecraft to leave Earth's gravity and reach the Moon. The mission achieved operational experience and tested the Apollo command module systems, including communications, tracking, and life-support, in cis-lunar space and lunar orbit, and allowed evaluation of crew performance on a lunar orbiting mission. The crew photographed the lunar surface, both farside and nearside, obtaining information on topography and landmarks as well as other scientific information necessary for future Apollo landings. See Fig. 24. Additionally, six live television transmission sessions were done by the crew during the mission, including the famous Christmas Eve broadcast in which the astronauts read from the book of Genesis. All systems operated within allowable parameters and all objectives of the mission were achieved. The flight carried a three man crew: Commander Frank Borman, Command Module Pilot James A. Lovell, Jr., and Lunar Module Pilot William A. Anders.

The Apollo 8 spacecraft consisted of a command module similar to Apollo 7 except that the forward pressure and ablative hatches were replaced by a combined forward hatch, which would be used for transfer to the Lunar Module on later missions. The spacecraft mass of 28,817 kg (63,531 lbs) is the mass of the CSM including propellants and expendables. A Lunar Module was not used on the Apollo 8 mission but a Lunar Module Test Article which was equivalent in mass (9027 kg) to a Lunar Module was mounted in the spacecraft/launch vehicle adapter as ballast for mass loading purposes.

The spacecraft was launched on December 21, 1968 at 12:51:00 UT (7:51 a.m. EST), and was placed in a 190.6 km × 183.2 km (118.4 miles × 113.8 miles) Earth parking orbit with a period of 88.2 minutes and an inclination of 32.51 degrees. At 15:41:37 UT a third-stage burn injected the Apollo spacecraft into translunar trajectory. Orbit insertion took place on 24 December at 09:59:20 UT into an elliptical 310.6 km by 111.2 km lunar orbit. Two orbits later a second burn placed Apollo 8 into a near-circular 110.4 by 112.3 km (68.6 by 69.8 miles) orbit for eight orbits. The transearth injection burn took place on 25 December at 06:10:16 UT after a total of 10 lunar orbits.

Apollo 8 splashed down in the Pacific Ocean on 27 December 1968 at 15:51:42 UT (10:51:42 a.m. EST) after a mission elapsed time of 147 hrs, 0 mins, 42 secs. The splashdown point was 8 deg 7.5 min N, 165 deg 1.2 min W, 1,000 miles SSW of Hawaii and 5 km (3 mi) from the recovery ship USS Yorktown. The Apollo 8 Command Module is on display at the Chicago Museum of Science and Industry, Chicago, Illinois.

Fig. 24.   The rising Earth is about five degrees above the lunar horizon in this telephoto view taken from the Apollo 8 spacecraft near 110 degrees east longitude. The horizon, about 570 km (250 statute miles) from the spacecraft, is near the eastern limb of the Moon as viewed from the Earth. This is one of the more famous images of the Earth from the Apollo program, taken by the Apollo 8 astronauts as they became the first humans to circumnavigate the Moon. (*Image courtesy of NSSDC*).

See also **Spaceflight U.S. Manned (Mercury, Gemini, Apollo, Skylab, Apollo-Soyuz Project and Space Shuttle Program)**.

The Apollo program included a large number of uncrewed test missions and 12 crewed missions: three Earth orbiting missions (Apollo 7, 9 and Apollo-Soyuz), two lunar orbiting missions (Apollo 8 and 10), a lunar swingby (Apollo 13), and six Moon landing missions (Apollo 11, 12, 14, 15, 16, and 17). Two astronauts from each of these six missions walked on the Moon (Neil Armstrong, Edwin Aldrin, Charles Conrad, Alan Bean, Alan Shepard, Edgar Mitchell, David Scott, James Irwin, John Young, Charles Duke, Gene Cernan, and Harrison Schmitt), the only humans to have set foot on another solar system body. Total funding for the Apollo program was approximately $20,443,600,000.

Apollo 10. The Apollo 10 spacecraft was launched from Cape Kennedy at 12:49 p.m., EDT, on May 18, 1969. This spacecraft was the second Apollo mission to orbit the Moon, and the first to travel to the Moon with the full Apollo spacecraft, consisting of the Command and Service Module (CSM-106, "Charlie Brown") and the Lunar Module (LM-4, "Snoopy"). The spacecraft mass of 28,834 kg (63,568 lbs) is the mass of the CSM including propellants and expendables. The LM mass including propellants was 13,941 kg (30735 lbs). The primary objectives of the mission were to demonstrate crew, space vehicle, and mission support facilities during a manned lunar mission and to evaluate LM performance in cislunar and lunar environment. The mission was a full "dry run" for the Apollo 11 mission, in which all operations except the actual lunar landing were performed. The flight carried a three man crew: Commander Thomas P. Stafford, Command Module (CM) Pilot John W. Young, and Lunar Module (LM) Pilot Eugene A. Cernan.

After launch, the spacecraft was inserted into a 189.9 km × 184.4 km (118 miles × 115 miles) Earth parking orbit at 17:00:54 UT, followed by translunar injection after 1 1/2 orbits at 19:28:21 UT. The CSM separated from the Saturn V 3rd stage (S-IVB) at 19:51:42 UT, transposed, and docked with the LM at 20:06:37. After a three day cruise, Apollo 10 entered an initial 315.5 km × 110.4 km (196 miles × 69 miles) lunar orbit on 21 May 1969 at 20:44:54 UT, using a 356 sec. SPS burn. A second SPS burn lasting 19.3 seconds circularized the orbit to 113.9 km × 109.1 km (71 miles × 68 miles).

On 22 May Stafford and Cernan entered the LM and fired the SM reaction control thrusters to separate the LM from the CSM at 19:36:17 UT.

The LM was put into an orbit to allow low altitude passes over the lunar surface, the closest approach bringing it to within 14 km (8.7 miles) of the Moon. All systems on the LM were tested during the separation including communications, propulsion, attitude control, and radar. Numerous close-up photographs of the Moon's surface, in particular the planned Apollo landing sites, were taken. The LM descent stage was jettisoned into lunar orbit. The LM and CSM rendezvous and redocking occurred 8 hours after separation at 03:22 UT on 23 May.

Later on May 23 the LM ascent stage was jettisoned into solar orbit, and on 24 May at 10:25:29 UT after 31 lunar orbits the CSM rockets fired for trans-earth injection. CM-SM separation took place on 26 May at 16:22:26 UT and Apollo 10 splashed down in the Pacific Ocean on 26 May 1969 at 16:52:23 UT (12:52:23 p.m. EDT) after a mission elapsed time of 192 hrs, 3 mins, 23 secs. The splashdown point was 15 deg 2 min S, 164 deg 39 min W, 400 miles east of American Samoa and 5.5 km (3.4 mi) from the recovery ship USS Princeton.

All systems on both spacecraft functioned nominally, the only exception being an anomaly in the automatic abort guidance system aboard the LM. In addition to extensive photography of the lunar surface from both the LM and CSM, television images were taken and transmitted to Earth. The Apollo 10 Command Module "Charlie Brown" is on display at the Science Museum, London, England.

Luna 15. Launch Date/Time: July 13, 1969 at 02:54:42 UT from Tyuratam (Baikonur Cosmodrome), U.S.S.R. Luna 15 was placed in an intermediate earth orbit after launch and was then sent toward the Moon. The spacecraft was capable of studying circumlunar space, the lunar gravitational field, and the chemical composition of lunar rocks. It was also capable of providing lunar surface photography. After completing 86 communications sessions and 52 orbits of the Moon at various inclinations and altitudes, the spacecraft impacted the lunar surface on July 21, 1969.

Apollo 11. Launch Date/Time: July 16, 1969; 09:32:00 am EST. Launch Complex 39-A Kennedy Space Center, FL. Apollo 11 was the first mission in which humans walked on the lunar surface and returned to Earth. On 20 July 1969 two astronauts (Apollo 11 Commander Neil A. Armstrong and LM pilot Edwin E. "Buzz" Aldrin Jr.) landed in Mare Tranquilitatis (the Sea of Tranquility) on the Moon in the Lunar Module (LM) "Eagle" while the Command and Service Module (CSM) (with CM pilot Michael Collins) continued in lunar orbit. During their stay on the Moon, the astronauts set up scientific experiments, took photographs, and collected lunar samples. The LM took off from the Moon on 21 July and the astronauts returned to Earth on 24 July.

After launch on Saturn V SA-504, Apollo 11 entered Earth orbit. After 1 1/2 Earth orbits, the S-IVB stage was re-ignited at 16:16:16 UT for a translunar injection burn of 5 minutes, 48 seconds putting the spacecraft on course for the Moon. The CSM separated from the S-IVB stage containing the LM 33 minutes later, turned around and docked with the LM at 16:56:03 UT. About an hour and 15 minutes later the S-IVB stage was injected into heliocentric orbit. During translunar coast a color TV transmission was made from Apollo 11 and on 17 July a 3-second mid-course correction burn of the main engine was performed. Lunar orbit insertion was achieved on 19 July at 17:21:50 UT by a retrograde firing of the main engine for 357.5 seconds while the spacecraft was behind the Moon and out of contact with Earth. A later 17 second burn circularized the orbit. On 20 July Armstrong and Aldrin entered the LM for final checkout. At 18:11:53 the LM and CSM separated. After a visual inspection by Collins, the LM descent engine fired for 30 seconds at 19:08 UT, putting the craft into a descent orbit with a closest approach 14.5 km (9 miles) above the Moon's surface. At 20:05 the LM descent engine fired for 756.3 seconds and descent to the lunar surface began.

The LM landed at 20:17:40 UT (4:17:40 p.m. EDT) in Mare Tranquilitatis (the Sea of Tranquility), (Fig. 25) Armstrong reporting, "Houston, Tranquility Base here—the Eagle has landed." Armstrong stepped onto the lunar surface at 02:56:15 UT on 21 July (10:56:15 p.m. July 20 EDT) stating, "That's one small step for man, one giant leap for mankind". He then collected a small contingency sample of lunar material. Aldrin followed 19 minutes later, calling the lunar surface "Magnificent desolation". The astronauts then unveiled the plaque mounted on a strut behind the ladder and read the inscription aloud: "Here men from the planet Earth first set foot on the Moon July 1969, A.D. We came in peace for all mankind." They put up an American flag and talked to President Nixon by radiotelephone.

Fig. 25.   This picture was taken from the Apollo 11 LM window during the descent to the lunar surface shortly before landing. It shows the area of the Moon near the touchdown point in the Sea of Tranquility. Landing occurred on July 20 at 20:18 UT (4:18 p.m. EDT) at 00.57 S, 23.49 E. The view is to the north. (*Image courtesy of NSSDC*).

The astronauts deployed the EASEP (Fig. 26) and other instruments, took photographs, and collected 21.7 kg (48 lbs) of lunar rock and soil. The astronauts traversed a total distance of about 250 meters (820 ft), both ranging up to about 100 meters (328 ft) from the LM. They took two core tube samples of lunar soil and packed these along with the lunar samples and the solar wind experiment into the sample boxes. Aldrin returned to the LM first, after 1 hour 41 minutes on the lunar surface, Armstrong followed about 12 minutes later, at 05:09:32 UT, after transferring the sample boxes up to Aldrin and placing a packet of memorial items on the ground. The EVA ended at 5:11:13 UT when the astronauts returned to the LM and closed the hatch.

The LM lifted off from the Moon at 17:54:01 UT on 21 July after 21 hours, 36 minutes on the lunar surface. After docking with the CSM at 21:34:00 UT, the LM was jettisoned into lunar orbit at 00:01:01 UT on 22 July. Transearth injection began at 04:54:42 UT on 22 July with a 2 1/2 minute firing of the CSM main engine. A mid-course correction was made later on 22 July. The CM "Columbia" separated from the SM at 16:21:13 UT on 24 July. Apollo 11 splashed down in the Pacific Ocean on 24 July 1969 at 16:50:35 UT (12:50:35 p.m. EDT) after a mission elapsed time of 195 hrs, 18 mins, 35 secs. The splashdown point was 13 deg 19 min N, 169 deg 9 min W, 400 miles SSW of Wake Island and 24 km (15 mi) from the recovery ship USS Hornet.

The performance of the spacecraft was excellent throughout the mission. The primary mission goal of landing astronauts on the Moon and returning them to Earth was achieved. Armstrong was a civilian on his second spaceflight (he'd previously flown on Gemini 8), Aldrin was a USAF Colonel on his second spaceflight (Gemini 12), Collins was a USAF Lt. Colonel also on his second flight (Gemini 10). The backup crew for this mission was Jim Lovell, Fred Haise, and William Anders. The Apollo 11 Command Module "Columbia" is on display at the National Air and Space Museum in Washington, D.C.

Zond 7. Launch Date/Time: August 7, 1969 at 23:48:06 UT from Tyuratam (Baikonur Cosmodrome), U.S.S.R. Zond 7 was launched towards the moon from a mother spacecraft (69-067B) on a mission of further studies of the moon and circumlunar space, to obtain color photography of the earth and the moon from varying distances, and to flight test the spacecraft systems. Earth photos were obtained on August 9, 1969. On August 11, 1969, the spacecraft flew past the moon at a distance of

Fig. 26. Apollo 11 astronaut Edwin Aldrin deploying the Early Apollo Science Experiments Package (EASEP). The package consists of a seismometer on the ground in front of Aldrin powered by a solar panel to the left. The white rod jutting out of the top of the instrument is the antenna to send the results back to Earth. Partially hidden behind Aldrin's right hand is the laser ranging retroreflector. (*Image courtesy of NSSDC*).

1,984.6 km (1,233 miles) and conducted two picture taking sessions. Zond 7 reentered the earth's atmosphere on August 14, 1969, and achieved a soft landing in a preset region south of Kustanai.

Apollo 12. Launch Date/Time: November 14, 1969 at 16:22:00 UT (11:22:00 a.m. EST). Apollo 12 was the second mission in which humans walked on the lunar surface and returned to Earth. On 19 November 1969 two astronauts (Apollo 12 Commander Charles P. "Pete" Conrad and LM Pilot Alan L. Bean) landed in Oceanus Procellarum (Ocean of Storms) on the Moon in the Lunar Module (LM) while the Command and Service Module (CSM) (with CM pilot Richard F. Gordon) continued in lunar orbit. During their stay on the Moon, the astronauts set up scientific experiments, took photographs, examined the nearby Surveyor 3 spacecraft which had landed on the Moon 2 1/2 years earlier and removed pieces for later examination on Earth, and collected lunar samples on two moonwalk EVA's. The LM took off from the Moon on 20 November and the astronauts returned to Earth on 24 November.

Launch took place under cloudy, rain-swept skies on Saturn V SA-507 from Pad 39A at Kennedy Space Center. The spacecraft was struck by lightning 36 seconds after launch and again 52 seconds after launch, which momentarily shut off electrical power and cut out telemetry contact. Power was automatically switched to battery backup while the crew resored the primary power system. There were no further problems with the power system and the spacecraft entered planned Earth parking orbit at 11 minutes 44 seconds after liftoff. After 1 1/2 orbits the S-IVB stage was re-ignited at 19:15:14 UT for a translunar injection burn of 5 min. 45 sec. putting the spacecraft on course for the Moon. The CSM separated from the S-IVB stage containing the LM 25 minutes later, turned around and docked with the LM at 19:48:53 UT. The S-IVB stage was then jettisoned into Earth orbit instead of planned heliocentric orbit due to error in the instrument unit. During lunar coast, the LM was checked out to ensure no electrical damage had been caused by the lightning. A midcourse correction was made on 16 November at 02:15 UT. A six minute SPS burn on 18 November at 03:47:23 UT put the Apollo 12 into lunar orbit. Two orbits later a second burn circularized the orbit. Conrad and Bean entered the LM "Intrepid" and it separated from the CSM at 04:16:03 UT on 19 November. The LM descent engine fired for 29 seconds at 05:47 UT, and the LM landed in Oceanus Procellarum near the rim of Surveyor crater at 06:54:35 UT. Conrad and Bean performed two surface EVA's, one on 19

November and one on 20 November, during which an Apollo lunar surface experiments package (ALSEP) was placed on the lunar surface, 34.4 kg (76 lbs) of samples of the lunar terrain were acquired, various photographs were exposed by the astronauts during lunar surface activities, and parts were taken from the Surveyor 3 spacecraft for examination.

The LM lifted off from the Moon on 20 November at 14:25:47 UT after 31 hours 31 minutes on the lunar surface. After docking with the CSM at 17:58:22 UT, the LM was jettisoned at 20:21:30 and intentionally crashed into the Moon creating the first recorded artificial moonquake. Transearth injection began at 20:49:16 UT on 21 November with a firing of the CSM main engine. A mid-course correction was made on 22 November. The CM "Yankee Clipper" separated from the SM on 24 November at 20:29:21. Apollo 12 splashed down in the Pacific Ocean on 24 November 1969 at 20:58:24 UT (3:58:24 p.m. EST) after a mission elapsed time of 244 hrs, 36 mins, 24 secs. The splashdown point was 15 deg 47 min S, 165 deg 9 min W, near American Samoa and 6.9 km (4.3 mi) from the recovery ship USS Hornet.

Performance of the spacecraft, the first of the Apollo H-series missions, was very good for all aspects of the mission. The primary mission goals of an extensive series of lunar exploration tasks, deployment of the ALSEP, and demonstration of the ability to remain and work on the surface of the Moon for an extended period were achieved. Conrad was a Navy Commander on his third spaceflight (previously on Gemini's 5 and 11, later to fly on Skylab 2), Bean was a Navy Lt. Commander on his first flight (he later flew on Skylab 3), and Gordon was a Navy Commander on his second flight (Gemini 11). The backup crew for this mission was David Scott, Alfred Worden, and James Irwin. The Apollo 12 Command Module "Yankee Clipper" is on display at the Virginia Air and Space Center in Hampton, Virginia. The returned Surveyor 3 camera is on display at the Smithsonian Air and Space Museum in Washington, D.C.

Apollo 13. Launch Date/Time: April 11, 1970 at 19:13:00 UT (02:13:00 p.m. EST). Apollo 13 was intended to be the third mission to carry humans to the surface of the Moon, but an explosion of one of the oxygen tanks and resulting damage to other systems resulted in the mission being aborted before the planned lunar landing could take place. The crew, commander James A. Lovell, Jr., command module pilot John L. Swigert, Jr., and lunar module pilot Fred W. Haise Jr., were returned safely to Earth on 17 April 1970.

Apollo 13 was launched on Saturn V SA-508 from pad 39A at Kennedy Space Center. During second stage boost the center engine of the S-II stage cut off 132 seconds early, causing the remaining four engines to burn 34 seconds longer than normal. The velocity after S-II burn was still lower than planned by 68 m/sec, so the S-IVB orbital insertion burn at 19:25:40 was 9 seconds longer than planned. Translunar injection took place at 21:54:47 UT, CSM/S-IVB separation at 22:19:39 UT, and CSM-LM docking at 22:32:09 UT. The S-IVB auxilliary propulsive system burned at 01:13 UT on 12 April for 217 seconds to put the S-IVB into a lunar impact trajectory. (It impacted the lunar surface on 14 April at 01:09:41.0 at 2.75 S, 27.86 W with a velocity of 2.58 km/s at a 76 degrees angle from horizontal.) A 3.4 second mid-course correction was made at 01:27 UT on 13 April.

A television broadcast was made from Apollo 13 from 02:24 UT to 02:59 UT on 14 April and a few minutes later, at 03:06:18 UT Jack Swigert turned the fans on to stir oxygen tanks 1 and 2 in the service module. The Accident Review Board concluded that wires which had been damaged during pre-flight testing in oxygen tank no. 2 shorted and the teflon insulation caught fire. The fire spread within the tank, raising the pressure until at 3:07:53 UT on 14 April (10:07:53 EST 13 April; 55:54:53 mission elapsed time) oxygen tank no. 2 exploded, damaging oxygen tank no. 1 and the interior of the service module and blowing off the bay no. 4 cover. With the oxygen stores depleted, the command module was unusable, the mission had to be aborted, and the crew transferred to the lunar module "Aquarius" and powered down the command module "Odyssey".

The service module, which had been kept attached to the command module to protect the heat shield, was jettisoned on 17 April at 13:15:06 UT and the crew took photographs of the damage. The command module was powered up and lunar module was jettisoned at 16:43:02 UT. Any parts of the lunar module which survived atmospheric re-entry, including the SNAP-27 generator, planned to power the ALSEP apparatus on the lunar surface and containing 3.9 kg (8.6 lbs) of plutonium, fell into the Pacific Ocean northeast of New Zealand. Apollo 13 splashed down in the Pacific Ocean on 17 April 1970 at 18:07:41 UT (1:07:41 p.m. EST) after a mission elapsed time of 142 hrs, 54 mins, 41 secs. The splashdown point was 21 deg 38 min S, 165 deg 22 min W, SE of American Samoa and 6.5 km (4 mi) from the recovery ship USS Iwo Jima.

Although the planned mission objectives were not realized, a limited amount of photographic data was obtained. Lovell was a Navy captain on his fourth spaceflight (he'd flown previously on Gemini 7, Gemini 12, and Apollo 8), Haise and Swigert were both civilians on their first spaceflights. The backup crew was John Young, Charles Duke, and John Swigert (who replaced Thomas Mattingly on the prime crew after the crew was exposed to German measles). The Apollo 13 Command Module "Odyssey" is now at the Kansas Cosmosphere and Space Center, Hutchinson, Kansas. It was originally on display at the Musee de l'Air, Paris, France.

Luna 16. Launch Date/Time: September 12, 1970 at 13:25:53 UT from Tyuratam (Baikonur Cosmodrome), U.S.S.R. Luna 16 was the first robotic probe to land on the Moon and return a sample to Earth and represented the first lunar sample return mission by the Soviet Union and the third overall, following the Apollo 11 and 12 missions. The spacecraft consisted of two attached stages, an ascent stage mounted on top of a descent stage. The descent stage was a cylindrical body with four protruding landing legs, fuel tanks, a landing radar, and a dual descent engine complex. See Fig. 27. A main descent engine was used to slow the craft until it reached a cutoff point which was determined by the onboard computer based on altitude and velocity. After cutoff a bank of lower thrust jets was used for the final landing. The descent stage also acted as a launch pad for the ascent stage. The ascent stage was a smaller cylinder with a rounded top. It carried a cylindrical hermetically sealed soil sample container inside a re-entry capsule. The spacecraft descent stage was equipped with a television camera, radiation and temperature monitors, telecommunications equipment, and an extendable arm with a drilling rig for the collection of a lunar soil sample.

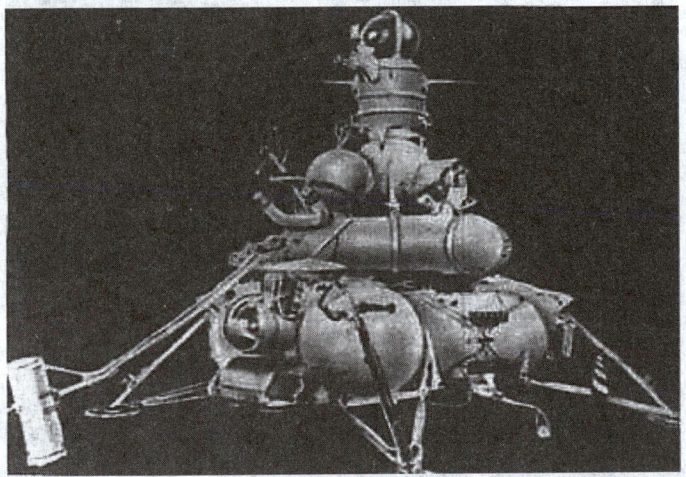

Fig. 27.   Luna 16 Spacecraft. (*Image courtesy of NASA*).

The Luna 16 automatic station was launched toward the Moon from a preliminary Earth orbit and after one mid-course correction on 13 September it entered a circular 111 km (69 miles) lunar orbit on 17 September 1970. The lunar gravity was studied from this orbit, and then the spacecraft was fired into an elliptical orbit with a perilune of 15.1 km (9 miles). The main braking engine was fired on 20 September, initiating the descent to the lunar surface. The main descent engine cut off at an altitude of 20 m (66 ft) and the landing jets cut off at 2 m (7 ft) height at a velocity less than 2.4 m/s, followed by vertical free-fall. At 05:18 UT, the spacecraft soft landed on the lunar surface in Mare Foecunditatis (the Sea of Fertility) as planned, approximately 100 km (62 miles) west of Webb crater. This was the first landing made in the dark on the Moon, as the Sun had set about 60 hours earlier. According to the Bochum Radio Space Observatory in the Federal Republic of Germany, strong and good quality television pictures

were returned by the spacecraft. However, such pictures were not made available to the U.S. by any sources so there is question as to the reliability of the Bochum report. The drill was deployed and penetrated to a depth of 35 cm (14 inches) before encountering hard rock or large fragments of rock. The column of regolith in the drill tube was then transferred to the soil sample container. After 26 hours and 25 minutes on the lunar surface, the ascent stage, with the hermetically sealed soil sample container, lifted off from the Moon carrying 101 grams (3.6 ounces) of collected material at 07:43 UT on 21 September. The lower stage of Luna 16 remained on the lunar surface and continued transmission of lunar temperature and radiation data. The Luna 16 re-entry capsule returned directly to Earth without any mid-course corrections, made a ballistic entry into the Earth's atmosphere on 24 September and deployed parachutes. The capsule landed approximately 80 km SE of the city of Dzhezkazgan in Kazakhstan at 03:26 UT.

Zond 8. Launch Date/Time: October 20, 1970 at 19:55:39 UT. Zond 8 was launched from an earth orbiting platform, Tyazheliy Sputnik (70-088B), towards the moon. The announced objectives were investigations of the moon and circumlunar space and testing of onboard systems and units. The spacecraft obtained photographs of the earth on October 21 from a distance of 64,480 km. The spacecraft transmitted flight images of the earth for three days. Zond 8 flew past the moon on October 24, 1970, at a distance of 1110.4 km and obtained both black and white and color photographs of the lunar surface. Scientific measurements were also obtained during the flight. Zond 8 reentered the earth's atmosphere and splashed down in the Indian Ocean on October 27, 1970.

Luna 17/Lunokhod 1. Launch Date/Time: November 10, 1970 at 14:44:01 UT. Luna 17 was launched from an earth parking orbit towards the Moon and entered lunar orbit on November 15, 1970. The spacecraft soft landed on the Moon in the Sea of Rains. The spacecraft had dual ramps by which the payload, Lunokhod 1, descended to the lunar surface. Lunokhod 1 was a lunar vehicle formed of a tub-like compartment with a large convex lid. The rover stood 135 cm (4.43 ft) high and had a mass of 840 kg (1,852 lbs). It was about 170 cm (5.6 ft) long and 160 cm (5.25 ft) wide and had 8 wheels each with an independent suspension, motor and brake. The rover had two speeds, ~1 km/hr (0.6 mi/hr and ~2 km/hr (1.2 mi/hr). Lunokhod was equipped with a cone-shaped antenna, a highly directional helical antenna, four television cameras, and special extendable devices to impact the lunar soil for soil density and mechanical property tests. See Fig. 28. An x-ray spectrometer, an x-ray telescope, cosmic-ray detectors, and a laser device were also included. The vehicle was powered by a solar cell array mounted on the underside of the lid. Lunokhod was intended to operate through three lunar days but actually operated for eleven lunar days. The operations of Lunokhod officially ceased on October 4, 1971, the anniversary of Sputnik 1. Lunokhod had traveled 10,540 m (6.5 miles) and had transmitted more than 20,000 TV pictures

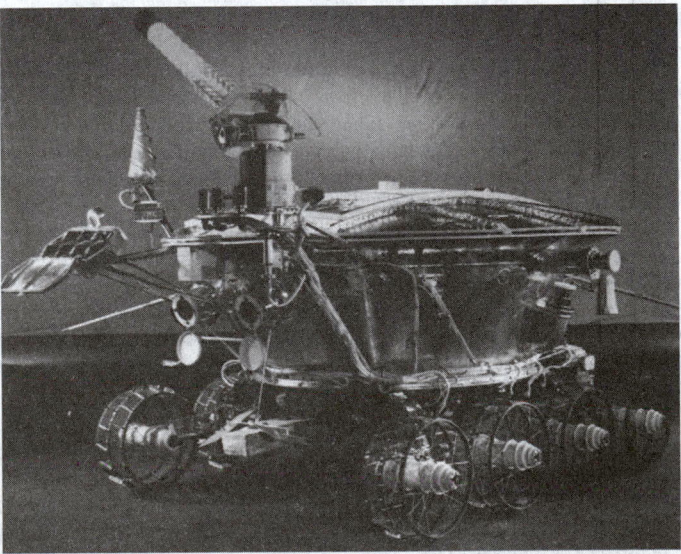

Fig. 28.   Lunokhod 1 lunar vehicle. (*Image courtesy of NSSDC*).

and more than 200 TV panoramas. It had also conducted more than 500 lunar soil tests.

Apollo 14. Launch Date/Time: January 31, 1971 at 21:03:02 UT (04:03:02 p.m. EST). Apollo 14 was the third mission in which humans walked on the lunar surface and returned to Earth. On 5 February 1971 two astronauts (Apollo 14 Commander Alan B. Shepard, Jr. and LM pilot Edgar D. Mitchell) landed near Fra Mauro crater on the Moon in the Lunar Module (LM) while the Command and Service Module (CSM) (with CM pilot Stuart A. Roosa) continued in lunar orbit.

After a delay of 40 minutes, 2 seconds due to clouds and rain, Apollo 14 was launched into Earth parking orbit from pad 39A of Kennedy Space Center on Saturn V SA-509. Earth orbit insertion occurred at 21:14:51 UT followed by translunar injection at 23:37:34. An early first mid-course correction was made to make up for the launch delay so the spacecraft would arrive at the Moon on schedule. The CSM separated from the S-IVB stage containing the LM at 00:05:31 UT on 1 February. Five attempts were made to dock the CSM and the LM, all unsuccessful because the catches on the docking ring did not release. The sixth attempt, at 02:00:02 UT, was successful and no further problems with the docking mechanism occurred. The S-IVB stage was released into a lunar impact trajectory. (It impacted the lunar surface on 4 February at 07:40:55.4 UT at 8.09 S, 26.02 W with a velocity of 2.54 km/s at a 69 degree angle from the horizontal.) A second mid-course correction was made on 2 February and a third on 4 February. Lunar orbit insertion occurred at 06:59:43 UT on 4 February.

The LM "Antares", with Shepard and Mitchell aboard, separated from the CSM, piloted by Roosa, at 04:50:44 UT on 5 February and landed at 09:18:11 UT in the hilly upland region 24 km (15 miles) north of the rim of Fra Mauro crater at 3.6 S, 17.5 W. The astronauts made two moonwalk EVA's totaling 9 hours, 23 minutes, one on 5 February and one on 6 February, during which the Apollo lunar surface experiments package (ALSEP) was placed on the surface of the moon, 42.9 kg (95 lbs) of lunar samples were acquired, and photographs were taken. At the end of the second EVA Shepard hit two golf balls. Experiments were also performed from the CSM in equatorial orbit.

The LM lifted off from the Moon at 18:48:42 UT on 6 February after 33 hours, 31 minutes on the lunar surface. After the LM docked with the CSM at 20:35:53 UT the lunar samples and other equipment were transferred from the LM and the LM was jettisoned at 22:48:00 UT. It impacted the Moon on 8 February 00:45:25.7 UT at 3.42 S, 19.67 W. Transearth injection began at 01:39:04 UT on 7 February. One small mid-course correction was made on 8 February during transearth coast. The CM "Kitty Hawk" separated from the SM at 20:35:44 UT on 9 February. Apollo 14 splashed down in the Pacific Ocean on 9 February 1971 at 21:05:00 UT (4:05:00 p.m. EST) after a mission elapsed time of 216 hrs, 1 min, 58 secs. The splashdown point was 27 deg 1 min S, 172 deg 39 min W, 765 nautical miles south of American Samoa. The astronauts and capsule were picked up by the recovery ship USS New Orleans. This was the last Apollo mission in which the astronauts were put in quarantine after their return.

Performance of the spacecraft, the third of the Apollo H-series missions, was good for most aspects of the mission. The primary mission goals of deployment of the ALSEP and other scientific experiments, collection of lunar samples, surface photography, and photography, radio science and other scientific experiments from orbit were achieved with the exception of the full coverage planned for the Hycon camera. Shepard, 47, was a Navy captain on his second spaceflight (he'd flown previously as the first American in space on Mercury Redstone 3), Roosa, 37, was an Air Force major on his first spaceflight, and Mitchell, 40, was a Navy commander also on his first spaceflight. The backup crew for this mission was Eugene Cernan, Ronald Evans, and Joe Engle. The Apollo 14 command module "Kitty Hawk" is currently on display at the Astronaut Hall of Fame in Titusville, Florida.

Apollo 15. Launch Date/Time: July 26, 1971 at 13:34:00 UT (9:34:00 a.m. EST). Apollo 15 was the fourth mission in which humans walked on the lunar surface and returned to Earth. On 30 July 1971 two astronauts (Apollo 15 Commander David R. Scott and LM pilot James B. Irwin) landed in the Hadley Rille/Apennines region of the Moon in the Lunar Module (LM) while the Command and Service Module (CSM) (with CM pilot Alfred M. Worden) continued in lunar orbit.

Apollo 15 was launched on a Saturn V SA-510 from Pad 39A at Kennedy Space Center. The spacecraft was inserted into Earth orbit at 13:45:44 UT and translunar injection took place at 16:30:03 UT. The CSM separated from the S-IVB stage at 16:56:24 UT and docked with the LM at 17:07:49 UT. The S-IVB stage was released and burns at 19:22 UT and 23:34 UT sent the stage into a lunar impact trajectory. (It impacted the lunar surface on 29 July at 20:58:42.9 UT at 1.51 S, 11.81 W with a velocity of 2.58 km/s at a 62 degree angle from the horizontal.) A short was discovered in the service propulsion system and contingency procedures were developed for using the engine. A mid-course correction was performed on 27 July at 18:14:22 UT and another on 29 July at 15:05:15. During cruise it was discovered that the LM range/range-rate exterior glass cover had broken and a small water leak had developed in the CM "Endeavour" requiring repair and clean-up. The SIM door was jettisoned at 15:40 UT and lunar orbit insertion took place at 20:05:47 UT. The descent orbit maneuver was executed at 00:13:49 UT on 30 July.

Scott and Irwin entered the LM "Falcon" and the LM-CSM undocking maneuver was initiated at 17:48 UT but undocking did not take place. Worden found a loose umbilical plug and reconnected it, allowing the LM to separate from the CSM at 18:13:30 UT. The LM fired its descent engine at 22:04:09 UT and landed at 22:16:29 UT on 30 July 1971 in the Mare Imbrium region at the foot of the Apennine mountain range at 26.1 N, 3.6 E. Scott and Irwin made three moonwalk EVAs totaling 18 hours, 35 minutes. See Fig. 29. During this time they covered 27.9 km (17.3 miles), collected 76.8 kg (169 lbs) of rock and soil samples, took photographs, and set up the ALSEP and performed other scientific experiments. This was the first mission which employed the Lunar Roving Vehicle which was used to explore regions within 5 km (3 miles) of the LM landing site. After the final EVA Scott performed a televised demonstration of a hammer and feather falling at the same rate in the lunar vacuum. The CSM remained in a slightly elliptical orbit from which Worden performed scientific experiments.

Fig. 29. Apollo 15 astronaut James Irwin and the rover at the edge of Hadley Rille on the Moon. David Scott took this Hasselblad picture looking northwestward from the flank of St. George crater. The rille has been widened and made shallower over time by wasting from the walls. Talus blocks are visible on the walls and on the floor of the rille. (*Image courtesy of NSSDC*).

The LM lifted off from the Moon at 17:11:22 UT on 2 August after 66 hours, 55 minutes on the lunar surface. After the LM docked with the CSM at 19:09:47 UT the lunar samples and other equipment were transferred from the LM. The LM was jettisoned at 01:04:14 UT on 3 August, after a one orbit delay to ensure LM and CSM hatches were completely sealed. The LM impacted the Moon on 3 August 03:03:37.0 UT at 26.36 N, 0.25 E, 93 km west of the Apollo 15 ALSEP site, with an estimated impact velocity of 1.7 km/s at an angle of ~3.2 degrees from horizontal. Experiments were performed from orbit over the next day. On

5 August, Worden carried out the first deep space EVA when he exited the CM and made three trips to the SIM bay at the rear of the SM to retrieve film cannisters and check the equipment. Total EVA time was 38 minutes, 12 seconds. The CM separated from the SM at 20:18:00 UT on 7 August. During descent, one of the three main parachutes failed to open fully, resulting in a descent velocity of 35 km/hr (21.8 mph), 4.5 km/hr (2.8 mph) faster than planned. Apollo 15 splashed down in the Pacific Ocean on 7 August 1971 at 20:45:53 UT (4:45:53 p.m. EDT) after a mission elapsed time of 295 hrs, 11 mins, 53 secs. The splashdown point was 26 deg 7 min N, 158 deg, 8 min W, 330 miles north of Honolulu, Hawaii and 9.8 km (6.1 mi) from the recovery ship USS Okinawa.

Performance of the spacecraft, the first of the Apollo J-series missions, was excellent for most aspects of the mission. The primary mission goals of exploration of the Hadley-Appenine region, deployment of the ALSEP and other scientific experiments, collection of lunar samples, surface photography, and photography and other scientific experiments from orbit, and engineering evaluation of new Apollo equipment, particularly the rover, were achieved. Scott, 39, was an Air Force Colonel on his third spaceflight (he'd flown previously on Gemini 8 and Apollo 9), Worden, 39, was an Air Force major on his first spaceflight, and Irwin, 41, was an Air Force lt. colonel also on his first spaceflight. The backup crew for this mission was Richard Gordon, Vance Brand, and Harrison Schmitt. The Apollo 15 command module "Endeavor" is on display at the USAF Museum at Wright-Patterson Air Force Base, Dayton, Ohio.

Luna 18. Launch Date/Time: September 2, 1971 at 13:40:40 UT from Tyuratam (Baikonur Cosmodrome), U.S.S.R. Luna 18 was placed in an earth parking orbit after it was launched and was then sent towards the Moon. On September 7, 1971, it entered lunar orbit. The spacecraft completed 85 communications sessions and 54 lunar orbits before it was sent towards the lunar surface by use of braking rockets. It impacted the Moon on September 11, 1971, at 3 degrees 34 minutes N, 56 degrees 30 minutes E (selenographic coordinates) in a rugged mountainous terrain. Signals ceased at the moment of impact.

Luna 19. Launch Date/Time: September 28, 1971 at 10:00:22 UT from Tyuratam (Baikonur Cosmodrome), U.S.S.R. Luna 19 was placed in an intermediate earth parking orbit and, from this orbit, was sent toward the Moon. It was placed in a lunar orbit on October 3, 1971. Luna 19 extended the systematic study of lunar gravitational fields and location of mascons (mass concentrations). It also studied the lunar radiation environment, the gamma-active lunar surface, and the solar wind. Photographic coverage via a television system was also obtained.

Luna 20. Launch Date/Time: February 14, 1972 at 03:27:59 UT from Tyuratam (Baikonur Cosmodrome), U.S.S.R. Luna 20 was placed in an intermediate earth parking orbit and from this orbit was sent towards the Moon. It entered lunar orbit on February 18, 1972. On 21 February 1972, Luna 20 soft landed on the Moon in a mountainous area known as the Apollonius highlands near Mare Foecunditatis (Sea of Fertility), 120 km (75 miles) from where Luna 16 had impacted. While on the lunar surface, the panoramic television system was operated. Lunar samples were obtained by means of an extendable drilling apparatus. The ascent stage of Luna 20 was launched from the lunar surface on 22 February 1972 carrying 30 grams (1.06 ounces) of collected lunar samples in a sealed capsule. It landed in the Soviet Union on 25 February 1972. The lunar samples were recovered the following day.

Apollo 16. Launch Date/Time: April 16, 1972 at 17:54:00 UT (12:54:00 p.m. EST). Apollo 16 was the fifth mission in which humans walked on the lunar surface and returned to Earth. On 21 April 1972 two astronauts (Apollo 16 Commander John W. Young and LM pilot Charles M. Duke, Jr.) landed in the Descartes region of the Moon in the Lunar Module (LM) while the Command and Service Module (CSM) (with CM pilot Thomas K. Mattingly, II) continued in lunar orbit.

Apollo 16 was launched on a Saturn V SA-511 from Pad 39A at Kennedy Space Center. (The launch was postponed from the originally scheduled date, March 17, because of a docking ring jettison malfunction.) The spacecraft entered Earth parking orbit at 18:05:56 UT and translunar injection took place at 20:27:37 UT. The CSM and S-IVB stage separated at 20:58:59 UT and CSM-LM docking was achieved at 21:15:53 UT. The S-IVB stage was released into a lunar impact trajectory, but due to an earlier problem with the auxiliary propulsion system (APS) helium regulators, which resulted in continuous venting and loss of helium, the second APS burn could not be made. Tracking of the S-IVB was lost on 17 April at 21:03 UT due to a transponder failure. (The S-IVB stage impacted the Moon on 19 April at 21:02:04 UT at 1.3 N, 23.8 W with a velocity of 2.5 to 2.6 km/s at a 79 degree angle from the horizontal, as estimated from the Apollo 12, 14 and 16 seismic station data.) A mid-course correction was performed at 00:33:01 UT on 18 April. During translunar coast a CSM navigation problem was discovered in which a false indication would cause loss of inertial reference, this was solved by a real-time change in the computer program. The SIM door was jettisoned on 19 April at 15:57:00 UT and lunar orbit insertion took place at 20:22:28 UT. Two revolutions later the orbit was lowered to one with a perilune of 20 km (12.4 miles).

At 15:24 UT on 20 April Young and Duke entered the LM "Orion". The LM separated from the CSM at 18:08:00 UT, but the LM descent was delayed almost 6 hours due to a malfunction in the yaw gimbal servo loop on the CSM which caused oscillations in the service propulsion system (SPS). Engineers determined that the problem would not seriously affect CSM steering and the mission was allowed to continue with the LM descent. The LM landed at 02:23:35 UT on 21 April in the Descartes highland region just north of the crater Dolland at 9.0 S, 15.5 E. Young and Duke made three moonwalk EVAs totaling 20 hours, 14 minutes. During this time they covered 27 km (16.8 miles) using the Lunar Roving Vehicle, collected 94.7 kg (209 lbs) of rock and soil samples, took photographs, and set up the ALSEP and other scientific experiments. Other experiments were also performed from orbit in the CSM during this time.

The LM lifted off from the Moon at 01:25:48 UT on 24 April after 71 hours, 2 minutes on the lunar surface. After the LM docked with the CSM at 03:35:18 UT the lunar samples and other equipment were transferred from the LM and the LM was jettisoned at 20:54:12 UT on 24 April. The LM began tumbling, apparently due to an open circuit breaker in the guidance and navigation system. As a result the planned deorbit and lunar impact could not be attempted. The LM remained in lunar orbit with an estimated lifetime of one year. The instrument boom which carried the orbital mass spectrometer would not retract and was jettisoned. On 25 April at 20:43 UT Mattingly began a cislunar EVA to retrieve camera film from the SIM bay and inspect instruments, two trips taking a total of 1 hour, 24 minutes. The CM "Casper" separated from the SM on 27 April at 19:16:33 UT. Apollo 16 splashed down in the Pacific Ocean on 27 April 1972 at 19:45:05 UT (2:45:05 p.m. EST) after a mission elapsed time of 265 hrs, 51 mins, 5 secs. The splashdown point was 0 deg 43 min S, 156 deg 13 min W, 215 miles southeast of Christmas Island and 5 km (3 mi) from the recovery ship USS Ticonderoga.

Performance of the spacecraft, the second of the Apollo J-series missions, was good for most aspects of the mission. The primary mission goals of inspecting, surveying, and sampling materials in the Descartes region, emplacement and activation of surface experiments, conducting inflight experiments and photographic tasks from lunar orbit, engineering evaluation of spacecraft and equipment, and performance of zero-gravity experiments were achieved despite the mission being shortened by one day. Young, 41, was a Navy Captain who had flown on three previous spaceflights (Gemini 3, Gemini 10, and Apollo 10; he later flew on STS-1 and STS-9), Mattingly, 36, was a Navy lt. commander on his first spaceflight (he later flew STS-4 and STS-51 C), and Duke, 36, was an Air Force lt. colonel also on his first spaceflight. The backup crew for this mission was Fred Haise, Stuart Roosa, and Edger Mitchell. The Apollo 16 Command Module "Casper" is on display at the Alabama Space and Rocket Center in Huntsville, Alabama.

Apollo 17. Launch Date/Time: December 7, 1972 at 05:33:00 UT (12:33:00 a.m. EST). Apollo 17 was the sixth and last Apollo mission in which humans walked on the lunar surface. On 11 December 1972 two astronauts (Commander Eugene A. Cernan and LM pilot Harrison H. Schmitt, the first scientist on the Moon) landed in the Taurus-Littrow region of the Moon in the Lunar Module (LM) while the Command and Service Module (CSM) (with CM pilot Ronald E. Evans) continued in lunar orbit.

Apollo 17 lifted off after a 2 hour, 40 minute delay due to a malfunction of a launch sequencer. Launch was on Saturn V SA-512 from Pad 39A at Kennedy Space Center and was the first nighttime launch of an Apollo. The spacecraft began Earth parking orbit at 05:44:53 UT and translunar injection took place at 08:45:37 UT. The CSM separated from the S-IVB at 09:15:29 UT and CSM-LM docking took place at 09:29:45 UT. The S-IVB was released at 10:18 UT into a lunar impact trajectory. (It impacted the lunar surface on 10 December at 20:32:42.3 UT at 4.21 S, 12.31 W with a velocity of 2.55 km/s (1.58 miles/s) at a 55 degree angle from the horizontal.) A single mid-course correction requiring a 1.6 second burn of the Service Propulsion System (SPS) was made at 17:03:00 UT on 8 December. On December 10 at 15:05:40 UT the SIM bay door was jettisoned and a 398 second burn of the SPS was initiated at 19:47:23 UT to insert Apollo 17 into lunar orbit. Approximately 4 hours 20 minutes later another maneuver lowered the orbit to a perilune of 28 km (17.4 miles). At 14:35 UT on 11 December Cernan and Schmitt entered the LM "Challenger".

The LM separated from the CSM at 17:20:56 UT on 11 December 1972 and reduced its orbit to 11.5 km (17.4 miles) perilune at 18:55:42 UT. The descent burn took place at 19:43 UT and the LM landed at 19:54:57 UT on the southeastern rim of Mare Serenitatis in a valley at Taurus-Littrow, at 20.2 N, 30.8 E. Cernan and Schmitt made three moonwalk extra-vehicular activities (EVAs) totaling 22 hours, 4 minutes. See Fig. 30. During this time they covered 30 km (18.64 miles) using the Lunar Roving Vehicle, collected 110.5 kg (243.6 lbs) of lunar samples, took photographs, and set up the ALSEP and performed other scientific experiments. Evans performed experiments from orbit in the CSM during this time.

Fig. 30. Photograph taken by Eugene Cernan, commander of Apollo 17, of lunar module pilot Harrison Schmitt standing in front of a large split boulder on the Moon. The lunar rover is in the foreground at left. The picture was taken during the third EVA at the Taurus-Littnow site on the Moon at Station 6, the base of the North Massif. Apollo 17 was launched on 7 December 1972 and landed on the Moon 11 December. It was the last of the Apollo Moon landing missions. (*Image courtesy of NSSDC*).

The LM lifted off from the Moon at 22:54:37 UT on 14 December after 75 hours on the lunar surface. After the LM docked with the CSM at 01:10:15 UT on 15 December the lunar samples and other equipment were transferred from the LM and the LM was jettisoned at 04:51:31 UT. The LM impacted the Moon at 06:50:20.8 UT at 19.96 N, 30.50 E, approximately 15 km from the Apollo 17 landing site, with an estimated impact velocity of 1.67 km/s at an angle ~4.9 degrees from horizontal. After another 1 1/2 days in lunar orbit, transearth injection took place at 23:35:09 UT on 16 December. On 17 December at 20:27 UT Evans began a cislunar spacewalk EVA consisting of three trips to the SM SIM bay to collect camera and lunar sounder film over a period of 67 minutes. The CM "America" and SM separated at 18:56:49 UT on 19 December. Apollo 17 splashed down in the Pacific Ocean on 19 December 1972 at 19:24:59

UT (2:24:59 p.m. EST) after a mission elapsed time of 301 hrs, 51 mins, 59 secs. The splashdown point was 17 deg 53 min S, 166 deg 7 min W, 350 nautical miles SE of the Samoan Islands and 6.5 km (4 mi) from the recovery ship USS Ticonderoga.

Performance of the spacecraft, the third of the Apollo J-series missions, was excellent for all aspects of the mission. The primary mission goals of investigating the lunar surface and environment in the Taurus-Littrow region, emplacing and activating surface experiments, performing experiments in lunar orbit, obtaining and returning lunar surface samples, and enhancing the capability for future astronaut lunar exploration were achieved. Cernan, 38, was a Navy captain with two previous spaceflights (Gemini 9, Apollo 10), Evans, 39, was a Navy commander making his first spaceflight, and Schmitt, 37, was a civilian also making his first spaceflight. The backup crew for this mission was John Young, Stuart Roosa, and Charles Duke. The Apollo 17 command module capsule "America" is on display at the Johnson Space Center in Houston, Texas.

Luna 21/Lunokhod 2. Launch Date/Time: January 8, 1973 at 06:55:38 UTC from Tyuratam (Baikonur Cosmodrome), U.S.S.R. The Luna 21 spacecraft landed on the Moon and deployed the second Soviet lunar rover (Lunokhod 2). The primary objectives of the mission were to collect images of the lunar surface, examine ambient light levels to determine the feasibility of astronomical observations from the Moon, perform laser ranging experiments from Earth, observe solar X-rays, measure local magnetic fields, and study mechanical properties of the lunar surface material.

The SL-12/D-1-e launcher put the spacecraft into Earth parking orbit followed by translunar injection. On 12 January 1973, Luna 21 was braked into a 90 × 100 km (56 × 62 miles) orbit about the Moon. On 13 and 14 January, the perilune was lowered to 16 km (10 miles) altitude. On 15 January after 40 orbits, the braking rocket was fired at 16 km (10 miles) altitude, and the craft went into free fall. At an altitude of 750 meters (2,461 ft) the main thrusters began firing, slowing the fall until a height of 22 meters (72 ft) was reached. At this point the main thrusters shut down and the secondary thrusters ignited, slowing the fall until the lander was 1.5 meters (4.92 ft) above the surface, where the engine was cut off. Landing occurred at 23:35 UT in LeMonnier crater at 25.85 degrees N, 30.45 degrees E. The lander carried a bas relief of Lenin and the Soviet coat-of-arms.

After landing, the Lunokhod 2 took TV images of the surrounding area, then rolled down a ramp to the surface at 01:14 UT on 16 January and took pictures of the Luna 21 lander and landing site. It stopped and charged batteries until 18 January, took more images of the lander and landing site, and then set out over the Moon. The rover would run during the lunar day, stopping occasionally to recharge its batteries via the solar panels. At night the rover would hibernate until the next sunrise, heated by the radioactive source. Lunokhod 2 operated for about 4 months, covered 37 km (23 miles) of terrain including hilly upland areas and ridges, and sent back 86 panoramic images and over 80,000 TV pictures. Many mechanical tests of the surface, laser ranging measurements, and other experiments were completed during this time. On June 4 it was announced that the program was completed, leading to speculation that the vehicle probably failed in mid-May or could not be revived after the lunar night of May-June. The Lunokhod was not left in a position such that the laser retroreflector could be used indicating that the failure may have happened suddenly.

Luna 22. Launch Date/Time: May 29, 1974 at 08:57:00 UT from Tyuratam (Baikonur Cosmodrome), U.S.S.R. Luna 22 was a lunar orbiter mission. The spacecraft carried imaging cameras and also had the objectives of studying the Moon's magnetic field, surface gamma ray emissions and composition of lunar surface rocks, and the gravitational field, as well as micrometeorites and cosmic rays. Luna 22 was launched into Earth parking orbit and then to the moon. It was inserted into a circular lunar orbit on 2 June 1974. The spacecraft made many orbit adjustments over its 18 month lifetime in order to optimize the operation of various experiments, lowering the perilune to as little as 25 km (15.5 miles). Maneuvering fuel was exhausted on 2 September and the mission was ended in early November.

Luna 23. Launch Date/Time: October 28, 1974 at 14:30:32 UT from Tyuratam (Baikonur Cosmodrome), U.S.S.R. Luna 23 was a Moon lander mission which was intended to return a lunar sample to Earth. Launched to the Moon by a Proton SL-12/D-1-e booster, the spacecraft was damaged during landing in Mare Crisium (Sea of Crises). The sample collecting apparatus could not operate and no samples were returned. The lander continued transmissions for 3 days after landing.

Luna 24. Launch Date/Time: August 9, 1976 at 15:04:12 UT from Tyuratam (Baikonur Cosmodrome), U.S.S.R. The last of the Luna series of spacecraft, the mission of the Luna 24 probe was the third Soviet mission to retrieve lunar ground samples (the first two were returned by Luna 16 and 20). After entering a 115 × 115 km (71.5 × 71.5 miles) lunar orbit with an inclination of 120 degrees, the probe landed in the area known as Mare Crisium (Sea of Crisis) at 12.75 N, 62.2 E on 18 August 1976. Using a sample arm and drill, the mission successfully collected 170.1 grams (6 ounces) of lunar samples and deposited them into a collection capsule. The capsule was launched from the Moon at 5:25 UT on 19 August and landed at 5:55 UT on 22 August in western Siberia, about 200 km (124 miles) southeast of the town of Surgut, where the samples were collected for scientific study.

Lunar Missions 1990-2008

Hiten. Launch Date/Time: January 24, 1990 at 11:46:00 UT from Uchinoura Space Center, Japan. Hiten (originally called Muses-A) was an ISAS (Japanese Space Agency) Earth orbiting satellite designed primarily to test and verify technologies for future lunar and planetary missions. The spacecraft carried a small satellite named Hagoromo which was released in the vicinity of the Moon. Hiten itself was put into a highly elliptical Earth orbit which passed by the Moon ten times during the mission, which ended when Hiten was intentionally crashed into the Moon on 10 April 1993. Hiten was named after a flying, music-playing Buddhist angel. Hagoromo was named for the veil worn by Hiten. This mission included Japan's first-ever lunar flyby, lunar orbiter, and lunar surface impact, making Japan only the third nation to achieve each of these goals. The primary objectives of the mission were to: 1) test trajectory control utilizing gravity assist double lunar swingbys; 2) insert a sub-satellite into lunar orbit; 3) conduct optical navigation experiments on a spin-stabilized spacecraft; 4) test fault tolerant onboard computer and packet telemetry; 5) conduct cis-lunar aerobraking experiments; and 6) detect and measure mass and velocity of micro-meteorite particles. Three follow-on objectives were also added later in the mission: excursion to the L4 and L5 Lagrangian points of the Earth-Moon system, orbit of the Hiten spacecraft around the Moon, and hard landing on the lunar surface.

Clementine. Clementine was a joint project between the Ballistic Missile Defense Organization (BMDO), formerly the Strategic Defense Initiative Organization and NASA. The objective of the mission was to test sensors and spacecraft components under extended exposure to the space environment and to make scientific observations of the Moon and the near-Earth asteroid 1620 Geographos. The observations included imaging at various wavelengths including ultraviolet and infrared, laser ranging altimetry, and charged particle measurements. These observations were originally for the purposes of assessing the surface mineralogy of the Moon and Geographos, obtaining lunar altimetry from 60 N to 60 S latitude, and determining the size, shape, rotational characteristics, surface properties, and cratering statistics of Geographos.

Clementine was launched on 25 January 1994 at 16:34:00 UT (12:34:00 p.m. EST) from Vandenberg AFB aboard a Titan IIG rocket. After two Earth flybys, lunar insertion was achieved on February 21. Lunar mapping took place over approximately two months, in two parts. The first part consisted of a 5 hour elliptical polar orbit with a perilune of about 400 km at 28 degrees S latitude. After one month of mapping the orbit was rotated to a perilune of 29 degrees N latitude, where it remained for one more month. This allowed global imaging as well as altimetry coverage from 60 degrees S to 60 degrees N.

During 71 days in lunar orbit, Clementine systematically mapped the 38 million square kilometers (23,612,105 million square miles) of the Moon in 11 colors in the visible and near-infrared parts of the spectrum. In addition, the spacecraft took tens of thousands of high-resolution and mid-infrared thermal images, mapped the topography of the Moon with a laser ranging experiment, improved our knowledge of the surface gravity field of the Moon through radio tracking, and carried a charged particle telescope to characterize the solar and magnetospheric energetic particle environment. See also **Clementine Mission**.

After leaving lunar orbit, a malfunction in one of the on-board computers on May 7 at 14:39 UT (9:39 a.m. EST) caused a thruster to fire until it had used up all of its fuel, leaving the spacecraft spinning at about 80 RPM with no spin control. This made the planned continuation of the mission, a flyby of the near-Earth asteroid Geographos, impossible. The spacecraft remained in geocentric orbit and continued testing the spacecraft components until the end of mission.

AsiaSat 3/HGS-1. Launch Date/Time: December 24, 1997 at 23:19:00 UT from Tyuratam (Baikonur Cosmodrome), Kazakhstan. Asiasat 3 was a communications satellite launched by Hong Kong, People's Republic of China. It was to be used primarily for television distribution and telecommunications services throughout Asia, the Middle East, and Australasia, with multiple spot beams for selected areas.

The body-stabilized satellite was 26.2 m (86 ft) tip-to-tip along the axis of the solar arrays and 10 m (33 ft) across the axis of the antennas. The bus was essentially a cube, roughly 4 m (13 ft) on a side.

Power to the spacecraft was generated using two sun-tracking, four-panel solar wings covered with Ga-As solar cells, providing up to 9900 watts A 29-cell Ni-H battery provided power to the spacecraft during eclipse operations. A bipropellant propulsion system, consisting of twelve conventional bi-propellant thrusters, was used for stationkeeping.

Two 2.72 m (8.92 ft) diameter Gregorian shaped-surface antennas were mounted on opposing sides of the bus, perpendicular to the axis along which the solar arrays were mounted. One of these antennas operated in C-band, the other in Ku-band. Focused area coverage was provided by a 1.3 m (4.3 ft) diameter, dual-gridded shaped reflector operating in the Ku-band. A 1 m (3.28 ft) diameter Ku-band steerable spot-beam antenna provided the spacecraft with the ability to direct 5 degree coverage of any area on the Earth's surface visible to the spacecraft from its orbit location. Both of these antennas were mounted on the nadir side of the spacecraft.

The satellite was to be placed into a geosynchronous orbit, but a malfunction in the fourth (DM 3) stage resulted in a short life and a dysfunctional orbit. Later investigation revealed that the DM 3 may have been designed only for a maximum payload of 2.4 metric tonnes and had previously mislaunched two earlier payloads which exceeded this limit (as did Asiasat 3).

Following its failure to achieve the proper orbit, the manufacturer (Hughes Global Services) purchased the spacecraft back from the insurers and renamed it HGS 1. HGS 1 was then successfully maneuvered into two successive flybys of the Moon to place it into geosynchronous orbit, the first time such a maneuver was performed by a commercial satellite.

Lunar Prospector. Launch Date/Time: January 7, 1998 at 02:28:44 UT (January 6, 9:26:44 p.m. EST) aboard a three-stage Athena 2 rocket, the Lunar Prospector a 105 hour cruise to the Moon. The Lunar Prospector (LP), was the first of NASA's new, cheaper, "Discovery"–class mission. The Lunar Prospector was designed for a low polar orbit investigation of the Moon, including mapping of surface composition and possible deposits of polar ice, measurements of magnetic and gravity fields, and study of lunar outgassing events. Data from the 19 month mission allowed construction of a detailed map of the surface composition of the Moon, and improved our understanding of the origin, evolution, current state, and resources of the Moon.

The spacecraft carried 6 experiments: a Gamma Ray Spectrometer (GRS), a Neutron Spectrometer (NS), a Magnetometer (MAG), an Electron Reflectometer (ER), an Alpha Particle Spectrometer (APS), and a Doppler Gravity Experiment (DGE). The instruments are omnidirectional and require no sequencing. The normal observation sequence is to record and downlink data continuously.

The spacecraft is a graphite-epoxy drum, 1.37 meters (4.5 ft) in diameter and 1.3 meters (4.25 ft) high with three radial 2.5 m (8.2 ft) instrument booms. A 1.1 m (3.6 ft) extension boom at the end of one of the 2.5 m (8.2 ft) booms holds the magnetometer. Total initial mass (fully fueled) was 296 kg (652.5 lbs). It is spin-stabilized (nominal spin rate 12 rpm) with its spin axis normal to the ecliptic plane. The spacecraft is controlled by 6 hydrazine monopropellant 22-Newton thrusters, two aft, two forward, and two tangential. Three fuel tanks mounted inside the drum hold 138 kg (304 lbs) of hydrazine pressurized by helium. The power system consists of body mounted solar cells which produce an average of 186 W and a 4.8 amp-hr rechargeable NiCd battery. Communications are through two S-band transponders, a slotted, phased-array medium gain antenna for downlink, and an omnidirectional low-gain antenna for downlink and uplink. There is no on-board computer, all control is from the ground, commanding a single on-board command and data handling unit. Data are downlinked directly and also stored on a solid-state recorder and downlinked after 53 minutes, to ensure all data collected during communications blackout periods are received.

The mission ended on 31 July 1999 at 9:52:02 UT (5:52:02 EST) when Lunar Prospector was deliberately targeted to impact in a permanently shadowed area of a crater near the lunar south pole. It was hoped that the impact would liberate water vapor from the suspected ice deposits in the crater and that the plume would be detectable from Earth, however, no plume was observed.

Total cost for the mission was $62.8 million including development ($34 million), launch vehicle (~$25 million) and operations (~ $4 million). See also **Lunar Prospector Mission**.

Smart 1. Launch Date/Time: September 27, 2003 at 23:14 UT. The European Space Agency's Science Programme encompasses, in addition to the ambitious 'Cornerstone' and medium-sized missions, recently dubbed 'flexi-missions', small relatively low-cost missions. These have been given the generic name SMART — 'Small Missions for Advanced Research in Technology'. Their purpose is to test new technologies that will eventually be used on bigger projects.

SMART-1 is the first in this programme. Its primary objective is to flight test Solar Electric Primary Propulsion ion drive as the key technology for future Cornerstones in a mission representative of a deep-space one. ESA's BepiColombo mission to explore the planet Mercury could be the first to benefit from SMART-1's demonstration of electric propulsion. Another objective is to test new technologies for spacecraft and instruments.

The primary scientific objectives of the mission are to return data on the geology, morphology, topography, mineralogy, geochemistry, and exospheric environment of the Moon in order to answer questions about planetary formation accretional processes, origin of the Earth-Moon system, the lunar near/far side dichotomy, long-term volcanic and tectonic activity, thermal and dynamical processes involved in lunar evolution, and water ice and external processes on the surface.

The planetary objective selected for the SMART-1 mission is to orbit the Moon for a nominal period of six months. It is the first time that Europe has sent a spacecraft to the Moon. In addition to the use of solar electric primary propulsion to reach Earth's natural satellite, the spacecraft will carry out a complete programme of scientific observations in lunar orbit.

The SMART-1 spacecraft launched on 27 September 2003 from Kourou, French Guiana as an auxiliary passenger on an Ariane-5 Cyclade which launched two other large satellites as its primary payload. It was put into a geostationary transfer orbit, 742 × 36,016 km (461 × 22,379 miles), inclined at 7 degrees to the equator. The spacecraft used its ion drive over a period of 14 months to elongate its Earth orbit and utilized three lunar resonance maneuvers in August, September, and October 2004 to minimize propellant use. Its final continuous thrust maneuver took place over 100 hours from 10 to 14 October 2004. Lunar orbit capture occurred on 13 November 2004 at a distance of 60,000 km (37,282 miles) from the lunar surface. The ion engine began firing in orbit at 05:24 UT (12:24 a.m. EST) on 15 November to start a 4.5 day period of thrust to lower the orbit. The first perilune took place on 15 November at 17:48 UTC (12:48 p.m. EST) at an altitude of about 5,000 km (3,107 miles) above the lunar surface. The engine was then used to lower the initial 4,962 × 51,477 km (3,083 × 31,986 miles) altitude, 5 day, 9 hour period, 81 degree inclination orbit, putting SMART-1 into a 300 × 3,000 km (186 × 1,864 miles) polar orbit. Lunar commissioning began in mid-January 2005 and lunar science operations in February 2005. The mission has been extended from its originally planned 6-month lifetime by a year, so it will now conduct mapping of the Moon's surface and evaluating the new technologies onboard from lunar orbit until August 2006. The xenon-ion engine was shut down in September 2005 after exhausting its fuel supply. It operated for almost 5000 hours and underwent 843 starts and stops. SMART-1 is expected to crash into the Moon on 17 August 2006. The total cost of the spacecraft is estimated at 100 million euros in 2001 economic conditions (~ $90 million U.S.). See also Smart 1 Mission.

Lunar-A. The scientific objectives of the Lunar-A mission to the Moon are to image the surface of the Moon, to monitor moonquakes, measure the near-surface thermal properties and heat flux, and to study the lunar

core and interior structure. To achieve these objectives, Lunar-A will carry a mapping camera and two surface penetrators. The surface penetrators are equipped with seismometers and devices to measure heat flow. The seismometers will monitor moonquake activity over the course of a year and this information will be used to learn about the structure of the Moon's interior and the size of the core. The heat flow measurements will provide information on the thermal state and evolution of the Moon.

The launch has been delayed a number of times for technical and financial reasons, it is now being re-evaluated. There is now no launch schedule for LUNAR-A, it will certainly launch no earlier than 2006. After launch, Lunar A will go into Earth parking orbit. The spacecraft will then be injected into an orbit around the Earth and Moon. After four and a half of these orbits, Lunar-A will swing out into a wide single orbit with an apogee of 1,185,000 km (737,324 miles). At the end of this orbit the spacecraft will re-encounter the Moon and be inserted into lunar orbit. This orbit will have an inclination of 30 degrees and bring Lunar-A within 40 km (24.85 miles) of the Moon's surface. The spacecraft will deploy two 13 kg (28.7 lb) penetrators over the course of a month. They will be individually released and impact the Moon at 250 to 300 m/s, burrowing 1 to 3 meters (3.28 to 9.84 ft) into the surface. One penetrator will be targeted at the equatorial area of the near side (in the vicinity of the Apollo 12 and 14 landing sites) and one at the equatorial far side. After deploying the penetrators, the orbiter will move up to a 200 to 300 km (124 to 186 miles) near circular mapping orbit. Data will be stored in memory in the penetrators and transmitted to the orbiter when it transits over each penetrator every 15 days.

Chandrayaan-1. Proposed launch date: September 1, 2007 from Sriharikota, India. Chandrayaan-1 is an Indian Space Research Organization (ISRO) mission designed to orbit the Moon over a two year period with the objectives of upgrading and testing India's technological capabilities in space and returning scientific information on the lunar surface. The spacecraft bus is roughly a 1.5 meter (4.92 ft) cube with a dry weight of 523 kg (1,153 lbs). See Fig. 31. It is based on the Kalpansat meteorological satellite. It will also carry a 30 kg (66 lb) probe designed to be released from the spacecraft and penetrate the lunar surface. Power is provided by a solar array which generates 750 W and charges lithium ion batteries. A bipropellant propulsion system is used to transfer Chandrayaan-1 into lunar orbit and maintain attitude. The spacecraft is 3-axis stabilized using attitude control thrusters and reaction wheels. Knowledge is provided by star sensors, accelerometers, and an inertial reference unit. Telecommand communications will be in S-band and science data transmission in X-band.

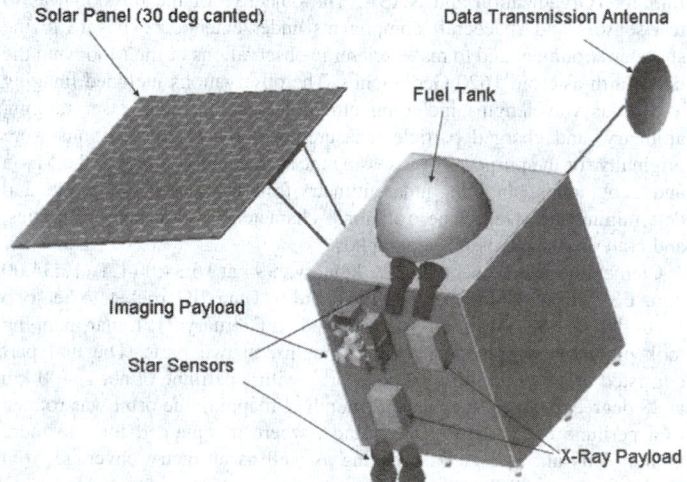

Fig. 31. Chandrayaan-1 Lunar Orbiter. (*Image courtesy of NSSDC*).

The scientific payload has a mass of 55 kg (121 lbs) and contains three Indian instruments. The Terrain Mapping stereo Camera (TMC) has 5 meter (16 ft) resolution and a 40 km (24.85 miles) (swath in the panchromatic band and will be used to produce a high-resolution map of the Moon. The Hyper Spectral Imager (HySI) will perform mineralogical mapping in the 400-900 nm band with a spectral resolution of 15 nm and a spatial resolution of 80 m (262 ft). The Lunar Laser Ranging Instrument (LLRI) will determine the surface topography. A fourth instrument, an X-ray flourescence spectrometer, will be built by the British. It will have three components: a Low Energy X-ray spectrometer (LEX) covering 0.5–10 keV with a ground resolution of 10 km (6.2 miles), a High Energy X-ray/gamma ray spectromenter (HEX) for 10–200 keV measurements with ground resolution of about 20 km (12 miles), and a Solar X-ray Monitor (SXM) to detect solar flux in the 2–10 keV range. LEX will be used to map the abundance of Si, Al, Mg, Ca, Fe, and Ti at the surface, the HEX will measure U, Th, 210Pb, 222Rn degassing, and other radioactive elements, and the SXM will monitor the solar flux to normalize the results of LEX and HEX. Ten kilograms of payload mass has been reserved for a proposed foreign instrument. The five finalists are being evaluated for this final slot on the mission and a final decision will be made in 2005.

The spacecraft will be launched on a PSLV C5 (Polar Satellite Launch Vehicle) from the Satish Dhawan Space Center in Sriharikota on the southeast coast of India in September of 2007 at the earliest. The PSLV will inject Chandrayaan-1 into a 240 × 36,000 km (149 × 22,369 mi) geosynchronous transfer orbit. After a 5.5 day lunar transfer trajectory the spacecraft will be captured into an initial 1,000 km (621 mi) (near circular orbit which will be lowered to a 200 km (124 mi) checkout orbit and finally into a 100 km (62 mi) circular polar orbit. It will stay in orbit and return data for at least two years. Chandrayaan is Hindi for "Voyage to the Moon". Total cost of the mission is about $100 million U.S.

Chang'e 1. Chang'e 1 is planned to be the first of a series of Chinese missions to the Moon. The spacecraft will launch in late 2007 on a CZ-3A booster and orbit the Moon for a year to test the technology for future missions and to study the lunar environment and surface regolith. The orbiter is based on the DFH-3 Comsat bus and will have a mass of roughly 2,000 kg (4,409 lbs), 150 kg (331 lbs) of which will be the scientific payload. The payload will include a stereo camera system to map the lunar surface, an altimeter to measure the distance between the spacecraft and the surface, a gamma/X-ray spectrometer to study the overall composition and radioactive components of the Moon, a microwave radiometer to map the thickness of the lunar regolith, and a system of space environment monitors to collect data on the solar wind and near-lunar region. The Chang'e program is named for a Chinese legend about a young fairy who flies to the Moon.

Selene. SELENE (SELenological and ENgineering Explorer) is a Japanese Space Agency (JAXA) lunar orbiter mission. The primary objective of the mission is a global survey of the Moon, obtaining data on elemental abundance, mineralogical composition, topography, geology, gravity, and the lunar and solar-terrestrial plasma environments and to develop critical technologies for future lunar exploration, such as lunar polar orbit injection, three-axis attitude stabilization, and thermal control. The mission consists of three satellites, an orbiter containing most of the scientific equipment, a VLBI (Very Long Baseline Interferometry) Radio (VRAD) satellite, and a relay satellite designed to receive a doppler ranging signal from the orbiter when it is around the far side out of direct contact wth the Earth and transmit the signal to Earth to estimate the far-side gravitational field.

The orbiter main bus is box-shaped, roughly 2.1 × 2.1 × 4.8 m (6.9 × 6.9 × 15.75 ft), divided into a 2.8 m (9.2 ft) long upper, or mission module which contains most of the scientific instruments and a 1.2 m (3.9 ft) long lower, or propulsion module. A solar array wing is mounted on one side of the spacecraft. A 1.3 meter (4.3 ft) high-gain antenna is mounted on one side 90 degrees from the solar panel. A 12 m (39 ft) magnetometer boom juts out of the top of the spacecraft and four 15 m (49 ft) radar sounder antennas protrude from the top and bottom corners of the mission module. The total dry mass of the spacecraft is 1,720 kg (3,792 lbs), with 795 kg (1,753 lbs) of propellant. See Fig. 32.

Power is supplied by the solar array, consisting of 22 square meters (72 square feet) of GaAs/Ge cells which can generate up to 3486 W and charges four NiH2 batteries of 50 V, 35 Ah capacity. Communications are via S- and X-band through the high-gain antenna with a data rate up to 10 Mbps downlink to a 60 m ground dish in X-band, and 40 or 2 kbps S-band downlink. Four S-band omnidirectional antennas are used for command uplink at 1 kbps. Onboard data recording capacity is 10 Gbytes. Thermal control is achieved by radiators, louvers, and heaters.

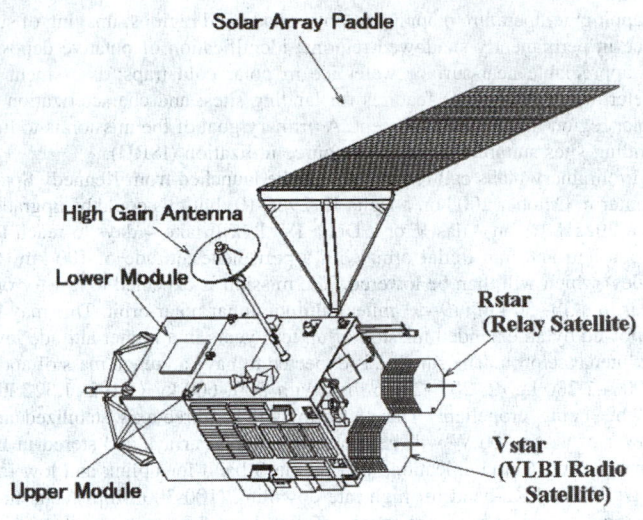

Fig. 32.   Selene.

A 500 N bipropellant (NTO/N2H4) main engine is mounted in the propulsion module. Orbital maintenance and yaw-pitch attitude are controlled by twelve bipropellant 20 N thrusters. Roll attitude is controlled by eight monopropellant 1 N thrusters. The spacecraft is three-axis stabilized with attitude control provided by four Sun sensors, two IMUs (Inertial Measurement Units), two star trackers, four 20 Nms reaction wheels, and the thrusters. The mission module carries 13 instruments for use in science investigations: a multi-band imager, terrain camera, high definition TV camera, spectral profiler, x-ray spectrometer, gamma-ray spectrometer, radar sounder, laser altimeter, magnetometer, plasma imager, charged particle spectrometer, plasma analyzer, and radio science equipment.

The VRAD subsatellite for SELENE is an octagonal cylinder 0.99 × 0.99 × 0.65 m (3.25 × 3.25 × 2.13 ft) in size with a mass of 53 kg (117 lbs). A dipole antenna protrudes from the top center of the subsatellite. The spacecraft is spin-stabilized at 10 rpm and has no propulsion units. Power is provided by a 70 W Si solar cell array covering the sides of the satellite which charges a 13 AH, 26 V NiMH battery. It holds one X-band and three S-band radio sources. The satellite in conjunction with the relay satellite will enable differential VLBI observations from the ground. The spacecraft will start in a 100 × 800 km (62 × 497 ft) polar orbit and is expected to survive in orbit for over one year. The VRAD and Relay satellites are mounted on top of the mission module before release.

The SELENE launch has been postponed to the summer of 2007 due to budgetary considerations. SELENE will be launched on an H-IIA rocket from Tanegashima Space Center into a 270 km (168 miles) Earth parking orbit with an inclination of 30.4 degrees. From this orbit, a lunar transfer trajectory injection burn will be conducted. Two mid-course correction maneuvers are planned, one about 24 hours after launch and another 103 hours after launch. SELENE will be injected into lunar 100 km × 13,000 km (62 × 8,078 miles), 16-hour polar orbit, 127 hours after launch. The spacecraft will make 6 orbit-transfer maneuvers to lower the orbit to a 2 hour, 100 km (62 miles) circular polar science orbit. During the transition to lower orbit, the relay satellite will be released into a 100 km × 2,400 km (62 × 1,491 miles) polar orbit and the VRAD satellite will be released into a 100 × 800 km (62 × 497 miles) orbit. The main orbiter will maintain the circular orbit for one year of science operations, using correction burns roughly every two months to maintain the orbit within 30 km (19 miles) of the 100 km (62 miles) nominal orbit. An option to lower the orbit to 40–70 km after one year is being considered.

Lunar Reconnaissance Orbiter (LRO). The Lunar Reconnaissance Orbiter (LRO) is a Moon orbiting mission scheduled to launch in the fall of 2008. The first mission of NASA's Robotic Lunar Exploration Program, it is designed to map the surface of the Moon and characterize future landing sites in terms of terrain roughness, usable resources, and radiation environment with the ultimate goal of facilitating the return of humans to the Moon. The following measurements are listed as having the highest priority: characterization of deep space radiation environment in lunar orbit; geodetic global topography; high spatial resolution hydrogen mapping; temperature mapping in polar shadowed regions; imaging of surface in permanently shadowed regions; identification of putative deposits of appreciable near-surface water ice in polar cold traps; assessment of meter and smaller scale features for landing sites; and characterization of polar region lighting environment. A primary goal of the mission is to find landing sites suitable for in-situ resource utilization (ISRU).

Preliminary plans call for the LRO to be launched from Kennedy Space Center in October 2008 on a Delta II (2925-10) but this could be upgraded to a 2925H-10, an Atlas V or a Delta IV. It will take 4 days to reach the Moon and enter an initial orbit with a periselene altitude of 100 km (62 miles) which will then be lowered. The mission is expected to last for one year in a 30–50 km (19–31 miles) altitude lunar polar orbit. This may be followed by an extended mission of up to 5 years in a higher altitude low-maintenance orbit. The satellite is expected to have a launch mass of about 1,000–1,200 kg (2, 205–2,646 lbs), with 500–600 kg (1,102–1,322 lbs) of this being propellant. The platform will be three-axis stabilized and power of about 400 W will be provided by solar arrays and stored in Li-ion batteries. Communications will be via S-band for uplink and low rate downlink and Ka-band for high rate downlink (100-300 Mbps). The final design is expected soon, the spacecraft will have the capability of carrying about 100 kg (220.5 lbs) of scientific payload which will be composed of: a high resolution (one meter (3.28 ft) or better) camera to acquire images of small scale landing site hazards and document lighting conditions at the lunar poles; a laser altimeter to measure landing site slopes and search for polar ices; a neutron detector to search for water ice and characterize the space radiation environment; a radiometer to map the temperature of the lunar surface to identify cold traps and possible lunar ice deposits; a Lyman-alpha mapper to observe the lunar surface in ultraviolet, looking for surface ices and frosts and imaging permanently shadowed regions; and a cosmic ray telescope to investigate background space radiation. NASA has also signed an agreement with the U.S. National Reconnaissance Office to cooperate on the development of a miniature synthetic aperture radar sensor to map the Moon's surface.

See also **Apollo Missions**; **Lunar Roving Vehicle**; and **Moon (Earths)**.

Additional Reading

Beattie, D.A., and R.D. Launius: *Taking Science to the Moon: Lunar Experiments and the Apollo Program*, Johns Hopkins University Press, Baltimore, MD, 2003.

Bussey, B., and P.D. Spudis: *The Clementine Atlas of the Moon*, Cambridge University Press, New York, NY, 2003.

Chaikin, A.L., with Foreword by Tom Hanks: "Man on the Moon," *The Voyages of the Apollo Astronauts*, Penguin Groug (USA), New York, NY, 1998.

Eckart, P., and B. Aldrin: *The Lunar Base Handbook*, The McGraw-Hill Companies, Inc., New York, NY, 1999.

Schmitt, H.J.: *Return to the Moon: Exploration, Enterprise, and Energy in the Human Settlement of Space*, Springer-Verlag New York, LLC, New York, NY, 2005.

Scott, D., and A. Leonov: *Two Sides of the Moon: Our Story of the Cold War Space Race*, St. Martin's Press, New York, NY, 2004.

Staff: "Committee on Planetary and Lunar Exploration Space Studies Board," *Quarantine and Certification of Martian Samples*, National Academy Press, Washington, D.C., 2002.

Ulivi, P., and D.M. Harland: *Lunar Exploration-Human Pioneers, Robotic Surveyors*, Springer-Verlag New York, LLC, New York, NY, 2004.

Web References

A History of the Lunar Orbiter Program: http://www.hq.nasa.gov/office/pao/History/TM-3487/top.htm

A History of Project Ranger: http://history.nasa.gov/SP-4210/pages/Cover.htm

Apollo 8 Mission: http://nssdc.gsfc.nasa.gov/planetary/lunar/apollo8info.html

Chandrayaan-1 Mission: http://www.isro.org/chandrayaan-1/announcement.htm

Lunar Prospector Mission: http://lunar.arc.nasa.gov/

Lunar Reconnaissance Orbiter (LRO) Mission: http://lunar.gsfc.nasa.gov/missions/

NSSDC Image Catalog; The Moon: http://nssdc.gsfc.nasa.gov/imgcat/html/group_page/EM.html

SELENE Mission: http://www.isas.ac.jp/e/enterp/missions/selene/index.shtml

Smart 1 Mission: http://smart.esa.int/science-e/www/area/index.cfm?fareaid=10

Unmanned Space Project Management; Surveyor and Lunar Orbiter: http://www.hq.nasa.gov/office/pao/History/SP-4901/table.htm

DAVID R. WILLIAMS, NASA Goddard Space Flight Center

LUNAR PROSPECTOR MISSION. The *Lunar Prospector*, the first dedicated lunar mission in 25 years, was a tremendous success. Following

a near flawless launch on Jan 6, 1998, a four-day journey to the Moon and entry into lunar orbit, the tiny spin-stabilized spacecraft sent data back to Earth. Lunar data from the circular polar-mapping orbit started arriving January 15.

On March 5, 1998 *Prospector* scientists captured the public's imagination by announcing the discovery of a definitive signal for water ice at both of the lunar poles. At that time, a conservative analysis of the available data indicated that a significant quantity of water ice, possibly as much as 300 million metric tons, was mixed into the regolith (lunar soil) at each pole, with a greater quantity existing at the north pole. The first competitively selected Discovery class mission had conclusively demonstrated that, not only could a cost-capped, fast-development mission succeed, it could do ground-breaking science in the process.

The first operational gravity map of the Moon was announced at the same time. Since then, Lunar Prospector engineer's have taken advantage of the missions own science results and the gravity data have been used to facilitate orbit maintenance.

Mission Profile

At 9:28 p.m. (EST) on January 6, 1998, *Lunar Prospector* (LP) blasted off to the Moon aboard a Lockheed Martin solid-fuel, three-stage rocket called Athena II. It was successfully on its way to the Moon for a one-year, polar orbit, primary mission dedicated to globally mapping lunar resources, gravity, and magnetic fields, and even outgassing events. About 13 minutes after launch, the Athena II placed the Lunar Prospector payload into a "parking orbit" 115 miles above the Earth. Following a 42-minute coast in the parking orbit, *Prospector's* Trans Lunar Injection (TLI) stage successfully completed a 64-second burn, releasing the spacecraft from Earth orbit and setting it on course to the Moon, a 105-hour coast. See Fig. 1. The official mission timeline began when the spacecraft switched on 56 minutes, 30 seconds after liftoff. Shortly after turning the vehicle on, mission controllers deployed the spacecraft's three extendible masts, or booms. Finally, the spacecraft's five instruments—the gamma-ray spectrometer, alpha particle spectrometer, neutron spectrometer, magnetometer and electron reflectometer—were turned on. On Sunday, January 11, at 7:20 a.m. (EST), *Lunar Prospector* was successfully captured into lunar orbit, and a few days later began its mission to globally map the Moon. See Fig. 2.

The *Lunar Prospector* spacecraft is shaped like a drum, 4.25 feet (1.3 meters) high, with a diameter of 4.5 feet (1.4 meters). When full of fuel, Prospector weighed 650 pounds (295 kilograms). The three science masts carrying its five science instruments and isolating them from the spacecraft's electronics are each 8 feet (2.4 meters) long. The spacecraft was built by Lockheed Martin Missiles & Space, of Sunnyvale, California. See Fig. 3.

The *Lunar Prospecter* was a small spin-stabilized spacecraft in a polar orbit with a period of 118 minutes at a nominal altitude of 100 km (63 miles). Since the Moon rotates a full turn beneath the spacecraft every lunar cycle (~27.3 days) as it zips around the Moon every 2 hours, Prospector visited a polar region every hour and completely covered the lunar surface twice a month. *Prospector's* one-year-long primary mission with an optional extended mission of a further 6 months at an even lower altitude enabled large amounts of data to collect over time. For some science instruments, a significant amount of time was required to obtain high quality usable data. Thus, *Prospector's* polar orbit and long-mission time rendered it ideal from the standpoint of globally mapping the Moon.

*(1.3 m in diameter X 1.4 m tall bus with three 2.5 meter science masts carrying its five science instruments and isolating them from the spacecraft's electronics)

Lunar Prospector Scientific Goals

As a Discovery-class mission, Prospector's scientific goals were carefully chosen to address outstanding questions of lunar science both efficiently and effectively. In the Post-Apollo era, NASA convened the Lunar Exploration Science Working Group (LExSWG) to draft a list of the most pressing, unanswered scientific riddles still facing the lunar-science community. In 1992, LExSWG produced a document, entitled "A Planetary Science Strategy for the Moon." The following lunar science objectives were listed: How did the Earth-Moon system form? How did the Moon evolve? What is the impact history of the Moon's crust? What constitutes

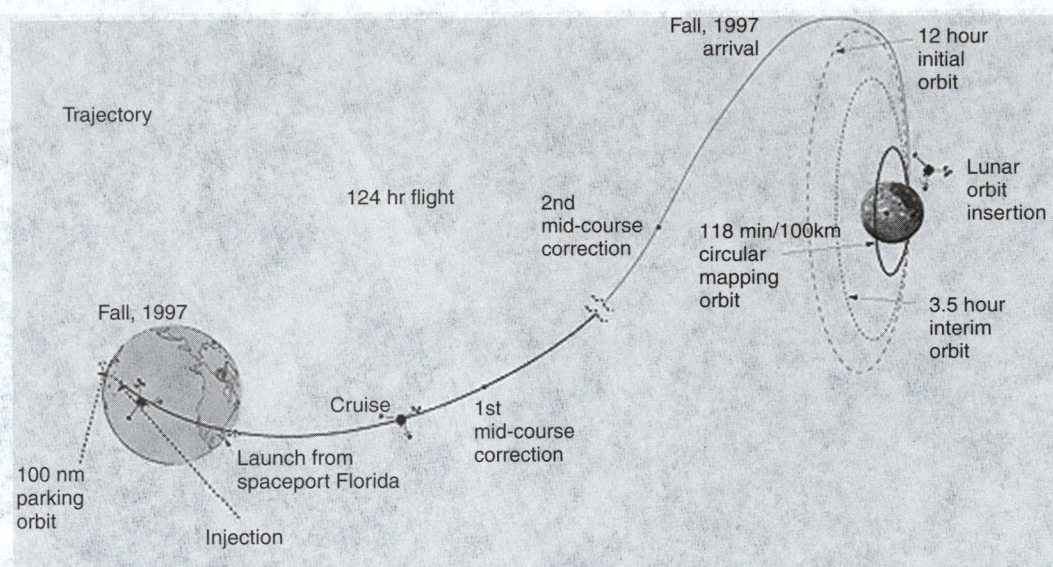

Fig. 1.   This diagram shows Prospector's flight path to the Moon. The spacecraft circles the Earth before achieving injection over Australia. There will be at least two mid-course corrections before reaching the Moon. Prospector is inserted in an initial large elliptical orbit around the Moon and goes through an interim orbit before achieving its final circular mapping orbit. This method of gradual orbit insertion assures the fuel efficiency vital to Prospector's low-cost design. (*NASA, Goddard Space Flight Center.*) http://lunar.arc.nasa.gov/lunar/ Archives-Images Choose a Mission (Lunar Prospector) Select A Category: (Mission) Fig. 2.

the lunar atmosphere? What can the Moon tell us about the history of the Sun and other planets in the Solar System?

Lunar Prospector mission designers carefully selected a set of objectives and a payload of scientific instruments which would address as many of LExSWG's priorities as possible, while remaining within the tight budget confines of NASA's "faster, better, cheaper" Discovery Program.

Lunar Prospector's identified critical science objectives were:

- "Prospect" the lunar crust and atmosphere for potential resources, including minerals, water ice and certain gases,
- Map the Moon's gravitational and magnetic fields, and
- Learn more about the size and content of the Moon's core.

The six experiments (five science instruments) which address these objectives were:

- Neutron Spectrometer (NS)—Map hydrogen using neutrons having several different energy ranges and thereby infer the presence or absence of water.
- Gamma Ray Spectrometer (GRS)—Map 10 key elemental abundances, several of which offer clues to lunar formation and evolution.
- Magnetometer and Electron Reflectometer (Mag/ER)—These two experiments combine to measure lunar magnetic field strength at the surface and at the altitude of the spacecraft and thereby greatly enhance understanding of lunar magnetic anomalies.
- Doppler Gravity Experiment (DGE)—Make an operational gravity map of the Moon for use by future missions as well as LP by mapping gravity field measurements from changes in the spacecraft's orbital speed and position.
- Alpha Particle Spectrometer (APS)—Map out-gassing events by detecting Radon gas (current outgassing events) and Polonium (tracer of recent, i.e. 50 years).

The Basics of Lunar Selenology (Geology)

Formation The current most widely held theory of how the Moon was formed is the impact theory. This theory suggests that a large body, possibly the size of Mars impacted the Earth some 4 billion years ago and the material, which was thrown off, a combination of Earth material and the impactor, coalesced into Earth's Moon. The body was originally molten. As it cooled, light elements such as calcium and aluminum and silica crystallized into feldspar and other minerals and floated upwards. Heavier minerals such as iron and magnesium-rich olivine and pyroxene formed and sank downward forming the lunar mantle. Incompatible elements such as KREEP (Phosphorous-K, Rare Earth Elements, and Potassium) were

trapped between the layers. See also **Rare-Earth Elements and Metals**. Repeated large impacts on the lunar surface itself, around 4.1 to 3.9 billion years ago, formed craters, which gradually filled with basaltic lava. These include the large dark areas that can be seen from Earth, the maria such as Mare Tranquilitatus-the Sea of Tranquillity and Mare Imbrium-the Sea of Rains. Over time, the Moon has continued to be bombarded from space by crater forming impacts. As the Moon cooled, however, lava ceased to fill the craters, and the impacts began to "garden" or break up the brittle lunar surface, creating the powdery dust, which is known as regolith or lunar "soil."

Lunar topography can be roughly divided into two main types—the maria with their distinctive dark color, and the highlands which are significantly different in composition as shown in Fig. 4.

Photographic Experiments

Ground-based astronomy provided early views of the Moon's near side, the side that perpetually faces the Earth. In addition, thirty-eight previous lunar missions have photographed the Moon, supplying scientists with a plethora of still imagery and television footage of the lunar surface. In 1959, the Soviet's *Luna 3* spacecraft was the very first to capture a full composite view of the far side of the Moon. During the 1960s and 1970s, NASA successfully flew 22 missions to the Moon, including the historic Apollo missions. The Soviets sent another 20. Throughout the 1960s and 1970s, these missions continued to amass images of the Moon and, in 1965, the American public watched for the first time, live pictures transmitted from NASA's *Ranger 9* spacecraft. Cameras aboard NASA's *Lunar Orbiter 1* spacecraft took the first pictures of Earth from the Moon. All of the Apollo missions returned lunar photographs to Earth; even the failed *Apollo 13* mission yielded a limited amount of photographic data. The U.S. Department of Defense *Clementine* spacecraft, which briefly orbited the Moon in 1994, obtained a significant set of lunar imagery. The polar orbit was an ellipse, roughly 400 kilometers above the surface at the closest point. The payload aboard *Clementine* was a camera and a topographic mapper.

Lunar Prospector's Design Philosophy

Lunar Prospector was never intended to carry a camera, nor did it require a true onboard computer for two key reasons. *Prospector*, from its inception, was designed to be a simple, cost-efficient spacecraft. Cameras require pointing in a given direction. The *Lunar Prospector* spacecraft was spin-stabilized, an engineering strategy that is well understood, highly successful and simple and cheap to achieve. With a camera, Prospector would have

Fig. 2. This artist's conception of Lunar Prospector shows the Spacecraft in lunar orbit with its instrument masts fully deployed. Prospector's primary mission will keep the spacecraft in a 100 km polar mapping orbit for a full year or more. This orbit will provide higher quality science data than has previously been obtained. Prospector has an Extended Mission option of a 10 km, very high resolution orbit for a brief period of time. (*NASA, Goddard Space Flight Center.*) http://lunar.arc.nasa.gov/lunar/ Archives-Images Choose a Mission (Lunar Prospector) Select a Category (Spacecraft) (Fig. 13).

TABLE 1. APPROACHES TO LUNAR SCIENCE AND PREVIOUS MISSIONS

Experiment Type (* = aboard LP)	Previous lunar missions
Photographic Studies	U.S. Missions: Lunar Orbiter 1–5, Surveyor 1, Surveyor 3, Surveyor 5–7, Ranger 7–9, Apollo 8, Apollo 10–17, Clementine Soviet missions: Luna 3, Luna 9, Luna 12–13, Luna 16–17, Luna19–22, Zond 3, Zond 6–8
Surface Sampling/Soil Analysis	U.S Missions: Surveyor 3, Surveyor 5–7, Apollo 11–12, Apollo 14–17 Soviet Missions: Luna 16–17, Luna 20–21, Luna 24
Magnetic Studies*	U.S. Missions: Apollo 12, Apollo 14–17 Soviet Missions: Luna 10, Luna 21–22
Gravity Studies*	U.S. Missions: Apollo 12, Apollo 14–17, Clementine Soviet Missions: Luna 10–11, Luna 14, Luna 19, Luna 22
Alpha Particle Spectroscopy*	U.S. Missions: Apollo 15–16
Gamma-Ray Spectroscopy*	U.S. Missions: Apollo 15–16 Soviet Missions: Luna 10–11, Luna 22
Neutron Spectroscopy*	U.S. Missions: none, although Apollo 17 had a "neutron probe"

required a whole different engineering strategy, as well as a computer to control the camera. The result would have been a far more complex, heavier, costlier spacecraft that may well not have survived the Discovery selection process. Secondly, *Lunar Prospector* was a science-focused mission. That is, *Prospector* was designed to answer the highest-priority outstanding questions the planetary science community still has about the Moon. The spacecraft's five instruments were chosen according to that

criterion. Previous Moon missions have, in fact, taken thousands of pictures of the Moon. We already have quite detailed knowledge of what the Moon looks like. What we did not know were details about its composition, volcanic activity, magnetic and gravitational fields and/or its resources.

Lunar Sampling/Soil Analysis Experiments

The first spacecraft to dig into the soil of an extraterrestrial body (which happened to be the Moon) was NASA's *Surveyor 3*, which landed on a geographic region of the Moon called the *Oceanus Procellarum* in April 1967. Over the next several months, each of the *Surveyor landers* (5–7) probed the lunar surface with instruments called alpha-scattering surface analyzers and soil mechanics surface samplers. Each was designed to physically probe the lunar soil and gather measurements about the abundances of the various elements that constitute lunar rocks and soil. Data gathered with these two instruments was valuable, but limited in the sense that only specific regions of the Moon were analyzed (the landing sites). Some of the Soviet missions also excavated lunar samples for further analyses on Earth.

All of the *Apollo landers* (11–17), with the exception of the failed *Apollo 13*, performed experiments with lunar surface samples. With tools such as hammers, scoops, rakes and tongs, the Apollo astronauts collected many lunar rock samples and performed a series of tests on the mechanics of the lunar soil. In addition, they returned over 800 ponds of lunar rock to Earth to enable further detailed study.

Lunar Prospector's **Approach**

Lunar Prospector, a polar-orbiting spacecraft, did not touch down on the lunar surface during either its primary (first year) or extended mission (up to an additional six months after that, until its fuel supply runs out). While previous lunar landing missions have returned an immense amount of physical data, the main limitation of the experiments thus far had

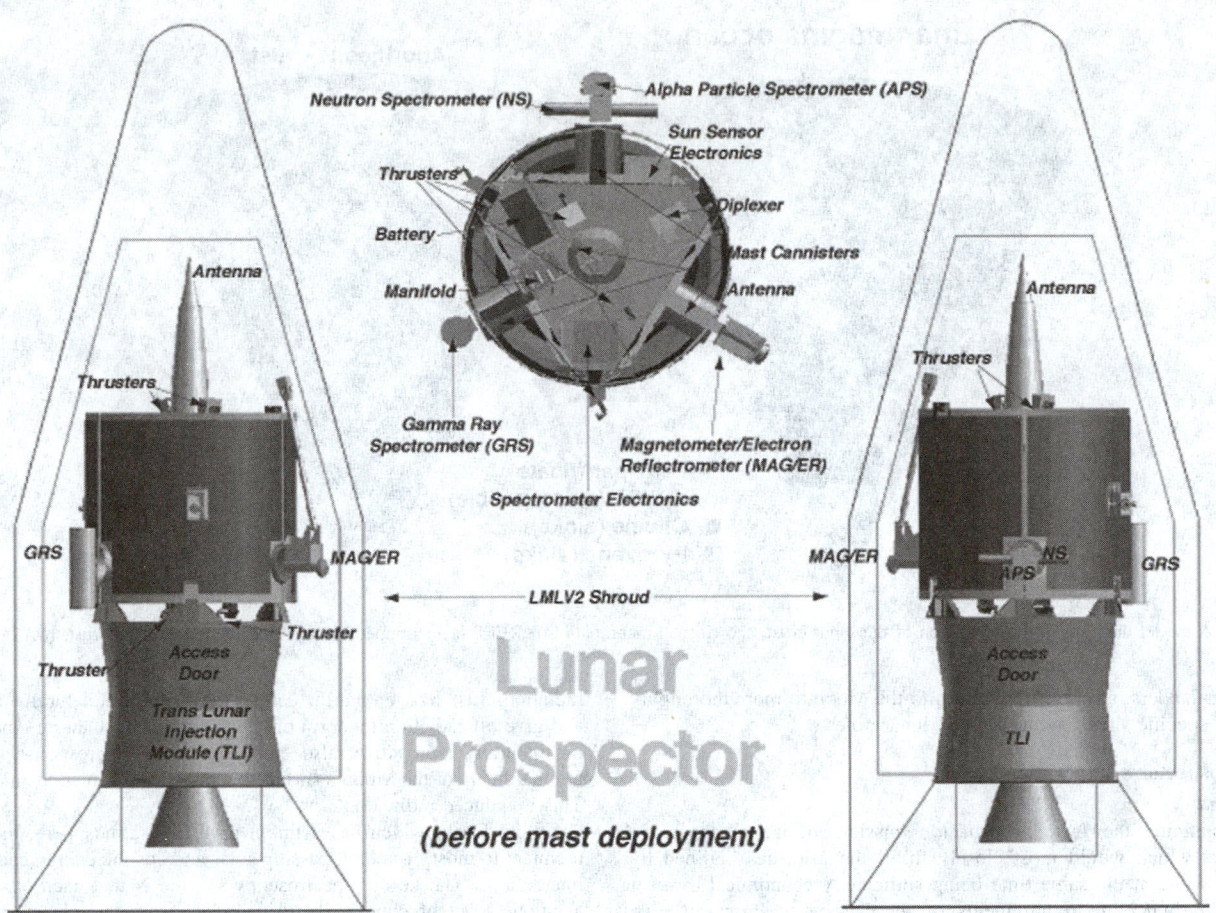

Fig. 3.   This diagram of the Lunar Prospector shows the Spacecraft with its instrument masts stowed. The three sets of science instruments: (1) Gamma Ray Spectrometer (GRS), (2) Neutron Spectrometer (NS) with Alpha Particle Spectrometer (APS), and (3)Magnetometer with Electron Reflectometer (Mag/ER) are identified. The antenna which will facilitate the Doppler gravity experiment, as well as command and control is identified. The major electronics boxes are also identified. (*NASA, Goddard Space Flight Center*.) http://lunar.arc.nasa.gov/lunar/ Archiver-Images Choose a Mission (Lunar Spacecraft) Select a Category (Spacecraft) Fig. 2.

applicability to the Moon as a whole. Lunar soil and rocks at the Moon's equator, where the majority of such experiments have been conducted tell a limited story. Global elemental abundances need to be determined. Until Prospector's confirmation of polar water ice, that valuable resource and potential sample was not seriously considered as the basis for a sample return mission. Before a future sample return mission could be contemplated, it was necessary to have an understanding of the entire global situation in order to profitably plan such a mission. Hence, over time, scientists rely on data from both orbiters, such as LP and landers, to completely understand a planet and develop an approach to its study.

Remote Sensing Experiments

In contrast to direct geologic studies of actual lunar rocks and soil, remote sensing experiments analyze the lunar surface and atmosphere from a distance. Photography is, of course, a kind of remote sensing study, one with which we are familiar. Spectroscopy, however, provides information not discernible to the human eye. Using gamma-ray spectroscopy, *Apollo 15* and *16* were the first missions to attempt to measure the concentrations of elements that make up the lunar crust using spectroscopy. (A primitive version of the Apollo GRS was flown aboard the Soviet's *Luna 10* spacecraft in 1966, but this instrument gathered only limited data). The general findings from the *Apollo 15/16* GRS experiment indicated that there are two distinct geographic regions of the Moon — the highlands and the mare — with chemical compositions that are, in fact, markedly different (for example, iron and magnesium being relatively enriched in the mare). Those data also indicated that the concentrations of certain other elements, such as thorium and titanium, are non-uniformly distributed and that the far side of the Moon differs from the near side.

The *Apollo 15/16* GRS acquired data on the abundance of four elements — thorium, potassium, iron, and titanium — near the equatorial region of the Moon. In concert, the landing modules for those missions collected samples from the same region, permitting comparison of chemical composition data acquired via remote sensing techniques (GRS) and direct sampling (mass spectrometry of soil and rock samples). *Lunar Prospector* mission scientists, of course, would not be able to conduct similar studies comparing in situ samples with remotely acquired data, however, global coverage offers insights not achievable with single point investigations. In addition, because we do have an existing store of lunar rock, LP data is being compared to current samples and may reveal important information.

Lunar Prospector's Role

Since *Lunar Prospector* was an orbiting spacecraft, its entire science payload consists of remote sensing instruments, three spectrometers and the two magnetic field instruments. As with any form of spectroscopy, the quality of the data improves with the time of sampling (number of counts detected). Statistically, it is very important for mission scientists to continually gather spectral information over the same lunar surface regions in order to subtract-out background "noise" caused by not only the cosmic environment but also by the elements which make up the materials on the spacecraft and the instruments themselves. However, with LP, interference from the spacecraft and its electronics were kept to a minimum by the fact that the instruments were separated from the body of the vehicle by eight-foot booms.

Using spectroscopy and other remote sensing techniques, the Prospector mission can perform global-mapping studies not possible with landing craft, such as were typical of the Apollo series of missions. For the scientific community, such an approach represented the logical next step in lunar exploration. Based on the scientific findings of the *Lunar Prospector*,

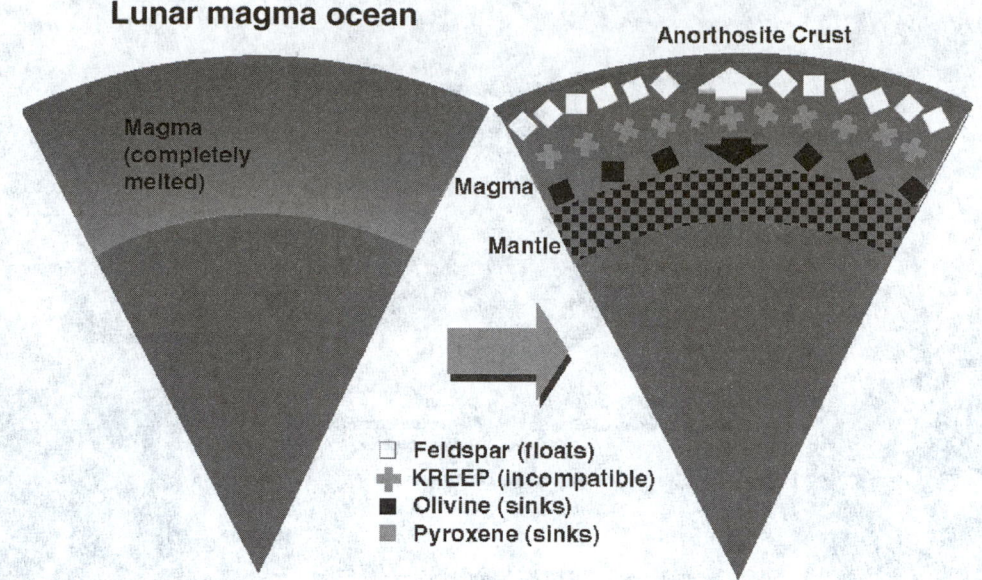

Fig. 4.   A model diagram of the formation of the lunar crust, showing an incompatible KREEP layer trapped between the mantle and the crust. (*NASA, GSFC.*)

future lunar landers may someday return to the Moon to more thoroughly investigate specific sites, say at the 'icy' lunar poles.

Lunar Prospector's Experiments
Background

When planning the *Lunar Prospector* mission, designers chose six experiments which would most closely fulfill the priorities outlined by LExSWG, while at the same time being sufficiently economical to meet the stringent budgeting requirements of the Discovery Program. Five instruments aboard Prospector are conducting six experiments. Three spectrometers (gamma-ray, alpha-particle and neutron) were globally mapping the Moon's surface and tenuous atmosphere (consisting mainly of gas release events) to determine which minerals and gases are present, and in what abundancy. Two instruments, the magnetometer and electron reflectometer, were measuring the Moon's anomalous magnetic properties, and in doing so, began to characterize the nature and size of a possible lunar core. Finally, the vehicle's own telemetry (communication system) served as the "instrument" for *Lunar Prospector's* Doppler Gravity Experiment, which globally mapped the Moon's non-uniform gravitational fields.

Why Use Spectroscopy?

Physicists use spectroscopy for a variety of research pursuits. A rather general term, spectroscopy is nothing more than the process of visualizing a substance by splitting its light and other emissions into their constituent parts. One of the simplest spectra to consider is a rainbow: visible light is dispersed into its different energies (called wavelengths) when it passes through droplets of rainwater in the atmosphere. Each wavelength absorbs (and thus reflects) a different color, which is readily apparent to the naked eye. This is so because each wavelength has a characteristic energy level. Spectrometers are instruments, which record such dispersions of energy, called spectra.

Since it is no trivial matter to collect rocks and soil samples from other planets, break them up into their individual elements and thereby determine their chemical makeup, spectrometers play a major role in planetary exploration. Knowing what planets, asteroids and moons are made of helps scientists piece together the history and evolution of the Solar System. One type of spectroscopy, gamma-ray spectroscopy, is a technique in which scientists can measure (from orbit) the composition of surface material (down to several inches) around any celestial body possessing little or no atmosphere (such as is the case for the Moon or for Mars, for instance).

The way all of *Lunar Prospector's* three spectrometers worked is by detecting (remotely, from orbit) "signature" energies emitted by various elements in the lunar soil and atmosphere. Such energy emitted by the Moon comes from two sources: "natural" radiation and "induced"

radiation. Just like on Earth, certain elements are naturally radioactive and give off radiation (energy) on a constant basis. Other elements, while not naturally radioactive, also emit energy, but in response to constant bombardment of the Moon's surface by galactic and solar cosmic radiation. This is induced radiation.

Unlike Earth, which has a thick, protective atmosphere, the Moon is recipient to most, if not all, passing solar and cosmic energy causing many interactions. The key to spectroscopy's value is that each element emits a unique level of energy. By using a spectrometer, planetary scientists can discern the chemical makeup of the surface of a planet by looking for signature "peaks" (in energy emission) of an energy output plot—an identifying barcode of sorts for the actual elements, which make up the crust and atmosphere. Because solids and gases emit vastly different levels of energy, spectrometers are usually calibrated to detect ranges of energies. Each of *Prospector's* three spectrometers "sees" a characteristic energy range, permitting the detection of energy particles such as fairly-low-energy neutrons (the neutron spectrometer's range of detection is from less than 0.3 eV [electron volts] to hundreds of keV [thousand electron volts]), high-energy gamma rays (the gamma-ray spectrometer's range of detection is ~0.3 MeV [million electron volts] to 9 MeV), higher energy neutrons (a portion of the gamma-ray spectrometer is used to detect neutrons with energies from 0.5 MeV to 9.5 MeV) and alpha particles (the alpha particle spectrometer's range of detection is ~4.1 MeV to 6.6 MeV).

Gamma-Ray Spectroscopy

Lunar Prospector's gamma-ray spectrometer (GRS) mapped the abundances of ten elements on the Moon's surface:

thorium (Th)	silicon (Si)
potassium (K)	aluminum (Al)
uranium (U)	calcium (Ca)
iron (Fe)	magnesium (Mg)
oxygen (O)	titanium (Ti)

The GRS is especially sensitive to the heavy, radioactive element thorium and the lighter element potassium. These elements are particularly plentiful in the part of the crust that is last to solidify. Thus, mission scientists are able to determine the global distribution of KREEP (K-potassium, Rare Earth Elements, and P-phosphorous), a chemical "tracer" of sorts which helps to tell the story of the Moon's volcanic and impact history. The data produced by the GRS are helping scientists to understand the origins of the lunar landscape, and may also tell future explorers where to find useful metals like aluminum and titanium.

How *Lunar Prospector's* GRS Works

A gamma ray is a very energetic photon (a tiny parcel of light)—more energetic than a visible light ray or an X-ray. When gamma rays reached the orbiting *Lunar Prospector* spacecraft, they passed through a crystal of bismuth germanate (BGO crystal) in the GRS.

The various atoms inside this detector give off a flash of light when the radiation hits them. Gamma-ray photons with high energies produce brighter flashes than gamma-ray photons with low energies. The light produced by the gamma-ray is then measured by a photomultiplier tube (PMT) which converts the light signal into an amplified electronic signal. Finally, this electronic signal is sent back to Earth for scientists to analyze. The energy of a given gamma ray tells scientists exactly which kind of atom emitted it.

To fully appreciate the potential of LP's GRS, a useful comparison can be drawn with the earlier Apollo GRS experiments. In contrast to *Lunar Prospector*, which mapped the elemental composition of the entire lunar surface, the mapping performed by *Apollo 15* and *16* only covered about 20 percent of the lunar surface—specifically, the region around the Moon's mid-portion or equator. Another difference between the Apollo-era and the *Lunar Prospector* gamma-ray spectroscopy studies is the detecting crystal inside the GRS instrument itself. The *Apollo 15/16* instrument used a sodium iodide (NaI) crystal, whereas *Lunar Prospector's* crystal was composed of bismuth, germanium, and oxygen atoms (BGO crystal).

Since the combined atomic weight of bismuth, germanium and oxygen exceeds that of sodium and iodine, a BGO crystal is significantly denser than an NaI crystal. As a result, a BGO crystal is better able to stop gamma rays in their tracks and, as such, offers greater detection sensitivity—on the order of two- to eight-fold higher—than an NaI crystal. What that means is that energy spectra measured with a BGO crystal can be more cleanly separated (lines can be more easily distinguished from one another on an energy plot) than spectra measured with a NaI crystal as illustrated in Fig. 5. In addition, certain elements, which were unmappable by the Apollo GRS, such as uranium, aluminum, calcium and magnesium, were detected with Lunar Prospector's more sensitive instrument. This offered mission scientists more opportunities to distinguish subtle features of the lunar landscape. The ability to measure concentrations of aluminum and calcium, for instance, may unearth clues as to the makeup of certain types of highland rock formations. Similarly, the distribution of titanium serves as a useful "probe" for mare regions. Other geochemical clues to planetary evolution include the presence of iron stores, the ratio of iron oxide (FeO) to magnesium oxide (MgO), the ratio of potassium to uranium (which hints at remelting rates of primordial condensates), and the ratio of thorium (Th) to uranium, which serves as a marker for the relative abundance of volatile compounds.

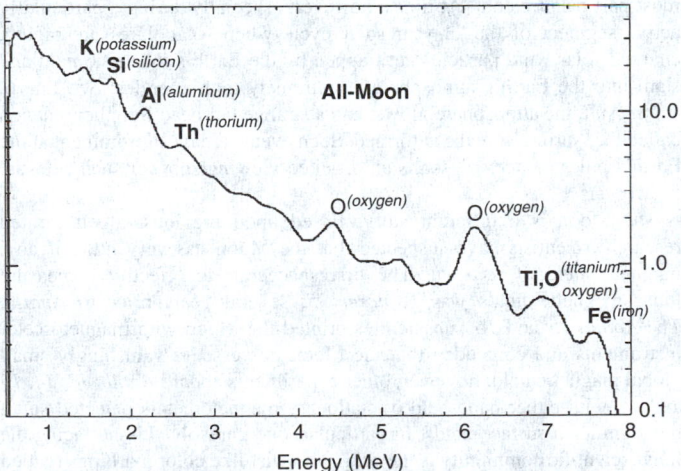

Fig. 5. Sample data plot from Lunar Prospector's gamma-ray spectrometer. (*NASA, GSFC*.)

While gamma-ray data is highly informative, relatively few gamma-rays leave the Moon's surface and escape into space. Much as a camera working in conditions of low light can compensate by increasing exposure time, gamma ray spectroscopy generally benefits from a significantly long detection period, allowing spectral lines to fill in over time. One sweep—or even a few sweeps—over the Moon's surface would not give

mission scientists enough information to determine the concentration of radioactive elements. In addition, the stable (non-radioactive) elements do not emit gamma rays as readily as do naturally radioactive ones.

The GRS also contributed indirectly to the search for water on the Moon. The bismuth germanate crystal is surrounded by a shield of borated plastic (anti-coincidence shield) that detects high-energy (fast) neutrons. The main purpose of this shield is to allow correction for background signal caused by solar and galactic cosmic rays. It differentiates between gamma rays and the cosmic ray background. In addition, because it is borated, it also measures fast (high energy) and epithermal (medium-energy) neutron fluxes. Mission scientists are using this information, in concert with the lower-energy (thermal and epithermal) neutron counts detected by Lunar Prospector's neutron spectrometer, to detect water ice at the lunar poles.

Alpha-Particle Spectroscopy

The Apollo series of missions revealed that the Moon had not been perpetually cold and dead, as once believed, but rather was host to a series of dramatic volcanic eruptions in which vast seas of molten lava flooded much of the lunar surface. While the majority of such activity most likely occurred very early in the Moon's history over three billion years ago, the Moon is thought to still harbor some remnant volcanic and tectonic activity. Outgassing events, in which alpha-particle emissions of radon leak out from the lunar interior, are scientific evidence of such activity. Polonium, a natural-decay product of the heavier element radon, itself a natural-decay product of the still heavier element uranium, collects around vents and provides keys to their recent history. This is due to the fact that the entire decay chain takes over 21 years, so that Polonium is evidence of vent activity over the last half century. Radon gas and polonium were likely to be detected because of their relatively long half-lives (3.8 days for [222] radon and 138 days for [210] polonium; one half-life is the amount of time it takes for half of a given radioactive sample to decay into another substance).

Ancient volcanic vents, seismic fractures, impacts and pore openings in the lunar soil all provide paths for radon to find its way to the lunar surface. Actually, radon itself is present in very small quantities, but thought to be mixed in with other gases, such as nitrogen, carbon monoxide and carbon dioxide. Determining where and when such gas-release events take place can tell scientists just how active the Moon actually is, as well as help to identify the source(s) of the Moon's small and tenuous atmosphere.

An alpha particle is the nucleus of a helium atom: two protons and two neutrons bound together. Like gamma rays, alpha particles escape from radioactive elements as part of their natural-decay process. The alpha particles are emitted with a precise energy that serves as a fingerprint for the atom from which they came. *Lunar Prospector's* Alpha-Particle Spectrometer (APS) detected these events.

Housed inside the APS instrument are ten separate wafers of silicon. Silicon is a semiconductive material. When an alpha particle hits a silicon wafer, it creates a small track of charge. When a 25-volt bias is applied to the silicon wafer, the alpha particle's charge is funneled into an amplifier which then increases the charge. Since that pulse of charge is directly proportional to the signature energy of the alpha particle, scientists can infer the identity of the element, which emitted the alpha particle. The APS contains ten such silicon detectors, each sandwiched between gold and aluminum disks, and arranged on five out of six sides of a cube, enabling nearly a complete field of detection.

Before Lunar Prospector

Apollo 15 and *16* scientists, using an APS instrument for the first time, found evidence for a spatially variable distribution of radon and polonium. Those studies identified a striking correlation between polonium and the edges of lunar maria, especially Mare Fecunditatûs, but also at nearly all maria investigated.

Earlier astronomical studies noted that the Aristarchus crater region was the site of phenomena dubbed "transient optical events," in which regions of the lunar surface glowed and changed color for short periods of time. Some scientists believe that the light flashes may represent transient venting of volatile materials.

Factors in Analyzing APS Data

Data acquired by the APS aboard Prospector appeared in the form of counts—very similar to the way the GRS instrument works. As with the

GRS data, the number of counts accumulated (and thereby the time of sampling), were the key determinants of the sensitivity of the data. The *Apollo 15* and *16* missions gathered only about 8 days' worth of data around the Moon's equator. The Lunar Prospector gathered 18 months of mapping time.

One issue mission planners had to take into account when designing alpha-particle spectroscopic experiments (and in fact any type of experiment in which a spectrometer measures cosmic radiation) was the timing of the solar cycle. Repeating every 11 years, the solar cycle is a periodic phenomenon in which the overall extent of radiation (in the form of solar particles called *protons* and *alpha particles*) produced by the Sun varies in a predictable manner. A new solar cycle began in 1997, at which time sunspot activity was at a minimum and galactic cosmic rays (GCR) were at their 11-year maximum. Since more GCR protons imply more induced planetary gamma-ray emission, it was an optimal time for performing global spectrographic mapping experiments, such as gamma-ray and neutron spectroscopy, because the inherent signal to be detected would be higher than usual. However, getting closer to solar maximum, solar activity and its associated solar energetic-particle population increases, leading to higher background radiation "noise", so mission scientists had to take into account such stray counts and subtract them from the overall data.

Neutron Spectroscopy

Lunar Prospector mission scientists devised the neutron spectroscopy experiment to search for water ice at the poles of the Moon. As the world found out on March 5, 1998, at the mission's first science data return press conference, preliminary results from the experiment were indeed positive: water ice does exist on the Moon, and there appears to be more of it at the North than at the South pole. *Lunar Prospector* had detected a significant amount of hydrogen, which is inferred to be in the form of water. This was the first direct evidence of the presence of water ice at the Moon's frigid poles. *Lunar Prospector* is also the first interplanetary mission ever to use the neutron spectroscopy technique to detect water. *Prospector's* neutron spectrometer (NS) works by detecting hydrogen, by way of subatomic particles called *neutrons*.

Neutron Science

The materials we use every day are made up of molecules, which are made up of atoms. Inside the atoms are even tinier pieces of matter called subatomic particles. Neutrons are one type of subatomic particle—they are present in every atom of every molecule in our bodies as well as in all of the synthetic and natural substances in our environment. Besides being a basic building block of matter, neutrons serve as a useful experimental tool for physicists. Materials scientists, for instance, are interested in how atoms are packed within the molecules of different materials. How the individual atoms of a given material stack up against each other in large part determines the properties (strength, plasticity, etc.) of that particular material. One way scientists study molecular structure is to bombard atoms with high-energy neutrons and then wait and see where and how fast the neutrons scatter.

The same thing happens naturally in space. When cosmic rays collide with atoms in the lunar crust, they violently dislodge neutrons and other subatomic particles. Some of the neutrons escape directly into space—essentially unchanged—as "fast" neutrons. Other neutrons shoot off into the crust, where they slam into other atoms, bouncing around like pinballs. If they only run into heavy atoms, they do not lose very much energy in the collisions, and are still traveling at close to their original speed when they finally bounce off into outer space. But if a neutron encounters a hydrogen atom—which is the same size as a neutron—it will slow greatly or even stop, much like a speeding billiard ball running into a stationary one. If the Moon's crust contains an abundance of hydrogen at a certain location—say, a crater with water ice in it—any neutron that bounces around in the crust before heading out to space will cool off (slow down) rapidly. When *Lunar Prospector* flew over such a crater, the NS detected a definitive dropoff in the number of these ("epithermal" or medium energy) neutrons.

How Lunar Prospector's NS Works

The NS has two different counters—a cadmium-wrapped canister of [3]helium and a tin-wrapped canister of [3]helium. When a neutron collides with an atom of [3]helium, a nuclear reaction takes place, producing a burst of energy. That burst of energy tells mission scientists that they have detected a neutron. Except for the outside wrapping, the two counters are nearly identical. The cadmium-wrapped counter filters out all but the epithermal neutrons, because cadmium is good at screening away the slow-moving thermal neutrons, whereas the tin-wrapped counter lets all of the neutrons through. Since the two counters are otherwise identical, counts can be subtracted, and any difference between the two must be attributable to thermal neutrons.

Lunar Prospector measured "fast" and thermal plus epithermal neutron flux with a separate instrument (the anti-coincidence shield of the GRS). The respective count rates of the different types of neutron fluxes are an indicator of hydrogen, and hence the presence of water ice, embedded within the lunar soil.

Mission scientists received data from the spacecraft every 32 seconds. Since the data contained random noise, several passes over the surface and careful statistical analysis were required to analyze the data. However, since the spacecraft passed over the poles every orbit (whereas it passed over any given region on the equator only a few times a month), the NS produced the most accurate data in the polar regions.

Magnetometer/Electron Reflectometer Studies

The magnetometer and electron reflectometer aboard the *Lunar Prospector* collected valuable data to help unravel puzzles that have intrigued scientists for more than a quarter of a century. What kind of magnetic field(s) exists on the Moon? What kind of natural resources are buried in the Moon's crust and is there a core? If so, what are its characteristics? Can we build a lunar base? How did the Moon form and evolve—what is its history?

The MAG/ER experiment aboard *Lunar Prospector* was designed with two primary goals in mind: scanning the lunar crust for signs of permanent magnetization, and searching for electrical currents flowing deep within the lunar interior—the sign of a conductive metallic lunar core. The two instruments combined to calibrate the Moon's global magnetic field strength: the magnetometer measures the field surrounding the spacecraft, and the electron reflectometer surveys the lunar surface.

How Much Do We Already Know about the Moon's Magnetic Field?

Scientists have known for years that the Earth is magnetic—there is a strong magnetic field surrounding our own planet, originating with electrical currents swirling inside the Earth's iron-rich metallic core. The boundary between the Earth's magnetosphere (a dipole field) and the influence of the Sun's charged particle activity (from the solar wind) for the most part balances out. At times, however, especially during a particularly active segment of the 11-year solar cycle when solar flares are raging, charged solar wind particles get trapped by the Earth's magnetic field and slam into the Earth's atmosphere at extremely high energies. See Fig. 6. As a result, the atmosphere glows, and a strange but beautiful phenomenon called the Aurora Borealis is formed. Such events are most prominent at the Earth's poles; hence, Alaska is an oft-cited viewing spot for such celestial fireworks.

The Moon is a different story. Based upon previous, albeit limited research, scientists have suspected that the Moon has very little, if any, magnetic field of its own. The first spacecraft to effectively measure lunar magnetic fields was *Explorer 35*, several years prior to *Apollo*. The *Apollo 15* and *16* subsatellites orbited the Moon with magnetometer instruments and concluded that the Moon possessed a vanishingly small global magnetic field; however, other experiments aboard *Apollo 12, 14, 15* and *16* with either hand-held or stationary magnetometers detected small but significant surface fields. In particular, one unresolved issue facing the lunar scientific community is the origin of swirl-like color markings (called "albedos") visible from orbit on the surface of the Moon that are thought to be due to magnetic anomalies.

Why Are We Interested in the Magnetic Field?

The presence of such non-uniformly distributed magnetic regions has led planetary scientists to conjecture about the Moon's impacts (which may have imparted some of these locally distributed magnetic properties) and the possible presence of a small, iron-laden core (roughly estimated to a maximum of about 500 km in radius). From the perspective of planetary science, better understanding the size and nature of the Moon's core can

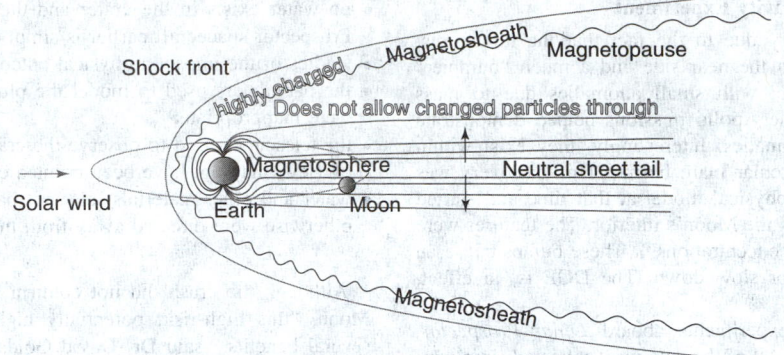

Fig. 6.   Illustration showing the Earth's magnetosheath and the shock front it creates. (*NASA, GSFC.*)

tell a lot about its evolution, history and beginnings—such as, did the Moon really arise as the result of a cataclysmic collision between the Earth and a Mars-sized body? In addition, metals discovered in the crust may be an extremely valuable resource. Now that mission scientists have found a potential source of water on the Moon, the presence (and relative abundance) of other materials may become paramount to the eventual establishment of a future possible lunar base.

How it Works

Lunar Prospector's magnetometer/electron reflectometer (MAG/ER) measurements were affected by three sources: the Earth's magnetic field (35,000 nano Teslas), the relatively weak field of the Moon (0 to 300 nano Teslas) and the very weak field carried from the Sun by the solar wind (approximately 10 nano Teslas). Mission scientists were able to compare these measurements to determine which magnetic field variations are caused by surface features (local deposits of magnetic material) or alternatively, by the Moon's core. Both instruments were copies of detectors housed aboard the *Mars Global Surveyor* spacecraft, launched in December 1996, with some modifications to adapt them for a spinning spacecraft.

The triaxial fluxgate MAG is a standard device that is also used to measure magnetic fields on Earth. "Triaxial" means that it includes sensors to measure the strength of the field in three different directions. This enables scientists to determine not only the maximum strength of the field but also the direction in which it points. The "fluxgate" is an electric coil through which the magnetic field passes. By measuring the variation of the current passing through the coil, the MAG determines the strength of the magnetic field. It can measure magnetic fields as weak as one-millionth the strength of the Earth's magnetic field.

Prospector's ER measured the magnetic field at the surface of the Moon. The Moon, like every other body in the Solar System, is constantly barraged by electrons from the solar wind. Unlike the Earth, the Moon was not thought to have a magnetic field strong enough to repel these tiny charged particles, and they would be expected to spiral toward the surface in giant loops, typically several miles wide. The electrons that descend in a tighter spiral make it to the surface of the Moon and are absorbed there. But if there is magnetic material on the Moon, it will reflect some of the electrons back into space. When those reflected electrons reached the *Lunar Prospector* spacecraft, the ER measured their pitch angle (the angle at which they bounce). Scientists saw an abrupt cutoff above a certain angle—because all the electrons with a larger pitch angle were absorbed by the Moon. That cutoff tells mission scientists just how intense the magnetic field is at the lunar surface.

Together, the two instruments detected local variations in the Moon's magnetic field that arise from selenological features on the lunar surface. *Lunar Prospector's* MAG instrument not only measured lunar magnetic fields, but also external fields present in space plasmas (a plasma is an electrically conductive gas comprised of both neutral and ionized particles and free electrons). As a result, only a small fraction (about 10 percent) of the MAG's data was pertinent to lunar scientists. What that means is that culling relevant data posed a significant challenge for *Prospector's* MAG/ER team. Essentially, this involved careful pruning of the data as well as paying attention to what times of the month the data was gathered, as certain times (and thereby location relative to the Earth and its own magnetic field) were better than others for measuring lunar fields.

The interaction between charged particles in the solar wind and the Earth's magnetic field (about 220 miles above the surface of the Earth) is a region scientists call the magnetosphere. Within the magnetosphere are two separate regions: the magnetosheath (which is magnetically "noisy") and the geomagnetic tail (which, by comparison, is "quiet"). It was important that mission scientists take lunar magnetic measurements while the spacecraft passed through the quieter geomagnetic tail. The data was expressed in units called "gammas" (which are the same as nano Teslas, or one-billionth of a Tesla) and were tagged with spacecraft positional data to produce computer files which demonstrate magnetic field components as a function of location relative to the Moon as well as spacecraft altitude.

As is the case with all of *Prospector's* scientific experiments, the spacecraft's polar orbiting pattern permitted complete global analysis of the Moon. So, for the first time, scientists had a handle on all of the Moon's magnetic features, not just those associated with isolated geologic regions. Such analyses also permitted scientists to correlate magnetic properties with lunar surface features—the MAG data was essential component in analyzing the data from the ER instrument. Data from the ER was plotted as electron flux (number of particles per unit area per unit time) as a function of time. The ER "probes" (from orbit, of course) the lunar surface by recording the behavior of space-born electrons after they bounce off of the lunar crust. But while the ER was capable of measuring the strength of electron paths ("bent" by tiny, locally dispersed lunar magnetic fields), this instrument is incapable of determining where on the Moon it was measuring electrons at any given time. Positional data from the MAG put the ER data into perspective, allowing mission scientists to match up electron flux strengths with the magnetic fields along which they travel. Thus, the two instruments (which were housed together on one of Prospector's eight-foot booms) provided mission scientists with a complete picture of the Moon's magnetic and electrical properties.

Gravity Studies

The gravity field of the Moon strongly influences the altitude of a spacecraft in low-circular orbit. The most dramatic example is the *Apollo 16* subsatellite. After being deployed in a near-circular orbit from the command and service module, the eccentricity increased quickly and the spacecraft impacted the lunar surface 35 days after the release strictly due to the force of the gravity field. Understanding the precise nature of a planet's gravity field is vital to all exploration and experimentation.

As presented at the March 5, 1998 science return press conference, *Lunar Prospector's* Doppler Gravity Experiment (DGE) has provided the first polar low-altitude measurement of the lunar gravity field. This provided the spacecraft with the first truly operational gravity map of the Moon and immediately improved orbit and fuel efficiency. Improved gravity information, will not only help scientists build better models of the role of impact processes on the history and evolution of the Moon, but will also help in estimating the lunar core size and metallic iron content. A more practical benefit of the new lunar gravity data provided by *Prospector's* DGE experiment is that a more precise gravity map of the Moon will inevitably aid future mission planners in planning fuel-efficient journeys to the Moon, and may even help identify potential resources.

Lunar Prospector's Doppler Gravity Experiment

The Moon has a large asymmetry due to the fact that the lunar crust is thicker on the far side than on the near side and a much "bumpier" gravitational field than the Earth, with small anomalies due to mass concentrations on the surface. The Apollo missions helped demonstrate such sizable positive gravity anomalies. Interestingly, they exist within the topographically low, large circular mare basins. This discovery was unexpected and opposite of any physical model at that time and started the development of new models of the Moon's interior. The features were called *mascons* (short for "mass concentrations"). These bumps cause an orbiting spacecraft to speed up or slow down. The DGE is, in effect, drawing a map of the bumps.

The DGE, unlike the other experiments aboard *Lunar Prospector*, required no extra instrumentation. All of the data was, collected simply by communicating with the spacecraft. As the spacecraft orbited the Moon, its speed was determined by the Doppler effect, the same effect that causes a police siren to sound higher when the police car is moving toward you and lower when it is moving away from you. The "siren," in this case, is the spacecraft's radio signal, whose frequency shifts slightly as it moves toward Earth or away from it. Relative to the near side, lunar farside gravity is poorly determined because the spacecraft is not in view from the Earth when over the lunar far side. However, some information is obtained by observing changes in the LP orbit due to the accumulated acceleration of the farside gravity as the spacecraft comes out of occultation (back into view).

By tracking the velocity of the spacecraft, mission scientists were able to infer the forces acting upon it. For over 99 percent of the duration of the mission (excepting only periods when the engines are being fired) the only force on *Lunar Prospector* was gravity. Thus, by simply circling the Moon and sending signals back to Earth, *Lunar Prospector* has mapped the Moon's global gravitational field. *Lunar Prospector* completed this gravitational map in the first two months of the mission. However, the results of the DGE, were greatly improved with data from the extended, low-altitude phase of the mission. At this low altitude of 6 miles (10 km), the precision of the gravity data was improved by a factor of over 100.

Lunar Prospector End of Mission Profile

The controlled crash of NASA's *Lunar Prospector* spacecraft into a crater near the south pole of the Moon on July 31, 1999 produced no observable signature of water, according to scientists digging through data from Earth-based observatories and spacecraft such as the *Hubble Space Telescope*.

This lack of physical evidence leaves open the question of whether ancient cometary impacts delivered ice that remains buried in permanently shadowed regions of the Moon, as suggested by the large amounts of hydrogen measured indirectly from lunar orbit by *Lunar Prospector* during its main mapping mission.

In a low-budget attempt to wring one last bit of scientific productivity from the low-cost *Lunar Prospector* mission, NASA worked with engineers and astronomers at the University of Texas to precisely crash the barrel-shaped spacecraft into a specific shadowed crater. NASA accepted the team's proposal based on successful scientific peer review of the idea and the pending end of the spacecraft's useful life, although the chances of positive detection of water were judged to be less than 10 percent.

Worldwide observations of the crash were focused primarily on using sensitive spectrometers tuned to look for the ultraviolet emission lines expected from the hydroxyl (OH) molecules that should be a by-product of any icy rock and dust kicked up by the impact of the 354-pound spacecraft.

"There are several possible explanations why we did not detect any water signature, and none of them can really be discounted at this time," said Dr. Ed Barker, assistant director of the university's McDonald Observatory at UT Austin, who coordinated the observing campaign. These explanations include:

- the spacecraft might have missed the target area;
- the spacecraft might have hit a rock or dry soil at the target site;
- water molecules may have been firmly bound in rocks as hydrated mineral as opposed to existing as free ice crystals, and the crash lacked enough energy to separate water from hydrated minerals;
- no water exists in the crater and the hydrogen detected by the Lunar Prospector spacecraft earlier is simply pure hydrogen;
- studies of the impact's physical outcome were inadequate;
- the parameters used to model the plume that resulted from the impact were inappropriate;
- the telescopes used to observe the crash, which have a very small field of view, may not have been pointed correctly;
- water and other materials may not have risen above the crater wall or otherwise were directed away from the telescopes' view.

Although the crash did not confirm the existence of water ice on the Moon, "this high-risk, potentially high-payoff experiment did produce several benefits," said Dr. David Goldstein, the aerospace engineer who led the UT Austin team. "We now have experience building a remarkably complex, coordinated observing program with astronomers across the world, we established useful upper limits on the properties of the Moon's natural atmosphere, and we tested a possible means of true 'lunar prospecting' using direct impacts." See also **Moon (Earth's)**.

Web References

Lunar Prospector: http://lunar.arc.nasa.gov/
Lunar Prospector Archives: http://lunarprospector.arc.nasa.gov/
Lunar Prospector Mission: http://nssdc.gsfc.nasa.gov/planetary/lunarprosp.html

LUNAR ROVING VEHICLE.   The Lunar Roving Vehicle (LRV) was an electric vehicle designed to operate in the low-gravity vacuum of the Moon and to be capable of traversing the lunar surface, allowing the Apollo astronauts to extend the range of their surface extravehicular activities. Three LRVs were driven on the Moon, one on Apollo 15 by astronauts David Scott and Jim Irwin, one on Apollo 16 by John Young and Charles Duke, and one on Apollo 17 by Gene Cernan and Harrison Schmitt. Each rover was used on three traverses, one per day over the three day course of each mission. On Apollo 15 the LRV was driven a total of 27.8 km (17.3 miles) in 3 hours, 2 minutes of driving time. The longest single traverse was 12.5 km (7.8 miles) and the maximum range from the Lunar Module (LM) was 5.0 km (3.1 miles). On Apollo 16 the vehicle traversed 26.7 km (16.6 miles) in 3 hours 26 minutes of driving. The longest traverse was 11.6 km (7.2 miles) and the LRV reached a distance of 4.5 km (2.8 miles) from the LM. On Apollo 17 the rover went 35.9 km (22.3 miles) in 4 hours 26 minutes total drive time. The longest traverse was 20.1 km (12.5 miles) and the greatest range from the LM was 7.6 km (4.7 miles).

The Lunar Roving Vehicle had a mass of 210 kg (463 lbs) and was designed to hold a payload of an additional 490 kg (1,080 lbs) on the lunar surface. The frame was 3.1 meters (10.17 ft) long with a wheelbase of 2.3 meters (7.55 ft). The maximum height was 1.14 meters (3.74 ft). The frame was made of aluminum alloy 2219 tubing welded assemblies and consisted of a 3 part chassis which was hinged in the center so it could be folded up and hung in the Lunar Module quad 1 bay. It had two side-by-side foldable seats made of tubular aluminum with nylon webbing and aluminum floor panels. An armrest was mounted between the seats, and each seat had adjustable footrests and a velcro seatbelt. A large mesh dish antenna was mounted on a mast on the front center of the rover. The suspension consisted of a double horizontal wishbone with upper and lower torsion bars and a damper unit between the chassis and upper wishbone. Fully loaded the LRV had a ground clearance of 36 cm (14.2 inches). See Fig. 1.

The wheels consisted of a spun aluminum hub and an 81.8 cm (32.2 inches) diameter, 23 cm (9.06 in) wide tire made of zinc coated woven 0.083 cm (0.03 in) diameter steel strands attached to the rim and discs of formed aluminum. Titanium chevrons covered 50% of the contact area to provide traction. Inside the tire was a 64.8 cm diameter bump stop frame to protect the hub. Dust guards were mounted above the wheels. See Fig. 2. Each wheel had its own electric drive, a DC series wound 0.25 hp motor capable of 10,000 rpm, attached to the wheel via an 80:1 harmonic drive, and a mechanical brake unit. Maneuvering capability was provided through the use of front and rear steering motors. Each series wound DC steering motor was capable of 0.1 hp. Both sets of wheels would turn in opposite

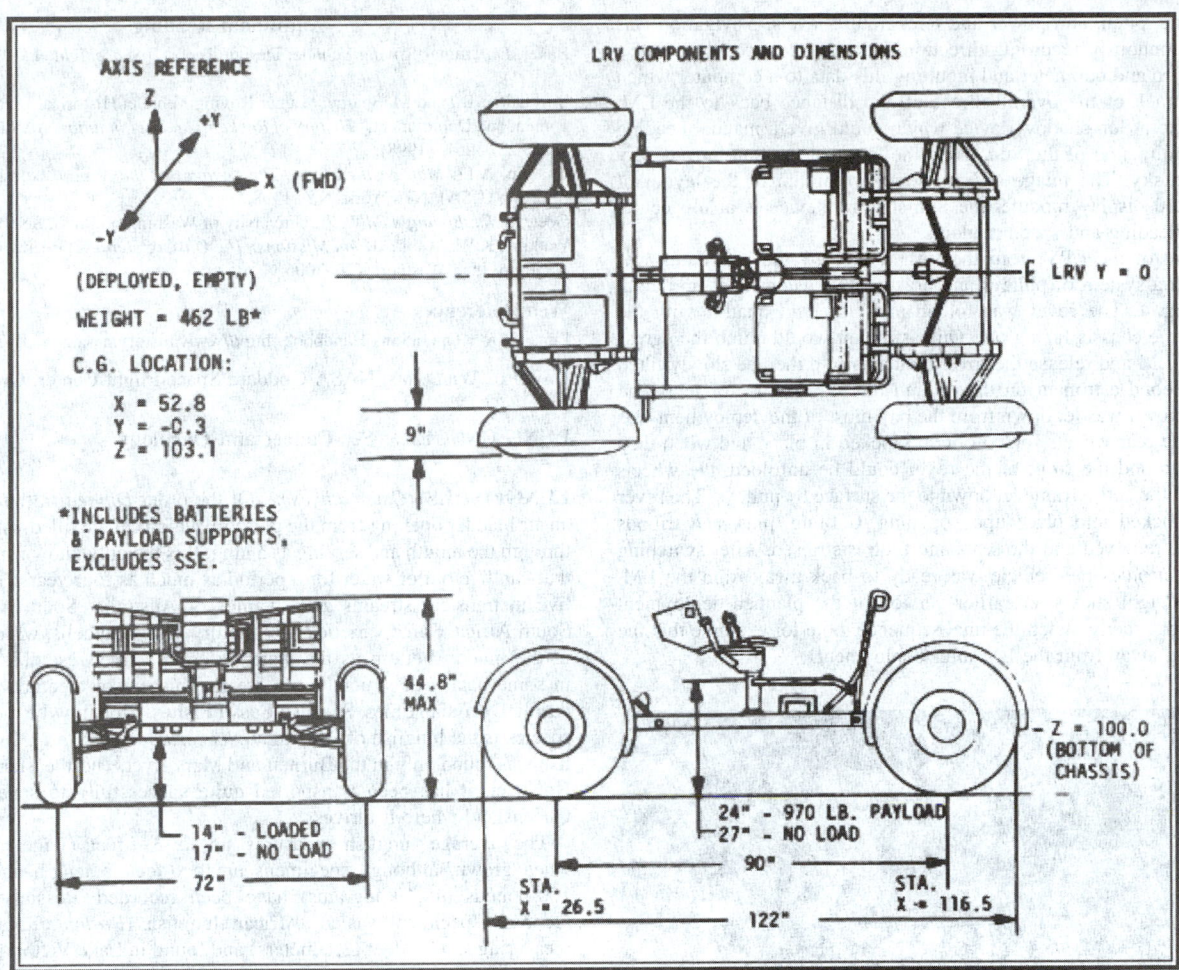

Fig. 1.   Schematic of Lunar Roving Vehicle used on Apollo missions 15, 16, and 17. (*image courtesy of NSSDC*).

directions, giving a steering radius of 3.1 meters (10.2 ft), or could be decoupled so only one set would be used for steering. Power was provided by two 36-volt silver-zinc potassium hydroxide non-rechargeable batteries with a capacity of 121 amp-hr. These were used to power the drive and steering motors and also a 36 volt utility outlet mounted on front of the LRV to power the communications relay unit or the TV camera. Passive thermal controls kept the batteries within an optimal temperature range. See Fig. 3.

Fig. 3.   Pictured on the front of the LVR is a 36 volt utility outlet to power the communications relay unit or the TV camera. (*image courtesy of NSSDC*).

A T-shaped hand controller situated between the two seats controlled the four drive motors, two steering motors and brakes. Moving the stick forward powered the LRV forward, left and right turned the vehicle left or right, pulling backwards activated the brakes. Activating a switch on the handle before pulling back would put the LRV into reverse. Pulling the handle all the way back activated a parking brake. The control and display modules were situated in front of the handle and gave information on the

Fig. 2.   Close up of Lunar Roving Vehicle (*image courtesy of NSSDC*).

speed, heading, pitch, and power and temperature levels. Navigation was based on continuously recording direction and distance through use of a directional gyro and odometer and inputting this data to a computer which would keep track of the overall direction and distance back to the LM. There was also a Sun-shadow device which could give a manual heading based on the direction of the Sun, using the fact that the Sun moved very slowly in the sky. The image at left shows a diagram of the layout of the control and display module, the Sun-shadow device is at top center between the heading and speed readouts.

Deployment of the LRV from the LM quad 1 by the astronauts was achieved with a system of pulleys and braked reels using ropes and cloth tapes. See Fig. 4. The rover was folded and stored in quad 1 with the underside of the chassis facing out. One astronaut would climb the egress ladder on the LM and release the rover, which would then be slowly tilted out by the second astronaut on the ground through the use of reels and tapes. As the rover was let down from the bay most of the deployment was automatic. The rear wheels folded out and locked in place and when they touched the ground the front of the rover could be unfolded, the wheels deployed, and the entire frame let down to the surface by pulleys. The rover components locked into place upon opening. Cabling, pins, and tripods would then be removed and the seats and footrests raised. After switching on all the electronics the vehicle was ready to back away from the LM. The image at right shows an earlier version of the planned deployment which does not exactly match the final sequence, note for example that the rover is facing away from the LM after deployment.

Fig. 4.   Deployment of the LRV from the Lunar Module. (*image courtesy of NSSDC*).

The original cost-plus-incentive-fee contract to Boeing (with Delco as a major sub-contractor) was for $19 million and called for delivery of the first LRV by 1 April 1971, but cost overruns led to a final cost of $38 million. Four lunar rovers were built, one each for Apollos 15, 16, and 17, and one that was used for spare parts after the cancellation of further Apollo missions. There were other LRV models built: a static model to assist with human factors design, an engineering model to design and integrate the subsystems, two 1/6 gravity models for testing the deployment mechanism, a 1-gravity trainer to give the astronauts instruction in the operation of the rover and allow them to practice driving it, a mass model to test the effect of the rover on the LM structure, balance and handling, a vibration test unit to study the LRV's durability and handling of launch stresses, and a qualification test unit to study integration of all LRV subsystems. The LRV was developed in only 17 months and yet performed all its functions on the Moon with no major anomalies. Harrison Schmitt of Apollo 17 said, "....the Lunar Rover proved to be the reliable, safe and flexible lunar exploration vehicle we expected it to be. Without it, the major scientific discoveries of Apollo 15, 16, and 17 would not have been possible; and our current understanding of lunar evolution would not have been possible." See also **Apollo Lunar Missions**; **Lunar Exploration**; **Moon (Earth's)** and **Spaceflight U.S. Manned**.

Additional Reading

Baker, D.: "Lunar Roving Vehicle: Design Report," *Spaceflight*, **13**, 234–240, (July 1971).
Burkhalter, B., and M. Sharpe: "Lunar Roving Vehicle: Historical Origins, Development and Deployment," *History of Rocketry and Aeronautics*, AAS History Series, **22**, 227–261, (1998).
Chaikin, A.L.: *Man on the Moon: The Voyages of the Apollo Astronauts*, Penguin Group (USA), New York, NY, 1998.
Geer, M.W.: *Boeing's Ed Wells*, University of Washington Press, Seattle, WA, 1992.
Watkins, K.W.: *Apollo Moon Missions: The Unsung Heroes*, Greenwood Publishing Group, Inc., Westport, CT, 2006.

Web References

Lunar Rover Operations Handbook: http://www.history.nasa.gov/alsj/lrvhand.html

DAVID R. WILLIAMS, NASA Goddard Space Flight Center, Greenbelt, MD

LUNG CANCER.   See **Cancer and Oncology**.

LUNGFISHES *(Osteichthyes)*.   Of the order *Dipneusti*, the lungfish has an air bladder opening from the pharynx which can be filled with air gulped through the mouth and serving as a lung. It is the only known species of fish that can live out of water for a period as much as four years. The dipnoids live in transient streams and swamps of Australia, South America, and South Africa. Some varieties pass the dry season in cells, which they form in the muddy bottom as the water dries up. They resemble amphibians in some details of structure and are reminiscent of creatures that existed during Devonian times. See also **Fossil Fishes**. Probably the most primitive species is the lungfish of Australia (*Neoceratodus forsteri*). Originally, the fish was found only in the Burnett and Mary rivers (northeastern Australia). However, it has been transported quite successfully to several lakes in Queensland where it thrives.

The average lungfish measures up to $3\frac{1}{2}$ feet (1 meter) in length when grown, although specimens up to 6 feet in length and weight of 100 pounds (45.4 kilograms) have been recorded. Possibly the largest recorded specimen was an African lungfish (*Protopterus aethiopicus*), measuring some 7 feet (2.1 meters) and found in Lake Victoria. See Fig. 1.

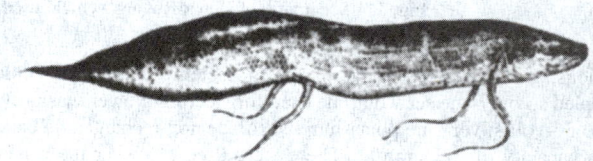

Fig. 1.   African lungfish.

It is interesting to note that similarities between the African and South American lungfishes have contributed to the hypothesis that there may have been a land connection between South America and Africa at an earlier period. See also **Earth Tectonics and Earthquakes**.

Lungfishes form mud-ball cocoons within which they encase themselves. This is a mucous cocoon, which becomes quite leathery after hardening. During estivation in their cocoons, lungfishes lose both weight and length, factors that are quickly recovered in about a month after coming out of the cocoon.

LUPUS.   See **Systemic Lupus Erythematosus**.

LUPUS (the wolf).   A southern constellation located near Libra.

LUSTER.   This term is used by mineralogists to describe the appearance of the surface of a mineral, usually a crystal face, in reflected light. The principal types of luster are: metallic, adamantine, vitreous, resinous, greasy, pearly. The degrees of luster may be defined as: splendid, shining, glistening, or dull. Schillerization is a peculiar form of submetallic luster observed in different directions in certain minerals such as schillerspar, diallage, hypersthene, etc.

LUTEIN.   Lutein is an antioxidant found in several areas of the body including the skin and eyes. In recent years, researchers have discovered that age-related eye diseases such as cataracts and macular degeneration might benefit from the antioxidant effects found in vitamins and minerals from fruits or vegetables.

Lutein belongs to a chemical class of compounds called carotenoids and is found in the central area of the retina called the macula. The macula is responsible for acute central vision, and damage to this portion of the retina severely limits a patient's ability to read, recognize faces, and perform any other task that requires straight-ahead vision. Some research suggests that lutein may protect the macula from potentially damaging forms of light.

Antioxidants act as scavengers, preventing the formation of and neutralizing the damage from free radicals-substances thought to be responsible for acceleration of the aging process and causing damage such as lowered immune system responses, heart attacks, arthritis, eye diseases, cancers, and other disorders. Free radicals are produced by the body in response to environmental pollutants such as cigarette smoke, pesticides, smog, radiation, and many drugs. Although free radicals do have certain beneficial effects, there is extensive evidence to indicate that they can damage both the structure and function of cell membranes, causing degenerative diseases and conditions.

Lutein deposited in the retina and lens of the eye is thought to protect the macula from damaging oxidation by filtering blue light. The substance apparently absorbs and dissipates dangerous ultraviolet light rays that are associated with age-related macular degeneration, the leading cause of irreversible blindness among those 65 years of age and beyond.

The body does not make its own lutein and must depend on dietary sources to get an adequate supply. Nutritionists suggest 6 milligrams per day, which would be two salad bowls of spinach, for instance. Generally, a diet rich in yellow and orange fruits and leafy green vegetables will provide an adequate supply, although most people do not get enough to reach the 6 milligram level. The foods richest in lutein are spinach, collards, kale, mustard greens, and turnip greens. Other fruits and vegetables that contain lutein are: Romaine lettuce, leeks, pears, broccoli, brussels sprouts, butternut squash, cabbage, carrots, celery, corn, cornmeal, cucumbers, green beans, honeydew, kiwi, mango, okra, oranges and orange juice, peppers, (green and orange bell), persimmons, pumpkins, red grapes, yellow squash and zucchini squash.

Cooked vegetables appear to be a better source of lutein because cooking unlocks the cell walls and releases the substance making it more available. Lutein also appears to be more easily absorbed when vegetables are served with a source of fat, such as cooking oil or butter.

Lutein has been the subject of much interest over the past few years, with the general conclusion being that a diet rich in lutein may reduce the risk of developing macular degeneration. In any event, lutein-rich foods should be a part of any balanced diet. See also Carotenoids; and Macula.

Vision Rx, Inc., Elmsford, NY

LUTETIUM. [CAS: 7439-94-3]. Chemical element symbol Lu, at. no. 71, at. wt. 174.97, fourteenth and last element in the Lanthanide Series in the periodic table, mp. 1,663°C (3,025°F), bp 3,402°C (6,156°F), density 9.841 g/cm³ (20°C (68°F). Elemental lutetium has a close-packed hexagonal crystal structure at 25°C (77°F). The pure metallic lutetium is silver-gray in color and retains its luster at room temperature indefinitely. Although not as extensively studied as most of the other lanthanides, most of the basic properties of lutetium are known, and it behaves as a normal trivalent metal with no magnetic transitions because its $4f$ levels are completely filled. There are two natural isotopes ^{175}Lu and ^{176}Lu. The latter isotope is radioactive with a half-life of 2.2×10^{10} years. Fourteen artificial isotopes are known. The element was first identified by G. Urban in 1907 and independently by Carl Auer von Welsbach in 1908. See also Chemical Elements: History of the Origin. Although not investigated fully, lutetium is classified with a low acute-toxicity rating. Lutetium is the least abundant of the Lanthanide elements, estimated as present on the average of 0.5 ppm in the earth's crust. Potentially, however, it is more plentiful than mercury, cadmium, or any of the precious metals. Electronic configuration

$$1s^2 2s^2 2p^6 3s^2 2p^6 3d^{10} 4s^2 4p^6 4d^{10} 4f^{14} 5s^2 5p^6 5d^1 6s^2$$

Ionic radius Lu³⁺ 0.848 Å. Metallic radius 1.735 Å. Other important physical properties of lutetium are given under Rare-Earth Elements and Metals.

The source of lutetium to date has been the processing of the other heavy rare-earth metals. Because of very limited availability, little research was conducted on lutetium until the mid-1960s. Most of these studies now are concentrating on prospective uses in phosphors, semiconductor, and other electronic circuitry components. A lutetium dithalocyanine complex has received much consideration recently for application in large, thin screens for television projection.

See references listed at ends of entries on **Chemical Elements**; and **Rare-Earth Elements and Metals**.

K.A. GSCHNEIDNER, Jr., B. EVANS, Iowa State University, Ames, IA

LUTH (*Reptilia, Chelonia*). Also called the leathery turtle. A large marine turtle, *Dermochelys coriacea*, which reaches a length of 6 feet (1.8 meters). Its carapace is formed of bony plates connected together but not joined to the spinal column or ribs. Its flesh is not palatable.

LUX. A photometric unit of illuminance or illumination equal to one lumen per square meter. A level of illumination between 200 and 1000 lux is generally considered to be adequate for homes and offices.

See also **Foot-Candle**.

LWOFF, ANDRÉ MICHEL (1902–1994). Lwoff was a French microbiologist who studied the phenomenon of lysogeny in bacterial cultures, leading to important contributions to the understanding of the biology of viruses.

When André Lwoff joined Felix Mesnil's laboratory at the Institut Pasteur in 1921, the institute was beginning to deviate from its traditional focus on infectious disease, immunology and epidemiology—microbiology in the service of medicine. Bacterial physiology was emerging as a valid field of research. Lwoff himself was experiencing an analogous shift: though he had started his studies in medicine (largely due to pressure from his father), his real interest was in laboratory biology. He had just graduated with a bachelor's degree in the natural sciences, and was continuing with his medical studies even as he became engrossed in the work at the laboratory. He earned the medical degree in 1927, and followed this with a doctorate in natural science in 1932. See also **Bacteriology (The History)**.

Lwoff's research began with ciliates and other simple organisms, identifying specific nutritional requirements or 'growth factors' for different species. This research led to entirely new fields of inquiry in the biochemistry of bacterial metabolism. It was during these studies that he began his collaboration with his wife, Marguerite, which would last until her death in 1979.

In the 1930s, Lwoff's friends Eugéne and Elisabeth Wollman were studying lysogenic bacteria, or bacteria that appeared to produce bacteriophage (viruses that prey on bacteria). During World War II, the Wollmans were lost to Auschwitz, and Lwoff was active in the French Resistance. In the 1940s lysogeny became less fashionable, with Max Delbrück's phage group denying that a bacterium could spontaneously produce bacteriophage without first having been infected. After the war, Lwoff picked up the Wollmans' research subject, putting his energies to lysogeny. Lwoff employed a painstaking technique whereby individual bacterial cells are isolated, thus allowing a lineage to be followed, individual by individual. Lwoff showed that a bacterium can indeed produce bacteriophage in a culture initially free of bacteriophage, that this trait is hereditary, and that in lysogenic bacteria the genetic material of the phage is incorporated with the genetic material of the bacterium. In 1965 Lwoff received the Nobel Prize in Physiology or Medicine for these studies, an award shared with Jacques Monod and François Jacob. See also **Delbrück, Max Ludwig Henning (1960–1981); Jacob, Francois (1920–Present); Monod, Jacques Lucien (1910–1976); and Virology (The History)**.

SUZANNE BOARD, Toronto, Ontario, Canada

LYASES. See **Enzyme**.

LYCOPSIDA. A group of plants (Club Mosses) that contains about 500 species, most of which are included in two genera, *Lycopodium* and *Selaginella*. The species of *Lycopodium* are trailing plants often called ground pines, or ground hemlock, as well as club mosses. Many of them are common plants of dry open places in the temperate zone. The plants have long creeping stems growing on the surface of the ground or several inches beneath it. From this prostrate stem short dichotomously branched roots extend down into the ground, and erect branches grow upward. The stems are covered with many small, pointed, dark green leaves. In the more primitive species the reproductive structures or sporangia are found in the axils of ordinary leaves. In other species the sporangia are borne in the axils of modified leaves aggregated at the tip of an erect branch, forming a slender cone or strobilus. The many spores borne within the sporangia are all alike and for this reason Lycopodium species are said to be homosporous. The spores are disseminated by the wind, and in time develop into gametophytes. The gametophytes of *Lycopodium* species are extremely small tuberous bodies, which grow slowly and reach maturity only if they are invaded by an endophytic fungus. Generally the gametophyte or prothallus develops underground. In the upper surface of the prothallus both antheridia and archegonia are found. Each antheridium contains many straight, biciliate sperms. These swim to the egg, with which one unites, forming a zygote. From this the new sporophyte develops. At first the sporophyte depends on the gametophyte for its food substances. Thus it obtains nutriment by means of a special absorbing structure called a suspensor which grows into the tissue of the gametophyte.

Lycopodium plants are widely used as material from which to make Christmas wreaths. For this purpose the entire plant is often ripped from the ground. The spores of *Lycopodium* are also gathered and sold under the name of Lycopodium powder. Formerly these spores were used in making explosive mixtures and for flashlights.

The genus *Selaginella*, containing some 400 species, is most abundant in the tropics. A few species of small plants are found in temperate regions. In the tropics there are both terrestrial and epiphytic species. The general habit of the plant is much like that of *Lycopodium*. The sporangia are formed in the axils of leaves at the tips of the branches, forming terminal strobili or cones. At the base of the sporophylls there is also a small scale, called a ligule, of unknown function. The sporangia are of two kinds, one, a megasporangium, containing four large spores, called megaspores; the other a microsporangium containing many small spores or microspores. Species of *Selaginella* are therefore heterosporous, a character which distinguishes them from *Lycopodium* species.

LYELL, CHARLES (1797–1875).

Charles Lyell was a British geologist, well known for his *uniformitarian* geology, his arguments on geological time and the age of the earth, his subdivisions of the Tertiary, his influence on Darwin, and application of Darwinian theory to humans. Lyell came from a landowning family in Forfarshire, Scotland, but was largely brought up in Hampshire in southern England. He studied at Oxford, intending a legal career, but became interested in geology by attending William Buckland's lectures. Following tours on the Continent in the 1820s, Lyell decided to devote himself to geology. After publishing *Principles of Geology* (3 vols, 1830–33), Lyell was appointed Professor of Geology at King's College, London, but soon resigned, thereafter making a living by his writings, which included *Elements of Geology* (1838) and *Antiquity of Man* (1863).

Lyell's *uniformitarianism*, developed from James Hutton's Theory of the Earth (1795), involved both methodological and substantive elements. He thought that, to be scientific, geologists should observe present phenomena and processes and use these to understand the past. Also, he believed that, within broad limits, past and present conditions on earth were similar, but conditions at any given locality varied due to slow, chiefly vertical, movements of the earth's surface. Geology should not be concerned with the earth's origins. The earth was very ancient, as could be demonstrated by, for example, observations of Mount Etna. This grew at a rate determinable from records of lava flows; so, knowing the mountain's overall size, its age could be estimated. Yet this ancient mountain overlay recent shell deposits. Thus the earth as a whole must be extremely ancient. Such arguments later warranted Darwin's assumption of ample time for "the origin of species."

In Volume 2 of *Principles* Lyell examined Lamarck's theory but ended up rejecting it. Lyell's model to account for the observed changes in fossil forms in strata was that new species were created from time to time by an unknown natural agency, and died out if changed environmental circumstances became unfavorable. Thus there was a slow changeover of forms, with species 'tracking' changing environmental conditions. Lyell supposed that, if particular past conditions returned, then past life forms such as the iguanodon might reappear.

Given a slow changeover of species, the older the strata the smaller the proportion of extant forms. Thus Tertiary strata were subdivided: Eocene (3% fossils resembling recent forms); Miocene (17%); Older Pliocene (35–50%); Newer Pliocene (96%). A large faunal change (as between Secondary and Tertiary rocks) represented a long period of nondeposition, during which a complete cycle of species change had presumably occurred. The theory was unsatisfactory in that if old types recurred due to renewal of past conditions the regular replacement of old by new forms would be disturbed and the envisaged stratigraphic method might fail. But Lyell's subdivision of the Tertiary was accepted.

Lyell's contemporaries approved his uniformitarian methodology and his *gradualism*, but many, including Darwin, could not accept Lyell's contention that the principal animal groups (humans excepted) had always been on earth. His fellows mostly believed in some kind of progressive development, either continuous or saltatory, which was an anathema to Lyell.

The young Darwin admired Lyell's geology, but had difficulty in reconciling the worldwide evidence collected during the Beagle voyage with the expectation of organisms being specifically related to their environmental circumstances. Darwin's evolutionary theory was, in part, a response to this problem. Lyell slowly and reluctantly accepted Darwin's theory after Darwin expounded it to him in 1856—when Lyell was himself making voluminous notes on the 'species question'. In *Antiquity of Man* Lyell deployed Darwinian theory, accepted human evolution, and incorporated mid-century ideas about palaeoclimates (glacial theory) and the recent discoveries of Neanderthal and other paleolithic remains. But for Lyell humankind was always a "special creation."

Additional Reading

Bartholomew, M.: "Lyell and Evolution: An Account of Lyell's Response to the Prospect of an Evolutionary Ancestry for Man," *British Journal for the History of Science*, **6**, 261–303 (1973).

Gould, S.J.: *Time's Arrow, Time's Cycle: Myth and Metaphor in the Discovery of Geological Time*, Harvard University Press, Cambridge, MA. 1987.

Hodge, J.: "Darwin Studies at Work: A Re-examination of Three Decisive Years (1835–37)," in: Levere, T., and W. Shea, eds., *Nature, Experiment, and the Sciences*, Kluwer Academic Publishers, Dordrecht, Netherlands, 1990, pp. 249–274.

Rudwick, M.J.S.: "Lyell on Etna and the Antiquity of the Earth," in: Schneer, C.J., ed., *Toward a History of Geology*, MIT Press, Cambridge, MA, 1969, pp. 288–304.

Rudwick, M.J.S.: "The Strategy of Lyell's Principles of Geology," *Isis*, **61**, 4–33 (1970).

Rudwick, M.J.S.: "Charles Lyell's Dream of a Statistical Paleontology," *Paleontology*, **21**, 225–244 (1978).

Secord, J.A.: *Charles Lyell: Principles of Geology*, Penguin Books, London, UK, 1997.

Wilson, L.G.: *Sir Charles Lyell's Journals on the Species Question*, Yale University Press, New Haven, CT, 1970.

Wilson, L.G.: *Charles Lyell: The Years to 1841: The Revolution in Geology*, Yale University Press, New Haven, 1972.

DAVID OLDROYD, The University of New South Wales,
Sydney, New South Wales, Australia

LYMAN ALPHA EMISSION LINE.

Feature in the emission spectrum of solar radiation, identified with neutral hydrogen, which occurs at a wavelength of 121.567 nm. Because of the extremely low absorption cross section of molecular oxygen at this particular wavelength, this radiation penetrates deep into the earth's atmosphere. It is important to the dissociation of trace gases in the upper mesosphere and to the formation of the lower ionospheric layers by ionizing certain minor atmospheric constituent gases, especially nitric oxide.

LYMAN-ALPHA HYGROMETER.

A hygrometer based on the absorption of radiation by water vapor at the Lyman-alpha line, which is an emission line of atomic hydrogen at 121.567 nm.

Lyman-alpha radiation can be generated by a glow discharge in hydrogen, and detection is normally accomplished by a nitric oxide ion chamber. Two magnesium fluoride windows both at the radiation source and the detector bound the absorption path. Lyman-alpha hygrometers are used on aircraft and on meteorological towers for high-frequency humidity measurements. Inconveniences of the method, like drift of the source intensity or contamination of the windows, are overcome by special calibration techniques or by baselining the high-frequency output to the humidity values provided by a slower, but stable, hygrometer.

LYMAN-ALPHA RADIATION.   The radiation emitted by hydrogen at 1216 angstrom, first observed in the solar spectrum by rocket-borne spectrographs. Lyman-alpha radiation is very important in the heating of the upper atmosphere thus affecting other atmospheric phenomena.

LYME DISEASE.   Lyme disease is a multi-stage, multi-system infection caused by *Borrelia burgdorferi*, a member of the family of spirochetes, or corkscrew-shaped bacteria. See Fig. 1. *Borrelia burgdorferi* is transmitted to humans by bites from the ticks colloquially referred to as "deer ticks" in the north-eastern and north central regions of the United States. *Ixodes scapularis* (black-legged tick), *Ixodes pacificus* (western black-legged tick), *Ixodes dammini* and *Amblyomma americanus* are the primary vectors for Lyme disease. (A Vector is an organism (as an insect) that transmits a pathogen (or virus) that causes a disease).

Both of the latter ticks are known to infest white-tailed deer (*Odocoileus virginianus*) and the whitefooted mouse (*Peromysus leucopus*). Ixodes ticks are significantly smaller than common dog and cattle ticks. They are no bigger than a pinhead in their larval and nymphal stages, and are only slightly larger as adults. See Figs. 2, 3, and 4. Lyme disease was first identified in 1975 in the town of Lyme, in southeastern Connecticut, USA. There, a clustering of cases with arthritis-like symptoms were observed to have distinctive skin lesions, which led to the recognition of a new tick-borne disease.

The earliest stage of the illness is commonly characterized by the appearance of a distinctive skin rash called *erythema migrans*. Approximately 3 to 20 days following a tick bite, a red *macule* or papule appears at the bite site. It then expands to a large, annular erythematous lesion (*erythema chronicum migrans*) about 6 to 16 centimeters or larger in diameter. The lesion sometimes shows central clearing, and secondary concentric rings may develop within the original ring. The lesions are not pruritic and may be multiple. Other symptoms that often occur at the onset of early Lyme disease may include fever, chills, malaise, headache, aching in the muscles (*myalgias*) and neuralgic pain in the joints (*arthralgias*). In addition, early symptoms can include secondary skin lesions, facial nerve palsy, and *lymphocytic meningitis*. Early treatment with an appropriate antibiotic therapy is almost always successful. If left untreated, or inadequately treated, the

early Lyme disease progresses to late Lyme disease, which is characterized by distinctive arthritic, neurologic, and cardiac manifestations. These symptoms require a more intensive therapy, and permanent after-effects may occur.

The incidence of Lyme disease has been steadily increasing, making Lyme disease an important public health problem in some areas of the United States. The total of cases reported each year increased 25-fold between 1982 and 1997, with a total of more than 103,000 cases reported in that 15-year time-span. See Fig. 5. As a rapidly emerging infectious disease, Lyme disease accounts for more than 90% of all vector-borne illness reported in the United States. Cases are most concentrated in the northeastern, north central and pacific coastal regions of the United States, although 48 states and the District of Columbia, parts of Europe and Australia have reported Lyme disease. The geographic distribution of *Ixodes dammini* and *Amblyomma americanus*, the principal vectors of *Borrelia burgdorferi*, continues to spread, and with it, there has been an increase in the reported incidence of Lyme disease.

The diagnosis and confirmation of Lyme disease is difficult. Clinical manifestations can be confused with those of other illnesses and diseases,

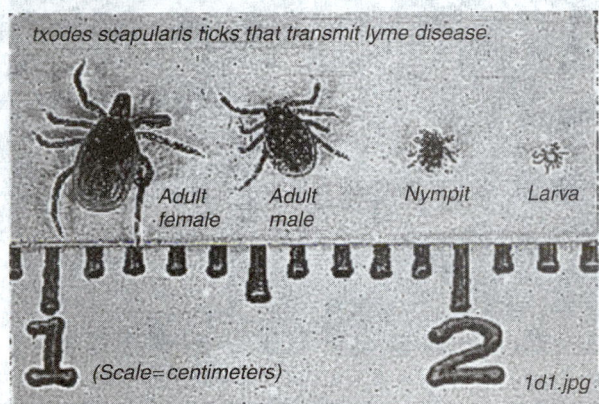

Fig. 2.   The tick vectors of Lyme disease have four life stages: egg, larva, nymph, and adult. Compare the differences in size of the three later stages on a centimeter scale. (Center for Disease Control, National Center for Infectious Diseases, Division of Vector-Borne Infectious Diseases, Atlanta, Georgia.)

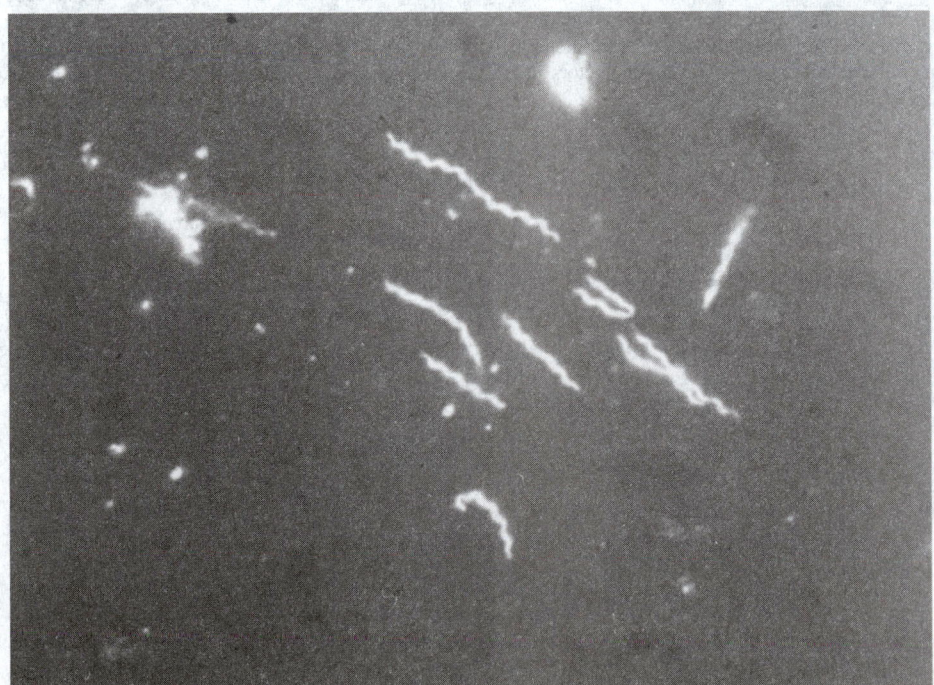

Fig. 1.   The corkscrew shaped bacteria, Borrelia burgdorferi, are the cause of Lyme disease. (Center for Disease Control, National Center for Infectious Diseases, Division of Vector-Borne Infectious Diseases, Atlanta, Georgia.)

Ixodes scapularis on a grass stem.

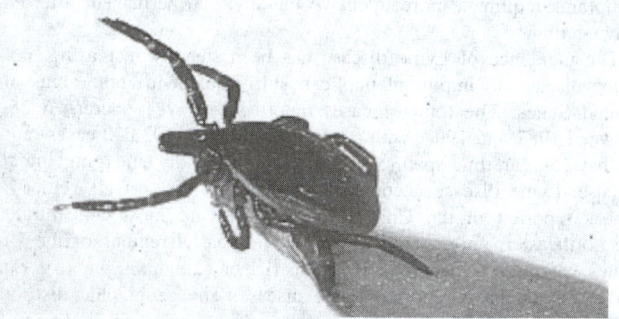

Fig. 3.   An adult female is shown on a grass stem. (Center for Disease Control, National Center for Infectious Diseases, Division of Vector-Borne Infectious Diseases, Atlanta, Georgia.)

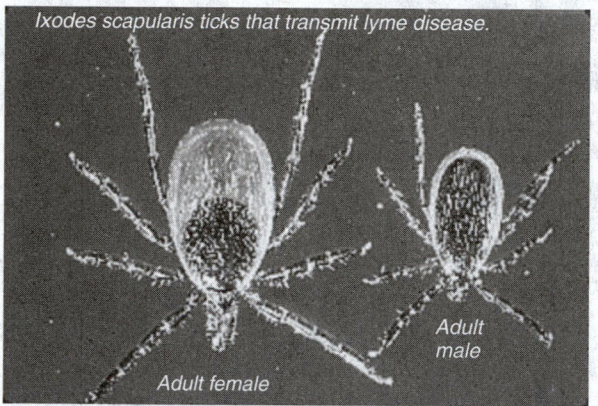

Ixodes scapularis ticks that transmit lyme disease.

Adult male

Adult female

Fig. 4.   Adult female and male ticks are different in size and coloring. Note that the adult female is almost two times larger than the adult male. The female has black on red coloring, whereas the male is all black in appearance. (Center for Disease Control, National Center for Infectious Diseases, Division of Vector-Borne Infectious Diseases, Atlanta, Georgia.)

especially in late Lyme disease. Researchers continue to work to identify antigens as well as new methods that can improve upon the current diagnostic tools for Lyme disease. As the disease is more intensively studied, the recognized clinical spectrum widens, yet major factors in the origination and development of the disease remain unknown.

LYMPH.   A clear fluid that circulates in the tissue spaces of vertebrates and passes into the venous system by way of a tubular lymphatic system. It is derived from the liquid plasma of the blood but is more watery and contains no red corpuscles. It serves as an intermediary between the blood itself and the tissues of the body.

Lymph is found lying free in the serous sac cavities of the body, i.e., the peritoneum, pleura, and the spaces in the brain filled with cerebrospinal fluid, which may also be classified as lymph, although it differs in composition from the fluid found in the lymph vessels.

Lymph is derived from the plasma of the blood either by filtration, diffusion, or osmosis through the capillary walls or by active secretion of endothelial cells making up capillary walls. Its composition is very similar to that of the blood plasma.

The function of lymph is to bring nourishment to the tissue cells and to return waste matter and other toxic material to the blood stream by way of the lymphatic vessels, or directly into the blood stream through the capillary walls. Lymph has been compared with the body fluids of invertebrates, commonly designated as hydrolymph and hemolymph.

The small tubules of the lymphatic system resemble capillaries and the larger ducts, called lymphatics, are similar to veins, although of a more delicate structure in relation to their size. Like veins, they have valves that aid in promoting flow through the movements of the surrounding muscles. They are irregular in diameter, forming reservoirs at some points and dilating in the amphibians, reptiles, and birds to form lymph hearts whose pulsations propel the lymph toward the heart. The smaller vessels converge like blood vessels to form larger trunks. In humans, the chief vessels are the thoracic duct, and the right lymphatic duct, which empty into the large veins at the sides of the neck. Along the course of the lymph vessels are groups of lymph nodes. They serve as filters, which localize and retard the spread of toxic and infectious elements that are being returned to the blood stream. The lymph nodes also serve as centers for formation of lymphocytes, which form one of the main divisions of blood cells.

LYMPHATIC SYSTEM.   The lymphatic system in mammals has evolved in parallel with the blood circulation from the single hemolymph systems of lower orders. Lymphatic pathways form a distinct circulation of fluid and cells which is separate from, but closely integrated with, the blood circulation, although the two circulations differ considerably in the composition of their fluids, their cellular content and their hydrodynamics.

The lymphatic system consists of a network of vessels lined by endothelial cells which start as endothelial tubes called *initial lymphatics* in the tissue spaces and end in the thoracic duct, which anastomoses with the great veins in the root of the neck. Lymphatic vessels thus serve as a tissue drainage system which provides an accessory route for the return of fluid, protein and cells to the bloodstream. An intact lymphatic system is

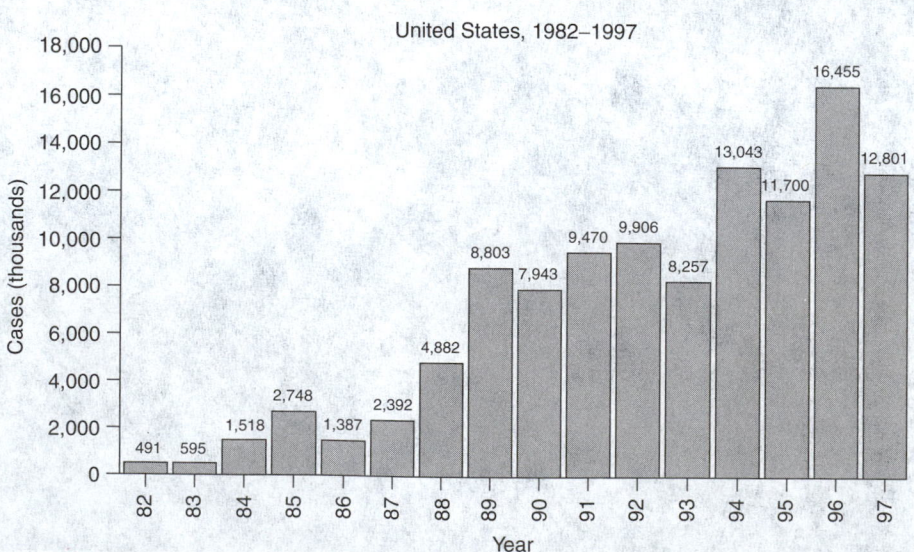

United States, 1982–1997

Fig. 5.   The reports of Lyme disease in the United States have increased dramatically since the first cases were recognized in southeastern Connecticut in 1975. (Center for Disease Control, National Center for Infectious Diseases, Division of Vector-Borne Infectious Diseases, Atlanta, Georgia.)

thus mandatory for the maintenance of normal blood volume, the control of interstitial fluid volume and the prevention of tissue oedema.

Recent studies now indicate that the lymphatic system also plays a major role in the metabolism and turnover of extracellular matrix (ECM) constituents such as hyaluronan and other glycosaminoglycans which form an important part of the tissue "glue" which binds cells together. Lymphatic vessels in the wall of the gut provide the main pathway for the absorption of fat in the small intestine.

Finally the lymphatic system is the key integrating element in the global functioning of the immune system, providing the pathway by which the cells of the immune system circulate between the blood, peripheral tissues and lymph nodes. The lymphatic system is essential for the amplification of the immune response to foreign antigen and for the dissemination of immunological memory to antigen throughout the whole body. See also **Immune System and Immunology** and **Lymph**.

Structure

Almost all tissues of the body except the central nervous system (CNS) have lymphatic vessels that drain fluid from the interstitial spaces. The lymphatic system in mammals consists of a well-defined network of vessels that collect tissue fluid from most organs and tissues in the body and transport it, initially with little modification, as lymph. The lymph passes through one or more lymph nodes to be ultimately discharged into the bloodstream, mainly by one large vessel, the thoracic duct, which drains into the large central veins (Fig. 1). Lymph from the right side of the head and neck, the right arm and part of the thorax enters the right lymphatic duct, which joins the venous system at the confluence of the right internal jugular and subclavian veins. Virtually all the lymph draining the rest of the body enters the right lymphatic duct, which joins the venous system at the confluence of the left internal jugular and subclavian veins.

The lymphatic system starts in the interstitial space as blind-ending endothelial channels called initial lymphatics. These microlymphatics lie in the connective tissues of the body and provide a framework for parenchymal cells and a space for the distribution of blood vessels and nerves. The interstitial space also acts as the site where fluids, nutrients and metabolites are exchanged between parenchymal cells and blood.

Microlymphatics and the Extrinsic Lymphatic Pump. The anatomy of microlymphatics is important in understanding how fluid is moved from the tissue spaces into the initial lymphatics (i.e., how lymph is formed). The endothelial cells that form the initial lymphatics are positioned on a fine but dense collagen network. The cells are attached to the collagen by anchoring filaments which are fixed to the centre of the endothelial cells, thus leaving the unattached edges of the cell to function as inlet valves to the lymphatic lumen. These valves are open when the initial lymphatic is expanded, thus allowing unrestricted movement of fluid, macromolecules and cells into the initial lymphatic, but they are closed when the lymphatic is compressed when they seal the endothelial barrier and prevent lymph from leaking back into the interstitial space.

Intrinsic Lymphatic Pump. Several initial lymphatics join to form a collecting duct, which has a smooth muscle coat and is contractile. Rhythmic contraction of this lymphatic smooth muscle is referred to as the intrinsic lymphatic pump which propels lymph downstream from the initial lymphatics. The presence of bicuspid valves in the collecting ducts prevents retrograde flow when the collecting lymphatic is contracting. Collecting lymphatic segments with an upstream and a downstream valve and intrinsic smooth muscle to compress the lymphatic lumen are known as lymphangions and are believed to be important in sucking fluid out of initial lymphatics.

In many organs, for example skeletal muscle and the intestines, the transition from initial lymphatics to collecting lymphatics occurs only outside the organ so that the entire intraorgan lymphatic system consists of initial lymphatics. Initial lymphatics, however, have no smooth muscle coat and are not contractile. The mechanical propulsion of lymph seems to be dependent on extrinsic tissue deformation of initial lymphatics to open and close lymphatic inlet valves and to massage lymph downstream into the collecting ducts. Important factors in this extrinsic lymphatic pump which affect the flow of lymph into initial lymphatics include vasomotion and pulse pressure changes in neighboring arterioles and, in the case of the intestines, the rhythmic smooth muscle contraction that occurs during peristalsis. Additional mechanisms contributing to the extrinsic pump include skeletal muscle contraction, respiration, walking, skin tension and external tissue compression.

Formation of Lymph. After their filtration from blood capillaries, fluid and macromolecules must be transported across the tissue spaces into initial lymphatics (lymph formation), from where they are propelled along collecting ducts, ultimately to enter the bloodstream via the thoracic and right lymphatic ducts. In humans between 3 and 4 L of lymph per day is returned to the blood via the thoracic and right lymphatic ducts. This lymph flow is associated with the return to the blood of an amount of protein approximately equal to the total plasma protein content of the blood. The interstitial fluid volume, which makes up 15–25% of body weight (60–70% of the interstitial fluid volume is in skin and skeletal muscle), is kept fairly constant, largely because any increase in interstitial pressure results in increased lymph flow. Starling forces operating across the blood capillary result, however, in a net filtration of fluid and leakage of protein into the interstitium, which is sucked into the initial lymphatics through

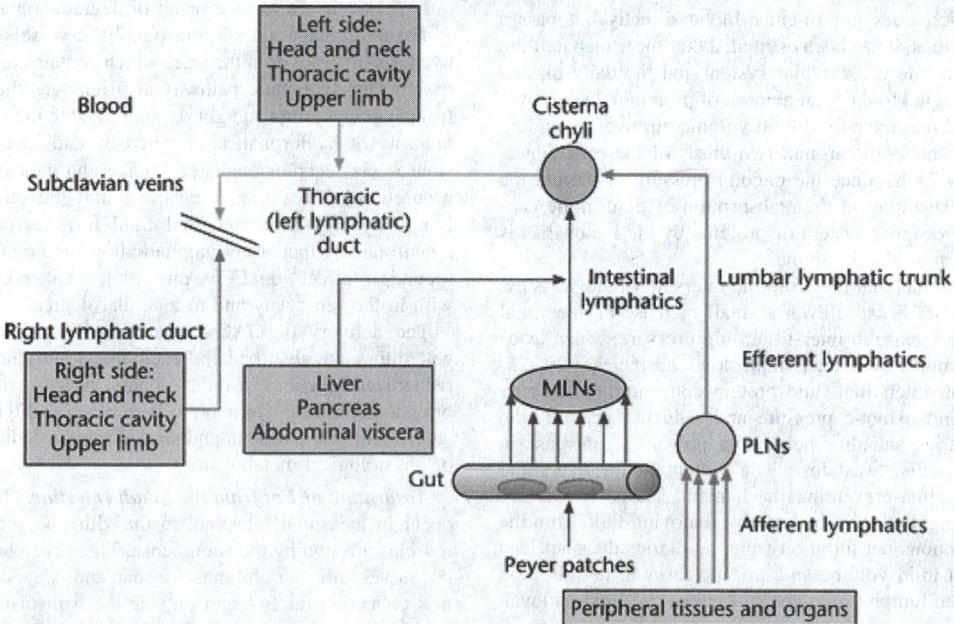

Fig. 1.   Scheme showing the principal pathways of lymph flow in mammals. PLN, peripheral lymph node; MLN, mesenteric lymph node.

their inlet valves as a result of the rhythmic contraction and expansion of initial lymphatics produced by the extrinsic lymphatic pump. In addition, both the extrinsic and intrinsic lymphatic pumps operating in the collecting duct system may create a suction effect on the initial lymphatics, thus facilitating the emptying of these vessels and promoting lymph flow. In some tissues, such as kidney and intestines, transendothelial passage of fluids and macromolecules by vesicular transport may also occur.

Prelymphatics. A further factor to be considered in the transport of substances across the interstitial space from blood to lymph is the presence of prelymphatics, which are nonendothelialized channels in the ECM which may direct fluid and macromolecules towards the initial lymphatics. Glycosaminoglycans such as hyaluronan in the ECM exclude some protein, so that the protein concentration in the interstitium is not uniform and prelymphatic channels may reflect differences in the configuration of the extracellular matrix which offer preferential pathways for fluid and macromolecules to cross the tissue spaces into lymph.

Pathways of Lymph Through Lymph Nodes. Lymph nodes are aggregations of lymphocytes and other accessory cells of the immune system which lie at intervals along the collecting vessels of the lymphatic system. Lymph nodes are positioned outside the individual organs they drain, and they divide lymph into two discrete compartments that differ in their cellular and molecular composition. Afferent or prenodal lymphatics deliver peripheral lymph from the tissues to lymph nodes, whereas efferent or central lymph channels transport lymph to nodes further along the lymphatic chain or, in the case of the thoracic duct, directly into the bloodstream (Fig. 1). See also **Lymphocytes**.

Afferent lymphatics branch sequentially as they approach a lymph node to form a radiating pattern which is distributed over a large area of the surface of the node. These terminal afferents then describe one of two different pathways. Some enter the subcapsular sinus, and from there lymph flows to trabecular sinuses in the cortex and medulla, while other penetrating lymphatics deliver lymph flows through the medullary sinuses that surround the medullary cords. The complex arrangement of lymph pathways provides a mechanism for increased contact between lymph and the nodal circulation for fluid exchange and for interactions between the cells of the immune system. Lymph leaves the node through initial lymphatics that lie deep in the medulla or just under the capsule, and emerges through many small efferent vessels over a large area of the node which then join to form one or more major efferent vessels, which then transport the lymph downstream to the next lymph node in the chain.

Function

Regulation of Interstitial Fluid Volume and Blood Volume. The lymphatic system is a key segment in the massive daily circulation of extracellular fluid and protein. Because there is a continuous leakage of protein from the capillaries which does not re-enter blood directly but passes instead into initial lymphatics, it has been estimated that more than half the plasma protein resides outside the vascular system and the daily plasma protein turnover is about equal to the total amount of protein in the plasma. It has also been estimated that interstitial fluid volume turnover (including that reabsorbed in lymph nodes) is around two-thirds of the entire interstitial fluid volume every 24 h. Since the oncotic pressure of the plasma proteins is the key force resulting in the reabsorption of fluid in the capillaries, the return to the vascular system of proteins by the lymphatics is vital in maintaining a normal blood volume.

The interstitial fluid volume, and therefore the interstitial fluid pressure, is kept within narrow limits because even a small increase in interstitial fluid volume leads to an increase in interstitial fluid pressure, which leads to a greatly increased lymph flow rate. Lymph flow can increase 20–25-fold, resulting in a fall in interstitial fluid protein concentration, thereby lowering the tissue colloid osmotic pressure and reducing net filtration of fluid from the capillaries and thus helping to prevent an increase in interstitial fluid volume. This "washdown" of protein in the interstitial fluid is an important oedema-preventing mechanism. Tissue lymphatics play a major role in the prevention of oedema by removing fluid from the interstitium so that continuous net filtration from capillaries does not lead to an increased intestinal fluid volume and any increases in net filtration are countered by increased lymph flow rates and interstitial fluid removal.

Lymphatic Drainage of the Central Nervous System. The CNS, unlike most other tissues, does not contain lymphatic vessels. What then are the implications for the return of extravasated fluid? Cerebral oedema in mammals is a condition seen after head trauma and it can be a major contributor to irreversible brain damage. The brain has evolved structures called arachnoid granulations which have been thought to provide the major exit route for cerebral spinal fluid (CSF) drainage. However, numerous studies now suggest a physiological link between cerebral interstitial fluid, CSF and extracranial lymph, thereby raising the possibility that the lymphatic system may play a significant role in the fluid balance in the CNS. Recent experiments suggest that one-third to one-half of all CSF drainage occurs via the lymphatic pathway into lymphatics and lymph nodes in the neck. The precise anatomical structures are being debated, but it seems that fluid passes through perineural extensions of the arachnoid space. The olfactory nerve tract seems to be most prevalent, but exit follows other cranial nerves as well. Currently it appears that, although the brain contains no lymphatic vessels, the lymphatic drainage of the CNS may have an important function not only in normal CSF balance but also under conditions of raised intracranial pressure, cerebral oedema and subarachnoid hemorrhage. See also **Central and Peripheral Nervous Systems**.

Turnover of Extracellular Matrix Constituents Such as Glycosaminoglycans. The lymphatic system is intricately involved in the turnover of ECM components such as hyaluronic acid (HA) and proteoglycans, which are collectively called glycosaminoglycans (GAGs). A large part of the volume of all tissues consists of extracellular space which is filled with an intricate network of molecules called the extracellular matrix (ECM). The ECM contains various protein fibers interwoven in a hydrated gel composed of a network of GAG chains. The GAGs are a heterogeneous group of negatively charged polysaccharide chains which, except for HA, are covalently linked to protein to form proteoglycan molecules. They occupy a large volume and form hydrated gels in the extracellular space. The ECM not only functions as a scaffold to support surrounding tissues and preserve their physical integrity, but also plays an important part in regulating the development, migration, proliferation, shape and function of cells in contact with it.

HA is the simplest of the GAGs. It is found in variable amounts in all tissues and is especially abundant in early embryos. HA is produced in large quantities during wound repair and is an important constituent of joint fluid, where it serves as a lubricant. It plays a part in resisting compressive forces in tissues and is an important space filler through which blood leucocytes continuously migrate as they patrol the tissues in their role as scavengers and immune cells. The degradation of ECM components is tightly controlled and their regulated turnover is critical to a variety of important processes, for example during involution of the uterus after childbirth and in wound repair, and in all tissues there is a continuous turnover of the ECM as a result of degradation and resynthesis.

Tissue drainage via afferent lymphatic vessels accounts for 70–80% of hyaluronan turnover in the skin, which contains the most hyaluronan of any tissue. This lymphatic pathway also removes the majority of hyaluronan from synovial joints and gut. Upon arrival at the draining lymph nodes, the majority of hyaluronan is captured by endothelial cells lining lymphatic sinuses via receptor-mediated uptake and catabolized, with only a small proportion of hyaluronan escaping into the efferent lymph leaving the lymph node. When interstitial fluid flow increases, depletion of tissue hyaluronan by increased lymphatic flow, and hyaluronan uptake may have secondary effects on ECM physiology. Other GAGs appear to be dealt with in the same way and to a similar degree.

The delivery of GAGs in afferent lymph to draining lymph nodes where they are absorbed and catabolized introduces a new category to the recognized functions of the lymphatic system, whereby specialized groups of endothelial cells common to both blood and lymphatic vessels represent an element of the reticuloendothelial system dedicated to the catabolic arm of physiological metabolism.

Absorption of Fat from the Small Intestine. Three layers of lymphatics occur in the small intestine: in the villi, the submucosa and the smooth muscle surrounding the submucosa. These lymphatic vessel networks have no valves, no smooth muscle coat and are noncontractile, so that the first collecting ducts begin outside the wall of the small intestine. Initial lymphatics in the villi are called *lacteals* and these vessels are responsible for the absorption of fat from the small intestine.

Intestinal lymph flow is considerably increased when water or fat is absorbed during digestion, whereas proteins and carbohydrates elicit little or no increase in flow. Cholesterol together with triglycerides and phospholipids absorbed by the intestinal mucosa are taken up entirely, or almost entirely, by the lacteals in the villi. In the intestinal lymph these fats exist in the form of lipoprotein complexes or chylomicrons. Chylomicrons consist of a core of triglyceride with some cholesterol ester, with a surface membrane consisting of a mosaic of protein, free cholesterol and triglyceride in a monolayer of phospholipid. See also **Lipids**.

Fat-soluble vitamins A, D, E and K are also absorbed from the gut into intestinal lymph. Lymph flow from the small intestine is considerable and forms around 80% of total thoracic duct lymph flow. The majority of protein in intestinal lymph is derived from circulating plasma proteins. An exception is the absorption by lacteals of maternal antibodies from ingested colostrum, which occurs in neonates from many species including ruminants, rats, mice, dogs and cats.

The Lymphatic System and Immunity. Most naturally occurring immune responses result from the entry of infectious microorganisms through breaches in skin or mucous membranes. As a result, immune responses to invading pathogens are almost always initiated in the lymph node draining their site of entry. The lymphatic system is uniquely structured to collect antigen draining from sites of invasion and, after filtration and capture of antigen, to present it to specific antigen-reactive lymphocytes which then start the immune response to that antigen. Microorganisms are carried to the lymph node in afferent lymphatic vessels, either free in lymph or after phagocytosis by macrophages and dendritic cells. The arrival of antigenic material at the lymph node then triggers a remarkable series of events by which the lymphatic system not only distributes throughout the body the effector cells and antibodies that lead to the eventual elimination of invading pathogens, but it also disperses the memory lymphocytes which continue to circulate between the bloodstream and the lymphatic system, providing immunity against any subsequent infection by the same pathogen. The transport of cells through single lymph nodes is considered in more detail in a later section.

Another important function of the lymphatic system in immunity lies in its role in the recirculation of lymphocytes. Unlike the cells of most organs which are fixed in highly organized structures, lymphocytes are mobile cells that are continuously migrating from the blood through almost all tissues of the body before being returned to the bloodstream via lymph. This continuous migration of lymphocytes is a fundamental feature of the immune system because it allows a very wide repertoire of antigen-specific cells to exist at very low concentrations. The migration of lymphocytes through tissues, including lymph nodes, not only allows tissues to be continuously screened for invading microorganisms but it also provides a mechanism for immune surveillance and killing of any cells, such as cancer cells, that express abnormal proteins such as tumour antigens on their cell surface.

Composition of Lymph

Molecular Composition. Afferent or prenodal lymph drains individual tissues and organs and, in the absence of any experimental data to the contrary, is believed to have the same composition as the interstitial fluid that it collects. There have been relatively few studies on the fluid composition of afferent lymph, but it has been suggested by some investigators that considerable volumes of fluid are directly reabsorbed into the bloodstream in lymph nodes. If this proposition proves correct, it has considerable implications for the magnitude of fluid volume turnover and for the concentration of protein in afferent and efferent lymph. It has been suggested, for example, that total afferent lymph flow in a 65-kg human may be around 12 L per day at a protein concentration of 20 g L^{-1}, with a total efferent lymph flow of about 4 L per day and a protein concentration of 60 g L^{-1}. This would mean that total daily fluid volume turnover would approach two-thirds of the entire interstitial fluid volume per day.

The concentration of protein in efferent lymph varies in lymph draining different tissue beds and is much higher in lymph coming from tissues such as liver and ovaries where capillary permeability to plasma proteins is high. Lymph also contains all the coagulation factors, although it clots more slowly than plasma. Lymph draining tissues such as testis and ovaries may also acquire significant amounts of hormones. Efferent lymph has a similar electrolyte composition to plasma. Efferent lymph can also acquire a range of cytokines as well as immunoglobulins which are released from lymph nodes undergoing an immune response. Caution should, however, be exercised in interpreting the fluid and molecular composition of efferent lymph because the concentration of substances such as proteins in lymph may be changed considerably if large volumes of fluid are reabsorbed into the blood in lymph nodes. See also **Cytokines**.

Cellular Composition. Enormous numbers of lymphocytes migrate from the bloodstream into the tissues and lymph nodes every day. The lymphatic system has the essential function of returning these wandering cells to the blood and of seeding any cells generated during immune responses throughout the entire body. In fulfilling this function the lymphatic system plays a key role in integrating the widely dispersed organs and tissues that make up the immune system so that it can function, in effect, as a single organ.

Afferent lymph contains populations of dendritic cells which are important antigen-presenting cells, and macrophages which engulf any particulate matter such as cell debris and any microorganisms that may have entered tissues during episodes of infection. During bouts of inflammation, afferent lymph also contains high concentrations of granulocytes which have exited the blood in response to a variety of chemotactic agents. Efferent lymph contains much higher concentrations of lymphocytes than afferent lymph. Although lymphocytes leave the bloodstream in most tissues, they do so in vastly greater numbers in lymph nodes (see Fig. 2). Lymphocytes exit the blood in lymph nodes at specialized traffic sites in nodes that lie in the deep cortex. Specialized postcapillary venules, known as high endothelial venules (HEVs) because of their cuboid-shaped endothelium, transmit vastly greater numbers of lymphocytes than do other tissues.

Transport of Cells and Molecules

Transport of Fluid and Molecules. Previous sections have considered how the lymphatic system provides the critical transport pathway which allows the continuous recycling of plasma protein and interstitial fluid between blood and the interstitium. In so doing, the flow of lymph makes a vital contribution to extracellular fluid homeostasis and ensures that the effects of capillary exchange of fluids and macromolecules on the volume of interstitial fluid and on the blood volume are tightly controlled. The following sections deal with the role of the lymphatic system in the transport of lymphocytes and other cells around the body.

Transport of Cells. There are roughly 10^{12} lymphocytes in the lymphoid tissues of a young adult 70-kg (154 lb) human and, after allowing for size and age, there are comparable numbers of lymphocytes present in the lymphoid tissues of other mammals. The majority of lymphocytes in primary lymphoid organs such as thymus and bone marrow are fixed and do not leave their sites of production in these organs. There are also a large number of lymphocytes present in secondary lymphoid organs such as spleen and lymph nodes which do not recirculate. The remaining lymphocytes belong to a population of lymphocytes that are continuously recirculating between blood, tissues and lymph. This population has been estimated to be about 10% of the lymphocytes present in young adult animals. Most studies on lymphocytes circulating in lymph have been conducted in sheep, and the data provided in the following sections have been obtained in this species, although they probably apply to other mammals including humans. A notable exception is the pig, which has very few cells present in lymph because most recirculating lymphocytes enter and leave lymph nodes via the nodal blood supply in this species.

Recirculation of Lymphocytes Through an Individual Lymph Node. The passage of lymphocytes through a sheep popliteal lymph node weighing 1 g (0.04 ounce) is shown in Fig. 2. Cells can enter lymph nodes either from afferent lymph or from blood via HEVs in the node itself. Having entered a lymph node, they can leave only by passing into efferent lymph. In a resting lymph node (one that is unstimulated by antigen) fewer than 2% of lymphocytes are generated within the node itself, so that the efferent lymph output of lymphocytes is a direct measure of the input of cells into the node.

The output of cells from a wide range of lymphatics has been investigated in sheep by inserting soft plastic cannulae into lymphatic vessels and collecting cells over periods of days or weeks in conscious animals. Around 90–95% of lymphocytes enter lymph nodes from the blood directly via HEVs and the rest via afferent lymph from the tissues.

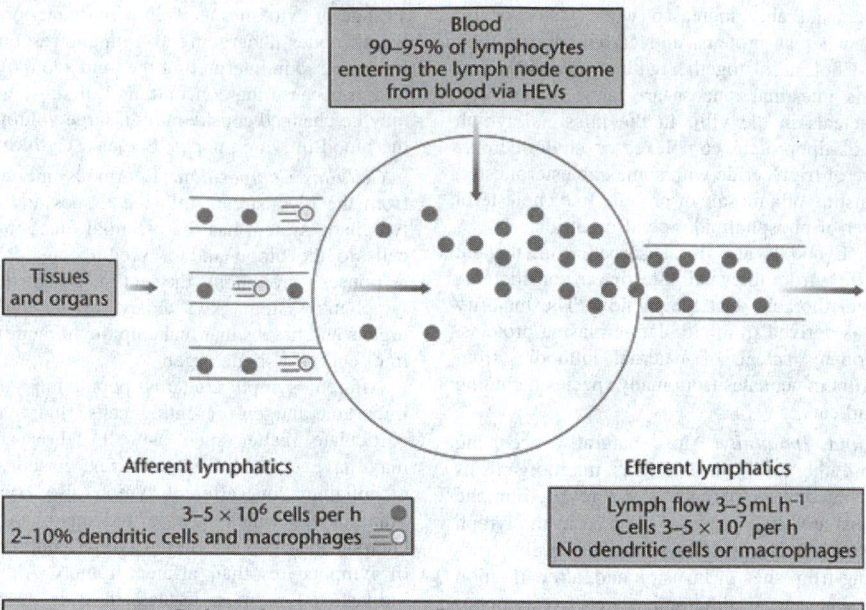

Fig. 2.   Scheme of lymphocyte recirculation through a sheep popliteal lymph node weighing 1 g. Collectively, about $1–1.5 \times 10^9$ lymphocytes per hour are returned in this way to the bloodstream via lymph. HEV, high endothelial venule.

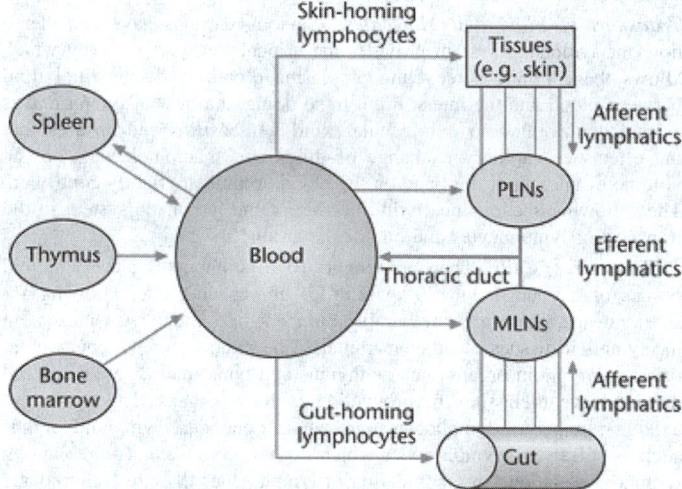

Fig. 3.   Scheme showing pathways of lymphocyte recirculation. The journey from blood through lymph nodes and back into lymph takes about 1 day. Tissue-specific streams direct gut-homing cells to the intestines and skin-homing cells to skin. Lymphocytes are added to the pool from their sites of production in thymus and bone marrow. PLN, peripheral lymph node; MLN, mesenteric lymph node.

The relationship of cell output per gram of lymph node weight holds for lymph nodes generally, except the mesenteric lymph node where the afferent input from the intestines is higher than that in other nodes. During an immune response, lymphocytes are recruited from the blood at up to 5–10 times greater rates, and large numbers of memory and effector cells are produced within the lymph node and released into lymph and subsequently seeded throughout the body via the lymphatic and blood circulation. Cytokines such as interferon γ, interleukins 1, 6, 8, and granulocyte–macrophage colony-stimulating factor are released into efferent lymph, and antibody-secreting cells and immunoglobulins also appear in the efferent lymph.

Tissue-specific Streams of Recirculating Lymphocytes in Lymph. For many years after lymphocyte recirculation was first described by Sir James Gowans, this process was believed to be completely random, although it was known that activated or blast cells homed preferentially to their tissue of origin. Since the discovery some 20 years ago that a large population of T lymphocytes (as distinct from blast cells) were preferentially recirculating between the small intestine, lymph and blood, it has become apparent that the tissue-specific migration or homing of lymphocytes targeting various tissues and organs is a key factor in the development and regulation of global immune responses. It is now recognized, for example, that intestinal lymph contains lymphocytes that preferentially home to gut, lymph node-homing lymphocytes target lymph nodes, while populations of lymphocytes in lymph coming from skin recirculate preferentially to skin and subcutaneous tissues (Fig. 3). It is also thought that antigen-experienced or memory lymphocytes show altered migration patterns from naive lymphocytes, which recirculate preferentially, but not exclusively, to lymph nodes. During immune responses newly generated cells can be released into lymph, so that what appears to be a homogeneous population of lymphocytes in lymph is a complex mixture of subpopulations with differing recirculation properties and functions which are being constantly reassorted as they journey between blood and lymph.

Additional Reading

Aukland, K., and R.K. Reid: "Interstitial-lymphatic Mechanisms in the Control of Extracellular Fluid Volume," *Physiological Reviews*, **73**, 1–78 (1993).

Boulton, M., M. Flessner, D. Armstrong, J. Hay, and M. Johnston: "Determination of Volumetric Cerebrospinal Fluid Absorption into Extracranial Lymphatics in Sheep," *American Journal of Physiology*, **274**, R88–R96 (1998).

Cahill, R.N.P., H. Frost, and Z. Trnka: "The Effects of Antigen on the Migration of Recirculating Lymphocytes Through Single Lymph Nodes," *Journal of Experimental Medicine*, **143**, 870–879 (1976).

Cahill, R.N.P., D.C. Poskitt, H. Frost, and Z. Trnka: "Two Distinct Pools of Recirculating Lymphocytes: Migratory Characteristics of Nodal and Intestinal T Lymphocytes," *Journal of Experimental Medicine*, **145**, 420–428 (1977).

Fraser, J.R.E., R.N.P. Cahill, W.G. Kimpton, and T.C. Laurent: "Lymphatic System," in: Comper, W.D.: *Extracellular Matrix*, Vol. 1, *Tissue Function*, Harwood Academic Publishers, Newark, NJ, 1996.

Guyton, A.C., A.E. Taylor, and H.J. Granger: *Circulatory Physiology II: Dynamics and Control of the Body Fluids*, W.B. Saunders, Philadelphia, PA, (1975).

McDowell, J., and M. Windelspecht: *Lymphatic System*, Greenwood Publishing Group, Inc., Westport, CT, 2004.

Reed, R.K., N.G. McHale, J.L. Bert, C.P. Winlove, and G.A. Laine: *Interstitium, Connective Tissue and Lymphatics*, Portland Press, Ltd., London, UK, 1995.

Schmid-Schönbein, G.W.: "Microlymphatics and Lymph Flow," *Physiological Reviews*, **70**, 987–1028 (1990).

Yoffey, J.M., and F.C. Courtice: *Lymphatics, Lymph and the Lymphomyeloid Complex*, Academic Press, London, UK, 1970.

Ross N.P. Cahill, University of Melbourne, Melbourne, Australia
Wayne G. Kimpton, John B. Hay, University of Toronto, Toronto, Canada

LYMPHOCYTES.   Lymphocytes are cells found in the bloodstream and lymphoid organs that make specific, inducible immune responses against infectious agents.

Introduction

Lymphocytes are the cells that mediate specific, inducible (adaptive) and generally long-lasting immune responses following infection or the administration of a vaccine. In addition, exposure to most vaccines or natural infections leads to accelerated and enhanced adaptive immune responses following subsequent re-exposure to the pathogen, a phenomenon known as immunological memory. Lymphocytes have the capacity to recognize substances foreign to the body and then to mount responses that result in their neutralization and or elimination. The earliest mediators of adaptive immunity to be recognized were antibodies. These are circulating proteins (collectively called immunoglobulins) that bind to components of microorganisms (e.g. viral proteins or capsular polysaccharides of bacteria) collectively called antigens. See also **Antibody**.

Although lymphocytes can be found in virtually any tissue, most are in specialized lymphoid organs. They are produced in central lymphoid organs, comprising the thymus and bone marrow, which, together with the peripheral lymphoid organs (spleen, tonsils and lymph nodes, and the gut-associated lymphoid tissues, such as the Peyer's patches and the appendix), constitute the lymphoid system (Fig. 1). Lymphocytes are the major cell type found in the lymph, the fluid found in the lymphatics. These are a network of vessels that drain tissue fluids via the lymph nodes, which are highly organized aggregates of lymphocytes and other cells (such as macrophages and dendritic cells), into the bloodstream, via the thoracic duct. Most lymphocytes therefore recirculate continuously from blood to lymph, thereby maximizing their chances of encountering antigen. An adult human contains some 2×10^{12} lymphocytes, and the total mass of the lymphoid system is comparable to that of the liver. The peripheral lymphoid organs fulfill multiple functions in immunity: (1) they act as effective filtering devices for microorganisms which may have entered the circulation and (2) they are the major sites at which lymphocyte responses to antigens occur. See also **Antigen**; and **Lymphatic System**.

The adaptive immune system has evolved two mechanisms for responding to antigens: (1) humoral immunity, which is mediated by antibodies

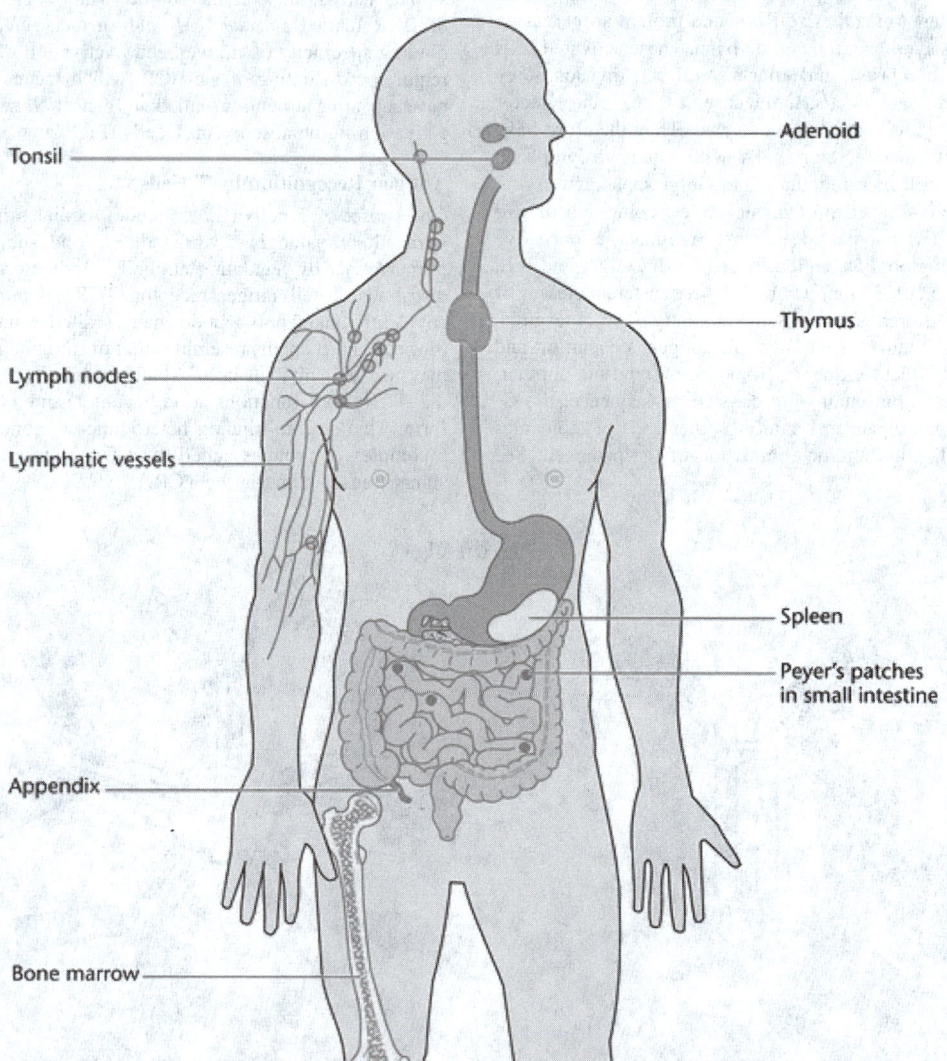

Fig. 1.   Schematic depiction of the human lymphoid system. B lymphocytes develop in the bone marrow and T cells in the thymus. These cells then migrate to the peripheral (secondary) lymphoid organs, such as the lymph nodes, spleen and gut-associated lymphoid organs, which are the sites where they respond to antigens. The lymphatic vessels drain tissue fluids, including lymphocytes, into the lymph nodes and ultimately into a vessel called the thoracic duct, which joins the bloodstream at the left subclavian vein (not shown). Lymphocytes in the blood re-enter the lymphoid organs. The recirculation of lymphocytes therefore provides a mechanism whereby these cells can monitor most tissues in the body for the presence of antigens.

and (2) cell-mediated immunity, involving the elimination of infected host cells by specialized "killer" cells, for example. Antibodies are produced by B lymphocytes (so-called because they arise in the bone marrow), which have encountered antigen. T lymphocytes (which develop functional maturity within the thymus) execute cell-mediated immune responses.

Morphology

Lymphocytes are small (6–10 μm in diameter) cells, with a dense nucleus, very little cytoplasm and little in the way of organelles. In a healthy individual circulating lymphocytes are quiescent cells (i.e., in the G_0 stage of the cell cycle); they do not synthesize DNA and synthesize only minimal amounts of RNA and protein. Although under conventional light microscopy lymphocytes all look pretty similar, more sophisticated analytical approaches have revealed that they can be divided into a variety of highly specialized subpopulations, which have evolved to fulfill specific functions in immunity. A particularly illuminating analytical approach has been the use of monoclonal antibodies, which specifically recognize cell surface proteins (called *markers*), characteristic of different subsets of lymphocytes. These markers now include a wide variety of proteins, such as CD4 and CD8, which are only found on T cells, and molecules such as CD19, which are characteristic of B cells.

Following encounter with antigen, quiescent T or B cells enter the cell cycle: they enlarge and start to synthesize RNA and protein at increasing rates, preparatory to DNA synthesis and cell division. These activated cells are called *lymphoblasts* (or blasts), and after several cell divisions they differentiate into so-called *effector cells*. In the case of B cells, the effector cells (plasma cells) secrete antibodies at a high rate. They therefore have the typical morphology of other secretory cells, with extensive cytoplasm and highly developed endoplasmic reticulum and Golgi apparatus.

Antibodies are flexible Y-shaped protein molecules, composed of two heavy (H) and two light (L) polypeptide chains. Mammals produce five distinct classes of immunoglobulins, called IgM, IgD, IgG, IgA and IgE, each having a different type of H chain (Fig. 2). These different classes of immunoglobulins fulfill different functions in immunity. The N-terminal portions of both H and L chains are involved in antigen recognition and consequently vary in amino acid sequence from one antibody to another. The C-terminal portions of both chains are constant in sequence; the C-regions of the H chains participate in recruiting other cells or molecules which lead to the neutralization and/or elimination of the pathogen. See also **Antibody Classes**.

Specificity

The immune system is capable of responding to the huge array of antigens that an individual may encounter in their environment. Both T and B lymphocytes recognize and respond to antigens by means of cell surface proteins (antigen receptors), which have the capacity to (1) bind to foreign antigens and (2) transduce biochemical signals across the cell membrane that lead to the activation of the appropriate clone(s). During the development of lymphocytes millions of unique cells are produced, each with an antigen receptor with a different specificity. At this time clones of cells that react with components of the body (self antigens) are eliminated in a process known as immune tolerance. The remaining cells comprise the primary immune repertoire and colonize the peripheral lymphoid organs (Fig. 3). The introduction of a foreign antigen leads to the marked expansion of those clones of cells that recognize antigenic components of the pathogen; this process is known as clonal selection.

Antigen Recognition by B Cells

B cells can recognize foreign materials in the circulation or in tissues, such as intact microbial proteins or polysaccharides. They do this by means of a clonally distributed antigen receptor, the B-cell antigen receptor (BCR). These receptors are composed of membrane-anchored forms of antibody molecules, associated with two additional polypeptides, which transduce signals across the cell membrane. Mature B cells express two classes of BCR (called *surface IgM* and *surface IgD*), with identical antigen-binding specificity on a given cell. Activation of B cells by antigen alone requires crosslinking of the BCR, which means that only antigens which have repeating structures (called *antigenic determinants*) will induce B-cell activation in the absence of T cells (Fig. 4).

Antigen Recognition by T Cells

One subset of T cells (CD8 T cells) mediates immune responses against intracellular pathogens, whilst the second subpopulation (CD4 T cells) primarily act by regulating antibody production by B cells. The clonally distributed T-cell antigen receptor (TCR) of most T cells consists of two covalently linked polypeptide chains (called α and β). The overall structure of the α and β chains resembles that of immunoglobulin molecules, in that they are also divided into V-(antigen-binding) and C-regions. However, the TCR only functions as a receptor and is not found in a soluble form. This antigen-binding heterodimer is noncovalently associated with a complex of proteins called the *CD3 complex*, which is responsible for signal transduction via the TCR.

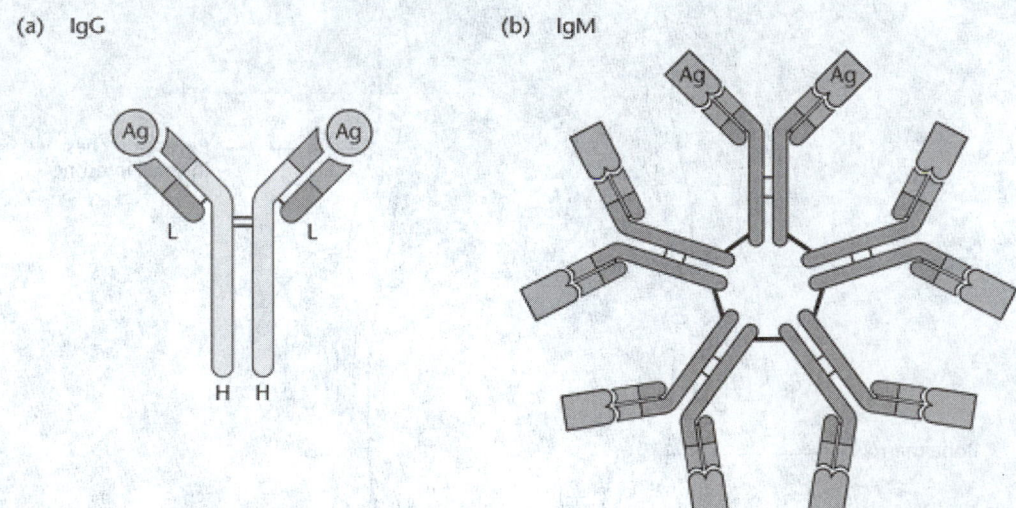

(a) IgG   (b) IgM

Fig. 2.   Structure of antibody molecules. (a) The basic structure of an antibody molecule, such as IgG, composed of two heavy (H) chains (in IgG called the γ chain) and two light (L) chains joined to each other by disulfide bridges. The N-terminal half of each L chain and the corresponding 25% of the H chain are variable in sequence (V regions, green) and together comprise the antigen-binding site of the molecule. The remainder of the H chain and L chain (C regions) are constant in sequence and, in the case of the H chain, are responsible for carrying out effector functions (such as opsonization) of the molecule once it has bound antigen. (b) The structure of a secreted IgM antibody. It is composed of a pentamer of the basic four-chain structure, in this case containing μH chains, and again divided into V and C regions. Secretory IgM therefore contains 10 antigen-binding sites.

1. Establishment of the primary repertoire

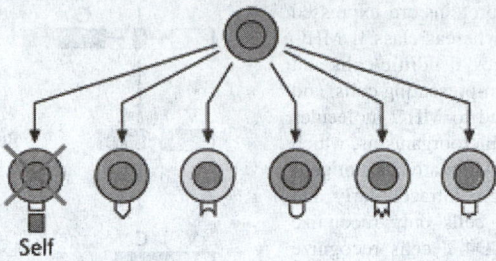

Self

2. Generation of effector cells

3. Generation of memory cells

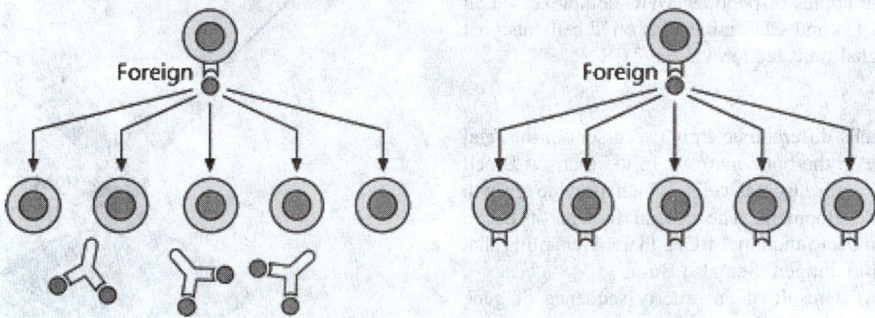

Foreign

Foreign

Fig. 3. The clonal selection hypothesis. (1) During the development of T and B lymphocytes thousands of clones of cells bearing antigen receptors of differing specificities are generated in the central lymphoid organs. Any that bind to self antigens with sufficiently high affinity are deleted early in their development. The remainder colonize the secondary lymphoid organs and comprise the primary immune repertoire. (2) The introduction of a foreign antigen induces activation, proliferation and differentiation of those clones which are specific for that antigen. This leads to the emergence of a markedly expanded population of specific effector cells (shown here as plasma cells, secreting specific antibodies), which neutralize and eliminate the antigen. (3) At the same time, a proportion of these antigen-specific cells enter a distinct differentiation pathway, which also involves substantial cell proliferation. However, these cells then return to a quiescent state, when they are called memory cells. Memory cells are responsible for the capacity of immunized individuals to mount more rapid and effective immune responses following re-exposure to the same antigen.

(a) B cell   (b) CD4 T cell   (c) CD8 T cell

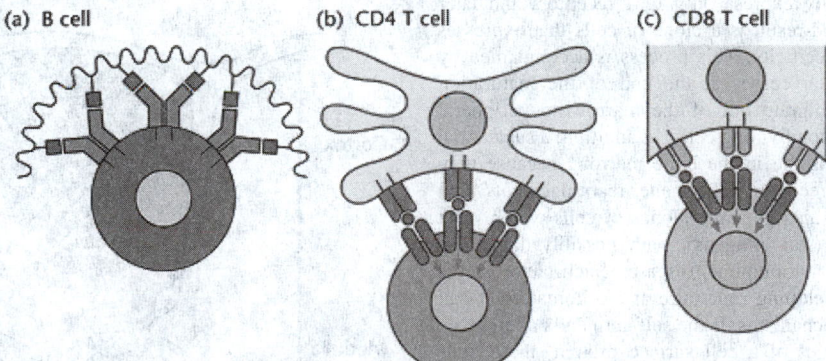

Fig. 4. Antigen recognition by T and B cells. (a) The B-cell receptor (BCR) can recognize native components of pathogens (illustrated here schematically as a long-chain bacterial capsular polysaccharide). This type of antigen can crosslink a substantial number of receptors, leading to the delivery of activating signals across the cell membrane (arrows) and subsequent expansion of that clone of cells. Most protein antigens require additional signals from CD4 helper T cells to induce B-cell activation. T cells, in contrast, only recognize peptide fragments of foreign protein antigens, associated with MHC proteins on other cells. (b) CD4 T cells recognize peptides associated with class II MHC proteins (shown in green): these derive from proteins which have been taken up by antigen-presenting cells (such as a dendritic cell depicted here) and degraded intracellularly. (c) CD8 T cells also recognize peptides, but only those associated with class I MHC proteins (grey). In this case, the peptides are derived from components of intracellular pathogens (e.g., from a virus-infected cell).

T cells recognize peptide fragments of foreign proteins when these are bound to grooves on the proteins encoded by genes of the major histocompatibility complex (MHC). Class I MHC proteins are expressed on the membranes of virtually all nucleated cells, whereas class II MHC proteins are only found on cells such as macrophages, dendritic cells and B cells. These class II-bearing cells are called antigen-presenting cells, and their importance is explained below. Peptides bound to MHC molecules may derive from degraded proteins of intracellular microorganisms, which are then presented on the surface of infected cells. Alternatively, antigen-presenting cells may take up proteins, degrade them intracellularly and present the peptides on MHC molecules. CD8 T cells only recognize peptides associated with class I molecules, and CD4 T cells recognize peptides associated with class II MHC proteins (Fig. 4). Again, the delivery of activating signals via the TCR requires crosslinking of the receptor, i.e. the recognition of multiple copies of peptide–MHC complexes on an antigen-presenting cell. The CD4 and CD8 molecules on T cells also act as coreceptors, amplifying signal transduction via the TCR.

Development

Development of B Cells. B cells differentiate from precursors in the fetal liver and throughout adult life in the bone marrow. In the marrow B cell precursors interact physically with stromal cells, which provide growth factors essential for B-cell development. The crucial feature of B-cell development is the production of a functional BCR, thereby enabling that cell to recognize and respond to antigen. See also **Bone**.

Expression of the BCR is the result of an orderly sequence of gene rearrangements in both the H and L chain loci: in essence, a functional H chain is the result of recombination between four gene segments, the V (variable), D (diversity) and J (joining) segments, which together form the antigen-binding region of the chain, and the constant region, encoded by a separate segment (C). Similarly, a complete L chain is the product of V, J and C gene segments (Fig. 5). The gene segments which form the V regions exist as multiple copies in the genome (e.g. there are some 50 V_H segments). These recombine in a combinatorial fashion: this plus the combinatorial association of different H chains with different L chains generates the millions of B-cell receptors found in the immune repertoire.

The earliest cell in the B-cell lineage is the pro-B cell, in which rearrangement of the H chain gene takes place. This matures into a pre-B cell, which rearranges the L chain gene. Finally, these cells develop into immature B cells, which initially express only IgM receptors and later on express IgD as well. The end-result is a clone of cells that expresses antigen receptors of a single specificity. This process is accompanied by considerable cell division, which ceases at the end of the maturation process, so that B cells that migrate out of the marrow to peripheral lymphoid organs are quiescent, nondividing cells. In addition, a substantial proportion of developing B cells die in the bone marrow, because they fail to produce a functional receptor. The gene rearrangements that accompany B-cell development also generate clones of cells which react with components of the body (self antigens), with possibly dangerous consequences, culminating in autoimmune disease. Such autoreactive clones are purged from the developing repertoire at the immature B-cell stage of development by two mechanisms. If the self antigen is a repeating structure (such as multiple copies of a cell surface protein) these cells are eliminated by a process known as apoptosis. Alternatively, if the self antigen is a soluble protein, which induces minimal crosslinking of the BCR, the cells are functionally silenced by a process known as clonal anergy.

Development of T Cells. Precursors of T cells migrate from the bone marrow and colonize the thymus, an organ in the upper thorax. The thymus is composed of lobules, each one consisting of an outer cortical region and an inner medullary region (Fig. 6). Most of the cells in the thymus (especially in the cortex) are lymphoid cells (called *thymocytes*), which are closely associated with epithelial cells, comprising the thymic stroma. The rate of T-cell production in the thymus is highest in young individuals, and the thymus shrinks during adult life, thereby indicating that the complete T-cell repertoire is laid down before puberty in humans. A young mouse produces about 5 million thymocytes per day, but only about 2% of these leave the thymus to colonize peripheral lymphoid organs. The remainder die in the thymus (also by apoptosis). The process of T-cell development commences in the thymic cortex and mature cells are found in the medulla.

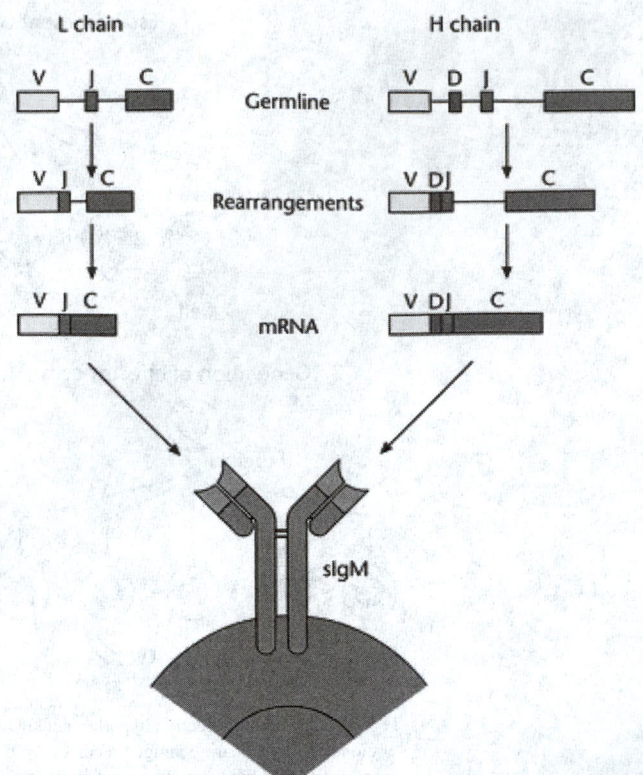

Fig. 5. Generation of the BCR during B-cell development by gene rearrangements. In all cells of the body except B cell, the genes encoding the V, J and D segments of the H chain and V and J of the L chain are separated from each other (germline configuration). During the development of B cells recombination leads to the deletion of intervening stretches of DNA, generating a complete V-region gene, consisting of fused VDJ segments in the case of the H chain and VJ in the case of the L chain. These become spliced to the C-region genes and then transcribed into the complete polypeptide chains, which form the surface IgM receptor.

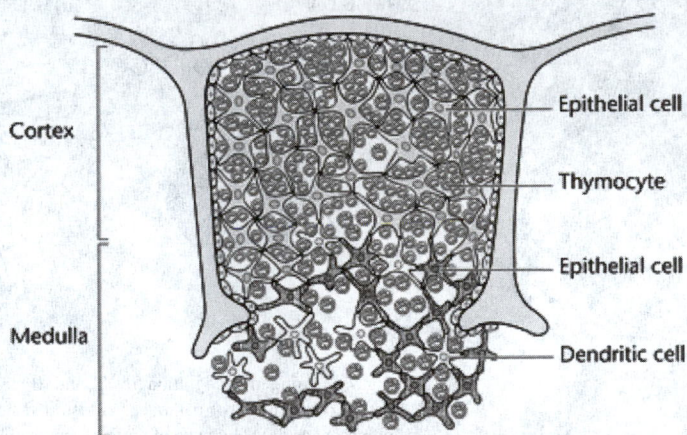

Fig. 6. Schematic structure of the thymus. The thymus is composed of lobules, each of which is divided into an outer (cortical) and inner (medullary) region. The cortex contains dense aggregates of immature developing T cells (thymocytes), interspersed with epithelial cells. The medulla consists of mature thymocytes, epithelial cells and dendritic cells.

Again, the critical event that occurs during T-cell development is the expression of a functional TCR. See also **Thymus Gland**.

The most immature thymocytes lack all of the markers characteristic of peripheral T cells: hence they do not express the TCR, or CD4 and CD8. During the maturation process these cells, termed *double-negative cells*, first express both the CD4 and CD8 molecules (*double-positive cells*) then differentiate into mature T cells which express either of these markers (*single-positive thymocytes*). This is accompanied by gene rearrangements that lead to the expression of the clonally distributed TCR. The α and β chains of the TCR are generated via mechanisms similar to those described above for Ig H and L chains. Hence, the β chain is the product of four gene segments: V, D and J (forming the variable region) and C forming the constant region. Similarly, the α chain is the product of V, J and C gene segments.

Once thymocytes express a functional TCR they undergo a process of selection, which shapes the T-cell repertoire. This occurs at the CD4/CD8 double-positive stage of development, principally at the corticomedullary junction of thymic lobules, by mechanisms that are not wholly understood. In the thymus, double-positive cells encounter a wide array of peptides derived from self proteins displayed on MHC molecules of epithelial cells and thymic dendritic cells. At least two models have been proposed to explain how T cells are selected to enter the peripheral repertoire (positive selection). The first hypothesis (the instructional model) postulates that recognition of peptide on MHC class II leads to the shut-off of CD8 synthesis, generating a CD4 T cell, while recognition of peptide on class I has the converse effect. The second hypothesis (stochastic model) proposes that lineage commitment to single-positive cells occurs before positive selection. In any event, it is likely that the strength of interaction (affinity) of the TCR for a particular peptide–MHC complex determines the fate of that cell. Cells whose receptors do not bind will die, while those that bind with moderate affinity will be positively selected to become part of the repertoire. Finally, thymocytes that interact with self peptide–MHC complexes with high affinity will be induced to die (negative selection). The thymus therefore plays a key role in the elimination of self-reactive cells from the T-cell repertoire.

Function

T Lymphocytes. T lymphocytes generally first encounter antigen in the form of peptide–MHC complexes in the peripheral lymphoid organs on the surfaces of antigen-presenting cells such as dendritic cells and macrophages (Fig. 4). This encounter leads to their activation, proliferation and the generation of effector cells. The proliferation of activated T cells is mediated by growth factors, most notably interleukin 2, a protein produced by these cells themselves. See also **Interleukins**.

CD4 T Cells. CD4 T cells, also called *helper T cells*, recognize peptides in association with class II MHC proteins. A major function of CD4 effector cells is to interact with B cells and deliver helper signals to enable the B cell to mount antibody responses to protein antigens. B cells are very efficient at specifically taking up antigens, by virtue of their antigen receptors. These are degraded and presented on class II MHC molecules on their surfaces. If a CD4 effector cell encounters its specific peptide on the surface of a B cell this leads to the formation of an antigen-specific conjugate between the two cells (T cell–B cell collaboration): during this interaction both contact-mediated and soluble stimuli are produced by T cells, which result in B-cell activation. The contact-mediated signal emanates from a protein on CD4 effector T cells called CD154, which binds to a stimulatory receptor on B cells called CD40. This, in conjunction with soluble growth and differentiation factors (called *cytokines*) induces the production of antibodies by activated B cells (Fig. 7). The nature of the cytokines produced by CD4 helper cells also determines the classes of antibodies that B cell produce. Another important function of CD4 T cells is to activate macrophages; this again is mediated by cytokines. Activated macrophages play an important role in the elimination of certain intracellular pathogens, such as the leprosy bacillus. See also **Cytokines**.

CD8 T Cells. CD8 T cells are much more specialized than CD4 T cells and are precommitted to differentiate into cytotoxic T cells after encountering peptides in association with class I MHC molecules. The principal function of these cells is to eliminate virus-infected target cells. When a CD8 T cell recognizes peptide fragments of viral proteins associated with MHC class I molecules on the target this causes the release of proteins called *cytotoxins*, stored in intracellular granules. These penetrate the membrane of the target cell and initiate an intrinsic death program within the target. In addition, CD8 T cells (and some CD4 T cells) can express a protein (called the *Fas ligand*) following their activation, which leads to the lysis of target cells expressing the Fas protein. Fas/Fas ligand interactions are thought to be principally involved in controlling the

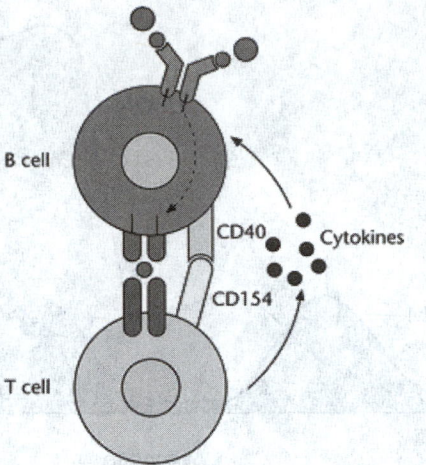

Fig. 7. T cell–B cell cooperation. CD4 T cells that have encountered antigen on dendritic cells (see Fig. 4) become activated and express a protein called CD154. B cells bind and internalize protein antigens (dashed arrow) and present peptides on their class II MHC molecules (dark green). When these cells encounter T cells specific for that antigen, the two cells form conjugates. Signals delivered via the interaction of CD154 with CD40, in conjunction with cytokines produced by T cells, lead to B-cell activation.

magnitude of clonal proliferation of lymphocytes and hence in maintaining homeostasis within the lymphoid system.

B Lymphocytes. As discussed above, the induction of antibody responses to protein antigens requires the interaction of B cells with CD4 T cells. In consequence, these antigens are called *T cell-dependent antigens*. However, other antigens, as exemplified by bacterial capsular polysaccharides, are able to induce antibody responses without T-cell help. This is because these T-independent antigens are large, polymeric molecules with a repeating structure and can therefore induce extensive crosslinking of the BCR, which is in itself sufficient to induce B-cell activation and the generation of plasma cells (Fig. 4).

The first class of antibodies to be formed in response to any antigen is IgM (Fig. 8). Later in the response, some B-cell blasts undergo a process called *class switching*, so that their progeny subsequently produce antibodies of other classes, such as IgG or IgE, for example. Class switching occurs as a result of DNA recombination, whereby the C-gene segment of the IgM H-chain gene is deleted and the V-region gene recombines with a C segment from another H chain class (such as IgG). The purpose of this process is to produce antibodies of different classes, but with the same antigenic specificity. See also **Antibody Classes**.

Antibodies exert their protective effects against infectious agents in several different ways.

1. They can neutralize the toxic effects of bacterial toxins, for example, or prevent viruses or bacteria from invading host cells.
2. Coating the surface of microorganisms with antibodies facilitates their removal by phagocytic cells, a process known as opsonization.
3. Binding of antibodies to antigen may activate the proteins of the complement cascade: certain complement components are also involved in opsonizing microorganisms. Others induce the recruitment of phagocytic cells to the area of infection, or when bound to antibodies, lead to the destruction of bacteria, by creating holes in their cell walls.

It is clear that the different classes of immunoglobulin evolved because they fulfill different functions in adaptive immunity and often in different locations in the body. IgM, for example, is extremely potent at activating the complement system, whereas IgA is the principal class of antibodies found in secretions, such as tears, saliva and the gut.

Primary exposure to a vaccine or an infection induces a burst of clonal expansion of specific lymphocytes, which differentiate into effector cells, such as plasma cells, over a period of some 5 days. Most of these cells then die by apoptosis. However, a proportion of these cells enter a different differentiation pathway and become long-lived, nonproliferating cells called memory cells (Fig. 3). Both T and B cells can become memory

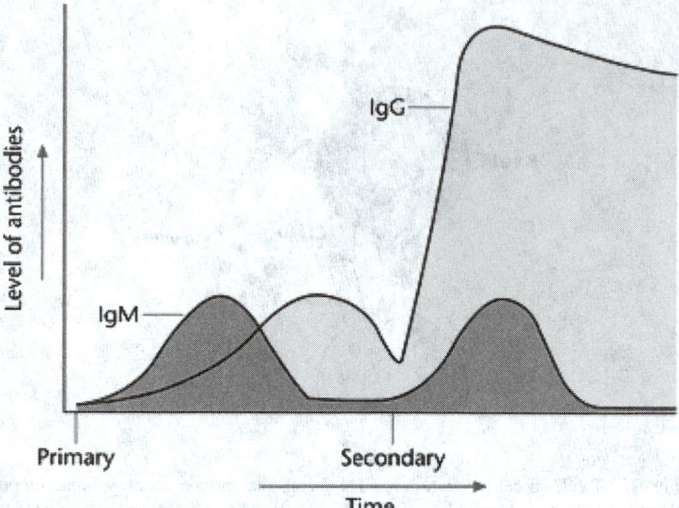

Fig. 8. Immunological memory. The graph illustrates the levels (titres) of antibodies elicited after primary and secondary (booster) immunization with a typical T-dependent antigen. The primary response is relatively slow in onset, composed initially of IgM antibodies, followed by low levels of IgG. Following secondary immunization, antibodies appear more rapidly and reach higher titres. They are principally of the IgG class and have higher affinity for the antigen than those produced in the primary response.

cells, although the mechanisms involved are better understood in the B-cell lineage. The existence of immunological memory can be demonstrated by measuring antibody titres after a primary and secondary (booster) injection of say, a protein antigen: the primary response is often slow and modest in magnitude. Secondary responses after boosting occur more rapidly and are much larger in magnitude. Primary responses are generally dominated by IgM antibodies, which tend to have low affinity for antigen (Fig. 8). Secondary responses are typically mostly IgG, with much higher affinity (a process called *affinity maturation*).

Memory B cells arise in foci called *germinal centers in peripheral lymphoid organs*: here B-cell blasts proliferate rapidly and their V-region genes undergo a process of mutation. These mutations may increase or decrease the affinity of the cell's BCR for antigen. B cells in germinal centers are exposed to a depot of antigen retained on specialized presenting cells called *follicular dendritic cells*. Only cells whose receptors have high affinity for the antigen are selected to enter the peripheral pool as memory cells. These cells remain quiescent until they re-encounter antigen. They then become reactivated (with the help of T cells, as above) and generate the secondary antibody response. The induction of memory is a T cell-dependent process, so that T-independent antigens do not elicit immunological memory, but instead evoke primarily an IgM antibody response, even following repeated exposure to the antigen.

See also **Immune System and Immunology**.

Additional Reading

Herbert, W.J., P.C. Wilkinson, and D.I. Stott: *Dictionary of Immunology*, 4th Edition, Elsevier Science & Technology Books, New York, NY, 1995.

Janeway, C., P. Travers, M. Walport, and M. Shlomchik: *Immunobiology: The Immune System in Health and Disease*, 6th Edition, Taylor & Francis, Inc., Philadelphia, PA, 2004.

Roitt, I.M., and S.J. Martin: *Roitt's Essential Immunology*, 11th Edition, Blackwell Publishers, Malden, MA, 2006.

Singh, H., and R. Grosschedl: *Molecular Analysis of B Lymphocyte Development and Activation*, Springer-Verlag New York, LLC, New York, NY, 2005.

Virella, G.: *Medical Immunology*, 6th Edition, CRC Press, LLC, Boca Raton, FL, 2007.

GERRY G.B. KLAUS, National Institute for Medical Research, London, UK

LYMPHOCYTIC CHORIOMENINGITIS (LCM). Lymphocytic choriomeningitis, or LCM, is a rodent-borne viral infectious disease that presents as aseptic meningitis (inflammation of the membrane, or meninges, that surrounds the brain and spinal cord), encephalitis (inflammation of the brain), or meningoencephalitis (inflammation of both the brain and meninges). Its causative agent is the lymphocytic choriomeningitis virus (LCMV), a member of the family Arenaviridae, that was initially isolated in 1933. Although LCMV is most commonly recognized as causing neurological disease, as its name implies, asymptomatic infection or mild febrile illnesses are common clinical manifestations. Additionally, pregnancy-related infection has been associated with abortion, congenital hydrocephalus and chorioretinitis, and mental retardation.

LCM and milder LCMV infections have been reported in Europe, the Americas, Australia, and Japan, and may occur wherever infected rodent hosts of the virus are found. However, the disease has historically been under reported, often making it difficult to determine incidence rates or estimates of prevalence by geographic region. Several serologic studies conducted in urban areas have shown that the prevalence of LCMV infection among humans ranges from 2% to 10%.

LCMV is naturally spread by the common house mouse, *Mus musculus*. Once infected, these mice can become chronically infected by maintaining virus in their blood and/or persistently shedding virus in their urine, a common characteristic of other arenavirus infections in rodents. Chronically infected female mice usually transmit infection to their offspring, which in turn become chronically infected.

Humans become infected by inhaling infectious aerosolized particles of rodent urine, feces, or saliva, by ingesting food contaminated with virus, by contamination of mucus membranes with infected body fluids, or by directly exposing cuts or other open wounds to virus-infected blood. LCMV infection has also been documented among staff handling infected hamsters. Person-to-person transmission has not been reported, with the exception of vertical transmission from an infected mother to fetus.

The incubation period of LCMV infection is usually between 8 and 13 days. A characteristic biphasic febrile illness then follows. The initial phase, which may last as long as a week, typically begins with any or all of the following symptoms: fever, malaise, anorexia, muscle aches, headache, nausea, and vomiting. Other symptoms that appear less frequently include sore throat, cough, joint pain, chest pain, testicular pain, and parotid (salivary gland) pain. Following a few days of remission, the second phase of the disease occurs, consisting of symptoms of meningitis (for example, fever, headache, and a stiff neck) or characteristics of encephalitis (for example, drowsiness, confusion, sensory disturbances, and/or motor abnormalities, such as paralysis). LCMV has also been known to cause acute hydrocephalus, which often requires surgical shunting to relieve increased intracranial pressure. In rare instances, infection results in myelitis (inflammation of the spinal cord) and presents with symptoms such as muscle weakness, paralysis, or changes in body sensation. An association between LCMV infection and myocarditis (inflammation of the heart muscles) has been suggested.

During the first phase of the disease, the most common laboratory abnormalities are a low white blood cell count (leukopenia) and a low platelet count (thrombocytopenia). Liver enzymes in the serum may also be mildly elevated. After the onset of neurological disease during the second phase, an increase in protein levels, an increase in the number of white blood cells or a decrease in the glucose levels in the cerebrospinal fluid (CSF) is usually found.

Previous observations have shown that most patients who develop aseptic meningitis or encephalitis due to LCMV recover completely. No chronic infection has been described in humans, and after the acute phase the virus is cleared. However, as in all infections of the central nervous system, particularly encephalitis, temporary or permanent neurological damage is possible. Nerve deafness and arthritis have been reported. Infection of the human fetus during the early states of pregnancy may lead to developmental deficits that are permanent. LCM is usually not fatal. In general, mortality is less than 1%.

Aseptic meningitis, encephalitis, or meningoencephalitis requires hospitalization and supportive treatment based on severity. There is no specific drug therapy for LCM. Anti-inflammatory drugs, such as corticosteroids, may be considered under specific circumstances. Although studies have shown that ribavirin, a drug used to treat several other viral diseases, is effective against LCMV *in vitro*, there is no established evidence to support its use for treatment of LCM in humans.

Individuals of all ages who come into contact with urine, feces, saliva, or blood of the house mouse are potentially at risk for infection. Laboratory workers who themselves handle infected animals are also at risk. However, this risk can be minimized by utilizing animals from sources that regularly test for the virus, wearing proper protective laboratory gear, and following appropriate safety precautions. Owners of pet mice or hamsters may be at risk for infection if these animals originate from colonies with circulating LCMV, or if the animals become infected from other wild mice. Human fetuses are at risk of acquiring infection vertically from infected maternal blood.

Like many other rodent-borne infectious diseases, LCMV infection can be prevented by avoiding or minimizing direct physical contact with rodents or exposure to their excreta. Adequate ventilation should be provided to any heavily infested, previously unventilated enclosed room or dwelling prior to cleanup. A liquid disinfectant, such as a diluted household bleach solution, should be applied to visible rodent droppings and their immediate surroundings. Gloves should be worn when disinfecting and cleaning up rodent excreta. Rodent spring traps may be set up in households or dwellings where rodent infestations are a concern.

The geographic distributions of the rodent hosts are widespread both domestically and abroad. However, infrequent recognition and diagnosis, and therefore under reporting, of LCM, have limited scientists' ability to estimate incidence rates and prevalence of disease among humans. Understanding the epidemiology of LCM and LCMV infections will help to further delineate risk factors for infection and develop effective preventive strategies. Increasing physician awareness will improve disease recognition and reporting, which may lead to better characterization of the natural history and the underlying immunopathological mechanisms of disease, and stimulate future therapeutic research and development. See also **Arenaviruses**; and **Viral Hemorrhagic Fevers**.

Additional Reading

Galasso, G.J., T.C. Merigan, and R.J. Whitley: *Antiviral Agents and Human Viral Diseases*, 4th ed., Lippincott Williams & Wilkins, Philadelphia, PA, 1997.

Jahrling, P.B. and C.J. Peters: "Lymphocytic Choriomeningitis Virus: A Neglected Pathogen of Man," *Arch. Pathol. Lab Med.*, **1 16**, 486–488 (1992).

Leland, S.S. and S. Ozmat: *Clinical Virology*, W.B. Saunders Company, Philadelphia, PA, 1996.

Love C.B. and P.B. Jahrling: *Viral Hemorrhagic Fever*, DIANE Publishing Company, Collingdale, PA, 1996.

Pattison, J.R., J.E. Banatvala, and A.J. Zuckerman: *Principles and Practice of Clinical Virology*, 4th Ed., John Wiley & Sons, Inc., New York, NY, 2000.

Peters, C.J., M. Buchmeier, P.E. Rollin, and T.G. Ksiazek: "Arenaviruses," in R.B. Belshe, ed., *Textbook of Human Virology*, 2nd ed., Mosby-Year Book, Inc., St. Louis, MO, 1991.

Peters, C.J. et al.: "Arenaviridae: Biology of Viruses," in: B.N., Fields, D.M. Knipe, and P.M. Howley, eds., *Fields Virology*, 3rd ed., Lippincott Williams & Wilkins, Philadelphia, PA, 1996.

Richman, D.D., R.J. Whitley, and F.G. Hayden: *Clinical Virology*, Harcourt Brace & Company, San Diego, CA, 1998.

Centers for Disease Control and Prevention (CDC), Atlanta, GA

LYMPHOGRANULOMA VENEREUM. Caused by strains of *Chlamydia trachomatis*, this is a venereal disease which features the appearance of a transitory primary genital lesion, a vesicle or a papule, followed by other stages. In the male, the primary genital lesion, which occurs within 5 to 21 days after the implicating sexual exposure, usually takes the form of a painless vesicle or papule, or as a chancriform lesion (*chancroid*). In the male, it is commonly located on the coronal sulcus of the penis; in females, on the labia or posterior vaginal wall. The situation is usually self-limiting and healing occurs within a few days. Extragenital infections occur in persons who deviate from normal sex profiles. The infection may continue in a second and a third stage, where the disease is extended to include regional lymph nodes. These nodes enlarge, produce pus, and ultimately form buboes. In a third stage, lymphatic obstruction may occur. Lymphatic obstruction may infrequently produce dilation of lymphatic channels and hypertrophy of subcutaneous connective tissue and of skin (*genital elephantiasis*).

Treatment may include oral tetracycline or sulfonamides, such as sulfisoxazole. It is a good practice to test all patients with lymphogranuloma venereum for syphilis because the two infections are frequent companions.

LYMPHOKINES. These are soluble substances (factors) produced by lymphocytes which aid in regulating a variety of immune responses. See also **Immune System and Immunology**. They are not immunoglobulins, but are synthesized by lymphocytes of undetermined structure and are classified according to the target cells they affect.

Chemotactic factors are lymphokines that attract certain cells, e.g., monocytes, neutrophils, etc., to a particular site. They are synthesized and released within 24 hours of lymphocyte stimulation.

Migration inhibition factor (MIF), the first lymphokine discovered, inhibits macrophage migration and is produced by certain sensitized lymphocytes stimulated by an exquisite antigen. Mitogens and antigen-antibody complexes can, however, nonspecifically trigger lymphocytes to produce MIF and antigen need not be present for MIF to inhibit macrophage migration.

Macrophage-activating factor (MAF) induces morphologic, metabolic, and functional changes in macrophages, which enhance the cell's ability to kill microorganisms and tumor cells.

Leucocyte-inhibitory factor (LIF) inhibits neutrophil (but not monocyte) migration in vitro.

Interleukin. Macrophages produce a monokine termed Interleukin-1 (lymphocyte activating factor) which combines with an antigen or mitogen to stimulate T lymphocytes to produce Interleukin-2 (T-cell growth factor) which in turn causes the proliferation of other T cells, such as helper, suppressor, or cytotoxic cells.

Cytotoxic and Cystostatic Factors. Lymphocytes from several species produce lymphokines, which kill or inhibit the growth of susceptible target cells. Among these lymphokines are several proliferation-inhibiting or colony-inhibiting factors and inhibitors of DNA synthesis.

Tissue Factor. Lymphocytes stimulated in vitro by an antigen or mitogen produce a procoagulant material that is biologically similar to tissue thromboplastin, but is antigenically distinct from it. The pathophysiologic importance of tissue factor is undetermined.

Lymphokines that enhance antibody production are termed helper factors and are produced by helper T lymphocytes. Lymphokines that suppress antibody production are termed suppressor factors and are produced by suppressor T lymphocytes.

Colony-stimulating activity is a lymphokine that induces differentiation of bone marrow cells in vitro into granulocytes and agranular mononuclear cells.

Osteoclast activating factor (OAF) is a lymphokine which causes bone resorption by activating osteoclasts.

Lymph node lymphocytes, from rats immunized with myelin basic protein, release immunoglobulin-binding factor that combines with antigen-antibody complexes, causing IgG-sensitized sheep erythrocytes to agglutinate.

Interferons. These are a family of glycoproteins that act against a wide range of viruses. These types are known and can be distinguished on the basis of their physicochemical characteristics, the cell types producing them, and the types of stimuli causing their production. They do not however act directly upon viruses, but render cells resistant to viral infection.

Additional Reading

Cohen, S.: *Lymphokines & the Immune Response*, CRC Press, LLC., Boca Raton, FL, 1989.

Gills, S.: *Recombinant Lymphokines and Their Receptors*, Marcel Dekker, Inc., New York, NY, 1987.

Goldstein, A.L.: *Thymic Hormones and Lymphokines: Basic Chemistry and Clinical Applications*, Perseus Books, Boulder, CO, 1984.

ANN C. DE BALDO, Ph.D., University of South Florida, Tampa, FL

LYMPHOMA. A lymphoma is defined as a tumor of lymphoid tissue. Some diseases are also referred to as lymphomas and comprise a group of malignant disorders that usually arise in the lymph nodes. These diseases present a broad spectrum of clinical features. Sometimes the term non-Hodgkin's lymphomas is used to distinguish other lymphomas from Hodgkin's disease because they do not have the characteristic giant cells of Hodgkin's disease. The causes of the lymphomas remain poorly understood. One suspected cause is viral infection. The epidemiology of the lymphomas may provide evidence as to the etiology of the diseases. Some lymphomas occur with much higher frequency in some geographical regions than in others. Diagnosis of a lymphoma is by biopsy. Treatment of the non-Hodgkin's disease lymphomas tends to parallel the therapy used in treating Hodgkin's disease. See also **Hodgkin's Disease**; and **Lymphosarcoma**.

LYMPHOSARCOMA. A disease, similar to leukemia and Hodgkin's disease in its symptoms. Although still generally used, the term lymphosarcoma is pathologically obsolete; it basically refers to what are now known as lymphocytic lymphomas and includes what were formerly known as lymphocytic lymphosarcomas, the so-called lymphoblastic lymphosarcoma, follicular lymphoma, and Burkitt's tumor.

It is important to realize that the distinction between chronic lymphocytic leukemia and those cases of lymphocytic lymphoma in which some of the tumor cells enter the circulating blood stream is not made easily and indeed may be impossible.

The lymphocytic lymphomas may arise in any group of lymph nodes in a tonsil, the spleen, or thymus, but lymph nodes are by far the most common site of their origin. The invasive growth of the sarcoma leads to fusion and coalescence of affected nodes and their adherence to adjacent tissues—these latter may be deeply penetrated by the tumor. Microscopically the normal structure of the affected tissue is obliterated by masses of closely packed tumor cells and in many cases the cells cannot be distinguished from normal lymphocytes. The cells of any given tumor are, however, usually uniform in type—either small or large. Occasionally a well differentiated tumor that consists predominantly of small lymphocytes may be found to include areas in which the cells are larger lymphocytes; more rarely, pleomorphic areas are seen. Whatever the type of cell, the lymphomas with mixed cytological constitution trend to be more rapidly progressive. An early indication of developing malignancy by cell differentiation is enlargement of the nucleolus. There is seldom any tendency for necrosis to occur in the lymphomas unless they have become anaplastic.

Acute lymphatic leukemia of childhood may be lymphosarcoma with the abnormal cells overflowing into the circulating blood. Whereas usually this disease is fatal if untreated, when patients receive early and proper treatment, both blood and bone marrow may revert to normal condition with full restoration of health for variable periods. In lymphosarcoma, x-ray therapy is the major and most universally effective form of treatment. Chlorambucil and nitrogen mustard have been effective chemotherapeutic agents in earlier stages of the disease; cyclophosphamide and vinblastine sulfate, in later stages. Prednisone may prove beneficial in patients with fever and hematologic disturbances no longer suitable for treatment with x-ray or other drugs.

Additional Reading

Cabanillas, F. and M. Rodriguez: *Advances in Lymphoma Research*, Vol. 85, Kluwer Academic Publishers, Norwell, MA, 1997.

Crocker, J.: *Advances in Lymphoma Research*, Vol. 85, Blackwell Science, Inc., Malden, MA, 1993.

Katz, R.: *Current Issues in Lymphoma*, S. Karger AG, Basel, Switzerland, 1994.

R. C. V.

LYNX. See **Cats**.

LYOPHILIC. Characterizing a material that readily goes into colloidal suspension in a liquid; if into water, it is called *hydrophilic*. The colloid is stabilized by the formation of an adsorbed layer of molecules of the dispersing medium about the suspended particles. Systems of this type are said to be lyocratic. Examples include glue, gelatin, and milk-fat particles.

LYOPHOBIC. Characterizing a material that exists in the colloidal state but with a tendency to repel liquids: if the liquid is water, the material is called *hydrophobic*. Such colloids are generally stabilized by the adsorption of ions and coagulate when the charge is neutralized. Examples include colloidal gold, and colloidal arsenic sulfide.

LYRA (the harp). One of the small constellations, which contains a number of most interesting objects for viewing through a small or large telescope. Lyra is most easily distinguished by an equilateral triangle having the star Vega at one of its apexes. This star is the brightest in the northern celestial hemisphere. It lies almost in the direction in which the sun and all the planets are moving, due to solar motion. And although it is a long way from the pole of rotation of the celestial sphere at present, it will be the pole star about 12,000 years hence, due to precession. The star Epsilon Lyrae is one of the most famous multiple stars in the entire sky. It can be resolved into two components through a field glass and, on a clear night, into four components through a 6-inch ($\approx$15-centimeter) telescope. Also to be found in this constellation are several other double stars, and the famous ring nebula, an interesting object when viewed through a 6-inch telescope. (See Fig. 1, and map accompanying entry on **Constellations**.)

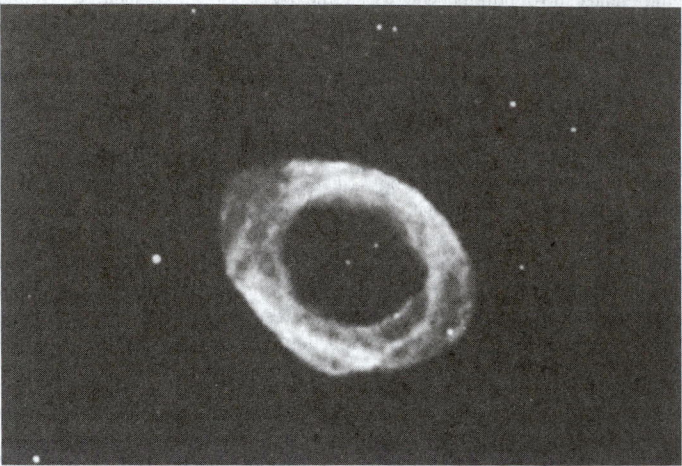

Fig. 1. Ring nebula in Lyra. This nebula is gradually expanding, and its edges are quite red. (*Lick Observatory*.)

LYRE BIRD (*Aves, Passeriformes*). This bird ranges from Queensland to Victoria and is highly regarded in Australia, sometimes pictured on stamps and official seals. The male lyre bird is well known for its display of plumage and performance to attract females of the species. The "lyre frame" feathers are something like those of a bird of paradise or peacock when displayed. Of several species, there is the superb lyre bird (*Menura novaehollandiae*) and Albert's lyre bird (*M. alberti*). The tail feathers range from $1\frac{1}{2}$ to $2\frac{1}{2}$ feet (0.5 to 0.8 meter) in length and usually are white and brown. The head is small, the legs are long, the claws are strong. The male is known for its incredible mimicking of other birds. There is one pale-purple, thick-shelled egg of about the size of a chicken egg. The incubation period is 6 weeks. The young do not leave the nest for 12 weeks after hatching.

LYRIDS. A name given to certain meteor showers that are observed about April 20 of each year. The orbit of the radiant point was definitely associated with the orbit of comet 1861 I. by Weiss. Records of showers from this radiant are found as far back as 687 B.C. A report written by the Chinese in 15 B.C. indicates that during the Lyrid shower of that year, "after the middle of the night, stars fell like rain." Several other accounts of striking showers during April are on record; in particular, we find many newspaper accounts of the Lyrid shower of 1803, which was observed over the United States from North Carolina to New Hampshire. The Richmond, Va., *Gazette* of April 23, 1803, gives a long and vivid account of the shower occurring on the morning of April 20, stating that "from one until three those starry meteors seemed to fall from every point of the heavens, in such numbers as to resemble a shower of sky rockets."

A few scattered members of this shower are observed coming from the radiant point in the constellation of Lyra every year, but there has not been any very striking display since 1803. Because there have been striking showers in the past, the assumption is that a large swarm of meteors exists at some undetermined point along the orbit, and that we may be treated to another brilliant display during some April in the future.

LYSIS (Bacteriology).   The dissolution of cells, e.g., bacteria, or red blood cells; their breakdown from structural form to a structureless fluid.

LYSIS (Physiology).   The gradual decline of the symptoms of disease, referring especially to the gradual abatement of fever.

LYSOSOMES.   Subcellular organelles believed to contain digestive enzymes capable of breaking down many of the cellular constituents.

Disruption of the lysosomes and liberation of these enzymes may occur under certain conditions and can lead to lysis of the cell. See also **Cell (Biology)**.

LZ77 AND LZ78 ALGORITHMS.   See **Data Compression**.

LZW (LEMPEL-ZIV-WELCH).   See **Data Compression**.

M

MAAR. A volcanic explosion crater without a prominent volcanic cone.

MACADAMIA TREE. Of the family *Proteaceae*, the macadamia tree (*Macadonia ternifolia*) is native to New South Wales and South Queensland. It was introduced into the United States (Hawaii, California, and Florida), South Africa, the Mediterranean countries, and the West Indies in the late-1900s. The tree is best known for its tasty and rich nut, now highly valued as a confection in many parts of the world. The tree must have drained rich soil, a warm climate, and considerable care, particularly in removing suckers that grow around the roots. The tree can attain a height of from 35 to 50 feet (10.5 to 15 meters) and the trunk may measure a foot or more in circumference in a fully-grown specimen. The leaf is dark green, glossy, and deciduous. The most highly regarded varieties, out of a total of over 80, in Hawaii are the Pahau, Keauhou, Kapea, Kohala, and Nuuomir. The nut of the tree was first popularized by Brisbane nurserymen in the early 1900s. Many varieties are of no significant commercial value.

MACAQUE. See **Monkeys and Baboons**.

MACE. See **Spices**.

MACERATE. (1) To soften or break up a fibrous substance by long soaking in water at or near room temperature, often accompanied by mechanical action, as in the preparation of paper stock in the beater. (2) In the plastics industry, to comminute a fabric so that it can be used as a filler in a plastics composition. (3) The term is also used in pharmacy to describe a method of preparing medicinal compositions.

MACH ANGLE. The angle a shock wave makes with the direction of motion as determined by the velocity of the object and the velocity of shock propagation.

MACH, ERNST (1838–1916). Mach was born in Chirlitz-Turas, Moravia (now Chrlice-Turany, Czech Republic). He earned his Ph.D. in physics from the University of Vienna. Beginning in 1867, he worked as a professor in experimental physics at Charles University in Prague. He did research in projectile motion. He is best known for his theory of science called positivism. The basis of this theory is that no concept can enter into a scientific hypothesis unless it can be explained completely in terms of observations. His most important published works include *History and Root of the Principle of the Conservation of Energy* and *The Science of Mechanics*.

See also **Mach Angle**.

J. M. I.

MACHINE. A mechanical device wherein mechanical energy is applied at one point and delivered in a more useful form at another point. The term machine applies traditionally in physics to any one of the basic devices of an elementary nature. These fall into two classes: (1) those dependent upon the vector resolution of forces, such as the inclined plane, wedge, screw, and toggle joint; and (2) those in which there is an equilibrium of torques, such as the lever, pulley, and wheel-and-axle.

MACHINE LANGUAGE. 1. A language, occurring within a computer, ordinarily not perceptible or intelligible to persons without special equipment or training.

2. A translation or transliteration of sense 1 into more conventional characters, but frequently still not intelligible to persons without special training.

MACHINE (Simple). Examples of simple machines include levers, pulleys, the inclined plane, gears and gear trains, and the screw. Most of these concepts have been known almost since antiquity.

Levers. A lever consists of a bar of nearly rigid material, either straight or bent, a fulcrum F, a weight W, and a force P. The components F, W, and P can be applied, in any position relative to each other, to the bar as shown in Fig. 1. Levers are used to move large forces (weights) by means of smaller forces, or are used to either amplify or diminish arc motion (Fig. 2). The arms A and B must be perpendicular to the lines of actions (directions) of their respective forces P and W. Then, by a balanced moment equation about F,

$$PA = WB$$

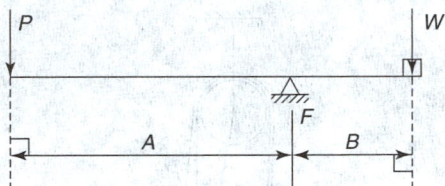

Fig. 1.   Simple first-class lever.

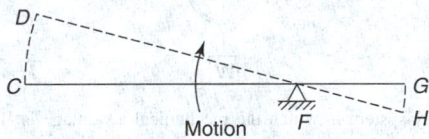

Fig. 2.   Simple lever in motion.

The mechanical advantage of a lever, which defines the ability of an available force P to overcome a resisting force W, is given by the ratio W/P or A/B and results in the expression

$$\text{Mechanical advantage} = \frac{W}{P} = \frac{A}{B}$$

The use of a lever to amplify (or decrease) motion is shown in Fig. 2. When bar CG rotates about F, point G moves through the arc GH, and C will move through the arc CD, and

$$\frac{CD}{CF} = \frac{GH}{FG}$$

Levers continue to be used in many modern kinds of machinery and are notably visible in large weighing scales, conveyors and other automated equipment, and earth-moving equipment. The crowbar is an excellent example of a simple lever-type tool.

In a *first-class lever*, the fulcrum is always between the power and the load and thus the mechanical advantage can be greater than one, one, or less than one. In a *second-class lever*, the power is always less than the load and the mechanical advantage is greater than one. In a *third-class lever*, the power is always greater than the load and the mechanical advantage is less than one. In second-class and third-class levers, the fulcrum is always to one side of the load and power—never in between the application of these two forces.

Pulleys. In pulley systems, forces are transmitted by ropes, chains, etc. in conjunction with pulley wheels and axles. As noted from the lever examples, a wheel and axle is an adaptation of a lever rotating about its fulcrum.

Considering a theoretical frictionless system of pulleys, the force (pull) in any part of a continuous rope is constant and equal to P. Then, by establishing the number of supporting forces (ropes) and the weight W which is being moved, $nP = W$, where n is the number of supporting ropes. In Fig. 3, two forces (ropes) support the weight W, hence $2P = W$ and $P = W/2$. In Fig. 4, four forces (ropes) support the weight W, and $4P = W$ and $P = W/4$. In Fig. 5, five forces (ropes) support W, and $5P = W$ and $P = W/5$.

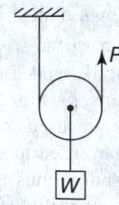

Fig. 3. Simple rope pulley.

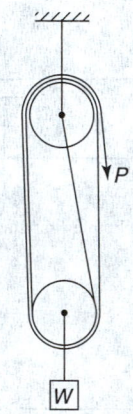

Fig. 4. Pulley system in which the mechanical advantage is $W/(W/4) = 4$.

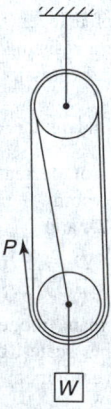

Fig. 5. Pulley system in which the mechanical advantage is $W/(W/5) = 5$.

The mechanical advantage of a pulley system is the ratio of the weight to be moved to the applied pull in the rope, or $W/P = n$, the same as the number of supporting ropes. In Fig. 3, $W/(W/2) = 2$, or for Fig. 4, the mechanical advantage is $W/(W/4) = 4$. In Fig. 5, the mechanical advantage is $W/(W/5) = 5$.

Pulley systems can be analyzed by the use of work done by the force P and its relation to the work done by W. The displacement (Fig. 4) of P is four times that of W. Thus the work done by P is $4PS$, where $4S$ is the distance that P moves. The work done by W is WS. The mechanical advantage is $4S/S = 4$ and $4P = W$, as given above.

Differential Pulley. This device makes use of two pulleys of different radii, r_1 and r_2, attached to each other and rotating about a common axle. An endless chain connects the dual pulley to a second free pulley wheel, as shown in Fig. 6. The chain and the corresponding teeth on the dual pulleys prevent slipping between the chain and pulleys.

From the previous analysis of pulleys, a force equal to $W/2$ acts in the chain at points A and B. A moment equation about axle C then gives

$$Pr_1 + \frac{W}{2}r_2 = \frac{W}{2}r_1$$

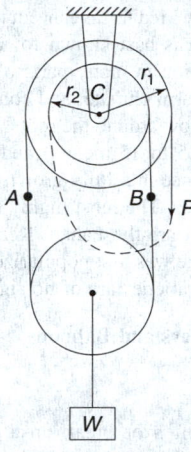

Fig. 6. An endless chain connects a dual pulley to a second free pulley wheel.

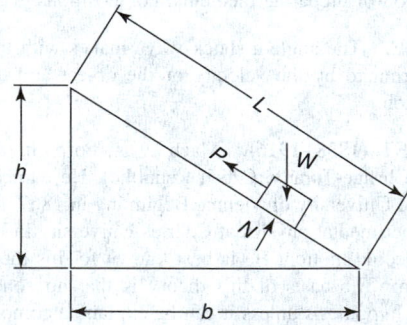

Fig. 7. Inclined plane.

from which

$$P = \frac{W(r_1 - r_2)}{2r_1}$$

The mechanical advantage of the differential pulley is the ratio W/P.

Inclined Plane. As a simple machine, the inclined plane is presumed to be rigid and smooth. The weight W which moves along the incline is a vertically downward force partially supported (N) by the frictionless plane assumed for this example. If the weight W is to be at rest on the incline (Fig. 7), the force system must be balanced in directions normal and parallel to the inclined plane. Balancing the forces parallel to the incline,

$$P - W\frac{h}{L} \text{ or } PL + Wh$$

This is equivalent to a work equation, where P is displaced a distance L, and W is lifted a distance h. When the force P acts to the left in a horizontal direction, P moves an equivalent distance b, and then

$$PB = WH$$

The mechanical advantage is the ratio W/P or L/h, where the force is parallel to the incline. A *wedge* is equivalent in its analysis to an inclined plane. When t is the thickness of a wedge, $P/W = t/L$.

The Screw. This device is an inclined plane wrapped around a cylinder in such a way that the height h is parallel to the axis of the cylinder. If p is the height of travel in one circumference of the screw thread, r is the radius of the thread, and friction is neglected, by the work method of analysis one obtains

$$P 2\pi r = Wp$$

The mechanical advantage is

$$\frac{W}{P} = \frac{2\pi r}{p}$$

Gears and Gear Trains. A gear is a wheel with projections uniformly spaced around its circumference. It is usually meshed with a second similar wheel of a different diameter so that the circumferential forces and rotational speeds are different. Two meshed gear wheels (Fig. 8), the teeth of which are not shown are used as an example. From a balanced moment equation,

$$Pr_1 = Wr_1$$

and the mechanical advantage is the ratio W/P or R_1/r_1.

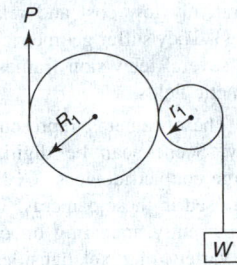

Fig. 8.   Two meshed gear wheels (teeth not shown.)

Gear trains consisting of more than two gear wheels have the same relationship. In Fig. 9, letting the wheels be represented by letters, one has

$$PR_1R_2R_3 = Wr_1r_2r_3$$

and the mechanical advantage is

$$\frac{W}{P} = \frac{R_1R_2R_3}{r_1r_2r_3}$$

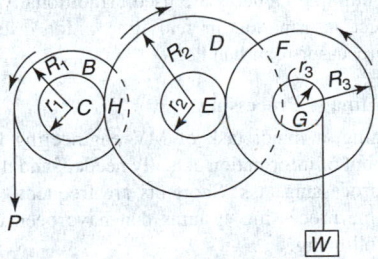

Fig. 9.   Simple gear train.

By inspection, and understanding that the points of contact (such as point H) have a common speed, the directional relations can be determined. If wheels B and C are rotating counterclockwise, D and E rotate clockwise, and F and G rotate counterclockwise.

Spur gears are those which have their teeth cut parallel to parallel axes of rotation. Bevel gears are used when the axes of rotation intersect; their teeth are cut on conical surfaces with the apex at the point of intersection of the axes. Speed ratios are inversely proportional to pitch diameters, number of teeth, or number of revolutions. Referring to Fig. 9,

$$\frac{\text{rpm of } E}{\text{rpm of } F} = \frac{2R_3}{2r_2} = \frac{R_3}{r_2}$$

When a screw meshes with a cogged wheel, it is known as a *worm* and *worm wheel*. When the worm has a single continuous thread, one revolution of the worm will cause the wheel to rotate through a circumferential distance equal to the distance between two consecutive teeth. Thus, a worm must rotate 48 revolutions if a worm wheel with 48 teeth is to rotate one revolution. The same relationship is true if speeds are considered.

Efficiency. The efficiency of a machine is defined as the ratio of the output to the input. The efficiency of an inclined plane, for example, is given by the expression:

$$\text{Efficiency } (\%) = \frac{h}{h + fb} 100$$

for the case where the applied pull tends to move the block up the incline (Fig. 7), and where f is the coefficient of friction between the incline and the flock. See also entry on **Efficiency**.

J. W. Breneman (deceased)

MACHINE VISION (Recognition and Applications).

Since its initial serious recognition in the late 1970s, machine vision (MV) has been variously defined:

> MV is the process of extracting information from visual sensors for the purpose of enabling machines to make intelligent decisions.
> or
> MV is part of the larger technology of artificial visual perception that substitutes (partially or totally) the human visual capability by instruments that are backed up by electronic data processing (notably complex computations).

As will be noted from the foregoing, the real and practical objectives of MV tend to be nebulous and of far-reaching expectations.

MV became a "buzz word" of the 1980s—to the extent that MV, combined with robots, would bring about the most ambitious goals in terms of production automation. The fact is that, by the early 1990s, MV had lost much of its earlier luster. With relatively few exceptions, most contemporary MV systems are confined to sophisticated and demanding object-detection problems, with the majority of such detection problems currently being solved by the less exotic types of sensors previously described in this article. Because total MV systems in general still remain overly complex and costly, other nonvisual sensors have been greatly improved and applied more imaginatively to achieve many of the objectives that once appeared to lie within the province of MV. Cost, once again, has been the traditional motivating factor.

"Seeing" robots, once a major goal of MV, has proved disappointing from the standpoint of MV. To be true, in practice, robots, once programmed, do perform in many instances as though they could "see," but this has been brought about more by other instrumental detectors (tactical, for example) than by literal "viewing" of parts, pieces, machines, and entire manufacturing scenes. For example, in the late 1980s a large automotive manufacturer originally invested rather heavily in MV systems in connection with robotic operations, but after a year or so of trials, abandoned them for possible consideration at some future date.

To be sure, MV developments will continue, and systems will be installed where other simpler, lower-cost approaches do not suffice. Coupled with trends toward reducing computation costs, MV may become more competitive.

Artificial Visual Perception

Industrial MV techniques initially stemmed from the interests of the military in a technology known as pattern recognition. This may be defined as seeing, analyzing, and interpreting patterns, as of scenery, juxtaposition, dimensional magnitudes, color, and other characteristics of the visual environment.

Human visual communication (information transfer) with the outside environment depends on the interactions (predominantly absorption, reflection, and refraction) of light (visual) radiation emanating from physical objects (in point, two, or three dimensions), thus enabling the human observer to cope with the surrounding environment in a safe and efficient manner. The vision activity (human or machine) must be complemented by some form of processor (human brain or electronic counterpart) to make a final identification of what has been seen. This process, defined by a few words, is pattern recognition.

The means that are applied to perform pattern recognition is known as the pattern processor. The human brain, in processing a pattern, does an amazing job of sorting out extraneous information in the input to quickly identify what really is present (the objects of interest that are in the pattern).

Interest in pattern recognition dates back some 70 years. The more the process is studied, the more one finds how fundamental the process really is to the human brain function. The amount of information (inputs) that the human system can observe and process is tremendous, but nevertheless remains poorly understood. This explains why MV, from a technological viewpoint, is closely associated with the study of artificial intelligence.

Elements of Pattern Recognition

Any pattern recognition system contains the same three basic elements: (1) sensing, (2) processing, and (3) implementing actions based upon input data.

The primary objective of pattern recognition is to classify a given unknown pattern as belonging to one of several classes of patterns. The applications of pattern recognition are many and varied. In the case of character recognition, for example, the patterns are easily generated and recognized by humans, and the basic goal has been to improve the human-machine communication. In other situations, the patterns are difficult for humans to recognize rapidly, as, for example, the very rapid interpretation of an electrocardiogram. The wide range of applications, as well as the relationship to diverse disciplines, including machine vision, communication and control, and the area of linguistics (word and sentence reading), have broadened the interest basis in pattern recognition.

Information Content of an Image

In terms of MV as used in industrial production situations, the information content of the image falls into three basic categories:

1. *Geometry*, which in turn portrays shape, position, dimension, and a number of other associated properties, including density and texture (which can be inferred from the known geometry in most cases).
2. *Color*, which is very helpful when present. There are, of course, color-blind persons and instruments that preclude its use.
3. *Movement*, which is present in two or more images of a dynamic process.

Extracting Information from Images

As with other forms of industrial instrumentation, the MV system senses (reads may be a better term) the image, but before the data obtained from the image can become meaningful, the sensed information must be compared with some form of standard, or prelearned, pattern. In more familiar terms, the combined actions of sensing and comparing constitute measurement.

As compared with the usual type of industrial sensor, such as a thermocouple, where measured data are conveniently available in the form of a ready-made electrical signal, in MV one deals not with just one or a very few points or locations of emanating data, but with thousands-plus bits of information—because what is being observed (measured) is comprised of a multitude of points (picture pixels[1]). In an artificial vision system, image data, as may be gathered by an electronic camera, must be compared electronically with information, that is, in some fashion with data that are stored in electronic memory.

Initially the application of a general-purpose computer to MV was the only choice available. The problem immediately encountered was the fact that the computer was designed to process computational data, *not* data patterns. The data from a video camera in an MV system are pattern data and, thus, are very different from computational data as exemplified by financial balance sheets or linear regression analysis. The situation is summarized as follows:

Image data quantity 484 × 320 pixel density
Hence, 154,880 pixels per frame

[1] A pixel is a picture element, a small region of a scene within which variations of brightness are ignored.

At 6 bits of gray-level data per pixel, 929,280 pixel values per frame
At a 30 times a second refresh rate, 27,878,400 bits per second

Since the early days of MV, a number of innovations have contributed to the simplification of this problem, including improvements in gray-scale systems, better algorithms, the pipelined image processing engine, the geometric arithmetic parallel processor, and associative pattern processing.

Machine Vision Sensors

Factors that usually must be considered are (1) optics and lighting, (2) field of view, (3) resolution, (4) signal-to-noise ratio, (5) time and temperature stability, and (6) cost. MV sensors that have been or are currently in use include the following.

Line Scanners. These include solid-state arrays, flying-spot scanners, and prism, mirror, or holographically deflected laser cameras. These scanners are fast high-resolution devices, which are relatively free of geometric distortion in one dimension. In order to capture a complete two-dimensional scene, the second dimension is obtained either by motion of the object past the scanner or by mirror or prism deflectors. Mechanical motion tends to slow down data acquisition and, in some cases, produces geometric distortion.

Area-Type Scanners. These scanners were used in early systems utilizing closed-circuit television cameras with vidicon image sensors. Advantages were comparatively low cost and relatively high resolution (300 to 500 television lines). They suffer geometric distortion, temperature instability, lag (requiring several television frames for full erasure), and sensitivity to nearby magnetic fields.

Solid-State Cameras. These cameras represent a major advance in image-sensing technology. Scenes can be digitized onto an array of photosensitive cells. Charge-coupled devices (CCDs) or charge injection devices (CIDs) have been used in these cameras. The arrays form a pixel grid containing the data currently appearing on camera. The solid-state camera is available in a variety of pixel densities, notably, 128 × 128, 256 × 256, 512 × 512, 1024 × 1024, and so on.

The limiting factor of cameras for imaging systems is not necessarily the resolution or speed. Camera technology essentially has kept up with the ability of computationally based processes to handle information, especially when multiple cameras are used. Traditionally, in MV systems sensors can detect information in real time a lot faster and in larger quantities than processors can handle.

Machine Vision Image Processing

From the beginning, a major task of MV engineering has been that of reducing the amount of information actually needed, and thus lessening and simplifying the processing task. Shortcuts are frequently taken to reduce data volume. Some processing systems that have been or are in current use include the following.

Binary MV System. The representation of all combinations possible in every pixel of every scene is unrealistic. Conversion of each pixel point into a binary value reduces the pixel data, but it also reduces the accuracy of the analysis. Binary systems evaluate each pixel as black or white. A threshold-adjusting capability can allow users to select what intensity of signal is to be the black-white border.

Gray-Level System. This system can interpret each pixel's value as a specific gray tone. These systems vary in precision, with each pixel being evaluated as 16, 64, or even 256 different values.

Windowing the Scene. This system can reduce the total data system load. Since analysis of an entire scene often is unnecessary for proper recognition, the use of a window can eliminate unneeded image data.

Segmentation. A common means of data reduction, this technique divides the image data into areas of interest and then interpolates surrounding pixels into those areas. There are several ways to segment an image, but all detract from accuracy. Four procedural options for segmentation may be considered:

1. *Algorithms*. In general, algorithms are a set of mathematical models that can be used to describe an image. A few years ago, the SRI algorithms were developed by the Stanford Research Institute. These consisted of about 50 different features that are extracted from a

binary image, such as the size of blank areas (holes) or objects (blobs) and their *centroids* or *perimeters*.

2. *Neighborhood processing.* This is an averaging method, achieved by treating each pixel value as if it were part of a group or neighborhood. This gray-level data-reduction technique can change an otherwise complicated image into areas with well-defined lines. Each pixel's point value is calculated by considering the value of a neighboring pixel. The process effectively averages the scene into regions or areas.

3. *Convolution.* If an image is represented by the rate of light change per pixel instead of light density, the image will look like a line drawing because the greatest intensity of a pixel change is at the boundaries.

4. *String encoding.* In run-length, string, or connectivity encoding, the values of the first and last pixel positions of each scan line are compared to see if they are equal and, therefore, belong to the same region. The Tables generated by this process also consider the vertical changes in pixel state and, as may be expected, are rather short because most simple binary images contain relatively few transition borders.

5. *Associative pattern processing.* Rather than increasing the capability of processing large amounts of data in a shorter time span through the route of improving computer and data-handling techniques (hardware and software) as just described, a fundamentally different approach has three functional sections (as developed by APP (pattern processing technologies). (1) An image analyzer contains all the pixel data for the current image on the camera. (2) It converts the data into a unique statistical fingerprint, which is a unique set of response memory addresses. The response memory contains the statistical fingerprints for all trained images. (3) A matching histogram compares the fingerprints of the current image with those stored in the response memory. The theory of operation is based on a statistical phenomenon that reduces an image's data into a precise fingerprint, which is a set of values for an image that differs from other images and is a minute fraction of the total data of the image. Instead of one image being trained, all the images needed are trained and no programming is involved. One simply shows an image to the camera, allows the fingerprint memory locations to drive the response memory, and loads in a label for each image.

The Recognition Process

Once the user describes the image mathematically, the second half of the process step can occur. Data processing has been mentioned previously. The programming of a general-purpose computer requires programming expertise that is costly. Using this computational approach may entail performance levels that are too slow for real-time operation, unless the application only involves very basic procedures. Gains in processing power can be taken up quickly by application complexity with its increasing data demands.

Machine Vision Applications

Some form of actuation or control is the last step in implementing MV. Reasonably standardized methods of communication usually can be used with such common interfaces as RS-232 or IEEE-488. These are networking protocols developed by the Institute of Electrical and Electronics Engineers.

MV suppliers in recent years have turned to specialized products oriented to specific industries. The advent of generic software has been possible because of numerous application similarities.

MV applications essentially fall into four categories: (1) inspection, (2) location, (3) measurement, and (4) recognition.

MV can inspect to determine if parts are acceptable. Inspection tasks can be further categorized into one of four functions: (1) surface flaw detection, (2) presence or absence of certain features, (3) dimensional tolerance verification, and (4) shape verification. Three examples that use a gray scale are shown in Figs. 1, 2, 3, and 4.

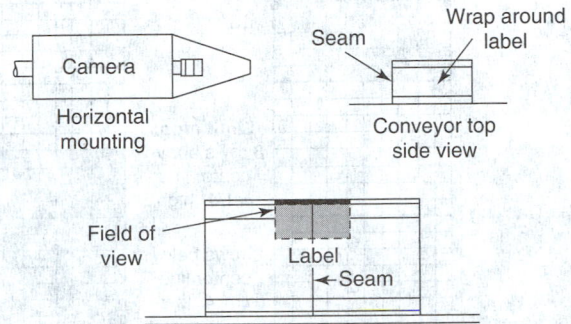

Fig. 1.   Use of a template or pattern match algorithm to check the accuracy of a label on a can of food. A label is wrapped (360°) around a tuna fish sized can (upper diagram). The camera is placed so that the field of view is located horizontally across the area where the label ends touch. The camera's field of view should be located as shown in the lower diagram. To set up the reference template, the operator adjusts the horizontal/vertical axis of the camera by viewing a display on the controller, during which time the operator also adjusts the mounted camera to obtain the view shown in Fig. 2. (*Cutler-Hammer Products, Eaton Corporation.*)

Discrete-piece Identification — Bar Coding

The initial attempts to codify pieces, such as bank check numbers and addresses (zip codes) on mail, took the form of optical or magnetic character reading, and numerous systems of this type remain in place today. See also **Magnetic Ink Character Recognition (MICR)**; and **Optical Character Recognition (OCR)**. In principle, these systems now are a part of the overall concept of machine vision. However, more closely allied

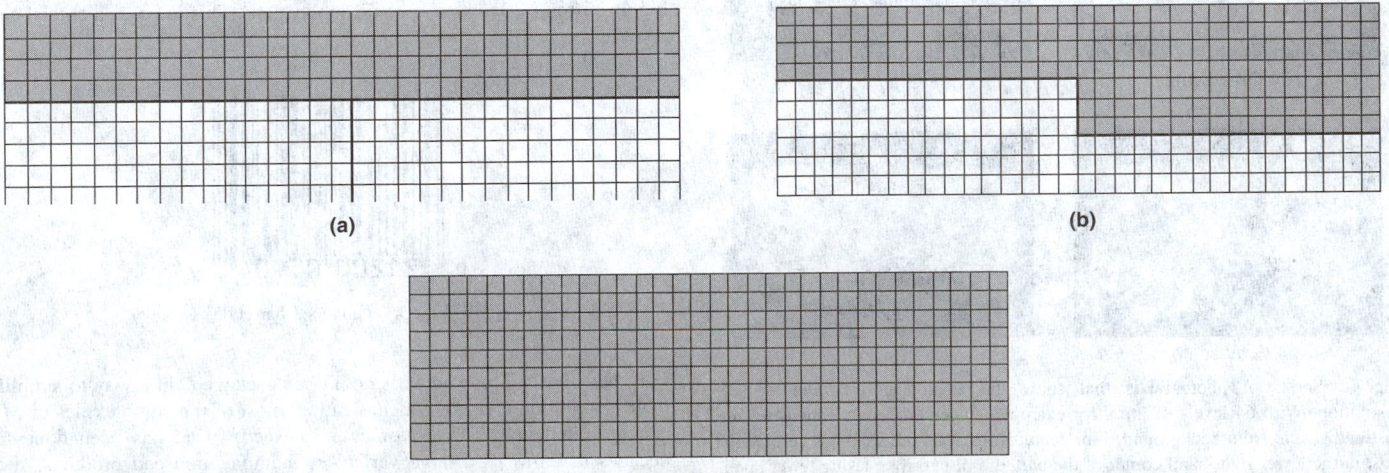

Fig. 2.   Displays under three possible conditions: (**a**) Label is on can and is properly placed (can appears black; label, white); (**b**) a poorly placed label, indicating a mismatch area of 3 pixel rows; and (**c**) no label on can (screen appears black). (*Cutler-Hammer Products, Eaton Corporation.*)

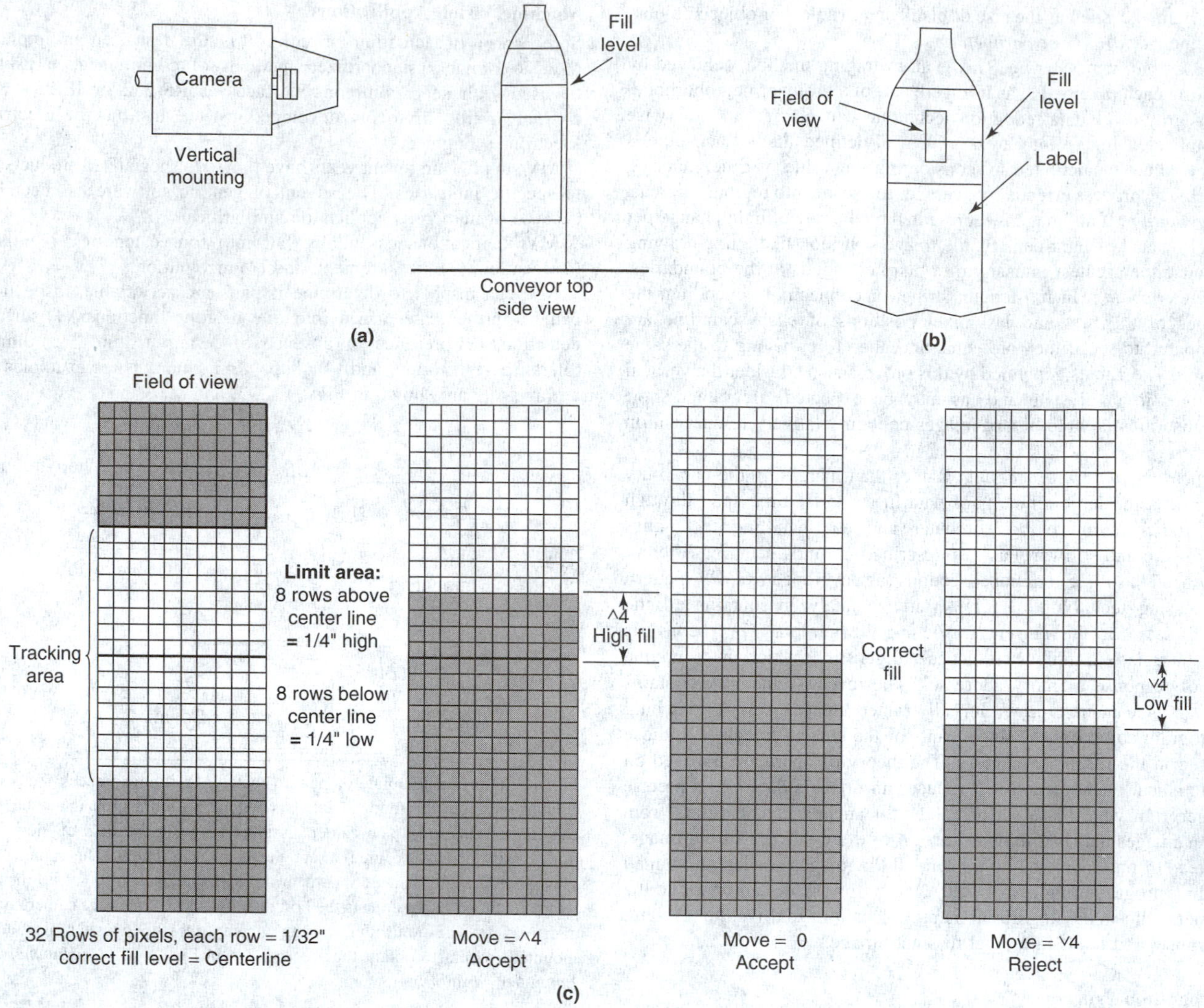

Camera

Vertical mounting

Fill level

Conveyor top side view

(a)

Field of view

Fill level

Label

(b)

Field of view

Tracking area

Limit area:
8 rows above center line = 1/4" high

8 rows below center line = 1/4" low

32 Rows of pixels, each row = 1/32" correct fill level = Centerline

High fill

Move = ^4
Accept

Correct fill

Move = 0
Accept

Low fill

Move = ∨4
Reject

(c)

Fig. 3.   Use of a Y-axis algorithm to check accuracy of filling: (**a**) arrangement of camera and conveyor juxtapositions; (**b**) location of field of view for given high and low levels; (**c**) reference images, showing high fill, correct fill, and low fill. (*Cutler-Hammer Products, Eaton Corporation.*)

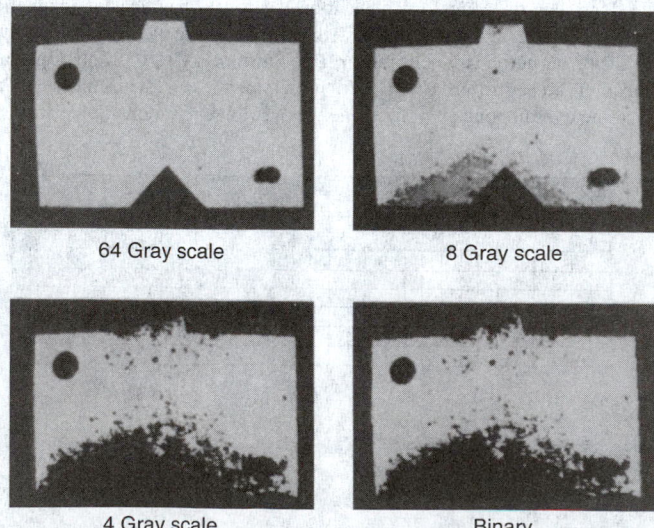

64 Gray scale

8 Gray scale

4 Gray scale

Binary

Fig. 4.   Series of photographs that show the advantages of high-intensity resolution. With 64 levels of gray, for example (upper left), part boundaries are distinguishable from background. Fine features, such as the shadows cast by the post in the lower right-hand corner of the part, are discernible. Using fewer levels of gray (8,4, and 2 as shown) renders the scene more subject to the effects of shading. The slightly shaded portion of the part has deteriorated, as have other subtle features. (*Analog Devices, Inc.*)

with machine vision is the machine-readable bar code, which depends only upon the line width and line location of printed bars, as shown in Fig. 5.

The lack of uniformity of parts, pieces, subassemblies, and so on, that persists in the discrete-piece manufacturing industries imposes a myriad of complexities, as just alluded to in the preceding description of machine vision. Thus MV applications, as generally defined, must be highly customized to very specific applications. This imposes a crucial cost restraint.

0   21200 05904   9

Fig. 5.   Universal bar code.

If, however, pieces and parts could be "packaged" in a way to simplify their recognition and exact identification, most of the complexities of MV can be eliminated. That is, of course, exactly what has been done for decades in terms of a large variety of manufactured end products, such as individual boxes of nuts and bolts, screws, cases of canned goods, boxed machine parts, containerized detergents, paints, hardware, plastic-encapsulated items, and even palletized groupings of identical merchandise.

The general concept of containerization, of course, originally was developed to simplify shipping and distribution to the end user. Later these packaging concepts became the basis for vastly improved inventory control and for automated warehousing by manufacturers, distributors, wholesalers, and, indeed, the final point of purchase.

With many millions of items produced, it became obvious that an extensive codification system was needed if piece identification were ever to be automated. Early trials with color coding proved that color-based systems were grossly inadequate. Entering from another direction (commercial and financial), the concepts of magnetic and optical coding were developed and adopted rather quickly by financial institutions. However, early magnetic and optical systems required a certain closeness of the object with the detecting means, as, for example, in identifying a check or reading the address on an envelope. Specially designed alphanumeric characters have served financial and commercial needs, but it is expected that bar coding ultimately may be used.

In the late 1960s, the first supermarket checkout stand, developed by the Hughes Aircraft Company, was installed in a Los Angeles supermarket and operated successfully. The system did not get underway nationally for several years, however, because of the unwillingness of grocery manufacturers to print a bar code on each product package. By the late 1980s, however, the system had become a favorite of manufacturers and store operators. Uses also were found in automated warehousing and inventory accounting. Bar code readers became an important segment of the electronics industry.

Invention of the laser scanner has made it possible to read bar codes on boxes in a warehouse as far as 10 feet (3 meters) away. Well-designed bar code readers are considered to be very close to error free. Replacement of older style code readers with the hand-held wand has increased the acceptance and versatility of the system. Bar-coded tags for unpackaged materials, such as fabrics, have received wide acceptance. See also **Artificial Intelligence: Machine Vision**.

Additional Reading

Burke, M.: *Handbook of Machine Vision Engineering: Image Processing*, Vol. 2, Chapman & Hall, New York, NY, 1997.

Cipolle, R., and A. Pentland: *Computer Vision and Human-Computer Interaction*, Cambridge University Press, New York, NY, 1998.

Davies, E.R.R.: *Machine Vision: Theory, Algorithms Practicalities*, 2nd Edition, Academic Press, Inc., San Diego, CA, 1996.

Jain, R.C.: *Introduction to Machine Vision*, The McGraw-Hill Companies, Inc., New York, NY, 2000.

Kanellopoulos, I., G.G. Wilkinson, and T. Moons: *Machine Vision and Advanced Image Processing in Remote Sensing*, Springer-Verlag, Inc., New York, NY, 1999.

Myler, H.R.: *Fundamentals of Machine Vision, SPIE International Society for Optical Engineering*," Bellingham, WA.1999.

Parker, J.R.: *Algorithms for Image Processing and Computer Vision*, John Wiley & Sons, Inc., New York, NY, 1996.

Pietikainen, M.K.: *Texture Analysis in Machine Vision*, World Scientific Publishing Company, Inc., Riveredge, NJ, 2000.

Sanz, J.L.C.: *Image Technology: Advances in Image Processing, Multimedia and Machine Vision*, Spriner-Verlag, Inc., New York, NY, 1996.

Sonka, M., V. Hlavac, and R. Boyle: *Image Processing, Analysis, and Machine Vision*, Brooks/Cole Publishing Company, Pacific Grove, CA, 1999.

Zuech, N.: *Understanding and Applying Machine Vision*, Marcel Dekker, Inc., New York, NY, 1999.

MACHINE WORD.   For a given computer, the number of information characters handled in each transfer. This number is usually fixed, but may be variable in some computers.

MACH (M or Ma).   See **Units and Standards**.

MACH NUMBER.   See **Aerodynamics and Aerostatics**; **Airplane**; and **Supersonic Aerodynamics**.

MACH PRINCIPLE.   The inertia of any system arises from the interaction of that system and the rest of the universe, including distant parts thereof. Postulated by applying the Mach criterion to the concept of absolute space. Applied by Einstein to the hypothesis that the metric of space-time is determined by the distribution of matter and energy.

MACKERAL BREEZE.   See **Winds and Air Movement**.

MACKERELS (*Osteichthyes*).   Of the order *Scombridae*, mackerels have an elongated, fusiform (spindlelike) body, which is only slightly compressed. The tail typically has one to three longitudinal keels; it is long and rather thin. Scales are absent or tiny, sometimes forming a corselet on the front of the body. A wavy lateral line is present. The large head is tapered. The wide mouth opening extends at least to beneath the eyes; the jaws have large or small, sharp teeth. The wide gill opening has four gill arches. There are 31 to 61 vertebrae, and the vertebral column is well ossified.

There are 33 genera with a large number of species. All mackerels are epipelagic (high-sea) fishes. Many of them undertake extensive feeding migrations; a few species swim to coastal waters to spawn, while others are never in shallow water. The anatomy of these extremely fast swimmers reflects their great maneuvering ability. The pectoral and pelvic fins are recessed into shallow grooves. They click into place with a jerk of the fin base. The dorsal fin has a similar recess. The large majority of mackerels live in tropical and subtropical regions.

Atlantic Mackerel. This fish (*Scomberomorius scombrus*) is the best-known species. Its mature length ranges between about 10 and 20 inches (25 and 50 centimeters). This species is distributed from the Mediterranean and Black Sea along Europe's Atlantic coast to the arctic, and from there across the Atlantic Ocean to Labrador and the American East Coast to Cape Hatteras. The back is grass green, with numerous irregular dark stripes; the sides and underside have a mother-of-pearl hue with a reddish shimmer. The first and second dorsal fins are widely separated. The tail shaft lacks a median keel. The bony eye rings are well developed. There is no swim bladder. See Fig. 1.

Fig. 1.   Atlantic mackerel.

Like many other mackerels, the Atlantic mackerel occurs in dense schools just beneath the water surface, and sometimes there are so many of them together that they churn up the water at the surface. Since they lack a swim bladder, they can dive quickly to evade predators, such as sharks, tuna, and dolphins. They feed chiefly on small crustaceans, juvenile herring, sardines, and anchovies, as well as sand lances.

After the winter rest in deep waters, sexually mature mackerel seek coastal waters in April and May, where they spawn during the early summer months (March to April in the Mediterranean). They always migrate in schools. One female can lay up to 500,000 eggs, each of which has an oil bubble and floats on the water surface. The young hatch after about six days, and grow very rapidly. After 2 years, they are already over 7.8 inches (20 centimeters) long and at the end of the third year, they are about 11.8 inches (30 centimeters) long and reach sexual maturity.

Since they are highly valued because of their fine-tasting meat, Atlantic mackerels are commercially fished just about wherever they occur. Of all the numerous salt water food fishes, the Atlantic mackerel has presented marine researchers with one of the most fascinating and elusive puzzles. The curious periodic cycle of the fish, from scarcity to overabundance, has been known since the time of the American colonists in the mid-1600s. At that time, the fish had been established as an important staple commodity. Unpredictable fluctuations in the catches occurred from year to year and, as early as 1670, American settlers had enacted laws to prevent overfishing. In many parts of the world, changes in the mackerel's migration habits bring these same periodic fluctuations and thus affect the economy of fisheries and nearby ports.

Out of every million eggs one fish must survive to keep the population constant. To be considered a successful year, survival of more than four fish at the 2-inch (5-centimeter) size per million eggs is needed. In the egg and larval stages, many factors influence mortality. Adverse winds may push the waters in an unfavorable direction during the time when the fry lack sufficient motility and prevent them from reaching suitable nursery grounds. A lack of desirable zooplankton when needed reduces survival. Both conditions occurring together in one season are disastrous. Besides contending with the caprices of nature, mackerel are preyed upon by many forms of sea life.

As an example of poor survival, Atlantic mackerel mortality from the fertilized egg to a 2-inch (5-centimeter) stage is 99.9996%. Hence, a fluctuation in survival of several thousandths of 1% may make the difference between a dominant and a weak year. The mackerel catch in the Gulf of Maine may vary by a factor of 100 between a good year and a poor one.

Fresh mackerel is considered by many to be one of the choicest of food fishes. In the 1800s, mackerel were caught close inshore, dressed, and placed in tubs of salt water, which was changed frequently to keep the fish cool. The object was to catch the mackerel and get them to market before daylight, where they were sold in the cool of the morning.

Chub Mackerel. This fish (*Scomber japonicus*) is closely related to the Atlantic mackerel and occurs worldwide, but in less abundance. Chub mackerel are smaller and not as valuable commercially, except between the Yellow Sea and Sakhalin Island, where the fish has large commercial importance. It is the most fertile of the mackerel species, in that a single female can lay over 1 million eggs.

Rastrelliger mackerels are closely related to *S. japonicus*. For some Indian fisheries, the *Rastrelliger kanagurta* is practically as important as the oil sardine. Although not restricted to the west coast of India, like the sardine this species of mackerel forms a major fishery along that coast, where the fishing grounds and seasons of the two species overlap. This species moves in dense shoals into coastal waters toward the end of the southwest monsoon and is captured in shore seine (*rampani*), drift net, boat seine, and purse seine. When the catch of mackerel in the rampani is very large, the fish are impounded for up to a week by staking the net in a semicircle from the shore; the fish are then bailed out as per demand.

Spanish Mackerel. This fish (*Scomberomorius maculatus*) is actually a tuna and probably is the most elongated of tunas. These fishes often appear in huge schools in the Atlantic and Pacific Oceans. See Fig. 2. Other tunas are described in entry on **Tuna.**

See also **Fishes**.

Fig. 2.   Spanish mackerel.

Additional Reading

Bond, C.E.: *Biology of Fishes*, 2nd Edition, Harcourt Brace College Publishers, San Diego, CA, 1996.

Eschmeyer, W.N., C.J. Ferraris, M.D. Hoang, and D.J. Long: *Catalog of Fishes*, California Academy of Sciences, San Francisco, CA, 1998.

Evans, D.H.: *The Physiology of Fishes*, 2nd Edition, CRC Press, LLC., Boca Raton, FL, 1998.

Paxton, J.R. and W.N. Eschmeyer: *Encyclopedia of Fishes*, 2nd Edition, Academic Press, Inc., San Diego, CA, 1998.

MACLAURIN SERIES.   A special case of the Taylor series. If $f(x)$ and all of its derivatives remain finite at $x = 0$, then $f(x)$ may be expanded to

$$f(x) = f(0) + f'(0)x + f''(0)\frac{x^2}{2!} + \cdots + f^{(n-1)}(0)\frac{x^{n-1}}{(n-1)!} + R_n$$

where R_n is the remainder after n terms. When the remainder converges towards zero as the number of terms increases, the result is the Maclaurin series for $f(x)$ at $x = 0$.

Equations for R_n are:

$$\frac{x^n}{n!}f^{(n)}(\theta x), 0 < \theta < 1$$

$$\frac{1}{(n-1)!}\int_0^x f^{(n)}(x-t)t^{n-1}dt$$

See also **Series**.

MACLEOD EQUATION.   A constant relationship between the surface tension of a liquid, its density, and the density of its vapor, of the form:

$$\frac{\gamma^{1/4}}{\rho - \rho'} = \text{constant}$$

where γ is the surface tension of the liquid, ρ is its density, and ρ' is the density of its vapor.

MACLEOD, JOHN JAMES RICHARD (1876–1935).   John Macleod was a Scottish physiologist who helped to isolate insulin. Macleod graduated in medicine from the University of Aberdeen in 1898, and following a year of study in Leipzig was appointed a demonstrator of physiology at the London Hospital, then a lecturer in biochemistry in 1902. From 1903 to 1918 he was Professor of Physiology at the Western Reserve University, in Cleveland, Ohio. He then moved to the University of Toronto, where he was appointed Professor of Physiology.

Much of Macleod's career was devoted to the study of carbohydrate metabolism. He is best remembered for his part in the discovery and isolation of the pancreatic hormone insulin in 1921, for which he and Frederick Banting were awarded the Nobel Prize for Physiology or Medicine in 1923. His interest in diabetes began in 1905, and as early as 1913 he published *Diabetes: Its Pathological Physiology*. In Toronto Macleod provided Banting with the facilities of his laboratory. Although the idea upon which the research was based was very much Banting's, the discovery was a product of teamwork. In this spirit of collaboration Banting shared his half of the prize with his research assistant C. H. Best, whilst Macleod shared his with the chemist J. B. Collip, who assisted in producing an extract of high purity. See also **Banting, Frederick Grant (1891–1941)**.

In 1928 Macleod returned to Scotland as Regius Professor of Physiology at the University of Aberdeen. At the end of his career, influenced by the ideas of Claude Bernard, he undertook work on the possible existence of a diabetogenic center in the brain.

Additional Reading

Bliss, M.: *Discovery of Insulin*, University of Chicago Press, Chicago, IL. 2007.

Gillispie, C.C.: *Dictionary of Scientific Biography* (1970–1980), Vol. 8, pp. 614–615, Charles Scribner's Sons, New York, NY, (1970–1980).

Stevenson, L.G.: *Nobel Prize Winners in Medicine and Physiology, 1901–50*, Henry Schuman, New York, NY, 1953, pp. 109–114.

Claire E.J. Herrick London, UK

MACROASSEMBLER.   An assembler which permits the user to define pseudo computer instructions, which may generate multiple computer instructions when assembled. The source statements that may generate multiple computer instructions are termed *macrostatements* or *macroinstructions*. On a process-control digital computer, a macroassembler can be a most significant tool. By defining a set of macrostatements, for example, a process-control engineer can define a process-control programming language that is specifically oriented to the process.

Macrolibraries. This is comprised of a set of defined macrostatements and the associated computer instructions. A macrolibrary results for example, when a process-control engineer defines a language for a system. The effectiveness of a macroassembler depends largely on the ease of creating, manipulating, modifying, and linking various macrolibraries.

See also **Assembler (Computer System)**.

MACROCLIMATE.   The general large-scale climate of a large area or country, as distinguished from the mesoclimate and microclimate. See also **Mesoclimate**; and **Microclimate**.

MACROECOLOGY.   Macroecology is the science of examining ecological systems at large spatial scales, focusing on the statistical distributions of species among ecologically important variables such as body size, geographic range size, energy use and phylogeny.

Introduction

Ecological science since its inception around the beginning of the twentieth century focused attention on relationships among plants, animals and the nonbiological environment within relatively small spatial units, roughly corresponding to the area that a single human could realistically traverse while collecting data in the field. The area encompassed by such studies rarely exceeded 100 hectares (247 acres), and often was much smaller. A few laboratory studies of microbial communities focused on very small spatial scales. As more data accumulated, it became clear that many ecologicalphenomena operated on much larger scales. Parallel progress in biogeography suggested that many properties of the geographic distributions of speciesand biomes must be understood within an ecological framework. Within paleontology, the evolutionary dynamics of groups of species in the fossil record required an understanding of the ecological context of evolutionary processes. These disparate fields of study required a new conceptual construct in order to establish the nature of the links among them.

Macroecology is one approach to providing a set of conceptual links among sciences that attempt to provide an understanding of biological diversity. Operationally, macroecology examines the statistical distributions of species among variables likely to be affected by ecological and evolutionary processes shaping diversity. Patterns in these distributions are then used to suggest theoretical insights into the nature of the processes that shaped them. The theoretical insights obtained often require developing models that integrate processes occurring at different spatial and temporal scales.

Empirical Patterns

A major portion of the research done in macroecology is devoted to documenting patterns in the statistical distributions of species of variables that reflect the operation of ecological and evolutionary processes. To do so, it is necessary to show that a particular pattern is not the result of sampling biases and random processes that do not involve biological mechanisms. This is particularly challenging because much of the data analyzed inmacroecological investigations are gleaned from studies conducted for reasons other than documenting macroecological patterns. Variables often examined in macroecological studies include body mass, geographic range size and shape, local abundance, phylogeny, resource use behavior and energy flux.

Statistical distributions of species with respect to several variables often lead to patterns among variables that are not well characterized by simple linear models. Many patterns are subtle and consist of variation of the probability density of species across two or more dimensions that consist of the ecological variables being examined. For this reason, it is often difficult to use conventional correlation and regression techniques that implicitly assume the relationship between variables is linear. Mathematical transformations of variables (most commonly logarithmic) may not linearize relationships among variables, and more complicated statistical treatments are sometimes necessary.

The following discussion examines several of the more common patterns documented in macroecology to provide examples of the general empirical approach and to illustrate some of the statistical complications that arise.

Body Mass Distributions. In examining the distribution of body mass among species, it is a common practice to take the logarithms of masses prior to comparisons in order to reduce the orders of magnitude differences among species. When plotted on a logarithmic scale, histograms of body masses are typically skewed, most commonly towards larger body masses [Brown, et al.:], but are left-skewed for some tax a [Roy, et al.]. The skewness of body mass distributions varies among taxa within a larger group. For example, log body mass distributions of different orders of birds have different degrees of skewness.

Body mass distributions appear to change shape when examined across different spatial extents [Brown and Nicoletto ref.]. For entire continents, body mass distributions among species within higher level taxa (e.g., class) appear to be peaked and skewed. As the spatial extent examined decreases, the distribution of body masses among species appears to approach a uniform distribution. This appears to be a consequence of the fact that assemblages of species within small areas are missing many of the modal-sized species that accumulate when different small areas are aggregated together into larger assemblages.

Distribution and Abundance. It has long been known that species found in many localities tend to be more common in those localities than species

that are narrowly distributed in space. This pattern has only recently been quantified [Brown ref.]. Interestingly, the relationship appears to hold across scales. That is, species that are common and distributed widely across a set of sites located within a single geographic region, also tend to have high average abundance within their geographic range and have ranges that cover a relatively large area. However, the geographic range of a species does not necessarily correlate with the abundance of a species at a single location. This is because abundance varies considerably across the range of a species. Most species achieve their highest abundance within a relatively small geographic region, or many times, in several regions. In places further removed from these regions, abundance is much lower. The distribution — abundance relationship only holds when abundance is averaged across a relatively large suite of local sites.

Two different methods are often used to quantify distribution, and these are not necessarily correlated. First, distribution is measured using the proportion of individual locations or censuses that a species is observed on. This is straightforward to do when locations being examined are discrete (e.g., oceanic islands). It is somewhat more problematic if abundance varies continuously across the area being studied. In such cases, there is an area effect that may confound the relationship between distribution and abundance: larger census areas have a greater probability of containing species that are rare or only incidental within the region being studied. A second method of measuring distribution is to count the total area occupied by a species within the region being studied. Again, there are complications with the meaning of the measurement, because if species are discontinuously distributed within their ranges, the total area covered by the range may include regions within it that are not occupied by the species [Gaston ref.].

Diversity and Energy Availability. In any local ecological region, the amount of solar radiation will determine the rate of primary productivity. All else being equal, higher rates of primary production should result in a greater biomass available to be apportioned among different species of plants. This should increase the amount of energy available to primary consumers. If there are multiple sources of energy within an ecosystem, then the number of species should increase as the number of those energy sources increases. Consequently, species richness should increase with productivity [Wright ref.]. Although this suggestion seems reasonable in principle, in practice it is very difficult to determine precisely what units of energy are relevant to species richness. Incident solar radiation in a region is not sufficient as a measure of energy availability since there is a very large number of climatic, topographical and physical variables that might modify the effect of solar radiation on primary productivity and its consequent allocation among species. Correlation of measures of primary productivity and climate with species richness of trees and vertebrates indicated that no single variable was consistently correlated with the number of species in these groups on the North American continent [Currie ref.]. However, every group of species was correlated with at least one variable related to energy availability.

Over very broad geographic regions, there appears to be a latitudinal gradient in species richness. Most groups of plants and animals have more species that are found in tropical latitudes. Consequently, when a longitudinal transect is taken from the equator to the poles, there will be more species at low latitudes along this transect than at high latitudes. The trend can be reversed within taxa that are composed primarily of species that are adapted to high latitudes, but such taxa are relatively rare. Presumably, the increase in species richness near the equator is ultimately caused by the greater amount of sunlight that hits the Earth at those latitudes, but exactly how this translates into increased species numbers is not clear. If species at tropical latitudes were more specialized and experienced greater interspecific competition than those further north, then one would expect that geographic ranges of tropical species would be smaller [Stevens ref.]. Because of statistical complications, this has been difficult to demonstrate empirically.

Theoretical Work

Perhaps the most important function of macroecology is its focus on explanation of patterns across spatial and temporal scales. Such explanations should eventually lead to new insights into how biological and physical systems interact with one another at the scale of the Earth's surface. Such interactions are resolved over long temporal scales, and consequently are difficult to study. Macroecology shares this problem with geology and astronomy. In the latter fields, progress is made through use of sophisticated theoretical models that predict or explain observed patterns. Although macroecology is still in its early stages of development, such theoretical models are beginning to be developed. In particular, the important role that body size plays in macroecology is emerging as a theme in the theoretical development of the field.

A recurring theme in many macroecological studies is the importance of allometric relationships. Such relationships are obtained from plots of ecological, morphological or physiological variables with body size. Such plots are often best fit by a power function of the form $y = a\,M^b$, where y is the variable of interest, and M is some measure of body size (e.g. mass, volume, length). For example, population density, D, scales as $D = a_d\,M^{-0.75}$ for a number of mammal species assemblages [Damuth ref.], while basal metabolic rate, B, for mammals scales as $B = a_b\,M^{0.75}$. Allometric relationships often imply subtle, non-intuitive consequences. In the example just cited, the total basal metabolic rate of a population $(D \times B)$ is implied to be independent of body mass. That is, a population of mice requires the same energy for maintenance as a population of elephants. See also **Morphometrics**.

The existence of similar allometric exponents for a wide variety of ecological and physiological processes has led to the realization that there is a single, general explanation [West, et al.:]. If natural selection maximizes the rate of energy delivery to the structures of which an organism or cell is composed, then the networks that deliver energy to those structures must be fractal-like branching networks that are maximally space filling. That is, these networks, given constraints on their physical structure that arise from the kinds of materials they are built from, must have the maximal amount of surface area across which energy exchanges take place. Since the materials that are being exchanged are molecules, the network essentially spreads its surface area out within the volume of the organism to fill that volume completely. From a macroecological perspective, this means that the number of organisms that can be packed into a given amount of space will be profoundly affected by body mass. Given a fixed amount of energy in a particular environment, the energy can be used by either a few large organisms or many small ones. Since organisms are constructed of fractal-like branching networks, there is a regular decline in the number of organisms that can use a given amount of energy as body mass increases. Consequently, the scaling of density with body mass is the inverse of the scaling of energy use with body mass.

Directions of Future Work

From an empirical perspective, one of the major challenges for macroecology is developing methods to compare data from disparate sources. Many of the studies conducted up to this point have used data collected from a variety of sources. This potentially introduces biases and unknown sources of variation into observed patterns. Biases can give rise to explanations that assume the pattern results from a biological process rather than a sampling artefact. Unknown sources of variation inflate estimates of sampling variances and therefore decrease statistical power. Furthermore, measures such as body size, geographic range size, and even species number may not be directly comparable among different groups of species. For example, some organisms continue to grow larger until they die, while others reach an adult weight and then cease to grow. Organisms that do not sexually reproduce may not be easily broken into species. This may be particularly true in microbes where genetic exchange regularly occurs among taxonomically distinct organisms. Conceptual principles applied to species assemblages from distantly related taxa will continue to be a problem in macroecology. Potential solutions to these problems include statistical techniques for estimating measurement error, theoretical models with robust definitions of macroecological variables, and innovative techniques for measuring and estimating such variables.

Macroecology has the potential to provide conceptual bridges among many different facets of ecology and evolutionary biology. Since it deals with species assemblages, many aspects of population and community ecology might be incorporated in explanations of macroecological patterns. Geographical scale spatial patterns documented in macroecological studies bridge the gap between ecology and biogeography. Evolutionary explanations for macroecological phenomena often appeal to nonrandom patterns

of origination and extinction of taxa, hence data and concepts from paleontology and macro evolution are relevant. Most macroecological studies are comparative in nature; hence, information on the phylogenies of species assemblages is often relevant to the interpretation of demonstrated patterns.

In summary, macroecology is based on the assumption that much can be learned about the causes underlying broad patterns in the natural world by examining and understanding the properties of assemblages of species inhabiting large geographic regions. In order to do so, however, a number of empirical issues will need to be addressed, such as establishing the meaning of conceptually important variables that may have fundamentally different biological causes in different groups of organisms. Principles derived from the study of large, geographic-scale species assemblages may provide links among a number of different biological disciplines.

Additional Reading

Blackburn, T.M., and J. Gaston: *Macroecology: Concepts and Consequences*, Cambridge University Press, Cambridge, UK, 2004.

Brown, J.H.: "On the Relationship between Abundance and Distribution among Species," *American Naturalist*, **124**, 255–279 (1984).

Brown, J.H., and P.F. Nicoletto: "Spatial Scaling of Species Composition: Body Masses of North American Land Mammals," *American Naturalist*, **138**, 1478–1512 (1991).

Brown, J.H., P.A. Marquet, and M.L. Taper: "Evolution of Body Size: Consequences of an Energetic Definition of Fitness," *American Naturalist*, **142**, 573–584 (1993).

Brown, J.H.: *Macroecology*, University of Chicago Press, Chicago, IL, 2000.

Currie, D.J.: "Energy and Large-scale Patterns of Animal- and Plant-Species Richness," *American Naturalist*, **137**, 27–49 (1991).

Damuth, J.: "Population Density and Body Size in Animals," *Nature*, **290**, 699–700 (1981).

Gaston, K.J.: *Rarity*, Chapman and Hall, New York, NY, 1994.

Gaston, K.J., and T.M. Blackburn: *Pattern and Process in Macroecology*, Blackwell Science, Inc., Malden, MA, 2000.

Maurer, B.A.: "Macroecology and Consilience," *Global Ecology and Biogeography*, **9**, 275–280 (2000).

Roy, K., D. Jablonski, and K.K. Martien: "Invariant Size-frequency Distributions along a Latitudinal Gradient in Marine Bivalves," *Proceedings of the National Academy of Sciences of the USA*, **97**, 13150–13155 (2000).

Stevens, G.C.: "The Latitudinal Gradient in Geographic Range: How So Many Species Coexist in the Tropics," *American Naturalist*, **133**, 240–256 (1989).

West, G.B., J.H. Brown, and B.J. Enquist: "The Fourth Dimension of Life: Fractal Geometry and Allometric Scaling of Organisms," *Science*, **284**, 1677–1679 (1999).

Wright, D.H.: "Species-energy Theory: an Extension of Species-area Theory," *Oikos*, **41**, 496–506 (1983).

BRIAN A. MAURER, Michigan State University, East Lansing, MI

MACROMEDIA FLASH®.   See **Data Compression**.

MACROMETEOROLOGY.   The study of the largest-scale aspects of the atmosphere, such as the general circulation and weather types. There is a wide gap between this scale and the relatively small scale of mesometeorology. The gap is bridged by those atmospheric characteristics, referred to as cyclonic scale, that are commonly the subject of synoptic chart analysis. See also **Micrometeroology**.

MACROMOLECULAR SCIENCE.   Macromolecules or polymers are now such a common feature of modern life that it is sometimes forgotten that the rapid development of such materials in industry has taken place in the last 60 years. In fact, the idea that such molecules exist was proposed by Staudinger, Mark, and coworkers in the 1920s and it was some time before these concepts were accepted by the scientific community.

Polymers are long-chain molecules synthesized by the linking together of a large number of identical or similar small units termed *monomers*. The large size of macromolecules leads to their many useful properties. Macromolecular science is the study of these properties in relation to their chemical and physical structure. Naturally occurring or biological macromolecules, such as proteins, carbohydrates, nucleic acid, and natural rubber, are studied by the same methods as synthetic polymers, resulting in a number of industrial and biomedical applications.

In response to the needs of industry, a number of centers for graduate study in macromolecular science have developed. The first such center was at Brooklyn Polytechnic Institute (now Polytechnic University) and large centers now exist at Case Western Reserve University in Cleveland Ohio (Fig. 1), University of Massachusetts (Amherst) and University of Akron (Ohio). The first accredited engineering undergraduate degree program in polymer science was developed in the Department of Macromolecular Science at Case Western Reserve University.

In the past, research emphasis has centered on the synthesis of new macromolecules. This is still a major emphasis, especially as applied to the production of new polymers or composite systems having unique electrical and/or mechanical properties and to the synthesis and fabrication of unique transport and barrier membranes. Such research areas relate directly to the current emphasis on high tech, high price products rather than on bulk commodity polymers.

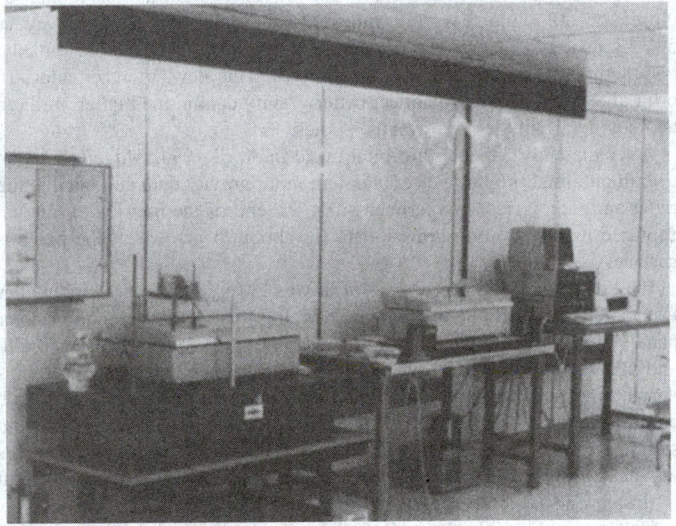

Fig. 1. Automated Langmuir-Blodgett deposition station used for production of ultrathin films and located in a clean room at the Polymer Microdevice Laboratory, Case Western Reserve University, Cleveland, Ohio.

At least equally important in aiding in the solution of current problems of society has been the development, primarily over the past ten years, of the fields of polymer physics and engineering. The general thrust of this development has been to gain an understanding of the physical structure and morphology of macromolecules and to develop methods to modify these parameters to produce useful properties in both polymer and composite systems. See accompanying illustration. See also **Molecular and Supermolecular Electronics**; and **Colloid System**.

JOHN BLACKWELL, Case Western Reserve University, Cleveland, OH

MACROMOLECULE. A molecule, usually organic comprised of an aggregation of hundreds or thousands of atoms. Such giant molecules are in general of two types: (1) Individual entities (compounds) that cannot be subdivided without losing their chemical identity. Typical of these are proteins, many of which have a molecular weights running into the millions. (2) Combinations of repeating chemical units (monomers) linked together into chain or network structures called polymers. Each monomer has the same chemical constitution as the polymer, e.g., isoprene (C_5H_8) and polyisoprene (C_5H_8)$_x$. Synthetic elastomers (plastics) are typical of this kind of macromolecule. Cellulose is the most common example found in nature. Most macromolecules are in the colloidal size range.

See also **Colloid System**; and **Polymerization**.

Additional Reading

McPherson, A.: *Introduction to Macromolecular Crystallography*, John Wiley & Sons, Inc., New York, NY, 2002.

Munk, P., and T.M. Aminabhavi: *Introduction to Macromolecular Science*, 2nd Edition, John Wiley & Sons, Inc., New York, NY, 2002.

Pittman, C.U., M. Zeldin, and J.E. Sheats: *Macromolecules Containing Metal and Metal-like Elements: Biomedical Applications*, Vol. 3, John Wiley & Sons, Inc., Hoboken, NJ, 2004.

Sun, S.F.: *Physical Chemistry of Macromolecules: Basic Principles and Issues*, 2nd Edition, John Wiley & Sons, Inc., Hoboken, NJ, 2004.

Wohrle, D., and A.D. Pomogailo: *Metal Complexes and Metals in Macromolecules: Synthesis, Structure and Properties*, John Wiley & Sons, Inc., New York, NY, 2003.

MACROPHAGE. See **Immune System and Immunology**.

MACROSCOPIC. Large enough to be visible to the naked eye or under low order of magnification.

MACULA. The macula is a small area, less than $\frac{1}{4}$ inch, in the center of the retina at the back of the eye. It is responsible for sharp, clear central vision and the ability to perceive color.

Like the film in a camera, the retina receives light rays from the front of the eye and transmits those light rays through the optic nerve to the brain where the rays are converted into images. The densely packed photoreceptor (light-sensitive) cells in the macula control all of the eye's central vision and are responsible for the ability to read, drive a car, watch television, see faces, and distinguish detail. The rest of the retina handles peripheral vision that enables your eyes to see objects off to the side while you are looking forward.

There are two types of photoreceptor cells in the cornea: rods and cones. The rods provide vision at low light levels, while the cones provide sharp vision and discrimination. Because the macula contains a high concentration of cones, straight-ahead vision is in sharp focus, particularly in bright light. Most of the rods are located in the periphery of the retina, so faint objects are more visible if you do not look directly at them. A dim star, for instance, is best seen when your eyes are not aimed directly at it.

The most common cause of functional blindness in people over the age of 60 is macular degeneration, a deterioration or breakdown of the macula. Damage to the macula results in the loss, either partial or complete, of ability to see objects clearly in the center of vision. Although not totally blind, the person has difficulty performing tasks that require "straight-on" vision, such as driving a car, reading, or watching television. Because peripheral vision is not affected, the person can adapt somewhat to the loss of central vision and continue to pursue some normal daily activities, such as walking, without assistance.

There are two types of macular degeneration. The "dry" form is usually the result of aging and thinning of the macula's layers, and the "wet" form occurs when abnormal blood vessels under the retina leak fluid and blood, causing scarring. Vision loss with dry macular degeneration occurs gradually over a number of years, and the affected person may not be aware of any problem. Dry macular degeneration is the less serious of the two forms. With the wet form of this disease, central vision capabilities can be damaged rapidly. Early detection usually results in more successful treatment. See also **Retina**.

Vision Rx, Inc., Elmsford, NY

MADAR. See **Silk Cotton Trees**.

MADDEN-JULIAN OSCILLATION. A quasi periodic oscillation of the near-equatorial troposphere, most noticeable in the zonal wind component in the boundary layer and in the upper troposphere, particularly over the Indian Ocean and the western equatorial Pacific.

This phenomenon is named after the co-discoverers. The oscillation can be detected globally in winds near the tropopause. The period of the oscillation varies between about 30 and 50 days, and appears to represent an eastward-propagating disturbance with the structure of a Kelvin wave with a vertical half-wavelength of the depth of the troposphere, but with a phase speed of only about 8 m s^{-1}, much less than that of an adiabatic Kelvin wave. The disturbance is accompanied by strong fluctuations of deep convection, easily detectable using satellite observations, and is a major contributor to intraseasonal weather variability in equatorial regions from eastern Africa eastward to the central Pacific.

AMS

MADDER. See **Rubiaceae**.

MADRONE TREE. See **Heather Shrubs and Trees**.

MADUROMYCOSIS. A disease caused by various fungi (*Madurella mycetomi*) or actinomycetes (*Nocardia brasiliensis*). It usually affects the foot, hand, and legs with tissues becoming necrosed and swollen after infection.

See also **Antiparasitic Agents, Antimycotics**.

MAFIC. Said of an igneous rock composed chiefly of one or more ferromagnesian, dark-colored minerals in its mode; also, said of those minerals.

MAGELLAN MISSION TO VENUS. The *Magellan* spacecraft, named after the sixteenth-century Portuguese-born explorer whose expedition first circumnavigated the Earth. NASA's *Magellan* spacecraft used sophisticated imaging radar to make the most highly detailed maps of Venus ever captured during its four years in orbit around Earth's sister planet from 1990 to 1994.

After concluding its radar mapping, *Magellan* made global maps of Venus's gravity field. Flight controllers also tested a new maneuvering technique called aerobraking, which uses a planet's atmosphere to slow or steer a spacecraft.

Craters shown in the radar images that *Magellan* sent to Earth tell scientists that Venus's surface appears relatively young—resurfaced about 500 million years ago by widespread volcanic eruptions.

The planet's present harsh environment has persisted at least since then, with no features detected suggesting the presence of oceans or lakes at any time in the planet's past.

Scientists also found no evidence of plate tectonics, the movements of huge crustal masses on Earth that cause earthquakes and result in the drifting of continents over time spans of hundreds of millions of years.

Magellan's mission ended with a dramatic plunge to the planet's surface, the first time an operating planetary spacecraft has ever been intentionally crashed. Contact was lost with the spacecraft October 12, 1994, at 10:02 Universal Time (3:02 a.m. Pacific Daylight Time). The purpose of the maneuver was for Magellan to gather data on Venus's atmosphere before it ceased to function during its fiery descent.

Mission Overview

Magellan was the first planetary spacecraft to be launched by a space shuttle. The shuttle *Atlantis* from Kennedy Space Center in Florida carried aloft *Magellan* on May 4, 1989. *Atlantis* took *Magellan* into low Earth orbit, where it was released in the shuttle's cargo bay. A solid-fuel motor called the Inertial Upper Stage (IUS) then fired, sending *Magellan* on a 15-month cruise looping around the Sun 1-1/2 times before it arrived at Venus on August 10, 1990. A solid-fuel motor on *Magellan* then fired, placing the spacecraft in orbit around Venus.

Magellan's initial orbit was highly elliptical, taking it as close as 294 kilometers (182 miles) from Venus and as far away as 8,543 kilometers (5,296 miles). The orbit was a polar one, meaning that the spacecraft moved from south to north or vice versa during each looping pass, flying over Venus's north and south poles. *Magellan* completed one orbit every 3 hours, 15 minutes.

During the part of its orbit closest to Venus, *Magellan's* radar mapper imaged a swath of the planet's surface approximately 17 to 28 kilometers (10 to 17 miles) wide. At the end of each orbit, the spacecraft radioed back to Earth a map of a long ribbon-like strip of the planet's surface captured on that orbit. Venus itself rotates once every 243 Earth days. As the planet rotated under the spacecraft, *Magellan* collected strip after strip of radar image data, eventually covering the entire globe at the end of the 243-day orbital cycle.

By the end of its first such eight-month orbital cycle between September 1990 and May 1991, *Magellan* had sent to Earth detailed images of 84 per-cent of Venus's surface. The spacecraft then conducted radar mapping on two more eight- months cycles from May 1991 to September 1992. This allowed it to capture detailed maps of 98 percent of the planet's surface. The follow-on cycles also allowed scientists to look for any changes in the surface from one year to the next. In addition, because the "look angle" of the radar was slightly different from one cycle to the next, scientists could construct three-dimensional views of Venus's surface.

During *Magellan's* fourth eight-month orbital cycle at Venus from September 1992 to May 1993, the spacecraft collected data on the planet's gravity field. During this cycle, *Magellan* did not use its radar mapper but instead transmitted a constant radio signal to Earth. If it passed over an area of Venus with higher than normal gravity, the spacecraft would slightly speed up in its orbit. This would cause the frequency of *Magellan's* radio signal to change very slightly due to the Doppler effect—much like the pitch of a siren changes as an ambulance passes. Thanks to the ability of radio receivers in the NASA/JPL Deep Space Network to measure frequencies extremely accurately, scientists could build up a detailed gravity map of Venus.

At the end of *Magellan's* fourth orbital cycle in May 1993, flight controllers lowered the spacecraft's orbit using a then-untried technique called aerobraking. This maneuver sent *Magellan* dipping into Venus's atmosphere once every orbit; the atmospheric drag on the spacecraft slowed down *Magellan* and lowered its orbit. After the aerobraking was completed between May 25 and August 3, 1993, *Magellan's* orbit then took it as close as 180 kilometers (112 miles) from Venus and as far away as 541 kilometers (336 miles). *Magellan* also circled Venus more quickly, completing an orbit once every 94 minutes. This new, more circularized orbit allowed *Magellan* to collect better gravity data in the higher northern and southern latitudes near Venus's poles.

After the end of that fifth orbital cycle in April 1994, *Magellan* began a sixth and final orbital cycle, collecting more gravity data and conducting radar and radio science experiments. By the end of the mission, *Magellan* captured high-resolution gravity data for about 95 percent of the planet's surface.

In September 1994, *Magellan's* orbit was lowered once more in another test called a "windmill experiment." In this test, the spacecraft's solar panels were turned to a configuration resembling the blades of a windmill, and *Magellan's* orbit was lowered into the thin outer reaches of Venus's dense atmosphere. Flight controllers then measured the amount of torque control required to maintain *Magellan's* orientation and keep it from spinning. This experiment gave scientists data on the behavior of molecules in Venus's upper atmosphere, and lent engineers new information useful in designing spacecraft.

On October 11, 1994, *Magellan's* orbit was lowered a final time, causing the spacecraft to become caught in the atmosphere and plunge to the surface; contact was lost the following day. Although much of Magellan was believed to be vaporized, some sections probably hit the planet's surface intact.

Venus

One of the handful of planets known to the ancients, Venus is often called Earth's sister planet because of its similar size and distance from the sun. Earth is 12,756 kilometers (7,926 miles) in diameter, compared to Venus at 12,103 kilometers (7,520 miles); Earth orbits the sun at an average 149.6 million kilometers (93 million miles), compared to Venus at 108.2 kilometers (67.2 million miles). The two planets' densities are also similar—5.52 grams per cubic centimeter for Earth, compared to 5.24 grams per cc for Venus. Because Venus is closer to the sun than Earth is, it always appears close to the sun from our point of view as either a glistening, bright evening or morning "star." See also **Venus**.

Despite the similarities, however, in other ways Venus is very much unlike Earth. Venus has a surface temperature of about 470 degrees Celsius (about 900 degrees Fahrenheit); the atmospheric pressure at the surface is 90 times greater than Earth's. Venus's atmosphere is nearly devoid of water, made up of 97 percent carbon dioxide; its upper clouds contain sulfuric acid. Venus has no moons, and no magnetic field has been detected. It rotates on its axis in a retrograde direction—that is, opposite that of Earth and most of the other planets—very slowly, once every 243 Earth days.

The first spacecraft mission to another planet was the NASA/JPL spacecraft *Mariner 2*, which executed a fly by of Venus in December 1962. Other U.S. spacecraft to visit Venus have included *Mariner 10*, which flew by Venus in 1974 on its way to Mercury in the first mission to more than a single planet; and *Pioneer Venus*, a 1978 mission that included an orbiter with an altimeter and imaging radar that functioned at lower resolution than *Magellan's*, as well as multiple probes that descended into Venus's atmosphere. The then-Soviet Union also sent a number of spacecraft to Venus, including four—*Venera 9, 10, 13* and *14*—that landed on the surface and made closeup pictures of the rocky terrain briefly before the searing heat caused them to stop functioning. Two other Soviet missions, *Venera 15* and *16*, used orbiting imaging radar similar to *Magellan's* but at a lower resolution.

The Magellan Spacecraft

Built partially with spare parts from other missions, the *Magellan* spacecraft was 4.6 meters (15.4 feet) long, topped with a 3.7-meter (12-foot) high-gain antenna. Mated to its retrorocket and fully tanked with propellants, the spacecraft weighed a total of 3,460 kilograms (7,612 pounds) at launch.

The high-gain antenna, used for both communication and radar imaging, was a spare from the NASA/JPL *Voyager* mission to the outer planets, as were *Magellan's* 10-sided main structure and a set of thrusters. The command data computer system, attitude control computer and power distribution units are spares from the *Galileo* mission to Jupiter. *Magellan's* medium-gain antenna is from the NASA/JPL *Mariner 9* project. Martin Marietta Corp. was the prime contractor for the *Magellan* spacecraft, while Hughes Aircraft Co. was the prime contractor for the radar system. See Fig. 1.

Magellan was powered by two square solar panels, each measuring 2.5 meters (8.2 feet) on a side; together they supplied 1,200 watts of power. Over the course of the mission the solar panels gradually degraded, as expected; by the end of the mission in the fall of 1994 it was necessary to manage power usage carefully to keep the spacecraft operating. See Fig. 2.

Imaging Radar

Because Venus is shrouded by a dense, opaque atmosphere, conventional optical cameras cannot be used to image its surface. Instead, Magellan's imaging radar uses bursts of microwave energy somewhat like a camera flash to illuminate the planet's surface.

Magellan's high-gain antenna sends out millions of pulses each second toward the planet; the antenna then collects the echoes returned to the spacecraft when the radar pulses bounce off Venus's surface. The radar pulses are not sent directly downward but rather at a slight angle to the side of the spacecraft, the radar is thus sometimes called "side-looking radar." In addition, special processing techniques are used on the radar data to result in higher resolution as if the radar had a larger antenna, or "aperture"; the technique is thus often called "synthetic aperture radar," or SAR.

Synthetic aperture radar was first used by NASA, on JPL's Seasat oceanographic satellite in 1978; it was later developed more extensively on the Spaceborne Imaging Radar (SIR) missions on the space shuttle in 1981, 1984 and 1994. Another imaging radar mission is also planned as part of the NASA/JPL *Cassini* mission to Saturn in 1997 to map the surface of the ringed planet's major moon Titan. See also **Cassini Mission to Saturn**; and **Saturn**.

Besides its use in imaging, *Magellan's* radar system was also used to collect altimetry data showing the elevations of various surface features. In this mode, pulses were sent directly downward and Magellan measured the time it took a radar pulse to reach Venus and return in order to determine the distance between the spacecraft and the planet.

Science Results

Magellan returned maps of Venus's surface and its gravity field in unprecedented detail that will be a resource for many years for scientists studying the planet. The mission held many surprises for scientists, and resulted in a number of theories about the planet being revised.

From the craters visible in Magellan's Venus maps, scientists believe they are looking at a relatively young planetary surface, perhaps about 500 million years old. Since Venus formed at the same time as Earth 4.6 billion years ago, some event or events 500 million years ago must have resurfaced the planet. Scientists believe that this may have been the work of massive outpourings of lava from planet-wide volcanic eruptions. Although Venus may still have active volcanoes, no visible outpourings of lava have yet been detected in comparisons of *Magellan* images between one eight-month orbital cycle and another.

Although some scientists speculate that Venus may have once been a temperate planet that fell victim to a runaway greenhouse effect creating enormously high temperatures, *Magellan's* maps show no telltale signs of past major water bodies such as shorelines or ocean basins. Also, surface features show no evidence of being eroded by water—although there is evidence of wind erosion in the form of numerous sand dunes and wind streaks.

One of the hopes that scientists had for *Magellan* was to find out if Venus, like Earth, has plate tectonics—movements of crustal masses that

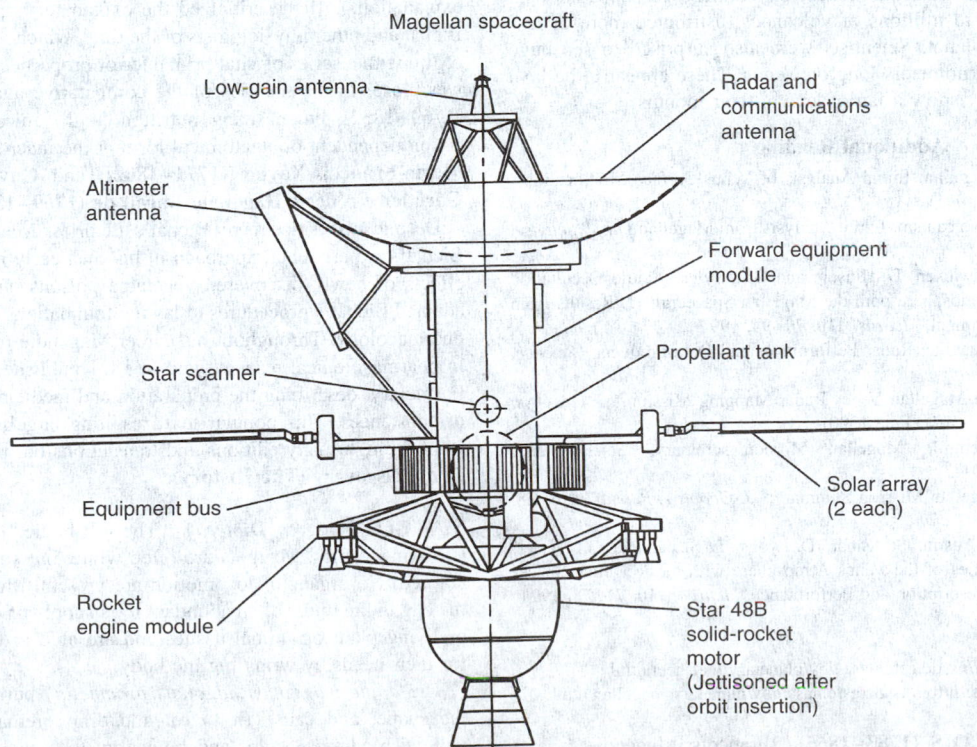

Magellan spacecraft

Low-gain antenna

Radar and communications antenna

Altimeter antenna

Forward equipment module

Star scanner

Propellant tank

Equipment bus

Solar array (2 each)

Rocket engine module

Star 48B solid-rocket motor (Jettisoned after orbit insertion)

Fig. 1. *Magellan* spacecraft (*NASA, GSFC*).

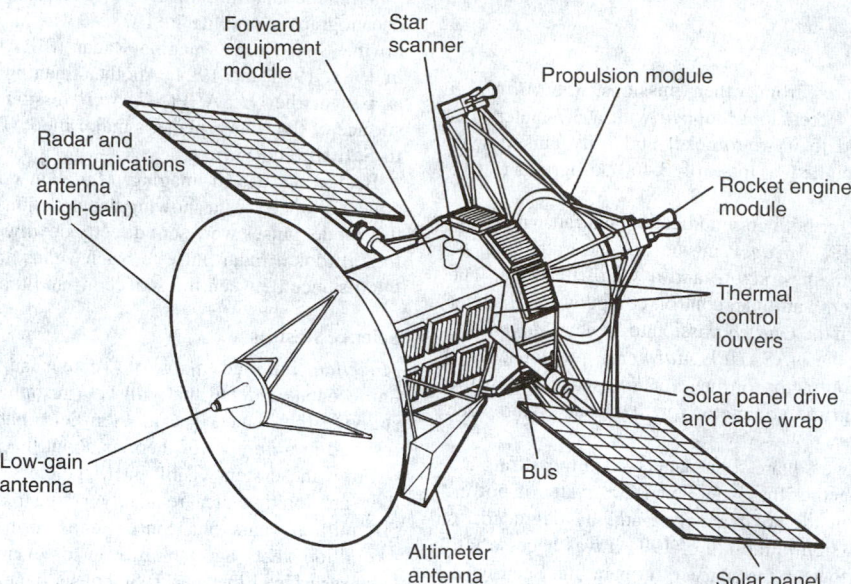

Magellan spacecraft

Forward equipment module

Star scanner

Propulsion module

Radar and communications antenna (high-gain)

Rocket engine module

Thermal control louvers

Solar panel drive and cable wrap

Low-gain antenna

Bus

Altimeter antenna

Solar panel

Fig. 2. *Magellan* spacecraft with solar panels extended (*NASA, GSFC*).

on Earth cause earthquakes and result in the drifting of continents over time periods of hundreds of millions of years. They in fact found no evidence of plate tectonics in the data returned by the mission. Scientists suspect that, although Venus is very similar in size to Earth, its interior is probably different in major ways. In particular, Venus seems to lack an "asthenosphere," a buffer layer within Earth between the outer part of the planet and the mantle beneath. As a result, the gravity signature of features on Venus closely reflect surface topography, whereas on Earth such a correspondence does not always occur.

Scientists are also intrigued by the distribution of volcanoes around Venus. On Earth, volcanoes occur in groups such as the so-called "Ring of Fire" around the Pacific Rim. Venus, by contrast, is peppered with hundreds of thousands to millions of volcanoes distributed more or less randomly around the planet. Scientists were also surprised to see huge channels thousands of kilometers long on Venus. These appear to be lava channels, and frequently show a fan of lava at their mouths.

Additional Reading

Arvidson, R.E. et al.: "Magellan: Initial Analysis of Venus Surface Modification," *Science*, 270 (April 12, 1991).

Head, J.W. et al.: "Venus Volcanism: Initial Analysis from Magellan Data," *Science*, 276 (April 12, 1991).

Jenkins, J., P. Steffes, J. Twicken, D. Hinson, and G.L. Tyler: "Radio Occultation Studies of the Venus Atmosphere with the Magellan Spacecraft: 2. Results from the October 1991 Experiment," *Icarus*, 110, 79–94, 1994.

Pettengill, G.H. et al.: "Magellan Radar Performance and Data Products," *Science*, 260 (April 12, 1991).

Saunders, R.S. et al.: "The Magellan Venus Radar Mapping Mission," *J. Geophys. Res.*, 95, No. B6, 8339–8355, (June 1990).

Saunders, R.S., G.H. Pettengill: "Magellan: Mission summary," *Science*, 252, 247–249, (April 12, 1991).

Saunders, R.S. et al.: "Magellan Mission Summary," *J. Geophys. Res.*, 97, No. E8, 13067–13090, (Aug. 1992.)

Steffes, P., J. Jenkins, R. Austin, S. Asmar, D. Lyons, E. Seale, and G.L. Tyler: "Radio Occultation Studies of the Venus Atmosphere with the Magellan Spacecraft: 1. Experimental Description and Performance," *Icarus*, 110, 71–78, 1994.

Web References

Magellan Fact Sheet: http://nssdc.gsfc.nasa.gov/planetary/factsheet.html

Magellan Mission to Venus: http://nssdc.gsfc.nasa.gov/planetary/magellan.html

MAGENDIE, FRANÇOIS (1783–1855). François Magendie was a French medical Figure of the early nineteenth century who was one of the founders of experimental physiology and pharmacology. François

Magendie was born in Bordeaux, the elder son of a surgeon who moved his family to Paris in 1791. Applying the educational principles of Jean-Jacques Rousseau (1712–1778), Magendie senior provided no formal schooling for his two sons, so that François had not learned to read or write by his tenth birthday. He insisted on entering school shortly afterward, however, and by the age of fourteen had won the grand prize in a major essay competition. In 1799 he became an apprentice to the surgeon Alexis Boyer (1757–1833) and then entered the new Parisian medical school, graduating in 1808.

Although remembered primarily as an experimentalist, Magendie's first scientific publication (1809) was a brief theoretical article on physiological explanation. In it he criticized the explanatory approach used by Xavier Bichat and other physiologists of the time, which identified the phenomena of life with a series of vital principles or properties. Magendie's alternative was closer to the theories of the comparative anatomist Georges Cuvier (1769–1832), emphasizing nutrition at the micro level and functional action dependent on anatomical form at the macro level. See also **Bichat, Marie-François-Xavier (1771–1802);** and **Cuvier, Georges Léopold Chrétien Frédéric Dagobert Baron de (1769–1832)**.

Despite his criticism of Bichat's theories, Magendie closely followed Bichat's experimental approach in his own early researches. Studying the effects of newly discovered vegetable poisons on mammals in 1809, he adapted Bichat's procedures to lay the foundations of modern experimental pharmacology. Throughout his career Magendie retained a strong interest in pharmacological research and in 1821 published the first edition of his Formulary, describing the preparation and medicinal uses of a wide range of substances. This popular text, resulting largely from his own studies, went through many editions and translations during his lifetime. See also **Drug Discovery (The History)**.

MAGGOT (*Insecta, Diptera*). The soft-bodied larva of many species of two-winged flies. Maggots are often white, but some species are brightly colored. No organs of locomotion are present. Most maggots hatch from the egg in the midst of an abundant food supply of decaying organic matter or living plant or animal tissues and are able to move about sufficiently for their needs by wriggling the body.

The *apple maggot* (*Rhagoletis promnella*) burrows into fruit, distorts the shape, and causes it to rot and drop prematurely. The maggot is yellowish-white in color and ranges up to $\frac{3}{8}$ inch (9–10 millimeters) in length. The adult version is the black fly. The apple maggot is found in the United States from the Dakotas to New England and from the southeastern

Canadian border into Arkansas, Ohio, and Georgia. However, the pest is uncommon in the southern parts of this range.

This maggot, one of the most serious insect pests of apple, either ruins the fruit entirely, or makes it unappetizing for consumption. Heavily infested fruit will be reduced to a brown, rotted mass, filled with yellowish, legless maggots. When the fruit is slightly infested, there is no external indication of maggots within. However, when the fruit ripens, burrows made by the maggots show as dark lines under the skin of the fruit. Larvae in prematurely dropped fruit can continue to live and become adult flies that will reinfest the fruit on the trees. Thus, it is important to remove and burn such fruit immediately after it has dropped. The adult flies appear in late June and early July, at which time they insert eggs under the skin of the fruit. Hibernation occurs in small puparia located just below the fruit surface. The sweet and subacid varieties of apple are the most frequently attacked varieties.

The *artichoke stem maggot* (*Straussia longipennis*) is a small, yellow-colored maggot which bores into the pith of the stems. The adults are yellow flies with two banded wings.

The *cabbage root maggot* (*Pegomya brassicae*) is a headless, legless, white maggot, about $\frac{1}{4}$ inch (6 millimeters) long, which destroys seed in the soil while also attacking the underground parts of plants which have germinated. For control, a suitable chemical insecticide, such as diazinon or chlordane, should be applied to the soil at the base of the plants when the leaves appear; the treatment should be repeated soon after transplanting or thinning. Such chemicals should not be applied to the edible parts once they are formed on the cabbage. This maggot also attacks radish.

The *onion maggot* (*Phorbia cepetorum*) is a legless, white, root-eating maggot which attains a length of about $\frac{1}{3}$ inch (8 millimeters). Distribution is in the northern United States. A dust spray containing malathion provides effective control, but this should not be applied within three days of harvesting.

The *orange maggot* (*Trypeta ludens*) is a dirty-white maggot that attains a length of about $\frac{1}{2}$ inch (12–13 millimeters). The maggot burrows into the pulp of the fruit and up to 20 maggots may be found in a single orange. The adult version is a light-yellow fly with brown markings and bands on the wings. The orange maggot is particularly serious in the Mexican citrus groves. Control is essentially by picking infested fruit and immediately destroying to prevent reinfestation.

The *mushroom maggot* (*Sciara sp.*) is a small maggot, white to yellow in color, with a black head. Treatment is mainly by prevention, keeping flies out of the mushroom growing area, fumigating regularly, and sterilizing the manure growing medium by heating to at least 150 °F (66 °C).

The *raspberry cane maggot* (*Phorbia rubivora*) is a small maggot, white, that burrows into new canes and girdles the shoot. During April and May, a fly deposits the eggs.

The *seed-corn maggot* (*Pegomya fusciceps*) attacks the germinating seeds and roots of many plants, notably bean and pea. This maggot is headless, legless, whitish in color. The best prevention is to use seed that have been commercially treated for seed-corn maggot control.

Other maggots that are quite destructive to food crops include the rice-stem maggot and the seed maggot.

MAGMA.   The term for molten material. A natural, complex, liquid, high-temperature, silicate solution ancestral to all igneous rocks, both intrusive and effusive. The locus of a magma is within the lithosphere (crust) under great pressure and an impenetrable cover which helps the magma to retain its original gases and water vapor in solution. The origin of magma is not known but it is generally assumed that separate magma chambers may exist within the lithosphere.

MAGNESITE.   The mineral magnesite is carbonate of magnesium, $MgCO_3$. It is a hexagonal mineral, but usually found massive. It has a rhombohedral cleavage; conchoidal fracture; brittle; hardness, 4–4.5; specific gravity is approximately 3.0 (average); luster, vitreous to dull; color, white, gray, yellow, or brown; transparent to opaque. Most magnesite is believed to have been derived from the action of carbonated waters upon rocks rich in magnesium. Magnesite-bearing waters, on the other hand, may have in some cases acted upon calcite or dolomite. Magnesite deposits are known in Greece, Austria, Norway, India, Australia, and the Republic of South Africa. In the United States, magnesite is found in California

and Nevada, some of which deposits seem to be of original sedimentary character. Magnesite is in demand for the manufacture of refractories and various compounds of magnesium.

MAGNESIUM.   [CAS: 7439-95-4]. Chemical element, symbol Mg, at. no. 12, at. wt. 24.305, periodic table group 2, mp 649 °C, bp 1,090 °C, critical temperature (calculated) 1,867 °C, density 1.74 g/cm^3 (20 °C), 1.64 g/cm^3 (solid at 650 °C), 1.57 g/cm^3 (liquid at 650 °C). Elemental magnesium has a close-packed hexagonal crystal structure, as do the common alloys of magnesium except those that contain lithium in excess of 11%.

Magnesium is a silver-white metal, malleable and ductile when heated; unattacked by dry oxygen, by H_2O or alkalis at room temperature; when heated to about 800 °C reacts in air or steam and emits a brilliant white light of high actinic power; reactive with acids including carbonic at room temperature; reactive upon heating with nitrogen, phosphorus, arsenic, sulfur, in some cases with such vigor as to constitute a hazard.

Magnesium occurs extensively in the earth's crust, ranking 8th among the chemical elements in terrestrial abundance. An average composition of igneous rocks contains 2.09% magnesium. Of the elements present in seawater, magnesium ranks 5th with an estimated 6,125,000 tons of magnesium per cubic mile (1,323,000 metric tons per cubic kilometer) of seawater, its content exceeded only by hydrogen, oxygen, sodium, and chlorine. Magnesium is a constituent of over 150 minerals and also is found in bitterns and subterranean brines and salt beds. Only a few magnesium minerals are important commercially, notably dolomite [CAS: 17069-72-6], CaO·MgO·2CO$_2$, magnesite [CAS: 13717-00-5], brucite [CAS: 1317-43-7], carnallite [CAS: 1318-27-0], and olivine [CAS: 1317-71-1] as a source of magnesium. See also **Dolomite**. More than half of metallic magnesium produced is extracted from seawater. There are three naturally occurring isotopes, ^{24}Mg through ^{26}Mg; and three radioactive isotopes have been identified, ^{23}Mg, ^{27}Mg, and ^{28}Mg, all with comparatively short half-lives measured in seconds, minutes, or hours. The first known magnesium compound to be isolated was Epsom salt, $MgSO_4$, which Nehemiah Grew obtained in 1695 by evaporating the mineral waters at Epsom, England. In 1754, Joseph Black demonstrated that magnesia and lime were two different substances, but the exact identify of magnesia was not reported until 1808 by Sir Humphrey Davy who demonstrated that magnesia was an oxide of a heretofore unknown element. He first termed the element magnium. Metallic magnesium was first isolated by A. Bussy in 1828 when he fused magnesium chloride with potassium. Michael Faraday produced the first magnesium metal electrolytically in 1883. First ionization potential 7.64 eV; second, 14.97 eV. Oxidation potential Mg → Mg^{2+} + 2e$^-$, 2.375 V; Mg + 2OH$^-$ → $Mg(OH)_2$ + 2e$^-$, 2.67 V. Other important physical properties of magnesium are given under **Chemical Elements**.

Production

There are two principal magnesium production processes: (1) electrolytic, and (2) metallothermic reduction. Electrolytic processes account for 80% of commercial production. In this process, seawater is pumped into large settling tanks where it is treated with lime. Roasted oyster shells sometimes are used if a convenient source is nearby. The lime precipitates the magnesium as the insoluble hydroxide. The hydroxide is filtered and then converted into a slurry with fresh H_2O. Subsequent treatment with HCl converts the $Mg(OH)_2$ into $MgCl_2$. The latter compound is dried and then electrolyzed in the fused state to produce molten magnesium and chlorine gas. The latter is recycled. The magnesium is cast into ingots. In the thermal or ferrosilicon process, used in some European countries, a mixture of magnesium oxide and powdered ferrosilicon (an iron-silicon alloy) is fed into a retort and heated under vacuum to about 1,200 °C. The magnesium is freed in the form of vapor and condenses into crystals at the cool end of the retort. The crystals then are remelted and cast into pigs.

Uses of Magnesium

Magnesium finds principal uses as a primary metal to which other metals are added in various alloying amounts to enhance the properties of magnesium. Magnesium is the lightest of all structural metals and consequently the metal has enjoyed much attention over the years in connection with the transportation industry, notably for applications in the aircraft, aerospace, and automotive industries. Vehicle designers constantly are aware of the additional power requirements for simply moving "dead weight" that wastes fuel and contributes to air pollution.

In addition to its use as a structural metal, magnesium is an important metallurgical chemical in the form of a deoxidizer and desulfurizer and as the constituent of numerous industrial and laboratory chemical compounds.

Magnesium Alloys. Even prior to the use of magnesium as a structural metal in the aerospace field, in 1921, Louis Chevrolet put a set of magnesium-alloy pistons in the Ford racing car that won the Indy 500 for him that year. The magnesium pistons gave racing and sports cars faster acceleration and deceleration. This application of magnesium was not intended so much as a dead-weight savings feature for the car, but rather more in terms of inertia (obviously also relative to weight). Although the magnesium pistons provided better acceleration/deceleration because of smaller inertia, the early designers encountered what is known as piston slap, which results when the piston material has a considerably higher coefficient of thermal expansion than the cylinder material does.

The use of magnesium castings for auto wheels was introduced a few years later and also serves the principal purpose of reducing inertia. With wheels, it is not just faster acceleration/deceleration that can be achieved, but also minimizing the amount of unsprung weight for a smoother and easier-to-control ride and minimizing the problem of gyroscopic action of the rapidly spinning wheels. Designers of racing cars switched from wire-spoke wheels to magnesium-alloy wheels in the early 1950s. The use of magnesium has increased not only in racing cars, but some passenger cars, both for the purpose of reducing inertia and weight. Today, magnesium is used for transmission and differential housings and a variety of other racing car parts. Serious attention continues to be given to major engine components, such as the cylinder block, head, and oil sump, all of which are candidates for reducing dead weight and increasing fuel economy.

The most extensive use of magnesium castings in automobiles commenced in 1936, with the introduction of the Volkswagen Beetle. Each Beetle used from 40 to 50 pounds (18 to 23 kg) of primary magnesium ingot plus scrap metal.

Magnesium has a density only $\frac{2}{3}$ that of aluminum, $\frac{1}{4}$ that of zinc, and about $\frac{1}{5}$ that of irons and steels. In addition to the obvious aerospace and automotive applications, other applications include hand trucks, containers, materials-handling equipment, portable electric and pneumatic tools (such as chain saws), hand tools, luggage, sporting goods, dockboards, and tooling jigs and fixtures. It has been found that lighter-weight equipment significantly reduces accidents and lost time due to injuries. On an arbitrary scale, where the power required to machine magnesium alloys is 1.0, the Figures for other metals are: aluminum alloys, 1.8; brass, 2.3; cast iron, 3.5; mild steel, 6.3; and nickel alloys, 10.0.

Some magnesium alloys are listed in Tables 1 and 2.

Magnesium in Other Metal Alloys. Magnesium is an important alloying ingredient in the production of other base metal alloys. When added during metallurgical processing, magnesium in small amounts has a marked effect on final properties of the metals:

Aluminum — Magnesium increases resistance to corrosion, facilitates heat treatment, and increases most mechanical properties. If magnesium-containing aluminum is remelted, the magnesium may be lost and should be replaced by adding pure magnesium to the casting ladle or pot.

Copper — Magnesium improves tensile strength and allows age hardening. Magnesium is used mainly as a deoxidizer, notably in copper-nickel-zinc alloys and in leaded brasses and bronzes. The magnesium is added during melting.

Lead — Magnesium increases hardness, strength, and resistance to creep. Magnesium also is used as a debismuthizer in refining primary lead.

Nickel — Magnesium, in combination with carbon, forms an age-hardenable alloy. The main use of magnesium is to deoxidize and

TABLE 1. REPRESENTATIVE MAGNESIUM ALLOYS

Alloy Designation	Elements Added	Tensile Strength 1,000 psi	Brinell Hardness	Melting Point °C	Forms Available	Features
AZ31B	3% Al 1% Zn	29	49	627	Sheet, plate, extrusions, forgings.	Moderate strength, good formability, general-purpose alloy. Dent resistant, weldable.
AZ91B	9% Al 0.6% Zn	33	67	596	Die casting alloy.	Good strength and castability. Popular for portable tools, business machines, vehicles.
AZ91C	8.7% Al 0.7% Zn	40	53	596	General-purpose sand and permanent-mold casting alloy.	Good castability, pressure tightness, and weldability. Moderate strength.
HK31A	3% Th 0.7% Zn 0.7% Zr	38	57	649	Sheet and plate for aerospace uses. (200–370 °C). Sand and permanent-mold castings.	Good short-time, elevated temperature characteristics. Weldable without stress relief. Low microporosity in cast form.
HM21A	0.6% Mn 2% Th	35	56	650	Sheet, plate, forgings for aerospace uses. (200–425 °C)	Very stable at elevated temperatures. Good creep strength and formability. Weldable without stress relief.
HM31A	1.2% Mn (min) 3% Th	44	63	605	Extrusions for aerospace uses. (200–425 °C)	Excellent elevated temperature properties. Weldable without stress relief.
QE22A	2% Pr 0.7% Zr 2.5% Ag	40	78	549	Castings for aerospace uses. (up to 260 °C)	Superior tensile strength plus excellent creep and fatigue strength.
ZK60A	5.7% Zn 0.5% Zr	47	—	635	Highly stressed parts of aerospace and military uses. Used as a forging alloy.	High strength, good toughness, good spot-weldability. Limited arc-weldability.

Note: 1 psi (pounds/square inch) = 0.0069 megapascal (MPa).
Designation of Magnesium Alloys (an ASTM system now accepted by the SAE). A four-part system is used:

1. Letters indicate the two principal alloying elements: A, Aluminum; E, Rare-Earth; H, Thorium; K, Zirconium; M, Manganese; Q, Silver, S, Silicon; T, Tin; Z, Zinc. Thus HK signifies a thorium-zirconium magnesium alloy.
2. The approximate amounts (percent, wt) of the two principal alloying materials follow to the immediate right of the alloying element letters. Thus HK31 indicates approximately 3% thorium, and 1% zirconium.
3. The next two letter symbols to the right are used to distinguish two different alloys of the same chemical composition. Any letter may be used except I and O.
4. A fourth part of the designation (not indicated in this table) is separated by a dash from the foregoing parts and is used to indicate temper and other characteristics, such as F (as fabricated), 0 (annealed), H10 and H11 (slightly strain hardened), H23, H24, and H26 (strain hardened and partially annealed), T4 (solution heat treated), T5 (artificially aged only), T6 (solution heat treated and artificially aged), and T8 (solution heat treated, cold worked, and artificially aged). Thus, the complete designation may appear as: AZ91C-T6 for an aluminum-zinc-magnesium alloy containing 9% Al, 1% zinc, C indicating that this is the third alloy standardized with the same percentages of Al and Zn and T6 indicating that the alloy is solution treated and artificially aged.

TABLE 2. MAGNESIUM CASTING ALLOYS FOR AUTOMOTIVE APPLICATIONS

AM60B[2]	Die-casting alloy for uses needing toughness and ductility.		
	5.5–6.5% Al	0.25% Mn (min)	0.002% Ni (max)
	0.010% Cu max	0.22% Zn (max)	0.10% Si (max)
		0.005% Fe (max)	
AZ91D[2]	Provides an optimum combination of properties with die castability.		
	8.3–9.7% Al	0.15% Mn (min)	0.02% Ni (max)
	0.030% Cu (max)	0.35–1.0% Zn	0.10% Si (max)
		0.005% Fe (max)	
AZ91E[2]	A sand and permanent-mold casting alloy with properties and castability similar to AZ91B.		
	8.1–9.3% Al	0.17–0.5% Mn	0.0010% Ni (max)
	0.015% Cu (max)	0.40–1.0% Zn	0.20% Si (max)
		0.005% Fe (max)	
ZE41A	A sand and permanent mold casting alloy for applications to 175°C (350°F). Low microporosity and good pressure tightness.		
	0.0% Al	0.15% Mn (max)	0.40–1.0% Zr
	0.010% Cu (max)	0.75–1.75% Re	0.01% Ni (max)
		3.5–5.0% Zn	
ZC63	A proprietary sand and permanent-mold casting with properties similar to ZE41A, but less expensive.		
	0.0% Al	0.25–0.75% Mn	
	2.4–3.0% Cu	5.5–6.5% Zn	

desulfurize the melts, including pure nickel, nickel-chrome, and nickel-copper alloys.

Tin — Magnesium increases hardness and tensile strength. The effect of magnesium on tin can be dramatic. However, too much magnesium will reduce corrosion resistance and ductility.

Zinc — Magnesium improves dimensional stability and reduces the intergranular corrosion of zinc die castings. Magnesium refines the grain and increases hardness and creep strength of zinc sheet. Magnesium also is used in zinc-base bearing metals and in zinc alloy metalworking dies.

Magnesium alloy extrusions have become very popular for numerous items in recent years. Extrusion is particularly attractive as a parts making method — when extruded parts and sheet can be easily joined to form an assembly, where the desired shapes are too costly to machine from castings, and where pieces cut from extrusions can replace individually cast or forged parts. Final products with outstanding performance qualities coupled with light weight include concrete hand finishing tools, tennis racquets, portable shelters for the military, snowshoes, and improved luggage, among others.

The use of magnesium composites has become popular for rotary engine parts. Rotary engines remain attractive for business aircraft, boats, industrial equipment and compressors, and well over a million rotary-engine-powered cars have been built. In a research program (NASA Lewis Research Center) rotary engine parts are made from graphite-fiber-reinforced magnesium. An AZ91 C magnesium alloy is reinforced by 30% (vol) graphite fibers.

Progress has been in the early 1990s toward the development of metal-matrix composites (MMCs) that blend liquid magnesium alloys with ceramic particles, such as silicon carbide (SiC) and alumina (Al_2O_3). The method is similar to methods that have been developed for aluminum composites in that blending is accomplished by way of a high-shear process. Major differences of the new process result from increased general reactivity of magnesium and the difference in surface chemistry between the Al-SiC and Mg-SiC systems.

The particulate-reinforced MMCs are lightweight and demonstrate a significant increase in modulus and tensile strength at both ambient and elevated temperatures of the unreinforced material. The process was announced in late 1992 by Magnesium Elektron Ltd., Manchester, U.K.

Chemistry and Compounds

The behavior of magnesium is intermediate between that of beryllium and the higher alkaline earths. While it reacts readily with halogens, oxygen, and sulfur to form halides, oxide, and sulfide, it reacts with cold water only when the formation of protective oxide is prevented by amalgamation. All its compounds are divalent. Its oxide does not react with water to form the hydroxide, and it does not normally form a peroxide. Its major difference from the higher elements of the group is its much greater number of complexes. Anhydrous magnesium halides, especially, combine easily with many oxygen-functional organic compounds to form addition compounds. These reactions usually suggest covalent or dative bonding (both electrons from the oxygen) of the magnesium. Magnesium salts often form amines and amine complexes, though these are less stable than beryllium complexes. Magnesium also forms some basic salts, and many more of its salts are hydrated than are those of the higher alkaline earths. The metal reacts with alkyl and aryl halides to form the Grignard reagents, through which many organic reactions are conducted. The Grignard reagents themselves form complexes with ethers, tertiary amines, tertiary phosphines and many other type compounds. See also **Grignard Reactions.**

Important compounds of magnesium include the following:

Magnesium Acetate. [CAS: 142-72-3]. Anhydrous magnesium acetate, a white, crystalline, deliquescent solid, occurs in two forms: α-$Mg(C_2H_3O_2)_2$, formed by the reaction of MgO and concentrated acetic acid (13–33%) in boiling ethyl acetate, and β-$Mg(C_2H_3O_2)_2$ which is formed using 5–6% acetic acid. Of commercial interest is magnesium acetate tetrahydrate [CAS: 16674-78-5], $Mg(C_2H_3O_2)_2 \cdot 4H_2O$, a colorless to white crystalline solid obtained from aqueous solution. The tetrahydrate is the only stable phase below 68°C, the transition point of the anhydrous salt. A monohydrate [CAS:60582-92-5], $Mg(C_2H_3O_2)_2 \cdot H_2O$, can be prepared from the reaction of MgO and acetic acid in slightly hydrated isobutyl alcohol.

Magnesium acetate is hygroscopic and should be stored in a cool, dry place. Personal protective equipment to be used when handling magnesium acetate includes chemical safety goggles, chemical resistant gloves, and a NIOSH/MSHA approved respirator. To keep exposure to respirable dust to a minimum, mechanical exhaust is required. Although magnesium acetate is a relatively low hazard chemical, intravenous poisoning can occur if this material is not handled properly. Magnesium acetate is incompatible with strong oxidizers. When heated to decomposition, acrid smoke and irritating fumes may evolve.

The largest use for magnesium acetate is in the production of rayon fiber, which is used for cigarette filter tow. Magnesium acetate also has uses as a dye fixative in textile printing, as a deodorant, disinfectant, an antiseptic in medicine, and as a reagent chemical.

Magnesium Acetylacetonate. [CAS: 14024-56-7], $Mg(C_5H_7O_2)_2$, crystalline powder, slightly soluble in water, resistant to hydrolysis, a chelating nonionizing compound.

Magnesium Alkyls. Magnesium alkyl compounds RMg, RMgR, or RMgR', along with other compounds are useful as polymerization catalysts. These compounds should not be confused with alkyl magnesium halides or the much discussed ether solvated Grignard reagents. Magnesium alkyls may, however, be prepared from Grignard reagents. See also **Grignard Reactions.**

Magnesium alkyls are white, crystalline, pyrophoric solids that react vigorously with water, alcohols, and other compounds containing an active hydrogen. Magnesium alkyls, soluble in ether solutions but insoluble in benzene and some alkane solutions, decompose at 170–200°C. The molecular weights of unsolvated compounds fall in the range of 100–200, but the molecular weights in solution, as determined by cryoscopic methods, are in the range of 1,000–10,000. The low solubility and high molecular weights in solution are attributed to extensive association resulting from the electron-deficiency of the magnesium.

Magnesium alkyls are used as polymerization catalysts for alpha-alkenes and dienes, such as the polymerization of ethylene, and in combination with aluminum alkyls and the transition-metal halides. Magnesium alkyls have been used in conjunction with other compounds in the polymerization of alkene oxides, alkene sulfides, acrylonitrile(qv) and polar vinyl monomers. Magnesium alkyls can be used as a liquid detergents. Also, magnesium alkyls have been used as fuel additives and for the suppression of soot in combustion of residual furnace oil.

Magnesium Amide. $Mg(NH_2)_2$, whitish to gray crystals, d1.40, decomposes when heated, formed by reaction of magnesium and ammonia under elevated pressure. Use: Catalyst for polymerization.

Magnesium Arsenate. (arsenic acid, magnesium salt), [CAS: 10103-50-1], $Mg_3(AsO_4)_2 \cdot xH_2O$, white powder, when pure it is insoluble in water

Magnesium Ammonium Arsenate. $MgNH_4AsO_4$, white precipitate, solubility 0.0013 molar, formed by reaction of soluble magnesium salt solution and sodium arsenate in the presence of excess ammonium hydroxide, and upon igniting yields magnesium pyroarsenate, $Mg_2As_2O_7$, white solid.

Magnesium Benzoate. $Mg(C_7H_5O_2)_2 \cdot 3H_2O$, white crystalline powder, loses $3H_2O$ at 110C, mp approximately 200C, soluble in water and alcohol.

Magnesium Borate. $3MgO \cdot B_2O_3$ (orthoborate) or $Mg(BO_2)_2 \cdot 8H_2O$ (metaborate) transparent, colorless crystals or white powder, soluble in alcohol, acetic acid and inorganic acids, slightly soluble in water, formed by heating magnesium oxide and boric anhydride. Magnesium borate is used as a preservative, antiseptic in medicine, and fungicide.

Magnesium Boride. [CAS: 12007-25-9], Mg_3B_2, brown solid, by reaction of boron oxide and magnesium powder ignited.

Magnesium Bromide. [CAS: 7789-48-2], $MgBr_2 \cdot 6H_2O$, white solid, soluble, formed by reaction of magnesium carbonate and hydrobromic acid. Magnesium bromide is found in seawater, some mineral springs, natural brines, inland seas and lakes such as the Dead Sea and the Great Salt Lake, and salt deposits such as the Stassfurt deposits. In seawater, it is the primary source of bromine. By the action of chlorine gas upon seawater or seawater bitterns, bromine is formed.

Magnesium Bromide Hexahydrate. [CAS: 13446-53-2], $MgBr_2 \cdot 6H_2O$, which crystallizes from an aqueous solution at temperatures above $0\,°C$, is highly hygroscopic and isomorphous with $MgCl_2 \cdot 6H_2O$. It is also formed by the reaction of magnesium carbonate and hydrobromic acid. The solubility of magnesium bromide is 101 g/ 100 mL of water at $20\,°C$; the solubility of the hexahydrate is 160 g/ 100 mL of 95% ethanol at $20\,°C$.

Magnesium bromide is soluble in alcohols and forms addition compounds with numerous organic substances such as alcohols. The hexamethanolate, $MgBr_2 \cdot 6CH_3OH$, and the ethanolate, $MgBr_2 \cdot C_2H_5OH$, both exist. By gradually adding bromine to a cold mixture of magnesium powder and dry ether, the dietherate of magnesium bromide diethylether [CAS: 17950-53-7], $MgBr_2 \cdot 2C_4H_{10}O$, is formed. Other compounds form with ammonia. For example, the compound magnesium bromide hexammoniate [CAS: 75198-46-8], $MgBr_2 \cdot 6NH3$, is easily prepared from anhydrous $MgBr_2$ and ammonia gas. Thermal decomposition of the hexammoniate yields the diammoniate [CAS: 75198-47-9], $MgBr_2 \cdot 2NH_3$, and monoammoniate [CAS: 75198-48-0], $MgBr_2 \cdot NH_3$.

Magnusium bromide is used in medicine as a sedative in treatment of nervous disorders, in electrolyte paste for magnesium dry cells, and as a reagent in organic synthesis reactions.

Magnesium Carbonate. $MgCO_3$, white solid, K_{sp} 4.0×10^{-5}, formed by reaction of soluble magnesium salt solution and sodium carbonate or bicarbonate solution. Present in carbonate minerals and rocks, magnesite (more or less pure magnesium carbonate), dolomite (magnesium-calcium carbonate mixtures), dolomitic limestone. When ignited yields magnesium oxide and CO_2; when treated with acids yields the corresponding magnesium salt and CO_2, but with carbonic acid yields soluble magnesium bicarbonate. Magnesium bicarbonate, $Mg(HCO_3)_2$, colorless solution, by reaction of magnesium carbonate and carbonic acid, yields, upon boiling, magnesium carbonate and CO_2; magnesium ammonium carbonate, $(MgCO_3 \cdot NH_4)_2CO_3 \cdot 4H_2O$, white precipitate (soluble in ammonium chloride solution) by reaction of soluble magnesium salt solution and excess ammonium carbonate. Uses: Magnesium salts, heat insulation and refractory, rubber reinforcing agent, inks, glass, pharmaceuticals, dentrifice and cosmetics, free-running table salts, antacid, making magnesium citrate, filtering medium.

Magnesium Chlorate. [CAS: 10326-21-3]. $Mg(ClO_3)_2 \cdot 6H_2O$, white powder; very hygroscopic. D 1.8, mp 35C (decomposes at 120C), soluble in water; slightly soluble in alcohol. Uses: Defoliant, desiccant.

Magnesium Chloride. [CAS 7786-30-3]. $MgCl_2$, is one of the primary constituents of seawater and occurs in most natural brines and salt deposits formed from the evaporation of seawater. It occurs infrequently in nature as the mineral bischofite [CAS: 13778-96-6], $MgCl_2 \cdot 6H_2O$. Large deposits of oceanic origin contain the mineral carnallite [CAS: 1318-27-0], $KCl \cdot MgCl_2 \cdot 6H_2O$. Magnesium chloride, one of the most commercially important magnesium compounds, is available in the anhydrous and

hexahydrate forms. Both are deliquescent and form saturated solutions on standing in a moist atmosphere.

Anhydrous magnesium chloride is soluble in lower alcohols. In 100 g of methanol, its solubility is 15.5 g at $0\,°C$ and 20.4 g at $60\,°C$. In ethanol, the solubility is 3.61 g at $0\,°C$ and 15.89 g at $60\,°C$. Upon cooling, anhydrous $MgCl_2$ forms addition compounds with alcohols of crystallization such as magnesium chloride hexamethanolate [CAS: 57467-93-0], $MgCl_2 \cdot 6CH_3OH$, and magnesium chloride hexaethanolate [CAS: 16693-00-8], $MgCl_2 \cdot 6C_2H_5OH$. Both of these alcoholates are deliquescent.

Magnesium chloride forms double salts with potassium and ammonium chlorides. Carnallite is an important source of $MgCl_2 \cdot 6H_2O$. Magnesium chloride hexammoniate [CAS: 68374-23-2], $MgCl_2 \cdot 6NH_3$, can be formed by the reaction of anhydrous magnesium chloride and ammonia gas in a closed system, or by the action of ammonia on an aqueous solution of $MgCl_2$ and ammonium chloride, NH_4Cl, upon subsequent cooling to. Thermal decomposition of $MgCl_2 \cdot 6NH_3$ yields the diammoniate [CAS: 68374-24-3], $MgCl_2 \cdot 2NH_3$, and the monoammoniate [CAS: 68374-25-4], $MgCl_2 \cdot NH_4$.

The largest use for magnesium chloride brine is as a suppressant for dust on dirt roads, construction sites, unpaved parking lots, mines, and quarries. A corrosion inhibitor may be added to the brine to reduce corrosion on structures, such as steel surfaces, that are associated with the sites where the brine is used. The inhibitor forms a protective coating so that the brine will not corrode metal surfaces. Magnesium chloride brine also may be used to melt ice on road surfaces, sometimes in conjunction with an abrasive such as sand. Brines also have applications in oil-well completion fluids, as a component of some herbicides, and in regeneration of ion-exchange resins.

Magnesium chloride hexahydrate is almost exclusively used for melting ice. It is used in conjunction with, or in place of, salt for removal of ice and snow from sidewalks and roadways. Magnesium chloride has a lower freezing point than salt and is generally less corrosive to asphalt and cement, but it is significantly more expensive. In most cases, salt is used as the major melting agent for ice and snow, but if the surface that is being treated is expensive to maintain, the additional cost for magnesium chloride can be justified.

Another important use of magnesium chloride is in the preparation of oxychloride cements, $5Mg(OH)_2 \cdot MgCl_2 \cdot 8H_2O$, for flooring (nonsparking), wall plaster compositions, fire-resistant panels, fireproofing of steel beams, and grinding wheels. These cements (known as Sorel cements) are vermin resistant flooring cements used in industrial buildings. The cements are produced on-site by adding a 20% solution of $MgCl_2$ in water and a dry mix consisting of magnesium oxide, fillers (eg, wood fiber), and fine aggregates. After a few hours of setting, the cements form a dense but smooth-textured stonelike product

Magnesium chloride is also used in the processing of sugar beets and textiles, in water treatment, as a fireproofing agent for wood, as a dust control agent in mines and on haul roads, as an ingredient of floor-sweeping compounds, refrigeration brines, and fire-extinguishing agents.

Anhydrous magnesium chloride, along with magnesium bromide and magnesium iodide, is used in a process for producing organometallic compositions such as alkyllithium compounds used as reagents in the preparation of pharmaceuticals and special chemicals. Molten magnesium chloride has been used in the preparation of pure crystalline ceramic powders such as crystalline cordierite, forsterite, enstatite, and spinel. The introduction of $MgCl_2$-supported $TiCl3$ (Ziegler-Natta) catalysts has changed the manufacturing technology of polypropylene because of greatly enhanced catalytic productivity. Iso as a catalyst, $MgCl_2 \cdot 6H_2O$ is used in finishing to increase strength properties of cotton fabric

Magnesium Chromate. [CAS: 13423-61-5], $MgCrO_4 \cdot 5H_2O$, small readily soluble, yellow crystals, formed by reaction of magnesium carbonate and chromic acid solution. Use: Since it does not produce a fusible alkaline residue when thermally decomposed, it is used as a corrosion inhibitor in the water coolant of gas turbine engines. Insoluble basic magnesium chromates also are available. Their potential applications are in the treatment of light metal surfaces.

Magnesium Citrate. [CAS: 144-23-0] $Mg_3(C_6H_5O_7)_2.$, white solid, soluble, formed by reaction of magnesium carbonate and citric acid.

Magnesium Fluoride. [CAS: 7783-40-6]. MgF_2, is a fine white crystalline powder with low chemical reactivity. This relative inertness makes possible some of its uses, eg, stable permanent films to alter light transmission properties of optical and electronic materials. The reaction with sulfuric acid is so sluggish and incomplete that magnesium fluoride is not a suitable substitute for calcium fluoride in manufacturing hydrogen fluoride. Magnesium fluoride resists hydrolysis to hydrogen fluoride up to 750°C (61). Bimetallic fluorides, such as $KMgF_3$ [CAS: 28042-61-7], are formed on fusion of MgF_2 alkali metal and ammonium fluorides. MgF_2 is birefringent and only mildly affected by high energy radiation, making possible optics for the uv region.

Established uses of magnesium fluoride are as fluxes in magnesium metallurgy and in the ceramics industry. A proposed use is the extraction of aluminum from arc-furnace alloys with Fe, Si, Ti, and C. The molten alloy in reacting with magnesium fluoride volatilizes the aluminum and magnesium, which are later separated above the melting point of MgF_2. A welding flux for aluminum as well as fluxes for steel contains MgF_2.

Optical windows of highly purified magnesium fluoride which transmit light from the vacuum ultraviolet (140 nm) into the infrared are recommended for use as ultraviolet optical components for use in space exploration.

Magnesium Formate. $Mg(CHO_2)_2 \cdot 2H_2O$, colorless crystals, soluble in water; insoluble in alcohol and ether; combustible. Used in Analytical chemistry.

Magnesium Gluconate. $Mg(C_6H_{11}O_7)_2 \cdot 2H_2O$, white powder or fine needles, odorless, almost tasteless, soluble in water, combustible, formed by reaction of magnesia or magnesium carbonate dissolved in gluconic acid. Uses: Medicine, and vitamin tablets.

Magnesium Hydroxide. [CAS: 1309-42-8] $Mg(OH)_2$, occurs naturally as the mineral brucite [CAS: 1317-43-7]. Brucite, usually found as a low temperature, hydrothermal vein mineral associated with calcite, aragonite, talc, or magnesite, appears as a decomposition product of magnesium silicates associated with serpentine, dolomite, magnesite, and chromite. Brucite also occurs as a hydrated form of periclase, and is found in serpentine, marble, chlorite schists, and in crystalline limestone. At one time brucite was recovered commercially from deposits at Wakefield, Quebec and Nye County, Nevada; both operations have since ceased.

Magnesium hydroxide is produced from aqueous solutions of magnesium salts. To precipitate and recover magnesium hydroxide from solutions of magnesium salts, a strong base is added. The more commonly used base is calcium hydroxide [CAS: 1305-62-0], derived from lime [CAS: 1305-78-8], CaO, or dolime [CAS: 50933-69-2], CaO·MgO. Lime and dolime are calcination products of limestone and dolomite, respectively. See also **Lime and Limestone**.

The principal use of magnesium hydroxide is in the pulp and paper industries. The main captive use is in the production of magnesium oxide, chloride, and sulfate. Other uses include, ceramics, sugar refining, pharmaceuticals (antacid, laxative), plastics, flame retardants/smoke suppressants, residual fuel oil additive, sulfate pulp, uranium processing, dentrifrices, in foods as frying agent, color retention agent, frozen desserts, and the expanding environmental markets for wastewater treatment and So_x removal from waste gases.

Magnesium Hypophosphite. $Mg(H_2PO_2)_2 \cdot 6H_2O$, white solid, soluble, formed by reaction of magnesium carbonate and hypophosphorous acid.

Magnesium Iodide. [CAS: 10377-58-9], can exist as two deliquescent and heat-sensitive compounds: the octahydrate [CAS: 7790-31-0], $MgI_2 \cdot 8H_2O$, and the hexahydrate [CAS: 75535-11-4], $MgI_2 \cdot 6H_2O$. Soluble in alcohols and many other organic solvents, and forms numerous addition compounds with alcohols, esters, aldehydes, esters, and amines. One example is magnesium iodide dietherate [CAS: 29964-67-8], $MgI_2 \cdot 2C_4H_{10}O$, prepared by gradual addition of iodine to a mixture of magnesium and dry ether. Magnesium iodide dietherate, which occurs as white, needle-like crystals, is very hygroscopic and becomes yellowish after several hours, and then brown after a day because of separation of iodine. The action of water upon magnesium iodide dietherate leads to the formation of the octahydrate salt, $MgI_2 \cdot 8H_2O$.

Magnesium iodide is used in the deoxygenation of oxiranes into olefins and iodine. Anhydrous MgI2 is used in a process for producing organometallic and organobimetallic compositions, which are important in the preparation of pharmaceutical and special chemicals.

Magnesium Methoxide. (magnesium methylate), [CAS: 27428-49-5] $(CH_3O)_2Mg$, colorless, crystalline solid, decomposes on warming, formed

by reaction of magnesium and methanol. Uses: Dielectric coatings, a cross-linking agent to form stable gels, and catalyst.

Magnesium Lactate. $Mg(C_3H_5O_3)_2 \cdot 3H_2O$, white solid, soluble, formed by reaction of magnesium carbonate and lactic acid.

Magnesium Molybdate. [CAS: 13767-03-8] $MgMoO_4$, crystalline powder, soluble in water. Use: Electronic and optical applications.

Magnesium Nitrate. [CAS: 10377-60-3]. Anhydrous magnesium nitrate $Mg(NO_3)_2$, is very difficult to isolate, white crystals, D1.45, mp 95-100C, decomposes at 330C, soluble in water and alcohol, deliquescent. The commercial product is the deliquescent hexahydrate [CAS: 13446-18-9], $Mg(NO_3)_2 \cdot 6H_2O$.

Magnesium nitrate is prepared by dissolving magnesium oxide, hydroxide, or carbonate in nitric acid, followed by evaporation and crystallization at room temperature. Impurities such as calcium, iron, and aluminum are precipitated by pretreatment of the solution with slight excess of magnesium oxide, followed by filtration. Most magnesium nitrate is manufactured and used on site in other processes.

A soluble form of magnesium nitrate is used as a fertilizer in states such as Florida, where drainage through the porous, sandy soil depletes the magnesium. Used as a prilling aid in the manufacture of ammonium nitrate and in Pyrotechnics. Another use is as an alternative to sulfuric acid in the purification of nitric acid.

Magnesium Nitride. [CAS: 60195-15-5]. Mg_3N_2, yellow solid, with moist air or water yields ammonia and magnesium hydroxide, formed by heating magnesium to a high temperature in nitrogen or NH_3 (hydrogen gas evolved).

Magnesium Oleate. $Mg(C_{18}H_{33}O_2)_2$, yellowish mass, soluble in linseed oil, hydrocarbons, alcohol, and ether, insoluble in water, combustible, formed by reaction of soluble magnesium salt solution and sodium oleate. Uses: Varnish driers, in dry-cleaning solvents (to prevent spontaneous ignition), emulsifying agent, and lubricant for plasticizers.

Magnesium Oxalate. [CAS: 547-66-0]. $MgC_2O_4.2H_2O$, white solid, insoluble, K_{sp} 8.6×10^{-5}, formed by reaction of soluble magnesium salt solution and ammonium oxalate solution.

Magnesium Oxide. [CAS: 1309-48-4]. MgO, also known as magnesia, occurs in nature only infrequently as mineral periclase, most commonly as groups of crystals in marble. The principal commercial forms of magnesia are dead-burned magnesia (periclase), caustic-calcined (light-burned magnesia), hard-burned magnesia, and calcined dolomite. These materials are usually formed by the thermal decomposition or chemical reaction of various magnesium compounds including magnesite ore, magnesium hydroxide, magnesium chloride, and synthetic magnesium carbonate.

There are many processes for producing magnesium oxide. Martin Marietta Magnesia Specialties, Inc.,http://www.magspecialties.com/Default.htm, mines dolomitic limestone in Woodville, Ohio. The limestone is calcined at a high temperature under controlled conditions to produce calcined dolomite or dolime [CAS: 50933-69-2] which upon reaction with magnesium chloride-rich brine produces magnesium hydroxide and calcium chloride. The insoluble magnesium hydroxide is then separated from the liquid calcium chloride carrier and calcined under controlled conditions. The various grades of magnesia range from very reactive light-burned to nonreactive dead-burned.

Another process, in use globally, involves the mining, crushing, sizing, and subsequent calcination of natural magnesite. The chemical purity of the magnesia produced is dependent on the mineralogical composition of the natural magnesite. This magnesia is often less pure than magnesia produced by other processes.

The seawater process used by American Premier, National Magnesia Chemicals, and others, involves decarbonating limestone or dolomite to the point where all CO_2 is removed without converting the resulting magnesia to a chemically inactive form. Reaction of filtered seawater, treated to remove bicarbonate and/or sulfate, and dolime is followed by seeding with magnesium hydroxide to promote crystal growth. Upon formation of magnesium hydroxide, flocculants are added and the magnesium hydroxide precipitate is allowed to settle while the spent seawater is disposed to the sea. The precipitate is washed, filtered, and dried to obtain magnesium hydroxide, which is calcined to produce light-burned, hard-burned, or dead-burned magnesium oxide.

Dead Sea Periclase Ltd., on the Dead Sea in Israel, http://www.periclase. com/ uses yet another process to produce magnesium oxide. A concentrated magnesium chloride brine processed from the Dead Sea is sprayed into a reactor at about 1700 °C. The brine is thermally decomposed into magnesium oxide and hydrochloric acid. To further process the magnesia, the product is slaked to form magnesium hydroxide which is then washed, filtered, and calcined under controlled conditions to produce a variety of MgO reactivity grades.

Uses: Refractories, especially for steel furnace linings, polycrystalline ceramic for aircraft windshields, electrical insulation, pharmaceuticals and cosmetics, inorganic rubber accelerator, oxychloride and oxysulfate cements, paper manufacture, fertilizers, removal of sulfur dioxide from stack gases, adsorption and catalysis, semiconductors, and food and feed additive.

Magnesium Peroxide. [CAS: 14452-57-4]. MgO_2, white solid, insoluble in water, soluble in dilute acids with formation of hydrogen peroxide, formed by reaction of soluble magnesium salt solution and sodium or barium peroxide.

Magnesium peroxide is used mainly in medicine for treating hyperacidity in the gastric intestinal tract, and in the treatment of metabolic diseases such as diabetes and ketonuria. It is also used in the preparation of toothpaste and antiseptic ointments. All of these uses involve a mixture of magnesium peroxide, magnesium oxide, magnesium hydroxide, and an admixture of magnesium carbonate. Magnesium peroxide is also used in bleaching and agricultural applications. See also **Bleaching Agents**.

Magnesium Ammonium Phosphate. $MgNH_4PO_4$, white precipitate, K_{sp} 2.5×10^{-12}, by reaction of soluble salt solution and sodium phosphate in the presence of excess ammonium hydroxide, upon igniting yields magnesium pyrophosphate, $Mg_2P_2O_7$, white solid.

Magnesium Phosphate, Dibasic. (dimagnesium orthophosphate; dimagnesium phosphate; magnesium phosphate, secondary; magnesium hydrogen phosphate), [CAS: 7782-75-4]. $MgHPO_4 \cdot 3H_2O$, white, crystalline powder, D 2.13, loses water at 205C, decomposes at 550-650C, decomposes to pyrophosphate on heating, soluble in dilute acids, slightly soluble in water. Uses: Stabilizer for plastics, food additive, and medicine (laxative).

Magnesium Phosphate, Monobasic. (magnesium biphosphate; acid magnesium phosphate; magnesium tetrahydrogen phosphate), $MgH_4(PO_4)_2 \cdot 2H_2O$, white, hygroscopic, crystalline powder, decomposes to metaphosphate on heating, soluble in water and acids, insoluble in alcohol. Uses: Fireproofing wood, and as a stabilizer for plastics.

Mangnesium Phosphate, Tribasic. (Mangnesium phosphate, neutral; trimagnesium phosphate), $Mg_3(PO_4)_2 \cdot 8H_2O$ or $4H_2O$, soft, bulky, white powder, odorless, tasteless, loses all water at 400 C, soluble in acids, insoluble in water, formed by reaction of magnesium oxide and phosphoric acid at high temperatures. Uses: Dentifrice polishing agent, pharmaceutical antacid, adsorbent, stabilizer for plastics, food additive and dietary supplement.

Magnesium Salicylate. $Mg(C_7H_5O_3)_2 \cdot 4H_2O$, white solid, soluble in water and alcohol, formed by action of salicylic acid on magnesium hydroxide. Use: Medicine (antiinfective).

Magnesium Silicide. [CAS: 39404-03-0]. Mg_2Si, bluish crystals, mp 1085 C, d 1.9, decomposes on heating above 500 C, also by water and hydrochloric acid, formed by heating magnesium powder with silicon in ratio of 20:6. Uses: Semiconductor technology, and electrical equipment.

Magnesium Stannate. $MgSnO_3 \cdot 3H_2O$, white crystalline powder, decomposes at 340C, soluble in water. Use: Additive in ceramic capacitors.

Magnesium Stannide. Mg_2Sn, blue-white crystals, Mp 775C, soluble in water and dilute hydrochloric acid, has electrical and magnetic properties. Use: Semiconductor technology, magnetochemistry, thermoelectric research.

Magnesium Stearate. [CAS: 557-04-0]. $Mg(C_{18}H_{35}O_2)_2$ or with one H_2O, soft, white, light powder, tasteless, odorless, insoluble in water and alcohol. Uses: Dusting powder, lubricant in making tablets, drier in paints and varnishes, flatting agent, in medicines, stabilizer and lubricant for plastics, emulsifying agent in cosmetics, and dietary supplement.

Magnesium Sulfate. [CAS: 7487-88-9]. $MgSO_4$, is found widely in nature as either a double salt or as a hydrate, colorless crystals, very soluble in water, soluble in glycerol, sparingly soluble in alcohol. The more important mineral forms are: kieserite [CAS: 14168-73-1] $MgSO_4 \cdot H_2O$; starkeyite [CAS: 24378-31-2] $MgSO_4 \cdot 4H_2O$; pentahydrite [CAS: 15553-21-6] $MgSO_4 \cdot 5H_2O$; hexahydrite [CAS: 13778-97-7] $MgSO_4 \cdot 6H_2O$; epsomite [CAS: 10034-99-8] $MgSO_4 \cdot 7H_2O$; vanthoffite [CAS: 15557-33-2] $3Na_2SO_4 \cdot MgSO_4$; bloedite [CAS: 15083-77-9] $Na_2SO_4 \cdot MgSO_4 \cdot 4H_2O$; langbeinite [CAS: 13826-56-7] $K_2SO_4 \cdot 2MgSO_4$; leonite [CAS: 15226-80-9] $K_2SO_4 \cdot MgSO_4 \cdot 4H_2O$; schoenite [CAS: 15491-86-8] $K_2SO_4 \cdot MgSO_4 \cdot 6H_2O$; kainite [CAS: 67145-93-1] $4KCl \cdot 4MgSO_4 \cdot 11H_2O$; polyhalite [CAS: 15278-29-2] $K_2SO_4 \cdot MgSO_4 \cdot 2CaSO_4 \cdot 2H_2O$.

Magnesium sulfate forms many double salts, including naturally occurring minerals. The sulfuric acid double salts $MgSO_4 \cdot H_2SO_4$ [CAS: 10028-26-9], $MgSO_4 \cdot H_2SO_4 \cdot 3H_2O$ [CAS: 75198-53-7], and $MgSO_4 \cdot 3H_2SO_4$ [CAS: 39994-66-6] are crystallized from solutions of $MgSO_4$ in H_2SO_4. The amine double salts $MgSO_4 \cdot NH_3 \cdot 3H_2O$ [CAS: 75198-54-8], $MgSO_4 \cdot 2NH_3 \cdot 4H_2O$ [CAS: 75198-56-0], and $MgSO_4 \cdot 2NH_3 \cdot 2H_2O$ [CAS: 75198-55-9] are products of $MgSO_4 \cdot 7H_2O$ and gaseous ammonia.

Manufacture and Processing. Anhydrous $MgSO_4$ can be prepared only by dehydration of a hydrate. Crystallization from aqueous solution is not possible. Aqueous solutions of $MgSO_4$ can be prepared by dissolving MgO, $Mg(OH)_2$, or $MgCO3$ in sulfuric acid; or absorbing SO_2 using a $Mg(OH)_2$ slurry to form the soluble bisulfite, $Mg(HSO_3)2$, followed by air oxidation to SO^{-2}_4.

Technical-grade epsom salt is prepared by dissolving MgO, $Mg(OH)_2$, or $MgCO_3$ in sulfuric acid. The reaction mixture is crystallized to separate the product. In one process $MgSO_4$ solution is recycled from crystallizers to a reaction vessel containing sulfuric acid and low reactivity MgO. After pH adjustment to slightly acidic conditions and a 4–5 h reaction time, a 34% $MgSO_4$ mother liquor at 82 °C is produced. Iron is precipitated and insolubles are filtered from the mother liquor. Epsom salt is crystallized at 15 °C and screened; the 24% $MgSO_4$ filtrate is recycled. The epsom salt crystals are dried at low temperature in a rotary oven. Following filtration, the 34% mother liquor can be diluted to 24% and sold as a solution. The theoretical yield is 1 t of epsom salt per ton MgO. The actual yield depends on particle size, reactivity, and purity of the MgO. The heat of reaction is often the determining factor for using MgO or $Mg(OH)_2$ as the reagent. The $Mg(OH)_2$ reaction generates only 65–75% of the heat that the MgO reaction does.

To prepare a USP-grade epsom salt, higher purity MgO or $Mg(OH)_2$ is used. USP and food grades require low chloride levels, limiting allowable chloride content of the MgO to 0.08 wt %. Trace impurities including iron and aluminum are precipitated using excess MgO. Following crystallization, the epsom salt is washed free of mother liquor.

Natural and synthetic magnesium sulfate have a wide array of uses. The largest use for magnesium sulfate in all forms is for consumer goods. About 30% of magnesium sulfate is used in food additives and pharmaceuticals. Magnesium sulfate heptahydrate, epsom salts, is used for mineral baths and in medicine as a cathartic and analgesic soaking agent for bruises, sprains, localized inflammations, and insect bites. Magnesium sulfate is used as a micronutrient in some food products, and it is used in the production of high fructose corn syrup (HFCS). In the early 1980s, replacement of most or all of the sugar by HFCS in soft drinks was expected to increase the market for HFCS dramatically, and as a consequence, boost magnesium sulfate consumption by as much as 5% per year. Although some sugar has been replaced, the high estimates of growth for magnesium sulfate in this application did not materialize.

Animal feeds and fertilizers represent about 22% of the U.S. market for magnesium sulfate. Most applications for magnesium sulfate use synthetically produced material because of its higher purity. Purity requirements for animal feeds and fertilizers are not as stringent, so they use mainly the natural minerals, which are imported into the United States. The most effective way of preventing grass tetany is to provide magnesium to the pasture through fertilization. Magnesium also may be supplied in the form of epsom salts or kieserite that is added to the feed or drinking water.

Pulp and paper processing accounts for about 14% of magnesium sulfate use in the United States. Magnesium sulfate is used by kraft pulp mills that use oxygen delignification on soft woods, but it is also used in conjunction with sodium silicate to increase the life of hydrogen peroxide in oxygen-based bleaching processes. Miscellaneous use, which represent about 9% of magnesium sulfate demand, include textiles, matches, photographic solutions, rubber coagulation, refractory bonding agent in bricks, and oxysulfate cements.

Magnesium Sulfide. [CAS: 12032-36-9] MgS, red-brown, crystalline solid, decomposes above 2000C, decomposes in water. Use: Source of hydrogen sulfide, laboratory reagent.

Magnesium Sulfite. White, crystalline powder, D 1.725, mp loses $6H_2O$ at 200C, bp (decomposes), slightly soluble in water, insoluble in alcohol.

The white hexahydrate [CAS: 13446-29-2], $MgSO_3 \cdot 6H_2O$, is prepared by adding an excess of sulfur dioxide, SO_2, to a suspension of magnesium hydroxide, $Mg(OH)_2$, or basic magnesium carbonate, [CAS: 12306-51-3], $5MgO \cdot 4CO_2 \cdot 5H_2O$. The formation of magnesium bisulfite (magnesium hydrogen sulfite), $MgHSO_3$, unisolable in solid form, in the presence of excess SO_2 increases the solubility of magnesium sulfite in the liquid phase. In dilute solutions of both magnesium sulfite and magnesium bisulfite, the solubility of magnesium sulfite increases with increasing temperature independent of $MgHSO_3$ concentration. The basic salt $11MgSO_3 \cdot 2Mg(OH)_2 \cdot 22H_2O$ forms as dilute solutions of magnesium sulfite are heated.

Use: Manufacture of paper pulp (as bisulfite). See also **Pulp (Wood) Production and Processing**.

Magnesium Sulfonates. Magnesium sulfonates are detergents containing magnesium carbonate or magnesium complexes as the metallic portion, and an oil-soluble magnesium-based substrate, dispersed as a colloid in petroleum oil. By definition a soap is commonly the sodium or potassium salt of a high molecular weight fatty acid. The term metallic soap refers to substitution of another metal for the sodium, in this case, magnesium. Classification of detergents reflects their alkalinity. Magnesium sulfonates may be either neutral or overbased.

Principal uses of magnesium sulfonates are as additives to engine oils, automatic transmission fluids, gear oils and industrial oils. See also **Hydraulic Fluid**. In engine lubricating oils, in concentrations of 1–2%, the primary function is as a sludge dispersant and neutralizer of acidic contaminants from partially oxidized fuels, oil degradation products, and NOx. The noncarbonated forms may be used also in corrosion-resistant coatings for metals, and as liquid fuel additives, in smoke suppression and in vanadium scavenging

Mangnesium Tungstate. [CAS: 13573-11-0], (magnesium wolframate) $MgWO_4$, white crystals, D 5.66, soluble in acids, insoluble in water and alcohol, formed by interaction of solutions of magnesium sulfate and ammonium tungstate. Use: Fluorescent screens for X-rays, luminescent paint.

Magnesium Zirconium Silicate. $MgZrSiO_5$, or $MgO \cdot ZrO_2 \cdot SiO_2$, white solid, mp 1760 C, d 80 lb/ft^3, insoluble in water and alkalies, slightly soluble in acids. Use: Electrical resistor, ceramics, glaze opacifier.

Additional Reading

Avedesian, M.M. and H. Baker: *Magnesium and Magnesium Alloys*, ASM International, Materials Park, OH, 1999.

Davis, J.R.: *Metals Handbook*, 2nd Edition, ASM International, Materials Park, OH, 1998.

Greenwood, N.N. and A. Earnshaw: *Chemistry of the Elements*, 2nd Edition, Butterworth-Heiemann, Inc., Woburn, MA, 1997.

Kainer, K.U.: *Magnesium Alloys and Their Applications*, John Wiley & Sons, Inc., New York, NY, 2000.

Kainer, K.U.: *Magnesium Alloys and Technologies*, John Wiley & Sons, Inc., New York, NY, 2003.

Kainer, K.U.: *Magnesium: Proceedings of the 6th International Conference Magnesium Alloys and their Applications*, John Wiley & Sons, Inc., New York, NY, 2004.

Kaplan, H.I., J. Hryn, and B. Clow: *Magnesium Technology 2000: Proceedings of the Symposium Sponsored by the Light Metals Division of the Minerals, Metals and Materials Society (TMS) and the International Magnesium*, Warrendale, PA, 2000.

Krebs, R.E.: *The History and Use of Our Earth's Chemical Elements*, Greenwood Publishing Group, Inc., Westport, CT, 1998.

Lide, D.R.: *CRC Handbook of Chemistry and Physics*, 88th, Edition, CRC Press, LLC., Boca Raton, FL, 2007.

MAGNESIUM (In Biological Systems). Magnesium is an integral part of the molecule of chlorophyll, the green pigment in plants that absorbs solar energy. See also **Chlorophylls**. Magnesium deficiency is a fairly common cause of poor crop yields, especially among crops produced on sandy soils. Magnesium is a prosthetic ion in enzymes that hydrolyze and transfer phosphate groups. Hence it is essential for energy-requiring biological functions, such as membrane transport, generation and transmission of nerve impulses, contraction of muscles, and oxidative phosphorylation. See also **Phosphorylation (Oxidative)**. Magnesium is essential for the maintenance of ribosomal structure and thus protein synthesis. Magnesium may be related to the incidence of ischemic heart disease among Western populations.

The accumulation of magnesium from the soil by plants is strongly affected by the species of plant. The leguminous plants, such as clovers, beans, and peas, usually contain more magnesium than grasses, tomatoes, corn (maize), and other nonleguminous plants, regardless of the level of available magnesium in the soil where they grow.

A very high level of available potassium in the soil interferes with the uptake of magnesium by plants, and magnesium deficiency in plants is often found in soils that are very high in available potassium. High levels of available potassium may occur naturally, especially in soils of subhumid and semiarid regions; or they may be caused by heavy applications of certain commercial fertilizers or animal manure. On sandy and loamy soils, applications of magnesium fertilizers are often effective in increasing crop yields and the concentration of magnesium in the crop, but on fine-textured, clay-containing soils, especially those with substantial reserves of potassium, the application of a magnesium fertilizer may not cause higher magnesium concentration in crops. Since magnesium is not a highly toxic element in either plants or animals, precautions against its overuse are rarely necessary. When animals are fed diets primarily of grains, a proper balance among magnesium, calcium, and phosphorus should be maintained to minimize danger from urinary calculi.

The biological functions of magnesium, such as its essential role as a nutrier, its activation of enzyme systems, and its pharmacological properties, have been widely investigated. Nevertheless, some aspects of its critical physiological role remain obscure.

Distribution in System. Magnesium, primarily an intracellular ion, is distributed among all tissues. It constitutes about 0.05% of the animal body and, of this, 60% occurs in the skeleton and only 1% in extracellular fluids.

Reported serum magnesium values for most species range from 1.0 to 3.5 meg/liter, with a mean value of about 2. Between 65% and 80% of the plasma magnesium is ultrafilterable, and most of this exists as the free ion. The nonfilterable portion is reversibly bound to plasma protein. Cerebrospinal fluid contains slightly more than plasma. Interstitial fluid is similar to plasma ultrafiltrate.

The magnesium content of soft tissues varies from 0.06 to 0.13% of dry weight and remains remarkably constant regardless of the magnesium status of the animal. Normally, the intracellular concentration is more than 20 times that of the interstitial fluid, and the highest concentration occurs in the cell nucleus. Maintenance of such a large concentration gradient across the cell membrane suggests an active transport mechanism.

In late 1990's, R.R. Preston (University of Wisconsin–Madison) reported that recent reappraisals of the role of ionized magnesium in cell function suggests that many cells maintain intracellular free Mg^{2+} at low concentrations and that external agents can influence cell functions via changes in intracellular Mg^{2+} concentration. There is considerable evidence to suggest that intracellular free magnesium ions may be a key physiological regulator of cell activity.

The relatively large proportion of magnesium found in the skeleton, which amounts to about 0.6% of dry, fat-free bone, serves in part as a body reserve. It occurs largely as Mg^{2+} and $MgOH^+$ ions held by electrostatic attraction to the apatite crystal surface. During deficiency in young animals, 30% or more of bone magnesium can be mobilized for metabolic functions. Calcium ions appear to replace the magnesium that occupied the original adsorption sites.

Metabolism. The rate of absorption from the intestine exerts an important role in magnesium metabolism. Whereas in vitro studies show that magnesium absorption is positively correlated with the concentration of magnesium, it does not appear to be a purely passive process. Magnesium absorbed in excess of body needs is excreted primarily by way of the kidney. Urinary excretion is controlled primarily by a filtration-reabsorption mechanism so that magnesium appears in the urine only when glomerular filtration exceeds tubular reabsorption. Acute renal failure is accompanied by hypermagnesemia. In some species, considerable endogenous magnesium is lost by way of the feces, the amount depending upon the magnesium status of the animal and upon other dietary factors, such as the digestibility of the diet. The endogenous fecal magnesium in calves has been estimated at 3.5 milligrams/kilogram of body weight.

In contrast with the metabolism of calcium, no one endocrine gland exerts a primary regulatory function on magnesium. Thyro-parathyroidectomy in dogs causes only a temporary lowering of plasma magnesium. Adrenalectomy causes a rise, whereas hyper-aldosteronism produces a fall in the plasma level. Administrative of deoxycorticosterone or aldosterone to sheep lowers the magnesium concentration in plasma. Magnesium-deficient animals exhibit a higher metabolic rate than normal, and the toxic effect of excess thyroxine is partially overcome by increasing the dietary level of magnesium.

Function. Although magnesium activates isolated enzymes, in most cases an absolute requirement is difficult to establish because the enzymes are partially active without added magnesium. The stimulating effect is not always specific for magnesium. In some cases, manganese or calcium will also activate the system.

Magnesium is particularly concerned with enzyme-catalyzed reactions involving the cleavage of phosphate esters and the transfer of phosphate groups. Magnesium ions activate phosphatases and the phosphorylation reactions involving adenosine triphosphate (ATP). Among the latter group may be mentioned glucokinase, phosphoglucokinase, phosphofructokinase, myokinase, creatine transphosphorylase, arginine transphosphorylase, and flavokinase. It has been suggested that an ATP-Mg complex is the active substrate inasmuch as ATP forms a 1 : 1 complex with magnesium and maximal activation occurs when the ATP: Magnesium ration is 1. Alkaline phosphatases, pyrophosphateses, and ATPase are activated by magnesium, as are enolase, certain peptidases, and pyruvic oxidase. Since magnesium is tied to ATP utilization, it follows that magnesium plays a role in important metabolic processes, including the synthesis of protein, fat, and nucleic acids, and in the trapping and utilization of energy derived from catabolism of carbohydrate and fat.

There is little change in magnesium concentration of soft tissues from deficient animals even at the point of expiration. This does not preclude the possibility that a small component of the cell, such as the nucleus or a cell particulate, is deprived of its critical level, but the dramatic drop in extracellular magnesium suggests that a function outside the cell is of greatest significance. It appears that tetany and convulsions in deficient animals result from a derangement of neuromuscular transmission. Magnesium ion possesses strong pharmacological properties, depressing both the central and peripheral nervous systems. These effects are counteracted by calcium. In the presence of normal calcium levels, a reduction of extra-cellular magnesium is believed to increase the release of acetylcholine and to decrease the rate of its hydrolysis. Such effects would increase the irritability of the neuromuscular system.

Magnesium generally has not been considered a major factor in bone formation and strength, but recent studies suggest closer attention be given to dietary levels of magnesium in this regard. Because of the close interrelationship with calcium, it is not surprising to see research findings of magnesium interfering with calcium entry into cells of the islets of the pancreas in studies of diabetes. The recognized presence of magnesium as part of numerous enzyme systems has led to observations of the reduction in carbohydrate metabolism associated with a deficiency and to beneficial effects in reducing blood cholesterol and lipids associated with other dietary agents when supplemental magnesium is added to the diet. The relationship to calcium also shows up in a study showing that adding magnesium to the rations of laying hens causes an increase in shell thickness, with a consequent reduction in the number of broken eggs.

Magnesium-Induced Diarrhea. It is well established that an excessive intake of magnesium causes diarrhea. This source of diarrhea is difficult to differentiate from other causes of diarrhea. Consequently, the diagnosis may be long and costly unless the physician questions a patient on possible excessive magnesium intake, which may result from large dosages of antacids or off-the-shelf food supplements.

Pathology of Magnesium Deficiency. Although there are numerous clinical symptoms, two cardinal aspects of pathology have been observed in all species of higher animals. These are hyper irritability and soft tissue calcification. While there are species differences as to the dominating syndrome, this is determined in part by the severity of the deficiency. Metastatic calcification is more likely to occur in a chronic deficiency in which the animal does not succumb at an early age. Hyper irritability, terminating in convulsions and death, has been observed in rat, rabbit, pig, calf, chick, and duck. Magnesium deficiency in humans is characterized by muscle tremors and twitching, often accompanied by delirium and occasionally by convulsions. The guinea pig, calf, dog, and cotton rat are prone to metastatic calcification and develop grossly visible deposits in and around joints, along the muscles of the rib cage, and also in the heart, great vessels, and other critical organs. Most soft tissues show an elevated ash content and marked histopathology. Some researchers have hypothesized that long-term intakes of marginal dietary levels of magnesium may be related to the incidence of ischemic heart disease.

The first clinical symptom of magnesium deficiency is a hypomagnesemia which occurs in cattle and less frequently in sheep and is described by such names as grass tetany, grass staggers, lactation tetany, and wheat pasture poisoning. It is observed most frequently when animals are first grazed on lush grass or wheat pastures. The disease is characterized by irritability, tetany, and convulsions, and all animals have subnormal plasma magnesium. Symptoms can be relieved by administration of magnesium salts and cam be prevented by providing extra magnesium in the diet.

Hypomagnesemia, often associated with hypocalcemia, is frequently encountered in heavy users of alcohol. Alcoholism sometimes is ascribed to impaired intestinal calcium absorption.

Nutritional Requirements and Dietary Supplementation. As is true of many mineral nutrients, the requirement for magnesium is affected by other dietary constituents, by the age and species of the animal, and by the criterion of adequacy applied. An allowance for magnesium has been included in the Recommended Dietary Allowance since 1968. Calcium and magnesium have an important effect upon magnesium availability. Either of these ions in excess increases the requirements for magnesium, and their effects are additive. Since calcium is known to compete with magnesium pharmacologically, it is reasonable to believe that it also competes with

TABLE 1. DISTRIBUTION OF MAGNESIUM IN VARIOUS FOOD GROUPS

Group (Types of Samples Tested)	Magnesium Concentration (Milligrams/100 grams (wet))	Magnesium-Calorie Ratio (Micrograms/Kilocalorie)
Milk products (cheeses, ice cream, milk, puddings)	6.8–25.7	18–198
Meat and meat alternates (chicken, dried beef, eggs, fish, sausage)	9.8–37.6	20–353
Vegetables (cabbage, carrot, onion, turnip)	6.7–20.6	196–1000
Breads and cereals (buns, cereals, cornbread, crackers, croutons, English muffins, pasta, taco shells)	10.6–126.0	27–325
Baked desserts (cakes, cookies, doughnuts, pastries, sweet rolls)	4.6–53.2	18–307
Candies	21.8–89.9	63–225

magnesium for absorption sites in the intestine. It is believed that phosphate decreases magnesium absorption by formation of insoluble magnesium phosphates, and excess of calcium aggravates the effect of creating a more alkaline intestinal medium. Excess magnesium can be considered toxic, but this effect is largely due to the induction of a calcium deficiency. Magnesium deficiency in humans generally has not been fully documented except in cases of predisposing and complicating disease states.

Distribution of magnesium in food classes is summarized in Table 1.

Additional Reading

Sigel, H. and A. Sigel: *Metal Ions in Biological Systems: Compendium on Magnesium and Its Role in Biology, Nutrition, and Physiology*, Marcel Dekker, Inc., New York, NY, 1990.

Theophanides, T.M. and J. Anastassopoulou: *Magnesium, Current Status and New Developments: Current Status and New Developments: Theoretical, Biological, and Medical Aspects*, Kluwer Academic Publishers, Norwell, MA, 1997.

Tsang, R.C.: *Calcium and Magnesium Metabolism in Early Life*, CRC Press, LLC., Boca Raton, FL, 1995.

Vedral, J.L.: *Dietary Reference Intakes: For Calcium, Phosphorus, Magnesium, Vitamin D, and Fluoride*, National Academy Press, Washington, DC, 1997.

Web References

NIH Clinical Center National Institutes of Health: http://www.cc.nih.gov/ccc/supplements/magn.html

International Magnesium Association: http://www.intlmag.org/

Magnesium Update Information: http://www.krispin.com/magnes.html

The Magnesium Home Page: http://www.members.tripod.com/Mg/

MAGNET (Superconductivity).   See **Superconductivity**.

MAGNETIC DECLINATION.   See **Isogonic Line**.

MAGNETIC DIPOLE.   In geomagnetism, either of the two points on the earth's surface where a free- swinging magnetic needle points in a vertical direction.

The line connecting these two points does not pass through the center of the earth. These two points constantly move at a slow rate. They are presently in northern Canada and in the Antarctic south of Australia.

See also **Dipole**.

MAGNETIC DOUBLE REFRACTION.   The splitting, into two components, of a radio wave traveling in a region of free electrons. This is due to the interaction of the Earth's magnetic field and the alternating field of the radio wave. Except for waves near the gyrofrequency, the components of the split wave, the ordinary ray and the extraordinary ray, will travel with slightly different velocities and be reflected at different heights. See **Magnetoionic Theory**.

MAGNETIC EQUATOR.   See **Aclinic Line**.

MAGNETIC FIELD.   A region of space wherein any magnetic dipole would experience a magnetic force or torque; often represented as the geometric array of the imaginary magnetic lines of force that exist in relation to magnetic poles.

MAGNETIC FIELD INTENSITY.   The magnetic force exerted on an imaginary unit magnetic pole placed at any specified point of space. It is a vector quantity. Its direction is taken as the direction toward which a north magnetic pole would tend to move under the influence of the field. If the force is measured in dynes and the unit pole is a cgs unit pole, the field intensity is given in *oersteds*. Also called *magnetic intensity, magnetic field*, or *magnetic field strength*. Prior to 1932 the *oersted* was called the *gauss*; but the latter term is now used to measure magnetic induction (within magnetic materials), whereas *oersted* is reserved for magnetic force.

By definition, one magnetic line of force per square centimeter (in air) represents the field intensity of 1 *oersted*.

MAGNETIC FLOWMETER.   See **Flow Measurement (Liquids and Gases)**.

MAGNETIC INDUCTION.   A vector field, usually denoted by **B**, defined as follows. The torque **N** experienced by a magnetic dipole with magnetic dipole moment **m** is

$$\mathbf{N} = \mathbf{m} \times \mathbf{B}.$$

Thus by measuring **N** for **m** oriented in two orthogonal directions, the magnetic induction components are obtained as torque components divided by the magnitude of **m**. The fundamental relation linking electric field **E** and magnetic induction **B** to the force on a charge q with velocity **v** is the Lorentz force equation

$$\mathbf{F} = q(\mathbf{E} + \mathbf{v} \times \mathbf{B}).$$

Magnetic induction is sometimes called magnetic field, a term usually applied to a different field **H**, related to **B** but different from it. In free space, **B** and **H** are proportional:

$$\mathbf{H} = \frac{\mathbf{B}}{\mu_0},$$

where μ_0, the permeability of free space, is a universal constant. **B** is the primitive field, whereas **H** is secondary, not strictly needed but convenient. Care must be exercised in deciding if, by magnetic field, **B** or **H** is meant. What is usually meant by the electric and magnetic fields (or the electromagnetic field) are **E** and **H**, although according to the Lorentz force equation **E** and **B** are the fundamental fields. Moreover, the Lorentz transformation preserves the (**E**, **B**) structure but not the (**E**, **H**) structure.

AMS

MAGNETIC INK CHARACTER RECOGNITION (MICR).   Developed primarily as the common machine language for bank check handling, the early work on this system dates back to the early 1950s, when the American Bankers Association commenced writing specifications for a suitable system to be used with the then rapidly developing electronic business machines. The system is used by nearly all banking institutions.

As shown in Fig. 1, specifically designed symbols are used for all numerals and zero. A few additional symbols for special coding are also shown. To take full advantage of the system, it was necessary to modify to some degree the conventional shapes of the numbers, but generally the numbers can be easily read with the naked eye, thus serving two purposes. The waveform for each number is shown below each numeral in the diagram and it will be noted that there is a marked contrast between each waveform, thus assuring reliability of reading accuracy.

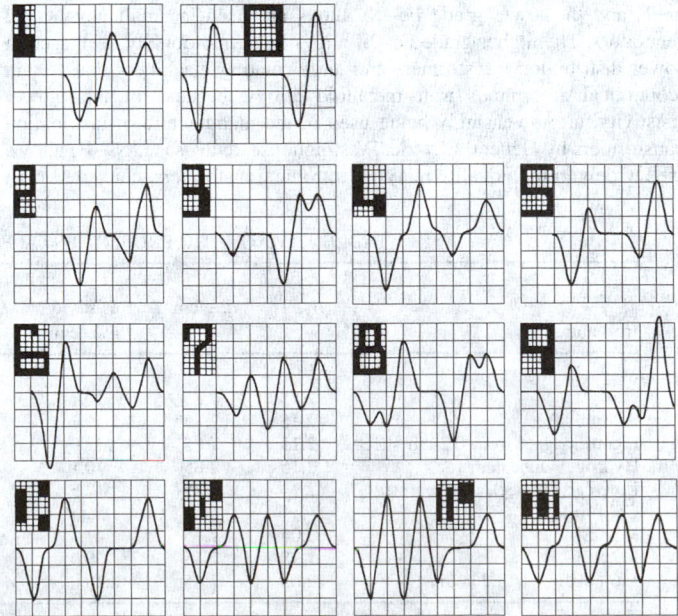

Fig. 1.   Magnetic ink numerals and associated electron waveforms. The symbols in lower row are used for special coding purposes.

As the MICR-coded documents pass a special reading head, the symbol is converted to its relevant waveform and with further electronic translation can be introduced into a data processing system.

Other systems seriously considered by the American Bankers Association before final selection of MICR included binary or bar codes, spot code (decimal system, fluorescent ink, magnetic bar code, perforations, and notches). The principal criteria against which each system was judged included accuracy, tolerances, printing practicability, customer acceptance, verification, cost, format, and resistance to mutilation and obliteration.

For some applications, optical character recognition (OCR) is preferred over MICR. Each approach has its advantages and disadvantages and sometimes selection of the most effective approach is quite difficult. See also **Optical Character Recognition (OCR).**

MAGNETIC K-INDICES.

An approximately logarithmic measure of geomagnetic disturbance activity based on the range of the most disturbed magnetic element during each 3-hour interval of the day. The K-indices are assigned integers from 0 to 9. The K-indices averaged over the observatories of the earth are called planetary indices K_p and divided into 28 grades.

MAGNETIC LUNAR DAILY VARIATION (symbol L).

A periodic variation of the Earth's magnetic field that is in phase with the transit of the moon. This variation is essentially a tidal effect. The amplitude of this variation changes with the phase of the moon, the seasons, and the sunspot cycle.

MAGNETIC MATERIALS: BULK.

All materials that are magnetized by, i.e., exhibit a response in magnetic field are magnetic materials and are classified according to the nature of the response, e.g., as ferromagnetic or ferrimagnetic, the latter typified by the ferrites. Most commercially important magnetic materials are ferromagnets and ferrimagnets. See also **Ferrimagnetism**; and **Ferrite**.

Soft Magnetic Materials

Soft magnetic materials are characterized by high permeability and low coercivity. There are six principal groups of commercially important soft magnetic materials: iron and low carbon steels, iron−silicon alloys, iron−aluminum and iron−aluminum−silicon alloys, nickel−iron alloys, iron−cobalt alloys, and ferrites. In addition, iron−boron-based amorphous soft magnetic alloys are commercially available. Table 1 summarizes the properties of some of these materials. Table 2 summarizes properties of some ferrites.

Because of low coercivity and high magnetic permeability, iron and low carbon steels tend to be used in static applications. Low carbon steels and the lower grade Fe−Si alloys are used in small motors and generators. The higher grade Fe−Si alloys have traditionally been used in power distribution transformers and large rotating machinery, but certain economical amorphous iron−metalloid alloys, because of their lower resistivity, are increasingly being used in the manufacture of distribution transformers by General Electric, Westinghouse, and Osaka. Ni−Fe alloys, used widely in high quality relays, electronic transformers, converters, and

inverters in the electronics industry, have much higher permeability and much lower resistivity than Fe−Si alloys. Soft ferrites (oxides) are suitable for high frequency applications. The Co−Fe alloys are used because of their higher saturation polarization (flux density) electrical resistivity and Curie temperature compared to the iron−nickel alloys, but have the disadvantage of poorer workability and higher cost. Thus, they are used in special applications.

Soft magnetic ferrites are oxides and they are electrical insulators. Because of their exceptionally higher resistivities, ferrites are particularly suitable for high frequency applications, of about 100,000 cycles (10 kHz).

Hard Magnetic Materials

Hard or permanent magnetic materials are characterized by high coercivity and high energy product. The important commercial hard magnetic materials are hard ferrites such as ferroxdure (Table 3), rare-earth (R)-cobalt alloys, and the ternary alloys based on $Nd_2Fe_{14}B$. The last exhibits the highest coercivities and energy products. The use of Alnico and the binary R−Co alloys has continually decreased because of the high cost of cobalt. These are being replaced by the ternary NdFeB materials, including $Nd_2Fe_{14}B$.

Hard ferrites are used widely in electromechanical devices, e.g., generators, relays, motors, and magnetos; electronic applications, e.g., loudspeakers, traveling-wave tubes, and telephone ringers and receivers; antitheft tags, holding devices such as door closers, seals, and latches; and are perennial favorites in various toy designs. Loudspeakers are the largest use of permanent magnets (ca 50%). Strontum ferrites exhibit higher coercivities and are increasingly being produced.

The commercial development of magnets based on $Nd_2Fe_{14}B$ boride proceeded rapidly, and they are now being used in many diverse applications, for example, servo devices for machine tools, for over 30 dc motors for fully equipped automobiles (windshield wipers, cooling fans, window and antenna lift motors, etc), magnetic resonance imaging (mri), computer disk drives, and medical device applications. Their largest use is in positioning motors for computer hard disk drives. New designs of electrical machines are now taking place.

See also **Magnetism.**

Additional Reading

Buschow, K.H.: *Handbook of Magnetic Materials*, Vol. 15, Elsevier Science, New York, NY, 2003.

Buschow, K.H. and F.R. De Boer: *Physics of Magnetism and Magnetic Materials*, Kluwer Academic Publishers, Norwell, MA, 2002.

Buschow, K.H. and E.P. Wohlfarth: *Handbook of Magnetic Materials: A Handbook on the Properties of Magnetically Ordered Substances*, Vol. 5, Elsevier Science, New York, NY, 1990.

Chatterji, T.: *Neutron Scattering from Magnetic Materials*, Elsevier Science & Technology Books, New York, NY, 2006.

Chen, C.W.: *Magnetism and Metallurgy of Soft Magnetic Materials*, North-Holland, New York, NY, 1977.

Herbst, J.F.: "Permanent Magnets," *Am. Sci.*, 251 (May−June 1993).

Heirich, B. and J.A.C. Bland, eds.: *Ultrathin Magnetic Structures*, Springer, Berlin, 1994.

Luborsky, F.E. ed.: *Amorphous Metallic Alloys*, Butterworths, London, 1983.

TABLE 1. MAGNETIC PROPERTIES OF FULLY ANNEALED IRON AND IRON ALLOYS

Iron and Alloys	B_s, T[b]	d, g/cm³	Resistivity, $\mu\Omega \cdot cm$	$H_c(B_m = 1$ T)[b] A/cm[a]	Permeability, A/cm[a] H = 0.8	H = 8	Core Loss (1.5 T[b], 60 Hz), W/kg[c] 0.35 mm	0.46 mm	0.64 mm
Magnetic ingot iron									
cast	2.15	7.85	10.7	0.68	3,500	1,500			
0.2-cm sheet	2.15	7.85	10.7	0.88	1,800	1,575			13.20
Electromagnet iron, 0.2-cm sheet	2.15	7.85	12.0	0.81	2,750	1,575			
Hydrogen-annealed iron	2.15	7.85	10.1	0.04	14,000	1,580			
Low carbon steel, decarburized	2.14	7.85	12.5	0.70	2,000	1,530	8.10	9.2	11.44
Cold-rolled									
M36 Si−Fe	2.04	7.75	41.0	0.36	7,400	1,485		3.85	4.73
M22 Si−Fe	1.98	7.65	49.0	0.31	8,100	1,450		3.63	4.29
M6 (110)[001] 3.2% Si−Fe	2.03	7.65	48.0	0.06	16,000	1,820	1.45		

[a] To convert A/cm to Oe, divide by 0.7958.
[b] To convert T to G, multiply by 10^4.
[c] At thickness shown.

TABLE 2. CHARACTERISTICS OF FERRITES

Property code	MnZn ferrites									NiZn ferrites		
	H5A	H5B	H5C2	H5E	H6F	H6H3	H6K	H7C1	H7C2	K5	K6A	K8
Practical frequency, MHz	<0.2	<0.1	<0.1	<0.01	0.2–2.0	0.01–0.8	0.01–0.3	<0.3	<0.2	<8	1–50	<200
Initial permeability, μ_0	3,300	5,000	10,000	18,000	800	1,300	2,200	2,500	3,900	290	25	16
Relative loss factor, $\tan\delta/\mu_i \times 10^6$, at (kHz)	<2.5 (10)	<6.5 (10)	<7.0 (10)		<17 (1,000)	<1.2 (100)	<3.5 (100)			<28 (1,000)	<150 (10,000)	<250 (100,000)
Temperature coefficient of $\mu_i \times 10^6$ from −30 to 20°C, $(\mu_2 - \mu_1/\mu_1^2(T_2 - T_1))$	−0.5 to 2.0	−0.5 to 2.0	−0.5 to 1.5	−0.5 to 2.0		0.3 to 2.0			0.4 to 1.2	−4.0 to 2.0		
Curie temperature, °C	>130	>130	>120	>115	>200	>200	>130	>230	>200	>280	>450	>500
Saturation flux density, T^a	0.41	0.42	0.40	0.44	0.40	0.47	0.39	0.51	0.48	0.33	0.30	0.27
Disaccommodation factor, $D \times 10^6$ (from 1–10 min), $(\mu_1 - \mu_2/\mu_1^2\log(t_2/t_1))$ where t = time	<3	<3	<1	<1	<12	<5	<2			<30	<20	
Resistivity, $\Omega \cdot m$ Applications	1	1	0.15 Transformers	0.05	4	25 Inductors	8	10	2 Power supplies	20×10^5	2.5×10^5 Inductors	1.0×10^5

^{a}To convert T to G, multiply by 10^4.

TABLE 3. MAGNETIC PROPERTIES OF COMMERCIAL PERMANENT MAGNET MATERIALS

Material	T_C, °C	$(BH)_{max}$, kJ/m^{3a}	B_r, T^b	H_c, kA/m^c
Ferroxdure (SrFe$_{12}$O$_{19}$)	450	36	0.42	250
Alnico 9	850	72	1.05	120
SmCo$_5$	724	144	0.87	600
Sm(Co$_{0.68}$Cu$_{0.10}$Fe$_{0.21}$Zr$_{0.01}$)$_{7.4}$	800	240	1.10	510
Nd$_2$Fe$_{14}$B	312	290	1.23	880

^{a}To convert kJ/m^3 to G · Oe, multiply by 12.57×10^4.
^{b}To convert T to G, multiply by 1×10^4.
^{c}To convert kA/m to Oe, divide by 7.958×10^{-2}.

O'Handley, R.C.: *Modern Magnetic Materials: Principles and Applications*, John Wiley & Sons, Inc., New York, NY, 1999.

Sharrock, M.P.: "Particulate Magnetic Recording Media: A Review," *IEEE Trans. Magn.*, **25**(6), 4374 (1989).

Spaldin, N.: *Magnetic Materials: Fundamentals and Device Applications*, Cambridge University Press, New York, NY, 2003.

Wolfarth, E.P. ed.: *Ferromagnetic Materials—A Handbook on the Properties of Magnetically Ordered Substances*, Vols. 1 and 2, Elsevier, New York, NY, 1980; E.P. Wolfarth and K.H.-J. Buschow, eds., Vols. 3 and 4, Elsevier, 1988.

MAGNETIC MATERIALS: THIN FILMS AND PARTICLES. The largest use of magnetic films and particles, in the form of tapes and disks for recording and retention of audio, visual, and digital information, is in memory and storage technologies. Price per bit of information, including the cost of the peripheral electronics, and performance, as denoted by access time, generally are used to characterize the various memory technologies. Power, modular capacity, reliability, nonvolatility, etc, are also factors describing the efficacy of memories. See also **Magnetic Tape Storage (Computer)**.

Magnetic Properties and Structure

The static or low frequency magnetic properties pertinent to thin-film materials generally are utilized to characterize magnetic materials. As a first approximation, these properties serve to suggest utility for device applications. Saturation magnetization M_s and Curie temperature T_C are intrinsic (structure insensitive) properties and are equal to the bulk values when thick films are made properly. For very thin highly paramagnetic films, such as those of platinum, sandwiched between ferromagnetic, e.g., Co, or antiferromagnetic, e.g., Cr, thin films, in the form of multilayered structures being developed for recording heads, a magnetization can be induced in the normally paramagnetic material. See also **Magnetic Materials: Bulk**. The surface area-to-volume ratio of the individual layers is so large that the atomic moments at the interfaces play an important role.

Fabrication

Fabrication methods include thermal evaporation, sputtering, magnetron sputtering, pulsed laser evaporation, molecular beam epitaxy, chemical vapor deposition, electrolytic and electroless deposition, and growth from solution.

Materials

Magnetic storage materials for storage of audio and video information as well as of digital data are in the form of tape and disks. There are two states of remanent magnetization for recording: longitudinal, in which the magnetization is in the plane of the recording medium; and perpendicular, in which the magnetization is normal to the plane. For particulate media in which the acicular submicronic particles are single domain and embedded in plastic, the magnetization is confined in the direction of the long dimension.

Multilayer materials exhibiting high magnetization and permeability are undergoing considerable research and development for advanced recording heads. The discovery of giant magnetoresistance in multilayered nano-thick magnetic materials is expected to become important for advanced read heads.

Particulate Materials. There are three principal classes of particulate magnetic materials: γ-ferric oxide, γ-Fe$_2$O$_3$, and its modifications;

TABLE 1. MAGNETIC PROPERTIES OF COMMON MAGNETIC RECORDING MEDIA

Material	B_r, T^a	H_c, kA/m^b
γ-Fe$_2$O$_3$	0.11	26
Fe$_2$O$_3$–Fe$_3$O$_4$	0.15	37
Co–γ-Fe$_2$O$_3$	0.15	52
CrO$_2$	0.15	45
BaFe$_{12}$O$_{19}$	0.12	64
Fe	0.30	120

^{a}To convert T to G, multiply by 10^4.
^{b}To convert kA/m to Oe, divide by 7.958×10^{-2}.

chromium dioxide, CrO$_2$; and iron. A comparison of the remanent magnetization, B_r, and coercivity, H_c, for several material systems is shown in Table 1.

Recording Heads. Materials that are suitable for read/write recording heads for tapes and disks are characterized by high saturation flux density, low remanent induction to avoid erasure of information when the writing current ceases, and low hysteresis and low eddy-current loss, particularly for high data rates or high frequency operation. In addition, because of the small air gap between the head and recording medium, the head material should be abrasion resistant. Dust particles and the magnetic attraction between head and tape can lead to abrasion.

For general-purpose audio recording, laminated Ni–Fe alloys exhibit the required high saturation and low remanence and eddy-current losses; moreover, abrasion is low. Head wear is improved by use of precipitation hardened material. The spinel structure oxides, manganese–zinc and nickel–zinc ferrites, exhibit good abrasion resistance and high frequency characteristics and in some cases are the preferred material despite their relatively low saturation.

For high quality audio and video recording where the recording medium is CrO$_2$ or Co impregnated γ-Fe$_2$O$_3$, sputtered Sendust alloy films (9.6 wt % Si, 5.4 wt % Al, balance Fe) and ferrites are used as head materials.

Thin-Film Magnetic Metallic Media. Advanced magnetic recording media are in the form of thin films. The metallic media are typically sputtered films having carbon overcoats for protection. Cobalt-based alloys have been developed for use as longitudinal, i.e., *c*-axis of the crystalline Co-alloy parallel to the plane of the substrate, magnetic recording media. See also **Cobalt**. Magnetic disks are presently fabricated alloys on NiP coated aluminum alloy disk substrates.

Magnetooptic Materials. The application of magnetooptic effects to optical memory systems, such as for laser beam writing and magnetooptic read, has been the subject of much research.

Memory systems based on laser writing and reading through the interaction of electromagnetic radiation, either through reflection utilizing the Kerr effect or by transmission utilizing the Faraday effect, have begun to appear in the marketplace.

The magnetic storage media being employed are ternary amorphous alloys composed of the rare-earth elements gadolinium, Gd, and terbium, Tb, with Fe and Co for use in the near infrared. These materials are compatible with GaAs-based lasers. These alloys are ferrimagnetic.

The rare-earth (R) garnets, R$_3$Fe$_5$O$_{12}$, which are ferrimagnetic, are being investigated for magnetooptic recording.

Amorphous single-domain CoTaZr cores having Al$_2$O$_3$ interlayers where the CoTaZr thickness is from 0.23–0.9 µm, depending on the number of layers, and Al$_2$O$_3$ is 0.01-µm thick, were evaluated for use as thin-film heads. This material combination is attractive for low noise heads operating at frequencies up to 40 MHz.

Magnetic Superlattices. The discovery in the late 1980s of giant magnetoresistance (GMR) in antiferromagnetically coupled Fe/Cr superlattices stimulated great interest. Properties of metallic superlattices consisting of thin alternate single-crystal layers of different magnetic materials as well as alternate layers of magnetic and nonmagnetic materials were examined.

Magnetoresistive recording heads offer much more sensitivity than inductive heads and there is strong evidence as of this writing that such heads will be used exclusively by the year 2000. The trend in the development of head materials is toward thin-film media. Although

Permalloy films ($Ni_{18}Fe_{19}$) are used for magnetoresistive sensors, the change in resistance is only about 2.5%. Higher magnetoresistive materials are needed.

Magnetic Fluids. Magnetic fluids are stable colloidal suspensions of ferromagnetic particles, such as Fe_3O_4, and of subdomain size (ca 10 nm) in aqueous or organic bases. The fluid behaves as a homogeneous Newtonian liquid and reacts to a magnetic field. These materials are used in bearings, rotary-shaft seals, and feedthroughs.

See also **Magnetism**.

Additional Reading

Buschow, K.H.: *Handbook of Magnetic Materials*, Vol. 15, Elsevier Science, New York, NY, 2003.

Buschow, K.H. and F.R. De Boer: *Physics of Magnetism and Magnetic Materials*, Kluwer Academic Publishers, Norwell, MA, 2002.

Buschow, K.H. and E.P. Wohlfarth: *Handbook of Magnetic Materials: A Handbook on the Properties of Magnetically Ordered Substances*, Vol. 5, Elsevier Science, New York, NY, 1990.

Chatterji, T.: *Neutron Scattering from Magnetic Materials*, Elsevier Science & Technology Books, New York, NY, 2006.

Chen, C.W.: *Magnetism and Metallurgy of Soft Magnetic Materials*, North-Holland, New York, NY, 1977.

Freund, B. and S. Suresh: *Thin Film Materials: Stress, Defect Formation and Surface Evolution*, Cambridge University Press, New York, NY, 2003.

Harper, J.M.E.: "Ion Beam Techniques in Thin Film Deposition," *Solid State Technol.*, **129** (Apr. 1987).

Herbst, J.F.: "Permanent Magnets," *Am. Sci.*, **251** (May–June 1993).

Heirich, B. and J.A.C. Bland, eds.: *Ultrathin Magnetic Structures*, Springer, Berlin, 1994.

Kryder, M.H.: "Data Storage in 2000: Trends in Data Storage Technologies," *IEEE Trans. Magn.*, **25**(6), 4358 (1989).

Luborsky, F.E. ed.: *Amorphous Metallic Alloys*, Butterworths, London, 1983.

Nalwa, H.S.: *Handbook of Thin Films*, Five-Volume Set, Elsevier Science, New York, NY, 2001.

O'Handley, R.C.: *Modern Magnetic Materials: Principles and Applications*, John Wiley & Sons, Inc., New York, NY, 1999.

Ohring, M.: *Materials Science of Thin Films*, 2nd Edition, Elsevier Science, New York, NY, 2001.

Sharrock, M.P.: "Particulate Magnetic Recording Media: A Review," *IEEE Trans. Magn.*, **25**(6), 4374 (1989).

Spaldin, N.: *Magnetic Materials: Fundamentals and Device Applications*, Cambridge University Press, New York, NY, 2003.

Venables, J.A.: *Introduction to Surface and Thin Film Processes*, Cambridge University Press, New York, NY, 2000.

Wolfarth, E.P. ed.: *Ferromagnetic Materials—A Handbook on the Properties of Magnetically Ordered Substances*, Vols. 1 and 2, Elsevier, New York, NY, 1980; E.P. Wolfarth and K.H.-J. Buschow, eds., Vols. 3 and 4, Elsevier, 1988.

MAGNETIC MOMENT.   1.   The quantity obtained by multiplying the distance between two magnetic poles by the average strength of the poles.

2.   A measure of the magnetic flux set up by the gyration of an electric charge in a magnetic field. The moment is negative, indicating it is diamagnetic, and equal to the energy of rotation divided by the magnetic field.

3.   (symbol m). In atomic and nuclear physics, a moment, measured in Bohr magnetons, associated with the intrinsic spin of the particle and with the orbital motion of the particle in a system. Also called *magnetic dipole moment.*

MAGNETIC MOMENT (Particle).   Use of the term magnetic moment for a nuclear or atomic particle or system of particles usually denotes the magnetic dipole moment. For a particle or system in a magnetic field, the interaction energy is the negative of the product of the field strength H by the component of the magnetic dipole moment μ_H of the particle in the direction of the field ($\mu_H H$). A magnetic moment is associated with the intrinsic spin of a particle and with the orbital motion of a particle in a system, e.g., nuclei with finite spins have finite magnetic moments between about -2 and $+6$ nuclear magnetons.

MAGNETIC NORTH.   See **North**.

MAGNETIC POLES (Earth).   See **Earth**.

MAGNETIC PRESSURE.   The energy density associated with a magnetic field. In a very real sense, there is energy stored in a magnetic field, and since energy per unit volume is equivalent to force per unit area or pressure, one may speak of the pressure exerted by a magnetic field. For plasma containment in a thermonuclear device, the magnetic pressure must be greater than the kinetic pressure of the plasma. A pressure of 1 atmosphere corresponds approximately to 5,000 *gausses*, and the pressure is proportional to the square of the field.

MAGNETIC RESONANCE IMAGING (MRI).   Magnetic resonance imaging (MRI) is a noninvasive method of mapping the internal structure of the body. It employs radiofrequency radiation (RF) together with carefully controlled magnetic fields to produce high-quality images of the body which display the underlying anatomic substrate in terms of the spatial distribution of hydrogen nuclei and parameters relating to their motion, in cellular water and lipids [Smith and Ranallo; and Horowitz, refs.]. The first applications of MRI were in the nervous system [Hawkes, et al.:] but the method is now widely applied throughout the body and an extensive knowledge base has been established to guide its use in clinical practice, either alone or in conjunction with other techniques. See also **Imaging (Medical)**.

Basic Imaging Technique

Magnetic resonance (MR) refers to the phenomenon whereby the nuclei of certain atoms, which have an odd number of protons or neutrons and so possess a net charge, absorb and emit energy at a specific frequency when placed in a magnetic field. Hydrogen nuclei (protons) are particularly favorable for imaging because of their great abundance in biological material. Because of their charge and angular momentum, protons behave as tiny bar magnets spinning about their magnetic axis. When a group of protons is placed in a strong uniform magnetic field, their magnetic moments experience a couple, tending to turn them parallel to the field. More of the magnets align with the applied field than against it, so that there is a net magnetization in the direction of the field. This is a vector quantity, having both magnitude and direction. Because the nuclei are spinning, they respond to the magnetic couple like a gyroscope, so that their axes are both tilted and caused to rotate at exactly the same frequency around the magnetic field direction in a movement known as precession. There is a linear relationship between the applied magnetic field and the precessional frequency (f).

If a collection of protons in a magnetic field is exposed to an oscillating RF field at their precessional frequency, a strong interaction or resonance effect occurs. This is referred to as magnetic resonance, a phenomenon described independently by Felix Bloch and E. M. Purcell in 1946, for which they received the Nobel Prize in 1952. This interaction leads to absorption of energy by the precessing protons, and their motion is disturbed so that the direction of the net nuclear magnetization is tilted away from the direction of the main field through an angle that depends on the strength and duration of the RF pulse. Most commonly 90° and 180° pulses are used which rotate the net magnetization through 90° and 180° respectively. After the disturbance induced by one or more pulses of RF, the net magnetization vector returns to its equilibrium position along the direction of the main field. As it does so, the changing magnetization induces a small voltage in a receiver coil (which is placed around the part to be imaged) in accordance with Faraday's law of electromagnetic induction.

The electrical signal is known as the free induction decay (FID), and its magnitude and duration reflect complex molecular interactions in the tissue environment which determines the so-called *relaxation times of the protons*. The T_1 relaxation time refers to the time taken for the system to return to thermal equilibrium after disturbance, and the T_2 relaxation time indicates the characteristic decay of the FID due to irreversible dephasing of the initially coherent precession of protons following disturbance by an RF pulse.

By altering the pattern of RF pulses applied during imaging in a so-called *pulse sequence*, the dependence of the resulting signal can be altered to weight the resulting image contrast to emphasize a particular characteristic of signal behavior (Fig. 1). T_1- and T_2-weighted images represent the mainstay of conventional imaging. Because MR is an inherently complex phenomenon, however, methods of modulating the signal to reflect other

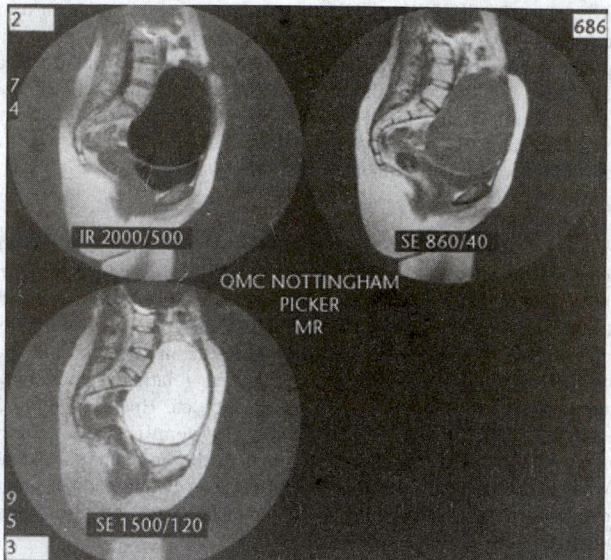

Fig. 1. Midline sagittal images of the lower abdomen and pelvis, showing a large ovarian tumor situated above the bladder. Note the change in signal returned by the fluid within the tumor and the urine in the bladder with different pulse sequences.

tissue interactions at the molecular level, such as diffusion, have been devised.

To produce an image of a particular body part, spatial information is required so as to localize the signal emanating from a particular volume element (voxel). Because the resonant frequency is proportional to the strength of the magnetic field, a linear variation of the field in one direction will result in the resonant frequency of protons to vary in like fashion. By appropriate use of so-called *field gradients*, different volume elements will be labeled by having a different resonant frequency for the protons contained within them. In practice, the complex resulting FID is digitized and then the amplitude of the component frequencies is determined by analysis in a computer using a mathematical algorithm called *Fourier analysis*. If an additional gradient field is applied before detection of the signal, this will induce differences in the phase of the local nuclear signals when compared to a reference. In practice, a series of phase-encoding magnetic gradients of different strengths is applied so as to obtain the spatial distribution of signal in a direction perpendicular to the gradient applied for frequency encoding. This is the basis of the now almost universally applied two-dimensional Fourier transform method of imaging. During the imaging process a matrix of numbers is generated, and this array with K_x and K_y axes is known as K space. Fourier transformation of the data in K space give rise to an MR image. The majority of images are acquired on a 256×256 matrix of picture elements (pixels) distributed over the relevant organ of the body. For the same field of view, a larger matrix (e.g., 512×512 pixels) will provide a higher spatial resolution.

In order to define the imaging plane, current scanners use a method known as selective excitation. In this a gradient field is applied during excitation with an RF pulse that contains a preselected narrow frequency band so that only those protons whose resonant frequency falls within this band will be excited. By this means, imaging can be restricted to a particular slice of chosen thickness. As an extension of this method, multiple slices are usually collected during each data acquisition.

Fast Imaging

Until recently, conventional MRI was relatively slow. An examination requiring multiple pulse sequences and interrogation in several planes could take as long as an hour to complete. Furthermore, images were often contaminated by obtrusive artifacts arising from movement of body organs. The development of rapid pulse sequences [Haase, et al.:] now allows examination of the abdomen in a single breath-hold and the acquisition of large three-dimensional data sets in a few minutes, which can subsequently be interrogated in any chosen plane. The fastest imaging strategy, echo-planar imaging, [Mansfield, ref.] allows the acquisition of a complete

two-dimensional image in typically 64–128 ms, and this has provided the opportunity to study dynamic events in the body, such as movement of the heart, with high temporal resolution as well as the kinetics of administered contrast media.

The principal component of any MR scanner is the magnet, which may be either permanent or superconducting, constructed to provide a high degree of field homogeneity within the main bore. The higher the field strength, the greater the intrinsic signal to noise ratio of the MR signal. Within the magnet bore is situated a gradient coil assembly, and the maximum amplitude of field gradients generated determines the minimum slice thickness and field of view, and thereby the spatial resolution of the system. A single coil surrounding the body part to be imaged may both transmit the RF and receive the MR signal but, where these are separate, appropriately configured coils can be placed in close apposition to the part to be imaged with a consequent improvement in the signal to noise ratio.

Gradient amplifiers are required to generate the magnetic field gradient required for spatial encoding, and a large computer is necessary to control the system and carry out image reconstruction. These may be displayed on the basic console or on a remote console, which can be operated independently. See also **Digital Processing and Information Technology in Imaging**.

Acquisition of Appropriate Image Data

The images that can be created from a particular part of the body represent the set of mappings of the spatial distribution of one or more of its properties. The outstanding attribute of MRI is its ability to provide a remarkable range of tissue contrast, and this reflects the multiplicity of intrinsic tissue parameters that can be exploited. The machine variables that can be controlled by the user have a complex interdependence, which impacts on the final signal to noise ratio, spatial resolution, image contrast and total examination time. Before selecting a protocol for a given examination, careful thought must be given to the possible prospective diagnosis so that the appropriate imaging planes, pulse sequences, matrix size, and number and thickness of slices are chosen to accomplish the best demonstration of suspected pathology in an acceptable time frame.

There is typically no single set of parameters that allows this to be accomplished, and so multiple data acquisitions are usually required. The repertoire of possibilities for any given clinical problem is continually being extended and refined as the knowledge base of imaging expands. The safe conduct of MRI is an important consideration and, in particular, appropriate exclusions from imaging must be made (e.g. the presence of a cardiac pacemaker) and it must be ensured that all equipment for anesthesia and monitoring is compatible with the scanner environment [Shellock and Kanal, ref.].

Contrast Agents

The principal reason for developing contrast agents for MRI was to provide a means of discriminating between magnetically similar tissues so as to highlight pathological changes. Despite a wide range of different structures, these agents produce their effect by altering either the proton density or the relaxation times of a given tissue. When translated to their effect on the signal intensity returned by a particular tissue, this may be either a signal enhancement or signal loss. Contrast agents have been developed that have an appropriate distribution in the body and a sufficiently low toxicity to make them acceptable for clinical use. There are two major groups of contrast agent, whose effect on signal intensity is indirect, being achieved through alterations induced in proton relaxation rates.

Paramagnetic Contrast Agents

These are based on a paramagnetic ion which produces its own local fluctuating magnetic field, the size of which is dependent on the number of unpaired electrons in the outer orbital shells. This produces an effective enhancement of the relaxation of local water protons. Gadolinium is highly paramagnetic, having seven such unpaired electrons, and its intrinsic toxicity is substantially reduced by binding it to a suitable chelating agent such as diethylenetriamine pentaacetic acid (DTPA). A number of gadolinium chelates are now available for use as contrast media and, because their effect is greater on T_1 than on T_2 relaxation rates in human tissues, the enhancement produced is best shown on T_1-weighted images [Gadian, et al.:]. The agents are given by intravenous injection and their use is now well established in the nervous system (Fig. 2), cardiovascular system and musculoskeletal system. A number of contrast agents have also been developed for hepatobiliary imaging by chelating a paramagnetic ion (manganese or gadolinium) to a ligand that has an affinity for liver cells.

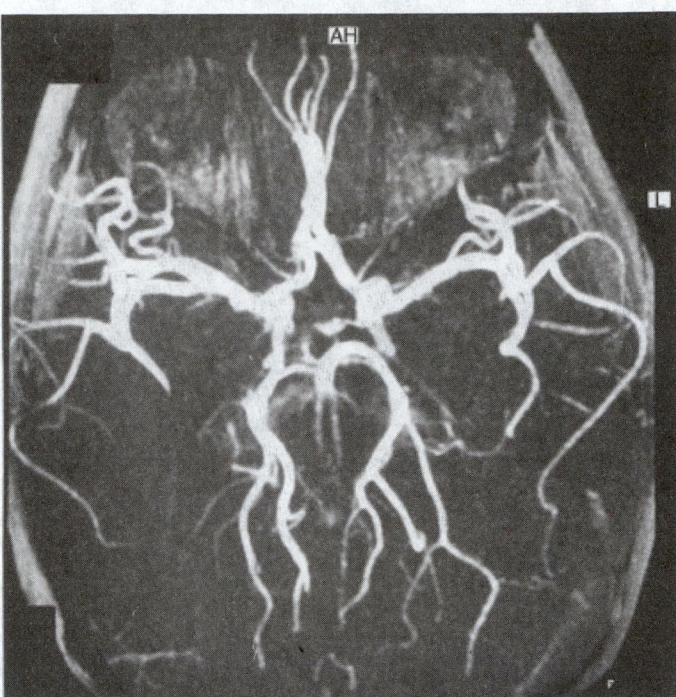

Fig. 3.   "Time of flight" magnetic resonance angiogram, showing the major blood vessels that supply the brain.

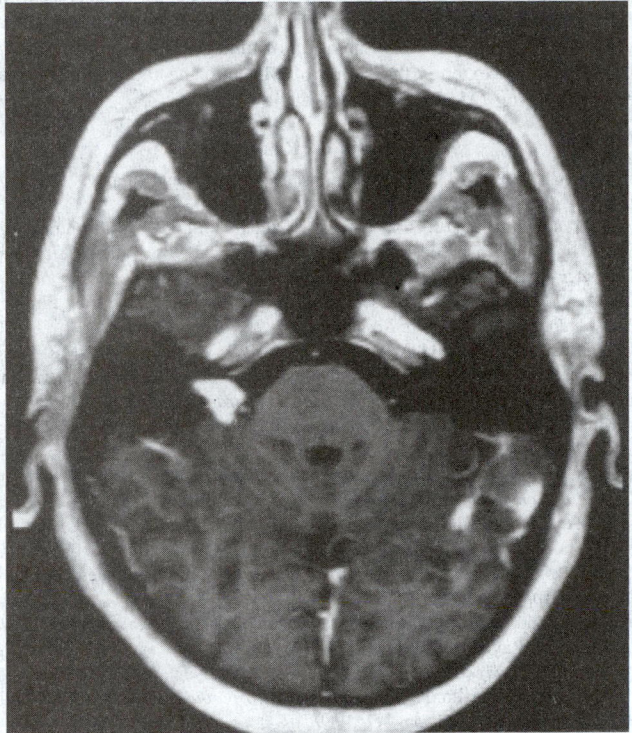

Fig. 2.   Transverse axial magnetic resonance scan of the brain following administration of intravenous contrast agent, showing a small enhancing tumor adjacent to the left side of the brainstem.

Super Paramagnetic Contrast Agents. These consist of tiny particles of iron oxide crystals which can become transiently magnetized when placed in a magnetic field. The associated induced local field produces rapid spin dephasing of protons in the vicinity of such particles and thus a reduction in T_2 values with a fall in signal intensity on T_2-weighted images. These agents have been introduced for liver imaging following their uptake in reticuloendothelial cells in the liver after intravenous injection of an appropriate formulation.

Magnetic Resonance Angiography

Flowing blood can return either a high or low signal depending on its direction, its velocity and the pulse sequence employed. An understanding of the phenomenology underlying this observation has led to the development of MR angiography, which portrays vascular architecture by mapping the protons in flowing blood in such a way that they return a high signal against a background of little or no signal from surrounding stationary tissues.

In the "time of flight" technique [Laub and Kaiser, ref.], a selected volume of tissue receives multiple RF pulses which saturate the protons in stationary tissues so that they have no net magnetization, whereas the fully magnetized spins contained in flowing blood entering the volume from outside will return a high signal during data collection (Fig. 3.)

In the "phase contrast" technique [Dumoulin and Hart, ref.], different velocity-dependent phase shifts are produced in two consecutive images such that, on subtraction of the first from the second, there is cancellation of the phase shift from stationary tissues whereas a positive signal is returned from flowing blood. With this method, direction of blood flow as well as its velocity can also be determined.

In both methods there may be signal loss or regions of low signal where there is turbulent flow, and also difficulty in resolving small vessels. By combining a rapid bolus injection of an intravascular contrast agent (which provides a magnetic label) in a time of flight acquisition, these limitations

can be reduced significantly. Rapid acquisition techniques and careful bolus timing are necessary, however, to ensure that the peak concentration of contrast is correctly positioned within the data acquisition period.

Image Display

In conventional MRI the acquired data consists of a matrix of numbers that represent the relative signal intensities, and these can be stored on magnetic tape or optical disks. They are subsequently presented as an analogue display of the voxel values in an anatomical cross-section. The dynamic range of the data set is so wide that it cannot be partitioned into a single grey scale that can be appreciated by the human visual system. To interrogate the whole data set, therefore, this must be viewed at different window widths around a variable window level.

The enormous number of data generated during MR examinations can be displayed in a wide variety of formats, including three-dimensional projections and multicolor rendering. The rapid cinematic presentation of contiguous transverse axial sections allows information across several sections to be integrated more readily. Cine presentations are now a standard mode of display for cardiac studies. Software developments have provided the facility to interrogate a three-dimensional data set and produce secondary image sets in any chosen plane. In MR angiography a special computer algorithm (maximum intensity projection) allows a selection of all bright pixels above a threshhold in either contiguous two-dimensional sections or a three-dimensional data set, and these are then displayed in continuity to provide a portrayal of the relevant vascular architecture.

Clinical Applications of Magnetic Resonance Imaging

MRI was shown at the outset to have outstanding qualities as an imaging technique for examination of the nervous system. It has no planar restriction, thus allowing direct coronal and sagittal imaging; it provides a wide range of tissue contrast and avoids the use of ionizing radiation. MRI has been shown to be more sensitive than computed tomography to the presence of intracranial pathology, and the multiplanar facility permits assessment of the relationship of important structures to tumors and other pathology before operation. The sensitivity of MRI has been increased through the development of contrast agents which highlight vascular lesions and tumors. Studies of the extracranial and intracranial blood vessels can be made, allowing the identification of malformations and areas of pathological narrowing. MRI is now the preferred technique for studying the spinal axis, giving excellent delineation of the spinal cord and vertebral column. It is used extensively in the evaluation of degenerative disc disease and acute disc protrusions (Fig. 4). MRI is a highly cost-effective method of evaluating disorders of the major joints of the body, especially the knee and shoulder, and has replaced all other techniques. It has become the method of choice for staging the extent of tumors of bone, marrow and soft tissue.

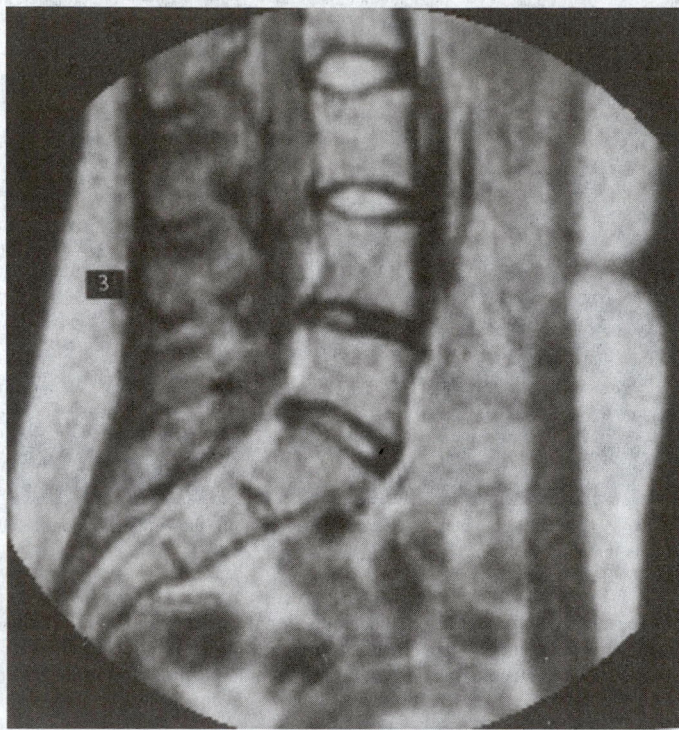

Fig. 4.   Midline sagittal magnetic resonance scan of the lower spine, showing the vertebral bodies and intervertebral discs. Note the narrowing of the two lower discs and their backward protrusion into the spinal canal.

MRI has made a significant contribution to the assessment of pelvic disease, particularly in defining the extent of tumors. The advent of fast imaging strategies has permitted imaging of the abdomen to be carried out within a single breath-hold, so that high-quality images of the liver, pancreas and kidneys can now be made. The excellent spatial resolution over a wide field of view, lack of planar restriction and inherent contrast between the blood pool and vascular wall commended MRI early on as a method with promise for imaging the heart and great vessels. This potential became fully realized only with the advent of fast imaging strategies. The development of rapid acquisition contrast enhanced MR angiography has allowed the generation of high-quality vascular images of the main blood vessels of the abdomen, thorax and limbs, and this noninvasive technique is rapidly replacing conventional diagnostic angiography.

Recent Developments

Functional Brain Imaging. The human brain is known to contain anatomically distinct processing regions whose location and organization together provide the basis for a functional model of its activity. It is now well established that there is a tight coupling between neuronal activity, energy consumption and regional blood flow. Activation studies using imaging techniques, by assessing the local response of brain metabolism or hemodynamic induced by internal or external stimuli, can map the underlying anatomy. High-speed functional MR techniques have been developed which are sensitive to blood volume, tissue perfusion and changes in blood oxygenation, specifically the relative amount of paramagnetic deoxyhemoglobin. Besides fundamental contributions to cognitive neuroscience, such studies have clinical value, for instance the mapping of the relationship of the motor cortex of the brain to structural pathology before surgery.

Interventional Procedures. X-ray computed tomography has proved to be a useful method for guiding stereotactic and needle biopsy procedures.

The development of nonmagnetic biopsy needles has meant that MR can be harnessed in the same way, and stereotactic frames are now available that are compatible with the MR environment. This has allowed neurosurgeons to capitalize on the excellent tissue contrast in the portrayal of intracranial pathology. The development of open-design magnet systems, together with fast scanning methods which provide MR fluoroscopy at near real-time rates, has allowed the direction and monitoring of both interstitial laser therapy and focused ultrasound ablation of tumors. This is a growing field of activity which extends the scope of MR into the therapeutic domain. See also **Imaging (Medical): An Overview**.

Additional Reading

Dumoulin, C.L., and H.J. Hart: "Magnetic Resonance Angiography," *Radiology*, **161**, 717–720. (1986).
Gadian, D., J.A. Payne, D.J. Bryant, et al.: "Gadolinium DTPA as a Contrast Agent in MR Imaging: Theoretical Projections and Practical Observations," *Journal of Computer Assisted Tomography*, **9**(2), L242–251 (1985).
Haase, A., J. Frahm, and D. Mathaei: "Rapid NMR Imaging Using Low Flip-angle Pulses," *Journal of Magnetic Resonance*, **67**, 258–266 (1986).
Hawkes, R.C., G.N. Holland, W.S. Moore, and B.S. Worthington: "NMR Tomography of the Brain: A Preliminary Clinical Assessment with Demonstration of Pathology," *Journal of Computer Assisted Tomography*, **4**: 577–586 (1980).
Horowitz, A.: *MRI Physics for Radiologists*, 3rd Edition, Springer-Verlag New York, LLC, New York, NY, 1995.
Laub, G.A., and W.A. Kaiser: "MR Angiography with Gradient Motion Refocusing," *Journal of Computer Assisted Tomography*, **12**, 377–382 (1988).
Liney, G.P.: *MRI in Clinical Practice*, Springer-Verlag New York, LLC, New York, NY, 2006.
Mansfield, P.: "Multiplanar Image Formation Using NMR Spin Echoes," *Journal of Physics*, **C10**, L55–L58. 1977.
Shellock, F., and E. Kanal: *Magnetic Resonance—Bioeffects, Safety and Patient Management*, Raven Press, New York, NY, 1994.
Siegelman, E.S.: *Body MRI*, Elsevier Health Sciences, New York, NY, 2004.
Smith, H., and F. Ranallo: *A Non-mathematical Approach to Basic MRI*, Medical Physics Publishing, Madison, WI, 1989.
Stark, D.D., and W.G. Bradley: *Magnetic Resonance Imaging*, 3rd Edition, Mosby, Inc., St Louis, MO, 1999.
Weishaupt, D., and V. Koechli: *How Does MRI Work?: An Introduction to the Physics and Function of Magnetic Resonance Imaging*, 2nd Edition, Springer-Verlag New York, LLC, New York, NY, 2006.
Westbrook, C., J. Talbot, and C.K. Roth: *MRI in Practice*, 3rd Edition, Blackwell Publishers, Malden, MA, 2005.

BRIAN S. WORTHINGTON, University of Nottingham, Nottingham, UK

MAGNETIC SEPARATION.   Use of a magnetic field to remove unwanted magnetic particulates from solid or liquid mixtures of nonmagnetic materials, e.g., sands, and mineral processing. Low-gradient fields are suitable for separation of strongly magnetic materials, whereas high-gradient fields can separate particles of materials that are weakly magnetic, such as coliform bacteria from municipal wastes and sulfur from coal. Removal of magnetic impurities from industrial wastewater is called magnetic filtration, e.g., reconditioning of boiler water and regeneration of condensate in power plants. See also **Electromagnetic Separation**; **Magnetic Materials**; **Mass Spectrometry**; and **Separation: Magnetic Separation**.

MAGNETIC STORM.   A worldwide disturbance of the Earth's magnetic field. Magnetic storms are frequently characterized by a sudden onset, in which the magnetic field undergoes marked changes in the course of an hour or less, followed by a very gradual return to normality, which may take several days. Magnetic storms are caused by solar disturbances, though the exact nature of the link between the solar and terrestrial disturbances is not understood. They are more frequent during years of high sunspot number. Sometimes a magnetic storm can be linked to a particular solar disturbance. In these cases, the time between solar flare and onset of the magnetic storm is about 1 or 2 days, suggesting that the disturbance is carried to the Earth by a cloud of particles thrown out by the sun. When these disturbances are observable only in the *auroral zones*, they may be termed polar magnetic storms.

MAGNETIC TAPE STORAGE (Computer).   Information is recorded on tape by magnetizing narrow (lengthwise) strips (termed tracks) in a pattern corresponding to a sequence of binary states (1s and 0s). Binary data, often corresponding to a single byte, are stored in column form across the width of the tape. A read/write head is usually associated with each row of magnetized material, so that one column can be read or written at a time as the tape traverses the head. The density of recorded data is most commonly 800, 1600, or 6250 bits per linear inch (approximately 315, 630, or 2461 bits per linear centimeter). A nine-track, 2400-foot (~732-meter) tape will carry from 20 to 156 million characters, dependent upon the density. Magnetic tape storage is most often used to provide offline archival storage of data. A drawback to tape storage is the fact that it must be read serially. Retrieval of randomly distributed information can be time consuming, compared, for example, with disk storage.

Tape material usually is polyester plastic, one side of which is coated with a suspension of ferrite or magnetic oxide particles. The common tape size is $\frac{1}{2}$ inch (12.7 millimeters) in width, $1\frac{1}{2}$ mils (0.0015 inch; 0.4 millimeter) in thickness, and 2400 feet (~732 meters) in length. Tape widths of $\frac{1}{4}$, $\frac{3}{4}$, and 1 inch (0.64, 1.91, and 2.54 centimeters); and lengths of 200, 300, 600, 1200, and 3600 feet (~61, 91, 183, 366, and 1097 meters) also are available for use in the computer field.

Recorded material can be removed from the tape by passing the tape through a strong, constant magnetic field (dc erase), or through a high-frequency alternating magnetic field (ac erase). Information is stored on the tape by magnetizing the magnetic film in one direction or the other in a pattern determined by the binary data. The information is recorded on the tape in a parallel by bit, serial by digit, format, i.e., the bits which comprise a digit are recorded across the width of the tape and the digits are recorded sequentially along the length of the tape.

The magnetic tape is stored on reels and is transferred from one reel to the other past two-gap read/write heads in the course of reading or writing data. Vacuum columns or other tension mechanisms are provided so that the tape can be moved rapidly a few inches at a time without waiting for the movement of the reels. The two-gap head allows automatic error checking of the tape data while it is being written. The first gap is used for writing, the second gap for reading.

The reading or writing of data on the tape is controlled by clock circuits within the logic of the tape unit. In writing, the frequency at which the clock is stepped is a function of the tape speed and the recording density. In reading, the oscillator that drives the read clock is also gated by the first bit being read. To compensate for the time it takes for the tape motion to get up to speed, read delay and write delay counters are designed into the logic of the tape unit.

On a tape read operation, the character is stored in a buffer register as it is read from the tape and is then transferred to the digital computer while the next character is being read. On a write operation, the next character is fetched from the computer while the previous character is being written. Since data are not written on the tape until the tape gets up to speed, the information to be written is blocked into records to minimize the number of gaps caused by the write delay.

See also **Storage (Computer)**.

Thomas J. Harrison, IBM Corporation, Boca Raton, FL

MAGNETISM. A magnet is a body possessing the property of attracting magnetic substances. The so-called permanent magnet should be used where a constant magnetic field is to be produced, inasmuch as a well-made permanent magnet loses its magnetism very slowly, and then only up to a certain value, after which it is said to be aged. Therefore, it essentially maintains a constant degree of magnetism unless subjected to strong demagnetizing effects. Permanent magnets for precision apparatus are artificially aged during manufacture.

A bar that has been magnetized is found to have poles, which are centers where magnetic attraction is strongest. If the magnet is free to turn, the pole which points northerly is aptly termed the north pole; the other, the south pole. Like poles repel; unlike poles attract. Thus, because of the magnetic properties of the earth, a magnet can serve as a compass. The poles, of course, have no physical reality, but provide a convenient concept for describing certain magnetic phenomena.

Magnetic Field. The region surrounding a magnet (or an electric current) is endowed with specific properties. The most familiar is the torque experienced by a small magnet when placed in such a region. For any point of the field, there is only one direction in which the small magnet

will reach stable equilibrium. The direction in which the north pole of the magnet points when in equilibrium is termed the direction of the field. By moving a small magnet in a magnetic field, it is found that the field direction will follow curved lines of force. If the field is due to the current flowing in a conductor, the lines form completely closed curves enclosing the conductor. If the field is due to a magnet, they appear to enter the iron at the south pole and emerge at the north pole, inferring that they complete themselves through the iron as they do through a coreless, current-carrying helix. A magnetic field may be defined as a vector function field described by the magnetic induction. The term magnetic field is used interchangeably to refer to magnetic induction and magnetizing force.

The difference in the magnetic potential at two points in a magnetic field is measured by the work necessary to move a unit magnetic pole against the field from one point to the other. This difference is sometimes called a magnetomotive force, in analogy to electromotive force.

The lines of magnetic intensity around a current-carrying wire are circular, having the plane of the circle perpendicular to the axis of the wire. The direction of the lines of force is determined by the "right-hand rule." When the wire is grasped by the right hand, the fingers encircling the wire, and the thumb pointing along the wire in the direction of the current, the fingers encircle the wire in the direction of the lines of force. This rule is used to determine in which direction the north pole would lie in a helix or solenoid of wire, for, instead of having a straight wire encircled by magnetic lines, the wire itself is bent into a circular form by being wound in a solenoid or helix. The lines of force will then produce an axial magnetic field, so that one end of the solenoid is equivalent to a north pole, the other to a south pole.

In the vicinity of magnets, electric currents, or time-varying electric fields, it is found that a small current-carrying wire loop, if free to turn, will come to an equilibrium orientation. This behavior is attributed to an auxiliary field vector **B** called the magnetic induction or the magnetic flux density. The magnetic moment of a current loop or a magnetized body is a measure of the magnetizing force **H**, produced by the current or magnetized body. The magnetic moment of a plane current loop is a vector (**m**), normal to the plane of the loop and directed so that the current has a clockwise rotation around **m**. The magnitude of **m** is the product of the current and loop area. The magnetic moment of a magnetized body is the vector summation of the magnetic moments of the internal current loops and spins of the body. The magnetizing force produced at a displacement **r** from a small source of moment **m**

$$H = -\nabla \frac{\mathbf{m} \cdot \mathbf{r}}{r^3}$$

where ∇ is del. the vector differential operator.

The ratio of the magnetic moment of a magnetized body to its volume, expressed by the equation $\mathbf{B} = \mu_0(\mathbf{H} + 4\pi\mathbf{M})$, where μ_0 is the permeability of free space, **B** is magnetic induction, **H** is the external magnetizing force, and **M** is the magnetization. (The symbol I is sometimes used for intensity of magnetization.) The application of an increasing magnetizing force to a ferromagnetic substance yields a resulting intrinsic induction that asymptotically approaches a constant value known as the saturation magnetization. Spontaneous magnetism is the magnetic saturation of the domains of a ferromagnetic material, even in the absence of an applied magnetizing force. Each domain is magnetized to saturation at all temperatures below the Curie point, although the material as a whole may be unmagnetized because of the differing orientations of the various domains. See also **Ferromagnetism**.

The intensity of magnetization I induced at any point in a body is proportional to the strength of the applied field H:

$$I = \kappa H \, (\text{or } \kappa = I/H)$$

where κ is a constant of proportionality depending on the material of the body. It is called the magnetic susceptibility per unit volume, and may be defined qualitatively as the extent to which a material is susceptible to induced magnetism. For an isotropic body, the susceptibility is the same in all directions. However, for anisotropic crystals, the susceptibilities along the three principle magnetic axes are different, and measurements on their powder samples give the average of the three values.

Magnetic susceptibility is obviously related to magnetic permeability, and the following relationships may be derived:

$$B = \mu H = 4\pi I + H$$

Therefore,

$$\mu = 4\pi I/H + 1 = 4\pi \kappa + 1$$

or

$$\kappa = (\mu - 1)/4\pi$$

Magnetic Flux. The magnetic flux through any closed figure, such as a circle, a rectangle, or a loop of wire, is the product of the area of the Figure by the average component of magnetic induction normal to that area. Thus, if a rectangle 5 cm $\times$ 8 cm is placed in a region where there is a uniform magnetic induction of 2,500 gauss (other units defined shortly), and at an angle of 30° with the lines of induction, the magnetic flux through it is 2,500 gauss °40 cm^2 $\times$ sin 30° = 50,000 gauss-cm^2 or "maxwells." The magnitude of this quantity is often conventionally represented by imagining the lines of induction to be so spaced that the number of them through a given area is equal to the number of gauss-cm^2 or maxwells of flux through that area. The flux in the above example would be commonly expressed as 50,000 "lines." When a coil has several (n) turns and each turn has approximately the same flux (ϕ) through it. This product, which is called the "linkage," is expressed in "maxwell-turns" or "line-turns."

The magnetic flux or the linkage through a loop or a coil may be measured by putting into the circuit a ballistic (undamped) galvanometer and then suddenly removing the flux (or the coil). If the resistance of the whole circuit and the constant of the galvanometer are known, the flux may be calculated from the "throw" of the galvanometer.

The *gauss* (G) is the electromagnetic CGS unit of magnetic flux density. The *tesla* (T) is the **SI** unit. The *maxwell* (**Mx**) is the electromagnetic CGS unit of magnetic flux. The weber (Wb) is the SI unit.

Magnetic Circuit. The flux from a bar magnet or from a straight electromagnet issues from one end of the magnet or coil, bends around, and reenters at the other end. As mentioned before, this can be exhibited by exploring the region with a small magnet or compass needle. If there is an iron frame or ring extending from one pole of the magnet or coil around to the other, and in the case of the coil, running clear through it, the magnetic flux is not only concentrated largely in the iron, but is much greater in total amount than if the induction is entirely in the air. Even a short gap in the iron reduces the flux considerably.

The analogy of such a magnetic path to an electric circuit is easily seen. The magnetic flux corresponds to a current. The magnet or coil corresponds to a battery, and provides magnetomotive force just as a battery supplies electromotive force. The amount of flux produced by a given magnetomotive force depends upon the dimensions and material of the "magnetic circuit," e.g., the length and cross section of the iron ring followed by the flux and the permeability of the iron; just as the dimensions and material of the electric conductor determine its resistance. The attribute of the magnetic circuit (corresponding to resistance) is called its reluctance.

It must be remembered that this analogy is purely mathematical, not physical. In magnetism, there is no flow of charge, as in electricity.

For a single magnetic circuit consisting of two parts, magnetic material having a permeability μ_1, length l_1, and A_1, and an air gap of permeability μ_0, length l_2 and area A_2, Ampere's circuital law gives $\int H \cos \theta \, dl = H_1 l_1 + H_2 l_2 = NI$, N is the number of turns and I the current through the coil producing the magnetic flux Φ_m. Now as

$$H_1 l_1 = B_1 l_1/\mu_1 = \Phi_m l_1/A_1\mu_1$$

and

$$H_2 l_2 = B_2 l_2/\mu_0 = \Phi_m l_2/A_2\mu_0$$

So

$$\Phi_m[l_1/A_1\mu_1 + l_2/A_2\mu_0] = NI$$

or

$$\Phi_m = \frac{NI}{[l_1A_1\mu_1 + l_2/A_2\mu_0]} = \frac{F}{R}$$

where F is the magnetomotive force in ampere-turns and R, the reluctance of the magnetic circuit. This relation is a magnetic analogy of Ohm's law known as Bosanquet's law.

The *gilbert* (Gb) is the electromagnetic CGS unit of magnetomotive force. The *ampere* (*or ampere turn*) (A) is the SI unit. The *oersted* (Oe) is the electromagnetic CGS unit of magnetic field strength. The *ampere per meter* (A/m) is the SI unit.

Magnetic Materials. The electrical industry is dependent upon four basic types of materials: (1) good conductors of electricity; (2) high-resistivity conductors capable of withstanding high temperatures; (3) insulators; and (4) magnetic materials. The development of better magnetic materials has contributed to marked improvements in motors, generators, transformers, and instruments of all kinds.

The ferromagnetic elements are iron, nickel, and cobalt. Of these, iron is the only one that has important magnetic applications in pure, or commercially pure form. All three elements, and many others which themselves are nonmagnetic, are used in special alloys in which certain magnetic characteristics are developed to a high degree. It is even possible, by alloying, to develop magnetic material from certain nonmagnetic elements.

There are two distinct groups of ferromagnetic materials, those that are easily demagnetized, and those that are not. These are often designated as soft and hard magnetic materials. In many respects, iron is an excellent soft, magnetic material. It attains the highest saturation value of magnetic induction found in any material, with the exception of certain cobalt alloys. However, its maximum permeability (maximum ratio of magnetic induction to magnetizing force) is surpassed by other materials. When used as a core in a field induced by ac, the so-called "iron losses" are relatively high. Iron losses consist of energy losses related to the area of the hysteresis loop, and of eddy-current losses caused by induced current circuits within the metal. High values of residual induction (B_r on the typical hysteresis curve) and coercive force (H_c) are indications of high hysteresis loss. See also **Hysteresis**. A low value of electrical resistivity results in high eddy current losses. However, other factors such as the thickness of the sheets in laminated cores, and to a lesser degree grain size also affect eddy-current losses. By special heat treatments and by the use of specially purified iron, much higher values of permeability and lower hysteresis losses can be obtained.

With a good grade of silicon electrical steel, the hysteresis loss can be cut in half and the eddy-current loss reduced even more because of the high resistivity compared with iron. This results in much more efficient operation of alternating-current equipment.

Certain high-nickel alloys of the Permalloy type have very high permeabilities at low and moderate inductions and give low hysteresis losses. For many special applications in instruments and communications equipment, these higher cost alloys are economically justified. Power transformers, motors and generators, etc., are designed for silicon electrical steels.

In the use of any of these materials for ac applications, it is desirable to reduce the thickness of individual laminations to reduce eddy-current losses. For special applications at higher than usual power frequencies, extremely thin strips are used, approaching $\frac{1}{1000}$ of an inch (0.025 millimeter) in thickness. The most common thickness at 60-cycle power frequencies is about $\frac{14}{1000}$ inch (0.35 millimeter). In the communications field, where the use of very high frequency current greatly aggravates the problem of eddy-current losses, alloys of the Permalloy type are used. They are produced in powdered form, given a very thin insulating film, and compacted under high pressure into a suitable core shape.

In the production of magnetically soft materials, it is important that elements such as carbon and sulfur, which disturb the continuity of the base-metal crystal structure, be reduced to low residual amounts, and that the material be used in a strain-free condition. High annealing temperatures resulting in coarse grain structures are used. These conditions are, for the most part, reversed when magnetically hard materials are sought. These materials must be capable of reaching high inductions, but upon removal of the direct magnetizing force the magnetic flux remaining in the circuit should be high, therefore a hysteresis loop of large area is desired. A good overall measure of the quality of a permanent magnet is the "maximum external energy." This is the maximum value which can be obtained for the product of the coordinates of a point on the curve between B_r and H_c on the hysteresis curve. See also **Magnetic Materials: Bulk**; and **Magnetic Materials: Thin Films And Particles**.

Progress in Development of Magnetic Alloys. During the last 100 years of experience in producing and using permanent magnets, alloy magnets as of the early 1980s are the most powerful by a factor of 30. Until the late 1930s, magnets were produced from steel. Then considerable research was devoted to producing and testing various magnetic alloys. Alnico magnets became available in the mid-1940s. representative present-day Alnico magnets contain a number of other metals. For example, Alnico 5 contains 8% aluminum, 14% nickel, 24% cobalt, and 3% copper, the remainder being iron. Alnico 9 contains 7% aluminum, 15% nickel, 35% cobalt, 4% copper, 5% titanium, the remainder being iron. Work on the development or magnetic alloys containing various rare-earth elements commenced in the late 1950s. A number of magnetic alloys containing rare-earth elements and cobalt are available today. As pointed out by Chin (1986), the magnetism of the R-Co compounds is due to the interatomic exchange between the spins of the two sublattices plus the spin-orbit coupling within the rare-earth atoms. The spins of the lighter rare-earth elements, such as cerium, praseodymium, neodymium, and samarium, align parallel with the cobalt atoms, resulting in high saturation magnetization values at room temperature. Other rare-earth elements align antiparallel with the cobalt atoms and thus the values are lower. In the mid-1970s, the alloy $Sm_2(Co, Fe, Cu)_{17}$ was introduced. See also **Rare-Earth Elements and Metals.**

The importance of improved magnetic alloys is sometimes underestimated. Magnets are used widely in telecommunications and in scores of electrical and electronic instruments and appliances, and it has been estimated that the demand for permanent magnets has increased by a factor of three since 1970. Chromium-cobalt-iron alloys have increased markedly in application. It is interesting to note that these magnets have essentially about the same coercive force and maximum energy product as Alnico 5, but with only half the cobalt content.

See also **Magnetostriction.**

Molecular Magnets. Ordinary iron and alloy magnets are made of atomic constituents that are difficult to modify. By contrast, molecular magnets may be customized much more easily, and they account for the past decade of research into formulating superior molecular magnets. In mid-1991, A.I. Epstein (Ohio State University) and J.S. Miller (DuPont Central Research and Development Facility) appear to have synthesized an organic-based magnet that retains its magnetism at room temperature and up to and including an elevated temperature of 350°K. At the latter temperature, the polymeric material commences to break down chemically. M. Hoffman (Northwestern University) observes that, if this is a molecular magnet, it represents an enormous leap. No researcher in the field has come even close to 350°K.

This particular magnet, although of exceeding interest in the laboratory, is not considered for practical applications because the material deteriorates rapidly unless it is contained within an inert atmosphere. According to a paper released by J.M. Manriquez, et al., the magnet is the result of a reaction of bis (benzene) vanadium with tetrocyanoethylene (TCNE). The material is an amorphous black solid that exhibits field-dependent magnetism and hysteresis at room temperature.

Quantum Effects and Tiny Magnets. Since early computer designs of the 1950s, there has been tremendous pressure exerted to microminiaturize magnetic information systems. This marked trend continues as designers strive for increased information density. But is there a limit? Some scientists now are showing serious concern that, as magnetic devices become smaller and smaller, the point may be reached where quantum effects could cause upheavals in the magnetic storage system. It is interesting to note that, in the computers of the 1950s, an estimated 100 billion atoms were needed to store one bit of information. In the early 1990s, this Figure had dropped to an estimated 1 billion atoms. Experts then predicted that, by the year 2000, the storage of one bit of information would require only 100,000 or fewer atoms. Researchers at the IBM Thomas J. Watson Research Center and other researchers in the computer field have pioneered techniques for studying magnets composed of 100,000 atoms at temperatures approaching absolute zero. Some of these researchers, who have instrumentation that can measure a magnetic field with a million times the sensitivity of traditional instruments, have observed what they believe may be evidence of tunneling and other effects of quantum mechanics in experimental magnetic systems. One group claims that, when a tiny magnet is cooled to absolute zero, the

north and south poles of this magnets can be reversed effortlessly, causing havoc to enter the magnetic storage system. Another group clings to the concept of classical mechanics and the idea that "spins" cannot occur without receiving energy — thus a so-called "energy barrier." By contrast, consistent with quantum mechanics, there is a "chance" that spins will be capable of breaking through the energy barrier. As pointed out by R. Ruthen, "This phenomenon is analogous to the ability of an electron to tunnel through an energy barrier."

Thus, a profound difference of opinion remains and will require continued research to find the truth.

An excellent description of the early investigations of electricity and magnetism is given by L. Pearce Williams (reference listed).

Additional Reading

Aharoni, A.: *Introduction to the Theory of Ferromagnetism*, Oxford University Press, Inc., New York, NY, 2000.

Amato, I.: "Some Molecular Magnets Like It Hot," *Science*, **1379** (June 7, 1991).

Beeteson, J.S.: *Visualising Magnetic Fields*, Academic Press, Inc., San Diego, CA, 2000.

Campbell, W.H.: *Earth Magnetism*, Academic Press, Inc., San Diego, CA, 2000.

Chatterji, T.: *Neutron Scattering from Magnetic Materials*, Elsevier Science & Technology Books, New York, NY, 2006.

Chin, G.Y.: "Magnetic Materials," in *Encyclopedia of Materials Science and Engineering*, MIT Press, Cambridge, MA, 1986.

Comstock, R.L.: *Magnetic Recording*, John Wiley & Sons, Inc., New York, NY, 1999.

Craik, D.J.: *Electricity, Relativity and Magnetism*, John Wiley & Sons, Inc., New York, NY, 1999.

Edmonds, D.: *Electricity and Magnetism in Biological Systems*, Oxford University Press, Inc., New York, NY, 2001.

Guimarces, A.P.: *From Lodestone to Supermagnets: Understanding Magnetic Phenomena*, John Wiley & Sons, Inc., Hoboken, NJ, 2005.

Hamilton, D.P.: "A Reprieve for MIT's Magnet Lab," *Science*, **850** (August 23, 1991).

Kubler, J.: *Theory of Itinerant Electron Magnetism*, Oxford University Press, Inc., New York, NY, 2000.

Majils, N.: *The Quantum Theory of Magnetism*, World Scientific Publishing Company, Inc., Riveredge, NJ, 2000.

Makarova, T., and F. Palacio: *Carbon-Based Magnetism: An Overview of the Magnetism of Metal Free Carbon-Based Compounds and Materials*, Elsevier Science & Technology Books, New York, NY, 2006.

Manriquez, J.M. et al.: "A Room-Temperature Molecular/Organic-Based Magnet," *Science*, **1415** (June 7, 1991).

Miller, J.S., A.J. Epstein, and W.M. Reiff: "Molecular/Organic Ferromagnets" *Science*, **40** (April 1, 1988).

Miller, J.S., M. Drillon: *Magnetism: Molecules to Materials V*, John Wiley & Sons, Inc., Hoboken, NJ, 2005.

Narlikar, A.V.: *Frontiers in Magnetic Materials*, Springer-Verlag New York, LLC, New York, NY, 2005.

Prinz, G.A.: "Hybrid Ferromagnetic-Semiconductor Structures," *Science*, **1092** (November 23, 1990).

Sellmyer, D.J., Y. Liu, and D. Shindo: *Handbook of Advanced Magnetic Materials*, Springer-Verlag New York, LLC, New York, NY, 2005.

Ruthen, R.: "Quantum Magnets," *Sci. Amer.*, **28** (July 1991).

Staff: *Glencoe Science: Electricity and Magnetism*, 2nd Edition, Glenco/McGraw-Hill, New York, NY, 2004.

Ulrich, S., R.F. Bishop, J. Richter, and J.J. Farnell Damian: *Quantum Magnetism*, Springer-Verlag New York, LLC, New York, NY, 2004.

Williams, L.P.: "AndrÉ-Marie AmpÈre," *Sci. Amer.*, **90** (January 1989).

Winters, A.J., et al.: "Large-Scale Superconducting Separator for Kaolin Processing," *Chem. Eng. Progress*, **36** (January 1990).

MAGNETITE. [CAS: 1309-38-2]. The mineral magnetite, ferroferric oxide, Fe_3O_4, is isometric, commonly occurring in octahedrons, dodecahedrons, and massive, granular, and laminated forms. It is brittle with an uneven fracture; cleavage is not distinct, but with pressure an octahedral parting may develop; hardness, 5.5–6.5; specific gravity, 5.18; luster, metallic to dull; color, iron black, streak, black. It is opaque and strongly magnetic; when possessing polarity it is known as lodestone. Important large ore bodies are products of magmatic segregations, with titanium a prominent constituent of such deposits. Magnetite is a common mineral in the igneous rocks, especially those of the ferromagnesian varieties, and is found in many metamorphic types. It is associated with corundum in emery.

In northern Sweden are located what may be the largest magnetic deposits in the world, believed to have been formed by segregation in the magma. Magnetite is also found in Norway, in the Urals, Italy, Switzerland, Australia, and Brazil. In the United States, the Precambrian rocks of the Adirondacks contain large beds of magnetite, as well as extensive deposits of titaniferous magnetite, and the mineral is found also in New Jersey, Arkansas, and Utah. In Canada it is found in Quebec and Ontario. The Iodestone or natural magnet is found in Siberia, the Harz Mountains, Germany; the Island of Elba; and at Magnet Cove Arkansas. The name magnetite is said to be derived from the district of Magnesia, near Macedonia. There is, however, a fable that it was named for a shepherd, Magnes, whose iron-bound staff and shoes with iron nails struck to the ground in which magnetite was present.

This mineral is an important ore of iron, 72% being metallic iron. Magnetite sometimes is referred to as magnetic iron ore.

See also **Ocean Resources (Mineral)**.

ELMER B. ROWLEY, Union College, Schenectady, NY

MAGNETO. An alternating-current generator that uses one or more permanent magnets to produce its magnetic field. Also called a magneto-electric generator.

MAGNETOCHEMISTRY. A subdivision of chemistry concerned with the effect of magnetic fields on chemical compounds; analysis and measurement of these effects, (e.g., magnetic moment and magnetic susceptibility) are important tools in crystallographic research and determination of molecular structures. Substances that are repelled by a magnetic field are diamagnetic (water, benzene); those that are attracted are paramagnetic (oxygen, transition element compounds). Diamagnetic materials have only induced magnetic moment; paramagnetic materials have permanent magnetic moment. Magnetochemistry has been useful in detection of free radicals, elucidation of molecular configurations of highly complex compounds, and in its application to catalytic and chemisorption phenomena. See also **Nuclear Magnetic Resonance (NMR) and Magnetic Resonance Imaging (MRI)**.

MAGNETOELECTRIC. Of or pertaining to electricity produced by or associated with magnetism. Electromagnetic pertains to magnetism produced by or associated with electricity.

MAGNETOHYDRODYNAMIC GENERATOR. The greatest concentration of research, development, and design effort in the technology of converting sources of energy into electricity (for use by the electric utility field) has been concentrated on rotating equipment. However, since about the mid-1970s, there also has been considerable research and experimentation with new kinds of batteries, fuel cells, thermionic converters, and magnetohydrodynamic generators. The possible use of magnetohydrodynamic generators in times of crisis that affect vast, interconnected electrical networks and which have resulted in wide regional blackouts in a few instances has received considerable attention.

The magnetohydrodynamic (MHD) generator is one in which a thermally ionized gas is forced at high temperature, pressure, and velocity through a duct situated in a transverse magnetic field. An induced voltage appears in the third mutually perpendicular direction (the Hall effect), and this voltage may be tapped by electrodes within the duct. If the exhaust gas from the MHD generator is used to heat steam for a conventional generator, a larger portion of the thermal spectrum will be utilized and the system efficiency may be raised from the present approximately 40% to possibly 50 or 55%. Heat for the system may come from the use of fossil fuel or nuclear reactors. An additional advantage of the MHD generator would be the absence of rugged moving parts and close tolerances in the MHD portion of the cycle.

The comparative conceptual simplicity of the MHD generator is evident from Fig. 1. The steady-state operation of this unit is governed approximately by the following one-dimensional flow equations:

Force/Unit Volume

$$\rho v \frac{dv}{dx} + \frac{dp}{dx} - (\underline{J} \times \underline{B})_x + F = 0$$

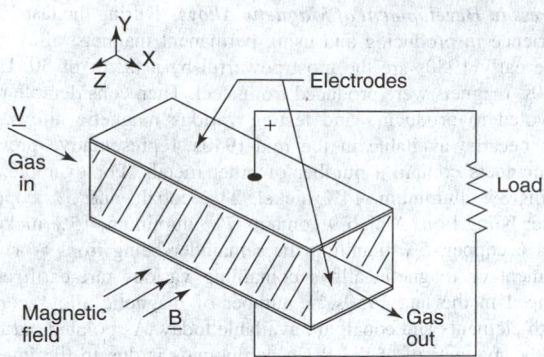

Fig. 1.   Highly simplified diagram of MHD generator.

Power/Unit Volume

$$\rho v \left(\frac{dH}{dx} + \frac{dv}{dx} \right) - \underline{J} \cdot \underline{E} + G = 0$$

Mass Flow

$$\rho v A = \text{constant}$$

and by Ohm's law, modified to account for magnetic fields and Hall effect:

$$J + \mu(\underline{J} \times \underline{B}) = \sigma[E + (\underline{v} \times B)]$$

where ρ = gas density
$\underline{v}$ = gas velocity factor
ρ = gas pressure
$\underline{J}$ = current density vector
$\underline{B}$ = applied magnetic field
F = wall friction force/unit volume
H = gas enthalpy
$\underline{E}$ = electric field vector
G = heat loss through walls/unit volume
A = cross sectional area of duct
μ = electron mobility in gas
σ = electrical conductivity in gas

The last equation is of most interest in working with the electrical load, since it greatly influences the way in which power is drawn from the unit. The axial component of $(\underline{J} \times \underline{B})$ produces an axial field, called the Hall field, which may be several thousand volts/meter. This causes currents to flow along the electrodes, creating high ohmic losses. To minimize these losses, the electrodes are segmented and separated by strips of insulating material, as shown in Fig. 2. A large MHD unit would typically have over 100 segments.

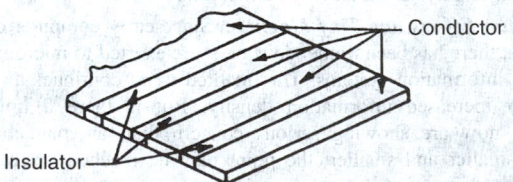

Fig. 2.   Section of segmented electrode.

For a duct with segmented electrodes, there are several ways to connect the load. Two of these connections are shown in Fig. 3. In the Faraday connection (a), the total load is divided among the electrodes; each electrode segment then behaves like a separate MHD generator. The electrical properties can be derived from the last equation by setting J_x (the axial component of current) = 0 because of the segmenting, and $J_y = -\alpha E_y$, where α represents the load admittance. The Faraday connection makes most efficient use of the thermal energy of the gas, but the requirement of multiple load connections is a drawback. This shortcoming can be overcome, however, by paralleling inverter-transformers at the ac line. The Hall connection uses the induced Hall potential to drive the load. The top and bottom of each electrode segment are joined, so that no transverse electric field exists.

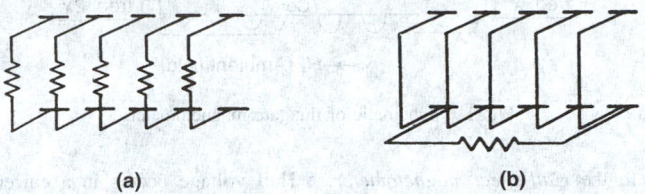

(a)   **(b)**

Fig. 3. Load configuration for segmented electrode generator: (a) Faraday connector; (b) Hall connection.

Principal Applications to Date. Particularly in connection with the investigation of the effectiveness of MHD units to supplement or reduce required system load shedding in a situation where an island is formed with insufficient generation and with resulting declining frequency, it is believed that the use of quick-responding MHD units can significantly reduce the amount of required underfrequency load shedding, mainly in situations where the sum of the rating of the MHD units plus fast-acting system reserve is approximately equal to the load/generation imbalance at the time of separation. The benefits appear to be less pronounced, but are still significant even if this sum is only one-half of the imbalance. The MHD unit seems most effective when all of the bootstrap power output of the unit is used to build up the unit magnetization to rate level before providing power to the system. Bus voltage oscillations seem less severe for this procedure than for the approach where the MHD magnet is pre-energized, or where the unit magnetization is allowed to build up more slowly (by using only part of the bootstrap startup power for energizing purposes, with the remainder allowed to flow into the power system).

MHD units (as proposed for quick-start, short-service devices) do not seem particularly appropriate for subduing short-term (0–3 second) power system swings for two reasons. (1) In order to provide the response time required to be effective, 15 megawatts of continuous excitation would be required to obtain quick availability (20–40 milliseconds) of 200 megawatts for each unit. (2) Quite a few units would have to be scattered throughout the system to provide coverage for even a relatively small number of possible contingencies. It is obvious that the costs associated with the required continuous excitation power and the number of units needed for reasonable system coverage would not make this application of MHD units attractive.

It is generally concluded that quick-start MHD units should not be considered as an alternative to underfrequency load-shedding practices, but rather as a supplement to these practices. Underfrequency load shedding is attractive for combating rapidly dropping frequency because it is effective, has comparatively low installation costs, can be implemented independently at many scattered substations throughout the system with resulting high reliability, and can be tailored rather easily to the power system's requirements. However, if a particular power system operator did not choose to incorporate underfrequency load shedding, a sufficient number of quick-responding MHD units could be used to boost drooping frequency during the time period (up to several minutes), when the reserve capability of the other prime movers is being realized. Hence, this would potentially allow time for centrally controlled manual load shedding, if required. In this situation, the rating of MHD units would be more effective than the spinning reserve of conventional prime movers. This is because the MHD units are much faster to respond and do not suffer from such problems as the steam depletion problem inherent to many steam units after about one minute of significantly increased output.

The potential use of MHD units as infrequently-used peaking units may be considered, but other types of peaking units, such as gas turbines and diesel units, become very competitive for usages exceeding 100 hours/year.

MAGNETOHYDRODYNAMICS (MHD).

The behavior of high-temperature ionized gases passed through a magnetic field. A power-generating method using MHD involves an open cycle in which hot combustible gases from coal, seeded with cerium or potassium to increase electrical conductivity, constitute the working fluid. These are sent through a nozzle surrounded by a magnet; the electricity induced by movement of the ionized gas through the magnetic field is passed to electrodes and the gas sent to a steam generator. Efficiency is rated at 50–60% compared with 40% for conventional fossil fuel plants and 33% for plants using nuclear fuels. Two-phase liquid-metal systems are being studied as auxiliary units for a number of energy converters. MHD is an important field of expansion of research activity on new sources of energy; its high efficiency and low pollution factor indicate that it may have a significant future in electric power supply.

MAGNETOIONIC THEORY.

The theory of propagation of electromagnetic radiation through a medium containing ions in the presence of an external magnetic field. It applies to the propagation of radio waves in the ionosphere, and provides theoretical relationships among such aspects of the subject as the index of refraction, radiofrequency, free-electron density, electron collision frequency, the earth's magnetic field (components relative to the direction of propagation), the nature of polarization, etc. See **Magnetic Double Refraction**.

MAGNETOMETER.

An instrument for measuring the magnitude and sometimes the direction of a magnetic field. Specific areas of application include: (1) determining the magnetic moment of a specimen and the magnetization of materials (2) calibrating electromagnets and permanent magnets; and (3) measuring the strength of magnetic fields and their components on or near the surface of the earth, as well as in space. Numerous physical principles are applied in a variety of magnetometers.

The *classical astatic magnetometer* is used to determine the magnetic moment of rod-shaped samples. The specimen can be exposed to the controllable homogeneous magnetic field produced in the center region of a solenoid that is appreciably longer than the specimen. Its magnetic moment is characterized by strength and direction of the field at a known distance from the sample. Two equivalent permanent magnet needles horizontally placed and rigidly linked by a nonmagnetic rod in a vertical position comprise the measuring system. Arranged in antiparallel alignment, the needles have a wide distance between them, as compared to length. This astatic system, unaffected by the earth's field, is suspended on a calibrated torsion wire. The axis of the test specimen is placed perpendicular to the axis of the lower needle and in its plane of rotation. It is possible to cancel the field of the magnetizing solenoid at the location of the sensing needle by means of a magnetically opposing solenoid located in the same plane. The magnetic moment of the specimen can be derived from the angular deviation of the needle occurring from the action of the static field. With a modified astatic magnetometer, the magnetic properties of very small specimens in the form of fine wires or films have been measured. The specimen is arranged parallel to the axis of the astatic system and a short distance from the closely spaced short needles, in order that each needle will sense the pole of the neighboring sample. With this method, films as thin as 10^{-5} centimeter and weighing less than 1 milligram have been studied.

A more recent instrument for measurement of very small samples is based upon the relative periodic displacement of the dipole field of the specimen against pickup coils and subsequent amplification of the small, induced alternating current voltages.

The *vibrating sample magnetometer* makes use of a mechanical oscillator that can be either a loudspeaker or a motor. The sample, attached to the lower end of a nonmagnetic shaft, moves up and down, between and parallel to the pole faces of an electromagnet. Amplitude of the oscillation is approximately 2 millimeters, the frequency is approximately 80 Hz with loudspeaker drive. The sample stays within the homogeneous region of the magnetizing field. Two series opposing signal coils with approximately 20,000 turns each are located on each side of the sample, with axes parallel to its motion and perpendicular to the exciting field. The magnetic moment can be derived directly by comparison with the voltage excited by a standard sample of known magnetization, such as nickel.

In the *vibrating coil magnetometer*, the sample is kept in a fixed position in the homogeneous region of the magnetic field between the pole faces. The signal coil oscillates at approximately 40 Hz, its axis and velocity vector being collinear with the dipole axis of the specimen. The distance between coil and sample is large enough to allow for installation of temperature- and pressure-generating apparatus to enclose the sample.

Special precautions have to be taken to eliminate the signal produced by the curvature of the magnetizing field. The sensitivity of this method is considered excellent.

The *pendulum magnetometer* is a rather simple apparatus, which utilizes the ponderomotive force that a sample experiences in an inhomogeneous magnetic field. The device is based on the concept that this force is, for small deviations from the position of maximum field strength, proportional to the displacement perpendicular to the field lines between the hemispherical pole pieces of an electromagnet. This condition creates a simple harmonic motion. The specimen is fastened to a lightweight bar whose movement is constrained only in the direction of its length and without rotation by a quinquefilar suspension. The magnetization is determined from measurements of the periods of oscillation with and without magnetic field.

In the *vibrating reed magnetometer*, the pendulum is replaced by a metallic reed of nonmagnetic material, and the sample is attached to one end. The vibration of this spring is excited by a piezoelectric transducer driven from an oscillator. The resonance frequency of the reed is observed with and without field. The magnetization can be calculated by comparison with a reference sample.

The *gaussmeter* or *fluxmeter* is a laboratory instrument used for calibrating electromagnets. It consists of a small direct-current generator. A small generator coil is wound on a nonmagnetic core, placed at the end of a 3- to 4-foot long axis, and driven by an alternating current motor with constant speed of revolution. The induced sinusoidal voltage is rectified by a commutator and read on a voltmeter calibrated in magnetic field units. The device is inherently linear.

The *earth inductor* is used for measurement of the inclination (magnetic dip) of the earth field. A coil, connected through commutator and brushes to a sensitive galvanometer, is rotated about its diameter by a hand-powered flexible shaft. When the rotation axis of the coil is brought in line with the earth field, the galvanometer will read zero. The inclination of the axis against the horizontal plane can be read with an accuracy of approximately 0.1 minute.

The *classical magnetometer of Gauss* can be used to determine the horizontal field intensity of the earth field in stationary observatories. Two measurements are required: (1) the measurement of the period of a permanent magnet, which is vertically suspended on a torsion fiber, when it oscillates in the horizontal plane about the magnetic meridian; and (2) the measurement of the deflection angle, which a magnetic needle experience attached to the same suspension, when the permanent magnet acts at a present distance in a preferred position together with the earth field upon the needle. This is an absolute method.

The *sine galvanometer* is an absolute instrument for determination of horizontal intensity. With it, the suspended detector magnet is acted upon by a field produced by a calibrated coil and the horizontal field component. Accurate measurements of the coil current and deflection angle of the needle are required.

The *flux-gate magnetometer* is well suited to airborne applications and is used for detecting magnetic anomalies and for exploring the earth's surface in search of mineral deposits. The instrument also has been used in space probes, and in underwater research.

The operation is based upon the change of permeability of a highly sensitive material in weak fields. The device incorporates two permalloy cores in parallel position. A coil is wound on each of the cores. The two windings are opposed in their polarities and connected in an impedance bridge circuit in such a manner that an alternating current voltage supplying the bridge will not produce a diagonal voltage. The bridge is balanced in the absence of an external field. See Fig. 1. If a component of the earth's field parallels the axis of the core, the bridge is unbalanced due to the opposing magnetic biases. In use, three mutually perpendicular flux gates are mounted on a platform that can be rotated by servomotors in two perpendicular planes. Activated by the diagonal voltage from one flux gate, each servomotor rotates its twin core, decreasing the unbalance and holding the cores in zero position. The combined action of two gates then brings the cores of the third gate into the direction of the field. An additional field winding encloses both cores of the third gate. A servomechanism controlled by the diagonal voltage of this gate provides a current through the field winding, in order to annul the earth field. The magnitude of this current is a measure of the strength to the earth's field. The instrument is sensitive to about 10^{-5} oersted.

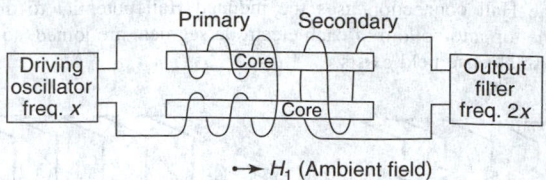

Fig. 1.   Schematic of fluxgate magnetometer.

In the *Hall-effect magnetometer*, a Hall voltage occurs in a current-carrying sample of semiconducting material perpendicular to the current and perpendicular to an applied static magnetic field. The magnitude of this voltage is proportional to the field. The Hall voltage amounts to about 100 millivolts at 10,000 gauss, with a 100 milliampere current. This is primarily a laboratory device.

Nuclear magnetometers are also well adapted to airborne measurements. They utilize the magnetic properties of atomic structures. The two principal types are: (1) Proton precession and (2) alkali vapor. Both types have the advantage of being insensitive to the direction of the field and of producing absolute measurements in terms of frequency, the physical quantity most easily measured with high accuracy. See Fig. 2. The *alkali-vapor magnetometer* makes use of the fact that if circularly polarized light is shone through vapor of an alkali metal that is being excited by a varying radio-frequency magnetic field, light is absorbed at certain frequencies related to the energy levels in the atoms of a steady external magnetic field. By measuring the absorption frequencies, it is possible to determine continuously the external magnetic field.

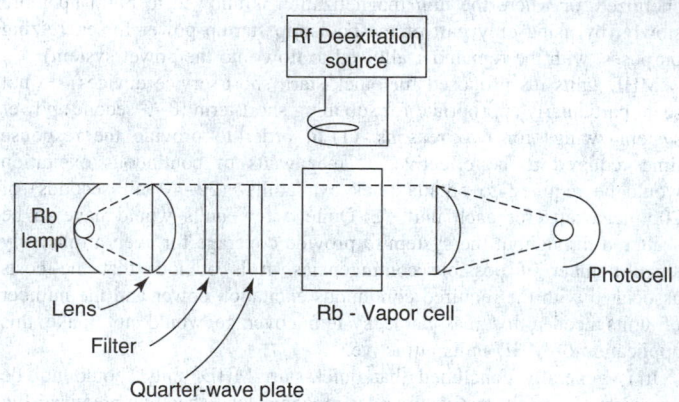

Fig. 2.   Schematic of rubidium-vapor magnetometer with no signal feedback.

The *metastable-helium magnetometer* was developed for spaceborne applications, which require measurement of changes in planetary magnetic fields accurate to 1 part in 5 million and a sensitivity of 0.01 ores;d. In operation, the system measures the absorption of infrared radiation by helium in a metastable state. Mechanically, a helium spectral-emission lamp and a helium-filled absorption cell are excited by a radio frequency source. A sensitive infrared detector then measures the amount of infrared radiation absorbed by the metastable helium. This amount is proportional to variations in planetary magnetic fields. The system also is used in antisubmarine warfare detection systems.

MAGNETO-OPTICAL ROTATION. Some substances are in themselves optically active, that is, they rotate the polarization plane of polarized light passed through them. In 1845, Faraday discovered that glass and other substances devoid of this property acquire it when placed in a strong magnetic field. This is the so-called Faraday effect. The light must traverse the substance along the lines of force. The direction of the rotation is reversed if the field is reversed, but is the same with respect to the observer whether the light is going or coming, so that a beam passing one way and reflected back has its rotation thereby doubled (which is not the case with natural activity). Some substances produce right-handed, some left-handed, rotation in light traveling in the direction of the field.

MAGNETOPAUSE.   The sharp boundary between the earth's magnetosphere and the solar wind of interplanetary space. The magnetopause represents the outer termination of the earth's atmosphere and extends from a distance of several earth-radii in the sunward direction to a much larger and rather indefinite distance in the anti-sunward direction, resulting in a cometlike shape with a long tail directed away from the sun. See also **Atmosphere (Earth)**; and **Interplanetary Medium**.

MAGNETOSPHERE.   The region of the Earth's atmosphere where ionized gas plays an important part in the dynamics of the atmosphere and where the geomagnetic field, therefore, plays an important role. The magnetosphere begins, by convention, at the maximum of the *F layer* at about 350 kilometers and extends to 10 or 15 Earth radii to the boundary between the atmosphere and the interplanetary plasma. See also **Earth**.

MAGNETOSPHERIC MULTISCALE (MMS) MISSION.   See **Space Science Missions: Sun**.

MAGNETOSTATIC THEORY.   See **Electromagnetic Phenomena**.

MAGNETOSTRICTION.   When a polycrystalline nickel sample is placed in a magnetic field, it contracts along the field direction by about 30 parts per million and elongates in the transverse direction by about half that amount. There is also a small volume change. Such changes in dimension of magnetic materials with variation of magnetic field strength or direction are termed *magnetostriction*. See also **Magnetic Materials: Bulk**. They are measured by strain gages, optical dilatometers, capacitance variation, and x-ray analysis. Below the Curie Temperature, magnetostriction in weak fields is caused by domain rotation, becoming appreciable at fields near the knee of the *B-H* curve. See also **Hysteresis**. The saturation magnetostriction of a single crystal depends upon the direction of the (sublattice) magnetization and the direction of measurement with respect to the crystal axes. Magnetostriction coefficients vary greatly, depending upon the material, temperature, and the magnetization state. The source of magnetostriction is the dependence of magnetic energy on strain. Because the elastic energy is quadratic in strain while the magnetoelastic energy is linear in strain, the minimum free energy occurs at nonzero strain.

Magnetostriction has been put to practical use in the magnetostrictive resonator. This is essentially an iron rod maintained in longitudinal elastic vibration by a high-frequency current in a helix wound upon it, and used, through the joint operation of the Joule and Villari effects, to control the frequency of the current, somewhat after the manner of the familiar piezoelectric (crystal) resonator. It is also used in band-pass electrical wave filters.

MAGNETOSTRICTIVE DELAY LINE.   In electronic computers, a device in which a wave is induced by the characteristic, possessed by nickel and certain other materials, of shortening in length when placed in a magnetic field. The wave travels at the speed of sound through the material.

MAGNETRON.   A self-excited oscillator used as a radar transmitter tube. Magnetrons are characterized by high peak power, small size, efficient operation, and low operating voltage. Emitted electrons interact with an electric field and a strong magnetic field to generate microwave energy. Because the direction of the electric field that accelerates the electron beam is perpendicular to the axis of the magnetic field, magnetrons are sometimes referred to as crossed-field tubes. Unlike a klystron, a magnetron is not a coherent transmission source, but has a randomly changing phase from pulse to pulse. A coaxial magnetron uses a different architecture and has better stability, higher reliability, and longer life. Magnetrons are used in inexpensive radars and microwave ovens.

See also **Klystron**; and **Microwave Tubes**.

MAGNIFYING POWER.   Crudely defined, the magnifying power of an optical instrument is the ratio of the apparent size of an object as seen through the instrument to the apparent size of the same object as seen without the instrument. For a telescope, the magnifying power may be defined as the ratio of the size of the retinal image obtained with the instrument to the size of the retinal image obtained without any optical aid. When a positive eyepiece is used, the magnifying power may be shown to be directly equal to the ratio, of the focal length of the object glass of the telescope, to the focal length of the eyepiece. For example, a telescope with an object glass of 10-foot (3-meter) focal length will have a magnifying power of 60 when used with a positive eyepiece of 2-inch (~13-millimeter) focal length. It should be noted that changing the eyepiece will change the magnifying power of the telescope. That is, if an eyepiece with 0.5-inch (5-centimeter) focal length were used with the above telescope, the magnifying power of a telescope with interchangeable eyepieces is meaningless unless the particular eyepiece is specified for the particular telescope.

MAGNITUDE (Stellar).   See **Stellar Magnitude**.

MAGNOLIA TREES.   Members of the magnolia family (*Magnoliaceae*), there are some 35 species, nearly all of which prefer the warm climates found in the southern United States and in parts of India, China, and Japan. However, the evergreen form (*Magnolia grandiflora*) ranges from New Brunswick in Canada south in the eastern United States to Florida. The tree requires rich soil and considerable water. Adult trees may attain a height of over 100 feet (30 meters). The general form is pyramidal. The bark is gray, rough, and has thin scales. Because of the wide spread of the tree, it is a favorite for shade and landscaping. The leaf is large, elliptical, and from 5 to 8 inches (12.7 to 20 centimeters) in length. The leaf is smooth, of a lustrous green color, and has a leathery feel. The flower is creamy white, very showy, fragrant, and from 6 to 8 inches (15.2 to 20 centimeters) across. Depending upon area, the trees bloom from April to June. The fruit is ovoid, 3 to 4 inches (7.6 to 10 centimeters) in length, and of a brown color when ripe. The wood is soft, light, satiny, close-grained, and of light ocher color and of no commercial value. Several record magnolias have been selected by American Forests. See Table 1. The tulip tree, with tulip-shaped, bluish-green leaves, greenish-yellow flowers, and a long, conelike fruit, is also a member of *Magnoliaceae*.

Breakthrough in Magnolia's DNA. The Clarkia shale deposit near Moscow, Idaho, differs from most rich fossil sources by the manner in which ancient fossils are preserved. Most fossils do not contain DNA because they are completely mineralized over long periods of time. The fossils at Clarkia are classified as compression fossils — that is, cold, oxygen-free sediments in a lake bottom squeezed a magnolia leaf for an estimated 20 million years. Fossils from this region contain biomolecules and subcellular structures, like those of modern plants. Researchers from the University of California at Riverside and the University of Georgia found the leaf unchanged. D.E. Giiannasi observed, "When we first cracked open the sediment, the leaf was still dark green." Analysis of the magnolia leaf has yielded some important genetic information. Scientists now are directing their attention to other trees. This research represents an early step in the melding of molecular biology and paleontology to shed light on plant evolution.

MAGPIE (*Aves, Passeriformes*).   Moderately large long-tailed birds related to the crows. The common North American species, *Pica pica*, ranges from Alaska to Arizona and eastward into Iowa. It is a black and white bird. The yellow-billed magpie, *P. nuttalli*, flies only in California. The European and Asiatic species of magpies are more brightly colored.

The magpies build very large untidy nests and are noted for their curiosity, adaptability, and noisiness.

MAHOGANY TREES.   Members of the mahogany family (*Meliaceae*), these trees are found in Central America and the West Indies. A few related species are found in tropical Africa. Species of the genus *Swietenia*, are large trees with pinnately compound leaves like those of ash trees and small flowers in panicles in the leaf axils.

The trees grow in a variety of habitats, often in most inaccessible places, so that it is very difficult to get the cut logs to the market. Mahogany is usually classified according to the region from which it comes, as Cuban mahogany, Honduras mahogany, etc.

The first mahogany to appear in Europe was brought in as ballast in a ship, the heavy logs being very suitable for that purpose. In port it

TABLE 1. RECORD MAGNOLIAS IN THE UNITED STATES[1]

Specimen	Circumference[2]		Height		Spread		Location
	Inches	Centimeters	Feet	Meters	Feet	Meters	
Ashe magnolia (1993) (*Magnolia ashei*)	55	140	52	15.8	37	11.3	Pennsylvania
Bigleaf magnolia (1999) (*Magnolia macrophylla*)	55	140	67	20.4	32	9.8	Georgia
Cucumbertree magnolia (1985) (*Magnolia acuminata*)	293	744	75	22.9	83	25.3	Iowa
Frasier magnolia (1998) (*Magnolia acuminata*)	118	300	121	36.9	33	10.1	Tennessee
Pyramid magnolia (1988) (*Magnolia pyramidata*)	62	157	65	19.8	32	9.8	Florida
Pyramid magnolia (1999) (*Magnolia pyramidata*)	46	117	84	25.6	30	9.1	Florida
Southern magnolia (1994) (*Magnolia grandiflora*)	268	681	98	29.9	90	27.4	Mississippi
Sweetbay magnolia (1991) (*Magnolia virginiana*)	173	439	92	28	52	15.8	Arkansas
Umbrella magnolia (1993) (*Magnolia tripetala*)	122	310	50	15.2	50	15.2	Pennsylvania

[1] From the "National Register of Big Trees," American Forests (by permission).
[2] At 4.5 feet (1.4 meters).

was necessary to remove these in order to get in more cargo. So it was offered to English woodworkers. At first it was rejected as too hard to work and of little use, being heavy and dark-colored. Gradually it found favor, until it became a highly prized wood for fine furniture making. The first mahogany used in cabinet work was Spanish mahogany, *Swietenia mahagoni*. For a long time mahogany from Santo Domingo held first rank and was eagerly sought after. It was a very hard dark wood, which could be given a very high polish, and was very durable. The supply is now nearly exhausted. Cuban mahogany is another variety which gives a dark red wood, and which finishes with a very fine glossy surface. In this, as in some other varieties, the wood is frequently marked with very small white pores of chalk-like substance. With age the wood of these varieties gradually darkens; it does not, however, lose its beautiful smooth finish. Other species of *Swietenia* are shipped from Panama, from Mexico, and from South American countries.

The heavy logs of mahogany are removed from their native forests and shipped to American or foreign markets. Especially valuable are those logs which, when cut, show a wavy grain or other irregularities. Even more valuable are blocks of mahogany that come from a large forking of the stem; from these the beautifully grained crotch mahogany is obtained. This is usually cut into thin veneers. In drying, mahogany shrinks very little, and once dry, it is very durable, twisting or warping very little.

The record West Indies mahogany (*Swietenia mahagoni*) growing in the United States is located in Everglades National Park, Florida. See Table 1.

African mahogany is becoming increasingly valuable. It is obtained from large trees of the genera *Khaya* and *Entandrophragma*, both members of the same family as mahogany, and yielding woods very similar to it. Numerous other tropical woods are imported under the name of mahogany, because of a similarity in texture or color. Some of these are valuable, but many are definitely inferior substitutes.

The so-called Tree of Heaven or Tree of the Gods, commonly known as ailanthus, is also a genus of trees in the *Meliaceae* family. They have large pinnate leaves, small flowers, and winged fruit. One species, *Ailanthus glandulosa*, is a native of China and Japan. It has been introduced into the eastern United States, largely because of its excellent resistance to atmospheric pollutants. One objectionable feature is the unpleasant odor of the male flowers. One specimen singled out by American Forests is located in Hamilton Tennessee. See Table 1.

The lauan tree, which grows in the Philippines, Malaya, and Sarawak, is also of the *Meliaceae* family. Several genera of the trees grow in these areas. On the American market, the wood is known as Philippine mahogany, obtained chiefly from *G. shorea*. Dark-red Philippine mahogany, called tangil, is obtained from *S. polysperma* and most closely resembles true mahogany. It is also called Bataan mahogany. The wood from numerous varieties of the lauan tree ranges widely in color and other characteristics — so that when specifying mahogany, one must be quite exacting as to source.

The Chinatree (*Melia azedarach*) is also of the *Meliaceae* family. The tree is native to Asia, but is found in the southern parts of the United States and in Mexico. The tree generally is considered an attractive, ornamental shade tree. The top is nearly flat. The leaves are deciduous, alternate, and bi-pinnately compound. They are about 15 in. (38 cm) long, pointed, and light green. The tree grows rapidly, but has a rather short life span. The fruit is a fleshy, yellow berry, less than 1 in. (2.5 cm) in diameter. The flower is purple and fragrant and hangs in long clusters.

The tree is commonly referred to in the southern United States as the chinaberry tree, although other names include pride of India tree, umbrella tree, Cape lilac, Indian lilac, Persian lilac, China tree, and paradise or paraiso tree. In an article by M.D. Hodgins (American Forests, 22–61, May 1979), the very extensive plantings of this tree along the streets of Savannah, Georgia are described. The pale lilac blossoms of the tree perfume the air. Although still proliferating in suburban and rural areas throughout the southeastern United States, urban use was progressively discouraged because, as pointed out by H.S. Traub, "the 'smelly, gooey

TABLE 1. RECORD MAHOGANY FAMILY TREES IN THE UNITED STATES[1]

Specimen	Circumference[2]		Height		Spread		Location
	Inches	Centimeters	Feet	Meters	Feet	Meters	
Tree-of-Heaven (1999) (*Ailanthus altissima*)	248	630	67	20.4	64	19.5	Tennessee
Chinaberry (1967) (*Melia azedarach*)	222	564	75	22.9	96	29.3	Hawaii
West Indies mahogany (1992) (*Swietena mahagoni*)	175	445	79	24.1	96	29.3	Florida

[1] From the "National Register of Big Trees," American Forests (by permission).
[2] At 4.5 feet (1.4 meters).

mess' the trees made when they dropped their berries led to their removal from the historic district of Savannah."

The record chinaberry tree (*melia azedarach*) growing in the United States is located in Kaobe, South Kona, Hawaii. See Table 1.

MAIDENHAIR TREE. The sole surviving member of an order of *Gymnosperms* which in Mesozoic times was very abundant and widely distributed. Also called the Ginkgo tree, it is of a single-species family *Ginkgoaceae* (Ginkgo family). There is uncertainty as to whether the tree still may be growing wild in any region today. Occasional reports are received indicating that the tree has been seen on the mountain slopes of China. The tree has been cultivated in the temple gardens of China and Japan for centuries. In late years, the Ginkgo has been widely planted in Europe, America, and other parts of the world. Because the tree dates back some 200 million years, some authorities observe that the tree has outlived its natural enemies. But even in modern times, the tree seems quite resistant to human-created environments, including the fumes from buses and cars along streets where it occasionally will be seen planted in a masonry tub or opening in the sidewalk. The normal height of an adult tree is about 70 to 80 feet (21 to 24 meters). One specimen in Milan, Italy was reported to have reached 125 feet (37.5 meters). Planters like to avoid female trees because the covering over the nut (edible) radiates a most unpleasant odor' when it decomposes. Although the tree does not have needles or cones, it frequently is classified with the conifers because of its assumed ancestral relationship to the conifers.

The tree often has a tall slender pyramidal shape when young; others, and especially older specimens, are wide spreading. The branches are of two kinds; a long shoot that grows rapidly in length, composed mostly of woody tissue; and short shoots or spurs which elongate very slowly. These short shoots have a large pith, a thick cortex and very little wood. A short shoot may sometimes (especially in the case of injury to the long shoot) become a long shoot. The leaves of the ginkgo tree are somewhat variable in shape. Those on the long shoot are wedge-shaped and deeply notched, those of the short shoot broadly wedge-shaped and little or not at all notched. The veins are furcate, forked, as in the ferns. It is to the leaves that the tree owes its common name, Maidenhair tree, since their shape suggests that of the maidenhair fern. The trees are dioecious, the two types of reproductive organs being borne on different trees. The male strobilus is composed of many sporophylls, each with two sporangia. The female counterpart consists of a long slender peduncle or stalk bearing two ovules. In most cases, one of those aborts early. The pollen grains, which consist of three small disk-shaped cells and one relatively large one, are carried to the ovule by the wind. There each forms a pollen tube, which digests its way through the tissue surrounding the gametophyte. A pollen tube that has nearly reached the gametophyte contains two large sperms, each of which has a spiral coil of cilia at its anterior end. One of these sperms passes to one of the large eggs contained in the female gametophyte, and joins with it. The nucleus of the sperm unites with that of the egg, which is then said to be fertilized. The ovule containing the fertilized egg enlarges. When mature it is about an inch in diameter and green. Its outer covering is fleshy and has a curious rancid odor, which is very noticeable when this coating is crushed. Within this fleshy coat is a dry covering surrounding the gametophyte and the embryo plant. The seeds germinate readily, forming a long tap root and a short erect shoot. Young plants are rather susceptible to low temperatures, requiring some protection in the northern states.

MAILLARD REACTION.   See **Amino Acids**.

MAIN STANDARD TIME.   See **Meteorology**.

MAIZE.   See **Corn (Maize)**.

MAKO.   See **Sharks**.

MAKSUTOV-BOUWERS TELESCOPE. A system, independently devised by Maksutov and Bouwers, wherein a negative meniscus lens is used to introduce spherical aberration, causing a small chromatic effect with very little focusing effect. The primary mirror is then made spheroidal, which cancels the spherical aberration introduced by the lens. This catadioptic system is finding wide applications because of the ease with which one can compute the surface points of a spheroid. The telescope can be designed for use at different foci.

See also **Telescope (Astronomical-Optical)**.

MALACHITE. The mineral malachite [CAS: 1319-53-5] is a basic carbonate of copper [CAS: 12069-69-1] corresponding to the formula $Cu_2(CO_3)(OH)_2$. It is monoclinic, crystals tending to be acicular, but usually found massive. It is a brittle mineral; hardness, 3.5–4; specific gravity, 4.05; luster, vitreous to silky or dull; color, green; streak, green; translucent to opaque. Malachite is an alternation product found associated with other copper-bearing minerals. It is a rather common mineral and is found quite widely distributed. Large quantities have been found in the Ural Mountains; it is also found in Germany, France, England, Zaire, Rhodesia, and Australia. In the United States, beautiful radiated masses of fibrous crystals have been found in Berks County, Pennsylvania, as well as in Tennessee at Ducktown, and in Arizona, Nevada, and Utah. Malachite, besides being an ore of copper, has been used for various ornamental purposes. The word malachite is derived from the Greek, meaning a *mallow*, because of its green color.

MALAPRADE REACTION. Compounds containing two hydroxyl groups or a hydroxyl and an amino group attached to adjacent carbon atoms undergo cleavage of the carbon-to-carbon bond when treated with periodic acid.

MALARIA. Essentially a disease of the tropics, malaria is known to infect 200 to 400 million people of whom 10% will probably die of the disease in adulthood. The mortality in children is about 50% and some 2 billion people (about one-third of the world's population) are threatened by the disease. Distribution of the disease is roughly indicated in Fig. 1.

Malaria was recognized as a human disease more than 5000 years ago. Enlarged spleens, presumably due to malaria, have been found in Egyptian mummies more than 3000 years old, and the Ebers Papyrus (c. 1570 bc) mentions splenomegaly, fevers and a variety of cures for such ailments. Malaria antigen was detected in samples of skin and lung from mummies dating from 3204 bc and 1304 bc. Malaria probably came to southern Europe from Africa via the Nile valley, but infections possibly resulted from the closer contact between the peoples of Europe and Asia Minor. The clear discussion of quartan and tertian fevers by the physician Hippocrates (460–370 BC) leaves little doubt that by the fifth century bc P. malariae (quartan malaria) and P. vivax (tertian malaria) were present in Greece. Hippocrates also observed that quartan malaria was the less dangerous of the two types, and noted the relationship between enlarged spleens and marshes. Since there is no mention in any of the writings of Hippocrates of severe, malignant, tertian fevers, it appears that P. falciparum infections were rare or nonexistent. Some historians have suggested that Alexander the Great (356–323 BC) did not extend his conquests into the subcontinent of India because, in part, he died from malaria. It is presumed that he became infected in India; however, since 20 months elapsed between the time of infection and his death in Babylon at the age of 33, his death was unlikely to have been due to P. falciparum. See also **Hippocrates of Cos (460BC-370BC);** and **Parasitology (The History)**.

The cause of malaria was first discovered in 1880 by a French Army surgeon, Charles Laveran, who identified the malaria parasite while examining with the aid of a microscope the fresh blood of a patient infected with falciparum malaria. When Laveran made this discovery in Constantine, Algeria, the leading European medical professionals were under the spell, so to speak, of Louis Pasteur. The bold concept that malaria could be caused by the presence of millions of minute animal parasites in the blood, and not by bacteria, was difficult to accept. Thus six years were required to convince the sceptical medical profession of the validity of Laveran's discovery. By 1886, Camillo Golgi in Pavia, Italy identified two specific human malaria parasites, namely, P. vivax and P. malariae. A third species, P. falciparum, was identified by Ettore Marchiafava in Rome in 1889. An English physician, Patrick Manson, first suggested (1894) that mosquitoes may be the means for transmission of the disease — by drawing out the malaria parasites from human blood — and that transmission would occur by ingestion of water contaminated by infected dead mosquitoes. Manson was prompted to this observation by prior experience in noting that mosquitoes could suck up the microscopic threadlike worms from the blood of patients infected with filariasis. Of course, Manson's theory of transmission accurately implicated the mosquito, but erred in the details of transmission. Manson desired to prove his theory, but realized that England was not a suitable location. Manson motivated a British Army surgeon, Ronald Ross, to carry on the research in India. For a few years, Ross

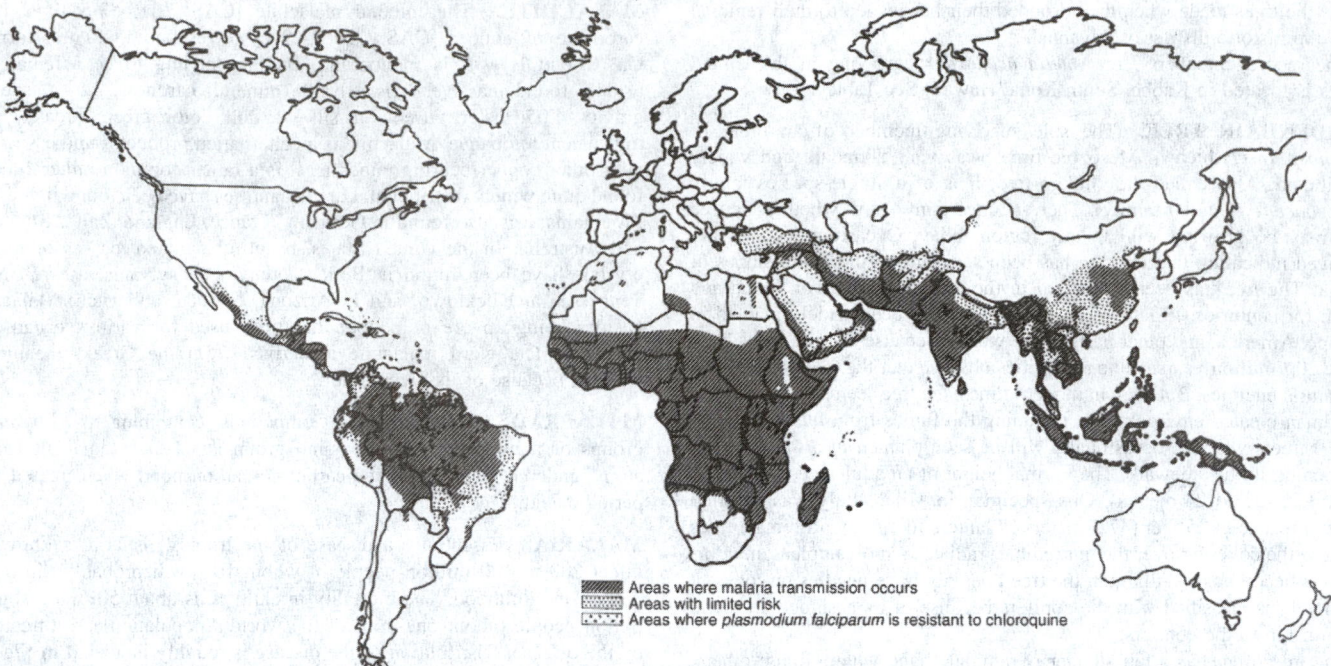

Fig. 1.   World malaria risk chart, including geographical distribution of principal vectors, geographical distribution of falciparum malaria, and areas where Plasmodium falciparum is resistant to chloroquine. Chart prepared by International Association for Medical Assistance to Travellers (IAMAT). Among the several laudable purposes of IAMAT is "To provide world wide medical assistance ... to travelers who, while absent from their home country, may find themselves in need of medical or surgical care or of any form of medical treatment by providing these travelers with the names of centers in countries other than that to which they are native, the names of locally licensed medical practitioners who have command of the native language spoken by the traveler and who agreed to a stated and standard list of medical fees for their services." IAMAT headquarters is located in Guelph, Ontario, Canada, N1H 7L5.

experimented with mosquitoes at random, but it was not until 1897 that his attention was brought to the *Anopheles* mosquito. Late in that year, while making a microscopic examination of the gut of mosquitoes that had fed on a patient with malignant malaria, he noted protozoa growing only in the Anopheles. However, because of a transfer from Madras to Calcutta, Ross commenced to work with the malaria parasite of birds, which is transmitted by a *Culex* species. Ross proved that the spindle-shaped malaria organisms (sporozites) freed by rupturing of the fertilized egg, migrate from the gut of the mosquito to the salivary glands to be injected into the victim by the bite of the mosquito. To Ross goes the credit for the discovery that malaria is transmitted by the bite. In the same year (1898), a Canadian pathologist, William McCallum, also working with birds, was able to interpret and describe the fertilization process of the parasite, taking place in the gut of the mosquito.

During the period 1886–1899 a group of four Italian scientists worked in Rome on the problem of transmission of malaria in humans. In 1898, a breakthrough came with the observations of Grassi, a physician with a keen interest in zoology and mosquitoes in particular. Grassi noticed that when malaria was present there was always a large population of *Anopheles*, while in areas of large *Culex* populations there was no malaria in humans. From the Campagna Romana near Rome, Grassi collected *Anopheles* mosquitoes which his colleague, Amico Bignami, let feed on a volunteer patient of the Santo Spirito Hospital near Saint Peter's Basilica. On November 1, 1898, the patient had developed the classical symptoms of falciparum malaria. In other volunteers, they proved that only Anopheles mosquitoes transmit malaria in humans.

Even with the significance of this discovery, there remained a link missing in the cycle of the malaria parasite. Still unexplained was the time lapse between the introduction of the parasites through the bite of the mosquito and the appearance of the symptoms of malaria. It was not until 1936 that an Italian scientist. Giulio Raffaele, discovered, while working with birds, that the malaria parasites entering the host first undergo a cycle of transformation within the blood-forming cells of the liver. Later, in 1948, British researchers H.E. Shortt and P.C.C. Garnham demonstrated the liver cycle of the malaria parasite in humans. Following a period of extensive trials on monkeys, a human volunteer was bitten during three days by nearly 800 *Anopheles* infected with *P. falciparum*. On the fifth

day, a surgeon removed a small piece of tissue from his liver (biopsy), which when examined under the microscope, demonstrated the growth of the parasites in the liver cells.

More than fifty different species of *Plasmodium* can cause malaria in humans, monkeys, birds, fish, and cattle. However, only four species attack humans: *P. falciparum, P. vivax, P. malariae,* and *P. ovale*.

Vivax is the mildest form of malaria and is characterized by periodic chills and fever, an enlarged spleen, anemia, severe abdominal pain and headaches, and extreme lethargy. If left untreated, the disease tends to be self-limiting within a period of 10 to 30 days, but will recur periodically. Although the fatality rate of vivax malaria is low, the disease is highly debilitating and makes the patient more vulnerable to other diseases. In addition to the aforementioned symptoms, falciparum malaria presents edema of the brain and lungs and blockage of the kidneys. Unless treated promptly, the fatality rate of falciparum malaria is high.

The life cycle of the parasite and its course in the human body proceeds in the following way. The saliva of the mosquito contains the *Plasmodium* at the lance-shaped sporozoite stage of its life cycle. Upon inoculation of the host by biting, the sporozoites quickly migrate to the liver where they divide and develop into multinucleated schizonts. Within 6 to 12 days, the schizonts disrupt and release into the blood the form known as *merozoites*. Each liver cell infected by one sporozoite releases into the blood stream from 5000 to 10,000 merozoites. These later invade the host's erythrocytes where they grow and form more schizonts which, in turn, again divide, releasing more merozoites into the blood stream to repeat the cycle. The principal symptoms of malaria are associated with the rupture of the schizonts, the periodic lysis of the blood cells with release of merozoites and toxic wastes, which cause the regular fevers and chills of malaria.

The effects of insect and parasite control chemicals in West Africa and Southeast Asia appear to be losing much of their former ability, particularly concerning *Plasmodium falciparum*, the deadliest form of malaria. For many years, despite some of their adverse effects, insecticides such as DDT, dieldrin, and HCK have been used to control the Anopheles mosquitoes. There is considerable evidence that shows that these chemicals no longer suffice as the principal control measures. This situation precipitates an even greater interest in malaria vaccine(s), which are mentioned later in this article.

As briefly outlined earlier, the quest to prevent and cure malaria and a number of similar African fevers is now in its second century of trial-and-error efforts. D.J. Wyler (Tufts University School of Medicine) places past efforts in perspective, citing a remarkable variety of substances, such as cinchona bark extracts, infusions from wormwood, quinine, and cogeners. The author describes as 40-year drug-discovery program conducted at the Walter Reed Army Institute of Research, during which time one-quarter million compounds were tested as potential antimalarial drugs. Of these, only two, mefloquine and halfantrine were licensed. The details provided by Wyler provide valuable and interesting reading for the person who is interested in tropical diseases, such as malaria.

The American Association for the Advancement of Science (AAAS) has established a Sub-Saharan Africa Program, under a cooperative agreement with the Agency for International Development, to develop and evaluate strategies to combat malaria in Africa. The AAAS is using knowledge within its affiliated societies to review sociocultural, economic, and behavioral factors; environmental and urbanization issues; health care delivery systems; and natural science applications for malaria prevention and control. Each year 80 percent of the 100 million cases of malaria worldwide occur in Africa. Included will be studies of resistance of parasites to drugs and lack of eradication efforts.

Course of the Disease. This varies with the causative protozoan.

1. *Vivax* or benign *tertian malaria* is the most common type. The incubation period ranges from 10 days to 4 weeks. Generally, paroxysms of chills and fever appear on the 14th day after the bite of an infected female anopheline mosquito. During this time the parasite has been multiplying in the liver cells of the patient. Paroxysms continue to recur every other day, as the parasite completes its 48-hour cycle of development, now in the blood. During the paroxysm, the patient first goes through a "cold stage" during which he has chilly sensations, his skin is blue, his teeth chatter and there is violent shaking. After an hour, the "hot stage" is ushered in, with a rapid rise in temperature to as high as 107 °F (41.7 °C); the skin is hot and dry and the patient complains of severe headache. The fever lasts about 2 hours, and is followed by the "sweating stage," during which there is profuse perspiration, the temperature falls to normal, the headache disappears, and although weak and drowsy, the patient feels well.

2. *P. ovale* produces a disease very similar to tertian malaria.

3. *Quartan malaria*, produced by *P. malariae*, has an incubation period of 18–40 days. The paroxysms occur every 72 hours, and are longer and somewhat more severe than those accompanying tertian malaria.

4. *Falciparum, malignant tertian* or *estivo-autumnal malaria*, is the severest form of malaria and causes most of the fatalities. The paroxysms occur irregularly after a 12-day incubation period. They are severe, and accompanied by high temperatures. The so-called cerebral, algid, hemorrhagic and pernicious types of malaria represent forms of falciparum malaria with different localizations of the parasite. In the cerebral type, the onset is rapid with delirium and coma, and death may occur in several hours without return to consciousness. "Black-water fever" or hemorrhagic malaria is a type in which hemolysis or dissolution of the red cells occurs, and dark urine due to the presence of hemoglobin is an outstanding feature. In the algid form, there are vomiting, diarrhea, and subnormal temperature. Diagnosis of malaria is made by examination of blood films taken during episodes of fever, when the parasites may be seen.

Microscopy is, however, a time-consuming procedure. Immunological methods, although more rapid, cannot distinguish between past and present infections. Another procedure has now been developed. This uses a DNA probe which enables a technician in the field to process 1000 samples per day, as compared with a microscopist's 60 samples.

A.E. Greenberg (Centers for Disease Control, Atlanta, Georgia) and a group of researchers reported in July 1991 on a possible relationship between *Plasmodium falciparum* malaria and the acquired human immunodeficiency virus type 1 infection, both infections that commonly occur in Zaire. Since the cellular immune system is critical to protection against malaria, it is biologically plausible that *P. falciparum* could occur more frequently or be more severe in HIV-infected persons with profound CD4 lymphocyte depletion. Further, it was postulated that repeated malarial infection may accelerate the progression of HIV-related disease. At the

end of a 13-month study involving several hundred people, the conclusion was, "In this study malaria was not more frequent or more severe in children with progressive HIV-1 infection and malaria did not appear to accelerate the rate of progression of HIV-1 disease."

Attempts to Control Malaria. A threefold approach has been taken in attempting to control malaria: the elimination of the mosquito, antimalarial drugs, and vaccines.

Mosquito Control. Attempts to destroy the breeding grounds of the mosquito, as was done in the marshes around Rome, or spraying the infected areas with an insecticide, have had limited success. Whereas area drainage has yielded good results, the use of insecticides (especially DDT) has ultimately not been so effective. After an initial period of success, *Anopheles* developed a resistance to the chemical and proceeded to reoccupy its old habitats. In recent years, the use of DDT has been banned by many countries.

Antimalarials

Antimalarials can be categorized according to their mode of action.

Tissue Schizonticides. These eradicate the liver stages of the parasite and thereby prevent their entry into the blood. As a class, therefore, they are useful for prophylaxis. Some tissue schizonticides can act on the long-lived tissue forms (hypnozoites) of *P. vivax* and *P. ovale* and thus can cure the latter infections by preventing relapses.

Blood Schizonticides. These destroy the erythrocytic stages of the parasites and are useful for the clinical cure of falciparum malaria or suppression of relapsing infections.

Gametocytocides. These annihilate the sexual forms of the plasmodia (gametocytes) and also destroy the stages of the parasites in the *Anopheles* mosquito.

Sporontocides. These act against the sporozoites and oocysts in the mosquito and thereby prevent the transmission of the disease.

It should be noted that drugs may operate by more than one mechanism, and may possess a specific mode of action against one species of plasmodium but lack efficacy against others. In addition, antimalarial drugs may be classified according to their structural types.

Antimalarial Drugs

Quinine (**1**), [CAS: 804-63-7] ($C_{20}H_{24}N_2O_2 \cdot \frac{1}{2}H_2O_4S$), the first known antimalarial, is a 4-quinolinemethanol that has served as a model for the design of numerous antiplasmodial drugs.

(1)

Its history began in mid-seventeenth century Peru when the Incas told the Jesuits of the medicinal properties of the bark of an evergreen mountain tree they called quina-quina (later called cinchona). The bark, when made into an aqueous infusion, was capable of curing malaria. The use of cinchona spread to Europe and the alkaloid from it, quinine, was isolated in 1820. Quinine, an extremely bitter substance, has been used by millions of malaria sufferers. In recent times, it has been employed successfully to treat chloroquine-resistant strains of *P. falciparum* but it frequently fails to provide a complete cure of the infection. Quinine acts on the asexual blood forms of the plasmodia in a manner slower than many synthetic drugs. Overdoses cause tinnitus and visual disturbances, side effects that disappear on withdrawal of the drug. It can also cause premature contractions in women in late stages of pregnancy. Although quinine is a favored antimalarial for parenteral administration, it is nevertheless hazardous by this route. Quinidine (**2**), has been shown to be even more effective in combatting the disease (Table 1). However, it has undesirable cardiac side effects that reduce its suitability as an antimalarial. Mixtures of cinchona alkaloids, known as totaquine, are easier to produce and have

TABLE 1. MALARIA ANTIPROTOZOAL AGENTS[a]

Structure number	Compound name	CAS Registry Number	Molecular formula	Structure
		4-Quinolinemethanols		
(2)	Quinidine (sulfate)	[50-54-4]	$C_{20}H_{24}N_2O_2 \cdot \frac{1}{2}H_2O_4S$	
(3)	Mefloquine	[53230-10-7]	$C_{17}H_{16}F_6N_2O$	
		Phenanthrene-methanol		
(4)	Halofantrine	[69756-53-2]	$C_{26}H_{30}Cl_2F_3NO$	
		4-Aminoquinoline		
(5)	Hydroxy-chloroquine	[118-42-3]	$C_{18}H_{26}ClN_3O$	
		8-Aminoquinolines		
(6)	Pamaquine[b]	[491-92-9]	$C_{19}H_{29}N_3O$	
(7)	Primaquine (phosphate)[c,d]	[63-45-6]	$C_{15}H_{21}N_3O \cdot 2H_3O_4P$	
		Antifolates		
(8)	Dapsone	[80-08-0]	$C_{12}H_{12}N_2O_2S$	
(9)	Sulfadoxine[c]	[2447-57-6]	$C_{12}H_{14}N_4O_4S$	
(10)	Sulfalene[e]	[152-47-6]	$C_{11}H_{12}N_4O_3S$	

TABLE 1. (*Continued*)

Structure number	Compound name	CAS Registry Number	Molecular formula	Structure
(11)	Trimethoprim[c,e]	[738-70-5]	$C_{14}H_{18}N_4O_3$	
(12)	Chlorguanide	[500-92-5]	$C_{11}H_{16}ClN_5$	
(13)	Cycloguanil	[516-21-2]	$C_{11}H_{14}ClN_5$	

Others

(14)	Menoctone[b]	[14561-42-3]	$C_{24}H_{32}O_3$	
(15)	Pyronaridine	[74847-35-1]	$C_{29}H_{32}ClN_5O_2$	
(16)	Doxycycline	[564-25-0]	$C_{22}H_{24}N_2O_8$	
(17)	Chloram-phenicol	[56-75-7]	$C_{11}H_{12}Cl_2N_2O_5$	
(18)	Febrifugine	[24159-07-7]	$C_{16}H_{19}N_3O_3$	
(19)	Artemisinin	[63968-64-9]	$C_{15}H_{22}O_5$	

(continued)

TABLE 1. (*Continued*)

Structure number	Compound name	CAS Registry Number	Molecular formula	Structure
(20)	Artemether	[71963-77-4]	$C_{16}H_{26}O_5$	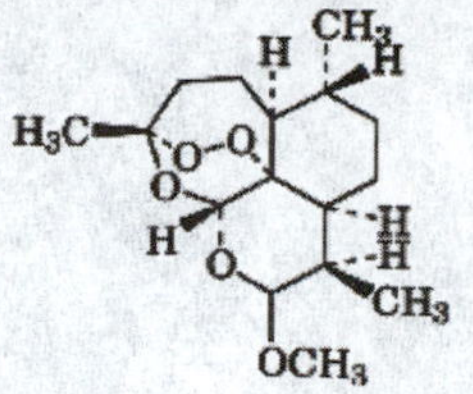
(21)	Arteether	[75887-54-6]	$C_{17}H_{28}O_5$	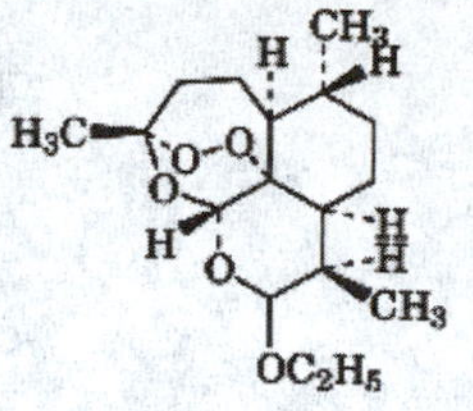
(22)	Artesunic acid	[88495-63-0]	$C_{19}H_{28}O_8$	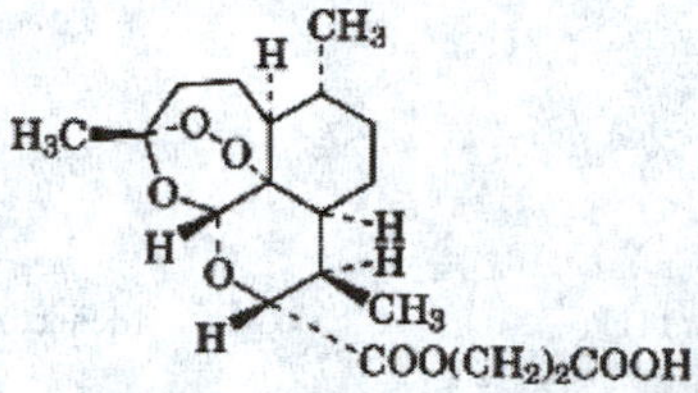
(23)	Artelinic acid	[109637-83-4]	$C_{23}H_{30}O_7$	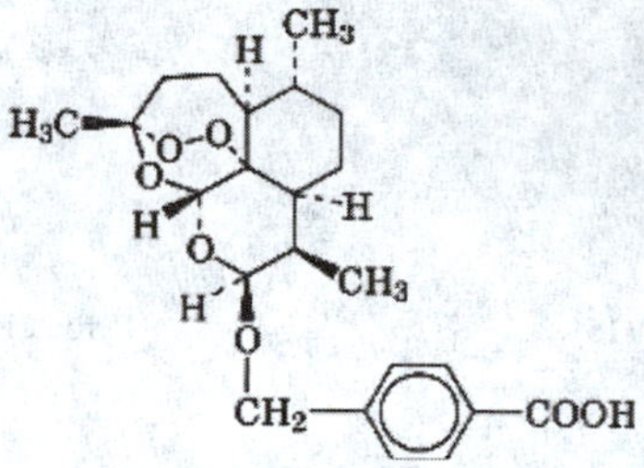

[a] Other applications are indicated in footnotes.
[b] Theileriasis.
[c] Pneumocystosis.
[d] American trypanosomiasis.
[e] Toxoplasmosis.

been employed in treatment. Totaquine has been standardized to contain a minimum of 15% quinine.

The success of quinine inspired the search for other antimalarials. The greatest impetus for the development of synthetic drugs came this century when the two World Wars interrupted the supply of cinchona bark to the combatants. A structurally related 4-quinolinemethanol is mefloquine (**3**, Lariam® [CAS: 51773-92-3]), which now serves as an effective alternative agent for chloroquine-resistant *P. falciparum*. This is a potent substance that requires less than one-tenth the dose of quinine to effect cures. There are some untoward side effects associated with this drug such as gastrointestinal upset and dizziness, but they tend to be transient. Mefloquine is **NOT recommended** for use by those using beta-blockers, those whose job requires fine coordination and spatial discrimination, or those with a history of epilepsy or psychiatric disorders. A combination of mefloquine with Fansidar (a mixture of pyrimethamine and sulfadoxine) is known as Fansimef but its use is not recommended. Resistance to mefloquine has been reported even though the compound has not been in wide use.

The best example of the class of phenanthrene-methanols is halofantrine (**4**, Halfan® [CAS: 36167-63-2]), a drug that is effective against chloroquine-resistant malaria and is now being evaluated in Africa. It produces temporary gastrointestinal disturbances.

Chloroquine (**24**), [CAS: 50-63-5], $C_8H_{26}ClN_3 \cdot 2H_3O_4P$ is the most effective of the hundreds of synthesized 4-aminoquinolines and was the antimalarial of choice until resistance developed to it in Colombia and Southeast Asia in 1962.

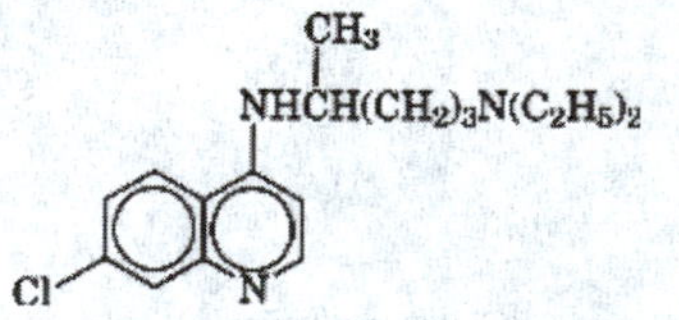

(**24**)

It is still one of the most important antimalarial agents in use for suppression (prophylaxis) in endemic areas where there remains sensitivity to the drug. It acts as a blood schizonticide and rarely produces serious side effects in individuals taking it prophylactically. Ironically, it was the widespread prophylactic use that most likely led to drug resistance. For treatment of the disease, the phosphate salt is administered orally, or the chloride, parenterally. The minor side effects include gastrointestinal upset, headache, dizziness, blurred vision, and pruritus (itch), but discontinuance of the drug is seldom necessary. Those who have used chloroquine continually for more than 6 years, either for prophylaxis or in the treatment of arthritis, are recommended to have ophthalmologic exams to check for possible retinal damage. It was demonstrated that chloroquine resistance can be reversed *in vitro* by treating *P. falciparum* parasitized erythrocytes with the calcium channel blocking drug verapamil. Subsequently, other calcium channel blockers and antihistamines have been found to be effective in overcoming chloroquine resistance *in vitro* and in rodents. Desipramine (Norpramin) and other antidepressants have been shown to reverse chloroquine resistance in monkeys. Alternatives to chloroquine for treatment are amodiaquine (**25**), [CAS: 86-42-0], $C_{20}H_{22}ClN_3O$ and hydroxychloroquine (**5**, Table 1), a chloroquine modification in which one of the ethyl groups on the tertiary nitrogen atom of the side chain is replaced with hydroxyethyl. These compounds are less effective, and amodiaquine is more toxic than chloroquine. The 4-aminoquinolines also act as gametocytocidal agents against *P. vivax*, *P. ovale*, *P. malariae* and immature gametocytes of *P. falciparum*.

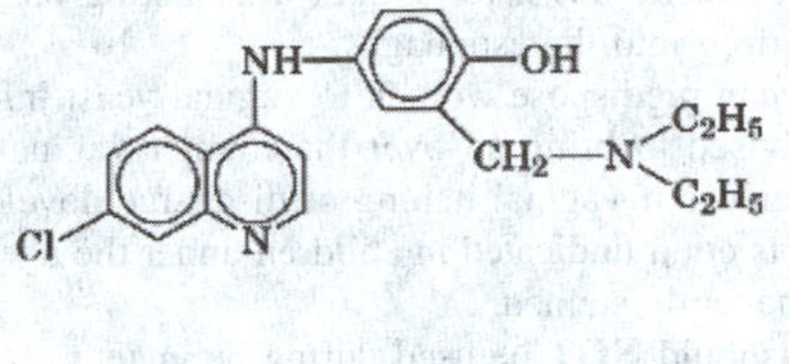

(**25**)

The 8-aminoquinolines are effective against both the primary and secondary tissue forms and sexual blood forms of the parasites. At toxic levels they are also active against asexual blood forms in humans. Pamaquine (**6**, plasmochin, plasmoquine), the oldest useful member of the class of 8-aminoquinolines, was synthesized in Germany in the mid-1920s. Primaquine [90-34-6] (**7**), which is less toxic and more effective, is the most widely used of the 8-aminoquinolines. It is gametocytocidal against all species of human malaria parasites and is an antirelapsing drug, but has the undesirable ability to cause severe hemolysis in glucose-6-phosphate dehydrogenase (G6PD) deficient individuals. This accounts for its recommended administration in limited doses over 1–2 weeks and its low utilization. During pregnancy, it is considered inadvisable for primaquine to be taken as it could be passed on to a G6PD-deficient fetus, thereby causing hemolytic anemia. Primaquine is generally used in conjunction with a blood schizonticide such as chloroquine (**24**), amodiaquine (**25**), or pyrimethamine (**26**, [CAS: 58-14-0] $C_{12}H_{13}ClN_4$) and is used for the prevention of relapses only against *P. vivax* and *P. ovale*.

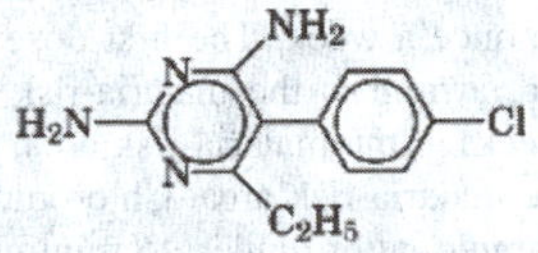

(**26**)

The drugs known as antifolates act as effective blood schizonticides. Unfortunately, the parasites readily develop resistance to them. Most antifolates show poor oral tolerance, absorption, and host toxicity. They fall into two types depending on the mechanisms by which they operate.

The first are competitors of PABA (*p*-aminobenzoic acid) and thus interrupt host *de novo* formation of the tetrahydrofolic acid required for nucleic acid synthesis. Examples of drugs that fall into this group are the sulfones and sulfonamides. The most well-known of the sulfones is dapsone (**8**, 4,4'-diaminodiphenyl sulfone, DDS), whose toxicity has discouraged its use. Production of folic acid, which consists of PABA, a pteridine unit, and glutamate, is disturbed by the substitution of a sulfonamide (structurally similar to PABA). The antimalarial sulfonamides include sulfadoxine (**9**, Fanasil [2447-57-6]), sulfadiazine (**27**, [CAS: 68-35-9] $C_{10}H_{10}N_4O_2S$) and sulfalene (**10**, sulfamethoxypyrazine [152-47-6], Kelfizina). Compounds of this group are rapidly absorbed but are cleared slowly.

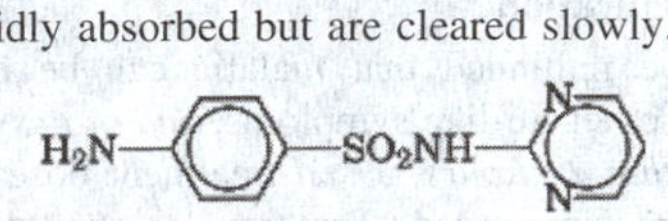

(**27**)

The second type of antifolates bind preferentially with, and thus selectively inhibit, the enzyme dihydrofolate reductase contained in the plasmodia. This interferes with the ability of the malaria parasites to convert dihydrofolate to tetrahydrofolic acid. In the erythrocyte host, however, dihydrofolate reductase is considerably less sensitive to these drugs and the blood cells are capable of utilizing exogenous folate.

Dihydrofolate reductase inhibitors are potent blood schizonticides that act on the asexual blood stages. Members of this class include the pyrimidines, pyrimethamine (**24**), and trimethoprim (**11** Table 1). Pyrimethamine is a slow-acting drug that is not recommended for use in acute attacks and is potentiated by combination with sulfadoxine (**9**, Table 1) (mixture is called Fansidar), dapsone (**8**, Table 1) (mixture is called Maloprim), sulfalene (**10**, Table 1) (mixture is called Kelfimeta or Metakelfin), or chloroquine (**24**) (mixture is called Darachlor). It is an effective suppressant against *P. falciparum*. The mode of action of trimethoprim (**11**, Table 1) is similar to that of pyrimethamine (**24**). Because it is also slow-acting, it is generally used in combination with the fast-acting sulfonamide, sulfalene (**10**, Table 1), to give an effective treatment against *P. falciparum*. Related structurally and functionally are the so-called open and cyclic triazines, exemplified by chlorguanide (**12**, Table 1, proguanil [CAS: 500-92-5], or the hydrochloride, Paludrine), a biguanide, and cycloguanil (**13**, Table 1, Camolar [CAS: 609-78-9]), respectively. The compounds have low toxicity and are also effective as causal prophylactics, ie, they destroy parasites before they enter the red blood cells.

Quinones and naphthoquinones were explored during the World War II Antimalarial Drug Program. Now that chloroquine resistance is a serious problem, compounds of this group such as menoctone (**14**, Table 1) are being reinvestigated.

Quinacrine (**28**, [CAS: 83-89-6] $C_{23}H_{30}ClN_3O$) is an acridine that was used extensively from the mid-1920s to the end of World War II. It acts much like chloroquine and is reasonably effective. Because it causes the skin to turn yellow and, in high doses, causes yellow vision, the drug is no longer in use as an antimalarial. Pyronaridine (15, Table 1), a 1-azaacridine developed in China, appears to be effective against mefloquine-resistant, but not entirely against chloroquine-resistant, strains of *P. falciparum*.

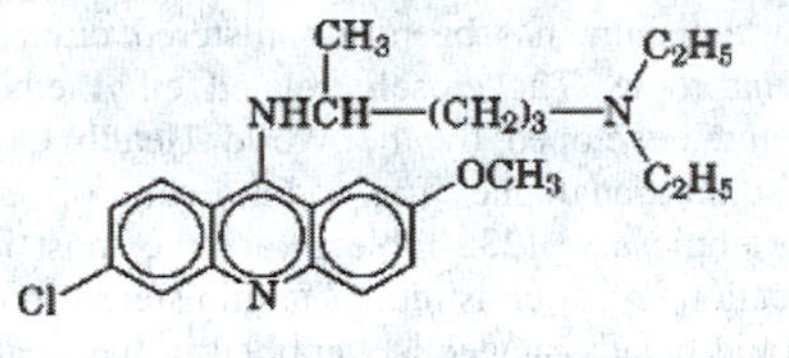

(**28**)

Some antibiotics, such as the tetracyclines, tetracycline (**29**, [CAS: 60-54-8] $C_{22}H_{24}N_2O_8$), doxycycline (**16**, Table 1), and rifampin (**30**, [CAS: 13292-46-1] $C_{43}H_{58}N_4O_{12}$), chloramphenicol (**17**, Table 1), and clindamycin (**31**, [CAS: 18323-44-9] $C_{18}H_{33}ClN_2O_5S$) have modest antimalarial properties, but are slow-acting.

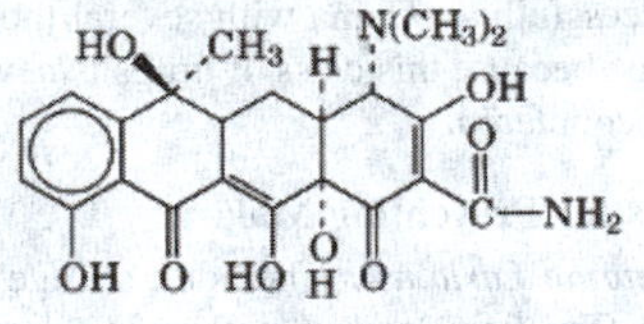

(**29**)

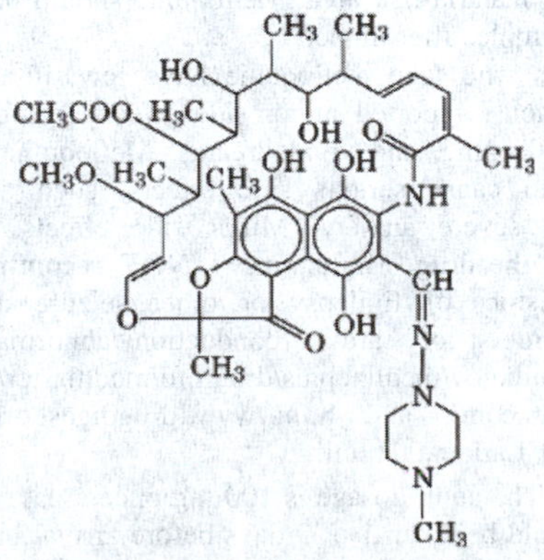

(**30**)

CH₃CH₂CH₂ (structure diagram)

(31)

A Chinese traditional herbal treatment for malaria obtained from the roots of *Dichroa febrifuga* is called Ch'ang Shan and was investigated in the 1940s. Febrifugine (**18**, Table 1), the alkaloid responsible for its activity, was isolated and found to be considerably more active than quinine in experimental infections. Unfortunately, the drug caused nausea and vomiting in humans. Synthesized analogues were generally less effective than the parent.

The most promising group of new antimalarials is based on the compound artemisinin (**19**, Table 1), an unusual sesquiterpene lactone endoperoxide first isolated in China in 1972 and then in the United States from the weed *Artemisia annua*. Artemisinin, also called qinghaosu, meaning in Chinese "extract of green herb," is a rapid-acting and relatively nontoxic therapeutic agent. Its low oil and water solubility has been increased by making certain structural modifications to the molecule. Oil-soluble artemether (**20**, Table 1), the methyl ether of the lactol dihydroartemisinin, has been administered clinically in China by the intramuscular route. The closely related ethyl ether, arteether (**21**, Table 1), is being developed by the World Health Organization as an alternative oil-soluble analogue. The sodium salts of artesunic acid (**22**, Table 1) and artelinic acid (**23**, Table 1) are the most interesting water-soluble modifications of artemisinin. If administered intravenously, these sodium salts could be advantageously applied in the treatment of cerebral malaria where rapid return to consciousness of the comatose patient and reduction of the parasitemia are the primary clinical goals. Although the water-soluble compound sodium artelinate is more stable in aqueous solution than sodium artesunate, only the latter has been tested clinically. Artelinic acid and related compounds effectively eliminated or prevented the establishment of parasitemia in *P. berghei*-infected mice when they were administered by the transdermal route. Artemisinin and its derivatives have been used successfully in China with several thousand *P. falciparum* and *P. vivax* patients; because this class of drugs is new, parasite resistance has not yet been encountered.

Prescription Drugs for Preventing Malaria

Mefloquine/brand name Lariam®. The adult dosage is 250 Mg salt (one tablet) once a week. The first dose of mefloquine should be taken 1 week before arrival in the malaria risk area, and once a week, on the same day of the week, in the malaria-risk area, and once a week for 4 weeks after leaving the malaria-risk area. Mefloquine should be taken on a full stomach, for example, after dinner.

Most travelers who take mefloquine have few, if any, side effects. The most commonly reported minor side effects include nausea, dizziness, difficulty sleeping, and vivid dreams. Mefloquine has very rarely been reported to cause serious side effects, such as seizures, hallucinations, and severe anxiety. Minor side effects, usually do not require stopping the drug. Mefloquine is **NOT recommended** for travelers with a history of Epilepsy or other seizure disorders; severe psychiatric disorders; or cardiac conduction abnormalities. See also http://www.nlm.nih.gov/medlineplus/druginfo/medmaster/a603030.html; http://www.lariam.com/; and http://www.drugdigest.org/DD/DVH/Uses/0,3915,6178%7CLariam,00.html.

Doxycycline. The adult dosage is 100 mg once a day. The first dose of doxycycline should be taken 1 or 2 days before arrival in the malaria-risk area, once a day, at the same time each day, in the malaria-risk area, and once a day for 4 weeks after leaving the malaria-risk area.

See also http://www.drugs.com/doxycycline.html; and http://www.medicinenet.com/doxycycline/article.htm.

Doxycycline Side Effects and Contraindications

• Doxycycline may cause photosensitivity. Travelers should be advised to avoid midday sun, use a sunscreen with SPF of at least 15, wear long-sleeved shirts, long pants, and a hat.
• Patients are advised to take doxycycline on a full stomach to minimize nausea and to not lie down for 1 hour after taking the drug to prevent reflux of the drug into the esophagus.
• Doxycycline can predispose women to vaginal yeast infections. Women should be advised to bring an over the-counter vaginal yeast infection medication for use if vaginal itching or discharge develops.
• Doxycycline is contraindicated in children under the age of 8; teeth may become permanently stained.
• Doxycycline should **NOT** be used during pregnancy.

Malarone™. Malarone is a new antimalarial drug in the United States. Malarone is a combination of two drugs (atovaquone and proguanil) and is an effective alternative for travelers who cannot or choose not to take mefloquine or doxycycline.

The adult dosage is one adult tablet (250 mg atovaquone/100 mg proguanil) once a day. The first dose of Malarone should be taken 1 to 2 days before travel in the malaria-risk area, once a day in the malaria-risk area and once a day for 7 days after leaving the malaria-risk area. The dose should be taken at the same time each day with food or milk.

Although side effects are rare, abdominal pain, nausea, vomiting, and headache can occur. Malarone **should not** be taken by patients with severe renal impairment (creatinine clearance <30 ml/min); pregnant women or women breast-feeding infants weighing less than 11 kg (24 pounds) should not take Malarone to prevent malaria; and infants weighing less than 11 kg should not be given Malarone. See also http://www.malarone.com/aboutmalarone.html; http://www.fda.gov/bbs/topics/ANSWERS/ANS01026.html; and http://www.healthlink.mcw.edu/article/979237802.html.

Chloroquine/brand name Aralen®. The adult dosage is 500 mg (salt) chloroquine phosphate once a week. The first dose of chloroquine should be taken 1 week before arrival in the malaria-risk area, once a week, on the same day of the week, in the malaria-risk area, and once a week for 4 weeks after leaving the malaria-risk area. Chloroquine should be taken on a full stomach, for example, after dinner, to minimize nausea.

Although side effects are rare, nausea and vomiting, headache, dizziness, blurred vision, and itching have been reported. Chloroquine may worsen the symptoms of psoriasis. See also http://www.drugs.com/cons/Aralen.html.

Hydroxychloroquine sulfate/brand name Plaquenil®.

The adult dosage is 400 mg (salt) once a week. The first dose of hydroxychloroquine sulfate should be taken 1 week before arrival in the malaria-risk area, once a week, on the same day of the week, in the malaria-risk area, and once a week for 4 weeks after leaving the malaria-risk area. Hydroxychloroquine sulfate should be taken on a full stomach, for example, after dinner, to minimize nausea. Hydroxychloroquine sulfate may be better tolerated than chloroquine.

Although side effects are rare, nausea and vomiting, headache, dizziness, blurred vision, and itching have been reported. Hydroxychloroquine sulfate may worsen the symptoms of psoriasis.

See also http://www.drugs.com/cons/Plaquenil.html.

Self-Treatment Medication

Travelers should be reminded that malaria can be fatal. If a traveler develops a fever or other flu-like symptoms, *and professional medical care is not available within 24 hours*, a self-treatment dose of either Fansidar or Malarone™ is recommended. The traveler should seek professional medical care as soon as possible after self-treatment. Malaria symptoms will occur, at least seven to nine days after being bitten by an infected mosquito. Fever in the first week of travel is unlikely to be malaria; however, travelers should be advised to have any fever promptly evaluated.

Fansidar® may be used for presumptive self-treatment for travelers if: they are **NOT** allergic to sulfa drugs and their travel itinerary does **NOT** include the Amazon basin of South America, Southeast Asia, and certain countries in eastern and southern Africa (Kenya, Malawi, Mozambique,

South Africa, Tanzania, and Uganda). These countries have documented Fansidar®-resistant *Plasmodium falciparum* malaria.

See also http://www.drugs.com/cons/Fansidar_Systemic.html.

Malarone™ may be used for presumptive self-treatment for travelers not taking Malarone for prophylaxis. Travelers on Malarone prophylaxis who take presumptive self-treatment should use Fansidar if they are traveling to an area without Fansidar resistance.

Those travelers who cannot take Fansidar® or Malarone™ for presumptive self-treatment should consult the CDC Malaria Hotline.

Note: The foregoing information on prescription drugs for preventing Malaria was furnished by the Centers for Disease Control and Prevention, Atlanta, GA.

Malaria Vaccines. Development of a malaria vaccine has been hindered mainly by a lack of a suitable source of parasites from which a vaccine could be prepared. Despite the development of a means of culturing the parasite, however, very little further progress has appeared. A large problem is the low immunogenecity of malaria parasites, which means that inducing immunity against them is difficult. In the face of the changing envelope of the trypanosome it appears unlikely that a suitable antibody against malaria will be found in the near future. Antibodies against the sporozoite sheath have been claimed to provide a lasting protection against infection by *Plasmodium*, but this claim has been strongly questioned and it is more probable that the monoclonal antibody approach will have more chance of success.

The parasite, *Plasmodium*, takes on several forms during its sojourn in the human host. At each phase, the parasite possesses a distinct protein coat. Obviously, then, an effective vaccine must induce antibodies to at least several of these, which is a very large order for one vaccine.

See also **Antiparasitic Agents, Antiprotozoals**.

Additional Reading

CDC: *Morbidity and Mortality Report*, Center for Disease Control, Atlanta, Georgia (Issued weekly).
Doolan, D.L.: *Malaria Methods and Protocols*, Humana Press, Totowa, NJ, 2000.
Dronamraju, K.R., and P. Arese: *Malaria*, Springer-Verlag New York, LLC, New York, NY, 2006.
Good, M.F. and A.J. Saul: *Molecular Immunological Considerations in Malaria Vaccine Development*, CRC Press, LLC., Boca Raton, FL, 1994.
Greenberg, A.E., et al.: "Plasmodium Falciparum Malaria and Perinatally Acquired Human Immunodeficiency Virus Type 1 Infections in Kinshasa, Zaire," *N. Eng. J. Med.*, 105 (July 11, 1991).
Greene, L. and M. Danubio: *Adaptation to Malaria: The Interaction of Biology and Culture*, Gordon and Breach Publishing Group, Newark, NJ, 1998.
Hoffman, S.L.: *Malaria Vaccine Development: A Multi-Immune Response Approach*, ASM Press, Washington, DC, 1996.
Humar, A., S. Sharma, D. Zoutman, et al.: "Fatal Falciparum Malaria in Canadian Travelers," *Can Med Assoc J.*, **156**, 1165–1167 (1997).
Khusmith, S., et al.: "Protection Against Malaria by Vaccination with Sporozoite Surface Protein 2 Plus CS Protein," *Science*, 715 (May 5, 1991).
Kinoshita, J.: "Malaria Vaccines," *Sci. Amer.*, 34 (May 1988).
Laird, M.: *Avian Malaria in the Asian Tropical Subregion*, Springer-Verlag Inc., New York, NY, 1998.
Langhorne, J.: *Immunology and Immunopathogenesis of Malaria*, Springer-Verlag New York, LLC, New York, NY, 2005.
Miller, K.D., A.E. Greenberg, and C.C. Campbell: "Treatment of Severe Malaria in the United States with a Continuous Infusion of Quinidine Gluconate and Exchange Transfusion," *N. Eng. J. Med.*, **321**, 65–70 (1989).
Nagel, R.L.: "Red-Cell Cytoskeletal Abnormalities—Implications for Malaria," *N. Eng. J. Med.*, 1558 (November 29, 1990).
Poser, C.M. and G.W. Bruyn: *An Illustrated History of Malaria*, Parthenon Publishing Group, New York, NY, 1999.
Rennie, J.: "Birds of a Fever: A Lethal Malaria May Have An Avian Origin," *Sci. Amer.*, 25 (July 1991).
Rosenthal, P.J., C. Peterson, F.R. Geertsma, et al.: "Availability of Intravenous Quinidine for Falciparum Malaria [Letter]," *N. Eng. J. Med.*, **335**, 138 (1996).
Schecter, J.: "Parasite Pacification," *Technology Review (MIT)*, 10 (October 1987).
Sherman, I.W.: *Malaria: Parasite Biology, Pathogenesis, and Protection*, ASM Press, Washington, DC, 1998.
Sherman, I.W.: *Molecular Approaches to Malaria*, ASM Press, Washington, DC, 2005.
Snowden, F.: *The Conquest of Malaria: 1900–1962*, Yale University Press, New Haven, CT, 2006.
Staff: *Malaria Strategies for Africa*, AAAS Sub-Saharan Africa Program, American Association for the Advancement of Science, Washington, DC, 1990.
Staff: "Treatment with Quinidine Gluconate of Persons with Severe Plasmodium falciparum infection: Discontinuation of Parenteral Quinine from CDC Drug Service," *MMWR*, **40** (RR-4) 21–23 (1991).
Staff: "Availability of Parenteral Quinidine Gluconate for Treatment of Severe or Complicated Malaria," *MMWR*, **45**, 494–495 (1996).
Staff: *Quinidine Gluconate Injection [Package Insert]*, Eli Lilly Company, Indianapolis, IN, February, 2000.
Staff: *Malaria—a Medical Dictionary, Bibliography, and Annotated Research Guide to Internet References*, ICON Health Publications, San Diego, CA, 2003.
Sullivan, D., and S. Krishna: *Malaria: Drugs, Disease and Post-genomic Biology*, Springer-Verlag New York, LLC, New York, NY, 2005.
Udomsangpetch, R., et al.: "Human Monoclonal Antibodies Against *P. falciparum*," *Science*, **231**, 57 (1986).
Valkiunas, G.: *Avian Malarial Parasites and Other Haemosporidia*, Routledge, New York, NY, 2004.
Wahlgren, M. and P. Perlmann: *Malaria: Molecular and Clinical Aspects*, Gordon & Breach Publishing Group, Newark, NJ, 2000.
Wyler, D.J.: "Bark, Weeds, and Iron Chelators—Drugs for Malaria," *N. Eng. J. Med.*, 1519 (November 19, 1992).
Zucker, J.R. and C.C. Campbell: "Malaria: Principles of Prevention and Treatment," *Infect. Dis. Clin. No. Am.*, **7**, 547–567, (1993).

Web References

Centers for Disease Control and Prevention (CDC), United States: http://www.cdc.gov/malaria/
Malaria Foundation International: http://www.malaria.org/
Malaria Journal: http://www.malariajournal.com/
Malaria Vaccine Initiative: http://www.malariavaccine.org/
National Center for Biotechnology Information: http://www.ncbi.nlm.nih.gov/Malaria/
World Health Organization: http://www.who.int/topics/malaria/en/
WHO/TDR Malaria Database: http://www.wehi.edu.au/MalDB-www/who.html

R.C. VICKERY, Blanton, Dade City, FL

DANIEL L.KLAYMAN, Walter Reed Army Institute of Research

MALEO *(Aves, Galliformes)*. A peculiar bird of Celebes and neighboring East Indian islands. The head and neck are covered with naked red skin and the crown bears a black prominence resembling a helmet. The plumage is mostly black but that of the breast and belly is salmon colored. The large eggs are buried in hot sand. See also **Galliformes**; **Megapode**; and **Mound Birds**.

MALIC ACID. [CAS: 6915-15-7], $C_4H_6O_5$, is a tart-tasting organic dicarboxylic acid that plays a role in many sour or tart foods. It is sometimes referred to as a fruit acid. This is because malic acid is found in apples and other fruits. It is also found in plants and animals, including humans. In its ionised form it is malate, an intermediate of the TCA cycle along with fumarate. See also **TCA Cycle**. It can also be formed from pyruvate as one of the anaplerotic reactions.

Malic acid is a chiral molecule. The naturally occurring stereoisomer is the L-form. The L-form is also the biologically active one. There is some preliminary evidence that malic acid, in combination with magnesium, may be helpful for some with fibromyalgia. Malic acid sold as a supplement is mainly derived from apples and, therefore, is the L-form. Malic acid is a naturally occurring compound that plays a role in the complex process of deriving adenosine triphosphate (ATP; the energy currency that runs the body) from food.

Malic acid is absorbed from the gastrointestinal tract from whence it is transported via the portal circulation to the liver. There are a few enzymes that metabolize malic acid. Malic enzyme catalyzes the oxidative decarboxylation of L-malate to pyruvate with concomitant reduction of the cofactor NAD+ (oxidized form of nicotinamide adenine dinucleotide) or NADP+ (oxidized form of nicotinamide adenine dinucleotide phosphate). These reactions require the divalent cations magnesium or manganese. Three isoforms of malic enzyme have been identified in mammals: a cytosolic NADP+-dependent malic enzyme, a mitochondrial NADP+-dependent malic enzyme and a mitochondrial NAD(P)+-dependent malic enzyme. The latter can use either NAD+ or NADP+ as the cofactor but prefers NAD+. Pyruvate formed from malate can itself be metabolized in a number of ways, including metabolism via a number of metabolic steps to glucose. Malate can also be metabolized to oxaloacetate via the citric acid cycle. The mitochondrial malic enzyme, particularly in brain cells, may play a key role in the pyruvate recycling pathway, which utilizes dicarboxylic acids and substrates, such as glutamine, to provide pyruvate to maintain the citric acid cycle activity when glucose and lactate are low.

Malic acid is also a natural constituent of many fruits and vegetables that are preserved by fermentation. This acid may be broken down during fermentation by certain bacteria into lactic acid and carbon dioxide. This reaction is desired to reduce the acidity in certain types of wines, and is undesired in the fermentation of cucumbers because of gaseous spoilage from carbon dioxide accumulation inside the fruit.

Many traditional fresh and fermented ready-to-eat foods are dependent on their acidity as the primary means for controlling the presence of disease- causing bacteria. In such foods, how quickly pathogenic microorganisms are inactivated is dependent on both the level of acidity (pH) and the identity and amount of the specific acid associated with the food. Recent concern about the survival of acid resistant pathogens in apple cider produced a need for better information on how malic acid, the principal acid in apples, affects bacteria. The current study helps address that need by providing information on how malic acid and pH interact to inactivate the foodborne pathogen, *Listeria monocytogenes*. These results demonstrate that malic acid is one of the gentler food acids.

Malic acid is both derived from food sources and synthesized in the body through the citric acid cycle. Its importance to the production of energy in the body during both aerobic and anaerobic conditions is well established. Under aerobic conditions, the oxidation of malate to oxaloacetate provides reducing equivalents to the mitochondria through the malate-aspartate redox shuttle. During anaerobic conditions, where a buildup of excess of reducing equivalents inhibits glycolysis, malic acid's simultaneous reduction to succinate and oxidation to oxaloacetate is capable of removing the accumulating reducing equivalents. This allows malic acid to reverse hypoxia's inhibition of glycolysis and energy production. This may allow malic acid to improve energy production in Primary fibromyalgia (FM), reversing the negative effect of the relative hypoxia that has been found in these patients.

Because of its obvious relationship to energy depletion during exercise, malic acid may be of benefit to healthy individuals interested in maximizing their energy production, as well as those with FM. In the rat it has been found that only tissue malate is depleted following exhaustive physical activity. Other key metabolites from the citric acid cycle needed for energy production were found to be unchanged. Because of this, a deficiency of malic acid has been hypothesized to be a major cause of physical exhaustion. The administration of malic acid to rats has been shown to elevate mitochondrial malate and increase mitochondrial respiration and energy production. Surprisingly, relatively small amounts of exogenous malic acid were required to increase mitochondrial energy production and ATP formation. Under hypoxic conditions there is an increased demand and utilization of malic acid, and this demand is normally met by increasing the synthesis of malic acid through gluconeogenesis and muscle protein breakdown. This ultimately results in muscle breakdown and damage. In a study on the effect of the oral administration of malic acid to rats, a significant increase in anaerobic endurance was found. Interestingly, the improvement in endurance was not accompanied by an increase in carbohydrate and oxygen utilization, suggesting that malic acid has carbohydrate and oxygen-sparing effects. In addition, malic acid is the only metabolite of the citric acid cycle positively correlated with physical activity. It has also been demonstrated that exercise-induced mitochondrial respiration is associated with an accumulation of malic acid. In humans, endurance training is associated with a significant increase in the enzymes involved with malic acid metabolism.

Because of the compelling evidence that malic acid plays a central role in energy production, especially during hypoxic conditions, malic acid supplements have been examined for their effects on FM. Subjective improvement in pain was observed within 48 hours of supplementation with 1200–2400 milligrams of malic acid, and this improvement was lost following the discontinuation of malic acid for 48 hours. While these studies also used magnesium supplements, due to the fact that magnesium is often low in FM patients, the rapid improvement following malic acid, as well as the rapid deterioration after discontinuation, suggests that malic acid is the most important component. This interesting theory of localized hypoxia in FM, and the ability of malic acid to overcome the block in energy production that this causes, should provide hope for those afflicted with FM. The potential for malic acid supplements, however, reaches much farther than FM. In light of malic acid's ability to improve animal exercise performance, its potential for human athletes is particularly exciting.

Additionally, many hypoxia related conditions, such as respiratory and circulatory insufficiency, are associated with deficient energy production. Therefore, malic acid supplements may be of benefit in these conditions. Chronic Fatigue Syndrome has also been found to be associated with FM, and malic acid supplementation may be of use in improving energy production in this condition as well. Lastly, malic acid may be of use as a general supplement aimed at ensuring an optimal level of malic acid within the cells, and thus, maintaining an optimal level of energy production. See also **Food Additives**.

Two other dicarboxylic acids have similar names and should not be confused with malic acid; Maleic acid, and Malonic acid.

MALIGNANCY.   See **Cancer and Oncology**.

MALLEUS.   See **Hearing and the Ear**.

MALLOPHAGA.   The bird lice or biting lice, constituting a small order of insects. They have flattened bodies with many spines directed backward, short legs, and biting mouths. Wings are lacking. Most species live as external parasites on birds but a few are found on mammals.

MALLOW FAMILY.   See **Malvaceae (Mallow Family)**.

MALNUTRITION.   Malnutrition is one of the greatest contributors to morbidity and mortality in the world, taking its greatest toll on the poor and developing segments of the population. In recent decades, much has been learned about the complex interrelationships between nutritional status and disease. Nutritional deficits in prenatal life are now known to have a far-reaching impact on health in adult life [Ravelli, et al.:]. Isolated micronutrient deficiencies have been linked to the pathophysiology of multiple disease processes, and malnutrition has a profound impact upon normal immunological function [Kubena and McMurray]. Increased awareness of the pathophysiological processes associated with malnutrition can guide therapies to reduce the associated morbidity and mortality, but improved access to food providing a balanced combination of energy and nutrients continues to be the key to prevention of malnutrition. See also **Nutritional Science (The History)**.

Pathophysiology of Malnutrition

Metabolism. In protein or energy deficiency, the human body has a versatile array of compensatory mechanisms to maintain energy substrates to vital tissues. Short periods of starvation lead to adaptive response in the kidneys, heart and skeletal muscle. These tissues switch from using glucose to fatty acids as their primary source of metabolic fuel. Preferential utilization of fuel from adipose tissue allows the preservation of glucose availability for organs such as the brain, while minimizing protein degradation. After limited glycogen stores have been used, amino acids derived from skeletal muscle become the primary sources of glucose production for the brain. During prolonged starvation, the brain adapts and begins to utilize keto acids as a protein-sparing alternative source of fuel.

Even with multiple compensatory mechanisms activated to minimize protein losses during starvation, there appears to be constitutive proteolysis and aminogenolysis to provide amino acids as a source of carbon skeletons for glucose synthesis [Owen, et al.:]. After 18 days of starvation, obese individuals use protein as a fuel source for approximately 7% of their daily resting energy requirements. However, the provision of as little as 7.5 g oral carbohydrate can reduce urinary nitrogen excretion, a sensitive marker of protein loss, by 50% in the setting of prolonged starvation [Sapir, et al.:]. There are no body stores of protein, and therefore protein losses in starvation lead to decreased lean body mass and decreased muscle mass. This occurs in parallel with depletion of fat stores, but has greater physiological consequences.

Adaptation to prolonged protein energy malnutrition, in addition to those stated, include a compensatory decline in resting metabolic requirements, decreased activity and, in children, diminished-somatic growth. When measured as a function of oxygen consumption, resting metabolic rate (RMR) decreases in parallel to decreases in body mass such that the RMR per kilogram of body mass is essentially constant in acute starvation. There is evidence, however, that more protracted starvation is associated with a decline in RMR independent of weight loss [Keys, et al.:]. The goal of each of these adaptive changes is to provide adequate fuels to body tissues while minimizing additional protein loss.

Endocrinology in Malnutrition. Endocrinological responses to malnutrition play an integral role in the body's adaptation to states of energy and protein deficiency. After a brief period of starvation there is increased activation of the hypothalamic–pituitary–adrenal axis. Corticotrophin-releasing hormone, through pituitary activation, stimulates the release of corticotrophins and subsequent adrenal production of glucocorticoids. The increase in glucocorticoids, in combination with decreased levels of circulating insulin and leptin during weight loss, leads to increased hypothalamic production of neuropeptide Y. Neuropeptide Y has been demonstrated to result in increased food intake and to suppress sympathetic nervous system outflow, thereby decreasing energy expenditure — both important responses to starvation.

During periods of protein energy deficiency, plasma insulin concentrations are reduced. In children with kwashiorkor (i.e. severe energy and protein malnutrition), abnormal responses to oral glucose have been noted and may be attributed to insulinopenia. While this condition corrects with restoration of adequate nutrition *in utero* has been linked to increased risk of long-term abnormalities of insulin and glucose regulation [Ravelli, et al.:]. Prenatal exposure to famine, in addition to resulting in lowered birthweight, is associated with impaired glucose tolerance in adulthood, a potential precursor to diabetes.

Another important regulatory hormone affected by starvation is insulin-like growth factor (IGF) I (also known as somatomedin C). IGF-I has insulin-like functions on many tissues, and promotes protein synthesis and cell proliferation. IGF-I is secreted by many tissues, but its production is extremely sensitive to nutritional supply. Therefore, starvation and malnutrition are associated with a dramatic reduction in IGF-I levels, which returnto normal with restoration of adequate protein and calories. Diminished concentrations of IGF-I may, for example, contribute to the osteoporosis associated with energy and protein malnutrition in the elderly [Bonjour, et al.:] since IGF-I is known to be a potent stimulator of bone remodeling. The administration of human recombinant IGF-I is being investigated in states of chronic malnutrition such as anorexia nervosa [Grinspoon, et al.:].

Immunology in Malnutrition. The immune system is sensitive to impaired nutrition, and the loss of normal immune function in the setting of malnutrition often exacerbates energy deficits in the form of increased infections. Protein energy malnutrition is associated with impaired lymphocytic function and depressed or delayed hypersensitivity skin test reactions. Deficits in specific nutrient substrates such as lipids and antioxidant vitamins are thought to alter cell membrane structure and fluidity, and consequently may interfere with cytokine and antigen receptor function on lymphocytes and phagocytes. See also **Immunodeficiency**.

Certain micronutrients have also been implicated in normal immune function. For example, zinc deficiency is associated with impaired T-cell proliferation and function. Similarly, magnesium is critical in the development of humoral immunity and B-cell function [Kubena and McMurray].

Impaired nutrition and protein deficiency contributes to the breakdown of mechanical barriers which are a necessary component of human immune defenses. For example, prolonged malnutrition leads to intestinal villus atrophy and disrupts the immunological integrity of the gut. This injury, in addition to being permissive to enteric pathogens, further exacerbates malnutrition by causing protein energy malabsorption. Skin integrity and woundhealing are also compromised in states of chronic malnutrition. Restoration of adequate energy and nutrition can reverse these pathological processes and restore normal immune function.

Frequency and Clinical Implications of Malnutrition

Malnutrition remains one of the major contributing factors to disease and death in the world. While its impact is greatest in poor and underdeveloped portions of the world, where protein energy malnutrition accounts for 49% of deaths in children under the age of 5 years, malnutrition and micronutrient deficiency have a significant deleterious effect on health in developed nations. Suboptimal nutrition among certain vulnerable segments of the population, for example young children, pregnant women, the elderly andthe chronically ill, has a major influence on health and comorbid disease processes.

Protein Energy Malnutrition. Rates of protein energy malnutrition remain high throughout the world despite efforts by the World Health Organization (WHO) and many others to reduce malnutrition in severely affected areas. The prevalence of malnutrition is particularly high in Asia, where two-thirds of the world's malnourished children live, and in regions of Africa and Latin America. In additionto poverty, diarrhoeal diseases, and long-term issues of inadequate food supplies, acute episode of famine associated with natural disasters, population displacement and war, continue to plague these areas. Somalia in 1992 and the recent crisis in Sudan represent the modern-day challenges to world hunger and starvation which kill and injure humans on a massive scale.

Children are uniformly more vulnerable to protein energy malnutrition than adults. Current estimates suggest that 167.9 million children under the age of 5 years are malnourished according to measurements of appropriate weight for age. The first important exposure to malnutrition for many children is during prenatal life. Intrauterine growth retardation (IUGR) is present in as many as 30 million births per year (23.8%) in developing countries and is associated with increased rates of stillbirth, death in infancy, childhood malnutrition, growth failure, and impaired cognitive and neurological development. IUGR has also been associated with an increased risk in adulthood of cardiovascular disease, chronic lung disease, diabetes and high blood pressure. Throughout infancy and childhood, protein energy malnutrition may contribute to developmental delay, growth retardation, increased risk of infection and death.

Iron Deficiency Anaemia. There is a broad spectrum of possible aetiologies of anaemia, but iron deficiency due to nutritional inadequacies is the number one cause of anaemia worldwide. Estimates provided by the WHO indicate that the prevalence of iron deficiency is 34–98% and an associated prevalence of anaemia of 18–53% in developing countries. Approximately 58% of pregnant women in developing countries suffer from anaemia. Low intake of dietary iron, poor bioavailability of dietary iron [Tatala, et al.:] and infectious illnesses combine to make iron deficiency a major health issue in many nations. Complications of anaemia include increased risk of fetal death, low birthweight, increased mortality rate, developmental and cognitive delay, poor growth and impaired work performance. See also **Anemias**.

Vitamin A Deficiency. Elimination of vitamin A deficiency by the year 2000 was identified as one of the major goals of the Food and Agriculture Organization of the United Nations/World Health Organization (FAO/WHO) International Conference on Nutrition, held in Rome in 1992. Vitamin A deficiency in developing countries is a leading cause of blindness and visual impairment, and has been linked to increased risk of premature death as well as increased risk of maternal–fetal transmission of human immunodeficiency virus infection invitamin A-deficient mothers [Darnton-Hill].

Obesity. Over nutrition or obesity is another form of malnutrition which carries with it impressive levels of comorbidity and mortality. The prevalence of obesity has been rising steadily in Western developed nations over the past decade and estimates of adulthood clinical obesity, defined as a body mass index greater than 30 kg m^{-2} are as high as 22.5% in the United States. Similarly alarming trends have been found in U.S. children. The rates of obesity among children between the ages of 6 and 11 years have jumped from 15% in 1980 to current rates of 22%. Complications of obesity include insulin resistance, type II diabetes, dyslipidemia, steatohepatitis, increase risk of cardiovascular disease, arthritis, and increased risks of breast, endometrial and colonic cancer. Obesity, its growing prevalence among children and adults and its impact on health are becoming increasingly recognized as a major component of malnutrition affecting the developed world today. See also **Obesity**.

Approaches to the Management of Malnutrition

The treatment of malnutrition must begin with efforts directed towards worldwide prevention. Complex societal, economic and political factors must be overcome in order to provide effective coordinated programs targeted at improving access to nutrition. Initiatives by organizations such the WHO, OMNI (Opportunities for Micronutrient Intervention) and UNICEF (United Nations International Children's Emergency Fund) have made advances in the reduction of protein energy malnutrition and micronutrient deficiencies by creating partnerships between government, industry, and academic and scientific communities. For example, protein energy malnutrition among children under 5 years of age has been reduced from 36.4% in 1975 to the current rate of 27.4%. This reflects improvement in the world food supply, as well as improved health and social services to needy areas. National food fortification programs have been implemented in several countries to overcome vitamin A deficiency by fortifying flour, sugar and margarine with supplemental vitamin A.

The treatment of severe malnutrition has benefited from advances in understanding of pathophysiology and metabolism during refeeding. The increasing availability of appropriate fluid and electrolyte solutions for resuscitation in severe dehydration and the refinement of other oral refeeding formulas has led to improved clinical outcome and decreased complications of refeeding. Recent experience has demonstrated that the use of a high-protein oral formula (16.4% of energy derived from protein), which was previously believed to be optimal in severe protein energy malnutrition, was associated with a 3-fold increase in the mortality rate and delayed recovery compared with a lower protein formula (8.5% of energy derived from protein) among patients with advanced starvation and oedema at the time of presentation [Collins, et al.:]. There was no advantage, however, to the lower protein formula among patients with marasmus (nonoedematous protein energy malnutrition).

Summary

Malnutrition and micronutrient deficiencies remain a major public health concern around the world. Increased appreciation of the metabolic and physiological processes associated with malnutrition will guide the development of strategies to prevent and treat the comorbidity and mortality linked to nutrient deficiencies. If progress towards the elimination of malnutrition is to be made in the next century, the provision of adequate and appropriate nutrition must remain a priority of governments, public health organizations, and healthcare providers worldwide.

Additional Reading

Bonjour, J.P., M.A. Schurch, and R. Rizzoli: "Nutritional Aspects of Hip Fractures," *Bone,* **18**(supplement 3), 139S–144S (1996).

Collins, S., M. Myatt, B. Golden: "Dietary Treatment of Severe Malnutrition in Adults," *American Journal of Clinical Nutrition,* **68**, 193–199 (1998).

Darnton-Hill, I.: "Developing Industrial–governmental–academic Partnerships to Address Micronutrient Malnutrition," *Nutrition Reviews,* **55**(3), 776–781 (1997).

Grinspoon, S.K., H. Baum, K. Lee, et al.: "Effects of Short-term Recombinant Human Insulin-like Growth Factor I Administration on Bone Turnover in Osteopenic Women with Anorexia Nervosa," *Journal of Clinical Endocrineand Metabolism,* **81**(1), 3864–3870 (1996).

Keys, A., J. Brozek, A. Henschel, O. Mickelsen, and H.L. Taylor: "Biochemistry: Nature of the Biochemical Problems," *The Biology of Human Starvation,* Vol. 1, pp. 289–339. University of Minnesota Press, Minneapolis, MN, 1950.

Kubena, K.S., and D.N. McMurray: "Nutrition and the Immune System: A Review of Nutrient–nutrient Interactions," *Journal of the American Dietetic Association,* **96**, 1156–1164 (1996).

Owen, O.E., K.J. Smalley, D.A. D'Alessio, M.A. Mozzoli, and E.K. Dawson:

Ravelli, A.C.J., J.H.P. van der Meulen, R.J.P. Michels, et at.: "Glucose Tolerance in Adults after Prenatal Exposure to Famine," *Lancet,* **351**, 173–177.

Sapir, D.G., O.E. Owen, J.T. Cheng, et al.: "The Effect of Carbohydrates on Ammonium and Ketoacid Excretion during Starvation," *Journal of Clinical Investigation,* **51**, 2093–2102 (1972).

Staff, Icon Health Publications: *Malnutrition: A Medical Dictionary, Bibliography, and Annotated Research Guide to Internet References,* ICON Health Publications, San Diego, CA, 2004.

Tatala, S., U. Svanberg, and B. Mduma: "Low Dietary Iron Availability is a Major Cause of Anemia: A Nutrition Survey in the Lindi District of Tanzania," *American Journal of Clinical Nutrition,* **68**, 171–178 (1998).

Taube, F.: "As Obesity Rates Rise, Experts Struggle to Explain Why," *Science,* **280**, 1367–1368 (1998).

World Health Organization: Malnutrition—the Global Picture. http://www.who.int/en/

W. Allan Walker
Colleen Hadigan
Children's Hospital, Boston, MA

MALOJA WIND. See **Winds and Air Movement.**

MALPIGHI, MARCELLO (1628–1694). Marcello Malpighi, the son of Marcantonio Malpighi and Maria Cremonini, was born and baptized in Crevalcore outside Bologna, Italy on 10 March 1628. He entered the University of Bologna in 1646, and with the encouragement of his tutor Francesco Natali, began to study medicine in 1649. Four years later, he received doctorates in both medicine and philosophy. In 1656 Malpighi was appointed to the chair of theoretical medicine at the University of Pisa. However, he eventually tired of Pisa and in 1659 he returned to Bologna. He later would spend four years in Messina, as principal chair of medicine (1662–1666), before returning again to Bologna. In 1667, Malpighi married Francesca Massari, who was 19 years his senior and the sister of his former tutor Bartolomeo Massari. Malpighi was called to be chief physician to Pope Innocent XII in 1691 and died in Rome on 29 November 1694. His body was dissected in a Roman church.

Malpighi is perhaps most renowned as a microscopist, using various lenses of different powers and both reflected and transmitted light in his investigations. He was also skilled in preparing microscopical samples, utilizing such methods as boiling, drying and deaeration. His work also relied on the so-called "microscope of nature"—whereby the investigation of anatomical structures and functions in lower animals serve to reveal the nature of their human parallels. This approach was grounded in Malpighi's belief that nature usually "undertake[s] its great works only after a series of attempts at lower levels, and to outline in imperfect animals the plan of perfect animals" (qtd. DSB IX, 63).

His first work, *De Pulmonibus* (1661), greatly increased the understanding of lungs. It describes how, using a microscope, he discovered that the lungs are surrounded by a capillary network. This finding confirmed the theory of the blood's circulation, and thus *De Pulmonibus* is hailed by some scholars as the beginning of anatomical microscopy.

Another significant work of Malpighi's was his treatise on the tongue (*De Lingua,* 1665). He peeled two layers off the tongue—one of which now bears his name—and thus revealed the papillae (sensory receptors), which he grouped into three separate orders. *De Polypo Cordis* (On Heart Polyps, 1666) was Malpighi's primary hematological text, in which he theorized that polyps developed from the coagulation of the blood. He also described how one could break blood down into its parts by repeatedly washing a polyp with water. Furthermore, he was able to observe red blood corpuscles by microscopically examining the water washed off a polyp. Malpighi also made significant embryological discoveries; his *De Formatione* (1675) describes many such discoveries about chick embryos. *Anatome Plantarum* (1675), Malpighi's work on plant anatomy, was one of the earliest microscopic investigations of plant anatomy.

A doctor as well as a scientist, Malpighi was a defender of rational approaches to medicine, i.e. the use of anatomical and physiological theories in determining medical treatments. Thus, he wrestled with the problem of reconciling the medical theories of old with the anatomical discoveries of the day.

Additional Reading

Adelmann, H.: *Marcello Malpighi and the Evolution of Embryology,* 5 Vols. Cornell University Press, Ithaca, NY, 1966.

Belloni, L.: "Marcello Malpighi and the Founding of Anatomical Microscopy," In: Bonelli, M., and W. Shea: *Reason, Experiment and Mysticism in the Scientific Revolution,* Science History Publications, New York, NY, 1975, pp. 95–110.

Bertoloni, M.D.: *Marcello Malpighi: Anatomist and Physician,* Leo S. Olschki Publishers, Florence, Italy, 1997.

C. Christopher Smith, Indiana University, Bloomington, IN

MALT. Unless otherwise specified, malt usually connotes barley malt. Malt, however, can be prepared from other cereal grains. Between 75 and 80% of the malt produced in the United States goes into the manufacture of beer and associated beverages. Nearly 15% of the production goes to the manufacture of distilled alcohol products; and the remainder (about 5–6%) goes into the preparation of malt syrups, breakfast foods, malted milk concentrates, and coffee substitutes.

Barley malt is barley that has been germinated by moisture under controlled conditions and for a specified time, after which the germinated plants are carefully dried under controlled conditions. The drying or kilning operation and other operations in the total malting process are customized to the final product in which the malt will be used. The principal stages of the malting process are diagrammed in Fig. 1.

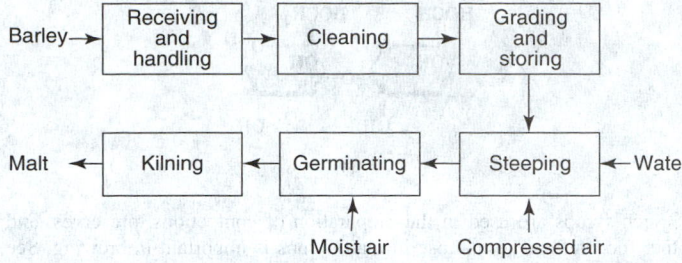

Fig. 1.   Simplified flowsheet of operations involved in preparing malt.

Upon receipt, the barley must be inspected carefully to make certain that it fully meets the minimum acceptable specifications established by the malt manufacturer. The qualities desired in grain for malting are described in the entry on Barley.

Prior to processing, the barley is stored for up to 6 weeks to permit any further ripening (after-ripening) to take place. This enhances later germination. The cleaning operation that follows removes impurities, unwanted foreign seeds, and damaged and broken kernels. Because the size of the kernels affects handling during the germinating operation, the barley is graded into 2 or 3 size ranges, each of which is malted separately.

The role served by malt in the later production of beer and distilled spirits is that of furnishing enzymes, which convert starches and other ingredients during the brewer's and distiller's mashing operations. J. de Clerck states: "Quantitatively the most significant chemical constituent of the grain is starch which constitutes some two-thirds of the dry weight of barley. Apart from a comparatively small contribution from other sugars in barley it is the starch which eventually furnishes some 85–90% of malt extracts, of which some 70% is fermented in brewing and sometimes nearly all in distilling. Starch, therefore, occupies a key position in relation to the brewing and distilling industries and its chemistry is, apart from its intrinsic biochemical interest as the final product of photosynthesis, of outstanding importance." The starch granules of barleys are first seen in the cells of the endosperm a few days after development of the seed begins as small spheres later developing into the bean-shaped and lenticular forms characteristic of mature starch. Luers observed that barley starch undergoes enzymic degradation during malting and that malt starch is different from barley starch. This was recognized as early as 1902. The loss in starch that occurs during malting has been estimated by various researchers. Luhder (1908) first established that the starch content of barley declined considerably during malting. For example, Moravian barley before malting contains 65.57% starch and after malting, 54.89%. Of the two starch constituents amylopectin and amylose, the former is more susceptible to enzymolysis during malting. Several theories have been proposed to explain this difference. The principal factors that influence the amount of starch in barley are environmental during growth and also relate to the variety planted.

Brewers' and Distillers' Malts. Barleys of large grain size are desired for preparing brewers' malt. The larger sizes usually have a greater percentage of starch and relatively little protein. The grains, however, do contain adequate amyloytic enzymes for solubilizing native starch and also some of the other ingredients of the mashing operation at the brewery.

In contrast, the primary objective of distillers' malt is to furnish enzymes in larger quantities for later use in converting grains and starch present in other substances. Barley best suited for this purpose is of a relatively high nitrogen content, a factor that relates to the ability to produce amylases. In the United States, certain barley varieties, such as *Kindred, Manchuria, Montcalm*, and *Odessa*, are specifically selected for making distillers' malt.

Steeping. Once size-graded, the barley is transferred to large steeping tanks equipped with conical or funnel-shaped bottoms. Cool, clean water is added to partially fill the tank. Air agitation assists in cleaning the barley. Any light-weight kernels present automatically rise to the surface of the water and are skimmed off. During a period of from 45 to 60 hours, the barley soaks up water (steeps). At the end of the steeping period, the moisture content of the barley will range between 45 and 48%, representing sufficient moisture to commence germination. Distillers' malt usually requires a somewhat longer period of steeping in order to bring the moisture content up to a minimum of 50%. The process that takes place during steeping is, not surprisingly, essentially the same as the changes that occur when the seed is planted in moist soil.

Germinating. During the next process step, germination, enzymes are produced or liberated in a structure situated between the germ and endosperm of the barley kernel. During germination, the cell wall is made more permeable by the action of the enzyme cytase. Mellowness and friability of the finished malt is determined by this enzymatic action. In barley malt, the amylase enzymes are the most important. These enzymes convert starch into maltose sugars and dextrins in later brewing processes. During germination, only a small part of the starch present should be converted to maltose or other sugars, recalling the prior mention of starch losses. The principal conversion of starch occurs later in mashing operations at the brewery or distillery. Other enzymes that transform proteins also are produced or activated during malting.

Over the years, three principal malting procedures have been developed. The oldest of these is the germinating floor or floor malting, still practiced in some European countries. The steeped barley, once removed from the steeping tanks, is spread in heaps about 1 foot (0.3 meter) in height. The first spreading occupies perhaps only 40% of the available floor area. The barley commences to dry out and sprout, producing small hairlike fibers (rootlets). It is necessary to aerate the early-sprouting barley (known as *green malt*), which is accomplished by turning the heap over and over with forks. The barley is spread over a larger floor area. These actions, plus the use of forced aeration, cool the green malt. This procedure is repeated at regular intervals over a period of 5 to 7 days. The process is confined to the cooler seasons of the year because germinating temperature should be maintained within a span between 60° and 70°F (15.6° and 21.1°C). Excessive temperature accelerates germination and generates waste, increasing starch losses and excessively large rootlets. Technically, the optimum cut-off time for germination is when the plumule or acrospire reaches the length of the kernel. An extension of the floor malting process to avoid any seasonal limitations on the process is pneumatic floor or compartment malting. Instead of placing the barley on the floor in heaps, it is placed in box-shaped compartments, which hold the grain during the entire germinating period. Purified air is circulated through these compartments. The air is both temperature and humidity controlled. The green malt is turned by mechanical screws within the compartments. Such compartments will contain up to 5000 bushels (108 metric tons) of green malt.

In *drum malting*, widely used in modern malting plants, there are two concentric hollow metal cylinders. Grain is placed in the spaces between the cylinders, which are perforated for introduction and circulation of humidity-and-temperature-controlled air. To permit maximum movement of the grain, the equipment is only partially filled. The revolving action of the cylinders keeps the grain tumbling, constantly exposing new surfaces of the grain to the air stream. Capacity of these drums is up to 650 bushels (about 13.5 metric tons).

Drying or Kilning. These are not high-temperature or rotary kilns as one may visualize in connection with a cement or ore plant, but they are usually 2-story buildings. Drying is commenced on the upper story by spreading the green malt in layers of 2 to 3 feet (0.6 to 0.9 meter) in depth. The primary purpose of the kilning operation is to halt further germination, although the conditions of kilning also affect the final end-use properties of the malt. Hot air is drawn through the malt at a relatively low temperature for the first 24 hours to accomplish a partial drying. The malt is then transferred to the lower story, where the air temperature is increased. In the case of brewers' malt, the final kilning temperature is in the range of 160° to 180°F (71° to 82°C), and the malt is retained in the kiln until the moisture has reached a content of about 4%. The entire process requires from 48 to 72 hours. In the case of distillers' malt, the final kilning temperature is lower—in the range of 120° to 140°F (49° to 60°C) and the final moisture content is from 5 to 7%. The lower temperature preserves higher enzymatic activities. The higher temperature for brewers' malt introduces a more intense malt flavor and aroma, usually desired for the brewing process. In the case of porter, bock beer, and stout, the kilning temperature is higher and a caramel malt is produced. This imparts a dark color and distinctive flavor during the brewing of these malted beverages.

After kilning, the malt is cleaned to remove rootlets and any broken kernels that may remain in the batch.

Malting Process Innovations. In recent years, several steps to modernize and improve the operations just described have been made. Sophisticated control and conveying systems have assisted in more exacting control of processing conditions and reducing hand labor costs. There is a growing trend toward continuous operations. Some consideration has been given to the use of gibberellic acid as part of the steepwater or in the form of a spray during the germinating process. The objective of these steps would be that of increasing yield. It also has been found that potassium bromate in the steepwater will depress respiration and rootlet formation.

A nutritional profile is given in Table 1.

TABLE 1. NUTRITIONAL PROFILE OF MALT (100-gram samples)

	Dry Malt	Dried Malt Extract
Water	18.7 g	11.5 g
Food energy	374	374 cal
Protein	1.3 g	1.7 g
Fat	1.8 g	trace g
Carbohydrate	78.8 g	91.1 g
Calcium	—	50 mg
Phosphorus	—	299 mg
Iron	4.0	9.0 mg
Sodium	—	83 mg
Potassium	—	234 mg
Vitamin A	—	—
Thiamine	0.50	0.36 mg
Riboflavin	0.32	0.47 mg
Niacin	9.4	10.1 mg
Vitamin C (ascorbic acid)	—	—

Source: U.S. Dept. of Agriculture, Washington, DC.

Malt from Other Grains. Sometimes wheat malt is used as a flour supplement for bakery products. The malt provides a source of a-amylase, which degrades starch to sugars, the fermentation of which causes rising of the dough. Of course, barley malt is also used in some bakery product flours. Wheat malt imparts a characteristic flavor to beer. Most consumers find this undesirable. However, *Weissbier*, made with wheat malt, is quite popular in certain regions of Germany. In producing wheat malt, kilning temperatures are more moderate.

Additional Reading

Hough, J.S.: *The Biotechnology of Malting and Brewing*, Cambridge University Press, New York, NY, 1992.

Hough, J.S., D.E. Briggs, R. Stevens, and T.W. Young: *Malting and Brewing Science*, Vol. 2, Chapman & Hall, New York, NY, 1999.

Hough, J.S. and T.W. Young: *Malting and Brewing Science*, Vol. 1, Kluwer Academic Publishers, Norwell, MA, 1999.

Staff, Briggs Corporation: *Malts and Malting*, Aspen Publishers, Inc., Gaithersburg, MD, 2000.

MALTHA. A black, viscous, natural bitumen consisting of a complex mixture of hydrocarbons. Its viscosity and rheological properties lie between those of crude oil and semisolid asphalt. It is the chief component of Athabaska oil sands.

MALTITOL. See **Sugar Alcohols**.

MALTOSE. Maltose [CAS: 69-79-4] (malt sugar) is a disaccharide, 4-*O*-α-D-glucopyranosyl-D-glucose (**1**), comprising two molecules of glucose (dextrose). Although occurring in some plants and fruits, it is more frequently recognized as a structural component of starch. Pure maltose is isolated with difficulty from a directed starch hydrolysate, ie, high maltose corn syrup, by precipitation with ethanol. Purification can be achieved by way of the β-maltose octaacetate. Removal of the acetate groups allows crystallization of the monohydrate of β-maltose. Commercial maltose typically contains 5–6 wt % of the trisaccharide maltotriose with traces of glucose. High maltose syrups from starch typically contain ca 8–9 wt % glucose, 40–80 wt % maltose, with higher saccharides as the remainder.

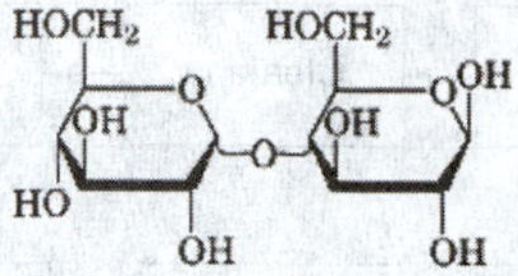

(1)

Such syrups are used in the preparation of confections, preserves, and other foodstuffs. The maltose in malt syrups is important in brewing. See also **Beer and Other Malt Beverages**. Intravenous feeding (primarily in Europe and Japan) and sports beverage formulations take advantage of the fact that energy release from maltose becomes accessible to the body at a slower rate than energy supplied by monosaccharides.

Important physical and functional properties of maltose and maltose syrups include sweetness, viscosity, color stability, humectancy, freezing point depression, and promotion of beneficial human intestinal microflora growth. Maltose possesses ca 30–40% of the sweetness of sucrose in the pure state.

Hydrogenation of high maltose syrups gives a mixture of sugar alcohols, from which maltitol [CAS: 585-88-6] (**2**) can be isolated in crystalline form. Maltitol is almost as sweet as sucrose (0.9 times) and has been promoted as a sweetener in various food applications. See also **Carbohydrates**; and **Sucrose**.

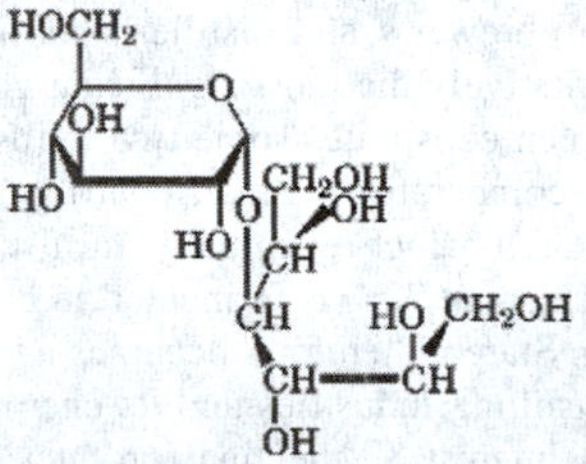

(2)

MALTOSE SYRUP. See **Carbohydrates**; **Syrups**; and **Sweeteners**.

MALUS COSINE-SQUARED LAW. A law applying to the intensity of polarized light as affected by the polarizing apparatus. If a beam of plane-polarized light is passed through a Nicol prism, for example, the intensity (flux density) of the emergent beam falls off, as the prism is rotated, from a maximum value when the transmission plane of the prism coincides with the plane of vibration of the light, to zero when it is at right angles to that direction. The intensity varies as the square of the cosine of the angle through which the prism has been thus rotated. The same law applies to the effect of a glass reflector, reflecting always at the polarizing angle, as the plane of reflection is rotated around the stationary, polarized incident beam.

MALVACEAE (Mallow Family). The plants of this family include herbs, shrubs and trees (the latter tropical), and are rich in mucilaginous substance. The leaves are alternate, and in most cases palmately lobed and veined, with small deciduous stipules. The flowers are regular and perfect, often large and showy, and variously borne. They have five (or rarely, fewer) more or less united sepals, five petals, and numerous stamens, which characterized the family by having their filaments joined to form a tube which surrounds the styles.

The most important member of this family is the cotton plant, whose fibers outrank in commercial importance all others. See also **Cotton**. Okra or Gumbo, *Hibiscus esculentus*, a native of tropical Africa, is another member of importance. See Fig. 1. It is a coarse annual plant with large veiny leaves and showy axillary flowers. The slender 5-ribbed pods are used in soups, or when young, are cooked and used in salads. Okra has also been used as a source of fibers for paper manufacture. Another member of some importance is *Althaea officinalis*, the Marsh Mallow. The underground rootstock of this plant is not only rich in mucilage, but is also used medicinally in ground form, the bark being removed before grinding.

Many members of the family are grown as ornamental plants, among them being the Hollyhock, *Althaea rosea*, species of *Malva*, the true

Fig. 1.   Okra plant (Hibiscus esulentus) showing pods ready to harvest. (USDA photo.)

Mallows, and the Rose of Sharon, *Hibiscus syriacus*, which becomes a large bush or even a small tree, with showy pink or white flowers.

MALVINAS CURRENT.   See **Ocean Current.**

MAMMA.   See **Clouds and Cloud Formation.**

MAMMALIA.   The Latin word *mamma* (meaning breast) provides the key to that large class of animals designated *Mammalia*. There is an exceedingly great variety of characteristics among mammals. Their size ranges from the tiny shrew of some 2 inches (5 centimeters) in length, to the whales, which may attain 100 feet (30 meters) in length and a weight of 130 tons (117 metric tons) or more. The principal point of commonality is that all are vertebrates, the females of which possess mammary glands, milk-secreting glands for feeding the young. Within the framework of this definition, humans are mammals. There are 19 orders and some 5,000 species of mammals, as shown in Table 1. Several examples of each order and reference to specific entries in this volume on mammals are given in the Table 1.

Because of the variety among mammals, generalizations are difficult and can be misleading, but with these factors in mind, the following observations may be made. (1) The young of most species develop in the body of the mother and all are nourished by milk secreted by special glands known as the mammary glands. (2) There are typically two sets of teeth, the so-called milk teeth or temporary teeth which the young retain for a varying length of time — followed by permanent teeth set in sockets in the jawbones. The teeth are of four kinds, i.e., incisors, canines, premolars, and molars. (3) Mammals are one of the more dominant forms of animal life of the phylum *Chordata*, this dominance greatly aided by superior intelligence, locomotion, and well-developed sensory organs in most cases. (4) Most mammals have a covering of skin, out of which grows hair. (5) Although predominantly creatures of the land, both above and below the surface, species of mammals are found in the seas, and a few species can take to the air. (6) Mammals are found practically everywhere on earth — if not in their native habitat, often in areas of

TABLE 1. MAMMALS (Alphabetical Listing of Orders)

Order and Examples	Major Entries in this Encyclopedia
ARTIODACTYLA	Artiodactyla
(Even-toed Hoofed Animals)	
Antelopines	Antelope
(Antelopes and gazelles)	
Antilocaprines	Pronghorn Antelope
Bovines	Bovines
(Oxen, cattle, buffalo, bison, and duikers)	
Camelines	Camels and Llamas
Caprines	Goats and Sheep
Cervines	Deer
(Deer, muntjacs, moose, and reindeer)	
Giraffines	Giraffe and Okapi
Hippopotamines	Hippopotamus
Suines	Suines
(Pigs and peccaries)	
Tragulines	Tragulines
(Chevrotains)	
CARNIVORA	Carnivora
(Flesh-eating Mammals)	
Felines	Cats
(Lions, tigers, leopards, jaguars, ocelots, domestic cats, lynxes, servals, jaguarondis, and cheetahs)	
Viverrines	Viverrines
(Civets, hemigales, and mongooses)	
Hyaenines	Hyena
Procyonines	Raccoons
Canines	Canines
(Wolves, jackals, foxes, and dogs)	
Ursines	Bears
Mustelines	Mustelines
(Weasels, badgers, skunks, and otters)	
Ailuridae	Pandas
CETACEA	Whales, Dolphins, and Porpoises
(Hairless, fish-like Water Mammals)	
CHIROPTERA	Bats
(Flying Mammals)	
DERMOPTERA	Dermoptera
(Gliding Mammals)	
(Flying Lemur or Kobego)	
EDENTATA	Edentata
(Anteaters, sloths, and armadillos)	
HYRACOIDEA	Hyraxes
INSECTIVORA	Moles and Shrews
(Insect-eating Mammals)	
LOGOMORPHA	Rabbits and Hares
(Leaping Mammals)	
MARSUPIALIA	Marsupialia
(Pouched Mammals)	
(Opossums, bandicoots, phasogales, phalangers, koalas, wombats, and kangaroos)	
MONOTREMATA	Monotremata
(Egg-laying Mammals)	
(Duckbills and spiny anteaters)	
PERISSODACTYLA	Perissodactyla
(Odd-toed Hoofed Mammals)	
Equines	Horses, Asses, and Zebras
Tapirines	Tapir
Rhinocerotines	Rhinoceros
PHOLIDOTA	Pholidota
(Scaly Mammals)	
(Pangolins)	
PINNIPEDIA	Sea-Lions and Seals
(Fin-footed Mammals)	
PRIMATES	Primates
(Top Mammals)	
Tupaioids	Moles and Shrews
(Tree-shrews)	
Lorisoids	Lorisoids
(Lorises and bush-babies)	
Lemuroids	Lemur
Tarsioids	Tarsioids
(Tarsiers)	

(continued)

TABLE 1. (Continued)

Order and Examples	Major Entries in this Encyclopedia
Hapaloids	Marmoset
Ceboids	Monkeys and Baboons
(Half-monkeys and hand-tailed monkeys)	
Simioids	Monkeys and Baboons
(Colobine monkeys, long-tailed monkeys, dog-faced monkeys, the black ape, baboons, the gelada, and drills)	
Anthropoids	Anthropoids
(Lesser apes, gibbons, Greater apes, gorillas, chimpanzees, and orangutans)	
PROBOSCIDEA	Elephant
RODENTIA	Rodentia
(Gnawing Mammals)	
Sciuromorphs	Rodentia
Squirrels	Squirrels and Other Sciuromorphs
Beavers	Beaver
Myomorphs	Rodentia
(Mice, rats, and jerboas)	
Hystricomorphs	Rodentia
Porcupines	Rodentia
SIRENIA	Sea-Cows
(Manatees and dugongs)	
TUBULIDENTATA	Aard-Vark

extreme climates, where their great adaptability after migration often assures survival. Humans create artificial shelters and, in essence, often create their own environment through the application of energy for heating and cooling, as well as applying technology for raising and processing a food supply. They can exist for long periods of time in the most severe of climates and, of course, in recent years have proved their ability to survive in the extremely hostile environment of interplanetary space.

Much knowledge pertaining to the adaptation of mammals has been learned from the study of fossils. See also **Paleontology**.

Additional Reading

Balog, J.: "A Personal Vision of Vanishing Wildlife," *Nat'l. Geographic*, 84 (April 1990).

Dixson, A.F.: *The Natural History of the Gorilla*, Columbia University Press, New York, NY, 1981.

Dutcher, K. "The Secret Life of America's Ghost Cat," *Nat'l. Geographic*, 38 (July 1992).

Gould, E., G. McKay: *Encyclopedia of Mammals*, 2nd Edition, Academic Press, Inc., San Diego, CA, 1998.

Grzimek, B.: *Grzimek's Animal Life Encyclopedia*, Vols. 10–13, Van Nostrand Reinhold, New York, NY, 1975. (A Classic Reference.)

Heinz-Georg, K., E.M. Lang: *Handbook of Zoo Medicine*, John Wiley & Sons, Inc., New York, NY, 1982.

Linden, E., F. Lanting: "Bonobos, Chimpanzees With a Difference," *Nat'l. Geographic*, 46 (March 1992).

Linden, E., M. Nichols: "A Curious Kinship: Apes and Humans," *Nat'l. Geographic*, 2 (March 1992).

Macdonald, D.: *Encyclopedia of Mammals*, Barnes & Noble Books, New York, NY, 1999.

Maple, T.L., M.P. Hoff: *Gorilla Behavior*, John Wiley & Sons, Inc., New York, NY 1981.

Napier, J.R., P.H. Napier: *The Natural History of the Primates*, MIT Press, Cambridge MA, 1994.

Preuschoft, H., et al.: *The Lesser Apes*, Edinburgh University Press, Edinburgh, U.K., 1984.

Schaller, G.B., et al.: *The Giant Pandas of Wolong*, University of Chicago Press, Chicago, Illinois, 1985.

Staff: *Encyclopedia of Mammals*, Marshall Cavendish, Inc., Tarrytown, NY, 1997.

Walther, F.R., et al.: *Gazelles and Their Relatives*, Noyes, Park Ridge, New Jersey, 1983.

Zhi, Lu: "Newborn Panda in the Wild," *Nat'l. Geographic*, 60 (February 1993).

MAMMARY GLAND. A large gland, which secretes milk for the nourishment of the young of mammals.

MAMMOTH. See **Elephant**.

MANAKIN (*Aves, Passeriformes*). Brightly colored, small wren-size bird of Central and tropical South America. There are several species. The males are known for their singing and dancing to exhibit their beautiful plumage to the usually drab, green-colored females. Much like the bowerbird, the male clears away an area for his demonstrations. Most species are polygamous. The female builds the nest, incubates and hatches the eggs, and rears the young. An open-type nest is constructed in the fork of a tree. There are usually two pastel eggs with dark markings. Incubation period is from 19 to 21 days. Manakins feed mostly on insects.

MANDIBLE. See **Fishes**.

MANGANESE. [CAS: 7439-96-5]. Chemical element, symbol Mn, at. no. 25, at. wt. 54.9380, periodic table group 7, mp 1,244 $\pm$ 3°C, bp 1,962°C, density 7.3 g/cm^3 (solid), 7.21 (single crystal) (20°C). Manganese has a cubic (complex) crystal structure.

Manganese is a little-known element other than to a small circle of technical specialists who are predominantly metallurgists and chemists. Yet it is the fourth most used metal in terms of tonnage, being ranked behind iron, aluminum and copper, with in the order of 20 million tons of ore being mined annually (2000).

Manganese has numerous applications which impact on our daily lives, whether it be as consumers of objects made of steel, of portable batteries, or of beverage cans based on aluminum. In each case, manganese plays a vital role in improving the properties of the alloys and compounds involved in each specific application.

In the mid-17th century, the German chemist Glauber obtained permanganate, the first usable manganese salt. Nearly a century later, manganese oxide became the basis for the manufacture of chlorine. Yet manganese was only recognized as an element in 1771, by the Swedish chemist Carl Wilhelm Scheele. It was isolated in 1774 by one of his collaborators, J.G. Gahn. At the beginning of the 19th century, both British and French scientists began considering the use of manganese in steelmaking, with patents granted in the U.K. in 1799 and 1808. In 1816, a German researcher observed that manganese increased the hardness of iron, without reducing its malleability or toughness.

In 1826 Prieger in Germany produced a ferromanganese containing 80% manganese in a crucible. J.M. Heath produced metallic manganese in England in about 1840. The following year, Pourcel began industrial-scale production of "spiegeleisen", a pig-iron containing a high percentage of manganese, and in 1875 he started the commercial production of ferromanganese with a 65% manganese content. The major breakthrough in the use of manganese occurred in 1860. At that time, Sir Henry Bessemer was trying to develop the steelmaking process which was to bear his name. But he experienced difficulty with an excess of residual oxygen and sulphur in the steel. The problems were overcome thanks to the beneficial effect of manganese, disclosed in a patent granted to Robert Mushet in 1856. Mushet suggested adding "spiegeleisen" after the blow to introduce both manganese and carbon and remove oxygen. This procedure made the Bessemer process possible, and thus paved the way for the modern steel industry. Ten years later, in 1866, Sir William Siemens patented the use of ferro-manganese in steelmaking so as to control the levels of phosphorus and sulphur.

Subsequently, and in contrast to all the early work involving manganese and steelmaking, Leclanché in 1868 developed the dry cell battery. This uses manganese dioxide as a depolariser in a simple yet effective dry cell and the battery market today is the second largest consumer of manganese. The history of manganese in the 20th century has been a stream of new processes and metallurgical/chemical applications developed with a significant impact on markets as diverse as beverage cans, agricultural pesticides and fungicides and electronic circuitry used in consumer products. Details of these applications are analysed later.

Manganese is a silver-white metal, not notably hard (becomes hard on alloying with carbon), brittle, capable of taking a brilliant polish but readily oxidized upon heating, reacts with water upon boiling, soluble in dilute acids. In terms of abundance, manganese is present in igneous rocks to an average extent of 0.10% (weight). In terms of cosmic abundance, in the estimate by Harold C. Urey (1952), using a base Figure of 10,000 for silicon, the Figure for manganese is 75. Manganese is estimated as 34th among the elements in its content in seawater, an estimated 9.5 tons per cubic mile of seawater. There are eight isotopes of manganese, ^{50}Mn through ^{57}Mn, all radioactive with exception of ^{55}Mn. Half-lives range from a fraction of a second for ^{50}Mn to approximately 140 years for ^{53}Mn. Electronic configuration $1s^2 2s^2 2p^6 3s^2 3p^6 3d^5 4s^2$. Ionic radius Mn^{2+} 0.83 Å. Metallic radius 1.365 Å. First ionization potential 7.32 eV; second, 15.7 eV. Oxidation potentials $Mn \rightarrow Mn^{++} + 2e^-$, 1.18 V; $Mn^{2+} + 2H_2O \rightarrow MnO_2 + 4H^+ + 2e^-$, −1.28 V; $Mn^{2+} \rightarrow Mn^{3+} + e^-$, 1.51 V; $Mn^{2+} + 4H_2O \rightarrow MnO_4^- + 8H^+ + 5e^-$, −1.52 V; $MnO_2 + 2H_2O \rightarrow MnO_4^- + 4H^+ + 3e^-$, −0.168 V; $Mn(OH)_2 + OH^- \rightarrow Mn(OH)_3 + e^-$, 0.40 V; $MnO_4^- \rightleftharpoons MnO_4^- + e^-$, −0.54 V; $MnO_2 + 4OH^- \rightarrow MnO_4^- + 2H_2O + 3e^-$, −0.58 V. Other important physical properties are given under **Chemical Elements**.

Occurrence. The most common manganese ore is pyrolusite, MnO_2. Other commercial ores include braunite, Mn_2O_3; hausmannite, Mn_3O_4; and rhodochrosite, $MnCO_3$. Although not of industrial value, manganese also exists in nature as the silicate, sulfate, sulfite, and tungstate. See also **Pyrolusite**; and **Rhodochrosite**.

Manganese Nodules. These are rocks composed largely of ferromanganese oxides formed by precipitation at the bottom of lakes and the oceans. They range in size from micrometers to meters. Their morphology is highly variable. They contain up to 55% manganese, 35% iron, and 2% nickel, cobalt, and copper. Manganese nodules were first discovered in the open ocean by Thompson, Murray, and Renard during the *Challenger* expedition (1873–1876). Buchanan reported the occurrence of nodules in the Firth of Clyde, a shallow-water area, and by the end of the century at least five additional occurrences of manganese nodules in shallow marine environments had been discovered. Early workers chemically analyzed about a score of manganese nodules and hypothesized about their mechanism of growth. Two principal concepts emerged: (1) they grow by the slow precipitation of manganese from seawater; and (2) they are formed by the rapid precipitation of manganese released in submarine volcanism.

Until the 1950s, little additional work was done except for some early measurements of manganese nodule growth rates. During recent years, however, there has been a strong revival of interest in manganese nodules, stimulated both by the expansion of oceanographic facilities and the realization of the economic importance of the nodules as ores. It has been found that in large areas of the ocean floor, manganese nodules may be absent. In other areas, they may cover nearly 100% of the area. In all of the Pacific Ocean, nodules have been estimated to cover approximately 10% of the ocean floor. The estimated coverage in the Indian and Atlantic Oceans is less. The local variability in manganese nodule concentration is large. Two ocean bottom photographs only a few meters apart may show very different nodule concentrations. In some locations, the weight concentration of nodules ranges up to 5 g/cm^2.

Manganese nodules are composed of cryptocrystalline minerals. They are known to consist of three major manganese phases: (1) δ MnO_2 (birnessite); (2) 10-Å manganite; and (3) 7-Å manganite. The first is the most highly oxidized form, and has a chemical composition of about $Mn_{1.9}$. Barnes (1967) examined the depth dependence of the mineralogy in nodules taken from the Pacific. His data indicate that above 3,500 m in depth, the only important manganese phase is δ MnO_2, but, below the 3,500 m depth, both 10-Å manganite and 7-Å manganite coexist with the δ MnO_2. The observed phase changes may be pressure induced.

During recent years, the growth rates of manganese nodules have been determined by various methods. Results all indicate that the nodules measured grow at a rate of a few millimeters per million years. This does not exclude the possibility that nodules in certain areas evolve more rapidly, but it appears that most deep-sea nodules grow slowly. There is some belief that the nodules are primarily the result of bacterial fixation of manganese. Other investigators believe that the nodules are formed by inorganic precipitation of metals supersaturated in sea water. There is some experimental and theoretically tenable evidence to support both concepts.

Research gathered during the Deep Sea Drilling Project (DSDP) and the International Decade of Ocean Exploration (IDOE) programs is described in the entry **Ocean Resources (Mineral)**.

Hypotheses continue to evolve concerning the manner in which manganese nodules develop. In the early 1980s, in an effort to sweep away some of the mysteries concerning the Mn nodules, a consortium of American researchers participated in the MANOP (Manganese Nodule Project) program. The program involved the creation of mathematical models, not a simple task because it has been estimated that nodule attrition is about one atomic layer of the Mn-O structure per year. As reported by J. Dymond and colleagues (Oregon State University), two different processes operate within sediments, the particular process depending on whether any oxygen remains below the sediment surface. The amount of oxygen present, of course, is a function of the biological productivity of the overlying surface waters. Where winds and currents mix the sea in the needed manner, microscopic plants and animals do well and part of their inorganic skeletons and probably about one percent of their organic tissues will sink to the bottom—along with clay washed from the land. Regardless of how deep the sea floor is, bacteria and animals dwelling on and in the sediment will oxidize organic matter. However, a small percentage of their organic matter does reach the bottom. If the falling organic matter is light, not all of the sediment oxygen will be consumed. It is proposed that toxic chemical alterations (oxic diagenesis) of the sediment can then supply metals to nodules. It is further reasoned that oxic sediments must be altered chemically to produce nodules because under such oxidizing conditions, Mn and Fe are tied up as insoluble oxides, which cannot move and thus cannot be incorporated in nodules. Several possible diagenetic reactions, such as those involving volcanic ash and skeletal opal, have been suggested. As reported by Kerr (1984), the Oregon State University team concluded that the nodule composition most typical of growth under oxic diagenesis is that of the nodule bottom most rich in trace metals from a siliceous sediment in the tropical North Pacific. In terms of the overall picture, some scientists have observed that manganese nodules should not be there, but are! Obviously Mn nodule research will require many more years in supplying answers. See also **Ocean Resources (Mineral)**.

Todorokites may be defined as calcium-bearing manganese oxides, which are found in terrestrial Mn ore deposits, in weathering products of Mn-bearing rocks, and in some Mn nodules. In some cases, todorokites are the principal constituents of Mn nodules. Knowledge collected concerning todorokites has contributed and will continue to contribute to a better understanding of Mn nodule formation in ocean waters. See also **Todorokites**.

Processing. Manganese metal can be obtained from oxide ores by reduction with carbon, aluminum, magnesium, or sodium in an electric furnace. The main form in which manganese is used is *ferromanganese*. This material contains approximately 80% manganese and 20% iron. Ferromanganese is generally made in a blast furnace or an electric-arc furnace. Usually a mixture of ores is used, proportioned to yield the final desired specifications of the alloy. To reduce slag volume, low-silica ores are preferred. It is also desirable to maintain a low phosphorus content in the alloy. The charge to the electric furnace process for making ferromanganese is the manganese ore, coke, and limestone. The loss of metal to the slag is determined by the silica present. Usually about 75% of the metal is recovered. Where high-purity manganese is produced, the ore is first roasted to MnO, then leached with H_2SO_4 to form the sulfate. The solution is then neutralized to precipitate iron and aluminum. Other impurities are removed as the sulfides. Electrolysis of the resulting solution yields a 99.94% pure manganese metal.

The high-purity (electrolytic) manganese is used as a deoxidizing agent and sometimes as a constituent of nonferrous metals where it improves strength, ductility, and hot-rolling properties. Because of their very high temperature thermal coefficient of expansion, manganese-base alloys with 72% manganese (balance is copper and nickel) are used in bimetals for switching applications. Manganese (60–80% Mn) and copper alloys find application because of their vibration-damping properties.

The standard ferromanganese (7% carbon; 74–78% Mn) is used both to produce a manganese alloy steel and as a deoxidant. As early as 1856, Robert Mushed used *spiegeleisen* (10–23% Mn; 4–5% C) in alloys. Where the carbon content of steel is critical, low-carbon ferromanganese is added. Silicomanganese is used as a blocking agent to stop the reaction of carbon and oxygen in steel. Developed in 1888, Hadfield steel contains about 13% manganese. It finds use where a very hard material is needed and it has the interesting property of increasing its hardness when subjected to repeated impacts. In the 200 series of stainless steels, manganese is replacing nickel in order to achieve more economical austenitic materials.

Manganese Inorganics. A number of chemical processes have been developed to upgrade Mn ore which produce an intermediate Mn compound. These intermediates usually are free of most siliceous matter. Although these processes were designed to convert the compound to an oxide for use in metallurgical applications, the purity of the compounds often renders them suitable for commercial use.

The ammonium carbonate process (developed by Manganese Chemicals Corp.) is the first such upgrading process that has reached commercial application. The high-grade manganese carbonate produced is sold to the chemical industry. The process involves reducing the ore to MnO by roasting with gases rich in CO as the initial step. The calcine is then ground and leached in an aqueous solution containing 18 moles of NH_3 and 3 moles of CO_2. The resulting product is decomposed to yield $MnCO_3$ and NH_3.

The manganese nitrate process is the second upgrading process that has reached commercial application. The high-purity manganese oxides produced are sold to the chemical and ferrite industries. The process involves the reaction between NO_2 and manganese ore to form manganese nitrate solution. The resulting aqueous solution is then thermally decomposed to produce MnO_2 and nitrogen oxides. The nitrogen oxides are recycled to the leaching step, while the MnO_2 is recovered and processed by reduction to Mn_2O_3, Mn_3O_4, or MnO. Processes of lesser importance include the chloride and sulfur oxide processes and bacterial leaching.

Chloride Process. In 1985, investigators at the Argonne National Laboratory reported on the success of a two-step process that extracts cobalt and manganese from low- and medium-grade ores that are mined mainly for other metals. Inasmuch as cobalt and manganese are strategically critical minerals for the United States and several other industrial nations, a viable process for secondary sources of Co and Mn is attractive. A molten salt is used to dissolve more than 90% of the Co and Mn found in common nickel- and copper-bearing ores. The salt mixture contains the chlorides of sodium, potassium, and magnesium (mp 750°C; 1382°F). Approximately one part (wt) of the ore requires four parts of the chloride mixture. The latter is recyclable. The desired metals are subsequently separated electrolytically.

Chemistry of Manganese. Manganese has a $3d^5 4s^2$ electron configuration, and compounds in all oxidation states from 0 to 7+ are known, although those of 1+ and 5+ are uncommon. The reducing power of the manganese atom (Mn→ Mn^{2+}, 1.18 V) is less than that of magnesium, although the first and second ionization potentials are closely similar, due to the higher heat of sublimation of manganese. However, manganese is oxidized by the halogens, H+ or even H_2O to the dipositive state.

Like so many other metals manganese forms compounds with nitrogen, carbon and even oxygen that exhibit unusual valences, or are even of nonstoichiometric character. With nitrogen manganese combines with unusual valence of 5+ to form Mn_3N_5; with carbon it forms Mn_3C, while with free oxygen it forms first MnO, then Mn_3O_4, and finally Mn_2O_3. An exception to this rule is the MnO_2 produced by thermodecomposition of concentrated manganese nitrate solutions where the oxygen-to-manganese ratio is 1.99+.

Manganese (0) compounds are exemplified by the carbonyls discussed below.

Manganese(I) is found chiefly in the few complex ions, such as the hexacyanomanganate(I) ion $[Mn(CN)_6]^{5-}$, produced by vigorous reduction (e.g., by aluminum in alkaline solution) of the corresponding manganese(II) ion $[Mn(CN)_6]^{4-}$, or in isocyanide complexes (formed by reduction of the diiodide with alkyl isocyanides) $[Mn(RNC)_6]^+$ where R is an alkyl radical.

Manganese(II) (manganous) compounds are obtained, as stated above, by action of water, halogens (except fluorine) or acids upon the metal, or by reduction of more highly oxidized compounds in acid solution. Many salts of Mn^{2+} are known, including all four of the common halides, the nitrate, the sulfate, the sulfite, various phosphates, the arsenate, and many salts of organic acids, e.g., the acetate, butyrate, citrate, lactate, oleate, and tartrate. The manganese(II) compounds are in general relatively resistant to oxidation, due to the stability of the half-filled $3d$ subshell. However, the oxide, MnO, and the hydroxide, $Mn(OH)_2$, are rather easily oxidized by air.

This stability of the Mn(II) state is also reflected in the relatively strong oxidation potential of Mn^{3+} (manganic) ion (the value for Mn^{3+}/Mn^{2+} being -1.51 V), and the readiness with which Mn(III) compounds disproportionate. Manganese(III) fluoride, produced by the action of fluorine on lower compounds, reacts with H_2O to produce the difluoride, hydrogen fluoride, and MnO_2. In general, however, the manganic compounds such as dimanganese trisulfate and manganese triacetate, decompose in H_2O to divalent manganese ions and Mn_2O_3, forming the MnO_2 only if the pH is definitely below 7. The phosphate, $MnPO_4$, is easily formed by action of nitric acid on manganese(II) phosphate in concentrated phosphoric acid. The fluoro salt K_3MnF_6 is formed by reduction of potassium permanganate, $KMnO_4$, in 40% hydrofluoric acid by an excess of diethyl ether or manganese(II) salt. Manganese(III) also forms a variety of complexes with chelating agents, e.g., oxalate, glycine, acetylacetone, and the like. Like other tripositive transition metal ions it forms alums. The cyanide $K_3Mn(CN)_6$ is stable. All Mn(III) compounds undergo hydrolysis except the complexes.

In addition to the dioxide and the manganites, formed by fusion of MnO_2 with alkali, manganese(IV) forms a number of complexes, such as K_2MnF_6 by reduction of potassium permanganate in 40% hydrofluoric acid with a limited amount of diethyl ether or manganese(II) salts; and Cs_2MnCl_6, by action of cold concentrated HCl containing cesium chloride on MnO_2. Complex iodates are known, e.g., $M_2^1[Mn(IO_3)_6]$, as are cyanides, formed by the action of potassium cyanide on potassium permanganate and said to be $K_4Mn(CN)_8$ (cf. $K_4Mo(CN)_8$ and $K_4Re(CN)_8$.)

Manganese rarely occurs with an oxidation number of 5+. In addition to the nitride, there is another compound of interest, in that it can be formed in solution. It is an oxyanion of pentavalent manganese, MnO_4^{3-}, which occurs in the compound, $Na_3MnO_4 \bullet 7H_2O$, formed by reduction of the manganate in strongly alkaline formate or sulfite solutions or by heating MnO_2 in alkali hydroxide at very high temperature. Upon neutralization, it disproportionates to the manganate (and MnO_2).

The manganates, containing MnO_4^{2-}, and produced by alkaline oxidation of MnO_2, are the principal known compounds of hexavalent manganese. They are unstable in neutral or acidic solution, undergoing disproportionation to permanganate (MnO_4^-) and MnO_2. In basic solution, the reaction is reversible. The equilibrium is displaced toward the MnO_4^- by the action of strong oxidants.

The permanganates are strong oxidizing agents, and are usually reduced down to Mn^{2+} under acidic conditions, but to MnO_2, manganate, MnO_4^{2-}, or even hypomanganate, MnO_4^{3-}, under progressively more alkaline conditions. Permanganic acid, $HMnO_4$, and its anhydride, Mn_2O_7, can be obtained at lower temperatures, but they are unstable, decomposing above 0°C. Permanganyl fluoride, MnO_3F, formed by the action of liquid hydrogen fluoride on potassium permanganate, decomposes at about 0°C. In strongly acidic media, such as 100% H_2SO_4, manganese (VII) appears to exist as permanganyl ion, MnO_3^+. The sigma bond hybridization in MnO_4^-, MnO_4^{2-} and MnO_4^{3-} is best represented as d^3s.

The only compound of manganese with carbon monoxide alone is the decacarbonyl dimanganese, $(CO)_5MnMn(CO)_5$, but several hydrogen-containing carbonyls, such as $HMn(CO)_5$, halogen-containing carbonyls, such as $Mn(CO)_5Br$, alkyl carbonyls, such as $C_2H_5Mn(CO)_5$ and oxygen-function organometallic compounds, such as

$$[CH_3C(= O)O]_3Mn$$

are known. With the exception of the dicyclopentadienyl compounds, $C_5H_5MnC_5H_5$ manganese does not combine with unsubstituted hydrocarbons or their radicals.

It is interesting to note that some bacteria found near manganese ore plants have the ability to dissolve manganese oxides in solutions of pH 5–6 by the slow addition of H_2SO_4. The only requirement other than the organisms is a nutrient solution. The extraction of manganese as a sulfate is on the order of 71.7–99.9%, depending on the ore. The action of the bacteria is not fully understood.

See also **Manganese (In Biological Systems)**.

Health and Safety Factors

Health and Environment. Manganese in trace amounts is an essential element for both plants and animals and is among the trace elements least toxic to mammals, including humans. Exposure to abnormally high concentrations of manganese, particularly in the form of dust and fumes, is, however, known to have resulted in adverse effects to humans. Two kinds of disease owing to manganese are known in humans: manganic pneumonia and manganism.

Airborn manganese concentrations in the U.S. range from 0.02 to 0.57 $\mu g/m^3$ in urban areas and 0.0017–0.047 $\mu g/m^3$ in non-urban areas. The ACGIH (American Conference of Governmental Industrial Hygienists: http://www.acgih.org/) recommends a TLV (threshold limit values) of 5 mg/m^3.

Plant Safety. Of the many ferroalloy products produced in electric furnaces, ferromanganese has the greatest potential for furnace eruptions or the more serious furnace explosions.

Most of the serious eruptions of manganese furnaces can be traced to a set of conditions that cause bridging or hang-up of the charge materials so that the normal downward movement through the furnace is disrupted or retarded. Safe operation of ferromanganese furnaces requires careful control of raw material particle size, oxygen content of the ore blend, and charge stoichiometry.

Additional Reading

Davis, J.R.: *Metals Handbook*, 2nd Edition, ASM International, Materials Park, OH, 1998.

Greenwood, N.N. and A. Earnshaw: *Chemistry of the Elements*, 2nd Edition, Butterworth-Heinemann, Inc., Woburn, MA, 1997.

Kerr, R.A.: "Manganese Nodules Grow by Rain from Above," *Science*, **223**, 576–577 (1984).

Klimas-Tavantzis, D.: *Manganese in Health and Disease*, CRC Press, LLC., Boca Raton, FL, 1994.

Krebs, R.E.: *The History and Use of Our Earth's Chemical Elements*, Greenwood Publishing Group, Inc., Westport, CT, 1998.

Lewis, R. and N. Sax: *Sax's Dangerous Properties of Industrial Materials*, 10th Edition, John Wiley & Sons, Inc., New York, NY, 2000.

Lide, D.R.: *CRC Handbook of Chemistry and Physics*, 88th Edition, CRC Press, Boca Raton, FL, 2007.

Varentsov, I.: *Manganese Ores of Supergene Zone: Geochemistry of Formation*, Kluwer Academic Publishers, Norwell, MA, 1996.

J.Y. WELSH
D.F. DE CRAENE
Chemicals Corporation Baltimore, MD

MANGANESE (In Biological Systems). Manganese is required by both plants and animals. Although its deficiency is normally a problem in small areas of fields, it has caused economic losses in the production of cereal small grains on some alkaline soils. In acid soils, manganese is more soluble and plants may be damaged by excessive uptake of the element. Reduced crop yields due to manganese toxicity on acid soils are probably responsible for greater economic losses in a number of regions of the world than are reduced crop yields as the result of manganese deficiency. Measurement of the total manganese concentration in any soil is of little value for predicting possible manganese deficiency or toxicity. The amounts of soluble manganese are more directly related to the level of manganese in plants, but soluble manganese in the soil may fluctuate over short periods because of flooding or drying of the soil or the addition of fresh organic matter. The concentration of manganese in food and in feed plants varies widely; it is more dependent upon the acidity or alkalinity of the soil than on the amount of manganese used in fertilizers.

Although established as an essential trace element, manganese is less well understood than many of the other trace elements. The evidence for its essentiality rests extensively on the consequences of limiting or curtailing the supply of the element of various organisms. Manganese deficiency has induced in most organisms studied a diminished life expectancy. The element is associated with reproductive processes. Additional manifestations depend upon the kind of organism under observation, its age, the degree and duration of manganese deficiency, as well as the coexistence of still another deficiency. For example, in plants, a striking manifestation is chlorosis in which the leaves become pale or yellow, while the veins of the leaves remain green. Manganese plays a significant role in photosynthesis by plants, but it also participates in the regulation of several other enzymic processes.

In poultry, manganese deficiency causes a different clinical picture when it affects the egg than when it affects the hatched bird. In the case of the egg, the embryo become swollen and deformed, and their skeletons become defective and fragile ("chondrodystrophy"). Adult birds develop

perosis (slipped tendon) which is an enlargement and malfunction of the tibial metatarsal joint, followed by slipping of the Achilles tendon from the condyles. The bone deformities seen in poultry also can be induced in mammals.

Manganese deficiency also results in the birth of "crooked calves" that are born with enlarged joints, stiffness, and twisted legs.

If the deficiency is induced prior to birth, there is a high intrauterine mortality and whatever young are born alive tend to suffer from an inability to coordinate their muscles (ataxia). These young also have convulsions, delayed growth, and defective bone formation. Adequate manganese intake after birth will correct many of these anomalies, but not the ataxia. If the deficiency is imposed on adult female mammals, ataxia develops infrequently. Instead there appear anemia, defective bone formation, infertility, a tendency to miscarry, and a tendency to absorb the embryo which die within the uterus. The sickly offspring are jeopardized after birth by a disinterest on the part of the manganese-deficient mothers. These avoid nursing their young even when they produce adequate milk. In males, in addition to poor growth, bone deformities and anemia, impotence and infertility develop. Adult animals also develop defects in metabolizing body fat, which are reflected in abnormal amounts and abnormal distribution of body fat. This liptropic effect of manganese extends also to the metabolism of cholesterol and, in this particular role, it can be antagonized by vanadium. The bone deformities are ascribed primarily to poor synthesis of the mucopolysaccharides that make up the matrix of the bones. The infertility is a consequence of death of the testicle's germ cells.

On the other hand, manganese in excessive amounts can cause manganese toxicity. In the past, this disease has mainly affected miners who work either in manganese mines or in ore crushing mills. The manganese ore enters the body by inhalation of the dust. Among the many miners exposed throughout the world, some develop brain symptoms. Involvement of the brain manifests itself first in mental aberrations. Later, neurological changes occur in the form of trembling, rigidity, salivation, mask-like face, and a general appearance of a person afflicted with Parkinson's disease. Chronic manganese poisoning has occurred in epidemics. The condition is incurable, but not necessarily life-limiting.

It is believed that high manganese diets in cattle will decrease fiber digestion. Manganese interferes with iron metabolism by antagonizing the enzyme system that oxidizes or reduces iron at the absorption site, thus affecting iron availability. It is also suggested that manganese may cause a condition in ruminant cattle (hypomagnesia). Most of the foregoing observations can be explained on the basis of manganese activating various cellular enzymes. Much importance has been given to the particular enzyme systems responsible for oxidative phosphorylation. These systems determine the generation and utilization of energy from foodstuffs by the cells. Additionally, manganese appears to activate many other enzymes (arginase, enolase, peroxidases). It also appears to participate in the structure of the nucleic acids responsible for the manufacture of enzymes and other proteins. Manganese probably plays a number of unique roles. No other metal replacement for the element in biological functions has been identified to date.

Manganese occurs in the liver of the animal body. Even though the amount of manganese present in mammalian tissues is very small, its concentration seems to be accurately controlled by elaborate mechanisms. These mechanisms function primarily by promoting the excretion of excesses of the element from the body rather than by regulating the amounts of manganese the body absorbs. The mechanisms are located in the liver and on the mucosa of the gut. In cases of manganese toxicity, it is assumed that these mechanisms become saturated.

One vital feature of manganese, which is not widely appreciated, is its role as an essential element in maintaining human health. Recommended daily dietary intake levels have been established by US regulatory authorities in an effort to ensure the maintenance of good health.

The exact role of manganese is not fully understood, but complex cellular reactions involving metallo-enzymes have been identified. Humans have well-developed homeostatic control mechanisms whereby manganese levels are regulated to keep them in the desired range. Medical research into conditions arising from an excess or deficit of body manganese is being carried out in a number of institutions.

Cereals and pulses (peas and beans) are the major sources of manganese in human diets, and diets containing these foods can be expected to provide adequate manganese. Dietary supplements for manganese (feeds and foodstuffs) include: manganese chloride, manganese gluconate, manganese glycophosphate, manganese hypophosphate, and manganese sulfate. The usual intake of this mineral is 2 to 5 mg/day, and absorption is 5 to 10%.

Additional Reading

Adriano, D.C.C.: *Biogeochemistry of Trace Metals*, Lewis Publishers, Boca Raton, FL, 1992.

O'Dell, B.L., R.A. Sunde: *Handbook of Nutritionally Essential Minerals*, Vol. 2, Marcel Dekker, Inc., New York, NY, 1997.

Klimas-Tavantzis, D.: *Manganese in Health and Disease*, CRC Press, LLC., Boca Raton, FL, 1994.

Staff: *Handbook of Inorganic and Organometallic Chemistry*, Gmelin Institute Series, Springer-Verlag Inc., New York, NY, 1997.

Staff: *Dietary Reference Intakes: Vitamin A, Vitamin K, Arsenic, Boron, Choromium, Copper, Iodine, Iron, Manganese, Molybdenum, Nickel, Silicon, Vanadiu*, National Academy Press, Washington, DC, 2001.

Underwood, E.J. and W. Mertz: *Trace Elements in Human and Animal Nutrition*, 5th Edition, Academic Press, Inc., San Diego, CA, 1990.

Web Reference

The Linus Pauling Institute: Micronutrient Information Center: http://lpi.oregonstate.edu/infocenter/minerals/manganese/index.html

MANGANITE.   The mineral manganite is a hydrous oxide of manganese corresponding to the formula $MnO(OH)$. It occurs in prismatic monoclinic crystals, sometimes in massive columnar forms, granular, concretionary, and stalactitic. It is a brittle mineral, with perfect prismatic cleavage; hardness, 4; specific gravity, 4.33; luster, submetallic; color, steel gray to iron black; streak, red-brown to almost black; opaque. Manganite is of secondary origin and it may itself alter to pyrolusite. It is usually associated with other manganese minerals. It is found in the Harz Mountains, Germany; Sweden; Cornwall and Cumberland, England; and in the United States in Michigan. It is an ore of manganese. See also **Pyrolusite**.

MANGO TREE.   See **Cashew and Sumac Trees**.

MANGROVE TREE.   Of the family *Rhizophoraceae*, the mangrove is a moderate-sized tree which grows on low, often submerged, coastal lands. It is found, for instance, in all tropical American coasts. The leaves of the plant are opposite, entire, dark green, and rather tough. The flowers are borne in small clusters and are perfect, with four sepals, four pale yellow linear petals, four to twelve stamens and single two-celled inferior ovary. Only one ovule develops. The seed usually germinates while the fruit is still attached to the tree. A long thick hypocotyl grows from the fruit, and attains a length of 5–10 inches (12.7 to 25 centimeters) and a diameter of less than 3/4 inch (1.9 centimeters). Eventually an abscission layer develops, so that the fruit, in which the young seedling is well advanced in germination, falls to the soft muddy ground, in which the new tree will grow. In a favorable location the hypocotyl puts out many roots, which anchor the young plant; then the epicotyl quickly grows. It is characteristic of the mangrove that from the stem and branches there grow out arching prop roots which soon form an intricate mass in which is deposited silt and all sorts of debris floating in the water. Because of this the mangrove causes a gradual building up of the land around it, until eventually the black slimy mud in which it grows gives place to a low coastal land which gradually becomes usable by man.

In addition to its land-forming function, the mangrove has other uses. The wood is dark red or reddish-brown, fine-grained, and hard; it is used in charcoal making. The bark contains tannin and so is employed in tanning hides. From the young shoots, a reddish dye may be obtained, which, however, is of little value.

Several other species of similar habitat and growth pattern are also called mangroves.

Two record mangrove trees selected by American Forests are described in the Table 1.

MANHATTAN PROJECT (The).   More than 55 years ago, work at Los Alamos and elsewhere in the world set in motion developments in military and civil applications of nuclear science and technology. Over the years these ongoing developments have shaped history. The resulting "Nuclear Age" has had a significant impact on many aspects of society—nationally and internationally.

The Manhattan Project of the "Second World War" represents the most remarkable congregation of scientific minds in human history. New scientific ground was broken which helped to produce numerous additional discoveries. Modern computer theory largely grew from bomb-related research with the first huge mainframe computers being used mainly for bomb design.

On August 2nd 1939, just before the beginning of World War II, Leo Szilard persuaded Albert Einstein to write to then President Franklin D. Roosevelt. Einstein and several other scientists told Roosevelt of efforts in Nazi Germany to purify U-235 with which might in turn be used to build an atomic bomb. It was shortly thereafter that the United States Government began the serious undertaking known only then as the Manhattan Project. Simply put, the Manhattan Project was committed to expedient research and production that would produce a viable atomic bomb.

General Leslie R. Groves, Deputy Chief of Construction of the U.S. Army Corps of Engineers, was appointed to direct this top-secret project. Groves established three large engineering and production centers at remote U.S. sites: the Clinton Engineer Works at Oak Ridge, Tenn.; the Hanford Engineer Works in eastern Washington State; and Project Y, a code-named site 100 miles north of Albuquerque at Los Alamos, N.M. All three sites still exist and contribute to America's nuclear arsenal.

The most complicated issue to be addressed was the production of ample amounts of "enriched" uranium to sustain a chain reaction. At the time, Uranium-235 was very hard to extract. In fact, the ratio of conversion from Uranium ore to Uranium metal is 500:1. An additional drawback is that the 1 part of Uranium that is finally refined from the ore consists of over 99% Uranium-238, which is practically useless for an atomic bomb. To make it even more difficult, U-235 and U-238 are precisely similar in their chemical makeup. This proved to be as much of a challenge as separating a solution of sucrose from a solution of glucose. No ordinary chemical extraction could separate the two isotopes. Only mechanical methods could effectively separate U-235 from U-238. Several scientists at Columbia University managed to solve this dilemma.

A massive enrichment laboratory/plant was constructed at Oak Ridge, Tennessee. H.C. Urey, along with his associates and colleagues at Columbia University, devised a system that worked on the principle of gaseous diffusion. Following this process, Ernest O. Lawrence (inventor of the Cyclotron) at the University of California in Berkeley implemented a process involving magnetic separation of the two isotopes.

TABLE 1. RECORD MANGROVE TREES IN THE UNITED STATES[1]

Specimen	Circumference[2]		Height		Spread		Location
	Inches	Centimeters	Feet	Meters	Feet	Meters	
Black mangrove (1996) *(Avicennia germinans)*	101	257	43	13.1	57	17.4	Florida
Red mangrove (1995) *(Rhizophora mangle)*	47	119	58	17.7	42	12.8	Florida

[1]From the "National Register of Big Trees," American Forests (by permission).
[2]At 4.5 feet (1.4 meters).

Following the first two processes, a gas centrifuge was used to further separate the lighter U-235 from the heavier non-fissionable U-238 by their mass. Once all of these procedures had been completed, all that needed to be done was to put to the test the entire concept behind atomic fission.

Over the course of six years, ranging from 1939 to 1945, more than 2 billion dollars were spent on the Manhattan Project. The formulas for refining Uranium and putting together a working bomb were created and seen to their logical ends by some of the greatest minds of our time. Among these people who unleashed the power of the atomic bomb was J. Robert Oppenheimer.

Oppenheimer was the major force behind the Manhattan Project. He literally ran the show and saw to it that all of the great minds working on this project made their brainstorms work. He oversaw the entire project from its conception to its completion.

Finally the day came when all at Los Alamos would find out whether or not *The Gadget* (code-named as such during its development) was either going to be the colossal dud of the century or perhaps end the war. It all came down to fateful morning of midsummer, 1945. The first nuclear explosion in history took place in New Mexico, at the Alamogordo Test Range, on the Jornada del Muerto (Journey of Death) desert, in the test named Trinity. Trinity, was the conclusion of the Manhattan Project to build the bomb in a frantic race with Adolf Hitler's scientists.

This test was intended to prove the radical new implosion weapon design that had been developed at Los Alamos during the previous year. This design, embodied in the test device called *Gadget*, involved a new technology that could not be adequately evaluated without a full scale test. The gun-type uranium bomb, in contrast, was certain to be effective and did not merit testing. In addition, since no nuclear explosion had ever occurred on Earth, it seemed advisable that at least one should be set off with careful monitoring to test whether all of the theoretical predictions held.

At 5:29:45 (Mountain War Time) on July 16th, 1945, in a white blaze that stretched from the basin of the Jemez Mountains in northern New Mexico to the still-dark skies, *The Gadget* ushered in the Atomic Age. The light of the explosion then turned orange as the atomic fireball began shooting upwards at 360 feet per second, reddening and pulsing as it cooled. The characteristic mushroom cloud of radioactive vapor materialized at 30,000 feet. Beneath the cloud, all that remained of the soil at the blast site were fragments of jade green radioactive glass. All of this caused by the heat of the reaction.

The brilliant light from the detonation pierced the early morning skies with such intensity, that residents from a faraway neighboring community would swear that the sun came up twice that day. Even more astonishing is that a blind girl saw the flash 120 miles away.

Upon witnessing the explosion, reactions among the people who created it were mixed. Isidor Rabi felt that the equilibrium in nature had been upset—as if humankind had become a threat to the world it inhabited. J. Robert Oppenheimer, though ecstatic about the success of the project, quoted a remembered fragment from Bhagavad Gita. "I am become Death," he said, "the destroyer of worlds."

Several participants, shortly after viewing the results, signed petitions against loosing the monster they had created, but their protests fell on deaf ears. As it later turned out, the Jornada del Muerto of New Mexico was not the last site on planet Earth to experience an atomic explosion.

Detonation

As many know, atomic bombs have been used only twice in warfare. The first and foremost blast site of the atomic bomb is Hiroshima. A Uranium bomb (which weighed in at over 4 1/2 tons) nicknamed "Little Boy" was dropped on Hiroshima August 6th, 1945. The Aioi Bridge, one of 81 bridges connecting the seven-branched delta of the Ota River, was the aiming point of the bomb. Ground Zero was set at 1,980 feet. At 0815 hours, the bomb was dropped from the Enola Gay. It missed by only 800 feet. At 0816 hours, in the flash of an instant, 66,000 people were killed and 69,000 people were injured by a 10 kiloton atomic explosion.

The point of total vaporization from the blast measured one half of a mile in diameter. Total destruction ranged at one mile in diameter. Severe blast damage carried as far as two miles in diameter. At two and a half miles, everything flammable in the area burned. The remaining area of the blast zone was riddled with serious blazes that stretched out to the final edge at a little over three miles in diameter.

On August 9th 1945, Nagasaki fell to the same treatment as Hiroshima. Only this time, a Plutonium bomb nicknamed "Fat Man" was dropped on the city. Even though the "Fat Man" missed by over a mile and a half, it still leveled nearly half the city. Nagasaki's population dropped in one split-second from 422,000 to 383,000. 39,000 were killed, over 25,000 were injured. That blast was less than 10 kilotons as well. Estimates from physicists who have studied each atomic explosion state that the bombs that were used had utilized only 1/10th of 1 percent of their respective explosive capabilities.

While the mere explosion from an atomic bomb is deadly enough, its destructive ability doesn't stop there. Atomic fallout creates another hazard as well. The rain that follows any atomic detonation is laden with radioactive particles. Many survivors of the Hiroshima and Nagasaki blasts succumbed to radiation poisoning due to this occurrence.

The atomic detonation also has the hidden lethal surprise of affecting the future generations of those who live through it. Leukemia is among the greatest of afflictions that are passed on to the offspring of survivors.

While the main purpose behind the atomic bomb is obvious, there are many by-products that have been brought into consideration in the use of all weapons atomic. With one small atomic bomb, a massive area's communications, travel and machinery will grind to a dead halt due to the EMP (Electro-Magnetic Pulse) that is radiated from a high-altitude atomic detonation. These high-level detonations are hardly lethal, yet they deliver a serious enough EMP to scramble any and all things electronic ranging from copper wires all the way up to a computer's CPU within a 50 mile radius.

At one time, during the early days of The Atomic Age, it was a popular notion that one day atomic bombs would one day be used in mining operations and perhaps aid in the construction of another Panama Canal. Needless to say, it never came about. Instead, the military applications of atomic destruction increased. Atomic tests off of the Bikini Atoll and several other sites were common up until the Nuclear Test Ban Treaty was introduced. Photos of nuclear test sites here in the United States can be obtained through the Freedom of Information Act.

See also **Nuclear Fission**; and **Oppenheimer, J. Robert (1904–1967)**.

Additional Reading

Gonzales, D.: *The Manhattan Project and the Atomic Bomb in American History,* Enslow Publishers, Inc., Berkeley Heights, NJ, 2000.

Groueff, S.: "Manhattan Project: The Untold Story of the Making of the Atomic Bomb," iUniverse.com, Inc., New York, NY, 2000.

Sparks, R.C., B.G. Storms: *Twilight Time: A Soldier's Role in the Manhattan Project in Los Alamos,* Los Alamos Historical Society, Los Alamos, NM, 2000.

Web References

Hanford Site: http://www.hanford.gov/doe/culres/historic/index.htm

History of the Plutonium Production Facilities at the Hanford Site Historic District, 1943–1990, Manhattan Project: 1943-1946, Cold War Era: 1947–1990: http://www.hanford.gov/docs/rl-97-1047/index.htm

Los Alamos National Laboratory: http://lib-www.lanl.gov/infores/history/history.htm

Nuclear History Site: http://geocities.com/RainForest/Andes/6180/

The US National Atomic Museum's virtual tour of the Manhattan Project: http://www.atomicmuseum.com/tour/manhattanproject.cfm

MANIC-DEPRESSIVE (Bipolar) ILLNESS. This is a major mental illness, estimated as of the 1990s to affect 5 persons per 100,000 population, a number that is considerably less than for depression without mania. Researchers at Washington University have established six criteria for the diagnosis of manic-depressive illness. (1) Hyperactivity and aggressiveness. This applies to some or all personal activities—motor, social, and sexual. (2) Incessant talking—with no toleration for interruptions, resulting in a "pushy" behavior that can lead to much frustration in persons attempting to communicate with the manic-depressive patient. (3) Rapid transfer and alteration of thought patterns, resulting in racing from one idea to the next, often precipitated by mention of a key word. (4) An air of grandeur—promotion of schemes requiring large sums of money, with little basis in fact, dramatized by excessive use of credit cards, telephone calls, etc. (5) Reduced need to sleep. (6) Overreaction to numerous stimuli—the patient, easily distracted by unimportant events. In using the foregoing criteria, the examiner must make certain that there have been no preexisting psychiatric disturbances as may result from alcoholism or schizophrenia. The etiology of this illness appears to be connected with genetic factors as well as environment. Frequently, familial connections are apparent, but no genetic connection has been revealed thus far.

In 1987, a group of researchers announced that a statistical analysis of members of the Amish community, using a genetic technique (linkage analysis), indicated that a gene located at the top of the short arm of chromosome 11 seemed to be responsible for the heritability of certain forms of manic-depressive illness. This series of experiments later was challenged and refuted. The Amish community had been selected because of its meticulous family records, dating back to the early 18th century when 30 Amish couples increased in numbers to about 15,000 persons today.

Role of Lithium. In 1949, a researcher, J.F.J. Cade (Australian psychiatrist), accidentally discovered that lithium had a calming effect on laboratory animals. This led to the investigation of lithium as a possible drug for use in treating manic-depressive illness. Subsequent research indicated that lithium antagonizes synaptic transmission of cathecholamines (specifically by inhibiting norepinephrine and dopamine release). See also **Central and Peripheral Nervous Systems**. Lithium carbonate is the drug of choice and is widely used. Because of several side effects of lithium, considerable clinical evaluation of the patient is required prior to its administration. Inasmuch as many cases of mania require immediate or very early hospitalization, various antipsychotic drugs may be used as an interim measure. Because manic-depressive patients are also subject to periods of depression (thus a biopolar situation), it is fortunate for many patients that lithium also is an effective prophylaxis for depression. In less responsive patients, more dramatic therapy, such as electroconvulsive therapy, may be required. Complications of prolonged lithium therapy include benign diffuse nontoxic goiter, usually treatable with L-thyroxine. The physician must monitor all factors which would increase sodium loss (increases lithium retention). Lithium has been shown to cause serious renal damage when administered over long periods. See also **Schizophrenia**.

The bidirectional therapeutic effect of lithium carbonate has puzzled investigators for several years. In 1972, D.S. Janowski (Vanderbilt University) proposed that adrenergic-cholinergic unbalances are the root cause of bipolar illness. A group of researchers (Ben Gurion University of the Negev, Israel) suggested in 1988 that lithium blocks the activation of G proteins by neurotransmitters binding to both adrenergic and cholinergic receptors. It had been established previously that lithium cripples a messenger system linked to cholinergic receptors. In studies of rats, the Israeli team reported that lithium also is effective in modulating adrenergic receptors. This, at least, assists in explaining why lithium functions bidirectionally. But other researchers, who are aware of the wide distribution of G proteins, wonder why lithium's effectiveness is limited to the central nervous system. This part of the puzzle, however, has not been solved because researchers in the United Kingdom observe, "At therapeutic doses, the side effects of lithium are negligible." W.R. Sherman (Washington University) has observed, "Studies of lithium and manic depression, like all studies of drug action in psychiatry, are disarmed by their reliance on subjective information. In spite of the fact that thousands of Americans lead normal lives because of lithium therapy, it's almost impossible to get real evidence of what's going on!"

Additional Reading

Gabbard, G.O.: *Treatments of Psychiatric Disorders*, 3rd Edition, American Psychiatric Press, Inc., Washington, DC, 2001.

Goodwin, F.K. and K.R. Jamison: *Manic-Depression Illness*, Oxford University Press, Inc., New York, NY, 1990.

Halleck, S.L.: *Evaluation of the Psychiatric Patient: A Primer*, Kluwer Academic Publishers, Norwell, MA, 1991.

Hawton, K. and P. Cowen: *Dilemmas and Difficulties in the Management of Psychiatric Patients*, Oxford University Press, Inc., New York, NY, 1991.

Lam, D.H., P. Hayward, S.H. Jones, and J.A. Bright: *Cognitive Therapy for Biopolar Disorder*, John Wiley & Sons, Inc., New York, NY, 1999.

Mirin, S.M., J.T. Gossett, and M.C. Grob: *Psychiatric Treatment: Advances in Outcome Research*, American Psychiatric Press, Inc., Washington, DC, 1991.

Mondimore, F.M.: *Biopolar Disorder: A Guide for Patients and Families*, Johns Hopkins University Press, Baltimore, MD, 2000.

Pearlman, T.: *The Threatened Medical Identity of Psychiatry: The Winds of Change*, Charles C. Thomas, Springfield, IL, 1992.

Reid, W.H., G.U. Balis, and B.J. Sutton: *The Treatment of Psychiatric Disorders*, 3rd Edition, Brunner/Mazel, New York, NY, 1997.

Staff: *Practice Guideline for the Treatment of Patients with Bipolar Disorders*, American Psychiatric Press, Inc., Washington, DC, 2000.

Swartz, C.M.: "Serum Lithium During Treatment of Bipolar Disorders," *N. Eng. J. Med.*, 1159 (April 19, 1990).

Walden, J. and H. Grunze: *Bioplar Illnesses: New Ways of Treatment*, S. Karger Publishers, Inc., Farmington, CT, 2000.

Weller, M. and M. Eysenck: *The Scientific Basis of Psychiatry*, 2nd Edition, W.B. Saunders, Philadelphia, PA, 1992.

Yonkers, K. and B.B. Little: *Management of Psychiatric Disorders during Pregnancy*, Oxford University Press, Inc., New York, NY, 2001.

Web Reference

American Psychiatric Association: http://www.psych.org/

MANNICH REACTION. Reaction of active methylene compounds with formaldehyde and ammonia of primary or secondary amines to give β-aminocarbonyl compounds.

MANNING EQUATION. An equation relating the mean velocity of flow to channel characteristics. It is expressed as

$$V = \left(\frac{1}{n}\right) R^{2/3} S_f^{1/2},$$

where, at a given section, V is mean velocity, n is the Manning's roughness coefficient (indicative of the resistance to the flow), S_f is the gradient of the total head line, and R is the hydraulic radius.

MANNITOL. See **Sugar Alcohols**.

MANOMETER. A relatively simple instrument that provides direct measurement of positive pressure, vacuum, and differential pressure. The manometer is also used indirectly for the measurement of flow by sensing the output of a pressure-differential producing device, such as a venturi or orifice plate. The manometer operates on the fundamental principle of displacing a liquid column by the unknown pressure force to be measured. Two types of manometer are illustrated in Fig. 1. See also **Barometer**.

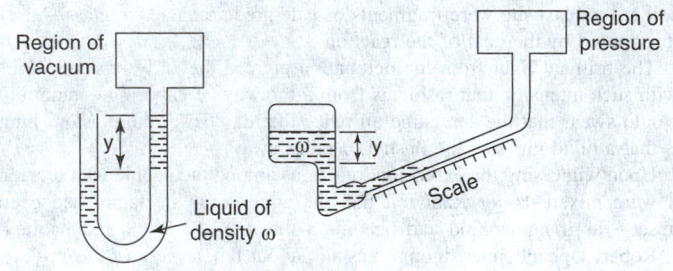

Fig. 1.   Types of manometer: U-tube manometer at left; inclined-tube manometer at right.

MANSON, PATRICK (1844–1922). Patrick Manson was a British doctor who established the role of arthropods as secondary hosts in parasitic diseases and who founded the discipline of tropical medicine in Britain. Manson was born in Aberdeenshire, Scotland. He joined the family engineering firm but ill-health ended his apprenticeship. Interested in natural history since youth, he attended Aberdeen Medical School, graduated in 1866, and worked initially at Durham Lunatic Asylum. After a year, he joined the Chinese Imperial Maritime Customs service. He was stationed on Formosa (Taiwan), but transferred to Amoy in 1871. Here he continued his interest in filariasis, particularly surgical removal of excess tissue. See also **Filariasis**. He also sought literature on filariasis, Timothy Lewis's work on *Filaria sanguinis hominis* proving inspirational. Lewis postulated that microfilaria, immature forms of larger worms, caused filariasis. Manson speculated on the means of transmission and, returning to Amoy, investigated his mosquito theory. Harnessing early skills in natural history with clinical observations, his experiments successfully implicated the mosquito and its blood meal in the transmission of microfilaria.

In 1890, Manson established a London consulting practice in conjunction with an appointment as Physician to the Seaman's Hospital Society. He lobbied for improved teaching of tropical medicine and following his appointment as Medical Adviser to the Colonial Office he founded the London School of Tropical Medicine in 1899. During this period, Manson assisted Ronald Ross in his discovery of the mosquito transmission of malaria. Manson's classic textbook, Tropical Diseases. A Manual of the Diseases of Warm Climates, first edition 1898, ran to five further editions in his lifetime. He was elected a Fellow of the Royal Society in 1900 and received a knighthood in 1903.

Additional Reading

Bynum, W.F., and C. Overy: *The Beast in the Mosquito: The Correspondence of Ronald Ross and Patrick Manson.* Rodopi, Amsterdam, Netherlands. 1998.

Gillispie, C.C.: *Concise Dictionary of Scientific Biography*, 2nd Edition, Charles Scribner's Sons, New York, NY, 2000.

Manson-Bahr, P.H.: *The Life and Work of Sir Patrick Manson*, Cassell, London, UK, 1927.

HELEN J. POWER, Wellcome Trust Centre for the History of Medicine at UCL, London, UK

MANTIS FLY *(Insecta, Neuroptera).* Small predacious insects superficially like the mantis in form. They make up the family *Mantispidae* and are also called mantispas.

MANTIS *(Insecta, Orthoptera).* A large insect of predatory habits. The body is moderately broad and bears four wings. The first segment of the thorax is long and rather slender, adding to the reach of the powerful raptorial front legs. The head is prominent and has large eyes.

From the fancied suppliant air of these voracious insects as they await their prey with the forelegs uplifted they are called praying mantises, and their owlish expression has given them the rarer name of soothsayers. The common species is *Mantis religiosa.*

The mantis is found principally in America and southern Europe. There are about 800 species. The common species (*Mantis religiosa*) measures about 2 inches in length, but in South America the species is larger and sometimes these insects will attack small birds, frogs, and lizards. Sometimes, a mantis will devour its young and the female may eat the male.

The mantis is a great help in destroying flies, grasshoppers, caterpillars, and other insects. They are harmless to human beings.

See Fig. 1.

MANTISSA. See **Logarithm**.

MANTLE (Earth). See **Earth**.

MANUFACTURING MESSAGE SPECIFICATION (MMS). MMS (Manufacturing Message Specification) is an internationally standardized application layer messaging service for exchanging real-time data and supervisory control information between networked intelligent electronic devices (IED) and/or computer applications. MMS is used in manufacturing automation for communications between programmable logic controllers (PLC), robots, computer numerical controls (CNC), and computers. In the electric utility industry, MMS is used for communications between utility control centers, substations, and intelligent electronic devices (IED) such as remote terminal units (RTU), protection relays, and meters.

History of MMS

Work on MMS was originally begun during the Manufacturing Automation Protocol (MAP) effort sponsored by the General Motors Corporation in the 1980s. The MAP effort was undertaken to develop a standardized automation protocol for use across a range of industrial applications in

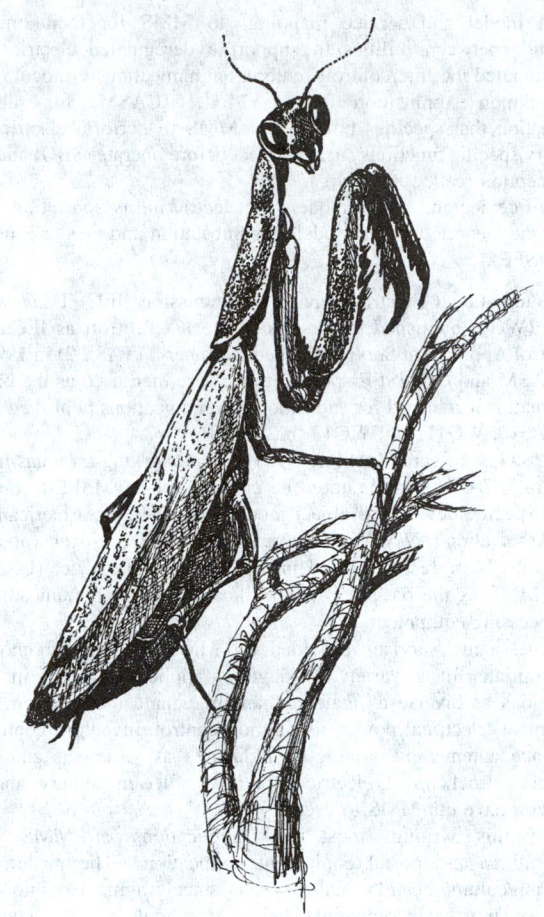

Fig. 1.   Praying mantis (mantis religiosa). Length ranges up to 7.5 cm.

both the discrete[1], continuous[2], and batch[3] processing industries. A broad group of engineers from the programmable logic control (PLC), computer numerical control (CNC), robotic, automotive, chemical, oil processing, and communications industries came together under the auspices of the International Organization for Standardization's (ISO) technical committee 184 (TC184) to develop an application layer protocol suitable to satisfy this broad range of interests. The result was the MMS standard published as ISO9506 part 1 (Services) and part 2 (Protocol) in December of 1988.

In 1990 the Electric Power Research Institute (EPRI) began an effort to develop a standard for electric utilities to use for building a modern real-time communications architecture called the Utility Communications Architecture (UCA™). One of the objectives of the UCA effort was to leverage existing technology and adapt it to the needs of electric utilities. The UCA effort selected the use of Ethernet, fiber optics, TCP/IP-Internet networking, and MMS as an application layer protocol for real-time communications. In addition to the use of MMS at the application level, UCA also defines:

- A set of profiles for running MMS over networks (TCP/IP and OSI) and serial links for spread spectrum (SS) or multiple address system (MAS) radios.

[1] Discrete manufacturing processes are characterized as producing products that are measured in discrete units (e.g. automotive parts and assembly, appliances, toys, and tools).

[2] Continuous manufacturing processes are characterized by producing products that are measured in bulk units (e.g. oil refining, power generation, fertilize, and chemicals).

[3] Batch manufacturing processes are characterized by producing discrete units of continuously produced products (e.g. food processing).

™ UCA is trademark of EPRI.

- Object model and service mappings to MMS for facilitating data exchange between utilities to support a deregulated electric energy market called the Intercontrol Center Communications Protocol (ICCP).
- A Common Application Service Model (CASM) for substation automation that specifies how to use MMS to perform electric utility industry specific functions such as select before operate (SBO) and report by exception (called *reporting*).
- Defined device and object models for electric utility substation devices called the Generic Object Models for Substation and Feeder Equipment (GOMSFE).

The International Electro-Technical Commission (IEC) TC57 working group 6 (WG6) published the results of the ICCP effort as IEC60870-6 Telecontrol Application Service Element Number 2 (TASE.2) in 1996. The UCA CASM and GOMSFE specifications were then used as the basis for the international standard for substation communications published by IEC TC57 WG10, WG11, and WG12 in 2001.

In 1998 Gas Research Institute (GRI) began working on a gas industry version of UCA that builds upon the concepts of GOMSFE to build gas industry specific device and object models. In 1999, the American Water Works Association (AWWA) developed a version of UCA for water utility applications. More recently, the United States Postal Service (USPS) has utilized MMS as the basis for its specifications for communicating with mail processing equipment.

The messaging services provided by MMS are generic enough to be appropriate for a variety of devices, applications, and industries. Applications as diverse as material handling, fault annunciation, energy management, electrical power distribution control, inventory control, and deep space antenna positioning in industries as varied as automotive, aerospace, petrochemical, electric/gas utility, office machinery, and space exploration have put MMS to useful work.

As of this writing, most new applications of MMS are in electric utility and postal equipment applications. The manufacturing networking/communications industry has since fragmented into a large number of mostly incompatible fieldbus technologies for connecting controls to I/O systems and numerous TCP/IP based application protocols for communications between automation controllers and business information systems. These systems tend to be either very simplistic, making them unsuitable for the more complex interactions required by a typical MMS application, or are specific to a very narrow application niche such as motion control and distributed I/O. MMS remains the only internationally standardized application level protocol that has a proven track record of being effective across a broad range of industries.

The VMD Model

The primary goal of MMS was to specify a standard communications mechanism for devices and computer applications that would achieve a high level of interoperability[4]. In order to achieve this goal, it would be necessary for MMS to define much more than just the format of the messages to be exchanged. A common message format, or protocol, is only one aspect of achieving interoperability. In addition to protocol, the MMS standard also provides definitions for:

- **Objects**. MMS defines a set of common objects (e.g., variables) and defines the network visible attributes of those objects (e.g., name, value, type).
- **Services**. MMS defines a set of communications *services* (e.g., read, write) for accessing and managing these objects in a networked environment.
- **Behavior**. MMS defines the network visible behavior that a device should exhibit when processing these services.

This definition of objects, services, and behavior comprises a comprehensive definition of how devices and applications communicate

---

[4] Interoperability is the ability of two or more networked applications to exchange useful information between them without the user of the applications having to create the communications environment. Most communication protocols can provide some level of interoperability. However, many of them are too specific to work outside of a very narrow appliction niche. Others are not specific enough and offer too many choices for developers. This results in incompatible systems.

in which MMS calls the *Virtual Manufacturing Device* (VMD) model. The VMD model is the key feature of MMS. The VMD model specifies how MMS devices, also called servers, behave as viewed from an external MMS client application point of view. The VMD model only specifies the network visible aspects of communications. The internal detail of how a real device implements the VMD model (i.e., the programming language, operating system, CPU type, input/output (I/O) systems) are not specified by MMS. By focusing only on the network visible aspects of a device, the VMD model is specific enough to provide a high level of interoperability. At the same time, the VMD model is still general enough to allow innovation in application/device implementation and making MMS suitable for applications across a range of industries and device types. See Fig. 1.

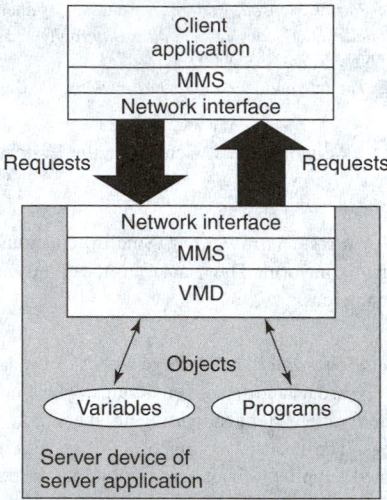

Fig. 1. The VMD model provides a consistent and well defined view for client applications of the objects contained in the VMD. Clients use MMS services to access and manipulate those objects. MMS requires that all servers behave according to the VMD model.

Client/Server Relationship. MMS is a client/server based protocol A *server* is a device or application that contains data and executes commands. A *client* is a networked application or device that accesses and manipulates data or issues command requests to a server (Fig. 1). While MMS defines the services for both clients and servers, the VMD model defines the network visible behavior of servers only. See Fig. 2. Many MMS applications and devices can provide both MMS client and server functions. The VMD model would only define the behavior of the server functions of those applications. Any MMS application or device that provides MMS server functions must follow the VMD model for all the network visible aspects of the server application or device. MMS clients are only required to conform to rules governing message format or construction and sequencing of messages (the protocol).

Real and Virtual Devices and Objects

There is a distinction between a real device and a real object (e.g., a PLC with a part counter) and the *virtual* device and objects (e.g. VMD, domain, variable, etc.) defined by the VMD model. Real devices and objects have peculiarities (a.k.a. product features) associated with them that are unique to each brand of device or application. Virtual devices and objects conform to the VMD model and are independent of brand, language, operating system, etc. Each developer of a MMS server device or MMS server application is responsible for "hiding" the details of their real devices and objects, by providing an *executive function*. The executive function translates the real devices and objects into the virtual ones defined by the VMD model when communicating with MMS client applications and devices. See Fig. 3.

Because MMS clients always interact with the virtual device and objects defined by the VMD model, the client applications are isolated from the specifics of the real devices and objects. A properly designed MMS client application can communicate with many different brands and types of devices in the exact same manner. This is because the details of the real

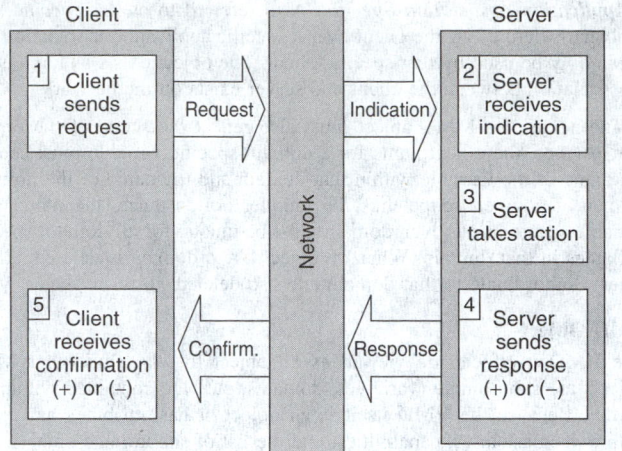

Fig. 2.   Client/Server Interactions. MMS clients and servers interact with each other by sending/receiving request, indication, response, and confirmation service primitives over the network. The figure depicts the interactions between a client and server for a MMS confirmed service where (1) the client sends a request, (2) the server receives an indication, (3) the server performs the desired action, (4) the server sends a positive (+) response if the action was successful or a negative (−) response if there was an error, and (5) the client receives the confirmation (+) or (−). An Unconfirmed Service is send by the server and has only the request and indication service.

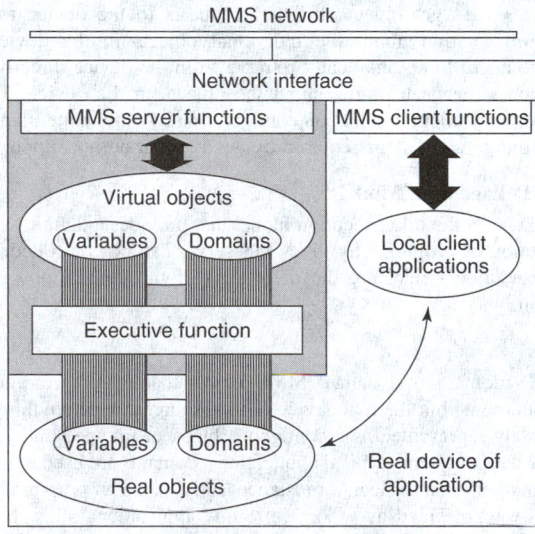

The MMS VMD model

Fig. 3.   Real and Virtual Objects. The executive function provides a translation, or "mapping" between the MMS defined virtual objects and the real objects used by the real device. Applications local to the VMD, and the objects contained in them, are only accessible to a remote MMS client application if the executive function provides the mapping function for those objects and applications. Client applications local to the VMD may access and manipulate the real objects without using MMS.

devices and objects are hidden from the MMS client by the executive function in each VMD. This virtual approach to describing server behavior does not constrain the development of innovative devices and product features and improvements. The MMS VMD model places constraints only on the network visible aspects of the virtual devices and objects, not the real ones.

MMS Device and Object Modeling

The implementor of the executive function (the application or device developer) must decide how to "model" the real objects as virtual objects. The manner in which these objects are modeled is critical to achieving interoperability between clients and servers among many different developers. Inappropriate or incorrect modeling can lead to an implementation that is difficult to use or difficult to interoperate with. One of the key benefit of using UCA and ICCP is the additional object modeling definitions these standards provide for electric utility specific applications.

Objects

MMS defines objects that are found in many typical devices and applications requiring real-time communications. For each object MMS defines a set of properties or *attributes* that describe various aspects of the object such as its name, status/state, value/contents, etc. The objects defined by MMS are[5]:

- *VMD*. The device itself is an object.
- *Domain*. A resource (e.g. memory) represented by a block of untyped data.
- *Program Invocation*. A runnable program consisting of one or more domains.
- *Variable*. An named (or unnamed) element of typed data.
- *Type*. A description of the format of a variable's data.
- *Named Variable List*. A list of variables that is named as a list.
- *Event Condition, Event Enrollment, Event Action*. These are objects that are related to the control, processing and notification of events.
- *Semaphore*. An object used to control access to a shared resource.
- *Journal*. An object used to keep a time-based record of variables and events.
- *Operator Station*. A display and keyboard for use by an operator.
- *File*. Data stored in files on a file server.

MMS Services

The MMS client uses MMS services to access and manipulate the objects and their attributes. Each class of object has a unique set of services available to the client. In general, these services support the following actions on the objects[5]:

- *Create Object*. Many MMS objects can be created clients using MMS services. Examples of object creation services are: CreateNamedVariableList, InitiateDownloadSequence (for domains), CreateProgramInvocation, etc.
- *Delete Object*. Objects that are created by clients can also be deleted by clients. Some examples are DeleteNamedVariableList, DeleteDomain, DeleteProgramInvocation, etc.
- *Get Attributes*. The MMS client can determine the attributes of a given MMS object are by using MMS services. Examples of MMS services used to obtain object attributes include GetVariableAccessAttributes (retrieves the definition of a variable, Read (retrieves the value of a variable), GetDomainAttributes, etc.
- *Change Attributes*. The MMS client can also modify the attributes of an MMS object by using MMS services. The MMS services that can be used to change an object's attributes include Write (change the value of a variable), Start/Stop (changes the state of program invocations), Write Journal, etc.

ICCP-TASE.2 Objects

The ICCP (IEC60870-6 TASE.2) standard defines the following additional objects that are useful for inter-utility data exchange:

- *Bilateral Table*. An object that represents an agreement between a local and remote node regarding the data to be exchanged and how the data exchanges will be controlled. The use of the bilateral table allows both sides of a data exchange to carefully control the information to be exchanged.
- *Data Set*. An object, represented by an MMS Named Variable List that is used to collect data into logical groups.
- *Transfer Set*. The object, represented by one or more data sets, that is sent in a transfer report that the server sends to the client on a periodic basis or when the data changes (called "report by exception") using the MMS unconfirmed service of InformationReport.
- *Account*. A special type of transfer set that contains information, stored in a row/column format, related to energy exchange schedules and device outages of high-voltage electrical transmission systems.

[5] More detailed descriptions of MMS objects, attributes, and services is available on the Internet at: http://www.sisconet.com/techinfo.htm

UCA Objects

The UCA (IEC61850) standard defines additional objects and services for SCADA, distribution and substation automation applications:

- *Select Before Operate*. The SBO object is used as an interlock to coordinate control commands issued by multiple UCA clients to the same point. The SBO object is mapped to a structured MMS variable.
- *Log Object*. The UCA log is an object that is used to store a time based record of events and variables as a sequence of events (SOE) log. UCA clients can then use the log to retrieve the data for archiving. The UCA log is mapped to a MMS journal object.
- *UCA Report*. The UCA report allows UCA clients to control how unsolicited report by exception (RBE) data is sent to them by an IED. The UCA report is mapped to a structured MMS variable using MMS named variable lists for the actual reports.
- *GOMSFE Objects*. UCA defines numerous object models for many common IED functions used in the electric utility. GOMSFE object models are mapped to MMS domains and structured MMS variables with standardized names.
- *Generic Object Oriented Substation Event*. A GOOSE object is broadcast over a local area network (LAN) by protection relays to exchange protection signals as a replacement for individually hard-wired signals. See Fig. 4.

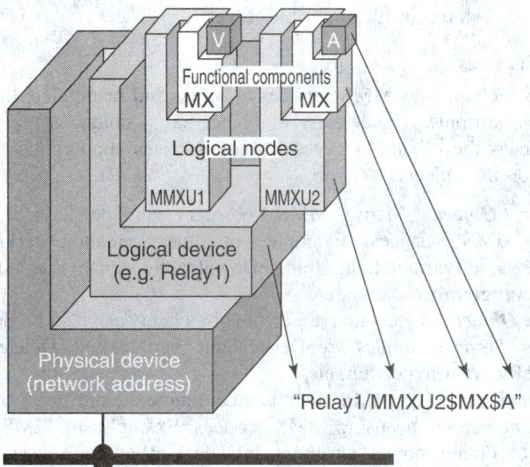

Fig. 4. A GOMSFE name preserves the hierarchy of the date. Logical Devices, modeled as MMS domains, are broken down into functional nodes for measurement functions (MMXU), basic protection relay functions (PBRO), and numerous other device specific models. Each logical node is further broken down into functional components such as measurements (MX), status (ST), descriptions (DC), etc. Each functional component consists of one or more groupings of variables such as amps (A), volts (V), watts (W), vars., (VA), etc. A MMS client specify the entire name to access a specific element or can specify just the logical node designation (e.g. MMXU2), to get all date associated with that logical node.

Object Attributes and Scope

Associated with each object are a set of attributes that describe that object. MMS objects have a name attribute and other attributes that vary from object to object. Variables have attributes such as, name, value, type. Other objects, program invocations for instance, have attributes like name and current state.

Subordinate objects exist only within the scope of another object. For instance, all other objects are subordinate to, or contained within, the VMD itself. Some objects, such as the *operator station* object, may be subordinate only to the VMD. Some objects may be contained within other objects, such as variables contained within a *domain*. This attribute of an object is called its *scope*. The object's scope also reflects the lifetime of an object. An object's scope may be defined to be:

- *VMD-Specific*. The object has meaning and exists across the entire VMD (is subordinate to the VMD). The object exists as long as the VMD exists.
- *Domain-Specific*. The object is defined to be subordinate to a particular domain. The object will exist only as long as the domain exists.

- *Application-Association-Specific*. Also referred to as *AA-Specific*. The object is defined by the client over a specific application association and can only be used by that specific client. The object exists as long as the association between the client and server exists on the network.

The name of a MMS object must also reflect the scope of the object. For instance, the object name for a domain-specific variable must specify the name of the variable within that domain and the name of the domain. Names of a given scope must be unique. For instance, the name of a variable specific to a given domain must be unique for all domain specific variables in that domain. When an object like a domain is deleted, all the objects subordinate to that domain are also deleted.

VMD Object

The VMD itself can be viewed as an object to which all other MMS objects are subordinate (variables, domains, etc., are contained within the VMD). Because the VMD itself is an object, it has attributes associated with it such as status, capabilities, and the list of subordinate objects. The MMS client has several services available to interact with the VMD object. These services include Status, Unsolicited Status, Identify, GetNameList, and others.

Self Describing Devices

One of the unique capabilities of MMS devices is their ability to support self-description. This means that the device can completely describe all the objects contained within it without the client having to be configured in advance. Self description is accomplished by a MMS client using the GetNameList service to first obtain a list of the named objects that are defined in the device such as domains, variables, journals, etc. Then the client can issue "get object attribute" requests to the device to obtain detailed information about the individual objects in the device. This allows the client to automatically retrieve from the device directly all the information it needs to know about the objects in the device. This can greatly simplify client application configuration by eliminating many of the manual configuration steps required by devices that do not support MMS.

The VMD Execution Model

The VMD has a flexible execution model that provides a definition of how the execution of programs by the MMS server can be controlled. Central to this execution model are the definitions of the domain and program invocation objects.

Domains

The MMS domain is a named MMS object that is a representation of some resource within the real device. This resource can be anything that is appropriately represented as a contiguous block of untyped data (referred to as load data). In many typical applications, domains are used to represent areas of memory in a device. For instance, a PLC's ladder program memory is typically represented as a domain. Some applications allow blocks of variable data to be represented as both domains and variables. MMS provides no definition for, and imposes no constraints on, the content of a domain. To do so would be equivalent to defining a "real" object (i.e., the ladder program). The content of the domain is left to the implementor of the VMD.

MMS provides a set of services that allow domains to be uploaded from the device or downloaded to the device. The MMS domain services do not provide access to subordinate objects individually within the domain. The domain can only be uploaded or downloaded as a single contiguous block of untyped data.

UCA Usage of Domains. Within UCA a domain is treated as a representation of a logical device. This allows a single VMD to represent multiple devices each containing their own domain-specific objects, particularly variable access objects. This is important for UCA because of the extremely large number of devices in a typical utility. For instance, a single substation could have many thousands of customers connected to it and therefore many thousands of meters. To require client applications to directly address each device using a unique network would require that the network infrastructure support many millions of addresses to accommodate all the potential devices. Although this is possible to do using current networking technology, it would require higher levels of performance on the network, thereby increasing its cost. The utility industry has addressed this concern historically by using Remote Terminal Units (RTU) as data concentrators and gateways. Modeling data in devices as domain-specific objects allows UCA to easily support a data concentrator architecture while still conforming to the VMD model. By remaining within the bounds of the VMD model, generic non-UCA MMS clients are still able to interoperate with a UCA device without any prior knowledge of the UCA specific object models (GOMSFE) and application service models (CASM).

Program Invocations

It is through the manipulation of *program invocations* that a MMS client controls the execution of programs in a VMD. A program invocation is an execution thread which consists of a collection of one or more domains. Simple devices with simple execution structures may only support a single program invocation containing only one domain. More sophisticated devices and applications may support multiple program invocations containing several domains each.

As an example, consider how the MMS execution model could be applied to a personal computer (PC). When the PC powers up, it downloads a domain called the operating system into memory. When you type the name of the program you want to run and hit the <return> key, the computer downloads another domain (the executable program) from a file and then creates and runs a program invocation consisting of the program and the operating system. The program by itself cannot be executed until it is bound to the operating system. MMS services are provided that allow the client to create, delete, get the attributes, and control the execution of program invocations. See Fig. 5.

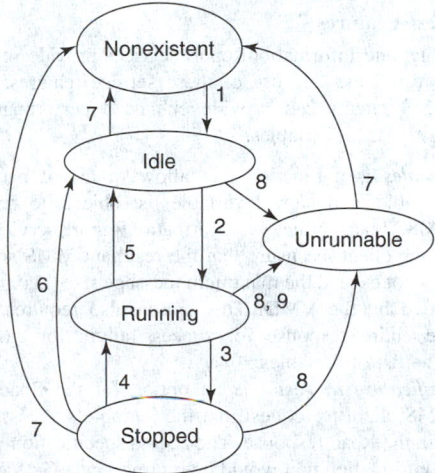

Fig. 5. MMS Clients use MMS services to cause state transitions in the program invocation as follows: 1. CreateProgramInvocation, 2. Start, 3. Stop or program stop, 4. Resume, 5. End of program and reusable = true, 6. Reset, 7. DeleteProgramInv., 8. Kill or error condition, and 9. End of program and reusable = false.

Example Batch Controller. As an example of how the MMS execution model can be applied to a typical device, let us look at a VMD model for a simple batch controller. See Fig. 6. The example in the figure depicts how the VMD model could be applied to define a set of objects (e.g., domains, program invocations, variables) appropriate for a batch controller. This model will provide clients an appropriate method of controlling the batch process using MMS services. In order to startup and control these two processes, a MMS client using this controller would perform the following actions:

1.   Initiate and complete a domain download sequence for each domain: recipe and data domains A and B, I/O domains A and B, and the control program domain.

2.   Create program invocation A consisting of I/O domain A, the control program domain, and recipe and data domain A.

3.   Start program invocation A. Create program invocation B consisting of I/O domain B, the control program domain, and recipe and data domain B.

4.   Start program invocation B.

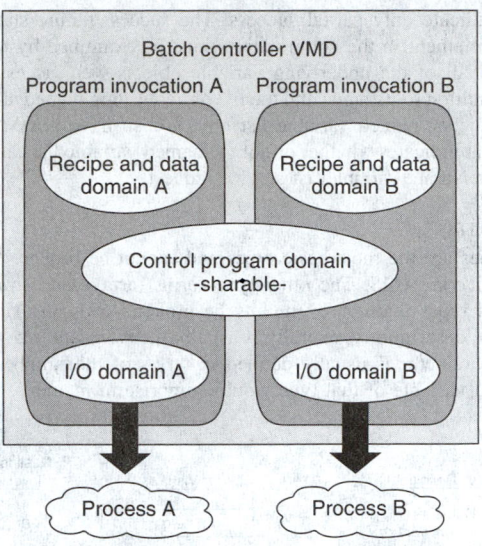

Fig. 6.   Our example batch controller allows us to control two identical batch oriented processes. An I/O domain, specific to each process ties the data in the corresponding recipe/data domain to the process. The control program domain in sharable and contains the control algorithm. The MMS client can create a separate program invocation to control each process. This allows the client to control the recipe and perform supervisory control (start, stop, etc.) of both processes independently from each other.

The example above demonstrates the flexibility of the VMD execution model to accommodate a wide variety of real world situations. Further examples might be a loop controller where each loop is represented by a single domain and where the control loop algorithm (e.g., PID) is represented by a separate but common *sharable* domain. Process variables, setpoints, alarm thresholds, etc. could be represented by domain-specific (control loop) variables. A program invocation would consist of the control loop domains and their algorithm domains needed to control the process.

Variable Access Model

MMS provides a comprehensive and flexible framework for exchanging variable information over a network. The MMS variable access model includes capabilities for *named, unnamed* (addressed), and named lists of variables. MMS also allows the *type description* of the variables to be manipulated as a separate MMS object (named type object). MMS variables can be simple (e.g., integer, boolean, floating point, string) or complex (e.g., arrays and structures). The services available to access and manage MMS variable objects support a wide range of data access methods from simple to complex.

MMS Variables

A real variable is an element of typed data that is contained within a VMD. A MMS variable is a virtual object that represents a mechanism for MMS clients to access the real variable. The distinction between the real variable (which contains the value) and the virtual variable is that the virtual variable represents the access path to the variable, not the underlying real variable itself.

The named variable object describes the access to the real variable by using a MMS object name. MMS clients need only know the name of the object in order to access it. Remember that the name of a MMS variable must also specify the *scope* of the variable (the scope can be VMD, domain, including the domain name, or association specific).

MMS also defines a *named variable list* object that provides an access mechanism for grouping both named and unnamed variable objects into a single object for easier access. A named variable list is accessed by a MMS client by specifying the name (which also specifies its scope) of the named variable list. When the VMD receives a Read service request from a client, it reads all the individual objects in the list and returns their value within the individual elements of the named variable list.

Because the named variable list object contains independent subordinate objects, a positive confirmation to a Read request for a named variable list may indicate only partial success. The success/failure status of each individual element in the confirmation must be examined by the client to ensure that all of the underlying variable objects were accessed without error. In addition to its name and the list of underlying named and unnamed variable objects, named variable list objects also have a MMS deletable attribute that indicates whether or not the named variable list can be deleted via a DeleteNamedVariableList service request.

Variable Type

Simple types are the most basic types and cannot be broken down into a smaller unit via MMS. The other type forms (arrays and structures) are constructed types that can eventually be broken down into simple types. Simple type descriptions generally consist of the *class* and *size* of the type. The size parameter is usually defined in terms of the number of bits or bytes that a variable of that type would comprise in memory. See Fig. 7.

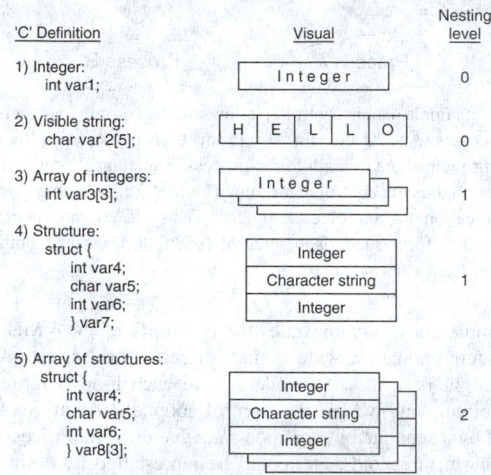

Fig. 7.   The level of complexity that a VMD can support is defined by its nesting level. A simple variable has a nesting level of zero. An array of simple variables (e.g., an array of Integers) requires a nesting level of one. An array of structured variables, each consisting of simple variables, requires a nesting level of two and so forth.

The various classes of simple types defined by MMS consists of:

- **Boolean**. Variables of this type can only have the values of *true* (value = non-zero) or *false* (value = zero). There is no size parameter for Boolean types and are generally represented by a single byte or *octet*.
- **Bit String**. A Bit String is a sequence of bits. The size of a Bit String indicates the number of bits in the Bit String. The most significant bit of the most significant byte in the string is defined as Bit0 in MMS terminology.
- **Boolean Array**. A Boolean Array is also a sequence of bits where each bit represents a *true* or *false*. It differs from an array of Booleans in that each element in a Boolean Array is represented by a single bit while each element in an array of Booleans is represented by a single byte. The size parameter specifies the number of Booleans (number of bits) in the Boolean Array.
- **Integer**. MMS Integer's are signed integers. The size parameter specifies the number of bits of the integer in 2's complement form.
- **Unsigned**. The Unsigned type is identical to the Integer type except that it is not allowed to take on a negative value. Because the most significant bit of an Integer is essentially a sign bit, an Unsigned with a size of 16 bits can only represent 15-bits of values or values of 0 through 32,767.
- **Floating Point**. The MMS definition for floating point can accommodate any number of bits for the format and exponent width including the IEEE 754 single and double precision floating point formats commonly in use today.
- **Octet String**. An Octet String is a sequence of bytes (*octet* in ISO terminology) with no constraint on the value of the individual bytes. The size of an Octet String is the number of bytes in the string.

- **Visible String**. The Visible String type only allows each byte to contain a printable character. The character set is defined by ISO 10646 and is compatible with the common ASCII character set. The size of the Visible String is the number of characters in the string.
- **Generalized Time**. This is a representation of time specified by ISO 8824. It provides millisecond resolution of date and time.
- **Binary Time**. This is a time format that represents time as either (1) time of day by a value which is equal to the number of milliseconds from midnight or (2) date and time by a value that is equal to the time of day and the number of days since January 1, 1984.
- **BCD**. Binary Coded Decimal format is where four-bits are used to hold a binary value of a single digit of zero to ten. The size parameter is the number of decimal digits that the BCD value can represent.
- **Object Identifier**. This is a special class of object defined by ISO 8824 that is used to define ISO registered network objects.

An *array* type defines a variable that consists of a sequence of multiple identical (in format not value) elements. Each element in an array can also be an array or even a structured or simple variable. MMS allows for arbitrarily complex nesting of arrays and structures.

A *structured* type defines a variable that consists of a sequence of multiple, but not necessarily identical, elements. Each individual element in a structure can be of a simple type, an array, or another structure. MMS allows for arbitrarily complex nesting of structures and arrays. A structured variable consisting of individual simple elements requires a nesting level of 1. A structured variable consisting of one or more arrays of structures containing simple variables requires a nesting level of 3.

Variable Access Features

The Read, Write, and InformationReport services provide several features for accessing Variables. The use of these service features, as described below, by MMS clients can provide enhanced performance and very flexible access to MMS Variables.

- **List of variables** is a function that allows a list of named variable, unnamed variable, and named variable list objects to be accessed in a single MMS Read, Write, or InformationReport service. Care must be taken by the client to ensure that the resultant MMS service request message does not exceed the maximum message size (maximum segment size) supported by the VMD. This option also requires that a client examine the entire response for success/failure for each individual element in the list of variables.
- **Access specification in result** is an option for the Read service that allows a MMS client to request that the variable's access specification be returned in the Read response. The access specification would consist of the same information that would be returned by a GetVariableAccess-Attributes service request.
- **Alternate Access** allows a MMS client to 1) partially access only specified elements contained in a larger arrayed and/or structured variable, and 2) rearrange the ordering of the elements contained in structured variables.

Event Management Model

In a real sense, an event or an alarm is easy to define. Most people have an intuitive feel for what can comprise an event within their own area of expertise. For instance, in a process control application, it is common for a control system to generate an alarm when the process variable (e.g., temperature) exceeds a certain preset limit called the high alarm threshold.

The MMS event management model provides a framework for accessing and managing the network communication aspects of these kinds of events. This is accomplished by defining three named objects that represent 1) the state of an event (*event condition*), 2) who to notify about the occurrence of an event (*event enrollment*), and 3) the action that the VMD should take upon the occurrence of an event (*event action*).

For many applications, the communication of alarms can be implemented by using MMS services other than the event management services. For instance, a system can notify a MMS client about the fact that a process variable has exceeded some preset limit by sending the process variable's value to a MMS client using the InformationReport service. The InformationReport service is how the typical ICCP-TASE.2 and UCA applications communicate event information. Other schemes using other

MMS services are also possible. When the application is more complex and requires a more rigorous definition of the event environment the MMS event management model can be used.

Event Condition Object

A MMS *event condition* object is a named object that represents the current state of some real condition within the VMD. It is important to note that MMS does not define the VMD action (or programming) that causes a change in state of the event condition. In the process control example given above, an event condition might reflect an *idle* state when the process variable was not exceeding the value of the high alarm threshold and an *active* state when the process variable did exceed the limit. MMS does not explicitly define the mapping between the high alarm limit and the state of the event condition. Even if the high alarm limit is represented by a MMS variable, MMS does not define the necessary configuration or programming needed to create the mapping between the high alarm limit and the state of the event condition. From the MMS point of view, the change in state of the event condition is caused by some autonomous action on the part of the VMD that is not defined by MMS.

Event Actions

An *event action* is a named MMS object that represents the action that the VMD will take when the state of an *event condition* changes. An event action is optional. When omitted, the VMD would execute its event notification procedures without processing an event action. An event action, when used, is always defined as a confirmed MMS service request. The event action is linked to an event condition when an *event enrollment* is defined. For example, an event action might be a MMS Read request. If this event action is attached to an event condition (by being referenced in an event enrollment), when the event condition changes state and is enabled, the VMD would execute this Read service request just as if it had been received from a client. Except that the Read response (either positive or negative) is included in the EventNotification service request that is sent to the MMS client enrolled for that event. A confirmed service request must be used (i.e., Start, Stop, Read). Unconfirmed services (e.g., InformationReport, UnsolicitedStatus, and EventNotification) and services that must be used in conjunction with other services (e.g., domain upload-download sequences), cannot be used as event actions.

Event Enrollments

The *event enrollment* is a named MMS object that ties all the elements of the MMS event management model together. The event enrollment represents a request on the part of a MMS client to be notified about changes in the state of an *event condition*. When an event enrollment is defined, references are made to an event condition, an *event action* (optionally), and the MMS client to which EventNotification should be sent. See Fig. 8.

Semaphore Management Model

In many real-time systems there is a need for a mechanism by which an application can control access to a system resource. An example might be a workspace that is physically accessible to several robots. Some means to control which robot (or robots) can access the workspace is needed. MMS *semaphores* are named objects that can be used to control access to other resources and objects within the VMD. For instance, a VMD that controls access to a setpoint (a variable) for a control loop could use semaphores to only allow one client at a time to be able to change the setpoint (e.g., with the MMS Write service). The MMS semaphore model defines two kinds of semaphores. *Token* semaphores are used to represent a specific resource within the control of the VMD. *Pool* semaphores consist of one or more named tokens each representing a set of similar but distinct resources under the control of the VMD.

Because semaphores are used solely for the purpose of coordinating activities between multiple MMS clients, the scope of a semaphore cannot be *AA-specific* where the object exists only an association between a single VMD and a single MMS client.

Token Semaphores

A token semaphore is a named MMS object that can be a representation of some resource, within the control of the VMD, to which access must

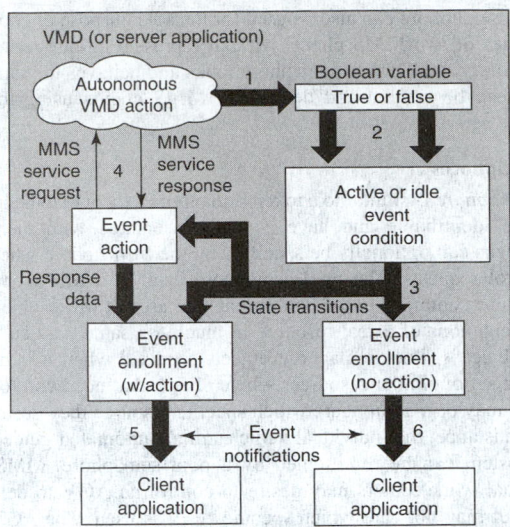

Fig. 8.   A Monitored event condition has a Boolean variable associated with it that the VMD sets (1) via some form of local autonomous action. The VMD periodically evaluates this variable (2) to determine the state of the event condition. MMS clients "enroll" to be notified of event condition state transitions (3) by defining an event enrollment. If an event action is defined for the event enrollment, the VMD obtains the response (4) to the event action (a MMS service request) and inserts the response data into the event notification (5) that is sent to the client. Event enrollments without an event action have their event notifications sent to the client (6) without any event action response data.

be controlled. A token semaphore is modeled as a collection of tokens that MMS clients take and relinquish control of using MMS services. When a client *owns* a token, the client may access the resource that the token represents. This allows both multiple or exclusive ownership of the semaphore. An example of a token semaphore might be where two users want to change a setpoint for the same control loop at the same time. In order to coordinate their access to the setpoint a token semaphore, containing only one token, can be used to represent the control loop. When a user "owns" the token, they can change the setpoint. The other user would have to wait until ownership is relinquished. See Fig. 9.

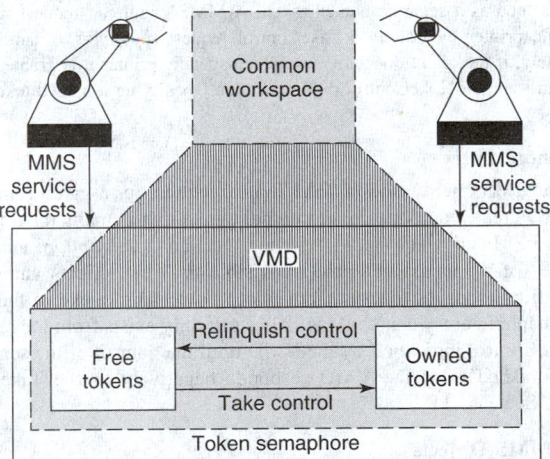

Fig. 9.   A token semaphore is modeled as a collection of free tokens and owned tokens. When a robot (a MMS client in this example) wants to access the common workspace, it will issue a TakeControl request to the VMD controlling the workspace. If there is a free token available, the VMD will grant control by moving a token from the free state to the owned state and then responding positively to the TakeControl request. If the other robot had already owned the token, then the VMD would respond negatively to the TakeControl. The token representing the common workspace would remain under the control of the robot until control was released with a RelinquishControl request or upon a control time out by the VMD. The total number of tokens availables how many simultaneous owners can exist for the same semaphore.

A token semaphore can also be used for the sole purpose of coordinating the activities of two MMS clients without representing any real resource. This kind of "virtual" token semaphore looks and behaves the same except that they can be created and deleted by MMS clients using the Define Semaphore service.

Pool Semaphores

A *pool semaphore* is similar to a token semaphore except that the individual tokens are identifiable and have a name associated with them. These named tokens can optionally be specified by the MMS client when issuing TakeControl requests. The pool semaphore itself is a MMS object. The named tokens contained in the pool semaphore are not MMS objects. They are representations of a real resource in much the same way an unnamed variable object is. Pool semaphore objects are used when it is desired to represent a set of similar resources where clients that need control of such a resource may or may not care which specific resource they desire control over. For instance, the individual vehicles in an automated guided vehicle (AGV) system can be represented by a pool semaphore. MMS Clients at individual work centers may desire to control an AGV to deliver new material but may not care which specific AGV is used. The AGV system VMD would decide which specific AGV, represented by a single named token, would be assigned to a given MMS client. The name of the pool semaphore is independent of the names of the named tokens (see Fig. 10). Pool semaphores can only be used to represent some real resource within the VMD. Therefore, pool semaphores cannot be created or deleted using MMS service requests and cannot be *AA-specific* in scope.

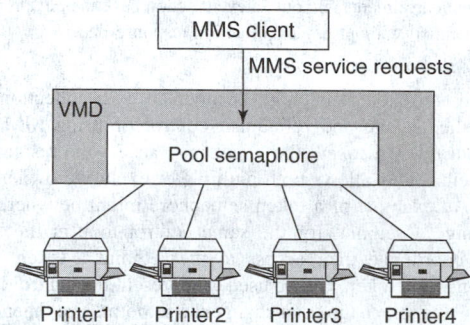

Fig. 10. A pool semaphore can be useful to control access to similar but distinguishable resources, such as a printer pool. In the example, each printer is represented as a separate named token. A MMS client can request control of a specific printer by issuing a TakeControl request specifying a named token. Alternately, if the client does not care which specific printer it is granted control of, it can issue the TakeControl request without specifying a named token.

Semaphore Entry

When a MMS client issues a TakeControl request for a given semaphore, the VMD creates an entry in an internal queue that is maintained for each semaphore. Each entry in this queue is called a *semaphore entry*. The attributes of a semaphore entry are visible to MMS clients and provide information about the internal semaphore processing queue in the VMD. The semaphore entry is not a MMS object. It only exists from the receipt of the TakeControl indication by the VMD until the control of the semaphore is relinquished or if the VMD responds negatively to the TakeControl request. See Fig. 11.

Other MMS Objects

Journal Objects. A MMS journal represents a log file that contains a collection of records (called a *journal entry*) that are organized by time stamps. Journals are used to store time based records of tagged variable data, user generated comments (called *annotation*), or a combination of events and tagged variable data. Journal entries contain a time stamp that indicates when the data in the entry was produced, not when the journal entry was made. This allows MMS journals to be used for applications where a sample of a manufactured product is taken at one time, analyzed in a laboratory off-line, and then placed into the journal at a later time. In this case, the journal entry time stamp would indicate when the sample was taken.

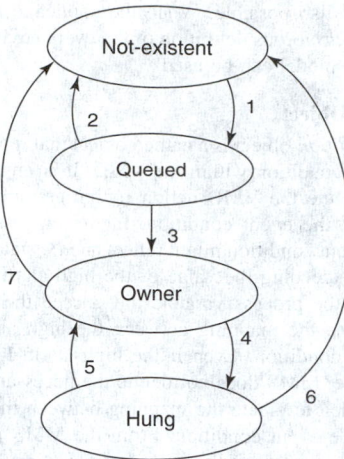

Fig. 11. A semaphore entry is created each time a client attempts to take control of a semaphore. The semaphore entry reflects the state of the relationship between the client and the semaphore. State transitions can be caused by local action by the VMD or by the following events: 1. TakeControl request received, 2. Timeout, Cancel request, or abort; 3. Semaphore available (token free), 4. Application association aborted, 5. TakeControl with preempt, 6. Control time out, or 7. RelinquishControl request received or time out.

MMS clients read the journal entries by specifying the name of the journal (which can be *VMD-specific* or *AA-specific* only) and either 1) the date/time range of entries that the client wishes to read or 2) by referring to the *entry ID* of a particular entry. The entry ID is a unique binary identifier assigned by the VMD to the journal entry when it is placed into the journal. Each entry in a journal can be one of the following types:

- *Annotation*. This type of entry contains a textual comment. This is typically used to enter a comment regarding some event or condition that had occurred in the system.
- *Data*. This type of entry would contain a list of variable tags and the data associated with those tags at the time indicated by the time stamp. Each variable tag is a 32-character name that does not necessarily refer to a MMS variable, although it might.
- *Event-Data*. This type of entry contains both variable tag data and event data. Each entry of this type would include the same list of variable tags and associated data as described above, along with a single event condition name and the state of that event condition at the time indicated by the time stamp.

Operator Station Object

The *operator station* is an object that represents a means of communicating with the operator of the VMD via a keyboard and display. An operator station is modeled as character based input and output devices that may be attached to the VMD for the purpose of communicating with an operator local to the VMD. Because the operator station is a representation of a physical feature of the VMD, it exists beyond the scope of any domain or application association. Therefore, MMS clients access the operator station by name without scope. There can be multiple operator stations for a given VMD.

Files

MMS also provides a set of simple file transfer services for devices that have a local file store but do not support a full set of file services via some other means. For instance, an electric meter may use the file services for transferring oscillography (waveform) files to an interested MMS client. The MMS file services support file transfer only, not file access. Although these file services are defined in an annex within the MMS standard, they are widely supported by many commercial MMS implementations.

Context Management

MMS provides services for managing the context of communications between two MMS nodes on a network. These services are used to establish and terminate application *associations* and for handling protocol errors between two MMS nodes. The terms *association* and *connection* are sometimes used interchangeably although there is a distinction from a network technology point of view. The node that initiates the association with another node is referred to as the *calling* node. The responding node is referred to as the *called* node.

In a MMS environment, two MMS applications establish an application association between themselves using the MMS Initiate service. This process of establishing an application association consists of an exchange of some parameters and a negotiation of other parameters. The exchanged parameters include information about restrictions that pertain to each node that are determined solely by that node (e.g. which MMS services are supported). The negotiated parameters are items where the called node either accepts the parameter proposed by the calling node or adjusts it downward as it requires (e.g., the maximum message size).

The calling application issues an Initiate service request that contains information about the calling node's restrictions and a proposed set of the negotiated parameters. The called node examines the negotiated parameters and adjusts them as necessary to meet its requirements. It then returns the results of this negotiation and the information about it's restrictions in the Initiate response. Once the calling node receives the Initiate confirmation, the application association is established and other MMS service requests can then be exchanged between the applications.

Once an application association is established, either node can assume the role of *client* or *server* independent of which node was the calling or called node. For any given set of MMS services, one application assumes the client role while the other assumes the role of server or VMD. Whether or not a particular MMS application is a client, server (VMD), or both is determined solely by the developer of the application.

Associations vs. Connections

Although many people may refer to network connections and application associations interchangeably, there is a distinct difference. A connection is an attribute of the underlying network layers that represents a virtual circuit between two nodes. For instance, telephone networks require that two parties establish a connection between themselves (by dialing and answering) before they can communicate. An application association is an agreement between two networked applications governing their communications. It is analogous to the two telephone parties agreeing to use a particular language and to not speak about religion or politics over the telephone. Application associations exist independent of any underlying network connections (or lack thereof).

In a connection oriented environment (e.g. TCP/IP) the MMS Initiate service is used to signal to the lower layers that a connection must be established. The Initiate service request is carried by the network through the layers as each layer goes through its connection establishment procedure until the Initiate indication is received by the called node. The connection does not exist until after all the layers in both nodes have completed their connection establishment procedures and the calling node has received the Initiate confirmation. Because of this, the association and the connection are created concurrently in a connection oriented environment.

In a connectionless environment (e.g. 3-layer), it is not strictly necessary to send the Initiate request before two nodes can actually communicate. In an environment where the Initiate service request is not used before other service requests are issued by a MMS client to a VMD, each application must have prior knowledge of the other application's exchanged and negotiated parameters via some local means (e.g., a configuration file). This foreknowledge of the other MMS application's restrictions is the application association from a MMS perspective. Whether an Initiate service request is used or not, application associations between two MMS applications must exist before communications can take place. In some connectionless environments such as the UCA 3-layer for serial link communications, MMS nodes still use the Initiate service to do the application association negotiation before communicating data or control information.

Summary

MMS provides a very flexible real-time messaging architecture for a wide range of applications. The concept of the VMD model of MMS is used in many lower level networking systems such as Foundation Fieldbus and Profibus where a more limited subset of MMS, called the Fieldbus Messaging System (FMS), is used. MMS has been used for years as a messaging system for Ethernet networks in automotive manufacturing, pulp-paper, aerospace, and other large and complex material handling systems. Recent technological innovations have allowed MMS to be applied into smaller and more resource limited devices like small PLCs, RTUs, meters, breakers, and relays. With the backing of EPRI, GRI, U.S. Postal Service, and a large number of utility users, the use of MMS is gaining momentum in these industries as a considerable number of equipment suppliers now support MMS and UCA. MMS is an effective bridge between the plant floor and the management information systems (MIS) as well as between the process control systems of the power plant and the distribution systems of the utility. With the additional refinements of GOMSFE and CASM, the level of interoperability between dissimilar equipment that is being achieved is welcome relief to an industry long plagued with numerous incompatible proprietary communication methods.

RALPH E. MACKIEWICZ, SISCO, Inc., Sterling Heights, MI

MANY-BODY FORCE. An interaction between two particles that becomes modified when a third particle is present, e.g., the forces between polarizable molecules. A large part of the experimental data of physics is concerned with natural objects that may be looked upon as being made up from smaller bodies. Thus, results the so-called many-body problem; there are several approaches to its solution. This highly theoretical topic is beyond the scope of this volume, but reference is suggested to the "Many-Body Problem," in *The Encyclopedia of Physics* (R.M. Besancon, editor), 3rd Edition, Van Nostrand Reinhold, New York, 1985.

MA (or Ma). Millions of years.

MAPLE SYRUP. Maple syrup is prepared by concentrating (evaporation or reverse osmosis) sap from the maple tree to a concentrated solution containing predominantly sucrose. Its characteristic flavor and color are formed during evaporation.

Maple syrup and sugar production predate the establishment of both the United States of America and the Canadian Federation. Production of this unique product is one of the few agricultural endeavors not brought to this continent by European settlers. The production process and practices have survived to the present and represent an interesting developmental and historical account. Native Americans were the first to discover the fact that sap from maple trees could be processed into maple syrup. While there are no authenticated accounts of how this process was discovered there are several interesting legends. Undoubtedly these have been modified over time, but it is likely this discovery was accidental.

Maple syrup comes from eastern Canada, particularly Quebec, and the northern United States, especially New England, New York State and the Great Lake states. It is made as far south as Virginia and as far west as Indiana and Minnesota (see Fig. 1). Sixteen states and four Canadian provinces produce maple syrup. However, it can be made wherever maples grow. Most maple trees, even boxelders, can be used as a source of sap, but the sugar maple (*Acer saccharum*) and black maple (*A. nigrum*) are the most favored.

There are thirteen native maple species in North America. While most of these species are probably tapped to some extent, at least by hobbyists, sugar maple (Fig. 2) and black maple, along with red maple (*Acer rubrum*), provide most of the commercial sap. A fourth maple species, silver maple (*Acer saccharinum*), is sometimes tapped, particularly in roadside operations, and is often confused with red maple. See also **Maple Trees.**

The following is used with the gracious permission of Ohio State University.

Physiology of Sap Flow

Maple sap represents the raw material from which pure maple products including maple syrup, maple sugar and various confection products are made. It is a totally natural product which is obtained from the sapwood (light colored, outer xylem tissue) of several species of maple trees.

The physiology of sap flow in maple trees is a complex biological process which is not completely understood. While it has been the subject of several investigations, the specific nature of the process remains unclear. Several aspects of the process are understood, however, and this knowledge can be helpful to maple producers. It is interesting to note that while

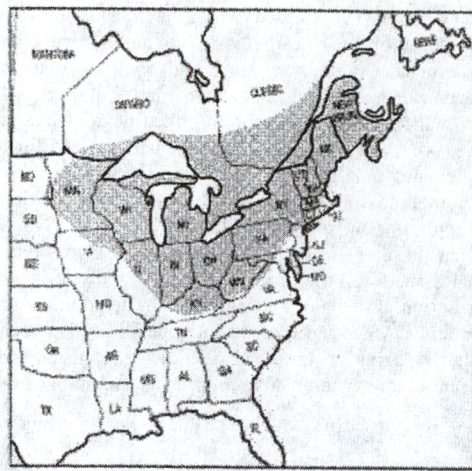

Fig. 1.   Area of commercial maple syrup production in North America. (courtesy of the Massachusetts Maple Producers Association and Ohio State University).

Fig. 2.   Excellent specimen of sugar maple growing in Pisgah National Forest, North Carolina. (U.S. Forest Service photo).

the production of maple sap by maple trees is a natural physiological phenomenon, the flow and collection of sap is not; rather, it occurs as a result of wounding the tree and "harvesting" the sap which flows in response to this wounding.

Maple sap flow may occur anytime during the "winter" dormant season when air temperatures fluctuate above and below the freezing point. In the northern United States and adjacent Canada this can occur from early October through the following April. However, the largest flows take place in late winter and early spring during the months of February, March and April. During this period, winter begins to lose its grip in many parts of the maple region. Colder days and nights give way to periods when temperatures during the day frequently rise above the freezing point. When these warmer daytime temperatures are followed by below freezing periods, usually at night, strong sap flows can be expected.

Fluctuations in air temperatures above and below freezing during the dormant season cause maples to develop strong positive sap pressure (well above atmospheric pressure) provided there is adequate available soil moisture. Current theory suggest that when temperatures fall below

freezing, negative pressure (suction) is created within the sapwood of maple trees as a result of sap freezing, carbon dioxide dissolving in cooled sap, and gas contraction during cooling. As a result of this negative pressure, water moves from the soil into the tree increasing the sap volume. When the temperature then rises above freezing and the frozen sap thaws, forces (including pressure from released gases, osmotic<caused by the presence of sugar and other substances dissolved in the sap, previous gas compression, and gravity) act on the increased sap volume to create a positive pressure. This pressure develops first in the twigs, then in the trunk, and finally in the roots of the maple tree. The positive pressures that develop can be considerable, rising to 40 or more pounds per square inch in untapped trees. When temperatures fall, the process is reversed and pressures are reduced. Below freezing temperatures are required for strong negative pressures to again develop. Much weaker sap pressures develop if below freezing temperatures are not reached and sustained long enough for ice to form within the tree. The presence of sucrose in sap may also be involved in the maple sap flow mechanism.

Sap flow occurs when a wound is made in the sapwood of a maple tree which has positive sap pressure. Sap flow from the wound will continue as long as the pressures inside the tree are greater than atmospheric pressure outside the tree. This period of sap flow (pressure dissipation) will vary from a few (1 or 2) to several (15 to 20) hours in length. Length of the sap flow period and the amount of sap produced appear to be affected by several environmental and tree metabolic factors including the minimum and maximum temperatures experienced, sap sugar concentration, the duration of the freezing and thawing cycles, and the availability of soil moisture. For strong sap flows to be repeated, a suitable temperature cycle above and below freezing must again occur to allow strong positive sap pressure to develop. Sap flow temporarily ceases during the "maple season" when suitable temperature cycles do not occur (cold or warm periods) and ceases entirely when suitable temperature cycles end for the year.

The sugar in maple sap is the product of photosynthesis which occurred during the previous growing season. Photosynthesis produces carbohydrates which are predominately stored in the tree in the form of starch, although some may be stored as sugars (primarily sucrose). During the winter, in maple trees, some of this starch is converted to sucrose and is dissolved in the sap. The amount of sugar in the sap will depend on many factors including the genetics of the individual tree, the quality of the site on which the tree is growing, tree health, environmental conditions the previous growing season and time in tapping season. Sap sugar contents are sometimes lower at the beginning of the maple season, rise rapidly and then slowly decline as the season progresses. However, this may vary depending on weather conditions present the previous growing season.

Maple Sap Production: Tapping, Collection and Storage

The Tapping Process. Tap in such a way that you obtain at minimum cost, the maximum volume of high-quality sap of the highest sugar content without reducing the health and sap-producing ability of the trees.

While somewhat complicated, the above sentence summarizes tapping objectives and emphasizes the importance of tapping decisions to the success of a maple operation. While tapping is a seemingly simple task, the selection of trees to tap and the tapping techniques used significantly affect the quality of syrup produced and the economic success of the operation.

Correct tapping will (1) minimize taphole damage to the trees resulting in healthier trees, and (2) minimize contamination of the sap by microorganisms resulting in a higher-quality syrup. Microorganism contamination of sap is a major cause of low-quality syrup, causing darker grades and often causing undesirable flavors, and can also shorten taphole life by blocking sap flow. Proper tapping technique includes: use of proper equipment; tapping at the proper time; proper number, size, and placement of tapholes; taphole cleanliness, sanitation, and retapping; and proper seating of spouts.

Tapping Equipment. The basic equipment required for tapping maple trees is a drill bit, some type of tapper, spouts (spiles), a small mallet, and a clean bucket.

Drill Bit. A clean, sharp 7/16-inch diameter drill bit is required. When using a power drill use a fast cutting tapping bit, not a carpenter's speed bit. Dull bits are more difficult to use; create fuzzy, ragged tapholes which impede sap flow and provide additional surfaces within the taphole for microorganism growth; are more likely to produce oval holes which allow sap leakage and contamination by microorganisms; and often result in slower healing tapholes. Conversely, sharp bits are easier to use and cut clean round holes which close more quickly.

Drill bits should be as clean as possible to minimize contamination of the taphole. An effective way to accomplish this is to wash drill bits with detergent (if needed), triple rinse them with water, and then soak them for at least 15 minutes in a bleach solution (1 cup commercial household bleach in 1 gallon of water) or in alcohol. If rotten wood is encountered while tapping, bits should be rewashed with bleach to prevent transferring the contamination to other trees.

Tapper. A variety of tappers are used by sugar makers including a carpenter's brace, battery powered drill, gasoline powered tapper, and chainsaw engine with tapping adapter. The primary differences among these tappers are cost and speed of tapping. Power tappers are more economical for larger producers.

Spouts (Spiles). Spouts have three important functions: (1) transfer sap from taphole into container or tubing, (2) provide hanging support for container or connection for tubing, and (3) provide seal to minimize microorganism contamination of sap and taphole. Spouts have evolved from the early days of using hollowed out reeds or wooden shoots to the metal or plastic (for tubing) spouts of today. A number of different spout designs for both tubing and bucket/bag operations are currently in use. Each has its perceived advantages and disadvantages, but all commercially available spouts are satisfactory. All spouts are tapered to form a tight seal against the bark and outer sapwood when seated in the taphole.

Mallet. A small mallet with a wooden, rubber, plastic, or leather head is ideal for seating spouts.

Taphole Number, Size and Location. Proper tapping requires using the correct number of taps in a tree and separating the taps as much as possible to minimize the chance of tapping into old taphole compartments or discolored wood. Over the years, various tapping guidelines have related the number of allowable taps to tree diameter. One of the more common traditionally accepted tapping guidelines suggests that trees larger than 10 inches (25 centimeters) in diameter can be tapped and up to 4 taps can be put in larger trees (Table 1).

TABLE 1. TRADITIONAL TAPPING GUIDELINE

Tree Diameter		Tree Circumference		Number of Taps
Inches	Centimeters	Inches	Centimeters	
10–15	25.4–38.1	31–47	78.7–119.4	1
15–20	38.1–50.8	47–63	119.4–160	2
20–25	50.8–63.5	63–78	160–198.1	3
25+	63.5+	78+	198.1+	4

Historically, researchers and producers were confident that healthy, vigorous maple trees could support the tapping intensity outlined in Table 1. However, many maples are not in optimum health, but suffer to some degree from a variety of stress causing factors including wounds, insects, diseases, weather factors (such as drought), soil compaction, and air pollution. In recent years, maple trees throughout much of the northeastern United States have been subjected to severe stresses from gypsy moth, pear thrips, pollution, and weather. As a result, tapping guidelines have become more conservative, with one commonly recommended guideline suggesting 12 inches as the smallest tree to tap, and no more than 1 or 2 taps in any tree (Table 2).

TABLE 2. MORE CONSERVATIVE TAPPING GUIDELINE

Tree Diameter		Tree Circumference		Number of Taps
Inches	Centimeters	Inches	Centimeters	
12–18	30.5–45.7	38–57	96.5–144.8	1
18+	45.7+	57+	144.9+	2

The fact is that wounding a tree by tapping has a greater potential impact on tree health than removing too much sap. This is a strong argument for conservative tapping. Furthermore, research and field observations show

that the volume yield of sap per taphole can increase substantially when fewer tapholes per tree are used. In summary, using more taps per tree does not necessarily mean proportionally more sap.

The importance of accurately determining tree size and vigor when tapping cannot be overemphasized. Tree diameters can be estimated accurately with experience. They can be easily measured using a diameter or circumference tape (tree diameter = circumference divided by 3.14) or, most quickly, with a Biltmore stick. A Biltmore stick is a yard-stick-like scale stick which is placed against the trunk of the tree and sighted across to read tree diameter.

Tapholes can be located anywhere on the tree trunk but for ease of tapping and collecting they are generally located between 2 and 4 feet (0.6 and 1.2 meters) above the ground (or snow). Certainly this tapping band can and often should be widened, particularly upward, to achieve proper taphole spacing. Tapholes are made by drilling round 7/16-inch or 5/16-inch diameter holes 11/2 to 3 inches into the sapwood (Fig. 3). Because bark thickness of tapable maples varies between 3/8 inch and 1 inch, adjustments to total tapping depth should be made so that tapholes extend the desired depth into the sapwood.

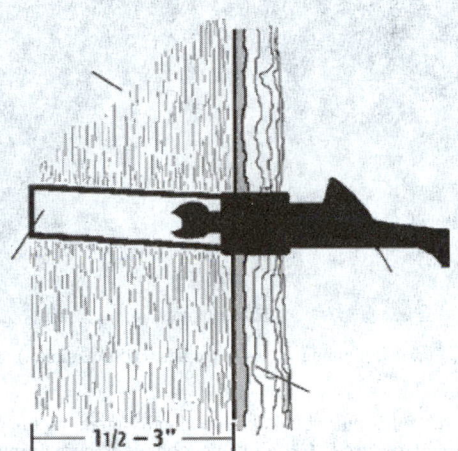

Fig. 3. Installation of spile (spout) into maple tree.

If the trees have been tapped before, new tapholes should be at least 6 inches to the side of old tapholes. Tapholes drilled in successive years should not be placed in a straight line around the tree. A useful pattern is to drill each new taphole at least 6 inches (15.2 centimeters) to the side *and* slightly above or below old tapholes. This results in a spiral pattern of tapping around the tree which, over the years, utilizes the entire tapping face of the trunk.

Immediately after drilling, a spout should be gently but firmly seated in the taphole. In a bucket/bag operation spouts must be seated firmly enough to prevent sap leakage and to support the weight of the bucket or bag. In a tubing operation, spouts must be seated firmly enough to prevent sap or air leakage. Care must be exercised, however, not to seat spouts too tightly. Spouts seated with excessive force may split the bark and wood around the taphole resulting in leakage and slow healing, and may block the sap flow into the taphole from the outer sapwood.

The danger of splitting tapholes when seating spouts is far greater when trees are frozen. When possible, trees should be tapped when they are not frozen. Producers with large operations who must tap frozen trees in order to complete tapping need to exercise extreme care when seating spouts in frozen wood.

Spouts must be gently removed from the tapholes immediately following the sugaring season before the tree begins growth. Care in spout removal will reduce injury to the cambium and facilitate taphole closure. Spouts left in tapholes during the growing season will interfere with closure and cause more injury when removed. Plugging tapholes after spout removal is not recommended as this practice could interfere with taphole closure.

Sap Collection and Transfer. It is in the method of collecting and transferring sap to the sugar house that some of the greatest differences exist among commercial maple operations. Historically, sap was collected in containers, such as wooden buckets, located below each spout.

Periodically the sap was dumped from the buckets into larger containers and transferred to the sugarhouse. Many modern maple operations still rely on this bucket or bag technology for collecting and transferring sap. Others have converted all or part of their sugarbush to a tubing collection and transfer system. There are certainly advantages and disadvantages to each system, the choice between them depending on a great many factors including size of operation, capital resources and planning horizon of the producer, characteristics of the sugarbush, and personal preference of the producer.

Buckets and Bags. Buckets are the most common collecting container and 16 quart galvanized buckets are most commonly used (Fig. 4). However, sap buckets made from other materials, particularly aluminum, and of different sizes are available.

Fig. 4.   Tapping of sugar maples. (USDA photo.)

As a general rule, only buckets specifically manufactured for maple collection should be used. Certainly no container containing lead, lead-containing paint, or lead solder should be used. No container capable of rusting should be used, and containers with thin galvanized coatings should be avoided as the coatings may quickly wear exposing a surface which will rust.

Each bucket should have a cover to keep out rain and falling debris. Rain dilutes sap resulting in a lower sugar content requiring more sap, time, and fuel to produce syrup. Debris not only is a physical contaminant that must be filtered, but may introduce off-flavors or microorganisms. Bucket covers come in two general types, those that attach with a wire pin to the spout and those that clamp to the bucket top.

While not as popular as buckets, plastic bags provide an alternative sap collecting container. Often cited advantages of sap bags over buckets are the following: (1) Because of their small bulk and weight they require less storage space and are easier to transport and hang. (2) They provide the spout and taphole greater protection from microorganisms. (3) Emptying bags is easier; they can be rotated on the spout and dumped with one hand. (4) They do not require a separate cover. (5) Ice during cold spells as well as sunlight may reduce microbial development. Frequently cited disadvantages of sap bags include the following: (1) They may break at the seams, particularly if sap in a full bag freezes. (2) They may tear or be damaged by rodents. (3) They are difficult to empty when filled with ice. (4) The bag may be too small to hold a large run. (5) Washing and rinsing bags may be difficult. (6) Sap may warm rapidly in the sun.

Plastic Tubing. While buckets and bags are effective in collecting sap, their use is labor intensive since frequent emptying and collection must be done. Oftentimes this is a daily task which becomes laborious as well as costly. To reduce the labor involved in sap collection, other sap collection schemes have been considered. In the past, wood troughs or metal conduit systems were used by some; producers although these too had their problems and limitations. In the 1950s relatively inexpensive plastic tubing became available for use in the sugarbush. Plastic tubing with accompanying spouts permitted sap from all tapholes in a given sugarbush or section of a sugarbush to flow into a common collection

tubing (conduits) network. This system could deliver sap directly to a collection tank, which in a few fortunate situations, was located at the sugarhouse. For others, sap flowed to a storage collecting tank from which it was transferred to the sugarhouse for processing. Since its initial introduction, refinements and improvements in the composition, construction and application of plastic tubing systems have been made such that today plastic tubing networks are considered to be state-of-the-art with respect to maple sap collection systems.

In order for a plastic tubing system to operate efficiently it is essential that a few basic considerations be observed. These are relatively simple but can lead to installation difficulties and operation inefficiencies if not observed. These basic considerations include:

1. All systems should be designed so gravity will help sap move through the system.
2. Tubing capacity and therefore size is a function of the topographic slope and number of taps.
3. Keep the system as simple and direct as possible to maximize sap movement.
4. Maintain the tightness and grade of all tubing lines as uniform as possible to facilitate sap movement.
5. Tubing systems must be inspected and maintained on a regular basis to minimize leaks and sap loss.

A properly designed, installed and operating tubing system will substantially reduce the amount of labor associated with sap collection. Additionally, it is possible to obtain larger sap yields from properly installed systems.

Vacuum Systems. The laborious task of gathering and transporting maple sap from the tree to the sugarhouse was dramatically reduced in the 1950s with the advent of plastic tubing. Although maple producers widely accepted the use of plastic tubing for sap collection, the levels of success they experienced varied. The use of vacuum on tubing sap collection systems was developed in the 1960s. Vacuum pumping was first conceived as a means of overcoming problems normally associated with plastic tubing networks, but was soon found to increase sap yield substantially. Thus, a profit can be expected from the investment in vacuum pumping equipment. Even though it was demonstrated that higher yields of sap were associated with higher levels of vacuum some maple producers have been reluctant to adopt vacuum pumping for fear of possible negative effects to trees or sap quality. However, research found that even with an increase in sap volume, no significant difference existed in sugar or other solutes between sap collected by gravity and sap collected at a vacuum level of 15 inches of mercury (abbreviated Hg) measured at the vacuum pump. Researchers found no difference in the biochemical composition of sap collected by gravity without vacuum (control) and with several maintained levels of vacuum at the taphole (10, 15, 20-in. Hg). However, significant increases of sap volume yield and a slight decrease in sugar concentration (more water released) occurred at each level of vacuum over the treatment without vacuum.

In order for more producers to benefit from the use of vacuum in sap collection systems, the basics of why vacuum can increase sap flow should be understood. Maple sap cannot be pulled or sucked from a taphole. Increasing sap flow using vacuum is more complex. Sap flow is generally associated with rising temperatures (above 32 °F (0 °C)) following freezing temperatures (below 32 °F (0 °C)) along with increasing pressure within the tree. When tree pressure is greater than atmospheric pressure, i.e. is positive, sap will flow freely when the tree is punctured as by a taphole. As temperatures drop below freezing, sap pressure changes from positive to negative causing sap flow to cease. Conditions often exist during the sap production season when the internal tree pressure is low or equal to atmospheric pressure creating less than optimal or "weeping flows." Vacuum pumping in effect lowers the atmospheric pressure in the tubing system, thus creating a pressure differential that is greater inside the tree than atmospheric pressure, thereby allowing sap to flow more freely from the taphole.

Several types of vacuum units are commercially available for maple producers and innovative producers have created their own vacuum systems. Some form of power is required to operate all vacuum systems. A general rule of thumb is that one cubic foot per minute (CFM) will deliver sufficient vacuum for 100 taps (example: 15 CFM for 1500 tapholes). Regardless of the type of vacuum system used, the benefits of vacuum can only be realized when tubing systems are properly designed and free of air leaks. Maintaining tubing systems can be accomplished more efficiently with the use of vacuum because leaks can be detected more easily by observing sap movement in the tubing, hearing the hiss associated with leaks and noting decreases in vacuum levels using a vacuum gauge.

Sap Storage. Sap storage is an important step in the production of quality maple products. Adequate sap storage is necessary not only to store sap until it is processed, but also to supply a constant flow of sap to the evaporator or other sap processing equipment. Proper sap storage involves proper sizing of the storage facilities, cleanliness, and managing temperature and light.

To avoid losing sap during good runs, producers should generally have a sap storage capacity equivalent to two good days of sap flow. On the average, this is interpreted as 2-3 gallons of storage per tap. In locations where back-to-back large flows are not uncommon, some producers have as much as 4 gallons of sap storage per tap. This large storage capacity, however, is the exception rather than the rule, and certainly is influenced by many factors including how quickly sap can be processed. Estimates of total storage capacity should include the storage tanks at the sugarhouse plus any secondary storage tanks (transfer or dump tanks) in the woods. Producers collecting with buckets/bags should also recognize that buckets/bags provide additional storage. While bucket/bag storage does not replace adequate tank storage at the sugarhouse, it is important in estimating storage needs to handle periods of large sap flow.

A wide variety of tanks made from a variety of materials are used for sap storage. Tanks should be made from an easily cleaned, nonporous material such as plastic, fiberglass, glass, stainless steel, galvanized steel, painted steel, etc. Porous tanks, such as unsealed concrete, are very difficult to clean and are less suitable as storage tanks. Commercial sap storage tanks are available from maple equipment vendors. Glass or stainless steel lined tanks from the dairy industry are often used. Tanks, both new and used, from a wide variety of other sources have been adapted by producers for sap storage. If used tanks are used for sap storage, their previous contents should have been only food-grade materials.

Maple Syrup Production

Pure maple syrup is made by concentrating the slightly sweet sap of the sugar maple tree. The basics needed for making maple syrup therefore are some sugar maple trees and a method of concentrating the sap into syrup. The group of maple trees that is used is called *sugarbush*, or *maple orchard*.

Maple sap, as it comes from the tree is a clear, slightly sweet liquid. The sugar content rages from one to four percent. A device called a "hydrometer" can be floated in the sap to determine the exact sugar content (for instructions on the use of a hydrometer, consult the *North American Maple Syrup Producers Manual* referenced at the end of this article). Sweeter sap is favored because less water will have to be evaporated to make maple syrup. The sap must be evaporated as soon as possible because the freshest sap makes the best quality syrup.

Maple syrup is traditionally made in a building called a "sugarhouse"—the name of the building comes from the time when most sap was actually turned into sugar. Sugarhouses vary in size and shape, each with its own character. Some may be rustic wood buildings out in the woods with poor access and no electricity, full of old tools and memories of grandfather's sugar seasons of the past. Still others might remind you of a modern food processing plant, brightly lit and streamlined. Each sugarhouse will have vent at the top, a cupola—which is opened to allow the steam of the boiling syrup to escape the building.

Antique or modern, each sugarhouse will contain an evaporator used to boil down the sap into syrup. Evaporators are made up of one or more flat pans which sit on an "arch," a type of firebox. Wood or oil, and sometimes gas or coal is burned at the front end, and the flames are drawn along the underside of the pan, heating and boiling the sap as they travel towards the rear. It commonly takes about one cord of wood or sixty gallons of oil to boil down 800 gallons of sap into maple syrup. Depending on the size of the evaporator and the number of trees tapped, this may represent anywhere from two hours to two whole days of boiling. The basic design of maple syrup evaporators has changed little over the years, although sugarmakers are always tinkering with new designs to make the process faster or more fuel efficient. The size of the evaporator depends on the

number of trees a producer has tapped. Modern two-section evaporators with a syrup and sap pan are commonly available in sizes ranging from 2 feet wide by 6 feet long (referred to as a 2 by 6 or 2×6) to 6 feet wide by 18 feet long (6×18). The larger the evaporator, the more water it will evaporate per hour, the greater its productivity. Many backyard and hobby sugarmakers use smaller arrangements, or boil down their sap on the kitchen stove.

An evaporator pan is divided into partitions, so that the sap is continuously flowing through the pan. Fresh sap enters at the back of the pan, where a float valve keeps the sap about an inch deep. As the water is boiled off, two things happen: First, the liquid becomes sweeter, and begins to move towards the front of the pan, traveling through the partitions. Secondly, more fresh sap is allowed into the rear of the pan. In this way the water is constantly being evaporated away, the liquid is becoming sweeter as it moves towards the front of the pan, and the float valve in the rear is always allowing more sap to be added to keep the level about an inch deep. It takes about forty gallons of this slightly sweet sap, boiled down, to make one gallon of pure maple syrup.

The sugarmaker concentrates his attention to the front of the evaporator where the boiling sap is turning a golden color as it approaches being maple syrup. From time to time he will check the temperature of the boiling liquid. When it reaches seven and a half degrees above the boiling point of water, it has reached proper density and has become maple syrup. Another way of checking for the proper density or sugar content is to place a scoop into the boiling syrup. If the drops along the bottom edge of the scoop begin to hold together like a sheet or apron, then the sugarmaker knows the syrup is done. Coming from the tree, maple sap is approximately 98% water and 2% sugar. When the syrup is finished, it is only 33% water and 67% sugar. At this stage a valve on the front of the pan is opened and some of the finished boiling syrup is drawn off the pan and is filtered. After filtering, the syrup is bottled and is ready for sale.

The length of the sugaring season is totally dependent upon the weather. It may last only a few weeks, or as long as six or eight weeks. As the days become increasingly warmer, and the nights rarely get below freezing, the buds on the branches of the maple trees begin to swell, marking the end of the season. Chemical changes take place within the tree as baby leaves begin to form within the buds. At this time the sap is no longer suitable for boiling down into syrup.

The Evaporator and its Function

The process of converting maple sap to syrup is essentially one of boiling the sap for a period of time to evaporate water and concentrate the sugar in the sap. During this heating process, chemical changes occur which impart the unique color and flavor to maple syrup. While the evaporation process is simple in concept, proper evaporation equipment and technique are critical to economically produce quality maple syrup.

Maple syrup evaporating equipment has undergone a great evolution in design. The first boiling equipment might have been a hollowed-out log in which water was evaporated from the sap by adding hot stones. The next evaporator was a metal kettle hung over an open fire. These early "evaporators" were all examples of batch-type evaporators in which the entire evaporation process was done in one container. Sap was continually added to the evaporator until the desired volume of finished syrup was obtained. Many hours of boiling were required resulting in a dark, strong flavored syrup.

A major improvement in the evaporation process was the use of multiple kettles. These were the forerunner of today's continuous evaporators. The sap was partly evaporated in the first kettle, transferred to the second kettle for further concentration, and then finally transferred to a third and sometimes a fourth kettle where evaporation was completed. The multiple kettle method was a semicontinuous operation which resulted in a higher-quality syrup (lighter colored) because the time of heating at near-syrup density was shortened. The next important change in the design of evaporating equipment was the introduction of the flat-bottom pan and the enclosed firebox. Both the increased heating surface of the pan and the confined fire increased the efficiency of the fuel. This design was quickly followed by partitioned pans, which were the forerunner of today's flue-type evaporators.

The modern flue-type evaporator, developed about 1900, was the last major change in design. The use of flues (deep channels in the sap pan) and altering the firebox so that it arched the hot gases between the flues, caused the hot gases and luminous flames to pass between the flues before escaping up the chimney. Fuel economy was increased. Also, the rate of evaporation was increased which shortened the evaporation time, improved the quality of the syrup, and lowered the cost of production.

Early evaporators were heated exclusively by wood fires. Modern evaporators are available which use a variety of heat sources including wood, fuel oil, gas, and steam.

The modern flue-type evaporator consists basically of two sections: (1) the sap pan, in which the flues are located, and (2) the syrup pan. The sections are separated to facilitate their removal from the arch for cleaning and repair. A semirigid pipe connects the pans. Partitions (baffles) are built in the pans to form channels through which the sap flows as the sugar concentration increases during evaporation. These baffles and the channels they form allow the evaporator to be operated in a continuous or semicontinuous manner, with sap being introduced into the sap pan and finished or near-finished syrup being removed from the syrup pan.

The sap pan, often referred to as the flue pan and normally the rear pan, is commonly made with narrow, deep channels to increase the pan surface and improve heat exchange. This construction is possible because sap in the sap pan never becomes concentrated enough to become viscous (thick), so there is little danger of burning. Fresh sap is admitted to the sap pan through a float valve that can be adjusted to maintain the desired depth of liquid in the evaporator.

The syrup pan, often called the front pan, is normally located over the firebox. Evaporation of the sap to syrup may be completed in this pan, or, in some instances, evaporation is stopped before finished syrup is produced. If this is done, finishing is accomplished in a separate finishing pan. The syrup pan has a flat bottom to permit evaporation of shallow layers of syrup without danger of burning and to facilitate cleaning.

Table 3 provides estimates of the evaporating capacity of conventional evaporators of different sizes. These estimates were developed by averaging the capacities published by several manufacturers. They should be used only as an approximate guide since there is considerable variation in capacities among manufacturers.

TABLE 3. ESTIMATES OF CONVENTIONAL EVAPORA-TOR CAPACITY IN GALLONS PER HOUR

Evaporator Size (Length × Width)	Evaporator Capacity (Gallons Per Hour)
2′ × 6′	25
2′ × 8′	35
30″ × 8′	50
30″ × 10′	60
3′ × 8′	70
3′ × 10″	85
3′ × 12′	100
40″ × 10′	110
40″ × 12′	120
4′ × 10′	120
4′ × 12′	140
4′ × 14′	170
5′ × 12′	205
5′ × 14′	230
5′ × 16′	280
6′ × 14′	310
6′ × 16′	340
6′ × 18′	380

Changes in Color (and Flavor) During Evaporation. During the evaporation process, sap is concentrated to the desired density and flavor and color develops as a result of chemical reactions that occur during heating. The extent and character of the color and flavor are determined, in part, by the length of time the sap is boiled. The longer sap is boiled, the darker it becomes. Making high-quality, light-colored syrup requires evaporation time be kept to a minimum. Anything that slows the evaporation process (uneven fire, weak fire, excessive sap depth in evaporator, etc.) will produce darker, usually stronger flavored syrup. At low sugar concentrations, relatively little color is produced by a given length of boiling, while at higher concentrations more color develops in the same length of boiling time.

How Much Sap? How Much Syrup?-The Rule of 86. Because the conversion of sap to syrup is essentially a concentration process, the sugar content (°Brix) of the sap determines the amount of sap required to produce a gallon of syrup. This relationship, termed the "Rule of 86", may be used to estimate the number of gallons of sap of a specific density required to produce a gallon of syrup. The relationship is as follows:

$$S = 86/X$$

where S is the number of gallons of sap required to produce one gallon of syrup, X is the Brix value of the sap, and 86 is a mathematical constant representing the percentage of solids on a weight-volume basis that is in a gallon of syrup (see following discussion).

It follows, then, that the number of gallons of water that must be evaporated from sap to obtain one gallon of syrup can be calculated by subtracting 1 from the above equation:

$$W = (86/X) - 1$$

As an example, using sap with a density of 2° Brix requires 43 gallons of sap to produce 1 gallon of syrup:

$$S = 86/X = 86/2 = 43 \text{gallons sap}$$

and 42 gallons of water must be evaporated to produce one gallon of syrup:

$$W = (86/X) - 1 = (86/2) - 1 = 43 - 1 = 42 \text{gallons water}$$

The "Rule of 86" calculates a good "working" estimate of the number of gallons of sap required or water that must be evaporated to produce a gallon of syrup. It is not exact because the mathematical formula was developed when "standard density syrup" had a density of 65.5° Brix. Syrup of this density contains 86.3 percent solids, hence, the "Rule of 86." Current "standard density syrup" has a density of 66° Brix and contains 87.2 percent solids. Further, geographic regions and individual producers differ slightly in the desired finish density of syrup. Nonetheless, the "Rule of 86" is a tradition in the maple industry and is quite satisfactory for practical purposes.

Fuel Systems for Evaporators. The fuel to be used to heat the evaporator should not be chosen as an afterthought. It should be considered early in the maple planning process and the entire heating system designed and assembled with components that enable it to operate effectively and efficiency. Wood and fuel oil are the two most commonly used fuels to provide heat for evaporating; natural gas and propane are far less common. A few producers use steam as a heat source to the evaporator, generating the steam in a wood, coal, oil, or gas boiler. Some producers have explored less traditional heat sources such as wood chips.

Wood Fuel. Wood is the traditional fuel for heating maple evaporators, and it is still used by a high proportion of producers. A sugarhouse with a wood—fired evaporator can be easily recognized by its ample storage space for fuelwood near the evaporator doors, the tall smokestack (usually more than twice as tall as the length of the evaporator), and the characteristic wood-burning arch (much more vertical firewall and much shallower under the sap pan than most oil-fired arches) (Figs. 5 and 6). Evaporating with wood fuel creates a pleasant, nostalgic atmosphere in the sugarhouse unmatched by any other fuel.

Fig. 5.   Wood burning evaporator with characteristic shallow arch below sap pan.

Fig. 6. Sugarhouse with readily accessible wood supply and relatively tall smokestack characteristic of wood-burning evaporator.

Producers report producing approximately 25 gallons of syrup from 1 full cord of firewood. A cord is a stack of wood occupying 128 cubic feet of volume, often represented as a stack of wood 4 feet high, 8 feet long, and 4 feet deep. Actual production will depend on the efficiency of the evaporator, as well as the species of wood burned. Table 4 presents a comparison of the total available heat in common firewood species. It should be stressed that Table 4 provides total available heat values. Most wood-burning evaporators will use less than half of this heat in evaporation. Obviously, all species are not the same when it comes to providing heat. Other things being equal, the heavier (denser) the wood, the higher its available heat.

TABLE 4. APPROXIMATE AVAILABLE HEAT IN ONE CORD OF WOOD AT 20% MOISTURE CONTENT, ASSUMING 5780 BTUS PER POUND OF WOOD

Species	BTUs Available per Cord (millions)
Locust, Black	24.9
Hickory, Shagbark	24.9
Ironwood	24.6
Apple	23.9
Oak, White	23.2
Beech	21.7
Maple, Sugar	21.7
Birch, Yellow	21.3
Oak, Red	21.3
Ash, White	21.0
Ash, Green	20.0
Walnut, Black	19.5
Birch, Paper (White)	18.7
Maple, Red	18.7
Hackberry	18.7
Cherry, Black	18.7
Elm, Slippery (Red)	18.4
Sycamore	17.9
Elm, American	17.6
Ash, Black	17.2
Maple, Silver	16.5
Boxelder	16.2
Cottonwood	14.0
Willow, Black	13.6
Aspen	13.3
Basswood	12.5

Wood used in an evaporator may be round or split, depending on its size, but should be thoroughly seasoned (dried). Most hardwood species will require from 9 to 12 months to thoroughly season. This means that wood used in the evaporator should be cut the previous late spring or early summer. Wood cut later than this will have higher moisture contents and will burn less efficiently because a portion of the heat will be used to evaporate the excess moisture content.

Most traditional wood burning evaporators are less than 50 percent efficient in using the available heat for evaporation. Some may be substantially less than 50 percent in efficiency. The importance of efficiency cannot be overemphasized. As an example, suppose a producer is using wood with 20,000,000 BTUs available heat per cord, and an evaporator with an efficiency of 30 percent. If the sap is 2.5° Brix, the temperature of sap in the storage tank is 35°F (2°C), and the boiling temperature of standard density syrup is 219°F (104°C), approximately 19 gallons of syrup will be made per cord[1]. If the efficiency of the evaporator can be increased to 40 percent, an easily attainable figure, approximately 25 gallons of syrup can be made from the same cord of wood.

Oil Fuel. Experiments with oil as a fuel for maple evaporators date back to the early 1950s; recommendations for installations have been available since the late 1950s or early 1960s. Interest in the development of the oil-fired evaporator came, at least in part, from the growth in size of many maple operations during that time and the accompanying need to boil for extended periods of time (often around the clock) with minimum labor. Today oil is a common fuel for heating evaporators, second only to wood (Fig. 7).

Fig. 7. Oil burning evaporator with two oil burners.

Commonly cited advantages of using oil as a heat source for evaporating maple sap include the following:

- Oil-fired evaporators require less labor to operate than wood fired units.
- No farm labor is required to harvest and process the fuel as is required with wood.
- With oil-fired evaporators it is easier to maintain a constant, more uniform heat source than wood. This may result in the production of higher-quality syrup.
- Oil as a fuel is generally considered cleaner since little or no soot or ashes are produced.

Commonly cited disadvantages of using oil to fire evaporators include:

- Requires more capital investment in equipment (oil burners).
- Requires a different arch than wood fired evaporators, though in new installations this is not necessarily more expensive.
- Fuel must be purchased rather than produced by labor.
- Does not use wood material generated by sugarbush management.
- Requires mastery of a new technology.

Number 2 fuel oil is normally used to fire one or more oil burners in an oil-fired evaporator. Number 2 fuel oil contains approximately 139,000 BTUs of available heat per gallon. Using the same example used above with wood, if the evaporator were 100 percent efficient, it would require slightly more than 2.2 gallons of fuel oil to produce a gallon of syrup from 2.5° Brix sap, at 35°F, if standard density syrup boiled at 219°F. Traditional oil-fired evaporators are commonly 65 to 75 percent efficient, and therefore usually require in excess of 3 gallons of fuel oil to produce a gallon of syrup. Currently the most efficient evaporators on the market require approximately 2.5 gallons of fuel oil per gallon of syrup produced.

Arch design for evaporators reflects the fact that oil and wood burn differently. Wood burns with a luminous flame over essentially the entire

[1] BTU is required to raise 1 pound of water 1°F up to the boiling temperature, and 970 BTUs are required to convert 1 pound of water to steam.

length of the evaporator. Under the flue pan, much of this flame results from burning gases released from the wood and moving in the draft toward the stack. Since most of the heat transferred from the fire to the evaporator is radiant (about 80 percent), an evaporator design which encourages movement of the flame along the bottom of the pans is desirable when burning wood. Wood evaporators, therefore, have nearly vertical firewalls and a relatively shallow depth to the arch floor beyond the firewall.

Oil, in contrast, burns in a relatively compact ball; its flame does not move along the bottom of the pans. In traditionally—designed, oil-fired evaporators, to efficiently transfer heat from the oil fireball to the bottom of the evaporator pans, the burning ball must illuminate (by radiation) the entire area under the surface of the pans. This has been accomplished by tilting the firewall (at least the upper portion) back under the flue pan, essentially lowering the arch floor below the front of the flue pan. This provides a direct line for radiant heating from the fire ball to the underside of the pans. However, a number of fairly new technologies deviate markedly from traditional designs, and appear to achieve comparatively high efficiencies. Among these are the use of relatively small V—shaped combustion chambers and the use of rear—mounted burners.

Other Fuels. Natural gas and propane are clean and efficient fuels for firing evaporators. Natural gas may be particularly attractive to producers with access to an inexpensive supply, such as a well. Propane is sometimes the fuel of choice of hobby producers to fire small evaporators.

Historically, wood fired arches were converted to use natural gas. Today, gas-fired arches, similar in appearance to basic oil fired arches, along with natural gas and propane burners, are available from maple equipment suppliers.

Because gas heat can be easily adjusted and shut off when the desired temperature is attained, propane gas (L.P.) is commonly used to fire finishing pans and canning units. Its cleanliness also makes it an attractive fuel to use adjacent to canning or other maple product making locations in the sugarhouse.

The evaporation of maple sap by steam has been practiced by a few producers for many years although its use has never been widespread.

The principle behind a steam evaporator is relatively simple. When water vaporizes into steam it requires energy (970 BTUs per pound). When steam condenses it releases this heat energy. Steam is formed in the boiler and transferred by pipe to the evaporator where it condenses in pipes running through the sap in the evaporator pan. The energy that is released when the steam condenses boils the sap.

Both high- and low-pressure steam systems have been used to evaporate maple sap. Low-pressure systems are favored by many producers because they are safer and usually less expensive to install.

Wood chips may be an attractive fuel to maple producers who have access to large quantities of chips or wood suitable for chipping. Commercial evaporators which use wood chips are available. Some units have automated chip loading which incorporates a chip hopper, an agitator to separate the chips, and an auguring mechanism to load the chips into the evaporator firebox. Some wood-chip evaporators require forced-air draft both below and above the grate to support effective combustion. Fuel systems using a wood-chip evaporator require chip drying and storage facilities under cover to prevent moisture deterioration and freezing, as well as chip handling equipment such as a blower, augur conveyer, or bucket loader. If wood is being converted to chips, chipping equipment will also be needed.

Wood chips can also be processed through gasification to a clean, efficiently burning fuel, with the added advantage that gasification units can burn green hardwood chips. Producers can purchase green chips, or buy or rent a chipper and produce their own.

Increasing Evaporation Efficiency. *Reverse Osmosis (RO)*, originally used for desalinating salt water, is a process in which the water content of maple sap is reduced by forcing sap under pressure across membranes that contain pores large enough to pass water molecules, but too small to permit the passage of sugar and other large molecules (Fig. 8). As the sap passes over these semipermeable membranes some of the water passes out of the sap through the membranes leaving behind a more concentrated sap. Repeated passes of the sap through the membranes result in additional water loss and increasingly higher sugar concentrations.

Reverse osmosis has been used commercially in the maple business since the mid 1970s, though less efficient technology dates back to the late 1950s.

Fig. 8. Reverse osmosis machine used to concentrate sap and reduce evaporation time.

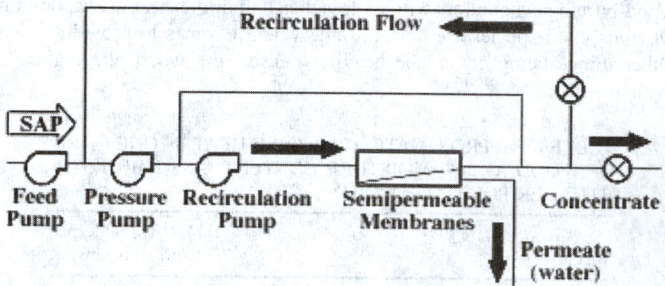

Fig. 9. Schematic of operation of reverse osmosis machine.

The objective of using reverse osmosis is to remove a substantial portion (approximately 75%) of the water from the sap, concentrating it to between 7° and 10° Brix before it enters the evaporator, thereby reducing evaporator fuel costs and boiling time. Other advantages frequently cited for reverse osmosis include shortening the holding time for unprocessed sap (less spoilage) and shortening the time sap is processed at high temperatures.

Figure 9 illustrates how a reverse osmosis unit works. Sap ("feed") is pumped into the machine, placed under pressure, and recirculated across the semipermeable membranes. "Concentrate" is the concentrated sap drawn off. "Permeate" is the water that has been removed from the sap by passing through the membrane. The actual operation of a reverse osmosis unit primarily involves the adjustment of valves to control the flow rate of sap as it flows through the unit. Maintaining the proper pressures and flow rates through the unit and the correct relationship between the amount of liquid moving across the membrane (the recirculation flow), the amount of water being removed (permeate), and the amount of concentrated sap being drawn off (concentrate) are critical to effective operation.

Preheaters offer a relatively inexpensive method of increasing the efficiency and profitability of a conventional evaporator. In a conventional open evaporator, the heat in the water vapor (steam) generated in the boiling process is lost. A preheater uses a portion of this heat to increase the temperature of incoming sap, thereby reducing the amount of fuel required to heat the sap in the evaporator. Most preheaters consist of a series of parallel tubes (usually copper) located in the hood of the sap pan (Fig. 10). Properly sized preheaters commonly increase the efficiency of a conventional evaporator by about 15 percent. In fact, some of the newer, more sophisticated evaporators, such as the "piggyback" and "steam—away," incorporate the preheater concept into their design. As an additional benefit, preheaters provide a source of clean, hot water.

Forced-Draft Units. Using forced air to supply supplemental air to the fire to increase the rate and completeness of combustion of wood and

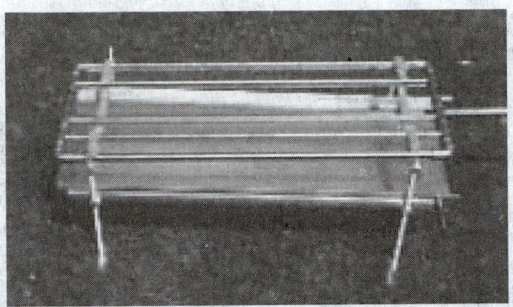

Fig. 10. A relatively simple preheater assembled from copper tubing. (U.S.F.S.).

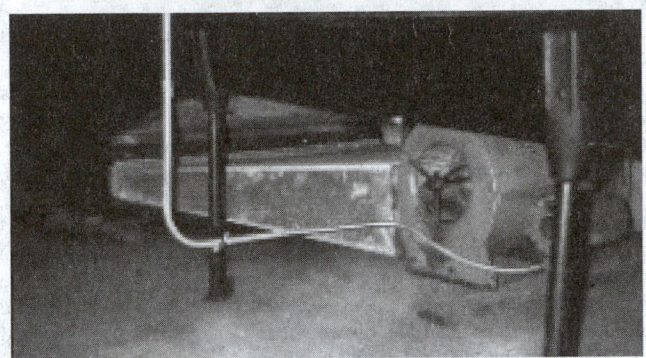

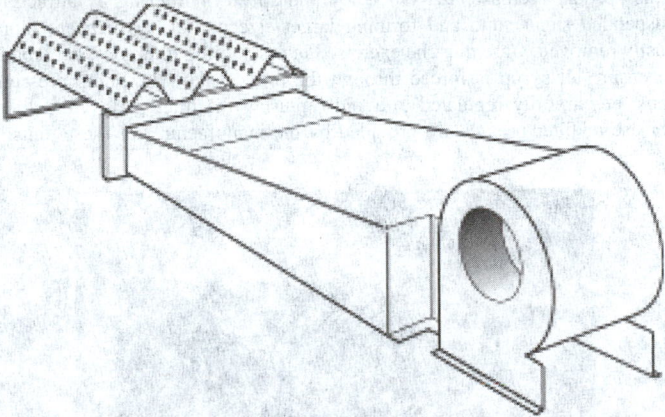

Fig. 11. A forced-draft blower unit can increase fuel efficiency in traditional wood burning evaporators.

Fig. 12. The piggyback evaporator incorporates the preheater concept into the evaporator design. Unit slightly modified for photographic purposes. (W. Brown).

wood gases can substantially increase it's efficiency. In a traditional wood arch this can be achieved using a commercially available forced-draft unit (Fig. 11). Basically, most of these units consist of a specially designed hollow grate with holes through which air is forced, one or more squirrel-cage blowers (fans), electric motor(s), and a rheostat to control fan speed. Some manufacturers offer kits which also provide forced air near the top of the firebox to more completely burn gases.

Other Evaporator Units. Piggyback® (Fig. 12), STEAM AWAY® and other similar units are relatively recent developments in evaporator design that combine the preheater concept with that of tandem sap pans. With these units, an evaporator pan is mounted on top of the sap pan on a tight drip trap suspension collar. This pan acts as a preheater, absorbing the heat in the steam produced in the sap pan. In addition, air is forced through the sap in the top pan creating turbulence which enhances the evaporation of water. As a result, sap being transferred from this pan is not only heated, as it is in a preheater, but becomes more concentrated.

Syrup Filtration, Grading, Packing and Handling

Maple syrup is produced by concentrating the sugars (principally sucrose) in maple sap through evaporation. Concentration results from boiling the sap and reducing the water content. In the course of evaporation the characteristic flavor and color associated with pure maple syrup is produced. In the process of evaporation not only is the sugar concentration increased but other compounds present in the sap increase in concentration as well. While these non-sucrose components are present in extremely small quantities in sap, evaporation can increase their content to a significant level.

Syrup Filtration. Maple syrup as it is taken off the evaporator contains suspended solids, commonly referred to as *niter* or *sugar sand*. The composition of sugar sand includes the calcium and magnesium salts of malic acid. Malic acid is one of the organic acids present in maple sap. These salts form during evaporation when they are precipitated out of the sap solution because they become less soluble as the concentration of sucrose increases due to boiling. The precipitates, in suspension, impart a gritty texture to unfiltered maple syrup. Sugar sand can occur in various forms ranging from a nondescript oily substance which is dark in color to a fine-grained crystalline material which is light in color. The color of the suspended sugar sand can cause syrup to appear cloudy or darker in color than it actually is.

The amount of sugar sand in various syrups is not the same. It will vary from year to year as well as during the season. It will also vary from one sugarbush to another even within the same season. Factors responsible for the amount of mineral compounds in maple sap are not completely understood. Apparently soil and site factors upon which the maple trees are growing contribute. It is also probable that local weather factors are also responsible, at least in part, although the relationships have not been proven.

Syrup which is to be marketed commercially must have all suspended material and other precipitates removed. The clarification process is necessary to meet some individual state and/or federal grading standards. Two methods of removing suspended materials are available to most producers. These involve sedimentation and filtration systems.

Sedimentation. In the past, many producers used a sedimentation or settling process to clarify maple syrup. Although this process is simple, it has several disadvantages and is not commonly used by commercial producers. Sedimentation consists of placing finished syrup (of correct density) in a large container and allowing the suspended material to settle out. This process may require a long amount of time ranging from a few days to a week or more. Sedimentation is effective for removing most suspended material, however, some very fine particles may remain suspended. When samples of the syrup in the settling tank indicate the syrup is clear, the syrup is carefully removed. Before packaging, it must be reheated to above 180 °F (82 °C) to ensure a sterile pack. This reheating may darken the syrup thereby resulting in a reduction of grade.

Because of the difficulties associated with sedimentation, most producers use a filtering process to clarify finished syrup. Filtration involves pouring or forcing finished syrup through a filter or series of filters to remove all suspended particles. Syrup may be filtered either through a gravity system or a forced pressure system. When done correctly, either process will result in effective clarification.

Gravity Systems. Gravity filtration systems are of two types, cone and flat filters. The so called "felt-hat" system uses a cone-shaped wool or Orlon felt bag which is suspended over an appropriately sized and shaped container (Fig. 13). Hot syrup from the evaporator is poured into the bag and allowed to flow into the collection container. Most commonly, two or three bags will be suspended over a single collection container. Syrup is placed in each bag as space allows. Filtering receptacles or tanks are frequently equipped with canning spigots so syrup can be directly packaged following filtering. If this is done some provision should be made to keep filtered syrup hot (180 °F) before packing.

Fig. 15.   Cloth cone-filter with paper prefilters.

Fig. 13.   A cone-filter, syrup-filtering system.

Flat filters consist of a wool or Orlon felt sheet which is placed over a hardware cloth screen (Fig. 14). As with filter bags, two or three adjacent filtering units may be constructed. Hot syrup is poured on the filter and allowed to pass through to the collection tank below. Once in the collection tank, syrup may be packaged directly if it is kept hot (above 180 °F) and a packing-valve filling assembly is present.

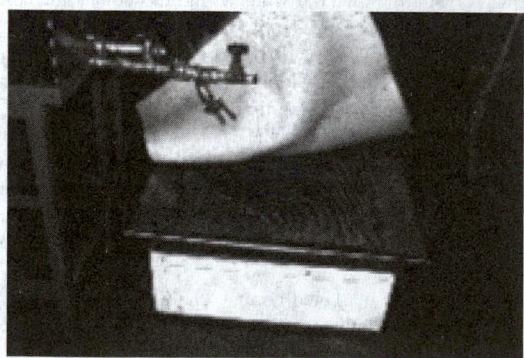

Fig. 14.   A flat-filter, syrup-filtering system.

As hot syrup passes through the felt filter, either in a cone-shaped or flat configuration, suspended sugar sand or other particles collect on the filter and are removed from the syrup. This collected material accumulates on the filter surface and will reduce the rate of flow of syrup through the filter. Eventually the filter will become clogged and must be cleaned. To prolong the period of use between washings it is recommended that a prefilter be used. Paper or nylon prefilters are available which can be placed on top of the felt filter (Fig. 15). When accumulations of sugar sand build up, the prefilter is either replaced or is removed, washed and reinstalled. This process will extend the life of the felt filter and will increase the rate at which filtration occurs.

Pressure Filters. For producers who process relatively large volumes of syrup or for those who do not want the bother associated with gravity filters, pressure filtering units are available. Although more expensive, pressure units are capable of filtering finished syrup quite rapidly and, at the same time, producing a product of the highest clarity. Pressure filters consist of a mechanical pump (gear pumps are traditionally used although diaphragm pumps are gaining in popularity because of less wear) which forces the syrup through a series of filter plates and disposable filter pads (Fig. 16). Before pumping finished syrup through the filter a small amount of filter aid (diatomaceous earth) is added to the finished syrup. (Follow manufacturer's recommendations as to amount to use.) This product increases the efficiency and speed of filtering by attracting suspended sugar sand and forming larger size particles which are more easily removed. Care must be exercised in assembling filter plates and pads to ensure all syrup is forced through the press. Likewise, pump pressure must be carefully regulated to avoid rupturing the filter pads. Guidelines for use of filter presses are provided by the manufacturer of these units.

Fig. 16.   A pressure-filter syrup-filtering system.

Maintaining and Adjusting Density. Using a refractometer to determine the density of syrup by measuring its refractive index is an accurate, yet simple method. A refractometer works by measuring the refractive index of a solution (syrup) which is directly related to the amount of dissolved solids (sugar) present in the solution. Refractometers are precise instruments and are particularly well suited for determining the density of syrup in Brix units at room temperatures. They are not well suited for measuring the density of hot syrup (180 °F and above) but are very convenient for larger operations which buy and sell syrup. They are also commonly used to determine the density of syrups entered in various competitions.

Hydrometers are the instruments of choice for most maple producers in determining and checking syrup density. They are well adapted for use in the sugarhouse and can provide accurate density determinations. Hydrometry is based on a physics principle which states the density of a liquid (syrup in this case) can be measured by evaluating the amount of displacement of a floating body. All that is required to determine syrup density is a calibrated hydrometer, a hydrometer cup and a thermometer. If a temperature compensated hydrometer is available (hydrotherm) and the temperature of the syrup solution being evaluated is within the indicated range, a separate thermometer is not required. The hydrometer is carefully placed in the syrup in the hydrometer cup. It will come to rest at a certain level at which time the density value of the syrup can be read directly from a scale sealed in the stem of the hydrometer (Fig. 17). Care must be observed to note the exact point on the hydrometer stem which is in contact with the surface of the syrup.

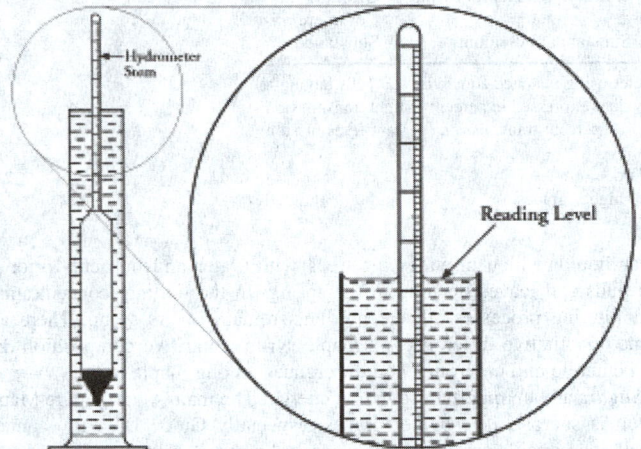

Fig. 17. An hydrometer can be used to measure syrup density (a-top). The hydrometer cup should be filled to the top and the reading should be taken on the hydrometer stem at the point shown in (b-bottom).

Brix Scale. The Brix scale relates the density of syrup to sugar solutions of the same density and known percentages of sugar. The Brix value does not express the true percentage of sugar in a solution containing sugar plus other dissolved solids; rather, it indicates what the percentage of sugar would be if the density of the solution were due only to dissolved sugar. The Brix scale is particularly well suited for measuring the density of maple syrup because 98 percent of the dissolved solids are sugar. For practical purposes, the Brix value equals the percentage of sugar in the syrup.

A good approximation of the weight of sugar in any lot of maple syrup, whether or not it is standard-density syrup, can be found by multiplying the weight of the syrup by its density (° Brix) and dividing by 100. This information is important to the producer who sells their syrup wholesale, since the price is based on its solids (sugar) content or weight. Thus, 100 pounds of syrup at 65° Brix contains 65 pounds of sugar, whereas 100 pounds of standard-density syrup (66.0° Brix) contains 66.0 pounds of sugar. Therefore, 100 pounds of the low-density syrup has a lesser value than 100 pounds of standard-density syrup. Likewise, 100 pounds of syrup with a density of 66.8° Brix contain 66.8 pounds of sugar, which is more than that contained in 100 pounds of standard-density syrup, and it has a greater value.

Baumé Scale. The Baumé Scale relates the density of a liquid to that of a salt solution. It does not express directly the solids content of maple syrup and does not, therefore, have the same utility for calculating syrup sugar content. Its use, however, is established in the maple industry with many producers using it exclusively. See also **Specific Gravity**. Table 5 presents Brix/Baumé equivalents for converting between the two scales. Standard density syrup, the equivalent of 66° Brix syrup, is defined on the Baumé scale as 35.6°.

Syrup Grading. In both the United States and Canada, federal grading standards and guidelines exist. Additionally, several states have recognized grading standards applicable to some or all of their production. While terminology among the different regulating units will vary, the grading standards are generally very similar.

Grading Standards. The primary characteristic of maple syrup around which grades are determined is color. Syrup which is the lightest in color (assuming minimum light transmittance standards are met) is assigned the "highest" grade while syrup of darker colors is designated differently. It is perhaps unfortunate that grading standards are based on color in that flavor characteristics are the feature most desired by consumers. To be sure, color and flavor are often related, yet it seems unfortunate that full-bodied syrups with stronger maple flavors are usually assigned a lower grade.

TABLE 5. COMPARISON OF BRIX WITH BAUMÉ SCALES FOR VALUES BETWEEN 30° AND 70° BRIX AT 68°F

°Brix	°Baumé	°Brix	°Baumé	°Brix	°Baumé	°Brix	°Baumé	°Brix	°Baumé	°Brix	°Baumé
30.0	16.6	43.5	23.8	57.0	30.9	62.0	33.5	64.7	34.9	67.4	36.3
30.5	16.8	44.0	24.1	□	□	62.1	33.6	64.8	34.9	□	□
31.0	17.1	44.5	24.4	57.5	31.2	62.2	33.6	64.9	35.0	67.5	36.3
31.5	17.4	□	□	58.0	31.5	62.3	33.7	□	□	67.6	36.4
32.0	17.7	45.0	24.6	58.5	31.7	62.4	33.7	65.0	35.0	67.7	36.4
□	□	45.5	24.9	59.0	32.0	□	□	65.1	35.1	67.8	36.5
32.5	17.9	46.0	25.2	59.5	32.2	62.5	33.8	65.2	35.2	67.9	36.5
33.0	18.2	46.5	25.4	□	□	62.6	33.8	65.3	35.2	□	□
33.5	18.5	47.0	25.7	60.0	32.5	62.7	33.9	65.4	35.2	68.0	36.6
34.0	18.7	□	□	60.1	32.5	62.8	33.9	□	□	68.1	36.6
34.5	19.0	47.5	26.0	60.2	32.6	62.9	34.0	65.5	35.3	68.2	36.7
□	□	48.0	26.2	60.3	32.6	□	□	65.6	35.3	68.3	36.7
35.0	19.3	48.5	26.5	60.4	32.7	63.0	34.0	65.7	35.4	68.4	36.8
35.5	19.6	49.0	26.8	□	□	63.1	34.1	65.8	35.5	□	□
36.0	19.8	49.5	27.0	60.5	32.7	63.2	34.1	65.9	35.5	68.5	36.8
36.5	20.1	□	□	60.6	32.8	63.3	34.2	□	□	68.6	36.9
37.0	20.4	50.0	27.3	60.7	32.9	63.4	34.2	66.0	35.6	68.7	36.9
□	□	50.5	27.5	60.8	32.9	□	□	66.1	35.6	68.8	37.0
37.5	20.6	51.0	27.8	60.9	33.0	63.5	34.3	66.2	35.7	68.9	37.0
38.0	20.9	51.5	28.1	□	□	63.6	34.3	66.3	35.7	□	□
38.5	21.2	52.0	28.3	61.0	33.0	63.7	34.4	66.4	35.8	69.0	37.1
39.0	21.4	□	□	61.1	33.1	63.8	34.4	□	□	69.1	37.1
39.5	21.7	52.5	28.6	61.2	33.1	63.9	34.5	66.5	35.8	69.2	37.2
□	□	53.0	28.9	61.3	33.2	□	□	66.6	35.9	69.3	37.2
40.0	22.0	53.5	29.1	61.4	33.2	64.0	34.5	66.7	35.9	69.4	37.3
40.5	22.2	54.0	29.4	□	□	64.1	34.6	66.8	36.0	□	□
41.0	22.5	54.5	29.6	61.5	33.3	64.2	34.6	66.9	36.0	69.5	37.3
41.5	22.8	□	□	61.6	33.3	64.3	34.7	□	□	69.6	37.4
42.0	23.0	55.0	29.9	61.7	33.4	64.4	34.7	67.0	36.1	69.7	37.4
□	□	55.5	30.2	61.8	33.4	□	□	67.1	36.1	69.8	37.5
42.5	23.3	56.0	30.4	61.9	33.5	64.5	34.8	67.2	36.2	69.9	37.5
43.0	23.6	56.5	30.7	□	□	64.6	34.8	67.3	36.2	70.0	37.6

TABLE 6. MAPLE SYRUP GRADE DESINGATION BASED ON PERCENT OF LIGHT TRANSMITTANCE, UNITED STATES AND CANADA

% Light Transmittance[a]	Canada Federal and Quebec	United States[b] USDA	Vermont State
Not less than 75%Tc	No. 1 Extra Light	Grade A Light Amber	Fancy
Between 60.5 & 74.9%Tc	No. 1 LightGrade A	Medium Amber	Same as U.S.
Between 44.0 & 60.4%Tc	No. 1 MediumGrade A	Dark Amber	Same as U.S.
Between 27.0 & 43.9%Tc	No. 2 Amber	Grade B for Reprocessing	Grade B
Less than 27%Tc	No. 3 Dark	□	Commercial
□	□	SubstandardVermont[c]	Substandard[c]

[a] Percent light transmission measured with a spectrophotometer using matched square optical cells having a 10 mm light path at a wavelength of 560 nm, with the color values expressed in percent of light transmission as compared to analytical reagent glycerol fixed at one hundred percent transmission. Percent transmission determined in this way is symbolized "%Tc".
[b] Color actually determined by comparison with glass standard.
[c] Does not meet density, clarity, color, or flavor standards of higher grades.

Grading standards of all regulating governments are based on color in relation to light transmittance. A comparison of U.S. and Canadian grades is shown in Table 6. While color is the primary characteristic used in differentiations among grades it is understood all syrup will be at minimum density, 66.0° Brix at 68 °F, and will not possess any off flavors or other undesirable characteristics not identified with pure maple syrup.

Grade A Light Amber, is very light and has a mild, more delicate maple flavor. It is usually made earlier in the season when the weather is colder. This is the best grade for making maple candy and maple cream.

Grade A Medium Amber, is a bit darker, and has a bit more maple flavor. It is the most popular grade of table syrup, and is usually made after the sugaring season begins to warm, about mid-season.

Grade A Dark Amber, is darker yet, with a stronger maple flavor. It is usually made later in the season as the days get longer and warmer.

Grade B, sometimes called Cooking Syrup, is made late in the season, and is very dark, with a very strong maple flavor, as well as some caramel flavor. Although many people use this for table syrup; because of its strong flavor, it's often used for cooking, baking, and flavoring in special foods.

Grading Kits. Standardized grading kits for both U.S. and Canadian grades are available. These consist of permanent glass standards which correspond to the minimum percent light transmittance standards for each grade (Fig. 18). In addition to permanent standards, colored glycerine solution grading kits are also available (Fig. 18). Temporary glycerine solution kits may be accurate when they are new. However, there is a tendency for some of the solutions to change color and become inaccurate with age. If glycerine kits are used, they should be periodically compared to permanent glass standards to evaluate their accuracy.

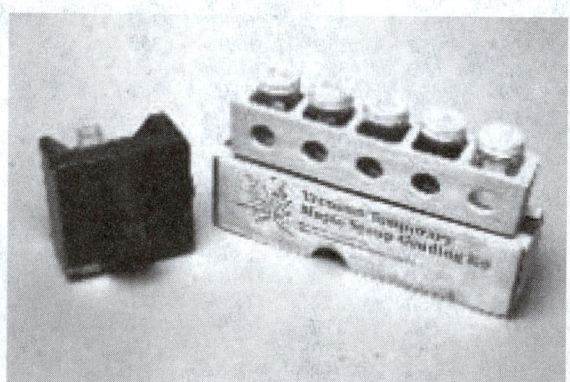

Fig. 18. Permanent glass standard grading kits, such as the Lovibond on the left, and temporary glycerine grading kits, such as the Vermont temporary kit shown on the right, are available for color grading maple syrup.

When using a grading kit, syrup must be placed in the standardized bottles supplied with the kit. Using a different sized or shaped bottle as well as a bottle with thicker or thinner walls may give erroneous results.

Nutritional Information. Pure maple syrup is a 100% natural food, processed by heat concentration of pure maple sap. This sap is a sterile, clear liquid, which provides the trees with water and nutrients prior to the buds and leaves opening in the spring. In the boiling, concentrating, and filtering processes, all the nutrients remain in the syrup. There are some quantitative differences in maple syrup's nutritive composition due to metabolic and environmental differences among maple trees.

Sugars are an important source of energy. The main sugar in pure Maple syrup is sucrose. The darker grades, especially Grade B syrup, contain small and variable amounts of fructose and glucose. In order of sweetness, sucrose is less sweet than fructose, and sweeter than glucose.

Minerals have specific and nonspecific nutritional functions in the body's metabolism. In pure filtered maple syrup the main minerals present are: calcium, potassium, manganese, magnesium, phosphorus, and iron.

Vitamins are essential to maintain health; they cannot be manufactured by the body (except vitamin D) so they must be acquired through food or taken separately. In pure maple syrup trace amounts of vitamins are present, mainly B_2 (riboflavin), B_5 (pantothenic Acid), B_6 (pyridoxine), PP (niacin, B_1), biotin, and folic acid.

Amino Acids are the building blocks of the proteins. In pure maple syrup many amino acids are present in trace amounts.

Total Solids. The total solid's in pure maple syrup amount to 66.5%, with the remaining 33.5% consisting of water. Table 7 represents the main elements within the total solids.

Syrup Packing and Handling. Once maple sap has been processed into maple syrup and the correct density obtained, it is ready for packing. It does not require further treatment beyond the normal filtering process.

TABLE 7. TOTAL SOLIDS 66.5%

Carbohydrates	Percent	Organic Acids	Percent
Sucrose	88–89	Malic	0.141
Hexoses, Fructose,		Citric	0.015
Other Sugars	Trace	Fumaric	0.006
		Unidentified	Trace
Amino Acids	**Parts per Million**	**Vitamins**	**Micrograms/Liter**
Amino Nitrogens	50–280	Niacin (PP, B_1)	276
Phenols	450–1440	Pantothenic Acid (B_5)	600
		Riboflavin (B_2)	60
		Folic Acid	Trace
		Pyridoxine (B_6)	Trace
		Biotin	Trace
		Vitamin A	Trace
Minerals	**Parts per Million**	**Calories**	
Potassium	1300–3900	Maple Syrup	50 per Tablespoon
Calcium	400–2100		
Magnesium	12–360	Compares to:	
Manganese	2–220	Karo Corn Syrup	60 per Tablespoon
Sodium	0–6	Honey	60 per Tablespoon
Phosphorus	79–183	Molasses	40 per Tablespoon
Iron	0–36		
Copper	0–2		

Hot Packing. To prevent contamination of finished syrup by yeast or mold growth, finished syrup should be hot packed. If the temperature of the syrup following filtering is 180 °F (82 °C) or higher it can be packaged immediately. However, if the temperature has fallen below 180 °F it should be reheated to this temperature or slightly higher. It is not recommended to heat the syrup to temperatures greater than 200 °F (93 °C) since some darkening and accompanying loss of grade may occur. By maintaining temperatures above 180 °F when packing, most mold spores are killed, thereby eliminating possible contamination sources. It is important that the entire inside of the container be exposed to 180 °F. This is commonly accomplished by inverting the containers immediately after filling and sealing.

Syrup can be packed directly into containers intended for retail sale, or it can be placed in larger cans or drums (ranging from 5 to 55 gallons or more). These larger size containers may either be sold at wholesale or can be used as sources of syrup for filling retail containers later in year. Most maple producers will find it advantageous to pack some syrup in retail containers during the producing season; however, it is recommended that at least part of the crop be packed in bulk containers for filling later orders. This allows individual orders to be filled on a customized basis while making certain syrup quality at the time of filling is high. When syrup is to be stored for longer periods of time, storage in bulk as opposed to retail containers is preferred. It is very helpful to keep a small container of syrup from each batch in a freezer. This can be used to check the characteristics of the syrup batch without unsealing the drums.

Glass, Plastic and Metal Containers. All syrup offered for retail sale is packed either in glass, plastic or metal containers. Each has advantages as well as disadvantages associated with its use. While the handling and filling of each is generally similar, there are some guidelines regarding the use of each which can be offered.

Metal containers represent maple syrup packaged in the most traditional manner. They are well suited for packing in the 1 quart to 1 gallon size. When metal containers are clean and dry they are very satisfactory. Some producers prefer to rinse all cans with hot water before filling. If this is done they should be thoroughly and rapidly dried before use. If cans are not rinsed, check for loose metal fragments or rust before filling. Once filled, securely install the cap insert and screw on the cap.

Glass containers are preferred by some because they permit the natural color of the syrup to be seen. Glass is most common for 1-quart and smaller sizes. Glass, like all containers, should be stored in an inverted position before filling to avoid possible contamination.

In the past few years, containers made of various food-grade plastics have become the containers of choice for many. They are available in a variety of sizes and offer distinct advantages, particularly the absence of rust and increased resistance to breakage. However, color changes may occur in syrup stored in standard plastic containers for a prolonged period. For this reason, to avoid possible grade change, syrup offered for sale in these plastic containers should not be packed for periods longer than 3 months before the anticipated sale date.

The potential grade change associated with standard plastic containers appears to be associated with the fact that they are somewhat porous to air. Coated plastic containers are now being manufactured and marketed which are nonporous to air.

Stack Burn. The color of pure maple syrup results from a browning reaction which occurs in the latter stages of evaporation. If syrup is packaged hot, this same browning reaction can continue. This may result in changing the grade of medium amber syrup to dark amber or even darker. This darkening of color is known as stack burn. It is usually not a problem with light amber grade syrup.

To avoid stack burn, it is recommended that filled containers be allowed to cool before they are boxed or packed close together. If adequate space is available, this is commonly achieved by separating the filled containers to allow air to freely circulate around them and hasten cooling. Once the syrup and its container has reached room temperature, all containers can be stacked more closely. Some producers have hastened the cooling process by using a fan or by packing syrup in a cool room.

Other Maple Product

A non-table grade of syrup called "commercial" is also produced. This is very dark, with a very strong flavor, sometimes also with off-flavors (metabolism, buddy, ferment). Commercial maple syrup is generally used as a flavoring agent in other products.

Maple syrup and its artificial imitations are the preferred toppings for crêpes, pancakes, waffles, and French toast in North America. Maple syrup can also be used for a variety of uses, including: biscuits, fresh donuts, fried dough, fritters, ice cream, hot cereal, and fresh fruit (especially grapefruit).

It is also used as sweetener for apple sauce, baked beans, candied sweet potatoes, winter squash, cakes, pies, breads, fudge and other candy, milkshakes, tea, coffee and hot toddys.

Most "maple-flavored" syrups on the market today in the United States are imitation maple syrups (table syrups), usually with little (for advertising purposes) or no real maple content. They are usually thickened far beyond the viscosity of real maple syrup, as well. They are less expensive than real maple syrup. US labeling laws prohibit these products from being labeled "Maple Syrup", many simply calling the imitation "Syrup" or "Pancake Syrup". Québécois often refer to these cheap imitations as *Sirop de poteau* ("Pole Syrup"), implying the syrup has been made by tapping telephone poles.

Maple Sugar. Maple sugar is perhaps the most common product produced from maple syrup. In the earlier years of maple syrup production, most maple syrup was processed into maple sugar. This form allowed easier transport and storage while also providing a source of sugar which was used in cooking and baking. Maple syrup could, in essence, be reconstituted by adding water to maple sugar and allowing it to again become soluble.

Maple sugar is produced by heating maple syrup until a boiling temperature of 248 °F (120 °C) is reached. This is 36 °F (2 °C) degrees above the normal boiling point of water (212 °F (100 °C). As soon as the cooked syrup reaches this temperature it is removed from the heat and stirred. Stirring is continued until the solution begins to crystalize and stiffen. At this time it can be poured into molds, either of a small size or large blocks. The sugar will solidify in the cooking container if stirring is continued for too long or the contents of the cooking container are not transferred to molds. Maple sugar is highly hygroscopic (absorbs moisture) and should quickly be stored in dry, sealable containers.

Granulated Maple Sugar. Granulated maple sugar is prepared by heating maple syrup until the boiling point is 45 to 50 °F (7 to 10 °C) above the boiling point of water. It is then stirred either in the cooking vessel or in an appropriately sized container until granulation is achieved. This stirring can be done by hand or by using a mechanical stirring machine. Stirring continues until all moisture is essentially removed from the cooked syrup and crumbly, granulated sugar remains. At this point the product is sieved through a coarse screen (1/8-inch hardware cloth is commonly used) to make a uniform sized product. Granular sugar is hygroscopic (absorbs moisture) and should quickly be stored in dry, sealable containers.

Maple Spread (Maple Cream, Maple Butter). Maple spread, also called *maple cream* or *maple butter* (some believe these latter terms should be discouraged to avoid confusion with dairy products) is an increasingly popular product. Many producers have developed extensive markets for the product and now process much of their annual maple syrup crop into maple spread.

Maple spread, a fondant-type confection, is prepared by elevating the boiling point of maple syrup to a prescribed level, then rapidly cooling the cooked syrup followed by stirring. This procedure results in the formation of very small sugar crystals which together have a "peanut butter consistency." Maple spread is a delectable topping for toast, muffins or other similar products. For best results, the syrup from which maple spread is prepared should be U.S. Grade A Medium Amber or lighter. However, other grades of syrup can be used if they contain less than 4 percent invert sugar.

Maple Candy (Fondant). Maple candy or fondant is a nougat-type candy produced in some areas of the maple region. In a few locations it is locally referred to as maple cream because of its very fine crystalline structure. Maple candy is made in the same manner as maple spread except that the syrup is heated to a higher boiling point (27 °F (-2.8 °C) above the boiling point of water (212 °F (100 °C). The thickened syrup is cooled to 50 °F (10 °C) degrees and stirred as for making spread. Since there is less syrup left in the fondant, it will set up to a soft solid at room temperatures. Small amounts can be dropped on a marble slab, waxed paper or a metal sheet;

or it can be packed into molds. It is customarily packaged and sold in small boxes or canisters.

Molded Sugar. Like maple spread, the popularity of molded sugar has increased. It is a popular confection with a concentrated maple flavor. Like maple spread, approximately eight pounds of molded sugar can be prepared from one gallon of maple syrup.

Molded maple sugar contains little or no free syrup, thus it is stiffer than maple spread. The crystals in molded sugar are larger than in maple spread and are palpable to the tongue, but they should not be large enough to produce an unpleasant sandy effect. Molded sugar can be made from any of the top three grades of syrup. Unlike maple spread, a small amount of invert sugar is desirable because it reduces the tendency to produce large crystals that give the sugar a grainy texture. The invert sugar content can be increased by adding a small amount of dark amber syrup or 1/2 teaspoon of cream of tartar to one gallon of low invert syrup.

See also **Syrups**.

Note: Those interested in a more comprehensive discussion of maple syrup production may wish to obtain a copy of the *North American Maple Syrup Producers Manual,* a 178-page manual dealing with all aspects of maple product production from sugarbush management to marketing. This manual may be purchased through the Ohio State University Extension website. http://ohioline.osu.edu/lines/ebull.html.

Web References

"All About Maple Sugaring" by the Massachusetts Maple Producers Association: http://www.massmaple.org/
Hobby Maple Syrup Production: http://ohioline.osu.edu/for-fact/0036.html
Leader Evaporator Company: http://www.leaderevaporator.com/index.php
North American Maple Syrup Produces Manual: http://ohioline.osu.edu/b856/

MAPLE TREES. Of the family *Aceraceae* (maple family), the maple tree is native to the United States and Canada. There are 13 or more species of these trees in the United States (Table 1). The sugar maple is found in large numbers throughout most of the central parts of the United States. However, the species, *A. saccharum*, the major source of maple syrup is found mainly in Vermont, New Hampshire, New York, and Ohio, and to a lesser extent in the southern states. See also **Maple Syrup**.

TABLE 1. MAPLE SPECIES NATIVE TO THE UNITED STATES

Common Name	Scientific Name	Geographic Distribution
Sugar Maple	*Acer saccharum*	Northeast United States and Southern Canada
Black Maple	*Acer nigrum*	Northeast United States and Southeast Canada
Red Maple	*Acer rubrum*	Eastern United States and Southeast Canada
Silver Maple	*Acer saccharinum*	Eastern United States and Southeast Canada
Boxelder	*Acer negundo*	Eastern and Central United States and Canada
Mountain Maple	*Acer spicatum*	Northeast United States and Southeast Canada
Striped Maple	*Acer pensylvanicum*	Northeast United States and Southeast Canada
Bigleaf Maple	*Acer macrophyllum*	Pacific Coast United States and Canada
Chalk Maple	*Acer leucoderme*	Southeast United States
Canyon Maple	*Acer grandidentatum*	U.S. Rocky Mountains
Rocky Mountain Maple	*Acer glabrum*	Western United States
Vine Maple	*Acer circinatum*	Pacific Coast of United States and Canada
Florida Maple	*Acer barbatum*	Southeast United States Coastal Plain and Piedmont

Table 2 contains a descriptive comparison and Figures 1-4 illustrate characteristic leaves, bark, twigs, and fruits of sugar, black, red and silver maple. These four species share several characteristics in common. All have leaves of similar shape: a single leaf blade with the characteristic maple shape, 3-5 lobes radiating out like fingers from the palm of a hand (palmately lobed) with notches (called sinuses) between the lobes. Like all

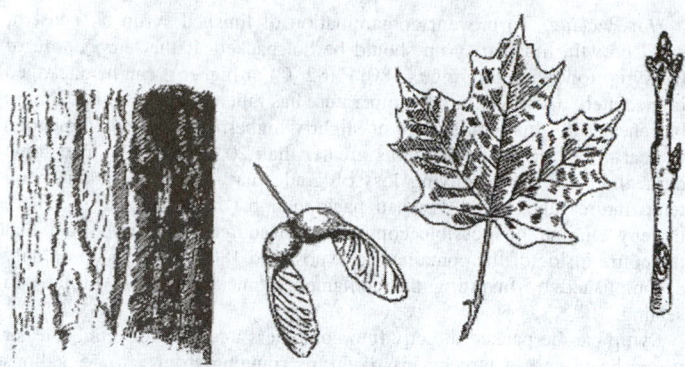

Fig. 1. Sugar maple bark, fruit, leaf, and twig. (Courtesy of the Massachusetts Maple Producers Association and Ohio State University).

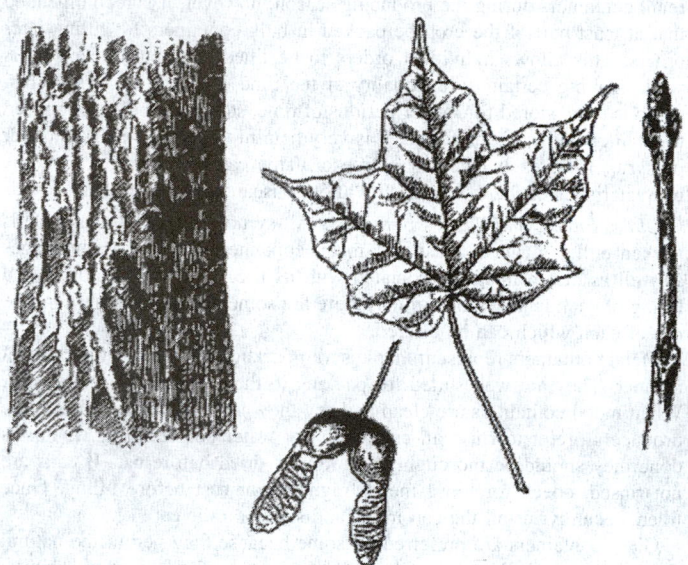

Fig. 2. Black maple bark, fruit, leaf, and twig. (courtesy of the Massachusetts Maple Producers Association and Ohio State University).

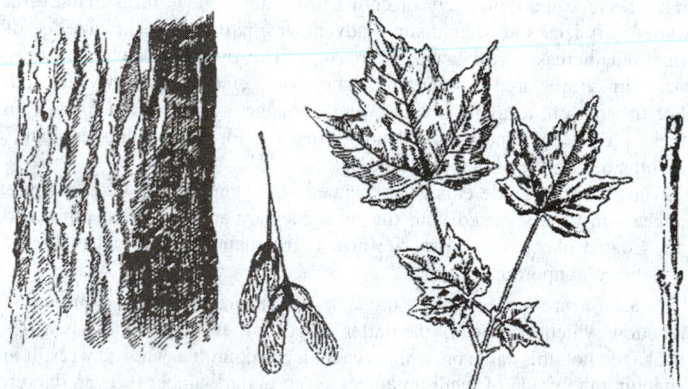

Fig. 3. Red maple bark, fruit, leaf, and twig. (courtesy of the Massachusetts Maple Producers Association and Ohio State University).

maples, the leaves, buds and twigs of all four are attached in pairs opposite each other along the branches. Also, all four produce a fruit called a samara (or double samara), which is a pair of connected, winged seeds.

Species Suitable for Maple Product Production

Sugar and Block Maple. Sugar (*Acer saccharum*) and Black Maple (*Acer nigrum*) are very similar species and unquestionably the most preferred species for producing maple products, primarily because of their high sugar content. Sugar maple occurs naturally throughout most of the northeastern

TABLE 2. IDENTIFYING CHARACTERISTICS OF SUGAR, BLACK, RED, AND SILVER MAPLE

Species	Leaf	Bark	Twig	Fruit
Sugar Maple	3-5 inches wide; 5lobed (rarely 3-lobed); bright green upper surface and a paler green lower surface; leaf margin without fine teeth (compare with red and silver maple).	Young trees up to 4-8 inches with smooth gray bark. Older trees developing furrows and ultimately long, irregular, thick vertical plates that appear to peal from the trunk in a vertical direction.	A somewhat shiny, brownish, slender, relatively smooth twig with 1/4-3/8 inch long sharply pointed terminal bud.	Horseshoe-shaped double-winged fruit with parallel or slightly divergent wings. Winged seed approximately 1″ long. Fruits mature in fall.
Black Maple	Similar to sugar maple but usually 3-lobed (sometimes five); often appears to be drooping; often with a thicker leaf and lear stem (petiole) than sugar maple; usually with two winglike or leaflike growths at the base of the petiole (stipules).	Similar to sugar maple but usually darker and more deeply grooved or furrowed.	Similar to sugar maple but twig surface with small warty growths (lenticels, which are not raised much above the bark surface in sugar maple) and often more hairy buds.	Similar to sugar maple with, perhaps, a slightly larger seed.
Red Maple	2-6 inches wide; 3lobed (occasionally weakly 5-lobed); sharply V-shaped sinuses; small sharp teeth along margin. Mature leaves have a whitish appearing underside.	Young trees up to 4-8 inches with a smooth light gray bark, developing into gray or black ridges and ultimately narrow scaly plates.	Slender, shiny, usually reddish in color; terminal buds 1/8-1/4 inch long, blunt, red; odorless if bark bruised or scraped.	V-shaped, double-winged fruit about 1/2-1 inch long. Fruit matures in spring.
Silver Maple	5-7 inches wide; deeply clefted; 5-lobed with the sides of the terminal lobe diverging toward the tip; light green upper surface and a silvery white underside; leaf margin with fine teeth (but not the inner edges of the sinuses).	Silvery gray on young trees breaking into long thin scaly plates that give the trunks of older trees a very shaggy appearance. Considerable red is seen in bark pattern as scales develop.	Similar to red maple but bruised or scraped bark has a very fetid or foul odor.	V-shaped, double-winged fruit 1 1/2 to 2 inches long, with widely divergent wings. One of two seeds present is often poorly developed or aborted. Fruit matures in spring.

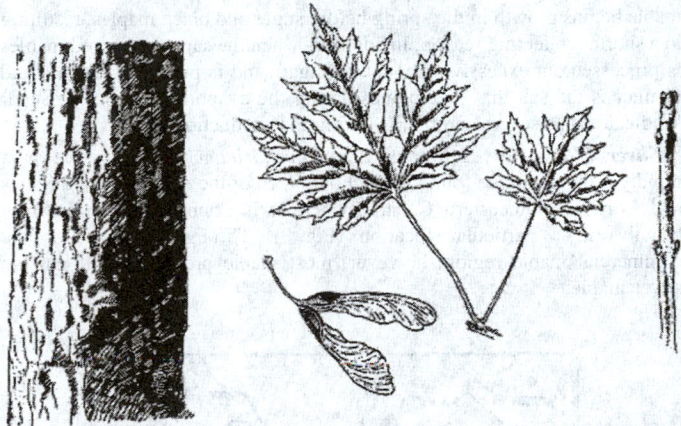

Fig. 4. Silver maple bark, fruit, leaf, and twig. (Courtesy of the Massachusetts Maple Producers Association and Ohio State University).

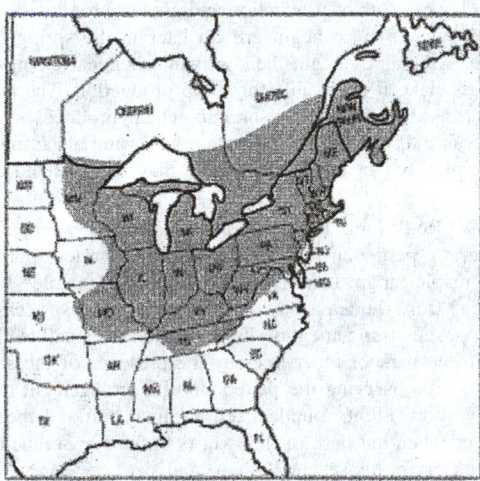

Fig. 5. Geographic range of sugar maple in North America. (Courtesy of the Massachusetts Maple Producers Association and Ohio State University).

United States and southeastern Canada (Fig. 5). Black maple, on the other hand, occupies a much smaller natural range (Fig. 6). Distinguishing between them may be more of an academic exercise than one useful in sugarbush management because (1) they are essentially identical in quality as sugartrees, and (2) they often hybridize producing trees with a range of characteristics, making it difficult to clearly distinguish between them.

Identifying a tree as a sugar or black maple (Table 2, and Fig. 1 and 2) is easily done from the leaves by observing 5-lobed leaves, the paired opposite attachment of the leaves along the stem and the lack of teeth along the leaf margin; from the bark of older trees by observing the long plates that remain attached on one side; from the twigs by observing the opposite arrangement of buds and the relatively long, pointed, brownish terminal bud; and from the seed by observing its horseshoe shape and size. Distinguishing between sugar and black maple is best done by comparing the leaf structure (particularly the number of lobes, droopiness and presence or absence of stipules along base of petiole) and by the degree of bumpiness of the twigs.

Sugar and black maples are found on a variety of soils and site conditions, but neither tolerates excessively wet or dry sites, and both grow best on moist, deep, well-drained soils. Black maple is more likely to be found along moist river bottoms. Both species can be found growing in pure stands, with each other, or with a wide variety of other hardwood species including American beech, American basswood, yellow birch, black cherry, northern red oak, yellow poplar and black walnut. Both species have been planted extensively as roadside trees which are often tapped as part of a sugaring operation. Plantations of sugar maple have also been established with the intent of developing efficient, productive sugarbushes. Both species are relatively long lived, capable of living well beyond 200 years, with trunk diameters greater than 30 inches (76 centimeters) and heights greater than 100 feet (30.5 meters).

Sugar and black maple both grow in the shade of other trees (they are shade tolerant), and trees of many different ages (sizes) are often found in a forest. Both species are also found in stands composed of trees that are essentially all the same age (size). Healthy sugar and black maple trees growing in overstocked uneven-aged or even-aged stands can be expected to achieve tapable size in 40 to 60 years, depending on overall site quality. Thinning or release cutting dramatically reduces this age-to-tapable-size.

Sugar and black maple are particularly attractive as sugartrees because of their high sap sugar content and the late date at which they begin growth in the spring. Sugar and black maple have the highest sap sugar content of

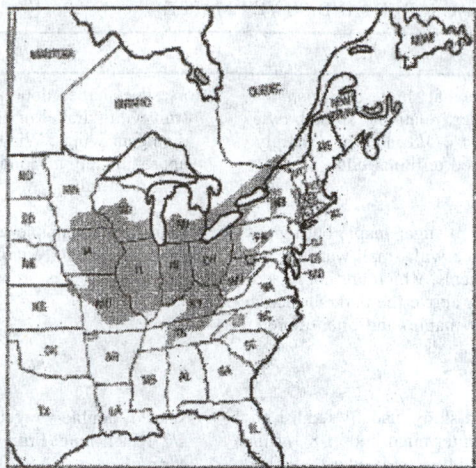

Fig. 6.   Geographic range of black maple in North America. (Courtesy of the Massachusetts Maple Producers Association and Ohio State University).

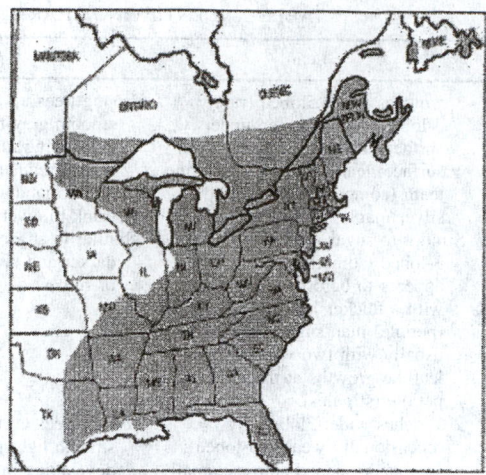

Fig. 7.   Geographic range of red maple in North America. (Courtesy of the Massachusetts Maple Producers Association and Ohio State University).

any of the native maples. While the exact sap sugar content of a tree will vary depending on many factors including genetics, site and weather, sugar and black maples generally average between 2.0 and 2.5 percent sap sugar content. It is not unusual to find many trees in a sugarbush well in excess of 3 percent, and occasionally higher. Genetic research on sugar maple suggests that the sap sugar content of planted seedlings can be increased by controlled breeding. Other things being equal, higher sap sugar content translates to lower costs of production and greater profits.

Black and sugar maples begin growth later in the spring than red or silver maple. As maples begin their growth, chemical changes occur in the sap which make it unsuitable for syrup production. The term "buddy sap" is often applied to late season sap which produces syrup with a very disagreeable flavor and odor. Because sugar and black maple resume growth later than red or silver maple, sap may be collected later in the spring.

Red Maple. Red Maple (*Acer rubrum*) is commonly tapped in certain geographic areas, particularly in the southern and western portions of the commercial maple range. Identifying a tree as a red maple (Table 2, and Fig. 3) is done from the leaves by observing the 3 lobes (occasionally 5), the paired opposite arrangement of the leaves and the small teeth along the margin; from the bark of older trees by the presence of the scaly plates; from the twig by observing the pairedopposite arrangement of the buds, the relatively short, blunt, rounded, red terminal bud and the lack of an offensive odor when the bark of the twig is bruised or scraped; and from the fruit by observing its severe V-shape and size.

Red maple is one of the most abundant and widespread hardwood trees in North America (Fig. 7). Probably no other species of forest tree, certainly no hardwood, can thrive on a wider variety of soil types and sites. Although it develops best on moderately well-drained to well-drained, moist soils, it commonly grows in conditions ranging from dry ridges to swamps. Because of the wide variety of sites on which red maple will grow, it is found growing naturally in pure stands and with an enormous variety of other tree species ranging from gray birch and paper birch, to yellow poplar and black cherry, and including sugar and black maple. Its rapid growth and ability to thrive on a wide variety of sites have resulted in its widespread planting as ornamental and street trees which are often tapped as part of a sugaring operation.

Compared to sugar and black maple, red maple is a relatively short-lived tree, rarely living longer than 150 years. Mature trees commonly average between 20 and 30 inches (51 and 76 centimeters) in diameter and 60 and 90 feet tall (18 and 27 meters). Like sugar and black maple, red maple is shade tolerant and is found in both even-aged and uneven-aged forests. Thinning or release cutting will substantially shorten the age-to-tapable-size.

From the perspective of producing maple syrup, red maple's most attractive characteristic is its ability to thrive on a wide variety of site conditions. In some areas of the commercial maple range, red maple is the only maple present on many sites. One either taps red maple or they do

not sugar. In other areas, red maple may be tapped along with sugar and black maples.

It is important to emphasize that good, high-quality maple syrup can be made from red maple sap. However, for sugaring, red maple does have three important weaknesses. First, the sap sugar content of red maple will be less, on the average, than that of nearby comparable sugar or black maples, perhaps by 1/2 percent or more. This lower sap sugar content translates to higher costs of production and lower profits. Secondly, red maple begins growth in the spring before sugar and black maples, resulting in a shorter collecting season. In addition, when the sap of some red maples is processed, an excessive amount of sugar sand is produced. Sugar sand or niter is the salt that precipitates during the evaporation process. Sugar sand can cause several problems during the production process.

Silver Maple. Silver Maple (*Acer saccharinum*) Silver maple is a rapidly growing maple found throughout much of the eastern United States and extreme southeastern Canada, where it is often tapped (sometimes heavily) in a particular location (Fig. 8). Throughout much of the commercial maple region, however, most maple producers will not tap silver maple.

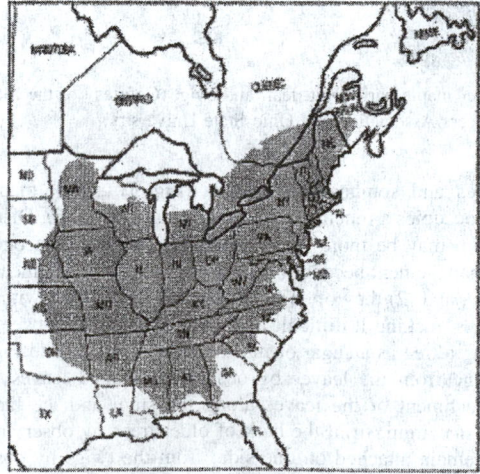

Fig. 8.   Geographic range of silver maple in North America. (Courtesy of the Massachusetts Maple Producers Association and Ohio State University).

Identifying a silver maple (Table 2, Fig. 4) is done from the leaves by observing the 5 lobes with the sides of the terminal lobe diverging toward the tip, the paired opposite arrangement of the leaves, the presence of fine teeth along the margin but not on the inner sides of the sinuses and the silvery white underside; from the bark of older trees by the trunk's shaggy appearance; from the twigs by observing the paired opposite arrangement of the buds, the relatively short blunt, rounded, red terminal bud and the presence of a fetid or foul odor when the twig is bruised or scraped; and from the fruit by observing its V-shape and size.

Under natural conditions, silver maple is primarily a bottom land and flood plain species, where it may occur in pure stands but is more commonly found associated with other bottom species such as American elm, sweetgum, pin oak, swamp white oak, eastern cottonwood, sycamore, and/or green ash. Silver maple is among the fastest growing hardwood species commonly planted in eastern North America, certainly the fastest growing maple. For this reason, it has been widely planted as an ornamental and street tree. Its use as an ornamental and street tree, at least in urban areas, has been discontinued in recent years because the wood of silver maple is very brittle and often breaks in severe wind, snow or ice storms. Nevertheless, large silver maple street trees are numerous in many areas and these are sometimes tapped as part of a sugaring operation.

Like the red maple, silver maple is a relatively short-lived tree when compared to the sugar or black maple, living perhaps 130-150 years. Because of its fast growth rate, however, mature trees can achieve diameters in excess of 3 feet (0.9 meter) and heights in excess of 100 feet (30.5 meters). On good sites with little competition from other trees, silver maple diameter growth may approach 1/2 inch (1.3 centimeters) per year (rates as high as 1 inch (2.5 centimeters) per year have been recorded). Silver maple's growth rate often responds dramatically to thinning or release cutting.

When compared to sugar, black and red maple, silver maple is a distinctly fourth choice for sugaring for several reasons. First, its sugar content is usually lower than red maple's, perhaps as much as 1/2 percent or more, which means even higher production costs and lower profits. Second, like red maple, it begins growth in the spring, earlier than sugar and black maple, resulting in a shorter collecting season. Third, like red maple, the evaporation of sap from some silver maples produces an excessive amount of sugar sand.

Other Maples

The maple when fully grown is quite a large tree, spreading, pendulous, and graceful. The branches are long, trim, and narrow. The leaf is about 3 to 4 inches (7.6 to 10 centimeters) across, 3- or 5-lobed, and relatively dense. Distinctive of shape, the leaf is beautifully portrayed on the flag of Canada. The leaf may be simple or compound, depending upon the species. The fruit is green, winged, and full centered. The wood is close-grained, hard, and light in color. It weighs nearly 40 pounds per cubic foot (640 kilograms per cubic meter) and is strong. The wood is used for numerous objects that require a durable and high-quality material. The species *A. macrophyllum* (Bigleaf Maple) is native to the western United States and is found in abundance in the Columbia River area. Several million board feet are shipped annually. The trees are quite large. The leaf is very long, from 8 to 12 inches (20.3 to 30.4 centimeters) wide, with stalks about 10 to 12 inches (25 to 30.4 centimeters) long. Along the Pacific coast, moisture from winter rains and summer fogs usually assures full, healthy growth.

The box elder (*Acer negundo*), is a fast-growing tree that drops its leaves early, is also a member of the family Aceraceae.

In recent years, tree scientists have become increasingly concerned with a condition that is sometimes referred to as *maple decline*, an insidious disease, which seems to prefer street and landscape plantings. Dr. Robert Norgren, a plant pathologist with the Wisconsin Department of Agriculture who compiles statistics on diseased trees has observed, "Maple decline will surpass Dutch elm disease, in both complexity and severity, for many of these [North-Central and Northeastern] states." Diseased trees show a decay of the trunk near the base, frequently below the soil line. Inspection has shown that the affected areas are frequently located on the root collar and sometimes on the large roots extending from the trunk. The decay is sometimes characterized by bark discoloration and looseness. Underneath healthy-appearing bark will be found infected wood and cambium that is dark brown or red in contrast with the normal healthy white. Laboratory analyses of infected trunk samples have shown two fungi to be involved, *Phytophthora* and *fusarium*. As pointed out by R. Dries (*American Forests*, 49–63, May 1982), for trees that show no decline symptoms, treatment may include carefully removing the soil from the root-collar area, leaving this area exposed, or filling it with a compost of hardwood bark. During dry periods, the tree should be watered. Further details are given by Dries in his article, "The Battle Against Maple Decline."

Striped maple (*Acer pensylvanicum*) and Mountain Maple (*Acer spicatum*) are two other native maples that are found growing within the commercial maple range (Figs. 9 and 10). Neither of these species is commonly tapped. Striped maple is a small slender tree which rarely attains tapable size. It is most easily identified by the opposite paired arrangement of its leaves and branches, its 3-lobed leaf with fine teeth on the margin, and striping on the branches and young trunks. Mountain maple is essentially a shrub. It is most easily identified by the opposite paired arrangement of its leaves and branches and its 3lobed leaf with coarse teeth. If these species occur in a sugarbush it is important to be able to identify them. They should not be confused with the desirable maple species when performing management practices such as thinning or release cuts.

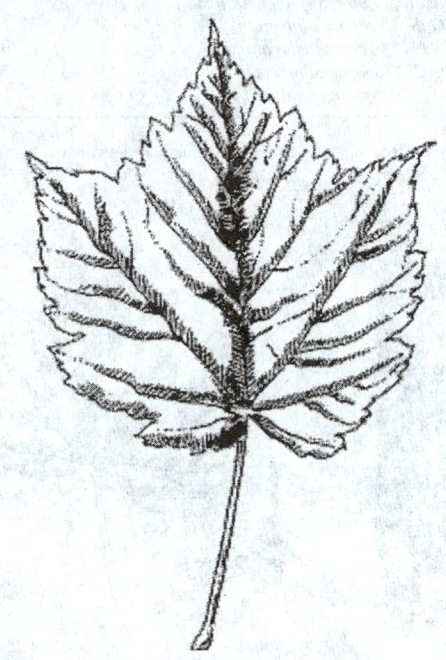

Fig. 9. Striped maple leaf. (Courtesy of the Massachusetts Maple Producers Association and Ohio State University).

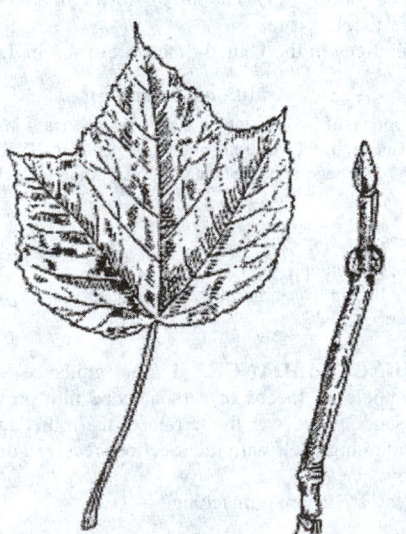

Fig. 10. Mountain maple leaf and twig. (courtesy of the Massachusetts Maple Producers Association and Ohio State University).

One exotic maple, Norway maple (*Acer platanoides*), is commonly planted as an ornamental and street tree and will attain tapable size. It is recognized by the opposite paired arrangements of its leaves and branches,

TABLE 3. RECORD MAPLE TREES IN THE UNITED STATES[1]

Specimen	Circumference[2] Inches	Circumference[2] Centimeters	Height Feet	Height Meters	Spread Feet	Spread Meters	Location
Bigleaf maple (1995) (*Acer macrophyllum*)	419	1064	101	30.8	90	27.4	Clatsop County, OR.
Black maple (1987) (*Acer nigrum*)	198	503	118	36	127	38.7	Allegan County, MI.
Canyon maple (2005) (*Acer grandidentatum*)	140	356	80	24.4	58	17.7	Tonto National Forest, AZ.
Chalk maple (1999) (*Acer leucoderme*)	34	86	54	16.5	50	15.2	Thompson Mills Forest, GA.
Chalk maple (2003) (*Acer leucoderme*)	38	97	50	15.2	42	12.8	Mount Berry, GA.
Florida maple (2003) (*Acer barbatum*)	146	371	91	27.7	75	22.9	Mount Berry, GA.
Mountain maple (1982) (*Acer spicatum*)	33	84	58	17.7	31	9.4	Houghton County, MI.
Mountain maple (2001) (*Acer spicatum*)	34	86	62	18.9	20	6.1	Tray Mountain, Georgia
Norway maple(2001) (*Acer platanoides*)	212	538.5	79	24.1	100	30.5	Kenyon College, OH.
Red maple (1997) (*Acer rubrum*)	276	701	141	43	88	26.8	Great Smoky Mountains National Park, TN.
Rocky Mountain maple (1996) (*Acer glabrum*)	107	272	67	20.4	55	16.8	Island County, WA.
Silver maple (2003) (*Acer saccharinum*)	347	881.4	115	35.1	61	18.6	Newberry, MI.
Striped maple (1997) (*Acer pensylvanicum*)	44	112	77	23.5	31	9.4	Great Smoky Mountains National Park, TN.
Sugar maple (2003) (*Acer saccharum*)	223	566.4	115	35.1	89	27.1	Lime, CT.
Vine maple (2003) (*Acer circinatum*)	38	96.5	64	19.5	37	11.3	Olympic National Park, WA.

[1]From the "National Register of Big Trees," American Forests (by permission).
[2]At 4.5 feet (1.4 meters).

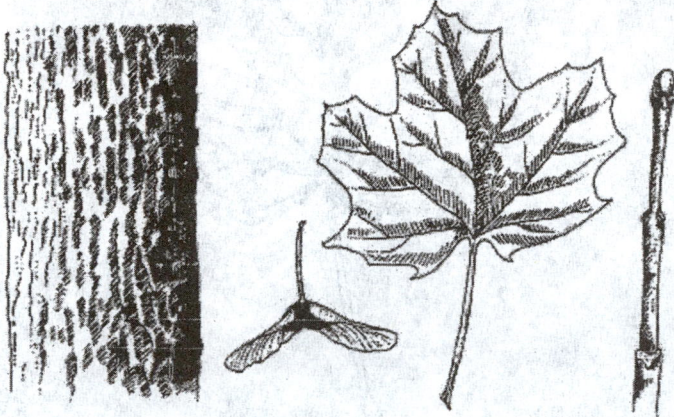

Fig. 11.   Norway maple bark, fruit, leaf, and twig.

its 7 lobed leaf without marginal teeth, and its 1 1/2 to 2 inch long samara with divergent wings (Fig. 11). The sap of Norway maple is not commonly used to produce maple syrup.

Record maple trees in the United States are listed in Table 3.

Additional Reading

The best overall source of information on maple tree species is *Maples of the World*, by D.M. van Gelderen, P.C. de Jong and H.J. Oterdoom, Timber Press, Portland OR, 1994. This 500 page book has data on every known species and cultivar of maple trees.

Web References

Maple-Trees.com: http://maple-trees.com/
National Register of Big Trees: http://www.americanforests.org/resources/bigtrees/register.php?sort=

MAP-MATCHING GUIDANCE. 1. The guidance of a rocket or aerodynamic vehicle by means of a radarscope film previously obtained by a reconnaissance flight over the terrain of the route, and used to direct the vehicle by aligning itself with radar echoes received during flight from the terrain below.

2. Guidance by stellar map matching.

MAPPING (Mathematics). A mapping f from a set D to a set R is a correspondence which associates with each point x in the set D a unique point $f(x)$ in R called the *image* of x. The set D is said to be the *domain* of f, while R is said to be the range of f. The mapping f is said to be *onto* or *subjective* if every point y in R is the image of some point x in D, and *one-to-one* or *injective* if for x_1 and x_2 in D with $x_1 \neq x_2$ it follows that $f(x_1) \neq f(x_2)$. See also **Isomorphism**.

MAP PROJECTIONS. The numerous methods for representing the surface of the earth on a plane. The term originated from the geometric-projection methods of drawing lines from some point out through the surface of the earth to a developable surface. The surfaces most commonly used in the geometric projections were the plane, the cylinder, and the cone. In most cases, the distortions introduced by the strictly geometric projections are intolerable for modern purposes, and the term map projection has been expanded to include all sorts of purely mathematical methods for representing the surface of the earth on a plane. The pattern of lines on the map or chart representing meridians of longitude and parallels of latitude form what is known as the graticule of the projection.

Additional Reading

Pearson, F.: *Map Projection: Theory and Applications*, CRC Press, LLC., Boca Raton, FL, 1999.
Synder, J.P.: *Flattening the Earth: Two Thousand Years of Map Projections*, University of Chicago Press, Chicago, IL, 1997.
Yang, O.H. and J.P. Snyder: *Map Projection Transformation: Principles and Applications*, Taylor & Francis, Inc., Philadelphia, PA, 1999.

MARANGONI EFFECT. See **Foam**.

MARATHON-HOWARD PROCESS. A treatment of waste sulfite liquor from sulfite pulp manufacture to recover chemicals and reduce steam pollution. The waste sulfite is treated with line and precipitates. (1) calcium sulfite for use in preparing fresh cooking acid for the sulfite pulp process, and (2) a basic calcium salt of lignin sulfonic acid (lignin sulfonates) that can be pressed and used as a fuel of used as raw material for vanillin, lignin plastics, and other chemicals. The remaining liquor with its BOD reduced 80% is the effluent.

MARBLE. A metamorphic form of calcium carbonate, usually containing admixtures of iron and other minerals that impart variegated color patterns. Marble chips are often used as source of carbon dioxide in laboratory experiments.

MARBURG-EBOLA VIRUS DISEASE. First discovered in Marburg, West Germany, this highly fatal disease is caused by the Marburg and Ebola viruses carried by vervet monkeys. The disease was recognized in 1967 among German animal handlers and laboratory personnel, as well as by hospital workers who treated victims of the disease. Cases also appeared among similar workers in Yugoslavia. Investigation indicated that the disease was spread among persons who were in contact with monkeys shipped from Uganda in 1967. In 1976, 500 cases of the disease were reported in Zaire and the Sudan, particularly in persons who lived in the Ebola River valley. Mortality with the Marburg agent is estimated to be 30%; with the Ebola virus, 85%.

The onset of the disease is abrupt with severe headache, sore throat, high fever and myalgia, followed by rapid prostration, and dehydration from diarrhea and vomiting. A generalized rash appears about the fifth day and lasts for two to three days. Severe bleeding occurs from the gastrointestinal tract and lungs. Death usually eventuates between the 7th and 16th days. Autopsy shows necrotic cellular lesions in the liver and other organs, including the kidneys.

Although monkey organs were the source of the first infection, primates are not thought to be the natural reservoirs since no infection developed in people who had handled whole monkeys. Experimental infection of several species of primates with body fluids from patients with the disease produced a uniformly fatal infection and the course of the illness was similar to that in humans. No natural reservoir of the virus has yet been discovered.

Care is primarily supportive, although antiviral agents may have a place in therapy. Barrier nursing and infection-control precautions are required; it is most important to handle blood and other body fluids and contaminated needles with great care.

Travel into endemic areas during periods of disease activity should be minimized and quarantine of vervet monkeys coming from such areas is mandatory.

R. C. V.

MARBURG HEMORRHAGIC FEVER.

Marburg hemorrhagic fever is a rare, severe type of hemorrhagic fever which affects both humans and nonhuman primates. Caused by a genetically unique zoonotic (that is, animal-borne) RNA virus of the filovirus family, its recognition led to the creation of this virus family. The four species of Ebola virus are the only other known members of the filovirus family.

The Marburg virus was first recognized in 1967, when outbreaks of hemorrhagic fever occurred simultaneously in laboratories in Marburg and Frankfurt, Germany and in Belgrade, Yugoslavia (now Serbia). A total of 37 people became ill; they included laboratory workers as well as several medical personnel and family members who had cared for them. The first people infected had been exposed to African green monkeys or their tissues. In Marburg, the monkeys had been imported for research and to prepare polio vaccine.

Recorded cases of the disease are rare, and have appeared in only a few locations. Although the 1967 outbreak occurred in Europe, the disease agent had arrived with imported monkeys from Uganda. No other case was recorded until 1975, when a traveler, most likely exposed in Zimbabwe, became ill in Johannesburg, South Africa and passed the virus to his travelling companion and a nurse. Two other cases were seen in 1980. One in Western Kenya not far from the Ugandan source of the monkeys implicated in the 1967 outbreak. This patient's attending physician in Nairobi became the second case. Another human Marburg infection was recognized in 1987 when a young man who had traveled extensively in Kenya, including western Kenya, became ill and later died.

The Marburg virus is indigenous to Africa. While the geographic area to which it is native is unknown, this area appears to include at least parts of Uganda and Western Kenya, and perhaps Zimbabwe. As with the Ebola virus, the actual animal host for the Marburg virus also remains a mystery. Both of the men infected in 1980 in western Kenya had traveled extensively, including making a visit to a cave, in that region. The cave was investigated by placing sentinels animals inside to see if they would become infected, and by taking samples from numerous animals and arthropods trapped during the investigation. The investigation yielded no virus. The sentinel animals remained healthy and no virus isolations from the samples obtained have been reported.

Just how the animal host first transmits Marburg virus to humans is unknown. However, as with some other viruses which cause viral hemorrhagic fever, humans who become ill with Marburg hemorrhagic fever may spread the virus to other people. This may happen in several ways. Persons handling infected monkeys, who come into direct contact with them or their fluids or cell cultures, have become infected. Spread of the virus between humans has occurred in a setting of close contact, often in a hospital. Droplets of body fluids, or direct contact with persons, equipment, or other objects contaminated with infectious blood or tissues are all highly suspect as sources of disease.

After an incubation period of 5–10 days, the onset of the disease is sudden and is marked by fever, chills, headache, and myalgia. Around the fifth day after the onset of symptoms, a maculopapular rash, most prominent on the trunk (chest, back, stomach), may occur. Nausea, vomiting, chest pain, a sore throat, abdominal pain, and diarrhea then may appear. Symptoms become increasingly severe and may include jaundice, inflammation of the pancreas, severe weight loss, delirium, shock, liver failure, massive hemorrhaging, and multiorgan dysfunction.

Because many of the signs and symptoms of Marburg hemorrhagic fever are similar to those of other infectious diseases, such as malaria or typhoid fever, diagnosis of the disease can be difficult, especially if only a single case is involved.

Antigen-capture enzyme-linked immunosorbent assay (ELISA) testing, IgM-capture ELISA, polymerase chain reaction (PCR), and virus isolation can be used to confirm a case of Marburg hemorrhagic fever within a few days of the onset of symptoms. The IgG-capture ELISA is appropriate for testing persons later in the course of disease or after recovery. The disease is readily diagnosed by immunohistochemistry, virus isolation, or PCR of blood or tissue specimens from deceased patients.

Recovery from the Marburg hemorrhagic fever may be prolonged and accompanied by orchitis, recurrent hepatitis, transverse myelitis or uvetis. Other possible complications include inflammation of the testis, spinal cord, eye, parotid gland, or by prolonged hepatitis. The case fatality rate for the Marburg hemorrhagic fever is between 23–25%.

A specific treatment for this disease is unknown. However, supportive hospital therapy should be utilized. This includes balancing the patient's fluids and electrolytes, maintaining their oxygen status and blood pressure, replacing lost blood and clotting factors and treating them for any complicating infections.

Sometimes, treatment also used has been transfusion of fresh-frozen plasma and other preparations to replace the blood proteins important in clotting. One controversial treatment is the use of heparin (which blocks clotting) to prevent the consumption of clotting factors. Some researchers believe the consumption of clotting factors is part of the disease process.

People who have close contact with a human or nonhuman primate infected with the virus are at risk. Such persons include laboratory or quarantine facility workers who handle nonhuman primates that have been associated with the disease. In addition, hospital staff and family members who care for patients with the disease are at risk if they do not use proper barrier nursin-Q techniques.

Due to our limited knowledge of the disease, preventive measures against transmission from the original animal host have not yet been established. Measures for prevention of secondary transmission are similar to those used for other hemorrhagic fevers. If a patient is either suspected or confirmed to have Marburg hemorrhagic fever, barrier nursing techniques should be used to prevent direct physical contact with the patient. These precautions include wearing of protective gowns, gloves, and masks; placing the infected individual in strict isolation; and sterilization or proper disposal of needles, equipment, and patient excretions.

In conjunction with the World Health Organization, the CDC has developed practical, hospital-based guidelines, titled *Infection Control for Viral Haemorrhagic Fevers in the African Health Care Setting*. The manual can help health-care facilities recognize cases and prevent further hospital-based disease transmission using locally available materials and few financial resources.

Marburg hemorrhagic fever is a very rare human disease. However, when it does occur, it has the potential to spread to other people, especially health care staff and family members who care for the patient. Therefore, increasing awareness among health-care providers of clinical symptoms in patients that suggest Marburg hemorrhagic fever is critical. Better awareness can help lead to taking precautions against the spread of virus infection to family members or health-care providers. Improving the use of diagnostic tools is another priority. With modern means of transportation that give access even to remote areas, it is possible to obtain rapid testing of samples in disease control centers equipped with Biosafety Level 4 laboratories in order to confirm or rule out Marburg virus infection.

A fuller understanding of the Marburg hemorrhagic fever will not be possible until the ecology and identity of the virus reservoir are established. In addition, the impact of the disease will remain unknown until the actual incidence of the disease and its endemic areas are determined. See also **Ebola Hemorrhagic Fever**; **Filoviruses**; and **Viral Hemorrhagic Fevers**.

Additional Reading

Fields, M.B.N., D.M. Knipe, P.M. Howley, and R.M. Chanock: *Fields Virology*, 3rd Edition, Lippincott Williams & Wilkins, Philadelphia, PA, 1996.

Galasso, G.J., T.C. Merigan, and R.J. Whitley: *Antiviral Agents and Human Viral Diseases*, 4th Edition, Lippincott Williams & Wilkins, Philadelphia, PA, 1997.

Gear, J.H.S.: *Handbook of Viral and Rickettsial Hemorrhagic Fever*, CRC Press, LLC, Boca Raton, FL, 1988.

Love, C.B. and P.B. Jahrling: *Viral Hemorrhagic Fever*, DIANE Publishing Company, Collingdale, PA, 1996.

Pattison, J.R., J.E. Banatvala, and A.J. Zuckerman: *Principles and Practice of Clinical Virology*, 4th Edition, John Wiley & Sons, Inc., New York, NY, 2000.

Richman, D.D., R.J. Whitley, and F.G. Hayden: *Clinical Virology*, Harcourt Brace & Company, San Diego, CA, 1998.

Voyles, B.A.: *The Biology of Viruses*, Mosby-Year Book, Inc., St Louis, MO, 1993.

Centers for Disease Control and Prevention (CDC), Atlanta, GA

MARCASITE. The mineral marcasite, sometimes called white iron pyrites, is, like ordinary pyrites, disulfide of iron corresponding to the same formula, FeS_2. Marcasite, however, crystallizes in the orthorhombic system often yielding serrate, spear-shaped twins, hence the name "cock's comb pyrites." It is a brittle mineral; hardness, 6–6.5; specific gravity, 4.92; luster, metallic; color, light bronze-yellow; streak, greenish-black; opaque. Marcasite alters very easily and may disintegrate with the formation of sulfuric acid and iron sulfate. Fossils replaced by marcasite are therefore often destroyed after being placed in collections. Marcasite is found in numerous places in Europe, notably in Czechoslovakia, France, and England; in Mexico; and in the United States in the lead districts of Illinois, Wisconsin, and Missouri. The name marcasite is believed to be of Arabic origin and formerly was applied to common pyrite.

MARCHING PROBLEM. A differential equation with initial conditions solved numerically by computing the values of the dependent variable recursively for systematically increasing values of the independent variable. For example, the wave equation is solved at each time-step before advancing to the next time-step. Hyperbolic equations may be formulated as marching problems. See also **Jury Problem**.

MARCONI, GUGLIELMO (1874–1935). Marconi was born in Bologna, Italy. He was educated privately at Bologna, Florence and Leghorn. Even as a boy he took a keen interest in physical and electrical science and studied the works of Maxwell, Hertz, Righi, Lodge and others. He began his scientific experiments in 1895, by setting up an electrical laboratory in the attic of his father's country estate, where he succeeded in sending wireless signals over a distance of one and a half miles, thus becoming the inventor of the first practical system of wireless telegraphy.

In 1896 Marconi took his apparatus to England where he was introduced to Mr. (later Sir) William Preece, Engineer-in-Chief of the Post Office, and later that year was granted the world's first patent for a system of wireless telegraphy. He demonstrated his system successfully in London, on Salisbury Plain and across the Bristol Channel, and in July 1897 formed The Wireless Telegraph & Signal Company Limited (in 1900 re-named Marconi's Wireless Telegraph Company Limited). In 1899 he established wireless communication between France and England across the English Channel. He erected permanent wireless stations at The Needles, Isle of Wight, at Bournemouth and later at the Haven Hotel, Poole, Dorset.

In 1900 he took out his famous patent No. 7777 for "tuned or syntonic telegraphy." In December 1901 determined to prove that wireless waves were not affected by the curvature of the Earth, he used his system for transmitting the first wireless signals across the Atlantic between Poldhu, Cornwall, and St. John's, Newfoundland, a distance of 2100 miles.

Between 1902 and 1912 he patented several new inventions. In 1902 the "magnetic detector" which then became the standard wireless receiver for many years. In 1905 he patented his "horizontal directional aerial" and in 1912 a "timed spark" system for generating continuous waves.

He received the Nobel Prize in Physics in 1909, which he shared with Professor Carl Ferdinand Braun. He is remembered best for his development of radio wave communication.

See also **Electronics**.

J. M. I.

MARCUS, RUDOLPH A. (1923–). Professor Marcus from the California Institute of Technology, Pasadena, California, won the Nobel Prize for chemistry in 1992 for his contributions to the theory of electron transfer reactions in chemical systems.

The processes Marcus has studied, the transfer of electrons between molecules in solution, underlie a number of exceptionally important chemical phenomena, and the practical consequences of his theory extend over all areas of chemistry. The Marcus theory describes, and makes predictions concerning, such widely differing phenomena as the fixation of light energy by green plants, photochemical production of fuel, chemiluminescence ("cold light"), the conductivity of electrically conducting polymers, corrosion, the methodology of electrochemical synthesis and analysis, and more.

From 1956 to 1965 Marcus published a series of papers on electron transfer reactions. His work led to the solution of the problem of greatly varying reaction rates.

MARE. (Plural: maria.) 1. One of several dark, low-lying, level, relatively smooth, plains-like areas of considerable extent on the surface of the earth's moon, having fewer large craters than the highlands, and composed of mafic or ultramafic volcanic rocks, e.g., Mare Imbrium (a circular mare) and Mare Tranquillitatis (a mare with an irregular outline). Lunar maria are completely waterless.

2. By extension, a dark area of the surface of Mars or other planets and satellites whose origin is not known. See also **Mars**; and **Moon (Earth's)**.

MARGARINE. See **Vegetable Oils (Edible)**.

MARGAY. See **Cats**.

MARGULES'S EQUATION. An equation for the equilibrium inclination of an interface separating two homogeneous air masses in a steady geostrophic motion parallel to the interface,

$$\tan\alpha = \frac{f}{g}\frac{(T_2 v_1 - T_1 v_2)}{(T_2 - T_1)},$$

where α is the angle of inclination of the surface to the horizontal, f the Coriolis parameter, g the gravitational acceleration, and T_1 and T_2 the absolute temperatures of the colder and warmer air masses, respectively, with speeds v_1 and v_2.

This equilibrium condition has been used to calculate the slope of atmospheric frontal surfaces.

AMS

MARIAN. See **Winds and Air Movement**.

MARICULTURE. See **Aquaculture**.

MARIJUANA. A drug derived from *Cannabis indica*, a variety of common hemp. The much more potent hashish is also derived from this plant. The active substance is located in the glandular hairs of the leaves and stems. From the pistillate flowers and fruits is obtained a resinous substance, which is smoked under the name of marijuana, hashish, or bhang, depending upon its concentration and particular mode of preparation. Synthetic cannabis, or synhexyl, is a pyrahexyl with an action more severe than the natural material.

Marijuana has been used as an agent for achieving euphoria since ancient times; it was described in a Chinese medical compendium traditionally considered to date from 2737 B.C. Its use spread from China to India and then to N Africa and reached Europe at least as early as A.D. 500.

The first direct reference to a cannabis product as a psychoactive agent dates from 2737 BC, in the writings of the Chinese emperor Shen Nung. The focus was on its powers as a medication for rheumatism, gout, malaria, and oddly enough, absent-mindedness. Mention was made of the intoxicating properties, but the medicinal value was considered more important. In India though it was clearly used recreationally. It was the Muslims who introduced hashish, whose popularity spread quickly throughout 12th century Persia (Iran) and North Africa.

In 1545 the Spanish brought marijuana to the New World. The English introduced it in Jamestown in 1611 where it became a major commercial crop alongside tobacco and was grown as a source of fiber.

By 1890, hemp had been replaced by cotton as a major cash crop in southern states. Some patent medicines during this era contained marijuana, but it was a small percentage compared to the number containing opium or cocaine. It was in the 1920's that marijuana began to catch on. Some historians say its emergence was brought about by Prohibition. Its recreational use was restricted to jazz musicians and people in show business. "Reefer songs" became the rage of the jazz world. Marijuana clubs, called tea pads, sprang up in every major city. These marijuana establishments were tolerated by the authorities because marijuana was not illegal and patrons showed no evidence of making a nuisance of themselves or disturbing the community. Marijuana was not considered a social threat.

Marijuana was listed in the United States Pharmacopoeia from 1850 until 1942 and was prescribed for various conditions including labor pains, nausea, and rheumatism. Its use as an intoxicant was also commonplace from the 1850s to the 1930s. A campaign conducted in the 1930s by the U.S. Federal Bureau of Narcotics (now the Bureau of Narcotics and Dangerous Drugs) sought to portray marijuana as a powerful, addicting substance that would lead users into narcotics addiction. It is still considered a "gateway" drug by some authorities. In the 1950s it was an accessory of the beat generation; in the 1960s it was used by college students and "hippies" and became a symbol of rebellion against authority.

The Controlled Substances Act of 1970 classified marijuana along with heroin and LSD as a Schedule I drug, i.e., having the relatively highest abuse potential and no accepted medical use. Most marijuana at that time came from Mexico, but in 1975 the Mexican government agreed to eradicate the crop by spraying it with the herbicide paraquat, raising fears of toxic side effects. Colombia then became the main supplier. The "zero tolerance" climate of the Reagan and Bush administrations resulted in passage of strict laws and mandatory sentences for possession of marijuana and in heightened vigilance against smuggling at the southern borders. The "war on drugs" thus brought with it a shift from reliance on imported supplies to domestic cultivation (particularly in Hawaii and California). Beginning in 1982 the Drug Enforcement Administration turned increased attention to marijuana farms in the United States, and there was a shift to the indoor growing of plants specially developed for small size and high yield. After over a decade of decreasing use, marijuana smoking began an upward trend once more in the early 1990s, especially among teenagers.

The use of marijuana is one of numerous problem interfaces between science and society. Solutions are for society to devise. Science assists society in constructing enlightened public policy.

Many marijuana users are unaware of several fundamental factors concerning the drug. This is particularly true of adolescents. For example: (1) the smoking of marijuana produces over 2000 separately identifiable chemicals. Many of these compounds remain in fat stores for several weeks, currently with unknown consequences. The half-life in humans of a single marijuana cigarette containing 2% concentration of THC (delta-9-tetrahydrocannabinol) ranges from 3 to 7 days, meaning that several weeks are required for the THC to exit the body completely. (2) On occasion, marijuana is adulterated with various substances. One of these compounds is PCP (phencyclidine, an established dangerous drug). (3) Other adulterants may include insect sprays (some with known carcinogenic properties); and dried shredded cow manure. The latter may contain salmonella, a major cause of food poisoning. Each of the fundamental ingredients of marijuana, as well as adulterants, has its own intrinsic toxicity.

Marijuana, when smoked or otherwise ingested, is reported by users to produce a dreamlike state of relaxation, a sense of contentment, improved social interaction, loss of inhibitions, and feelings of heightened self-awareness. To achieve such effects, marijuana acts upon the brain, including the initiation of chemical and electrophysiologic changes.

As pointed out by Schwartz, there are also the adverse effects of acute panic, flashback phenomena, and acute toxic psychosis, such as excitement, disorientation, confusion, delusions, depersonalization, delirium, and visual hallucinations, which can occur unpredictably in some individuals. Acute panic reactions may be accompanied by abdominal discomfort, headache, anxiety, depression, fear of dying, restlessness, uncontrollable hostility, anxiety, and paranoia. Also, accurately intoxicated individuals may have impaired reflexes, decreased attention, altered depth perception, and reduced long-term memory.

The drug causes acute changes in the heart and circulation which are characteristic of stress. Acute exposure to marijuana smoke causes bronchodilation. In chronic heavy smoking of the drug, there is inflammation and preneoplastic change in the airways, not unlike those produced by heavy tobacco smoking. Thus, there is a strong possibility that prolonged heavy smoking of the drug will lead to cancer of the respiratory tract and to serious impairment of lung function.

Professionals who treat marijuana users generally focus on the disturbed interpersonal relationships of the abuser. For example, adolescents may use marijuana as an act of defiance or to conceal and cover up particular difficulties in their lives. Fortunately, many chronic users ultimately outgrow their habit, often when they enter into family formation and make an increasing number of friendships with drug-free persons.

Additional Reading

Abel, E.L.: *Marijuana: The First Twelve Thousand Years*, Perseus Publishing, Boulder, CO, 2000.

Grinspoon, L. and J.B. Bakalar: *Marijuana: The Forbidden Medicine*, Yale University Press, New Haven, CT, 1997.

Iversen, L.L.: *The Science of Marijuana*, Oxford University Press, Inc., New York, NY, 2000.

Mack, A. and J. Joy: *Marijuana as Medicine? The Science beyond the Controversy*, National Academy Press, Washington, DC, 2000.

Marx, J.: "Marijuana Receptor," *Science*, 624 (August 10, 1990).

Musto, D.F.: "Opium, Cocaine, and Marijuana in American History," *Sci. Amer.*, 40 (July 1991).

Nahas, G.G., S. Agurell, K.M. Sutin, and D.J. Harvey: *Marijuana and Medicine*, Humana Press, Totowa, NJ, 1999.

Schwartz, R.H.: "Marijuana: A Crude Drug with a Spectrum of Underappreciated Toxicity," *Pediatrics*, **73**, 455 (1984).

Schwenk, C.R. and S.L. Rhodes: *Marijuana and the Work Place*, Greenwood Publishing Group, Inc., Westport, CT, 1999.

Smith, A. and E. Tanner: *Highlights: An Illustrated History of Cannabis*, Ten Speed Press, Berkeley, CA, 1999.

Somdahl, G.L.: *Marijuana Drug Dangers*, Enslow Publishers, Inc., Berkeley Heights, NJ, 1999.

Staff: *Facts About Marijuana and Smoking*, American Lung Association, New York, NY (revised periodically).

Stanley, D.: *Marijuana and Your Lungs*, Rosen Publishing Group, Inc., New York, NY, 2000.

Watson, J., J.E. Joy, and J.A. Benson: *Medical Use of Marijuana: Assessment of the Science Base*, National Academy Press, Washington, DC, 1999.

MARINE BAROMETER. A mercury barometer designed for use aboard ship. The instrument is of the fixed-cistern type. See also **Kew Barometer**. The mercury tube is constructed with a wide bore for its upper portion and with a capillary bore for its lower portion. This is done to increase the time constant of the instrument and thus prevent the motion of the ship from affecting the reading. The instrument is suspended in gimbals to reduce the effects of pitch and roll of the ship.

MARINE BIOLOGY. See **Ocean**.

MARINE CLIMATE. See **Climate**.

MARINE METEOROLOGY. See **Meteorology**.

MARINE RAINBOW. A rainbow seen in sea spray. Also called *sea rainbow*. It is optically the same as the ordinary rainbow, although the slightly different index of refraction of saltwater results in a shift in the angular radius of the bow, which is apparent if accompanied by a bow formed in raindrops.

See also **Atmospheric Optical Phenomena**.

MARITIME AEROSOL. Aerosol produced by processes taking place near the ocean surface. Air entrained by breaking ocean waves (visible as whitecaps) forms bubbles that break at the water surface. The whitecaps produce a fine aerosol; the Rayleigh jet produces coarser particles that are carried aloft by the turbulent eddies of the winds. Aerosol also forms from chemical reaction of ocean-produced gases.

MARITIME CLOUD.   See **Clouds and Cloud Formation**.

MARJORAM (Sweet Marjoram).   See **Spices**.

MARKHOR.   See **Goats and Sheep**.

MARK-HOUWINK EQUATION.   Defines the relationship between the intrinsic viscosity and molecular weight for homogeneous linear polymers.

MARKOV, ANDREI ANDREYVICH (1856–1922).   Markov was a renowned Russian mathematician working in number theory and probability theory. His name is best remembered for the concept of Markov chain, a series of events in which the probability of a given event depends only on the immediately previous event.
    See also **Markov Process**.

                                                                        J. M. I.

MARKOV CHAIN.   This expression is used in two different senses, both relating to a Markov process. In one sense, a process (x_i) is called a chain if the time parameter is discontinuous. In the other it is called a chain if the values of x are discontinuous. The former appears preferable.

MARKOV PROCESS.   A stochastic process such that the conditional probability distribution for the state at any future instant, given the present state, is unaffected by any additional knowledge of the past history of the system. See also **Autoregression**.

MARKOWNIFOFF RULE.   When a halogen acid adds to an asymmetrical ethylenic compound, the halogen usually appears on the carbon atom carrying the smaller number of hydrogen atoms; this order of addition is frequently reversed with hydrogen bromide if peroxides are present (peroxide effect).

MARLIN.   See **Billfishes**.

MARMOSET (*Mammalia, Primates*).   Small monkeys of Central and South America. They have only thirty-two teeth, four less than the other American monkeys, and the thumb is not opposable to the fingers. All digits but the great toe bear claws instead of nails. They constitute the family of Hapaloids.
    Most of these monkeys are called marmosets but the common Brazilian species, *Hapale jacchus*, is also known as the ouistiti and the group of long-tusked marmosets, *Mystax*, are called tamarins. One species, *Midas aedipus*, of the Isthmus of Panama, bears the French name pinché.

MARMOT.   See **Squirrels and Other Sciuromorphs**.

MARS.   The fourth planet from the Sun, Mars is the first planet in the solar system beyond Earth. Compared with Earth, Mars is a small planet. With a value of 1.00 for Earth, the mass of Mars is 0.107; the density, 0.719; volume, 0.149. The equatorial diameter of Mars is 4,226 miles (6,772 kilometers) as compared with that of earth of 7,960 miles (12,757 kilometers). Although Mars has been observed since early times, reliable and detailed data did not become available until commencement of the *Mariner* exploratory program in the 1960's. With the *Viking* programs of the last half of the 1970s, some important *Mariner* data had to be revised. With the present Mars Exploration Programs (1996-Present), which includes; Mars Reconnaissance Orbiter; Mars Express; 2001 Mars Odyssey; Mars Global Surveyor; and the Mars Exploration Rovers (Spirit and Opportunity) Missions, some important *Viking* data will have to be revised. See also **Mars Exploration Program**.
    The orbit of Mars is noticeably eccentric (0.093). The distance from Mars to the sun varies from a minimum of 129 million miles (208 million kilometers) at perihelion to a maximum of 155.3 million miles (249 million kilometers) at aphelion. Earth-bound observations of Mars are best when the planet is within a few months of opposition (when the Earth lies between the planet and the sun). During the remainder of the present century, Mars will be closest to earth on February 11, 1995, March 20, 1997, and May 1, 1999.

Mars has two satellites, or moons, Phobos and Deimos, both discovered in 1877 by Hall. Spacecraft have shown these bodies to be cratered, rocky, and chunky, and in recent years there has been serious speculation that these may not be moons in the usual sense, but rather captive asteroids. See also **Asteroid**. Phobos is quite small, with dimensions of approximately 12.4×17.4 miles (20×28 kilometers) and Deimos even smaller, 6.2×9.9 miles (10×16 kilometers).

Missions to Mars

The twin *Viking* missions to Mars, each with its own lander, represented a very sophisticated and successful venture. Among some scientists there remains perplexity regarding some of the main features of the planet, notably numerous channels and rifts at one time called "canals" by Earth-bound observers several decades ago. Knowledge of how these features look (including full-color) and their dimensions have been greatly enhanced, but the mysteries of their origins remains unknown. Earth-bound estimates and *Mariner's* measurements of Mars comparatively thin atmosphere were confirmed, a factor which detracted from the possibility of organisms living on the planet. The polar ice cap once thought to be frozen carbon dioxide has been found to be ice with possibly some frozen carbon dioxide with it. Biological experiments designed to detect living organism proved negative, but the apparently oxidizing characteristic of Martian soil has introduced new puzzles.
    An interesting view of Mars taken by *Viking Orbiter 1* showing the huge Mariner Valley (Valles Marineris) is given in Fig. 1. A close-up of this extremely impressive Martian feature is given in Fig. 2.

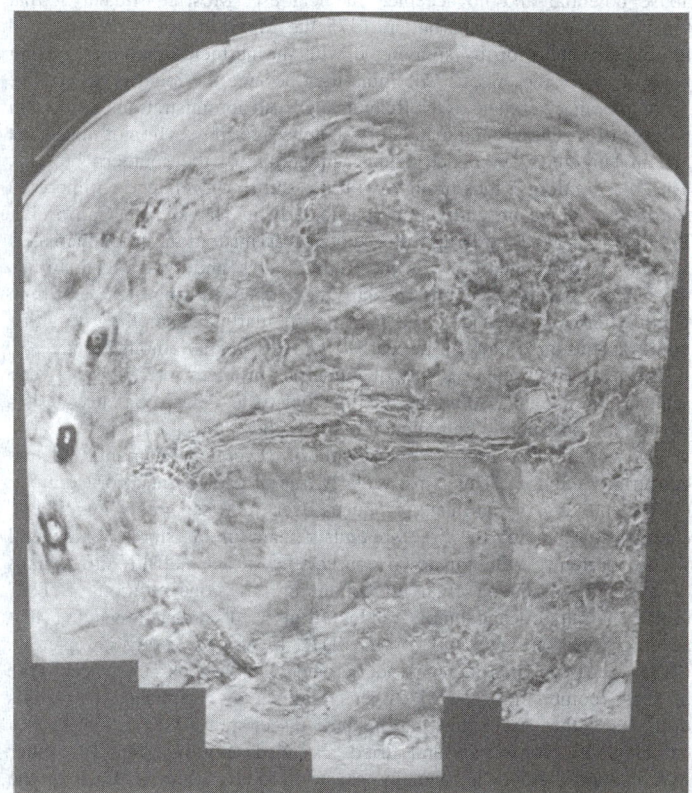

Fig. 1.   Mosaic of 102 photos of Mars taken on February 22, 1980 by *Viking Orbiter 1*. Several prominent Martian features and at least two unusual weather phenomena are visible. Valles Marineris (Mariner Valley), as long as the North American continent from coast to coast, stretches across the center. Three huge volcanoes of the Tharsis Ridge are visible at the left: Arsia Mons, Pavonis Mons, and Ascraeus Mons, proceeding from south to north. A sharp line, either a weather front or an atmospheric shock wave, curves north and east from Arsia Mons. This is the first time a feature like this had been seen on a planet. Four tiny clouds can be seen in the southernmost frame, just north of a large crater named Lowell. While the clouds are too close together to be resolved, even under high magnification, their shadows can be separated easily. The largest cloud is nearly 32 kilometers (20 miles) long. Measurements show the elevation of the clouds at nearly 28 kilometers (91,000 feet). Such distinct cloud-shadow patterns apparently are quite rare on Mars. (*NASA; Jet Propulsion Laboratory, Pasadena, California.*)

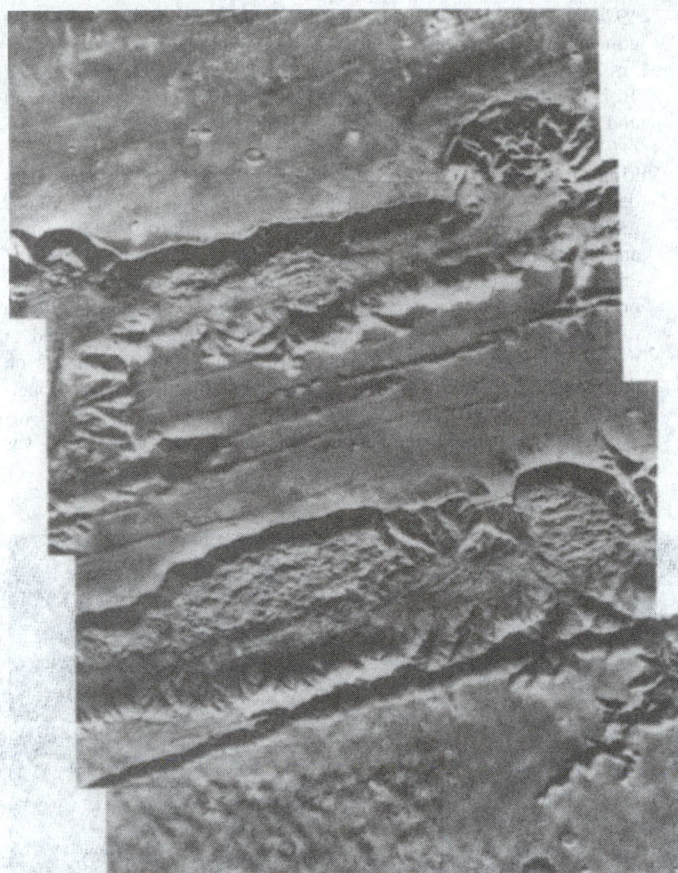

Fig. 2. Mosaic of the surface of Mars showing the west end of the Valles Marineris (all of which is shown in Fig. 1) from a range of about 4300 kilometers (2700 miles). These two canyons, running east-west across the picture, are each about 60 kilometers (37 miles) wide and more than 1 kilometer (0.6 mile) deep. Some scientists suggest that the canyons were originally formed by downfaulting of the crust along parallel faults. Other faults and collapsed depressions with the same trend are seen between the two canyons. After they were formed, it is suggested that the canyons were modified by erosion that formed great slumps on the walls and also cut side valleys to the main canyons. A few comparatively recent impact craters will be noted, particularly at the bottom right of the view. (*NASA; Jet Propulsion Laboratory, Pasadena, California.*)

The *Viking* missions are discussed in greater detail in a later section of this entry. Many missions preceded the *Viking* missions to Mars, and several have followed. The list below presents the chronology of missions to Mars:

Mars 1960A — USSR Mars Probe was launched on October 10, 1960, however, it failed to reach Earth orbit.

Mars 1960B — USSR Mars Probe. Launched on October 14, 1960. It also failed to reach Earth Orbit.

Mars 1962A — USSR Mars Flyby. Launched on October 24, 1962, this spacecraft failed to leave Earth orbit after the final rocket stage exploded.

Mars 1 — USSR Mars Flyby. Launched on November 1, 1962. The spacecraft weighed 1,969 pounds (893 kilograms). This mission was not successful due to communications failure.

Mars 1962B — USSR Mars lander. Launched on November 4, 1962. This spacecraft also failed to leave Earth orbit.

Mariner 3 — Launched on November 5, 1964 at a weight of 572 pounds (260 kilograms) by the USA, the solar panels did not open, preventing a successful flyby. *Mariner 3* remains in solar orbit.

Mariner 4 — Launched on November 28, 1964 at a weight of 572 pounds (260 kilograms) by the USA, *Mariner 4* reached Mars on July 14, 1965. It passed within 5,952 miles (9,920 kilometers) and returned data confirming that the atmosphere was composed of carbon dioxide,

Fig. 3. *Mariner 4* was the fourth in a series of spacecraft used for planetary exploration in a flyby mode. It was designed to conduct close-up scientific observations of the planet Mars and to transmit these observations to Earth. (*Courtesy of the Jet Propulsion Laboratory and NASA's National Space Science Data Center.*)

and identifying a small magnetic field. *Mariner 4* obtained 22 close-up photos of the surface of Mars clearly showing surface features, notably craters. *Mariner 4* remains in solar orbit. See Fig. 3.

Zond 2 — USSR Mars Flyby launched on November 30, 1964, which was unsuccessful. Contact with the spacecraft was lost and its fate is unknown.

Mariner 6 — USA Mars Flyby launched at a weight of 910 pounds (413 kilograms), the spacecraft reached Mars on July 31, 1969. It passed within 2,062 miles (3,437 kilometers) of the surface of the planet. *Mariner 6* remains in solar orbit.

Mariner 7 — USA Mars Flyby launched at a weight of 910 pounds (413 kilograms), the spacecraft reached Mars on August 5, 1969. It passed within 2,131 miles (3,551 kilometers) of the surface of Mars at the south pole region. Both *Mariner 6* and *Mariner 7* obtained data related to the atmosphere and surface composition. Over 200 photos were obtained during these two missions. *Mariner 7* remains in solar orbit. See Fig. 4.

Fig. 4. *Mariner 6* and *7* were designed to fly over the equator and southern hemisphere of the planet Mars. They were solar powered and capable of continuous telemetry transmission. Each spacecraft weighed 910 pounds (413 kilograms) and measured 11 feet (3.35 meters) from the scan platform to the top of the low-gain antenna. The width across the solar panels was 19 feet (5.8 meters). The eight-sided body of the spacecraft carried seven electronic compartments. A small rocket engine, used for trajectory corrections, protruded through one of the sides. The planetary experiments aboard the spacecraft were two television cameras, an infrared radiometer, and infrared spectrometer and as ultraviolet spectrometer. (*Courtesy NASA.*)

Mariner 8 — USA Mars Flyby launched May 8, 1971, this mission was unsuccessful as it failed to reach Earth's orbit.

Kosmos419 — Launched by the USSR May 10, 1971, this mission was unsuccessful as it failed to reach Earth's orbit.

Mars 2 — This spacecraft was a USSR Mars Orbiter/Soft Lander launched May 19, 1971 that weighed 10,230 pounds (4,650 kilograms). It failed in its landing mission as the *Mars 2 Lander*, which was released from the Orbiter on November 27, 1971, crash-landed on the surface of the planet. It is known that the breaking rockets failed, but no data was returned. The *Mars 2 Orbiter* returned data until 1972.

Mars 3 — This spacecraft was another USSR Mars Orbiter/Soft Lander that weighed 10,215 pounds (4,643 kilograms). It reached Mars on December 2, 1971, and successfully released the lander to the surface of the planet. It was the first successful landing on the surface of Mars, but the *Mars 3* failed to record and transmit more than 20 seconds of data to the orbiter. The *Mars 3 orbiter* collected data related to the surface temperature and atmospheric conditions until August 1972.

Mariner 9 — Launched by the USA May 30, 1971, the spacecraft weighed 1,116 pounds (506 kilograms). *Mariner 9* was the first US spacecraft to enter orbit around a body other than the Earth's moon, and it entered this orbit on November 24, 1971. Among the data obtained were the first high-resolution images of the Martian moons, Phobos and Deimos, and surface data detailing river and channel-like features. *Mariner 9* remains in Martian orbit. See Fig. 5.

Fig. 5.   The *Mariner 9* spacecraft was built on octagonal magnesium frame 18 inches (45.7 centimeters) deep and 54.5 inches (138.4 centimeters) across a diagonal. Four solar panel each 85 × 35 inches (215 × 90 centimeters), extended out from the top of the frame. Each set of two solar panels spanned 23 feet (6.89 meters) from tip to tip. Also mounted on the top of the frame were two propulsion tanks, the maneuver engine, a 5-foot (1.44 meters) long low gain antenna mast and a parabolic high gain antenna. A scan platform was mounted on the bottom of the frame, on which were attached the mutually bore-sighted science instruments (wide-and narrow-angle TV cameras, infrared radiometer, ultraviolet spectrometer, and infrared interferometer spectrometer). The overall height of the spacecraft was 7.5 feet (2.28 meters). (*Courtesy of NASA's National Space Science Data Center.*)

Mars 4 — Another of the USSR Mars Orbiter/Soft Lander vehicles, this mission was not wholly successful. Although it arrived at Mars in February 1974, it failed to enter orbit due to failure of the breaking rockets. A flyby at a distance of 1,320 miles (2,200 kilometers) returned limited data.

Mars 5 — A USSR Mars Orbiter/Soft Lander vehicle, the spacecraft weighed 10,230 pounds (4,650 kilograms) and entered Martian orbit in February 1974. Data obtained during this mission set the stage for the *Mars 6* and *Mars 7* missions.

Mars 6 — This USSR Mars Orbiter/Soft Lander vehicle, which weighed 10,230 pounds (4,650 kilograms) entered Martian orbit on March 12,

1974 and launched its lander. The lander successfully transmitted atmospheric data during its descent, but failed prior to landing.

Mars 7 — Another USSR Mars Orbiter/Soft Lander vehicle that weighed 10,230 pounds (4,650 kilograms), failed both to enter Martian orbit and to set the lander vehicle on the Martian surface. The *Mars 7 orbiter* and lander remain in solar orbit.

Viking 1 — Designed after the *Mariner* spacecraft, the USA Mars Lander/Orbiter was launched on August 20, 1975 weighing 7,478 pounds (3,399 kilograms). The orbiter weighed 1,980 pounds (900 kilograms) and the lander weighed 1,320 pounds (600 kilograms). *Viking 1* entered Martian orbit June 19, 1976, and its lander successfully set on the surface one day later on July 20, 1976 on the western slopes of Chryse Planitia. The lander and orbiter obtained data related to the weather on Mars, the Martian terrain, and microorganisms on the planet. The *Viking 1 orbiter* ran out of altitude control propellant August 7, 1980 and was deactivated. The *Viking 1 lander* was accidentally shut down and neither communication nor activation was ever regained. See Fig. 6.

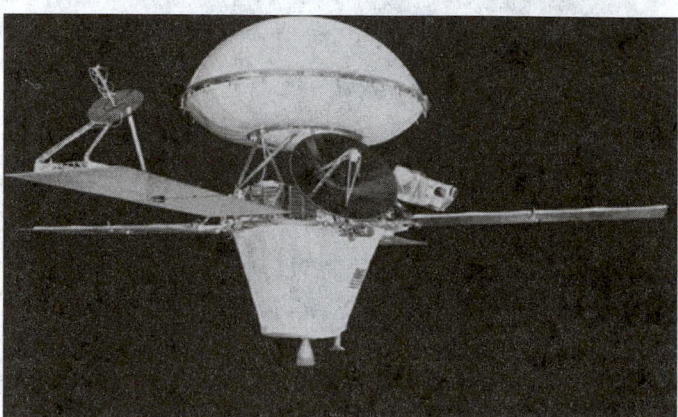

Fig. 6.   This image shows a model of one of the *Viking* spacecraft, which were made of two parts: an orbiter and a lander. The orbiter's initial job was to survey the planet for a suitable landing site. Later the orbiter's instruments studied the planet and its atmosphere, while the orbiter acted as a radio relay station for transmitting lander data. (*Courtesy NASA/JPL.*)

Viking 2 — Also designed after the *Mariner* spacecraft, the USA Mars Lander/Orbiter was launched on September 9, 1975 weighing 7,478 pounds (3,399 kilograms). See Fig. 7. The orbiter weighed 1,980 pounds (900 kilograms) and the lander weighed 1,320 pounds (600 kilograms). Viking 2 entered Martian orbit on July 24, 1976 and its lander set down at Utopia Planitia on August 7, 1976. See Fig. 8. While both landers had experiments to search for and identify microorganisms on Mars, the results of the experiments are still subject to debate. Both landers together obtained over 52,000 images while mapping the planet's surface. The *Viking 2 orbiter* ran out of altitude control propellant July 25, 1978 and was deactivated. Because the *Viking 2 lander* used the *Viking 1 orbiter* for communications, it had to be shut down the same time the *Viking 1 orbiter* was deactivated on August 7, 1980.

Phobos 1 — USSR Mars Orbiter/Lander weighing 11,000 pounds (5,000 kilograms) that was launched on July 7, 1988. The spacecraft was lost on the way to Mars due to a command error on September 2, 1988. See Fig. 9.

Phobos 2 — USSR Phobos Flyby/Lander, which weighed 11,000 pounds (5,000 kilograms), was launched on July 12, 1988. The spacecraft entered Martian orbit January 30, 1989, but failed at a distance of 480 miles (800 kilometers) from the Martian moon Phobos.

Mars Observer — USA Mars Orbiter was launched September 25, 1992 but failed to enter Martian orbit. Communication with the Mars Observer was lost August 21, 1993.

Mars Global Surveyor (MGS) — USA Mars Orbiter was launched November 7, 1996 to complete the mission of the Mars Observer. See Fig. 10. See also **Mars Global Surveyor Mission**.

Fig. 7. Launch of the *Viking 2* spacecraft from Cape Canaveral, Florida. (*Courtesy NASA/JPL.*)

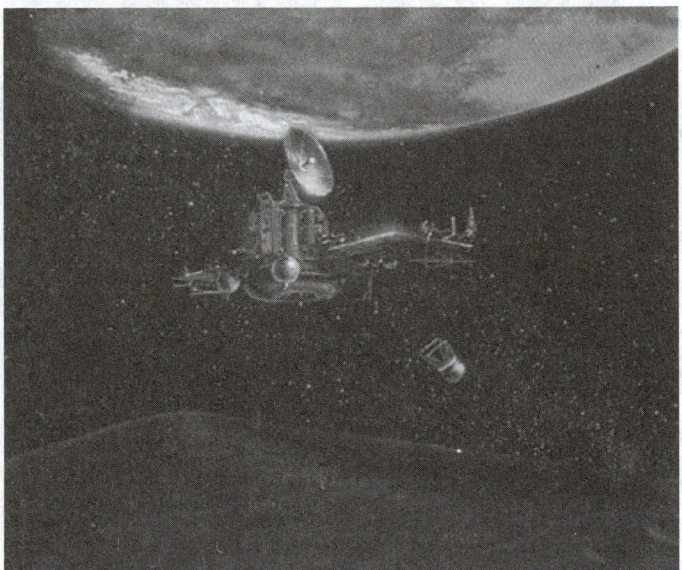

Fig. 9. This artist's concept depicts the *Phobos 1 & 2* spacecraft destined for Mars. They were the next-generation in the Venera-type planetary missions, succeeding those last used during the Vega 1 and 2 missions to comet P/Halley. (*Courtesy of NASA's National Space Science Data Center.*)

Fig. 8. Captured in this rendering is a *Viking* lander just before it touched down on the Martian surface. The parachute and upper aero shell can be seen in the upper left corner of the image. At this stage of the descent, the lander's terminal descent propulsion system (three retro-engines) had slowed the craft down so that velocity at landing was about 7 miles per hour (2 meters per second). Seconds after the lander reached the surface it began transmitting images back to the orbiter for relay to Earth. (*Courtesy NASA/JPL.*)

Mars 96 — Russia Orbiter and Lander which was launched November 16, 1996, consisted of an orbiter, two landers and two soil penetrators. The fourth stage of the rocket that launched the Mars 96 spacecraft ignited prematurely as the vehicle entered orbit. The spacecraft crashed into the ocean and sank between the coast of Chile and Easter Islands.

Mars Pathfinder — USA Lander and Surface Rover launched on December 4, 1996. The lander weighed 581 pounds (264 kilograms) and the rover vehicle weighed only 23 pounds (10.5 kilograms). Mars Pathfinder reached Mars July 4, 1997 and impacted the surface at a velocity of approximately 40 miles per hour (18 meters per second). It bounced into the air about 50 feet (15 meters), bounced another 15 times, and rolled to a stop approximately 2.5 minutes after impact about one-half mile (about 1 kilometer) from the site of initial impact.

Fig. 10. Captured in this rendering the *Mars Global Surveyor (MGS)* is designed to orbit Mars over a two year period and collect data on the surface morphology, topography, composition, gravity, atmospheric dynamics, and magnetic field. This data will be used to investigate the surface processes, geology, distribution of material, internal properties, evolution of the magnetic field, and the weather and climate of Mars. (*Courtesy of NSSDC.*)

The landing site, in the Ares Valley region at 19.33 °N, 33.55 °W, was named the Sagan Memorial Station. The rover, a six-wheeled vehicle named Sojourner, hit the Martian surface July 6. See Figs. 11

Fig. 11.   Artist view of Pathfinder on Mars. (Courtesy of NASA's National Space Science Data Center.)

Fig. 12.   The rover *Sojourner* is a six-wheeled vehicle launched with the Mars Pathfinder mission. It is controlled by an Earth-based operator, who uses images obtained by both the rover and lander systems. Note that the time delay is about 10 minutes, requiring some autonomous control by the rover. The primary objectives were scheduled for the first seven sols (1sol = 1martian day = ~24.7) (*Courtesy of NASA's National Space Science Data Center.*)

and 12. This highly successful mission returned 2.6 billion bits of information including over 16,000 images from the lander, 550 images from Sojourner, 15 chemical analyses of rocks, and extensive data on climatic conditions. See also **Pathfinder Mission to Mars**.

Planet B—Japan Mars Orbiter launched August 1998 by Japan's Institute of Space and Astronautical Science (ISAS) will be the first Japanese spacecraft to reach another planet.

Mars Surveyor '98 Orbiter—The Mars Surveyor '98 program is comprised of two spacecraft launched separately, the ***Mars Climate Orbiter*** (formerly the Mars Surveyor '98 Orbiter) and the Mars Polar Lander (formerly the ***Mars Surveyor '98 Lander***). The two missions were to study the Martian weather, climate, and water and carbon dioxide budget, in order to understand the reservoirs, behavior, and atmospheric role of volatiles and to search for evidence of long-term and episodic climate changes. The Mars Climate Orbiter was destroyed when a navigation error caused it to miss its target altitude

at Mars by 80 to 90 kilometers (50 to 56 miles), instead entering the Martian atmosphere at an altitude of 57 kilometers (35 miles) during the orbit insertion maneuver. Launched December 11, 1998, this spacecraft was to study the planet from polar orbit for one to two years using a variety of highly sophisticated instruments. See Fig. 13.

Fig. 13.   The *Mars Climate Orbiter* spacecraft was launched from the Cape Canaveral Air Force Station (CCAFS) Space Launch Complex 17 (SLC-17) on December 11, 1998. The Mars Orbit Insertion (MOI) propulsive maneuver will occur in September 1999 and will place the orbiter into a highly elliptical, near polar orbit around Mars. Peripapse will be lowered approximately 110 kilometers (68 miles) altitude to initiate the aerobraking maneuvers. Successive passes of the orbiter through the upper atmosphere of Mars will slow the vehicle and lower the apoapse of the orbit to 450 kilometers (280 miles) over the course of the 2 month aerobraking phase. The orbit then will circularized using the orbiter's onboard propulsion resulting in the design 400 kilometers (249 miles) altitude, circular, polar science mapping orbit. Science operation of the PMIRR and MARCI instruments will be conducted over the course of the one Martian year (687 Earth day) mapping mission. The orbiter will continue operations in a relay only mode following the science mission in support of any future U.S. or international Mars surface missions ending December 1, 2004. (*Courtesy of NASA/JPL.*)

Mars Surveyor '98 Lander—The companion vehicle to the Mars Surveyor'98 Orbiter launched January 1999. It was to study the Martian environment at the south pole region specifically climate, atmospheric conditions, and soil. It will be equipped with advanced meteorological equipment and a robotic arm for digging into the soil. See Fig. 14. The Mars Polar Lander (formerly Mars Surveyor '98 Lander) was lost most. The most likely cause of the Lander's failure, investigators decided, was that a spurious sensor signal associated with the craft's legs falsely indicated that the craft had touched down when in fact it was some 40 meters (131 feet) above the surface. When the landing legs unfolded they made a bouncing motion that accidentally set of the landing sensors, thus caused the descent engines to shut down prematurely and the Lander to free-fall out of the Martian sky. Another possible reason for failure was inadequate preheating of catalysis beds for the pulsing rocket thrusters. Hydrazine fuel decomposes on the beds to make hot gases that throttle out the rocket nozzles; cold catalysis beds caused misfiring and instability in crash review tests.

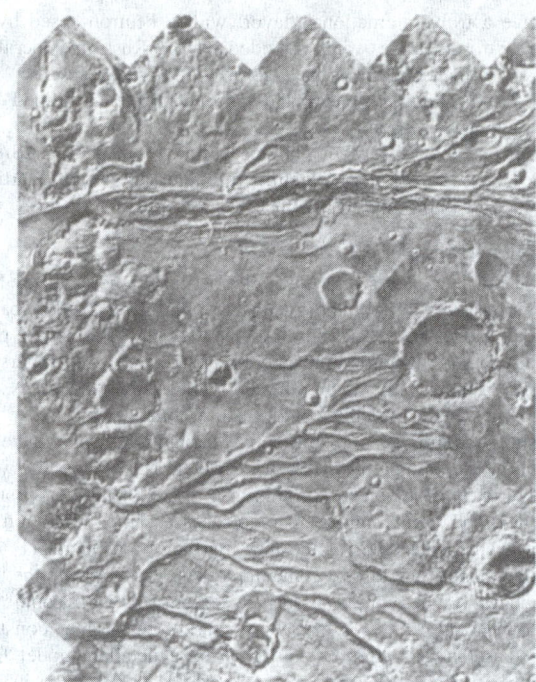

Fig. 14.   The *Mars Polar Lander* spacecraft was launched from Cape Canaveral Air Force Station (CCAFS) Space Launch Complex 17 (SLC-17) during a 25 day launch period beginning on January 3, 1999. Dimensions of the spacecraft 1.06 meters (3.5 feet) tall by 3.6 meters (12 feet) wide. The total spacecraft mass is 576 kilograms (1,270 pounds). The Lander weighs 290 kilograms (639 pounds). Propellant 64 kilograms (141 pounds). Cruise Stage 82 kilograms (181 pounds). Aeroshell and Heat Shield 140 kilograms (309 pounds). Mars Landing is scheduled on Dec. 3, 1999 with the end of the primary mission on March 1, 2000. The science payload includes Deep Space 2 Microprobes, Mars Descent Imager (MARDI), Light Detection and Ranging (LIDAR), New Millennium Microprobes, Mars Volatiles and Climate Surveyor (MVACS), Stereo Surface Imager (SSI), Robotic Arm and Camera, Meteorological Package (MET), and Thermal and Evolved Gas Analyzer (TEGA). (*Courtesy of NASA/JPL.*) See Fig. 26, same picture taken by different mission.

Fig. 15.   The *Mars Surveyor 2001 Orbiter* was scheduled for launch on April 18, 2001. It will arrive at Mars on Oct. 27, 2001, if launched on schedule. The 2000 Orbiter will be the first to use the atmosphere of Mars to slow down and directly capture a spacecraft into orbit in one step, using a technique called aerocapture. It will then reach a circular mapping orbit within about 1 week after arrival. The Orbiter will carry 2 main science instruments, the Thermal Emission Imaging System (THEMIS) and the Gamma Ray Spectrometer (GRS). THEMIS will map the mineralogy and morphology of the Martian surface using a high resolution camera and a thermal infrared imaging spectrometer. The GRS will achieve global mapping of the elemental composition of the surface and the abundance of hydrogen in the shallow subsurface. The gamma ray spectrometer was inherited from the lost Mars Observer mission. The 2001 Orbiter will also support communication with the Lander and Rover scheduled to arrive on Jan. 22, 2002. (*Courtesy of NASA/JPL, 2001 Artwork by Corby J. Waste.*)

Mars Surveyor 2001 — USA Mars Probe scheduled to be launched in 2001 as part of NASA's ten-year program to launch probes to Mars as favorable launch opportunities arise. See Figs. 15, 16 and 17. This mission was cancelled as part of the review and restructuring of NASA's Mars Exploration Program. See also **Phoenix Mars Mission**.

NASA Unveils its 21st Century Mars Campaigns

The seven-month retooling of its Mars campaign was prompted by the back-to-back loss last year of two spacecraft at the Red Planet. Subsequent investigative reports, including one authored by retired Lockheed Martin executive Tom Young, found bad management, a lack of training and an inadequate system of checks and balances, as well as too-tight budgets, doomed the Mars Climate Orbiter and Polar Lander missions, a $300 million-plus loss.

NASA will halt ambitious plans to send a lander/orbiter pair to Mars every 26 months, when the Earth and the Red Planet are closely aligned. Instead, it will now stagger the pace dispatching just one of each at the roughly two-year intervals.

The revised program also looks out beyond returning a sample of Martian soil to Earth for study. That goal has been pushed back to 2011 or later.

This program will represent a long-term strategy. It won't just end with Mars sample return like the previous one did. Officials, said the new program, allow for NASA to respond to any new discoveries on Mars, like the evidence that suggests water may have flowed on the planet's surface in the recent past, as well as accommodate the prospect of any of the planned missions failing.

What are missing from the equation are humans. NASA has already scrapped plans to launch in 2001, a package of experiments that would have laid some of the groundwork for future human missions to Mars, including experiments to produce oxygen from the Martian atmosphere and to assess the threat of its dust and radiation environments. Now, similar experiments might not make to Mars until 2005 at the earliest.

The agency plans on six major Mars missions for this decade alone, spending as much as $450 million a year on its near-term efforts. The missions include:

2001 — ***The 2001 Mars Odyssey Orbiter***, for high-resolution mapping and imaging.

2003 — Two water-sniffing ***Mars Exploration Rovers*** (Spirit & Opportunity), for detailed field geological work.

2005 — A ***Mars Reconnaissance Orbiter***: an orbiter modeled on NASA's successful Mars Global Surveyor, but capable of imaging objects as small as 30 centimeters (a foot) in diameter. Jim Garvin, NASA's Mars exploration program scientist, likened it to putting a microscope in orbit.

2007 — A "smart" surface lander equipped with a hazard avoidance system, precision landing capability and designed to deliver a rover laden with up to 270 kilograms (600 pounds) of scientific instruments; also in 2007, a "Scout" mission. The Phoenix mission is the first chosen for NASA's Scout program, an initiative for smaller, lower-cost, competed spacecraft. Named for the resilient mythological bird, Phoenix uses a lander that was intended for use by 2001's Mars Surveyor lander prior to its cancellation. It also carries a complex suite of instruments that are improved variations of those that flew on the lost Mars Polar Lander. The Mars Pathfinder landing in 1997 was within a 100×300-kilometer (60×200-mile) landing ellipse. "Where we want to be by 2007 is down to something that's 1 kilometer by 3 kilometers (0.62 by 2 miles) — a reduction by a factor of 100," Lavery said. The eventual goal is to land spacecraft

Fig. 16.   The *Mars Surveyor 2001 Lander* was scheduled for launch on April 10, 2001. It will land on Mars on Jan. 22, 2002, if launched on schedule. The 2001 Lander will carry an imager to take pictures of the surrounding terrain during its' rocket-assisted descent to the surface. The descent imaging camera will provide images of the landing site for geologic analyses, and will aid planning for initial operations and traverses by the rover. The 2001 Lander will also be a platform for instruments and technology experiments designed to provide key insights to decisions regarding successful and cost-effective human missions to Mars. Hardware on the Lander will be used for an in-situ demonstration test of rocket propellant production using gases in the Martian atmosphere. Other equipment will characterize the Martian soil proper and surface radiation environment. (*Courtesy of NASA/JPL, 2001 Artwork by Corby J. Waste.*)

Fig. 17.   As part of the 2001 mission redesign, the plans for the 2001 Rover have changed significantly. Current plans are to send the Marie Curie rover to Mars on the 2001 lander. This rover is very similar to the Pathfinder Sojourner Rover, and in fact is the same rover that was used within the Pathfinder "sandbox" test bed pictured above. (*Courtesy of NASA/JPL.*)

on the equivalent of a Martian dime — within a tight ellipse just a few hundreds of yards (meters) across, he said.

2009 — NASA building on the success of the two rover geologists that arrived at Mars in January, 2004, NASA's next rover mission is being planned for travel to Mars before the end of the decade. Twice as long and three times as heavy as the Mars Exploration Rovers Spirit and Opportunity, the **Mars Science Laboratory** would collect martian soil samples and rock cores and analyze them for organic compounds and environmental conditions that could have supported microbial life now or in the past. The mission is anticipated

to have a truly international flavor, with a neutron-based hydrogen detector for locating water provided by the Russian Federal Space Agency, a meteorological package provided by the Spanish Ministry of Education and Science, and a spectrometer provided by the Canadian Space Agency.

2011 — As early as 2011, but perhaps slipping to 2014, NASA could start a long-term project to return multiple samples of Martian soil and rock to Earth. See also **Mars Exploration Program**.

The Martian Atmosphere

As pointed out by Anders and Owen 1977, the thinness of the Martian atmosphere has been one of the great disappointments of the space age. At one time, Lowell, Dollfus, and others had suggested a surface pressure near 80 millibars (1 standard atmosphere on earth = 1013 millibars). Even with an atmosphere only about one-tenth the value of earth's, it was envisioned that Mars possibly could sustain some forms of life. *Viking 1* established a figure of 7.65 millibars. Substantiation of this relatively low pressure, indicating that Mars has only about 3% of the volatiles found on earth, revised scientific thinking in terms of the development of Mars. Anders and Owen suggested five processes, in combination, which may have been responsible for the thin martian atmosphere: (1) a small initial endowment of volatiles; (2) incomplete outgassing from the interior; (3) recondensation or trapping in surface regions; (4) catastrophic loss of an early atmosphere; and (5) gradual escape of the lighter constituents. Although these investigators do not describe a detailed model for how the Martian atmosphere passed from an early, dense state to its present condition, they offer a schematic scenario:

> The basic notion is that the atmosphere gradually decreased in density as a result of the deposition of carbon dioxide in the form of carbonates and the escape of nitrogen from the upper atmosphere. While the latter process was critical for the ultimate nitrogen abundance and isotope ratio, it should have played a small role in determining the total atmospheric pressure, since carbon dioxide was probably the most abundant gas. The depositional process (which may have included formation of nitrates or nitrites) was most active during the time when liquid water was most abundant — the cutting of the sinuous channels was thus a premonition of the end of the dense atmosphere. The apparent absence of an active Martian biota has prevented the recycling of volatiles through biological processes. Moreover, there is evidence that carbonates may form even under the present arid conditions on Mars.

Anders and Owen proceed in their interesting paper to make comparisons between the large and the small planets, with earth and Mars as paradigms.

Water and Water Vapor.   An infrared spectrometer operating at the 1.38 micrometer region, mounted on the scan platform, was used to detect water vapor in the Martian atmosphere.[1] This scanning device was used to measure the latitudinal variations and diurnal variations. By operating over a complete martian year, the instrument was able to measure the seasonal changes. The southern hemisphere, which was at the onset of winter, was found to have very little water vapor (0 to 0.3 precipitable micrometer). In contrast, the northern hemisphere showed a significant amount of water (up to 75 micrometers at 70–80 °N), a range of almost three orders of magnitude. The north polar region showed a slight drop in water vapor abundance. A strong diurnal repetitive cycle in certain regions, peaking out in the local mid-afternoon, was also found. Negative correlation existed between the elevation and the water vapor abundance, as would be expected. On the basis of the abundance of water vapor in the polar region, a lower limit can be put on the atmospheric temperature, namely 205 K (−68 °C; −90 °F). This value indicates that the permanent polar cap consists of water ice, a factor that was confirmed by the infrared thermal mapper (IRTM) also part of the *Viking* instrumentation package.

Observations of the latitude dependence of water vapor made from the *Viking 2 Orbiter* showed peak abundances in the latitude band 70–80 °N in the northern midsummer season. Total column abundances in the polar regions appeared to require near-surface atmospheric temperatures in excess of 200 K (−73 °C; −99 °F) and are incompatible with the survival of a frozen carbon dioxide cap at martian pressures. The remnant

[1] Information extracted from official NASA Langley Research Center report.

or residual on the north polar cap and the outlying patches of ice at lower latitudes are thus believed to be predominantly water ice whose thickness can be established within widely spaced limits, between 1 meter and 1 kilometer (~3 and 3280 feet). Broadband thermal and reflectance observations of the Martian north polar region in late summer yielded temperatures for the residual polar cap near 205 K (−68 °C; −90 °F). Residual cap and several outlying smaller deposits appeared to be water ice with included dirt. No evidence was found for a permanent carbon dioxide polar cap.

The first evidence of the direct visible exchange of water between the Martian surface and atmosphere was obtained by the *Viking 1 Orbiter* on July 24, 1976, as shown by Fig. 18.

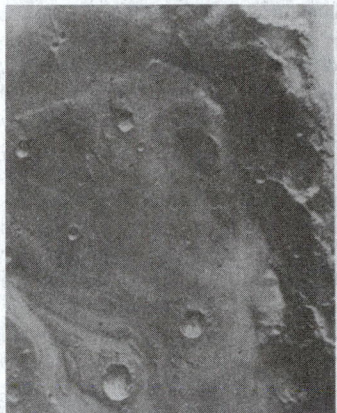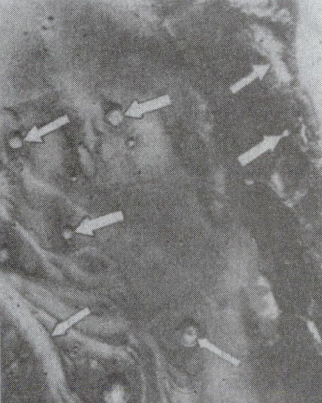

Fig. 18.   Two pictures taken a half-hour apart by *Viking 1 Orbiter* shows the development on Mars of early morning fog in low spots, such as crater and channel bottoms (see arrows on view at right). The scene at left was photographed shortly after martian dawn on July 24, 1976 from 12,400 kilometers (7700 miles) and, at right, 30 minutes later from 9800 kilometers (6100 miles). Slight warming of the sub-zero surface by the rising sun evidently drove off a small amount of water vapor which recondensed in the colder air just above the surface. Brightness measurements of the resulting fog patches indicated that a film of water about one micrometer thick had condensed. These fog patches were the first direct, visible evidence as to where the exchange of water between the martian surface and atmosphere may occur. (*NASA; Jet Propulsion Laboratory, Pasadena, California.*)

Ground Ice on Mars. As reported by Squyres (NASA Ames Research Center) and Carr (U.S. Geological Survey), many Martian landforms suggest the former presence of ground ice or water, including fretted and chaotic terrain, valley systems, outflow channels, and, with less certainty, various types of patterned ground and rampart craters. If sufficient ice is present now, the regolith should undergo quasi-viscous flow due to creep deformation of the ice. Accordingly, to determine where ice may be present, the researchers examined approximately 24,000 *Viking Orbiter* images taken with 5,000 kilometers (3,107 miles) of the surface and mapped the distribution of three types of features—lobate debris aprons, concentric crater fill, and terrain softening—that may indicate creep of near-surface materials.[2] In summary, the researchers observe that the origin of ice in the martian regolith is unclear at this time. The many lines of evidence implying that ice was common in the cratered uplands early in Martian history suggest that the ice was emplaced during an early period of intense outgassing. An alternative scenario would be the continuous outgassing throughout the planet's history at a rate substantially lower than the low-latitude depletion rates in order to keep the low latitudes ice-free. In either case, intense early meteoritic brecciation was probably largely responsible for the apparent capability of the deep regolith to hold large amounts of water.

[2] *Lobate debris aprons*—accumulations of erosion debris at the base of steep escarpments. *Concentric crater fill*—develops where debris aprons are confined within impact crates, and inward flow of material produces a pattern of concentric ridges. *Terrain softening*—a distinctive style of landform degradation apparent in high-resolution images. Definitions as given by Squyres and Carr.

Juvenile Water from Volcanism. In studying *Viking* data. Greeley (Arizona State University) reports that volcanism played a dominant role in the evolution of the Martian surface and environment. It is estimated that volcanism has occurred from at least the close of the period of heavy impact cratering (~3.9 billion years ago) to the age of the youngest rocks visible on the planet. Materials that are considered to be of volcanic origin cover more than half the surface of Mars. Perhaps the assumption can be made that, as with the Earth, juvenile water[3] was released on Mars in association with the eruption of these volcanic materials. By determining the volumes and ages of volcanic units and inferring the volatile content for the magmas, the amounts and timing of associated water release can be estimated. Tentative conclusions indicate the amount of juvenile water release on Mars would equal a layer some 46 meters (151 feet) deep over the entire planet. Most of this water was released in the first 2 billion years of Martian history. There are several uncertainties in estimates made to date, including lack of knowledge of volatile contents for magmas; even terrestrial values, as used, have large uncertainties, and extrapolation to Martian values is difficult. Uncertainties that stem from estimates of volcanic unit volumes can be reduced through more detailed mapping and determination of flow thicknesses, which will include additional new data, obtained from the *Mars Observer* in the 1990s.

Carbon Dioxide. At both *Viking* landing sites, it was found that the temperature was appreciably above the carbon dioxide condensation boundary, thus precluding the occurrence of carbon dioxide hazes in northern summer at latitudes at least 50 °N. Thus, the ground level mists seen in these latitudes would appear to be condensed water vapor. Neutral mass spectrometers carried on the aeroshells of both *Viking* spacecraft indicated that carbon dioxide is the major constituent of the Martian atmosphere over the height range of 120 to 200 kilometers (74 to 124 miles). Densities for carbon dioxide measured by the upper atmospheric mass spectrometers on both *Viking* spacecraft were analyzed to yield height profiles for the temperature of the Martian atmosphere between the aforementioned range. The upper atmosphere of Mars was found to be surprisingly cold with an average temperature of less than 200 K (−73 °C; −99 °F). The atmosphere contains detectable concentrations of nitrogen, argon, carbon dioxide, molecular oxygen, atomic oxygen, and nitric oxide. The upper atmosphere exhibits a complex and variable thermal structure and is well mixed to heights in excess of 120 kilometers (74 miles).

Carbon Dioxide and Heat Balance at Polar Caps. As reported by Paige and Ingersoll (California Institute of Technology), the Infrared Thermal Mappers (IRTM) aboard the two *Viking* orbiters obtained solar reflectance and IR emission measurements of the Martian north and south Polar Regions during an entire martian year. The observations were used to determine annual radiation budgets and to infer annual carbon dioxide frost budgets. The results provide further confirmation of the presence of permanent CO_2 frost deposits near the South Pole and show that the stability of these deposits can be explained by their high reflectivities. In the north, the observed absence of solid CO_2 during summer was primarily the result of enhanced CO_2 sublimation rates due to lower frost reflectivities during spring. The results suggest that the present asymmetric behavior of CO_2 frost at the Martian poles is caused by preferential contamination of the north seasonal polar cap by atmospheric dust. The investigators emphasize that the *Viking* results have made it clear that the annual heat balance at the Martian poles is not purely a local phenomenon, but may be strongly influenced by the complex, global scale geologic and atmospheric processes that bring dust to the Polar Regions each year. The forthcoming *Mars Observer* mission, which will follow a polar orbit about the planet, will furnish much needed additional data.

Nitrogen. Results from the neutral mass spectrometer carried on the aeroshell of *Viking 1* spacecraft showed evidence for NO in the upper

[3] In terms of Earth, *juvenile water* refers to water which has been derived from the crystal rocks or from the interior of the Earth, and at the time of its appearance in the circulating water of the hydrosphere represents an accretion to the available water supply. Juvenile water is, therefore, water which has not previously been a part of the hydrosphere. Although many investigators have tried to device means for identifying juvenile water, no satisfactory method has been found. It is difficult to be certain that a particular water, such as that charged by hot springs, geysers, furaroles, etc., is of juvenile origin. Magmatic water is derived from a magma or included in a magma (rock melt).

atmosphere of Mars and indicated that the isotopic composition of carbon and oxygen is similar to that of earth. Mars is enriched in ^{15}N relative to earth by about 75%, a consequence of escape that implies an initial abundance of nitrogen equivalent to a partial pressure of at least 2 millibars. The initial abundance of oxygen present either as carbon dioxide or water must be equivalent to an exchangeable atmospheric pressure of at least 2 bars in order to inhibit escape-related enrichment of ^{18}O. McElroy, Yung, and Nier (1976) constructed models for the past history of nitrogen on Mars based upon *Viking* measurements showing that the atmosphere is enriched in ^{15}N. The enrichment is attributed to selective escape, with fast atoms formed in the exosphere by electron impact dissociation of N_2 and by dissociative recombination of N_2^+. The initial partial pressure of N_2 should have been at least as large as several millibars and could have been as large as 30 millibars if surface processes were to represent an important sink for atmosphere HNO_2 and HNO_3.

Krypton and Xenon. These gases were discovered in the Martian atmosphere with the mass spectrometer on *Viking Lander* 2. Krypton is more abundant than xenon. The relative abundances of the krypton isotopes appear normal, but the ratio of xenon-129 to xenon-132 is enhanced on Mars relative to the earth value for this ratio. The mass spectrometer on *Viking Lander 1* had previously reported the detection of ^{36}Ar and the establishment of upper limits on Ne, Kr, and Xe in the atmosphere. The upper limit of krypton was close the value that would be predicted if the $^{36}Ar/Kr$ ratio on Mars were identical to that on earth. As pointed out by Owen et al. 1976, the Earth's atmosphere is deficient in xenon compared with the primordial gas in meteorites, and this is exactly the situation found on Mars. The xenon deficiency on earth has been attributed to the preferential adsorption of xenon in shale's and other sedimentary material after it was outgassed. One is thus led to the tentative conclusion that similar processes have been active on Mars, perhaps in association with the epochs of fluvial erosion that have left their imprint on the planet's surface. Owen et al. suggest as an alternative or supplementary suggestion that some of the xenon could be absorbed in the regolith.

Weather

The atmosphere of Mars appears to favor stability, much unlike earth in that regard. The annual temperature range for the Martian surface at the *Viking* landing sites was computed on the basis of thermal parameters derived from observations made with infrared thermal mappers. Viking Lander 1 site showed small annual variations in temperature, whereas VL-2 site showed larger annual changes. (Locations of the sites are described in the latter portion of this article.) At both sites, daily temperature ranges at the top of the soil were 183 to 268 K (-90 to $-5\,°C$; -130 to $+23\,°F$). Diurnal variations decreased with depth in an exponential manner. The maximum temperature of soil sampled beneath rocks at the VL-2 site was computed to be 230 K ($-43\,°C$; $-45\,°F$). Daily patterns of temperature, wind, and pressure were highly repetitive at both sites during the early summer period. Wind was found to have a vector mean of 0.7 meter/second from the southeast with diurnal amplitude of 3 meters/second. Pressure exhibited both diurnal and semidiurnal oscillations, although of substantially smaller amplitude than those of VL-1. It should be mentioned that Mars does not have an ozone layer in its atmosphere as a shield against ultraviolet radiation and the absence of it, of course, has some effect on its climate.

It will be obvious to the reader that a satisfactory model of Martian weather cannot be formulated with so many unanswered questions as thus far indicated in this article. One popular concept, based upon incomplete data, suggests that the Earth and Mars commenced largely with similar initial conditions, but that over the course of some 4 billion years, the two planets have evolved differently. Both of these premises seem reasonable and logical. The images from *Viking* certainly prove that the two planets are distinctly different, yet when compared with other planets in the solar system, Earth and Mars present more similarities. Fundamental differences (not a direct function of the passage of time) between the two planets include: (1) *size* (Mars is only a little more than half the size of Earth); (2) the much greater distance from the sun of Mars than of Earth (hence less radiation received); and (3) the orbit of Mars is more eccentric, or elliptical, than that of Earth. These factors all have a fundamental bearing on a planet's atmosphere and weather system.

It has been generally established that a planet's size determines the strength of the internal heat sources, coming mainly from radioactive decay processes and gravitational energy released during accretion, that drive tectonic and volcanic activity. Even though, as a smaller planet Mars sustained volcanic activity (as evident from *Viking* images), it was not imbued with the volcanic potential of Earth, which continues apace. Plate tectonics on Earth permits frequent access to internal heat sources, whereas *Viking* images indicate no recent evidence of plate tectonics; the entire crust appears to be a single plate. (Not full agreement among observers regarding this point.)

The greater distance of Mars from the sun obviously is a factor that must be built into any model of Mars. However, it may not be a major determining factor. With essentially general agreement that liquid water once existed on Mars and with suspicions that it may exist as ice in the regolith today (still very speculative), it follows that Mars received significant radiation from the sun in its early, formative periods, prior to losing its primitive atmosphere, to maintain water in the liquid state. (It has been theorized that once volcanic activity essentially abated on the planet, the atmosphere CO_2 level and hence greenhouse effect declined, causing cooling to the point where liquid water could no longer exist.) Kahn (Washington University) suggests that the surface pressure on the planet is so low today because CO_2 continued to be removed from the atmosphere and stored as carbonates by transitory pockets of liquid water. As interpreted by Haberle (NASA Ames Research Center), such pockets could have existed long after the global mean temperature had dropped below the freezing point; specifically, they could form as long as the surface pressure was sufficiently high to limit evaporation. Through the action of transitory water pockets the pressure was gradually reduced to its present value of 6.1 millibars. Haberle further observes that the small size of Mars probably had at least as much influence on its climate as has its distance from the sun. Moreover, the size of Mars has determined the fate not only of water and CO_2, but also of nitrogen, which is relatively scarce in the planet's atmosphere. The lower level of volcanic activity on Mars meant that less nitrogen was outgassed than on Earth. The smaller gravitational pull of Mars also made it easier for nitrogen to escape. (Although nitrogen does not have enough thermal energy to escape, it can acquire the necessary energy by dissociative recombination.)

Because of the eccentricity of Mars' orbit, the seasons on the planet are of unequal duration and intensity. (The Martian year is 687 Earth days long.) Perihelion (closest approach to the sun) occurs late in the southern hemisphere, making it 52 Earth days shorter than in the north. When at perihelion, Mars receives some 40% more solar radiation than when at aphelion. In terms of the Earth, this difference is only 3%. This asymmetry of seasons markedly affects the weather on Mars as it is known today, particularly influencing the cycles of CO_2, of water, and of dust (an important factor in the planet's current climate.) These cycles have been plotted by several researchers and are contained in the Haberle reference listed.

Obvious major differences in the weather of the two planets cast doubts on attempting to compare Earth models with those of Mars. The thin atmosphere of Mars today eliminates any consideration of a greenhouse effect. The absence of oceans eliminates the Earth's familiar hydrologic cycle. As pointed out later, these major differences in the way the martian atmosphere and "weather" system has evolved helps to explain some of the very unusual geologic features found in the Viking images.

Soil and Rocks

Nergal, Ares, and Mars were legendary names for a pinpoint of reddish light in the night sky, observed to move relative to the star field even in ancient times. Because of its color, Mars was an important part of the mythology of early civilizations, serving as an abode for gods of fire, war and terror in the minds of many populaces through the centuries. The ancients would not have been disappointed in the coloring (reddish-brown) of most of the Martian soil and rocks.

Steps taken in the operation of the surface sampler on the *Viking* landers is shown in Fig. 19. As pointed out by the x-ray analysis team[4] elemental analyses of fines in the martian regolith at the two widely separated landing sites were remarkably similar. At both sites, the uppermost

[4] Scientists with Martin-Marietta Aerospace Corp., NASA Langley Research Center, Pomona, College, the University of New Mexico, and the U.S. Geological Survey.

Fig. 19.   Operation of the surface sampler in obtaining martian soil was closely monitored by one of the Lander cameras because of the precision required in trenching a small area (8 × 10 inches; 20 × 25 centimeters) surrounded by rocks. The exposure of thin crust appeared in unique contrast with surrounding materials and became a prime target for organic analysis in spite of potential hazards. The large rock in the foreground is only 8 inches (20 centimeters) high. At left, the sampler scoop has touched the surface, missing the rock at upper left by a comforTable 6 inches (15 centimeters), and the backhoe had penetrated the surface about 0.5 inch (13 millimeters). The scoop was then pulled back to sample the desired point and (second view) the backhoe furrowed the surface, pulling a piece of thin crust toward the spacecraft. The initial touchdown and retraction sequence was used to avoid a collision between a rock (in the shadow of the arm) and a plate joining the arm and scoop. The rock was cleared by 2 to 3 inches (5 to 7.5 centimeters). The third picture was taken 8 minutes after the scoop touched the surface and shows that the collector head has acquired a quantity of soil. With the surface sampler withdrawn (right), the foot-long (0.3-meter) trench is seen between the rocks. The trench is 3 inches (7.5 centimeters) wide and about 1.5 to 2 inches (3.8 to 5 centimeters) deep. The scoop reached to within 3 inches (7.5 centimeters) of the rock at the far end of the trench. Penetration appears to have left a cavernous opening roofed by the crust and only about one inch (2.5 centimeters) of undisturbed crust separates the deformed surface and the rock. (*NASA; Jet Propulsion Laboratory, Pasadena, California.*)

regolith was found to contain abundant silicon and iron, with significant concentrations of magnesium, aluminum, sulfur, calcium, and titanium. The sulfur concentration is one to two orders of magnitude higher, and potassium (<0.25% by weight) is at least five times lower than the average for the earth's crust. The trace elements strontium, yttrium, and possibly zirconium, were detected at concentrations near or below 100 parts per million. Pebble-sized fragments at VL-1 site were found to contain more sulfur than the bulk fines, and were thought to be pieces of a sulfate-cemented duricrust. It is interesting to note that no phosphorus was found on Mars.

Each *Viking* lander carried an energy-dispersive X-ray fluorescence spectrometer for elemental analysis of samples of the Martian surface. This composition is best interpreted as representing the weathering products of mafic igneous rocks. A mineralogic model, derived from computer mixing studies and laboratory analog preparations, has suggested that the martian fines could be an intimate mixture of about 80% iron-rich clay; about 10% magnesium sulfate (kieserite perhaps), about 5% carbonate (calcite?) and about 5% iron oxides (hematite, magnetite, maghemite, goethite?). The mafic nature of the fines, which appear to be distributed globally, and their probable source rocks seem to preclude large-scale planetary differentiation of an earthly nature. The samples were characterized by abundant red-colored fine material and scattered blocks of generally angular rocks. More diversity was found at VL-1 than at VL-2.

Terrain

The Martian terrain in the vicinity of the two *Viking* lander sites is well illustrated in latter part of this article. Of course, these sites represented only a portion of the planet. There are ten or more volcanoes prominent on Mars, with scientific estimates of their age ranging from 100 million years to 1 billion years. See Figs. 20 and 21.

Evidence of erosion, including dry channels resembling riverbeds and tributaries, has led many analysts to conclude that Mars may have had a warmer, more water-rich climate in the past. Photographic evidence from spacecraft indicates that the once-reported "canals" are mostly illusory and that the dark patchy markings once suspected to be vegetation along these canals, varying with the seasons, are in reality simply deposits of

wind-blown dust that may be altered from time to time. That alterations can and do occur on Mars is shown by Fig. 22. With the benefit of a few years to assimilate observational data from *Mariner* and the *Viking* missions, the findings of Pieri et al. (1980), are exceptionally interesting. As explained by Pieri, *Mariner 9* orbital reconnaissance mission discovered ubiquitous valley networks in heavily cratered terrain. Branching valley networks throughout the heavily cratered terrain of Mars exhibit no compelling evidence for formation by rainfall-fed erosion (one of the popular hypotheses). Rather, the networks are diffuse and inefficient, with irregular tributary junction angles and large, undissected intervalley areas. The deeply entrenched canyons, with blunt amphitheater terminations, cliff-bench wall topography, lack of evidence of interior erosion of flow, and clear structural control, suggest headward extension by basal sapping.[5] The size-frequency distributions of impact craters in these valleys and in the heavily cratered terrain that surrounds them are statistically indistinguishable, suggesting that valley formation has not occurred on Mars for billions of years.

Pieri observes that the branching and coalescent character of the channels provoked immediate comparison to terrestrial riverine networks produced by fluvial[6] erosion, a process driven primarily by rainfall. Liquid water, however, cannot now exist at the surface of Mars for more than a few minutes owing to a very low atmospheric pressure and very cold temperatures during most of the year at most latitudes. It has been suggested that these features, if formed by processes similar to those that operate in the formation of earthly river systems, are relics of a more clement epoch. Thus, a major problem in the study of Mars is whether the valley networks could have evolved under current surface conditions, or whether a major shift in Martian climate occurred.

There are certain unifying characteristics of Martian valleys. (Valleys are distinguished from channels by the absence in the former of direct

---

[5] The undercutting, or breaking away of rock fragments, along the headwall of a cirque (semicircular, amphitheater-like, or armchair-shaped hollow of nonglacial origin), due to frost action at the bottom of the bergschrund (a deep and often wide gap or crevasse).

[6] Of or pertaining to a river or rivers — produced by the action of streams.

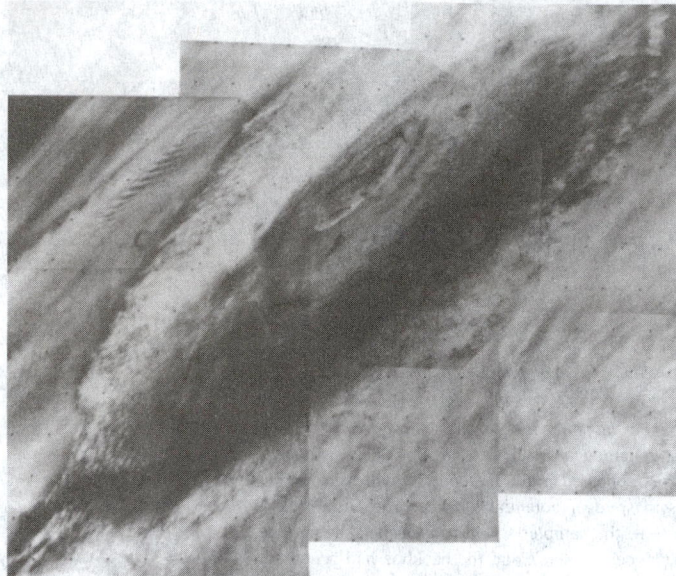

Fig. 20. The great Martian volcano Olympus Mons was photographed by the *Viking 1 Orbiter* on July 31, 1976 from a distance of 8000 kilometers (5000 miles). The 24-kilometer-high (15-miles) mountain is seen in mid-morning, wreathed in clouds that extend up the flanks to an altitude of about 19 kilometers (12 miles). The multi-ringed caldera (volcanic crater), some 80 kilometers (50 miles) across, pushes up into the stratosphere and appears cloud-free at this time. The cloud cover is most intense on the far western side of the mountain. A well-defined wave cloud train extends several hundred miles beyond the mountain (upper left). The planet's limb can be seen at the upper left-hand corner of the view. It also shows extensive stratified hazes. The clouds are thought to be composed mainly of water ice condensed from the atmosphere as it cools while moving up the slopes of the volcano. In the martian afternoon, the clouds develop sufficiently to be seen from earth and it is known that they are a seasonal phenomenon largely limited to spring and summer in the northern hemisphere of the planet. Olympus Mons is about 600 kilometers (375 miles) across at the base and would extend from San Francisco to Los Angeles. (*NASA; Jet Propulsion Laboratory, Pasadena, California.*)

Fig. 21. Fine detail in the interior of a martian crater can be seen in this photo taken by *Viking 1* of an area near the *Viking 2* landing site. The crater (on the left margin of the view) is about 40 kilometers (25 miles) in diameter and shows many features found in lunar craters. The central portion is crossed by numerous cracks. Similar features are seen at the huge lunar impact basin, Orientale. Their origin is unknown, but it has been suggested that the cracks were formed either by consolidation of lava that filled the crater after it formed, or by fallback from the impact process. Alternatively, the cracks may have formed long after the impact event by uplift of the crater floor. Between the cracked terrain and the crater rim is a region of chaotic debris. Beyond the rim there is no evidence of an ejecta blanket (rock material which is blasted from the crater by the shock of the impacting meteorite). The ejecta blanket is presumably overcovered by later deposits. (*NASA; Jet Propulsion Laboratory, Pasadena, California.*)

evidence of fluid erosion often found in the latter.) There is no clear evidence (streamlined obstacles, interior channels) of direct fluid erosion in any Martian valley. It is possible that such features are too small to be observed by present instrumentation, although *Viking Orbiter* images as small as 100 meters (328 feet) can be resolved. Walls of the valleys are typically rugged and cliff like, with some debris accumulation and talus, and the floors are generally flat. Mantling by materials of eolian and volcanic origin is common. Some valleys display cliff-bench interior topography, similar in character and scale to features in the Grand Canyon of the Colorado River. The most striking morphological characteristics, however, are the presence of steep-walled, cuspate terminations at the heads of the smallest tributary valleys. These steep-walled, amphitheater terminations suggest headward extension (sapping) by basal undermining and wall collapse, as in the predominant mode of headward extension for many earthly canyons. A variety of Martian terrain is shown in Figs. 23, 24, 25 and 26.

Martian valley networks lack the dendritic pattern so common to terrestrial streams. The Martian valley patterns show remarkable parallelism and lack of tributary competition for undissected intervalley terrain, and thus appear diffuse compared to terrestrial systems. Viewed from spacecraft, terrestrial drainage systems have a fine, filigreed texture, whereas Martian systems appear coarse.

Pieri has suggested that the valleys were formed on Mars during an ancient epoch by erosional processes involving not rainfall, but the movement of groundwater and its participation as a liquid or a solid in the undermining of less competent strata, causing progressive headward collapse. These processes, combined with modification by impact and eolian (wind) processes, have produced the degraded valleys seen on Mars today.

Even a brief description of the physical features of Mars is not complete without mention of the so-called "Spokane Flood" concept. As summarized by Baker 1978, in a series of papers published between 1923 and 1932, J.H. Bretz described an enormous plexus of proglacial stream channels that eroded into the loess and basalt of the Columbia Plateau in eastern Washington state. Bretz argued that this region (which he termed the Channeled Scabland) was the product of a cataclysmic flood, which he called the Spokane flood. Considering the nature and vehemence of the opposition to his hypothesis, which at one time was considered highly imaginative, its eventual scientific verification constitutes a fascinating episode in the history of modern science. The discovery of possible catastrophic flood channels on Mars has given new relevance to Bretz's insights. The connection between Bretz's proposal and parts of the Martian surface is well developed by Baker.

Volcanism. As readily apparent from *Viking* images, volcanism on Mars was widespread. According to Lucchitta (U.S. Geological Survey), volcanism has formed enormous shields, large composite cones, lobate lava flows, and possibly small cones and pseudocraters. Flood basalts similar to those filling lunar maria may have resurfaced ridged highland plateaus. Large deposits of pyroclastic material may also exist, although their presence is controversial. Dark patches are common on the planet as they are on Earth's moon, where they have been interpreted as pyroclastic materials. The association of dark patches with pyroclastic volcanism on Mars has been largely overlooked, because most dark patches are inside craters and were obviously accumulated by wind; the possibility was

Fig. 22. Changes observed by the *Viking Landers* over a period of time included water-ice snow seen by VL-2 during the winter at Utopia Planitia, and a thin dust layer deposited at both sites during the dust storms of 1977. As shown here, a change occurred by Chryse Planitia over a 4-day period in September 1978. Top photo is the "before" and bottom photo is the "after" view. Change (A) appears as a small circle-like formation on the side of a drift in the lee, or downward, side of "Whale Rock." This is believed to have been a small-scale landslide of an unstable dust layer which had accumulated behind the rock. Interpretation of this feature would be difficult without an earlier change (B) near "Big Joe," a slump. The new slump is observed approximately 25–35 meters (82–115 feet) from the lander craft and just under 1 meter (3.3 feet) across. This slumping was probably initiated by the daily heating and cooling of the surface by solar radiation. More importantly, it is now believed that, based upon the repeated occurrence of such slumping features, a dust layer which overlies the surface may, in fact, be redistributed fairly regularly during periods of high wind activity. (*NASA; Jet Propulsion Laboratory, Pasadena, California.*)

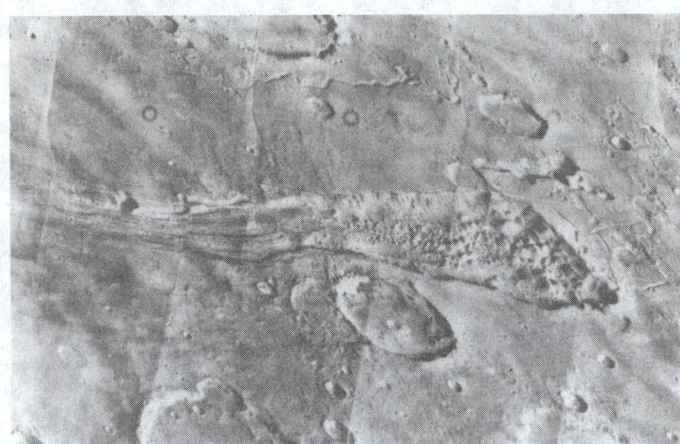

Fig. 23. The sinous rille (a relatively long, narrow, trench- or cracklike valley) at the top of this mosaic of 8 photos is believed by some scientists as indicative of flooding of the high plateau in the vicinity of an alternative landing site (known as Capri) for *Viking Lander 2*. In the foreground is a valley that may have been caused by downfaulting of the martian crust. The hummocks (rounded or conical knolls or mounds of comparatively small elevation) on the valley floor look like chaotic terrain. Some scientists have suggested that the subsidence may be partially caused by melting of the subsurface ice. The large areas of the collapsed terrain show the regional extent of this phenomenon. These views were taken by *Viking 1* on July 3, 1976 from a range of 2300 kilometers (1400 miles) and cover an area of about 300 × 300 kilometers (180 × 180 miles). South is toward the top as seen from the spacecraft. (*NASA; Jet Propulsion Laboratory, Pasadena, California.*)

neglected that some Martian dark patches, like lunar ones, may reflect pyroclastic vents. Lucchitta describes dark patches in Valles Marineris that may be such vents and may reflect young mafic volcanism. The evidence for past volcanism on Mars is commonly accepted, but none has been documented in the Valles Marineris equatorial rift system. A recent survey of the troughs in this valley revealed dark patches that are interpreted to be volcanic vents. The configuration and association of these patches with tectonic structures suggest that they are of internal origin; their albedo and color ratios indicate a mafic composition; and their stratigraphic position, crispness of morphologic detail, and low albedo imply that they are young, perhaps even recent. If this volcanism is indeed as young as it seems, Mars has been an active planet throughout most of its history.

Case for an Early "Wet" Mars. Certain features of the Martian surface, as observed by the Viking missions, have continued to intrigue and confuse analysts of the data returned from Mars. Included are ancient valleys, channels, and what appear from images to be tributary systems. What appear to be numerous extinct volcanoes and meteoritic craters over which more recent geologic features have been superposed are found in the images. For many of these features, the presence of water in relatively large quantities on the planet during its earlier phases offers the most tempting solution. Many scenarios have been developed. For example, some scientists at the Jet Propulsion Laboratory (Pasadena, California) suggested, at a symposium of the Lunar and Planetary Institute (see reference listed), that lakes, or a shallow sea or ocean, may have encompassed as much as 10 to 15% of the Martian surface and of a generous portion of the northern plains from 2 to 3 billion years ago.

Fig. 24. View taken by *Viking 1* on July 3, 1976 from a range of 2000 kilometers (1240 miles), looking southward across Valles Marineris. This huge equatorial canyon is about 2 kilometers (1.2 mile) deep. The area shown is 70 kilometers (43 miles) by 150 kilometers (94 miles). Aprons of debris on the canyon floor indicate how the canyon may have enlarged itself. The walls appear to collapse at intervals to form huge landslides that flow down and across the canyon floor. Linear striations on the landslide surface show the direction of flow. On the canyon's far wall in this view, one landslide appears to have ridden over a previous one. Streaks on the canyon floor, aligned parallel to the length of the canyon, probably are evidence of wind action. Layers in the canyon wall indicate that the walls are made up of alternate layers of lava and ash or wind-blown deposits. (*NASA; Jet Propulsion Laboratory, Pasadena, California.*)

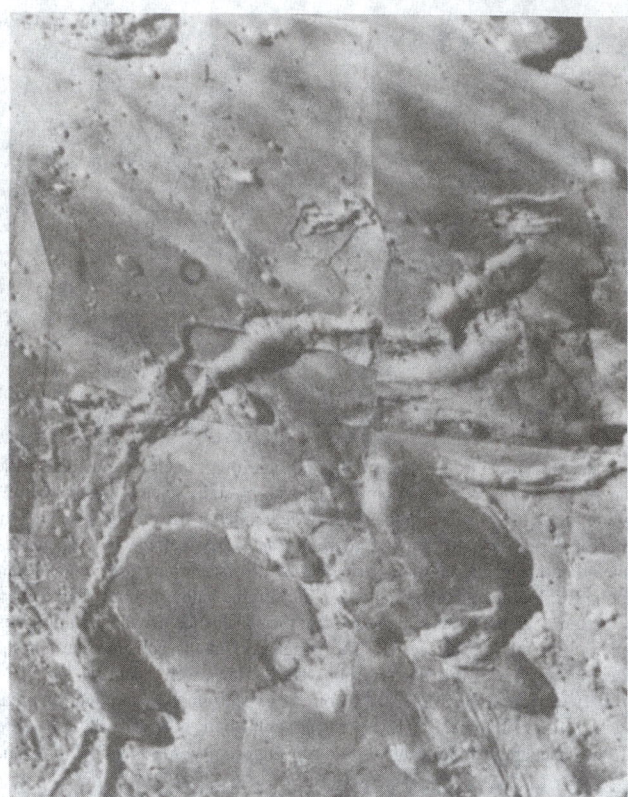

Fig. 25.   A view obtained by *Viking 1* on July 8, 1976 showing what appear to be fault zones in the martian crust in an area two degrees south of the equator. The fault valleys are widened by mass wasting and collapse. Mass wasting is the downslope movement of rocks due to gravity (possibly hastened by seismic shaking if present). (*NASA; Jet Propulsion Laboratory, Pasadena, California.*)

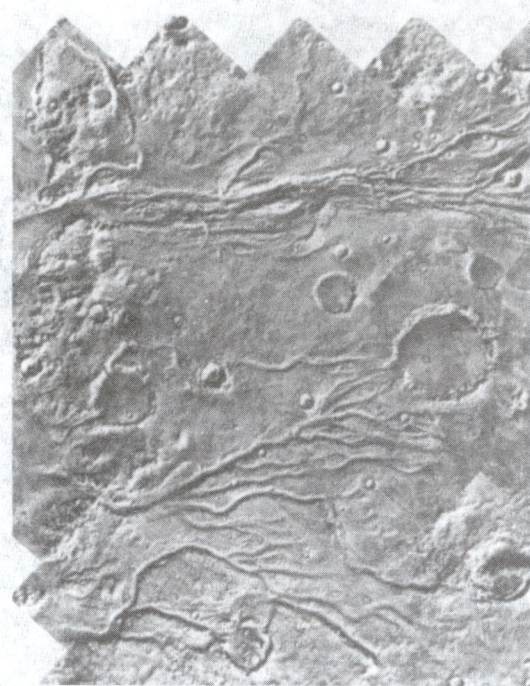

Fig. 26.   Mosaic of Martian surface made by *Viking Orbiter 1* over a period between August 4 and 9, 1976. The area is centered at 17°N, 55°W, to the west of the Viking 1 Lander site in Chryse Planitia. Just to the west of this area are the plains of Lunae Planum. The terrain shown in this view slopes from west to east with a drop of about 3 kilometers (1.9 mile). The channels are a continuation of those to the west of the VL-1 landing site and, to some scientists, suggest a massive flood of waters from Lunae Planum, across this intervening cratered terrain, and into the general region of the VL-1 landing site. In several cases, it will be noted that channels cut through craters; in others, the craters are clearly later than the assumed flood and are superimposed in the channels. (*NASA; Jet Propulsion Laboratory, Pasadena, California.*) See Fig. 14.

Timothy Parker (JPL) estimates that the water would have been about 100 meters deep (or less), making the sea's volume equivalent to a layer of about 10 meters deep covering the globe. This volume of water in a hypothetical sea would equal all the water that some geochemists have allowed for the entire planet. But, all do not agree. Michael Carr (U.S. Geological Survey) reported that the latest estimates of the amount of water hidden beneath the surface may be several times greater. Some investigators believe that the surface of Mars, something comparable to Earth's moon, is made up of rubble and porous soil well capable of storing ice, water, or brine. Stephen Clifford (Lunar and Planetary Science Institute) estimates that this megaregolith has the capacity to hold water equivalent to a global layer 200 to 500 meters deep. The main remaining question, of course, is how much of that capacity is actually filled? Certainly, the unanswered question of the amount of water that was and still may be trapped on the planet is central to preparing a satisfactory model of the planet.

Age Determination. Mars has been mapped extensively by *Mariner 9* and later by the *Viking* missions. One major goal in planetary science is to determine the chronology of development of the surfaces of the terrestrial planets, particularly Mars. As indicated in an excellent paper by Neukum and Wise 1976, cratering links to lunar time suggest that Mars died long ago. Fortunately, for the purpose of age determination from photographs, Mars is impact-cratered. Differences in impact crater frequencies at different sites reflect differences in age. Two attempts have been made to determine absolute age for Mars from its measured crater frequencies, based on extrapolations from the cratering chronology of the lunar surface (Hartman, 1973; Soderblom et al., 1974). Unfortunately, a straightforward comparison of Martian and lunar crater frequencies does not necessarily yield true ages; relative impact rates and the time dependence of the martian cratering rate are not known; and it is not certain whether the same meteoroid population bombarded both planets.

At *Mariner 9* resolution, the impact crater production size-frequency distribution of Mars is generally similar to that of the moon for crater diameters in the range 0.8 to 50 kilometers (0.5 to 31 miles), and it appears to have been relatively stable through time. The lunar and Martian crater curves can be brought into near coincidence by a diameter shift appropriate to reasonable impact velocity differences between bodies hitting Mars and the moon. This indicates that a common population of bodies impacted both planets and suggests the same or a very similar time dependence of impact flux. Constraints on relative lunar and Martian fluxes can be obtained by comparing crater frequency data for the lunar and Martian highlands and for Mars' satellite Phobos.

These cratering constraints, as pointed out by Neukum and Wise, provide the basis for a tentative Martian time scale derived from lunar data. Previous time scales have painted a picture of a disorderly planetary evolution of Mars, punctuated by a strange pulse of Tharsis Ridge tectonic and volcanic activity late in geological history. The new scale suggests a much more orderly evolution with Mars, like the moon, winding down most of its major planetary tectonic and volcanic disturbances in the first 1.5 billion years of its history. By 2.5 billion years ago the volcanic-tectonic era on Mars had ended.

Other Physical Characteristics of Mars

Doppler radio-tracking data have provided detailed measurements for a Martian gravity map extending from 30°S to 65°N latitude and through 360° of longitude. The feature resolution is approximately 500 kilometers (310 miles), revealing a huge anomaly associated with Olympus Mons, a mascon in Insidis Panitia, and other anomalies correlated with volcanic structures. Olympus Mons has been modeled as a disk of 600-kilometer (372-mile) surface area, having a mass of 9.7×10^{21} grams. The very large Olympus Mons anomaly should have a very significant impact on geophysical modeling of the planet. Similarly, the Elysium anomaly and the Insidis mascon should place constraints on the internal structure. Gravity in the southern hemisphere remains poorly resolved.

A three-axis short-period seismometer was delivered to the surface of Mars by *Viking Lander 2* on September 3, 1976. Noise background correlated well with wind gusts. Data returned to earth indicated that Mars is a very quiet body.

The amounts of magnetic particles held on the reference test chart and backhoe magnets on *Viking Landers 1* and *2* were comparable, indicating the presence of an estimated 3–7% (weight) of relatively pure, strongly magnetic particles in the soil. It is argued that the results indicate the presence, now or originally, of magnetite, which may be titaniferous.

Dust Devils on Mars. Several scientists, after studying Viking data, have reported the existence of dust devils (columnar, cone-shaped, and funnel-shaped clouds rising 1 to 6 kilometers (0.6 to 3.7 miles) above the surface) on Mars. Dust devils result from atmospheric conditions that occur close to the ground and are, therefore, sensitive to surface topography. Dust devils on Mars may be responsible for the initiation of large dust storms on the planet and for increasing the general atmospheric dust content.

Dust devils, as observed by Thomas and Gierasch (Cornell University), have meteorological as well as geological significance. Fluid motions in an atmospheric boundary layer can be driven either by stresses due to the mean wind (forced convection) or by buoyancy due to heating of the gas adjacent to the surface (free convection). Dust devils are an example of the latter. On Earth, large-scale eolian transport is generally due to forced convection. The investigators report that moderate to high winds characterize forced convection, and on Mars, where the atmospheric density is only about 1% of that on Earth, it is estimated that winds must exceed about 25 to 40 meters sec^{-1} to initiate soil movement.

One of the major geologic processes on Mars is the entrainment and transportation of dust by winds. Observations on the genesis and development of local and global dust storms on Mars are sparse.

Tornadolike Tracks on Mars. Some images from the *Viking Orbiter* reveal well defined, dark filamentary lineations in numerous locations on the Martian surface. On Earth, tornadic-intensity vortices commonly leave distinctive tracks whose appearance is similar to that of the Martian lineations. A high-resolution imaging system, as proposed for the *Mars Observer* mission, could resolve these ground tracks. The filamentary lineations, as reported by Grant and Schultz (Brown University) are from 2 kilometers (1.2 miles) to at least 75 kilometers (47 miles) long and less than 1 kilometer (0.6 mile) wide. Most are straight to curvilinear, and some have obvious nontopographically initiated gaps in their path. The visible occurrence of the lineations appears to be seasonal. In the southern hemisphere, they were visible (from *Mariner 9*) only from midsummer into early fall. After formation, they were rapidly modified and were no longer visible by midfall. In the northern hemisphere, lineations appear from early to midsummer. By late summer, these lineations also become smeared and faint.

Natural Laser Phenomenon Noted on Mars. Based upon observations made with the Goddard infrared heterodyne spectrometer during the period of January to April 1980, when the planet was near opposition, astronomers M.J. Mumma and colleagues (NASA-Goddard Space Flight Center) and D. Zipoy (University of Maryland) noted natural gain amplification in the mesosphere of Mars, probably representing the first definite identification of a natural infrared laser. Natural microwave amplifiers (masers) are abundant in interstellar clouds and some circumstellar shells, primarily among the rotational level populations of certain molecules, such as OH, SiO, and H_2O, but no optical lasers in nature had previously been observed. As reported by Mumma et al., many examples of natural nonthermal optical emission have been found, such as the infrared and ultraviolet auroras or the day glows of Earth, Jupiter, Mars, and Venus. Details are reported in the reference listed.

Pole Wandering and Crustal Shifts on Mars

Careful study of *Viking* images has revealed a number of features of the planet, which are very difficult to explain. For example, regions at the planet's equator seem once to have been near a pole. As observed by Schultz (see reference), in certain areas of the surface, erosion appears to have occurred at a very low rate (perhaps less than a millimeter in a million years). But, in other areas, at the same latitude, there are regions that have been heavily stripped and etched by the wind. Also, very old networks of narrow valleys, once cut into the surface by flowing water, suggest a warm climate, although such networks are seen within 10 degrees of the southern polar ice cap. While many details remain to be worked out, Schultz suggests that one hypothesis may explain all or most of the

contradictions: the orientation in space of the Martian crust has not always remained the same throughout geologic time, but rather, it has shifted with respect to the planet's axis of spin. This would require that the spin axis, which intersects the planet's surface at the north and south poles, would appear to have wandered over the planet's crust. This would indicate that certain regions of the crust, presently far from the poles, may have been at some time in the past within the Polar Regions. In introducing a detailed paper on this topic, Schultz observes that if Mars had undergone polar wandering, then Martian geology may have to be viewed in the context both of a dynamically changing planet like the Earth and of a stable, rigid body like the Earth's moon. In this sense, the Martian equivalent of plate tectonics might simply be the movement of the entire lithosphere, the solid outer portion of the planet, as one plate.

Martian Satellites

Mars has two satellites, Phobos, the inner and larger of the two moons, and Deimos. These satellites were visited during the *Viking* missions.

Phobos. This satellite revolves around Mars in an orbit of about 9,330 kilometers (5,800 miles) from the center of Mars (some 5,950 kilometers; 3,700 miles above the planet's surface). The diameter was stated in the introductory section of this entry. Its orbital period is 7 hours, 40 minutes. Because its orbital period is in the same direction as, but is less than, that of Mars, it rises in the west and sets in the east as seen from Mars. Phobos is heavily cratered and dark in color, of a material resembling carbonaceous chondrite meteorites. A system of grooves, possibly marking fractures, is associated with the largest crater, Stickner, which is about 10 kilometers (6.2 miles) across.

Viking Orbiter 1 flew within 480 kilometers (300 miles) of Phobos to obtain the view given in Fig. 27. A view much closer to the satellite is given in Fig. 28. A considerably later view, made in 1978, is given in Fig. 29.

Deimos. This satellite revolves around Mars in an orbit about 23,000 kilometers (14,260 miles) from the center of Mars. Five close flybys, within 1000 kilometers (620 miles) of Deimos, were made in October 1977. The closest encounter was on October 5, 1977 when the spacecraft passed within 50 kilometers (30 miles) of the moon's surface. Images indicated that the surface of Diemos differs considerably from that of Phobos. Deimos has many craters, but appears smoother than Phobos. See Fig. 30. With reference to the peculiar blocks observed on Deimos, which are visible in the illustration, Duxbury and Veverka 1978 suggest: "If the bright patches and blocks represent ejecta, then it is puzzling why apparently so much of it was retained by such a small satellite and why the process seems to be so much more efficient on Deimos than on Phobos. It is conceivable that the very close proximity of Phobos to Mars makes it easier for impact ejecta to escape from the inner satellite, but the mechanics of such a preferential process remain to be worked out."

Illustrations of Phobos and Diemos are also given in the entry on **Asteroid**.

Additional Post-Viking Mission Studies and Hypotheses

Further studies of the Viking information and observations made from Earth in recent years have posed interesting new questions pertaining to Mars.

Radar Images of Mars. In late 1991, D.O. Muhleman and a team of researchers (California Institute of Technology) conducted aperture synthesis mapping of Mars by using the Very Large Array (VLA) in New Mexico as the imaging instrument to detect continuous wave signals transmitted at 9.5 GHz (3.5 cm) from the Jet Propulsion Laboratory (JPL) 70-meter antenna in Goldstone, California. See also **Antenna (Communications)**. Summary of the project: "The surface of Mars was illuminated with continuous wave radiation. The reflected energy was mapped in individual 12-minute snapshots with the VLA in its largest configuration; fringe spacing's as small as 67 kilometers (42 miles) were obtained. The images reveal near-surface features, including a region in the Tharsis volcano area, over 2,000 kilometers (1,243 miles) in east-west extent, which displayed no echo to the very low level of the radar system noise. This feature (called *Stealth*) is interpreted as a deposit of dust or ash with a density less than about 0.5 grams/cubic centimeter and free rocks larger than 1 centimeter (0.4 inch) across. The deposit is envisioned to be several meters thick and may be much deeper. The strongest reflecting

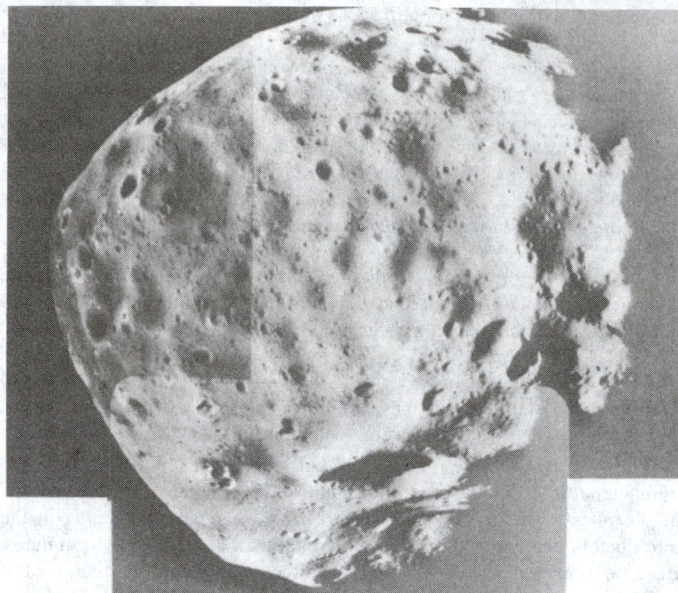

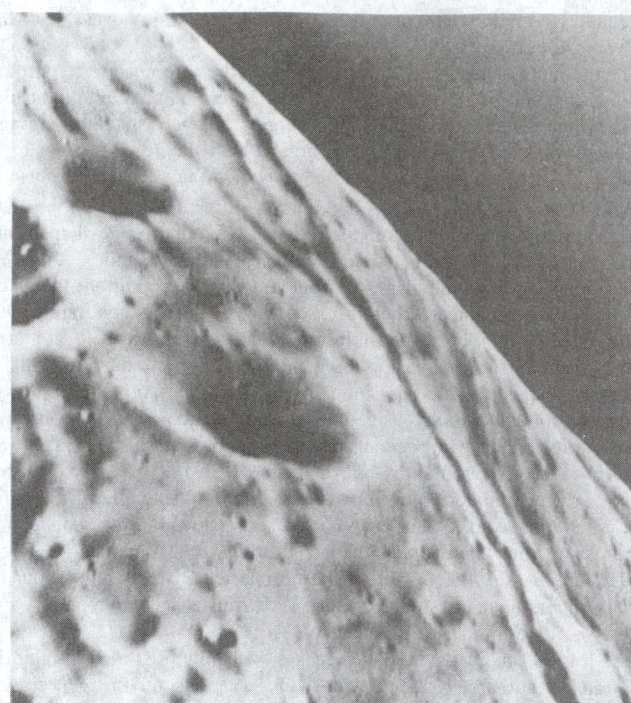

Fig. 27. View of Phobos taken by *Viking Orbiter 1* from a distance of 480 kilometers (300 miles). This mosaic of 3 pictures was made in February 1977. As seen here, Phobos is about 75% illuminated and is about 21 kilometers (13 miles) across and 19 kilometers (11.8 miles) from top to bottom. North is at top. The south pole is within the large crater (Hall) with a diameter of 5 kilometers (3.1 miles) and will be noted at bottom center where the pictures overlap. Some features as small 20 meters (65 feet) across can be seen. Remarkable features include striations, crater chains, a linear ridge, and small positive features which appear to be resting on the surface. A long linear ridge is seen starting near the south pole and extending to the upper right. A very sharp wall at the intersection of two craters (about 1 kilometer; 0.6 mile across) is seen along this ridge at right. A series of craters runs horizontally in the picture which is parallel to the orbit plane of Phobos. These crater chains are commonly associated with secondary cratering by ejecta from larger impacts. A surprising discovery has been made of what apparently resembles hummocks or small positive features. These features, primarily seen near the terminator (right), are about 50 meters (165 feet) in size and may be surface debris from previous impacts. (*NASA; Jet Propulsion Laboratory, Pasadena, California.*)

Fig. 28. *Viking Orbiter 1* took this close-up picture of Phobos from a range of 120 kilometers (75 miles) on February 20, 1977. This is the closest range at which a spacecraft has photographed the tiny satellite. At that range, Phobos is too large to be captured in a single frame. This picture covers an area 3×3.5 kilometers (1.86×2.17 miles). A single picture element is about 3 meters (7.5 feet) across. However, the high relative speed of Orbiter 1 and Phobos caused some image smear so that the smallest surface feature identifiable is between 10 and 15 meters (32 and 49 feet). The picture shows a region in the northern hemisphere of Phobos that has striations and is heavily cratered. The striations, which appear to be grooves rather than crater chains, are about 100 to 200 meters (328 to 656 feet) wide and tens of kilometers long. Craters range in size from 10 meters (32 feet) to 1.2 kilometers (0.75 mile) in diameter. The surface of Phobos appears similar to the highland regions of the earth's moon, which also is heavily cratered and an ancient terrain. The dark region above the limb of Phobos is an artifact of processing and does not indicate an atmosphere. (*NASA; Jet Propulsion Laboratory, Pasadena, California.*)

geological feature was the south polar ice cap, which was reduced in size to the residual south polar ice cap at the season of observation. The cap image is interpreted as arising from nearly pure carbon dioxide or water ice with a small amount of Martian dust (less than 2 percent by volume) and a depth greater than 2 to 5 meters (6.5 to 16 feet). Only one anomalous reflecting feature was identified outside of the Tharsis region, although the Elysium region was poorly sampled in this experiment and the North Pole was not visible from Earth." More detail is given in reference listed.

Radar Detection of Phobos. During the exceptionally close approach of Mars to Earth in the autumn of 1988, the Goldstone 70-meter (230 foot) antenna was used as a radar telescope to observe Phobos. See also **Deep Space Network**. A total of 117 transmit/receive cycles were completed. Radar echoes from the Martian satellite provided information about the object's surface properties at scales near the 3.5-centimeter (1.4 inch) observing wavelength. In summary, "Phobos's surface apparently resembles those of many (if not most) large, C-class asteroids in terms of bulk density, small-scale roughness, and large-scale topographic character, but differs from the surfaces of the moon and at least some small, Earth-approaching objects. Additional 3.5-centimeter (1.4 inch) and 13-centimeter (5 inch) radar observations of asteroids, comets, and the martian satellites can clarify these relations."

Simulating the Surface of Phobos. Researchers (University of Arizona, Lunar and Planetary Laboratory) in recent years have been attempting to simulate how certain distinct features of Phobos may have been formed and as imaged by the Viking orbiters. Lines (rows) of comparatively small craters resemble a beaded chain unlike other features found in the planetary system thus far. An initial hypothesis described the features as being formed out of secondary ejecta—that is, debris resulting in a

crater-causing impact. In attempting to duplicate the unusual feature in the laboratory, the researchers have developed an apparatus consisting of a pair of narrow, rigid glass plates and a variety of materials, including expanded vermiculite, silica sand, and small glass spheres. Intense interest in unusual surface conditions is not new in connection with planetary space explorations. Ponder, for example, the variety of scientific opinions that were expressed prior to touchdown of *Surveyor*, the first unmanned spacecraft to land on the Moon. From their work to date, the researchers have suggested that Phobos may be covered with a regolith some 300 meters thick!

A Peopled Mission to Mars. Although, in early 1990s, it was delayed for want of funding and other political considerations (due in part to some lack of interest on the part of the public) an *Apollo*-type mission to Mars is in advanced planning phase. Conservative scientists have suggested, however, that another *Viking*-type venture may be the safest and most sensible step prior to putting human lives at risk. Probably the most intense interest in a fully blown venture stemmed from Russian scientists, who have been preparing for a "Soviet" Phobos mission. A major concern has been and continues to be that of prolonged crew interest and mental and physical reactions to a sojourn in space that would require a currently calculated minimum of 15 months from launch to return on Earth. This has been referred to as the "Sprint" mission. For example, if it were assumed that the mission would leave Earth on 19 November 2004, it would reach at Mars on 30 July 2005 and depart Mars on 20 August 2005, returning to Earth on 2 February 2006. Time of stay on Mars would be less than 1 month and fuel costs would be at a maximum. A much longer

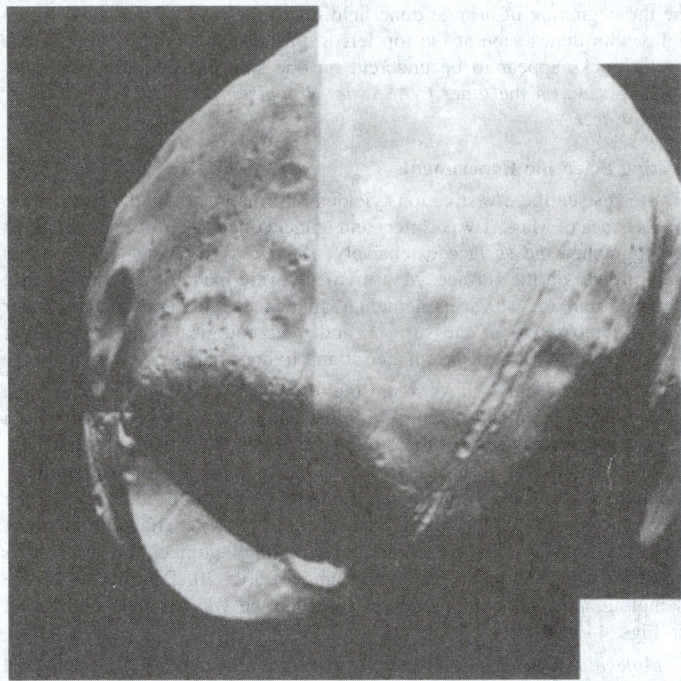

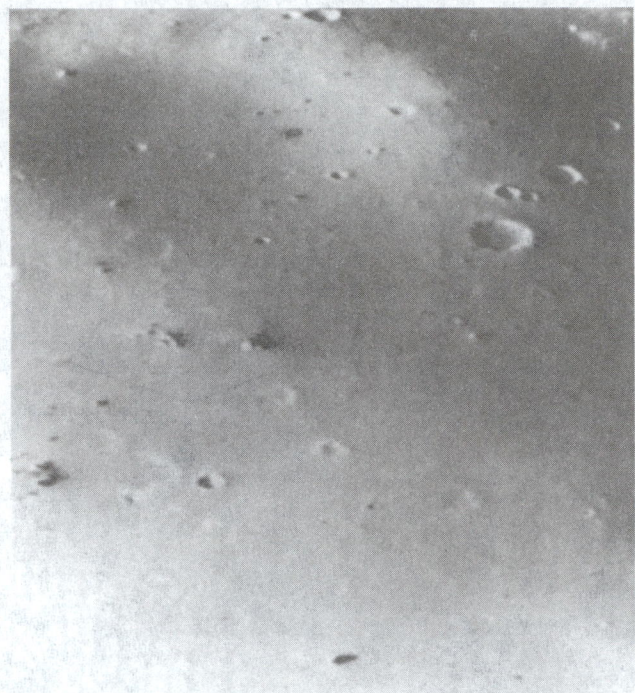

Fig. 29. This view of Phobos was made by *Viking Orbiter 1* on October 19, 1978 at a range of 612 kilometers (379 miles) during the spacecraft's 854th revolution of Mars. This view was made just before Phobos entered the shadow of the planet. The photomosaic shows the front side of Phobos which always faces Mars—from about 10° below the equator with north at the top. Stickney, the largest crater on Phobos (10 kilometers; 6.2 miles across) is at the left near the morning terminator. Linear grooves coming from and passing through Stickney appear to be fractures in the surface caused by the impact which formed the crater. Two earlier new encounters with Phobos brought Viking Orbiter 1 within close range of the satellite, but had not provided scientists with good opportunities to observe Stickney as well. This view provides new high-resolution coverage of the front side of Phobos (approximately 19 × 22 kilometers; 11.8 × 13.6 miles as seen here) as well as the highest resolution yet achieved of the western wall of Stickney. Kepler Ridge is casting a shadow in the southern hemisphere which partially covers the large crater (Hall) at the bottom. (*NASA; Jet Propulsion Laboratory, Pasadena, California.*)

Fig. 30. View of the Martian moon Deimos taken on October 15, 1977 when *Viking Orbiter* 2 passed within 50 kilometers (30 miles) of the satellite. The picture covers an area 1.2 × 1.5 kilometers (0.74 × 0.93 mile) and shows features as small as 3 meters (10 feet). Deimos is saturated with craters, but a layer of dust appears to cover craters smaller than 50 meters (165 feet) in diameter, making Deimos look smoother than the other Martian moon, Phobos. Boulders as large as houses (10 to 30 meters; 33 to 100 feet) across are strewn about the face. It is suggested that these objects may be blocks ejected from nearby craters. The spacecraft would have been clearly visible to an observer standing on the surface of Deimos. (*NASA; Jet Propulsion Laboratory, Pasadena, California.*)

(31 months) mission could be much more fuel efficient, and the stay on Mars could be considerably longer.

The Viking Missions to Mars

Two identical spacecraft, *Viking 1* and *Viking 2*, were launched in 1975 to explore Mars. In actuality, there were four spacecraft in all—a *Viking 1* orbiter and lander and a *Viking 2* orbiter and lander. Each orbiter and lander traveled together as one unit to rendezvous with Mars. The principal subsystems of the Viking spacecraft and separation, deorbit, entry, and landing sequences are shown in Fig. 31. The principal *Viking* events occurred as follows:

	Viking 1	*Viking 2*
Date of launch	August 20, 1975	September 10, 1975
Placed in elliptical orbit around Mars		June 6, 1976
	August 7, 1976	
Touchdown of Lander	July 20, 1976	September 3, 1976

Viking 1 traveled nearly 676 million kilometers (420 million miles) and *Viking 2* nearly 713 million kilometers (443 million miles) in their heliocentric Mars intercept trajectories prior to their respective insertions into elliptical orbits around the planet. See Figs. 32 and 33. Timing of the Viking missions was planned to achieve the trajectory situation shown. The Mars orbit insertion maneuvers for the Vikings require significant engine burns—in the case of *Viking 1*, for example, 38 minutes, consuming 2330 pounds (1057 kilograms) of propellant. Once in the Martian vicinity, radio signals required 22 minutes in either direction between earth and Mars,

thus a total of 44 minutes to execute a command and receive confirmation of that command. The general plan successfully followed in the *Viking* program involved orbiting the spacecraft in their respective orbits around Mars for several days, not only to commence imaging of the planet and certain scientific experiments, but to reconfirm earlier decisions concerning the best landing sites to finally elect for the two landers. Then, much as the *Surveyor* soft lunar landing craft had been placed on the earth's moon several years before, the landers were released from the aeroshells on the orbiting spacecraft, using parachute deployment at an elevation above the Martian surface of about 19,400 feet (5,913 meters) and the firing of terminal-descent engines at about 4,600 feet (1,402 meters) above the planet's surface. Both landers touched down successfully.

Later, adjustments were made to the *Viking* orbiters to provide better imaging of additional areas of the planet. Transmissions from the orbiters and the landers extending over an extensive period and information from one of the spacecraft was still being received during the early 1980s.

The Viking Landers

To assist in familiarizing many scientists at the control center in the detailed operation of the complex *Viking* landers, a test lander was installed in the auditorium at NASA's Jet Propulsion Laboratory in Pasadena, California. See Fig. 34. A diagram of one of the Landers is given in Fig. 35. Locations on Mars of the final landing sites are shown in Fig. 36. A view of the landing site for *Viking Lander 1*, taken from the orbiting *Viking 1*, prior to the landing is shown in Fig. 37.

The first photograph ever taken on the surface of Mars is shown in Fig. 38. This picture was taken just minutes after *Viking Lander 1* touched down successfully at Chryse Planitia. The center of the image is about 1.4 meters (5 feet) from camera No. 2 of the spacecraft. A similar view of the Martian surface taken by *Viking Lander 2* shortly after touchdown at Utopia Planitia is shown in Fig. 39.

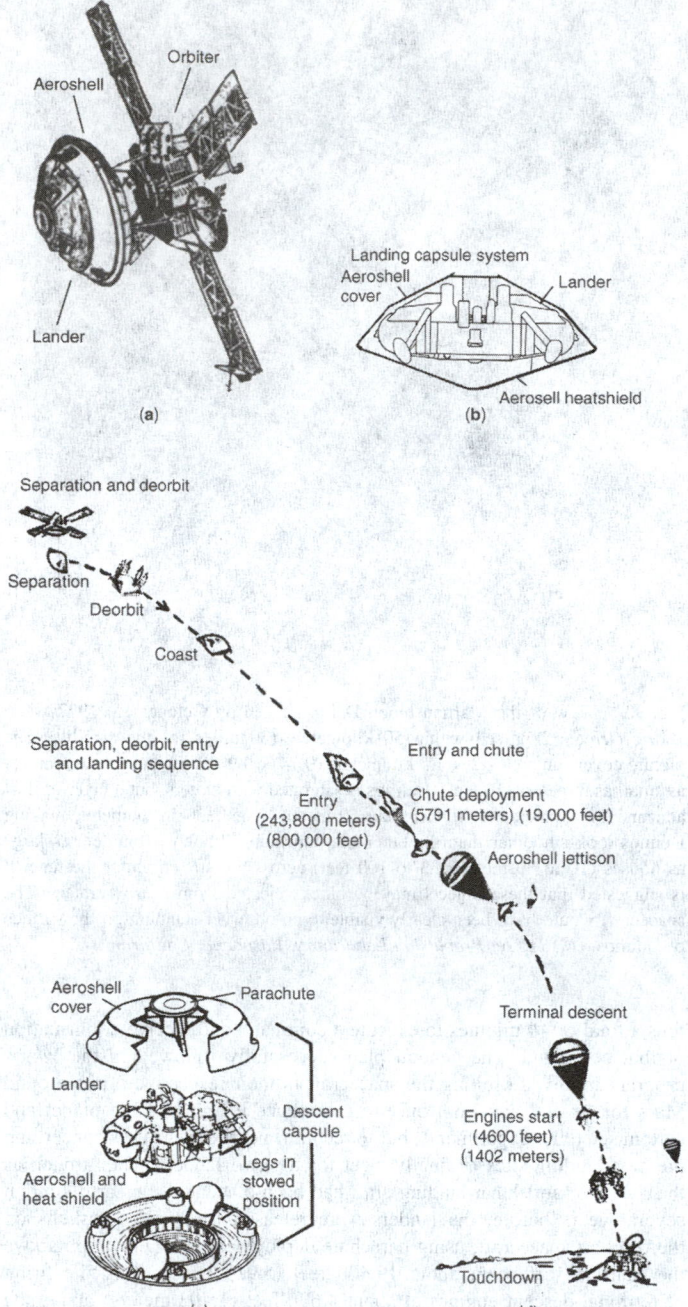

Fig. 31. Principal subsystems of the *Viking* spacecraft: (**a**) Orbiter and Lander linked together as they travel through space; (**b**) landing capsule system; (**c**) aeroshell cover, parachute, and descent capsule; and (**d**) separation, deorbit, entry, and landing sequence. (*NASA; Jet Propulsion Laboratory, Pasadena, California.*)

The first photograph of the Martian landscape (Chryse Planitia site) is shown in Fig. 40. In real color, this view is predominantly reddish brown. A diagram of the *Viking 1 Lander* and showing field of view with reference to equipment components is given in Fig. 41. Another striking view of the Martian landscape of Chryse Planitia is shown Fig. 42. The Martian landscape at the Utopia Planitia site is shown in Fig. 43.

At the center is seen the low-grain antenna for receipt of commands from earth. The projections on or near the horizon may represent the rims of distant impact craters. In the right foreground are color charts for Lander camera calibration, a mirror for the Viking magnetic properties experiment and part of a grid on the top of the Lander body. At upper right is the high-gain dish antenna for direct communication between the landed spacecraft and earth. Toward the right edge is an array of smooth, fine-grained material which shows some hint of ripple structure and may

be the beginning of a large dune field off to the right of the view, which joins with dunes seen at the top left in this 300° panoramic view. Some of the rocks appear to be undercut on one side and partially buried by drifting sand on the other (*NASA; Jet Propulsion Laboratory, Pasadena, California.*)

Viking Scientific Experiments

Thirteen scientific investigations yielded information about the atmosphere and surface of Mars. Two orbiters and landers operating for several months photographed the surface extensively from 1500 kilometers (930 miles) and directly on the surface. Measurements were made of the atmospheric composition, the surface, elemental abundance, the atmospheric water vapor, temperature of the surface, and meteorological conditions: direct tests were made for organic material and living organisms.[7]

Inorganic Chemistry. An x-ray fluorescence spectrometer was used to determine the elemental composition of samples at each lander site. Both sites yielded analyses of the fine-particle materials that are strikingly similar. Silica and iron in large amounts and magnesium, aluminum, calcium, and sulfur in significant amounts. More detail pertaining to the inorganic aspects of the Martian surface and rock materials are given in the entry on Mars. The trenching and sampling equipment on the Lander spacecraft are shown in the foreground of Fig. 20. Plan views of the sampling apparatus and procedures followed the *Viking* Landers are given in Figs. 44 and 45.

Molecular Analysis. Two samples from each lander site were analyzed for organic material with successive use of volatilization, pyrolysis, and detection by gas chromatography-mass spectrometry (GCMS). The sensitivity of the method is of the order of parts per billion. No organic compounds were detected at that level. The instrument was not designed to detect life—neither the quality or sensitivity permitted detecting biomass directly. The absence of organics in the sample was somewhat surprising considering the likelihood of carbonaceous chondrites reaching the Martian surface, or the possibility of de novo synthesis. Explanations involving dilution in the regolith and destruction by ultraviolet light or oxidation are all plausible. The GCMS was also used to measure the Martian atmosphere. It was ideally suited to measure isotopic ratios. Based upon measurements of nitrogen, argon, xenon, and krypton and their isotopic abundances, it is believed unlikely that a history of Mars outgassing will emerge.

Biology Experiments. Three experiments were conducted to test directly for life on Mars. The tests revealed a surprisingly chemically active surface—very likely oxidizing—but no evidence concerning the existence of life on the planet. The biological experiments were conducted with fully programmed and automated miniature laboratory equipment installed in each *Viking* lander. In the pyrolytic release (PR) experiment, Martian soil was placed in a chamber, after which carbon dioxide and carbon monoxide were added. These compounds were traceable because of the addition of radioactive carbon-14. The soil was incubated beneath a lamp that simulated Martian sunlight, but with no ultraviolet radiation present, as is the actual case on the planet today. If microorganisms were present, they would take up the radioactive gases. The chamber was heated to pyrolyze (decompose) any microbes present in the organic gases. The gases then were forced into an organic vapor trap, allowing other gases to pass to a radiation detector for "first count." With additional heating, the "organic vapors" were released and, if radioactive, they would indicate that living organisms were present. Results were negative.

In the labeled release (LR) experiment radioactive nutrient was added to a soil sample. Microorganisms present would digest the nutrient and release radioactive carbon dioxide. The soil is permitted to "incubate" for a period of a week or more, with further additions of nutrient. Results were negative.

In the gas exchange (GEX) experiment, scientists were looking for changes that Martian microbes might cause in gas levels over a long period. Soil was placed in a chamber, which was sealed to prevent gas leakage. Just sufficient nutrient flows admixed with water vapor would awaken spores or seeds, changing the gas level in the experiment. The results were negative. Considerable production of oxygen was noted from the GEX experiment, more or less explained as unusual Martian exotic chemistry.

---

[7] Information extracted from official NASA Langley Research Center report.

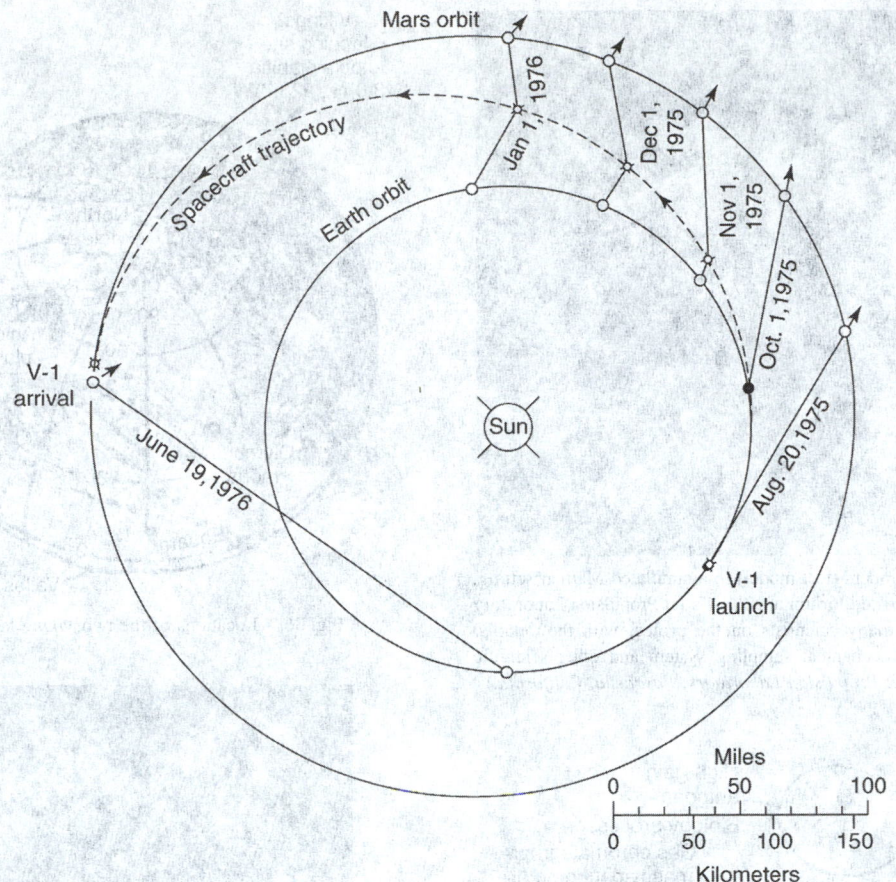

Fig. 32. Trajectory followed by *Viking 1*. Dates given show relationship of earth, spacecraft, Mars, and the sun at specific times.

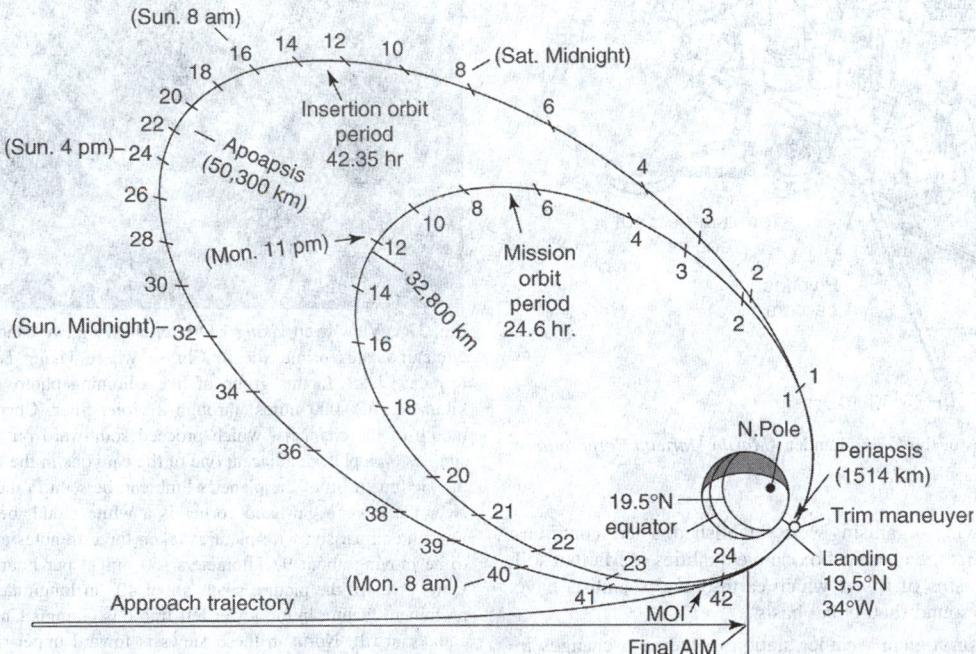

Fig. 33. Orbit geometry for the insertion and mission orbits. *Viking 1* completed only one revolution on the insertion orbit before the trim maneuver placed it on the mission orbit. The tick marks indicate spacecraft flight hours with periapsis as the zero point. (Periapsis = the orbital point nearest the focus of attraction; apoapsis = the farthest point.) Additional information at selected points along the insertion orbit indicate where the spacecraft was that day relative to Earth Pacific Daylight Time. A complete revolution of Mars on the mission orbit required 24.6 hours, the length of a Martian day. The orbit was synchronized with the landing site in that the spacecraft passed over the site once each day near periapsis, allowing maximum resolution orbital photography of that region for site certification and surface-data (after landing) correlation.

Fig. 34. One of the *Viking* Landers (test model) in a simulated Martian setting. This spacecraft was set up in the auditorium at NASA's Jet Propulsion Laboratory to thoroughly familiarize the many scientists on the project with the detailed operation of the spacecraft's mechanical sampling system and other scientific experiments aboard. (*NASA; Jet Propulsion Laboratory, Pasadena, California.*)

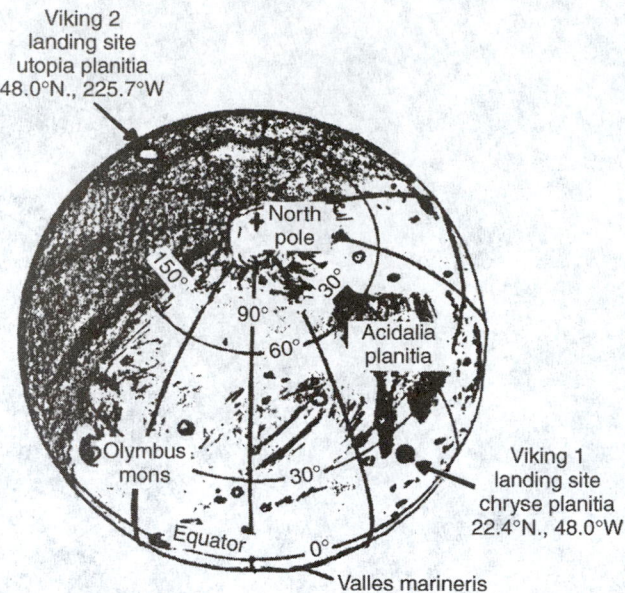

Fig. 36. Locations of the two *Viking* landing sites on Mars.

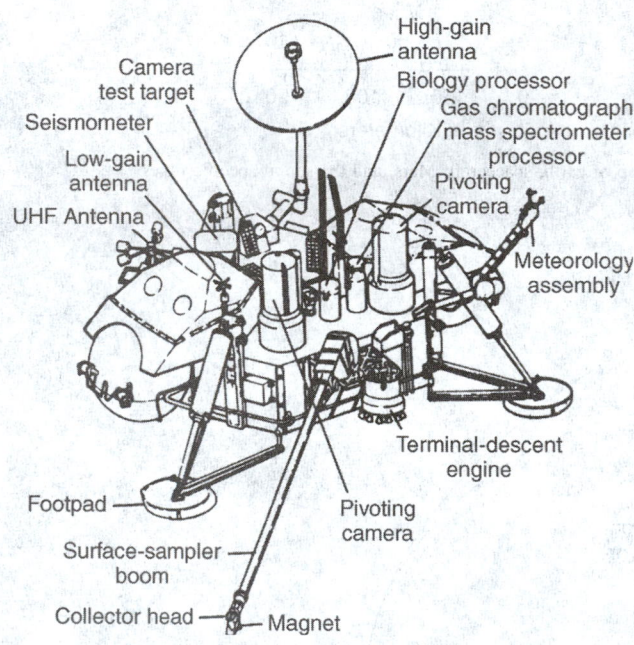

Fig. 35. Principal features of the *Viking* Lander. (*Martin Marietta Corporation.*)

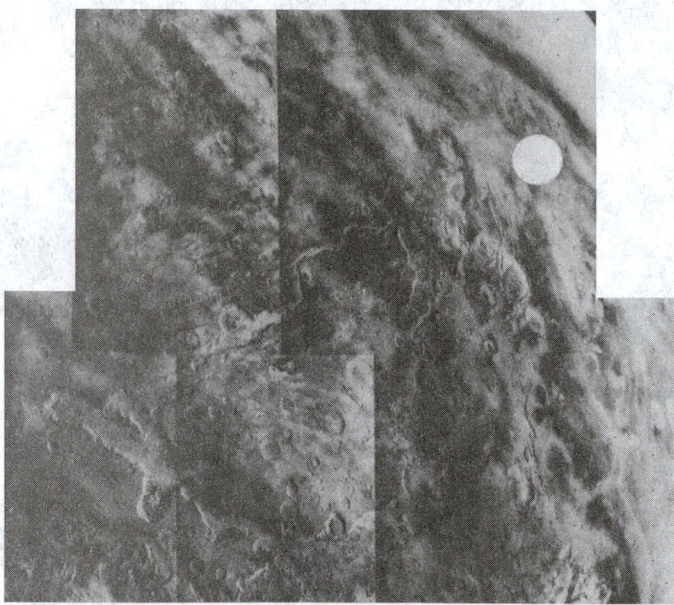

Fig. 37. View from *Viking* 1 Orbiter showing two candidate landing sites. White circle indicates prime site in Chryse where *Viking Lander 1* touched down a few days later. In this group of five adjoining photos taken from about 32,000 kilometers (20,000 miles) through a violet filter, Chryse is shown lying at the mouth of the channels, which proceed southward on the planet. An alternative site lies on a plateau adjacent one of the canyons in the lower (foreground) part of the picture. A bit of the planet's limb can be seen in the upper right-hand corner. Near the lower right-hand corner is a white cloud, believed to be ice crystals. From a comparison with pictures taken three minutes apart, the cloud was found to be moving about 97 kilometers (60 miles) per hour toward the upper left of view. Overall, the picture spans about 40° in longitude and 35° in latitude. The prominent feature in the lower left frame is Grangis Chasma, an arm of the great equatorial rift. North in these views is toward upper-right-hand corner. (*NASA; Jet Propulsion Laboratory, Pasadena, California.*)

No plausible ties with living organisms were established. There continues to be some speculation that the chemical oxidative qualities of Martian soil may support microorganisms of a sort which earth-bound scientists have not described even on a sound theoretical basis.

Meteorology. A meteorological weather station to measure changes in pressure, temperature, wind speed and direction operated well on both landers. Generally, the weather at both sites was repetitive, with a daily temperature variation of between 190 and 240K, (−83 °C (−118 °F) and (−33 °C (−28 °F) the peak usually occurring in mid-afternoon Martian time. The pressure at each site was in the range of 7 to 8 millibars. Daily pressure variations were about 0.3 millibar.

Seismology. The seismometer on Viking Lander 1 failed to be uncaged, but the Viking 2 functioned normally. Little or no quake activity was detected.

Atmospheric Water Detector. An infrared spectrometer operating at the 1.38 micrometer region was mounted on the scan platform of each lander. The device was used to measure the latitudinal variations and diurnal variations and, by operating over a complete Martian year, it was able to indicate seasonal changes. More moisture was found in the northern hemisphere than in the southern portion of the planet. These measurements helped to confirm that the permanent polar cap of Mars consists of water ice.

Fig. 38.   First photograph ever taken on surface of Mars, obtained by *Viking 1 Lander* just a few minutes after its successful touchdown on Chryse Planitia. The center of the image is about 1.4 meters (5 feet) from camera #2 of the spacecraft. Both rocks and finely granulated material—sand and dust—are observed. Many of the small foreground rocks are flat with angular facets. Several larger rocks exhibit irregular surfaces with pits and the large rock at the top left shows intersecting linear cracks. Extending from that rock toward the camera is a vertical linear dark band, which may be due to a one-minute partial obscuration of the landscape due to clouds or dust intervening between the sun and the surface. Associated with several of the rocks are apparent signs of wind transport of granular material. The large rock in the center is about 10 centimeters (4 inches) across and shows three rough facets. To its lower right is a rock near a smooth portion of the Martian surface, probably composed of very fine-grained material. It is possible that the rock was moved during the *Viking 1* descent maneuvers, revealing the finer-grained basement substratum; or that the fine-grained material had accumulated adjacent to the rock. There are numerous other furrows and depressions and places with fine-grained material elsewhere in the view. At right is a portion of footpad #2. Small quantities of fine-grained sand and dust are seen at the center of the footpad near the strut and were deposited at landing. The shadow to the left of the foodpad clearly exhibits detail, due to scattering of light either from the Martain atmosphere or from the spacecraft, observable because the Martian sky scatters light into shadowed areas. (*NASA; Jet Propulsion Laboratory, Pasadena, California.*)

Fig. 39.   First photograph of Martian surface taken by *Viking Lander* 2 at the Utopia Planitia site. The scene reveals a wide variety of rocks littering a surface of fine-grained deposits. Boulders in the 10–20 centimeter (4–8 inch) size range—somewhat vesicular (holes) and some apparently fluted by wind—are common. Many of the pebbles have tubular or platy shapes, suggesting that they may be derived from layered strata. The fluted boulder just above the Lander's footpad displays a dust-covered or scraped surface, suggesting it was overturned or altered by the foot at touchdown. Brightness variations at the beginning of the picture scan (left edge) probably are due to dust settling after landing. A substantial amount of fine-grained material kicked up by the descent engines has accumulated in the concave interior of the footpad. Center of the image is about 1.4 meters (5 feet) from the camera. Field of view extends 70° from left to right and 20° from top to bottom. This second landing location is in the northern latitudes about 7500 kilometers (4600 miles) northeast of the Viking 1 Lander site, where touchdown occurred 45 days earlier. (*NASA; Jet Propulsion Laboratory, Pasadena, California.*)

Fig. 40.   First panoramic view of Martian surface taking by *Viking 1 Lander*. The out-of-focus spacecraft component toward left center is the housing for the Viking sample arm, which is not yet deployed. Parallel lines faintly seen in the sky are an artifact and are not real features. However, the change of brightness from horizon toward zenith and toward the right (west) is accurately reflected in this picture, which was taken in the late Martian afternoon. At the horizon to the left is a plateau like prominence much brighter than the foreground material between the rocks. The horizon features are approximately three kilometers (1.8 miles) away. At left is a collection of fine-grained material reminiscent of sand dunes. The dark sinous markings in the left foreground are of unknown origin. Some unidentified shapes can be perceived on the hilly eminence at the horizon toward left center of view. The horizontal cloud stratum can be made out halfway from the horizon to the top of the picture.

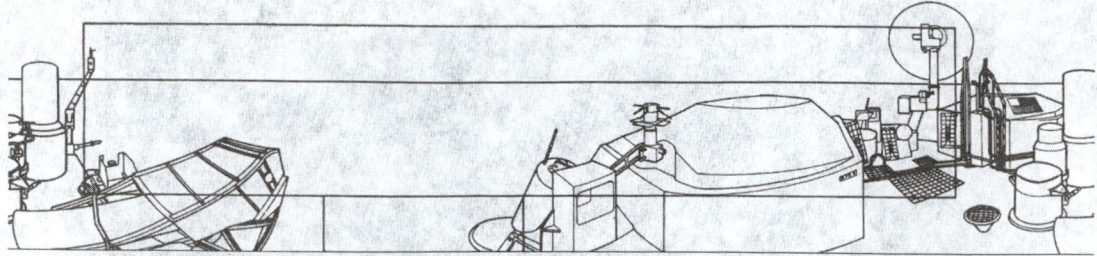

Fig. 41. This diagram illustrates a full 360° image from camera #2 on the *Viking Lander* spacecraft. The outlined image areas represent the Viking 1 Lander's first two pictures (Figs. 10 and 12). This type of diagram helped analysts at the mission control center to identify photo orientation in terms of parts of the Lander spacecraft components when they appeared in various views. (*NASA; Jet Propulsion Laboratory, Pasadena, California.*)

Fig. 42. Martian landscape as viewed by *Viking 1 Lander*, showing a dune field with features remarkably similar to many seen in the deserts on earth. The early morning lighting (7:30 A.M. local Mars time) reveals subtle details and shading. The picture covers 100°, looking northeast at left and southeast at right. Viking scientists observed that this area is reminiscent of regions in Mexico, California, and Arizona (Kelso, Death Valley, Yuma). The sharp dune crests indicate the most recent wind storms capable of moving sand over the dunes in the general direction from upper left to lower right. Small deposits downwind of rocks also indicate this wind direction. Large boulder at left is about 8 meters (25 feet) from the spacecraft and measures about 1 × 3 meters (3 × 10 feet). The meteorology boom, which supports the spacecraft's miniature weather station, cuts through the center of the picture. The sun rose two hours earlier and is about 30° above the horizon near the center of the view. In real color, the landscape is predominantly reddish brown. (*NASA; Jet Propulsion Laboratory, Pasadena, California.*)

Fig. 43. High-resolution photo of the Martian surface taken by Viking Lander 2 at the Utopia Planitia landing site. View was made on May 18, 1979 and relayed to earth by Orbiter 1 on June 7. The "rolling hill" Marscape is an artifact of the Lander's 8-degree tilt; the horizon is generally flat. There is some relief of the flat terrain, however, at lower scales locally and at greater scales on the horizon. In stereo, a few gullies and depressions take on considerable depth and dimension. A thin coating of water ice on the rocks and soil is visible. The time the frost appeared corresponded almost exactly with the build-up of frost one Martian year (23 earth months) earlier when a similar view had been taken. The frost remained on the surface for about one hundred days. Some scientists believe dust particles in the atmosphere may pick up bits of solid water. But carbon dioxide present in the Martian atmosphere also may freeze and adhere to particles, becoming sufficiently heavy to settle. Warmed by the sun, the surface evaporates the carbon dioxide and returns it to the atmosphere, leaving behind water and dust. The ice seen in this picture, like that which formed during the earlier martian year, is extremely thin, perhaps no more than 1/1000 inch (0.03 millimeter) thick. (*NASA; Jet Propulsion Laboratory, Pasadena, California.*)

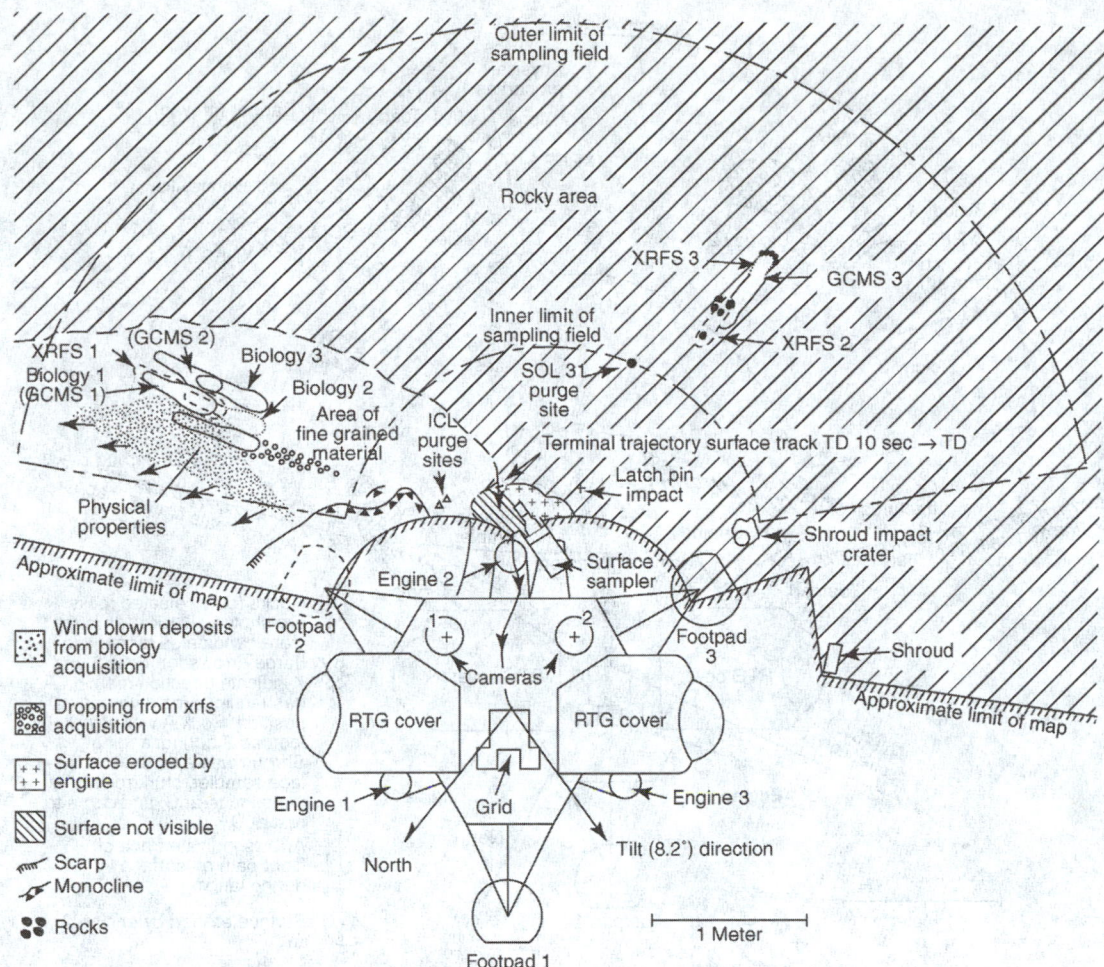

Fig. 44.   Plan view of *Viking Lander* 1, showing the spacecraft and its orientation, location of sample sites, locations of selected rocks for analytical experiments. 1Sol = 1 complete Martian day and night; XRFS = x-ray fluorescence spectrometer, GCMS = gas chromatograph-mass spectrometer, ICL = initial rock to be investigated. (*NASA; Jet Propulsion Laboratory, Pasadena, California.*)

Infrared Thermal Mapper. An infrared radiometer measured thermal emission of the surface and atmosphere. It was found that the atmospheric temperature above 20 kilometers (about 12 miles) varies from 165 K (near dawn) to 185 K at about 2:15 in the afternoon Martian time. This variation is believed to be initiated at the lower levels and radiatively propagated by dust in the atmosphere. The temperature of the surface was found to be highly variable.

Physical and Magnetic Properties. Cameras were used to determine certain physical and magnetic properties of the soil. Pictures of stroke gages, sample digging, footpad movement, areas underneath the lander, and rock movement were used to determine the bulk density, particle size, angle of internal friction, cohesion, adhesion, and penetration resistance of the Martian soil. Small permanent magnets on the sampler collected material indicating that the surface contains a few percent of magnetic material, very likely magnetite.

See also **2001 Mars Odyssey Mission**; **Pathfinder Mission to Mars**; **Mars Exploration Program**; **Mars Exploration Rover (Spirit & Opportunity) Mission**; **Mars Global Surveyor Mission**; **Phoenix Mars Mission**; and **Viking Mission to Mars**.

Classic References to Mars and Missions to Mars

The *Viking* and other missions to Mars by the United States and the former U.S.S.R. were quite thoroughly documented and represent a wealth of resource information. Information collected between 1970 and 1990, which appeared in the 7th Edition of this encyclopedia, is re-listed here for the serious scholar of Mars and its exploration.

Additional Reading (1976–1987)

Anders, E. and T. Owen: "Mars and Earth: Origin and Abundance of Volatiles," *Science*, **198**, 453–465 (1977).

Arvidson, R.E. et al.: "Three Mars Years: *Viking Lander 1* Imaging Observations," **222**, 463–478 (1983).

Baker, Victor R.: "The Spokane Flood Controversy and the Martian Outflow Channels," *Science*, **202**, 1249–1257 (1978).

Baker, V.R.: *The Channels of Mars*, Univ. of Texas Press, Austin, Texas, 1982.

Bogard, D.D. and P. Johnson: "Martian Gases in an Antarctic Meteorite?" *Science*, **221**, 651–654 (1983).

Carr, M.H.: *The Surface of Mars*, Yale University Press, New Haven, CT, 1981.

Carr, M.H., R.S. Saunders, R.G. Strom, and D.E. Wilhelms: *The Geology of the Terrestrial Planets*, Spec. Pubn. SP-469, National Aeronautics and Space Administration, Washington, DC, 1985.

Duxbury, T.C. and J. Veverka: "Deimos Encounter by Viking: Preliminary Imaging Results," *Science*, **201**, 812–814 (1978).

Ezell, E.C. and L.N. Ezell: *On Mars*, National Aeronautics and Space Administration, Washington, DC, 1984.

Greeley, R.: "Release of Juvenile Water on Mars," *Science*, **236**, 1363–1364 (1987).

Haberle, R.M.: "The Climate of Mars," *Sci. Amer.*, 54–62 (May 1986).

Hartman, W.K.: *J. Geophys. Res.*, **78**, 4096 (1973).

LPI: Mars: Evolution of Its Climate and Atmosphere, Proceedings of Symposium, Washington, DC, Lunar and Planetary Institute, Houston, TX (July 17–19, 1986).

Lucchitta, B.K.: "Recent Mafic Volcanism on Mars," *Science*, **235**, 565–567 (1987).

McElroy, M.B. et al.: "Composition and Structure of the Martian Upper Atmosphere: Analysis of Results from Viking," *Science*, **194**, 1295–1298 (1976).

Mumma, M.J. et al.: "Discovery of Natural Gain Amplification in the 10-Micrometer Carbon Dioxide Laser Bands on Mars: A Natural Laser," *Science*, **212**, 45–49 (1981).

Mutch, R.A. et al.: *The Geology of Mars*, Princeton Univ. Press, Princeton, NJ, 1977.

Neukum, G. and D.U. Wise: "Mars: A Standard Crater Curve and Possible New Time Scale," *Science*, **194**, 1381–1387 (1976).

NEWS: "Launch of *Mars Observer*—1992," *Science*, **235**, 743 (1987).

Wind blown deposits from biology acquisition

Dropping from xrfs acquisition

Surface eroded by engine

Surface not visible

Scarp

Monocline

Rocks

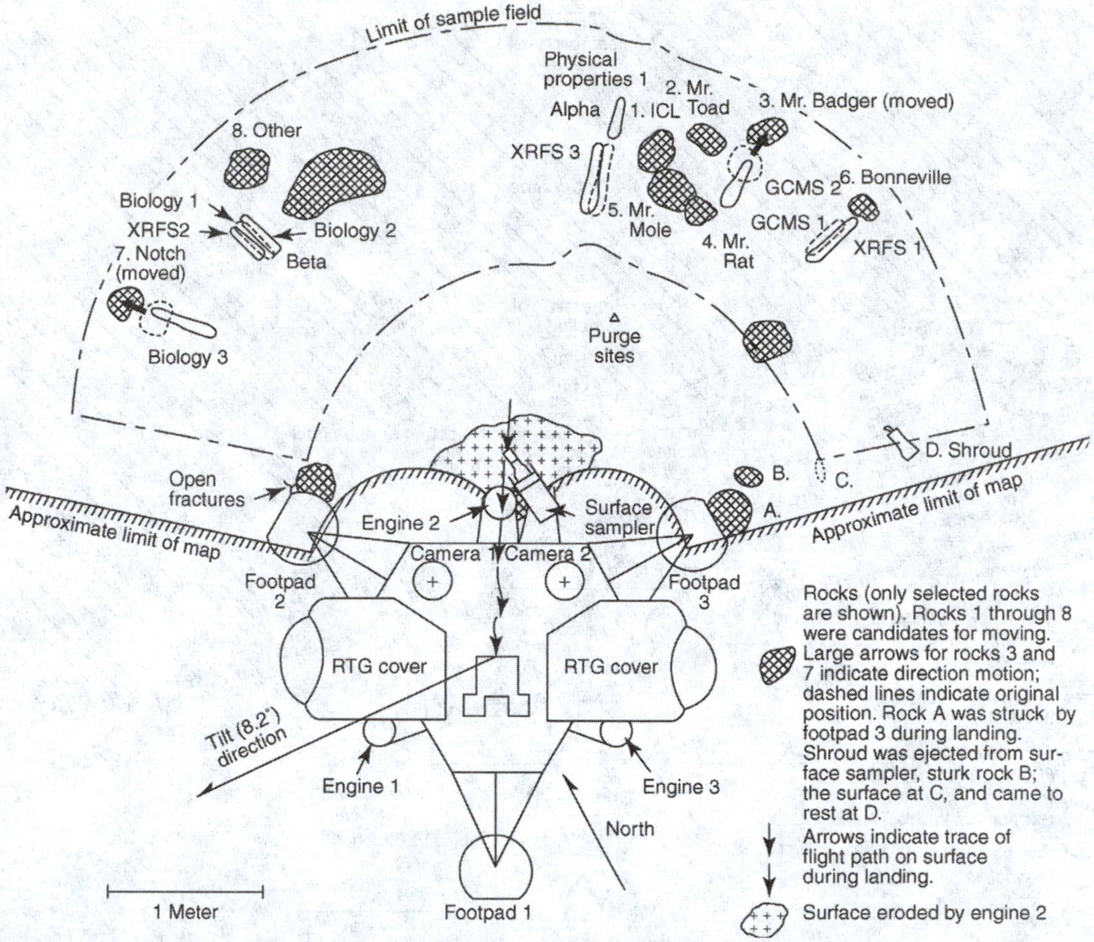

Fig. 45.   Plan view of Viking Lander 2, similar to that of Fig. 29. (NASA Jet Propulsion Laboratory, Pasadena, California.)

Owen, T. et al.: "The Atmosphere of Mars: Detection of Krypton and Xenon," *Science*, **194**, 1293–1295 (1976).

Paige, D.A. and A.P. Ingersoll: "Annual Heat Balance of Martian Polar Caps: *Viking* Observations," *Science*, **228**, 1160–1168 (1985).

Pieri, D.C.: "Martian Valleys: Morphology, Distribution, Age, and Origin," *Science*, **210**, 895–897 (1980).

Soffen, G.A. and C.W. Snyder: "The First *Viking* Mission to Mars, (contains 13 papers relating to *Viking 1 Lander* on Mars)," *Science*, **193**, 759–815 (1976).

Soffen, G.A.: "Scientific Results of the Viking Missions, (contains 20 papers relating to Viking missions to Mars)," *Science*, **194**, 1274–1353 (1976).

Soffen, G.A.: "Status of the *Viking* Missions, (contains 15 papers relating to the Viking mission to Mars)," *Science*, **194**, 57–105 (1976).

Squyres, S.W.: "The History of Water on Mars," *Ann. Review of Earth and Planetary Sciences*, **12**, 83–106 (1984).

Squyres, S.W. and M.H. Carr: "Geomorphic Evidence for the Distribution of Ground Ice on Mars," *Science*, 231–252 (1986).

Thomas, P. and P.J. Gierasch: "Dust Devils on Mars," *Science*, **230**, 175–177 (1985).

Additional Reading (1988–Present)

Beardsley, T.M.: "U.S.–Soviet Collaboration in Space Science Is Improving," *Sci. Amer.*, **21** (August 1988).

Beardsley, T.: "Slow Boat to Mars," *Sci. Amer.*, **14** (April 1990).

Bergreen, L.: *Voyage to Mars: NASA's Search for Life Beyond Earth*, The Putnam Publishing Group, New York, NY, 2000.

Boyce, J.M.: *Smithsonian Book of Mars*, Smithsonian Institution Press, Washington, DC. 2003.

Carr, M.H.: *Surface of Mars*, Cambridge University Press, New York, NY. 2007.

Chapman, M.G.: *Geology of Mars: Evidence from Earth-Based Analogs*, Cambridge University Press, New York, NY. 2007.

Collins, M.: "Mission to Mars," *Nat'l. Geographic*, **732** (November 1988).

Cornell, J.: "Red Weather (Mars)," *Technology Review (MIT)*, **19** (February/March 1990).

Eberhart, J.: "Soviet Findings from Phobos and Mars," *Science News*, **286** (October 28, 1989).

Eberhart, J.: "Phobos: Moonlet of the Pits," *Science News*, **301** (November 4, 1989).

Eberhart, J.: "Powerful Appeal of Mars' Missing Field," *Science News*, **150** (March 10, 1990).

Eberhart, J.: "Episodic Oceans: Mars," *Science News*, **283** (May 5, 1990).

Eberhart, J.: "The Sandy Face of Mars," *Science News*, **268** (October 27, 1990).

Eberhart, J.: "Mars: Let It Snow, let it snow…," *Science News*, **286** (November 3, 1990).

Flam, F.: "Swarms of Mini-Robots Set to Take on Mars Terrain," *Science*, **1621** (September 18, 1992).

Greeley, R. and B.D. Schneid: "Magma Generation on Mars: Amounts, Rates, and Comparisons with Earth, Moon, and Venus," *Science*, **996** (November 15, 1991).

Hamilton, D.P.: "NASA to Explore Three Possible Mars Missions," *Science*, **863** (February 22, 1991).

Harland, D.M.: *Water and the Search for Life on Mars*, Springer-Verlag New York, LLC, New York, NY. 2006.

Keating, G.M.M.: *Exploration of Venus and Mars Atmospheres*, Elsevier Science, New York, NY, 1995.

Kerr, R.A.: "Soviet Failure at Mars a Reminder of Risks," *Science*, **26** (April 7, 1989).

Kerr, R.A.: "Planetary Science Funds Cut," *Science*, **282** (January 19, 1990).

Kieffer, H.H., C. Snyder, B.M. Jakosky, and M.S. Matthews: *Mars*, University of Arizona Press, Phoenix, AZ, 1997.

Kiernan, V.: "Reactor Project Hitches onto Moon-Mars Effort," *Science*, **1482** (June 22, 1990).

Kiernan, V.: "Sailing to Mars," *Technology Review (MIT)*, **20** (November/December 1990).

McKay, C.P. and R.H. Haynes: "Should We Implant Life on Mars?" *Sci. Amer.*, **144** (December 1990).

Moore, P: *Patrick Moore on Mars, Cassell Illustrated*, London UK. 2007.

Muhleman, D.O. et al.: "Radar Images of Mars," *Science*, **1508** (September 27, 1991).

Ostro, S.J. et al.: "Radar Detection of Phobos," *Science*, **1584** (March 24, 1989).

Owens, T. et al.: "Deuterium on Mars: The Abundance of HDO and the Value of D/H," *Science*, **1767** (June 24, 1988).

Raeburn, P.: *Mars: Uncovering the Secrets of the Red Planet*, National Geographic Society, Washington, DC, 2000.

Rubincam, D.P.: "Mars: Change in Axial Tilt Due to Climate," *Science*, **720** (May 11, 1990).

Russell, C.T.: *2001 Mars Odyssey*, Springer-Verlag New York, LLC, New York, NY. 2004.

Schwartz, B.D.: "Muddy Evidence," *Sci. Amer.*, **28** (June 1989).

Sheehan, W.: *The Planet Mars: A History of Observation and Discovery*, University of Arizona Press, Phoenix, AZ, 1996.

Squyres, S.D., and A.H. Knoll: *Sedimentary Geology at Meridiani Planum, Mars*, Elsevier Science & Technology Books, New York, NY, 2005.

Staff: "Mars Mission," *Technology Review (MIT)*, **19** (May/June 1989).

Staff: "Mars Magnetism: A Moot Question?" *Science News*, **31** (July 14, 1990).

Staff: "Martian Atmosphere Eyed by Hubble Space Telescope," *Hughes News*, **5** (May 17, 1991).

Staff: "Instruments Help Unlock Mars' Secrets," *Hughes News*, **1** (October 2, 1992).

Tokano, T.: *Water and Life on Mars*, Springer-Verlag New York, LLC, New York, NY. 2004.

Touma, J. and J. Wisdom: "The Chaotic Obliquity of Mars," *Science*, **1294** (February 26, 1993).

Waldrop, M.M.: "Jet Propulsion Lab Looks to Life After Voyager," *Science*, **1037** (September 8, 1989).

Waldrop, M.M.: "Phobos at Mars: A Dramatic View—and Then Failure," *Science*, **1042** (September 8, 1989).

Waldrop, M.M.: "Asking for the Moon," *Science*, **637** (February 9, 1990).

Walker, M.: *Evolution of Hydrothermal Ecosystems on Earth and Mars*, John Wiley & Sons, Inc., New York, NY, 1996.

Walters, M.: *The Search for Life on Mars*, Perseus Publishing, Boulder, CO, 1999.

Web References

Mars Exploration: http://Mars.jpl.nasa.gov/

National Aeronautics and Space Administration: http://sse.jpl.nasa.gov/missions/mars_missions/mgs.html

MARSDEN CHART. A system introduced by Marsden early in the nineteenth century for showing the distribution of meteorological data on a chart, especially over the oceans. A Mercator map projection is used, the world between 80°N and 70°S latitudes being divided into Marsden squares each of 10° longitude. These squares are systematically numbered to indicate position. Each square may be divided into quarter squares, or into 100 one-degree subsquares numbered from 00 to 99 to give the position to the nearest degree.

AMS

MARSDEN SQUARE. A system that divides a Mercator chart of the world into squares of 10° latitude by 10° longitude. Each square is numbered and then subdivided into 100 one-degree squares numbered from 00 to 99. These 100 one-degree squares are applied to the eight octants of the globe. Marsden squares are mainly used for identifying the geographic position of meteorological data over the oceans.

MARS EXPLORATION MISSIONS (By the USSR). Mars research began long before the Space Age. Ground-based astronomical observations using photometry (I.K. Koval', N.P. Barbashev, et al.), polarimetry (A.V. Morozhenko et al.), and infrared spectrometry (V.I. Moroz) were performed in the Soviet Union throughout the 1950s and 1960s. G.A. Tikhov attempted to find evidence of life on Mars using spectroscopic techniques. Theoretical models for the internal structure of the planet began to be developed (V.N. Zharkov et al.). An overview of this period may be found in the Moroz Ref.

Early Soviet planetary spacecraft were developed by Special Design Bureau 1 (OKB-1, now NPO Energia) under the leadership of Chief Designer S.P. Korolev (1907–1966). B.E. Chertok has brilliantly described this period of history in his books. One of the primary motivations for the development of long-term Martian research programs was a desire to find life on the planet. It was generally understood that this project had to begin elsewhere—learning more about the planet Mars itself. Nevertheless, early Soviet spacecraft carried a spectrophotometer for observations of the so-called Sinton bands, which were, at that time, presumed to indicate the presence of organic material on Mars.

Eight attempts were made to launch a spacecraft to Mars, starting in 1962. Only the fourth spacecraft, the Mars-1 (1962) executed a flyby, but communications were lost prior to the flyby. Jet Propulsion Laboratory made two attempts (1964) during this period, one of which was successful: Mariner-4 performed a Mars flyby and transmitted an image of the planet.

In 1965, S.P. Korolev transferred all unmanned interplanetary space flight projects to another facility, the Design Bureau and Plant named after S.A. Lavochkin. The Design Bureau ("KB") was managed by Chief Designer G.N. Babakin (1914–1971). Later (1974) this facility was re-named to Nauchno-Proizvodstvennoe Obyedinenie imeni S.A. Lavochkina (NPO Lavochkin or NPOL), in translation Reserach-Industrial Association S.A. Lavochkin. For simplicity we will use the designation "NPO Lavochkin" for all periods covered below.

Many successful missions were carried out by NPO Lavochkin to the Moon, Venus, and eventually to Halley's Comet, but, the USSR's development of Mars projects was unsuccessful. Between 1969 and 1973, six Soviet spacecraft were launched, two in 1971 and four in 1973; however, virtually all of the new knowledge about Mars obtained by the early 1980s came from the NASA Mariner 9 and Viking missions. The Soviet contribution to research concerning the planet itself turned out to be quite meager, except for various issues relating to the interaction of the planet and its plasma sphere with the solar wind, where the Soviet Mars-3, Mars-5, and Phobos-2 spacecraft provided the basic results. It should also be noted that the Soviet Union was responsible for the first successful landing on Mars (Mars-2) and the first direct measurements performed in the Martian atmosphere (Mars-6). A NASA-published history of the NPO Lavochkin Mars efforts may be found in the Perminov Ref. The first papers discussing the scientific results of the 1973 Soviet missions were published in a special volume of the journal *Kosmicheskie Issledovaniya* [*Space Research*]. A description of the major scientific research results related to the Martian atmosphere and surface obtained during both Soviet and American missions of that era may be found in the Moroz Ref. *Physics of the Planet Mars*.

Mars-2 and Mars-3 Spacecraft

The Mars-2 and Mars-3 unmanned interplanetary spacecraft were launched on 19 and 28 May 1971. Each spacecraft included an orbiter (Fig. 1) and a lander. Upon arrival at Mars (on 27 November and 2 December 1971, respectively), the landers separated from the Mars-2 and Mars-3 spacecraft and reached the surface of the planet, thereby becoming the first spacecraft to attempt a landing on the surface of Mars. Each 1,000-kg (2,205 lbs) lander carried a pennant with the seal of the Soviet Union, a camera, and scientific instrumentation for studying soil samples. These were the first items made on Earth to land on Mars. The Mars-2 lander broke up at coordinates 44.2°S 313.2°W because of inaccurate targeting, which led to an error in the entry angle and caused the lander to impact on the surface before the parachute opened. The Mars-3 lander successfully landed on the surface at coordinates 45°S 158°W. The Mars-2 and Mars-3 orbiters (each weighing 2,265 kg (4,993 lbs)) were inserted into Mars orbit and performed a research program on the composition of the Martian atmosphere, surface photometry and IR radiometry, the magnetic field of Mars, and the interaction between the solar wind and the Martian plasma environment for more than 8 months. Both orbiters had cameras but no scientifically important images were transmitted. A global dust storm prevented imaging during first months after orbit insertion, and cameras failed by the time when dust dissipated. The Mars-2 and Mars-3 orbiters ceased operations in August 1972.

Mars-4 and Mars-5 Spacecraft

The Mars-4 and Mars-5 spacecraft for planetary studies of Mars from Mars orbit were launched on 21 and 25 July 1973 (Fig. 2). Mars-4 reached Mars on 10 February 1974 and passed within 2,200 km (1,367 miles) of the planet's surface without entering Mars orbit; however, it did transmit photographs of the planet to Earth. The Mars-5 spacecraft entered Mars orbit on 12 February 1974 (the orbiter had a mass of 2,200 kg (4,850 lbs). The spacecraft continued operating once it reached Mars orbit; data were obtained on the atmosphere and surface temperature patterns, and approximately 120 surface photographs and panoramas of the Martian Southern Hemisphere were transmitted to Earth.

Fig. 1.   General view of Mars-2 and Mars-3 spacecraft.

Fig. 2.   Mars-4 and Mars-5 orbiter.

Fig. 3.   Mars-6 and Mars-7 spacecraft.

Mars-6 and Mars-7 Spacecraft

The Mars-6 and Mars-7 Mars spacecraft were launched on 5 and 9 August 1973. (Figs. 3 and 4). Mars-6 reached the planet on 12 March 1974. The descent module separated from the orbiter, landed on the surface of Mars, and performed direct measurements of the Martian atmosphere during its descent. Mars-7 reached the planet on 9 March 1974. The descent module separated from the spacecraft and performed a flyby 1,300 km (807.8 miles) above the Martian surface.

After 1973, there were no USSR flights to Mars for many years. At that time, it would have been difficult to mount a standard mission (e.g., with a satellite and several lander modules) that could have obtained new scientific results at a level higher than that already achieved by the United States. There was a desire to undertake a project that would be more significant on a fundamental level. The project that was eventually selected involved the return of a sample of Martian material to Earth. NPO Lavochkin worked on this mission for several years, but the project turned out to be too risky and was eventually cancelled.

Fig. 4.   Mars-6 and Mars-7 descent module.

During this period, the USSR successfully continued its flights to Venus. A new direction for research was pioneered by R.Z. Sagdeev. These were research projects related to two solar-system bodies, Phobos and Halley's Comet.

Phobos-1 and Phobes-2 Spacecraft

The primary purpose of the Phobos spacecraft was to study not Mars itself, but Phobos. However, this mission was the next major post-Viking step in Mars research and marked the first mapping of the planet in the thermal infrared (the spatial resolution was approximately 1 km at a wavelength of approximately 10 μm). Approximately 40,000 near-infrared (1–3 μm) images were obtained using a mapping spectrometer. At that time, no instruments of this type had yet been used, even in remote sensing observations of Earth. This mission also marked the first spectroscopic remote sensing of the Martian atmosphere via measurements of the solar spectrum at various altitudes above the limb in several wavelength bands; these data were used to determine the vertical water vapor and aerosol distributions. The first measurements of the dissipative flux in the upper atmosphere of Mars were made. Several interesting new results were also obtained for Phobos. Precise mass and density values were determined, and brightness was measured as a function of wavelength. The brightness as a function of wavelength did not conform to previously held views regarding the composition of Phobos. A large amount of unique data was obtained on the plasma environment of Mars and its interaction with the solar wind. Papers describing the scientific results of the Phobos mission were published in the journals *Nature*, *Geophysical Research Letters*, and *Planetary and Space Science*. Despite the fact that Phobos-1and 2 was a short-term mission (transmitted data from orbit for approximately 2 months), the resulting scientific data exceeded the scope and quality of that obtained in all of the other Soviet Mars expeditions taken together.

The Phobos-1 and Phobos-2 new-generation spacecraft (Figs. 5 and 6) were launched on 7 and 12 July 1988, respectively (Table 1). These spacecraft performed research on Phobos, Mars, and outer space (Fig. 7). Thirteen countries, plus the European Space Agency (ESA), were involved in developing the scientific instrumentation (Table 2).

TABLE 1. BASIC SPECIFICATIONS FOR PHOBOS-2 SPACECRAFT

Launch date	12 July 1988
Mars-orbit entry date	29 January 1989
Duration of Earth–Mars flight	200 days
Time spent in Mars space prior to Phobos flyby	120 days
Duration of Phobos flyby	25 minutes
Orbiter mass	2600 kg (5,732 lbs)
Mass of scientific instrumentation	370 kg (816 lbs)

After the end of the mission of the two Phobos spacecraft, an intense controversy erupted over whether another mission of the same type or a new mission should be developed whose primary goal would be to study the planet Mars itself. The latter point of view won out and led to

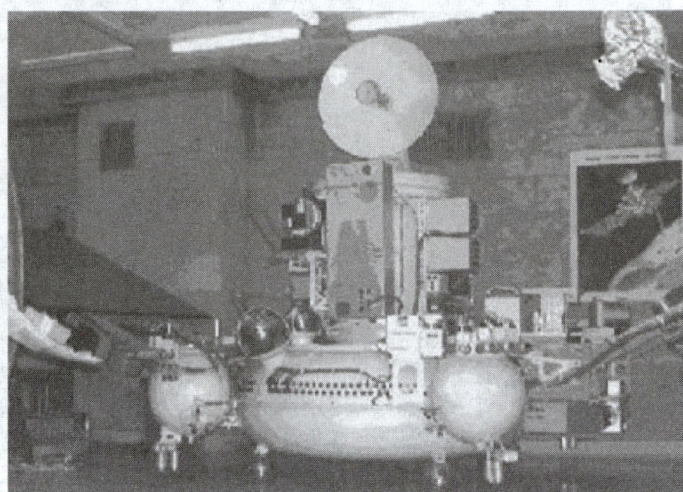

Fig. 5.   General view of the Phobos-1/2 spacecraft.

TABLE 2. BASIC SPECIFICATIONS OF PHOBOS LONG-DURATION AUTOMATIC STATION

Total mass	67 kg (148 lbs)
Mass of scientific payload	20.6 kg (45.4 lbs)
Height	1.8 m (6 ft)
Solar-panel length	2.5 m (8.2 ft)

approval of a very complicated project with the following five elements: a satellite in Mars orbit, small stations (landers), penetrators, a Mars rover, and an aerostat. The science experiments were largely developed within the framework of an international cooperative effort involving 20 countries. Launch was scheduled for 1994 but was then rescheduled for 1996. The situation was dramatic: Work on the project had begun in one country — the Soviet Union — and was continued and finally completed in another — the Russian Federation. The Russian Space Agency was established during this period, and Mars-96 became its first large scientific project. It soon became clear that it needed to be simplified. The Mars rover and aerostat were eliminated. However, there was a lack of funding, the project was continuously delayed, and the infrastructure crumbled. For the first time ever, the two-launch scheme had to be abandoned due to lack of funding.

Mars-96 was launched on 17 November 1996 but was unable to enter the transfer orbit because of a failure in the Block D stage. This was a catastrophe of a magnitude different from previous disasters. Many were convinced that this meant an end to the overall space-science research strategy inherited from the Soviet Union, with virtually unlimited funding, use of expensive launch vehicles, etc. However, it would have been very difficult for this strategy to continue, even had the Mars-96 launch been

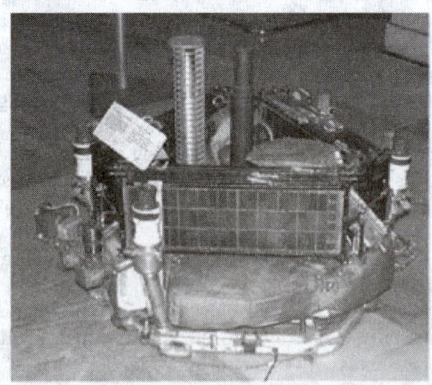

Transfer configuration

Deployed configuration

Fig. 6.   General view of Phobos long-duration automatic station.

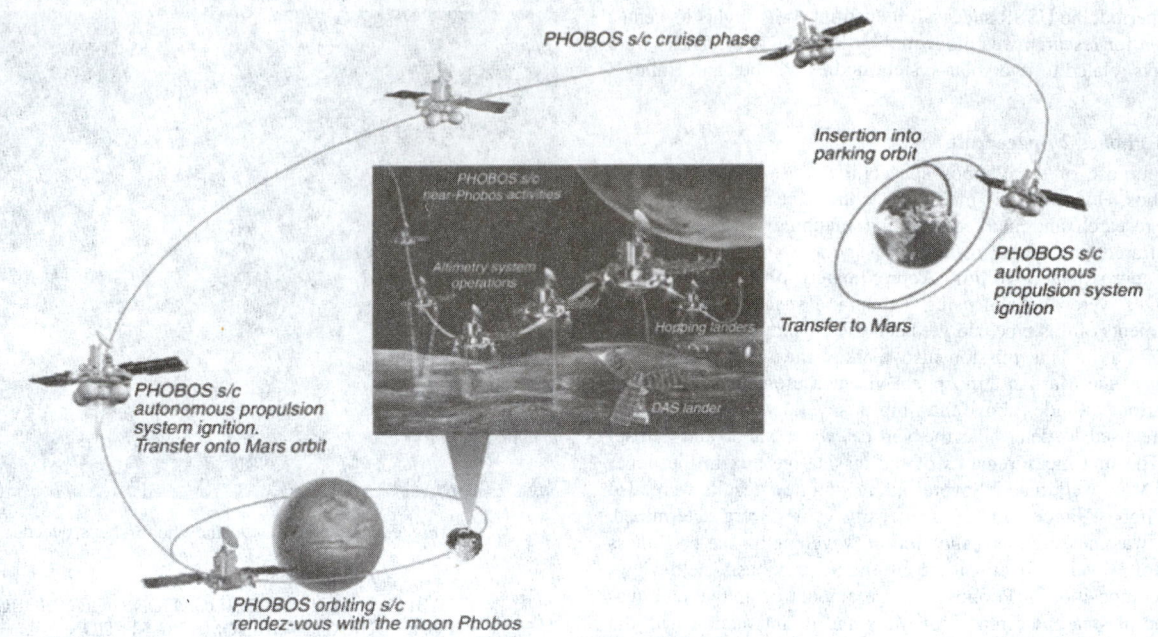

Fig. 7.   Phobos mission flight plan.

successful. Basic data on the Mars-94/96 spacecraft and its scientific instrumentation may be found in two *Institute for Space Research* preprints.

While work on the Mars-96 project was under way, an attempt was made to organize bilateral cooperation between the United States and Russia in space-science research; this attempt was initiated by W. Huntress, who at that time was NASA Deputy Administrator for Space Science. The Cold War was now over, and it was logical to combine efforts in Mars research, as well as several other areas. Several options were generated for a joint flight to Mars under the title "Together to Mars": A Russian launch vehicle, an American orbiter, and a Russian descent module. For various reasons, this effort never proceeded beyond the preliminary discussion phase. This cooperative effort eventually merely amounted to participation of Russian scientists in three American Mars missions. Two of them (Mars Climate Orbiter and Mars Polar Lander) failed but the third (Mars Odyssey 2001) was successful. Important observations of ice in the subsurface ground layer were made by Mars Odyssey with the gamma-spectrometer package that includes a Russian instrument (high energy neutron detector HEND).

Despite all of the lack of success in Mars research, Mars and its satellite, Phobos, remain highly attractive targets for the current Russian planetary program.

See also **Mars**; and **Mars Odyssey Mission**.

Additional Reading

Moroz, V.I.: *Fizika Planet.* Nauka, Moscow, 1967 [*Physics of planets*, NASA TTF-515, 1968].

Perminov, V.G.: *The Difficult Road to Mars. A Brief History of Mars Exploration in the Soviet Union*, NASA, Washington, DC, 1999.

Sagdeev, R.Z., V.M. Balebanov, and A.V. Zakharov.: "The Phobos Project," *Astrophys. Space Phys. Rev.*, **6** (part 1), 3–62 (1988).

Sagdeev, R.Z., and A.V. Zakharov: "Brief History of the Phobos Mission," *Nature*, **341**, 581–585 (1989).

Snyder, C.W., and V.I. Moroz: "Spacecraft Exploration of Mars," In: H.H. Kieffer, B.M. Jakosky, C.W. Snyder, and M.S. Mattheus (eds), *Mars*. The University of Arizona Press, Tucson, AZ, 1992.

Staff: "Series of initial papers describing the results of the Phobos mission," *Nature*, 341 (1989).

Staff: "Series of papers describing the results of the Phobos mission," *Geophys. Res. Lett.*, **17** (6) (1990).

Staff: "Series of papers describing the results of the Phobos mission," *Planetary and Space Sci.*, **39** (1): (1991).

Zakharov, A.V.: "The Plasma Environment of Mars: Phobos Mission Results," In: J.G. Luhmann, M. Tatrallyay, and R.O. Pepin (eds), *Venus and Mars: Atmospheres, Ionospheres, and Solar Wind Interactions*. Geophysical Monograph Series #66, 1992,pp. 327–344.

Zakharov, A.V.: "Mars-94," *Inst. Space Res.*, Preprint 1994.

Zakharov, A.V.: "Mars-96," *Inst. Space Res.*, Preprint 1996.

ALEXANDER V. ZAKHAROV, V.I. MOROZ, Russian Academy of Sciences, Moscow, Russian Federation

MARS EXPLORATION PROGRAM.   Since our first close-up picture of Mars in 1965, spacecraft voyages to the Red Planet have revealed a world strangely familiar, yet different enough to challenge our perceptions of what makes a planet work. Every time we feel close to understanding Mars, new discoveries send us straight back to the drawing board to revise existing theories.

You'd think Mars would be easier to understand. Like Earth, Mars has polar ice caps and clouds in its atmosphere, seasonal weather patterns, volcanoes, canyons and other recognizable features. However, conditions on Mars vary wildly from what we know on our own planet.

Over the past three decades, spacecraft have shown us that Mars is rocky, cold, and sterile beneath its hazy, pink sky. We've discovered that today's Martian wasteland hints at a formerly volatile world where volcanoes once raged, meteors plowed deep craters, and flash floods rushed over the land. And Mars continues to throw out new enticements with each landing or orbital pass made by our spacecraft.

Among our discoveries about Mars, one stands out above all others: the possible presence of liquid water on Mars, either in its ancient past or preserved in the subsurface today. Water is a key factor, because almost everywhere we find water on Earth, we find life. If Mars once had liquid water, or still does today, it's compelling to ask whether any microscopic life forms could have developed on its surface. Is there any evidence of life in the planet's past? If so, could any of these tiny living creatures still exist today? Imagine how exciting it would be to answer, "Yes!!"

Even if Mars is devoid of past or present life, however, there's still much excitement on the horizon. We ourselves might become the "life on Mars" should humans choose to travel there one day. Meanwhile, we still have a lot to learn about this amazing planet and its extreme environments.

To discover the possibilities for life on Mars–past, present or our own in the future–the Mars Program has developed an exploration strategy known as *Follow the Water*.

Following the water begins with an understanding of the current environment on Mars. We want to explore observed features like dry riverbeds, ice in the polar caps and rock types that only form when water is present. We want to look for hot springs, hydrothermal vents or subsurface water reserves. We want to understand if ancient Mars once held a vast ocean in the northern hemisphere as some scientists believe and how Mars may have transition from a more watery environment to the dry and dusty climate it has today. Searching for these answers means delving into the planet's geologic and climate history to find out how, when and why Mars underwent dramatic changes to become the forbidding, yet promising, planet we observe today.

To pursue these goals, all of our future missions will be driven by rigorous scientific questions that will continuously evolve as we make new discoveries.

Brand new technologies will enable us to explore Mars in ways we never have before, resulting in higher-resolution images, precision landings, longer-ranging surface mobility and even the return of Martian soil and rock samples for studies in laboratories here on Earth.

Science

The Mars Exploration Program is a science-driven program that seeks to understand whether Mars was, is, or can be, a habitable world. To find out, we need to understand how geologic, climatic, and other processes have worked to shape Mars and its environment over time, as well as how they interact today.

Mars is similar to Earth in many ways, having many of the same "systems" that characterize our home world. Like Earth, Mars has an atmosphere, a hydrosphere, a cryosphere and a lithosphere. In other words, Mars has systems of air, water, ice, and geology that all interact to produce the Martian environment. See also **Mars**.

What we do not know yet is whether Mars ever developed or maintained a biosphere–an environment in which life could thrive.

The key to understanding the past, present or future potential for life on Mars can be found in the four broad, overarching goals for Mars Exploration.

Goal 1: Determine if Life Ever Arose On Mars. During the next two decades, NASA will conduct several missions to address whether life ever arose on Mars. The search begins with determining whether the Martian environment was ever suitable for life.

On Earth, all forms of life need water to survive. It is likely, though not certain, that if life ever evolved on Mars, it did so in the presence of a long-standing supply of water. On Mars, we will therefore search for evidence of life in areas where liquid water was once stable, and below the surface where it still might exist today. Perhaps there might also be some current "hot spots" on Mars where hydrothermal pools (like those at Yellowstone) provide places for life. Recent data from Mars Global Surveyor suggest that liquid water may exist just below the surface in rare places on the planet, and the 2001 Mars Odyssey will be mapping subsurface water reservoirs on a global scale. We know that water ice is present at the Martian poles, and these areas will be good places to search for evidence of life as well.

In addition to liquid water, life also needs energy. Therefore, future missions will also be on the lookout for energy sources other than sunlight, since life on the surface of mars is unlikely given the presence of *superoxides* that break down organic (carbon-based) molecules on which life is based. Here on Earth, we find life in many places where sunlight never reaches—at dark ocean depths, inside rocks, and deep below the surface. Chemical and geothermal energy, for example, are also energy sources used by life forms on Earth. Perhaps tiny, subsurface microbes on Mars could use such energy sources too.

NASA will also look for life on Mars by searching for telltale markers, or biosignatures, of current and past life. The element carbon, for instance, is a fundamental building block of life. Knowing where carbon is present and in what form would tell us a lot about where life might have developed.

We know that most of the current Martian atmosphere consists of carbon dioxide. If carbonate minerals were formed on the Martian surface by chemical reactions between water and the atmosphere, the presence of these minerals would be a clue that water had been present for a long time—perhaps long enough for life to have developed.

On Earth, fossils in sedimentary rock leave a record of past life. Based on studies of the fossil record on Earth, we know that only certain environments and types of deposits provide good places for fossil preservation. On Mars, searches are already underway to locate lakes or streams that may have left behind similar deposits.

So far, however, the kinds of biosignatures we know how to identify are those found on Earth. It's possible that life on another planet might be very different. The challenge is to be able to differentiate life from nonlife no matter where one finds it, no matter what its varying chemistry,

structure, and other characteristics might be. Life detection technologies under development will help us define life in non-Earth-centric terms so that we are able to detect in all the forms it might take.

Goal 2: Characterize the Climate of Mars. A top priority in our exploration of Mars is understanding its present climate, what its climate was like in the distant past, and the causes of climate change over time.

The current Martian climate is regulated by seasonal changes of the carbon dioxide ice caps, the movement of large amounts of dust by the atmosphere and the exchange of water vapor between the surface and the atmosphere. One of the most dynamic weather patterns on Mars is the generation of dust storms that generally occur in the southern spring and summer. These storms can grow to encompass the whole planet. Understanding how these storms develop and grow is one goal of future climatic studies.

A better understanding of Mars' current climate will help scientists more effectively model its past climatic behavior. To do that, we'll need detailed weather maps of the planet and information about how much dust and water vapor are in the atmosphere.

Monitoring the planet for this information over one full Martian year (687 Earth days) will help us understand how Mars behaves over its seasonal cycle and guide us toward understanding how the planet changes over millions of years.

The layered terrain of the Martian Polar Regions also holds clues about the planet's past, much like the rings of a tree provide a record of its history. When and how were these polar layers deposited? Was the climate of Mars ever like that of Earth? And if so, what happened to change the planet into the dry, cold, barren desert it is today? Those are the questions that our missions still have to answer.

Goal 3: Characterize the Geology of Mars. How did Mars become the planet we see today? What accounts for the differences and similarities between Earth and Mars? These questions will be addressed by studying Mars' geology. As part of the Mars Exploration Program, we want to understand how the relative roles of wind, water, volcanism, tectonics, cratering and other processes have acted to form and modify the Martian surface.

For example, Mars is home to incredibly large volcanoes, which can be 10 to 100 times larger than those on Earth. One reason for this difference is that the crust on Mars doesn't move the way it does on Earth. That means the total volume of lava piles up into one, very large volcano.

A recent discovery by the Mars Global Surveyor spacecraft of large areas of magnetic materials on Mars indicates that the planet once had a magnetic field, much like Earth does today. Because magnetic fields in general act to shield planets from many forms of cosmic radiation, this discovery has important implications for the prospects for finding evidence of past life on the Martian surface. Study of the ancient magnetic field also provides important information about the interior structure, temperature and composition of Mars in the past. The presence of magnetic fields also suggests that Mars was once more of a dynamic Earth-like planet than it is today.

Of fundamental importance are the age and composition of different types of rocks on the Martian surface. Geologists use the age of rocks to determine the sequence of events in a planet's history. Composition information tells them what happened over time. Particularly important is the identification of rocks and minerals formed in the presence of water. Water is one of the keys to whether life might have started on Mars.

What other materials might be trapped in those rocks with information about the planet's history? How are the different rock types distributed across the surface? Future orbiting and landed missions will carry special tools designed to help answer these questions.

Thirty four Mars meteorites discovered on Earth can also give us clues about the planet. See also http://www2.jpl.nasa.gov/snc/.

Goal 4: Prepare for the Human Exploration of Mars. Eventually, humans will most likely journey to Mars. Getting astronauts to the Martian surface and returning them safely to Earth, however, is an extremely difficult engineering challenge. A thorough understanding of the Martian environment is critical to the safe operation of equipment and to human health, so the Mars Exploration Program will begin to look at these challenges in the coming decade.

The safety of astronauts is of paramount importance to NASA. Mars lacks an ozone layer, which on Earth shields us from lethal doses of solar ultraviolet radiation. We do not have good information about the amount of ultraviolet radiation that reaches the Martian surface.

A more detailed understanding of the radiation environment will provide the information necessary to assess the effects of UV radiation on astronauts, as well as help engineers design protective space suits and habitats.

We do know that the Martian soil contains "superoxides." In the presence of ultraviolet radiation, superoxides break down organic molecules. While superoxides' effect on astronauts is probably not serious, their impact and that of any other unique chemical aspects of the Martian soil must be assessed before human exploration of Mars can begin.

To pave the way for human exploration, 2001 Mars Odyssey will begin to analyze the radiation environment on Mars. This mission and the Mars Reconnaissance Orbiter will continue to search for water resources that, if discovered, could be used to support future human explorers. Eventually, robotic spacecraft, rovers, and drills could be used to access water resources in advance of, and during, human exploration.

Advanced entry, descent and landing techniques that reduce the G-forces on landers will also be developed for spacecraft and astronaut safety.

While robotic exploration will pave the way for the long-term possibility of human missions to Mars, much of the necessary scientific and technological work for this goal is carried out by NASA's Johnson Space Center. For more information about eventual human missions to the Red Planet, visit NASA's Human Spaceflight Web section on Mars Exploration and Beyond; http://spaceflight.nasa.gov/mars/.

Technology

Technology development makes missions possible. Each Mars mission is part of a continuing chain of innovation: each relies on past missions for new technologies and contributes its own innovations to future missions. This chain allows NASA to continue to push the boundaries of what is currently possible, while relying on proven technologies as well.

NASA is pursuing an aggressive, science-driven agenda of robotic exploration of Mars with a series of orbiters and landers. These missions carry science instruments selected to answer questions the planetary science community has posed to better characterize the planet (See Mars Exploration Program Analysis Group, MEPAG, http://mepag.jpl.nasa.gov/). The overarching objective is increased understanding with regard to Life, Climate, Geology, and Preparation for Human Exploration.

Many new technologies need to be developed and infused into future Mars missions, which demand the following capabilities:

- Better landing accuracy, with active hazard-detection-and-avoidance capability.
- Access to high-priority sites with terrain too complex for landing current rovers.
- Increased mobility to sample diverse geological sites and reach targets of interest.
- Longer-lived, more robust and higher-output energy systems to allow year and longer surface operations in a greater range of adverse conditions.
- Technologies to access the subsurface and acquire samples for in situ analysis.
- New and improved science instruments.
- In situ sample acquisition, preparation, and distribution systems.
- Increased autonomy to enable increased return of high-priority science.
- Planetary protection techniques.
- Technologies for possible return of samples to Earth for analysis.

The Mars Technology Program (MTP) http://marstech.jpl.nasa.gov/, is responsible for technology-development plans that are consistent with NASA's Mars Exploration vision, and implementing and infusing those technologies into future missions.

Technologies are selected for development funding by competitions via NASA Research Announcements (NRA) http://www.hq.nasa.gov/office/procurement/nraguidebook/ and by direct funding when appropriate.

Exploration of Mars

So far, the exploration of Mars has occurred in three stages:

- **Flybys** When we were just starting out in solar system exploration, the very first missions simply flew by Mars, taking as many pictures

as possible on their way past. Flyby missions include: Mariner 3-4 | Mariner 6-7.

- **Orbiters** As our knowledge and technologies grew, we began putting spacecraft in orbit around Mars for longer term, global studies. Orbital Missions include: Mariner 8-9 | Viking 1-2 | Mars Observer | Mars Climate Orbiter | Mars Global Surveyor | 2001 Mars Odyssey | Mars Express | Mars Reconnaissance Orbiter.
- **Landers and Rovers** Then, with even more capabilities over the years, we began to land on the surface. Today, we're not only landing in one place, but have shown that we can be mobile on the surface of Mars. Landed missions include: Viking 1-2 | Pathfinder | Polar Lander/Deep Space 2 | Mars Exploration Rovers | Mars Express.

In the future, Mars exploration may bring:

- **Airplanes and Balloons** We may send craft that can study the planet from a perspective we've never achieved before: soaring views from the Martian sky.
- **Subsurface Explorers** Going below the ground will tell us more about the geology of the planet, the presence of water, and maybe even clues about whether Mars was ever a habitat for life.
- **Sample Returns** Going below the ground will tell us more about the geology of the planet, the presence of water, and maybe even clues about whether Mars was ever a habitat for life.

Mission Types

Orbiters. Orbiters circling Mars have shown us that Mars, though hostile, has many similarities to Earth: canyons, volcanoes, craters, gullies and runoff channels, clouds, weather patterns, rocks, hills, polar ice caps, eclipses, and more. They've revealed a wealth of information about the red planet's atmosphere, landforms, gravity, magnetic fields, elemental and mineral composition, internal structure, and weather.

Orbiters on future missions will examine the planet's environment even more precisely. They will search for evidence of water's existence both now and in the past, on the surface and underground. The history of water is important to all four science themes of the Mars Exploration Program: climate, geology, life, and preparation for human exploration previously discussed.

Orbiters will also help identify scientifically interesting sites on the surface for further investigation and provide knowledge of large rocks and other hazards at landing sites to be avoided.

Orbiters will also play a key role as communications relays for rovers, landers, balloons, and airplanes. For instance, it won't be practical for rovers, busy navigating hazardous terrain, to point their small antennas towards faraway Earth and transmit information. Instead, it will be easier for the rover to send data to a Mars orbiter, which in turn will precisely point a large antenna at Earth and communicate large volumes of images and other data.

Orbiters can also assist with navigation of other spacecraft approaching Mars. For example, Mars Reconnaissance Orbiter will carry a test camera that pinpoints the position of approaching spacecraft, helping navigators back on Earth know exactly how to direct the spacecraft for precision orbit insertion and landing.

Orbiters may also play a vital role in returning samples from the surface of Mars. Just as it is not efficient to communicate large volumes of science data directly from the Martian surface to Earth, it is not practical to launch a rocket containing samples from the Martian surface and expect it to navigate itself back to Earth and land safely. Instead, the preferred solution is to launch a small canister containing samples into orbit around Mars, have an orbiter rendezvous with and capture the canister, and then eject the spacecraft from Mars orbit back towards Earth.

Landers. Landing on a planet hundreds of millions of miles away from Earth is incredibly challenging. The surface of Mars has many hazards. There are deep canyons, volcanic mountains and craters of all sizes formed by meteorite impact. Some areas are strewn with boulders; others are shrouded in sand dunes. In addition, landers must rapidly decelerate from 13,000 miles per hour in a matter of minutes.

Landers therefore need systems to ensure a safe landing. For example, Mars Pathfinder and the Mars Exploration Rovers both airbag landing systems. When hitting the surface, the lander is protected inside something that looks like a huge, bouncing ball. Future landers using airbags might include new technologies that would prevent the *ball* from rolling down steep terrain once it lands. See also **Mars Exploration Rovers**; **Pathfinder Mission to Mars**.

Another new landing gear design would use a structure that acts as a shock absorber to protect the main body of the lander. This shock absorber would also have legs to help stabilize the craft upon landing. Both the shock absorber and legs could crumble or be destroyed without jeopardizing the main lander.

Even better than mere protective systems, "smart landers" of the future will be able to achieve safe landings by scanning the terrain about half-a-mile above the surface and guiding themselves to much smaller target zones—down to a few hundred feet in size. Then, they will be able to dodge large hazards like boulders and crevices at the very last moment. In case they aren't aware of last-minute hazards, these landers will carry impact protection systems that will be lighter than those on our present landers.

These sophisticated approaches to landing safely will also allow future landers to carry more equipment. This greater *load* capability will enable missions that are not feasible with current technology, such as sample returns of rock and soil to Earth or heavy drilling equipment for exploring deep beneath the Martian surface in a search for liquid water and possible past or present signs of life.

Rovers. Even with the development of increasingly precise landers, we don't want to be stuck in just one spot on Mars since we never know what exciting discovery might be just around the next bend. Rovers give us the ability to explore an area beyond the place where landers touch down.

Our first rover to visit Mars, Sojourner, traveled about 100 meters (109 yards) over an entire month. See also **Mars**. The next two rovers to visit Mars, the Mars Exploration Rovers, will land on different parts of the planet. Both twins are also expected to have the ability to rove approximately 1 kilometer (about six-tenths of a mile) over the duration of their mission. Future rovers may travel over several miles on the surface, allowing us to move from a safe landing site to investigate surface features that are currently too difficult to access.

One approach to long-range mobility is inflatable rovers, which would use very large, inflatable wheels to climb over rocks, instead of traveling around them. This would allow inflatable rovers to travel much farther and faster than current rovers. These rovers are only inflated after arrival on the Mars surface. Prior to inflation, they are lighter and, when deflated, can be packed in a much smaller volume than a conventional rover of the same physical size. When deployed, the inflatable rover is approximately 50 times its packed volume, whereas a conventional rover is only two or three times bigger than when packed.

Rovers will also increasingly rely on smart technologies to know where they are, where they want to go, and which soil and rock samples are worth studying and collecting.

Future rovers may also have adjustable shoulders that allow them to drop low to the ground or elevate themselves to navigate through a gully or crater. They will no doubt conduct more sophisticated experiments, such as using ground-penetrating radar studies, to search for evidence of water.

Future rovers will also travel locations where they can rendezvous with other surface vehicles. Eventually, teams of robotic rovers might even work together to build an infrastructure of robotic colonies, laying the groundwork for human visits and human bases. Thus, if human exploration is indeed possible on the Red Planet one day, it will be robotic rovers that pave the way.

Balloons. Balloons provide a unique vantage point for scientific observation. Balloons can fly one hundred times closer to the surface of Mars than orbiters and can travel a thousand times further than rovers in a comparable period, thus providing views of much broader areas of the surface.

Balloons have been flying for decades in Earth's stratosphere, which has an atmosphere as thin as that on the surface of Mars. Conventional stratospheric balloons have lifetimes limited to a few days because of the daily heating and cooling of the balloon. Helium super-pressure balloons, currently under development for the Ultra Long Duration Balloon (ULDB), will fly more than 100 days and perhaps as long as a year here on Earth. Smaller super-pressure balloons carrying payloads of only a few pounds have already flown for as long as a year. See also **Scientific Balloons**.

Using this technology, a future Mars balloon would deploy soon after a spacecraft entered the Mars atmosphere and would then rapidly inflate from a helium tank as it descended beneath a parachute. After inflation, the parachute and tanks would detach and the balloon would fly at a nearly constant altitude both day and night. The balloon's internal pressure would be higher during the day than at night, although the balloon volume would remain the same. Strong, lightweight, leak-proof materials are under development to permit large tool kits of science instruments to be flown on such a balloon. Tests of balloon deployment in the Earth's atmosphere are currently underway as well.

Another kind of lightweight balloon that might be useful on Mars is called a *solar Montgolfiere balloon*, named after the French brothers who flew the first hot air balloon. It does not have to be inflated with a light gas such as helium. Instead, the balloon would deploy upon entering the Martian atmosphere. An opening at the bottom of the balloon would fill up with Martian "air" while falling to the surface. The balloon would then be quickly heated by the sun, which provides buoyancy. Montgolfieres are attractive because they are not vulnerable to leaks, as any leaking Martian *air* would be quickly replaced and re-heated by the sun. However, the balloon lifetime is limited to a few hours, because it is only buoyant until the sun goes down. However, the Montgolfiere balloon can play two important roles in exploration: provide a soft, slow landing for small craft on potentially hazardous terrain, with greater control than a parachute-assisted or rocket landing system; and go back up into the atmosphere after dropping off the landed craft so that it could image the surface further and gather more science data before nightfall.

Subsurface Explorers. Even if rovers and balloons, continuously move around and near the surface of Mars one day, we should never judge a planet by its cover. Today's desert-like Martian surface likely hides the presence of water below ground. To "follow the water" to where it is today, we must go beneath the surface of the planet with subsurface explorers.

The subsurface of Mars may resemble some of the colder parts of Earth. For example, in Antarctica or Iceland, we know that water is stored in a layer of permafrost and beneath that, as liquid groundwater. Even if the ancient surface water on Mars evaporated, there may still be substantial reservoirs of water, in either liquid or frozen form, in the subsurface.

The very first subsurface exploration of Mars for NASA will be in partnership with the European Space Agency (ESA) in their Mars Express mission. This spacecraft carries a subsurface radar instrument that will use a 40-meter (130-foot) antenna to detect and map subsurface water. Electric signals will be sent down the antenna, creating low-frequency radar waves. The radar waves will penetrate the Martian surface as deep as five kilometers (three miles) and will be reflected back to the spacecraft by different subsurface features, including water. This data will give us a three-dimensional understanding of where and how much water may be distributed in the Martian subsurface.

A lander on Mars Express called Beagle 2 will also carry the first robotic mole. Mimicking the behavior of the small furry earth-bound creatures that burrow into the ground, robotic moles will drill underground by pulverizing rock and soil, avoiding the need for a complex drill stem. Beagle 2's mole will only have the ability to penetrate less than a meter (less than 3 feet) below the surface.

A much more capable mole is under development in NASA's technology program. Weighing about 20 kilograms (44 pounds), it will be capable of drilling hundreds of meters (hundreds of yards) into the ground and possibly deeper at a rate of 10–20 meters (33–66 feet) a day. Excavated soil would be moved to the back of the mole and a small tube leading to the surface would help alleviate the pressure from the growing mounds of soil. The tube would also send soil samples back to the surface and carry power to the robotic mole. The samples sent up to the surface would be studied for scientific data such as mineral content and oxidation levels of subsurface soil. A mole drilling at the polar cap would study the layers of ice that tell the story of its history, much like the rings of a tree reveal many things from its past. All of this data would provide clues in the search for ancient, or possibly current, life.

Once we know in more detail where the water lies, the next step is to drill in those locations. To get to the zone where frozen water—and possible dormant life—might be present, we will probably need to drill to a depth of 200 meters (656 feet). Liquid groundwater will be even deeper. That's no easy feat, but it's critical for understanding the possibility of past or present life on Mars and for confirming that water resources are available for future human explorers.

Deep subsurface access on Mars will have unique challenges. First of all, unlike on Earth, we will not be able to use a drill to go through mud, water, or probably even gas pressure to carry the cuttings away from the bit. We will need new systems for fluidless drilling. Second, we will need an effective means of keeping the hole open while the drilling proceeds. On Earth, this task is normally done with steel casing, which is very heavy. Engineers are actively seeking alternative ways that don't require us to send heavy equipment to Mars given the expense. Finally, we will have to develop systems that allow the drill to make operational decisions for itself. On Earth, drills can get stuck very quickly, so a Mars robotic drill or subsurface explorer must know how to recognize, avoid, and solve problems on its own.

Sample Returns. Rovers and other space vehicles do a great job studying Mars. However, the most exhaustive studies of rock, soil, and atmospheric particles can only be conducted in laboratories here on Earth. After all, nothing beats the hands-on expertise of scientists. However, bringing samples back is challenging since it requires rockets that can ascend from the surface of Mars to orbit and may require vehicles that can rendezvous and capture the sample for delivery to Earth.

With all of the rock and soil samples that are available on Mars, we need the ability to determine which samples are the most scientifically interesting. Scientists and engineers are currently developing many tools and instruments to make the right choices. The initial identification of interesting rocks will probably be done in the same way that a field geologist studies rocks on Earth, by using visual information such as color and texture. For this job, electronic imaging systems are essential. We need some systems to take high-quality images of rocks from a distance of several meters (several feet), while we need others to look at rock features on microscopic scales of millimeters or less. These abilities along with spectroscopy, a technology using ultraviolet, visible or infrared light to analyze a rock's chemical composition, are also being developed. This chemical information will give clues to a rock's origin and history.

After a scientifically interesting rock has been selected based on its chemical composition and other factors, we must obtain a sample of it that is small enough to be brought back to Earth, yet large enough to preserve important texture and structure. Instruments have already been designed to drill into rocks and retrieve cores from the inside. These interior rock samples should be better preserved than the outside of the rock, which will have been exposed to, and chemically altered by, the Martian atmosphere.

Since searching for evidence of present or past life is a key objective, the sampling system carried on the rover must not contaminate the sample with any organisms brought from Earth. The coring apparatus must be thoroughly cleaned before launch so the samples won't interact with dust or biological material from Earth. After all, we wouldn't want to bring a sample all the way from Mars and study its features, only to discover that we're studying Earth materials along with it. We want "pure" Martian samples, straight from the source!

Once the rover has its samples, they will be placed in a small spherical container weighing a few kilograms. To increase our ability to bring back samples untainted with Earth materials, samples must be sealed in a capsule for launch. This capsule must be able to seal completely in order to prevent contamination of the sample by the Earth's atmosphere or biosphere upon landing on Earth. Technologies for remotely welding metal to make clean airtight seals are needed to protect the returned samples. The sealing process must also assure that material of Martian origin remains on the outside of the container to avoid inadvertent release of the material on Earth. Once sealed, a small rocket called a Mars Ascent Vehicle will launch the capsule from the surface of Mars.

From this point, there are several possible approaches to bringing the sample to Earth. The most practical of these appears to be using an orbiter to capture the sample container while it is in Mars orbit. Methods are being studied for finding a small canister in Mars orbit; navigating the orbiter to rendezvous with the canister and capturing the canister, all with commands initiated 100 million kilometers away. Although traveling at the speed of light, the commands will take almost half an hour to reach the spacecraft.

The journey back to Earth involves special precautions to ensure safe containment of the sample. The samples may be delivered directly to Earth, but could be returned via the space shuttle. Although it is highly

unlikely that living organisms will be found on the samples, NASA will implement a wide range of precautions to preclude inadvertent release. This protocol will analyze the samples in containment to determine if they are hazardous. The samples will be released for scientific analysis only when it is determined that they are non-hazardous.

Past Missions

For a complete chronology of past missions see Table 1.

See also **Mars**.

Present Missions

Mars Global Surveyor. Launched November 7, 1996, Mars Global Surveyor was the first successful U.S. mission launched to Mars since the Viking mission in 1976. After a 20-year absence at the planet, Mars Global Surveyor ushered in a new era of Mars exploration. See Fig. 1.

Fig. 1. Mars Global Surveyor accomplished all of its science goals, and is currently in an extended mission phase (*NASA*).

The mission has studied the entire Martian surface, atmosphere, and interior. One of the most exciting observations of the spacecraft's wide-angle camera system, known as the Mars Orbital Camera, is that the red planet has very repeatable weather patterns. Each day the camera operates, it collects images that are used to build up a daily global map. These maps provide a record of changing meteorological conditions on Mars. Weather patterns observed by the spacecraft include some dust storms that repeat in the same location within a week or two of the time they occurred in the previous year. In addition, local disturbances and dust devils may start up at any time after the first day of spring and continue until Martian autumn.

An impressive and magnificent display of high-resolution images from the Mars Global Surveyor has documented gullies and debris flows suggesting that occasional sources of liquid water, similar to an aquifer, were once present at or near the surface of the planet. Magnetometer readings have shown that the planet does not have a global magnetic field but has localized magnetic fields in particular areas of the crust. Temperature data and close-up images of the Martian moon Phobos have determined that the moon is covered by a layer of powdery material — the pulverized output of millions of years of meteoroid impacts — at least 1 meter (3.28 feet) thick.

By studying Mars for several Martian years (a Mars year is about twice as long as an Earth year), Mars Global Surveyor has observed gully formation, new boulder tracks, recently formed impact craters, and

TABLE 1. HISTORICAL LOG OF PAST MARS MISSIONS

Launch Date	Name	Country	Result	Reason
1960	Korabl 4	USSR (flyby)	Failure	Did not reach Earth orbit
1960	Korabl 5	USSR (flyby)	Failure	Did not reach Earth orbit
1962	Korabl 11	USSR (flyby)	Failure	Earth orbit only; spacecraft broke apart
1962	Mars 1	USSR (flyby)	Failure	Radio Failed
1962	Korabl 13	USSR (flyby)	Failure	Earth orbit only; spacecraft broke apart
1964	Mariner 3	US (flyby)	Failure	Shroud failed to jettison
1964	Mariner 4	US (flyby)	Success	Returned 21 images
1964	Zond 2	USSR (flyby)	Failure	Radio failed
1969	Mars 1969A	USSR	Failure	Launch vehicle failure
1969	Mars 1969B	USSR	Failure	Launch vehicle failure
1969	Mariner 6	US (flyby)	Success	Returned 75 images
1969	Mariner 7	US (flyby)	Success	Returned 126 images
1971	Mariner 8	US	Failure	Launch failure
1971	Kosmos 419	USSR	Failure	Achieved Earth orbit only
1971	Mars 2 Orbiter/Lander	USSR	Failure	Orbiter arrived, but no useful data and Lander destroyed
1971	Mars 3 Orbiter/Lander	USSR	Success	Orbiter obtained approximately 8 months of data and lander landed safely, but only 20 seconds of data
1971	Mariner 9	US	Success	Returned 7,329 images
1973	Mars 4	USSR	Failure	Flew past Mars
1973	Mars 5	USSR	Success	Returned 60 images; only lasted 9 days
1973	Mars 6 Orbiter/Lander	USSR	Success/Failure	Occultation experiment produced data and Lander failure on descent
1973	Mars 7 Lander	USSR	Failure	Missed planet; now in solar orbit.
1975	Viking 1 Orbiter/Lander	US	Success	Located landing site for Lander and first successful landing on Mars
1975	Viking 2 Orbiter/Lander	US	Success	Returned 16,000 images and extensive atmospheric data and soil experiments
1988	Phobos 1 Orbiter	USSR	Failure	Lost en route to Mars
1988	Phobos 2 Orbiter/Lander	USSR	Failure	Lost near Phobos
1992	Mars Observer	US	Failure	Lost prior to Mars arrival
1996	Mars Global Surveyor	US	Success	More images than all Mars Missions
1996	Mars 96	USSR	Failure	Launch vehicle failure
1996	Mars Pathfinder	US	Success	Technology experiment lasting 5 times longer than warranty
1998	Nozomi	Japan	Failure	No orbit insertion; fuel problems
1998	Mars Climate Orbiter	US	Failure	Lost on arrival
1999	Mars Polar Lander	US	Failure	Lost on arrival
1999	Deep Space 2 Probes (2)	US	Failure	Lost on arrival (carried on Mars Polar Lander)
2001	Mars Odyssey	US	Success	High resolution images of Mars
2003	Mars Express Orbiter/Beagle 2 Lander	ESA	Success/Failure	Orbiter imaging Mars in detail and lander lost on arrival
2003	Mars Exploration Rover — Spirit	US	Success	Over 70,000 images lasting 8 times longer than warranty
2003	Mars Exploration Rover — Opportunity	US	Success	Over 58,000 images lasting 8 times longer than warranty
2005	Mars Reconnaissance Orbiter	US	TBD	

diminishing amounts of carbon dioxide ice within the south polar cap. Data from the spacecraft's laser altimeter has given scientists their first 3-D views of Mars' north polar ice cap. Changes in radio transmissions as they are refracted by the Martian atmosphere have enabled scientists to create vertical profiles of atmospheric temperature and pressure. Spacecraft accelerations due to gravity have given scientists a better understanding of the interior of Mars. Findings such as these have shown that Mars is a dynamic planet with a history of seasonal and long-term change recorded in the planet's surface. See also **Mars Global Surveyor Mission**; and **Space Science Missions: Solar System**.

2001 Mars Odyssey. Mars Odyssey was launched April 7, 2001 on a Delta II rocket from Cape Canaveral, Florida, and reached Mars on October 24, 2001, 0230 Universal Time (October 23, 7:30 pm PDT/ 10:30 EDT). Still in orbit around Mars, NASA's 2001 Mars Odyssey spacecraft has collected more than 130,000 images and continues to send information to Earth about Martian geology, climate, and mineralogy. See Fig. 2.

Measurements by Odyssey have enabled scientists to create maps of minerals and chemical elements and identify regions with buried water ice. Images that measure the surface temperature have provided spectacular views of Martian topography. Odyssey is currently supporting landing site selection for the Phoenix Scout Mission, to be launched in 2007.

Early in the mission, Odyssey determined that radiation in low-Mars orbit—an essential piece of information for eventual human exploration because of its potential health effects—is twice that in low-Earth orbit. Odyssey has provided vital support to ongoing exploration of Mars by relaying data from the Mars rovers to Earth via the spacecraft's UHF antenna. See also **2001 Mars Odyssey Mission**.

Mars Express. NASA is participating in a mission of the European Space Agency and the Italian Space Agency called *Mars Express*, which has been exploring the atmosphere and surface of Mars from polar orbit

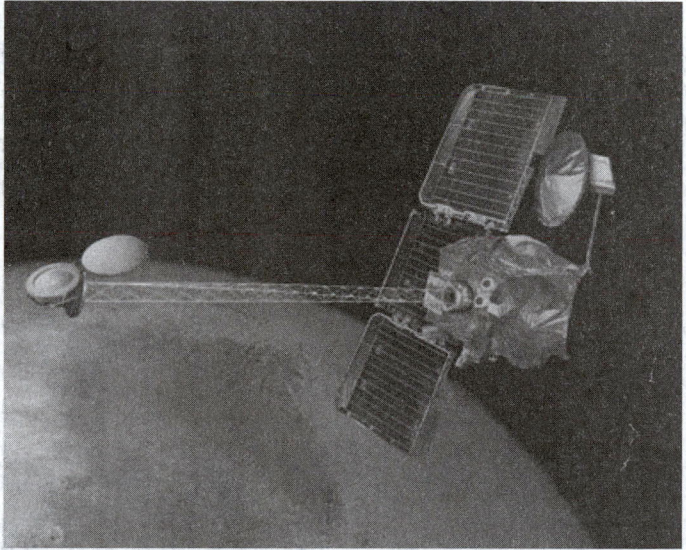

Fig. 2.   2001 Mars Odyssey over Mars (*NASA*).

since arriving at the red planet in 2003. The spacecraft carries a science payload derived in part from European instruments lost on the ill-fated Russian Mars '96 mission, as well as a communications relay to support lander missions. Mars Express was successfully launched on June 2, 2003 atop a Soyuz-Fregat launcher from the Baikonur Cosmodrome in Kazakhstan.

The mission's main objective is to search for sub-surface water from orbit. Seven scientific instruments on the orbiting spacecraft have conducted rigorous investigations to help answer fundamental questions about the geology, atmosphere, surface environment, history of water, and potential for life on Mars. Examples of discoveries—still debated by scientists–by Mars Express are evidence of recent glacial activity, explosive volcanism, and methane gas.

Initially, Mars Express also carried a small lander called Beagle 2, named for the ship in which Charles Darwin set sail to explore unchartered areas of the Earth in 1831. The lander was lost on arrival in December, 2003.

NASA's involvement with the mission includes joint development of a radar instrument called MARSIS—short for the Mars Advanced Radar for Subsurface and Ionospheric Sounding—with the Italian Space Agency. MARSIS has already provided information about features beneath the Martian surface, including buried impact craters, layered deposits, and hints of deep underground water ice.

NASA's involvement also includes coordination of radio relay systems to make sure that different spacecraft operate together; a hardware contribution to the energetic neutral atoms analyzer instrument; and backup tracking support from NASA's Deep Space Network during critical mission phases.

Spacecraft. The Mars Express spacecraft has been designed to take a payload of seven state-of-the-art scientific instruments and one lander to the red planet and allow them to record data for at least one Martian year, or 687 Earth days. The spacecraft also carries a data relay system for communicating with Earth.

The mission is a test case for new working methods to speed up spacecraft production and minimize mission costs. These new methods have had two major impacts on spacecraft design. Weight was kept to an absolute minimum: 116 kilograms (256 pounds) was allowed for the seven instruments and 60 kilograms (132 pounds) for the lander. And off-the-shelf technology, or technology developed for the Rosetta mission to a comet, was used wherever possible.

The instruments sit inside the spacecraft bus which is a honeycomb aluminium box just 1.5 meters (5 feet) long by 1.8 meters (6 feet) wide by 1.4 meters (4.5 feet) high. See Fig. 3. The lander, Beagle 2, was attached to the outside of the bus. Payload, lander, spacecraft and on-board fuel weighed 1,223 kilograms (2,696 pounds) at launch.

signals when the spacecraft is a long way from Earth. When it is close to Earth at the beginning of its journey, communication is via a low gain antenna which is a 40 centimeter (15.75 inch) aerial protruding from the spacecraft bus.

For more information, see the U.S. participation in Mars Express http://marsprogram.jpl.nasa.gov/express/ and the Mars Express home page at the European Space Agency http://sci.esa.int/science-e/www/area/index.cfm?fareaid=9.

Mars Exploration Rovers (Spirit and Opportunity). NASA's twin robot geologists, the Mars Exploration Rovers, launched toward Mars on June 10 and July 7, 2003, in search of answers about the history of water on Mars. Spirit and Opportunity landed on opposite sides of the red planet on January 3 and January 24 PST (January 4 and January 25 UTC). With far greater mobility than the 1997 Mars Pathfinder rover, these robotic explorers have trekked for miles across the Martian surface, conducting field geology and making atmospheric observations. Carrying identical, sophisticated sets of science instruments, both rovers have found evidence of ancient Martian environments where intermittently wet and habitable conditions existed. See also **Pathfinder Mission to Mars**.

During the rovers' landings, parachutes deployed to slow the descending spacecraft, rockets fired to slow them still more just before impact, and airbags inflated to cushion their landing. After bouncing and rolling to a halt, a protective structure of petals opened, brought the landers to an upright position, and provided a platform from which the rovers drove onto the Martian surface.

Since leaving their landing sites, the twin rovers have sent more than 100,000 spectacular, high-resolution, full-color images of Martian terrain as well as detailed microscopic images of rocks and soil surfaces to Earth. Four different spectrometers have amassed unparalleled information about the chemical and mineralogical makeup of Martian rocks and soil. Special rock abrasion tools, never before sent to another planet, have enabled scientists to peer beneath the dusty and weathered surfaces of rocks to examine their interiors.

Each rover weighs nearly 180 kilograms (about 400 pounds). Two and a half years after landing, both rovers are still working and have far exceeded their initial 90-day warranties on Mars. See Fig. 4.

Fig. 3.   Mars Express Orbiter on top of the Soyuz launcher upper stage (Fregat). (*Image courtesy of ESA.*)

Fig. 4.   Mars Exploration Rover (*NASA/JPL*).

The circular dish attached to one face of the spacecraft bus is a 1.6 meter (5.25 foot) diameter high gain antenna for receiving and transmitting radio

Opportunity's study of *Eagle* and *Endurance* craters revealed evidence for past inter-dune playa lakes that evaporated to form sulfate-rich sands. The sands were reworked by water and wind, solidified into rock, and soaked by groundwater. Opportunity is examining more sedimentary bedrock exposures along a route leading from *Endurance* to *Victoria Crater*, where an even broader, deeper section of layered rock is likely exposed that could reveal new aspects of Martian geologic history in Meridiani Planum.

While Spirit's initial travels in Gusev Crater revealed a more basaltic setting, after reaching the *Columbia Hills* the rover found a variety of rocks indicating that early Mars was characterized by impacts, explosive volcanism, and subsurface water. Unusual-looking bright patches of soil turned out to be extremely salty and affected by past water. At *Home Plate*, a circular feature in the *Inner Basin* of the *Columbia Hills*, Spirit discovered finely layered rocks that are as geologically compelling as those found by Opportunity and that may hold clues to a history of past water in Gusev Crater. See also **Mars Exploration Rovers (Spirit and Opportunity)**.

Mars Reconnaissance Orbiter (MRO). NASA's Mars Reconnaissance Orbiter, launched in August, 2005, after a seven-month cruise to Mars and six months of *aerobraking* to reach its science orbit, Mars Reconnaissance Orbiter will seek to find out about the history of water on Mars with its science instruments. See Fig. 5. They will zoom in for extreme close-up photography of the Martian surface, analyze minerals, look for subsurface water, trace how much dust and water are distributed in the atmosphere, and monitor daily global weather.

Fig. 5.   Mars Reconnaissance Orbiter (MRO) (*NASA/JPL*).

The MRO carries the most powerful camera ever flown on a planetary exploration mission for homing in on details of Martian terrain with extraordinary clarity. While previous cameras on other Mars orbiters were able to identify objects no smaller than a dinner table, this camera will be able to spot something as small as a dinner plate. This capability will not only provide an astoundingly detailed view of the geology and structure of Mars, but will help identify obstacles that could jeopardize the safety of future landers and rovers. Calibration tests during the spacecraft's aerobraking phase have provided an impressive glimpse of what the camera will reveal as the mission continues.

The Mars Reconnaissance Orbiter also carries a sounder to find subsurface water, an important consideration in selecting scientifically worthy landing sites for future exploration.

Other science instruments on this multitasking, multipurpose spacecraft will identify surface minerals and study how dust and water are transported in the Martian atmosphere. A second camera will provide medium-resolution images to provide a broader geological and meteorological context for more detailed observations from higher-resolution instruments.

The Mars Reconnaissance Orbiter will also serve as the first installment of an *interplanetary Internet*, a crucial service for future spacecraft. As the first link in a communications bridge back to Earth, the Mars Reconnaissance Orbiter will be used by several international spacecraft in coming years.

Since beginning its primary science phase in November 2006, the orbiter has returned enough data to fill nearly 1,000 CD-ROMs. This ties the record for Mars data sent back between 1997 and 2006 by NASA's Mars Global Surveyor mission.

The rate of data return is expected to increase over the coming months as the relative motions of Earth and Mars in their orbits around the sun shrink the distance between the planets. By the conclusion of its first science phase in 2008, the mission is expected to have returned more than 30 terabits of science data, enough to fill more than 5,000 CD-ROMs. Observations will be used to evaluate potential landing sites for future missions and to increase our understanding of Mars and how planets change over time.

The orbiter's primary mission ends about five-and-a-half years after launch, on December 31, 2010.

Mars Exploration Beyond 2005

NASA has been developing a long-term Mars exploration program that charts a course for the next two decades. The new program incorporates the lessons learned from previous mission successes and failures, and builds on scientific discoveries from past missions. International participation, especially from Italy and France, will add significantly to the plan.

Smart Lander and Long-Range Rover. NASA proposes to develop and to launch a roving long-range, long-duration science laboratory that will be a major leap in surface measurements and pave the way for a future sample return mission. NASA is studying options to launch this mobile science laboratory mission as early as 2009. This capability will also demonstrate the technology for "smart landers" with accurate landing and hazard avoidance in order to reach what may be very promising but difficult-to-reach scientific sites.

Scout Missions. NASA also proposes to create a new line of small *Scout* missions which would be selected from proposals from the science community, and might involve airborne vehicles (e.g., airplanes or balloons) or small landers, as an investigation platform. Exciting new vistas could be opened up by this approach either through the airborne scale of observation or by increasing the number of sites visited. The first Scout mission launch is planned for 2007.

Sample Return and Other Missions. In the second decade of the century, NASA plans additional science orbiters, rovers and landers, and the first mission to return samples of Martian rock and soil to Earth. Current plans call for the first sample return mission to be launched no earlier than 2014. Options that would significantly increase the rate of mission launch and/or accelerate the schedule of exploration are under study. Technology development for advanced capabilities such as miniaturized surface science instruments and deep drilling to hundreds of meters will also be carried out in this period.

The program envisions significant international participation, particularly by France and Italy. In cooperation with NASA, the French and Italian space agencies plan to conduct collaborative scientific orbital and surface investigations and to make other major contributions to sample collection/return systems, telecommunications assets and launch services. Other nations also have expressed interest in participating in the program.

Future Missions

Phoenix. Scheduled for launch in August 2007, the Phoenix mission is the first chosen for NASA's Scout program, an initiative for smaller, lower-cost, competed spacecraft. Named for the resilient mythological bird, Phoenix uses a lander that was intended for use by 2001's Mars Surveyor lander prior to its cancellation. See Fig. 6. It also carries a complex suite of instruments that are improved variations of those that flew on the lost Mars Polar Lander.

In the continuing pursuit of water on Mars, the poles are a good place to probe, as water ice is found there. Phoenix will land on the icy northern pole of Mars between 65 and 75-north latitude. During the course of the 150 Martian day mission, Phoenix will deploy its robotic arm and dig trenches up to half a meter (1.6 feet) into the layers of water ice. These

Fig. 6.   Phoenix Scout Lander (*NASA*).

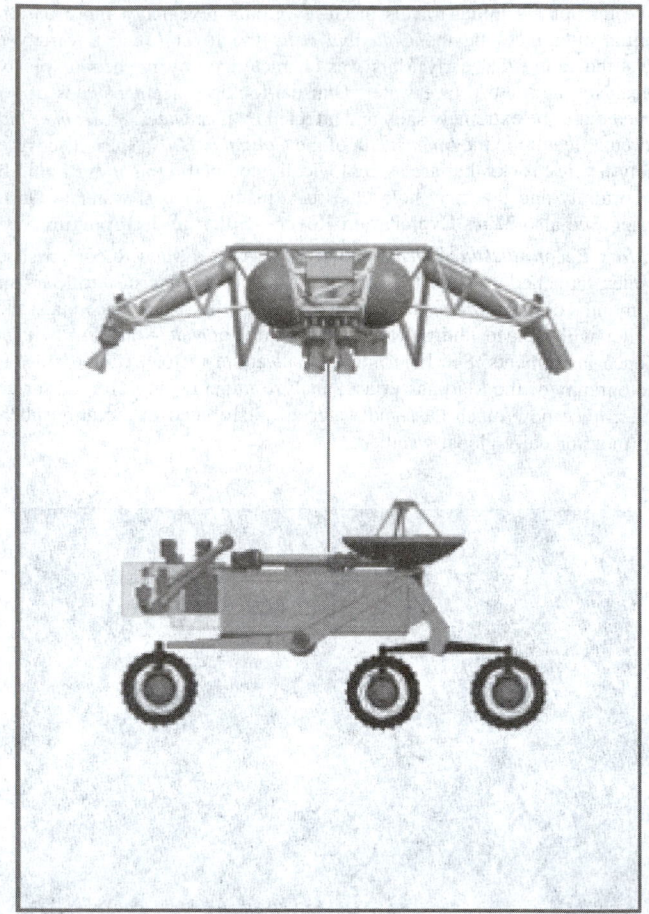

Fig. 7.   CAD/CAM Drawing of Sky Crane and Rover. This drawing shows how the rover would be lowered on a tether during entry, descent, and landing (EDL). (*Image courtesy of NASA/JPL.*)

Fig. 8.   Mars Science Laboratory (*NASA*).

layers, thought to be affected by seasonal climate changes, could contain organic compounds that are necessary for life.

To analyze soil samples collected by the robotic arm, Phoenix will carry an "oven" and a "portable laboratory." Selected samples will be heated to release volatiles that can be examined for their chemical composition and other characteristics.

Imaging technology inherited from both the Pathfinder and Mars Exploration Rover missions will also be implemented in Phoenix's stereo camera, located on its 2-meter (6.6-foot) mast. The camera's two "eyes" will reveal a high-resolution perspective of the landing site's geology, and will also provide range maps that will enable the team to choose ideal digging locations. Multi-spectral capability will enable the identification of local minerals.

To update our understanding of Martian atmospheric processes, Phoenix will also scan the Martian atmosphere up to 20 kilometers (12.4 miles) in altitude, obtaining data about the formation, duration and movement of clouds, fog, and dust plumes. It will also carry temperature and pressure sensors.

See also **Phoenix Mars Mission**.

Mars Science Laboratory (MSL). The Mars Science Laboratory is scheduled to launch in the fall of 2009 with an arrival date of October, 2010.

Building on the success of the two rover geologists Spirit & Opportunity that arrived at Mars in January, 2004, NASA's next rover mission is being planned for travel to Mars before the end of the decade. Twice as long and three times as heavy as the Mars Exploration Rovers Spirit and Opportunity, the Mars Science Laboratory would collect martian soil samples and rock cores and analyze them for organic compounds and environmental conditions that could have supported microbial life now or in the past. The mission is anticipated to have a truly international flavor, with a neutron-based hydrogen detector for locating water provided by the Russian Federal Space Agency, a meteorological package provided by the Spanish Ministry of Education and Science, and a spectrometer provided by the Canadian Space Agency. See also **Mars Exploration Rover (Spirit and Opportunity) Mission**.

Mars Science Laboratory is intended to be the first planetary mission to use precision landing techniques, steering itself toward the martian surface similar to the way the space shuttle controls its entry through the Earth's upper atmosphere. In this way, the spacecraft would fly to a desired location above the surface of Mars before deploying its parachute for the final landing. As currently envisioned, in the final minutes before touchdown, the spacecraft would activate its parachute and retro rockets before lowering the rover package to the surface on a tether (similar to the way a skycrane helicopter moves a large object). See Fig. 7. This landing method would enable the rover to land in an area 20 to 40 kilometers (12

to 24 miles) long, about the size of a small crater or wide canyon and three to five times smaller than previous landing zones on Mars.

Like the twin rovers now on the surface of Mars, Mars Science Laboratory would have six wheels and cameras mounted on a mast. See Fig. 8. Unlike the twin rovers, it would carry a laser for vaporizing a thin layer from the surface of a rock and analyzing the elemental composition of the underlying materials. It would then be able to collect and crush rock and soil samples and distribute them to on-board test chambers for chemical analysis. Its design includes a suite of scientific instruments for identifying organic compounds such as proteins, amino acids, and other acids and bases that attach themselves to carbon backbones and are essential to life as we know it. It could also identify features such as atmospheric gases that may be associated with biological activity.

Using these tools, Mars Science Laboratory would examine martian rocks and soils in greater detail than ever before to determine the geologic processes that formed them; study the martian atmosphere; and determine the distribution and circulation of water and carbon dioxide, whether frozen, liquid, or gaseous.

NASA plans to select a landing site on the basis of highly detailed images sent to Earth by the Mars Reconnaissance Orbiter beginning in 2006, in addition to data from earlier missions.

NASA is considering nuclear energy for powering the Mars Science Laboratory. The rover would carry a U.S. Department of Energy radioisotope power supply that would generate electricity from the heat of plutonium's radioactive decay. This type of power supply could give the mission an operating life span on Mars' surface of a full Martian year (687 Earth days) or more. NASA is also considering solar power alternatives that could meet the mission's science and mobility objectives.

Technology. Technology development makes missions possible.

Mars Science Laboratory and other Mars missions are all part of a continuing chain of innovation: each relies on past missions for technology and contributes new technologies to future missions.

This chain of innovation allows NASA to continue to push the boundaries of what is currently possible, while relying on proven technologies as well.

Mars Science Laboratory would inherit some design elements from the Mars Exploration Rover mission, including: a six-wheel drive; a rocker-bogie suspension system; and scene-scanning instruments mounted to a mast to assist the mission team to select exploration targets and driving routes (see Fig. 9).

Fig. 9. Rover Wheel Sizes. Showing heritage, the relative sizes of Pathfinder (far left), Mars Exploration Rovers Spirit & Opportunity (center), and Mars Science Laboratory (far right) wheels keep growing. (*Image courtesy of NASA/JPL.*)

Mars Science Laboratory would also advance many new technologies, perhaps the most important of which are those for entry, descent, and landing. New, innovative systems include those for:

- landing a large mass of 900–1100 kilograms (about 1,984–2,425 pounds)
- accessing more of the globe, including landing sites with slopes as great as 30 degrees and with rocks as large as 0.75 meters (about 2.5 feet)
- providing precision guided entry to improve landing accuracy from a range of hundreds of kilometers to only 20 kilometers (12 miles)
- creating advanced terrain-sensing techniques for the detection and avoidance of surface hazards during descent through the martian atmosphere
- creating a robust and efficient touchdown system
- collecting and preparing rock and soil samples for scientific analysis
- providing a robust software architecture with reusable components
- increasing the capability to survive the extreme temperature ranges on Mars (−120 degrees Celsius to 85 degrees Celsius, or −184 to 185 degrees Fahrenheit) for years as opposed to months.

The spacecraft would land without the use of airbags and would run on a rebuilt engine inherited from the Viking landers sent to Mars in the 1970s. (The rover itself is being designed to be larger than both Viking landers and equipped with wheels.) The new engine would be constantly throttleable, meaning it would not need to be pulsed on and off during landing, allowing the rover to touch-down gently, using its wheels as a landing gear.

The Mars Science Laboratory rover would carry instruments for drilling rock cores, grinding the rock into samples with individual grains as small as 150 micrometers (a tad thicker than a strand of human hair), and dividing and transporting the samples to the individual scientific instruments on board.

Science Instruments

Mast Camera (MastCam). The Mast Camera, or MastCam for short, would take color images, three-dimensional stereo images, and color video footage of the Martian terrain. See Fig. 10.

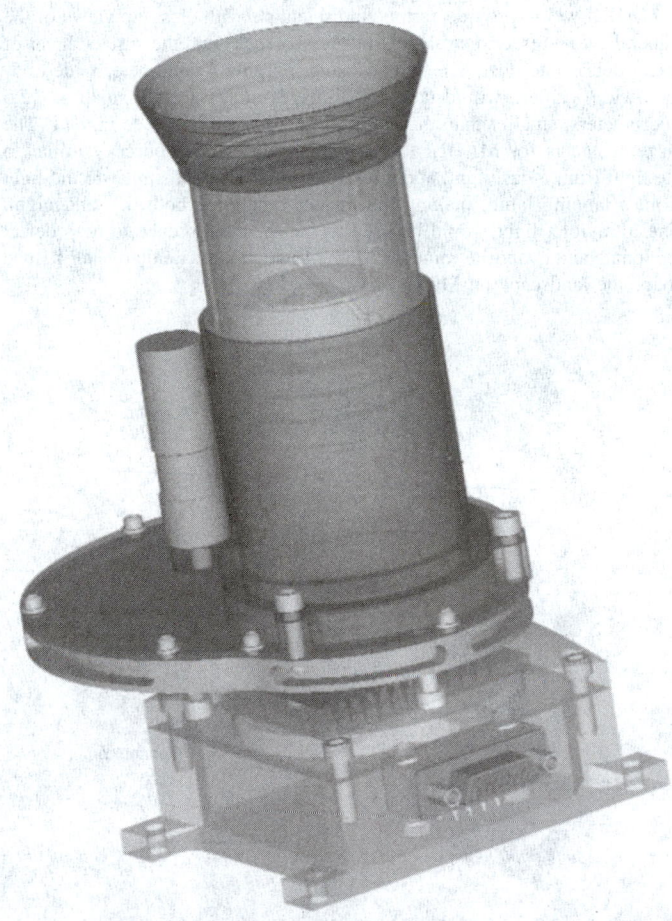

Fig. 10. MastCam Zoom Camera. The mast camera on the Mars Science Laboratory would capture the martian terrain in color photos, three-dimensional images, and high-definition video. (*Image courtesy of NASA/JPL-Caltech/MSSS.*)

Like the cameras on the Mars Exploration Rovers that landed on the red planet in 2004, the MastCam design consists of two duplicate camera systems mounted on a mast extending upward from the Mars Science Laboratory rover deck. The cameras would function much like human eyes, producing three-dimensional stereo images by combining two side-by-side images taken from slightly different positions.

Several new features on the MastCam would make it more versatile than previous rover cameras.

- The MastCam would be able to take high-definition video at 10 frames per second.
- A 10:1 zoom lens would allow scientists to view features of the landscape as if the rover were much closer to the item of interest than it actually

is. By zooming in on an object of interest, the object would fill up more of the field of view.

- The MastCam is being designed to take single-exposure, color snapshots similar to those taken with a consumer digital camera on Earth. In addition, it would have multiple filters for taking sets of monochromatic (single-color) images. These images would be used to analyze patterns of light absorption in different portions of the electromagnetic spectrum.
- Electronics on the MastCam would process images independently of the rover's central processing unit.
- The MastCam would have an internal data buffer for storing thousands of images or several hours of high-definition video footage for transmission to Earth.

Mars Hand Lens Imager (MAHLI). Second only to the rock hammer, the hand lens is an essential tool of human geologists. Usually carried on a string around the person's neck, the hand lens helps a geologist in the field identify the minerals in a rock. The robotic geologist, Mars Science Laboratory, is being designed to carry its own equivalent of the geologist's hand lens, the Mars Hand Lens Imager (MAHLI).

MAHLI would provide earthbound scientists with close-up views of the minerals, textures, and structures in Martian rocks and the surface layer of rocky debris and dust. The self-focusing, roughly 4-centimeter-wide (1.5-inch-wide) camera would take color images of features as small as 12.5 micrometers, smaller than the diameter of a human hair. See Fig. 11. The current plan is for MAHLI to carry both white light sources, similar to the light from a flashlight, and ultraviolet light sources, similar to the light from a tanning lamp, making the imager functional both day and night. The ultraviolet light would be used to induce fluorescence to help detect carbonate and evaporite minerals, both of which indicate that water helped shape the landscape on Mars.

Fig. 12.   Mars Descent Imager. The Mars Descent Imager would provide color video footage of the landing site and the surrounding terrain during the Mars Science Laboratory's descent to the surface of the red planet. (*Image courtesy of NASA/JPL-Caltech/MSSS.*)

Fig. 11.   Mars Hand Lens Imager. The Mars Hand Lens Imager would provide close-up views of microbial-size features. (*Image courtesy of NASA/JPL/MSSS.*)

MAHLI's main objective would be to help the Mars Science Laboratory science team understand the geologic history of the landing site on Mars. MAHLI would also help researchers select samples for further investigation.

Mars Descent Imager (MARDI). Knowing the location of loose debris, boulders, cliffs, and other features of the terrain would be vital for planning the path of exploration after the Mars Science Laboratory rover arrives on the red planet. The Mars Descent Imager would take color video during the rover's descent toward the surface, providing an "astronaut's view" of the local environment. See Fig. 12.

As soon as the rover jettisoned its heat shield several kilometers above the surface, the Mars Descent Imager would begin producing a five-frames-per-second video stream of high-resolution, overhead views of the landing site. It would continue acquiring images until the rover lands, storing the

video data in digital memory. After landing safely on Mars, the rover would transfer the data to Earth.

In addition to helping earthbound planners select an optimum path of exploration, the Mars Descent Imager would provide information about the larger geologic context surrounding the landing site. It would also enable mappers to determine the spacecraft's precise location after landing.

Alpha Particle X-Ray Spectrometer (APXS). The Alpha Particle X-Ray Spectrometer would measure the abundance of chemical elements in rocks and soils. See Fig. 13. Funded by the Canadian Space Agency, the APXS would be placed in contact with rock and soil samples on Mars and would expose the material to alpha particles and X-rays emitted during the radioactive decay of the element curium. X-rays are a type of electromagnetic radiation, like light and microwaves.

Alpha particles are helium nuclei, consisting of 2 protons and 2 neutrons. When X-rays and alpha particles collide with atoms in the surface material, they knock electrons out of their orbits, producing an energy release by emitting X-rays that can be measured with detectors. The X-ray energies enable scientists to identify all important rock-forming elements heavier than sodium.

The APXS would take measurements both day and night. Its sensor head is designed to be smaller than a soda can and would contain a highly sensitive X-ray detector in the middle of an array of curium sources. The longer the instrument would be held in place on the surface of a rock or soil sample, the more clearly the signal from the sample could be determined. Most APXS measurements would take two to three hours to reveal all elements, including small amounts of trace elements. Ten minutes of operation would be sufficient for a quick look at major elements.

As a contact instrument, the APXS is being designed to work in concert with other payload elements on the instrument arm and in the body of the Mars Science Laboratory rover, such as the CheMin instrument and the Rock Abrasion Tool. Scientists would use the APXS to help characterize and select rock and soil samples and then examine the interiors of the rocks following abrasion. By analyzing the elemental composition of rocks and

Fig. 13.   This picture shows the Alpha Particle X-Ray Spectrometer used on the Mars Exploration Rovers. The improved APXS instrument on the Mars Science Laboratory rover would be able to detect elemental composition more quickly and work both day and night. (*Image courtesy of NASA/JPL-Caltech/Cornell/Max Planck Institut für Chemie/University of Guelph.*)

Fig. 14.   The Laser-Induced Remote Sensing for Chemistry and Micro-Imaging instrument will identify atomic elements in martian rocks. (*Image courtesy of NASA/JPL-Caltech/LANL/J.-L. Lacour, CEA.*)

soils, scientists would seek to understand how the material formed and if it was later altered by wind, water, or ice. The APXS on NASA's two Mars Exploration Rovers has already helped scientists conclude that water once played a major role in Mars' geologic past.

Two earlier missions to Mars carried previous versions of the Alpha Particle X-Ray Spectrometer. The first was the Alpha Proton X-Ray Spectrometer, launched to Mars on the Pathfinder mission in late 1996. The second was the APXS, on board both the Mars Exploration Rovers that arrived on the red planet in January, 2004.

In addition to the Canadian Space Agency and NASA, major organizations involved in developing the APXS include the University of Guelph; MDA Space Missions; the University of California, San Diego; and Cornell University.

Laser-Induced Remote Sensing for Chemistry and Micro-Imaging (Chem-Cam). Looking at rocks and soils from a distance, ChemCam would fire a laser and analyze the elemental composition of vaporized materials from areas smaller than 1 millimeter on the surface of Martian rocks and soils. An on-board spectrograph would provide unprecedented detail about minerals and microstructures in rocks by measuring the composition of the resulting plasma, a visible brief glow of ionized material.

ChemCam would also use the laser to clear away dust from Martian rocks and use a remote camera to produce extremely detailed images. See Fig. 14. The camera would be able to resolve features 5 to 10 times smaller than those visible with cameras on NASA's two Mars Exploration Rovers that began exploring the red planet in January 2004. In the event the Mars Science Laboratory rover could not reach a rock or outcrop of interest, ChemCam would have the capability to analyze it from a distance.

From a distance of 1 to 9 meters (3.3 to 30 feet), ChemCam would be able to:

- rapidly identify the kind of rock being studied (for example, whether it is volcanic or sedimentary);
- determine the composition of soils and pebbles;
- measure the abundance of all chemical elements, including trace elements and those that might be hazardous to humans;
- recognize ice and minerals with water molecules in their crystal structures;
- measure the depth and composition of weathering rinds on rocks; and,
- provide visual assistance during drilling of rock cores.

The ChemCam instrument has two parts: a mast package and a body unit. On the mast would be a telescope to focus the laser and the camera, a

laser for vaporizing surfaces, and a remote micro-imager. The mast package would be tilted or rotated as needed for optimum viewing of the rock.

A fiber-optic link would feed light from the telescope on the mast to a body unit inside the rover. The body unit would carry three spectrographs that would divide the plasma light into its constituent wavelengths for chemical analysis. The body unit would also have its own power supply and an electronic interface to the rover's central computer system.

Developing the ChemCam instruments for NASA are the Los Alamos National Laboratory (LANL) and the Centre d'Etude Spatiale des Rayonnements (CESR), with major contributions from JPL, Ocean Optics Inc., and the Commissariat a l'Energie Atomique (CEA).

Chemistry & Mineralogy X-Ray Diffraction/X-Ray Fluorescence Instrument (CheMin). The Chemistry and Mineralogy instrument, or CheMin for short, would identify and measure the abundances of various minerals on Mars. See Fig. 15. Examples of minerals found on Mars so far are olivine, pyroxenes, gypsum, hematite, goethite, and magnetite.

Minerals are indicative of environmental conditions that existed when they formed. For example, olivine and pyroxene, two primary minerals in basalt, form when lava solidifies. Jarosite, found in sedimentary rocks by NASA's Opportunity rover on Mars, is a mineral salt that precipitates out of water.

Using CheMin, scientists would be able to further study the role that water, an essential ingredient for life as we know it, played in forming minerals on Mars. For example, kieserite is a mineral that contains magnesium, sulfur, and water. Anhydrite is a calcium and sulfur mineral with no water in its crystal structure. CheMin would be able to distinguish the two. Scientists would also use CheMin to search for mineral clues indicative of a past martian environment that might have supported life. For instance, certain microorganisms can use hydrogen released by water-altered minerals as an energy source. CheMin would be able to identify the particular minerals involved in such reactions.

To prepare rock or soil samples for analysis, a rock crusher on the rover would grind them into a powder. A sieve would allow only very fine grains of the sample to fill a reusable sample holder. A vibration system would mix the grains to evenly distribute them.

CheMin would then direct a beam of X-rays as fine as a human hair through the powered material. X-rays, like visible light, are a form of electromagnetic radiation. They have a much shorter wavelength that cannot be seen with the naked eye. When the X-ray beam interacts with the rock or soil sample, some of the X-rays would be absorbed by atoms in

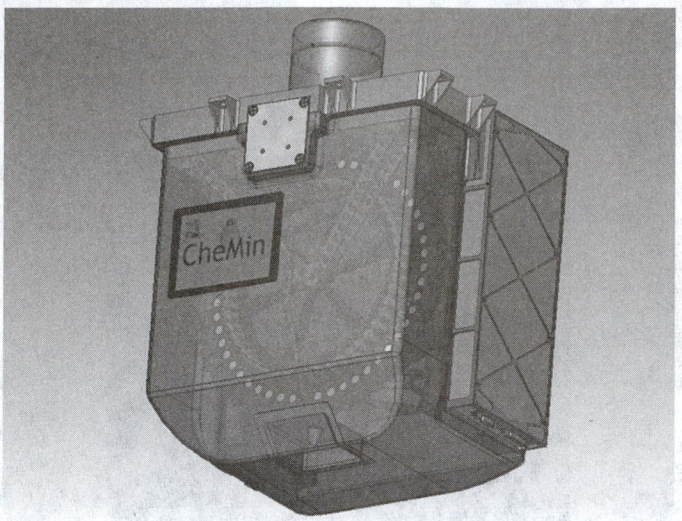

Fig. 15. Chemistry & Mineralogy X-Ray Diffraction/X-Ray Fluorescence Instrument. Designed to be about the size of a laptop computer inside a carrying case, the Chemistry and Mineralogy Instrument would identify and measure the abundances of minerals on Mars. A rotating wheel in the center of the rectangular housing would carry individual rock and soil samples for chemical analysis. (*Image courtesy of NASA/JPL-Caltech.*)

the sample and re-emitted or fluoresced at energies that are characteristic of the particular atoms present.

In X-ray diffraction, some X-rays bounce away at the same angle from parallel surfaces of atoms in the sample. When this happens, they mutually reinforce each other and produce a distinctive signal. Scientists can measure the angle at which X-rays are diffracted toward the detector and use that to identify minerals. For example, if the mineral halite (common table salt, or NaCl), were placed in CheMin, the instrument would produce a specific diffraction pattern that would identify the structure of halite. The instrument would also provide a compositional analysis showing that sodium and chlorine were present in equal amounts.

Because all minerals diffract X-rays in a characteristic pattern and all elements emit X-rays with a unique set of energy levels, scientists would use the information from X-ray diffraction and X-ray fluorescence to identify the crystalline structure and elemental composition of materials the rover encounters on Mars. A Charge-Coupled Device (CCD) would collect both diffraction and fluorescence information. An X-ray-sensitive silicon diode would collect a redundant set of fluorescence information.

Sample Analysis at Mars Instrument Suite (SAM). The Sample Analysis at Mars instrument suite will take up more than half the science payload on board the Mars Science Laboratory rover and feature chemical equipment found in many scientific laboratories on Earth. See Fig. 16. Provided by NASA's Goddard Space Flight Center, Sample Analysis at Mars would search for compounds of the element carbon, including methane, which are associated with life and explore ways in which they are generated and destroyed in the martian ecosphere.

Actually a suite of three instruments, including a mass spectrometer, gas chromatograph, and tunable laser spectrometer, Sample Analysis at Mars would also look for and measure the abundances of other light elements, such as hydrogen, oxygen, and nitrogen, associated with life.

The mass spectrometer would separate elements and compounds by mass for identification and measurement. The gas chromatograph would heat soil and rock samples until they would vaporize, and would then separate the resulting gases into various components for analysis. The laser spectrometer would measure the abundance of various isotopes of carbon, hydrogen, oxygen, and nitrogen in atmospheric gases such as methane, water vapor, carbon dioxide, nitrous oxides, and hydrogen peroxide. These measurements would be accurate to within 10 parts per thousand.

Because these compounds are essential to life as we know it, their relative abundances would be an essential piece of information for evaluating whether Mars could have supported life in the past or present.

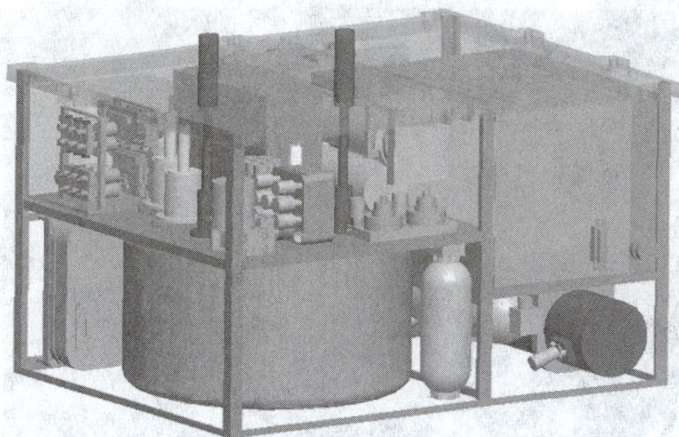

Fig. 16. The Sample Analysis at Mars instrument suite would weigh about 83 pounds (38 kilograms) and make up about half the science payload of the Mars Science Laboratory mission. It is a suite of three instruments that would search for carbon-based compounds associated with life. (*Image courtesy of NASA/JPL-Caltech.*)

Radiation Assessment Detector (RAD). The Radiation Assessment Detector (RAD) is designed to be one of the first instruments sent to Mars specifically to prepare for future human exploration. Smaller than a milk carton in size, RAD would measure and identify all high-energy radiation on the martian surface, such as protons, energetic ions of various elements, neutrons, and gamma rays. That includes not only direct radiation from space, but also secondary radiation produced by the interaction of space radiation with the martian atmosphere and surface rocks and soils. See Fig. 17.

Fig. 17. Radiation Assessment Detector Smaller than a milk carton in size, the Radiation Assessment Detector would point its telescope skyward and use an array of silicon detectors and a crystal of cesium iodide to measure galactic cosmic rays and solar particles that pass through the martian atmosphere. (*Image courtesy of NASA/JPL-Caltech/SRI.*)

To prepare for future human exploration, RAD would collect data that would allow scientists to calculate the equivalent dose (a measure of the effect radiation has on humans) to which people would be exposed on the surface of Mars. RAD would also assess the hazard presented by radiation to potential microbial life, past and present, both on and beneath the martian surface. In addition, RAD would investigate how radiation has affected the chemical and isotopic composition of martian rocks and soils. (Isotopes are atoms of the same element having the same number of protons but a different number of neutrons.)

A stack of paper-thin, silicon detectors and a small block of cesium iodide would be used to measure high-energy charged particles coming through the martian atmosphere. As the particles passed through each of the detectors, they would lose energy, producing electron or light pulses. An internal signal processor would analyze the pulses to identify each high-energy particle and determine its energy. In addition to identifying neutrons, gamma rays, protons, and alpha particles (subatomic fragments consisting of 2 protons and 2 neutrons, identical to helium nuclei), RAD would identify heavy ions up to iron on the periodic table. The RAD is designed to be lightweight and energy efficient so as to use as little of the Mars Science Laboratory's available mass and energy resources as possible.

Dynamic of Albedo Neutrons (DAN). One way to look for water on Mars is to look for neutrons escaping from the planet's surface. Cosmic rays from space constantly bombard the surface of Mars, knocking neutrons in surface soils and rocks out of their atomic orbits. If liquid or frozen water happens to be present, hydrogen atoms in water molecules slow the neutrons down. In this way, some of the neutrons escaping into space have less energy and move more slowly. These slower particles can be measured with a neutron detector. See Fig. 18.

Scientists expect to find hydrogen on Mars in two forms: water ice and minerals that have molecules of water in their crystal structures. At the request of the Russian Federal Space Agency, the Mars Science Laboratory rover would carry a pulsing neutron generator called the Dynamic of Albedo Neutrons, or DAN for short that would be sensitive enough to detect water content as low as one-tenth of 1 percent and resolve layers of water and ice beneath the surface. Albedo is a scientific word for the reflection or scattering of light. Funded by the Russian government, the instrument would focus a beam of neutrons onto the martian surface from a height of 2.6 feet. The neutrons are expected to travel 1 to 2 meters (3 to 6 feet) below the surface before being absorbed by hydrogen atoms in subsurface ice.

Scientists estimate that, near the martian poles, water ice makes up 30 percent to 50 percent of shallow subsurface deposits. If the beam of neutrons encountered a layer of water ice beneath the surface, DAN would detect a relatively greater amount of slower neutrons reflected at the surface. If there are no ice layers or water-logged minerals beneath the surface, DAN would detect a relatively greater amount of faster neutrons reflected at the surface.

Rover Environmental Monitoring Station (REMS). The Mars Science Laboratory rover would carry a weather monitoring station provided by the government of Spain on behalf of investigators at the Centro de Astrobiología (INTA-CSIC). The Rover Environmental Monitoring Station would provide a daily report of atmospheric weather conditions on Mars. Attached to the vertical mast on the Mars Science Laboratory deck, the station would measure atmospheric pressure, humidity, ultraviolet radiation from the sun, wind speed, wind direction, ground temperature, and air temperature.

A sensor in the electronics box would measure atmospheric pressure. Separate sensors would track humidity and ultraviolet radiation. Two wind sensors would detect and characterize atmospheric circulation. An XY wind sensor would measure wind direction moving horizontally, similar to breezes blowing across the surface on Earth, and a Z wind sensor would measure vertical atmospheric movement, such as thermal currents moving up or down. See Fig. 19.

Mars Exploration Beyond 2009. NASA is developing a long-term Mars exploration program that charts a course for the next two decades. This visionary program will build on scientific discoveries from past missions and incorporate the lessons learned from previous mission successes and failures.

NASA remains committed to creating additional *Scout* missions, such as the Phoenix lander, which would be selected from proposals submitted by members of the science community. Such missions might involve airborne vehicles, such as airplanes or balloons, or small landers that serve as investigation platforms. This approach could open up exciting new vistas by increasing the number of martian sites visited. The next Mars Scout is planned for launch in 2011.

In the second decade of the 21st century, NASA plans additional science orbiters, rovers, and landers. One proposal is for a Mars Sample Return mission that would use robotic systems and a Mars ascent rocket to collect and send samples of martian rocks, soils, and atmosphere to Earth for detailed chemical and physical analysis. Researchers on Earth could measure chemical and physical characteristics much more precisely than they could by remote control. On Earth, they would have the flexibility to

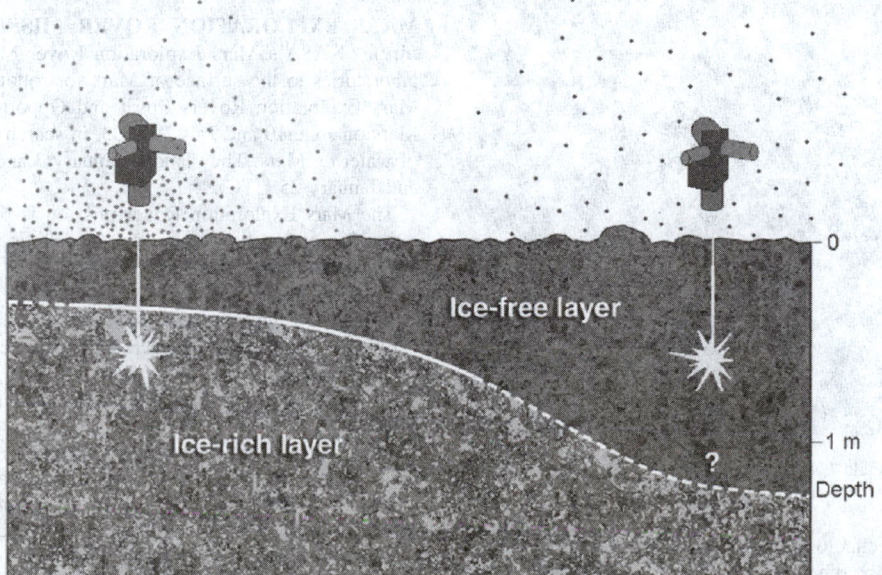

Fig. 18. Water, whether liquid or frozen, absorbs neutrons more than other substances. The Detector of Albedo Neutrons on the Mars Science Laboratory rover would use this characteristic to search for subsurface ice on Mars. (*Image courtesy of NASA/JPL-Caltech/Russian Federal Space Agency.*)

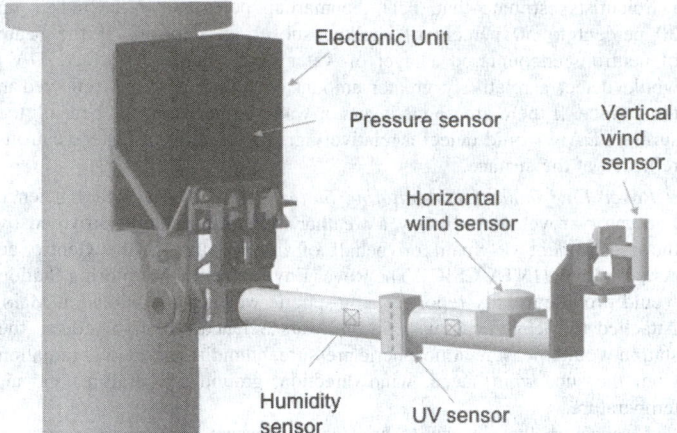

Fig. 19. The Rover Environmental Monitoring Station would monitor atmospheric pressure, humidity, wind currents, and ultraviolet radiation from the sun. (*Image courtesy of NASA/JPL-Caltech/INTA (Instituto Nacional de Tecnica Aeroespacial.)*)

make changes as needed for intricate sample preparation, instrumentation, and analysis if they encountered unexpected results. In addition, for decades to come, the collected Mars rocks could yield new discoveries as future generations of researchers apply new technologies in studying them.

Another proposal is for an Astrobiology Field Lab that would conduct a robotic search for life. It would be the first mission since Viking in the 1970s to look specifically for evidence of past or present life. The robotic lab would carry instruments for identifying and measuring the chemical building blocks for life (as we know it), including thousands of carbon-carrying compounds, elements such as sulfur and nitrogen, and oxidation states of trace metals associated with life. See Fig. 20. It would conduct detailed analysis of geologic environments identified by the 2009 Mars Science Laboratory as being conducive to life. Such environments might include fine-grained sedimentary layers, hot spring mineral deposits, icy layers near the poles, or sites such as gullies where liquid water once flowed or may continue to seep into soils from melting ice packs.

Fig. 20.   Astrobiology-Field-Lab. (Image courtesy of NASA/JPL)

NASA is interested in technologies that would increase the rate of mission launch or accelerate the schedule of exploration. The agency plans to invest in advanced capabilities such as miniaturized surface science instruments and deep drilling systems that can extend hundreds of meters beneath the surface. See Fig. 21.

Fig. 21.   Deep-drill Lander. (*Image courtesy of NASA/JPL*)

Web References

Mars Exploration Rovers: http://marsrovers.jpl.nasa.gov/home/index.html
Mars Express: http://marsrovers.jpl.nasa.gov/home/index.html
2001 Mars Odyssey: http://marsprogram.jpl.nasa.gov/odyssey/
Mars Reconnaissance Orbiter: http://mars.jpl.nasa.gov/mro/
Mars Global Surveyor: http://marsprogram.jpl.nasa.gov/mgs/
Phoenix: http://phoenix.lpl.arizona.edu/
Mars Science Laboratory: http://marsprogram.jpl.nasa.gov/msl/index.html

National Aeronautics and Space Administration (NASA Staff)

MARS EXPLORATION ROVER MISSION (Spirit and Opportunity). NASA's Mars Exploration Rover Mission delivered two mobile laboratories to the surface of Mars for robotic geological fieldwork. The Mars Exploration Rovers Spirit and Opportunity were launched toward Mars on June 10 and July 7, 2003, in search of answers about the history of water on Mars. They landed January 3 and January 24 PST (January 4 and January 25 UTC).

The Mars Exploration Rover mission is part of NASA's Mars Exploration Program, a long-term effort of robotic exploration of the red planet. See also **Mars Exploration Program**.

Primary among the mission's scientific goals is to search for and characterize a wide range of rocks and soils that hold clues to past water activity on Mars. The spacecraft are targeted to sites on opposite sides of Mars that appear to have been affected by liquid water in the past. The landing sites were Gusev Crater, and Meridiani Planum. See **Landing Sites**.

After the airbag-protected landing craft settled onto the surface and opened, the rovers rolled out and took panoramic images. This give scientists the information they needed to select promising geological targets that tell part of the story of water in Mars' past. Then, the rovers drove to those locations to perform on-site scientific investigations.

These are the primary science instruments carried by the rovers:

Panoramic Camera (Pancam): for determining the mineralogy, texture, and structure of the local terrain.

Miniature Thermal Emission Spectrometer (Mini-TES): for identifying promising rocks and soils for closer examination and for determining the processes that formed Martian rocks. The instrument will also look skyward to provide temperature profiles of the Martian atmosphere.

Mössbauer Spectrometer (MB): for close-up investigations of the mineralogy of iron-bearing rocks and soils.

Alpha Particle X-Ray Spectrometer (APXS): for close-up analysis of the abundances of elements that make up rocks and soils.

Magnets: for collecting magnetic dust particles. The Mössbauer Spectrometer and the Alpha Particle X-ray Spectrometer will analyze the particles collected and help determine the ratio of magnetic particles to non-magnetic particles. They will also analyze the composition of magnetic minerals in airborne dust and rocks that have been ground by the Rock Abrasion Tool.

Microscopic Imager (MI): for obtaining close-up, high-resolution images of rocks and soils.

Rock Abrasion Tool (RAT): for removing dusty and weathered rock surfaces and exposing fresh material for examination by instruments onboard.

Before landing, the goal for each rover was to drive up to 40 meters (about 44 yards) in a single day, for a total of up to one 1 kilometer (about three-quarters of a mile). Both goals have been far exceeded! As of sol 1102 (February 7, 2007), Spirit's total odometry was 6,926.42 meters or 6.9 kilometers (22,724.5 feet or 4.3 miles); on February 6, 2007 Opportunity's odometer rolled past 10 kilometers or 10,000 meters (6.2 miles or 32,808 feet). The rovers have worked on Mars for more than 10 times their originally planned three-month missions.

Moving from place to place, the rovers perform on-site geological investigations. Each rover is sort of the mechanical equivalent of a geologist walking the surface of Mars. The mast-mounted cameras are mounted 1.5 meters (5 feet) high and provide 360-degree, stereoscopic, humanlike views of the terrain. The robotic arm is capable of movement in much the same way as a human arm with an elbow and wrist, and can place instruments directly up against rock and soil targets of interest. In the mechanical *fist* of the arm is a microscopic camera that serves the same purpose as a geologist's handheld magnifying lens. The Rock Abrasion Tool serves the purpose of a geologist's rock hammer to expose the insides of rocks.

The scientific objectives of the Mars Exploration Rover mission are to:

1. Search for and characterize a variety of rocks and soils that hold clues to past water activity. In particular, samples sought will include those that have minerals deposited by water-related processes such as precipitation, evaporation, sedimentary cementation, or hydrothermal activity.
2. Determine the distribution and composition of minerals, rocks, and soils surrounding the landing sites.
3. Determine what geologic processes have shaped the local terrain and influenced the chemistry. Such processes could include water or wind erosion, sedimentation, hydrothermal mechanisms, volcanism, and cratering.
4. Perform "ground truth"—calibration and validation—of surface observations made by Mars orbiter instruments. This will help determine the accuracy and effectiveness of various instruments that survey Martian geology from orbit.
5. Search for iron-containing minerals, identify and quantify relative amounts of specific mineral types that contain water or were formed in water, such as iron-bearing carbonates.
6. Characterize the mineralogy and textures of rocks and soils and determine the processes that created them.
7. Search for geological clues to the environmental conditions that existed when liquid water was present. Assess whether those environments were conducive to life.

Landing Sites

Selection of the landing sites for the two Mars Exploration Rovers involved over two years of intensive study by more than 100 scientists and engineers. Their job was to find sites that offered both excellent chances for a safe landing and outstanding science after the landings are achieved.

To qualify for consideration, candidate sites had to be near Mars' equator, low enough in elevation (so the spacecraft would pass through enough atmosphere to slow them), not too rugged, no too rocky and not too dusty. In all, 155 potential sites met the initial safety constraints. The two that made the final cut satisfied all of the safety criteria; they also show powerful evidence of past liquid water, but in two very different ways:

Gusev Crater, named after the 19th-century Russian astronomer Matvei Gusev, is an impact crater about 150 kilometers (95 miles) in diameter and about 15 degrees south of Mars' equator. It lies near the transition between the planet's ancient highlands to the south and smoother plains to the north. See Fig. 1.

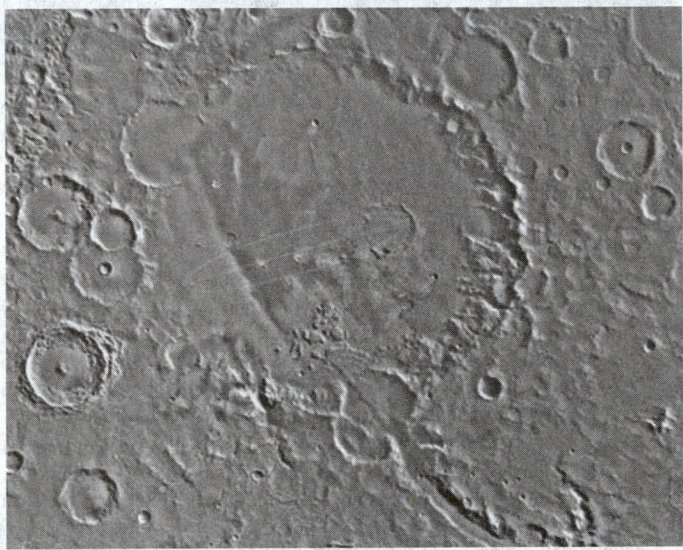

Fig. 1.   The designated landing site for the first Mars Exploration Rover mission is Gusev Crater, seen here in its geological context from NASA Viking images. (*Image courtesy of JPL.*)

What makes Gusev an attractive landing side is a 900-kilometer=long (550-mile) meandering valley that enters the crater from the southeast. Called *Ma'adima Vallis* (from the Hebrew name for Mars), this valley is believed to have been eroded long ago by flowing water. The water likely cut through the crater's rim and filled much of the crater, creating a large lake not unlike current crater lakes here on Earth such as Lake Bosumtwi in Ghana. The lake is gone now, but the floor of Gusev Crater may contain water-laid sediments that still preserve a record of what conditions were like in the lake when the sediments were deposited.

Are lake sediments still preserved at Gusev Crater, or have been buried by younger geologic materials? If sediments can be found, what do they reveal about the conditions that existed in the lake? Did the lake create an environment that would have been suitable for life? Are there other clues at Gusev that can reveal more about whether Mars had a warmer, wetter past?

Meridiani Planum is near the Martian equator, halfway around the planet from Gusev. The region of the planet in which it lies has been known as Meridiani since the earliest days of telescopic study of Mars, because it lies near the planet's meridian, or line of zero longitude. *Planum* means plains, and the name fits: Meridiani Planum is one of the smoothest, flattest places on Mars. See Fig. 2. The scientific appeal of Meridiani comes not from its smooth landscape, but from it strange mineral composition. Looking down from orbit, the thermal emission spectrometer instrument on the Mars Global Surveyor spacecraft has shown that Meridiani Planum is rich in an iron oxide mineral called gray hematite. Gray hematite is found on Earth, where it usually—through not always—forms in association with liquid water. Did the formation of the telltale mineral hematite at Meridiani involve liquid water? If it did, what was the process? Was the water in a lake? Was it percolating through rocks? If water was present, were the conditions at Meridiani favorable for life? Or did the hematite form by some other process that did not involve water at all? And what other clues does Meridiani Planum hold regarding past conditions on Mars?

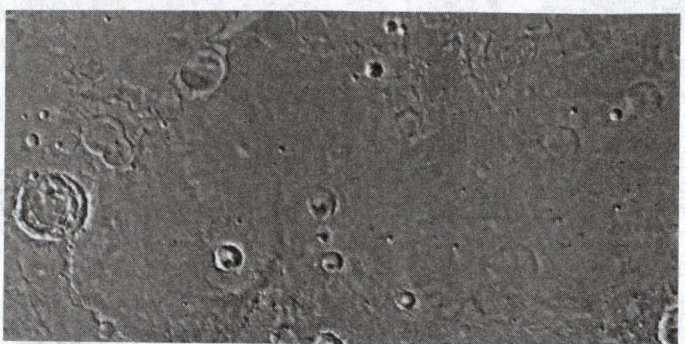

Fig. 2. The designated landing site for the second Mars Exploration Rover mission is Meridiani Planum, seen here in its geological context from NASA Viking images. (*Image courtesy of JPL*.)

Mission Timeline

Sequences of launch, cruise, and arrival operations will dispatch each rover to a different area of the planet three weeks apart to explore those areas.

While the rovers and the instruments they carry are the centerpieces of the project, each rover mission also depends on the performance of other components: the launch vehicle; a cruise stage; a system for entering Mars' atmosphere, descending through it and landing; a versatile system for deep-space communications; see **Spacecraft**; Earth facilities for data processing; and an international team of engineers, scientists and others.

Launch. Launch phases begin when the spacecraft transfers to internal power on the launch pad and ends when the spacecraft is declared stable, healthy, and ready to accept commands and the launch telemetry has been played back.

A launch vehicle provides the velocity needed by a spacecraft to escape Earth's gravity and set it on its course for Mars.

When mission planners are considering different launch vehicles, what they take into consideration is how much mass each launch vehicle can lift into space. The Boeing Delta II launch vehicle was selected for the Mars Exploration Rover mission because it has the right liftoff capability for the weight requirements and because it's extremely reliable.

Rover A (*Spirit*) launched using a Delta II 7925 launch vehicle from Space Launch Complex 17A (SLC-17A) at the Cape Canaveral Air Force Station in Florida. Rover B (*Opportunity*) launched using a Delta II 7925H launch vehicle from SLC-17B at the Cape.

The Delta II family of launch vehicles has been in service for over 10 years and has successfully launched 90 projects including the last five NASA missions to Mars: *Mars Global Surveyor* and *Mars Pathfinder* in 1996, *Mars Climate Orbiter* in 1998, *Mars Polar Lander* in 1999, and *2001 Mars Odyssey* in 2001. For more information on the Delta II family of launch vehicles; see http://marsrovers.jpl.nasa.gov/mission/launch vehicle.html.

The launch periods for each rover were as follows:

Mission	Open	Close	Arrival
Rover A (Spirit)	June 5, 2003	June 24, 2003	January 4, 2004
Rover B (Opportunity)	June 25, 2003	July 15, 2003	January 25, 2004

To allow changeover of ground equipment at the launch pads, the two missions must be launched at least 10 days apart. There were two daily launch opportunities throughout both launch periods, providing a high probability of liftoff within the back-to-back MER launch periods. A constant arrival date is used for both rovers. Spirit: the first rover, launched on June 10. Opportunity: the second rover, launched on July 7.

The major activities in this mission phase include: the liftoff and boost phase of the launch; insertion into a circular parking orbit; a coast period followed by additional launch vehicle burns necessary to inject the spacecraft onto a trajectory to Mars; separation of the spacecraft from the launch vehicle; initial acquisition by the Deep Space Network; verification of the initial spacecraft health and operating conditions; and the verified execution of a minimal set of post-launch commands.

Cruise. The Cruise phase begins soon after separation from the third stage of the Delta-II launch vehicle when the spacecraft completes the Launch phase. Cruise ends when the spacecraft is 45 days from entry into the Martian atmosphere. Since the arrival of each rover is held constant no matter when they launch within the launch period, the end of the cruise phase also has specific dates set: November 20, 2003 for Rover A (Spirit) and December 11, 2003 for Rover B (Opportunity).

Both rover missions will take "type I" interplanetary transfers to Mars. Type I transfer trajectories are the fastest possible. Since each spacecraft will travel millions of miles to reach Mars, engineers will perform three trajectory correction maneuvers during cruise to "tweak" the position of the spacecraft to ensure it arrives at Mars and lands at the planned location. (Three more trajectory correction maneuvers are also planned during the approach phase.)

During cruise, the spacecraft will spin at 2 rpm and will require periodic spin axis pointing updates to maintain the antenna pointing toward Earth and the solar panels pointed to the sun as their relative positions change during flight.

Due to a wide separation between Sun and Earth relative to the early trajectories, both spacecraft will use a cruise low-gain antenna (CLGA) for early cruise communications. After the first couple of months of cruise, as the Earth and Sun come closer together as viewed from the spacecraft, the spacecraft will switch to a medium-gain antenna (MGA) for communications during the rest of cruise.

In short, the activities during this phase include:

- health checks and maintenance of the spacecraft in its cruise configuration
- monitoring and calibrating the spacecraft and subsystems
- attitude correction turns (aforementioned spins to maintain the antenna pointing toward Earth for communications and the solar panels pointed toward the sun for power)
- navigation activities, including trajectory correction maneuvers, for determining and correcting the vehicle's flight path and for training navigators prior to approach
- preparation for entry, descent, and landing and surface operations, including X-band communication tests used during entry, descent, and landing

Approach. To ensure a successful entry, descent, and landing, engineers begin intensive preparations 45 days before the spacecraft enters the Martian atmosphere. They call this preparation period the approach phase. It lasts until the spacecraft enters the Martian atmosphere, which extends 3,522.2 kilometers (2,113 miles) as measured from the center of the red planet.

The mission team will switch to "Mars time" during this phase. (The Martian day, or sol, is about 40 minutes longer than an Earth day, so the team will need to synchronize their schedules to the rovers.) The activities that engineers are focused on during the approach phase include:

- the final three trajectory correction maneuvers, which are used to make final adjustments to the spacecraft's incoming trajectory at Mars
- attitude pointing updates, as necessary, for communications and power
- frequent "Delta DOR" measurements that monitor the spacecraft's position and ensure accurate delivery
- start of the entry, descent, and landing behavior software, which automatically executes commands during that phase
- entry, descent, and landing parameter updates
- spacecraft activities leading up to the final turn to the entry attitude and separation from the cruise stage
- the loading of surface sequences and communication windows needed for the first several sols (a "sol" is a martian day)

During the approach phase, the amount of requested tracking by the Deep Space Network is substantially increased to allow engineers to determine more accurate trajectory solutions in the final weeks before arrival at Mars. This tracking will support the safe delivery of the spacecraft to the surface of Mars. The Deep Space Network's 34-meter (111.5 foot) and 70-meter (230 foot) antennas will be able to provide tracking coverage of the spacecraft during the approach phase. See also **Deep Space Network**.

Entry, Descent, and Landing (EDL). The entry, descent, and landing (EDL) phase begins when the spacecraft reaches the Mars atmospheric entry interface point (3522.2 kilometers or about 2,113 miles from the center of Mars) and ends with the lander on the surface of Mars in a safe state.

The rovers arrived during the latter half of the northern winter/southern summer on Mars. Rover A will land at approximately 2:00 p.m. local time on Mars (with Earth set an hour after landing), whereas Rover B will land at around 1:15 p.m. local time on Mars (with Earth set as long as two-and-a-half hours after landing). That means that both rovers will land in the Martian afternoon while the Earth is still in view, allowing the Earth to receive the landing signal if the lander is on the base petal.

Entry, descent, and landing for the Mars Exploration Rover mission is an adaptation of the Mars Pathfinder method:

- An aeroshell and a parachute decelerate the lander through the Martian atmosphere.
- Prior to surface impact, retro-rockets are fired to slow the lander's speed of descent, and airbags are inflated to cushion the lander at surface impact.
- After its initial impact, the lander bounces along the Martian surface until it rolls to a stop.
- The airbags are then deflated and retracted, and the lander petals and rover egress aids are deployed.
- Once the petals have opened, the rover deploys its solar arrays, and places the system in a safe state.

See also **Pathfinder Mission to Mars**.

Communications during entry, descent, and landing will occur through a pair of low-gain antennas, one mounted on the backshell and the other on the rover itself. About 36 ten-second radio tones will be transmitted to Earth during descent through the atmosphere, which takes approximately six minutes. These tones are coded to indicate the accomplishment of critical steps in the entry, descent, and landing timeline.

To learn more about the Steps in the entry, descent, and landing phase; see http://marsrovers.jpl.nasa.gov/mission/tl entry1.html.

Rover Egress. The rover egress phase is defined as the period of time following placement of the rover into a "safe state" after landing to driving the rover off of the lander onto the Martian surface.

The egress phase will likely last through Sol 4 or 5 (the landing sol is designated Sol 1). The major deployment and egress activities are power intensive, and must occur during the Martian day when the solar arrays can charge the rover battery, and when the rover and the landing site are naturally illuminated. The actual rate for egress activities is largely dictated by the availability of power, the duration and number of ground decision points, and the availability of a communications link.

The lander has no further function or capability after egress is complete, so there will be no pictures of the rovers themselves in operation (in contrast to the Mars Pathfinder lander, which was active and imaged the rover Sojourner throughout its surface operations). Pictures taken from the Mars Exploration Rover will be rather like those you might take from the inside of your car, as we'll see parts of the Rover in each picture.

Egress consists of a carefully choreographed series of steps by the rover; each initiated by the ground only after verification that the previous step has been successfully executed. Because there is limited knowledge of the terrain at the landing sites before the rovers arrive, the flight team must interactively assess the rover data, command the rover, and then assess the resulting data before proceeding to subsequent steps. Consequently, many possible egress scenarios exist that depend on the outcome of the events leading up to egress.

The principal steps during this phase include:

- deployment of the Pancam mast and the high-gain antenna
- characterization of the lander, the landing site, and the surrounding terrain
- rover stand up
- calibration of science instruments
- selection of a suitable egress path
- egress from the lander onto the martian surface

To learn more about the Rover Egress phase; see http://marsrovers.jpl. nasa.gov/mission/tl rover egress steps.html.

Surface Operations. Surface Operations begin once the rover has completed its egress. The rovers were designed to last for 90 days on the Martian surface.

Surface operations includes two highly interconnected efforts: rover navigation, which allows the rover to reach places scientists identify as interesting for further study; and science investigations, which use the rover's science instruments to discover more about the martian environment.

Engineers responsible for rover navigation and science team members must work closely together to achieve mission goals. What the rover will actually do on the surface will depend on complex calculations from the science team on which rock, soil, and other targets are high-priority and then intense discussion with the engineering team on whether the rover can actually move toward those targets safely and quickly.

Before the rovers get to Mars, the science and engineering teams practice surface operations here on Earth through *field tests*. In preparation for this mission, a rover called *FIDO* was taken out to a remote desert location, while the scientists and engineers worked together to move it toward interesting science targets. To see what this experience was like; see the FIDO Field Test site http://mars.jpl.nasa.gov/mer/fido/.

Toward the end of the surface phase for both missions, both power and telecom capabilities will be decreasing, as the Earth and the Sun become more distant from Mars, dust falls on the solar panels, the batteries lose capacity, and the Sun moves further North past the landing site latitude. Eventually, it is expected that the rover will be unable to store up enough thermal or battery energy to prevent its components' overnight temperatures from falling below flight allowable levels. That will sooner or later result in failure of one or more of those components, silencing the rover forever.

Spacecraft

The spacecraft is separate from the launch vehicle. It is the protective "spaceship" that enables the precise cargo (that is, the rover!) to travel between Earth and Mars once the launch vehicle has already projected it outside of Earth's atmosphere and gravity pull.

The spacecraft includes the mechanical units that safely carry and maneuver the rover as it enters the Martian atmosphere and lands on Mars. Once on the surface, like a semi-truck carrying a racecar cross-country, it will lower a ramp to let the rover drive out.

The spacecraft consists of: (1) **Cruise Stage:** Configuration for travel between Earth and Mars. (2) **Entry, Descent, and Landing System:** Configuration for entry into the Martian atmosphere. Includes the aeroshell (the heatshield and backshell), the parachute, the airbags, and a lander structure. (3) **Rover:** A wheeled vehicle with science instruments for discoveries on the Martian surface.

The spacecraft design for the Mars Exploration Rover mission is largely based on the successful Mars Pathfinder system for entry, descent, and landing. The rover design is based on the Athena Rover on the previously cancelled Mars 2001 lander mission.

Each system has a total launch mass of 1,063 kilograms (2,343 pounds). The mass of each primary part of the spacecraft are listed in Table 1.

TABLE 1. MASS OF EACH PRIMARY PART OF SPACECRAFT

	Allocated Mass in kg (lbs)	Cumulative Mass in kg (lbs)
Rover	185 kg (408 lbs)	185 kg (408 lbs)
Lander	348 kg (767 lbs)	533 kg (1,175 lbs)
Backshell/Parachute	209 kg (742 lbs)	742 kg (1,636 lbs)
Heat Shield	78 kg (172 lbs)	820 kg (1,808 lbs)
Cruise Stage	193 kg (425 lbs)	1,013 kg (2,233 lbs)
Propellant	50 kg (110 lbs)	1,63 kg (2,343 lbs)

Cruise Configuration. The cruise stage is the configuration of the spacecraft for travel between Earth and Mars. The cruise stage is very similar to the Mars Pathfinder design and is approximately 2.65 meters (8.7 feet) in diameter and 1.6 meters (5.2 feet) tall when attached to aeroshell. It has a launch mass of 1063 kilograms (about 2,344 pounds). See Figs. 3, 4, 5.

The primary structure is aluminum with an outer ring of ribs covered by the solar panels, which are about 2.65 m (8.7 feet) in diameter. Divided into five sections, the solar arrays can provide up to 600 Watts of power near Earth and 300 Watts at Mars.

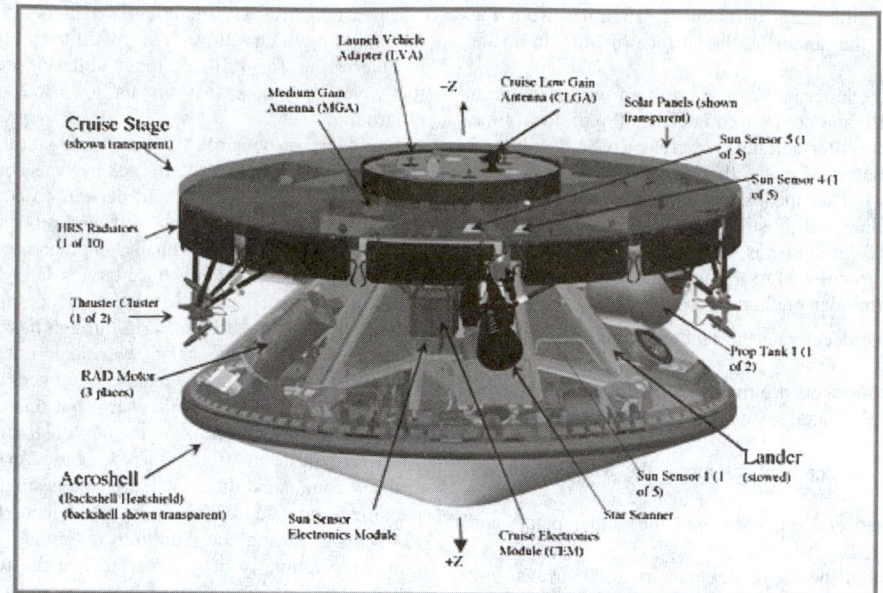

Fig. 3. Cruise stage configuration of the spacecraft. (Image courtesy of NASA/JPL)

Fig. 4. The cruise stage provides power, communication and propulsion for the rover and lander within its cocoon made up of the heat shield and backshell for the cruise to Mars. See also Fig. 3. (*Image courtesy of JPL.*)

Heaters and multi-layer insulation keep the spacecraft electronics *warm*. In contrast, there is also a freon system used to remove heat from the flight computer and telecommunications hardware inside the rover so they don't get overheated. Cruise avionics systems allow the flight computer in the rover to interface with other electronics such as the sun sensors, the star scanner, and the heaters.

Entry, Descent, and Landing Configuration. The entry, descent, and landing system consists of:

Aeroshell. The aeroshell forms a protective covering during the seven month voyage to Mars. The aeroshell, together with the lander and the rover, constitute what engineers call the *entry vehicle.* The aeroshell's main purpose is to protect the lander with the rover stowed safely inside from

Fig. 5. One of two Mars Exploration Rovers sits inside its cruise stage waiting to undergo environmental testing at NASA's Jet Propulsion Laboratory. In this photo, engineers are preparing the rover for vibration testing to ensure that it can undergo the rigors of launch and entry into the Martian atmosphere.

the intense heating of entry into the thin Martian atmosphere on landing day.

The aeroshell for the Mars Exploration Rovers is based on the Mars Pathfinder and Mars Viking designs.

The aeroshell is made of two principle parts: the heat shield (flat, brownish half) and the backshell (large, white-painted, cone-shaped half) See Fig. 6.

Fig. 6.   Aeroshell. (*Image courtesy of JPL.*)

Fig. 7.   Suspended by an overhead crane in the Payload Hazardous Servicing Facility, the Mars Exploration Rover (MER) aeroshell is guided by workers as it moves to a rotation stand. (*Image courtesy of JPL.*)

The heat shield protects the lander and rover from the intense heat from entry into the Martian atmosphere and aerodynamically acts as the first "brake" for the spacecraft.

The backshell carries the parachute and several components used during later stages of entry, descent, and landing, including:

- a parachute (stowed at the top of the backshell)
- the backshell electronics and batteries that fire off pyrotechnic devices like separation nuts, rockets and the parachute motor
- a Litton LN-200 Inertial Measurement Unit (IMU), which is a device that, like your inner ear, monitors and reports the orientation of the backshell as it swings under the parachute
- three large solid rocket motors called RAD rockets (Rocket Assisted Descent), each providing about a ton of force for over 2 seconds
- three small solid rockets called TIRS (mounted so that they aim horizontally out the sides of the backshell) that provide a small horizontal kick to the backshell to help orient the backshell more vertically during the main RAD rocket burn

Built by the Lockheed Martin Astronautics Co. in Denver, CO., the aeroshell is made out of an aluminum honeycomb structure sandwiched between graphite-epoxy face sheets. See Fig. 7. The outside of the aeroshell is covered with a layer of phenolic honeycomb. A phenolic compound is made from benzene and is typically used in various plastics, disinfectants, and pharmaceuticals. This phenolic honeycomb is filled with an ablative material (also called an "ablator"), which dissipates heat generated by atmospheric friction.

The ablator itself is a unique blend of cork wood, binder and many tiny silica glass spheres. It was invented for the heat shields flown on the Viking Mars lander missions 25 years ago. A similar technology was used in the first US manned space missions Mercury, Gemini and Apollo. It is specially formulated to react chemically with the Martian atmosphere

during entry and essentially take heat away, leaving a hot wake of gas behind the vehicle. (Normal friction without an ablator would cause the heat shield to burn up.) The heat loss to the Martian atmosphere lowers the kinetic energy of the entry vehicle, thereby it slowing it down … a lot … fast! The vehicle will slow from 10,000 mph to about 1000 mph in about a minute, producing about 10 "Earth gees" of acceleration on the lander and rover.

Both the backshell and heat shield are made of the same materials, but the heat shield has a thicker (1/2 in) layer of the ablator. Also, instead of being painted, the backshell will be covered with a very thin aluminized Mylar blanket to protect it from the cold of deep space. The blanket will vaporize during Mars atmospheric entry.

Parachute. The design of the parachute is driven by "loads" (the forces the parachute experiences as it fully inflates). Loads are calculated by using atmospheric density, velocity, parachute drag area, and mass. The 2003 parachute design is part of a long-term Mars parachute technology development effort and is based on the designs and experience of the Viking and Pathfinder missions. See Fig. 8. The parachute for this mission is 40% larger than Pathfinder's because the largest load for the Mars Exploration Rover is between 18,000 and 19,000 pounds (80,100–84,600 N*) when the parachute fully inflates. By comparison, Pathfinder's inflation loads were approximately 8,000 pounds (35,600 N). (*N stands for "Newton," the unit of force required to accelerate a mass of one kilogram one meter per second.)

Fig. 8.   The parachute will help slow the spacecraft down during entry, descent, and landing. (*Image courtesy of JPL.*)

The parachute is made out of two durable, lightweight fabrics: polyester and nylon. The parachute has a triple bridle (the tethers that connect the

Fig. 9. Mars Exploration Rover parachute deployment testing in the world's largest wind tunnel at NASA's Ames Research Center, Moffet Field, CA. (Image courtesy of JPL.)

parachute to the backshell). This bridle is made out of Kevlar, the same material used in bullet-proof vests. See Fig. 9.

The materials used to make the parachute must be strong, yet lightweight enough to fit inside a very small area and to prevent excess weight within the backshell. The amount of space available on the spacecraft for the parachute is so small that the parachute must be pressure packed. Before launch, a team must tightly fold together the 48 suspension lines, three bridle lines, and the parachute. The parachute team loads the parachute in a special structure that then applies a heavy weight to the parachute package several times (like sitting on a suitcase to pack down clothes to fit inside). Before placing the parachute into the backshell, the parachute is heat set to sterilize it.

After the parachute is deployed at an altitude of about 10 km (6 miles) above the surface, the heat shield is released using 6 separation nuts and push-off springs. The lander then separates from the backshell and "rappels" down a metal tape on a centrifugal braking system built into one of the lander petals. The slow descent down the metal tape places the lander in position at the end of another bridle (tether), which is made of a nearly 20-meter-long (65-foot-long) braided Zylon.

Zylon is an advanced fiber material similar to Kevlar that is sewn specifically in a webbing pattern (like shoelace material) to make it stronger. (Zylon is often used in lines for sailing here on Earth.) The Zylon bridle provides space for airbag deployment, distance from the solid rocket motor exhaust stream, and increased stability. The bridle incorporates an electrical harness that allows the firing of the solid rockets from the backshell as well as provides data from the backshell inertial measurement unit (which measures rate and tilt of the spacecraft) to the flight computer in the rover.

Because the atmospheric density of Mars is less than 1% of Earth's, the parachute alone cannot slow down the Mars Exploration Rover enough to ensure a safe, low landing speed. The spacecraft descent is assisted by rockets that bring the spacecraft to a dead stop 10–15 meters (30–50 feet) above the Martian surface.

A radar altimeter unit is used to determine the distance to the Martian surface. The radar's antenna is mounted at one of the lower corners of the lander tetrahedron. When the radar measurement shows the lander is a few meters (feet) above the surface, the Zylon bridle will be cut, releasing the lander from the parachute and backshell so that it is free and clear for landing. The radar data will also enable the timing sequence on airbag inflation and backshell RAD rocket firing.

Airbags. Airbags used in the Mars Exploration Rover mission are the same type that Mars Pathfinder used in 1997. See also **Pathfinder Mission to Mars**. Airbags must be strong enough to cushion the spacecraft if it lands on rocks or rough terrain and allow it to bounce across Mars' surface at freeway speeds after landing. To add to the complexity, the airbags must be inflated seconds before touchdown and deflated once safely on the ground.

While most new automobiles now come with airbags, spacecraft don't. The fabric being used for the new Mars airbags is a synthetic material called *Vectran* that was also used on Mars Pathfinder. Vectran has almost twice the strength of other synthetic materials, such as Kevlar, and performs better at cold temperatures.

Denier is a term that measures the diameter of the thread used in the product. There will be six 100-denier layers of the light but tough Vectran protecting one or two inner bladders of the same material in 200-denier. Using the 100-denier means there is more actual fabric in the outer layers where it is needed, because there are more threads in the weave.

Each rover uses four airbags with six lobes each, which are all connected. Connection is important, since it helps abate some of the landing forces by keeping the bag system flexible and responsive to ground pressure. The fabric of the airbags is not attached directly to the rover; ropes that crisscross the bags hold the fabric to the rover. The ropes give the bags shape, which makes inflation easier. While in flight, the bags are stowed along with three gas generators that are used for inflation.

Lander. The spacecraft lander is a protective *shell* that houses the rover and protects it, along with the airbags, from the forces of impact.

The lander is a strong, lightweight structure, consisting of a base and three sides "petals" in the shape of a tetrahedron (pyramid-shaped). See Fig. 10. The Lander structure consists of beams and sheets that are made from a "composite" material. Composites such as fiberglass are made of strong fibers or fabrics that are stiffened with a glue, or "matrix". The lander beams are made out of carbon-based layers of graphite fiber woven into a fabric, creating a material that is lighter than aluminum and more rigid than steel. Titanium fittings are bonded (glued and fitted) onto the lander beams to allow it to be bolted together. The Rover is held inside the lander with bolts and special nuts that are released after landing with small explosives.

The three petals are connected to the base of the tetrahedron with hinges. Each petal hinge has a powerful motor that is strong enough to lift the weight of the entire lander (The Rover plus Lander is about 530 kilograms, which would weigh almost 1,200 pounds on Earth but only 437 pounds on Mars. The Rover alone is 170 kilograms, weighing 375 pounds on Earth and 140 pounds on Mars.) Having a motor on each petal ensures that the lander can place the rover in an upright position no matter which side the lander comes to rest on after the bouncing and rolling subsides on the surface of Mars. See Fig. 11.

The Rover contains sensors called accelerometers that can detect which way is down (toward the surface of Mars) by measuring the pull of gravity. The Rover computer, knowing which way is down, commands the correct lander petal to open to place the rover upright. Once the base petal is down and the rover is upright, the other two petals are opened.

The petals will initially open to an equally flat position, so all sides of the lander are straight and level. The petal motors are strong enough that if the two outer petals land on rocks, the middle petal with the rover will be held in place like a bridge above the surface of Mars. The middle petal will hold at a level even with the height of the petals resting on rocks, making a straight flat surface throughout the length of the open, flattened lander. The flight team on Earth may then send commands to the rover to adjust the petals to create a better pathway for the rover to drive off of the lander and safely move onto the Martian surface without dropping off a steep rock.

The process of the rover moving off of the lander is called the egress phase of the mission. The rover must be able to safely drive off of the lander without getting its wheels caught up in the airbag material or without dropping off a sharp incline.

To aid in the egress process, the lander petals contain a retraction system that will slowly drag the airbags toward the lander to get them out of the path of the rover. (This step is performed before the Lander petals are opened.) Small ramps or *ramplets* are also connected to the petals, which fan out and create "driving surfaces" that fill in large spaces between the lander petals. These ramplets, nicknamed *Batwings*, are made out of a strong Vectran cloth, which is similar to the Kevlar used in bullet-proof vests. See Fig. 11. The *batwings* help cover dangerous, uneven terrain, rock obstacles, and leftover airbag material that could get entangled in the rover wheels. These Vectran cloth surfaces make a circular area from

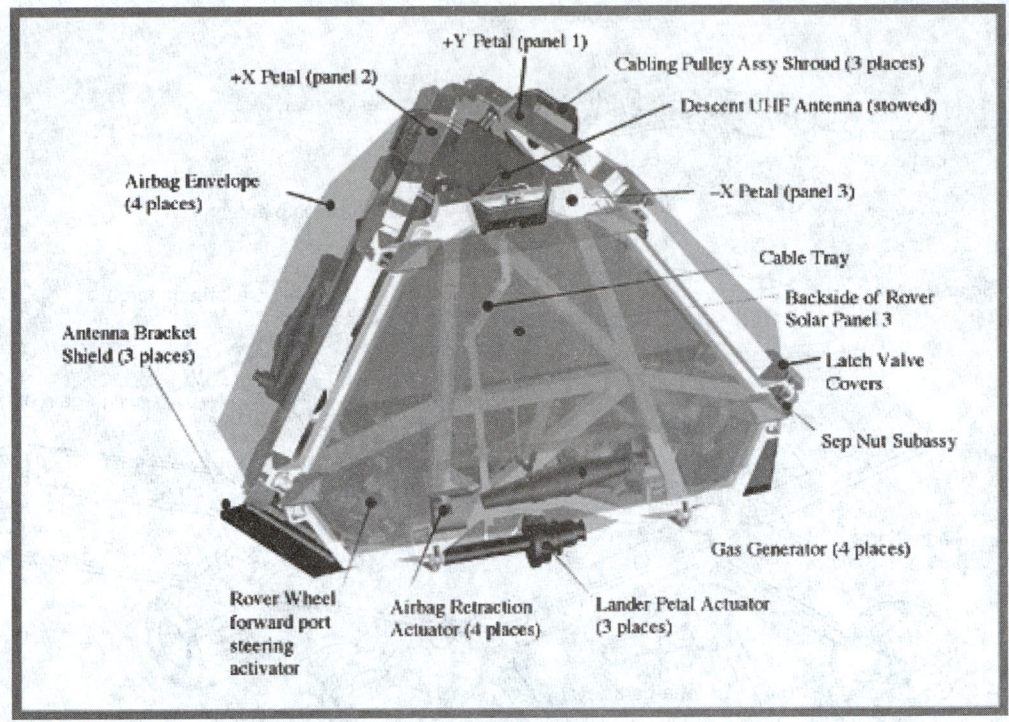

Fig. 10.   The Lander Structure. (Image courtesy of JPL.)

Fig. 11.   Rover and Lander. (*Image courtesy of JPL.*)

which the rover can roll off the lander, providing additional directions the rover can leave the lander. The ramplets also lower the height of the *step* that the rover must take off of the lander, preventing possible death of the rover. If the rover banged its belly on a rock or smashed into the ground as it was moving off the lander, the entire mission could be lost.

About 3 hours is allotted to retract the airbags and deploy the lander petals.

Surface Operations Configuration. Originally each rover was designed to conduct science investigations on the surface of Mars for 90 sols (90 Martian days). These goals have been far exceeded!

Science investigations are enabled by: the rover and the science instruments it carries.

Rover. The Mars Exploration Rovers will act as robot geologists while they are on the surface of Mars. In some senses, the rover's parts are similar to what any living creature would need to keep it *alive* and able to explore. See Fig. 12. The rover has a:

- a body: a structure that protects the rovers' "vital organs"
- brains: computers to process information
- temperature controls: internal heaters, a layer of insulation, and more

- a "neck and head": a mast for the cameras to give the rovers a human-scale view
- eyes and other "senses": cameras and instruments that give the rovers information about their environment
- arm: a way to extend its reach
- wheels and "legs": parts for mobility
- energy: batteries and solar panels
- communications: antennas for "speaking" and "listening"

Rover Body. The rover body is called the *warm electronics box*, or WEB for short. Like a car body, the rover body is a strong, outer layer that protects the rover's computer, electronics, and batteries (which are basically the equivalent of the rover's brains and heart). The rover body thus keeps the rover's vital organs protected and temperature-controlled.

The warm electronics box is closed on the top by a triangular piece called the Rover Equipment Deck (RED). See Fig. 13. The Rover Equipment Deck makes the rover like a convertible car, allowing a place for the rover mast and cameras to sit out in the Martian air, taking pictures and clearly observing the Martian terrain as it travels.

The gold-painted, insulated walls of the rover body also keep heat in when the night temperatures on Mars can drop to −96 °C (−140 °F).

Brains. Unlike people and animals, the rover brains are in its body. The rover computer (its "brains") is inside a module called The *Rover Electronics Module* (REM) inside the rover body. See Fig. 13. The communication interface that enables the main computer to exchange data with the rover's instruments and sensors is called a *bus* (a VME or Versa Module Europa bus to be exact). This VME bus is an industry standard interface bus to communicate with and control all of the rover motors, science instruments, and communication functions.

The computer is composed of equipment comparable to a high-end, powerful laptop computer. It contains special memory to tolerate the extreme radiation environment from space and to safeguard against power-off cycles so the programs and data will remain and will not accidentally erase when the rover shuts down at night.

On-board memory includes 128 MB of DRAM with error detection and correction and 3 MB of EEPROM. That's roughly the equivalent memory of a standard home computer. This onboard memory is roughly 1000 more than the Sojourner rover from the Pathfinder mission had.

The rover carries an Inertial Measurement Unit (IMU) (Fig. 13) that provides 3-axis information on its position, which enables the rover to make precise vertical, horizontal, and side-to-side (yaw) movements. The

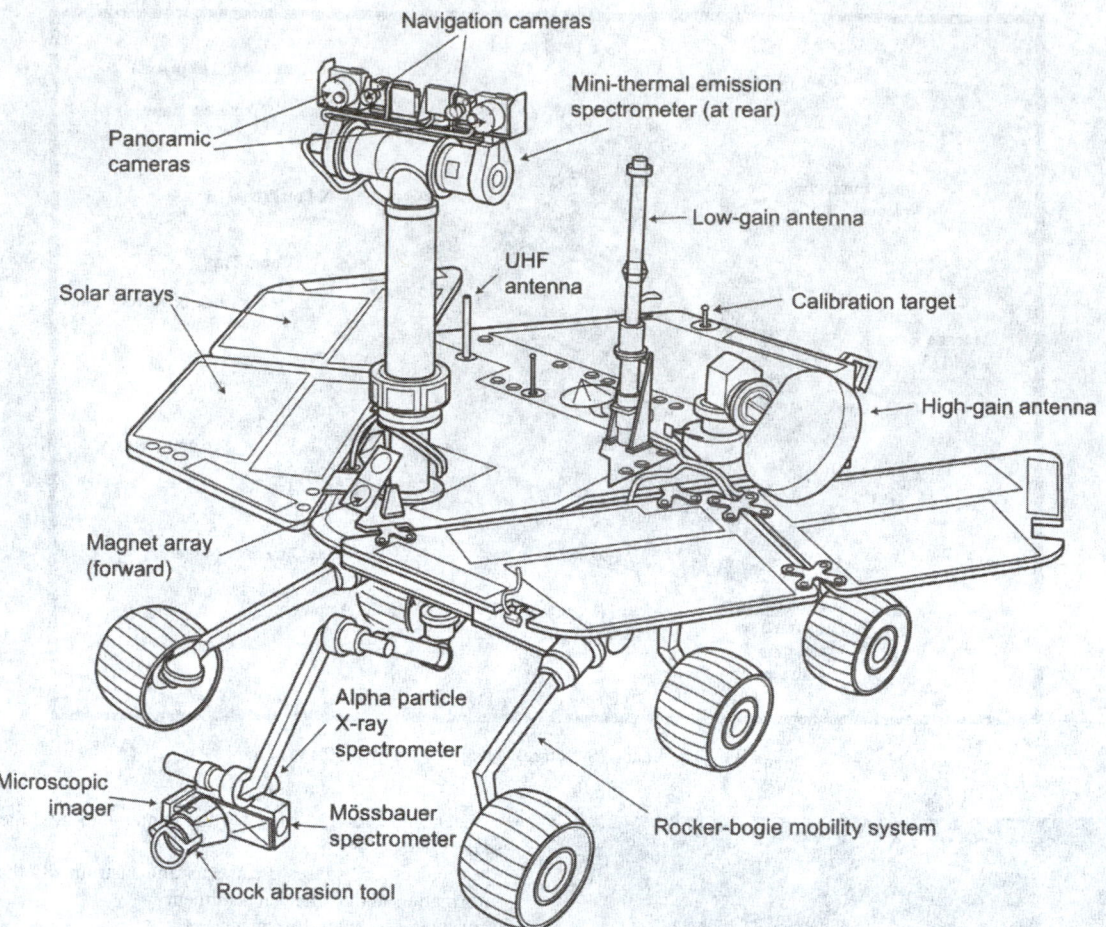

Fig. 12. Mars Rover. (*JPL.*)

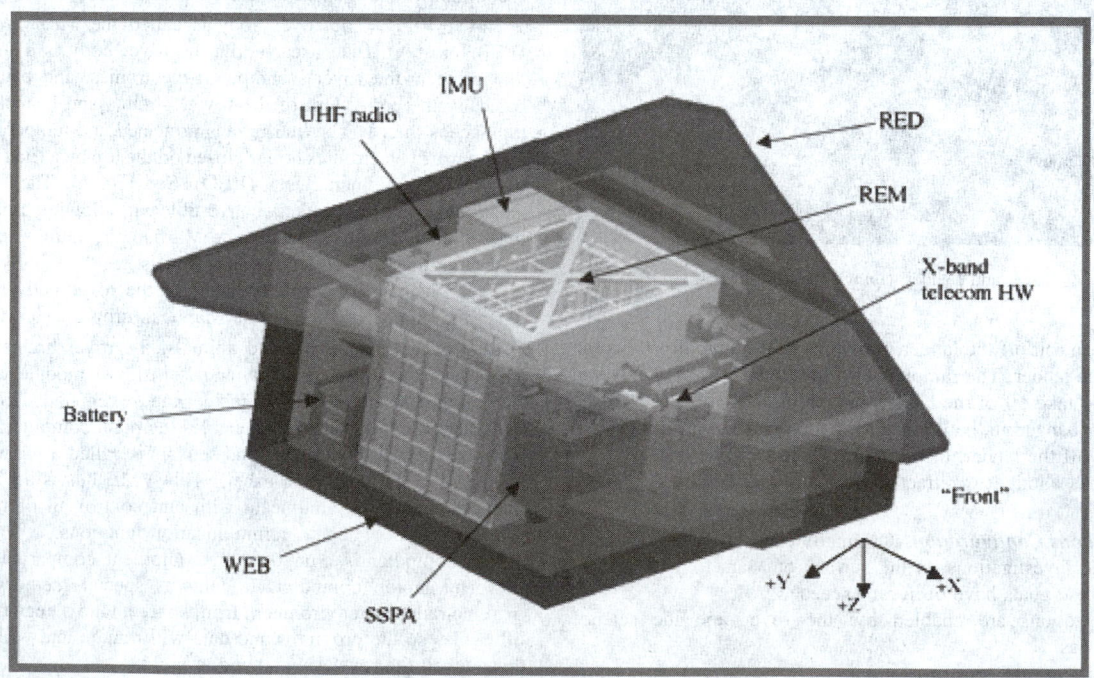

Fig. 13. The warm electronics box illustrated. (*JPL.*)

device is used in rover navigation to support safe traverses and to estimate the degree of tilt the rover is experiencing on the surface of Mars.

Just like the human brain, the rover computers register signs of health, temperature, and other features that keep the rovers "alive."

The software in the main computer of the rover changes modes once the cruise portion of the mission is complete and the spacecraft begins to enter the Martian atmosphere. Upon entry into the Martian atmosphere, the software executes a control loop that monitors the *health* and status of the vehicle. It checks for the presence of commands to execute, performs communication functions, and checks the overall status of the rover. The software does similar health checks in a third mode once the rover emerges from the lander.

This main control loop essentially keeps the rover *alive* by constantly checking itself to ensure that it is both able to communicate throughout the surface mission and that it remains thermally stable (not too hot or too cold) at all times. It does so by periodically checking temperatures, particularly in the rover body, and responding to potential overheating conditions, recording power generation and power storage data throughout the Mars sol (a Martian day)[1], and scheduling and preparing for communication sessions.

Activities such as taking pictures, driving, and operating the instruments are performed under commands transmitted in a command sequence to the rover from the flight team.

The rover generates constant engineering, housekeeping and analysis telemetry and periodic event reports that are stored for eventual transmission once the flight team requests the information from the rover.

Temperature Controls. Like the human body, the Mars Exploration Rover cannot function well under excessively hot or cold temperatures. In order to survive during all of the various mission phases, the rover's "vital organs" must not exceed extreme temperatures of $-40\,°C$ to $+40\,°C$ ($-40\,°F$ to $104\,°F$).

The rover's essentials, such as the batteries, electronics, and computer, which are basically the rover's heart and brains, stay safe inside a Warm Electronics Box (WEB), commonly called the *rover body*. Heaters are packed inside the rover body, and like a warm coat, the WEB (Fig. 13) walls help keep heat in when the night temperatures on Mars can drop to $-96\,°C$ ($-140\,°F$). Just as an athlete sweats to release heat after an intense workout, the rover's body can also release excess heat through its radiators, similar to ones used in car engines.

There are several methods engineers used to keep the rover at the right temperature:

- Preventing heat escape through gold paint
- Preventing heat escape through insulation called "aerogel"
- Keeping the rover warm through heaters
- Making sure the rover is not too hot or cold through thermostats and heat switches
- Making sure the rover doesn't get too hot through the heat rejection system

Many of these methods are very important to making sure the rover doesn't *freeze to death* in the cold of deep space or on Mars. Many people often assume that Mars is hot, but it is farther away from the sun and has a much thinner atmosphere than Earth, so any heat it does get during the day dissipates at night. In fact, the ground temperatures at the rover landing sites will swing up during the day and down again during the night, varying by up to $113\,°C$ (or $235\,°F$) per Mars day. That's quite a temperature swing, when you consider that Earth temperatures typically vary by tens of degrees on average between night and day.

At the landing sites, an expected daytime high on the ground might be around $22°$ Celsius ($71°$ Fahrenheit). An expected nighttime lows might be $-99°$ Celsius ($-146°$ Fahrenheit). Atmospheric temperatures, by contrast, can vary up to $83°$ Celsius ($181°$ Fahrenheit). An atmospheric daytime high might be $-3°$ Celsius ($26°$ Fahrenheit), while a nighttime low might be $-96°$ Celsius ($-140°$ Fahrenheit).

Neck and Head. What looks like the rover *neck and head* is called the *Pancam Mast Assembly*. It will stand from the base of the rover wheel 1.4

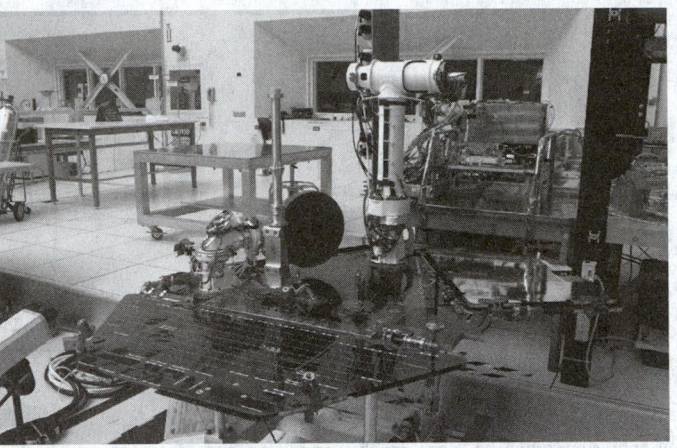

Fig. 14.   Mast Assembly (*JPL*).

meters tall (about 5 feet). See Fig. 14. This height will give the cameras a special *human geologist's* perspective and wide field of view.

The pancam mast assembly serves two purposes: (1) to act as a periscope for the Mini-TES science instrument that is housed inside the rover body for thermal reasons; (2) to provide height and a better point of view for the Pancams and the Navcams

Essentially, the pancam mast assembly enables the rover to see in the distance. The higher one stands, the more one can see. You can test this out for yourself by lying on the ground and observing as much as possible, then standing up and seeing the difference in the amount of greater detail you can observe about the world with a broader field of view.

One motor for the entire Pancam Mast Assembly head turns the cameras and Mini-TES $360°$ in the horizontal plane. A separate elevation motor can point the cameras $90°$ above the horizon and $90°$ below the horizon. A third motor for the Mini-TES elevation, enables the Mini-TES to point up to $30°$ over the horizon and $50°$ below the horizon.

During cruise, the Pancam Mast Assembly lays flat against the top of the rover equipment deck in a stowed configuration. After the lander opens on the surface of Mars, pyros release the bolts holding it down. Pyros are solid mechanical devices that contain as much power as a bullet release mechanism in a gun. Pyros were designed in the 1960s for the Apollo mission as a safe way to release bolts and strong devices on spacecraft while humans were in the vicinity.

The Pancam Mast Assembly rises from the rover equipment deck by driving a motor that moves the Pancam upward, in the shape of a helix. It sweeps out in a cone-like manner as it deploys. Once the Pancam Mast Assembly is in its full-upright position, it will not stow again, but will stay upright for the entire duration of the mission.

Eyes and Other Senses. Each rover has nine *eyes*. See Fig. 15. Six engineering cameras aid in rover navigation and three cameras perform science investigations.

Each camera has an application-specific set of optics.

Four Engineering Hazcams (Hazard Avoidance Cameras). Mounted on the lower portion of the front and rear of the rover, these black-and-white cameras use visible light to capture three-dimensional (3-D) imagery. This imagery safeguards against the rover getting lost or inadvertently crashing into unexpected obstacles, and works in tandem with software that allows the rover make its own safety choices and to "think on its own."

The cameras each have a wide field of view of about 120 degrees. The rover uses pairs of Hazcam images to map out the shape of the terrain as far as 3 meters (10 feet) in front of it, in a *wedge* shape that is over 4 meters wide at the farthest distance. It needs to see far too either side because unlike human eyes, the Hazcam cameras cannot move independently; they're mounted directly to the rover body.

Two Engineering Navcams (Navigation Cameras). Mounted on the mast (the rover *neck and head*), these black-and-white cameras use visible light to gather panoramic, three-dimensional (3D) imagery. The Navcam is a stereo pair of cameras, each with a 45-degree field of view to support ground navigation planning by scientists and engineers. They work in

---

[1] Each Martian day is longer than an Earth day, lasting 24 hours, 39 minutes, 35 seconds.

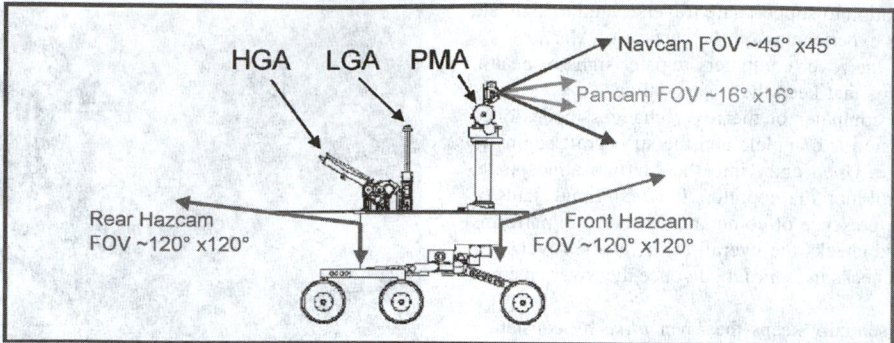

Fig. 15.   The rover's *eyes* and other *senses* (*JPL.*).

cooperation with the Hazcams by providing a complementary view of the terrain.

Two Science Pancams (Panoramic Cameras). This color, stereo pair of cameras is mounted on the rover mast and delivers three-dimensional panoramas of the Martian surface. As well as science panoramas, the narrow field of view and height of the cameras basically mimic the resolution of the human eye (0.3 milliradians), giving the world a view similar to what a human geologist might see if she or he were standing on the surface of Mars. Also, the Pancam detectors have 8 filters per *eye* and between the two *eyes* there are 11 total unique color filters plus two-color, solar-imaging filters to take multispectral images. The Pancam is also part of the rover's navigation system. With the solar filter in place, the Pancam will be pointed at the Sun and therefore will be used as an absolute heading sensor. Like a sophisticated compass, the direction of the Sun combined with the time of day tells the flight team exactly which way the rover is facing.

One Science Microscopic Imager. This monochromatic science camera is mounted on the robotic arm to take extreme close-up pictures of rocks and soil. Some of its studies of the rocks and soil will help engineers understand the properties of the smaller rocks soil that can impact rover mobility (how much resistance it has against the rover wheels, how far they will sink).

Arm. The rover arm (also called the *instrument deployment device* or IDD) holds and maneuvers the instruments that help scientists get up-close and personal with Martian rocks and soil.

Much like a human arm, the robotic arm has flexibility through three joints: the rover's shoulder, elbow, and wrist. The arm enables a tool belt of scientists' instruments to extend, bend, and angle precisely against a rock to work as a human geologist would: grinding away layers, taking microscopic images, and analyzing the elemental composition of the rocks and soil. See Fig. 16.

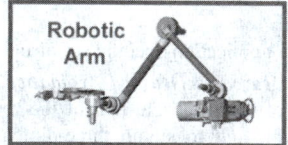

Fig. 16.   The rover's arm (*JPL*).

At the end of the arm is a turret, shaped like a cross. This turret, a hand-like structure, holds various tools that can spin through a 350-degree turning range.

The four tools, or science instruments, (Fig. 17) on the robotic arm are: (1) The Microscopic Imager provides close-up images of rocks and soil. (2) The Mössbauser Spectrometer analyzes the iron in rocks and soil. (3) The Alpha Particle X-Ray Spectrometer analyzes the elemental composition of rocks and soil. (4) The Rock Abrasion Tool (RAT)grinds away the outer surface of rock to expose fresh material

The forearm also holds a small brush so that the Rock Abrasion Tool can spin against it to *brush its teeth* and rid the grinding tool of any leftover pieces of rock before its next bite.

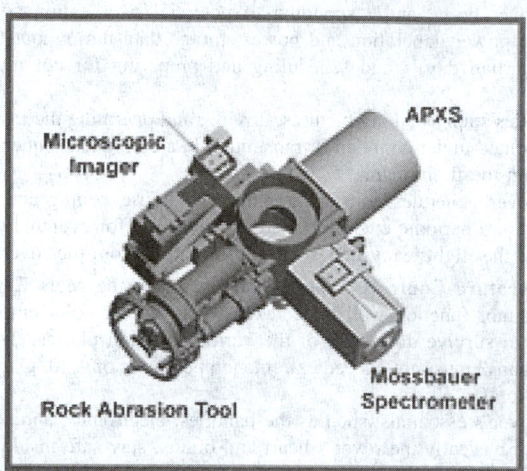

Fig. 17.   Science instruments on the rover's arm (*JPL*).

Thirty percent of the mass of the titanium robotic arm comes from the four instruments it holds at the end of the arm. This weight makes maneuvering the lightweight arm a bit of a challenge — like controlling a bowling ball at the end of a fishing rod. The arm must be as lightweight as possible for the overall health of the mission, and holes are even cut out in places where there is no need for solid titanium.

Once the arm and instruments have succeeded in one location; and before the rover begins another traverse, the arm stows itself underneath the *front porch* of the rover body. The elbow hooks itself back onto a pin, and the turret has a T-bar that slides back into a slotted ramp. The fit is almost as tight as a necklace clasp, and it can withstand shocks of 6 G's while roving along the rocky terrain. "Six G's" is roughly equivalent to dropping a box onto a hard floor from a height of 20 centimeters (almost 8 inches). During launch and landing, the arm is restrained by a retractable pin restraint, and can withstand even higher loads of 42 G's.

Wheels and Legs. The Mars Exploration Rover has six wheels, each with its own individual motor.

The two front and two rear wheels also have individual steering motors (1 each). This steering capability allows the vehicle to turn in place, a full 360 degrees. The 4-wheel steering also allows the rover to swerve and curve, making arching turns.

The design of the suspension system for the wheels is similar to the Sojourner rover *rocker-bogie* system on the Pathfinder mission. The suspension system is how the wheels are connected to and interact with the rover body.

The term *bogie* comes from old railroad systems. A bogie is a train undercarriage with six wheels that can swivel to curve along a track.

The term *rocker* comes from the design of the differential, which keeps the rover body balanced, enabling it to *rock* up or down depending on the various positions of the multiple wheels. Of most importance when creating a suspension system is how to prevent the rover from suddenly and dramatically changing positions while cruising over rocky terrain. If one side of the rover were to travel over a rock, the rover body would go out of balance without a *differential* or *rocker*, which helps balance the angle the rover is in at any given time. When one side of the rover goes up, the differential or rocker in the rover suspension system automatically makes the other side go down to even out the weight load on the six wheels. This system causes the rover body to go through only half of the range of motion that the *legs* and wheels could potentially experience without a *rocker-bogie* suspension system.

The rover is designed to withstand a tilt of 45 degrees in any direction without overturning. However, the rover is programmed through its *fault protection limits* in its hazard avoidance software to avoid exceeding tilts of 30 degrees during its traverses.

The rover rocker-bogie design allows the rover to go over obstacles (such as rocks) or through holes that are more than a wheel diameter (26 centimeters or a little over 10 inches) in size. Each wheel also has cleats, providing grip for climbing in soft sand and scrambling over rocks. See Fig. 18.

Fig. 18.   This up-close photo shows the spiral flectures which act as shock absorbers and the orange Solimide that fills the flectures, preventing rocks and debris from interfering with the driving and steering actuators (*JPL*).

The rover has a top speed on flat hard ground of 5 centimeters (2 inches) per second. However, in order to ensure a safe drive, the rover is equipped with hazard avoidance software that causes the rover to stop and reassess its location every few seconds. So, over time, the vehicle achieves an average speed of 1 centimeter per second. The rover is programmed to drive for roughly 10 seconds, then stop to observe and understand the terrain it has driven into for 20 seconds, before moving safely onward for another 10 seconds.

Energy. The rover requires power to operate. Without power, it cannot move, use its science instruments, or communicate with Earth. The main source of power for each rover comes from a multi-panel solar array. They look almost like *wings*, but their purpose is to provide energy, not fly.

When fully illuminated, the rover solar arrays generate about 140 watts of power for up to four hours per sol (a Martian day). The rover needs about 100 watts (equivalent to a standard light bulb in a home) to drive. Comparatively, the Sojourner rover's solar arrays provided the 1997 Pathfinder mission with around 16 watts of power at noon on Mars. That's equivalent to the power used by an oven light. This extra power will potentially enable the rovers to conduct more science.

The power system for the Mars Exploration Rover includes two rechargeable batteries that provide energy for the rover when the sun is not shining, especially at night. Over time, the batteries will degrade and will not be able to recharge to full power capacity. Also, by the end of the 90-sol mission, the capability of the solar arrays to generate power will likely be reduced to about 50 watts of power due to anticipated dust coverage on the solar arrays (as seen on Sojourner/Mars Pathfinder), as well as the change in season. Mars will drift farther from the sun as it continues on its yearly elliptical orbit, and because of the distance, the sun will not shine as brightly onto the solar arrays. Additionally, Mars is tilted

on its axis just like Earth is, giving Mars seasonal changes. Later in the mission, the seasonal changes at the landing site and the lower position of the Sun in the sky at noon than in the beginning of the mission will mean less energy on the solar panels.

Communications. The rover has both a low-gain and high-gain antenna that serves as both its *voice* and its *ears*. They are located on the rover equipment deck (its *back*).

The low-gain antenna sends and receives information in every direction; that is, it is *omni-directional*. The antenna transmits radio waves at a low rate to the Deep Space Network (DSN) antennas on Earth. The high-gain antenna can send a *beam* of information in a specific direction and it is steerable, so the antenna can move to point itself directly to any antenna on Earth. The benefit of having a steerable antenna is that the entire rover doesn't necessarily have to change positions to talk to Earth. Like turning your neck to talk to someone beside you rather than turning your entire body, the rover can save energy by moving only the antenna.

Not only can the rovers send messages directly to Earth, but they can uplink information to other spacecraft orbiting Mars, utilizing the 2001 Mars Odyssey and Mars Global Surveyor orbiters as messengers who can pass along news to Earth for the rovers. The orbiters can also send messages to the rovers. See also **2001 Mars Odyssey Mission; and Mars Global Surveyor Mission**.

The benefits of using the orbiting spacecraft are that the orbiters are closer to the rovers than the Deep Space Network antennas on Earth and the orbiters have Earth in their field of view for much longer time periods than the rovers on the ground.

The radio waves to and from the rover are sent through the orbiters using UHF antennas, which are close-range antennas which are like walky-talkies compared to the long range of the low-gain and high-gain antennas. One UHF antenna is on the rover and one is on the petal of the lander to aid in gaining information during the critical landing event. The Mars Global Surveyor will be in the appropriate location above Mars to track the landing process. (2001 Mars Odyssey will not be in the vicinity.)

Instruments. Science instruments are state-of-the-art tools that will acquire information about the Martian environment. They include:

Cameras: The Panoramic Camera (Pancam). Pancam is a high-resolution color stereo pair of CCD cameras that will be used to image the surface and sky of Mars. The cameras are located on a *camera bar* that sits on top of the mast of the rover. See Fig. 19.

The Pancam Mast Assembly (PMA) allows the cameras to rotate a full 360° to obtain a panoramic view of the Martian landscape. The camera bar itself can swing up or down through 180° of elevation. Scientists will use Pancam to scan the horizon of Mars for landforms that may indicate a past history of water. They will also use the instrument to create a map of the area where the rover lands, as well as search for interesting rocks and soils to study.

The Pancam cameras are small enough to fit in the palm of your hand (270 grams or about 9 ounces), but can generate panoramic image mosaics as large as 4,000 pixels high and 24,000 pixels around. Pancam detectors are CCDs (charge coupled devices). These devices form the image, just as film does in a film camera.

Each *eye* of the Pancam carries a filter wheel that gives Pancam its multispectral imaging capabilities. Images taken at various wavelengths can help scientists learn more about the minerals found in Martian rocks and soils. Blue and infrared solar filters allow the camera to image the sun. These data, along with images of the sky at a variety of wavelengths, will help to determine the orientation of the rover and will provide information about the dust in the atmosphere of Mars. The Pancam color imaging system has, by far, the best capability of any camera ever sent to the surface of another planet.

The Microscopic Imager (MI). The Microscopic Imager is a combination of a microscope and a CCD camera that will provide information on the small-scale features of Martian rocks and soils. It will complement the findings of other science instruments by producing close-up views of surface materials. Some of those materials will be in their natural state, while others may be views of fresh surfaces exposed by the Rock Abrasion Tool.

Microscopic imaging will be used to analyze the size and shape of grains in sedimentary rocks, which is important for identifying whether water may have existed in the planet's past.

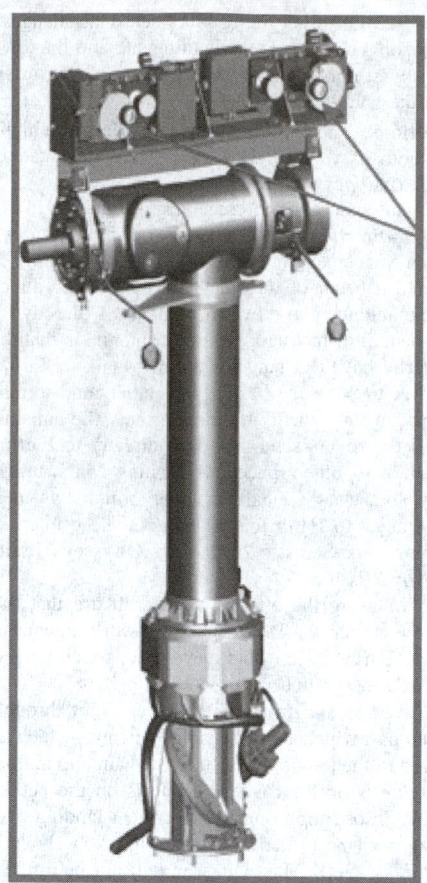

Fig. 19.   The Pancam mast assembly. (*JPL*)

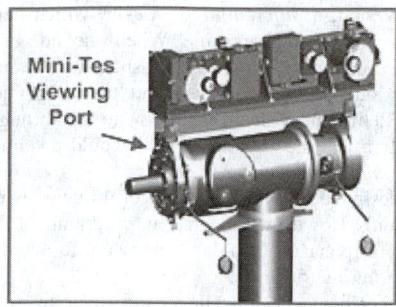

Fig. 20.   Mini-TES locations. (*JPL*)

The Microscopic Imager is located on the arm of the rover. Its field of view is 1024 x 1024 pixels in size and it has a single, broad-band filter so imaging is in black and white.

Spectrometers

Miniature Thermal Emission Spectrometer (Mini-TES). Mini-TES is an infrared spectrometer that can determine the mineralogy of rocks and soils from a distance by detecting their patterns of thermal radiation. All warm objects emit heat, but different objects emit heat differently. This variation in thermal radiation can help scientists identify the minerals on Mars. Mini-TES will record the spectra of various rocks and soils. These spectra can be studied to determine the type of minerals and their abundances at selected locations. One particular goal will be to search for minerals that were formed by the action of water, such as carbonates and clays. Mini-TES will also look at the atmosphere of Mars and gather data on temperature, water vapor, and the abundance of dust.

Mini-TES weighs 2.1 kg (almost 5 lbs) and is located in the body of the rover at the bottom of the *rover neck*, known as the Pancam Mast Assembly (PMA). See Fig. 20. Scanning mirrors located in the Pancam Mast Assembly act like a periscope to send light down to the instrument. This structure allows Mini-TES to see the terrain around the rover from the same vantage point as Pancam. Mini-TES looks one way, and the Pancams looks the other way. To make observations of the same location from both of the instruments, the Pancam Mast Assembly (the rover's neck) must be commanded to swivel.

Mössbauer Spectrometer (MB). Many of the minerals that formed rocks on Mars contain iron, and the soil is iron-rich. The Mössbauer Spectrometer is an instrument that was specially designed to study iron-bearing minerals. Because this science instrument is so specialized, it can determine the composition and abundance of these minerals to a high level of accuracy. This ability can also help us understand the magnetic properties of surface materials.

The Mössbauer Spectrometer sensor head is small enough to fit in the palm of your hand. It is one of four instruments mounted on the turret at the end of the rover arm. Its electronics are housed inside the body of the rover (in the Warm Electronics Box, or WEB). See Fig. 13. Measurements are taken by placing the instrument's sensor head directly against a rock or soil sample. One Mössbauer measurement takes about 12 hours.

Alpha Particle X-Ray Spectrometer (APXS). The APXS determines the elemental chemistry of rocks and soils using alpha particles and X-rays. Alpha particles are emitted during radioactive decay and X-rays are a type of electromagnetic radiation, like light and microwaves. The APXS carries a small alpha particle source. The alphas are emitted and bounce back from a science target into a detector in the APXS, along with some X-rays that are excited from the target in the process. The energy distribution of the alphas and X-rays measured by the detectors is analyzed to determine elemental composition. The elemental composition of a rock describes the amounts of different chemical elements that have come together to form all of the minerals within the rock. Knowing the elemental composition of Martian rocks provides scientists with information about the formation of the planet's crust, as well as any weathering that has taken place.

As with the other instruments on the arm of the rover, the APXS sensor head is small enough to hold in your hand. Its electronics are housed in the warm electronics box (WEB) (Fig. 13) located in the body of the rover. Most APXS measurements are taken at night and require at least 10 hours of accumulation time, although just X-ray alone will only require a few hours.

Grinder Rock Abrasion Tool (RAT). The Rock Abrasion Tool is a powerful grinder, able to create a hole 45 millimeters (about 2 inches) in diameter and 5 millimeters (0.2 inches) deep into a rock on the Martian surface.

The RAT is located on the arm of the rover and weighs less than 720 grams (about 1.6 lbs). It uses three electric motors to drive rotating grinding teeth into the surface of a rock. Two grinding wheels rotate at high speeds. These wheels also rotate around each other at a much slower speed so that the two grinding wheels sweep the entire cutting area. The RAT is able to grind through hard volcanic rock in about two hours.

Once a fresh surface is exposed, scientists can examine the abraded area in detail using the rover's other science instruments. This means that the interior of a rock may be very different from its exterior. That difference is important to scientists as it may reveal how the rock was formed and the environmental conditions in which it was altered. A rock sitting on the surface of Mars may become covered with dust and will weather, or change in chemical composition from contact with the atmosphere. See also http://www.honeybeerobotics.com/rat.html.

Magnets Magnet Arrays. Mars is a dusty place and some of that dust is highly magnetic. Magnetic minerals carried in dust grains may be freeze-dried remnants of the planet's watery past. A periodic examination of these particles and their patterns of accumulation on magnets of varying strength can reveal clues about their mineralogy and the planet's geologic history.

Each rover has three sets of magnetic targets that will collect airborne dust for analysis by the science instruments. One set of magnets will be carried by the Rock Abrasion Tool. As it grinds into Martian rocks, scientists will have the opportunity to study the properties of dust from these outer rock surfaces.

A second set of two magnets is mounted on the front of the rover at an angle so that non-magnetic particles will tend to fall off. These magnets will be reachable for analysis by the Mössbauer and APXS instruments. A third magnet is mounted on the top of the rover deck in view of the

Pancam. This magnet is strong enough to deflect the paths of wind-carried, magnetic dust.

The instruments use calibration targets, including a sundial, to determine accurate colors, brightness's, and other information collected by the instruments.

Calibration Targets. When you adjust the color on your television set, you do so by picking something on the screen that you know should be a certain color (such as grass should be green) and you adjust your set accordingly. This is a form of calibration. You used the color of the grass as a reference point. Instruments that go to Mars also need to be calibrated so that scientists receive accurate information. There has to be a known reference—a calibration target.

The rover's calibration targets are objects with known properties. For example, the Mössbauer Spectrometer's calibration target is a thin slab of rock that is rich in magnetite. The APXS also uses another reference point on the inside of its dust doors. When these doors are closed, they protect the APXS sensor head from Martian dust and offer a calibration target on their interior surfaces. Mini-TES has an internal target located in the Pancam Mast Assembly as well as an external target on the deck of the rover.

The Pancam calibration target is, by far, the most unique the rover carries. It is in the shape of a sundial and is mounted on the rover deck. See Fig. 21. Pancam will image the sundial many times during the mission so that scientists can adjust the images they receive from Mars. They will use the colored blocks in the corners of the sundial to calibrate the color in images of the Martian landscape. Pictures of the shadows that are cast by the sundial's center post will allow scientists to properly adjust the brightness of each Pancam image. Children provided artwork for the sides of the base of the sundial.

Fig. 21. Rover sundial. (*JPL*)

Spirit Mission Highlights

A traveling robotic geologist from NASA has landed on Mars and returned stunning images of the area around its landing site in Gusev Crater. See Figs. 22, 23, 24

Mars Exploration Rover Spirit successfully sent a radio signal after the spacecraft had bounced and rolled for several minutes following its initial impact at 11:35 p.m. EST (8:35 p.m. Pacific Standard Time) on January 3.

NASA chose Spirit's landing site, within Gusev Crater, based on evidence from Mars orbiters that this crater may have held a lake long ago. A long, deep valley, apparently carved by ancient flows of water, leads into Gusev. The crater itself is basin the size of Connecticut created by an asteroid or comet impact early in Mars' history. Spirit's task is to spend the next three months exploring for clues in rocks and soil about whether the past environment at this part of Mars was ever watery and suitable to sustain life.

Spirit traveled 487 million kilometers (302.6 million) miles to reach Mars after its launch from Cape Canaveral Air Force Station, Fla., on June 10, 2003. Its twin, Mars Exploration Rover Opportunity, was launched July 7, 2003, and is on course for a landing on the opposite side of Mars on Jan. 25 (Universal Time and EST; 9:05 p.m. on Jan. 24, PST).

More than two weeks after landing, team members waited for news of the first hole to be drilled on Mars. Spirit was to use its rock abrasion tool to grind a hole in a rock dubbed "Adirondack." See Fig. 25.

Technicians at NASA's Deep Space Network antenna outside Canberra, Australia reported that they had received no confirmation from the rover that it had gotten instructions from Earth. Because rain was affecting communications, they didn't worry too much at first.

The next day passed with no communication either. When a smaller antenna failed to make contact with NASA's 2001 Odyssey spacecraft orbiting Mars, the rover team became concerned.

"This is the way it starts when you lose a spacecraft, you lose communication," said Adler. "So we were definitely in the *Pasadena, we have a problem* situation."

The team told the rover to send a simple beep to Earth—a five-minute blast of radio communication to let them know it was alive and listening. No beep. Again, no beep. Mission controllers requested help from antenna stations around the world. No beep. Hours passed.

Finally they decided to send a command at a different rate, similar to talking more slowly. This time, they got a response. They knew then the rover was potentially alive.

The team was working on Mars time, in a building at JPL isolated from the California time of day by blackout shades on the windows. It was nighttime for the Spirit rover and nighttime for Spirit's handlers.

"Sleeping and eating were optional," said Adler. "There were cots we could sleep on in our offices. This was our one objective, our primary objective in our lives, was to get our spacecraft back."

The next transmissions from Spirit were garbled. The rover was babbling, doing things like sending nonsensical communications that it date-stamped as being from the year 2038.

In the midst of the gibberish, the team got a 30-minute session of data that actually made sense. A key packet of engineering and health data told them the rover's temperature was way up and its battery power was way down. The rover's health was failing because it was constantly rebooting its computer.

"So we knew then we were in a race," said Adler. "The battery was going down and it wasn't going to sleep the way it was supposed to. The rover had been up constantly, apparently doing resets. Somehow we had to get this rover to go to sleep and recharge its batteries."

Spirit had a fever and it was dying. Like a child pretending to obey, the rover even went so far as to say "OK" when team members told it to go to

Fig. 22. This mosaic image taken on January 4, 2004 by the navigation camera on the Mars Exploration Rover Spirit shows a 360 degree panoramic view of the rover on the surface of Mars. (*Image courtesy of NASA/JPL*)

Fig. 23. This image taken by the hazard avoidance camera on the Mars Exploration Rover Spirit shows the rover's rear lander petal and, in the background, the Martian horizon. Spirit took the picture right after successfully landing on the surface of Mars.

Fig. 25. Scientists lost contact with Spirit as it prepared to drill a hole into a rock dubbed "Adirondack." (*Image courtesy of NASA/JPL/Cornell.*)

"So now we had won the race," Adler said. But their patient was far from well. Each morning the rover would wake up confused. Engineers spent four days tracing all of the tasks it performed, deleting its files, reformatting its computer hard drive, and watching its every move.

In the end, they saved not only the Spirit rover, but the Opportunity rover as well, which was carrying all the same software and hardware as Spirit. While Spirit was in a fight for its life, its rover twin, Opportunity, had arrived on the other side of Mars.

Fig. 24. This is one of the first images beamed back to Earth shortly after the Mars Exploration Rover Spirit landed on the red planet.

sleep. By now, engineers were getting wary and radioed another message to Mars. Wide awake, the rover answered. The team figured Spirit had reset its systems more than 100 times trying to get itself back into shape. It was depleting its own battery trying to get well.

The team finally decided to try a different approach. They put the rover in "crippled" mode, said Adler, which was actually far better than the condition it was in. They cut off the flash file memory, which turned out to be the site of the malfunction. When they finally got the rover to listen, the first thing they told it to do was go to sleep.

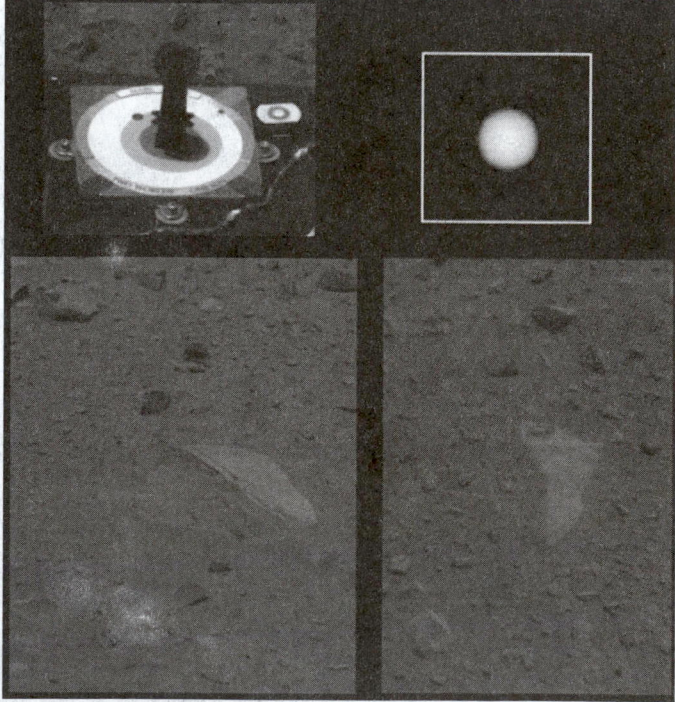

Fig. 26. Spirit's potential target rocks *Cake* on the left and *Blanco* on the right. (Image courtesy of NASA/JPL/Cornell)

Two of Spirit's potential target rocks can be seen on the lower left and right in Fig. 26. The rock on the left has been named *Cake*, and the white rock on the right has been named *Blanco*. These are the first images sent back from the Pancam panoramic camera on Spirit since the rover experienced communications problems on the 18th sol, or Martian day, of its mission. They were acquired at Gusev Crater on Sol 26 (Jan. 29, 2004), showing that the camera's health remains excellent during Spirit's recovery. In the upper left is the MarsDial, Pancams calibration target. The monochrome image in the upper right shows the sun, magnified five times, as part of a routine sequence of images designed to monitor the dust abundance in the Martian atmosphere. The dust abundance appears to be decreasing slowly with time, consistent with the atmosphere continuing to clear after the large dust storm of last December.

The round, shallow depression in Fig. 27 resulted from history's first grinding of a rock on Mars. Spirit's rock abrasion tool ground off the surface of a patch 45.5 millimeters (1.8 inches) in diameter on a rock nicknamed Adirondack during the rover's 34th sol on Mars, Feb. 6, 2004. The hole is 2.65 millimeters (0.1 inch) deep, exposing fresh interior material for close inspection with the Microscopic Imager and two spectrometers on the robotic arm. This image was taken by Spirit's Pancam panoramic camera.

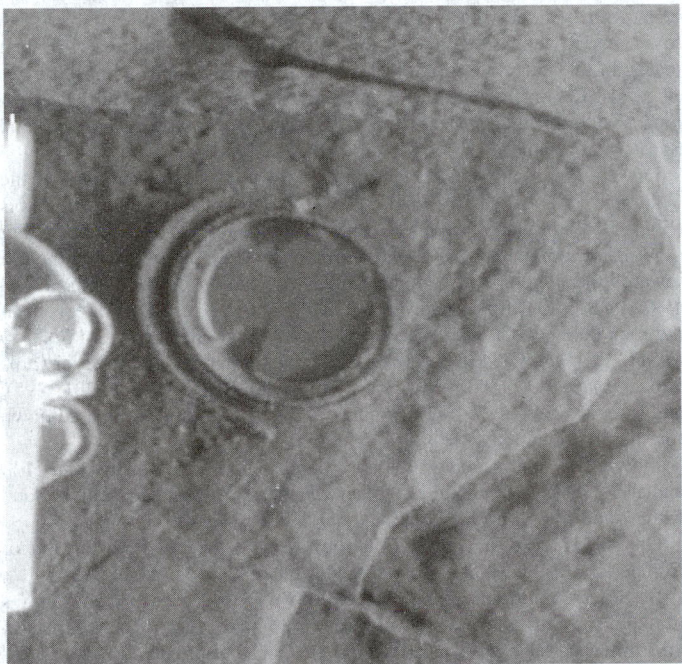

Fig. 27.   Shallow depression left after history's first grinding of a rock on Mars. (*Image courtesy of NASA/JPL/Cornell/Honeybee Robotics*)

In Fig. 28, Spirit's Pancam camera captured this image of the shallow depression dubbed *Laguna Hollow* before the rover drove into it to sample its bed of fine sediments. Scientists will use the hollow to study the atmospheric processes that shaped Mars because, in contrast to surrounding rocky terrain, it contains windblown dust and possibly salty clumps of soil. This image was taken on the 45th sol of Spirit's mission (Feb. 18, 2004).

To dig its first trench Spirit used its left front wheel to dig into the soil at *Laguna Hollow* and create a trench 7 centimeters (3 inches) deep. See Fig. 29. The soil at this location is more cohesive than the material where Spirit's twin, Opportunity, dug its first trench at Meridiani Planum. Scientists and engineers plan to begin up-close inspection of the soil in this trench (dubbed *Road Cut*) by placing the Microscopic Imager on the floor and the walls before conducting readings with the rover's Mössbauer and alpha particle X-ray spectrometers. This view is from Spirit's front Hazcam (hazard-avoidance camera). (Feb. 19, 2004).

Fig. 29.   Trench dug by Spirit. (Image courtesy of NASA/JPL.)

On February 26, 2004 Spirit's Navcam (navigational camera) took this image of a rock named *Humphrey* that stands approximately .6 meters (about 2 feet) tall and is one of the largest blocks of what scientists believe is ejected material from the crater dubbed *Bonneville*. It is likely a basaltic rock and its fractures are thought to have been caused by the impact as it was hurled from the crater to its current location. The name *Humphrey* was inspired by Humphries Peak — the tallest peak in Arizona. See Fig. 30.

On March 1, 2004 another interplanetary *first*, for the rock abrasion tool (RAT). The three dark circular areas on the rock *Humphrey* make up a rock abrasion tool mosaic, created by the tool's stainless steel brushes in about 15 minutes. See Fig. 31. Even though a triple brushing was never conducted in a testbed, the RAT's previous performance on the

Fig. 28.   Laguna Hollow. (*Image courtesy of NASA/JPL/Cornell.*)

Fig. 30.   A rock named *Humphrey*. (*Image courtesy of NASA/JPL.*)

Fig. 31.   Abrasion tool mosaic made by RAT on the rock dubbed "Humphrey." (*Image courtesy of NASA/JPL/Honeybee Robotics.*)

rock "Adirondack" convinced the science and engineering teams that it was fully capable of such an operation. A larger, dust-free area is needed so that the Mini-TES can conduct an assessment of the rock. This is an image from Spirit's Navcam (navigation camera) taken during the rover's 56th sol on Mars.

The first-ever microscopic look inside a drift on Mars was provided by Spirit's Microscopic Imager after the rover successfully dug into the side of the drift nicknamed *Serpent*. This view of the scuffed interior of the drift is dominated by larger, pea-shaped particles. These grains are not natural to the inside of the drift, but are crust particles that have tumbled into the scuffed area as a result of the digging. They were covered with a fine layer of dust when they were on the surface of *Serpent*, but lost their

dust cover in the process of falling into the scuff, giving scientists clues about the strength—or lack of strength—of the bond between the dust and sand particles at Gusev Crater.

On March 5, 2004, NASA announced that Spirit had found hints of water history on Mars in a rock dubbed "Humphrey" (Fig. 31). Dr. Ray Arvidson of Washington University, St. Louis, reported during a NASA press conference: "If we found this rock on Earth, we would say it is a volcanic rock that had a little fluid moving through it." In contrast to the rocks found by the twin rover Opportunity, this one was formed from magma and then acquired bright material in small crevices, which look like crystallized minerals. If this interpretation holds true, the minerals were most likely dissolved in water, which was either carried inside the rock or interacted with it at a later stage, after it formed.

On March 11, 2004, the rover reached Bonneville crater (Fig. 32) after a 400 yard journey. This crater is 150 yards across and about 30 yards deep. Scientists decided that it would be a bad idea to send the rover down into the crater, as they saw no targets of interest inside. Spirit drove along the southern rim, tore up some sand dunes, and continued to the southwest towards the Columbia Hills.

Spirit reached Missoula crater on Sol 105. The crater is roughly 100 yards across and 20 yards deep. Missoula crater was not considered a high priority target due to the older rocks it contained (Fig. 33). The rover skirted the northern rim, and continued to the southeast. It then reached Lahonten crater on Sol 118, and drove along the rim until Sol 120 (Fig. 34). Lahonten is about 60 yards across and about 10 yards deep. A long, snaking sand dune stretches away from its southwestern side, and Spirit went around it.

On Sol 159, Spirit reached the first of many targets at the base of the Columbia Hills called *West Spur*. Hank's Hollow was studied for 23 sols (Fig. 35). Within Hank's Hollow was the strange looking rock dubbed "Pot of Gold". From here, Spirit took a northerly path along the base of the hill towards the target Wooly Patch, which was studied from Sol 192 to Sol 199. By Sol 203, Spirit had driven southward up the hill and arrived at the rock dubbed "Clovis". Clovis was ground and analyzed from Sol 210 to Sol 225 (Fig. 36). Following Clovis came the targets of Ebenezer (Sols 226–235), Tetl (Sol 270) (Fig. 37), Uchben and Palinque (Sols 281–295), and Lutefisk (Sols 296–303). From Sols 244 to 255, Spirit powered down for solar conjunction.

Slowly, Spirit has made its way around the summit of Husband Hill, and at Sol 344 was ready to climb over the newly designated "Cumberland Ridge" and into "Larry's Lookout" and "Tennessee Valley".

On Sol 371, Spirit arrived at a rock named "Peace" near the top of Cumberland Ridge. Spirit ground it with the RAT tool on Sol 373.

By Sol 390 (Mid-February 2005), Spirit was advancing towards "Larry's Lookout", by driving up the hill backwards in reverse. The scientists at this time were trying to conserve as much energy as possible for the climb.

Spirit also investigated some targets along the way, including the soil target, "Paso Robles", which contained the highest amount of salt found on the red planet. The soil also contained a high amount of phosphorus in its composition, however not nearly as high as another rock sampled by Spirit, "Wishstone". Squyres said of the discovery, "We're still trying to work out what this means, but clearly, with this much salt around, water had a hand here".

On 9 March 2005 (probably during the Martian night), the rover's solar panel efficiency jumped from around 60% of what it had originally been to 93%, followed on 10 March by the sighting of dust devils. NASA scientists speculate a dust devil must have swept the solar panels clean, possibly significantly extending the duration of the mission. This also marks the first time dust devils had been spotted by either Spirit or Opportunity, easily one of the top highlights of the mission to date. Dust devils had previously been photographed by only the Pathfinder probe.

Spirit has provided substantial evidence in support of the mission's primary scientific goals: to search for and characterize a wide range of rocks and soils that hold clues to past water activity on Mars. Spirit has also obtained astronomical observations and atmospheric data.

For the latest updates on Spirit; See http://marsrovers.jpl.nasa.gov/mission/status spiritAll.html#sol1159

Opportunity Mission Highlights

On Jan. 24, 2004, halfway around the planet from Spirit's landing site, the airbags cushioning the Opportunity rover slammed into the surface,

Fig. 32.   Bonneville Crater. (Image courtesy of JPL/NASA)

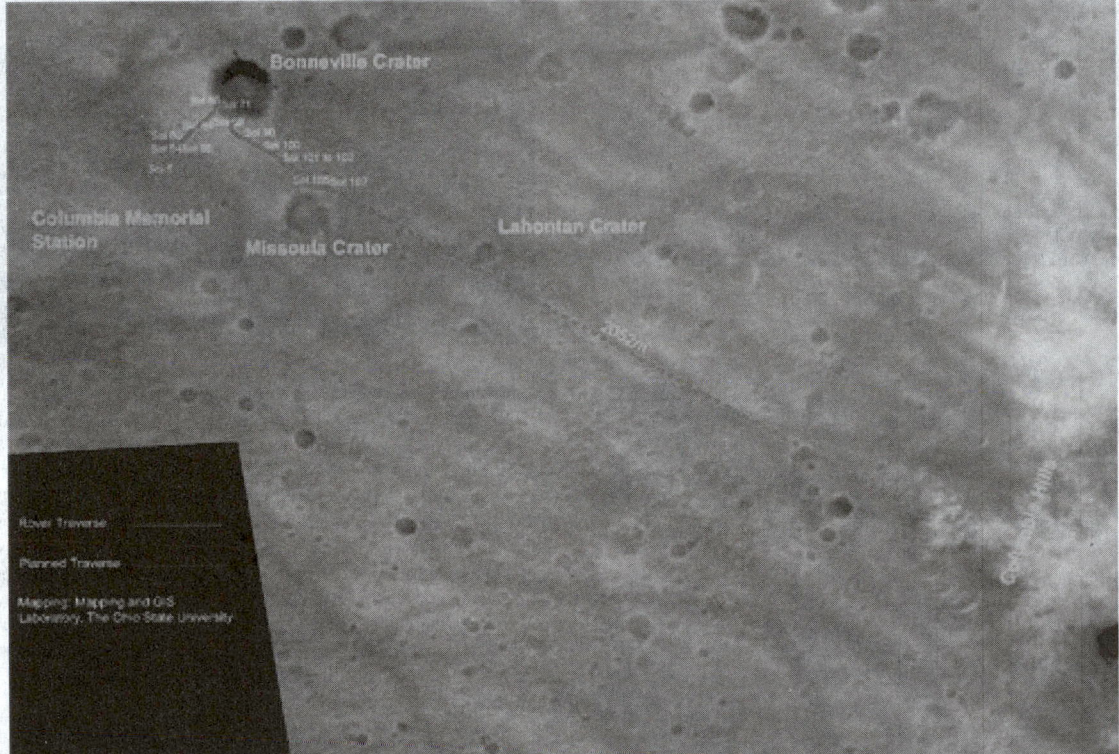

Fig. 33.   This overview map made from Mars Orbiter camera images illustrates the path that the Mars Exploration Rover Spirit has taken from its first sol on the red planet through its 107th sol. As of sol 112 (April 26, 2004), Spirit has passed "Missoula" crater and sits approximately 1,900 meters (1.18 miles) away from its destination at the western base of the "Columbia Hills." While most of Spirit's journey has been over the very angular rocks that make up the ejecta fields surrounding "Bonneville" crater, the rover's next 50 or so sols will be spent traversing over Martian plains that are dominated by rounder, vesicle-filled rocks (NASA).

bounced high into the atmosphere, hit the ground again, and continued bouncing and landing and bouncing and landing before slowly rolling to a halt. Mission engineers at NASA's Jet Propulsion Laboratory, Pasadena, Calif., received the first signal from Opportunity on the ground at 9:05 p.m. Pacific Standard Time Saturday via the NASA Deep Space Network, which was listening with antennas in California and Australia.

Opportunity landed in a region called Meridiani Planum (Fig. 37), halfway around the planet from the Gusev Crater site where its twin rover, Spirit, landed three weeks ago. By initial estimates, Opportunity landed about 24 kilometers (15 miles) down range from the center of the target landing area. Although Meridiani is a flat plain, without the rock fields seen at previous Mars landing sites, Opportunity rolled into an impact crater approximately 20 meters (65.5 feet) in diameter, with the rim of the crater approximately 10 meters (32 feet) from the rover. NASA Scientists were so excited about landing in a crater that they called the landing a "hole in one." Later, the crater was named Eagle crater. This was the darkest landing site ever visited by a spacecraft on Mars (Fig. 38). It would be two weeks before she was able to get a better look of her surroundings.

Mission scientists were intrigued by the abundance of rock outcrops

dispersed throughout the crater, as well as the crater's soil, which appeared to be a mixture of coarse gray grains and fine reddish grains. This sweeping look at the unusual rock outcropping near Opportunity was captured by the rover's panoramic camera. Scientists believe the seemingly layered rocks are either volcanic ash deposits or sediments laid down by wind or water. It was given the name Opportunity Ledge.

Geologists say that the layers — some no thicker than a finger — indicate the rocks likely originated either from sediments carried by water or wind, or from falling volcanic ash. "We should be able to distinguish between those two hypotheses", said Dr. Andrew Knoll of Harvard University, Cambridge, a member of the science team for Opportunity and her twin, Spirit. If the rocks are sedimentary, water is a more likely source than wind, he said. See also http://marsrovers.nasa.gov/newsroom/pressreleases/20040127a.html.

On February 19, 2004 the survey of "Opportunity Ledge" was declared successful. A specific target in the outcrop (dubbed "El Capitan"), whose upper and lower portions appeared to differ in layering and weathering characteristics, was selected for further investigation (Fig. 39). El Capitan, about 10 centimeters (4 inches) high, was named after a mountain in Texas.

Fig. 34.   Lahonten carter on Sol 120. (NASA)

Fig. 36.   At a rock called "Clovis," the rock abrasion tool on NASA's Mars Exploration Rover Spirit cut a 9-millimeter (0.35-inch) hole during the rover's 216th martian day, or sol (Aug. 11, 2004). The hole is the deepest drilled in a rock on Mars so far. This false-color view was made from images taken by Spirit's panoramic camera on sol 226 (Aug. 21, 2004) at around 12:50 p.m. local true solar time — early afternoon in Gusev Crater on Mars. To the right is a "brush flower" of circles produced by scrubbing the surface of the rock with the abrasion tool's wire brush. Scientists used rover's Mössbauer spectrometer and alpha particle X-ray spectrometer to look for iron-bearing minerals and determine the elemental chemical composition of the rock. This composite combines images taken with the camera's 750-, 530-, and 430-nanometer filters. The grayish-blue hue in this image suggests that the interior of the rock contains iron minerals that are less oxidized than minerals on the surface. The diameter of the hole cut into the rock is 4.5 centimeters (1.8 inches) (NASA).

Fig. 35.   This false-color composite panoramic camera image highlights mysterious and sparkly dust-like material that is created when the soil in this region is disturbed. NASA's Mars Exploration Rover Spirit took this image on sol 165 (June 20, 2004) in "Hank's Hollow," using filters L2, L5 and L7 (NASA).

Opportunity reached "*El Capitan*" on Sol 27, and took first picture of the rocks with her panoramic camera.

During a press conference on March 2 mission scientists discussed their conclusions about the bedrock and the evidence for the presence of liquid water during their formation. They presented the following reasoning to explain the small, elongated voids in the rock visible on the surface and after grinding into it (Fig. 40 and 41).

These voids are consistent with features known to geologists as "vugs". These are formed when crystals form inside a rock matrix and are later removed through erosive processes, leaving behind voids. Some of the features in this picture are "disk-like", which is consistent with certain types of crystals, notably sulfate minerals.

Scientists used the rover's microscopic imager and two spectrometers to look at the details of the freshly exposed, clean surfaces created by the rover's rock abrasion tool. Seeing beyond the veil of dust and coatings on the surface of the rock, scientists obtained the best views of the chemical composition of the areas. These data indicated that the rocks are made up of types of sulfate that could have only been created by interaction between water and martian rock.

The chemical make-up of the two holes is slightly different, giving scientists an inkling into the geologic history of this area. This history may help to explain the origin of the granular hematite found around the small crater cradling Opportunity and the "El Capitan" rock region.

The sulfates and the other chemicals found in the rocks at this location on Mars also occur on Earth, but only rarely. In places like Rio Tinto, Spain, similar minerals are forming today, and microorganisms live and thrive there.

Analyzing these two clean surfaces created by the rock abrasion tool proves that Mars had interactions between water and rock over extended amounts of time. Life on Earth is sustained by extended interaction

Fig. 37.   This image shows one of the Mars Exploration Rover Opportunity's first breathtaking views of the Martian landscape after its successful landing at Meridiani Planum on Mars. On the left, the rover's mast can be seen in a stowed position. Opportunity landed Saturday night at approximately 9:05 PST. The image was taken by the rover's navigation camera (NASA).

between water and the environment. The fact that scientists have now found evidence of a similar relationship between water and rock on Mars does not necessarily mean that life did develop on Mars, but it does bring the possibility one step closer to reality.

Opportunity's wheel tracks can be seen at the bottom left and right sides of this image. The tracks extend to the center of the image, indicating where Opportunity sat when it analyzed the rocks with the instruments on its robotic arm.

Additionally, mission members presented first data from the Mossbauer spectrometer taken at the bedrock site. The iron spectrum obtained from the rock *El Capitan* shows strong evidence for the mineral jarosite. This mineral contains hydroxyl radicals, which indicates the presence of water when the minerals were formed. Mini-TES data from the same rock showed that it consists of a considerable amount of sulfates.

On April 30, 2004 *Opportunity* reached Endurance crater (Fig. 42), which was known to have many layers of rocks. In May the rover circumnavigated the crater, and made observations with Mini-TES and the panoramic camera. The rock "Lion Stone" was investigated and found to be similar in composition to the layers found in Eagle crater.

On June 4, 2004 mission members announced their intention to drive Opportunity into Endurance, even if it should turn out to be impossible to get back out, targeting the various rock layers that were identified in the pictures from the crater rim. "This is a crucial and careful decision for the Mars Exploration Rovers' extended mission", said Dr. Edward Weiler, NASA's associate administrator for space science. Dr Squyres, principal investigator from Cornell University said: "Answering the question of what came before the evaporates is the most significant scientific issue we can address with Opportunity at this time."

A first drive into the crater was executed June 8 and Opportunity backed out again the same day. It was found that the angle of the surface was well inside the safety margin (about 18 degrees), and the full excursion towards the rock layer of interest was started. During Sols 134 (June 12), 135, and 137 the rover drove deeper and deeper into the crater, each time executing the drive exactly as planned.

Wispy clouds, similar to Earth's cirrus clouds, were spotted.

After exiting Endurance crater, in January 2005 Opportunity went to examine its own discarded heat shield (Fig. 43). While in the vicinity of the heat shield, it happened to come upon an object which was immediately suspected and soon confirmed to be a meteorite. The meteorite was promptly named Heat Shield Rock, and is the first meteorite identified on another planet or moon (Fig. 44).

After about 25 Sols of observations Opportunity headed south for a crater named Argo, nearly 300 m from the heat shield.

The rover was commanded to dig another trench on the vast plains of Meridiani Planum, on Sol 366, and observations continued until Sol 373 (February 10, 2005). The rover then passed the craters "Alvin" and "Jason", and by Sol 387, approached a "crater triplet" on its way to Vostok Crater. Along the way, Opportunity set a distance record for one-day travel by either rover: 177.5 meters (582 feet), on February 19, 2005. On Sol 389 (February 26, 2005), the rover approached one of the three craters, dubbed Naturaliste (Fig. 45). A rock target named "Normandy" was chosen for investigation on Sol 392, and Opportunity remained there until Sol 395.

Opportunity reached Vostok Crater on Sol 399, finding it mostly filled with sand and lined by outcrops. It was then ordered south into what has been called "etched terrain", to search for more bedrock.

On April 26 (Sol 446) Opportunity inadvertently dug itself into a sand dune: Mission scientists reported that images indicated all four corner wheels were dug in by more than a wheel radius, just as the rover attempted to climb over a dune about 30 centimeters (12 inches) tall. The sand dune was designated "Purgatory Dune" by mission planners.

Fig. 38.   This 360-degree panorama is one of the first images beamed back to Earth from the Mars Exploration Rover Opportunity shortly after it touched down at Meridiani Planum, Mars. The image was captured by the rover's navigation camera on Sol 1 of the mission, showing the interior of Eagle crater (NASA).

Fig. 39.   Close up of "El Capitan" in the "Opportunity Ledge" region on Mars. (JPL/NASA/Cornell.)

Fig. 41.   This navigation camera image taken by the Mars Exploration Rover Opportunity on the 36th Martian day, or sol, of its mission (March 1, 2004) shows the layered rocks of the "El Capitan" area near the rover's landing site at Meridiani Planum, Mars. Visible on two of the rocks are the holes drilled by the rover, which provided scientists with a window to this part of the red planet's water-soaked past.

Fig. 40.   Voids on bedrock on Mars. (JPL/NASA/Cornell/USGS.)

The rover's condition was simulated on Earth prior to any attempt to move, out of concern that the rover might become permanently immobilized. After various simulations intended to mimic the properties and behavior of Martian sand were completed, the rover executed its first wheel movements on May 13 (Sol 463), intentionally advancing only a few centimeters, after which mission members evaluated the results.

During Sol 465 and 466 more drive commands were executed, and with each trial the rover moved another couple of centimeters. At the end of each movement, panoramic images were acquired to investigate the atmosphere and the surrounding dune field (Fig. 46). The sand dune escape maneuver was successfully completed on June 4 (Sol 484), and all six wheels of Opportunity were on firmer ground. After studying "Purgatory" from sol 498 to sol 510, Opportunity proceeded southwards towards "Erebus crater".

Opportunity studied Erebus crater (Fig. 47), a large, shallow, partially buried crater and a stopover on the way south towards Victoria crater, between October 2005 and March 2006.

New programming to measure the percentage of slip in the wheels was successful in preventing the rover from getting stuck. Another

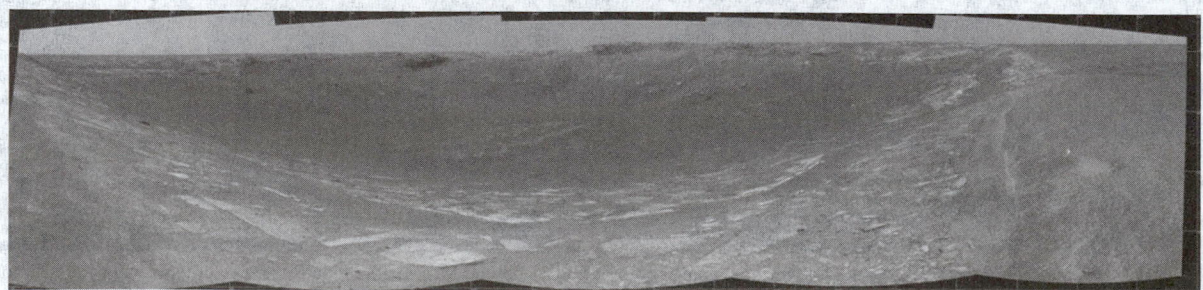

Fig. 42.   Opportunity's view of "Endurance Crater." The crater is about 130 meters (about 430 feet) in diameter. (NASA/JPL)

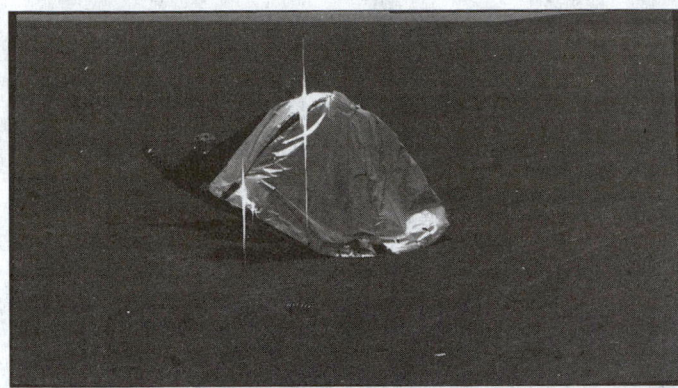

Fig. 43. Discarded heat shield from Mars rover Opportunity on the surface of Mars, imaged on Sol 335 of the mission, or January 2, 2005. Engineers were surprised to see that the heat shield had turned itself inside out (inserted). Note Heat Shield Rock is visible in the background just to the upper left of the heat shield, and two metal springs are also visible on the ground. The vertical streaks are imager artifacts caused by sunlight glinting off aluminium. (NASA/JPL/Cornell)

Fig. 46. Picture of the dug in back wheels of the Opportunity rover. Taken on Sol 468. (JPL/NASA)

"Purgatory" — like incident was averted on sol 603, when onboard slip check software stopped a drive after slip reached 44.5%. It proceeded over many ripples and 'half-pipes' taking photographs after each sol's journey.

On sol 628 (November 3, 2005) Opportunity woke up in the midst of a mild dust storm that lasted three days. The rover was able to drive in self protective auto-mode during the storm but could not take any post drive images. Less than three weeks later, another cleaning event cleared the dust off of the solar array so as to produce around 720 watt-hours (80% of max). On sol 649 (December 1, 2005), it was discovered the motor used to stow the robotic arm for travel was stalling. This problem took nearly two weeks to fix. Since then, the arm is only stowed for travel and is extended at night to save the arm from getting stuck.

Opportunity observed numerous outcroppings around "Erebus Crater." It also collaborated with ESA's Mars Express by using the miniature thermal emission spectrometer and panoramic camera, and took images of a transit across the sun by Phobos. On sol 760 (March 22, 2006), Opportunity began the journey to its next destination, "Victoria Crater."

"Victoria Crater" is a massive impact crater approximately 7 kilometers (4.4 miles) from the original landing site. Victoria's diameter is six times larger than Endurance crater. Scientists believe that rock outcrops along the walls of Victoria will yield more information about the geologic history of Mars, if the rover survives long enough to investigate them.

On Sol 951 (September 26, 2006) Opportunity reached the rim of "Victoria Crater" and transmitted the first substantial views of "Victoria Crater," including the dune field at the bottom of the crater (Fig. 48). The Mars Reconnaissance Orbiter recently photographed "Victoria Crater," (Fig. 49).

Fig. 44. NASA's Mars Exploration Rover Opportunity has found an iron meteorite on Mars, the first meteorite of any type ever identified on another planet. The pitted, basketball-size object is mostly made of iron and nickel. Readings from spectrometers on the rover determined that composition. Opportunity used its panoramic camera to take the images used in this approximately true-color composite on the rover's 339th Martian day, or sol (Jan. 6, 2005). This composite combines images taken through the panoramic camera's 600-nanometer (red), 530-nanometer (green), and 480-nanometer (blue) filters (NASA/JPL/Cornell).

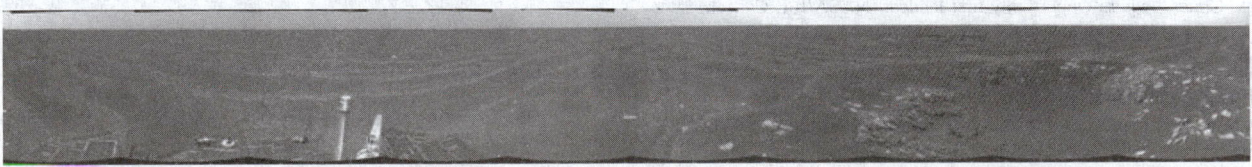

Fig. 45. Naturaliste Crater in Meridiani Planum (JPL/NASA)

Fig. 47.   This is a mosaic assembled from some of the images taken by the panoramic camera on NASA's Mars Exploration Rover Opportunity during the rover's 590th sol (Sept. 21, 2005). The view is toward the south and includes rock exposures north of "Erebus Crater," with the crater in the background. The rover will investigate the exposed rocks in the foreground and will take additional panoramic-camera images of Erebus Crater, which is about 300 meters (about 984 feet) across.

Erebus Crater dwarfs the landing-site crater, "Eagle Crater," which measures about 22 meters (72 feet) in diameter. And, it is nearly twice the diameter of "Endurance Crater," which, at 130 meters (430 feet) wide, has been compared to a stadium.

The camera's red filter was used for taking the images in this mosaic. It admits light with a wavelength of 750 nanometers. (NASA/JPL-Caltech/Cornell)

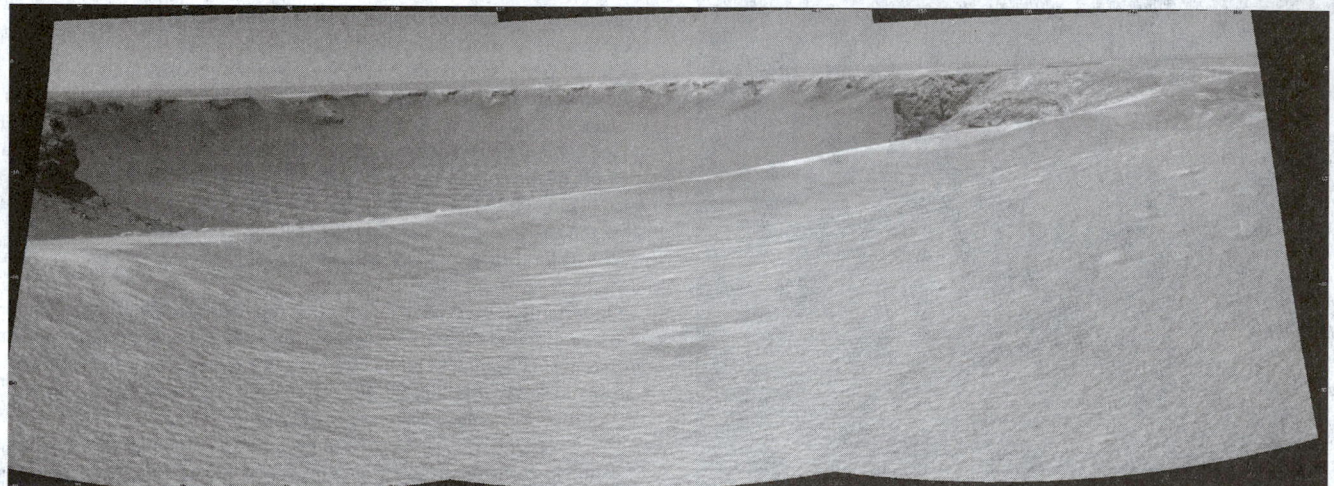

Fig. 48.   NASA's Mars rover Opportunity reached the rim of "Victoria Crater" in Mars' Meridiani Planum region with a 26-meter (85-foot) drive during the rover's 951st Martian day, or sol (Sept. 26, 2006). After the drive, the rover's navigation camera took the three exposures combined into this view of the crater's interior. This crater has been the mission's long-term destination for the past 21 Earth months.

A half mile in the distance one can see about 20 percent of the far side of the crater framed by the rocky cliffs in the foreground to the left and right of the image. The rim of the crater is composed of alternating promontories, rocky points towering approximately 70 meters (230 feet) above the crater floor, and recessed alcoves. The bottom of the crater is covered by sand that has been shaped into ripples by the Martian wind.

The position at the end of the sol 951 drive is about six meters from the lip of an alcove called "Duck Bay." The rover team planned a drive for sol 952 that would move a few more meters forward, plus more imaging of the near and far walls of the crater.

Victoria Crater is about five times wider than "Endurance Crater," which Opportunity spent six months examining in 2004, and about 40 times wider than "Eagle Crater," where Opportunity first landed. (NASA/JPL-Caltech)

The crater is about 800 meters (half a mile) in diameter. North is up. (NASA/JPL-Caltech/MSSS)

Opportunity has provided substantial evidence in support of the mission's primary scientific goals: to search for and characterize a wide range of rocks and soils that hold clues to past water activity on Mars. In addition to investigating the "water hypothesis", Opportunity has also obtained astronomical observations and atmospheric data.

For a month by month review of Opportunity; See http://marsrovers.jpl.nasa.gov/mission/wir/index.html

The Final TKO

So, when answering the frequently asked question, "when will the rovers stop working?" engineers and scientists can truly respond that it's up to the sun and the rovers. Spirit and Opportunity will likely never reach their "golden years," but a little creative "leapfrog" is helping them stay in the ring and stave off retirement.

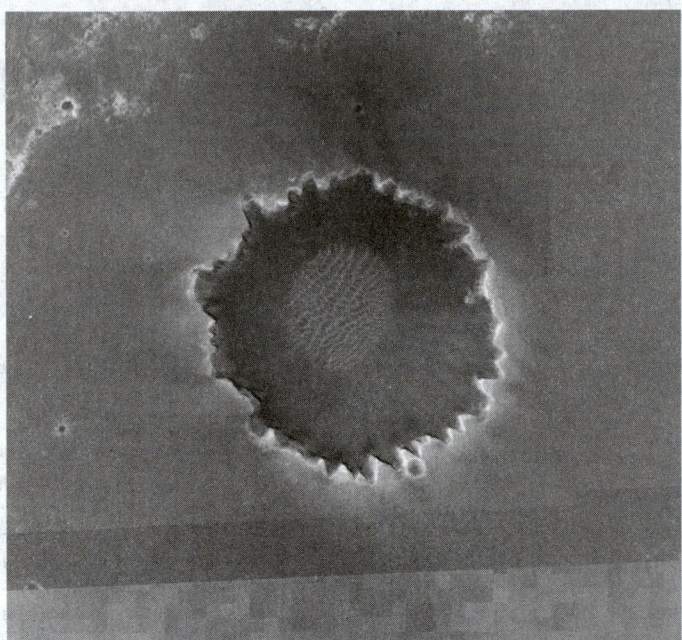

Fig. 49. This image from the Mars Orbiter Camera aboard NASA's Mars Global Surveyor spacecraft shows an overview of "Victoria Crater" and a portion of the area NASA's Mars Exploration Rover Opportunity has covered to reach the enormous depression.

Web References

Mars Exploration Rover Mission: http://marsrovers.jpl.nasa.gov/home/index.html

List of All Features 2001-2002: http://marsrovers.jpl.nasa.gov/spotlight/features-2002.html

List of All Features 2003: http://marsrovers.jpl.nasa.gov/spotlight/features-2003.html

List of All Features 2004: http://marsrovers.jpl.nasa.gov/spotlight/features-2004.html

List of All Features 2005: http://marsrovers.jpl.nasa.gov/spotlight/features-2005.html

List of All Features 2006: http://marsrovers.jpl.nasa.gov/spotlight/index.html

Press Release Images Spirit 2004: http://marsrovers.jpl.nasa.gov/gallery/press/spirit/2004.html

Press Release Images Spirit 2005: http://marsrovers.jpl.nasa.gov/gallery/press/spirit/2005.html

Press Release Images Spirit 2006: http://marsrovers.jpl.nasa.gov/gallery/press/spirit/2006.html

Press Release Images Spirit 2007: http://marsrovers.jpl.nasa.gov/gallery/press/spirit/index.html

Press Release Images Opportunity 2004: http://marsrovers.jpl.nasa.gov/gallery/press/opportunity/2004.html

Press Release Images Opportunity 2005: http://marsrovers.jpl.nasa.gov/gallery/press/opportunity/2005.html

Press Release Images Opportunity 2006: http://marsrovers.jpl.nasa.gov/gallery/press/opportunity/2006.html

Press Release Images Opportunity 2007: http://marsrovers.jpl.nasa.gov/gallery/press/opportunity/index.html

Jet Propulsion Laboratory (JPL), Pasadena, CA

National Aeronautics And Space Administration (NASA),
Washington, DC

MARS EXPRESS. See **Mars Exploration Program**; and **Space Science Missions: Solar System**.

MARS GLOBAL SURVEYOR (MGS) MISSION. NASA's Mars Global Surveyor (MGS) orbiter is the oldest Mars spacecraft currently in operation, (circa 2006) and it has been studying the red planet for nearly a decade. The Mars Global Surveyor was the first successful U.S. mission launched to Mars since the Viking mission in 1976. After a 20-year absence at the planet, the Mars Global Surveyor ushered in a new era of Mars exploration with its five science investigations. The Mars Global Surveyor orbiter launched on November 7, 1996, and began orbiting Mars on September 11, 1997 (September 12, UTC), and has since returned more scientific data about the evolution of the red planet than all other previous Mars missions combined. Its camera has returned more than 240,000 images to Earth.

The Mars Global Surveyor pioneered the use of *aerobraking*, using careful dips into the atmosphere for friction to shrink a long elliptical orbit into a nearly circular one. The mission then started its primary mapping phase in April 1999. The original plan was to examine the planet for one Mars year, nearly two Earth years. Based on the value of the science returned by the spacecraft, NASA has extended its mission four times.

The spacecraft's longevity is a testament to innovative engineering, and it has enabled scientists to develop an unprecedented understanding of how the Martian landscape and seasons have been changing over time. When Mars Global Surveyor launched, it carried with it a limited amount of propellant, and the spacecraft was expected to run out of fuel in April of 2003, ending the capability to steer the ship with small engine thrusters and point the science instruments accurately.

In August of 2001, the Mars Global Surveyor team implemented a clever *angular momentum management* strategy that minimized the need for thrusters to help stabilize and balance the spacecraft and keep it on target over viewing areas on the surface of Mars. Part of the angular momentum management scheme tips the spacecraft at an angle 16 degrees backwards (instead of pointing straight down at the surface), which effectively cuts the use of fuel by 800% on a daily basis.

Mars Global Surveyor circles in a polar orbit (traveling over the North Pole to the South Pole and back to the North Pole) once every two hours, twelve times a day, collecting global "snapshots" from 400 kilometers (249 miles) above the Martian surface. Long past Mars Global Surveyor's life expectancy, scientists are continuing to glean valuable new information on daily and seasonal weather patterns, geological features and the migration of water vapor from hemisphere to hemisphere over time. The spacecraft's laser altimeter has given scientists their first 3-D views of the striking topography of Mars. Mars Global Surveyor's suite of science instruments includes a high-resolution camera and a mineral detector that have helped engineers and scientists select safe landing sites rich in various minerals like hematite, a mineral often formed in liquid water, that help rover and landed missions search for more clues about the history of water on Mars.

In a landmark discovery, imaging scientists have seen gullies and debris flow features that suggest there could be current sources of liquid water at or near the surface of Mars. While all current Mars missions aim to understand the role of water on Mars to unlock the mystery of whether life could have developed on Mars in the past or may survive in the subsurface now or in the future, Mars Global Surveyor also has been able to characterize the topography, gravity, magnetic fields, thermal properties, surface composition, and atmosphere of Mars.

Mar Global Surveyor has fulfilled all of its science objectives and continues to evolve with the Mars Exploration Program. See also **Mars Exploration Program**. Principal goals for the orbiter's latest mission extension include continued weather monitoring to form a continuous set of observations with NASA's Mars Reconnaissance Orbiter, scheduled to reach the red planet in March 2006; imaging of possible landing sites for the Phoenix 2007 Mars Scout lander and 2009 Mars Science Laboratory rover; continued mapping and analysis of key sedimentary-rock-outcrop sites; and continued monitoring of changes on the surface due to wind and ice. Because the narrow-angle camera has imaged only a small fraction of the surface, new discoveries about surface features are likely to come at any time.

The orbiter's current mission extension has a budget of $7.5 million per year. The Mars Global Surveyor mission cost $150 million to develop and build and $65 million to launch.

Technology

Technology makes space exploration possible. Each Mars mission is part of a continuing chain of innovation: each relies on past missions to identify needed new technologies and each contributes its own innovations to benefit future missions. This chain allows NASA to continue to push the boundaries of what is currently possible, while relying on proven technologies.

The Mars Global Surveyor team implemented four innovative procedures that have enabled the Mars Global Surveyor spacecraft to return more science data than all previous Mars missions combined, including the highest-resolution images from Mars thus far, and the longest-lived active Mars mission.

Propulsion. Two innovations have minimized propulsion usage, allowing Mars GlobalSurveyor to successfully operate at Mars for more than eight years, conducting long-term studies all the while.

Angular Momentum Management Plan. As Mars Global Surveyor orbits the red planet, the force of gravity from the uneven terrain below tugs and pushes the spacecraft. Mountains and valleys cause the spacecraft to fly in a non-circular orbit at slightly varying altitudes. The effect is similar to turbulence in an airplane.

To stabilize the spacecraft above the Martian atmosphere (but within the pull of gravity), the spacecraft uses three reaction wheels, or "flywheels." The reaction wheels spin much like the whirling "tea cups" on a child's amusement ride. If a girl inside the teacup wants to make the teacupspin clockwise, she grabs the wheel inside the teacup and pulls counterclockwise. If a boy inside the same teacup wants to slow the teacup down, he grabs the wheel and pulls it clockwise. On the spacecraft, if the gravity of a mountain pulls the spacecraft, the reaction wheels spin and pull the spacecraft back, keeping it balanced.

Early in the mission, Mars Global Surveyor used small thrusters to adjust the reaction wheels and keep the spacecraft from drifting off course or off balance. Now, with the angular momentum management plan, the spacecraft balances itself naturally, without the need to use as much fuel to fight the external forces of gravity. For example, instead of pointing straight down, the spacecraft is tilted backward 16 degrees, which helps distribute the gravitational pull of the planet more evenly on the body of the spacecraft and cuts fuel consumption by 800 percent.

Aerobraking. To reduce the mass and expense of fuel needed for the mission, the Mars Global Surveyor team used a braking technique called aerobraking to trimthe spacecraft's initial, highly elliptical orbit into a nearly circular orbit after arriving at Mars. The aerobraking technique eliminated the need for 1500 kilograms (3,300 pounds) of braking propellant during the 700-million-kilometer (435-million-mile) interplanetary journey to Mars. The lighter weight reduced the size of the launch vehicle required from a Titan III to a Delta rocket, saving an additional $250 million.

The Magellan spacecraft at Venus was the first planetary spacecraft to use aerobraking, in a demonstration in the summer of 1993. Its success cleared the way for the use of aerobraking by Mars Global Surveyor. And the success of Mars Global Surveyor's aerobraking paved the way for 2001 Mars Odyssey and Mars Reconnaissance Orbiter to use the same technique.

Remote Science Instrumentation. Two other innovations changed how the spacecraft is flown to increase the ability to collect science data from Mars.

Image Motion Compensation. Since 2003, the camera and spacecraft teams for Mars Global Surveyor have perfected a technique that allows the entire spacecraft to roll so that the camera can scan surface details at three times higher resolution than if the spacecraft did not roll. The image motion compensation technique adjusts the rotation rate of the spacecraft to match the ground speed under the camera.

The Mars Orbiter Camera acquires the highest-resolution images ever obtained from a Mars-orbiting spacecraft. During normal operating conditions, when the spacecraft does not roll, the smallest objects that can be resolved on the Martian surface are about 4 to 5 meters (13 to 16 feet) across. With the adjusted-rotation technique, called "compensated pitch and roll targeted observation" or "CPROTO," objects as small as 1.5 meters (4.9 feet) are visible in images from the same camera. Resolution capability of 1.4 meters (4.6 feet) per pixel is improved to one-half meter (1.7 feet) per pixel. Because the maneuvers are complex and the amount of data that can be acquired is limited, only 150 images of this new type have been taken so far.

Beta Supplement. A vital component of collecting science data is the ability to send the information to Earth. The High Gain Antenna is a dish antenna that send sand receives data at high rates. To communicate with NASA's Deep Space Network antennas, it must point toward Earth. See also **Deep Space Network**.

In 1997, after Mars Global Surveyor got into orbit around Mars, investigators realized that the High Gain Antenna's range of motion was surprisingly limited. When the spacecraft was in certain positions, they were unable to adjust the antenna to point it directly toward Earth. The team discovered that the problem stemmed from a loose screw that most likely vibrated out of position during launch.

In order to maximize science return, the team figured out a way to flip-flop the antenna during every 2-hour orbit. This flip-flop, officially named the "Beta Supplement," provides two 25-minute intervals for communicating with Earth during each orbit. The flip-flop or rewind of the antenna is a complex operation, as can be seen in the Beta Supplement Animation.

While the antenna must point toward Earth for communications, the solar panels must point toward the Sun for energy, and the science instruments must point down toward Mars to collect data. The Mars Global Surveyor team continues to creatively twist and turn the spacecraft during each revolution about Mars to accommodate all of these needs.

For more information on Mars Technology; see http://marstech.jpl.nasa.gov/.

Spacecraft

Mars Global Surveyor is a well-designed machine that has been successfully operating since launch on November 7, 1996. The Mars Global Surveyor team has been working creatively to conserve fuel and invigorate the aging spacecraft. The engineering and science teams have taken many preventative measures to minimize risk, limit use of resource consumables, and keep the spacecraft in good health for gathering data that yield clues into the mysteries of Mars.

The spacecraft, fabricated at the Lockheed Martin Space Systems plantin Denver, Colorado, looks like a rectangular-shaped box with wing-like projections extending from opposite sides. The body (or bus) houses the computers, radio system, solid-state data recorders, fuel tanks, and other equipment. Attached to the outside of the bus are several rocket thrusters, which were fired to adjust the spacecraft's path during cruise to Mars and to modify the spacecraft orbit around the planet.

When fully loaded with propellant at the time of launch, the vehicle weighed 1,060-kilograms (2,342 pounds). The spacecraft is about 3 meters (10 feet) tall with its braking engine and instruments. The bus or main body of the spacecraft measures 1.2 by 1.2 meters (4 by 4 feet) and is 12 meters (40 feet) across from tip to tip when the solar panels are fully unfolded. The high-gain antenna is deployed on a 2-meter-long (6-1/2-foot) boom. See Fig. 1.

To minimize costs, spare units left over from the [1]Mars Observer mission were used in portions of the spacecraft's electronics and for some of the science instruments. The spacecraft design also incorporated new hardware—the radio transmitters, solid-state recorders, propulsion system, and composite material bus structure-and retains many backup and redundant features of the original Mars Observer design in case of failure of critical elements such as the primary processors, recorders or transmitters.

The solar arrays, which always point toward the Sun (when the spacecraft isn't behind the planet), provide 980 watts of electricity for operating the electronic equipment and for charging nickel hydrogen batteries. The batteries provide electricity when the spacecraft is mapping the dark side of Mars. To maintain appropriate operating temperatures, most of the outer exposed parts of the spacecraft, including the science instruments, are wrapped in thermal blankets.

Spacecraft communications with Earth utilize X-band frequencies for radio tracking, return of science and engineering telemetry, commanding, and the radio science experiments. Primary communications to and from the spacecraft occur through the 1.5-meter-diameter (4.9-foot) high-gain

---

[1] After a 17-year gap since its last mission to the red planet, the United States launched Mars Observer on September 25, 1992. The spacecraft was based on a commercial Earth-orbiting communications satellite that had been converted into an orbiter for Mars. The payload of science instruments was designed to study the geology, geophysics and climate of Mars.

The mission ended with disappointment on August 22, 1993, when contact was lost with the spacecraft shortly before it was to enter orbit around Mars. Science instruments from Mars Observer are being reflown on two other orbiters, Mars Global Surveyor and 2001 Mars Odyssey.

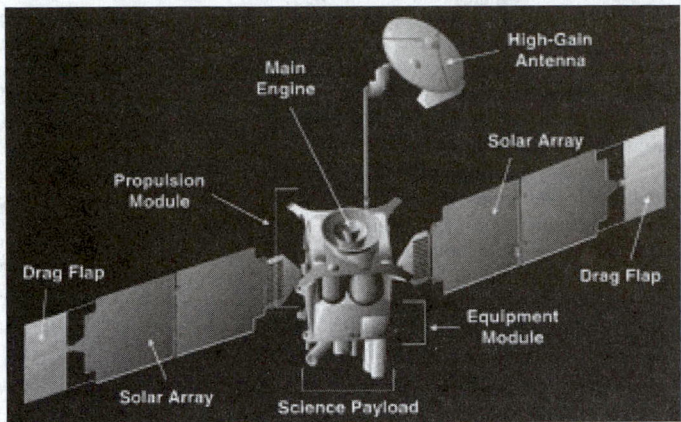

Fig. 1.   Drawing of Mars Global Surveyor. (*Image courtesy of NASA/JPL.*)

antenna. Mars Global Surveyor can receive instructions from Earth at a maximum rate of 12.5 commands per second.

Subsystems. The spacecraft has several major subsystems.

Attitude and Articulation Control Subsystem (AACS). Provide Pointing Control of The Spacecraft during All Mission Phases, both Nominal and Off Nominal, and Attitude Knowledge of the Spacecraft Orientation Relative to Fixed Reference Frames. Also provides for articulation of the Solar Arrays and the High Gain Antenna.

Design Features

- 3-Axis Stabilized
- Celestial Sensor, Horizon Sensor, 4-pi Sun Sensors, and Inertial Measurement Unit Used in Various Combinations to Determine Attitude in Desired Reference Frames
- Fine Fine Pointing Control Using Reaction Wheels
- Thrusters Used for Large Torque Capability, Reaction Wheel Momentum Unloading, Orbit Maintenance, Backup Attitude Control and Safe Mode
- Autonomous Periapsis Raise Maneuver Implemented in Safe Mode for Aerobraking
- Extensive Reuse of MO Hardware and Algorithms

Command and Data Handling. Executes Real-time Hardware Commands, Provides for S/C Command Measurement Interfaces, Provides for On-board Processing, Provides for Engineering and Science Data Storage Playback

Design Features

- Marconi 1750A Processors for SCP and EDF, 8086 Processor for PDS
- SCP RAM Increased to 128K
- 20 K PROM for Safe Mode
- 4 –.75 Gb Solid State Recorders Based on Clementine Design to Save Mass
- Extensive Reuse of Mars Observer Spares

Electrical Power Subsystem. Two solar arrays, each 3.53 meters (11.6 feet) long by 1.85 meters (6.1 feet) wide will gather power from the Sun to generate electricity. Each array mounts to the main body of the spacecraft close to the attachment point between the equipment and propulsion modules. Rectangular-shaped, metal flaps attached at the ends of both arrays add another 81.3 centimeters (32 inches) to the overall structure. These "drag flaps" serve no purpose other than to increase the spacecraft's susceptibility to air resistance when it flies through the Martian atmosphere to lower its orbit.

Each array consists of two panels, an inner and outer panel comprised of gallium arsenide and silicon solar cells, respectively. During mapping operations at Mars, the amount of power produced by the arrays will vary from a high of 980 Watts when Mars is closest to the Sun, to a low of about 660 Watts when Mars is farthest from the Sun. In order to understand the difficulty of designing and operating an interplanetary spacecraft, consider the fact that 980 Watts is less power than that used by an ordinary hair dryer.

While in orbit around Mars, the solar arrays will provide power as Surveyor flies over the day side of the planet. When the spacecraft passes over the night side, energy will flow from two nickel-hydrogen (NiH2) batteries to compensate for the temporary loss of power from the solar arrays. These two batteries can provide power for over an hour before requiring a recharge from the solar arrays.

Propulsion Subsystem. Surveyor's main rocket engine nozzle sits outside the propulsion module on the end opposite the equipment module. When fired during major maneuvers, this engine will provide a thrust of 659 Newtons. If applied at sea level, that amount of force would be strong enough to suspend a 148 pound person in mid-air.

The main engine derives its power by burning a combination of nitrogen tetroxide (N2O4) and hydrazine (a derivative of N2H4). Engineers working on the project call this chemical combination "hypergolic." This term refers to the fact that the two chemicals will spontaneously combust when they come into contact with each other. In other words, no spark is needed to ignite the engine.

In addition to the main engine, Surveyor also carries small "attitude-control" thrusters attached to each corner of the propulsion module. Each one of these tiny rocket engines will only provide 4.45 Newtons of thrust. Their main purpose involves performing small course correction changes and keeping the spacecraft from wobbling out of control during main-engine burns. In total, Surveyor will carry 385 kilograms (849 pounds) of propellant. Nearly 75% of that amount will be expended during the MOI burn to allow the spacecraft to slow down and enter orbit around Mars.

Relay Antenna. Of the six main instruments that will fly to Mars on Surveyor, the Mars Relay is the only one not designed to take scientific measurements. Instead, this cylindrical-shaped antenna will focus its efforts on collecting data transmitted to Surveyor from landers on the Martian surface. After collecting the data, Surveyor will transmit the data back to Earth. The advantage of using Surveyor as a relay satellite for Mars landers is that the lander spacecraft will not need to carry a large antenna to talk with the Earth. Instead, the reduction in the weight of the lander from not carrying a large antenna can be used for more scientific instruments.

The relay operates at a UHF frequency of 437.1 megaHertz and can listen to stations on the Martian surface up to 5,000 km (3,125 miles) away from Surveyor. In December 1999, NASA will use the relay antenna for the first time. That mission is scheduled to land a scientific package near the South Pole of Mars for a three-month study.

The Centre Nationale d'Estudies Spatiales in France provided the relay antenna to NASA as part of an international cooperation toward the study of Mars.

Telecommunications Subsystem. In order to communicate with the Earth, Surveyor will use a 1.5-meter (59-inch) diameter high-gain antenna dish that sits at the end of a 2.0 meter (6.6 foot) long boom attached to the propulsion module. Two rotating joints, called gimbals, hold the antenna to the boom. The gimbals will allow the antenna to automatically track and point at the Earth while the science instruments observe Mars.

One of the high-gain antenna's two main functions will involve receiving command programs sent by the flight operations team on Earth. These programs, called command sequences, will contain instructions to tell Surveyor what functions to perform for time periods up to seven weeks in duration. During commanding periods, data will flow to Surveyor at a rate up to 750 commands per minute (500 bits per second).

The antenna's other main function involves sending data back to Earth. All transmissions broadcast by Surveyor will utilize X- band radio signals near 8.4 gigaHertz. That frequency would be equivalent to 8,400 on an FM receiver if ordinary radio dials reached that high. To complicate matters, Surveyor's radio transmitter will broadcast with a power of only 25 Watts. By the time the signal crosses millions of kilometers of space to reach the Earth, the signal strength will diminish to less than one millionth of one billionth of a Watt. How small is that number? If one could gather energy at that rate to charge a battery, it would take 30 million years to store enough charge to run a wrist watch for one second.

In order to receive the faint signal transmissions sent from Mars by Surveyor, NASA will utilize the gigantic tracking antennas of the Deep Space Network. These 34-meter (112-foot) diameter antennas are located in the Mojave Desert, Spain, and Australia. The large size of these Earth-based tracking antennas will allow Surveyor to transmit scientific data at rates up to 85,333 bits per second. In contrast, the average home computer modem functions at 14,400 bits per second.

Science Instruments

Mars Global Surveyor is a global mapping mission, carrying a suite of science instruments designed to study the entire Martian surface, atmosphere, and interior. Mars Global Surveyor provides a global portrait of how Mars looks today, and helps scientists better understand the history of Mars' evolution. With this information, humans have a better understanding of the history of all the inner planets of the solar system, including Earth.

MOC (Mars Orbiter Camera). This camera produces a daily wide-angle image of Mars similar to weather photographs of the Earth, and takes narrow angle images. See Fig. 2.

Fig. 2. The Mars Global Surveyor (MGS) Mars Orbiter Camera (MOC). MOC consists of 3 cameras: a narrow angle camera that obtains grayscale (black-and-white) high resolution images (typically 1.5 to 12 m/pixel) and red and blue wide angle cameras for context (240 m/pixel) and daily global imaging (7.5 km/pixel).

- The Mars Orbiter Camera provides a continuous record of Martian weather and supports the Mars Reconnaissance Orbiter as it flies toward Mars and enters the aerobraking phase.
- The narrow-angle lens captures images of objects as small as 0.5 meters (1.7 feet) across. These ultra-high-resolution images have been achieved using a technique that adjusts the rotation of the spacecraft to match the ground speed under the camera. These observations look for changes in surface features over time and help characterize the properties of fluids involved in gullies identified on Mars.
- The Mars Orbiter Camera continues to monitor the dynamic seasonal processes and will continue to support the landing site selection process for current and future rover and lander missions.
- In addition to semi-annual releases of large collections of archived pictures (250,000 images so far), the Mars Orbiter Camera team posts a new image daily, and solicits public suggestions for camera targets on Mars.

Principal investigator is Dr. Michael Malin, Malin Space Science Systems Inc., San Deigo, CA. See also http://www.msss.com/.

MOLA (Mars Orbiter Laser Altimeter). This experiment measured the height of Martian surface features like mountains and depths of valleys.

The Mars Orbiter Laser Altimeter has created the most accurate global topographic map of any planet in the solar system, giving scientists elevation maps precise to within about 30 centimeters (1 foot) in the vertical dimension. Data from the laser altimeter establishes pathways for the flow of past water and the locations, sizes, and volumes of watersheds. It also detects the heights of clouds and identifies dynamical features in the atmosphere, such as gravity waves. The Mars Orbiter Laser Altimeter shows seasonal changes in the height of the martian surface (aka, snow depth) that represented the first direct global measurement of the amount and distribution of condensed carbon dioxide.

In June 2001, part of the laser reached the end of its life, but a sensor continues to detect changes in surface brightness in the near infra-red part of the spectrum. These data provide evidence of cloud coverage and atmospheric variations.

Principal investigator is Dr. David Smith, NASA Goddard Space Flight Center. See also http://ltpwww.gsfc.nasa.gov/tharsis/mola.html.

TES (Thermal Emission Spectrometer). This instrument studies the atmosphere and maps the mineral composition of the surface by analyzing infrared radiation, which scans for heat emitted from the surface of Mars.

The Thermal Emission Spectrometer discovered new mineralogical and topographic evidence that suggest Mars had abundant water and thermal activity in its early history. Data indicate clear evidence of an ancient hydrothermal system, implying that water was stable at or near the surface and that a thicker atmosphere existed in Mars' early history. Measurements show an accumulation of the mineral hematite, a mineral that typically originates in standing bodies of water, near the Martian equator. This deposit of hematite, and knowledge of other minerals on the surface of Mars helped scientists direct the Mars Exploration Rovers to the Meridiani Planum and Gusev Crater.

Principal investigator is Dr. Phil Christensen, Arizona State University. See also http://tes.asu.edu/.

Magnetometer (Electron Reflectometer). The magnetometer studies the magnetic properties of Mars to gain insight into the interior of the planet and better understand the early history and evolution of Mars.

The magnetometer discovered the planet's magnetic field is not globally generated in the planet's core, but is localized in small, particular areas of the crust. Multiple magnetic differences were detected at various points on the planet's surface, indicating that magma solidified as it came up through the crust and cooled very early in Mars' evolution. The localized magnetic fields create a new model for the way in which the solar wind interacts with the planet (in contrast to the way the solar wind interacts with planets like Earth and Jupiter, which have strong magnetospheres).

Principal investigator is Dr. Mario Acuna, NASA Goddard Space Flight Center. See also http://mpfwww.jpl.nasa.gov/mgs/sci/mag/data1/mag_first.html.

Radio Science (Gravity Field Experiment). This radio science investigation uses data provided by the spacecraft's telecommunications system, high-gain antenna, and onboard ultra-stable oscillator (an ultra precise clock) to map variations in the gravity field. These measurements also enable scientists to determine the atmospheric pressure at specific locations as the spacecraft sends its signal through the atmosphere while disappearing behind the planet and re-emerging every orbit.

Dr. G. Leonard Tyler, of Stanford University, is the radio science team leader.

See also http://mpfwww.jpl.nasa.gov/mgs/sci/uso/uso.html.

NASA's Mars Global Surveyor May Be at Mission's End

On November 22, 2006 NASA's Mars Exploration Rover Opportunity did not detect any signal from the Mars Global Surveyor orbiter on Wednesday during an attempt to get the orbiter to transmit to the rover.

NASA's Mars Global Surveyor has likely finished its operating career. The spacecraft has served the longest and been the most productive of any mission ever sent to the red planet.

"Mars Global Surveyor has surpassed all expectations," said Michael Meyer, NASA's lead scientist for Mars exploration at NASA Headquarters, Washington. "It has already been the most productive science mission to Mars, and it will yield more discoveries as the treasury of observations it has made continues to be analyzed for years to come." Its camera has returned more than 240,000 images to Earth.

The orbiter has not communicated with Earth since November 2, 2006. Preliminary indications are that a solar panel became difficult to pivot, raising the possibility that the spacecraft may no longer be able to generate enough power to communicate. Engineers are also exploring other possible explanations for the radio silence.

"Realistically, we have run through the most likely possibilities for re-establishing communication, and we are facing the likelihood that the amazing flow of scientific observations from Mars Global Surveyor is over," said Fuk Li, Mars Exploration Program manager at NASA's Jet Propulsion Laboratory, Pasadena, Calif. "We are not giving up hope, though."

Efforts to regain contact with the spacecraft and determine what has happened to it will continue. NASA's newest Mars spacecraft, the Mars Reconnaissance Orbiter, pointed its cameras toward Mars Global Surveyor on Monday. "We have looked for Mars Global Surveyor with the star tracker, the context camera and the high-resolution camera on Mars Reconnaissance Orbiter," said Doug McCuistion, Mars Exploration

Program director at NASA Headquarters. "Preliminary analysis of the images did not show any definitive sightings of a spacecraft."

Mars Global Surveyor launched on Nov. 7, 1996, and began orbiting Mars on Sept. 11, 1997. It pioneered the use of *aerobraking* at Mars, using careful dips into the atmosphere for friction to shrink a long elliptical orbit into a nearly circular one. The mission then started its primary mapping phase in April 1999. The original plan was to examine the planet for one Mars year, nearly two Earth years. Based on the value of the science returned by the spacecraft, NASA extended its mission four times.

"It is an extraordinary machine that has done things the designers never envisioned despite a broken wing, a failed gyro and a worn-out reaction wheel. The builders and operating staff can be proud of their legacy of scientific discoveries and key support for subsequent missions," said Tom Thorpe, project manager for Mars Global Surveyor at JPL.

The spacecraft evaluated landing sites for the twin NASA rovers that landed in 2004 and sites for future landings of the Phoenix and Mars Science Laboratory missions. It monitored atmospheric conditions during aerobraking by newer orbiters. It served as a relay link for the rovers and provided mapping information about their surroundings.

"When we watched the launch 10 years ago, we wondered if we would make the specified mission length. We certainly were not thinking of a 10-year operating life," said JPL retiree Glenn Cunningham, who managed the Global Surveyor project through development and launch.

A few of the mission's many important discoveries about Mars include:

- The spacecraft's camera found gullies cut into many slopes that have few, if any, impact craters. This indicates the gullies are geologically young. Scientists interpret this as evidence of action by liquid water, essentially in modern times.
- The mineral-mapping infrared spectrometer found concentrations of a mineral that often forms under wet conditions, fine-grained hematite. This discovery led to selection of a hematite-rich region as the landing site for NASA's Mars Exploration Rover Opportunity.
- Laser altimeter measurements have produced an unprecedented global topographic map of Mars. The instrument revealed a multitude of highly eroded or buried craters too subtle for previous observation, and mapped canyons within the polar ice caps.
- The magnetometer found localized remnant magnetic fields, indicating that Mars once had a global magnetic field like Earth's, shielding the surface from deadly cosmic rays.
- The camera found a fan-shaped area of interweaving, curved ridges interpreted as evidence of an ancient river delta resulting from persistent flow of water over an extended period in the planet's ancient past.
- A long life allowed Global Surveyor to track changes through repeated annual cycles. For three Martian summers in a row, deposits of carbon-dioxide ice near Mars' South Pole shrunk from the previous year's size, suggesting a climate change in progress.

On January 10, 2007 NASA has formed an internal review board to look more in-depth into why NASA's Mars Global Surveyor went silent in November 2, 2006 and recommend any processes or procedures that could increase safety for other spacecraft.

The preliminary report by the internal review board was released on April, 13, 2007 and is as follows. "The loss of the spacecraft was the result of a series of events linked to a computer error made five months before the likely battery failure," said board Chairperson Dolly Perkins, deputy director-technical of NASA Goddard Space Flight Center, Greenbelt, Md.

On Nov. 2, after the spacecraft was ordered to perform a routine adjustment of its solar panels, the spacecraft reported a series of alarms, but indicated that it had stabilized. That was its final transmission. Subsequently, the spacecraft reoriented to an angle that exposed one of two batteries carried on the spacecraft to direct sunlight. This caused the battery to over heat and ultimately led to the depletion of both batteries. Incorrect antenna pointing prevented the orbiter from telling controllers its status, and its programmed safety response did not include making sure the spacecraft orientation was thermally safe.

The board also concluded that the Mars Global Surveyor team followed existing procedures, but that procedures were insufficient to catch the errors that occurred. The board is finalizing recommendations to apply to other missions, such as conducting more thorough reviews of all non-routine changes to stored data before they are uploaded and to evaluate spacecraft contingency modes for risks of overheating.

"We are making an end-to-end review of all our missions to be sure that we apply the lessons learned from Mars Global Surveyor to all our ongoing missions," said Fuk Li, Mars Exploration Program manager at NASA's Jet Propulsion Laboratory, Pasadena, Calif.

Mars Global Surveyor, launched in 1996, operated longer at Mars than any other spacecraft in history, and for more than four times as long as the prime mission originally planned. The spacecraft returned detailed information that has overhauled understanding about Mars. Major findings include dramatic evidence that water still flows in short bursts down hillside gullies, and identification of deposits of water-related minerals leading to selection of a Mars rover landing site.

JPL manages Mars Global Surveyor for the NASA Science Mission Directorate, Washington.

Web Reference

Mars Global Surveyor: http://marsprogram.jpl.nasa.gov/mgs/.

Jet Propulsion Laboratory (JPL)

MARSHALL-PALMER RELATION. The Z–R relationship developed by J. S. Marshall and W. M. Palmer (1948) consistent with an exponential drop-size distribution.

The relationship is $Z = 200R^{1.6}$, where $Z(\mathrm{mm}^6\ \mathrm{m}^{-3})$ is the reflectivity factor and $R(\mathrm{mm}\ \mathrm{h}^{-1})$ is the rainfall rate. The relationship is sometimes generalized to the form $Z = aR^b$, where a and b are adjustable parameters.

See also **Drop-Size Distribution**.

Additional Reading

Marshall, J.S., and W. Mck. Palmer: "The Distribution of Raindrops with Size," *J. Meteor.*, **5**, 165–166 (1948).

MARSH GAS. See **Natural Gas**.

MARSH, OTHNIEL CHARLES (1831–1899). Othniel Marsh was an American vertebrate paleontologist whose fossil excavations shaped the popular understanding and scientific study of dinosaurs and early mammals.

Born in Lockport, New York, the son of a farmer and businessman, Marsh graduated from Philips Academy in Andover, Massachusetts in 1856. Support from his wealthy maternal uncle, George Peabody, allowed Marsh to attend Yale University, where he studied with geologist James Dwight Dana and chemist Benjamin Silliman Jr. Following his graduation from Yale in 1860, Marsh spent two years doing postgraduate work at Yale's Sheffield Science School. After persuading George Peabody to fund an endowed chair and Museum of Natural History at Yale, Marsh became the first Chair of Paleontology in the United States, a position he held from 1866 until his death.

Marsh gained fame and scientific acclaim from the vast fossil-digging enterprises in the western United States, directed by Marsh from his base at the Peabody Museum of Natural History. Marsh hired dozens of fossil collectors to work in the Cretaceous and Jurassic era beds found between Kansas and Wyoming. Shipped by the thousand to Marsh, these fossils were then organized, described and reconstructed. From 1870 to 1896, Marsh is credited with describing over 500 new vertebrate species, including the dinosaurs *Allosaurus*, *Stegosaurus*, *Apatosaurus* and *Diplodocus*. He advanced Darwinian evolution by describing the Odontornithes or birds with teeth as a link between birds and reptiles. In addition, his careful reconstruction of the evolution of the horse from Eocene to Pleistocene became a model for Darwinian evolution and remained a paradigm for generations thereafter. See also **Fossil Record**.

He was a hero in the American press, which published accounts of the bone collectors and of Marsh's specimens. His influence over American science was felt in his work as chief paleontologist for the US Geological Survey (1882–1892), and as President of the National Academy of Sciences (1883–1895). His *Dinosaurs of North America* (1896) became seminal in organizing all the relevant information then known about dinosaurs. His reputation suffered when his feud with fellow paleontologist Edward Drinker Cope became public in 1890. His last years were spent in financial difficulty and poor health. See also **Cope, Edward Drinker (1840–1897)**; **Dinosauria (Dinosaurs)**; and **Paleontology (The History)**.

Additional Reading

Gillispie, C.C.: *(1970–1980) Dictionary of Scientific Biography*, Charles Scribner's Sons, New York, NY, 1981.

McCarren, M.J.: *The Scientific Contributions of Othniel Charles Marsh*, Peabody Museum, New Haven, CT, 1993.

Wallace, D.R.: *The Bonehunter's Revenge*, Houghton Mifflin, Boston, MA, 1999.

TIMOTHY W. KNEELAND, Nazareth College, Rochester, NY

MARS RECONNAISSANCE ORBITER (MRO) MISSION. See **Mars Exploration Program**.

MARS SCIENCE LABORATORY (MSL) MISSION. See **Mars Exploration Program**.

MARSUPIALIA. Pouched mammals, including the kangaroos, opossums, koala, and many less familiar forms. The order is characterized by the presence of a marsupium or pouch on the abdomen of the female in which the young complete their development. They are born in a very early stage. With few exceptions, such as the opossums of the Americas, marsupials occur in the Australian region. The general organization of Marsupialia is given in Table 1.

Didelphids. Opossums of various species make up this family. They are of moderate size, comparable to a large domestic cat or small dog. They are of a gray color with deep fur. The tail is long and scaly; the nose is sharp. See Fig. 1. There usually are from 6 to 16 young, usually more than the number of teats provided for in the mother's pouch. Hence, some of the young frequently die of starvation. The weaker and smaller young (approximately $\frac{1}{2}$ inch; 1.3 centimeters in length) are carefully placed into the pouch by the mother immediately after birth. As they grow larger and stronger, the mother will carry them on her back. The young use their prehensile tail very early for hanging on to the mother's tail or fur coat.

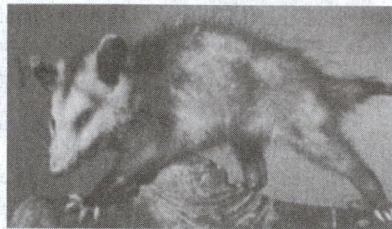

Fig. 1.   Opossum. (*A.M. Winchester.*)

The common opossum, *Didelphys virginiana*, ranges from the Great Lakes to the Gulf and westward into Oklahoma. Other species range from Mexico to southern Brazil. With exception of the water opossum, they are arboreal animals which eat insects, small birds, and fruit. The water opossum or yapock, *Chironectes*, of South America is a swimming animal and lives on fishes and other aquatic life, thus in these habits resembling the behavior of the mink and otter. The hind toes are webbed.

Dasyurids. The phascogales generally are rodent-like creatures. The Tasmanian Devil, *Sarcophilus ursinus*, resembles the badger in its stout build and large head. It is nocturnal in habits, hiding during the day in a burrow or in natural crevices. The animal eats all kinds of living creatures, even killing forms much larger than itself. The Tasmanian Wolf, *Thylacinus*, is a pouched animal closely resembling the wolves and of similar habits. Like the wolves of the northern hemisphere, this species has been killed in large numbers for its attacks on domestic animals and is now restricted to the wilder mountainous parts of Tasmania. The pouched mole is a rare mammal of the Australian deserts. The species, *Notoryctes typhlops*, is known for its highly specialized burrowing. Like the true moles, the pouched mole has enormous claws and rudimentary eyes.

Peramelids. Various species of the bandicoot comprise this family. The bandicoot is a medium-to-small burrowing animal of the Australian region, stoutly built and quadrupedal.

Phalangerids. Although there are exceptions, the phalangers are comprised generally of small-to-medium-size animals with thick woolly fur. Some resemble mice or squirrels superficially and all are arboreal. With

TABLE 1. GENERAL ORGANIZATION OF MARSUPIALIA Pouched Mammals

DIDELPHIDS

American Opossums (*Didelphidae*)
  Common Opossums (*Didelphis*)
  Four-eyed Opossums (*Metacheirus....*)
  Woolly Opossums (*Philander....*)
  Mouse-Opossums (*Marmosa*)
  Shrew-Opossums (*Monodelphis....*)
  Water Opossums (*Chironectes*)

DASYURIDS

Phascogales (*Phascogalinae*)
  Broad-footed Phascogales (*Antechinus*)
  Flat-headed Phascogales (*Planigale*)
  Brush-tailed Phascogales (*Phascogale*)
  Crest-tailed Marsupial Mice (*Dasycercus*)
  Crest-tailed Marsupial Rats (*Dasyuroides*)
  Narrow-footed Phascogales (*Sminthopsis*)
  Pouched Jerboas (*Antechinomys*)
  Native Cats (*Dasyurus*)
  Tasmanian Devil (*Sarcophilus*)
Pouched Wolf (*Thylacininae*)
Numbats (*Myrmecobiinae*)
Marsupial Moles (*Notoryctidae*)

CAENOLESTIDS

PERAMELIDS

Long-nosed Bandicoots (*Perameles*)
Short-nosed Bandicoots (*Thylacis*)
New Guinea Bandicoots (*Echymipera*)
Rabbit-Bandicoots (*Marcrotis*)
Pig-footed Bandicoots (*Choeropus*)

PHALANGERIDS

Honey-Suckers (*Tarsipedinae*)
Phalangers (*Phalangerinae*)
  Dormouse-Phalangers (*Cercoertus*)
  Striped Phalangers (*Dactylopsila*)
  Feather-tailed Phalangers (*Distoechurus*)
  Gliding Feather-tails (*Acrobates*)
  Leadbeater's Phalanger (*Gymnobelideus*)
  Flying-Phalangers (*Petaurus*)
  Ring-tailed Phalangers (*Pseudocheirus....*)
  Great Gliders (*Schoinobates*)
  Brush-tailed Phalangers (*Trichosurus*)
  Scaly-tailed Phalanger (*Wyulda*)
  Cuscuses (*Phalanger*)
The Koala (*Phascolarctidae*)
Wombats (*Wombatidae*)
Musk Rat-Kangaroos (*Hypsiprymnodontinae*)
Rat-Kangaroos (*Potoroinae*)
  Long-nosed Rat-Kangaroos (*Potorous*)
  Short-nosed Rat-Kangaroos (*Bettongia*)
  Desert Rat-Kangaroos (*Caloprymnus*)
  Rufous Rat-Kangaroos (*Aepyprymnus*)
Kangaroos (*Macropodinae*)
  Tree-Kangaroos (*Dendrolagus*)
  Hare-Wallabies (*Lagorchestes*)
  Rock-Wallabies (*Petrogale* and *Peradorcus*)
  Nail-tailed Wallabies (*Onychogale*)
  Pademelons (*Setonyx* and *Thylogale*)
  True Wallabies (*Wallabia*)
  Wallaroos (*Osphranter*)
  True Kangaroos (*Macropus*)

exception of the koala, they have prehensile tails. In addition to the true phalangers, the group includes the cuscuses, the flying phalangers, and long-snouted phalangers. The koala, shown in Fig. 2, is a curious pouched marsupial of eastern Australia. Because of its resemblance to a teddy bear, it is sometimes called the native bear. The animal is robust and achieves a length of about 2 feet (0.6 meters). An arboreal animal, the koala feeds on eucalyptus leaves, preferring those leaves with the greatest oil content. The animal can digest from 2 to 3 pounds (0.9 to 1.4 kilograms) of this foliage per day. The koala matures at 4 years of age and lives for about 20 years. Breeding occurs every other year. The gestation period is 35 days. After a short time in the pouch, the mother carries the young on her back. Highly regarded by Australians, the animal is protected by law.

Fig. 2. Female koala with baby.

Fig. 3. Kangaroo.

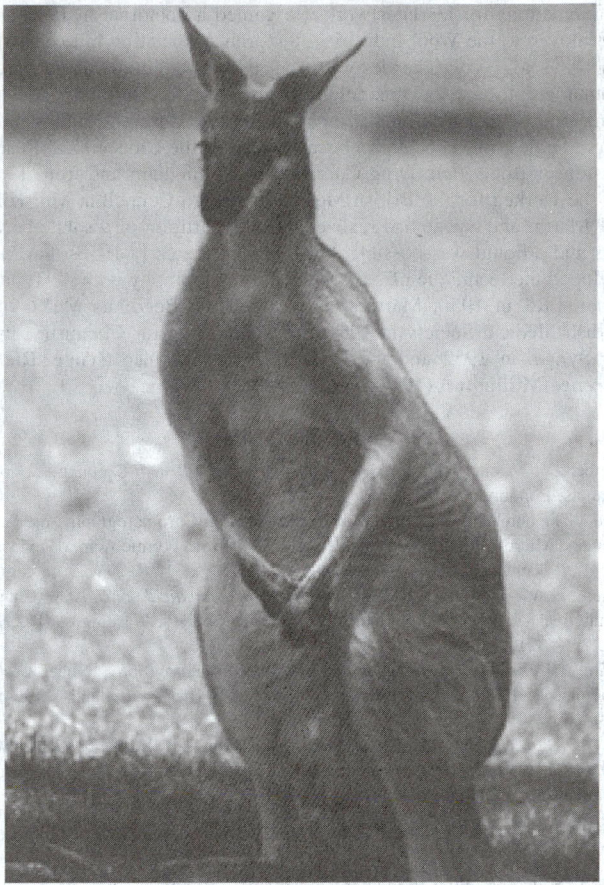

Fig. 4. Wallaby.

Additional Reading

Aitkin, L.: *Hearing-the Brain and Auditory Communication in Marsupials*, Springer-Verlag, Inc., New York, NY, 1998.

Austad, W.N.: "The Adaptable Opossum," *Sci. Amer.*, 98–105 (February 1988).

Dawson, T.J.: *Kangaroos: The Biology of the Largest Marsupials*, Cornell University Press, Ithaca, NY, 1995.

Degabriele, R.: "The Physiology of the Koala," *Sci. Amer.*, **243**, 1, 110–117 (July 1980).

Grzimek, B.: *Grzimek's Animal Life Encyclopedia*, Vol. 10, Mammals 1, Van Nostrand Reinhold, New York, NY, 1975.

Lee, A.K. and A. Cockburn: *Evolutionary Ecology of Marsupials*, Cambridge University Press, New York, NY, 1985.

Sharman, G.: *National Geographic*, **155**, 2, 192, 209 (1979).

Szalay, F.S.: *Evolutionary History of the Marsupials and an Analysis of Osteological Characters*, Cambridge University Press, New York, NY, 1994.

MARTEN. See **Mustelines**.

MARTENSITE. The chief constituent of hardened carbon tool steels. It is a solution of carbon or Fe_3C in β-iron, or and exceedingly fine-grained α-iron with carbon or Fe_3C in atomic or molecular dispersion. Carbon content up to 1%; easily obtained by quenching small bodies of hypereutectoid steel in cold water; more difficult to obtain in low-carbon steels.

MARTIN, ARCHER JOHN PORTER (1910–2002). Archer John Porter Martin was born in London on 1 March 1910. He entered Peterhouse, Cambridge in 1927 and, influenced by J. B. S. Haldane, specialized in biochemistry. After graduation in 1932 Martin spent a year in the physical chemistry laboratory and then joined the Dunn Nutritional Laboratory under L. J. Harris and Sir Charles Martin. Here he gained valuable experience in separating and isolating closely related, biologically active compounds using fractional separation techniques; he also studied the pathological effects of vitamin E deficiency and attempted to isolate the vitamin by solvent extraction and chromatographic techniques, which laid

The wombat, also a phalangerid, is a stoutly built pouched animal, broad of body and with short thick legs. The wombat has a pair of chisel-like incisor teeth in each jaw, like the rodents, and displays similar feeding habits. The wombat lives in burrows or in rock crevices. There are several species.

Possibly best known of all marsupials is the kangaroo. The typical kangaroo, *Macropus*, has large hind legs and a strong tail. The animal sits upright and moves by spring leaps, not touching the front feet to the ground. The largest kangaroos reach a height of more than 6 feet (1.8 meters) and a weight of 200 pounds (91 kilograms). See Fig. 3. Kangaroos usually travel in groups. They are easy to frighten and tend to panic under tension. However, they are known to bite rather severely and to box with their forefeet. The feeding habits are much like deer, foraging on grasses and tender shoots of plants. Some species can cause considerable crop damage.

Moderate and small-size members of the kangaroo family, which resemble the true kangaroo in form, are called wallabies. See Fig. 4. The distinction between kangaroos and wallabies is not rigidly scientific, the large species being called kangaroos and the smaller species wallabies, with a transition in the larger wallabies which are also known as brush kangaroos. Wallabies, like their large relatives, have powerful hind legs and small forelegs, and are bipedal in locomotion. They vary from the hare wallaby, *Lagorchestes*, less than 2 feet (0.6 meter) long, to the rednecked wallaby whose body is $3\frac{1}{2}$ feet (1 meter) long, exclusive of tail. The spur-tailed wallabies, *Onychogale*, are peculiar in having the tail tipped with a horny spur.

the foundations for his later work. He gained a doctorate in 1936 and in 1938 moved to the Wool Industries Research Association in Leeds where, with R. L. M. Synge, he developed a method of liquid–liquid partition chromatography for the separation of amino acids. They also developed the method of paper chromatography, using two different liquids moving at right angles. In 1946 Martin became head of the Biochemical Research Division of Boots Pure Drug Company in Nottingham and from 1948 to 1959 he worked for the British Medical Research Council at Mill Hill. In 1941 Martin and Synge had realized that the partition of a solute between a gas and a liquid was possible, but it was in the early 1950s that Martin developed the technique of gas–liquid chromatography with A. T. James. Elected FRS in 1950, Martin was awarded the Berzelius Medal of the Swedish Medical Society in 1951, the Nobel Prize for Chemistry, jointly with Synge, in 1952 and the CBE in 1960. See also **Synge, Richard Laurence Millington (1914–1994)**.

Additional Reading

Campbell, W.A., and N.N. Greenwood: *Contemporary British Chemists*, Taylor and Francis, London, UK. 1971.

James, A.T., and A.J.P. Martin: "Gas–liquid Partition Chromatography: the Separation and Micro-estimation of Volatile Fatty Acids from Formic Acid to Dodecanoic Acid" *Biochemical Journal*, **50**, 679–690 (1952).

Laitnen, H.A., and G.W. Ewing: *A History of Analytical Chemistry*, Chap. V, Analytical Division of the American Chemistry Society, Washington, DC. 1977, pp. 296–321.

Martin, A.J.P.: *The Development of Partition Chromatography*, Nobel Lectures: Chemistry, 1942–1962, Elsevier, Amsterdam, The Netherlands. 1964, pp. 359–371.

Shetty, P.H.: "Archer John Porter Martin." In: James, L.K.: *Nobel Laureates in Chemistry 1901–1992*, pp. American Chemical Society and the Chemical Heritage Foundation, Washington DC, 1993, pp. 352–355.

NOEL G. COLEY, The Open University, Milton, Keynes, UK

MARTIN *(Aves, Passeriformes)*. Any of several species of swallows, represented on all continents except Australia. They have the short beak and wide mouth of the swallows and either a forked or a square tail. They nest about houses, on cliffs, and in burrows and hollow trees. North America has one species, the purple martin, *Progne subis*. The male is glossy blue-black and the female somewhat duller. These birds commonly nest in the cornices of buildings or in bird houses where they live in colonies. See also **Swallow**.

MARTINGALE. Originally, a process known to gamblers, under which the loser at a fair game doubled his stakes for the next, and so on at each loss. The paradox is that in the long run he appeared certain to win sooner or later and at that point would have a net gain. More recently, the term has been given a precise significance in the theory of stochastic processes. A stochastic process $\{x_t\}$ is called a martingale if $E\{|x_t|\}$ is finite for all t, and

$$E\{x_{t_{n+1}}|x_{t_1}, \ldots, x_{t_n}\} = x_{t_n}$$

with probability unity for all $n \geq 1$ and $t_1 < \cdots t_{n+1}$.

MARVIN SUNSHINE RECORDER. A sunshine recorder of the type in which the timescale is supplied by a chronograph. It consists of two bulbs, one of which is blackened, that communicate through a glass tube of small diameter. The tube is partially filled with mercury and contains two electrical contacts. When the instrument is exposed to sunshine, the air in the blackened bulb is warmed more than that in the clear bulb. The warmed air expands and forces the mercury through the connecting tube to a point where the electrical contacts are shorted by the mercury. This completes the electrical circuit to the pen on the chronograph. The Marvin sunshine recorder is equally sensitive to the direct rays of the sun and to diffuse radiation from the sky (the heat from the latter at midday in overcast may be more than that from direct sunshine in the early morning); thus, the instrument is not without ambiguities. It is standard equipment at National Weather Service stations.

MASCON. A large-scale, high-density lunar (or planetary, as Martian) mass concentration below a ringed mare. See also **Mare**.

MASER. An acronym for microwave amplification by stimulated emission of radiation. The device is identical in theory of operation to the laser except that it operates at frequencies in the microwave region of the electromagnetic spectrum, rather than in the light range. See also **Laser**.

Consider a stream of atoms in equilibrium for a given energy transition, that is, some are emitting radiation at the frequency of the transition and others are absorbing it. This process can be represented by the equation

$$h v_{21} = E_2 - E_1$$

where E_2 represents the energy of an atom in an excited state, E_1 represents its energy in a lower state, such as the ground state, h is Planck's constant, v_{21} represents the frequency of a photon which would excite the atom from the state E_1 to state E_2, or conversely, the frequency of a photon that would be emitted in the reverse transition. Now consider a large number of atoms in thermal equilibrium in a closed box. The radiation inside the box will be black body radiation; in other words, the number of photons of given energy will be uniquely determined by the temperature. In addition, the temperature will fix the proportion of atoms in the excited state, which will be given by the Boltzmann distribution

$$\frac{N(E_2)}{N(E_1)} = \exp\left(\frac{-(E_2 - E_1)}{kT}\right)$$

where $N(E_2)$ and $N(E_2)$ are the populations of the states E_1, E_2, respectively. These two assertions, governing the number of excited atoms and the number of photons, respectively, are both fundamental thermodynamic principles and should fit together into a consistent picture. To bring about the energy balance necessary for equilibrium, an extra term is introduced, corresponding to a second process of radiation emission. The second term is quite different from the first in that it represents a process whose rate (or probability) is proportional to the intensity of radiation of frequency v falling on the atom. The radiation field stimulates the excited atom to emit, and we can contrast this process of stimulated emission with the more familiar process of spontaneous (random) emission. The latter is more familiar because it is overwhelmingly dominant so long as hv/kT is large, and this is so in the optical and infrared regions of the spectrum, where most of our accumulated experience lies. On the other hand, if hv/kT is small (as it is in the microwave region), stimulated emission is the more important; however, stimulated emission in the visible region may be significant under certain conditions.

In terms of photons, stimulated emission appears as the interaction of the photon with an excited atom, which leads to the emission by the atom of a second photon, identical with the primary. In other words the photon has multiplied.

To accomplish amplification it is only necessary to produce more photons by stimulated emission than are lost by absorption. The details of the balance at equilibrium show that the probability of an atom in the ground state absorbing a photon, and of one in the excited state emitting one by stimulation, are equal. Hence, if amplification is to be possible, the number of atoms in the excited state must be greater than the number in the ground state. A look at the Boltzmann equation shows that this cannot be attained at any (positive) value of temperature, which means that it cannot be attained in thermal equilibrium. The problem then is essentially one of disturbing equilibrium so as to bring about the population inversion required.

Although having a number of limitations, the ammonia maser can be used to describe the principles involved. What is involved is picking out the excited atoms from the unexcited ones, and segregating them. This implausible procedure is in certain cases possible, the most notable involving the ammonia molecule, NH_3. The structure of this molecule is pyramidal, with the nitrogen atom at the vertex and the three hydrogen atoms forming the base. It is possible for the molecule to execute vibrations in which the nitrogen atom vibrates back and forth through the plane of the hydrogen atoms. The energy difference between this excited state and the ground state (no vibration) corresponds to a wave-length in the microwave region, i.e., about 3 centimeters. Because of its lack of symmetry, the molecule in the ground state will have a dipole moment, but the average dipole moment of the molecule executing the vibrations is described as zero. If ammonia molecules are formed into a beam by allowing the gas to stream out of a collimating tube into a vacuum, and this beam is passed through a nonuniform electric field, the separation can be effected. The ground state molecules will be deflected by the nonuniform electric field, and lost to the beam; the excited molecules, however, by virtue of their zero dipole moment, experience no deflecting force, and are introduced into a resonant cavity resonating at the appropriate frequency. The cavity will then contain a preponderance of excited molecules, provided the lifetime of the excited state is long enough (as it will be at microwave frequencies). See Fig. 1.

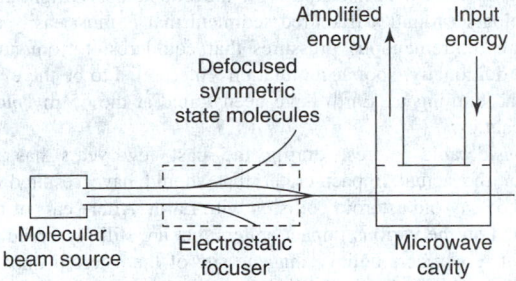

Fig. 1.   Schematic representation of an ammonia beam maser. (*After Gordon, Zeiger, Townes.*)

If now an input signal of the resonant frequency is fed into the cavity, it can bring about spontaneous emission, and amplification will occur. The shortcoming of this arrangement is that the ammonia resonance cannot be tuned and it represents an impracticably narrow bandwidth. On the other hand, like any other amplifier, it can be made to oscillate, so that it can be used as a very stable frequency standard (ammonia clock).

The three-level maser provides a more versatile arrangement for achieving inversion. Three energy levels are associated with the same atomic or molecular system. See Fig. 2. A strong microwave signal of frequency v_{31} corresponding to $E_3 - E_1$ raises some of the atoms to E_3. A limit is reached when the populations of E_3 and E_1 are the same, since then the radiation absorbed is just balanced by stimulated emission, a state of affairs known as "saturation." Not all the atoms in state E_3 will return directly to E_1 — some will return via E_2. By the choice of states of suitable lifetimes ($\tau_2 > \tau_3$) it is possible to arrange that the number of atoms in E_2 exceeds that in E_1 because the "pump" frequency maintains a "head" of atoms in E_3. Thus E_2 and E_1 are inverted and amplification is possible.

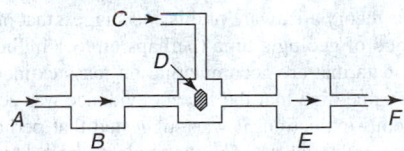

Fig. 2.   Three-level maser amplifier, where A is signal input; B is the first isolator; C is the power input for pumping; D is the crystal; E is the second isolator, usually operated at low temperature as a noise reflector; and F is the signal output.

Maser amplifiers are used where the requirement for a very low noise amplifier outweighs the technological problems of cooling to low temperatures. They have been used in passive and active radiostronomical work, in satellite communications, and as preamplifiers for microwave spectrometry. The ammonia and the atomic hydrogen masers have been studied as frequency standards and have been used in accurate tests of special relativity.

Additional Reading

Arecchi, F. et al.: *Instabilities and Chaos in Quantum Optics II*, Perseus Publishing, Cambridge, MA, 1988.

Bertolotti, M.: *Masers and Lasers: An Historical Approach*, Institute of Physics (IoP), London, UK, 1988.

Elitzur, M.: *Astronomical Masers*, Kluwer Academic Publishers, Norwell, MI, 1992.

MASK (Computer System).   A pattern of digits used to control the retention or elimination of portions of another pattern of digits. Also, the

use of such a pattern. For example, an 8-bit mask having a single i bit in the ith position, when added with another 8-bit pattern, can be used to determine if the ith bit is a 1 or a 0; that is, the ith bit in the pattern will be retained and all other bits will be 0s.

As another example, there are situations where it is desirable to delay the recognition of a process interrupt by the digital computer. A mask instruction permits the recognition of specific interrupts to be inhibited until it is convenient to service them.

MASS.   The physical measure of the principal inertial property of a body, i.e., its resistance to change of motion. At speeds small compared with the speed of light, the mass of a body is independent of its speed. Under these circumstances, the masses m_1 and m_2 of two bodies may be compared by allowing the two bodies to interact. Then

$$m_1/m_2 = |a_2|/|a_1|$$

where $|a_1|$ and $|a_2|$ are the magnitudes of the respective accelerations of the two bodies as a result of the interaction. This permits the measurement of the mass of any particle with respect to a standard particle (for example, the standard kilogram). At higher speeds, the mass of a body depends on its speed relative to the observer according to the relation:

$$m = m_0/\sqrt{1 - v^2/c^2}$$

where m_0 is the mass of the body as found by an observer at rest with respect to the body, v is the speed of the body relative to the observer who finds its mass to be m, and c is the speed of light in empty space as prescribed by the theory of relativity.

When relativistic mechanics is appropriate, e.g., when speeds comparable to the speed of light are involved, mass may be converted into energy and vice versa, hence the energy of the system must be converted into mass through the Einstein equation

$$E = mc^2$$

where c is the speed of light in empty space, before the conservation law may be applied. See also **Gravitation**; and **Inertia**.

MASS (Center of).   The center of mass is that point in a collection of mass-particles which moves as if the total mass of the collection were concentrated there and the resultant of all the external forces were acting there. The position vector of such a point is given by

$$\mathbf{r} = \frac{\sum_{i=1}^{n} m_i \mathbf{r}_i}{\sum_{i=1}^{n} m_i}$$

where m_i is the mass of ith discrete particle, $\mathbf{r}_i$ is the position vector of ith discrete particle.

MASS DEFECT.   The difference Δ between the atomic number A and the atomic mass M of a nuclide, $\Delta = A - M$. The negative of the mass defect, $-\Delta$, is known as the mass excess.

MASS-ENERGY EQUIVALENCE.   The formula $E = mc^2$, which equates a quantity of mass m to a quantity of energy E. The conversion factor c^2 is the square of the speed of light. The relationship was developed from the relativity theory (special), but has been experimentally confirmed. See also **Relativity and Relativity Theory.**

MASS EXTINCTIONS.   Estimates of the number of species of plants and animals that have lived on Earth during the past 600 million years (my) range from hundreds of millions to several billion, yet fewer than two million are known to live on Earth today. It is apparent that most species that have lived are extinct. Paleontologists have calculated extinction rates for most of the groups of geologically important organisms that have lived during the past 600 my, and these data suggest some predictability of rates of extinction and evolution. However, the normal or "background" extinction rates of Earth's inhabitants are punctuated by a few intervals when extinction rates were much greater. These were times of drastic reduction of biologic diversity and resulted in biologic catastrophes that are called *mass extinctions*. The occurrence of mass extinctions is well known, and the major divisions of geologic time for the past 600 my are largely based on the intervals of time between mass extinctions. Thus, the Paleozoic Era ended approximately 245 my ago at a mass extinction event that may have affected between 77 and 96% of all marine invertebrate species while significantly reducing Earth's terrestrial fauna. The younger Mesozoic Era ended approximately 65 my ago when approximately 60 to 75% of marine species became extinct; this mass extinction is thought to have included dinosaurs and many other reptiles. There were other major extinction events during the past 600 my, the most important of which are noted on the Fig. 1. In addition to the events noted, different paleontologists believe that there may have been 1 to 3 additional times of major mass extinction and perhaps 8 or more times of minor mass extinction. The events noted at 245 my and 65 my are the most severe and dramatic of all mass extinctions. Most of the data available are based on marine organisms, but include information on terrestrial vertebrates as well. Although the data are small, it appears that plants may have been more resistant to mass extinctions than animals.

Documentation of extinction is produced from specialized actuarial Tables that are calculated to demonstrate background extinction levels for genera and families of organisms. These calculations are convincing and suggest that extinction of any group of organisms may be the expected part of the evolutionary process. The explanations for the extinction of a large number of groups during a very brief interval of geologic time are much more complicated. Background extinction is the result of assorted kinds of ecologic interactions and biologic changes, and the extinction of an entire family comprised of multiple species of organisms may or may not be due to some shared deficiency. In contrast, mass extinctions may be the result of a single event that, together with feedback, amplifies the catastrophe and cuts across ecologic boundaries in the elimination of a diversity of unrelated organisms.

One of the most profound hypotheses of the past 10 years is that the Cretaceous mass extinction (an event that occurred approximately 65 my ago) may have resulted from the impact of an asteroid of approximately 10 km in diameter. It is proposed that the direct damage caused by such an impact fractured Earth's crust (and perhaps the upper mantle), damaging an area of several hundred or thousand km. Heat generated and released from the impact would affect a large part of Earth, perhaps burning large vegetated regions. If the initial impact occurred in the ocean, this would result in the generation of tsunamis that could destroy life living in low-lying coastal regions. Release of crust/mantle chemicals following impact may have poisoned surface waters, affecting additional forms of marine life. It is theorized that debris from the impact would result in the formation of a gigantic dust cloud consisting of pulverized crustal particles ejected from the impact and that the cloud probably would be so dense that it could effectively block solar radiation and plunge Earth into darkness. Calculations on the period of darkness resulting from the dust cloud range from a few days to a few years. Just how long would depend on poorly understood factors of atmospheric circulation 65 my ago. Much smaller volcanic eruptions during historic times have produced atmospheric dust that has affected solar radiation and altered normal Earth weather patterns, sometimes for a period of several years. Reduction or even elimination of photosynthesis could occur if the duration of such a cloud was prolonged. Any reduction of photosynthesis, in turn, would affect the food chain both in the oceans and on land. The demise of autotrophic organisms would have a chain reaction on heterotrophic organisms at several different levels. Organisms that survived the immediate effects of heat and/or tsunamis would be affected by such a massive disruption of the food chain. An eventual cleansing of the atmosphere by an acid rain of Armageddon proportions might be the final straw for survivors of the earlier sequence of events. It is proposed that Earth's climate 65 my ago might even have been changed by such a chain of events, shifting from what is known to have been a mild climate to a worldwide cooling that is also known to have occurred.

But other than the general Earth cooling known to have occurred at this time, what is the evidence to support such a scenario? There appears to be general agreement that between 60 and 75% of marine species living during the Cretaceous Period became extinct during a very short period of time at the close of the Cretaceous, generally dated as 65 my ago.

Terrestrial organisms, including the dinosaurs, flying reptiles, and a few other types, became extinct at approximately the same time. Most important for this theory is the 1980 discovery of the chemical element iridium (Ir) in abundances well above that of normal Earth crustal levels concentrated in the sediment that is believed to have accumulated during the time of the 65 my mass extinction event. Additional investigations showed similar non-Earth abundances of Ir worldwide, all occurring in rocks dated at 65 my. Such concentrations of Ir are most common in asteroids, but not in Earth's crust. Associated with the anomalously high Ir at many sites, geologists found a fractured sediment that is most easily explained as a result of tremendous pressures that could result from an asteroid impact. Additionally, soot accumulations interpreted to be the evidence of widespread burning on Earth have been found at the 65 my old layer at many localities.

Of considerable interest during the past few years has been the search for the actual impact crater that should have resulted from the theorized 65-my-old asteroid collision with Earth. After years of tabulating information on the various impact craters that are still preserved on Earth, a group of researchers believe that the site of the asteroid impact 65 my ago is located in the Gulf of Mexico, adjacent to the north coast of the Yucatan peninsula centered at Chicxulub. This 180-km wide submarine fossil impact scar is difficult to study, but, fortuitously, it had been drilled by oil companies so that a great deal of the geologic framework for the structure is known.

Evidently the Chicxulub crater is the right size and age to accommodate almost everything that has been proposed by the proponents of an asteroid impact as the cause of the mass extinction 65 my ago. The evidence from this crater structure includes the right geologic age for the impacted sediment (Late Cretaceous), believable age melt rocks within and surrounding the crater, and the remarkable similarity in the age of the crater melt rocks and ejecta that have been found great distances away. In fact, all of the stratigraphic, sedimentologic, petrographic, geochemical, and geophysical data that are studied at the Chicxulub crater are consistent with the idea that this is the site of an asteroid impact 65 my ago and may be the "smoking gun" that proves the asteroid impact/mass extinction theory.

However, it is extremely important to note that the imprecision of dating events occurring millions of years ago casts a degree of uncertainty on the precise contemporaneity of all the data considered to be critical for the theory. If higher resolution of geologic time is accomplished, it is possible that the 65-my-old event (or other mass extinction events that are considered contemporaneous) may turn out to have occurred over a longer time interval than could be explained by the single impact theory. Proponents of the theory are aware of this and suggest that multiple impacts over a short period of geologic time (perhaps up to 1 million years) could also explain the data that are accumulating on mass extinction.

At almost the same time that the impact evidence was accumulating for the Late Cretaceous extinction, it was suggested that perhaps all or most of the major mass extinctions during the geologic past may have occurred as the result of similar asteroid impacts. See Fig. 1. These were produced by periodic perturbations of our solar systems by cyclic encounters with an asteroid swarm or by some planet whose orbit has been obscured by the sun. Periods of between 26 and 30 my, were calculated for the known times of mass extinctions (at least for the last 100 my), and this periodicity is assumed to be that of the source of the perturbations. In fact, the reality of periodicity is still debated, and the arguments for periodicity of asteroid impacts are tempered by the fact that a worldwide Ir anomaly as definitive as that found in the sediment of the Late Cretaceous (65 my) extinction has not been found at any of the other mass extinction levels of the Paleozoic or Mesozoic.

Even though most of the publicity for mass extinctions has featured the Late Cretaceous event (most likely because of the extinction of dinosaurs at that time), this event was not as significant as the mass extinction that marks the close of the Paleozoic Era, 245 my ago. This event, some 180 my earlier than the Cretaceous event, is considered to be the most profound of all of Earth's mass extinctions. It is reported to have resulted in the extinction of up to 96% of all marine species, and many terrestrial species were affected as well. There is no Ir anomaly for this event as there is for that of the Late Cretaceous, and some evidence supports the idea of multiple causes for the mass extinction at this time. A great deal of evidence points to widespread volcanic eruptions 245 my ago. The volcanic

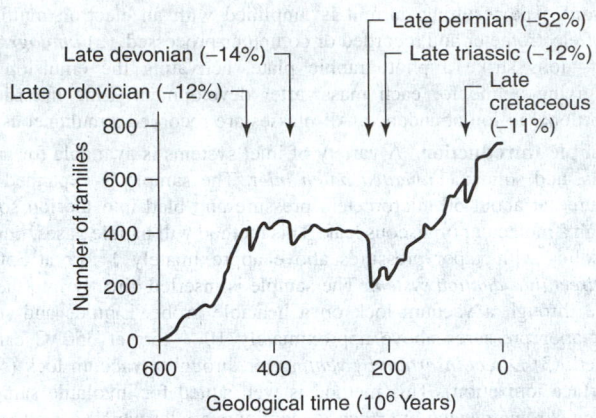

Fig. 1. Chart indicating five of the six to eight major mass extinction events, with the relative magnitudes of these events shown in terms of percentage of families of marine invertebrates and vertebrates affected. (After Raup and Sepkoski, 1982.)

theory is supported by recent dating of Siberian basaltic flows and volcanic tuff that are within the theoretical time range of the Late Permian (245 my) extinction event. According to this theory of mass extinction, widespread volcanism may have injected large amounts of sulfur dioxide into the atmosphere, resulting in a decrease in solar radiation and consequent global cooling that affected marine and terrestrial life. This extinction event of 245 my ago is not as well studied, nor are the data as well controlled as that of the mass extinction of 65 my ago. In addition to the known volcanic activity of this time, other specialists point out that there may have been catastrophic ecosystem devastation produced by reorganization of Earth's continental shelves during this time that also was a time of major plate tectonic activity.

An even older major mass extinction is that of the Late Devonian (~367 my). Recent reports of glass spherules, interpreted to represent material similar to the micro tektites (known from the 65-my-old extinction event), and the tentative identification of a crater in the Baltic Shield of approximately the same age, suggest that an asteroid impact may be responsible for this mass extinction. This is perhaps the closest match to date for the Late Cretaceous event.

The interest generated by these kinds of studies has stimulated a great amount of research for each of the major and minor mass extinction events. Paleontologists and geologists continue to study possible causes, and, although the reality of multiple mass extinctions of Earth's organisms at intervals during the past 600 my is fairly well established, certainly there is no consensus on the actual mechanism of extinction. The asteroid impact proposed to explain the Late Cretaceous event is attractive, but has a few problems that bother some specialists. The volcanic theory for the Late Permian event is supported by a variety of data, but even these can be interpreted differently. The precise timing of the extinction of different groups of organisms is difficult to establish, and the older the extinction event, the more difficult it is to unequivocally establish synchroneity of extinction of different species. Understanding and organizing all of the components involved in mass extinction is analogous to the creation of a picture from pieces of a jigsaw puzzle. Because some of the pieces of the mass extinction puzzle have been lost during the millions of years since the occurrence of the event, we may never be able to complete the whole picture. But we are beginning to understand the picture of mass extinction to a much greater extent than was possible even a few years ago.

Additional Reading

Alvarez, L.W., W. Alvarez, F. Asaro, and H.V. Michel: "Extraterrestrial Cause for the Cretaceous-Tertiary Extinction," *Science*, **208**, 1095–1105 (1980).
Bernhard, T.: *Extinction*, University of Chicago Press, Chicago, IL, 1996.
Campbell, I.H., G.K. Czamamske, V.A. Fedorenko, R.I. Hill, and V. Stepanov: "Synchronism of the Siberian Traps and the Permian-Triassic Boundary," *Science*, **258**, 1760–1763 (1992).
Claeys, P., J-G Casier, and S.V. Margolis: "Microtektites and Mass Extinctions: Evidence for a Late Devonian Asteroid Impact," *Science*, **257**, 1102–1104 (1992).
Clark, D.L., C.Y. Wang, C.J. Orth and J.S. Gilmore: "Conodont Survival and Low Iridium Abundances Across the Permian-Triassic Boundary in South China," *Science*, **223**, 984–989 (1986).
Courtillot, V.: *Evolutionary Catastrophes: The Science of Mass Extinction*, Cambridge University Press, New York, NY, 1999.
Eldredge, N.: "The Miners Canary: Unraveling the Mysteries of Extinction," Princeton University Press, Princeton, NJ, 1994.
Friedman, G. et al. (Editors): *Extinction Events in Earth History: Proceedings of the Project 216: Global Biological Events in Earth History*, Springer-Verlag New York, Inc., New York, NY, 1990.
Kitchell, J.A. and D. Pena: "Periodicity of Extinctions in the Geologic Past: Deterministic Versus Stochastic Explanations," *Science*, **226**, 689–692 (1984).
Ramino, M.R. and R.B. Stothers: "Geological Rhythms and Cometary Impacts," *Science*, **226**, 1427–1431 (1984).
Raup, D.M. and J.J. Sepkoski, Jr.: "Mass Extinctions in the Marine Fossil Record," *Science*, **215**, 1501–1503 (1982).
Swisher, C.C.III, J.M. Grajales-Nishimura, A. Montanari, S.V. Margolis, P. Claeys, W. Alvarez, P. Renne, E. Cedillo-Pardo, F.J-M.R. Maurrasse, G.H. Curtis, J. Smit, and M.O. McWilliams: "Coeval ⁴⁰Ar/³⁹Ar Ages of 65.0 Million Years Ago from Chicxulub Crater Melt Rock and Cretaceous-Tertiary Boundary Tektites," *Science*, **257**, 954–958 (1992).

DAVID L. CLARK, Ph.D., University of Wisconsin–Madison

MASS-LUMINOSITY RELATION. Observationally, it is found that a plot of absolute bolometric magnitude against mass for giant and main sequence stars reveals a definite correlation. The observational diagram is shown in Fig. 1. Using only visual binaries, Eddington observed this trend and then deduced the relation theoretically. One may conveniently use the approximation

$$\log L = 3.3 \log M$$

where L and M are in solar units.

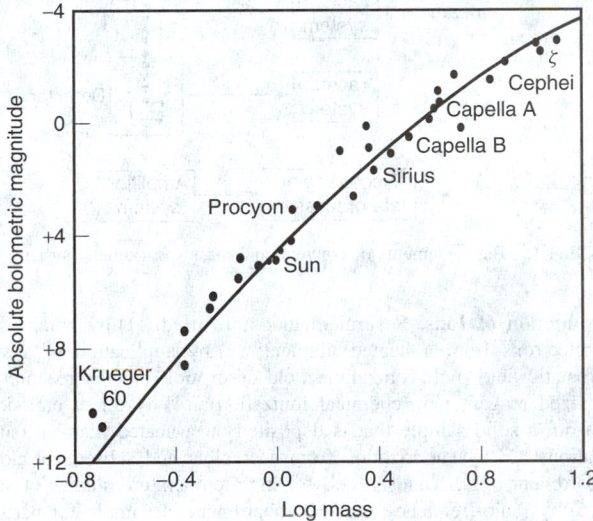

Fig. 1. Mass-luminosity curve. (*Eddington.*)

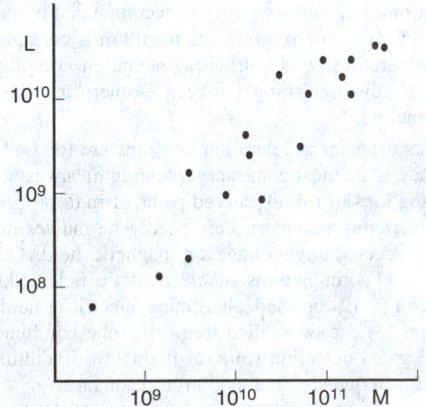

Fig. 2. Mass-luminosity relation for selected galaxies.

A similar diagram can be constructed for galaxies, and is shown in Fig. 2. Again, a clear relationship is exhibited, but there is no theoretical explanation for this fact.

MASS NUMBER.   The total number of nucleons in the nucleus of an atomic species is its mass number, which then is numerically equal to the sum of the atomic number and the neutron number of the species. See also **Chemical Elements**.

MASS SPECTROMETRY.   A general quantitative and qualitative analyzer for most components in all types of samples — gas, liquid, or solid, but with some volatility limitations. A complete analysis is obtained from nanogram samples in a few seconds or minutes. The range is from parts per billion to 100% purity. Accuracy is ±1%; the specificity is good. Conventional methodology is first described after which recent trends are presented.

Operating Principle.   With reference to Fig. 1, the sample to be studied is introduced into an evacuated area (ion source), where it is ionized, accelerated by an electrostatic field, and separated according to mass. The various masses are collected and measured.

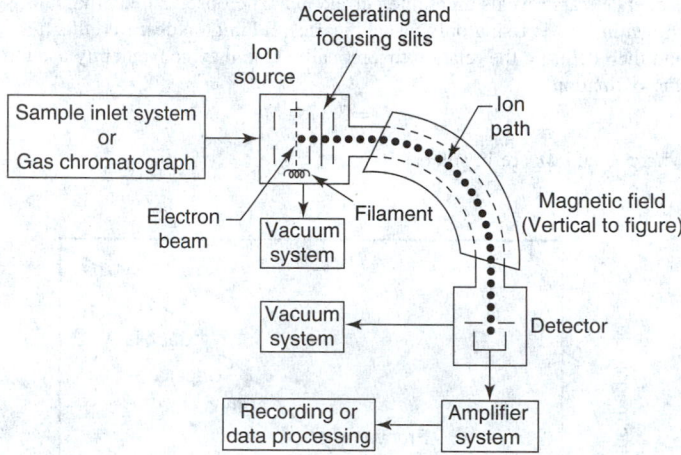

Fig. 1.   Basic elements of conventional mass spectrometer system.

Production of Ions.   Several methods are used: (1) by bombardment with electrons from a heated filament; (2) by application of a strong electrostatic field (field ionization, field desorption); (3) by reaction with an ionized reagent gas (chemical ionization); (4) by direct emission of ions from a solid sample that is deposited on a heated filament (surface ionization); (5) by vaporization from a crucible and subsequent electron bombardment (e.g., Knudsen cell for high-temperature studies of solids; and (6) by radio-frequency spark bombardment of sample for parts-per-billion (ppb) elemental analysis of solids as encountered in metallurgical, semiconductor, ceramics, and geological studies. Ions also are produced by photoionization and laser ionization.

Fragmentation.   Ionization usually is accompanied by partial fragmentation of the molecule. The fragmentation pattern is constant for a specific molecule and operating conditions. Fragmentation complicates computation, but permits distinguishing between isomers and gives molecular-structure information.

Ion Separation.   After acceleration, the ions are focused and separated according to mass. The most common separating means is a magnetic field, which causes the ions to follow curved paths of radii proportional to their masses. Many different geometries are used. The masses may be scanned by varying the accelerating voltage or magnetic field. Other separation means include: (1) combinations of electrostatic fields (double focusing cycloidal focusing); (2) crossed alternating electrical fields (quadrupole mass filter); and (3) use of a filled-free drift tube combined with pulsed ion source and gated detection (time-of-flight). Still additional means are omegatron, radio frequency, and cyclotron resonance.

Detection.   Commonly used detection means are: (1) *electrical* — ion beams are successively scanned across a collector where they pick up

electrons. The resulting current is amplified with an electron multiplier and/or electrometer and recorded or computer-processed. (2) *photographic plate* — ions strike a photographic plate, activating the emulsion and thus giving a line for each mass, after development. Line intensity is proportional to ion abundance. All masses are recorded simultaneously.

Sample Introduction.   A variety of inlet systems is available for gases, liquids, and solids: (1) *heated batch inlet*. The sample is expanded into a volume at about 50 micrometers pressure and bled into the ion source through a molecular or viscous leak. This method will handle gases, liquids, and solids with vapor pressures above approximately 1 torr at 350°C. (2) *Direct introduction system*. The sample is inserted directly into the ion source through a vacuum lock on a heatable probe. Liquids and solids with vapor pressures above approximately 10^{-7} torr at 350°C can be handled. (3) *Direct insertion by venting*, or through a vacuum lock (spark or surface ionization). This method is well suited for involatile samples. (4) *Gas chromatograph interface*. A continuous inlet which permits mass spectrometric analysis of the separated components as they emerge from the gas chromatograph.

The combination of gas chromatograph and mass spectrometer provides a separating and identifying capability not achievable by other means, particularly for very small samples.

Data Processing.   The high data output rate of many mass spectrometer systems requires data processing to fully utilize the capability of the instrument. The present trend is toward systems using small dedicated computers with digital tape, core, or disk memory, and printer, plotter, and cathode-ray tube output. Typical outputs available are: (1) mass and abundance printouts; (2) mass versus abundance plots; (3) elemental composition printouts (high-resolution mass spectrometer); (4) "total ionization chromatogram" plot (summed ion plot, similar to gas chromatogram); (5) mass chromatogram (similar to total ionization chromatogram, but for a selected mass or masses only); and (6) quantitative analysis printout.

Outputs are available from raw data, data with background or other spectra subtracted, and normalized data. A number of other options of data selection, manipulation, and output are usually available from the instrument manufacturers.

Uses and Applications.   Common uses for mass spectrometers include: (1) compound identification; (2) elemental formula determination; (3) molecular-structure determination; (4) quantitative mixture analysis; (5) ppb solids elemental analysis; (6) isotope ratio determination; (7) leak detection (helium tracer); (8) residual-gas analysis in vacuum systems; and (9) age dating.

Mass spectrometers are widely used for studies and determinations in: organic chemistry; petroleum and biological laboratories; nuclear investigations; geochemistry and cosmochemistry; metallurgical, semiconductor, and ceramics investigations; pollution control; space programs; agricultural and pesticide research and manufacture; flavors and fragrances chemistry; and as a basic research tool in studies of ion-molecule reactions, high-temperature chemistry kinetics, free radicals, and thin films. Broad classes of instruments are summarized in Table 1.

TABLE 1. TYPES AND CHARACTERISTICS OF MASS SPECTROMETERS

General Class	Type	Major Uses
Leak detector	Magnetic analyzer (helium only)	Detects small leaks, using helium gas tracer.
Residual-gas	Magnetic or quadrupole analyzer	Analysis of gases in vacuum systems
Low-resolution	Magnetic, cycloidal, or quadrupole analyzer	Analysis of gases or light liquids
Medium-resolution	Magnetic analyzer	Identification, molecular-structure studies, mixture analysis, isotope ratios.
High-resolution	Magnetic and electric (mass and energy) focusing	Molecular-structure determinations, ppb solids analysis, isotope ratios of solids, high temperature chemistry

Advancements in Mass Spectrometers

The great strides made by mass spectrometry since its inception several years ago are aptly put forth by Delgass and Cooks (see reference), who describe the status of mass spectrometry as of the late 1980s. The applications of mass spectrometry have penetrated into physical chemistry (bond dissociation energies, ion enthalpies, proton affinities); organic chemistry (structure studies, organic ion structure and fragmentation); biology (drug metabolism, stable isotope tracer work, modifications in biopolymers); the earth sciences (chronology/dating of geological and life extinction events); and environmental science (trace organic analysis).

Mass spectrometry is undergoing a rapid development that shows little indication of abating either in the areas of instrument refinement or of extended applications. Mass spectrometry is expected to play a major role in the revitalized science of materials and surface phenomena.

Tandem Mass Spectrometry. Coupling mass spectrometers in series has many advantages for the analysis of specific organic compounds in complex mixtures. Sensitivity to picograms of targeted compounds can be achieved with high specificity and almost instantaneous response. As reported by McLafferty (see reference), the targeted compound is selectively ionized, and its characteristic ions are separated from most others of the mixture in the *first* mass spectrometer. The selected primary ions then are decomposed by collision and, from the resulting products, the *final* (or second) mass analyzer selects secondary ions characteristic of the targeted compound. Tandem mass spectrometry (MS-MS) can achieve specificities and sensitivities equivalent to those of methods such as radioimmunoassay and gas chromatography/mass spectrometry, while performing analyses in much shorter times. Just a few of the materials and systems successfully studied by tandem mass spectrometry include: polynuclear aromatics, DNA pyrolysis, steroid mixtures, ion structures, stereoisomers, pyrolysis of bacteria, alkaloids in plants, penicillins, polychlorodibenzodioxins, petrochemicals, drug metabolites, enkephalins, peptide mixtures, diesel exhaust, odors in the air, concealed drugs, parathion in lettuce, and ion plasmas.

Tandem mass spectrometry has been particularly effective in molecular structure determinations. To increase the number and absolute abundance of peaks in the secondary mass spectrum, it is necessary to add energy to the separated primary ions. Collisionally activated dissociation (CAD) is frequently used.

Fourier Transform Mass Spectrometry (FTMS). This technique enables chemists to use mass spectrometry in expanding and new ways. The technique enables the measurement of high molecular masses and, by application of ion manipulation (MS-MS), molecules can be degraded into more manageable pieces for which accurate mass measurement is feasible. As observed by Gross and Rempel, the unique applications of FTMS result from its ability to store ions. Ions in the trap can be reacted in very specific ways (chemical ionization) or activated by using lasers to give photodissociation spectra. Instead of operation at 1 torr of pressure as in the case of conventional mass spectrometers, FTMS chemical ionization must be conducted at 10^{-6} torr of reagent gas and 10^{-8} torr of sample.

Coupling Lasers with Mass Spectrometers. The analysis of inorganic atomic species can be facilitated by using lasers with mass spectrometers. A tunable dye laser, by itself or in combination with a pump laser, ionizes atoms by resonant excitation processes. The ions are then analyzed in the mass spectrometer. This combination of techniques has much potential for overcoming traditional limits of sensitivity and selectivity and will lead to increasing applications in analytical chemistry. As observed by Fassett, et al. (see reference), a large potential exists for the application of *multiphoton resonance ionization mass spectrometry*, ranging from basic spectroscopic studies of atoms and molecules to the detection of solar neutrinos and quarks. Discovery of the optogalvanic effect (Green, et al.) and earlier work (Hurst, et al.) led to the development of *laser-enhanced ionization* (*LEI*). It has been determined that elements suitable for resonance ionization by the one-photon-resonant, two-photon ionization scheme (wavelengths between 260 and 355 nm) include: Na, Ca, Ba, Cr, Mn, Sc, Cr, Mn, Ru, Rh, Pd, Pt, Au, Ga, Ge, Sn, Bi, Eu, Gd, Tb, Ho, Tm, and Yb. Elements for which resonance ionization has been demonstrated in the laboratory include: Li, Mg, Sr, Y, Ti, Zr, Hf, V, Ta, Mo, W, Re, Fe, Os, Co, Ni, Al, In, Ge, Sn, Pb, Dy, Er, Lu, Th, and U. Some elements also have been ionized by the two-photon-resonant, three-photon ionization scheme, including: Be, C, and I.

It is reported by Fassett, et al. that the resonance ionization process has an inherently high elemental selectivity. Mass spectrometric detection provides an increased selectivity that is a practical necessity for analytical problems in which nonspecific background ionization must be characterized and differentiated. Resonance ionization is ideally suited to mass spectrometry; ionization is well defined in both time and space, and only a small excess of translational kinetic energy is added to the atom by the process.

Nuclear Accelerators and Mass Spectrometers. An interesting area is that of using nuclear accelerators as high-sensitivity mass spectrometers. As recently as a decade ago, the Grenoble cyclotron was used as a mass spectrometer to measure ratios of $^{10}Be/^9Be$ of $10^{-8}, 10^{-9}$, and 10^{-10} in standardized beryllium oxide samples. Measurements of this type also can be used to determine cosmogenic ^{10}Be profiles in various geophysical reservoirs, such as sea sediments and polar ice. See also **Particles (Subatomic)**.

Mass Spectrometers as Process Analyzers. The use of mass spectrometers in process analysis has not been widespread because of its perceived complexity and high cost. Technological advancements over the past decade or two have reduced costs and simplified mass spectrometry to the point where it is suitable for a number of process analytical applications in place of infrared absorption or gas chromatography (GC) techniques. The *quadrupole mass spectrometer* is well accepted in such applications because of low cost, good reliability, and ease of computer-controlled interfaces.

Simply, a quadrupole mass spectrometer can be divided into three parts: (1) the ionizer, (2) the mass filter, and (3) the detector, all of which are contained in a vacuum chamber maintained at a low pressure. When a gaseous sample is introduced into the system's ionizer, it is bombarded with a stream of electrons, producing positively charged parent ions (ions with the same molecular weight as the neutral molecule), and fragment ions. The ionizer has a series of lenses that serve to collimate the cloud of sample molecules toward the mass filter.

A quadrupole mass filter is a set of four rods disposed parallel and symmetrically with one another; opposite rods are electrically connected. An rf and dc voltage of equal potential, but opposite charge, is applied to each set of rods (Fig. 2). By varying the absolute potential applied to the rods, it is possible to stop all ions except those of a given m/e (mass-to-charge) ratio.

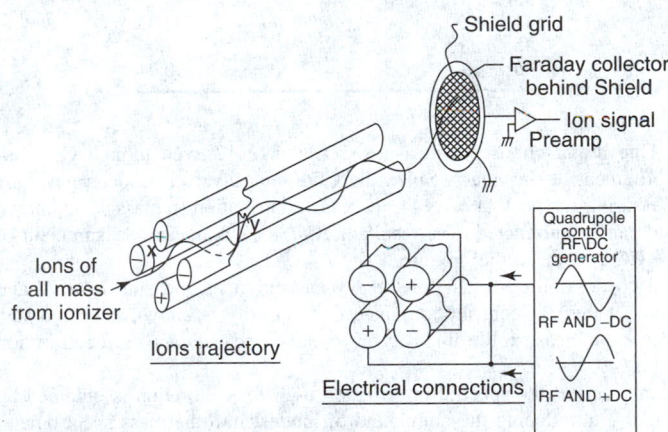

Fig. 2. Schematic diagram of quadrupole mass spectrometer of type used for process analysis. (*Extranuclear Laboratories.*)

Finally, the ions flowing down the quadrupole strike the Faraday plate detector. In some cases, the signal is amplified further by an electron multiplier. Thus, there is obtained a spectrum of signal intensity versus m/e value. Each molecule has a unique fragmentation pattern so that a spectrum can be used as a fingerprint for compound identification. In addition, it is possible to quantitate the amount of a particular compound by comparing sample signal intensity with the intensity produced by a known amount of the compound.

Many applications of the quadrupole mass spectrometer use a gas or liquid chromatograph to introduce the sample into the ionizer. When the spectrometer is used in this manner, it is most common to scan a wide mass range (50–1000 amu) at rates on the order of 1000 amu/second for compound identification. For process analyses, it is most common to introduce the sample directly into the ionizer and scan a shorter mass range. For both applications, computer systems are needed to collect the enormous amounts of data produced.

The computer is also used to control the potential applied to the quadrupole in order to scan the mass range of interest. Alternatively, the computer can command the appropriate potential required to focus a particular mass, a technique referred to as Selected Ion Monitoring (SIM). Process analysis is essentially continuous SIM to achieve online quantitation of the components in the process stream.

Once the stream components to be analyzed have been identified, the mass spectrum of each component must be compared to the spectra of all other components to find unique mass peaks. Not all constituents of the mixture will have unique peaks; some may have intense peaks that might be useful to measure that compound except that this mass peak also exists in another compound. Considering a stream containing N_2, CO, H_2O), and CO_2, Table 2 shows the mass peaks in each component's spectrum and their relative intensities using the largest peak from each compound as 100% in each case. Inspection of the table shows that m/e 18 and m/e 44 are unique peaks for H_2O and CO_2, respectively. However, the parent molecular ions of both N_2 and CO appear at m/e 28. Thus, some other method of measuring these components must be applied.

TABLE 2. MASS PEAKS IN EACH COMPONENT'S SPECTRUM AND THEIR RELATIVE INTENSITIES. (Largest Peak from each compound is used as 100% in each case)

Mass	H_2O	CO	N_2	CO_2
1	6.00			
12		4.00		9.00
14			5.00	
15		0.03		
16	4.00	1.00		10.00
17	28.00			
18	100.00			
28		100.00	100.00	11.00
29		1.00	0.73	0.09
30			0.02	
44				100.00
45				1.00
46				0.04

One might choose to measure CO at m/e 12 even though CO_2 has a fragment at this mass. Since the CO_2 intensity can be determined by measuring the peak at m/e 44, the CO_2 contribution to mass 12 is known and can be subtracted. The resultant at m/e 12 is then a measurement of the CO present in the mixture.

Two possibilities exist for the N_2 measurement; measure m/e 28 and subtract the CO contribution since CO intensity is known from the step above, or measure the intensity at m/e 14 and subtract the contribution from CO.

In practice, the spectral information used for calibration is obtained by actually introducing the compound of interest into the mass spectrometer and measuring the spectrum. This is usually done by using a known binary mixture of the particular compound in nitrogen or argon. The ratio of each mass peak compared to some standard ion, for example, m/e 28 of nitrogen, is measured. The fragmentation of the peaks of interest are then programmed into a fragmentation matrix. For each compound to be measured, only one mass peak is programmed.

For the analysis of N_2, CO, H_2O, and CO_2, the fragmentation matrix becomes:

	12	18	28	44
CO	4.00*	0.00	100.00	0.00
H_2O	0.00	100.00*	0.00	0.00
N_2	0.00	0.00	100.00*	0.00
CO_2	9.00	0.00	11.00	100.00*

The peaks to be measured for each component are indicated by an asterisk. The subtraction of interfering mass peaks is accomplished by inverting the matrix to yield:

	12	18	28	44
CO	25	0	0	−2.25
H_2O	0	1	0	0
N_2	−25	0	1	2.14
CO_2	0	0	0	1

In the actual analysis, the amplitudes of peaks at masses 12, 18, 28 and 44 are measured and multiplied by the inverted matrix. For example:

	12	18	28	44	
CO	25	0	0	−2.25	I(12)
H_2O	0	1	0	0	I(18)
N_2	−25	0	1	2.14	I(29)
CO_2	0	0	0	1	I(44)

Therefore, CO intensity is determined by:

$$25[I(12)] + 0[I(18)] + 0[I(28)] - 2.25[I(44)]$$

H_2O, N_2, and CO_2 are similarly calculated. Intensities may be converted to percent concentration by dividing by the component's sensitivity and normalizing to 100%. The formulae for these calculations are therefore:

$$INT(J) = IBKG(J) - BKG(J)$$

$$CONC(J) = \text{Sum over J } [INVER(I,J) * INT(J)]/SENS(J)$$

$$\%CONC(J) = CONC(J) * 100/\text{Sum over J } [CONC(J)]$$

A modern mass spectrometric process analyzer should be designed for rapid response time, wide dynamic range, accuracy, and ease of computer operation. Because the instrument is a nonspecific detector, it is useful in detecting almost any gaseous compound that produces ions in the mass spectrometer's range (typically 200 amu). The instrument is adaptable to process analysis because it is easily and rapidly directed to monitor any mass ion peak within its range. One modern instrument is controlled by a computer system, which commands masses, gathers and reduces data, and presents useful information almost instantly to the operator. The instrument enables the operator to design analysis, calibration, and sampling sequence methods to optimize conditions for a particular application, An analysis method is designed by specifying the molecular weight and the mass to be monitored for each of the components to be analyzed. The data system then searches a library (electronically stored) for the spectra and sensitivity of the compounds of interest. The fragmentation matrix used for the analysis is thereby constructed. Changing the parameters for any or all compounds can be easily accomplished by the operator. Complex streams may require more than one calibration gas to properly calibrate all stream components.

The analyzer has three modes of operation: (1) *Sweep Mode* — the system can be commanded through the CRT keyboard to sweep the quadrupole over a specified range with the data displayed on the oscilloscope. This mode is useful for tuning or to determine the masses present in a particular stream. (2) *Manual Mode* — during real-time operation, the operator can display and change parameters in order to fine-tune a particular analysis. Data can be presented as intensity (raw signal voltages), percent concentration, or taken as a ratio to a specific component. Optionally, data can be printed or electronically stored. (3) *Automatic Analysis*?

Additional Reading

Adams, F., R. Gibbels, and R. Van Grieken: *Inorganic Mass Spectrometry*, 2nd Edition, John Wiley & Sons, Inc., New York, NY, 2001.

Barker, J.: *Mass Spectrometry*, 2nd Edition, John Wiley & Sons, Inc., New York, NY, 1997.

Chait, B.: *The Application of Matrix-Associated Laser Desorption Time-of-Flight Mass Spectrometry to the Analysis of Proteins*, Pittcon, Pittsburgh, PA, (March 1992).

Conzemius, R.J.: *Prospects for Radically New TOF-MS Instrumentation; Making Single-Ion Concepts Provide Higher Performance*, Pittcon, Pittsburgh, PA, (March 1992).

Cotter, R.J.: *Designing Time-of-Flight Instruments to Solve Biological Problems*, Pittcon, Pittsburgh, PA, (March 1992).

Dass, C.: *Principles and Practice of Biological Mass Spectrometry*, John Wiley & Sons, Inc., New York, NY, 2000.

Delgass, W.N. and R.G. Cooks: "Focal Points in Mass Spectrometry," *Science*, **235**, 545–552 (1987).

Enke, C.G.: *Time-of-Flight Mass Spectrometry with Time-Array Detection*, Pittcon, Pittsburgh, PA, (March 1992).

Fassett, J.D. et al.: "Laser Resonance Ionization Mass Spectrometry," *Science*, **230**, 262–267 (1985).

Fenn, J.B. et al.: "Electrospray Ionization for Mass Spectrometry of Large Biomolecules," *Science*, 64 (October 6, 1989).

Goeringer, D.E., G.L. Glish, and S.A. McLuckey: *Fixed-Wavelength Laser Ionization/Tandem Mass Spectrometry for Mixture Analysis in the Quadrupole Ion Trap*, Pittcon, Pittsburgh, PA, (March 1992).

Grant, E.R. and R.G. Cooks: "Mass Spectrometry and Its Use in Tandem with Laser Spectroscopy," *Science*, 61 (October 5, 1990).

Green, R.B. et al.: *J. Am. Chem. Soc.*, **98**, 8517 (1976).

Hoffmann, E., De, V. Stroobant, and J. Charette: *Mass Spectrometry: Principles and Applications*, John Wiley & Sons, Inc., New York, NY, 1996.

Hurst, G.S., J.H. Nayfeh, and J.P. Young: *Phys. Rev. A*, **15**, 2283 (1977).

McLafferty, F.W.: "Studies of Unusual Simple Molecules by Neutralization-Reionization Mass Spectrometry," *Science*, 925 (February 23, 1990).

Siuzdak, G.: *Mass Spectrometry for Biotechnology*, Academic Press, Inc., San Diego, CA, 1996.

Watson, J.T.: *Introduction to Mass Spectrometry*, Lippincott-Raven Publishers, Philadelphia, PA, 1997.

MASTICATORY SUBSTANCES. The property of chewiness is one of several components that comprise so-called mouth feel experienced by the consumer of a food product. Whereas chewiness may be highly undesirable in a cut of roast beef, this property is the predominant rewarding factor of some food products, notably certain kinds of novelties, such as chewing gum. Biting and deformation resistance can be created or improved by the use of a number of essentially rubber-like substances. These are commonly termed *masticatory substances*. Chewiness is the main advantage contributed by such substances and thus other ingredients, such as sweeteners, flavorings, colorants, etc. are admixed with them to result in an overall attractive product for particular consumers.

Masticatory substances are of (1) vegetable origin, or (2) the products of organic synthesis. In the case of anhydrous lanolin, the source is fat from the wool of sheep.

Vegetable Substances. Masticatory substances derived from vegetables are gums from various plants and trees of the families *Apocynaceae* (dogbane family), *Euphorbiacea* (spurge family), *Moraceae* (mulberry family), and *Sapotaceae* (sapodilla family). Most of the naturally derived substances have unfamiliar names, usually known well only by persons in the trade. For example, from the *Apocynaceae* family are obtained jelutong, leche caspi or sorva, pendare, and perillo. From the *Euphorbiaceae* family, there are candelilla wax, chilte, and natural rubber (latex solids). From the *Moraceae* family, there are leche de vaca, Niger gutta, and tunu or tuno. From the *Sapotaceae* family, there are chicle, chiquibul, crown gum, gutta hang kang, gutta katiau, massaranduba balata, nispero, rosidinha, and Venezuelan chicle. Other naturally derived substances include lanolin, petroleum wax, rice bran wax, and natural terpene resin.

Synthetic Substances. During the last several decades, naturally derived masticatory substances have been displaced to a considerable degree by synthetic materials—for reasons of availability, economics, and, frequently, better control over purity. These developments essentially paralleled the development of the synthetic rubbers for industrial uses. Some of the synthetic substances now used and listed in the "Food Chemicals Codex," published by the National Academy of Sciences (Washington, DC) include: butadiene-styrene 75/25 rubber; butadiene-styrene 50/50 rubber; glycerol ester of partially dimerized rosin; glycerol ester of partially hydrogenated wood rosin; glycerol ester of tall oil rosin; isobutylene-isoprene copolymer (butyl rubber); methyl ester of rosin (partially hydrogenated); paraffin (synthetic, by Fischer-Tropsch process); pentaerythritol ester of partially hydrogenated wood rosin; polyethylene; polyisobutylene, polyvinyl acetate; and terpene resin (synthetic).

MASTITIS. Acute infection of the breast. The breast becomes swollen, red, and tender. Fever and malaise generally accompany the infection. Lactation is halted and antibiotics are administered. Further relief may be provided by a breast support and cooling with ice bags. In some cases, formation of a breast abscess will require incision and drainage.

Staphylococcal mastitis is most frequently seen in nursing mothers in the early postpartum period, but may occur in neonates or women who are not lactating, or have not been recently pregnant. Group B streptococci, once considered exclusively as animal pathogens causing bovine mastitis, have been clearly established as important human pathogens.

Bovine Mastitis. This is an inflammation of the udder and is the most costly disease to dairypeople.

Most of the economic loss from the disease results: (1) when cows produce less milk, because of permanent udder damage; (2) when cows are culled, because they do not respond to therapy or do not produce enough milk; (3) when milk must be discarded, from cows showing signs of inflammation and from those undergoing antibiotic treatment; and (4) when treatment with antibiotics or other drugs is expensive. Market milk is examined regularly to assure more uniform compliance with regulations prohibiting the sale of milk from diseased cows. When milk is found to contain an excessive number of leucocytes, the producer is warned to take corrective action. If the excessive cell count continues, the producer will be shut off from the market. It is estimated that, from all effects, the disease costs the American dairy industry over $1/2 billion per year.

Over 20 different organisms cause mastitis. All of these can be transmitted from cow to cow. Two types of bacteria most frequently found in any herd are *Streptococcus agalactiae* and *Staphylococcus aureus*. The former bacteria live only in the cow's udder, but *S. aureus* can be harbored in a variety of places in the cow's environment.

With this disease, the cow's udder becomes inflamed, a condition that can be acute or chronic. Acute mastitis is easy to recognize and is dangerous. This condition is usually fatal. In the acute form, the entire body of the cow reacts to the infection. Effects are depression, rough coat, dull eyes, loss of appetite, constipation, fever, and, eventually, death.

In the chronic form, the affected quarter is hot, very hard, and tender. The hardness of the udder is caused by an influx of white blood cells and fluids from the blood to fight the bacterial infection. If the action of the bacteria goes unchecked, some of the normally functioning tissue will be destroyed. It will then be replaced by scar tissue, which can be felt as hard lumps or knots. Eventually, the entire gland atrophies and milk production may stop completely.

MASU SALMON. See **Salmon**.

MATAMATA. See **Turtles**.

MATHEMATICAL CLIMATE. See **Climate**.

MATHEMATICAL PHYSICS (Equations of). The name is sometimes given to a set of partial differential equations of second order, of which the following are the most commonly met with:

(1) the Laplace equation,

$$\nabla^2 \phi = 0$$

and its inhomogeneous analogue, the Poisson equation;

(2) the equation of wave motion,

$$c^2 \nabla^2 \phi = \partial^2 \phi / \partial t^2;$$

(3) the diffusion equation, which also applies to thermal conduction,

$$a^2 \nabla^2 \phi = \partial \phi / \partial t;$$

(4) the equation of telegraphy,

$$a \partial^2 \phi / \partial t^2 + b \partial \phi / \partial t = \partial^2 \phi / \partial x^2.$$

The parameters are observable quantities and t is the time. In modern theoretical physics, the differential equations of quantum mechanics, particularly the Schrödinger wave equation, must be included.

Many special functions (described elsewhere) owe their importance to the fact that they are useful in constructing solutions of these equations. In fact, the name "equations of mathematical physics" is sometimes also given to the linear ordinary differential equations, which arise when the above partial differential equations are solved by the method of separation of variables. These ordinary equations are all specializations arising from confluence from the generalized Lamé equation.

See also **Bessel Function**; **Lamé Equation (Generalized)**; **Laplace Equation**; and **Poisson Equation**.

MATHEMATICS.   Several score entries pertaining to mathematics and related topics appear in this encyclopedia. Consult the alphabetical index. Particularly, check the following key words and phrases: angles, triangles, squares, and polygons; arithmetic and algebra; charts, graphs, tree maps; coordinates; curves; sections and solids; determinants and matrices; differential equations; equations; functions; groups; integrals and integration; interpolation and approximation; series; space and topology; statistics, distribution, and probability; tensors; and vectors.

Mathematical Symbols. The symbolism of mathematics dates back at least three centuries. To a large degree, there is standardization of these symbols. However, one may find optional symbolic devices, each equally acceptable. Simply because of limitations imposed by the English and Greek alphabets, there is some duplication of symbol usage between mathematics and other scientific fields. An abridged list of some of the more commonly used mathematical symbols is given in the Abridged List of Mathematical Symbols.

MATHEMATICS (State of the Art Reviews).   During the summer of 1985, a group of state-of-the-art reviews was initiated by the National Research Council (NRC) at the request of the National Science Foundation (NSF). The purpose of these reviews was to assess and monitor world trends, relative strengths, and competitiveness of the United States in rapidly evolving areas of science and technology. Particular emphasis was placed on developments that influence the rate at which these scientific fields evolve. *Applied Mathematics* was one of three such studies undertaken.[1] The study on mathematics was conducted by the Panel on Mathematical Science under the auspices of the Board of Mathematical Sciences of the NRC's Commission on Physical Sciences, Mathematics, and Resources. In its final report the panel described major trends in modern mathematics, illustrating these trends with a few examples. The report, in full, was published in 1986. The study was funded by U.S. Government resources. Chairman of the Panel on Mathematical Sciences for the report was Phillip A. Griffiths, Duke University.[2]

Current Summary.   In this 9th Edition of the encyclopedia, an attempt is made to update the 1985 findings.

The executive summary of the 1985 report stresses the following trends in the mathematical sciences:

1. Mathematical sciences research is strong worldwide; the United States is maintaining the leading role.
2. Mathematics is unifying internally.
3. Applications of mathematics in both traditional and new areas are flourishing and involving more central areas of mathematics.
4. Mathematics is the driving force behind new areas of computational science and, in turn, is profoundly influenced by high-speed computing.

There are several corollaries of the trends, which are especially notable in view of a declining number of Ph.D. degrees in the mathematical sciences earned in the United States. The expanding number and sophistication of tools needed for successful research as areas of mathematics become intertwined now require protracted study often beyond the Ph.D. degree. This corollary development is especially critical for those working at high levels of applications of mathematics. The difficulty in reaching mathematical research frontiers with the requisite deep, broad range of knowledge, without the opportunity for extended study, may partially

[1] The other two studies: Cell Biology; Materials Science.

[2] Other panel members: Hyman Bass (Columbia University); Peter Bickel and Alexandre Chorin (University of California, Berkeley); Richard Dudley (Massachusetts Institute of Technology); Wendell Fleming (Brown University); Ronald L. Graham (AT&T Bell Laboratories); David Kazhdan (Harvard University); Cathleen S. Morawetz (New York University); Richard Schoen (University of California, Los Angeles); and Michael E. Taylor (State University of New York at Stony Brook).

explain the decreasing Ph.D. production. Continued concern should be expressed over the decline in the number of Ph.D.s in the mathematical sciences in the U.S., and in view of the critical role of mathematics, this trend must be reversed. The competition for students among the various fields of science is keen. Thus it is critical that mathematics be able to provide sufficient inducements to candidates to maintain its vitality. That the U.S. maintains mathematical preeminence is due in part to the commitment and investment made in the past to an outstanding group of researchers. Unfortunately, this commitment has been considerably weakened over the past decade, as was documented in the 1984 NRC report, Renewing U.S. Mathematics: Critical Resource for the Future, released by the National Academy in 1984.

Unity of Mathematics

The unification that is taking place within mathematics is obvious to people in the field and will be apparent in the examples (vignettes) given in this article, particularly with reference to D-modules, computational complexity, the Yang-Mills equations, and operator algebras. Unification occurs when there is a confluence of seemingly independent phenomena, motivating cooperative study of the significant underlying patterns. One symptom of this accelerating unification is the increasing difficulty that agencies are having in assigning proposals to their discipline programs, which now substantially overlap. The trend burdens young investigators with a need to pursue increasingly broad training, as is noted in several of the examples given later. In mathematics, it is becoming critical to lengthen the training period substantially.

An example of this confluence of areas is the Korteweg-de Vries (KdV) equation $u_t = uu_x = u_{xxx} = 0$ [where u is an unknown function $u(x, t)$ in one space dimension and time], which arises both as the simplest nonlinear dispersive equation in shallow-water wave theory and as the equation of isospectral evolution of the potential in the Schrödinger equation $\psi_{xx} = (k^2 - u)\psi = 0$ of quantum mechanics. The intensive study of the KdV equation during the past quarter century has affected many major areas of mathematics. For example, recently a young Japanese mathematician, using a development in the study of KdV equations initiated a decade before by the Moscow school, solved a major problem in algebraic geometry that was first discussed more than a century ago by Riemann, a German.

Mathematics and Other Sciences

The symbiotic relationship between mathematics and its areas of applications is ever deepening as more areas of science and engineering become almost indistinguishable from subareas of mathematics, and this relationship is producing exciting and intriguing new mathematics. Cross-disciplinary collaboration between mathematicians and professionals in other fields is accelerating and deserves encouragement. An important number of interactions between mathematics and science and engineering are not exactly interdisciplinary, but might be more accurately described as resonance phenomena in that advances in one field spur development in another. Examples of this important trend are described in the examples on Yang-Mills equations and operator algebras described later in this article. The broadening and deepening of these applications, as noted in the example on nonlinear hyperbolic conservation laws, create pressure for mathematicians to pursue significant postdoctoral study.

See Table 1 for a list of mathematical symbols.

New Opportunities for Mathematics

There is a growing trend for mathematics to be incorporated into the language, not only in the physical sciences and technology, but also into other fields, such as the social sciences. Mathematical models (descriptions of real-world events that use mathematical language) form the basis of econometrics and health policy analysis. The survival analysis example, described later, demonstrates this. It examines statistical and mathematical methods used to provide a realistic analysis for problems in medical research, reliability theory, actuarial computations, and demographic studies. Mathematical analyses contribute substantially to decisions about economic and health policies, which in turn have enormous financial and social consequences.

TABLE 1. ABRIDGED LIST OF MATHEMATICAL SYMBOLS

Symbol	Definition
$\times$ or $\cdot$	Multiplied by; or times
$\div$ or;	Divided by
$+$	Plus; added to; positive
$-$	Minus; subtracted from; negative
$\pm$	Plus or minus; positive or negative
$\mp$	Minus or plus; negative or positive
$=$ or $::$	Equals; is equivalent to
$\equiv$	Identical with
$\cong$	Is approximately equal to; approximates
$\neq$	Unequal; does not equal; not equivalent to
$>$	Greater than
$\gg$	Much less than
$<$	Less than
$\ll$	Much greater than
$\geq$	Greater than or equal to
$\leq$	Less than or equal to
$)$	Therefore
$\angle$	Angle
Δ	Increment or decrement
Δx	Change in x
Δf	Change in value of f
$\perp$	Perpendicular to
$\parallel$	Parallel to
$a \in A$	a is an element of set A
$\{\}$	Notation of a set
$\{3\}$	The set of which 3 is the only element
$\{2, 7, 15, 36\}$	The set whose elements are 2,7,15,36
$X = \{x \mid x \text{ is a real number}\}$	The set of all real numbers
$X \times Y$	The set of all ordered pairs (x, y) of real numbers
$\emptyset$	The empty set
$\cup$	The universal set
$A \subseteq B$	Set inclusion
$A \subset B$	Proper set inclusion
$A \cap B$	Intersection of A and B
$A \cup B$	Union of A and B
A'	Complement of A
$|a|$	Absolute value
$a + bi$	Complex number in conventional notation
(a, b)	Complex number in ordered pair notation
$r(\cos \theta + i \sin \theta) = r \operatorname{cis} \theta$	Complex number in polar form
$a \equiv b, \bmod m$	a congruent to b modulo m
$1.\overline{142857}$	Repeating decimal
$n!$	factorial
f	Function
$f(x)$	Value of f at x
$f : X \to Y; f : (x, y)$	Notations for a function
f^{-1}	Inverse of f
$e^x = \exp x$	Exponential function
$\log x$	Logarithmic function
$\sin x$	Restricted sine function
$\arcsin x = \sin^{-1} x$	Inverse sine function
(r, θ)	Polar coordinates
$\lim\limits_{x \to \infty} f(x)$	Limit of the function f as x approaches ∞
$\lim\limits_{x \to \alpha} f(x)$	Limit of the function f as x approaches a
$\lim\limits_{x \to \infty} S_n$	Limit of a sequence as n approaches infinity
p, q, r	Propositions
p_x, q_x, r_x	Open sentences
P, Q, R	Truth sets of p_x, q_x, r_x
$\forall x$	For all x
$\exists x$	For some x
$p \wedge q$	Conjunction of p and q
$\sim p$	Negation of p
$p \to q$	The implication, "If p then q"
$p \leftrightarrow q$	p is equivalent to q
$[\,]$	Greatest integer function
$g \circ f g(f(x))$	Composite of g and f
$\sum$	Summation
$A_a^b = \lim\limits_{n \to \infty} \sum f(x_i) \Delta x$	Area under $y = f(x)$ from $x = a$ to $x = b$
$\int_a^b f(x)\,dx$	Definite integral
$D_x f = f'(x) = df/dx$	Derivative of f
$P(A)$	Probability of the event A
$P(A \mid B)$	Probability of A, given B
$P_{n,r}$	Number of permutations of n things, taken r at a time

Symbol	Definition
$P_{n,n}^{r_1, r_2}$	Number of permutations of n things, n at a time, of which $r_1 r_2 \ldots$ are alike
$C_{n,r}$	Number of combinations of n things, taken r at a time
$\binom{n}{r}$	Binomial coefficient
$\begin{pmatrix} a & b \\ c & d \end{pmatrix}$	2×2 matrix
$\circ$	Group operation
$\oplus$	Sum of vectors
$\begin{pmatrix} 1 & 2 & 3 \\ 3 & 1 & 2 \end{pmatrix}$	Permutation
$P(x)$	Polynomial
(x, y)	Rectangular coordinates of a point in the plane
$\sqrt{x}$	Square root of x
x^n	x raise to a power of n. n is referred to as the exponent

There is no longer any question as to whether mathematical analysis will substantially influence discussions of public policy, but only whether it will be used appropriately and effectively. It is essential that those making the decisions understand and influence the assumptions used to form the mathematical model and that mathematicians comprehend the applications sufficiently well that they address and solve the *correct* problem. In fields where mathematical models are not subject to experimental verification—such as those dealing with the most drastic consequences—it is especially essential that the mathematics be critically scrutinized.

The Role of Computation

Computational mathematics is an integral part of the mathematics discussed in the examples given later about complexity and nonlinear hyperbolic equations, and there are two important observations worthy of emphasis: (1) Computational methods pervade almost all aspects of science, and mathematics is the foundation of these methods. Today's complex problems, involving computational solutions, range from the design of computer architecture itself through mathematical modeling of physical, chemical, biological, and engineering processes. Mathematics, the intellectual basis of computational science, has been and will continue to be the key to the dynamic revolution being created by the computer in science and engineering. (2) Computational results provide the insight for the development of mathematical theory. For example, the behavior of the solutions to the KdV equation previously mentioned was first discovered numerically. The mathematical theory, in turn, provides a deeper understanding of the models, revealing phenomena that enable people to analyze and test previous computational results and conceive of new computations that will facilitate further theory.

Core Mathematics now consist of three basic operations—*computation, abstraction*, and *generalization*. Raw information leading to mathematical discovery comes from concrete examples, and it is increasingly the role of computation to provide such examples, although they are frequently formulated mathematically. Abstraction is the process of distilling the essential features from such examples. Number and space are, respectively, abstractions of the process of counting and of our experience in the physical world, and the mathematical idea of functions similarly abstracts human ideas of measurement and motion. In these contexts, ideas from one manifestation of the abstraction are often relevant in solving problems in seemingly dissimilar situations. Generalization uncovers hidden analogies between abstract patterns of mathematics and frequently extends the range of applicability of such patterns.

Thus, in the development of mathematics, periods of computation often alternate with periods of theorizing. During the former, new raw data are generated and horizons expanded. Eventually, there is a plethora of information in need of an intellectual framework on the basis of which masses of material can be comprehended simultaneously. As reunification occurs, seemingly disparate examples are often revealed as different aspects of the same phenomenon.

Computation, abstraction, and generalization need each other to be meaningful. Recent mathematics has focused more on concrete problems than on abstractions as it revels in the computing power suddenly available to it, and evidence of a new unification is now appearing. At the same time, our computing power has matured to the point where it can be an enormous asset in the investigation of the still more complex mathematical examples that will inevitably be suggested by pending research.

Organization of Mathematics

More and more, the mathematical community is now organizing itself by areas of interest instead of fields of traditional study. A variety of mathematicians traditionally educated in the disparate fields of topology, algebra, differential geometry, algebraic geometry, complex variables, and partial differential equations now must cooperate in solving some of the newer, exciting equations. In numerous organizations, research is accomplished by teams frequently gathered around major pieces of scientific equipment or laboratories. Each team consists of at least one senior scientist and many, sometimes a great many, junior investigators working on related problems. Mathematicians likewise are often informally grouped around a common research interest. Such mathematical groups, however, are usually geographically separate, and the analogue of access to a common major piece of scientific equipment is their ability to gather together for sustained periods of collaboration. (Possibly this type of process in broadening the student mathematicians can be targeted in educational institutions.)

Language of Mathematics

At least three centuries ago, mathematics developed a language of its own, which has become thoroughly distinctive and international. Just as it takes years for students from one country to become fluent in the spoken language of another, any aspiring student, discounting his/her nationality, spends years studying the *common language* of mathematics. As a result, each can open the other's mathematics books and recognize the topic under discussion (even if ignorant of the other's verbal language).

Language Problem for the Nonmathematician. The complex and dynamic language of mathematics that has developed over the past brings great satisfaction in terms of international understanding for those who are fluent in it, but the professional language does present substantial obstacles to those who have not devoted years to its study. The history of mathematics is one of progressively less translatability for nonmathematicians. In contrast, over a long period, there has been an ever-broadening use of mathematics in what might be termed nontraditional mathematical areas. (What may be the solution to this problem?)

A Sampler of Interesting New Topics in Mathematical Science

Traditionally, core mathematics has been broadly divided into analysis, algebra, and geometry/topology, although all three subfields include extensive applied mathematics components. (Geometry and topology are not synonymous, but each term is often used for areas in their broad confluence.) However, as the interplay of analysis, algebra, and geometry/topology becomes more complicated, even the division of core mathematics into subfields (not to mention its distinction from applied mathematics) seems artificial. Indeed, the essential unity of mathematics is vivid as one reviews some important recently discovered relationships among these traditional subfields.

The Concept of Chaos — A Fresh Perspective

Since the early formative years of science, the precepts of classical mechanics were entrenched firmly in the pursuit of dynamic systems and guided by the unwavering notion that the behavior of complex systems could be predicted accurately, provided that one had enough information and intelligence. The concept (or theory) of chaos has challenged this historic approach. The ground rules are changing!

The "sufficient information" doctrine first was challenged at the atomic level by quantum mechanics in the 1920s; then, during the 1980s, prior tenets received another blow with the emergence of chaos theory. This theory holds that, for microscopic or macroscopic systems, tiny variations in initial conditions sometimes may create unexpected, radically different outcomes, seemingly making it impossible to predict fully the behavior of some systems. Perhaps most startling of all, such behavior can arise in relatively simple systems governed by a few uncomplicated equations. Thus, relatively simple or highly complex systems can exhibit chaos.

During the course of the first score of years of its existence, chaos theory has generated wide interest in academia, and, although relatively few practical applications have emerged, research is intensifying. As pointed out by a number of researchers in the field, the science of system dynamics is entering a new era, one that is comparable to the time frame when quantum mechanics was "fleshing out." A period of intellectual ferment transpired years before the transistor made its debut. Presently, chaos is being searched for its rightful scientific underpinnings.

Numerous researchers in recent years have stressed the ubiquity of chaos, ranging from fluid dynamics to electric power networks to physiology — ad infinitum. Before describing, within the next few pages, the searches underway for the practical application of chaos theory, a variety of general observations made by some of the leaders who are pursuing this strange blend of mathematics and physics may serve as an introduction.

John Dorning, University of Virginia: "For decades, engineers, scientists, and mathematicians alike, for the most part, when confronted with nonlinearity looked the other way or looked and shrugged their shoulders, or worse yet, looked and saw nothing at all beyond which their intellectual tunnel vision allowed."

Jong Kim, Electric Power Research Institute: "With chaos, we're on the brink of a new classical dynamics, and people thought that classical physics was dead."

John Stringer, Electric Power Research Institute: "It's called the curse of dimensionality — the amount of data you need to understand a system rises exponentially with the system's dimensionality, that is, the number of independent variables or degrees of freedom needed to describe it. Some of the projects involving what we thought would be simple questions have turned out to be very difficult. And, of course, there's the problem of noise. In many cases, it may be very hard to get data sets that are sufficiently tidy for understanding chaos. On the other hand, chaos theory can help us learn the limits of predictability for very complex systems, such as the weather, and may even give us new tools for controlling these systems."

Chaos in Electric Power Generation and Distribution. John Douglas (Electric Power Research Institute, EPRI) observes, "The implications of chaos theory for electric power equipment and networks are both disturbing and exciting. On the one hand, an unsuspected potential for instability may lurk among the operating conditions of systems thought to be well understood."

Sudden voltage collapses on power grids, for example, may indicate the presence of underlying chaotic dynamics. On the other hand, understanding chaos may provide unprecedented control over some of the most complex and elusive natural processes, such as combustion, corrosion, and superconductivity.

Along these lines, EPRI (Palo Alto, California) established a workshop consisting of specialists selected from a variety of disciplines, ranging from physics and engineering to physiology and computer science, in an effort to bring chaos theory to bear on practical problems of the power industry.

Jong Kim and John Stringer, both leaders of this workshop, astutely observe, "Is chaotic behavior impossible to understand? Not necessarily, according to current theory, which describes an underlying order in seemingly random phenomena. Using the tenets of deterministic chaos, EPRI is doing exploratory research on several dynamic processes of importance to the utility industry — searching for points of departure from linear behavior and for the reasons why predictable dynamics become chaotic. For power delivery systems, this research may help define the difference between a stable network and a system failure. For combustion processes, on the other hand, chaotic behavior may actually be encouraged in order to optimize the turbulent mixing of fuel and air that leads to higher combustion efficiency. Convection and metal passivation also may have chaotic aspects. Understanding chaotic behavior in such processes — learning how to control it and, if desirable, reverse it — could lead researchers to a better grasp of complex natural phenomena and to very practical fixes as well."

The EPRI researchers stress that the application process probably will be long and complex. The problem is how to distinguish "deterministic chaos" from stochastic, or totally random, behavior. John Stringer emphasizes, "Chaos has an underlying order, a pattern that's not periodic, but isn't completely random either. In any real system, however, some stochastic processes are also likely to be present as noise. It's like looking for a fuzzy

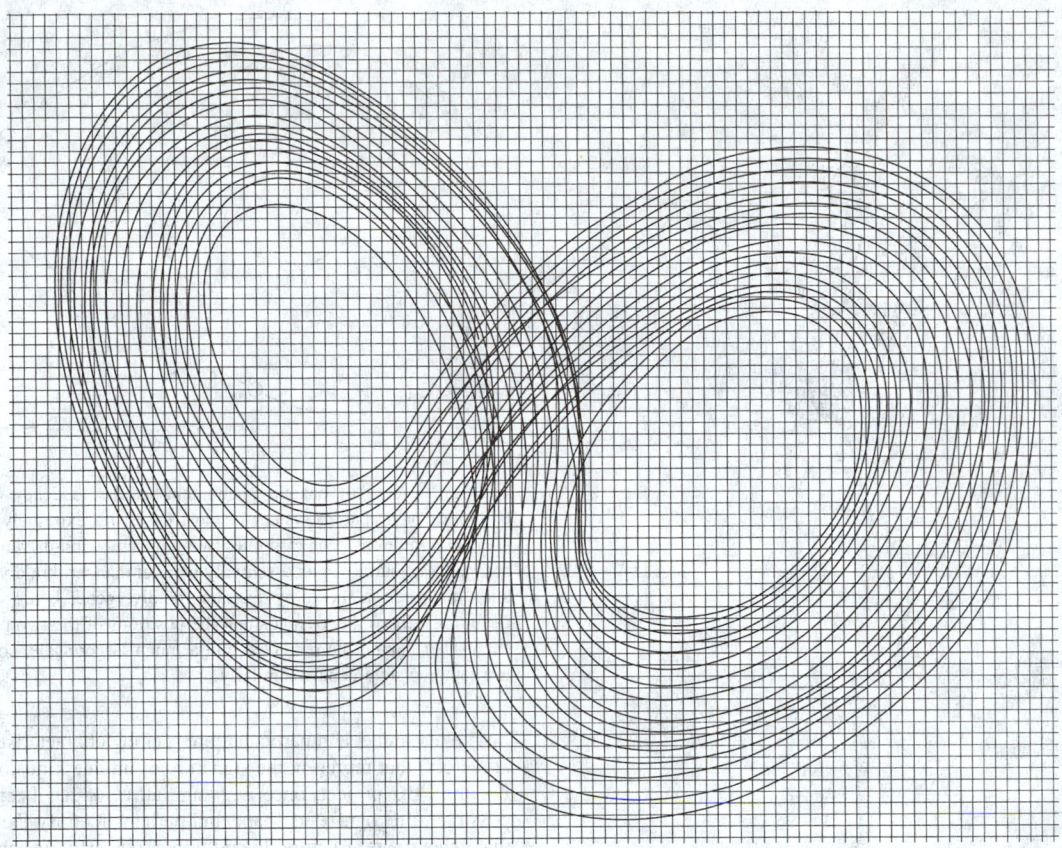

Fig. 1.   Reasonable facsimile of a computer-generated butterfly pattern. This is a representation of the Edward Lorenz butterfly pattern (1961) that displays an "orderly disorder." The pattern exhibits deterministic chaos. The pattern was created by Lorenz as he attempted to develop a simple computer model of weather based on convection currents in the atmosphere. In an elegantly created computer graphic of the pattern, it would be noted that the lines in the shape never quite touch each other, thus imputing the pattern as an infinitely complex microstructure.

When, in the investigation of a chaotic phenomenon, a recognizable pattern of data emerges, the pattern may be indicative of the types of equations that generate the pattern's structure. Such patterns frequently are referred to in the literature of chaos as *strange attractors*.

A powerful tool of chaos theory, known as the Takens embedding theorem, may create an identifiable pattern simply by analysis of how any one of the key variables evolves.

pattern through a fog!" Indeed, what eventually led to the current revolution in the science of dynamic systems was the slow realization that chaotic behavior is ubiquitous. Today, chaotic behavior is found in the fluctuations in predator-prey populations, fibrillation of the heart, the dripping of a faucet, and trends in the price of cotton; these all demonstrate telltale signs of chaotic behavior. They are amenable to the same mathematical formalities.

In commenting on the EPRI programs targeted for study, researchers observe that the equations that describe potentially chaotic systems often have at least one thing in common: *nonlinearity*. The proportionality and easy-to-solve aspects of linear equations are *not* present. In addition to lacking proportionality, nonlinear equations differ qualitatively as conditions change.

Currently, EPRI is exploring four specific chaos-focused research projects: (1) convective flows, which have potential application in nuclear safety and thermal power plant operating efficiency; (2) fluidized beds, which are an attractive model for understanding and potentially improving control over chaotic flow processes, keeping in mind that chaos can be a desirable condition because it leads to good mixing; (3) power grids — learning how to avoid any combination of conditions that could lead to chaos; and (4) the kinetics of metal passivation, in an effort to reveal how chaotic processes influence corrosion.

Convection (The Problem of Nonlinearity). As early as 1899, the French mathematician Jules-Henri Poincaré first recognized the possibility for chaotic behavior in dynamic systems. However, it was not until 1961 that the meteorologist Edward Lorenz observed the phenomenon when he was attempting to construct a simple computer model of weather on the basis of convection currents in the earth's atmosphere. Lorenz was puzzled by the sensitivity of the model to what appeared to be insignificant differences

in starting conditions. Even after Lorenz developed a very abstract version of convection using only three variables and three equations, he still encountered unexpected complex behavior. Because the variables changed in a very complex manner, Lorenz found it impossible to forecast their values with any degree of certainty over long periods of time. As Lorenz mapped their long-term trends, he found that the variables produced a three-dimensional pattern, as shown in Fig. 1. Because in three dimensions it appeared something like the outstretched wings of a butterfly, this type of pattern is now dubbed the "butterfly" and is frequently mentioned in the current literature on chaos.

The development of chaos science to where it is today is exquisitely summarized by John Douglas: "It took more than a decade and a half for this phenomenological pattern (Lorenz butterfly) — globally organized but locally unpredictable — to gain enough recognition to be named and it took even longer for investigations of chaos to earn scientific respectability."

As pointed out by Douglas, "Convection starts out as a smooth flow that speeds up as the temperature difference between the top and bottom of a fluid increases. But, beyond a certain point, instabilities begin to appear and at great enough temperature differences, the flow becomes turbulent — that is, *chaotic*." See Figs. 2 and 3.

The EPRI program on chaos in convective systems, now underway at the University of Virginia, is targeting on the development of a generic model of nonlinear flow in critical electric power devices, such as transformers, heat exchangers, and boiling water (nuclear) reactors. The three original Lorenz equations, previously mentioned, are being used as a starting point. Researchers are attempting to determine what conditions lead to chaotic flow and what effect this transition has on the heat-removal efficiency of the devices in question. Researchers also will investigate the feasibility of

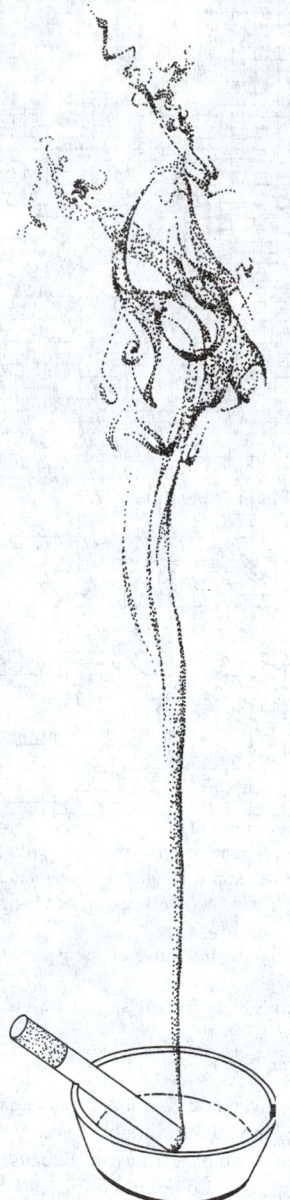

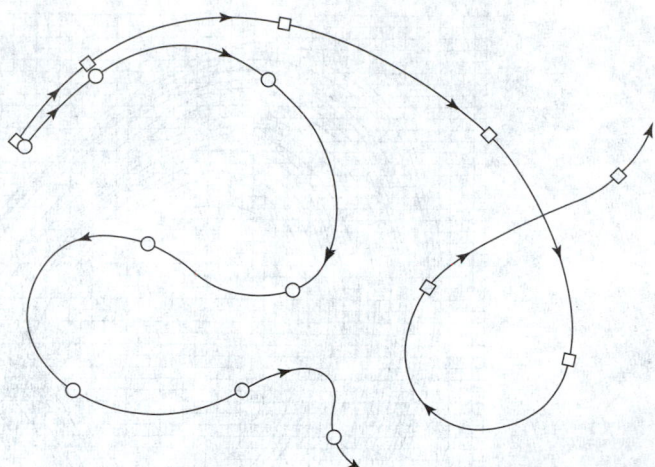

Fig. 3. A tenet of chaotic phenomena is the tendency for divergence. The dynamics of a chaotic system in most cases are extremely sensitive to starting (initial) conditions. For example, two adjacent wind-driven particles may commence their journey in an essentially parallel manner, but diverge radically after a short distance. As shown by diagram, the divergence is not slight, but is a radical departure. Chaos theory holds that the positions of the two particles, as related to each other, cannot be predicted. On a much larger scale, convection currents of a weather system behave similarly.

The researchers, in particular, are seeking characteristic patterns (strange attractors) that represent the signature of unique kinds of chaotic behavior. For the FBC project, once a strange attractor has been found, it may be possible to recognize its shape and perhaps identify the types of equations that generate its structure. In addition, a powerful principle of chaos theory (the Takens embedding theorem) implies that, at least theoretically, the form of a strange attractor should be identifiable simply by examining how any of the key variables evolves over time. This is regarded as pertinent to the FBC project because the simultaneous measurement of multiple variables, such as pressure, particle velocity, et al., in a particular small region of interest is extremely difficult, if not impractical, to attempt.

Also, there is the ever-present problem of noise in a system. The question is posed: "How can one detect a recognizable chaotic pattern when it may be obscured by the random meanderings of stochastic events?"

Investigators at Oak Ridge National Laboratory believe that they have found a way to separate major dynamic effects from noise, producing a series of somewhat messy, but nevertheless recognizable, strange attractors.

Chaos in Power Grids. As emphasized by John Douglas (EPRI), "The very idea that chaos may occur in electric power grids is about as welcome to utility planners as a heart flutter — and for many of the same reasons. If confirmed, it would at least prompt a fundamental rethinking of the analytical methods used to ensure network stability. At worst, it could mean that power systems harbor an unappreciated potential for voltage oscillations and collapse — the network equivalent of a heart attack."

EPRI research reports that a study commenced in 1989 focuses on the chain of events that might lead to fully chaotic behavior. Simple network models indicate that the onset of chaos is preceded by system bifurcations, in which operating characteristics suddenly can oscillate between two sets of conditions. Problems multiply if one or both of the new operating conditions lead to instability. Also, successive bifurcations (dividing into two or more branches) can so disrupt a system that eventually the number of possible operating states becomes infinite — that is, chaos ensues. See Fig. 4.

A common type of bifurcation observed in utility power systems is *period doubling*, in which line frequency jumps between 60 and 30 Hz. Mark Lauby (EPRI power system engineering manager) observes, "It's like plucking the D string on a violin and watching the G string start to vibrate in response. As long as the violin is constructed to withstand both kinds of vibration, there's no problem. But some power systems might be hypersensitive to the new frequency and collapse after a bifurcation — which would be like having the violin fall apart in your hands. With chaos there are an infinite number of possible frequencies, and the system could collapse from experiencing any of these outside its

Fig. 2. For a considerable height, the smoke from a lighted cigarette at rest initially rises in typical laminar flow. But then the flow of smoke takes on a rather violent, turbulent flow behavior. The fine structure of the turbulent flow pattern currently is beyond exact prediction. This is an example of chaos as related to fluids.

reversing this process — that is, driving the system back from chaotic to periodic flow.

Jong Kim, project leader, comments, "We need to understand what conditions can result in chaotic behavior in major power systems. It's not always to be avoided, of course. Chaotic flow is an aid when you want more fluid mixing, for example. But, we do need to learn how to control chaos, including how to reverse it. Such research will be particularly important for ensuring the stability of the next generation of so-called passively safe reactors, which rely on natural convection to provide emergency cooling capability."

Chaos in Fluidized-Bed Combustion (FBC). This program has at least two important targets: (1) instabilities in low-nitrogen oxide (NOx) burners, and (2) the formation of air pollutants at various stages in conventional boilers. The EPRI plan notes that an FBC unit makes an ideal model for studying chaos, since coal and limestone particles are lifted by upward-rushing gases to form a suspended bed of material that acts like a turbulent fluid.

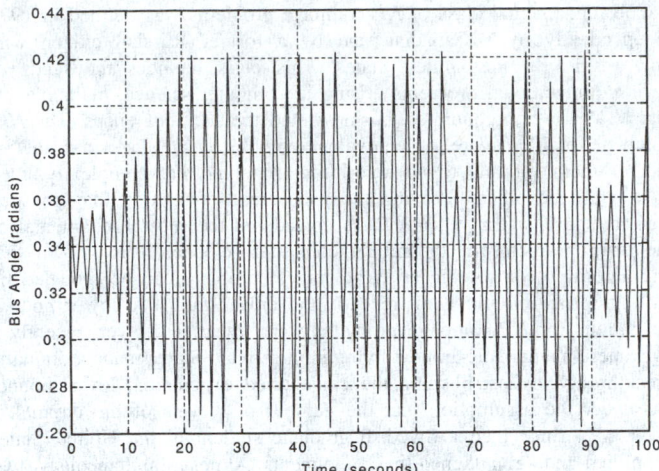

Fig. 4. Power system stability. Shown here is the relative generator bus angle on a simulated system, driven to an unstable, chaotic state by increasing the reactive power by only about 8%, again demonstrating the characteristic of chaos in which small divergences result in what appear to be disproportionately large changes. (*Electric Power Research Institute.*)

design limits. What we hope to do is to learn how to eliminate conditions that could set up network bifurcations and thus block the path to chaos."

Research along these lines is being conducted at Cornell University and the University of California at Berkeley. Investigators are using highly simplified power system models, from which they have identified various kinds of bifurcations as well as chaos, among the myriad operating conditions that affect the behavior of variables in the "state space" of the system. Although the researchers consider these observations rather tentative, the problems observed could conceivably result from modeling inadequacies, but nevertheless they raise important warning "flags." The studies do establish the presumption that chaotic behavior will exist in most power system models; however, it is not yet clear if chaos occurs for parameters in regions sufficiently near (ordinary, conventional) operating regimes to affect the stability region of utility power systems to a significant extent.

Utility engineers long have recognized that power systems have nonlinearities. Now that chaos and bifurcations exist in these systems, ways must be developed to reliably identify their presence. Then remedial measures for controlling systems and steering them away from these conditions must be tackled. Analysis tools must be developed to enable the design of power systems that will be free of the threats of chaotic behavior.

Exploring Chaos at the Microscopic Level. It is well-established knowledge that the corrosion of metals exposed to reactive gases and liquids can be prevented by improving *metal passivation* (that is, by reducing the chemical activity of a metal surface as its electric potential increases). This is extremely important, not only to the power industry, but to other industrial applications. It is estimated, for example, that 4% of the gross national product of the industrialized nations is lost to corrosion.

Researchers have found that the sets of equations used by various models to describe metal passivation are all nonlinear and thus pose an inherent possibility for chaotic phenomena. Work underway through EPRI sponsorship at Batelle Memorial Institute and Ohio University has led to the development of a new model of metal passivation. The model predicts that chaos will occur during the passivation process and identifies several different bifurcation routes to full chaotic behavior. To date, the physical implications of the model remain unclear because of the surprisingly complex ways in which chaos appears and disappears again in response to small changes in key parameters. This type of research, of course, requires extremely precise control of electrochemical conditions at the metal surface.

Chaos and Fractal Geometry. The concepts of chaos and fractal geometry have been researched essentially during the same time frame. The interrelationship of the two disciplines has been observed by numerous scientists during the past few years. Fractional dimensionality contributes an additional complexity in one's understanding of chaos. See also article on **Fractal Geometry**.

This connection is being explored by numerous researchers, including the EPRI group.

As pointed out by John Douglas, "The idea of multiple and fractional dimensionality is not as unfamiliar as it might seem." In addition to the traditional three geometric dimensions used to describe the size or location of an object in space, other dimensions usually are required to describe the behavior of a dynamic system.

For example, in considering the several independent variables that determine the flight of a baseball, to the three aforementioned (x, y, z) data values, one must add the three directions of linear velocity along with three spin coordinates. Thus, as pointed out by Douglas, there is a total of nine dimensions, if each variable is assigned its own axis for graphing purposes. In analyzing the flight of a *curve ball*, Douglas observes that "dimensions" researchers attempt to explain that real-world phenomena are not simply the dimensions of physical length, width, and height, but rather the parameters that drive the dynamics of a system, such as pressure, velocity, and temperature.

D-Modules

Algebraic geometry has been one of the most lively areas of research in algebra during recent decades. It is the study of geometric objects that are the loci of points satisfying polynomial equations in two or more variables, such as the familiar conics from classical geometry. Meanwhile, algebraic topology has become a leading area of geometry. Considerably more general geometric objects than loci of polynomial equations are studied in algebraic topology; in the 1950s and 1960s this was an especially active field and was discussed on a number of occasions in the popular scientific literature.

Any geometric object has a group of symmetries. For example, a cube is invariant under a finite set of rotations. Similarly, a sphere is symmetric under (the infinite set of) all rotations and reflections around its center. A continuous symmetry group, such as the latter example, is called a Lie group. Lie groups can also be viewed as certain groups of matrices with their usual matrix multiplication. This multiplication is not commutative; that is, in general $XY \neq YX$. The set of derivatives along curves through the identity matrix of a particular Lie group can be viewed as an additive set of matrices and is called a Lie algebra. The theory of Lie groups and algebras, one of the great achievements of modern mathematics, originated from questions in differential equations, which has always been central in analysis. But "group" and (of course) "algebra" are quintessentially algebraic notions. Thus Lie theory, based on geometric symmetries and increasingly useful for physicists, incorporates aspects of algebra, analysis, and geometry.

Another place where these three fields meet is in *D*-modules, which have been developed recently in Japan. A module (with or without the *D*) is an algebraic structure consisting of a group such as a vector space whose elements can be multiplied by another set of mathematical objects such as matrices. *D*-modules are modules whose vectors can be multiplied by partial differential operators with analytic coefficients. One motivation for their study was to focus attention on the equations themselves rather than solutions to differential equations. They were investigated using methods of algebraic geometry invented in France in the early 1960s.

The theory of *D*-modules is also related to the Riemann-Hilbert problem, posed in Germany around the turn of the century. Suppose we have a linear, homogeneous system of n first-order ordinary differential equations for n functions $f_1, \ldots f_n$ of one complex variable z. It might be written $f' = Af$, where the coefficient A is an n by n matrix whose elements are analytic functions of z. In general the system will have n independent solutions $(f_{11}, \ldots, f_{1n}), (f_{21}, \ldots, f_{2n}), \ldots, (f_{n1}, \ldots, f_{nn})$ such that each solution $(g_i, \ldots, g_n)$ can be written uniquely as a linear combination $g_i = c_1 f_{1i} + c_2 f_{2i} + \cdots + c_n f_{ni}$ for some constants $c_1, \ldots, c_n$. But there may be singularities, points at which one or more of the elements in $A(z)$ tend to infinity or are otherwise irregular. Not only may solutions be undefined at the singularity, but when we follow a curve around the singularity, one solution may be transformed into a different one. Going around the given singularity (and no others) once in a clockwise direction has the effect of transforming a solution by multiplication by an invertible matrix, depending only on the system and not the path traversed. The

Riemann-Hilbert problem, a question in analysis, was to show a converse for the theorem just presented — that for any admissible map from a set of singularities into invertible matrices, there is a system of n differential equations, as described, such that encircling each singularity changes the solution by the corresponding matrix. The solution, now several decades old, has been extensively generalized.

Mathematicians working in this country and France were investigating ostensibly a quite different area — an area at the intersection of algebraic geometry and algebraic topology that focuses on the concept of duality in geometry. In 1977 researchers in the United States conjectured the existence of a relation between this geometric theory and a fundamental problem from Lie groups. Their conjecture was then proved independently by two pairs of mathematicians, one in France and the other in the Soviet Union. Their proof surprised the mathematical world by using the generalized Riemann-Hilbert results, thus again linking algebra with geometry through differential equations.

Papers about *D*-modules are impossible to classify into the three traditional fields of analysis, algebra, and geometry/topology, a problem for the editors of *Mathematical Reviews*. It also suggests how very tentative any division of mathematics into subfields may be. See also **Lie Group**.

Computational Complexity — Algorithms

An algorithm is a procedure for solving a given class of problems with a specified set of mathematical tools. In classical geometry the tools were straight edge and compass, and the ancient Greeks provided a simple algorithm for trisecting any segment using only these. Their extensive attempts to trisect a general angle were proved futile in the nineteenth century, when it was shown that there can be no such algorithm.

If the tools in algebra are addition, multiplication, division, and taking radicals (square roots, cube roots, etc.), then there is a familiar algorithm for solving any quadratic equation, $ax^2 + bx + c = 0$. However, a question posed by Renaissance Italians was answered when it was proved by Abel, a Norwegian in the early nineteenth century, that there can be no such algorithm for solving equations of degree five or more. The tenth problem posed by Hilbert at the 1900 International Congress of Mathematicians asked whether there is an algorithm for deciding if a general polynomial equation, $f(x_1, \ldots, x_n) = 0$, with integer coefficients has a solution in the positive integers. In the 1960s a Soviet mathematician, strongly incorporating the work of two Americans, proved that there is no such algorithm.

Such decision questions were formerly addressed primarily by logicians, but computers have given new urgency to algorithmic questions. Computers after all operate with only a few primitive tools, and the programs that instruct them are essentially algorithms. For most problems that computers are asked to solve the existence of some algorithm is usually evident; instead, the problem is to find algorithms that are efficient and reliable. The mathematical analysis of these practical considerations has spawned the field now called "complexity theory."

This field measures the complexity of an algorithm by the maximum number $f(n)$ of basic steps that the algorithm needs to solve those cases of the problem requiring n digits in their statement. The theoretically tractable problems, called type P, are those for which there is an algorithm with $f(n)$ growing no faster than some polynomial in n. There are adjustments to this assertion, most notably in the well-publicized linear programming problem discussed below, where the simplex algorithm, though exponentially long in the worst cases, is remarkably efficient in the vast majority of its application.

Each generation of computer scientists has quickly encountered pressing problems that overwhelm available computational resources. Unable to obtain exact answers, they resort to simulation, approximation, and sampling, perhaps by using Monte Carlo techniques. Such approaches are not always appropriate, as when a massive computation is used to make a momentous yes/no decision. In the development of antiballistic missile software during the early 1960s scientists tried to cope with such situations by the simultaneous use of many processors in a multiprocessor environment. It was found that this could result in unpredictable and often serious deterioration in the performance of the system as a whole. This discovery generated a serious study of such anomalies. The resulting work in the mid-1960s provided rigorous bounds for these deleterious effects. Performance guarantees of this type, usually called "worst-case bounds," remain a major focus of algorithm analysis.

The fundamental class of *NP*-complete problems was defined in 1971 independently by a Canadian and by a former Russian citizen who now resides in the United States. This class includes thousands of basic computational problems arising in computer science, mathematics, physics, biology, economics, business, and the social sciences. The *NP*-complete problems are of equivalent complexity in the sense that if one of them submits to an algorithm of polynomial complexity, then so can all the others. Proving the widespread conviction that no such algorithm exists (that $P \neq NP$) is considered the most fundamental of the open problems in theoretical computer science. It is known that any algorithm for solving an *NP*-complete problem can be modeled by an appropriate circuit. Until recently the only established lower bounds on circuits for *NP*-complete problems were linear. However, recently a Taiwanese mathematician now living in the United States made a dramatic breakthrough by establishing the conjectured exponential lower bound, but under the assumption that the number of "levels of the circuits is bounded". Since then a Swedish graduate student in the United States simplified and strengthened these arguments. Almost simultaneously, two Soviet mathematicians independently established an exponential lower bound assuming that the circuit is "monotone," a related but distinct development. This work has been extended by an Israeli doing postdoctoral work in the United States, and many complexity specialists now believe that the tools are finally becoming available to prove the corresponding results for unrestricted circuits, thereby finally settling this fundamental problem.

As the difficulty of *NP*-complete problems was realized, attention shifted to other approaches. These included approximation algorithms, average-case instead of worst-case performance analysis, and randomized algorithms that give a confident guess rather than a firm answer. Timely examples of the latter are the twin problems of deciding if a large integer n is prime or, if it is not, of factoring n. One such algorithm randomly selects an integer k less than n, performs a simple test, and announces either that n is definitely composite or that the problem is still undecided. About three quarters of the possible k's will establish that a composite n is not prime. Thus, performing the test with 100 independent k's that do not prove n to be composite justifies the conclusion that n is prime with a mere one in 4^{100} chance of error.

The study of primes has long been central to number theory, a field that has been pursued for its own splendid beauty but traditionally was considered to have few pragmatic consequences. Recent innovations in cryptography have completely reversed the latter perception. The security of important cryptosystems depends crucially on the belief that the problem of finding the prime factors of a random number with a thousand decimal digits or more is, and will be for decades, a computationally infeasible problem. However, faith in this belief is beginning to erode because of some recent unexpected advances in primarily testing and factoring. Essentially overnight, a Dutch mathematician produced the currently best factoring algorithm by using the theory of elliptic curves from algebraic geometry, a field mentioned in the previous vignette. It is not yet clear whether this will lead to even more effective algorithms that will undermine the security of these cryptosystems.

This vignette concludes with a discussion of linear programming (LP), a subject that occurs widely in discrete optimization. For example, a linear profit function, possibly depending on a large number of variables, is to be maximized in a region defined by linear constraints. This question is geometrically equivalent to finding the highest point in some high-dimensional convex solid or polyhedron in n-dimensional space, where n can be quite large. The importance of these problems, which have many applications in such varied areas as airline scheduling, meteorology, portfolio management, and telephone traffic routing, was first observed more than 40 years ago by Dantzig in the United States and Kantorovitch in Russia. Dantzig developed the very effective "simplex algorithm" for solving LP problems, which examines first one vertex and then another, moving along the outside edges of the high-dimensional polyhedron in such a way as to improve the function that is to be optimized at each step, until the optimal vertex is reached. Even though these n-dimensional polyhedrons, which typically arise in practical problems, can have exponentially many vertices, over 40 years of experience with the simplex algorithm indicates that only rarely are more than $4n$ or $5n$ vertices tested before the optimum point is attained, despite the fact that pathological examples can be constructed that do indeed require that all vertices be tested.

After the concept of *NP*-completeness was introduced in 1971, researchers struggled without success to find either a polynomial-time algorithm for LP or a proof that it was *NP*-complete. In 1979 the polynomial "ellipsoid algorithm" was produced in the Soviet Union, but the bound on this method grows as n^6, rendering it impractical for large problems, e.g., those having hundreds of thousands of variables. Subsequently, however, the ellipsoid algorithm has had a significant impact in the theory of combinatorial optimization.

In 1984 a young Indian mathematician in the United States made a striking breakthrough when he discovered an iterative method that plunges through the interior of the polyhedron, transforming it nonlinearly at each step so as to stay as far as possible from the boundaries of the changing solid. The number of steps required by this method is on the order of $n^{3.5}$, a significant improvement over the earlier n^6, and a variety of applications have shown that when implemented cleverly, this algorithm seems to perform significantly faster than the simplex algorithm. See also **Linear Programming**.

Very recent efforts in understanding this new method indicate that an *n*-dimensional polyhedron can be equipped with a coordinate system that transforms its interior into a certain quasi-hyperbolic geometry so that the trajectories to the optimal vertex form geodesics in this space. Obviously, this area is extremely active worldwide, and much more work is needed before a full understanding can be achieved. Nevertheless, it is already apparent that complexity theory addresses many practical problems and that sophisticated core mathematics, including algebraic geometry, number theory, and geometry, is vital to complexity theory.

Nonlinear Hyperbolic Conservation Laws

The development of both theory and numerical methods for solving nonlinear hyperbolic conservation laws (NLHCLs) has been an exciting area of recent mathematical research. There are many applications of NLHCLs because they describe many important physical systems, including some in aerodynamics, meteorology, water waves, plasma physics, and combustion. In gas dynamics these are the laws of conservation of mass, momentum, and energy.

The major technical obstacle to both solving and analyzing these systems is the fact that their solutions are not smooth, that is, they do not have derivatives of all orders. Many standard approximation procedures require smoothness, and, furthermore, solutions tend to be unique only if certain physical constraints (called entropy conditions) are satisfied. Since there are other important equations in hydrodynamics, relativity, and optics that do not have smooth solutions, NLHCLs are providing a testing ground for innovations with potentially wide applicability.

The first systematic computational attempts to solve NLHCLs, motivated by problems of jet propulsion and in the Manhattan Project, were made in the United States during World War II. John von Neumann and others provided a clear formulation of the problem and introduced several crucial ideas, in particular artificial viscosity (a justifiable technique for smoothing the problem) and a first analysis of stability. These ideas were greatly extended after the war and gave rise to an elegant theory of weak solutions that treated the lack of smoothness without smoothing. In particular, the Lax-Wendroff scheme and its many variants yielded acceptable solutions for many practical problems. However, many other problems remained unsolved, and the theory was incomplete.

In the 1950s the relevance of the Riemann problem, well known to chemists and engineers working with the Riemann shock tube, became fully recognized. The Riemann problem contains the pathology of the general NLHCL problem but is in a more tractable form. An American mathematician gave a mathematical analysis of the Riemann problem, and then a Russian mathematician incorporated the American's analysis into a numerical construction that described the misbehavior of the solutions in a natural way. These developments led to a proof, developed in the United States, that solutions exist for one-dimensional NLHCLs subject to some technical restrictions. This result gave rise to practical algorithms that in turn generated sharper existence results.

The Russian's construction has been generalized in several ways, and dramatic progress has arisen in the last three years from a combination of these ideas. In particular, reliable and efficient solutions of the equations

of gas dynamics in any number of space dimensions are now available, and they reveal and explain intricate physical phenomena that previously had been only dimly comprehended. Examples include the disclosure of unexpected complexity in flows involving interacting discontinuities, the discovery of transition criteria from regular to Mach reflection for waves impinging on a surface, and the revelation of instability mechanisms for supersonic jets. Both theory and experiments in NLHCL have been aided by elaborate computations from experiments using new laser technologies.

Recently, much computational activity has been directed toward solving systems of NLHCLs that are structurally more complex than those of gas dynamics, in particular those that arise in combustion theory or in the analysis of flow of porous media—a subject of great relevancy for oil recovery. New methodologies have appeared involving front tracking, mesh refinement, and piecewise parabolic approximations. Some of these are related to higher-order versions of the aforementioned Russian construction.

Many important practical problems remain open. For example, there is as yet no reliable numerical method for solving the equations of combustion theory in more than one space dimension except in the low-Mach-number limit. The newer numerical methods are so complex that computer science questions regarding their implementation, similar to those discussed in the previous vignette, have become crucial. Also, perturbations of NLHCLs, for example by boundary layers, are beginning to be considered.

Why numerical methods fail and why they sometimes succeed so spectacularly are questions that have been studied successfully in recent years, especially through the reexamination of the precise role and possible forms of artificial viscosity. New theoretical tools for understanding practical algorithms and new ideas, such as the notion of variation diminishing schemes, are producing a slow confluence of algorithms rooted in disparate a priori notions of what is important for practical calculations.

New techniques of functional analysis, in particular the compensated compactness method that originated in France, have given new impetus to existence theory, removing some of the earlier limitations. The compensated compactness analysis is significant to the broader context of homogenization and order/disorder phenomena, two other areas in which the French have been involved. In addition, partial existence and stability results have appeared for nonsmooth solutions in more than one space dimension for both convex and nonconvex systems.

The strong interaction between mathematics and practical applications in NLHCL is clear from this account. All the major advances in computation have been anchored in theoretical developments. Indeed, most of the more innovative practical algorithms are due to mathematicians, and more mathematicians have been involved than is apparent. An explosion of knowledge could at this time be safely forecast if there were more high-caliber people active in the field.

However, there are too few people with a combined understanding of the abstruse physical, mathematical, computer science, and related aspects of NLHCLs. Such an understanding requires a broad education in several fields that is not easily available in the United States. One explanation for the strong role played by Israelis in this field may be that in Israel many mathematicians are exposed to engineering problems and to programming during a lengthy military service. In other countries, such as Japan, China, Russia, and in much of Europe, most students complete algebra by the seventh grade and soon begin calculus, leaving time during secondary school for those with mathematical and technical talent to pursue advanced topics. Given that changes in precollegiate education will take a long time to evolve, the most immediate solutions in the United States would seem to be an extension of the graduate student years through supporting young investigators with adequate postdoctoral fellowships and providing enrichment for talented undergraduate students.

Yang-Mills Equations

In 1954, Yang in the United States and Mills in England constructed a nonlinear version of Maxwell's equations that incorporated a non-Abelian group, typically SU(2), the group of two by two unitary complex matrices with a determinant one. [SU(2) is a three-dimensional Lie group.] This was first conceived as a classical theory transplanted to Lorentz space-time; but when it is used in quantum theory, it is convenient to use Euclidean space-time. The theory has been incorporated into nearly every model of particle physics since the construction in 1975 of instant on solutions and the more recent construction of multi-instant on solutions in four-dimensional space. Shortly thereafter, the Penrose twist or theory was shown to transform an apparently very messy nonlinear system of partial differential equations into an elegant problem in algebraic geometry. Then the equations themselves were noticed to be natural geometric objects.

The Yang-Mills equations depict the curvatures (fields) of connections (potentials) as a principal bundle over a Riemannian manifold. By now, Yang-Mills theory is prominent in pure mathematics. It has already affected subjects as diverse as differential geometry, algebraic geometry, the topology of four-dimensional manifolds, the calculus of variations, nonlinear partial differential equations, index theory (or anomalies), and even the representation of infinite-dimensional groups, and it remains fertile research ground.

The extended impact of Yang-Mills theory does not yet involve the complete equations but concentrates on their role as nonlinear extensions of Laplace's equations, which are well known to be fundamental to earlier mathematics, physics, and engineering. In two variables Laplace's equation is related to the Cauchy-Riemann equations, part of the foundation of complex analysis. One form of the Yang-Mills equations, called the self-dual Yang-Mills equations (SDYM), is a four-variable analogue of the Cauchy-Riemann equations. The instantons are solutions of these equations, and they appear to have properties nearly as basic as solutions of the Cauchy-Riemann equations. The SDYM equations are in turn important in algebra, geometry/topology, and analysis, respectively.

At first glance, the SDYM equations seem absolutely intractable for writing explicit solutions. Even for SU(2), a small essentially non-Abelian Lie group, there are nine first-order equations with twelve unknown dependent variables as well as the four independent variables on the space. There are three extra degrees of freedom, due to gauge symmetries, so it is not surprising that insight from algebraic geometry had to precede progress in topology. The Penrose twist or methods were used to transform these equations in four-dimensional space to a problem concerning holomorphic bundles on a six-dimensional manifold. These methods from algebraic geometry are surprisingly general and can lead in many directions, for example, toward Kac-Moody Lie algebras and models with loop groups.

Learning about the structure of the space of instantons over four-dimensional manifolds has generated profound insight into the topological structure of general four-dimensional manifolds, demonstrating the fundamental value of an equation specific to a low dimension like four. This was helpful to physicists, who have since developed the fundamental intuition of instantons as solitons, the wavelike solutions of KdV and Sine-Gordon that superimpose nonlinearly.

It is interesting to note that mathematicians had been able to deal with the structure of manifolds in dimensions two, three, five, and greater. The Cauchy-Riemann equations are used in two dimensions, geometric methods are employed in three, and mathematicians find five and more dimensions amenable to standard methods of algebraic topology. The gap of the fourth dimension appears to be filled by SDYM theory.

In analysis, one of the key properties of Yang-Mills theory is its conformal invariance. Some of the basic equations about instantons can be formulated in the context of the calculus of variations. The conformal invariance of the theory implies that the variational problem does not satisfy the conditions that are required in order to use the method of direct steepest descent, so helpful in three-dimensional work. The attempt to understand the failure of the steepest descent method for the Yang-Mills problem has led to the development of new variational techniques, which are useful on a variety of problems. It is interesting, and possibly significant, that the three fundamental scale-invariant geometric problems coincide with three basic models of quantum fields: the Yamabe problem (phi-four theory), harmonic maps (sigma models), and the Yang-Mills equations.

In any case, Yang-Mills theory is a beautiful example of the intense bonds between current theoretical physics and all subfields of core mathematics. Yang-Mills theory is a young discipline that will undoubtedly attract many more mathematicians in the near future. Its results to date are primarily due to the efforts of English, American, and Russian researchers. Although Americans cannot claim to dominate the field, they have certainly contributed significantly to its development. The necessity of extremely broad training, mentioned in the final paragraph of the preceding vignette, also applies to successful research in this field. Increasingly, postdoctoral or protracted study is necessary in order to become a successful researcher in many areas of mathematics.

Operator Algebra

The area of operator algebras is currently very active and provides another excellent example of unexpected interactions between areas of core mathematics and the natural sciences. Quantum physics originated the Heisenberg Uncertainty Principle, which forces consideration of quantities P and Q satisfying $PQ - QP = h/2\pi i$. It motivated a search for appropriate mathematical systems containing infinitely many such noncommuting variables. In the 1920s, M.H. Stone and J. von Neumann demonstrated that such systems require a general theory of algebras of operators on Hilbert spaces, generalizations of finite-dimensional vector spaces. These were studied extensively in the 1930s by F. Murray and von Neumann.

The mathematics of operator algebras is characterized by a profound blend of noncommutative algebra and infinite-dimensional analysis. Although operator algebras exhibit a very rich structure, no serious work following that of Murray and von Neumann appeared until after World War II.

The simplest example of an operator algebra is the entire set of n by n dimensional matrices for a specific n. A general operator algebra is a subset of the bounded linear transformations on a (generally infinite-dimensional) Hilbert space that is closed under addition, multiplication, an adjoint operation, and a suitable limiting process. If the system is closed only for the strongest limiting process, it is called a C^*-algebra. If it also contains the most general limits, it is said to be a von Neumann algebra. The full algebras of all linear transformations on Hilbert spaces of arbitrary dimensions are the building blocks of the simplest operator algebras. However, consideration of proper sub algebras of the universal operator algebra over some Hilbert space reveals far more complex situations. This complexity can be slightly relieved and the study reduced to three basic types of von Neumann algebras simply called Types I, II, and III.

During the early postwar years, both American and French mathematicians made substantial progress in operator algebras, including some important applications to the theory of infinite-dimensional group representations. The French school became relatively inactive by the early 1960s and reemerged in the mid-1970s when a young mathematician, subsequently a Fields Medalist, began working in the area. During the late 1950s and early 1960s, some Japanese mathematicians entered the field. Simultaneously, numerous theoretical physicists became involved and provided valuable insights derived from their physical applications. It was shown that there exist infinitely many distinct Type II and Type III factors. The latter, especially, remained mysterious because they did not possess the special functionals called traces that were so fundamental to studying the structure of the other two types. When it was discovered that the physically interesting algebras are of Type III and certain classes were explicitly parameterized by physicists, parts of the mystery began to unravel.

Major international conferences began having a crucial influence on operator algebra theory in the mid-1960s. In the late 1960s, a conference held in the United States removed a key obstacle to analyzing Type III factors by displaying results proved independently by a Japanese mathematician and some Dutch and German physicists. At a subsequent conference, the French Fields Medalist, then a student, was motivated to work on the subject. He opened entirely new vistas by investigations that combined algebra and analysis in new ways and led to the classification of the algebraic structure of Type III factors. In the 1970s, researchers turned toward the geometric aspects of the subject. A new extension theory of C^*-algebras generated a successful synthesis of geometry and algebra, resulting in a unified view of features from these two subjects. At about the same time, the fundamental Atiyah-Singer Index Theorem was extended from locally trivial families to the more general foliations by using operator algebras. Additionally, a Soviet mathematician solved specific cases of a long-standing problem in differential topology concerning smooth deformations by developing powerful new techniques in C^*-algebras.

Recently, an American mathematician, born in New Zealand and educated in Switzerland, has found a totally unexpected connection between three apparently diverse fields — knot theory, the classification of sub factors of Type II factors, and the theory of Hecke algebras.

This activity has occurred between 1984 and 1986, a catalyst being the fortuitous meeting of the Mathematical Science Research Institute at Berkeley in 1985, sponsored by the National Science Foundation, where experts in operator algebras were meeting concurrently with those in low-dimensional topology, the field that contains knot theory. The excitement of the connection between von Neumann algebras and knot theory may be overshadowed by investigations of its utility to biologists in describing large-scale structures of DNA. A knot is a closed curve in three-space, and a link is a (possibly interlocking) system of knots. Although they can be surprisingly complicated, knots and links can be adequately represented by a projection onto the plane. Two relatively simple examples are shown in Fig. 5.

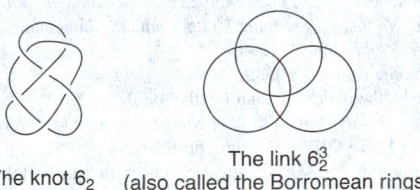

The knot 6_2 The link 6_2^3
(also called the Borromean rings)

Fig. 5.   Examples of relatively simple knots. (*National Research Council.*)

Although people probably have always used knots, the first known attempt to list and classify them mathematically (as opposed to mechanically) resulted from an erroneous hypothesis of Lord Kelvin in the late nineteenth century that atoms were knotted vortices in the ether. His vain hope of deriving the periodic table by classifying knots stimulated scientific advances entirely different from his vision — in the characteristic but unpredictable manner that pure research, stimulated by attractive ideas, often yields unanticipated harvests quite different from their original intent.

Knot theory considers two knots or links to be the same if one may be deformed without crossing strings until it is identical to the other. It is exceedingly difficult (and obviously fundamental) to decide when two given links are the same. Trial and error methods are rarely satisfactory. One approach to classifying two mathematical entities in some class is to assign each entity in the class some mathematical label (called an invariant) that coincides on the two entities considered to be the same. Thus a search for appropriate invariants for links began. In the 1920s polynomial invariants were developed by studying the topology of the space remaining when the link is removed from ordinary three-space. The corresponding polynomial for the knot 62 in Fig. 5 is $1 - 3x + 3x^2 - 3x^3 + x^4$.

Another approach to knots first studied in the 1920s was a mathematical structure called the braid group. Braids can be spliced together thus forming a group. However, splicing the top of a braid to its bottom forms a knot or link, and indeed all knots and links may be so formed (in possibly many ways). See Fig. 6.

Fig. 6.   Examples of knots of the braid group. (*National Research Council.*)

Braids and their invariants have rewarded investigators with considerable valuable insight over the years. However, their connection with operator algebras remained unnoticed until the proof of a deep result about sub factors of Type II factors required an analysis of one representation of the braid group. Coincidentally, almost the same representation, arising from a special case of the Hecke algebras, had been discovered by mathematical physicists in the 1960s as they partially solved the Potts model of statistical mechanics. The connection with operator algebras generated a trace function for the braids that could be used to recover numerical information. The trace of a braid then provided a new (Laurent) polynomial invariant for the associated knot or link. It was more sensitive

than the earlier polynomial in that it separated knots that were previously indistinguishable. It could even distinguish knots from their mirror images. Computers can be used to compute this polynomial for many links. For the knot 6_2, it is $x^{-1} - 1 + 2x - 2x^3 - 2x^4 + x^5$ and, for the link 6_2^3, it is $-x^{-3} + 3x^{-2} - 2x^{-1} + 4 - 2x + 3x^2 - x^3$.

Both operator algebra and topology already have produced substantial generalizations of this new invariant that have been used to solve many venerable problems of knot theory. Similar work is expected to shed new light soon on the Potts model, von Neumann algebras, knot theory, statistical mechanics, quantum physics, and possibly even basic structures of life via the DNA application. In any case, these developments emphasize again the eternity of a good mathematical result, the harmony of all mathematics, the unpredictable relationships between fields, and the many bridges across the humanly created gap between core mathematics and basic science.

At present, only a few institutions in the United States offer training in operator algebras. In this area, as with other areas of mathematics, the United States has become increasingly dependent on hiring people from other countries. In addition, some related areas with potential for interaction with engineering have been underemphasized in the United States. This is an ideal time to put a greater effort into providing the correct conditions for new young researchers and strong leadership in the United States in the subject.

Survival Analysis

Since problems in collecting, analyzing, and interpreting data are universal, it is not surprising that people in many countries (including England, France, Scandinavia, Russia, and the United States) have made significant contributions both to statistics and also to probability, the branch of core mathematics that has until recently provided the major theoretical support for statistics. Expansions in technology have both motivated and enabled the most striking progress in statistics during the past decade. Advances in instrumentation and communication have generated enormously complex sets of data, and the growth of computing power permits the collection and management of such data. The United States has been the unquestioned world leader in statistical computing, mainly because the availability and sophistication of its equipment has been unmatched elsewhere.

However, other countries have made significant contributions; a notable example from England is David Cox's proportional hazards model for life history data. Life history analysis is a body of statistical techniques useful in medical research, reliability theory, actuarial computations, and demographic studies. John Graunt initiated this field in 1662 with his invention of life Tables for analyzing English mortality data, but recent clinical studies are far more complex. Typically, these begin with patients who have an unpleasant disease (possibly at different stages) being assigned randomly to two or more different treatments; they are then followed until they die or disappear or the study ends. Observers record variables, called covariates, that might affect the survival of the patients, including some that do not change, such as sex and age at diagnosis, and some that do, such as blood pressure and glucose level. Statisticians use these observations to study the relationships between the covariates and the survival time of patients after contracting the disease, and especially the comparative merits of the treatments. One goal may be to predict the effect of different treatment on life expectancy of future sufferers of the disease, another may be to determine which factors prolong the patient's normal functioning as long as possible. Subtleties such as the role of individuals who are still alive at the end of the study and those who withdraw or disappear complicate the analysis.

One model for the life history of a patient (or piece of equipment) includes a vector $Z(t)$ of covariates that vary (possibly randomly) with time, from zero, when the study beings, to the time T when the study is terminated, and a finite-valued process $Y(t)$ designating the state of the dependent variable. These states might be "alive-and-functioning," "alive-not-functioning," "dead," or "lost." The treatments appear as coordinates of $Z(t)$. Analysis of these models concentrates on the intensity of changes; these intensities correspond to a stochastic process $J(t, y_0, y)$, where $j(t, y_0, y)dt$ represents the probability that Y changes from y_0 to y between time t and $t + dt$ assuming a past of $Z(t)$. The life histories of the n patients are regarded as a set of independent observations that can be used to estimate J.

Until the early 1970s, research concentrated on the extremes of either relatively simple situations, with few assumptions, or on vastly more complicated parametric models, with heavy assumptions on j. An example of the former is the setting for the product limit estimator for the probability of surviving beyond time t, when no covariants are measured and only questions of life expectancy are of interest. The assumptions of the parametric models were found to be too unrealistic by experts in biostatistics.

In 1973 Cox proposed his "proportional hazards model" for survival-time data, basing his work in part on a model developed during the 1960s at the National Cancer Institute. In this model

$$J(t, \text{Alive, Dead}) = \exp[\theta^T Z(t)]\lambda(t)$$

where $\lambda \neq 0$ is a (nonrandom) function and θ is a vector of unknown parameters. $J(t, \text{Alive, Dead})$ can be an essentially arbitrary function of $Z(s), 0 \leq s \leq t$.

The attractive features of the Cox model are easily seen if we specialize to fixed covariates. The model permits an arbitrary lifetime distribution for a control population and postulates a linear approximation for the log of the ratio of intensities corresponding to two different values of Z. The effects of the covariates can then be measured in terms of the components of θ.

The inferential procedures proposed by Cox were quickly applied to heart transplant and other data by statisticians and biomedical scientists in the United States. Evidence of these applications appears in numerous citations of the model, mostly in the medical literature. Although the model seemed to give sensible answers in applications, its highly nonlinear nature and the complex probabilistic structure of the data prevented rigorous theoretical analysis for some time.

The theory for the case of time-independent covariates was independently developed in the United States and Denmark in the mid- to late 1970s, but these methods could not handle the much more difficult case of time-varying covariates. In 1975, the thesis of a Norwegian student studying in the United States with a United States statistician who emigrated from France showed how to attack the analytic problems in this area. He applied a multivariate counting process framework to both survival analysis models and more general life history models. Most significantly, he exhibited applications to these models of the deep results in continuous time martingales and stochastic integrals, which were introduced by researchers in Japan, the United States, and France. Then others, primarily in The Netherlands, Norway, Denmark, England, and the United States, used these techniques, both to analyze the heuristic suggestions of Cox, and to devise and investigate new inferential procedures in more complicated life history models. Despite their flexibility, the Cox models are still burdened by the questionable assumption that the ratio of intensities for two individuals can be modeled parametrically. It is not clear that the counting process techniques will prove adequate for the analysis of the newer, more flexible, semiparametric models that have been proposed, but they should provide a good starting point.

The analysis of life history data is a rapidly expanding field widely used in a variety of disciplines, including biomedical science, demography, and sociology, fields that reciprocate by continually presenting statisticians with data for which previous methods are inadequate. Although the interaction of theory and application and the international nature of this work are hardly new features of statistics, they have become more prominent recently.

The United States and England are the primary world centers of statistical activity. An important pattern in the field, which is illustrated here, is that foreign scientists and students come to study, lecture, and meet in the United States, and subsequently many of the best remain as citizens or permanent residents.

Note: Some portions of this article also appeared in the publicly funded report, "Mathematical Sciences: A Unifying and Dynamic Resource," prepared by the National Research Council, Washington, DC.

Additional Reading

Papers and Reports, Board of Mathematical Sciences, Washington, DC.
"A Challenge of Numbers: People in the Mathematical Sciences," 1990.
"Actions for Renewing U.S. Mathematical Sciences Departments," 1990.

"Applications of the Mathematical Sciences to Materials Science," 1991.
"Calculus for a New Century: A Pump, Not a Filter," 1987.
"Chairing the Mathematical Sciences Department of the 1990s," 1990.
"Combining Information: Statistical Issues and Opportunities for Research," 1992.
"Discrimination Analysis and Clustering," 1988.
"Educating Mathematical Scientists: Doctoral Study and the Postdoctoral Experience in the United States," 1992.
"Everybody Counts: A Report to the Nation on the Future of Mathematics Education," 1989.
"Mathematical Foundations of High-Performance Computing and Communications," 1991.
"Mathematical Opportunities in Nonlinear Optics," 1992.
"Mathematical Sciences: A Unifying and Dynamic Resource," 1986.
"Mathematical Sciences: Some Research Trends," 1988.
"Mathematical Sciences, Technology, and Economic Competitiveness," 1991.
"Moving Beyond Myths: Revitalizing Undergraduate Mathematics," 1991.
"Number Theory, Proceedings," 1990.
"Probability and Algorithms," 1992.
"Renewing U.S. Mathematics: A Plan for the 1990s," 1990.
"Report of the Advisory Panel to the Mathematical and Information Sciences Directorate, Air Force Office of Scientific Research," 1987.
"Selected Opportunities for Mathematical Sciences Research Related to the Navy Mission: An Update," 1990.
"Select Opportunities for Mathematical Sciences Research Related to the Navy Mission," 1987.
"Spatial Statistics and Digital Image Analysis," 1991.
"Statistical Models and Analysis in Auditing," 1988.
"Statistics — A Guide to Assessing Societal Risk, Proceedings," 1991.
"Symposium on Statistics in Science, Industry, and Public Policy, Proceedings," 1989.
"The Future of Statistical Software," 1991.
"The Impact of Mathematics: Nonlinear Mathematics, Chaos, and Fractals in Science, Proceedings," 1990.

Papers Pertaining to Chaos

Bennett, C.H.: "Quantum Cryptography: Uncertainty in the Service of Privacy," *Science*, 752 (August 7, 1992).
Campbell, A.: "That Uncertain Something," *Case Western Reserve University Magazine*, 18 (November 1992).
Cipra, B.: "Cross-Disciplinary Artists Know Good Math When They See It," *Science*, 748 (August 7, 1992).
Cipra, B.: "Putting the Pedal to the Metal in a Controlled Chaotic Laser," *Science*, 1309 (November 20, 1992).
Douglas, J.: "Seeking Order in Chaos," *Electric Power Research Institute Journal*, 5 (June 1992).
Fischer, P. and W.R. Smith: "Chaos, Fractals, and Dynamics," Marcel Dekker, New York, NY, 1985.
Flam, F.: "The Quest for a Theory of Everything Hits Some Snags," *Science*, 1718 (June 12, 1992).
Garfinkel, A. et al.: "Controlling Cardiac Chaos," *Science*, 1230 (August 28, 1992).
Gleick, J.: "New Images of Chaos That Are Stirring a Science Revolution," *Smithsonian*, 122 (December 1987).
Goldberger, A.L., D.R. Rigney, and B.J. West: "Chaos and Fractals in Human Physiology," *Sci. Amer.*, 42 (February 1990).
Gutzwiller, M.C.: "Quantum Chaos," *Sci. Amer.*, 78 (January 1992).
Horgan, J.: "Nonlinear Thinking," *Sci. Amer.*, 26 (June 1989).
Kerr, R.A.: "Does Chaos Permeate the Solar System?" *Science*, 144 (April 14, 1989).
Krasner, S., Editor: "The Ubiquity of Chaos," American Association for the Advancement of Science, Waldorf, MD, 1990.
Lundqvist, S., N.H. March, and M.P. Tosi, Editors: "Order and Chaos in Nonlinear Physical Systems," Plenum, New York, NY, 1988.
McCauley, J.L.: *Chaos, Dynamics, and Fractals: An Algorithmic Approach to Deterministic Chaos*, Cambridge University Press, New York, NY, 1994.
Pool, R.: "Chaos Theory: How Big an Advance?" *Science*, 26 (July 7, 1989).
Pool, R.: "Putting Chaos to Work," *Science*, 626 (November 2, 1990).
Staff: "Nonlinear Mathematics, Chaos, and Fractals: Symposia Proceedings," Board of Mathematical Sciences, Washington, DC (April 28, 1988).
Stewart, I.: "Does God Play Dice? — The Mathematics of Chaos," Science News Books, Washington, DC, 1989.
Sussman, G.J., and J. Wisdom: "Chaotic Evolution of the Solar System," *Science*, 56 (July 2, 1992).

Papers Pertaining to Strings and Knots

Cipra, B.A.: "To Have and Have Knot: When are Knots Alike?" *Science*, 1291 (September 9, 1988).
Cipra, B.A.: "From Real Numbers to Strings of Zeros," *Science*, 1140 (March 3, 1989).

Cipra, B.: "Knotty Problems — and Real-World Solutions," *Science*, 403 (January 24, 1992).

Horgan, J.: "The Pied Piper of Superstrings," *Sci. Amer.*, 42 (November 1991).

Jones, V.F.R.: "Knot Theory and Statistical Mechanics," *Sci. Amer.*, 98 (November 1990).

Papers Pertaining to Factoring

Cipra, B.A.: "Mathematicians Reach Factoring Milestone," *Science*, 374 (October 21, 1988).

Cipra, B.A.: "PCs Factor a Most Wanted Number," *Science*, 1634 (December 23, 1988).

Cipra, B.A.: "Math Team Vaults Over Prime Record," *Science*, 815 (August 25, 1989).

Cipra, B.A.: "Big Number Breakdown," *Science*, 1608 (June 29, 1990).

Fackelmann, K.A.: "Fermat-Number Factors," *Sci. News*, 389 (June 23, 1990).

Ruthern, R.: "Factoring Googols," *Sci. Amer.*, 22 (December 1988).

Papers Pertaining to Geometry

Banavar, J.R., A. Maritan, and A. Stella: "Geometry, Topology, and Universality of Random Surfaces," *Science*, 825 (May 10, 1991).

Berry, M.: "The Geometric Phase," *Sci. Amer.*, 46 (December 1988).

Borwein, J.M., and P.B. Borwein: "Ramanujan and Pi," *Sci. Amer.*, 112 (February 1988).

Cipra, B.A.: "Getting a Grip on Elliptic Curves," *Science*, 30 (January 6, 1989).

Cipra, B.A.: "The Circle Can Be Squared!" *Science*, 528 (May 5, 1989).

Cipra, B.A.: "Another Piece of 3.14159," *Science*, 1260 (June 16, 1989).

Cipra, B.A.: "Packing Your n-Dimensional Marbles," *Science*, 1035 (March 2, 1990).

Cipra, B.A.: "A Speedier Way to Decompose Polygons," *Science*, 270 (July 19, 1991).

Corcoran, E.: "Geometry Decrees a New Dome," *Sci. Amer.*, 102 (September 1989).

Dewdney, A.K.: "The Theory of Rigidity," *Sci. Amer.*, 126 (May 1991).

Hey, A.J.G. et al.: "Topological Solutions in Gauge Theory and Their Computer Graphic Representation," *Science*, 1163 (May 27, 1988).

Horgan, J.: "Pythagoras's Bells," *Sci. Amer.*, 25 (July 1990).

Peterson, I.: "The Color of Geometry," *Sci. News*, 408 (December 23/30, 1989).

Peterson, I.: "Curves for a Tighter Fit," *Science News*, 316 (May 19, 1990).

Strauss, S.: "Impossible Matter," *Technology Review (MIT)*, 19 (January 1991).

Wallich, P.: "Simple Geometry Brings Supercomputers to Their Knees," *Sci. Amer.*, 126 (December 1990).

Papers Pertaining to Computer Mathematics, Algorithms, Networks

Bauschlicher, C.W., Jr. and S.R. Langhoff: "Quantum Mechanical Calculations to Chemical Accuracy," *Science*, 394 (October 18, 1991).

Bern, M.W. and R.L. Fraham: "The Shortest-Network Problem," *Sci. Amer.*, 84 (January 1989).

Cipra, B.A.: "Do Mathematicians Still Do Math?" *Science*, 760 (May 19, 1989).

Cipra, B.A.: "Combinatorial Arguments," *Science*, 595 (August 11, 1989).

Cipra, B.A.: "Algebra: A Hotbed of Radicalism," *Science*, 1190 (September 15, 1989).

Cipra, B.A.: "In Math, Less is More — Up to a Point," *Science*, 1081 (November 23, 1990).

Cipra, B.A.: "The Breaking of a Mathematical Case," *Science*, 165 (January 11, 1991).

Dewdney, A.K.: "Old and New Three-Dimensional Mazes," *Sci. Amer.*, 136 (September 1988).

Hau, Feng-hsiuung et al.: "A Grandmaster Chess Machine" *Sci. Amer.*, 44 (October 1990).

Peterson, I.: "Natural Selection for Computers," *Science*, 346 (November 25, 1989).

Peterson, I.: "Little Fermat: Scheme for Multiplication Leads to A Unique Computer," *Sci. News*, 222 (October 6, 1990).

Pool, R.: "Computing in Science," *Science*, 44 (April 3, 1992).

Rowe, D.E. and J. McCleary: "The History of Modern Mathematics," Academic Press, San Diego, CA, 1994.

Singh, J. and M. McBride: "Successfully Model Complex Hazard Scenarios," *Chem. Eng. Progress*, 71 (October 1990).

Wallich, P. "Computers Are Changing the Spirit of Mathematics," *Sci. Amer.*, 24 (March 1989).

Yam, P.: "Math Exorcise: A New Algorithm Lifts the Curse of Dimensionality," *Sci. Amer.*, 29 (July 1991).

MATINAL. See **Winds and Air Movement**.

MATRIX (Adjacency). Let G be a nonoriented graph possessing e elements $\varepsilon_1, \varepsilon_2, \ldots, \varepsilon_e$ and v vertices $\beta_1, \beta_2, \ldots, \beta_v$. The adjacency matrix A of G is a square matrix of order v in which $a_{ij} = 1$ if β_i and β_j are adjacent (see also **Vertex**) and zero otherwise; a_{ij} denotes the entry in A located in the ith row and jth column of A. Note that $a_{ji} = 0 (j = 1, 2, \ldots, v)$, i.e., the diagonal of A is composed solely of zeros.

Let B be the degree matrix of G. The *Matrix Tree Theorem* states that all the primary cofactors of the matrix $M = B - A$ of a connected nonoriented graph G are equal to the number of trees in G. A graph having the associated matrices A, B, M below is shown in Fig. 1.

$$A = \begin{bmatrix} 0 & 1 & 1 & 1 \\ 1 & 0 & 1 & 1 \\ 1 & 1 & 0 & 1 \\ 1 & 1 & 1 & 0 \end{bmatrix}$$

$$B = \begin{bmatrix} 3 & 0 & 0 & 0 \\ 0 & 3 & 0 & 0 \\ 0 & 0 & 3 & 0 \\ 0 & 0 & 0 & 3 \end{bmatrix}$$

$$M = \begin{bmatrix} 3 & -1 & -1 & -1 \\ -1 & 3 & -1 & -1 \\ -1 & -1 & 3 & -1 \\ -1 & -1 & -1 & -3 \end{bmatrix}$$

Observe that the (1,1) cofactor of M is precisely equal to 16, the number of trees in G. See also **Digraph**; and **Tree (Mathematics)**.

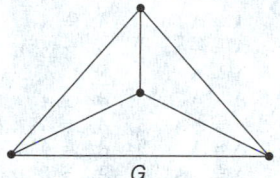

G

Fig. 1.   Concept of adjacency matrix.

MATRIX INVERSION. The determination for a given matrix A of its inverse matrix A^{-1} such that the product $AA^{-1} = A^{-1}A = E$, the unit matrix. The unit matrix is also symbolized by I. All its elements are zero except those with equal subscripts, which are unity. Inversion is defined only for square matrices.

MATRIX (Mathematics).   Consider a set of elements, finite in number, which may be arranged in rows and columns. If A_{ij}, B_{ij} are the elements in ith row and jth column of two such arrays and if these arrays combine to form a product with elements $C_{ij} = \Sigma A_{ik} B_{kj}$, then they are called matrices. A convenient symbolism is $A, B, C = AB$.

A matrix containing m rows and n columns is of order $(m \times n)$ and it has mn elements. If $m \neq n$, the matrix is rectangular; if $m = n$, it is square; if either m or n is unity, the matrix is called a vector, for its $n$elements can be regarded as vector components in an n-dimensional space. A row vector, of order $(1 \times n)$, will be denoted by $[x]$; a column vector, of order $(n \times 1)$ by $\{x\}$.

Matrix elements may be real, complex, or pure imaginary quantities of a very general nature. The concepts may be extended to matrices of infinite order.

If two matrices are equal, $A = B$, and $A_{jk} = B_{ij}$. The sum or difference of two matrices is defined by $A \pm B$, and $A_{ij} = B_{ij}$. The sum or difference of two matrices is defined by $A \pm B = C$, $A_{ij} \pm B_{ij} = C_{ij}$. Also A, where c is a constant, has elements cA_{ij}.

Two matrices A and B can form a product AB only if the number of columns in A equals the number of rows in B. Thus, if A is of order $(n \times h)$ and B of order $(h \times m)$; the product $C = AB$ is of order $(n \times m)$ and A, B are conformable. The following cases can also occur: $Ax = y$; $[s]A = [y]$; $[x]y = $ a scalar; $x[y] = B$, where B is square with rows or columns proportional to each other. Matrix multiplication is not necessarily commutative, but the distributive and associative laws hold. If $AB \neq BA$, the quantity $(AB - BA)$ is called the commutator. A special combination called the direct product can be defined. It is of interest in group theory.

Matrices with special properties are named as follows: (a) Null, indicated by 0, with zero for all its elements. In this case, $A + 0 = A$; $A0 = 0A = 0$.

(b) Unit, **E** (for German, *Einheit*). Its elements are δ_{ij}, the Kronecker delta, and $\mathbf{EA} = \mathbf{AE} = \mathbf{A}$. (c) Diagonal, all elements are zero except those with equal subscripts; thus, if **D** is diagonal, $D_{ij} = D_i \delta_{ij}$ or in another symbolism $\mathbf{D} = \text{diag}(D_1, D_2, \ldots, D_n)$. (d) Triangular, all elements above or below the main diagonal are zero ($T_{ij} = 0$, $i > j$ or $i < j$). (e) Transposed, given **A** with elements A_{ij}, its transposed matrix is $\tilde{\mathbf{A}}$ or $\mathbf{A}'$ with elements A_{ji}. If $\mathbf{ABC} \cdots \mathbf{X} = \mathbf{Y}$, then $\tilde{\mathbf{Y}} = \tilde{\mathbf{X}} \cdots \tilde{\mathbf{C}}\tilde{\mathbf{B}}\tilde{\mathbf{A}}$. (f) Singular, the determinant of **A** vanishes. All rectangular matrices are singular since their determinant, by definition, does not exist. (g) Adjoint, find the cofactor A_{ij} of A_{ij} in $|A|$ and transpose to get $\hat{\mathbf{A}}$ with elements A_{ji}. It follows from the properties of determinants that $\mathbf{A}\hat{\mathbf{A}} = \hat{\mathbf{A}}\mathbf{A} = |A| \; \mathbf{E}$. (h) Reciprocal or inverse, $\mathbf{A}^{-1} = \hat{\mathbf{A}}/|A|$, as defined from (g). Only a square matrix, thus a nonsingular one, has a reciprocal. If $\mathbf{ABC} \cdots \mathbf{X} = \mathbf{Y}$, then $\mathbf{Y}^{-1} = \mathbf{X}^{-1} \cdots \mathbf{C}^{-1}\mathbf{B}^{-1}\mathbf{A}^{-1}$, as for (e). (i) Complex conjugate, only if **A** has complex elements. Using an asterisk for this operation, the complex conjugate to **A** is $\mathbf{A}^*$ with elements A_{ij}^*. Unlike some previous cases, $\mathbf{A}^*\mathbf{B}^*\mathbf{C}^* \cdots \mathbf{X}^* - \mathbf{Y}^*$. (j) Associate (also called adjoint or Hermitian conjugate) $\mathbf{A}\dagger = (\mathbf{A}^*) = (\tilde{\mathbf{A}})^*$. In this case, $\mathbf{Y}\dagger = \mathbf{Y}\dagger \cdots \mathbf{C}\dagger \mathbf{B}\dagger \mathbf{A}\dagger$.

Certain other special matrices lead to names and relations as shown in Table 1.

TABLE 1. SPECIAL MATRICES

Name	Relation	Matrix Elements
Symmetric	$\mathbf{A} = \tilde{\mathbf{A}}$	$A_{ij} = A_{ji}$
Skew symmetric	$\mathbf{A} = -\tilde{\mathbf{A}}$	$A_{ii} = 0$; $A_{ij} = -A_{ji}$
Orthogonal	$\mathbf{A} = \tilde{\mathbf{A}}^{-1}$	$\sum A_{ij} A_{kj} = \sum A_{ji} A_{jk} = \delta_{ik}$
Real	$\mathbf{A} = \mathbf{A}^*$	$A_{ij} = A_{ij}^*$
Pure imaginary	$\mathbf{A} = -\mathbf{A}^*$	$A_{ij} = i B_{ij}$; B_{ij} real
Hermitian	$\mathbf{A} = \mathbf{A}\dagger$	$A_{ij} = A_{ji}^*$
Skew Hermitian	$\mathbf{A} = -\mathbf{A}\dagger$	$A_{ii} = 0$; $A_{ij} = -A_{ji}^*$
Unitary	$\mathbf{A} = (\mathbf{A}\dagger)^{-1}$	$\sum A_{ij} A_{kj}^* = \sum A_{ji} A_{jk}^* = \delta_{ik}$

MATRIX (Triangular). A matrix in which all elements are zero above the diagonal (a lower triangular matrix), or else below the diagonal (an upper triangular matrix). It is a unit upper or lower triangular matrix when also every diagonal element = 1. The product of two triangular matrices of the same type is again of that type, i.e., (unit) upper or lower; and the reciprocal of a triangular matrix (when it exists) is also of the same type. A triangular matrix is properly triangular if the diagonal is null.

MATTER-ANTIMATTER SYMMETRY. See **Quantum Mechanics**.

MAXILLOFACIAL PROSTHETIC MATERIALS. See **Polymeric Dental Materials**.

MAXIMUM AND MINIMUM (Mathematics). If is often of interest to study the behavior of a plane curve, described by the equation $y = f(x)$, as the independent variable x is continually increased. The possibilities are: (1) the curve continuously rises or falls; (2) the curve reaches a maximum and then falls from it or reaches a minimum and then rises; (3) an abrupt break in the curve appears at a certain pair of values (x_0, y_0); (4) some peculiar change in the slope or direction of the curve occurs. This article is concerned with the conditions for a maximum or a minimum; for cases (1) and (3), see also **Continuous Function**; for case (4), see also **Singular Point of a Curve**.

If the curve $y = f(x)$ is plotted, it is easy to see that the curve rises or increases if its tangent with the X-axis is acute or if the slope, dy/dx, is a positive number. Conversely, if the curve decreases or falls, the tangent or dy/dx is negative. In order for a maximum to occur, the derivative must change its sign and at the maximum point (x_0, y_0), $dy/dx = 0$. Similar considerations show that a minimum also occurs at $f'(x) = 0$ but the sign of the derivative must there change from minus to plus.

A more careful investigation of the situation shows that the derivative can also vanish when neither a maximum nor a minimum occurs, for the curve could momentarily become parallel to the X-axis and then continue to rise or fall without passing through either a maximum or a minimum. One is thus led to consider the second derivative and one then finds that a rule, generally applicable, is: $f( )$ is a maximum at $x = x_0$ if $dy/dx = 0$; $dy/dx < 0$ for $x > x_0$; $dy/dx > 0$ for $x < x_0$; $d^2y/dx^2 < 0$ for $x = x_0$.

The corresponding rule for a minimum is $dy/dx = 0$ at $x = x_0$; $dy/dx < 0$ for $x > x_0$; $dy/dx > 0$ for $x > x_0$; $d^2y/dx^2 > 0$ for $x = x_0$.

A third possibility arises when both dy/dx and d^2y/dx^2 vanish at $x = x_0$. Before consideration of this case, let us refer to the concavity or convexity of a curve. These terms are to be interpreted in their usual non-technical way with respect to the X-axis but technically *concavity* means that the curve lies above its tangent and *convexity* means that the tangent is above the curve. Hence, if d^2y/dx^2 is positive (negative) at $x = x_0$, the curve has passed from *concavity* to *convexity*, or the reverse. Such a situation is called a point of inflection.

These arguments can be extended to the case of several independent variables and, in fact, both a maximum and a minimum, called a saddle point or a minimax, can occur simultaneously. A more refined treatment of such problems is the subject matter of the calculus of variations. A neutral term to designate either a maximum or a minimum is *extremal*.

See also **Variations (Calculus of)**.

MAXIMUM THERMOMETER. A thermometer so designed that it registers the maximum temperature attained during an interval of time. The liquid-in-glass type of maximum thermometer has a bore that is constricted between the bulb and graduated portion of the stem. As the temperature rises, a portion of the mercury is forced past the constriction and into the graduated section. This mercury is retained when the temperature falls and serves to indicate the highest temperature reached. Bimetallic thermometers with a circular scale are also used as maximum thermometers. A free index mounted concentrically with and driven by the thermometer index is held by friction at the maximum temperature.

See also **Minimum Thermometer**; and **Townsend Support**.

AMS

MAXIMUM UNAMBIGUOUS RANGE. The maximum range from which a transmitted radar, lidar, or sodar pulse can be reflected and received before the next pulse is transmitted. This range, $r_{\max}$, is given by $r_{\max} = cT/2$, where T is the interpulse period and c is the speed of light (or speed of sound in the case of sodar).

Range is measured by the time delay between pulse transmission and reception, ordinarily assuming that the received pulse is associated with the most recent transmitted pulse. Targets at ranges beyond $r_{\max}$ therefore appear at ranges closer than $r_{\max}$ because of range folding. Special coding of the pulses permits discrimination between echoes from the most recent transmitted pulse and earlier ones, enabling the measurement of ranges beyond $r_{\max}$.

AMS

MAXIMUM UNAMBIGUOUS VELOCITY. The maximum range of radial velocity that can be observed without ambiguity by a Doppler radar, sodar, or lidar.

Velocities outside this interval are folded into the interval. For a pulsed radar or sodar operating with an interpulse period T, or a lidar with an A/D converter sampling interval T, the unambiguous velocity interval is $-\lambda/4T$ to $+\lambda/4T$, where λ is the operating wavelength. This interval can be extended by the use of dual PRF and other techniques.

AMS

MAXWELL EQUATIONS. See **Electromagnetic Phenomena**.

MAXWELL, JAMES CLERK (1831–1879). A Scottish physicist, Maxwell worked in electromagnetism and the kinetic theory of gases. Maxwell studied molecules of gases in rapid motion. He treated them statistically and is remembered for the Maxwell-Boltzmann kinetic theory of gases. Maxwell's most important achievement was showing that a few simple mathematical equations could express the behavior of electric and magnetic fields and their interrelated nature. His work is known as Maxwell's equations and is a great achievement for 19th century physics. His important publications include, *Theory of Heat, Treatise on Electricity and Magnetism, and Matter and Motion*.

See also **Boltzmann's Distribution Law**; **Bridge Circuits (Electrical/Electronic)**; **Color**; **Electromagnetic Phenomena**; **Electron Theory**; **Equilibrium**; **Gravitation**; **Kinetic Theory**; **Photography and Imagery**; **Quantum Mechanics**; and **Statistical Mechanics**.

J. M. I.

MAYER, MARIA GOEPPERT (1906–1972). Mayer was born in Germany and immigrated in 1930 to the United States. Her early work is in molecular spectra and is important for laser physics. She also proposed, with Teller, a theory for the origin of the elements. In 1948, she developed the nuclear shell model. She won the Nobel Prize in Physics in 1963.

J. M. I.

MAYFLY *(Insecta, Ephemeroptera)*. Also known as shad flies, salmonflies, and lake flies. The adults are sluggish insects with slender filaments at the caudal end of the body and large, triangular front wings. The hind wings are much smaller, in some species rudimentary. The immature insect is aquatic and in most species feeds on decaying vegetable matter. It may live for several years, while the adult stage lasts only a few days.

Mayflies of one species emerge as adults in large numbers within a short period and are sometimes very abundant near favorable bodies of water. They fly at twilight and can sometimes be seen in large gray clouds at a distance of more than a mile over the islands of Lake Erie, where they are especially abundant. In the cities bordering the lake they are attracted to lights and their dead bodies are sometimes swept up in bushels after a heavy flight. Under such conditions they are a nuisance but not a serious pest. Their value as food for fishes more than offsets what little harm they do. See Fig. 1.

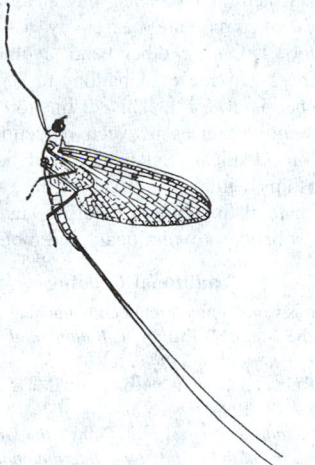

Fig. 1.   Mayfly. (*USDA.*)

MAYR, ERNST WALTER (1904–2005). Ernst Mayr was one of the 20th century's leading evolutionary biologists. He was also a renowned taxonomist, tropical explorer, ornithologist, historian of science, and naturalist. His work contributed to the conceptual revolution that led to the modern evolutionary synthesis of Mendelian genetics, systematics, and Darwinian evolution, and to the development of the biological species concept.

Born 5 July 1904 in Kempten (Allgäu, Bavaria) and died February 3, 2005 in Bedford, Massachusetts U. S. Ernst was the second of three sons born to Otto and Helene Mayr. His father was a judge in the Bavarian court system, and the family had a long tradition of practice in medicine. Following his father's career, the family moved to Würzburg in 1908, then to Munich in 1914. Otto Mayr was an enthusiastic naturalist, and the family took frequent excursions into the countryside. In Würzburg, Ernst began his interest in natural history, especially birds. See also **Evolution (History)**; and **Taxonomy (The History)**.

Otto Mayr died unexpectedly in July 1917. Afterwards his wife and sons moved to Dresden. Ernst left for university in 1923. Considerable family savings were lost during the hyperinflation of the early 1920s. Each son attended university on scholarships and savings from their mother's pension. Following family tradition, Mayr began medical training at the University of Griefswald in 1923. A professional career was expected by family pressure (his two brothers studied engineering and law, respectively). Mayr later claimed his choice of schools was dictated by a desire to situate himself in an ornithologically interesting region of Europe. Despite high marks in his medical examinations at university,

Mayr switched to zoology after earning a candidate in medicine degree in 1925.

In 1923 Mayr observed an occurrence of the migratory duck *Netta rufina* well outside its expected range. Confident this was important, Mayr visited the Museum of Natural History at the University of Berlin to report his results to the curator of birds, Erwin Stresemann, who was also the editor of *Ornithologische Monatsberichte*. This first publication fuelled Mayr's interest in ornithology. Stresemann offered him a volunteer position at the Museum during university holidays. This encouragement, including the promise to place him as an expedition-based collector, drew Mayr into the subject.

Mayr completed his PhD in 1926, 16 months after moving to Berlin. With Stresemann as his supervisor, Mayr studied the European distribution and biogeography of the serin finch, *Serinus canaria serinus*. On completion, he received an assistantship at the Museum.

Several attempts failed to secure a place for Mayr on a collecting expedition, but Stresemann convinced several prominent competitors to hire Mayr in a cooperative venture. In February 1928 Mayr left Berlin on two separate projects. First, he was to collect in Dutch New Guinea jointly for the Rothschild Museum at Tring (UK) and the American Museum of Natural History (AMNH) (USA). Second, he was to collect in the former German Mandated New Guinea for Stresemann. Before returning home Mayr was asked to join the Whitney Expedition in the Solomon Islands, collecting again for the AMNH. This added a year to his travels. Returning to Berlin in 1930, Mayr set to work studying the New Guinea material. His 1932 article "A Tenderfoot Explorer in New Guinea" provides a taste of this experience.

On his return, Mayr was thought to be a rising star in ornithology. At Tring, he was considered to replace the curator Ernst Hartert. Financial problems meant this post was never offered. Instead, Mayr was recruited by the AMNH to process its backlog of ornithological materials from the South Seas. Mayr began in January 1931.

In 1932, Walter Rothschild sold his ornithological collection to the AMNH. Numbering 280,000 skins and other materials, this accession had to be assimilated into the AMNH collection. Mayr accepted this expansion of duties, becoming associate curator of the Whitney–Rothschild Collection. For twenty years, Mayr's principal responsibility was the curation of this collection.

Mayr worked at the AMNH until 1953, rising to the rank of curator. He resigned to accept an Agassiz Professorship at the Museum of Comparative Zoology (MCZ) at Harvard University. He served as Director of the museum from 1961 to 1970. The Harvard appointment meant an important change in status to Mayr: from museum curator to full professor. Mayr became emeritus professor at Harvard in 1975.

Impact on the Life Sciences

Ornithology. Mayr's taxonomic specialty is birds of the southwestern Pacific, authoring both technical publications and field guides, such as his (1945) *Birds of the Southwest Pacific*, co-authored by Jaques. In the 1930s and 1940s Mayr was active in the Linnaean Society of New York, which mixed Manhattan's birding enthusiasts with professional ornithologists centered at the AMNH. He had a role in cultivating numerous PhD students in the subject. From 1931, Mayr contributed substantially to several editions of James Peters' *Check-list of Birds of the World*. When Peters died unexpectedly in 1952, Mayr oversaw the completion of the *Check-list*'s next edition.

Speciation and Evolution. In the late 1930s, Mayr turned his interest to the process of speciation. His (1942) *Systematics and the Origin of Species* earned Mayr an international reputation and the Leidy Medal (1946). Mayr was part of a small group of biologists confident that recent developments in genetics, biogeography, ecology and behavior made possible a new and rigorous understanding of speciation. Through the 1940s, Mayr cultivated this subject as his own specialty. See also **Biogeography (The History)**.

Mayr's approach to speciation emphasized geographic isolation and polytypic species, which collected geographically distinct populations together as varieties within a single species provided these populations interbreed. Mayr became a strong advocate of allopatric speciation, which required physical, geographical isolation of a population prior to their divergence into a new species. For Mayr, geographically distinct populations represented more than mere varieties; they were "incipient species." He also championed the "biological" or "genetical" species concept, which emphasized the role of interbreeding and gene flow in speciation. As gene flow dwindled between separate populations, Mayr argued, a new speciation event was underway. This was "evolution in action." Its study emphasized isolating mechanisms and barriers to gene flow. Though Mayr admitted many isolating mechanisms as plausible, he emphasized the importance of geographic isolation in the process. Physical isolation was necessary, he argued, for evolving barriers to accumulate within populations.

Mayr's emphasis on polytypic species and geographic isolation imported ideas from his Berlin mentors, Stresemann and Berhnard Rensch. He was also influenced by many strands of American biology, including the population genetics research of Theodosius Dobzhansky. *Animal Species and Evolution* (1963) represents the maturation of Mayr's evolutionary theorizing. See also **Dobzhansky, Theodosius (1900–1975)**.

Mayr played an important role in renovating evolutionary studies during the 1930s and 1940s. Intellectually, this involved a multidisciplinary synthesis of data and methodologies. It also involved a shift from documenting evolutionary patterns to examining its mechanisms. Mayr worked hard to convince skeptical naturalists this was a legitimate area for inquiry and to convince skeptical experimental biologists that naturalists could make unique contributions to the subject. By the early 1940s, Mayr was community building among biologists interested in speciation processes. After World War II, this expanded into the Society for the Study of Evolution, and the journal *Evolution*, which Mayr founded and edited from 1947 to 1949.

Systematics. Mayr's impact on systematics involved several levels. First, he worked to discipline systematics and to establish its place within the life sciences. Pressing for status meant transforming the perception of systematics from diffuse to explicit roles as a cornerstone in the subject. Disciplining the subject meant reform from within: convincing colleagues they had a responsibility to keep up-to-date on developments elsewhere in the life sciences, and convincing them they ought to use their specialized knowledge to speak to general biological problems (such as evolution).

Mayr first asserted himself in these roles during the 1930s within the American Ornithologists Union. He was one of a dozen self-proclaimed "bird biologists" striving to shift the focus of American ornithology away from birding *per se* and towards general biological subjects: behavior, physiology, migration and evolution. This also manifested during Mayr's role with the AMNH exhibit "biology of birds," which opened in 1948. The same desires to assert systematics drove Mayr towards interdisciplinary research in the 1940s.

In the 1950s and 1960s Mayr worked at the national and international level through his role as Director of the MCZ and his position within the National Academy of Sciences. Mayr was key in establishing "systematic biology" as a field within the life sciences. His co-authored (1953) book *Methods and Principles of Systematic Zoology* became a standard text in the discipline.

Among systematists, Mayr led a faction self-described as "evolutionary systematics." George Gaylord Simpson and Arthur Cain were two other leading proponents of this approach. The basic principle of evolutionary systematics is to use classification to express evolutionary relationships, with a special focus on groups close to the species level and a formal recognition of subspecies using trinomial nomenclature. See also **Simpson, George Gaylord (1902–1984)**.

Evolutionary theory sets the principles guiding the systematists' formulation of these groups. For example, because gene flow became the currency of evolutionary studies, species became the fundamental unit of attention. Geographic and ecological varieties were recognized as incipient species. Species clusters or super species were set neatly within a genus. Variation within populations was understood to be an essential element of population structure. Characters thought to be adaptations to local conditions were highlighted within taxonomic distinctions as the features making a group distinct. The presence of geographic barriers was deemed sufficient to draw boundaries in classifications. In the overall scheme of classification, taxonomic groups were built around a common ancestor and its descendants, following Darwin's emphasis on continuity and descent with modification. See also **Systematics: Historical Overview**.

Not all taxonomists accepted evolutionary systematics. Some preferred artificial classification systems for convenience. The rise of numerical taxonomy and cladistics in the 1960s promoted vigorous debate about the nature of systematics and its relations to evolution. This debate continues today.

History and Philosophy of Science. Since the 1950s, Mayr has written extensively on the history and philosophy of biology, as with his monumental (1982) *Growth of Biological Thought* and his additions to Stresemann's (1975) *Ornithology from Aristotle to the Present*. In the early 1970s, Mayr organized a conference on evolutionary studies in the twentieth century, with proceedings published as *Evolutionary Synthesis* (1980), co-edited with Provine. He also has written extensively on Darwin and leading evolutionists in the nineteenth and twentieth centuries. In addition to his scholarship, Mayr has played an important role as patron and advocate for these subjects. See also **Evolution (History)**; and **Taxonomy (The History)**.

As a historian, Mayr's work revolves around three themes. First, he works to organize the history of biology to explain how interests in his generation arose. Second, he works to credit those in the past who advocated the positions he defends. Third, he defends the historical importance of naturalist traditions in the history of biology when emphasis seems to shift elsewhere, such as towards molecular biology.

In philosophy, Mayr contributed to the notion of causation with a distinction between proximate and ultimate causes. Functional biology, he argued, focuses on proximate causes by investigating the operation and interaction of structural elements. Explanations in physiology are paradigm examples of proximate causes by asking "how" questions (How does this work?). On the other hand, evolutionary biology seeks ultimate causes. These provide explanations for "why" questions (Why does this exist; what is it for?). This distinction nicely describes the division within the life sciences between reductionistic specialties that emphasize physico-mechanical explanations and holistic specialties that emphasize relationships within the broad expanse of evolutionary time. The proximate/ultimate distinction arose during Mayr's role in conflicts in the 1950s and 1960s between evolutionary and molecular biologists.

Additional Reading

Cain, J.: "Ernst Mayr as Community Architect: Launching the Society for the Study of Evolution and the Journal," *Evolution, Biology and Philosophy*, **9**, 387–427 (1994).

Greene, J., and M. Ruse: "Special Issue on Ernst Mayr at Ninety," *Biology and Philosophy*, **9**, 263–427 (1994).

Haffer, J.: *'We Must Lead the Way on New Paths': the Work and Correspondence of Hartert, Stresemann, and Ernst Mayr—International Ornithologists*, Jochen Hölzinger, Ludwigsburg, Germany, 1997.

Hey, J., F.J. Ayala, and W.M. Fitch: *Systematics and the Origin of Species: On Ernst May's 100th Anniversary*, National Academies Press, Washington, DC, 2005.

Mayr, E.: "A Tenderfoot Explorer in New Guinea: Reminiscences of an Expedition for birds in the Primeval Forests of the Arfak Mountains," *Natural History*, **32**, 83–97 (1932).

Mayr, E.: *Systematics and the Origin of Species, from the Viewpoint of a Zoologist*, Columbia University Press, New York, NY, 1942.

Mayr, E.: *Animal Species and Evolution*, Belknap Press of Harvard University Press, Cambridge, MA, 1963.

Mayr, E.: *The Growth of Biological Thought: Diversity, Evolution, and Inheritance*, Belknap Press of Harvard University Press, Cambridge, MA, 1982.

Mayr, E., and F.L. Jaques: *Birds of the Southwest Pacific, a Field Guide to the Birds of the Area Between Samoa, New Caledonia, and Micronesia*, Macmillan, New York, NY, 1945.

Mayr, E., E.G. Linsley, and R.L. Usinger: *Methods and Principles of Systematic Zoology*, McGraw-Hill, Companies, Inc., New York, NY, 1953.

JOE CAIN, University College London, London, UK

McCARTY, MACLYN (1911–2005). Maclyn McCarty was an American medical microbiologist whose work on the transforming principle of the pneumonia-causing bacteria led to the discovery that DNA was the chemical basis of genetic information.

Growing up as the son of a sales manager in the American Midwest and West, Maclyn McCarty decided very early in life that he would be a medical researcher and also, to study at Johns Hopkins University, the country's most renowned medical school of the day. After obtaining a bachelor's degree in 1933 from Stanford University where he majored in biochemistry, McCarty followed through with his intentions and went to Hopkins. He completed his medical training and a pediatric internship, and then moved to a bacteriology research laboratory in New York City in 1940. After one year at New York University's medical school, he moved to the then Rockefeller Institute of Medical Research where he stayed for the remainder of his career.

Upon his arrival at the Rockefeller Institute, McCarty joined the laboratory of Oswald Avery where the primary interest was to find the basis of pathogenicity of *Streptococcus pneumoniae* (called pneumococci), the causative agent of pneumonia, a very serious disease during that period. Specifically, McCarty was to carry forward a line of investigation begun earlier by Colin MacLeod, aimed at pinpointing the source of what was dubbed the "transforming principle," of the bacteria. At the time he joined Avery's laboratory, all that was known about this transforming principle was that it was present in virulent bacteria (whether dead or alive) and that it was capable of imparting pathogenicity or virulence to nonpathogenic forms of the same organism, presumably by affecting the latter's ability to produce capsular materials. Although the chemical identity of the transforming principle (i.e. the gene for the capsule) was unknown, the scientific community had assumed that this molecule was a protein, because at the time proteins were the only molecules known to contain "information." However, as the Avery laboratory reported in 1944, the ability to transform bacteria actually lay in the DNA fraction of the bacteria. Soon after reporting these findings, McCarty provided the confirming evidence for the chemical identity of the transforming principle by demonstrating that it lost the ability to transform avirulent bacteria when treated with the DNA-degrading enzyme deoxyribonuclease but retained activity in the presence of proteinases. He published this important sequel to the 1944 paper in 1946. See also **Avery, Oswald Theodore (1877–1855)**.

Additional Reading

McCarty, M.: *The Transforming Principle: Discovering that Genes are made of DNA*, WW Norton, New York, NY, 1985.

NEERAJA SANKARAN, Yale University, New Haven, CT

McCOLLUM, ELMER VERNER (1879–1967).

Elmer McCollum was born and raised on a pioneer farm in Kansas, McCollum's education began in a one-room school. But he went on to graduate from the University of Kansas and gain a PhD in organic chemistry at Yale. After a few months' study of biochemistry with Professor Mendel, he obtained a position with Professor Hart at the University of Wisconsin. He was needed to help explain why cows fed solely with wheat products had dead calves whilst those fed on corn products had healthy calves, when the chemical analyses in use at the time showed no significant differences between the diets. McCollum's problem was to discover what was lacking in the wheat. He chose, against his Dean's wishes, to use the rat, with its rapid growth, small food consumption and short lifespan, as a small animal model. He worked with diets of relatively purified ingredients: protein, carbohydrates, fat and minerals. This led to the finding that some, but not all, fats contained trace amounts of a factor (to be named vitamin A) that was essential for the normal growth of young rats. It was found later that shortage of this same factor in "wheat" rations had been responsible for their failure to support healthy reproduction in cattle. He also showed that, to make a complete diet, products from cereals needed to be supplemented with "protective foods" such as milk and leafy vegetables. After giving the prestigious Harvey Lectures in 1917, he was appointed to a professorship at the School of Hygiene and Public Health at the Johns Hopkins University in Baltimore. His work continued to show the complexity of nutritional requirements, for example that the requirement for vitamin D by the rat, in order to prevent rickets, depended on the ratio of calcium-to-phosphorus in the diet. Unfortunately, his reliance on the rat led him astray in asserting that the human disease of scurvy could not be a vitamin deficiency. (It was later realized that rats, unlike humans, are able to synthesize vitamin C.) In 1918, he published *The Newer Knowledge of Nutrition*, which was to go through five editions over the next 21 years, and he was much in demand for service on national and international committees concerned with nutrition. When the national fortification of white flour with iron and three vitamins was proposed in 1941, he was against the idea. But his alternative, adding non-fat milk solids, yeast, and wheat and corn germs, to provide a wider spread of micronutrients, was rejected.

After his formal retirement in 1946, he continued to work and write on the history of nutrition. See also **Nutritional Science (The History)**.

Additional Reading

McCollum, E.V.: *A History of Nutrition*, Houghton Mifflin, Boston, MA, 1957.
McCollum, E.V.: *From Kansas Farm Boy to Scientist*, University of Kansas Press, Lawrence, KS, 1964.
Rider, A.A.: "Elmer Verner McCollum—a Biographical Sketch," *Journal of Nutrition*, **100**, 1–10 (1970).

KENNETH J. CARPENTER, University of California, Berkeley, CA

MCLEOD GAGE.

A liquid-level vacuum gage in which a known volume of gas, at the pressure to be measured, is compressed by the movement of a liquid column to a much smaller known volume, at which the resulting higher pressure is measured.

MCMILLAN, EDWIN M. (1907–1991).

An American physicist who won the Nobel Prize in chemistry in 1951 along with Glenn T. Seaborg for their discoveries in the chemistry of the transuranium elements. His work included research in nuclear physics and particle accelerator development as well as microwave radar and sonar. He and his colleagues discovered neptunium and plutonium. He was the recipient of the Atoms for Peace prize in 1963. His Ph.D. in Physics was awarded from Princeton University.

M COMPONENT.   See **Lightning**.

M-DISPLAY.

In radar, a display in which target distance is determined by moving an adjustable blip along the baseline until it coincides with the horizontal position of the target signal deflections. The control which moves the blip is calibrated in distance. Also called *M-scan, M-scope, or M-indicator*.

MEADOW-BROWN (*Insecta, Lepidoptera*).

Butterflies of dull gray-brown color marked with eye-like spots, in some species set in a yellow patch on the fore wings. Family *Satyridae*.

MEADOWLARK (*Aves, Passeriformes*).

A common North American bird (Aves) more closely related to the blackbirds and orioles than to the true larks. The eastern meadowlark, *Sturnella magna*, is less attractive than the western, *S. neglecta*, which has a brief but glorious song. Both have the characteristic yellow breast with a black chevron at the throat.

MEALYBUG (*Insecta, Homoptera*).

A small, very damaging insect on citrus and in glasshouse operations. The insects, of several species as indicated below, are distributed worldwide.

Citrophilus mealybug (Pseudococcus gahani (Green)).
Citrus mealybug (Pseudococcus citri (Risso)).
Coconut mealybug (Pseudococcus nipae (Maskell)).
Grape mealybug (Pseudococcus maritimus (Enrohorn)).
Long-tailed mealybug (Pseudococcus adonidum (Linne)).
Mexican mealybug (Phenacoccus gossypii (Townsend and Cockerell)).

In the United States, they are notably destructive of citrus crops in California. All stages of development of the insect can be found throughout the year.

The life cycle of all species is about the same. The appearance differs slightly. The citrus mealybug has a dense white powder over its back. The long-tailed mealybug is ovoviviparous. Adult females of the other species, ranging from $\frac{1}{8}$ to $\frac{1}{3}$ inch (3 to 8 millimeters) in length, deposit from 300 to 600 eggs. The females secrete a waxy, cottony mass into which the eggs are placed. From 1 to 3 weeks are required for hatching. The mealybug young feed on the sap and juices from fruit and leaves. Their movement is slow and limited. It requires from 1 to 4 months to complete their growth. The female passes through a pupa stage before becoming an adult. The females are wingless. Usually there are from 2 to 4 generations per year.

Biological methods have been quite successful in controlling mealybugs. In the California citrus orchards, great numbers (millions) of the coccinellid beetle (*Cryptolaemus montrouzieri*), have been very effective in controlling the mealybug. Several hymenopterous parasites have been imported and are effective: *Coccophagus gurneyi* and *Tetracnemus pretiosus* against the citrophilus mealybug; *Leptomastidea abnormis* against the citrus mealybug; *Anaphopus sydneyensis* and *T. peregrinus* against the long-tailed mealybug.

MEAN DEVIATION. The mean deviation of n sample values x about a point a is defined as $\Sigma|x - a|/n$. Used as a measure of dispersion, a is usually chosen to be the sample mean, or the sample median about which the mean deviation is a minimum.

MEAN FREE PATH. 1. Average distance a molecule travels in a gas between collisions. This concept has a meaning only to the extent that the paths of molecules are mostly straight lines interrupted by changes in direction (collisions) over comparatively shorter distances of order the molecular size. Molecules in a liquid are never free in this sense. The mean free path L in a gas of a single molecular species with a Maxwell–Boltzmann distribution of speeds is

$$L = \frac{1}{\sqrt{2}nS},$$

where n is the number density of molecules and S is the mutual collision cross section. To the extent that a molecule can be considered a hard sphere with diameter d, $S = \pi d^2$. The concept of molecular diameter is fuzzy and each method for determining it yields different results. At sea level the mean free path in air is of order 0.1 μm.

2. Average distance a photon travels in a turbid medium between scattering (scattering mean free path) events, or the average distance a photon travels before being absorbed (absorption mean free path). The scattering mean free path in a medium is the inverse of its scattering coefficient; the absorption mean free path is the inverse absorption coefficient; and the total mean free path is the inverse of the sum of scattering and absorption coefficients. At visible wavelengths, the scattering mean free path in clouds is of order 10 m.

AMS

MEAN LIFE. For any unstable system that decays in accordance with the laws of probability, such as either excited states of atoms or radioactive nuclides, the number N of individual units of the system in existence at any time t is $N = N_0 e^{-\lambda t}$, in which N_0 is the number of units existing at time $t = 0$ and λ is the decay constant. Then the number of units that have a lifetime between t and $t + dt$ is $N\lambda dt = N_0\lambda e^{-\lambda t}dt$ and the total lifetime of all N_0 nuclides is

$$L = \int_0^\infty tN\lambda\,dt = \int_0^\infty tN_0\lambda e^{-\lambda t}\,dt = N_0/\lambda$$

The average lifetime, or mean life τ, of a single nuclide is $\tau = 1/\lambda$. Another commonly used unit of lifetime is the half life, $t_{1/2}$ which is less than the mean life by a factor $\ln 2 = 0.693$, such that $\tau = t_{1/2}/0.693$. The mean life is used as a measure of the lifetime of many other systems besides radioactive nuclides, such as metastable states, mesons, hyperons, the recombination of carriers of opposite sign in semiconductors, and the rate of absorption of thermal neutrons in matter. See also **Decay Constant**; and **Radioactivity**.

MEAN SEA LEVEL. The average height of the sea surface, based upon hourly observation of the tide height on the open coast or in adjacent waters that have free access to the sea. These observations are to have been made over a "considerable" period of time. In the United States, mean sea level is defined as the mean height of the surface of the sea for all stages of the tide over a 19-year period. Selected values of mean sea level serve as the sea level datum for all elevation surveys in the United States. In meteorology, mean sea level is used as the reference surface for all altitudes in upper-atmospheric work; in aviation it is the level above which altitude is measured by a pressure altimeter. Along with mean high water, mean low water, and mean lower low water, mean sea level is a type of tidal datum. See also **Altimetry**.

MEAN SEA LEVEL PRESSURE. See **Altimetry**.

MEAN SIDEREAL TIME. See **Time**.

MEAN SOLAR DAY. The duration of one rotation of the Earth on its axis, with respect to the mean sun. The length of the mean solar days is 24 hours of mean solar time or 24 hours 3 minutes 56.555 seconds of mean sidereal time. A mean solar day beginning at midnight is called a civil day; and one beginning at noon, 12 hours later, is called an astronomical day.

MEAN SOLAR SECOND. Prior to 1960 the fundamental unit of time, equal to 1/86,400 of the mean solar day. Now replaced by the *ephemeris second*. In radar, a display in which target distance is determined by moving an adjustable blip along the baseline.

MEAN SQUARE ERROR. If a statistic t estimates a population parameter θ, the mean square error of t is defined as $E(t - \theta)^2$, when E denotes expectation. It is equal to the sum of the variance of t plus the square of the bias.

MEAN TEMPERATURE. The average temperature of the air as indicated by a properly exposed thermometer during a given time period, usually a day, a month, or a year. For climatological tables, the mean temperature is generally calculated for each month and for the year. For charts, the observed mean values at station level are reduced to sea level by adding a correction for elevation, usually taken as 0.5 °C for each 100 m (1 °F for 360 ft), but in some mountainous countries different rates are used, based on local observations. See also **True Mean Temperature**.

MEAN VALUE THEOREMS. The first law of the mean for integrals is

$$\int_a^b f(x)\,dx = (b - a)f(z)$$

where $a \le z \le b$ and $f(x)$ is a continuous function.

The second law of the mean is

$$\int_a^b f(x)\phi(x)\,dx = \phi(a)\int_a^z f(x)\,dx$$

with z and $f(x)$ restricted as before; $\phi(x)$ is also continuous and a positive monotonic decreasing function in the interval (a, b). Another form of the second law is

$$\int_a^b f(x)\phi(x)\,dx = \phi(a)\int_a^z f(x)\,dx + \phi(b)\int_z^b f(x)\,dx$$

where $\phi(x)$ is not necessarily always positive.

There are similar formulas for the case where $\phi(x)$ is an increasing function. The two forms of the second theorem are known as the forms of *Bonnet* and of *DuBois-Reymond*, respectively.

A mean value theorem also exists for a derivative. Let $f(x)$ be a function which has a finite derivative at all points of the interval (a, b). Then there exists a value of z between a and b such that $f(b) - f(a) = f'(z)(b - a)$. The theorem may be interpreted geometrically, for it states that the tangent to a smooth curve is parallel to an intermediate point on a chord of the curve. The procedure can be generalized to give the extended mean value theorem

$$f(b) = f(a) + (b - a)f'(a) + \frac{(b - a)^2}{2!}f''(a) + \cdots$$

$$+ \frac{(b - a)^{n-1}}{(n - 1)!}f^{(n-1)}(a) + \frac{(b - a)^n}{n!}f^{(n)}(a)$$

See also **Taylor Series**.

MEASLES. A viral infection that remains a major cause of childhood morbidity and mortality in developing countries. A World Health Organization estimate (1990) indicates that more than 2 million children die of measles each year. Measles also is an important cause of blindness, diarrhea, and malnutrition in many children who survive the acute illness. Vaccines are effective in reducing the incidence of measles in many countries. Infants in areas endemic for measles, however, lose maternal antibody before they are 9 months old and thus remain at high risk for measles and account for 20 to 30% of all patients with measles in some large urban areas.

In the advanced and industrialized nations, the incidence of measles has decreased markedly over the last few decades, mainly because of the increase in the number of vaccinations and, in particular, because of the increased effectiveness of vaccines. For example, with the introduction of live measles immunization, the number of cases of measles in the United States declined from 450,000 in 1964 to 22,400 in 1967. With this decline, the serious complications of the disease and resulting fatalities from encephalitis also dropped in 1979. The decline continued throughout the 1980s, and professionals set a target to eliminate indigenous measles from the United States. As of the early 1990s, measles remained a notifiable disease in the United States, with approximately 2100 cases reported in 1992. In contrast, the incidence of measles in many parts of the world remains one of major concern.

As reported by Georges Peter (Rhode Island Hospital and Brown University School of Medicine), "The marked decline in the incidence of vaccine-preventable diseases in the United States has correlated with rates of immunization of approximately 95% or more in school-age children. These rates can be attributed in part to the enactment and enforcement of school immunization laws in each state. Among children in the first two years of life, however, rates of immunization in some areas are substantially below the national goal of 90 percent for completion of the recommended immunizations by the second birthday. The gap is especially prominent in some intercity populations." In a study by Cutts, et al. (see reference), "The principal cause of the measles epidemic of 1989 through 1990 was failure to vaccinate children at the recommended age." The study notes that this deficiency also caused concern over possible outbreaks of other vaccine-preventable diseases in the United States. In 1990, there was an outbreak of 25 cases of congenital rubella infection in Southern California. In 57% of the women who delivered infants with the congenital rubella syndrome, one or more missed opportunities for rubella-susceptibility testing and vaccination were identified. See also **Immune System and Immunology**.

The etiologic agent is a single-stranded RNA paramyxovirus of which only one strain is known; symptomatic variations in the disease course are related to local environmental factors rather than the causative agent. The virus is spread by droplet infection and gains access through the respiratory tract and conjunctivae. It replicates mainly in the pharynx and regional lymph nodes. After 2 or 3 days of infection, primary viremia (virus in blood) develops. In the fifth to sixth day, a secondary viremia occurs. This produces a rash, fever, and conjunctivitis, with coryza (acute head cold) present in the tenth to fourteenth day after incubation. The incubation period runs from 7 to 14 days. Characteristically, the rash begins at the hairline and spreads down over face, neck, trunk, and eventually over the entire body. The lesions start as tiny flat red spots; they enlarge and spread to become confluent in many areas. At the height of the disease, the temperature may be as high as 105 °F (40.6 °C). The patient suffers from itching and burning of the skin. Marked sensitivity of the eyes to light, and cough, are present at this stage. Koplik's spots appear in the oral mucosa as whitish-blue centers in red erythematous backgrounds. After the eruption reaches its peak, it begins to fade, usually in the order of its appearance and, during this period, the patient improves dramatically.

Therapy is mainly supportive. Although usually benign, measles may develop some untoward complications, otitis media, giant cell interstitial pneumonia, and encephalitis being the most common. These should be treated as separate disease entities, with penicillin or other antibiotics as necessary.

In July 1990, G.D. Hussey (University of Cape Town, South Africa) and M. Klein (Red Cross War Memorial Children's Hospital, Rondebosch, South Africa) reported on a study of 189 children in an attempt to determine the possible effectiveness of vitamin A therapy in the treatment of measles. This therapy first was proposed in the mid-1980s. Conclusions of the report: "Treatment with vitamin A reduces morbidity and mortality in measles, and all children with severe measles should be given vitamin A supplements, whether or not they are thought to have a nutritional deficiency." Other than vaccination for prevention, this is one of the few therapies recommended other than careful supportive care.

Additional Reading

Cliff, A.D., M. Smallman-Raynor, and P. Haggett: *Measles: A History*, Blackwell Science, Inc., Malden, MA, 1993.

Cutts, F.T. et al.: "Monitoring Progress Toward U.S. Preschool Immunization Goals," *J. Amer. Med. Assn.*, 1952 (May 1992).

Diaz-Ortega, J. et al.: "Immunization of Six-Month Old Infants with Different Doses of Edmonston-Zagreb and Schwarz Measles Vaccines," *N. Eng. J. Med.*, 580 (March 1, 1990).

Eobbins, A. and P. Freeman: "Obstacles to Developing Vaccines for the Third World," *Sci. Amer.*, 126 (November 1988).

Hussey, G.D. and M. Klein: "A Randomized Controlled Trial of Vitamin A in Children with Severe Measles," *N. Eng. J. Med.*, 160 (July 19, 1990).

Kurstak, E.: *Measles and Poliomyelitis: Vaccines, Immunization, and Control*, Springer-Verlag, Inc., New York, NY, 1993.

Oehen, S., H. Hengartner, and R.M. Zinkernagel: "Vaccinations for Disease," *Science*, 195 (January 11, 1991).

Peter, G.: "Childhood Immunizations," *N. Eng. J. Med.*, 1794 (December 17, 1992).

Web Reference

Center for Disease Control and Prevention: http://www.cdc.gov/health/diseases.htm

MEASURED CEILING.   See **Meteorology**.

MEASURE OF LOCATION.   A quantity calculated from a frequency distribution intended to indicate the position of the distribution on the scale of measurement. The commonest measure of location is the arithmetic mean.

MEASURING WORM (*Insecta, Lepidoptera*).   A caterpillar that loops the body by drawing the hind legs up close to the front legs as it crawls. The movement is associated with the lack of most of the legs near the middle of the body. Caterpillars of the family *Geometridae* and a few species of *Noctuidae* are of this type. The adult moths of the *Geometridae* family are relatively large, having a wingspan of one inch (2.5 centimeters) on the average, but some species exceeding 2 inches (5 centimeters). Their bodies are slender and they have a rather delicate appearance, something like a butterfly. However, as with other moths, their wings are spread when they are at rest. Closely related are the spring and fall cankerworms, the currant spanworm, and the snow-white linden moth.

MECHANICAL EQUIVALENT OF HEAT.   The conservation of heat per se is observed only for systems not involving the performance of mechanical or electrical work. Count Rumford (ca. 1800) was the first to establish this fact in his well-known cannon-boring experiments carried out in the arsenal of the Duchy of Bavaria in Munich. He observed that when his drills became dull, heat was produced in great quantities limited only by the amount of work done against friction. He concluded that the large-scale mechanical energy used in overcoming friction could only be converted into the motions of the ultimate particles of matter, a motion not directly observable, but detected by our senses as heat. His results were confirmed and extended by the late work of Joule and Helmholtz, in particular, and also provided a more reliable value for the so-called *mechanical equivalent of heat*.

This is taken as the amount of mechanical (or electrical) energy which when converted into heat is equivalent to exactly 1 calorie. The value for this important constant is 4.185 joules per 15° calorie. Here the joule is the work performed when power is expended at the rate of 1 watt for 1 second. Thus, a 100-watt lamp bulb converts 100 joules of electrical energy to thermal energy each second; this amounts to 100/4.185, or about 24 calories.

The foregoing experiments emphasized that heat is merely another form of the universal quantity *energy*. Its transformation always occurs at the rate of 4.185 joules per calorie whether heat goes into external work or work is dissipated through friction into heat.

MECHANICS.   The science that deals with the effects of forces upon bodies at rest or in motion. The laws and phenomena of gases and liquids and solid bodies have a part in this subject, and it is one of the basic studies of engineering, physics and astronomy. It is customary to subdivide mechanics into the study of fluids and the study of particles or bodies of solid materials. It is to the latter field that the term mechanics is frequently restricted. For convenience it is subdivided into statics and dynamics. Dynamics is usually further subdivided into kinematics and kinetics. Statics deals with bodies at rest, in equilibrium under the action of forces or of moments; kinematics deals with abstract motion and kinetics treats of the effect of forces or of moments upon the motions of material bodies.

Fluid mechanics is that branch of mechanics that deals with those fundamental laws that apply to all fluids (liquids or gases) at rest or in motion.

See also **Dynamics**; **Kinematics**; **Kinetics**; and **Rotation**.

MECOPTERA. Insects with four narrow wings with numerous veins. The head is prolonged downward in a beak bearing biting mouth parts at the tip. The relatively few species inhabit moist woods and are not commonly known.

This order includes two chief forms, the scorpion flies and a group of slender long-legged insects usually known by their generic name, *Bittacus*. In the former the tip of the abdomen is modified so that it resembles that of a scorpion slightly. *Bittacus* has the general appearance of the crane flies but for its four wings, and is chiefly remarkable for the grasping joints at the tips of the legs.

MEDAWAR, PETER BRIAN (1915–1987). Sir Peter Medawar was born on 28 February 1915 in Rio de Janeiro. He was the second son of an English mother, Edith Muriel Dowling, and Nicholas Agnatius Medawar, a naturalized British businessman from Lebanon who went to Brazil as the agent of a dental supplies manufacturer. Peter Medawar, a member of the Athenaeum Club, disliked portentousness and what he called the *snobbisms* of the English. He was knowledgeable about music, especially opera which he loved, and he enjoyed village cricket.

After attending indifferent preparatory schools in southern England with his older brother Philip, Medawar entered Marlborough College (a large public school in Wiltshire), which he greatly disliked, especially its belief in the notion of "untutored brilliance" in both academic work and sport. He did though pay tribute to his biology teacher Dr A. G. Lowndes. He went up to Magdalen College Oxford and graduated with a first-class degree in zoology in 1936. He had also developed his taste for philosophy and some enthusiasm for mathematics at Oxford. Although he decided not to complete the final procedures — to supplicate — for the award of his DPhil. degree he was awarded a DSc in 1947. In 1937 he married Jean Shinglewood Taylor, a fellow zoology student whom he described as the most beautiful woman in Oxford. Initially they both worked on tissue culture under Professor (later Sir) Howard Florey. Thus began the research that led to Medawar's introduction of the new science of immunology, to which he devoted his professional life, and for particular research in which he was awarded his Nobel Prize in 1960. Following scholarships and Fellowships at St John's and Magdalen Colleges Oxford, Medawar was appointed to the Mason Professorship in Zoology in the University of Birmingham at the age of 32. He attributed this success to the help of Dr (later Sir) Solly Zuckerman who had left Oxford to take up the Chair of Anatomy in Birmingham. In Birmingham, Medawar was joined by his first and esteemed graduate student at Oxford, Rupert Everett Billingham. Their early work concerned the use of skin grafts to investigate the mechanism of pigment spread in the coats of parti-colored animals such as guinea-pigs and cattle. At the end of two years Medawar had to admit that his hypothesis on which they had been working was mistaken. In Medawar's view all scientists who are in the least imaginative will sometimes take a wrong view — a hazard of the job. See also **Florey, Howard Walter (1910–1985)**; and **Immunology (The History)**.

In 1951 he moved to London to take up the Chair of Zoology and Comparative Anatomy at University College London (UCL), where his Oxford tutor J. Z. Young had already been appointed to a Chair. Medawar, who always washed his own glassware for his highly sensitive immunological work on transplants, remained at UCL until 1962, when he became Director of the prestigious National Institute for Medical Research (NIMR), at Mill Hill in north London. Following serious illness in 1969 he was, from 1971 to 1986, head of the transplantation section of the Clinical Research Centre (CRC), which like the NIMR was government funded through the Medical Research Council (MRC). Throughout his life Medawar received a prodigious number of awards and honors both at home and abroad. These included Companion of the British Empire (CBE) 1958, Companion of Honor (CH) 1972 and the Order of Merit (OM) 1981. In 1949 he was awarded his Fellowship of the Royal Society (FRS) and received its Royal Medal in 1959 and its Copely Medal in 1969. He was knighted in 1965.

Medawar was a skilled experimentalist who believed that collaborative research worked well in science and technology; he was himself a good team player. The potential relationship between his tissue culture work and transplantation had been highlighted by the predicament in wartime Oxford of a crashed airman with 60% of his body covered with third degree burns. This incident led to Medawar spending a short time in the Burns Unit of the Glasgow Royal Infirmary where, with Tom Gibson, he gained experience in comparing the results of grafting a patient's own skin — an autograft — with that from a donor, a homograft. Whilst the autografts were accepted the homografts were rejected. A second series of homografts from the same donor were, as Gibson speculated, rejected even faster. They noticed that the homografts were invaded by lymphocytes which set up the inflammatory reaction of rejection.

In 1945 the American geneticist R. D. Owen had discovered that in cattle most fraternal twins (developed from two fertilized eggs) contain and may retain throughout life a stable mixture of each other's red cells. Medawar and Billingham in Birmingham showed that most such fraternal twins would accept skin grafts, that is homografts, from each other. This phenomenon, which did not apply to other nonfraternal siblings, whose grafts were rejected, was called *immunological tolerance*. Sir Frank McFarlane Burnet had explained this phenomenon in terms of "self and nonself" substances, and postulated that this capacity was not inherited but acquired during fetal life. At UCL Medawar and Leslie Brent set out to induce immunological tolerance in other organisms that is, to abolish their natural capacity to recognize and destroy the genetically foreign graft tissue of homograft transplants. Working on mouse embryos they succeeded in establishing what they called "acquired immunological tolerance." For this work Mcfarlane Burnet and Medawar shared the 1960 Nobel Prize for medicine.

Medawar wrote and spoke eloquently, as exemplified by his 1959 BBC Reith Lectures on "The Future of Man" and by his numerous publications, some co-authored with his wife. These include real gems which explained and popularized his views both on scientific discovery and on the recording of scientific work, in which he showed the influence on his thinking about scientific method — not The Scientific Method, a notion he dismissed — of the philosopher Karl Popper. They also present aspects of his own innovative work on immunology and its relevance to human transplantation.

As part of his role as President in 1969 of the British Association for the Advancement of Science, Medawar was reading the lesson in Exeter Cathedral when he suffered a severe stroke. This paralyzed his left arm and leg and severely impaired his vision. Further strokes eventually led to the loss of his left eye. He faced these setbacks in his inimitable manner requesting, for instance, that his replacement eye should be beer bottle brown with just a hint of sparkle. Although his condition came to preclude laboratory work, he continued to publish until his death in 1987. He left behind his wife and four children. He also left behind an unforgettable legacy to those who were taught by him, to so many of those who worked with him and to the countless recipients of transplant procedures who have benefited from his pioneering work in immunology. See also **Stroke**.

Additional Reading

Lance, E.M., P.B. Medawar, and E. Simpson: *An Introduction to Immunology*. Wildwood House, London, UK, 1977.

Medawar, P.B.: *An Unsolved Problem of Biology: An Inaugural Lecture delivered at University College London*, HK Lewis, London, UK, 1951.

Medawar, P.B.: *The Uniqueness of the Individual*, Methuen, London, UK, 1957.

Medawar, P.B.: *Advice to a Young Scientist*, Harper and Row, London, UK, 1979.

Medawar, P.B.: *The Limits of Science*, Harper and Row, New York, NY, 1984.

Medawar, P.B.: *Memoir of a Thinking Radish. An Autobiography*, Oxford University Press, Oxford, UK, 1986.

Mitchison, N.A.: "Peter Brian Medawar 28 February 1915–2 October 1987," *Biographical Memoirs of Fellows of the Royal Society*, Vol. 35, The Royal Society, London, UK, 1990, pp. 281–301.

DIANA E. MANUEL, Wellcome Institute for the History of Medicine, London, UK

MEDDIES. First identified in 1978, Meddies are so named because they flow out of the Mediterranean Sea. Typical Meddy scales and properties include diameters of $\sim$50–100 km (31–62 miles), a vertical extent of $\sim$600–1,400 m (1,969–4,593 ft), a drift velocity of $\sim$2–3 cm s^{-1} (with occasional stalls), a rotation velocity of $\sim$20–30 cm s^{-1}, a rotation period of $\sim$4–10 days, a lifetime of months to two years, and a salinity core contrast of 0.2–1 psu (practical salinity units). Meddies contain more than 900 billion kilograms (a billion tons) of salt.

Meddies are salt lenses containing high amounts of original Gibraltar Outflow Water in their interior. With spatial scales smaller than the internal Rossby radius, they belong to the energetic class of submesoscale coherent vortices. They rotate clockwise (anticyclonically) like solid bodies and are encapsulated by strong contrasts (gradients) of water masses (properties) and a sharp vorticity front at their periphery. Meddies interact with partner vortices, depending on their geographical position and eddy population. Spontaneous Meddy release represents random salt sources within the Mediterranean salt tongue and questions a large-scale advection–diffusion salt balance in the North Atlantic.

Using sensor data from several U.S. and European satellites, researchers from NASA's Jet Propulsion Laboratory, Pasadena, Calif.; the University of Delaware, Newark; and Ocean University of China, Qingdao; have developed a method to detect salty, submerged eddies called "Meddies" that occur in the Atlantic Ocean off Spain and Portugal at depths of more than 1,000 meters (one-half mile).

These warm, deep-water whirlpools, part of the ocean's complex circulatory system, help drive the ocean currents that moderate Earth's climate. The research marks the first time scientists have detected this phenomenon from space, and the first use of a new multi-sensor technique that can track changes in ocean salinity. Results are reported in the April issue of the American Meteorological Society's *Journal of Physical Oceanography*.

"Since Meddies play a significant role in carrying salty water from the Mediterranean into the Atlantic, new knowledge about their trajectories, transport and life histories is important to understanding their mixing and interaction with North Atlantic water," said Professor Xiao-Hai Yan of the University of Delaware http://www.ocean.udel.edu/cms/xyan/, lead author of the study and co-director of the university's Center for Remote Sensing. "Ultimately, we hope this will lead to a better understanding of their impact on global ocean circulation and global climate change."

While warm water ordinarily resides at the ocean's surface, the warm water flowing out of the Mediterranean Sea has such a high salt concentration that when it enters the Atlantic Ocean at the Strait of Gibraltar, it sinks to depths of more than 1,000 meters (one-half mile) along the continental shelf. This underwater river then separates into clockwise-flowing Meddies that may continue to spin westward for more than two years, often coalescing with other Meddies to form giant, salty whirlpools that may stretch for hundreds of miles.

"Since the Mediterranean Sea is much saltier than the Atlantic Ocean, the Meddies constantly add salt to the Atlantic," Yan said. Without this steady salt-shaker effect, he notes, the conveyor belt of ocean currents that help distribute heat from the tropics toward the North Pole might be diminished, resulting in colder temperatures in regions such as New England and northwestern Europe that currently experience more temperate climates.

"There is concern about global climate change shutting down the ocean currents that warm the Atlantic Ocean," Yan says. "The melting of sea ice at the North Pole could add enormous amounts of fresh water to the Atlantic, reducing its salinity enough to slow the sinking of cooler water, which would shut down the conveyor belt of ocean currents that help warm major regions of the planet."

Yan and his team drew on data from several satellite sensors that can read an important signal of a Meddy's presence Fig. 1. Altimeters flying aboard NASA's Topex/Poseidon and Jason http://www.nasa.gov/centers/jpl/missions/jason.html, satellites and the European Space Agency's European Remote Sensing and Environment (Envisat) satellites http://www.esa.int/esaEO/SEMWYN2VQUD_index_0_m.html, measured the height of the sea surface compared to average sea level, revealing the difference in altitude where a Meddy entered the Atlantic.

Specialized microwave radars called scatterometers, including the former NASA Scatterometer (Nscat) on Japan's Midori-1 spacecraft and the current SeaWinds instrument on NASA's QuikScat spacecraft http://winds.jpl.nasa.gov/missions/quikscat/index.cfm, measured the surface wind over the ocean, providing data needed to remove the surface variability "noise" caused by the wind blowing over the ocean's surface.

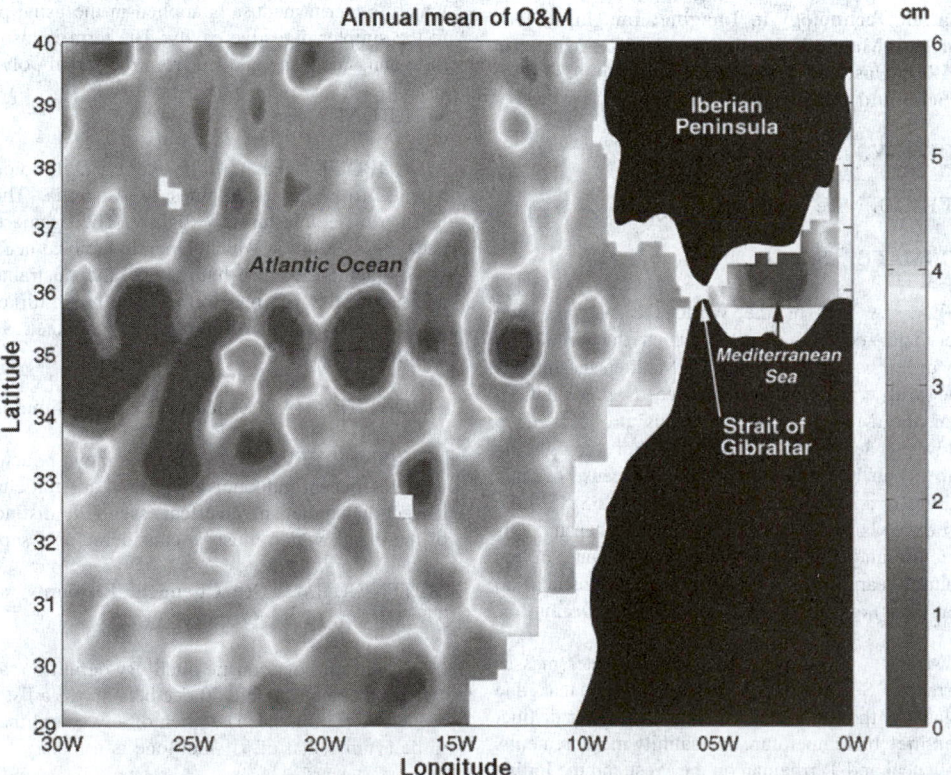

Fig. 1. Coupling data collected by several different satellite-borne sensors, researchers from the University of Delaware, NASA's Jet Propulsion Laboratory, and the Ocean University of China have been able to "break through" the ocean's surface to detect "Meddies" — super-salty warm-water eddies that originate in the Mediterranean Sea and then sink more than a half-mile underwater in the Atlantic Ocean. The Meddies are shown in red in this scientific figure. (*University of Delaware/NASA*).

"By carefully removing the stronger surface signatures of upper ocean processes, we were able to unveil the surface signatures of deeper ocean processes, such as the Meddies, to these space-based sensors," said Dr. W. Timothy Liu, QuikScat project scientist at JPL.

The scientists also analyzed data provided by an infrared spectrometer known as the Advanced Very High Resolution Radiometer http://noaasis.noaa.gov/NOAASIS/ml/avhrr.html, which flies aboard National Oceanic and Atmospheric Administration satellites. This instrument maps heat emitted by the ocean's top layer and showed the increase in temperature from a warm Meddy before it began sinking.

While the technique is not yet 100 percent accurate, Yan and his colleagues are continuing to refine it, and are exploring its application to other coastal regions of the world. They are currently examining salinity variations in the East China Sea before and after the building of the Three Gorges Dam, the largest dam in the world. The data will help researchers assess the dam's impacts on the ecosystem and on water circulation patterns.

See also **Eddy**; and **Whirlpool**.

ALAN BUIS, Jet Propulsion Laboratory, Pasadena, CA

MEDIAN. One of several measures of central tendency.1) Pertaining to a series of numbers, the median is the middle term when the numbers are arranged in algebraic order. If the number of terms is even, the median is taken as halfway between the two middle terms. 2) Pertaining to a continuous random variable x, the median is that value that divides the probability distribution into two equal areas. Hence, in terms of the distribution function $F(x)$, the median is that value of x for which $F(x) = 1/2$. In case $F(x)$ is discontinuous, the median is defined in such a way as to yield consistent results when the area is cumulated from either end of the distribution.

See also **Quantile**.

MEDIATOR. See **Immune System and Immunology**.

MEDICAL IMAGING. See **Computed Tomography (CT); Digital Processing and Information Technology in Imaging; Imaging (Medical): An Overview; Nuclear Magnetic Resonance (NMR); Magnetic Resonance Imaging (MRI); Positron Emission Tomography (PET); Ultrasound; and X-ray Scan and Other Medical Imagery**.

MEDICAL MICROBIOLOGY. See **Microbiology**.

MEDIEVAL WARM PERIOD. See **Climate**.

MEDITERRANEAN CLIMATE. See **Climate**.

MEDITERRANEAN SEA. A part of the World Ocean with little communication with the major ocean basins, in which the circulation is controlled by density differences. Two types of Mediterranean Seas are distinguished. In the arid type, evaporation exceeds precipitation, increasing the density of the Mediterranean Sea; this produces deep convection and an outflow of dense Mediterranean Water through the connection with the main ocean below an inflow of less dense oceanic water. Examples are the Eurafrican Mediterranean and the Red Sea. In the humid type, precipitation exceeds evaporation, lowering the density; this produces a two-layer structure and inflow of denser oceanic water below water of less dense Mediterranean Water. Examples are the Australasian Mediterranean (Indonesian Seas) and the Baltic Sea. See also **Ocean**.

MEDITERRANEAN WATER. A study of the oceanic water masses contained in the Mediterranean Sea is complicated by the fact that, due to the numerous peninsulas and the high rate of evaporation, four distinct bodies of water can be identified by temperature and salinity measurements. These are the Algiers-Provencal and Tyrrhenian on the West and the Ionian and Levantine on the East. Moreover, in each area all three types, surface water, intermediate and bottom water, can be identified. The oxygen content is also a valuable differential property, since the salinity is generally consistently over 36%. In fact, as the Mediterranean water leaves the Straits of Gibraltar its salinity averages over 37%, and its temperature over 18 °C

(64.4 °F). It forms currents, which, because of these properties, can be traced far into the South and North Atlantic Oceans.

MEDIUM LOW EARTH ORBIT. This orbit takes place at an altitude up to 1,400 km (870 miles) and is particularly suited for constellations of satellites mainly used for telecommunications. A satellite in this orbit travels at approximately 7.3 km (4.5 miles) per second.

See also **Earth-Orbiting Satellites (Data Receiving And Handling Facilities); Orbit (Astronomy); Satellites (Communications and Navigation); and Satellites (Scientific and Reconnaissance)**.

MEDIUM-RESOLUTION INFRARED RADIOMETER (MRIR). See **Nimbus Satellite Program**.

MEDLAR TREES. See **Rose Family**.

MEDULLA OBLONGATA. See **Central and Peripheral Nervous Systems**.

MEDUSA (or Hydromedusa). A form of coelenterate in which the body is shortened on its principal axis and broadened, sometimes greatly, in contrast with the hydroid or polyp. It varies from bell-shaped to a thin disk, scarcely convex above and only slightly concave below. The upper or aboral surface is called the ex-umbrella and the lower surface, on which the mouth opens, the subumbrella. The latter may be partly closed by a membrane extending inward from the margin. This structure is called the velum. The digestive cavity consists of a central chamber, the stomach, and radiating canals, which extend toward the margin. These canals may be simple or branching and few or many. The margin of the dish bears tentacles and sensory organs.

In the class *Hydrozoa*, medusae are the sexual individuals of many species, alternating in the life cycle with asexual polyps; but in the *Scyphozoa* or jellyfish proper, the medusa alone is well developed.

MEDUSOID. The medusa of certain coelenterates of the class *Hydrozoa*. A hydromedusa. Medusoids differ from the free-swimming jellyfishes to which the term medusa is applied in the usual presence of the velum and in the simpler digestive cavity. The term is also applied in some cases to the young medusae budded from the larval polyp stage of the jellyfishes.

MEERKAT. See **Viverrines**.

MEGAPODE (*Aves, Galliformes*). Dull-colored birds of the Pacific islands, from the Philippines to Australia. They have strong legs and feet and resemble turkeys slightly. The eggs are deposited in mounds of decaying vegetation which generate the heat necessary for incubation. The family includes the brush turkeys of the Australian region and the maleo in addition to the true megapodes. The brush turkey also is called the zebra bird, shell bird, and warbling grass parakeet. See also **Galliformes**; and **Mound Birds**.

MEGATHERMAL CLIMATE. See **Climate**.

MEISSNER EFFECT. When a superconductor is cooled in a magnetic field the lines of induction are pushed out at the transition, as if it exhibited perfect diamagnetism, an effect essentially distinct from the zero resistivity of the metal, which must be considered as a separate phenomenon.

MEITNERIUM. See **Chemical Elements**; and **Chemical Elements: The History of the Origin**.

MEL. A unit of acoustic pitch. By definition, a simple tone of frequency 1000 cycles per second, 40 decibels above a listener's threshold, produces a pitch of 1000 mels. The pitch of any sound that is judged by the listener to be n times that of a 1-mel tone is n mels.

MELALEUCA TREE. In an effort to afforest the Everglades at the turn of the century, the melaleuca tree was introduced from Australia in 1906. The effort was stepped up in the 1930s when melaleuca seeds were scattered over the region by aircraft. In recent years, the effort has backfired because the melaleuca has damaged and continues to threaten the diversity of life that once characterized the Everglades. The melaleuca is a large, bushy tree that can reach 80 feet (24+ meters) in height and have a diameter of 40 in. (102 cm). The tree is well adapted to damp areas. Like the casuarina (see also **Casuarina Tree**), the melaleuca is extremely rugged and very difficult to eradicate. One biologist has observed, "Cut one down and you can wind up instead with three or four—every cut piece sprouts back. Bulldoze one, and it will grow back from the roots. Disturb one in any way and its seed pods open up." The melaleuca is equipped with mechanisms that make it a natural marvel. During flooding, it produces new roots up the stems as the water level rises; seedlings produce new leaves and continue to grow underwater. during fires, the pods burst from the heat and disperse their seeds, while a thick, punky bark acts as excellent insulation. During high winds, seeds are released and broken branches quickly sprout and take root. Frost seems to be the tree's only principal adversary.

The melaleuca is not without some benefits. For example, the trees provide abundant pollen and nectar over much of the year and it is estimated that beekeepers maintain some 200,000 bee colonies on the melaleucas. Some authorities believe that melaleuca wood may have potential as a hardwood raw material. The heartwood is resistant to decay and termites. The wood can be seasoned and finished to rival the attractiveness of cherry, black walnut, or mahogany.

The record melaleuca tree, melaleuca quinquenervia, in American Forests is located in new Ft. Denaud, Florida. Also referred to as the cajeput tree, this record holder is 62 feet (18.9 meters) high, has a spread of 28 feet (8.58 meters), and a circumference of 231 rule (554.5 centimeter).

MELAMINE. [CAS: 108-78-1]. $(N \equiv C - NH_2)_3$, formula weight 126.12, white solid, mp $355\,°C$, sp gr 1.56. The compound may be considered the trimer of cyanamide, [CAS: 420-04-2] or as the triamide of cyanuric acid [CAS: 108-80-5]. Melamine resembles an amide more than an amine. Liebig first prepared melamine in 1834. In early production methods, melamine was prepared from calcium cyanamide [CAS: 156-62-7] through conversion to the cyanodiamide and then to the trimer, melamine. The compound now is synthesized from urea. The production of melamine exceeded 450 metric tons annually in the early 1970s and has been growing at a rate of about 5% annually. Most of the melamine made is condensed with formaldehyde or other aldehydes to form resins. These resins possess particularly outstanding resistance to heat, water, and many chemicals. The electrical properties and surface hardness also are rated high. The consumption of melamine for these resins is: (1) protective and decorative laminates, 45%; (2) molding compounds, 30%; (3) textile resins, 9%; (4) coatings, 7%; (5) paper-treating resins and various adhesives, 9%. Typically, in a modern synthesis process, (1) urea is thermally decomposed into a gas mixture of cyanic acid and NH_3 : $H_2N \cdot CO \cdot NH_2 \rightarrow HCNO + NH_3$; (2) cyanic gas is thermally decomposed into a melamine-CO_2 vapor:

$$6HCNO \rightarrow (N \equiv C - NH_2)_3 + 3CO_2.$$

Step 1 is endothermic; step 2 is exothermic; the overall reaction is endothermic. Because of the large quantities of CO_2 and HN_3 generated, the process often is undertaken in connection with urea manufacture, which permits the off-gases to be recycled usefully. The melamine synthesis may be carried out at low or medium pressures with the assistance of a catalyst; or at higher pressures without a catalyst.

MELANIN.   See Dermatitis and Dermatosis.

MELANOMA. Melanoma arises from specialized cells in the skin called *melanocytes*. They are situated as single cells within the basal layer of epidermis, eye and inner ear. Melanocytes synthesize and distribute the pigment melanin, which imparts skin color, to neighboring keratinocytes. Melanin in melanosomes provides protective coloration against the damaging effects of ultraviolet radiation. Developmentally, melanocytes arise from pluripotent cells of the neural crest. Their survival and migration from the neural crest, and their differentiation, depend on spatial and temporal expression of a variety of proteins including growth factors and adhesion receptors. See also **Dermatitis and Dermatosis**.

Pathobiology of Melanoma

Melanoma progression from normal melanocytes is divided into distinct steps characterized by histolopathological and experimental features (Fig. 1). Least is known about human melanocyte precursor cells, which can also be found in the adult. Potentially, these cells are more likely to give rise to naevi and melanomas than mature, fully differentiated cells. Common naevi can progress to dysplastic naevi and melanomas, but more likely regress after four or five decades. Their development is potentially due to errors occurring in the growth regulation between melanocytes and keratinocytes. Both cell types in normal skin exist in a fine balance (homeostasis) of growth and differentiation, and we begin to understand that potential perturbance in adhesion and gap junction formation disrupts the control that keratinocytes exert over melanocytes. Dysplastic naevi are direct precursors of melanoma. However, their diagnosis and that of radial growth phase (RGP) primary melanomas is difficult and controversial because of the current absence of molecular markers. Surgical resection of RGP melanomas leads invariably to cure, suggesting that these lesions have not yet developed competence for tumorigenic growth and metastasis [Satyamoorthy, et al.:]. Vertical growth phase (VGP) primary melanomas, on the other hand, contain multiple genetic abnormalities and have metastatic competence by growing invasively and independently of exogenous growth factors. They also produce tumors in immunodeficient mice [Herlyn, et al.:].

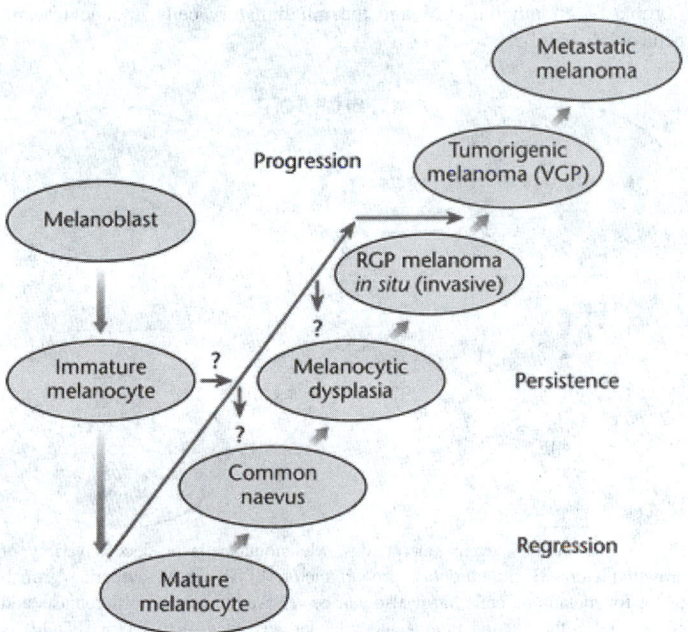

Fig. 1. Melanoma development and progression. The model implies that melanoma commonly develops and progresses in a sequence of steps from naevic lesions, which can be identified histologically in approximately 35% of cases. Thus, melanoma may also develop directly from normal cells. The roles of melanomablasts (immature melanocytes) in melanomagenesis remains poorly defined. Cells from naevic lesions show persistence, but the nontumorigenic lesions tend to disappear through apoptotic or differentiation pathways which still need to be defined at the molecular level. RGP, radial growth phase; VGP, vertical growth phase.

The process of metastasis in melanoma consists of a series of interlinked and orderly steps. Metastasis is a consequence of survival and growth of cell subpopulations of clonal origin that exist within the primary neoplasm. Different metastases can originate from the proliferation of cells derived from the original clone, suggesting that the metastatic phenotype is genetically unstable. It is also reversible. The outcome of metastasis depends on the interactions of metastatic cells with the host, particularly how functions of the organ are affected by the expanding malignant cells. Organ-specific metastases have been demonstrated in melanoma and may even be specific to a particular site within one organ. Melanoma metastases to distant skin, brain, gastrointestinal tract, lung and liver are most common, and those to the kidney and pancreas are less common.

Dissemination of melanoma cells from the site of origin includes a cascade of steps, such as: (1) detachment of tumor cells from the primary tumor; (2) traversation of the basement membrane (mostly of lymphatic vessels); (3) migration into the draining (sentinel) lymph node, where the cells are trapped but from where they can easily dislodge again to migrate to additional lymph nodes until they finally reach the blood vessels; (4) circulation in the bloodstream; (5) extravasation from the vessel; and (6) invasion, survival and growth in the distant organ [Juhasz, et al.:]. In these processes, several important categories of proteins are involved, including growth factors and their receptors, proteases and their substrates, adhesion receptors and their ligands.

Growth Factor Nexus

Autocrine and paracrine regulatory networks exist in the epidermis that affect the survival, proliferation and pigmentation of human melanocytes. These regulatory networks induce distinct biological and histopathological features characterizing each stage of melanocyte development and melanoma progression. The members of this network include neuropeptides, endothelin 1, 2 and 3, α-melanocyte stimulating hormone, basic fibroblast growth factor (bFGF), insulin-like growth factor (IGF) I, platelet-derived growth factor (PDGF) A or B, stem cell factor, hepatocyte growth factor/scatter factor, vascular endothelial growth factor, interleukin (IL) 8, and monocyte chemotactic protein 1, which represent the most significant participants in an intricate crosstalk with cells in the stroma, particularly fibroblasts, endothelial cells and the inflammatory cells, monocytes and neutrophils (Fig. 2.).

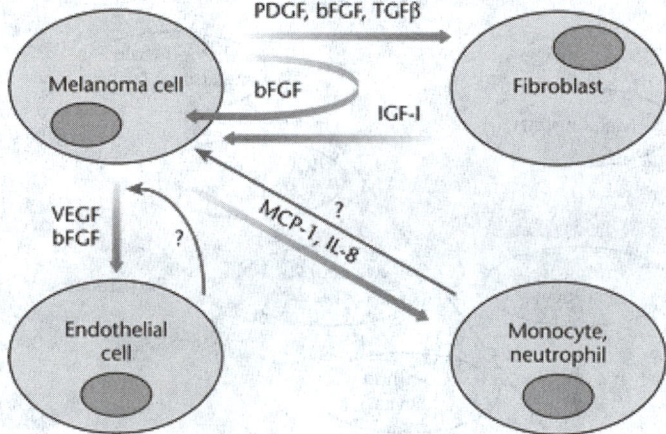

Fig. 2.   Melanoma–stroma interactions. Melanoma cells produce a variety of growth factors. Basic fibroblast growth factor (bFGF) is an autocrine growth factor for melanoma cells but it also can be released to stimulate fibroblasts and endothelial cells. On the other hand, platelet-derived growth factor (PDGF) is produced only for fibroblast stimulation. Activated fibroblasts, in turn, can secrete insulin-like growth factor I (IGF-I) which is a strong mitogen for melanocytic cells from all stages of tumor progression. Endothelial cells are activated by melanoma-derived vascular endothelial growth factor (VEGF) and bFGF. Melanoma cells also produce chemokines for monocyte attraction (monocyte chemotactic protein 1 (MCP-1)) or neutrophil attraction (interleukin 8 (IL-8)). TGF β, transforming growth factor β.

One of the most significant aspects of melanoma progression is the acquisition of growth autonomy and the expression of multiple growth factors and receptors by tumor cells but not by normal melanocytes. Among the growth factors produced by melanoma cells, bFGF, transforming growth factor α and β, PDGF-A and B chains, melanoma growth stimulatory activity and interleukins have been most extensively characterized [Mattei, et al.:]. The complex signaling networks mediated by these melanoma-derived factors are responsible for autocrine growth stimulation of melanoma cells and for paracrine actions of growth factors in the generation of a microenvironment favorable for tumor survival and invasion. Foremost among them is bFGF, which plays a key role in autonomous growth of melanoma cells, imparting both autocrine and paracrine effects: bFGF is produced for self stimulation and for stimulation of cells in the stroma, for example fibroblasts and endothelial cells. PDGF, on the other

hand, is not produced for autocrine growth because the melanoma cells do not express the PDGF-B receptor [Forsberg-Nilsson, et al.:]. Instead PDGF is produced solely for activation of fibroblasts, which in turn produce IGF-I for stimulation of the malignant cells, which express the IGF-I receptor. Here, we describe the roles and effects of bFGF and IGF-I in melanoma because both growth factors appear to be central for tumor progression.

Basic Fibroblast Growth Factor Pathway

bFGF is the most important autocrine growth factor in melanoma. Inhibition of bFGF with antibodies and antisense deoxyoligonucleotides inhibits melanoma growth *in vitro* and *in vivo*. Similarly, inhibition of the receptor with antisense deoxyoligonucleotides abolishes tumor growth [Becker, et al.:]. Virtually all melanoma cells tested produce bFGF. bFGF may already be found in naevi to allow cells to leave the epidermis, become independent from the stimulation by keratinocytes and migrate into the dermis (Fig 3a). In melanoma cells, bFGF acts as a motility factor, induces angiogenesis and is responsible for activation of the invasion machinery, particularly by activating the plasminogen activator and collagenase cascades. bFGF lacks a signal peptide and is released by melanoma cells by an unknown mechanism.

bFGF messenger ribonucleic acid contains multiple alternative CUG start sites rather than AUG (Fig. 3b). However, the biological properties of the different molecular weight proteins translated from these start sites are not very well characterized. All the translated products are present in the nucleus, whereas only the 18-kDa product is secreted. While the nuclear localization–retention signal is present in the high molecular weight forms of bFGF, extracellular bFGF is apparently transported to the nucleus in a receptor-dependent manner.

The biological activity of bFGF is mediated through high-affinity cell surface receptors (FGFRs), ranging in size between 110 and 150 kDa (Fig. 3c). There are four related genes encoding high-affinity FGFRs (FGFR-1 to FGFR-4); structural variants of these receptors are additionally generated due to alternative splicing. This leads to different ligand-binding specificities and affinities. FGFRs consist of three immunoglobulin-like domain loops in the extracellular domain, a transmembrane domain and two intracellular tyrosine kinase domains; the first is for adenosine triphosphate (ATP) binding, and the second is the catalytic domain. The signal sequence in the N-terminus is followed by the first immunoglobulin loop, which is not required for binding of FGF or heparan sulfate but which affects binding affinity. This is followed by an "acid box" which is acidic and rich in threonine/serine residues of unknown function. The second immunoglobulin loop is thought to be involved in the functional activity of FGFR and binds heparan sulfate and FGF. The sequences between immunoglobulin loops 2 and 3 and immunoglobulin loop 3 are the chief source of alternative splicing that affects the interaction of bFGF to its receptor. The immunoglobulin domain 3 is encoded by two separate exons. Exon IIIa encodes the N-terminal half while two alternative exons, IIIb and IIIc, code for the C-terminal half. This region is responsible for homeo-interactions and it acts as a secondary FGF-binding domain. FGFR-1 and FGFR-2 exon IIIc also have high affinity for bFGF, whereas FGFR-4 contains only exon IIIc of the immunoglobulin domain 3 and therefore acts as a low-affinity receptor for bFGF. The transmembrane domain of FGFR not only serves as an anchor to the membrane but it also facilitates oligomerization of the receptor. Receptor dimerization and protein tyrosine kinase activation results after binding of bFGF to the receptor and leads to the recruitment of SH2-containing molecules, triggering signal transduction and leading to activation of the mitogen-activated protein (MAP) kinase pathway which results in activation of transcription factors that modulate target genes for proliferation. At the same time, association of phospholipase C γ with the FGFR tyrosine kinase domain leads to production of second messengers, such as diacylglycerol or inositol phosphates, which are able to mimic or induce protein kinase C in cellular growth control.

Insulin-like Growth Factor I Pathway

IGF-I is one of the few known growth factors not produced by melanoma cells. Because it is essential for growth of normal melanocytes, nevus cells, and RGP and early VGP primary melanoma cell lines, it has major biological significance for tumor progression. Thus, IGF-I is the prototype

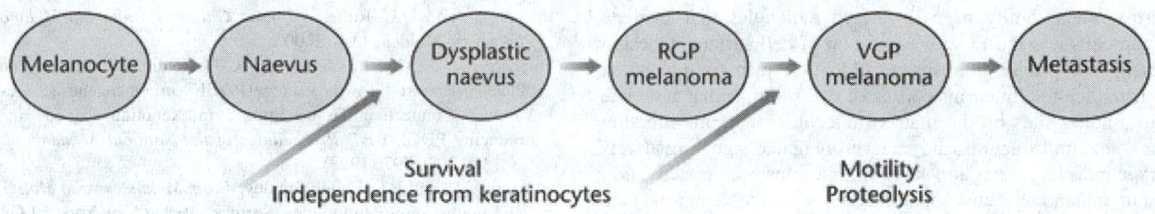

(a) bFGF in melanoma progression

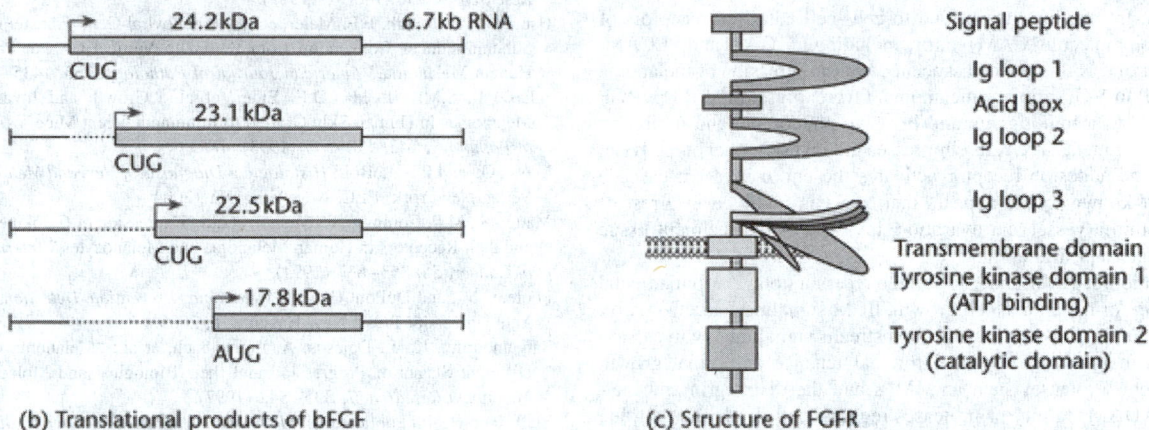

(b) Translational products of bFGF

(c) Structure of FGFR

Fig. 3. Significance of basic fibroblast growth factor (bFGF) and structures of bFGF and the FGF receptor (FGFR). (a) bFGF is produced by nevus cells for survival when cells are migrating from the epidermis to the dermis. Melanoma cells produce bFGF for autocrine growth stimulation, but bFGF also stimulates angiogenesis and stroma formation, and plays a major role in invasion by acting as a motility factor and by activating the expression of proteases important for invasion. (b) Four translational products of bFGF are produced by melanoma cells but apparently only the 17.8-kDa product is released by cells for paracrine functions, whereas the other products are produced for autocrine growth stimulation and can also translocate to the nucleus. & Nonoptimal Kozak sequence. RNA, ribonucleic acid. (c) Melanoma cells express predominantly FGFR-1, which transmits signals after ligand binding through the tyrosine kinase domains. For signal transduction, two pathways are apparently activated concomitantly, the mitogen-activated protein kinase pathways and the phospholipase C γ pathway, but a melanoma-specific signaling is not known. Ig, immunoglobulin; ATP, adenosine triphosphate.

paracrine growth factor. Nearly all cells of the melanocytic lineage express IGF-I receptor. The mitogenic effect of IGF-I was first established in 3 T3 cells when the cells were also treated with PDGF. IGF-I stimulates cells to enter the S phase of the cell cycle. The 70-amino-acid single-chain polypeptide consists of the N-terminal B domain of 29 amino acids, followed by C, A and D domains of 12, 21 and eight residues respectively. The A and B domains of IGF-I are homologous to insulin A and B chains, and IGF-I shares 70% homology with IGF-II. In humans IGF-I is mostly produced by activated fibroblasts, liver cells, adipose and ovarian cells. Some tumors such as breast carcinoma also secrete IGF-I. See also **Insulin**.

The IGF-IR is a heterotetrameric cell surface glycoprotein consisting of 1367 amino acids. It is synthesized as a single-chain polypeptide with a 30-residue signal peptide and is proteolytically cleaved at amino acid position 707–710 into a and b subunits. The heterotetrameric form of IGF-IR is composed of two each of a and b subunits joined by several disulfide bonds. The IGF-I-binding domain is located in a subunit that is rich in cysteine residues and consists of several N-glycosylation sites. The transmembrane domain in the b subunit anchors the complex to the cell surface. The b subunit also contains ATP and tyrosine kinase-binding domains, which transmit the signals after IGF-I binding to the receptor. The IGF-IR binds not only IGF-I, but also IGF-II and insulin, although with greatly reduced affinity. Receptor aggregation and internalization also appears to be important for the mitogenic activity of IGF-I. Signaling occurs through MAP kinase similar to the bFGF signaling pathway, but IGF-I appears to activate additional pathways such as the antiapoptotic pathways involving Bad (Bcl2/Bcl-X$_L$ associated death promoter). However, a full assessment of the multifunctional role of IGF-I in melanoma has not been done.

Extracellular Matrix and Adhesion Interplay

Human melanoma cells produce a variety of matrix-degrading enzymes. Expression in normal cells is absent unless the cells are activated, and expression increases with progression. Invasive and metastatic melanoma cells express the highest levels when compared with noninvasive and nontumorigenic melanoma cells. The biological activity of proteolytic enzymes is widely regulated. Serine and metalloproteases are secreted as inactive proenzymes, often bound to an inhibitor, and require proteolytic cleavage for activation. Thus, the biological activity of invasion-related proteases is regulated by the presence of: (1) an activator, (2) an inhibitor and (3) a binding protein or receptor. The reaction partners (inhibitor or receptor) are specific for each enzyme and may not necessarily be expressed by the melanoma cells: malignant cells may express the tissue plasminogen activator but not its inhibitor, which instead can be found on normal fibroblasts in the stroma. Melanoma cells can activate the enzymes through an intracellular (furin) pathway, or they are activated through an extracellular (plasmin) pathway.

Proteolytic enzymes have substrate specificity for different components of the extracellular matrix (ECM). The degradation of complex matrices requires the cooperation of proteases with specificity for different glycoproteins and collagens. Matrix metalloproteinases (MMPs) are a family of zinc-dependent endopeptidases collectively capable of degrading essentially all ECM components. These enzymes can be produced by different cell types in normal skin, such as fibroblasts, keratinocytes, macrophages, endothelial cells, mast cells and eosinophils, and their activity is specifically inhibited by tissue inhibitors of metalloproteinases, which bind to active MMPs with a 1:1 stoichiometry. MMPs are not expressed constitutively in skin, but are induced temporarily in response to exogenous signals such as cytokines, growth factors, cell matrix interactions and altered cell–cell contacts. In contrast, melanoma cells express the enzymes constitutively. However, little is known about how the melanoma cells direct enzyme activation during cell growth, tumor expansion and invasion.

Invasion of malignant cells requires altered cellular interaction with ECM. Integrin-type cell adhesion receptors play an important role in this process. Integrins are a family of cell surface molecules that mediate adhesion between cells and the ECM. Expression of cell surface adhesion molecules changes with progression from RGP to VGP primary melanoma and to metastatic melanoma. Integrins consist of two subunits that associate to bind specific binding sites on the matrix molecules. Integrins function as receptors for signal transduction, and integrin-mediated signaling affects cell growth, tissue morphogenesis and survival. Integrins require activation for adhesion, but melanoma cells apparently express the integrins in a constitutively active form. Whereas most integrins have specific ECM-binding partners, one integrin, the $\alpha v\beta 3$ vitronectin receptor which has a broad specificity, can bind more than five different ECM proteins, including vitronectin, von Willebrand factor, fibronectin, thrombospondin and fibrinogen. In addition, it can bind to cell–cell adhesion receptors of the cell adhesion molecule (CAM) family including L1-CAM and PECAM. The $\beta 3$ subunit of $\alpha v\beta 3$ is highly associated with progression of melanoma cells from RGP to VGP primary melanoma. Overexpression of $\beta 3$ in RGP melanoma leads to a tumorigenic and invasive phenotype, and inhibition of the molecule in metastatic cells suppresses the invasive phenotype [Hsu, et al.:]. How one adhesion receptor activates the entire invasive cascade for cells is not known, but it appears that $\alpha v\beta 3$ is involved in arrest of tumor cells within a vessel, extravasation, invasion of the recipient tissue bed, and ultimately proliferation.

The cooperation of ECM and cell surface receptors plays a fundamental role in signaling for melanoma cell growth. ECM signaling for cell–matrix and cell–cell interactions leading to downstream signaling involves several molecules that are also essential for maintenance of normal growth and development. Proteases such as MMPs and the disintegrin and metalloprotease (ADAM) family of proteases regulate cell death and cell fate specifications. Upon integrin binding to ECM ligands, multiple signaling pathways become activated and cytoskeletal rearrangements occur. A number of protein kinases act downstream of the integrin–ligand interaction, including the tyrosine kinases (syk, Src and FAK), the serine threonine kinases (Raf, mitogen-activated/extracellular receptor-regulated kinase (MEK) and extracellular signal regulated kinase (ErK)), phosphatidylinositol 3-kinase and Akt. The crosstalk of the signaling pathways activated by integrins is important for the regulation of melanocyte proliferation and melanoma tumor progression, but little is known about whether this occurs in a cell type-specific manner.

Emerging Concepts

It has become clear that melanoma is a consequence of multifactorial perturbations and that melanoma cells control the surrounding fibroblasts, keratinocytes, endothelial cells and inflammatory cells for their own benefit, and utilize them for survival. The ability of melanoma cells to invade and metastasize to distant organs in the body suggests their adaptability to very different microenvironments. Therefore, there is a need to develop suitable model systems to study cells in their three-dimensional environment, instead of the typical monolayer-grown cell. The mechanisms by which the host factors interact with the tumor remains to be elucidated. A clear understanding of the signal transduction pathways between growth factors and their receptors, adhesion molecules and their ligands, cytokines, chemokines and their receptors should lead to the identification of new targets for therapy. Crosstalk between several signal transduction pathways and their redundancies suggests the importance of understanding the precise role of complex biological processes. To date there is only a limited understanding of gene regulation in melanoma. Transcription factors that are specific for melanocytic lineage have been identified but their impact on melanoma progression remains to be seen.

See also **Skin Cancer**.

Additional Reading

Agarwala, S.: "Melanoma," *Translational Research and Emerging Therapies*, CRC Press, LLC, Boca Raton, FL, 2006.

Atillasoy, E.S., J.T. Seykora, P.W. Soballe, et al.: "UVB Induces Atypical Melanocytic Lesions and Melanoma in Human Skin," *American Journal of Pathology*, **152**, 1179–1186 (1998).

Barnhill, R.L.: *Pathology of Malignant Melanoma*, Springer-Verlag New York, LLC, New York, NY, 2004.

Becker, D., D.E. Johnson, P.L. Lee, U. Rodeck, and M. Herlyn: "Inhibition of the FGF Receptor 1 (FGFR-1) Gene in Human Melanocytes and Malignant Melanomas Leads to Inhibition of Proliferation and Signs Indicative of Differentiation," *Oncogene*, **7**, 2303–2313 (1992).

Buchan, J., and D.L. Roberts: *Pocket Guide to Malignant Melanoma*, Blackwell, Publishers, Malden, MA, 2000.

Forsberg-Nilsson, K., I. Valyi-Nagy, C-H Heldin, M. Herlyn, and B. Westmark: "Platelet-derived Growth Factor (PDGF) in Oncogenesis: Development of a Vascular Connective Tissue Stroma in Xenotransplanted Human Melanoma Producing PDGF-BB," *Proceedings of the National Academy of Sciences of the USA*, **90**, 393–397 (1993).

Hearing, V.J., and P.L. Leong Stanley: *From Melanocytes to Malignant Melanoma: The Progression to Malignancy*, Springer-Verlag New York, LLC, New York, NY, 2006.

Herlyn, M., J. Thurin, G. Balaban, et al.: "Characteristics of Cultured Human Melanocytes Isolated from Different Stages of Tumor Progression," *Cancer Research*, **45**, 5670–5676 (1985).

Hsu, M-Y, D-T Shih, F.E. Meier, et al.: "Adenoviral Gene Transfer of $\beta 3$ Integrin Subunit Induces Conversion from Radial to Vertical Growth Phase in Primary Human Melanoma," *American Journal of Pathology*, **153**, 1435–1442 (1998).

Juhasz. I., S.M. Albelda, D.E. Elder, et al.: "Growth and Invasion of Human Melanomas in Human Skin Grafted to Immunodeficient Mice," *American Journal of Pathology*, **143**, 528–537 (1993).

Massi, G., and P.E. LeBoit: *Histological Diagnosis of Nevi and Melanoma*, Springer-Verlag New York, LLC, New York, NY, 2004.

Mattei, S., M.P. Colombo, C. Melani, et al.: "Expression of Cytokine/growth Factors and their Receptors in Human Melanoma and Melanocytes," *International Journal of Cancer*, **56**, 853–857 (1994).

Poole, C.M., and DuPont Guerry: *Melanoma: Prevention, Detection, and Treatment*, Yale University Press, New Haven, CT, 2005.

Satyamoorthy, K., E. Dejesus, A. Linnenbach, et al.: "Melanoma Cell Lines from Different Stages of Progression and their Biological and Molecular Analyses," *Melanoma Research*, **7**, S35–S42 (1997).

Staff, Icon Health Publications: *Melanoma: A Medical Dictionary, Bibliography, and Annotated Research Guide to Internet References*, ICON Health Publications, San Diego, CA, 2004.

Thompson, J.F., D.L. Morton, and B.K. Kroon: *Textbook of Melanoma*, Taylor & Francis, Inc., Philadelphia, PA, 2003.

Web References

MayoClinic.com: www.mayoclinic.com/health/melanoma/DS00439
Melanoma.com: www.melanoma.com/
National Cancer Institute: http://www.cancer.gov/cancertopics/types/melanoma
The Skin Cancer Foundation: http://www.skincancer.org/melanoma/index.php

KAPAETTU SATYAMOORTHY,
MEENHARD HERLYN,
Wistar Institute, Philadelphia, PA

MELIOIDOSIS. The unusual causative agent of this disease, *Pseudomonas mallei*, is a free-living organism widely distributed in stagnant ponds, streams, rice paddies and soils in endemic areas of the tropics. Most infections have emanated from southeast Asia, but indigenously acquired disease has been reported from India, Korea, the Philippines, Australia, Panama, Turkey, and the United States. The organism is a small, pleomorphic, Gram-negative, strictly anaerobic bacillus, which shows a prominent bipolar pattern on staining.

Human infection usually results from contact of the broken skin with infected water, or by inhalation of contaminated dust. Five patterns of the disease have been observed. The *subclinical* pattern is generally asymptomatic and manifest only by elevated antibody levels. The *acute localized infection* takes the form of a localized pustule at the site of the skin injury and may be self-limited, develop into lymphangitis and regional adenopathy, or progress to acute septicemia. *Pulmonary infection* is the most common clinical presentation, ranging from a mild pneumonitis to a fulminant necrotizing pneumonia. *Acute septicemia* may derive from any manifestation of the disease and is the most mortal of the variations. Metastatic abscesses may be seen in the liver, spleen, lungs, lymph nodes, bone, and brain. Hepatosplenomegaly is felt and rales and friction rubs are heard. *Chronic suppurative infection* characterizes the fifth form, in which chronic cavitary lesions in the upper lungs may develop within months or years after the primary infection. Chronic abscesses can also develop and a protracted intermittently febrile wasting illness may result.

Melioidosis should be considered in any patient who has ever resided in an endemic area and who has an acute or chronic illness fitting one of the aforementioned patterns. Chloramphenicol, tetracycline, trimethoprim-sulfamethoxazole and kanamycin are the preferred therapeutic agents.

MELLOR-YAMADA PARAMETERIZATION. A parameterization of a complex model for turbulent flows in the planetary boundary layer. See also **Parameterization**.

A series of simplifications of the full turbulence model to remove complex terms and form a closed set of equations leads to a hierarchy of so-called closure models of decreasing complexity labeled level 4 through level 1. The level 2.5 model is widely used; it incorporates only one additional equation and produces conventional turbulent fluxes.

AMS

MELTING LAYER. See **Meteorology**.

MELTING LEVEL. The altitude at which ice crystals and snowflakes begin to melt as they descend through the atmosphere. In cloud physics and in radar meteorology, this is the accepted term for the 0 °C (32 °F) constant-temperature surface. See also **Bright Band**.

It is physically more apt than the corresponding operational term, freezing level, for melting of pure ice must begin very near 0 °C, but freezing of liquid water can occur over a broad range of temperatures (between 0° and −40 °C (32 °F and −40 °F; see supercooling).

See also **Freezing Point**;and **Melting Point**.

MELTING POINT. The temperature at which a solid substance undergoes fusion, that is, melts, changes from solid to liquid form. The melting point of a substance should be considered a property of its crystalline form only. At the melting point the liquid and solid forms of a substance exist in equilibrium. All substances of crystalline nature have their characteristic melting points. For very pure substances the temperature range over which the process of fusion occurs is very small. The melting point of a pure crystalline solid is a function of pressure; it increases with increasing pressure for most substances. However, in the case of ice (and a few other substances) the melting point decreases with increasing pressure. Under a pressure of one standard atmosphere, the melting point of pure ice is the same as the ice point, that is, 0 °C (32 °F).

MEMBRANE (Semipermeable). See **Semipermeable Membrane**.

MEMBRANE SEPARATIONS TECHNOLOGY. The separation of materials (solids from liquids; liquids from other liquids; solids of one size from solids of another size; gases from liquids; etc.) is one of the fundamental unit operations needed by the chemical, food, and related fluid processing industries. Such operations include adsorption, absorption, distillation, evaporation, extraction, filtration, ion exchange, settling, and preparative chromatography, among others. In recent years, membrane technology has made important inroads into several of these more traditional unit operations, and on adsorption, distillation, and filtration in particular. Either membrane technology has replaced traditional separation operations or is used in connection with them—with resulting improvement of the separation (in terms of product quality) and in operating efficiency as well. The membranes used are synthetic polymers, the pores of which are made in a number of interesting ways. Membrane separations are not a single technology, but rather they differ in the methodology used and in the degree of separation that can be effected. Membrane-using subtechnologies include electrodialysis, electrolysis, microfiltration, ultrafiltration, and reverse osmosis.

The general design of the equipment in which membranes are used is commonly (1) *perpendicular flow*, i.e., where the flow of unprocessed material approaches the "filtering" medium in an effrontal manner, passing through the medium, the processed material exiting the medium on the opposite side, and the material removed remaining on the surface of the medium. This is the common figuration that applies to traditional filters, such as cartridge, bag, diatomaceous-earth precoated, backwashable sand, and backwashable mixed-media filters. Some of these configurations allow regeneration by way of backflushing; some do not—the used media must be discarded. See also **Filtration**. (2) In *crossflow*, the influent unprocessed stream is separated into two effluent streams, known as the *permeate* and *concentrate*, respectively. The permeate is that fraction which has passed through a semipermeable membrane; the concentrate is that stream which has been enriched with the solutes or suspended solids, i.e., those materials which have not passed through the membrane. This design permits the membrane medium to operate continuously in a self-cleaning mode, with solutes and solids swept away by the concentrate stream which is running parallel to the membrane (hence the term "crossflow"). As with conventional filtration, sometimes the trapped material (filter cake) is the principal desired end product; in other cases, the desired product is clarified effluent. However, in contrast with conventional filtering, membrane, methodologies not only separate solids (or gases) from liquids (or gases), but proper selection of the membrane will allow separation of solids (particles) by size range. Thus, the permeate (what passes through the membrane) will contain much smaller particle sizes (in the molecular range) than will the concentrate. Again, depending upon the objective of the process, the permeate or the concentrate will be the principal product of interest. In some instances, both products may be of vital interest. See Fig. 1.

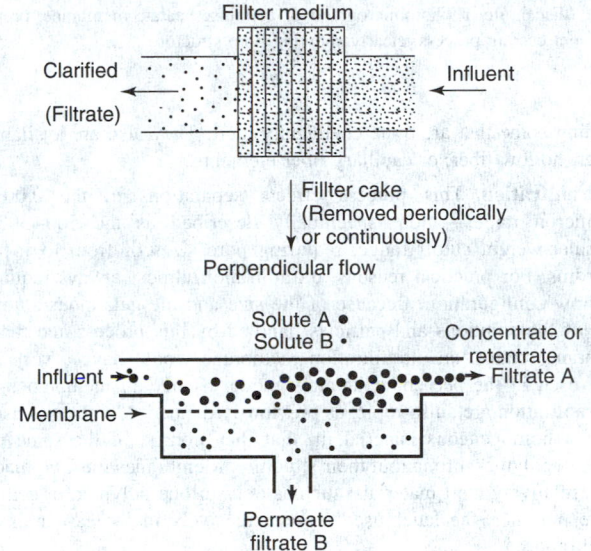

Fig. 1.   Perpendicular flow contrasted with crossflow. Perpendicular flow is shown here in the familiar terms of conventional filtering, but the principle is the same if a membrane is used. There is one influent and a single effluent. In crossflow, there are two effluents, a concentrate (or retentate) and a permeate, usually both liquids of different concentrations in terms of particle size. However, membranes are also used for separating acid gases and hydrocarbons.

Membranes may be *isotropic*, i.e., their pore structure and material are the same throughout the membrane; or they may be *anisotropic*, i.e., they have a dense skin layer on top which defines the degree of separation effected, with a spongy support layer underneath. The dimensions of the pores range widely as detailed later in this article. Membranes are made by a number of processes, including solvent casting and mechanical stretching to form pores in an otherwise impervious film. Irradiation, followed by acid etching, has been used to create pores. The cellulose acetate membrane (Fig. 2) is made by casting thin sheets of polymer dissolved in a water-miscible solvent on a flat plate, usually glass. Shortly after casting, the cast solution is immersed in water. The water diffuses into the solution and causes the polymer to coagulate at a rate that is a complex function of the polymer and solvent properties. These membranes are porous throughout, possessing a thin, relatively dense skin near one surface.

The range of small particles, molecules, and ions dealt with by membrane separation technology, particularly as encountered in the biochemical field are as follows.

Microfiltration. This process effects separations in the 0.02–2.0 micron range and historically has been run in the perpendicular flow mode, requiring disposal of the membrane medium as a result of binding by the retained material. Crossflow technology is increasing, as it proves practical. Microfiltration membranes are of an isotropic and homogeneous morphology, i.e., the pore structure is consistent throughout. There is some movement, however, toward the use of "skinned" anisotropic membranes. Microfiltration membranes are available in a wide variety of polymers,

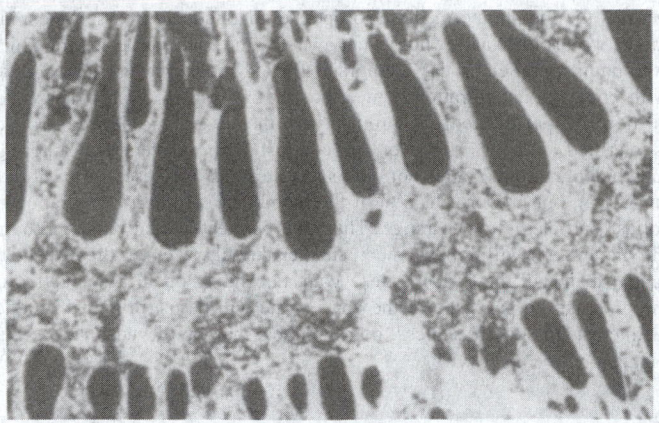

Fig. 2. Facsimile of photomicrograph of cellulose acetate membrane, prepared by solvent casting process, clearly showing pore structure.

including some that are quite chemically inert. They also are available as tubular, hollow fiber, or capillary fiber elements.

Ultrafiltration. This process effects separations in the 0.002 to 0.2 micron range—more specifically described as the 500–300,000 molecular-weight cutoff range, requiring pore sizes of from 15 to 1000 angstroms. For practical reasons, ultrafiltration almost always requires a crossflow configuration. Because of the size and the gelatinous nature of many of the solutions and particles handled by this process and that the membrane retains, an ultrafiltration membrane would have a very short life if used in the perpendicular flow mode. Nearly all membranes used in ultrafiltration are anisotropic, as previously defined. Membranes usually are of a homogeneous material, in that they consist of the same polymer or copolymer throughout their structure. Membranes must be made of tough, relatively inert materials, such as polysulfone polymer, or cellulose acetate polymer, the latter used more extensively in the earlier days of ultrafiltration.

Reverse Osmosis. Sometimes called *hyperfiltration*, this is the most technically complex class of membrane technology. Reverse osmosis effects separations both at the micromolecular and ionic size ranges. Pore sizes range from 5 to 15 angstroms. These membranes can effect separation of solutions down to a molecular weight of 150 and sometimes lower. Reverse osmosis membranes are anisotropic. Cellulose acetate has long been the favorite, especially for many industrial and medical applications. However, homogeneous polyamide-type membranes are finding a large share of the market and are increasingly favored for seawater desalting purposes.

Applications of Membrane Separations Technology. The food, biochemical, and petrochemical industries are the largest users. In the food processing industry, there are three main categories of use—processing, waste treatment, and pure water make-up. Processing applications include milk concentration and fractionation, numerous fermentation operations, and the production of colorants, among many others. Waste handling operations include corn processing byproducts, soybean protein reclamation, and handling meat processing oils and fish processing proteins and oils.

Membrane separations technology is also widely used in the pharmaceutical industry in connection with recovering antibiotic products. The list of references given at the end of this article is rather long. There the reader will find a wealth of information on applications, including the growing recognition by the petroleum and petrochemical industries of the viability of membrane technology as a replacement or companion for the more traditional separation processes.

Membranes Configurations for Use in Separation Equipment. For insertion into separation equipment, frequently with large throughputs, membranes are available in several configurations: (1) the tubular form ($\frac{1}{2}$ in; 12.5 mm) is common; (2) hollow-fiber; (3) plate-and-frame; and (4) spiral wound. The Paulson reference covers membrane formats in considerable detail.

See also **Desalination**.

Additional Reading

Abelson, P.H.: "Synthetic Membranes," *Science*, 1421 (June 23, 1989).

Beaudry, E.G. and K.A. Lampi: "Membrane Technology for Direct-Osmosis Concentration of Fruit Juices," *Food Technology*, 121 (June 1990).

Bedzyk, M.J. et al.: "Diffuse-Double Layer at a Membrane-Aqueous Interface Measured with X-ray Standing Waves," *Science*, 52 (April 6, 1990).

Carroll, L.E.: "New Process Concentrates Juices, Preserving 'Fresh Notes,' ' *Food Technology*, 148 (October 1989).

Dziezak, J.D.: "Membrane Separation Technology Offers Processors Unlimited Potential," *Food Technology*, 107 (September 1990).

Friedman, R.: "Seawater to Drink," *Technology Review (MIT)*, 14 (August/September 1989).

Hsieh, H.P.: *Inorganic Membranes for Separation and Reaction*, Elsevier Science, New York, NY, 1996.

Kosenoglu, S.S., J.T. Lawhon, and E.W. Lusas: "Use of Membranes in Citrus Juice Processing," *Food Technology*, 90 (December 1990).

Kosenoglu, S.S., J.T. Lawhon, and E.W. Lucas: "Vegetable Juices Produced with Membrane Technology," *Food Technology*, 124 (January 1991).

Matsurra, T.: *Synthetic Membranes and Membrane Separation Processes*, CRC Press, LLC., Boca Raton, FL, 1994.

Noble, R.D., C.A. Koval, and J.J. Pellegrino: "Facilitated Transport Membrane Systems," *Chem. Eng. Progress*, 58 (March 1989).

Noble, R.D. and S.A. Stern: *Membrane Separations Technology: Principles and Applications*, Elsevier Science, New York, NY, 1995.

Paulson, D.J., R.L. Wilson, and D.D. Spatz: "Crossflow Membrane Technology and Its Applications," *Food Technology*, 77–87 (December 1984).

Rousseau, R.W.: "Handbook of Separation Process Technology," John Wiley & Sons, New York, NY, 1987.

Singh, R.: "Surface Properties in Membrane Filtration," *Chem. Eng. Progress*, 59 (June 1989).

Spillman, R.W.: "Economics of Gas Separation Membranes," *Chem. Eng. Progress*, 41 (January 1989).

MEMBRANE TECHNOLOGY. See **Desalination**; and **Ocean Resources (Energy)**.

MEMORY, ELECTRONIC—CHRONOLOGY TO 1990. The need to remember information—to store and later to retrieve data—is one of the principal chores of electronic information processing equipment (computers, etc.). The time interval during which information must be retained (stored) ranges from extremely short time spans to very long periods of time. The time interval permitted to store and retrieve information also varies. Costs of memory systems also vary. Consequently, over a number of years, several information storage formats have emerged. In addition to memory system capacity and speed, protection of memory from physical destruction has been of major concern. Also, in numerous cases, a memory must be safeguarded against unauthorized access.

Considering a digital data processing system (Fig. 1), the basic functional elements consist of the central processing unit (CPU), main memory, input/output interface (I/O), and the associated peripherals for the input and output of information. Some, but not always all of these functional elements may be used in a given system. In the case of some industrial control applications, the I/O simply may be a display panel with switches and light-emitting diode (LED) display. *All* of the elements in the diagram use memory, including the peripherals. The characteristics and type of memory used varies greatly for each element.

The relative features and physical residences of those memory types used in each of the identified functional elements are shown in Fig. 2, commonly referred to as the *memory hierarchy*. Note that the memory characteristics used in each system element differ considerably.

Register Memory. For example, the register memory is used for the *temporary storage* of data being modified by the CPU. Thus, to insure fast execution of operations, the register storage should be as fast as the logic used to implement the CPU. The size of storage needed is usually small because only a few bits of data can be utilized on each execution cycle of the CPU.

Local Storage. This is similar in usage to that of the registers, here the emphasis is on the *temporary storage* of small amounts of data being moved between elements, such as the peripherals noted in Fig. 1. Generally, the performance needed for these applications does not require

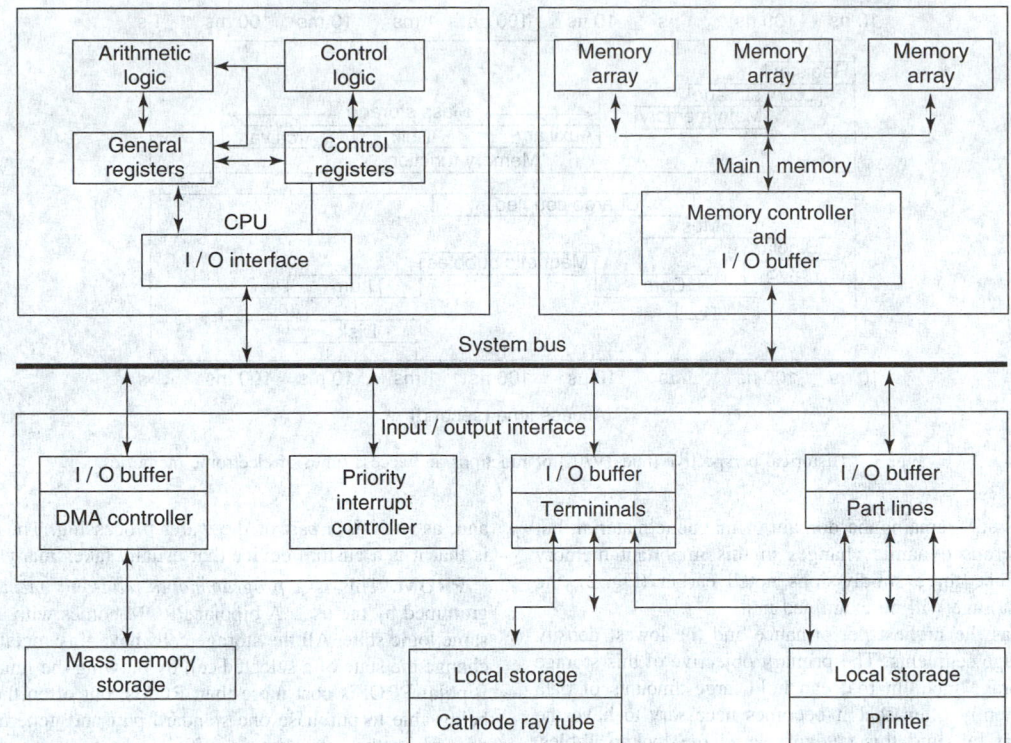

Fig. 1.   All elements shown in this diagram require electronic memory.

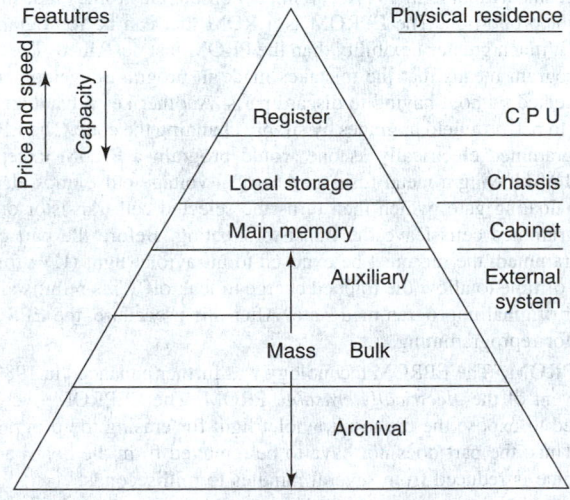

Fig. 2.   Memory system hierarchy.

the high speed of register storage. However, the memory size will be two to three orders of magnitude larger than that of the register storage memory.

Main Memory. As the name implies, this is the main working data storage device of the system. This random access read/write memory acts as a buffer between the CPU and the bulk or mass storage and is used to hold current resident programs, subroutines, and data. Hence, this memory is usually quite large (from thousands to millions of words). Since the current instructions and data of the programs being executed are stored in this memory, high speed is also essential. Thus, in most systems, the size and performance of the main memory will determine the overall performance of the system.

Mass Memory. This memory is used to store data and programs when they are not being used. In general, this data storage medium is the least expensive; it is also of very high density and slow. These low-cost memories employ *sequential address accessing*, which results in variable access times. The access time is variable because the time needed to get from one address to the next depends on the location of the last

address versus the location of the next address. With the exception of Mass Memory, other memory technologies use the faster *random access* method, which permits all addresses to be selected in the same time regardless of location. As noted in Fig. 2, there are three categories of mass memory: (1) auxiliary, (2) bulk, and (3) archival. Both bulk and archival memories are used for permanent data storage, whereas an auxiliary memory is referred to as *semipermanent*.

Historical Perspective

Prior to the mid-1970s, magnetic storage media predominated. Functions and access times of electronic memories of that period are given in Fig. 3. The most dramatic changes in memory technology occurred in connection with mainframe applications. The dominant main memory technology had switched to semiconductors by 1980. The rapid changeover to semiconductor memory started with the introduction of the 1024 bit *dynamic* random access memory (DRAM) in 1970. Dynamic storage is accomplished by storing charge on a capacitor and, therefore it is necessary to periodically recharge to maintain data. Reasons for quick acceptance of the DRAM for main memory applications were the increase in density and performance with lower cost. Since 1970, the density of these memories has increased manyfold.

Bipolar Technology. Bipolar static RAM (random access memory) technology historically has been used for register storage. Static RAM storage is accomplished through cross-coupled gates. Thus data as such does not require recharging to retain the data. Although the bipolar technology has gone through dramatic changes in density, speed, and power, the MOS (metal-oxide semiconductor) technology is strongly competitive. Through clever design techniques and process changes, such as high-performance *N*-channel metal oxide semiconductors (NMOS) and complementary MOS (CMOS), MOS static RAMs (SRAMs) approached the performance of bipolar units. Hence, with their higher density and lower power, they replaced bipolar RAMs in many system applications.

Local Storage. In a similar manner, the local storage memory applications went the route of register storage, but much faster. Since the performance of local storage is much less critical than that for register, static MOS memories quickly replaced bipolar RAM for these applications. Due to the high density, low power, and relatively fast access, the local storage applications widely use NMOS and CMOS SRAMs (static RAMs).

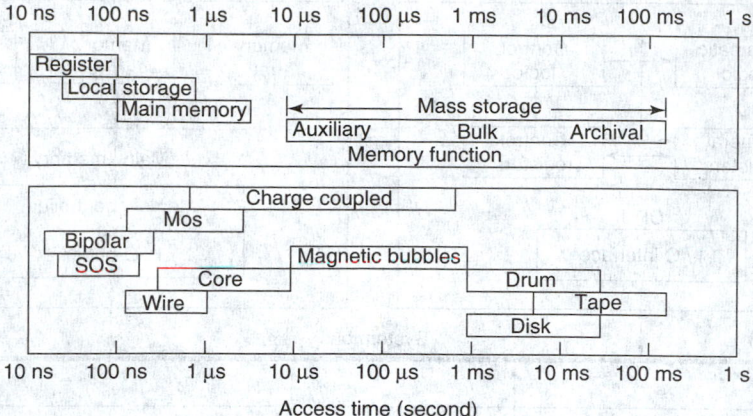

Fig. 3.   Historical perspective (mid-1970s) of functions and access times of electronic memories.

Mass Storage. This still remains the domain of magnetic material, but there have been numerous dynamic changes in this important memory segment. Considering the three subdivisions noted earlier (Fig. 2), the progress of auxiliary storage will be examined first.

Auxiliary storage has the highest performance and the lowest density of the three mass storage segments. The primary objective of this storage is to have a large storage medium that can hold large amounts of data that are needed sufficiently often that it becomes necessary to have the data on line. Data that fall into this category are large lookup Tables, data files, and backup operating systems with diagnostics, among others. Since auxiliary memory is always on line, its performance requirements differ from those of bulk and archival memory (Fig. 3). In the 1960s and early 1970s, memory technologies used for this application were magnetic bubbles and charge coupled devices (CCDs). Inasmuch as the CCD did not retain the density and lower cost advantages over DRAMs as had been expected, the CCDs eventually gave way to high-density DRAMs for this application. Thus, by the 1980s, the popular DRAMs were being used in many auxiliary storage applications. The use of DRAMs for this application developed rapidly. The system is sometimes referred to as semiconductor DISK.

Bubble memory can be used for auxiliary storage, but it does not have performance comparable to DRAM. Magnetic bubble memory does have one valuable advantage over DRAM in mass storage applications, namely, *nonvolatility*. When power is lost, bubble memory retains data, whereas DRAMs lose data and must be reloaded on powerup. Although bubble is magnetic, it is not mechanical in nature like magnetic tape or disk. These properties (nonvolatile and nonmechanical) make the bubble memory ideal for mass storage in harsh environments, such as encountered in military, industrial control, and portable systems.

For several decades, the primary storage device for bulk and archival storage have been magnetic — tape or disk. Fixed- and movable-head disks are used for bulk applications, whereas both movable-head disks and tape are used for offline archival storage.

Read-Only Memories (ROMs)

The memories discussed thus far have been read/write memories. There are other memory types. Their performance, density, and cost characteristics follow the memory hierarchy (Fig. 2). Read-only (ROM) memories utilize both the MOS and bipolar process technologies. These memories are used to hold code data that will *not change* with time. They also can be used for virtually all of the functional building blocks (Fig. 1). The ROM is used in a variety of applications, including the microcodes in a CPU (instruct CPU what to do), code converters, and frequently used Tables and constants, among many other functions. ROMs can be used as a substitute for combinational logic elements. In numerous instances, because of its bit density, the ROM can replace a controller design using many logic devices with just one or two ROMs.

A major advantage of a ROM over other semiconductor memories is that the ROM has the smallest storage cell of any memory device and thus the smallest die area of any memory of identical bit capacity. The programming of a ROM is accomplished with a metal mask interconnect and, as such, is a part of the wafer processing. Thus, its major drawback is that it is a custom device that usually takes many weeks to produce.

PROM. This is a *programmable read-only memory* that can be programmed by the user. A bipolar PROM comes with all storage cells in the same logic state. All the storage cells have tiny metal links so the user can change the state of a selected cell by blowing the link with a current pulse. Bipolar PROMs cost more than ROMs, but often the added flexibility of being able to purchase one standard part and generating the code pattern desired on the same day is sufficient in many cases to offset the higher cost over ROM.

EPROM. This is an *electrically programmable read-only memory* that first became available in 1971. The impact on the electronics industry was of major proportion. The EPROM is a ROM that can be reprogrammed, giving it much greater flexibility than the PROM just described. The ability to reprogram means that the mistakes made in program development can be corrected without having to discard parts. Another key advantage is the ability to perform field upgrades by simply changing the code. The EPROM is programmed electrically as one would program a PROM except that instead of blowing a metal fuse, an electrical voltage pulse traps electrons onto a floating gate, which then turns the selected cell transistor on. The unprogrammed cells leave the transistor floating. Before the part can be reprogrammed, the die must be exposed to ultraviolet light (UV) for a set period of time to allow the trapped charge to leak off. This returns all cells to their original unprogrammed state. After this procedure, the EPROM is ready for reprogramming.

EEPROM. The EPROM technology was further enhanced in 1980 with the advent of the *electrically erasable* PROM. The EEPROM overcomes the need to expose the die to ultraviolet light for erasing the programmed cells. Thus, the part does not have to be removed from the board and the erase time is reduced from several minutes to milliseconds.

The Groundwork for Creating the Electronic Memories of the 1990s
Transferring Data Between Functional Modules

For many years, the trend was to refine the existing memory technologies by increasing the density and speed and, where possible, to reduce the power. Each new generation was essentially a carbon copy of the existing one, but with enhanced performance and density. Generally, the memory cost performance improvements had a significant impact on the overall system cost performance and reliability. Unfortunately, each new memory component generation yielded a smaller increment of cost performance at the system level. A major reason for diminished return of the cost performance at the system level was the bottleneck of transferring data between the functional modules of the computer (Fig. 1). As indicated in that figure, the modules communicate with each other over a single data bus. With a single bus system, only two modules can use the bus at any one time. Due to long lines and heavy loading, the bus data bandwidth of most systems is limited to the range of 2 to 5 MHz. Widening the data bus is one obvious way to improve overall system performance.

An extension of adding intelligence to the modules is to build a multi-processing system with multiple buses that can have both distributed and common shared memory so that many instructions can be processed simultaneously. A key problem (limitation) of these system schemes is the development of software that will take advantage of all the processing power. Among other system concepts used to enhance performance are pipeline architecture, cache, and interleaving.

Enhancing Main Memory Performance. As previously mentioned, the main memory performance has the largest impact on the overall performance of a computer system. Thus, it is logical that one would expect system designers to look at ways to modify this memory technology. An evolutionary process essentially commencing in the mid-1970s.

The initial DRAMs had only one addressing mode, but the number increased markedly. The first special addressing mode to appear was *page mode* on the 4 K DRAM. This dramatically reduced the access and cycle times of the DRAM (Fig. 4). This simple feature reduced the time required to transfer large blocks of contiguous data between disk storage and main memory. Other peripherals, such as graphic terminals, benefited from this feature by providing a greater bandwidth needed for screen refresh.

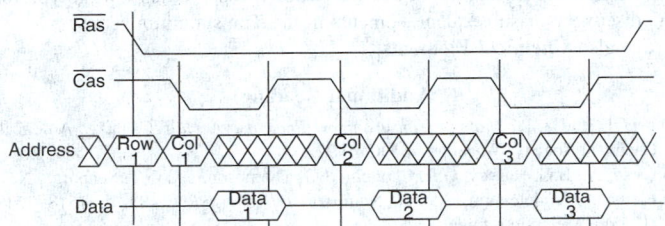

Fig. 4.   Early page addressing mode. (Motorola Inc.)

Since the appearance of page mode, a variety of other popular addressing modes and cycles (bearing some resemblance to page mode) were added. These include *ripple mode, static column, nibble, CAS* (column-address strobe) before *RAS* (row-address strobe), and *hidden refresh*, among others. Although the foregoing features improve system performance, the limit to what can be accomplished is determined by the bus, which is a major bottleneck to system improvement.

Bus Limitations and Improvements. An obvious way to overcome the bus problem is to have more than one bus per system. This approach has been slow in acceptance, however, because of cost for all but the very high end high performance computers. Until relatively recently, the bit density of semiconductor memory was not sufficient to allow for special features that would require a lot of silicon to implement (a cost restraint). With low-bit-density memory, even small systems will require many memory components for main memory, i.e., into the hundreds for many systems.

With this situation, the support electronics required to interface to the bus and control the operation of the memories can be amortized over many memory components and thus make this a feasibly cost-effective approach.

However, with the fourfold increase in density with each new generation of semiconductor memory, more memory becomes available per system with fewer components required. For example, with 64 K, one needs 32 memory components for a fourth of a megabyte, while at the 256 K bit density, only eight memory components are needed. It follows that with 1 M bit density, only two memory components are required, etc. With the high bit density and low package count, alternate solutions at the silicon level become economically feasible, which only a few years ago would have been impossible.

The logical solution to an economical multibus system lies in the ability to make a cost-effective multiport semiconductor memory. The ability to implement such a memory is no longer visionary, but is a reality as dual-port memories with higher and higher densities become available. This is exemplified by Fig. 5. In this case, the new memory components are actually two memories in one. One is a 256 K DRAM organized as 64 K × 4; the other is a 256 × 4 random access serial port. The DRAM port has performance characteristics similar to standard DRAMs. The serial port is very high performance, with cycle times in the 25 ns range. Although these two memories can operate totally independent of each other, with a special cycle, the DRAM port can be connected to the serial port and 1024 bits of data can be transferred between the two ports in one DRAM cycle time.

A representative system block diagram using this type of memory is shown in Fig. 6. To take full advantage of the high-speed data bus, it is advantageous to also employ a cache memory. To accomplish this requires a dual-port static RAM. Since both ports of this memory operate independently, the real advantage to the system architecture is that it is possible to transfer data between the disk and the main memory serial port or the serial port and cache at the same time the CPU is executing a program from data out of the DRAM port. The performance cost tradeoffs of this system are far superior to that of the conventional single-bus system. It is interesting to note that the dual port DRAM was not designed specifically for this application, but rather for the high-speed data storage required for high-resolution graphics.

Memory with On-Chip Control. Also with densities of 1 M bit and beyond, it is possible to design memories with on-chip support logic so that the memory requires little if any support logic to function in the system. Examples of this are *intelligent* DRAMs, which include the refresh function on chip. As the chip density increases, it is more economically feasible to add the control logic that does the clock timing for the memory chip. Thus, the memory will require no support electronics whatever and will interface directly to the CPU bus. With this level of integration, the word width of the memory component is not ×1 or ×4, but rather ×$\frac{8}{9}$

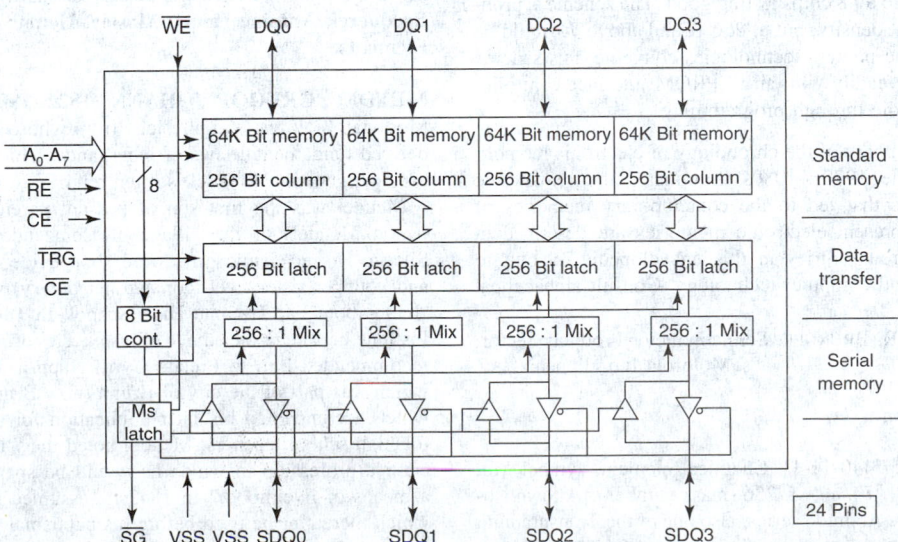

Fig. 5.   Early dual-port system block diagram. (Motorola Inc.)

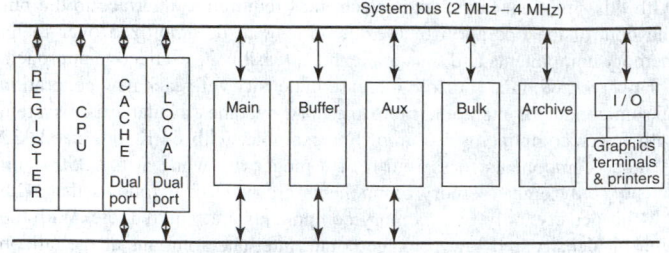

Fig. 6.   Early multiport dynamic RAM (64 K × 4). (Motorola Inc.)

or × $\frac{16}{18}$. Thus, a complete memory system with a half to one megabyte can be accomplished with a half-dozen chips or less.

Similar changes at the other end of the semiconductor spectrum emphasize high performance. In both the bipolar and high-speed HCMOS and GaAs SRAMs, dramatic changes are occurring. In addition to such architectural changes as dual port, the support electronics, which includes the address latching, internal timing generation, such as the write pulse, can accommodate memories with access and cycle times below 10 ns. Other features include comparator logic so that the SRAM can function as a content-addressable RAM.

Chipmaking technology is also changing. In order to meet the higher density, with ever increasing speed and manageable power, designers are turning to GaAs (gallium arsenide) and BiCMOS. With GaAs, it is possible to achieve sub-nanosecond speeds. This places even greater emphasis on having on-chip latching and control features in order to take advantage of these speeds at the system level.

Gate-Array with SRAM.   Another relatively recent development is the use of small on-chip SRAM storage with large gate arrays and a form of wafer scale integration that will make very large storage arrays possible in one package. Some of the latest very large gate arrays (1 K to 20 K range) have an on-chip SRAM array. Since these gate arrays are used in the design of very high-performance custom CPUs and controllers that require the need for temporary storage of processed data, having the SRAM storage on-chip eliminates the costly time required to store the data off-chip as is traditional.

Cluster of Memory Chips.   Some semiconductor makers are investigating methods of using a cluster of memory chips that test good at wafer probe and then interconnect them at assembly into one package. As an example, take the entire wafer and segment it into eight adjacent 64 K DRAM chips and test each before scribing. Next, interconnect the good chips into a variety of organizations. Examples of possible memory organizations are: For two good chips, you would have only 128 K × 1; 3 good chips would yield 192 × 1; four good chips could be 64 K × 4, 256 K × 1, or 128 K × 2, and so on, up to all 8 chips testing good. The scheme is probably marginal at the 64 K bit density, but at 256 K and above, it would be much more attractive. As the process technologies converge, this scheme may be accomplished more readily with an EEPROM interface controller that interconnects the good die through programming.

Editor's Note: This summary of the chronology of electronic memory development during the 1970s–1980s illuminates the remarkable engineering capabilities that led to the contemporary memories of the early 1990s. The present electronic memories are described in connection with numerous entries in this encyclopedia relating to electronic equipment and computer technology. Consult alphabetical index.

R. BRUNNER, Semiconductor Products Sector,
Motorola Inc., Phoenix, AZ

MEMORY LOSS.   See **Amnesia**.

MENDELEVIUM.   [CAS: 7440-11-1]. Chemical element, symbol Md, at. no. 101, mp 827 °C (1,521 °F), at. wt. 256 (mass number of known isotope), radioactive metal of the Actinide series, also one of the Transuranium elements. Mendelevium was named after Dimitri "*Mendeleev*," the Russian chemist who contributed so much to the development of the periodic

table. The element was produced synthetically and first identified by A. Ghiorso, B.G. Harvey, G.R. Choppin, S.G. Thompson, and G.T. Seaborg at the University of California at Berkeley in 1955. See also **Chemical Elements: History of the Origin**. ^{256}Md was produced by the bombardment of ^{253}Es on gold foil with 48-MeV alpha particles in the 60-inch cyclotron at Berkeley. By ion exchange treatment of the dissolved gold foil, only one or two atoms of ^{256}Md were obtained, which decayed (half-life 1.3 hours) by K-electron capture to ^{256}Fm, which underwent its characteristic spontaneous fission. Probable electronic configuration

$$1s^2 2s^2 2p^6 3s^2 3p^6 3d^{10} 4s^2 4p^6 4d^{10} 4f^{14} 5s^2 5p^6 5d^{10} 5f^{13} 6s^2 6p^6 7s^2.$$

Ionic radius Md^{3+} 0.96 Å.

Another isotope, ^{255}Md, is also formed during the bombardment of ^{253}Es by alpha particles. It also decays by electron capture, and has a half-life of 30 minutes.

Regarding the first identification, scientists considered it notable in that only in the order of 1 to 3 atoms per experiment were produced, thus making Md the first element to be discovered on an atom-at-a-time basis. The techniques developed in the search for Md served as a prototype for the discovery of subsequent elements in the Transuranium series.

See also **Chemical Elements**.

Additional Reading

Fuger, J. and L.R. Morss: *Transuranium Elements: A Half Century*, American Chemical Society, Washington, DC, 1992.

Ghiorso, A., B.G. Harvey, G.R. Choppin, S.G. Thompson, and G.T. Seaborg: "New Element Mendelevium, Atomic Number 101," *Phys. Rev.*, **98**, 5, 1518–1519 (1955). (A classic reference).

Hulet, E.K. et al.: "Mendelevium: Divalency and Other Chemical Properties," *Science*, **158**, 486–488 (1967).

Lide, D.R.: *CRC Handbook of Chemistry and Physics*, 88th Edition, CRC Press, LLC, Boca Raton, FL, 2007.

Seaborg, G.T.: *Transuranium Elements*, Dowden, Hutchinson & Ross, Stroudsburg, PA, 1978 (A classic reference.)

Seaborg, G.T. and W.D. Loveland: *The Elements beyond Uranium*, John Wiley & Sons, Inc., New York, NY, 1990.

MENDELEYEV, DIMITRI (1834–1907).   Born in Siberia, Mendeleyev made a fundamental contribution to chemistry in 1869 by establishing the principle of periodicity of the elements. His first periodic table recognized the regular variation in the chemical and physical properties of the elements and classified the 63 elements known at the time into groups placing the elements in ascending order of atomic weight and grouping them by similarity of properties. So accurate was Mendeleyev's thinking that he predicted the existence and atomic weights of several elements that were not actually discovered until years later. The original table has been modified and corrected several times, notably by Moseley, but it has accommodated the discovery of isotopes, rare gases, etc. Its importance in the development of chemical theory can hardly be overestimated. See also **Becquerel, Antoine Henri**; **Mosley, Henry**; and **Periodic Table of the Elements**.

MENDEL, GREGOR JOHANN (1822–1884).   Gregor Mendel was a Moravian biologist and physicist from whose plant breeding studies were derived fundamental laws of inheritance, which founded a new branch of biology, genetics. Mendel is often referred to as the father of genetics.

Mendel was the first son of peasant parents (Rosine Schwirtlich and Anton Mendel) of the village of Heinzendorf near Odrau in Austrian Silesia. His academic gifts were soon evident, but although his parents and younger sister (who donated her dowry) were supportive, he was on "half rations" at Troppau High School. In 1841 at Olmütz Philosophical Institute he sought the advice of his physics professor, Friedrich Franz, who recommended him to Prelate Cyrill Napp at the monastery at Altbrunn, which was part of the city of Brünn (now Brno). In a brief autobiography which accompanied his later application for entrance to an examination for high school teachers, Mendel noted that "it was incumbent on him to enter a profession in which he would be spared perpetual anxiety about a means of livelihood." In 1843 he assumed his monastic name Gregor, which thereafter he used before his baptismal name, Johann.

Under Napp, the monastery was a major intellectual and cultural center. Most monks were engaged in artistic or scientific activities. From 1845 to

1848 Mendel attended Brünn Theological College. Here the Professor of Agriculture, Franz Diebl, was teaching that a nonmystical, but "unknown force" might cause new characters emerging in hybrids to remain constant (i.e., to "breed true" so that when hybrids of the same type were crossed uniform offspring with the hybrid characters would be produced). Such hybrids would then constitute new varieties, from which new species might emerge.

Mendel was ordained priest in 1847, and Napp recommended him as a supply teacher, writing that "he is very diligent in the study of the sciences; but he is much less fitted for work as a parish priest, the reason being that he is seized by an unconquerable timidity when he has to visit a sick-bed or to see anyone ill and in pain." Mendel taught mathematics and Greek at Znaim High School with much success. Despite a quip that a visiting school inspector had "more fat than understanding," Mendel was well liked, and was encouraged to take examinations for a formal degree. His Vienna examiners noted 'that he was devoid neither of industry nor talent. It would seem, however, that he can have had no opportunity for acquiring exhaustive knowledge, and must have lacked access to the necessary means of study.'

Napp then arranged for him to spend two years at Vienna University studying mathematics, physics and biology. Teachers such as the botanist Franz Unger and the physicist Christian Doppler (discoverer of the "Doppler effect") gave Mendel what he needed, but not what the examiners needed. When he returned from an examination in physics and natural history in 1856, he "had a sore head" after a difference of opinion with the botany examiner. Although his teaching continued to receive high praise, he never reattempted the examination.

His studies of heredity in peas (1856–1863) were reported to the Brünn Society for the Study of Natural Science in 1865, and published in the Society's *Transactions* (1866), which were distributed to academic centers throughout Europe. His abstract units (later known as genes) were constant, but by their reassortment only variable hybrids were generated. Less clear results were obtained with hawkweed, published in 1870. Again Mendel concluded that he had not produced constant hybrids but, in contrast to his pea experiments, each character under study depended on multiple units. There were also studies on mice and bees, meteorology, and sun-spots. The German translation of Charles Darwin's *The Origin of Species by Natural Selection* (1864) and many other scientific books in the monastery library, contain annotations in Mendel's neat handwriting.

Prior to Darwin's theory, many biologists regarded the property of being a member of a species as due to a stable "harmony of the inner nature" or "essence" distinct from the character differences (e.g. tallness, smallness) which were commonly observed between species subgroups ("varieties"). Mendel's failure to find constant hybrids, suggested that some "other force" than hybridization *perse* would usually be needed for a new species to emerge. Could this force be natural selection?

We do not know when (and if) they first met, but correspondence dating from 1866 to 1871 survives between Mendel and the Professor of Botany at Munich, Carl von Nä;geli. Opposing Darwin's natural selection, Nä;geli considered that: "the formation of . . . constant varieties or races is not the consequence and the expression of outer agencies, but is brought about through inner causes." The outer agency, natural selection, worked like a gardener clipping the exuberant branching of a tree, but was powerless to bring about the initial branching. It is possible that these views were close to Mendel's. Regarding natural selection Mendel is quoted as saying: "This much already seems clear to me that nature does not modify species in any such way, so some other force must be at work." He also wrote "a factor that so far has received little attention is involved in the variability of cultivated plants."

In 1868 Mendel succeeded Napp as Prelate, noting in a letter to Nä;geli that "all at once, from my modest sphere as teacher of experimental physics, I am translated into one in which a good deal seems strange to me." He became increasingly involved in monastic and community affairs, which included membership and eventually chairmanship of the Moravian Mortgage Bank. Although his political leanings were liberal, from 1874 until his death in 1884 he was much engaged in a bitter dispute with the German Liberal Party, which had imposed severe taxes on monasteries. His advocacy on behalf of the monasteries was politically naive and counterproductive. Although he traveled widely on the continent, there is no evidence that Mendel visited other scientists or attempted to increase awareness of his own work.

Mendel's studies were briefly mentioned by Wilhelm Focke in 1881, but their significance was not appreciated. Nä;geli did not mention the studies in his published work. It was not until 1900 that botanists on the continent of Europe drew attention to Mendel's 1866 paper. One of these, Hugo de Vries, found constant hybrids appearing among varieties of the large flowered evening primrose, in the absence of natural selection; many of these were later found to be due to polyploidy, which can be a cause, but not the usual cause, of the origin of species.

William Bateson argued Mendel's case to the English-speaking world. Refusing to accept natural selection as an originator of species, and perhaps influenced by Nä;geli's "inner cause" and Mendel's "other force," he attributed the origin to a "residue," which was inherited like, but was distinct from, the Mendelian units (genes). In Bateson's view, the reproductive isolation that was critical for the origin of species resulted from failure of the different residues in the gametes of the two parents to complement "like sword with scabbard" when their offspring were forming the gametes for the next generation (i.e., during meiosis). Recent studies of the origin of species in fruit fly (the phenomenon known as Haldane's rule) are consistent with this. See also **Bateson, William (18261–1926)**.

There is a continuing debate as to the extent to which Mendel appreciated the significance of his observations, and why they were initially neglected. Perhaps the key to understanding this nineteenth century monk is that, like some of the most innovative "molecular biologists" of the twentieth century, he approached problems of biology from the perspective of a physicist. He who can write of the "unity in the plan of development of organic life," that "distinguishing traits . . . can only be caused by differences in the composition and grouping of . . . elements," and that "development must necessarily proceed in accord with a law," has some sense of what we now call a "genetic program." See also **Evolution (History)**.

Additional Reading

Forsdyke, D.: *William Bateson: Two Levels of Genetic Information*, http://post.queensu.ca/~forsdyke/bateson1.htm.

Forsdyke, D.R.: "Haldane's rule: Hybrid Sterility Affects the Heterogametic Sex First Because Sexual Differentiation is on the Path to Species Differentiation," *Journal of Theoretical Biology*, **204**, 443–452 (2000).

Mawer, S.: *Gregor Mendel: Planting the Seeds of Genetics*, Abrams, Harry N. Inc., New York, NY, 2006.

Olby, R.: *The Origins of Mendelism*, 2nd Edition, Chicago University Press, Chicago, IL, 1985.

Orel, V.: *Gregor Mendel, the First Geneticist*, Oxford University Press, Oxford, UK, 1996.

Orel, V.: "Constant Hybrids in Mendel's Research," *History and Philosophy of the Life Sciences*, **22**, 291–299 (2000).

Sandler, I.: "Development: Mendel's Legacy to Genetics," *Genetics*, **154**, 7–11 (2000).

Web Reference

Donald R. Forsdyke: http://post.queensu.ca/~forsdyke/homepage.htm#Homepage

DONALD R. FORSDYKE, Queen's University, Kingston, Ontario, Canada

MENHADEN. Of the genus *Brevoortia*, the menhaden is one of the most important commercial fishes in North America. Of the four species of menhaden occurring along the coasts of the United States, two contribute nearly all of the commercial catch. The Atlantic coast catch consists principally of the *Atlantic menhaden* (*B. tyrannus*), which ranges from Nova Scotia to northern Florida. The *Gulf menhaden* (*B. patronus*), which ranges from the west coast of Florida to Mexico, contributes most of the catch from that area. Two other species, the *yellowfin menhaden* (*B. smithi*), which occurs mainly along the east and west coasts of Florida; and the *finescale menhaden* (*B. gunteri*), which occurs from the western Gulf of Mexico to Florida, are of relatively minor importance. Other species are known.

Menhadens are small, oily, herring-like fishes closely related to shad, alewife, herring, and sardine. The largest authenticated specimen was about 19 inches (48 centimeters) long, but most of those caught are less than 12 inches (30.5 centimeters) in length and weigh somewhat less than 1 pound (0.45 kilogram). The average size of Gulf menhaden at any particular age is considerably less than that of the Atlantic species.

Menhaden are prolific spawners. The number of eggs produced by a single female varies, depending on the age and size of the fish. Individual estimates of the number of eggs spawned by Atlantic menhaden range from 38,000 for a medium-size fish to hundreds of thousands for a larger fish. Adult menhaden feed on microscopic plants and animals rather than on other fish. As the fish swims through the water, the planktonic particles are effectively strained from the water by sieve-like gills.

The abundance of menhaden fluctuates greatly. Although these fluctuations in abundance in certain local areas most likely are controlled to a large degree by such environmental factors as water temperature and food availability, variations in abundance of the entire resource appear to be largely caused by changing survival of individual yearly spawnings (year classes). For example, records show that the highly successful 1958 spawning of Atlantic menhaden yielded about seven times as many fish to the fishery as did the 1964 spawning. Since the late 1950s, the catch of menhaden has been decreasing.

In the United States, over the years, the menhaden catch has accounted for a large percentage of the marine oils and fish solubles produced. The fish is caught for its reduction products. Processing plants are usually located quite close to the fishing areas.

See also **Fishes**.

MENINGITIS. The meninges are the covering membranes of the brain and spinal chord. Inflammation of these membranes, especially of the *piamater* and the *arachnoid mater*, is called *meningitis*. See also **Central and Peripheral Nervous Systems**. Meningitis takes a number of forms, including: (1) *aseptic meningitis*, (2) *bacterial meningitis*, (3) *fungal* or *cryptococcal meningitis*, and (4) a number of less common forms, such as *Mollaret's meningitis*.

Aseptic Meningitis. The virus or viruses that cause this disease are positively identified in only about one-quarter of cases. However, increasing emphasis is being placed on more thorough diagnosis and clinical testing for the treatment of specific viral infections.

It is generally believed that enteroviruses cause the majority of cases, but numerous other viruses are seen in this disease. Without identifying the specific virus, aseptic meningitis is relatively simple to diagnose particularly during a spring or summer enteroviral epidemic and where the patient has a rash, a symptom of enterovirus infection. Positive determination of viral etiology is a relatively complex procedure and requires, among other procedures, the inoculation of suckling mice with pharyngeal washings, a stool specimen, and cerebrospinal fluid from the patient.

Symptomatic of aseptic meningitis are fever, headache, and rigidity or stiffness of the neck. Further confirmation can be obtained from examination of the cerebrospinal fluid. Upon completion of the initial diagnosis, treatment and support must be commenced because a full identification of the causative virus requires several days. Statistics indicate that in about three-fourths of the cases of enteroviral meningitis, the virus has been spread by the fecal-oral route and that coxsackieviruses A and B and echoviruses are implicated. See also **Coxsackie Virus**; and **Virus**. The next most frequent causative agents are mumps and varicella zoster viruses. Because of the presence of parotitis (inflammation of salivary glands), the diagnosis is simplified. A less frequent agent, herpes simplex type 2 virus, is seen particularly in women, often with an accompanying primary herpes simplex genital infection. An infrequent agent is lymphocytic choriomeningitis virus, an arenavirus. This is found most often in young adults.

Mollaret's meningitis is a relatively uncommon disease characterized by recurring episodes (self-limited) of aseptic meningitis. Episodes persist for 2 to 5 days and remit spontaneously. The patient may be free of symptoms for weeks or even months. As reported by L.J. Yamamoto (University of Colorado) and a research team, measures have been taken to establish a microbiologic cause, but thus far this has been elusive. However, in at least one case, herpes simplex virus (HSV) has been found in cerebrospinal fluid.

Bacterial Meningitis. Three microorganisms account for over three-fourths of the cases of bacterial meningitis. These are influenza bacillus, meningococcus, and pneumococcus. Meningitis is most common among infants less than one year old. Epidemics of meningococcal meningitis are not uncommon. Areas where many people live in close proximity, as in crowded urban neighborhoods or in military barracks, are particularly productive of this disease. The meningococci reside in the nose and throat

(present in about 5% of population) and spread of the agent via the respiratory route is believed to be the principal pathway of transmission from human to human. The highest incidence of the disease occurs during the winter and spring. The majority of cases among adolescents and young adults is caused by meningococcus, whereas in adults, pneumococcus accounts for most cases. Influenza bacillus rarely causes meningitis in adults except in cases of head injury.

In the United States, approximately 2000 cases were reported (mandatorily) in 1992 to the Centers for Disease Control in Atlanta, Georgia.

Symptoms of bacterial meningitis include headache, fever, and stiff neck. Infection of the respiratory tract and/or middle ear may also be present. With progress of the disease, more dramatic features, including stupor, coma, and seizures, may be presented. Symptoms vary in degree among patients. Although clinical features may be quite indicative of the disease to the physician, positive diagnosis requires examination of the cerebrospinal fluid. At one time, sulfonamides were extensively used in the treatment of meningococcal meningitis, but the organisms have become resistant to these drugs and hence penicillin is the usual drug of choice.

Much less frequent, but of growing concern is *listerial meningitis*, caused by a thin gram-positive bacillus found extensively in soil and in the intestinal tract of many birds and animals. Infection of humans with the bacillus is called *listerosis* and the most common form (60–70% of cases) of listerosis is listerial meningitis. The symptoms of this disease are sometimes less pronounced than with other types of bacterial meningitis, frequently with only fever and headache presented. The usual neck stiffness is absent. Onset of the disease is frequently subacute, extending over a period of several weeks. Antibiotic therapy (ampicillin or penicillin G) is the usual course of treatment followed. Alternative drugs are tetracycline or erythromycin.

Meningitis frequently develops as a consequence of gonococcemia. See also **Gonorrhea and Gonococcemia**.

Cryptococcal meningitis is a common *opportunistic infection* found in patients in the United States who have the human immunodeficiency virus (HIV). This occurs in 5–10% percent of patients with the acquired immunodeficiency syndrome (AIDS). A few years ago, it was believed that a full cure of cryptococcal disease was highly improbable and that most AIDS patients who had completed primary therapy for cryptococcal meningitis remained at risk for relapse. In 1992, W.G. Powderly (Washington University School of Medicine) and a team from other hospitals and medical departments reported on a controlled trial of fluconazole or amphotericin B in preventing relapse in such cases. Conclusions of the study: "Fluconazole taken by mouth is superior to weekly intravenous therapy with amphotericin B to prevent relapse in patients with AIDS-associated cryptococcus meningitis after primary treatment with amphotericin B."

Additional Reading

Berkow, R. and M.H. Beers: *The Merck Manual*, 17th Edition, Merck & Company, Inc., Whitehouse Station, NJ, 1999.

Cartwright, K.: *Meningococcal Disease*, John Wiley & Sons, Inc., New York, NY, 1995.

Davies, P.A. and P.T. Rudd: *Neonatal Meningitis*, Cambridge University Press, New York, NY, 1995.

Lange, J.M.A. and P. Reiss: "Suppressive Therapy for Cryptococcal Meningitis," *N. Eng. J. Med.*, 565 (August 20, 1992).

Powderly, W.G. et al.: "A Controlled Trial of Fluconazole or Amphotericin B to Prevent Relapse of Cryptococcal Meningitis in Patients with the Acquired Immunodeficiency Syndrome," *N. Eng. J. Med.*, 793 (March 19, 1992).

Roos, K.L.: *Meningitis: 100 Maxims in Neurology*, Vol. 4, Oxford University Press, Inc., New York, NY, 1996.

SchFonfeld, H. and H. Helwig: *Bacterial Meningitis*, S. Karger Publishers, Inc., Farmington, CT, 1992.

Staff: *Morbidity and Mortality Weekly Report*, Massachusetts Medical Society, Waltham, Massachusetts (issued weekly).

Taylor, H.G. et al.: "The Sequelae of Haemophilus Influenzae Meningitis in School-Age Children," *N. Eng. J. Med.*, 1657 (December 13, 1990).

Tunkel, A.R.: *Bacterial Meningitis*, Lippincott Williams & Wilkins, Philadelphia, PA, 2001.

Yamamoto, L.J. et al.: "Herpes Simplex Virus Type 1 DNA in Cerebrospinal Fluid of a Patient with Mollaret's Meningitis," *N. Eng. J. Med.*, 1082 (October 10, 1991).

Web Reference

Center for Disease Control and Prevention: http://www.cdc.gov/health/diseases.htm

MENISCUS. The curved surface of a liquid, particularly noticeable in vessels of tubes of small diameter and due to the surface tension of the liquid. If the liquid wets the containing vessel, the meniscus is concave; otherwise it is convex. The meniscus of mercury in glass is convex.

MENOPAUSE. See **Gonads**.

MENTAL DISORDERS. See **Brain Disorders**.

MERCAPTANS. Hydrogen sulfide yields two classes of organic compounds: (1) hydrosulfides, and (2) sulfides. The hydrosulfides are termed mercaptans, a name derived from *mercurium captans*, because of their ability to react with mercuric oxide to form crystalline compounds. Mercaptans also are termed thioalcohols and sulfur-alcohols. A more general term *thiols* also is used. This term not only embraces mercaptans, but also covers thioethers, sulfhydrates, and thiophenols.

Ethyl mercaptan [CAS: 75-08-1], C_2H_5SH, (legal label name for ethanethiol) one of the better known mercaptans, is an odorous liquid, mp $-121\,°C$, bp $36-37\,°C$, sp gr 0.839. The compound is very slightly soluble in H_2O; soluble in alcohol and ether. It is prepared by distilling ethyl potassium sulfate with potassium hydrogen sulfide. Additional mercaptans can be prepared in a similar manner with the corresponding proper ingredients. All mercaptans have unpleasant garlic-type odors; when oxidized with HNO_3 they yield sulfonic acids.

Some formulations for styrene-butadiene rubber (GR-S) contain dodecyl mercaptan which plays the role of a chain-transfer agent used to control the molecular weight of the final synthetic product.

MERCATOR PROJECTION. The mercator method of projecting the surface of the earth on a plane was invented by Gerard Mercator, who lived in Flanders during the latter part of the seventeenth century. At the present time, 90% of the chart work of the deep-sea navigator is done on the mercator chart. To realize the general outline of the method, say that we have a terrestrial globe and wish to peel off the surface. One method of procedure is to make cuts through the surface along meridians of longitude and remove the sectors of the surface thus obtained. If the material of the surface construction has sufficient elasticity, these segments, with a little stretching, can be placed on a plane. The segments will be tangent along the equator, but will separate as the poles are approached, thus forming a discontinuous map. To make the map continuous requires stretching the segments along parallels of latitude, the stretching increasing as higher latitudes are reached. This east-west stretching will introduce distortion in the shape of objects, e.g., a circular object will be distorted into an ellipse. To maintain the shape of objects, the same amount of stretching must be introduced in the north-south direction as is necessary in the east-west direction, in any latitude, to make the segments tangent. This will result, finally, in objects retaining their true shape, but also, in considerable alteration of the relative sizes of surface features in different latitudes.

On the completed mercator chart, meridians of longitude and parallels of latitude appear as perpendicular straight lines. The meridians are equally spaced, but the distance between successive parallels becomes greater as we proceed away from the equator. This distance becomes so great that mercator projection is not used beyond 70° latitude. The actual computation of the distances of the successive parallels from the equator, taking into account the ellipticity of the earth, is a complicated mathematical task.

MERCATOR SAILING. If the distance run by a ship does not exceed 300 nautical miles, the various problems of dead reckoning and the sailings, may be solved without sensible error by methods that assume the surface of the earth to be a plane. The distances run by modern steamships and aircraft during a single day frequently exceed this limit, and the true shape of the earth's surface must be taken into consideration. The method of solving the various problems connected with course and distance in such cases is usually one that depends for its theory upon the mercator projection, and is known as mercator sailing.

In Fig. 1, we have the representation of the general problem, drawn on a section of a mercator chart. The vessel is proceeding from A in latitude ϕ_1 and longitude λ_1, to B in latitude ϕ_2 and longitude λ_2. The course is extended to cross the equator, EE', at the point E. The angle C is approximately the rhumb-line course between the two points. M_1 and

M_2 represent the stretched distances of the parallels ϕ_1 and ϕ_2 from the equator, the values being taken from tables of meridional parts. D_1 and D_2 represent the difference of longitude of the points A and B from the point of equator crossing of the extended course.

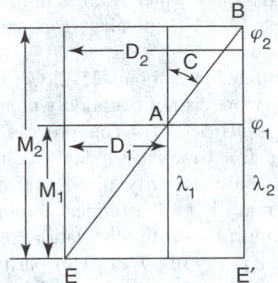

Fig. 1.   Mercator sailing.

Now call $M_2 - M_1 = m$ and $D_2 - D_1 = \delta\lambda$ (the longitude difference between A and B), and we have at once, from the figure, $\delta\lambda = m\tan C$. The value of C thus computed, taking into account the mercator stretching, is frequently referred to as the mercator course between A and B. The distance along AB is a stretched distance and, if calculated from the above figure, or obtained graphically, would be greater than the distance actually traveled between A and B. To avoid this discrepancy, the distance is computed by the methods of plane sailing; the plane triangle is solved using the mercator course as computed above for the vertex angle and the difference of latitude as one leg. The so-called mercator distance is obtained as $d = (\phi_1 - \phi_2)$ secant C, the value of $(\phi_2 - \phi_1)$, the difference of latitude being expressed in minutes of arc, which is practically identical with nautical miles.

See also **Course**; **Navigation**; and **Rhumb Line**.

MERCURY. [CAS: 7439-97-6] Chemical element, symbol Hg, at. no. 80, at. wt. 200.59, periodic table group 12, mp $-38.84\,°C$, bp $356.58\,°C$, density 13.546 g/cm^3 (liquid), 14.193 g/cm^3 (solid). Solid mercury has a rhombohedral crystal structure. The element, sometimes referred to as quicksilver, is a silver-white liquid metal at standard conditions. There are seven stable isotopes of mercury, ^{196}Hg, ^{198}Hg through ^{202}Hg, and ^{204}Hg, and seven radioactive isotopes, ^{192}Hg through ^{195}Hg, ^{197}Hg, ^{203}Hg, and ^{205}Hg. With exception of ^{194}Hg (half-life of approximately 130 days) and ^{203}Hg (half-life of approximately 46 days), the half-lives of the radioactive isotopes are short, measured in terms of minutes or hours.

First ionization potential 10.434 eV; second, 18.65 eV; third, 34.3 eV. Oxidation potentials Hg $\rightarrow \frac{1}{2}Hg_2^{2+} + e^-$, -0.7986 V; Hg $\rightarrow Hg^{2+} + 2e^-$, -0.852 V; $Hg_2^{2+} \rightarrow 2Hg^{2+} + 2e^-$, -0.905 V. Hg $+ 2OH^- \rightarrow HgO + 2H_2O + 2e^-$, -0.098 V; $2Hg + 2OH^- \rightarrow Hg_2O + H_2O + 2e^-$, -0.123 V. Other important physical properties of mercury are given under **Chemical Elements**.

Mercury forms alloys, called amalgams, with most metals, but not with iron or platinum; does not wet glass but forms a convex surface when in a glass container; is slightly volatile at ordinary temperatures and a health hazard due to its poisonous effect; slowly tarnishes in moist air; upon heating in air or oxygen, somewhat below its boiling temperature of $357\,°C$, forms mercuric oxide slowly, as in the classical experiment by Lavoisier on the composition of air; may be purified by distillation and condensation (health hazard); unattacked by dilute HCl or H_2SO_4, but dissolved by dilute or concentrated HNO_3 with the formation of mercurous and mercuric nitrates, respectively, and by hot concentrated H_2SO_4 with the formation of both mercurous and mercuric sulfates; unattacked by alkalis. Discovery ancient.

When cooled to sufficiently low temperatures, mercury becomes *superconducting*, virtually conducting electricity with no resistance. See also **Superconductivity**.

Mercury was mined as early as 500 B.C. and currently ranks tenth in worldwide production of nonferrous metals. The unusual combination of physical properties possessed by the element give it an importance exceeding its production rating. The chief source of mercury is cinnabar, HgS, the red sulfide that contains 86.2% mercury. See also **Cinnabar**. Although the mineral occurs widely throughout the world, relatively few deposits are of commercial importance, notably those in China, Italy, Mexico, the Philippines, Peru, Spain, the former Soviet Union, the former Yugoslavia, and the United States. World reserves of mercury currently are estimated at 4 million flasks, of which sources in the United States account for about 300,000 flasks. A flask contains 76 pounds (34.5 kilograms) of the liquid metal.

The ore is concentrated to about 25–50% mercury by flotation. Beneficiation of mercury ores is not commonly practiced. The concentrate is roasted: $HgS + O_2 \rightarrow Hg + SO_2$. The process is essentially one of distillation because the freed mercury quickly volatilizes, after which it is condensed with a resulting purity of 95% (furnace plants) to 98% (retort plants). The mercury is further refined, through filtering, oxidation, acid leaching of the impurities, or by distillation, to yield prime or virgin mercury with an average purity of 99.9%. This purity is satisfactory for all but the most exacting requirements. Some special mercury chemicals may require that the virgin mercury be triple distilled. Significant quantities of secondary mercury are recovered from waste products, such as dental amalgams, sludges, used batteries, used instruments, and other mercury-bearing materials. Secondary recovery accounts for close to 20% of the total domestic production of mercury in the United States.

Toxicity: Mercury and its compounds, with few exceptions, are highly poisonous to living organisms. Particularly in connection with finely divided mercury metal, extreme care must be taken to avoid inhalation of and contact with the element. The chances of poisoning are increased because awareness of the presence of the metal is reduced when it is finely divided. The fine gray mercury powder is easily generated when liquid mercury is rubbed against or agitated with grease, chalk, sugar, ether, and numerous other substances.

Mercury can cause acute renal failure and nephrotic syndrome. Chronic exposure to or ingestion of mercury may lead to polyneuropathy. Confirmation of poisoning is sometimes made by analysis for the presence of mercury in hair, fingernails, serum, and urine. Removal of the metal may be hastened by the oral administration several times a day for a limited period of D-penicillamine. As pointed out by Beary (1979), with reference to the relatively high mercury content of whale meat, when humans consume contaminated whale meat, the lipid-soluble methylmercury is concentrated in the cells of the nervous system and very slowly eliminated from the body, even when all intake is stopped.

One of the early examples of mercury poisoning may have involved the so-called "dark year" (1693) in the life of Sir Isaac Newton, when it was reported that Newton broke away from friends and associates, accusing them of plotting against him, when he slept very little, and when he reported conversations that actually did not occur. Initial explanations of Newton's erratic behavior included his failure to obtain certain appointive government positions, overwork, and the traumatic fire that destroyed a number of his valuable manuscripts. This evidence was not convincing, however, to investigators P.E. Spargo and C.A. Pounds. Through contacts with Newton's descendants, this team obtained four hairs taken from the head of the master and subjected them to laboratory tests. In the very interesting account by Broad (1981), the investigators concluded that Newton's madness was "due principally to poisoning by the metals which he used so frequently and with such cavalier disregard for his own safety." Locks of hair were located at Trinity College, Cambridge. One hair was found to contain 197 ppm Hg (even quite high in terms of modern-day Hg poison cases). Broad concludes his article, "Perhaps poisoning by mercury was not only the cause of Newton's brief lunacy, but was also the pivotal event that nudged the superstitious genius away from his researches in the laboratory to the seemingly less dangerous ways of the world." There are several doubters of the Spargo-Pounds hypothesis and the controversy is likely to persist.

Uses of Mercury. Because of the poisonous nature of Hg, wherever possible the industrial uses of the element have been reduced and research to find suitable substitutes for Hg for many applications has accelerated during recent years. Traditionally, mercury's largest usage has been in connection with electrical apparatus (switching, etc.), the electrolytic preparation of chlorine and caustic soda (mercury cells), in antifouling and mildew-proofing agents and paints, in industrial and clinical thermometers,

in pharmaceutical preparations, in agricultural herbicides and pesticides, in dental preparations, in the preparation of amalgams, and in use of Hg as a catalyst for certain chemical reactions.

The consumption of mercury by the chlor-alkali industry may range from 15% to as high as 35% of the total in any given year, depending upon new construction. Large amounts of mercury are required for the start-up of mercury cell operations, whereas replacement requirements are quite low. Several areas of mercury use are declining gradually, particularly in the pharmaceutical field where sulfa drugs, iodine, and various antiseptics and disinfectants have made inroads on mercury chemicals. Mercury compounds, used for many years in the treatment of syphilis, for example, largely have been displaced by antibiotics and other treatments. Because of the fundamentally toxic nature of mercury and its compounds, the agricultural uses of mercury formulations are being deemphasized, with constant research for substitute materials. In the dental field, a number of metal powders, porcelain, and plastic materials have displaced mercury amalgams in many dental applications. In the explosives field, several compounds, such as lead azide, diazodinitrophenol, and other organic initiators, are serving the same function as mercury fulminate. The use of mercury as a heat-transfer medium in boilers was essentially abandoned a number of years ago. On the other hand, mercury-base catalysts are increasing in application and, to date, suitable substitutes for mercury in the antifouling and mildew-proofing formulations area have not been found. The essential properties of mercury that will be difficult to replace are its high specific gravity, fluidity at room temperatures, and excellent electrical conductivity.

In addition to some diminishment of mercury usage for various products because of increasing awareness of its toxicity potential, conservation-minded technologists also have pointed to the relatively limited world resources of the metal. Considerable ingenuity has been used to replace mercury. For example, diaphragm cells can be used in caustic-chlorine production, organic biocides are replacing mercury-containing compounds, gold recovery is accomplished by the cyanide process rather than by the amalgamation process, and plastic paints and copper oxide paints can be used in place of mercuric biocidal paints. The use of the diaphragm cell for chlorine-alkali production possibly is the most dramatic substitution in terms of mercury conservation. Diaphragm cells require no mercury, whereas the traditional mercury cells once accounted for as much as 35% of the mercury use in some years.

Chemistry and Compounds. The apparent anomaly between mercury and the lighter elements of transition group 2, in that mercury regularly forms both univalent and divalent compounds, while zinc and cadmium do so very rarely, is partly understood from the observation that mercury(I) salts ionize even in the gaseous state to Hg_2^{2+}, rather than Hg^+. Evidence for this double ion is provided by its Raman spectral line, by the lineal Cl—Hg—Hg—Cl units in crystals of mercury(I) chloride, and by the emf of mercury(I) nitrate concentration cells. The anomaly is further removed by the observation that cadmium also forms a (much less stable) diatomic ion Cd_2^{2+}, e.g., in $Cd_2(AlCl_4)_2$.

Oxides. Heating of mercury in air yields the (divalent) oxide HgO, which at higher temperatures decomposes into its elements. Mercury(II) oxide is also precipitated from solutions of mercury(II) salts by alkaline solutions. Alkalies precipitate a yellow form, while alkali carbonates give a red one. The yellow is apparently a finely divided form of the red, since they are crystallographically identical, but differ slightly in certain chemical and physical properties, including solubility. Mercury(II) oxide exhibits solubility in solutions of alkali salts, which is attributed to formation of complex ions such as $[Hg(OH)_2NO_3]^-$ and $[Hg(OH)_2SO_4]^{2-}$.

Halogen Compounds. All eight compounds of univalent and divalent mercury with the single halogens are known, as well as several compounds of mercury(II) with two halogens, such as HgBrI and HgClI. The mercury(I) halides are insoluble in water, with the exception of the fluoride which, like mercury(II) fluoride, is hydrolyzed by water. Like the zinc and cadmium halides, mercury(II) halides behave anomalously in aqueous solution, and for similar reasons, i.e., the presence of complex ions and un-ionized molecules. In the case of mercury(II) halides, with their more covalent character than zinc or cadmium halides, the ionization is somewhat less, and the concentration of Hg^{2+} relatively low. Thus, in aqueous solution, mercury(II) chloride, $HgCl_2$, is present largely as un-ionized molecules, but also ionizes $HgCl^+$ and Cl^-, and only secondarily and to a slight extent to give Hg^{2+}. In the presence of added Cl^-, an $HgCl_2$ solution is a complex system involving equilibria between $HgCl_2$, $HgCl^+$, Cl^-, Hg^{2+}, and the complex ions $HgCl_3^{-}$ and $HgCl_4^{2-}$. Similarly, the hydrolysis of $HgCl_2$, though slight, involves several equilibria whose relative importance varies with the concentration of the solution. In more concentrated solutions the hydrolysis of $HgCl_2$ to $HgOHCl$, Cl^- and H^+ is prominent, while in more dilute solutions the most important equilibria involve the ionization $HgCl_2$ to $HgCl^+$ and Cl^-, and the hydrolysis of $HgCl_2$ to $[HgOHCl_2]_2{}^{2-}$ and H^+, and that of $HgCl^+$ to $HgOHCl$ or $[HgOHCl]_2$ and H^+. Finally, oxyhalides, such as $HgBr_2 \cdot 3HgO$, $HgCl_2 \cdot 2HgO$, $HgCl_2 \cdot 3HgO$ and $HgCl_2 \cdot 4HgO$ are also obtainable, usually by action of alkali hydroxides upon mercury halides. Mercury(II) iodide, like the oxide, is polymorphic. It has three forms, yellow, red, and white, the second being the most stable up to $127\,^\circ C$, where it undergoes a definite transition to the yellow. The colorless $HgI_4{}^{2-}$ is very stable, especially to alkalies, and is used in Nessler's reagent.

Salts. Mercury forms many salts, both of mercury(I) and mercury(II). In general, action of oxidizing acids upon the metal yields the latter, while the former requires either a limited amount of the oxidant or indirect methods. Mercury(I) salts are made by treating a solution of a soluble mercury(II) salt with metallic mercury. Thus, heating mercury with H_2SO_4 or HNO_3 yields mercury(II) sulfate or nitrate, $HgSO_4$ or $Hg(NO_3)_2$, respectively, crystallizing as hydrates, while mercuric(I) nitrate results from the use of cold acid in limited amount, and mercuric(I) sulfate is produced by the last method as well as from mercury(I) nitrate and sulfuric acid. The other salts of both univalent and divalent mercury include the acetates, antimonates, arsenates, bromates, carbonates, chlorates, chromates, fluorosilicates, iodates, oxalates, perchlorates, periodates, phosphates, tartrates, thiocyanates, tungstates, uranates, and vanadates. Also known in both valences are the arsenides (from arsine and the mercury solutions), azides (from hydrozoic acid and the mercury solutions), nitrides and phosphides. Only mercury(II) selenide and telluride exist. Hg_2S_2 has been reported to be obtained as a black powder, but is believed to be a mixture of Hg and HgS. The latter exists in two forms, the black form that is usually precipitated by hydrogen sulfide, and the red, cinnabar, precipitated by H_2S from a solution of mercury(II) acetate and ammonium thiocyanate. The black changes to the red in liquid H_2S, and the red to the black on heating to $386\,^\circ C$. Cinnabar is the thermodynamically stable form at room temperature.

Compounds with Nitrogen and Sulfur. The reactions of mercury(I) compounds and NH_3 are complex, and published results vary. Recent (x-ray) studies show that this reaction, modified by the presence of ammonium chloride, NH_4Cl, yields three aminobasic compounds containing divalent mercury: $Hg_2NCl \cdot H_2O$, $HgNH_2Cl$, and $Hg(NH_3)_2Cl_2$. The first of these is the chloride of Millon's base, $Hg_2NOH \cdot 2H_2O$, which is produced by warming HgO with aqueous ammonia.

Mercury is the least active of the elements of its group as an electron acceptor from oxygen; however, with sulfur it is more active, the mercury halides forming dialkyl sulfide addition products, $R_2S \cdot 2HgX_2$, and HgS dissolves in alkali sulfides forming $[HgS_2]^{2-}$ or $Hg(SH)_4{}^{2-}$. Like zinc and cadmium, mercury forms a series of ammines, which with the mercury halides are principally the diammines, $[Hg(NH_3)_2]X_2$, where X is a covalently bonded halogen atom, and with more ionic mercury compounds, e.g., the nitrate and sulfate, especially in the presence of high concentrations of ammonium salts, the tetrammines, e.g., $[Hg(NH_3)_4](NO_3)_2$. These complexes also form with amines and diamines, e.g., ethylenediamine, which contributes three molecules per Hg^{2+} ion. Besides forming univalent and divalent cyanides, mercury forms complex cyanide ions, $[Hg(CN)_3]^-$ and $[Hg(CN)_4]^{2-}$, as well as additional compounds with the mercury(II) halides of the structure HgCNX. Mercury forms insoluble thiocyanates and complex ions, e.g., $Hg(SCN)_3{}^-$, $Hg(SCN)_4{}^{2-}$. Mercury cyanates and fulminates are insoluble. Mercury(II) mercaptides, $Hg(SR)_2$, decompose on heating to give HgS and R_2S. Mercury(II) hydrogen sulfite, $Hg(HSO_3)_2$, is actually mercuridisulfonic acid, $Hg(SO_3H)_2$, with Hg–S bonds. Like the halides, compounds such as $Hg[C(NO_2)_3]_2 \cdot Hg[C(CN)_3]_2$, $Hg(NO_2)_2$ (dinitromercury), $Hg(CF_3)_2$ are noteworthy for their lack of ionic dissociation.

Organometallic Compounds. Mercury also forms a large number of organometallic compounds of the type HgR_2, where R may be not only an alkyl radical, but an alkoxy radical, an acyl radical, a halogenated alkyl

radical, an alkylthio radical, an aryl radical or a perfluoroalkyl radical. In addition, mercury also forms numerous organometallic compounds of structure RHgX, where X is a halogen atom, and R one of the foregoing organic radicals.

The organic compounds containing mercury that are used as disinfectants, germicides, and antiseptics are known as mercurials. Among these are Merthiolate, Mercurochrome, and Metaphen. Merthiolate is the sodium salt of ethylmercurithiosalicylic acid, $C_2H_5HgS \cdot C_6H_4 \cdot COONa$. It contains 49.5% of mercury. It is a crystalline, cream-colored powder which is very soluble in water, for about 1 gram dissolves in 1 milliliter of water. It is much less soluble in alcohol, 1 gram in 8 milliliters. It is insoluble in organic solvents like benzene and ether. It is used as an antiseptic for tissues in concentrations of the order of 1:1,000 to 1:30,000. It is commonly used as an antiseptic in biologics.

Mercurochrome, also known by the name of *Merbromin* and by many other trade names, is the disodium salt of 2,7-dibromo-4-hydroxymercurifluorescein, $C_{20}H_8Br_2HgNa_2O_6$. It forms green, iridescent scales or granules, which are freely soluble in water, yielding a bright red solution with dilute solutions having a yellow-green fluorescence. It is generally used in a 2% aqueous solution as a mild antiseptic. It is nearly insoluble in alcohol and is insoluble in organic solvents like acetone and ether. Most other common mercurials and iodine solution are considered to be better antiseptics.

Additional Reading

Agocs, M.M. et al.: "Mercury Exposure from Interior Latex Paint," *N. Eng. J. Med.*, 1096 (October 18, 1990).

Broad, W.J.: "Sir Isaac Newton: Mad as a Hatter," *Science*, **213**, 1341–1344 (1981).

Clarkson, T.W.: "Mercury—An Element of Mystery," *N. Eng. J. Med.*, 1137 (October 18, 1990).

Fackelmann, K.A.: "Painting a Perilous Picture of Mercury," *Science News*, 244 (October 20, 1990).

Greenwood, N.N. and A. Earnshaw: *Chemistry of the Elements Revised and Updated*, 2nd Edition, Butterworth-Heinemann, Inc., Woburn, MA, 1997.

Krebs, R.E.: *The History and Use of Our Earth's Chemical Elements: A Reference Guide*, Greenwood Publishing Group, Inc., Westport, CT, 1998.

Lide, D.R.: "*CRC Handbook of Chemistry and Physics*," 88th Edition, CRC Press, LLC., Boca Raton, FL, 2007.

Raloff, J.: "Mercurial Risks from Acid's Reign," *Science News*, 154 (March 8, 1991).

Stone, R.: "Mercury's Metabolic Fingerprint," *Science*, 29 (April 3, 1992).

Stwertka, A. and E. Stwertka: *A Guide to the Elements*, Oxford University Press, Inc, New York, NY, 1998.

MERCURY BAROMETER. (Or mercurial barometer; formerly called *Torricelli's tube*.) A glass manometer, employing mercury in its vertical column, that is used to measure atmospheric pressure.

The basic construction, unchanged since Torricelli's experiment in 1643, is a glass tube about three feet long, closed at one end, filled with mercury, and inverted with the open end immersed in a cistern of mercury. With the cistern surface exposed to atmospheric pressure, the height of the mercury column varies with that pressure. Mercury barometers may be classified into three groups according to their construction: cistern barometers, siphon barometers, and weight barometers.

See also **Aneroid Barometer**; **Barometer**; **Cistern Barometer**; and **Siphon Barometer**.

MERCURY (Planet). The first of the planets in order of distance from the Sun. At maximum elongation, Mercury is not more than 28 degrees away from the Sun as seen from earth. Thus, the planet does not rise above the horizon for more than 2 hours before sunrise, or remain above the horizon for more than 2 hours after sunset. Consequently, observations of Mercury from earth are less than ideal. This behavior of Mercury also led early observers to believe that Mercury was actually two planets — Apollo observed in the early morning and Mercury as seen in the early evening.

The mean distance of Mercury from the Sun is about 58 million kilometers (40 million miles). The planet's orbit is inclined some $7°$ to the plane of the ecliptic and has a large eccentricity of 0.206. Because of this eccentricity, the distance of Mercury from the sun varies from a minimum of 46 million kilometers (28.5 million miles) at perihelion to a maximum of 70 million kilometers (43.4 million miles) at aphelion. The orbital velocity of Mercury varies accordingly, ranging from 39 kilometers

(24.2 miles) per second at aphelion to 57 kilometers (35.2 miles) per second, the fastest of all of the planets.

Orbits computed for Mercury on the basis of Newtonian mechanics, and allowing for all perturbations, revealed a progressive eastward advance of the longitude of perihelion of 531 seconds per century. Later observations of the actual advance of perihelion of 574 seconds per century were made. The discrepancy of 43 seconds led to a search for an intra-mercurial planet. However, the application of the theory of relativity, or relativistic mechanics, to the problem showed that the added advance of perihelion is due to a relativistic mass effect, and was one of the early, classical proofs of Einstein's theory of relativity.

The albedo of Mercury is quite low, on the order of 0.056. This is much like the albedo of the moon, which led to the conclusion for many years that the two surfaces are similar and, indeed, recent observations also indicate this to some degree. The brightness of Mercury is on the order of −0.4 apparent visual magnitude, and its color is somewhat redder than that of the Sun. Like Venus and the moon, Mercury shows changes of phase.

Exploration of Mercury. The absence of special features, such as satellites and rings, coupled with the limiting viewing time of Mercury from Earth, have contributed to a better understanding and interest in Mercury as compared with the outer planets. The most dramatic and thorough exploration occurred in 1974 by way of the *Mariner 10* spacecraft. Between the time of that encounter and the early 1990s, additional findings of Mercury have been made from Earth-based instrumentation, including spectroscopy and radar imaging. In terms of long-range studies, Mercury is not high on the list of exploratory priorities. However, prior to the breakup of the U.S.S.R., Soviet scientists had proposed in 1989 that a Mercury "lander" be planned for launching sometime during the period, 1998 to 2000. Of course, as of mid-1993, the extent to which Russia may participate in space exploration programs is vague. At the Conference on Solar System Exploration held at California Institute of Technology in August 1989, Valery Barsukov, a leading Soviet space scientist, pointed out that Mercury holds interest for exploration on several points. Mercury is the closest planet to the sun and thus presents great extremes of daytime heat and nighttime cold. Further, Mercury is the only one of the inner planets whose surface has not been probed by a landing craft.

Mariner 10 Mission. The principal scientific objective of the *Mariner 10* mission was the exploratory study of Mercury. Thus, the instruments selected were chosen to serve these objectives. However, *Mariner 10* encountered with Venus on February 5, 1974 because of the use of Venus' gravity field as a "third stage" so to speak to achieve the encounter with Mercury. Some very useful information was obtained concerning Venus. See also **Venus**.

The first pass (closest encounter) of *Mariner 10* with Mercury occurred on March 29, 1974, after a flight of 146 days, but photographing commenced on March 23 from a distance of 3.36 million miles (5.4 million kilometers). Periodic photographic operations continued until April 3 when the probe was 2.17 million miles (3.5 million kilometers) beyond the planet. Over 2,000 television frames were transmitted from twin, high-resolution cameras. Because the study of the interaction between Mercury and the solar wind was given a high scientific priority, a darkside passage was selected for the first flyby. Nevertheless, excellent computer-enhanced pictures were taken at about 145,000 miles (234,000 kilometers) away from the planet, about 6 hours before the closest approach.

Mariner 10's second encounter, after a long voyage around the sun, with Mercury occurred on September 21, 1974. Much additional information and many photographs of surface features of the planet were gained on the second flyby. See Figures 1, 2, and 3.

Landforms. The principal landforms on Mercury are basins, scarps, ridges, and plains that resemble analogous forms on the moon. Where the plains are absent, overlapping craters and basins form rugged terrain. The plains materials have many of the characteristics of the lunar maria and have been cratered to approximately the same degree. This twofold division of the surface morphology of Mercury is strikingly similar to that of the moon. Prominent structural features of the planet include irregular scarps which are up to 1 kilometer high (0.6 mile), extend for hundreds of kilometers, and cut across large craters and intercrater areas. Similar features are absent on the moon. No features suggestive of either earthlike plate tectonics or large-scale tensional faulting in the crust have been recognized. Mariner spacecraft observations included: (1) extensive

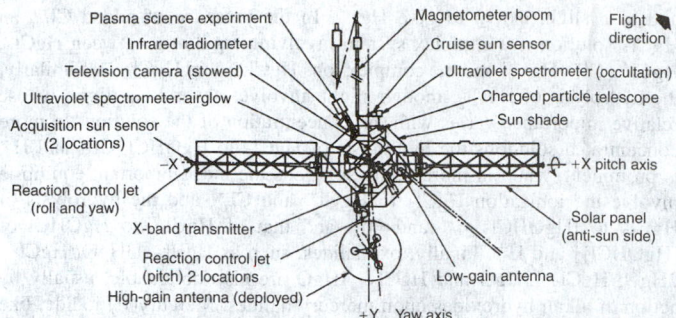

Fig. 1.   *Mariner 10* spacecraft looking sunward along the Z axis (normal to plane of diagram at the intersection of the X and Y axes). (*Jet Propulsion Laboratory.*)

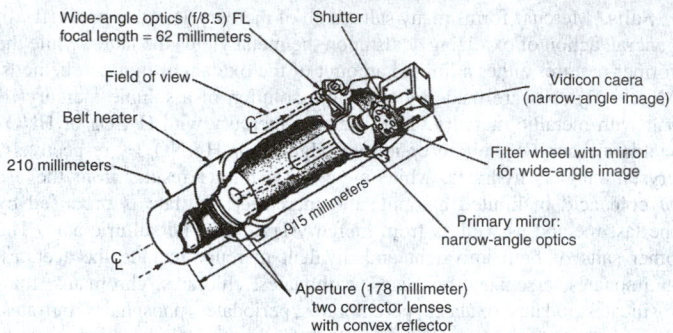

Fig. 2.   Schematic diagram of television camera used on *Mariner 10*. (*Jet Propulsion Laboratory.*)

flooding by rock materials at least grossly similar to those of the lunar maria has occurred on Mercury; (2) the heavily cratered surfaces on Mercury record the final periods of heavy impact bombardment at Mercury; (3) in the half of the planet observed by Mariner 10, Mercury, like the moon, seems to exhibit a hemispherically nonuniform distribution of flooded basins; and (4) mare-like surfaces now have been formed on the moon, Mars, and Mercury which show a surprising similarity in accumulated impacts, although only those of the moon have been dated radiometrically.

Volcanism on Mercury? In addition to *Mariner* data, an estimated one-third of the planet (mainly the equatorial region) has been mapped from radar data in recent years. Many of these areas were not visible by Mariner. Initially, based mainly on *Mariner* data, some of the surface features of Mercury definitely appeared to some astronomers as being of volcanic origin. However, during the interim, counterarguments have developed, and it is highly likely that a final answer must await the next encounter by a spacecraft with the planet. One argument that has been proposed is that the plains surrounding Caloris (See Fig. 3) were the result of ejecta from an impact—that is, the plains were formed from debris flows.

During the course of diagnosing Mercury's past history and current appearance, frequent comparisons of the planet with earth's moon have been made. More recently, some scientists point out that this analogy may have been overworked.

Atmosphere of Mercury. It appears that there is virtually no atmosphere on Mercury. Findings of the *Mariner* program showed that the atmosphere of the planet, like that of earth's moon, is maintained in an extremely tenuous minimum state by weak solar wind accretion and radioactive decay processes, and depleted by strong removal mechanisms. Unlike the moon, Mercury has a high daytime surface temperature in the range of 100–325 °C (212–617 °F) that promotes the production of water vapor, which may be the dominant constituent derived from solar wind protons.

In 1985, Andrew Potter (NASA Johnson Space Center) and Thomas Morgan (Southwestern University) reported that sodium had been observed in the atmosphere of Mercury. The spectrum of Mercury at the Fraunhofer sodium D lines shows strong emission features that are attributed to resonant scattering of sunlight from sodium vapor in the atmosphere of the planet. The total column abundance of Na was estimated to be 8.1×10^{11} atoms per square centimeter, which corresponds to a surface density at

Fig. 3. Mosaic of 18 pictures, selected from over 2,300 photographs of Mercury taken by *Mariner 10*, shows the largest structural feature on the planet discovered to date. In the left half of this photomosaic taken on March 20, 1974, is a ring basin 1300 kilometers (800 miles) in diameter and bounded by mountains which rise as high as 2 kilometers (1.2 miles). The structure is centered at 190°W. longitude and 30°N latitude and is intensely disrupted with many fractures and ridges. The feature is similar in size and appearance to the Imbrium Basin on Earth's moon and most likely originated by impact of a body at least tens of kilometers in diameter. The huge basin has provisionally been named *Caloris*, or hot basin, for its position near one of the subsolar points when the planet is nearest the sun. (*NASA; Jet Propulsion Laboratory, Pasadena, California.*)

the subsolar point of about 1.5×10^5 atoms per cubic centimeter. The most abundant atmospheric species found by the *Mariner 10* mission to Mercury was helium, with a surface density of 4.5×10^3 atoms per cubic centimeter. It now appears that Na vapor is a major constituent of Mercury's atmosphere. The atomic weight of sodium is sufficiently high so that the thermal escape of Na from Mercury should be negligible. However, it has been suggested that the solar wind can sweep away ions produced by photoionization of Mercury's atmosphere. This may be a major loss process for atmospheric gases that otherwise would not escape and may account for the manner in which the planet originally lost most of its atmosphere. It has been further suggested that Mercury's atmosphere may resemble a cometary coma rather than an Earthlike planetary atmosphere. More information is needed before the sources and sinks of Na in Mercury's atmosphere are understood. By measuring the variation of Na emission

with phase, time, and distance of the planet from the sun, it may be possible to determine what processes control it.

In May 1990, A.E. Potter and T.H. Morgan (NASA), using a Bowen image slicer and a charge-coupled device (CCD), probed the spectra of Mercury. Their report concluded, "Monochromatic images of Mercury at the sodium D_2 emission line showed excess emission in localized regions at high northern and southern latitudes and day-to-day global variations in the distribution of sodium emission. These phenomena support the suggestion that magnetospheric effects could be the cause. Sputtering of surface minerals could produce sodium vapor in polar regions during magnetic substorms, when magnetospheric ions directly collide with the surface. Another important process may be the transport of sodium ions along magnetic field lines toward polar regions, where they impact directly on the surface of Mercury and are neutralized to regenerate neutral sodium atoms. Day-to-day variations in planetary sodium distributions could result from changing solar activity, which can alter the magnetosphere in time scales of a few hours. Observations of the sodium exosphere may provide a tool for remote monitoring of the magnetosphere of Mercury."

During the period between June 1986 and January 1988, A.L. Sprague, R.W.H. Kozlowski, and D.M. Hunten (University of Arizona), using a 1.5-meter Cassegrain reflecting telescope and spectrographic instrumentation, observed enhanced abundances of neutral potassium (K) in the atmosphere of Mercury above the longitude range containing the Caloris Basin. Results of large data sets showed typical K column abundances of $\sim 5.4 \times 10^8$ K atoms/cm^2. The scientists observe that this enhancement is consistent with an increased source of K from the well-fractured crust and regolith associated with this large impact basin. The phenomenon is localized because, at most solar angles, thermal alkali atoms cannot move more than a few hundred kilometers from their source before being lost to ionization by solar ultraviolet radiation.

Thermal Properties. The infrared radiometer on *Mariner 10* measured the thermal emission from Mercury with a spatial resolution element as small as 40 kilometers (25 miles) in a broad wavelength band centered at 45 micrometers. The minimum brightness temperature (near local midnight) in these near-equatorial scans was 100 K. Along the track observed, the temperature declined steadily from local sunset to near midnight, behaving as would be expected for a homogeneous, porous material with a thermal inertia of 0.0017 cal^{-2}sec$^{-1/2}$ degrees K^{-1}, a value only slightly larger than that of the moon. From near midnight to dawn, however, the temperature fluctuated over a range of about 10 degrees Kelvin, implying the presence of regions having thermal inertia as high as 0.003 cal^{-2}sec$^{-1/2}$ degrees K^{-1}.

Polar Ice Caps? M.A. Slade, B.J. Butler, and D.O. Muhleman (Jet Propulsion Laboratory) conducted the first unambiguous full-disk radar mapping of Mercury at 3.5-centimeter wavelength with the Goldstone 70-meter antenna transmitting and 26 antennas of the Very Large Array receiving. This study is reported to provide evidence for the presence of polar ice. The radar experiments were designed to image that half of the planet that was not photographed by *Mariner 10*. The investigators reported, "The orbital geometry allowed viewing beyond the north pole of Mercury, a highly reflective region clearly visible on the north pole during the experiments. The team initially was surprised to find that ice could exist on Mercury (planet nearest the sun). The radar data, however, were consistent with reflections from ice. Later, the existence of the bright north polar feature was confirmed at Arecibo, and a compact feature near the south pole also was discovered. See Fig. 4. The plausibility of ice at the poles, not apparent in conventional telescopic images, is supported by the fact that radar can penetrate beneath the surface, where ice can exist because it is well protected from the baking hot surface temperatures. The concept of ice at the poles of Mercury still is somewhat debatable. For example, why haven't the ices evaporated over billions of years?

Magnetic Properties. Data from the spacecraft indicated that fluxes of protons with energies of about 550 keV and electrons with energies of about 300 keV, which exceed approximately 104 and 105 cm^{-2} sec^{-1}, respectively, have been discovered in the magnetosphere of Mercury. Electron fluxes less than 103 cm^{-2} sec^{-1} also were observed in the outbound pass of the spacecraft through the magnetosheath. On the basis of the experimental evidence and a knowledge of the general magnetic field intensities and directions along the trajectory of *Mariner 10* provided by the magnetic field observations, it is shown that the radiation events

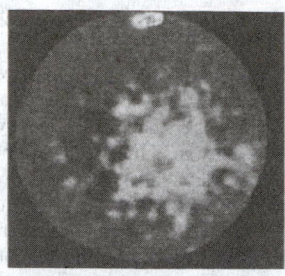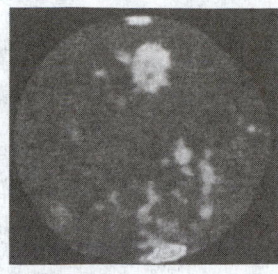

Fig. 4.   Radar images of Mercury made in 1992. Suspected ice cap is shown at Mercury's north pole in both images. In a view taken at a somewhat later date (right), a bright image also is evident at the south pole of the planet. Some scientists propose that this region also may be made up of ice.

observed in the magnetosphere and magnetosheath are transient and are not interpretable in terms of stable trapped particle populations. Furthermore, experimental evidence strongly supports the view that the particles are impulsively accelerated and that the acceleration source is not more distant from the point of observation along lines of force than about 8×10^3 to 16×10^3 kilometers (3 to 6.5 units of Mercury's radius). The phenomena discovered at Mercury will place more stringent conditions on allowed models for electron and proton acceleration than have heretofore been possible in studies within the earth's magnetosphere.

One scientist with the *Mariner* program reported on magnetic field observations. Rather unexpectedly, a very well-developed, detached bow shock wave, which develops as the super-Alfvénic solar wind interacts with the planet, was observed. Also, a magnetosphere-like region, with a maximum field strength of 98 gammas at closest approach (704 kilometers altitude), was observed and contained within boundaries similar to the terrestrial magnetosphere. The obstacle deflecting the solar wind flow is global in size, but the origin of the enhanced magnetic field was not uniquely established. The field may be intrinsic to the planet and distorted by interaction with the solar wind. It may also be associated with a complex induction process whereby the planetary interior-atmosphere-ionosphere interacts with the solar wind flow to generate the observed field by a dynamo action.

A magnetic field about 1% as strong as the earth's, together with the planet's high mean density of 5.4 grams per cubic centimeter, suggests that Mercury may have an iron core similar to earth's core, in which the magnetism is generated.

The concept of an intrinsic magnetic field and, in fact, the nature of the core of Mercury, essentially remain an enigma. A group of scientists, including J.O. Burns, at the University of New Mexico have used radio emissions to probe Mercury's subsurface. These investigators have observed, "There does not appear to be any excess heat arising from the core of Mercury. The temperature map can be explained by reradiation of solar energy." J.O. Burns (New Mexico State University), in combining the data of the mapping studies with Mercury's slow rotation, observes, "This indicates that Mercury does not have a large, hot molten core that many believe is needed to produce the strong magnetic field via the dynamo model."

Classic References to Mariner 10 Mission (1974) to Mercury

Broadfoot, A.L. et al.: "Mercury's Atmosphere from Mariner 10: Preliminary Results," *Science*, **185** (4146), 166–169 (1974).

Chase, S.C. et al.: "Preliminary Infrared Radiometry of the Night Side of Mercury from Mariner 10," *Science*, **185** (4146), 142–145 (1974).

Dunne, J.A.: "Mariner 10 Mercury Encounter," *Science*, **185** (4146), 141–142 (1974).

Howard, H.T. et al.: "Mercury, Results on Mass, Radius, Ionosphere, and Atmosphere from Mariner 10 Dual-Frequency Radio Signals," *Science*, **185** (4146), 179–180.

Metz, W.D.: "Mercury: More Surprises in the Second Assessment," *Science*, **185** (4146), 132 (1974).

Murray, B.C. et al.: "Venus: Atmospheric Motion and Structure from Mariner 10 Pictures," *Science*, **183** (4131), 1307–1315 (1974). (Describes Mariner 10 cameras and photographic techniques.)

Murray, B.C. et al.: "Mariner 10 Pictures of Mercury: First Results," *Science*, **184** (4135), 459–461 (1974).

Murray, B.C.: "Mercury's Surface: Preliminary Description and Interpretation from Mariner 10 Pictures," *Science*, **185** (4146), 169–179 (1974).

Ness, N.F. et al.: "Magnetic Field Observations near Mercury: Preliminary Results from Mariner 10," *Science*, **185** (4146), 151–159 (1974).

Ogilvie, K.W. and J.D. Scudder, et al.: "Observations at Mercury Encounter by the Plasma Science Experiment on Mariner 10," *Science*, **185** (4146), 145–151 (1974).

Simpson, J.A. et al.: "Electrons and Protons Accelerated in Mercury's Magnetic Field," *Science*, **185** (4146), 160–169 (1974).

Additional Reading

Eberhart, J.: "Cold Message from Mercury's 'Hot Poles'," *Science News*, 375 (June 16, 1990).

Holden, C.: "An Ice Cap on the Hottest Planet," *Science*, 935 (November 15, 1991).

Potter, A.E. and T.H. Morgan: "Evidence for Magnetospheric Effects on the Sodium Atmosphere of Mercury," *Science*, 835 (May 18, 1990).

Slade, M.A., B.J. Butler, and D.O. Muhleman: "Mercury Radar Imagine: Evidence for Polar Ice," *Science*, 635 (October 23, 1992).

Sprague, A.L. et al.: "Caloris Basin: An Enhanced Source for Potassium in Mercury's Atmosphere," *Science*, 1140 (September 7, 1990).

Sprague, A.L., R.W.H. Kozlowski, and D.M. surHunter: "Detecting Potassium on Mercury," *Science*, 974 (May 17, 1991).

Staff: Time-Life Books, *The Near Planets*, Time-Life, Inc., New York, NY, 1992.

Strom, R.: *Mercury: The Elusive Planet*, Smithsonian Institution Press, Washington, DC, 1987.

Vilas, F. et al. (Editors): *Mercury*, University of Arizona Press, Phoenix, AZ, 1997.

Waldrop, M.M.: "A New Soviet Plan for Exploring the Planets," *Science*, 211 (October 13, 1989).

Web References

http://sse.jpl.nasa.gov/features/planets/mercury/mercury.html (National Aeronautics and Space Administration, United States).

http://sse.jpl.nasa.gov/missions/merc_missns/m10.html(National Aeronautics and Space Administration, United States)

MERCURY SURFACE, SPACE ENVIRONMENT, GEOCHEMISTRY, AND RANGING (MESSENGER) MISSION.   See **Discovery Program**.

MERGANSER.   See **Waterfowl**.

MERIDIAN.   On the celestial sphere, a great circle that passes through the poles of rotation and, hence, is perpendicular to the celestial equator. The *local meridian* is the great circle passing through both the poles of rotation and, also, through the zenith; it is both a vertical circle and an hour circle. However, the local meridian differs from the ordinary circles on the sphere in that it apparently remains fixed as the celestial sphere apparently rotates once each day. An object on the local meridian is said to be in culmination. If it is on that section of the meridian between the horizon and above the pole, it is said to be in upper culmination; if below the horizon, or below the pole and still above the horizon, it is said to be in lower culmination. The spherical coordinates both of hour angle and astronomical azimuth are measured from the local meridian in the direction of apparent rotation of the celestial sphere.

A *terrestrial meridian* is a curve cut on the surface of the earth by a plane containing the earth's axis of rotation. The local meridian of a point on the surface of the earth is the meridian passing through that point. The plane of the local terrestrial meridian coincides with the plane of the local celestial meridian. Terrestrial longitude is measured from the local meridian of Greenwich, England.

See also **Bearing (Navigation)**; and **Celestial Sphere and Astronomical Triangle**.

MERIDIAN CIRCLE.   A telescope, adjusted so that the collimation plane of the instrument is in the plane of the local meridian, and which may be rotated about a horizontal axis. The instrument is usually fitted with a circle, which is accurately graduated in degrees, minutes, and seconds, and is perpendicular to the axis of rotation, hence, in the plane of the meridian. If the instrument does not carry the circle in the meridian, it is known as a transit circle.

At the principal focus of the telescope is placed a reticle with an odd number of vertical wires, the middle one of which is in the collimation plane. One horizontal wire through the optic axis of the telescope is usually present.

The instrument is used to determine the equatorial coordinates of the stars, when the local sidereal time and terrestrial latitude are known; or, conversely, to determine accurate local sidereal time by observation of stars of known right ascension. The local sidereal time of the instant of passage of a star across the middle wire of the reticle must be the right ascension of the star. The declination is obtained from the readings of the graduated circles when the star passes through the field of view along the horizontal wire. The instrument is also used for the accurate determination of terrestrial longitude by determining the local sidereal time and knowing the corresponding Greenwich time. See also **Telescope (Astronomical-Optical)**.

MERIDIAN LOSSLESS PACKING (or MLP). See **Data Compression**.

MEROMORPHIC FUNCTION. A function of a complex variable for which every point in the finite plane is either a regular point or a pole. For example,

$$f(z) = \frac{P(z)}{Q(z)}$$

where P and Q are polynomials, without common factor, is a meromorphic function.

MERRIFIELD, R. BRUCE (1921–). An American chemist who won the Nobel Prize for chemistry in 1984. Merrifield was cited for work on the use of solid matrix as an aid to chemical synthesis of complex peptides and proteins. His synthesis techniques have been used in the development of solid matrix-bound inorganic and organic agents. Awarded doctorate from U.C.L.A.

MESA. A flat-topped, steep-sided, table-like mountain capped with a formation or stratum, which is relatively horizontal and resistant to erosion. When such a topographic feature is less than 1 square mile (2.6 square kilometers) in area it is usually called a butte.

MESOANALYSIS. The representation of temperature, moisture, pressure, and wind variations on horizontal scales of 10–100 kilometers (6–62 miles). The analysis seeks to define mesoscale features of the observed temperature, pressure, moisture, and wind fields that can be related to important local and regional circulations that in turn may have a significant impact on local and regional weather systems. The mesoanalysis differs from the more conventional synoptic-scale representation of the wind and pressure features in that smaller-scale features inherent in the wind, pressure, and moisture fields are retained in the analysis.

AMS

MESOCLIMATE. The climate of a natural region of small extent, for example, valley, forest, plantation, and park. Because of subtle differences in elevation and exposure, the climate may not be representative of the general climate of the region.

AMS

MESOCLIMATOLOGY. The study of mesoclimates; the climatology of relatively small areas that may not be climatically representative of the general region. The data used in mesoclimatology are mostly standard observations. The size of the area involved is rather indefinite and may include topographic or landscape features from a few acres to a few square miles, such as a small valley, a forest clearing, a beach, or a village site.

AMS

MESOCYCLONE. A cyclonically rotating vortex, around 2–10 kilometers (1.2–6.2 miles) in diameter, in a convective storm. The vorticity associated with a mesocyclone is often on the order of $10^{-2}s^{-1}$ or greater. (It should be noted that a mesocyclone is not just any cyclone on the mesoscale; it refers specifically to cyclones within convective storms.) Mesocyclones are frequently found in conjunction with updrafts in supercells. Tornadoes sometimes form in mesocyclones. Persistent mesocyclones that have significant vertical extent are detected by Doppler radar as mesocyclone signatures. Tornado warnings may be issued when a mesocyclone signature is detected.

AMS

MESOCYCLONE SIGNATURE. The Doppler velocity pattern of a mesocyclone within a severe thunderstorm. In a storm-relative reference frame, the idealized signature is symmetric about the radar viewing direction with marked azimuthal shear across the core region between peak Doppler velocity values of opposite sign. Typical signatures consist of Doppler velocity differences of 25–75 m s^{-1} across core diameters of 2–8 kilometers (1.2–5 miles), with resulting azimuthal shear values of 5×10^{-3} s^{-1} to 2×10^{-2} s^{-1}.

AMS

MESODESMIC STRUCTURE. A type of ionic crystal in which one of the cation-anion bonds is equal in strength to all the bonds from the cation to the other anions. The silicates are important members of this class.

MESOJET. A mesoscale wind maximum. It typically may have an along-flow length scale of tens to hundreds of kilometers (miles) and a cross- flow length scale of <100 kilometers (<62 miles). Mesojets differ from planetary-scale jets, which can have length scales of several thousand kilometers, and synoptic-scale jets, which may have length scales of 1,000–2,000 kilometers (621–1,242 miles) and are commonly found in association with progressive synoptic-scale troughs and ridges. Larger mesojets may also sometimes be known as jet streaks. Mesojets can form adjacent to prominent orographic features in association with terrain-channeled flow. Mesojets are also seen in association with organized mesocale convective systems as typified by the evaporatively driven rear-inflow jet commonly found behind active squall lines lines. Mesojets may also be found in conjunction with prominent lower-tropospheric stable layers where the airflow can become decoupled from the planetary boundary layer, especially at night. An exceptionally well organized lower-tropospheric mesojet extending over hundreds of kilometers might be known as a low-level jet.

MESONS. The *mesons* are subatomic particles of the hadron family. See also **Hadrons**. Fermionic hadrons are called *baryons*; the others are called *mesons*. The meson family consists of eight members which fall into a triplet of *pions*, a singlet *eta*, a doublet of *kaons*, and a doublet of *antikaons*. (They are all pseudoscalar (spin zero and odd parity) and exhibit strong interactions.) The charged particles are coupled to the photon, but even the neutral members can participate in electromagnetic interaction by virtue of the large probability for virtual dissociation into charged particles. They participate in a variety of weak interactions including the nuclear beta decay interaction.

It is found that the kaons, the hyperons (baryons other than the neutron and proton) and their antiparticles, collectively known as *strange particles*, can decay by weak interactions not involving leptons or photons, with a lifetime which is large compared to the natural periods appropriate to strong interactions. On the other hand, these particles are produced copiously in high-energy nuclear collisions. These two circumstances can be understood in terms of the existence of another additive quantum number (*hypercharge*) which is conserved in strong and electromagnetic interactions, but violated in weak interactions.

The meson-baryon system exhibits further regularities as far as strong interactions are concerned. The neutron and the proton have very nearly the same mass and similar nuclear interactions although their electromagnetic properties are quite different. The three *pions* have different electric charges, but again they have approximately equal masses and similar nuclear interactions. This kind of multiplet structure is evident for other strongly interacting particles; the *kaons* form a doublet, the *sigma hyperons* form a triplet, the *xi hyperons* form a doublet, and the *lambda hyperon* remains a singlet. See also **Particles (Subatomic)**.

The pion (*pi meson*) was first recognized in 1947 in photographic films made by C.F. Powell, P.S. Occhialini, and their collaborators of cloud chamber tracks made by cosmic rays high in the Andes. The masses of these pions were greater than those of the previously discovered muons, corresponding more closely to those predicted by K. Yukawa in his theory of the nuclear structure of the atom. The positive or negative pion has a mass 273 times that of the electron, and a charge equal in magnitude to that of the electron. Both positively charged and negatively charged pions are found in cosmic rays. Neutral pions may also be present in cosmic rays, but are produced in much greater abundance by high-energy particle accelerators and are therefore more easily detected in these laboratories.

The first artificially produced pions were made in 1948 by the impact of 380 MeV alpha particles, from the Berkeley synchrocyclotron, on a target of carbon or certain metals. These were charged pions. Others were produced later by beams of protons and deuterons. The first evidence of neutral pions were the gamma-rays produced in 1950 by the impact of 175 MeV protons upon similar targets (carbon, beryllium, etc.). These gamma-rays had a minimum energy of about 140 MeV, which would be expected from the decay of a (neutral) pion into two gamma-rays. The same method is used today to produce beams of charged pions. The life of the neutral pion is so much shorter (about 10^{-16} seconds against 2.6×10^{-8} seconds) that beams cannot be produced. Unless it is captured by an atom, or reacts with another particle, a charged pion decays into a muon of the same sign and a neutrino or antineutrino. Like the other mesons the pion is a boson. It has zero spin.

Evidence of the *kaon* was found in 1944 by L. Laprince-Ringuet and M. LhÉritier in a cloud chamber photograph of a cosmic ray event. It was found again in 1947 by Rochester and Butler as a V-shaped track in a cloud chamber, the particle forming the other side of the V being probably a pion. For that reason it was first called a charged V-particle, which has been superseded by kaon.

See also **Quarks**.

Additional Reading

Drell, P.S., and D.L. Rubin: *Heavy Quark Physics*, Springer-Verlag New York, LLC, New York, NY, 1997.

Ejiri, H., and H. Toki: *Nucleon-Hadron Many-Body Systems: From Hadron-Meson to Quark-Lepton Nuclear Physics*, Oxford University Press, New York, NY, 1999.

Henley, E.M., and H.W. Pauchy: *Lectures on Quarks, Mesons and Nuclei*, World Scientific Publishing Company, Inc., Riveredge, NJ, 1989.

Jarczyk, L., H. Machner, A. Maqiera, and C. Guaraldo: *Meson 2002: Production, Properties and Interaction of Mesons*, World Scientific Publishing Company, Inc., Riveredge, NJ, 2002.

MESOPAUSE.   The top of the mesosphere and the base of the thermosphere. The mesopause is usually located at heights of 85–95 km (53–59 miles), and is the site of the coldest temperatures in the atmosphere. Temperatures as low as 100 K ($-173\,°C$ ($-279\,°F$)) have been measured at the mesopause by rockets. See also **Atmospheric Shell**.

AMS

MESOSCALE.   Pertaining to atmospheric phenomena having horizontal scales ranging from a few to several hundred kilometers, including thunderstorms, squall lines, fronts, precipitation bands in tropical and extratropical cyclones, and topographically generated weather systems such as mountain waves and sea and land breezes. From a dynamical perspective, this term pertains to processes with timescales ranging from the inverse of the Brunt–Väisälä frequency to a pendulum day, encompassing deep moist convection and the full spectrum of inertiogravity waves but stopping short of synoptic-scale phenomena, which have Rossby numbers less than 1.

AMS

MESOSCALE CELLULAR CONVECTION (MCC).   A regular pattern of convective cells that can develop in an atmospheric boundary layer heated from below or radiatively cooled from cloud top. This phenomenon is readily observed in satellite imagery during cold air outbreaks when continental air passes over the relatively warm coastal ocean. Cloud lines, marking horizontal roll vortices, form initially in the developing marine atmospheric boundary layer. These lines evolve into open cells, which are defined by clouds in the upward motion along the edges of honeycomb-shaped cells, with less cloudy subsiding air in their centers. The convective structure further evolves into closed cells, which have cloudy centers and cloud-free edges.

AMS

MESOSCALE CONVECTIVE COMPLEX (MCC).   A subset of mesoscale convective systems (MCS) that exhibit a large, circular (as observed by satellite), long-lived, cold cloud shield.

The cold cloud shield must exhibit the following physical characteristics.

- Size: **A**—Cloud shield with continuously low infrared (IR) temperature $\leq -32\,°C$ must have an area $\geq 10^5$ km^2; and **B**—Interior cold cloud region with temperature $\leq -52\,°C$ ($\leq -62\,°F$) must have an area $\geq 0.5 \times 10^5$ km^2.
- Initiate: Size definitions **A** and **B** are first satisfied
- Duration: Size definitions **A** and **B** must be met for a period ≥ 6 h.
- Maximum extent: Contiguous cold cloud shield (IR temperature $\leq -33\,°C$ ($\leq -27\,°F$) reaches maximum size.
- Shape: Eccentricity (minor axis/major axis) ≥ 0.7 at time of maximum extent.
- Terminate: Size definitions **A** and **B** no longer satisfied.

Alternatively, a dynamical definition of an MCC requires that the system have a Rossby number of order 1 and exhibit a horizontal scale comparable to the Rossby radius of deformation. In midlatitude MCS environments, the Rossby radius of deformation is about 300 kilometers (186 miles).

AMS

MESOSCALE CONVECTIVE SYSTEM (MCS).   A cloud system that occurs in connection with an ensemble of thunderstorms and produces a contiguous precipitation area on the order of 100 km or more in horizontal scale in at least one direction. An MCS exhibits deep, moist convective overturning contiguous with or embedded within a mesoscale vertical circulation that is at least partially driven by the convective overturning. See also **Mesoscale**.

AMS

MESOSPHERE.   The region of the atmosphere lying above the stratosphere and extending from the stratopause at about 50 km (31 miles) height to the mesopause at 85–95 km (53–59 miles). See also **Mesopause**.

The mesosphere is characterized by decreasing temperature with increasing height, reflecting the decreasing absorption of solar ultraviolet radiation by ozone. Many features of the mesosphere remain poorly understood since in situ measurements are difficult. The region is too high for balloon operations and too low for satellites to orbit. Rockets, while useful, usually travel too rapidly through the region to produce reliable measurements. See also **Atmospheric Shell**.

AMS

MESOSPHERIC JET.   There are two zonal jet streams in the layer known as the mesosphere. In January, there is a westerly (blowing from the west) jet located at 70 kilometers (43.5 miles) between 25° and 45 °N with a maximum wind of 60 m s^{-1}. There is also an easterly jet of comparable intensity located between 30° and 50 °S. In July, the direction of the mesospheric jet in each hemisphere is reversed. The westerly jet in the southern hemisphere in July descends to a somewhat lower level (about 50 kilometers (31 miles) and has a stronger intensity (about 100 m s^{-1})

AMS

MESOTHERMAL CLIMATE.   See **Climate**.

MESOZOIC.   A major subdivision of the geologic time-scale. The era of "Middle" life, or the age of reptiles. Subdivided from the base up, into the following periods: Triassic, Jurassic, Comanchean, Cretaceous. The era was characterized by: the rise of dinosaurs (Triassic); rise of birds and flying reptiles (Jurassic); rise of flowering plants (Comanchean); great development of ammonites which became extinct at the end of the Cretaceous; culmination and extinction of most reptiles, and rise of the archaic mammals between the Cretaceous and the Tertiary. The Mesozoic era began 200 million years ago and lasted 140 million years.

MESQUITE TREE.   Commonly found in the southwestern United States, Mexico, other parts of Central America, the West Indies, and parts of South America, the mesquite (*Prosopis L.*) is a very hardy, small-to-medium-size tree. Examples of mesquites growing in the United States, as reported by American Forests are shown in Table 1.

Some beholders of the tree consider it scraggly and unkempt; to others, the dropping habit of its branches is attractive. The tree does not offer full shade. Its general appearance is very suggestive of its location in a desert and drought-stricken locale. The leaves are small, compound pinnately, and drought and heat resistant. Flowers are small and greenish white and occur in 2-to-5 in. (5–13 cm) spikes. Seeds, which develop in large 6 to 12 in. (15–30 cm) flat pods, are dropped to the ground between June and August. Succulent and containing sugar, the seed pods are welcomed by livestock.

TABLE 1. RECORD MESQUITE TREES IN THE UNITED STATES[1]

Specimen	Circumference[2]		Height		Spread		Location
	Inches	Centimeters	Feet	Meters	Feet	Meters	
Honey mesquite (typ.) (1997)	172	437	57	17.4	87	26.5	Texas
(*Prosopis glandulosa var. glandulosa*)							
Screwbean mesquite (1983)	39	99	30	9.1	36	11	Texas
(*Prosopis pubescens*)							
Screwbean mesquite (1983)	35	89	28	8.5	40	12.2	Texas
(*Prosopis pubescens*)							
Velvet mesquite (1993)	196	498	46	14	60	18.3	Arizona
(*Prosopis velutina*)							
Western honey mesquite (1997)	76	193	47	14.3	39	11.9	California
(*Prosopis glandulosa var. torreyana*)							

[1] From the "National Register of Big Trees," American Forests (by permission).
[2] At 4.5 feet (1.4 meters).

In Mexico, the pods are used to form a type of tamale (*mesquitamales*), which are sun dried and can be preserved as food for long periods.

The trunk of the mesquite generally is short, often twisted, and features limbs that are gnarled at the base. Thorns (11/2 in; 4 cm) long are found among the leaves and along the branches, but rarely on the trunk. The silhouette of the mesquite varies immensely—from that of a weeping willow to unlimited grotesque configurations. In a tree that is well developed and balanced above ground, the root system will be large, but not abnormally so. But where there is limited above ground development, the root system will be huge. Some taproots will reach out for water down to 60 feet (18 meters) below ground level. It has been observed by some foresters in the arid American Southwest, that the mesquite is one of the region's most interesting trees because of its heat and drought resistance resulting from its huge root system.

Along with its other survivability characteristics, the mesquite is relatively free from insects, pests, and fungus diseases. Its natural enemy is mistletoe, which hangs in heavy masses from the branches and causes deformation and atrophy of the branches, as well as asymmetrical burls on the trunk and major limbs. The tree is notably resistant to heartwood decay. Formation of a natural resin (closely related to acacia gum) covers over pruning cuts and other wounds, contributing to the longevity of the tree amidst very adverse conditions. Very important, as a legume, the mesquite can fix nitrogen in the soil. Further descriptions of the properties and lore of the mesquite tree can be found in "The Indomitable Mesquite," by J.H. Haller, American Forests, 20–24, August 1980.

MESSENGER RNA.   See **Cell (Biology)**; and **Human Genome Project**.

METABOLISM.   The chemical transformations occurring in an organism, from the time a nutrient substance enters it until it has been utilized and the waste products eliminated. In animals and humans, digestion and absorption are primary steps, followed by complicated series of degradations, syntheses, hydrolyses, and oxidations, in which agents such as enzymes, bile acids, and hydrochloric acid take part. These transformations are often localized with respect to organs, tissues, and types of cells involved.

Basal metabolism is the rate of total heat production of an individual who is awake but in complete mental and physical repose, at comfortable temperature and without having had food for at least 12 hours. See also **Basal Metabolism**. Under these conditions, oxidation of stored nutrients provides the sole source of energy expended and heat is measurable by calorimetry. See also **Digestion**.

METACHROSIS.   The change of colors in animals by the expansion or contraction of pigment cells.

METALLIC COATINGS.   Metallic coatings provide an inexpensive way to modify or control the properties of a base material. Coatings may be functional or decorative, permanent or temporary, sacrificial, or noble. The metallic coating may be continuous, or it may be patterned into discontinuous functional or decorative areas. The base material may be metallic, nonmetallic inorganic, or organic such as plastic, paper, or fiber. The common criterion for the purposes herein is that the metallic coating is functionally bonded to the base material. Galvanized zinc items, gold clad jewelry, electroplated materials, and semiconductor chips are among the most common types of materials having metallic coatings. See also **Conversion Coatings**; **Electroplating**; and **Semiconductors**.

Metallic coatings have been used since ancient times, initially for jewelry and decorative applications. One of the earliest techniques, used by the ancient Egyptians, was the use of gold foil. Gold foil, produced by beating gold to thicknesses of <0.1 μm can be easily bonded to all types of metals, paper, leather, and other materials. Many early artifacts originally thought to have been fabricated by means of a chemical displacement coating of a precious metal, were probably made by selective dissolution of the less noble components of an alloy. Glass mirrors formerly were made by amalgamating a thin tin sheet with mercury, pressing it to a glass sheet, and heating to vaporize the mercury and form a direct metal-to-glass bond. Soldering is an ancient metal coating technique used for joining two metals, used as of the 1990s in highest volume in the electronics industry.

Some of the first electroplating applications in the early 1800s involved attempts to coat inexpensive base metals with thin layers of pure gold, or even more cheaply, gold-colored brass. Economic reasons are almost, always foremost in the decision of coating selection. Exceptions, are where environmental or legislative measures require more expensive, less toxic coatings, as in replacement of cadmium [CAS: 7440-67-7]. The use of metallic coatings often conserves rare metals, allowing the more abundant materials to serve as substrates. Bulk corrosion-resistant materials such as stainless steel or Monel use large amounts of scarce nickel and chromium. One metal may serve many different functions, depending on the application. Thus gold [CAS: 7440-57-5] may be the permanent decorative coating on jewelry, a permanent functional coating on electrical connectors or anodes, or a temporary functional and protective coating for solder joint connections on printed circuit boards in which the thin gold layer protects the underlying nickel from oxidation then dissolves in the molten solder to allow solder bonding to the nickel.

Metallic coatings are most often selected for protective function. Decorative ability is a common secondary function. Most coatings applied by hot dipping, such as galvanizing or aluminizing, use a film of a more chemically reactive metal over a less reactive material such as iron alloy. These are sacrificial coatings because the zinc or aluminum slowly dissolves instead of the underlying steel.

Decorative coatings must be sufficiently corrosion resistant to maintain an attractive appearance during the anticipated life of the composite product. Composites of multiple metal layers or mixtures of metallic and nonmetallic coatings are becoming increasingly common as material properties are selectively tailored. Aluminum [CAS: 7429-90-5] coatings on mirrors, laser disks, and compact disks are given a protective sealer coating of transparent plastic to inhibit the sacrificial oxidation of the shiny aluminum to the dulled aluminum oxide. Thin transparent rhodium [CAS: 7440-16-6] coatings are used to make tarnish-resistant silver jewelry. Chromium [CAS: 7440-47-3] coatings combine both decoration and corrosion resistance.

Another important function of metallic coatings is to provide wear resistance. The most technically demanding types of coatings are those grown in precisely defined multiple layers.

Plating and galvanizing are the most common methods of applying metallic coatings, but many other processes, such as hot dipping, cementation, thermal spraying, sputtering, and chemical vapor disposition, have been developed. All metallizing techniques are potential sources of pollution, a subject of increasing ecological, economic, legislative, and public health interest.

There are a plethora of commercially useful methods for applying a metallic coating. Many more techniques have been demonstrated in the laboratory. Each method has different critical parameters, i.e., maximum and minimum coating thickness producible, substrate temperature, bulk or imagewise deposition, coating adhesion value, type and cost of coating equipment, labor requirement, scrap rate, rework capability, and safety and waste disposal aspects. For convenience, metallic coating methods may be divided into classes defined by the general way in which the metal is applied, e.g., liquid-phase, gas-phase, and vacuum-phase metallizing or by direct physical or thermal bonding.

The two largest-volume processes, in terms of amount of metal used and amount of surface area coated, are the liquid-phase metallizing processes hot dip coating and electroplating.

Many of the newer coating methods give metallized coatings that are quite distinct from simple single or multiple coatings of metals or alloys. Many of these newer coatings, which cannot be made except by one or a small family of processes, are defined as much by process method as physical and chemical structure. Examples of these coatings include ion-plated surfaces, laser heat-treated surfaces, modifications of existing surfaces to predetermined depths by selective additions of atoms or heat; and Teflon–electroless nickel composite coatings.

Liquid-Phase Metallizing Techniques

Hot Dip Galvanizing. The largest single type of metallic coating process in terms of amount of metal used is hot dip galvanizing. Half of all galvanized zinc is used on steel coils, one-third is employed for coating of parts after fabrication, and the remainder for wires and tubes. The other uses, in approximate order, are electroplating, zinc-filled paints, zinc spray processes, zinc foils having conductive adhesive, and mechanical zinc plating.

In hot galvanizing, zinc is applied to iron and steel parts by immersing the parts into a bath of molten zinc. Whereas, in principle, almost any metal could be coated with molten zinc, this coating serves no worthwhile purpose on most metals. The combination of zinc and ferrous materials are almost uniquely suited to each other. Aluminum and cadmium are the only other similar combinations. Zinc provides iron parts with better corrosion protection by developing a coating of zinc and zinc compounds on the base metal surface.

Hot Dipping Using Other Metals. Processes include coating with aluminum, coating with 55% aluminum–zinc alloy, hot dip tin coating of steel and cast iron, hot dip terne coating, and solder coatings.

Electroplating. Many developments in electroplating have been the direct result of increased coating functionality and economy, and others of the increasing need for environmental and legislative compliance. The field of electroplating has expanded rapidly.

The most common electroplated metal is zinc, followed by nickel, copper, and chromium. Many other metals and metaloids can be electroplated, including manganese, iron, cobalt, gallium, germanium, arsenic, selenium, ruthenium, rhodium, palladium, silver, cadmium, indium, tin, lead, bismuth, mercury, antimony, gold, iridium, and platinum. Types of electroplating include fused salt plating, plating from nonaqueous solvents, and brush plating.

Chromium metal seems to be biologically harmless owing to inertness. Trivalent chromium Cr(III), one of the essential mineral nutrients, has not been shown to be harmful. Hexavalent chromium, Cr(VI), found in chromic acid and used in electroplating baths, is very toxic as well as a suspected carcinogen. Two basic types of chromium are plated: hard chromium up to 7100 μm or more, and decorative chromium at 0.25–1.0 μm. No replacement coating has been found for thick hard chromium deposits for wear resistance and parts salvage, although electroless nickel can partially substitute for chromium plating baths. Many decorative applications are being converted to Cr(III). Products from the newest Cr(III) baths are essentially equivalent to those from the older decorative Cr(VI) baths, although some cosmetic color differences exist which have prevented complete conversion to the Cr(III) baths.

Cyanide solutions formerly were ubiquitous in electroplating shops. The best cleaners contained sodium cyanide. The removal of cyanides from all

plating baths has been a general goal since the latter part of the 1980s. As of this writing, cyanide-containing cleaners are rare and are only used for special purposes. Several noncyanide copper strike baths have been introduced.

The greatest tonnage decrease in cyanide plating has occurred for zinc plating. As of this writing, less than one-third of all zinc is plated from cyanide baths. The remainder is plated from alkaline noncyanide, zinc potassium chloride, and zinc ammonium chloride baths.

Chlorinated and fluorinated cleaners have been widely used as degreasers during surface preparation prior to plating. Recognition of the role of many of these solvents as either global warming gases or ozone depletion agents has led to prohibitions against their continued use. See also **Pollution (Air)**. Many new and improved cleaning systems are being developed, such as alcohol-based cleaners, emulsion cleaners, and cleaners based on natural bioproducts such as limonene. During this process a massive reformulation of many aqueous cleaners has also occurred.

Cadmium usage, illegal in most of Europe, is being discouraged elsewhere. The U.S. military has cadmium specifications for electronic, fastener, and marine equipment, which requires only cadmium. Tin is being substituted for tin–lead as a metallic etch resist during printed circuit board production.

Electroless Plating. The metallizing process known as electroless plating is mainly used for deposition of copper on plastics and for nickel–phosphorus alloy on plastics and metals. Formaldehyde is the most common copper reducing agent, giving pure coatings. Sodium hypophosphite is the agent mainly used for electroless nickel. An unusual nickel–phosphorus alloy or solid solution is formed. Unlike electroplating, electroless plating can be used on almost any substrate, metallic or nonmetallic. Often electroless plating is used as the first coating to make glass ceramic, or plastic conductive, followed by conventional electrolytic plating.

Immersion Plating. A simplified aqueous metal deposition process which does not use electric current, immersion plating works only when a metal of higher electromotive force, such as copper, is deposited on a metal of lower electromotive force, such as iron or aluminum. The coatings are typically very thin and porous. One important application is immersion plating of copper sulfate bath onto steel wire, to use as a drawing agent. Very thin immersion gold deposits are used for inexpensive jewelry.

Miscellaneous Techniques. Lasers have been used for both electroless and electrolytic plating; selective dissolution has been used from ancient times to give the appearance of a thin plated coating of precious metal; and mercury layers plated onto the surface of analytical electrodes serve as liquid metal coatings.

Gas-Phase Metallizing Techniques

Metal or Thermal Spray Coatings. These specific processes include plasma arc spray, flame spray, laser spray, and electric arc spray, depending on the energy input source. Rods, wires, and powders are used as coating material sources. Thermal spray metal and ceramic coatings have diverse properties suitable for numerous applications, including corrosion resistance, e.g., zinc and aluminum, especially against oxidation or salt water corrosion; high temperature oxidation, e.g., nickel, cobalt, and chromium alloys; electrical conductivity, e.g., radio frequency interface (RFI) shielding by zinc or tin on nonconductors; electrical resistance, e.g., insulating layers in induction heating coils and high temperature strain gauges; wear resistance, e.g., chromium–nickel–boron alloys and carbide-containing coatings; catalytic surfaces; nuclear moderators; and dimensional buildup for salvaging worn metal parts.

A great advantage of zinc arc spray is that it can be applied to almost any plastic. Zinc arc spray, also suitable for prototypes and small lots of materials, is less suited for very small parts and parts having blind holes or complex interior surfaces, or where warpage is a problem.

Zinc arc spraying or flame spray equipment is hardly more hazardous than a welding torch, and only safety goggles and gloves are required. Safety aspects emphasize reduction of noise and vapor inhalation.

Thermal spray processes can be used to give coatings of chromium carbide or nickel chromium for erosion resistance, copper nickel indium for fretting resistance, tungsten carbide cobalt for wear and abrasion resistance, and even aluminum silicon polyester mixtures for abradability.

Carburizing and Nitriding. Several commonly used metallurgical surface treatments are applied by gas-phase reactions in a reducing atmosphere for carburizing, and in a nitrogen atmosphere for nitriding. These treatments are used to increase the surface hardness of ferrous alloys by diffusion of carbon and nitrogen at high temperature.

Pack Diffusion. Pack diffusion or cementation processes are similar to pack carburizing, and are used to coat iron, nickel, cobalt, and copper with chromium, boron, zinc (Sheradizing), aluminum, silicon, titanium, molybdenum, and other metals.

Chromizing and Related Diffusion Processes. Chromizing is similar to aluminizing. A thin corrosion and wear resistant coating is applied to low cost steels such as mild steel, or to a nickel-based alloy. In the related boronizing process, a thin boron alloy is produced for extreme hardness, wear, and corrosion resistance. Siliconizing is yet another process used especially for coating of the refractory metals Ti, Nb, Ta, Cr, Mo, and W.

Metals. Aircraft and space vehicles, turbine generators, and other such applications require high strength at high temperatures along with excellent oxidation resistance. Superalloys, i.e., complex nickel and cobalt-based alloys, and refractory metals, e.g., niobium, tungsten, molybdenum, tantalum, and their alloys, are used for applications at temperatures above 1000 °C. In many case the coatings must be resistant both to oxidation and to hot corrosion by sulfidation from sulfur-bearing gases.

Two types of coatings have been used for superalloys: diffusion coatings, in which a layer of nickel, cobalt, platinum, or palladium aluminide, i.e., NiAl, CoAl, PtAl, or PdAl, is formed on the surface by diffusion; and overlay coatings, in which a complex coating material such as nickel–cobalt–chromium–aluminum–yttrium, NiCoCrAlY, is applied to the surface. Pack cementation is the most widely used process for applying diffusion coatings to superalloys.

Vacuum-Phase Metallizing Techniques

Vacuum-phase metallizing techniques all depend on the use of a vacuum as part of the metallizing process. The tonnage of metal deposited by these techniques is insignificant compared to hot dip galvanizing or plating, and the total surface area metallized is much less than that done by plating. However, the economic added value of vacuum metallizing probably exceeds that of either plating or galvanizing because of the extremely rapid growth and deep market penetration of semiconductor devices. Projections call for at least a 10% compounded annual growth rate into the twenty-first century. The state of the art in vacuum metallizing is in semiconductors processing.

Thermal Evaporation. Thermal evaporation is done in a high vacuum to minimize chemical side reactions of the evaporated active metal. Thermal evaporation is inexpensive and efficient. It is used for low cost items such as aluminized plastic sheet and decorative Christmas tinsel, second surface coating of transparent plastic auto parts, and metallization of glass and ceramics. Aluminum is the predominant metal used. Color effects such as gold or brass are achieved by applying a dyed translucent organic protective over the aluminum.

Sputtering or Glow Discharge. Sputtering can be done using both conductive and nonconductive items. A low pressure atmosphere of argon is used in the deposition vessel. This method gives excellent coating adhesion and more consistent coatings than simple vacuum deposition. It can also be applied to a wider range of materials, including complex metal alloys and oxides. The equipment and operating costs are higher than vacuum deposition, and deposition rates are lower. Reactive sputtering is a variation in which two or more deposition sources are used. This process can be used to give compounds such as silicon carbide on the surface.

Chemical Vapor Deposition. This process is distinct from simple thermal evaporation because chemical vapor deposition (CVD) depends on a chemical reaction at the surface of the part and in that way is analogous to electroless plating. This process uses a gas of one or more chemical species, which react at a heated substrate to form an appropriate film. The film can be metallic, nonmetallic, single element, or compound. Among the coatings available are copper and aluminum conductors, diamond and tantalum oxide dielectrics, lead zirconium titanate piezoelectrics, and bismuth strontium calcium cuprate superconductors. Some of the largest applications are for coating TiC and TiN on cutting tools.

Ion Implantation. Also known as ion plating, ion implantation is a high vacuum process for modifying the surface properties of any material. This is a surface modification technique rather than a surface coating technique. Semiconductors are commonly modified using this technique, because ion implantation can be done using total or imagewise scanning of the beam over the surface. Nitrogen, chromium, and phosphorus can be used to harden metals and increase the corrosion resistance. Metastable alloys can be produced that are difficult to make by other processes.

Laser Hardening and Modification. Lasers are used to surface harden ductile steels and improve the toughness to a depth of 0.35 mm or more. Lasers can also be used to bond solid or powder coatings to a surface. Typical coatings are nickel or titanium carbide on iron, and nickel, cobalt, manganese, and titanium carbide, TiC, on aluminum. Use of lasers with other specialized coating methods is common.

Metallizing by Direct Physical or Thermal Bonding

Direct bonding techniques are among the oldest types of metallizing, and the most versatile. Many methods depend on heat or pressure and an adhesive layer to glue the coating to the substrate. Methods for metallizing on a metallic surface often depend on removal or displacement of a preexisting surface oxide layer. Many metals form intermetallic alloys or self-diffuse into one another even at room temperature, but the surfaces in contact must be clean and oxide-free. Processes include lamination, mechanical plating, slurry coatings, and roll bonding or strip roll welding.

Environmental Concerns

Each type of metallic coating process has some sort of hazard, whether it is thermal energy, the reactivity of molten salt or metal baths, particulates in the air from spray processes, poisonous gases from pack cementation and diffusion, or electrical hazards associated with arc spray or ion implantation. Vacuum or inert gas operations can produce flammable dusts or powders when opened for cleaning. Most of the hazards are confined to the operator and immediate environs in the operating plant. OSHA is the primary regulator of these hazards in the United States, although many local and state agencies, especially fire departments, also regulate coatings plants. Adequate training, documentation, and protective equipment are the minimum requirements. The principal regulatory burden falls on wastes and discharges which leave the plant.

The U.S. EPA regards metallizers such as platers, surface finishers, and printed circuit board producers as among the most important source polluters for metals. Much production of electroplated items has, however, shifted from the United States to less environmentally stringent countries.

The surviving U.S. plants have embraced all types of waste treatment processes. See also **Wastes and Pollution**. The most desired pollution prevention processes are those which reduce the total amount of waste discharged. Treatment and disposal are less strongly emphasized options. Zero wastewater discharge facilities and water recycling processes are becoming more common.

Discharge limits vary between localities and among plants. Table 1 shows federal EPA maximum discharge limits for a number of metals for a new metal-finishing installation.

Newer federal limits on metals contents of sewage sludge combined with laws on fuller treatment of the sewage sludge and its allowed disposal methods have affected limits. The metallic content of sludges from municipal waste treatment facilities is becoming of great concern.

Air pollution is recognized as a significant problem for coating facilities.

Many types of waste treatment and waste minimization processes are in common use in the metallization industry. Air scrubbers are commonly required, even for general acid fume removal from plating shops. Additionally, many newer technologies have been adapted for use in metallizing operations. These include air stripping, antimisting agents, biological destruction, carbon absorption, countercurrent rinsing, crystallization, distillation, Donnan dialysis, electrodialysis, electrowinning, evaporation, filtration, fluocculation, hydrolysis, incineration, ion exchange, metallic replacement, neutralization, oxidation, pH adjustment, photolysis, precipitation, process modification, reduction, reverse osmosis, salt splitting, sedimentation, solidification, and spray rinsing.

Additional Reading

Durney, L.J. ed.: *Electroplating Engineering Handbook*, 4th Edition, Van Nostrand Reinhold Co., Inc., New York, NY, 1984.

Metal Finishing Guidebook and Directory, Vol. 92, No. 1A, Elsevier Publishing, New York, 1944; collected vols., Publications, Inc., Hackensack, NJ, updated yearly.

Plating and Surface Finishing, Semiconductor International, and *Metal Finishing* provide some of the best information on innovative metallizing methods.

Safranek, W.H.: *The Properties of Electrodeposited Metals and Alloys*, 2nd Edition, American Electroplaters and Surface Finishers Society, Orlando, FL, 1986.

Schweitzer, P.A.: *Corrosion Resistant Linings and Coatings*, Marcel Dekker, Inc., New York, NY, 2001.

Sedlacek, V.: *Metallic Surfaces, Films and Coatings*, Elsevier Science & Technology Books, New York, NY, 1992.

METALLOBIOMOLECULES. Natural products, the biologically active forms of which contain one or more metallic elements. Metallobiomolecules may be transport and storage proteins, such as cytochromes (Fe), ferritin (Fe), transferrin (Fe), cruloplasmin (Cu), myoglobin (Fe), or hemoglobin (Fe); or they may be enzymes, such as carboxypeptidases (Zn), aminopeptidases (Mg, Mn), phosphatases (Mg, Zn, Cu), hydroxylases (Fe, Cu, Mo), or isomerases and synthetases, such as coenzymes (Co). Metallobiomolecules also may be nonproteins, such as siderophores (Fe) or chlorophyll (Mg). Ibers and Holm provide an excellent review of metallobiomolecules in *Science*, **209**, 223–235 (1980).

METALLOGRAPHY. Study of the structure and properties of metals and alloys, principally by microscopic and x-ray diffraction methods. The term is also used in a broader sense to include the processing of metals by mechanical and heat treatments and the fabrication and testing of finished products. In this usage it is synonymous with physical metallurgy.

METALLOID. A chemical element that may exhibit physical and chemical properties both of a metal and a nonmetal sometimes is referred to as a metalloid. Antimony, arsenic, and tellurium are examples. Less frequently, metalloid refers to elements, such as carbon, silicon, phosphorus, and sulfur, which are added in small amounts in the manufacture of iron and steel.

METALLOPROTEINS. Proteins, especially in solution, readily participate in a greater variety of chemical reactions than any other class of compounds of biological interest. This reactivity is a function primarily of the many polar side chains containing $-OH$, $-COOH$, $-NH_2$, $-SH$, and other groups, all of which can, to varying extents, interact with metal ions. Proteins can bind metals, some of them very tightly. However, relatively specific and nonspecific binding should be differentiated.

A negatively charged protein molecule exerting a nonspecific electrostatic attraction on metal ions would not qualify as metalloprotein. The term metalloprotein is restricted to compounds in which under natural conditions a metal ion is relatively specifically and strongly bound to a protein molecule in such a way that the compound can be isolated and shown to contain a stoichiometric amount of metal.

A variety of metal ions are found in biologically important metalloproteins. Metalloproteins occur in a wide range of biological systems. The function of the metalloproteins, as indicated in Table 1, varies widely from one compound to another.

The chemical properties of the metal in these compounds may be greatly affected by bonding to a protein ligand. The bound metal can play one of many roles. Thus, in an enzyme, the metal ion may permit the formation of a ternary complex between protein, metal and substrate or coenzyme. An instance of this role is provided by the enzyme enolase, which is unable to catalyze the equilibrium between 2-phosphoglycerate and 2-phosphopyruvate in the absence of Mg ions. In other enzymes, the metal may actually participate in electron transport by cyclic oxidation and reduction. Such is probably the case with the Cu in polyphenol oxidase. The metal may serve primarily for the maintenance of a specific spatial folding of the polypeptide chains in the protein molecule.

The strength of metal-protein bonds in metalloproteins may vary from relatively loose association to very tight binding. When the metal ion is able to dissociate with some ease from the protein, it is usually a single ligand responsible for the metal binding. Such ligand groups are mainly found in the amino acid side chains of the protein molecule (e.g., $-NH_2$ or $-OH$ groups). The interaction of metal and ligand may exhibit strong pH dependence because of competition between metal and hydrogen ions.

TABLE 1. REPRESENTATIVE METALLOPROTEINS AND THEIR FUNCTIONS

Name	Metal	Source	Function
Hemocuprein	Cu	Erythrocytes	Unknown
Ceruloplasmin	Cu	Serum	Oxidase (probable)
Hepatocuprein	Cu	Liver	Unknown
Polyphenol oxidase	Cu	Mushroom	Enzyme
Hemocyanin	Cu	Mollusks	Respiratory pigment
Tyrosinase	Cu	Mushroom	Enzyme
Metallothionein	Cd + Zn	Kidney	Na Reabsorption (probable)
Xanthine oxidase	Mo	Liver	Enzyme
Carbonic anhydrase	Zn	Erythrocytes	Enzyme
Alcohol dehydrogenase	Zn	Yeast	Enzyme
Ferritin	Fe	Spleen	Fe storage
Transferrin	Fe	Plasma	Fe transport
Conalbumin	Fe	Eggs	Fe storage
Ferredoxin	Fe	Bacteria	Electron transport
DPNH-cytochrome *c* reductase	Fe	Heart muscle	Electron transport
Hemovanadin	V	Tunicates	Respiratory pigment

Of the single ligand groups, by far the strongest is the $-SH$ group in the amino acid cysteine. Even stronger metal bonding to protein may be observed when a divalent or trivalent metal forms chelate complexes with the protein. Chelation is often indicated not only because of the strength of the bond, but also because of the specificity of the reacting site on the protein molecule for one particular metal. Such a specificity may reflect the coordination requirements of the various metals. The preferred electron donor in the formation of protein-metal coordination compounds is N, such as that of the imidazole nucleus of histidine, but S and O may also participate in this process. If the protein contains carboxyl or phosphoryl groups, strong ionic bonds between metal and protein may be formed. A completely different type of protein-metal interaction is illustrated by the Fe-containing protein ferritin. Basically, this compound consists of a coat of protein (apoferritin) surrounding a micelle of hydrated iron hydroxide. The metal can be readily and reversibly removed from the apoprotein.

See also **Chelates and Chelation**.

METALLOTHIONEINS. These are low-molecular weight, cysteine-rich proteins that bind metal ions. As reported by Furey, et al.(*Science*, **231**, 704, 1986), metallothioneins and their genes have several potential kinds of physiological activity, including: (1) the genes are induced by metal ions and glucocorticoid hormones; (2) transcription is modulated during embryonic development; (3) the genes may be involved in control of cell differentiation and proliferation; (4) the proteins may function to activate Zn requiring apo-enzymes and regulate cellular metabolism; and (5) the proteins may act as free-radical scavengers. It appears that metallothioneins are synthesized in response to ultraviolet radiation. Cadmium, in addition to zinc, is found in metallothioneins.

METALLURGY. The science and technology of metals and alloys. Process metallurgy is concerned with the extraction of metals from their ores and with the refining of metals; physical metallurgy, with the physical and mechanical properties of metals as affected by composition, processing, and environmental conditions; and mechanical metallurgy, with the response of metals to applied forces. (*From Glossary of Metallurgical Terms and Engineering Tables, American Society for Metals, with permission.*)

Early sources define metallurgy as the process of extracting metal from ores. For many metals, the primary source materials as of the 1990s are still crude metalliferous ores. For some metals however, recycled materials contribute significantly to total metal production. For example, in the United States the recycling (qv) rate of all-aluminum used beverage cans is over 50%. For an energy-intensive metal such as aluminum, this represents a substantial energy saving. Recycled aluminum requires only 5% of the energy needed to make aluminum from bauxite ore.

Metallurgy includes not only the treatment of crude ore and scrap, but also the processing of intermediates, ie, concentrates, and wastes such as slags, tailings, etc, for contained metal values.

Definitions

The field of metallurgy has a unique and frequently very specialize vocabulary. Understanding this language helps to clarify certain concepts and processing steps. Definitions of key terms follow.

To *concentrate* is to take an action to intensify in strength or purity by the removal of valueless or unneeded constituents.

Electrometallurgy covers the various electrical processes for the working metals.

Flotation is the method of mineral separation in which a froth created in water by a variety of reagents floats some finely crushed minerals.

Gangue consists of the undesired minerals associated with ore, mostly nonmetallic.

Hydrometallurgy refers to the treatment of ores, concentrates, and other metal-bearing materials by wet processes.

Leaching is the extracting of a soluble metallic compound from an ore by selectively dissolving it in a suitable solvent.

A *mineral* is an inorganic substance occurring in nature and having a definite chemical composition or a characteristic range of chemical composition, and distinctive physical properties or molecular structure.

Mineral dressing is the physical or chemical concentration of raw ore into a product form which a metal can be recovered at a profit.

Ore is a mineral or aggregate of minerals from which a valuable constituent, expecially a metal, can be profitably extracted.

Pyrometallurgy is metallurgy involved in winning and refining metals where heat is used, as in roasting and smelting.

Rosating is the heating of solids, frequently to promote a reaction with a gaseous constituent in the furnace atmosphere.

Smelting is any metallurgical operation in which metal is separated by fusion from those impurities with which it may be chemically combined or physically mixed. Scores of entries in this encyclopedia deal directly or indirectly with various aspects of metallurgy. Check the alphabetical index. In addition to each of the individual metals (aluminum, cadmium, iron, vanadium, etc.) described in separate articles in this encyclopedia, also check the following key words and phrases: alloys; amalgam; annealing; brazing; brittle fracture; calorizing (and other metal treating processes); corrosion; creep; metallography; phase diagram; powder metallurgy; temper; welding; wire drawing; wrought iron; etc.

Additional Reading

Gupta, C.K., and N. Krishnamurthy: *Extractive Metallurgy of Rare Earths*, CRC Press, LLC., Boca Raton, FL, 2004.

Hosford, W.: *Physical Metallurgy*, CRC Press, LLC., Boca Raton, FL, 2005.

Russel, A., and K. Loong Lee: *Structure Property Relations in Nonferrous Metals*, John Wiley & Sons, Inc., Hoboken, NJ, 2005.

Seetharaman, S.: *Fundamentals of Metallurgy*, CRC Press, LLC., Boca Raton, FL, 2005.

Walsh, R.A., and D. Cormier: *McGraw-Hill Machining and Metalworking Handbook*, 3rd Edition, The McGraw-Hill Companies, Inc., New York, NY, 2005.

METAL-MARK (*Insecta, Lepidoptera*). Butterflies of small or moderate size, in many species marked with metallic spots and dashes. They constitute the family *Riodinidae* (*Erycinidae*). Relatively few species occur in temperature climates but in the tropics they are numerous and varied.

METALS (The). In terms of classification, several of the chemical elements are referred to as metals, principally because of the metallic qualities they exhibit. The new group designating system is used here. There are several subclassifications:

Group 11: In order of increasing atomic number, these are copper, gold, and silver. Sometimes, these metals also are referred to as noble metals, principally because they sometimes occur in nature in elemental form. Gold and silver also are frequently referred to as "coinage" metals. The elements of this group are characterized by the presence of one electron in an outer shell. Although copper and gold also have other valences, all of the elements in this group have a 1+ valence in common.

Group 12: In order of increasing atomic number, these are zinc, cadmium, and mercury. The elements of this group are characterized by the presence of two electrons in an outer shell. Although mercury also has a valence of 1+, all of the elements in this group have a 2+ valence in common.

Group 4: In order of increasing atomic number, these are titanium, zirconium, and hafnium. The elements of this group are characterized by the presence of two electrons in an outer shell. Although titanium and zirconium also have other valences, all of the elements in this group have a 4+ valence in common.

Group 5: In order of increasing atomic number, these are vanadium, niobium (sometimes called columbium), and tantalum. Vanadium and tantalum have two electrons in an outer shell; niobium has one electron in its outer shell. Although niobium and vanadium also have other valences, all of the elements in this group have a 5+ valence in common.

Group 6: In order of increasing atomic number, these are chromium, molybdenum, and tungsten. Chromium and molybdenum have one electron in their outer shells; tungsten has two electrons in its outer shell. Although chromium and molybdenum also have other valences, all of the elements in this group have a 6+ valence in common.

Group 7: In order of increasing atomic number, these are manganese, technetium, and rhenium. Manganese and rhenium have two electrons in their outer shells; technetium has one electron in its outer shell. Although manganese and rhenium also have other valences, all of the elements in this group have a 7+ valence in common.

Groups 8, 9, 10: In order of increasing atomic number, these are iron, cobalt, nickel, ruthenium, rhodium, palladium, osmium, iridium, and platinum. Ruthenium, rhodium, and platinum have one electron in their outer shells; iron, osmium, cobalt, and nickel have two electrons in their outer shells; iridium has 17 outer electrons and palladium 18 outer electrons. Although all of these elements fall into one group, they appear in the classification in three subgroupings (hence the sometimes-used term *triads*): (1) iron, cobalt, and nickel each have valences of 2+ and 3+; (2) ruthenium, rhodium, and palladium each have valences of 4+, in addition to other valences; (3) osmium, iridium, and palladium each have valences of 4+, in addition to other valences.

In terms of the periodic classification, all elements here designated as the metals fall between highly alkaline elements (alkali metals and alkaline earths) at the left end of the table and the acidic elements, ending with the halogens at the right end of the table. Thus, the term "transition elements" sometimes is used to describe these in-between elements. Actually, the term transition can be applied to the differences between any series of elements within the overall classification, or between individual elements within a group—because of the gradual alteration in chemical behavior that takes place between groups and between elements.

METAMERE. A division of the animal body occurring as one of a series along the principal axis. Segments of this type are well developed in the earthworms. As a general rule, each segment of such a body contains similar internal organs but a certain amount of specialization of the segments is evident both internally and externally in different parts of the body.

METAMORPHIC ROCKS. Metamorphism of rock material, whatever its original nature and origin, may produce such profound changes that the resulting mass has distinctive characters which warrant a new classification. The chief agents of metamorphism are pressure and heat, and circulating liquids and gases, and usually a long time interval is required for their operation. Metamorphic changes affect mineral composition, texture, and structure of rocks. Minerals that compose most rocks are definite chemical combinations, which commonly are stable only under definite conditions. If these minerals are subjected to radically new conditions, they will tend to change slowly into new chemical combinations, stable under the new conditions. For example, bituminous coal that has been subjected to long-continued pressure and increased temperature in a belt of deformation becomes changed into anthracite coal through loss of volatile constituents and concentration of fixed carbon.

METAMORPHOSIS. Development of the individual after birth or hatching, involving marked change in form as well as growth and differentiation.

Metamorphosis usually accompanies change of habitat or of habits. In some forms, however, it is merely development through a succession of forms, which probably represent ancestral stages in the evolution of the species. The first type is illustrated by many insects and by amphibians.

Immature dragon flies are aquatic although the adults are flying insects, and frogs undergo a transition from the aquatic tadpole to the air-breathing, if not entirely *terrestrial*, adult. Change of habits is illustrated by the transformation of free-swimming young of many aquatic invertebrates into sessile adults, and by the development of adult butterflies and moths with sucking mouths from caterpillars which eat solid food. The crustaceans afford many examples of transformation through several immature forms without conspicuous change of habits or of habitat.

The immature stages are wholly or partly designated by the term larva. In the complex metamorphosis of insects, however, only the first stage is called the larva and even it sometimes bears a different name. The distinction depends on the nature of the metamorphosis.

Some insects hatch from the egg with the general form of the adult and the attainment of the adult stage is marked chiefly by the completion of the wings. This type of metamorphosis is said to be gradual and in its early stages the insect is called a nymph. The orders that develop in this way are grouped as the Paurometabola. A few orders are aquatic in early life and are then called naiads. They transform directly into the terrestrial adult and are known as the Hemimetabola, or insects with incomplete metamorphosis. The Holometabola, with complete metamorphosis, pass through a larval stage, then enter an inactive stage known as the pupa, and finally become the conspicuously different adult. A few beetles undergo hypermetamorphosis with a sequence of different larval forms preceding pupation.

METASTABLE NUCLEI.   Nuclei in excited nuclear states that have measurable lifetimes (exceeding 10^{-10}–10^{-9} second).

METASTABLE STATE.   Three common uses of this term denote: (1) A peculiar state of pseudo-equilibrium, in which the system has acquired energy beyond that for its most stable state, yet has not been rendered unstable. Thus, by using great care, water at 760 millimeters pressure may be heated several degrees above its normal boiling point, say to 105 °C, yet not boil. In this condition it has received heat energy beyond that normally required for liquid-vapor equilibrium, energy which it might be expected to release by spontaneously exploding into steam; and only a slight disturbance will precipitate that change, but the disturbance must come from some external source.

(2) The term has been used in atomic physics for various excited states, but its most general usage today is for an excited state from which all possible quantum transitions to lower states are forbidden transitions by the appropriate selection rules.

(3) In nuclear physics, the term is used to denote the states in which metastable nuclei are found.

METASTABLE SYSTEM.   A system that has a measurable lifetime in an energy state that is not its lowest.

METASTASIS.   See **Cancer and Oncology**.

METCHNIKOFF, ELIE (IIYA) (1845–1916).   Metchnifoff was a Russian biologist who discovered phagocytes, developed a cellular theory of immunity, demonstrated the cells of acute inflammation and studied ageing.

Metchnikoff, also spelled Mechnikov, was born at Ivanovska, Kharkov Province, Russia in 1845. The son of an officer in the Imperial Guard, he attended the University of Kharkov, graduating in 1864.

He was fascinated by science even from childhood, translating a German textbook of physics at the age of 15 and submitting for publication his first article on zoological research at 17. At 18, his second zoological paper appeared in translation in a British microscopy journal. He studied with prominent German zoologists and anatomists as a graduate student, but received his master's degree at the University of St Petersburg in 1867. *The Origin of Species*, by Charles Darwin, acquired during a visit to Germany, exerted a profound influence on his scientific career. He engaged in zoological studies under Tschelkoff at Kharkov. Metchnikoff was greatly influenced by reading Müller's book *Für Darwin*, which purportedly confirmed Darwin's theory by using embryonic structures to trace genealogies. His research during this period demonstrated alternation of sexual and asexual generations in nematodes. Metchnikoff considered the organism to be intrinsically disharmonious and its biology based

on integration of disharmonious elements by active processes. He wrote extensively about this throughout his life and applied it to social and natural philosophy. Metchnikoff and his colleague, Kowalevsky, conducted embryological investigations in Göttingen (1865) and completed them in Naples (1867) before returning to St Petersburg for their doctorates (1868), where they shared the prestigious von Baer prize.

After pursuing studies in Germany, Metchnikoff returned to Russia, becoming Professor of Zoology and Comparative Anatomy at the University of Odessa (1870), and married. He was plagued with eye trouble and bouts of depression and lost his young wife to tuberculosis despite his devoted efforts to save her life. His second marriage, to a much younger neighbor, was happy and the world is indebted to her for a biography of her famous husband.

Metchnikoff related his embryological research to evolution. Besides his extensive investigations (1877) on intracellular digestion of invertebrates, he sought the basic theoretical principles of comparative embryology. This led to his interest in inflammation and infectious disease. By tracing mesodermal amoeboid cell function through the phyla, he merged his studies in embryology with his research on inflammation. Inheriting enough money in 1882 to render him financially independent, he moved to Messina, Italy, where he continued his research as he was not temperamentally suited to a teaching career. While there he observed the digestive powers of the mesodermal cells in transparent larvae of starfish and connected the process with the cause of immunity against infectious disease. He returned to Odessa and made further studies on phagocytosis in the water flea, *Daphnia*, infected by a parasite.

He was appointed Director of the Bacteriological Institute in Odessa in 1886 where he was charged with the manufacture of vaccines against rabies, fowl cholera and other diseases. Disagreements with the clinical staff and disputes with members of the local medical society and conservative newspapers together with other problems caused him to move to St Petersburg, but the situation was no better. Thereafter, he turned his eyes to Western Europe, first visiting the Hygiene Institute of Robert Koch, who showed no interest. Fortunately, Louis Pasteur invited him to join his new Institute, which was under construction. His familiarity with Pasteur's work on silkworm diseases and microbial interactions attracted Metchnikoff to accept the post in Paris. He began in a relatively small laboratory with a few Russian students but his activities soon increased. See also **Koch, Heinrich Hermann Robert (1843–1910)**.

He demonstrated phagocytes in higher animals, finding that many of the white blood cells or leucocytes are phagocytic, defending the body against acute infection by engulfing invading microorganisms. Metchnikoff related phagocytes to active defense against infectious disease agents which led to a revolutionary concept of cellular immunity. The ensuing debate between the proponents of humoral immunity and champions of the new cellular immunity theory stimulated much valuable research. Metchnikoff spent the remainder of his scientific career at the Institut Pasteur, becoming a Sub-director and Chief of Service. He shared the Nobel Prize in Medicine or Physiology in 1908 with Paul Ehrlich for his work in immunology and made many more contributions to immunity and bacteriology. He authored several important books including one entitled *Immunity in Infective Diseases* and another on the *Comparative Pathology of Inflammation*. Metchnikoff's phagocytic cell theory of immunity led Sir Almroth Wright and others to develop vaccine therapy. His research between 1892 and 1901 established the cells and function of the reticuloendothelial system. In 1895, he demonstrated the *in vitro* lysis of cholera vibrios, which Pfeiffer had shown *in vivo* in immunized guinea-pigs. Together with Roux (1903–1904), he successfully inoculated higher apes with syphilis. See also **Wright, Almroth Edward (1861–1947)**.

Metchnikoff extrapolated his studies in embryology, and subsequently immunology, to broad concepts of the biological underpinnings of ageing, human behaviour and organization. These views were expressed in his books entitled *The Nature of Man* and *The Prolongation of Life*.

He became interested in longevity and ageing in his later years and advocated the consumption of yogurt. He believed that lactic acid producing bacteria in the gut prolonged lifespan. His views on intestinal autointoxication aroused much interest and notoriety.

In addition to immunology and microbiology, Metchnikoff had eclectic interests that included research on zoology, embryology, pathology, microbial pesticides, epidemiology, physical anthropology, chemotherapy and gerontology. He attracted wide media attention, especially with his pronouncements on health.

At the Institut Pasteur, students, scholars, musicians, actors, philosophers and statesmen often visited his laboratory. He was a friendly and generous individual and was liberal in his views on education and social issues championing the educational rights of women.

He devoted much of his subsequent career to elaborating and championing his cellular theories of immunity. He attracted a number of students to his laboratory, including Jules Bordet, and published many papers and a series of lectures on the comparative pathology of inflammation. A colorful and influential personality in the early attempts to understand the mechanisms of immunity, he had a long, happy and distinguished career at the Institut Pasteur, where he died in 1916.

J.M. CRUSE
R.H. LEWIS
University of Mississippi Medical Center, Jackson, MS

METEOR 3M-1 (SAGE III) MISSION.   See **Earth Observing System (EOS)**.

METEOROGRAPH.   See **Meteorology**.

METEOROIDS AND METEORITES.   A *meteoroid* is one of countless small, solid particles existing in interplanetary space, which, when encountering the Earth's atmosphere, burn up. During the short instant of ablation, a brilliant track of light often can be observed, commonly called "falling star," "shooting star," or "fireball." When ablation is accompanied by an explosion, the meteoroid may be called a bolide. Sometimes a loud detonation following the explosion may be heard. Considering the large number of meteoroids entering the Earth's atmosphere, the *bolides* are rare. The brilliance of a meteoroid trail normally does not exceed that of zero stellar magnitude, i.e., no brighter than the planet Jupiter. Most debris from meteoroids ultimately falls to the Earth as dust. Sometimes meteoroids enter the Earth's atmosphere in swarms at given times each year when the Earth is passing an area of space where there is a high population of meteoroids. An example is the Leonid Meteor Show, which occurs in November each year. See also **Leonids**.

A *meteorite* is that portion of a meteoroid that survives passage through the Earth's atmosphere and lands on or penetrates the surface of the Earth. Many thousands of meteorites have been identified over several decades. These represent but a small portion of the total number estimated to have struck the Earth. Very little research has been conducted to determine at what average rate meteorites may strike the Earth.[1] Meteorites may strike the earth as isolated instances, or they may fall in swarms. Evidence of several very large swarms has been recorded. The Tungusta fall in Russia in 1908 leveled trees within a radius of nearly 48 kilometers (30 miles). In 1947, also in Russia, the Sikhote-Aline fall caused heavy damage. In the Holbrook, Arizona area, a meteorite shower (1912) produced more than 20,000 stones, many of which were recovered. Large meteorites can produce immense craters, as exemplified by the large Barringer meteorite crater in Arizona. See Fig. 1.

Information returned from planet-orbiting satellites has shown extensive cratering on Mercury, Mars, and several other planets and their moons. Telescopic observation of craters on the Earth's moon were first made a few centuries ago. The comparative absence of large craters on earth is most likely due to the earth's atmosphere, which causes meteoroids to "burn up" before they can impact the earth's crust, but other factors also contribute. Among these factors is the location of the earth within the solar system, which may be unfavorable insofar as the potential for impacts is concerned. The geological and tectonic history of the earth may have contributed to covering over or otherwise obliterating the existence of craters caused during the earlier periods of the developing earth. See also **Astrobleme**.

[1] In 1984, Halliday, Blackwell, and Griffin (Herzberg Institute of Astrophysics, National Research Council of Canada) conducted a study involving a network of 60 cameras in western Canada to determine the influx rate of meteorites. By way of calculations too complex to describe here, the researchers derived the following figures: 19,000 impacts per year of meteorites having a mass of about 0.1 kg; 4100 impacts (1 kg); and 830 impacts (10 kg). See reference listed.

Fig. 1. Barringer meteorite crater located near Canyon Diablo in northern Arizona. It is a classic example of a huge meteor crater, approximately three-fourths of a mile (1.2 km) in diameter and more than 600 feet (183 m) deep. (*Meteor Crater Society.*)

Meteorites found on earth range in size from that of a pea (and probably smaller, considering the difficulty in noting them) up to tons. The Hoba West meteorite, discovered in South Africa, weighs 70 tons (63 metric tons) and measures about 3×3 meters (10×10 feet). The ahnighito meteorite, one of the largest, was found in Greenland and brought to the American Museum of Natural History (New York) by Admiral Peary many years ago. A large meteorite weighing about 1050 kilograms (2300 pounds) fell on Kansas in 1948. In addition the large meteorite crater in northern Arizona, previously mentioned, the Ries Basin in Germany has a notable meteorite crater, originally measuring about 27 kilometers (17 miles) in diameter, and estimated to have been formed between 15 and 20 million years ago.

There is no really characteristic shape to meteorites except that many look like stones and could be casually observed without comment. Many fragment during passage through the atmosphere, producing a variety of shapes. Some meteorites exhibit intense sculpturing caused by the intense heat when the object passes through the earth's atmosphere.

Many larger and much older craters have been located in recent years. In particular, the Chubb Crater in northern Canada is an example of an old, filled-in, meteor impact crater.

The 2-ton (1.8-metric-ton) Allende meteorite, which fell in northern Mexico in 1969, has been of particular fascination to scientists. From these studies, several concepts pertaining to the origin of the solar system have evolved. Central to most of these studies of the Allende and other meteorites is the determination of isotopic ratios of various elements, such as oxygen-16 to oxygen-18; of magnesium-26 to magnesium-25 to magnesium-24.

It has been estimated that meteorites impact the earth's surface with a velocity ranging from about 11 kilometers (7 miles) per second up to a theoretical velocity (maximum if object arrives from within the solar system) of nearly 72 kilometers (45 miles) per second. It is believed that the average velocity, however, is about 16 kilometers (10 miles) per second.

When Antarctic exploration revealed the presence of numerous meteorites, this broadened the database for scientists who are pursuing this special discipline. A connection with oceanographic exploration may yield additional rich evidence of meteor impacts and craters. Such studies, of course, also interlock with the scientific exploration of past mass extinctions on Earth.

In 1990, researchers with the Geological Survey of Canada serendipitously located an exceptionally large crater buried beneath the waters of Lake Huron at a point that straddles the Canadian–United States border. The researchers captured an excellent magnetic image of what may be largest meteorite crater on record. It has been estimated that a hole the size of the "Can-Am" would have required a projectile about 3 miles (4.5 kilometers) in diameter. In observing this discovery, J. Melosh

(University of Arizona) commented, "We would expect 100 craters this size or greater during the last 500 million years. We know of about six." Melosh further observed that impact data to date have been too scattered to prove any correlation between the collisions and species extinctions. In summary, Melosh observed, "With more craters like Can-Am to study, we can test those correlations."

Over a period of several years, some scientists have expected to find a "killer" crater near Cuba. A number of geologists visited the suspected southwestern corner of Cuba and reported at the annual meeting of the Geological Society of America that investigators may best conduct their research elsewhere. However, some scientists remain unconvinced that a massive crater does not exist in the Caribbean region.

Classes of Meteorites

There are three broad classes of meteorites that strike the Earth:[2]

1. *Iron meteorites* (sometimes called siderites) account for about 6% of the meteorites found. These objects are essentially a nickel-iron alloy containing from 4% to 20% Ni (rarely greater), together with several accessory minerals, of which troilite (FeS), schreiberside (Fe, Ni, Co)$_3$P, and graphites are important. Iron meteorites tend to weather more rapidly than other types once they impact the earth. The cross section of an iron meteorite exhibits a thin layer of vitreous material, caused by fusing of the substance during entry, which may be up to about 2 millimeters ($\frac{1}{16}$ inch) in thickness. Iron meteorites, when viewed sectionally and particularly after they have been ground flat, polished, and etched, will show intersecting bands which are known as Widmanstatten lines. The Widmanstatten structure has been defined as a triangular pattern of iron meteorites composed of parallel bands or plates of kamacite bordered by taenite and intersecting one another in two, three, or four directions. The kamacite bands, arranged parallel to the octahedral planes in the host taenite, are produced by exsolution from an originally homogeneous taenite crystal. As the bands become finer (thinner), the nickel content increases. The structure is named after A.B. Widmanstatten, the Austrian mineralogist who discovered the structure in 1808.

2. *Stony-iron meteorites* (sometimes called siderolites) account for 1% to 2% of the meteorites found. These objects contain at least 25% and approximately equal amounts (by weight) of nickel-iron and heavy basic silicates, such as pyroxene and olivine. Geological and geochemical analysis indicates that these meteorites formed inside parent bodies resembling asteroids, typically some hundreds of kilometers in diameter, about 4.5 billion years ago. Some scientists consider them to be fragments of these bodies broken apart by collisions among themselves during the formation of the solar system. Further evidence indicates that they are related to the asteroids observed today. Studies show that their spectra resemble those of certain asteroids, especially the Apollos, whose orbits cross that of the earth and theoretically could at some time intersect the Earth. Orbits observed for two or three meteorites resemble Apollo-type orbits and reach into or across the asteroid belt, suggesting that some of these bodies may have somehow been deflected out of the belt, eventually falling to Earth. See also **Asteroid**.

3. *Stony meteorites* constitute over 90% of the meteorites known to strike Earth. These objects consist largely or entirely of silicate minerals, chiefly olivine, pyroxene, and plagioclase. Stony meteorites are sometimes called *chondrites*. See also Chondrite. Most of the meteorite material has been modified from primitive chemical forms by geological processes, such as heating, exposure to pressure and radiation; some have been entirely melted and resolidified. Approximately 6% of the stony meteorites are *carbonaceous chondrites*. They are characterized by the presence of hydrated, clay-type silicate minerals (usually fine-grained serpentine or chlorite); by considerable amounts and a great variety of organic compounds (hydrocarbons, fatty and aromatic acids, porphyrins, etc.) believed to be of extraterrestrial origin; by an absence or almost total absence of free nickel-iron; and by abnormally high contents of inert gases, particularly xenon. Much of the organic matter is a black insoluble complex of compounds of high molecular weight; the water content (usually water of hydration) may be up to 20% (weight). Carbonaceous chondrites are

grouped in three types, each characterized by the amount of organic material they contain and by other compositional features:

Type I   contain the greatest amount of water and organic matter (3% of 5% combined carbon; 24% to 30% ignition loss)

Type II   are of an intermediate composition (12% to 24% ignition loss)

Type III contain high-temperature minerals and some metallic components (2% to 12% ignition loss)

Possibly such objects may represent surviving material from the matter that formed in the ancient solar system and aggregated into planets.

Nearly all meteorites contain small amounts of *troilite* (FeS). It is a variety of pyrrhotite with almost no ferrous-iron deficiency.

Diamond in Meteorites. S.S. Russell (Oxford University) and C.T. Pillinger reported in 1992 that the presence of diamond in meteorites has been known for at least a century. Scientists have theorized that the mechanism for diamond formation in meteorites probably does not compare with the manner in which diamond was formed on Earth, but that meteoritic diamond probably is associated with *shock processing* — that is, upon impact. Analysis of material from the iron meteorite crater in Canyon Diablo, Arizona, tends to confirm this hypothesis.

In contrast, however, diamond polymorphs found in an Antarctic iron meteorite shows no evidence of terrestrial shock. This, then, would suggest that the meteoritic diamond had been present prior to impact. An alternative mechanism, chemical vapor deposition (CVD), also has been suggested. In studying a particular meteorite (Abbe, an enstatite chondrite), Russell and co-researchers observe, "Because the Abee diamonds have typical solar system isotopic compositions for carbon, nitrogen, and xenon, they are presumably nebular in origin rather than presolar. Their discovery in an unshocked meteorite eliminates the possibility of origins normally invoked to account for diamonds in ureilites and iron meteorites and suggests a low-pressure synthesis. The diamond crystals are "100 nanometers in size, of an unusual lath shape, and represent "100 ppm of Abee by mass."

Analysis of Meteoritic Materials. As pointed out in 1989 by R. Zenobi (Stanford University) and a research team, "Meteorites are highly heterogenmous mineralogically and chemically, even on a submillimeter scale. Techniques for elemental analysis are well established for the investigation of microgram amounts of solids and the investigation of surfaces with micrometer-scale resolution. In contrast, analysis of organic molecules has been impossible even in a few milligrams of carbonaceous chondrites, the meteorite group with the highest carbon content." Reasons stated for this difficulty in analysis include: (1) very low concentrations of organic materials (ppm range) and (2) the large number of different organic compounds found in meteorites. Usually, multiple extraction and purification must precede analysis. The variety of organic compounds is exemplified by naphthalene, 2-methylnapthalene, 2,3-dimethylnaphthalene, phenanthrene, anthracene, 2-methylanthracene, 9-methylantracene, 2-tertbutylacene, fluorene, fluoranthene, m poyrene, perylene, and coronene.

In analyzing the Allende meteorite, the aforementioned team, the technique used was that of two-step laser desorption spectrometry. The analysis of each meteorite, of course, presents special problems, and thus numerous meteorite analytical techniques have been devised and customized for the problem.

SNC Meteorites. In addition to the three main classes of meteorites just described, a relatively new group (very small in number at present) is attracting considerable scientific attention, mainly because they are considered of lunar or Martian origin. Currently attention is being given to about eight-odd rocks that have been collected around the world. Three of these rocks are known to be meteorites and they had been named Shergotty, Nakhla, and Chassigny. Thus, the acronym is SNC.

These meteorites are youthful geologically speaking (1.3 billion years old), whereas 4.0 billion years is considered a minimum age for other meteorites. Thus far, the age determination tends to disqualify them as coming from an asteroid, although not all experts agree. The SNC meteorites resemble certain basaltic rocks found on Earth, also tending to disqualify asteroids as a source. Further, their features, including their oxygen isotope composition, indicate that indeed they are meteorites and not Earth rocks.

In 1992, H.R. Karlsson (NASA) and a team of researchers reported that recent evidence suggests that the planet Mars once had a water-rich atmosphere and flowing water on its surface. The paper observes,

---

[2] Some of the definition material is from the "Glossary of Geology," American Geological Institute, Washington, DC (by permission).

"The SNC meteorites, purportedly of Martian origin, contain 0.04 to 0.4 percent water by weight. Oxygen isotopic analysis can be used to determine whether this water is extraterrestrial or terrestrial. Such analysis reveals that a portion of the water is extraterrestrial and furthermore was not in oxygen isotopic equilibrium with the host rock. Lack of equilibrium between water and host rock implies that the lithosphere and hydrosphere of the SNC parent body formed two distinct oxygen isotopic reservoirs. If Mars was the parent body, the maintenance of two distinct reservoirs may result from the absence of plate tectonics on the planet." See also **Mars**.

Antarctic Meteorites. A meteorite, now called Allan Hills 81005 (only 31 grams and about 3 cm across), was found in the Antarctic region. From studies of lunar rocks returned by the *Apollo* mission, experts felt from the start that the meteorite was of lunar origin even though the concept of lunar meteorites had not been popular for a number of years. When the meteorite's oxygen isotope composition was checked, it was found to be like that of moon rocks and unlike most meteorites. But, upon further analyses and hypothesizing, lunar origin was questioned. One problem was that of timing. The age of the meteorite is about 2 billion years more recent than when the moon was volcanically active. Further, there were suspicions about certain characteristics which indicated that the object had to come from a body of sufficient size to permit the kind of melting and crystallization processes that produced Earth's basalt lavas.

J.T. Wasson (University of California, Los Angeles) reported in 1990, "Eighty-five percent of the iron meteorites collected *outside* Antarctica are assigned to 13 compositionally and structurally defined groups; the remaining 15 percent are ungrouped. Of the 31 iron meteorites recovered from Antarctica, 39 percent are ungrouped. This major difference in the two sets is almost certainly not a stochastic variation, a latitudinal effect, or an effect associated with differences in terrestrial ages." Wasson suggests that, during impacts on asteroids, smaller fragments tend to be ejected into space at higher velocities than larger fragments, and thus, on the average, small meteroids undergo more changes in orbital velocity than the larger ones. As a result, the set of asteroids that contributes small meteoroids to Earth-crossing orbits is larger than the set that contributes larger meteroids. It is suggested that most small iron meteorites may escape from the asteroid belt as a result of impact-induced changes in velocity that reduce their perihelia to values less than the aphelion of Mars.

A.J.T. Jull (University of Arizona) and a team of researchers reported in 1988 that weathering products have been observed on many Antarctic meteorites, but their consequences for cosmochemical analysis have remained poorly understood. The time of formation has not been quantitatively evaluated. The research team observed, "Weathering products are typically in the form of 'rust' or hydrated iron oxides and hydroxides; however, sulfate and carbonates have also been observed. Most workers have presumed that weathering proceeds slowly in the cold, dry Antarctic environment. Most Antarctic meteorites have been exposed to terrestrial conditions for 10^4 to 10^5 years and small increments of weathering can accumulate into measurable effects. Many specimens, however, are remarkably well preserved."

The Jull team reported, in particular, on studies of the Antarctic meteorite LEW 85320 (H5 chondrite). The research paper concludes that "Results from carbon-14 dating suggest that, although the meteorite has been in Antarctica for at least 3.2×10^4 to 3.3×10^4 years, nesquehonite[3] formed after A.D. 1950."

Meteoric Communications. It has been estimated that tiny meteors enter Earth's atmosphere, producing an ionized residue at the rate of over a billion events per day. The practical use of this phenomenon as a communications medium for bounding very high frequency radio signals has been under investigation during the past 15–20 years. A number of firms are commercializing the method, which has much appeal for trucking firms. The system competes directly with satellite communications for such purposes. The Conservation Service of the U.S. Department of Agriculture has used meteoric communications for monitoring snowpack and other weather data collected by instruments in remote areas. The system also has been used in connection with military communications.

Ionospheric Interference. In late 1989, P. Kaufmann (Universidade de Sao Paulo) reported on the findings of a study of the effects that the large June 1975 meteoroid storm had on communications worldwide. A brief summary of the report: "The June 1975 meteoroid storm detected on the moon by the Apollo seismometers was the largest ever observed. Reexamination of radio data taken at that time showed that the storm also produced pronounced disturbances on Earth, which were recorded as unique phase anomalies on very low frequency (VLF) radio propagation paths in the low terrestrial ionosphere. Persistent effects were observed for the major storm period (20–30 June 1975), including reductions in the diurnal phase variation, advances in the nighttime and daytime phase levels, and reductions in the sunset phase delay rate. Large nighttime phase advances, lasting a few hours, were detected on some days at all VLF transmission, and for the shorter propagation path, they were comparable to solar Lyman alpha daytime ionization. Ion production rates attributable to the meteor storm were estimated to be about 0.6 to 3.0 ions per centimeter cubed per second at the E and D regions, respectively. The storm was a sporadic one with a radiant (that is, the point of apparent origin in the sky) located in the Southern Hemisphere, with a right ascension 1 to 2 hours larger than the sun's right ascension.

K. Brecher (Boston University) and J. Crouchley (University of Queensland, Australia) corroborated in the study.

Additional Reading

Benoit, P.H. and D.W.G. Sears: "The Breakup of a Meteorite Parent Body and the Delivery of Meteorites to Earth," *Science*, 1685 (March 27, 1992).

Eugster, O.: "History of Meteorites from the Moon Collected in Antarctica," *Science*, 1197 (September 15, 1989).

Flam, F.: "Seeing Stars in a Handful of Dust," *Science*, 580 (July 26, 1991).

Halliday, I. et al.: "The Frequency of Meteorite Falls on the Earth," *Science*, **223**, 1405–1407 (1984).

Hodge, P.: *Meteorite Craters and Impact Structure*, Cambridge University Press, New York, NY, 1994.

Hoon, J. et al.: "Application of Two-Step Laser Mass Spectrometry to Cosmogeochemistry: Direct Analysis of Meteorites," *Science*, 1523 (March 25, 1988).

Horan, M.F. et al.: "Rhenium-Osmium Isotope Constraints on the Age of Iron Meteorites," *Science*, 1118 (February 28, 1992).

Hutchinson, R. et al.: *Catalogue of Meteorites*, Cambridge University Press, New York, NY, 1999.

Jull, A.J. et al.: "Rapid Growth of Magnesium-Carbonate Weathering Products in a Stony Meteorite from Antarctica," *Science*, 417 (October 21, 1988).

Jurewicz, A.J.G., D.W. Mittlefehldt, and J.H. Jones: "Partial Melting of the Allende (CV3) Meteorite: Implications for Origins of Basaltic Meteorites," *Science*, 695 (May 3, 1991).

Karlsson, H.R. et al.: "Water in SNC Meteorites: Evidence for a Martian Hydrosphere," *Science*, 1409 (March 13, 1992).

Kaufmann, P. et al.: "Effects of the Large June 1975 Meteoroid Storm on Earth's Ionosphere," *Science*, 787 (November 11, 1989).

Lowe, D.R. et al.: "Geological and Geochemical Record of 3400-Million-Year-Old Terrestrial Meteorite Impacts," *Science*, 959 (September 1, 1989).

Margolis, S.V., P. Claeys, and F.T. Kyte: "Microtektites, Microkrystites, and Spinels from a Late Pliocene Asteroid Impact in the Southern Ocean," *Science*, 1594 (March 29, 1991).

Mark, K.: *Meteorite Craters*, University of Arizona Press, Phoenix, AZ, 1995.

McSween, H.Y. Jr.: *Meteorites and Their Parent Planets*, Cambridge University Press, New York, NY, 1987.

Melosh, H.J.: *Impact Cratering: A Geologic Process*, Oxford University Press, Inc., New York, NY, 1996.

Russell, S.S., J.W. Arden, and C.T. Pillinger: "Evidence for Multiple Sources of Diamond from Primitive Chondrites," *Science*, 1188 (November 22, 1991).

Russell, S.S. et al.: "A New Type of Meteoritic Diamond in the Enstatite Chondrite Abee," *Science*, 206 (April 10, 1992).

Stix, G.: "Meteoric Messages," *Sci. Amer.*, 167 (September 1990).

Stolzenburg, W.: "Impact Crater May Lie Beneath Lake Huron," *Science News*, 133 (September 1, 1990).

Walker, R.J. and J.W. Morgan: "Rhenium-Osmium Isotope Systematics of Carbonaceous Chondrites," *Science*, 519 (January 27, 1989).

Wasson, J.T.: "Ungrouped Iron Meteorites in Antarctica: Origin of Anomalously High Abundance," *Science*, 900 (August 24, 1990).

Weaver, K.F.: "Meteorites," *National Geographic*, 390 (September 1986).

Zenobi, R. et al.: "Spatially Resolved Organic Analysis of the Allende Meteorite," *Science*, 1026 (November 24, 1989).

Web Reference

Ames research Center, National Aeronautics and Space Agency, United States. http://www.arc.nasa.gov/index.html

---

[3] A hydrogen magnesium carbonate that occurs as a weathering product on the surface of Antarctic meteorite LEW 85320.

METEOROLOGICAL EQUATOR. See **Meteorology**.

METEOROLOGY. The study dealing with the phenomena of the atmosphere. This includes not only the physics, chemistry, and dynamics of the atmosphere, but is extended to include many of the direct effects of the atmosphere upon the earth's surface, the oceans, and life in general. The goals often ascribed to meteorology are the complete understanding, accurate prediction, and artificial control of atmospheric phenomena. In a more restricted sense, meteorology is the science of weather, being particularly concerned with the physics of the elements that make the weather. A distinction can be drawn between meteorology, in this sense, and climatology, the latter being primarily concerned with average, not actual, weather conditions. See also **Bjerknes, Vilhelm Frimann Koren (1862–1951)**.

Meteorology may be subdivided, according to the methods of approach and the applications to human activities, into a large number of specialized sciences. For example, the aeronautical, agricultural and industrial branches of meteorology are concerned with the application of meteorological information and techniques within their respective fields. *Hydrometeorology* is the branch directly concerned with hydrologic problems, particularly flood control, hydroelectric power, irrigation and water resources. *Radio meteorology* (also known as *radioelectric meteorology*) is concerned with the propagation of radio energy through the atmosphere and with the use of radio equipment in meteorological studies; while radar meteorology embraces all meteorological matters connected with radar.

There are specializations in terms of atmospheric regions to which study is devoted. *Aerology* is the study of the free atmosphere throughout its vertical extent, as distinguished from studies confined to the layer of the atmosphere adjacent to the earth's surface. *Aeronomy* is a recently introduced term denoting, basically, the physics of the upper atmosphere. It is concerned with upper-atmospheric composition (i.e., nature of constituents, density, temperature, etc.) and chemical reactions. Studies of atmospheric phenomena are designated from the largest-scale aspects to the smallest, respectively, as *macrometeorology, mesometeorology,* and *micrometeorology*.

Still other specializations include: *synoptic meteorolgy*, the study and analysis of weather information obtained from synoptic reports, charts, and weather observations; *applied meteorology*, the application of current weather data, analyses, and/or forecasts to specific practical problems; *dynamic meteorology*, the study of atmospheric motions as solutions of the fundamental equations of hydrodynamics or other systems of equations appropriate to special situations, as in the statistical theory of turbulence; and *physical meteorology* (also called *atmospheric physics*), dealing with optical, electrical, acoustical, and thermodynamic phenomena of the atmosphere, its chemical composition, the laws of radiation, and the explanation of clouds and precipitation. Mathematical theory of the motions of the atmosphere and the forces responsible falls in the field of dynamic meteorology. Atmospheric thermodynamics lies so near the borderline of physical and dynamical meteorology that it is treated as often in one as in the other branch.

There are numerous entries in this encyclopedia relating directly to meteorology. Check the alphabetical index. Particularly check such topical terms as: **Air Mass; Altimetry; Atmosphere (Earth); Atmosphere-Ocean Interface; Aurora; Barometer; Climate; Clouds and Cloud Formation; Fog and Fog Clearing; Fronts and Storms; Gust Front; Jet Streams; Polar Front Theory; Precipitation and Hydrometeors; Psychrometric Chart; Sky Cover; Visibility; Weather Technology; Winds and Air Movement;** and **Wind Shear**.

Glossary

An abridged glossary of Meteorological terms would include:

Abatement — the lessening or reduction of an atmospheric condition that is considered detrimental to animals, plants, humans, or structures. In air pollution meteorology, abatement refers to a reduction in peak intensity, duration, average concentration, or exposure to those chemicals in the air that are considered pollutants.

Ablation — the removal of snow, ice, or water from a glacier, snow-field, etc.; in this sense, the opposite of accumulation. These processes include melting, evaporation, wind erosion, and an avalanche. Air temperature is the dominant factor in controlling ablation, precipitation amounts exercising only secondary control.

Ablation Area — that portion of a glacier surface below the firn line where ablation exceeds accumulation; the opposite of accumulation area. See also **Ablation (Glaciology)**.

Abrolhos Squalls — (or *abroholos*.) Rain or thundersqualls of the frontal type experienced near the Abrolhos Islands (18°S) off the coast of Brazil from May through August.

Absolute Angular Momentum — in the atmosphere the absolute angular momentum M per unit mass of air is equal to the sum of the angular momentum relative to the earth and the angular momentum due to the rotation of the earth:

$$M = ur\cos\phi + \Omega r^2\cos^2\phi,$$

where r is the distance from the center of the earth to the particle, u the relative eastward component of velocity, φ the latitude, and Ω the angular rotation rate of the earth. Since the earth's atmosphere is shallow, the variable r is often replaced by the constant a, defined as the radius of the earth. The absolute angular momentum per unit mass is then approximated by

$$M = ua\cos\phi + \Omega a^2\cos^2\phi.$$

Absolute Annual Range of Temperature — the difference between the highest and lowest temperature observed at a location.

Absolute Coordinate System — (or absolute reference frame.) an inertial coordinate system that has its origin on the axis of the earth and is fixed with respect to the stars. Thus, any mechanical quantities in meteorology defined with respect to this frame take into account the rotation of the earth. See also **Coriolis Effect**.

Absolute Humidity — the mass of water vapor per unit volume of air containing the water vapor; usually expressed as grams of water vapor per cubic meter of air.

Absolute Instability — property of an ambient air layer that is unstable for both saturated (cloudy) and unsaturated (clear) air parcels.

Absolute Stability — property of an ambient air layer that is stable for both saturated (cloudy) and unsaturated (clear) air parcels.

Absolute Zero — the theoretical temperature at which a body does not emit electromagnetic radiation and all molecular activity ceases (usually some subatomic activity takes place); 0 K. See also **Absolute Zero**.

Absorption — the process whereby a portion of the radiation incident on an object is converted to heat. See also **Absorption Spectrum**.
A further process always results from absorption, that is, the irreversible conversion of the absorbed radiation into some other form of energy within and according to the nature of the absorbing medium. The absorbing medium itself may emit radiation, but only after an energy conversion has occurred. See also **Attenuation**.

Absorption Band — a connected series of closely spaced or overlapping absorption lines. Absorption bands are common features in the absorption spectra of polyatomic gases. See also **Absorption Band**.

Absorption Hygrometer — (sometimes called *chemical hygrometer*.) A type of hygrometer that measures the water vapor content of the atmosphere by means of the absorption of vapor by a hygroscopic chemical. See also **Hygrometry and Psychrometry**.

ACCAS — (usually pronounced ACK-kis) — AltoCumulus CAStellanus; mid-level clouds (bases generally 8 to 15 thousand feet), of which at least a fraction of their upper parts show cumulus-type development. These clouds often are taller than they are wide, giving them a turret-shaped appearance. ACCAS clouds are a sign of instability aloft, and may precede the rapid development of thunderstorms. See also **Clouds and Cloud Formation**.

Acceleration — the rate of change with time of the velocity vector of a particle. See also **Acceleration**.

Acceleration of Gravity — See also **Acceleration (Due to Gravity)**.

Accelerometer — an instrument that measures acceleration. See also **Accelerometer**.

Accessory Cloud — a cloud which is dependent on a larger cloud system for development and continuance. Roll clouds, shelf clouds, and wall clouds are examples of accessory clouds.

Accumulation Area — the portion of the glacier surface above the firn line where the accumulation exceeds ablation; the opposite of ablation area. See Ablation Area.

Accumulation Rain Gauge — (or *accumulative rain gauge*) a particular group of rain gauges in which the precipitation is accumulated over time. The depth of accumulated precipitation can be determined by the level of a float, by weighing, or by manual direct measurement of water depth. For long-term, unattended operation a known amount of liquid that prevents evaporation is placed in the collection container.

Acid Deposition — the accumulation of an acidic chemical from the atmosphere to the surface of the earth, or to plants and structures at the surface. Acids have high concentrations of hydrogen ions when dissolved in water, indicated by a pH less than 7. Acids can corrode metals, dissolve some types of rocks such as limestone, injure plants, and exacerbate some conditions in humans and animals. Acid deposition can occur in two forms: 1) wet deposition including acid rain, acid snow, acid hail, acid dew, acid frost, and acid fog; and 2) dry deposition including fallout of heavy particles, gravitational settling of lighter particles, and interception by and reaction with plant surfaces. Sometimes all forms of acid deposition are loosely called acid rain, although literally acid rain refers only to the liquid form. See also **Acid Rain**.

Actinometer — the general name for any instrument used to measure the intensity of radiant energy, particularly that of the sun. See also **Actinometer**.

Actinometry — the science of measurement of radiation, originally more general, but now used mainly to describe photochemical techniques of measuring ultraviolet radiation by chemical actinometers.

Active Cloud — this category of cumulus cloud distributes air pollutants from the atmospheric boundary layer to the free atmosphere. These clouds have reached their level of free convection, allowing latent heat released during water-vapor condensation to contribute to the positive buoyancy of the cloudy air. These types of clouds are usually produced by thermals, but can eventually decay into passive clouds before disappearing completely. Corresponding morphological species include cumulus mediocris, cumulus congestus, and cumulonimbus. See also **Clouds and Cloud Formation**.

Active Front — a front, or portion thereof, that produces appreciable cloudiness and, usually, precipitation.

Actual Elevation — the vertical distance above mean sea level of the ground at the meteorological station. This term is denoted by the symbol H in international usage.

Adaptation Luminance — (or adaptation brightness; also called *adaptation level, adaptation illuminance, brightness level, field brightness*, and *field luminance*). See also **Adaptation Luminance**.

Adaptive Grid — a grid on which the number or geometric distribution of points changes in response to the characteristics of the evolving flow that is being described. Adaptive grids are most commonly used to place higher resolution in regions where error is likely to be large. Adaptive grids are most commonly used to place higher resolution in regions where error is likely to be large, for instance, in areas where the gradient (or Laplacian operator) is large.

Adfreezing — the process by which one object becomes adhered to another by the binding action of ice.

Adhesion Efficiency — in cloud physics, the fraction of ice particles colliding with a collector ice particle that actually adheres to, or aggregates with, the collector; or, in the case of the riming process, the fraction of water drops colliding with a collector ice particle that actually adheres (freezes) to the collector.

Adhesive Water — water retained by soil constituents as a result of the molecular attraction between the water and the soil.

Adiabatic Atmosphere See also **Adiabatic Atmosphere**.

Adiabatic Process — a process in which a system does not interact with its surroundings by virtue of a temperature difference between them. See also **Adiabatic Process**.

Adiabatic Temperature Changes — Temperature changes related to changes of pressure without external gain or loss of heat. In a compressible fluid, such as seawater, temperature rises as the fluid is compressed and adiabatic cooling occurs during expansion. The latter is of practical concern when water samples are taken in thermally insulated water bottles are taken at great depth and raised to the surface.

Adiabatic Temperature Gradient — the rate of change of temperature due to pressure under adiabatic conditions. In practice, since in the sea the pressure changes can be considered proportional to depth changes, the adiabatic temperature gradient is usually given as rate of change per unit depth, instead of per unit pressure. For practical purposes, the unit of depth is often chosen as 1000 m.

Adsorption Isotherm — a boundary on a phase diagram that expresses the partitioning of a compound between solid and aqueous phases. It is the isothermal equilibrium relationship between the concentration of a compound sorbed to the solid phase and the concentration of the same compound in the aqueous solution in contact with the solid phase.

Advection Fog — 1. A type of fog caused by the advection of moist air over a cold surface, and the consequent cooling of that air to below its dewpoint. A very common advection fog is that caused by moist air over a cold body of water (sea fog).

2. Sometimes applied to steam fog. See also **Advection**; **Dew Point**; and **Fog and Fog Clearing**.

Advective Change of Temperature — the contribution to local temperature change that is caused by the horizontal or vertical advection of air. The horizontal component of change, usually the most important in the troposphere, is proportional to the horizontal temperature gradient and the magnitude of the component of the wind in the direction of the gradient. The vertical component is proportional to the vertical velocity and the static stability and depends also on whether the air is saturated.

Advective-gravity Flows — a type of cold-air downslope flow at the bottom of the boundary layer where the two dominant processes are advection and buoyancy. Once fully developed, this idealized flow is constant in time as long as cold air is supplied or produced. Wind speed increases with the square root of downslope distance, and flow depth increases linearly with distance. Typical depths are 2–10 m (6.5–33 ft). See also **Katabatic Wind**.

Advisory Area — area over which a meteorological advisory forecast applies.

Advisory — statements issued by a weather service that discuss weather situations of inconvenience that do not carry the danger of warning criteria, but, if not observed, could lead to hazardous situations. Some examples include snow advisories stating possible slick streets, or fog advisories for patchy fog condition causing temporary restrictions to visibility. See **Special Weather Statements**.

Aerodynamics — the study of the forces exerted on and the flow around bodies, especially aircraft, moving relative to a gas, especially the atmosphere. Aerodynamics is sometimes used as a synonym for the science of flight. See also **Aerodynamics and Aerostatics**.

Aerograph — in general, any self-recording instrument carried aloft by any means to obtain meteorological data. See **Meteorograph**.

Aerologation — (also called *single-grid heading*.) A method utilizing grid navigation to simplify flight problems attendant to pressure-pattern flight. A map projection is used in which the great circle course approximates a straight line, and a rectangular grid overlay is superimposed on the map projection and oriented along the central meridian of the projection. Then the great circle course makes equal angles with all north–south lines on the grid. When the net drift-correction angle is applied to correct for wind effect, the entire course may be flown on a single heading with respect to the rectangular grid overlay.

Aerometeorograph — a self-recording instrument used on aircraft for the simultaneous recording of atmospheric pressure, temperature, and humidity. See **Meteorogram**.

Aeronautical Climatology — the application of the data and techniques of climatology to aviation meteorological problems.

Aerosol — a colloidal system in which the dispersed phase is composed of either solid or liquid particles, and in which the dispersion medium is some gas, usually air. There is no clear-cut upper limit to the size of particles composing the dispersed phase in an aerosol, but as in all other colloidal systems, it is rather commonly set at 1 μm. Haze, most smokes, and some fogs and clouds may thus be regarded as aerosols. However, it is not good usage to apply the term to ordinary clouds with drops so large as to rule out the usual concept of colloidal stability. It is also poor usage to apply the term to the dispersed particles alone; an aerosol is a system of dispersed phase and dispersing medium taken together. See also **Aerosol**.

Agricultural Drought — Conditions that result in adverse crop responses, usually because plants cannot meet potential transpiration as a result of high atmospheric demand and/or limited soil moisture. Drought severity may be defined according to the Palmer Drought Severity Index or functionally expressed as yield-reducing water stress. See also **Drought Indices**.

Agricultural Meteorology — in general, meteorology and micrometeorology as applied to specific agricultural systems and of agriculture as applied to specific atmospheric conditions. This discipline may emphasize atmospheric transport of insects, pathogens, etc., that impact agriculture as well as energy and mass exchange of plants and animals with the atmospheric environment. The effect of soils and vegetation on the ratio of sensible and latent energy exchange is representative of the impact of agriculture on meteorology. See also **Micrometeorology**.

Air Drainage — the general term for gravity-induced, downslope flow of relatively cold air. Winds thus produced are called gravity winds, slope winds, katabatic winds, or drainage winds. See also **Katabatic Wind**; and **Winds and Air Movement**.

Air Mass — 1. A widespread body of air, the properties of which can be identified as 1) having been established while that air was situated over a particular region of the earth's surface (airmass source region), and 2) undergoing specific modifications while in transit away from the source region. An air mass is often defined as a widespread body of air that is approximately homogeneous in its horizontal extent, particularly with reference to temperature and moisture distribution; in addition, the vertical temperature and moisture variations are approximately the same over its horizontal extent. The stagnation or long-continued motion of air over a source region permits the vertical temperature and moisture distribution of the air to reach relative equilibrium with the underlying surface.

2. In radiation, the ratio of the actual path length taken by the direct solar beam to the analogous path when the sun is overhead from the top of the atmosphere to the surface. Extrapolation of surface measurements to zero air mass was the original method for estimating the value of solar irradiance at the top of the atmosphere. See also **Air Mass**.

Air Pocket — a local downdraft or an abrupt reduction of headwind or increase in tailwind that causes an airplane to drop suddenly.

Air Pollution Meteorology — the subdiscipline of meteorology devoted to the study of air pollution. Topics include sources of pollutants, emission rates, plume rise, fallout, dry and wet deposition, chemistry, precipitation scavenging, dispersion (molecular diffusion and turbulent transport), short- and long-range transport (advection), trapping, venting by cumulus clouds, complex terrain and mesoscale circulations, receptors, impact on society, alerts and episodes, policy and regulation, modeling, prediction, control, and climate change.

Air Sampler — a device used on a moving platform such as an aircraft that, when quickly opened and closed, captures a representative sample of the atmosphere.

Air Toxins — are hazardous air pollutants that are known or suspected to cause cancer or other serious health effects (such as birth or developmental defects). The U.S. Clean Air Act Amendments of 1990 require emission reductions for 188 hazardous air pollutants from industrial factories and other sources. As of 1996, the U.S. Environmental Protection Agency has issued standards for 47 source categories, such as chemical plants, oil refineries, aerospace manufacturers, and steel mills, as well as dry cleaners, commercial sterilizers, secondary lead smelters, and chromium electroplating processors. See also **Clean Air Act**.

Aircraft Ceiling — 1. After U.S. weather observing practice, the *ceiling classification* applied when the reported ceiling value has been determined by a pilot while in flight within one and one-half nautical miles of any runway of the airport. Aircraft ceilings may refer to vertical visibility or obscuring phenomena aloft as well as to clouds, and are designated A in aviation weather observations.

2. The maximum altitude at which any given aircraft can be operated safely. See **Ceiling Classification**.

Aircraft Thermometry — is the art of measuring the temperature of the environment from aircraft.

Airspeed — the speed on an exposed (usually airborne) object, relative to the atmosphere. In a calm atmosphere, airspeed equals ground speed.

Alaska Current — the eastern part of the North Pacific subpolar gyre. It is a shallow current carrying relatively warm water northward and thus has a climate influence similar to that exercised by the North Atlantic and Norwegian Currents on the climates of northwestern Europe, though on a smaller scale. It flows cyclonically around the Gulf of Alaska, feeding into the Alaskan Stream. Freshwater from the many rivers of Canada and Alaska reduces the water density near the coast; the result is a pressure gradient normal to the coast that constrains the current geostrophically to the coastal region and increases its speed to 0.3 m s^{-1}. See also **Gyres**.

Alaskan Stream See **Ocean Currents**.

Aleutian Low — the low pressure center located near the Aleutian Islands on mean charts of sea level pressure. It represents one of the main centers of action in the atmospheric circulation of the Northern Hemisphere. The Aleutian low is most intense in the winter months; in summer it is displaced toward the North Pole and is almost nonexistent. On a daily basis, the area of the Aleutian low is marked by alternating high and low pressure centers, moving generally to the eastward; it is not the scene of an intense stationary low. Normally the depth of intensity of the low pressure areas exceeds the intensity of the high pressure areas, so that the region is one of low pressure on the average. The travelling cyclones of subpolar latitudes usually reach maximum intensity in the area of the Aleutian low. The Aleutian low and its counterpart in the Atlantic Ocean, the Icelandic low, compose the Northern Hemisphere's subpolar low pressure belt. See **Icelandic Low**.

Alexander's Dark Band — the dark region of the sky seen between the primary and secondary rainbows. It is named after Alexander of Alphrodisias, the first person known to have commented upon it. He was the head of the Lyceum from 198 to 211 A.D.

Alidade — a stationary instrument, mounted on a stand, that measures the angle subtended at the stand by the horizon and an object in space. The clinometer is a portable form of alidade frequently used with a ceiling projector to determine the height of clouds. An alidade usually measures the elevation angle only; a theodolite measures azimuth as well.

Almeria–Oran Front — a frontal region located in the western Mediterranean Sea between Spain and Algeria that separates salty Mediterranean surface water from inflowing Atlantic water. It marks the transition from hydraulic control of the Atlantic inflow through the Straits of Gibraltar to geostrophic control of the flow. The front is associated with a strong current that feeds into the Algerian Current.

Alti-electrograph — a balloon-borne instrument for recording the value of the atmospheric electric field strength within active thunderstorms. A disk of specially treated paper (pole-finding paper), slowly rotated by an aneroid element as the balloon ascends, lies between two iron electrodes, the upper one being electrically connected to a discharge point attached to the balloon and the lower one being electrically connected to a long trailing wire. Currents passing through the paper under the influence of the potential difference set up in the two electrodes discolor the treated paper in a pattern with a size roughly related to the ambient electric field strength. A thermal element and a humidity element that record in aerograph fashion complete the device. See **Aerograph**.

Ambient Air — 1. Background, environmental, or surrounding air. When studying the dynamic and thermodynamic processes acting on an individual element such as an air parcel, cloud, smoke plume, raindrop, or ice crystal, ambient air represents the atmosphere outside of that element. The ambient air is often assumed to be static and of relatively large domain, within which the element resides.

2. The air that surrounds us, within which we live. When air pollutants of high concentration from exhaust or stack gases are emitted into cleaner air, the resulting polluted mixture is called the ambient air. National Ambient Air Quality Standards (NAAQS) http://www.epa.gov/air/criteria.html, apply to this final mixture, not to the undiluted emission gases.

Ambient Pressure—the air pressure that is characteristic of the atmosphere surrounding a small- scale feature such as a cumulus cloud.

Ambient Temperature—the temperature that is characteristic of the atmosphere surrounding a small-scale feature such as a cumulus cloud.

Amorphous Frost—Hoarfrost that possesses no apparent simple crystalline structure; opposite of crystalline frost. The lack of distinct crystal structure in forms of amorphous frost, however, is only a matter of scale. Such frost is built up of a multitude of units each of which has its own crystal structure, although no unit fits compatibly with its neighboring unit. See also **Precipitation and Hydrometeors**.

Anelastic Approximation—an approximate system of equations for deep and shallow atmospheric convection. The equations are derived under the assumptions that the percentage range in potential temperature is small and that the timescale is set by the Brunt–Väisälä frequency. Acoustic waves are thereby filtered—hence the term anelastic, meaning elastic energy is not allowed. If the vertical scale of motion is small compared to the depth of an adiabatic atmosphere, the anelastic equations reduce to the Boussinesq equations for shallow convection.

Angel—a radar echo caused by a physical phenomenon not discernible by eye at the radar site. Angels may appear as coherent or incoherent echoes. When diffuse and incoherent appearing, they are sometimes called ghost echoes. Angel echoes observed by radars with wavelengths of about 10 cm and less are usually caused by birds or insects. Radars with longer wavelengths and radar wind profilers, which operate in the UHF and VHF radio frequency bands, regularly detect echoes from the optically clear air that are caused by spatial fluctuations of the atmospheric refractive index. See also **Refractive Index**.

Antarctic Front—the semipermanent, semicontinuous front between the antarctic air of Antarctica and the polar air of the southern oceans; generally comparable to the arctic front of the Northern Hemisphere.

Anthropogenic Heat—heat released to the atmosphere as a result of human activities, often involving combustion of fuels. Sources include industrial plants, space heating and cooling, human metabolism, and vehicle exhausts. In cities this source typically contributes 15–50 W m^{-2} to the local heat balance, and several hundred W m^{-2} in the center of large cities in cold climates and industrial areas.

Antihail Rocket—a type of cloud seeding device that is fired into a cloud region where it releases a seeding agent. Such rockets have been launched from the ground and from aircraft. The seeding agent may be generated as smoke from the rocket motor, or it may be part of the rocket payload. Hail suppression programs in Italy, Russia, Yugoslavia, and Switzerland have used rockets to deliver cloud seeding agents into regions of hail-producing thunderstorms where there are large amounts of supercooled liquid water. See also **Clouds and Cloud Formation**; and **Cloud Seeding**.

Apob—an observation of pressure, temperature, and relative humidity taken in the free atmosphere by means of an aerometeorograph (mounted under the wing of an aircraft). This term was abbreviated from "airplane observation," and is rarely (if ever) used except in the above restricted sense. See **Aerometeorograph**.

Apocenter—the point on any orbit farthest from the center of attraction; the opposite of pericenter (see **Pericenter**). See also **Aphelion**.

Apogee—the point in an orbit at which any orbiting object, for example, planet, moon, artificial satellite, is farthest from the central object; the opposite of perigee (see **Perigee**). See also **Perihelion**.

Applications Technology Satellite (ATS)—See also **Geostationary Operational Environmental Satellite (GOES)**.

Aquiclude—a geologic formation that may contain water but is incapable of transmitting water in significant quantities.

Aquifer System—a group of two or more aquifers that are separated by aquitards or aquicludes. See also **Aquifers**.

Aquifer Test—a test to determine hydrologic properties of the aquifer involving the withdrawal of measured quantities of water from, or addition of water to, a well and the measurement of resulting changes in head in the aquifer both during and after the period of discharge or additions. See also **Aquifers**.

Aquifuge—a geologic formation that has no interconnected openings and hence cannot receive or transmit water.

Aquitard—a geologic formation that is not permeable enough to yield significant quantities of water to wells, but on a regional scale can contribute significant water to the underlying or overlaying aquifers. See also **Aquifers**.

Arago Distance—the angular distance from the antisolar point to the Arago point. The Arago distance is sensitive to the presence of foreign scattering particles in the atmosphere since these increase the contribution of the negative (horizontal) component of skylight polarization and hence shift the location of the point where the negative component is just equalled by the positive component. Thus the Arago distance is a useful measure of atmospheric turbidity. Its value is generally close to 20°, and is a function of solar elevation angle and of the wavelength of the light with which polarization is studied [Sekera Ref.].

Arago Point—one of the three commonly detectable points along the vertical circle through the sun at which the degree of polarization of skylight goes to zero; a neutral point. The Arago point, so named for its discoverer, is customarily located at about 20° above the antisolar point, but it lies at higher elevations in turbid air. The latter property makes the Arago distance a useful measure of atmospheric turbidity. Measurements of the location of this neutral point are typically more easily carried out than measurements of the Babinet point and the Brewster point, both of which lie so close to the sun (about 20° above and below the sun, respectively) that glare problems become serious. [Neuberger Ref.].

Arched Squall—the name applied to a squall in the Tropics when the squall cloud features a well- developed arcus (or roll cloud). It is usually a relatively violent storm. See **Sumatra**.

Arctic Air—a type of air mass with characteristics developed mostly in winter over arctic surfaces of ice and snow. Arctic air is cold aloft and extends to great heights, but the surface temperatures are often higher than those of polar air. For two or three months in summer arctic air masses are shallow and rapidly lose their characteristics as they move southward. See **Polar Air**.

Arctic Front—the semipermanent, semicontinuous front between the deep, cold arctic air and the shallower, basically less cold polar air of northern latitudes; generally comparable to the antarctic front of the Southern Hemisphere.

Arctic Zone—1. (Formerly called North Frigid Zone.) Geographically, the area north of the Arctic Circle (66°34′N).

2. (Same as tundra.) Biogeographically, the area extending northward from the arctic tree line to the "limit of life." It is also used for the level above the timber line in mountains. See also **Tundra**.

Aridity—the degree to which a climate lacks effective, life-promoting moisture; the opposite of humidity, in the climate sense of the term. The overall concept of aridity versus humidity is coming to be known as precipitation effectiveness. Two basic approaches have been made. The first, used by W. Köppen and modified by Bailey, does not openly define aridity, but rather assigns delimiting values of annual precipitation (treated with regard to distribution and temperature) to separate a dry climate from other types. The second approach actually prescribes a measure of aridity or precipitation effectiveness and uses these values as a primary parameter of classification. Of this type are Thornthwaite's precipitation- effectiveness index and moisture index, E. de Martonne's index of aridity, W. Gorczyński's aridity coefficient, Lang's moisture factor, and Ångström's humidity coefficient.

Aridity Coefficient—a function of precipitation and temperature designed by W. Gorczyński to represent the relative lack of effective moisture (the aridity) of a place. It is given by (latitude factor)×(temperature range)×(precipitation ratio). The latitude factor is the cosecant of the latitude (taken as 3.0 for 0°–4°). The temperature range is the difference (°F) between the means of the hottest and coldest months. The precipitation ratio is the difference between the highest and lowest annual totals (adjusted to a 50-year record) divided by the average. The value of this coefficient is about 100 in the middle of the Sahara; in the United States it ranges from 70 at Bagdad, California, to 2 at Eureka, California.

Aridity Index — 1.   As used by C. W. Thornthwaite in his 1948 climatic classification: an index of the degree of water deficit below water need at any given station; a measure of aridity. It is calculated, independently of the opposite humidity index, as follows:

$$\text{aridity index} = 100d/n,$$

where d (the water deficit) is the sum of the monthly differences between precipitation and potential evapotranspiration for those months when the normal precipitation is less than the normal potential evapotranspiration; and where n is the sum of monthly values of potential evapotranspiration for the deficient months. Thornthwaite puts the aridity index to two uses: 1) as a component of the moisture index; 2) as a basis for the more detailed classification of moist climates (perhumid, humid, and moist subhumid climates). See **Humidity Index**.

2.   See **Index of Aridity**.

Astronomical Twilight — the twilight stage during which the sun's unrefracted center is at elevation angles $-12° > h_0 > -18°$. During a clear evening's astronomical twilight, horizontal illuminance due to scattered sunlight decreases from 0.008 lux to $\square 6 \times 10^{-4}$ lux. At $h_0 = -18°$, 1) no horizon glow is visible at the sun's azimuth (the bright segment's upper boundary is at the observer's astronomical horizon), 2) sixth-magnitude stars can be seen near the zenith, and 3) scattered sunlight's residual illuminance is less than that from starlight and airglow. See also **Atmospheric Optical Phenomena**; and **Aurora and Airglow**.

Atmometer — (Also called *evaporimeter*, *evaporation gauge*, *atmidometer*.) The general name for an instrument that measures the evaporation rate of water into the atmosphere. Four main classes of atmometers may be distinguished: 1) large evaporation tanks sunk into the ground or floating in water; 2) small evaporation pans; 3) porous porcelain bodies; and 4) porous paper wick devices. The evaporation from a surface depends greatly upon the nature of the surface and the exposure of the surface to the atmosphere. Measured evaporation rates should be compared only between identical instruments.

Atmoradiograph — a device for measuring the frequency of occurrence of atmospherics, the intensity of which is greater than a predetermined level.

Atmospheric Radiation — See also **Atmosphere (Earth)**.

Atmospheric Shell — See also **Atmosphere (Earth)**.

Aurora — See also **Aurora and Airglow**.

Autan — See also **Winds and Air Movement**.

Aviation Weather Observation — (Also called airways observation.) An evaluation, according to set procedure, of those weather elements that are most important for aircraft operations. It always includes the cloud height or vertical visibility, sky cover, visibility, obstructions to vision, certain atmospheric phenomena, and wind speed and direction that prevail at the time of the observation. Complete observations include the sea level pressure, temperature, dewpoint temperature, and altimeter setting. Aviation weather observations are further classified as record, special, check, and local extra observations. The first two types are encoded and transmitted as reports on communications circuits.

Azores High — the semipermanent subtropical high over the North Atlantic Ocean, so named especially when it is located over the eastern part of the ocean. The same high, when displaced to the western part of the Atlantic, or when it develops a separate cell there, is known as the *Bermuda high*. On mean charts of sea level pressure, this high is one of the principal centers of action in northern latitudes. See **Bermuda High**.

Backdoor Cold Front — A cold front that leads a cold air mass toward the south and southwest along the Atlantic seaboard of the United States. This is one of the occurrences to which New Englanders give the name sea turn, for the cold wind following a backdoor cold front blows from the northeast quadrant.

Back-Sheared Anvil — Colloquial expression for a cumulonimbus anvil that spreads upwind into relatively strong winds aloft. A back-sheared anvil implies strong divergent flow near the summit of a high-speed convective storm updraft. These anvils often exhibit a crisp appearance with sharp, distinctive edges.

Bai — (Also called sand mist.) In China and Japan, a "mist" that occurs in spring and fall, when loose earth is churned up by the wind so that clouds of dust rise to great heights, afterward collecting moisture and falling as colored mist that produces a thick coating of very fine yellow dust.

Bai-u — (Also called tsuyu.) In southern Japan and in parts of China, the name of the season of heaviest rainfall. The bai-u season (June and early July in Japan, May to July in China) is the most important period for the cultivation and transplanting of rice. The bai-u rains are also called plum rains or mold rains, with reference to the season when plums ripen and to the effects of continued dampness.

Balance Year — the period between minima in net balance on a glacier. It corresponds roughly to the year between the end of one summer and the end of the following summer. Because the dates defining the balance year are meteorologically dependent, in practice glaciologists usually adhere to a "measurement year" defined by specific calendar dates.

Balloon Ceiling — After United States weather observing practice, the *ceiling classification* that is applied when the ceiling height is determined by timing the ascent and disappearance of a ceiling balloon or pilot balloon. Balloon ceilings are designated B in an aviation weather observation, and they may pertain only to clouds or to obscuring phenomena aloft. See **Ceiling Classification**.

Barber — 1.   A severe storm at sea during which spray and precipitation freeze onto the decks and rigging of boats.

2.   (Also spelled berber.) In the Gulf of St. Lawrence, a local form of blizzard in which the wind-borne ice particles almost cut the skin from the face.

3.   Same as frost smoke.

Barometry — The study of the measurement of atmospheric pressure, with particular reference to ascertaining and correcting the errors of the different types of barometer. See also **Barometer**.

Base Map — A map designed for the presentation and analysis of data; it usually includes only the coordinates, geographical and major political outlines, and sometimes the larger lakes and rivers. Many modifications exist for specific uses throughout the geophysical sciences, such as the frequent inclusion of fixed reference points (or station positions). Mountains and contour lines are generally omitted, but high ground may be indicated by a single contour line and shading.

Bay Ice — 1.   Any recently formed sea ice that is sufficiently thick to impede navigation.

2.   In Labrador, one-year ice that forms in bays and inlets.

3.   In the Antarctic, sometimes applied to heavy floes recently broken away from an ice shelf.

Bermuda High — The semipermanent subtropical high of the North Atlantic Ocean, so named especially when it is located in the western part of the ocean. This same high, when displaced toward the eastern part of the Atlantic, is known as the *Azores high*. On mean charts of sea level pressure, this high is a principal center of action. Warm and humid conditions prevail over the eastern United States, particularly in summer, when the Bermuda high is well developed and extends westward. See **Azores High**.

Bifilar Electrometer — An electrometer of the electrostatic type in which the potential to be measured is applied to two metal-coated quartz fibers, and the deflection due to their mutual repulsion is observed through a low-power microscope. The bifilar electrometer is used for potential measurements in atmospheric electricity studies. See also **Electrometer**.

Bio-Assay — Method for testing toxic effects of water with the help of living organisms; specifically: 1) the use of a change in biological activity as an indicator of a sample's response to biological treatment; 2) determining toxicity of wastewaters by using viable organisms as test organisms.

Biogeochemical Cycle The transformation and transport of substances within and among the atmosphere, biosphere, hydrosphere, and lithosphere via biological, geological, and chemical processes that are often cyclical in nature.

Bubble Bursting Process of bursting of air bubbles, rising to the ocean surface after a breaking wave, leading to breaking of a thin film cap and, for bubbles greater than a few millimeters in diameter, subsequent ejection of a Raleigh jet. These particles evaporate to leave residues that may act as cloud condensation nuclei in air of maritime origin.

Bumpiness Rapid variation of the vertical component of air motion causing an aircraft to jolt alternatively upward and downward. Bumpiness is associated generally with either convection currents in an unstable atmosphere or a flow of air across surface irregularities or both. It is more common and intense over land than over the sea. It is most marked in the lowest kilometer of the atmosphere but may extend to much higher levels, especially over mountainous terrain. Different types of aircraft may experience different types and intensities of bumpiness when flying through identical atmospheric conditions.

Burning Index A relative number related to the contribution that fire behavior makes to the amount of effort needed to contain a fire of a specified fuel type. The calculated burning index falls on a scale of 1–100: 1–11 is no fire danger; 12–35 is medium danger; 40–100 is high danger.

Cap (Also called lid) — a region of negative buoyancy below an existing level of free convection (LFC) where energy must be supplied to the parcel to maintain its ascent. This tends to inhibit the development of convection until some physical mechanism can lift a parcel to its LFC. The intensity of the cap is measured by its convective inhibition. The term capping inversion is sometimes used, but an inversion is not necessary for the conditions producing convective inhibition to exist.

Ceiling Classification — in aviation weather observation, a description or explanation of the manner in which the height of the ceiling is determined. The different types of ceilings according to this classification are *aircraft ceiling*, *balloon ceiling*, *estimated ceiling*, *indefinite ceiling*, *measured ceiling*, and *precipitation ceiling*.

Ceiling Light — (Also called ceiling projector) A type of cloud-height indicator that uses a searchlight to project vertically a narrow beam of light onto a cloud base. The height of the cloud base is determined by using a clinometer, located at a known distance from the ceiling light, to measure the angle included by the illuminated spot on the cloud, the observer, and the ceiling light.

Ceilometer — an automatic, active remote-sensing instrument for detecting the presence of clouds overhead and measuring the height of their bases. For optically thin clouds, such as most cirrus, more than one layer may be detected, but when optically thick clouds, such as liquid water stratus, are present, the light beam is unlikely to penetrate much beyond the base of the lowest liquid layer. Laser ceilometers use intense pulses of light in a very narrowly collimated, vertically directed beam, and have collocated transmitter and receiver systems. The cloud base heights may be displayed in a variety of time-height section images or backscatter intensity profile plots. Some older ceilometers use separated transmitter and receiver units. The instruments are designed to work during the day or night.

Celestial Sphere — The apparent sphere of infinite radius, having Earth as its center. It is upon the "inner surface" of this sphere that all heavenly bodies, the ecliptic, and the celestial equator appear. Disregarding the effects of topography and refraction near the horizon, for practical purposes half of this sphere may be considered visible from any point on the earth's surface at any time. See also **Celestial Sphere and Astronomical Triangle**.

Central Water — The water mass of the permanent or oceanic thermocline, which is located at a depth of between 150 and 800 meters (492 and 2, 625 feet). Central water is formed by subduction in the subtropics between 25° and 45° latitude in both hemispheres. It therefore spans a wide temperature and salinity range, with temperature and salinity both decreasing with depth. Each ocean has its own central water with its own specific temperature–salinity relationship depending on the atmospheric conditions in the formation region. These are distinguished by appropriate names, for example, South Atlantic Central Water, Western North Pacific Central Water, etc.

Change of Phase — A process in which a substance that can exist in two or more phases is converted from one phase to another. The most important meteorological examples are the evaporation, condensation, freezing, melting, deposition, and sublimation of water.

Clear Ice (Or clear icing) — Smooth compact rime, usually transparent, fairly amorphous, with a ragged surface, and morphologically resembling glaze. This term has two different major applications. 1) Most commonly, it is used as a synonym for glaze, particularly with respect to aircraft icing. Factors that favor clear ice (or glaze) formation are large drop size, rapid accretion of liquid water, slight supercooling, and slow dissipation of latent heat of fusion. Thus, an aircraft flight through supercooled rain at an air temperature of 0° to −4 °C is most conducive to clear icing. This type of icing does not seriously distort airfoil shape, but it does add appreciably to the weight of the craft. 2) The term may also be applied to homogeneous bodies of glacier ice and lake ice.

Clinometer — An instrument for measuring angles of inclination. In meteorology, it is used in conjunction with a ceiling light to measure cloud height at night. With it the observer measures the angle included by the illuminated spot on the clouds, the observer, and the ceiling light.

Closed Low — A low that may be completely encircled by an isobar or contour line. (This means an isobar or contour line of any value, not necessarily restricted to those arbitrarily chosen for the analysis of the chart.) Strictly, all lows are closed. However, in weather-map analysis terminology, this designation is used commonly in two respects: 1) on surface charts, to distinguish a low from a trough, especially as a low develops within the trough; and 2) on upper-level charts, to accentuate the fact that the circulation is closed, especially at levels and over latitudes where such an occurrence is unusual. The definition of closed high is analogous.

Coffin Corner — A term used to describe the range of Mach numbers between the buffeting Mach number and the stalling Mach number within which an aircraft must be operated. The buffeting and stalling Mach numbers approach each other with altitude; when they become the same, the ceiling of the aircraft is reached.

Cold-air Pool — A topographic depression, such as a valley or basin, filled with cold air. The cold air is heavy, and settles to the bottom of the depression. This air can remain stagnant, trapped by the surrounding higher terrain, resulting in long periods of poor air quality and fog, depending on the sources of pollution and amount of moisture in the air, respectively.

Cold Type Occlusion — (Also called cold occlusion, cold occluded front, cold-front-type occlusion.) According to the Norwegian cyclone model, the situation where a cold front catches up with a portion of a warm front above the cold frontal surface. This conceptualization of cold type occlusion development is not often observed in nature; however, it is undisputed that in strong cyclones the low center often retreats toward the cold air separating itself from the cold and warm fronts. A trough in sea level pressure is found between the cyclone center and the wave on the front, and this trough is the occluded front. Regardless of the formation processes, characteristics of a cold type occlusion are 1) a warm temperature or thickness ridge along the occluded front; 2) a trough in the sea level pressure field along the occluded front; 3) relatively colder air behind the front; and 4) an increase in lower-tropospheric static stability behind the front.

Colored Rain — Rain that leaves a colored stain on the ground and on exposed objects, often red or rusty in hue. The coloration is usually the result of rain picking up particles as it falls through a dust-filled subcloud layer. The subcloud layer, usually rich in iron oxide, may originate from an area far from the observed colored rain event. This phenomenon has been observed frequently in Italy with particles advected northward from the Sahara.

Complex Terrain — A region having irregular topography, such as mountains or coastlines. Complex terrain can also include variations in land use, such as urban, rural, irrigated, and unirrigated. Complex terrain often generates local circulations, or modifies ambient synoptic weather features, to create unique local weather characteristics such as katabatic winds, anabatic clouds, and sea breezes. In regions of complex terrain, weather forecast models must have high resolution to reproduce numerically the terrain-induced weather features.

Conformal Map — (Also called isogonal map, orthomorphic map.) A map that preserves angles; that is, a map such that if two curves intersect at a given angle, the images of the two curves on the map also intersect at the same angle. On such a map, at each point, the scale is the same in every direction. Shapes of small regions are preserved, but areas are only approximately preserved (the property of area conservation is peculiar to the equal-area map). The most commonly used conformal map is probably the Lambert conic projection, with standard latitudes at 30° and 60° N. On the standard latitudes, the scale is exact; between them, it is decreased by not more than about 1%; outside them, distortion increases rapidly. The Mercator and stereographic projections are also conformal maps.

Constant-Pressure Chart — (Also called isobaric chart, isobaric contour chart.) The synoptic chart for any constant-pressure surface, usually containing plotted data and analyses of the distribution of, for example, height of the surface, wind, temperature, and humidity. Constant-pressure charts are most commonly known by their pressure value; for example, the 1000-mb chart (which closely corresponds to the surface chart), the 850-mb chart, 700-mb chart, 500-mb chart, etc.

Cooling Power — In the study of human bioclimatology, one of several parameters devised to measure the cooling effect of the air upon a human body. Essentially, cooling power is determined by the amount of applied heat required by a device to maintain it at a constant temperature (usually 34 °C (93 °F); the entire system should be made to correspond, as closely as possible, to the external heat exchange mechanism of the human body.

Instruments used in applying this principle include the katathermometer, the frigorimeter, and the coolometer.

Cooling Temperature — In the study of human bioclimatology, one of several parameters devised to measure the cooling effect of the air upon the human body. It is the skin temperature of a sphere that is in thermal equilibrium with the surrounding air; the sphere is supplied heat at a constant rate, and its heat exchange characteristics correspond to those of a blond, fair-skinned human.

Crop Moisture Index (CMI) — The Crop Moisture Index (CMI) uses a meteorological approach to monitor week-to-week crop conditions. It was developed by Palmer in 1968 from procedures within the calculation of the PDSI. Whereas the PDSI monitors long-term meteorological wet and dry spells, the CMI was designed to evaluate short-term moisture conditions across major crop-producing regions. It is based on the mean temperature and total precipitation for each week within a climate division, as well as the CMI value from the previous week. The CMI responds rapidly to changing conditions, and it is weighted by location and time so that maps, which commonly display the weekly CMI across the United States, can be used to compare moisture conditions at different locations. Weekly maps of the CMI are available as part of the USDA/JAWF Weekly Weather and Crop Bulletin http://www.usda.gov/oce/weather/.

Cross Section — In weather analysis and forecasting, a graphic representation of a vertical surface in the atmosphere, along a given horizontal line or path, and extending from the earth's surface to a given altitude. The type of data and analysis presented on such a cross-sectional chart depends upon its purpose. In meteorology, a synoptic cross section is prepared from synoptic weather data. In aviation, a flight cross section (or route cross section) is a graphic forecast of conditions expected to be encountered along the proposed flight route; therefore, time varies along the horizontal axis of the chart.

Deepening — A decrease in the central pressure of a pressure system as depicted on a constant-height chart, or an analogous decrease in height on a constant-pressure chart; the opposite of filling. The term is usually applied to low pressure rather than to high pressure, although technically it is acceptable in either sense. The deepening of a low is commonly accompanied by the intensification of its cyclonic circulation, and the term is frequently used to imply the process of cyclogenesis. Deepening can be quantitatively expressed in at least two ways: either 1) as the time rate of central-pressure decreases; or 2) as that component of the pressure tendency at any fixed point that is attributable neither to the motion of the pressure system relative to that point nor to the diurnal influence of atmospheric tides.

Dependent Meteorological Office (DMO) — An office that provides meteorological service for international air navigation in accordance with International Civil Aviation Organization specifications. Its functions are to 1) prepare forecasts under the guidance of a main meteorological office; 2) supply meteorological information and briefings to aeronautical personnel; 3) supply meteorological information required by an associated supplementary meteorological office. See **Meteorological Watch Office**.

Differential Chart — General term for a chart showing the amount and direction of change of a meteorological quantity in time or in space, for example, change chart, vertical differential chart.

Diffusion Hygrometer — A hygrometer based upon the diffusion of water vapor through a porous membrane. In its simplest form, it consists of a closed chamber having porous walls and containing a hygroscopic compound. The absorption of water vapor by the hygroscopic compound causes a pressure drop within the chamber that is measured by a manometer. See also **Manometer**.

Disdrometer — An instrument that measures and records the sizes of raindrops. A common type of disdrometer consists of a sensitive transponder that measures the momentum of individual drops as they fall onto an exposed horizontal surface. Size is determined from momentum through calibration, and the drop-size distribution is obtained by keeping a tally of the number of drops in different size categories that fall onto the surface in a given period of time.

District Forecast — In U.S. Weather Bureau historical records usage, a general weather forecast for conditions over an established geographic "forecast district." Forecast districts are relatively large areas of the order of tens of thousands or hundreds of thousands of square miles.

Diurnal — Daily, especially pertaining to actions that are completed within 24 hours and that recur every 24 hours; thus, most reference is made to diurnal cycles, variations, ranges, maxima, etc. The diurnal variability of nearly all of the meteorological elements is one of the most striking and consistent features of the study of weather. The diurnal variations of important elements at the earth's surface can be summarized as follows: 1) temperature maximum occurs after local noon and minimum near sunrise; 2) relative humidity and fog are the reverse of temperature; 3) wind generally increases and veers by day and decreases and backs by night; 4) cloudiness and precipitation over a land surface increase by day and decrease at night; over water the reverse is true, but to a lesser extent; 5) evaporation is markedly greater by day; 6) condensation is much greater at night; 7) atmospheric pressure varies diurnally or semidiurnally according to the effects of atmospheric tides.

Dropsonde Observation — An evaluation of meteorological data received from a descending dropsonde. The dropsonde is a small expendable instrument package that is released from an aircraft. As it descends, it radio transmits pressure, temperature, and relative humidity data back to the aircraft. By tracking the position of the dropsonde using radio-navigation techniques [e.g., Global Positioning System (GPS)], wind speed and direction data can be obtained. The processed data are usually presented in terms of height, temperature, dewpoint, and winds at mandatory and significant pressure levels. A dropsonde observation is comparable to a rawinsonde observation. Data from the dropsonde are usually received and processed in the aircraft that dropped the instrument.

Dropsonde — (Also called parachute radiosonde.) A radiosonde with a parachute dropped from an airplane carrying receiving equipment for the purpose of obtaining an upper-air sounding during descent. See also **Radiosonde**.

Dry Deposition — The process by which atmospheric gases and particles are transferred to the surface as a result of random turbulent air motions. See **Wet Deposition**.

Dry Freeze — The freezing of the soil and terrestrial objects caused by a reduction of temperature when the adjacent air does not contain sufficient moisture for the formation of hoarfrost on exposed surfaces. With respect to vegetation alone, this is termed a *black frost*. A dry freeze is usually considered to be a more local and short-period (probably radiative) phenomenon than a freeze.

Dryline — A low-level mesoscale boundary or transition zone hundreds of kilometers in length and up to tens of kilometers in width separating dry air from moist air. The length of the dryline is related to large-scale terrain or large-scale weather system features, whereas its width is related to mesoscale processes. In its quiescent state, the dryline may be considered the intersection of the top of a low-level moist layer with large-scale features of sloping terrain. In this state the shallow layer of moisture near the higher terrain is eroded by turbulent mixing with daytime heating. Moisture gradients are additionally strengthened by horizontal convergence resulting from downward transport of horizontal momentum in the dry air. In a more dynamically active state the dryline often advances away from the higher terrain as an integral component of an extratropical cyclone or frontal wave. In such cases it extends equatorward from the cyclone or wave. In this state moisture gradients and boundary motion are largely influenced by downward transport of horizontal momentum resulting from larger-scale sinking in the dry air. The dryline is found all over the world. In the United States the dryline, which marks the boundary between moist air from the Gulf of Mexico and dry continental air from the west, is found in the Plains region. It is most often present during the spring, where it is often the site of thunderstorm development. Typically the dryline in the United States advances eastward during the day and retreats westward at night.

Dynamic Meteorology — The study of atmospheric motions as solutions of the fundamental equations of hydrodynamics or other systems of equations appropriate to special situations, as in the statistical theory of turbulence. The restrictions of this definition suffice to distinguish dynamic meteorology from other fields, for example, physical meteorology or synoptic meteorology, such distinctions being a function of the state of the science rather than of the subject matter itself.

Encroachment Method — A technique used in boundary layer meteorology to denote a type of mixed layer growth rate that is proportional to the rate of warming of the mixed layer divided by the ambient lapse rate immediately above the top of the mixed layer. In other words, heating from the surface causes the depth of the mixed layer to increase. Also known as the thermodynamic method, the mixed-layer top can be found by finding the area under the early morning sounding (between the sounding and an adiabat for warmer air) that is equal to the area under a plot of the surface heat flux versus time up to the time of interest.

Equatorial Trough — 1. The quasi-continuous belt of low pressure lying between the subtropical high pressure belts of the Northern and Southern Hemispheres. This entire region is one of very homogeneous air, probably the most ideally barotropic region of the atmosphere. Yet humidity is so high that slight variations in stability cause major variations in weather. The position of the equatorial trough is fairly constant in the eastern portions of the Atlantic and Pacific, but it varies greatly with season in the western portions of those oceans and in southern Asia and the Indian Ocean. It moves into or toward the summer hemisphere. It has been suggested that this name be adopted as the one general term for this region of the atmosphere. Thus, the equatorial trough would be said to contain regions of doldrums; portions of it could be described as intertropical convergence zones; and within it there might be detected intertropical fronts. However, one weakness of this nomenclature is that it alludes specifically and only to the existence of a trough of low pressure. Perhaps an even more general term might be preferable, for example, atmospheric equator.

2. Same as meteorological equator.

Estimated Ceiling — After U.S. weather observing practice, the *ceiling classification* applied to a ceiling height that is determined in any of the following ways: 1) by means of a convective-cloud height diagram or dew point formula; 2) from the known heights of unobscured portions of natural landmarks, or objects more than one and one-half nautical miles from any runway of the airport; 3) on the basis of observational experience, provided the sky is not obscured by surface-based hydrometeors or lithometeors, and other guides are lacking or considered unreliable; or 4) determined by ceilometer or ceiling light when the penetration of the light beam is in excess of normal for the particular height and type of layer, or when the elevation angle of the clinometer or ceilometer-detector scanner exceeds 84°. Such a ceiling is denoted E in aviation weather reports. See **Ceilometer**; and **Clinometer**.

Fair — With respect to weather, generally descriptive of pleasant weather conditions, with due regard for location and time of year. It is subject to popular misinterpretation, for it is a purely subjective description. When this term is used in weather forecasts (National Weather Service), it is meant to imply 1) no precipitation; 2) less than 0.4 sky cover of low clouds; 3) no other extreme conditions of cloudiness, visibility, or wind; 4) unrestricted visibility; and 5) light winds of generally less than 10 knots (5 m s^{-1}).

Fetch — 1. The distance upstream of a measurement site, receptor site, or region of meteorological interest, that is relatively uniform. If a measurement site is located in the middle of a farm field with homogeneous land use, and if there are no changes to the land use and no obstructions such as trees or buildings immediately upstream of the site, then the site is said to have "large fetch". Large fetch is usually considered good if the measurements are to be representative of the atmosphere over the farm field. Similarly, measurements over a homogeneous forest could also have large fetch if there are no clearcuts or changes in the tree characteristics upstream of the measurement site.

2. (Also called generating area.) An ocean area where waves are generated by a wind having a constant direction and speed.

3. The length of the fetch area, measured in the direction of the wind in which ocean waves are generated. In many cases, the fetch is limited by the upwind distance to the coast.

Firn — Old snow that has become granular and compacted (dense) as the result of various surface metamorphoses, mainly melting and refreezing but also including sublimation. The resulting particles are generally spherical and rather uniform. Firnification, the process of firn formation, is the first step in the transformation of snow into land ice (usually glacier ice). Some authorities restrict the use of firn to snow that has lasted through one summer, thereby distinguishing it from spring snow. Originally, the French term, "névé," was equivalent to the German term, "firn," but there is a growing tendency, especially among British glaciologists, to use "névés" for an area of firn, that is, generally for the accumulation area above or at the head of a glacier

First-Order Climatological Station — A meteorological station at which autographic records or hourly readings of atmospheric pressure, temperature, humidity, wind, sunshine, and precipitation are made, together with observations at fixed hours of the amount and form of clouds and notes on the weather.

First-Order Station — After U.S. National Weather Service practice, any meteorological station that is staffed in whole or in part by National Weather Service (Civil Service) personnel, regardless of the type or extent of work required of that station.

Footprint — 1. In micrometeorology, the region of ground that affects a turbulent flux measurement above the surface. Also known as the source-weight distribution function, the footprint can figuratively be described as the ensemble average field-of-view of a turbulent flux measurement. The footprint function is derived from a suitable model of turbulent transport. Alternately, an analogous footprint can be defined for scalar concentrations or for radiative fluxes.

2. In remote sensing, the instantaneous field-of-view of an airborne remote sensing instrument.

Footprint Modeling — A modeling approach to determine the relative importance of upwind source areas on the value of an atmospheric variable at a given height. The footprint is dependent on atmospheric stability and the surface type and is different for scalars than for higher-order turbulence statistics (such as turbulent fluxes). Footprint models use either Eulerian or Lagrangian dispersion theory. The former is usually based on idealized assumptions (e.g., horizontal homogeneity, restriction of the vertical range), whereas the latter (stochastic particle modeling) may take into account more realistic situations but is computationally much more expensive.

Frost Hazard — The risk of damage by frost. It may be expressed as the probability or frequency of killing frost on different dates during the growing season, or as the distribution of dates of the last killing frost of spring or the first of autumn. A strict application of the concept would take into account the actual species or group of plants that might potentially be killed, as different plants sustain frost damage at different temperatures. Wind-chill factors should also be considered.

Frost Heaving (or Frost Heave) — The upward or outward movement of the ground surface (or objects on, or in, the ground) caused by the formation of ice in the soil. This process can damage plant roots through breaking or desiccation, cause cracks in pavement, and damage the foundations of buildings, even below the frost line. Moist, fine-grained soil at certain temperatures is most susceptible to frost heaving.

Originally, frost heaving was thought to occur due simply to the freezing of water in soil. However, the vertical displacement of soil in frost heaving can be significantly greater than the expansion that occurs when ice freezes. In the 1960s, frost heaving was demonstrated in soil saturated in benzene and nitrobenzene, which contract when they freeze.

The current understanding is that certain soil particles have a high affinity for liquid water. As the liquid water around them freezes, these soils draw in liquid water from the unfrozen soils around them. If the air temperature is below freezing but relatively stable, the heat of fusion from the water that freezes can cause the temperature gradient in the soil to remain constant. The soil at the point where freezing is occurring continues to draw in liquid water from the soils below it, which then freezes and builds up into an "ice lens". Depending on the soil's affinity for moisture and amount of moisture available, a significant amount of soil displacement can result.

The earliest known documentation of frost heaving came in the 1600s.

Frost Line — Maximum depth of frozen ground during the winter. The term may refer to an individual winter, to an average over a number of years, or to the greatest depth since observations began. The frost line varies with the nature of soil and the protection afforded by vegetal ground cover and snow cover, as well as with the amount of seasonal cooling.

Building codes must take this into account, as foundations must be dug down to or below this point. Failure to do so will cause frost heaving to dislodge the building at least slightly, causing damage which may become a serious threat to the building's structural integrity.

Frost Smoke — (Sometimes called barber.) A rare type of fog formed in the same manner as a steam fog, but at colder temperatures so that it is composed of ice particles instead of water droplets.

General Circulation — 1. (Also called planetary circulation.) In its broadest sense, the complete statistical description of large scale atmospheric motions. These statistics are generated from the ensemble of daily data and include not only the temporal and spatial mean flows (e.g. zonal westerlies and easterlies) but also all other mean properties of the atmosphere that are linked to these flows (e.g., semipermanent waves and meridional cells) that together form the general circulation. The general circulation also includes higher-order statistics that measure the spatial and temporal variability of the atmosphere necessary to understand the large-scale temporal and spatial mean state of the atmosphere (e.g., seasonal changes and the effects of transient cyclones).

General Circulation Model (GCM) — A time-dependent numerical model of the atmosphere. The governing equations are the conservation laws of physics expressed in finite-difference form, spectral form, or finite-element form. Evolution of the model circulation is computed by time integration of those equations starting from an initial condition. The GCM can be used for weather prediction or for climate studies.

Geomagnetism — (Also called terrestrial magnetism, geomagnetic field.)

1. The earth's magnetic field or the geophysical phenomena caused or affected by this field.

2. The scientific study of the earth's magnetic field, including its variation in space and time, and its relation to other geophysical phenomena (e.g., aurora). Geomagnetism belongs to the same family of earth sciences as geodesy and geomorphology.

Gill Anemometer — A lightweight version of a propeller anemometer sometimes used instead of the more expensive sonic anemometer for the measurement of turbulent velocity fluctuations. The "traditional" Gill propeller anemometer is a three-axis arrangement, which is particularly suited for the measurement of the vertical wind. The so-called K-Gill propeller vane has two propellers, mounted at an angle of 90°, that are aligned with the mean wind by a vane. One particular type of sonic anemometer is sometimes referred to as "Gill sonic anemometer" in the literature of the early 1990s.

Green Flash — A flash of green light seen on or (seemingly) adjacent to the upper rim of the low sun (at either sunrise or sunset). The green flash is a mirage, but the image formed in this case is of a portion of the sun rather than of an earthbound object. In addition to the displacement and distortion that is characteristic of mirages, there is also significant dispersion. The upper edge of the low sun normally has a thin green rim (occasionally blue) that is too narrow to be seen by the naked eye unless the rest of the sun is obstructed, say, by the horizon. It is often asserted that the green flash is seen in this way: a mere transient view of the green rim between obscuration by the rest of the sun and obstruction by the horizon. Yet such a sequence produces a singularly poor flash. Rather, the remarkable flashes always seem to involve multiple and magnified images of the green rim. Indeed, the presence of such multiple images of a small portion of the sun is a good indicator of a forthcoming flash. The optical signature of multiple images is a serrated edge to the sun. The refraction that displaces the image of the low sun up from the position it would occupy in the absence of an atmosphere does so more strongly for shorter wavelengths. This leads to a red rim on the bottom of the sun and a blue or green rim on the top.

Green Rim (Also called *green segment*) — The upper rim of the low sun usually displays a green rim. The transient sighting of this rim, between its obscuration by the rest of the sun and its obstruction by the horizon, is sometimes credited with being the origin of the green flash. But, while the color of the flash is provided by the green rim, the size and transient nature has more to do with multiple-image mirages.

Haar — A name applied to a wet sea fog or very fine drizzle that drifts in from the sea in coastal districts of eastern Scotland and northeastern England. It occurs most frequently in summer.

Heat Wave — (Also called hot wave, warm wave.) A period of abnormally and uncomfortably hot and usually humid weather. To be a heat wave such a period should last at least one day, but conventionally it lasts from several days to several weeks. In 1900, A. T. Burrows more rigidly defined a "hot wave" as a spell of three or more days on each of which the maximum shade temperature reaches or exceeds 90 °F. More realistically, the comfort criteria for any one region are dependent upon the normal conditions of that region. In the eastern United States, heat waves generally build up with southerly winds on the western flank of an anticyclone centered over the southeastern states, the air being warmed by passage over a land surface heated by the sun.

Helioseismology — Study of the the internal structure (i.e., density, temperature, abundance, etc.) and dynamics (i.e., mixing, rotation rate, etc.) of the sun from measurements of its surface oscillations.

Hemispheric Model — A numerical model constructed assuming a boundary condition along the equator. In many applications, the boundary condition is based upon vanishing meridional velocity along the equator. In some spectral formulations, the boundary condition may be formulated using symmetric or antisymmetric expansion (basis) functions.

High — In meteorology, an area of high pressure, referring to a maximum of atmospheric pressure in two dimensions (closed isobars) in the synoptic surface chart, or a maximum of height (closed contours) in the constant-pressure chart. Since a high is, on the synoptic chart, always associated with anticyclonic circulation, the term is used interchangeably with anticyclone.

Homogeneous Atmosphere — 1. A hypothetical atmosphere in which the density is constant with height. The lapse rate of temperature in such an atmosphere is known as the autoconvective lapse rate and is equal to g/R (or approximately 3.4 °C/100 meters (38 °F/328 feet), where g is the acceleration of gravity and R is the gas constant for air. A homogeneous atmosphere has a finite total thickness that is given by $R_d T_v/g$, where R_d is the gas constant for dry air and T_v is the virtual temperature (K) at the surface. For a surface temperature of 273 K (32 °F), the vertical extent of the homogeneous atmosphere is approximately 8,000 meters (26,247 feet). At the top of such an atmosphere both the pressure and absolute temperature vanish.

2. Same as adiabatic atmosphere.

Humidity Index — as used by C. W. Thornthwaite in his 1948 climatic classification, an index of the degree of water surplus over water need at any given station. It is calculated, independent of the opposing aridity index, as follows:

$$\text{humidity index} = 100 s / n$$

where s (the water surplus) is the sum of the monthly differences between precipitation and potential evapotranspiration for those months when the normal precipitation exceeds the latter, and n (the water need) is the sum of monthly potential evapotranspiration for those months of surplus. The humidity index has two uses in Thornthwaite's classification: 1) as a component of the moisture index; and 2) as a basis for detailed classification of dry climates [Thornthwaite, C.W. Ref.].

Hydrography — The measurement and study of depths and currents in open seas, lakes, estuaries, and rivers.

Hydrometeorology — 1. Study of the atmospheric and terrestrial phases of the hydrological cycle with emphasis on the interrelationship between them.

2. Meteorology plus hydrology. Many countries use the word in this sense to name the official service charged with the dual responsibility of weather and hydrologic functions.

3. (Rare.) That branch of meteorology that deals with the hydrometeors.

Hydrostatic Model — An atmospheric model in which the hydrostatic approximation replaces the vertical momentum equation. This implies that vertical acceleration is negligible compared to vertical pressure gradients and vertical buoyancy forces, a good approximation for synoptic and subsynoptic scales of motion. Hydrostatic models have been successfully applied with horizontal resolutions as small as about 10 km, resolving even some mesoscale circulations. Global and regional weather prediction models have traditionally been hydrostatic models.

Hyetal Equator — A line (or transition zone) that encircles the earth and lies between two belts that typify the annual time distribution of rainfall in the lower latitudes of each hemisphere; a form of meteorological equator. It lies slightly north of the geographic equator, reaching its most northerly position at about 10 °N latitude near the mouth of the Orinoco River in South America. The hyetal equator is more or less centrally situated in the belt of tropical rainfall, which has two rainy seasons and generally one main dry season, the latter occurring in the winter of the corresponding hemisphere.

Hyetal Region — A region in which the amount and seasonal variation of rainfall are of a given type. According to Köppen in his climatic classification, the main types of hyetal regions are 1) desert, with rainfall rare and irregular; 2) winter-dry, with main rainy season in summer, characteristic of a monsoon climate; 3) summer-dry, with rainy season in winter, such as a Mediterranean climate; 4) rain at all seasons, but not evenly distributed through the year; 5) rain on more than half the days in every month of the year. Minor types may be distinguished in which the season of rainfall maximum is either broken by a short dry season or is displaced toward spring or autumn. *Compare* humidity province.

Hythergraph — A type of climatic diagram where the coordinates are some form of temperature versus a form of humidity or precipitation. A common, specific use is to show the annual "march" of mean monthly values of temperature and precipitation at a given station. Also, a comfort chart may be considered a hythergraph. There are many other possibilities.

Ice Forecast — Describes the predicted position of ice boundaries and expected ice phenomena (ice concentration, distribution, stage of development, thickness and direction of drift, number and size of icebergs) for a specified period and for a specified locality, based on forecast meteorological and oceanographic conditions and the regional ice climatology. An ice forecast is often issued to cover the period between the current ice analysis and the next scheduled ice analysis.

Icelandic Low — 1. A low pressure center located near Iceland; (mainly between Iceland and southern Greenland) on mean charts of sea level pressure. It is a principal center of action in the atmospheric circulation of the Northern Hemisphere. It is most intense during winter, having a January central pressure below 996 mb. In summer, it not only weakens but also tends to split into two centers, one near Davis Strait and the other west of Iceland. Like its Pacific counterpart, the Aleutian low, its daily position and intensity vary greatly so that it is best regarded as a region where migratory lows tend to slow up and deepen. See **Aleutian Low**.

2. Any low, on a synoptic chart, centered near Iceland.

Indefinite Ceiling (Formerly called ragged ceiling) — After U.S. weather observing practice, the ceiling classification applied when the reported ceiling value represents the vertical visibility upward into surface-based atmospheric phenomena (except precipitation). Such phenomena include fog, blowing snow, and all of the lithometeors. All indefinite ceilings are estimations, but one of the following must be used as a guide: 1) the distance an observer can see vertically into the obstruction; 2) the height corresponding to the top of a ceiling-light beam; 3) the height at which a ceiling balloon completely disappears; 4) the height determined by the sensor algorithm at automated stations. The letters "VV" (vertical visibility) are used to designate an indefinite ceiling.

Index of Aridity — a measure of the precipitation effectiveness or aridity of a region, proposed by De Martonne 1925, given by the following relationship:

$$\text{index of aridity} = \frac{P}{T + 10},$$

where P (cm) is the annual precipitation and T (°C) the annual mean temperature (De Martonne)

Indian Equatorial Jet — An intense eastward flow of about four-week duration found at the equator in the Indian Ocean during the transition periods between the monsoon seasons (April–June and October–December). Speeds at the equator can exceed 1 m s^{-1} but fall off to less than 0.2 m s^{-1} at 2 °S and 2 °N. The jet is the result of sea level adjustment to the change of wind direction from one monsoon season to the next.

Indicated Airspeed (IAS) — The airspeed read or recorded directly from an airspeed indicator. Indicated airspeed is usually lower than the actual airspeed and must be corrected for both temperature and density to yield true airspeed. The correction is quickly accomplished on an ordinary navigational computer.

Indicated Altitude — The altitude read directly from a pressure altimeter when set to the prescribed altimeter setting. This value differs from the corrected altitude as a function of the difference between the actual density of the underlying air and that of the standard atmosphere. The vertical separation of aircraft on airways is based on indicated altitude, and in general, standard aircraft operating procedure calls for the use of indicated altitude. See also **Altimetry**.

Industrial Climatology — A type of applied climatology that studies the effect of climate and weather on industry's operations. The goal of industrial climatology is to provide industry with a sound statistical basis for all administrative and operational decisions that involve a weather factor.

Inland Sea Breeze — A circulation similar to a sea breeze, except not at a shore. The inland sea breeze is a very weak thermal circulation caused by temperature contrast between different land surfaces and is sometimes observed between cool irrigated farm land and neighboring dry desert land. This phenomenon is observed only when the synoptic-scale winds are very light.

Integral Depth Scale — An average height that is weighted by some other characteristic of the vertical profile. For example, the potential temperature profile in the stable boundary layer at night often has an exponential shape because the greatest cooling has occurred nearest the ground and the temperature change decreases with height. By finding the area under the potential temperature change curve and dividing by the temperature change at the surface, a height scale is obtained that is an integral measure of the depth of the stable boundary layer. For this particular example of an exponential profile, the integral depth corresponds to the *e*-folding depth of the profile.

Integral Length Scales — Of the three standard *turbulence length scales*, the ones that are measures of the largest separation distance over which components of the eddy velocities at two distinct points are correlated. They characterize the energy-containing range of eddy length scales. In the most general form, the integral scales (expressed here as a tensor) are functions of position and are defined in terms of the normalized two-point velocity correlations.

Internal Tides — Tidal waves that propagate at density differences within the ocean. They travel slowly compared with surface gravity waves and have wavelengths of only a few tens of kilometers, but they can have amplitudes of tens of meters. The associated internal currents are termed baroclinic motions.

Internal Water Circulation — The conceptual hydrological cycle for a specified continental surface that comprises water evaporating from the surface, condensing in the overlying atmosphere, and falling back to the surface. In reality, some of the evaporated water leaves the region in the wind and is replaced with water vapor brought into the region by wind. See also **Hydrological Cycle**.

Internal Wave — A wave in fluid motion having its maximum amplitude within the fluid or at an internal boundary (interface). The concepts of internal and external waves originated in the study of gravity waves in homogeneous incompressible fluids, and it makes no difference in the dynamics of the wave whether the static stability of the fluid is concentrated in a free surface or in an interface. However, internal waves in a fluid with continually varying density have maximum amplitudes and nodal surfaces within the fluid itself, so that these are properly distinguished from external waves.

International Index Numbers — A system of designating meteorological observing stations by number, established and administered by the World Meteorological Organization. Under this scheme, specified areas of the world are divided into blocks, each bearing a two-number designator; stations within each block have an additional unique three-number designator, the numbers generally increasing from east to west and from south to north. The international language of this system facilitates quick identification of the source of any meteorological report.

Interstellar Dust (Also called cosmic dust) — Small, solid particles in the supposedly empty space between stars. Dark blotches in the Milky Way are not due to the absence of stars but rather to attenuation of starlight by interstellar dust. See also **Interplanetary Medium**.

Intertropical Convergence Zone — 1. (Also called ITCZ, equatorial convergence zone.) The axis, or a portion thereof, of the broad trade-wind current of the Tropics. This axis is the dividing line between the southeast trades and the northeast trades (of the Southern and Northern Hemispheres, respectively). It is collocated with the ascending branch of the Hadley cell. At one time it was held that this was a convergence line along its entire extent. It is now recognized that actual convergence occurs only along portions of this line. For further discussion, see **Equatorial Trough**; and **Intertropical Front**.

2. Same as meteorological equator.

Ionospheric Storm — Term used to denote the major changes that take place in the F-region as a result of solar activity. Ionospheric storms are closely associated with magnetic storms and can lead to severe disruptions of radio-wave propagation, particularly at high latitudes. See also **Ionosphere**.

Ionospheric Tides — Term denoting the system of electrical currents and fields generated in the ionosphere by tidal motions in the background atmosphere. These tidal motions are forced mainly by solar heating of atmospheric water vapor and ozone, and reach large amplitudes at ionospheric heights.

Ionospheric Trough — A portion of the F-region, centered on the magnetic dip equator, in which electron densities are anomalously low with peaks at $15°-20°$ latitude on both sides. The *equatorial trough* appears during daytime and is absent at night. It is attributed to diffusion of ionospheric plasma down the magnetic field lines from the magnetic equator as the F-region rises in response to an eastward electric field generated by dynamo action in the lower ionosphere. One

or more troughs in electron density are also found at high latitudes and are magnetically linked to the plasmapause located at a radial distance of several earth radii in the outer magnetosphere. See also **Ionosphere**.

Isallobaric — Of equal or constant pressure change; this may refer either to the distribution of equal pressure tendency in space or to the constancy of pressure tendency with time. The term allobaric, or simply, "pressure change," could be used instead. This term, preferably, should not be used to mean "of isallobars."

Isallobaric Chart — Chart that presents analyses of the changes of atmospheric pressure during a specific time interval.

Isentropic Analysis — Analysis of processes within the atmosphere on the basis of the location and configuration of various isentropic surfaces, distribution of atmospheric processes, and motion on them.

Isentropic Chart — A constant-entropy chart; a synoptic chart presenting the distribution of meteorological elements in the atmosphere on a surface of constant potential temperature (equivalent to an isentropic surface). It usually contains the plotted data and analysis of such elements as pressure (or height), wind, temperature, and moisture at that surface.

Köppen Classification — A climatic classification scheme developed by Wladimir Köppen (1846–1940), a German climatologist. This scheme is based upon annual and monthly means of temperature and precipitation and also takes into account the vegetation limits. It is a tool for presenting the world pattern of climate and for identifying important deviations from this pattern.

Lake Effect — Generally, the effect of any lake in modifying the weather about its shore and for some distance downwind. In the United States, this term is applied specifically to the region about the Great Lakes or the Great Salt Lake. More specifically, lake effect often refers to the generation of sometimes spectacular snowfall amounts to the lee of the Great Lakes as cold air passes over the lake surface, extracting heat and moisture, resulting in cloud formation and snowfall downwind of the lake shore.

Lake-Effect Snow — Localized, convective snow bands that occur in the lee of lakes when relatively cold airflows over warm water. In the United States this phenomenon is most noted along the south and east shores of the Great Lakes during arctic cold-air outbreaks.

Land Breeze — A coastal breeze blowing from land to sea, caused by the temperature difference when the sea surface is warmer than the adjacent land. Therefore, it usually blows by night and alternates with sea breeze, which blows in the opposite direction by day.

Large Scale — In meteorology, a scale in which the curvature of the earth is not negligible. This is the scale of the high tropospheric long-wave patterns, with four or five waves around the hemisphere in middle latitudes. These waves are within the province of both the general circulation and synoptic meteorology, but the terminology should distinguish this scale from that of the migratory high and low pressure systems of the lower troposphere. Rossby waves and other long barotropic waves are large-scale disturbances.

Lee-Wave Separation — The production of small wavelength mountain waves near a mountaintop under conditions of very strong static stability. When the air is very stable and wind speeds are slow, the natural wavelength of air is often much shorter than the width of the mountain, as indicated by a very small Froude number. For this situation, the buoyant restoring force in the air is so strong that the air resists vertical displacement to get over the mountaintop, and instead most of the air flows around the sides of the mountain. The shallow layer of air near the mountaintop that is able to be displaced upward over the mountain will continue in vertical oscillation as it blows downstream, or separates, from the mountain. See also **Downslope Windstorm** in the article **Winds and Air Movement**.

Length of Record — The period during which observations have been maintained at a meteorological station, and which serves as the frame of reference for climatic data at that station. The standard length of record for the purpose of a normal has been fixed by the World Meteorological Organization as 30 years (i.e., three consecutive 10-year periods), which is a reasonable average for the length of a homogeneous record desirable for most of the meteorological elements. Homogeneous records as long as 50 years are rare due to breaks or gradual changes being introduced by changes in the hours of observation, in the observational practices, in the site or instruments used, or by a gradual change in the character of the surrounding country, such as the growth of a city. It is often possible, however, to account for these changes and to construct a composite record that may cover a century or more.

Lithometeor — The general term for dry substances suspended in the atmosphere, including dust, haze, smoke, and sand.

Lowest Astronomical Tide — The lowest level of tide that can be predicted to occur under average meteorological conditions and under any combination of astronomical conditions; often used to define chart datum where the tides are semidiurnal.

Main Meteorological Office (MMO) — An office that provides meteorological service for international air navigation in accordance with International Civil Aviation Organization specifications. MMOs 1) prepare forecasts; 2) supply meteorological information and briefings to aeronautical personnel; 3) supply meteorological information required by an associated dependent meteorological office or supplementary meteorological office.

Main Standard Time — Synoptic hour when meteorological stations make surface synoptic observations that are broadcast on a regional or worldwide scale. The main standard times are 0000, 0600, 1200, and 1800 UTC. See also **Universal Time Coordinated (UTC)**.

Marine Meteorology — The part of meteorology that deals mainly with the study of oceanic areas, including island and coastal regions. In particular it serves the practical needs of surface and air navigation over the oceans. Since there is a close interaction between ocean and atmosphere, and oceanic influences upon weather and climate can be traced far inland over the continents, modern meteorology uses this name mainly for making regional or administrative distinctions.

Marine Weather Observation — The weather as observed from a ship at sea, usually taken in accordance with procedures specified by the World Meteorological Organization. The following elements are usually included: total cloud amount; wind direction and speed; visibility; current weather; pressure; temperature; selected cloud-layer data, that is, amount, type, and height; pressure tendency; seawater temperature; dewpoint temperature; state of the sea (waves); sea ice; and icing onboard ship. Also included are the date and time, and the name, position, course, and speed of the ship. The encoded and transmitted marine observations are known as ship reports.

Mean Temperature — The average temperature of the air as indicated by a properly exposed thermometer during a given time period, usually a day, a month, or a year.

For climatological tables, the mean temperature is generally calculated for each month and for the year. For charts, the observed mean values at station level are reduced to sea level by adding a correction for elevation, usually taken as $0.5\,°C$ ($32.9\,°F$) for each 100 meters (328 feet), but in some mountainous countries different rates are used, based on local observations. See **True Mean Temperature**.

Measured Ceiling — After U.S. weather observing practice, the ceiling classification that is applied when the ceiling value has been determined by means of 1) a ceiling light or ceilometer, provided that penetration of the beam is not in excess of that normally experienced for the height and type of layer and that the elevation angle indicated by the clinometer or ceilometer detector does not exceed 84°; 2) the timed disappearance of a radiosonde balloon with its height computed; 3) the known heights of unobscured portions of objects, other than natural landmarks, within $1\frac{1}{2}$ nautical miles of any runway of the airport. A measured ceiling pertains only to clouds or to obscuring phenomena aloft. It is designated M in aviation weather observations. See **Ceilometer**.

Melting Layer — The altitude interval throughout which ice-phase precipitation melts as it descends. The top of the melting layer is the melting level. The melting layer may be several hundred meters deep, reflecting the time it takes for all the hydrometeors to undergo the transition from solid to liquid phase. The temperature of the melting layer is typically $0\,°C$ ($32\,°F$) or slightly warmer.

Melting Level — The altitude at which ice crystals and snowflakes begin to melt as they descend through the atmosphere. In cloud physics and in radar meteorology, this is the accepted term for the $0\,°C$

($32\,°F$) constant-temperature surface. It is physically more apt than the corresponding operational term, freezing level, for melting of pure ice must begin very near $0\,°C$ ($32\,°F$), but freezing of liquid water can occur over a broad range of temperatures (between $0°$ and $-40\,°C$ ($32\,°F$ and $-40\,°F$).

METAR — 1. Abbreviation for Meteorological Terminal Air Report; also known as Aviation Routine Weather Reports.

2. The international standard code for hourly and special observations that took effect on 1 July 1996. See **Record Observation**.

Meteorograph — an instrument that automatically records the measurement of two or more meteorological elements. See **Aerograph**.

Meteorological Equator — 1. The parallel of latitude of $5\,°N$; so named because this is the annual mean latitude of the equatorial trough.

2. (Also called *equatorial trough*). The axis of the barotropic current that characterizes the low troposphere in equatorial regions. This axis is marked by the presence of a convergence line (the *intertropical convergence zone*). See **Equatorial Trough**; and **Intertropical Convergence Zone**

Meteorological Watch Office — A meteorological office specified under International Civil Aviation Organization procedures to maintain watch over meteorological conditions within a defined area or along designated routes or portions thereof for the purpose of supplying meteorological information, in particular, meteorological warnings. A meteorological watch office may be an independent office or may be part of a main meteorological office (MMO) or dependent meteorological office. In the United States, this office is either a National Center for Environmental Prediction (an independent office) or a Warning and Forecast Office (an MMO).

Micrometeorology — A part of meteorology that deals with observations and processes in the smallest scales of time and space, approximately smaller than 1 km and less than a day (i.e., local processes). Micrometeorological processes are limited to shallow layers of frictional influence (slightly larger-scale phenomena like convective thermals are not part of micrometeorology). Therefore, the subject of micrometeorology is the bottom of the atmospheric boundary layer; namely, the surface layer. Exchange processes of energy, gases, etc., between the atmosphere and the surface (water, land, plants) are important topics. Therefore, micrometeorology is closely connected with most of the human activities in the atmosphere. Microclimatology describes time averaged (long- term) micrometeorological processes, and micrometeorologists are interested in their fluctuations.

Mountain Observation — A collection of simultaneous meteorological measurements taken and recorded in a mountainous location. The harshness of the high-mountain environment, the inaccessibility of sites, and the remoteness of these regions are special problems that have limited the availability of long-term records of mountain weather. These difficulties are compounded by the issue of representativeness of a measurement. Over flatter, simpler terrain, care is taken to place instrumentation in exposed locations where the measurement can be considered as representing a larger area. Barry (1922) defines at least three types of situations in the mountains: "summit, slope, and valley bottom — apart from considerations of slope orientation, slope angle, topographic screening, and irregularities of small-scale relief." Thus it is very difficult, perhaps inappropriate, to claim representativeness for a single observation, and one must interpret mountain observations with great caution. Barry further states, "These factors necessitate either a very dense network of stations or some other approach to determining mountain climate. In the future, the use of ground-based and satellite remote sensors combined with intensive case studies of particular phenomena, may provide the best solution."

Multiyear Ice — Sea ice that has survived more than two summer melt seasons. Such ice is typically 3 meters (10 feet) or more thick, is less saline, and has smoother hummocks and ridges than does younger ice. Undeformed multiyear ice is distinguished by its undular surface (remnants of drained or refrozen melt ponds). Multiyear ridges are distinct from first-year ridges in that they are typically smaller, more rounded, nearly solid ice and are therefore a serious impediment to surface ships.

Nautical System—A system for expressing distance, speed, and acceleration in which 1) the distance of one minute of arc along a meridian or great circle is one nautical mile; 2) a nautical mile per hour is a knot; 3) a nautical mile per hour per hour is the acceleration in knots per hour. Although the nautical system originated with marine operations, it has been adopted to report winds and aircraft speeds.

Neoglacial—A period of general expansion of glaciers variously defined as spanning from approximately 3000 to 2000 years ago or covering the last 4000–5000 years. See also **Glacier**.

Nephology—The study of clouds.

Nova Zemlya—This phenomenon owes its name to an event in 1596 when explorers (in search of the northeast passage) wintering on the island of Nova Zemlya saw a distorted image of the sun two weeks before astronomical calculation would have had it rise. Since that time, the term has been used generically for any such observation of an image of the sun when the actual sun was substantially below the horizon. In the original case, the angular difference between the image and the object was 5°. This is explained by the ducting of the sunlight between the surface and a lifted inversion.

Oasis Effect—Evaporative cooling effect due to heat advection when a source of water exists in an otherwise arid area. In addition to true desert oases, the oasis effect is also characteristic of natural bodies of water in arid surroundings, melting snow patches, irrigated fields in arid areas, or irrigated urban lawns and parks. Latent heat flux from such an oasis can exceed the locally available radiative flux twofold; advection of sensible heat from surrounding warmer surfaces and airmass subsidence over the cooler area provides the remainder. Evaporation also exceeds the local precipitation, the extra water coming from wells, river flow, and irrigation.

Obscuration—1.   (Also called obscured sky cover.) In U.S. weather observing practice, the designation for the sky cover when the sky is completely hidden by surface-based obscuring phenomena. It is encoded "X" in aviation weather observations; it always constitutes a ceiling, the height of which is the value of vertical visibility into the obscuring phenomenon.

2.   A surface-based obscuring phenomenon.

Obscuring Phenomenon—(Also called obscuration.) Any collection of particles, aloft or in contact with the earth's surface, dense enough to be discernible to the observer. Examples are haze, dust, smoke, fog or ice fog, spray or mist, drifting or blowing snow, duststorms or sandstorms, dust whirls or sand whirls, and volcanic ash. Potentially, all hydrometeors and lithometeors may be obscuring phenomena.

Opaque Sky Cover—In U.S. weather observing practice, the amount (in tenths) of sky cover that completely hides all that might be above it; opposed to transparent sky cover. In the case of an obscuration or partial obscuration, the total sky cover attributed thereto must be opaque.

Operational Weather Limits—(Also called minimums.) The limiting values of ceiling, visibility and wind, or runway visual range, established as safety minimums for aircraft landings and takeoffs. Civil aircraft operate under limits stated in Civil Air Regulations and military aircraft operate under limits established by the respective military organizations. Limits for day and night operations usually differ. Also, the limits vary according to airport environment, navigational aids, and type of aircraft.

Orographic—Relating to mountains and mountain effects. Often refers to influences of mountains or mountain ranges on airflow, but also used to describe effects on other meteorological quantities such as temperature, humidity, or precipitation distribution. A major effect is orographic lifting. See also **Clouds and Cloud Formation**; **Orographic Lifting**; **Orographic Precipitation**; **Orographic Vortex**; and **Winds and Air Movement**.

Pericenter—the point of any orbit nearest to the center of attraction; the opposite of apocenter. See also **Perihelion**.

Perigee—the point in an orbit at which any orbiting object, for example, planet, moon, artificial satellite, is closest to the attracting body; the opposite of apogee. See also **Perihelion**.

Physical Meteorology—A subfield of meteorology generally restricted to that part of meteorology not explicitly devoted to atmospheric motions. There is no real distinction between atmospheric physics and physical meteorology. Physical meteorology usually deals with optical, electrical, acoustic, and thermodynamic phenomena of the troposphere, its chemical composition, the laws of radiation, and the physics of clouds and precipitation.

Polar Air—A type of air mass with characteristics developed over high latitudes, especially within the subpolar highs. Continental polar air (cP) has low surface temperature, low moisture content, and, especially in its source regions, great stability in the lower layers. It is shallow in comparison with arctic air. Maritime polar air (mP) initially possesses similar properties to those of continental polar air, but in passing over warmer water it becomes unstable with a higher moisture content. See **Arctic Air**.

Polar Trough—In tropical meteorology, a wave trough in the circumpolar westerlies having sufficient amplitude to reach the Tropics in the upper air. At the surface it is manifest as a trough in the tropical easterlies, but at moderate elevations it is characterized by westerly winds. It moves generally from west to east, accompanied by considerable cloudiness. Cumulus congestus and cumulonimbus clouds are usually found in and around the trough lines. Early- and late-season (June and October) hurricanes of the western Caribbean frequently form in polar troughs.

Portable Mesonet Stations—Portable observation systems that can be set up and operated unattended at field sites to measure, for example, temperature, winds, humidity, surface pressure, solar and infrared radiation, rainfall, and sometimes more specialized variables such as sensible and latent heat fluxes, momentum flux, surface radiation temperature, net radiation, soil moisture, and soil heat flux. They are often self-contained, powered by solar cells and batteries, and able to transmit data to a base station from remote locations. Typically they are deployed in a network from a few kilometers to tens of kilometers apart to provide baseline data for mesoscale multifaceted observational studies.

Precipitation Ceiling—After U.S. weather observing practice, a ceiling classification applied when the ceiling value is the vertical visibility upward into precipitation. This is necessary when precipitation obscures the cloud base and prevents a determination of its height. All precipitation ceilings are estimations, but the following are used as guides: the height corresponding to the upper limit of a ceilometer reaction; the top of a ceiling-light projector beam; or the height at which a ceiling balloon or pilot balloon completely disappears. These guides usually indicate values that are lower than the actual vertical visibility. Precipitation ceilings are designated P in aviation weather observations.

Precipitation Station—A station where only observations of precipitation are made. See **Third-Order Climatological Station**.

Racoon—A zero-pressure balloon flying high above a very cold tropopause in tropical or summer midlatitudes. When the balloon cools and descends at night, its radiation temperature does not change, but its lift increases as it descends to the colder levels. The lost lift is overcome and the balloon floats at a lower altitude without the need for ballast. The flight altitude of the balloon is radiation-controlled.

Radar Meteorological Observation—(Or radar weather observation.) An evaluation of the echoes that appear on the display of a weather radar in terms of the orientation, intensity, tendency of intensity, height, movement, and unique characteristics of echoes, which may be indicative of certain types of severe storms (such as hurricanes, tornadoes, or thunderstorms).

Radar Meteorology—The study of the atmosphere and weather using radar as the means of observation and measurement. One of the branches of physical meteorology, it shares much in common with cloud physics and, more generally, atmospheric remote sensing. For essays on the history of radar meteorology and research reviews, see Atlas (1990) Ref.

Radio Meteorology—That branch of meteorology embracing the propagation of radio waves in the atmosphere and the use of such waves for the remote sensing of clouds, storms, precipitation, turbulence, winds, and various physical properties of the atmosphere.

Record Observation—(Commonly called hourly observation.) A type of aviation weather observation; the most complete of all such observations, usually taken at regularly specified and equal intervals (hourly, usually on the hour). This type of observation has been replaced by the METAR. See **METAR**.

Regional Basic Synoptic Network — A WMO network consisting of synoptic stations with a specified observational program satisfying minimum regional requirements that enable members to fulfill their responsibilities within World Weather Watch and in the application of meteorology. See **Synoptic Station**.

Regional Meteorological Center — A center for the Global Data Processing System that has the primary purpose of issuing meteorological analyses and prognoses on a regional scale for a specified geographic area.

Regional Meteorological Telecommunications Network (RMTN) — A broadcast system for distributing various National Weather Service weather charts and information to an area that is smaller than an entire hemisphere.

Rotoscope — 1. A device used in experimental meteorology for viewing the relative motions in a rotating fluid system. It consists of a Dove prism mounted in a rotating barrel and aligned along the axis of rotation of the vessel containing the fluid. Since the image in the prism rotates twice during each revolution of the prism itself, the rotating vessel will appear stationary if the rate of rotation of the barrel is one-half the rotation of the vessel, and in the correct sense.

2. An instrument that measures the flow velocity of a gas through a tube. It consists of a lightweight piston, externally threaded, that fits loosely inside a vertically mounted glass tube. The gas flows in at the bottom of the tube, lifting the piston and imparting a rotary motion to it as air flows through the threaded section. The height to which the piston rises is a measure of the flow velocity of the particular gas.

Satellite Meteorology — The use of artificial earth-orbiting satellites for the purposes of imaging the atmospheric, land, and oceanic systems; providing atmospheric profiling; and collecting and relaying environmental data. Satellite meteorology involves the sampling of weather and climate features in time and space, as well as the development of new algorithms and interpretation methods, satellite sensors, and products for weather and weather applications.

Second-Order Climatological Station — A station at which observations of atmospheric pressure, temperature, humidity, winds, clouds, and weather are made at least twice daily at fixed hours, and at which the daily maximum and minimum of temperature, the daily amount of precipitation, and the duration of bright sunshine are observed.

Second-Order Station — After U.S. National Weather Service practice, a station staffed by personnel certified to make aviation weather observations and/or synoptic weather observations.

Sonora Weather — A term used by old-timers living in the coastal plain of Southern California to describe summer episodes of hot, humid weather associated with widespread middle and high cloudiness and thunderstorm activity, sometimes intense, over the mountains and deserts to the east. Summer thunderstorms, unusual over the coastal plain of southern California, occur during such episodes. Sonora weather is a particular manifestation of the Southwest monsoon, occurring when the more typical summer monsoonal flow of midlevel moisture into Arizona and New Mexico (from Sonora State, Mexico, and points south) is shunted westward.

Special Weather Statements — Statements issued by a weather service that discuss weather situations including long-term events that describe phenomena of interest or concern to a number of users. These are used when a warning is in progress to provide additional detail, describe events, and provide appropriate response recommendations. These statements may refer to inclement or hazardous weather, but are not so limited.

Standard Atmosphere — 1. A hypothetical vertical distribution of atmospheric temperature, pressure, and density that, by international agreement, is taken to be representative of the atmosphere for purposes of pressure altimeter calibrations, aircraft performance calculations, aircraft and missile design, ballistic tables, etc. The air is assumed to obey the perfect gas law and the hydrostatic equation, which, taken together, relate temperature, pressure, and density variations in the vertical. It is further assumed that the air contains no water vapor and that the acceleration of gravity does not change with height. This last assumption is tantamount to adopting a

particular unit of geopotential height in place of a unit of geometric height for representing the measure of vertical displacement, for the two units are numerically equivalent in both the metric and English systems, as defined in connection with the standard atmosphere. The current standard atmosphere is that adopted in 1976 and is a slight modification of one adopted in 1952 by the International Civil Aeronautical Organization (ICAO), which, in turn, supplanted the NACA Standard Atmosphere (or U.S. Standard Atmosphere) prepared in 1925. It assumes sea level values as follows:

- —Temperature 288.15 K (15 °C)
- —Pressure 101 325 Pa (1013.25 mb, 760 mm of Hg, or 29.92 in. of Hg)
- —Density 1225 g m^{-2} (1.225 g L^{-1})
- —Mean molar mass — 28.964 g mole^{-1}.

The parametric assumptions and physical constants used in preparing the current standard atmosphere are as follows.

1. Zero pressure altitude corresponds to that pressure that will support a column of mercury 760 mm high. This pressure is taken to be 1.013250×10^6 dynes cm^{-2}, or 1013.250 mb, and is known as one standard atmosphere or one atmosphere.
2. The gas constant for dry air is 2.8704×10^6 ergs gm^{-1}K^{-1}.
3. The ice point at one standard atmosphere pressure is 273.16 K.
4. The acceleration of gravity is 980.665 cm s^{-2}.
5. The temperature at zero pressure altitude is 15 °C or 288.15 K.
6. The density at zero pressure altitude is 0.0012250 gm cm^{-3}
7. The lapse rate of temperature in the tropopause is 6.5 °C km^{-1}.
8. The pressure altitude of the tropopause is 11 km.
9. The temperature at the tropopause is −56.5 °C.

The ARDC Model Atmosphere, 1959, extended the above standard approximately as follows:

1. The lapse rate from 11 to 25 km is 0 °C km^{-1}.
2. The lapse rate from 25 to 47 km is +3.0 °C km^{-1}; temperature at 47 km is +9.5 °C.
3. The lapse rate from 47 to 53 km is 0 °C km^{-1}.
4. The lapse rate from 53 to 75 km is −3.9 °C km^{-1}; temperature at 75 km is −76.3 °C.
5. The lapse rate from 75 to 90 km is 0 °C km^{-1}.
6. The lapse rate from 90 to 126 km is +3.5 °C km^{-1}; temperature at 126 km is +49.7 °C (molecular-scale temperatures).
7. The lapse rate from 126 to 175 km is +10.0 °C km^{-1}; temperature at 175 km is 539.7 °C (molecular-scale temperatures).
8. The lapse rate from 175 to 500 km is +5.8 °C km^{-1}; temperature at 500 km is 2424.7 °C (molecular-scale temperatures).

The U.S. Extension to the ICAO Standard Atmosphere is essentially a recomputation of the above data from the surface to 300 kilometers (186 miles).

2. A standard unit of atmospheric pressure, the 45° atmosphere, defined as that pressure exerted by a 760-mm column of mercury at 45° latitude at sea level at temperature 0 °C (acceleration of gravity = 980.616 cm s^{-2}). One 45° atmosphere = 760 mm Hg (45°) = 29.9213 in Hg (45°) = 1013.200 mb = 101.320 kPa.

3. A standard unit of atmospheric pressure, defined as that pressure exerted by a 760-mm column of mercury at standard gravity (980.665 cm s^{-2} at temperature 0 °C). This is a unit recommended for meteorological use. One standard atmosphere = 760 mm Hg = 29.9213 in Hg = 1013.250 mb = 101.325 kPa.

4. With respect to radio propagation, that hypothetical atmosphere in which standard propagation exists, that is, one in which the index of refraction decreases with height at a rate of 12 N-units per 1000 ft. See also **Atmosphere (Earth)**.

Standard Precipitation Index — An index developed by McKee et al. [1993] to quantify precipitation deficit at a given location for multiple timescales. Standardized precipitation is the difference of precipitation from the mean for a specified time divided by the standard deviation, where the mean and standard deviation are determined from the climatological record. The fact that precipitation is not normally distributed is overcome by applying a transformation (i.e., gamma function) to the distribution.

Storm Warning—1. A warning of sustained winds of 48 knots (55 mph) or more either predicted or occurring, not associated with tropical cyclones. The storm-warning signals for this condition are 1) one square red flag with black center by day and 2) two red lanterns at night.

2. Prior to 1 January 1958, a warning, for marine interests, of impending winds of from 28 to 63 knots (32 to 72 mph).

3. A message in plain language broadcast to all Navy ships and merchant ships over appropriate Fleet Broadcasts. The warning gives information on the position, movement, intensity, etc., of a low pressure center.

Sumatra—A squall, with wind speeds occasionally exceeding 13 m s^{-1} (30 mph), in the Malacca Strait between Malay and Sumatra during the southwest monsoon (April through November). It usually blows from the southwest, sometimes from the west or northwest, raising a heavy sea on the Malay coast. The wind veers and strengthens and a heavy bank or arch of cumulonimbus arcus passes overhead (arched squall) with heavy rain and often thunder. Sumatras usually occur at night; they bring a sudden drop in temperature and are generally due to the descent of air cooled by radiation on the high ground of northern Sumatra. In a few cases they mark an air mass boundary during the advance of the monsoon. They are said to occur simultaneously along a line of 320 km (200 miles) or more that advances in a direction between southeast and northeast at about 9 m s^{-1} (20 mph).

Supercell—An often dangerous convective storm that consists primarily of a single, quasi-steady rotating updraft, which persists for a period of time much longer than it takes an air parcel to rise from the base of the updraft to its summit (often much longer than 10–20 min). Most rotating updrafts are characterized by cyclonic vorticity (*see* mesocyclone). The supercell typically has a very organized internal structure that enables it to propagate continuously. It may exist for several hours and usually forms in an environment with strong vertical wind shear. Supercells often propagate in a direction and with a speed other than indicated by the mean wind in the environment. Such storms sometimes evolve through a splitting process, which produces a cyclonic, right-moving (with respect to the mean wind), and anticyclonic, left-moving, pair of supercells. Severe weather often accompanies supercells, which are capable of producing high winds, large hail, and strong, long-lived tornadoes.

Supplementary Meteorological Office (SMO)—An office that, according to International Civil Aviation Organization specifications, is competent to supply aeronautical personnel with 1) meteorological information received from a main meteorological office or dependent meteorological office; 2) meteorological reports otherwise available.

Surface Chart (Also called *surface map, sea level chart, sea level pressure chart*)—An analyzed chart of surface weather observations. Essentially, a surface chart shows the distribution of sea level pressure, including the positions of highs, lows, ridges, and troughs and the location and character of fronts and various boundaries such as drylines, outflow boundaries, sea-breeze fronts, and convergence lines. Often added to this are symbols of occurring weather phenomena, analysis of pressure tendency (isallobars), indications of the movement of pressure systems and fronts, and perhaps others, depending upon the intended use of the chart. Although the pressure is referred to mean sea level, all other elements on this chart are presented as they occur at the surface point of observation. A chart in this general form is the one commonly referred to as the weather map. When the surface chart is used in conjunction with constant-pressure charts of the upper atmosphere (e.g., in differential analysis), sea level pressure is usually converted to the height of the 1000-mb surface. The chart is then usually called the 1000-mb chart.

Surface Weather Observation—An evaluation of the state of the atmosphere as observed from a point at the surface of the earth, as opposed to an upper-air observation. This term is applied mainly to observations that are taken for the primary purpose of preparing surface synoptic charts. Major types of surface observation are synoptic weather observation, aviation weather observation, and marine weather observation. Climatological observations may also be included.

Synoptic Analysis—The analysis of synoptic charts, or the body of techniques so employed.

Synoptic Chart—In meteorology, any chart or map on which data and analyses are presented that describe the state of the atmosphere over a large area at a given moment in time. The possible variety of such charts is almost limitless, but in meteorological history there has been a more or less standard set of synoptic charts, including surface charts and the constant- pressure charts of the upper air. Other synoptic charts include isentropic charts and constant- height charts, both used for upper-air analysis. There are a number of auxiliary and special-purpose synoptic charts, including thickness charts, tropopause charts, stability charts, change charts, continuity charts, etc., that have useful applications for preparing forecasts of weather events at various locations.

Synoptic Station—A station at which meteorological observations are made for the purposes of synoptic analysis. The observations are made at the main synoptic times of 0000, 0600, 1200, 1800 UTC and normally at the intermediate synoptic hours of 0300, 0900, 1500, 2100 UTC and are entered into a coded format for dissemination.

Synoptic Weather Observation—A surface weather observation, made at periodic times (usually at three-hourly and six-hourly intervals specified by the World Meteorological Organization), of sky cover, state of the sky, cloud height, atmospheric pressure reduced to sea level, temperature, dewpoint, wind speed and direction, amount of precipitation, hydrometeors and lithometeors, and special phenomena that prevail at the time of the observation or have been observed since the previous specified observation.

Temperature-Efficiency Index (T-E Index)—For a given location, a measure of the long- range effectiveness of temperature (thermal efficiency) in promoting plant growth. Numerically, the T-E index is equal to the sum of the 12 monthly temperature-efficiency ratios (T-E ratio).

Temperature-Efficiency Ratio (T-E Ratio)—For a given location and month, a measure of thermal efficiency equal to the departure of the normal monthly temperature above 32°F divided by 4:

$$\text{T-E ratio} = \frac{T - 32}{4}$$

where T is the normal monthly temperature in degrees Fahrenheit, except that all values below 32° are counted as 32°. No dimensions are assigned to this ratio.

Temperature-Humidity Index (THI)—(Also known as discomfort index, effective temperature). An index to determine the effect of summer conditions on human comfort, combining temperature and humidity.

Several equations have been used to calculate the index, dependent on the data availability:

$$\text{THI} = 0.4(T_d + T_w) + 15$$

$$\text{THI} = 0.55 T_d + 0.2 T_{dp} + 17.5$$

$$\text{THI} = T_d - (0.55 - 0.55 RH)(T_d - 58)$$

where T_d is the dry-bulb temperature in °F, T_w is the wet-bulb temperature in °F, T_{dp} is the dewpoint temperature in °F, and RH is the relative humidity in percent. In the equation, RH is used as a decimal; in other words, 50% relative humidity is indicated as 0.50. Studies have shown that relatively few people in the summer will be uncomfortable from heat and humidity while THI is 70 or below; about half will be uncomfortable when THI reaches 75; and almost everyone will be uncomfortable when THI reaches 79. There are portions of the United States in which the THI has reached values around 90.

Temperature–Moisture Index — An index indicating the initiation of convective cloud. This is defined as

$$\text{temperature–moisture index} = [150.0 - \text{sfc}/100 - 2(T850 + T700 + T500)]$$

where sfc is the surface elevation of the station in m; T850, T700, and T500 are the 850-, 700-, and 500-hPa temperatures in °C; and PW700 is the precipitable water in cm between the surface and the 700-hPa level. The more negative the index, the greater the probability of convective precipitation.

Thermoneutral Zone — The range of ambient temperature in which normal metabolism provides enough heat to maintain an essentially constant body temperature in homeothermic animals. The limits of the zone depend on the species and breed of an animal and its age, sex, degree of acclimatization, how it is fed, and even the time of day. The zone is narrow for the young of a species (e.g., 29°–30°C for a chick) and wide for a well- fed large adult (e.g., −30° to +25°C for a cow).

Thermotropic Model — A model atmosphere used in numerical forecasting in which the parameters to be forecast are the height of one constant-pressure surface (usually 500 mb) and one temperature (usually the mean temperature between 1000 and 500 mb). Thus, a surface prognostic chart can also be constructed. The quasigeostrophic approximation is employed and the thermal wind is assumed constant with height.

Thickness Chart — A type of synoptic chart showing the thickness of a certain physically defined layer in the atmosphere. Currently it almost always refers to an isobaric thickness chart, that is, a chart of vertical distance between two constant-pressure surfaces and is often proportional to the temperature of that layer. This chart consists of a pattern of thickness lines either drawn directly to data plotted on the chart or, more commonly, drawn by the single graphical process of differential analysis.

Third-Order Climatological Station — As defined by the World Meteorological Organization in 1956, a station, other than a precipitation station, at which the observations are of the same kind as at a second-order climatological station, but are 1) not so comprehensive; or 2) made once a day only; or 3) made at other than the specified hours. This designation is not used officially in the United States, but types of stations that would fit under this category include climatological substations and certain aeronautical weather reporting stations.

Tower of the Winds — An octagonal marble building in Athens erected not later than 35 BC and still standing. The sides face the points of the Athenian compass and carry a frieze of male personifications of the winds from those directions: Boreas (N), Kaikias (NE), Apheliotes (E), Euros (SE), Notos (S), Lips (SW), Zephyros (W), and Skiron (NW). These figures are reproduced on the tower of the Radcliffe Observatory, Oxford, United Kingdom, and in the Library of the Blue Hill Observatory near Boston, Massachusetts. The Tower was not a meteorological observatory, though it originally carried a wind vane on the roof, but was built to measure time, the walls bearing sundials, with a water clock inside for use during cloudy weather.

Transosonde — The flight of a constant-level balloon, the trajectory of which is determined by ground tracking equipment.

Thus, it is a form of upper-air, quasi-horizontal "sounding." The most usual observations are of successive positions of the balloon located by radio direction-finding or radio-navigation equipment, giving trajectory, wind speed, and wind direction. Instrumentation can also be added to the balloon to sense and transmit pressure, temperature, relative humidity, and other meteorological elements.

Tropical Meteorology — The study of the tropical atmosphere. The dividing lines, in each hemisphere, between the tropical easterlies and the midlatitude westerlies in the middle troposphere roughly define the poleward boundaries of this region. Whereas many circulation systems in middle and high latitudes are nearly adiabatic and quasigeostrophic, tropical systems are often strongly influenced by cumulus convection and surface heating, and can be less often dealt with using quasigeostrophic techniques. Many tropical circulations are driven or strongly influenced by coupling with the ocean. Examples of important tropical systems include the Hadley and Walker circulations, monsoons, tropical cyclones, the Madden–Julian oscillation, easterly waves, and El Niño–Southern Oscillation. The stratospheric circulation is dominated by the quasi-biennial oscillation and also contains the ascent branch of the Dobson circulation. Although tropical meteorology may be said to be a distinct endeavor, there are strong interactions between tropical and extratropical circulation systems.

True Mean Temperature — As adopted by the International Meteorological Organization, a monthly or annual mean of air temperature based upon hourly observations at a given place, or on some combination of less frequent observations designed to represent this mean as nearly as possible.

Turbulence Length Scales — Measures of the eddy scale sizes in turbulent flow. The separation between the largest and smallest sizes is determined by the Reynolds number. The largest length scales are usually imposed by the flow geometry, for example, the boundary layer depth. Because turbulence kinetic energy is extracted from the mean flow at the largest scales, they are often referred to as the "energy-containing" range. The smallest scales are set by the viscosity and the rate at which energy is supplied by the largest-scale eddies. Intermediate between these scales are the inertial subrange scales for which turbulence kinetic energy is neither generated nor destroyed but is transferred from larger to smaller scales. Smaller-scale eddies are generated from the larger eddies through the nonlinear process of vortex stretching. Typically, energy is transferred from the largest eddies to the smallest ones on a timescale of about one large- eddy turnover. There are standard turbulence length scales for each of the eddy scale sizes; integral length scales for the energy-containing eddies, Taylor microscale for the inertial subrange eddies, and Kolmogorov microscale for the dissipation range eddies.

Weather Modification — (Also called *weather control*.) In general, any effort to alter artificially the natural phenomena of the atmosphere. The term usually refers to cloud seeding activities, but can also include constructing windbreaks, dissipating fog by the forceful addition of heat or water spray, or preventing frost formation on crops by cloud spray, heating, or mixing processes. Inadvertent weather modification refers to accidental weather effects resulting from the release of greenhouse gases, aerosols, and dust, or changes in albedo or surface properties of the earth associated with urban, industrial, or agricultural activity. See also **Cloud Seeding**.

Wet Deposition — is the removal of atmospheric gases or particles through their incorporation into hydrometeors, which are then lost by precipitation. See **Dry Deposition**.

Wind-Chill Index — A means of quantifying the threat of rapid cooling during breezy or windy conditions that may result in hypothermia in cold conditions. The index is used to remind the public to minimize exposure when outdoors and to take precautionary actions. In the late 1940s, Antarctic explorers Siple and Passel experimented with measuring the time it took to freeze 250 grams (9 ounces) of water in different temperature and wind conditions. They developed empirical formulas relating these data to the rate of heat loss from exposed human skin. They developed the following formula which was used to determine the wind-chill index. At wind speeds of 4 mph (6 kilometers per hour) or less, the wind chill temperature is the same as the actual air temperature:

$$T_{WC} = 0.0817(3.71V^{0.5} + 5.81 - 0.25V)(T - 91.4) + 91.4$$

where V is wind speed in mph and T is temperature in °F.

Winds-Aloft Observation — The measurement and computation of wind speeds and directions at various levels above the surface of the earth. Among the methods employed are 1) the visual tracking, by theodolite, of ascending balloons (pilot-balloon observation or rabal); 2) the use of a radio direction finder to track the radio signals emitted by an ascending radiosonde or other type of transponder (rawinsonde observation or rawin observation); 3) the use of radar to track a balloon-borne radar target, sometimes in combination with a radiosonde (rawin observation). The tracking of high-altitude, constant- level balloons (transosonde) may be considered to fall within this group. Winds-aloft data are included in many aircraft observations, particularly in aircraft weather reconnaissance flights.

Zonal— In meteorology, latitudinal, that is, easterly or westerly; opposed to meridional.

Special Note: Because the subjects described in this article are related to numerous other articles in this encyclopedia, the reader is referred to "Additional Reading" lists which accompany articles on **Atmosphere (Earth)**; **Atmosphere-Ocean Interface**; **Atmospheric Pressure**; **Atmospheric Turbulence**; **Climate**; **Clouds and Cloud Formation**; **Fog and Fog Clearing**; **Gust Front**; **Jet Streams**; **Lightning**; **Ocean**; **Precipitation and Hydrometeors**; **Tsunami**; **Weather Technology**; and **Wind and Air Velocity Measurements**.

Additional Reading

Ackerman, S.A., and J.A. Knox: *Meteorology: Understanding the Atmosphere (with 1pass for MeteorologyNOW)*, 2nd Edition, Thomson Brooks/Cole, Farmington Hills, MI, 2006.

Aquado, E., and J. Burt: *Understanding Weather and Climate*, 4the Edition, Pearson Education, Upper Saddle River, NJ, 2006.

Atlas, D.: *Radar in Meteorology*, American Meteorological Society, Boston, MA, 1990.

Barry, R.G.: *Mountain Weather and Climate*, 2nd Edition, Routledge, London, UK, 1992.

Bean, B.R., and E.J. Dutton: *Radio Meteorology*, National Bureau of Standards Monograph No. 92, U.S. Government Printing Office, Washington, DC, 1996.

De Martonne, E.: *Index of Aridity*, In *Traité de Géographie Physique*, Paris. 1925.

Dunlop, S.: *Dictionary of Weather (Paperback Reference Series)*, Oxford University Press, New York, NY, 2005.

Emanuel, K.A., and S. Emanuel: *Divine Wind: The History and Science of Hurricanes*, Oxford University Press, New York, NY, 2005.

Grotjahn, R.: *Global Atmospheric Circulations: Observations and Theories*, Oxford University Press, New York, NY, 1993.

Holton, J.R.: *An Introduction to Dynamic Meteorology*, 4th Edition, Elsevier Science & Technology Books, New York, NY, 2004.

Jacobs, W.C.: "Aerobiology," In *Compendium of Meteorology*, American Meteorological Society, Boston, MA, 1951, pp. 1103–1111.

Kokhanovsky, A.: *Cloud Optics*, Springer-Verlag New York, LLC, New York, NY, 2005.

Lamb, H.H., and K. Frydendahl: *Historic Storms of the North Sea, British Isles and Northwest Europe*, Cambridge University Press, New York, NY, 2005.

Lutgens, F.K., and E.J. Tarbuck: *Atmosphere: An Introduction to Meteorology*, 10th Edition, Pearson Education, Upper Saddle River, NJ, 2006.

Lynch, A., and J. Cassano: *Applied Atmospheric Dynamics*, John Wiley & Sons, Inc., Hoboken, NJ, 2006.

McKee, T.B., N.J. Doesken, and J. Kleist: "The Relationship of Drought Frequency and Duration to Time Scales," Preprints, *Eighth Conference on Applied Climatology*, Anaheim, CA, American Meteorology Society, 1993, pp. 179–184.

Middleton, W.E.K.: *Invention of the Meteorological Instruments*, Johns Hopkins Press, Baltimore, MD, 1969.

Neuberger, H.: *Arago Point*, In *Introduction to Physical Meteorology*, 1951, pp. 196–204.

Ochoa, G., and T. Tin: *Climate: The Force That Shapes Our World and the Future of Life on Earth*, Rodale Press, Inc., Emmaus, PA, 2005.

Oliver, J.E.: *Encyclopedia of World Climatology*, Springer-Verlag New York, LLC, New York, NY, 2005.

Peixoto, J.P., and A.H. Oort: *Physics of Climate*, Springer-Verlag New York, LLC, New York, NY, 1992.

Saucier, W.J.: *Principles of Meteorological Analysis*, Dover Publications, Mineola, NY, 2003.

Sekera, Z.: *Arago Distance*, In *Compendium of Meteorology*, 1951, pp. 79–90.

Smith, J.: *Facts on File Dictionary of Weather and Climate*, 2nd Edition, Facts on File, Inc., New York, NY, 2006.

Taylor, F.W.: *Elementary Climate Physics*, Oxford University Press, New York, NY, 2005.

Thornthwaite, C.W.: "An Approach toward a Rational Classification of Climate," *Geogr. Rev.*, **38**, 55–94 (1948).

Wallace, J., and P.V. Hobbs: *Atmospheric Science: An Introductory Survey*, 2nd Edition, Elsevier Science & Technology Books, New York, NY, 2006.

American Meteorological Society (AMS), Boston, MA

Peter E. Kraght, Certified Consulting Meteorologist, Mabank, TX

METEOR SHOWER. A term applied to indicate a number of meteoroids coming from the same general part of the sky known as a radiant point. It has been theorized that these large concentrations of falling meteoroids are the remains of disintegrated comets. According to this theory, the earth's orbit may intersect the orbits of any one of several comets, and, due to the attraction of earth's gravity, the meteoroid particles would be attracted and fall into the atmosphere.

Multitudes of these meteoroid particles revolve together around the sun; if clustered together, they are known as "meteor swarms"; if more evenly distributed throughout the orbit, they are called "meteor streams." The orbits of meteor swarms or streams and the orbit of the earth may intersect; and it is then that a meteor shower may be observed from earth, the intensity and duration of which depends upon the distribution of meteoroids within the intersecting portion of the orbit. Some meteor showers occur annually (or biannually), and some periodically but at longer intervals. The showers are commonly named after the constellations or stars in which their radiant points occur. Thus, the most conspicuous and dependable annual display, with trails visible for 2 or 3 weeks, is known as the Perseids (after the constellation Perseus). The Orionids (after the constellation Orion) and the Geminids (after the constellation Gemini) also occur annually. The Draconids, on the other hand, occur only every 13 years, and the Leonids at 33-year intervals. Because meteor showers so often occur in the orbits of comets, they are commonly named after their associated comet as well as after a star or constellation.

Certain meteor showers observed in the daytime (and sometimes only then) have been observed in the course of systematic surveys of meteor activity by means of radar echoes.

See also **Leonids**; and **Perseids**.

METER. The origins of the meter go back to at least the 18th century. At that time, there were two competing approaches to the definition of a standard unit of length. Some suggested defining the meter as the length of a pendulum having a half-period of one second; others suggested defining the meter as one ten-millionth of the length of the earth's meridian along a quadrant (one fourth the circumference of the earth). In 1791, soon after the French Revolution, the French Academy of Sciences chose the meridian definition over the pendulum definition because the force of gravity varies slightly over the surface of the earth, affecting the period of the pendulum.

Thus, the meter was intended to equal 10^{-7} or one ten-millionth of the length of the meridian through Paris from pole to the equator. However, the first prototype was short by 0.2 millimeters because researchers miscalculated the flattening of the earth due to its rotation. Still this length became the standard. (The engraving at the right shows the casting of the platinum-iridium alloy called the "1874 Alloy.") In 1889, a new international prototype was made of an alloy of platinum with 10 percent iridium, to within 0.0001, that was to be measured at the melting point of ice. In 1927, the meter was more precisely defined as the distance, at 0°, between the axes of the two central lines marked on the bar of platinum-iridium kept at the BIPM, and declared Prototype of the meter by the 1st CGPM, this bar being subject to standard atmospheric pressure and supported on two cylinders of at least one centimeter diameter, symmetrically placed in the same horizontal plane at a distance of 571 mm from each other.

The 1889 definition of the meter, based upon the artifact international prototype of platinum-iridium, was replaced by the CGPM in 1960 using a definition based upon a wavelength of krypton-86 radiation. This definition was adopted in order to reduce the uncertainty with which the meter may be realized. In turn, to further reduce the uncertainty, in 1983 the CGPM replaced this latter definition by the following definition:

The meter is the length of the path travelled by light in vacuum during a time interval of 1/299 792 458 of a second.

Note that the effect of this definition is to fix the speed of light in vacuum at exactly 299 792 458 m·s^{-1}. The original international prototype of the meter, which was sanctioned by the 1st CGPM in 1889, is still kept at the BIPM under the conditions specified in 1889.

See also **Units and Standards**.

METHANE. [CAS: 74-82-8]. CH_4, formula weight 16.04, colorless, odorless (when pure) gas, mp $-182.6\,°C$, bp $-161.4\,°C$, sp gr 0.415 (at $-164\,°C$).

Sometimes referred to as *marsh gas* or *fire damp*, methane is practically insoluble in H_2O, and moderately soluble in alcohol or ether. The gas burns when ignited in air with a pale, faintly luminous flame, forming an explosive mixture with air between gas concentrations of 5% and 13%.

Methane is the principal constituent of natural gas, averaging 75% by weight. Natural gas from the Pennsylvania fields is almost 99% methane, but some gas from Kentucky fields contains as little as 23% methane. See also **Natural Gas**. Pipeline gas from several fields typically will contain about 78% methane, 13% ethane, 6% propane, 1.7% butane, and 0.6% pentane. The remaining fraction consists of gases higher in the alkyl series. While generally not referred to as such, methane can be classified as a major fuel. The heating value of pure methane is 995Btu/ft^3 (8856 Calories per cubic meter).

Methane, as the major constituent of natural gas, is an extremely important raw material for numerous synthetic products. For most processes, it is not required to isolate and purify the methane, but the natural gas as received may be used. The high percentage of CH_4 in various feedstocks makes possible the formation of synthesis gas: $CH_4 + H_2O \rightarrow CO + 3H_2$. The percentages of CO and H_2 in synthesis gas vary depending on the end product to be made. Synthesis gas is used widely in the manufacture of NH_3, oxo-chemicals, and methyl alcohol. See also **Synthesis Gas**.

In addition to the preparation of synthesis gas, which is used so widely in various organic syntheses, methane is reacted with NH_3 in the presence of a platinum catalyst at a temperature of about 1,250 °C to form hydrogen cyanide: $CH_4 + NH_3 \rightarrow HCN + 3H_2$. Methane also is used in the production of olefins on a large scale. In a controlled-oxidation process, methane is used as a raw material in the production of acetylene.

Most artificial gases, such as producer gas, coal gas, water gas, manufactured gas, and town gas contain a high content of methane. In addition to its use as a basic chemical and fuel, methane is of notable interest because of its role as the anchor compound of the *alkanes* (paraffin or aliphatic hydrocarbons). All of these compounds may be considered derivatives of methane.

Some methane is manufactured by the distillation of coal. Coal is a combustible rock formed from the remains of decayed vegetation. It is the only rock containing significant amounts of carbon. The elemental composition of coal varies between 60% and 95% carbon. Coal also contains hydrogen and oxygen, with small concentrations of nitrogen, chlorine, sulfur, and several metals. Coals are classified by the amount of volatile material they contain, that is, by how much of the mass is vaporized when the coal is heated to about 900 °C in the absence of air. Coal that contains more than 15% volatile material is called bituminous coal. Substances released from bituminous coal when it is distilled, in addition to methane, include water, carbon dioxide, ammonia, benzene, toluene, naphthalene, and anthracene. In addition, the distillation also yields oils, tars, and sulfur-containing products. The non-volatile component of coal, which remains after distillation, is coke. Coke is almost pure carbon and is an excellent fuel. However, it may contain metals, such as arsenic and lead, which can be serious pollutants if the combustion products are released into the atmosphere.

Carbon monoxide and hydrogen react to form CH_4 in the presence of a nickel catalyst. Methane also is formed by reaction of magnesium methyl iodide in anhydrous ether (Grignard's reagent) with substances containing the hydroxyl group. See also **Grignard Reactions**. Methyl iodide (bromide, chloride) is preferably made by reaction of methyl alcohol and phosphorus iodide (bromide, chloride).

Additional Reading

Buxton, S.: *Guide to Organic Stereochemistry: From Methane to Macromolecules*, Pearson Education, Boston, MA, 1997.

Clever, H. et al.: *Methane*, Elsevier Science, New York, NY, 1987.

Lee, S.: *Methane and Its Derivatives*, Marcel Dekker, Inc., New York, NY, 1996.

Liu, Chang-Jun, M. Aresta, and R.G. Mallinson: *Utilization of Greenhouse Gases*, American Chemical Society, Washington, DC, 2003.

Mastalerz, M. et al.: *Coalbed Methane: Scientific, Environmental, and Economic Evaluation*, Kluwer Academic Publishers, Norwell, MA, 1999.

METHANOGENS. Cells that resemble bacteria in a superficial way, but that have unique genetic and metabolic characteristics. Methanogens are anaerobic, methane-producing microorganisms that occur in a wide variety of places—the gastrointestinal tract of animals, including humans, in the sediments of natural waters, in sewage treatment plant vessels and piping, and in natural hot springs. As proposed by Woese (University of Illinois) and Fox (University of Houston), the methanogens probably make up a third line of descent of cells in addition to the prokaryotes (bacteria and blue-green algae cells which do not have a well-defined nucleus) and eukaryotes (more complex cells with a nucleus). See also **Cell (Biology)**. These researchers also have suggested that there may be still other kinds of cells that do not meet the criteria set down for prokaryotes and eukaryotes.

Methanogens are distinguished from bacteria on at least three counts: (1) The cell walls do not contain muramic acid, the characteristic constituent of the peptidoglycans that form bacterial cell walls. (2) Their metabolism differs markedly from bacteria. A number of coenzymes apparently unique to methanogens have been identified. Some of these enzymes are involved in methyl transfer reactions, including the formation of methane. One of the coenzymes is possibly the smallest coenzyme yet to be discovered. The methanogens also differ in the manner in which carbon dioxide is fixed into cellular carbon. However, the pathway has not been clearly identified. (3) The RNA sequences of methanogens differ from those of other organisms. These observations have indicated to Woese and Fox that although the methanogens share a common ancestor with prokaryotes and eukaryotes, an independent line of descent branched off at possibly about the same time the other cell types diverged.

Although not fully understood, the methanogens place new challenges to the evolutionary biologists for further explanation in terms of the development of early life on earth and may be very valuable toward understanding life on extraterrestrial bodies as these may be explored over future years.

Barker (University of California at Berkeley) and Huntgate (University of California at Davis) as early as the mid-1950s noted that methanogens differ radically from bacteria.

Methanogens take part in the terminal stages of organic matter degradation and survive on carbon dioxide and hydrogen yielded by anaerobic bacteria and converting them to methane.

METHANOL. See **Methyl Alcohol**.

METHOD OF COMPONENTS. A trigonometrical procedure for adding forces in which the components of each force along a chosen set of orthogonal coordinates axes generally symbolized by x, y, and z are determined. The components along the x, y, and z axes are then added up separately to give the components of the resultant force. The magnitude and direction of the resultant force can then be determined from its components. These procedures can be used to add any vector quantities.

METHYL ALCOHOL. [CAS: 67-56-1]. CH_3OH, formula weight 32.04, colorless, liquid at ambient temperatures with mild characteristic alcohol odor, mp −97.6 °C, bp 64.6 °C, sp gr 0.792. Also known as *methanol*, the compound is miscible in all proportions with H_2O, ethyl alcohol, or ether. When ignited, methyl alcohol burns in air with a pale blue, transparent flame, producing H_2O and CO_2. The vapor forms an explosive mixture with air. The upper explosive limit (% by volume in air) is 36.5 and the lower limit is 6.0.

Methyl alcohol possesses distinct narcotic properties. It is also a slight irritant to the mucous membranes. The principal toxic effect is exerted on the nervous system, particularly the optic nerves and possibly the retinae. The effect upon the eyes has been attributed to optic neuritis, which subsides, but is followed by atrophy of the optic nerve. Once absorbed, methyl alcohol is only very slowly eliminated. Coma resulting from massive exposures may last as long as 2 to 4 days. In the body, the products formed by its oxidation are formaldehyde and formic acid, both of which are toxic.

Chemical Properties

Methyl alcohol is a versatile material, reacting (1) with sodium metal, forming sodium methylate, [CAS: 124-41-4], sodium methoxide CH_3ONa plus hydrogen gas, (2) with phosphorus chloride, bromide, iodide, forming methyl chloride, bromide, iodide, respectively, (3) with H_2SO_4 concentrated, forming dimethyl ether $(CH_3)_2O$, (4) with organic acids, warmed in the presence of H_2SO_4, forming esters, e.g., methyl acetate CH_3COOCH_3, [CAS: 79-20-9], methyl salicylate $C_6H_4(OH) \cdot COOCH_3$, [CAS: 119-36-8] possessing characteristic odors, (5) with magnesium methyl iodide in anhydrous ether (Grignard's solution), forming methane as in the case of primary alcohols, (6) with calcium chloride, forming a solid addition compound $4CH_3OH \cdot CaCl_2$, which is decomposed by H_2O, (7) with oxygen,

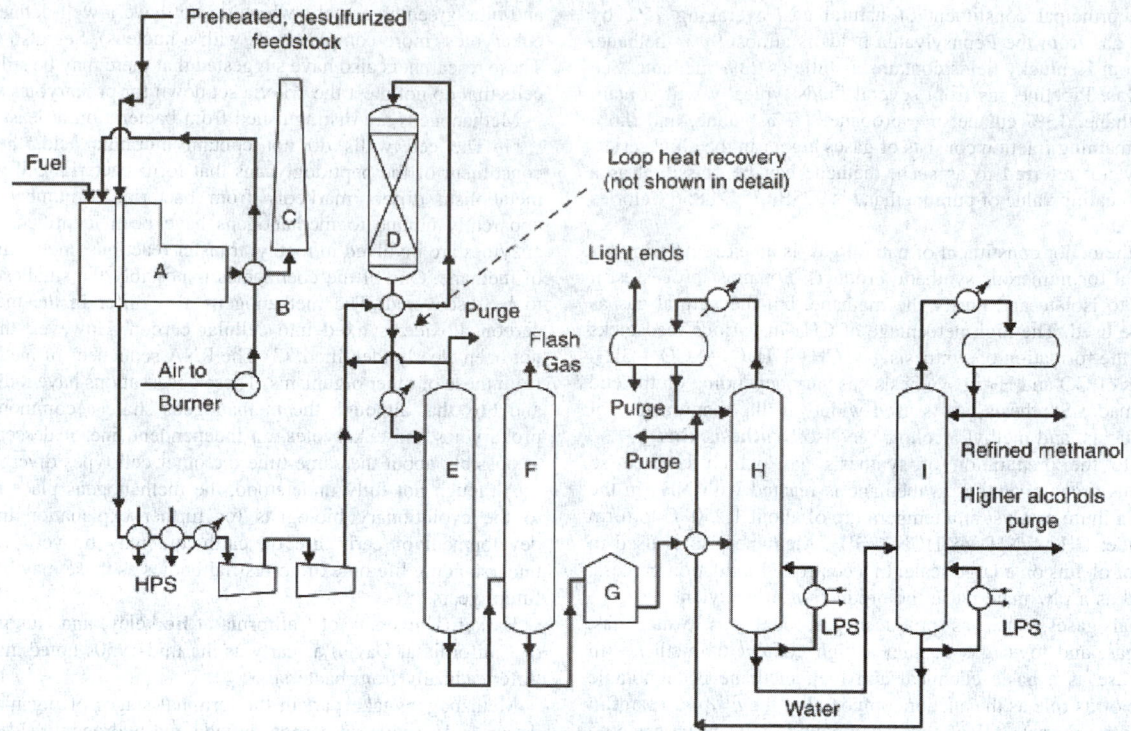

Fig. 1.   Low-pressure methanol production. (A) Burner and superheater; (B) air preheater; (C) stack; (D) methanol converter; (E) separator; (F) flash vessel; (G) crude storage; (H) topping column; (I) refining column. HPS = high-pressure steam; LPS = low-pressure steam. (*Imperial Chemical Industries, Ltd.*)

in the presence of heated smooth copper or silver forming formaldehyde. The density of pure methyl alcohol is 0.792 at 20 °C compared with H_2O at 4 °C (the corresponding figure for ethyl alcohol is 0.789), and the percentage of methyl alcohol present in a methyl alcohol-water solution may be determined from the density of the sample.

A common test for methyl alcohol is by its oxidation in air with a hot copper wire to form formaldehyde.

At one time, most methyl alcohol was obtained by the destruction distillation of hardwoods (hence the name *wood alcohol*) at about 350 °C, along with a yield of acetic acid and small percentages of acetone in the water condensate. Interest in returning to wood as a source has revived because of fossil fuel shortages.

Production of Methyl Alcohol[1]

Synthetic methanol is one of the major raw materials of the organic chemical industry. Methanol has economic stability and a steady growth rate owing to the low costs of production and diversity of applications. Nearly all the methanol producers also make formaldehyde, which is the main end use (more than 50%) of methanol. The other main end uses are dimethyl terephthalate, methacrylates, methylamines (for resins, herbicides, and fungicides), methyl halides (for silicones, tetramethyl lead, butyl rubbers, paint removers, photographic films, aerosol propellants, (diminishing use), and degreasing compounds), acetic acid, and solvents.

An important process for production of synthetic protein uses methanol as feedstock. The use of methanol as a fuel, either as pure methanol, as a mixture (approximately 15%) with gasoline, or as a feedstock for synthetic gasoline is envisaged for possible large-scale application; as well as use in gas turbines for electricity generation. See also **Wastes as Energy Sources**.

There are three principal commercial grades of methanol (as defined in U.S. Federal Specification O-M-232f: June 5, 1975): *Grade A*, synthetic, 99.85% by weight (solvent use); *Grade AA*, synthetic, 99.85% by weight (hydrogen and carbon dioxide generation use); and *Grade C*, wood alcohol (denaturing use).

The most recent advances in methanol synthesis are the low- and intermediate-pressure processes of the type shown in Fig. 1. The synthesis

step of this process developed by Imperial Chemical Industries, Ltd. relies upon a copper-based catalyst, which gives good yields of methanol at pressures of 50 and 100 atmospheres. These pressures are substantially below those of the 250–350 atmospheres required by earlier processes. The high catalyst activity allows the synthesis reaction to take place at a relatively low temperature of 250–270 °C. As a result, methanation is avoided, and byproduct formation is lower, giving increased process efficiency.

The development of this low-pressure technology has caused a major reassessment of the economics of methanol production. The energy required to compress the synthesis gas from its production pressure to the synthesis unit is reduced by a factor between 2 and 3. The lower synthesis pressure allows the exclusive use of centrifugal compressors in plants with capacities as low as 15 million gallons (0.57 million hectoliters) per year. Small producers find attractive the savings in investment, operating, and maintenance costs made possible by low-pressure operation. Plants range in capacity from 15 million gallons (0.57 million hectoliters) to 250 million gallons (9.46 million hectoliters) per year.

Synthesis gas is prepared by the steam reforming or partial oxidation of a liquid or gaseous hydrocarbon feedstock, or by direct combination of carbon dioxide with purified hydrogen-rich gases. Economic considerations usually favor the steam-reforming route for a naphtha or natural gas feedstock. In this instance, desulfurized feedstock is preheated, mixed with superheated steam, and reacted over a conventional catalyst (normally nickel-based) in multitubular reformer. The reformer usually is operated at between 15 and 30 atmospheres and at a tube outlet temperature of 840–900 °C. The reforming conditions are chosen to give the most economic overall production costs. Methane slip (amount of unconverted methane) usually is greater than for conventional high-pressure synthesis processes, since the cost of compressing the additional methane is less significant with the low-pressure process. With a naphtha feedstock, an almost exact stoichiometric ratio of carbon oxides to hydrogen in the synthesis gas is achieved, but when natural gas is the feedstock, there is an inherent deficiency of carbon. Established practice for many years has been to add carbon dioxide from an external source in preparing a stoichiometric synthesis gas. Development of the low-pressure process has shown that this addition of carbon dioxide is not required and that, depending upon the cost

[1] Remainder of this entry prepared by J.R. Masson, Process Engineering Consultant, Davy McKee (Oil & Chemicals) Ltd., London, England.

of carbon dioxide production, the production of methanol from natural-gas feedstock alone is economic.

After heat recovery and cooling, the synthesis gas is compressed to the required synthesis pressure and passed into the synthesis loop at the suction of a circulator. The circulator, which boosts the pressure of the circulating gases to make up the total loop pressure drop, also is a centrifugal machine. Feed-gas preheating is carried out by heat exchange with the hot gases leaving the converter. Heat recovery is incorporated into the loop to recover the heat of reaction of methanol synthesis.

Synthesis takes place in a hot-wall converter over the low-pressure methanol-synthesis catalyst at 250–270 °C. Temperature control of the converter is effected by injecting cold gas at appropriate levels in the catalyst bed, using specially developed distributors that provide excellent gas mixing while allowing free passage of the catalyst for easy charging and discharging. After leaving the converter and passing through the feed-gas preheater the converted gases are cooled, and crude methanol is condensed and separated from the uncondensed gases, which are recycled with makeup synthesis gas to the converter. A continuous gas purge is taken from the synthesis loop in order to remove an accumulation of inert gases. This purge is recycled to the synthesis-gas preparation section as reformer fuel. The crude methanol is reduced in pressure before passing forward to the methanol-purification section, where methanol of the required purity is produced by conventional distillation methods.

Economics in fuel gas consumption are achieved by use of recovered heat in reboiling in the distillation columns. In addition, distillation schemes involving three or four columns have also been developed with reduced reboil heat requirements.

Health and Safety

Methanol is not classified as carcinogenic, but can be acutely toxic if ingested: 100-250 mL may be fatal or result in blindness. The principal physiological effect is acidosis resulting from oxidation of methanol to formic acid. See also **Acidosis**. Methanol is a general irritant to the skin and mucous membranes. Methanol vapor can cause eye and respiratory mucous membranes. Methanol vapor can cause eye and respiratory tract irritation, nausea, headaches, and dizziness.

Methanol does not pose an undue toxicity hazard if handled in well-ventilated areas, and is rated as a slight health hazard by the National Fire Protection Association (NFPA): http://www.nfpa.org/.

Storage and Handling

Menthanol is stable under normal storage conditions. Methanol is not subject to hazardous polymerization reactions, but can react violently with strong oxidizing agents. The greatest hazard involved in handling methanol is the danger of fire or explosion. The NFPA classifies methanol as a serious fire hazard.

Additional Reading

Chang, C.: *Hydrocarbons from Methanol*, Marcel Dekker, Inc., New York, NY, 1983.

Cheng, W. and H. Kung: *Methanol Production and Use*, Marcel Dekker, Inc., New York, NY, 1994.

Murrell, J. and H. Dalton: *Methane and Methanol Utilizers*, Kluwer Academic Publishers, Norwell, MI, 1992.

Web Reference

Methanol Institute: http://www.methanol.org/

METHYL BROMIDE. CH_3Br, (bromomethane) [CAS: 74-83-9], is a colorless liquid or gas with practically no odor. Its physical properties are mp −93.7 °C (−137 °F); bp 3.56 °C (38 °F); d^{20}_4 1.6755 kg/m^3; 3.974 kg/m^3; n_D^{20} 1.4218; vapor pressure at 20 °C (68 °F), 189.3 kPa (1420 mmHg); viscosity at −20, 0, and 25 °C (−4, 32, and 77 °F): 0.475, 0.397, and 0.324 mPa·s, respectively. Heat capacity of the liquid at −13 °C (9 °F) and of the vapor at 25 °C (77 °F), 824 (197) and 448 (107) J/kg·K, (cal/kg·K), respectively; heat of vaporization at 3.6 °C (38.5 °F), 252 J/g (60.2 cal/g); critical temperature (calculated) 194 °C (381 °F); expansion coefficient −15 to 3 °C (5 to 37 °F), 0.00163/K; dielectric constant at 0 °C (32 °F) and 0.001–0.01 MHz, 9.77; dipole moment gas, 1.81D. Methyl bromide is miscible with most organic solvents and forms a bulky, crystalline hydrate below 4 °C (39 °F). Its solubility in water varies

with pressure: at normal pressure, methyl bromide plus water vapor, the solubility is 1.75 g/100-g solution (20 °C (68 °F).

Methyl bromide is present in the atmosphere as the result of both natural (oceanic production) and anthropogenic (use as a soil fumigant) sources; this compound is the largest single source of bromine to the stratosphere, with a tropospheric mixing ratio of about 10 parts per trillion (by volume).

Methyl bromide reacts with several nucleophiles and is a useful methylation agent for the preparation of ethers, sulfides, amines, etc. Tertiary amines are methylated by methyl bromide to form quaternary ammonium bromides. The reactivity of methyl bromide is summarized in Figure 1.

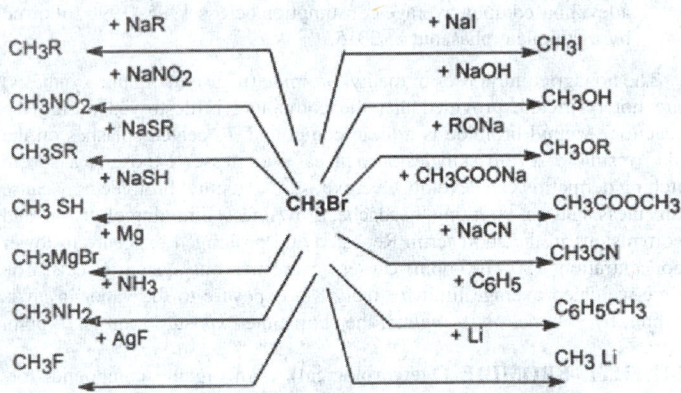

Fig. 1.   The reactivity of alkyl bromides.

Methyl bromide, when dry (<100 ppm water), is inert toward most materials of construction. Carbon steel is recommended for storage vessels, piping, pumps, valves, and fittings. Copper, brass, nickel, and their alloys are sometimes used. Aluminum, magnesium, zinc, and alloys of these metals should not be used, because under some conditions dangerous pyrophoric Grignard-type compounds may be formed. A severe explosion due to the ignition of a methyl bromideair mixture by pyrophoric methylaluminum bromides produced by the corrosion of an aluminum fitting has been reported. Nylon and poly(vinyl chloride) (PVC) should also be avoided for handling methyl bromide.

Methyl bromide is nonflammable over a wide range of concentration in air at atmospheric pressure, and offers practically no fire hazards. With an intense source of ignition its explosive limits are from 13.5 to 14.5% by volume.

The commercial manufacture of methyl bromide is based on the reaction of hydrogen bromide [CAS: 10035-10-6] with methanol [CAS: 67-56-1]. The hydrogen bromide used could be generated in situ from bromine and a reducing agent. The uses of sulfur [CAS: 63705-05-5] or hydrogen sulfide [CAS: 7783-06-4] as reducing agents are described, the latter process having the advantage. A new continuous process for the production of methyl bromide from methanol and aqueous HBr in the presence of a silica supported heteropolyacid catalyst has recently been described. Methyl bromide can also be coproduced with other organic bromine compounds by the reaction of the methanol solvent with hydrogen bromide formed as a by-product. The processes include coproduction of methyl bromide with bromostyrenes, tribromophenol, potassium and sodium bromide, and especially tetrabromo bisphenol A.

Worldwide, methyl bromide is used principally as a space fumigant used for killing soil parasites (nematodes, fungi, weeds, insects, and rodents) in agriculture and for the sanitation of cereal and other crops under storage and before shipment. Methyl bromide is also used as an intermediate for the manufacture of pharmaceuticals (clidinium bromide [CAS: 3485-62-9], clobazam [CAS: 22316-47-8], glycopyrrolate [CAS: 596-51-0], mepenzolate bromide [CAS: 25990-43-6], mepivacaine hydrochloride [CAS: 1722-62-9], methscopolamine bromide [CAS: 155-41-9], pancuronium bromide [CAS: 15500-66-0], propantheline bromide [CAS: 50-34-0], pyridostigmine bromide [CAS: 101-26-8]), biocides (CTAB [CAS: 57-09-0]), insecticides (pirimicarb [CAS: 23103-98-2]), and chemical reagents (methylmagnesium bromide [CAS: 75-16-1] and tetramethylammonium bromide [CAS: 64-20-0]). The current world consumption of methyl bromide as an intermediate is ~1000 t/a (~1.5% of total world consumption).

Due to its role in the depletion of the ozone layer, an international agreement has been reached calling for its reduced consumption and complete phasing out in the developed countries. In September 1997 at the Ninth Meeting of the Parties of Montreal Protocol, members agreed to a schedule for a reduction in the use of the fumigant. See also **Montreal Protocol**; and **Ozone Depletion (Science of)**. The final agreements include

1. For industrial nations, a 25% reduction in use in 1999 followed by a 50% reduction in 2001 and a complete phaseout in 2005 with an allowance for critical exemptions.
2. For developing nations, a 20% reduction in consumption in 2005 at a level based upon average consumption across 1995–1998 followed by a complete phaseout in 2015.

The nonagricultural uses of methyl bromide (its use in organic synthesis) are not restricted, provided that the compound is destroyed during the reaction. Methyl bromide is a toxic compound. Repeated splashes on the skin result in severe skin lesions. In cases of lesser exposure, a severe itching dermatitis can develop. Overexposure to methyl bromide may cause dizziness, nausea, vomiting, headache, drowsiness, dimming of vision, and convulsions in the short term. Repeated and prolonged exposure to lower concentrations (30–100 ppm) causes severe nervous system effects. The time-weighted average limit for daily 8-h exposure to the vapor in air is 5 ppm by volume, or 19 mg/m^3; the short-time exposure limit is 15 ppm.

METHYL BROMIDE (Meteorological). An organic compound, formula CH_3Br, present in the atmosphere as the result of both natural (oceanic production) and anthropogenic (use as a soil fumigant) sources; this compound is the largest single source of bromine to the stratosphere, with a tropospheric mixing ratio of about 10 parts per trillion (by volume).

See also **Ozone Depletion (Science of)**.

METHYL CHLORIDE. [CAS: 74-87-3] (chloromethane, mono chloromethane), at ordinary temperatures and pressures is a colorless gas with a very mild odor and sweet taste. Methyl chloride is the most abundant single halocarbon, formula CH_3Cl, found in the atmosphere, with a mixing ratio of about 600 parts per trillion (by volume) in the troposphere. This compound is mostly of natural origin, as a result of production in the oceans. It provides the natural background amount of chlorine that was believed to be present in the preindustrial stratosphere and that will likely be present in the future following the phaseout of other chlorine source compounds (chlorofluorocarbons, etc.). Methyl chloride is handled commercially as a liquid. It is miscible with the principal organic solvents and only slightly soluble in water. The dry liquid is stable and noncorrosive; however, in the presence of moisture, the liquid slowly decomposes and becomes corrosive to metals, particularly aluminum, zinc, and magnesium. Gaseous methyl chloride is moderately flammable. Prolonged exposure to high concentrations of the vapor can produce severe toxic effects. Methyl chloride is used mainly in the manufacture of silicones, synthetic rubber, and as a general methylating agent. Impure methyl chloride was produced in the laboratory as early as 1835 by Dumas and Peligot, who heated wood spirit, i.e., crude methyl alcohol, with a mixture of sulfuric acid and common salt. It was later made by Schiff and by Walker and Johnson by the reaction of phosphorus chlorides with methyl alcohol. One of the first preparations of pure methyl chloride was probably that of Groves in 1874. Groves passed hydrogen chloride into a boiling solution of zinc chloride in twice its weight of wood spirit. Berthelot obtained the compound by chlorinating methane.

During the last quarter of the nineteenth century, methyl chloride was manufactured on a small scale in Europe for use as a refrigerant and in the synthesis of dyes. Large-scale production began in the United States about 1920, chiefly to meet refrigerant requirements. Manufacture in the United Kingdom began in 1930. Production increased greatly after 1943 when methyl chloride was required as the starting material for methyl silicones and fluorinated refrigerants. After World War II, methyl chloride production in the United States increased more than tenfold and in the late 1980s was between 3.6 and 4.5×10^8 kg per year.

Methyl chloride uses break down into the following categories: intermediates for silicons (89%); methyl cellulose ethers (3%); quaternary ammonium compounds (3%), herbicides (2%); butyl rubber (1%); and miscellaneous uses (2%).

Most methyl chloride is used as an intermediate feedstock in silicone fluids, elastomers, and resins. Silicon fluids are used in processing aids, such as surfactants, release agents, and lubricants. They are also used in cosmetics, polishes, paper coating, and in electrical, pharmaceutical, and medicinal applications. Silicone elastomers are used by the construction industry in sealants and adhesives. Silicone resins are used in high-temperature duty and weather resistant coatings

See also **Chlorinated Organics**.

METHYL CHLOROFORM. [CAS: 71-55-6]. A chlorine-containing organic compound, formula CH_3CCl_3. It is used in industrial applications as a solvent and a degreasing agent. The production of this compound is now forbidden as a result of the Montreal Protocol. See also **Montreal Protocol**. Because its sources to the atmosphere (anthropogenic activity) and its losses (mostly through reaction with the hydroxyl radical) are well understood, the atmospheric abundance of this compound has been used to infer the average hydroxyl concentration in the troposphere.

METRIC SPACE. A generalization of distance and related concepts to abstract sets.

A metric space is a set X such that to each pair of points x and y, there is associated a nonnegative real number $\rho(x, y)$, called the *distance* from x to y. This distance satisfies the conditions: (1) $\rho(x, x) = 0$, (2) $\rho(x, y) = 0$ implies $x = y$, (3) $\rho(x, y) = \rho(y, x)$, and (4) $\rho(x, y) + \rho(y, z) \geq \rho(x, z)$. The latter is the familiar triangular inequality.

The function $\rho(x, y)$ is said to be a metric for X. Examples of metric spaces are the unit circle in the plane and three-dimensional space with the usual distances. Less familiar examples are Hilbert space and any set X with the distance function $\rho(x, y) = 1$ for $x \neq y$ and $\rho(x, x) = 0$.

A metric can be used to define a topology for X called the *metric topology*. The set $S(x, \varepsilon)$ of points in X at distance less than ε from the point x is said to be the *open ball* of radius ε centered at x. A subset U of X is then said to be *open* in the metric topology if for each x in U there is an $\varepsilon > 0$ so that $S(x, \varepsilon)$ is contained in U. A topological space (X, τ) is said to be *metrizable* if a metric can be defined on X in such a way that open sets in τ are open in the metric topology and, conversely, most familiar topological spaces are metrizable.

A sequence of points, $x_1, x_2, x_3, \ldots$ in the metric space X is said to *converge* to the point x in X if the sequence of real numbers $\rho(x, x_1), \rho(x, x_2), \rho(x, x_3), \ldots$ converges to 0. A sequence converges in this sense if, and only if, it converges in the metric topology.

A sequence of points $x_1, x_2, x_3, \ldots$ in X is said to be a *Cauchy sequence* if for $\varepsilon > 0$ there exists an integer N so that $n, m \geq N$ implies $\rho(x_n, y_m) < \varepsilon$. A sequence that converges to a point in X can be shown to be a Cauchy sequence. A metric space is said to be *complete* if every Cauchy sequence converges to some point in X. The real number line is an example of a complete metric space, while the set of points between 0 and 1 with the usual distance is not complete.

See also **Topology**.

METROLOGY. Metrology usually is described as the field of knowledge concerned with measurement. However, this rather general description can be slightly misleading, as it does not sufficiently emphasize that any metrological consideration and operation ultimately focus on the notions of the accuracy and uniformity of measurement. In fact, as metrology concentrates on these two aspects of measurement, it does not encompass the bulk of experimental physics, but rather can be viewed as a branch of applied physics. As such, it provides the scientific basis of measurement with special regard to the accuracy of measurement as well as to methods and means to ensure uniformity of measurements. Accuracy and uniformity of measurement contribute to the growth of scientific knowledge, bring about reliability and confidence, support the quality of products, protect the consumer, and add to the capability of economy to successfully compete in the markets. Therefore, metrology is important to all fields of science, technology, and commercial or social transactions.

In practice, metrological tools and operations can be expensive. This is particularly true for measurements on the highest possible level of accuracy such as measurements of the fundamental constants of physics.

They usually are performed in national metrology laboratories or academic institutions. But even at somewhat lower levels of accuracy, such as in manufacturing industries, adequate metrology may still cause significant cost. Nonetheless, appropriate on-line measurements of critical parameters of a manufacturing process, acknowledged by many as introducing "non-value-added" expense, can often be shown to be ultimately less expensive than the application of only poor production metrology (see e.g. Kudva and Potter, 1992). In fact, the present level of development of some technologies, such as microelectronics, would not have been possible without paralleling advances in production metrology.

Metrology may be divided into three partially overlapping fields:

1. Scientific metrology, dealing with the theory of systems of units; the theory of measurement; methods of evaluation of results of measurement and uncertainty of measurement (numerical expression of the accuracy of measurement); terminology; realization of the units on the highest possible accuracy level; and very accurate measurements of, for instance, fundamental constants, or measurements to test or compare physical theories.

2. Applied scientific metrology, dealing with all components of practical importance that influence the accuracy and uniformity of measurements. Such components are the closeness between the actual reference quantity used in measurement and the defined unit the reference quantity is supposed to represent; the properties of the measuring instruments; and the choice of the measurement procedure. Thus, applied metrology includes the representation of the units or of their multiples and submultiples by standards arranged in steps of uncertainty, the maintenance of the standards, the dissemination of the units by comparison of standards; and calibration of measuring instruments.

3. Legal metrology, dealing with the inspection examination of properties of measuring instruments intended to be used in the public interest. The properties must meet legally specified requirements (mostly in the form of specified maximum permissible errors of the instrument). Regulations are set up, for instance, for instruments used in consumer protection or in medicine, or for instruments monitoring environmental pollution. Legal metrology is not intimately connected to scientific metrology and its direct applications.

Historically, up to the last few decades, metrology was perceived by much of the scientific community as a science of auxiliary and custodial character, not contributing essential elements to scientific progress. This view simply disregards important scientific discoveries by metrologists such as the establishment of the laws of blackbody radiation in 1900. After development of a method to realize accurately the concept of the blackbody, it was the subsequent accurate measurement of its radiation's spectral energy distribution that enabled Planck to initiate quantum physics.

But undoubtedly the tasks of metrology for a long time were predominantly understood as to provide, maintain, and disseminate a set of units—in particular, of length and mass—in order to support uniformity of measurements in commercial and social transactions. But the units of length and mass in former times were only locally defined by arbitrary man-made artifacts, and these were reproduced as accurately as possible to yield reliable secondary material measures valid in the geographical region for which the units were defined. In Europe, for example, this local restriction often made trade difficult even between different cities of the same country, as the cities usually insisted on having their own units of length and mass.

It was only in 1875 that a first step to national and international uniformity was made. In this year, because of the apparent growing need for common standards and accurate measurements in the emerging industrial society, the Convention du Métre (agreement on the metric system) was signed by 17 countries, including the United States. A decimal system of weights and measures was established and the agreed-on units, the prototypes, of length and mass were decided to be kept in Paris. This event can be viewed as the birth of modern metrology. Additionally, to propagate and further to develop the metric system, the Conférence Générale des Poids et Mesures (CGPM), the Bureau International des Poids et Mesures (BIPM), and the Comité International des Poids et Mesures (CIPM) were founded at that time. See also **Unit**.

In the sequel, national metrological institutions were installed that, on the basis of the above-mentioned agreement, were in charge of scientific metrological research and of the national uniformity of measurement in technology and industry by realizing, maintaining, and disseminating national standards in concordance with the internationally agreedon units. The first of such institutes were the Physikalisch-Technische Reichsanstalt (now Physikalisch-Technische Bundesanstalt, PTB), founded in 1887 in Germany, http://www.ptb.de/index_en.html, the National Physical Laboratory, NPL, founded in 1900 in England, http://www.npl.co.uk/, and the National Bureau of Standards (now National Institute of Standards and Technology, NIST), founded in 1901 in the U.S., http://www.nist.gov/.

In the course of time, progress in measurement showed that particularly the accuracy of a length measurement was nearly exclusively determined by the accuracy with which the actual reference quantity in measurement could be compared with the prototype in Paris. For example, whereas the lattice parameter of a crystal could very accurately be measured in terms of a specific x-ray wavelength, on expressing the lattice parameter in the length unit meter the experimental accuracy immediately was lost.

It was only when improved interferometric measurement techniques suggested redefining the meter as a multiple of the wavelength of a characteristic electromagnetic radiation of krypton atoms (^{86}Kr) that the accuracy in length measurements was increased. The multiple was chosen in such a way that for all practical purposes there was no change to the original prototype definition of the meter (concept of the "continuity of the units").

The unit of length with its above new definition was one member of the set of seven base units of the coherent International System of Units (SI) that in 1960 replaced all of the various earlier unit systems. The definitions of these base units, however, should not be viewed as fixed once and for all within the SI. For example, according to the needs of even higher accuracy and taking into account the principle of continuity, a redefinition of the length unit was agreed on in 1983, tracing it to the definition of the unit of time, the second, which can be realized with a higher accuracy than any other of the base units. Also, in 1979, the 1967 definition of the unit of luminous intensity, the candela, was changed.

The least satisfying definition in the SI is that of the unit of mass, the kilogram, which since 1889 is given by an artifact maintained in Paris at the BIPM. Any change with time of the mass of the prototype results in the change of the unit. Great efforts are being made in modern metrology to replace this anachronism by a very accurate determination of the Avogadro constant.

However, whereas the increasing sophistication of measuring instruments and techniques allow for improvement of the utmost accuracy of measurement by appropriate definition of the SI units, the corresponding primary or even secondary standards as well as the methods of unit transfer may differ markedly from their prototype predecessors. An example is the standard of the meter, represented by the wavelength of a specific visible line of the He-Ne laser. The line is linked experimentally to the frequency (in the infrared) of the radiation corresponding to a well-defined transition of the cesium atom, i.e., it is derived from the radiation used in defining the second. This standard meets scientific needs and is known to be of a high accuracy, but it is far from the much more practical line scale, which allows for meter comparisons by a simple physical operation. Elaborate interferometry must be used to compare the wavelength with a line scale accurately.

Therefore, quite a large part of metrological work in the national institutions consists of providing a hierarchy of standards of decreasing accuracy. The chain starts from the first transfers from the unit and, as in the case of the line scale, ends in an appropriate physical embodiment of the unit to be used in practice. The hierarchy of standards is of special importance in quality management of measuring equipment and standards. Here, for calibration purposes measurement standards must have the property of "traceability," which means that the accuracy of the standard in use must be traceable to that of a corresponding national or international standard through a documented unbroken chain of comparison.

The basic idea underlying the SI is to define, whenever possible, a unit not by an arbitrary artifact, but by a property of a prescribed specific natural phenomenon that can be realized very accurately at any place and time. The impact of this concept on metrology resulted in a change of its role and scope in science. Modern metrology is acknowledged to be a universal discipline intimately linked to nearly all frontier developments of the physical sciences and of technology.

See also **International System Of Units (SI) Unit**; and **Units and Standards**.

Additional Reading

Bowen, D.K., and B.K. Tanner: *X-Ray Metrology*, CRC Press LLC, Boca Raton, FL, 2006.

Bucher, J.L.: *Metrology Handbook*, ASQ Quality Press, Milwaukee, WI. 2004.

Ciarlini, P., M.G. Cox, G.B. Rossi, and F. Pavese: *Advanced Mathematical and Computational Tools in Metrology VI*, World Scientific Publishing Company, Inc., Hackensack, NJ, 2004.

de Silva, G.M.: *Basic Metrology for ISO 9000 Certification*, Butterworth-Heinemann, Woburn, MA, 2002.

Kudva, S.M., and R.W. Potter: In M.T. Postek (Ed.), 6th Annual Conference on Integrated Circuit Metrology, Inspection and Process Control, SPIE Proceedings Vol. 1673, SPIE. Bellingham, WA, 1992.

Mercer, C.R.: *Optical Metrology for Fluids, Combustion, and Solids*, Kluwer Academic Publishers, Norwell, MA, 2003.

Pennella, C.R.: *Managing the Metrology System*, ASQ Quality Press, Milwaukee, WI, 2004.

Smith, G.T.: *Industrial Metrology: Surfaces and Roundness*, Springer-Verlag New York, LLC, New York, NY, 2002.

Staff: United Nations, *Road Map for Quality: Guidelines for the Review of the Standardization, Quality Management, Accreditation and Metrology (SQAM) Infrastructure at National Level*, International Trade Centre, Geneva–Switzerland, 2005.

Staff: French College of Metrology, *Metrology in Industry: The Key for Quality*, Iste Publishing Company, Washington, DC, 2006.

Trusler, J.P.: *Physical Acoustics and Metrology of Fluids*, 2nd Edition, Taylor & Francis, Inc., Philadelphia, PA, 2007.

WOLFGANG WÖGER
SIGMAR GERMAN
Physikalisch-Technische Bundesanstalt, Braunschweig, Germany

METRONIC CYCLE. A period of 19 years, after which the various phases of the moon fall on approximately the same days of the year as in the previous cycle. The Metonic cycle is the basis for the golden numbers used to determine the data of Easter. Four such cycles form a Callippic cycle.

MEYER REACTION. Preparation of alkylstannonic acids by reacting alkali stannite with an alkyl iodide. When applied to alkali arsenites or plumbites, the reaction yields alkylarsonic and alkylplumbonic acids, respectively.

MEYER-SCHUSTER REARRANGEMENT. Acid catalyzed rearrangement of secondary and tertiary α-acetylenic alcohols to α,β-unsaturated carbonyl compounds: aldehydes when the acetylenic group is terminal, ketones when it is internal.

MICA. [CAS: 12001-26-2]. Mica is a generic term that applies to a wide range of hydrous aluminum silicate minerals characterized by sheet or plant-like structure, and possessing to varying degrees, depending on composition and weathering, flexibility, elasticity, hardness, and the ability to be split into thin (1 μm) sheets. All micas form flat six-sided monochromic crystals, and possess cleavage parallel to the basal plane.

Mica exists in nature in a wide variety of compositions. Muscovite and phlogopite are the only natural micas of commercial importance. Vermiculite, although not considered a true mica by most mineralogists, is a micaceous mineral formed from the weathering of phlogopite or biotite and is also of commercial importance. Fluorphlogopite, $K_2Mg_6(Al_2Si_6O_{20})F_4$, is a synthetic mica made from pure chemical oxides.

Mica has been classified into three groups: (1) the mica group proper, (2) the clintonite or brittle micas group, and (3) the chlorite group. Supplementary to these are the vermiculites, which are hydrated compounds that result from the alteration of any one of the micas, but usually biotite. All minerals in these groups belong to the monoclinic crystal system, and all show plane angles of 60 and $120°$ on the basal section. The crystals usually form in hexagonal or rhombohedral-shaped scales, prisms, or plates. The basic structural unit of mica is a layer composed of two silicon tetrahedral sheets with a central octahedral sheet.

Muscovite is dioctahedral, having a theoretical composition of 11.8% K_2O, 45.2% SiO_2, 38.5% Al_2O_3, and 4.5% H_2O. Muscovite mica formed as a primary mineral in pegmatites and granodiorite differs in physical properties compared to muscovite mica formed by secondary alteration (mica schist). The main differences are in flexibility and ability to be delaminated. Primary muscovite is not as brittle and delaminates much easier than muscovite formed as a secondary mineral.

Mining

Flake mica is mined from weathered and hard rock pegmatites, granodiorite, and schist and gneiss by conventional openpit methods. In soft, residual material, dozers, shovels, scrapers, and front-end loaders are used to mine the ore. Often kaolin, quartz, and feldspar are recovered along with the mica. See also **Clays**; and **Silicon**.

Hard rock mining of these ore bodies requires drilling and blasting with ammonium nitrate and dynamite. After blasting, the ore is reduced in size with a drop ball and then loaded on trucks for transportation to the processing plant. Mica, quartz, and feldspar concentrates are separated, recovered, and sold from the hard rock ore.

Pockets of mica crystals are found in pegmatite stills and dikes or grandonite ore bodies. Sheet mica is mined by both underground and open-pit mining procedures. Underground mining is accomplished by driving a shaft, formed with tungsten carbide-tipped air drills, hoists, and explosives. After blasting, the mica is placed in boxes or bags for transporting to the trimming shed where it is graded, split, and cut to various specified sizes for sale.

Sheet mica is no longer mined in the U.S. Most sheet mica is mined in India.

Beneficiation Processes

In the early to mid-1900s, flake or scrap mica was mainly processed by a jigging procedure which consists of hydraulically washing a pile of bulldozed ore across a series of roll crushers and Trommel screens gaped at different size openings.

The grade of mica produced by jigging is very poor, usually about 75% concentrate, and recovery of available mica low (50%). Specifications on mica have become more stringent, therefore a more efficient processing method has been devised that provides higher quality mica, as well as more efficient recovery.

Because of improved mica processing operations, low cost earthen waste impoundment ponds have been built to store solid waste and thereby provide for a relatively cheap means of meeting new federal and state environmental laws. There are several methods of preparing ore for beneficiation after it arrives at the plant site weathered granodiorite ore. An alkaline–cationic circuit may be used by inserting a second conditioner containing lignin sulfonate, adjusting the pH to 8.0, and adding NaOH and DRL (distilled tall oil) fatty acid to the first conditioner.

Flake mica is also produced as a by-product from processing feldspar ore (hard granodiorite) from mica schist which normally contains from 30–60% recrystallized muscovite mica along with quartz and iron minerals. The quartz is usually not suitable for glass sand or high purity material, however.

The preparation of sheet mica for feedstock for various punching and machining operations involves cobbing mica blocks or books to remove dirt, rock, and defective mica, trimming and splitting into sizes and thicknesses suitable for punching and milling to desired shapes, and grading the finished mica sheets according to size and quality. The waste mica resulting from cobbing and trimming (scrap mica) is often mixed with flake mica for processing by dry or wet ground procedures.

The grade determines whether the mica can be used in high technology electronic instruments, e.g., computer-aided tomography (CAT) scan, or in low technology devices, e.g., a toaster. Many types of insulators, as well as the base for electronic circuits, are formed from the high quality sheets of mica by a punch pressing operation.

Procedures for Production

The general pieces of equipment used in grinding flake mica or mica concentrate into saleable mica products are hammer mills of various types, fluid energy mills, Chaser or Muller mills for wet grinding, and Raymond or Williams high side roller mills. Another method is being developed, called a Duncan mill (J. M. Huber, Inc.), that is similar in many respects to an attrition mill. All of these mills are used in conjunction with sieves, and all but some types of hammer mills-incorporate air classifiers as a part of the circuit.

This constitutes by far the largest commercial use for mica. It is largely produced from the beneficiation of weathered and unweathered pegmatites, granodiorite, and metamorphic schists, although some higher grades are produced from trimmings of sheet mica or Type A (low quality) mica blocks.

Wet ground mica products account for approximately 15% of the total mica market. Exact sizing of mica products coupled with surface treatment procedures have led to a greater use for wet ground mica in plastic compositions, particularly automobile bodies. These quality products demand a high dollar value.

Dry ground mica concentrate is processed into usable products by several different grinding methods. Relatively coarse particle sizes (1.651–0.147 mm (10–100 mesh)) are used in oil-well drilling muds, some types of welding rod coatings, asphalt roofing shingles, and some other types of fillers. These products are ground on a hammer mill in closed circuit with a sieve. Roofing micas produced from mica schist are often ground in a Raymond or Williams high side roll mill in closed circuit with an air classifier and a sieve. The finer particle-size micas ≤0.147–0.044 mm (−100 to −325 mesh), used mainly in textured paints and joint cement compounds, are ground on several types of fluid energy mills, but generally a mill of the Majacs type. The finest dry ground mica product is ground with superheated steam (Micronized, KMG Minerals).

Testing of Mica

There are several conventional tests required by consumers of ground mica. They include screening and the determination of bulk density, true specific gravity, chemical analysis, moisture, free silica, refraction index, oil absorption, brightness, grit content, and aspect ratio.

By-Products of Mica

The main by-products of mica processing plants are kaolin, quartz, and feldspar. Some plants produce all of these products for sale.

Uses

Good quality sheet mica is widely used for many industrial applications, particularly in the electrical and electronic industries, because of its high dielectric strength, uniform dielectric constant, low power loss (high power factor), high electrical resistivity, and low temperature coefficient. Mica also resists temperatures of 600–900°C, and can be easily machined into strong parts of different sizes and shapes.

When the primary property needed for a particular application is insulation, built-up mica made by binding layered mica splittings together serves as a substitute for the more expensive sheet mica. The principal uses for built-up mica are segment plate, molding plate, flexible plate, heater plate, and tape.

Wet ground mica is used because of its unique properties, i.e., luster, slip and sheen, and high aspect ratio. It is used in wallpaper and coated paper, nacreous pigments, as a coating for rubber, in outdoor house paint, and in aluminum paints. Mica is used in all types of sealers for porous surfaces, such as wallboard masonry, and concrete blocks, to reduce penetration and improve holdout and as a filler in plastics to improve its electrical and thermal resistance and its insulating qualities. See also **Sealants**.

Dry ground mica produced by hammer milling and screening is used in oil-well drilling, coatings for roofing shingles, roofing felt, and for some types of welding rod flux.

The largest use for fine, dry ground mica is in the manufacture of wallboard joint cements. Ground mica that is essentially ≤0.147 mm-100 mesh and ~70% passing a 0.044 mm (325 mesh) Tyler sieve is used in the joint compound mixture as a filler and extender. These compounds are used to fill joints between panels of gypsum plaster board. Mica contributes to making a nonabsorbing smooth surface that reduces shrinkage and eliminates cracks. It is also used in the finished coating on ceilings and to prepare thermal insulation and acoustical qualities of ceiling tile and prefabricated concrete. See also **Calcium**.

Fine particle-size dry ground mica is also used as an extender and filler in certain texture and traffic paints. Mica particles are stronger than iron and not brittle like other inerts. It is an antifriction, antifouling, antisettling, anticorrosive, antitarnish, and antisiege agent. It is a superior reinforcing pigment that acts as a sealer over porous surfaces and reduces penetration and flushing; moreover, it improves the moisture resistance of protective coatings and adhesion to all types of surfaces.

Micronized mica is a trade name (KMG Minerals, formerly English Mica Co.) for a very fine particle-size dry ground product, usually ground with superheated steam in a special fluid energy mill and used as a replacement for wet ground mica in certain types of paints. Micronized mica, preferably calcined, is also used in cosmetic applications, i.e., nail varnishes, lipsticks, eyeshadows, and barrier cream, because is has the advantages of high ultraviolet light stability, excellent lubricity, skin adhesion, and compressibility. Some of these micas are coated with oxides like titanium and iron.

Environmental and Health Regulations

Mica mining is subjected to local, state, and federal laws. The Mining, Safety and Health Administration (MSHA) regularly monitors mica mining operations for safety violations.

Both state air and water environmental departments together with the U.S. EPA regulate and oversee air and water quality associated with mica mining operations. Most states have land management departments that regulate dam safety, erosion, sedimentation, and reclamation. The mica mines must control erosion and sedimentation and restore the mined out areas. This is accomplished either by backfilling or contouring and seeding operations, or in cases where this is impractical or undesirable, lakes for water-related recreation may be built. The Corps of Engineers have jurisdiction over laws governing wetlands.

Health regulations are supervised by county and state health departments. There are no known health problems caused by the mica crystal, however, most industrial mica products contain some free silica particles that can cause silicosis and some states require employees who work in mica plants to receive an annual x-ray.

Additional Reading

Davis, L.L.: *Minerals Yearbook*, U.S. Bureau of Mines, Washington, DC, 1991–1993, p. 4, 5, 7–9.
Grim, R.W.: *Clay Mineralogy*, McGraw-Hill Book Co., Inc., New York, NY, 1968, 596 pp.
Madhukar, B.B.L., and S.N.P. Srivastava: *Mica and Mica Industry*, Taylor & Francis, Inc., Philadelphia, PA, 1995.
Preston, J.B.: "Mica," *Pigment Handbook*, John Wiley & Sons, Inc., New York, NY, 1971, 30 pp.
Rajgarhia, M.L.: *Ground Mica*, Mica Manufacturing Co., Private Ltd., Calcutta, India, 1987, p. 30; *British Standards*, British Standards Institute, London.

MICELLE. An electrically charged colloidal particle, usually organic in nature, composed of aggregates of large molecules, such as found in surfactants and soaps. The term is also applied to the casein complex in milk.

See also **Colloid Systems**.

MICHEL, HARTMUT (1948–). Awarded Nobel Prize for chemistry in 1988, along with Johann Deisenhofer and Michel Huber, for work that revealed the three-dimensional structure of closely-linked proteins that are essential to photosynthesis. Doctorate awarded in 1977 by the University of Wurtzburg, Germany.

MICROBIOLOGY. Microbiology is the study of agents too small to be seen with the unaided eye. These agents (and their fields of study) are bacteria (bacteriology), fungi (mycology), protozoa (protozoology), algae (phycology) and viruses (virology).

The study of microorganisms would not be possible, and microbiology would not have developed as a discipline, without the aid of the light microscope, an instrument that makes it possible to magnify minute specimens up to two thousand fold. Some of the earliest microscopes were made by Zacharias Janssen (1580–c. 1638), a Dutch optician who, as a young boy, combined two lenses to construct crude microscopes. These early instruments subsequently were improved and by the late 1600s microscopes were capable of magnifying objects by 50–300 diameters. Anton van Leeuwenhoek (1632–1723), a Dutch linen draper, made hundreds of these microscopesduring his lifetime and observed small, living particles, which he called *animalcules*, through them. Leeuwenhoek is considered to be the 'fatherof microbiology' because of his work with these early microscopes. See also **Bacteriology (The History)**; **Leeuwenhoek, Antoni van (1632–1723)**; and **Light Microscopy**.

Louis Pasteur, a French chemist, played a significant role in the development of microbiology with his experiments to disprove the theory of spontaneous generation. This theory, which became popular during the eighteenth century, proposed that spoilage organisms arose spontaneously in putrefied food. Pasteur captured airborne microorganisms on guncotton filters to show that these airborne microbes were responsible for food spoilage and did not arise spontaneously. See also **Pasteur, Louis (1822–1895)**.

In 1876 a German country physician named Robert Koch discovered that the lethal, contagious cattle disease anthrax could be transmitted from animal to animal by injecting blood from an infected cow into a healthy cow. His experiments were formulated into a set of criteria, known as Koch's postulates, which have become the cornerstone for associating specific microorganisms with specific infectious diseases. Koch's postulates initiated an era known as the Golden Age of Microbiology (1876–1906), during which the causes of many microbial diseases were discovered. See also **Koch, Heinrich Hermann Robert (1843–1910)**.

Through the accomplishments of these and many other individuals, microbiology developed into a distinct discipline. Today, microbiology continues to grow as a discipline and has entered into a new golden age closely affiliated with molecular biology and biotechnology.

Aspects of Prokaryology, Bacteriology, Mycology, Protozoology, Phycology, and Virology

The five agents (bacteria, fungi, algae, protozoa and viruses) associated with microbiology each have their own unique characteristics. Bacteria, which do not have a well-defined membrane-enclosed nucleus, are considered prokaryotes (Latin *pro*, before, Greek *karyon*, nucleus), whereas fungi, algae and protozoa have a well-defined membrane-enclosed nucleusand, therefore, are eukaryotes. Viruses, which are not cellular but are intracellular parasites containing either deoxyribonucleic acid (DNA) or ribonucleic acid (RNA), are not considered either prokaryotes or eukaryotes.

Classification of Microorganisms. The separation between prokaryotes and eukaryotes has become less distinct in recent years with the emergence of molecular techniques and the discovery in the 1980s that some prokaryotes have features that more closely resemble the eukaryotes. In 1981, Carl Woese proposed that all living organisms be separated into three domains: Archaea (Greek *Archaios*, ancient, as in the Archaean era, a geological period approximately 3.9–2.6 billion years ago), Bacteria and Eukarya (Figure 1). This separation is based on the unique ribosomal RNA sequences for members of each domain, as well as other characteristics such as cell wall and plasma membrane composition, complexity of RNA polymerases and mechanism of protein synthesis. Ribosomal RNA sequences have especially been important in determining the evolutionary, or phylogenetic, relationships among organisms in the three domains and indicate that the Archaea are the most primitive organisms. See also **Taxonomy (The History)**.

The Archaea are a diverse group of prokaryotes that inhabit extreme environments. These organisms include methanogens (organisms found in anaerobic swamps, marshes and animal intestinal tracts that produce methane as a product of metabolism), sulfate reducers (organisms occurring in marine hydrothermal vents and that use sulfate as an electron acceptor), extreme halophiles (organisms found in high-salt environments such as the Great Salt Lake and the Dead Sea), extremely thermophilic sulfur metabolizers (organisms that exist in high-temperature environments and are capable of metabolizingsulfur), and cell wall-less thermophiles (organisms without walls that livein high-temperature environments).

The remaining prokaryotes are classified under the domain Bacteria. The domain Bacteria is an extensive group of microorganisms, consisting of hundred of thousands of species that are ubiquitous and found in various environments, including soil, water, plants and animals. Both the Archaea and the Bacteria typically are grouped into taxonomic families, genera and species, primarily on the basis of structural and morphological characteristics, such as cell shape, size and appendages, and secondarily on the basis of biochemical and physiological traits, such as growth factor requirements, range of carbohydrates used as carbon and energy sources, and end products of metabolism. This classical approach to taxonomy is used to organize prokaryotes in a reference manual called *Bergey's Manual of Systematic Bacteriology*. This classification of prokaryotes is rapidly undergoing revision and modification as newer molecular techniques provide more accurate information on phylogenetic relationships of living organisms through ribosomal RNA and DNA sequence analyses. Molecular analyses are also important in characterizing the vast majority of prokaryotes that are nonculturable and can only be classified by RNA or DNA composition. See also **Bacteriology**.

Prokaryotes

The typical prokaryote is 1–2 µm long and 0.2–0.5 µm in diameter, although the smallest prokaryotes (*Mycoplasma*) have diameters of 125–250 nm and the largest prokaryotes (*Epulopiscium*) have dimensions of 600 µm by 80 µm. Prokaryotes exist in three basic forms: rod (bacillus), spherical (coccus) and spiral (spirillum) (Figure 2).

Prokaryotic DNA is found in a single, circular chromosome, although some prokaryotes have small, extrachromosomal, circular pieces of DNA called *plasmids*, which carry genes for antibiotic resistance, toxins and resistance to heavy metals. The cytoplasm and nuclear region of prokaryotes is typically surrounded by a phospholipid bilayer plasma membrane

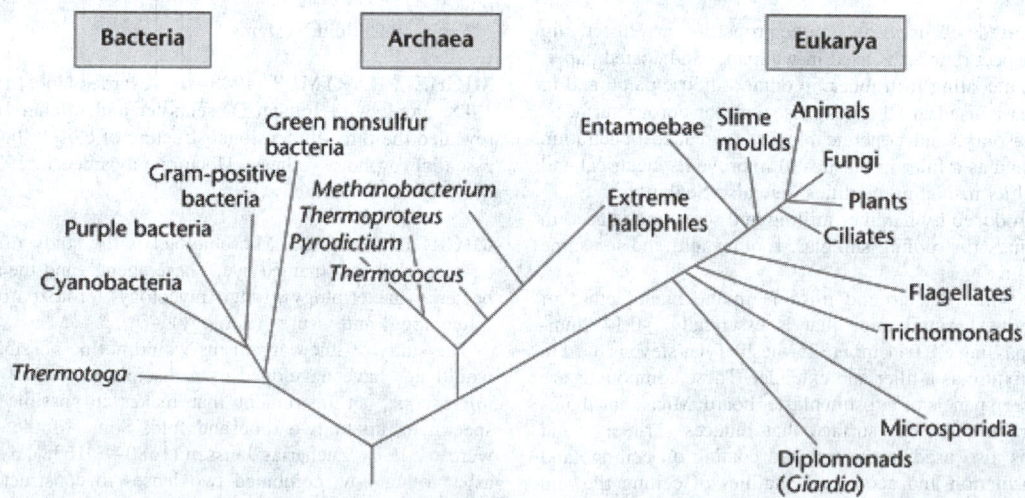

Fig. 1.   Universal phylogenetic tree. The phylogenetic relationships among organisms in the three domains Bacteria, Archaea and Eukarya are shown. Separations are based on the unique ribosomal RNA sequences for members of each domain.

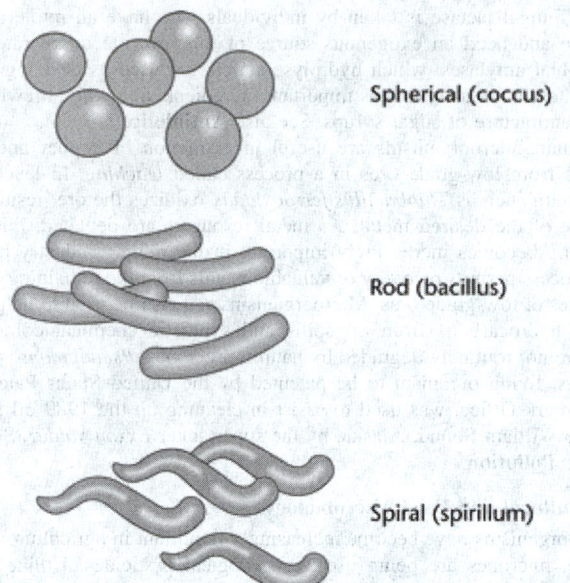

Spherical (coccus)

Rod (bacillus)

Spiral (spirillum)

Fig. 2. Basic forms of prokaryotes. Prokaryotes exist in three fundamental forms: spherical (coccus), rod (bacillus) and spiral (spirillum).

and a cellwall consisting of the repeating carbohydrate derivatives N-acetylglucosamine and N-acetylmuramic acid complexed with amino acids in a rigid structure called peptidoglycan.

A common staining procedure used in microbiology, the Gram stain (named after Hans Christian Gram (1853–1938), who first developed this stain in 1884), separates most prokaryotes into two groups: Gram-positive prokaryotes, which have a thick peptidoglycan cell wall layer; and Gram-negative prokaryotes, which have a thin peptidoglycan cell wall layer and an additional external membrane called the outer membrane. Gram-positive prokaryotes, which include bacteria such as *Staphylococcus* and *Streptococcus*, stain deep violet with the Gram stain, whereas Gram-negative prokaryotes, which include bacteria such as *Escherichia coli* and *Pseudomonas*, stain pink or red.

Many prokaryotes have long, whip like appendages called *flagella* which aid in motility. Some prokaryotes, particularly those associated with disease, have outer layers of material called capsules which protect them from phagocytosis. Certain prokaryotes are capable of producing, under unfavorable conditions, a thick, multilayered structure called an *endospore*, which is resistant to high temperatures, desiccation, many disinfectants and other adverse environmental conditions. The endospore is a dormantform of the bacterium in which there is little metabolism. When conditionsbecome favorable again, the endospore germinates into a normal, vegetative cell. *Bacillus* and *Clostridium* are examples of prokaryotes that form endospores.

Prokaryotes generally divide by binary fission, a process in which one cell divides into two equal cells. The most rapid-growing prokaryotes can divide every 15–20 min under ideal growth conditions. Because prokaryotes double in number with every generation, growth occurs rapidly and the number of prokaryotes in a population of cells can increase exponentially from one cell to several billion cells in a few hours. This rapidgrowth of prokaryotes and their binary fission into equal daughter cells makes these microorganisms ideal tools for genetic and metabolic studies.

Fungi

Fungi are eukaryotic microorganisms that, like plants, have rigid cell walls and are nonmotile but, unlike plants, lack chlorophyll and are nonphotosynthetic. Fungi are as diverse as the prokaryotes and can be found in various environments ranging from arid deserts to tropical forests. Fungi live primarily off dead, decaying organic material and are therefore important in the decomposition and recycling of organic matter.

Fungi can exist as two fundamental morphological forms: molds, which consist of filaments called *hyphae* that can branch into masses known as mycelia, and yeasts, which are unicellular, oval forms. Some fungi,

such as *Candida*, are dimorphic and alternate between the mould and yeastforms, depending on environmental conditions. Fungi are capable of sexual and asexual reproduction. Sexual reproduction can involve an alternation ofgenerations between a sporophyte that produces spores and a gametophyte that produces gametes. Examples of asexual reproduction are budding, in whicha new daughter cell arises as a bud on the surface of the mother cell, and formation of spores at the tips of hyphae.

Fungi are economically important. Yeasts (for example, *Saccharomyces cerevisiae*), which ferment carbohydrates to alcohol and carbon dioxide, are extensively used in the making of bread, wine and beer. Mushrooms are a type of fungus and certain cheeses such as Roquefort (*Penicillium roqueforti*) and brie (*Penicillium camemberti*) are produced from fungi. Many antibiotics (for example, penicillin, cephalosporin and griseofulvin) and microbial enzymes (for example, amylases, pectinases and proteases) are produced by fungi. However, fungi can cause serious human diseases, such as histoplasmosis, coccidioidomycosis, cryptococcal meningoencephalitis, dermatophytosis (for example, athlete's foot and ringworm) and sick building syndrome; animal diseases, such as poultry haemorrhagic syndrome and ich (fish dermatitis); and plant diseases, such as apple scab, potato blightand corn brown spot. Fungi also are responsible for bread mould, mildew, rot and other destructive processes. See also **Fungus**; and **Yeasts and Molds**.

Protozoa

Protozoa are unicellular, nonphotosynthetic eukaryotic microorganismsthat lack cell walls. Some protozoa are large enough to be seen with the unaided eye, although most are microscopic. Protozoa ingest materials by pinocytosis, a process in which liquids are surrounded by invagination of the plasma membrane and brought into the cell, and phagocytosis, a similar process in which larger particulate matter is engulfed. See also **Protozoa**.

Protozoa reproduce sexually or asexually. In asexual reproduction, the parent cell mitotically divides either into two equal daughter cells (binary fission) or into several daughter cells (multiple fission). Some protozoa asexually reproduce by budding or by spore formation. Sexual reproduction can occur by several methods, including conjugation, in which two cellsjoin, exchange nuclei and produce progeny by budding or fission.

Protozoa are diverse in their habitats and distribution. Some cause human diseases, such as malaria (*Plasmodium*), toxoplasmosis (*Toxoplasma*), amoebic dysentery (*Entamoeba*), cryptosporidiosis (*Cryptosporidium*) and trichomoniasis (*Trichomonas*). Others (for example, *Paramecium*) are innocuous members of the biosphere, where they exist as important initial links in the food chain.

Algae

Algae are plant-like eukaryotic microorganisms that are distinguished from the fungi and protozoa by their chlorophylls and photosynthetic abilities. They contribute significantly to the aquatic food chain in marine and freshwater environments. Algae range in size from single-celled microscopic organisms to large, complex cell aggregates that attain lengths of more than 30 meters (98 feet). The colorful algal hues often visible in aquatic environments result from different chlorophylls and other photosynthetic pigments. Diatoms, elaborate shells that have two, thin overlapping halves, are among the simplest algae. Certain algae exist in association with microscopic animals as plankton. Other algae exist as multicellular aquatic structures known as seaweed.

Most algae reproduce asexually by mitotic division, whereas others form spores or reproduce by fragmentation of cells from larger aggregates. Some algae reproduce sexually by forming diploid zygotes from haploid gametes.

Over the years, algae have become increasingly important as food and food additives for humans. The red alga *Porphyra*, known as nori, is popular in sushi and as a toasted soup additive. Other red algae are also eaten as vegetables or as sweet jellies. Seaweed, which has high iodine and vitamin value, is used as a diet supplement. Algae can also produce economic loss when they overgrow, or bloom, because of excess nutrients in the water, which are often the result of human pollution or the release of sewage to bodies of water. Outbreaks of red tide are caused by algal blooms. The algae release toxic products that can kill fish and other marine life. See also **Algae**; and **Ocean Resources (Living)**.

Viruses

Viruses are considered microorganisms, but occupy a unique place in the microbial world because they possess only a single type of nucleic acid (single-stranded or double-stranded DNA or RNA) contained within a simple protein coat. Furthermore, viruses are intracellular parasites that rely upon an infected host for metabolism and reproduction. Viruses that infect bacteria are called *bacteriophages*. Most viruses are too small (less than 200 nm in diameter) to be seen with the light microscope and can only be seen with an electron microscope. Because the viral particle is so small, its nucleic acid can code for only a few genes. Consequently, viruses are among the simplest forms of life. They often have repeating chemical building blocks in their structures and assume only a few symmetrical forms. See also **Virus**.

Viruses replicate within an infected host cell, using host protein-synthesizing machinery such as ribosome's to assist in replication. Following replication, progeny viruses are released as the host cell is lysed and dies. Successful viral infections typically lead to physical cell damage inprokaryotes and eukaryotes. Viruses are responsible for diseases such as acquired immune deficiency syndrome (AIDS), influenza, measles, poliomyelitis, rabies and encephalitis. A unique infectious agent that is simpler in structure than the virus is the prion. Prions (proteinaceous infectious particles), infectious agents that contain only protein and no nucleic acid, cause neurological diseases such as Creutzfeldt–Jakob disease and bovinespongiform encephalopathy (mad cow disease). See also **Creutzfeldt-Jakob Disease and Related Diseases**.

Applied Microbiology, Agricultural and Food Microbiology, Molecular Microbiology, Biotechnology

It often is believed that most microorganisms are harmful. In reality the vast majority of microorganisms are beneficial. In fact, life as we know it could not exist without microbes. Microorganisms play critical, indispensable roles in the recycling of chemical elements in the biosphere. As animals and plants die, their dead tissues are decomposed by microorganisms, which recycle the tissue components as elements into the biosphere. Organic compounds are degraded into carbon dioxide, which is released into the atmosphere to be fixed once again into organic material by photosynthetic plants, algae and prokaryotes. Proteins, nucleic acids and other organicnitrogenous compounds are decomposed by microorganisms to ammonia, nitrate or nitrogen gas. These products are assimilated into new nitrogenous compounds or, in the case of nitrogen gas, released into the atmosphere. Other chemical elements are recycled in similar fashion by microorganisms.

The decomposition of organic matter by microorganisms is also the basis of sewage treatment. Sewage is the liquid human, animal, plant, industrial and agricultural waste that is carried by a system of pipes and other conduits called *sewers* to a central discharge point. Sewage must be treated to remove hazardous chemical waste and pathogenic microorganisms before the water can be reused. Chemical wastes in sewage are degraded by microorganisms during a multistage process. The processed sewage is then chlorinated, filtered and/or treated by other means to kill pathogens prior to reuse as water. The effectiveness of such treatment to remove pathogens is generally measured by the presence or absence of indicator organisms. Indicator organisms, although not necessarily pathogenic, provide a relative index of faecal contamination of water. Coliforms and enterococci, which arebacteria that often are associated with human faeces, are two common indicator organisms that are used to monitor water quality. Water that has no indicator organisms and is suitable for drinking is said to be potable. See also **Bioremediation**; and **Wastes and Pollution**.

Although pathogen-free, potable water is not sterile, as evident from the colonization and attachment of microorganisms to the interior surfaces of water pipes. These biofilms, which consist of metabolically active communities of microbes in a matrix of polysaccharides, slow the flow of water and contribute to pipe corrosion. Biofilms are also significant in oral hygiene, where acid-producing bacteria (for example, *Streptococcus mutans*) embedded in a polysaccharide/glycoprotein matrix develop as dental plaque on teeth.

Applied Microbiology

Industry depends heavily on microorganisms for the production of antibiotics, vitamins, enzymes and various other commercial products. Streptokinases produced by *Streptococcus* are used in medicine to dissolve blood clots. Fungal lactase is taken by individuals who have an intolerance to lactose and need an exogenous source of this enzyme to tolerate milk. Microbial amylases, which hydrolyse starch, are used as desizing agents in the textile industry and as important ingredients in baking, brewing and the manufacture of sugar syrups. See also **Antibiotic**.

Certain microorganisms are useful in extraction of copper and other metals from low-grade ores in a process called *leaching*. In leaching, a bacterium such as *Thiobacillus ferrooxidans* oxidizes the ore, resulting in release of the desired metal. As metal resources are depleted, microbial leaching becomes increasingly important in the mining industry because the process permits recovery of valuable metals from the remaining natural supplies of low-grade ores. Microorganisms can also be used to help break down hydrocarbons from oil spills and synthetic chemical compounds that are not routinely degraded by natural processes. *Pseudomonas putida*, the first living organism to be patented by the United States Patent and Trademark Office, was used to assist in cleaning up the 1989 oil spill in Prince William Sound, Alaska, by the supertanker *Exxon Valdez*. See also **Water Pollution**.

Agricultural and Food Microbiology

Microorganisms have become increasingly important in agriculture; forexample, microbes are being used as biological pesticides. Unlike chemicalpesticides that have broad lethal effects, are recalcitrant and persist as harmful chemicals in soil for many years, biological pesticides selectivelykill specific insects, are not harmful to other living organisms, and regenerate as the microbe grows. *Bacillus thuringiensis* is one popular bacterial pesticide that is effective for the control of caterpillars, bollworms, cabbage worms and gypsy moths.

Microorganisms have also been used in agriculture to introduce genetic material into plants. The Ti plasmid of the bacterium *Agrobacterium tumefaciens* is often used as a vehicle in genetic engineering to transferdesired genes into plants to produce transgenic crops that are more resistant to infection, disease and spoilage.

Microbes are the only living organisms that are capable of reducing atmospheric nitrogen, which constitutes 80% of the gases in the earth's atmosphere, to ammonia (a process called nitrogen fixation) for utilization by plants and animals. In some instances, prokaryotes fix nitrogen via symbiotic associations with plants; in others, microbial nitrogen fixation occursnonsymbiotically. It is estimated that prokaryotes are responsible for fixing over 135 million metric tons of nitrogen annually. Were it not for prokaryotic nitrogen fixation, soils would be nitrogen-deficient and much more chemical fertilization would be required.

The food industry relies heavily on microorganisms for the production of fermented milks and milk products, fermented foods and alcoholic beverages. Milk and milk products, such as acidophilus milk (*Lactobacillus acidophilus*), yogurt (*Lactobacillus bulgaricus*, *Streptococcus thermophilus*), butter (*Streptococcus*) and cheese (for example, *Lactobacillus* and *Streptococcus*), are produced by the direct action of microorganisms on milk. Many important foods, including sausage (*Pediococcus*), sauerkraut (*Lactobacillus*, *Leuconostoc*) and olives (*Lactobacillus*), are made by microbial fermentations. Yeasts (*Saccharomyces cerevisiae*) are used in the baking industry to raise, or leaven, the dough and in the brewing industry to ferment barley grains. Carbon dioxide produced by these yeasts causes the dough to rise, and the ethanol produced by the yeasts contributes to the alcoholic content of beer. See also **Fermentation**; and **Yeasts and Molds**.

Molecular Microbiology and Biotechnology

No other living organism has been studied as extensively as *E. coli*. This bacterium has been used for so many studies because it is easy to grow and has a small chromosome. Under ideal growth conditions of optimum temperature and ample nutrients, *E. coli* doubles in number every 15–20 min and within a few hours produces billions of identical cells in a small laboratory flask. The chromosome of *E. coli* contains 4.7 million base pairs — approximately 0.1% of the amount of DNA found in a eukaryotic cell. Some of the earliest observations of DNA replication, RNA synthesis and protein synthesis were made with *E. coli* cells. Our knowledge of the control of gene expression is based primarily on landmark experiments performed with prokaryotes. Microorganisms are adept in regulating the types and quantities of proteins formed. Without these precise molecular regulatory mechanisms, microbial cells would needlessly expend their energy synthesizing unnecessary proteins. See also **Molecular Biology**.

Microorganisms have also served as laboratory models to study the transfer of genetic material between microbial cells. Although prokaryotes do not reproduce sexually, they can transfer genetic material by several mechanisms, including transformation (uptake of naked DNA), transduction (transfer of DNA via a bacteriophage from one cell to another cell) and conjugation (direct transfer of DNA between two cells). Genetic exchange among prokaryotes is an important mechanism that has led to genetic diversity and evolution among these microorganisms, and has formed the basis for genetic engineering. See also **Genetic Engineering**; and **Genetics and Gene Science (Classical)**.

Through genetic engineering, or the deliberate modification of genetic material in a cell or organism, it is now possible to manipulate DNA and develop products for science and commerce. Prokaryotes play an important role in genetic engineering. They produce enzymes called restriction endonucleases that cleave specific areas of DNA. Using restriction enzymes, one can isolate a gene of interest from a prokaryotic or eukaryotic chromosome and insert this gene into a DNA vehicle called a cloning vector. The cloning vector can then be inserted into a host, where the desired gene is replicated and expressed. Insulin, human growth hormone, interferon and streptokinase are examples of products available through cloned genes. These advances in biotechnology have significantly improved human life through purer and less expensive products made possible by genetic engineering.

Medical Microbiology

Infectious microorganisms capable of causing disease live in delicate balance with a host. When such microbes successfully cause damage to the host, disease results. There are two main mechanisms by which microbes damage a host. Pathogens can produce toxins, which are harmful to the host. These toxins can be secreted by the pathogen (exotoxin) or can be part of the outer membranes of Gram-negative bacteria (endotoxin). Exotoxins are among the most potent toxins. For example, 1 mg of botulinum exotoxin could kill 1000 human beings. Microbial pathogens can also invade and harm host tissue by physically damaging the tissue or by depriving it of nutrients. Microbial invasion is assisted by prokaryotic cell appendages (pili) that aid in attachment to host cells, enzymes (leucocidins) that degrade white blood cells, enzymes (collagenases) that degrade collagen in muscle tissue, and thick outer cell coverings (capsules) that protect the invading microbe from phagocytosis. The host responds to microbial invasion and toxins by producing protective antibodies that bind to and inactivate the microbe or toxin, and white blood cells that attack and phagocytose the invaders.

A human host acquires infectious diseases by different routes, including the respiratory tract, the oral cavity and digestive system, and the skin and genitourinary system. In certain instances, the microbial pathogen may be transmitted directly to the host, as in the cases of sexually-transmitted diseases, such as gonorrhoea (*Neisseria gonorrhoeae*) and syphilis (*Treponema pallidum*), or respiratory diseases, such as tuberculosis (*Mycobacterium tuberculosis*) and influenza (influenza virus). In other instances, the pathogen may be transmitted indirectly by inanimate objects, such as towels, or by other living organisms, such as insects or rodents. Rocky Mountain spotted fever (*Rickettsia rickettsii*) and malaria (*Plasmodium*) are examples of diseases that are transmitted indirectly by an insect vector to the human host. After a pathogen successfully infects a host, there usually is an incubation period during which the pathogen multiplies and establishes itself before visible symptoms and disease occurs. See also **Malaria**; **Sexually Transmitted Diseases**; **Syphilis**; and **Tuberculosis**.

Antibiotics are often administered to the host to help in the fight against infectious prokaryotic and eukaryotic microorganisms. Antibiotics kill or inhibit pathogens by inhibiting nucleic acid or protein synthesis, damaging the plasma membrane, preventing cell wall synthesis or interfering with cell metabolism. Viruses, which are unable to reproduce independently and replicate only inside a host cell, are not affected by antibiotics and must therefore be treated by other, limited therapies, such as chemical compounds that specifically block viral enzymes or viral nucleic acid synthesis. The indiscriminate use of antibiotics in recent years has led to the development of microorganisms that are resistant to many common drugs. These antibiotic-resistant microbes represent a challenge to conventional chemotherapy and are of major concern to the

medical profession. Microbial infections can also be prevented by the use of vaccines. Vaccines, which prompt the host to produce protective antibodies against infectious agents and other foreign antigens, have especially been helpful in protection against childhood diseases such as diphtheria (*Corynebacterium diphtheriae*), whooping cough (*Bordetella pertussis*), poliomyelitis (poliovirus) and measles (rubeola virus). See also **Antibiotics**; **Antiviral Drugs**; and **Vaccines**.

Importance of Microbiology as a Scientific Discipline

Microorganisms are intimately and intricately interwoven into our daily lives. These ubiquitous microscopic forms of life are found wherever other life forms occur and, in some instances, where no other forms of life exist. These unseen organisms, first observed over three centuries ago by Leeuwenhoek with his crude microscopes, are the basis of all life and play indispensable roles in our biosphere. Life and all that it encompasses could not exist without microbes. The study and understanding of these microorganisms would not be possible without the development of microbiology as a scientific discipline.

See also **Virology (The History)**.

Additional Reading

Black, J.G.: *Microbiology: Principles and Explorations*, 6th Edition, John Wiley & Sons, Inc., New York, NY, 2004.
Black, J.G., and L.M. Lewis: *Microbiology: Principles and Explorations SG*, 6th Edition, John Wiley & Sons, Inc., Hoboken, NJ, 2005.
Cann, A.J.: *Principles of Molecular Virology*, 4th Edition, Elsevier Science & Technology Books, New York, NY, 2005.
Cappuccino, J., and N. Sherman: *Microbiology: A Laboratory Manual*, 8th Edition, Benjamin Cummings, San Francisco, CA, 2007.
Carter, J., and V. Saunders: *Virology: Principles and Applications*, John Wiley & Sons, Inc., Hoboken, NJ, 2007.
Flint, S.J., V.R. Racaniello, L.W. Enquist, and A.M. Skalka: *Principles of Virology: Molecular Biology, Pathogenesis, and Control of Animal Viruses*, 2nd Edition, ASM Press, Washington, DC, 2003.
Kun, L.Y.: *Microbial Biotechnology: Principles and Applications*, 2nd Edition, World Scientific Publishing Company, Inc., Hackensack, NJ, 2006.
Madigan, M.T., J.M. Martinko, and J. Parker: *Brock Biology of Microorganisms*, 9th Edition, Prentice-Hall, Inc, Upper Saddle River, NJ, 2000.
Murray, P.R., K.S. Rosenthal, and M.A. Pfaller: *Medical Microbiology: with Student Consult Access*, Elsevier Health Sciences, New York, NY, 2005.
Talaro, K.P.: *Foundations in Microbiology: Basic Principles*, 6th Edition, The McGraw-Hill Companies, Inc., New York, NY, 2006.
Tortora, G.J., B.R. Funke, and C.L. Case: *Microbiology: An Introduction*, 7th Edition, Benjamin Cummings, San Francisco, CA, 2001.

DANIEL V. LIM, University of South Florida, Tampa, FL

MICROCHEMISTRY. A branch of analytical chemistry that involves procedures that require handling of very small quantities of materials. Specifically, it refers to carrying out various chemical operations (weighing, purification, quantitative and qualitative analysis) on samples ranging from 0.1 to 10 mg; this often involves use of a microscope, and still more often chromatography. See also **Microscope (Chemical)**.

MICROCLIMATE. The fine climatic structure of the air space that extends from the very surface of the earth, to a height where the effects of the immediate character of the underlying surface no longer can be distinguished from the general local climate (mesoclimate or macroclimate). The microclimate varies with and in turn is superimposed upon the larger-scale conditions. While some rigid limits have been placed on the thickness of the layer concerned, it is more realistic to consider variable thicknesses. (Observe the microclimate of a putting green versus that of a redwood forest.) Generally, four times the height of surface growth or structures defines the level where microclimatic overtones disappear. Microclimate can be subdivided into as many different classes as there are types of underlying surface. With sufficient detail, this could be almost limitless. Currently, the most studied broad types are the "urban microclimate," affected by pavement, buildings, air pollution, dense inhabitation, etc., the "vegetation microclimate," concerned with the complex nature of the air space occupied by vegetation, and its effects upon the vegetation and the microclimate of confined spaces (the cryptoclimate) of houses, greenhouses, caves, etc.

See also **Cryptoclimate**; **Macroclimate**; **Mesoclimate**; and **Phytoclimatology**.

Additional Reading

Geiger, R.: "Micoclimate," In *Compendium of Meteorology*, American Meteorological Society, Boston, MA. 1951, pp. 993–1003.

MICROCRYSTALLINE. A form in which a number of high-polymeric substances have been prepared. They include cellulose, chrysotile asbestos, amylose (starch), collagen, nylon, and certain mineral waxes. On the microsopic level, these substances are composed of colloidal microcrystals connected by molecular chains. The process involves breaking up the network of microcrystals (by acid hydrolysis in the case of cellulose) and separating them by mechanical agitation. The size range of the microcrystals is from 2.5 to 500 nanometers (millimicrons). The products form extremely stable gels that have a number of commercial use possibilities. Petroleum-derived waxes of high molecular weight have been available in microcrystalline form for many years. Chlorophyll has a naturally microcrystalline structure.

MICRODISPLAYS. Microdisplays are very small displays that are viewed through the use of optics. Although no formal definition exists for the size of a microdisplay, most in the industry would agree on a diagonal measurement of one inch or less. The two most competitive features of microdisplay technology are the ability to display large numbers of pixels and to do so in a lightweight package that occupies a small volume. Traditional display technologies tend to get larger, heavier, and more expensive as image size and pixel count increase. The fact that microdisplays can be highly compact without sacrificing image quality gives them an advantage over existing products. In addition, these features of microdisplays enable a number of new products, such as lightweight headsets, that could not be served by existing technologies.

High-density information displays currently have from 100,000 to more than 1 million pixels. Clearly, a one-inch display with such pixel counts must be coupled to an optical system to create reasonable viewing conditions. There are two broad methods used to implement microdisplays: *projection* and *near-eye*. Projection systems are designed to magnify a small, real image onto a screen for viewing by one or more users. In contrast, near-eye applications use an optical system to project a virtual image intended to be viewed by a single person (hence, they are also called personal viewers). These systems are best suited for compact handheld or head-mounted devices.

Within the category of head-mounted displays (HMDs), there are a number of divisions. The display may be monocular (presented to only one eye) or binocular (presented to both eyes). A binocular display can be stereo or biocular, where the first consists of a single image that the two eyes naturally see from different perspectives (as in the real world), and the second presents separate, slightly different images to each eye. Both monocular and binocular HMDs can be immersive (blocking the viewer from seeing anything but the display) or transparent (allowing the viewer to see both the displayed image and the real world).

Most microdisplays are fabricated on silicon (or quartz) substrates rather than glass, as in the production of direct-view flat panel displays, a quality that has both technical and business consequences. By building the display directly on semiconductor substrates, designers can integrate electronic components—such as row and column drivers, digital and video interface, and control circuits—directly alongside the display on the same substrate. This can lead to higher performance with lower manufacturing costs. On the business side, this has led to production arrangements that are a departure from those typically seen in the display industry, more closely resembling those of the semiconductor industry. Rather than require new investments in production facilities that use increasingly large area glass substrates, microdisplays can be fabricated on existing semiconductor equipment, leading to the possibility of flexible arrangements between display developers and semiconductor firms.

There are four types of microdisplays: *transmissive*, which modulate an external light source as it is transmitted through the device; *reflective*, which modulate an external light source by varying the properties of a reflecting surface; *emissive*, which produce light internally; and *scanning*, which write images directly onto the retina.

Transmissive Microdisplays

Direct-view displays based on the light-modulating properties of liquid crystals are ubiquitous. Over the past decade, new materials and manufacturing techniques have evolved to allow for very small (10–20 micron) pixel sizes, resulting in the development of transmissive microdisplays that function much as their direct-view counterparts. In this section, detailed descriptions are given only for twisted-nematic (TN) displays, but many of the same principles apply to other types, such as ferroelectric.

Amorphous silicon (a-Si) is currently the most widely used semiconductor channel material for direct-view active matrix LCD applications. It is especially convenient because it is deposited at a low temperature around 300 °C (572 °F), meaning that relatively inexpensive glass substrates can be used. In addition, its low mobility (although still high enough to provide adequate *on* current) makes for TFTs with very small *off* state currents, meaning supplementary storage capacitors are rarely necessary. However, this low mobility makes a-Si incompatible with integrated driver architecture, which is required for the compactness of LCD microdisplays. Integrated drivers are made possible by using the higher-mobility polycrystalline silicon (poly-Si, or polysilicon) as the channel material.

All transmissive LCD microdisplays, with the exception of those from Sharp and Kopin Corporation http://www.mdreport.com/summaries/october99.html (see the following sections on continuous-grain silicon and single-crystal silicon), use poly-Si TFTs. As noted previously, the systems called microdisplays are near one inch in diagonal size and are viewed indirectly through external optics, disqualifying very small LCDs that are viewed directly (such as in some viewfinders).

Poly-Si microdisplays are used in two major applications: video/digital camera viewfinders, which benefit from very small pixels, and projection LCDs (particularly for front projection, where poly-Si devices will dominate for the next five years). To date, nearly all poly-Si TFT-LCDs, of all sizes, have been fabricated using high-temperature techniques on quartz substrates, but there has been a great deal of effort applied to low-temperature manufacturing on glass, because of reduced costs.

Reflective Microdisplays

Liquid Crystal on Silicon. Of all the types of AMLCDs, liquid-crystal-on-silicon (LCOS) devices most directly use semiconductor manufacturing techniques, as the active matrix array and associated driver circuits are designed and built on a silicon wafer, using complementary metal oxide semiconductor (CMOS) processing techniques. The LCOS displays are the most heavily researched microdisplays.

Instead of thin-film transistors on glass, LCOS displays use bulk silicon devices to control the pixels from outside the optical path. See Fig. 1. The silicon-based CMOS chip serves as both the active matrix and the reflective layer, on top of which a thin layer of liquid crystal, glass plate, and polarizer are deposited. Instead of the a-Si or poly-Si TFTs used in transmissive LCD cells, LCOS devices use the high-speed switching capability provided by single-crystal silicon. LCOS devices also have a simplified LCD structure because the device is reflective, not transmissive, like most LCDs. For example, only one polarizer is required, and the thickness of the layer of liquid crystal can be reduced, allowing for faster switching.

Two important design choices are made when developing an LCOS device: the type of liquid crystal material and the type of control circuit to be used. The nematic and ferroelectric liquid crystal modes are the most popular types for such a display. The nematic types provide a good contrast ratio and are typically coupled with a DRAM switching array. Ferroelectric liquid crystals provide very fast switching speeds, but they do not allow for grayscale; the pixel is either black or white. Grayscale can be created temporally by taking advantage of the fast switching speed to dither the binary value within a frame period. Ferroelectric liquid crystals are typically coupled with SRAM devices, which provide the fast frame transfer rates needed to provide temporally dithered grayscales and which are also binary devices.

Liquid crystal microdisplays can also be configured to modulate unpolarized light, an approach that has the advantage of much greater light transmission by eliminating the significant absorption of polarizers. There are two ways to use unpolarized light. In one approach, light is scattered when the liquid crystal molecules are arranged in one fashion and reflected when the molecules are rearranged, a process that is controlled

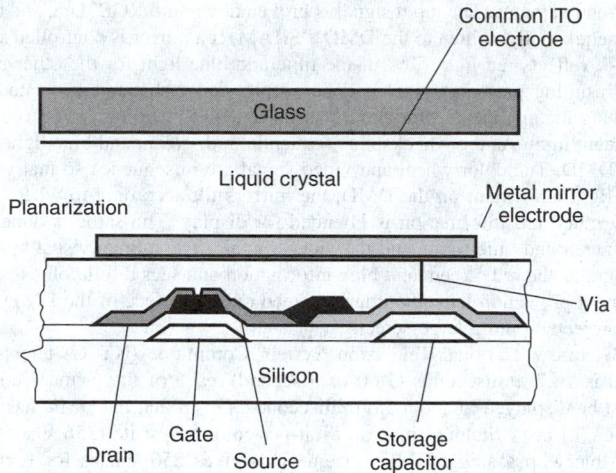

Fig. 1.   Structure of a LCOS Microdisplay (*Courtesy of CMD*).

by the application of a voltage across the cell. Scattering displays typically use polymer-dispersed liquid crystal materials. The second approach uses the principle of diffraction, in which a periodic arrangement of *on* or *off* pixels (forming a grating) causes light to constructively or destructively interfere. This interference pattern can then be filtered by a Schlieren stop to pass light when in certain grating modes; by altering the grating structure, amplitude and color may be controlled. Twisted-nematic liquid crystal material is the most common type used in diffractive systems.

Color can be achieved in LCOS devices in several ways. The most popular approach for projection applications is to use three LCOS chips and dichroic mirrors to separate red, green, and blue components. Color-filter wheels have also been used to present sequential color to a single chip. For personal viewers, red, green, and blue light sources (typically light-emitting diodes) are used to illuminate sequential frames of data at three times normal video rates.

A number of engineering issues remain with LCOS. First is the light source, which should be as close to a point source as possible because the display is so tiny. Also, improvements in the optics to handle the fine pixel pitch would increase image quality, particularly in the rear-projection screen and the polarizing beamsplitter. The manufacturing of LCOS microdisplays is also not yet a mature practice; manufacturing lines for common LCDs and TFTs are generally not appropriate for LCOS displays. These reflective displays require a thin cell gap (1–3 microns), with a tolerance of 100–200 nm, as well as specialized filling and sealing procedures. Pixel pitches in microdisplays are as much as five times smaller than in direct-view displays, adding to the engineering challenge. Current yields are far below acceptable levels. Applications for LCOS devices are grouped into four main types: Front projectors; viewfinders (camera and camcorder) and viewers integrated into other handheld devices (cellular telephones and PDAs); head-mounted displays; and rear-projection monitors and televisions.

Most companies use TN liquid crystals, although two use ferroelectric liquid crystals (FLCs).

The first devices reached the market in the form of front projectors for business applications. The LCOS front projectors developed so far are not very reflective and thus require bright illumination to function. These illumination requirements restrict LCOS projectors from competing in the fast-growing market segment of ultralight projectors, due to the need for a hefty lamp and optics and a specialized cooling system.

LCOS displays for viewfinders (in camcorders or digital cameras, for instance) have the prospect of reaching the market in the near term.

LCOS projectors are the first successful application of this type of microdisplay, providing credibility for further development.

The HMD (which could be implemented with a variety of display types, including LCOS) is widely touted as an up-and-coming method for viewing information. Many start-up microdisplay companies are developing HMDs for games, personal movie viewing, wearable computers, and screens for cellular telephones and PDAs. Some of these systems require total attention (such as movie viewers), but others strive for integration with normal vision, either by showing the display only to one eye or by superimposing the image over the normal view.

Interest in rear-projection desktop monitors is now increasing after several companies rejected the concept in 1999, and the LCOS display could be a contender in this market. IBM has demonstrated a 28-inch-diagonal monitor with three $2,048 \times 2,048$–pixel TN microdisplays. It is yet to be demonstrated that LCOS-based rear projectors can compete with direct-view CRT or LCD in the desktop monitor market or with the rear-projection CRT or plasma display panel (PDP) in the television market.

These activities indicate that there is currently great interest in (and resources devoted to) microdisplays on the supply side. So far, performance has been adequate but not stellar, and the price of products is still quite high. Both of these factors must improve in order to generate significant demand.

LCOS devices are just beginning to take hold in the display market. Although there is potential for high growth in the coming years, market acceptance will be contingent on improvements in price, performance, and ergonomics of these products.

Microelectromechanical systems (MEMS). Devices use semiconductor fabrication equipment and processes to build miniature systems that have both mechanical and electrical components. They rely on physical movement of a display element (pixel or line) to modulate the amplitude or phase of an external light source. The most common type of MEMS device used in the display industry is the Digital Micromirror Device™ (DMD), but two other types the Grating Light Valve (GLV) and the Actuated Mirror Array (AMA) are also relevant.

Digital Micromirror Device (DMD). The DMD is a monolithic, micromechanical spatial light modulator integrating discrete, tilting mirror elements with CMOS addressing circuits at each pixel location. The device was invented by Larry J. Hornbeck of Texas Instruments in 1987. Devices with 800×600 pixels and $1,024 \times 768$ pixels were available in 1998, and systems with $1,280 \times 1,024$ pixels were introduced in 1999. The technology is capable of providing projected images from all currently operational or proposed high-definition standards; demonstration systems have been built with $1,920 \times 1,080$ pixels. Hence, the DMD is a potential technology for HDTV projection systems. The devices can be made in high volume at relatively low cost in a specialized semiconductor-manufacturing environment. This DMD is sold by Texas Instruments to numerous projection-system vendors who use it as the core of their systems. The subsystem driving the products built around the DMD chip is called a Digital Light Processing™ (DLP) engine.

The DLP circuit board accepts an incoming signal from a computer and, using an array of digital-signal processors, converts the signal to digital sequences. This digital system, controlled by software and algorithms, converts text, graphics, and video into a digital signal that yields tones and color values that rival the image quality of photographic film. No analog signals are used in the control circuit.

After processing, the digital signal moves to the heart of the DLP system, the DMD. The DMD functions like an array of light switches. It consists of an array of 17 micron $\times$ 17 micron mirrors, fabricated on hinges on top of a static RAM (SRAM) chip. The DMD decodes the incoming signal, providing each mirror with its own unique instruction. Then each mirror is responsible for creating a single pixel in the projected image and receives a new instruction approximately every millisecond.

Depending on the state of the SRAM cell, the mirror is electrostatically attracted to either one or the other address electrodes by a combination of bias and address voltages. Thus, the mirror rotates until its tip touches a landing electrode, which is held at the same potential as the mirror. A "1" in the memory cell causes the mirror to rotate $+10$, while a "0" in the memory cell allows the mirror to rotate -10. Figure 2 shows an array of micromirrors in three positions: flat, on, and off. The mirrors tilt into the *on* or *off* states when 5 volts are applied to one of the two offset electrodes while the other is grounded. Although the DMD can be operated in the analog mode, the mirrors are biased in such a way that only the digital landing sites of $+10$ or -10 are possible. The digital mode of operation permits operation within the low-voltage CMOS framework and ensures uniform deflection angles.

Because the process is completely digital, DLP projection systems based on one-, two-, or three-DMD configurations can serve as the digital display interface for not only presentations but also computers, televisions, or any

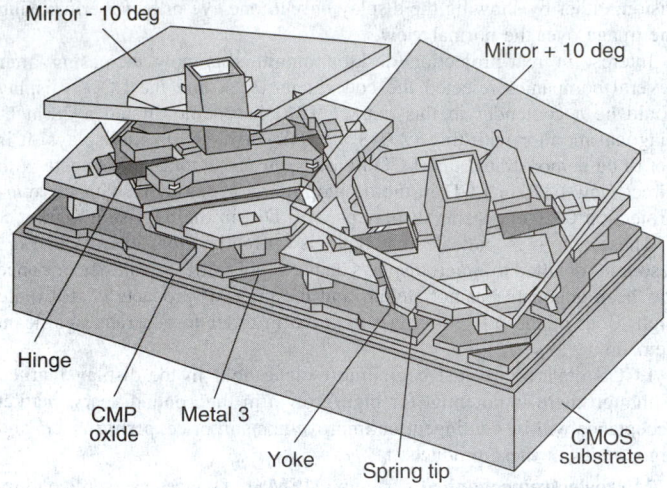

Fig. 2.   DMD Pixel Structure. (*Courtesy of Texas Instruments.*)

future application where digital light can be harnessed and effectively used. In the three-chip system, one chip is used for each of the primary colors (red, green, and blue), similar to other projection systems. But, because of the DMD's high speed and light throughput (optical efficiency of the DMD pixels have been measured to be 61% by Texas Instruments, and no polarizers are needed), it is also practical to create a single-chip system, which is less costly and is self-converged. Diagrams of the three system configurations are shown in Fig. 3.

In a one-chip system, color is added to the image by filtering light through a color system. Typically, a white light source is focused onto a color wheel through the use of condensing optics. The color wheel is a red, green, and blue (RGB) filter system that spins at 60 Hz to give 180 color fields per second. In this configuration, DLP operates in a sequential color mode. As white light hits the red filter, the red light is transmitted and the blue and green lights are absorbed. The same holds true for the blue and green filters: The blue filter transmits blue and absorbs red and green, while the green filter transmits green and absorbs red and blue. Thus, when a color wheel is used, two-thirds of the light is blocked at any given time.

The light that passes through the color wheel is then imaged onto the surface of the DMD. The integrator rod shown in Figure 3 serves to match the beam cross section to the shape of the DMD, and the total internal reflection (TIR) prism simplifies the optical layout; both are optional in

low-end versions. The input signal is broken down into RGB data, and data are sequentially written to the DMD's SRAM. If a mirror is controlled such that it reflects red light 50% of the time and blue light for the other 50% the resulting projected pixel will be purple. Color shades are created by varying the amount of time the mirror is rotated toward or away from the incident light. As the wheel spins, sequential red, green, and blue lights hit the DMD. The color wheel and video signal are in sequence so that when red light is incident on the DMD, the mirrors tilt according to where and how much red information is intended for display. The same is done for the green and blue lights and the video signals. The human visual system integrates the red, green, and blue information and sees a full-color image. Using a projection lens, the image formed on the surface of the DMD can be projected onto a large screen.

Because a National Television System Committee (NTSC) television field is 16.7 milliseconds (1/60 of a second), each of the primary colors must be displayed in about 5.6 milliseconds. Given that the DMD has less than a 15 ms switching time, an 8-bit-per-color grayscale (256 shades) is possible with a single DMD system. This gives 256 shades for each of the primary colors, or 2,5633 (16.7 million) possible colors that can be generated.

The two-chip DMD system uses a color-filter wheel that passes two of the secondary colors, magenta and yellow. The magenta segment of the color wheel allows both red and blue to pass through, while the yellow segment passes red and green. The result is that red light is passing through the filter system at all times. Blue and green alternate with the rotation of the magenta-yellow color wheel and are each essentially on half of the time.

By biasing light transmission toward the red, the two-chip subsystem compensates for the red-light deficiency in long-lifetime metal halide lamps. Compared to a one-chip system, the two-chip system yields roughly three times the amount of red-light output. Because the color wheel is made up of only two filters instead of three, the blue and green light output is increased by roughly 50%.

Once through the color wheel, the light is directed to a system consisting of a TIR and dichroic prisms. At this point, the constant red light is split off and directed to a DMD dedicated to handling red light and red component-video signals. The sequential blue and green light is directed to another DMD that is driven by blue and green component-video signals and is configured to handle the sequential colors.

The three-chip system uses one chip for each of the primary colors: red, green, and blue. Light is separated solely by dichroic prisms. No color-filter wheel is required, so light output is maximized. In addition to the higher brightness achieved by the three-chip system, the image quality — a function of not only sophisticated signal processing but also higher-bit

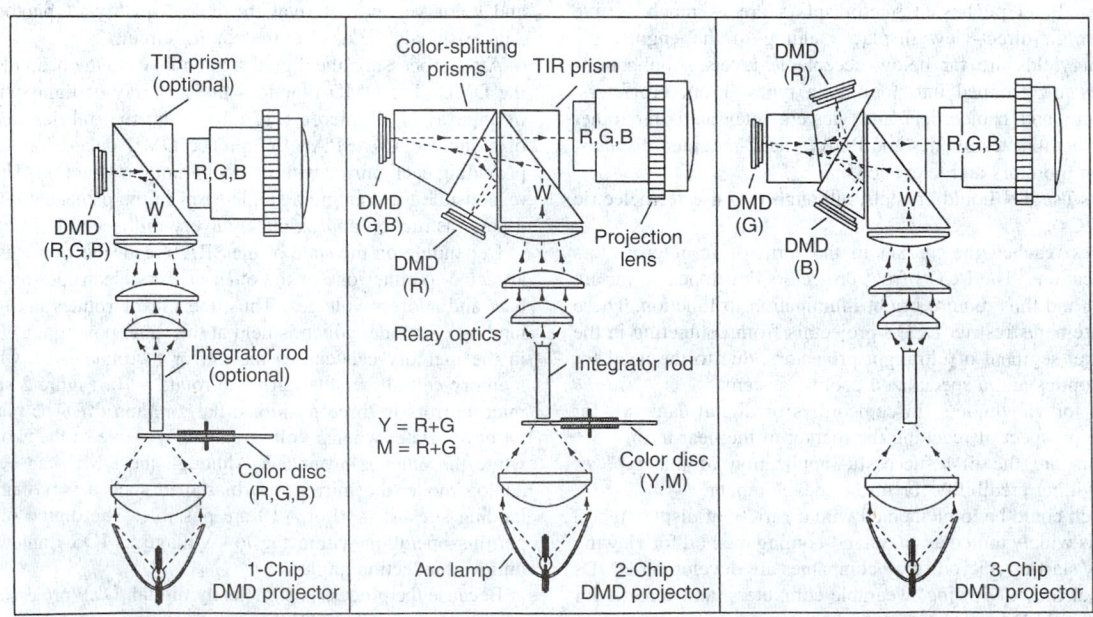

Fig. 3.   Three DMD Systems. (*Courtesy of Texas Instruments.*)

color depth — is also increased. Because light is directed to each DMD for the whole field, 10-bit grayscale per color is possible, allowing three-chip systems to reproduce 1,024 shades of gray compared with the one-chip system's 8-bit color and 256 shades of gray.

The grating light valve (GLV), invented at Stanford University and being developed by Silicon Light Machines (Sunnyvale, California), is a silicon-based integrated circuit with a micromachine reflection phase grating. The structure consists of a series of suspended silicon-nitride beams, referred to as ribbons, which are coated with a reflective aluminum layer. The ribbons are approximately 3 microns wide, 100 microns long, and 100 nanometers thick, and they are arranged in parallel. Each ribbon is supported at both ends and is suspended above an air gap of approximately 650 nanometers. When a potential difference is applied between the aluminum coating on a ribbon and a conductive layer below the air gap, the ribbon is pulled down by electrostatic force. This effect is very fast; the switching time can be as short as 20 nanoseconds.

A typical pixel consists of a set of six ribbons. If no voltages are applied, the set of ribbons functions as a mirror; when a pixel is addressed, alternate ribbons are deflected, forming a grating surface, and incoming light is diffracted. All that is required is the application of a voltage to the selected pixel; no transistors are required to switch and hold a pixel. The hysteresis effect inherent in the device also allows the deflection to be maintained at a voltage level in between the turn-on and turn-off voltages. The other three ribbons remain in a fixed position, and need only a fixed bias voltage. The amount of diffraction from a pixel is related to the depth of the grating at that pixel, which is proportional to the applied voltage. A Schlieren optics system is used to allow only the diffracted light to pass.

The distinctive feature of a GLV projection system is that it consists of linear arrays of pixels, which are scanned perpendicular to the array to form an image rather than a typical two-dimensional pixel array. The fast switching time of the GLV allows for the horizontal dimension to be accomplished through scanning. A projection system using a white-light laser source, separated into red, green, and blue components has been demonstrated. The primary colors illuminate three GLV arrays with 1,080 pixels and are scanned with a galvanometric mirror to produce an image with 1,920 × 1,080 pixels. The company reported that this system had a contrast ratio of 200:1, showed 16.7 million colors, and had a maximum refresh rate of 96 Hz. The pixel elements can be fabricated in as few as 7 mask steps, promising low tooling and manufacturing costs relative to more complex MEMS devices.

The actuated mirror array (AMA) was developed by Aura Systems, El Segundo, California, in the early 1990s, and licensed to Daewoo Electronics, which has been developing the concept further under the name thin-film micromirror array-actuated (TMA). The device is similar to the DMD in concept and system design, although the TMA requires a Schlieren stop to operate properly, whereas the DMD's larger rotation angles allow the display to be coupled directly to the projection lens. The key difference is that the TMA uses piezoelectric materials, rather than electrostatic forces, to tilt the mirror/pixels. Piezoelectric materials constrict or expand in response to an applied voltage; each mirror sits on a pair of piezoelectric posts, and when one post is constricted and the other is expanded, the mirror tilts toward the constricted post.

A linear response to an applied voltage was reported, up to a tilt of 4 to 10 volts, and a frequency response up to 30 kHz, or approximately 33 ms switching time, allowing for a gamut of 16.7 million colors using three chips. A device with 640 × 640 pixels, each 97 × 97 microns was also reported, subsequently announced a 1,032 × 774–pixel device with 50-micron square pixels was announced. Due to the large chip size (approximately 3 inches in diagonal), three-chip demonstration systems have achieved 5,000 ANSI lumens.

Emissive Microdisplays

Emissive microdisplays produce their own light, obviating external light sources or reliance on reflection. All emissive microdisplays are based on technologies that were first developed for direct-view displays and then applied to chip-based systems later. There are three main types of emissive displays: thin-film electroluminescent (TFEL) displays; vacuum fluorescent displays (VFDs), which are called Vacuum Fluorescent on Silicon (VFOS) when they are in microdisplay form; and organic light-emitting diode (OLED) displays. The microdisplay version of the field emission display (FED) was briefly pursued, but is no longer being developed.

Thin-Film Electroluminescent Microdisplays

Active matrix thin-film electroluminescent (AM-TFEL) displays are miniature AC-TFEL structures constructed on an integrated circuit backplane. By using active matrix addressing, AM-TFEL displays can provide higher information content than direct-view TFEL displays; microdisplays with up to 1,280 × 1,024 pixels have been demonstrated, using a 0.7-inch-diagonal substrate and a resolution of 2,000 lines per inch. The high voltages (80 volts AC) used in such devices require the use of silicon-on-insulator substrates that isolate the high voltage signals from nonselected pixels and logic. Color AMEL devices have been produced using filtered white phosphor and field-sequential liquid crystal shutters.

In 1996, Planar announced the first AM-TFEL head-mounted display, which has since evolved into a line of microdisplays called MicroBrite™. These devices are designed for such applications as hands-free maintenance computers, military targeting and communications systems, portable personal communications and computing, medical imaging and virtual reality for commercial applications, and consumer entertainment. The 2000 model has an active area of 0.61 inch × 0.45 inch, is 1.7 mm thick, and weighs less than 3 grams. The monochrome (amber) image has a VGA pixel format (640 × 480 pixels) with 32 gray levels and is sold as a developer's kit with interface electronics for digital LCD VGA, analog VGA, or video. Integrators such as Liteye and Kaiser Electro-Optics offer products based on Planar's technology. A color version was expected to be available in late 2000. See also **Electroluminescent Displays**.

Vacuum Fluorescent on Silicon (VFOS) Microdisplay

The VFOS display is a blend of semiconductor technology and conventional VFD technology. The display is a phosphor dot matrix on top of a 5.4 mm × 6 mm semiconductor chip that integrates memory function and display driver circuits. The basic structure of this display device is similar to the conventional VFD except for the structure of the anode. In a typical VFD, wiring, insulation, and phosphor layers are fabricated directly on the glass plate and no semiconductor devices are arranged internally. With VFOS, phosphor matrices are fabricated on a semiconductor chip and arranged on a glass plate. The chip and the outer leads are connected by wire bonding. These types of graphic displays contain semiconductor driver ICs with embedded memory.

Display Research Laboratories (Los Altos, California) has pioneered VFOS. Claimed benefits of VFOS displays over conventional VFDs include lower noise, longer life, higher brightness, and more stable operation under varied environmental conditions. See also **Vacuum Fluorescent Displays**.

Organic Light-Emitting Microdisplays

A new direction in OLED development is to build the device on a silicon backplane for use as a microdisplay. This technology is being pursued by eMagin Corporation, which has licensed Kodak's small-molecule patents. The company's active matrix OLED display has 12-micron pixels in a 1,280 × 1,024 (SXGA) pixel format on a 0.77-inch-diagonal chip. It is capable of showing full-motion video. The prototype shown in May 2000 was monochrome, but there are plans to add color filters (which will reduce the pixel format). The device is intended for consumer products such as cellular telephones and movie viewers and also for use as a virtual computer monitor.

Because OLEDs emit light perpendicularly from an areal source (that is, the emission is Lambertian), they are very well suited for near-eye applications. No special optics are needed to make the light resemble the light that normally strikes the eyes. The flip side of this property is that OLED microdisplays are poorly suited for projection applications, because they are not bright enough.

Many challenges remain in commercializing OLED microdisplays. As with other microdisplays, the manufacturing process is not yet fully automated, but OLEDs have the additional burden of the lack of large-scale manufacturing even for the direct-view version of the technology. Equipment has not been standardized or refined for large-area OLED displays as it has been for LCDs, VFDs, and TFEL panels. Also, the organic light-emitting materials have limited lifetimes and are sensitive to environmental factors like moisture, heat, and oxygen. Also, there are interface and driver issues; standard chips and connectors are not yet established for this technology. See also **Organic Light-Emitting Diodes (OLEDs)**.

Scanning Microdisplays

Microdisplays for projection applications produce a *real* image on a screen, and those discussed so far for near-eye applications produce a *virtual* image that appears to float before the viewer's eyes. The virtual image is a magnification of a tiny real image. But there is an alternative method for near-eye applications: *scanning microdisplays*. These devices "paint" the image directly on the retina, so in some sense, there is no image.

Scanning microdisplays work by sweeping a light beam to create an image, much as a CRT sweeps an electron beam. However, unlike electrons, which can be directed with electromagnetic fields, photons must be directed with mirrors, because they have no charge. By coordinating the sweep and some form of light modulation, the image is produced.

Only four components are needed: a controllable light source, a scanner, drive electronics to control the scanning, and viewing optics. The light source should be compact and low power; a laser diode, laser, or LED is a standard choice (note that the intensity of the light used cannot harm the eye). The brightness is controlled by one of several types of optical attenuator. In the case of color displays, three light sources are used (red, green, and blue). The light mix is controlled separately from the brightness control by an acousto-optic or electro-optic modulator.

There are two kinds of devices that sweep light beams: line scanners and pixel scanners. Line scanners are as wide as the image they produce, sweeping the light beam only from top to bottom. These simple scanners are widely used in fax machines and document scanners. Although common line scanners are too slow for display applications, faster ones for displays are being developed.

Pixel scanners, which are analogous to CRTs and are the only kinds used as microdisplays so far, control pixels one by one, or in groups of a few, using drive electronics that change the mixture of red, green, and blue light appropriately. The beam sweeps in a two-dimensional raster pattern to create the image. The horizontal sweep must be much faster than the vertical sweep and is typically accomplished with a magnetically controlled mechanical resonant scanner or a MEMS device. A galvanometer or a MEMS device can perform the slower vertical sweep. See also **Cathode-Ray Tube**.

A mechanical resonant scanner consists of a reflective plane surface that can tilt like a seesaw under the influence of magnets on either side. Varying the fields controls the position continuously within a range that defines the sweep angle. Operating frequencies between 15 and 20 kHz can be achieved. Since this is a uniaxial scanner, it must be combined with another scanner (usually a galvanometer) to produce the two-dimensional raster motion.

Using a MEMS device allows for a much more sophisticated implementation of the same concept. In one arrangement, a silicon micromirror is etched away from the wafer surface, suspended by thin flexures on all four sides. About one axis (the fast, horizontal sweep), the mirror vibrates in a torsional oscillation mode, while in the other direction (the slow, vertical sweep), the flexures are moved capacitively. The scanner is biaxial and also operates between 15 and 20 kHz. (Note that the other type of micromirror device commonly used in display technology flips between two distinct states, rather than sweeping continuously).

As in all display systems, image quality in a scanning microdisplay results from a complex trade-off of many variables. The two factors with the greatest influence on the system's image properties are the size of the micromirrors and their deflection angle. The product of these two parameters (at a given wavelength) is directly proportional to the resolution of the image. Clearly, there is a trade-off: the bigger the mirror, the less deflection is necessary. There are also manufacturing trade-offs that limit the design space. Large mirrors carry the penalty of large inertia and hence slow response. Increasing the stiffness of the flexure mechanism (the lever that moves the mirror) to speed the mirror's response results in higher stress during mirror motion, which reduces the acceptable deflection angle (which was small anyway, because the mirror is large). Large mirrors are also less likely to be adequately flat and in general take up more space in the microdisplay system, which should be as compact as possible.

Small mirrors are fast and have larger deflection angles, but they also have limitations. If the deflection angle gets too large, there are problems with excessive stress in the mirror, which can lead to reduced lifetime. Changing the shape of the flexure to a long, thin profile helps mitigate stress, but can result in coupling to undesirable modes of the mirror. The ultimate effect of all these problems is a blurry, shifting, or flickering image.

Many other factors play a role in the MEMS system design also. The optimization of mirror size and deflection angle must be tailored to each wavelength, so color displays must have three different micromirror setups. Other important variables include device materials, mirror thickness, and flexure dimensions. Together with the mirror properties discussed above, these parameters interact to determine the overall cost and performance of the microdisplay.

To finally produce the image, the scanned beam must be directed into the viewer's eye using optics. The beam size and sweep angle must be defined to correctly enter the pupil while preserving the diffraction-limited nature of the original beam. There are a number of optical arrangements that can be employed, each with trade-offs in size/weight, complexity, field of view, and distortion effects.

Microvision produces scanning microdisplays based on both magnetically controlled scanners and its proprietary MEMS technology, which uses silicon micromirrors less than one millimeter on a side. The company has demonstrated horizontal scanners that operate at up to 19 kHz. Each half cycle of oscillation sweeps one line. If the scan is performed *bidirectionally*, meaning that the first line is scanned left to right, the second is scanned right to left, and so on, then the effective horizontal sweep frequency is 38 kHz. Painting a complete SVGA image (600 lines) thus requires 15.8 ms, for a frame rate of 63 Hz. The scanners can be improved to operate above 30 kHz, allowing for at least SXGA format (1,024 lines).

The compact version of Microvision's display, intended for consumer use as a wearable computer or gaming unit, is a monochrome, monocular VGA (640 × 400 pixels) HMD weighing about 1.5 pounds. The company is also engaged in designing and building high-performance binocular (and necessarily biocular) versions for surgery or military applications. These systems are more substantial helmet-type units and will incorporate full color and SVGA (800 × 600) or higher pixel formats. Handheld units for the wireless market are planned; studies show that retinal scanning displays indeed work when simply held near the eye. These products use either LEDs or lasers as the light source.

MicroOptical Corporation, http://www.microopticalcorp.com/ an integrator that makes glasses-mounted display systems based on liquid crystal microdisplay technology, is also developing MEMS scanning technology through a grant from DARPA. It will use lasers as the light source. The company claims that MEMS display systems can outperform LCD systems in terms of brightness, power efficiency, and reduced cabling.

Research work on scanning displays is also going on at numerous universities. The founders of Microvision originated at the Human Interface Technology Lab (HIT) at the University of Washington, Seattle, http://www.hitl.washington.edu/ where topics ranging from interface software design to human factors in wearing HMDs are investigated. There are also studies and design work proceeding at the University of California, San Diego, and Delft University, the Netherlands.

The experience of using a scanning microdisplay is different from viewing the virtual image that is seen in most HMDs. Everyone who has used a direct-view computer monitor (or looked at photograph) knows that even the most accurate representation of three-dimensional space on a flat surface cannot be mistaken for the real world. This experience has to do with a multitude of visual cues and physiological phenomena that people use to judge distance. Flat, finite-sized surfaces cannot reproduce all the cues people need to perceive an image as three dimensional. Even though a scanning microdisplay writes directly on the retina (as the real world does), it too cannot provide the full experience of reality; however, sophisticated optical processing of the scanning beam can satisfy more of the brain's criteria for three-dimensionality than a flat virtual image. Thus, scanning microdisplays can potentially produce more realistic stereoscopic images than any other type of microdisplay.

One key feature of scanning microdisplays is that the brightness and contrast are adjustable over a wide range by changing the intensity of the light source (a feature that is also available on CRTs but not reflective displays, other emissive displays, or LCDs). Hence, a person with low vision could use a scanning display to read or to watch television. A monochrome VGA display that is modified for patients with low-vision by has been produced. Although it is still being refined, this application is the subject of continued research at medical centers.

Although much research and development has focused on producing scanning microdisplays for playing games, performing field work, or for military use, a significant portion is devoted to the medical use of these displays for patients with low vision.

In summary, scanning microdisplays, like other types of microdisplays, are finally finding applications as interest in high-resolution near-eye displays is increasing. As with most new technologies, price is an issue; scanning microdisplays are very expensive and face the usual problem of needing volume production to achieve lower prices but needing lower prices to achieve high demand.

However, the crucial concern is whether consumers will like using near-eye displays. Exploring this issue means looking at the ergonomic factors of near-eye microdisplays, an area in which there have been few significant studies. The companies that succeed in developing the near-eye microdisplay market will be those that attend to human factors first. See also **Television (TV)**; **Optical Character Recognition (OCR)**; and **Semiconductor**.

For additional reading, refer to Flat Panel Display Technology entry.

Stanford Resources, Inc., San Jose, CA

MICROELECTRONICS. A constantly evolving technology, microelectronics is concerned with the design and manufacture of semiconductor electronic circuitry–containing components measured in microscopic dimensions. Also referred to as *chips, microchips*, and *integrated circuits*, microelectronic circuits usually are subsystems of larger systems, such as computers, telecommunications switches and transmission systems, guidance systems for aircraft and missiles, and satellites. The tiny chips also are found in commonplace items, such as personal computers, automobiles, household appliances, digital wristwatches, calculators, toys, and video and audio entertainment systems.

Brief Chronology. The birth of microelectronics can be traced to the invention of the *transistor* in 1947 at Bell Laboratories in Murray Hill,

New Jersey. With a name coined from its ability to *transfer* electric signals across a *resistor*, the transistor initially replaced the hot and bulky vacuum tube. The scope of transistor applications later expanded dramatically. In 1958, the integrated circuit (IC) was invented. In addition to combining numerous electrical components into one device, integrated circuits or microchips offered several advantages over discrete diodes and transistors, including smaller size, greater reliability, improved performance, lower cost, and reduced power requirements. See also **Diode**; and **Transistor (Invention and Development)**.

As of the mid-1990s, electronic systems were composed of many different microelectronic components formed from various types of semiconductor materials. Each component is selected because of a specific ability, such as amplifying, switching, or blocking electrical current flow. Based on its ability, an electrical component is classified as either *passive* (resistors, inductors, and capacitors) or *active* (diodes and transistors). As the word implies, semiconductor materials are not good or bad conductors of electrical current, but fall somewhere in between.

The two most commonly used semiconductor materials are silicon and gallium arsenide. Silicon, which is the second most abundant element in the earth's crust and the primary ingredient of sand, has become to the electronics revolution what steel was to the Industrial Revolution. Gallium arsenide, which is more expensive to manufacture than silicon, is useful for a variety of high-frequency, low-power applications. GaAs is faster than Si, can convert electronic signals to laser light and back again, and is radiation hardened, meaning it is less susceptible to radiation, a trait that is useful in satellite applications.

Function of a Chip. A chip may function as a *memory storage* device or a *processor*. There are several types of memory, with the two most common being random access memory (RAM) and read-only memory (ROM). RAM is the basic memory storage unit in a computer. It stores information temporarily in binary form and can be changed by the user. ROM, on the other hand, cannot be changed. It permanently stores binary information that is used repeatedly. See Fig. 1.

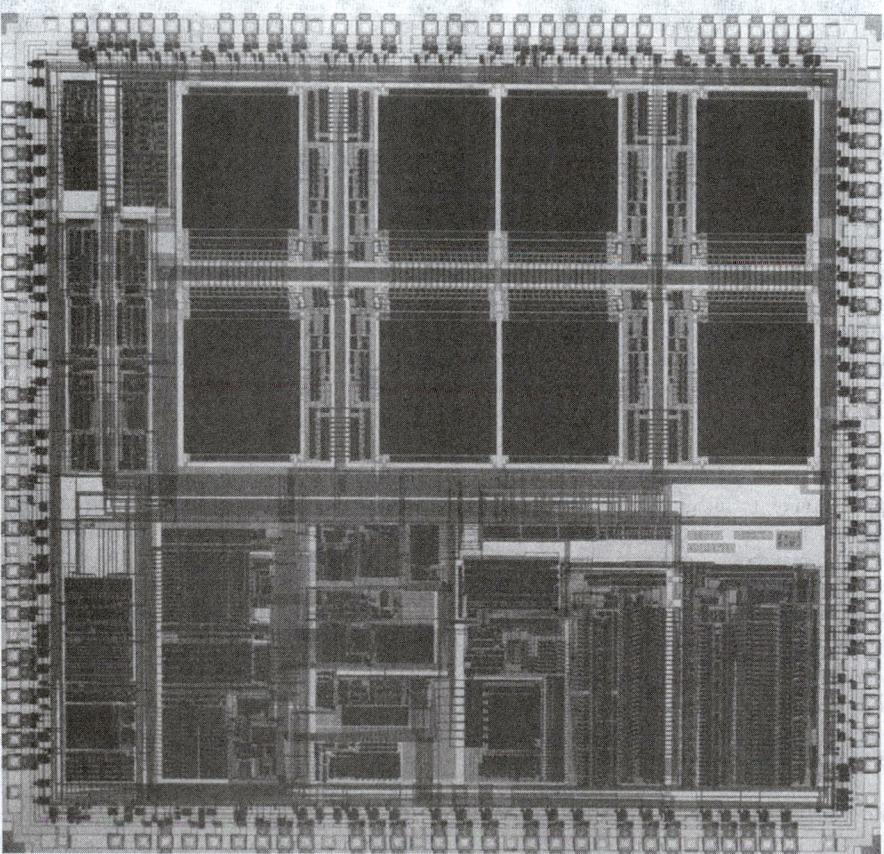

Fig. 1. AT&T's Digital Signal Processor (DSP) 1610, one of the world's fastest and most easily customized chips, operates at speeds up to 40 million instructions per second. Specifically designed to process speech for transmission and reception in digital cellular telephone networks, the DSP 1610's modular architecture gives designers the ability to customize the chip for different applications with relative ease and speed, without sacrificing performance. Actual size of the DSP 1610 is slightly less than a square centimeter. (*AT&T Bell Laboratories*.)

Fig. 2.   Nestled in the heart of 44 conductor leads is a tiny AT&T UNITE® semiconductor chip, only $\frac{1}{4}$-inch (0.6 cm) square. This chip is a key component in the digital telephone, personal computers, workstations, and other terminals used with the new digital telecommunications network (the Integrated Services Digital Netork [ISDN] now being implemented throughout the world. The UNITE® chip formats (organizes) digitized voice and data signals so that they can share the same four-wire line for ISDN operations. The network will simplify the world's public communications systems by eliminating the need for separate networks to handle voice, data, and images. (*AT&T Bell Laboratories. UNITE is a registered trademark of AT&T.*)

The chip may also be a processor or microcomputer, which in reality is a *computer-on-a-chip*. The application of the integrated circuit (IC) may be continuous, as in a digital wristwatch, or occasional, as in a stereophonic audio amplifier. The area of a chip may be as large as a thumbnail or as small as a grain of salt. The chip may be purely electronic, or it may be an electro-optical IC that combines electronic circuitry with photonics (i.e., which involves processing information with light pulses).

Manufacture of a Chip.   To be useful, a microchip must be mounted in a "carrier" or "package" (i.e., a housing that protects the delicate chip while connecting its circuit inputs, outputs, power supply, and ground to conductive pins that are pressure-fitted or soldered to other conductors). See Fig. 2. With the gradual shift from analog to digital technology, which represents information as a superfast stream of digital on-off pulses coded in binary notation, a single IC may now contain a million or more logic gates (on and off switches composed of combinations of transistors, diodes, and their connecting circuits).

The replacement of discrete diodes and transistors with integrated circuits based on more advanced microelectronics technology has significantly lowered the costs of sophisticated electronic functions. A logic gate that cost several dollars in 1950 now costs a fraction of a penny, and the cost is still declining. In addition, reliability of electronic circuits has improved by a factor of about 100,000 from the changeover of vacuum tubes to discrete transistors to integrated circuits.

The development of microelectronic devices is made possible by computer-based systems that design a circuit and its implementation in physical structures and simulate its operation. This process, which uses software on personal computers or workstations, is called computer-aided design (CAD). Essentially, it allows the chip designer to "construct" a chip step by step, function by function, using powerful software to simulate the operation and speed of an IC. Some of the power of CAD comes from the use of libraries of previously designed and tested circuit modules. The explosive growth in chip complexity over the years has been largely fueled by increasingly powerful software that not only facilitates chip design, but vastly shortens the interval needed to bring a new chip to the market.

The manufacturing process for integrated circuits typically consists of more than a hundred steps, during which dozens of copies of an integrated circuit are formed on a single wafer of silicon. Generally, each copy of the circuit is constructed by the creation of 8 to 20 or more patterned layers on the wafer, also known as a substrate. This layering process, which uses a printing technique called lithography, creates electrically active structures on the semiconductor wafer surface, along with the circuit patterns interconnecting them.

Fabrication of microelectronic circuits relies on complex machinery housed in "clean rooms." The air in these rooms is constantly filtered to remove particles and gases that might contaminate the circuits during the manufacturing process. See Fig. 3 on p. 2338. Clean rooms are rated according to the maximum number and size of particles allowed per cubic foot of air; a Class 100 room has only 100 particles per cubic foot that are 0.5 micrometers in size and is regarded as cleaner than a hospital operating room. (A micrometer is a millionth of a meter. Human hairs average 75 micrometers in width.) New chip fabrication systems that are now being built for ever-more complex integrated circuits require Class 1 clean rooms with particle sizes of 0.2 micrometers to prevent contamination that leads to defective chips. Such rooms and the fabrication lines they contain are extremely expensive to construct, with costs reaching $500 million in the early 1990s. As a result, only the largest chip makers will be able to afford an investment of that magnitude.

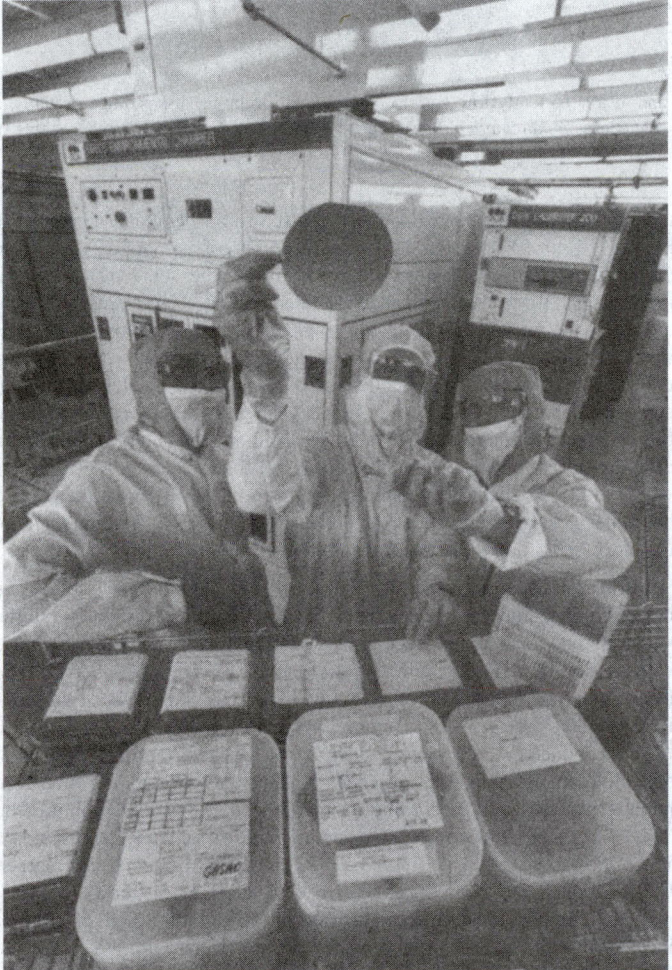

Fig. 3. More complex and powerful integrated circuits are possible using a photolithographic resist that relies on deep ultraviolet radiation in the wavelength range of 200 to 300 nanometers. The resist, a thin film of which is coated onto the surface of a semiconductor wafer, can be used to produce highly complex chips with features as small as 0.3 micrometer, or roughly $\frac{1}{400}$ the diameter of a human hair. Three AT&T Bell Laboratories researchers in Murray Hill, New Jersey, inspect a wafer made using the advanced resist. (*AT&T Bell Laboratories.*)

The fabrication sequence for a microelectronic circuit starts with molten purified polycrystalline silicon that is used to form a single crystal ingot.

The next step involves slicing the silicon ingot into thin wafers, up to 8 inches (20 cm) across. These are polished to a mirror-like luster.

Fabrication of a chip is accomplished through a series of repeated steps involving patterning and processing. In the first step, called thermal oxidation, the wafers are heated and exposed to ultra-pure oxygen in diffusion furnaces under carefully controlled conditions. This forms a silicon dioxide film of uniform thickness on the surface of the wafer. A photoresist or light-sensitive film is applied to the wafer. Next a pattern mask containing a distinctive design is used to expose a portion of the silicon to light while covering other areas. A mask resembles a photographic negative on a glass sheet, usually about 6×6 inches (15×15 cm). A wafer is mounted into a step-and-repeat machine that holds a mask above a special reducing lens. Because of a reduction ratio as large as 10 to 1, the machine exposes the mask to only a small portion of the wafer and then moves to the next position and exposes it again. Each time, the mask pattern is duplicated on the surface of the wafer. This process is repeated many times until that pattern has filled the wafer's surface area, thus creating multiple copies of the integrated circuit being fabricated.

The next step is called etching. It involves dissolving a portion of the silicon dioxide insulating layer on the surface of the silicon wafer and washing away the unexposed resist material. This leaves the desired circuit features on the wafer. Another step, called doping, exposes the surface of the wafer to impurities that alter the electrical characteristics of the silicon. The thermal oxidation, masking, etching, and doping steps are then successively repeated for each new pattern, or layer. A given chip may have 20 or more layers precisely registered with each other for component and conductor locations.

After all the layers of circuits have been constructed, a process called metallization imprints metal wiring patterns on the wafer so that the individual devices are connected as needed. Finally, each chip on the wafer is tested electrically, and a diamond saw cuts the wafer into single chips. The chips that perform correctly are then assembled into a package that provides the contact leads for the chip. A wire bonding machine attaches wires—each a fraction of the width of a human hair—to the leads of the package. Encapsulated with a plastic coating for protection, the chip is tested again prior to delivery to the customer.

Variety of ICs Produced. The semiconductor industry currently manufactures a wide variety of ICs that can be used in a virtually limitless array of products. The most common ICs include:

Microprocessor/central processing unit of a computer on a single chip. A chip with processing capability, but without extensive random access memory (ROM). A new architectural concept for microprocessors uses reduced instruction set computing (RISC), which speeds the processing of data by limiting the number of instructions the processor has to execute. The fastest RISC microprocessor processes information in chunks of 64 bits, each at a speed of 200 megahertz.

Digital signal processors (DSPs)—These devices represent the fastest-growing segment of the semiconductor industry. DSPs are specialized chips that perform high-speed addition and multiplication, digital filtering, data decoding, and formatting. One of the DSPs developed by AT&T Microelectronics performs about 40 million instructions a second, with up to five basic operations within each instruction. This is substantially faster than a typical desktop computer. The power of DSPs to do digital computations on analog information—both sounds and images—ideally suits them for consumer electronic products, cellular telephones, and laptop computer modems. In addition, they also can be used in large products like high-definition televisions.

Telecommunications integrated circuits—A broad line of specialty chips used for managing signal traffic in telecommunications equipment and networks. Telecommunications ICs usually perform dedicated functions, such as converting voice signals into digital format for faster transmission and then, at the receiving end, back into voice signals again.

Hybrid integrated circuits and multichip modules—Essentially a different way to package ICs, hybrids and multichip modules are complete electronic circuits in which numerous, separately manufactured items are arrayed on a suitable passive substrate, such as ceramic or polymer, and interconnected by metallization patterns, very fine wires, or both. Hybrids and multichip modules afford faster communications

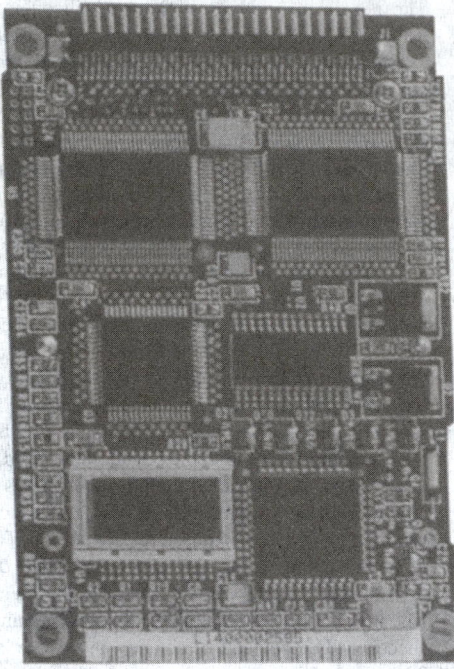

Fig. 4.   At some future date when chips will be able to store data in large volumes and at a sufficiently low cost to compete with rotating-disk storage systems, a flash memory may be used. This type of memory has no moving parts and, due to its ruggedness, is already able to compete with electromechanical drive systems in products that are used in demanding environments. (*AT&T Bell Microelectronics.*)

between chips and faster system performance compared to printed circuit boards.

Application specific integrated circuits (ASICs) — Semicustom chips designed to be tailored by individual customers to meet specific requirements, thus saving time and cost. One type of ASIC, called standard cell, is constructed from basic circuit modules that can be selected and wired together by the designer. Another type, called gate array, consists of identical circuit elements constructed on the wafer and later wired together according to customer requirements. An alternative approach, called field-programmable gate arrays, allows customers to program their own connections among circuit elements.

Memory chips — In addition to read-only memory, memory chips include two principal types: dynamic random access memory (DRAM) and static random access memory (SRAM). DRAMs store information in the form of binary digits, represented by the presence or absence of an electrical charge in capacitors. The stored charge must be refreshed periodically by circuitry to counteract the leakage of the capacitors. SRAMs hold binary information in a digital logic circuit, whose charge does not have to be periodically refreshed. Like DRAMs, SRAMs lose memory when power is turned off. See Fig. 4.

One problem facing researchers is that ever-increasing circuit density and performance depend on ever-shrinking lines, or features, that define circuit elements. These now measure less than 1 micrometer. Soon the lines won't be visible with optical microscopes because they will be smaller than the wavelength of light. As a result, light-based lithography, or photolithography, likely will give way to new forms of lithography based on new energy sources. Among the options being examined at Bell Laboratories and elsewhere are the use of X-rays and electron beams to carry on the lithography process. In the meantime, scientists have continued to extend the life of silicon technology by shrinking the size of features using photolithography. As of early 1992, the microelectronics industry was manufacturing the most advanced chips, with feature sizes of 0.5 micrometers. Researchers estimate that, by about the year 2010, chips may have line sizes of 0.05 micrometers, well below the 0.125-micrometer level that is now under research.

Developers double the number of components on a microchip every few years, which means two or three new generations of chips were produced during the 1990s. That trend will have matured by the year 2010, with chips containing up to a few billion components each for highly packed

circuits, such as memory. For typical custom logic circuits, the range is in the 10^8 to 10^9 region. The actual number of components per chip produced may be constrained by economics to be somewhat less than the physical limit. With advanced semiconductor technology, complete electronic systems will be put on a single chip, even systems as complex as today's supercomputers. Ideally, researchers look for ways to lengthen the lifespan of the existing technology, an approach consistent with good economics and good engineering. Besides the physical processing to make the chips, chip designers need additional automated tools to help them design and test advanced chips. High-level, modular design capabilities will continue to allow developers to focus on what they want the chip to do, rather than worrying about the details of the chip's 50 million or more transistors, an impossible task for humans.

Future of Microelectronics. The familiar "bulk-effect" solid-state devices may mature in the early part of the next century. At that point, the smallest functional bulk-effect transistor operating at room temperature will measure about 400 atoms by 400 atoms. The next frontier may be based on single-electron devices and ballistic transistors, whose behavior is described by the laws of quantum physics as applied to single particles.

Semiconductors will play a central role in the future of the U.S. electronics industry — chips provide the fundamental building blocks in hundreds of products. According to the National Advisory Committee on Semiconductors (NACS), in 1990 semiconductors annually represented a \$21 billion industry in the United States and a \$63 billion world market. U.S. market share, which has declined steadily in the past 25 years, was 37%. Semiconductors are critical to the \$384 billion U.S. electronics industry and the \$751 billion world market for electronics. According to NACS, by the year 2000, world semiconductor sales are expected to rise to \$200 billion, and the world electronics industry to \$2 trillion. Thus, electronics, already the largest employer in the United States, holds the promise for strong growth in the decade ahead. And a key linkage is the dependence of all electronic systems on semiconductors.

In the "Information Age," the amount of information transmitted in communications and stored and processed in computers continues to expand at a phenomenal rate. To cope with this growth, faster, more powerful microchips are essential for the advancement of information products during the 1990s and beyond. The increasing computing power of ICs also promises to deliver a wide range of easy-to-use consumer and

Fig. 5. As chips become smaller and incorporate more functions, fewer of them are required to accomplish a task, such as sending and receiving digital information over a telephone line. Such is the job of a modem, and the few chips here are able to do that task in a package the size of a credit card and only 5 millimeters thick. Just a few years ago, modems were the size of a hardcover novel. (*AT&T Microelectronics*.)

hand-held information products that may, for example, respond to speech and handwritten input. See Fig. 5.

Advanced chips are needed for national and global high-speed networking of voice, data, images, and video for business and commerce, academia, research, and entertainment. New applications will become possible, such as intelligent systems for toll booths and highways that will read chips built into vehicles. By shortly after the turn of the century, it is likely that people will be able to talk directly with their computers as an adjunct to using keyboards. This will be possible, thanks to digital signal processors that can help understand human speech and generate a synthesized voice. Video, speech recognition and synthesis, and character-recognition capabilities will be relatively easy just by adding a few chips to a product such as a laptop computer. See Fig. 6.

Fig. 6. The AVP1000 three-chip video codec makes it cost-effective for manufacturers to bring advanced video telephony and multimedia features, such as full-motion video and video-conferencing, to desktop personal computers and workstations. (Codec refers to "compression/decompression," a technique required to process image and sound information in a manageable digital form.) The chip set shown here compresses image information with sufficient rapidity to enable video to be sent through digital communication lines. The set, which replaces printed circuit boards (PCBs), also decompresses stored or transmitted video information to recreate a motion display. (*AT&T Microelectronics*.)

Products using such capabilities will range from individual electronic student tutors to high-performance military systems. Speech recognition and speech synthesis will mature in a variety of service capabilities based on the ability of intelligent machines to talk and listen much as people do. For example, by the year 2010 we should have the ability to transfer unique personal speech characteristics across languages, to make possible customized speech translators. Speech in one language would be automatically translated into a second language, which might then be synthesized with the voice characteristics of the original speaker. And this translation would be done in real time.

During its lifetime, microelectronics has been one of the most spectacular forces in the history of science and engineering. It has been the most powerful driver of progress in the communications field and also was at the forefront of breakthroughs in many other industries, including aerospace, computers, and consumer electronics. Microelectronics promises to continue as a major force well through its maturity during the coming decade, reaffirming that "smaller is better."

Additional Reading

Jaeger, R.: *KC's Problems and Solutions for Microelectronic Circuits*, Addison Wesley Longman, Inc., Reading, MA, 1988.
Sedra, A. and K. Smith: *Microelectronic Circuits*, Oxford University Press, Inc., New York, NY, 1997.
Smith, K.: *KC's Problems and Solutions for Microelectronic Circuits*, Oxford University Press, Inc., New York, NY, 1997.

THOMAS K. LANDERS, AT&T Bell Laboratories, Short Hills, NJ

MICROENCAPSULATION. Microencapsulation is the coating of small solid particles, liquid droplets, or gas bubbles with a thin film of coating or shell material. Here, the term microcapsule is used to describe particles with diameters between 1 and 1000 µm. Particles smaller than 1 µm are called nanoparticles; particles greater than 1000 µm can be called microgranules or macrocapsules.

Many terms have been used to describe the contents of a microcapsule: active agent, actives, core material, fill, internal phase (IP), nucleus, and payload. Many terms have also been used to describe the material from which the capsule is formed: carrier, coating, membrane, shell, or wall. In this article the material being encapsulated is called the core material; the material from which the capsule is formed is called the shell material.

Table 1 lists representative examples of capsule shell materials used to produce commercial microcapsules along with preferred applications.

Microcapsules can have a wide range of geometries and structures. Figure 1 illustrates three possible capsule structures. Parameters used to characterize microcapsules include particle size, size distribution, geometry, actives content, storage stability, and core material release rate.

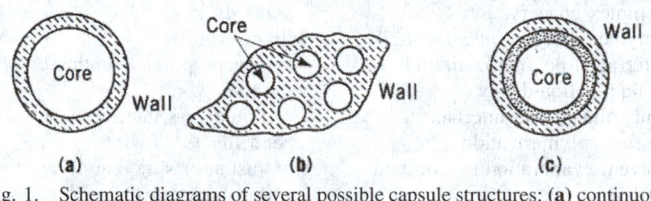

Fig. 1. Schematic diagrams of several possible capsule structures: (**a**) continuous core/shell microcapsule in which a single continuous shell surrounds a continuous region of core material; (**b**) multinuclear microcapsule in which a number of small domains of core material are distributed uniformly throughout a matrix of shell material; and (**c**) continuous core capsule with two different shells.

Encapsulation Process

Classification of the many different encapsulation processes is useful. Previous schemes employing the categories chemical or physical are unsatisfactory because many so-called chemical processes involve exclusively physical phenomena, whereas so-called physical processes can utilize chemical phenomena. An alternative approach is to classify all encapsulation processes as either Type A or Type B processes. Type A processes are defined as those in which capsule formation occurs entirely in a liquid-filled stirred tank or tubular reactor. Emulsion and dispersion stability play a key role in determining the success of such processes. Type B processes are

TABLE 1. SHELL MATERIALS USED TO PRODUCE COMMERCIALLY SIGNIFICANT MICROCAPSULES

Shell Material	Regulatory Status	Chemical Class	Encapsulation Process	Applications
Gum arabic	Edible	Polysaccharide	Spray drying	Food flavors
Gelatin	Edible	Protein	Spray drying	Vitamins
Gelatin-gum arabic[a]	Nonedible[b]	Protein-polysac-charide complex	Complex coacervation	Carbonless paper
Ethylcellulose	Edible	Cellulose ether	Wurster process or polymer—polymer incompatibility	Oral pharmaceuticals
Polyurea or polyamide	Nonedible	Cross-linked polymer	Interfacial polymerization	Agrochemicals and carbonless paper
Aminoplasts	Nonedible	Cross-linked polymer	*in situ* polymerization	Carbonless paper, fragrances, and adhesives
Maltodextrins	Edible	Low molecular weight carbohydrate	Spray drying and desolvation	Food flavors
Hydrogenated vegetable oils	Edible	Glycerides	Fluidized bed	Assorted food ingredients

[a] Treated with glutaraldehyde.
[b] For intended application, i.e., carbonless paper.

processes in which capsule formation occurs because a coating is sprayed or deposited in some manner onto the surface of a liquid or solid core material dispersed in a gas phase or vacuum. This category also includes processes in which liquid droplets containing core material are sprayed into a gas phase and subsequently solidified to produce microcapsules. Emulsion and dispersion stabilization can play a key role in the success of Type B processes also.

Many Type A and Type B processes are similar. For example, solvent evaporation is a key step in most spray dry encapsulation protocols (Type B) and protocols involving solvent evaporation from an emulsion (Type A). The difference in these protocols is that evaporation in the former case occurs directly from a liquid to a gas phase, whereas in the latter case evaporation involves transfer of a volatile liquid from a dispersed phase to a continuous liquid phase from which it is subsequently evaporated. Another example is encapsulation by gelation. In Type A gelation processes, the droplets that are gelled and become microcapsules are formed by dispersion in a liquid phase and are gelled in this phase. In Type B gelation processes, droplets formed by atomization or extrusion into a gas phase are subsequently gelled either in the gas phase or a liquid gelling bath.

Most Type A processes might be classified as chemical processes, whereas most Type B processes are classified as mechanical processes. Representative examples of both types of processes follow. Type B processes tend to be promoted by organizations that sell and service equipment for producing microcapsules. Most Type A processes are not promoted by equipment manufacturers, but are developed and used by organizations that produce microcapsules.

Type A processes	Type B processes
complex coacervation	spray drying
polymer–polymer incompatibility	fluidized bed
interfacial polymerization at liquid–liquid	interfacial polymerization at solid–gas
and solid–liquid interfaces	or liquid–gas interfaces
in situ polymerization	centrifugal extrusion
solvent evaporation or in-liquid drying	extrusion or spraying into a desolvation bath
submerged nozzle extrusion	rotational suspension separation (spinning disk)

Applications

Microcapsules are used in a number of pharmaceutical, graphic arts, food, agrochemical, cosmetic, and adhesive products. Other specialty products also exist, thus the concept of microencapsulation has been accepted by a wide range of industries. In order to illustrate how microcapsules are used commercially, it is appropriate to describe a number of commercial microcapsule-based products and the role that microcapsules play in these products.

Carbonless copy paper is by far the largest single commercial application of microcapsules. This product consumes thousands of tons of capsules annually.

Success of all carbonless paper products depends on the microcapsules, leuco dyes, and reactive coating. A number of leuco dyes are available.

The concept of microencapsulation has intrigued the pharmaceutical industry for many years, because it offers the possibility of providing a number of important new oral and parenteral dosage forms. Microcapsules in oral dosage forms could conceptually taste-mask bitter pharmaceuticals, provide extended release *in vivo*, provide enteric release, improve the stability of incompatible drug mixtures, provide resistance to oxidation, reduce volatility, and distribute a drug in many small carrier particles so that effects of the drug on the sensitive walls of the stomach are minimized. Microencapsulated parenteral formulations could provide prolonged delivery of drugs with short half-lives *in vivo* and perhaps even achieve targeted drug delivery. For these reasons, microencapsulation has received much attention by pharmaceutical scientists. Several microcapsule-based oral pharmaceutical formulations which offer some of these features are available.

The use of microcapsules for a variety of biomedical and biological applications has been promoted for many years. Several biomedical microcapsule applications are in clinical use or have approached clinical use. One application is the use of air-filled human albumin microcapsules as ultrasound contrast agents. Another biomedical application of microcapsules is the encapsulation of live mammalian cells for transplantation into humans. The purpose of encapsulation is to protect the transplanted cells or organisms from rejection by the host.

A number of food ingredients or additives have been encapsulated and are available commercially. Solid ingredients encapsulated are typically water-soluble and are encapsulated with a hydrophobic or hydrophilic coating material usually applied by the Wurster process. Both types of coating materials are well-accepted food-grade products. See also **Food Additives**.

The microencapsulation of pesticides and herbicides has been an active area of development that has produced several commercial products. The function of the microcapsules is to prolong activity while reducing mammalian toxicity, volatilization losses, phytotoxicity, environmental degradation, and movement in the soil. Ideally, encapsulation would also reduce the amount of agrochemical needed. See also **Pesticides**; and **Herbicides**.

Advertising inserts that utilize encapsulated perfumes and flavors contain a coating of scent-filled capsules which break and release scent when the insert is torn open are widely used as a marketing tool, primarily for new perfumes. Children's crayons loaded with encapsulated scents are appearing on the market. The capsules break during the drawing process thereby releasing a scent characteristic of the drawn object.

Microcapsules are used in several film coatings other than carbonless paper. Encapsulated liquid crystal formulations coated on polyester film are used to produce a variety of display products including thermometers. Polyester film coated with capsules loaded with leuco dyes analogous to those used in carbonless copy paper is used as a means of measuring line and force pressures. Encapsulated deodorants that release their core contents as a function of moisture developed because of sweating represent another commercial application. Microcapsules are incorporated in several cosmetic creams, powders, and cleansing products.

A majority of the fasteners used in automobiles in the U.S. are coated with microcapsules loaded with an adhesive. Other uses include encapsulated ammonium polyphosphate incorporated in plastics that acts as a fire-retardant and microencapsulated oil-field chemicals for use by the oil industry.

Additional Readings

Bakan, J.A.: in L. Lachman, H.A. Lieberman, and J.L. Kanig, eds., *The Theory and Practice of Industrial Pharmacy*, 3rd Edition, Marcel Dekker, New York, 1986.

Cohen, S. and H. Bernstein: *Microparticulate Systems for the Delivery of Proteins and Vaccines*, Marcel Dekker, Inc., New York, NY, 1996.

Deasy, P.B.: *Microencapsulation and Related Drug Processes*, Marcel Dekker, Inc., New York, 1984.

Kondo, A.: *Microcapsule Processing and Technology*, Marcel Dekker, Inc., New York, 1979.

Thies, C.: *How-to-Make Microcapsules: Lecture and Lab Manual*, Thies Technology, St. Louis, Mo., 1994.

Whateley, T.L.: *Microencapsulation of Drugs*, Taylor & Francis, Inc., Philadelphia, PA, 1992.

MICROFILM AND MICROFORM. Prior to the advent of convenient computer systems for storing long-term information, microfilm (photographic reduction on 16- or 35 mm film) was widely used. Microform, microfiche, and microcard were variations of this technique. For some applications, some users still use these techniques.

MICROGRAVITY AND MATERIALS PROCESSING. The development of space transportation systems during the past several years has drawn attention to materials processing in a reduced gravity environment. Actually, exploratory work in this area has been proceeding since the early *Apollo* missions to the moon. For example, materials processing experiments were carried out during the *Skylab* program and on the *Apollo-Soyuz* test program fight. More recently, tests have been conducted on some of the *Space Shuttle* projects.

Effects of Reduced Gravity. The reduced gravity of space offers a unique environment for materials processing. It is untrue, of course, to claim that a state of zero gravity is achieved as, for example, on space flights as currently experienced. Rather, the gravity is greatly reduced — to about 10^{-6} g. Hence, the term *micro* rather than *zero* gravity is more appropriate to use here. It is interesting to note that spacecraft in low earth orbit may experience changes in the gravitational field, referred to as *g-jitter*, which are caused by such factors as maneuvering the craft while in orbit, atmospheric drag, and even movement of the crew within the spacecraft. The *g-jitter* phenomenon can cause "spikes" in the gravitational field, ranging from 10^{-2} g to 10^{-1} g. These may have an adverse effect upon on-board experiments and processes.

Manned or unmanned orbital spaceflight is not the only possibility for reduced-gravity studies of materials processing. Drop tubes and towers, aircraft, and sounding rockets also offer opportunities for varying levels and time periods of reduced gravity. Time spent in low gravity and payload size are the limiting factors for earth-based facilities. Drop tube experiments provide 2.5 to 4.5 seconds at 10^{-8} g to 10^{-9} g for free falling droplets, whereas a drop tower allows an entire experimental package, weighing 100 kg, to be tested. By flying in parabolic flight paths, aircraft ranging in size from a KC-135 to a single-seat F-104 can provide 10^{-1} g to 10^{-2} g for 15 to 60 seconds for payloads ranging from 10 to 35 kg. Sounding rockets provide up to 300 seconds at 10^{-5} g. When compared with the duration time provided by an orbiting space platform, the earthbound methods are quite brief.

One effect of the reduction of gravitational forces experienced in space is the virtual elimination of *buoyancy-driven convection*. Convection within a fluid medium arises when the medium is subjected to a nonvertical thermal gradient. Temperature differences create density gradients as gases or liquids expand upon heating. In the presence of a gravitational field, the less dense volume of the medium is displaced by a denser, cooler volume, resulting in the circulation of gases or liquids, commonly known as convective flow.

A more unstable form of convection occurs when a denser fluid lies above a less dense fluid, corresponding to a situation where the medium is heated from below. If viscous forces outweigh the buoyant forces within the medium, this unstable condition can be maintained. If the buoyancy is greater than the viscosity, however, the volume element rises too quickly, resulting in spontaneous flow, which takes the form of cells or vortex rolls.

Concentration gradients caused by chemical reaction within a fluid medium may also cause convection. In this case, density gradients occur when a particular chemical component is consumed or produced by reaction.

Convective flow is mathematically characterized by the Grashof number Gr, which represents the ratio of buoyant to viscous forces. The Grashof number is given by the expression

$$Gr = \frac{gl^3 \beta \Delta T}{\mu^2}$$

where g is the gravitational acceleration; β the coefficient of expansion; ΔT the temperature gradient; and μ the kinematic viscosity. In an earth-based experiment, convection can be reduced somewhat within practical limits by altering the geometry of a given system or minimizing temperature gradients. Conducting a similar experiment in low earth orbit can reduce the value of Gr by six orders of magnitude.

Because convection and diffusion occur simultaneously in an earth-based process, corresponding studies in a microgravity environment can help identify the effects of these two phenomena on a given process such as crystal growth.

The reduced gravity of space can also be used to process materials in a container-free environment. This feature of microgravity processing is particularly advantageous when a material of high purity is desired, or for achieving a high degree of supercooling in a sample. Containerless processing is especially useful for obtaining glasses and alloys.

The presence of a gravitational field causes substances of differing densities to separate out. In microgravity, however, this gravity-induced separation is eliminated, thereby producing a more uniform mixture. This effect is useful in the processing of alloys and organic polymers.

Representative Microgravity Experiments

Biological Materials. The degree of purity of biological materials severely limits their usefulness. Electrophoresis is a commonly used method of separation and purification of substances such as cells, enzymes, and proteins. This technique relies upon the fact that surface charge distribution, and thus mobility in an electric field, vary from one material to another. The degree of separation, product yield, and purity are limited by convection which is caused by concentration gradients within the process medium.

A continuous flow electrophoresis (CFE) process has been used to effect separation of biological materials in microgravity. The absence of convection permits continuous processing of relatively large volumes of material, higher yields, finer separation, and higher product purity than are possible on earth. Erythropoietin, which is produced by the kidneys and controls red blood cell production in the body, has been produced by CFE on board the *Shuttle*. The first CFE experiment performed in 1982 in *Spacelab* yielded 463 times more material than comparable earth-based processes. Separation rates were boosted in later flights to yield 700 times more material having a fourfold increase in purity over products obtained on earth.

Polymers. Research efforts in the area of organic polymer growth in space seek to take advantage of the absence of phase separation due to density differences. On earth, density differences in nonhomogeneous mixtures of organic liquids produce buoyancy-driven convection and cause immiscible liquids to separate. These phenomena affect the growth of organic polymers, causing flaws in the final product. In the absence of phase separation and convection, more uniform mixtures can be produced. Secondary effects such as surface tension can also be utilized to obtain more perfect polymers and organic compounds.

One type of experiment performed in space was the diffusive mixing of organic solutions (DMOS) study conducted by the 3M Corporation. The DMOS experiments mixed different types of organic solutions to yield crystalline material. The purpose of the study was to determine the effect of microgravity upon the ordering of organic molecules upon crystallization. The crystals grown in the experiment were not only significantly larger than similar crystals grown on earth, but possessed much better optical and electrical properties as well.

The dominance of surface tension, due to the lack of convection in microgravity, has been used to produce perfectly round spheres of polystyrene-latex. The spheres are grown in space by the coalescence of an emulsion. Under conditions where surface tension controls the process, droplets do not readily break up, thereby allowing large spheres to coalesce. As a result, large, perfect spheres, having a diameter of up to 30 mm, can be produced in space. In comparison, a maximum diameter of 5 mm can be produced on earth. These spheres are offered commercially for reference and calibration applications. As such, they are the first commercial products to be made in space.

Physical Metallurgy. One obstacle to the processing of alloys on earth is that components of a given mixture are often immiscible. As a result, density differences cause the components to separate as the bulk melt cools. By eliminating this gravity-driven separation, the manufacture of alloys can benefit from a microgravity environment.

Several alloy systems have been studied in space. In general, these experiments have yielded promising results, showing that finer, more homogeneous mixtures of components can be obtained in microgravity. In space, reduced convection in the melt apparently reduces microsegregation and heat transfer. This allows materials possessing highly directional physical properties, such as magnetic coercivity and microstructure, to be produced.

Containerless processing is also of interest in physical metallurgy, as it provides opportunities to study thermophysical properties of high temperature metals and alloys, avoid sample contamination due to contact with container walls, and observe the solidification of materials that have been rapidly cooled from the melt.

Containerless processing is accomplished on earth under the influence of gravity by using electrostatic, acoustic, or electromagnetic energy to levitate a sample. Sample size is limited, however, by power requirements for levitation. The application of the forces necessary to levitate a substance also induces a certain amount of mixing and heating. Gravity-driven convection is also present in this situation and can cause unwanted mixing of liquid samples.

To date, containerless processing in near zero gravity has been limited to drop tube experiments, which provide a few seconds of low gravity for small drops of material. Alloys studied in this fashion have been undercooled as much as 500 °C, which corresponds to a cooling rate of greater than 10^6 K/sec. Samples obtained in these experiments have exhibited metastable or peritectic phases which are extremely difficult to obtain under normal conditions.

Some levitation of samples will still be necessary to carry out containerless processing in space, although the magnitude of the forces necessary to do so will be small relative to those required in earth-based work. Levitation would be necessary only to avoid contact between sample and container, so larger samples could be used in space-based experiments.

Other metallurgy experiments scheduled for space will examine the role of macrosegregation in the processing of metals, the feasibility of using directional solidification in the processing of different classes of alloys, and the manufacture of alloys which cannot be produced on earth due to density induced separation.

Glasses. The manufacture of glasses also benefits from containerless processing in space. As with alloys, the purity of glasses can be affected by contact and subsequent reaction with container walls. In the case of glass processing, however, contact between sample and reactor wall also causes crystallization, and hence loss of the amorphous glassy state. In addition, less viscous glasses require high cooling rates in order to prevent crystallization. By avoiding contact-induced nucleation during cooling, containerless processing may be used to obtain larger, high purity samples of such glasses.

As mentioned previously, some levitation is required in microgravity to maintain sample positioning. Several designs, some of which are capable of processing temperatures up to 1600 °C, have been developed for use in space experiments.

Glass processing experiments that have already been flown on the *Shuttle* have been concerned with melt homogenization, bubble behavior in molten glass spheres, preparation of glass microballoons, and comparing properties of space-produced glasses with those manufactured on earth. Results to date indicate that glasses having different microstructures than those of glasses processed on earth can be obtained. Galliacalcia, sodium-borate, and lead-silica glasses have been selected for the above experiments. The list of glasses to be studied will be expanded to include materials that are particularly difficult to produce on earth. Among these materials are heavy cation (Zr, Hf, Th) glasses that tend to react with containers, and silica glasses that must be processed at high temperatures.

Crystal Growth. Single crystals of both organic and inorganic substances can be grown from either the vapor or liquid phase, using several different experimental techniques. Gravity-driven convection affects the motion of these fluid media, greatly affecting the mixing and transport of individual chemical components. Experiments aimed at examining the effects of eliminating buoyancy-driven convection upon different crystal growth techniques have been performed on *Skylab*, the *Apollo-Soyuz* mission, and several *Space Shuttle* flights.

Growth of crystals from the vapor relies upon the presence of a temperature gradient. Concentration gradients of vapor species and subsequent migration from a source region to a seed crystal, substrate, or deposition region are caused by this temperature difference. In the case of physical vapor transport (PVT), solid source material vaporizes at one temperature. The gaseous vaporization products migrate, usually through an atmosphere of inert gas, to another temperature where solid material condenses. In space, the PVT method has been used to grow highly ordered organic thin films onto silicon wafer substrate and large single crystals of germanium selenide.

Another PVT experiment examined the growth of HgI_2 onto a seed crystal. This substance has potential for use as a radiation detector. Due to its high density, however, the HgI_2 crystal structure readily deforms during earth-based processing. Large crystals of HgI_2 have been grown by the PVT method aboard *Spacelab 3*. Growth times in space were considerably less than those normally required on earth. Performance of the space-grown crystals as radiation sensors is matched by only the very best crystals obtained on earth.

Unlike PVT, chemical vapor transport (CVT) utilizes a highly reactive gaseous substance — such as a halogen or metal halide — to transport source material to a region of the reaction container where single crystals condense from the vapor. In earth-based CVT studies, under conditions where buoyancy-driven convection drives the overall transport process, crystal size is generally small. The morphology of crystals grown under these conditions is often poor; surfaces are marked by large numbers of defects and irregular growth steps. In contrast, space-grown crystals grown by chemical vapor transport are much larger, have smoother growth steps, and fewer defects. Chemical homogeneity within these crystals is also considerably better than in similar crystals grown on earth.

The presence of convection also affects crystal growth from the melt. Single crystals of Te-doped InSb were grown from the melt on *Skylab*. The crystals obtained in space were free of striations caused by convection-driven growth rate fluctuations that are normally seen on earth. Future space experiments will examine the growth of electronic materials such as GaAs from a solution subjected to an electric current.

R‌OBERT P. S‌ANTANDREA, Ph.D., Los Alamos National Laboratory, Los Alamos, NM

MICROMANOMETER. A manometer capable of measuring very small pressure changes or differences.

MICROMETEORITE. A very small meteorite or meteoritic particle with a diameter in general less than a millimeter. See also **Moon (Earth's)**; and **Poynting-Robertson Effect**.

MICROMETEOROLOGY. The part of meteorology that deals with observations and processes in the smallest scales of time and space, approximately smaller than 1 km (0.6 mi) and less than a day (i.e., local processes). Micrometeorological processes are limited to shallow layers of frictional influence (slightly larger-scale phenomena like convective thermals are not part of micrometeorology). Therefore, the subject of micrometeorology is the bottom of the atmospheric boundary layer; namely, the surface layer. Exchange processes of energy, gases, etc., between the atmosphere and the surface (water, land, plants) are important topics. Therefore, micrometeorology is closely connected with most of the human activities in the atmosphere. Microclimatology describes time averaged (long-term) micrometeorological processes, and micrometeorologists are interested in their fluctuations. See also **Macrometeorology**.

MICROMETER. The micrometer represents a general principle of physical measurements, used on various instruments such as comparators; spherometers, compensators, interferometers, etc. It is essentially a screw of accurately known, uniform pitch (commonly 1 millimeter or 0.5 millimeter), provided with a large head whose periphery is divided into equal parts, forming a scale. Turning the screw through a given number of these parts causes the shaft to travel through a distance which is a proportionate fraction of the pitch. For example, if the pitch is 0.5 millimeter and the head is divided into fiftieths, each scale division corresponds to a travel of 0.01 millimeter.

Micrometer also is a unit of length. See also **Units and Standards**.

MICRONEWTON THRUSTERS. See **Space Technology 7 (ST7) Mission.**

MICROPHONE. An electroacoustic transducer that responds to sound waves and delivers essentially equivalent electric waves. Making this conversion as faithfully as possible is one of the most important criteria in microphone design. The transducer should be linear for minimum waveform distortion and should have the greatest possible dynamic range. The transducer should have minimal effect on the dynamic phenomenon and whatever effect it does have should be determined by calibration or computation so that data can be appropriately corrected.

Microphones are of two basic types: (1) *pressure-sensing*, used to observe the pressure waveform of sound; and (2) *velocity-sensing*, used to observe the velocity waveform. For measurement purposes, the pressure-sensing microphone usually is preferred because of its superior frequency response and dynamic range.

Pressure-sensing microphones used for measurement purposes operate on the capacitive, piezoelectric, or moving-coil principle. In a pressure microphone, the electric output substantially corresponds to the instantaneous sound pressure of the impressed sound wave. When stability and a broad flat frequency response are required for accurate measurement, a condenser (capacitive) microphone is used.

Condenser microphone. In a capacitive transducer, mechanical-to-electrical conversion is effected by mechanically induced changes in electrical capacitance. The capacitance between two conducting plates, separated by an insulator, is inversely proportional to the distance between the plates, provided that the plates are large enough relative to their separation so that edge effects can be ignored.

When the metal diaphragm of a condenser microphone is displaced by acoustic pressures, the diaphragm produces a capacitance change between the diaphragm and a rear electrode. Air is the insulator between the conductors. The components of a simplified condenser microphone are shown in Fig. 1. The small hole in the rear cavity behind the diaphragm is to equalize static pressure across the diaphragm. The most common method of sensing capacitance change in this type of microphone is to apply, between the diaphragm and the rear electrode, a large dc voltage. This is called the polarizing voltage and should come from a high-resistance source. The voltage decreases as the distance between the conductors decreases and vice versa. This voltage change can be amplified and observed on a display instrument as the electric analog of the acoustic pressure waveform.

input-output characteristic. Condenser microphone sensitivities fall in the range from -50 dB to -100 dB re 1 volt per dyne/cm^2. The flat response range, wherein sensitivity is relatively constant, generally is from less than 100 cycles/sec (Hz) to several thousand Hz. While the polarizing voltage technique is simple and practical, it requires very high electrical resistance for adequate low-frequency response and to ensure that the charge does not change as the diaphragm moves.

The resonant circuit technique also has been used with capacitive transducers. This technique does not require high-resistance circuitry and responds to static pressures, but the response is less linear than the polarizing voltage technique and requires more complex electronic circuits.

Piezoelectric Microphones. These can be substituted when it is not convenient to use the peripheral electrical circuitry essential to the capacitive type—and when some loss of accuracy can be tolerated; or when the sound pressure is high enough to overload the condenser microphone. A piezoelectric transducer normally used is an accelerometer. Voltage sensitivities of piezoelectric accelerometers typically range from 1 to 100 mV/gram. The piezoelectric accelerometer is used widely because it is simple, rugged, and requires no outside power supply.

Carbon-button Microphones. These employ a packet of carbon granules, which the sound field acts upon through a linkage to vary the resistance of the packet. With a source of direct current, this device produces a dc voltage proportional to the pressure.

Moving-coil Microphone. This device operates on the same principle as an electric motor/generator and the loudspeaker. That is, when an electrical conductor is moved through a magnetic field, a voltage is induced in the conductor that is proportional to (1) the strength of the magnetic field; (2) the length of exposed conductor; and (3) the velocity of motion. Loudspeakers sometimes are used in reverse as receivers of sound, but their size and uneven frequency response limits their utility. See also **Loudspeaker**. Microphones designed on the moving-coil principle are used widely on sound stages and in recording studios. Moving-coil (dynamic) microphones work well with low-impedance electrical circuitry, but since their electrical output is proportional to the velocity of coil motion, the devices are unresponsive to very low frequency or quasi-static pressure changes.

Ribbon Velocity Microphone. Also termed the pressure gradient microphone, this device employs a metal ribbon suspended between the poles of a magnet. The pressure gradient in the acoustic field, acting on both sides of the ribbon, induces a voltage across the ribbon that is proportional to the sound pressure.

Electrostatic Microphone. This is a capacitor type microphone wherein variations in sound pressure, acting on a stretched metal diaphragm, effect variations in capacity. This microphone normally is used in a dc circuit with a voltage source and a resistor to develop a voltage across the resistor; or in an electronic-tube circuit, to modulate the frequency of an oscillator. The output is proportional to pressure in either application.

Directivity of Microphones. This is the relative sensitivity of the microphone to the angle of arrival of a plane wave perpendicular to a specified axis. Most commonly, microphones are omnidirectional; bipolar; or cardioid. See Fig. 2. Omnidirectional microphones usually are of the pressure type. Velocity microphones have bipolar response, showing high sensitivity to the front and rear, but little to the sides. Cardioid microphones are unidirectional, highly indifferent to sounds from outside the cone of prime sensitivity.

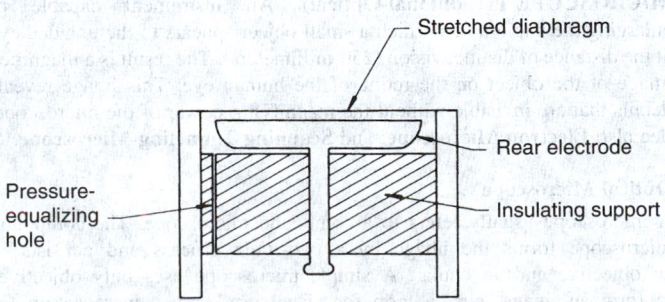

Fig. 1.   Components of a simplified condenser microphone.

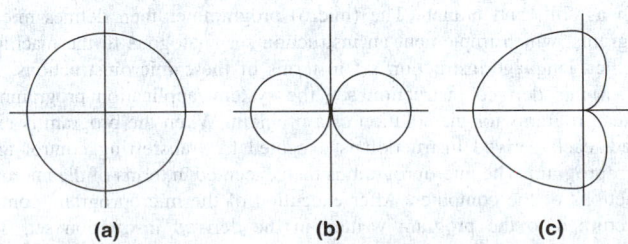

Fig. 2.   Directivity of microphone: (**a**) omnidirectional; (**b**) bipolar; (**c**) cardioid.

The sensitivity of the transducer is the ratio of its electric output to its mechanical input. Thus, the sensitivity is the slope of the linear range of the

See also **Acoustics** for description of electret and other recent developments in microphone design.

MICROPROCESSOR (Chronology).

The microprocessor, in terms of concept and reduction to practical applications in thousands of end products, will be recorded in the history of the electronics industry as a truly major achievement. The microprocessor also has had a profound effect upon the design of nearly all forms of electronic equipment, including *smart* or *intelligent* sensors, transducers, and controllers.

A microprocessing unit (MPU) is a computer central processing unit (CPU) built as a single semiconductor chip. (In some early designs, an MPU consisted of a small number of chips.) The difference between a microprocessor and a microcomputer has progressively grown fuzzy. Traditionally, the microcomputer is somewhat more complete than a MPU. This is because the microcomputer contains not only the CPU logic, but also memory for storing programs and I/O (input/output) data interfaces for exchanging data with peripheral devices, and timing circuits to control the flow of data within the computer, and the Advanced Micro Devices *Am29000*, among others. Several giant firms entered the microprocessor manufacturing arena, producing a highly competitive market during the early 1990s. The microprocessor has continued to advance in capacity and performance.

A microcomputer is frequently a single-board computer containing several to many MPUs. Marked enhancement of the MPU in the relatively recent past has further muddied the distinction.

History records that the first commercial microprocessor (the *4004*) was introduced by Intel in 1971. Attention to standardized architecture lowered the cost of the chip design and development, permitting amortization over many more units. Intel introduced the *4040* and *8008* shortly thereafter, and the *8080* in 1974. By 1976, several manufacturers were supplying 8-bit (basic word length) microprocessors in large volumes, significantly reducing unit cost. In 1980, more powerful 16-bit MPUs were developed, examples of which were Texas Instruments' *9900*; Intel's *8086*, Zilog's *Z8000*, and Motorola's *MC 68000*. One of the first 32-bit MPUs was introduced in 1984, the Motorola *MC 68020*, with its emphasis on higher execution speed, larger address space (4 billion locations) and more instructions relative to the 16-bit *MC 68000* chip. The *MC 68020* used the equivalent of about 200,000 transistors on an integrated circuit chip measuring about 9.5×8.9 mm. In 1986 and early 1987, a number of other manufacturers announced the availability of 32-bit MPUs, including the Fairchild *Clipper*, the Motorola *68030*. See Fig. 1.

See also **Microelectronics**.

MICROPROGRAM.

Microprogramming is a technique of using a special set of instructions for an automatic computer. Elementary logical and control commands are used to simulate a higher-level instruction set for a digital computer. The basic machine-oriented instruction set in many computers is comprised of commands, such as add, divide, subtract, and multiply. These are executed directly by the hardware. The hardware actually implements each function as a combination of elementary logical functions, such as AND, OR, and EXCLUSIVE OR. The manner of exact implementation usually is not of concern to the programmer. Compared with the elementary logical functions, the add, subtract, multiply, and divide commands are a higher-level language set in the same sense that a macrostatement is a higher-level instruction set when compared with the machine-oriented language-instruction set. See also **Macroassembler**.

In a microprogrammed computer, the executable (micro) instructions which may be used by the (micro) programmer, are comprised of a logical function, such as AND and OR and some elementary control functions, such as shift and branch. The (micro) programmer then defines microprograms, which implement an instruction set analogous to the machine-oriented language-instruction set in terms of these microinstructions. By use of this derived instruction set, the systems/application programmer writes programs for the solution of a problem. When the program is executed, each derived instruction is executed by transferring control to a microprogram. The microprogram is then executed in terms of the microinstructions of the computer. After execution of the microprogram, control is returned to the program written in the derived instruction set. The microprogram typically is stored in a read-only storage (ROS) and thus is permanent and cannot be changed without physical replacement.

An advantage of microprogramming is increased flexibility. This can be realized by adapting the derived, machine-oriented instruction set to a particular application. The technique enables the programmer to

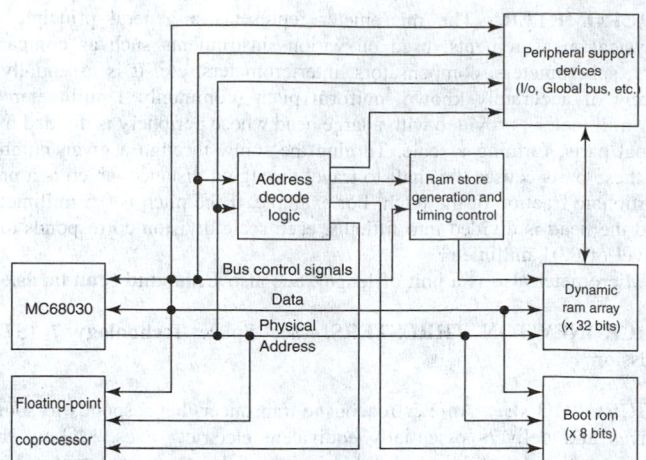

Fig. 1. Circuit organized around a microprocessor—in this case, the Motorola 68030 introduced in 1986 (32-bit). Block diagram illustrates a representative example, one that is useful in numerous cost-sensitive, performance-oriented applications that may range from real-time process control to the execution of multitasking virtual-memory operating systems. The system may consist of some address decode logic, a RAM control section, a RAM array, a group of standard input/output devices and interface modules, and optional elements, such as a floating-point coprocessor and boot ROM.

The address code logic monitors the physical address outputs of the 68030. During a bus cycle, it generates a select signal for the specific device being accessed by the 68030. The logic unit could also provide timing control for simple devices with fixed access times, as in the case of the boot ROM. The on-chip cache memory of the 68030 and paged memory-management unit do much of the work that, in another system, would be handled by the decode logic. The peripheral section shown is the same type found in prior existing systems. It may contain serial and parallel I/O devices, DMA controllers, an interface to a global or extension bus, and other functions. The RAM control section is the standard control logic required to maintain and access the RAM array as modified to support the 68030's burst-mode bus operations, this being a feature of the microprocessor's efficient on-chip bus-interface unit. The burst mode takes advantage of dynamic RAMs with a nibble mode or equivalent to shorten access time to the array and to burst-fill the 68030's two on-chip caches with fresh data or instructions. Modification to the RAM control increases the cost slightly as well as complexity, but yields a significant increase in overall performance. Delivery of the 68030 commenced in quantity in 1987. (*High-End Microprocessor Products Group, Motorola, Inc.*)

implement a function, such as square root, directly without subroutines or macrostatements. Thus significant programming and execution efficiency is realized if the square-root function is commonly required. Also, the instruction set of a character-oriented computer can be implemented on a word-oriented machine where adequate microprogramming is provided.

See also **Program (Computer)**.

THOMAS J. HARRISON, IBM Corporation, Boca Raton, FL

MICROSCOPE (Traditional-Optical).

An instrument capable of enlarging the angle under which a small object appears to the unaided eye at the distance of distinct vision (250 millimeters). The result is a magnified image of the object on the retina of the human eye. This image reveals details that are invisible without the magnifying power of the microscope. See also **Electron Microscope**; and **Scanning Tunneling Microscope**.

Optical Microscope

A microscope usually refers to a compound microscope. The compound microscope forms the image by a two-stage process and consists of an objective and an ocular. A simple microscope uses only objectives to form an image, e.g., a loop for visual use, or a camera set-up for photomicrography.

The total magnification of a compound microscope is

$$M = \frac{250}{f_m} = M_{\text{objective}} \times M_{\text{ocular}} = \frac{t}{f_{\text{objective}}} \times \frac{250}{f_{\text{ocular}}}$$

where f_m = focal length of the microscope, millimeters
$M_{\text{objective}}$ = magnification of the objective
M_{ocular} = magnification of the ocular (eyepiece)
t = optical tube length, millimeters
$f_{\text{objective}}$ = focal length of the objective, millimeters
f_{ocular} = focal length of the ocular, millimeters

Any magnification changer, e.g., a system with continuously variable focal length (zoom system) between the objective and eyepiece increases the total magnification by the tube factor of this system. With objectives ranging in power from about $1\times$ to $100\times$ (and even higher) and eyepieces ranging in power from about $4\times$ to $20\times$ and higher, it is possible to cover a magnification range from approximately $5\times$ to $2,000\times$. The total magnification M should be in a range given by $500 \times N.A. \leq M \leq 1,000 \times N.A.$ to obtain best results. $N.A.$ is the numerical aperture of the objective. Since the largest $N.A.$ of an objective is approximately 1.4, the upper limit of M is in the neighborhood of $1,400\times$. As M exceeds this figure, one enters the range of "empty magnification" without gaining additional resolution (or resolving power). The latter factor is expressed for incoherent light by

$$d = \frac{\lambda}{2 \times N.A.}$$

where d = smallest resolvable distance or detail,
$N.A. = n \sin \alpha$ (n = efractive index of medium between object and objective; α = half-angle of cone of light entering the objective).

This formula indicates several ways to improve the resolving power, either by reducing the wavelength of light, or by increasing the $N.A.$ (immersion objective), or both at the same time.

Besides sufficient resolution as one prerequisite for a satisfactory image, the object must be imaged with an optimum of contrast. To enhance the contrast, the specimen is either stained (amplitude object, contrast generated by absorption of light), or, if this is not possible (e.g., a living specimen, a phase object with negligible absorption, contrast near zero), the contrast is optically created. Common methods, for example, are phase contrast and interference contrast which utilize the interference phenomenon of light. Another method is darkfield illumination (illumination of the specimen by inclined pencils of light, thus permitting only light scattered by the object to enter the objective). The object appears bright on a dark background. Oblique illumination was often used before these methods became available.

Types of Microscopes. These instruments can be divided into two large groups: (1) microscopes for transmitted light, and (2) microscopes for incident (reflected) light. Within these groups, microscopes are commonly distinguished either according to their application (e.g., polarizing microscope, fluorescence microscope, and so on), or to their illumination system (e.g., brightfield microscope or darkfield microscope), or even by a combination of the foregoing factors (e.g., darkfield-fluorescence microscope).

Function of Microscope. The cross section of a microscope for transmitted light is shown in Fig. 1. The light path at the left explains the formation of the image of the specimen. The path at the right is that of the rays for formation of the images of the light sources and the pupils of condenser and objective. In reality, both paths have to be superimposed to obtain the true conditions of the optical system and its participation in image formation.

The light source is normally built into the base of the microscope. Koehler illumination is commonly used. This system utilizes the uniformly illuminated area of the lamp condenser and the field diaphragm as a light source. This diaphragm is imaged by the condenser into the specimen plane and by the objective into the intermediate image plane. The filament of the bulb is imaged by the lamp condenser into the plane of the condenser diaphragm. Thus, one illuminates only the field of the specimen imaged by the objective and protects the specimen from excessive heat and reduces stray light. Reduced stray light results in improved contrast. An advantage of the Koehler illumination system is its capability of fully illuminating the exit pupil of the objective. Only then is its maximum resolving power available. Optimum contrast occurs with the aperture diaphragm closed to approximately $\frac{2}{3}$ to $\frac{3}{4}$ of its entire diameter. The microscopist, therefore, has to make a compromise between required resolution and contrast.

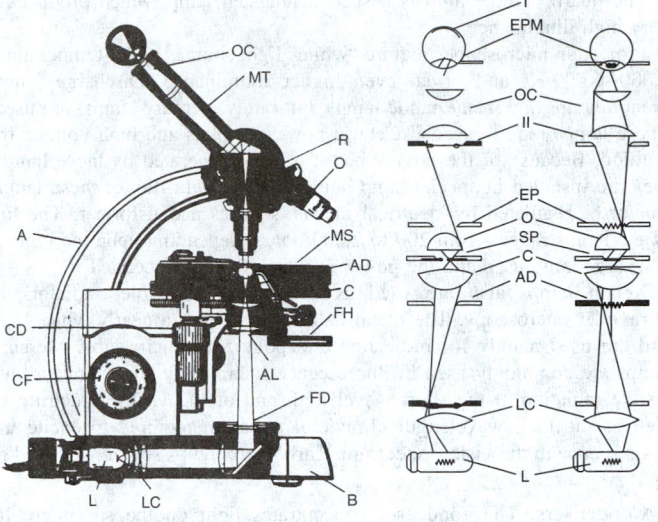

Legend:

A	Arm of microscope	FD	Field diaphragm (iris)
CD	Coaxial stage drive	AL	Auxiliary lens
CF	Coaxial coarse and fine motion knobs	FH	Swing-out filter holder, diaphragm (iris)
L	Tungsten light source	B	Base of microscope
LC	Lamp condenser	I	Image of object
OC	Ocular	EPM	Exit pupil of microscope
MT	Inclined monocular tube	II	Intermediate image plane
R	Revolving nose piece	EP	Exit pupil of objective
O	Objectives	SP	Specimen
MS	Microscope stage with built-in mechanical stage	SP	Specimen
AD	Aperture	C	Abbe condenser

Fig. 1. Microscope for transmitted light showing two main light paths. (*Carl Zeiss.*)

The critical illumination system images the filament of the light source into the specimen, thus creating there a higher intensity. However, it is very difficult to homogeneously illuminate the specimen (photomicrography). Critical illumination, therefore, is seldom used.

The light traverses the condenser, which illuminates a small field of the specimen. The objective forms an inverted and reversed magnified image of the object in the intermediate image plane of the microscope. The ocular further magnifies this image by the magnification factor of the eyepiece.

The human eye can be replaced by a photographic camera with or without an automatic exposure device. Several microscopes are available, with built-in cameras with automatic exposure control.

Microscopes usually feature a wide variety of interchangeable condensers, objectives, tubes, eyepieces, and other components to permit a rapid change from one application to the next.

Light Sources. The common light source of conventional microscopes for transmitted or reflected light is the low-voltage tungsten lamp. Its electric rating varies from 6 volts, 15 watts to 12 volts, 100 watts. For visual work in brightfield, 15 watt or 30 watt bulbs provide a sufficient light flux and brightness. For photomicrography, polarizing microscopy, darkfield observation, and interference microscopes, incandescent lamps of a higher wattage are preferred.

Tungsten lamps permit a variation of their light flux and the brightness of the image by a regulating transformer. In black-and-white photomicrography, this method is not applicable because the color temperature of the lamp is a function of the voltage and amperage through the filament. The color temperature must remain constant after it has been matched with the color temperature of the film by inserting various color-balancing filters in the beam. If this is not done, the color rendition of the film will deteriorate. The color temperature of tungsten lamps ranges from 2,850 K to 2,400 K. Special types reach 3,600 K.

Tungsten lamps are sensitive to overload. If the voltage increases by 10%, the illuminance will rise by 45%, but the life of the bulb will decrease by about 70%. In the other direction, if the voltage is 10% less than the rated voltage, the illuminance is reduced by 35%, but the life of the bulb is increased by 450%. The average life of a tungsten microscope bulb at rated power is approximately 100 hours.

The quartz iodide lamp is a special tungsten lamp, which produces a very high illuminance.

Projection microscopes require "white" light sources (color temperature, 5,000–6,000 K) and of an even higher illuminance. Discharge lamps (xenon lamps and metal halide lamps, but rarely mercury lamps) are used. These lamps require special electric power supplies and high voltage for ignition. Because of the large amount of heat generated by these lamps, they are installed in special lamp housings. The light flux of these lamps cannot be regulated by electrical means, such as a transformer. The life of the lamps ranges from 200 to 2,000 hours, depending upon the type of discharge lamp used and the power at which they are operated.

Xenon lamps also are used as continuous ultraviolet radiators in ultraviolet microscopy. The metal halide lamps emit a nearly white light and are used mainly for projection and polarizing microscopy. Mercury lamps are commonly used in fluorescence microscopy. They emit a blue visible radiation at the short wavelength end of the visible spectrum as well as the long wavelength ultraviolet radiation required to excite the fluorescence in the visible spectrum. Carbon arc lamps seldom are used as light sources.

Condensers. The condenser concentrates light on the specimen. Its numerical aperture must be variable so that it may be adjusted to that of the objective in order to obtain optimum resolution and contrast. The condenser images the field diaphragm into the specimen plane. Each method of illumination requires its own condenser. Condensers in modern microscopes are, in most cases, multipurpose units, e.g., one condenser for brightfield, phase contrast, and darkfield; or for brightfield, phase contrast, and interference contrast. Classes of condensers include: (1) condensers for transmitted light, (2) condensers for reflected light, and (3) special condensers for either transmitted or reflected light (for phase contrast, interference contrast, phase contrast-fluorescence, ultraviolet microscopy, interference microscopy, pancratic condensers).

The simplest condenser for transmitted light brightfield is the Abbe condenser, consisting of one lens for numerical apertures up to 0.6 and of three lenses for numerical apertures up to 1.3. Since neither configuration is chromatically nor spherically corrected, they should be supplemented by highly corrected achromatic-aplanatic condensers (*N.A.* up to 1.4) for high quality microscopy, e.g., color photography with apochromatic objectives.

Incident light illuminates opaque specimens. The reflected fraction of the incident light forms the image according to the same principle as previously described. The objective (either for brightfield or for brightfield and darkfield) is its own condenser, as shown in Fig. 2. The reflector either is a thin plane parallel glass plate mounted at 45°, or a prism, the latter used mainly for polarizing microscopy.

Objectives. The microscope objective is the most important component of the optical system of the microscope inasmuch as it essentially determines the image quality. Typical objectives for one-stage imagery in photomicrography are *Luminars, Micro Tessars, Macro objectives, Summars, Milars, Photars,* and so on.

The majority of microscopes use the two-stage magnification previously described. The objectives are divided into different categories according to their optical design and type of correction.

Achromats. These are the simplest microscope objectives, corrected for equal back focal length of one blue and one red wavelength falling in the green region of the visual spectrum. Spherical aberration is fully corrected for one wavelength in the green region. They are ample for all routine work, although they show some field curvature. Planachromats, Flatfield Achromats, Plano Objectives form a "flat" or plane intermediate image, important for photomicrography.

Fluorite Objectives. When compared with achromats, the fluorite objectives incorporate one or two fluorite lenses to obtain an improved chromatic correction (reduced chromatic difference of magnification). The result is a higher *N.A.* than that of an achromat of the same magnification. Fluorite objectives form an image with a contrast superior to the achromats and their image quality is very close to that of apochromats. They are, therefore, often called *semiapochromats.* Synthetic fluorspar is used mainly for fluorite objectives. Fluorite objectives still show a slight field of curvature. They are rarely available as flatfield objectives.

Apochromats. These are the best and most expensive objectives. The back focal length is identical for three specific wavelengths in the blue, green, and red regions of the spectrum. Further, the apochromats are aplanatic for at least two wavelengths. A consequence of the excellent correction is the high *N.A.* obtainable with these objectives as compared with fluorites of the same magnification. Apochromats also are available as flatfield objectives.

Objectives for Reflected Light. This group of objectives is usually provided as special flatfield achromats. They are corrected for use with specimens without cover glass. Any objective with *N.A.* > 0.3 forms a poor image due to increased spherical aberration when it is used for an uncovered specimen, but has been corrected for use with standard cover glasses of 0.17 millimeter (Europe) to 0.18 millimeter (United States) thickness.

Special Objectives. These systems are for use in the ultraviolet and consist of either quartz lens systems corrected for a wide wavelength (Carl Zeiss *Ultrafluars,* 250–600 nanometers), or a single wavelength (Vickers *Monochromats*); or they are mirror objectives (Leitz, Bausch and Lomb, and so on). To this group also belong objectives for interference

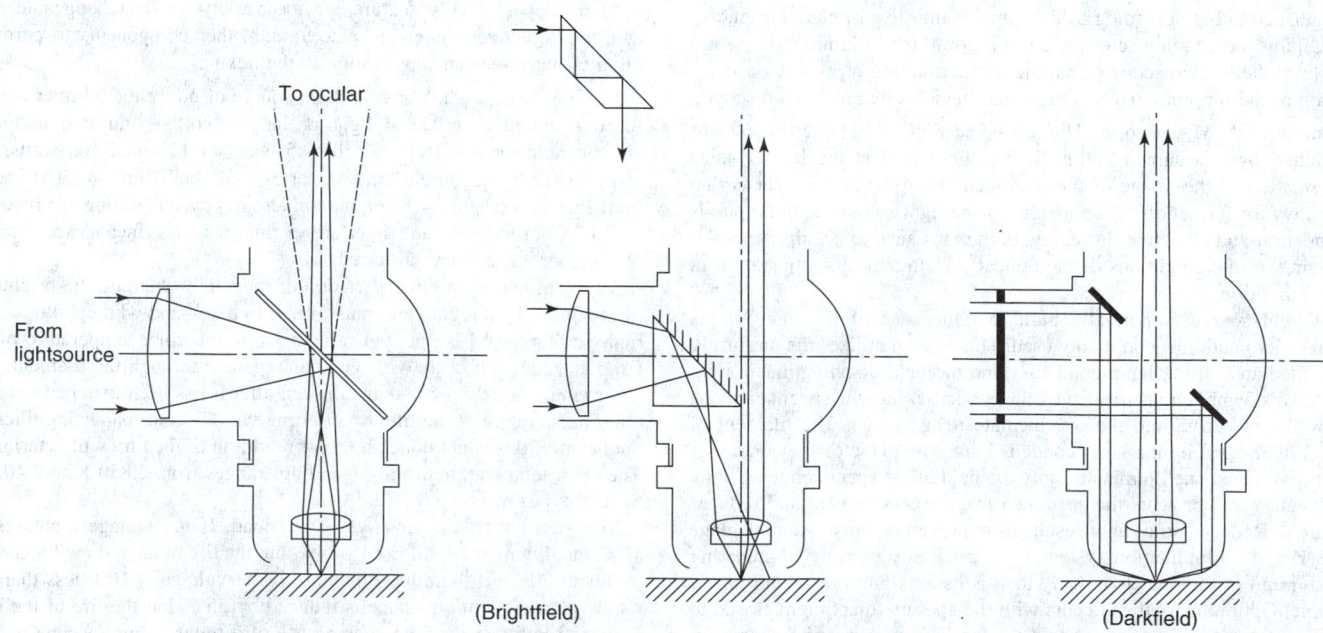

Fig. 2. Illuminators for incident light.

microscopes, and special immersion objectives (water, methylene iodide, glycerine, and so on).

The described objectives also are available as phase contrast objectives.

Any of these objectives is corrected not only for use with or without cover glass, but also for a certain mechanical tube length (distance between shoulder of objective and shoulder of eyepiece—upper end of straight microscope with no optics in between objective and eyepiece). The commonly applied distance is 160 millimeters or infinity (American Optical Company) for transmitted light. For reflected light, the mechanical tube length varies between 180 millimeters and 250 millimeters and infinity, depending upon the manufacturer of the instrument. The objectives usually are designated by their type (except achromats), magnification, numerical aperture, tube length, and cover glass correction.

Oculars. The ocular at the upper end of the microscope tube enlarges the immediate image formed by the objective. Together with an attachment camera, it can also form the final image on a film plane. All oculars are derived either from the Huygens or Ramsden type.

Due to the low numerical aperture of oculars, only astigmatism, field curvature, distortion and the chromatic difference of magnification need be corrected. The commonly used type of oculars is the compensating eyepiece, whose chromatic difference of magnification is equal but opposite to that of the objective. For use with flatfield objectives, differently corrected oculars are used to fully utilize the performance of the flatfield objectives. Many eyepieces have a high eyepoint to permit the microscopist to wear his corrective glasses.

Special eyepieces are available for measurements (filar eyepieces, micrometer eyepieces, interference eyepieces, image-splitting eyepieces, and so on) and for teaching purposes (pointer eyepieces and demonstration eyepieces). The magnifications range from $5\times$ to $25\times$, of which $8\times$, $10\times$, $12.5\times$, and $15\times$ eyepieces are mainly used.

Photographic Equipment. Photography, both black-and-white and color, through a microscope plays an ever increasing role in modern microscopy. Attachment cameras are either manual, or semiautomatic, or fully automatic devices for all common film formats. Standard sizes are 35 millimeters, $2\frac{1}{4}$ inches $\times 4\frac{1}{4}$ inches, 4 inches $\times$ 5 inches. In addition in Europe sizes include 56×72 millimeters, 6.5×9 centimeters, and 9×12 centimeters. In a modern 35-millimeter attachment camera, a photomultiplier tube measures the light flux, an exposure device keeps the shutter open for the exact exposure time, depending upon the brightness of the image and the speed of the film emulsion. After the shutter is closed, an electric motor advances the film to the next frame. Filter factors, coverage of the field, and other factors can be compensated. In some cases, the automatic camera is built into the microscope stand. Cutaway view of a photomicroscope is shown in Fig. 3.

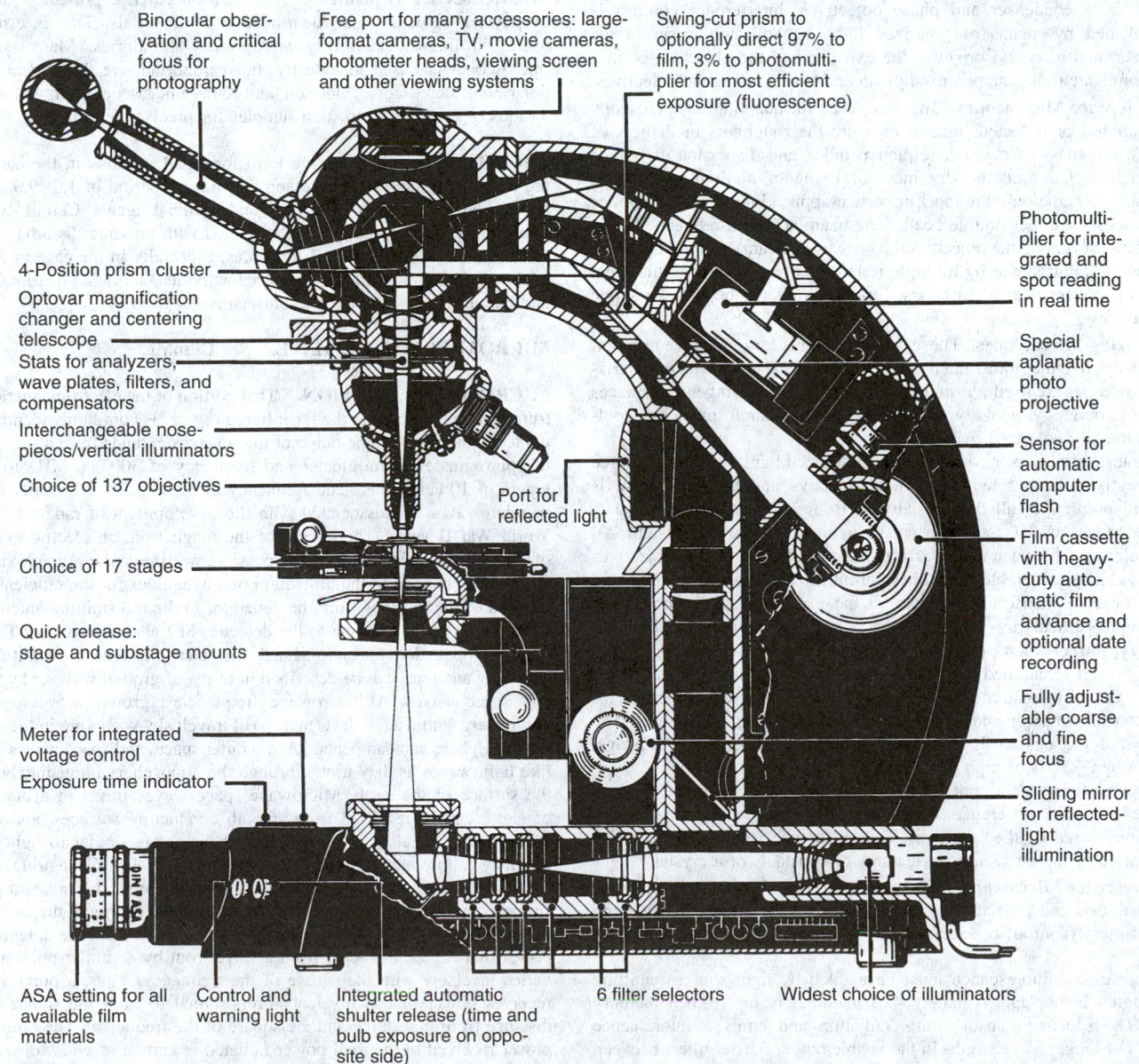

Binocular observation and critical focus for photography

Free port for many accessories: large-format cameras, TV, movie cameras, photometer heads, viewing screen and other viewing systems

Swing-cut prism to optionally direct 97% to film, 3% to photomultiplier for most efficient exposure (fluorescence)

4-Position prism cluster

Optovar magnification changer and centering telescope

Stats for analyzers, wave plates, filters, and compensators

Interchangeable nose-piecos/vertical illuminators

Choice of 137 objectives

Choice of 17 stages

Quick release: stage and substage mounts

Meter for integrated voltage control

Exposure time indicator

Port for reflected light

Photomultiplier for integrated and spot reading in real time

Special aplanatic photo projective

Sensor for automatic computer flash

Film cassette with heavy-duty automatic film advance and optional date recording

Fully adjustable coarse and fine focus

Sliding mirror for reflected-light illumination

ASA setting for all available film materials

Control and warning light

Integrated automatic shutter release (time and bulb exposure on opposite side)

6 filter selectors

Widest choice of illuminators

Fig. 3.   Cutaway view of Zeiss Photomicroscope III.

Special Microscopes

The basics of the transmitted light microscope just described apply, with few exceptions, to most special microscopes.

Metal Microscopes for Incident Light. The classic instrument of this type utilizes the principles of Le Chatelier. It is an inverted microscope with the objective underneath the specimen, which is placed with its polished surface facing down on the microscope stage. The size of these instruments varies from simple routine laboratory microscopes to the big bench-type (Reichert, E. Leitz, American Optical Company, Vickers, Unitron, and so on) to the console type (Bausch and Lomb) research instruments. Also, *upright* microscopes with incident illumination are used as metallographs (Zeiss), a term often used for research camera microscopes.

Plankton and tissue culture microscopes form the second group of inverted microscopes.

Phase contrast microscopes (after Zernicke) and *interference contrast microscopes* (after Nomarski, Françon, and so on). These utilize the interference of light to yield an image of high contrast for nonabsorbing phase objects. These objects alter only the phase of light according to the differences in optical thickness of the specimen. Since the human eye and the photographic emulsion are insensitive to phase variations, the phase contrast and the interference contrast system convert these phase variations into visible contrast.

The Phase Contrast Microscope. This microscope requires both a special phase condenser and phase objectives. Interference contrast is accomplished by means of polarized light, a Wollaston prism in the condenser, and a second prism in the exit pupil of the objective, or in a plane conjugate to it. The prisms are matched to certain regular objectives.

Interference Microscopes. These are used for quantitative microscopy in transmitted or reflected light to measure the roughness or flatness of surfaces, thickness of coating, refractive index and dispersion of liquids or solid materials, and the dry mass of living or nonliving biological material — to name only the most important applications. The interference microscopes are either double beam (one beam acts as a reference beam; the second, traversing the object, as the measuring beam), or multiple beam instruments (interference by multiple reflection of rays). Due to the great variety of interference microscopes, further details are beyond the scope of this review.

Polarizing Microscopes. These are instruments for qualitative or quantitative work in either transmitted or reflected light. The majority of polarizing microscopes are used for quantitative studies of birefringent substances, (e.g., in mineralogy, geology, metallurgy, fiber research, medicine, zoology) in linearly polarized light.

The substage polarizer converts the unpolarized light of the light source into linearly polarized light. Between objective and eyepiece, there is a second *polarizer*, called the analyzer. Both are in calibrated mounts rotatable to 180° or 360°, and they can be swung out of the path of light. All optical elements between the polarizers must be strain free. The objectives are individually centerable. The tubes permit the use of an Amici-Bertrand lens to observe interference figures in the exit pupil of the objectives (conoscopic path of rays), because the conoscopic image (e.g., together with a gypsum or quartz plate red first order) renders information on the type of crystal (examined e.g., uniaxial or biaxial, positive or negative). Most of the investigations require the polarizers crossed (directions of oscillation of polarizer and analyzer oriented at 90° to each other). The crosshair of the oculars is congruent with these directions. The stage, rotatable by 360°, is of high precision.

Compensators (Senarmont, Berek, Ehringhaus, Brace-Koehler). These measure the path difference between the ordinary and extraordinary beams introduced by the specimen. This difference value is the basis for computation of the characteristic features or constants of a crystal.

Fluorescence Microscopes. These are routine and research instruments for transmitted and reflected light and used to study specimens showing either primary (natural) or secondary (induced by fluorochromes) fluorescence.

In both cases, fluorescence must be excited by light sources emitting wavelengths below approximately 420 nanometers (as from a mercury lamp). The specimen absorbs this radiation and emits a fluorescence radiation of longer wavelengths in the visible range. Barrier filters between objective and eyepiece prevent light of the light source from hitting the observer's eye and allow only the fluorescence light to pass.

High aperture brightfield, or darkfield condensers illuminate the specimen, which is imaged through objectives having a numerical aperture as high as possible because the fluorescence is normally relatively weak. The image brightness increases with the square of the *N.A.*

Fluorescence microscopy is a very sensitive method. Fluorochromes diluted to 1 ppm can be detected without difficulty.

Ultraviolet Microscopes. These are equipped with quartz optics and work in the wavelength range from approximately 240 nanometers through the visible range, unless objectives corrected for a fixed wavelength are used. Unstained material on quartz slides showing no absorption in the visible often absorb considerably in the ultraviolet. Strong xenon light sources provide the ultraviolet light and ultraviolet-image converters present a visible image of the otherwise invisible specimen to the observer.

Additional Reading

Bradbury, S.: *An Introduction to the Optical Microscope*, Oxford University Press, New York, NY, 1999.

Marmasse, C.: *Microscopes and Their Uses*, Gordon & Breach Publishing Group, Newark, NJ, 1980.

Wilson, C.: *The Invisible World: Early Modern Philosophy and the Invention of the Microscope*, Princeton University Press, Princeton, NJ, 1997.

W.G. HAVEMANN, Carl Zeiss, Inc., New York, NY

MICROSCOPY (Chemical). Use of a microscope primarily for study of physical structure and identification of materials. This is especially useful in forensic chemistry and police laboratories. Many types of microscopes are used in industry; most important are the optical, ultra-, polarizing, stereoscopic, electron, and X-ray microscopes. Organic dyes of various types are used to stain samples for precise identification.

MICROSEISM. A collective term for small motions in the earth that are unrelated to an earthquake and that have a period of 1.0–9.0 s. They are caused by a variety of natural and artificial agents. Certain types of microseisms seem to be closely correlated with pressure disturbances and can be used to locate such disturbances, especially in the case of tropical cyclones. In addition, traffic, industrial activities, and wind flexure of trees and tall structures can create microseisms.

MICROTHERMAL CLIMATE. See **Climate**.

MICROWAVE RADIATION. That portion of the electromagnetic spectrum that lies adjacent to the far-infrared region, is commonly identified by the term microwave. The range of microwaves extends from a wavelength of approximately 1 millimeter and frequency of 300,000 MHz to wavelength of 10 centimeters and frequency of 3,000 MHz. The early interest in microwaves was associated with the development of radar during the World War II years. The advent of the magnetron, an electric generator of high-power microwaves, made possible wartime radar at approximately 3,000 MHz and led to the utilization of waveguides for the efficient transmission of microwaves from the generator to the transmitting antenna and from the receiving antenna to the detector. See also **Microwave Tubes**.

The propagation of radio signals in space between transmitting and receiving antennas can be described in terms of ground waves, sky waves, and space waves. At microwave frequencies, ground waves attenuate completely within a few feet (meters) of travel; sky waves are influenced by the ionosphere and can penetrate into outer space, and space waves behave like light waves as they travel through the atmosphere immediately above the surface of the earth. Microwave space waves travel in a direct line of sight, can be reflected from smooth conducting surfaces, and can be focussed by reflectors or lenses. Their behavior is similar to light waves and they follow many of the rules of optics. See also **Waveguide**.

If a space wave is radiated from a point antenna, the radiated energy spreads out like an ever expanding sphere, and the energy of the wave front decreases inversely with the square of the distance from the antenna. The power that can be extracted from a wave front by a similar point antenna varies inversely with the square of the frequency. Thus, a point antenna receives power that is inversely proportional to both the square of the distance from the source and the square of the frequency. The ratio of the power received to the total power radiated is known as *path attenuation*.

When the receiving antenna is a parabola-shaped dish, the power extracted from the wave front is greatly increased. The ratio of the power received by such an antenna to the power received by a theoretical point antenna is defined as antenna gain. The gain of a parabolic antenna increases with the antenna area and the operating frequency. Thus, for a given microwave path with fixed-size antennas, the path attenuation increases with frequency, the antenna gain increases with frequency, and the over-all result is that one tends to offset the other.

Because microwave transmission follows essentially a straight line, reflectors are often used to redirect a beam over or around an obstruction. The simplest and most common reflector system consists of a parabolic antenna mounted at ground level that focuses a beam on a reflector mounted at the top of a tower. This reflector inclined at 45° redirects the beam horizontally to a distant site where a similar "periscope" reflector system may be used to reflect the beam down to another ground level. If two sites are separated by a hill, it may be necessary to use a large, flat-surface reflector referred to as a "billboard" reflector. In a typical system, a "billboard" reflector might be located at a turn in a valley, effectively bending the beam to follow the valley. Many arrangements are possible which, in effect, resemble huge mirror systems.

Microwaves are ideally suited for communication systems where a broad frequency bandwidth of the order of several megacycles is required for the rapid transmission of signals that contain a large amount of information, such as television. See also **Radar**; and **Telephony (Telecommunications)**.

MICROWAVE REGION.   That portion of the electromagnetic spectrum lying between the far infrared and the conventional radiofrequency spectrum. The microwave region is commonly regarded as extending from 300,000 MHz to 1,000 MHz (1 millimeter to 30 centimeters in wavelength). See also **Electromagnetic Phenomena**.

MICROWAVE SPECTROSCOPY.   A type of adsorption spectroscopy used in instrumental chemical analysis that involves use of that portion of the electromagnetic spectrum having wavelengths in the range between the far infrared and the radiofrequencies, i.e., between 1 mm and 30 cm. Substances to be analyzed are usually in the gaseous state. Klystron tubes are used as microwave source.

MICROWAVE TUBES.   To avoid the difficulties resulting from the short transit time of electrons as well as the effects of interelectrode capacitances, microwave tubes have been designed to make use of velocity-modulated beams of electrons. Traveling-wave tubes, magnetrons, and klystrons are special-purpose tubes designed to operate in the microwave region.

A *traveling-wave tube* is a broad-band microwave tube which depends for its characteristics upon the interaction between the field of a wave propagated along a waveguide and a beam of electrons traveling with the wave. In this tube, the electrons in the beam travel with velocities slightly greater than that of the wave, and on the average are slowed down by the field of the wave. The loss in kinetic energy of the electrons appears as increased energy conveyed by the field of the wave. The traveling-wave tube may, therefore, be used as an amplifier or as an oscillator. See Fig. 1.

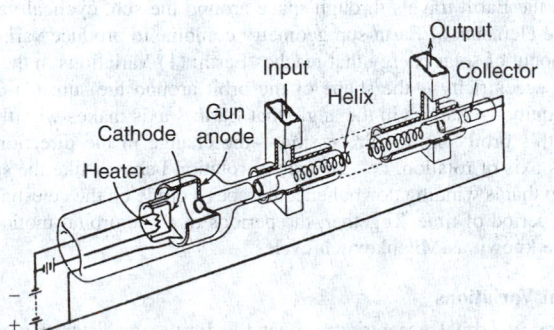

Fig. 1.   Schematic diagram of early traveling-wave amplifier demonstrates basic principles of device.

A *magnetron* is a vacuum tube, which functions under the joint action of an externally applied magnetic field and the electric field between its anode and cathode. In one form, it consists of a cylindrical cathode and a coaxial

anode structure. The anode may be a single cylinder or be split lengthwise for all or part of its length. These tubes were originally designed for use in ultra high-frequency oscillator circuits where conventional vacuum tubes could not operate effectively. The electric field is created by applying a high, direct-current potential between the filament and anode structure, while the magnetic field is applied longitudinally by external permanent or electromagnets. When the tube is properly connected to a resonant line it can be made to operate as an oscillator for certain values of the applied fields. A magnetron of suitable dimensions can be made to generate frequencies measured in thousands of MHz (wavelengths of a few centimeters).

A *klystron* is an electron tube of the velocity-modulated type used in ultra-high frequency circuits. At these very high frequencies (hundreds or thousands of MHz), conventional vacuum tubes proved useless because of lead and electrode inductance and capacitance, and transit-time effects. The klystron was a solution to that problem. The tube may consist of a cathode, grid, and perforated anode somewhat like the electron gun of a cathode ray tube, followed by two cavity resonators separated by a calculated distance, and finally a collector. Except for the collector, all electrodes have grid-like surfaces, so that electrons can pass on through them. The beam of random-velocity electrons passing through the grid is accelerated by the positive potential applied to the first resonant cavity structure, causing this structure to serve as an anode. These electrons pass through the grids into the cavity (buncher). The standing waves in the cavity act on the electrons and cause them to change speed so that they arrive at the second cavity (catcher) in bunches, having passed out of the first into a field-free space, and then into the second, through the grids in the sides of the cavities. Here the energy of the electrons is absorbed by the field and contributes to the useful output and normally supplies also the driving energy for the buncher. The electrons then pass on to the collector and return to the cathode. By proper adjustment of the voltages and spacings of the cavities, the circuit may be made to oscillate or amplify as desired. The circuit of a modulated klystron oscillator is shown in Fig. 2.

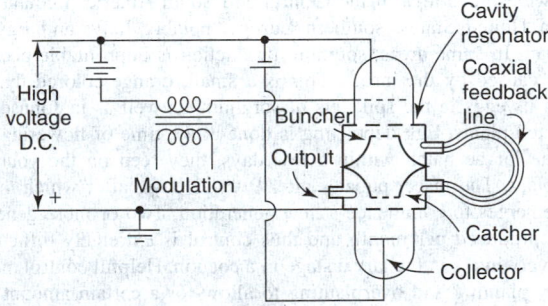

Fig. 2.   Modulated klystron oscillator.

MIDAS.   A two-object trajectory measuring system whereby two complete Cotar antenna systems and two sets of receivers at each station, with the multiplexing done after phase comparison, are utilized in tracking more than one object at a time.

MIDGE (*Insecta, Diptera*).   Of the family *Chironomidae*, several species of very small insects (from $\frac{1}{10}$ to $\frac{1}{5}$ inch; 2.5 to 5 millimeters long) that resemble in build and habit the much larger mosquitoes and crane flies. Generally, they are scavengers, but a few species are of serious economic interest in food production.

Clover seed midge (Dasyneura leguminicola, Lintner). This insect, when present in relatively large numbers, can devastate a clover seed crop, but is not damaging if the clover is produced for hay. The maggots infest the clover seed. Red clover is the principal target of attack, while alsike, mammoth, crimson, white, and sweet clover are relatively immune to attack. A related species, *D. gentneri* (Pritchard), however, does attack Ladino and alsike clover. Distribution of the insect is throughout the United States and in the southern part of Canada.

Pear midge (Contarinia pyrivora, Riley). This insect causes blotched, deformed, and prematurely dropping fruit. The maggots are white to-orange in color and only about $\frac{1}{7}$ inch (3 to 4 millimeters) in length. On occasion over 100 maggots per fruit may be found, with almost complete consumption of the interior of the fruit.

Wheat midge (Sitodiplosis mosellana, Gehin). The maggots are pink-to-reddish in color and only about $\frac{1}{12}$ inch (2 millimeters) long. They are found among the bracts and feed on the kernels of the wheat. Cultural methods usually are adequate protection. These include crop rotation, fall plowing to bury and destroy the larvae, and destruction of all debris from infested fields. The control chemical etrimfos is also effective.

The punkie (Culicoides spp.). Also called sand fly or no-see-um, the punkie is a small biting midge, found in abundance at certain times along streams in the eastern mountains of the United States and at some parts of the seashore. Although not injurious to food crops, the insects can be extremely annoying to persons who work in the fields. See Fig. 1.

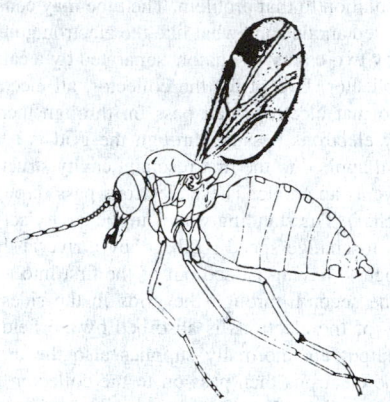

Fig. 1.   A punkie or "no-see-um."

Sorghum midge (Centarinia sorghicola). This insect affects sorghum over a widespread area of the Central and South American countries, the southern United States, southern Europe, notably Italy, and Egypt and the Sudan. In some areas, sorghum production is unprofitable because of damage caused by the insect. This is a small, orange-colored fly, which deposits its eggs in the spikelets of sorghum, as well as in Johnson grass and related plants. Egg depositing is done at the time of flowering. When the white larvae hatch within a few days, they feed on the young seed of the plant. The insect pupates after 9 to 12 days, after which the adult midge emerges to commence a new generation. Two or more generations may be produced per month and thus control is extremely difficult. The insect overwinters in the larval stage in a cocoon. Helpful control measures are early planting and overplanting to allow for a certain amount of this damage, as well as the usual good practice of destroying plant residues and debris. Grass species such as Johnson grass and related weeds that act as hosts for the midge also must be controlled.

MIDGLEY, THOMAS, JR. (1889–1944).   An American chemist and inventor. One of the most creative and brilliant chemists of his era. Midgley's early work was in the field of rubber chemistry and technology, especially in the development of synthetic and substitute rubbers that were being introduced in the 1930s. He worked with Kettering at General Motors and then became vice president of Ethyl Corporation, as well as of the Ohio State University Research Foundation. His innovative genius was responsible for the development of organic lead compounds for antiknock gasoline and later for the discovery of fluorcarbon refrigerants for which he did the basic research. He was recipient of many of chemistry's highest honors including the Nichols medal, the Perkin medal, and the Priestly medal.

MID-OCEAN RIDGES.   The global mid-ocean ridge system is the largest single volcanic feature on the Earth, encircling it like the seams of a baseball. Here the Earth's crust is spreading, creating new ocean floor and literally renewing the surface of our planet. Older crust is recycled back into the mantle elsewhere on the globe, typically where plates collide. The mid-ocean ridge consists of thousands of individual volcanoes or volcanic ridge segments which periodically erupt.

Beneath a typical mid-ocean ridge, mantle material partially melts as it rises in response to reduced pressure. This melted rock, or "magma", may collect in a reservoir a few kilometers (mile) below the seafloor, awaiting eruption. Much of the magma eventually freezes in place there within the crust, forming the bulk of the new oceanic crust without erupting at all. Average oceanic crust is about 10 kilometers (6 miles) thick, but only the upper 1 to 3 kilometers (0.6 to 1.9 miles) are formed by eruption processes. When magma pressure builds up enough to force its way out to the seafloor, eruption occurs. "Dikes" are magma-filled cracks and are the conduits that magmas flow through to reach the surface. A typical ridge eruption leaves behind a dike up to 2 meters (6.5 feet) in width, extending between the crustal magma chamber and the eruptive fissure at the surface. Lavas pour from the fissure across the surface of the volcanic seafloor, adding a thin coat of new lava (typically <10 m (<33 feet) thick) with each eruption. This process of magma ascending and lavas erupting is on-going and perpetual. At the Juan de Fuca Ridge, the spreading process creates an average width of ~6 meters (~20 feet) of new crust in 100 years.

Because the ridge is the site of focused volcanic activity, seawater circulates actively to cool the new crust. This heated water reacts with the volcanic rock, dissolving out metals and depositing them around seafloor hot springs. Within the volcanic upper crust, subterranean chambers and fractures filled with heated water act as incubators for microbes that live in some of the harshest conditions ever discovered to support life. These microbes are the foundation for a rich ecosystem that thrives only at these hydrothermal vents. See also **Ocean**.

MIESCHER DEGRADATION.   Adaptation of the Barbier-Wieland carboxylic acid degradation to permit simultaneous elimination of three carbon atoms, as in degradation of the bile acid side chain to the methyl ketone stage. Conversion of the methyl ester of the bile acid to the tertiary alcohol, followed by dehydration, bromination, dehydrohalogenation, and oxidation of the diene yields the required degraded ketone.

MIGNONAC REACTION.   Formation of amines by catalytic hydrogenation of aldehydes and ketones in liquid ammonia and absolute ethanol in the presence of a nickel catalyst.

MILANKOVITCH, MILUTIN (1879–1958).   The Serbian astrophysicist Milutin Milankovitch is best known for developing one of the most significant theories relating Earth motions and long-term climate change. Born in the rural village of Dalj, Serbia, Milankovitch attended the Vienna Institute of Technology and graduated in 1904, with a doctorate in technical sciences. After a brief stint as the chief engineer for a construction company, he accepted a faculty position in applied mathematics at the University of Belgrade in 1909, a position he held for the remainder of his life.

Milankovitch dedicated his career to developing mathematical theory of climate based on the seasonal and latitudinal variations of solar radiation received by the Earth. Now known as the Milankovitch Theory, it states that as the Earth travels through space around the sun, cyclical variations in three elements of Earth-sun geometry combine to produce variations in the amount of solar energy that reaches Earth: (1) Variations in the Earth's orbital eccentricity—the shape of the orbit around the sun: (2) Changes in obliquity—changes in the angle that Earth's axis makes with the plane of Earth's orbit: and (3) Precession—the change in the direction of the Earth's axis of rotation, i.e., the axis of rotation behaves like the spin axis of a top that is winding down; hence it traces a circle on the celestial sphere over a period of time. Together, the periods of these orbital motions have become known as Milankovitch cycles.

Orbital Variations

Changes in orbital eccentricity affect the Earth-sun distance. Currently, a difference of only 3% (5 million kilometers) exists between closest approach (perihelion), which occurs on or about January 3, and furthest departure (aphelion), which occurs on or about July 4. This difference in distance amounts to about a 6% increase in incoming solar radiation (insolation) from July to January. The shape of the Earth's orbit changes from being elliptical (high eccentricity) to being nearly circular (low eccentricity) in a cycle that takes between 90,000 and 100,000 years. When the orbit is highly elliptical, the amount of insolation received at perihelion would be on the order of 20 to 30% greater than at aphelion, resulting in a substantially different climate from what is experienced today. See Figs 1 and 2.

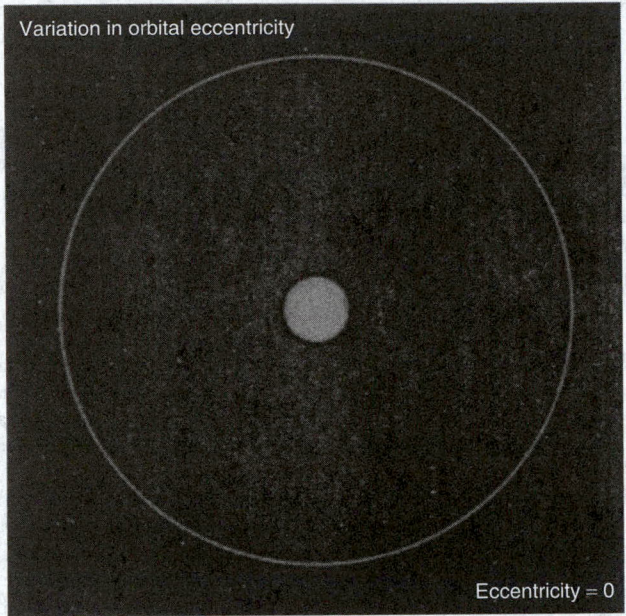

Fig. 1. The eccentricity of the Earth's orbit changes slowly over time, from nearly zero to 0.07. As the orbit gets more eccentric (oval) the difference between the distance from the Sun to the Earth at perihelion (closest approach) and aphelion (furthest away) becomes greater and greater. Note that the Sun is not at the center of the Earth's orbital ellipse, rather it is at one of focal points. (Image by Robert Simmon, *NASA GSFC*.)

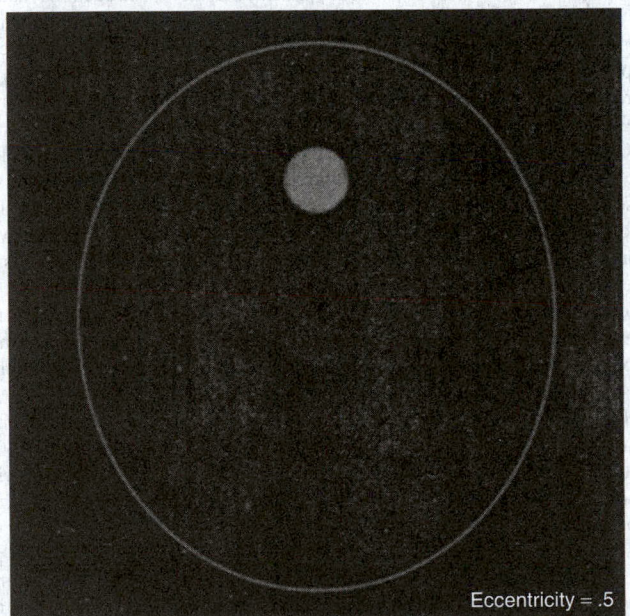

Fig. 2. The eccentricty of the orbit shown in this image is a highly exaggerated 0.5. Even the maximum eccentricity of the Earth's orbit (0.07) it would be impossible to show at the resolution of a web page. Even so, at the current eccentricity of .017, the Earth is 5 million kilometers closer to Sun at perihelion than at aphelion. (Image by Robert Simmon, *NASA GSFC*.)

Obliquity Change in Axial Tilt

As the axial tilt increases, the seasonal contrast increases so that winters are colder and summers are warmer in both hemispheres. Today, the Earth's axis is tilted 23.5° from the plane of its orbit around the sun. But this tilt changes. During a cycle that averages about 40,000 years, the tilt of the axis varies between 22.1 and 24.5°. Because this tilt changes, the seasons as we know them can become exaggerated. More tilt means more severe seasons — warmer summers and colder winters; less tilt means less severe seasons — cooler summers and milder winters. It is the cool summers that are thought to allow snow and ice to last from year-to-year in high latitudes, eventually building up into massive ice sheets. There are positive feedbacks in the climate system as well, because an Earth covered with more snow reflects more of the sun's energy into space, causing additional cooling. See Fig. 3.

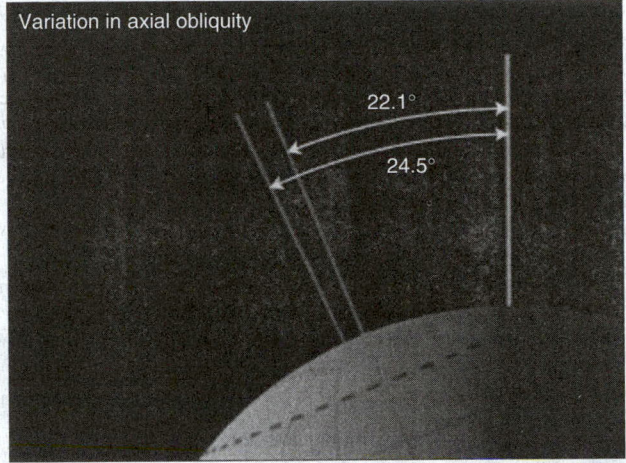

Fig. 3. The change in the tilt of the Earth's axis (obliquity) effects the magnitude of seasonal change. At higher tilts the seasons are more extreme, and at lower tilts they are milder. The current axial tilt is 23.5°. (Image by Robert Simmon, *NASA GSFC*.)

Precession

The change in orientation of the Earth's rotational axis — alters the orientation of the Earth with respect to perihelion and aphelion. If a hemisphere is pointed towards the sun at perihelion, that hemisphere will be pointing away at aphelion, and the difference in seasons will be more extreme. This seasonal effect is reversed for the opposite hemisphere. Currently, northern summer occurs near aphelion. See Fig. 4.

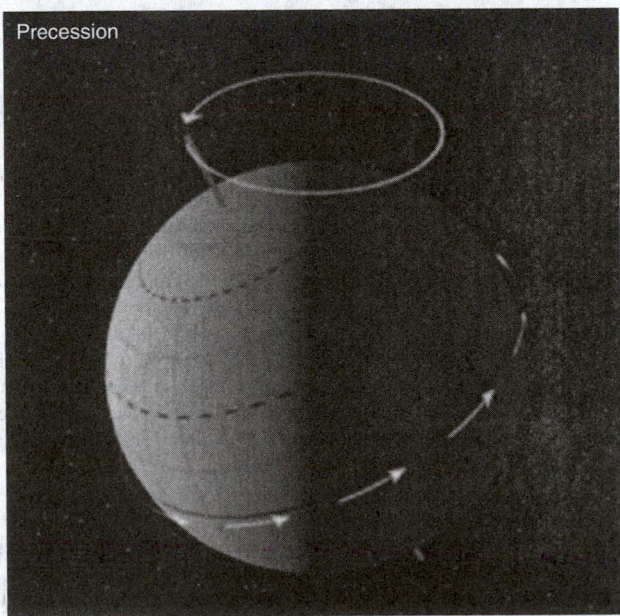

Fig. 4. (Image by Robert Simmon, *NASA GSFC*).

Using these three orbital variations, Milankovitch was able to formulate a comprehensive mathematical model that calculated latitudinal differences in insolation and the corresponding surface temperature for 600,000 years prior to the year 1800. He then attempted to correlate these changes with the growth and retreat of the Ice Ages. To do this, Milankovitch assumed that radiation changes in some latitudes and seasons are more important to ice sheet growth and decay than those in others. Then, at the suggestion of German Climatologist Vladimir Koppen, he chose summer insolation at 65° North as the most important latitude and season to model, reasoning that great ice sheets grew near this latitude and that cooler summers might reduce summer snowmelt, leading to a positive annual snow budget and ice sheet growth.

But, for about 50 years, Milankovitch's theory was largely ignored. Then, in 1976, a study published in the journal *Science* examined deep-sea sediment cores and found that Milankovitch's theory did in fact correspond to periods of climate change (Hays et al.). Specifically, the authors were able to extract the record of temperature change going back 450,000 years and found that major variations in climate were closely associated with changes in the geometry (eccentricity, obliquity, and precession) of Earth's orbit. Indeed, ice ages had occurred when the Earth was going through different stages of orbital variation.

Since this study, the National Research Council of the U.S. National Academy of Sciences has embraced the Milankovitch Cycle model.

...orbital variations remain the most thoroughly examined mechanism of climatic change on time scales of tens of thousands of years and are by far the clearest case of a direct effect of changing insolation on the lower atmosphere of Earth (National Research Council, 1982).

See also **Climate**.

Additional Reading

Hays, J.D., J. Imbrie, and N.J. Shackleton: "Variations in the Earth's Orbit: Pacemaker of the Ice Ages," *Science*, **194**(4270), 1121–1132 (1976).

Hays, J.D.: *Encyclopedia of Weather and Climate*, S.H. Schneider, Editor, Oxford University Press, New York, NY, 1996.

Lutgens, F.K. and E.J. Tarbuck: *The Atmosphere*, Prentice-Hall, Inc., Upper Saddle River N.J., 1998.

Staff: National Research Council, *Solar Variability, Weather and Climate*, National Academy Press, Washington, DC, 1982.

Web References

About.com: Milankovitch Cycles—Changes in Earth-sun Interaction: http://geography.about.com/science/geography/library/weekly/aa121498.htm?once=true&

Alaska Science Forum: The Earth's Changing Orbit: http://www.gi.alaska.edu/ScienceForum/ASF8/825.html

MILDEW. See **Fungus**.

MILK AND MILK PRODUCTS. Milk has been a source for food for humans since the beginning of recorded history. Although the use of fresh milk has increased with economic development, the majority of consumption occurs after milk has been heated, processed, or made into butter. The milk industry became a commercial enterprise when methods for preservation of fluid milk were introduced. The successful evolution of the dairy industry from small to large units of production, ie, the farm to the dairy plant, depended on sanitation of animals, products, and equipment; cooling facilities; health standards for animals and workers; transportation systems; construction materials for process machinery and product containers; pasteurization and sterilization methods; containers for distribution; and refrigeration for products in stores and homes.

Intricate chemical and microbiological problems arise in the production, processing, and distribution of milk products. Milk is a complex mixture of fat (4%), protein (3.5%), carbohydrate (4.8%) and mineral components (0.7%) and is an excellent bacterial growth medium; hence the need for care and cleanliness in handling.

Milk Fat

Milk fat or butterfat is a mixture of triglycerides of various fatty acids. Milk fat also contains a small percentage of cholesterol (0.37%), a substance characteristic of fats and oils of animal origin in contrast to those of plant origin which contain plant sterols. The phospholipids, lecithin, caphalin, and sphingomyclin are present in milk within the range of 0.03–0.04%. These are fatlike substances containing phosphorus and nitrogen. They have emulsifying properties and are associated with the fat-globule surfaces: hence their tendency to concentrate in butter and buttermilk.

Milk fat is distinguished from all other fats in that it is the only one containing butyric acid (C_4) as a component of glycerides. This acid occurs in the fat in a concentration of approximately 10 mole % and the total of C_6, C_8, and C_{10} fatty acids accounts for another 10 mole % of the component fatty acids. These acids (C_4–C_{10}) are often called the volatile fatty acids of milk fat. In the free form, they have a pungent characteristic flavor which is important in many types of cheese.

The deterioration of milk fat is an important cause of off-flavor development in dairy products and its control requires technical understanding of the processes involved. Three major types of deterioration associated with milk fat are recognized:

1. *Rancidity*, due to free volatile fatty acids liberated from the glycerides by enzymic (lipase) hydrolysis. Lipases are normal components of raw milk, and are inactivated by the heat of pasteurization.
2. *Tallowiness or oxidation*, due to autoxidation of unsaturated fatty acids with the production of flavorful unsaturated aldehydes. These reactions are accelerated by oxygen, high storage temperatures, and copper catalysts. Oxidation is usually the primary cause for spoilage of dried whole milk, cream, butter, and butteroil.
3. *Heat-generated flavors*, due to the formation of lactones and methyl ketones from hydroxy and keto acid precursors, which occur in trace quantities in milk fat. These flavors are considered to be desirable in fried and baked goods and are partly responsible for the unique condiment properties of butter in food preparation. However, they are undesirable in dried whole milk and evaporated milk where the objective is to make a bland product as much like fresh milk as possible.

Proteins

The proteins of milk fall into two groups: casein, precipitable by both acid and proteolytic enzymes such as rennin; and whey proteins, which are acid soluble but heat-denaturable. There is about 3% casein in milk, the removal of which leaves a whey of approximately 1.0% nitrogenous matter. Of this 0.6% is heat-coagulable protein and 0.4% is nonheat-coagulable. The latter fraction is composed of protein-like fragments of a proteose or peptone nature plus other nitrogenous substances. Among these are small percentages of urea, creatin, creatinine, uric acid, and various forms of amino nitrogen.

Casein exists in milk as a calcium caseinate-calcium phosphate complex; the ratio of these components is approximately 95.2 to 4.8. The dispersed casein particles appear to be spherical in shape and of various sizes. The size distribution of the casein micelles is not constant, but varies with aging, heating, concentration, and other processing treatments. Processing alters the water-binding of casein and this in turn affects the apparent viscosity of products that contain casein. Changes in hydration have not been measured quantitatively although the casein particles of raw milk appear to consist of one volume of water-free protein and three volumes of solvate liquid.

The whey or serum proteins have been partially resolved into three relatively homogeneous, crystallizable proteins: (1) β-lactoglobulin (50% of total serum protein), (2) an albumin resembling the albumin of bovine blood (5% of the serum protein), and (3) α-lactalbumin (12% of the serum protein).

On heat denaturation, the serum proteins show decreased solubility at pH 4.7 and in concentrated salt solutions. There is some variability in response toward heat treatment, but complete denaturation will occur during heating within the range of 60 to 80 °C for periods of time up to two hours. There is practically no denaturation during normal pasteurization. Heat increases the activity of the sulfhydryl groups and the sulfhydryl titer can be employed as a measure of denaturation. The $-SH$ groups are readily oxidized in liquid systems and consequently they appear to act as antioxidants to protect milk fat in dairy products. The fat of fluid milk, heated to produce a high $-SH$ titer, shows increased resistance to oxidation and this carries through to the dried product which exhibits superior storage stability, if it is made from high heat milk. The high sterilization temperature to which evaporated milk is subjected and the low oxygen content in the can protect this product from development of oxidized and tallowy flavors during storage. See also **Proteins**.

Milk is widely used as an ingredient for bread and other baked goods to which it adds substantial nutritional value. Milk is heated when used in bread to avoid softening of the dough and reduction of loaf volume. Why heat improves the baking properties of milk is not clear, but good baking properties have long been associated with low whey-protein-nitrogen values.

Lactose

The sugar of milk, lactose ($C_{12}H_{22}O_{11}$), occurs in the milk of all mammals. It is mildly sweet with a final solubility in water of 10.6% at 0 °C, 17.8% at 25°, 29.8% at 49°, 58.2% at 89°. Lactose, on hydrolysis by acid or the enzyme lactase, yields a mixture of approximately equal parts of glucose and galactose, together with a small but variable quantity of oligosaccharides. The products of lactose hydrolysis are much more soluble than the original disaccharide. Lactose is a reducing sugar which is converted to lactobionic acid on mild oxidation. Two forms which differ in solubility and optical rotation are known. Alpha-lactose hydrate crystallizes at ordinary temperatures with one molecule of water, but this is lost with the formation of the anhydrous form during heating to a temperature between 149 and 200.3 °F. Anhydrous β lactose, more soluble than alpha, crystallizes from supersaturated lactose solutions above 200.3 °F. Solid beta-hydrate has never been prepared. The crystalline alpha-hydrate is stable in dry air at room temperatures, but both anhydrous forms readily absorb moisture and change to alpha-hydrate at ordinary temperatures. Alpha-lactose crystallizes out in some dairy products and because of the hardness of its crystals and their slow and limited solubility, "sandy" products may result.

The crystallization of lactose in frozen concentrated milk has been associated with a denaturation of casein which ultimately appears as a gel structure in the thawed product. Gelation in frozen milk can be retarded by enzymic hydrolysis of part of the lactose before freezing or by addition of a polyphosphate salt.

Lactose, when fermented by lactic bacteria, is the source of the lactic acid formed in sour milk and whey. Lactose is helpful in establishing a slightly acid reaction in the intestine, which assists in calcium assimilation.

Mineral Components

When milk is heated to a temperature high enough to volatilize the water and oxidize the organic constituents, the residue of inorganic oxides that remains is called the milk ash; its major components are: K_2O, CaO, Na_2O, MgO, Fe_2O_3, P_2O_5, Cl, and SO_3. The calcium and phosphorus of the ash are of special interest because of their nutritional importance and because calcium phosphate is part of the casein micelle, influencing its physiochemical behavior toward coagulation with rennin, acid, and heat. Minor inorganic constituents are present in milk in trace amounts, i.e., iron, copper, zinc, aluminum, manganese, iodine, and cobalt.

Miscellaneous Components

The hydrogen-ion concentration of milk increases slightly with age, after milking, as natural carbon dioxide escapes. Most samples of cow's milk vary within the range of pH 6.5–6.7. Titratable acidity of fresh milk which may vary from 0.13 to 0.16%, expressed as lactic acid is an arbitrary measurement influenced by the protein and salt-buffer systems present in the particular sample. Citrates, phosphates, and carbonates are the principal buffers in milk.

Milk contains some important vitamins. The vitamin D content may vary from 30 I.U. per quart in summer to 6 in winter, depending upon the feed and the sunlight which reach the cow. Both pasteurized and evaporated milk are often fortified by the addition, on a fluid basis, of 400 I.U. of vitamin D per quart. Vitamins A, D, and E (alpha-tocopherol) are fat-soluble and stable at the heat treatments used in processing milk and milk products. The remaining vitamins are water-soluble and of varying stability. Vitamins B_1 (thiamine) and C (ascorbic acid) are partially destroyed by heat, while B_6 and B_{12}, are relatively heat-stable. Vitamin B_2 (riboflavin) is heat-stable but it is quickly destroyed by light. In spite of the varying sensitivity of the water-soluble vitamins toward heat, pasteurized milk is a good source of all the milk vitamins except C.

Two types of enzymes in milk are important: those useful as an index of heat treatment and those responsible for bad flavors. Phosphatase is destroyed by the heat treatments used to pasteurize milk; hence its inactivation is an indication of adequate pasteurization. Lipase catalyzes the hydrolysis of milk fat which produces rancid flavors. It must be inactivated by pasteurization or more severe heat treatment to safeguard the product against off-flavor development. Other enzymes reported to have been found in milk include catalase, peroxidase, protease, diastase, amylase, oleinase, reductase, aldehydrase, and lactase.

Additional Reading

Considine, D.M., Editor: *Foods and Food Production Encyclopedia*, Van Nostrand Reinhold, New York, 1982.

Early, R.: *Technology of Dairy Products*, 2nd Edition, Blackie Academic & Professional, London, UK, 1999.

Ernstrom, C.A. and N.P. Wong: *Fundamentals of Dairy Chemistry*, AVI, Westport, Connecticut, 1974.

Fox, P.F. and P. McSweeney: *Dairy Chemistry and Biochemistry*, Chapman & Hall, New York, NY, 1999.

Harper, W.J. and C.W. Hall: *Dairy Technology and Engineering*, AVI, Westport, CT, 1976.

Lewis, M.J. and N.J. Heppell: *Continuous Thermal Processing of Foods: Pasteurization and Uht Sterilization*, Aspen Publishers, Inc., Gaithersburg, MD, 2000.

Marth, E.H., Editor: *Standard Methods for Examination of Dairy Products*, American Public Health Association, Washington, DC, (Revised periodically).

Mazza, G.: *Functional Foods: Biochemical and Processing Aspects*, CRC Press LLC, Boca Raton, FL, 1998.

Robinson, R.K.: *Dairy Microbiology Handbook*, 3rd Edition, John Wiley & Sons, Inc., New York, NY, 2002.

Selitzer, R., Editor: "The Dairy Industry in America," *Dairy and Ice Cream Field* (publishers), New York, 1977.

Spreer, E. and A. Mixa: *Milk and Dairy Product Technology*, Marcel Dekker, Inc., New York, NY, 1998.

Varnam, A.H. and J.P. Sutherland: *Milk and Milk Products: Technology, Chemistry, and Microbiology*, Vol. 1, Aspen Publishers, Inc., Gaithersburg, MD, 2001.

Wong, N.P.: *Fundamentals of Dairy Chemistry*, 3rd Edition, Chapman & Hall, New York, NY, 1999.

BYRON H. WEBB, U.S. Department of Agriculture, Washington, DC

MILKFISH (*Osteichthyes*).   Of the order *Isospondyli*, family *Chanidae*, the milkfish occurs in the tropical waters of America and Africa. It can be described as a silvery-appearing fish which attains a length of about 5 feet (1.5 meters). The milkfish is considered a valuable food fish, notably in the Philippines. The fish is known for its ability to withstand very hot water, as may be found in shallow-water areas.

MILKWEEDS (*Asclepiadaceae*).   A family of some 325 genera, with over 1,700 species of shrubs, woody vines, and perennial herbs. All contain a milky juice from which rubber may be made. The family is particularly abundant in the tropics, especially in Africa. The forms are extreme: many are lianas of great length; others have leaves modified into pitcher-like forms; some are epiphytes; while many, especially in Africa, are very much like Cacti in appearance. Indeed; many species are sold under the name of cactus. The uses made of various milkweeds are many and varied. The young shoots of many species are eaten as greens; other species yield dyes. The juices of several are violent poisons, as *Gonolobus*, from which an arrow-poison is obtained, and *Cynanchum*, which is used to stupefy fish. Many are grown for their weird shape, or because of their beauty, as the Waxplant, *Hoya carnosa*.

MILLER INDICES.   In mineral crystallography the identity of a crystal face consists of a series of whole numbers which are the products of their parameters relating to that face by their inversion, and where required the clearing of fractional values. A parameter is the relative intercept of a crystallographic axis on a given crystal face.

Assuming parameter values on a given crystal face to be $1a$, $1b$, $\frac{1}{2}c$ would on inversion yield 1, 1, $\frac{2}{1}$; parameters $1a$, $1b$, $2c$ would on inversion yield 1, 1, $\frac{1}{2}$; and parameters of $3a$, $2b$, $6c$ would on inversion yield $\frac{1}{3}$, $\frac{1}{2}$, $\frac{1}{6}$. Clearing the fractions in each instance would yield Miller indices of (112), (221) and (231) respectively.

The three Miller indices for a crystal face in all systems except the hexagonal, which requires four indices, are always given in the same order as their crystallographic axes, a, b, c, respectively; a^1, a^2, a^3, in the isometric system; a^1, a^2, c, in the tetragonal system; and a^1, a^2, a^3, c, in the hexagonal.

If the parameter intercepts for a given face are unknown, general indices (hkl) may be used if that face intercepts all three axes; four in the hexagonal, with general indices ($hkil$). If a crystal face cuts two axes and parallels the third, general indices would be identified as ($h0l$), ($0kl$), or ($hk0$) as applicable to that face; in the hexagonal system as ($h0hl$), etc.

See also **Crystal**; and **Mineralogy**.

MILLERITE. The mineral millerite [CAS: 1314-04-1] is nickel sulfide, NiS, whose slender hexagonal interwoven crystals so suggestive of hairs has led to the application of the name "capillary pyrites." It occurs also as radiated masses and coatings. It is brittle; hardness, 3–3.5; specific gravity, 5.48–5.52; luster, metallic; color, brass-yellow, often with an iridescent tarnish. Millerite is found in association with other nickel-bearing minerals and other sulfides. European localities are Bohemia, Westphalia, Wales, etc.; and in the United States at Antwerp, New York; with pyrrhotite in Lancaster County, Pennsylvania; at St. Louis, Missouri; Keokuk, Iowa; and Milwaukee, Wisconsin. In Canada millerite occurs in Oxford, Quebec, and in the famous Sudbury District, Ontario. It is used as an ore of nickel. Millerite was named for the English mineralogist, W.H. Miller.

MILLET (*Gramineae*). Cereal and forage grasses of several different genera are included in the term *millet*. All have a fibrous root system, ample foliage, and rather small grains. They are grown extensively in the occidental countries as forage crops and in some of the oriental countries for human food as well. Of the world's cereal crops, millet ranks sixth in terms of quantity produced.

Millets are members of *Gramineae* (grass family). Compared with other major cereal crops, the millets produce better yields under highly adverse conditions, such as those brought on by inadequate fertility of soils, intensity of heat, and drought. Millets are grown in numerous regions of the world where severe environmental conditions preclude the successful planting of other cereal crops. Parts of Russia, China, India, and Africa are examples. From the standpoint of consumption, the millets tend to be less desirable than other cereals, but there often is no other choice, giving rise to the term "poor man's cereal" for the millets. To a considerable extent, over a long period of years, the millets have become less important in those areas where genetic research and plant-breeding experiments have improved the performance of maize (corn) and wheat when planted in areas previously considered marginal. On the other hand, research of the millets, considered inadequate for many years, is increasing and is pointing toward new areas of interest in millets.

Finger Millet. (*Eleusine coracana*). Also known as African millet, birdsfood millet, coracana millet, nagli, and ragi, this plant is grown in Africa (Ethiopia, Somalia, the Sudan) and in southern Asia (notably India's states of Madras and Mysore) for human consumption. The plant is hardy and tolerates poor soils. Unlike millets in general, finger millet is tolerant of moisture. In India it is frequently found near rice-growing areas. The plant does not do well, however, in frequent heavy rains. Moist mountainous slopes (the foothills of the Himalayas, for example) are particularly favorable to the plant.

Ditch Millet. (*Paspalum scrobiculatum*). Also known as koda millet (India), this is a forage grass principally grown in India and New Zealand. Several grasses (genus Paspalum) are found in the southeastern United States, such as Bahia grass and Dallis grass.

Browntop Millet. (*Panicum ramosum*). This plant was introduced into the United States from India in 1915. There are relatively limited plantings in the southeastern states (Alabama, Florida, Georgia) for hay and pasture. As a dividend, the seed supports wild birds (quail, doves, etc.). The plant is self-seeding, a disadvantage for use in rotation with other crops. At one time, the plant was known as German hay grass. The plant is not to be confused with *browntop panicum* (*Panicum culatum*), which is a native of Central and South America.

Foxtail Millet. (*Setaria italica*). Also called German millet, Italian millet, and Hungarian grass, foxtail millet is widely grown in China, Japan, India, and Manchuria. In the United States, it is planted in limited quantities in the Great Plains states as a forage crop. In some areas, it is sown as an emergency crop when all other plants have failed. In China, this millet

is next to rice and wheat as a cereal for human consumption. In some regions, such as Armenia and Turkey, this millet is used as a substitute for proso millet in areas of very poor soils and during extended hot, dry periods. Foxtail millet is frequently mixed with proso millet in such areas. Foxtail millet is an annual grass and is slender and erect, reaching a height of from 1 to 6 feet (0.3 to 1.8 meters).

Japanese Millet. (*Enchinochloa crusgalli* var. *frumentacea*). Also called billion dollar grass (United States), Japanese barnyard millet, and Sawan millet (India), this plant is often present (as a weed) where other millets are cultivated. It is purposely planted as a catch crop in parts of Australia (New South Wales and Victoria) for hay and pasture. Some limited plantings have been made in the United States (New York, Pennsylvania, Pacific Northwest) as a green feed.

Most Japanese millet grown in India is consumed by low-income populations. It is often boiled with milk and sugar. In Japan and eastern India, the millet is mixed with rice and used for making beer. It also serves well as a forage grass. As many as eight forage crops may be produced in a single year.

Pearl Millet. (*Pennisetum glaucum* or *P. typhoideum*). This is an annual, warm-weather grass that has been cultivated in Africa and Asia since antiquity. The plant is also known as cattail millet, penicillaria, and Mands forage plant (in the United States); as bullrush millet or Dukha (in Africa); as candle millet or dark millet (in Europe); and as baira, cumbo, or sajja (in India).

Pearl millet was probably domesticated somewhere in the dry savanna fringing the southern Sahara desert between western Sudan and Senegal-Mauritania. The plant is extremely drought-resistant and can be grown at the limits of agriculture near the deserts of Africa and India, where millions of people depend upon it for survival. It is one of the most nutritious

Fig. 1.   Pearl millet (*Pennisetum glaucum*). (*USDA diagram.*)

cereals, containing good quantities of phosphorus, minerals, and vitamins A and E.

In Africa, pearl millet serves as a substitute for sorghum on sandy soils and in dry regions. It is estimated that about 20.5 million hectares (50.6 million acres) are planted in India, and in the Sudan, at the edge of the Sahara, 1.2 million hectares (3 million acres) are planted, making millet second only to sorghum as the major cereal crop. Pearl millet is well adapted to the extensive dry periods found in the western Ghats and the plains of Rajputana in India. Limited quantities of pearl millet are grown in the sandy soils of the eastern coastal plain of the United States. Research tests have shown that pearl millet produces more beef per area planted than any other grazing crop used in that area.

Pearl millet has coarse, pithy stems that grow from 6 to 12 feet (1.8 to 3.6 meters) in height. See Fig. 1. The stems are about 1 inch (2.5 centimeters) in thickness. Blades are up to 3 feet (0.9 meters) in length and up to 3 inches (2.5 centimeters) in width. The spike ranges from 8 to 18 inches (20 to 46 centimeters) in length and about 0.5 inches (12 millimeters) in thickness. The heads, as illustrated, appear as beads glued to a round stick. The seeds of pearl millet are somewhat larger than those of other millets, ranging from 3 to 4 millimeters long and about 2.25 millimeters wide. They are of a gray-yellow color and of an obovoid shape. The plant is cross-pollinated.

The *Starr* variety was developed in Georgia and made available in 1950. A later hybrid, the. Gahi (Georgia Hybrid No. 1) became available. This yields about 50% more than the *Starr* variety.

In Africa and Asia, pearl millet is ground into meal, from which gruel, porridge, and baked products are prepared.

Proso Millet. (*Panicum milliaceum*). Also called broom-corn millet hog millet, and Hershey millet (in the United States); and common millet (in Europe). There are three main subspecies of the plant: (1) *P. miliaceum effusion* (characterized by broad panicles that spread out in all directions; (2) *P. miliaceum contractum* (one-sided panicles that have a limited spread); and (3) *P. miliaceum compactum* (compact, thick, erect panicles). Most varieties of proso millet grown in the Middle Eastern countries are derived from southeastern Russian varieties.

Additional Reading

Buerkert, B. et al. (Editors): *Wind Erosion in Niger: Implications and Control Measures in a Millet-Based Farming System*, Kluwer Academic Publishers, New York, NY, 1996.

Khairwal, I. and C. Ram: *Pearl Millet, Seed Production and Technology*, South Asia Books, Colombia, MO, 1990.

Leslie, J. and R. Frederikson (Editors): *Disease Analysis through Genetics and Biotechnology: Interdisciplinary Bridges to Improved Sorghum and Millet Crops*, Iowa State University Press, Ames, IA, 1995.

Web Reference

United States Department of Agriculture website. http://www.usda.gov/

MILLIKAN, ROBERT (1863–1953).

Millikan was an American experimental physicist who earned his doctorate at Columbia University (1895) for research on the polarization of light emitted by incandescent surfaces using for this purpose molten gold and silver at the U.S. Mint.

Millikan made numerous momentous discoveries, chiefly in the fields of electricity, optics, and molecular physics. During an oil-drop experiment, he devised a way of measuring the size of an electric charge and confirmed that electric charge comes in identical chunks, or quanta. His earliest major success was the accurate determination of the charge carried by an electron, using the elegant "falling-drop method"; he also proved that this quantity was a constant for all electrons (1910), thus demonstrating the atomic structure of electricity. From 1912–1915, Millikan verified experimentally Einstein's all-important photoelectric equation, and made the first direct photoelectric determination of Planck's constant h. In addition his studies of the Brownian movements in gases put an end to all opposition to the atomic and kinetic theories of matter. During 1920–1923, Millikan occupied himself with work concerning the hot-spark spectroscopy of the elements (which explored the region of the spectrum between the ultraviolet and X-radiation), thereby extending the ultraviolet spectrum downwards far beyond the then known limit. The discovery of his law of motion of a particle falling towards the earth after entering the earth's atmosphere, together with his other investigations on electrical phenomena, ultimately led him to his significant studies of cosmic radiation (particularly with ionization chambers).

Throughout his life Millikan remained a prolific author, making numerous contributions to scientific journals. He held honorary doctor's degrees from some twenty-five universities. Millikan received the Nobel Prize for physics in 1923.

See also **Electron Theory.**

J. M. I.

MILLIPEDE (*Diplopoda*).

Also known as the thousand-legged worm, there are several species. Millipedes are damaging to some plants in much the same way as certain wireworms and white grubs. Vegetable crops attacked include bean, beet, carrot, cauliflower, corn (maize), cucumber, lettuce, muskmelon, parsnip, pea, potato, radish, squash, and tomato. The pest is not an insect, but a member of the class *Diplopoda*. Millipedes are widely distributed. They appear to be more of a nuisance in New York, the Great Lakes area, especially Ohio, and the Pacific coastal states. The species *Julus impressus* is damaging to vegetable crops; the species *Orthomorpha gracilis* (Koch) is injurious to plants raised in glasshouses.

In addition to tender plant parts, millipedes consume all kinds of decayed vegetable materials, manure, decayed leaves, and seeds. Millipedes are egg-laying and the generations develop slowly, producing about one complete generation per year. Control measures are similar to those used for sowbug and wireworms.

MILSTEIN, CESAR (1927–2002).

Cesar Milstein was born in 1927 in Bahia Blanca, Argentina. He attended the Collegio Nacional, Bahia Blanca from 1939 to 1944, and then the University of Buenos Aires. After graduating in 1952, he continued studies for the doctorate, which he received in 1957, and worked as a staff member inthe Institute of Microbiology at the university until 1963. During a leaveof absence, Milstein worked in the Department of Biochemistry at Cambridge University in England, receiving a second doctorate there in 1960. Returning to Cambridge in 1963, he joined the staff of the Medical Research Council Laboratory of Molecular Biology, later directing its Protein Chemistry Division together with F. Sanger. See also **Sanger, Frederick (1918–Present)**.

Milstein and Associates established the complete sequence of the light chain portion of an immunoglobulin molecule. He determined the nucleotide sequence of a large segment of the light chain messenger RNA, discovering that there is only one type of messenger RNA for both domains within that chain. The separate domains of light and heavy chains are designated constant and variable. Milstein reasoned that the constant domain genes may be separate in the germline but that they must come together in antibody-synthesizing cells. Milstein and Köhler proceeded to develop a method for preparing monoclonal antibodies from hybridomas in 1975 for which they subsequently shared the Nobel Prize in Medicine or Physiology with N. K. Jerne in 1984. They fused antibody-forming spleen cells with mutant myeloma cells to form hybridomas to study the genetic basis of antibody diversity. The mutant myeloma cells conferred immortality on the fused cell line, whereas the spleen cells conferred antibody specificity. Thus, the hybrid cell or hybridoma cell line was endowed with the ability for long-term survival in culture, producing an unlimited quantity of monoclonal antibodies. The success of the technique depended upon a selective method to recover only fused cells. Therefore, they employed a mutant myeloma cell line that was deficient in the enzyme hypoxanthine phosphoribosyltransferase. Cells deficient in this enzyme would perish in a medium containing hypoxanthine, aminopterin and thymidine (HAT). However, hybrid cells would survive and could be isolated since the antibody-synthesizing cell incorporated into the hybridoma would furnish the enzyme. Thus, the immortal hybridoma clone would produce abundant quantities of monoclonal antibodies with specificity for a specific antigenic determinant. Milstein wrote the report and received most of the initial honors and credit. See also **Jerne, Niels Kaj (1911–1994)**.

This revolutionary technique permitted the development of permanent cultures from a single clone that could be maintained indefinitely, providing an unlimited amount of specific monoclonal antibody. The technique developed with mouse cells was later extended to perfect human hybridomas and could even be used to synthesize purified monoclonal antibodies against impure antigens.

Additional Reading

Clark, W.R.: *Experimental Foundations of Modern Immunology*, 4th Edition, John Wiley & Sons, Inc., New York, NY, 1991.

Eichmann, K.: *Kohler's Invention*, Springer-Verlag New York, LLC, New York, NY, 2005.

Goding, J.W.: *Monoclonal Antibodies: Principles and Practice*, 3rd Edition, Elsevier Science & Technology Books, New York, NY, 1996.

Hurrell, J.G.R.: *Monoclonal Hybridoma Antibodies: Techniques and Applications*, CRC Press, LLC, Boca Raton, FL, 1982.

Köhler, G.F., and C. Milstein: "From Antibody Structure to Immunological Diversification of Immune Response," *Science*, **231**, 1261–1268 (1975).

McMichael, A.J., and J. Fabre: *Monoclonal Antibodies in Clinical Medicine*, Elsevier Science & Technology Books, New York, NY, 1982.

Milstein, C.: "Monoclonal Antibodies from Hybrid Myelomas: Theoretical Aspects and some General Comments," In: McMichael, A.J., and J.W. Febre: *Monoclonal Antibodies in Clinical Medicine*, Academic Press, London, UK. 1982, pp. 3–13.

Wade, N.: "Hybridomas: The Making of a Revolution," *Science*, **215**, 1073–1075. (1982).

J.M. CRUSE,
R.E. LEWIS,
University of Mississippi Medical Center, Jackson, MS

MILT. See **Fishes**.

MIMETITE. The mineral mimetite is a chloro-arsenate of lead corresponding to the formula $Pb_5(AsO_4)_3Cl$. It is monoclinic (pseudohexagonal); brittle; hardness, 3.5; specific gravity, 7.0–7.25; luster, resinous; color, usually yellow to brown but may be colorless or white; translucent. Mimetite is a rather rare secondary mineral occurring in altered lead deposits. Found in Bohemia; Saxony; Cornwall and Cumberland, England; South West Africa; Mexico; and in the United States, in Pennsylvania and Utah. The name mimetite is derived from the Greek word meaning imitator, because of the similarity of mimetite and pyromorphite. See also **Pyromorphite**.

MIMICRY. The resemblance of an animal to some other living thing or inanimate object. Mimicry may involve both color and form, hence it is closely related to coloration. It is supposed to benefit the mimic either by concealing it from its enemies, by causing them to mistake it for something undesirable, or by enabling it to approach its prey without giving alarm.

MIMOSA TREE. See **Acacia Trees**.

MINDANAO CURRENT. See **Ocean Currents**.

MINDANAO TRENCH. See **Ocean**.

MINERAL NUTRIENTS. Minerals that are essential to life are the source of metals and other inorganic elements involved in the most fundamental processes. For example, oxygen, required by the cells of animals, is utilized with the aid of metal complexes. In humans both iron-containing hemoglobin and zinc-containing carbonic anhydrase play pivotal roles in binding oxygen and delivering it to the cells. Moreover, enzymes developed to protect cells from high levels of oxygen also contain metals. One such class of protective enzymes is known as the superoxide dismutases (SODs). These contain metals such as manganese, copper, zinc, and iron. Mutations in the copper- and zinc-containing superoxide dismutase gene have been linked to amyotrophic lateral sclerosis.

As for other biological substances, states of dynamic equilibrium exist for the various mineral nutrients as well as mechanisms whereby a system can adjust to varying amounts of these minerals in the diet. In forms usually found in foods, and under circumstances of normal human metabolism, most nutrient minerals are not toxic when ingested orally. Amounts considerably greater than the recommended dietary allowances (RDAs) can generally be eaten without concern for safety (Table 1).

Some elements found in body tissues have no apparent physiological role, but have not been shown to be toxic. Examples are rubidium, strontium, titanium, niobium, germanium, and lanthanum. Other elements are toxic when found in greater than trace amounts, and sometimes in trace amounts. These latter elements include arsenic, mercury, lead, cadmium, silver, zirconium, beryllium, and thallium. Numerous other elements are used in medicine in non-nutrient roles. These include lithium,

TABLE 1. ESSENTIAL MINERAL NUTRIENTS

Element[a]	Body Content, mg/kg Body wt	Daily Requirement, mg
PRINCIPAL ELEMENTS		
Calcium	14,000–20,000	800–1,200[b,c]
Phosphorus	11,000–12,000	800–1,200[b,c]
Sulfur	1,600–2,500	[d]
Potassium	2,000–3,500	2,000
Sodium	1,500–1,600	500
Chlorine	1,200–1,500	750
Magnesium	270–500	280[b,c,e];350[b,f]
TRACE AND ULTRATRACE ELEMENTS		
Iron	60–66	10[b,f];15[b,e]
Fluorine	37	1.5–4.0[b]
Zinc	33–50	12[b,c,e];15[b,f]
Silicon	15–16	5–20
Copper	1.0–2.5	1.5–3.0
Boron	0.69	0.5–1.0
Selenium	0.2–0.3	0.055[b,c,e]; 0.07[b,f]
Iodine	0.2–0.4	0.15[b,c]
Manganese	0.2–4.0	2.0–5.0[b]
Molybdenum	0.1–0.5	0.075–0.25[b]
Chromium	0.06–0.2	0.05–0.2[b]
Cobalt	0.02	0.003[g]
Tin	0.2	
Vanadium	0.14	<0.01
Nickel	0.07–0.14	<0.10

[a] Generally not ingested in elemental form.
[b] Values are for adults.
[c] Increased amounts are required during pregnancy and lactation.
[d] Adequate intake with adequate intake of protein.
[e] Value for females.
[f] Value for males.
[g] As vitamin B_{12}.

bismuth, antimony, bromine, platinum, and gold. The interactions of mineral nutrients with carbohydrates, fats, and proteins, minerals with vitamins, and mineral nutrients with toxic elements are areas of active investigation.

The amount of each element required in daily dietary intake varies with the individual bioavailability of the mineral nutrient.

The Principal Elements

Calcium. Calcium [CAS: 7440-70-2], the most abundant mineral element in mammals, comprises 1.5–2.0 wt % of the adult human body, over 99 wt % of which is present in bones and teeth. About 48% of serum calcium is ionic, ca 46% is bound to blood proteins, the rest is present as diffusible complexes, e.g., of citrate. The calcium ion level must be maintained within definite limits.

Bones act as a reservoir of certain ions, in particular Ca^{2+} and PO_4^{3-}, which readily exchange between bones and blood. Bone structure comprises a strong organic matrix combined with an inorganic phase which is principally hydroxyapatite, $3Ca_3(PO_4)_2 \cdot Ca(OH)_2$. Bones contain two forms of hydroxyapatite. The less soluble crystalline form contributes to the rigidity of the structure. The crystals are quite stable, but because of the small size present a very large surface area available for rapid exchange of ions and molecules with other tissues. There is also a more soluble intercrystalline fraction. Bone salts also contain small amounts of magnesium, sodium, carbonate, citrate, chloride, and fluoride. Osteoporosis is reported to result when bone resorption is relatively faster than bone formation. The calcium ion, necessary for blood-clot formation, stimulates release of bloodclotting factors from platelets. See also **Blood**; and **Anticoagulants**.

In normal adults, the blood Ca_{2+} level is established by an equilibrium between blood Ca^{2+} and the more soluble intercrystalline calcium salts of the bone. Additionally, a subtle and intricate feedback mechanism responsive to the Ca^{2+} concentration of the blood that involves the less soluble crystalline hydroxyapatite comes into play. The thyroid and parathyroid glands, the liver, kidney, and intestine also participate in Ca^{2+} control.

In addition to hypocalcemia, tremors, osteoporosis, and muscle spasms (tetary), calcium deficiency can lead to rickets, osteomalacia, and possibly heart disease. These, as well as Paget's disease, can also result from faulty utilization of calcium. Calcium excess can lead to excess secretion of calcitonin, possible calcification of soft tissues, and kidney stones when combined with magnesium deficiency.

Phosphorus. Eighty-five percent of the phosphorus [CAS: 7723-14-0], the second most abundant element in the human body, is located in bones and teeth. Whereas there is constant exchange of calcium and phosphorus between bones and blood, there is very little turnover in teeth. The Ca:P ratio in bones is constant at about 2:1. Every tissue and cell contains phosphorus, generally as a salt or ester of mono-, di-, or tribasic phosphoric acid, as phospholipids, or as phosphorylated sugars. Phosphorus is involved in a large number and wide variety of metabolic functions. Examples are carbohydrate metabolism, adenosine triphosphate (ATP) from fatty acid metabolism, and oxidative phosphorylation.

The formation of phosphate esters is the essential initial process in carbohydrate metabolism. See also **Carbohydrates**. The glycolytic, i.e., anaerobic or Embden-Meyerhof pathway comprises a series of nine such esters. The phosphogluconate pathway, starting with glucose, comprises a succession of 12 phosphate esters. Cyclic adenosine monophosphate (cAMP), produced from ATP, is involved in a large number of cellular reactions including glycogenolysis, lipolysis, active transport of amino acids, and synthesis of protein. Inorganic phosphate ions are involved in controlling the pH of blood. The principal anion of intercellular fluid is HPO_4^{2-}.

Phospholipids, components of every cell membrane, are active determinants of membrane permeability. They are sources of energy, components of certain enzyme systems, and involved in lipid transport in plasma. Because of their polar nature, phospholipids can act as emulsifying agents. The structure of most phospholipids resembles that of triglycerides except that one fatty acid radical has been replaced by a radical derived from phosphoric acid and a nitrogen base, e.g., choline or serine.

Phosphorus is an essential component of nucleic acids, polymers consisting of chains of nucleosides, a sugar plus a nitrogenous base, and joined by phosphate groups. In ribonucleic acid (RNA), the sugar is D-ribose; in deoxyribonucleic acid (DNA), the sugar is 2-deoxy-D-ribose.

Phosphorus nutrient deficiency can lead to rickets, osteomalacia, and osteoporosis, whereas an excess can produce hypocalcemia. Faulty utilization of phosphorus results in rickets, osteomalacia, osteoporosis, and Paget's disease, and renal or vitamin D-resistant rickets.

Sulfur. Sulfur [CAS: 63705-05-5] is present in every cell in the body, primarily in proteins containing the amino acids methionine, cystine, and cysteine. Inorganic sulfates and sulfides occur in small amounts relative to total body sulfur, but the compounds that contain them are important to metabolism. Sulfur intake is thought to be adequate if protein intake is adequate and sulfur deficiency has not been reported.

Although sulfur is in the same group of the Periodic Table, Group 16(VIA), as oxygen, sulfur functions much more like phosphorus, Group 15(VA), in biological systems. In fat metabolism, sulfur plays a key role analogous to that of phosphorus in carbohydrate metabolism. Fatty acid synthesis and degradation begin and end with the same compound, acetyl-S coenzyme A (acetyl–SCoA).

Detoxification systems in the human body often involve reactions that utilize sulfur-containing compounds. For example, reactions in which sulfate esters of potentially toxic compounds are formed, rendering these less toxic or nontoxic, are common as are acetylation reactions involving acetyl–SCoA. Another important compound is *S*-adenosylmethionine (SAM), the active form of methionine. SAM acts as a methylating agent, e.g., in detoxification reactions such as the methylation of pyridine derivatives, and in the formation of choline (qv), creatine, carnitine, and epinephrine. Sulfur nutrient deficiency results in retarded growth, and faulty utilization in homocystinuria.

Sodium and Potassium. [Sodium CAS: 7440-23-5], [Potassium CAS: 7440-09-7]. Whereas sodium ion is the most abundant cation in the extracellular fluid, potassium ion is the most abundant in the intracellular fluid. Small amounts of K^+ are required in the extracellular fluid to maintain normal muscle activity. Some sodium ion is also present in intracellular fluid.

Sodium ion acts in concert with other electrolytes, in particular K^+, to regulate the osmotic pressure and to maintain the appropriate water and pH balance of the body. Homeostatic control of these functions is accomplished by the lungs and kidneys interacting by way of the blood. Sodium is essential for glucose absorption and transport of other substances across cell membranes. It is also involved, as is K^+, in transmitting nerve impulses and in muscle relaxation. Potassium ion acts as a catalyst in the intracellular fluid, in energy metabolism, and is required for carbohydrate and protein metabolism.

Maintenance of the appropriate concentrations of K^+ and Na^+ in the intra- and extracellular fluids involves active transport, i.e., a process requiring energy. Sodium ion in the extracellular fluid (0.136–0.145 M Na^+) diffuses passively and continuously into the intracellular fluid (<0.01 M Na^+) and must be removed. This sodium ion is pumped from the intracellular to the extracellular fluid, while K^+ is pumped from the extracellular (ca 0.004 M K^+) to the intracellular fluid (ca 0.14 M K^+). The energy for these processes is provided by hydrolysis of adenosine triphosphate (ATP) and requires the enzyme Na^+–K^+ ATPase, a membrane-bound enzyme which is widely distributed in the body. In some cells, e.g., brain and kidney, 60–70 wt % of the ATP is used to maintain the required Na^+–K^+ distribution.

Sodium and potassium ions are actively absorbed from the intestine. As a consequence of the electrical potential caused by transport of these ions, an equivalent quantity of Cl^- is absorbed. The resulting osmotic effect causes absorption of water.

Selective excretion and reabsorption of Na^+ and K^+ are accomplished by means of the kidney tubular cell membranes. The volume of extracellular fluid is directly related to the Na^+ concentration which is closely controlled by the kidneys. Homeostatic control of Na^+ concentration depends on the hormone aldosterone. The kidney secretes a proteolytic enzyme, rennin, which is essential in the first of a series of reactions leading to aldosterone. In response to a decrease in plasma volume and Na^+ concentration, the secretion of rennin stimulates the production of aldosterone resulting in increased sodium retention and increased volume of extracellular fluid.

Salt-free or low salt diets often are prescribed for hypertensive patients. However, sodium chloride increases the blood pressure in some individuals but not in others. Conversely, restriction of dietary NaCl lowers the blood pressure of some hypertensives, but not of others. Genetic factors and other nutrients, e.g., Ca^{2+} and K^+, may be involved. The optimal intakes of Na^+ and K^+ remain to be established. See also **Diet**.

Potassium and/or sodium deficiency can lead to muscle weakness and sodium deficiency to nausea. Hyperkalemia resulting in cardiac arrest is possible from 18 g/d of potassium combined with inadequate kidney function. Faulty utilization of K^+ and/or Na^+ can lead to Addison's or Cushing's disease.

Chlorine. The chlorides are essential in the homeostatic processes maintaining fluid volume, osmotic pressure, and acid–base equilibria. Most chloride is present in body fluids; a little is in bone salts. Chloride [CAS: 16887-00-6] is the principal anion accompanying Na^+ in the extracellular fluid. Less than 15 wt % of the Cl^- is associated with K^+ in the intracellular fluid. Chloride passively and freely diffuses between intra- and extracellular fluids through the cell membrane. If chloride diffuses freely, but most Cl^- remains in the extracellular fluid, it follows that there is some restriction on the diffusion of phosphate.

Some of the blood Cl^- is used for formation in the gastric glands of hydrochloric acid, HCl, required for digestion. Hydrochloric acid is secreted into the stomach where it acts with gastric enzymes in the digestive processes. The chloride is then reabsorbed with other nutrients into the blood stream. Chloride is actively transported in gastric and intestinal mucosa. In the kidney, chloride is passively reabsorbed in the thin ascending loop of Henle and actively reabsorbed in the thick segment of the ascending loop, i.e., the distal tubule. In the chloride shift, Cl^- plays an important role in the transport of carbon dioxide.

Numerous neurotransmitter receptors, e.g., glutamate, γ-aminobutyric acid (GABA), and benzodiazepine (called the valium receptor), have been identified as chloride channel proteins. The genetic defect in cystic fibrosis involves defective-functioning chloride channel proteins with excessive Cl^- loss. Deficient Cl^- during development adversely affects language skills in humans, as well as impaired growth in infants and metabolic alkalosis.

Fruit and vegetable juices high in potassium have been recommended to correct hypokalemic alkalosis in patients on diuretic therapy. Apparently the efficacy of this treatment is questionable. A possible reason for ineffectiveness is the low Cl^- content of most of these juices. Because Cl^- is high only in juices in which Na^+ is high, these have to be excluded.

Magnesium. In the adult human, 50–70% of the magnesium [CAS: 7439-95-4] is in the bones associated with calcium and phosphorus. The rest is widely distributed in the soft tissues and body fluids. Most of the nonbone Mg^{2+}, like K^+, is located in the intracellular fluid where it is the most abundant divalent cation. Magnesium ion is efficiently retained by the kidney when the plasma concentration of Mg^{2-} falls; in this respect it resembles Na^+. The functions of Na^+, K^+, Mg^{2+}, and Ca^{2+} are interrelated so that a deficiency of Mg^{2+} affects the metabolism of the other three ions.

Magnesium is essential in numerous metabolic processes. It is the activator of many enzymes, e.g., adenyl cyclase, alkaline phosphatases, and the phosphokinases, pyrophosphatases, and thiokinases. Because the phosphokinases are required for the hydrolysis and transfer of phosphate groups, magnesium is essential in glycolysis and in oxidative phosphorylation. The thiokinases are required for the initiation of fatty acid degradation. Magnesium is also required in systems in which thiamine pyrophosphate is a coenzyme.

As an activator of the phosphokinases, magnesium is essential in energy-requiring biological processes, such as activation of amino acids, acetate, and succinate; synthesis of proteins, fats, coenzymes, and nucleic acids; generation and transmission of nerve impulses; and muscle contraction.

Regulation of serum Mg^{2+} appears to result from a balance among intestinal absorption, renal reabsorption, and excretion. The controlling factor is probably the renal threshold.

A severe magnesium deficiency in humans is seldom encountered except as a secondary effect resulting from numerous disease states, e.g., chronic alcoholism with malnutrition, acute or chronic renal disease, long-term Mg^{2+}-free parenteral feeding, protein–calorie malnutrition, and hyperthyroidism. In these situations, it is difficult to attribute specific clinical manifestations to magnesium deficiency. The specific role of magnesium in cardiovascular disease, e.g., arrythmia, spasms, or ischemia, remains a subject of conflicting research findings.

Neuromuscular irritability, convulsions, muscle tremors, mental changes such as confusion, disorientation, and hallucinations, heart disease, and kidney stones have all been attributed to magnesium deficiency. Excess Mg^{2+} can lead to intoxication exemplified by drowsiness, stupor, and eventually coma.

Trace Elements and Ultratrace Elements

Iron. The total body content of iron [CAS: 7439-89-6], i.e., 3–5 g, is recycled more efficiently than other metals. There is no mechanism for excretion of iron and what little iron is lost daily, i.e., ca 1 mg in the male and 1.5 mg in the menstruating female, is lost mainly through exfoliated mucosal, skin, or hair cells, and menstrual blood.

A large percentage of the iron in the human body is in hemoglobin: 85 wt % in the adult female, 60 wt % in the adult male. The remainder is present in other iron-containing compounds involved in basic metabolic functions, or in iron transport or storage compounds.

Absorption of iron from food to maintain homeostasis, tightly controlled, increases in instances of increased demands, such as during pregnancy and lactation, and iron-deficiency states which are the result of blood loss or iron-deficiency anemia resulting from inadequate iron intake. Iron absorption is greatly reduced in the normal individual when iron stores are adequate or excessive. Absorption is enhanced by acid conditions and reducing agents. Heme iron from animal sources is absorbed more readily than nonheme iron from cereals and vegetables.

A system of internal iron-exchange exists which is dominated by the iron required for hemoglobin synthesis. For formation of red blood cells, iron stores can furnish 10–40 mg/d of iron, as compared to 1–3 mg from dietary sources. Only ca 10 wt % of ingested iron actually is absorbed. Transferrin is essential for movement of iron and without it, as in genetic absence of transferrin, iron overload occurs in tissues. This hereditary atransferrinemia is coupled with iron-deficiency anemia. The

iron overload in hereditary or acquired hemochromatosis results in fully saturated transferrin and is treated by phlebotomy.

Iron deficiency is a significant worldwide nutritional problem and cause of anemia which can also lead to a decreased resistance to infection. Insufficient dietary iron intake; iron losses, e.g., bleeding and parasite infestation; and malabsorption of iron are the principal causes. The groups at greatest risk for developing iron-deficiency anemia are menstruating females, pregnant or nursing females, and young children. Children can experience impaired psychomotor development and intellectual performance.

Iron toxicity resulting from excess absorbable iron ingestion is rare except in Africa where fermented beverages made in large iron pots have levels of iron approaching 80 mg/L in a brew where the pH is very low. This results in Bantu siderosis which can result in hemochromatosis, i.e., damage to various organs from excessive storage of iron. This condition can cause numerous disease states, e.g., hepatic fibrosis and diabetes in 80% of the cases of idiopathic hemochromatosis patients. Iron overload is frequently a complication of repeated blood transfusions in anemias, e.g., thalassemia. The lethal dose of ferrous sulfate for a two-year old is 2 g; for an adult the lethal dose is from 200–300 g.

Fluorine. Fluoride [CAS: 16984-48-8] is present in the bones and teeth in very small quantities. Human ingestion is from 0.7–3.4 mg/d from food and water.

Fluoridation of public water supplies, a common practice throughout much of the United States, may be an effective means of significantly reducing the incidence of dental caries. See also **Fluorine**. Concern regarding the narrow range of safety between effective and toxic fluoride concentrations has been expressed and poisoning from excessive fluoride, fluorosis, added to public water has been reported. Assertions that fluoridation of water supplies increases the incidence of cancer have not been substantiated.

Excess fluoride ingestion damages developing teeth, causing mottling, chalky-white coloration, and pitting.

Zinc. The 2–3 g of zinc [CAS: 7440-66-6] in the human body are widely distributed in every tissue and tissue fluid. About 90 wt % is in muscle and bone; unusually high concentrations are in the choroid of the eye and in the prostate gland. Almost all of the zinc in the blood is associated with carbonic anhydrase in the erythrocytes. Zinc is concentrated in nucleic acids, and found in the nuclear, mitochondrial, and supernatant fractions of all cells.

Zinc is essential for the function of many enzymes, either in the active site, i.e., as a nondialyzable component, of numerous metalloenzymes or as a dialyzable activator in various other enzyme systems.

Zinc-hormone interactions include hormonal influence on absorption, distribution, transport, and excretion of zinc and zinc influence on synthesis, secretion, receptor binding, and function of numerous hormones. Zinc enhances pituitary activity by increasing circulating levels of growth hormone, thyroid-stimulating hormone, luteinizing hormone, follicle-stimulating hormone, and adrenocorticotropin. The role of zinc in insulin action is recognized but not well understood. Zinc is required for maintenance of normal plasma concentrations of vitamin A and for normal mobilization of vitamin A from the liver.

Zinc was confirmed as essential for humans in 1956 and deficiency symptoms were reported in 1961. The size of the human fetus is correlated with zinc concentration in the amniotic fluid and habitual low zinc intake in the pregnant female is thought to be related to several congenital anomalies in humans. Low zinc intakes result in hypogonadism, dwarfism, mental retardation, low serum and red blood cell zinc in humans and animals, and retarded growth and teratogenic effects on the nervous system in rats.

In children suffering from marginal zinc deficiency, impaired taste acuity, poor appetite, and suboptimal growth can be reversed upon zinc supplementation. Accelerated wound healing occurs in humans upon zinc supplementation, suggesting that marginal zinc deficiency in humans may be more widespread than has been thought. Zinc supplementation has also been effective in alleviating symptoms of active rheumatoid arthritis in clinical trials. Acrodermatitis enteropathica, a hereditary disease that involves aberrant zinc metabolism, responds to oral zinc supplementation. Excessive zinc intake may interfere with copper metabolism.

Silicon. Silicon [CAS: 7440-21-3] comes mainly from ingestion of silicates, primarily from vegetables. It is found in the serum as silicic acid [CAS: 7699-41-4], $Si(OH)_4$, and normal blood serum levels are ca

1 mg/100 mL regardless of intake because of efficient kidney excretion of excess. Silicon is necessary for calcification, growth, and as cross-linking material in mucopolysaccharide formation. Silicon is especially helpful in situations where the diet is low in calcium or high in aluminum, or thyroid function is inadequate. The human requirement may be 5–20 mg/d. Silicon deficiency may lead to altered metabolism of connective tissue and bone and/or aluminum accumulation in the brain.

Copper. All human tissues contain copper [CAS: 7440-50-8]. The highest amounts are found in the liver, brain, heart, and kidney. In blood, plasma and erythrocytes contain almost equal amounts of copper, i.e., ca 110 and 115 mg/100 mL, respectively.

In plasma, ca 90 wt % of copper is in the metalloprotein ceruloplasmin, also known as a_2-globulin, mol wt 151,000, which contains 8 atoms of copper per molecule. Ceruloplasmin has been identified as a ferroxidase(I) which catalyses the oxidation of aromatic amines and of Fe^{2+} to Fe^{3+}. The ferric ion is then incorporated into transferrin which is necessary for the transport of iron to tissues involved in the synthesis of iron-containing compounds, e.g., hemoglobin. Lowered levels of ceruloplasmin interfere with hemoglobin synthesis.

Copper deficiency is characterized by poorly formed collagen which leads to bone fragility and spontaneous bone fractures in animals, and also results in cardiac hypertrophy. Abnormal electrocardiographs have been noted when low copper diets were fed to humans. Anemia, neutropenia, and bone disease have been reported in children having protein calorie malnutrition (PCM) and accompanying hypocupremia. At least two genetic diseases involving copper are known: Wilson's disease, an autosomal recessive disease, usually detected in adulthood, and Menke's kinky-hair syndrome.

Analytical data indicate that many diets contain less than the RDA for copper. Excessive copper has been reported to be fatal for oral dose levels of copper sulfate of 200 mg/kg body weight for a child and 50 mg/kg for adults.

Boron. The essentiality of boron [CAS: 7440-42-8], first accepted for higher plants in 1923, then for animals, was recognized in 1981 for human metabolism. Boron is reported to help maintain function or stability of cell membranes and is thought to be involved with hormone reception and transmembrane signaling. Enhanced need for boron may develop with nutritional or metabolic stress involving other nutrients, e.g., magnesium deprivation or physiological changes in calcium metabolism. Boron depletion impairs cognitive function. Organs known to contain the highest levels of boron are bone, spleen, and thyroid. An excess of boron can, however, cause seizures in infants, riboflavinuria, and gastrointestinal upset.

Selenium. Selenium [CAS: 7782-49-2], thought to be widely distributed throughout body tissues, is present mostly as selenocysteine in seleno-proteins or as selenomethionine. Animal experiments suggest that greater concentrations are in the kidney, liver, and pancreas and lesser amounts are in the lungs, heart, spleen, skin, brain, and carcass.

The most clearly documented role for selenium is as a necessary component of glutathione peroxidase. Selenium is also involved in the functions of additional enzymes, e.g., type 1 iodothyronine deiodinase, leukocyte acid phosphatase, and glucuronidases. A role for selenium in electron transfer has been suggested as has involvement in nonheme iron proteins. Selenium and vitamin E appear to be necessary for proper functioning of lysosomal membranes. A role for selenium in metabolism of thyroid hormone has been confirmed.

Alkali disease and blind staggers of grazing livestock in the western U.S. were reported as the result of selenium poisoning. There are unusually high concentrations of selenium in certain plants, because of selenium-accumulating properties, or in ordinary plants growing on highly seleniferous soils. No treatment for this type of poisoning is known, thus excess selenium in the animal diet must be avoided. Prolonged ingestion of up to 600 mcg/d of selenium did not produce toxic effects in humans. Toxic effects in humans have been reported, however, from chronic ingestion of food in China supplying 5 mg/d of Se, and from supplements in the U.S. of 27–2387 mg/d of Se. Effects include hair loss, changes in the nails, gastrointestinal upset, and peripheral neuropathy.

Pure selenium deficiency, without concurrent vitamin E deficiency, is not generally seen except in animals on experimental diets. In China, selenium deficiency in humans has been associated with Keshan disease, a cardiomyopathy seen in children and in women of child-bearing ages, and Kashin-Beck disease, an endemic osteoarthritis in adolescents. Selenium may have anticarcinogenic effects possibly because of the antioxidant properties of selenium compounds.

Iodine. Of the 10–20 mg of iodine [CAS: 7553-56-2] in the adult body, 70–80 wt % is in the thyroid gland. See also **Iodine (In Biological Systems)**. The essentiality of iodine, present in all tissues, depends solely on utilization by the thyroid gland to produce thyroxine and related compounds. Well-known consequences of faulty thyroid function are hypothyroidism, hyperthyroidism, and goiter. Dietary iodine is obtained from eating seafoods and kelp and from using iodized salt.

The functions of the thyroid hormones and thus of iodine are control of energy transductions. These hormones increase oxygen consumption and basal metabolic rate by accelerating reactions in nearly all cells of the body. A part of this effect is attributed to increase in activity of many enzymes. Additionally, protein synthesis is affected by the thyroid hormones.

In many parts of the world, simple goiter is endemic and usually results from dietary iodine deficiency or goiterogens in foods which bind iodine. Iodine deficiency disorders (IDD) include cretinism, myxedema, hypothyroidism, and goiter. In technologically advanced countries, the problem of iodine deficiency has been minimized by the use of iodized salt. Faulty utilization of iodine can lead to Grave's disease. A sodium iodide excess can also produce goiter and an excess of 500 mg/kg body weight can be fatal.

Manganese. The adult human body contains ca 10–20 mg of manganese [CAS: 7439-96-5] widely distributed throughout the body. The largest Mn^{2+} concentration is in the mitochondria of the soft tissues, especially in the liver, pancreas, and kidneys. Manganese concentration in bone varies widely with dietary intake.

Manganese is essential for normal body structure, reproduction, normal functioning of the central nervous system, and activation of numerous enzymes. An excess of manganese can lead to neural damage and possible impaired insulin production.

In animals, manganese deficiency results in wide-ranging disorders, e.g., impaired growth, abnormal skeletal structure, disturbances of reproduction, and defective lipid and carbohydrate metabolism. Although overt manganese deficiency has not been induced in humans, some forms of epilepsy in humans and animals and a decrease in glucose tolerance in animals have been linked to low levels of manganese in the tissue.

Molybdenum. Molybdenum [CAS: 7439-98-7] is a component of the metalloenzymes xanthine oxidase, aldehyde oxidase, and sulfite oxidase in mammals. Two other molybdenum metalloenzymes present in nitrifying bacteria have been characterized: nitrogenase and nitrate reductase. The molybdenum in the oxidases, is involved in redox reactions. The heme iron in sulfite oxidase also is involved in electron transfer. Foods rich in molybdenum include legumes, dark green vegetables, liver, whole-grain cereals, and milk. Xanthine oxidase, mol wt ca 275,000, present in milk, liver, and intestinal mucosa, is required in the catabolism of nucleotides. Xanthine oxidase is also involved in iron metabolism.

A copper–molybdenum antagonism involving sulfate occurs in animals, i.e., large amounts of molybdenum and sulfate can depress copper absorption. Cattle grazing on pasturage of high Mo content succumb to teart or peat scours, characterized by diarrhea and general wasting. Control involves increasing copper intake. The Cu–Mo antagonism has been observed in humans. Significant increases in urinary copper excretion have been observed with increasing Mo intake.

Molybdenum deficiency in humans results in deranged metabolism of sulfur and purines and symptoms of mental disturbances. Toxic levels produce elevated uric acid in blood, gout, anemia, and growth depression. Faulty utilization results in sulfite oxidase deficiency, a lethal inborn error.

Chromium. Chromium(III) [CAS: 7440-47-3] potentiates the action of insulin and may be considered a cofactor for insulin. Chromium is thought to form a complex with insulin and insulin receptors.

Studies of elderly people and mildly diabetic patients showed significant improvement in the glucose tolerance test (GTT) when chromium supplementation of 150–200 μg/d was given. In other tests, these positive results were not obtained. It is possible that not all subjects are capable of utilizing inorganic chromium to the same extent. Some may require a preformed GTF (glucose tolerance factor). Chromium chloride supplementation has been effective in normalizing impaired glucose tolerance in malnourished children and in patients receiving total parenteral nutrition for a long time. The most available form of chromium is GTF obtained from brewer's yeast. Chromium deficiency may also lead to atherosclerosis and peripheral neuropathy.

Cobalt. Cobalt [CAS: 7440-48-4] is nutritionally available only as vitamin B_{12}. Although Co^{2+} can function as a replacement *in vitro* for other divalent cations, in particular Zn^{2+}, no *in vivo* function for inorganic cobalt is known for humans. In ruminant animals, B_{12} is synthesized by bacteria in the rumen.

In pernicious anemia, the bone marrow fails to produce mature erythrocytes as a result of defective cell division, a consequence of impaired DNA synthesis which requires vitamin B_{12}. If the disease goes untreated, extensive neurological damage, e.g., irreversible degeneration of the spinal cord by demyelinization, may occur because of faulty fatty acid metabolism.

Vitamin B_{12} deficiency commonly is caused by inadequate absorption resulting from a lack or insufficient intrinsic factor (IF). Intrinsic factor is a glycoprotein, mol wt ca 50,000, which binds vitamin B_{12} in a 1:1 molar ratio. The B_{12}–IF complex, formed in the stomach, is absorbed in the ileum. Absorption in this part of the intestine occurs because of the specific characteristics of the cells of the microvilli (brush border) of the ileum. The IF remains in the intestine attached to the epithelial cells. Transport of B_{12} into the blood stream requires Ca^{2+}. In the blood, B_{12} is bound to transcobalamin II (transport protein). Whatever bound B_{12} is not utilized immediately is stored in the liver. With increasing quantities of dietary B_{12}, the fraction that is absorbed decreases. Generally, vitamin B_{12} is excreted in the urine, but with large intake, some is excreted in the bile. A nutritional excess of cobalt can lead to polycythemia.

Tin. The widespread use of canned foods results in a daily intake of tin [CAS: 7440-31-5] that is ca 1–17 mg for an adult male. At this level it has not been shown to be toxic. Some grains also contain tin. Too much tin can adversely affect zinc balance and iron metabolism. Essentiality has not been confirmed for humans. It has been shown for the rat. An enhanced growth rate results from tin supplementation of low tin diets. Animals on deficient diets exhibit poor growth and decreased feed efficiency.

Vanadium. Vanadium [CAS: 7440-62-2] is essential in rats and chicks. Estimated human intake is less than 4 mg/d. In animals, deficiency results in impaired growth, reproduction, and lipid metabolism, and altered thyroid peroxidase activities. Vanadium may play a role in the regulation of (NaK)–ATPase, phosphoryl transferases, adenylate cyclase, and protein kinases.

Nickel. There is considerable evidence for the essentiality of nickel [CAS: 7440-02-0] in animals. Various pathological manifestations of nickel deficiencies have been observed in chicks, cows, goats, pigs, rats, and sheep. Average intake is reported to be about 60–260 µg/d, and a dietary requirement for humans of less than 100 µg/d has been suggested. *In vitro* studies have shown nickel to be an activator of several enzymes.

Arsenic. Arsenic [CAS: 7440-38-2] is under consideration for inclusion as an essential element. No clear role has been established, but arsenic, long thought to be a poison, may be involved in methylation of macromolecules and as an effector of methionine metabolism.

Health and Safety Factors

Under unusual circumstances, toxicity may arise from ingestion of excess amounts of minerals. This is uncommon except in the cases of fluorine, molybdenum, selenium, copper, iron, vanadium, and arsenic. Toxicosis may also result from exposure to industrial compounds containing various chemical forms of some of the minerals. Aspects of toxicity of essential elements have been published.

Efficient homeostatic controls of mammalians generally prevent serious toxicity from ingestion of the mineral nutrients. Toxicity may occur under conditions far removed from those of nutritional significance or for individuals suffering from some pathological conditions. Because of very low concentrations in foods, the trace elements are not toxic under normal nutritional conditions. Exceptions are selenium and iron.

See also **Calcium (In Biological Systems); Cobalt (In Biological Systems); Copper (In Biological Systems); Magnesium (In Biological Systems); Molybdenum (In Biological Systems); Nickel (In Biological Systems); Phosphorus (In Biological Systems); Potassium and Sodium (In Biological Systems);** and **Zinc (In Biological Systems).**

Additional Reading

Considine, D.M., and G.D. Considine: *Foods and Food Production Encyclopedia*, Van Nostrand Reinhold Company Inc., New York, NY, 1982.

Ensminger, A.H., M.E. Ensminger, J. Konlande, and J.R.K. Robson: *Food and Nutrition Encyclopedia*, 2nd Edition, CRC Press, Boca Raton, FL, 1994.

Linder, M.C., ed.: *Nutritional Biochemistry and Metabolism with Clinical Applications*, 2nd Edition, Appleton & Lange, Norwalk, CT, 1991.

Mertz, W., ed.: *Trace Elements in Human and Animal Nutrition*, Vols. 1 and 2, Academic Press, San Diego, CA, 1987.

Seelig, M.S., and A. Rosanoff: *The Magnesium Factor*, Avery Publishing Group, Inc., Garden City Park, NY, 2003.

Shils, M.E., J.A. Olson, and M. Shike, eds.: *Modern Nutrition in Health and Disease*, 8th Edition, Vols. 1 and 2, Lea & Febigher, Philadelphia, PA, 1993.

MINERALOGY. The science of mineralogy is concerned with the formation, occurrence, properties, composition, and classification of minerals. Various definitions of a mineral have been proposed. Possibly, the most acceptable may be, "a naturally occurring inorganic substance, usually crystalline, possessing a relatively definite chemical composition and physical characteristics." It should be pointed out that some naturally formed organic substances, particularly of an economic resource nature, are sometimes classified as minerals.

Although in its broadest application, mineralogy is as ancient as human civilization, mineralogy is a modern science, the mineralogist taking full advantage of all modern tools and instruments for exploration, analysis, testing, and study of minerals. Several major scientific advances in the materials field have stemmed from the study of minerals as will be pointed out shortly.

Presumably in the early ages man used minerals as weapons. Through the passage of time and attainment of knowledge regarding certain mineral characteristics man learned, notably initially by accident but later by design, that the content of minerals provided essential materials for his expanding needs. Very early, the natural form and beauty of certain minerals became objects for personal adornment. Later it was found that both the form and the innate beauty of minerals were enhanced by cutting and polishing them. Although the science of mineralogy touches the life of every person, a fundamental understanding of minerals is not common.

Modern mineralogy is the product of research and discovery by many persons. Robert Hooke (1665) foretold the atomic theory by constructing models of alum crystals out of leaden musket balls. Nicolaus Steno (1669) discovered the constancy of interfacial angles between corresponding faces of quartz crystals from many localities. This was later formalized by Rome de l'Isle (1764) under his *law of constancy of crystal interfacial angles*. In 1784, René Just Haüy proposed the theory of *integral molecules* by stacking calcite rhombs to show that structural units could produce exact external facial planes of various forms of calcite crystals. Haüy is known as the "father" of geometrical crystallography. A great advance in mineral studies was made in 1828 when Nicol invented the nicol prism for investigating the behavior of polarized light in crystallized minerals. In 1912, Max von Laue, a student of Roentgen, theorized that the wavelengths of x-rays and atomic spacing of crystals may be of the same magnitude. Laue found that the diffracted rays, when passed through a crystal, substantiated his theory. This discovery opened up an entirely new field of mineral research, i.e., *crystal chemistry*.

Origin of Minerals

Minerals are products of formation and deposition from Earth's natural open solution systems, as opposed to substances formed as products of closely controlled laboratory systems, and thus may vary in many instances from an exact chemical content and formula. Minerals crystallizing from such open solution systems utilize effectively the ions required to form the mineral, but may incorporate within that structure other nonessential ions foreign to it. The incorporation of those foreign ions may, under favorable conditions of chemical affinity, replace certain specific elements named in the given formula. Or there may be defects in the space lattice of the mineral, e.g., vacancies in the crystalline structure comparable to a classroom with X number of seats orderly arranged where, under ideal conditions, each seat would be occupied by a student. Owing to some extraneous circumstance one or more students may be absent, yet the total seating capacity and orderly arrangement would remain constant. Deviations of the first type may be a product of included impurities in the structure; of the second type, by occupancy of a foreign compatible element in the vacant space(s) of the host crystal structure. See "**Isomorphism and Diadochy**" later in this article.

In other minerals, the formula may vary within restricted limits, such as (Zn, Fe) S for sphalerite, where ferrous iron substitutes for zinc within the sphalerite structure. Two factors prevail in such chemical substitution within a given mineral structure: (1) reasonably comparable ionic radii (approximately, but not strictly limited to ($\pm 15\%$); and (2) maintenance of electrical neutrality of the compound. In the event the substituting ions have a different valence or charge, electrical neutrality may be obtained by an accompanying substitution elsewhere within the crystal structure. Basically, a mineral is a homogeneous inorganic solid with an ordered atomic arrangement, which places it in the category of a crystalline material and possessing a definite, though not a fixed chemical formula.

Natural systems strive toward a state of equilibrium when all component units attain their lowest energy level. The respective energy level of the elements within any mineral is dependent upon the physical environment, principally the temperature, pressure, and chemical substances present, at the time and place of its formation. Any later change in its environment may cause a change in the mineral's composition and form. Whatever the primary or intermediate environmental conditions may have been, the mineral as observed, represents its present equilibrium energy state or crystal structure.

Matter exists in three states — gaseous, liquid, and solid.* In the gaseous state, the elements move freely about within their environment, their only contact being haphazard collisions. Elements in the liquid state are in closer contact with each other, but still retain freedom of mobility. The solid state is characterized by the chemical elements combining under atomic bonds of various types and strengths into a structured system. For any element within the structure, there are equivalent elements in a definite three-dimensional crystalline pattern. Under restricted circumstances, two or more minerals may form in contact with one another in such a manner as to preclude their complete development as single crystals. Those minerals which develop fully and which exhibit well-formed polyhedral (many sides) facial planes are referred to as being *euhedral*. Those minerals that exhibit no facial planes are referred to as *anhedral*. Imperfectly formed crystals are referred to as *subhedral*. Crystalline minerals, where the crystallinity can be determined only by aid of a microscope, are said to be *microcrystalline*. Where the materials are so finely divided as to be discernible only by x-ray analysis they are referred to as *cryptocrystalline*. The rate of crystallization plays an important part in the resultant crystalline character of minerals.

Crystalline solids are bonded together by electrical forces, which originate in the constituent atoms. The position of the respective unit particles within the crystal structure are determined by geometric factors, along with considerations of electrical neutrality and lattice energy. One type of bond is produced by the lending or borrowing of electrons in an attempt to complete the outer electron shell of the atom. A stable compound is formed when the outer shell is complete and the mutual attraction is termed an *ionic bond*. For example, sodium with one electron in its outer electron shell lends that electron to a neutral chlorine atom with seven electrons in its outer electron shell. The result is a stable electron shell for each ion (charged atom) which then join by electrostatic attraction in a crystal structure to form sodium chloride, NaCl. Minerals so bonded are of moderate hardness, readily soluble, and poor conductors of electricity and heat.

The bond produced by the sharing of electrons by which a stable configuration is achieved is termed a *covalent bond*. For example, chlorine with seven electrons in its outer shell needs one additional electron to achieve stability. If its nearest neighbor is another chlorine atom, the two atoms combine in such a way that the one electron serves in the outer shell of each atom, thus achieving a stable configuration. Diamond is another example of this type. Carbon with four electrons in its outer shell mutually shares the four electrons with four adjoining carbon atoms, thus achieving

a stable configuration by completing the shell with eight electrons. This type of electron-sharing or covalent bonding is the strongest of all chemical bonds. Minerals so bonded are generally insoluble, possess a high degree of stability, and are nonconducting.

Metallic bonds consist of a structure of positive ions through which free electrons can drift. The structure of metals may be envisioned as a mesh of electrons that surround the atomic nuclei and bind them together. This electron mobility produces minerals of generally a low hardness and of high electrical and thermal conductivity.

The *Van der Waals* bond is a very weak attraction between atoms, usually between essentially neutral atoms or groups of atoms. These neutral and essentially uncharged structure units are held together within the crystal lattice by virtue of small residual charges on their surfaces. This type of bonding is rare in minerals. Minerals so bonded often display ready cleavage and low hardness. For example, the foliated scales of graphite are linked weakly together by Van der Waals bonds.

Every mineral is a product of the redistribution or recombination of its component chemical elements to form a stable substance. The process is known as *crystallization*. The process may involve *precipitation* of chemical elements from aqueous solutions at the earth's surface; or from siliceous melts (magmas) from the earth's interior. In either situation, the process is dependent upon the degree of concentration of the constituent chemical elements present and the temperature/pressure conditions. Precipitation from vapor also is possible. An example is the hot vapor, rich in sulfur dioxide, which is emitted from vents associated with volcanoes. Upon becoming exposed to the cooler atmosphere, crystal sulfur is deposited around those vents. Snow crystals are another example of precipitation from vapor.

Crystal Structure

Associated chemical units become systematically arranged in the crystal structure, which is constructed from a single motif that develops repetitively. The resulting three-dimensional array is called the *space lattice* of the crystal. The lattice or framework is defined by three directions and by the distance along those directions where the motif repeats itself. Because the units within the structure adhere to a strict arrangement, the external facial planes of a crystal represent the limiting surfaces of that growth and are an external expression of its internal atomic order. Crystals are formed, therefore, where constituent atoms or ions are free to combine in constant chemical proportions and are an expression of the environmental conditions that promote their formation.

Periodic repetitions of a space lattice cell in three dimensions from the original cell will completely partition space without overlapping or omissions. It is possible to develop a limited number of such three-dimensional patterns. Bravais, in 1848, demonstrated geometrically that there were but fourteen types of space lattice cells possible, and that these fourteen types could be subdivided into six groups called *systems*. Each system may be distinguished by symmetry features, which can be related to four symmetry elements:

1. *Symmetry with respect to a point.* If through a central point in a geometric figure, lines are drawn from a point on one side of the figure to a similar point equidistant on the other side, the figure is symmetrical to a point.

2. *Symmetry with respect to a plane.* A geometrical figure is symmetrical with respect to a plane when for each edge, solid angle, or face on one side, there is a corresponding edge, solid angle, or face on the other side of that plane. One side is, in fact, a mirror image of the opposite side. That plane is called a *plane of symmetry*.

3. *Symmetry with respect to a line.* If during a complete revolution of 360° about a given axis, a geometrical figure repeats itself in appearance two or more times, it is said to be symmetrical with respect to a line, or to an *axis of symmetry*. Possible axes of symmetry are twofold, threefold, fourfold, and sixfold. A rotation of onefold or 360° is equivalent to no rotation at all.

4. When a crystal is rotated about an axis and inverted about the central point and at that point repeats itself, it is said to have an *axis of rotary inversion*. It is a twofold axis of rotary inversion if the geometrical figure is rotated 180° and then inverted. Additionally, there are threefold, fourfold, and sixfold axes of rotary inversion possible.

---

* Traditional concepts. Details on more advanced concepts on the states of matter are given elsewhere in this volume.

Within each of the six crystal systems, there are specific *crystal classes*. Each class displays distinctive symmetry elements. There are 32 possible classes distributed among the six crystal systems. One of the crystal classes within each system possesses all of the symmetry elements that are characteristics of its space lattice cell. These are called the *holohedral* class of that system. Other classes within each system possess somewhat fewer symmetry elements and are called *merohedral* classes.

It is significant to note that in most geometrical situations regarding crystal study, the distinguishing characteristic is symmetry, not geometry. This is especially true in the case of a mineral such as pyrite (FeS_2) which assumes a cubic shape, but its symmetry, controlled by the three sets of opposing striations on its crystal faces, identify it as belonging to a much lower order symmetry than that of a true cube.

The six crystal systems are identified by hypothetical lines of reference known as *crystallographic axes*, and their angular relationship to each other. Crystal orientation and axial order are given as: (1) front to back; (2) right side to left side; and (3) top side to bottom side. The front, right, and top side axial ends represent the positive ends of those respective axes. The back, left, and bottom side ends are designated as the negative ends. The relative intensity of crystal growth along those axes gives to each crystal face a distinctive identifying character. This character is evidenced by the physical similarity of equivalent facial planes on the crystal and their relationship to the crystallographic axes of that form. The six systems, their crystallographic axes, and interaxial angular relationship are defined and illustrated in the Fig. 1.

Crystallographic identification of facial planes on a crystal become possible through assignment of numerical values to a face which represents its relationship to the crystal axes. *Parameters*, or relative intercepts, are obtained by plotting coordinates of crystal faces with respect to their crystallographic axes. The actual distances of axial intercepts of the crystal face are determined and expressed as a unit of measurement. The product of these values is known as *Miller indices*.

A Miller indices face on the front face of a cube would be (100), signifying that face intersects the a^1 axis at 1 unit length from the center of the crystal, and is parallel to axes a^2 and a^3, or intersects those axes at infinity. In this system, zero (0) is the numerical substitute for infinity. A (111) Miller indices face identifies that facial plane as intersecting each of the three crystallographic axes of that form at 1 unit length from the crystal center.

The Miller Indices identify the orientation of a face in relationship to its axes of reference regardless of its size and position on the crystal. Haüy first proposed this basic law of crystallography — "crystal faces make simple rational intercepts on suitable crystal axes." Inasmuch as the intercepts are simple, it follows that the Miller indices should likewise be simple whole numbers.

See also **Crystal**; and **Miller Indices**.

Crystal Forms

Form is used here to designate the general outward appearance of a crystal, specifically to a group of crystal faces that bear identical relationships to the crystal's symmetry elements. It is essentially a geometric form with

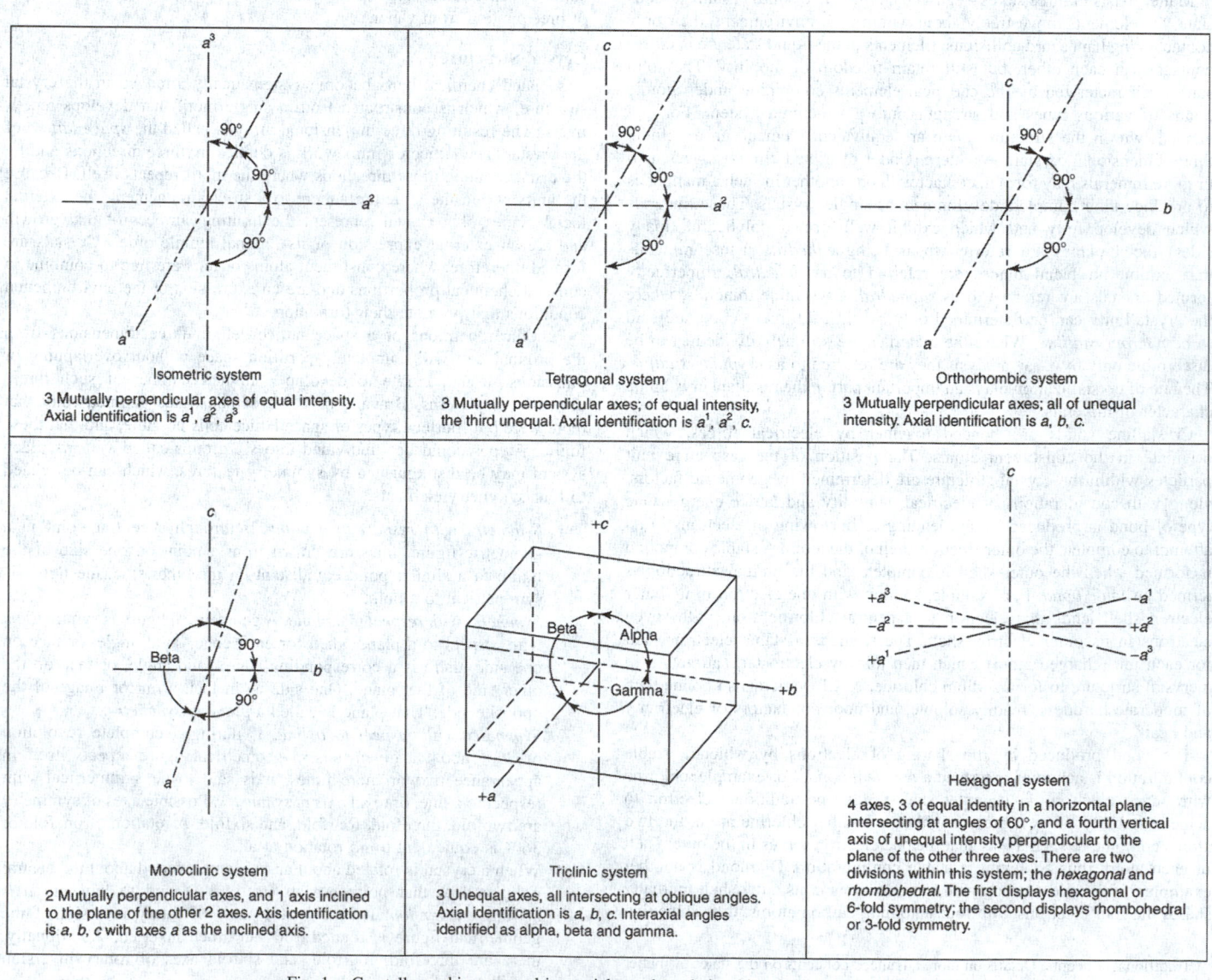

Fig. 1.   Crystallographic axes and interaxial angular relationship of six crystal systems

equivalent facial planes in their relationship to the crystal lattice symmetry elements. In this regard, an octahedron in the isometric system would be identified as a "closed form," inasmuch as its eight equivalent faces totally enclose the crystal space; its form identification would be {111}, enclosed in braces. Even though the crystal faces may be equivalent they may vary widely in size and distance from the crystal center, owing to irregular or distorted development during their formation. Braces shown around the Miller indices, e.g., {111} signify form identification, as opposed to Miller indices shown in parentheses, e.g. (111), which identify a specific crystal face only. In crystals possessing different lengths of their crystallographic axes general form {*hkl*} would be used. The hexagonal system with four axes requires a four-unit form identification.

In the isometric system, crystals are recognized by their geometrical forms, e.g., cube, octahedron, trapezohedron, pyriteohedron, and so on. In all other systems, crystal faces are given specific names, which refer to their relationship with their respective crystallographic axes. The most common facial forms in these systems are *pedion, pinacoid, dome, prism,* and *pyramid.* Crystal forms bear specific relationships to symmetry axes or planes, and not to the general shape of the crystal.

Physical Mineralogy

This aspect of mineralogy is concerned with several observable physical characteristics, such as color, hardness, lustre, fracture, cleavage, magnetic properties, radioactivity, fluorescence, and specific gravity.

Color. The color of a mineral is a product of selective absorption of certain wave lengths of visible white light by atoms within the mineral structure. A mineral color may be indicative of a species, but more often is of descriptive value only. There are two broad classification categories of color in minerals—*allochromatic* and *idiochromatic.* Allochromatic minerals are those which occur with variable colors, such as quartz, SiO_2; corundum, Al_2O_3; and calcite, $CaCO_3$. Color in such minerals may be a product of included foreign elements, rate of crystallization, or a defective lattice structure. Idiochromatic minerals are those in which the color is a characteristic constant, such as the green of malachite, $Cu_2CO_3(OH)_2$; the blue of azurite, $Cu_3(CO_3)_2(OH)_2$; the black of magnetite, Fe_3O_4; and the red of cinnabar, HgS. The actual color of most minerals can be obtained by rubbing the mineral on an unglazed porcelain plate and observing the powder (streak) of that mineral left on the plate. This test is essential in certain instances inasmuch as many minerals display a physical color foreign to their actual color, e.g., hematite may appear black or blue, but its streak color is always red.

Hardness. The hardness of a mineral is its resistance to scratching. Testing for hardness is based upon the premise that all minerals possess such resistance to a lesser or greater degree. The Mohs scale of hardness has been universally adopted to test this physical property. In the list below, the ascending numeric order of the mineral named will scratch all of those of lower order.

1 Talc
2 Gypsum
3 Calcite
4 Fluorite
5 Apatite
6 Orthoclase
7 Quartz
8 Topaz
9 Corundum
10 Diamond

In a relative sense, minerals from 2 to 2.5 hardness can be scratched by a fingernail; 4 by a penny, and 5 to 6 by a knife blade or piece of glass.

This scale is strictly relative and nonlinear because there is a much wider differential, for example, between corundum and diamond than between topaz and corundum. Hardness also varies according to crystallographic direction in certain minerals. This test should always be made on a freshly broken area because certain minerals are subject to surface alteration. A hardness test on an altered area can be misleading.

TABLE 1. HARDNESS OF REPRESENTATIVE MINERALS

(Mohs Scale)			
Agate	6–7	Galena	2.5
Alabaster	1.7	Garnet	6.5–7
Andalusite	6.5–7.5	Graphite	0.5–1
Aragonite	3.5	Kaolinite	2.0–2.5
Asbestos	5.0	Magnetite	6
Barite	3–3.5	Marble	3–4
Beryl	7.8	Mica	2.8
Corundum	9	Opal	5.5–6.5
Diatomaceous earth	1–1.5	Pumice	6
Dolomite	3.5–4	Pyrite	6–6.5
Emery	7–9	Serpentine	3–4
Flint	7	Tourmaline	7–7.5

The hardness (Mohs scale) of several minerals is given in Table 1.

LUSTRE. The lustre of a mineral is a product of both light reflection and refraction. Reflection is the governing factor in translucent to opaque minerals; refraction in transparent minerals. There are two broad categories in describing lustre—*metallic* and *nonmetallic.* Minerals of nonmetallic lustre are further subdivided into categories, such as vitreous (glass); adamantine (diamond); resinous (sphalerite); silky (asbestos); waxy (chalcedony and opal); greasy (some quartz and diamonds); and pearly (talc or mica).

Fracture. The product of irregular breaking of a mineral is termed fracture. Categories include conchoidal or conch-like (quartz); uneven (serpentine); and hackly (copper). Fracture in a mineral is unrelated to its crystal structure.

Cleavage. The cleavage of a mineral is the product of regular breaking of the mineral along specific surfaces related to the mineral's internal structure. The cleavage plane is always parallel to a possible crystal face and, therefore, is a reflection of the atomic structure of the mineral. Three planes of cleavage are present in the cube and rhombohedron. Four cleavage planes produce an octahedron, six planes a dodecahedron, both of which fall within the isometric system.

Magnetic Properties. A few minerals possess the property of being attracted by a magnet and they are known as *ferromagnetic* minerals. The more common ferromagnetic minerals are magnetite, Fe_3O_4, and pyrrhotite, $Fe_{2-x}S$. Those minerals which are natural magnets are known as *lodestones.* Most minerals are affected to some extent in a magnetic field. The minerals that are repelled are known as *diamagnetic*; those that are weakly attracted are known as *paramagnetic.*

Radioactivity. Basically, this involves the spontaneous disintegration of uranium and thorium minerals. Both of these mineral species disintegrate at a steady rate that is completely unaffected by external chemical or mechanical conditions. The end-product of this disintegration is lead within the mineral structure. Geophysicists are able to determine the geologic age of a specific uranium or thorium mineral and its host environment by measuring the amount of lead present and computing the known time required to produce this amount. With the impetus of atomic energy, many previously unknown uranium minerals were uncovered and identified.

Fluorescence. This is the property of certain minerals to absorb ultraviolet radiation and convert that energy into visible light. The rays energize certain elements within the mineral, causing the excitation of electrons in the orbital shell. If the activating source is removed and visible light continues for a period of time following, the phenomenon is known as *phosphorescence.* The visible light continues until the electrons return to their normal orbital electron shell.

Specific Gravity. The density or specific gravity of a substance is the amount by which that substance decreases in apparent weight when first weighed in air, followed by weighing in water. The density is obtained by dividing the weight of the substance in air by the loss of its weight in water.

Isomorphism and Diadochy. *Isomorphism* is the substitution of an atom, ion, or radical for another within a mineral structure. The degree of substitution is controlled largely by two factors: (1) temperature, and (2) ionic size. Only ions of similar size will readily substitute for one another.

Under certain conditions, isomorphism may be either partial or complete. The spinel, garnet, and amphibole mineral groups represent a complete isomorphous series. Partial isomorphism is exemplified by chemical components displaying extensive ionic substitution at high temperature levels, as evidenced by the nearly complete random mixing of potassium and sodium in microcline $KAlSi_3O_8$ and albite $NaAlSi_3O_8$ feldspars. Reducing temperature causes segregation of their component content according to their respective radii, which then group together, forming two separate mineral phases. This phenomenon is known as *exsolution* and is shown by the mixed feldspar known as perthite, common in igneous and high-grade metamorphic rocks. Other isomorphous examples include calcite and siderite; magnesite and siderite.

Diadochy is a term applied to the substitution of foreign atoms/ions within the same space lattice of a crystal atomic structure. Diadochic substitution refers only to specific crystal structures, as substitution elements may be diadochic in one structure and not in another. For example, excess Al^{3+} (0.57kX) may substitute diadochically for Fe^{3+} (0.64kX) in the epidote structure, thereby producing clinozoisite; when within the dolomite structure, Fe^{2+} (0.83kX) substitutes for Mg^{2+} (0.78kX), ankerite is the product; marmatite (ferroan sphalerite) results when Fe^{2+} (0.83kX) replaces Zn^{2+} (0.83kX) within the sphalerite structure. Diadochy involves the actual substitution/replacement of a given compatible element for another within a crystal structure, which, as in isomorphism, may be either partial or complete. In like manner, high temperature contributes materially to such substitution.

Polymorphism

Polymorphism is the capability of a substance to exist in more than one crystal form. The basic controlling factors appear to be temperature-pressure conditions at the time of formation, which controls the type of atomic packing within the structure. Examples of polymorphism are given in Table 2.

TABLE 2. REPRESENTATIVE POLYMORPHIC FORMS>

Chemical Composition	Mineral	Crystal System	Specific Gravity
Carbon	Diamond	Isometric	3.5
Carbon	Graphite	Hexagonal	2.2
Al_2SiO_5	Sillimanite	Orthorhombic	3.2
Al_2SiO_5	Kyanite	Triclinic	3.7
FeS_2	Pyrite	Isometric	5.0
FeS_2	Marcasite	Orthorhombic	4.9

Pseudomorphism

Pseudomorphism is the change of the original chemical composition of a substance into some other equally definite compound by the action of natural agencies. Pseudomorphism exists when the external crystalline form of a mineral is inconsistent with its internal chemical composition and atomic structural arrangement. It is always a secondary process. The altered substance is known as a *pseudomorph*.

Types of pseudomorphic alteration include:

Substitution This is a process whereby silica replaces wood fiber to form silicified (petrified) wood. Quartz replacing fluorite is another example. In the latter instance, the original fluorine in the fluorite is removed by silica-rich solutions that first remove the fluorine and then substitute silica in its place.

Incrustation This is a process whereby one mineral forms a crust over another mineral, e.g., prehnite over anhydrite crystals. Later solutions may remove the anhydrite, but the space occupied by the anhydrite remains as a cast surrounded by the prehnite.

Alteration results from the addition of new material or partial removal of original material from a mineral, e.g., anglesite after galena; or gypsum after anhydrite.

Paramorphism results when the internal crystal structure of a mineral is changed to a polymorphous form, yet retaining the external crystal form of the original mineral, e.g., aragonite after calcite.

Twinning

Twinning in crystals results from the intergrowth of two or more individuals in such a way as to yield parallelism in the case of certain parts of the different individuals and, at the same time, other parts of the different individuals are in reverse positions in respect to each other. For example, an octahedral crystal of magnetite is twinned when one-half of the crystal is rotated 180° parallel to an octahedral facial plane. This type of twinning is known as *spinel twinning*, owing to its common occurrence in the spinel group of minerals.

Mineral aggregates are often grouped together to form compound crystal structures. Such groupings may be a product of irregular and accidental growth which do not conform to basic twinning laws. Two or more intergrown crystals should not arbitrarily be labeled as twins unless their twin relationship can be established.

Rock Types, Associated Minerals and Their Uses

Minerals are the basic building blocks of all rock types found in the earth's crust. Rocks are classified into three broad categories: *igneous, sedimentary*, and *metamorphic*.

Igneous rocks have their origin deep within the earth's interior from molten magmas of siliceous (silica-rich) melts. The rocks resulting from deep-seated solidification are known as *plutonic*; while those that have been extruded on the surface as lava flows are termed *extrusives*. Plutonic rocks are products of slow cooling and possess well-formed crystalline structure, e.g., granite. Extrusive rocks are products of fast cooling and occur as glassy or microcrystalline masses, e.g., obsidian and basalt.

Minerals occurring in igneous rocks possess crystalline character, but their rate of precipitation from the parent melt prevented their development as euhedral crystals and thus they occur as granular (anhedral) aggregates within these rocks. The more prominent component minerals include quartz, feldspar, and feldspathoid family members, micas, pyroxenes, amphiboles, and olivine. Zircon, magnetite, ilmenite, hematite, apatite, pyrite, and garnet are commonly associated in these rock types.

Pegmatites represent a residual phase of igneous depositions, characterized by extremely coarse crystalline material, that results from the presence of associated volatiles, e.g., water vapor, carbon dioxide, sulfur dioxide, and others, which decrease the viscosity and facilitate crystallization. Quartz, feldspar, and mica are the more common minerals found in this environment, but such bodies are also hosts for many rare minerals and several types of gem stones, e.g., beryl, tourmaline, and topaz.

Sedimentary rocks are products of deposition from either the mechanical or chemical breakdown of all preexisting rock types. Precipitation from aqueous bodies rich in soluble salts produces economically valuable beds of halite and gypsum. Weathering of iron-bearing rocks has produced extensive deposits of hematite. Limestone, chalk, and diatomite are products of biochemical precipitation. Minerals commonly associated with sedimentary rocks include calcite, galena, sphalerite, pyrite, marcasite, fluorite, barite, celestite, and quartz.

Metamorphic rocks are those that have undergone a reconstitution or redistribution of the chemical elements contained in the original formations to new mineral species. The process of change involves attaining a state of equilibrium of the constituent elements with the newly imposed environment. Chlorite, biotite, garnet, staurolite, andalusite, kyanite, and sillimanite, with ubiquitous quartz, are a few of the more common minerals associated with metamorphic rocks.

Minerals of Primary Industrial and Economic Importance

Economically valuable mineral occurrences are the products of particular types of worldwide geological formations that are, or have been, hosts to such minerals. In each occurrence, the minerals found represent the products of geological processes, which include:

1. The character and concentration of the mineral components within the source magmas from which the minerals were formed.

2. Secondary deposition from either percolating solutions or gaseous emanations from intrusive formations, causing chemical reactions within the intruded formations.

3. Precipitation from chemically supersaturated solutions.

4. Alluvial deposits resulting from the erosion weathering of the original host rocks.

Minerals that are a product of the first type of formation include the basic native metals, gold, silver, and copper.

Gold is primarily a product of deposition from ascending hydrothermal solutions associated genetically with siliceous-rich igneous rocks. Pyrite and other sulfide minerals are common associates within which the gold is often physically admixed. Surficial weathering of such deposits removes the sulfides, leaving free gold as a residual deposit. Erosion of these deposits results in alluvial deposits of placer gold, both as flakes and nuggets. Other characteristics and the uses for gold are described under **Gold**.

Silver occurs both as native ore and in combination with various silver sulfide minerals. Native silver is predominantly a product of primary deposition from hydrothermal solutions. Minor occurrences are products of oxidation of silver sulfide minerals with which the native ore is secondarily associated. See also **Silver**.

Native copper commonly occurs in the oxidized zones of copper deposits in association with cuprite, malachite, and azurite. The native copper deposits on the Michigan Keeweenaw Peninsula represent an exceptional occurrence. The copper occurs there as veins within igneous trap rocks interbedded with conglomerates. See also **Copper**.

Similar and valuable minerals include cinnabar (mercury); antimony (for type metal and battery plates); galena (lead and silver); argentite, pyrargyrite, and proustite (silver); sphalerite (zinc); chalcocite, chalcopyrite, bornite, malachite, azurite, and cuprite (copper); nickeline and pentlandite (nickel); bauxite (aluminum); magnetite, hematite, and goethite (iron).

Diamonds were found originally as loose crystals in geologically ancient alluvial stream beds. Later, their host formations were found to be a basic igneous rock (kimberlite) in the Republic of South Africa. Diamonds are the products of extremely high-temperature, high-pressure environment and are composed of pure carbon. See also **Diamond**.

Trap rocks (basalts) are products of volcanic action, either as extensive lava flows, or as intrusive dikes in preexisting rocks. Secondary mineralization within such rocks from circulating waters produces interesting suites of zeolitic minerals, such as analcime, heulandite, natrolite, stilbite, mesolite, and others.

These minerals possess the ability to exchange ions contained in the mineral structure for those in solutions. This facility promotes the use of zeolitic minerals (or their synthetic counterparts) as water softeners. Water rich in calcium (hard water), when passed in solution through a tank containing zeolites, loses the calcium ions by absorption in the zeolite structure, with substitution of calcium ions by sodium ions. A reverse process may be initiated with the sodium ions replacing calcium ions in the structure, thereby reconstituting the original zeolite composition. See also **Zeolite Group**.

Halite, gypsum, and anhydrite are products of precipitation from large bodies of supersaturated salt water. The salt and gypsum of commerce are derived from such deposits. See also **Gypsum**; and **Sodium Chloride**.

Landlocked inland seas and lakes become enriched with various soluble elements from waters draining into those basins. Sylvite and carnallite are valuable for their potassium content. They represent the final evaporation products of landlocked bodies of supersaturated sea water. Two famous localities are Stassfurt, Germany and near Carlsbad, New Mexico. Minerals formed from the evaporation of boron-rich waters include borax, kernite, colemanite, and ulexite. See also separate alphabetical entries for these minerals. The only known locality for kernite is the Mohave Desert in California, where a deposit of great extent exists — with potential reserves of millions of tons beneath the desert floor.

Pegmatites are valuable mineral sources. These formations represent the residual phase of igneous crystallization from magmas rich in siliceous content. As crystallization of their component elements proceeds, these magmas become increasingly enriched with volatile substances (mineralizers), such as water vapor, carbon dioxide, chlorine, fluorine, phosphorus,

and others. The volatiles reduce the viscosity of the residual magmas and facilitate crystallization, as previously mentioned. When the residual liquids are injected into cooler rocks, they crystallize from their peripheral borders inward. The great mobility of the constituents enhances the growth of large mineral crystals — a characteristic feature of pegmatite bodies. Beryl and spodumene crystals from pegmatites attain sizes in terms of feet and tons. Feldspar, quartz, and mica crystals of comparable character are not uncommon.

There are two genetic pegmatite types — *simple* and *complex*. Simple pegmatites are recognized by their coarse texture and normal granite components, e.g., quartz, feldspar, and mica. Pegmatites produce the feldspar of commerce and mica for industrial and commercial uses. See also **Feldspar**; and **Mica**. Complex pegmatites are characterized by the presence of rare elements, in addition to the normal feldspar, quartz, and mica. Such bodies are also hosts for many semiprecious gem stones, such as amethyst, rose quartz, topaz, tourmaline, beryl, and chrysoberyl. Many rare-earth minerals obtained from complex pegmatite minerals include columbite/tantalite (columbium or niobium; tantalum), lepidolite, triphylite, spodumene, ambylgonite (lithium), zircon (zirconium), and monazite (thorium oxide).

A most unusual pegmatite occurs near Ivigtut, Greenland. This consists of a cryolite with subordinate siderite, chalcopyrite, galena, and sphalerite. Cryolite is a fluoride of sodium and aluminum. For many years, cryolite was mined from this single occurrence for use as a flux in the electrolytic recovery of aluminum from bauxite, the major ore source of aluminum. Synthetic sodium aluminum fluoride essentially has replaced the need for natural cryolite. See also **Aluminum**; **Bauxite**; **Cryolite** and **Mica**

Quartz and tourmaline crystals once were commercially important for their piezoelectric properties as radio oscillation wafers and other electronic and instrumental uses. Synthesized quartz crystals have largely replaced the need for natural quartz for such applications.

Nuclear fission reactors are supplied with materials from uranium-bearing minerals of primary origin, e.g., uraninite/pitchblende, and other uranium-bearing minerals of secondary origin, e.g., carnotite, tyuyamunite, torbernite, and autunite.

Minerals of economic importance within sedimentary formations include, but are not limited to fluorite, barite, phosphorite, and oolitic hematite. Fluorite is utilized as a flux in steelmaking and when of high quality as lenses and prisms in the optical industry. Barite is an essential mineral used in gas- and oil-well drilling. Phosphorite, a product of chemical precipitation from seawater, when treated with sulfuric acid, produces superphosphate fertilizer. Oolitic hematite deposits of extensive size are important sources of iron ore.

Garnet is a common mineral component in metamorphic rocks. A major occurrence of this type is at the summit of Gore Mountain, North River, in Warren County, New York. The garnet is a composite of almandine and pyrope with a hardness exceeding that of most world garnets. Gore Mountain garnet retains sharp cutting edges even when crushed to submicron size, making it an outstanding abrasive. It is used extensively as an abrasive (garnet paper) and as a glass-polishing agent in the optical industry.

Titaniferous iron ores, represented by the mineral ilmenite, occur within crystalline metamorphic environments. These ores are the major source of titanium. See also **Titanium**.

Major sources of industrial (non-fuel) minerals are shown in Table 3.

Ocean Sources of Minerals

Mineral requirements for future world needs has focused increased attention to potential ocean resources. Major attention has been directed to petroleum resources. Associated with the petroleum are salt domes or bedded salt deposits, often with anhydrite and sulfur. Their potential is dependent upon development of economically feasible recovery methods. Not enough is known at this time about the origin of sulfur to satisfactorily predict precise occurrences. The Frasch process is presently being utilized in certain offshore deposits of this type. The economics of sulfur also may be affected by availability of large quantities of the element from Claus recovery units used in connection with the desulfurization of flue gases.

TABLE 3. MAJOR SOURCES OF INDUSTRIAL (NONFUEL) MINERALS

ALUMINUM
   Major source is bauxite (gibbsite and boehmite). Deposits are found worldwide, except in Antarctica. Major producers: Australia: Caribbean countries; Venezuela; Brazil; Indonesia.
CHROMIUM
   Nearly all chromium ores are found in the Republic of South Africa and Zimbabwe.
COBALT
   Zaire, Zambia, and Russia.
COPPER
   Chile, United States, Russia, Canada, Zambia, and Zaire.
GOLD
   Republic of South Africa: Russia, Brazil, and the United States (not extensive).
IRON
   Russia, the United States, Brazil, Australia, and China.
LEAD
   United States, Russia, Australia, and Canada
MANGANESE
   United States, Japan, and Western Europe. (Major reserves are in South Africa and Russia)
NIOBIUM (Columbium)
   Brazil and Canada
NICKEL
   Russia, Canada, Australia, and Indonesia.
PLATINUM-GROUP METALS
   Russia and Republic of South Africa.
SILVER
   United States. (Silver is mined in more than 53 countries.)
TANTALUM
   Thailand, Australia, and Brazil.
TITANIUM
   Russia, Japan, United States
ZINC
   Canada, Russia, Australia, Peru, and United States.

Source: U.S. BUREAU OF MINES.

Valuable deposits of detrital sands and lime muds occur on the continental shelves of many world areas. These can be recovered by dredging operations. Diamonds are presently being recovered by means of vacuum suction tubes from detrital subsea sands adjacent to the Orange River section off the south-western African coast. Dependent upon the nearshore geology, it is known that iron, copper, and coal deposits extend into the subsea areas. In several world areas (Scotland and Japan for coal; Finland and Canada for iron ore; the English coast for tin and copper) deposits have been mined from underground entrances from the adjacent land areas. Sphene and zircon, plus other heavy minerals, have been noted in Texas offshore sediments.

Phosphorite, a major source of phosphorus, is known to occur both as nodular masses and crusts on rocks in subsea areas. Although enormous amounts of phosphorite are accessible in relatively shallow water, marine phosphorites have not been economically competitive with terrestrial supplies.

Metallic sulfides of copper, zinc, and iron have been found in central oceanic rocks and muds under conditions that indicate their deposition from hydrothermal solutions. Such solutions, rich in carbon dioxide, leached metallic elements from both basic rock masses and sedimentary formations with which they came in contact. When such solutions ascended, with concomitant cooling, the minerals were precipitated in the overlying sediments.

Manganese and iron oxides occur as nodular masses in many subsea world areas. They presently are of more interest for their copper, nickel, and cobalt content than for the manganese. Most extensive occurrences are at great ocean depths, as much as 3,500 to 4,500 meters. Fullest exploitation of these deposits and the metallic sulfides will require not only additional technical knowledge for their initial recovery, but also for their refinement to a marketable form. Again, the economics depends

upon future demands for the metals as the continental deposits become depleted. Beyond these considerations, the persistent problem of ownership of oceanic resources must be solved as a condition of large-scale recovery. See also **Manganese**.

Lunar Rocks

Geological specimens collected on the *Apollo* lunar missions are indicative of an anhydrous igneous origin. There are three major rock types: (1) a potassium-rich basalt; (2) anorthosite; and (3) an iron, titanium-rich basalt. The first two types are prevalent in the highland areas; the latter in the maria terrain. They occur as crystalline vesicular masses, breccias, and regolithic mantle dust. The absence of an atmosphere and weathering processes on the moon has left the rocks and their component minerals unchanged through eons of time since their formation. Secondary mineralization, therefore, is generally absent and the rocks exhibit a rather limited mineralogy.

Lunar rocks differ in their chemical content rather than type of rock from their terrestrial counterparts. They consistently contain more titanium and chromium, less sodium, and most are richer in iron content. Lunar plagioclase, the major mineral component of anorthosite, is almost always the calcium-rich anorthite, $CaAl_2Si_2O_8$, indicating extreme magmatic differentiation. Lunar basalts are olivine-rich and have been found to be from 3 to 10 times richer in ilmenite (titaniferous iron) as compared with terrestrial basalts.

Clinopyroxene materials, which are common in terrestrial basalts, are well represented in lunar rocks. They include diopside, hedenbergite, johannsenite, aegerine-augite, spodumene, jadeite, augite, pigeonite, omphacite, and fassaite. The more prominent lunar mineral species noted include ilmenite, with rutile intergrowths in certain subfloor maria basalts, cristobalite/tridymite, and pyroxferroite, a mineral closely related in both structure and composition to terrestrial pyroxmangite. Accessory minor minerals include troilite, chromite, ulvospinel, apatite-whitlockite, potash feldspar, quartz, hafnium-rich baddeleyite, and perovskite. Two newly classified species have been recorded—armalcolite, an iron-magnesium titanate, named after *Apollo* astronauts *Arm* strong, *Al* drin, and *Col* lins; and zirkelite, an oxide of calcium, iron, thorium, uranium, with titanium, niobium, and zirconium. Euhedral iron crystals in a pyroxene-rich vug of recrystallized breccia were recovered on the *Apollo 15* mission.

Tiny translucent-to-opaque glassy spherules are prominent in the lunar regolith, within which ilmenite (as thin plates) with minor olivine are present.

Lunar mineralogy is generally analogous to that of terrestrial basalts, the major difference being the lack of oxygen during crystallization which has resulted in the presence of free iron and the exotic minerals, such as troilite, pyroxferroite, armalcolite, and zirkelite. The bulk mineralogy, however, is quite similar to terrestrial rocks with pyroxene, plagioclase, ilmenite, and olivine as the dominant minerals.

Phase equilibrium research of mineral solids has revealed vital information regarding their molecular structure. Application of knowledge gained from this research has extended into the fields of metallurgy, glass and ceramics, and a more adequate interpretation of mineral geology.

Mariner 9 fly-by of Mars revealed a surface terrain of massive blocks of tumbled character cut by ridges and graben-type troughs. Huge volcanic peaks dominate a pock-marked landscape. Extensive channels characteristic of concentrated erosive powers of torrential floods were also evident, as were braided stream systems emanating from what were resolved to be plateau-type elevations.

Viking spacecraft equipped with a hoe-type scoop and spectrometer analyzed the surface soil to be of a character suggestive of an igneous mafic rock origin, rich in magnesium and iron. The *Viking Landers'* spacecraft analysis of the surface soil chemical composition by x-ray fluorescence revealed a low SiO_2 concentration ($\sim 45\%$) with iron as Fe_2O_3 near 20%. Further analysis revealed that the regolith mantle soil consisted essentially of iron-rich clay mineral with iron hydroxide, and minerals of sulfate and carbonate content with approximately 1% water by weight. Magnets attached to *Viking's* scoop attracted magnetic material aggregates. It is quite probable that the magnetic material represents a component part of that regolith and possibly the soil is enriched by both magnetite (Fe_3O_4, color black) and maghemite (γ-Fe_3O_4, color yellowish-brown). The yellowish-brown surface color may be the product of thin coating of hydrated iron oxides, with nontronite/montmorillionite as the host soil.

Much remains to be resolved before final definitive answers can be given in this area of planetary investigation and evaluation.

Classification of Minerals

Minerals are classified in groups, according to their chemical composition, based upon the dominant anion or anionic group. The system works well in various ways. Generally, the dominant anion or anionic group brackets minerals of corresponding characteristics which tend to occur in quite similar environments. The dominant chemical subdivisions are:

Elements Minerals composed of uncombined chemical elements, e.g., Au, gold; Ag, silver; Cu, copper; although minor impurities may be present within the structure.

Sulfides Minerals composed of compounds of metals with sulfur.

Sulfosalts Minerals composed of compounds of semimetals with sulfur.

Halides Minerals composed of compounds of metals with fluorine, chlorine, bromine, and iodine.

Oxides and Hydroxides Minerals composed of compounds of the metallic elements with oxygen.

Carbonates Minerals composed of compounds of a metal with the carbonate radical CO_3.

Borates Minerals composed of compounds of a metal with the borate radical, BO_3.

Nitrates Minerals composed of compounds of a metal with the nitrate radical, NO_3.

Sulfates Minerals containing the sulfate radical, SO_4.

Chromates Minerals containing the chromate radical, CrO_4.

Molybdates Minerals containing the molybdate radical, MoO_4.

Tungstates Minerals containing the tungstate radical, WO_4.

Phosphates Minerals containing the phosphate radical, PO_4.

Arsenates Minerals containing the arsenate radical, AsO_4.

Vanadates Minerals containing the vanadate radical, VO_4.

Silicates This mineral classification encompasses the largest group of mineral species and includes most of the important rock-forming minerals, such as the feldspars, feldspathoids, pyroxenes, amphiboles, micas, olivine, and quartz. Silicon is the basic chemical element, as the name implies. The small silicon cation combines with four oxygens to form an SiO_4 tetrahedral structure. The SiO_4 formula leaves a net negative charge, which requires additional combinations with other tetrahedra or anions to effect a neutral balance. The type and degree of such tetrahedral combinations control the final structural character and act as a convenient classification of the silicate mineral family.

Subclassification of silicates are:

Nesosilicates, with each tetrahedron existing within the structure as isolated SiO_4 units.

Sorosilicates involve the pairing of SiO_4 tetrahedra. The shared-oxygen anion represents the link between these tetrahedra.

Cyclosilicates involve two oxygens from each SiO_4 tetrahedron combining with oxygen in adjacent tetrahedral units to form *ring* structures.

Inosilicates are the product of oxygen sharing between adjacent tetrahedra to form *single* or *double* chains. In the single-chain structure, two oxygens from each tetrahedra combine with adjacent tetrahedra. In the double chain, half of the tetrahedra share three oxygens, while the other half share only two.

Phyllosilicates involve the sharing of three oxygens in each tetrahedron with adjacent tetrahedrons to form *sheet* structures. Minerals in this classification are usually flaky in character and relatively soft.

Tectosilicates involve the sharing of all four oxygens in each tetrahedral unit with adjacent tetrahedrons to form a three-dimensional *framework* of SiO_4 units linked together. The product is a strongly bonded structure with a silicon-oxygen ratio of 1:2. The greater portion of the earth's crust is composed of minerals found within this classification.

Most minerals exhibit a variation in their chemical composition, with the exception of the elements (see preceding list). The substitution of one ion for another is common, since minerals crystallize in solutions of complex composition.

A full listing of all minerals described separately in this encyclopedia is given in Table 4.

Additional Reading

Arem, J.E.: *Color Encyclopedia of Gemstones*, 3rd Edition, Chapman & Hall, New York, NY, 1994.

Boyd, F.R. and J.J. Gurney: "Diamonds and the African Lithosphere," *Science*, **232**, 472–477 (1986).

Boyle, R.W.: *Gold: History and Genesis of Deposits*, Chapman & Hall, New York, NY, 1990.

Brierley, C.L.: "Microbiological Mining," *Sci. Amer.*, **44–53** (August 1982).

Brown, W.L., Editor: *Feldspars and Feldspathoids*, Reidel, Boston, 1984.

Campbell, A.N. et al.: "Recognition of a Hidden Mineral Deposit by an Artificial Intelligence Program," *Science*, **217**, 927–929 (1982).

Carmichael, R.S. and S. Robert: *Practical Handbook of Physical Properties of Rocks and Minerals*, CRC Press, LLC., Boca Raton, FL, 1990.

Cornelis Klein, C. and C.S. Hurlbut: *Manual of Mineralogy*, 21st Edition, John Wiley & Sons, Inc., New York, NY, 1998.

Crowson, P.: *Minerals Handbook: 1996–1997*, Groves Dictionaries, Inc., New York, NY, 1996.

Derry, D.R.: *A Concise World Atlas of Geology and Mineral Deposits*, John Wiley & Sons, Inc., New York, NY, 1981.

Dietrich, R.V., B.J. Skinner, and R. Vincent: *Cambridge University Press Gems, Granites, and Gravels: Knowing and Using Rocks and Minerals*, Cambridge University Press, New York, NY, 1990.

Glusker, J.P.: *Structural Crystallography in Chemistry and Biology*, John Wiley & Sons, Inc., New York, NY, 1982.

Goeller, H.E. and A. Zucker: "Infinite Resources: The Ultimate Strategy," *Science*, **223**, 456–462 (1984).

Golden, D.C., C.C. Chen, and J.B. Dixon: "Synthesis of Todorokite," *Science*, **231**, 717–719 (1986).

Hein, J.R.: *Siliceous Sedimentary Rock-Hosted Ores and Petroleum*, John Wiley & Sons, Inc., New York, NY, 1987.

Holland, H.D. and M. Schidlowski: *Mineral Deposits and the Evolution of the Biosphere*, Springer-Verlag, Inc., New York, NY, 1982.

Kahle, A.B. and A.F.H. Goetz: "Mineralogic Information from a New Airborne Thermal Infrared Multispectral Scanner," *Science*, **222**, 24–27 (1983).

Kelly, E.G. and D.J. Spottiswood: *Introduction to Mineral Processing*, John Wiley & Sons, Inc., New York, NY, 1982.

Lide, D.R.: *CRC Handbook of Chemistry and Physics 86th Edition*, CRC Press, LLC., Boca Raton, FL, 2005.

Meyer, C.: "Ore Metals Through Geologic History," *Science*, **227**, 1421–1428 (1985).

Nancollas, G.H.: *Biological Mineralization and Demineralization*, Springer-Verlag, Inc., New York, NY, 1982.

Nesse, W.D.D.: *Introduction to Mineralogy*, Oxford University Press, Inc., New York, NY, 1999.

Newton, R.C., A. Navrotsky, and B.J. Wood: *Thermodynamics of Minerals and Melts*, Springer-Verlag, Inc., New York, NY, 1981.

O'Reilly, W.: *Rock and Mineral Magnetism*, Chapman and Hall, New York, NY, 1984.

Ozima, M. and S. Zashu: "Primitive Helium in Diamonds," *Science*, **219**, 1067–1068 (1983).

Park, C.F.: *The Geology of Ore Deposits*, 4th Edition, W.H. Freeman Company, New York, NY, 1998.

Parker, S.P.: *McGraw-Hill Dictionary of Geology and Mineralogy*, The McGraw-Hill Companies, Inc., New York, NY, 1997.

Robinson, E.S.: *Basic Physical Geology*, 3rd Edition, John Wiley & Sons, Inc., New York, NY, 1991.

Rona, P.A.: "Mineral Deposits from Sea-Floor Hot Springs," *Sci. Amer.*, **84–92** (January 1986).

Sawkins, F.J.: *Metal Deposits in Relation to Plate Tectonics*, 2nd Edition, Springer-Verlag, Inc., New York, NY, 1989.

Sohn, H.Y. et al.: *Processing of Energy and Metallic Minerals*, American Inst. of Chemical Engineers, New York, NY, 1982.

Swanson, E.A., D.F. Strong, and J.G. Thurlow: *The Buchans Orebodies*, Geological Association of Canada, Toronto, Ontario, 1981.

Touloukian, Y.S. et al.: *Physical Properties of Rocks and Minerals*, Vol. **2**, The McGraw-Hill Companies, Inc., New York, NY, 1989.

TABLE 4. MINERALS DESCRIBED IN THIS ENCYCLOPEDIA

Arsenates	Oxides and Hydroxides	Silicates (Inosilicates)	Silicates (Sorosilicates)	Calaverite
Annabergite		Acmite-Aegerine	Allanite	Chalcopyrite
Erythrite	Alabandite	Actionolite	Clinozoisite	Chalcopyrite
Mimetite	Alexandrite	Aegerine	Epidote	Cinnabar
Scorodite	Anatase	Amphibole	Hemimorphite	Cobaltite
	Bauxite	Anthrophyllite	Lawsonite	Covellite
BORATES	Brookite	Augite	Prehnite	Galena
Boracite	Brucite	Babingtonite	Vesuvianite	Gersdorffite
Borax	Cassiterite	Bustamite	Zoisite	Greenockite
Colemanite	Cat's-Eye	Crocidolite		Hessite
Inyoite	Chromite	Cummingtonite	SILICATES (Tectosilicates)	Krennerite
Kernite	Chrysoberyl	Diallage	Agate	Mercasite
Ulexite	Columbite	Diopside	Amethyst	Millierite
	Corundum	Enstatite	Analcime	Molybdenite
CARBONATES	Cuprite	Glaucophane	Bloodstone	Nickeline
Aragonite	Diaspore	Hornblende	Cairngorm Stone	Orpiment
Azurite	Emery	Hypersthene	Cancrinite	Pentlandite
Barytocalcite	Fergusonite	Jade	Carnelian	Pyrite
Bastnasite	Franklinite	Jadeite	Chabazite	Pyrrhotite
Calcite	Gahnite-Zinc Spinel	Pyroxene	Chalcedony	Realgar
Cerussite	Geikielite	Rhodonite	Citrine	Skutterudite
Chalk	Goethite	Riebeckite	Danburite	Sperrylite
Dolomite	Hematite	Serandite	Desmine	Sphalerite (Biende)
Magnesite	Ilmenite	Spodumene	Feldspar	Stannite (Mineral)
Malachite	Limonite	Tremolite	Flint	Stibnite
Phosgenite	Magnetite	Uralite	Harmotome	Sylvanite
Rhodochrosite	Manganite	Wollastonite	Heulandite	Tetradymite
Siderite	Perovskite		Hyalite	Wurtzite
Smithsonite	Pitchblende	SILICATES (Nesosilicates)	Jasper	
Strontianite	Psilomelane	Andalusite	Lazurite	SULFOSALTS
Travertine	Pyrolusite	Chrondrodite	Leucite	Boulangerite
Witherite	Pyrophanite	Datolite	Natrolite	Bournonite
	Rutile	Dumortierite	Nepheline	Enargite
CHROMATES	Spinel	Fayalite	Opal	Geocronite
Crocoite	Tantalite	Forsterite	Perthite	Jamesonite
ELEMENTS	Tenorite	Garnet	Petalite	Polybasite
Amalgam	Thorianite	Kyanite	Phillipsite	Poustite
Antimony	Uraninite	Olivine	Pollucite	Pyrargyrite
Bismuth	Wad	Phenacite	Quartz	Stephanite
Copper	Zincite	Silimanite	Scolecite	Tetrahedrite
Diamond		Sphene	Sodalite	
Electrum	PHOSPHATES	Staurolite	Stilbite	TUNGSTATES
Gold	Amblygonite	Thorite	Tridymite	Scheelite
Graphite	Apatite	Topaz	Wernerite	Wolframite
Mercury	Autunite	Willemite	Zeolite Group	
Platinum	Lazulite	Zircon		VANADATES
Plumbago	Monazite		SULFATES	Carnotite
Quicksilver	Pyromorphite	SILICATES (Phyllosilicates)	Alabaster	Tyuyamunite
Silver	Torbernite		Alunite	Vanadinite
Sulfur	Triphylite	Apophyllite	Anglesite	
	Turquois	Asbestos	Anhydrite	OTHER MINERALOGICAL
HALIDES	Vivianite	Biotite	Antlerite	TERMS
Atacamite	Wavelite	Chlorite	Barite	Abrasion pH
Carnallite		Chloritoid	Brochantite	Carbonado
Chlorargyrite	SILICATES (Cyclosilicates)	Chrysotile	Celestite	Diamond
Cryolite	Axinite	Garnierite	Chalcanthite	Diatomite
Fluorite	Beryl	Glauconite	Epsomite	Clay
Halite	Chrysocolla	Kaolinite	Glauberite	Fuller's Earth
Sylvite	Cordierite	Lepidolite	Gypsum	Gangue
	Dioptase	Mica	Jarosite	Gem Stones
MOLYBDATES	Dravite	Muscovite	Polyhalite	Kimberlite
Wulfenite	Elbaite	Phlogopite		Peridotite
	Emerald	Pyrophyllite	SULFIDES	Tripolite
NITRATES	Euclase	Sepiolite	Argentite	Vitrophyre
Niter	Iolite	Serpentine	Arsenopyrite	
Soda-Niter	Liddicoatite	Talc	Bismuthinite	
	Tourmaline		Bornite	

Wills, B.A.: *Mineral Processing Technology: An Introduction to the Practical Aspects of Ore Treatment and Mineral Recovery*, Butterworth-Heinemann, Inc., Woburn, MA, 1997.

Web Reference

Mineralogical Society of America: http://www.minsocam.org/

ELMER B. ROWLEY F.M.S.A., Union College, Schenectady, NY

MINIMAX METHOD OF ESTIMATION. Suppose we wish to estimate a population parameter ϕ from a sample, and that we can specify the amount of loss we shall make if we adopt any value ϕ_i when the true value is ϕ_j. Denoting the sample values collectively by x, suppose also that we have some rule R that tells us which value of ϕ to adopt for any particular sample x. We can then calculate the expected loss or risk as a function of ϕ. For any given rule R we can find the maximum value of the risk, and the minimax estimate is provided by the particular rule R_0 that minimizes this maximum risk. (For application in **Game Theory**, see that entry.)

MINIMUM THERMOMETER. A thermometer that automatically registers the lowest temperature attained during an interval of time. The alcohol-in-glass minimum thermometer contains a dumbbell-shaped index that is kept on the bulb side of the meniscus by surface tension. The thermometer is installed in a horizontal mounting (*see* **Townsend Support**) so that as the temperature falls, the index is pulled toward the bulb and remains at the minimum point as the temperature rises. A bimetallic thermometer with a circular dial is also used as a minimum thermometer. A free index, mounted concentrically with and driven by the thermometer index, is held by friction at the minimum temperature.

See also **Maximum Thermometer**.

MINI-SUPERCELL. A convective storm that contains similar radar characteristics to those of a supercell (e.g., hook echo, weak echo region (WER), and bounded weak echo region (BWER), but is significantly smaller in height and width. See also **Hook Echo**. The diameter of the radar-detected rotation is 1–8 kilometers (0.6–5 miles). This is a relatively new storm type, the existence of which has been confirmed by data from the recently installed WSR–88D (Weather Surveillance Radar-1988 Doppler) radars in the United States. Mini-supercells occur in areas where the height of the equilibrium level is low, most often in the northern United States, but possibly under certain weather conditions in any area of the world. They are sometimes found in landfalling tropical cyclones.

See also **Weather Surveillance Radar-1988 Doppler (WSR-88D)**.

MINNOWS. See **Viviparous Topminnows**.

MINOR (Of a Matrix). If a matrix **B** is formed from a matrix **A** by striking out certain rows and columns, then **B** is called a minor of **A**. The matrix **C** formed by the deleted rows and columns is called the complementary minor of **B** in A. If **A, B, C** are all square, then $(-1)^k$ **C** is called the algebraic complement of **B**, where k is the sum of the indices of the rows and columns of **C**.

MINT FAMILY (*Labiatae*). The greater number of the 3,000 species of this family are herbaceous plants, widely distributed in the temperate regions. The family also includes a few small trees and shrubs, found mostly in the American tropics. The characters distinguishing these plants are so outstanding as to make identification easy. Commonly the stem is 4-angled, appearing square in cross section, with the leaves opposite, simple, and without stipules. The flowers are borne in racemes or more frequently in dense axillary cymes. The flowers are irregular, with the calyx composed of five sepals united to form a tube, and the corolla composed of fused petals which form a 2-lipped tube, the upper lip of three, the lower of two lobes. The stamens, either two or four in number, are inserted on the corolla tube. The pistil is composed of a bicarpellate ovary, each of whose carpels becomes constricted early in development to form a 4-parted ovary, from which a 2-lobed style arises. The fruit is commonly composed of four achenes or nutlets. The flowers of this family are mainly cross-pollinated by insects: in some species the corolla tube is short enough to allow bees to obtain the nectar located in a disk at the base of the ovary.

Pushing into the corolla tube to reach this nectar causes the pollen in the anthers to be shaken onto the insect's back, where it will be carried to the stigmas of another flower. Other species having longer corolla tubes are cross-pollinated by butterflies.

Many members of this family have volatile oils located in epidermal glands on the leaves, giving to the plants characteristic odors. Because of these volatile oils many of them are useful to man, some as condiments, some as perfumes, and some as drugs. Food products are rare in this family, the genus *Stachys* having species which form tubers, eaten in some European countries. *Salvia* species are widely grown because of the showy scarlet flowers, often accompanied by brightly colored bracts. The leaves of the garden sage, *Salvia officinalis*, are used as flavoring for poultry dressing.

The leaves of *Ocimum basilicum* (basil), *Thymus vulgaris* (thyme), and *Origanum vulgare* (majoram), as well as *Salvia*, are used for flavoring. *Rosamarinus officinalis* gives rosemary oil; *Lavandula vera* and other species, oil of lavender; and *Pogostemon patchouly*, oil of Patchouli—all these oils being used in making perfumes.

Certain species of Mints, commonly called nettles, are troublesome weeds, often with strong rank odors.

MIRAGE. An image formed when the atmosphere behaves as a lens. A mirage, derived from the Latin *mirari*, meaning 'to be astonished', is an optical phenomenon which often occurs naturally. Mirages have a very small angular extent compared with that of the sky: They are normally seen near the horizon and involve image displacements and distortions of less than half a degree. Consequently, even though visible with the naked eye, they are easiest to see with the aid of binoculars or when photographed with a telephoto lens. The name applied to a particular type of mirage is dependent upon the way in which the appearance of the image differs from that of the object. The simplest distinction for the observer is that between a mirage that exhibits but a single image and one showing multiple images. If there is only a single image, and if that image is displaced down from the position of the object, it is said that there is *sinking*; if up, *looming*. If the image appears vertically enlarged, there is *towering*; if vertically shrunken, there is *stooping*. Recognition of these states depends critically on one's knowledge or memory of the appearance of the scene in the absence of a mirage, because all that is seen is the image. However, the change is often so striking as to make classification fairly easy. Mirages are explained by the refraction of light through an atmosphere with a gradient of refractive index. The refractive index of air depends mainly on the molecular number density of air, but as the layer through which the majority of the refractive bending occurs is often fairly thin, this density variation is primarily dependent upon temperature. Indeed, it is a simple matter to associate a particular type of mirage with the shape of a temperature profile. Because the observer is located inside this atmospheric lens, the mirage can change its appearance markedly as a result of slight changes in position, say changing the height of the observer above a surface. See also **Atmospheric Optical Phenomena**; **Inferior Mirage**; **Lateral Mirage**; **Looming**; **Sinking**; **Stooping**; **Superior Mirage**; and **Towering**.

MIRA (Omicron Ceti). A star that has the distinction of being the first variable star initially announced as such. In 1596, the Dutch astronomer Fabricius noticed a star of about the third magnitude that had not previously been recorded. The star faded within a few weeks, but was again seen and recorded by Bayer, in 1603. In 1638, Holwarda, another Dutch astronomer, again observed the star, found it to disappear and to then return to visibility about 11 months later. At maximum brightness, the star is easily visible to the naked eye, having a magnitude of about 3.5; but at minimum, it can be seen only with a telescope of aperture greater than 1 inch (2.54 centimeters), because its magnitude is only about 9. In the near infrared, its magnitude changes very little.

Many determinations of the period of variability have been made, the time from maximum to minimum averaging about 330 days, with variations between times of successive maxima amounting to as long as one month. The diameter of the star is of the order of magnitude of 4.2×10^8 kilometers or large enough to contain the sun and all the members of the solar system in their orbits to beyond the planet Mars. Coupled with the variation in brightness is a variation in diameter amounting to about

5.1×10^7 kilometers, together with a temperature variation of roughly 500 K (900 °F).

The spectrum displays fascinating changes in the near infrared region, due primarily to the absorption bands of VO and TiO. See Fig. 1.

Fig. 1.   Spectrum of Mira. (*University of Michigan Observatory.*)

MIRROR NEPHOSCOPE.   A nephoscope in which the motion of the cloud is observed by its reflection in a mirror. Also called *reflecting nephoscope*, or *cloud mirror*. See also **Nephoscope**.

A representative instrument consists of a black mirror disk, engraved with special concentric circles calibrated in degrees, and mounted on a tripod stand fitted with leveling screws. An eyepiece is arranged so that it can be rotated about the center of the mirror and adjusted to various distances above the mirror surface. The observer orients the mirror so that its zero corresponds to true north, and then adjusts the eyepiece until the cloud is observed at the center of the mirror. The cloud's direction of motion is indicated by the azimuth at which the image leaves the mirror.

MIRROR NUCLIDES.   Pairs of nuclides, having their numbers of protons and neutrons so related that each member of the pair would be transformed into the other by exchanging all neutrons for protons and vice versa.

MIRRORS AND LENSES.   This article is chiefly concerned with spherical or plane reflecting and refracting surfaces. A spherical mirror may be treated as a 1-base zone of a spherical surface, the axis of which is the straight line passing through the center of curvature and the pole of the zone. When the diameter of the mirror is small compared with its radius of curvature, and when the rays make only small angles with the axis, so that spherical aberration may be neglected, such a mirror produces fairly sharp images which are easily calculated from the laws of reflection. Taking the pole O as origin and the mirror axis as X-axis (Fig. 1), and representing the radius OC by r, the image of a point $P_1(x_1, y_1)$ in the plane XY is the point P_2 whose coordinates, for either a concave or a convex mirror, are

$$\left. \begin{array}{l} x_2 = \dfrac{rx_1}{2x_1 - r} \\[3mm] y_2 = \dfrac{-ry_1}{2x_1 - r} \end{array} \right\}$$

r is $+$ or $-$ according as the mirror is convex or concave. If x_2 turns out positive, it means that the image is virtual.

In Fig. 1, r, x_1 and x_2 are all *negative*, as is y_2, so that the image is real and inverted. If the incident rays are parallel to the axis, the focus of the reflected rays, called the focal point of the mirror, is on the axis at $x = r/2$. For a plane mirror, $r = \infty$ and $x_2 = -x_1$, $y_2 = y_1$.

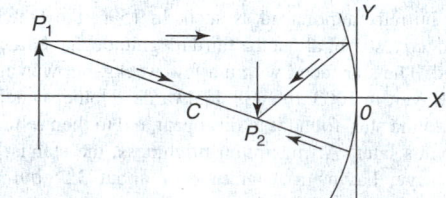

Fig. 1.   Formation of real image by concave mirror.

Spherical lenses have various combinations of convex, concave, or plane surfaces. There is always a point on the axis, called the optical center, such that if a ray in traversing the lens is in line with this point, the entering and emerging parts of the ray are parallel. For a very thin lens this point may be considered at the center of the lens and is a suitable origin. If the radii of curvature r_1, r_2 are large compared with the diameter of the lens,

and if the refractive index of the lens is n, the equations giving the image of the point x_1, y_1 made by a thin lens are

$$\left. \begin{array}{l} x_2 = \dfrac{r_1 r_2 x_1}{r_1 r_2 + (n-1)(r_2 - r_1)x_1} \\[3mm] y_2 = \dfrac{r_1 r_2 y_1}{r_1 r_2 + (n-1)(r_2 - r_1)x_1} \end{array} \right\}$$

r_1 and r_2 refer to the left and right surfaces, respectively, which is the order in which the light encounters them. P_1 and P_2 are called conjugate points of the system. The focal length is obtained by letting $x_1 + \infty$ and calculating x_2 from (2), which gives

$$f = \frac{r_1 r_2}{(n-1)(r_2 - r_1)}$$

This enables us to write (2) in simpler form:

$$\left. \begin{array}{l} x_2 = \dfrac{f x_1}{f + x_1} \\[3mm] y_2 = \dfrac{f y_1}{f + x_1} \end{array} \right\}$$

If x_2 turns out negative, the image is virtual; if y_2 is negative, it is inverted. For lenses of appreciable thickness or of strong curvature, the calculations are not so simple, and in general there are two unequal focal lengths, depending upon which way the rays pass through the lens.

The reciprocal of the focal length of a lens, called its "focal power," is a measure of the converging or diverging effect of the lens. It is commonly expressed in diopters or reciprocal meters; thus if the focal length is 50 centimeters or $\frac{1}{2}$ meter, the focal power is 2 diopters.

MISCIBILITY.   The ability of two or more substances to mix, and to form a single, homogeneous phase.

MISSILE PROPELLANTS.   See **Rocket Propellants**.

MISSILRY.   The art or science of designing, developing, building, launching, directing, and sometimes guiding a rocket missile; any phase or aspect of this art or science. This term is sometimes spelled missilery, but is then pronounced as a three-syllable word.

MISSISSIPPIAN PERIOD.   A geologic period in the Paleozoic Era. Type locality, Mississippi Valley. The period began about 280 million years ago and lasted for about 25 million years. The term Mississippian, first proposed by H.S. Williams in 1891, is roughly equivalent to the more general term, Lower Carboniferous. In Britain, the formations of this system are grouped under the terms Culm and Mountain Limestones, which, as in the United States, immediately succeed the Devonian and are followed by the upper Carboniferous, or Pennsylvania System (U.S.), and Coal Measures (Britain). The formations of this system are chiefly sandstones and shales in the Appalachian Geosyncline, representing delta and estuarine deposits of considerable thickness which pass Westward into thinner facies of marine shales and limestones. In the Rocky Mountain region occurs a great thickness of marine Mississippian called, locally, the Madison Limestone. The marine life of the Mississippian is chiefly characterized by echinoderms and foraminifera. Petroleum occurs in the Mississippian formations of Southeastern Ohio, West Virginia, Southeastern Pennsylvania and Eastern Kentucky.

MIST.   1.   A suspension in the air consisting of an aggregate of microscopic water droplets or wet hygroscopic particles (of diameter not less than 0.5 mm or 0.02 in.), reducing the visibility at the earth's surface to not less than 1 kilometer or 5/8 mi. The term mist is used in weather reports when there is such obscurity and the associated visibility is 1,000 meters (3,281 feet) or more, and the corresponding relative humidity is 95% or more, but is generally lower than 100%. These hydrometeors form a thin greyish veil that covers the landscape. It also reduces visibility, but to a lesser extent than fog.

2.   In popular usage in the United States, same as drizzle. See also **Drizzle**.

MISTRAL.   See **Winds and Air Movement**.

MITCHELL, PETER D. (1920–1992). A British biochemist who was the recipient of the Nobel Prize for chemistry in 1978 for his contribution to the understanding of biological energy transfer through the formulation of the chemiosmotic theory studies of cellular energy transfer. A graduate of Cambridge and recipient of many awards, he was the Director of Research, Glynn Research Institute.

MITE *(Arachnida, Acarina).* Minute animals related to spiders, ticks, and scorpions. Control chemicals used against these creatures are called *acaricides.* As is evident from the following descriptions, there are many species and varieties of the mites, with a remarkable degree of specialization exhibited. Also, many of the mites have shown outstanding capacity to develop genetic resistance to various control chemicals after long periods of exposure. Since the mites are so small, they are difficult to identify, but this is sometimes necessary to determine what chemicals may or may not be effective against a given infestation. Various species of mites are severely damaging to fruit crops, both deciduous and citrus, as well as certain field crops, such as alfalfa and clover. Further, the mites are serious pests to humans and domestic animals.

Citrus Mites

Among the most damaging mites on citrus are the red spider mite, the citrus rust mite, and the citrus bud mite.

Red Spider Mite (family Tetranychidae, many species). This is one of the most damaging pests on citrus and several other fruit crops. These insects are found in all stages all year long in the warmer citrus growing regions. Bright red eggs are deposited by females on the underside of leaves or in silken webs spun by the mite and located near twigs and fruit. The egg laying goes on for 2 weeks, the female laying two or three eggs per day. The total life cycle is from 3 to 5 weeks and is temperature-dependent. The larva has 6 legs and the protonymph and deutonymph have 8 legs. The adult mite ranges in color from purple to red and appears to be of a velvety texture. The six-spotted mite larvae are yellowish-green.

Citrus Rust Mite (Phyliocoptrula oleivora, Ashmead*); Citrus Bud Mite (Aceria sheldon,* Ewing*).* The citrus rust mite is the second (to purple scale insect) most severe economic pest on citrus in the Gulf States. The pest sucks sap from the leaves and fruit skins, with resulting russeting of oranges and silvering of lemons. This mite attacks orange, grapefruit, lemon, and lime. Sulfur is particularly effective in controlling the rust mite.

Citrus Bud Mite. This mite found in California is a more recent species. It attacks buds and blossoms, causing poorly shaped fruit. Oil emulsions are effective against bud mite.

Deciduous Fruit Mites

Several species of mite are injurious to deciduous fruit trees. They tend to specialize, but their life cycles are similar.

Brown Mite (Bryobia arborea, Morgan and Anderson*).* Particularly damaging to apple during dry seasons. Habilitates the foliage and sucks sap from buds and leaves, causing foliage to turn yellow. In some cases, the twigs of a tree may have so many tiny spherical red eggs attached that they will have a red aura about them. Until the late 1950s, this mite was confused with the *clover mite (Bryobia praetiosa,* Koch*).* The brown mite is found in Canada and in the northern and southwestern parts of the United States. Target trees of the mite include almond, apple, cherry, peach, pear, plum, prune, and walnut—as well as the raspberry plant.

European Red Mite (Panonychus ulmi, Koch*),* or *Paratetranychus pilosus).* A very significant pest of deciduous fruits in North America. The mite was first reported in the United States in 1911 and, although it occurs throughout the United States, it is most common in regions that lie north of 37 °N latitude (about San Francisco in the West; Saint Louis in the Midwest; and Richmond, Virginia in the East). Light invasions cause speckling of leaves. Heavy infestations cause paling and discoloration of foliage, causing leaves to drop. Fruit bud formation is difficult and fruit may be deformed, smaller than normal, and of poor color. All deciduous fruit trees can be affected; the mite is most severe against apple, pear, plum, and prune.

Pacific Mite (Tetranychus pacificus, McGregor*).* This species is often a severe economic pest of deciduous fruit, causing extensive webbing and discoloration of foliage. Fruit drops prematurely and is poorly colored. A thousand or more mites may be found on a single leaf. Fruits attacked include almond, apple, blackberry, grape, pear, plum, prune, and walnut. The mite also damages alfalfa, bean, clover, and cotton. The mite, about $\frac{1}{60}$ inch (0.4 millimeter) in length, occurs along the Pacific coast of North America, south to California from British Columbia. Adult females (summer) are of a green color and have brownish-black spots. Overwintering is by adult females (bright orange) that habitate in leaves, trash, or under pieces of bark.

Vegetable Plant Mites

Many species of spider mite cause extensive damage of certain vegetable crops. See Fig. 1. They are of the family *Tetranychidae.* Particularly active during hot, dry seasons, these mites cause damage to bean and a number of other vegetables. They tend to specialize. When infested, the leaves of a plant appear pale and progressively drop off. Prior to dropping, the leaves will appear yellow and red-brown in splotches of varying size. The undersides of the leaves take on a whitish powdery appearance. These whitened areas are made up of the wrinkled and empty skins of very tiny eggs, which are suspended by hardly visible silken threads. On the silk and on the leaf will be found many very small, eight-legged mites, ranging in color from white to green to red, and which are only about $\frac{1}{60}$ inch (0.4 millimeter) long. These mites pierce the leaves and consume the sap.

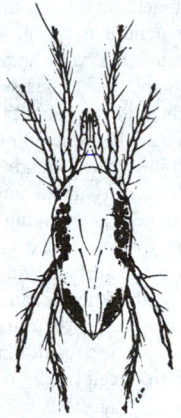

Fig. 1.   Adult female of the two-spotted spider mite. (*USDA.*)

The mites overwinter as females, usually of an orange color. Because there are so many species and the mites are so small, it requires an expert to identify a given species with certainty.

Control measures against the spider mite include mechanical removal of the intricate and continuously made webbing, spraying the plant with cool, clear water at frequent intervals, tasks that can be done more conveniently under glasshouse conditions and for plants that can withstand a wet soil. Control chemicals can be effective.

Processed Food Product Mites

These mites tend to specialize in certain substances as their sources of food.

Flour Mite (Acarus sino, Linne, or *Tyroglyphus farinae,* De Geer*).* Also known as the *grain mite,* this mite can be extremely troublesome to workers who handle flour, meal, and grain in various stages of processing. The itching resulting from bites is sometimes called "grocer's itch." Infestation in flour mills and grain-processing plants not only is a hazard to workers, but also severely damages product, sometimes requiring reprocessing or disposal.

Cheese Mite (T. catellanii, Hirst*).* Damaging to cheese in storage and often associated with the *cheese skipper,* description of which follows.

Cheese Skipper (Piophila casei, Linne*).* This insect infests and seriously damages or destroys cheese as well as smoked and cured meats. Distribution of the mite is believed to be worldwide. The adult skipper is a two-winged fly, about $\frac{1}{6}$ inch (4 millimeters) in length. Each female lays as many as 500 eggs, often in groupings of 40 to 50.

Ham Mite (family Tyroglyphus). This mite is very similar to the cheese mite except that it attacks smoked and cured meats. Curing temperature below a range of 30–36 °F (−1.1 to +2.2 °C) is required to prevent the presence of this mite in storage rooms. The mite is very adaptable in terms of its life cycle, and can inhabit a space for a long time without feeding. However, they do require a minimum relative humidity of 11%. Control is by fumigation with sulfur dioxide. Allethrin is also effective.

Mushroom Mite (Tyroglyphus lintneri, Osborn). Of an off-white coloration, this mite is about $\frac{1}{32}$ inch (1.5 millimeters) long. When present in large numbers, the mites produce a brownish powder which has an unpleasant, musty odor that is easily imparted to adjacent products. The mite is responsible for eating holes in the caps and stems of mushrooms and also for eating the spawn. Handlers also can be bothered with grocer's itch when handling infested products.

Animal Mange and Human Mites

Several species of mite produce serious problems for animals and humans. As with other mites, they tend to specialize.

Itch or Mange Mite (Sarcoptes scabiei, De Geer). This mite attacks humans, horses, hogs, and cattle. See Fig. 2. In animals, they cause *mange.* Several strains of this mite exist, each adapted best to a given host. The damage is caused by the female mite burrowing into the skin of its host and depositing eggs in the tunnels. A century or so ago, when the cause of the resulting itching was not known and treatments were not available, a number of generations of the mites could be produced in a given host, including a human being. This gave rise to the term "seven year itch." These mites are very small. The female is about $\frac{1}{60}$ inch (1.5 millimeters) long and the male is only half that length. The tunnels made by the female mite are about $\frac{1}{50}$ inch (0.2 millimeter) in diameter and about 1 inch (2.5 centimeters) long. In this tunnel, up to 24 eggs may be laid. The mite goes through an egg stage, a larva stage (6-legged creature), two 8-legged nymphal stages. Once the eggs are laid, the females prepare new tunnels and mating occurs within the tunnels. The main indication of the disease is extreme itching. Sometimes the tunnels can be seen just below the surface of the skin of the host. Common locations include the back of the knee, genitalia, and in between fingers and toes. Pimples and pustules form; they are subject to infections. The infested host can become quite ill as a result of the intense itching, concern, and stress, and the venom introduced by the presence of large numbers of mites.

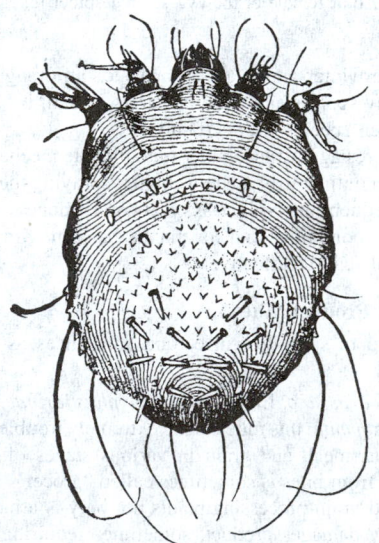

Fig. 2.   Female sarcoptic mange mite. (*USDA.*)

The infestation can be transmitted to another host because the creatures tend to crawl outside at night and a person occupying the same bed with an infested person has an excellent opportunity of also becoming infested. Until control methods take effect, isolation of the infested host is highly desirable. One control measure, the United States Army NBIN

formulation consists of: Benzyl benzoate, 68% (weight); DDT (or suitable substitute), 6%; Benzocaine (ethyl para-aminobenzoate), 12%; and sorbitan monooleate poloxyalkalene derivative (*Tween 80*), 14%. With exception of the head, about 2.5 ounces (75 cubic centimeters) of the diluted emulsion should be applied to the body. The compound should not be washed off for 24 hours. A second application may be required. Commercial preparations are also available.

In hogs, the mite can be diagnosed by first noting if hogs are scratching extensively and if their hair is standing erect. If there are no gray hog lice present, it is most likely that the hogs are infested with the mites. If so, inspection of the neck and back will show inflammation. The skin should be scraped to permit the mites to escape from the burrows. A close examination will show the 8-legged mites moving about.

The same mite attacks horses, hogs, mules, cattle, foxes, rabbits, squirrels, sheep and other common mammals. It is important to note that the resulting condition, mange, is highly contagious and transferable from one host to the next, regardless of species of host.

Poultry Mites

Poultry Mite or Chicken Mite (Dermanyssus gallinae, De Geer). This mite feeds on the blood of poultry during the night and usually leaves the fowl during the day. Since the presence of the mite cannot be detected by inspection during the day, the cause of an unhealthy flock may not be immediately traced to this mite. A telltale is the excrement of the mites, which appears like salt and pepper dusted about. Closer inspection will indicate the presence of many very tiny gray-and-brown mites, which appear as specks. The mite devitalizes its host, causing the fowl to be inactive, droopy, and listless. Some will cease laying eggs, and chicks and sitting hens may ultimately succumb to this persistent attack. Persons working in poultry houses also can be irritated by the mites. Control is essentially prevention — keeping the houses extremely clean. Where mites are suspected (or as prevention), the area can be sprayed with malathion. The birds can be sprayed with dilute formulations. Isolation of birds when moved from one house to the next is also suggested.

Northern Fowl Mite or Feather Mite (Ornithonyssus or Liponyssus sylviarum, Canestrini and Fanzago). This mite causes a drop in egg production and, if an infestation is severe, many fowl may perish. Eggs are laid in the fluff feathers of the bird.

MITSUNOBU REACTION.   Intermolecular dehydration reaction occurring between alcohols and acidic components on treatment with diethyl azodicarboxylate and triphenyl phosphine under mild neutral conditions. The reaction exhibits streospecificity and regional and functional selectivity.

MIXED CLOUD.   See **Clouds and Cloud Formation**.

MIXED LAYER (ML).   1.   Sometimes called *convective mixed layer, convective boundary layer,* or *mixing layer* in air-pollution meteorology. A type of atmospheric boundary layer characterized by vigorous turbulence tending to stir and uniformly mix, primarily in the vertical, quantities such as conservative tracer concentrations, potential temperature, and momentum or wind speed. Moisture is often not so well mixed, showing a slight decrease with height. The vigorous turbulence can be caused by either strong winds or wind shears that generate mechanical turbulence (called *forced convection*), or by buoyant turbulence (called *free convection*) associated with large thermals. The buoyantly generated mixed layers are usually statically unstable, caused by heating at the bottom boundary such as the earth's surface or radiative cooling at the tops of cloud or fog layers within the mixed layer. The terms mixed layer, convective mixed layer, and convective boundary layer commonly imply only the buoyantly stirred layer. During fair weather over land, mixed layers are usually daytime phenomena generated buoyantly, with growth caused by entrainment of free-atmosphere air into the mixed-layer top.

2.   In oceanography, a fully turbulent region of quasi-isopycnal water (i.e., virtually uniform potential density) that, in the case of the surface mixed layer, is bounded above by the air-sea interface and below by the transition layer.

Mixed layer depth is often defined as the depth at which potential density differs from that of the surface by 0.01 kg m^{-1}.

AMS

MIXED-LAYER VENTING.   Removal of pollutants out of the top of the atmospheric boundary layer through the mixed-layer capping inversion. Normally pollutants cannot escape through the capping inversion. However, penetrating cumulus clouds, thunderstorms, mountain circulations, and frontal circulations can force polluted air through the inversion to vent pollutants into the free atmosphere.

MIXER.   1. In a transmission, recording, or reproducing system, a device having two or more inputs, usually adjustable, and a common output, which operates to combine linearly, in a desired proportion, the separate input signals to produce an output signal. The term is sometimes applied to the operator of the above device.

2. In a superheterodyne receiver, the first detector (or transducer, heterodyne conversion).

MIXING.   1.   The result of irregular fluctuations in fluid motions on all scales from the molecular to large eddies. Gradients of conservative properties such as potential temperature, momentum, humidity, and concentrations of particles and gaseous constituents are reduced by mixing, tending toward a state of uniform distribution. See **Turbulence, Eddy flux, Diffusion**.

2.   In electronics, the nonlinear (nonadditive) combining of signals. The common mixing element is a diode or set of diodes. The common desired result of mixing two sinusoidal signals is the multiplicative product, with terms at the sum and difference frequencies. Mixing is used to shift signals to different carrier frequencies.

MIXING AND BLENDING.   These operations are important to chemical research and processing. In exploratory work in the laboratory the effects of mixing may be very great, and it is essential that the desired type and amount of mixing can be reproduced or can be varied by known amounts. The primary purpose of mixing is to distribute components as uniformly as possible; temperature distribution is frequently a major purpose. These may be followed by a chemical reaction or a transfer of matter between phases, and by a transfer of heat for temperature control. The mixer produces mechanical effects only. Molecules of themselves will diffuse, but mixing impellers produce flow which results in forced convection and mixing. Hence, reactants can be brought to an interface as rapidly as desired by controlling the fluid motion. Most fluid mixing is done by rotating impellers.

Both large scale (mass flow) motion and small scale (turbulent) motion are ordinarily required to bring about rapid mixing. The discharge stream from an impeller initiates the large scale flow pattern. Turbulence is generated mostly by the velocity discontinuities adjacent to the stream of fluid flowing from the impeller, and also by boundary and form separation effects. Turbulence spreads throughout the mass flow and is carried to all parts of the container. Some mixing operations require relatively large mass flows for best results, whereas others require relatively large amounts of turbulents. There is usually an optimum ratio of flow to turbulence for a desired mixing operation, whether it is a simple blending of immiscible liquids or a mass transfer followed by chemical reaction.

In the research laboratory it is important to recognize the effect of mixing on reaction rate or on other performance criteria. Energy must be supplied to produce fluid motion, thus, to compare mixing with different equipment or with different sizes of the same type impeller, it is essential that the comparisons be made on the basis of equal power input.

For the same power, the ratio of flow to turbulence from mixing impellers can be varied by changing the size and speed of the impeller. Figure 1 illustrates the differences in mass flow and turbulence which can be achieved for the same power input for dimensionally similar impellers. A large-diameter low-speed impeller produces a large ratio of flow to turbulence, whereas a small-diameter high-speed impeller will give a small ratio. Curve A, Figure 2, illustrates a reaction best accomplished by large flow and small turbulence. This curve, which is typical of blending operations, shows that the rate of blending increases to a maximum with

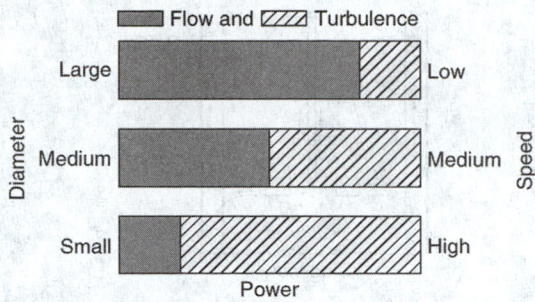

Fig. 1.   Constant power, effect of impeller size, and speed on flow and turbulence.

a large impeller as impeller diameter is increased (and impeller speed is decreased) with power input constant.

Curve B of Figure 2 is typical of gas-liquid contacting operations. Here the rate pf mass transfer between phases increases to a maximum at small impeller diameter and then decreases as impeller diameter is increased. The significance is that more turbulence is available with the small impeller and that turbulence is more important than flow in this operation.

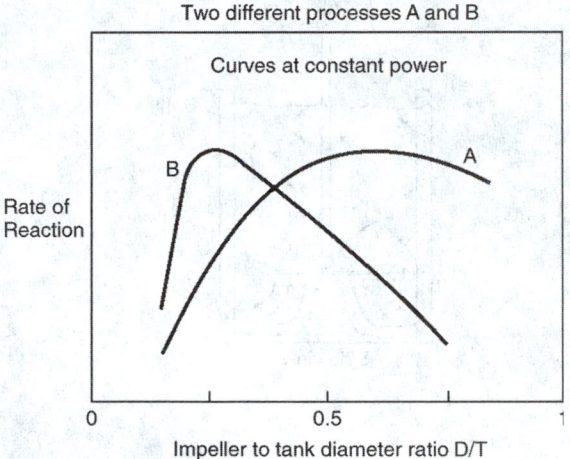

Fig. 2.   Effect of impeller size on reaction rate at equal power output

In all bench-scale and pilot-plant work where mixing is important, the effect of the impeller diameter-turbine diameter ratio should be determined so that the type of flow motion best suited to the operation can be found. If an optimum ratio is found, it becomes the basis for large scale design.

Mixing Vessels, Flow Patterns, and Impellers. Flow motion is dependent upon the shape and fitting of the container, the shape and position of the rotating impeller, and the physical properties of the fluid. The best mixing is usually one which produces lateral and vertical flow currents, and these currents must penetrate to all portions of the fluid; swirling motion should be avoided. Cylindrical vessels provide the best environment for mixing.

The most useful impellers are the simple flat paddle, the marine-type propeller, and the turbine. If any of these are on a vertical shaft rotating on the center line of a cylindrical vessel, the fluid motion will be one of rotation. A vortex forms around which the liquid swirls. A minimum of turbulence and of vertical and lateral flow motion will result. Very little power can be applied.

Rotary motion (and surface vortex) can always be stopped by inserting projections in the body of the fluid; when these are at the side if the tank they are called baffles, and this is the method most commonly used to obtain good mixing in large industrial equipment. The propeller with baffles will produce an axil flow pattern, Fig. 3, and the paddle and turbine will produce radial flow, Fig. 4.

Motionless Mixers. Mixing of molten polymers has been a problem in the plastics industry for many years mainly because of the high viscosity

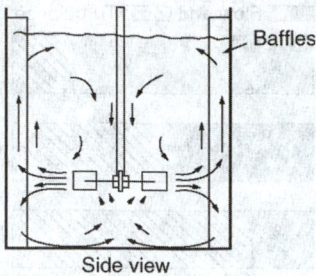

Side view

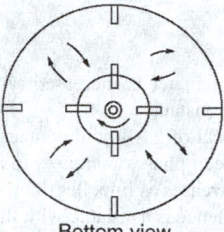

Bottom view

Fig. 3.   Typical flow patterns from axial flow impeller in baffled tank.

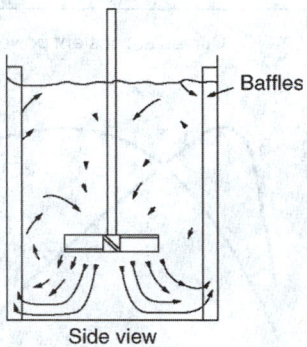

Side view

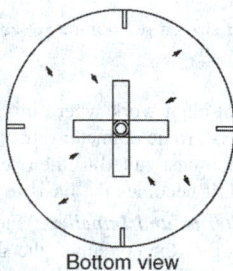

Bottom view

Fig. 4.   Radial flow pattern for flat blade turbine positioned on center in baffled tank.

of the melt. Blending of color concentrations, fillers, stabilizers, and other additives have sometimes been difficult to achieve efficiently with traditional mixing approaches, such as the extruder screw and other rotating devices. These problems have been partially solved through the application of so-called motionless mixers. A stationary baffle installed in a pipe can utilize the energy of the flowing fluid to produce mixing. Turbulent conditions are required for this simplistic approach. In recent years, more sophisticated motionless mixers have been developed. One design consists of a number of short right- and left-hand helices. The opposite hand helices are welded together so that their leading edges are 90 degrees to the trailing edge of the preceeding element. The mixing unit is housed in a tube or pipe for in-line installation. Materials entering the static mixer experiences flow division at the leading edge of each element. As the flow divides, it follows the semicircular channel of each element and repeatedly divides at succeeding junctions of additional elements, resulting in flow division and

radial mixing. Other approaches involving complex geometric stationary elements have been developed. Some of these units also can function for heating and cooling as well and mixing. [See "Motionless Mixers in Plastic Processing," Schott, Weinstein, and LaBombard, *Chem. Eng. Progress,* **71**(11) 52–58 (1975)].

Automatic Blending Systems. For a number of industrial applications, in-line blending of liquids and solids has replaced former batch-type operations. In these systems, all components flow together simultaneously to a central collection point where they combine to form the finished product. A modern in-line blending system is shown in Fig. 5. The blend controller will normally utilize microprocessor technology with a cathode-ray tube (CRT) display. Each fluid component is pumped from a storage tank, through a strainer, and then through a flowmeter, with the meter and valve carefully selected for prevailing process conditions (viscosity, temperature, pressure, flow rates). The signal from the flowmeter is fed to the blend controller which compares the actual flow rate to the desired flow rate. For minor components, such as dyes or additives, it is sometimes most practical to control the flow rate by means of proportioning pumps which inject a precise amount of the fluid when a pulse signal from the blend controller is received. This type of open-loop control is cost efficient, but some means for assuring flow (detecting any dry line) should be considered inasmuch as there is no direct fluid measurement device used. Other variations of measurement and control involve the use of variable-speed pump motor controllers (SCRs) for flow control; adding a flowmeter in series with an injection pump. Weigh-belt feeders with variable-feed/speed control and tachometer/load-cell outputs are frequently used for blending powders and aggregates.

A blend controller block diagram is shown in Fig. 6. A system for preparing bread and pastry dough is shown in Fig. 7. Applications for continuous blending systems are frequently found in the petroleum, petrochemical, food and beverage, building materials, pharmaceutical, automotive, and chemical industries, among others.

MIXING CLOUD.   See **Clouds and Cloud Formation.**

MIXING LENGTH.   1.   An average distance of air parcel turbulent movement toward a reference height, where the average is a root-mean-square distance. It is also known as Prandtl's mixing length, l, after Ludwig Prandtl who devised it in 1925 to explain turbulent fluxes such as the Reynolds stress, τ. Prandtl started with Boussinesq's first-order turbulence closure hypothesis that $\tau = \rho K(\overline{dU}/dz)$, where is density, $\overline{U}$ is average horizontal velocity, and K is kinematic eddy viscosity. He further recognized that exchange coefficient K has units of length times velocity, and proposed that $K = lw$, where w is a representative average turbulent vertical velocity. Prandtl also suggested that turbulent vertical motions are caused by the collision of air parcels moving horizontally at different speeds. This results in turbulent vertical velocity being proportional to turbulent horizontal velocity. From this, it can be shown that eddy viscosity can be approximated by $K = l^2|\overline{dU}/dz|$, which can be used in Boussinesq's first-order closure.

2.   A mean length of travel over which an air parcel maintains its identity before being mixed with the surrounding fluid; analogous to the mean free path of a molecule.

Additional Reading

Stull, R.B.: *Introduction to Boundary Layer Meteorology,* Springer-Verlag New York, LLC, New York, NY, 1988.

AMS

MIXING LINE.   A method of thermodynamic analysis where the conserved variables for two different states of air (i.e., air parcels) are plotted on a thermodynamic diagram, and the ending state of a mixture of the two parcels is found on the straight line connecting the two initial states, with relative distance along the line proportional to the relative amounts of each parcel in the mixture. Mixing line analysis can help determine the origin of air within clouds.

MIXING POTENTIAL.   The amount of turbulent mixing necessary to eliminate any static and dynamic instabilities in the atmosphere. Because atmospheric circulations that include turbulent eddies have finite velocities,

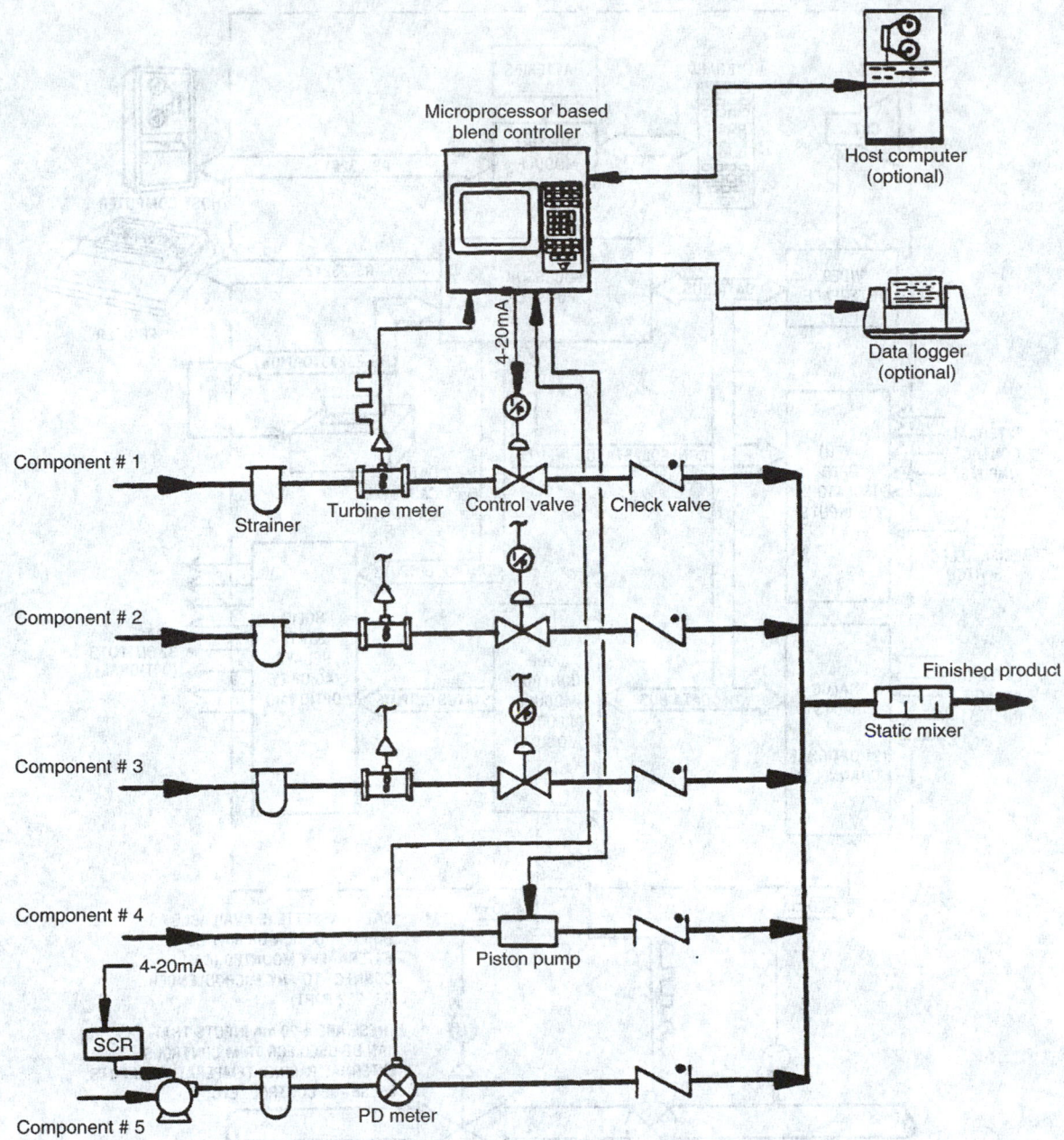

Fig. 5.   Typical blender configutaion. (*Waugh Controls Corporation.*)

the actual amount of mixing that can occur during a finite time interval might be less than the mixing potential. Some nonlocal turbulence models parameterize the amount of turbulence as proportional to the mixing potential, where local as well as nonlocal instabilities are considered.

MIXING RATIO. The ratio of the mass of a variable atmospheric constituent to the mass of dry air. If not otherwise indicated, the term normally refers to water vapor. For many purposes, the mixing ratio may be approximated by the specific humidity. Either r or w is commonly used to symbolize water vapor mixing ratio, with r used for thermodynamic terms in this glossary. In terms of the pressure p and vapor pressure e, the mixing ratio r is

$$r = \frac{0.622e}{p - e}$$

MIZAR (*ζ Ursae Majoris*). An interesting star in the Big Dipper, and probably the first double star ever observed. See also **Double Star**. Mizar forms with the fourth magnitude star Alcor a naked-eye double, and Mizar, itself, has a close companion that is telescopically visible. It was first observed as a double star by Riccioli, in 1650. Tradition says that observation of the pair Mizar-Alcor was considered a test of good eyesight among the American Indians. However, if this was a difficult pair for the Indians to separate, their eyesight could not have compared very favorably with that of modern times, because this is an easy double for most people.

As well as being the first visual double star to be discovered, Mizar also has the distinction of being the first spectroscopic binary discovered. In 1889, E.C. Pickering discovered that the spectral lines of this star were alternately double and single, a phenomenon that can be adequately explained only by the star being a close binary. In 1908, both the fainter companion of Mizar and the more distant, bright companion Alcor were found to be spectroscopic binaries.

M-JPEG (MOTION JPEG). See **Data Compression**.

MNG (MULTIPLE-IMAGE NETWORK GRAPHICS). See **Data Compression**.

MOAZAGOTL. See **Clouds and Cloud Formation**.

MOAZAGOTL WIND. See **Winds and Air Movement**.

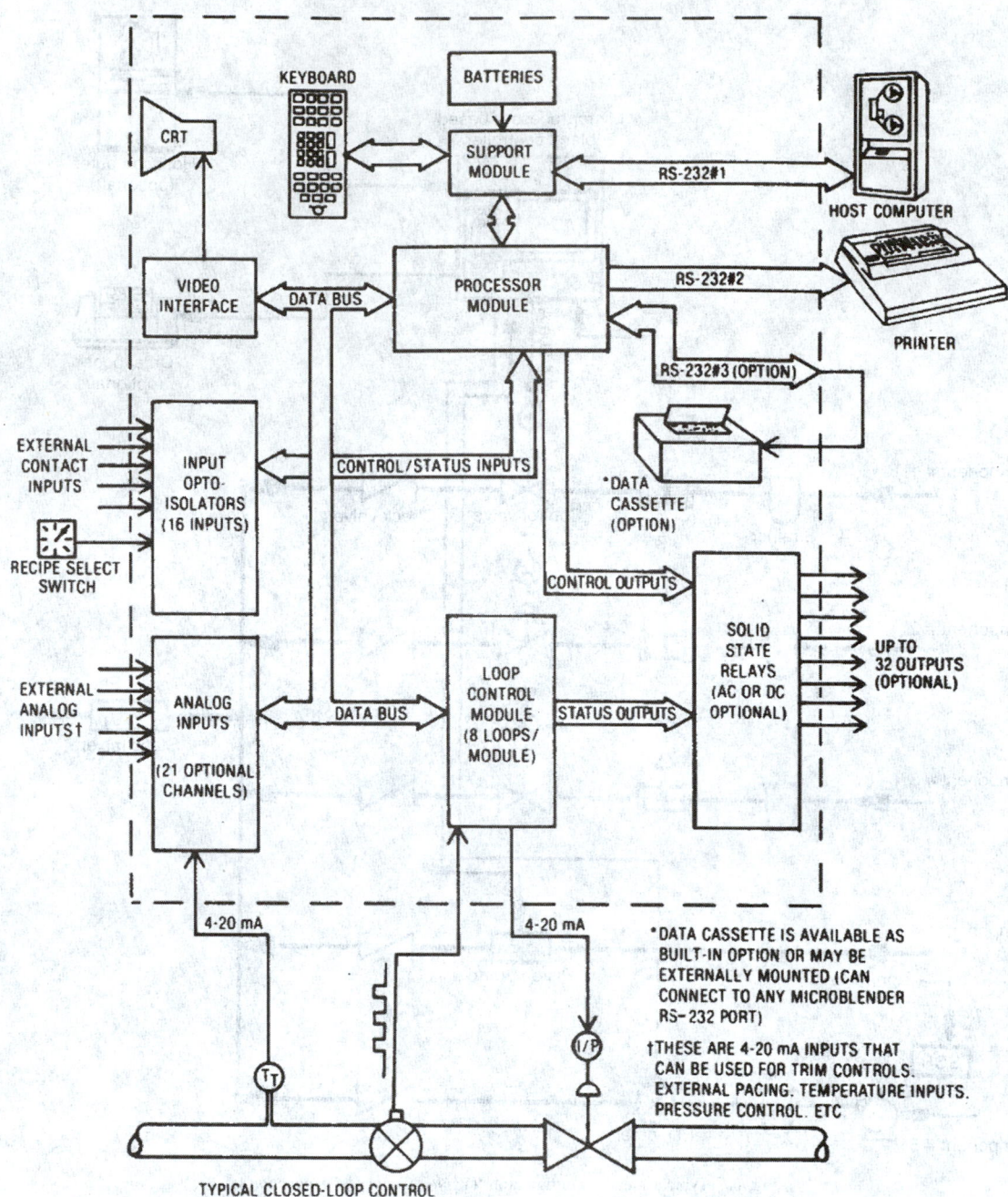

Fig. 6.   Blend controller block diagram.

MOBILE DEVICES AND PROTOCOLS.

The value of a mobile communication device is its ability to organize and access important information quickly. Mobile devices are often associated with the cutting-edge technology of the last 4 years. Although having gained an increasing popularity over recent years, mobile devices have a history of 15 years. The first generation of mobile devices was the analog mobile phone. They were large and cumbersome, resembling a small briefcase. With the launch of the second generation, mobile phone units became smaller and easier to handle. Currently, third-generation phones are starting to be released, advancing in design and dynamic capabilities. One area of mobile communications that has been growing exponentially is mobile computing.

Mobile Computing

Mobile computing merges cell phones and computer technology. Resulting from this merge are mobile devices that unite the mobile communications of the cell phone and the expanded functionality of a computer. One company specializing in this field is Symbol Technology, http://www.symbol.com/, which designs products that use radio frequency identification (RFID), infrared and wireless data transmissions. This article previews several devices and highlights the standard wireless communication protocols, so-called *smart phones*, which are beginning resemble PDAs and PocketPCs. Some of the IEEE wireless communication protocols that will be highlighted are the Institute of Electrical and Electronics Engineers (IEEE) Standards 802.11, 802.11b (Wi-Fi), 802.11a, and 802.11 g. A brief background on Bluetooth (a short range wireless connectivity standard) and wireless application protocol (WAP), as well as RFID and global positioning system (GPS), are also addressed in limited form.

It is important to distinguish the differences between mobile and nonmobile devices. The distinction is dictated by the use of a physical connection in the form of a cable, in which the device is connected to a network. Devices using a network cable are categorized as nonmobile. Conversely, devices that use wireless connectivity via a wireless modem are mobile.

Different Styles of Mobile Devices. The IEEE essentially defines a mobile device as a small portable electronic organizer, PDA, smart phone, clamshell, cell phone, laptop, or a PDA and phone combination that allows a

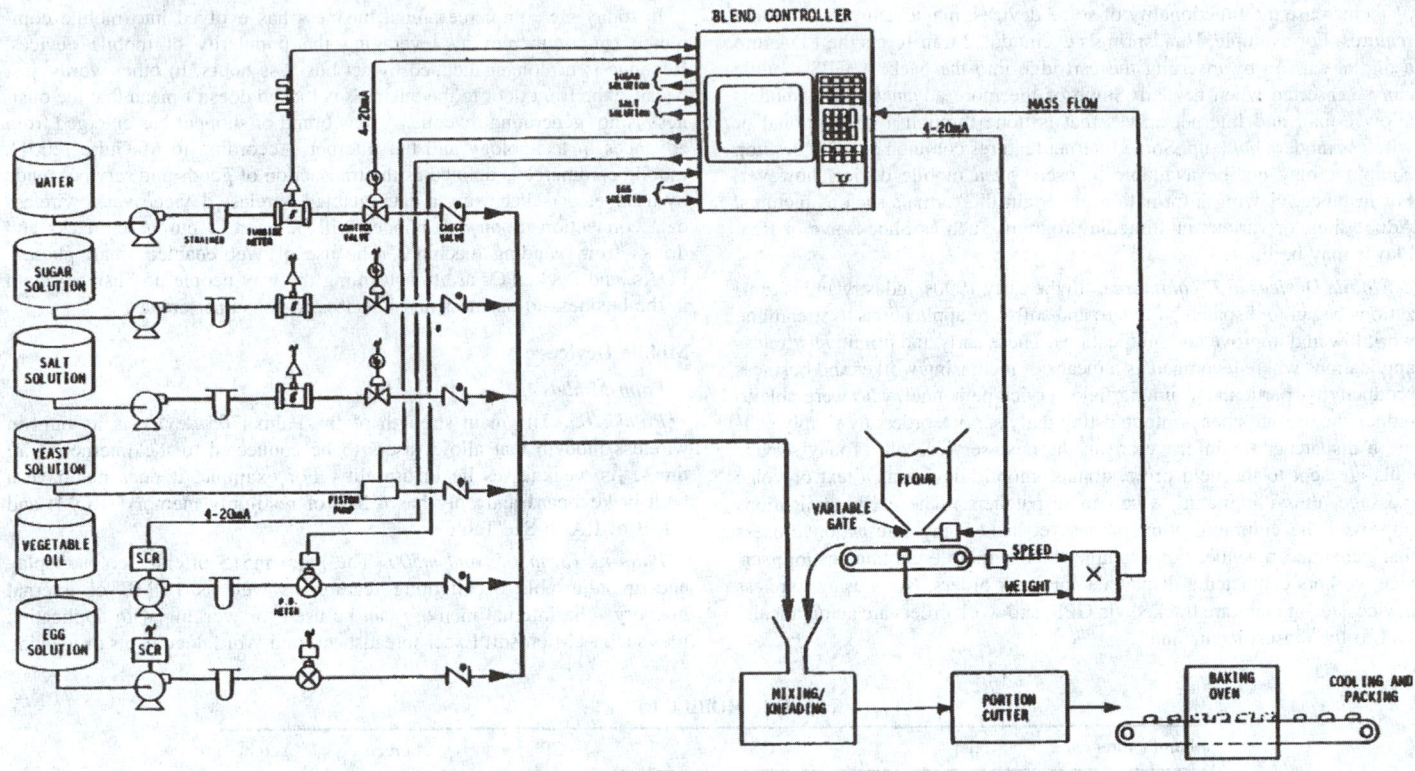

Fig. 7. Blending system for preparing bread and partry dough. (*Waugh Controls Corporation.*)

user to enter and manage data without a physical connection to a network. In addition, levels of information processing and data functionality are also used to classify mobile devices. Some high-end PDAs, like the Palm i705, have a color display. (The majority of PDAs have a monochrome or gray-scale display.) The current color range of the PDA and PocketPC models are anywhere from 256 to 64,000 colors. PDA screen displays are either active matrix or passive matrix. Active matrix (AM) liquid crystal display (LCD) uses transistors to control each pixel. AMLCDs are more responsive, faster, and can be viewed at larger angles than passive matrix displays, making them easier to read. Passive matrix displays use fine lines that intersect. The point of intersection is called an LCD element, which presumably either blocks or passes light through for viewing (IEEE, 2001). PDAs perform an information exchange or synchronization with a desktop computer by placing the PDA in a cradle, or docking station. The docking station connects directly to the desktop computer with a cable, and, using synchronization software, the data exchange is performed. Some models have infrared technology that can synchronize without the use of wires or a docking station. For infrared synchronization to occur, both devices must have the capability from a hardware and software standpoint. In general, the higher the central processing unit (CPU) speed, the faster the unit will execute tasks and the more expensive the unit. Nevertheless, higher speed processors require more battery power and may deplete batteries quickly. Some larger models use AAA alkaline batteries, and others use lithium-ion rechargeable batteries. Models typically have 2–32 MB of random access memory (RAM), and some models may have an expansion slot built into the device, allowing for a memory upgrade. In general, for PDAs,the degree of Web accessibility is not yet equal to what you gain from a full-fledged computer or PocketPC.

The Differences Between a PDA and a PocketPC.

Although both PDAs and PocketPCs are mobile, the pinnacle distinction can be found in the device's ability to engage software applications and manipulate data. Mobile computing devices allow you to store, organize, send, and receive messages as well as access information without a physical hard-wire connection. Mobile computing devices have a CPU, which allows them to receive data, perform a specific process with the data, and produce output. A PDA lacks general-purpose application software needed to perform data manipulation functions like that of a mobile computer. Although the early PDAs paved the way for mobile computing to reach its current level

of utility, today's PDAs are less functionally dynamic compared with PocketPCs.

Components and Data Input.

Entering data is done either through a scaled-down keyboard or through a penlike stylus. The stylus is primarily used in the writing area called the graffiti pad. Either rechargeable or disposable batteries power all devices, as dictated by the manufacture. PDAs primarily use a variation of the Palm OS, and a Motorola 33-MHz DragonBall CPU, whereas the PocketPCs use a StrongARM 206-MHz CPU and run a Windows OS. The amount of memory will vary with the make and model of the device. Optional are springboards and memory sticks. The average PDA screen size is 3.6 inches, with resolution ranging from 160 × 160 to 240 × 320 pixels. On PocketPC devices, the viewable screen averages 3.5 inches with resolution of 240 × 320−pixels. PDAs typically have between 2–32 MB of built-in memory. Often 2-MB memory is generally sufficient for maintaining address books, an active calendar, taking notes, and loading useful programs. More memory may be necessary for storing larger files, such as digital photos or audio recordings. Increasing the amount of memory can be accomplished with small flash-memory storage cards. These cards are inserted into slots in the PDA. The memory in most PocketPC models typically ranges from 32–64 MB. Adding smart memory cards in either PDAs or PocketPCs is advantageous when applications or files call for increased memory.

There are essentially three data-entry options available to both PDA and PocketPC users: the mini portable foldout keyboard, requiring a flat surface and thus limiting where the keypad can be used; the numeric keypad, which is identical to those on cell phones; and the stylus, which has similar functions to that of a mouse and is used for navigation.

Functions and Features.

Most PDAs operate on a Palm OS. Data input can be screen based, keyboard based, or both. More sophisticated devices, such as the PocketPC, usethe Windows-based OS (PocketPC). What distinguishes the PDA from the handheld or PocketPC is the lack of window's application software that is available and compatible for these devices. PocketPC is capable of running word-processing, spreadsheet, and money management applications and provides Internet and e-mail access. The Palm OS has some office management software, but it is not well-suited to Microsoft office applications. PDA devices aretypically less expensive than mobile computing models, because most lack advanced software applications.

To increase the functionality of some devices, manufactures add certain features. For example, Handspring's eyemodule2 transforms the PDA into a digital camera by inserting the cartridge into the back. A GPS module can be inserted when needed. Some of the more advanced PDA models allow e-mail and Internet access that is done through a conventional or wireless modem hook up. Some Internet features common on most desktop computer may not be available to users on a mobile device, however. For instance, viewing information in certain file formats such as pictures, Adobe files, or various multimedia programs such as Shockwave or Real Player may be limited.

Mobile Devices in E-commerce. In the early 1990s, industry and organizations began to dispatch hardware and software applications to streamline workflow and improve communications. These early and primitive wireless applications where leveraged as a means of increasing worker and business productivity, particularly among field service personnel, who were able to reduce the use of paper printoutslisting daily service orders by simply calling a dispatcher for information on the next service call. (Today service calls are sent to the field professionals' mobile device as a text or voice message, thus eliminating a call to dispatchers.) The early applications improved the communication processyet lacked any transaction process that generated revenue. For example, before mobile communication, service workers contacted a dispatcher for work orders. Now, using wireless devices, technicians are tracked via GPS, and work orders are automatically sent to the closest technician.

In today's e-commerce arena, business has evolved into mobile commerce (m-commerce) by leveraging the popularity of mobile devices. Revenue is no longer dictated by set business hours. In other words, just because the front door to the business is locked doesn't mean that the business is not generating revenue. A new brand of shopper has emerged from advances in technology and the Internet. According to McGuire [2001], mobile commerce is defined as the transaction of goods and services made by a buyer or seller using a data-enabled wireless device over a wireless data connection. Soon cell phones will be used to purchase snacks and drinks from vending machines. The use of Web-enabled smart phones, PDAs, and PocketPCs are transforming the way people do business, both in the business-to-business and business-to-consumer arenas.

Mobile Devices

Palm Models

Palm i705. The main strength of the Palm i705 device is its built-in wireless modem that allows users to be connected to the Internet at all times. Its weaknesses lie in usability. For example, it does not have a built-in keyboard and only has 4 MB of read-only memory (ROM) and 8 MB of RAM. See Table 1.

Palm m515, m505, and m500. The Palm m515 offers a color display and an adjustable backlighting feature, as well as 16 MB of internal memory. The internal memory can be used for working with application files such as Microsoft Excel spreadsheets and Word documents or viewing

TABLE 1. MOBILE DEVICE

Manufacturer	Model	OS	CPU	Screen	Weight
Casio	Cassiopeia *BE-300*	Windows CE	166 MHz/	3.5 in	5.9 ounces
	Cassiopeia E-200	Pocket PC 2002		240 × 320	5.9 ounces
Compaq Computer	IPAQ pocket PC H3835 H3850	Pocket PC 2002	206 MHz StrongARM	3.7 in 240 × 320	6.7 ounces
Handspring	Visor neo	PALM OS 3.5.2	33 MHz Motorola DragonBall	3.0 in	5.4 ounces
	Visor pro			160 × 160	5.7 ounces
	Visor edge				4.8 ounces
	Visor prism				5.4 ounces
Hewlett Packard	Jornada 565	Pocket PC 2002	206 MHz StrongARM	3.5 in 240 × 320	5.9 ounces
	Jornada 568				6.1 ounces
NEC	MobilePro P-300	Pocket PC 2002	206 MHz StrongARM	3.8 in 240 × 320	7.9 ounces
	MobilePro 790				2 lb
Palm	Pilot 1000	Palm OS 3.5	33 MHz Motorola DragonBall	3.6–3.8 in 160 × 160	5.7 ounces
	Pilot 5000	Uses faster seeming DragonBall			5.8 ounces
	Palm IIIxe				5.6 ounces
	Palm IIIc	DragonBall EZ processor and Palm OS 3.1+			6.8 ounces
	Palm V				5.4 ounces
	Palm VII				4.4 ounces
	Palm M 100	Palm OS 4.0			6.4 ounces
	Palm M500	Palm OS 4.1			4.0 ounces
	Palm M505				4.9 ounces
	Palm i705				5.9 ounces
PC-EPhone	EPhone	Pocket PC	206 MHz StrongARM	6.0 in 240 × 320	10.5 ounces
Blackberry	857	Proprietary	32bit 386	3.0 in 160 × 160	4.7 ounces
	957				4.8 ounces
	5810				4.7 ounces
Sharp	Zaurus	Linux 2.4 QT	206 MHz StrongARM	3.5 in 240 × 320	7.3 ounces
	SL-5500	Personal Java			
Sony	CLIE' PEG T-615 C	PALM OS 4.1	33 MHz Motorola	320 × 480	4.9 ounces
	CLIE' PEG T-760 C	PALM OS 4.1	DragonBall		5.7 ounces
Symbol	PPT 2800	Pocket PC 2002	206 MHz StrongARM	320 × 240	11.8 ounces
Toshiba	570	Pocket PC 2002	400 MHz	240 × 320	6.7 ounces

(Sources: Blackberry, Casio, Compaq, Handspring, Hewlett Packard, NEC, Nokia, Palm, PC-EPhone, Sharp, Sony, Symbol, and Toshiba.) See also **Web References**.

presentations and video clips. The m505 and m500 models extend PDA functionality with a secure digital and multimedia card expansion slot and a universal connector for memory. Additional features can be added that help with information backup, wireless modems, and camera options. The visible differences are found in the screen background; the Palm m500 handheld uses a high-contrast monochrome display on a white background and is limited to 8 MB of memory. The Palm m505 handheld has a 16-bit color display. The PalmVII is the first wireless Web-enabled device produced by Palm. This unit requires the user to uplink to receive e-mail, rather than synchronizing periodically throughout day.

Palm m100. In addition to the standard software applications, the m100 model has the date book, address book, to-do list, memo pad, note pad, and calculator. It has a monochrome display, requiring two AAA batteries and has 2 MB of memory.

Handspring Models

Handspring Treo. The Treo is the latest mobile device from Handspring. The Treo combines the dynamics of data organization in a PDA with the communication features of a cell phone. This mobile computing device can send and receive e-mail and text messaging, has wireless Internet access, and can place phone calls. The Treo is a dual-band phone, thus allowing three-way calling. It uses the 3.5 Palm OS and 33-MHz Motorola DragonBall processor. The device has 16 MB of memory and a 160×160−pixel monochrome touch screen. One advantage of this Handspring device is that it comes with as witch that enables easy toggling between the standard ringing tones or vibrate mode. It also has a dedicated message button allowing quick access to manage e-mail and short text messages.

Handspring Visor Pro. This PDA has 16 MB of ROM and 0 RAM. The Visor Pro runs on the Palm OS and weighs 5.4 oz. A weakness of this model is that it as a side-mounted infrared port, which can be a challenge when lining up infrared technology. The Visor Pro uses a 33-MHz DragonBall processor and has a 160×160-pixel touchscreen. The expansion slots can integrate a digital camera, GPS, and Visor-phone, which use Handspring's Internet browser, Blazer Browser.

Handspring Visor Neo. The Visor Neo uses the Palm OS and software applications. Using Motorola DragonBall VZ 33 MHz processor and 16 MB of memory, this unit can send and receive information with any infrared equipped Palm OS. The main issues challenging Handspring are primarily related to the business's production.

Sony Models

Sony Clié PEG-T 615 C. This device uses a 33-MHz Motorola DragonBall processor and runs the Palm desktop and Palm OS applications. This Clié has 16 MB of RAM and a 320×320−pixel high-resolution color screen. It weighs 4.9 oz and has a memory slot expansion slot.

Sony Clié 760 C. This model has 16 MB of memory and an additional slot for a memory stick, which is needed to view videos and listen to MP3s. The Sony Clié has a 320×320−pixel, high-resolution color screen and a center toggle switch, which is used to navigate through applications. A weakness with the 760 C is that additional software is needed work with Microsoft Word and .Excel files. Also, the screen is reduced in size to accommodate the graffiti pad.

Casio Models

Cassiopeia E-200. This device is a PocketPC running windows CE OS. With its 64 MB of RAM and a color-screen resolution of 240×320 pixels, the E-200 offers desirable functionality to the user. The only limitation is the lack of software applications.

Cassiopeia BE-300. This device is a PocketPC that runs 166-MHz processor and has 16 MB of RAM, a 320×240−pixel color screen, and weighs 5.9 oz. The E-200 has a Type II expansion slot available to a CF card. Data entry is done with a touchscreen and stylus. The lack of available software is the main drawback of this device.

HandERA Model

HandERA Model 300. The HandERA 330 weighs 5.9 oz and has 8 MB of RAM. It runs on the Palm OS. The gray-scale screen resolution is 240×320 pixels. The advantage of this device is that the entire screen can be used to work with data, and the graffiti window can be hidden or brought back on command. Minimal memory and the use of a slower and

more cumbersome serialcable (compared with a USB or infrared port) to synchronize data are its limitations.

Hewlett Packard Jornada and Compaq iPAQ Models

iPAQ Models. The iPAQ H3870 Pocket PC includes Microsoft's Pocket PC 2002 OS, 32 MB of ROM, 64 MB of RAM, Intel's 206-MHz StrongARM SA-110, and a USB cradle. There are three models to choose from. The latest version is the H3870 PocketPC. It has embedded Bluetooth technology. This 3870 weighs 6.6 ounces and has a 3.7-inch color screen with a resolution of 248×320 pixels. It also has an Secure Digital (SD) card option. The iPAQ H3760 has 64 MB of RAM and is identical to H3870 with one exception: it does not have Bluetooth. The iPAQ H3850 also has 64 MB of RAM and uses a dual USB or serial cradle. It also incorporates a card slot and the use of voice-command and voice-control software. It also does not have Bluetooth.

The Jornada Model. The Jornada 564 PocketPC features a robust 206-MHz, 32-bit StrongARM processor and 32-MB synchronous dynamic RAM (SDRAM). Its 240×320−pixel display supports up to 65,536 colors and features a 3.5-inch color LCD touchscreen. Jornada's 565 Pocket PC also has a Type I memory slot. The Jornada 564 is capable of wireless and Internet functionality. A weakness of the Jornada is its small keypad, which makes data entry difficult. The Jornada 568 weighs 6.1 oz has a 206-MHz, 32-bit StrongARM processor; it has 64 MB of SDRAM and 32 MB of Flash ROM. The screen has a resolution of 240×320 pixels.

NEC Models

NEC MobilePro P300. This device weighs 7.9 oz, making it heavier than the 5.7-oz average. It has a screen resolution of 240×320 pixels. The P300 also supports wireless connections and secure remote access capabilities. The P300 has two expansion slots supporting a 32-MB Compact Flash (CF) card and a 32-Mb SD card. A weakness of this device is the magnetic cover used to hold the device shut; magnetic fields can damage data. The 2002 model will phase out the magnetic cover and reduce the weight to 6.7 oz.

MobilePro 790. This is a clamshell-style device that weighs 2 pounds and uses a 168-MIPS (million instructions per second) processor. It operates using Microsoft Windows for Handheld PC 2000 and features 24 MB ROM, 32 MB RAM, and internal 16-MB Flash Memory that allocates 14 MB available for storage. The 790, like Hewlett Packard's Jornada, is weak in keypad functionality for data entry.

PC-EPhone Model

PC-EPhone. The PC-EPhone claims to be a Palm-sized PC that integrates cell phone functionality. It uses the 206 MHz Strong ARM processor and has a 4.0-inch display with 640×480−pixel resolution. It has 32 MB of RAM and 32 MB of ROM. Little is known about its durability and compatibility.

Research In Motion BlackBerry Models

RIM 857, RIM 957 and RIM 5810. The RIM models of the BlackBerry line of mobile devices are at the cutting edge of wireless technology and business connectivity. Once turned on, BlackBerry mobile devices remain connected to a wireless network. It does not require synchronizing of applications; e-mail messages find the device. The models differ in their 8 or 20 line graphical display. Currently, only the 5810 provides service that integrates customers' Internet service provider e-mail account to the device. Depending on the user, the device can be a distraction because of its high level of connectability.

Sharp

Sharp Zaurus SL-5500. This device incorporates a 206-MHz Strong ARM processor and a 3.5-inchthin film transistor (TFT) front-lit screen with 65,536 colors and a resolution of 240×320 pixels. The SL-5500 weighs 7.3 oz and has 16 MB of ROM and 64 MB of RAM. The two expansion slots house the Compact Flash and SD cards and can be handled simultaneously. The Zaurus has a built-in cursor key for simple navigation and runs Linux OS. One weakness is that this unit caters to more technical users with its customizable command line prompt option. Also, there is no technical support for Macintosh users, and because it uses the Linux OS, application functionality may be an issue.

Symbol Technologies

Symbol PPT 2800. The PPT 2800 is the "industrial strength" PocketPC, with 64 MB of RAM and 32 MB of ROM all enclosed in an outer casing that can withstand a 4-foot drop onto concrete. Its fundamental application environment is in large retail chains and in manufacturing industries. It weighs 10.3 oz and has 320×240–pixel resolution and an optional 16-level gray scale or 64-k color background. The PPT 2800 has the functional expandability to meet most applications. A definite benefit is the removable backup battery pack, eliminating the need to cradle the device to charge the battery. Also, the wireless Internal (WLAN), External (WWAN) antenna is built in, securing the modem from unexpected impact. The cost of the PPT 2280 is the main drawback. For the average user in the market for a PocketPC, the PPT 2880 may be cost prohibitive.

Toshiba

Toshiba 570 PocketPC. The 570 weighs 6.7 oz and is equipped with a 400-MHz Intel PXA250 processor. It also has a 3.5-inch TFT color display with 240×320–pixel resolution and 65,536 colors. The 570 has 64 MB of SDRAM and a slot for both the SD and CF card. Along with integrated Wi-Fi (IEEE Standard 802.11b), the 400-MHz processor and dual expansion slots offer several noteworthy benefits.

Smart Phones

Nokia

Nokia 7650. The Nokia 7650 weighs 5 oz and uses multimedia messaging, accomplished by merging image, audio and traditional voice calls. The 7650 alsouses the Symbian OS, has an integrated WAP browser, and is outfitted with Bluetooth. A limitation is the sliding keypad, which may make it challenging to use with one hand.

Nokia 9000i Communicator. The Nokia 9000 $\times$ Communicator series has been around since 1996and has evolved into a sophisticated palmtop computer, built in the form ofa global system for mobile communication (GSM) cellular phone. The Nokia 9000i has a clamshell design; when closed the 9000i operates as a normal cellular phone but with text messaging. When opened, the phone is transformedinto wireless PDA, the top half delivering a 640×200–pixel LCD screen and the bottom half offering a keyboard. The Communicator uses the GEOS 3.0 OS and has a 24-MHz Intel 386 processor with 8 MB of memory. The memory is divided into three individual partitions, 4 MB allocated to the OS and application software, 2 MB for program execution, and 2 MB for data storage. The pitfalls are the small keyboard and the fact that users must navigate through screens using buttons.

Nokia 9290. The Nokia 9290 addressed several of the challenging issues with the 9000i and increased the amount of memory to 56 MB and talk time to 10 from 3 hours. The memory is divided with 16 MB committed to user needs, 8 MB to execution, 16 MB to application, and 16 MB to the memory card. The 9290 uses a 32-bit ARM RISC CPU on the Symbian OS.

Ericsson

Ericsson T86 tri-mode GPRS. The Ericsson T86 weighs less than 3 oz and runs on a Symbian OS. This phone has a color screen and built-in Bluetooth capability. It also has optional modules for wireless earplugs, a video camera, and a four-way navigational joystick. A weakness is that the screen size is tiny and the buttons are small.

Ericsson T61 d. This device features voice-activated dialing and supports Bluetooth technology. With a calendar, calculator, and stopwatch, the T61 d is blurring the line between cell phones and PDAs. As with most smart phones, the main challenge is screen size. The seven-line graphic display limits the amount of information that the user can view.

Wireless Protocols

Institute of Electrical and Electronics Engineers (IEEE). The IEEE 802.11 standard is the foundation on which PDAs and PocketPCs operate. Products that employ this technology support a broad range of enterprises and individual users. With such explosive growth in wireless infrastructure and applications, understanding some basics about mobile computing technology and its limitations and variations can be helpful. This section touches on the protocols used for mobile communications and highlights some of the differences. This section of the article is a general overview, not a comprehensive analysis. See Table 2.

Over the past 15 years, the use of mobile computing devices has increased rapidly as cellular and digital services have become more affordable. Businesses are incorporating cell phones and PDAs as a way

TABLE 2. SMART PHONES

Manufacturer	Model	Screen	Weight
Nokia	9000i	Gray Scale	13.9 ounces
	9290	Color	8.6 ounces
Ericcson	T86 tri-mode	Color	3.0 ounces
	T61 d	Color	3.1 ounces

for employees to manage their work. This equipment also provides more efficient customer service. Unfortunately, as the popularity of mobile units increased, compatibility issues have become problematic. To remedy this, the IEEE partitioned the 2.4-GHz microwave band into two separate sections, frequency hopping spread spectrum (FHSS) and direct sequence spread spectrum (DSSS). The standards that emerged are 802.11 (FHSS), 802.11b (DSSS), 802.11a (DSSS), 802.11 g (DSSS), and 802.11e (DSSS). Their characteristics are as follows: 802.11 uses the 79 distinct FHSS channels for transmission of data; 802.11b uses a more exotic encoding technique and 14 fixed frequency channels up to 11 Mbps transmission in the license-free 2.4 GHz band. Because of current market share, 802.11b is the current wireless standard supporting all Ethernet network protocols. Data transmissions via 802.11a can reach up to 54 Mbps in the 5-GHz band using an orthogonal frequency division multiplexing (OFDM) encoding scheme. Standard 802.11 g addresses QoS side of wirelessnetworking and is currently in the development stages. Currently IEEE is investigating how best to standardize 802.11e as a wireless security protocol (IEEE,). The frequencies of 802.11b and 802.11 g share wavelength space with microwave ovens and cordless telephones, causing signal collision and interference.

Wireless Application Protocol (WAP). In 1997, Ericsson, Motorola, and Nokia formed the wireless application protocol forum. This forum was established to provide a worldwide open standard, enabling the delivery of Internet service to cell phones. WAP technology is aiming to be the global standard for mobile phones and wireless devices. The wireless application protocol supports standard data formatting and transmission for wireless devices. An important issue for WAP is that applications can be developed on any OS, thus providing interoperability among device families. This could cause an unnecessary increase in the number of interfaces that Web developers will need to design and maintain. We are in the earliest stages of wireless application development, and many changes will occur in the future. It will be important to reassess strategic goals before taking action on cutting-edge and nonstandardized wireless protocols. See Figure 1 for WAP infrastructure overview. See also **Wireless Application Protocol (WAP)**.

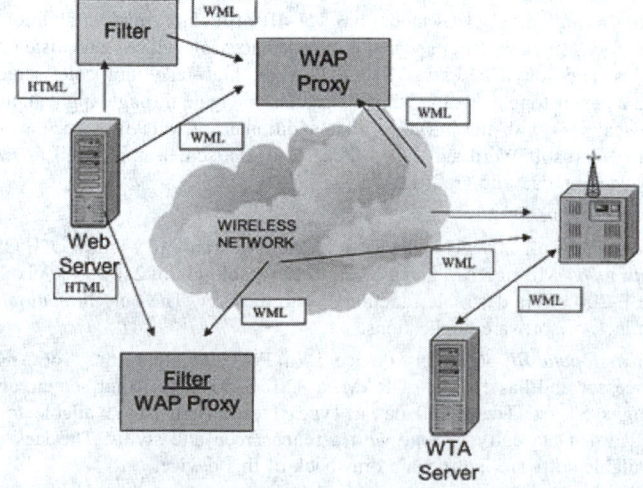

Fig. 1. WAP infrastructure overview.

Radio Frequency Identification. A basic RFID system consists of three components: the antenna, the transponder (tag), and the transceiver (tag reader). The antenna can be incorporated with the transceiver and decoder, becoming a reader. Readers can behandheld devices, such as the Symbol PPT 2280. RFID works as follows: the antenna emits radio signals to activate the tag; once the tag is activated, the reader either reads the tag or writes information to it (Table 2)). The type of tag dictates the functionality. Tags are either active (read and write) or passive (read only). The range of data transmission can reach more than 90 feet, subject to ideal working environments.

Global Positioning System. GPS is a method using satellites and radio frequencies to determine aspecific position without making observations of one's surroundings. The GPS satellites orbit at about 12,000 miles above the Earth (Table 3).GPS devices can process signals received from four or more satellites' transmissions simultaneously to calculate the receiver's location within 6 feet. See also **Global Positioning System (GPS)**; and **Navigation**.

TABLE 3. WIRELESS APPLICATION PROTOCOL INFRASTRUCTURE OVERVIEW

Applications Layer
Transport Protocol Layer
TCP/IP HID RF COMM
Logical Link Control Layer
Connections linking to devices are established and released
Link Manager Layer
Link Manager Protocol maintains connections
Baseband Layer
Coding/Decoding, packet handling and frequency hopping
Radio Frequency Layer
Frequency combination, convert bits to symbols

Bluetooth The Bluetooth Protocol Stack. Originally created by the Ericsson Company, Bluetooth is a short-range wireless technology similar to the 802.11 family. The use of Bluetooth in mobile computing is designed to link devices within a range of 30–35 feet. "Location independent" or ad hoc networks will benefit most from Bluetooth. Another feature is that Bluetooth performs authentication, payload encryption, and key handling. Because Bluetooth continuously hops between channels and switches between synchronous connection-oriented and asynchronous connectionless links, packet interference is primarily eliminated.With the multitude of compatibility options associated with Bluetooth, and with the increase of devices being manufactured with Bluetooth capabilities, it will likely surpass 802.11b in the arena of wireless communication within the next few years.

See also **Mobile Operating Systems and Applications**; and **Wireless Communications Applications**.

Additional Reading

Goldman, J., and P. Rawles: *Local Area Networks: A business-oriented Approach*, John Wiley & Sons, Inc., New York, NY, 2000.

McGuire, M.: Mobile Business Markets: What Can't Users Live Without, Paper presented at the Gartner Group Symposium, Lake Buena Vista, FL, October 2001, pp. 8–12.

Web References

Blackberry: http://na.blackberry.com/eng/devices/
Casio: http://www.casio.com/index.cfm
Compaq: http://h18000.www1.hp.com/
Handspring: http://www.handspring.com/
Hewlett Packard: http://www.hp.com/
NEC: http://www.necsolutions-am.com/products/psgateway.cfm
Nokia: http://www.nokiausa.com
Palm: http://www.palm.com/
PC-EPhone: http://www.pc-ephone.com/tech.html
Sharp: http://www.sharp.com
Sony: http://www.sony.com
Symbol: http://www.symbol.com/products/
Toshiba: http://www.toshiba.com

JULIE R. MARIGA
BENJAMIN R. POBANZ
Purdue University

MOBILE OPERATING SYSTEMS AND APPLICATIONS.

Mobile computing, defined as a generalization of all mobile computing devices including personal digital assistants (PDAs, e.g., Palm Pilots, Pocket PCs), smart phones, and other wireless communication devices, will to change dramatically in the next 2 to 5 years. There are a number of reasons for these changes, but two primary factors are the convergence of next-generation handhelds and high-speed wireless technology. The operating systems (OSs) found in today's handhelds will provide the foundation for future devices and applications. The world of technology is changing in many ways. Important considerations for companies are the challenges, trends, and opportunities of deploying mobile computing technologies. Companies need to ensure they have a mobile strategy in place to stay competitive and capitalize on new developments in the mobile computing area.

Many factors drive companies to grow their mobile computing infrastructure, including decentralized workforces, telecommuting, travel, device capabilities and proliferation, mobile infrastructure focus, networking choices, cost of ownership issues, mobile business to employee transactions, and companies' control of handheld devices. Which OS should companies or individuals implement? It depends on a number of factors. One important issue to consider is which application(s) need to be used on the device. Answering this question may help to eliminate certain operating systems and devices. Another important factor to consider is portability of applications, which is important because devices change rapidly. Portable applications can be used on new devices without having to be rewritten or upgraded. If applications are developed in a language that allows for portability, such as Java, then these can be deployed to a wide range of devices, including handhelds that support various operating systems, embedded Linux devices, and pure Java devices. Another important issue to consider in selecting an OS is what type of development tools are available as well as the number and strength of the programmers available to create and maintain applications. Currently, the Palm OS supports the largest number of packaged applications. Many of these, however, are better suited for individual rather than business use.

There are four main factors driving the mobile business phenomenon: (a) economics, (b) business needs, (c) social trends, and (d) technology. Economics includes the falling prices of mobile airtime and the inexpensive cost of devices. Jones stated that over the next 5 years, costs will continue to decrease, which will allow for new mobile applications to be developed and Bluetooth chip sets to cost under $5. This will enable electronic devices to be networked. Business needs include organizations requiring new types of mobile applications to increase customer service and allow for better supply chain management. With regard to social trends, in many countries, mobile devices have become a lifestyle accessory, mainly among younger adults. As young adults continue to want more functionality from their devices and applications, there will be a mix between the mobile technology and entertainment and fashion. Finally, new core technologies such as WAP, i-mode, Bluetooth, and 3G networks are enabling a new generation of mobile applications. As these four factors evolve, they will continue to push the growth of the mobile business arena.

Computing Architectures

It is important to understand the overall computing architecture when using mobile devices and developing mobile applications. There are two general architectures, the occasionally connected application model and the tiered computing architecture. This occasionally connected model would be used when the client is not permanently connected to the rest of the system. An important consideration with this model is that the developer should not make assumptions about data accessibility. The majority of business logic, processing, and will be done on the backend or server. The tiered computing architecture generally includes three types of architecture that fall into the tiered architecture model. The first is the one-tiered approach, in which the mobile device is used only as a display device, and the software, data validation, and business rule enforcement take place on the backend or server. In a two-tiered architecture, the business rules are divided, with some performed on the client-end application running on a dedicated device and the remainder on the server. An advantage to this type of architecture is that the network traffic can be reduced and overall performance improved. The disadvantage is that maintenance can be difficult. When a business rule changes, modifications need to be made to both the client device and the server. The three-tiered architecture model places a middle layer between the client device and the server. The middle layer implements a majority of the business processing, and the client device handles the display functions; the server provides data storage. The biggest advantage to this architecture is that scalability is extremely high. Any changes to business rules can be made on the middle layer. This type of architecture is also called n-tier architecture.

Mobile Computing Operating Systems

Before selecting a device, a few important decisions need to be made about what OS will run the device and what type of applications will run on the selected OS. There are three primary operating systems to choose from for handheld devices: the Palm, Pocket PC, and Symbian operating systems. The following sections look at each of these operating systems. For the next few years, information system managers will be forced to confront the various mobile platform and OS dependencies. A key for building success in the mobile computing arena will be for organizations to have a strategic plan in place for how mobile computing fits with the overall business plans.

Palm Operating System. The Palm OS is a compact operating system developed and licensed by PalmSource, Inc., http://www.access-company.com/home.html, for personal digital assistants (PDAs) manufactured by various licensees. It has been the most popular platform for handheld devices since 1997 when the first Palm Pilot was launched. Some advantages include its ease of use, market share, synchronization with many calendaring and scheduling applications, and its contact management systems. The Palm OS uses a simple handwriting recognition system called graffiti instead of a keyboard. A user can plug in a keyboard if desired. Since first introduced, the Palm OS has lost market share to devices running the Pocket PC OS and the Windows CE OS. Palm has created PalmSource as a Palm subsidiary, which will allow PalmSource the ability to concentrate on developing and enhancing the operating system. The Palm OS runs on the following devices: Palm, Aceeca, AlphaSmart, Garmin, Kyocera, Samsung, Sony, and Symbol. As of this writing, the latest version is Palm OS 6, which has created new opportunities for end users and developers.

With the Palm OS, software developers can build data applications to use with Palm devices, which can be implemented via wireless or synchronized data access to corporate data. The Palm OS has the following strengths:

- The number of partners working with Palm is extremely large.
- A large number of applications are available to run on the Palm OS.
- Palm has a healthy percentage of the market share.

Weaknesses with Palm and Palm OS are the following:

- The core OS functionality is limited (compared with Windows CE/Pocket PC).
- Palm as a company has undergone major changes, and some question its leadership and future business directions.

Primary Components. The Palm OS consists of five main components: 1. Palm OS software; 2. Reference hardware design; 3. Data synchronization technology for one-button synchronization; 4. Platform component tools including an application programming interface (API) that enables developers to write applications; and 5. Software interface capabilities to support hardware add-ons.

Figure 1 shows the overall architecture of the Palm OS.

The Palm OS is owned by Access Co. Ltd., a mobile software company in Japan, announced January 25, 2007 that is renaming Palm OS "Garnet OS" instead. The company has done so to differentiate Garnet OS from Palm Inc., manufacturers of the Treo handset and holders of the Palm trademark.

This isn't the first time the name "Garnet" has been associated with Palm OS. In February 2004, PalmSource introduced different names for the Palm operating system—Palm OS 5 became Palm OS Garnet, and Palm OS 6 became Palm OS Cobalt.

Besides Garnet OS, Access' specific focus is in developing a Linux-based platform for mobile phones and other devices. The Access Linux Platform, as the company calls it, will include a compatibility layer for Garnet OS.

Windows CE Operating System. Windows CE is an operating system developed by Microsoft for different types of mobile devices made by several leading consumer electronics manufacturers including Casio, Compaq, Hewlett Packard, and Symbol. Windows CE is a modular operating

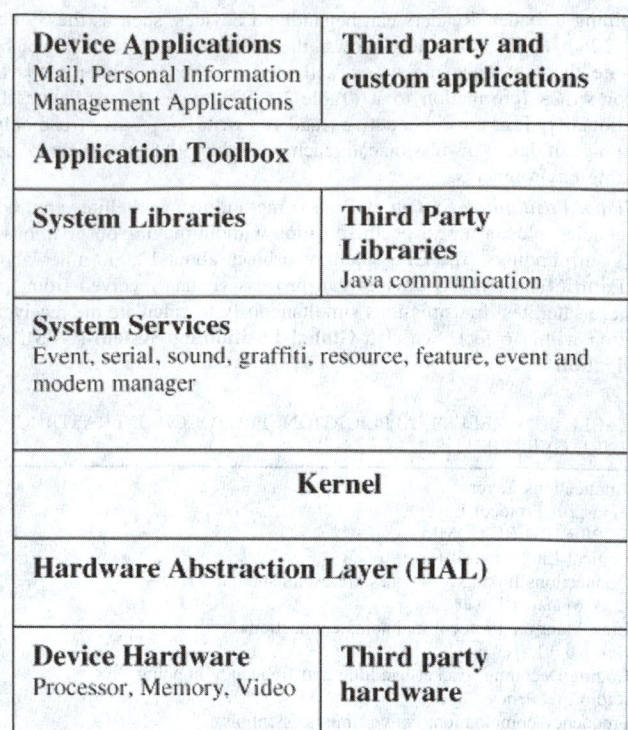

Device Applications Mail, Personal Information Management Applications	Third party and custom applications
Application Toolbox	
System Libraries	Third Party Libraries Java communication
System Services Event, serial, sound, graffiti, resource, feature, event and modem manager	
Kernel	
Hardware Abstraction Layer (HAL)	
Device Hardware Processor, Memory, Video	Third party hardware

Fig. 1.   Palm operating system architecture.

system that provides many building blocks for developers to choose from. It can be large or small, depending on the size and strength needed for a device. It provides users with the familiar look and feel of Microsoft Windows. Windows CE is a 32-bit OS designed to meet the needs of a broad range of devices, from enterprise tools (such as industrial controllers, communications hubs, and point-of-sale terminals) to consumer products (such as cameras, telephones, and handheld and home entertainment devices). A typical Windows CE-based embedded system is targeted for a specific use, and many analysts think that Windows CE should be used for any industry or business applications. Improved kernel services allow the OS to respond more quickly to various processing needs. The improved services make the OS ideally suited for industrial applications such as robotics, test and measurement devices, and programmable logic controllers. The OS interoperates easily with desktop environments that are based on Microsoft Windows NT operating system and Microsoft Windows 2000. This makes it the easier for overall enterprise system integration that combines small mobile systems with high-performance desktops servers and workstations. There are several applications that come with the Pocket PC including Pocket Word, Pocket Excel, and Pocket Internet Explorer, Microsoft money, active sync, note taker, and file explorer. Some of the features found in the OS include the following:

- *Increased Device Security:*—enhancements include encrypted passwords and programming interfaces that allow third-party developers to extend anti-virus software.
- *Windows Media Player:*—this enhancement delivers the ultimate digital media playback for mobile device users. With Media Player, a user can listen to digital music or watch a short video. The media player supports Windows media content, MP3 audio files, Windows media audio, and Windows media video.
- *Pocket Internet Explorer:*—this feature allows users to surf the Web online or download Web pages to read while offline. It supports HTML and WAP sites.
- *eReader:*—this feature allows users to read their favorite books on their mobile devices.
- *Pocket Outlook:*—this is the mobile companion to Microsoft Outlook. Users can schedule appointments, manage contacts, read, write, and send e-mail.
- *Industry Standard Expansion Slot:*—this feature allows users to expand the use of their device as their needs grow. Some users might use an

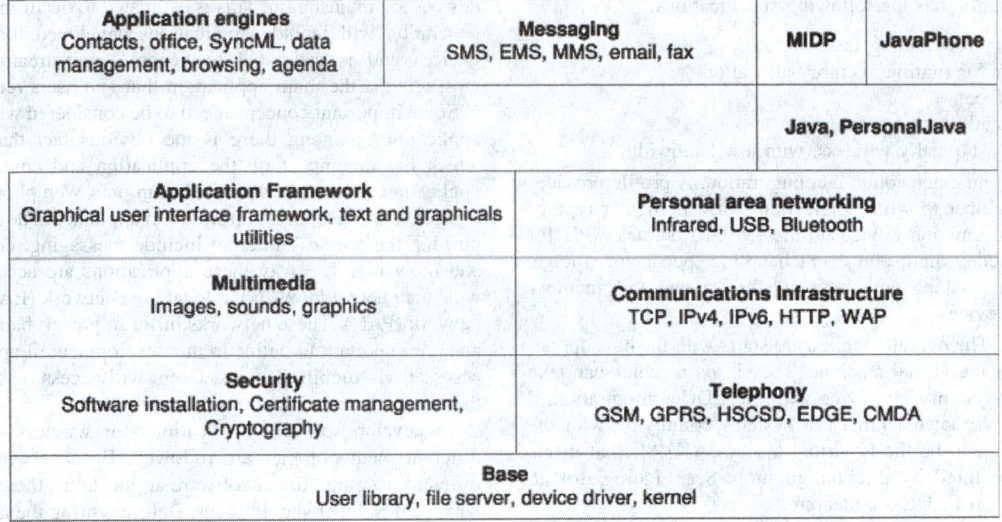

Application engines Contacts, office, SyncML, data management, browsing, agenda	Messaging SMS, EMS, MMS, email, fax	MIDP	JavaPhone
		Java, PersonalJava	

Application Framework Graphical user interface framework, text and graphicals utilities	Personal area networking Infrared, USB, Bluetooth
Multimedia Images, sounds, graphics	Communications Infrastructure TCP, IPv4, IPv6, HTTP, WAP
Security Software installation, Certificate management, Cryptography	Telephony GSM, GPRS, HSCSD, EDGE, CMDA
Base User library, file server, device driver, kernel	

Fig. 2.   Overview of the Symbian operating system.

expansion slot to add more memory or to add a modem, digital camera, or scanner.

Symbian Operating System. The Symbian OS is an advanced, open, standard system licensed by the world's leading mobile phone manufacturers. It is designed for the specific requirements of open and data-enabled 2G, 2.5G, and 3G mobile phones. The Symbian OS includes a multitasking kernel, integrated telephony support, communications protocols, advanced graphics support, data management, a low-level graphical user interface infrastructure, and a variety of application engines. Some of the key features of the operating system include the following:

- *Full Suite of Application Engines*: — Applications include a contact database, an alarm, an agenda, charts, a word processor, a database, and a help application.
- *Browsing*: — This operating system includes the Opera Web browser, which has the ability to browse over global system for mobile communications, general packet radio service, code-division multiple access, and transmission control protocol/Internet protocol connections and to browse local files.
- *Messaging*: — This allows users to create, send, and receive text, enhanced, and multimedia messages, as well as e-mail and faxes.
- *Multimedia*: — The Symbian OS includes a graphics subsystem that allows shared access to the screen, keyboard, and pointing devices. It also includes an interface that allows for drawing to any graphics device. In addition, it provides audio recording and playback.
- *Communication Protocols*: — The Symbian OS supports numerous protocols used for communication such as transmission control protocol, user datagram protocol, Internet protocol v4, Internet protocl v6, point-to-point protocol, domain name system, secure sockets layer, transport layer security, file transfer protocol, hypertext transport protocol, wireless application protocol, and Bluetooth.
- *Mobile Telephony*: — The telephony subsystem provides multiple APIs to its clients, which offers integration with the rest of the OS toaccomplish advanced data services. Some of the functionality included is phone and network information, access to the network names detected by the phone, information about the current network, and retrieval of signal and battery strengths and network registration changes.
- *Data Synchronization*: — The SyncML client is used for datasynchronization. It includes a contact and an agenda database adapter so that the two databases can be synchronized. It also includes a database adapter that allows the Symbian client to synchronize its data with the backend database.
- *Security*: — The Symbian OS security subsystem enables dataintegrity and authentication by supporting underlying secure communications protocols such as secure sockets layer and IP security. The security subsystem includes a cryptography module and certificate management module.

- *Software Development*: — The Symbian OS is delivered to its licensees and development partners in two products, the Symbian OS customization kit and the Symbian OS development kit. This allows other companies to develop software to is compatible with the Symbian OS.
- *Support for Multiple user Interfaces*.:

The features are shown in Figure 2.

J2ME Operating System. Java was introduced in 1995 with the idea of developing a language that would allow developers to write their code once and then run it on any platform supporting a Java Virtual Machine (JVM). In 1997, a new edition was released, Java 2 Enterprise Edition, to provide support for large-scale enterprise wide applications. The most recent addition to the family is the Micro Edition (Java ME), which targets "information appliances" ranging from Internet ready television to cellular phones. Following is a summary of the available Java platforms:

- *Standard Edition (J2SE)*: — This edition is designed to run on desktop and workstation computers.
- *Enterprise Edition (J2EE)*: — This edition provides built in support for servlets, Java server pages, and XML. J2EE is aimed at server-based applications.
- *Micro Edition (J2ME)*: — This edition is designed for devices with limited memory, display, and processing power.

J2ME is aimed at consumer devices with limited hardware capabilities. By introducing J2ME, devices no longer need to be static in nature. It allows users the option to browse, download, and install Java applications and content; devices running J2ME can access features inherent to the Java language and platform. J2ME capabilities vary greatly. Products may include cellular phones, PDAs, pagers, entertainment devices, and automotive navigation as well as others. An understanding of configurations and profiles is necessary to understand how J2ME can accommodate a broad range of electronics and devices.

Configurations. A configuration defines a Java platform for a broad range of devices and is closely linked to a JVM. It defines the Java language features and the core Java libraries of the JVM for that particular configuration. What a configuration applies to is primarily based on the memory, display, network connectivity, and processing power available on the device. On the Sun Microsystems Web site, the J2ME Frequently Asked Questions states, "The J2ME technology has two design centers, things you hold in your hand and things you plug into the wall." Two currently defined configurations are the connected device configuration (CDC) and the connected, limited device configuration (CLDC) (http://java.sun.com/javame/index.jsp). The CDC configuration has the following specifications:

- 512 kilobytes memory for running Java
- 256 kilobytes for runtime memory allocation
- Network connectivity, possibly persistent and high bandwidth

The CLDC configuration has the following specifications:

- 128 kilobytes memory for running Java
- 32 kilobytes memory for runtime memory allocation
- Restricted user interface
- Low power and typically battery powered
- Network connectivity, typically wireless, with low bandwidth

Profiles. A profile is an extension to a configuration. A profile provides the libraries for a developer to write applications for a particular type of device. The mobile information device profile (MIDP) defines APIs for user interface components, input and event handling, persistent storage, networking and timers, taking into account the screen and memory limitations of mobile devices.

J2ME Architectures. The overall architecture starts with the host operating system followed by the virtual machine. The virtual machine can take one of two forms: For systems complying with the CDC configuration, it will be the traditional virtual machine. For systems complying with the CLDC configuration, it will be the K virtual machine (KVM) that meets the specifications as required by the configuration. See Figure 3 for an overview of the CDC and CLDC architecture.

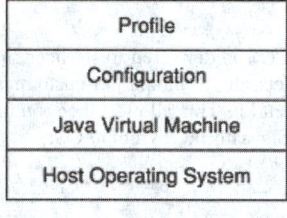

| Profile |
| Configuration |
| Java Virtual Machine |
| Host Operating System |

CDC Architecture

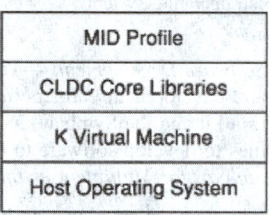

| MID Profile |
| CLDC Core Libraries |
| K Virtual Machine |
| Host Operating System |

CLDC Architecture

Fig. 3.   Overview of connected device configuration and the connected, limited device configuration architectures.

Conclusion

The future of mobile computing looks promising. Each of the operating systems described in this chapter has its strengths and weaknesses (Table 1).

TABLE 1. MOBILE PLATFORMS STRENGTHS AND WEAKNESSES

Platform	Keys
Palm	*Strengths:* Market share, number of partners, and number of available applications
	Weaknesses: Somewhat hardware dependent and core operating system functionality is limited
Symbian	*Strengths:* Telephony area, number of partners, Eurocentric
	Weaknesses: Lack of marketing compared with Palm and Microsoft and many companies involved in development
Microsoft	*Strengths:* Integration with existing infrastructure and applications and performance
	Weaknesses: Anti-Microsoft sentiment among the public and no true Java support
Linux	*Strengths:* Open source code, inexpensive, and popular in Asianmarkets
	Weaknesses: No wide base of applications available, many variations of the core language, no owner

As mobile technology continues to evolve, so do mobile applications. The first generation of mobile applications was primarily focused on the organizer, which consists of contact and appointment features. These applications are still used heavily in handheld devices. The second generation of applications began to expand their application needs beyond the embedded applications. Many users want access to office and PC-based applications, such as Microsoft Office and e-mail. The third generation of applications will involve the requirement for continuous wireless connectivity to various applications. During this generation, data synchronization will become much more common. The fourth generation of mobile applications will be built based on the assumption that handheld devices are mainstream. This is estimated to occur in 2004–2005. The fifth generation will include applications developed for managing numerous devices and is estimated to become mainstream in 2006. Currently, messaging is the main application that end users require.

Some important concepts need to be considered when developing mobile applications. Among these is the obvious fact that the user is mobile, which has an impact on the application and content delivered. Mobile applications will more than likely run on a Web phone, a handheld device, a pager, a Web PC, or an information appliance. The four main applications pushing the wireless Internet include messaging, conversing, interacting, and browsing. The way these applications are accessed could be over a wide area network (WAN), a local area network (LAN), or a personal area network (PAN). These networks differ in power, bandwidth, and data rates and are important to define in the development of applications that will be accessed via mobile devices. Users will access or browse sites that keep their content current and easy to navigate.

To develop software applications for wireless devices, a number of different methodologies are followed. Because wireless applications are different from traditional software applications, there are certain factors on which a developer should focus. Before writing the application, one should identify and study the intended audience. The three basic steps in studying the user are (a) developing a persona, (b) describing scenarios, and (c) creating storyboards. A persona is a single, concrete characterization of someone who uses an application. There can be more than one persona for each application, but developers should establish at least one. The creation of a persona involves the identification of the typical personality, writing a background for that personality, and describing characteristics and actions that define it and its typical activities. A persona should have a realistic personality and, at a minimum, should include a name, a title, a picture, biographical information, personality traits, and goals. The first paragraph of the persona's written information should introduce the user without reference to a specific technology. A scenario is a concise description of a persona using technology components to achieve a goal. The scenario considers how the user handles the hardware, how the application is operated, and how the content is used. A storyboard diagrams the entire story of application use, one screen at a time, and shows the display, navigation, and interaction for the wireless device. There are numerous programming languages available to develop applications, but the most popular wireless programming languages are Java, C, and C++.

A developer should also create a mobile database (with content) and design a logical application. The developer lays out a working model, builds wireless screens, and codes business logic to attach to real content. A mobile application should include back office administrative functions, keeping in mind that someone will have to manage the mobile user content on the server. Developing useful tools with good interfaces can help to simplify administrative tasks. Developers should also design a Web start page, which allows a user to set up an account and personalize the mobile application. A key to developing successful mobile applications is to allow users to personalize them to fit particular needs.

See also **Mobile Devices and Protocols**; **Wireless Application Protocol (WAP)**; and **Wireless Communications Applications**.

Note: Portions of this article originally appeared in the Internet Encyclopedia, John Wiley & Sons, Inc., New York, NY.

Additional Reading

Boling, D.: *Programming Microsoft Windows CE.NET*, Microsoft Press, Redmond, WA, 2003.

Carlson, J., and A. Schmitz: *Palm Organizers: Visual QuickStart Guide*, 4th Edition, Peachpit Press, Berkeley, CA, 2004.

Chen, Hsiao-Hwa: *Next Generation CDMA Technologies*, John Wiley & Sons, Inc., Hoboken, NJ, 2007.

Dyszel, B.: *Palm for Dummies*, 2nd Edition, John Wiley & Sons, Inc., New York, NY, 2002.

Jipping, M.J.: *Mobile Operating Systems Using Symbian OS: A Tutorial Guide*, John Wiley & Sons, Inc., Hoboken, NJ, 2007.

Keogh, J.: *J2ME: The Complete Reference*, The McGraw-Hill Companies, Inc., New York, NY, 2003.

Morris, B.: *The Symbian OS Architecture Sourcebook: Design and Evolution of a Mobile Phone OS*, John Wiley & Sons, Inc., Hoboken, NJ, 2007.

Muench, C.: *The Windows CE Technology Tutorial: Windows Posered Sollutions for the Developer*, Pearson Education, Upper Saddle River, NJ, 2000.

Mykland, R., J. Keogh, and M. Graham: *Palm OS Cobalt Programming from the Ground Up*, 2nd Edition, The McGraw-Hill Companies, Inc., New York, NY, 2005.

Wilding-Mcbride, D.: *Java Development on PDAs: Developing Application for PocketPC and Palm Devices*, Pearson Education, Upper Saddle River, NJ, 2003.

Web References

ACCESS: http://www.access-company.com/home.html
ACEECA: http://www.aceeca.com/home/
AlphaSmart: http://www.alphasmart.com/
Garmin: http://www.garmin.com/support/collection.jsp?product=010-00264-00
Kyocera: http://www.kyocera-wireless.com/pc-mobility/
Palm: http://www.palm.com/us/support/
Samsung: http://www.samsung.com/Products/MobilePhones/Sprint/SPH$\{-\}$
I500SSXAR.asp
Sony: http://esupport.sony.com/perl/select-system.pl
Symbian Operating System: http://www.symbian.com/
Symbol: http://software.symbol.com/palmos_directory.cfm
Windows CE Operating System: http://msdn2.microsoft.com/en-us/embedded/default.aspx

MOBILITY. Quantitative measure of the velocity of ions in an electric field of unit strength.

The mobility can be expressed as

$$\mu = \frac{v}{E},$$

where v is the velocity of the ion and E is the electric field strength. The units are square meters per volt per second. Heavier ions will have lower mobilities than lighter ones and values of mobility vary inversely with density, producing slower mobilities near the earth's surface.

MODEL PLASTERS. See **Dental Materials**.

MODEL (Scientific). Fundamentally, a model may be defined as a representation of some or all of the properties of a device, system, or object. The more properties represented by the model, the more complex the model becomes and thus one constantly faces tradeoffs in constructing a satisfactory model. There are three basic classes of models: (1) *mathematical models*, wherein the representation is comprised of procedures (algorithms), mathematical equations, and so on; (2) *physical models*, such as models of rivers and dam systems, airfoils and ship contours for use in wind tunnels and similar apparatus, and construction projects of extreme three-dimensional complexity (see Fig. 1); and (3) *logical models*. Simulation is closely associated with modeling, in that electrical circuits may be set up to correspond with mechanical, thermal, and fluid systems. See also alphabetical index.

MODEMS. Transmission of data between two devices requires the usage of a transmitter, a receiver, and a transmitting media that provides a path between the transmitter and the receiver. According to the manner in which data are transmitted, there are two fundamental modes of transmission: (1) parallel and (2) serial.

In a parallel mode of transmission, data are transmitted one byte (or character) at a time. This mode of transmission requires a minimum of 8 lines, with additional lines for control signaling, for transmitting the 8 bits (or one byte) of data from the transmitter to the receiver. This transmission method yields a very high data rate at the expense of increased costs due to the presence of a large number of cables between the communicating devices. Hence, it is used in communication between computers and peripheral units where cable distances are relatively short and data transfers must occur rapidly (such as between a printer and a computer).

The parallel mode of transmission becomes increasingly expensive as the distance between the two communicating devices increases relative to the increase in the cost of the cables. An alternative to parallel transmission is the serial mode of transmission, wherein the data are transmitted in sequence over one line, that is, one bit at a time. Instead of requiring additional lines for control signals, a preset sequence of bits can be used for a similar purpose, allowing a two-wire circuit with one wire serving as an electrical ground to be used for data transmission. Since the public switched telephone network (PSTN) already provides a two-wire facility for voice transmission, it is quite logical for serial transmission to utilize the available infrastructure for a cost-effective transmission mechanism.

Fig. 1.   Physical model ($\frac{1}{32}$ of full scale) of a large crude refining unit. (*Foster Wheeler Energy Corporation.*)

To communicate over serial lines, the terminal devices need to convert the parallel data into a serial datastream. A *universal asynchronous receiver/transmitter* (UART) on the terminal device usually handles this, and the resulting serial datastream is transmitted/received using a common serial interface. However, there is a basic incompatibility between the digital signals transmitted by a terminal device and the analog signals transmitted by a PSTN line since the PSTN was originally designed to carry only voice signals. Although digital signals can be transmitted over an analog telephone line, the digital pulse-distorting effects of resistance, inductance, and capacitance on the analog PSTN line limit their transmission distance. Further, the presence of analog amplifiers to boost analog voice signal levels in the PSTN pose additional problems with digital data since the analog amplifier would boost the digital signal along with the distortions and would, therefore, increase the distortion in the digital data transmission.

Because of the incompatibilities between the digital signals produced by terminal devices and the analog signals that telephone lines were designed to carry, a conversion device is required to enable digital signals to be carried on an analog transmission medium. This conversion device is a modem, a contraction of the term modulator–demodulator. The modulator portion of the device converts (or modulates) the digital signals into analog signals for transmission over the PSTN line, while the demodulator portion of the device converts (or demodulates) the analog signal into digital format. Therefore, the modulator portion of the modem can be considered to be the transmission component of a communication system, and the demodulator can be considered to be the receiver component of a communication system.

With the emergence of digital telephony, some portions of the PSTN are designed to carry voice signals in a digital format. Usage of services such as the Integrated Services Digital Network (ISDN) allows PSTN subscribers for end-to-end voice and data communication in digital format. These digital networks use a bipolar signaling scheme for transmission over twisted-pair cable. A digital modem is therefore utilized to convert the unipolar signals generated by the terminal devices to a bipolar format used by the digitized PSTN. See also **Telephony (Telecommunications)**.

Modems, depending on the type of datastreams they operate on, work in either an asynchronous or a synchronous mode. In an asynchronous mode of operation, often referred to as a *start/stop* transmission, each character is encoded into a series of pulses. The transmission is started by a start pulse followed by the encoded character (a series of pulses). The receiver is notified of the completion of the transmission of a character by the transmission of a stop pulse that may be equal to or longer than the start pulse depending on the transmission code being used.

In a synchronous mode of operation, a group of characters are transmitted in a continuous bitstream. Modems located at each end of the transmission medium normally provide a timing signal or clock to establish the data transmission rate and hence enable the devices attached to the modems to identify the appropriate characters as they are being transmitted or received. Before the data transmission is initiated, the transmitting and the receiving devices must establish synchronization between themselves. To keep the receiving clock in step with the transmitting clock for the duration of a bitstream representing a large number of consecutive characters, the data transmission is preceded by the transmission of a special set of synchronization characters. An error-free data transmission after the synchronization process is achieved by using an error detection scheme known as *cyclic redundancy check* (CRC). This mode of serial data transfer yields a much higher data rate at the expense of complex circuitry because the receiver must remain in phase with the transmitter for the duration of the transmitted group of characters.

Modem Operation

A modem connects a PC (or a computing device) with the outside world through a series of cables. The connection between the modem and the PC is usually accomplished by a serial cable (in case of an external modem) or through the system bus itself (in case of an internal modem). The modem speaks with the outside world through a telephone line.

In the common scenario of a PC speaking with the outside world through a modem, the PC is referred to as a DTE (data terminal equipment) while the modem is referred to as a DCE (data communication equipment). An exception is the case of an internal modem wherein the concept of a DTE does not exist since the definitions of DTE and DCE are plausible with respect to a RS-232 connection (and an internal modem does not employ any RS-232-type connections since it is a bus-connected device and not a serial-interface-connected device).

The speeds at which a DTE can transmit information to a modem are usually much larger than the speeds at which the modem can transmit that information to the outside world. This speed difference warrants the existence of a number of mechanisms to ensure the timely and accurate exchange of information.

In an effort to bridge the communication speed differences, various data compression schemes are employed. The data compression schemes currently in use are the MNP-5 and the V.42bis. MNP-5 is derived from Microcom Network Protocol standard (devised by Microcom, now acquired by Compaq) and yields a data compression rate of up to 2 : 1. However, MNP-5 accumulates a large amount of overhead while compressing data. Further, it is rather inefficient in transmitting precompressed files since it cannot sense the need for compression. Using this scheme with a precompressed file usually results in the transmission of a larger file as compared to the original compressed file.

V.42bis is a data compression standard approved by ITU-T (International Telecommunication Union — Telecommunication Sector) http://www.itu. int, which yields a compression ratio of up to 4 : 1. It builds on the advantages of the MNP-5 protocol and includes the ability to sense when compression is required. This makes it more efficient when transmitting precompressed files.

It should be noted that the two standards, MNP-5 and V.42bis, are both exclusive in nature; that is, they cannot be used at the same time. Further, for optimal performance, the DTE-DCE communication link (usually the terminal and the modem connection) should be able to sustain the data rate that these compression schemes afford. In the case of the MNP-5, since the compression rate is 2 : 1, for every bit that the modem sends out, the DTE is required to transmit 2 bits' worth of information. Therefore, the bit rate of the DTE should be 2 times that of the DCE. In the case of V.42bis, the DTE speed should be 4 times that of the DCE. For example, if your modem is operating at 14,400 bps and no data compression schemes are being used, the DTE speed would need to be set at 14,400 bps (bits per second). However, if MNP-5 were to be used, the DTE speed would need to be set at 28,000 bps (2 × DCE speed) since the MNP-5 supports a compression ratio of 2 : 1. In the case of V.42bis, the DTE speed would need to be 57,600 bps (4 * DCE speed) as V.42bis supports a compression ratio of 4 : 1.

A convenient way in which these speed differences are quoted in modem terminology is through the concept of a baud. A *baud* is defined as the rate at which information is transmitted. In contrast, *bit rate* is defined as the rate at which bits are transmitted. Put another way, a baud could be defined as one pulse (or signal) interval in a carrier signal while bit rate is the number of bits that are transmitted per second (or signal). The relation between baud and bit rate is given below:

$$\text{Bit rate(in bits per second)} = \text{baud} \times \text{bits per baud}$$

With increasing transmitting speeds, the emphasis on better error control mechanisms has been increasing for the accurate exchange of information. Two distinct error control schemes that are used in most modern modems are MNP-4 and V.42. MNP-4 includes the functionalities of MNP-2 and MNP-3 while V.42 employs LAP-M (Link Access Protocol — Modem) protocol as the primary error control mechanism and reverts to MNP-4 as a backup. These error control schemes retransmit corrupted data using 16–32-bit CRCs. V.42 is an ITU-T specification that yields slightly better performance than MNP-4. In V.42, the primary error control mechanism is LAP-M. Data are grouped together in terms of frames, and these frames are transmitted over the communication channel along with a CRC header for error control. In case of LAP-M, each frame has a size of 128 bytes, and up to 15 frames (by default) can be sent without waiting for an acknowledgment from the receiver. This translates to a storage requirement of $128 \times 15 = 1920$ bytes at the transmitting end for accomplishing error-free transmission since the transmitter would need to retransmit all 15 frames if the transmitter receives a negative acknowledgment from the receiver. V.42 employs 16- or 32-bit CRC fields (although 32-bit CRC is more common these days.)

Because of the presence of many operating speeds, it is quite possible that the operating speed of one modem may be more than that of another modem involved in a typical end-to-end connection between two communicating terminals. In such cases, the receiving modem needs to be able to inform the transmitting modem to pause before it can fully process the data it received. This is accomplished using flow control mechanisms. Flow control mechanisms can be broadly classified as either software-based or hardware-based.

In *software-based* flow control mechanisms (also referred to as XON/XOFF mechanisms), the receiver modem sends a special signal (usually a Control-S character) to the transmitting end requesting the transmitter modem to pause for a while. The transmitter modem stops sending any new information to the receiver modem until it receives another special signal from the receiver modem (which is usually a Control-Q character) informing the transmitter modem that it can resume sending data. The advantage of this scheme is that no additional hardware support is required since the pause/resume signals are handled by the communication software (or the firmware in the modem). The disadvantage of this scheme is that the presence of noise in the transmission media can result in the loss of the pause/resume signals and therefore affect the operation of the modems. If the pause signal were lost, the transmitter modem would keep transmitting data at a rate that the receiver modem cannot handle, resulting in over-runs at the receiving modem. If the resume signal is lost, the transmitter modem would never know when to resume transmission of data and therefore the transmission media would remain silent forever. Because of the transmission of special characters to signify pause/resume actions during transmission, XON/XOFF flow control mechanism should not be used for the transfer of binary data since the modems could falsely interpret the presence of Control-S character in the original binary file as a signal to pause data transmission.

In sharp contrast to software-based flow control schemes, *hardware-based* flow control schemes depend on special hardware support for ensuring proper control over the flow of data. This is also referred to as RTS/CTS (ready to send/clear to send) flow control mechanism. In the case of external modems, a specific wire in the serial cable (that is used to connect the terminal to the modem) is used for exchanging flow control information. In internal modems, in the absence of a serial cable, flow control functionality is built into the modem itself.

Data transfer (more commonly known as *file transfer*) protocols exist for the transmission of binary data between two modems over a communication channel (which is usually a telephone line). Commonly used data transfer protocols include Xmodem, Xmodem-CRC, Xmodem-1 K,

Ymodem, and Zmodem. In the *Xmodem* transfer protocol, binary data are transmitted in 128-byte chunks and a checksum is appended to these 128-byte blocks so that the receiver can verify the integrity of the received block and intimate the transmitter of the status of its reception. If the receiver determines any errors in the reception, the transmitter modem would resend the entire 128-byte block.

Xmodem-CRC adds CRC functionality to the basic Xmodem protocol, while Xmodem-1 K transfers data in terms of 1-kbyte data chunks as against the standard 128-byte data chunks used by the standard Xmodem protocol. The *Ymodem* protocol is quite similar to the *Xmodem-1 K* protocol and is seldom used on noisy communication channels (like telephone lines) due to its inability to perform efficiently in such environments. Ymodem-G, an improvement on the original Ymodem protocol, yields slightly faster data transfer rates by eliminating software error control mechanisms and relying on the underlying hardware to perform the required error control operations.

Zmodem is the most widely used data transfer protocol over dialup connections because of its improved resilience to noisy environments and the higher data rates. It employs 32-bit CRC fields and does not wait for an acknowledgment from the receiver before it transmits the next block of data.

Modulation Techniques and Modem Standards

Because of the comparatively smaller bandwidths that are offered by present-day dialup connections, modems employ a modulation scheme to transmit more information at the same bit rate. This is accomplished by converting data into *symbols* and transmitting the symbols over the communication channel. The conversion of a datastream into a symbol stream is carried out by using an appropriate modulation scheme. The usage of these modulation schemes leads to a much better utilization of bandwidth because more information (in terms of data bits) can be inserted into specific *symbols* that are transmitted over the communication channel.

The most common modulation scheme is AM (amplitude modulation). This forms the basis for a few more advanced modulation schemes like QAM (quadrature amplitude modulation). In AM, symbols are defined in terms of the amplitude of the original signal and these symbols are transmitted over a carrier through the analog communication channel. A major disadvantage of AM is the fact that as the amplitude of the signal decreases, it becomes increasingly difficult to separate the signal from noise in the communication channel.

QAM is an improvement over AM in which information is encoded on the basis of the deviations in the phase and amplitude of the carrier wave. This so-called *two-dimensional* encoding leads to a greater encoding efficiency and, therefore, a higher data rate. See also **Modulation**.

TCM (trellis-coded modulation) is based on the QAM scheme. In TCM, additional bits are added to each *symbol* to accomplish *forward correction*. This leads to a better error control ability and bit errors introduced into the communication process can be effectively reduced.

FM (frequency modulation) is the frequency counterpart for the AM scheme wherein the modulation is accomplished in terms of the frequency rather than the amplitude. While this modulation scheme is used more widely for radio broadcasts, its application in dialup connections is not widespread. A variation of FM modulation is FSK (frequency shift keying), which was designed primarily for transmission of data across a telephone line. In this scheme, the presence of bit "1" is represented by a specific frequency tone and the presence of bit "0" is represented by another specific frequency tone. To afford two-way communication, FSK allows the specification of two different sets of frequency tones.

PSK (phase shift keying) is similar in principle to FSK, with the sole difference that variations in phase of a constant frequency carrier are used to determine the bit values in the original datastream. A signal with unchanged phase is used to signify the presence of a bit whose value is the same as that of the previous bit. A 50% change in the phase is used to signify a bit value that is different from the previous bit. Differential PSK (DPSK) is a refinement of PSK wherein the changes in phase are determined by comparing the phase of the current state with that of its previous state.

PCM (pulse code modulation) is a modulation scheme wherein analog data are encoded into a specific number of bits (usually 8 bits) for transmission over a communication channel. This is, strictly speaking, not a true modulation scheme since a carrier is not employed at all. The encoding of analog information into digital format is accomplished using a quantizer and a sample-and-hold circuit.

Several modem standards have been introduced that allow modems to exchange data universally. While it is not possible to discuss every modem standard in this document, a few representative standards are discussed here.

Bell 103 is one of the older standards that allow data to be transmitted and received at a rate of 300 bps (bits per second). It employs FSK modulation and uses 1 bit to represent a baud (i.e., bit rate is equal to baud rate). Bell 202 is considered an improvement over the Bell 103 as it supports a 1200 bps data rate using the same FSK modulation technique as employed by Bell 103.

With the widespread usage of modems across the globe, the need arose for a set of modem standards that could facilitate modems throughout the world to communicate with each other. CCITT (later known as *ITU-T*) undertook to formulate a set of universally applicable standards to this effect. One of the earlier CCITT standards for data communication was the CCITT V.21, which allowed a data rate of 300 bps using FSK modulation (quite similar in operation to the Bell 103 standard). CCITT V.22 used DPSK to obtain data rates of 1200 bps at a baud rate of 600 baud (i.e., the number of bits per baud was set to be equal to 2). This standard is similar to the Bell 212 A.

More advanced standards were later released by CCITT (viz., V.32 and V.32bis) that allowed data rates of up to 14,400 bps using different modulation techniques such as TCM and QAM. ITU-T V.34 is a more recently established standard that allows modems to transfer data at rates up to 33,600 bps.

V.34 is currently the fastest end-to-end analog modem standard. Because of the dependence of the more advanced standards such as V.90 on the V.34 standard, let us take a closer look at the details of the V.34 analog telephony standard.

Modems supporting V.34 can sustain data transmission capacities of 2400–28,800 bps. A feature referred to as *line probing* was introduced in this standard to allow modems to identify the capacities and quality of the phone landline and adjust themselves to allow, for each individual connection, the most optimal data transmission rate. V.34 also supports a synchronous auxiliary channel with a data rate of 200 bps that could be used in tandem with the primary data channel for signaling information.

The bandwidth offered by a phone line is around 3–4 kHz, and the maximum symbol rate that is supported by V.34 is 3429 symbols per second. The operation of V.34 near the theoretical limits of the phone line spurred the design engineers to incorporate a mechanism within V.34 to autonegotiate the available bandwidth on a phone line and adjust the data transmission rates accordingly. A new handshake protocol, called as V.8 mode negotiation handshake, was introduced to enable two V.34 compatible modems to exchange feature and mode negotiation information via V.21 standards (300 bps FSK modulated communication). V.8 mode is used by the two V.34 modems to identify them with other telephone network equipment. It is also used to determine whether the call is destined for a data or facsimile operation. Negotiation of the available modes of modulation is also accomplished along with ability to support V.42 and V.42bis standards. During this handshake, the modems send a series of tones to each other, at specific frequencies and known signal levels. The received signal level is employed in the computation of the maximum possible available bandwidth for communication.

Line probing is employed immediately after V.8 handshake to determine parameters such as the optimal bandwidth and carrier frequency, preemphasis filters, and optimal output power level to be used during communication.

Fax Transmission

Facsimile (or fax) is defined as the process of sending a document from one terminal to another. In an effort to utilize the existing PSTN infrastructure, traditional fax transmission involved the usage of modems at either communicating terminals. The transmission of a document is preceded by an exchange of capabilities between the sender and the receiver and, at the end of the transmission, a confirmation of delivery sent from the receiver to the sender.

The earlier fax machines, also referred to as group 1 fax machines, were designed to handle fax transmission over an analog telephone network.

These conformed to ITU-T T.2 standard for fax transmission and yielded a transmission rate of 6 min per page. With improvements in fax devices, ITU-T later introduced the T.3 standard that allowed for the transmission of up to 3 min per page. Group 3 fax machines, the most widely deployed category of fax machines, were standardized in 1980 for digital facsimile devices to communicate over analog telephone lines. Group 3 machines are based on ITU-T standards T.30 (for fax transmission) and T.4 (for fax file formats) and yield a transmission rate of 6 to 30 seconds per page.

The procedures outlined in the T.30 recommendation comprise of 5 distinct phases: (1) call establishment, (2) control and capabilities exchange, (3) in-message processing and message transmission, (4) postmessage, processing, and (5) call release.

The call establishment phase consists primarily of establishing a connection between the calling and the called terminals and exchanging fax tones. The calling machine dials the telephone number of the called machine and the calling tone (referred to as CNG) is received at the called machine. The CNG tone beeps indicate the existence of a fax call as against a normal voice call. The called fax machine answers the ring signal by going off-hook. After a 1-s delay, the called fax machine sends a 3-s, 2100-Hz tone back to the calling machine.

In the premessage processing phase, the terminals carry out various identification procedures along with command procedures to establish a command set of capabilities for the successful transmission of facsimile data. During the identification phase, the terminals exchange information regarding, among others, the bit rate, page length, data compression format, telephone number, and name of the organization. The called machine sends its digital identification signal (DIS) at 300 bps identifying its capabilities (using V.21 protocol), including its optional features. For example, the called fax machine could send a DIS identifying its capability to support V.17 standard (14,400 bps data rate). On receipt of the DIS, the calling fax machine sends a digital command signal (DCS) locking the called unit into the selected capabilities. The calling machine sends a training check field (TCF) through the modem to ensure that the channel is suitable for transmission at the accepted data rate. The called fax machine sends a "confirmation to receive" (CTR) signal to confirm that the receiving modem is trained (adjusted for low-error operation).

The in-message processing phase takes place in parallel with the message transmission phase since the in-message processing phase handles the signaling required for the transmission of facsimile data. This includes control signals needed for error detection, error correction, and line supervision. The ITU-T T.4 recommendation governs the message transmission phase and addresses issues related to the dimension of the document, the transmission time per scanned line, the coding scheme, modulation/demodulation techniques, and similar. The modem standards that are supported for the transfer of facsimile data include V.27ter (4800/2400 bps), V.29 (9600/7200 bps), V.33 (14,400/12,000 bps), and V.17 (14,400/12,000/9600/7200 bps).

The postmessage processing phase consists of procedures to deal with tasks such as end-of-message signaling, multipage signaling, and confirmation signaling and is used as a precursor to the call release phase. The calling fax machine sends a "return to control" (RTC) command that effectively switches both modems to the 300 bps data rate condition (V.21 standard). The called fax machine sends a message confirmation (MCF) signal indicating the document was received successfully. If multiple pages exist, a multipage signal (MPS) is sent. The partial page signal (PPS) is sent for error correction of the transferred document. In the call release phase, the calling terminal transmits a disconnect (DCN) signal to the called terminal for the release of the call. It is important to note that no response is expected for the call release signal from the called terminal.

Other Modems

The typical bandwidth allocated to each individual user on a telephone line (also referred to as a *subscriber line*) is around 3400 Hz. This places an upper limit on the data rates that a voice band modem can achieve since the transmitting/receiving symbol rate cannot be higher than the available bandwidth. A digital subscriber line (DSL) overcomes this limitation by overlaying a data network onto the existing PSTN. This is accomplished by letting the data network use the same subscriber lines as the POTS (Plain Old Telephone System), with the only exception that the data signals use a different frequency band for data communication.

A device called a POTS splitter is responsible for splitting and recombining the two types of signals: voice signals and data signals, at both ends of the subscriber line. Since the data network and the voice network use the same subscriber line, the telephone companies need only upgrade their switching/terminal devices to handle the data signals. Therefore, the delivery of high-speed data services to customers without considerable investment in infrastructure is possible.

Depending on the data rates that are supported, there are different variations to DSL. ADSL (asymmetric DSL) is characterized by a different data rate from the service provider to the customer (the downstream direction) as compared to the data rate from the customer to the service provider (the upstream direction). The upstream data rates are typically 10 times slower than the downstream data rates and range from 100 to 800 kbps. In sharp contrast, SDSL (symmetric DSL) offers a symmetric data rate; that is, the upstream and downstream data rates are equal. VDSL (very-high-speed DSL) is a new addition to the DSL family and is being developed to provide data rates as high as 25 Mbps in either direction (upstream or downstream).

In ADSL, the transceivers (transmitter and receiver units) are designed to carry more than one logical channel on a single physical channel to support different data rates. In addition, the transceivers support an embedded operations channel (EOC), ADSL overhead channel (AOC) etc that are used mostly for synchronization purposes. The logical data channels on an ADSL link are grouped together as either downstream channels or upstream channels. The downstream channels are simplex in nature and are designated AS0, AS1, AS2, and AS3. Each channel has an allowable data rate up to 8, 4, 3, and 1.5 Mbps, respectively. The duplex channels are named LS0, LS1, and LS3. These support different data rates in upstream and downstream directions and are usually configured as upstream channels by setting the downstream channel data rate equal to 0.

The presence of many different logical channels enables ADSL to support a wide variety of applications since different channels operate at different data rates. However, the total physical bandwidth available in the channel should be greater than the sum of bandwidths of all the logical channels (since a portion of the bandwidth will be consumed by control/synchronization signals).

Since different carrier frequencies are employed for traditional voice transmission, upstream data and downstream data, ADSL uses frequency-division multiplexing (FDM) to multiplex these different signals onto one physical channel. Voice transmission is carried out by letting POTS use the lower-frequency band (0–3400 Hz), while the downstream data is transferred in the higher-frequency band (138 kHz–1.104 MHz). A guard band of approximately 26 kHz is placed between the POTS band and the upstream band to reduce the possibility of interference between voice conversations and data transmission operations.

In sharp contrast to voice band or DSL technologies, cable modems capitalize on the existence of a network of coaxial cables (that are used primarily for video applications) to transmit data. Since the coaxial cables support high bandwidths, cable modems could be used for high-speed data transfers. This is a prime reason for cable technology offering formidable competition to DSL.

Cable systems, however, have traditionally supported only downstream traffic and the available bandwidth is shared among several users since a single line serves many cable subscribers. Therefore, the cable system had to be reconditioned to support upstream traffic from the subscriber back to the coaxial distribution point.

A typical cable system consists of an uplink site that encodes and modulates video content obtained from various sources (tapes, DVDs, live feeds, etc.). The modulated video content is transmitted to downlink site via a satellite. The video content is transmitted from the downlink site to the cable subscribers through a network of coaxial cables. This transfer of video information from the downlink site to the cable subscribers is accomplished with the aid of headend (HE). An HE is responsible for modulating and scrambling each channel that is supported on the cable system and transmitting the scrambled information onto the local hybrid fiber coaxial cable) (HFC) distribution box. Subscribers connect to the HFC through a series of local taps (where each HFC could service 500–2000 homes per fiber node). Data communication is enabled into the cable system by incorporating routing/switching functionality in the HE. This is accomplished by using a broadband router (like a Cisco uBR 7200 series router) for data connectivity. At the customer premise, an additional device would be needed to allow users to utilize the data communication facilities that are available on the same coaxial line that delivers audiovideo content. This additional device is referred to as a *cable modem* (CM).

The downstream communication (from the cable company to the subscriber) is carried out in the frequency range of 54–860 MHz. The upstream communication (from the subscriber to the cable company) is carried out in the frequency range of 5–42 MHz. The manner in which data are transferred over a cable network is specified by DOCSIS (data over cable service interface specification). Typically, a CMTS (cable modem termination system) is employed at the HE to modulate (or demodulate) the signals sent (or received) from the CM. The CM is associated with the customer premise equipment (CPE) and is responsible for communicating with the CMTS for data communication. CMs support DOCSIS defined connectors, namely, RJ45 ports for Ethernet, RJ11 ports for voice, and F-connector for video.

DOCSIS-compliant devices require the presence of a few servers that provide information regarding IP addresses through the dynamic host configuration protocol (DHCP), time of day timestamps (as defined in RFC 868) and CM configuration files through TFTP (Trivial File Transfer Protocol).

When a CM is powered up, it scans the downstream channel for a synchronizing clock signal. The HE continues broadcasting information regarding upstream channel descriptors, upstream channel frequencies, downstream channel descriptors, and other data that the CM can use for determining the details of the channels it can use for upstream and downstream communication. Once the CM recognizes the upstream and downstream channels, it begins identifying the bandwidth that can be used for communication.

At this stage, the CM–HFC interface line protocol is considered to be up but the CM is not yet ready to start transferring data with other hosts on the global Internet since it does not yet have an IP address. A DHCP server provides the CM with an IP address, a default gateway, the address of the TFTP server, the CM configuration filename, and the address of a time-of-day (ToD) server that the CM can use for synchronizing its operations. The CM uses the address of the TFTP server to obtain the required files to configure itself as a network entity to handle data transfers across the cable network.

See also **Cable Modems**

Additional Reading

Abe, G., and A. Buckley: *Residential Broadband*, Cisco Press, Indianapolis, IN, 1999.

Biglieri, E., D. Divsalar, P.J. Mclane, and M.K. Simon: *Introduction to Trellis-Coded Modulation with Applications*, Macmillan, New York, NY, 1991.

Proakis, J.G., and M. Salehi: *Fundamentals of Communication Systems*, Pearson Education, New York, NY, 2004.

Raushmayer, D.J.: *ADSL/VDSL Principles*, Macmillan Technical Publishing, New York, NY, 1999.

Taub, H., and D.L. Schilling: *Principles of Communication Systems*, 2nd Edition, The McGraw-Hill, Companies, Inc., New York, NY, 1996.

Web References

Recommendation V.42bis, *Data Compression Procedures for Data Circuit Terminating Equipment (DCE) Using Error-Correcting Procedures*, ITU-T, http://www.itu.int

Recommendation V.21 (11/88) — *300 Bits per Second Duplex Modem Standardized for Use in the General Switched Telephone Network*, ITU-T, http://www.itu.int

Recommendation V.32 (03/93), *A Family of 2-Wire, Duplex Modems Operating at Data Signalling Rates of up to 9600 Bit/s for Use on the General Switched Telephone Network and on Leased Telephone-Type Circuits*, ITU-T, http://www.itu.int

Recommendation V.34 (02/98), *A Modem Operating at Data Signalling Rates of up to 33 600 Bit/s for Use on the General Switched Telephone Network and on Leased Point-to-Point 2-Wire Telephone-Type Circuits*, ITU-T, http://www.itu.int

Recommendation V.90 (09/98), *A Digital Modem and Analogue Modem Pair for Use on the Public Switched Telephone Network PSTN at Data Signalling Rates of up to 56,000 Bit/s Downstream and up to 33 600 Bit/s Upstream*, ITU-T, http://www.itu.int

Recommendation T.4 (04/99), *Standardization of Group 3 Facsimile Terminals for Document Transmission*, ITU-T, http://www.itu.int

Ravi Bhagavathula
Hyuck Kwon
Wichita State University, Wichita, KS

MODERATOR. A substance used to slow down neutrons by means of collisions. Moderators play an important part in the design and operation of nuclear reactors. Moderators *thermalize* neutrons to an energy of about 0.025 eV. See also **Nuclear Power Technology**.

MODE WATER. A term for water of exceptionally uniform properties over an extensive depth range, caused in most instances by convection. Mode waters represent regions of water mass formation; they are not necessarily water masses in their own right but contribute significant volumes of water to other water masses. Because they represent regions of deep sinking of surface water, mode water formation regions are atmospheric heat sources. Subantarctic Mode Water is formed during winter in the subantarctic zone just north of the subantarctic front and contributes to the lower temperature range of central water; only in the extreme eastern Pacific Ocean does it obtain a temperature low enough to contribute to Antarctic Intermediate Water. Subtropical Mode Water is mostly formed through enhanced subduction at selected locations of the subtropics and contributes to the upper temperature range of central water. Examples of Subtropical Mode Water are the 18 °C (64 °F) water formed in the Sargasso Sea, Madeira Mode Water formed at the same temperature but in the vicinity of Madeira, and 13 °C (55 °F) water formed not by surface processes but through mixing in Agulhas Current eddies as they enter the Benguela Current.

AMS

MODULAR ARITHMETIC. See **Encryption**.

MODULATION. The process, or the result of the process, whereby some characteristic of one wave is varied in accordance with some characteristic of another wave. Usually one of these waves is considered to be a carrier wave while the other is a modulating signal. The various types of modulation, such as amplitude, frequency, phase, pulse width, pulse time, and so on are designated in accordance with the parameter of the carrier which is being varied.

Amplitude modulation (AM) is easily accomplished and widely used. Inspection of Fig. 1 shows that the voltage of the amplitude modulated wave may be expressed by the following equation

$$v = V_c(1 + M\sin\omega_m t)\sin\omega_c t$$

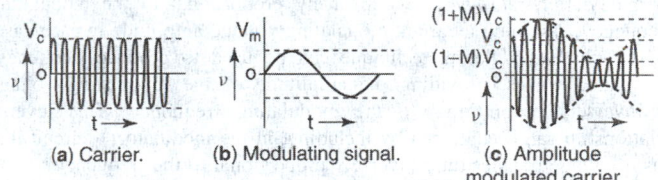

(a) Carrier.   (b) Modulating signal.   (c) Amplitude modulated carrier

Fig. 1.   Amplitude modulation.

where ω_c and ω_m are the radian frequencies of the carrier and modulating signals, respectively. The modulation index M may have values from zero to one. When the trigonometric identity $\sin a \sin b = \frac{1}{2}\cos(a - b) - \frac{1}{2}\cos(a + b)$ is used in the equation above, this equation becomes

$$v = V_c\sin\omega_c t + \frac{MV_c}{2}\cos(\omega_c - \omega_m)t - \frac{MV_c}{2}\cos(\omega_c + \omega_m)t$$

This equation shows that new frequencies, called side frequencies or side bands, are generated by the amplitude modulation process. These new frequencies are the sum and difference of the carrier and modulating frequencies.

Amplitude modulation is accomplished by mixing the carrier and modulating signals in a nonlinear device such as a vacuum tube or transistor amplifier operated in a nonlinear region of its characteristics. The nonlinear characteristic produces the new side-band frequencies. Frequency converters or translators and AM detectors are basically modulators. The various types of pulse modulation are actually special types of amplitude modulation.

Frequency modulation (FM) is illustrated by Fig. 2. The frequency variation, or deviation, is proportional to the amplitude of the modulating signal. The voltage equation for a frequency modulated wave follows.

$$v = V_c \sin(\omega_c t + M_f \sin\omega_m t)$$

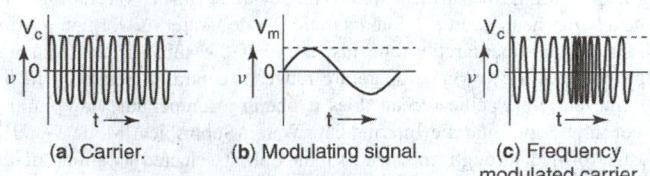

(a) Carrier.   (b) Modulating signal.   (c) Frequency modulated carrier

Fig. 2.   Frequency modulation.

The modulation index M_f is the ratio of maximum carrier frequency deviation to the modulating frequency. This ratio is known as the deviation ratio and may vary from zero to values of the order of 1,000. FM requires a broader transmission bandwidth than AM but may have superior noise and interference rejection capabilities. A large value of modulation index provides excellent interference rejection capability but requires a comparatively large bandwidth. The approximate bandwidth requirement for a frequency modulated wave may be obtained from the following relationship

$$\text{Bandwidth} = 2(\text{Modulating frequency})(M_f + 1)$$

The noise and interference characteristics of FM transmission are normally considered satisfactory when the modulation index or deviation ratio is five or greater.

Phase modulation is accomplished when the relative phase of the carrier is varied in accordance with the amplitude of the modulating signal. Since frequency is the time rate of change of phase, frequency modulation occurs when the phase-modulating technique is used and vice versa. In fact, the equation given for a frequency-modulated wave is equally applicable for a phase-modulated wave. However, the phase-modulating technique results in a deviation ratio, or modulation index, which is independent of the modulating frequency, while the frequency modulating technique results in a deviation ratio which is inversely proportional to the modulating frequency, assuming invariant modulating voltage amplitude in each case.

The phase-modulating technique can be used to produce frequency-modulated waves, providing the amplitude of the modulating voltage is inversely proportional to the modulating frequency. This inverse relationship can be obtained by including, in the modulator, a circuit that has a voltage transfer ratio inversely proportional to the frequency.

See also **Modems**.

MODULATORS. See **Sensors: Optical**.

MODULUS. 1. The absolute value of a complex number. It may be interpreted as the length of a vector representing the number in complex space. Thus, the modulus of $(a + ib)$ is $(a^2 + b^2)^{1/2}$.

2. The modulus of common logarithms is $\log e = 0.434294\ldots$, the factor which converts a natural logarithm to a common logarithm. Similarly, the modulus of natural logarithms is $\ln 10 = 2.302585\ldots$

3. A parameter which occurs in integrals or elliptic functions.

4. A formula, coefficient, or constant that expresses a measure of a property, force, or quality, such as of elasticity, efficiency, density, or strength.

MODULUS OF ELASTICITY. The ratio of the unit stress to the unit deformation of a structural material is a constant, as long as the unit stress is below the proportional limit, and is called the modulus of elasticity. The shearing modulus of elasticity is frequently called the modulus of rigidity. See also **Elastic Constants and Moduli**; and **Elasticity**.

MODULUS OF RUPTURE. The modulus of rupture in bending of a material is found by testing a transversely loaded beam of constant cross section to failure, and substituting the maximum bending moment, the moment of inertia of the cross section, and the distance from the neutral axis to the extreme fiber in the flexure formula:

$$S_M = \frac{Mc}{I}$$

The torsional modulus of rupture is obtained by testing a shaft of constant, circular cross section to failure and then substituting the maximum torque, polar moment of inertia of the cross section and the radius in the torsion formula:

$$S_s = \frac{Tc}{J}$$

The bending or torsional modulus of rupture may be used to predict the maximum bending or torsional moment which a member can resist.

MOHAIR. This very resilient hair is obtained from the Angora goat. The staple length ranges from 5 to 8 inches (12.5 to 20 centimeters), but Turkish fibers go up to 10 inches (25.5 centimeters). Mohair provides a characteristic crisp, resilient, and slightly scratchy hand to fabrics even when used in very low percentages with other fibers. See also **Fibers**; and **Goats and Sheep**.

MOHOROVICIC, ANDRIJA (1857–1936). Mohorovici was born in Volosko, Istria, Austrian Empire. He earned his degree in mathematics and physics at the University of Prague. He studied seismic waves and is best remembered for his discovery of the Mohorovicic Discontinuity, which led to the understanding that the Earth has a thin and brittle crust.

See also **Earth**.

J. M. I.

MOHS SCALE. An empirical scale of the hardness of mineral or mineral-like materials originally consisting of 10 values, ranging from talc, with a rating of 1, to diamond, with a rating of 10. The rating is based on the ability to each material to scratch the one directly below it in the series. The number of materials has been expanded from 10 to 15 with the addition of several synthetically produced substances (e.g., silicon carbide) between the original 9 to 10 positions. The scale is named after the German mineralogist, Friedrich Mohs (1773–1839). See also **Hardness**; and **Mineralogy**.

MOIRE PATTERN. If one draws a regular pattern, such as vertical lines of a given width and spacing, on a transparent sheet and then overlays this sheet onto another sheet that is ruled with lines, but of somewhat differing line widths and spacings—and then moves the two ruled surfaces horizontally with relation to each other, a shimmering effect will be noted. This is because of differences in the reinforcement of the lines and spaces one over the other. Walker (1978) suggests an interesting experiment with a comb and mirror. As one views the handheld comb against the mirror image of the comb (at just the right distances from the mirror), various periodic patterns will be noted along the length of the comb. This occurs because the teeth of the comb are "in step" in some locations and "out of step" in other locations. By changing the distance of the comb from the mirror, keeping the comb parallel to the mirror, different Moire patterns will be observed. Possibly this principle was first put to use in connection with various novel devices for Victorian entertainment. A present very useful application of the principle is in connection with dimension measurement and guidance systems for automated machine tools. A crisp scientific definition is "a pattern resulting from interference beats between two sets of periodic structures in an image."

Additional Reading

Cassin, C.: *Visual Illusions in Motion: With Three Different Moire Screens*, Dover Publications, Mineola, NY, 1997.

Grafton, C. Belanger: *Optical Designs in Motion: With Moire Overlays*, Dover Publications, Mineola, NY, 1990.

"Moire Effects, the Kaleidoscope and Other Victorian Diversions," by Jearl Walker, *Sci. Amer.*, **239**, 6, 182–186 (December 1978).

Patorski, K.: *Handbook of the Moire Fringe Technique*, Elsevier Science, New York, NY, 1993.

MOISSAN, HENRI (1852–1907). A Native of Paris, Moissan was a professor at the School of Pharmacy from 1886 to 1900 and at the Sorbonne from 1900 to 1907. At the former institution, he first isolated and liquefied fluorine in 1886 by the electrolysis of potassium acid fluroide and anhydrous hydrogen fluoride. His work with fluorine undoubtedly shortened his life as it did that of many other early experimenters in the field of fluorine chemistry. He won great fame by his development of the electric furnace and pioneered its use in the production of calcium carbide, making acetylene production and use commercially feasible in the preparation of pure metals, such as magnesium, chromium, urnaium, tungsten, etc., and in the production of many new compounds, e.g., silicides, carbides, and regractories. In 1906, he was awarded the Nobel Prize in chemistry.

MOIST-ADIABATIC LAPSE RATE (OR SATURATION-ADIABATIC LAPSE RATE). The rate of decrease of temperature with height along a moist adiabat.

It is given approximately by Γ_m in the following:

$$\Gamma_m = g \frac{1 + \dfrac{L_v r_v}{RT}}{c_{pd} + \dfrac{L_v^2 r_v \varepsilon}{RT^2}},$$

where g is gravitational acceleration, c_{pd} is the specific heat at constant pressure of dry air, r_v is the mixing ratio of water vapor, L_v is the latent heat of vaporization, R is the gas constant for dry air, ε is the ratio of the gas constants for dry air and water vapor, and T is temperature. This expression is an approximation to both the reversible moist adiabatic lapse rate and the pseudoadiabatic lapse rate, with more accurate expressions given under those definitions. When most of the condensed water is frozen, this may be replaced by a similar expression but with L_v replaced by the latent heat of sublimation.

See also **Pseudoadiabatic Lapse Rate**; and **Sublimation**.

AMS

MOIST CLIMATE. See **Climate**.

MOIST CONVECTION. Atmospheric convection in which the phase changes of water play an appreciable role. All cumuliform clouds are manifestations of moist convection. The enthalpy exchange between condensing water vapor or freezing liquid water and air (see **Latent Heat**) is a major contributor to the positive buoyancy of updrafts, while the reverse exchange between air and evaporating water or melting ice contributes strongly to the negative buoyancy of downdrafts.

See also **Clouds and Cloud Formation**.

MOIST STATIC ENERGY. A thermodynamic variable (analogous to equivalent potential temperature) calculated by hypothetically lifting air adiabatically to the top of the atmosphere and allowing all water vapor present in the air to condense and release latent heat:

$$s_e = C_p T + gz + L_v r,$$

where g is gravitational acceleration, L_v is the latent heat of vaporization, C_p is the specific heat at constant pressure for air, T is absolute temperature, z is height above some reference level (either the local surface at $z = 0$ or the height where the ambient pressure is 100 kPa), and r is the water vapor mixing ratio in the air.

AMS

MOISTURE-CONTINUITY EQUATION. The water vapor storage equation as applied to the atmosphere. The general form of the equation is written

$$\frac{dS}{dt} = I + E - O - P,$$

where I is the atmospheric moisture inflow, E the evapotranspiration from the ground, O the atmospheric moisture outflow, P precipitation, and dS/dt the time rate of change of moisture storage in the portion of the atmosphere under consideration. In practice the equation is more commonly applied to

a finite interval of time and the various terms become mean values in this interval.

AMS

MOISTURE FACTOR. One of the simplest measures of precipitation effectiveness, given by Lang as

$$\text{moisture factor} = \frac{P}{T},$$

where P (cm (in) is precipitation and T (°C (°F) mean temperature for the period in question.

This index recognizes only that as temperature increases, the effective moisture decreases due to greater evaporation. A number of greater refinements of this concept exist: De Martonne's index of aridity; Angström's humidity coefficient; Gorczyński's aridity coefficient; Thornthwaite's precipitation-effectiveness index and moisture index; and Köppen's formulas for outlining steppe climate and desert climate.

MOISTURE INDEX.

1. That portion of total precipitation used to satisfy plant (vegetation) needs.
2. As used by C. W. Thornthwaite in his 1948 climatic classification: an overall measure of precipitation effectiveness for plant growth that takes into consideration the weighted influence of water surplus and water deficiency as related to water need and as they vary according to season. For a given station, it is calculated by the formula

$$I_m = \text{humidity index} - 0.6(\text{aridity index}),$$

which becomes

$$I_m = \frac{100s - 60d}{n},$$

where I_m is the moisture index, s the water surplus, d the water deficiency, and n the water need. The calculation of s and d is made on a normal month-to-month basis, with s being the total surplus from all months having a water surplus, and d the total of all monthly deficiencies; each is represented by the difference between monthly precipitation and monthly potential evapotranspiration (in centimeters or inches). Here n is the annual potential evapotranspiration. The moisture index replaced Thornthwaite's previously used (1931) precipitation-effectiveness index.

See also **Palmer Drought Severity Index (PDSI)**; and **Precipitation-Effectiveness Index**.

Additional Reading

Thornthwaite, C.W.: "An Approach toward a Rational Classification of Climate," *Geographical Review*, **38**, 55–94 (1948).
Thornthwaite, C.W.: "Climates of North America According to a New Classification," *Geographical Review*, **21**, 633–655 (1931).

AMS

MOJAVE SPACEPORT. See **Spaceports U.S**.

MOLAL CONCENTRATION. A one molal solution contains one mole of a particular substance (the solute) in 1,000 grams of solvent. Thus, a 0.5 molal solution of potassium chloride in water contains $0.5 \times$ (gram-molecular weight of KCl = 74.555), or 37.278 grams of the salt in 1,000 grams of H_2O. See also **Molar Concentration**; and **Normal Concentration**.

MOLAR CONCENTRATION. A one molar solution contains one mole of a particular substance (the solute) in 1,000 milliliters of solution. Thus, a 0.5 molar solution of potassium chloride in water will be prepared by placing $0.5 \times$ (gram-molecular weight of KCl = 74.555), or 37.278 grams of the salt in a vessel and then adding H_2O, while thoroughly mixing to assure complete solution of the salt, until a total volume of 1,000 milliliters of solution is obtained. Molar is abbreviated M. Thus, the solution in the foregoing example would be $0.5M$ KCl. Molar solutions sometimes are referred to as *formal* solutions, not to be confused with normal solutions. See also **Molal Concentration**; and **Normal Concentration**.

MOLAR HEAT. The product of the gram-molecular weight of a compound and its specific heat. The result is the heat capacity per gram-molecular weight.

MOLASSES. [CAS: 68476-78-8]. A type of syrup, molasses is a by-product of the sugar industry. It is the mother liquor remaining after crystallization and removal of sucrose from the juices of sugar cane or sugar beet and is used in a variety of food and nonfood applications. Molasses, first produced from sugarcane in China and India centuries ago and later in Europe and Africa, was introduced as the by-product of cane-sugar production into Santo Domingo by Columbus in 1493. During Colonial times, molasses was very important to the American colonies for the production of rum. In 1733, the British Parliament passed the Molasses Act to tax molasses imported from foreign countries. This attempt to restrict trade was ignored by the colonies and was, in part, responsible for the American Revolution.

Manufacturing

Raw sugar is produced from sugarcane by a process that involves extraction of the sugar in water, treatment to remove impurities, concentration, and several crystallizations. After the first crystallization and removal of first sugar, the mother liquor is called first molasses. First molasses is recrystallized to obtain a second lower quality sucrose (second sugar) and a second molasses. After a third crystallization, the third molasses contains considerable nonsucrose material, and additional recovery of sucrose is not economically feasible. The third molasses is sold as blackstrap, final, or cane molasses. Raw sugar obtained from the above process is mixed with water to dissolve residual molasses and then separated by centrifugation. This process is called affination and the syrup is referred to as affination liquor. The sugar is dissolved in water, treated to remove color and impurities, and subjected to several crystallizations to obtain refined sugar. The mother liquor from the final crystallization is combined with affination liquor and crystallized to produce a dark sugar (remelts) which is recycled to raw sugar. The remaining mother liquor is called refiners molasses and is similar to final molasses but usually of better quality.

In beet sugar manufacture, the beet juice does not contain reducing sugars such as fructose and glucose, which are present in cane juice, but may contain raffinose. Because of the absence of reducing sugars, sucrose level in beet molasses is not reduced to the same extent as for cane. Final molasses from beet contains ca 60 wt % sucrose (dry basis) compared to 30 wt % sucrose (dry basis) in cane molasses. Treatment of diluted beet molasses with calcium oxide precipitates sucrose as tricalcium sucrate (Steffen process), which is recycled to the incoming hot beet juice. During recycling, raffinose accumulates in the final molasses and retards crystallization if not removed. Therefore, a portion of the final molasses, called discard molasses, is periodically removed.

Ion-exclusion chromatography is increasingly being used to remove sugar from beet molasses. As much as 90% sugar recovery is achieved by this technique, which involves passing diluted and clarified molasses through a column containing a strong acid cation exchange resin. Sucrose is absorbed on the resin and nonsucrose is recycled back to the sugar process. The resulting molasses contains 12–20% residual sugar compared to traditional molasses that contains sugar.

High test molasses (invert molasses) is produced from cane sugar when sucrose manufacture is restricted because of overproduction. The cane sugar at ca 55 wt % solids is enzymatically converted to invert syrup to prevent crystallization and evaporated to a syrup. The product is used in the same applications as blackstrap molasses.

Molasses from other sources include citrus and corn sugar (hydrol) molasses. Citrus molasses is produced from citrus waste and contains 60–75% sugars. Corn sugar molasses is the mother liquor remaining after dextrose crystallization and contains a minimum of 43% reducing sugars expressed as dextrose.

Molasses is shipped in drums, barrels, tank trucks, tank cars, barges, and sea vessels. Because of high viscosity, molasses must be heated in some situations to facilitate pumping. However, prolonged heating must be avoided to prevent caramelization. See also **Sugar** and **Syrups**.

Composition

Molasses composition depends on several factors, eg, locality, variety, soil, climate, and processing. Cane molasses is generally at pH 5.5–6.5 and contains 30–40 wt % sucrose and 15–20 wt % reducing sugars. Beet molasses is ca 7.5–8.6 pH, and contains ca 50–60 wt % sucrose, a trace of reducing sugars, and 0.5–2.0 wt % raffinose. Cane molasses contains less ash, less nitrogenous material, but considerably more vitamins than beet molasses.

Uses

The primary use of molasses is in animal feed. Molasses, which provides a carbohydrate source, salts, protein, vitamins, and palatability, may be used directly or mixed with other feeds. The carbohydrate content of 24.6 L (6.5 gal) of blackstrap molasses is considered to be equal to 0.035m³ (one bushel) of corn as measured by the energy produced from 0.035m³ of corn and the amount of molasses required to produce the same amount of energy. When molasses is less expensive than corn, sales increase; when the reverse is true, sales decrease.

Molasses is also used as an inexpensive source of carbohydrate in various fermentations for the production lactic acid, citric acid, monosodium glutamate, lysine, and yeast. Blackstrap molasses is used for the production of rum and other distilled spirits.

Food applications utilize first and second molasses in baking (bread, cakes, cookies) for the molasses flavor. Molasses is also used in curing of tobacco and meats, in confections such as toffees and caramels, and in baked beans and glazes.

MOLD. See **Fungus**; and **Yeasts and Molds**.

MOLD INHIBITORS. *(Antimycotics)* See **Bakery Processes, Yeast-Raised Products**.

MOLE CRICKET *(Insecta, Orthoptera)*. Burrowing crickets whose large forelegs give them a superficial resemblance to moles.

MOLECULAR AND SUPERMOLECULAR ELECTRONICS. Professor Gareth Roberts FRS, Director of Research, Thorn EMI plc, and Professor of Engineering Science at the University of Oxford.

The microelectronics and optoelectronics industries will continue to grow vigorously well into the 21st Century. Until now, they have relied largely on inorganic materials such as silicon and lithium niobate in single crystal form. However, as the perceived limitations inherent in these materials begin to restrict the realization of more complex system designs, more attention is being focussed on the *organic* solid state. The richness of the variety of organic molecular materials offers enormous potential compared with the relative paucity of structures achievable with inorganic compounds, even when due allowance is made for the recent exciting developments in inorganic quantum well semiconductors (Kelly and Weisbach 1986).

The ability to enlist the assistance of synthetic organic chemists to produce organic materials with tailored properties has, of course, already been used to advantage in several applications. The best known is that of liquid crystals and their use in displays and digital thermometers. New phenomena and types of molecule are still being discovered and seem likely to lead to successful large area displays for high definition television and to high density information stores. Other examples are piezoelectric polymers as very sensitive hydrophones for submarine detection, photoconducting polymers for electrocopying, and photochromic molecules for reversible high density optical storage and signal processing. Biosensors and chemical sensors for converting specific biochemical or chemical solute or gas interactions into electrical signals for use in industrial or medical diagnostics can also be mentioned. All are examples of 'Molecular Electronics,' that is, they are fields in which organic molecular materials perform an *active* function in the processing of information and its transmission and storage. This definition does not embrace their use in possible roles such as insulation, adhesion or encapsulation. Thus, molecular electronics is interpreted broadly and is not limited to phenomena concerning the movement of electrons only. Electromagnetic radiation, polarization phenomena, and various forms of electromechanical and electrochemical energy transfer are also included in the definition. A common feature of all the examples cited and of the area in general is that progress is achieved *via* 'molecular engineering' that is, using the ability to manipulate the architecture of a material to optimize a specific physical parameter.

An alternative definition exists for molecular electronics; this is formulated in terms of switching on a molecular scale and is aimed more at the long term problem of fabricating molecular electronic devices suitable for assembly into a computer (Carter 1986). It is interesting to note that only a modest diminution in the size of electronic circuit components is required before the scale of individual molecules is reached; in fact, many existing circuit elements could already be accommodated within the area occupied by a leukaemia virus. An illustration of the rapid evolution of silicon based microelectronics may be gained by studying Figure 1. If this systematic reduction in feature size suggested by the good log-linear graph is sustained, then the extrapolated line indicates device geometries with nanometer dimensions in approximately thirty years' time! The requirements of reliability and testing of complex structures suggest a system approach rather than the traditional one, which uses the properties of individual circuit elements. it appears likely that sequential designs, because of their vulnerability, will be abandoned in favor of supermolecular arrays acting as concurrent processor networks. For this reason, and to differentiate it from 'molecular electronics,' signal transport and control in nanometer scale assemblies is referred to as 'supermolecular electronics'.

precise thickness, coupled with the degree of control over their molecular architecture has now firmly established a role for such layers in thin film technology. It seems likely that an understanding of their physical properties and utilization in molecular based devices will assist generally in the transition from molecular to supermolecular electronics. The likely pattern for this evolutionary process is given in Figure 2. The plan is speculative and envisages three main stages before the advent of applied, complex supermolecular systems. In the near term it predicts that current research and development will result in organic materials ousting inorganic materials in existing applications, e.g. optoelectronics. Hybrid technologies, comprising a novel device partly based on conventional solid state materials and partly on organic compounds, could possibly follow in the mid-term, say 5 or 8 years. Thereafter, at some stage dictated by the emergence of reliable, stable supermolecular assemblies, a true watershed will occur. When this occurs, materials and process technologies of conventional solid state devices will be superceded by radically new types of devices. This era will be equivalent to that witnessed about forty years ago when inorganic semiconductor materials were developed. Just as then, novel effects should be discovered and these in turn will lead to the fabrication of novel devices that can be integrated in novel systems.

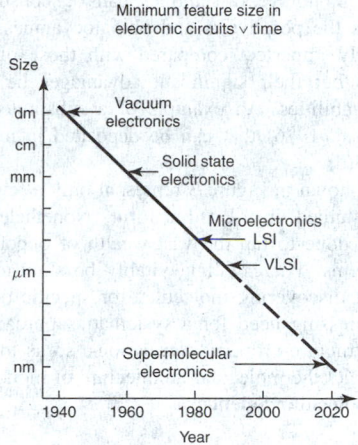

Fig. 1. Linear feature size of commercial electronic circuits versus time; the bottom arrow speculatively points to an era where switches on a molecular scale will have application in computer systems.

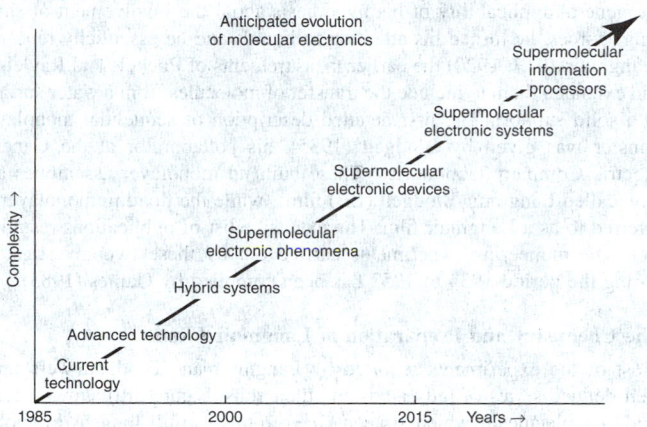

Fig. 2. The anticipated evolution of molecular electronics to supermolecular electronics.

Animals can solve with little apparent effort the tasks for which advanced supermolecular information processors are required; accordingly, some enthusiasts have speculated that the thirty-year time scale could be foreshortened using biological molecules. However, such thoughts are misplaced, for nature constructs organic materials for purposes other than those required for logic and memory functions. What can be learnt from studying biological systems are the scientific principles of organization and assembly; eventually, it may be possible to apply these concepts to construct synthetic supermolecular arrays. These will require nonlinear interactions between neighbors so that configurations of 'on' and 'off' elements propagate with time. However, it will be a difficult task to control the correlations between the parallel processing elements whether they be based on electronic, vibrational or magnetic effects. A great deal of fundamental research will be required to identify molecular systems with the necessary degree of cooperativity and to develop the routine analytical techniques designed for investigations in three dimensions with molecular scales of resolution.

Three-dimensional integration is extremely difficult using silicon technology while chemically nonspecific methods such as molecular beam epitaxy are relatively crude. A self-assembly technique is a far more attractive alternative if regular, three-dimensional, ordered structures are required. This involves the construction of unique assemblies whose architectures depend on the shapes and charge distributions of the units from which they are built, as distinct from the methods used to assemble them. There is a considerable degree of self assembly associated with organic monomolecular films deposited using the Langmuir trough technique (Roberts 1985). About half a century after the report of their discovery, intense interest is now being displayed in these so called Langmuir-Blodgett films. Their

Molecular and supermolecular electronics are broad and interesting subjects. Moreover, they require a multidisciplinary approach where collaboration between biologists, chemists, computer scientists, electronic engineers and physicists is of paramount importance. To illustrate these features, this article concentrates on Langmuir-Blodgett films. Equal emphasis is placed on their importance in basic science and on their potential applications, especially in the area of electronics.

Historical Review of Langmuir-Blodgett Films

According to Tabor (1980) the earliest written record of observations of the spread of oil on water is in cuneiform on clay tablets, dating from Hammurabi's period (18th century B.C.) in Babylonia. The earliest technical application of organic monolayer films is believed to be the Japanese printing art called subminagashi, involving a suspension of submicron carbon particles and protein molecules spread on the surface of water. The distinctive patterns so formed can be transferred by lowering a sheet of paper onto the water surface. There are also many references dating from the classical times of Plutarch. Aristotle and Pliny describing the ability of oil spread on water to dampen surface waves and ripples. It was this property which attracted Benjamin Franklin, the versatile American statesman, to the subject. During his frequent visits to Europe in the 18th century A.D. to negotiate the sovereignty of his country with the French and the British he carried out his famous 'teaspoonful of oil' experiment. Often quoted and picturesque, Franklin's (1774) account to the Royal Society included the following phrases: "At length, being at Clapham where there is, on the common, a large pond, which I observed to be one day very rough with wind, I fetched out a cruet of oil, and dropped a little of it on the water. I saw it spread itself with surprising swiftness upon the surface ... I then went to the windward side, where (the waves) began to form and there the oil, though not more than a teaspoonful, produced an instant calm over a space of several yards square, which spread amazingly, and extended itself gradually till it reached the lee side, making all that quarter of the pond, perhaps half an acre, as smooth as a looking glass. After this, I contrived to take with me, whenever I went into the country, a little oil in the upper hollow joint of my bamboo cane, with which I might repeat the experiment as opportunity should offer; and I found it constantly to succeed."

Franklin must have been too preoccupied with political affairs to place his observations on a quantitative basis. Had he done so he might well have calculated that a volume of one teaspoonful (approximately 2 ml) spread over an area of nearly half an acre (200 m^2) leads to a surface coating approximately 1 nm thick. However, it is Lord Rayleigh (1890) who is given the distinction of first suspecting that the maximum extension of an oil film on water represents a layer one molecule in thickness. For a direct measurement on molecular sizes he was indebted to Pockels (1891) whose simple apparatus later became the model for what is now called a Langmuir trough; using very simple equipment he calculated the precise thickness of a monomolecular layer of castor oil on water to be 1 nm. This significant observation was not fully followed up until the pioneering work by Langmuir on the adsorption of gases or solutes by solids. In order to test the general applicability of his hypothesis about the involvement of short range forces, he turned his attention to liquids and he essentially repeated (Langmuir 1917; 1920) the earlier measurements of Pockels and Rayleigh, and extended them to include the transfer of molecules from a water surface to a solid support. The first detailed description of sequential monolayer transfer was given by Blodgett (1935), his collaborator at the General Electric Company laboratories. These built up monolayer assemblies are now called Langmuir-Blodgett (LB) films, while the floating monolayer is referred to as a Langmuir film. The extensive list of publications resulting from the pioneering experiments carried out by these two investigators during the period 1934 to 1952 has been compiled by Gaines (1983).

The Chemistry and Preparation of Langmuir Films

Most of the experiments reported by Langmuir and Blodgett were on a well-defined series of fatty acids and their salts. Figure 3(a) shows stearic acid, a molecule in which sixteen CH$_2$ groups form a long hydrophobic chain; the other end of the molecule terminates in a hydrophilic carboxylic acid group. When dissolved in a suitable solvent and spread on the surface of water, molecules may be compressed with the aid of a barrier. Figure 4 shows a plot of the surface pressure (differential surface tension) versus area occupied per molecule for stearic acid. The monolayer undergoes a number of phase transformations during compression; the well-defined sequence can be viewed as the two-dimensional analogue of the classical transitions observed with pressure-volume isotherms. However, it should be emphasized that some materials, while forming acceptable quality LB films, do not display the well-defined break points shown in Fig. 4.

Generally speaking, the approach to the synthesis of suitable molecules for examination with a Langmuir trough has been an ad hoc one and has relied on the modification of known materials. For example, the alkyl group of fatty acids may be replaced by chains containing one or more double bonds. The ω-tricosenoic acid (Barraud 1983) molecule shown in Figure 3(b), which is similar to stearic acid but contains a terminal double bond, displays all the essential film-forming qualities including solubility in convenient organic solvents, stability at the surface of water, shear resistance, stability against collapse, and suitable orientation features. It is relatively straightforward to attach long aliphatic chains to a molecule and spread a monolayer. However, this may well dilute the desirable properties of the basic molecule; moreover, for stability reasons, the presence of long side groups will severely restrict their practical applicability. It has therefore been recognised that the scope of the Langmuir trough technique would be considerably enhanced if interesting materials containing only short, stable, side groups could be formed into LB films. A good example is provided by the anthracene derivative (Vincett et al. 1979) shown in Figure 3(c); multilayers of excellent quality can be obtained, even though the alkyl group contains only four aliphatic carbons and the hydrophobic group is attached to the ring structure *via* only two methylene groups. Extremely robust monolayer assemblies can be constructed using dye molecules such as the porphyrins and phthalocyanines. In general their quality is relatively imperfect compared with those of the classic film forming materials but their significant advantages lie in their thermal and mechanical stabilities. An example of a substituted phthalocyanine molecule (Hue et al. 1986) that can be deposited in monolayer form is shown in Figure 3(d).

The molecules shown in Figure 3 represent only a few of the materials that have been studied in LB film form. Nonetheless, a great deal more needs to be done to tap the vast wealth of opportunities available with organic systems. There will inevitably be short-term opportunistic attempts aimed at discovering molecules for specific devices. However, there is a more pressing need for a systematic approach that will yield rules governing structure-property correlations, so as to enable scientists confidently to predict the molecular architecture of monolayer assemblies. See also **Macromolecular Science**.

Langmuir-Blodgett Film Deposition

An LB film is formed by transferring a floating monolayer onto a solid substrate. The quality of the Langmuir film and the surface pressure at which 'dipping' occurs is established using the type of isotherm shown in Figure 4. The subphase is normally ultra-pure water, because it is readily available and it has an exceptionally high value of surface tension. The composition of the subphase, including its purity, pH, and ionic strength, can have a profound influence on factors such as the solubility of the monolayer and segregation effects resulting in molecular aggregates or domains.

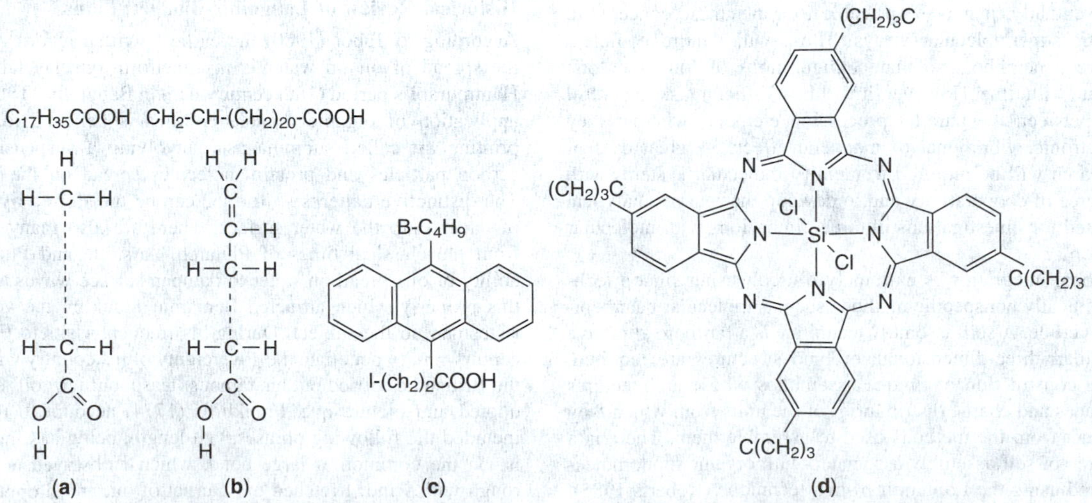

Fig. 3.   A selection of molecules used to form LB films: (a) fatty acid; (b) ω-tricosenoic acid; (c) 9-butyl-10-anthrylpropionic acid; (d) tetra-4-tert-butyl-10-phthalocyaninato silicon dichloride.

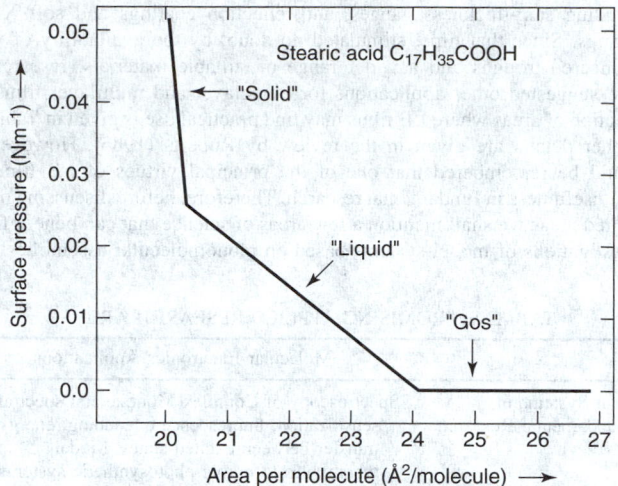

Fig. 4.   Surface pressure versus area characteristic for stearic acid.

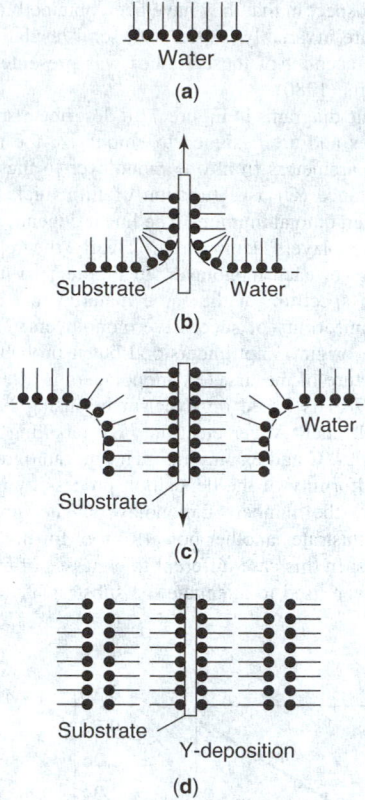

Fig. 5.   Langmuir-Blodgett film deposition (Y-type) on a hydrophilic substrate: (**a**) monolayer on the surface of water; (**b**) first layer on withdrawal; (**c**) second layer (second insertion); (**d**) substrate with three layers (after second removal).

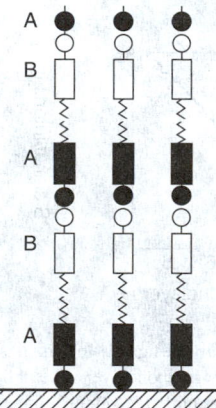

Fig. 6.   Alternating organic multilayer structure which enables a Y-type LB film to be produced of non-centrosymmetric character (molecular lego!).

Using conventional LB film technology, the substrate is raised and lowered vertically through a compact floating monolayer; the surface pressure at which this occurs is normally just above the 'knee' in the steeply rising sector of the isotherm indicating low compressibility in the monolayer. At this stage, if conditions are carefully controlled and appropriate molecules are used, one monolayer is transferred during each excursion through the subphase surface. The most common deposition mode (Y-type) is illustrated in Figure 5(a), where the molecules can be seen to stack in a head to head and tail to tail configuration. The floating molecules on a water surface are shown at the top left in this diagram. With a hydrophobic substrate (for example, a group III-V compound semiconductor), no pick-up occurs during the first immersion and the first monolayer is therefore deposited during the first withdrawal as shown in Figure 5(b). The surface is now hydrophobic and deposition does now occur during the next immersion into the water. Thus, one monolayer coverage is obtained on each traversal of the liquid surface. With a hydrophobic surface such as freshly etched silicon, pick up also occurs during the first insertion. Sometimes, the common deposition mode illustrated in Figure 5 is not followed and one of the other two possible configurations, X and Z-type, is observed, where transfer occurs only during immersion or withdrawal, respectively. The surface quality and chemical composition of the substrate is bound to control the nature of the deposited layer. When adhesion is poor, some researchers have resorted to the less satisfactory method of placing the substrate flat on the liquid surface, a technique first used by Langmuir and Schaefer.

Many modifications of the very early film balances have been described by Gaines (1983). However, the upsurge in interest in LB films has led to greater attention being placed on trough design and control systems to meet the stringent requirements of scientists and engineers. Therefore, modern instruments are relatively sophisticated and, for device-related work, need to be situated on anti-vibration Tables in clean environments. Although it is possible to automate most features, the primary benefit at the present time lies in efficient data collection and the ease with which data can be manipulated. For example, phase transitions are more apparent when the differential of the pressure-area isotherm is plotted. No difficulties are envisaged in scaling up the Langmuir trough or in the design of continuous fabrication arrangements. When an important practical application is discovered there will be a need to produce a specially designed trough capable, for example, of coating a moving belt or multiple wafers of silicon.

A recent development in trough design is worthy of special attention as it could have important commercial significance. It has arisen because of the need to produce non-centrosymmetric structures that display interesting non-linear physical effects. The conventional Y-type films are symmetrical in character and experience has shown that X and Y type layers, although non-symmetrical, are usually imperfect. Therefore, an alternative approach to producing noncentrosymmetric structures is to use alternate layers of two different materials where the contributions of adjacent molecules do not cancel. See Fig. 6. The additions of a fixed beam and a revolving center section to an automated constant perimeter barrier Langmuir trough enables the formation of an alternating Y-type structure of two different molecules spread in the two distinct areas of the subphase. The structural qualities of the LB films prepared in this way can be of high quality (Holcroft et al. 1985); another advantage of the rotating substrate arrangement, which is conducive to fast dipping, is that the meniscus, unlike that in the vertical dipping method, is always in the same direction.

Many different experimental techniques indicate that carefully prepared films of appropriate molecules do indeed possess a high degree of structural order. The reader is referred to the proceedings of the two international conferences on LB films for literature references describing the vast range of characterization experiments that have been employed (Roberts and Pitt 1983; Gaines 1985). These include ellipsometry, electron spin resonance, infrared dichroism, photoacoustic spectroscopy, secondary ion mass spectroscopy, surface potential, polarized X-ray and electron diffraction, neutron reflection and diffraction. Most of the electrical data for LB films are suspect in that they have been obtained for films deposited onto metals that are invariably coated with semi-insulating native oxides. A comprehensive account of these studies was presented in a review by Vincett and Roberts (1980).

The four separate diagrams in Figure 7 all describe results for fatty acids or their derivatives and are designed to emphasize the reproducibility of various physical parameters from one monolayer to the next. Figure 7(a) shows the capacitance (C) as a function of film thickness for cadmium arachidate deposited onto aluminium. The linear dependence of $C^{-'}$ versus the number of monolayers demonstrates clearly the repeatability of the dielectric thickness of each monolayer. In Figure 7(b) it is a band in the infrared reflection spectrum of the same material that has been used to demonstrate the uniformity of successive monolayers. The reason for the scatter around the origin is not understood but it probably reflects in this case that the structure of the first few monolayers is affected by the metal underlay. Figure 7(c) is based on experiments using barium stearate as the absorber for L shell Auger electrons. By labelling the molecules in these overlays with ^{14}C and examining their autoradiographs it is possible to confirm the uniformity of the deposition process by plotting the count of ^{14}C rays versus the number of monolayers. The final diagram in the set, Figure 7(d), illustrates another powerful tool for investigating organic coatings on metals. In this case different thicknesses of cadmium dimethyl arachidate have been used to attenuate the substrate X-ray photoemission signal.

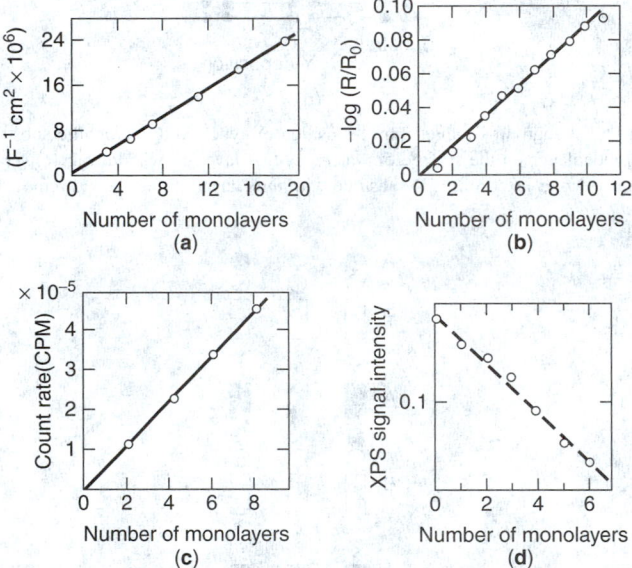

Fig. 7. These diagrams are designed to emphasize the reproducibility of various physical parameters in monolayer assemblies of different thicknesses: (**a**) reciprocal capacitance per unit area versus number of monolayers of cadmium arachidate on an aluminum substrate (Roberts et al., 1978); (**b**) absorption intensity versus number of monolayers for the symmetric carboxylate stretching mode of cadmium arachidate at 1432cm^{-1} (Allara and Swalen, 1982); (**c**) count rate of ^{14}C rays versus number of layers of barium stearate labelled with [14]C (Mori et al., 1980); (**d**) X-ray photoelectron signal intensity versus number of layers of cadmium dimethyl arachidate on silver (Brundle et al., 1979).

During the next few years many techniques-oriented scientists will be attracted to work on LB films because they provide interesting novel structures whose molecular architecture can be systematically controlled. The quality of the floating monolayer is also important and needs to be characterized as does the interface between the first deposited monolayer and the substrate. Fluorescence microscopy and Brillouin and Fourier transform infrared spectroscopies are currently being used to address these problems.

Applications of Langmuir-Blodgett Films

Following the pioneering work of their famous employees, the General Electric Company introduced several simple applications of LB films

including step-thickness gauges, anti-reflection coatings and soft X-ray gratings. Since that time, stimulated no doubt by the availability of well engineered troughs and a wider range of suitable materials, researchers have suggested other applications for monolayer and multilayer films. A selection of areas where LB films may find practical use is given in Table 1. Further details are given in the review by Roberts (1985). However, it should be remembered that one of the principal virtues of LB films is their usefulness in fundamental research. Therefore, before discussing more applied areas we shall mention a few areas of science that can benefit from investigations of model systems based on monomolecular assemblies.

TABLE 1. PROMISING APPLIED RESEARCH AREAS

Topic	Molecular Electronics Applications
Model Systems in Fundamental Research	Spectroscopy of Complex Monolayers: spectral sensitization, fluorescence quenching, energy transfer between excited states. Model membranes to mimic photosynthetic systems. Modification of solid surface properties. Examination of lipids, proteins and membrane phenomena; organic semiconductors.
Applied Chemistry	Surface chemistry and behavior of surfactants: catalysis; filtration/reverse osmosis membranes; adhesion; surface lubrication, e.g. magnetic tape; encapsulation.
Electron Beam Microlithography	Good sensitivity and contrast, acceptable plasma etching resistance; less scattering of electrons and therefore better resolution; negative and positive resists possible.
Integrated Optics and Storage Optics	Film thickness plus refractive index of film and hence guided wave velocity can be controlled with great precision; acceptable attenuation loss. Possible uses in conventional optics and optical data storage: photochromic and ablative systems. Optical sensors e.g. based on coated fibres.
Nonlinear Physics	Control of molecular architecture to produce asymmetric structures with high non-linear coefficients, e.g. in electro-optics, pyroelectric detectors, or acoustoelectric devices.
Dilute Radioactive Sources	Radioactive nuclide incorporated in conventional LB film; used to measure the ranges of low energy electrons.
Electronic Displays	Large area capability of LB films is an advantage; the monolayers can either be the active electroluminescent layer or used to enhance efficiency of an inorganic diode; passive application to align liquid crystal displays. Deposition of liquid-crystal type molecules also possible.
Photovoltaic Cells	Used as a tunnelling layer in an MIS solar cell or as an active layer in p-n junction diode, perhaps involving an inorganic/organic junction.
Two Dimensional Magnetic Arrays	Magnetic atoms e.g. Mn, periodically spaced in LB film; possible applications include magnetic control of superconducting junctions and bubble and magneto-optical devices.
Field Effect Devices	Accumulation, depletion and inversion regions possible with a variety of semiconductors; can therefore form the basis of several devices e.g. CCD, bistable switch, gas detector or pyro/piezo FET, is suitable LB films are used.
Biological Membranes	Attractive supporting membranes for commerical exploitation of biological material, e.g. immobilization of membrane bound enzymes in solid state sensors; ISFET type structures.
Supermolecular Structures	Speculative work aimed at superconductors, organic metals, 3D memory storage, molecular switches.

Model System in Basic Research

(a) **Energy transfer in complex monolayers:** The Langmuir trough technique provides a method of constructing simple artificial systems of co-operating molecules on a substrate. The pioneer in this field has been Kuhn (1983); the elegance of his and his colleagues' work is evident in their reviews of the subject. These describe the use of LB films to investigate intermolecular interactions and various photophysical and photochemical processes. Their supermolecular structures have mainly involved long chain fatty acids as matrices for appropriate synthetic dyes and have been ingeniously designed to clarify the different interactions that can occur between various molecules via photon, electron and proton transfer. An example of their research, designed to investigate the Förster type of energy transfer from a sensitising molecule, S, to acceptor molecule, A, is given in Figures 8 and 9. If S is a compound that absorbs in the ultraviolet part of the spectrum and fluoresces in the blue, while A absorbs in the blue and fluoresces in the yellow, then interesting effects are observed when the system is irradiated with ultraviolet light. If there is a sufficient distance between S and A, as in Figure 8(a), the fluorescence of S appears since A does not absorb UV radiation. However, below a certain threshold distance, as in Figure 8(b), the excitation energy of S is transferred to A and the yellow fluorescence of A is expected. Similar experiments based on fluorescence quenching indicate that the rate constant of the electron transfer decreases exponentially with increasing barrier thickness separating a donor chromophore and an electron acceptor. In the example shown in Figure 9, N,N'-dioctadecylthiacyanine has been used in conjunction with a viologen acceptor layer to observe the steady fluorescence intensities of the cyanine dye monolayer in the absence (I_0) and in the presence of the acceptor layer (I). The quantity, $[I_0/I) - 1]$ is proportional to the rate constant of the electron transfer; its linear dependence with d, the distance between the chromophores, is evidence of electron tunnelling. In a similar series of experiments it has been possible to investigate the energy transfer mechanism responsible for spectral sensitization.

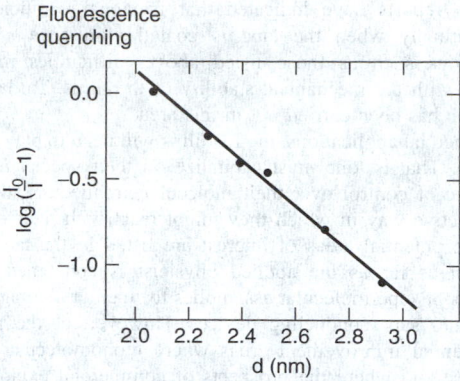

Fluorescence quenching

Fig. 9.   The fluorescence intensity I_0 of a donor dye is reduced to a value I in the presence of an acceptor dye. The logarithm of $[(I_0/I) - I]$ is shown as a function of d, the spacing between the donor and acceptor planes. (Mobius, 1981.)

the primary process in photosynthesis and for achieving an efficient photoinduced charge separation by appropriate modelling of potential profiles. Chlorophyll has been studied in this context. Some of the results may have relevance to solar photochemical conversion devices.

(c) Metal-ion incorporation: The addition of divalent ions into the liquid subphase in a Langmuir trough can increase both the shear resistance and cohesion of the monolayer. For this reason it is more common to find reports of studies on fatty acid salts than their acids. By adjusting the pH of the subphase it is possible to assemble multilayers containing metal ions separated by the width of an integral number of monolayers (for Y-type deposition, two monolayers). The ability to do this has been capitalized upon in several fundamental investigations, three of which will be mentioned here. The first of these relates to two-dimensional magnetic monolayers involving iron or manganese ions. Using electron spin resonance Pomerantz (1980) has demonstrated that at temperatures near 2 K, the resonance field and line-shapes were affected, thus signifying the rapid development of a large internal magnetic field. His results have been interpreted in terms of a predominantly anti-ferromagnetic state, but with a weak ferromagnetic component. Further experiments are required to clarify magnetic ordering in two-dimensional space.

In the area of surface science, X-ray photoemission is now used extensively to study organic materials on surfaces. In such experiments it is important to establish electron mean free path lengths as a function of kinetic energy. An example, of this type of investigation in LB films is illustrated in Figure 7(d); generally it is found that the mean free paths for ordered multilayers are significantly longer than those for conventionally produced polymers (Clark et al. 1981).

A third example showing the usefulness of using the Langmuir trough technique to provide matrices containing regularly spaced metal ions lies in radioactivity. Mori et al. (1980) used radioactive stearate monolayers in which some of the hydrogen atoms had been replaced by nuclides such as ^{51}Cr, ^{54}Mn, ^{55}Fe, ^{57}Co, ^{65}Zn and ^{109}Cd, to produce dilute and standard radioactive sources. By labelling the molecules with ^{14}C and examining autoradiographs he was able to confirm the uniformity of the deposition process as shown in Figure 7(c). Using conventional monolayers with well-controlled dimensions as overlays they were also able to demonstrate that Auger electrons from the L shell with an energy of approximately 0.5 keV are almost completely absorbed by fifteen monolayers of barium stearate. Experiments of this kind are of importance in fields such as medical physics and upper atmosphere science.

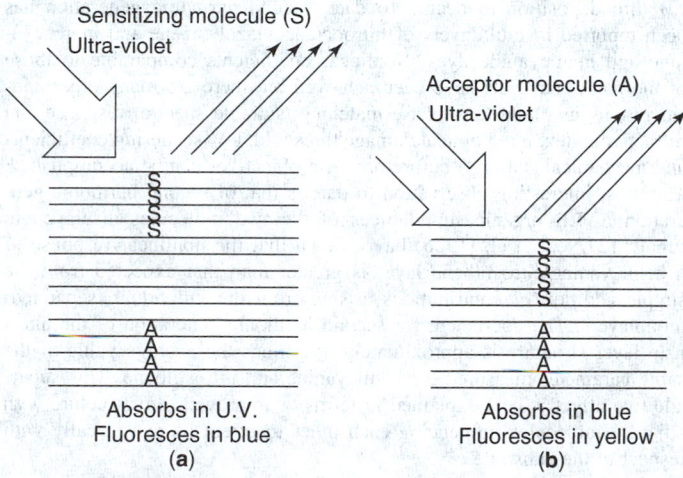

Sensitizing molecule (S)
Ultra-violet

Acceptor molecule (A)
Ultra-violet

Absorbs in U.V.
Fluoresces in blue
(a)

Absorbs in blue
Fluoresces in yellow
(b)

Fig. 8.   Schematic diagram showing basis of experiments designed to investigate energy transfer from a sensitizing molecule (S) to an acceptor molecule (A). The number of monolayers separating the two species governs the spectral response of the fluorescence spectrum. In **(b)** the separation distance is sufficiently small for the excitation energy of S to be transferred to A. In **(a)** the acceptor molecules does not absorb the ultraviolet radiation.

(b) Biological membranes: The physical structure and chemical nature of classical LB films gives them a close resemblance to naturally occurring biological membranes. For example, because the two ends of a lipid molecule have incompatible solubilities, they spontaneously organize themselves in the form of a bilayer, or essentially a two layer LB film. Scientists have suggested that they might provide a suitable model of the lipid membrane for probing the cooperative interactions between its constituents. However, caution must be exercised in assessing the biological relevance of this type of work and associated studies aimed at incorporating ionophores into phospholipid layers. Much of this research is targetted at novel integrated solid state devices incorporating biological molecules such as enzymes.

In some cases, LB films are useful to facilitate physical studies of biological molecules e.g. to measure the ionic permeability of reconstituted membranes. Supermolecular structures have also been designed to mimic

Promising Applied Research Areas

Langmuir-Blodgett films may have value in many applied areas of traditional interest to the industrial chemist such as adhesion, encapsulation and catalysis. The permeability characteristics of monolayer assemblies may also find application as synthetic membranes for ultra fine filtration, gas separation and reverse osmosis. For example, Albrecht et al. (1985) have proved the efficiency of polymeric diacetylene monolayers on semi-permeable supports in reducing the flow of CH_4. One interesting possibility lies in using LB monolayers as lubricants in magnetic tape technology.

Unpublished reports have indicated that frictional coefficients can be reduced markedly when the tape is coated with a few monolayers. In applications such as those listed above, difficulties may well be encountered with the mechanical stability of the films. To date relatively little research has been carried out in this area.

For commercial applications, the LB films will need to play an essential, integral role. That is, one must capitalize on their special features such as the degree of control over their molecular architecture, their thinness and the selective way in which they might react with their environment. Some of the potential areas of interest are listed in the table. The long term interest as far as the applied physicist is concerned, lies in the possible uses of supermolecular assemblies for memory storage, molecular switching, and superconducting devices. However, at the present time it is in potential improvement areas where monomolecular films show most promise and where the prospects of commercial exploitation seem reasonable in the medium term. A few of these areas are described below; most of the illustrations are based on work carried out in the author's research laboratory:

(a) Nonlinear physics: There is evidence that many organic molecules possess very high non-linear coefficients and therefore, if LB films with the required architecture can be formed, these could form the basis of novel devices. In order to avoid the symmetry inherent with conventional Y-type deposition, X and Z type films have been studied; some have displayed a permanent polarization with a strong component in a direction perpendicular to the substrate. However, as has already been mentioned, films produced in this way with their dipoles supposedly aligned in a common direction, are invariably of poorer quality than Y-type layers. a possible method of improving the structure is to use electric or magnetic fields to help align the molecules but efforts to orient films on the subphase and substrate have met with only limited success. The problem can be overcome by using organic superlattices based on alternating layers of two different materials (See Fig. 6). A good example is given in Figure 10 which shows a superlattice comprising acid and amine molecules whose dipole moments are in opposite senses but when deposited in Y-type LB film form are aligned in the same direction. Two areas of particular interest that would capitalize on this feature of organic superlattices are pyroelectricity and optoelectronics. Each of these will now be considered in turn.

crystals, such as triglycine sulfate, which possess a high pyroelectric coefficient, cannot be produced in thin film form. Thus the future development of relatively cheap thermal imaging systems with reasonable performance and an optimum thickness of approximately 0.5 μm, requires a materials breakthrough. A normal detector consists of a capacitor whose dielectric is an oriented pyroelectric material; using this type of device Christie et al. (1986a, b) have observed encouraging results using the alternate layer structure shown in Figure 10. The pyroelectric coefficient, p, has been determined using both dynamic and static detection techniques. For the simple fatty acid/fatty amine superlattice $p \simeq 1$ C cm^{-2} K^{-1} but more recent results involving a system where proton transfer occurs from the acid to the amine have yielded higher values, comparable with those observed for triglycine sulfate. The exploitation of this work will depend not only on the properties of the LB films but also on our ability to deposit the layers onto reticulated structures with low thermal mass.

Although many of the potential optical applications of LB films are in transmission optics employing the linear response properties of molecules, it is in the area of non-linear optics where the most exciting applications are perceived. Highly efficient nonlinear optic materials permit functions such as those illustrated in Figure 11 to be performed in a totally optical manner without the need for electron-photon conversion processes. Second harmonic generation and parametric amplification can be obtained using inorganic single crystal materials such as lithium niobate, but recently, organic crystals such as 3-methyl-4-nitroaniline have been shown to possess exceptionally large second order electrooptic coefficients (Zyss 1982). However, a thin film geometry is preferred; then it will be possible to integrate nonlinear interactions, linear filtering and transmission functions into one precision monolithic structure. Therefore, researchers are currently substituting such molecules with appropriate side groups to enable LB film deposition to occur. To date, second harmonic generation has been reported in multilayers of nitrooctadecylazobenzene and mercocyanine and hemicyanine dyes. Nonlinear coefficients comparable to those of inorganic materials have been achieved but there are many other considerations involved (e.g. phase matching, suitable spectral response and refractive index, good optical damage threshold, low scattering coefficients and mechanical stability) before practical objectives can be accomplished. The most interesting observation to date is that of second harmonic generation in LB organic superlattices of the two molecules displayed in Figure 12. Neal et al. (1986) have shown that the nonlinear response of a hemicyanine/nitrostilbene layer is greater than that expected from the simple addition of contributions arising from the individual (separated) monolayers. The coefficient for second harmonic generation of the alternate layer structure is approximately five times the average value of the same parameter measured for hemicyanine and nitrostilbene. This super-additive effect is best explained in terms of improved film structure with adjacent molecules influencing each other to orient more vertically with respect to the substrate.

Fatty acid + fatty amine
superlattice

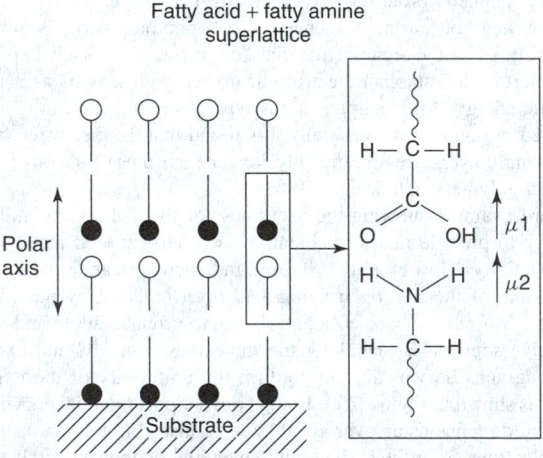

Fig. 10.   The left-hand diagram shows an organic superlattice with a unique polar axis. The two types of molecule involved could be a fatty acid and a fatty amine. The insert is designed to show that these two materials have dipole moments in opposite senses with respect to the hydrophobic chain. Thus, the Y-type film has a resultant dipole moment.

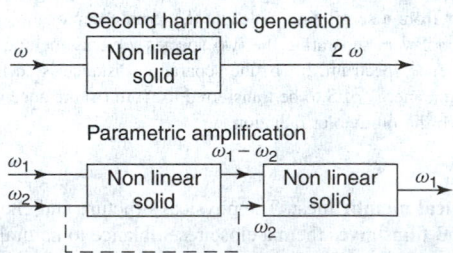

Fig. 11.   Schematic diagram illustrating the ability of nonlinear materials to double the frequency of incident laser radiation and also function as a parametric amplifier when two beams of different frequency are involved. Most of the non-centrosymmetric solids currently in use are inorganic, but they may be replaced by organic single crystal thin films.

Pyroelectric devices respond to a rate of change of temperature rather than to changes of temperature as in other, types of thermal detector. This gives them inherent advantages but their full potential has yet to be realized. For applications where both high speed and sensitivity are required, conventional materials have been unsuccessful. The desirable pyroelectric properties of inorganic materials appear to vanish for thicknesses less than 10 μm, and pyroelectric organic single

(b) Enhanced device processing: In integrated circuit technology the quest for faster speeds and larger memories has led to a gradual refinement of microlithographic methods for producing smaller and more closely spaced circuit elements. Sub-micron resolution is now required and this has necessitated a move away from conventional photolithography to techniques involving electron or ion beams and x-rays. The main disadvantage

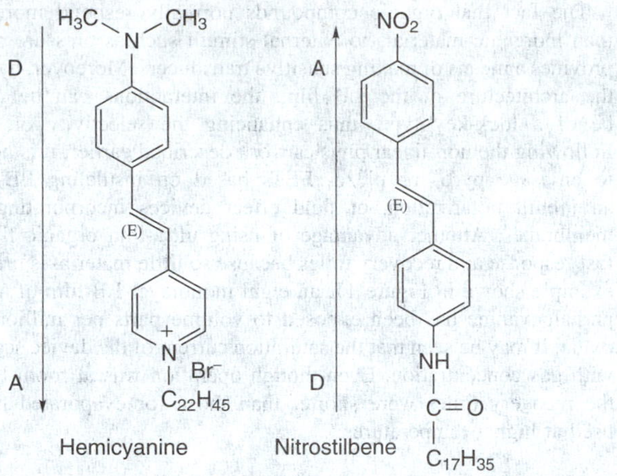

Fig. 12.   These two molecules, one a hemicyanine and the other a nitrostilbene dye, can be used to form an organic superlattice displaying a high coefficient for second harmonic generation.

of electron beam systems lies with their scattering characteristics; this enforces the requirement that the resist materials be pinhole free and less than 1 µm thick. Conventional spin-coated polymers display unacceptably large pin-hole densities and variations of thickness; however, LB films have already demonstrated their capability in this regard. There are good examples of both positive and negative resists but the best material reported to date is the ω-tricosenoic acid molecule shown in Figure 3(b). In purified form this material has adequate sensitivity, may be deposited at a rate of 0.5 cm s^{-1} to produce uniform coatings in the range 30–90 nm, and has been shown capable of a line resolution of 60 nm. The main property which requires some improvements is the etch resistance in plasma processing (Barraud et al. 1983).

Interconnects become increasingly important as the size of electronic circuits reduces. It has been suggested that protein layers might be useful in this regard. The method involves first depositing a synthetic protein and patterning it using a conventional resist. The NH$_2$ groups in the exposed protein layer can then be used to adsorb silver ions from a silver nitrate solution; this physical development stage converts the ions to metallic silver. An alternative approach is to fabricate supermolecular assemblies that conduct. The most successful attempt to date uses the molecule shown in Figure 13. Barraud et al. (1985) have succeeded in producing close packed multilayers of this TCNQ derivative, where the polar planes are separated by insulating lamellar regions. The resistivity of this N-stearylpyridinium$^+$1TCNQ$^-$ film is approximately 1 Ω cm. Of course, there are many other attractions in producing conducting LB films, including their uses.

There are several niches where LB films could have a useful role in semiconductor technology. Probably the most important is the ability of an oriented monolayer to change the effective barrier height at a semiconductor surface. Researchers at the University of Durham first demonstrated the effect on cadmium telluride and showed how it could be used to improve the efficiency of a photovoltaic diode (Dharmadasa et al. 1980). Similar effects have now been confirmed on a variety of materials including ZnSe, InP, GaP and GaAs and related III-V semiconductor alloys. The control afforded by the Langmuir-Blodgett technique permits the degree of band bending to be adjusted to suit the particular application, e.g. the increase in efficiency of an electroluminescent diode. In most applications it has been necessary to use robust monolayers of phthalocyanine which can withstand the large current densities involved. Figures 14 and 15 show data for gallium phosphide (Batey et al. 1983; Petty et al. 1985); for convenience the error bars have been removed. The LB thickness required to optimize the electroluminescence efficiency is found to be approximately 21 nm; this value is determined by the ability of minority carriers to cross the semi-insulating phthalocyanine film. Similar results have recently been achieved using zinc selenide layers grown using MOCVD. Blue electroluminescence is observed provided the organic film is present.

There are many methods of producing a surface layer on a semiconductor. However, experience has shown that when an energetic process such as evaporation, sputtering, or growth from a plasma, is used to deposit

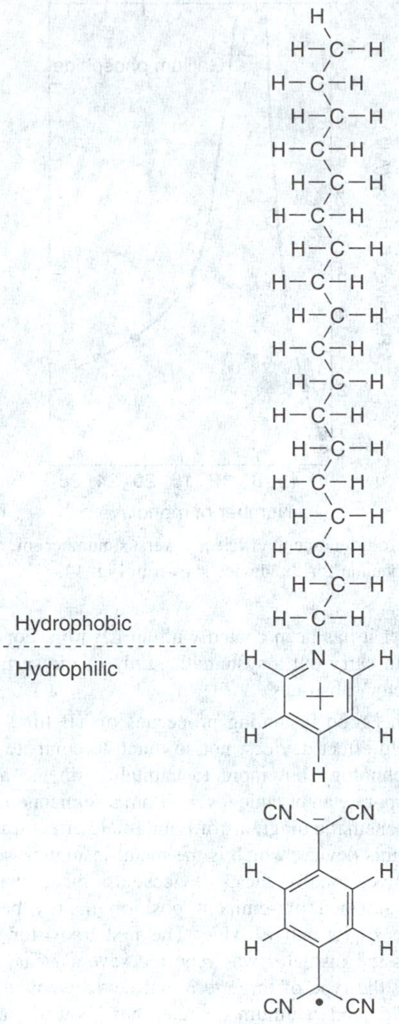

Hydrophobic
- - - - - - - - - - -
Hydrophilic

Fig. 13.   A substituent, pyridinium tetracyanoquinodimethane (TCNQ) molecular system.

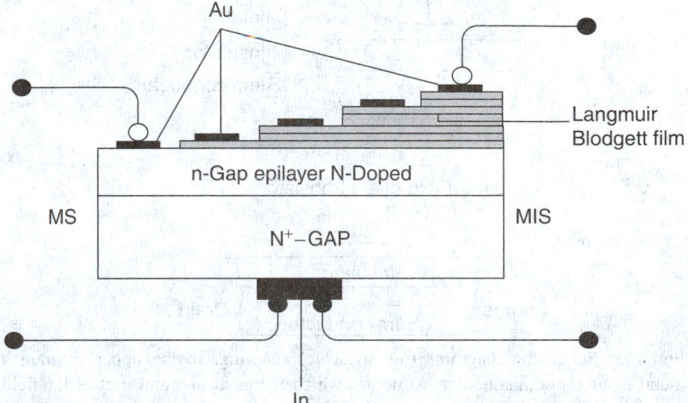

Fig. 14.   Schematic diagram (not to scale) of a Gold-LB film — Gallium Phosphide device structure.

a thin film onto a semiconductor, a surface damaged layer is produced which invariably dominates the electrical characteristics of the junction so formed. However, the Langmuir trough technique, being a low temperature deposition process, provides a means of circumventing this particular difficulty. On the other hand, it does mean that how the substrate is prepared before dipping is of considerable importance in determining the quality of the interface produced. That is, the nascent 'oxide' layer formed during the etching procedure remains relatively undisturbed and this can play a

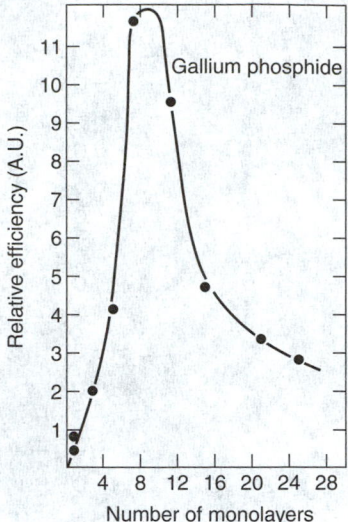

Fig. 15. The electroluminescent efficiency versus number of monolayers or substituted phthalocyanine for the device shown in Fig. 14.

vital role even after it has been coated with an LB film. For this reason it is important first to carry out a systematic study of the surface chemistry of the semiconductor substrates.

(c) Sensors: The good insulating properties of LB films suggest their possible use in field effect devices, not so much to compete with existing semiconductor technology but more to capitalize on the advantages of being able to incorporate an organic layer within a semiconductor structure. Figure 16 shows schematic diagrams of both a field effect transistor (FET) and the 'heart' of this device, which is the metal-insulating-semiconductor (MIS) diode. Conventionally these devices are made using inorganic materials; silicon holds a pre-eminent position mainly because of the insulating qualities of its native oxide. The first transistor incorporating LB monolayers as the insulator was reported several years ago (Roberts et al. 1978); using the type of three terminal device shown in Figure 16, on indium phosphide and cadmium stearate, they showed that the channel conductivity between the source and the drain could be modulated by the action of a gate electrode.

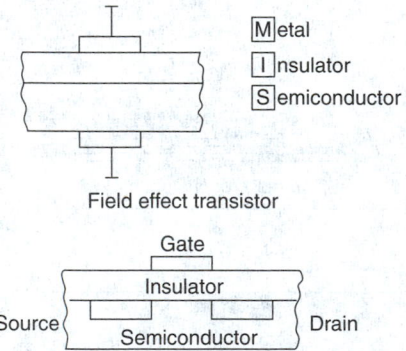

Fig. 16. Schematic diagrams (not to scale) showing, in the upper diagram a metal-insulator-semiconductor structure which forms an integral part of the field effect transistor shown in the lower diagram.

Subsequently, other semiconductors have been used and results have confirmed the ease with which a range of single crystal surfaces can be accumulated, depleted or inverted with an applied voltage. It is recognised that in all cases, the LB film is deposited on top of a nascent 'oxide' layer and that the insulation is provided essentially by a double dielectric structure. In the case of silicon, this can be used to advantage, in that a closely packed LB film layer can seal its surface from the atmosphere and thereby greatly retard the development of interface states. The organic film is also efficient in increasing the dielectric strength of a leaky silicon oxide film.

The fact that organic compounds normally respond more positively than inorganic materials to external stimuli such as pressure or radiation, provides a means of making sensitive transducers. Moreover, by controlling the architecture of the LB film, the interactions can be designed to be of a lock-key type, thus enhancing the selectivity of the device. Following the non-linear physics work described earlier, it is now possible to envisage pyro- or piezo- FETs based on insulating LB films with an inbuilt polarization, or field effect devices incorporating biological membranes. Another advantage of using ultra-thin organic films is their fast response and recovery times because so little material is present. In the example shown in Figure 17, an eight monolayer LB film of a substituted phthalocyanine has been exposed to volume parts per million of nitrous oxide. It may be seen that the saturation current of the device scales linearly with gas concentration. Even though operation was at room temperature, the recovery times were shorter than those for evaporated film devices used at higher temperatures.

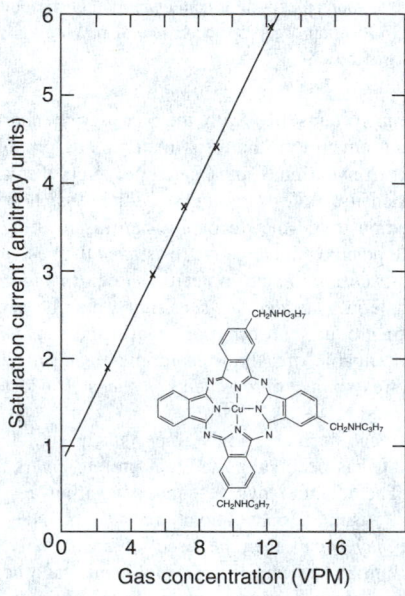

Fig. 17. Saturation current versus NO_2 gas concentration for a device incorporating eight monolayers of the asymmetrically substituted phthalocyanine molecule shown in the insert (Baker et al., 1983.)

It is not necessary to confine the discussion of microelectronic LB film based sensors to MIS or FET structures. For example, the switching voltage of a bistable switch or the characteristics of a gate-controlled diode could be very sensitive to an ambient. Moreover, optical and acoustic devices frequently show interesting threshold or resonance effects which could form the basis of a sensor. One area receiving particularly strong attention is that of surface plasmon resonance (SPR). The principle of this optical detection method is illustrated in Figure 18. A surface plasmon is a surface charge density wave at a metal surface. If the metal is sandwiched between two materials of different dielectric constant, then resonance can occur; this is observed as a very sharp minimum of the light reflectance when the angle of incidence is varied. The resonance angle is ultrasensitive to variations in the refractive index of the medium adjacent to the metal film. For example, the small change in an organic material due to gas absorption can easily be monitored even for concentrations in the part per billion range. In a practical situation, one normally selects an angle of incidence approximately half way down the reflectance minimum curve when no special gas is present; the change in intensity of the reflected light is then monitored at a constant angle.

There is widespread interest in the potential of LB films as biosensors as many believe that the incorporation of biological molecules such as enzymes will lead to novel devices. Some are exploring the deposition of biologically active molecules onto the gate electrodes or oxides of field effect transistors but optical sensors, probably based on fiber optics, are the most favored technique. In all cases the aim is to couple the specificity of interaction of chemicals or biochemicals with proteins or enzymes e.g.,

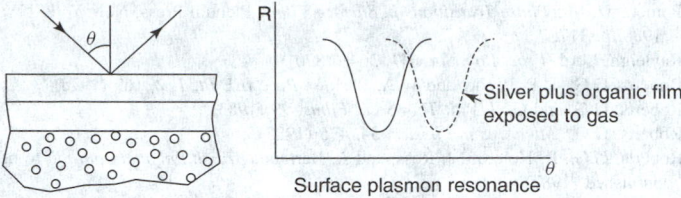

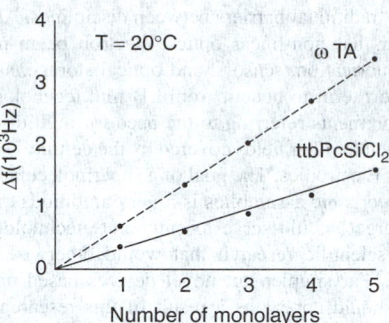

Fig. 18. Schematic diagrams illustrating the basis of the surface plasmon resonance technique. (Left) a beam of radiation striking the back surface of a glass prism coated with a metal film (usually evaporated silver). The reflected intensity and angle of reflection are extremely sensitive to variations in the dielectric on the metal surface. (Right) The shift in the R/θ plot when the organic film is exposed to a gas.

Fig. 19. The change in resonance frequency of 18 MHz piezoelectric quartz crystals coated with LB films of different thickness. The molecular structures of the organic films are shown in Figs. 3(b) and 3(d).

the change in their molecular conformation, with the sensitivity and signal transduction properties of the device. It is recognized that stability and lifetime may be problem areas and, for this reason, cross linked polymers are being explored as the hosts for the active species.

Some of the most convenient types of sensor are based on acoustoelectric devices; these can either be conventional bulk piezoelectric oscillators normally made of quartz, or surface acoustic wave (SAW) devices. The resulting change in the quiescent resonant frequency of a quartz oscillator coated with LB films of different thickness provides a very simple and elegant way of monitoring the reproducibility of monolayer deposition. Figure 19 illustrates how the quartz oscillator functions as a microgravimetric sensor; from the change in frequency it is possible to determine the density of the thin films. There is a further change when the organic films are exposed to minute concentrations of gas. The results presented in Figure 20 are for ω-tricosenoic acid in the presence of acetic acid, a species used for the detection of heroin. Greater effects can be obtained if the organic film is specially sensitised. Another way of increasing the sensitivity is to use acoustic surface waves. In such devices, input and output interdigitated electrodes are formed on a piezoelectric substrate usually made of quartz or lithium niobate. These perform the conversion between electric and acoustic energy. The single crystal surface region between the transducers serves as a propagation path for acoustic waves and thus forms a delay line which has application in electronic signal processing. Figure 20 shows a dual line configuration specially designed for sensing purposes. Basically, the device comprises two identical SAW (surface acoustic wave) oscillators positioned alongside each other. The hatched regions are earth shields to minimize reflections and cross talk between the two oscillators. The selective coating is placed in the propagation path of one of the oscillators thus affecting the delay time; the relative shift in frequency between the two oscillators is measured using a mixer circuit to obtain the difference frequency and then passing the resulting signal through a low pass filter. The change in frequency (Δf) between the two channels is then directly attributable to the sensor layer and other extraneous effects such as those due to temperature changes, are eliminated or very much reduced. Other geometries, for example, a surface acoustic wave resonator, are also possible. The device shown in Figure 21 has been constructed by Roberts et al. (1987) using quartz and interdigitated electrodes 24 μm wide separated by gaps of 25 μm; the operating frequency of the device is 98.4 MHz. Results have been obtained for both insulating (ω-tricosenoic acid) and conducting (pyridinium TCNQ) LB films as the sensor layer. In the case of the ω TA, only a mass loading effect is observed but with the TCNQ films, electric field effects are also apparent; these are associated with interactions between the surface acoustic wave and mobile charge carriers in the LB film. The combination of a device capable of measuring mass changes as low as femtograms per square centimeter, and the ability to detect minute changes in the electrical characteristics of a monolayer, augers well for sensors based on surface acoustic waves and monolayers.

Conclusions

We have discussed several research areas where there appears to be a tangible benefit in using monomolecular assemblies rather than organic or inorganic thin films deposited by other means. There are many ways of producing organic films and the onus will be on the Langmuir-Blodgett enthusiasts to demonstrate the special advantages to be gained using their

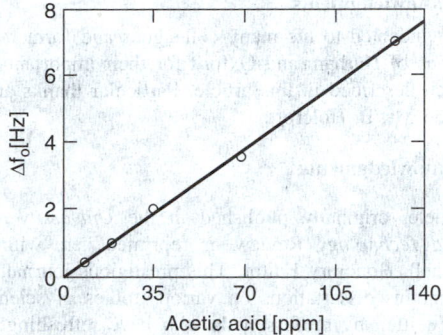

Fig. 20. Response characteristics of ω-tricosenoic acid coated quartz crystal oscillator at 22 °C exposed to acetic acid. A greater change in the resonance frequency can be obtained if the film is specially sensitized to detect acetic acid.

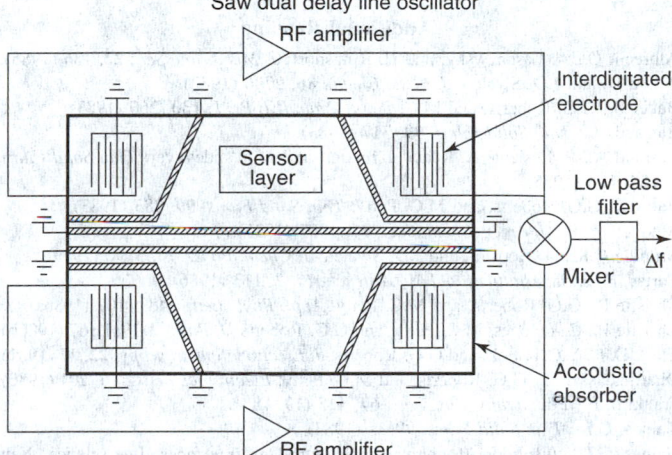

Fig. 21. A surface acoustic wave dual-delay line oscillator. The sensitive layer is placed in the propagation path of one of the two SAW devices. The difference in frequency (Δf) between the two channels provides a direct result of the mass loading and electric field effects associated with the sensor layer.

technique for a particular application. There is every likelihood that this will occur but it will require the combined efforts of high caliber teams with knowledge of physics, chemistry, electronics and biology. The ability of the synthetic chemist to manipulate the molecular architecture of a material to optimize a specific physical parameter or figure of merit will be vital. However, the role of the physicist or engineer in identifying the targets and guiding the main thrust of the research program will also be essential. Industry is already well able to organize interdisciplinary activities. The relatively inexpensive equipment requirements associated with the Langmuir trough technique coupled with the elegance of the fundamental science and the interesting applied prospects for LB films provides an excellent opportunity for the academic community to similarly break down the traditional barriers between disciplines. At the present time it would appear that non-linear optics, electron beam microlithography, magnetic tape lubrication, sensors, and optical storage, are the areas most likely in the short term to benefit from LB film technology.

The above comments referring to the need for multidisciplinary activity apply equally well to other fields covered by the definition of Molecular and Supermolecular Electronics. The goal of a supermolecular information processor based on organic assemblies is a very ambitious one. Nevertheless, pursuing this target should serve to enumerate technologies and identify areas of basic scientific research that would otherwise remain dormant and unexplored. Far simpler but novel devices based on organic molecular materials should appear as a result of this research effort. It seems likely that Langmuir-Blodgett films will play a key role in helping scientists identify basic physical phenomena in supermolecular assemblies and at the same time enable engineers to become more familiar with devices incorporating organic films.

Author's Acknowledgments

The author is indebted to his many colleagues and former colleagues at the Universities of Durham and Oxford for their important contributions to the research described in this article. Particular thanks are due to Dr. M.C. Petty and Mr. B. Holcroft.

Editor's Acknowledgments

This article, originally published in the *University of Wales Science and Technology* Review, is reprinted here with approval of J.H. Purnell, Honorary Editor. This prestigious journal, issued quarterly, contains dissertations on various topics of science and technology, written by professionals who have refreshingly new viewpoints. The *Review*, which commenced publication in March 1987, is a welcome contribution to the scientific community throughout the English-speaking world. Publication Office: MBN2 Cardiff, Marketing Department, 17th Floor, Pearl House, Greyfriars Road, Cardiff, Wales CF1 3XX, United Kingdom.

Additional Reading

Albrecht O., A. Laschewsky, and H. Ringsdorf: *J. Membrane Sci.*, **22**, 186 (1985).

Allara D. and J.D. Swalen: *J. Phys. Chem.*, **86**, 2700 (1982).

Baker S., G.G. Roberts, and M.C. Petty: *Proc. IEE Pt. 1*, **130**, 260 (1983).

Barraud, A.: *Thin Solid Films*, **99**, 317 (1983).

Barraud A., P. Lesieur, A. Ruaudel-Teixier, and M. Vandevyver: *Thin Solid Films*, **134**, 195 (1985).

Batey J., G.G. Roberts, and M.C. Petty: *Thin Solid Films*, **99**, 283 (1983).

Blodgett, K.B.: *J. Amer. Chem. Soc.*, **57**, 1007 (1935).

Brundle C.R., H. Hopster, and J.D. Swalen: *J. Chem. Phys.*, **70**, 5190 (1979).

Carter, F.L.: *Superlattices and Microstructures*, **2**, 113 (1986).

Christie P., G.G. Roberts, and M.C. Petty: *Appl. Phys. Letts.*, **48**, 1101 (1986a).

Christie P., C.A. Jones, M.C. Petty, and G.G. Roberts: *J. Phys.*, **D19**, L167 (1986b).

Clark D.T., Y.C.T. Fok, and G.G. Roberts: *J. Electron. Spectroscopy*, **22**, 17 (1981).

Dharmadasa I.M., G.G. Roberts, and M.C. Petty: *Electronics Letts.*, **16**, 201 (1980).

Franklin, B.: *Phil. Trans. Roy. Soc.*, **64**, 445 (1974).

Gaines, G.L.: *Thin Solid Films*, **99**, ix (1983).

Gaines, G.L.: *Insoluble Monolayers at Liquid-Gas Interfaces*, Interscience, New York, NY, 1966.

Gaines, G.L.: *Thin Solid Films*, **132, 133, 134** (1985).

Holcroft B., M.C. Petty, G.G. Roberts, and G.J. Russell: *Thin Solid Films*, **134**, 83 (1985).

Hua Y.L., G.G. Roberts, M.M. Ahmed, M.C. Petty, M. Hanack, and M. Rein: *Phil. Mag.*, **B53**, 105 (1986).

Kelly M.J. and Weisbach, C.: *The Physics and Fabrication of Microstructures and Microdevices*, Springer-Verlag, Berlin (1986).

Kuhn, H.: *Thin Solid Films*, **99**, 1 (1983).

Langmuir, I.: *J. Amer. Chem. Soc.*, **39**, 1848 (1917).

Langmuir, I.: *Trans. Faraday Soc.*, **15**, 62 (1920).

Mobius, D.: *Accts. Chem. Res.*, **14**, 63 (1981).

Mori C., H. Noguchi, M. Mizuno, and T. Watanabe: *Jap. J. Appl. Phys.*, **19**, 725 (1980).

Neal D., M.C. Petty, G.G. Roberts, M.M. Ahmed, W.J. Feast, I.R. Girling, N.A. Cade, P.V. Kolinsky, and I.R. Peterson: *Proc. Int. Symp. on the Applications of Ferroelectric Materials*, Philadelphia (1986).

Petty M.C., J. Batey, and G.G. Roberts: *IEE Proc. Pt. 1*, **132**, 133 (1985).

Pockels, A.: *Nature*, **43**, 437 (1891).

Pomerantz, M.: *Phase Transitions in Surface Films*, Plenum Press, New York, NY, 1980, p. 317.

Rayleigh, Lord: *Proc. Roy. Soc.*, **47**, 364 (1890).

Roberts, G.G., K.P. Pande, and W.A. Barlow: *Proc IEE Pt. 1*, **2**, 169 (1978).

Roberts, G.G. and C.W. Pitt: *Thin Solid Films*, **99** (1983).

Roberts, G.G.: *Advances in Physics*, **34**, 475 (1985).

Roberts, G.G., B. Holcroft, J. Ross, and A. Barraud: *British Polymer Journal*, to be published (1987).

Tabor, D.: *J. Colloid and Interface Science*, **75**, 240 (1980).

Vincett P.S., W.A. Barlow, F.T. Boyle, J.A. Finney, and G.G. Roberts: *Thin Solid Films*, **60**, 265 (1979).

Vincent, P.S. and G.G. Roberts: *Thin Solid Films*, **68**, 135 (1980).

Zyss, J.: *J. Non. Cryst. Solids*, **47**, 211, (1982).

GARETH ROBERTS, FRS, Thorn EMI plc and University of Oxford

MOLECULAR BEAM. A unidirectional stream of neutral molecules passing through a vacuum, generally with thermal velocity. Such a beam may be produced by emergence from a pinhole in a chamber containing low pressure gas or vapor, and it may be defined by a system of slits. By passing the beam through known electric or magnetic fields, quantities such as nuclear magnetic moments can be determined.

MOLECULAR BIOLOGY. A self-defining term — the study of biological substances and phenomena at the molecular level.

For many years, biologists, aided by numerous laboratory instruments and methodologies (X-ray crystallography, electron microscopy, chromatography, electrophoresis, etc.) constructed a vast databank of biochemistry. This has been and continues to be of inestimable value to those professionals who deal with health, medicine, the industrial use of biochemicals, among other areas that require such information. There remained, however, a dissatisfaction concerning the absence of knowledge pertaining to the manner in which biochemical processes are directed and take place at the molecular level. See also **Industrial Biotechnology**.

The first break occurred early in the 1950s when very illuminating findings were made concerning the complex biomolecule, DNA. The research of DNA, in essence, constituted the beginnings of *molecular biology*. A chronology of these early years is given in the article on **Genetics and Gene Science**. This pioneering work altered the thrust of biochemical research, as previously typified by traditional biochemistry, to a concentration and emphasis on studying biological events at the molecular level.

A second break occurred in the early 1970s when it was discovered that a strand of the DNA molecule can be cut by restriction enzymes and that the sticky ends can be reassembled by a new technology known as *genetic recombination* or *recombinant DNA*. This is frequently the starting point for projects engaged in biological research at the molecular level. It is no surprise that molecular biology has become a highly diverse, wide-spectrum, multifaceted discipline. The many hundreds of research projects in this area underway today obviously cannot be delineated here. The highlights of just a few research targets are given here to provide a representative view of projects underway and inroads made to date.

Cell Membrane Research. Cellular membranes serve as the interface between the cell and the organism of which it is a part. It has been established that the movement of cells, as directed during growth and development, must involve, in some way, the plasma membranes. It also has been observed that plasma membranes play some role in cancerous growth, where cell multiplication and migration proceed at an uncontrolled rate. Membranes are composed of lipids that interact with each other in a watery medium to form a closed and flexible compartment. This formation process occurs at a relatively rapid rate. Bretscher reports that an area of membrane equivalent to the area of the entire surface of the cell requires less than an hour to form. A good start has been made by a number of investigators toward understanding the detailed molecular structure of membranes. In retrospect, it is interesting to note that recent findings pertaining to the basic frame of membranes was qualitatively proposed as early as 1925 by Gorter and Grendel (University of Leiden). This framework is composed of a double layer of lipid molecules. One end of the molecule is hydrophilic, that is, soluble in water; the other end is hydrophobic, a hydrocarbon that is oily and insoluble in water. Most commonly, membrane lipids are phospholipids. See also **Lipids**.

Cytoplasm Research. The existence of fibers in cytoplasm was first observed microscopically about a century ago. It has been learned during the last few years as the result of biochemical and immunological studies, that a distinct set of proteins characterizes each filament system. Proteins in the cytoplasm make up a highly structured, yet changeable matrix. The matrix determines the cell shape, its division and motion, as well as the transport of vesicles and organelles. Research in molecular genetics is targeted to reveal the function of cytoskeletal proteins. Weber and Osborn report that current knowledge of the cell matrix is contributing to diagnosis and research in human pathology. For example, typing of intermediate filaments by way of immunofluorescence microscopy will distinguish the major tumor groups. Keratins are found in carcinomas. What is missing is detailed knowledge at the molecular level of why and how the protein molecules in the cytoplasm function.

Immune System Research. This is one of the most active research areas as of the late 1980s. It has been observed that proteins responsible for recognizing foreign invaders are the most diverse of all known proteins. Tonegawa observes that these proteins are encoded by hundreds of scattered gene fragments and these can be combined in millions, probably billions of ways. This gives some idea of how complex research in this area really is. It is one of the most difficult of tasks for molecular biologists, but is proceeding apace. See also **Immune System and Immunology**.

Cellular Intercommunications Research. It is well known that the majority of higher organisms utilize hormones or systems of neurons for intercellular communication. See also **Central and Peripheral Nervous Systems**; and **Hormones**. Snyder observes that chemical messengers mediate long-range communication via hormones; short-range communication via nerve cells. The two systems differ in directness, but some messenger molecules are common to both. Hormonal molecules are usually peptides and steroids. Communication between cells or groups of cells is mandatory for the survival of all multicellular organisms. Snyder also observed that as investigators come to know the properties and functions of highly specialized messenger molecules, it will be possible to develop better therapeutic agents for a number of diverse conditions, including hormonal abnormalities, heart disease, and mental illness.

Cellular Intracommunications Research. The communication within a cell is being investigated at the molecular level. The mechanism by which a cell receives external signals by way of receptors has been reasonably well understood for a number of years. But what comprises the cell's internal mechanism for reacting to received signals by way of the receptors? Berridge refers to receptors as "molecular antennas." Obviously, the barrier to the flow of information from receptor to mechanisms within the cell is the plasma membrane, previously mentioned. How are external signals acted upon internally within the cell to cause it to secrete, to contract, to metabolize, or to grow, among several other biological reactions? What are the "second messengers" within the cell? How do they function? A second messenger, cyclic adenosine monophosphate (cyclic AMP) has been identified. Other second messengers include Ca^+ ions, inositol triphosphate (IP_3), and diacylglycerol (DG). Berridge also reported that some evidence has been gained to the effect that the aforementioned substances are cannibalized from the plasma membrane itself. Obviously, the cell must include genes required for the synthesis of the proteins used in the internal pathways. Aberration of these gene functions could lead to abnormalities of cellular growth and hence to uncontrolled growth and structural transformations typical of cancer. Cells related in some way to tumor growth revealed thus far are called *oncogenes*. Berridge and other researchers are attempting to identify additional second messengers that may be present and how they function in internal communication.

Biological Development and Growth Research. For many years, the processes which take place within an organism to cause it to grow (the early embryonic phase), leading to maturity (adulthood) and later to initiate the aging phase have been among the least understood of biological phenomena. As aptly observed by Gehring: How is the basic architecture of an embryo laid down? How does the linear information of DNA generate a three-dimensional organism? One of the most recent and intriguing discoveries is that of the so-called *homeobox*, which is found as a short stretch of DNA. When the gene containing the homeobox is translated into a protein, the homeobox yields a string of amino acids that are believed to bind to the DNA double helix. By binding to the DNA of particular genes, the protein may be able to turn them on or off. The homeobox was first noted in studies of *Drosophila*; subsequent studies have revealed a homeobox within a range of organisms extending from worms to humans. It is well established, of course, that animals develop in numerous ways. Researchers now question whether or not the molecular mechanisms underlying development may be more universal than previously suspected. Again, it appears that the answers to still another of the biological secrets may lie in better understanding the process at the molecular level.

Adaptive and Evolutionary Processes. Wilson observes that in the past most biologists working in this area have concentrated their attention at the level of the whole organism. Mutation frequently has been suggested as the cause of adaptive changes. Prior to studies at the molecular level, however, it was *not* known that mutations may accumulate at steady rates over time in the genes of all lineages. Wilson, a proponent of the molecular evolution concept, reports on two assumptions: (1) Heritable differences among organisms result from differences in their DNAs, and (2) molecular evolutionists must not only measure differences in DNA, but also explain the origin of the differences and their relation to organismal differences. Wilson also points out that two critical elements of molecular evolution are (1) *point mutations*, specifically those occurring in the genes coding, and (2) *regulator mutations*. A point mutation is defined as a single replacement of a DNA base; a regulatory mutation is any change in a gene or within the vicinity of a gene that determines whether the gene is active or inactive. Research on point mutations led to the concept of a *molecular clock* and in the discovery of still another kind of genetic change, which Wilson calls a *neutral mutation*. This mutation is neither advantageous nor disadvantageous for an organism. Wilson also refers to the "pressure" to evolve, noting that this stems from geologic forces as well as from the brains of mammals and birds. It is obvious that the study of molecular evolution occupies a special position in contemporary biology.

See also **Biological Timing and Rhythmicity**; and **Evolution**.

Additional Reading

Ausubel, F. et al.: *Short Protocols in Molecular Biology: A Compendium of Methods from Current Protocols in Molecular Biology*, John Wiley & Sons, Inc., New York, NY, 1999.

Berridge, M.J.: "The Molecular Basis of Communication within the Cell," *Sci. Amer.*, 142–148 (October 1985).

Lackie, J. and J. Dow: *The Dictionary of Cell and Molecular Biology*, 3rd Edition, Academic Press, Inc., San Diego, CA, 1999.

Stahl, W. and A. Hershey: *We Can Sleep Later: Alfred D. Hershey and the Origins of Molecular Biology*, Cold Spring Harbor Laboratory Press, Cold Spring Harbor, NY, 2000.

Snyder, S.H.: "The Molecular Basis of Communication Between Cells," *Sci. Amer.*, 132–137 (October 1985).

Tonegawa, S.: "The Molecules of the Immune System," *Sci. Amer.*, 122–128 (October 1985).

Weber, K. and M. Osborn: "The Molecules of the Cell Matrix," *Sci. Amer.*, 110–118 (October 1985).

Wilson, A.C.: "The Molecular Basis of Evolution," *Sci. Amer.*, 164–170 (October 1985).

MOLECULAR DISTILLATION. A special form of distillation conducted at pressures of 1–7 micrometers in the laboratory and 3–30 micrometers in industrial applications. Compared with conventional laboratory vacuum distillations carried out between 1 and 10 millimeters of mercury pressure, this is a very high vacuum. One micrometer equals 0.001 millimeter of mercury pressure. The other feature of the molecular still is that the condenser is located within a distance less than the mean free path of the evaporating molecules from the evaporator portion of the apparatus. Thus, although a molecule may return to the distill and many hundred times before reaching the exit of a conventional vacuum still, 50% of the molecules in a properly functioning molecular still will reach the exit on their first try. Thus, efficiency is remarkably high.

Because of the absence of convection due to ebullition and because high viscosities and high molecular weights may impede diffusion within the distill and, the surface of the distill and in a molecular still may not always represent the total liquid. Therefore, efficient molecular distillation requires the mechanical renewal of the surface film. This is achieved by vigorous agitation, as in the *stirred-pot still*; by employing a *falling film*; or by using centrifugal force, as in the *centrifugal* molecular still. Commercial installations of falling-film stills achieve throughputs of many tens of liters per hour, whereas the centrifugal still is capable of several hundred liters per hour. Centrifugal stills usually are arranged in groups of from three to seven. This permits fractionation by multiple redistillation. Among the uses of molecular distillation are: the separation of mono and diglycerides for bread and paraffin wax for milk cartons; the distillation of plasticizers, fatty acid dimers, and synthetics; the distillation of vitamin A esters and intermediates; and the stripping of α, β, γ, and δ-tocopherols and sitosterols from vegetable oils.

See also **Distillation**.

MOLECULAR MICROBIOLOGY. See **Microbiology**.

MOLECULAR-SCALE TEMPERATURE. A temperature parameter T_M defined by the following relation:

$$T_M = T \frac{M_0}{M},$$

where T is the actual "kinetic" temperature, M_0 the molecular weight of air at sea level (28.966), and M the molecular weight of air at the point where the temperature is being specified.

Molecular-scale temperature has application in specifying temperatures at extremely high altitudes (see **Meteorology: standard atmosphere**). Below about 90 kilometers (56 miles), $T_M = T$, but above that level, T_M becomes increasingly greater than T.

AMS

MOLECULAR SIEVES. In this article, the term *molecular sieve* is restricted to inorganic materials that possess uniform pores with diameters in either the micro- (<2 nm) or meso- (2–20 nm) size range. The most technologically important molecular sieves are zeolites, i.e., crystalline silicate or aluminosilicate framework structures with channels of diameters <1.2 nm. Several of these topologies, with boron, gallium, or iron replacing aluminum, or germanium replacing silicon, have also been prepared. The chemical composition of microporous framework structures has been expanded considerably with the substitution of phosphorus for silicon, and new families of aluminophosphate and silicoaluminophosphate structures have been synthesized in the laboratory. Some of these frameworks have zeolite analogues, whereas others are unique. The addition of elements such as Mg, Ti, Mn, Co, Fe, or Zn into these structures has made it possible to generate metalloaluminophosphates, metallosilicoaluminophosphates, etc. Microporous sulfide-based framework structures are also possible.

Considerable synthesis effort has been devoted to developing frameworks with pore diameters within the mesoporous range; the largest synthesized are the phosphate-based AlPO-8 (14-membered ring), VPI-5 (18-MR), and cloverite (20-MR), which have pore diameters within the 0.8–1.3 nm range. A new family of mesoporous molecular sieves designated M41S has been discovered. Although not framework structures like zeolites, silicate, and aluminosilicate, M41S materials possess very uniform mesopores.

The technological applications of molecular sieves are as varied as their chemical makeup. Heterogeneous catalysis and adsorption processes make extensive use of molecular sieves. The utility of the latter materials lies in their microstructures, which allow access to large internal surfaces, and cavities that enhance catalytic activity and adsorptive capacity.

Zeolites

Molecular-sieve zeolites of the most important aluminosilicate variety can be represented by the chemical formula $M_{2/n}O \cdot Al_2O_3 \cdot ySiO_2 \cdot wH_2O$, where y is 2 or greater, M is the charge balancing cation, such as sodium, potassium, magnesium, and calcium, n is the cation valence, and w represents the moles of water contained in the zeolitic voids. The zeolite framework is made up of SiO_4 tetrahedra linked together by sharing of oxygen ions. Substitution of Al for Si generates a charge imbalance, necessitating the inclusion of a cation. The structures contain channels or interconnected voids that are occupied by the cations and water molecules. The water may be removed reversibly, generally by the application of heat, which leaves intact the crystalline host structure permeated with micropores that may account for >50% of the microcrystal's volume. In some zeolites, dehydration may produce some perturbation of the structure, such as cation movement, and some degree of framework distortion.

Zeolite minerals are formed over much of the earth's surface, including the sea bottom. Most zeolites occurring in cavities of basaltic and volcanic rocks are exceedingly rare. However, several zeolite minerals generated by the natural alteration of volcanic ash in alkaline environments over long periods of time occur in recoverable deposits. Although such minerals are rarely useful for catalytic application, mainly because of iron impurities (exception: Chabazite of Bowie, Arizona), more abundant zeolites, e.g., clinoptilolite, have found use as soil conditioners, additives to animal feed, as animal litter, in aquaculture to remove ammonia (phillipsite), and for ion exchange to remove heavy metals from industrial and mining effluents (chabazite).

Structure

Of the approximately 120 known framework aluminosilicates, ~50 occur naturally, the rest being synthetic. There are 56 structural types of zeolites known. Understanding the complexities of zeolite structures is made easier by recognizing three important structural keys: the basic arrangement of the individual structural units in space, which defines the framework topology; the location of the charge-balancing cations; and the channel-filling material, such as water or an organic template, which is incorporated as the zeolite is formed. After the channel-filling material is removed, the void space can be used for the adsorption of gases, liquids, salts, elements, metal complexes, etc. In turn, this void-filling property makes zeolites commercially useful in ion exchange, catalysis, etc.

There are two types of structures: one provides an internal pore system comprising interconnected cage-like voids; the second provides a system of uniform channels which, in some instances, are one-dimensional and in others intersect with similar channels to produce two- or three-dimensional channel systems. The preferred type has two- or three-dimensional channel systems to provide rapid intercrystalline diffusion in adsorption and catalytic applications.

In most zeolite structures, the primary structural units, tetrahedra, are assembled into secondary building units, which may be simple polyhedra such as cubes, hexagonal prisms, or truncated octahedra. The final framework structure consists of assemblages of the secondary units.

Crystal structures of zeolite minerals are illustrated by the zeolite chabazite. The structure of chabazite is hexagonal and the framework consists of double six-membered rings of $(Si,Al)O_4$ tetrahedra arranged in parallel layers in an AABBCC sequence. These tetrahedra are cross-linked by four-membered rings.

Many new crystalline zeolites have been synthesized and several fulfill important functions in the chemical and petroleum industries and in consumer products such as detergents. The structural formula of a zeolite is based on the crystal unit cell, the smallest unit of structure, represented by $M_{x/n}[(AlO_2)_x((SiO_2)_y] \cdot wH_2O$, where n is the valence of cation M, w is the number of water molecules per unit cell, x and y are, respectively, the number of AlO_4 and SiO_4 tetrahedra per unit cell, and y/x usually has values of 1–5.

The secondary structure unit in zeolites A, X, and Y is the truncated octahedron. These polyhedral units are linked in three-dimensional space through the four- or six-membered rings. The former linkage produces the zeolite A structure, and the latter the topology of zeolites X and Y and of the mineral faujasite.

The structure of the high silica zeolite ZSM-5 (y/x=10 to>5000) contains a high concentration of five-membered rings and has two intersecting channels. It has become a very important catalyst for petrochemical reactions.

Several types of structural defects or variants can occur which figure in adsorption and catalysis: (1) surface defects due to termination of the crystal surface and hydrolysis of surface cations; (2) structural defects due to imperfect stacking of the secondary units, which may result in blocked channels; (3) ionic species, e.g., OH^-, AlO_2^-, Na^+, SiO_4^-, may be left stranded in the structure during synthesis; (4) the cation form, acting as the salt of a weak acid, hydrolyzes in aqueous suspension to produce free hydroxide and cations in solution; and (5) hydroxyl groups in place of metal cations may be introduced by ammonium ion exchange, followed by thermal deammoniation.

Properties

Although several types of microporous solids are useful as adsorbents for the separation of vapor or liquid mixtures, the distribution of pore diameters does not enable separations based on the molecular-sieve effect. The most important molecular-sieve effects are shown by crystalline zeolites. The sieve effect may be total or partial.

Activated diffusion of the adsorbate is of interest in many cases. As the size of the diffusing molecule approaches that of the zeolite channels, the interaction energy becomes increasingly important. If the aperture is small relative to the molecular size, then the repulsive interaction is dominant and the diffusing species needs a specific activation energy to pass through the aperture. Similar shape-selective effects are shown in both catalysis and ion-exchange, two important applications of these materials.

In order to utilize the absorption properties of the synthetic zeolite crystals in processes, the commercial materials are prepared as pelleted aggregates combining a high percentage of the crystalline zeolite with an inert binder. The formation of these aggregates introduces macropores in the pellet which may result in some capillary condensation at high adsorbate concentrations. In commercial materials, the macropores contribute diffusion paths. However, the main part of the adsorption capacity is contained in the voids within the crystals.

Zeolites are high capacity, selective adsorbents capable of separating molecules based on the size and shape of the structure. They adsorb molecules, in particular those with a permanent dipole moment which show other interaction effects, with a selectivity that is not found in other solid adsorbents. Separation may be based on the molecular-sieve effect or may involve the preferential or selective adsorption of one molecular species over another. These separations are governed by several factors. The basic framework structure, or topology, of the zeolite determines the pore size and the void volume. The exchange cations, in terms of their specific location in the structure, their population density, their charge and size, affects the molecular-sieve behavior and adsorption selectivity of the zeolite. By changing the cation types and number, the selectivity of the zeolite in a given separation can be tailored or modified, within certain limits.

The high silica version of ZSM-5, also known as silicalite, is a hydrophobic adsorbent capable of adsorbing, e.g., ethanol from an aqueous solution.

In zeolites, catalysis takes place preferentially within the intracrystalline voids. Catalytic reactions are affected by aperture size and type of channel system, through which reactants and products must diffuse. Modification techniques include ion-exchange, variation of Si/Al ratio, hydrothermal dealumination or stabilization, which produces Lewis acidity, introduction of acidic groups such as bridging Si(OH)Al, which impart Brønsted acidity, and introducing dispersed metal phases such as noble metals. In addition, the zeolite framework structure determines shape-selective effects. Several types have been demonstrated, including reactant selectivity, product selectivity, and restricted transition-state selectivity. Nonshape-selective surface activity is observed on very small crystals, and it may be desirable to poison these sites selectively, with bulky heterocyclic compounds unable to penetrate the channel apertures, or by surface silation.

Some current and possible future zeolite catalyst applications are as follows: alkylation, cracking, hydrocracking, dewaxing, isomerization, hydrogenation and dehydrogenation, hydrodealkylation, methanation, shape-selective reforming, dehydration, methanol to gasoline, methanol to olefins, organic catalysis, inorganic reactions, H_2S oxidation, NH_3 reduction of NO, $H_2O \rightarrow 1/2O_2 + H_2$, and CO oxidation.

The exchange behavior of nonframework cations in zeolites, e.g., selectivity, and degree of exchange, depends on the nature of the cation, e.g., the size and charge of the hydrated cation, on the temperature, the concentration, and, to some degree, on the anion species. Cation exchange may produce considerable change in various other properties, such as thermal stability, adsorption behavior, and catalytic activity.

Framework Modification

The zeolite framework can be stabilized by hydrothermal treatment, which removes aluminum from the framework and forms aluminum cations. During this steaming process, the tetrahedral vacancies left behind in the framework are gradually refilled with silicon, which appears to migrate as a form of silicic acid from other parts of the framework and contributes to stabilization by repairing the damaged framework. Simultaneously, such other parts of the framework disappear under formation of mesopores. Cationic aluminum can be extracted with an acid, and a subsequent steaming causes further dealumination of the framework and migration

of silicon into the vacancies. Carefully controlled conditions can produce high silica forms of zeolites, e.g., zeolite Y. Since the $Si-O$ bond is shorter than the $Al-O$ bond, hydrothermal dealumination causes the unit cell parameter to decrease. Mesopores can be avoided by replacing the aluminum directly with external silicon, e.g., by treatment with silicon tetrachloride. In a reversal of the reaction with $SiCl_4$, aluminum can be introduced into the framework by reaction of the hydrogen or ammonium form with gaseous $AlCl_3$.

Manufacture

Zeolites are formed under hydrothermal conditions, defined here in a broad sense to include zeolite crystallization from aqueous systems containing various types of reactants. Most synthetic zeolites are produced under nonequilibrium conditions, and must be considered as metastable phases in a thermodynamic sense.

Many important types of zeolites have no natural mineral counterpart. Conversely, synthetic counterparts of many zeolite minerals are not yet known. The conditions generally used in synthesis are reactive starting materials such as freshly co-precipitated gels, or amorphous solids; relatively high pH introduced in the form of an alkali metal hydroxide or other strong base, including tetraalkylammonium hydroxides; low temperature hydrothermal conditions with concurrent low autogenous pressure at saturated water vapor pressure; and a high degree of supersaturation of the gel components, leading to nucleation of a large number of crystals.

A gel is defined as a hydrous metal aluminosilicate prepared from either aqueous solutions, reactive solids, colloidal sols, or reactive aluminosilicates such as the residue structure of metakaolin and glasses.

The gels are crystallized in a closed hydrothermal system at temperature varying from room temperature to about 200°C. The time required for crystallization varies from a few hours to several days. When prepared, the aluminosilicate gels differ in appearance, from stiff and translucent to opaque gelatinous precipitates and heterogenous mixtures of an amorphous solid dispersed in an aqueous solution. The alkali metals form soluble hydroxides, aluminates, and silicates. These materials are well suited for the preparation of homogeneous mixtures.

Gel preparation and crystallization is represented systematically using the $Na_2O-Al_2O_3-SiO_2-H_2O$ system as an example.

$$NaAl(OH)_4(aq) + Na_2SiO_3(aq) \xrightarrow{25°C} [Na_a(AlO_2)_b$$

$$(SiO_2)_c \cdot NaOH \cdot H_2O]gel \xrightarrow{25°-175°C} Na_x[(AlO_2)_x(SiO_2)_y] \cdot mH_2O$$

Typical gels are prepared from aqueous solutions of reactants such as sodium aluminate, NaOH, and sodium silicate; other reactants include alumina trihydrate ($Al_2O_3 \cdot 3H_2O$), colloidal silica, and silicic acid. The temperature strongly influences the crystallization time of even the most reactive gels.

Synthesis mechanisms of the typical low silica zeolites, such as A, X, and Y, are apparently different from the high silica zeolites such as ZSM-5. In the low silica zeolites, nuclei are formed consisting of alkali metal-ion complexes of the aluminosilicate species. Structural units consisting of four-membered rings, six-membered rings, and cages coordinated with cations are thought to be involved in the nucleation and crystallization. In the high silica zeolites, the mechanism appears to be a templating type where an alkylammonium cation complexes with silica by hydrogen bonding. These complexes cause the structures to replicate by hydrogen bonding of the organic cation with framework oxygen atoms.

Manufacturing processes for commercial molecular sieve products may be classified into three groups, hydrogel, clay conversion, and other.

Health and Safety Factors and Toxicology

Zeolites have applications in food, drugs, cosmetic products, and detergents. Thus, extensive toxicological and environmental studies have been carried out. Feeding of 5.0 g/kg of body weight (powder form of type 4A, 5A, 13X, and Y) for seven days produced no ill effect in rats. There is no contraindication to the use of zeolite A (Sasil) in detergents. No negative effect on biological wastewater treatment was found, and zeolite A showed no evidence of acute toxicity to four species of freshwater fish.

Uses

In most cases, the water content of the commercial product is below 1.5–2.5 wt %; certain products, however, are sold as fully hydrated crystalline powders. Molecular sieve products are used for adsorption (e.g., purification of water, carbon dioxide, and sulfur compounds); bulk separation of normal and isoparaffins, xylene, olefin, and oxygen from air; catalysis e.g., catalytic cracking, hydrocracking for fuels production, dewaxing of distillate fuel and lube basestocks, paraffin isomerization, catalysis of aromatic reactions (in selective toluene disproportionation, xylene isomerization, and ethylbenzene synthesis), synthesis of *p*-ethyltoluene, and the methanol-to-gasoline process; and ion exchange (e.g., cesium and strontium radioisotopes, ammonium ion removal, and detergent builders).

Additional Reading

Auerbach, S.M., K.A. Carrado, and P.K. Dutta: *Handbook of Zeolite Science and Technology*, Marcel Dekker, Inc., New York, NY, 2003.
Barrer, R.M.: *Zeolites and Clay Minerals as Sorbents and Molecular Sieves*, Academic Press, London, 1978.
Breck, D.W.: *Zeolite Molecular Sieves, Structure, Chemistry, and Use*, John Wiley & Sons, Inc., New York, NY, 1974.
Guisnet, M., and Jean-Pierre Gilson: *Zeolites for Cleaner Technologies*, Imperial College Press, London, UK, 2002.
Meier, W.M., D.H. Olson, and Ch. Baerlocher: *Atlas of Zeolite Structure Types*, 4th Revised Edition, Elsevier, New York, NY, 1996.
Occelli, M.L., and H. Kessler: *Synthesis of Porous Materials: Zeolites, Clays and Nanostuctures*, Vol. 69, Marcel Dekker, Inc., New York, NY, 1997.
Rabo, J.A. and G.J. Gajda: *Catal. Rev.-Sci. Eng.*, **31**, 385 (1989–1990).

MOLECULAR WEIGHT.   The sum of the atomic weights of the atoms in a molecule. That of methane (CH_4) is 16.043, the atomic weights being carbon=12.011, hydrogen=1.008. The chemical formula used in such a calculation must be the true molecular formula of the substance designated. For example, the molecular formula of oxygen is O_2 and its molecular weight is 31.998 (atomic weight of oxygen=15.999). For ozone the molecular formula is O_3 and the molecular weight is 47.997. The true molecular weight of a gas or vapor is found by calculating the weight of 22.4 L at 0 C and 760 mm Hg. The molecular weight of many complex organic molecules runs as high as a million or more (proteins and high polymers). See also **Avogadro Law**; and **Atomic Weight**.

MOLECULE.   In the traditional sense, a molecule is the smallest particle of a chemical substance capable of independent existence with retention of all its chemical properties. Molecules comprise one or more atoms which need not be of the same kind. Only the rare, or noble gases form single-atom or monatomic molecules. All other elements form bi-, tri-, quadri-, etc. atomic molecules, e.g., hydrogen, H_2; ozone, O_3; phosphorus, P_4; and sulfur, S_8; or hydrogen chloride, HCl; sodium sulfide, Na_2S, aluminum chloride, $AlCl_3$, carbon tetrachloride, CCl_4, and so on.

Structurally, a more specific definition would be that a molecule is a local assembly of atomic nuclei and electrons in a state of dynamic stability. The cohesive forces are electrostatic, but, in addition, relatively small electromagnetic interactions may occur between the spin and orbital motions of the electrons, especially in the neighborhood of heavy nuclei. The internuclear separations are of the order of $1-2 \times 10^{-10}$ meter, and the energies required to dissociate a stable molecule into smaller fragments fall into the 1–5 eV range. The simplest diatomic species is the hydrogen molecule, H_2^+, with two nuclei and one electron. At the other extreme, the protein ribonuclease contains 1876 nuclei and 7396 electrons per molecule.

Another form of molecule is known, however, and this is formed by atomic nuclei alone. Although these have not yet been found in nature, they may well play a role in stellar evolution. Under special conditions in high-energy interactions, they can be momentarily held together by effective bonds. Whether these bonds are the result of exchange or sharing of valence protons and neutrons is as yet moot. In these *nuclear molecules* a somewhat unstable balance is attained between long-range electrostatic repulsion of positively charged nuclei and the much stronger short-range nuclear force which determines the motions of protons and neutrons. Nuclear molecules are significant entities because they live much longer ($\sim 10^{-21}$ second) than the time usually taken for nuclei to collide ($\sim 10^{-23}$ second).

The molecular, or kinetic, theory of matter makes four assumptions: (1) that the molecules of which matter is composed are constantly in motion; (2) that their energy is increased by the addition of heat; (3) that they undergo elastic collision with each other and with the walls of a containing vessel; and (4) that they exert forces upon each other. As first developed by Heisenberg, Schrödinger, and Dirac, reduction of these theoretical assumptions to mathematical bases is somewhat inadequate and can relate only to interaction between hypothetical electron clouds. Except for the simplest of systems, the Schrödinger equation cannot be solved exactly. On the other hand, a more manageable understanding of atom interactions in molecules is afforded by the Valence Shell Electron Pair Repulsion Theory, independently enunciated by Nyholm and Gillespie. This theory proposes that both bonding and nonbonding pairs of outer atomic shell electrons in a molecule repel each other and establish themselves as far apart as possible.

Historically, molecules were regarded as being formed by the association of individual atoms. This led to the concept of *valency*, i.e., the number of individual chemical bonds or linkages with which a particular atom can attach itself to other atoms. When the electronic theory of the atom was developed, these bonds were interpreted in terms of the behavior of the valence, or outer shell, electrons of the combining atoms. Each atom with a partly-filled valence shell attempts to acquire a completed octet of outer electrons, either by electron transfer, as in (a) shown below, to give an electrovalent bond, resulting from coulombic attraction between the oppositely charged ions; or, as in (b) and (c) to give a covalent bond. The concept of (a) was proposed by Kossel in 1916; that of (b) and (c) by Lewis, also in 1916.

$$\text{Na}^+ \ [\ddot{\underset{\displaystyle ..}{\text{Cl}}}\!:]^- \qquad :\!\ddot{\underset{\displaystyle ..}{\text{Cl}}}\!:\!\ddot{\underset{\displaystyle ..}{\text{Cl}}}\!: \qquad \underset{\displaystyle R}{\overset{\displaystyle R}{R\!:\!\text{N}\!:\!\ddot{\text{O}}:}} \qquad R = CH_3$$

$$\text{(a)} \qquad\qquad \text{(b)} \qquad\qquad \text{(c)}$$

In (b), each chlorine atom donates one electron to form a *homopolar bond*, which is written Cl—Cl where the bar denotes on this theory one single bond, or shared electron pair. In (c), the nitrogen-oxygen bond is formed by two electrons donated by only the nitrogen atom, giving a *semipolar*, or *coordinate-covalent bond*, which is written $R_3N \rightarrow O$, and which is electrically polarized. Double or triple bonds result from the sharing of 4 to 6 electrons between adjacent atoms. More information on these bonding theories is given in the entries on **Chemical Composition; Chemical Elements**; and **Compound (Chemical)**.

However, difficulties arise in describing the structures of many molecules in this manner. For example, in benzene, C_6H_6, a typical aromatic compound, the carbon nuclei form a plane regular hexagon, but the electrons can only be conventionally written as forming alternate single and double bonds between them. Furthermore, an electron cannot be identified as coming specifically from any one of these bonds upon ionization. Such difficulties disappear in the quantum-mechanical theory of a polyatomic molecule, whose electronic wave function can be constructed from nonlocalized electron orbitals extending over all of the nuclei. The concept of valency is not basic to this theory, but is simply a convenient approximation by which the electron density distribution is partitioned in different regions in the molecule.

Molecular compounds consist of two or more stable species held together by weak forces. In *clathrates*, a gaseous substance, such as SO_2, HCl, CO_2, or a rare gas is held in the crystal lattice of a solid, such as beta-quinol, by Van der Waals-London dispersion forces. The gas hydrates, e.g., $Cl_2 \cdot 6H_2O$, contain halogen molecules similarly trapped in ice-like structures. The hydrogen bond, with energy ~ 0.25 eV, is responsible not only for the high degree of molecular association in liquids, such as water, but also for such molecules as the formic acid dimer, which contains two hydrogen bonds indicated by dashed lines.

$$
\begin{array}{ccc}
 & O\!-\!H\cdots\cdots O & \\
 & \diagup \qquad\qquad \diagdown\!\!\diagdown & \\
H\!-\!C & & C\!-\!H \\
 & \diagdown\!\!\diagdown \qquad\qquad \diagup & \\
 & O\cdots\cdots H\!-\!O &
\end{array}
$$

Molecular complexes vary greatly in their stability; in donor-acceptor complexes, electronic charge is transferred from the donor (e.g., NH_3) to the acceptor (e.g., BF_3), as in a semipolar bond. The $BF_3 \cdot NH_3$ complex has a binding energy with respect to dissociation into NH_3 and BF_3 of

1.8 eV. The bond here is relatively strong; the electron transfer can occur between the components in their electronic ground states. On the other hand, in weaker complexes, such as $C_6H_6 \cdot I_2$, with binding energy of about 0.06 eV, there is only a fractional transfer of charge from benzene to iodine. The actual ionic charge-transfer state lies at much higher energy than the ground state of the complex.

Complete pairing of all electrons present in a molecule and absence of any bonding orbitals was long taken to be a stable, unreactive state exemplified by the inert, or rare gases. In 1962, however, Bartlett unequivocally synthesized $XePtF_6$, and this was rapidly followed by the synthesis of other rare gas compounds whose existence was not predicted by classical valency theories. Compounds such as XeF_2, XeF_4, XeF_6, and $XeOF_4$ are quite stable, the average Xe-F bond energy in the square planar XeF_4 being 1.4 eV.

A molecule is characterized by (1) a stoichiometric formula; (2) the spatial distribution of the nuclei in their mean equilibrium or "rest" positions; and (3) the dynamical state.

The ratio $a: b: c: \ldots$ in a formula $A_aB_bC_c$, where $a, b, c, \ldots$ are the numbers of atoms of elements $A, B, C, \ldots$ that it contains is found by chemical analysis for these elements. The absolute values of $a, b, c, \ldots$ are then fixed by determination of the molecular weight. This principle is further described under the entry on **Compound (Chemical)**.

The spatial distribution of the nuclei in their mean equilibrium positions, at an elementary level, is described in geometrical language. For example, in carbon tetrachloride, CCl_4, the four chlorine nuclei are disposed at the corners of a regular tetrahedron, and the carbon nucleus is at the center. In the $[CoCl_4]^{2-}$ ion, the arrangement of the chlorine nuclei about the central metal nucleus is also tetrahedral, whereas in $[PdCl_2]^{2-}$, it is planar. For example, the pyramidal ammonia molecule NH_3 has a threefold rotation axis C_3 through the nitrogen nucleus and three reflection planes σ_v intersecting at the axis, and belongs to the $C_{3v}(3m)$ point group. Tetrahedral molecules CX_4 belong to the $T_d(\overline{4}3m)$ point group. Linear diatomic and polyatomic molecules belong to either of the continuous point groups $D_{\infty h}$ or $C_{\infty v}$ according to whether a center of symmetry is present or not.

The symmetry classification does not define the geometry of a molecule completely. The values of certain bond lengths or angles must also be described. In carbon tetrachloride, it is sufficient to give the C—Cl distance (1.77×10^{-10} meters), since classification under the T_d point group implies that all four of these bonds have equal length and the angle between them is 109°28″. In ammonia, both the N—H distance (1.015×10^{-10} meters) and the angle HNH (107°) must be specified. In general, the lower the molecular symmetry, the greater is the number of such independent parameters required to characterize the geometry. Information about the symmetry and internal dimensions of a molecule is obtained experimentally by spectroscopy, electron diffraction, neutron diffraction, and x-ray diffraction.

The dynamic state is defined by the values of certain observables associated with orbital and spin motions of the electrons and with vibration and rotation of the nuclei, and also by symmetry properties of the corresponding stationary-state wave functions. Except when heavy nuclei are present, the total electron spin angular momentum of a molecule is separately conserved with magnitude Sh, and molecular states are classified as singlet, doublet, triplet ... according to the value of the multiplicity ($2S + 1$). This is shown by a prefix superscript to the term symbol, as in atoms.

The Born-Oppenheimer approximation permits the molecular Hamiltonian H to be separated into a component H_e that depends only on the coordinates of the electrons relative to the nuclei, plus a component depending upon the nuclear coordinates. This in turn can be written as a sum $H_v + H_r$ of terms for vibrational and rotational motion of the nuclei, translation being ignored. The eigenfunctions Ψ of H may correspondingly be factorized as the product $\Psi_e\Psi_v\Psi_r$ of eigenfunctions of these three operators, and the eigenvalues of E decomposed as the sum $E_e + E_v + E_r$. In general, $E_e > E_v > E_r$.

Molecular Spectra. The spectra of substances in the molecular state, like atomic spectra, are made up of lines, although more complex. The transitions in a molecule which release the most energy (largest quanta) are due to electron changes, as in atoms, and the results of these changes are observed as lines in the ultraviolet region. But there are other ways in which a molecule can release or absorb energy. Thus, the component atoms oscillate with reference to each other within the molecule, and this motion apparently is "quantizied," i.e., changes abruptly from one state to another of different energy. But these "vibrational" energy changes are much less than the electronic, so that the resulting quanta and spectrum lines are of much lower frequency, and appear in the extreme red or near-infrared. Again, the molecule rotates, and the quantization of its rotational energy results in the emission of quanta of still lower frequency, appearing as lines in the far-infrared.

Molecular and laser spectroscopic approaches are making possible a deeper resolution of the dynamics of atomic and molecular motions and the potential energy surfaces governing energy transference within a molecule or group of molecules as chemical bonds are made and broken.

Integration of evolutionary biology with molecular biology, epitomized by the later work of Szent-Györgyi, is, in effect, translating morphogenesis into molecular language. Along metabolic pathways, intermediate molecules usually appear in such an order that thermodynamic stabilities increase progressively from starting materials to final products. Such stabilities are also associated with the influence of solvent water on chemical systems. The qualitative effects of this have been known for many years, but the quantitative data have so far been scarce. The solvent molecules tend to be reorganized in the neighborhood of the interacting groups, but not all molecules are attracted enough to overcome the self-cohesive properties of water.

Free energies of two or more moderately polar groups are often approximately additive and departures from this suggest special interactions between parts of the solute molecules and the surrounding solvent.

Polymerization. This term is used to designate a reaction in which a complex molecule of high molecular weight (or macromolecule) is formed from a number of simpler molecules. Thus the monomer formaldehyde, CHOH, can form the trimer trioxane $(CHOH)_3$, or the long-chain polymer paraformaldehyde, $HO(CHOH)_nH$, where $n = 8-100$. But the combining molecules may be of the same or different sorts. An *additional polymerization* is one in which like or unlike molecules combine without the elimination of any atoms or molecules. A *simple polymerization* involves only one species of molecule. *Copolymerization* is an addition polymerization in which two or more distinct molecular species are involved, each one of which is capable of polymerizing by itself. The high polymer formed contains each molecular constituent or an essential portion, as a distinct unit in the structure of the polymer. *Heteropolymerization* is an addition polymerization in which two or more molecular species are involved, one of which species will not polymerize by itself. It does, however, form distinct units in the high polymer.

A condensation polymerization is one in which the molecules undergoing polymerization react with the elimination of simple molecules like water, ammonia, and the like.

Polymerizations are also characterized by the state in which they are carried out such as: *gaseous polymerization* or those carried out in the vapor or gaseous phase; *mass polymerizations* in the liquid state; *solution polymerizations* carried out by first dissolving the material to be polymerized in an adequate solvent; *emulsion polymerizations* carried out in which one of the components is in an emulsion as in the case of rubber polymerizations; and *bulk polymerizations* in which the polymerization takes place without the use of a solvent or other medium. The wide variety of methods by which the process of polymerization can be used in the manufacture of plastics is exemplified by the production of polystyrene from styrene (vinylbenzene, $CH_2 \cdot CH \cdot C_6H_5$). One factor that conditions all these processes is the highly exothermic (high heat production) nature of this polymerization. This fact, together with the low heat conductivity of polystyrene, determines certain characteristics of an industrial process. This is particularly the case because the extent of the polymerization of styrene, like that of many other plastics, depends on the temperature. When the process is conducted at higher temperatures, the resulting polymer has a low molecular weight and low physical strength properties (i.e., it is weak and brittle). Extremely high molecular weight polymers, although mechanically tough, are more difficult to fabricate. Since the higher polymerization temperatures also result in faster polymerization rates, economic considerations dictate a compromise between faster production and better physical properties. Practical experience indicates a temperature range from 60 to 150°C.

The polymerization of styrene illustrates the application of several of the methods defined above.

In *batch mass polymerization* the reaction vessel is loaded with styrene monomer and heated to a temperature sufficient to initiate polymerization within a reasonable time. As polymerization proceeds, the temperature within the vessel rises, thus increasing the rate.

In *continuous mass polymerization* the reaction vessel contains monomer and polymer. As more polymer is formed it is drawn off the bottom of the reaction vessel while monomer is added to the top of the vessel. The temperature of the reaction is controlled by cooling coils within the vessel.

In *solution polymerization* the styrene monomer is diluted with a solvent. The solvent acts as a diluent, which decreases the rate of polymerization and also serves as a heat transfer medium for removing the excess heat developed by the reaction.

In *suspension polymerization* water is used as a diluent and as a heat transfer aid. Suspending agents such as starch and methylcellulose are used to keep the styrene monomer particles in suspension. The more efficient heat transfer of this process also allows for a narrower molecular weight distribution.

In *emulsion polymerization* the styrene monomer is emulsified with water by the addition of certain emulsifying agents. This results in very small particles and rapid polymerization rates. The heat of polymerization is dissipated by the water ingredient.

See also **Fibers**; and **Macromolecular Science**.

Additional Reading

Barondes, S.: *Molecules and Mental Illness*, Scientific American Library, New York, NY, 1999.

Eisberg, R. and R. Resnik: *Quantum Physics of Atoms, Molecules, Solids, Nuclei, and Particles*, 2nd Edition, John Wiley & Sons, Inc., New York, NY, 1990.

Parr, R. and Y. Weitao: *Density-Functional Theory of Atoms and Molecules*, Oxford University Press, New York, NY, 1994.

MOLE FRACTION.   As applied to a system, the *mole fraction* (sometimes spelled *mol fraction*) of a given substance in the system is found by dividing the number of moles of that substance by the total number of moles in the system. For a mixture of ideal gases, the mole fraction is equal numerically to the volume fraction. The *volume fraction* of a component in a mixture is found by dividing the volume of that component at the total pressure and at the temperature of the mixture by the volume of the mixture at the same pressure and temperature.

In a binary solution consisting of components X, and Y, the mole fractions, F_X and F_Y, respectively, are:

$$\text{Mole fraction of } X = F_X \frac{f_X}{f_X + f_Y}$$

$$\text{Mole fraction of } Y = F_Y \frac{f_Y}{f_X + f_Y}$$

where f = number of moles of specific component present.

It is apparent, of course, that the mole fraction of X plus the mole fraction of Y must equal unity, or, if expressed as a percentage (mole percent), must equal 100. In the instance of three or more components, the denominators of the prior expressions must reflect the additional moles present.

In considering a solution containing 50 grams of methyl alcohol (m.w., 32) in 1,000 grams of H_2O (m.w., 18), the mole percent of each component will be:

Moles of CH_3OH = 50/32 = 1.562

Moles of H_2O = 1,000/18 = $\underline{55.556}$

Total Moles   57.118

Mole percent CH_3OH = 1.562/57.118 × 100 = 2.735%

Mole percent H_2O = 55.556/57.118 × 100 = 97.265%

The same solution, expressed in terms of weight percentage, is:

Weight percent CH_3OH = 50/1,050 = 4.762%

Weight percent H_2O = 1,000/1,050 = 95.238%

In nuclear chemistry, the term *mole fraction* may be used to indicate the number of atoms of a given isotope in an isotopic mixture, as a fraction of the total number of atoms of that element in the mixture.

MOLES AND SHREWS.   Important members of the order of mammals known as *Insectivora*. Generally, members of this order are small animals, mostly nocturnal in habits. They live on insects and other small invertebrates and, in a few cases, on vegetation. The general organization of the *Insectivora* is given in Table 1.

Tenrecids. The tenrec is an animal of Madagascar. The species resembles the other members of the order in its compact body, short legs, and long sharp muzzle. Some specimens attain a length of 16 inches (41 centimeters) and are clothed with a mixture of spines, bristles, and hair. Tenrecs are nocturnal burrowing animals. The geogale is a watertenrec, a small animal of Madagascar resembling a mouse. The animal has webbed feet and a long tail. The geogale lives in and along streams and rivers. The solenodon is also a tenrecid. It is an animal of about the size of a rabbit, but with a long slender nose and a long naked tail. The claws are strong, those of the forefeet being much larger than those of the hind feet. There are two species, both of which occur in the West Indies.

Erinacids. The principal member of this family is the hedgehog. This mammal is small and compact, about 10 inches (25 centimeters) in length, with short legs and tail and a sharp nose. The entire upper surface is covered with sharp spines which protect the animal's body when it is curled up. However, a fox will roll the animal into a pool of water. To swim, the hedgehog must uncurl and thus lose much of its protection from the projecting spines. Then, the hedgehog is relatively easy for the fox

TABLE 1. GENERAL ORGANIZATION OF TENRECS, MOLES, AND SHREWS

INSECTIVORA	
TENRECIDS	SORICIDS
Tenrecs (*Tenrecidae*)	Shrews (*Soricidae*)
Common Tenrec (*Tenrec*)	Common Shrews (*Sorex* ...)
Striped Tenrecs (*Hemicentetes*)	Short-tailed Shrews (*Blarina* ...)
Hedgehog-Tenrecs (*Setifer*)	Common Water-Shrews (*Neomys*)
Rice-Tenrecs (*Oryzorictes*)	Musk-Shrews (*Crocidura*)
Long-tailed Tenrecs (*Microgale* ...)	Forest-Shrews (*Sylvisorex* ...)
Water-Tenrecs (*Limnogale* and *Geogale*)	Fat-tailed Shrews (*Suncus*)
Giant Water-Shrew (*Potamogalidae*)	Mole-Shrews (*Anourosorex*)
Solenodons (*Solenodontidae*)	Asiatic Water-Shrews (*Chimarrogale*)
Golden Moles (*Chrysochloridae*)	Web-footed Shrews (*Nectogale*)
	Girder-backed Shrew (*Scutisorex*)
ERINACIDS	MOLES
Hedgehogs (*Erinaceinae*)	Asiatic Shrew-Moles (*Uropsilus* and *Nasillus*)
Gymnures (*Echinosoricinae*)	Desmans (*Desmana* and *Galemys*)
	Eurasian Moles (*Talpa* ...)
MACROSCELIDS	Pacific Moles (*Scapanus* ...)
	Star-nosed Mole (*Condylura*)

to capture. Numerous species of hedgehogs live in Africa, Europe, and India; others are found in Asia north of the Himalayas. They are nocturnal animals. The European hedgehog, *Erinaceus europaeus*, is also called the urchin.

Soricids. This family of *Insectivora* is made up of several species of shrew. Shrews are small animals closely resembling mice in general appearance. They are common in Europe, Asia, Africa, and North America. Many species burrow. Some live near the water and have developed aquatic habits. A species found in Africa is characterized by the great development of the hind legs and is known as the jumping shrew. See Fig. 1. They also have very prolonged snouts, characteristic of many insect-eating mammals. The tree shrews or tupaias of the Oriental region constitute still another family of shrews. They have arboreal habits and are somewhat squirrel-like in appearance. All of these animals subsist on a diet of worms, insects, and plant shoots.

Fig. 1.   African jumping shrew. (*Painting by Charles R. Knight, American Museum of Natural History.*)

Fig. 2.   A mole. The hairless nose is sensitive to touch and is used as a guide. (*John H. Gerard from National Audubon Society.*)

Moles. Possibly the best known of the *Insectivora* is the mole. See Fig. 2. The mole is a burrowing animal of small size, and is highly specialized for life underground. The legs are short, but the front pair of legs is powerful, with broad feet and strong claws. The eyes are rudimentary and there are no external ears. The term *mole* is used in connection with mammals other than the true moles. In the order *Marsupialia*, a single rare Australian animal adapted for subterranean life is called the pouched mole or marsupial mole. The term also appears in mole voles, mole rats, Cape mole rats — all burrowing animals of the Old World belonging to the order *Rodentia*. These animals are much less extensively adapted (specialized) than the true moles. See also **Rodentia**.

It should be pointed out that one of the best known of insect-eating mammals is the anteater. However, the anteater is a member of the family *Edentata*. See also **Edentata**. The spiny anteater is a member of Monotremeta. See also **Monotremeta**.

MOLE (Stoichiometry).   Sometimes spelled *mol*, a mole is a quantity of a substance, expressed in specified mass units, that is equal to the molecular weight of the substance. For example, a *gram-mole* or *gram-molecular mass* of hydrogen H_2 will have a mass of $2 \times$ (atomic weight of hydrogen), or 2.016 grams. A gram-mole of carbon dioxide CO_2 will have a mass of $1 \times$ (atomic weight of carbon) plus $2 \times$ (atomic weight of oxygen), or 12.011 plus 31.998 = 44.009 grams. A pound-mole of ammonia gas NH_3 will have a mass of $1 \times$ (atomic weight of nitrogen) plus $3 \times$ (atomic weight of hydrogen), or 17.031 pounds. See also **Avogadro Constant**.

MOLE (UNIT OF AMOUNT OF SUBSTANCE).   Following the discovery of the fundamental laws of chemistry, units called, for example, "gram-atom" and "gram-molecule," were used to specify amounts of chemical elements or compounds. These units had a direct connection with "atomic weights" and "molecular weights," which were in fact relative masses. "Atomic weights" were originally referred to the atomic weight of oxygen, by general agreement taken as 16. But whereas physicists separated isotopes in the mass spectrograph and attributed the value 16 to one of the isotopes of oxygen, chemists attributed that same value to the (slightly variable) mixture of isotopes 16, 17, and 18, which was for them the naturally occurring element oxygen. Finally, an agreement between the International Union of Pure and Applied Physics (IUPAP) and the International Union of Pure and Applied Chemistry (IUPAC) brought this duality to an end in 1959/60. Physicists and chemists have ever since agreed to assign the value 12, exactly, to the "atomic weight," correctly the relative atomic mass, of the isotope of carbon with mass number 12 (carbon 12, ^{12}C). The unified scale thus obtained gives values of relative atomic mass.

It remained to define the unit of amount of substance by fixing the corresponding mass of carbon 12; by international agreement, this mass has been fixed at 0.012 kg, and the unit of the quantity "amount of substance" was given the name mole (symbol mol).

Following proposals of IUPAP, IUPAC, and the International Organization for Standardization (ISO), the CIPM gave in 1967, and confirmed in 1969, a definition of the mole, eventually adopted by the 14th CGPM (1971):

1. The mole is the amount of substance of a system which contains as many elementary entities as there are atoms in 0.012 kilogram of carbon 12; its symbol is "mol."
2. When the mole is used, the elementary entities must be specified and may be atoms, molecules, ions, electrons, other particles, or specified groups of such particles.

At its 1980 meeting, the CIPM approved the 1980 proposal by the Consultive Committee on Units of the CIPM specifying that in this definition, it is understood that unbound atoms of carbon 12, at rest and in their ground state, are referred to.

See also **The International System Of Units (SI)** and **Units and Standards**.

MOLE VOLUME.   A mole of gas will occupy a definite volume under definite conditions regardless of the nature of the gas. This definite volume is called the *mole volume*. Under a pressure of 760 torr and at a temperature of $0°C$, a gram-mole of gas will occupy 22.41 liters. This situation also applies to a mixture of gases. A pound-mole of gas will occupy 359 cubic feet at a pressure of 760 torr and at a temperature of $32°F$ ($0°C$).

Because the volume of one mole of gas at any specific pressure and temperature contains the same number of molecules even though there may be several different gases in the mixture, the percent by volume of any given gas is equal to the percent pressure exerted by that gas and is also equal to the mole percent of that gas. Mole percent equals volume percent equals pressure percent.

MOLINA, MARIO (1943–). A Mexican who won the Nobel Prize for chemistry along with Paul Crutzen and Frank Sherwood Rowland in 1995 for their work in atmospheric chemistry, particularly concerning the formation and decomposition of ozone. See also **Crutzen, Paul;** and **Rowland, Frank Sherwood.**

MOLLIER CHART. A thermodynamic diagram for a homogeneous system possessing two independent properties, in which enthalpy is the ordinate and entropy is the abscissa. Mollier charts are used widely in engine calculations, particularly those in which the working fluid is steam.

MOLLISOLS. See **Soil.**

MOLLUSCA. A major division of the animal kingdom containing the snails, oysters, clams, mussels, squids, octopus, nautilus and related forms. Mollusks are the most highly developed of the unsegmented invertebrates and are both diverse in form and numerous in species.

The phylum is characterized by the following structures. (1) The body is unsegmented. (2) A well-developed head is found in most species. (3) The body bears a ventral muscular protuberance, the foot. (4) A fold extends in most species from the dorsal wall, enclosing a cavity associated with respiration. The fold is the mantle and the cavity, the mantle cavity. (5) In many species the mantle secretes a shell. (6) The circulatory system consists of tubular vessels and open spaces, with a heart made up of a ventricle and two auricles.

Some mollusks are important as food. Clams, oysters, and scallops are the most familiar of the edible species but others are eaten. Pearls and mother-of-pearl are also molluscan products.

The phylum is divided into the following classes:

Class *Amphineura*. Without a distinct head. Shell absent or composed of a series of plates. Chitons.
Class *Gasteropoda*. With a distinct head. Shell absent, conical, or spiral. Snails and related forms.
Class *Scaphopoda*. Head indistinct. Shell cylindrical.
Class *Lamellibranchiata* (*Pelecypoda*). Head indistinct. Shell of two lateral parts (bivalve). Clams, mussels, oysters, etc.
Class *Cephalopoda*. Head distinct, with long tentacles. Squids, octopus, nautilus, etc.

Several mollusks are described under separate alphabetical entries in this encyclopedia. See illustration of mollusk shells in Fig. 1.

MOLLUSCUM CONTAGIOSUM. An orthopoxvirus infection characterized by multiple, painless, umbilicated nodules (2–5 mm diam) which can appear anywhere on the body except the palms and soles. The nodules, which are most commonly found in anogenital regions, rupture easily and may spread by sexual routes. Spread may also occur by close family contact under conditions of poor hygiene. Lesions may clear rapidly or persist for up to 18 months. When near the eye, lesions may be complicated by chronic conjunctivitis or superficial keratitis. Lesions resolve spontaneously without scarring and no antiviral therapy or vaccine is available.

MOLLUSKS. The mollusks are greatly diverse in structure and represented by many curious forms, from the highly active squids to the slow and sluggish snails. Yet all of these creatures have one or more basic morphological and embryological features that unite them into the phylum *Mollusca*. The mollusks are the second largest animal phylum, with thousands of species. Only a relatively few of these species are of importance as food substances—either as food for human consumption, or as bait for catching marine fishes, or for use as fertilizer. About 40% of the total world catch of mollusks is cephalopods, mainly squids, which are pelagic. The catches of the benthonic octopuses are less important. The gastropods only represent 1% (abalone and some small snails). The bivalves are by

Fig. 1.   Mollusk shells. (*A.M. Winchester.*)

far the most important resource of mollusks, with 59% of the total. In fact, most of the landings result from farming (oysters, mussels, clams), but the natural types are also important, for instance, scallops. The nomenclature and taxonomy of the mollusks tends to be complex, this accentuated by numerous changes in systematics over the years. Table 1 is offered to provide some assistance in deciphering the nomenclature.

Bivalves

From a commercial standpoint, the bivalves make up the largest economic class of mollusks. The bivalves are one of the most peculiar groups of animals. There is no recognizable head section and, therefore, it is difficult for a non-expert to differentiate between the front and back of the animal. Although most bivalves are able to move freely, one only rarely sees them changing their position or locality. The bivalve's body is surrounded by the laterally enlarged mantle folds and the two shell plates it secretes. An uncalcified ligament holds the two shell valves together without influencing their movement against each other. Along the hinge line, the two shell halves usually possess interlocking teeth. Muscular action is responsible for the tight and firm closure of the bivalve shell. A diagram is given in the entry on **Clam.**

Clams. The external appearance of the soft-shell clam is an irregular ellipsoid with the two valves or shells of nearly equal size. The right shell is usually slightly larger than the left. When closed, the two valves may touch along the ventral edge but gape widely in front, allowing passage of the foot, and in the rear, allowing extrusion of the siphons. The outer surface of the shell is covered with rough striations. The annular marks or "rings" result from the change in rate of growth during the year. Since cold winter temperatures restrict clam growth, the rings can indicate the age of the clam. The interior of the shell is relatively smooth, but has certain areas scarred at the points where major muscles are attached. Along the dorsal edge of each shell are the prominent hinge teeth; one valve carries a large, projecting, spoon-shaped structure, and the other valve has an

TABLE 1. ABRIDGED CLASSIFICATION OF EDIBLE MOLLUSKS

Common Name	Genus And Species	Family	Order	Subclass	Class
CLAMS					
Soft shell clams	*Mya arenaria*	*Myidae*	*Myoida*	*Heterodonta*	*Bivalvia* (The bivalves)
	Mya truncata				
Surf clams	*Spisula solidissima*	*Spisula*			
American quahog	*Mercenaria mercenaria*	*Veneridae* (Venus clams)	*Veneroida*		
Razor clam	*Solen marginatus* *Solen pellucidas*	*Solenidae* (Razor shells)			
Ocean quahog	*Arctica islandica*	*Arcticidae*			
Common edible cockle	*Cerostoderma edule*	*Cardiidae* (Cockles)			
Devon cockle	*Cerostoderma aculeatum*				
MUSSELS					
Mediterranean mussel	*Mytilus galloprovincialis*	*Mytilidae* (Marine mussels)	*Mytiloida*	*Anisomyaria*	
Blue mussel	*Mytilus edulis*				
California mussel	*Mytilus californianus*				
Asian mussel	*Mytilus crassitesta*				
Philippine mussel	*Mytilus smaragdinus*				
Indian mussel	*Mytilus viridis*				
New Zealand mussel	*Mytilus canaliculus*				
Brazilian mussel	*Mytilus perna*				
SCALLOPS					
American scallop	*Pecten magellanius*	*Pectinadae* (Scallops)	*Pteroidea*		
Pilgrim's scallop	*Pecten jacobaeus*				
Giant scallop	*Pecten maximus*				
OYSTERS					
Common European oyster	*Ostrea edulis*	*Ostreidae* (True oysters)			
	Ostrea lurida				
Japanese oyster	*Crassostrea gigas*				
Portuguese oyster	*Crassostrea angulata*				
American oyster	*Crassostrea virginica*				
	Crassostrea commenalis				
Chilean oyster	*Crassostrea chilensis*				
GASTROPODS					
Abalone	*Haliotis refuscens*	*Haliotidae* (Abalones)	*Archaeogastropoda* (Primitive univalves)	*Prosobranchia* (Front-gilled snails)	*Gastropoda* (The univalves)
	Haliotis gigantea				
	Haliotis tuberculata				
Giant conch	*Syrinx goliath*	*Melongenidae*	*Caenogastropoda* (Modern univalves)		
Queen conch	*Syrinx gigas*				
Crown conch	*Melongena melongena*				
Louisiana conch	*Thais haemastoma*	*Muricidae*			
Escargot snail	*Helix pomatia*	*Helicidae*	*Stylommatophora* (Land snails)		
CEPHALOPODS					
Common squid	*Loligo vulgaris*	*Loliginidae*	*Decabrachia*	*Dibranchiata*	*Cephalopoda* (Squids and Octopods)
No. American common squid	*Loligo pealei*				
No. European squid	*Loligo forbesi*				
Opalescent squid	*Loligo opalescens*				
Giant squid	*Architeuthis dux*	*Architeuthidae* (Giant squids)	*Teuthoidei*		
	Architeuthis harvey				
	Architeutis princeps				
Short-finned squid	*Illex illecebrosus*	*Ommatostrephidae*			
Common octopus	*Octopus vulgaris*	*Octopodidae* (Octopuses)	*Octobrachia*		
Common cuttlefish	*Sepia officinalis*	*Sepiidae* (The cuttlefishes)	*Decabrachia*		

inverted bowl. The bowl lies directly over the spoon. Between the two structures is a tough, rubber-like ligament that acts to keep the shells apart by opposing the closing action of the adductor muscles, much as a rubber eraser squeezed in a door hinge would force the door open.

Unlike the higher animals, the clam has blood circulating in an open system. The two-chambered heart pumps blood through a system of arteries that eventually empty into large, open lacunae. The blood is returned to the heart by a loose network of veins. Considerably more detail on the physiology of the clam can be found in the Hanks reference.

Life for a soft-shell clams begins as sperm and egg join in the coastal waters. The fertilized egg develops into a swimming trochophore larva in about 12 hours in cold New England water, probably sooner in warmer, southern waters. The larva moves by hair-like projections (*cilia*) arranged in distinct bands around the body. The larva has a mouth and a minute shell gland, which gives rise, within the following 24 to 36 hours, to the two calcified valves that envelop and protect the clam body throughout its life. After further physiological changes, the clam reaches a setting stage when metamorphosis is possible and the final adaptation to sedentary bottom existence can begin. This period is extremely critical in the life cycle for, once removed from transportation by water, the clam must establish itself within a relatively small area and its future success is influenced by the type of bottom material on which it settles.

Upon setting, the clam immediately attaches to sand grains, plants, or other materials, by a tough, horny thread (*byssus*). At the onset of cold weather, the young clam burrows and then becomes less active through the winter. Spring tides tend to move many clams from their winter burrows.

The soft-shell clam is found in bays, coves, estuaries, and other protected areas. It lives in a wide variety of sediments, ideally a mixture of mud and sand where it is less likely to be exposed to predation and climate. Clams live deeply buried in the bottom sediments, and the greatly extensible siphons may reach lengths of more than four times the length of the shell, drawing water and food from just above the bottom. Microscopic plants, animals, and other organic food substances are filtered from the water, possibly at a rate of about 45 liters (12 gallons) per day. Most soft-shell clam populations in New England are found in intertidal flats, exposed at low tide, whereas most clams in the Chesapeake Bay are in subtidal regions and are seldom exposed. In productive areas, clams occur in vast but erratically distributed beds.

Sexual maturity may be reached at one year of age (at a length of about 12 to 19 millimeters; 0.5 to 0.75-inch) in northern waters. The sexes are separate and can be determined only by microscopic examination. Each clam can produce millions of eggs or billions of sperm. Sex cells are exuded from the gonads through the siphon in successive puffs. Spawning is temperature-related and occurs from June to mid-August in northern areas. Clams north of Cape Cod have only one reproductive cycle per year, but south of Cape Cod, spawning may occur twice each year.

Southern clams may reach a shell length of 5 centimeters (2 inches), an acceptable commercial size, within 1.5 to 2 years. Maine clams require from 5 to 6 years to achieve this size. The green crab (*Carcinus maenas*) is the most serious predator (other than humans). Mass mortalities of some soft-shell clam populations have been reported from time to time, but causes are poorly understood.

Coastlines in the Northern Hemisphere on both sides of the Atlantic and Pacific Oceans support large numbers of soft-shell clams. The fishery on the Atlantic coast of North America is of the greatest importance, partly because of the heavy demands for these products in the nearby northeastern United States. The soft-shell clam (*Mya arenaria*) is found from the coast of Labrador to the region of Cape Hatteras, North Carolina. On the European coast, it has been recorded from northern Norway to the Bay of Biscay. In the late 1870s, young clams were accidentally introduced into San Francisco Bay, with shipments of seed oysters from the Atlantic coast. Since that time, the clams have spread from Monterey, California to Alaska. The soft-shell clam is also found along the western Pacific coast from the Kamchatka Peninsula to the southern regions of the Japanese islands. The southern range limits are well defined, but the northern limits are less marked because of the presence of a closely related species, *Mya truncata*, a northern form of which some populations exhibit similar external shell morphology.

Surf Clam. The Atlantic surf clam (*Spisula solidissima* Dillwyn) is estimated to have lived on the earth for many millions of years. The oldest known fossils are from deposits of the upper Miocene age in North Carolina. Surf clams are the largest bivalve mollusks in the region; some have a maximum length of over 20.3 centimeters (8 inches). Their shells are a familiar sight along ocean beaches after storms. Many names are in common use for the surf clam, such as *bar clam* (Canada); *hen clam* (Maine); *sea clam* (Massachusetts); and *beach clam* or *skimmer clam* (middle-Atlantic states). The surf clam industry has gained considerable importance since the 1950s.

The surf clam is found from the southern Gulf of Saint Lawrence to the northern Gulf of Mexico. In the northern part of its range, it is commonly abundant in the turbulent waters off outer beaches, just beyond the breaker zone, and is occasionally found to depths of 126 meters (420 feet) on Georges Bank. Surf clams have been taken at depths of 60 meters (200 feet) on the middle Atlantic continental shelf, but commercially useful concentrations are in depth of 11 to 55 meters (36 to 180 feet). South of Cape Hatteras, the clams are small and inhabit shallow water.

American Quahog. This hard-shell clam (*Mercenaria mercenaria*) should not be confused with the ocean quahog described later. The American quahog is an important commercial species along the northeastern Atlantic coast of the United States. Clams of the family *Veneriidae* are sometimes collectively referred to as the Venus clams. Many years ago, the American quahog was introduced to northwestern French waters from the eastern coast of North America. The range of the American quahog is from Cape Cod southward, although it is sometimes found as far north as the Maine coast (Casco Bay) and the Gulf of Saint Lawrence at Shediac. Young and tender quahogs are sometimes called *little-neck clams*.

Ocean Quahog. The natives of Iceland have eaten this hard-shell clam for at least two centuries, where it is called *ku-skioezl* and *krokfishur*. This use of ocean quahogs by Icelanders is the basis for its scientific name, *Arctica islandica* Linne. In North America, the clam is sometimes called *black quahog* and *mahogany clam*. The *Arcticidae* were abundant in earlier geological times; fossils of several genera and over 100 species have been found, largely in the North Atlantic region.

Mussels. Mussels (*Mytilidae*) enjoy a worldwide distribution and are extremely adaptable. Their meat is tasty and contains a high percentage of vitamins, protein, minerals, and other important nutritive substances. Mussels are popular throughout Europe, and are a particular favorite in France and the Netherlands. Certain countries cultivate edible mussels. In the Mediterranean region, the Mediterranean mussel (*Mytilus galloprovincialis*) is raised. Natural occurrence of these mussels in this tidal region greatly facilitates their cultivation. The related bearded mussel (*Modiolus barbatus*) is also a valued food item in the Mediterranean countries. A principal predator of the mussel is the starfish.

Mussel culture is required when natural beds are too small to support an expanding fishery. It is a two-part process: (1) The quality of natural stocks is improved by moving them to good growing areas; and (2) when supplied of seed mussels run short, new supplies sometimes can be obtained by bringing about the settlement of planktonic mussel spat, by various methods still being developed. As planktonic spat is quite abundant in European waters, the vast size of this resource, plus the fact that mussels live on phytoplankton, gives mussel culture an outstanding biological potential. See also **Aquaculture**.

The distribution of *Mytilus edulis* and *M. galloprovincialis* ranges from the Pacific and Atlantic coasts of North America to most of Europe and the African Mediterranean coast. *Mytilus perna* (or perna perna) is found in Brazil and Venezuela, while on the Pacific coasts of the Americas, there are *Mytilus californianus* in the north, and *Choromytilus choros* and *Aulocomya* species in Chile. Along Asiatic coasts and Japan is found *M. crassitesta*. *M. smaragdinus* is found in the Philippines as well as in India, where *M. viridis* and the Brown Mussel also occur. *Mytilus viridis* also has been reported in Malaysia, and *M. canaliculus* in New Zealand. In several of these locations, mussel fisheries exist, and the prospects of further development of them are worthy of examination. Potential mussel culture in Asia and South America appears particularly promising and important because of local protein shortages.

Data on the growth rates of mussels is not highly definitive. In the Philippines, *M. smaragdinus* is harvested at lengths of 5 to 10 centimeters (2 to 4 inches). In India, specimens of *M. viridis* of about 15 to 20 centimeters (6 to 8 inches) have been reported. Reports from Venezuela indicate that *Mytilus perna* will grow from 10 to 20 millimeters (0.4 to 0.8 inch) to 10 to 15 centimeters (4 to 6 inches) over a 5-month period, yielding 50% of flesh by weight. By comparison, *M. edulis* requires a year to grow from 10 to 75 millimeters (0.4 to 3 inches) and well over 3 years to reach the larger size in Wales. These figures suggest that mussel culture in Asia and South America has advantages.

Freshwater mussels are eaten but they are not an important source of food. They are taken in large numbers from the larger rivers of the United States for their shells, which are used in making buttons, and for the pearls they contain. As many as 50,000 tons of shells have been marketed in a single year, chiefly from the Mississippi River.

Scallops. Although the beautifully colored and sculptured shells of some scallops are much admired for their aesthetic qualities, scallops are important because of the economic value of the muscle, which is used for food. Populations of the largest American scallop (*Placopecten magellanius* Gmelin) occur from Labrador to New Jersey. Individual animals with an estimated age of 18 to 20 years and a diameter of 22.5 centimeters (9 inches) by 21 centimeters (8.25 inches) wide have been recorded in Maine. Some authorities believe that the greatest known sea scallop grounds are found between the 20- and 50-fathom curves on Georges Bank. Preceded by harvesting by American Indians, the commercial scallop fishery of Maine dates back to 1880. Scallop populations occur discretely in major estuaries and embayments from the Piscataqua River, separating Maine and New Hampshire, to the Saint Croix River, which forms the international boundary between Maine and New Brunswick, Canada. Vertical distribution in Maine waters ranges from mean low water in some areas to depths of over 100 meters in others. Concentrations of commercial importance are generally limited to the area from Penobscot Bay eastward. Sometimes, scuba divers are able to gather scallops in commercial quantities from rocky bottoms that are impossible to drag with conventional scallop gear.

Production of scallops tends to fluctuate widely from year to year. The principal lead for indication of why abundance fluctuations have taken place is the record of seawater temperatures taken at Boothbay Harbor, Maine since 1905. Since the offspring of any year's spawning—August to October—becomes of major importance to the fishery six years later, it appears from a study of temperature and production records that an association exists between seawater temperature six years earlier and highs and lows of scallop landings. Research indicates that an optimum temperature of 7.8–8.1 °C (46–46.5 °F) is required for peak landings and that temperature on either side of this narrow range seriously affects production. Some authorities, however, do not agree with this conclusion because a generally inverse relationship between Maine and New England (Georges Bank) landings suggest that seawater temperature is not an overriding influence.

In Maine inshore waters, sea scallops grow rapidly. At 6 months of age, a scallop may be only 2 millimeters (about $\frac{1}{16}$ inch) long. Within 4 years, it will have reached a size of 7.4 centimeters (2.9 inches); and by 10 years, 13 centimeters (5.1 inches).

Possibly the best-known scallop species worldwide is the pilgrim's scallop (*Pecten jacobaeus*), which reaches a size of 5 to 15 centimeters (1.9 to 5.7 inches). In decades past, many crusaders returning from the Mediterranean carried these shells on their hats or garments—thus the name for the scallop.

The giant scallop (*Pecten maximum*) is well known in the Mediterranean region. This scallop is also found in the North Sea.

In general, the scallop is mobile. A series of prominent, highly developed, deep-blue eyes are situated among the tentacles along the mantle margin. The location and degree of development of these visual organs can vary with species and within a species. The pilgrim's scallop detects the difference between light and dark as well as motion. The animal thus can perceive approaching starfishes or octopuses; in addition, it can recognize its enemies by their scent. When endangered, the scallop escapes by its thrusting manner of swimming. However, the visual acuity of these bivalves is not sufficient to help them orient toward a source of light during swimming.

Oysters. Oysters (*Ostreidae*) are found worldwide in a broad band between latitudes 64 °N and 44 °S. They are distributed from the intertidal zone down to approximately 39 meters (130 feet). Commercial edible oysters belong to two genera (*Ostrea* and *Crasostrea*). The more important

species are listed in the table. Pearl oysters are species of different families found only in warmer oceans, while the edible oysters include a European species (*Ostrea edulis*), a species of the Atlantic coast of North America (*O. virginica*), and the Pacific coast oyster (*O. lurida*). The common oyster of commerce was transplanted successfully to the Pacific coast many years ago. All are marine.

The soft body of an oyster is covered by two shells or valves. The right or top valve is flat, and the left or bottom valve is heavier and cupped. An oyster attaches and usually rests on its left valve. The two shells are joined together at the hinge by an elastic material known as the ligament. The shape of the shell is highly variable in *Crassostrea* and subcircular in *Ostrea*. More than 95% of the shell is calcium carbonate. Growth of the shell depends directly upon water temperature. Along the east coast of North America, the growing season is longer in the south than in the north. In Canada and New England, the season of shell growth is from 4 to 5 months; in the Chesapeake Bay, 6 to 7 months; and in the warm waters of Florida, oysters grow nearly all year. Oysters in northern waters require about five years to reach market size (10 centimeters; 4 inches). In Florida, this size is attained in some areas within 18 months.

Oyster meats are said to contain about every element found in seawater. The amount of meat yielded varies with geographic location and season. Meat quality drops during the spawning season. After spawning, oysters build up glycogen and are in their best condition at the height of this buildup. These seasonal changes are sometimes measured by calculation of percentage solids (dry weight of meat multiplied by 100 divided by weight of meat). In the spring, glycogen is converted into sex chemicals.

In American waters, the two principal predators are the Atlantic oyster drill (*Urosalpinx cinera*), and the thick-lipped drill (*Eupleura candata*). These are carnivorous gastropods. Certain species of snails are also detrimental to oysters. The type of predator is largely determined by water salinity. Numerous diseases affect oysters, as well as parasites.

For a successful industry, seed oysters must be readily available. Methods of collection vary with region. The Japanese catch seed oysters on shells suspended from rafts or on shells draped over wooden racks. In France, large quantities of seed are caught on lime-coated tiles, which are laid in rows in the intertidal zone. Australian seed oysters are caught on sticks. In the United States, most oysters are caught on shells scattered over the bottom. There is also some production of seed by suspending strings and bags of shells from rafts.

The development of off-bottom culture in Japan during the 1930s caused oyster production to increase at a rapid rate in that country. Presently, most areas in Japan, especially in protected bays, are being utilized. There is considerable potential for greater oyster production in South Atlantic waters.

Cockle. Of the family *Cardiidae*, the cockles are worldwide in distribution and contain some 200 species. The common edible cockle (*Cardium edule*) attains a length of 3.5 to 5 centimeters (1.4 to 1.9 inches). This is one of the most common and best-known bivalves in Europe. Shells of this form have been found in superabundance swept up on sandy beaches from Iceland to western Africa. The animals are characterized by whitish-yellow shells with regularly spaced ribs. There is an exhalent and inhalent siphon. The long foot is bent and tapers to a point. It enables the bivalve to move in a peculiar way. The animal extends the foot as far out of the shell as possible, up to nearly 5 centimeters (2 inches) and gropes for resistance from any object. When the bent portion of the foot is suddenly jerked into a straight position, which results in a push, the bivalve flings itself a distance of over 50 centimeters (20 inches). Some cockles possess eyes. The common cockle occurs along the coasts of the British Isles and is also found in the Baltic Sea. It is well regarded as food in various regions. A number of nonedible American varieties occur in the Cape Cod and Long Island waters.

Univalves

The univalves or *Gastropoda* are snails. Of this class of mollusks, probably the two most important species for food are the abalone, a marine species that inhabits the rocky shores of all continents and among many of the islands in the Pacific, Atlantic, and Indian Oceans; and the edible snail found in many parts of the world.

Abalone. All gastropods undergo a twisting, or torsion, early in their development, wherein the visceral mass rotates 180° counterclockwise and the gills, mantle cavity, and anus, which originally faced to the rear, come to lie behind the head. Abalone development typically proceeds from a short (3 to 6 days) pelagic free-swimming stage to a benthonic creeping phase and finally to a more sedentary and retiring adult existence.

The greatest concentrations of abalone populations, both in numbers of species and inhabitants, are off the coasts of Australia, Japan, and western North America. The largest species (*Haliotis rufescens*) occurs on the California coast; mature individuals average between 17.8 and 22.9 centimeters (7 and 9 inches). Some may exceed 28 centimeters (11 inches) in diameter. The next largest species (*H. gigantea*) occurs near Japan and attains a length of about 25.5 centimeters (10 inches). Other large haliotids live off south Australia, New Zealand, and South America.

Regional names for abalone vary widely — *abalone* in the United States; *Oreille de Mer* in France; *Ormer* or *Venus ear* in England; *mutton fish* in Australia; *paua* in New Zealand; *aulone* in Mexico; *awabi* in Japan.

Peeled abalones are sliced into steaks approximately 1.3 centimeters (0.5 inch) thick, usually with a slicing machine. The steaks have to be pounded to break down their connective fibers. Premium prices are received for large, white-meat steaks. Occasionally abalones will have grayish colored meat. They taste just as good, but bring a lower price.

Conch. The term *conch* is of Indo-European, origin and refers to the shell of mollusks. Presently, the term is predominantly used in reference to certain large prosobranch gastropids, particularly in the families *Strombidae, Cassidae*, and *Galeondidae*. The most important species from an economic standpoint belong to *Strombidae*.

Distributed throughout the world in tropical and subtropical waters, this group is represented by nearly 40 species in the Indo-Pacific region. The most extensive fishery, however, exists in the Caribbean area, where some seven species of *Strombus* thrive. The largest stromb, the rare and coveted *S. goliath*, lives off the coast of Brazil and attains a length of over 33 centimeters (13 inches). It is sometimes available in native markets near Recife. The queen conch (*S. gigas*), also known as pink-lipped conch, is the most important commercial species in the family and is used as an important source of protein by natives of Haiti and the Bahamas. The ornamental values of the conch shell are also highly regarded and exploited. Pearls also are formed from the mantle of the species.

The usual habitat of adult *S. gigas* is in beds of turtle grass on sandy bottoms. The animals have been found in depths to 60 meters (200 feet), but normally occur from the low-tide line to 30 meters (100 feet) because the plants in which they are found are limited to these depths. Almost exclusively herbivorous, *S. gigas* prefers a diet of algae and will move about to find better grazing areas — nearly 300 meters (1000 feet) in a relatively short period. *S. Gigas* reaches maturity in about 2 years. Some individuals live up to 10 or 25 years of age.

In Florida and along the Atlantic seaboard, *Busycon carica* and *B. canaliculatum* of the family *Galeodidae* are often referred to as conchs.

Edible Snail. Commonly referred to as escargot, the *Helix pomatia* inhabits vineyards and all regions that are not too moist, particularly brushy. With the onset of winter, the edible snail burrows into loose soil to a depth of nearly 30 centimeters (1 foot). The shell aperture becomes covered by an epiphragm, and thus the animal survives the cold season. *Helix pomatia* becomes active again with the warmth of spring, recuperates, and above all equalizes its loss of water. Crosscopulation occurs during the moist days of May or June. Edible snails produce sperm throughout the entire warm season, but eggs for only a limited time.

Even in ancient times, the *Helix* snail was a favorite food item, and the edible snail also played a role in folk medicine.

Cephalopods

The *Cephalopoda* represent the most complex of the classes of phylum *Mollusca*. One authority has stated, "From the standpoint of complexity of structure and behavior, cephalopods stand at the apex of invertebrate evolutionary development." The cephalopods include squid, octopus, and cuttlefish, all important either as direct food substances for human consumption, as bait, or as fertilizers for growing crops. Japan leads the world in the fishing of squid, and the great majority of the catch is a single species. Canada, especially Newfoundland and Labrador, is next in importance to the Orient. A single decapod species dominates, i.e., *Illex illecebros Illecebrosus* Leseur. Spain exceeds Canada in total catch of all cephalopods. See also **Cephalopoda**.

Squid. The short-finned squid (*Illex illecebrosus*) is fished commercially in Newfoundland, the entire catch being taken close to shore, particularly in Conception, Trinity, and Bonavista bays. In recent years, the squid have been used for bait in the local line fishery for cod, and exported to Portuguese, Norwegian, and Faroese ships longlining in the northwestern Atlantic for cod and sharks. Large quantities of dried squid were once exported to China for the human diet. Most of the world's commercial catch of squid is taken in Japan.

Octopuses. The octopod fishery is more specialized. Fishing is still largely a hand operation. In general, octopods are docile and timid in manner and behavior, as contrasted with the decapods (squid) which are aggressive and extremely active.

Additional Reading

Boon, D.D.: "Coloration in Bivalves," *J. Food Sci.*, **42**, 4, 1008–1015 (1977).

Boss, K.J.: "Conchs," in *The Encyclopedia of Marine Resources*, (F.E. Firth, editor), Van Nostrand Reinhold, New York, NY, 1969.

Davies, G.: "Mussels as a World Food Resource," in *The Encyclopedia of Marine Resources*, (F.E. Firth, editor), Van Nostrand Reinhold, New York, NY, 1969.

Ebert, E.E.: "Abalone," in *The Encyclopedia of Marine Resources*, (F.E. Firth, editor), Van Nostrand Reinhold, New York, NY, 1969.

Edmunds, W.J. and D.A. Lillard: "Sensory Comparison of Aroma Precursors in Marine and Terrestrial Animals," *J. Food Sci.*, **42**, 3, 843–844 (1977).

Grzimek, B.: *Grzimek's Animal Life Encyclopedia*, Vol. 3, *Mollusks and Echinoderms*, Van Nostrand Reinhold, New York, NY, 1974.

Hanks, R.W.: "Clams," in *The Encyclopedia of Marine Resources*, (F.E. Firth, editor), Van Nostrand Reinhold, New York, NY, 1969.

Sidwell, V.D. et al.: "Composition of Edible Portion of Raw (Fresh or Frozen) Crustaceans, Finfish, and Mollusks," *Marine Fisheries Review*, **36**, 3, 21 (1974).

Staff: *Atlas of the Living Resources of the Seas*, Food and Agriculture Organization (United Nations), Rome, 1976.

Staff: *Yearbook of Fishery Statistics*, Food and Agriculture Organization (United Nations), Rome (Issued annually).

MOLYBDENITE.   The mineral molybdenite is sulfide of molybdenum, MoS_2. Its hexagonal crystals are usually tabular to short prismatic, but if in massive form it may be foliated or granular. Has a perfect basal cleavage; is sectile; hardness, 1–1.5; specific gravity, 4.52–5.06; luster, metallic; color, very slightly bluish, lead gray; streak, greenish-gray; opaque. Molybdenite is one of the few minerals soft enough to give a distinctly greasy feel. Molybdenite is found as a contact mineral with cassiterite and wolframite, in granite pegmatites and sometimes in granites, syenites, or gneisses. It is found associated with tin ore in Saxony and Bohemia; in Norway; England; Australia; and in the United States in Colorado, Washington County and Oxford County, Maine, in New Hampshire, Connecticut, Pennsylvania, and Washington. Its name, derived from the Greek meaning lead, was formerly applied to minerals containing lead, to graphite and to molybdenite as well. Later the term was restricted to the latter mineral. It is an ore of molybdenum.

MOLYBDENUM.   [CAS: 7439-98-7]. Chemical element symbol Mo, at. no. 42, at. wt. 95.94, periodic table group 6b, mp 2,623 °C (4,753 °F), bp 4,639 °C (8,382 °F), density, 9.01 g/cm^3 (solid, 20 °C (68 °F), 10.2 g/cm^3 single crystal. Molybdenum has a body-centered cubic crystal structure. Molybdenum is silvery-white, tough, malleable, softer than glass, not oxidized by air at ordinary temperatures, but above 600 °C (1,112 °F) burns to form white molybdenum oxide. The metal is dissolved by dilute HNO_3 and aqua regia, but is made passive by concentrated HNO_3. Like chromium, molybdenum exhibits the phenomenon of passivity to a marked degree and even though it shows a strong reducing action when offering a fresh surface or in potentiometric determinations, it may be quite resistant to chemical action. The metal is attacked by fused alkalis. Seven isotopes occur naturally, ^{92}Mo, ^{94}Mo through ^{98}Mo, and ^{100}Mo; and five radioactive isotopes have been identified, ^{90}Mo, ^{91}Mo, ^{93}Mo, ^{99}Mo, and ^{101}Mo. With exception of ^{93}Mo, which has a half-life of approximately 10^4 years, the other radioactive isotopes have short half-lives, measured in terms of minutes and hours. The element does not appear on the list of the first 36 most frequently occurring elements in the earth's crust and thus may be considered a scarce element. Molybdenum is listed as the 42nd element in terms of estimated occurrence in the universe. The element ranks 25th among the elements occurring in seawater, there being an estimated 50 tons of molybdenum per cubic mile (10.8 metric tons per cubic kilometer) of seawater.

First ionization potential 7.18 eV. Oxidation potentials Mo → Mo^{3+} $+3e^-$, ca. 0.2 V; Mo(IV) ⟶ Mo(V)$+e^-$, (0.1 M KCl, pH 3)−0.01 V.

Other physical properties of molybdenum are given under **Chemical Elements**. See also summary of properties of refractory metals under **Niobium**.

Molybdenum (from the Greek *molybdos* meaning "lead-like") is not found free in nature, and the compounds that can be found were, until the late 18th century, confused with compounds of other elements, such as carbon or lead. Carl Wilhelm Scheele in 1778 was able to determine that molybdenum was separate from graphite and lead, and isolated the oxide of the metal from molybdenite. In 1781 Peter-Jacob Hjelm isolated an impure extract of the metal by reducing the oxide with carbon. Molybdenum was little used and remained in the laboratory until the late 19th century. Subsequently, a French company, Schneider and Co, tried molybdenum as an alloying agent in steel armor plate and noted its useful properties. See also **Chemical Elements: History of the Origin**.

The main sources of molybdenum are MoS_2 (molybdenite) [CAS: 1317-33-5], $CaMoO_4$ (powellite), and $PbMoO_4$ (wulfenite) [CAS: 10190-55-3], of which the first is by far the most important. See also **Molybdenite**; and **Wulfenite**. A representative molybdenite ore contains about 0.2% molybdenum. A concentrate containing about 90% MoS_2 is prepared through crushing, grinding and flotation operations. Most of the concentrate so produced is roasted in air to form technical molybdic oxide (about 90% MoO_3) [CAS: 1313-27-5], the product used to add molybdenum in most steelmaking processes. The prevalent industrial method for producing ferromolybdenum, the material used to add molybdenum to some cast irons and steels, is a thermit process wherein aluminum and silicon reduce a charge of iron oxide and molybdic oxide. Ferromolybdenum contains about 60% molybdenum.

Pure molybdic oxide (99.95% MoO_3) is prepared by sublimation of technical oxide or by calcining ammonium molybdate [CAS: 13106-76-8]. Metallic molybdenum powder is prepared commercially by hydrogen reduction of either pure molybdic oxide or ammonium molybdate in a two-step process. The first step, carried out at about 500 °C (932 °F) yields MoO_2. The second step, reduction of MoO_2 to Mo, is carried out at about 1,100 °C (2,012 °F).

Uses

Molybdenum is added to a number of alloy steels because its presence increases hardenability, toughness, cold-formability, and weldability. Even though the steels so treated may contain less than 1% Mo, these uses account for about 95 million pounds (~43 million kilograms) of the metal per year. One of the more recent special steels contains manganese [CAS: 7439-96-5], molybdenum, and niobium [CAS: 7440-03-1] for use in pipelines under arctic conditions. Stainless steels account for over 40 million pounds (~13.6 million kilograms) of Mo per year. In these steels, molybdenum increases corrosion resistance, elevated-temperature strength, and weldability. However, ferritic stainless steels that contain 18−26% chromium and 1−2% molybdenum are displacing the traditional 18−8 type stainless steels for many applications. For example, a steel containing 18% chromium and 2% molybdenum is finding wide use in solar energy panels.

The use of molybdenum in tool steels accounts for about 18 million pounds (~8.2 million kilograms) of Mo per year. In these steels, molybdenum provides better hot strength and improved resistance to softening and thermal cycling effects. A number of the earlier tungsten-grade tool steels have been replaced by molybdenum steels because of improved performance, lower metal density, and greater price stability. About 15 million pounds (~6.8 million kilograms) of Mo per year are consumed by the foundry industry where the metal improves the strength and abrasion resistance of cast iron. About 6 million pounds (~2.7 million kilograms) of Mo per year go into superalloys for high-temperature environments as required, for example, by jet engines. One of the later-developed superalloys is a nickel-base alloy that contains about 18% Mo and has the advantages of a very high melting point, low density, and low coefficient of thermal expansion. Uses of molybdenum compounds are described below.

Health and Safety

Molybdenum dusts and molybdenum compounds, such as molybdenum trioxide and water-soluble molybdates, may have slight toxicities if inhaled or ingested orally. Laboratory tests suggest, compared too many heavy metals, which molybdenum is of relatively low toxicity. Acute toxicity in humans is unlikely because the dose required would be exceptionally large. There is the potential for molybdenum exposure in mining and refining operations, as well as the chemical industry, but to date, no instance of harm from this exposure has been reported. Though water-soluble molybdenum compounds can have a slight toxicity, those that are insoluble, such as the lubricant molybdenum disulfide, are considered to be non-toxic.

The Occupational Safety and Health Administration (OSHA), http://www.osha.gov/, regulations specify maximum molybdenum exposure in an 8-hour day (40-hour week) to be 15 milligrams per cubic meter. The National Institute for Occupational Safety and Health (NIOSH), http://www.cdc.gov/Niosh/homepage.html, recommends exposure limit of 5000 mg per cubic meter.

In ruminants, the molybdenum toxicity occurs if the animals are let to graze on soil rich in molybdenum, but deficient in copper. The molybdenum causes excretion of copper reserves from the animal and cause copper deficiency. In young calves, the molybdenum toxicity is manifested as "teart" or shooting diarrhoea, where the dung is watery, full of air bubbles and with a fetid odor. In pigs and sheep, molybdenum toxicity combined with copper deficiency can lead to a condition called sway back or paralysis of hind quarters. In black coated animals, the toxicity of this metal is characterized by depigmentation of the skin surrounding the eyes, which is often referred to as "spectacled eyes"

Chemistry and Compounds

In keeping with its $4d^55s^1$ electron configuration, molybdenum forms many compounds in which its oxidation state is 6+, to an even greater extent than chromium. Also, like chromium, it forms compounds in which it is divalent and those in which it is trivalent; unlike chromium, it forms a number of pentavalent compounds, and a few more tetravalent compounds, especially complexes.

Among the divalent compounds of molybdenum are the dibromide, $MoBr_2$, [CAS: 13446-56-5], and the dichloride, $MoCl_2$, [CAS: 13478-17-6]. There is also a complex ion of divalent molybdenum $Mo_6Cl_8^{4+}$,, that is of particular interest because it does not yield Mo^{2+} ions. The chloride salt of the ion, Mo_6CL_{12} has been shown, by precipitation with Ag^+, to have two-thirds of its chlorine content present in a complex, so its structure is established as $[Mo_6Cl_8]Cl_4$. It is obtained by higher temperature dismutation of $MoCl_3$, [CAS: 13478-18-7], while the corresponding dibromide can be produced by direct reaction of the elements.

The relatively small number of trivalent molybdenum compounds generally exhibit the marked reducing action of the Cr^{3+} compounds, though not quite so strongly. They include the trichloride $MoCl_3$ and the tribromide $MoBr_3$, [CAS: 13446-57-6], as well as the sesquioxide Mo_2O_3, [CAS: 1313-29-7].

In addition to Mo_2O_3, the other oxides include MoO_2, [CAS: 18868-43-4], Mo_2O_5, [CAS: 12163-73-4], MoO_3, [CAS: 1313-27-5], the tetravalent and pentavalent oxides being obtainable from the hexoxide by hydrogen reduction, the MoO_3 being formed by direct combination of the elements.

In addition to the dioxide MoO_2 and the disulfide, MoS_2, example of tetravalent molybdenum compounds include the tetrachloride, $MoCl_4$, and tetrabromide, $MoBr_4$, [CAS: 13520-59-7], both of which are hydrolyzed in hot H_2O, [CAS: 7732-18-5]. Other tetravalent compounds include a few of the complexes.

The complexes also exist among the pentavalent molybdenum compounds, which include a number of simple compounds as well. In addition to the pentoxide, Mo_2O_5, [CAS: 12163-73-4], and pentasulfide, Mo_2S_5, pentahalides and oxyhalides, such as $MoCl_5$ [CAS: 10241-05-1] and $MoOCl_3$ are also known. Direct reaction with fluorine, yields MoF_6; [CAS: 7783-77-9], with chlorine, $MoCl_5$, [CAS: 10241-05-1] and with bromine, $MoBr_3$, [CAS: 13446-57-6].

As stated above, hexavalent compounds of molybdenum constitute the most numerous group. The hexavalent molybdenum oxyhalides include

$MoOF_4$, MoO_2F_2, $MoOCl_4$, MoO_2Cl_2, and $MoO_2Br_2.Mo_2O_3Cl_5$ may contain tetravalent and hexavalent molybdenum. Hexavalent molybdenum also exists in the sulfide, MoS_3, and in various oxyacid salts, and molybdates.

Molybdenum trioxide dissolves in alkaline solutions to yield, in more or less hydrated form, the molybdate ion, $MoOF_4^{2-}$.. However, the ionic species that exist in solutions of low pH are more complex than hydration of MoO_4^{2-} would indicate, and this is especially true of the compounds obtained from such solutions. Thus from strongly ammoniacal solution, $(NH_4)_2MoO_4$ is obtained, but from nearly neutral solutions containing NH_4^+ and MoO_4^{2-}, the complex $(NH_4)_6Mo_7O_{24}\cdot4H_2O$ crystallizes. The process of complex anion formation is considered to occur upon neutralization, or particularly upon acidification, of a solution, by the addition of a proton to an MO_4^{2-} ion, forming HMO_4^- ions, which condense with other oxyanions to form complexes, which can be crystallized as salts. Such salts are considered to be derived from "poly" acids. When such acids have one kind only of metal atom (e.g., Mo) in their anions, like the complex ammonium salt cited above, they are called *isopoly acids*; with more than one kind they are called *heteropolyacids*. The latter group comprises an entire field of molybdenum chemistry (as well as that of tungsten and related elements); the molybdophosphates are important in analysis and other applications. Other examples are the heteropolyacid salts formed by molybdates with oxyanions of boron, silicon, germanium, tin, arsenic, titanium, zirconium, and hafnium. In such compounds, one important group is the "12-group," containing 12 atoms of molybdenum, but many other proportions are known. Other interesting compounds are the "molybdenum blues," complex oxides of colloidal nature obtained by reduction of molybdates.

Molybdenum forms many other complexes. Of particular interest are the octacyano complexes, containing eight cyanide ions, CN^-, coordinated to a single tetravalent of pentavalent molybdenum ion, $Mo(CN)_8^{2-}$ and $Mo(CN)MO_8^{3-}$, the latter being exceptionally stable, and both form octacyanomolybdic acids $H_3[Mo(CN)_8]\cdot3H_2O$ and $H_4[Mo(CN)_8]\cdot6H_2O$. Molybdenum(III) forms salts of

$$H3MO(CN)^6$$

A fluoro complex of Mo(VI) has the structure, $[MoF_8]^{2-}$.

In liquid NH_3 solution, potassium amide reacts with MoO_3 to form the salt K_3MoO_3N, completely hydrolyzed by H_2O, in which the molybdenum atom is the center of a monomeric tetrahedral anion, being surrounded by three oxygen and one nitrogen atoms.

Like chromium, molybdenum forms a number of cyclopentadienyl compounds, many of which are carbonyls, e.g., $C_5H_5Mo(CO)_2NO$, $C_5H_5Mo(CO)_3X$ (where X may be Cl, Br, I, H, CH_3, C_2H_5 or C_3H_7). Molybdenum also forms a simple carbonyl, $Mo(CO)_6$.

Molybdates. Ammonium polymolybdate, also referred to as ammonium dimolybdate or molybdic acid (85%), is the highest-purity molybdenum compound commercially available (up to 99.97% purity). The compound is a source of very high-purity MoO_3 which is used as a catalyst in hydrogen-treating and hydrocracking processes. The compound also is used in electroplating baths and as an important laboratory reagent for determinations of phosphates, arsenates, and lead. Sodium molybdate Na_2MoO_4 [CAS: 7631-95-0] or $Na_2MoO_4 \cdot 2H_2O$ is made by dissolving MoO_3 in excess NaOH. The compound is widely used in the manufacture of molybdate-chromate orange pigments and several phosphomolybdic acid-organic pigments. The compound also is used as a condensation catalyst for phthalocyanine pigments, as a synthetic cutting fluid, and as a corrosion inhibitor in circulating cooling water systems and in glycol-based antifreeze formulations. Several of the soluble molybdates react with organic intermediates to form dyes that are used for furs and hair and have the advantage of being very colorfast. Zinc molybdate [CAS: 13767-32-3], because of its nontoxicity and excellent corrosion-inhibiting properties, is an excellent white pigment. Lithium molybdate [CAS: 13568-40-6] is an additive for porcelain enamel coatings. Iron, cobalt [CAS: 12640-46-9], and nickel molybdates are used as catalysts in hydrogenation, desulfurization, denitrification, and hydrocracking processes. Lead molybdate is used in connection with applying vitreous designs to glass bottles.

Sulfides. In addition to serving as the primary natural source of molybdenum, purified molybdenum disulfide MoS_2 [CAS: 1317-33-5] is an excellent lubricant when in the form of a dry film, or as an additive to oil or grease. The compound also is used as a filler in nylons, and as an effective catalyst for hydrogenation-dehydrogenation reactions. Molybdenum also combines with sulfur as the sesquisulfide Mo_2S_3 [CAS: 12033-33-9] and the trisulfide MoS_3, uses for which are under study.

Halides. Molybdenum pentachloride $MoCl_5$ [CAS: 10241-05-1] is used as a catalyst for several polymerization reactions involving olefins, vinyl monomers, trioxane, ethylene, vinylcyclohexane, cyclopentene, and butadiene. Vapor-phase coatings of molybdenum on metallic or ceramic substrates are prepared, using $MoCl_5$ as the starting compound.

Organomolybdenum Compounds. Soluble molybdates, molybdenum hexacarbonyl $Mo(CO)_6$, and several molybdenum halides form complex compounds with many organic oxygen, nitrogen, and sulfur compounds. Some of the oxygen-coordinated compounds include alkoxides, acetonates, oxalates, carboxylates, phenoxides, and organic chlorides. Nitrogen-coordinated compounds include several organic molybdates and chlorides. Sulfur-coordinated compounds include dialkyldithiophosphates, cysteine complexes, α-diketone complexes, and dialkyldithiocarbamates. Examples of industrial use of organomolybdenum compounds include pyrogalol-molybdate complexes in dyes, molybdenum oxalate in photochemicals, molybdenum dithiocarbamate as a lubricant additive, and molybdenum acetylacetonate as a catalyst for the polymerization of ethylene and the formation of polyurethane foam.

A discussion of the biological aspects of molybdenum is given in the next entry.

Additional Reading

Barry, H.F., and P.C.H. Mitchell, eds.: "Chemistry and Uses of Molybdenum," *Proceedings of Symposium*, sponsored by Climax Molybdenum Co., and University of Michigan, Ann Arbor, MI, August 1979.

Braithwaite, E.R.: "The Chemical Uses of Molybdenum and Its Compounds," *Chem. and Ind.*, **12**, 405–412 (1978).

Chianelli, R.R., et al.: "Molybdenum Disulfide in the Poorly Crystalline 'Rag' Structure," *Science*, **203**, 1105–1107 (1979).

Climax: Various product data sheets revised periodically, Climax Molybdenum Company, Greenwich, Connecticut 06830.

Gupta, C.K.: *Extractive Metallurgy of Molybdenum*, CRC Press, LLC, Boca Raton, FL, 1992.

Lander, H.N.: "Technological Progress with Molybdenum," *Molybdenum Mosaic*, **3**, 1, 2–11 (1978).

Lide, D.R.: *CRC Handbook of Chemistry and Physics*, 87th Edition, CRC Press, LLC, Boca Raton, FL, 2006.

Sutulov, A.: *International Molybdenum Encyclopedia*, Vol. III, Internet Publications, Santiago, Chile 1980.

Web References

Environmental Chemistry.com: http://environmentalchemistry.com/yogi/periodic/Mo.html

International Molybdenum Association: http://www.imoa.info/

Molybdenum Statistics and Information: http://minerals.usgs.gov/minerals/pubs/commodity/molybdenum/

ROBERT Q. BARR, Director Technical Information, Climax, Molybdenum Company, Greenwich, CT

MOLYBDENUM (In Biological Systems). Molybdenum is required in very low amounts by both plants and animals. Nutrient imbalances involving molybdenum and copper have caused serious problems in cattle and sheep production.

Molybdenum deficiencies are found in plants grown on certain acid soils, and sometimes the deficiency can be corrected by adding either small quantities of manganese compounds or larger quantities of limestone to the soil. The limestone makes the soil more alkaline and increases the availability of the native molybdenum in the soil. In certain parts of the world (including the eastern United States), small amounts of molybdenum fertilizer are used regularly for producing some vegetables, notably cauliflower. In Australia, large areas have been changed from near-desert conditions to productive agriculture through the application of molybdenized superphosphate.

In alkaline soils, molybdenum is more available to plants. Forage crops growing on some alkaline soils (as in the western United States) may take up high concentrations of molybdenum. The element is not toxic to the plants. They grow normally and may produce excellent yields. But cattle and sheep that eat these forages may suffer from molybdenum toxicity.

It is now well established that what appears to be molybdenum toxicity is actually a copper deficiency that is induced by the molybdenum. Thus, the symptoms of molybdenum toxicity are the same as those of copper deficiency and include fading of the hair and diarrhea. The condition may be prevented by supplementing the animal diet with extra copper, or by injecting copper compounds into the animal body, usually by an experienced veterinarian. Cattle are more susceptible to molybdenum-induced copper deficiency than other types of livestock. Horses and pigs are rather tolerant of high levels of dietary molybdenum.

High levels of molybdenum are generally considered to be 20 parts per million (ppm) or more in dry forage. Some symptoms of interference with copper metabolism in cattle may be evident when the forage contains as little as 5 ppm molybdenum if the forage is also low in copper. The effects of high-molybdenum forage in interfering with copper metabolism in animals are generally more severe if the animal diet is also high in sulfates.

In terms of humans, some research in New Zealand and the United Kingdom indicates that diets containing moderately high levels of molybdenum help to prevent dental decay. The high-molybdenum soils in the United States are seldom used for production of food crops and thus the effects of molybdenum toxicity from food substances are not well known.

Restriction of the molybdenum intake by young rats in a synthetic purified casein diet results in a decreased level of tissue, particularly small intestinal, xanthine oxidase. The enzyme levels are restored to normal by the inclusion of sodium molybdate and other molybdate compounds. Sodium tungstate is a competitive inhibitor of molybdate, and dietary intakes of tungstate greatly reduce the molybdenum and xanthine oxidase concentrations in tissues.

Legumes, cereal grains, and some green leafy vegetables are good sources of molybdenum, whereas fruits, berries, and most root or stem vegetables are poor sources. Vertebrate tissues are generally low in molybdenum, with concentrations in liver and kidney being higher than in other organs and cells. Excess molybdenum intake by cattle causes the disease known as "teart," characterized by severe diarrhea and degradation of general health.

For many years, it has been established that molybdenum is a catalyst for biological nitrogen fixation, that is, reducing molecular nitrogen to ammonia by a nitrogenase which exists in soil- and water-dwelling microorganisms. This was considered an exclusive process. However, in 1985, some British investigators (University of Sussex) confirmed that a second nitrogen-fixation scheme, one not involving molybdenum, also exists. Their experiments were conducted with *Azotobacter vinelandii*. As reported by Marx, the genes that code for the enzymes of the alternative system are being sought. Recent work has shown that the alternative system is activated in soils where there is a molybdenum deficiency. The detailed role of Mo in the principal pathway still is not well understood. Whether or not further research may lead to better growth of some plants in molybdenum-deficient soils remains unanswered.

Additional Reading

Allaway, W.H.: "The Effect of Soils and Fertilizers on Human and Animal Nutrition," *Cornell University Agricultural Experiment Station, Agriculture Information Bulletin* 378, U.S. Department of Agriculture, Washington, DC, 1975.

Braithwaite, E.R. and J. Haber: *Molybdenum: An Outline of Its Chemistry and Uses*, Elsevier Science, New York, NY, 1994.

Considine, D.M. and G.D. Considine: "Foods and Food Production Encyclopedia," 633–634, 674, 1307–1308, Appendix 2, Table 1, Van Nostrand Reinhold, New York, NY, 1982.

Kirchgessner, M. (editor): "Trace Element Metabolism in Man and Animals," *Institut für Ernährungsphysiologie, Technische Universität München*, Freising-Weihenstephan, Germany, 1978.

Lewis, R.J. and N.I. Sax: *Sax's Dangerous Properties of Industrial Materials*, 10th Edition, John Wiley & Sons, Inc., New York, NY, 1999.

Marx, J.L.: "Fixing Nitrogen without Molybdenum," *Science*, **229**, 956–957 (1985).

Mertz, W.: "The Essential Trace Elements," *Science*, **213**, 1332–1338 (1981).

Underwood, E.J.: "Trace Elements in Human and Animal Nutrition," 5th Edition, Academic Press, Inc., San Diego, CA, 1990.

MOMENT.   In physics and engineering, the term moment denotes the product of a quantity and a distance to some significant point connected with that quantity. The principal moments are moments of forces, moments of lines, moments of areas, and moments of masses. Two types of moments are statical moment and the moment of inertia. Unless specifically stated to be otherwise, the word moment would be taken to mean statical moment. A physical picture of moment may be obtained by considering the moment of a force (called torque). It is the magnitude of the force multiplied by the moment arm which is perpendicular dropped from the moment center to the line of action of the force. This moment is the turning effect on a body against which the force is applied. The moment of an area is the magnitude of the area multiplied by the perpendicular distance from the centroid of the area to the axis of moments. Similarly, the moment of a solid is its weight multiplied by the distance from its center of mass to the axis of moments. See also **Statics**.

MOMENTUM.   The momentum of a body is the product of its mass and linear velocity, while the moment of momentum (or angular momentum) of a body is the product of its moment of inertia and angular velocity. Thus linear momentum is MV, and angular momentum is $I\omega$. Both linear and angular momentum are vector quantities.

$$M = mass$$

$$I = moment\ of\ inertia$$

$$\omega = angular\ velocity$$

Because of the relation of momentum to force as set forth in the second of Newton's laws, this is a fundamental concept of dynamics.

A body tends to continue unchanged in momentum unless acted upon by external forces such as applied working forces, resistance, friction or air drag. The force F acting on a body for t seconds alters its momentum by Ft. The product Ft is known as the impulse, with a free body equal to change of momentum. In rotation, the corresponding quantity is the moment of impulse, that is, the product of the time by the applied torque or force moment.

The Law of Conservation of Momentum forms one of the basic cornerstones in physics and engineering. Its application is all-pervading, from the motion of the stars to the encounter and scattering of molecules, atoms, and electrons. The Law applies to either one particle or to a system of particles. See also **Conservation Laws and Symmetry**.

MONAZITE.   The mineral monazite is essentially a phosphate of the rare-earth metal cerium. (Ce, La, Nd, Th)PO$_4$; but other rare-earth metals are usually present. So constant is the presence of thorium that monazite is the chief source of thorium dioxide. It is monoclinic, but found ordinarily as translucent yellow to brown grains with a resinous luster, often as sand. Its hardness is 5.0–5.5; specific gravity, 4.6–5.4. Monazite is found in granites, pegmatites and similar rocks, but rarely in any concentration. The commercial deposits are residual sands. The Ilmen Mountains in the U.S.S.R., Norway, India, Madagascar, the Republic of South Africa, and Brazil are well known for their monazite deposits. In the United States monazite is known in Connecticut, New York, Virginia, North Carolina and Idaho. Monazite derives its name from the Greek word meaning solitary, in reference to the relative rarity of this mineral.

MONELLIN.   See **Sweeteners**.

MONGOOSE.   See **Viverrines**.

MONIN-OBUKHOW SIMILARITY THEORY.   A relationship describing the vertical behavior of nondimensionalized mean flow and turbulence properties within the atmospheric surface layer (the lowest 10% or so of the atmospheric boundary layer) as a function of the Monin–Obukhov key parameters. These key parameters are the height z above the surface, the buoyancy parameter ratio g/T_v of inertia and buoyancy forces, the kinematic surface stress $\tau_0/$, and the surface virtual temperature flux

$$Q_{v0} = H_{v0}/(\rho C_p) = \overline{w'T_{v0}},$$

where g is gravitational acceleration, T_v is virtual temperature, τ_0 is turbulent stress at the surface, is air density, Q_{v0} is a kinematic virtual heat flux at the surface, H_{v0} is a dynamic virtual heat flux at the surface, C_p is the specific heat of air at constant pressure, and $\overline{w'T_{v0}}$ is the covariance of vertical velocity w with virtual temperature near the surface. The key parameters can be used to define a set of four dimensional scales for the surface layer: 1) the friction velocity or shearing velocity, a velocity scale,

$$u_* = (\tau_0/\rho)^{1/2};$$

2) a surface-layer temperature scale,

$$T_{*SL} = Q_{v0}/u_*;$$

3) a length scale called the Obukhov length,

$$L = \frac{-u_*^3 T_v}{kg Q_{vo}},$$

where k is the von Kármán constant; and 4) the height above ground scale, z. These key scales can then be used in dimensional analysis to express all surface-layer flow properties as dimensionless universal functions of z/L. For example, the mean wind shear in any quasi-stationary, locally homogeneous surface layer can be written as

$$\frac{\partial U}{\partial z} = \frac{u_*}{z} f\left(\frac{z}{L}\right),$$

where f is a universal function of the dimensionless height z/L. The forms of the universal functions are not given by the Monin–Obukhov theory, but must be determined theoretically or empirically. Monin–Obukhov similarity theory is the basic similarity hypothesis for the horizontally homogeneous surface layer. With these equations and the hypothesis that the fluxes in the surface layer are uniform with height, the momentum flux, sensible heat flux, and fluxes of water vapor and other gases can be determined.

AMS

MONKEYS AND BABOONS. As shown by the organization of the monkeys and baboons in Table 1, there are two large families, the *Ceboids* and the *Simioids*. These mammals, in terms of position among the primates, rate just below the anthropoids, the latter including the lesser apes, gibbon, greater apes, gorillas, chimpanzees, and orangutans. The *ceboids* are also sometimes called the New World monkeys because they are found only in tropical America. The *simioids*, on the other hand, are sometimes called Old World monkeys because they are found in Africa and Asia and occur nowhere in the natural state in the New World. See Fig. 1. There is another group of small monkeys of Central and South America that constitute the family of *Hapaloids*, commonly known as marmosets. See also **Marmoset**.

General characteristics of monkeys, realizing that there are exceptions, include: (1) one pair of pectorally placed mammary glands, (2) frequent dilation of the cheeks into pouches for storing food; (3) the presence of two incisors on either side of each jaw, and well-developed, large canine teeth, the teeth being well adapted to crushing vegetables and fruits; (4) limbs that are of nearly equal length; (5) when present, the thumb (pollex) of

Fig. 1. Golden monkey (*Cercopithecus mitis kandri*), a rather rare subspecies of monkey that lives in the high-altitude forests near the Virunga volcanoes of east central Africa. The habitat is near that of the mountain gorilla. There are some 100 species of Old World monkeys.

the forelimbs is opposable to the other fingers; and (6) the great toe is opposable to other digits of the foot, permitting the feet, in essence, to perform as hands.

The *ceboids* generally have a broad partition in the nose; they possess an extra premolar on each side of the jaw; most species have a long, prehensile tail, providing much assistance in climbing and agile movement in tree tops. The face is free of hair; there are no callosities on the buttocks, the latter being a marked characteristic of the *simioids*. The thumb usually is not opposing, but resembles the other fingers. However, the big toe is more readily opposed to the other digits of the foot. The *ceboids* are essentially vegetarians and arboreal.

Probably no other mammal excels the spider monkey (*Ateles*) in arboreal life. This monkey was named because of the great length of its limbs in comparison with the rest of its body. The body is about 12 inches (30.5 centimeters) long and the tail is twice that length. The tail is used

TABLE 1. GENERAL ORGANIZATION OF MONKEYS AND BABOONS

Ceboids	Simioids
Half-Monkey (*Pithecinae*)	Colobine Monkeys (*Colobinae*)
Dourocoulis (*Aotes*)	Guerezas (*Colobus*)
Sakiwinkis (*Pithecia*)	Langurs (*Presbytis,...*)
Bearded Sakis (*Chiropotes*)	Snub-nosed Monkeys (*Rhinopithecus*)
Uacaris (*Cacajao*)	Proboscis Monkey (*Nasalis*)
Hand-tailed Monkeys (*Cebinae*)	Long-tailed Monkeys (*Cercopithecinae*)
Squirrel-Monkeys (*Saimiri*)	Guenons (*Cercopithecus*)
Capuchin Monkeys (*Cebus*)	Allen's Swamp Monkey (*Allenopithecus*)
Wooly Monkeys (*Lagothrix*)	Military Monkeys (*Erythrocebus*)
Wooly Spider Monkey (*Brachyteles*)	Mangabeys (*Cercocebus*)
Spider Monkeys (*Ateles*)	Dog-faced Monkeys (*Cynopithecinae*)
Howler Monkeys (*Alouatta*)	Macaques (*Macaca*)
	The Black Ape (*Cynopithecus*)
NOTE:	Baboons (*Papio*)
1. Ceboids are found in South America.	The Gelada (*Theropithecus*)
2. Simioids are found in Africa and Asia, south of the desert areas and the Himalayas in the east. One tribe of Barbary Apes, the only wild monkey in Europe (*Macaques*) lives on Gibraltar.	Drills (*Mancrillus*)

in assisting the legs when swinging from tree to tree. The fur usually is black, but is gray or brown in some species. Spider monkeys sometimes are used for pets and are considered intelligent. The *coaita* has a red face and is found in the Lower Amazon.

Capuchin monkeys are any of several species of South American monkeys of the genus *Cebus*, which are characterized by their moderately long prehensile tail. They differ from the other species with prehensile tails in having this organ fully covered with hair and not bare on the lower surface near the tip. This animal is also called the *sapajous*.

The Saki is a New World monkey of the genus *Pithecia*. Most of the species bear the name of the group, as the white-headed saki and the whiskered saki, but native names have been adopted for some. The hairy saki is called the parauacu and the black saki is also known as the cuxio. No members of this genus have prehensile tails.

The *simioids* are rated next to the anthropoid apes on the zoological scale. The nasal apertures are close together, the nasal septum being very narrow. With exception of the Colobine monkeys, all simioids have opposing thumbs and big toes. They have four incisors, two canines, four premolars, and six molars in each jaw. The canines are large and strong. In contrast to the ceboids, the simioids do not have a prehensile tail and, in fact, the tail may be quite rudimentary or essentially absent. They have cheek pouches. In some species, there may be no hair on the buttocks, but instead the presence of naked, hardened areas called callosities. In some species, the callosities are brilliantly colored. The simioids are considered to have superior intelligence and exhibit greater variety in food and living habits than the ceboids. The majority of pet and performing monkeys are simioids. Most species breed and otherwise do well in captivity, but are susceptible to respiratory diseases in northern climates unless suitable shelter precautions are taken.

The langurs represent a group of Old World monkeys characterized by slender build and extremely long tails. The legs are longer than the arms. They eat principally the leaves and young shoots of trees. Langurs live only in Asia and in some of the East Indian islands. Among the species of langurs that bear distinctive names are the hanuman, lutong, douc, negro monkey, leaf monkeys, bear monkey, white monkey, and purple-faced monkey. The latter species is found in Ceylon and is sometimes called the wanderoo.

Macaques are monkeys of several Asiatic and one African species. They are related to the mangabeys, but are stouter and have a slightly longer muzzle. The tail ranges from long to rudimentary. These monkeys make up the genus *Macacus* (Rhesus-like monkeys). Among the included species are the bonnet monkey, lion-tailed monkey, pig-tailed monkey, and magot. The Indian macaque is known as the rhesus monkey. A colony of macaques (also called Barbary apes) has inhabited the Rock of Gibraltar for many years. The crab-eating macaque *(kra)*, *Macacus cynomolgus*, is found in the Oriental region. The mangabey is an African monkey of a small group of species, also called the white-eyelid monkeys. They are slender animals with a fairly long muzzle and long tail. The proboscis monkey, *Nasalis larvatus*, is a moderately large monkey of Borneo, characterized by the long, fleshy, and somewhat drooping nose. This organ is largest in adult males, but is relatively small and upturned in the young. The nostrils open near each other on the lower surface. The species is closely related to the langurs. The thumbless monkeys constitute a group of African monkeys of the genus *Colobus*. They are named from the reduction of the thumb, which is either entirely absent or reduced to a small projection, with or without vestigial nail. Some of the included species are called colobs, one is a guereza, and one is the king monkey. See also **Marburg-Ebola Virus Disease**.

Baboons. The true baboons are characterized by a large head with the elongate, dog-like muzzles having nostrils at the tip. They walk on all fours. These animals constitute the genus *Pipio*, but the term baboon is also applied to the gelada baboon, *Theropithecus gelada*, of southern Ethiopia, also a dog-like species although the nostrils are some distance from the tip of the snout. Baboons sometimes are called dog-faced monkeys. Among the baboons, several species are known by special names, including the mandrill, *P. spinx*; the drill, *P. leucophaeus*; and the chacma, *P. porcarius*. All are found in the Ethiopian region. Baboons can be ferocious. Individually, they are able fighters and their defenses are augmented by their habit of living in bands. A group of males is reported to be a good match for some of the larger predators.

Additional Reading

Estes, R.: *The Behavior Guide to African Mammals*, The University of California Press, Berkeley, CA, 1991.

Smuts, B.: *Sex and Friendship in Baboons*, Harvard University Press, Cambridge, MA, 1999.

Strum, S.: *Almost Human: A Journey into the World of Baboons*, W.W. Norton and Company, Inc., New York, NY, 1990.

Web Reference

http://www.primates.com/welcome.htm (Photo gallery of monkeys and other primates).

MONKFISHES. See **Anglerfishes**.

MONOBASIC. Descriptive of acids having one displaceable hydrogen atom per molecule. Acids having two, three, or more displaceable hydrogen atoms are called dibasic, tribasic, and polybasic, respectively.

MONOCEROS. An equatorial constellation that lies between Canis Minor and Canis Major.

MONOCHROMATIC. 1. Having one color, strictly one frequency or wavelength of optical radiation. Actually, no finite amount of radiation will ever be strictly monochromatic. It will, at best, contain a narrow band of frequencies.

2. By analogy, a beam of particles, such as beta-particles or neutrons, is said to be monochromatic if all the particles have the same, or nearly the same, energies.

MONOCHROMATOR. An instrument used to supply a beam of light having some desired, narrow range of wavelengths. Although sometimes used in photochemistry experiments as a single instrument, it is usually part of a spectrophotometer. In that case, it has the following component parts: (1) an entrance slit producing a well-defined beam of heterochromatic radiation; (2) a prism or diffraction grating dispersing the incident radiation into a continuous spectrum; (3) some device to rotate the prism or grating so that the desired wavelengths of exit radiation are obtained; (4) an exit slit producing a narrow band of wavelengths. A monochromator becomes a spectrophotometer if it is preceded by a source of continuous radiation and followed by a sample holder, a detector, an amplifier, and a device for measuring the amplified output signal.

MONOCLONAL ANTIBODIES. Monoclonal antibodies are antibodies with a unique specificity, generally made by cloning cells containing a particular antibody gene to produce a population of identical cells derived from a single cell, which all produce the same antibody. This should be contrasted with polyclonal antibodies found in the serum of immunized animals, which consist of a very diverse mixture of antibodies against many different molecules.

Vertebrates have evolved an immune system that protects them from invading microorganisms. A major component of the immune response to such organisms is the production of antibody molecules. These molecules possess a binding site for structures on the surface of the invading organism. In order to be effective, the immune system must be capable of making antibodies that can bind the enormous diversity of molecular structures expressed by viruses, bacteria and other parasitic organisms, and must be capable of coping with mutations in these organisms. This challenge is met by the immune system in two ways. Firstly, B lymphocytes are capable of generating a diverse set of antibody structures (estimated at 10^{11}) by permutation and combination of a limited number of gene elements. Secondly, and uniquely in the body, the genetic elements that code for the antigen-binding structure of antibody are subjected to a high rate of mutation, coupled with a process that allows selection of cells that make antibodies that bind the antigen strongly. This process results in antibodies that bind their antigens with high specificity and high affinity. See also **Antibody** and **Lymphocytes**.

The potential of antibody as a tool in medicine has long been recognized. Antibodies made in animals have been used to assay for the presence of hormones (in pregnancy tests for example), and even to neutralize toxic substances (as in the use of antibodies against snake venom proteins, made in horses, to treat human victims of snake bite). Antibodies made in animals have a number of limitations. The product of a single B cell is multiple copies of antibody with a unique binding site. However, when we immunize an animal and subsequently bleed it we obtain, in the serum, a mixture of antibodies produced by the animal in response to the many antigenic molecules it encounters. Antisera made in animals are thus variable and of limited overall specificity.

In 1976, Georges Köhler and Cesar Milstein, working in Cambridge, England, developed a procedure to isolate and propagate the individual B cells making antibody against the antigen of interest. They fused cells obtained from the spleen of an immunized mouse with myeloma cells. Myeloma is a tumor of antibody-producing cells, and myeloma cell lines are available in which the cells multiply rapidly and produce large amounts of antibody—although not usually of a specificity that is of use to us. Some of Kohler and Milstein's fusion products ("hybridomas") retained these properties of indefinite propagation and high antibody secretion rates but made antibody coded for by the antibody genes of the mouse spleen cells. The mixture of hybridomas still would make a variety of antibodies, but this mixture could be separated out by cloning—that is, individual cells isolated and allowed to proliferate into separate populations, or clones. Then, it was a matter of screening the many clones to see which ones made antibody against the antigen of interest. See also **Köhler, Georges Jean Franz (1946-1995)** and **Milstein, Cesar (1927-2002)**.

This is the basis of the production of monoclonal antibodies. With relatively minor changes, this procedure has spawned an industry and revolutionized many aspects of medical diagnosis and research. Monoclonal antibodies are also being used increasingly as therapeutic agents.

Generation of Hybridomas

The process of generating hybridomas centers on the fusion reaction between immune spleen lymphocytes and myeloma cells, but the generation of these two components is critical and complex, involving several stages. The generation of hybridomas is illustrated schematically in Figure 1.

Strategies. Ideally, the target antigen, the molecular structure against which the antibody is to be made, will be available in pure form and in adequate quantity (a few milligrams). However, this ideal is often not achievable. Antibody against a complex multimolecular structure such as a virus or a blood cell may be desired, or the molecule may be known but it may not be possible to purify it in sufficient quantity. Monoclonal antibodies provide excellent reagents to use in purifying molecules from complex mixtures; thus, it may be that the antibody is wanted in order to be able to purify the antigen.

Monoclonal antibodies against individual components of a complex mixture can be made by devising a selection strategy that will identify hybridomas making antibody against the antigen in question. The most effective strategy depends on the situation. Differential screening of hybridoma clones, against the mixture containing the antigen of interest and against another mixture, lacking the antigen but otherwise as similar as possible, is widely used. For example, in making antibodies that will specifically identify a particular peptide growth factor produced in culture, a culture lacking the stimulus that elicited the production of the growth factor may be a useful control. In most cases, these strategies will not identify antibodies against the antigen with certainty, but will allow the selection of a group of hybridomas that are likely to include clones making antibody against the antigen of interest. Supplementary studies, for example Western blotting to characterize the molecular weight of the molecule detected, precipitation of the antigen followed by partial sequence analysis by mass spectrometry of peptide fragments and functional inhibition studies to neutralize biological activity, will then be needed to identify the hybridomas secreting the antibodies in question.

The principal consequence of these considerations is that a screening strategy must be in place before the immunization of mice.

Immunization. Virtually all hybridomas are made using immune cells from mice. Immunization protocols vary widely depending on the nature of the antigen, but generally involve a priming dose injected subcutaneously with adjuvant to provide a strong immune stimulus, followed about 4 weeks later by a booster dose, often given intravenously without adjuvant. It is sensible to use an immunization protocol, which has been described in the literature for a similar antigen.

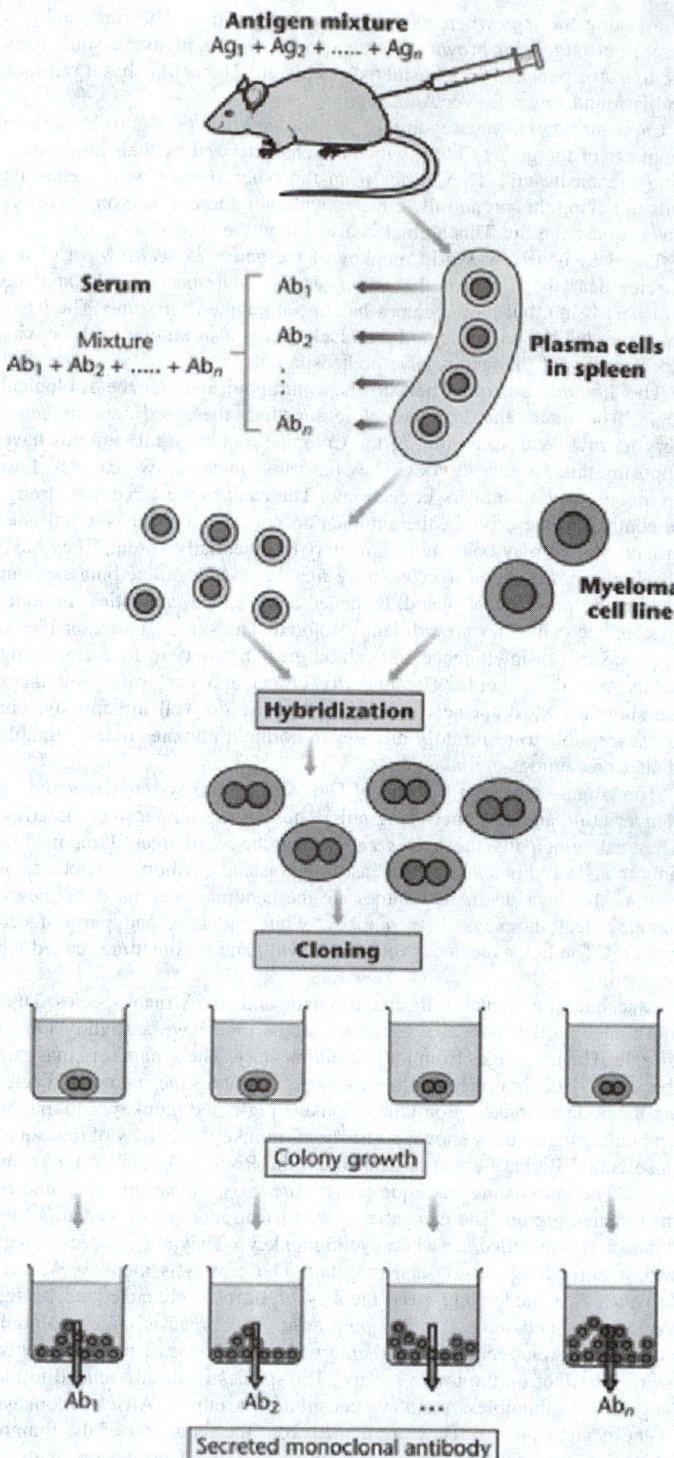

Fig. 1. Schematic representation of the process of immortalizing an antibody-producing clone by hybridization, cloning and selection of clones producing the desired antibodies.

Myeloma Cell Line. A small number of mouse myeloma lines are available. The earlier lines are able to make their own antibodies, so that the hybridoma can make the light and heavy chains of both the myeloma and the spleen cell fusion partner. The light and heavy chains are made independently and assemble in the cell; hence, such a hybridoma can make a variety of antibody molecules, only one of which will have the desired binding sites. To avoid this heterogeneity, myelomas have been selected that have lost the ability to make their own light and heavy chains.

Since myeloma cells grow continuously, a method is required to allow selective growth of hybridomas and suppress growth of the parent myeloma. The selection method used most widely depends on the use of myeloma cell lines that have lost the ability to make nucleotides by the salvage pathway. The main biosynthetic pathway can be blocked with the drug aminopterin, so that the myeloma cells cannot make DNA or RNA and die. Hybridomas, on the other hand, have the enzymes for the salvage pathway (derived from the spleen lymphocyte) and can grow, provided they are supplied with the substrates for the pathway, hypoxanthine and thymidine. The selective system is named after the three substances that are added to the culture medium, HAT (hypoxanthine, aminopterin and thymidine).

Myeloma cells are available for the production of hybridomas with rat, human and chicken lymphocytes, but most work has been carried out using the mouse, and extensive efforts to make human hybridomas for therapeutic purposes have yielded limited success. Heterohybridomas, using mouse myeloma and lymphocytes from a second species, have also had limited success. When human antibodies are required, interest has turned to the use of libraries of human antibody genes, genetic engineering methods and the use of mice genetically engineered to replace their mouse antibody repertoire by a human antibody repertoire.

Fusion Process. The reaction at the heart of the field of monoclonal antibodies, the fusion reaction, is surprisingly simple. Myeloma cells, taken from culture, are mixed with spleen cells isolated from an immunized mouse, at a ratio usually of one myeloma cell to 10 spleen lymphocytes. The cells are centrifuged together to form a pellet and resuspended in a small volume of a viscous solution of polyethylene glycol. After 1 min, the suspension is gradually diluted, care being taken not to break up small aggregates of cells, some of which will form the hybridomas. The cell suspension is washed and resuspended in culture medium. After a few hours in a tissue culture incubator (which provides a physiological pH and temperature), the cells are dispensed into small tissue culture wells at a concentration, which, from experience, is likely to yield single hybridoma colonies in the wells. The cells are then allowed to grow over the next 7–14 days, with occasional changes in medium.

There are many minor variations of this procedure, and many critical points. Polyethylene glycol is a rather inert chemical and probably stimulates fusion passively, by allowing cells to stick together and excluding water from the junction. The cells should be in good condition before the fusion—the myeloma cells should be growing exponentially and the lymphocytes should not be subjected to traumatic processes during preparation. This requires a gentle procedure for disaggregating the spleen and for removal of red cells. The fusion procedure itself is critical; inadequate or overvigorous resuspension of cells will reduce yields. Postfusion dilution, washing and plating out again need to be carried out with an understanding of the objective—at this stage, the cells are not yet fused stably; there are doublets with adhering and perhaps partly fused membranes, and these should not be disrupted. See also **Lymphocytes**.

Growth and Selection. Growth of hybridomas occurs gradually over the first 2 weeks after fusion. The yield may be improved by adding interleukin 6 (IL-6), which acts as a hybridoma growth factor. Other additives, found empirically to improve hybridoma yields, such as feeder cells or a variety of commercial supplements, are thought to act through IL-6. Cytokines have overlapping functions; thus, it is likely that other growth factors help hybridoma growth. See also **Interleukins**.

After a few days in culture, small colonies of hybridoma cells are seen using an inverted microscope and, if all goes well, these colonies expand to the point where they are visible by eye and begin to affect the pH of the medium, turning the indicator dye yellow. About 1 week after the initial fusion, the unfused myeloma cells should be dead, and it is then helpful to gradually supplement the cells with medium lacking aminopterin. Since hypoxanthine and thymidine are used up while aminopterin accumulates, the HAT medium should be replaced with HT medium.

Once the culture wells are showing visible colonies and the medium is turning yellow, it is time to test the supernatant for antibody. At this stage, there will be many colonies to test and maintain, perhaps several hundred, depending on the scale of the experiment. A priority is therefore to reduce the number of cultures by eliminating negative cultures. It is possible to test simply for immunoglobulin production, but this usually eliminates very few. A more useful test is to screen for binding to the antigen by a simple assay capable of being run daily on large numbers of samples.

Selected cultures must be cloned, because there is no guarantee that they arose from a single cell. Cloning is usually repeated several times over the first few weeks, because the fusion products are still genetically unstable and may produce loss mutants, which grow but do not produce antibody. Clones are grown into larger culture vessels to produce quantities of cells and antibody for more extensive evaluation.

At this stage cells should be cryopreserved, so that if anything goes wrong there is a seed culture to go back to. Many things can go wrong, including contamination with bacteria or fungi; thus, cells should be cryopreserved as soon as there is evidence that the colony may be producing a useful antibody. However, cryopreservation requires a few million cells, and hence cannot be performed until the culture has been sufficiently expanded.

Cryopreservation and Long-Term Maintenance. The amount of work required to establish a hybridoma is considerable; the final product is unique and may not be reproduced exactly in a subsequent fusion. It is therefore essential to establish a secure "bank" for hybridomas. This is achieved by storing ampoules of hybridomas in liquid nitrogen. The procedures for freezing down cells and for thawing them out to reestablish them in culture are straightforward; the critical issues are essentially administrative. It is important to freeze down at least 5–10 ampoules, to validate the "deposit" by reconstituting one ampoule, to maintain adequate records and to lay down more ampoules when stocks become low. It is wise to store a set of ampoules in a separate laboratory.

Hybridomas may be maintained in culture essentially indefinitely, by "splitting" the culture every 2–3 days, maintaining the cell concentration at between $2-3 \times 10^5$ and 10^6 cells mL^{-1}. However, it is more usual to grow them only when fresh antibody is required, and reestablish them from cryopreserved stocks when needed. If they are maintained for long periods, it is necessary to check antibody production regularly, because loss of production may occur.

Antibody Production. The amount of antibody secreted in small cultures such as microwells is adequate for screening. However, once a hybridoma has been found to make a useful antibody larger amounts will be needed. Cells may be grown in conventional culture flasks and supernatant may be harvested 2–3 times per week; such supernatants generally contain antibody at 1–5 µg mL^{-1}. This concentration is adequate for many assays; however, if the antibody needs to be purified, for example to conjugate it to an enzyme or fluorochrome, larger amounts and higher concentrations are needed. These have generally been prepared by growing the hybridoma as an ascitic tumour in mice, yielding ascitic fluid with antibody at concentrations of 1–5 mg mL^{-1}. This procedure is increasingly unacceptable for ethical reasons, as alternatives become available. Furthermore, the ascitic fluid contains, in addition to the antibody made by the hybridoma, antibodies made by the mouse against environmental antigens. The antibody is no longer truly monoclonal and monospecific and significant difficulties in interpretation may result. Fermenters and minifermenters are available for the production of monoclonal antibodies in culture at any scale from a few milligrams to the gram amounts required for clinical trials. While there are still significant difficulties and uncertainties, and production costs are high, these methods are gradually replacing ascites production in mice.

A number of methods are available for antibody purification in good yield and purity. For many purposes, however, conventional culture supernatant will work well, without the need for purification.

Applications

Monoclonal antibodies have a range of applications, which take advantage of the specific binding of their target antigen (Figs. 2 and 3).

Analytical Applications. Antibody may be used to detect an antigen, for example in forensic applications, in microbiological testing of foodstuffs and in diagnostic testing of blood samples for toxins or infectious organisms. The antibody-based test for the presence of antigen may be rendered quantitative, providing an assay for antigen. Antibody-based assays are widely used in medicine to determine levels of growth factors, hormones, blood cells or malignant cells; the applications are essentially unlimited. Still in analytical mode, monoclonal antibodies may be used to locate antigen. Antibodies are widely used in conjunction with color-forming labels and microscopy to localize antigens in tissue sections. This is known as immunohistochemistry.

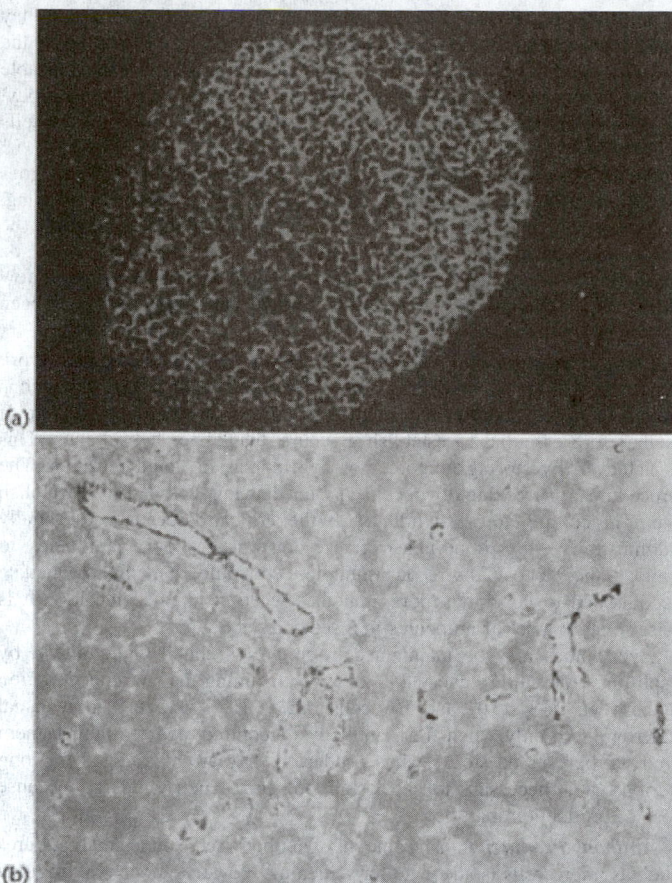

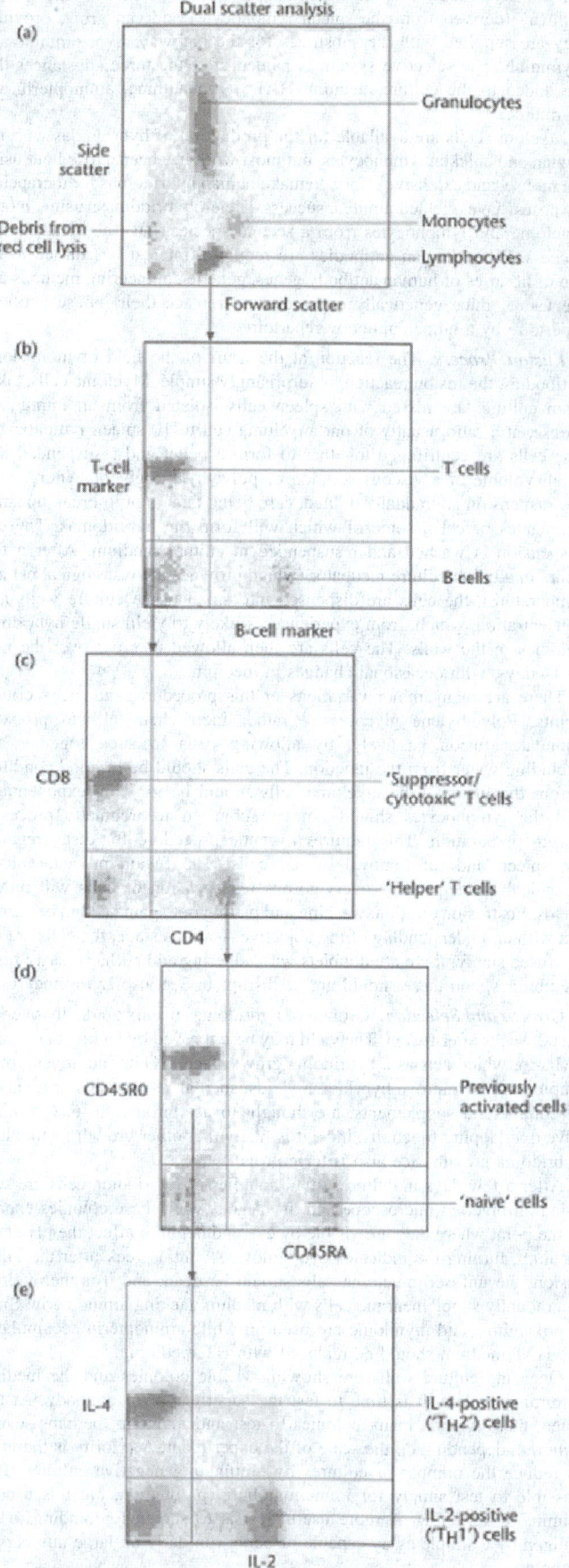

Fig. 2. Examples of the use of monoclonal antibodies to identify particular molecules and cell types in tissues. (a) Staining of human lymph node tissue with a monoclonal antibody against a protein called CD19. The protein is expressed on the surface of B lymphocytes, which are responsible for antibody production. The egg-shaped structure is a follicle of B lymphocytes, while the surrounding area (unstained) contains principally T lymphocytes, with a sprinkling of B lymphocytes. The follicle contains a germinal centre. Germinal centers develop in response to infection or other antigenic challenge. In the germinal centre, B lymphocytes divide rapidly and the genes coding for the antibody made by the cell are mutated and selected to give strong binding to antigen. (b) Human lymphoid tissue stained with an antibody against a protein called E selectin, or CD62E. This molecule is found on the endothelium of blood vessels and allows lymphocytes to attach to the endothelium and move from the blood into the tissue. The brown staining shows the distribution of E selectin on blood vessels, while the blue is a counterstain to reveal the background tissue. The varied shapes of the stained blood vessels are caused by the fact that the section cuts different blood vessels at different angles.

Preparative Applications. Antibodies may also be used preparatively to purify molecules or cells from crude mixtures. Immunoaffinity-based preparative techniques are very powerful compared to more traditional biochemical purification methods, although in general a successful purification procedure will combine both affinity-based and conventional methods. Antibody-based purification methods are useful from the laboratory scale, where the aim is to purify nanogram to milligram quantities of biological substances from complex mixtures such as serum, to production of therapeutic substances, such as the blood-clotting factor VIII from large volumes of blood. The use of antibody to identify particular cell types in complex mixtures has been extended to preparative methods to purify these cells. One approach is to link antibody to magnetic particles. A magnet is then used to physically separate cells that bear the antigen from cells that do not. A more powerful technique uses the fluorescence-activated cell sorter, in which a cell may be identified on the basis of antibody tagged with fluorescent dye and physically separated from the other cells (Fig. 2). Because the absolute differentiation of a particular cell type from all other cells may require several different markers, cell sorters can sort

Therapeutic Applications. Potential therapeutic applications of antibodies include the neutralization of toxins, the removal of infectious agents from the circulation and the destruction of body cells mediating disease, including autoimmune cells and cancer cells.

Polyclonal antisera have been used for many years in the treatment of snakebite and infections where the major threat to life is a toxin, such as tetanus. These conditions are relatively rare (bacterial infections such as tetanus generally being prevented by immunization), so that an individual is unlikely to need repeated treatment. The immune system recognizes antibody made in another species as a foreign protein and will make an antibody response against the protein. This means at best that second or subsequent treatment with antibody from the same species as the first treatment will be of limited effectiveness, because the protein is cleared rapidly; at worst, the resulting immune reaction can take the form of a life-threatening anaphylactic shock. See also **Venoms**.

Polyclonal antisera against human lymphocytes were developed in the 1970s to treat patients who were rejecting organ grafts. The principle was that the host immune response was responsible for the organ graft rejection; antibodies against key components of the immune system should suppress the rejection. These antisera have largely been superseded by a monoclonal antibody called OKT3, directed against a molecule expressed on human T cells and involved in T-cell function. OKT3 has been highly successful in reversing rejection episodes. The injected antibody is a mouse protein and is immunogenic, but the response is muted because the patients are immunosuppressed, both by the antibody itself and by other immunosuppressive therapy used to reduce the graft response. Nevertheless, OKT3, successful though it has been, has been perceived as limited in effectiveness partly by its immunogenicity.

There have been some successful results with other monoclonal antibodies in human therapy, but these must be set against a number of failed clinical trials and generally a lack of sustained success. This situation contrasts with the highly promising results in animal models, where antibodies can destroy established tumors, induce a state of permanent tolerance to transplanted organs, reverse autoimmune disease and rescue animals from acute toxicity caused by, for example, bacterial endotoxin. In comparison to the animal studies, therapeutic development of monoclonal antibodies for humans has yet to realize its full potential. However, in the last few years the number of successful trials has increased sharply and therapeutic antibodies now appear to be achieving their promise

Major factors that have limited the clinical applications of monoclonal antibodies have been:

- immunogenicity;
- difficulty and cost of production on an adequate scale;

- unwanted biological activity — due, for example, to direct effects on cells of the immune system;
- limited binding affinity, which necessitates the injection of large amounts of antibody in order to achieve a therapeutic effect;
- lack of direct functional action, requiring conjugation of drugs or other biologically active materials; and
- limited penetration into the target tissue — especially dense, poorly vascularized tumour tissue.

There is currently a mood of optimism that many of these difficulties are being overcome by antibody engineering to tailor the properties of the antibodies.

Antibody Engineering

Antibody is a protein that can be modified in a variety of ways: removing specific functional portions, reducing the overall size of the molecule, changing critical amino acids to increase affinity, linking to other functional molecules such as drugs, radioisotopes or toxins to add "teeth." It is generally easier in the long term to modify the DNA rather than the protein—this needs to be done once only, and the modified gene is then expressed to produce the modified protein. Antibody engineering is a relatively new field, which is still developing. Techniques that appear important include:

- preparation from existing hybridoma genes of genes coding for small proteins, which include the antigen-binding site but omit most of the rest of the molecule, including sequences responsible for some of the biological effects of antibodies;
- modification to increase antigen-binding affinity (affinity maturation);
- preparation of fusion proteins consisting of the antigen-binding site linked directly to, for example, a toxin, an enzyme, a sequence suitable for radioisotope labeling, another antibody sequence, to achieve increased or novel biological activities;
- modification to make the sequence more human-like and less immunogenic;
- generation of libraries of genes derived from human antibody genes, to avoid the immunogenicity associated with foreign protein;
- genetic modification of mice so that they produce human antibodies, and then immunizing and making hybridomas in the classical way; and
- preparation of antibody fragments in bacterial culture, to increase yield.

Products of these techniques are undergoing clinical trial and some have been approved for patient use. Several have become established therapeutic modalities.

Relative Merits of Polyclonal Antisera, Monoclonal Antibodies and Genetically Engineered Antibodies

The major characteristic of antisera, prepared by immunizing and subsequently bleeding animals, is the heterogeneity of the antibody preparation. This may be a strength or a weakness, depending on the application. An antiserum will contain a mixture of antibodies against the antigen of interest, together with antibodies against other antigens encountered by the animal. The latter may cause confusing reactions, and may need to be removed by absorption. The multiple antibodies against the antigen of interest, reacting with a multiplicity of epitopes and showing a spectrum of binding affinity, can be advantageous, giving a stronger reaction overall than is seen with a single monoclonal antibody.

There are important uses of antibodies where polyclonal antisera work very well. This is particularly true in radioimmunoassays, which have been established for many years; there is no pressing need to change. On the other hand, in immunohistochemistry, change was initially slow in coming, because polyclonal antisera apparently worked very well, until awareness of the increased specificity achievable by monoclonal antibodies became widespread. A polyclonal antibody may react with the antigen of interest in one tissue, but when applied to another tissue, it is not possible to know whether it is reacting with the same antigen. See also **Immunoassay**.

Monoclonal antibodies must always be preferred when specificity is important. However, monoclonal antibodies can be cross-reactive. A rabbit antiserum may react with the antigen of interest and additionally with a completely dissimilar molecule against which the rabbit made an antibody without being asked to. This type of nonspecificity is unlikely with a monoclonal antibody (unless it was made as an ascitic fluid). Cross-reactivity of antibodies results from their reaction with epitopes with a

Fig. 3. The analytical power of a panel of monoclonal antibodies against molecules on the surface of blood cells. (a) Light scatter distribution of blood cells analyzed in a flow cytometer, allowing the selection of the lymphocyte population for further analysis. If monoclonal antibodies against a T-lymphocyte marker and a B-lymphocyte marker, each attached to a different fluorescent dye, are added to the blood sample, the pattern seen in (b) can be obtained, allowing the selection of T cell for further analysis. (c) Resolution of T lymphocytes into two populations: lymphocytes marked with a monoclonal antibody against CD4, identifying a population containing "helper" cells, which provide positive signals in an immune response, and the CD8 population, which include suppressive activity. (d) If the CD4 lymphocytes are selected for further analysis, they may be further separated into cells (CD45RO-positive) that have previously been activated ("memory cells") and those that have not previously been activated ("naïve cells"). (e) These can be subdivided in turn according to the cytokines they make — cells that make IL-4 tend to favor antibody responses, while cells that make IL-2 tend to stimulate cell-mediated immunity. (Note that panels a–d show actual data, while panel e shows simulated data.) Technical limitations in flow cytometry instrumentation and the number of different fluorescent dyes available place limits on the extent of the analysis. Widely available instruments can analyze on the basis of three colors simultaneously (allowing, for example, the identification or separation of CD4 cells for further analysis), while more sophisticated research methods allow up to 11 simultaneous antibody markers, more than adequate to conduct the entire series shown in the figure. In practice, some steps can be left out; there is no need, for example, to include a B-cell marker or CD8 or CD45RA in identifying the IL-4-secreting helper T cells.

similar shape, which may be found on unrelated molecules. This type of cross-reactivity is found with monoclonal antibodies, and must be guarded against.

Apart from specificity, monoclonal antibodies possess a major advantage of reproducibility. Two rabbit antisera against a particular antigen will exhibit differences in affinity and in the mixture of antibodies against different epitopes of the antigen; they will also carry a different set of unrelated antibodies. Even two batches taken from the same rabbit may differ in activity and in cross-reactivity. The specificity of a monoclonal antibody, once established, should not vary. Differences in titre (amount of antibody) are relatively easy to adjust for. There remain subtle variations between batches of monoclonal antibody, depending on denaturation and aggregation during storage and processing, but these are minor compared to the variations between batches of rabbit antiserum.

In laboratory applications, the choice between polyclonal and monoclonal antibody may best be summed up as follows: if there is a choice use the monoclonal antibody. However, when there is no antibody available, it will sometimes be quicker to make a polyclonal antiserum; this should be replaced in due course with monoclonal antibody.

For therapeutic applications, the situation is sometimes radically different. A polyclonal antiserum against the venom of the Australian king brown snake will neutralize all of the toxic activities of this witches' brew—neurotoxins, anticoagulants, procoagulants and others we may not be aware of. To replace this well-tried preparation with a cocktail of monoclonals would be a major undertaking. Polyclonal antisera have an assured future in some therapeutic applications. In others, monoclonals have proved disappointing, but, as discussed above, antibody engineering is greatly extending the therapeutic scope of antibodies.

Additional Reading

Eichmann, K.: *Köhler's Invention*, Springer-Verlag New York, LLC, New York, NY, 2005.

Hudson, P.J., and C. Souriau: "Recombinant Antibodies for Cancer Diagnosis and Therapy," *Expert Opinion on Biological Therapy*, **3**, 305–318 (2003).

Kohler, G., and C. Milstein: "Continuous Cultures of Fused Cells Secreting Antibody of Predefined Specificity," *Nature*, **256**, 495–497 (1975).

Simmons, M.A.: *Monoclonal Antibodies: New Research*, Nova Science Publishers, Inc., Huntington, NY, 2005.

Subramanian, G.: *Antibodies: Novel Technologies and Therapeutic Use*, Springer-Verlag New York, LLC, New York, NY, 2004.

Winter, G., and C. Milstein: "Man-made Antibodies," *Nature*, **349**, 293–299 (1991).

Zola, H.: *Monoclonal Antibodies: A Manual of Techniques*, CRC Press, LLC, Boca Raton, FL, 1987.

Zola, H.: *Monoclonal Antibodies: The Second Generation*, BIOS Scientific Publishers, Oxford, UK, 1995.

Heddy Zola, Child Health Research Institute, Adelaide, Australia

MONOCOQUE. A type of construction, as of a rocket body, in which all or most of the stresses are carried by the skin. A monocoque may incorporate formers but not longitudinal members such as stringers.

MONOD, JACQUES LUCIEN (1910–1976). Monod was a French molecular biologist who, with F. Jacob, introduced the 'operon model' of the regulation of cellular activity. Jacques Monod was born in Paris to a Huguenot family. His father, Lucien, was an artist and art historian and his mother, Charlotte Todd McGregor, was American. Both parents had an important influence on young Jacques' education, pushing him toward science and music, though, with some hesitation, he finally chose to pursue a profession in biology.

After obtaining his degree from the Faculty of Science in Paris, Monod went to the Roscoff marine biology station, where he became acquainted with four scientists who shaped his conception of science and his scientific practice: Georges Teissier, who gave him a taste for quantitative description; André Lwoff, who introduced him to microbiology; Boris Ephrussi, whose field was physiological genetics; and Louis Rapkine, who taught him that only chemical and molecular descriptions could provide a complete interpretation of how living beings function. See also **Lwoff, André Michel (1902–1994)**.

In the autumn of 1931 Monod won a fellowship to study at the University of Strasbourg, in the laboratory of Edouard Chatton, the leading French protistologist of the time, where he became familiar with the techniques of microbiology. In 1932 Monod returned to Paris, spending two years in the Laboratory on the Evolution of Organized Beings, directed by Maurice Caullery, and in 1934 becoming an assistant in the zoology laboratory of the Sorbonne. In 1936 Monod spent a year at the California Institute of Technology, where the '*Drosophila* group' directed by T. H. Morgan was working. Here he learned not only genetics but the new scientific style based on collective effort, ease of personal relations between scientists, and freedom of critical discussion. See also **Microbiology**; and **Morgan, Thomas Hunt (1866–1945)**.

Back in Paris, Monod returned to the Sorbonne to prepare a doctoral dissertation (defended in 1941) on bacterial growth.

During World War II, Monod joined a Resistance group after the German occupation of France. In 1943 he joined the French Communist Party and the Franc-Tireurs Partisans, organizing the general strike that led to the liberation of Paris. In 1945 he resigned from the Communist Party, being in disagreement with its policy.

In late 1945 Monod joined the Pasteur Institute as laboratory director in the department headed by André Lwoff. He spent the rest of his scientific career in this institution. In collaboration with Alice Audureau, Monod continued his research on bacterial growth and enzyme adaptation, in particular studying the enzyme that became the classic subject of his later research, β-galactosidase.

In 1947 Monod produced a general report on enzymatic adaptation, assessing the respective roles of the hereditary and environmental factors (the substratum) in enzyme synthesis. From 1948 on, a fundamental contribution to the implementation of Monod's research program on enzyme adaptation was made by the American immunologist Melvin Cohn, who spent five years in Paris. The essential problem posed by the phenomenon of the induced biosynthesis of enzymes was to find out whether it consisted of the activation of a preexisting protein, a 'precursor', or whether, on the contrary, it involved the synthesis of a new protein. The linear kinetics ('Monod's plot') established by experiment showed that the synthesis of the enzyme after addition of the inductor is a constant fraction of the rate of total protein synthesis. Furthermore, the study of the incorporation of the radioactive isotope sulfur-35 into the protein in the course of its induced synthesis confirmed that the formation of β-galactosidase corresponded to the total synthesis of the protein on the basis of its elements, without the formation of precursors or intermediaries.

In 1953 Monod was assigned to create and direct the department of cellular biochemistry. The establishment of this department allowed Monod to develop a clear research program, based on the idea that fundamental chemical organization would be revealed at the level of cellular constituents rather than at the level of tissue or differentiated organs. Monod possessed a powerful tool for this study: the investigation of bacterial growth.

The new department had sufficient space to accept many students and visitors, primarily specialists in the physical and mathematical sciences. Beginning in 1957, Monod and François Jacob established a close collaboration. At the time, the development of research on inductible systems required the methods of crossing bacteria and of zygotic induction developed by Jacob in collaboration with Elie Wollman in André Lwoff's laboratory. This 'great collaboration' went beyond the solution to the problem of enzyme adaptation, uncovering the general mechanism of the regulation of protein synthesis and its genetic determination. The collaboration produced three theoretical models that proved fundamental in the development of molecular biology: the operon, messenger RNA, and allosteric interactions. See also **Jacob, François (1920–Present)**.

Together, Jacob and Monod conceived the famous Pa-Ja-Mo experiment, conducted in collaboration with Arthur Pardee, an American biochemist spending a sabbatical year in Paris. The conclusion to be drawn from this experiment was the existence of a double genetic determinism in protein synthesis. Two distinct genes intervened, one determining the structure of the synthesized molecule and the other controlling the expression of the first. Another conclusion was that different genes determining the structure of distinct proteins were subject to the same regulation system and this functional association was correlated with their genetic association.

Jacob and Monod suggested that there was a single structure sensitive to the repressor and controlling the activity of an entire group of genes. This is the concept of operon, a unit of coordinated expression made up of an operator and the group of structural genes that it coordinates.

The operon model opened the way to three research problems: (1) the nature of the repressor, (2) the mechanism of the repressor's chemical action and its relation to the target and the inductor, and (3) the molecular mechanisms of the transfer of genetic information for protein synthesis, the messenger.

Monod concentrated on the action mechanisms of the repressor and its chemical interactions with the target and the inductor and in 1961 Jacob and Monod generalized the concept of 'allosteric transition', a chemical interaction that allows complete freedom in the choice of chemical mechanisms, escaping all chemical constraints and obeying only the physiological constraints imposed by the system's consistency and submitted to the action of natural selection. Monod therefore considered the concept of allosteric interaction 'the second secret of life', the first being the double helix and the genetic code.

At the end of 1958, Monod was appointed Professor of Biochemistry at the Faculty of Science of the University of Paris. Later, in 1967, Monod was elected to the Collége de France, as the chair of molecular biology. His inaugural lecture in November 1967 was a solemn occasion for raising the philosophical implications of modern biology.

By interpreting the essential properties of organisms in terms of molecular structures, molecular biology had wrought a new definition of life, one that Monod summarized in three basic characteristics of biological objects: teleonomy, the existence of an 'internal program'; independent morphogenesis; and reproductive invariance. Particularly in his best-selling book *Chance and Necessity* (1970), Monod expounded his idea that modern biology stood in contradiction to any anthropomorphic interpretation of the universe or of life. The old ethical values no longer applied, and new ones had to be discerned. In a world in which science had demonstrated that human existence itself was contingent, acquisition of knowledge had to become the supreme value (the 'ethics of knowledge').

In 1965 Jacob and Monod were awarded the Nobel Prize in Physiology or Medicine for their research, together with André Lwoff. Monod then used his fame to demand university reform and to fight for the advancement of French science. During May 1968 Monod supported the student movement against the academic establishment.

Monod was an ardent defender of human rights. In the early 1950s he publicly protested about the repression of intellectuals in the USA; in the 1960s he aided intellectuals and Jews in Hungary and the Soviet Union and he came out against the French Secret Army Organization. He supported the activities of the French family planning movement, fighting for the legalization of abortion and in 1974 signed a plea in favor of beneficent euthanasia. The value and dignity of the individual was his fundamental ethical guideline.

In the 1960s Monod and his colleagues proposed thoroughgoing changes in the management of the Pasteur Institute and in 1971 Monod became the general director of the Institute. He assumed the post with a fully developed plan and with clear ideas—perhaps even dreams—of what an institution of biomedical research should be like. Monod improved the financial situation, reorganized research, eliminating or recasting many departments and creating others. Authoritarian and inflexible, he made decisions firmly and courageously. This created some difficult relations with many colleagues and Monod's years as head of the Pasteur Institute were difficult for personal reasons as well. In 1972 his wife died after a long illness. His administrative tasks made it impossible for him to continue his scientific activity, and he was compelled to resign from his position at the Collège de France.

In 1975 he sketched out a policy of scientific development for the Institute, centered on what he considered his fundamental vocation: the advance of the biological sciences in the service of humankind. But Monod had no time to implement this project. In October 1975 an incurable disease was diagnosed. He died in Cannes on 31 May 1976.

BERNARDINO FANTINI, University of Geneva, Geneva, Switzerland

MONOECIOUS PLANTS.   The flowers of many plants are unisexual; that is, they contain only stamens or pistils, but not both. When these two kinds of flowers are borne on the same plant, the plant is said to be monoecious. Familiar monoecious plants are oaks, corn, squash, begonia, and castor beans. The corresponding zoological term is hermaphrodite.

MONOMER.   A single molecule, or a substance consisting of single molecules. The term monomer is used in differentiation of dimer, trimer, etc., terms designating polymerized or associated molecules, or substances

composed of them, in which each free particle is composed of two, three, etc., molecules.

MONOMOLECULAR LAYER.   The early work of Rayleigh, Langmuir, Hardy and others has shown that it is possible to deposit on solid or liquid surfaces films which are one molecule thick. Any such layer is called a monomolecular layer, *unilayer* or *monolayer*. See also **Molecular and Supermolecular Electronics**.

MONOTREMETA.   A class of egg-laying mammals, containing the duck-billed platypus and the spiny anteaters of the Australian region. The duckbill, *Ornithorhynchus anatinus* (translated — "here is a creature with a bill like a duck"), also called platypus and duckmole, is about 18 inches (46 centimeters) long and has close fur something like that of the moles. The coat is glossy black hair with a waterproof undercoat of fur. The broad and flattened tail is covered with coarse hair, but is essentially bare underneath as in a beaver. The animal is aquatic. The muzzle is broad, flat, and naked, resembling the beak of a duck, but not horny. It is formed as part of the jaw bone and is very sensitive to touch. The animal also has a keen sense of smell. The legs are short, the web on the feet is loose, the claws are long and sharp and slightly curved. See Fig. 1. Duckbills are found in the streams of southern and eastern Australia. The animal nests in burrows in the banks and females deposit two small eggs at a time. The young, born naked, and blind, are nourished with milk secreted from glands on the mother's abdomen. There are no teats. The animal has pouches; like those of a squirrel, for collecting food. Spurs on the hind feet can inflict a poisonous venom, useful in capturing small animals, but not injurious to humans. The duckbill's voice sounds something like an angry puppy's growl. The diet consists of insects, crustaceans, and worms. The duckbill is considered shy by nature and, upon hearing the slightest noise, will disappear under water.

Fig. 1.   Duckbill or platypus.

The spiny anteater of Australia ranges from a few inches to 20 inches (51 centimeters) in length when fully grown, depending upon species. The body is covered with hair and spines. The animal has a slender snout, short legs, and strong claws. See Fig. 2. The spiny anteater is a burrowing animal. See also **Edentata**.

Fig. 2.   Australian echidna or spiny anteater.

MONSOON. Cyclic air masses accompanied by either much or little rain, carried by winds that reverse direction semiannually, are called monsoons. The word monsoon comes from the Arabic word "mausim," which is translated as "season." Monsoons occur over almost every continent, but they come in varying degrees of strength, as judged by rainfall production and wind speed. The most powerful and well-known monsoons occur in India and Southeast Asia.

Monsoons are caused by a difference in temperature, and thus air pressure, between landmasses and oceans. They are especially strong in Southeast Asia because the area is geographically conducive to monsoon weather. There is much more land in the Northern Hemisphere, namely, the Asian continent, and there is much more water in the Southern Hemisphere, namely the Indian Ocean. Due to the earth's tilt as it revolves around the Sun, the Northern Hemisphere receives more sunlight during the summer, and the Southern Hemisphere receives more sunlight during the winter. This means that there is always a marked difference between the air temperature over land and the air temperature over the sea in Southern Asia and India. The two variations that result are called wet and dry monsoons. Locally, they are called wet season (summer) and dry season (winter). Wet monsoons occur when the land in the Northern Hemisphere becomes heated by extensive sunlight. Due to convection, the warm air over land rises into the atmosphere, and cooler ocean air slides in to take its place. This cool air carries vast amounts of moisture, which is dropped on the landmass, in this case India and SE Asia.

Dry monsoons occur in the winter when the reverse occurs, and the oceans are heated by sunlight. The air over them rises, and cool air from the land leaves to take its place. This cold air takes moisture with it; thus, there is little rainfall over land.

The precipitation released during the wet season can fall quickly and unpredictably, sometimes resulting in floods, but usually resulting in enormous quantities of much needed rainfall for the farmers and residents of Southeast Asia. Due to this fact, monsoons are integral to the lives and well-being of virtually all of the people who live in India and Southeast Asia, as they have been for millennia. However, despite their importance to the area, they can cause great damage at times, as well. It is not uncommon for abnormal monsoon conditions to result in massive flooding or extended droughts. In fact, a 1987 Indian drought due to variations in the local monsoons was one of the worst of the twentieth century. And parts of India can receive 40 feet of rain or more in a span of less than 4 months.

The monsoon seasons are the distinctive feature of the Southeast Asian climate, but they do exist to lesser degrees elsewhere in the world. Africa, Australia, and the Southwestern United States all undergo minor monsoon seasons during the year. Generally, their monsoon weather is less pronounced than that of Southeast Asia because they do not have its unique geographical properties.

ARTHUR M. HOLST, Philadelphia Water Department, Philadelphia, PA

MONSOON CLIMATE.   See **Climate**.

MONSOON DEPRESSION.   See **Winds and Air Movement**.

MONSOON GYRE.   See **Winds and Air Movement**.

MONTAGNIER, LUC (1930–). Luc Montagnier is a native of Chabris, which is located close to Tours in France. After studying biology at the University of Poitiers, he received a license in Science in 1955 and a doctorate in medicine in 1960, both from the University of Paris. He continued to do research at both the institutions, and at the Pasteur Institute, mostly on viral disease and interferon, and, in 1974, become a Research Director of the Centre National de la Recherche Scientifique (CNRS). In 1983, his laboratory published the first paper in the scientific literature on the identification of a human retrovirus as the presumed cause of acquired immune deficiency syndrome (AIDS). In 1985, he was appointed to the prestigious position of Professor at the Pasteur Institute in Paris.

The discovery by the Montagnier laboratory of a virus was quickly confirmed by other groups in 1984. It was initially termed lymphadenopathy-associated virus (LAV) and later human immunodeficiency virus type 1 (HIV-1). This work quickly led to the establishment of culture techniques for efficient propagation of the virus and to an understanding of its protein and ribonucleic acid (RNA) structures and genetic composition. This,

in turn, led directly to the establishment of blood tests for the detection of HIV-1, a development that still represents a cornerstone of worldwide public health efforts to control the HIV/AIDS epidemic.

Professor Montagnier's laboratory is also credited with the identification of the Cluster of differentiation (CD)4 receptor as a point of attachment for HIV on to cells and the discovery of a related retrovirus termed HIV-2. He was also among the first to suggest that the aetiology of HIV disease might be multifactorial and that other microorganisms, including *Mycoplasma*, might be involved. The latter hypothesis is controversial.

Professor Montagnier went on to co-found the World Foundation for AIDS Research and Prevention in 1993. He has been a key spokesperson in support of worldwide research efforts towards the development of a safe and effective vaccine to protect against HIV infection as well as newer and less-expensive drugs that might be employed in developing countries, so as to ensure that HIV-infected individuals in resource-poor settings have the same opportunities to benefit from research advances as do people living in richer countries of the world. Professor Montagnier retired as a Research Director at the Pasteur Institute in 1998.

Professor Montagnier's research and humanitarian efforts toward the conquests of HIV/AIDS have also been recognized by world communities through multiple major awards and honors. These include the Lasker Award, the Japan Prize and the highest rank of the French Legion of Honor.

See also **Acquired Immune Deficiency Syndrome (AIDS)**; and **Human Immunodeficiency Viruses (HIV)**.

Additional Reading

Barre-Sinoussi, F., J.C. Chermann, F. Rey, et al.: "Isolation of a T-lymphotropic Retrovirus from a Patient at Risk for Acquired Immune Deficiency Syndrome (AIDS)," *Science*, **220**(4599), 868–871 (1983).

Gallo, R.C., and L. Montagnier: "The Discovery of HIV as the Cause of AIDS," *New England Journal of Medicine*, **349**, 2283–2285 (2003).

Klatzmann, D., E. Champagne, S. Chamaret, et al.: "T-lymphocyte T4 Molecule Behaves as the Receptor for Human Retrovirus IAV," *Nature*, **312**, 767–768 (1984).

Montagnier, L., J. Gruest, S. Chamaret, et al.: "Adaptation of Lymphadenopathy-Associated Virus (LAV) to Application in EBV-transformed B Lymphoblastoid Cell Lines," *Science*, **225**, 63–66 (1984).

Montagnier, L.: "Environmental Pathogenesis of Human Retroviruses," *AIDS Research and Human Retroviruses*, **12**, 357–359 (1996).

MARK A. WAINBERG, McGill University, Montreal, Canada

MONTE CARLO METHOD. A method for resolving problems in mathematics, statistics or operations research by the use of random sampling. For example, in the theory of integration it may be impossible to derive an explicit numerical value of a particular integral

$$I = \int_a^b f(x)\,dx$$

by classical mathematics. If the domain from a to b is sampled at random, that is to say n values x_i are chosen from the uniform distribution from a to b and $f(x_i)$ evaluated at each point, then the mean of the values of $f(x_i)$ multiplied by half the interval from x_{i-1} to x_{i+1} will tend to I as n increases.

In practice it is more usual to divide the range a to b into equal intervals and evaluate by quadrature, but if the integral is multiple a systematic lattice of points may involve too much computations and a smaller sample of random points may be easier to handle.

The method, however, has more obvious advantages in statistical situations where the integral or summatory processes cannot be explicitly written down. For example, many sampling distributions, say of a statistic t, are not derivable in tractable form; it is then possible to draw a random sample from the parent population, calculate t for each and hence to derive an empirical estimate of the distribution of t by repeating the operation a large number of times.

Analogous procedures apply to the study of a system in which the determinantal equations or inequalities can be written down but explicit solutions cannot be derived. The behavior of the system can be stimulated by feeding in values of the variables, and repeating the operation over different sets of values so as to explore the system under a variety of conditions. Such sets of values may be chosen systematically or by a random mechanism.

Usage varies, but it seems preferable to confine the term "Monte Carlo" to those cases where probabilistic sampling is employed.

SIR MAURICE KENDALL, International Statistics Institute, London

MONTREAL PROTOCOL. Following the discovery of the Antarctic ozone hole in late 1985, governments recognized the need for stronger measures to reduce the production and consumption of a number of CFCs (CFC 11, 12, 113, 114, and 115) and several Halons (1211, 1301, 2402). An international agreement known as the "Montreal Protocol on Substances that Deplete the Ozone Layer" was adopted on September 16, 1987 at the Headquarters of the International Civil Organization in Montreal; twenty-four countries and the European Economic Community signed a treaty that most observers at the time had thought would be impossible. Some years later, the Montreal Protocol on Substances That Deplete the Ozone Layer was characterized by the heads of the World Meteorological Organization (WMO) and the United Nations Environment Programme (UNEP) "as one of the great international achievements of the century."

The Protocol called for the Parties to phase down the use of CFCs, halons and other man-made ODCs. The Protocol was designed so that the phase out schedules could be revised on the basis of periodic scientific and technological assessments. See Table 1. Following such assessments, the Protocol was adjusted to accelerate the phase out schedules. It has also been amended to introduce other kinds of control measures and to add new controlled substances to the list. The Protocol came into force on January 1, 1989, when it was ratified by 29 countries and the EEC. Since then, it has undergone five revisions, in 1990 (London), 1992 (Copenhagen), 1995 (Vienna), 1997 (Montreal), and 1999 (Beijing) See also http://www.theozonehole.com/montext.htm. At present, 189 nations have become party to the Montreal Protocol. See also http://ozone.unep.org/Treaties_and_Ratification/2 C_ratificationTable.asp.

Historical and Scientific Background

Given the threats to life that have been averted through this landmark treaty, few would challenge their statement as hyperbole. Ozone, whose existence was unknown until 1840, has been characterized as "the single most important chemically active trace gas in the earth's atmosphere" [Albritton, et. al., 1987]. Without it, life as it currently exists on Earth could not have evolved. The Montreal Protocol, by phasing out certain chemicals, preserved the stratospheric ozone layer that absorbs harmful ultraviolet radiation from the sun. Depletion of this thin gaseous shield—which, if compressed to the planet's surface, would be no thicker than gauze—would have incalculable impacts on human, animal, and plant cells, as well as on climate and ecological systems. Recent research, for example, indicated that if anthropogenic ozone-depleting substances had continued their rapid accumulation in the upper atmosphere, there would have been a "runaway increase" in skin cancer over the next several decades [Slaper, et. al., 1996].

And yet, while the treaty was under negotiation, the science was still speculative, based on projections from evolving computer models of imperfectly understood atmospheric processes—models that yielded varying, sometimes contradictory predictions each time they were refined. Moreover, measurements revealed neither the theorized mid-latitude depletion of ozone nor any of the predicted impacts. The scientific, economic, technological and political issues involved in the negotiations were staggeringly complex. Chlorofluorocarbons (CFCs) and related substances seemed virtually synonymous with modern standards of living. They were ideal chemicals —nonflammable, nontoxic, noncorrosive. In the 1980's, they were finding new applications in thousands of products and processes across dozens of industries, from electronics, refrigeration, insulation, and plastics, to telecommunications, aerospace, pharmaceuticals, and agriculture. Powerful political and economic interests were aligned against meaningful controls.

Nevertheless, within less than six years after negotiations began in late 1986, the Montreal Protocol had been ratified by more than 100 (later over 160) nations and had undergone two major revisions that expanded the list of controlled substances from 8 to over 90 and that considerably strengthened timetables for reduction and phaseout of the dangerous chemicals [Benedick, 1998]. A veritable technological revolution was unleashed that in only a few years transformed entire industries. The protocol created the first-ever global environmental fund to assist developing nations, and promoted an unprecedented North-South collaboration in researching and diffusing new technologies that have now made ozone-depleting substances obsolete.

Even so, it was a near thing. For decades after their discovery in the 1930's, no one suspected that these "wonder-chemicals" could cause any harm, much less to the critical ozone layer. And, because the CFCs and their cousins have such long atmospheric lifetimes, their deleterious impacts will still be felt for decades, even after new emissions cease.

Unquestionably the indispensable element in the success of the Montreal Protocol was the role of science and scientists. Without the curiosity and courage of a handful of researchers in the mid-1970's, the world might have learned too late of the deadly, hidden dangers linked with rapidly expanding use of these substances. The initial, now legendary, hypotheses of Sherwood Rowland and Mario Molina at the University of California-Irvine unleashed a storm of criticism and controversy. They were vindicated by the 1995 Nobel Prize in Chemistry (together with Paul

TABLE 1. SUMMARY OF MONTREAL PROTOCOL CONTROL MEASURES

Ozone Depleting Substances	Developed Countries	Developing Countries
Chlorofluorocarbons (CFCs)	Phased out end of 1995[a]	Total phase out by 2010
Halons	Phased out end of 1993	Total phase out by 2010
Carbon tetrachloride	Phased out end of 1995[a]	Total phase out by 2010
Methyl chloroform	Phased out end of 1995[a]	Total phase out by 2015
Hydrochlorofluorocarbons (HCFCs)	Freeze from beginning of 1996[b]	
	35% reduction by 2004	
	65% reduction by 2010	Freeze in 2016
	90% reduction by 2015	At 2015 base level
	Total phase out by 2020[c]	Total phase out by 2040
Hydrobromofluorocarbons (HBFCs)	Phased out end of 1995	Phased out end of 1995
Methyl bromide	Freeze in 1995 at 1991 base level[d]	Freeze in 2002 at average
	25% reduction by 1999	1995–1998 base level
	50% reduction by 2001	20% reduction by 2005[e]
	70% reduction by 2000	Total phase out by 2015
	Total phase out by 2005	

[a] With the exception of a very small number of internationally agreed essential uses that are considered critical to human health and/or laboratory and analytical procedures.

[b] Based on 1989 HCFC consumption with an extra allowance (ODP weighted) equal to 2.8% of 1989 CFC consumption.

[c] Up to 0.5% of base level consumption can be used until 2030 for servicing existing equipment.

[d] All reductions include an exemption for pre-shipment and quarantine uses.

[e] Review in 2003 to decide on interim further reductions beyond 2005.

Crutzen of the Netherlands), but it is worth noting that the first popular book on this subject, published in 1978, was entitled The Ozone War [Dotto, and Schiff, 1978].

The complexity of the research effort was enormous. Ozone amounts to considerably less than one part per million of the total atmosphere, with 90 percent of it concentrated above six miles in altitude. The intrinsically unstable ozone molecules are continually being created and destroyed by complex natural forces involving solar radiation and interactions with even more minute quantities of other gases. Moreover, stratospheric ozone concentrations fluctuate wildly on a daily, seasonal, and solar-cyclical basis, and there are great geographical as well as altitudinal variations. Amidst all these fluxes, scientists faced a formidable challenge in predicting, and then detecting, the minuscule "signal" of a downturn in stratospheric ozone concentrations. This necessitated the development of ever more sophisticated computer models to simulate the stratospheric interplay among radiative, chemical, and dynamic processes such as wind and temperature for decades or centuries into the future. Intricate measuring devices had to be created and fitted onto aircraft, satellites, and rockets to monitor remote gases in quantities as minute as parts per trillion.

To understand the implications of a fading ozone layer, scientists had to venture far beyond atmospheric chemistry: they had to examine our planet as a system of interrelated physical, chemical and biological processes on land, in water, and in the atmosphere—processes that are themselves influenced by economic, political, and social forces. The Montreal Protocol became a truly multi- and interdisciplinary effort. Over the years, researching the dangers and solutions involved not only chemists and physicists, but also meteorologists, oceanographers, biologists, oncologists, economists, soil scientists, toxicologists, agronomists, pharmacologists, electrical, chemical, automotive and materials engineers, botanists, entomologists, and more.

It was not sufficient, moreover, for scientists merely to publish their findings. In order for the theories to be taken seriously and lead to concrete countermeasures, scientists had to interact with diplomatic negotiators and government policy makers. This meant that they occasionally had to leave the familiar atmosphere of their laboratories and assume an unaccustomed shared responsibility for the policy implications of their research. The history of the Montreal Protocol is filled with instances of scientists called upon to analyze the implications of alternative remedial strategies and policy measures.

International scientific consensus was also essential. The development of an accepted common body of data and analysis was prerequisite for a political solution among negotiating governments whose positions were initially very far apart. In 1984, a remarkable collaborative international research effort was launched, spearheaded by the National Aeronautics and Space Administration (NASA) and the National Oceanic and Atmospheric Administration (NOAA), in cooperation with the WMO, UNEP, the Federal Aviation Administration, the German Ministry for Research and Technology, and the Commission of the European Communities.

The Montreal Protocol later institutionalized this idea by establishing international expert panels to periodically assess scientific, technological, economic, and environmental knowledge and thereby guide the negotiators in the further evolution of the treaty. Over the years hundreds of scientific experts from dozens of countries participated in the effort to learn more about both the dangers and the possible technological solutions. This proved to be a central element in the protocol's success, facilitating agreement by negotiators on additional measures to protect the ozone layer. In effect, the Montreal Protocol was deliberately designed to be a dynamic process of narrowing the ranges of uncertainties, rather than a static solution based on the status quo.

The role of scientists in the ozone history provided useful lessons for the climate change issue. In the 1980's, scientific assessments on climate change appeared regularly, under the aegis of WMO and UNEP, from a small group of largely self-selected scientists called the Advisory Group on Greenhouse Gases. During the summer of 1987, while preparing for the conclusive final negotiation in Montreal, I recommended that the US take an initiative to establish a formalized international assessment body on climate change, similar to what we were doing on the ozone issue. My belief was that the findings would be more credible coming from a larger and diverse group of scientists under intergovernmental auspices.

This idea attracted unexpected allies and opponents. Anti-environmental officials within the Reagan Administration endorsed the concept, expecting it would provide governments more control over the science, while environmental groups feared that the process would become distorted by politics. My own feeling, grounded in the ozone experience, was that most scientists were unlikely to allow themselves to be subverted by political or commercial interests, and that governments would sooner be co-opted by the science than vice versa. The subsequent experience of the Intergovernmental Panel on Climate Change, founded in 1988, has confirmed this hope.

Another lesson from the protocol's success was the importance of public education: interpreting the continually evolving and sometimes confusing data, and communicating it intelligibly to the public and the media. This information flow mobilized public opinion on the potential dangers of a diminishing ozone layer, and thereby promoted political consensus for both policy measures and for funding research. The proponents of actions to protect the ozone layer generally avoided exaggerating their case as a means of capturing media and public attention. In this way, they maintained credibility and did not provide gratuitous ammunition to those interests that sought to minimize the problem. In 1976-78, US media interest, promoted and nurtured by some scientists, legislators, and environmental organizations, stimulated decisions by millions of individual consumers that led to the collapse of the domestic market for CFC aerosol sprays even before there was any government regulation. Later, UNEP and WMO played prominent roles, through workshops, publications, and electronic media, in disseminating relevant information, including the availability of new technologies, to officials, business, and public around the world.

The history of the Montreal Protocol also underscored the importance of having sufficient funding for all levels of science, from curiosity-driven basic research to applied engineering solutions. Initially, most funding came from government sources, in particular NASA and NOAA in connection with their space-related research. But this was not always the case. In 1985, when the U.K. Government was still opposed to strong controls over CFCs, it ceased financing British scientists in Antarctica who were coming up with disturbing evidence of stratospheric ozone losses. Interestingly, the gap was filled by the US Chemical Manufacturers Association which, although not favoring controls, was nevertheless more concerned about resolving the uncertainties, one way or the other.

Research investments by the private sector proved crucial in developing substitutes for the ubiquitous CFC family of chemicals. Unusual public-private partnerships were formed to find ozone-friendly solutions for products and processes where it had never been thought possible. In one case, Greenpeace teamed up with a former East German company to develop CFC-free refrigerators, which subsequently were adopted in European markets and promoted in China and India by the German and Swiss aid programs. The technological revolution had many novel aspects, ranging from collaboration by AT&T and a Florida citrus grower in developing new solvents for electronic circuit boards, to China's indigenous approach to replacing styrofoam with a biodegradable product of grass and straw. "Politics," stated Lord Kennet during early ozone debates in the House of Lords, "is the art of taking good decisions on insufficient evidence." The success of the Montreal Protocol stands as a beacon of how science can guide decision makers to overcome conflicting political and economic interests and reach solutions. The ozone history demonstrates that even in the real world of ambiguity and imperfect knowledge, the international community, with the assistance of science, is capable of undertaking difficult and far-reaching actions for the common good.

See also **Carbon Tetrachloride**; **Ozone**; **Ozone Depletion (Science of)**; and **Polar Research**.

Additional Reading

Albritton, D., et. al.: *Stratospheric Ozone: The State of the Science and NOAA's Current and Future Research*, National Oceanic and Atmospheric Administration, Washington, DC, 1987.

Benedick, R.E.: *Ozone Diplomacy: New Directions in Safeguarding the Planet*, Harvard University Press, Cambridge, MA, 1998.

Dotto, L., and H. Schiff: *The Ozone War*, Doubleday Publishing, Garden City, NY, 1978.

Obasi, P., and E. Dowdeswell: *The Changing Ozone Layer*, WMO/UNEP, Geneva, Switzerland, 1995.

Slaper, H., et. al.: "Estimates of Ozone Depletion and Skin Cancer Incidence to Examine the Vienna Convention Achievements," *Nature*, **384**, 256 (November 21, 1996).

Web Reference

The Montreal Protocol on Substances that Deplete the Ozone Layer: ozone.unep.org/pdfs/**Montreal-Protocol**2000.pdf

RICHARD E. BENEDICK, *Pacific Northwest National Laboratory, Richland, WA*

MOOD DISORDERS. The term "mood" can be described as the internal experience of a pervasive and sustained emotional state that colors our individual experience of the world around us, as well as our choice of how to respond to it. The study of mood has been viewed as essential to an understanding of mind and brain function since the end of the nineteenth century, when Charles Darwin described evolutionary components of emotional states, and William James the means by which bodily states could predispose to emotional experience. Only in the last two decades of the twentieth century, however, did technical advances in neuroscience allow a dissection of emotional processes in terms of cellular systems and neuroanatomical circuits. Although the terms are sometimes used interchangeably, mood is generally distinguished from emotion in having a longer duration and in being less intentional or object-related. Moods that are stable over time can be viewed as emotional traits that, together with nonaffective components, form the concept of temperament and represent consistent heritable and learned differences in the way particular emotions are experienced and responded to by the individual. Although understanding of the neuroanatomical circuitry underlying mood states is less clear than that involved in fear conditioning, a variety of neuroimaging procedures employing either animal models of "learned helplessness" or human studies of lesion deficit or empirical mood induction have localized mood regulation to sites in limbic, paralimbic and prefrontal cortical areas. Several recent studies have uncovered reciprocal changes in activation between subgenual cingulate and anterior insula regions and the right prefrontal and inferior parietal cortex in association with changes in negative mood in normal individuals experiencing transient sadness and in depressed patients recovering from chronic dysphoria.

From an evolutionary perspective, the phenomenology and physiology of depression have been conceptualized as an attempt to conserve resources and stop the organism from investing in activities that have a low likelihood of reward. Defeat and entrapment following social conflict may represent animal models that underlie the psychopathological state of human depression and may be relevant to understanding the behavioral strategies affected, even when the causation is genetic [Gilbert and Allan].

Disorders of Mood

The hallmark of mood disorders is a primary pervasive disturbance in emotional state that typically affects all aspects of the individual's life, causing changes not only in mood but also in psychomotor activity, cognition, appetite, libido and regulation of circadian rhythms. According to the International Statistical Classification of Diseases and Related Health Programs (ICD-10) and the Diagnostic and Statistical Manual of Mental Disorders (DSM-IV TR), the categorization of mood disorders includes individuals with either single or recurrent episodes of major depression (unipolar mood disorder), and those who, in addition to experiencing depressions, also have episodes in which an elevated or irritable mood is prominent (mania). The latter condition is also referred to as manic-depressive illness or, more recently, bipolar disorder. The nosology of mood disorders also includes conditions that exist on a continuum with both unipolar and bipolar disorders and differ principally in either the duration or intensity of the mood state or the frequency of mood change. In cyclothymic disorder, individuals experience numerous periods of hypomania (a state similar to mania but less severe in its symptomatology and not associated with marked impairment in social or occupational functioning), as well as numerous periods in which depressive symptoms are prominent. To meet the full criteria, the cycles must have persisted for two years and the depressive symptoms must not otherwise meet the criteria for a formal episode of major depression. Dysthymic disorder refers to a condition in which the depressed mood is less severe than in major depression but is persistent, often being present for many years and difficult to distinguish from the individual's state of usual function. When the depressive mood disorder is relatively mild and limited in duration and occurs shortly after an identifiable stressor or the death of a loved one,

the diagnosis of adjustment reaction with depressed mood or bereavement is appropriate. The diagnosis of a mood disorder is dependent upon its being viewed as primary and not secondary to the direct physiological effects of a substance (e.g. a drug of abuse) or a general medical condition (e.g., hypothyroidism). Within these categories there may be qualitative differences in the signs and symptoms observed; accordingly, specifiers describing the presence of psychotic features (delusions or hallucinations), of catatonia (marked motoric immobility or purposeless or peculiar voluntary movements) and of melancholia (the complete absence of capacity for pleasure in association with dramatic change in vegetative functions) are often appended to the primary diagnosis. It should be noted that all of these categorical distinctions are empirical in nature and that evidence for validity is lacking. The degree to which the various conditions relate to each other in causation or pathophysiology or to normal variations in mood state is currently unknown.

The diagnosis of major depression depends on the presence of a pervasive depressed mood or a loss of interest in usual activities and at least four of seven additional symptoms for a period of at least two weeks. Characteristic changes include a significant weight loss or loss of appetite, insomnia or increased need for sleep, psychomotor agitation or retardation, fatigue, feelings of worthlessness and guilt, impaired concentration, and recurrent thoughts of death or suicide. The mood change must be associated with a significant level of distress or impairment in usual functioning. Major depressions can be superimposed on dysthymic disorder, creating the clinical condition of "double depression." In individuals in whom the full symptomatic requirements are not met, the diagnosis of recurrent brief or minor depression is made.

Major depression has a lifetime prevalence of approximately 15%, with women experiencing a two times greater risk than men [Kessler, et al.:]. Until age 13, however, the gender ratio is equal. The average age of onset is in early adulthood, but there is increasing evidence over the last century that depression is becoming more prevalent and occurring at an earlier age in younger cohorts. Life events appear to play a role in precipitating episodes of depression and mania, but the relative degree of contribution appears inversely related to genetic loading and the total number of episodes experienced over time. A history of childhood maltreatment and negative parenting style increase vulnerability, and life events that involve loss of social support or that alter circadian rhythms seem to carry increased risk.

Most individuals recover from the initial episodes within six to nine months, but a minority will remain symptomatic over a period of years. Recurrence is frequent, with half of all patients experiencing another episode in three years, and three-quarters experiencing another in ten years. The lifetime risk for recurrence increases progressively with each successive episode; the total number of episodes and duration of the longest episode, together with thoughts of suicide and level of impairment, are the clinical features that best predict risk of depression in relatives.

Depression is associated with significant societal burden [Stewart, et al.:]. According to a study conducted by the World Health Organization, clinical depression is the fourth most disabling medical condition worldwide, and is expected to rise to the second position by the year 2020. Despite these figures, depression, and mood disorders in general, continue to be significantly undiagnosed and undertreated when identified, and associated with persistent social stigma [Kleinman, ref.]. Lifetime mortality due to suicide is approximately 3.5–4%, with depression accounting for the majority of the 30,000 deaths per year by suicide in the United States. Demographic risk factors for suicide include increased age, an absence of social support, unemployment, a family history of suicide, and the presence of alcohol or drug abuse. The best short-term predictors of suicidal action in the context of depression appear to be the presence of high psychic anxiety and agitation and extreme hopelessness.

The diagnosis of bipolar mood disorder requires a history of a manic episode in addition to that of major depression [Belmaker, ref.]. A manic episode is defined by a distinct period of an elevated, expansive or irritable mood lasting at least one week and the presence of at least three of seven possible associated symptoms. These include inflated self-esteem or grandiosity, a decreased need for sleep, being more talkative than usual, having racing thoughts, distractibility, increase in goal-directed activity or agitation, and evidence for poor judgment, impulsivity or excessive involvement in pleasurable activities. In some individuals, depressive episodes alternate with episodes of hypomania (bipolar II disorder) or present in a co-mingled fashion with mania (mixed episode).

Bipolar disorder occurs in 1–1.5% of the population, with an equal gender distribution. The age of onset is usually in late adolescence, but childhood onset is increasingly recognized, particularly in families where there is strong genetic loading. The course of illness is variable, with some individuals experiencing extended periods of normal mood between episodes and others rapid and frequent transitions between depression and mania. The majority will experience at least 4–6 episodes over their lifetime.

Investigations in the pathophysiology of depression have identified a number of neurobiological alterations, many of which parallel changes observed in animal models of stress-induced change [Nestler, et al.:]. Rats exposed to maternal deprivation or repetitive stresses in early postnatal life develop persistently exaggerated hypothalamic–pituitary–adrenal (HPA) hormonal responses to environmental demands. These in turn can influence noradrenergic and serotonergic functions, and damage hippocampal structure if sustained [Charney and Manji]. Basal elevations in cerebral spinal fluid (CSF) levels of cortisol and corticotrophin-releasing factor (CRF), as well as diminished cortisol suppression by dexamethasone, occur in a number of depressed patients and may be linked to associated findings of reduced dopamine turnover in CSF, alteration in growth hormone release and serotonin receptor abnormalities. Procedures inducing catecholamine and indoleamine depletion result in depressive relapses in remitted patients, but only in patients who have responded to the specific reuptake inhibitor in question, e.g. a noradrenaline (norepinephrine) reuptake inhibitor in the case of catecholamine depletion and an SSRI (selective serotonin reuptake inhibitor) in the case of tryptophan reduction. The finding that chronic stress can result in neuronal death and hippocampal atrophy raises the possibility that findings of neuronal and glial cell pathology and diminished spine density in postmortem brain tissue of depressed patients, and of hippocampal volume reductions on magnetic resonance imaging (MRI), are secondary to genetic and environmental alterations in stress response systems [Sheline, et al.:]. Other neuropathological changes reported in major depression include increased cortical blood flow and metabolism in amygdala, thalamus and orbital cortex, and decreased activation in dorsal prefrontal cortex and the anterior cingulate. Autopsy studies have also shown reduced glial density in prefrontal cortex in both major depression and bipolar disorder, implicating decreased neuronal connectivity with other brain regions [Coyle and Duman]. In patients with late-onset depression (onset after 60 years), cerebrovascular disease is likely to be more important as a causal factor than genetic or developmental influences. Many such individuals will have increased signal hyperintensities in periventricular and deep white matter on MRI. High rates of depression in conditions such as stroke, Parkinson disease, Huntington disease and multiple sclerosis indicate that changes in blood flow or metabolism in frontotemporal cortex, caudate, hypothalamus and brainstem may overlap to produce a clinical syndrome of depression that is indistinguishable clinically from that emerging out of genetic causation [Bogousslavsky and Cummings]. There is suggestive evidence that a lifetime history of depression increases risk for dementia and that late-onset depression is commonly a prodromal feature of Alzheimer disease, although such observations should not alter initial treatment recommendations.

Both stress and depression have been associated with alterations in immune function [O'Brien, et al.:]. Reported changes in depression include increases in acute-phase proteins, IgA, IgM and complement, as well as IL-1, INF α and tumor necrosis factor. In contrast, IL-2 levels appear to be reduced and numbers of natural killer cells decreased. Elevated adhesiveness and aggregation of peripheral blood leucocytes has also been reported. Such changes in autonomic and immune function may be relevant to epidemiological data that indicate that depression is associated with increased risk of cardiac morbidity and mortality and increased susceptibility to cancer [Lett, et al.:].

Major findings from sleep electroencephalographic (EEG) studies of depression include decreased sleep continuity, diminished delta-wave production and increased rates of rapid eye movement (REM), particularly in the first sleep period.

The pathophysiology of bipolar disorder is not as well understood, but a variety of reports indicate differences in signal transduction systems between patients and controls. Among the findings in the literature are reports of enhanced protein kinase C activity, increased G_α protein and phospholipase C-β levels, higher cAMP-dependent protein kinase levels and elevated intracellular calcium levels. Recurrent changes in these systems in association with mood cycling and their subsequent effects on gene expression have been linked to the clinical paradigm of kindling, in which intermittent subthreshold electrical or chemical stimuli produce increasing sensitization in the neuronal substrate to the point of eventually producing spontaneous and autonomous discharge activity. In the case of bipolar disorder, the experience of an increasing number of episodes over time may explain why cyclic frequency is increased and why the episodes themselves seem less dependent on environmental triggers.

Neuroanatomical changes have also been reported. Consistent findings include the presence of hyperintense lesions in subcortical white matter and grey nuclei, larger third and lateral ventricle volumes, and more sulcal prominence than are found in age-matched controls.

Genetics of Mood Disorders

Family studies of unipolar depression have documented that the risk of developing depression in a first-degree relative is 2-fold to 3-fold higher with a depressive than with an unaffected control proband. In bipolar disorder, the morbid risk to relatives is even greater (3-fold to 8-fold), in part because rates of both unipolar and bipolar disorders are increased. Because family studies cannot distinguish between genetic and environmental causation, researchers have investigated the concordance rate for illness in monozygotic (MZ) and dizygotic (DZ) twin pairs. The mean concordance rate in unipolar depression is 40% in MZ twins and 17% in DZ twins; in bipolar disorder the heritability quotient rises, with a concordance of 67% in MZ twins and 20% in DZ twins, using strict definitions for affected status. Pedigree analyses of both bipolar and unipolar disorders have indicated that the likely mode of inheritance is complex, and that multilocus genetic effects are likely in most families. In bipolar disorder there is limited evidence for Mendelian transmission of a single major locus in a few published pedigrees, but in the majority of cases the number of susceptibility loci and their relative contributions to risk are unknown. Molecular genetic studies have thus far concentrated on bipolar disorder because of evidence for greater genetic effect and because of greater reliability in determining phenotype [Mathews and Reus]. Although a number of reports of positive linkage and association have appeared, nonreplication of findings is the rule, in part owing to a lack of agreement regarding the threshold for statistical significance. Initial reports of linkage to Xq in Israeli families and to 11p in a large Amish kindred have been followed by further analyses and data that diminish the findings. Recently, more extensive and better-controlled studies, employing a combination of family and sib-pair linkage and association analyses, population isolates and full genome scans, have reported evidence for loci on 18p and 18q, as well as on 4p, 8q, 12q, 21q and 22q. Disappointment arising from the relative lack of progress thus far has led to a re-examination of the phenotypes chosen and to large-scale cooperative efforts to increase statistical power. Co-morbidity with substance abuse is common in mood disorders, for example, and increases the risk of inclusion of phenocopies, but may also represent pleiotropic effects of the genes involved. Attempts to utilize dimensional evaluations of core symptoms or "endophenotypes," trait markers that are heritable and stable, as alternative or additional determinants of affection have received increasing attention, but no current agreement exists as to which markers would best satisfy these criteria in genetic studies of mood disorders. Association studies of candidate genes have thus far suffered from the same criticisms as linkage studies. One of the most-studied polymorphisms, a candidate variant in the serotonin transporter promoter resulting in decreased protein expression, has been positively associated with both unipolar and bipolar mood disorders, and recently shown to interact with environmental stress in increasing risk for a depressive episode [Caspi, et al.:]. It is possible that this allele may also be relevant to an understanding of individual differences in response and tolerance to antidepressant agents.

Liability to suicidal behavior appears to be familially transmitted as a risk factor independently of formal psychiatric diagnosis. Reduced levels of a serotonin metabolite, 5-hydroxyindoleacetic acid (5-HIAA), have been reported in individuals who attempted suicide and in postmortem brain tissue of suicide victims, along with other alterations in serotonin function, including decreased brainstem levels, decreased transporter binding, and increased 5-HT_{1A} and postsynaptic 5-HT_{2A} receptor binding in prefrontal cortex.

Treatment of Mood Disorders

Empirical investigation has shown that the most efficacious treatment of mood disorders involves a combination of psychopharmacological and psychotherapeutic approaches. Mood disorders by their very nature exact a high toll on the social fabric of a patient's life and drug treatment alone is unlikely to repair the consequences of loss of self-esteem and confidence, guilt, or the continuing stresses associated with changes in financial status, employment or interpersonal relationships. The most successful psychotherapeutic strategies are time-limited and goal-directed and have utilized cognitive behavioral techniques to examine and change the erroneous self-perceptions and self-defeating actions that characterize the depressive state, or have focused on interpersonal interactions to identify supportive relationships and potential conflicts or deficits that increase risk for relapse [Pampallona, et al.:]. In the case of bipolar disorders, there is a particular benefit from a psychoeducational approach that educates the patient and family members about the course of illness and the need for prophylactic drug treatment and regularity in lifestyle to decrease risk of recurrence. Different treatment strategies and nontraditional interventions are often required to address the different phases of illness, i.e. prevention of onset, acute treatment, and maintenance, and to reach high-risk groups who traditionally underutilize services (low-income, uninsured and recent immigrant populations).

The pharmacological treatment of mood disorders emerged out of serendipity, but scientific evidence for efficacy and effectiveness is considerable none the less. Approximately 70–80% of patients with depression will respond beneficially to an initial or secondary antidepressant agent and a significant percentage of the remainder to a combination of medications [Bauer, et al.:]. Electroconvulsive treatment is even more predictably effective and is particularly useful in depressive syndromes with psychotic features. The monoamine theory of depression developed from the discovery that all clinically effective antidepressant drugs amplified either noradrenaline (norepinephrine) or serotonin function in the central nervous system, through either blockade of reuptake of the neurotransmitter from the synaptic cleft or inhibition of a relevant metabolic enzyme, monoamine oxidase. The first tricyclic antidepressant drugs (e.g. imipramine and amitriptyline) had potent effects on both noradrenaline and serotonin transporters and, in addition, a significant affinity for muscarinic cholinergic, α-adrenergic and histaminergic receptors. Accordingly, clinical benefit was often mitigated by the induction of side-effects such as dry mouth and constipation, hypotension and sedation. The newer generation of antidepressant agents, the selective serotonin reuptake inhibitors (SSRIs), the noradrenergic and serotonergic reuptake inhibitors (SNaRIs) and others, have been developed to exert a more specific primary action, without the side-effects associated with nonspecific receptor blockade. No available agent, however, is without limitations, and complaints of gastrointestinal upset, sexual dysfunction, agitation, sleep impairment and weight gain remain to a varying degree with each drug. All currently available antidepressant agents are equally effective across a cohort of patients, but individual response patterns may be unique. Selection of the most appropriate drug for a given individual depends on a review of family and personal history of drug response or intolerance, an evaluation of co-morbid medical factors or drug treatments that might make usage of a specific drug unwise or a drug interaction likely, and identification of individual preference as to which side-effects are likely to be most aversive in nature. A therapeutic response most typically occurs in 4–6 weeks, with objective changes in symptoms occurring before subjective relief is felt. Many patients with depression require indefinite treatment with antidepressant medications. Current guidelines for this recommendation include a history of three or more depressive episodes, or two episodes in which one has had psychotic or suicidal features or in which there has been only partial recovery. Treatment of any episode should extend for 6–9 months; the relapse rate for placebo is double that of active drug in the first year.

The mechanism of action of antidepressant drugs remains obscure [Schloss and Henn]. Although effects on reuptake occur within hours, therapeutic effects are delayed for some weeks. Accordingly, attention has focused on adaptations in intracellular signal transduction pathways as more relevant to therapeutic effect than changes in synthesis, reuptake, metabolism or receptor binding. Chronic administration of antidepressants upregulates cyclic adenosine monophosphate (cAMP) activity and leads to increased expression of the cAMP response element-binding protein (CREB). One of the target genes regulated by CREB produces a neurotrophin, brain-derived neurotrophic factor (BDNF), that is markedly decreased by environmental stress, particularly in the hippocampus, an area known to be sensitive to stress-induced insult and relevant to a number of the neuroanatomical, hormonal and clinical abnormalities observed in depressive disorders. Antidepressants may thus exert their effects through neural plasticity and changes in morphology and connectivity [Duman, ref.]. Other current hypotheses of drug action have addressed the transcriptional control of serotonin receptor genes and converging effects of antidepressants on neurosteroid regulation.

The treatment of bipolar disorder differs from that of unipolar disorder in that the preferred treatment for either the manic or depressive phases of the disorder is a mood stabilizer—lithium carbonate, sodium valproate or oxcarbazepine. An antidepressant is sometimes required when the depressive episode is particularly severe but generally should not be continued as a maintenance treatment because of the risks of precipitation of mania and acceleration of cycle frequency. Behavioral control of the acute manifestations of mania sometimes requires short-term adjunctive use of a benzodiazepine or an atypical neuroleptic agent, as mood stabilizers usually require several weeks to achieve full benefit. Lithium carbonate is the preferred first-line agent when mania is characterized by an elevated or expansive mood, but when the mood is irritable or dysphoric, or when rapid cycling or co-morbid substance abuse are present, valproate is more likely to be beneficial. Current treatment guidelines also indicate superior efficacy for lithium and, more recently, lamotrigine in the treatment of bipolar depressive episodes, in bipolar II disorder, and in the prophylactic treatment of both the manic and depressive phases of the syndrome [Keck, et al.:]. One of the most important clinical features of continued lithium maintenance, once acute benefit is achieved, is the dramatic effect it has on decreasing suicide risk over time. In cases of treatment resistance, combinations of mood stabilizers may be more effective than a single drug. In addition to the drugs mentioned, a variety of novel anticonvulsant agents, such as topiramate, zonisamide and tiagabine may possess mood-stabilizing properties, although definitive scientific trials are still lacking. Conflicting evidence for efficacy also exists for calcium channel blockers and for high doses of thyroid hormone.

The mechanism of action of mood stabilizers is unknown. As a simple cation, lithium influences nearly every neurotransmitter in some fashion. Converging effects on signal transduction systems led to the inositol depletion hypothesis, which was based on the finding that lithium was a potent inhibitor of inositol monophosphatase and thus could variably regulate cell signaling, depending on the initial state of activity. Competing hypotheses have focused on lithium's regulation of G protein function and on its decrease of glutamatergic activity over time. Additional effects on neuroplasticity have been uncovered, including modulation of glycogen synthase kinase-3 β, cyclic AMP-dependent kinase, protein kinase C and upregulation of the anti-apoptotic protein bcl-2. The multiplicity of actions makes it likely that more than one site is involved in lithium's ability to balance alternating fluctuations in neuronal activity.

Ongoing clinical investigations of agents that may exert antidepressant effect through novel mechanisms include compounds that block substance P, glutamate or corticosteroid receptors or that antagonize CRF [Manji, et al.:].

In addition to traditional prescription medications, several herbal and nutritional preparations appear to have at least modest antidepressant benefit. St John's wort, S-adenosylmethionine (SAM-E) and DHEA (dehydroepiandrosterone) have thus far received the greatest scientific attention.

Nonpharmacological treatments of depression include electroconvulsive therapy (ECT), transcranial magnetic stimulation, phototherapy, sleep deprivation and vagus nerve stimulation. The therapeutic efficacy of ECT is indisputable, but the treatment is underutilized, even though current practice carries minimal risk of side-effects or morbidity. There are few contraindications to its use and benefit is usually observed after a course of 6–8 sessions over a period of two weeks; the most common side-effects are transient short-term memory deficits and confusion. Transcranial magnetic stimulation is a recent procedure that is well tolerated and may have mood-enhancing effects in the absence of seizure induction. Phototherapy involves exposure to high-intensity, full-spectrum light, most commonly in the morning hours. It has demonstrable efficacy in depressions that have a seasonal component and is thought to ameliorate a depression related delay in circadian rhythms. Sleep deprivation is the only therapeutic intervention that has antidepressant benefit within 24 h, although the response is usually transient and occurs in only half of subjects. Vagus nerve stimulation is a technique that has proved effective in management of refractory partial complex seizures and, in open trials thus far, has resulted in significant mood improvement in treatment-resistant cases of depression. The mechanism of action is not known, and feasibility is limited by the need for surgical implantation.

Additional Reading

Aubry, Jean-Michel, N. Schaad, and F. Ferrero: *Pharmacotherapy of Bipolar Disorders*, John Wiley & Sons, Inc., Hoboken, NJ, 2007.

Bauer, M., P.C. Whybrow, J. Angst, et al.: "World Federation of Societies of Biological Psychiatry (WFSBP) Guidelines for Biological Treatment of Unipolar Depressive Disorders, Part 1: Acute and Continuation Treatment of Major Depressive Disorder," *World Journal of Biological Psychiatry*, **3**, 5–43 (2002).

Bauer, M., C. Whybrow, J. Angst, et al.: "World Federation of Societies of Biological Psychiatry (WFSBP) Guidelines for Biological Treatment of Unipolar Depressive Disorders, Part 2: Maintenance Treatment of Major Depressive Disorder and Treatment of Chronic Depressive Disorders and Subthreshold Depressions," *World Journal of Biological Psychiatry*, **3**, 69–86 (2002).

Belmaker, R.H.: "Bipolar Disorder," *New England Journal of Medicine*, **5**, 476–486 (2004).

Bogousslavsky, J., and J.L. Cummings: *Behavior and Mood Disorders in Focal Brain Lesions*, Cambridge University Press, Cambridge, UK, 2000.

Caspi, A., K. Sugden, T.E. Moffitt, et al.: "Influence of Life Stress on Depression: Moderation by a Polymorphism in the 5-HTT Gene," *Science*, **301**, 386–389 (2003).

Charney, D.S., H.K. Manji: "Life Stress, Genes, and Depression: Multiple Pathways Lead to Increased Risk and New Opportunities for Intervention," *Science's STKE*, **225**, re5: [DOI: 10.1126/stke.2252004re5] (2004).

Coyle, J.T., and R.S. Duman: "Finding the Intracellular Signaling Pathways Affected by Mood Disorder Treatments," *Neuron*, **38**, 157–160 (2003).

Daley, D.C., and G.A. Marlatt: *Addiction and Mood Disorders: A Guide for Clients and Families*, Oxford University Press, New York, NY, 2006.

Duman, R.S.: "Role of Neurotrophic Factors in the Etiology and Treatment of Mood Disorder," *Neuromolecular Medicine*, **1**, 11–25 (2004).

Gilbert, P., and S. Allan: "The Role of Defeat and Entrapment (arrested flight) in Depression: An Exploration of an Evolutionary View," *Psychological Medicine*, **28**, 585–598 (1998).

Keck, P.E. Jr. , E.B. Nelson, and S.L. McElroy: "Advances in the Pharmacologic Treatment of Bipolar Depression," *Biological Psychiatry*, **53**, 671–679 (2003).

Kessler, R.C., P. Berglund, O. Demler, et al.: "The Epidemiology of Major Depressive Disorder: Results from the National Comorbidity Survey Replication (NCSR)," *JAMA*, **289**, 3095–3105 (2003).

Kleinman, A.: "Culture and Depression," *New England Journal of Medicine*, **10**, 951–953 (2004).

Lett, H.S., J.A. Blumenthal, M.A. Babyak, et al.: "Depression as a Risk Factor for Coronary Artery Disease: Evidence, Mechanism and Treatment," *Psychosomatic Medicine*, **3**, 305–315 (2004).

Manji, H.K., J.A. Quiroz, J. Sporn, et al.: "Enhancing Neuronal Plasticity and Cellular Resilience to Develop Novel, Improved Therapeutics for Difficult-to-Treat Depression," *Biological Psychiatry*, **53**, 707–742 (2003).

Mathews, C.A., V.I. Reus: "Genetic Linkage in Bipolar Disorder," *CNS Spectrums*, **12**, 891–904 (2003).

Mondimore, F.M.: *Depression, the Mood Disease*, 3rd Edition, Johns Hopkins University Press, Baltimore, MD, 2006.

Nestler, E.J., M. Barrot, R.J. DiLeone, et al.: "Neurobiology of Depression," *Neuron*, **34**, 13–25, (2002).

O'Brien, S.M., L.V. Scott, T.G. Dinan, et al.: "Cytokines: Abnormalities in Major Depression and Implications for Pharmacological Treatment," *Human Psychopharmacology*, **6**, 397–403 (2004).

Pampallona, S., P. Bollini, G. Tibaldi, et al.: "Combined Pharmacotherapy and Psychological Treatment for Depression: A Systematic Review," *Archives of General Psychiatry*, **7**, 714–719 (2004).

Schloss, P., and F.A. Henn: "New Insights into the Mechanisms of Antidepressant Therapy," *Pharmacology & Therapeutics*, **1**, 47–60 (2004).

Sheline, Y.I., B.L. Mittler, and M.A. Mintun: "The Hippocampus and Depression," *European Psychiatry*, **3**, 300–305 (2002).

Stewart, W.F., J.A. Ricci, E. Chee, et al.: "Cost of Lost Productive Work Time among US Workers with Depression," *JAMA*, **289**, 3135–3144 (2003).

Sulser, F.: "The Role of CREB and Other Transcription Factors in the Pharmacotherapy and Etiology of Depression," *Annals of Medicine*, **34**, 348–356 (2002).

VICTOR I. REUS, University of California, San Francisco, CA

MOON (Earth's). The Moon the natural satellite of Earth, has positively affected our development in many profound ways. Its orbital presence helps stabilize Earth's axial precession and thus, prevents the alternating extremes of climate that some planets, such as Mars, experience. This equitable climate makes Earth a habitable world. The ocean tides induced by the Moon probably permitted vertebrate life to emerge from the sea more than 300 million years ago, leading to the development of land fauna and ultimately, to ourselves. Thus in a very real sense, we are here because the Moon exists.

Now, the Moon is exerting its beneficial influence on us once again by its existence. The Moon is the first milepost in humanity's movement into the solar system. We must pass by the Moon's orbit to go anywhere else in space. Thus, nature has provided us with a natural "way station," a place to learn how to live and work in space, refuel and refresh our spacecraft. In addition to these benefits, the Moon also happens to be a rather interesting place. Its surface contains a record of the important events that occurred in the early history of Earth. Moreover, it is an excellent platform to observe the Universe around us. In all of these ways, the Moon is an important part of our movement into space.

It has no formal name other than "The Moon", although in English it is occasionally called **Luna** (Latin for *moon*), or **Selene** (Greek for *moon*), to distinguish it from the generic "moon" (natural satellites of other planets are also called *moons*). Its symbol is a crescent. The terms *lunar*, *selene/seleno-*, and *cynthion* (from the Lunar deities Selene and Cynthia) refer to the Moon (aposelene, selenocentric, pericynthion, etc.).

The Moon is quite large in relation to the planet it orbits, about 1% of the mass of Earth and about one-fifth its radius. The diameter of the moon is 3476.6 ± 2.2 kilometers (2160 ± 1.4 miles). Other statistics of the moon are given in the entry on **Planets and the Solar System**. In surface area, the Moon is roughly the size of the continent of Africa, about 38 million square kilometers. Because the tenuous atmosphere of its surface is a near-perfect vacuum, no weather affects its terrain, and the sky is perpetually black. Stars are visible from the surface during daytime but are difficult to see because the glare reflected from the surface dilates your pupils. At high noon, the surface temperature can be more than $100\,°C$ ($212\,°F$) and at midnight, as low as $-150\,°C$ ($-238\,°F$). The lunar day (the time it takes to rotate once on its spin axis) is about 29 Earth days or 708 hours, and daylight hours on the Moon (sunrise to sunset) last almost 2 weeks. The Moon is famous for its low gravity, about one-sixth of Earth's. Thus, an astronaut who weighs 200 pounds on Earth weighs only 34 pounds on the Moon.

The first manmade object to land on the Moon was Luna 2 in 1959, the first photographs of the otherwise occluded far side of the Moon were made by Luna 3 that same year, and the first people to land on the Moon came aboard *Apollo 11* in 1969.

Data from the *Apollo* program indicate that the moon was formed 4.6 billion years ago and was strongly heated during its first few hundred million years, causing melting and formation of igneous rocks in its outer layers. Much of the outer crust, still preserved in the cratered uplands, formed from a low-density type of rock known as anorthosite or anorthositic gabbro. Large impact basins and other depressions are believed to have been flooded about 3.8 to 3.2 billion years ago by basaltic lavas generated beneath the surface, forming the dark mare plains that cover roughly 18% of the moon. Few geological processes appear to have occurred since that time, as there is virtually no atmosphere or water to permit erosive processes. The modest seismic activity of the moon tends to confirm the essential absence of geologic processes as found on earth. The rate of meteorite impact on the moon is low today (comparable to earth), but heavy cratering indicates that the rate was very high in the first half-billion years of lunar history. See Fig. 1.

Orbit and Phases of the Moon

The Moon moves in an elliptical path around Earth and completes its circuit once every 29 days. This time is equal to the amount of time it takes for

Fig. 1. The near-full moon as photographed shortly after trans-earth injection (TEI) by the Fairchild Metric Camera mounted in the Scientific Instrument Module (SIM) bay of the Apollo 16 Service Module. This view is looking generally westerly toward the large, circular Mare Crisium (Sea of Crises) on the horizon. Immediately east of Mare Crisium is Mare Marginis (Border Sea): and the more circular mare area south of Mare Marginis is Mare Smythii (Smyth's Sea). Most of the lunar area in this photo is on the far side of the moon. The most conspicuous crater, the smooth-floored one northeast of Mare Marginis, is Lomonosov. Neper is the large crater between Mare Marginis and Mare Smythii. The larger crater, Tsiolkovsky, is barely visible on the horizon at the southeastern edge of the Moon. (*National Aeronautics and Space Administration.*)

the Moon to rotate once on its axis (the lunar day). In consequence, the Moon shows the same hemisphere (called the *near side*) to Earth at all times. Conversely, one hemisphere is forever turned away from us (the *far side*).

The elliptical orbit of the Moon results in a variable distance between Earth and Moon. At perigee (when the Moon is closest to Earth), the Moon is a mere 356,410 km (221,463 miles) away; at apogee (the farthest position), it is 406,697 km (252,710 miles) away. This is different enough so that the apparent size of the Moon in the sky varies; its average apparent size is a little smaller than that of a dime held at arm's length. In works of art, a huge lunar disk looming above the horizon is often depicted, but such an appearance is an illusion. A Moon near the horizon can be compared in size with distant objects on the horizon, such as trees, making it seem large, and a Moon near zenith (overhead) cannot be compared easily with earthly objects, and hence, seems smaller. (See Fig. 2)

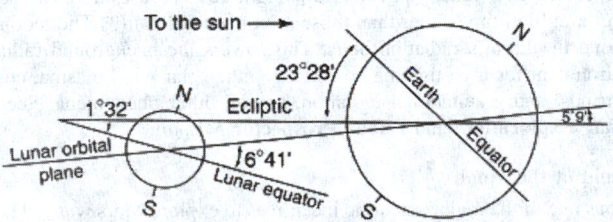

Fig. 2. Orbital planes and spin axes of Earth and Moon. Although Earth's axis is tilted 23° from the ecliptic, the Moon's is nearly perpendicular to it, resulting in grazing solar illumination near the poles.

The plane of the Moon's orbit lies neither in the equatorial plane of Earth nor in ecliptic plane, in which nearly all the planets orbit the Sun (Fig. 3). This relation poses some constraints on models of lunar origin. The spin axis of the Moon is nearly perpendicular to the ecliptic plane;

it has an inclination of about 1.5° from the vertical. This simple fact has some really significant consequences. Because its spin axis is vertical, the Moon experiences no "seasons", as does Earth, whose inclination is about 24°. So, as the Moon rotates on its axis, an observer at the pole would see the Sun hovering close to the horizon. A large peak near the pole might be in permanent sunlight, and a crater floor could exist in permanent shadow. In fact, we now know that such areas exist, particularly near the South Pole. The existence of such regions has important implications for a return to the Moon.

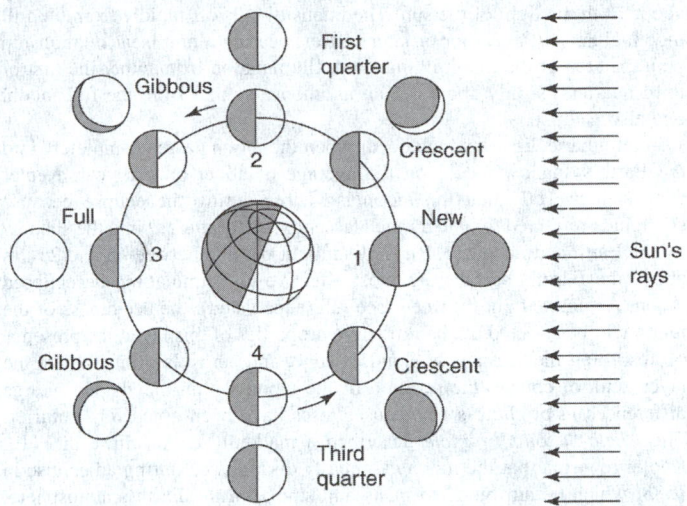

Fig. 3. Phases of the moon.

The moon, in orbiting around the earth, exhibits a strikingly interesting phase change. Since the moon shines by reflected light from the sun (and a very small amount of earth light), the phases are easily explained, as aided by Fig. 3. The moon is called *new moon* (position 1) when it is positioned between the earth and the sun, at which time the light of the sun does not reach the side of the moon that faces the earth. At full moon (position 3), the side of the moon facing the earth is completely flooded with sunlight. This is the position where the moon has half-completed a revolution about the earth. The various other phases are caused by this revolution about the earth, with a complete revolution occurring approximately once every $29\frac{1}{2}$ days. At position 2, essentially one-half of the surface is illuminated, although at that position, the phase is termed the *first quarter* (one-fourth of a revolution completed). Between positions 1 and 2, the lighted portion of the moon takes on a crescent shape and this is known as the *crescent phase*. Between positions 2 and 3, more than half the moon surface is illuminated and the term *gibbous phase* is used. At position 3, the moon is fully illuminated (*full moon*). Between positions 3 and 4, over half of the surface is illuminated and the term gibbous phase also applies here. At position 4, half of the surface is illuminated and this phase is termed the *third quarter* (three-quarters of a revolution completed). This phase is also commonly called the *last quarter*. Between position 4 and position 1, the crescent phase is resumed until a full return to position 1 occurs. If the plane of the moon's orbit were not inclined to the plane of the ecliptic (average value, 5°9′), there would be an eclipse of the sun at each new moon, and an eclipse of the moon at each full moon.

Also, because of the inclination of the moon's orbit, there is seasonal variation in the path in the sky followed by the moon. During wintertime in the Northern hemisphere, the moon will be higher in the sky than during the summertime. Soon after the new moon, a thin crescent of the moon will be brightly visible and, on a clear night, the remainder of the moon's surface will be noted to be very faintly lighted. This is due to reflection of sunlight from the earth back on to the moon and is called *earth-shine*.

The moon rotates slowly about its axis, which is tipped about 6.5° from the perpendicular to the plane of its orbit. This results in the remarkable circumstance that the period of rotation matches the moon's sidereal period of revolution and is in the same direction. Consequently, the moon always turns the same face toward the earth. The *sidereal month* is defined as the time required by the moon to complete one revolution about the earth with reference to the stars and is 27 days, 7 hours, 43 minutes, 11.5 seconds. The interval between new moon and new moon or full moon and full moon is called the *synodic month* and is 29 days, 12 hours, 44 minutes, and 2.8 seconds long.

The intensity of moonlight is quite weak even though the light of the full moon makes it possible for a person to move about on foot quite well on a clear night. The light from the full moon has an intensity of about 1/465,000th of sunlight on a clear day. It has been estimated that were the entire sky packed with full moons, limb to limb, they would give only about 1/6th the light of the sun. The intensity falls off rapidly from the full moon, either in the earlier or later phases, bearing a nonlinear relationship with the area of the moon illuminated. Illumination from either the first or third quarters is only about 1/10th as intense as that from the full moon. See also **Eclipse**.

Total lunar eclipses, which occur when the moon passes completely into the Earth's shadow, occur on the average of about once a year. Kepler suggested in 1604 that the moon is visible during the eclipse because sunlight is refracted (and to a much lesser extent, scattered) into the shadow by the Earth's atmosphere. Rays of sunlight passing between 5 and 25 km above the Earth's surface are most effective at illuminating the eclipsed moon. As pointed out by Keen (see reference listed), the brightness of the moon will be affected by the refractive properties of this layer, the presence of absorbing media such as clouds, ozone, and aerosols in and above the layer, and, of course, the position of the moon within the shadow. Large differences in brightness between eclipses have been noted for centuries. The *Anglo-Saxon Chronicle* described a markedly dark eclipse in 1110. Kepler reported that the moon essentially disappeared during an eclipse in 1588, which he attributed to mists and smoke in the Earth's atmosphere. It was not until 1883, however, that the relative brightness of the eclipsed moon was related to volcanic aerosols. During that year, the eruption of Krakatoa occurred. Researcher Keen and others in going back into historical records have been able to relate a number of "dark" eclipses with major eruptions. In the Keen report, details of these calculations are given. Observed brightnesses of 21 lunar eclipses during the 1960–1982 period were compared with theoretical calculations based on refraction by an aerosol-free atmosphere. Results indicate the global aerosol loading from the 1982 eruption of El ChicÓn is similar in magnitude to that from the 1963 Agung eruption.

Because of the tipping of the moon's axis, it is possible to see a bit past the one pole for half of the month and, conversely, to see past the other pole during the other half. The maximum that can be seen is about 6.5° of latitude on the moon. This is termed *libration in latitude*.

The gravitational attraction of the earth and sun cause the moon's orbit to change. The greatest influence is exerted by the sun and the second greatest by the earth's tidal bulge.

Laser altimeter data obtained from the *Apollo 15* and *16* missions provided two elevation cross sections of the moon separated by 35° of latitude. Prior to this technique, lunar topography was determined from earth by photography and radar techniques. Shadows were measured and stereoscopic effects were accounted for to determine surface elevations. Errors were large over distances of hundreds of kilometers. Results were very poor near the limbs. Numerous independent measurements of lunar elevations were made, including the elevations of Ranger impact point, Surveyor landing sites, *Apollo* survey sites, and *Saturn* booster impact points, in addition to elevations determined from velocity-height data of the *Lunar Orbiter*, and from earth-based radar. These results were obtained from front-side observations at random points with the exception of ten far-side landmark points obtained with the *Apollo* sextant. *Apollo 16* data put a 2-kilometer bulge toward the earth. However, combined data from this and other prior data are best fit by a sphere of radius 1,737 kilometers. The offset of the center of gravity from the optical center is about 2 kilometers toward the earth and 1 kilometer eastward.

Lunar Exploration

Probably the first major and highly successful step toward direct exploration of the moon was the soft lunar landing of the unmanned NASA spacecraft *Surveyor 1* on June 2, 1966. Prior to the findings of the first *Surveyor* spacecraft lander, scientists were uncertain about the moon's surface and its ability to support a landing craft. The impressive touchdown of *Surveyor* demonstrated that the lunar crust was strong and hard and that future spacecraft would not settle into deep layers of dust. Subsequent *Surveyor* missions collected important information preparatory to the launching of manned spacecraft. The next major aspect of the program was that of a manned lunar orbiter. During a two-day period (December 24–25, 1968), two U.S. astronauts made ten orbits around the moon and were the first persons to see the back side of the moon. On July 20, 1969, two U.S. astronauts (*Apollo 11* mission) set foot on the moon (Sea of Tranquility region). Five more manned missions, thus making a total of six, followed: November 18, 1969 (Ocean of Storms region); February 5, 1971 (Fra Mauro region); July 31–August 2, 1971 (Hadley Apennine region); April 20–23, 1972 (Descartes region): and December 11–14, 1972 (Taurus-Littrow region). Although the Soviet soft lunar lander *Luna 9* reached the moon on February 3, 1966, a few months prior to the landing of *Surveyor 1*, Russia did not follow through with a scientific program of lunar exploration comparable to the *Apollo* program. Costs of the U.S. lunar programs were approximately $25 billion. More than 2000 samples of lunar rocks and soil, weighing in the aggregate almost 400 kilograms (882 pounds), were returned to earth. These lunar materials were placed in storage chambers at the Manned Spacecraft Center near Houston, Texas under precisely controlled conditions. Samples for research by various investigators throughout the world were released. Elaborate care is taken not to lose track of the rocks' original orientations and shapes. The working assumption of persons in charge of the lunar samples is that they are essentially irreplaceable and that even more intelligent use of them will be made as the related sciences generally progress, but only if they are kept in a pristine state.

Lunar history turned out to be a much more subtle affair than most knowledgeable people had expected. As observed by Gillette (1972), the notion that the moon was a dead body made up of the primordial material from which the terrestrial planets coalesced now seems as quaint as the Ptolemaic idea that it was all shining crystal. It appears likely that if any bits of primordial crust do remain, they will have to be tracked down laboriously in the samples of soil, a task comparable to finding the proverbial needle in a haystack.

This process of image and rock analysis continues at various laboratories throughout the world, partly because these tasks are very demanding and partly because no new exploratory ventures to the moon are currently scheduled.

On numerous occasions researchers in the field have suggested that further data are needed from lunar exploration, in an effort to study the origin of the solar system, to build giant observatories on the far side of the satellite, and to commence further expeditions to Mars and the asteroids from the moon. More ambitious researchers are insisting that we also exploit lunar resources (minerals, etc.), as well as process certain materials under low-gravity conditions. A principal contention is using the moon as a base instead of building an entirely new space station. It has been suggested that using the moon as a platform for space astronomy would have at least three advantages. (1) Radio telescopes on the far side of the moon would be shielded from terrestrial radio emissions, greatly improving observational sensitivity and accuracy. (2) The moon provides a solid, seismically stable, high-vacuum platform for interferometric arrays. (For example, a lunar optical array might resolve astronomical details about a million times finer than those seen from Earth.) (3) The moon lies beyond the Earth's radiation belts. This lowers the background radiation environment for investigating cosmic rays, the solar wind plasma, cosmic neutrinos, and gravitational radiation, among other phenomena. See also **Lunar Exploration**; and **Lunar Prospector Mission**.

Origin of the Moon

Just as one of the major scientific incentives to explore Mars was to learn if any life existed on that planet, a major driving force for the earlier *Apollo* missions to the moon was that of attempting to determine the origin of the moon. Although the knowledge of the moon was greatly enhanced by the *Apollo* missions, the question of lunar origin was answered only very partially. While the moon is a specific situation, other planets have satellites, and it is strongly suspected that other "solar systems" exist and also most likely will be found to have satellites. Thus, the origin of the moon is best not approached as an isolated happening, but still in some unexplained way should relate to other like happenings in the solar system.

See also **Cosmology**. It is interesting to note that while there are similarities there also are inconsistencies with reference to the satellites of the other planets in the solar system. For example, Venus and Mercury have no moons; Mars has two large boulders as its only moons; the satellites of Earth and Pluto are of sufficient size to be considered sister planets; the satellites of Uranus revolve in a plane perpendicular to all others; Neptune's major satellite orbits in "reverse." The latest "giant impact" hypothesis of the moon's origin accommodates, to some degree, this variety.

Few areas of astronomy have nurtured so much controversy over such a long period of time as has been that dealing with the moon's origin. For many years, numerous investigators have researched this topic.[1]

Potpourri of Hypotheses. During the period ranging from the 1800s to the mid-1900s, three so-called classical hypotheses were discussed. During this period, there was marked vacillation among scientists regarding their support of this or that hypothesis. One or more of these classical hypotheses reigned until the more recent (circa 1975) concept that is currently called the *giant impact* was formulated. The three classical hypotheses were:

1. **Binary Planet or "Sister" Hypothesis** Scenario: The moon was formed by condensation from a cloud of material surrounding the Earth. A corollary (proposed by Gilbert); the moon was formed from a ring of small solid particles, the first stage of which produced the lunar craters.

2. **Rotational Fission or "Daughter" Hypothesis** This concept was based on the classically known dynamical instability of incompressible, inviscid bodies in uniform rotation. Scenario: Originally, Earth and moon were one. The outer layers of the proto-Earth were spinning too rapidly to achieve dynamic stability. Because of tidal dissipation, angular momentum was transferred from the Earth's rotation to the orbital motion of the moon. The moon commenced to recede from the Earth as the moon's angular speed (as seem from Earth) diminished. (G.H. Darwin, who conceived the hypothesis, "traced" the process back some 50 million years, at about which time he suggested the moon was only about 6000 miles (9654 km) from the surface of the Earth. At that time, the period of revolution of the moon and the Earth's period of rotation would have been about 5 hours.) The initiation of separation of the moon from the Earth was caused by solar influences, suggesting that the tides raised on the proto-Earth had a peaking time each 5-hour day. Any given place where this peaking occurred could resonate with the free oscillations, thus causing distortions sufficient to disrupt the body. In 1882, Osmond proposed a supplemental concept, suggesting that the scar left by the moon's initial separation did not fully "heal" — hence the Earth's ocean basins are the holes in the Earth left behind. This spawned the popular concept that the moon was "taken" out of the Pacific Ocean basin. This hypothesis was reasonably well received for a number of years, but tended to be replaced by the "capture" hypothesis proposed by Thomas See in the early 1900s. The fission concept was revived for a few years in the 1930s and, even today, probably lurks in the study halls. There are three convincing arguments against the fission hypothesis: (1) Modern calculations indicate that the spinning rate of the proto-Earth would have to be 2 hours or less in order to exceed the limit for dynamic stability. Currently, the achievement of such a high rate is considered improbable. (2) The amount of angular momentum needed for the fission instability would have to be approximately four times that of the present Earth-moon system. Boss explains this critical

problem in his paper (reference listed). (3) Because fission must occur through a dynamic instability, a fission origin of the moon would be apparently impossible unless the proto-Earth was nearly inviscid.

3. **Capture or "Wife" Hypothesis** This concept was first strongly advocated by Thomas See in the early 1900s and long supported by Harold Urey. Scenario: The moon was initially formed in the outer reaches of the solar system (in the vicinity of the orbit of Neptune). As this proto-moon moved through space it lost energy and, in some way, approached Earth. (This concept removed some of the objections to the other hypotheses, in that the capture theory would not necessarily conflict with certain inconsistencies, as previously mentioned, in connection with the satellites of other planets.) Upon reaching the vicinity of Earth, the proto-moon, by a combination of gravitational and other forces, reached a balance and formed its own orbit around the Earth. In a 1949 paper, Urey suggested that the moon represents more nearly the composition of the original dust cloud relative to the nonvolatile elements than does the Earth. Urey also argued that the fact that the moon does not have a shape corresponding to isostatic equilibrium indicates that it was frozen a long time ago. Urey prepared the way for the argument that exploration of the moon could provide information about the early history of the Earth. In lunar rock analysis, the moon's oxygen isotopic composition tends to support the simultaneous formation of the Earth and moon from the same portion of the solar nebular, perhaps while in orbit around each other. As found by researchers, the ratios of all three oxygen isotopes in lunar rocks and in terrestrial rocks are essentially identical. They do differ, however, from those ratios found in meteorites. But, the correlation tends to stop here because numerous other chemical differences between lunar and terrestrial materials exist.

Giant Impact Hypothesis. Proposed by A.G.W. Cameron (Harvard University) and researched by several other scientists who used computer modeling and simulation techniques over the past decade (Benz and Slattery of the Los Alamos Laboratory; Boss, Minzum, Peale, and Wetherill of the Carnegie Institution of Washington; and others), this relatively new hypothesis avoids many of the "hangups" that characterized prior hypotheses. As described in the Boss paper referenced, the scenario is about as follows: The origin of the moon is considered within the theory of formation of the terrestrial planets by accumulation of planetesimals. That theory predicts the occurrence of giant impacts, suggesting that the moon formed after a roughly Mars-size body impacted on the proto-Earth. The impact blasted portions of the proto-Earth and the impacting body into geocentric orbit, forming a prelunar disk from which the moon later accreted. Although other mechanisms for formation of the moon appear to be dynamically implausible, fundamental questions must be answered before a giant impact origin can be considered both possible and probable. The general theory of terrestrial planet formation provides the framework in which to consider lunar formation. The theory could change in the future, in which case certain revisions may have to be made in the giant impact hypothesis. A key element in the giant impact hypothesis of lunar origin is the mass distribution of the smaller bodies, called *planetesimals*, from which the terrestrial planets formed. Because of their physical and orbital similarities, Mercury, Venus, Earth-moon, and Mars each may have been formed by the same fundamental process. Much of the information from research on terrestrial planet formation compiled to date is based on the assumption that formation took place by accumulation of dust grains rather than by the gravitational collapse of gas in the solar nebula and subsequent removal of a gaseous envelope. See also **Cosmology**.

Boss estimates that in order to deposit the angular momentum of the Earth-moon system, a Mars-size body would have had to strike Earth nearly tangentially with a relative velocity of about 10 km sec^{-1}. This is consistent with the relative velocities of giant impacts now envisioned to have occurred in the late phases of accumulation.

Also, as observed by others, in the computer simulation, the collision shatters the impactor, and the rocky debris of the mantle subsequently become separated from the iron core. The gravitational fields of the three bodies — Earth, impactor core, and impactor mantle — through interacting forces keep the mantle debris in orbit. Within a very short period (hours), the gravitation of the debris itself draws it together into a roughly spherical moon. Thus, according to the giant impact scenario, the moon consists mainly of material from the impactor's mantle minus the volatile elements,

---

[1] Several well known scientists devoted much time and effort to the problem of the moon's origin and these included, in chronological order. Edouard Roche (1873); George Howard Darwin, son of Charles Darwin (1878); Osmond Fisher (1882); Grove Karl Gilbert (1890s); Thomas See (1909); H. Jeffreys (1930); Harold Urey (1951); J.A. O'Keefe (1963); T.C. Chamberlin (1963); J.F. Simpson (1964); D.U. Wise (1969); A.E. Ringwood (1970); A.G.W. Cameron (1973); Donald Davis (1975); W. Hartman (1975); A.P. Boss (1986); and S.J. Peale (1986); among many others. Dates given are the periods of early formulation or active support for a given hypothesis. Listings of major papers given by these and other scientists on the subject are listed at the ends of the articles by Brush (1982) and Boss (1986) cited for **Additional Reading** at the end of this encyclopedia article.

which are vaporized by heat from the impact. The impactor core is sacrificed and it crashes back into the Earth. Because it is heavy, it soon settles to the Earth's own iron core. The estimated date for the primeval cataclysm is about 4.6 billion years ago.

Lunar Topography

Craters. Very numerous on the moon are circular depressed structures (craters) with raised rims believed to be caused mostly by meteorite impacts, although some may be of volcanic origin. At one time, the largest craters were believed to be some 100 to 200 kilometers (62 to 124 miles) in diameter, but improved observations by earth-based instruments over the last few decades and the findings of the lunar orbiter and landing programs indicated that large mare-filled basins up to 1000 kilometers (620 miles) in diameter are also impact craters. Small craters are exceedingly numerous on the moon. The smallest are microscopic pits that dot the surfaces of exposed rocks. Ejected material blasted out of craters form bright rays (or radial streaks of bright material) and secondary impact craters (formed by ejected debris hitting the surface) approach the size of the parent primary crater in some cases.

Copernicus is a bright-rayed impact crater 95 km (60 miles) in diameter. It was recognized as stratigraphically important and features associated with it were used to define the most recent major time period of lunar history. Copernicus contains several prominent central peaks that rise about 800 meters above the floor of the crater; the largest is 12×5 km (3×7.5 miles) at the base. Images of these peaks taken from lunar orbit showed them to be massive and blocky. They were initially mapped as deep-seated bedrock. Subsequently, as observed by Pieters (Brown University), studies of crater dynamics indicated that the material of the central peaks in craters the size of Copernicus had likely been uplifted from an original depth of about 10 km ($6\frac{1}{4}$ miles) by dynamic rebound of local material in the terminal stages of the impact event.

The system for naming craters is generally accredited to Riccioli, (an Italian astronomer, who in 1651 proposed that the craters be named after famous scientists and other persons of prominence). Some craters on the moon so named include Copernicus, Archimedes, Newton, Plato, etc. Because the moon can be observed in considerable detail by earth-based instruments, detailed maps of crater locations as well as of other lunar features have existed for decades and were very helpful in the very early phases of planning the lunar exploratory missions.

Lunar Maria. Early astronomers thought that the lunar lowlands, which are dark, flat plains, were liquid bodies and thus the word sea (Latin = *mare*) was adopted to identify these features. The lunar maria are broad depressions filled with dark volcanic basalts, rich in iron and titanium. The maria were sampled at five sites during the *Apollo* program. The basalt ages fall into a surprisingly narrow time range, estimated between 3.15 and 3.85×10^9 years. It is reasoned that the mare basalts may have been generated by radioactive heating and partial melting in an iron-rich, plagioclase-poor region in the interior of the moon. Thus, they were not a product of the primary differentiation that gave rise to the lunar crust. That the basins were filled by a succession of flows is indicated by numerous overlapping flow fronts. Intervals between them indicate periods of cooling and solidification. Small dark steps or ledges are sometimes visible at the base of the highlands where the mare basalts lap against highland masses. This is particularly apparent on the western near sides. These have been called a form of "bath-tub ring," remains of the highest level attained by the lava as it was emplaced. The Planning Team report indicates that the most recent flows emanated from sources at the margins of the maria; they followed definite, but subdued axial channels, which may be collapsed lava tubes. Some of the flow units are as much as 350 kilometers in length. It is believed that older flows may have extended up to 1,200 kilometers north across the center of the basin (Mare Serenitatis). Prominent wrinkle ridges are found in the southern Imbrium region. These are believed to be compressional structures superimposed before complete solidification of the lavas.

It appears that the thicknesses of basalt are largest in the centers of maria and taper toward the margins, a condition possibly attributable to load-induced subsidence. In many small areas on the lunar nearside, usually at the margins of maria, a very dark material appears to have blanketed craters and other structures. Because of their darkness in comparison with the rest of the moon, they are apparent to the untrained eye when captured on a lunar photograph of good quality. Earlier it was conceived that the dark material comprised a layer of volcanic ash because of its mantling effect on other structures — as well as the fact that not many craters appear to postdate the dark material. This was considered as evidence of relatively recent volcanism on the moon.

Partly because the southeast margin of Mare Serenitatis includes a prominent deposit of dark mantling material, the *Apollo 17* landing site was selected at Taurus-Littrow. It turned out that the valley floor on which the *Challenger* landed is fully blanketed with this dark substance. However, no prominent component of ash-like material was found in these soil samples. It was learned that the blackness of the Taurus-Littrow soils derives from a high titanium and iron content in the underlying bedrock. The latter is a mare-type basalt similar chemically to that collected by *Apollo 11* in Mare Tranquillitatis. Apparently, the iron and titanium cause the bedrock and associated soil derived from it by comminution to have a high content of the black mineral ilmenite ($FeTiO_3$). They deeply color glasses generated by impacts on the local soil. Since the local basalts crystallized 3.8×10^9 years ago, the dark mantling material at the one place where it was sampled is attributed to early volcanic activity, i.e., in terms of eruptions that may have occurred at the beginning of the epoch of basin-filling volcanism, rather than later volcanism as has been anticipated.

The most unusual material discovered by the *Apollo 17* astronauts was an "orange soil," which constituted a 25-cm layer on the rim of Shorty Crater (110 m in diam. $\times$ 10 m deep). This was one of the few colorful spots observed on the moon. The orange color is due to a relatively high abundance of trivalent titanium. The soil also contains an excess of fission xenon isotopes attributable to neutron-induced fission of ^{235}U. It is estimated that these volcanic glasses were formed 3.63×10^9 years ago. Examination indicates that the glasses resided on the lunar surface for about 38 million years before they were deeply buried. The glass spherules were reexcavated by the impact that formed Shorty Crater some 17 million years ago and remained undisturbed until they were collected. As reported by Eugster and colleagues (Physikalisches Institut, University of Bern), the glass droplets from the bottom of the core were probably produced by lava fountaining a few tens of millions of years after the end of the lava flooding of Mare Serenitatis. They were then exposed to cosmic rays for about 38 million years.

Lunar Highlands. Many rocks obtained from the lunar highlands (See Fig. 4.) are feldspathic breccias.[2] Six or more categories of highland breccia rocks have been identified. Differences in these categories include: (1) variation in the composition of the fragments or clasts which they contain; (2) distribution of sizes; (3) the content of glass; and (4) the degree of thermal sintering or recrystallization that occurred since aggregation. However, some truly igneous rocks were also returned from the highlands. Such rocks may have resulted from melting in the interior of the moon, or from impact melting.

Lunar Crust and Rocks. It is generally believed that the moon's crust, estimated thickness in tens of kilometers, was produced by magmatic formation. This would have taken place when extensive melting occurred in the outermost layers of the moon (hundreds of kilometers deep). To date, the source of this heat has not been adequately explained. Concepts proposed include: (1) accretional heating during rapid formation of the moon; (2) electrical heating that would have resulted from an intense solar wind; or (3) tidal heating at the time the moon was in close proximity of the earth.

On a chemical basis, three broad classes of terra rocks have been recognized, as listed in Table 1. In elemental plots, these categories do not display well-separated clusters. Rather, they tend to grade into one another. No simple correlation of chemical class with petrographic texture has been formulated.

Considerations of phase equilibria and trace element partitioning make it most likely that the moon first separated a plagioclase-rich crust, presumably by upward flotation of plagioclase as it crystallized from an early surface magma system of monumental proportions, and that

---

[2] Feldspar is a group of minerals of the general formula, $Mal(Al, Si)_3O_8$, where $M = K$, Na, Ca, etc.: breccia is a coarse-grained clastic rock composed of particles greater than 2 millimeters in diameter, angular, and broken rock fragments that are cemented together in a fine-grained matrix.

Fig. 4.   A near vertical view of the Hadley-Apennine area, as photographed by the Fairchild metric mapping camera mounted in the Scientific Instrument Module (SIM) bay of the Apollo 17 Service Module in lunar orbit. Hadley-Rille meanders through the center of the picture. The Apennine Mountains dominate the picture. The Lunar Module touchdown point is on the right (east) side of the "chicken beak" in the lower center of the photograph. The craters Aristillus (top) and Autolycus are near the horizon at upper left. The small, distinct crater next to the rille is Hadley C. (*National Aeronautics and Space Administration*.)

subsequently the KREEP-rich and KREEP-poor noritic rock types were formed by partial melting of a feldspar-rich parent material (presumably the roots of the anorthsitic crust) and erupted onto the surface as lavas. See Table 1.

TABLE 1.   THREE CHEMICAL CLASSES OF TERRA ROCKS — SELECTED OXIDE AND ELEMENTAL ABUNDANCE (PARTS PER MILLION UNLESS OTHERWISE INDICATED)

Constituent	Anorthositic Rocks	KREEP-Poor Norites and Troctolites	Kreep-Rich Norites
Al$_2$O$_3$	>25%	20–25%	15–20%
FeO	0–5%	4–9%	8–10%
MgO	2–8%	8–16%	7–13%
P$_2$O$_5$	0–0.06%	0.1–0.3%	0.3–2%
K$_2$O	0.01–0.2%	0.05–0.1%	0.2–2.0%
Uranium	<0.4	0.4–1.0	2–6
Lanthanum	0.1–4.5	10–30	40–80
Europium	0.6–1.2	1–2	2–3
Europium (anomaly)	Positive	Negative	Negative
Hafnium	<0.01–1.5	4–10	10–30

Source: Lunar Sample Analysis Planning Team.

Note:

1. The use of the terrestrial rock names, anorthosite, norite, and troctolite refer to chemical similarities with no implication of an origin in deep-seated igneous bodies as in the case of the terrestrial counterparts.
2. Anorthositic rocks contain more than 65% plagioclase. Norites and troctolites contain roughly equal amounts of plagioclase and a mafic mineral. The latter is mainly orthopyroxene in the case of norites; olivine in the case of troctolites.
3. KREEP refers to potassium, rare-earth elements, and phosphorus.

Radiometric dating of terra rock samples has yielded interesting data. Ages derived from Rb-Sr internal isochrons and the ^{40}Ar-^{39}Ar technique consistently fall in the range of 3.85 to 4.05 × 10^9 years. The rocks are neither as old nor as scattered in age as was expected. Because it has been believed that the moon is of the same age as the rest of the solar system (about 4.6 × 10^9 years), it was expected that terra rocks, most of which suffered a series of shock brecciations and reheatings, would show a spectrum of apparent ages reaching back to that time. But, because of the consistency of the terra rocks in the 3.9 to 4.0 × 10^9 years age range, it is now believed that some major cataclysm affected the moon at that time, or perhaps had affected it continuously until about 3.9 × 10^9 years ago. Such events would have reset the clocks of the rocks throughout that part of the crust which to date has been accessible to exploration. It is visualized that the cataclysm may have been a highly intense and violent epoch of meteroid bombardment, possibly ending in the giant impacts that caused excavation of the Imbrium and Orientale basins. It is possible that the Orientale basin may have been the source of much of the highland material gathered at the *Apollo 16* and other landing sites. See also **Mineralogy**.

Previously, it had been apparent that debris from Orientale, the youngest of the major ringed basins, may have projected over the entire surface of the moon. However, it was felt that little more than a veneer of Orientale material would have been deposited at distances as far away from the basin as the Descartes landing site of *Apollo 16* (3,000 kilometers). As the Planning Team report indicates, the debris-moving capability of giant impacts has not been calibrated, and it is possible that such a thickness of Orientale ejecta was deposited at Descartes as to preclude substantial dilution by indigenous materials up to the present day. Depositions elsewhere on this scale would account for the remarkable similarity of terra rock types recovered from one site to the next.

Lunar Atmosphere

The *Apollo 17* station included a mass spectrometer for measuring the density and chemical composition of the lunar atmosphere. During the early observational period, ^{4}He, ^{40}Ar, and possibly Ne were detected. The daytime and nighttime ^{4}He concentrations at the lunar surface were found to be 3 × 10^3 and 6 × 10^4 atoms per cubic centimeter, respectively. This factor of 20 in the variation between daytime and nighttime concentrations corresponds with expected behavior of a noncondensable gas in the lunar atmosphere. However, the ^{40}Ar concentration exhibited unexpected behavior. This concentration dropped during the night to below 100 atoms per cubic centimeter, rising orders of magnitude a few hours before the sunrise terminator reached the measurement site. It has been proposed that ^{40}Ar is absorbed at the lunar surface during the night and released again at sunrise.

Fortunately, a record of past and present lunar atmospheres is preserved in the lunar fines and breccias. Solar ultraviolet radiation and charge exchange with the solar wind cause ionization of atmospheric atoms. Fine-grained lunar surface materials trap some of these accelerated ions. It has been recognized since *Apollo 11* that lunar soil particles contain a component of ^{40}Ar the abundance of which is propositional to the surface area of the particles and it is believed that this is retrapped gas from the lunar atmosphere. The amount of ^{40}Ar present in the lunar atmosphere is reflected by the ^{40}Ar/^{36}Ar ratio of trapped gas. Observation of these ratios indicate a variation that suggests that perhaps ^{40}Ar may have been more abundant in the ancient lunar atmosphere than prevails today.

Structure and Seismic Activity

Other than the fact that the moon is considerably less dense than other terrestrial planetary bodies (mean density of about 3.4 times that of water), very little was known pertaining to the interior of the moon prior to the *Apollo* explorations. Laser altimeter measurements from the *Apollo* orbiting command module indicated the lunar center of mass is displaced by about 2 kilometers (1.2 mile) earthward and 1 kilometer (0.6 mile) east of the center. Possibly this may be explained by the mean thickness of the low-density lunar crust being about a factor of 2 greater on the lunar farside than on the nearside. A crustal heterogeneity of this magnitude possibly also can account for the differences in the principal moments of inertia of the moon. These had been known before the *Apollo* missions from observations of the lunar librations. Previously, the moment differences had been attributed to a distorted, nonequilibrium form of the moon.

A network of four seismic stations deployed on the moon during the *Apollo* missions had recorded more than 1000 moonquakes between 1969 and 1976. The moonquakes are very small, with estimated magnitudes of less than about 2. The annual release of seismic energy is estimated at about 10^{15} ergs, or about ten orders of magnitude less than that of the earth. Except for a small number of shallow events, moonquakes occur deep inside the moon at depths of between 600 and 1000 kilometers (965 and 1610 miles). Moonquakes appear to occur repeatedly at distinct foci and events from each focus produce matching waveforms. During the aforementioned time span, nearly seventy such repeating foci were identified. The events at each focus appear to occur at 27-day intervals and there is evidence of a 206-day periodicity. These periodicities clearly relate the moonquake occurrences to the tidal forces acting on the moon. Further details are reported by Toksöz, Goins, and Cheng (1977).

The *Apollo* seismometer network has detected compressional seismic waves from meteoroid impacts and internal moonquakes on all sides of the moon. However, shear waves have been received only from the nearside events. Attenuation of shear waves by the interior of the moon indicate that it is probably hot and weak, possibly with a small fraction of the rock melted. The relatively rigid layer which transmits shear waves, the lunar lithosphere, is about 1,000 kilometers (621 miles) in thickness.

Magnetic Properties

An unexpected finding of the analysis of returned rocks, both crystalline and breccia, was their possession of a stable remanent magnetization. No dipole magnetic field in excess of 5 gammas (1 gamma = 10^{-5} gauss) had been detected by lunar orbiting satellites. In contrast, magnetometers landed during the *Apollo* missions indicated local magnetic field of up to 300 gammas. The *Apollo 15* and *16* subsatellites also had indicated complex magnetic anomalies up to 1 gamma. Since magnetization appears widespread on the lunar surface, it is estimated that a field of from 1,000 to 10,000 gammas was required when the rocks were formed some 3.2 to 4.1×10^9 years ago. Investigators do not fully agree on the mechanism that may have provided such an early magnetic field. It is believed that the rocks may have been heated above their Curie temperatures, either by volcanism or by cataclysm at the aforementioned time period. This would have taken place long after any transient magnetic field that may have been associated with the origin of the solar system would have passed. One proposal is that the moon had a small, fluid, metallic core and that this performed as a self-exciting dynamo at that time. Another proposal is that a primeval solar system field may have magnetized the cool interior of the moon while the surface layers were still molten and that later the field of the lunar interior magnetized the outer layers as they cooled, and that finally radioactive heating eliminated the interior magnetization.

Investigators Hood and Coleman (Univ. of California, Los Angeles) and Wilhelms (U.S. Geological Survey) reported in 1979 of additional studies of previously unmapped *Apollo 16* subsatellite magnetometer data collected at low altitudes over the lunar near side. The scientists found that medium-amplitude magnetic anomalies exist over the Fra Mauro and Cayley Formations (primary and secondary basin ejecta emplaced 3.8 to 4.0 billion years ago) but are nearly absent over the maria and over the craters Copernicus, Kepler, and Reiner and their encircling ejecta mantles. In their summary of this study, these investigators suggest that a single hypothesis that is consistent with the subsatellite data as well as with studies of returned samples and surface magnetic fields is that basin and crater ejecta deposits are major sources of the orbital anomalies. The level of magnetization is weak peripheral to the impact region inside the crater, but it rises beyond the rim and may rise sharply in the case of some ejected materials transported ballistically to large distances. Also, as suggested by Strangway et al. (1973), the ejected materials are probably also the materials most strongly shocked and heated by the impact event, a relative increase in the volume fraction of free iron grains capable of retaining a strong and stable magnetic remanence is reasonable to expect.

Meteorites from the Moon

Lunar meteorites, or lunaites, are meteorites from the Moon. In other words, they are rocks found on Earth that were ejected from the Moon by the impact of an asteroidal meteoroid or possibly a comet.

While impact cratering on the Moon can be a very destructive process, it also has the capacity to throw new samples of rock our way. As an impact crater is being excavated, some rocky material can be thrown out of the crater with enough velocity to escape the Moon's gravity and fall to Earth. Since impact craters occur at random locations, lunar meteorites provide a set of samples from all portions of the Moon, including the farside and Polar Regions, which Apollo astronauts and Luna spacecraft were unable to visit.

Meteoroids strike the Moon every day. Lunar escape velocity averages 2.38 km/s (1.48 miles per second), only a few times the muzzle velocity of a rifle (0.7–1.0 km/s). Any rock on the lunar surface that is accelerated by the impact of a meteoroid to lunar escape velocity or greater will leave the Moon's gravitational influence. Some ejected material becomes captured by the Earth's gravitational field and lands on Earth within a few hundred thousands of years (much shorter for some). Other ejected material, however, assumes an orbit around the Sun. Some of that material may eventually strike Earth. This can take a long time. Lunar meteorites Yamato-82192/82193/86032 and Dhofar 025 remained in space for 10 and 20 million years before finally landing on Earth.

The largest single stone is Kalahari 009 at 13.5 kg (30 lbs.). The rest are much smaller. The next biggest are Dar al Gani 400 (1425 grams = 3.1 lbs) and LAP 02205 (1226 grams = 2.7 lbs). Together, the five stones of the LAP 02xxx "pairing" are the second largest lunar meteorite (1875 grams = 4.1 lbs.). Several of the lunar meteorite fragments found in Antarctica and Oman only weigh a few grams (a U.S. nickel weighs 5 grams).

Of the >26,000 meteorites listed in the Catalogue of Meteorites, http://www.nhm.ac.uk/research-curation/projects/metcat/, only 1 in 1200 are lunar. Meteorites are very rare rocks; lunar meteorites are exceedingly rare. No lunar meteorites have yet been found in North America, South America, or Europe.

Nearly all lunar meteorites have been found in areas that are well known to be good places to find meteorites. All such places are dry deserts where there are geologic mechanisms for concentrating meteorites or where rocks of terrestrial origin are rare. Many lunar meteorites (and most meteorites overall) have been found in Antarctica (Fig. 5) by expeditions funded by the U.S. (**ANSMET**) http://geology.cwru.edu/%7Eansmet/, or Japanese (**NIPR**) governments, http://yamato.nipr.ac.jp/AMRC/index_en.html. The meteorites from Australia, Africa, and Oman were found by private collectors or purchased from local people by private collectors.

Allan Hills 81005 (ALHA81005), the first meteorite to be recognized as originating from the Moon, was found during the 1981—82 ANSMET collection season, on 18 January 1982. The three Yamato 79xxx meteorites were collected earlier, but not recognized to be of lunar origin until after 1982. The first lunar meteorite to be found appears to be Yamato 791197, on 20 November 1979.

Thus far, 36 lunar meteorites have been discovered. See http://epsc.wustl.edu/admin/resources/meteorites/moon_meteorites_list.html #DHO025. Many of these have been recovered in Antarctica, where they can be preserved in glacial ice for thousands of years after they fall. Some lunar meteorites, however, have also been found in other parts of the world. Generally these are hot desert regions where the meteorites are preserved because there is so little rain to affect them. See also **Meteroids and Meteorites**.

Future Human Activities on the Moon

The discovery of ice on the Moon has enormous implications for a permanent human return to the Moon. Water ice is made up of hydrogen and oxygen, two elements vital to human life and space operations. Lunar ice could be mined and disassociated into hydrogen and oxygen by electric power provided by solar panels deployed in nearby illuminated areas or by a nuclear generator. This hydrogen and oxygen is a prime rocket fuel, giving us the ability to refuel rockets at a lunar "filling station" and making transport to and from the Moon more economical by at least a factor of 10. Additionally, the water from lunar polar ice and oxygen generated from the ice could support a permanent facility or outpost on the Moon. The discovery of this material, rare on the Moon but so vital to human life and operations in space, will make our expansion into the solar system easier and reaffirms the immense value of our own Moon as the stepping-stone into the Universe.

The Moon as a Planetary Touchstone. Beyond obtaining new samples, emplacing a global network of geophysical stations would help us learn

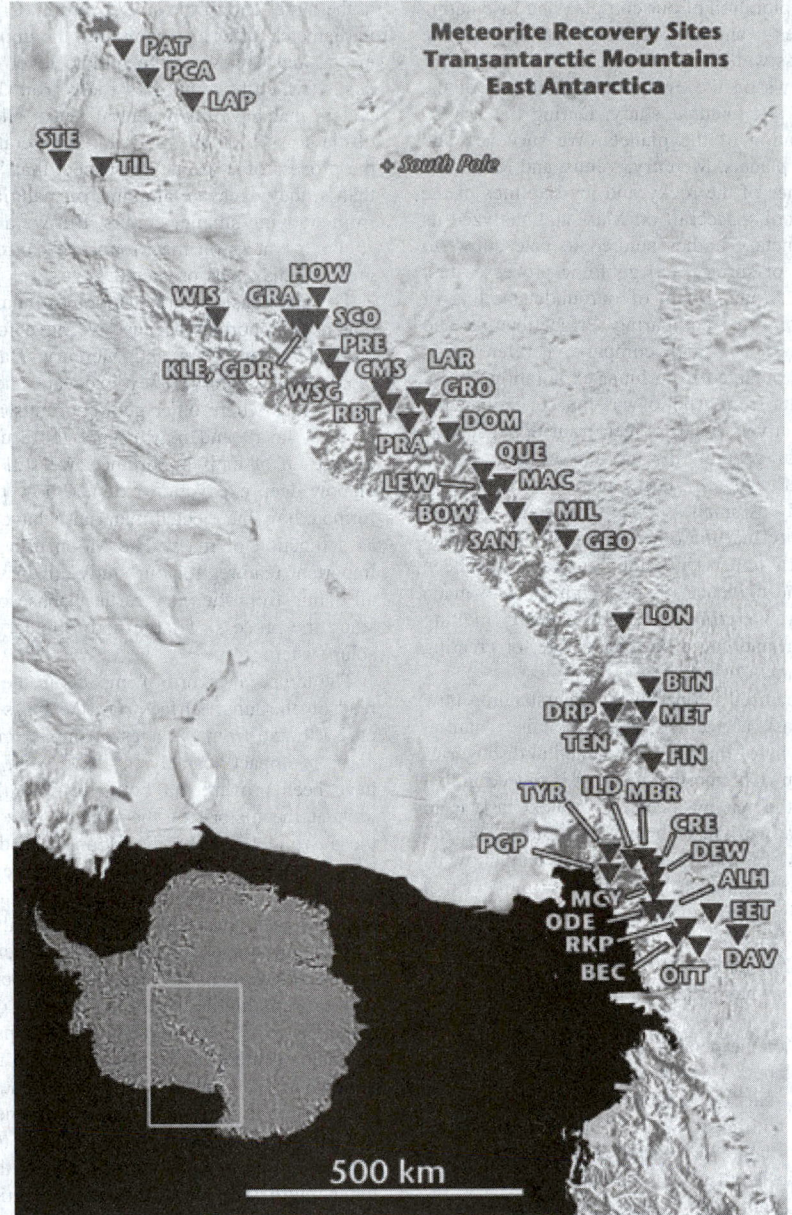

Fig. 5.   ANSMET Meteorite Locations. ALH (Allan Hills); BEC (Beckett Nunatak); BOW (Bowden Neve); BTN (Bates Nunataks); CMS (Cumulus Hills); CRE (Mt. Crean); DAV (David Glacier); DEW (Mt. DeWitt); DOM (Dominion Range); DRP (Derrick Peak); EET (Elephant Moraine); FIN (Finger Ridge); GDR (Gardner Ridge); GEO (Geologists Range); GRA (Graves Nunataks); GRO (Grosvenor Mountains); HOW (Mt. Howe); ILD (Inland Forts); KLE (Klein Ice Field); LAP (LaPaz Ice Field); LAR (Larkman Nunatak); LEW (Lewis Cliff); LON (Lonewold Nunataks); MAC (MacAlpine Hills); MBR (Mount Baldr); MCY (MacKay Glacier); MET (Meteorite Hills); MIL (Miller Range); ODE (Odell Glacier); OTT (Outpost Nunatak); PAT (Patuxent Range); PCA (Pecora Escarpment); PGP (Purgatory Peak); PRA (Mt. Pratt); PRE (Mt. Prestrud); QUE (Queen Alexandra Range); RBT (Roberts Massif); RKP (Reckling Peak); SAN (Sandford Cliffs); SCO (Scott Glacier); STE (Stewart Hills); TEN (Tentacle Ridge); TIL (Thiel Mountains); TYR (Taylor Glacier); WIS (Wisconsin Range); WSG (Mt. Wisting). (Antarctic Search for Meteorites) is a program funded by the United States government through the Office of Polar Programs of the National Science Foundation (NSF) and the Solar System Exploration Division of the National Aeronautics and Space Administration (NASA) in cooperation with the National Museum of Natural History (Smithsonian Institution). (*Map courtesy of NASA*).

more about the Moon's mantle and core structure, variations in its crustal thickness, and the enigmatic lunar paleomagnetism. It would also lead to a more accurate determination of the enrichment of the Moon in refractory elements by measuring lunar heat flow (because two of these elements, uranium and thorium, are radioactive and thus important sources of heat).

Eventually, humans will probably go to the Moon to live, and establishing a permanent presence there opens up scientific vistas that are difficult to foresee clearly. Each Apollo mission provided some geologic surprise within its sample collection. So there is little doubt that both the variety of rock types and geologic processes that have operated on the Moon exceed by far those that we have currently deciphered. From a permanent lunar base, we could begin a detailed exploration of our complex and fascinating satellite that could last for centuries — and uncover its secrets and also the early history of our home planet as well.

As fascinating as it is, our views on the evolution and history of the Moon have more relevance than just to lunar study. During the last 20 years, we made our first exploration of the planets. We surveyed and photographed all of the terrestrial planets, Mercury, Venus, and Mars, and conducted our initial reconnaissance of the rocky and icy satellites of the giant outer planets. We landed robot spacecraft on Mars and analyzed its surface materials. All of the planetary bodies studied to date show, to different degrees, the same kinds of surface and geologic processes first recognized and described on the Moon. Much of our understanding of planetary processes and history comes by comparing surface features and environments among the planets. In any such comparison, reference is inevitably made to knowledge we obtained from lunar exploration.

One of the most startling results from Apollo was the concept of the magma ocean. The Moon is a relatively small object, transitional in size between planets and asteroids. In general, the amount of heat that a planetary object contains is related to its size; larger planets contain more heat-producing elements. If a body as small as the Moon could undergo global melting, it is a near certainty that the other terrestrial planets also melted. The idea that early Earth underwent global melting has been bandied about for many years; the evidence of the magma ocean made such speculation respectable. Now, we think that early planetary melting may be a widespread phenomenon and could be responsible for creating all of the original crusts of the planets.

Knowing that global melting occurred is one thing, understanding how it operated in detail is another task altogether. The Moon is a natural laboratory to study this process. One of the most fundamental discoveries of Clementine is that the aluminum-rich, anorthosite crust is indeed global and provides strong support for the magma ocean. Our next task is to understand the complex processes at work in such an ocean. Did a "chilled" crust form and if so, are any pieces of it left? Such material would allow us to determine *directly* the bulk composition of the Moon, a parameter that is estimated indirectly (and very imprecisely) now. Are there any highly magnesian rocks in the highlands that are related to the magma ocean, not to the younger magnesium-rich suite of rocks? We have searched the Apollo collections for such rocks but have found none. They may exist at unvisited sites on the Moon.

Another process common to all of the planets is volcanism. The Moon is the premier locality to study planetary volcanism. The flood lavas of the maria span more than a billion years of planetary history and probably come from many different depths within the Moon. Thus, the lava flows are actually probes of the interior of the Moon, both laterally (across its face) and vertically (throughout the depths of the mantle). The inventory and study of the mare basalts will allow us to categorize both of these dimensions and the important additional dimension of *time*. By sampling, chemically analyzing, and dating many different samples of lava that cover the globe, we can piece together the changing conditions of the deep mantle across long periods.

The styles of eruption responsible for the maria appear to be typical of those on other planets. Flood volcanism — the very high rates of effusion responsible for the mare lavas — is seen on every terrestrial planet and appears to be especially widespread on Mars and Venus. What is largely unknown is the size and shape of the vents through which these lavas were extruded. On the Moon, eruptive vents might be exposed in several locations, including within the walls of grabens and irregular source craters. Detailed exploration and study of such features would help us understand a style of volcanism ubiquitous on the planets. Small, dome-like volcanoes, such as found in the Marius Hills, can be explored and examined to understand the styles and rates of lava eruption in creating these features. Small shield volcanoes are common on the surfaces of Mars and, particularly, on Venus. Relatively exotic processes, such as the erosion of terrain by flowing lava, has been proposed for the sinuous rilles of the Moon. Study of large rilles could help us decide whether this concept is correct.

From studying the Moon, we know that impact is one of the most fundamental of all geologic processes. Based on its population of craters of all sizes, where better to study and understand this important shaper of surfaces than on the Moon? Our ignorance is particularly vast for craters

at the larger end of the size spectrum. Craters such as Copernicus (100 km in diameter) offer a window into the upper crust through the study of their ejecta, and a view of the middle level of the crust through their central peaks (which have uplifted rocks from 15–20 km deep). The ubiquity of craters that have such central peaks allows us to reconstruct the nature of the crust in detail. Studying large craters will also clarify the nature of the process of impact. We suspect that large craters grow *proportionally*, that is, they excavate amounts of material to depths that can be predicted from studying smaller craters. But we are not certain that this is true. By studying craters on the Moon, we can determine whether this pattern of growth behaves as predicted.

The giant basins of the Moon pose many mysteries. Understanding such craters is important because we have found basins on the other planets, particularly on Mars and Mercury. Basins form in the earliest stages of planetary history. They excavate and redistribute the crust, serve as depressions where other geologic units may be deposited (such as stacks of thick lava), and may trigger the eruption of massive floods of lava. Yet for all of their importance, we still do not fully understand how far or how deep excavation extends, how the multiple, concentric rings are formed, how the ejected material behaves, and how far ejecta gets thrown as a function of mass. The Moon has preserved more than 40 of these important features for our study, all in various states of preservation and all dating from the very earliest phase of planetary history. Here we can study the process of large-body impact better than anywhere else in the solar system.

The temporal record of impacts in the Earth–Moon system can also be read on the lunar surface. On the basis of evidence of mass extinctions on Earth and from the ages of impact melts, cycles of bombardment and an early impact "cataclysm" have been proposed. Neither of these ideas have been proven, but both are potentially revolutionary and make us look at the history of the planets in a new way. The evidence to test these two ideas lies on the Moon. Episodic bombardment can be tested by sampling the melt sheets of many different craters and dating the samples. Episodes of intense cratering will be evident if groups of melt rocks have the same ages, spaced at constant intervals. On the other hand, a continuous distribution of ages would argue that such episodes of intense cratering do not occur. We cannot conduct this experiment on Earth or on any other planet — this highlights the uniqueness of the Moon for answering many questions in planetary science, questions that have application to a host of other scientific fields. The cataclysm is important because if the Moon underwent such an unusual bombardment history, Earth may also have experienced it, and applying the inferences made from lunar cratering to the other terrestrial planets would have to be reevaluated.

Studying the regolith will be one of the most important tasks during a lunar return. The regolith contains exotic samples flung from rock units hundreds of kilometers away. Using the regolith as a sampling tool, we can conduct a comprehensive inventory of the regional rock units by collecting many samples from the regolith at a single site. The regolith also contains a record of the output of the Sun during the last 3 billion years. To read this record, we must understand how the regolith grows and evolves. This knowledge will come only when we can study the regolith and its underlying bedrock in detail; to learn how its layers are formed; how the soil is exposed, buried, and re-exposed; and how volatile components might be mobilized and migrate through the soil. This knowledge is essential if we are to realize the goal of using the regolith as a recorder of the solar and galactic particles that have struck the Moon during its history.

Astronomy from the Moon. One day, the Moon will become humanity's premier astronomical observing facility. Consider its advantages. The Moon rotates very slowly (once every 29 days), so its "nighttime" is 2 weeks long. Moreover, because the Moon has no atmosphere, we can observe stars constantly, even in the daytime! The lack of an atmosphere also means that telescopes on the Moon will not be plagued by the "blurring" that a turbulent, thermally unstable air layer causes and that observations will not be degraded by "light pollution," the airglow that interferes with astronomy on Earth. The vacuum of the Moon also means that there are no "absorptions" to prevent certain wavelengths of radiation from being observed, such as the infamous "water absorption" in the atmosphere of Earth. The surface materials could be used as construction material for observatories.

Telescopes in Earth orbit or elsewhere in deep space also have many of these advantages. Why is the Moon better than these localities? The principal reason is that the Moon provides a quiet, stable platform. Seismic activity on the Moon is roughly one million times less than that of Earth. Because the Moon is a primitive, geologically dead world, it does not have the shifting, massive plates of our own dynamic Earth, along with its associated seismic trembling. Such stability of the surface would allow us to construct extremely sensitive instruments for observation, instruments that could not be constructed on Earth.

Likewise, a space-based telescope, such as the Hubble Space Telescope, must have its attitude carefully stabilized to achieve high-resolution observations. In addition, space telescopes have stringent pointing requirements and must not be pointed anywhere near the Sun. Both of these drawbacks mean that a telescope free-flying in space must carry attitude control fuel, precision gyroscopes, and equipment to protect the telescope's optics from solar burn damage. On the Moon, the quiet, stable base of the surface would alleviate such problems, allowing sensitive instruments to be erected and operated easily.

One such instrument, an interferometer, consists of an array of smaller telescopes. The smallest object that a telescope can see clearly is directly related to the size of its *aperture*, or the diameter of its mirror or lens. A telescope that has a larger aperture can resolve smaller or more distant features than a telescope that has a smaller aperture. However, there is a practical limit to the size that we can make telescopes. After a certain size is attained, such instruments become unwieldy and unstable. Interferometry is a technique whereby a series of small telescopes are operated as a larger aperture instrument. Each element of the array images some distant object. The light waves from this image are "added" in perfect phase and frequency to identical images obtained from other telescopes in the array, each separated by as much as several kilometers. The effect of this addition is to create an image of the same quality as if a telescope that had an aperture size equal to the separation distance had been used to image the star. This means that we can construct "telescopes" whose effective aperture sizes are *kilometers* across!

Even a small interferometer on the Moon would exceed the resolving capabilities of the very best telescopes on Earth and could even surpass the capabilities of the Hubble Space Telescope. Using such an instrument, we could resolve the disks of distant stars and observe and catalog "star spots," which are clues to the internal workings of stars. We could see individual stars in distant galaxies and catalog the stellar makeup of a variety of galaxy types. Optical interferometers could look into a variety of nebulae and observe the details of new stars and stellar systems in the very act of formation. Such a window onto the Universe is very likely to revolutionize astronomy in the same way it was changed when Galileo turned his crude "spyglass" toward the heavens in 1609.

The field of planetary astronomy would be completely changed by lunar observatories. The incredible resolving power of these instruments would allow us to examine deep sky systems, resolve the disks of extrasolar planets, and catalog the variety possible in other solar systems of our galaxy. All of our concepts of the way planets are created and evolve are derived from a single example, our own solar system. By observing the vast array of planetary systems circling nearby stars, we could see how they differ in such aspects as the number and spacing of planets, the ratio of giant gas planets to rocky "terrestrial" objects, and the evolution of those individual planets. Spectroscopic observations of these planets would allow us to determine the composition of their atmospheres, if any, and the surface composition, if visible. The compositions of planetary atmospheres could indicate the presence of life on these bodies. Carbon dioxide mixed with free oxygen in a planetary atmosphere would be a telltale indication of plant life, which "breathes" the former and manufactures the latter.

Astronomers look at the sky in many wavelengths other than the optical band. High-energy regions of the spectrum, such as X-ray and gamma-ray radiation, also contain important information about processes that occur in stars and galaxies. Supernovas (the sudden explosion of certain stars) produce copious amounts of high-energy radiation and energetic particles, such as cosmic rays. We think that certain particles derived from supernova explosions predate our solar system and that these stellar eruptions can induce planetary formation. We have already mentioned the use of the regolith as a recorder of energetic particle events. Using a high energy lunar observatory, we can watch the effusion of these particles and radiation as

they happen (supernovas are common and typically are occurring in some part of the sky at any given time). We have only begun to observe such stellar explosions from space and no doubt have much to learn.

At the other end of the spectrum, the Moon is an ideal place for observation in the thermal infrared and radio bands. Observing the sky in the long wavelengths of the thermal infrared (10 microns and longer wavelengths) is difficult because such detectors measure heat and they must be cooled to very low temperatures for use. Usually, this is done at great cost and difficulty by using cryogenic gases, such as very cold liquid helium ($-300\,°C$). Preserving such cold temperatures requires a lot of electrical power. The Moon is naturally cold. Surface temperatures during the lunar night may become as low as $-160\,°C$. In the shadowed areas near the South Pole, it may be as cold as $-230\,°C$, only 40 K above absolute zero! These temperatures would permit passive cooling of infrared detectors, allowing telescopes to be operated without costly and difficult-to-use cryogenic, cooling gas. Observing the thermal infrared sky would tell us much about dust clouds and nebulae in which new stars and planets are being formed.

The sky at certain radio wavelengths is almost completely unknown. Earth has an ionosphere, a layer of electrically charged atoms that makes certain radio waves bounce off it, and we cannot see the radio sky at certain frequencies. Moreover, the electrical din of Earth caused by radio stations, microwave cookers, automotive ignition systems, and the thousands of other static generators of modern civilization cause radio astronomers great vexation in their attempts to map the sky. The far side of the Moon is the *only* known place in the Universe that is permanently shielded from the radio noise of Earth. Locating a radio telescope on the far side would permanently place 3600 kilometers of solid rock between the observatory and the radio din of Earth. We will see sky for the first time at some radio wavelengths. History has shown us that whenever we look at the Universe with a new tool or through a new window of frequencies, we learn new things and reexamine present knowledge with new and sometimes startlingly different appreciation.

We can use the distinctive lunar terrain to our advantage. The small, bowl-shaped craters of the Moon are natural features that could be turned into gigantic "dish" radio antennas by laying conductive material (e.g., chicken wire) on their floors and hanging a receiver over and above the center of the crater at the "focus" of the dish. This technique has already been done on Earth at the famous Arecibo Radio Observatory in Puerto Rico, using a natural depression in the limestone bedrock to create a giant "dish" antenna. Interferometers could also be built at radio wavelengths, creating radio telescopes that have huge apertures. The large, flat mare plains would make an ideal site to lay out arrays of smaller telescopes. Manufacture of antenna elements from local resources could make constructing of extremely large instruments feasible.

Using the Moon as an astronomical observatory has great advantages, and many astronomers have taken up the banner for a return to the Moon. In the minds of some, astronomy is the principal reason for a lunar return. However, an observatory on the Moon also has its problems. The ubiquitous and highly abrasive dust must be very carefully controlled. Movements of people and machines will have to be minimized around telescope facilities because the slightest stirring up of dust could coat delicate optical surfaces. We would have to shield energetic detectors carefully from solar flares (this could be done by using the local regolith material.) We must guard against radio contamination of the far side because extensive operations of a base could ruin certain radio astronomical observations. Our task before a lunar return is to understand the impact of each problem fully and devise methods of working around it.

Despite these problems, the Moon offers unique opportunities for astronomy. Each time we see the sky more clearly or more completely, we obtain new insights into the way the universe works. A lunar window on the Universe around us will give us a new appreciation and understanding of both the Universe and of our place in it.

Additional Reading

Belton, M.J.S. et al.: "Lunar Impact Basins and Crustal Heterogeneity," *Science*, 570 (January 31, 1992).

Boss, A.P.: "The Origin of the Moon," *Science*, 341 (January 24, 1986).

Brush, S.G.: "Nickel for Your Thoughts: Urey and the Origin of the Moon," *Science*, **217**, 891–898 (1982).

Burke, B.F.: "Astrophysics from the Moon," *Science*, 1365 (December 7, 1990).

Burns, J.O. et al.: "Observations on the Moon," *Sci. Amer.*, 42 (March 1990).

Cohen, B.A.: *Lunar Meteorites and the Lunar Cataclysm*, Planetary Science Research Discoveries, Honolulu, HI, 2001. http://www.psrd.hawaii.edu/Jan01/lunarCataclysm.html

Eberhart, J.: "Going for the Whole Moon: The Lunar Observer Would Study What Other Spacecraft Missed," *Sci. News*, 264 (April 28, 1990).

Eugster, O.: "History of Meteorites from the Moon Collected in Antarctica," *Science*, 1197 (September 15, 1989).

Garber, M.: "New View of Universe to Come from Moon," *Hughes News*, 5 (May 3, 1991).

Gillette, R.: "The Aftermath of Apollo: Science on the Shelf?" *Science*, **178**, 1265–1268 (1972).

Gnos E., B.A. Hofmann, A. Al-Kathiri, S. Lorenzetti, O. Eugster, M.J. Whitehouse, I.M. Villa, A.J.T. Jull, J. Eikenberg, B. Spettle, U. Krahenbuhl, I.A. Franchi, and R.C. Greenwood: "Pinpointing the source of a lunar meteorite: Implications for the evolution of the Moon," *Science*, **305**, p. 657–659. (2004).

Gribbin, J.R. and S. Goodwin: *Empire of the Sun: Planets and Moons of the Solar System*, New York University Press, New York, NY, 1998.

Hartmann, W.K., R.J. Phillips, and G.J. Taylor: *Origin of the Moon*, Lunar and Planetary Institute Press, Houston, TX, 1986.

Hartmann, W.K.: *Moons and Planets*, 4th Edition, International Thomson Publishing, New York, NY, 1999.

Hood, L.L., P.J. Coleman, Jr., and D.W. Wilhelms: "The Moon: Sources of the Crustal Magnetic Anomalies," *Science*, **204**, 53–57 (1979).

Horgan, J.: "Blame It on the Moon," *Sci. Amer.*, 18 (February 1989).

Keen, R.A.: "Volcanic Aerosols and Lunar Eclipses," *Science*, **222**, 1011–1013 (1983).

Kerr, R.A.: "Making the Moon, Remaking Earth," *Science*, 1433 (March 17, 1989).

Kerr, R.A.: "Treasuring the Moon for 20 Years," *Science*, 1552 (March 24, 1989).

Kerr, R.A.: "A Rare View of the Moon," *Science*, 1651 (December 21, 1990).

Kerr, R.A.: "Did the Moon Suffer a Cataclysmic Bombardment?" *Science*, 1634 (June 19, 1992).

Marvin, U.B., J.W. Carey, and M.M. Lindstrom: "Cordierite-Spinal Troctolite, a New Magnesium-Rich Lithology from the Lunar Highlands," *Science*, 925 (February 17, 1989).

Morrison, D.C.: "An Unsung Legacy of the First Lunar Landing," *Science*, 447 (October 27, 1989).

Ottwell, G.: *The Understanding of Eclipses*, Furman University, Greenville, SC, 1991.

Porter, A.E. and T.H. Morgan: "Discovery of Sodium and Potassium Vapor in the Atmosphere of the Moon," *Science*, 675 (August 5, 1988).

Spudis, P.D.: *The Once and Future Moon*, Smithsonian Institution University Press, Washington, DC, 1996.

Strangway, D. et al.: "Magnetism and Magnetic Materials," (C. Graham, Jr., and editors), pp. 1178–1196, American Institute of Physics, New York, NY, 1973.

Taylor, G.J.: *A New Moon for the Twenty-First Century*, Planetary Science Research Discoveries, Honolulu, HI, 2000. http://www.psrd.hawaii.edu/Aug00/newMoon.html

Taylor, G.J.: *Uranus, Neptune and the Mountains of the Moon*, Planetary Science Research Discoveries, Honolulu, HI, 2001. http://www.psrd.hawaii.edu/Aug01/bombardment.html

Toksoz, M.N., N.R. Goins, and C.H. Cheng: *Science*, **196**, 979–981 (1977).

Waldrop, M.M.: "Asking for the Moon," *Science*, 637 (February 9, 1990).

Wilhelms, D.E.: *The Geologic History of the Moon*, U.S. Geological Survey, Denver, CO, 1988.

Wilhelms, D.E.: *To A Rocky Moon: A Geologist's History of Lunar Exploration*. University of Arizona Press, Tucson, AZ, 1993.

Web References

Chronology of Lunar and Planetary Exploration: http://nssdc.gsfc.nasa.gov/planetary/chrono.html

Exploring the Moon: http://www.space.edu/moon/

MOON ILLUSION. The moon seen near the horizon appears larger than the moon seen high in the sky. This difference is illusory, for there is no difference in the angular widths of moon from one situation to the other. (There is normally a small difference in the angular heights of the moon due to refraction in the atmosphere, but this serves to lessen the height on the horizon rather than increase it). Yet the illusion is sufficiently compelling to cause most observers to be convinced that the moon actually has a significantly larger angular size when near the horizon, and that this has a physical origin in the optics of the atmosphere. But the phenomenon is perceptual and its explanation lies in the realm of psychology. No single explanation has been found that accounts for all aspects of what people claim to see. One explanation that accounts for some aspects of

the phenomenon does relate to meteorological optics. The clear sky is not perceived to be a hemisphere, but a variety of shapes, for example, a flattened dome; the horizon being seen as significantly farther from the observer than the zenith. Further, the moon appears to be pasted on the horizon and so shares its distance. The perceptual phenomenon of size constancy will then cause something of fixed angular size but apparently varying in distance to appear larger at the greater distance of the horizon. Suffice it to say that there are aspects of the illusion that are consistent with this explanation and others that are at variance with it.

Additional Reading

Hershenson, M.: *The Moon Illusion*, Laurence Erlbaum Associates, Publisher, Hillsdale, NJ, 1989.

Ross, H.E., and C. Plug: *Mystery of the Moon Illusion*, Oxford University Press, New York, NY, 2002.

AMS

MOONQUAKES. Between 1969 and 1972, Apollo astronauts placed seismometers at their landing sites around the moon. The Apollo 12, 14, 15, and 16 instruments faithfully radioed data back to Earth until they were switched off in 1977. After reexamining Apollo data from the 1970s Clive R. Neal, associate professor of civil engineering and geological sciences at the University of Notre Dame and a team of 15 other planetary scientists concluded that the moon is seismically active. There findings were presented at a gathering of scientists at NASA's Lunar Exploration Analysis Group (LEAG) meeting in League City, Texas, October of 2005.

The Apollo instruments revealed that there are at least four different kinds of moonquakes: (1) deep moonquakes about 700 km (435 miles) below the surface, probably caused by tides; (2) vibrations from the impact of meteorites; (3) thermal quakes caused by the expansion of the frigid crust when first illuminated by the morning sun after two weeks of deep-freeze lunar night; and (4) shallow moonquakes only 20 or 30 kilometers (12 or 19 miles) below the surface.

The first three were generally mild and harmless. Shallow moonquakes on the other hand were doozies. Between 1972 and 1977, the Apollo seismic network saw twenty-eight of them; a few "registered up to 5.5 on the Richter scale," says Neal. A magnitude 5 quake on Earth is energetic enough to move heavy furniture and crack plaster.

Furthermore, shallow moonquakes lasted a remarkably long time. Once they got going, all continued more than 10 minutes. "The moon was ringing like a bell," Neal says.

On Earth, vibrations from quakes usually die away in only half a minute. The reason has to do with chemical weathering, Neal explains: "Water weakens stone, expanding the structure of different minerals. When energy propagates across such a compressible structure, it acts like a foam sponge — it deadens the vibrations." Even the biggest earthquakes stop shaking in less than 2 minutes.

The moon, however, is dry, cool and mostly rigid, like a chunk of stone or iron. So moonquakes set it vibrating like a tuning fork. Even if a moonquake isn't intense, "it just keeps going and going," Neal says. And for a lunar habitat, that persistence could be more significant than a moonquake's magnitude.

"Any habitat would have to be built of materials that are somewhat flexible," so no air-leaking cracks would develop. "We'd also need to know the fatigue threshold of building materials," that is, how much repeated bending and shaking they could withstand.

Representative lunar seismograms from the Apollo 16 station are shown in Figure 1.

What causes the shallow moonquakes? And where do they occur? "We're not sure," Neil says. "The Apollo seismometers were all in one relatively small region on the front side of the moon, so we can't pinpoint [the exact locations of these quakes]." He and his colleagues do have some good ideas, among them being the rims of large and relatively young craters that may occasionally slump.

"We're especially ignorant of the lunar poles," Neal continues. That's important, because one candidate location for a lunar base is on a permanently sunlit region on the rim of Shackleton Crater at the Moon's south pole.

Neal and his colleagues are developing a proposal to deploy a network of 10 to 12 seismometers around the entire moon, to gather data for at

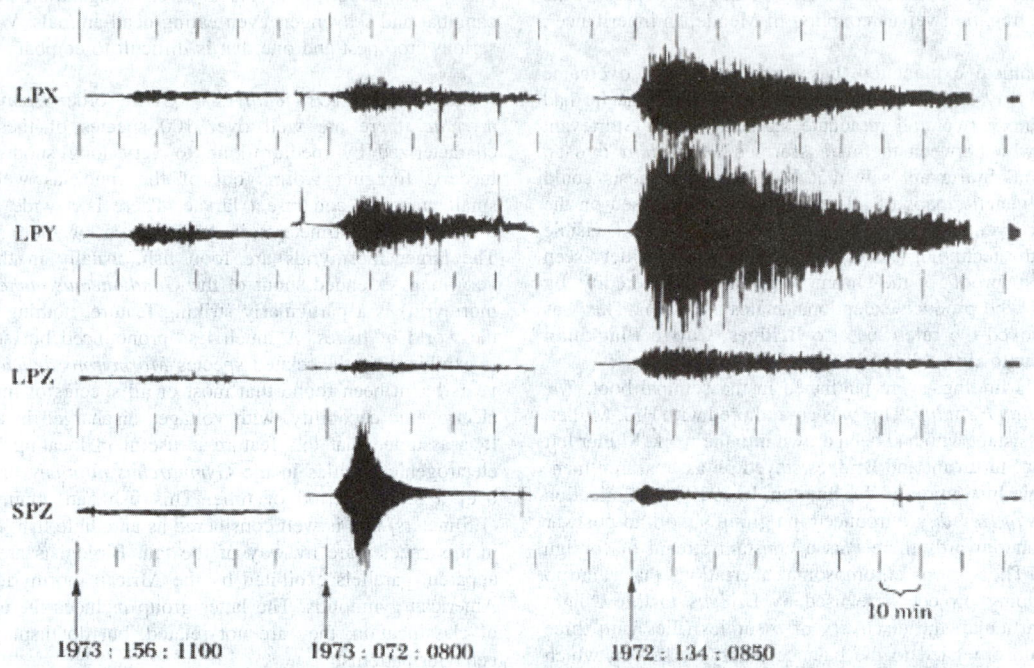

Fig. 1. Seismograms from three types of moonquakes recorded at the Apollo 16 station. LPX, LPY, and LPZ are the three long-period components and SPZ ist the short-period vertical component. The first column shows a deep-focus moonquake; the center column, a shallow moonquake; the third column shows records of the impact of meteoroid on the lunar surface. (*courtesy of NASA*).

least three to five years. This kind of work is necessary, Neal believes, to find the safest spots for permanent lunar bases.

And that's just the beginning, he says. Other planets may be shaking, too: "The moon is a technology test bed for establishing such networks on Mars and beyond."

Web References

Clive R. Neal: http://www.nd.edu/%7Ecneal/
Moonquakes: http://www.jsg.utexas.edu/news/feats/nakamura.html

TRUDY E. BELL, Science@NASA

MOORE, STANFORD (1913–1982). An American biochemist who won the Nobel Prize for chemistry in 1972, with Christian B. Anfinsen and William H. Stein, for enzyme studies. He was involved with the analysis of the action of the complex enzyme deoxyribonuclease. His Ph.D. was granted from the University of Wisconsin.

MOOSE. See **Deer.**

MORDANT. A substance capable of binding a dye to a textile fiber. The mordant forms an insoluble lake in the fiber, the color depending on the metal of the mordant. The most important mordants are trivalent chromium complexes, metallic hydroxides, tannic acid, etc. Mordants are used with acid dyes, basic dyes, direct dyes, and sulfur dyes. Premetallized dyes contain chromium in the dye molecule. A mordant dye is a dye requiring use of a mordant to be effective. See also **Dyes.**

MORELS. See **Ascomycetes.**

MORERA THEOREM. The converse of the Cauchy integral theorem. If $f(z)$ is a continuous function of the complex variable z, defined in finite simply connected domain,

$$\int_C f(z)\, dz = 0$$

for any closed contour on D, then $f(z)$ is an analytic function of z in D.

MORGAN, THOMAS HUNT (1866–1945). Morgan was an American zoologist and embryologist who established the chromosome theory of heredity.

Thomas Hunt Morgan was born a year after the end of the American Civil War into a family whose exploits as 'Morgan's raiders' had made them Confederate heroes. Two of his five uncles had been killed and his father, a former US consul to Messina, Sicily, was unable to find work.

At school Morgan's interests lay in natural history. He took a BS degree at the University of Kentucky and went on to Johns Hopkins University, Baltimore, at that time a pioneer in the biological sciences. His PhD dissertation used descriptive embryology to investigate the phylogeny of the sea spider but he rapidly moved into the new field of experimental embryology, working on the response of early embryos to artificial disruption and on regeneration in adults, seen as a parallel process. In 1904 he presented evidence that gradients are involved in regeneration. His books on experimental embryology and zoology were influential in changing biology from a descriptive to an experimental science.

From 1891 to 1904 Morgan was first Associate Professor and then Professor of Biology at Bryn Mawr, a women's college with close links to Johns Hopkins. In 1904 he was appointed to a new chair of Experimental Zoology at Columbia University, New York. The head of the department was E. B. Wilson, whose own research was in cytology. The idea that the segregation of the chromosomes during meiosis might provide a physical basis for Mendel's laws had been proposed in 1902 by Walter Sutton, one of Wilson's graduate students, and in 1905 Wilson and Nettie Stevens, herself a former graduate student of Morgan's, independently published a correlation in several insect genera between the number of 'X' chromosomes and sex. Females were XX; males XO or XY.

Morgan's embryological background made him skeptical about the role of the nucleus in development. However, in 1910 he found a sex-linked character in the fruit fly, *Drosophila melanogaster*. This was the famous white-eyed fly. Two other sex-linked characters were then found, one of which showed partial linkage (coupling) to white eyes. In 1909 the Belgian cytologist F. A. Janssens had described paired chromosomes twisted around each other in meiosis and suggested that they might exchange material. Morgan now proposed that the reason why some characters show coupling, and others do not, was related to 'the linear distance apart of the chromosomal materials that represent the factors' ('Random segregation versus coupling in Mendelian inheritance', in *Science*, 1911).

This simple mechanical explanation, based on his results, overcame Morgan's distrust of 'mystical' Mendelian factors. By this time he had taken into his laboratory two undergraduate students, A. E. Sturtevant and C. B. Bridges, who between them transformed *Drosophila* into an experimental organism. Sturtevant saw that Morgan's hypothesis could be used to construct genetic maps of all the chromosomes, based on the number of crossover events. Bridges built up the collection of visible markers; developed the technology; and provided what was widely seen at the time as the first 'proof' of the chromosome theory of heredity, by showing that females who produce exceptional males are XXY. This was the exception that proved the rule. See also **Bridges, Calvin Blackman (1889–1938)**; and **Sturtevant, Alfred Henry (1891–1970)**.

In 1915 the group's findings were published in the seminal book *The Mechanism of Mendelian Heredity*. This was co-authored with H. J. Muller, a graduate teaching assistant who had been drawn into the work. Muller left the group in 1915 but Sturtevant and Bridges stayed on as 'research men' funded by the Carnegie Institution of Washington. In 1919 the publication of *The Physical Basis of Heredity* introduced the group's work to postwar Europe and from then onwards there was a constant stream of foreign postdoctoral fellows. The various chromosomal aberrations that came to light during the mapping project were used by Bridges to throw light on gene action. In particular, the discovery of trisomics (flies with three copies of each chromosome) led to the balance theory of sex, in which the numerical ratio between the sex chromosomes and the autosomes was shown to be the determining factor.

In 1928 Morgan moved to the California Institute of Technology, Pasadena, to set up a Division of Biology, seeing this as an opportunity to establish biology in an institution where physics and chemistry were pre-eminent. His aim was 'to lay emphasis on the fundamental principles underlying the life processes in animals and plants'. There were to be no departmental divisions and close links with the Division of Chemistry. In 1933 Morgan was awarded the Nobel Prize in Medicine for his work in establishing the chromosome theory of heredity. He acknowledged this as a tribute to experimental biology and shared the prize money with Sturtevant and Bridges. In the same year the discovery of giant chromosomes in the salivary glands of insects transformed the cytology of *Drosophila* and made it possible to show a detailed correspondence between the genetic maps and the banding pattern visible on the salivary chromosomes.

After the move to California Morgan's own research returned to embryology. He never lost sight of the need to connect genetics to physiology and embryology. 'Between the characters that are used by the geneticist and the genes that his theory postulates lies the whole field of embryonic development.' A major problem lay in the fact that embryology was traditionally done in marine organisms, for which there was no genetic information. In 1935 George Beadle and Boris Ephrussi began their attempt to apply the techniques of experimental embryology to *Drosophila* at Pasadena; one of Morgan's last papers was on the genetics of the sea squirt *Ciona*. See also **Beadle, George Wells (1903–1989)**. Morgan's style of science influenced a generation of American scientists. His emphasis on experimental evidence and his distrust of 'metaphysics' was legendary, as was his dislike of spending money on equipment and his personal generosity. He was a fluent and prolific writer and had great charm.

Additional Reading

Allen, G.E.: *Thomas Hunt Morgan: The Man and His Science*, Princeton University Press, Princeton, NJ, 1978.

Horowitz, N.H.: "T. H. Morgan at Caltech: A Reminiscence," *Genetics*, **149**, 1629–1632 (1988).

Kohler, R.E.: *Lords of the Fly: Drosophila Genetics and the Experimental Life*, University of Chicago Press, Chicago, IL, 1994.

Shine, I., and S. Wrobel: *Thomas Hunt Morgan: Pioneer of Genetics*, University Press of Kentucky, Lexington, KY, 1976.

E.G. WINCHESTER, Wellcome Institute for the History of Medicine, London, UK

MORMON CRICKET (*Insecta, Orthoptera*).   A large wingless long-horned grasshopper of the western United States. It varies in color from pale green or yellow to black. It eats vegetation of all kinds and is a cannibal and scavenger, even eating dead animals. When abundant it is a serious crop pest and one that is difficult to combat.

MORMYRIDS (*Osteichthyes*).   Of the order *Isospondyli*, family *Mormyridae*, there are well over 100 species of these fishes. They are characterized by medium-long to very long snouts. They are bottom feeders, foraging worms out of the mud, as well as various other small creatures and insect larvae. There is a wide range of size, from 6 inches (15 centimeters) in length to 5 or more feet (1.5 + meters). The larger mormyrids are food fish, usually in the dried form. The very long, extended snout of the *Gnathonemus numerous* (elephant-nose mormyrid) is a particularly striking feature, nothing else quite like it in the world of fishes. A much less pronounced but still prominent snout is exhibited by the related species *Mormyrops deliciosus*. Only in recent years has it been found that most or all species of mormyrids possess an electrogenic capability, with voltages discharged in the microvolt range. It is assumed that this feature is useful in locating food sources. These electrogenic abilities in the *Gymnarchus niloticus* have been recognized over a longer period of time. This fish can attain a length of 5 feet (1.5 meters) and is well considered as an edible fish. Electrical discharges in this species are by way of the tail. Biologists are interested in some apparent parallels exhibited by the African mormyrids and the Southern American gymnotids. The latter group includes the electric eel. In terms of classification, they are not related, but do display somewhat similar behavior patterns.

MORNING GLORY.   See Clouds and Cloud Formation.

MORPHINE.   [CAS: 57-27-2]. About 10% of the weight of opium is morphine which was the first of the vegetable alkaloids to be isolated in 1805 by Sertürner. Since the source of the natural alkaloids is opium, all narcotics whose actions resemble those of morphine are sometimes referred to as opiates. Semisynthetic agents are usually made by altering the morphine molecule, and include such agents as diacetylmorphine (heroin), ethylmorphine (*Dionin*), dihydromorphinone (*Dilaudid*), and methyldihydromorphinone (metopon). Synthetic narcotics include agents with a wide variety of chemical structures. Some of the important synthetic agents are meperidine (piperidine type), levorphanol (morphinian type), methadone (aliphatic type), phenaxocine (benzmorphan type), and their derivatives. The structures of the various narcotics are given in Fig. 1.

Since morphine is responsible for the major actions of opium and the actions of all narcotics are qualitatively similar, morphine can be used as a model for discussing narcotic agents. The most prominent effects of morphine in the human body are on the central nervous system and the gastroenteric tract. The principal central action of morphine is the relief of pain, and this occurs in at least three ways: (1) morphine reduces central perception of pain probably at the thalamic level; (2) it alters the reactions to pain probably at the level of the cerebral cortex; and (3) it elevates the pain threshold by inducing sedation or sleep. In the medulla, morphine depresses the respiratory, cough and vasomotor centers and indirectly stimulates the vomiting center. The nuclei of the occulomotor (III) and vagus (X) nerves are stimulated by sufficient doses of morphine causing myosis (constriction of the pupils), bradycardia (slowing of the heart rate), and increased gastroenteric tone. The overall effect of morphine on the gastroenteric tract is spasmogenic and constipative. Morphine causes the constipative action by several means, including increased segmental movement of the large bowels, spastic tonus of the sphincters, decreased defecation reflex, and increased reabsorption of water in the large intestines to cause drying of feces.

The metabolic effects of morphine are not marked and are clinically unimportant. The metabolic rate may be decreased slightly due to the lowered activity and tone of the skeletal muscles resulting from the central depression. A rise in blood sugar may be observed after the injection of morphine. The hyperglycemia is due to glycogenolysis in the liver resulting from the release of epinephrine from the adrenal medulla. The lowering of urine production noted after the administration of the drug is due mainly to the release of antidiuretic hormone from the posterior pituitary gland.

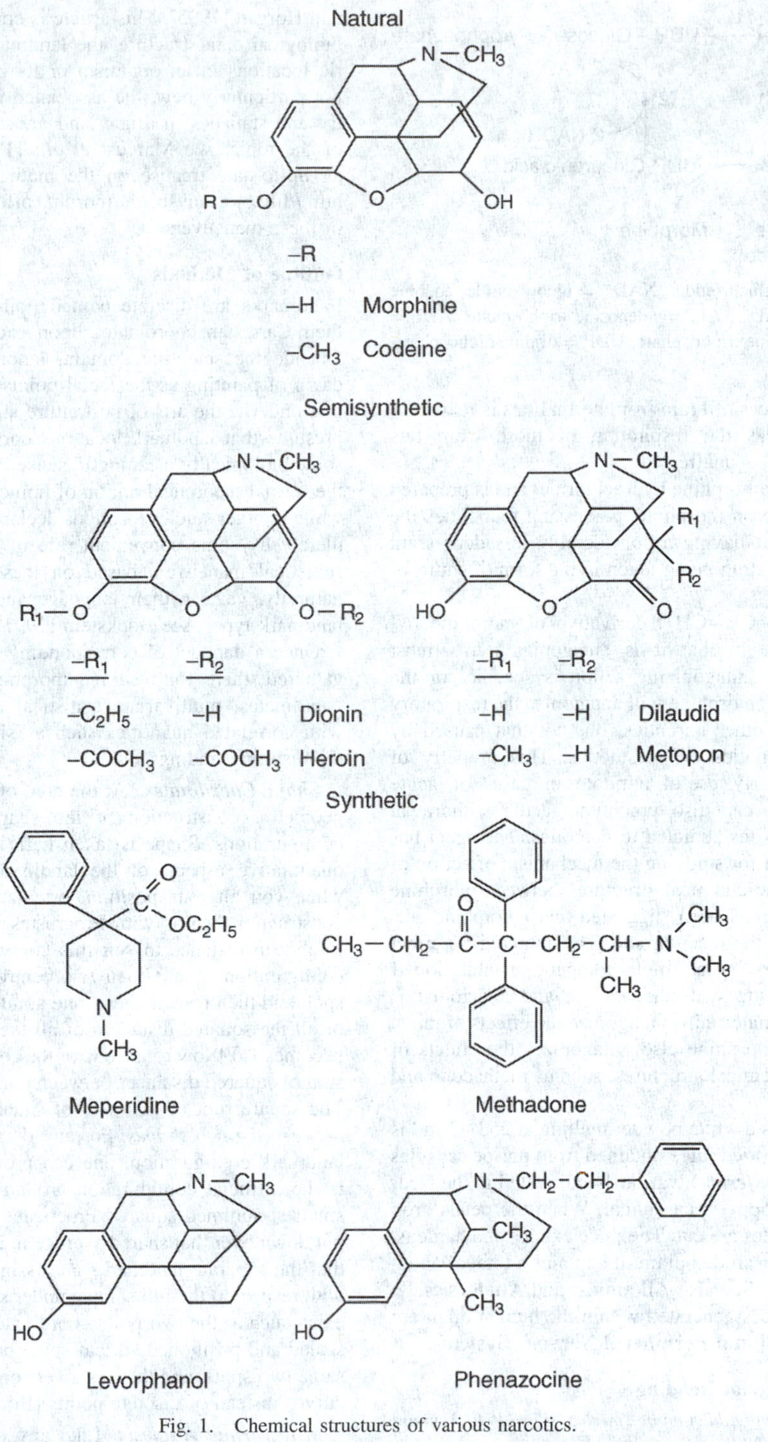

Natural

–R	
–H	Morphine
–CH$_3$	Codeine

Semisynthetic

–R$_1$	–R$_2$	
–C$_2$H$_5$	–H	Dionin
–COCH$_3$	–COCH$_3$	Heroin

–R$_1$	–R$_2$	
–H	–H	Dilaudid
–CH$_3$	–H	Metopon

Synthetic

Meperidine

Methadone

Levorphanol

Phenazocine

Fig. 1.   Chemical structures of various narcotics.

Morphine is detoxified or biotransformed mainly in the liver by conjugation with glucuronic acid. Morphine is conjugated by a series of reactions involving the formation of uridine diphosphoglucose (UDP-glucose), the oxidation of carbon-6 of glucose to form uridine diphosphoglucuronic acid (UDP-glucuronic acid) and the transfer of glucuronic acid to morphine to form the morphine glucuronide. This reaction is diagramed in Fig. 2. The following enzymes catalyze the sequential reactions; reaction (1), UDP-glucose pyrophosphorylase; reaction (2), UDP-glucose dehydrogenase; reaction (3), glucuronyl transferase; reaction (4) nucleoside diphosphokinase.

The most serious drawback in the use of morphine and other narcotic analgesics is their addictive potentiality. The characteristics of drug addiction include psychological need or habituation, tolerance and physical dependence. Habituation consists of an emotional and psychic dependence,

and in addiction, the habituation becomes an overpowering desire to take the drug. Tolerance is a phenomenon whereby the dosage of the drug must be continually increased to maintain equivalent pharmacological effects. Physical dependence develops when the tissues of the body become so adapted to the effects of the drug that the cells of the tissues cannot function normally without the drug in the environment. This is the most vicious characteristic of drug addiction.

The mechanisms underlying the development of tolerance are not fully understood. Biochemically, it may be attractive to explain tolerance by decreased absorption, altered distribution, increased biotransformation, and/or increased excretion of the drug. However, these processes have been shown to be unrelated to the development of tolerance. Thus, cellular adaptation offers the greatest likelihood for clarifying the phenomenon. Evidence for cellular adaptation is the finding that the respiration of

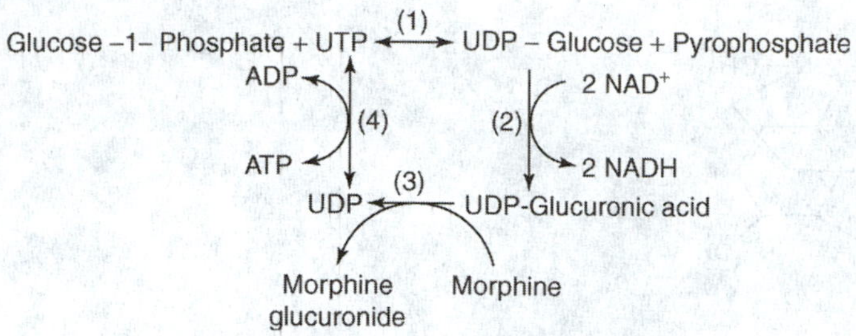

Fig. 2. Formation of morphine glucuronide. NAD^+ = nicotinamide adenine dinucleotide; NADH = reduced NAD^+; ATP = adenosine triphosphate; ADP = adenosine diphosphate; UTP = uridine triphosphate; UDP = uridine diphosphate.

chemically stimulated cortical slices of brain from normal rats is markedly depressed by morphine, whereas the respiration of those from rats chronically dosed with morphine is unaffected.

Heroin is diacetylmorphine (diamorphine hydrochloride) and is prepared by the action of acetic anhydride on morphine, possessing four times the analgesic affect of morphine, but having a considerably less depressant effect. Addiction is common, the drug being taken in the form of snuff, or by injection.

Nalorphine, the allyl ($-CH_2-CH=CH_2$) derivative of morphine (*N*-allylnormorphine) is remarkable in that it is antagonizing to almost all the effects of narcotics. The antagonizing action is specific for the narcotic analgesics. For instance, nalorphine will antagonize the respiratory depression due to morphine or other narcotics, but not that caused by other depressants, such as hypnotics or anesthetics. This property of nalorphine makes it a particularly useful antidote in cases of acute morphine poisoning. The agent can also precipitate acute withdrawal symptoms if administered to persons addicted to narcotics. The agent has become a useful biochemical tool for studying the mechanism of action of narcotics and tolerance. Since the chemical structures between morphine and nalorphine are so similar, it has been suggested that nalorphine acts by competing with morphine for the receptor site. The antagonistic effect of nalorphine cannot be explained by a simple competitive inhibition if equal affinity for the receptor site with the agonist and antagonist is assumed, because small doses of nalorphine antagonize the effects of much higher doses of the narcotic. Nalorphine also antagonizes the effects of synthetic narcotics of varying chemical structures, such as methadone and meperidine.

Morphine, $C_{17}H_{19}NO_3 \cdot H_2O$, is a white powder melting at 253 °C and is derived from opium which is the dried juice obtained from unripe capsules of the poppy plant (*Papaver somniferum*), variously cultivated in the Near East and Far East. The opium poppy is an annual. When the petals drop from the white flowers, the capsules are cut. The juice exudes and hardens, forming a brownish mass which is crude opium. It contains a total of about 20 narcotics, including morphine. See also **Alkaloids**; and **Analgesics**.

Natural morphine-like substances generated within the human brain are described in the entry on **Central and Peripheral Nervous Systems**.

Additional Reading

Courtwright, D.: *Dark Paradise: Opiate Addiction in America before 1940*, Harvard University Press, Cambridge, MA, 2001.

Gianino, J. et al.: *Intrathecal Drug Therapy for Spasticity and Pain: Practical Patient Management*, Springer-Verlag New York, Inc., New York, NY, 1995.

Web Reference

U.S. Food and Drug Administration Center for Drug Evaluation and Research. http://www.fda.gov/cder

MORPHOMETRICS. Morphometrics is the quantitative summary of size and shape differences among organisms or their components. Its methods vary according to the style of size or shape data under investigation. One familiar data type comprises measures of extent, such as lengths or volumes. For these, one typically uses principal component analysis of log-transformed measures or a related covariance-based technique [Reyment, 1991]. Another specialized toolkit, often used for the repeated tiny shapes or textures in microscopic images, is variously called *morphometry* or mathematical morphology, as surveyed, for instance, in Glasbey

and Horgan [1995]. This article is concerned with a separate, distinctively biological data structure, the landmark configuration of discrete geometric locations in an organism or its image. Although this type of data is not particularly new, the associated morphometric tools combine geometry and statistics in a new and unusual way. For an accessible overview of the topic, see Marcus, *et al.*: [1996]. Dryden and Mardia [1998] is an up-to-date treatise on the mathematical foundations, and a website http://life.bio.sunysb.edu/morph/, offers convenient PC software for most of these maneuvers.

Outline of Methods

Landmarks are discrete named points or curving structures located by their Cartesian coordinates upon each of a sample of biological forms. Outside the scientific domain, landmarks have been familiar since the dawn of painting as the facial points (bridge of the nose, tip of the chin) that underlie the arts of portraiture and caricature. Morphometric methods presume that landmark locations correspond across the entire data set in some biologically meaningful sense. Sometimes correspondence embodies the formal theoretical notion of homology from evolutionary morphology, while in other studies locations declared "the same" may have been aligned merely by some convenient rule of thumb. Biological interpretations of statistical maneuvers based on these different types of landmark will, naturally, vary in their language and their authority. For the variety of landmark types, see Bookstein [1991].

Once a data set of corresponding landmark points or curves has been gathered, there follow three morphometric steps: construction of shape coordinates, multivariate statistical analysis of these coordinates along with correlated quantities such as size or function, and visualization of the resulting findings.

Shape Coordinates. At the core of the landmark method is a canonical geometric construction of the shape variation in a set of landmark configurations. Shape is taken here in its ordinary English sense, as the quantitative aspects of the landmark configuration that do not change when you alter its position, orientation or scale. The formal geometric construction begins with Procrustes distance, a modification of ordinary Euclidean distance to suit this concept of shape. For any two landmark configurations, place their two centroids (centers of gravity) at the same spot, and then rescale them, one at a time, so that each Centroid Size (sum of all the squared distances of all the landmarks from this fixed location) becomes 1.0. Now rotate one of the configurations over the other so that the sum of squared distances between corresponding landmarks is a minimum. The square root of this sum of squares is called the *Procrustes distance between the two shapes*. For any data set of two or more specimens of a landmark configuration, one can produce their Procrustes average—this is the artificial configuration around which they have, as a sample, the smallest summed squared Procrustes distances—and when each form is put down over that shared average in its own Procrustes pose, the locations that the separate landmarks are assigned are called its *shape coordinates* with respect to the full sample under study. Other sets of shape coordinates exist, such as the two-point coordinates that result when configurations are scaled and positioned so that some pair of landmarks always lands in the same two spots, and there is a version of all this that applies to continuous curves instead of discrete points [Bookstein, 1997].

Multivariate Analysis. Like any other set of multiple measurements, shape coordinates can be averaged, regressed or correlated with other measurements, summarized by principal components, and the like. In studies of causes or effects of shape, such as group differences, one convenient summary quantity is the Procrustes variance accounted for by the causes or effects. This is the sum of the usual explained variances over all the shape coordinates when they are regressed on the claimed cause(s) or effect(s) of shape. Under certain assumptions, the ratio of explained to unexplained Procrustes variance can be tested against an F distribution [Dryden and Mardia, ref.]; but it is more realistic to examine it by a permutation test [Good, ref.]. Testing the Procrustes variance explained as a single sum means considering all possible patterns of shape difference on the given landmarks, not just the particular sort of difference (perhaps a length–width ratio) that the computation at hand happened to produce. For instance, to explore allometry (association of shape with size), one regresses each of the shape coordinates on the scaling factor called *Centroid Size* above; the significance test uses the sum of all the explained variances for all the shape coordinates. Other specific strategies apply to studies of bilateral symmetry, studies of geometrically uniform transformations, and large-scale growth gradients.

Visualization. Findings derived from the shape coordinates are intended to be drawn back in the picture plane or space where the landmark coordinates originated. For this purpose a powerful tool has been borrowed from continuum mechanics. The thin-plate spline extends any rearrangement of landmark locations into a deformation grid in the way that involves the least bending energy, which is the integrated squared second derivative of the grid mapping. A thin-plate spline deformation grid, in spite of its startlingly simple algebraic formula, attracts the scientist's eye to whatever gradient or localization of the form change may be implied by the joint displacements of the shape coordinates as a complete configuration. Group differences, regressions on causes or effects, and principal components of shape can all be visualized in this same way.

A Typical Application

In the following example, the landmarks are 13 points from the midsagittal plane (plane of symmetry) of the human head, as seen by magnetic resonance imagery, in 14 adults diagnosed with schizophrenia and 14 others. Figure 1 shows the landmarks *in situ* for one case. The first computational step results in the shape coordinate scatter shown in Figure 2: locations of 364 points (13 landmarks for 28 cases) after each form has been superimposed over the sample average in the way that minimizes summed squared distance. Figure 3 shows one simple multivariate finding in this data set, the contrast between average shapes of the schizophrenic and normal subsamples. The Procrustes variance explained by group is one-fourth of the sum of the squared distances between the paired dots here; it is significant (by permutation test) at about $P \sim 0.04$. Interpretation of the mean difference as a deformation, Figure 4, shows the schizophrenic mean to differ from the normal by a relatively focal expansion near the quadrigeminal cistern. Graphical enhancement of this expansion, Figure 5, shows it to be the only shape difference in the scene. (This Figure is an example of the method of creases [Bookstein, 2000], which extrapolates a transform forward or backward until some part of the grid is pinched off.) The expansion is situated between splenium point and colliculus, but is not quite aligned with the segment between them. See also **Magnetic Resonance Imaging (MRI)**.

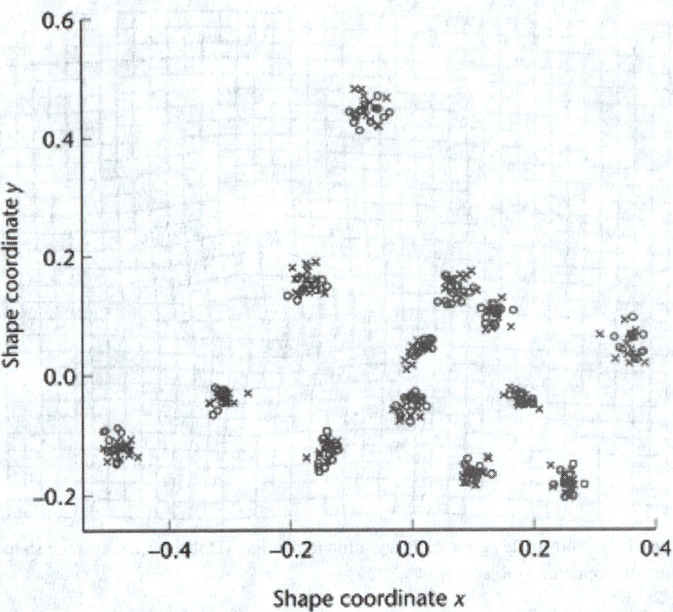

Fig. 2.   Procrustes shape coordinates for 28 cases; o, normal; ×, schizophrenic.

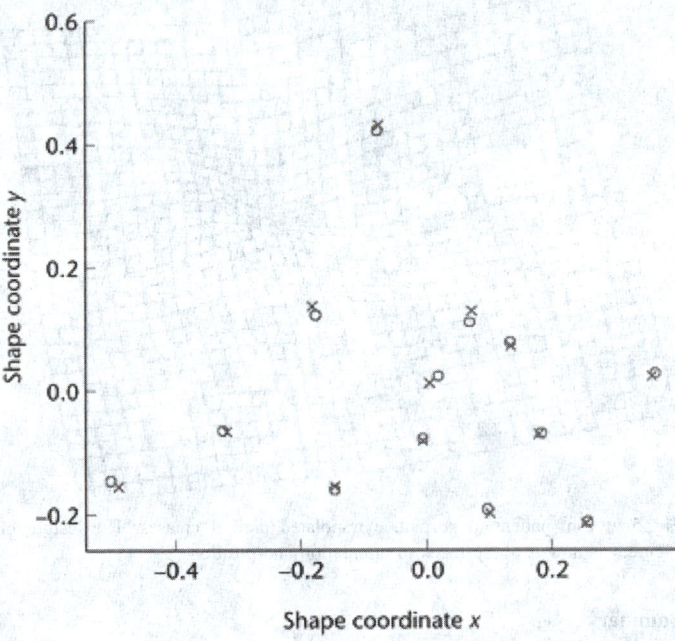

Fig. 3.   Group mean shapes, normal (o) and schizophrenic (×) subsamples.

[1991] and Marcus, et al.: [1996], they work to help summarize group differences, explore within-group modes of variability, predict form by its causes, and predict effects of form by form.

Future Developments

The range of applications of this relatively new methodology can be expected greatly to broaden over the next few years. There will be new visualization strategies for solid images that change over time, for instance, growth or disease processes as observed in a single organism. Other approaches yet to be designed will combine observations at different physical scales (from electron micrographs through histological slides to gross photographs), or from different biophysical modalities (e.g. electromagnetic activity along with biochemical concentration information in the animal brain), in one coherent spatially organized information hierarchy. The ultimate goal is to supply a quantitative foundation for rigorous study of interrelationships between form and function throughout a whole range of organismal phenomena.

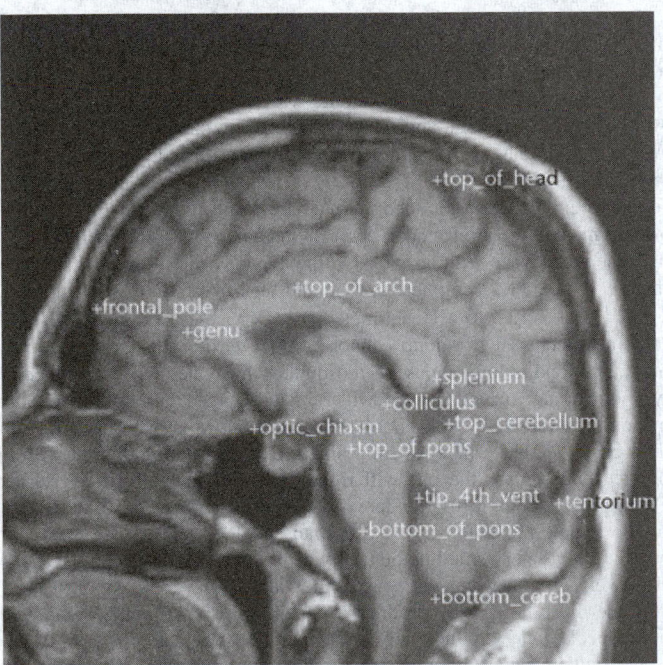

Fig. 1.   Thirteen landmarks on a typical midsagittal image.

This example is typical of the variety of applications of these tools. Across evolutionary and developmental biology, as surveyed by Bookstein

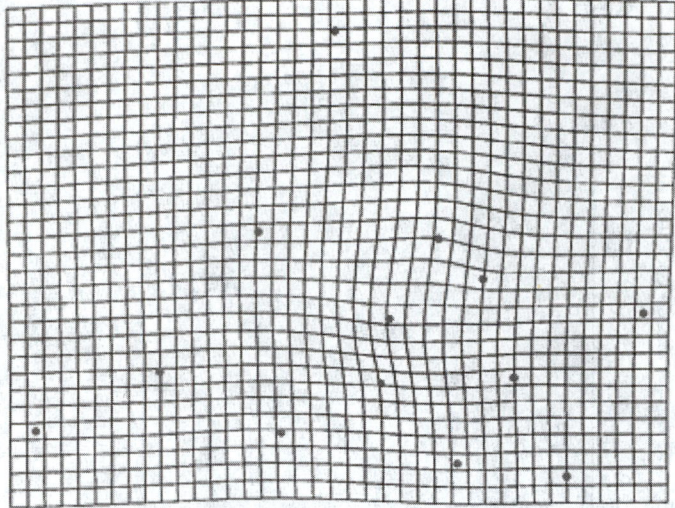

Fig. 4. A thin-plate spline calls attention to a bulge. (Dots indicate average shape for schizophrenic subjects.)

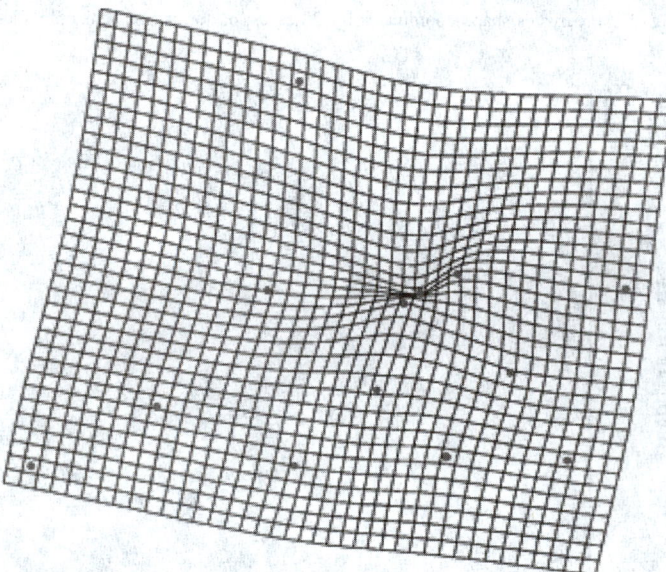

Fig. 5. From patient to normal, extrapolated until it creases. The feature of Figure 4 is now in sharp focus in orientation and position.

Summary

Morphometrics today emphasizes novel statistical techniques for landmark location data. Shape differences are summarized by multivariate analysis of shape coordinates, and findings are visualized by quantitative diagrams back in the plane or space where the landmark data originated. These analyses are good for summarizing variation within or between groups of organisms, for correlating their forms with causes or effects of form, and for focusing the scientist's attention on the specific regions of the form entailed in those group differences or correlations.

Additional Reading

Bookstein, F.L.: *Morphometric Tools for Landmark Data*, Cambridge University Press, New York, NY, 1997.

Bookstein F.L.: "Shape and the Information in Medical Images: a Decade of the Morphometric Synthesis," *Computer Vision and Image Understanding*, **66**, 97–118 (1997).

Bookstein, F.L.: "Creases as Local Features of Deformation Grids," *Medical Image Analysis*, **4**, 93–110 (2000).

Dryden, I.L., and K.V. Mardia: *Statistical Shape Analysis*, John Wiley & Sons, Inc., New York, NY, 1998.

Glasbey, C.A., and G.W. Horgan: *Image Analysis for the Biological Sciences*, John Wiley & Sons, Inc., New York, NY, 1995.

Good, P.I.: "Permutation Tests," *A Practical Guide to Resampling Methods for Testing Hypotheses*, 2nd Edition, Springer-Verlag New York, LLC, 2000.

Krim, H., and A. Yezzi, Jr.: *Statistics and Analysis of Shapes*, Springer-Verlag New York, LLC, New York, NY, 2005.

Marcus, L.A., M. Corti, A. Loy, G.J.P. Naylor, and D.E. Slice: *Advances in Morphometrics*, NATO Science Series A: Life Sciences, Vol. 284, Kluwer Academic Publishers, New York, NY, 1996.

Reyment, R.A.: *Multidimensional Palaeobiology*, Elsevier Science & Technology Books, New York, NY, 1991.

Small, C.G.: *The Statistical Theory of Shape*, Springer-Verlag New York, LLC, New York, NY, 1996.

FRED L. BOOKSTEIN, University of Michigan, Ann Arbor, MI

MOSAIC VISION. A theoretical interpretation of the action of compound eyes of arthropods. Each visual unit (ommatidium) of such an eye forms an image of part of the object toward which it is directed and the total image seen by the animal is supposed to be composed of many such partial images formed by the many units in the eye.

MOSELEY, HENRY (1887–1915). A British chemist who studied under Ernest Rutherford and brilliantly developed the application of X-ray spectra to the study of atomic structure; his discoveries resulted in a more accurate positioning of elements in the periodic table by closer determination of atomic numbers. Tragically for the development of science, Mosely was killed in action at Gallipoli in 1915.

MOSELEY'S LAW. The square root of the frequency of a given line of an element in the X-ray spectrum is directly proportional to the atomic number of the element.

MOSQUITO (*Insecta, Diptera*). A small two-winged fly with slender body, long legs, and narrow wings bearing scales along the veins. The larvae are aquatic. Male mosquitoes feed on plant juices and only the females suck blood, but in some species neither sex sucks blood.

Mosquitoes are well known as a nuisance and in warmer climates they are also dangerous because some species transmit disease. Malarial parasites are carried by mosquitoes of the genus *Anopheles* and yellow fever by species of *Aedes*. In regions where these diseases occur the destruction of mosquitoes by draining swampy areas where they may breed, and by applying oil to water that cannot be drained, is important.

The protection of patients from the attack of mosquitoes so that the disease cannot be carried to others is accomplished by adequate screening, a measure that is valuable both for comfort and safety in all dwellings.

Occurrence of the mosquito is quite widespread. They are found in the Americas, Europe, Asia, and Africa. They are abundant in Alaska and the Arctic regions. They also have been found in the Himalayas amid rock and snow at altitudes of 20,000 feet (6100 meters). Generally, mosquitoes are found wherever warm-blooded animals are to be found. For raising their young, mosquitoes prefer lakes, ponds, stagnant pools and puddles.

Dimensions vary with species. On the average, the mosquito is approximately $\frac{3}{4}$ inch (1.9 centimeters) long and from wing tip to wing tip about 1 to $1\frac{1}{2}$ inches (2.5 to 4.1 centimeters).

Some species of mosquito cannot produce fertile eggs without a prior meal of blood. Many types of mosquitoes have definite preferences in type of animal to feed on and will reluctantly feed upon another. Some mosquitoes excrete a sticky liquid and construct rafts on water by gluing their eggs together. The eggs are pointed. By constructing the raft with the pointed ends down, the raft will right itself automatically if overturned. The top part of the raft is rendered waterproof by the sticky exudate.

The female mosquito may lay as many as 40 to 200 eggs on water or near it. Only a very small amount of water is needed for the eggs to hatch. However, without water the eggs cannot survive.

Some mosquitoes may produce a generation every three weeks; while other species may produce only one generation per year.

Mosquitoes cannot breathe under water. The mosquito's breathing tube (sometimes called a snorkle) is kept dry by an oily secretion on it that repels water. If a thin film of petroleum is poured over the water, the petroleum enters the snorkle causing the mosquito larvae to smother. For many years, this has been an effective procedure in controlling mosquito population.

Some important species of mosquitoes include: *Anopheleini maculipennis*, common in Europe and the most numerous mosquito in the tropics; the *Aedes aegypti*, which transmits yellow fever in the Americas, Africa, and all tropical regions; the *A. albopictus*, which transmits dengue fever; the *A. obturbans*, which is found in the Pacific Islands and the coast of Asia; and the *Sabethes* and *Haemagogus*, which transmits jungle and yellow fever in jungle areas. See Fig. 1.

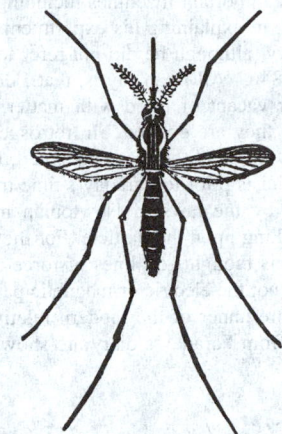

Fig. 1.   Yellow-fever mosquito, Stegomyic fasciata. (*F.B. Howard, USDA.*)

MÖSSBAUER EFFECT. The phenomenon of recoilless resonance fluorescence of gamma rays from nuclei bound in solids. It was first discovered in 1958 by R.L. Mössbauer. The extreme sharpness of the recoilless gamma transitions and the relative ease and accuracy in observing small energy differences make the effect an important tool in nuclear physics, solid-state physics, and chemistry.

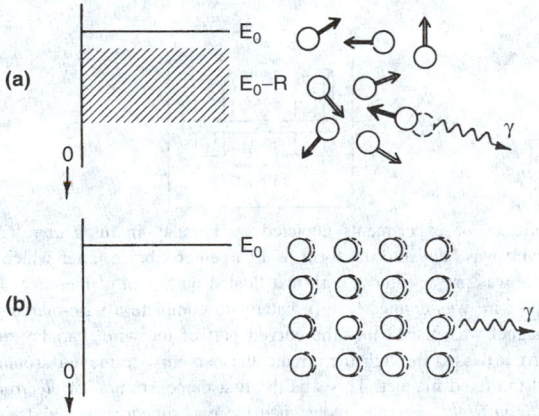

Fig. 1.   (**a**) Emission of a gamma-ray by a single atom moving randomly in a gas transfers appreciable energy to the emitting atom in the form of recoil kinetic energy, and reduces the energy of the gamma ray from the transition energy E_0 to some lower energy $E_0 - R$. (**b**) Emission of a gamma ray by an atom bound into a crystalline structure may sometimes cause recoil of the whole crystal, in which case, the loss of energy in the form if kinetic energy of the crystal is negligibly small and the gamma ray appear to be emitted with an energy E_0.

If a gamma ray is emitted by an atomic nucleus, the system to which the emitting atom belongs must recoil, in order to conserve momentum, in a direction opposite to that in which the gamma rays is emitted. Similarly, if an atomic nucleus absorbs a gamma ray, the system must continue to move, following absorption, in such a way that momentum is conserved. If the recoiling system is a single atom, such as in a gas and shown schematically in Fig. 1(a), the emitting atom carries away enough energy from the transition for the observed energy $E_o - R$ of the emitted gamma ray to be measurably less than the energy E_o of the nuclear transition that caused the gamma ray to be emitted, also indicated in (a) of the diagram.

Furthermore, a gamma ray that is absorbed by a single atom must transfer a measurable kinetic energy to that atom, as well as the energy of the nuclear transition. On the other hand, if the emitting or absorbing nucleus belongs to an atom that is bound into a crystalline structure, such that the structure as a whole can recoil, and as indicated schematically in Fig. 1(b), the kinetic energy that must be given to the crystalline system to conserve momentum is greatly reduced, compared to the energy that must be given to a single atom, because of the much larger mass of the system. The recoil energy is then so small that the gamma ray carries away essentially the full energy E_o of the transition in the case of emission, transferring such a small fraction of its energy to the absorbing system that emission and absorption appears to be recoil-free.

This process is observed, of course, in the analogous case of the resonance radiation in atomic transitions, in which case, the photons have energies in the range of light, commonly visible light. However, the protons of gamma radiation are so much more energetic that their energy loss by recoil of the nucleus emitting them is great enough, in the case of free atoms, for the resonance effect not to occur.

Mössbauer discovered, however, that in the case of atoms which are not free, but bound in a solid, the effect can often be observed. It is easily demonstrated when the normal, free-atom recoil energy is comparable to the energy of the quantized lattice vibrations. Under these conditions, zero-phonon processes are possible in which the entire energy of the nuclear transition goes into the gamma ray and the recoil momentum is taken by the solid as a whole. The resulting gamma rays then have the proper energy to be resonantly absorbed or scattered in an analogous zero-phonon process.

The Mössbauer effect is useful in determining nuclear level widths and Doppler effects. Another application is based upon the measurement of nuclear hyperfine structure, a measurement which is possible when the line-width of the gamma ray is smaller than the hyperfine interaction (that due to the coupling of the nuclear moments with external fields). In this application, the Mössbauer effect is almost unique because one obtains the splitting of both the ground and the first excited nuclear states. This effect, in turn, makes possible the determination of nuclear moments of the excited states, which can be important tests of nuclear models. Another important feature is the so-called isomer or chemical shift (terms used interchangeably) which measures the simple electrostatic interaction of the nucleus with its own s-electron and has given information about the difference in the nuclear radii of the ground and excited states.

MOSSBAUER, RUDOLF LUDWIG (1929).   Mossbauer was a German physicist who discovered and had named after him the Mossbauer effect. He received the Nobel Prize in Physics in 1961.

See also **Mössbauer Effect.**

J. M. I.

MOTH (*Insecta, Lepidoptera*).   Insects with the four wings at least partly scaly, the mouth parts formed for sucking, and the antennae rarely clubbed near the tips. No one character serves to distinguish all moths from the butterflies and skippers. Most moths are nocturnal but many are diurnal or crepuscular. Most of them have the antennae slender and tapering or broadened by setae or by processes from the segments, forming a comblike (pectinate) structure, but a few have an expansion just before the tip like that found in most skippers. Most of the caterpillars of moths pupate in a cocoon or a subterranean cell or in the tissues of plants but a few form a brightly colored naked pupa like those of the butterflies. The term does not apply to a principal division of the order but includes many families making up one entire suborder and most of the second. See Fig. 1.

There are at least 140,000 species of moth. The size varies considerably—the large Atlas moth of India measures about 10 to 12 inches (22.5 to 30 centimeters) (wing tip to wing tip) and the South American giant moth may be even larger. Many species have wing-tip spreads of only about $\frac{1}{2}$ inch (12 millimeters). Some of the numerous species of moth have been investigated quite intensively. Nerve impulses have been measured and habits studied in detail. Some moths go into a wing-beating ritual before taking to flight, warming up so to speak, much as a pilot may rev up an aircraft before taking off.

The moth larvae are soft and quite helpless when they emerge from their cocoon. However, the tiny parachute-like wings unfold and become useful wings within less than an hour. The moth achieves adult size, is firm,

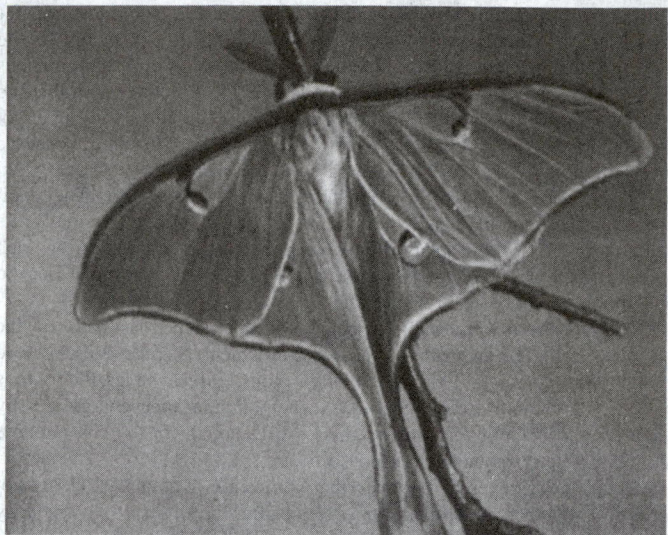

Fig. 1.   The luna moth. (*A.M. Winchester.*)

dry, and in full color within a very short period. Moth hormonal systems have been studied as well as habits. Some moths mimic other moths or even other insects, and may change coloration for purposes of camouflage almost instantly.

As in some other insects, the taste receptors of the moth are on the forelegs. The moth tastes prospective food by walking about on it. Each moth species has particular preferences in food and will follow the eating habits of its parents. In searching for a particular food, a moth may scan an entire garden within a short period without eating. The moth has a keen sense of smell and it is believed able to detect its mate over a distance of $\frac{1}{2}$ mile (0.9 kilometer). Although of the same order, moths differ basically from butterflies in the following major respects: When at rest, moths spread their wings, whereas the butterfly presses its wings close together over its head; the body of the moth is shorter and stouter than that of the butterfly.

Several species of moth are particularly damaging to food crops, this damage occurring mainly as the result of the larvae. Female moths commence the destructive activity when they find locations in which to place their eggs. Often, a thousand or more eggs will be deposited. The developing larvae are voracious eaters.

Some of the economic moths described in this book include: **Brown-Tail Moth**; **Bud Moth**; **Clear-Winged Moth**; **Codling Moth**; **Grain-Storage Insects**; **Gypsy Moth**; **Tussock Moth**; and **Wax Moth**. For food-destructive moths, see also **Dried-Fruit Insects**.

Additional Reading

Mitchell, R. et al.: *Butterflies and Moths: A Guide to the More Common American Species*, Western Publishing Company, Inc., North Platte, NE, 1987.

Tuskes, P. et al.: *The Wild Silk Moths of North America: A Natural History of the Saturniidae of the United States and Canada*, Cornell University Press, Ithaca, NY, 1995.

Young, M.: *The Natural History of Moths*, Morgan Kaufmann Publishers, San Mateo, CA, 1996.

MOTION (Gauss Principle of Least Constraint). 1. The motion of connected points is such that, for the elementary motion actually taken, the sum of the products of the mass of each particle into the square of the distance of its deviation from the position it would have reached if free, is a minimum.

2. The motion of a system of material points interconnected in any way and submitted to any influences, agrees at each instant as closely as possible with the motion the points would have if they were free. The actual motion takes place so that the constraints on the system are the least possible. For the measurement of the constraint, during any infinitesimal element of time, take the sum of the products of the mass of each point and the square of its deviation from the position the point would have occupied at the end of the element of time, if it had been free.

MOTMOT. See **Kingfishers and Other Coraciiformes**.

MOTOR (Electric). In an electric motor, electrical energy is converted into mechanical energy (torque in a rotary system). (Linear motors are described in the article on **Servomotors**). By way of a series of experiments conducted by Michael Faraday in the 1820s and culminating with his famous experiment of October 17, 1831, when he discovered the principle of electromagnetic induction, he had in essence invented the electric generator, among other important machines including the motor, dynamo, and transformer. Later, in explaining his experiments, Faraday developed the lines-of-force theory, although he did not refer to it by that name. He wrote in his diary in 1846, "All I can say is, that I do not perceive in any part of space, whether vacant or filled with matter, anything but forces and the lines in which they are exerted." In retrospect, going far beyond the immediate realm with which he was concerned, Faraday's vision was truly remarkable. Physicists prior to Faraday's time traditionally explained all physical processes by the laws of Newtonian motion and forces of mutual interaction working upon the particle. For the first time in physics, Faraday concentrated his thoughts on lines of force operating in space as of paramount concern, not the electric or magnetic particle. Thus, Faraday, unknowingly, was a forerunner of the modern relativistic era of physics. A few sketches taken from Faraday's diary are shown in Fig. 1.

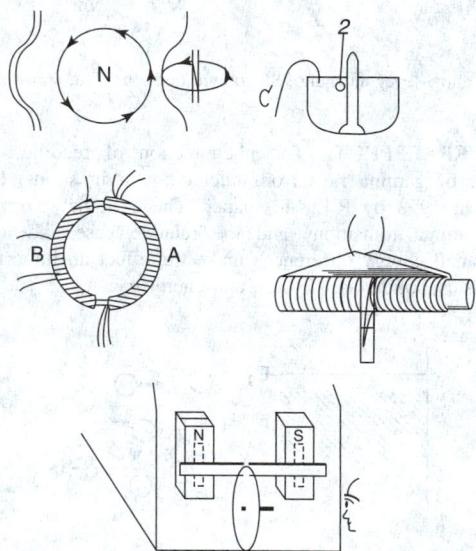

Fig. 1.   Series of experiments depicted by Faraday in his *Diary*. (*Top Left*) Copper wire was bent into the form of a carpenter's brace, after which one end of the wire was pressed into a cork that floated on a pool of mercury. The other end of the wire was connected to a battery to complete a wire-mercury circuit. A bar magnet was placed into the curved part of the wire. Faraday noted that when current passed through the circuit, the curved wire moved around until it contacted the fixed magnet. This was the first demonstration of *electromagnetic rotation*. (*Top Right*) Faraday redesigned the experiment, using a straight wire, again with one end impressed into a cork floating in the mercury container. Passage of current through the wire this time made the wire revolve continuously around the magnet and a reversal of current caused the wire to revolve in the opposite direction. This demonstrated the principle of the *electric motor*. (*Center Left*) In an effort to determine if a magnet would turn around a fixed conductor, Faraday wound two coils of wire (A and B) on an iron ring. Then he connected coil A to a battery and found that an intermittent current would flow in coil B. This demonstrated the principle of *electrical induction*. (*Center Right*) In another experiment, Faraday plunged a magnet in and out of a hollow cylinder and coil connected to a galvanometer. This showed that current can be induced as the result of relative motion of a conductor and a magnetic field. This demonstrated the principle of the *electric generator*. (*Bottom*) In another experiment, Faraday rotated a copper disk between the poles of a compound magnet, thus inducing a continuous current. This demonstrated the principle of the *dynamo*. (*Faraday's Diary, circa 1831.*)

From its modest start in 1831, the electric motor was years in development. The basic principles were understood, but creating practical operating machines required the ideas of many investigators. The commutator was developed by Henry, Pixii, and Wheatstone (1841); the drum armature by Siemens, Pacinotti, and von Alterneck (1867); ring armatures by Gramme (1870); disk armatures by Desroziers (1885); revolving magnetic fields and ac theory by Ferraris (1885); polyphase motors by Tesla (1888); the squirrel-cage rotor by Bradley (1889); and ac commutator motors by Eickmeyer, Thomson, and Atkinson (1889), among other important aspects of motor and generator design. During the decades that followed, refinement of electric machines (motors, etc.) continued apace. (Electricity on a large scale did not become available to major cities in the eastern United States until 1880s and 1890s, and somewhat later as the so-called electrical age moved westward.) Broad areas of motor improvements have included marked reductions in size and weight, greater efficiency, overload protection, and safety features. At present, much research is directed toward further improving efficiency by way of improved magnetic materials, conductors, thermal insulation, and even the refinement of relatively new concepts, such as linear and planetary motors. See also **Superconductivity**.

This article concentrates on the fundamentals of electric motors and on the major designs that have been traditionally available for a wide variety of uses. The somewhat specialized fields of servomotors and variable-speed motors used in industrial processes, robotics, etc. are discussed in the article on **Servomotors**. The principal classifications of electric motors are listed below and are diagrammed schematically in Fig. 2.

Electric motors are built in a range, varying from outputs of $\frac{1}{100}$ horsepower up to over 1,000 horsepower. A 50-horsepower motor is considered a large one, and the majority of electric motors used are in the range between $\frac{1}{4}$ and 10 horsepower. Standard motor sizes, above the flea power and fractional sizes, are $\frac{1}{4}$, $\frac{1}{3}$, $\frac{1}{2}$, $\frac{3}{4}$, 1, $1\frac{1}{2}$, 2, 3, 5, $7\frac{1}{2}$, 10, 15, 20, 25, 30, 40, and 50 horsepower. Sixty-cycle synchronous speeds are 3,600, 1,800, 1,200, 900, 720, 600, 514, and 450 rotations per minute. Full-load induction motor speeds are 2–5% less than these. The efficiency of the electric motor ranges from 75–95%. It is higher in large motors than in small. Induction motors are more efficient the higher the rated speed, but do motor efficiency is little affected by speed. Efficiency is often secondary to reliability; nevertheless, it is a factor to be considered, particularly if

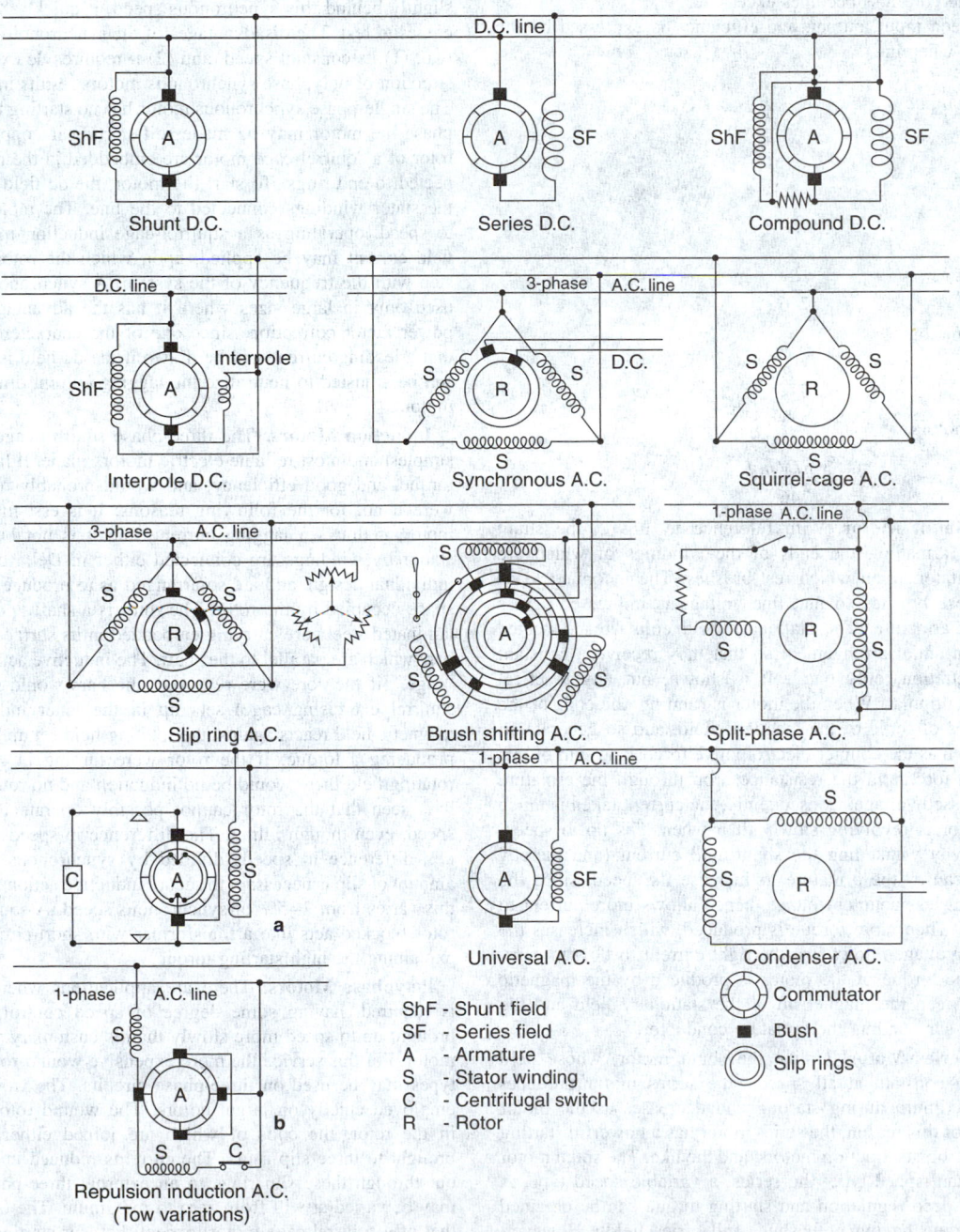

Fig. 2.   Fundamental circuits of some major types of electric motors.

the drive is heavy and the motor is well loaded over a considerable part of the time.

Direct-current motors are much less frequently employed than alternating current, because of the preponderance of ac over dc systems. However, speed control and starting torque are so excellent with dc that it is frequently used in ac territory where these characteristics are important. Dc power for motors is commonly obtained from the ac supply through motor-generators, mercury-arc tubes, or grid-controlled rectifiers, with voltages ranging from 110 to 600. The extra expense of the converter installation lays some handicap upon the employment of dc motors, and a number of methods have been devised to vary the speed of ac types, but, in the main, the latter are constant speed.

The losses sustained by a motor in converting electrical to mechanical power arise chiefly through the electrical and magnetic characteristics, but also, in some degree, through bearing friction and windage. The losses are, then, the resistance losses occasioned by current flowing through the conductors of the armature, the field, and the controller, the core losses of hysteresis and eddy currents, friction and windage. The cores of all motors must be built up of laminations insulated from one another by lacquer or enamel, otherwise the core loss becomes excessive.

The relation between input, output, and efficiency is expressed by the following equations, wherein

P = horse power output
η = efficiency
I = line current
V = lin voltage
$p.f.$ = power factor

For all dc motors,

$$P = \frac{\eta I V}{746}$$

For single-phase ac motors,

$$P = \frac{\eta I V p.f.}{746}$$

For three-phase ac motors,

$$P = \frac{\sqrt{3} \, \eta I V p.f.}{746}$$

Direct-Current Shunt Motor. With reference to Fig. 2, the shunt motor has a wound armature, the ends of the windings of which are brought to a commutator, upon which rest brushes. The incoming leads are connected to these brushes so that line voltage is impressed across the windings of the armature. The stationary field coils are connected across the brushes in shunt arrangement so that they receive a constant voltage. In the illustration, only one coil is shown, but any practical machine would be multipolar. When the motor is running, the coils of the armature cut the lines of force of the magnetic field, and so generate an internal voltage known as the counter-electromotive force. The sum of this counterelectromotive force and the resistance drop through the armature must equal the impressed voltage. Consequently, the current taken is much larger when the motor is revolving slowly than when it is up to speed. This also explains why weakening the shunt-field current (and thereby the magnetic field) causes the armature to increase its speed, since the weakened field causes less counter-voltage, hence allows more current to flow in the armature. Then more torque is produced, which increases the speed until the back voltage allows just the right current to flow to carry the existing load. The torque of the motor is produced by the magnetic reaction existing between the magnetism of the stationary field and the electromagnetic field surrounding the armature conductor.

Direct-Current Series Motor. Unlike the shunt motor, whose field current is practically constant at all speeds, the series motor produces a field, which is maximum during starting and decreases as the motor comes up to speed. For this reason, the series motor has a powerful starting torque, and is used for hoists, traction motors, and the like. The shunt motor is essentially a constant-speed type; the series, a variable speed type. A motor having better speed regulation and starting torque can be obtained by a compound winding having both shunt and series fields. However, the simplicity of the shunt-field motor, coupled with the possibility of

effecting a reasonable variation in speed by a variable resistance in the field circuit, has caused it to be widely used. It has been found, though, that any considerable weakening of the field is accompanied by sparking at the commutator, due to the demagnetizing armature reaction. Small poles, located between the shunt-field poles, and wound with series coils, will compensate for the distortion of the field flux, and such motors are known as interpole motors.

Alternating-Current Motors. The foregoing classification of motors indicates a primary division into synchronous and induction type alternating-current motors. Of these, the induction motor is the most important for many applications. However, the strictly constant-speed feature of the synchronous motor, dictates its selection for certain kinds of applications.

Synchronous Motors. The synchronous motor is practically an alternator operated inverted. It has a polyphase stator winding which carries the main line current. The field is wound on the rotor and is excited by dc brought to it by brushes resting on slip rings. The synchronous motor is stable only when operating at a synchronous speed corresponding to the frequency of the system, and if it is loaded to where it lags ever so slightly behind this synchronous speed, it quickly "falls out of step" and comes to rest. The disadvantages of the synchronous motor are principally two: (1) its constant speed, and (2) it requires dc excitation. Modern construction of polyphase synchronous motors results in good starting torque. The single-phase synchronous motor has no starting torque, but with three-phase the motor may be made self-starting if copper bars similar to the rotor of a squirrel-cage motor are embedded in the rotating field and connected to end rings. To start the motor, the dc field is open-circuited and the stator windings connected to the line. The motor will then come up to speed, operating as a squirrel-cage induction motor, after which the field current may be applied, upon which the rotor will lock itself into step with the frequency of the system. A synchronous motor is generally used only in large sizes where it has the advantage of providing some power factor correction, since one of the characteristics of this motor is that a leading current will be drawn if the dc field is overexcited, and this can be adjusted to neutralize the lagging current drawn by induction type motors.

Induction Motors. The three-phase squirrel-cage motor is one of the simplest and most reliable electric motors made. It has a powerful starting torque, and good efficiency, and would probably replace all other types were it not for the following reasons: It is essentially a constant speed motor, it draws a lagging current, and it is not built single-phase. The stationary windings are connected either in Delta or Y, as may suit the individual design, and are so arranged as to produce a rotating field in the space occupied by the rotor. The rotor is a shaft upon which is built up a laminated steel core carrying embedded in its surface copper or aluminum bars which are parallel to the shaft. The inductive action of the field on this "cage" (if the core were removed, the bars would resemble the familiar squirrel exercising cage) sets up in the latter induced currents whose magnetic field reacts against the rotating field set up by the stator winding, producing a torque. If the rotor were turning in synchronism with the rotating field there would be no induction and no rotor currents. Therefore it is seen that the rotor cannot possibly operate it at full synchronous speed, even though idling. The difference in speeds is expressed by slip, i.e., difference in speeds divided by synchronous speed, and a certain amount of slip is necessary to secure inductive action. As mentioned before, this varies from 2–5% of synchronous speed. A squirrel-cage motor with rotor blocked acts like a transformer with short-circuited secondary, thus explaining the high starting torque.

Polyphase Motors. There are applications where a polyphase motor is required, having some degree of speed control, and which may be brought up to speed more slowly than is customary with the squirrel-cage motor. For this service the more expensive wound-rotor and brush-shifting types may be used on three-phase circuits. The wound-rotor principle is employed chiefly on large motors. The wound rotor has polar windings in the rotor, the ends of which are joined either in Y or Delta, and brought to three slip rings. The currents induced in the rotor are brought out through these slip rings to an external three-phase resistance, which may be varied at will from zero to maximum. The operation is much like that of a squirrel-cage motor, except that, for starting, the rotor current is decreased by inserting the maximum of resistance in the external circuit.

This is gradually decreased as the motor comes up to speed, until all of the resistance is short circuited and the motor is operating inductively with a normal slip. Given constant torque, this motor may be varied in speed by varying the external resistance, but it is somewhat less efficient than the brush-shifting type, because of the energy consumed in the resistance. The brush-shifting motor is used where considerable speed variation is desired at good efficiency, as, for instance, when driving fans or pumps of large size. The brush-shifting motor has the primary winding on the rotating armature, similar to dc practice. This winding is connected to the three-phase line through slip rings. Another winding, called an adjusting winding, is also placed on the rotor; in fact, in the same slots, but is connected with a commutator, which is made fairly wide. The three-phase stator secondary windings are brought out individually to six brushes, which bear on the commutator, and are connected as shown in the diagram. Each set of three brushes is joined by a yoke, so that they may be moved simultaneously, and each pair is placed on opposite ends of the commutator. When these yokes are moved with respect to one another, they cause to be included a certain number of commutator segments in each secondary coil. When each pair of brushes is on a common commutator bar, the motor runs as a straight induction motor at slip frequency. By moving the brushes apart by rotating a yoke, the voltage induced in the commutator coil is added to that in the secondary, and the motor speeds up. These voltages may be subtracted by moving the brush in the opposite direction, resulting in slowing down the motor. Since the forces needed to move the yoke are very small, it may be readily operated by the light pressures produced in an automatic control system.

Single-Phase Motors. It is possible for a single-phase motor to operate inductively like the squirrel-cage motor, provided that it can be brought up to speed, but a single-phase squirrel-cage motor has no starting torque, so that numerous ways have been developed to overcome this limitation. In the split-phase motor, an inductance and resistance are used to displace the voltage at the mid-point so as to get an arrangement resembling a two-phase impressed voltage. Of course the starting torque obtainable is inferior to that of a polyphase motor, but is sufficient to start a motor attached to a drive requiring low starting torque. A fan illustrates this service. For heavy starting duty the starting torque of a single-phase motor is created by repulsion, which shifts over to induction as the motor comes up to speed. Several systems have been invented, two of which are illustrated. At *a*, in the lower left part of the diagram, the armature windings are brought out to a commutator, upon which rest two brushes connected externally by a low resistance conductor. A stator winding is connected across the line. The short-circuited armature has induced in it the large current necessary to secure starting torque. As the motor comes up beyond a certain speed, a centrifugally operated switch lifts the brushes from the commutator and applies to it a ring that short-circuits all the segments. When this is done the motor operates as a straight induction motor. This principle, known as repulsion-induction—i.e., repulsion starting and induction running—is employed in most small motors which are to produce large starting torques on single-phase supply. At *b* is another repulsion-induction principle, less complicated mechanically. Here, the switch operates during starting, and is closed for induction operation.

Universal Motor. This is a motor of the series type, which may be operated on either direct- or alternating- current. It is usually used in small sizes only, there being a compensating coil to prevent armature sparking and to improve the power factor.

Capacitor Motor. This is a split-phase motor, having the phase displaced by capacitance rather than by inductance. It is superior to the former in starting torque, efficiency, and power factor, but more expensive and slightly more bulky. The starting torque of the modern capacitor motor compares favorably with the repulsion-start motors, and, since the cost is less, the capacitor motor is replacing the repulsion type in many applications. In some of these motors the capacitor is disconnected by a centrifugal switch after starting; in others, it is left in the circuit to improve the operating characteristics and the power factor of the motor.

See also **Stepper Motors**.

Additional Reading

Gottlieb, I.: *Electric Motors and Control Techniques*, The McGraw-Hill Companies, Inc., New York, NY, 1994.

Hughes, A.: *Electric Motors and Drives: Fundamentals, Types and Applications*, Butterworth-Heinemann, Inc., Woburn, MA, 1993.

Kaiser, J.: *Electric Power: Motors, Controls, Generators, Transformers*, Goodheart-Willcox Publisher, Tinley Park, IL, 1998.

MOTOR FUELS. See **Petroleum**.

MOULT. 1. Ecdysis.

2. The shedding of old feathers by birds preparatory to the development of new. Most birds moult at least once a year, beginning just after the breeding season. Most flying birds shed the large flight feathers in pairs but a few shed them all together and temporarily lose the power of flight. The rest of the plumage is also shed and renewed little by little. Moulting is accompanied in many species by the seasonal changes in plumage which make some species so different in summer and winter. See also **Birds**; and **Poultry**.

3. The shedding of the outer layer of the skin by reptiles.

MOUND BIRDS *(Aves, Megapodiidae)*. This family of birds belongs to the suborder *Galli*. They are dark-colored and differ from all other birds in their particular method of incubating their eggs. They are between the size of domestic fowl and turkey. There are 7 genera with 12 species in the southwest of the Old World, with New Guinea as the focus. They are divisible into two tribes according to size.

The mound birds proper *(Megapodiini)* are small dark birds with short tails, often insular: They include (1) the Scrub Fowl *(Megapodius)* with the species Micronesian Scrub Fowl *(Megapodius laperouse)*, found on the Marianas, the Niuafoo Scrub Fowl *(Megapodius pritchardii)* found on Niuafoo (central Polynesia), and the Australian Scrub Fowl *(Megapodius freycinet)* in many subspecies from the Nicobar islands to northern Australia and Polynesia; (2) Maleo *(Macrocephalon maleo)* is found on Celebes; and (3) Wallace's Eulipoa *(Eulipoa wallacei)* is found on the Moluccas.

The large mound builders *(Alecturini)* are larger and are mostly tied to definite habitats. Their distribution is limited: (1) Latham's Brush Turkey *(Alectura lathami)* is found in eastern Australia; (2) Talegallas *(Talegalla)*, with three species on New Guinea; (3) the Combed Talegallas *(Aepypodius)*, with two species on New Guinea; and (4) the Mallee Fowl *(Leipoa ocellata)* inhabits dry areas in central, southern Australia. See Fig. 1.

Fig. 1. Mallee fowl, *Leipoa ocellata*. (Sketch by Glenn D. Considine.)

The ancient Egyptians first built ovens to incubate fowl eggs artificially and we do this today in electrically heated incubators. However, the mound birds, also called incubator birds, "discovered" this method much earlier. Many species lay their eggs near hot volcanic springs or even hotter lava. Others use the heat generated by rotting leaves and vegetation. Still others go to the seashore and lay their eggs in the sand where they can be warmed by the sun.

Mound builders use various means to incubate their eggs. The maleo *(Macrocephalon maleo)* and Wallace's eulipoa *(Eulipoa wallacei)* use only the sun's heat. They emerge from the dark forest and dig into their areas along the shore which are in sunshine. In the jungle fowl *(Megapodius)* the method of egg disposal varies greatly. The Australian scrub fowl *(Megapodius freycinet)* in particular, often selects places heated by volcanic action, such as New Britain and the Solomons. There it digs burrows, which may be up to a meter deep (3.3 feet), into warm soil. In other places these birds build large brood heaps, which have a diameter of 12 meters (39 feet) and a height of 5 meters (16 feet), of sandy soil and leaves. These are the largest structures built by birds. The incubation heat is provided partially by the sun and partially by fermentation of the leaves. In those places directly exposed to the sun, the heaps consist almost entirely of soil. In dense jungle they consist, however, almost entirely of leaves. Each egg is laid in the passage, which has been dug out at a place where there is a suitable incubation temperature. After egg-laying, the birds take no further steps to control the temperature. See also **Maleo**; and **Megapode**.

MOUNTAIN. Mountains may be classified in three chief groups: (1) mountains of accumulation (volcanoes); (2) mountains formed by crustal movements; (3) residual mountains (erosional remnants). Structural mountains are those whose form and relief have not, as yet, been particularly modified by erosion. The ridges are still anticlinal and the valleys synclinal. Later, in the cycle of erosion, the synclines may become mountains and the anticlines, valleys. If the folded mountainous region is ultimately reduced to a peneplain, and the peneplain is then lifted without further folding, a new cycle of erosion operating on the same, but base-leveled, structure will develop the type of topography now seen in the Appalachian mountains. See also **Antecedent Stream**; and **Earth**.

MOUNTAIN BUILDING. See **Earth**.

MOUNTAIN LION. See **Cats**.

MOUNTAIN OBSERVATION. See **Meteorology**.

MOUNTAIN-PLAINS WIND SYSTEMS. See **Winds and Air Movement**.

MOUNTAIN-VALLEY WIND SYSTEMS. See **Winds and Air Movement**.

MOUNTAIN WAVE. An atmospheric gravity wave, formed when stable air flow passes over a mountain or mountain barrier. Mountain waves are often standing or nearly so, at least to the extent that upstream environmental conditions (and diurnal forcing) are stationary. Two divisions of mountain wave are recognized, vertically propagating and trapped lee waves. Vertically propagating mountain waves over a barrier may have horizontal wavelengths of many tens of kilometers or more, usually extend upward into the lower stratosphere, and in pure form, tilt upwind with height. They can accompany foehn, chinook, or bora wind conditions. They have the capability to concentrate momentum on the lee slopes, sometimes in structures resembling a hydraulic jump, leading to occasionally violent downslope windstorms. When sufficient moisture is present in the upstream flow, vertically propagating mountain waves produce interesting cloud forms, including altocumulus standing lenticular (ACSL) and other foehn clouds. Intense waves can present a significant hazard to aviation by producing severe or even extreme clear air turbulence. Trapped lee waves generally have horizontal wavelengths of 5–35 kilometers (3–22 miles). They occur within or beneath a layer of high static stability and moderate wind speeds at low levels of the troposphere (the lowest 1–5 kilometers (0.6–3 miles) lying beneath a layer of low stability and strong winds in the middle and upper troposphere. These conditions are often diagnosed using a vertical profile of the Scorer parameter, a sharp decrease in midtroposphere indicating conditions favorable to trapped lee wave formation. Trapped lee waves assume the form of a series of waves running parallel to the ridges, and the crests of these waves often contain altocumulus, stratocumulus, wave clouds, or rotor clouds in parallel bands that can be very striking in satellite pictures. Because wave energy is trapped within the stable layer, these waves (and accompanying cloud bands) may dissipate only very slowly downwind, and they can continue downstream for many wavelengths spanning many tens of kilometers. Flow beneath the wave crests, occasionally made visible by rotor clouds, is often turbulent, thus presenting a significant hazard to low-level aviation. Vertically propagating mountain waves and trapped lee waves can coexist, and sometimes lee waves are incompletely trapped or

"leaky," leading to a variety of complex rotor interactions. This complexity of rotor patterns often produces interesting variations in cloud forms. As mountain waves propagate upward, the rotor's amplitude can grow to the point that the rotor "breaks," that is, the rotor becomes convectively unstable and overturns. Wave breaking can have an important role in vertically redistributing horizontal atmospheric momentum, as it slows the atmosphere by turbulent transport of the earth's momentum upward. See also **Atmosphere (Earth)**; **Clouds and Cloud Formation**; and **Winds and Air Movement**.

MOURNING CLOAK (*Insecta, Lepidoptera*).   A large butterfly, *Nymphalis* (*Euvanessa*) *iopa*, of Europe and North America. The wings are very deep maroon above, bordered with yellowish-white, and are slightly angular. The species hibernates in the adult stage and is often in flight on the first warm days of spring. See also **Butterflies**.

MOUSEBIRD (*Aves, Coliiformes*).   Peculiar bird of central Africa, also known as colies, and the only bird in its family. The mousebird is gray with dark tintings and, in a few species, spots of bright colors. The bird is small with a finch-like beak, short weak wings, and long tail. The feet are exceptionally strong. The mousebird only infrequently takes to the air, habituating the lower branches of trees and displaying mouse-like behavior—hence the name. They roost in very large groups, with their heads hanging downward. The diet is principally fruit. Cup-like nests are located in low shrubs and trees. See also **Coliiformes**.

MOUTH PROTECTORS.   See **Polymeric Dental Materials**.

MOVING AVERAGE.   A term used in time-series analysis. Given a series of values $u_1, u_2, \ldots, u_n$, the values

$$v_j = \sum_{k=0}^{m} u_{j+k} w_k$$

where w_k are a set of weights, are defined for $j = 1, 2, \ldots, n - m$. The average of a group of $m + 1$ consecutive terms moves, so to speak, along the series. The moving average is principally used to smooth the series, or to determine a trend in it.

MOZAMBIQUE CURRENT.   See **Ocean Currents**.

MP3 (MPEG-1 AUDIO LAYER 3).   See **Data Compression**.

MST (MESOSPHERE–STRATOSPHERE–TROPOSPHERE) RADAR.   A type of wind profiler designed to measure winds and other atmospheric parameters up to altitudes of 100 kilometers (62 miles) or more. In the troposphere and lower stratosphere (up to about 30 kilometers (19 miles) the radar signal is returned from refractive index fluctuations produced by turbulence in the neutral atmosphere. In the upper stratosphere and lower mesosphere (between about 60 and 100 kilometers (37 and 62 miles) refractive index variations are strengthened by the strong vertical gradient in electron density. Because the scale size of the refractive index fluctuations must be of the order of one-half the radar wavelength (Bragg scattering), and the minimum scale size of turbulence increases with height from a few centimeters in the lower troposphere to several meters in the upper stratosphere and lower mesosphere, most MST radars have operated in the VHF band (typically 30–60 MHz, 5–10 m wavelengths). Very sensitive UHF radars can detect echoes from incoherent scattering (thermal or Thomson scatter) by electrons above 60 kilometers (37 miles). MST radars are characterized by high-powered transmitters and large antennas. (VHF antennas range from 100 to 300 meters (328 to 984 feet) across.) Similar wind profilers that lack the transmitter power and antenna area to detect returns from the upper stratosphere and the mesosphere have often been called ST (stratosphere–troposphere) radars. See also **Radio Frequency Band**; and **Winds and Air Movement**.

MUCILAGES.   See **Gums and Mucilages**.

MUCORMYCOSIS.   Also called invasive phycomycosis or zygomycosis, this is a rare invasive disease caused by mucor-like fungi of the class

Phycomycetes, particularly the genera *Rhizopus* and *Mucor*, which are common airborne molds associated with decaying vegetable matter. They may invade the human body via lungs, sinuses, gastrointestinal tract, or through damaged skin, producing the disease of worldwide distribution. The characteristic trait of the infection is extensive localized necrosis followed by blood vessel invasion, possibly leading to infarction through thrombus formation.

In the compromised host, infection may lead to paranasal sinus destruction (often seen in diabetics with ketoacidosis), necrotic lung or skin lesions, and disseminated disease.

Basic symptoms of infection are fever and unilateral face pain with facial swelling, nasal obstruction, and proptosis. Palatal ulceration may occur, and the organisms may invade the brain.

Localized infections have been seen when surgical or burn wounds receive contaminated dressing packs.

Once infection has spread beyond the original site of invasion, phycomycosis is almost invariably fatal despite treatment. The most heroic measures are wide surgical debridement and amphotericidin B in maximum dosage.

MUCUS. A clear, slimy secretion secreted by animals where surfaces must be lubricated or moistened. It is produced at the surface of the body of fishes, and in some amphibians. Terrestrial vertebrates secrete it in the linings of the respiratory and digestive systems in abundance. A mucous membrane is a membrane composed of mucus-secreting epithelial cells. Mucous membranes line those canals, cavities, and tracts that communicate with the external air, such as the nose and throat, respiratory tract, generative and urinary passages, and the digestive system.

MUD DAUBER (*Insecta, Hymenoptera*). Any of several species of wasps that make their nests of mud.

The mud dauber has four wings, two antennae, two forelegs, with four additional legs attached to the hind part of the thorax. The insect is about $1\frac{1}{2}$ inches (4 cm) in length. The insects are referred to as "threadwaisted" because of the thin connection of the thorax and stomach. The mud dauber is a black-blue satin color and is not harmful. Small balls of mud are carried to the insect's nest which is often artistically designed. Separate cells are constructed. Each cell contains an adequate number of paralyzed spiders in which the mud dauber has deposited an egg. This is sufficient food for the larvae until its wings have developed.

MUELLER BRIDGE. See **Bridge Circuits (Electrical/Electronic)**.

MULBERRY FAMILY. Widely scattered in all but the coldest regions of the world, members of the mulberry family (*Moraceae*) include over 900 species of trees, shrubs, and herbs. A number are of economic importance. The many species of *Moraceae* contain a milky juice. The flowers are borne in axillary spikes or heads. Many members have dioecious flowers, i.e., the staminate and pistillate flowers are borne on different plants; while others are monoecious. The staminate flower has a variously three- to six-parted calyx, no petals, and one to four stamens with filiform filaments. The pistillate flower has a calyx of three to five, essentially united sepals and a single one- to two-celled superior ovary. The fruit varies greatly in different members of the family.

True Mulberries. Trees and shrubs of the genus *Morus* are widely distributed plants of temperate regions. The leaves are alternate. On a single tree one may observe interesting variations of the leaves; on one shoot they may all be entire, while on a nearby shoot they are variously and irregularly lobed or divided. The flowers develop early in the growing season. The plants are either monoecious or dioecious. The staminate flowers are borne in long catkins and soon fall from the tree. The calyx is divided into four lobes and there are four stamens inserted at its base. The pistillate inflorescence is a short dense catkin. The flowers have a four-lobed calyx and a single one-celled ovary. After pollination the calyx lobes become greatly swollen and fleshy, the individual fruits pressing together tightly to form a multiple fruit. These fruits may be white or pink in the White Mulberry, red in the Red Mulberry, and black in the Black Mulberry. The white mulberry, *Morus alba*, is grown in the Orient largely to supply food for silk worms. The roots of the tree yield a yellow dye, and the wood is used for various purposes. The black mulberry, *Morus nigra*, is grown largely for the fruits, which are greedily eaten by birds, including domestic poultry. The wood is also valuable. Red mulberry, *Morus rubra*, furnishes wood used in making shoe lasts, and for other purposes. Numerous species are planted as decorative trees.

Banyan Tree. The banyan (*Ficus benghalensis*) is found in tropical areas over most of the world. In the United States, the tree prefers parts of Florida and California along the warm coastal areas. The tree is evergreen with leaves from 4 to 8 inches (10 to 20 centimeters) long, alternate and ovate. The fruit occurs in pairs, is red and about $\frac{1}{2}$ inch (1.3 centimeters) in diameter. Banyan trees are considered sacred in parts of India. These trees have a habit of sending down from their spreading branches, aerial roots which enter the ground and increase in size until they resemble trunks. A single tree may have scores and, in some cases, hundreds of such "trunks," thus spreading the base of the tree over an unusually large area.

Breadfruit Tree. Native of the East Indian and Pacific islands, the breadfruit tree (*Artocarpus altilis*), is widely planted in tropical regions elsewhere. The tree not only is attractive and excellent for shade, but is also an important source of staple food, the fruit being used for making poi. When ripe, the fruit, weighing up to 30 pounds (13.6 kilograms), may be used in puddings and other dessert foods. The large ovoid fruits usually have a rough surface, but may be smooth in some varieties. Each fruit is composed of many achenes, each surrounded by a fleshy perianth and growing on a fleshy receptacle. Usually yellowish-brown when ripe, the fruit is roasted or boiled before being eaten. Some improved varieties of breadfruit are seedless. Whereas the fruit of the mulberry tree contains sugar, the breadfruit is rich in starch. The large leaves, from 1 to 2 feet (0.3 to 0.6 meter) in length, are deeply cut into pinnate lobes. Condensation dripping from the deeply-lobed leaves aids in maintaining the soil under the tree moist during dry periods. There are over 40 species of the breadfruit.

Fig Trees. The familiar fig is the fruit of a shrub or small tree (Ficus carica) which is probably native to southwestern Asia. Figs have been cultivated since earliest recorded times, being widely used by the Hebrews, greatly improved by the Greeks, and highly valued by the Romans. The fig tree is now grown in cultivation in nearly all tropical countries and many subtropical regions; in the United States it is grown to some extent in the southwestern states, notably in California.

The plants have alternate leaves, which are rather thick and rough surfaced above, but soft and hairy beneath. The leaves are deeply lobed in the cultivated varieties. The minute flowers are borne on the inside of the hollow receptacle, which develops into a pear-shaped body with a minute opening at its apex. The fruit developing from this is a synconium, composed of many small fruits inserted in the inner wall of a hollow fleshy receptacle. The narrow passage into this is partly closed by numerous small bracts.

Four kinds of flowers occur in fig trees: (1) Staminate flowers, each having four pollen-bearing stamens, occur in the wild "caprifig." A few cultivated forms have staminate flowers. Pollination is brought about by using pollen from caprifigs, and is called caprification. (2) Pistillate flowers, each with a single pistil which if pollinated produces a seed; these flowers are short-stalked. (3) The third flower type is the gall-flower, so-called because a small wasp, *Blastophaga grossorum*, lays its eggs in them. The developing larvae cause the ovaries to become swollen galls, incapable of developing seeds. This type of flower occurs only in caprifigs, in the basal portion of the synconium. (4) Lastly, in varieties of cultivated figs there are found sterile flowers, which will neither produce seeds nor become galls; these are called mule flowers. Caprifigs contain the first three types of flowers. If pollen is needed to insure fruit development, caprifigs must be planted, since they alone have pollen-bearing flowers. So, among Smyrna fig trees, the fruits of which fail to develop unless pollinated caprifigs must be planted.

In Mediterranean countries, where figs are grown in abundance, three crops are produced each year. The first fruits, known as profichi, are formed in the spring. In the pistillate flowers of these the female wasp lays her eggs, so that galls are formed in the profichi. When the young wasps emerge, these profichi are gathered and hung among Smyrna figs. Wasps escaping have to crawl past the staminate flowers near the aperture of the synconium and so are dusted with pollen. The wasp then enters and pollinates a flower of the second crop, thus insuring fruit development. This second crop is known as mammoni. The third crop, the mammae, remain on the trees. It is in these that the wasp passes the winter.

TABLE 1. RECORD TREES OF THE MULBERRY FAMILY IN THE UNITED STATES[1]

Specimen	Circumference[2]		Height		Spread		Location
	Inches	Centimeters	Feet	Meters	Feet	Meters	
MULBERRY							
Black mulberry (1999)	252	640	78	23.8	76	23.2	Maryland
(*Morus nigra*)							
Paper mulberry (1991)	157	399	75	22.9	55	16.8	Florida
(*Boussonetia papyrifera*)							
Red mulberry (1999)	301	765	52	15.8	52	15.8	Tennessee
(*Morus rubra*)							
Texas mulberry (1997)	36	91	29	8.8	31	9.4	Arizona
(*Morus microphylla*)							
White mulberry (1992)	292	742	59	18	73	22.3	Missouri
(*Morus alba*)							
FIG							
Florida strangler fig (1993)	360	914	63	19.2	72	21.9	Florida
(*Ficus aureo*)							
Shortleaf fig (1993)	248	630	41	12.5	57	17.4	Florida
(*Ficus citrifolia*)							

[1]From the "National Register of Big Trees," American Forests (by permission). [2]At 4.5 feet (1.4 meters).

Fig fruits are gathered and sometimes eaten fresh, but more frequently dried in the sun, then pressed together and shipped. Smyrna figs are often enlarged, by pulling during drying. The fruits produce a mild laxative effect, and so are sometimes prescribed in cases of chronic constipation. From them a wine is sometimes made, and alcohol produced. The wood of the fig is occasionally used in cabinet work.

Propagation is usually by stem cuttings, but sometimes by budding or grafting.

Strangler Fig (*Ficus aurea*). Closely related to the wild banyan, a single specimen has been known to spread its mass of foliage on stiltlike roots over several acres. The plant bears fruit, but the figs are not edible. The tree has a rough, leathery bark. It is also sometimes called *rubber tree*; this name arises from the white, sticky sap that exudes when the bark is bared. The leaves alternate on the branches with red fruit; the leaves are shiny and oval in shape. Propagation of the tree is unusual. To turn flowers into seed, an insect must be of minute size to penetrate a tiny hole in bottom of each fig. Since the strangler fig is a killer of other trees, Florida, for example, is fortunate in that it has no insects that will carry out the difficult pollination. Each of the many varieties of *F. aurea* requires its own kind of pollinating insects. The strangler fig proliferates in locations such as south Florida, because the seeds are dropped by birds, or are carried inland by ocean currents from nearby islands, or by fierce gales that come out of the Caribbean. A seed that falls on bare ground or porous limestone (plentiful in Florida) will germinate into a tree. The strangler fig is an epiphyte (can subsist on air only), but if seed sprouts in a host tree or plant, the early roots of the *F. aurea* work their way into the bark of the host, thus slowly but steadily depriving the host of light and moisture. As one observer (J. Fix) notes, "Woe betide the Florida tree that gets intimate with *F. aurea*." Trees frequently affected include the Sabal palm and the Florida oak. The strangler fig is extremely difficult to kill—because of its rampant root system that may reach hundreds of feet in all directions from the tree. This root system also damages water lines and residential drain fields.

Common Hop. The perennial climbing plant, *Humulus lupulus*, is another member of the mulberry family of economic importance. The plant has an extensive underground stem, or rhizome, from which rise the annual climbing stems, which twine in a clockwise direction around any supporting object. The hollow stem is ridged, with downward pointing hairs along each of the ridges. The opposite leaves are large and palmately veined. Hops are usually dioecious plants. The staminate inflorescence is a loose panicle; the pistillate, spike-like, with conspicuous bract-like structures subtending each branch. The staminate flowers have a five-parted calyx and five stamens; there is no corolla. The pistillate flower is a single ovary, partially surrounded by a small bract and having two long hairy stigmas; around the ovary is a cup-like perianth. Hops are wind-pollinated. After fertilization the bracts enlarge greatly. On their outer surface, and

also on the surface of the perianth and the subtending bract, yellow grains develop. These grains are called hop-meal and are multicellular cup-shaped bodies developing from single epidermal cells. The cells of these bodies secrete a yellow substance which fills the cup-shaped hollow and which contains the substances which makes hops valuable—an essential oil, resins, tannin, and a bitter substance, probably alkaloidal in nature. The resins are bitter and germicidal.

The principal use of hops is in the brewing of beer. Hops are prepared for the brewing process by drying and bleaching.

Some record trees of the mulberry family growing in the United States are listed in Table 1.

MULE. See **Horses, Asses, and Zebras.**

MULLER, HERMANN JOSEPH (1890–1967). Hermann Muller was an American geneticist and societal reformer whose studies in the United States, Russia and Scotland of mutations in fruit flies contributed to "the modern synthesis" of genetics in the 1930s.

"The arrival of this little man in Edinburgh with a large packing case full of synthetic stocks of *Drosophila* changed the whole atmosphere of the institute." Thus a colleague described H. J. Muller's entrance into the Institute of Animal Genetics in November 1937—the final stage of an eight-year scientific odyssey away from his native United States. Muller brought to the Institute not only superb analytical and technical skills, but also a Darwinist doctrine distilled from the contending views of three scientists—his thesis supervisor, the zoologist Thomas Hunt Morgan, and forever there, but in the shadows, William Bateson and Edmund B. Wilson. See also **Bateson, William (1861–1926); Morgan, Thomas Hunt (1866–1945);** and **Wilson, Edmund Beecher (1856–1939).**

When H. J. Muller was born in 1890 in New York City to second-generation immigrants of German and English origin, Morgan was studying developmental embryology in Naples. When Muller's father died suddenly in 1900, imposing a life of frugality on his family, William Bateson was introducing the work of Gregor Mendel to the English-speaking world. When Muller entered high school, Morgan was joining Edmund B. Wilson at Columbia College, New York, where the latter was exploring the relationship between the Mendelian character units, later known as "genes," and chromosomes. By the time Muller entered Columbia College at the age of 16, all three—Morgan, Bateson and Wilson—had written major books on Mendelism and evolution. While Morgan and Bateson disputed the role of Darwin's natural selection as an originator of species and were cautious in ascribing all inherited characteristics to genes, Wilson was less restrained. Most likely his lectures persuaded the young Muller of the correctness of the genic viewpoint and its compatibility with Darwinian dogma.

In 1908 Morgan decided to seek mutants among the progeny of a small, rapidly multiplying organism, the fruit fly (*Drosophila melanogaster*). In 1910 he identified the first mutant (white eye), while Muller, having graduated *cum laude* and now a graduate student in physiology, attended his biology lectures. Muller found Morgan:

Unduly loath to follow the facts to their logical conclusions. In fact, he really sought in Mendelism and mutation, substitutes for natural selection, and he long refused to adopt generalized, clear-cut ideas of chromosomal Mendelian inheritance.

In 1912 Muller began working in Morgan's "fly room" at Columbia with two other outstanding graduate students, Alfred H. Sturtevant and Calvin B. Bridges. By the time of Muller's doctorate in 1915, the group, under Morgan's "genial and democratic" supervision, had assigned genes to distinct chromosomes, mapped their linear order, and published the first edition of *The Mechanism of Mendelian Heredity*, which Bateson reviewed critically in the journal *Science*. See also **Bridges, Calvin Blackman (1889–1938)**; and **Sturtevant, Alfred Henry (1891–1970)**.

Whereas Sturtevant and Bridges remained with Morgan, in 1915 Muller went off to the Rice Institute in Texas with Julian Huxley. He returned to Columbia in 1918 as a temporary instructor and in 1922 joined the Department of Zoology in Austin, Texas. At the 1921 Toronto meeting of the American Association for the Advancement of Science, Bateson complimented the *Drosophila* group, but cautioned that although many genetic phenomena could be explained by genes, it did not follow that genes explained everything—there was still the question of the origin of species. See also **Huxley, Julian Sorrell (1887–1975)**.

At the same meeting Muller deemed it remarkable that a mutation could affect the functional expression of a gene, but not its replication. He suggested that the pairing of genes as parts of chromosomes undergoing meiotic synapsis might provide clues to gene structure and replication:

It is evident that the very same forces which cause the genes to grow should also cause like genes to attract each other, ⓒ If the two phenomena are thus dependent on a common principle in the make-up of the gene, progress made in the study of one of them should help in the solution of the other.

Decades later, on learning of the double-helix model for DNA, he set his students an essay, "How does the Watson–Crick model account for synapsis?" Crick took up the challenge in 1971 with his "impairing postulate," which Forsdyke later related (1996) to the question of the origin of species. See also **Crick, Francis Harry Compton (1916–2004)**.

In 1922 Muller married Jessie Jacobs of the Department of Mathematics, who was required to relinquish her position at the birth of their son in 1924. From this time Muller became increasingly active with respect to social causes, including minority rights, the teaching of evolution in schools, and positive eugenics. In 1927 he showed that X-rays would greatly increase the mutation rate. Arguing that with X-rays "there was no intensity below which mutation could not occur," Muller led a campaign to increase public and professional awareness of the dangers of radioactivity. He came under the surveillance of the Federal Bureau of Investigation, accumulating a 500 page dossier.

In 1928 Muller was at the peak of his scientific creativity. He had shown how disadvantageous mutations might accumulate in an asexual population by a process that became known as "Muller's rachet." He had pointed out that sexual reproduction would permit advantageous mutations arising in separate individuals to be incorporated into a single individual. Yet his professional relationships with his Texas colleagues were deteriorating, and his marriage was failing. Morgan in Pasadena was recruiting the world's talent to the Californian Institute of Technology, but, as far as we know, no invitation went to Muller. In 1932, depressed, he attempted suicide. Later that year, after addressing the 6th International Congress of Genetics at Ithaca, he sailed alone to Europe.

While working in a Berlin laboratory he witnessed for himself a raid by Hitler's storm troopers. The rise of Nazism, the social unrest accompanying the economic depression, and the seductive simplicity of the left-wing alternative, led to acceptance of an invitation from the geneticist N. I. Vavilov to work in Russia. His enthusiasm for the communist cause found expression in a 1934 article, "Lenin's doctrines in relation to genetics." Here he attacked Morgan's conservatism and Bateson's apparently metaphysical treatment of genes and chromosomes. However, as Stalin's purges took effect (eventually involving Vavilov

himself), disillusion followed. At the meeting in 1936 of the Lenin All-Union Academy of Agricultural Sciences, Muller declared the views of the biologist Lysenko as "irreconcilable with the fundamentals of genetics." Thereafter, he escaped "the closing jaws of Stalinism" by working briefly for the communists in the Spanish revolution, and then moving on to the Institute of Animal Genetics in Edinburgh. Here he met Charlotte Auerbach, S. P. Ray-Chaudhuri, Guido Pontecorvo, and his second wife, Dorothea Kantorowicz. He made a major contribution to *Evolution. The Modern Synthesis* (1942), edited by Julian Huxley.

In 1940 Muller fled from Europe at war to Amherst College in Massachusetts. However, his teaching was deemed unsatisfactory. In 1944, with a new daughter to support, Muller found himself at 54 an unemployed east-coast university reject. Rumors of his left-wing leaning, of his depression in the early 1930s, and of his Jewish background, all appeared to work against him.

By 1946 the entire position had changed. Muller had secured a position in the Zoology Department at Indiana University, obtained Rockefeller Foundation support, and been awarded the Nobel Prize in Physiology or Medicine. From this time on, he began to speak out more actively against Lysenkoism in the USSR and for social reform in the West. Young scientists who came under his influence included James Watson and Carl Sagan. See also **Watson, James Dewey (1928 — Present)**.

Additional Reading

Carlson, E.A.: *Genes, Radiation and Society: The Life and Work of H. J. Muller*, Cornell University Press, Ithaca, NY, 1981.

Crick, F.: "General Model for the Chromosomes of Higher Organisms," *Nature*, **234**, 25–27 (1971).

Forsdyke, D.R.: "Relationship of X chromosome Dosage Compensation to Intracellular Self/not-self Discrimination: a Resolution of Muller's Paradox," *Journal of Theoretical Biology*, **167**, 7–12 (1994).

Forsdyke, D.R.: "Different Biological Species "Broadcast" their DNAs at Different (C+G)% "Wavelengths"," *Journal of Theoretical Biology*, **178**, 405–417 (1996).

Forsdyke, D.R.: "Two Levels of Information in DNA: Relationship of Romanes' "Intrinsic" Variability of the Reproductive System, and Bateson's "Residue" to the Species-dependent Component of the Base Composition, (C+G)%," *Journal of Theoretical Biology*, **201**, 47–61 (1999).

Muller, H.J.: "Variation Due to Change in the Individual Gene," *American Naturalist*, **56**, 32–50 (1922).

Muller, H.J.: "Artificial Transmutation of the Gene," *Science*, **66**, 84–87 (1927).

Muller, H.J. "Evidence on the Precision of Genetic Adaptation," *Harvey Lectures*, **43**, 165–229 (1948).

Muller, H.J.: *Man's Future Birthright: Essays on Science and Humanity*, State University of New York Press, Albany, NY, 1973.

Muller, H.J., and E.A. Carlson: *Modern Concept of Nature: Essays on Theoretical Biology and Evolution*, State University of New York Press, Albany, NY, 1973,

Web Reference

Donald R. Forsdyke: http://post.queensu.ca/~forsdyke/homepage.htm#Homepage

DONALD R. FORSDYKE, Queen's University, Kingston, Ontario, Canada

MULLETS (*Osteichthyes*). Of the order *Mullus*, there are several species of mullet, distributed mainly in the Mediterranean and Black Seas and the Atlantic Ocean. From there, they move up to the Norwegian coast. The upper jaw has no teeth, but teeth are present on the vomer and the gums.

The *red mullet* (*Mullus ba rbatus*) has a very steep forehead and prefers muddy or sandy ground. The *striped mullet* (*Mullus surmuletus*), with a less steep forehead, is usually found above sandy ground. Both species attain sexual maturity in 2 to 3 years. Spawning generally occurs during summer, off the coast. The eggs float in the water because of their oil bubbles. In the fall, the young move into greater depths. Mullet were famous as food in ancient Rome, despite their small size, bringing a high price. They were brought into the banquet halls alive.

There is another *striped mullet* (*Mugil cephalus*); a much larger and important commercial fish. This species achieves a length up to nearly 3 feet (90 centimeters) and a weight of 31.7 pounds (7 kilograms). See Fig. 1. The color is ash gray with a dark blue shimmer. The sides of the body have 9 to 10 lighter longitudinal stripes. Distribution extends across all warmer seas, including the Mediterranean. Striped mullet are often found in lagoons or river mouths. Their meat is tasty and often, young striped mullet are kept in salt-water or brackish-water ponds. They are fed until autumn, when they reach a marketable size and are shipped off to be sold. There are several other species of mullets (suborder *Mugiloidei*; family *Mugilidae*). They are coastal fishes and have a high adaptability to salt water, brackish water, or fresh water. These lively and schooling fishes are generally found in tidal zones with a rich plant supply above soft ground. Their diet consists of plankton, snails, mussels, and other small organisms associated with algal colonies. They also feed on detritus and very small organisms found on the floor. This is the basis for the generic name *Mugil* (sucker).

See also **Fishes**.

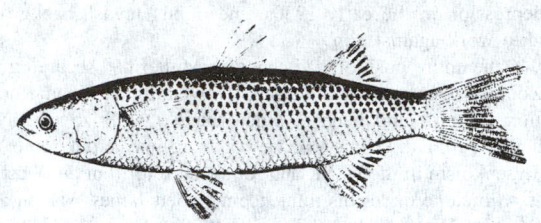

Fig. 1.   Striped mullet.

MULLIKEN, ROBERT S. (1896–1986).   An American chemist, physicist, and educator who won the Nobel Prize for chemistry in 1966 for his fundamental work concerning chemical bonds and the electronic structure of molecules by the molecular orbital method. Mulliken received is B.Sc. Degree in 1917 at the Massachusetts Institute of Technology, Cambridge, MA. and a Ph.D. degree at the University of Chicago, IL., in 1921.

MULLIS, KARY BANKS (1944–).   An American who won the Nobel Prize for Chemistry in 1993 for his invention of the polymerase chain reaction (PCR) method.

MULTIMEDIA FILES.   See **File Types**.

MULTIPARAMETER RADAR.   A radar capable of deriving more than one quantity from observations of a target. In meteorological applications, the term is usually applied to radars capable of measuring either 1) reflectivity and at least one polarization-dependent quantity, or 2) reflectivity and at least one wavelength-dependent quantity.

The term is not ordinarily applied to radars that operate with a single wavelength and measure only reflectivity and Doppler velocity.

MULTIPLE CORRELATION.   The correlation between a random variable and its regression function.

If Y denotes the regression function of a random variable (variate) y with respect to certain other variates $x_1, x_2 \ldots, x_n$ then the coefficient of multiple correlation between y and the x's is defined as the coefficient of simple, linear correlation between y and Y. However, the constants of the regression function automatically adjust for algebraic sign, with the result that the coefficient of correlation between y and Y cannot be negative; in fact, its value is precisely equal to the ratio of their two standard deviations, that is, $\sigma(Y)/\sigma(y)$. Therefore, the coefficient of multiple correlation ranges from 0 to 1, and the square of the coefficient of multiple correlation is equal to the relative reduction (or percent reduction), that is, the ratio of explained variance to total variance. Since, in practice, the true regression function Y is seldom known, it is ordinarily necessary to hypothesize its mathematical form and determine the constants by least squares, thus obtaining the approximation Y'. In that case, the conventional estimate of the multiple correlation is the sample value of the simple linear correlation (symbol R) between y and Y', although a better estimate is obtained by incorporating a correction for degrees of freedom. Such a corrected value R' is given as follows:

$$R' = \frac{[(N-1)R^2 - n]^{1/2}}{[N-(n+1)]^{1/2}},$$

where N denotes the sample size and $n + 1$ equals the total number of constants (including the absolute term) determined from the data. In case $(N-1)\,R^2 < n$, the value of R' is taken as zero.

See also **Regression**.

AMS

MULTIPLE ENDOCRINE NEOPLASIA.   See **Adrenal Disease**.

MULTIPLE-IMAGE NETWORK GRAPHICS (MNG).   See **Data Compression**.

MULTIPLE INTEGRAL.   The definition of an indefinite integral can be extended to cover the case of a function of several variables and then more than one integration will be required to evaluate the integral. The subject can be considered in various ways but one simple approach is that of partial integration as the inverse to partial differentiation. Thus, given the double integral

$$u = \iint f(x, y)\, dx\, dy$$

one wishes to determine u so that it will satisfy the partial differential equation $u_{xy} = f(x, y)$. The first integration is performed with respect to x, for example, holding y constant, and the second with respect to y, although the order of integrating does not matter. Constants of integration added to the result complete the work, although these are not really constants but arbitrary functions of the variables. Further generalization to triple, quadruple, etc., integrals offers no further difficulty in principle. Often, for ease in printing, a single integral sign is used for multiple integrals.

The definite multiple integral is commonly of more importance and it may be interpreted geometrically. For example, a function of three variables $f(x, y, z) = 0$ can be considered as a surface. Double integrals, with the appropriate limits, can then be formulated to give the volume of a solid bounded by two or more surfaces, the area of the surface itself, and the moment of inertia of a plane area. Similarly, a triple integral may be used to obtain the volume of a solid or of a closed surface. When given these geometrical interpretations double and triple integrals are often called surface and volume integrals. See also **Area**; **Vector Integral**; and **Volume (Geometry)**.

MULTIPLE INTERFEROMETER DETERMINATION OF TRAJECTORIES (MIDOT).   A trajectory measurement system with multiple-object-tracking capability utilizing two or more short-baseline stations and a data output consisting of a series of amplitude nulls that represent direction cosines at given times in the flight.

MULTIPLE MYELOMA.   See **Bone**.

MULTIPLE OBJECT PHASE TRACKING AND RANGING (MOPTAR).   A short-baseline continuous-wave phase comparison, trajectory measuring system, similar to the *Cotar* which consists of a crossed-baseline angle-measuring-equipment (AME) system and a distance-measuring-equipment (DME) system, wherein time sequencing of the ground station and transponders is used to track multiple targets.

MULTIPLE SCATTERING.   Scattering of radiation, usually electromagnetic but possibly acoustic, by an array of objects (e.g., atoms, molecules, particles) each of which is excited to scatter (radiate) not only by an external source but also by the scattered radiation from the other objects in the array.

Multiple scattering is distinguished from single scattering, an idealization strictly realized only with a single object excited by an infinitely distant source. Scattering as a consequence solely of excitation by the external source is sometimes referred to as primary scattering, the remaining scattering as secondary scattering, which is misleading in that it can be decomposed into an infinite series of primary, secondary, tertiary, and higher-order scattering. Multiple scattering can be classified according to the coherence properties of the array and the external source. For incoherent multiple scattering, phase differences of scattered waves are random, and scattered powers are additive. For coherent multiple scattering, phase differences of scattered waves are not random, and scattered fields are additive. Incoherent and coherent multiple scattering are idealizations. An example of (primarily) incoherent multiple scattering is scattering of sunlight by thick clouds. An example of (mostly) coherent multiple scattering is specular reflection by a glass of water. Scattering by a single cloud droplet is an example of scattering by a coherent array—the water molecules in the droplet stick together (cohere) in the sense that the phase differences between their individual scattered waves are fixed—whereas scattering by the entire cloud is incoherent in the sense that for droplets separated by random distances large compared with the wavelength, the phase differences between waves scattered by individual droplets are essentially random.

MULTIPLE SCLEROSIS. A *demyelinating* disease of the central nervous system presently of unknown cause, the onset of which seldom occurs before 15 or after 40 years of age. *Myelin* is the white, fatty substance that forms a sheath about certain nerve fibers. *Sclerosis* indicates the hardening (induration) of the sheath substance, forming zones of demyelination (*plaques*) that range in size and location. Multiple indicates *multiple sites*. Myelinated nerve fibers are minute in diameter, but may be very long. The myelin sheath conducts nervous impulses at a rate which enables muscles to make precise and delicate movements. It is the alteration or destruction of the sheath that interferes with these impulses and thus creates the features of multiple sclerosis (MS). The axon is not damaged. See also **Central and Peripheral Nervous Systems**.

Early symptoms of MS include clumsiness, slowness, stiffness, and weakness (42% of cases). There may be visual disturbances (hazy, misty, blurred vision), sometimes with pain associated with the eyeball and frequently a loss of central vision of one eye (34% of cases). Some patients describe tingling sensations, numbness, and a band-like tightness (18% of cases). There may be nausea, vomiting, and general light-headed feeling (7% of cases). In a minority of MS patients, there may be incontinence, loss of bladder sensation, and diminished or no sexual function (4% of cases). Following a general course, the disease can be highly variable among patients, particularly as to time and severity and mixture of symptoms.

Much has been learned about the nature and occurrence of MS, but the exact cause or causes remain elusive. Certain paths of research are converging on a probable profile of causes and the beginnings of an understanding of the pathophysiology of the disease. As of this time, the etiology can be summarized by the following superficial definition—MS is an exceptionally complex disease, resulting from poorly understood and intricate interaction of genetics, environment, geography, viruses, and the immune system.

Related Factors. For reasons not yet understood, there is a coarse relationship between climate and the incidence of MS. For example, in temperate latitudes, the disease is much more prevalent (50–100 cases per 100,000 population) than in the tropical latitudes, where the incidence is relatively low (5–10 cases per 100,000 population). This immediately suggests that the causative factors must be more prevalent in the temperate than in the tropical zones. But to date the relationship has not led researchers to the cause. Persons who emigrate from tropical to temperate zones apparently bring with them their low risk for the disease. This particularly applies to emigrants over 15 years of age. Multiple sclerosis also occurs more often among urban dwellers *not* from areas of poverty; this is in marked contrast to the pattern of so many other diseases, particularly of an infectious nature. These epidemiologic aspects of MS have tended to give support to a viral etiology. Also, a coarse relationship of a familial nature exists. It has been estimated that the risk of disease is about 15 times greater among those persons with a family history of the disease than for the population as a whole. The genetic connection, however, has not been worked out in a predictable, clear-cut manner.

Although a number of parallels have been observed among MS patients, such as an elevated level of IgG in the spinal fluid (80–90% of cases), the presence of antibodies to myelin and oligodendroglia, and alterations in lymphocyte distribution, among other factors, diagnosis of MS essentially must be made on the basis of clinical observations.

Although there is no specific treatment for MS, attention can be directed toward partially or temporarily alleviating some of its features. It has been generally believed that corticosteroid therapy is helpful, as in hastening the recovery of optic neuritis. General counsel to patients usually includes the desirability of avoiding excessive fatigue, emotional stress, and significant temperature changes. Pregnancy has not been found to carry a higher risk among MS patients, but usually is not advised because of the long-term prognosis for the mother. Elective surgical procedures are seldom indicated. A number of drugs have been used for controlling spasticity or painful muscle spasms, including diazepam, dantrolene, and baclofen. The physician will pay particular attention to monitoring bladder dysfunction.

Role of Viruses. There was reasonable consensus in the 1980s that probably several viruses play some causative role in MS. Until the mid-1970s, the evidence that a virus may be a causative factor was essentially epidemiological. Data patterns indicated that the disease may arise from a viral infection early in life. Research continues in an effort to prove the virus connection conclusively. Positive proof requires demonstrations such as the production of MS in a laboratory animal after injection with a suspect virus; or isolating a virus from a human with active MS. During the 1970s and continuing into the 1980s, some progress was made toward identifying several viruses in active MS patients. Some investigators found traces of measles virus at different sites in MS patients. Measles virus as a possible causative agent was suggested as early as 1962 when the virus was found in the blood and cerebrospinal fluid of an MS patient. Other investigators subsequently identified abnormal concentrations of antibodies to measles in the blood of many MS patients. More recently, such antibodies were found in the brain tissues of MS patients after death. But these findings are not considered clear-cut evidence of a singular connection with measles virus because abnormal concentrations of antibodies to other viruses, such as rubella, vaccinia, and herpes complex, among others, also have been found. There are also differences in the results of various investigators. Some investigators have found a substance called *blocking factor* in the blood of MS patients. This substance prevents leukocytes from destroying cells infected with measles virus. The substance is not found in normal individuals. Still other investigators have implicated a virus known as *MS-associated agent* (MSAA), which was first studied in 1972. Parainfluenza virus (called 6/94 virus) has been considered suspect by researchers in Germany and Japan. However, when the virus is injected into mice, while a chronic neurological disease is produced, it is quite unlike MS. The apparent persistence of evidence implicating several rather than one virus has bolstered the hypothesis that MS may be the result of several viral infections and that possibly one or two viruses, such as measles or 6/94 virus, may precipitate the disease. Findings also tend to support the hypothesis that MS may be caused by an inborn defect in the immune system of MS patients that permits viruses to proliferate in the central nervous system.

In 1986, a scientist at the University of Chicago observed that there is a serious derangement of immune function in multiple sclerosis. It appears to be an immune attack, but the mechanism is not understood. The immune attack may be against myelin basic protein, a component of the fatty myelin sheath. Other observers continue to associate an inherited predisposition in some persons toward the disease.

In 1985, workers at the National Cancer Institute detected in some MS patients traces of what may be a new viral relative of the human leukemia virus (HTLV-I—human T-cell lymphotropic virus-I). However, there is no evidence that the virus is a cause of MS. The high antibody concentrations found in MS patients may simply indicate their autoimmunity. However, there are precedents for suggesting that the HTLV viruses may produce neurological diseases. HTLV-III not only infects and destroys helper T cells, thereby causing the severe immune deficiency of AIDS, but also can be found in the brain. Attempts are being made to isolate the HTLV-I related viruses. Should the cloning or virus isolation be successful, this will provide researchers with a specific probe for future studies.

Additional Reading

Cook, S.D.: *Handbook of Multiple Sclerosis*, 3rd Edition, Marcel Dekker, Inc., New York, NY, 2001.

Donald W., D.W. Paty, and G.C. Ebers: *Multiple Sclerosis*, Vol. 51, Oxford University Press, Inc., New York, NY, 1997.

Feinstein, A.: *The Clinical Neuropsychiatry of Multiple Sclerosis*, Cambridge University Press, New York, NY, 1999.

Goodkin, D.E. and R.A. Rudick: *Multiple Sclerosis: Advances in Clinical Trial Design, Treatment and Future Perspectives*, Springer-Verlag, Inc., New York, NY, 1998.

Hawkins, C.P. and J.S. Wolinsky: *Principles of Treatment in Multiple Sclerosis*, Butterworth-Heinemann, Inc., Woburn, MA, 2000.

Staff: *Multiple Sclerosis: Current Status and Strategies for the Future*, National Academy Press, Washington, DC, 2001.

Web Reference

Multiple Sclerosis Foundation: http://www.msaa.com/

MULTIPLET. A complex energy level in an atom or molecule, which gives rise to a corresponding series of spectral lines. In an atom, energy levels with a given resultant electronic angular momentum $\mathbf{L}$ and nonzero resultant electron spin $\mathbf{S}$ split into a number of fine structure components with quantum numbers

$$\mathbf{J} = (\mathbf{L} + \mathbf{S}), (\mathbf{L} + \mathbf{S} - 1), \ldots, (\mathbf{L} - \mathbf{S})$$

The resulting multiplet has $2\mathbf{S} + 1$ components for $\mathbf{L} > \mathbf{S}$, and $2\mathbf{L} + 1$ components for $\mathbf{L} < \mathbf{S}$.

The term multiplet is applied also to the narrowly spaced groups of lines corresponding to transitions between the multiplet components of the same or of two different atomic energy levels. A line multiplet may have more components than either energy state involved (*compound multiplet*). See also **Atomic Spectra**.

MULTIPLICATION. An operation which is the inverse of division, used in arithmetic, algebra, and other branches of mathematics. The result of multiplying two or more numbers is a product. Sometimes, but not always, multiplication obeys the commutative and associative laws; in combination with addition, also the distributive law. The commutative law often fails for matrix or group multiplication.

Multiplication can also be considered as successive addition.

MULTIPLICITY. 1. The number $2S + 1$, representing the number of ways of vectorially coupling the orbital angular momentum vector L with the spin angular momentum vector S of an atom. This value represents the number of relatively closely spaced energy levels or terms in an atom which result from the coupling process. The value of the multiplicity is added as a left superscript to the term symbol, as ^3P (triplet P), ^4D (quartet D), etc. The multiplicity of molecules is analogous to that of atoms, and is expressed also by the number $2S + 1$. The multiplicity of an atomic or nuclear level is $2J + 1$, where J is the total angular momentum quantum number of the level.

2. The term is used in biology to indicate that in some cases of irradiation studies a number of targets must be inactivated before any effect is observed: for example, a small colony of bacteria must have hits on all members to prevent growth of the colony. However, the phrase *extrapolation number* has been suggested because factors that cannot affect the number of targets in an irradiated material can sometimes affect the multiplicity: for example, toxic agents released from one cell which may affect another. In either case, one obtains a dose-response curve where the initial low dose gives rise to nearly no observable effect followed by a region of exponential decline of activity with increasing dose. When one extrapolates this latter portion of the dose-response curve to zero dose, the point at which it intersects the biological activity coordinate is referred to as the extrapolation number of multiplicity.

MULTIPOSITION CONTROLLER. An automatic controller that has two or more discrete values of output. Curves indicating this type of control action are given under **Control Action.** In this action, the final controlling element is moved to one of two or more fixed positions, each corresponding to a definite range of values of the controlled variable. The controller can produce a three-position action if it is used with a final controlling element that takes a third position when the variable value is within the differential gap. Inasmuch as multiposition control is capable only of a limited number of corrections, this control action seldom produces an exact correction for any load condition and thus produces continuously cyclic control.

Two Position Controller. This is a special case of multiposition control wherein the controller has two discrete values of output. Curves indicating this type of control action are given under **Control Action**.

On-Off Controller. This, in turn, is a special case of two position control wherein the controller has two discrete values of output, i.e., either *fully on*, or *fully off*.

In the two position mode of control, the final controlling element is moved relatively quickly from one of two fixed positions to the other at a single value of the controlled variable. In the case of an actual controller, as in the instance of an electric temperature controller, when the temperature is at or above the setpoint value, the contact is closed and the valve closes; when the temperature is below the setpoint, the contact is opened and the valve opens. A two position controller cannot make an exact correction. The correction must be greater or less than exact. Thus, no stable balanced condition of input to output energy can be achieved. The controlled variable must cycle.

MULTIPROGRAMMING (Computer System). The essentials of a multiprogramming system in connection with digital computer operation are: (1) several programs are resident in main storage simultaneously, and (2) the central processing unit (CPU) is time shared among the programs. This makes for better utilization of a computer system. Where only one program may reside in the main storage at any given time, inefficient use of CPU time results when a program requests data from an input/output device. The operation is delayed until the requested information is received. In some applications, such delays can constitute a large portion of the program-execution time. In the multiprogramming approach, other programs resident in storage may use the CPU while a preceding program is awaiting new information. Multiprogramming practically eliminates CPU lost time due to input/output delays. Multiprogramming is particularly useful in process control or interactive applications which involve large amounts of data input and output.

Input/output delay-time control is a basic method for controlling the interplay between multiple programs. Where multiprogramming is controlled in this manner, the various programs resident in the storage are normally structured in a hierarchy. When a given program in the hierarchy initiates an input/output operation, that program is suspended until such time as the input/output operation is completed. A lower-priority program is permitted to execute during the delay time.

In a time-slice multiprogramming system, each program resident in the storage is given a certain fixed interval of the CPU time. Multiprogramming systems for applications where much more computation is done than input/output operation usually use the time-slice approach.

Multiprogramming systems allow multiple functions to be controlled simultaneously by a single process-control digital computer. A multiprogramming system allows a portion of the storage to be dedicated to each type of function required in the control of the process and thus eliminates interference between the various types of functions. In addition, it provides the means whereby asynchronous external interrupts can be effectively serviced on a timely basis.

See also **Program (Computer)**; **Time Sharing**; and **Time Slicing**.

THOMAS J. HARRISON, IBM Corporation, Boca Raton, FL

MULTIRESOLUTION SEAMLESS IMAGE DATA BASE (MRSID). See **Data Compression**.

MULTIVARIATE ANALYSIS. The analysis of data which are multivariate in the sense that each member bears the values of p variates. The principal techniques of multivariate analysis, beyond those admitting of straightforward generalism, e.g., regression correlation and the variance analysis, are cluster analysis, component analysis, factor analysis, and discriminatory analysis.

MULTIYEAR ICE. See **Meteorology**.

MUMPS. Also known as *epidemic parotitis*, mumps is a contagious disease characterized by swelling of the salivary glands located below the ear and below the angle of the jaw (parotid, submaxillary, and sublingual glands). The parotid glands, located just below and in front of the ears, are the ones principally affected and this swelling is often the first recognizable symptom of the disease. A sudden rise in temperature, to between 40 and 40.5 °C (104 and 105 °F), with or without vomiting and headache may also be a first symptom. The swelling of mumps is firm, and typically obliterates the angle of the jawbone, giving the face a pale, shiny, and somewhat bloated appearance. Enlargement may extend along the neck; the degree of swelling varies with the severity of the attack. Characteristically, swelling appears first on one side and then the other. The interval between enlargement of the opposite sides may be up to 12 days or more; in some cases, the second side never swells. The swelling in each side generally lasts from a week to 10 days; usually the swelling reaches its peak on the third day and gradually subsides thereafter. The early stages of the disease are marked by high fever, headache, pain in the back, reddened taste buds, and loss of appetite. There may be an excess of saliva, or the mouth and throat may be abnormally dry. The initial high temperature gradually subsides, but the patient usually has a mild fever so long as there is any swelling.

Mumps is primarily a disease of children and young adults. It occurs much less commonly later in life. The disease usually occurs in children between the ages of 5 and 15; children between the ages of 7 and 9 seem particularly susceptible. Infants appear to be immune for the first 8 to 10 months of life, and those under 2 years of age are considered just slightly susceptible. Although there can be serious complications of mumps in adults, particularly males, the disease and its complications are rarely fatal.

The causative agent, paramyxovirus, invades the salivary glands. The virus is transmitted mainly through direct contact with an infected person, although about 40% of exposed persons may not have apparent infection, but can infect others. The virus is present in the saliva and in the secretions of the nose. It may be present in secretions for up to 7 days before symptoms develop and for 9 days after swelling subsides. The incubation period of the disease is usually 13 to 21 days.

Firm diagnosis of mumps can be gained by detecting a rise in complement-fixing antibodies. Early in the disease, there is a rise in antibodies to S antigens and later an elevation of antibodies to V antigens. A mumps skin test antigen is not considered reliable. Management of the disease is essentially supportive.

The mumps virus is one of the most widely spread of all disease-causative agents and the disease is encountered throughout the world with exception of isolated locations, such as sparsely populated islands and essentially untraveled jungle communities. However, persons in those isolated areas, not having any immunity, are attacked in large numbers.

In from 5 to 20% of mumps-infected males, orchitis will occur. One-third of the males infected may have unilateral testicular atrophy following the disease. To relieve the pain of orchitis, an anesthetic block of the spermatic cord may be used. Ovarian involvement may occur in about 5% of adult women with mumps. Other organs, such as the pancreas, prostate, seminal vesicles, breasts, thyroid and thymus are less commonly involved. Central nervous system involvement is found in males at a rate 5 times that of women. One of the most serious complications is meningoencephalitis, which can be fatal. Involvement of the pancreas as a complication of mumps has suggested a tie between mumps and diabetes in some patients.

Mumps Vaccine. A live mumps vaccine has been available for several years. In many developed countries, such as the United States, the vaccine is administered almost routinely to children at about 15 months of age — along with vaccine for measles and rubella. The vaccine can be administered at any age. However, the vaccine will not prevent disease if a person has been exposed because the contact will have been shedding virus for several days before parotitis is evident. Unless contraindicated (for example, in persons who are immunosuppressed, immunodeficient, or pregnant), mumps vaccine is considered effective in almost 95% of persons vaccinated even though the degree of immunization induced is only about 20% as great as that resulting from natural infection. The vaccine has been instrumental in markedly reducing the occurrence of mumps as a general disease. Although mumps vaccine has been shown to reach the placenta, there is no evidence to date that the fetus will become infected.

Mumps virus is enveloped, containing single-stranded RNA. It averages 140 nanometers (0.14 micrometer) in diameter. Mumps hyperimmune globulin is sometimes administered to persons who have been exposed to the virus. However, there is little evidence that this is an effective preventive measure.

Web Reference

Center for Disease Control and Prevention: http://www.cdc.gov/health/diseases.htm

ANN C. DE BALDO, Ph.D., University of South Florida, Tampa, FL

MUON. The *muon* (μ^-) is an elementary particle of the lepton family. Properties include: Spin, $\frac{1}{2}$; mass (MeV), 105.66; lifetime, 2.20×10^{-6} second. The antiparticle is the positive muon (μ^+). The muon neutrino (ν) has spin, $\frac{1}{2}$; 0 mass; and is stable. The muon family appears to be simply a duplicate of the electron family except for a change in the unit of mass. See also **Particles (Subatomic)**.

The positive muon was discovered in cloud chamber photographs made by C.D. Anderson and S.H. Neddermyer on Pike's Peak in 1935, and the negative muon almost simultaneously in cloud chamber photographs made by J.C. Street and E.C. Stevenson. These particles have long been called mu-mesons, but since they are fermions (spin $\frac{1}{2}$) while all other mesons are bosons, the name *muon* is preferred, as is their classification with the leptons because of their small rest mass, which is about 206 m_e, where m_e is the mass of the electron. Another reason is their inability to interact with other particles through the nuclear forces.

Their charges are equal in magnitude to that of the electron. They are produced by the decay of pions (*pi mesons*) and (to a limited extent) by the decay of kaons and hyperons. Positive-negative muon pairs also can be generated by the action on matter of gamma-rays of energy greater than the rest masses of the particles, i.e., exceeding 211 MeV. Their lives are short, about 2.2×10^{-6} seconds in the free state, and the negative muon usually decays into an electron, a neutrino, and an antineutrino, while the positive muon usually gives a positron, as well as a neutrino and antineutrino. As explained in the entry on neutrino, there are two types of neutrinos and antineutrinos (ν_e or $\overline{\nu}_e$) like that produced in the decay of radionuclides, and a muon-associated neutrino or antineutrino (ν_μ or $\overline{\nu}_\mu$) so that these reactions would be written

$$\mu^- \rightarrow e^- + \nu_\mu + \overline{\nu}_e$$
$$\mu^+ \rightarrow e^+ = \overline{\nu}_e + \nu_e$$

Muons can easily penetrate many meters of iron and can sometimes cause problems in particle physics research. For example, the upsilon experiment at Fermilab in 1977, conducted by L.M. Lederman and others, required building a simple magnetic system that would remeasure each muon's energy after it emerged from the main detector. See also **Upsilon Particle**; and **Particles (Subatomic)**.

MUONIUM. The atom consisting of a positive muon and an electron. Thus, muonium may be regarded as a light isotope of hydrogen in which the positive muon replaces the proton. When a beam of positive muons is stopped in a gas (argon under such pressures as 50 atmospheres has been used in much of this research), muonium is formed directly in its ground state by the capture of an electron by a positive muon. The reaction is important because of its bearing upon the nature of the muon-electron interaction and the muon itself. The study has included measurement of the hyperfine structure interval in the ground state of muonium, and measurement of muon polarization as a function of time and impurity concentration. By adding such gases as oxygen (O_2) and nitric oxide (NO) as impurities to the argon, data on spin exchange of electron and muon is obtained, while with impurities such as nitrogen dioxide (NO_2) and ethylene (C_2H_2), evidence of such reactions as $NO_2 + M \rightarrow NO + OM$ and $C_2H_4 + M \rightarrow C_2H_4M$ is obtained.

MURINE.   Referring to rodents, notably mice and rats.

MUSA ANTENNA.   A multiple-unit steerable antenna consisting of a number of stationary antennas, the composite major lobe of which can be aimed electrically.

MUSCLE.   An organ formed of a bundle of contractile fibers attached to parts of the body which are moved in relation to each other when it shortens. Among the invertebrates, the fibers of a muscle are loosely associated, but in the vertebrates they are bound together and enveloped by special tissues.

The typical vertebrate muscle is surrounded by a connective tissue sheath, called the external perimysium, which continues into it a series of septa (septum), called the internal perimysium. Between the septa lie bundles of muscle fibers, each surrounded by a delicate continuation of the connective tissue closely joined to the surface of the fiber. Blood vessels and nerves supplying the muscle course through the perimysium.

Muscular tissue is made up of long slender cells specialized for contraction. It is usually mesodermal in origin. There are three kinds of muscular tissue: smooth, cardiac, and striated or skeletal. Among the invertebrates all muscle may be smooth, as in many worms, or striated, as in most arthropods. All three kinds are found in vertebrates and cardiac muscle is characteristic of the vertebrate heart. Muscle cells may contain a few scattered myofibrils, or a large number, occupying most of the cytoplasm. The myofibrils consist fundamentally of protein chains which are responsible for the contractile properties of muscles. See also **Contractility and Contractile Proteins**.

The smooth muscle tissue of vertebrates consists of slender tapering cells with a nucleus placed centrally. These cells are involuntary in action and are found in the walls of the alimentary tract excretory system, blood vessels, and other organs.

Striated muscle is so named because the many fibrils in its cytoplasm are made up of altering zones of different refractive quality which cause the fiber to show light and dark transverse bands. The fiber is long and cylindrical and contains many nuclei located at the periphery. It is surrounded by a delicate membrane, the sarcolemma. Most striated muscle is the foundation of voluntary movements and from its extensive association with the skeleton it receives the name skeletal muscle.

Cardiac muscle, like skeletal, is striated, although the striations are much finer. It differs in the general branching of its fibers and in its centrally placed nuclei. Cardiac muscle is rhythmically contractile, initiating the heartbeat of vertebrates. The nerves that innervate the vertebrate heart serve to modify but not to initiate the contraction of the heart muscle.

In the simplest animals having special muscular tissues, the worms, muscle cells pass across the loose tissues within the body and form layers in the body wall. Their disposal is often in a circular layer, with fibers running around the body, and a longitudinal layer with fibers parallel to the main axis. The contractions of these layers lengthen and shorten the body and so carry on the creeping movements characteristic of worms. Special groups of muscle fibers also govern such special structures as the setae of earthworms. These muscles have an attachment to the body wall called the origin, and an insertion in the tissue surrounding the part to be moved.

In animals having a rigid skeleton, an arrangement of muscle fibers, like that of the body wall of worms, persists in hollow organs like the alimentary tract. In organs like the heart and urinary bladder, a less regular arrangement provides for the uniform contraction of all parts of the wall. Locomotion and similar movements, however, are due to the action of muscles on skeletal supports that serve as systems of levers. Usually, the shortening of a single muscle regulates the movement of one part in relation to another; it is said to have its insertion in the part moved and its origin at the other point of attachment. When a muscle is used for slowing a movement, it is stretched; and when it is used for equilibrium, as in standing, the length does not change. These activities are all included under the term contraction for historical reasons.

The nature of movements varies greatly. Extension of jointed appendages, flexion, retraction, rotation, and other movements are carried out by opposed systems of muscles. Any movement is positive, due to the contraction of a specified muscle, and the return of the part to its former position results from the contraction of an opposed muscle, sometimes aided by the action of gravity. The anatomy of the muscular system and the relations of specified muscles have been worked out in great detail in the human body, in other vertebrates, and in some of the arthropods. See Fig. 1.

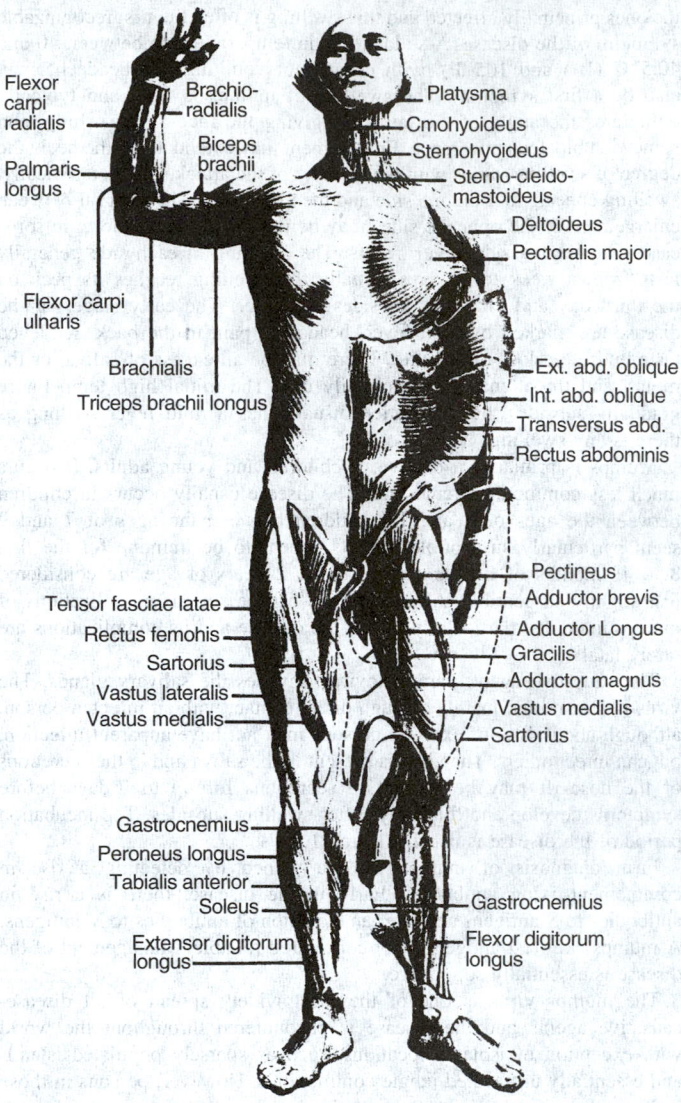

Fig. 1.   Ventral musculature of human.

Important disorders affecting muscle and the neuromuscular junction include alcoholic myopathy, dermatomyositis, drug- and endocrine-induced myopathies, glycogen storage disorder, malignant hyperthermia (a complication of anesthesia), mitochondrial and lipid storage myopathies, muscular dystrophy, myasthenia gravis, myastenic (Eaton-Lambert) syndrome, mycoblobinuria, and polymyositis, among others. See also **Muscular Dystrophy**; and **Myopathy**. Electromyography is an important diagnostic tool for differentiating muscle disorders. See also **Electromyography**.

Additional Reading

Burke, E.: *Optimal Muscle Recovery*, Avery Publishing Group, Inc., Garden City Park, NY, 1999.
Kendall, F. et al.: *Muscles: Testing and Function*, Williams and Wilkins, Philadelphia, PA, 1994.
Matthews, G.: *Cellular Physiology of Nerve and Muscle*, Blackwell Science, Inc., Malden, MA, 1997.

MUSCOVITE.   The mineral muscovite is a hydrated silicate of potassium and aluminum corresponding to formula $KAl_2(AlSi_3)O_{10}(OH)_2$, crystallizing in the monoclinic system although frequently hexagonal found in pseudohexagonal forms. Usually tabular in habit, the most prominent characteristic is the highly perfect basal cleavage yielding remarkably thin laminae, which are often highly elastic. Hardness, 2.5–3; specific gravity, 2.77–2.88; luster, vitreous to pearly; color, colorless through grays, browns, greens, yellows, and rarely violet or red; transparent to translucent.

Muscovite is the commonest mica, being found in granites, pegmatites, gneisses, and schists, and as a contact metamorphic mineral, or as a secondary mineral resulting from the alteration of topaz, feldspar, and kyanite. In pegmatites it is often found in immense sheets which are commercially valuable. A complete list of occurrences of muscovite would be impossible. In the United States, excellent specimens are found in the pegmatites of New England, where they are associated with rarer minerals like tourmaline, and beryl. Pennsylvania, Maryland, Virginia, North Carolina, Georgia, South Dakota, and New Mexico also furnish large and fine examples of this mineral. Foreign sources include Canada, Brazil, Norway, Sweden, the former U.S.S.R, and India. A single crystal weighting 85 tons, 10 feet (3 meters) in diameter, 15 feet (4.6 meters) in length was obtained from Inikurite Mine, Nellore, India.

The name muscovite comes from Muscovy-glass, a name formerly much used for this mineral because of its use by Russians for windows. It is in much demand for the manufacture of insulating and fireproofing materials, and to some extent as a lubricant. Muscovite sometimes is called *potash mica*.

MUSCULAR DYSTROPHY. One of a group of muscle disorders (myopathy). Dystrophy distinguishes this disease from other myopathies because of the heritable, progressive nature of the condition. In general terms, muscular dystrophy (MD) is a progressive weakness and wasting of muscles. There is no known effective treatment. Because of its genetic origin, genetic counseling prior to marriage and family formation is absolutely essential among thoughtful couples of childbearing age. Biochemically, the complex changes that occur during the progressive course of the disease are rather poorly understood even though much research has been devoted to such studies. As of the late 1980s, a number of studies are being directed toward membrane structure and function. Dystrophy is usually accompanied (probably preceded) by a number of abnormalities—changes in erythrocytes (red blood cells), alterations in membrane surface ultrastructure, as well as changes of phospholipids, membrane-associated enzymes, and the presence of higher concentrations of calcium (Ca^{2+}) in dystrophic muscle as compared to normal muscle. An alteration of membrane fluidity occurs, but the cause is not explained. Strongly suspected are the roles of proteases (enzymes associated with protein) and antigen-antibody complexes. Much more is understood concerning protein synthesis than protein catabolism (destruction-decomposition). See also **Contractility and Contractile Proteins**; and **Protein**. Investigators (State University of New York), working with chickens, showed that certain protease inhibitors (leupeptin and pepstatin) at least partially inhibit muscle degeneration, with success shown by these substances in delaying muscle atrophy. It also has been shown in studies at a number of laboratories that the symptoms of dystrophy can be partially alleviated by means of penicillamine, methysergide, and Dilantin. However, in the absence of detailed cell studies, the mechanism by which these drugs were effective is difficult to postulate.

Muscular dystrophy can be subcategorized into:

(1) **Duchenne's (Pseudohypertrophic) Dystrophy**. The sex and genetic linkage of this disorder is relatively well understood. Males between 2 and 5 years of age may develop hip-girdle weakness. This is characterized by a waddling gait and certain muscles, particularly of the calf, will become enlarged. The Duchenne form of MD is the most common type in children and occurs once in every 3000 to 4000 births. It is estimated that one-third of these cases arise from new mutation. By comparison, another X-linked disease, hemophilia, occurs only once in every 10,000 births. Although MD has been considered a disease of males, there are rare case where females also have the disease. The disease was diagnosed in 1977 (Lindenbaum, Oxford University) and it was determined that the child's X chromosome was broken at position Xp21. Later, a number of other girls were found to have similar broken X chromosomal genes. Other scientists who have made traditional linkage analyses trace to Xp21 as the site of the gene. (The general area of location constitutes about 10 million base pairs and represents about 20% of the short arm of the X chromosome.)

Traditionally, genetic researchers know in advance what protein products to seek. In the case of the MD gene, they are seeking the gene in the absence of an understanding of just how the gene functions. For several years, there has been evidence that indicates its approximate location. At least two approaches have been taken. Kunkel (Children's Hospital, Boston) and colleagues have been analyzing in great detail the region of the X chromosome in youths who have deletions of the MD gene. Worton and colleagues (Hospital for Sick Children, Toronto, Ontario) have been searching for the gene in children with translocations that interrupt the MD gene. It is beyond the scope of this encyclopedia to report on the fine details of these and other researches. Probes have now been developed which now can detect carriers with 98% accuracy, an invaluable achievement for assisting families with at least one child with the disease who want to know who carries the gene and who does not.

Currently, the prognosis is discouraging—confinement to a wheelchair during the teens, with fatalities often occurring prior to age 25. Not uncommon in association with this disorder are weakness of the heart muscles and a moderate mental retardation. Confirmation of the condition is found in the high serum creatine phosphokinase (CPK) levels of these individuals. A high percentage (up to 70%) of the female carriers of the disorder will also have high CPK levels.

(2) **Facioscapulohumeral Dystrophy**. This is a disorder that is not usually evident until adolescence. Because of the slow course of the disease, patients may live for many years. The disorder is characterized by a wasting of facial, pectoral, and shoulder-girdle muscles, making it difficult for the patient to smile, close the eyes, whistle, etc. In most cases, the lower limbs also will be involved. The heart muscle is not usually involved and CPK serum levels are not usually elevated.

(3) **Limb-Girdle Dystrophy**. This disorder involves the shoulder and/or hip girdle and varies in intensities among families. The onset usually occurs between the age of 20 and 50 years of age. There is no known effective treatment. The physician will be careful to differentiate this disorder from chronic polymyositis because it has been found that corticosteroids are sometimes helpful for the latter condition.

(4) **Ocular Dystrophy**. This is an uncommon disorder that is associated with disturbances of the retina, peripheral nerves, and central nervous system. Progressive external ophthalmoplegia may be the only feature of the disease. Like other dystrophies, there is considerable variation among families.

(5) **Oculopharyngeal Dystrophy**. This is also uncommon and usually commences during middle life. Symptoms include seeing, hearing, and swallowing difficulties.

(6) **Myotonic Dystrophy**. This is also heritable, usually becoming apparent during the teen years. Principal features are a wasting of facial muscles (usually producing facial features that are characteristic of the disorder), with accompanying difficulties in swallowing and cardiac arrhythmias. There are several other features which may develop as the result of neuromuscular degradation, including involvement of the muscles that serve the extremities, cataracts, premature balding, and sometimes various endocrine abnormalities, including diabetes mellitus and gonadal atrophy.

Additional Reading

Burnett, G. and S. Rioux: *Muscular Dystrophy*, Silver Burdett Press, Englewood Cliffs, NJ, 1995.

Emery, A.: *Muscular Dystrophy: The Facts*, Oxford University Press, Inc., New York, NY, 2000.

Siegel, I.: *Muscular Dystrophy in Children: A Guide for Families*, Demos Medical Publishing, New York, NY, 1999.

Web References

Muscular Dystrophy Association: http://www.mdausa.org/disease/als.html

United States Centers for Disease Control and Prevention http://www.cdc.gov/search.htm

MUSE PACK (or MPC). See **Data Compression**.

MUSHROOMS AND MUSHROOM CULTIVATION. Mushrooms have been found in fossilized wood 300 million years old [Editorial, 1997]. Almost certainly prehistoric people used mushrooms as food. The great early civilizations of the Greeks, Egyptians, Romans, Chinese and Mexicans prized mushrooms as a delicacy, appreciated their therapeutic nature and, in some cases, used them in religious rites. The association of mushrooms with thunderstorms was common in mythology, and it was formerly believed that mushrooms were formed by lightning and thunderstorms. Although some wild poisonous mushrooms became objects of fear and distrust, it is not surprising that the intentional cultivation of edible mushrooms had a very early beginning. Literature references indicate that *Auricularia auricula* was cultivated in China as early as ad 600 on wood logs. Other wood-rotting mushrooms such as *Flammulina velutipes* and *Lentinula edodes* were later grown in similar manner, but the biggest advance in mushroom cultivation came in France around 1600 when *Agaricus bisporus* was cultivated upon a composted substrate. In the Western world *A. bisporus* (champignon or button mushroom) has remained the mushroom that is produced in the greatest amounts, but now other species long popular in Asia (e.g. *Lentinula edodes*, *Pleurotus* spp.), and produced there in large numbers, are making inroads into Western markets. See also **Agarics**; **Basidiomycetes**; and **Fungus**.

Mushroom cultivation provides both nutritious protein rich food and medicinal products. Cultivated mushrooms have now become popular all over the world. The bioconversion of lignocellulosic biomass to food and useful products by mushrooms has already had an impact at national and regional levels, and the predictions are that this impact will continue to increase. Being without adverse legal, ethical or safety effects, this form of bioconversion technology has only favorable socioeconomic and employment impacts.

Mushrooms, like all other fungi, lack chlorophyll. They are unable to convert solar energy to organic matter like green plants, but they can convert the huge agricultural and forest waste materials into human food. The byproducts, spent substrates, can be used as animal feed and crop fertilizers. Therefore, sustainable development of mushroom cultivation can be called the *non-green lignocellulosic revolution*, because mushroom cultivation can generate equitable economic growth and protect and regenerate the environment.

Biodiversity of Mushrooms

Mushroom biology is the branch of mycology that deals with mushrooms. Fungi have been placed in a kingdom of their own called the *Myceteae*. The word mushroom may mean different things to different people in different countries. In this article, the term "mushroom" is defined [Chang and Miles] in the broad sense as "a macrofungus with a distinctive fruiting body which can be either epigeous or hypogeous and large enough to be seen with the naked eye and to be picked by hand." Thus, mushrooms need not be Basidiomycetes, nor aerial, nor fleshy, nor edible. Some mushrooms belong to the Ascomycetes, grow underground and have a nonfleshy texture. The most common type of mushrooms is umbrella shaped with pileus (cap) and stipe (stem), e.g. *Lentinula edodes* (Fig. 1), and some species additionally have a volva (cup), e.g. *Volvariella volvacea* (Fig. 2), or annulus (ring), e.g. *Agaricus campestris* (Fig. 3), or both, e.g. *Amanita muscaria* (Fig. 4). Furthermore, some mushrooms are in the form of pliable cups while others are round like golf balls. Some are in the shape of small clubs; some resemble coral; others are yellow or orange jelly-like globes; and some even resemble the human ear. In fact, there is a countless variety of forms. The structure that we call a mushroom is in reality only the fruiting body of the fungus. The vegetative part of the fungus, called the *mycelium*, comprises a system of branching threads and cord-like strands that branch out through soil, compost, wood log or other lignocellulosic material on which the fungus may be growing. After a period of growth and under favourable conditions, the established (matured) mycelium produces the fruit structure that we call the mushroom.

Mushrooms can be divided into four categories: (1) those that are edible and fleshy fall into the edible mushroom category, e.g. *Agaricus bisporus*; (2) those that are considered to have medicinal applications are referred to as medicinal mushrooms, e.g. *Ganoderma lucidum*; (3) those that are proven to be, or suspected of being, poisonous are called poisonous mushrooms or toadstools, e.g. *Amanita phalloides*; (4) a miscellaneous category that may tentatively be grouped together as "other mushrooms."

Fig. 1. *Lentinula edodes* (Shiitake mushroom). (Image courtesy of USDA).

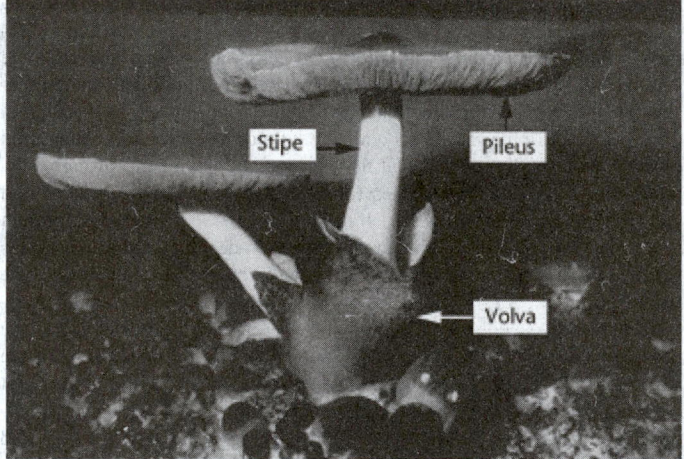

Fig. 2. *Volvariella volacea*, with volva.

This way of classifying mushrooms is by no means absolute since many kinds of mushrooms are not only edible but also possess tonic and medicinal properties. Of the fungi recognized so far, it has been suggested that over 10,000 species produce fruiting bodies of sufficient size and suitable texture to be considered as "macrofungi" [Kendrick]. Of these species, about 50% are considered to possess varying degrees of edibility and approximately 10% are poisonous, of which some 30 species are considered to be lethal [Miles and Chang].

As yet, no ascomycetous mushroom has been cultivated commercially with complete success. The truffle (genus *Tuber*) is successfully farmed, and patents have been issued for a cultivation method that produces fruiting bodies of the morel (genus *Morchella*) and commercial production is close at hand. In the class Basidiomycetes the commercially cultivated members

Fig. 3. *Agaricus campestris*. The gills are usually pink or silvery-grey at first, but are colored chocolate brown at maturity from the developing spores. The cap and stalk are usually some sort of white or greyish brown, but may have fibrils or scales that are darker (like the portobello).

Fig. 4. *Amanita muscaris*, with an annulus near the top and with the bulbous base adorned with several concentric zones of white scales representing the volva (Image courtesy of Mrs. Helena Pereira Lima Caruccio, Brazil).

belong to 14 genera (*Agaricus, Coprinus, Flammulina, Lentinula, Pleurotus, Hypholoma, Kurhneromyces, Pholiota, Stropharia, Volvariella, Dictyophora, Hericium, Auricularia* and *Tremella*) under 10 families (Agaricaceae, Coprinaceae, Tricholomataceae, Hypholomataceae, Strophariaceae, Pluteaceae, Phallaceae, Hericiaceae, Auriculariaceae and Tremellaceae) placed in five orders (Agaricales, Phallales, Aphyllophorales, Auriculariales and Tremellales) and two subclasses (Holobasidiomycetidae and Phragmobasidiomycetidae). See also **Basidiomycetes**.

Poisonous Mushrooms

Poisonous mushrooms are often called *toadstools*. However, like the term mushroom, toadstool also may mean different things to different people. To some it means any mushroom except the field mushroom (*Agaricus campestris*) or cultivated mushroom (*Agaricus bisporus*), to others it is any inedible or poisonous mushroom. Scientifically the term has no meaning at all. It is suggested that the term toadstool be dropped altogether in order to avoid confusion and the terms edible, medicinal and poisonous mushrooms used instead.

It has been estimated that the number of poisonous mushroom species is fairly large, i.e., about 1,000 species, of which some 30 species are known to be lethal. Since there is no known field test to determine if a mushroom is edible or not, people should never eat any mushroom unless they are absolutely certain of its identity.

The toxins contained in various species are very different in chemical composition, and thus the effects of poisoning differ considerably according to the species involved. In any case, suspected mushroom poisoning

should never be regarded lightly and medical assistance should be sought at once. The following summary of mushroom poisonings is based on Chilton [1978], and Shepherd and Totterdell [1988].

1. Amanita-type poisoning. Unquestionably the most dangerous type of mushroom poisoning is caused by the *Amanita phalloides* group. The toxins belong to the phallotoxin and amatoxin complexes. The majority of recorded deaths from mushroom poisoning have been caused by this group. The general symptoms of this type of poisoning are severe abdominal pains, violent vomiting, diarrhea, cold sweats and excessive thirst. These may last for 48 h, with dehydration, cramps and anuria.
2. Muscarine-type poisoning. Two toxins, muscarine and ibotenic acid, are involved. They occur in *Amanita muscaria, A. pantherina* and also in a number of *Inocybe* and *Clitocybe* species. The symptoms usually appear soon after eating the mushrooms, with vomiting, diarrhea and salivation. The most characteristic symptoms are nervous excitement, difficulties in breathing, shivering and a tendency to collapse.
3. Psychotropic or hallucinogenic poisoning. Several different toxins are involved, including psilocin and psilocybin, which are found in species of *Psilocybe, Conocybe* and *Stropharia*. The symptoms are varied, including vomiting, increased heartbeat and hallucinations.
4. Coprinus poisoning. Several *Coprinus* species, such as *C. micaceus* and *C. atramentarius*, if consumed with an alcoholic drink, are unpleasant, but not dangerous. The symptoms include reddening of the face, increased heartbeat, and in some cases, vomiting and diarrhea.
5. Poisoning from external sources. The poisoning is not caused by the mushrooms themselves but by toxic substances that have accumulated in the mushrooms. The principal causes are: (1) heavy metals due to polluting environmental conditions where the mushrooms are harvested; and (2) radioactive contaminants due to pollution by radioactive material; these are becoming a serious problem in certain areas for mushrooms hunting and consumption. The best publicized case was in the areas surrounding Chernobyl.

Edible Mushrooms

Of the 10,000 or more species of macrofungi (mushrooms), about 5,000 are considered to possess varying degrees of edibility [Miles and Chang] and more than 2,000 species from about 31 genera (Table 1) are regarded as prime edible mushrooms. Of these, only about 100 are experimentally grown, 50 economically cultivated, and around 30 commercially cultivated (Table 2); and only five or six species are cultivated on an industrial scale.

TABLE 1. GENERA OF PRIME EDIBLE MUSHROOMS

Basidiomycetes	
Agaricus	*Hypholoma*
Agrocybe	*Hypsizygus*
Amanita	*Lactarius*
Armillaria	*Lentinula*
Auricularia	*Lepista*
Boletus	*Lyophyllum*
Cantharellus	*Marasmius*
Calvatia	*Pleurotus*
Clitocybe	*Pholiota*
Coprinus	*Polyporus*
Cortinarius	*Russula*
Dictyophora	*Stropharia*
Flammulina	*Termitomyces*
Gloestereum	*Tremella*
Grifola	*Tricholoma*
Hericium	*Volvariella*
Ascomycetes	
Morchella	*Tuber*

The moisture content of fresh edible mushrooms varies in the range 70–95% depending upon the harvest time and environmental conditions,

TABLE 2. SPECIES OF COMMERCIALLY CULTIVATED EDIBLE MUSHROOMS

Agaricus bisporus[a]	*Pholiota nameko*
Agaricus bitorquis[a]	*Pleurotus citrinopileatus*
Agaricus blazei	*Pleurotus cornucopiae*
Auricularia auricula	*Pleurotus cystidiosus*
Auricularia fuscosuccinea	*Pleurotus djamor*
Auricularia polytricha	*Pleurotus eryngii*
Coprinus comatus	*Pleurotus florida*
Dictyophora duplicata	*Pleurotus ostreatus*[a]
Dictyophora indusiata	*Pleurotus sajor-caju*
Flammulina velutipes[a]	*Stropharia rugoso-annulata*
Grifola frondosa	*Tremella aurantia*
Hericium erinaceus	*Tremella fuciformis*
Hypsizygus marmoreus	*Volvariella diplasia*
Lentinula edodes	*Volvariella esculenta*
Lyophyllum ulmarium	*Volvariella volvacea*[a]

[a] Species have reached an industrial scale.

whereas it is about 10–13% in dried mushrooms. The protein content of cultivated species ranges from 1.75 to 5.9% of their fresh weight, with an average value of 3.5–4.0%. This means that the protein content of edible mushrooms, in general, is about twice that of onion (1.4%) and cabbage (1.4%) and four times and 12 times that of oranges (1.0%) and apples (0.3%), respectively. In comparison, the protein content of common meats is as follows: pork, 9–16%; beef, 12–20%; chicken, 18–20%; fish, 18–20%; and milk 2.9–3.3%. On a dry weight basis, mushrooms normally contain 19–35% protein as compared to 7.3% in rice, 12.7% in wheat, 9.4% in corn and 38.1% in soybean. Therefore, in terms of the amount of crude protein, edible mushrooms rank below most animal meats but well above most other foods, including milk, which is an animal product. Furthermore, the mushroom protein contains all nine amino acids that are essential for humans. In addition to their high quality protein, mushrooms are a relatively good source of the following individual nutrients: phosphorus, iron, and vitamins including thiamin, riboflavin, ascorbic acid, ergosterine and niacin. They have a low crude fat content with a high proportion of polyunsaturated fatty acids (72–85%) relative to total fatty acids. They are low in calories, carbohydrates and calcium.

The recent upsurge of interest in traditional remedies for various physiological disorders and the recognition of numerous biological response modifiers in mushrooms have led to the coining of the term "mushroom nutriceuticals" [Chang and Buswell]. A mushroom nutriceutical is a refined/partially defined mushroom extractive which is consumed in the form of capsules or tablets as a dietary supplement (not a food) and has potential therapeutic applications. A regular intake may enhance the immune responses of the human body, thereby increasing resistance to disease and, in some cases, causing regression of a disease state.

Mushroom Hunting

The fungi are found just about everywhere. Mushrooms are rather more selective than other fungi in that the size of the fruiting body requires the availability of more nutrients than are required for the production of asexual spores by microfungi. In damp places, such as tree-fern gullies and areas of rain forest, plentiful moisture leads to mushroom formation. They can be collected at most times of the year, but in drier regions they occur only after seasonal rains. In these areas there may be a particular flora of mushroom species associated with the seasons of autumn, summer and spring. Relatively few mushrooms are produced during the cold winter months although there are perennial fruiting bodies that persist during the winter. Formation of mushroom fruiting bodies depends very much on the pattern of rain and, in some years, there may be virtually a complete lack of fruiting.

Mushroom hunters, in addition to bringing basic equipment and field guide reference, which will vary depending on personal requirements and regional conditions, should record such items as date, time, location, smell, substrate (host) color, milk (if present), habitat and anything at all unusual about the specimen. Some important characteristics for identification disappear rapidly as the mushroom matures; these characteristics have to be recorded accurately at the time of collection. Ecologically, mushrooms can be classified into three groups: the saprophytes, the parasites and the mycorrhizae. There are only a few parasitic mushrooms. Most of the cultivated gourmet mushrooms are saprophytic fungi. Some are mycorrhizal mushrooms, e.g. Perigold black truffle, *Tuber melanosporum*, and matsutake mushroom, *Tricholoma matsutake*. It is difficult to bring these pricey wild gourmet species into cultivation because they are mycorrhizal. These mushroom species have a symbiotic relationship with some vegetation, particularly trees, i.e. there is a relationship of mutual need.

Whatever their motives (amateur, gourmet, scientific study), collectors should take some basic precautions to keep their materials in proper condition. The majority of fresh mushrooms are relatively fragile, and they should be protected from vibration and impact by careful packing. In general, plastic bags are unsuitable for collection of specimens, as they lead to "sweating" and rapid deterioration. Scientific collections may be made for the purpose of immediate examination and discarded, or they may be preserved for future reference. Although preservation is best achieved by drying, collections are often lost through the growth of moulds. Particularly in subtropical and tropical areas, collectors should be aware that it is essential to add paraformaldehyde or naphthalene to all collections. In addition, fleshy mushrooms should not be dried at temperatures above 35–40 °C (95–104 °F), since the hyphae and other microscopic structures become too strongly agglutinated, making later microscopical studies difficult.

Cultivation of Mushrooms

Cultivation of edible mushrooms had its beginning in China around ad 600 when *Auricularia auricula* was first cultivated. Mushroom cultivation can be relatively primitive or a highly sophisticated agricultural activity requiring a sizable capital outlay for mechanized equipment. The straw mushroom, *Volvariella volvacea*, is commonly grown in Southeast Asian countries on small, family-type farms. In contrast, in many Western nations, cultivation of the *Agaricus* mushrooms may be highly industrialized with a few farms producing a disproportionally large percentage of a country's output.

Although simple in concept, various intricacies associated with the process of mushroom cultivation must be understood for the enterprise to be successful. In every case, the ultimate aim is to obtain the maximum yield from a given surface area or a given weight of substrate per period of time by the use of high yielding strains, by shortening the cropping period, or by increasing the number of high yielding flushes. To achieve maximum yield requires an understanding of: (1) the nature of mushrooms, ecological habitats in wood and straw (mostly primary decomposers, directly breaking down a substrate for growth, e.g. *L. edodes* and *V. volvacea*) or dung (mostly secondary decomposers, relying on the previous activity of other microorganisms to partially degrade a substrate to a state wherein they can grow, e.g. *A. bisporus*); (2) the nature of substrate materials and their preparation; (3) appropriate control of physical, chemical and biological parameters (e.g. moisture content, pH, temperature, competitive microflora); and (4) proper management of mushroom beds, including pest and disease control.

The major phases of mushroom cultivation can be briefly described as follows.

1. Selection of an acceptable mushroom. There is considerable variation among edible mushroom species in the temperatures suitable for vegetative growth (spawn running) and fruiting body development.
2. Selection of a fruiting culture. A "fruiting culture" is defined as one with the genetic capacity (e.g. dikaryon) to form fruiting bodies under suitable growth conditions.
3. Development of spawn. A medium through which the mycelium of a fruiting culture has grown and which serves as the inoculum or "seed" for the substrate in mushroom cultivation is called the *mushroom spawn*. Substrates used in spawn production include various grains (rye, wheat and sorghum), rice straw cuttings, and cotton waste. A good spawn should be free of contamination, have vigorous growth, and survive storage well.
4. Preparation of substrate. Substrates for cultivating edible mushrooms normally require varying degrees of pretreatment in order to promote

growth of the mushroom mycelium to the practical exclusion of other microorganisms. Different substrate types for different mushrooms can be formulated according to the growing conditions.

5. Mycelial (spawn) running. Spawn running is the phase during which mycelium grows from the spawn and permeates the substrate. Good mycelial growth is essential for mushroom production and will depend on proper maintenance of the beds and mushroom house in terms of temperature, moisture content, humidity and aeration.

6. Mushroom development. The appearance of mushrooms normally occurs in rhythmic cycles called *flushes*. Suitable temperature, humidity and ventilation controls must be maintained during the cropping period since these factors will affect the number of flushes and the total yield obtained.

Cultivation techniques vary for different mushroom species in different countries. More information can be found in Chang and Miles [1989].

Additional Reading

Chang, S.T.: "Mushroom Biology: the Impact on Mushroom Production and Mushroom Products," In: Chang, S.T., J.A. Buswell, S.W. Chiu: *Mushroom Biology and Mushroom Products*, pp. 3–20. Chinese University Press, Hong Kong, China, 1993.

Chang, S.T., and J.A. Buswell: "Mushroom Nutriceuticals," *World Journal of Microbiology and Biotechnology*, **12**, 473–476 (1996).

Chang, S.T., and P.G. Miles: *Edible Mushrooms and Their Cultivation*, CRC Press, LLC, Boca Raton, FL, 1989.

Chang, S.T., and P.G. Miles: "Mushroom Biology—A New Discipline," *The Mycologist*, **6**, 64–65 (1992).

Chang, S.T., and P.G. Miles: *Cultivation, Nutritional Value, Medicinal Effect and Environmental Impact*, 2nd Edition, CRC Press, LLC, Boca Raton, FL, 2004.

Chilton, W.S.: "Chemistry and Mode of Action of Mushroom Toxins," In: Rumack, B.H., and E. Salzman: *Mushroom Poisoning: Diagnosis and Treatment*, pp. 88–117. CRC Press, LLC, Boca Raton, FL, 1978.

Danell, E., and F.J. Camacho: "Successful Cultivation of the Golden Chanterelle," *Nature*, **385**, 303 (1997).

Editorial: "The Magic of Mushroom," *Nature*, **388**, 340 (1997).

Hall, I.R., P. Buchanan, and S. Stephenson: *Edible and Poisonous Mushrooms of the World*, Timber Press, Inc., Portland, OR, 2003.

Kendrick, B.: The Fifth Kingdom, Mycologue Publications, Waterloo, Canada, 1985.

Miles, P.G., and S.T. Chang: *Mushroom Biology: Concise Basic and Current Developments*, World Scientific, Singapore, Southeast, Asia, 1997.

Shepherd, C.J., and C.J. Totterdell: *Mushrooms and Toadstools of Australia*, Inkata Press, Sydney, Australia, 1988.

Webster, J., and R.W.S. Weber: *Introduction to Fungi*, 3rd Edition, Cambridge University Press, New York, NY, 2007.

SHU-TING, CHANG, The Chinese University of Hong Kong, Hong Kong, China

MUSIC SCALE. See **Scale**.

MUSKEG. A swamp or bog occurring in depressions in poorly drained alluvial or glacial terrain in northern Canada or the United States. The depression usually accumulates a saturated, highly compressible mixture of mineral particles and decaying vegetal matter, topped by a hummocky surface of sphagnum moss, and incapable of supporting heavy loads or traffic. In the colder and wetter parts of Alaska, these accumulations spread widely over low-amplitude terrain and are not confined to depressions.

MUSKELLUNGE. See **Pike**.

MUSKMELON. See **Cucurbitaceae**.

MUSK OX. See **Goats and Sheep**.

MUSKRAT. See **Rodentia**.

MUSSEL. See **Mollusks**.

MUSTARD. See **Flavorings**.

MUSTARD FAMILY. See **Brassica**.

MUSTELINES (*Mammalia, Carnivora*). As shown by the organization of the *Mustelines* in Table 1, there are four main groupings — the weasels, the badgers, the skunks, and the otters.

TABLE 1. GENERAL ORGANIZATION OF THE WEASELS, BADGERS, SKUNKS, AND OTTERS

Mustelines	
Weasels (*Mustelinae*)	Skunks (*Mephitinae*)
True Weasels (*Mustela, …*)	Hog-nosed Skunks (*Conepatus*)
Polecats (*Putorius*)	Striped Skunks (*Mephitis*)
Minks (*Lutreola*)	Spotted Skunks (*Spilogale*)
Martens (*Martes, …*)	
Tayras (*Tayra*)	Otters (*Lutrinae*)
Grisons (*Grison, …*)	
Striped Weasels (*Poecilogale* and *Poecilictis*)	Common Otters (*Lutra*)
	Simung (*Lutrogale*)
Zorilles (*Zorilla*)	Clawless Otters (*Amblonyx*)
	Small-clawed Otters (*Aonyx* and *Paraonyx*)
Badgers (*Melinae, …*)	Saro (*Pteroneura*)
	Sea-Otter (*Enhydra*)
Wolverines (*Gulo*)	
Ratels (*Mellivora*)	
Eurasian Badgers (*Meles*)	
Sand-Badgers (*Arctonyx*)	
Teledu (*Mydaus*)	
American Badgers (*Taxidea*)	
Tree-Badgers (*Helictis*)	

The *weasel* is a small animal, slim and short-legged, related to the minks, ferrets, and martens. Several species of weasels occur in Eurasia and a dozen species in North America have been identified. The common or long-tailed weasel, *Mustela noveboracensis*, ranges from Illinois to the Carolinas and northward into Canada. The animal is brown above and yellowish below in the summer, becoming white in winter, with a black-tipped tail when in the northern part of its range. The winter phase also has been called ermine. The name long-tailed weasel applies also to another species, *M. longicauda*, found only on the plains from Kansas northward. The short-tailed weasel, *M. cicognanii*, resembles the common species, but is white below in summer. It ranges from the northern states to Alaska. The common weasel is shown in Fig. 1.

Fig. 1. Weasel. (*A.M. Winchester*.)

The *stoat*, *M. erminea*, is a slender, short-legged animal related to the weasels and sometimes called the greater weasel. It has been regarded by some zoologists as common to both hemispheres, living in northern latitudes, but authorities now consider the common North American weasel to be distinct, although a closely related species. Except in the extreme southern parts of their range, both of these animals turn white in winter.

The *polecat* is a long-bodied animal with short legs and is closely related to the weasels and minks. The polecat is known for its disagreeable odor, a characteristic responsible for the occasional use of this name for a skunk. The European polecat, also known as the foumart or foul marten and as the fitchet, fitcher, or fitcheu, is the source of the fur known on the market as fitch. Three other species occur in Europe and Asia. One, the black-footed polecat, commonly called the black-footed ferret, is found in the plains region of North America. The Cape polecat of South Africa is more closely related to the skunks than to the true polecats. The polecat feeds on small animals, birds, poultry, snakes, lizards, and frogs. The animal is considered bloodthirsty and hunts at night.

The *mink* is a slender semiaquatic animal with short legs, partially webbed toes, and a short, moderately bushy tail. It is related to the polecats. One species, *Mustela sibiricus*, occurs in Siberia; the *M. lutreola* is found in Eastern Europe; and the *M. vison* is found throughout North America. The fur varies from light to very dark brown and is thick and soft, mixed with long glossy hairs.

The *marten* is a slender animal with short legs and a moderately long bushy tail. Related to the weasels, the marten is larger. Several species live in the Northern Hemisphere. Some of the most valuable furs marketed are from martens, and all species bear fur of fine quality. The sable of northern Asia, *Martex zibellina*, and the marten of northern North America, *M. americana*, produce the most valuable fur, although that of the fisher, *M. pennanti*, is also good. The fisher formerly ranged over most of North America, but is now found only in the wilder northern areas. It is also called the pekan and the American martens, as well as the American sable.

The *tayra* is a South American animal of the weasel family. It is among the larger species, comparing with the otter in size, and has the characteristic long body and short legs of the weasels, with a long and rather heavily furred tail. The animals range from white to black in color. The grison is a small South American animal with long slender body and tail, short legs, and related to the weasels.

The *wolverine* is classified with the badgers (*Melinae*) and is a stoutly built short-legged animal, *Gulo luscus*, of northern North America, ranging from the mountain forests of Colorado and Pennsylvania to the Arctic region. The animal is over 3 feet (0.9 meter) long, with large feet, a short bushy tail, and shaggy fur. It is also known as the glutton. In French Canada, it is termed the carcajou. The wolverine is noted as a vicious killer and as a despoiler of camps. It not only damages foods that it cannot eat, but steals objects that it cannot use and so becomes a general nuisance to campers and trappers in the northern woods. Since it climbs, gnaws, and digs with great facility, protection of supplies is difficult. Wolverines eat dead animals as well as prey that they kill, and often they rob trap lines. A second species of wolverine, *G. luteus*, of slightly smaller size and different proportions, has been found as far south as the Sierra Nevadas in California.

The *ratel* is a short-legged animal with a broad depressed body and head, resembling the badger in form. The ratels (*Mellivora indica*) live in India and *M. capensis* in south and west Africa. They are nocturnal burrowing animals and eat other animals and insects. From their love of honey, they have been called the honey badger.

The *badger* is a stoutly built and short-legged burrowing animal. There are several species found throughout the Northern Hemisphere. The term is extended to the ferret-badgers, which are related to the skunks and to the ratels of India and Africa. Badgers are courageous and able fighters when attacked and are sometimes hunted for sport, but they are peaceful animals unless molested. The fur is moderately valuable and the long hairs, especially of the European species, *Meles taxus*, are used in making brushes. Badger-hair shaving brushes widely used for many years are an example. The badger measures about 28 inches (71 centimeters) long. It has a semiflat body, broad and flat tail. The fur is long, gray to red-brown. The feet and ears are black. The diet is mainly rodents. The ferret-badger occurs in several species in the forests of the Oriental region. They are intermediate between the ferret and the more stocky badger.

Skunks are animals of the Americas and widely known for their foul odor. The species are black and white, with a long bushy tail, under which are located the two glands that produce the defensive secretion. The skunks of North America have been divided into 15 species and a number of varieties. These species are grouped as the striped or common skunks and the small spotted skunks, belonging respectively to the two genera, *Mephitis* and *Spilogale*. Both genera are widely distributed. The fur, especially of the larger striped skunks, is of commercial value.

The *otter* is an elongate animal with short legs, broad head, and webbed toes. The otter is an excellent swimmer, sometimes appearing to catch fish simply for the pleasure of the pursuit. Otters are found on all continents except Australia. Different species vary from 2 to 3 feet (0.6 to 0.9 meter) in length, exclusive of the long tail, but there is relatively little difference in their general appearance. *Lutra vulgaris* is the common otter of Europe and the closely related *L. canadensis* is found in North America. Otter fur is thick, soft, and glossy, and among the valuable commercial furs. The fur of the sea otter, *Enhydris* (*Latax*) *lutris*, one of the largest species, is regarded as highly valuable. The species inhabits the northern Pacific, in both Asiatic and American waters, where it feeds on marine invertebrates. Once abundant, indiscriminate trapping has greatly reduced the population. Sea otters are known to travel under water for a half-mile (0.8 kilometers) without surfacing. They can travel on land faster than man can run. For a den, they prefer a dirt cave or a place dug out of the ground or old logs. They are reported to be very playful, sliding down mud or snow banks until the bank becomes smooth. It is also reported that they throw stones in water, diving in after them before they sink. The sea otter is readily trained, learning fast, exhibiting great curiosity, and displaying a friendly personality along with unusual intelligence.

MUTATION. A change in a gene which results in an altered effect of its expression. The great majority of mutations have effects that are harmful to the organism expressing them, but a very few may make the organism better adapted to its environment. Through selection, these favorable mutations are established in a population. The capacity to mutate is an important property of the genetic material since it is the origin of variation in the gene pool. See also **Evolution**; and **Genetics and Gene Science (Classical)**.

Mutations can be of several types. When the chromosome structure is involved, it is referred to as a chromosomal mutation or aberration. When the locus is restricted, the result is called a gene mutation. This type of mutation can be the result of a change in a single nucleotide by substitution of one purine or pyrimidine for another (transition), or of a purine by a pyrimidine or vice versa (transversion). Insertions and deletions of one or more nucleotides can also occur.

Although a gene consists of hundreds and even thousands of nucleotide bases, a change in just one of these bases can cause a change in the effect of the entire gene. One base alteration would cause a difference in one codon of the genetic code and a different amino acid might be inserted in the polypeptide chain formed. This could change the entire protein greatly. An example to illustrate is found in human hemoglobin. Normal hemoglobin consists of two pairs of polypeptide chains. Each chain contains about 150 amino acids. A change in one right of the DNA ladder causes the substitution of one amino acid, valine, for glutamic acid in the pair of beta chains. This small change is sufficient to alter the entire hemoglobin molecule and cause it to form in chains with other molecules when the oxygen tension is reduced. A severe anemia, sickle-cell anemia, results when this reaction takes place in the body. The red blood cells form into sickles and do not carry oxygen properly. Another change in the DNA ladder causes lysine to be substituted at this point on the polypeptide chain, resulting in hemoglobin C, which is also abnormal.

Spontaneous mutations are caused by unknown and omnipresent mutagenic agents (mutagens). Induced mutations are produced at will by subjecting the genetic material to a variety of known mutagens. The comparatively high mutation rates so obtained are superimposed on the spontaneous mutation rates which, as a rule, are quite low. Study of the action of these known mutagens on DNA (deoxyribonucleic acid—see also **Cell (Biology)**) furnishes information as to the chemical nature of gene action

TABLE 1. TYPICAL MUTAGENS AND THEIR POSSIBLE MUTAGENIC EFFECTS ON DNA

Mutagen	Possible Mutagenic Effect on DNA
X-rays	Largely unknown — probably caused by free radicals produced in aqueous solution
Ultraviolet radiation	Largely unknown — dimerization of thymine has been suggested
Auto-oxidizing agents, such as Fe^{2+}	Largely unknown — probably caused by free radicals produced in aqueous solution
Heat	Depurination or ionization of guanine at N-1, followed by a change in base pairing.
Alkylating and esterifying agents:	Alkylation of N-7 of guanine, possibly followed by depurination or ionization of guanine, and a change in base pairing
Mustards Dimethyl- and diethyl sulfate Beta-propiolactone Ethylmethane sulfonate	
Nitrous-acid	Deamination of adenine, guanine, and cytosine, followed by a change in base pairing
Hydroxylamine	Attachment to the C−C double bond of cytosine, followed by a change in base pairing
Base analogues:	Incorporation into DNA in place of thymine, followed by a change in base pairing
5-Bromouracil 5-Iodouracil 5-Chlorouracil	
Deuterium	Replacement of hydrogen, causing an unknown disturbance
Ionizing radiation	See also **Cancer Research**.

in general, and mutation in particular. Some typical mutagens and their possible effects on DNA are listed in Table 1.

ANN C. DeBALDO, Ph.D., University of South Florida, Tampa, FL

MUTATUS.   See **Clouds and Cloud Formation**.

MYALGIA.   Aches and pains in the muscles.

MYASTHENIA GRAVIS.   A disease characterized by excessive fatigability of the muscles after mild or moderate exertion. The muscles about the head and neck are the most seriously involved. In some instances, toward the end of a meal, the jaw muscles may be so tired that further chewing is impossible. After a few minutes of rest, there is sufficient strength to start again, but the fatigue returns rapidly. Difficulty in swallowing or keeping the eyes open may be noted. The disease affects about 5 persons per 100,000 population. In women, the onset usually is during the third decade, whereas for men, onset usually is in the seventh decade. The incidence is about 20% higher among females. The illness takes many forms, ranging from a mild and restricted disturbance of ocular muscles to that of a rapidly developing, generalized, and sometimes fatal, result. About 10% of myasthenia gravis cases are ultimately fatal. For persons who survive the first 3 years of the disease, the disease may stabilize, with a reasonable possibility for some recovery from symptoms.

The disease is currently poorly understood, but as of the early 1980s some scientists believe that monoclonal antibody techniques may ultimately lead to a cure. In the early stages, these techniques will lead to a better understanding of the mechanism of the disease.

Confirmation of the clinical diagnosis can be made with the edrophonium (Tensilon) test. Tensilon is a cholinesterase inhibitor with a brief (5 to 10 minute) duration of action. A marked improvement in power, which occurs in 30 to 60 seconds and lasts a few minutes, will confirm the diagnosis of myasthenia gravis in a majority of cases. Electromyography also may be an aid in diagnosing the disorder. Routine chest x-ray may yield helpful information as well as evaluation of thyroid function. The incidence of thyroid disorders is about 13% among myasthenic patients. Thymectomy has been used in a number of instances where myasthenia persists.

MYCOSES.   Mycoses are the infections and diseases caused by fungi.

Overview of Human Mycoses

Mycology is the branch of biology that deals with the study of fungi. Fungi play an important role in the health and wellbeing of humans. On the negative side, they cause allergies such as hypersensitivity pneumonitis and rhinitis in certain individuals, and also cause many infections in humans and animals.

Some of the fungal diseases of humans are life-threatening if not diagnosed quickly and treated aggressively. Understanding of the relationship of fungi to disease predates that for bacteria by several years with the work of Bassi in the early 1800s, who described the aetiology of muscardine, a fungal infection of silkworms, followed, from 1839 to 1853, with the works of Schoenlein, Remak and Gruby, who described the clinical entity known as *tinea favosa*, a form of ringworm of the scalp. These studies led to Robin's survey in 1853 in which he described several types of fungi that cause superficial infections of the skin and hair. During the 1890s and early 1900s seven important fungal infections were described: coccidioidomycosis (1892), "thrush" (a form of candidiasis; 1893), blastomycosis (1894), cryptococcosis (1895), aspergillosis (1897), histoplasmosis (1905) and paracoccidioidomycosis (1908). With the description of sporotrichosis in 1912, the major fungal diseases of humans had been described. However, from that period of time through to the 1940s fungal infections were considered to be rare and exotic diseases that were, for the most part, very difficult to treat and for the most part fatal. See also **Aspergillosis**; **Blastomycosis**; **Candidiasis**; **Coccidioidomycosis**; **Cryptocrystalline Histoplasmosis** and **Sporotrichosis**.

The epidemiological studies of Smith and others during World War II, who described subclinical and asymptomatic forms of coccidioidomycosis, a respiratory fungal disease that is endemic in parts of North, Central and South America, changed our thoughts about coccidioidomycosis and histoplasmosis, another respiratory fungal disease that is worldwide in distribution. In addition, the discovery of two cyclic polyene macrolides, nystatin in 1949 produced by *Streptomyces noursei* followed by amphotericin B in 1956 produced by *S. nodosus*, changed the clinical outcome of these life-threatening fungal infections by providing novel antibiotics that were useful in treating patients who had progressive life-threatening disease. Unfortunately, these compounds are extremely toxic to the patient and must be used with extreme caution. Recently, a class of compounds called the triazole derivatives have provided novel antifungal agents that are very useful in treating various systemic fungal infections because they can be taken orally are well tolerated, and less toxic than the polyenes. But they have disadvantages in that they are, for the most part, fungistatic, require prolonged treatment, are expensive to use, and there are growing numbers of fungi that are resistant to them. See also **Antifungal Agents**.

Fungi are ubiquitous in nature and, for the most part, free-living, with only a few being part of the normal flora of healthy individuals. Unlike other infectious organisms such as viruses, parasites and some bacteria, fungi do not need to infect human cells in order to survive and replicate. We are constantly exposed to their infectious propagules. In general, healthy and immunologically competent individuals have a high degree of innate resistance to fungal disease. When infections do occur most are either asymptomatic and subclinical or they are of short duration and of little clinical significance. This, however, is related to the number of infectious particles contained in the exposure.

There are approximately 100 000 recognized species of fungi. At the present time about 300 have been reported to cause disease in humans but the numbers are increasing. The escalating incidence of fungal diseases, particularly those due to the emergence of novel organisms and re-emergence of others, are attributable to advances and changes that have occurred in the modern practice of medicine. In particular, they include the use of potent broad-spectrum antibiotics to treat bacterial infections and the use of immunosuppressive therapies in organ transplantation and cancer chemotherapy. In addition, fungal infections have become a major problem because of the acquired immunodeficiency syndrome (AIDS) epidemic and the mere fact that humans are living longer. See also **Acquired Immune Deficiency Syndrome (AIDS)**.

In healthy individuals, only a small number of fungi are capable of causing significant infections. Once established these diseases can be classified according to the tissue levels that were the original sites of the infection. These include the superficial (hair, skin and nails), subcutaneous (dermis, muscle, connective tissues and occasionally bone), and the systemic (lungs). The term *opportunistic infections*, originally coined to describe fungal infections in general, is currently used to describe fungal infections caused by fungi with low pathogenic potential. Superficial fungal infections are caused by fungi that grow on tissues that contain keratin, a protein found in cells of the very superficial layers of the epidermis, hair and nails. The diseases caused by these fungi are usually cosmetic in nature and respond well to treatment, but in some cases they can be chronic and require prolonged courses of therapy. The subcutaneous infections are generally caused by fungi with low pathogenic potential and usually require some form of traumatic implantation into the deeper layers of tissue. The fungus survives and multiplies in these deeper tissues because of the tissue damage caused by trauma. In contrast to the superficial infections, these infections, with a few exceptions, are difficult to treat and on occasion require excision of the affected area or amputation of a limb. The systemic fungal infections are those that involve the lungs as the primary site of colonization.

There are five fungi included in this group and they all have the potential of causing disease in individuals who have no underlying immunological defect. Most patients experiencing these infections are usually asymptomatic but when symptomatic disease occurs, it is generally of short duration and does not require aggressive therapeutic intervention. In the case of opportunistic fungal diseases the infection can be attributed to the existence of an immunodepressed or debilitated state occurring in the patient. Patients suffering from these infections respond poorly to most forms of treatment until the conditions causing the immunosuppression or debilitated state are reversed or corrected.

Dimorphism, Virulence and Human Infections

Considerable information is available concerning the molecular bases of bacterial and viral virulence and factors that play a role in various plant diseases. There is, however, a lack of data concerning molecular genetic factors that contribute to and influence virulence of fungi that are implicated in diseases of humans. In general, healthy, immunologically competent individuals have a high innate resistance to fungi. Infections occur when certain protective physical barriers such as the skin and mucous membranes are breached or when the host's immune system becomes

defective. While most encounters with fungi are accidental, some fungi have developed mechanisms that facilitate their colonization or foster their survival within a hostile host environment; for example, the production of specific enzymes that hydrolyse keratin and factors that promote adherence to tissues, enabling fungi to gain access to and invade tissues. Several fungi that cause systemic infections and one that causes the subcutaneous disease, lymphocutaneous sporotrichosis, are dimorphic in that they are hyphal in nature but assume a unicellular morphology when they establish an infection in susceptible hosts. The parasitic phase of dimorphic fungal pathogens is generally limited to either unicellular structures such as yeast, endospores or sclerotic cells; the exceptions, *Candida albicans* and other species of *Candida*, undergo a yeast-to-hyphal transition (see below). The relationship between morphology and pathogenicity has served as a major incentive to elucidate the molecular basis of this adaptation process. Various environmental, nutritional and physical conditions appear to have a profound influence on this morphogenic process that leads to the pathogenic phase of these fungi.

Among the fungi that have been implicated in diseases of humans and animals, *Histoplasma capsulatum, Blastomyces dermatitidis, Paracoccidioides brasiliensis, Penicillium marneffei, Coccidioides immitis* and *Cryptococcus neoformans* have the intrinsic ability to cause infections and disease in healthy humans and animals. These organisms possess traits that contribute to their ability to survive at elevated temperatures of the host, elude or neutralize the hostile host environment they encounter and adapt in a manner that allows them to survive and multiply. Five of these fungi, *H. capsulatum, B. dermatitidis, P. brasiliensis, P. marneffei* and *C. immitis*, are dimorphic, changing from a hyphal organism in nature to a unicellular yeast when they invade tissues; *C. immitis* is different in that the single-celled arthroconidium undergoes various stages of development to form a large sporangium termed a spherule that is filled with endospores.

In the cases of *H. capsulatum, B. dermatitidis* and *P. brasiliensis*, three distinct stages, characterized by profound metabolic changes, distinguish the hypha-to-yeast transition when cells are transferred from an incubation temperature of 25 °C to 37 °C (77 °F to 98.5 °F). During stage 1 of the thermally induced morphogenesis there is a dramatic decrease in the respiratory rate and the transforming hyphae enter stage 2, which is characterized by a lowered respiratory rate followed by disappearance or reduction of cytochromes and a depletion of adenosine triphosphate (ATP) pools. Respiration in *H. capsulatum* yeasts and hyphae proceeds via a cyanide-sensitive cytochrome chain and salacyl hydroxamic acid (SHAM)-inhibited alternate oxidase. When both mitochondrial pathways are blocked by cyanide and SHAM, exogenous cysteine can stimulate oxygen consumption via a cytosolic oxidase present only in mature and developing yeasts. Furthermore, intracellular levels of cysteine decrease stage 2 and exogenous cysteine or other sulfhydryl compounds are required for passage through state 2 and entry into stage 3. In stage 2 cysteine is oxidized within mitochondria by an alternate shunt that bypasses the cyanide-sensitive and SHAM-sensitive respiration lost during stage 1. The role of cysteine in stage 2 is independent of its consumption by cysteine oxidase. Until stage 3, only this cysteine-stimulated respiration, which is constitutive in hyphae, still functions. In stage 3, cytochromes and respiration reappear and cysteine pools expand and distinct morphological changes consistent with yeast morphology begins during this period. Along with these changes, cysteine oxidase activity appears, with an absolute requirement for cysteine. In the absence of cysteine or other reducing compounds, a shift in temperature to 37 °C (98.5 °F) will initiate the differentiation process only through stage 1 into stage 2. In order to proceed to stage 3 and initiate the morphogenic change to yeast, a reducing environment, such as would be found within macrophages, must be present.

For fungi causing systemic infections the morphological transition is triggered by a significant temperature change [Maresca and Kobayashi; Matthews and Burnie, refs.]. In the case of dimorphic fungi, raising the temperature from 25 °C to 37 °C (77 °F to 98.5 °F) induces a hypha-to-yeast conversion, except for *C. albicans*, in which a yeast-to-hyphal transformation occurs. Homeostatic mechanisms that protect cells and entire organisms from the deleterious effects of such environmental stresses are termed the heat shock response. The products of this response are termed heat shock proteins (HSPs) and they play a major role in many cellular processes that affect host–parasite interactions. The heat shock response is a universal phenomenon in which a high temperature uncouples oxidative phosphorylation, hindering ATP production, elicits a nearly complete shutdown of RNA and protein synthesis, and stimulates the production of HSPs that have been highly conserved through evolution.

In *C. albicans* and *H. capsulatum* several proteins, between 6 and 110 kDa that are characteristic of the heat shock response appear quickly after the thermal shift. Furthermore, oxidative phosphorylation is uncoupled and ATP synthesis is immediately shut down in *H. capsulatum*. Thermally induced morphogenesis in this fungus has been described as a heat shock response followed by cellular adaptation to a higher temperature. The regulation of heat shock genes HSP70 and HSP82 in strains of *H. capsulatum* with varying thermotolerance and pathogenicity supports this notion. Temperatures at which HSPs and heat shock mRNA are not produced fail to induce the phase transition. Strains with increased virulence are more heat-resistant, pass through developmental stages 1 to 3 more quickly, and display a less severe heat shock response than less virulent strains. Less virulent strains can be made to mimic the response of the more virulent strains by making the temperature shift less extreme. Conversely, more virulent strains can be forced to resemble the less virulent strains by making the temperature shift more severe. Elucidation of the multiple heat-sensing mechanism regulating transcription of the heat shock genes may be important in elucidating the mechanisms that control dimorphic transitions and virulence in *H. capsulatum* and other systemic fungi that undergo this morphological change.

In concert with these changes, the rates of macromolecular synthesis change in characteristic patterns during dimorphic transitions of *C. albicans*, *H. capsulatum* and *P. brasiliensis*. Since all these fungi are induced to undergo morphogenesis by elevating the temperature, one might have anticipated increased rates of all macromolecular syntheses from simple thermodynamic considerations. Indeed, *C. albicans* hyphae increase mass twice as fast at 37 °C (98.5 °F) as yeast do at 24 °C (75 °F). However, by altering the carbon source, this organism can be cultivated in both forms within overlapping ranges of growth rate, implying that this alone is not critical to morphology. Similarly, when morphology is determined solely by pH, no differences in biosynthetic rates occur in the two forms. Thus it is interesting that a critical period of time spent in stationary phase or nutrient starvation is conducive to hyphal development when yeast cells are placed into fresh medium. Under these conditions there is an initial quiescent period (phase I), followed by a slow onset of RNA and protein synthesis (phase II), and ultimately, two major increases in the rates of RNA and protein synthesis after germ tube emergence (phase III).

In contrast, the rates of macromolecular synthesis are lower by about half in yeasts of *H. capsulatum* growing at 37 °C (98.5 °F) than in hyphae growing at 24 °C (75 °F). During thermally induced morphogenesis, the rates of RNA and protein parallel respiratory activity. In the temperature-susceptible strains they rapidly drop to practically nondetectable levels during stage 1, stay very low during stage 2, and begin to rise in stage 3. When yeast cells of *P. brasiliensis* are shifted from 37 °C (98.5 °F) to 22 °C (71.5 °F), growth and macromolecular synthesis cease for about 12 h and then resume at about half the initial rates as hyphae develop.

Of all the dimorphic systemic fungi, an interesting feature of infections caused by *H. capsulatum* is its propensity to parasitize macrophages. Furthermore, the yeast phase is able to survive and replicate within phagolysosomes of naive macrophages. That the hyphal and yeast phases of *H. capsulatum* differ in their calcium requirements for growth and in the production of a released calcium-binding protein has been demonstrated by Goldman and his colleagues [Batanghari, et al.:]. The hyphal phase of *H. capsulatum* requires calcium for growth whereas the yeast phase requires little free calcium. In addition, the yeast phase produces a calcium-binding protein whereas the hyphal phase does not. The mechanism and role of calcium binding protein is unknown but it has been speculated that it may provide yeast with the calcium they need to grow and replicate or may temper the harsh conditions within the macrophage.

A number of other molecules, including cAMP, have been attributed as "second messengers" of regulatory metabolites in biological processes. Internal cAMP concentrations are fivefold higher in hyphae than in yeasts of *H. capsulatum*. Similarly, cAMP levels are higher in hyphae of *B. dermatitidis*. Addition of phosphodiester (PDE) inhibitors to yeast cultures at 37 °C (98.5 °F) induces hyphal development. Addition of cAMP or PDE inhibitors to yeast cultures at 37 °C (98.5 °F) induces the development of hyphae. Cytochemical electron microscopy indicates a large concentration of PDE localized within vesicles in yeasts growing at 37 °C (98.5°). Most of this enzyme activity disappears when the temperature is lowered to 25 °C [77 °F] and hyphal development begins. Moreover, these substances induce the same physiological and biochemical changes that accompany temperature-induced morphogenesis.

That the shape of the cell wall ultimately determines fungal morphology is irrefutable: enzymatic digestion of this structure invariably generates spherical protoplasts, whether from yeasts or hyphae. Much work has been done to characterize the molecular components of yeast and hyphal walls from a variety of dimorphic fungi on the notion that form follows substance. Contrary to expectation, a great many dimorphic fungi were found to display essentially the same polymers in roughly the same relative amounts in both yeast and hyphal walls. Conversely, very different wall polymers were found to sustain quite similar cell morphologies in distantly related fungi. For example, *C. albicans*, *H. capsulatum*, *S. schenkii* and *Wangiella dermatitidis* have walls mainly composed of chitin fibrils within an amorphous matrix of β-glucans and mannoproteins regardless of the cell form. Yeast walls, which are invariably thicker than their hyphal counterpart, are generally made from the same building blocks as hyphal walls in a given organism. Two exceptions are *P. brasiliensis* and *B. dermatitidis*, in which the walls of the yeast forms contain mainly chitin, galactomannoproteins and β-glucans. The hyphal walls of *P. brasiliensis* has β-glucans in place of the α-glucans, while the hyphal walls of *B. dermatitidis* display 60% α-glucan and 40% β-glucan. The other major wall components remain unchanged in yeasts and hyphae of these species.

The dimorphic phase transition in fungi is a complex process effected at molecular, genetic and subcellular levels and ultimately manifested as a change in the pattern of cell wall assembly.

Morphogenesis in *Candida* spp

Fungi belonging to the genus *Candida*, with the exception of *C. glabrata*, possess interesting morphological potential in that they can exist as unicellular yeasts, as a pseudohyphal form or as septate hyphae, depending upon environmental conditions of growth. Various nutritional conditions will induce yeast-to-hypha morphogenesis. In *C. albicans*, specific sugars, amino sugars, amino acids, sulfhydryl compounds, proteins or peptides, and metal ions have all been described as critical inducers of morphogenesis and might be expected to influence cell physiology, including carbon and energy metabolism, each in their own way. In an attempt to minimize such influences, a defined medium was developed in which pH or temperature alone governs the morphological fate of yeast cells released from starvation. A number of physiological changes have been described during yeast-to-hypha morphogenesis that appear to be consistent irrespective of the carbon and energy source. The levels of several key metabolites and enzymes of carbohydrate metabolism do not systematically differ in growing yeast and hypha of *C. albicans*. Mitochondria from both phases possess cyanide-sensitive and SHAM-sensitive respiratory pathways. Respiratory quotients are comparable in lactate-induced yeasts and maltose-induced hyphae. Other studies suggest that yeasts of *C. albicans* have more respiratory activity and less fermentative activity than hyphae. Neither chloramphenicol nor cyanide inhibits morphogenesis in *C. albicans*. The role the alternative cyanide-resistant oxidative respiratory pathway plays in the morphological transition of *C. albicans* has not been resolved.

Polyamines interact with the biosynthetic machinery, implying regulatory potential. The levels of putrescine, spermadine and ornithine decarboxylase (ODC) activity increase significantly during the yeast-to-hypha morphogenesis in *C. albicans* and *C. lipolytica*. Several ODC inhibitors block hyphal development. This inhibition is reversed by exogenous putrescine or 5-azacytidine, a compound thought to prevent DNA methylation.

The relationship between cAMP levels and morphogenesis has been extensively investigated in *C. albicans*. Internal cAMP levels are higher in hyphae than in yeasts. Exogenous cAMP and PDE inhibitors induce a yeast-to-hypha transition at temperatures normally conducive to the yeast form. The patterns of flux in cAMP and response to the exogenous nucleotide are similar, if not identical, in all these studies although the morphogenic transitions are induced in different media, at different temperatures, and at different stages of growth. It has been reported that binding of calcium by calmodulin, an endogenous activator of PDE, influences dimorphism and cAMP in *Ceratocystis ulmi* and *C. albicans*.

Morphogenesis in *Wangiella dermatitidis*

Wangiella dermatitidis, one of the aetiological agents of phaeohyphomycosis, exhibits polarized growth processes leading to yeast budding or hyphal extension and nonpolarized processes leading to isotropically enlarged forms that may become unbudded, multinucleated, multicellular phenotypes that resemble sclerotic bodies of chromoblastomycosis [Peng, et al.:]. Budding yeast cells of this polymorphic fungus can be induced to form phenotypes resembling sclerotic cells by incubation of cells in pH 2.5 medium at 25 °C (77 °F). Further studies show that at this acidic pH, critical, but low concentrations of Ca^{2+} ion are crucial for regulating multicellular development and that higher concentrations of Ca^{2+} favor polarized growth. Studies with other fungi causing chromoblastomycosis indicate that Ca^{2+} concentrations and acidic conditions are important in regulating this morphological change. Whether Ca^{2+} deprivation *in vivo* induces this phenotypic expression in chromoblastomycosis is not known but the pathophysiological conditions existing in granulomatous tissue reactions involve processes that require Ca^{2+}. Several cellular and humoral processes such as the classical complement cascade, opsonization, phagocytosis, degranulation of eosinophils and mast cells and various inflammatory processes require Ca^{2+}. Thus in the milieu of tissue damage caused by traumatic implantation, the fungus may encounter competitive conditions of Ca^{2+} depletion.

Opportunistic Mycoses (Candidiasis, Cryptococcosis, Trichosporonosis, Aspergillosis, Zygomycosis, *Pneumocystis carinii* Infections)

Fungi are ubiquitous in nature and humans are constantly exposed to viable fungal elements. Most individuals tolerate these exposures with no sequelae, although some may develop allergies. Except for the systemic mycotic agents, all other fungi have a low inherent virulence and healthy immunologically competent individuals have a high degree of innate resistance to colonization by them. However, when protective barriers of the skin are breached or conditions of host debilitation exist, many people become susceptible to fungi. There is a growing list of exotic and rare fungi that have been implicated in infections, but the vast majority are caused by various species of *Candida, Aspergillus* and different zygomycetes. Although these organisms are called *opportunistic pathogens*, the term is based on the clinical setting and has no taxonomic significance. Three common ones, *C. albicans, A. fumigatus* and the zygomycetous fungi, have very different biological properties and host interactions. In addition, pathogenicity is not an all-or-nothing phenomenon. Individual species within *Candida, Aspergillus* and the zygomycetes exhibit a wide range of pathogenicity.

Various species of *Candida* are found as part of the normal flora of and are often isolated from healthy mucosal surfaces such as the mouth, vagina, rectal area and gastrointestinal tract. The clinical manifestations of candidiasis range from superficial infections of the skin and nails to life-threatening infections involving multiple organ systems. These organisms frequently colonize and cause problems of the skin in intertriginous, inframammary and the crural areas when these tissues become macerated or traumatized. A clinical condition often affecting infants and an important hallmark of human immunodeficiency virus (HIV) infections is an entity commonly known as "thrush." This is where various species of *Candida* colonize the superficial layers of mucous membranes of the oral cavity as creamy-white plaques. Untreated the condition may lead to erosive lesions that are extremely painful but not fatal. Infections also occur when intravenous lines such as catheters become contaminated. The disease can be life-threatening when haematogenous spread occurs in the immunosuppressed patient, particularly those who are neutropenic. See also **Human Immunodeficiency Viruses (HIV)**.

A striking feature of organisms in the genus *Candida*, with the exception of *C. glabrata*, is that they are able to multiply by forming blastospores, pseudohyphae and septate hyphae. *C. glabrata*, the aetiological agent of urinary tract infections and disseminated disease in patients who are immunologically defective, produces only blastospores. The spectrum of disease caused by these organisms includes: localized infections of the skin and nails; diseases of mucosal surface of the buccal cavity, oral pharynx, vagina, oesophagus, and bronchial tree; fungaemia and disseminated disease involving multiple organ systems, particularly in patients who are neutropenic.

Cryptococcosis is an infection caused by *Cryptococcus neoformans*, an encapsulated yeast that is a yeast in culture and in tissue. In contrast to the dimorphic fungal pathogens (see above), the asexual phase of *C. neoformans* is monomorphic. The most common clinical form of cryptococcosis involves infection of the brain and meninges. Pulmonary forms of the disease have many manifestations and the organism has been cultured on occasion from sputum of individuals in the absence of clinical disease. More importantly, cryptococcosis is an important cause of morbidity and mortality in AIDS patients and is a sign that an immune defect, such as AIDS, may be coexistent in the patient. There are several species of *Cryptococcus* but disease has been attributed only to *C. neoformans*. There are two varieties of *C. neoformans*: *C. neoformans* var. *neoformans* and *C. neoformans* var. *gattii*. Each variety has two serotypes: serotypes A and D for var. *neoformans* and B and C for var. *gattii*. As assayed by reactivity with antibody, glucuronoxylomannan is the predominant polysaccharide and determines serotype. The acidic mucopolysaccharide capsule is a virulence factor for this yeast. Encapsulated *C. neoformans* are not phagocytized whereas mutants lacking the capsule are readily phagocytized.

Most fungal infections caused by *Trichosporon beiglii* involve the hair in a clinical entity termed *white piedra*, but there are a number of recently described clinical situations in which this organism has caused systemic life-threatening infections in patients who are immunosuppressed. The aetiological agent is an organism that, at present, may genetically consist of several taxonomically distinct organisms, characterized morphologically by growing as either hypha or budding yeast. This conclusion comes from differences in gross colony morphology, isoenzyme analysis and restriction fragment length polymorphisms.

The clinical manifestations of infections caused by filamentous fungi vary depending on the aetiological agent and the immune status of the host. In contrast to most infections caused by the various species of *Candida*, which are endogenous in origin, infections caused by the various aspergilli are from an exogenous source where infectious propagules are numerous. Various species of *Aspergillus* are involved in animal diseases such as spontaneous abortion in sheep and cattle, pulmonary disease of birds, and certain mycotoxicoses. There are three forms of aspergillosis, i.e. allergic, secondary colonization and systemic disease.

Like species of *Aspergillus*, fungi belonging to the Zygomycetes are very common in the environment. Underlying factors leading to increased susceptibility to this group of fungi are conditions leading to metabolic acidosis, uncontrolled diabetes mellitus, leukemia and hyperglycemia. Rhinocerebral zygomycosis is the most common form of disease. It originates in the paranasal sinuses and involves the ocular orbit and palate with extensions into the brain. From a morphological viewpoint, organisms in this group are coenocytic and the hyphal elements frequently appear as ribbon-like structures in sections of diseased tissues. Several species of zygomycetes have been implicated in human disease but the organisms most frequently encountered are *Rhizopus arrhiza* and *Absidia corymbifera*.

Another interesting organism that fits into the category of opportunistic infections is *Pneumocystis carinii*, an agent that causes a severe interstitial plasma cell pneumonia. The patient population at risk of developing disease are the elderly, those with congenital and iatrogenically induced immunosuppression, and those with AIDS. Until recently this organism was classified as a protozoan. However, ultrastructural studies along with immunological and molecular genetic evidence indicate that this organism is a fungus and unique from others in that it possesses features that distinguish it from other fungi; in particular, its putative life cycle, different morphological forms found in tissues and its relative refractoriness towards most antifungal antibiotics. In tissues the organism may appear in a trophic form (1.5 to 5.0 μm) or as a uninucleated sporocyst (4 to 5.0 μm), or as a mature spore case (1.5 to 5.0 μm) containing eight oval to fusiform spores (1.0 to 3.0 μm). *P. carinii* appears to be very common in the environment and infection occurs at an early age. Subclinical infections in healthy individuals are probably frequent and the organism may remain dormant for long periods of time. See also **Fungal Infections (in Humans)**.

Additional Reading

Batanghari, J.W., G.S. Deepe Jr., E. Di Cera, and W.E. Goldman: "*Histoplasma* Acquisition of Calcium and Expression of *CBP1* during Intracellular Parasitism," *Molecular Microbiology*, **27**, 531–539 (1998).

Brock, D.L., and I.E. Alcamo: *Infectious Fungi*, Chelsea House Publishers, New York, NY, 2006.

Dismukes, W.E., J.D. Sobel, and P.G. Pappas: *Clinical Mycology*, Oxford University Press, New York, NY, 2003.

Heitman, J., J.E. Edwards, A. Mitchell, and S.G. Filler: *Molecular Principles of Fungal Pathogenesis*, ASM Press, Washington, DC, 2006.

Kavanagh, K.: "Medicial Mycology," *Cellular and Molecular Techniques*, John Wiley & Sons, Inc., Hoboken, NJ, 2007.

Maresca, B., and G.S. Kobayashi: "Heat Shock Proteins in Fungal Infections," In: van Eden, W., and D.B. Young: *Stress Proteins in Medicine*, Marcel Dekker, New York, NY, 1996 pp. 287–299.

Matthews, R., and J.P. Burnie: *Heat Shock Proteins in Fungal Infections*, RG Landes Company, Austin, TX, 1995.

Peng, M., C.R. Cooper, Jr., and P.J. Szaniszlor: "Genetic Transformation of the Dematiaceous Phaeohyphomycotic Fungus *Wangiella dermatitidis*," *Applied Microbiology and Biotechnology*, **44**, 440–450 (1995).

Richardson, M.D., and D.W. Warnock: *Fungal Infection: Diagnosis and Management*, 3rd Edition, Blackwell Publishers, Malden, MA, 2003.

GEORGE S. KOBAYASHI, Washington University School of Medicine, St Louis, MO

MYCOTOXINS.

Mycotoxins are natural products produced by fungi that evoke a toxic response in higher vertebrates and other animals when fed at low concentrations; some mycotoxins may also be phytotoxic or antimicrobial. Mycotoxins, however, are not grouped with antibiotics, compounds which also may be produced by fungi but have microorganisms as the target species. Also, poisonous metabolites produced by mushrooms and yeasts are not included in this group. Mycotoxins are not required for the growth of the producing fungus and, therefore, are considered secondary metabolites. Their production is associated with fungal development, e.g. sporulation (spore formation) and sclerotial formation. Sclerotia are specialized structures used by fungi for survival under adverse conditions. Presumably these compounds play some role in the ecology of the fungus, but their function has not been clearly defined. Many mycotoxins have been characterized and they show significant diversity in their chemical structures and biological activity. Further, many fungi produce multiple families of mycotoxins, including polyketides, modifications of amino acids and peptides, and terpenoids, etc.

Mycotoxicology

Mycotoxicology is the study of mycotoxins and the corresponding mycotoxicoses. The diversity of the discipline that encompasses mycotoxicology includes basic sciences, such as chemistry, biochemistry and molecular biology, and multidisciplinary subjects, such as plant and animal pathology, toxicology, pharmacology and immunology.

Natural Occurrence of Mycotoxins

The three genera *Aspergillus*, *Penicillium* and *Fusarium* comprise the largest number of mycotoxin-producing species (Table 1), but not all species within these genera produce toxins. Mycotoxins may be produced before harvest or during storage. Some species, such as *Aspergillus flavus*, *Alternaria alternata*, *Fusarium moniliforme* or *Fusarium graminearum*, are often found in the field, while other *Aspergillus* species and *Penicillium* species are found in storage.

The incidence and extent of mycotoxin contamination is dependent primarily on geographic and seasonal factors, as well as cultivation, harvesting, storage and transportation conditions. In the contaminated crops, mycotoxin production is also dependent on genetic, environmental and nutritional factors. Even a small amount of fungal growth can result in a significant amount of mycotoxin being produced. Mycotoxin production is dependent on well-defined ranges of temperature and water activity, whereas fungal growth can occur over a wide range of these environmental factors. Water activity (a_w) refers to water in food that is not bound to food molecules. For example, *A. flavus* and *A. parasiticus* can grow at temperatures from 12 to 48 °C (53.5 to 118 °F) (35–37 °C optimum (95–99 °F) and water activity as low as 0.80 (0.95 optimum); for

TABLE 1. NATURAL OCCURRENCE OF SELECTED MYCOTOXIGENIC FUNGI[a]

Fungi	Common crop substrate
Alternaria alternata	Cereal grains, tomato, animal feeds
Aspergillus flavus, A. parasiticus	Peanuts, corn, most treenuts, cottonseed
A. ochraceus	Figs. cereals, cheese
A. versicolor	Cereals, beans, coffee, feeds
A. nidulans	Corn, grains, cheese
A. clavatus	Corn, grains, cheese, apple, apple juice, beans, wheat
Fusarium graminearum	Wheat, corn, feeds, rice
F. moniliforme	Corn, sorghum, rice
F. sporotrichioides	Corn, feeds, hay
F. nivale	Corn, feeds, hay, peanuts, rice
Penicillium citrinium	Barley, corn, rice, walnuts
P. cyclopium	Peanut, corn, cheese
P. islandicum	Rice
P. rugulosum	Rice, sorghum
P. viridicatum	Barley, beans, cereals, coffee, feeds
P. verrucosum	
P. patulum	Apple, apple juice, beans, wheat
P. verticae	
P. puberulum	Barley, corn
P. rubrum	
P. purpurogenum	Feedstuffs, corn
P. politans	Cheese
P. roqueforti	

[a] Adapted from Chu [1998].

aflatoxin production these ranges are narrower–temperatures of 28–33 °C (82–91 °F) (31 °C optimum (88 °F) and water activity of 0.85–0.97 (optimum 0.90). Generally, the optimal temperature for production of mycotoxins is 24–28 °C (75–82 °F), with the exception that T-2 toxin is maximally produced at 15 °C (59 °F). Under storage conditions, each fungal species grows within fairly strict limits of temperature and water activity, and there is an ecological succession of different fungi as water activity and temperature of stored grain change.

Major research efforts throughout the world have resulted in the identification and characterization of several toxins produced by fungi growing on food and feed, quantification of these toxins and evaluation of their biological effects on human and animal health. The driving forces for these studies have been: (1) a food safety concern due to acute and chronic toxicity of these metabolites and the resulting human and animal illnesses; and (2) an economic issue due to the health costs associated with mycotoxicoses, and the inability to market toxin-contaminated field crops as well as stored food and feed.

Mycotoxicoses

Mycotoxicosis occurs when fungal toxins are ingested by animals or humans, and is frequently mediated through effects on various organs, most commonly liver, kidney and lungs and the nervous, endocrine and immune systems. Mycotoxicosis differs from "mycosis," a term that refers to diseases produced by direct pathogenic invasion by fungi. Mycotoxicosis is labeled acute or chronic on the basis of the amount of toxin ingested and the length of time the animal is exposed to the mycotoxin. Acute mycotoxicosis is caused by ingestion of extremely high levels of toxins in a short period, resulting in significant adverse physiological effects and even death. Long-term exposure to even low levels of toxins can lead to chronic mycotoxicoses resulting in reduced growth and increased susceptibility to other diseases. Chronic exposure of animals to some mycotoxins can result in cancer. The effect of a mycotoxin on an animal is dependent on many factors, such as: (1) the affected species; (2) intraspecies susceptibility depending on age, sex and nutritional status; and (3) the condition of the immune system of the affected animal. See also **Mycoses**.

At present more than 300 mycotoxins have been identified but only those implicated in human illnesses (mycotoxicoses) have been studied in detail. A list of human mycotoxicoses is summarized in Table 2; however, it should be emphasized that, although the listed mycotoxins have been

TABLE 2. HUMAN MYCOTOXICOSES AND THE IMPLICATED MYCOTOXINS[a]

Mycotoxin	Fungal source	Biological effect	Corresponding mycotoxicoses
Alternaria toxins (tenuazonic acid, alternariol)	*Alternaria alternata*	Apoptosis, mutagen	Onyalai disease
Aflatoxin	*Aspergillus flavus, A. parasiticus*	Apoptosis, mutagen, hepatotoxin, carcinogen, teratogen	Acute aflatoxicosis, hepatocarcinogenesis, Indian childhood cirrhosis, Reye's syndrome
Cyclopiazonic acid	*A. flavus, Penicillium cyclopium*	Nephrotoxin, cardiovascular lesion	Kodua poisoning
Ergot alkaloids	*Claviceps* spp., *A. fumigatus, P. chermesinum*	Neurotrophy	St Anthony's Fire, ergotism
Fumonisins	*Fusarium moniliforme, Fusarium* spp.	Apoptosis, hepatotoxin, neutrotoxin, respiratory illness	Equine leucoencephalopathy
Gliotoxin	*A. fumigatus, A. terreus, Trichoderma viride*	Respiratory illness	Lung disease
Moniliformin	*F. moniliforme*	Neurotoxin, cardiovascular lesion	Onyalai disease
Ochratoxin	*A. ochraceus, P. viridicatum, P. verrucosum*	Apoptosis, nephrotoxin, teratogen	Balkan nephropathy, renal tumours
Penicillic acid	*P. puberulum, A. ochraceus*	Neurotoxin, mutagen	—
Roquefortine	*P. roqueforti*	Neurotoxin	—
Rubratoxin	*P. rubrum, P. pupurogenium*	Apoptosis, hepatotoxin, teratogen	—
Sterigmatocystin	*A. versicolor, A. nidulans, A. parasiticus, A. flavus*	Hepatotoxin, carcinogen	Hepatocarcinogenesis
Trichothecenes (T-2, deoxynivalenol, etc.)	*F. sporotrichioides, F. nivale, Stachybotrys atra*	Apoptosis, dermatoxin, neurotoxin, keratogen	Alimentary toxic aleukia, Akakabi-byo disease, stachybotryotoxicosis, oesophageal cancer, red mould disease, yellow rain
Zearalenone	*F. graminearum*	Genitotoxin, oestrogenic effects, mutagen	Cervical cancer, premature menarche

[a] Adapted from Pohland and Wood [1987] and Chee [1998].

implicated in human illnesses, very few reports have demonstrated a direct connection between the mycotoxin and a corresponding mycotoxicosis. Only in the case of the aflatoxins have the relationships between exposure to the toxin and acute and chronic human mycotoxicoses been reasonably well established.

Economic Impact of Mycotoxins

The relevance of mycotoxins to human health has prompted the international trade and health organizations and governmental control authorities to regulate the levels of toxins permitted in commodities for commerce or food and feed for human and animal consumption. Over 50 countries of the world have developed such guidelines. These regulatory guidelines result in severe economic losses to farmers, producers and marketers, since food and feed contaminated with toxins above the legal limit must be destroyed. In severe outbreak years, several hundred million dollars worth of crop loss due to mycotoxin contamination has been estimated in the US alone. Worldwide losses in export markets for contaminated rice, corn, barley, cottonseed and peanuts may approach billions of dollars. In particular, economies of developing countries could be seriously affected when the presence of even the smallest amount of toxin in export commodities is not accepted by countries that strictly adhere to regulatory guidelines for levels of toxin in agricultural products for human or animal consumption.

Additional negative economic impact worldwide comes from human health costs, and animal death, as well as from a number of unseen losses, such as reduction in birth rate in certain animals, decline in milk production by dairy cattle and egg production in poultry, loss of quality of animal products, toxin residue problems, increased susceptibility to infection, losses due to reduced weight gain in cattle from toxin-contaminated animal feeds and poor grain quality of fungus-infected crops. Costs of chemical analyses, quality control and regulatory programs, research development, extension services and lawsuits must all be borne by the national economy.

Selected Mycotoxins

Some of the most prevalent naturally occurring mycotoxins, the toxin-producing fungi, and their major toxic effects in human and animals are discussed below.

Aflatoxins. Aflatoxins are by far the best characterized mycotoxins. They are a group of polyketide-derived, bis-furans produced primarily by and *A. parasiticus* on agricultural commodities, and infrequently by *A. tamarii* and *A. nomius*. At least 16 related toxins of this family have been identified (Fig. 1). The four major aflatoxins are B_1, B_2, G_1 and G_2. Other members of the aflatoxin family, aflatoxin M_1 and M_2 are metabolites of aflatoxin B_1 originally isolated from bovine milk. The chemical structure of aflatoxins involves fusion of dihydrofuranofuran and tetrahydrofuran rings with a substituted coumarin. These toxins fluoresce either blue (B_1 and B_2) or green (G_1 and G_2) under long wavelength ultraviolet radiation. The molecular genetics of the steps in the aflatoxin biosynthetic pathway have been studied extensively.

When water activity and temperature are favorable for fungal growth and aflatoxin production in stored commodities, these aflatoxigenic fungi produce toxins on virtually any food and feed. Aflatoxins thus contaminate a variety of agricultural commodities; however, the most predominant pre-harvest contamination has been found in groundnuts, tree nuts, cottonseed and corn.

The potential danger of aflatoxins to animal and human health was realized following the outbreak of turkey X disease in southeast England, which resulted in the deaths of tens of thousands of turkeys, ducklings and pheasants. Decades of studies have established that aflatoxins are mutagenic, teratogenic and hepatocarcinogenic in experimental animals. Aflatoxin B_1 (AFB$_1$) is the most toxic of this group and is one of the most carcinogenic natural compounds known; the order of toxicity of the four major aflatoxins is $B_1 > G_1 > B_2 > G_2$. Aflatoxin M_1 is 10-fold less toxic than B_1, but its presence in milk is of concern in human health.

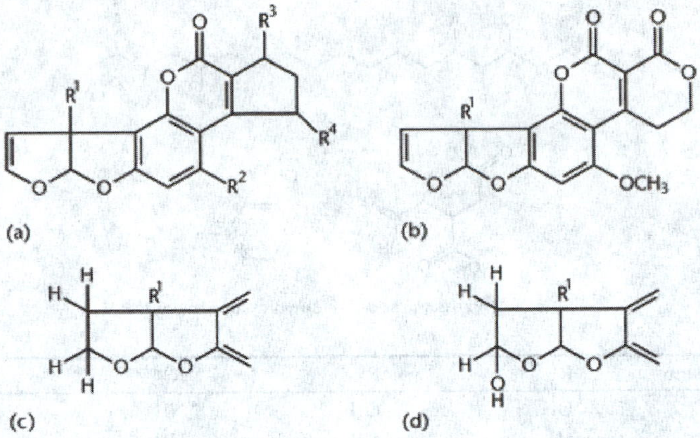

Fig. 1. Aflatoxins. Adapted from Chu [1998].

Aflatoxins	Structure	R^1	R^2	R^3	R^4
B$_1$	a	H	OCH$_3$	=O	H
M$_1$	a	OH	OCH$_3$	=O	H
P$_1$	a	H	OH	=O	H
Q$_1$	a	H	OCH$_3$	=O	OH
R$_0$	a	H	OCH$_3$	OH	H
R$_0$H$_1$	a	H	OCH$_3$	OH	OH
B$_2$	ac	H	OCH$_3$	=O	H
B$_{2a}$	ad	H	OCH$_3$	=O	H
M$_2$	ac	OH	OCH$_3$	=O	H
G$_1$	b	H	—	—	—
G$_2$	bc	H	—	—	—
G$_{2a}$	bd	H	—	—	—
GM	bc	OH	—	—	—

In animal models, activation of AFB$_1$ by microsomal cytochrome P-450 is required for its carcinogenic effect. The enzyme, cytochrome P-450 monooxygenase, in the liver converts AFB$_1$ to a variety of metabolites of increased polarity, including AFB$_1$-8,9-epoxide, the ultimate carcinogen that binds covalently to guanine residues in DNA, resulting in DNA damage, mutations and ultimately carcinomas.

Aflatoxin has been implicated in human hepatocellular carcinoma. An association of hepatocellular carcinoma and dietary exposure to aflatoxins was established from studies on patients living in high-risk areas of China and sub Saharan Africa. The correlation was demonstrated by a mutation hotspot, the transversion of guanine to thymine at the third base of codon 249 in the tumor suppressor gene p53, a transcription factor involved in the regulation of the cell cycle and which is commonly mutated in human cancers.

The US Food and Drug Administration prohibits interstate commerce of feed grain containing more than 20 ppb (μg kg^{-1}) aflatoxin B$_1$ and prohibits the sale of milk and eggs containing more than 0.5 ppb aflatoxin M$_1$. Levels varying from zero tolerance to 50 ppb are acceptable by at least 50 countries of the world.

Fumonisins. Fumonisins are a group of toxins produced by *Fusarium moniliforme* (proposed new name *F. verticillioides*), *Fusarium prolifera-tum* and other related species that readily colonize corn around the world. Nine structurally related fumonisins, including B$_1$, B$_2$, B$_3$, B$_4$, A$_1$ and A$_2$, have been described (Fig. 2). Chemically, fumonisin B$_1$ is a derivative of propane-1,2,3-tricarboxylic acid of 2-amino-12,16-dimethyl-3,5,10,14,15-pentahydroxy icosane; the other fumonisins result from removal of tricar-ballylic acid and other ester groups (Fig. 2). Fumonisins are chemically similar in structure to toxins (AAL) produced by *Alternaria alternata*. Production of fumonisins by *Alternaria* has been reported, and some fumonisin-producing fusaria have been known to produce AAL toxins.

Fumonisins B$_1$, B$_2$ and B$_3$ are the major compounds frequently found in corn and corn-based foods and other grains such as sorghum and rice. The level of fumonisins can be as high as 100 ppm (mg kg^{-1}), but is generally about 100-fold lower. Corn frequently contains detectable but low levels of fumonisins, even though there is no visible sign of fungal contamination. Fumonisins are quite stable; they cannot be removed from grains by changes in temperature, pH and salt concentration during food processing. Therefore, significant amounts of fumonisins are usually found in the end products of grain processing.

Fumonisins have been shown to have divergent biological and toxicolog-ical effects. Studies have shown that fumonisins are specific inhibitors of ceramide synthase; their toxic effect thus appears to be related to interfer-ence with sphingolipid biosynthesis in multiple organs, such as brain, lung, liver and kidney, of the susceptible animals. Fumonisin B$_1$ has also been shown to be a hepatotoxin and a carcinogen in rats. Reports have indicated a possible role of fumonisin B$_1$ in the aetiology of human oesophageal cancer in the regions of South Africa, China and northeastern Italy, where *Fusarium* species are common contaminants. It is also implicated in equine leucoencephalomalacia and porcine pulmonary oedema. Consequently, rec-ommendations have been made that permissible levels of fumonisins (B$_1$) in feeds be limited to 5 ppm for horses, 10 ppm for swine and 50 ppm for cattle and poultry. See also **Carcinogens**.

Trichothecenes. Several species of fusaria are parasites of corn, wheat, barley and rice. Under favorable conditions they produce a number of different types of tetracyclic sesquiterpenoid mycotoxins on corn, wheat, barley and rice (Fig. 3). Other fungal genera producing trichothecenes are *Myrothecium*, *Trichoderma*, *Trichothecium*, *Cephalosporium*, *Verti-cimonosporium* and *Stachybotrys*. The biosynthesis of the trichothecenes has been examined in detail.

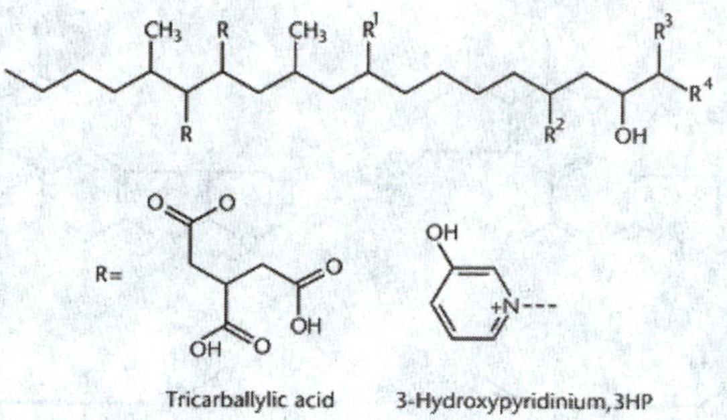

Fig. 2.   Fumonisins. Modified from Chu [1988].

Fumonisins	R^1	R^2	R^3	R^4
B_1	OH	OH	NH_2	CH_3
B_2	H	OH	NH_2	CH_3
B_3	OH	H	NH_2	CH_3
B_4	H	H	NH_2	CH_3
A_1	OH	OH	$NHCOCH_3$	CH_3
A_2	H	OH	$NHCOCH_3$	CH_3
C_1	OH	OH	H	H
C_4	H	H	H	H
P_1	OH	OH	3HP	CH_3
P_2	H	OH	3HP	CH_3
P_3	OH	H	3HP	CH_3

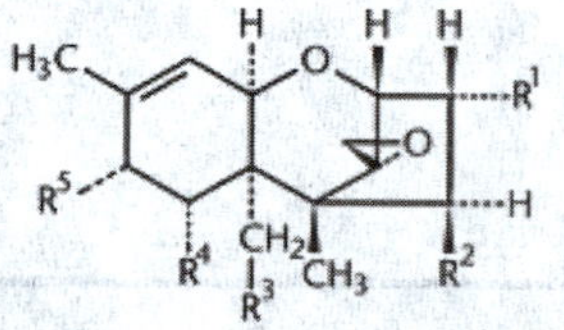

Trichothecenes	R^1	R^2	R^3	R^4	R^5
Diacetoxyscirpenol	OH	OAc^b	OAc	H	H
4-Monoacetoxyscirpenol	OH	OAc	OH	H	H
15-Monoacetoxyscirpenol[a]	OH	OH	OAc	H	H
Scirpentriol	OH	OH	OH	H	H
Deoxynivalenol	OH	H	OH	OH	=O
Nivalenol	OH	OH	OH	OH	=O
Fusarenon-x	OH	OAc	OH	OH	=O
T-2	OH	OAc	OAc	H	ISV^c
HT-2	OH	OH	OAc	H	ISV
T-2 triol	OH	OH	OH	H	ISV
3'-OH-T-2	OH	OAc	OAc	H	OH-ISV^d
T-2 tetraol	OH	OH	OH	H	OH
Neosolaniol	OH	OAc	OAc	H	OH
Roridin A, B, E, H, J	H	MC^a	MC	H	H
Roridin K	H	MC	MC	H	OH
Verrucarin A, B, J, K	H	MC	MC	H	H
Verrucarin L	H	MC	MC	H	OH
Satratoxin F, G, H	H	MC	MC	H	H
Verrucarol	H	OH	OH	H	H

[a]Macrocyclic.
[b]OAc, $OCOCH_3$.
[c]ISV, $OCOCH_2CH(CH_3)_2$.
[d]OH-ISV, $OCOCH_2C(OH)(CH_3)_2$.

Fig. 3.   Selected trichothecenes. Modified from Chu [1988].

The trichothecenes constitute a class of over 80 different members, but the ones causing the most concern are nivalenol, deoxynivalenol (DON) and T-2 toxin and the macrocyclic trichothecenes. The most potent trichothecenes are contact toxins and cause severe blistering and necrosis of the target tissue. Their cytotoxicity parallels their acute toxicity in animals, with T-2 toxin being more potent than nivalenol. Nivalenol is more potent than DON, which is almost never acutely toxic.

The human disease, alimentary toxic aleukia, in western Siberia prior to World War II, was attributed to ingestion of mouldy millet and wheat containing certain macrocyclic epoxytrichothecenes and the severe blistering agent T-2 toxin. Feed refusal by pigs exposed to mouldy corn containing deoxynivalenol (also called *vomitoxin*) was a severe economic problem in the 1970s in the Midwest United States. The major mycotoxin associated with head scab of wheat is deoxynivalenol. Red-mould disease caused by contamination of wheat and barley with *F. graminearum*, among other fusaria, was a sporadic problem in Japan and Korea in the 1970s. The combined result of *Fusarium* mycotoxin contamination is worldwide economic losses in the billions of dollars.

Another disease, termed stachybotryotoxicosis, caused the death of thousands of horses in the Soviet Union prior to World War II and was attributed to contamination of straw and hay with the fungus *Stachybotrys atra*. More recently, the growth of spores from *S. atra* on moist wood or building materials in houses has been associated with pulmonary hemosiderosis in young infants. Trichothecenes may be involved in this "sick building" syndrome in humans. Symptoms such as headaches, sore throats, hair loss, diarrhea, fatigue and general malaise have been reported in humans living in water-damaged houses.

The unusual 12,13-spiroepoxy ring of the trichothecenes has been shown to be necessary for their toxicity. Their toxicity is mainly due to the inhibition of initiation of polypeptide synthesis or polypeptide chain elongation during protein synthesis.

Contamination of cereal grains such as barley, corn, oats and wheat with T-2 toxin is less frequent than with DON. However, T-2 toxin is at least 20-fold more toxic to animals, and possibly to humans.

Alternaria Toxins. Alternaria has been known for centuries to cause various plant diseases such as early blight of potato and various leaf spots and fruit rot. Species of this fungus are widely distributed in soil and on aerial plant parts. Because *Alternaria* requires high moisture levels (28–34%) for growth, infection of seeds occurs when the seed moisture is high, either in early stages of development or after wetting of crops from rain.

More than 20 species of *Alternaria* are known to produce about 70 secondary metabolites belonging to a diverse chemical group including dibenzopyrones, tetramic acids, lactones, quinones, cyclic peptides, etc. However, only alternariol (Fig. 4), tenuazonic acid, altertoxin-I, alternariol methyl ether and altenuene are common natural contaminants of consumable items like fruits (apples), vegetables (tomato), cereals (sorghum, barely, oat) and other plant parts (such as leaves). The most common species of *Alternaria*, *Alternaria alternata* (formerly known as *Alternaria kikuchiana*) produces all important *Alternaria* toxins, including the five mentioned above and tentoxin, alteniusol, alteniric acid, altenusin and dehydroaltenusin. As mentioned previously, *Alternaria alternata* f. sp. *lycopersici* produces a group of host-specific toxins named AAL toxins which are similar to fumonisins.

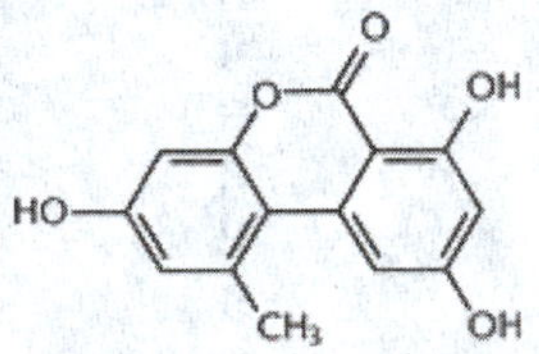

Fig. 4.   Alternariol. From Cole and Cox [1981].

Neurotropic Mycotoxins

Ergot Alkaloids and Related Toxins. Ergotism was one of the earliest documented incidences of a disease caused by mycotoxin consumption. Contamination of rye and millet with the "ergots," a name given to the hard,

dark-purple, sickle-shaped sclerotia from the parasitic fungus *Claviceps*, led to accumulation of a number of toxic alkaloids that affect the central and peripheral nervous systems and cause vasoconstrictive activity. Two types of ergot epidemics occurred, gangrenous and convulsive. In the gangrenous type the victim first experienced violent burning pains, termed "St Anthony's Fire," and then numbing and necrosis of the limbs, while the latter type caused severe convulsions and death. Epidemics of convulsive ergotism occurred from the Middle Ages in Europe, until as late as 1928 in Russia. In one epidemic in France in the 1700s it was estimated that 8,000 people died from eating grain containing as much as "25%" ergot. Acute human ergotism has essentially disappeared as a concern because the modern cleaning and milling processes for grains essentially remove most of the ergot sclerotia. However, ergot poisoning can still be a problem for livestock feeding on wild grasses infected with *Claviceps* spp.

Additional alkaloid-producing fungi are also of concern. Rye grass staggers and fescue toxicoses are caused by ingestion of grasses infected with endophytes *Neotyphodium* spp. and *Epichloe* spp. Representatives of these fungi can produce ergot alkaloids and indolediterpenes. The indoleterpenes can cause tremors in grazing livestock. The most biologically active form is the alkaloid whose diethylamide derivative, lysergic acid diethylamide (LSD), was made famous for its use as a recreational hallucinogen. Another class of ergot alkaloids includes the clavine alkaloids. At least 37 different types of clavine alkaloids are known and are made by *Claviceps* as well as unrelated fungi, such as *A. fumigatus* and *P. chermesinum*. See also **Hallucinogenic Drugs (Hallucinogene).**

The ergot alkaloids have three types of physiological effect: they cause contraction of smooth muscle; they block the action of serotonin and adrenaline; and they act on the hypothalmic–pituitary system to inhibit the secretion of prolactin. These properties have led to their being used to induce uterine contractions, to relieve migraine headaches and to treat prolactin-dependent disorders.

Other Neurotropic Mycotoxins. A number of other mycotoxins do not cause noticeable toxicity, but in some cases have neutrotoxic (tremorgenic) activity, as evidenced by uncontrollable shaking and disoriented behavior. In some instances animals will refuse food, and have lowered resistance to disease. Because the fungi that make some of these toxins, such as several strains of *Aspergillus* (*A. fumigatus*, *A. flavus*) and *Penicillium* (*P. oxalicum*) are widespread and contaminate a variety of different grains, their importance in causing mycological disease may be currently underestimated. Like the ergot alkaloids, some of these neutrotoxins are derived from tryptophan, for example the tremorgens, fumitremorgin, verruculogen, and roquefortine produced by certain strains of *Aspergillus* and *Penicillium* spp. *P. roqueforti*, used for the manufacture of roquefort cheese and other blue cheeses, produces roquefortine. The low levels of toxin in these cheeses are not considered to be problems for human health.

A. terreus, *A. flavus* and *A. fumigatus* produce tremorgenic toxins territrem A, B and C, and aflatrem. *A. flavus*, *A. wentii*, *A. oryzae* and *P. atrarovenetum* (as well as *Arthrinium sacchari*) can produce β-nitropropionic acid, which causes convulsions, congestion in lungs and liver damage. Citreoviridin and xanthomegnin, produced by *Penicillium* spp., are also frequently found in foods. Toxins causing tremorgenic effects in mice include penitrem (*P. crustosum* and *P. cyclopium*) and paspalitrem (also produced by *Penicillium* spp. and *C. paspali*).

Toxic Polyketides other than Aflatoxins. A number of polyketide metabolites are both toxic and carcinogenic. These include patulin and penicillic acid, citrinin and ochratoxin, zearalenone and cyclopiazonic acid.

Patulin. Patulin was originally considered desirable for its antibacterial properties; however, its toxicity to mammals precluded its use as an antibiotic. Patulin (Fig. 5), a metabolite of many species of *Penicillium* and *Aspergillus*, has been detected in apple juice made from partially decaying apples and in other fruit juices, as well as in wheat. Its toxic effects include contact oedema and hemorrhage upon ingestion. It appears to be carcinogenic in experimental animals but no instances of human cancers associated with ingestion of patulin are known.

Penicillic Acid. Penicillic acid (Fig. 6), a metabolite of *Penicillium* and *Aspergillus* spp., is a contaminant of corn and beans, but, even though it has been shown to be a liver toxin, no instances of animal or human toxicity have been reported. In spite of this the potential for these toxins to act synergistically with other more potent toxins is a concern, as the species that produce these toxins usually produce others.

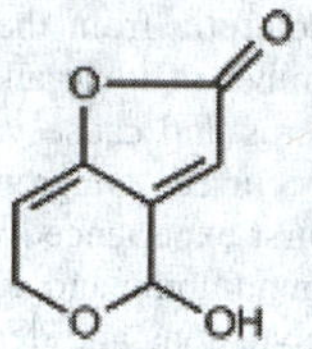

Fig. 5.   Patulin. From Cole and Cox [1981].

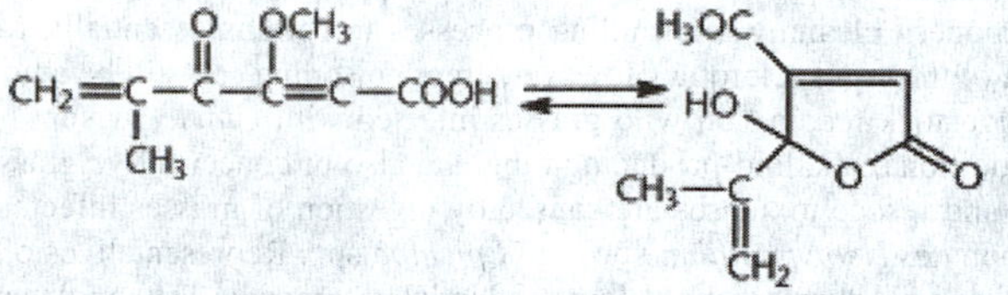

Fig. 6.   Penicillic acid. From Cole and Cox [1981].

Citrinin. As with patulin, citrinin was originally investigated for its antibacterial properties, but it has been reported to be a nephrotoxin in chickens fed corn contaminated with *P. citrinum.* This golden-yellow compound is easily detected as a free compound, but upon binding to proteins, particularly during the drying of corn, the color is lost, while the toxicity remains; therefore, some quantities of citrinin could be part of the diet, and lead to chronic, hard to diagnose kidney disease in susceptible individuals and animals fed products made from low-quality corn.

Ochratoxins. Although numerous *Penicillium* and *Aspergillus* species can make ochratoxins, the largest amounts are made by *A. ochraceus* and *P. cyclopium* (Fig. 7). This toxin can result in high incidence of renal tumors, genotoxicity and immunosuppressive effects. Ochratoxin is a food contaminant not only of cereals (wheat, maize, barley) but also of beans, dried fruits, coffee and cocoa. *In vitro* studies showed that ochratoxin is an inhibitor of transfer ribonucleic acid (tRNA) synthetase and protein synthesis. Ochratoxin has been considered to be the causative agent of kidney disease in pigs subjected to mouldy feed. Human exposure can come from ingestion of contaminated sausage meats. Outbreaks of kidney disease (Balkan endemic nephropathy) in rural populations in Bulgaria, Romania and the former Yugoslavia was associated with ochratoxin contamination. Administration of phenylalanine can reverse some of the toxic effects. Efforts to eliminate the presence of this toxin in food have not been successful.

Zearalenone. This toxin is produced by several *Fusarium spp.,* in particular *F. graminearum,* which also makes DON and has an unusual macrocyclic structure (Fig. 8). Zearalenone is found in the gluten of contaminated wet-milled wheat, and also in corn and corn products. It has oestrogenic properties and can cause premature onset of puberty in female animals. It has also been associated with increased incidence of cervical cancer. Zearalenone can be a phytotoxin; it can inhibit seed germination and embryo growth at low concentrations.

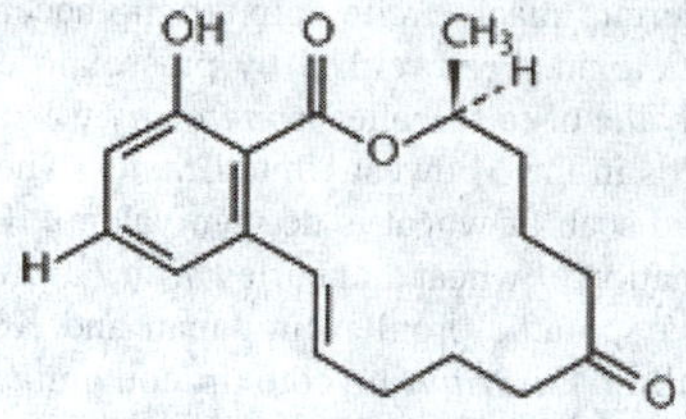

Fig. 8.   Zearalenone. From Cole and Cox [1981].

Cyclopiazonic Acid (CPA). This metabolite is produced on foods by several species of the genus *Penicillium,* but more significantly by *A. flavus* (Fig. 9). CPA is produced along with aflatoxins, and this association, together with a lack of well-developed analytical methods for determination of cyclopiazonic acid in foods, has prevented an assessment of the possible health effects of this toxin. Kodua poisoning, resulting from ingestion in India of kodo millet seeds contaminated with *Aspergillus,* has been attributed to cyclopiazonic acid.

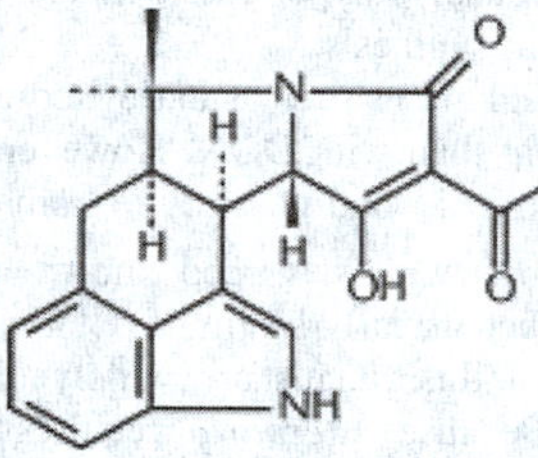

Fig. 9.   Cyclopiazonic acid. From Cole and Cox [1981].

Summary

While documented episodes of significant mycotoxin poisoning, particularly in the developed world, are rather uncommon and the food supply in

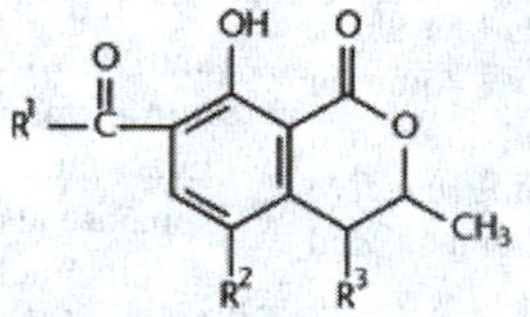

Ochratoxins	R^1	R^2	R^3
A	a	Cl	H
B	a	H	H
C	b	Cl	H
A Methyl ester	c	Cl	H
B Methyl ester	c	H	H
B Ethyl ester	b	H	H
4-Hydroxy ochratoxin A	a	Cl	OH
α	OH	Cl	H
β	OH	H	H

a, $C_6H_5CH_2CH(COOH)NH-$.
b, $C_6H_5CH_2CH(COOEt)NH-$.
c, $C_6H_5CH_2CH(COOMe)NH-$.

Fig. 7.   Ochratoxins. From Chu [1998].

most developed countries is generally considered safe, regulatory guidelines and strict enforcement against the sale of mycotoxin-contaminated commodities (particularly in the case of aflatoxin) has been instituted to ensure this safety. In less-developed parts of the world, mycotoxin-contaminated food and feed are utilized for consumption even though people are urged to discard mouldy food. Poor postharvest conditions and storage facilities in developing countries enhance the risk of exposure to mycotoxin-contaminated food; and the mycotoxin problem is therefore mainly an economic issue in developed countries, but a safety concern in developing countries. Rigorous government monitoring programs, better storage facilities and introduction of simple procedures to minimize toxin contamination, or detoxification of toxins, can reduce the risk of human and animal populations in the developing countries. Agricultural researchers are actively utilizing the latest tools in biotechnology to develop strategies to prevent fungal invasion of crop plants, or to prevent toxin production if the fungi do colonize the crops. These procedures will minimize the economic losses incurred in the developed countries.

Additional Reading

Bennett, J.W.: "Mycotoxins, Mycotoxicoses, Mycotoxicology and *Mycopathologia*," *Mycopathologia*, **100**, 3–5 (1987).

Bhatnagar, D., E.B. Lillehoj, and D.K. Arora: *Handbook of Applied Mycology.* Vol. 5: Mycotoxins in Ecological Systems, Marcel Dekker, New York, NY, 1991.

Chu, F.S.: "Mycotoxins–Occurrence and Toxic Effects," In: *Encyclopedia of Food and Nutrition*, Academic Press, New York, NY, 1998, pp. 858–869.

Cole, R.J., and R.H. Cox: *Handbook of Toxic Fungal Metabolites*, Academic Press, New York, NY, 1981.

Diaz, D.: *Mycotoxins Blue Book*, Blackwell Publishers, Malden, MA, 2005.

Eaton, D.L., and J.D. Groopman: *The Toxicology of Aflatoxins: Human Health, Veterinary, and Agricultural Significance*, Academic Press, New York, NY, 1993.

Eklund, M., J.L. Richard, and K. Mise: *Molecular Approaches to Food Safety: Issues Involving Toxic Microorganisms*, Alaken, Inc., Fort Collins, CO, 1995.

Magan, N., and M. Olsen: *Mycotoxins in Food: Detection and Control*, Woodhead Publishing Ltd., Cambridge, UK, 2004.

Miller, J.D., and H.L. Treholm: *Mycotoxins in Grain: Compounds other than Aflatoxin*, Eagan Press, St Paul, MN, 1994.

Pohland, A.E., and G.E. Wood: "Occurrence of Mycotoxins in Food," In: Krogh, P.: *Mycotoxins in Food*, Academic Press, New York, NY, 1987, pp. 35–64.

Riley, R.T., W.P. Norred, and C.W. Bacon: "Fungal Toxins in Foods," *Annual Review of Nutrition*, **13**, 167–189 (1993).

Sharma, R.P., and D.K. Salunkhe: *Mycotoxin and Phytoalexins in Human and Animal Health*, CRC Press, LLC, Boca Raton, FL, 1991.

Sinha, K.K., and D. Bhatnagar: *Mycotoxins in Agriculture and Food Safety*, Marcel Dekker, New York, NY, 1998.

Staff, CAST: *Mycotoxins, Economics and Health Risk.* Task Force Report No. 133(R), Council of Agricultural Science and Technology, Ames, IA, 1999.

Weidenborner, M.: *Encyclopedia of Food Mycotoxins*, Springer-Verlag New York, LLC, New York, NY, 2001.

DEEPAK BHATNAGAR
KENNETH C. EHRLICH
PERNG-KUANG CHANG
United States Department of Agriculture, New Orleans, LA

MYELOMA.   Essentially a tumor composed of cells normally found in the bone marrow (plasma cells). The stimulus for malignant conversion of the plasma cells in humans is not known, but the original mutation and much of the subsequent cell multiplication probably occurs at an early stage in the B cell maturation sequence. The cells grow relatively slowly with a doubling time of about four months. Electron microscopic studies of malignant plasma cells show megaloblastosis, that is, cytoplasmic maturity and nuclear immaturity.

Although myeloma may be classified, on the basis of the secreted protein, into a number of different immunochemical categories, there is little clinical difference between these varying groups. Myeloma resembles other malignancies in having its peak incidence in the seventh decade. The frequency is about five new cases per 100,000 of population, with commonality between sexes.

In about 10% of cases, the diagnosis is made by chance during blood chemistry analysis. Some 60% of patients present with skeletal pain, particularly affecting the back and ribs with deterioration of general health, fatigue, malaise, and anorexia. About 50% of myeloma patients also show

some evidence of renal impairment and bruising and epistaxis may be seen. A peripheral neuropathy may occur and about 10% of patients present with paraplegia.

Soft tissue plasmacytomas may be found but they are of little prognostic significance. Occasionally patients present with a solitary erosive bone deposit of myeloma.

In myeloma, there is an increased susceptibility to infection, associated with depression of normal serum immunoglobulin levels and a defective antibody response. In some patients, the underlying diagnosis may be made first at the time of an infection.

The diagnostic triad of myeloma includes the presence of osteolytic lesions in bones (well resolved by x-ray), atypical plasma cells in the bone marrow, and demonstration of an M protein band in the blood serum.

In treatment, efforts should be made to avoid hospitalization. A graded exercise program should be encouraged for increasing muscular tone and reducing osteoporosis. Solitary plasmacytomas may be treated by radiotherapy and a proportion of apparent cures is achieved. Where chemotherapy is indicated, melphalan and cyclophosphamide are the principal drugs used under careful monitoring. About 16% of patients fail to respond. The general management of patients is extremely important. Much can be done to relieve discomfort and pain with appropriate analgesics, but sooner or later the patient's condition escapes from control and death is ultimately due to renal disease, infection, hyperviscosity, or hemorrhage.

MYNA   *(Aves, Passeriformes).*   There are at least 110 species of this Old World family of birds, which are found in India and in many parts of Southeast Asia. They are related to the starlings. Most species have glossy, showy plumage, often with orange-yellow wattles. The face is partly naked. Most of the body is a purplish- or sooty-black. The myna feeds mostly on fruit, insects, and grain. Large groups numbering in the thousands may destroy crops and otherwise become a nuisance to communities, as do their relatives, the starlings. In some species of myna, the wings have odd notchings, which assist in making various mechanical noises characteristic of these birds. The Indian species (*Gracula acridotheres*) is quite adept at mimicking human speech and is frequently kept as a pet. Most species appear to like people. The species, *Artamus maximus*, sometimes called the Papuan wood swallow, is found in New Guinea. These are among the few species of birds that exhibit communal nesting and it is not uncommon to witness three adult birds sitting on a nest at one time. These birds are sooty black with white breasts. There remains much knowledge to be gained of some of the primitive species of myna which occur on many of the isolated Pacific islands of Melanesia and Micronesia.

MYOCARDIAL INFRACTION (MI).   See **Cardiovascular Devices; Ischemic Heart Disease; and Coronary Thrombosis**.

MYOPATHY.   Disease and degeneration of muscle and the neuromuscular junction make up a number of specific disorders, the majority of which are, at least in part, of genetic origin. Probably the best known of these is muscular dystrophy, which in itself has several subcategorizations. See also **Muscular Dystrophy**. *Myopathy* is a general term, taking in all disorders of muscle. *Dystrophy* refers to a specific kind of heritable, progressive myopathy.

Sometimes muscular weakness may be seen in association with various metabolic disorders, also usually found to be of genetic origin. It is sometimes difficult to associate the relationship—that is, which of the factors may be causative, which may be results, and which may simply occur together. There are inflammatory muscle diseases, such as *polymyositis* and *dermatomyositis*. *Myasthenia gravis* appears to be related to the function of the immune system. Some drugs, including a number of antibiotics, may contribute to muscle weakness, particularly in individuals who are prone to muscular disorders.

McArdle's disease is probably the better known of the glycogen storage diseases. In this disease, there is a deficiency of the enzyme *myophosphorylase*. This is a heritable disorder that usually appears during the late teen years. Commencing with pronounced muscle cramping and stiffness, which is accentuated by exercise, this disorder normally develops slowly into progressive muscular atrophy with consequent weakness of a permanent nature. *Pompe's disease* is due to a deficiency of the enzyme, α-1,4-glucosidase (*acid maltase*) and is usually fatal in infants. The disease usually involves the brain, liver, and heart in children. In an adult, there may be a chronic, indolent myopathy without involvement of other organs. In *Forbe's disease*, there is a deficiency of the enzyme amylo-1,6-glucosidase (*debrancher enzyme*) whereas in *Tarui's disease*, the deficiency is of the enzyme *phosphofructokinase*. These are uncommon diseases with definite genetic connections.

There are several disorders of energy metabolism in muscle, the causes of which are not fully understood. These include *mitochondrial* and *lipid storage* myopathies.

In certain individuals with a myopathic disorder, there appears to be a genetically determined susceptibility to *malignant hyperthermia*, an extremely rare and often fatal complication of anesthesia, particularly when halothane or succinylcholine are administered. Onset is sudden during anesthesia. A temperature as high as 110.3 °F (43.5 °C) may develop, frequently accompanied by acidosis, cardiac arrhythmias, circulatory collapse, coma, muscular rigidity, and frequently death. Muscular rigidity, although not the only major factor in the disorder, does interfere with procedures to assist the patient's ventilation. Prompt treatment measures, including immediate cessation of anesthesia, forced cooling of the body, oxygen therapy, and intravenous infusion with bicarbonate have been life saving in a number of cases.

Myopathy is sometimes associated with heavy consumption of alcohol over protracted periods. Certain drugs, such as colchicine, corticosteroids, some antimalarials, including chloroquine, and fluorinated compounds, such as triamcinolone, have been implicated in inducing myopathy.

Poorly understood are inflammatory muscle diseases. Most research has been concentrated on finding viral and immunologic mechanisms. In *polymyositis*, symptoms include muscle weakness as manifested by difficulties in rising from a chair or toilet, stair climbing, etc. Difficulty of swallowing may be present because of muscle involvement in the pharynx. Malaise and low-grade fever may be present. The disease is rarely fatal. *Dermatomyositis* is another inflammatory muscle disease which is accompanied by a skin rash, usually in the area of knuckles, elbows, and knees. The disease varies in intensity. There is a relatively high association with carcinoma in persons over 50 years of age (estimated at 25% of cases). Although the malignancy can appear in any organ, the lung is the most common site.

Myasthenia gravis is relatively uncommon, affecting an estimated 5 persons out of 100,000 population. See also **Myasthenia Gravis**.

Web References

Inflammatory and Immune Myopathies. http://www.neuro.wustl.edu/neuromuscular/antibody/infmyop.htm

Mitochondrial Myopathies United States National Institute of Neurological Disorders and Stroke. http://www.ninds.nih.gov/health_and_medical/disorders/mitochon_doc.htm

United States Centers for Disease Control and Prevention. http://www.cdc.gov/search.htm

MYOPIA (Nearsightedness).   Myopia is the ability of the eye to clearly see objects that are up close, but not far away. A nearsighted person can usually see well to read, but has trouble with distance vision, such as that used in driving. This is a common refractive eye condition created when the eyeball is more elongated than normal from front to back, or the cornea is too steep or dome-shaped. Myopia is an inherited condition that affects about one person in five.

The usual treatment for myopia is prescription eyeglasses with concave (inwardly curved) lenses or contact lenses that counteract the distortion created by corneas that are too outwardly curved in shape. A concave lens moves the image of a distant object backward onto the retina, thereby bringing it into proper focus.

Refractive eye surgery, which flattens the cornea, has also become a popular option for the correction of nearsightedness in recent years. The most popular of those procedures is Laser In-Situ Keratomileusis (LASIK), which uses an Excimer laser to reshape the cornea, often eliminating the need for corrective lenses. Another option that is becoming popular is Intrastromal Corneal Ring Segments (ICRSs), tiny plastic arcs that are implanted in the peripheral area of the cornea, causing the center of the cornea to flatten. These segments can be removed and/or replaced if needed. Refractive eye surgery is usually not recommended for people under 18 years of age. See also **Intrastromal Corneal Ring Segments (ICRS)**; **Laser In-Situ Keratomileusis (LASIK)**; **Refractive Eye Surgery**; **Retina**; and **Vision and the Eye**.

Vision Rx, Inc., Elmsford, NY

MYRISTIC ACID.   Also called tetradecanoic acid [CAS: 544-63-8], formula $CH_3(CH_2)COOH$. At room temperature, it is an oily, white crystalline solid. Soluble in alcohol and ether; insoluble in water. Specific gravity 0.8739 (80 °C); mp 54.4 °C; bp 326.2 °C. Combustible. The acid is derived by the fractional distillation of coconut oil. Myristic acid is used in soaps; cosmetics; in the synthesis of esters for flavorings and perfumes; and as a component of food-grade additives. Myristic acid is a constituent of several vegetable oils. See also **Vegetable Oils (Edible)**.

MYRMECOPHILE.   An animal that makes its home in the nests of ants. Among the insects that have become adapted to this mode of life are crickets, beetles, and larval flies. In most cases the myrmecophiles seem to feed at the expense of the ants, which may in turn eat secretions produced by their guests.

N

NACREOUS PIGMENT. A pigment, containing guanine crystals obtained from fish scales or skin, that produces a pearly luster. May be applied as surface coatings, as in simulated pearls, or incorporated into plastics. The pigment particle is generally a very thin platelet of high index of refraction. The crystals are readily oriented into parallel layers because of their shape. Being transparent, each crystal reflects only part of the incident light reaching it and transmits the remainder to the crystal below. The nacreous effect is obtained from the simultaneous reflection of light from the many parallel microscopic layers.

NADIR. 1. Satellite subpoint on the earth's surface that is centered directly below the satellite. This is also referred to as the point of zero nadir angle for a satellite in earth orbit.

2. The point on a given observer's celestial sphere diametrically opposite his zenith, that is, directly below the observer. See also **Celestial Sphere and Astronomical Triangle**.

NAIAD. An aquatic insect larva.

NAND (Circuit). A computer logical decision element which has the characteristic that the output F is 0 if, and only if, all the outputs are 1's. Conversely, if any of the input signals A or B or C of the three-input NAND element shown in Fig. 1 is not a 1, the output F is a binary 1. Although the NAND function can be achieved by inverting the output of an AND circuit, the specific NAND circuit requires fewer circuit elements. A two-input transistor NAND circuit is shown in Fig. 2. The output F is negative only when both transistors are cut off. This occurs when both inputs are positive. The number of inputs, or fan-in, is a function of the components and circuit design. See also **AND (Circuit)**. NAND is a contraction of NOT AND.

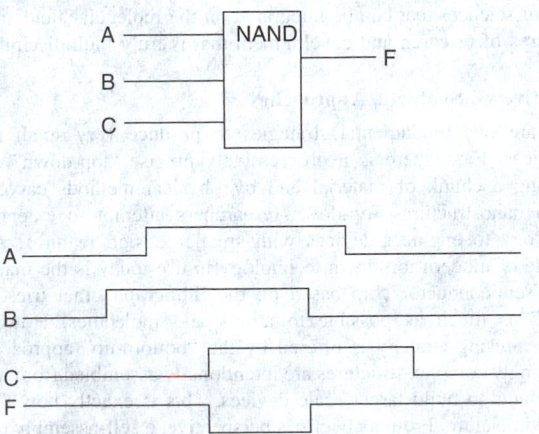

Fig. 1.   Schematic of a NAND circuit.

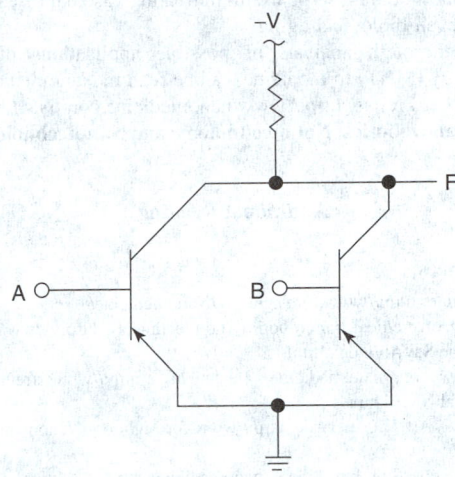

Fig. 2.   Transistor-type NAND circuit.

Thomas J. Harrison, IBM Corporation, Boca Raton, FL

NANNOPLANKTON. The portion of the floating and drifting aquatic animals (plankton), including minute species, which pass through ordinary nets and must be secured by centrifuging.

NANO (abbr. n). The prefix *nano-* comes from the Greek word *nanos*, or "dwarf," and means one-billionth (10^{-9}) of something: nanotechnology operates at the scale of a nanometer, about the width of six carbon atoms.

NANOMEDICINE. Despite advances in modern medicine over the last century, the state of current medicine has been limited both by its understanding and by its tools and in many ways is still more of an art than a science. Only within the last 50 years has medical science begun to examine disease pathology on a molecular level; thus, from a molecular viewpoint modern medicine remains crude. For example, today's drug is essentially a single molecule with an often sophisticated but always limited repertoire. With nanomedicine, tomorrow's "smart pharmaceuticals" could essentially be programmable machines with a range of "sensory," "decision-making," and "effector" capabilities.

So what exactly is nanomedicine? Technically, it is the application of nanotechnology (engineering of tiny machines) to prevention and treatment of disease in the human body. More specifically, it is the use of engineered nanodevices and nanostructures to monitor, repair, construct and control the human biological system on a molecular level. The most elementary of nanomedical devices will be used in the diagnosis of illnesses. A more advanced use of nanotechnology might involve implanted devices to dispense drugs or hormones as needed in people with chronic imbalance of deficiency states. Lastly, the most advanced nanomedicine involves the use of nanorobots as miniature surgeons, to repair or detect damages and infections. A typical blood borne medical nanorobot would be between 0.5-3 micrometres in size, because that is the maximum size possible due to capillary passage requirement. Carbon would be the primary element used to build these nanorobots due to the inherent strength and other characteristics of some forms of carbon (diamond/fullerene composites). Cancer can be treated very effectively, according to nanomedicine advocates. Nanorobots could counter the problem of identifying and isolating cancer cells as they could be introduced into the blood stream. These nanorobots would search out cancer affected cells using certain molecular markers. Medical nanorobots would then destroy these cells, and only these cells. This could be very helpful, since current treatments like radiation therapy and chemotherapy often end up destroying more healthy cells than cancerous ones. Nanorobots could also be useful in treating vascular disease, physical trauma, and even biological aging.

The first thorough analysis of possible applications of molecular nanotechnology (MNT) to medicine, can be read in Nanomedicine, a book series by Robert Freitas. http://www.nanomedicine.com/NMI.htm

See also **Nanorobotics**; **Nanotechnology**; and **Nanotechnology (Molecular)**.

Additional Reading

Web References

Foresight Institute, http://www.foresight.org/Nanomedicine/
General review of medical nanorobotics (non-technical), http://www.foresight.org/Nanomedicine/SayAh/index.html
General review of nanomedicine (technical), http://www.rfreitas.com/Nano/FutureNanofabNMed.htm
Nanomedicine Art Gallery, http://www.foresight.org/Nanomedicine/Gallery/index.html
Nanomedicine Book Site, http://www.nanomedicine.com/
Nanomedicine: Nanotechnology, Biology and Medicine, http://www.nanomedjournal.com/

NANOROBOTICS. The technology of creating machines or robots at or close to the scale of a millionth of a millimeter (10^{-9} meters). More specifically, nanorobotics refers to the still largely theoretical nanotechnology engineering discipline of designing and building nanorobots. Nanorobots are typically devices ranging in size from 0.1-10 micrometers and constructed of nanoscale or molecular components. As no nanorobots have so far been created, they remain a hypothetical concept at this time.

Another definition sometimes used is a robot, which allows precision interactions with nanoscale objects, or can manipulate with nanoscale resolution. Following this definition even a large apparatus such as an atomic force microscope can be considered a nanorobotic instrument when configured to perform nanomanipulation. Also, macroscale robots or microrobots which can move with nanoscale precision can also be considered nanorobots.

Nanomachines are largely in the research-and-development phase, but some primitive devices have been tested. An example is a sensor having a switch approximately 1.5 nanometers across, capable of counting specific molecules in a chemical sample. Initial uses of nanorobots to health care are likely to emerge within the next ten years with potentially broad biomedical applications. The ongoing developments of molecular-scale electronics, sensors and motors are expected to enable microscopic robots with dimensions comparable to bacteria. Recent developments on the field of biomolecular computing has demonstrated positively the feasibility of processing logic tasks by bio-computers, which is a promising first step to enable future nanoprocessors with increasingly complexity. Studies in the sense of building biosensors and nano-kinetic devices, which is required to enable nanorobots operation and locomotion, has been advanced recently too. Moreover, classical objections related to the real feasibility of nanotechnology, such as quantum mechanics, thermal motions and friction, has been considered and resolved and discussions about the manufacturing of nanodevises is growing up. Developing nanoscale robots presents difficult fabrication and control challenges. The control design and the development of complex nanosystems with high performance can be well analysed and addressed via simulation to help pave the way for future use of nanorobots in biomedical engineering problems. Another potential application is the detection of toxic chemicals, and the measurement of their concentrations, in the environment.

Additional Reading

Web References

Carnegie NanoRobotics Lab, http://www.me.cmu.edu/faculty1/sitti/nano/
Medical Nanorobotics, http://www.foresight.org/Nanomedicine/Nanorobotics.html
Nanorobotics in Medicine, http://www.nanomedicine.com/NMI.htm
Nanorobotics — Info Center ETHZ, http://www.infochembio.ethz.ch/links/en/werkstoffe_nanorob.html
Nanorobotics, NanoScience Today, September 2004, http://www.geocities.com/cbicpg/nanoscience/NST2004/nanorobots.htm

NANOSECOND (abbr nsec). 10E-9 second. Also called *millimicrosecond*.

NANOTECHNOLOGY. Nanoscience and nanotechnology are an extension of the field of materials science, and materials science departments at universities around the world in conjunction with physics, mechanical engineering, bioengineering, and chemical engineering departments are leading the breakthroughs in nanotechnology. Nanotechnology is the creation of functional materials, devices and systems through control of matter on the nanometer length scale (0.1–100 nm), and exploitation of novel phenomena and properties (physical, chemical, biological, mechanical, electrical...) at that length scale. One nanometer equals one thousandth of a micrometer or one millionth of a millimeter. For comparison, 10 nanometers is 1000 times smaller than the diameter of a human hair. A scientific and technical revolution has just begun based upon the ability to systematically organize and manipulate matter at nanoscale.

Properties at Nanoscale

Once the material sizes are reduced below 100 nm, they begin demonstrating an array of unique properties based on quantum mechanical effects, rather than the familiar Newtonian mechanics that operates in the macroscopic scale. The quantum effects may influence a variety of material properties such as conductivity, heat transfer, melting temperature, magnetization etc, without changing the chemical composition. That is, the same molecules are present, but their physical assembly is different. For example, colloidal structures immersed in liquid are subjected to constant attack from the liquid molecules, causing them to move about and bend in a random manner (Brownian motion). Although the combination of Brownian motion and strong surface forces is sometimes considered as a difficulty that nanoscience must overcome, these features at nanoscale converge to offer a remarkable opportunity to exploit new approaches to fabricating nanomachines. For example, some biological nanomachines such as molecular motors are not subject to these limitations because they actually depend in a profound way on Brownian motion. The Brownian motion can also lead to formation of rather complex nanoscale structures in liquid by self-assembly. The real power of nanoscience is the convergence of sciences that can be integrated on the molecular field and results in an area of research and development that is truly multidisciplinary.

Reductive vs. Synthetic Approaches

There are two fundamental strategies to produce very small machines or devices. First, there is a progressively precise "top-down" approach of taking a chunk of material and by physical methods carve out the desired nanostructure. Top-down researchers attempt to extend current technology to engineer devices with smaller design features. A typical example of modern top-down technology in use today is the manufacture of the semiconductor chip based on the lithography that tries to reach "down" as much as possible to atoms and molecules. However, the most promising strategy is offered by the "bottom-up" approach, where increasingly complex structures are intentionally assembled from nanoscale components to build larger scale devices. This is exactly how things are formed in Nature. From a chemist's perspective, a self-assembly technique is of greatest interest for developing nanoscale objects with complex compositions such as macromolecules and multicomponent systems. The macromolecules allow bottom-up creation of functionally designed devices economically. Although the future is in the bottom-up approach to nanotechnology, at present nanotechnology can be considered as a more primitive "top-down" stage.

Definitions and History

The birth of Nanotechnology is usually associated with a talk by Nobel Prize winner Richard Feynman. On December 29, 1959, Feynman gave a talk at the American Physical Society) meeting at the California Institute of Technology (CalTech) (http://www.caltech.edu/) entitled "There's Plenty of Room at the Bottom", during which he outlined the potential of having greater control of things at smaller dimensions. (The transcript of this classic talk has been made available on the web at http://www.zyvex.com/nanotech/feynman.html.) Indeed, scientists have been attempting to do so since then, before the term nanotechnology existed. The 'nano' catchphrase was coined by Dr. Eric Drexler, who is credited with introducing the scientific community to the modern potential of nano-scale phenomena and devices.

Nanotechnology has gained popularity in recent years due to the problems the semiconductor industry is already facing in upholding Moore's Law.[1] In the field of microelectronics, the drive towards miniaturization continues and transistor gate lengths of 65 nm are routinely fabricated in prototype circuits. The device density of modern computer electronics (i.e. the number of transistors per unit area) has grown exponentially, and this trend is expected to continue for some time. However, both economics and fundamental electronic limitations prevent this trend from continuing indefinitely. Thus, since technologies in use on chips in 2005 are already at the 65 nm scale and becoming more and more difficult to further miniaturize, it may require breakthroughs in nanotechnology to continue to see the constant increases in speed and decreases in price for computers that many take for granted. The problems facing the semiconductor industry are outlined in the "semiconductor roadmap," (http://www.cra.org/Activities/snowbird/2002/slides/rabaey.pdf) and many will ultimately require solutions which involve completely novel nanoscale devices and phenomena to achieve higher device densities semiconductor roadmap. Microchips have consistently gotten smaller, faster, and cheaper at once because creating smaller devices allows them to have a smaller capacitance, which allows greater switching speeds and thus processor clock speeds; in turn, the ability to pack more of these smaller transistors into a given area means greater economies of scale lead to cheaper chips.

More broadly, nanotechnology includes the many techniques used to create structures at a size scale below 100 nm, including those used for fabrication of nanowires, those used in semiconductor fabrication such as deep ultraviolet lithography, electron beam lithography, focused ion beam machining, atomic layer deposition, and molecular vapor deposition, and further including molecular self-assembly techniques such as those employing di-block copolymers. It should be noted, however, that all of these techniques preceeded the nanotech era, and are extensions in the development of scientific advancements rather than techniques which were devised with the sole purpose of creating nanotechnology or which were results of nanotechnology research.

The term nanoscience is used to describe the interdisciplinary fields of science devoted to the study of nanoscale phenomena employed in nanotechnology. This is the world of atoms, molecules, macromolecules, quantum dots, and macromolecular assemblies, and is dominated by surface effects such as Van der Waals force attraction, hydrogen bonding, electronic charge, ionic bonding, covalent bonding, hydrophobicity, hydrophilicity, and quantum mechanical tunneling, to the virtual exclusion of macro-scale effects such as turbulence and inertia. For example, the vastly increased ratio of surface area to volume opens new possibilities in surface-based science, such as catalysis.

The term nanotechnology is sometimes conflated with molecular nanotechnology (also known as "MNT"), a theoretical advanced form of nanotechnology believed by some to be achievable at some point in the future, based on productive nanosystems. Molecular nanotechnology would fabricate precise structures using mechanosynthesis to perform molecular manufacturing. Molecular nanotechnology, though not yet existent, is expected to have a great impact on society if realized. See also **Nanotechnology (Molecular)**.

---

[1] Moore's law is the empirical observation that at our rate of technological development, the complexity of an integrated circuit, with respect to minimum component cost will double in about 24 months.

Materials, Devices, and Technologies

As science becomes more sophisticated it naturally enters the realm of what is arbitrarily labeled nanotechnology. The essence of nanotechnology is that as we scale things down they start to take on novel characteristics. Nanoparticles (clusters at nanometre scale), for example, have very interesting properties and have proved useful as catalysts and in other uses since, for example when Charles Goodyear invented vulcanized rubber in 1839 or when the Mesoamericans achieved the same result some 2400 years earlier. If we ever do make nanobots, they will not be scaled down versions of contemporary robots. See also **Nanorobotics**. It is the same scaling effects that make nanodevices so special that prevent this. Nanoscaled devices will probably bear much stronger resemblance to nature's nanodevices: proteins, DNA, membranes etc. Supramolecular assemblies are a good example of this.

It was not until the 1980s that miniaturization of apparatus was developed, such as in scanning probe microscopy, includes several newly developed microscopy technologies. The most well-known are atomic force microscopy (AFM) and scanning tunneling microscopy (STM). In these techniques, images are obtained not by gathering reflected or refracted waves from a sample, as happens in conventional light or electron microcopy methods. These operate by employing a sharp tip (the probe) that scans across the surface at a distance of a few nanometers. In particular, the STM works by detecting small currents flows between the microscope probe and the sample being observed. When the probe is brought within about 1 nm of the sample, electrons from the sample begin to pass through the 1 nm gap into the tip or vice versa, depending upon the sign of the bias voltage. Just as per waves can pass over rocks without moving the rocks, so the electrons can pass over the energy barrier to the sample. In other words, this process invokes the wave properties of an electron to move across an energy barrier at lower energy than if it were a particle. The process is called tunneling. The resulting tunneling current varies with tip-to-sample spacing, and it is the signal used to create an STM image. For tunneling to take place, both the sample and the tip must be conductors or semiconductors. The STM was the first instrument to generate real-space images of surfaces with atomic scale resolution. See also **Scanning Tunneling Microscope**.

Several years after the invention of the STM, the atomic force microscope (AFM) was introduced. Instead of measuring the tunneling current, AFM detects the force between the probe and the sample surface. It has a very sharp tip on the end of a cantilever that moves over the surface of the sample. The variations of the surface topography cause bending of the cantilever which is detected by measuring the deflection of the laser beam.

The extension of a crystal (a piezo) is responsible for the movement of the tip across the surface. As the tip tracks the surface, the force between the tip and the surface causes the cantilever to bend. An optical lever measures the deflection of the cantilever. The optical lever of most machines consists of a laser beam reflected from the gold-coated back of the cantilever on to a positional sensitive diode. The positional sensitive diode can measure changes in position of the incident laser beam as small as 1 nm, thus, giving sub-nanometer resolution.

The AFM normally works in two modes: contact and noncontact modes. In contact mode the repulsive forces are measured as the sample cantilever pushes into the sample. These are interactions between the atoms in the tip and the atoms of the material's surface. Consequently the tip is in physical contact with the surface during the analysis and can cause physical damage to soft materials. However, a way around this is a variant of contact mode called 'tapping mode'. The cantilever is vibrated so that the tip contacts the surface intermittently. This intermittent contact serves to reduce the lateral forces incident on the soft sample, reducing possible surface damage or tip contamination but maintaining resolution. In noncontact mode the attractive van der Waals forces are measured by oscillating the cantilever with small amplitude from 5 to 10 nm from the surface of the sample. Because noncontact mode measures the weaker attractive forces, the lateral resolution is often said to be less than that achieved with contact mode. This is not correct, since noncontact AFM is used to demonstrate true atomic resolution. Apart from imaging, AFM can be used for sample manipulation at atomic or nanometer scale, as well as for nano lithographic processes.

The invention of scanning probe microscopes and a number of more specialized variants is perhaps the single most important development in nanoscale science and technology as a new discipline. This linked family of techniques was not the first form of microscopy that could provide information at the atomic and molecular scale, but a number of factors combine to make the impact of these techniques on nanotechnology particularly important.

One of the problems facing nanotechnology is how to assemble atoms and molecules into smart materials and working devices. Supramolecular chemistry is here a very important tool. Supramolecular chemistry is the chemistry beyond the molecule, and molecules are being designed to self-assemble into larger structures. In this case, biology is a place to find inspiration: cells and their pieces are made from self-assembling biopolymers such as proteins and protein complexes. One of the things being explored is synthesis of organic molecules by adding them to the ends of complementary DNA strands such as —A and —B, with molecules A and B attached to the end; when these are put together, the complementary DNA strands hydrogen bonds into a double helix, A=B, and the DNA molecule can be removed to isolate the product AB.

Conclusions and Outlook

Over the last twenty five years, microscale machines such as microelectromechanical systems (MEMS) have been developed, and their uses are growing rapidly. They currently represent the lowest size limits of commercially available machines. The nanotechnology equivalent of MEMS is nanoelectromechanical systems (NEMS) are in a very early stage of development in the laboratory and would involve using or mimicking naturally occurring biological motors, nanotubes, and related materials. In the development of nanodevices, problems start with localized nanoscale distortion or dislocation in materials at the atomic level, which may be a precursor to cracking and then total failure. Therefore, the role of nanosensors, eg, would be able to detect and send early warnings of these nanocracks, when repair or discard is easier and cost effective. These nanosensors may be active, ie, linked to an electrical or optical fiber circuit, or they may be passive and subject to external interrogation at periodic intervals using various types of radiation or by imaging which in turn will allow 'health checks' to be performed on all materials nondestructively. Repair at the local atomic level could be carried out with advanced laser and ion beam techniques, or atomic force nanoprobes, which can be used to supply energy locally to enable atoms to move back into correct, low energy positions. Although it will be a huge challenge to achieve localized repairs, the possibilities include apply, fill in, or "ion stitch" in, nanopatches, nanoparticles, or ions of the correct material. These are just a few of the potential nanotechnological applications. Much development is needed before any serious prospects emerge. Nanotechnology is very much at its infancy and as such it can be expected to give rise to a great deal of intellectually stimulating science before becoming a viable companion to the current silicon based technologies.

Additional Reading

Bhushan, B., and H. Fuchs: *Applied Scanning Probe Methods II: Scanning Probe Microscopy Techniques*, Springer-Verlag New York, LLC, New York, NY, 2005.

Bhushan, B.: *Nanotribology and Nanomechanics: An Introduction*, Springer-Verlag New York, LLC, New York, NY, 2005.

Borisenko, V.E., and S. Ossicini: *What Is What in the Nanoworld: A Handbook on Nanoscience and Nanotechnology*, John Wiley & Sons, Inc., Hoboken, NJ, 2004.

Challa, K.: *Biofunctionalization of Nanomaterials*, John Wiley & Sons, Inc., Hoboken, NJ, 2005.

Fisher, A.J., L. Kantorovich, W. Hofer, and A. Foster: *Scanning Probe Microscopy: Atomic Scale Engineering by Forces and Currents*, Springer-Verlag New York, LLC, New York, NY, 2005.

Foster, L.E.: *Nanotechnology: Science, Innovation, and Opportunity*, Prentice Hall Professional Technical Reference, Upper Saddle River, NJ, 2005.

Gad-el-Hak, M.: *MEMS Introduction and Fundamentals*. CRC Press LLC., Boca Raton, FL, 2005.

Grebel, H.: *Photonics with Nanoscale Structures*, Springer-Verlag New York, LLC, New York, NY, 2005.

Kelsall, R., I.W. Hamley, and M. Geoghegan: *Nanoscale Science and Technology*, John Wiley & Sons, Inc., Hoboken, NJ, 2005.

Leondes, C.T.: *Mems/Nems Handbook*, Springer-Verlag New York, LLC, New York, NY, 2005.

Malsch, N.H.: *Biomedical Nanotechnology*, CRC Press, LLC, Boca Raton, FL, 2005.

Mansoori, G.A.: *Principles of Nanotechnology: Molecular-Based Study of Condensed Matter in Small Systems*, World Scientific Publishing Company, Inc, Riveredge, NJ, 2005.

Minoli, D.: *Nanotechnology Applications to Telecommunications and Networking*, John Wiley & Sons, Inc., Hoboken, NJ, 2005.

Poole, C.P., and F.J. Owens: *Introduction to Nanotechnology*, John Wiley & Sons, Inc., New York, NY, 2003.

Schulte, J.: *Nanotechnology: Global Strategies, Industry Trends and Applications*, John Wiley & Sons, Inc., Hoboken, NJ, 2005.

Theodore, L.: *Nanotechnology: Basic Calculations for Engineers and Scientists*, John Wiley & Sons, Inc., Hoboken, NJ, 2005.

Theodore, L., and R.G. Kunz: *Nanotechnology: Environmental Implications and Solutions*, John Wiley & Sons, Inc., Hoboken, NJ, 2005.

Web References

Center for Biological and Environmental Nanotechnology, http://cben.rice.edu/

Center for Nano & Molecular Science & Technology- CNM at UT Austin, http://www.cnm.utexas.edu/

Center for Nanoscale Science and Technology at Rice University, http://cnst.rice.edu/

Center for Responsible Nanotechnology, http://crnano.org/

Cornell University Center for Nanoscale Systems, http://www.cns.cornell.edu/

Cornell NanoScale Science & Technology Facility, http://www.cnf.cornell.edu/

Foresight Institute, http://www.foresight.org/

International Council on Nanotechnology, http://icon.rice.edu/

Journal of Nanoscience and Nanotechnology, http://aspbs.com/jnn/

NanoTech Institute at the University of Texas at Dallas, http://nanotech.utdallas.edu/nn/index.asp

Nanotechnology and Nanomaterials A to Z, http://www.azonano.com/Materials.asp?Letter=

Nanotechnology basics, news, and general information, http://www.nanotech-now.com/

Nanotechnology, electronic journal since 1990, http://www.iop.org/EJ/journal/0957-4484

Nanotechnology by Dr.Ralph Merkle, http://www.zyvex.com/nano/

Nanotechnology Industries, http://www.nanoindustries.com/

National Nanotechnology Initiative, http://www.nano.gov/

Recent Developments In Nanotechnology, http://www.whatsnextnetwork.com/technology/index.php?cat=65

The California NanoSystems Institute, http://www.cnsi.ucla.edu/

The Kavli Institute of Nanoscience Delft, http://www.ns.tudelft.nl/live/pagina.jsp?id=a0626868-60 d3-4b20-a371-e7e96606e9a3&lang=en

The London Centre for Nanotechnology, http://www.london-nano.ucl.ac.uk/lcn/index2.htm

The MacDiarmid Institute for Advanced Materials and Nanotechnology, New Zealand, http://www.macdiarmid.ac.nz/

The MEMS and Nanotechnology Exchange / A repository of Nanotechnology fabrication information, http://www.mems-exchange.org/

The Smalley Group / Carbon Nanotechnology Laboratory, http://smalley.rice.edu/

NANOTECHNOLOGY (Molecular). Molecular nanotechnology is the production of functional materials and structures in the 0.1 to 100 nm range (the nanoscale) by any of a variety of physical and chemical methods. These methods include nanolithography, direct atomic and molecular manipulation with nanoscale probes, biotechnological selection and production of useful nanomaterials, and chemical synthesis and self-assembly of functional molecules and molecular aggregates. Nanoscale structures and devices may be constructed synthetically from their atomic or molecular constituents (the synthetic or "bottom-up" approach, also referred to as nanochemistry), or by fabrication techniques that use methods to form small structures from larger ones (the reductive or "top-down" approach). A natural analogy to nanotechnology can be found in the biosphere, wherein small and large molecules interact to form complex structures necessary for all the functions of living organisms.

Synthetic vs Reductive Technologies

Many of the devices that have thus far been envisioned as products of nanotechnology (e.g., nanoscale environmental sensors, information processors, and actuators) cannot be produced by the large-scale microfabrication techniques currently in use. The further development of nanotechnology hinges on the understanding and manipulation of physical laws and processes at the nanometer level, such as electronic, interatomic, and intermolecular interactions that can be manipulated to allow efficient assembly of nanostructures.

Biological systems employ a variety of synthetic strategies and processes that make efficient use of these interactions to form highly ordered molecular and supermolecular units that perform a wide variety of functions. The distribution and functional success of biological nanostructures in living organisms are an evolutionarily directed balance between the internal stability of the nanostructures (by atomic and molecular interactions) and their responses to the external environment. Thus, biological systems form important existence theorems and design criteria for the production of functional, nanoscale materials through nanotechnology employing the synthetic approach to nanostructures.

An important nature-mimicking methodology involves the use of covalent synthesis followed by molecular self-assembly of the synthesized molecules. These molecules are generally small mono- or oligomers that interact with each other and with other kinds of mono- or oligomers to form thermodynamically stable, nanoscale structures. See also **Molecular Recognition**.

Existence Theorem: Nanobiology

Biology is replete with complex, functional nanoscale structures formed by directed synthesis and subsequent self-assembly of the component molecules. Initially, the small molecules are produced by covalent synthesis, with the larger, functional structures resulting from many weak, noncovalent interactions that energetically overcome interactions with the solvent and the entropic advantages of disintegration of the ordered aggregates. The final structure represents a thermodynamic minimum, and incorrect subunits are rejected in the dynamic, equilibrium assembly. In this process, complementarity in shape and polarity provides the foundation for the association (binding) between components (e.g., phospholipids, polypeptide chains, proteins, nucleic acids). Shape-dependent association based on the nonspecific van der Waal's and hydrophobic interactions are made stronger and more specific by hydrogen bonds and electrostatic interactions, as well as, in some cases, covalent bonds (e.g., disulfides). Often, positive cooperativity is displayed, i.e., conformations of individual subunits change upon binding in such a manner that their affinity for other components of the final structure increases. Moreover, the amount of information required to execute the assembly of a particular structure (e.g., a protein) is minimized by use of only a few types of molecules and a limited number of binding interactions.

For example, a polypeptide is synthesized as a linear polymer derived from the 20 natural amino acids by translation of a nucleotide sequence present in a messenger RNA (mRNA). The mature protein exists as a well-defined three-dimensional structure. The information necessary to specify the final (tertiary) structure of the protein is present in the molecule itself, in the form of the specific sequence of amino acids that form the protein. This information is used in the form of myriad noncovalent interactions that first form relatively simple local structural motifs (helix and sheet structures associated through networks of hydrogen bonds); these motifs then tend to aggregate in ways that associate hydrophobic regions with one another and out of reach of water, and to place hydrophilic regions close to each other and to water. Thus, proteins self-assemble by two types of processes after synthesis: formation of relatively simple local structures from an unfolded polypeptide chain; and more complex structure-specific association (i.e., associations involving some form of molecular recognition) of these local structures.

In addition to self-assembly of protein structures, in living systems the complex maneuvers needed to achieve properly folded tertiary structures are facilitated by the function of a pre-existing protein machinery, of which, the molecular chaperones are an illustrative example. Chaperones are proteins that bind to and stabilize an otherwise unstable conformer of another protein, and by controlled binding and release, facilitate its correct fate *in vivo*. Molecular chaperones may be said to be the natural counterparts of assemblers and transporters envisaged as products of nanotechnology.

Among the many examples in biology of complex nanoscale structures and devices, one of the most ubiquitous and versatile classes is that of the membrane proteins. Membrane proteins, i.e., proteins that are associated with cellular and organellular membranes, serve myriad functions including recognition, adhesion, chemical triggering, ion and molecular transport, light harvesting, chemical and mechanical sensing, and actuation. Humans have attempted to mimic the function of some of these structures for a wide variety of uses, including chemical sensing, selective chemical transport (pumping), and tissue engineering.

An interesting class of proteins which may prove to be an illustrative archetype for future nanotechnological devices, if not integral components of such, are a class of molecular devices collectively known as motor proteins. Motor proteins are enzymatic protein complexes whose catalytic function results in a distinct mechanical function; a variety of examples are known, including those that perform functions analogous to levers, rotary motors, pumps and springs. The structures and mechanisms of action of a variety of motor proteins are being investigated.

Reductive Approaches

Conventional methods of microfabrication of integrated circuits and devices constitute the reductive (top-down) approach to the construction of micrometer and submicrometer-scale structures. The smallest features in commercial integrated circuits have measured 0.35 μm across, and technologies for further reduction in size have succeeded in forming feature sizes as small as 0.18 μm. Further miniaturization, however, will require major technological breakthroughs in the processes underlying microfabrication, especially photolithography, the heart of microfabrication. The technological barriers faced by the lithographers may not be completely insurmountable.

The number of transistors present on a chip has doubled approximately every 18 months since the integrated circuit was first developed. The main reason for this continuing decrease in the minimum feature sizes of transistors (and consequent increase in density of transistors on the chip) has been the development of photolithography. The most important limitation for further size reduction remains the development of new photo- and other lithographic techniques.

New light sources are currently being developed, such as the krypton — fluoride ultraviolet laser (wavelength 0.248 μm for features as small as 0.25 μm) and excimer lasers (wavelength 0.193 μm for features below 0.2 μm). But these technologies still need to overcome several obstacles before they can be implemented by the semiconductor industry. New photoresist materials, based on the deposition and patterning of monomolecular layers (e.g., self-assembled monolayers and Langmuir-Blodgett films), are also being developed to address the issues of further reduction in the limits of photolithography.

Further reduction in feature size to achieve nanoscale structures by photolithography will necessitate the use of ever smaller wavelengths of light.

Focused particle-based lithography, such as ion-, neutral atom-, and electron-beam lithography, are capable of achieving very high resolutions. However, the method of writing each circuit feature separately is serial and inherently slow so that the technology cannot be used for simultaneous fabrication of many chips. To speed up the process of electron-beam lithography, methods are being explored to scan a broad electron beam across the entire chip by projecting the beam through an appropriate mask (in mimicry of photolithography).

The most recent approach to reductive nanofabrication that can indeed construct nanoscale structures and devices uses microscopic tools (local probes) that can build the structures atom by atom, or molecule by molecule. Optical methods using laser cooling (optical molasses) are also being developed to manipulate nanoscale structures.

Atomic and Molecular Manipulation

The scanning tunneling microscope (STM) can image and manipulate matter on the atomic scale. In general, when a small conducting probe (the tip, consisting of one or a few atoms, or a metal) is placed close (less than 10 nm) to the surface of a conducting substrate, an electronic current results under a suitable bias due to the overlapping of the electronic wave functions of the probe and the surface. Because this tunneling current is exponentially dependent on the separation of the probe from the surface, imaging resolutions of fractions of an angstrom can be obtained. These images reflect both the topography and the electronic structure of the surface. The STM can also be used to modify surfaces locally. As a result, individual atoms and molecules can be manipulated with atomic-scale precision.

In parallel processes, e.g., field-assisted diffusion and sliding, the bond between the surface and the adatom is never completely broken. Field-assisted diffusion of an adatom on the surface occurs due to the presence of the intense, inhomogeneous electric field between the probe tip and the surface, which gives rise to a potential gradient.

The second class of atomic manipulations, the perpendicular processes, involves transfer of an adsorbate atom or molecule from the STM tip to the surface or vice versa. The tip is moved toward the surface until the adsorption potential wells on the tip and the surface coalesce, with the result that the adsorbate, which was previously bound either to the tip or the surface, may now be considered to be bound to both. For successful transfer, one of the adsorbate bonds (either with the tip or with the surface, depending on the desired direction of transfer) must be broken. The fate of the adsorbate depends on the nature of its interaction with the tip and the surface, and the materials of the tip and surface. Directional adatom transfer is possible with the application of suitable junction biases. See also **Scanning Tunneling Microscope**.

Laser beams provide another means of capturing individual atoms or molecules. When an atom is irradiated from both sides by laser light at a frequency slightly lower than the frequency at which the atom absorbs photons, then the atom loses some of its momentum. In particular, the laser beam propagating in a direction opposite that of the motion of the atom increases in frequency due to the Doppler effect, resulting in light absorption with subsequent isotropic emission (scattering). The light propagating in the same direction as the atom is not absorbed, so that the atom is pushed in a direction opposite its motion and slows down. By surrounding the atom with three sets of counterpropagating laser beams orthogonal to each other the atom can be cooled (i.e., slowed) in all three dimensions. Because the light field acts as a viscous drag force, the combination of laser beams is known as optical molasses.

Synthetic Approaches

The synthetic (bottom-up) approach offers a level of control over the selection and placement of atoms and molecules that is ultimately much higher than that offered by other methods of large-scale microfabrication (e.g., fabrication of integrated circuits). Such synthesis can employ a variety of chemical methods to produce nanoscale molecules. Most chemical synthetic reactions that produce large molecules (e.g., collections of atoms that can act as nanodevices) generate polydisperse materials. These materials are mixtures of oligo- or polymeric chains of varying molecular weights. For nanotechnological applications, it is important to have synthetic strategies that yield compounds of uniform length, size, and shape; inhomogeneities can be detrimental to the designated function of the molecule. Thus, the specific objectives of synthesis are to discover and develop rapid and efficient methods for the precise control of composition, molecular weight, stereochemistry, aggregation, and placement of functional molecules. Four strategies are currently in use (either separately or in combination with one another) for the fabrication of large molecules (substances with molecular weights of a few hundred to a few million): biotechnological synthesis, sequential covalent synthesis, covalent polymerization, and molecular self-assembly.

In nature, complex functional molecules are produced by using the chemical synthetic approach. The molecules are first formed by covalent synthesis. In particular, the polypeptides are formed by the directional joining of amino acids through amide bonds. The primary structure of the amino acids (i.e., sequence) during synthesis is specified by the mRNA (a transcript of the original gene encoded in the DNA). Once a polypeptide is produced, it then undergoes many conformational changes that reduce its size to a compact, native form that is the functional protein. These conformational changes are termed folding.

RNA is also capable of folding into specific shapes for ligand recognition and catalysis. Although there are only four naturally occurring bases available for the formation of tertiary structures through noncovalent interactions (while proteins have twenty different amino acids to choose from), myriad RNA shapes can be produced. Metal ions have been shown to confer extraordinary stability to RNA and RNA fragments.

Thus, protein and RNA folding studies form a fundamental paradigm for the design and synthesis of new, functional macromolecules with both final structure and function built into the primary structure of the macromolecule. Such synthetic strategies could be applied to the rational design of functional nanostructures, including drugs, sensing elements, photonic and electronic components, catalysts, and even mechanical devices.

Biotechnological Synthesis. Biotechnological synthesis of new nanomaterials (e.g., proteins) and biotechnological modification of living systems (e.g., conferring specific therapeutic properties, new genetic traits) exploit the many ways in which this protein manufacturing machinery can be modified. In particular, the DNA of an organism can be altered in a specific manner, resulting in a modified mRNA, and consequently, a modified or new protein.

Recombinant DNA technology provides a powerful tool for analysis, synthesis, and alteration of genes and proteins. It is based on the ability to rapidly synthesize polynucleotides with any sequence using nucleic acid enzymology. The unique base-pairing attributes of the constituents of the DNA and the ability to express the modified or synthetic DNA in microorganisms and eukaryotic cells result in a powerful tool for the production of synthetic molecules with specific biological functions.

Covalent Synthesis. The first strategy for chemical synthesis employs elaborate and sophisticated methods for assembling atoms into molecules based on the general strategy of sequential formation of covalent bonds. The atoms can also be assembled into subunits that are then reacted to form more complex, designed molecules (convergent synthesis).

Covalent synthesis of complex molecules involves the reactive assembly of many atoms into subunits with aid of reagents and established as well as innovative reaction pathways. These subunits are then subjected to various reactions that will assemble the target molecule. Very complex molecules can be synthesized in this manner.

Molecular Self-Assembly. Reductive techniques, such as those used in the microelectronics industry, can produce structural features smaller than about 200 nm. The use of proximal probes and other nanomanipulative techniques can be considered to be a hybrid of the reductive lithographic techniques and the synthetic strategies of assembling functional nanostructures atom by atom, or molecule by molecule. The organization of nanostructures and devices by the self-assembly of the component atoms and molecules, a ubiquitous phenomenon in biological systems, forms the noncovalent synthetic approach to nanotechnology.

In this approach, well-defined subunits (small molecules similar to, e.g., nucleotides) are first formed through covalent synthesis. Second, these subunits aggregate with themselves or with other subunits through covalent or noncovalent (or both) interactions to form large, stable, structurally defined assemblies. For the final supramolecular structure to be stable and to have a well-defined shape, the noncovalent connections must be collectively stable. Therefore, molecules must be stabilized by many noncovalent interactions.

Additional Reading

Ball, P.: *Designing the Molecular World: Chemistry at the Frontier*, Princeton University Press, Princeton, NJ, 1994.

Jones, F.J. and C.P. Poole, Jr.: *Introduction to Nanotechnology*, John Wiley & Sons, Inc., New York, NY, 2003.

Mulhall, D.: Our Molecular Future: How Nanotechnology, Robotics, Genetics, and Artificial Intelligence Will Transform Our World, Amherst, NY, 2002.

Ratner, M. and D. Ratner: *Nanotechnology: A Gentle Introduction to the Next Big Idea*, Prentice Hall Professional Technical Reference, Upper Saddle River, NJ, 2002.

Regis, E.: *Nano: The Emerging Science of Nanotechnology*, DIANE Publishing Company, Colllingdale, PA, 1998.

Rieth, M.: *Nano-Engineering in Science and Technology: An Introduction to the World of Nano-Design*, World Scientific Publishing Company, Inc., Riveredge, NJ, 2003.

Stroscio, J.A. and D.M. Eigler: *Science*, **254**, 1319–1326 (1991).

Tolles, W.M.: in G.M. Chow and K.E. Gonsalves, eds., *210th National Meeting of the ACS*, ACS, Chicago, Ill., 1996, pp. 1–15.

Whitesides, G.M. and co-workers: *Acc. Chem. Res.*, **28**, 37–44 (1995).

Wilson, M., K. Kannangara, and M. Wilson: *Nanotechnology: Basic Science and Emerging Technologies*, CRC Press LLC., Boca Raton, FL, 2002.

NANOTUBE. A nanotube (also known as a buckytube) is a member of the fullerene structural family, which also includes buckyballs. Whereas buckyballs are spherical in shape, a nanotube is cylindrical, with at least one end typically capped with a hemisphere of the buckyball structure. Their name is derived from their size, since the diameter of a nanotube is on the order of a few nanometers (approximately 50,000 times smaller than the width of a human hair), while they can be up to several centimeters in length. There are two main types of nanotubes: single-walled nanotubes (SWNTs) and multi-walled nanotubes (MWNTs).

Nanotubes are composed entirely of sp^2 bonds, similar to graphite. Stronger than the sp^3 bonds found in diamond, this bonding structure provides them with their unique strength. Nanotubes naturally align themselves into "ropes" held together by Van der Waals forces. Under high pressure, nanotubes can merge together, trading some sp^2 bonds for sp^3 bonds, giving great possibility for producing strong, unlimited-length wires through high-pressure nanotube linking.

While it has long been known that carbon fibers can be produced with a carbon arc, and patents were issued for the process, it was not until 1991 that Sumio Iijima, a researcher with the NEC Laboratory in Tsukuba, Japan, observed that these fibers were hollow. This feature of nanotubes is of great interest to physicists because it permits experiments in one-dimensional quantum physics.

The strength and flexibility of carbon nanotubes makes them of potential use in controlling other nanoscale structures, which suggests they will have an important role in nanotechnology engineering. The highest tensile strength an individual SWNT has been tested to be is 63 Gpa. (http://record.wustl.edu/archive/2000/02-03-00/articles/nanotube.html). In Earth's upper atmosphere, atomic oxygen erodes the carbon nanotubes, but other applications of carbon nanotubes rarely need the surface to be protected. Though it is debatable if nanotube materials can ever be made with a tensile strength approaching that of individual tubes, composites may still yield incredible strengths potentially sufficient to allow the building of such things as space elevators, artificial muscles, ultrahigh-speed flywheels, and more. MIT is working on combat jackets utilizing carbon nanotubes for ultrastrong fibers and for monitoring its wearer's condition.

Carbon nanotubes have already been used as composite fibers in polymers and concrete to improve the mechanical, thermal and electrical properties of the bulk product. Researchers have also found that adding them to polyethylene increases the polymer's elastic modulus by 30%. In concrete, they increase the tensile strength, and halt crack propagation.

Conductive carbon nanotubes have been used for several years in brushes for commercial electric motors. They replace traditional carbon black, which is mostly impure spherical carbon fullerenes. The nanotubes improve electrical and thermal conductivity because they stretch through the plastic matrix of the brush. This permits the carbon filler to be reduced from 30% down to 3.6%, so that more matrix is present in the brush. Nanotube composite motor brushes are better-lubricated (from the matrix), cooler-running (both from better lubrication and superior thermal conductivity), less brittle (more matrix, and fiber reinforcement), stronger and more accurately moldable (more matrix). Since brushes are a critical failure point in electric motors, and also don't need much material, they became economical before almost any other application.

Carbon nanotubes additionally can also be used to produce nanowires of other chemicals, such as gold or zinc oxide. These nanowires in turn can be used to cast nanotubes of other chemicals, such as gallium nitride. These can have very different properties from CNTs—for example, gallium nitride nanotubes are hydrophilic, while CNTs are hydrophobic, giving them possible uses in organic chemistry that CNTs could not be used for.

See also **Carbon**; and **Carbon Black**.

Additional Reading

Harris, P.: *Carbon Nanotubes and Related Structures: New Materials for the Twenty-First Century*, Cambridge University Press, New York, NY, 2001.

Kelly, L., and M. Meyyappan: *Carbon Nanotubes*, CRC Press, LLC, Boca Raton, FL, 2004.

Reich, S., C. Thomsen, and J. Maultzsch: *Carbon Nanotubes: Basic Concepts and Physical Properties*, John Wiley & Sons, Inc., Hoboken, NJ, 2004.

Rotkin, S.V., and S. Subramoney: *Applied Physics of Carbon Nanotubes: Fundamentals of Theory, Optics and Transport Devices*, Springer-Verlag New York, LLC, New York, NY, 2005.

Web References

The Nanotube site, http://www.pa.msu.edu/cmp/csc/NTSite/nanopage.html

The Wonderous World of Carbon Nanotubes, http://students.chem.tue.nl/ifp03/Wondrous%20World%20of%20Carbon%20Nanotubes_Final.pdf

NANOWIRE.   A nanowire is a wire of dimensions of the order of a nanometer (10^{-9} meters). At these scales, quantum mechanical effects are important—hence such wires are also known as "quantum wires".

The nanowires could be used, in a near feature, as components of nanotechnology to create electrical circuits out of compounds that are capable of being formed into extremely small circuits.

NANSEN BOTTLE.   A bottle used for collecting samples of seawater at any desired depth. After the bottle has been lowered to the desired depth, a weight is sent sliding down the wire to which the bottle is attached. This releases a catch when it strikes the bottle, which in turn closes the valves and traps the water inside. Special thermometers may be used to record the temperature at the desired depth.

NAPHTHALENE.   See **Coal Tar and Derivatives**; and **Organic Chemistry**.

NAPHTHENIC ACIDS.   The term *naphthenic acid*, as commonly used in the petroleum industry, refers collectively to all of the carboxylic acids present in crude oil. Naphthenic acids [CAS: 1338-24-5] are classified as monobasic carboxylic acids of the general formula RCOOH, where R represents the naphthene moiety consisting of cyclopentine and cyclohexane derivatives. Naphthenic acids are composed predominantly of alkyl-substituted cycloaliphatic carboxylic acids, with smaller amounts of acyclic aliphatic (paraffinic or fatty) acids. Aromatic, olefinic, hydroxy, and dibasic acids are considered to be minor components. Commercial naphthenic acids also contain varying amounts of unsaponifiable hydrocarbons, phenolic compounds, sulfur compounds, and water. The complex mixture of acids is derived from straight-run distillates of petroleum, mostly from kerosene and diesel fractions. See also **Petroleum**.

Chemical Structure

Naphthenic acids are based on saturated single or multicyclic condensed ring structures. The low molecular weight naphthenic acids contain alkylated cyclopentane carboxylic acids, with smaller amounts of cyclohexane derivatives occurring. The carboxyl group is usually attached to a side chain rather than directly attached to the cycloalkane. The simplest naphthenic acid is cyclopentane acetic acid [CAS: 1123-00-8], ($n = 1$).

Naphthenic acids are represented by a general formula $C_nH_{2n-z}O_2$, where n indicates the carbon number and z specifies a homologous series. The z is equal to 0 for saturated, acyclic acids and increases to 2 in monocyclic naphthenic acids, to 4 in bicyclic naphthenic acids, to 6 in tricyclic acids, and to 8 in tetracyclic acids.

Physical and Chemical Properties

Naphthenic acids are viscous liquids, with phenolic and sulfur impurities present that are largely responsible for their characteristic odor. Their colors range from pale yellow to dark amber. Naphthenic acids have wide boiling point ranges at high temperatures (250–350 °C). They are completely soluble in organic solvents and oils but are insoluble (50 mg/L) in water. Commercial naphthenic acids are available in various grades and are marketed by acid number, impurity level, and color. Chemically, naphthenic acids behave like typical carboxylic acids with similar acid strength as the higher fatty acids.

Naphthenic acid corrosion has been a problem in petroleum-refining operations since the early 1990s. Refineries processing highly naphthenic crudes must use steel alloys; 316 stainless steel [CAS: 11107-04-3] is the material of choice. Conversely, naphthenic acid derivatives find use as corrosion inhibitors in oil-well and petroleum refinery applications.

Occurrence

Not all crudes contain sufficient quantities of usable acids to make recovery an economic process. Heavy crudes from geologically young formations have the highest acid content, and paraffinic crudes usually have low acid content.

Manufacture

The commercial production of naphthenic acid from petroleum is based on the formation of sodium naphthenate. Naphthenic acids are recovered by caustic extraction of petroleum distillates rather than from crude petroleum. Crude naphthenic acid is obtained by acidulating the sodium naphthenate, and can be further refined to remove impurities.

Interest in synthetic naphthenic acid has grown as the supply of product has fluctuated. Oxidation of naphthene-based hydrocarbons, free-radical addition of carboxylic acids to olefins, and addition of unsaturated fatty acids to cycloparaffins have been studied but not commercialized.

Health and Safety Factors

Naphthenic acids are only slightly toxic to mammals but are toxic to fish, bacteria, and wood-destroying insects. The lethal oral dose for humans is approximately 1 L. Naphthenic acid is not listed as a carcinogen.

Uses

More than two-thirds of the naphthenic acids produced is used to make metal salts, with the largest volume being used for copper naphthenate consumed in the wood preservative industry. Oil field uses are primarily imidazolines for surfactant and corrosion inhibition. Besides the lubrication market for metals salts, the miscellaneous market is comprised of free acids used in concrete additives, motor oil lubricants, and asphalt-paving applications. See also **Lubricant**; and **Lubricating Agents**.

Naphthenic acid is ideal for synthesizing metal carboxylates that require a ligand with some oxidative stability, solubility in hydrocarbons and oils, and insolubility in water.

Another market application for naphthenic acid is the tire industry, where cobalt naphthenate is used as an adhesion promoter. Naphthenic acid esters have been repeatedly cited as surfactants, lubricants, and replacements for phthalates as plasticizers for PVC resins.

Naphthenyl alcohols are formed by reduction of the acids or their simple esters. They are valuable as surfactants, solvents, and components of lubricants. The acid halides are of value mainly as chemical intermediates.

Additional Reading

Lochte, H.L. and E.R. Littmann: *The Petroleum Acids and Bases*, Chemical Publishing Co., Inc., New York, NY, 1955.

Lower, E.S.: *Specialty Chem.*, **7**, 76 (1987); **7**, 282 (1987); **8**, 174 (1988); **9**, 135 (1989); **9**, 267 (1989).

Maass, W., E. Buchspiess-Paulentz, and F. Stinsky: *Naphthensäuren und Naphthenate*, Verlag für Chemische Industrie Ziolkowsky, H. Augsburg, Germany, 1961.

Narmetova, G., B. Khamidov, N. Ryabova, and E. Aripov: *Purification, Identification, and Use of Naphthenic Acid*, Fan, Tashkent, former USSR, 1983.

NAPIER'S RULES.   See **Spherical Trigonometry**.

NARBONNAIS.   See **Winds and Air Movement**.

NARWHAL.   See **Whales, Dolphins, and Porpoises**.

N'ASCHI.   See **Winds and Air Movement**.

NASOLACRIMAL.   The lacrimal gland located under the upper eyelid is what makes it possible for the eyes to produce tears, a necessity for keeping eyes moist and a functioning part of our emotions. Because the eyes produce fresh tears continuously, an efficient drainage system is necessary to drain the used tears from the eyes. This system is called the *nasolacrimal system*.

The nasolacrimal system starts with tiny openings on the brim of the upper and lower eyelids near the inner edge of the nose. These lead to the nasolacrimal tear ducts next to the bridge of the nose. The tears move from these ducts into the nasal cavity where they are either swallowed or discharged through the nose.

Occasionally, lacrimal or nasolacrimal functions become impaired, causing irritation and infection, and, sometimes, signifying other vision problems. One of the most common nasolacrimal problems is congenital nasolacrimal duct obstruction, where an infant's blocked tear duct becomes infected, causing matter to collect in the corner of the eyes and between the eyelids. In adults, excessive tearing and drainage problems can signify other areas of concern. It's important to distinguish between the two, because drainage and tearing are two separate issues that must be evaluated by an eye care professional.

The most common symptoms of lacrimal and nasolacrimal impairments are excess tearing (tears may run down the face) and mucous discharge. Eye infections, eyelashes, exposure to the wind, yawning, glaucoma, certain drugs, eyestrain, or even dry eyes can contribute to excessive tearing.

Treatment depends on the cause of the excessive tearing. Sometimes, simply removing an irritant or other environmental conditions contributes to a decrease in tear production. Other times, small plugs can be placed in the opening of the tear duct to decrease the amount of tears produced.

A blocked tear duct forces tears to build up in the eye and run down the cheeks. And because of the lack of drainage, leftover tears can remain within the eye and become infected. Injury, birth defects, consistent nasal infections, narrowing of the nasolacrimal system associated with age, and other infections can lead to improper tear drainage.

An ophthalmologist often instructs the patient to massage the lacrimal sac area several times a day to help unblock the duct. Congenital nasolacrimal duct obstruction almost always resolves itself as children grow, usually between 6 months and 1 year.

Probing and irrigating the nasolacrimal system may be sufficient to relieve the blockage. In some cases, an eye care professional will recommend surgery to open a blocked tear duct. See also **Tears**.

Vision Rx, Inc., Elmsford, NY

NATAMYCIN (PIMARICIN).   See **Food Additives**.

NATIONAL BUREAU OF STANDARDS (U.S.).   The NBS was established by the U.S. Congress in 1901. The name of the institution was changed **just a few years ago** to the National Institute of Standards and Technology (NIST). See also **NIST**.

NATIONAL CENTER FOR ATMOSPHERIC RESEARCH (NCAR). This organization was founded in 1960 by a consortium of 14 universities with doctoral programs in atmospheric sciences. Its staff numbers about 750 people. The headquarters is in Boulder, Colorado. By the early 1990s, the consortium had grown to more than 50 members in the United States and Canada. NCAR's chief source of funding is the U.S. National Science Foundation.

NCAR's research consists of four basic areas: (1) atmospheric chemistry; (2) climate and its links with other environmental systems; (3) solar and solar-terrestrial physics; and (4) microscale and mesoscale meteorology (the study of phenomena as small as the formation of ice crystals in clouds and as large as major thunderstorm systems). Additionally, one group of researchers looks at how societies and individuals respond to changes in their environment. All of the foregoing activities involve field and laboratory studies, theoretical work, and computer modeling.

Supercomputing power and observing facilities are among the services provided by NCAR because they are too costly and specialized for individual universities to acquire. NCAR also adapts research technology for practical applications, particularly in the area of aviation safety. In 1990, NCAR acquired a CRAYY-MPS/864 supercomputer, which, at the time, was the most powerful computing system in existence. With such computer power, researchers are able to construct models, or mock worlds, allowing them to explore speculative scenarios of such events as the fates of chemicals in the atmosphere, the formation of clouds, the global circulation of winds, and the movements of ocean eddies. These models are important for exploring phenomena that cannot be observed or studied firsthand.

Much of the information for NCAR's projects is gained from advanced observing systems. Included in such systems are advanced radar technology based on the Doppler effect and a fleet of research aircraft, which carry a variety of sophisticated instruments. Projects directed toward early practical application include improvement of aircraft flight safety and the efficient use of U.S. airspace. Surface wind shifts, microbursts, tornadoes, hail, lightning, and heavy rain pose hazards and compromise the efficiency of the aviation industry. So-called "Nowcasts" are designed to enable predictions within a few minutes to an hour, as contrasted with longer-term hazardous weather forecasts. A program known as Terminal Doppler Weather Radar detects and warns of wind shifts near the airport terminal, approach, and departure zones, and runways.

NATIONAL CLIMATIC DATA CENTER (NCDC).   This center is part of the Department of Commerce, National Oceanic and Atmospheric Administration (NOAA), and the National Environmental Satellite, Data and Information Service (NESDIS). The Center is responsible for archiving and distributing climate data. In November 1951, the Weather Bureau, Air Force and Navy Tabulation Units in New Orleans, LA were combined and formed into the National Weather Records Center in Asheville, NC. Authority to establish the joint Weather Records Center was granted under section 506(c) of the Federal Records Act of 1950 (Public Law 754, 81st Congress). The Center was eventually renamed the National Climatic Data Center.

The climate data that NCDC receives are from a wide variety of sources, including satellites, radar, remote sensing systems, NWS cooperative observers, aircraft, ships, radiosonde, wind profiler, rocketsonde, solar radiation networks, NWS Forecast/Warnings/Analyses Products, Military Services, Federal Aviation Administration, and the Coast Guard. NCDC archives 99 percent of all NOAA data, including over 320 million paper records; 2.5 million microfiche records; over 500,000 tape cartridges/magnetic tapes, and has satellite weather images back to 1960. As operator of the World Data Center-A for Meteorology, which provides for international data exchange, NCDC also collects data from around the globe.

NCDC supports many forms of data and information dissemination such as paper copies of original records, publications, atlases, computer printouts, microfiche, microfilm, movie loops, photographs, magnetic tape, floppy disks, CD-ROM, electronic mail, on-line dial-up, telephone, facsimile and personal visit. The National Archives and Records Administration has designated NCDC as the Commerce Department's only Agency Records Center.

The Center, which produces numerous climate publications and responds to requests from all over the world, provides historical perspectives on climate which are vital to studies on global climate change, the greenhouse effect, and other environmental issues. The Center stores information essential to industry, agriculture, science, agriculture, hydrology, transportation, recreation, and engineering. This information can mean tens of millions of dollars to concerned parties. NCDC annually publishes over 1.2 million copies of climate publications that are sent to individual users and 33,000 subscribers.

See also **Weather Technology**.

Web References

National Climatic Data Center (NCDC): http://www.ncdc.noaa.gov/oa/ncdc.html

National Environmental Satellite, Data and Information Service (NESDIS): http://www.nesdis.noaa.gov/

National Hurricane Center: http://www.nhc.noaa.gov/

NATIONAL FIRE DANGER RATING SYSTEM (NFDRS). A system that directly integrates the effects of fuels, topography, and weather into components that deal with fire occurrence and fire behavior potential. The system uses the components to derive indices that indicate the number of fires, difficulty of containment, and finally, the total fire control job in a rating area. The system is intended to provide guidance for short-range planning by evaluating the near upper limits of the behavior of fires that might occur in an area during the rating period. It is not designed to serve as a direct fire behavior forecast.

History

In 1954 there were eight different fire-danger rating systems in use across the county. Better communication and better transportation were beginning to make mutual assistance agreements between fire control agencies more practical than in the past. State compacts, and in the case of the Federal government, interagency and interregional agreements were bringing fire control teams together from widely separated areas of the county. It became necessary to establish a national system for estimating fire danger and fire behavior to improve and simplify communications among all people concerned with wildland fires.

Work on a national rating system began in 1959. By 1961, the basic structure for a four-phase rating system had been outlined and the fire phase (spread phase) was ready for field testing. However, since the remaining phases of the rating system — ignition, risk, and fuel energy — were not available, a number of fire control agencies preferred to remain with the systems then in use. Adaptations, interpretations, and additions to the spread phase quickly followed, making it obvious that the spread phase was not uniformly applicable across the country.

More research followed and in 1965 a research project headquartered in Seattle was established to provide a fresh look a the needs and requirements for a national, fire-danger, rating system. After canvassing

many fire control agencies across the country, the Seattle research group recommended new directions for research that would lead to the development of a complete, comprehensive, National Fire-Danger Rating System. A target date of 1972 was established for getting a complete system ready for operational use.

In 1970, a preliminary version of the system was tested at field sites in Arizona and New Mexico. In 1971, an improved version of the system was used operationally in the Southwest. Field trials were also conducted elsewhere across the country at stations from Maine to California and from Florida to Alaska. The system then became operational nationwide in 1972.

Philosophy of NFDRS

The philosophy of the National Fire Danger Rating System can be summarized as follows:

- The system only considers "initiating fires", or fires which are not behaving erratically, or spreading through downwind spotting or crowning. In other words, the fires are burning through continuous bed of fuels on the surface of the ground.
- The system measures that portion of containment which is directly attributable to fire behavior and thus limits its scope to predicting fire behavior at the head of the fire.
- The length of the flames at the head of the fire are directly related to fire behavior.
- The system evaluates the "worst" conditions on a rating area by 1) taking fuel and weather measurements when fire danger is normally the highest (mid- to late-afternoon), 2) measuring fire danger in the open, and 3) measuring fire danger on south to west exposures. This means that extrapolation of fire danger to other areas not in the immediate vicinity of the fire danger stations would involve scaling the fire-danger values down, not up.
- The system provides ratings and indices which are interpreted in terms of fire occurrence and fire behavior.
- Fire-danger ratings are relative, not absolute. In other words, when a component or index of the system doubles, a doubling of the fire activity or intensity should be expected.

Structure of NFDRS

The basic structure of the system provides three indexes. . .. (Occurrence Index, Burning Index, and the Fire Load Index). . .. designed to aid in planning fire control activities on a fire protection unit. These indexes are derived from three fire behavior components — Spread Component(SC), Energy Release Component (ERC), and the Ignition Component (IC). The scale for each index runs from 0 to 100.

Fire Danger Maps

Each day during the fire season, national maps of selected fire weather and fire danger components of the National Fire Danger Rating System are produced by the Wildland Fire Assessment System (WFAS-MAPS), located at the USDA Forest Service Rocky Mountain Research Station in Missoula, Montana. http://www.wfas.us/component/option,com_frontpage/Itemid,1/.

Current fire danger and forecasted fire danger maps are available (Figs. 1 and 2).

Web Reference

National Fire Danger Rating System: http://www.wrh.noaa.gov/sew/fire/olm/nfdrs.htm

NATIONAL HURRICANE CENTER (NHC). The National Hurricane Center (NHC) maintains a continuous watch on tropical cyclones over the Atlantic, Caribbean, Gulf of Mexico, and the Eastern Pacific from 15 May through November 30. The Center prepares and distributes hurricane watches and warnings for the general public, and also prepares and distributes marine and military advisories for other users. During the "off-season" NHC provides training for U.S. emergency managers and representatives from many other countries that are affected by tropical cyclones. NHC also conducts applied research to evaluate and improve hurricane forecasting techniques, and is involved in public awareness programs.

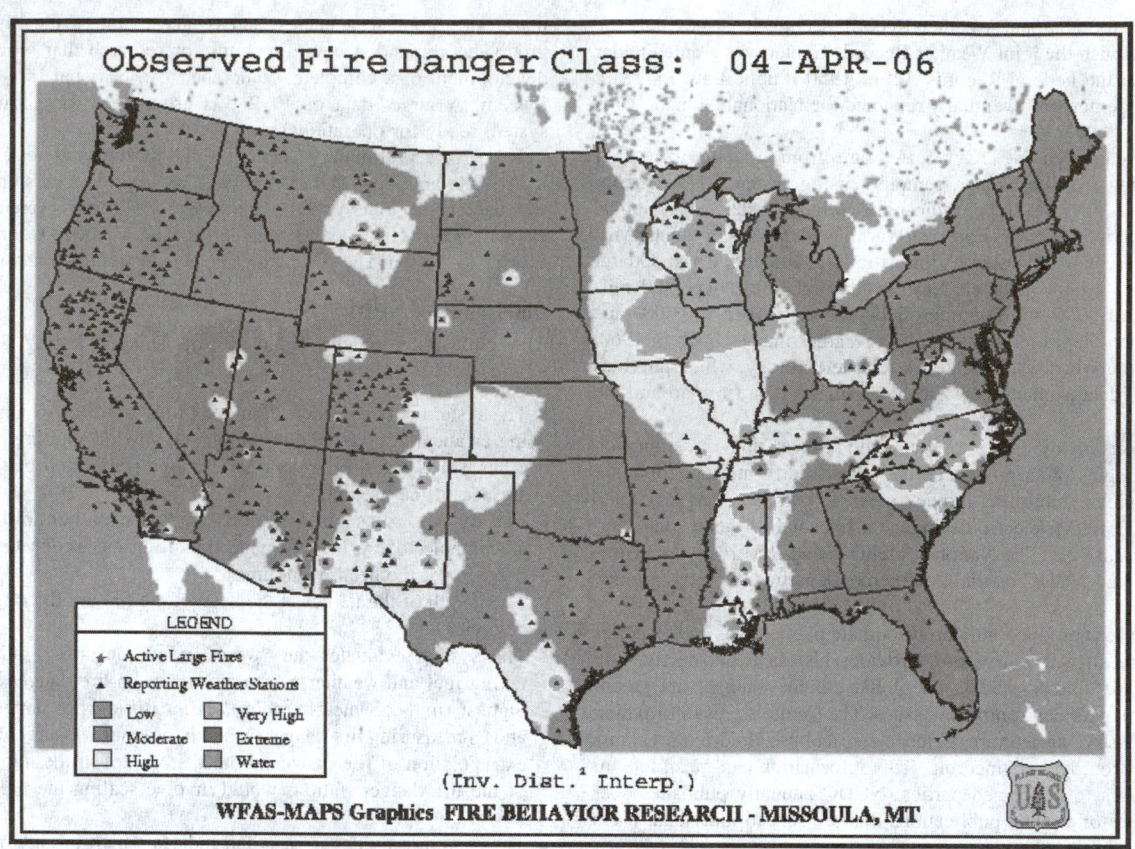

Fig. 1.   Observed Fire Danger Map.

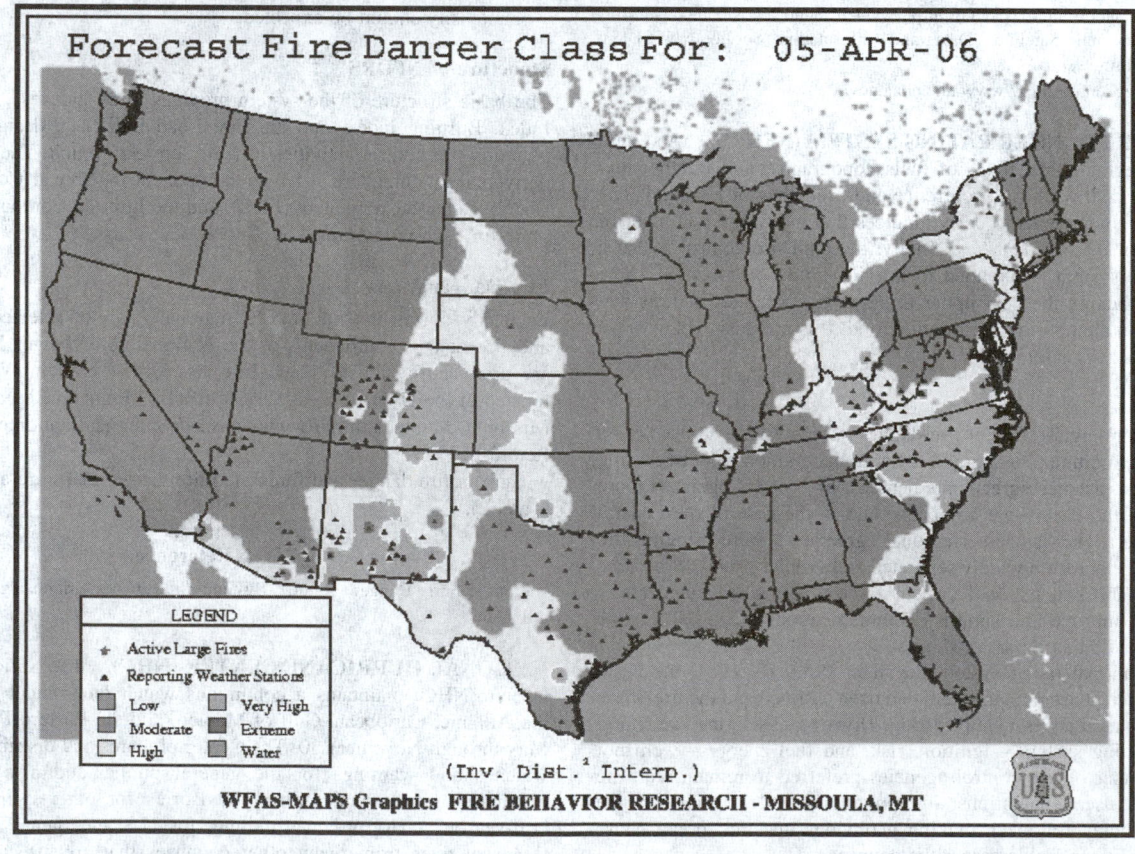

Fig. 2.   Forecast Fire Danger Map.

NHC also contains the Chief, Aerial Reconnaissance Coordination, All Hurricanes (CARCAH) unit. It is a small three person unit, an Operating Location of the 53rd Weather Reconnaissance Squadron (Hurricane Hunters) out of Keesler Air Force Base near Biloxi, Mississippi. CARCAH's mission is to coordinate all aerial reconnaissance requirements at NHC (Atlantic requirements) and at the Central Pacific Hurricane Center (Central Pacific requirements), then task the flying units to meet these requirements.

Data from the reconnaissance aircraft (normally a WC-130) is fed directly to CARCAH via satellite down link. It is quality controlled then provided directly to the hurricane specialist for use in the forecast and warning process. It is also entered into the world weather networks.

During the winter season, CARCAH also coordinates the aerial reconnaissance requirements in support of the National Winter Storms Operations Plan, which provides for flights off the U.S. east coast and over the Gulf of Mexico when severe winter storms are expected. The data is again received at CARCAH and quality controlled before it is submitted to the National Centers for Environmental Prediction. It is then included in their suite of computer weather prediction models.

The National Hurricane Center/Tropical Prediction Center is co-located with the National Weather Service Miami Forecast Office on the main campus of Florida International University at 11691 S.W. 17th Street, Miami, Florida. This location is about 12 miles west of downtown Miami and 8 miles southwest of Miami International Airport.

See also **Weather Technology**.

Web References

Central Pacific Hurricane Center (CPHC): http://www.prh.noaa.gov/hnl/cphc/
National Centers for Environmental Prediction (NCEP): http://www.ncep.noaa.gov/
National Winter Storms Operations Plan: http://www.ofcm.gov/nwsop/2000/nwsop.htm
The Hurricane Hunters: http://www.hurricanehunters.com/
Tropical Prediction Center (TPC): http://www.nhc.noaa.gov/aboutintro.shtml

NATIONAL INSTITUTES OF HEALTH (NIH). The National Institutes of Health (NIH), http://www.nih.gov/, founded in 1887, is the United States federal government's principal agency for the support of medical research both at its campus in Bethesda, Maryland, and through grants and contracts awarded to biomedical research scientists at institutions across the United States, and, to a more limited extent, in other countries. The NIH is a component of the U.S. Department of Health and Human Services. See also http://www.dhhs.gov/.

NIH Mission

NIH is the steward of medical and behavioral research for the Nation. Its mission is science in pursuit of fundamental knowledge about the nature and behavior of living systems and the application of that knowledge to extend healthy life and reduce the burdens of illness and disability.

The goals of the agency are as follows:

1. foster fundamental creative discoveries, innovative research strategies, and their applications as a basis to advance significantly the Nation's capacity to protect and improve health;
2. develop, maintain, and renew scientific human and physical resources that will assure the Nation's capability to prevent disease;
3. expand the knowledge base in medical and associated sciences in order to enhance the Nation's economic well-being and ensure a continued high return on the public investment in research; and
4. exemplify and promote the highest level of scientific integrity, public accountability, and social responsibility in the conduct of science.

In realizing these goals, the NIH provides leadership and direction to programs designed to improve the health of the Nation by conducting and supporting research:

- in the causes, diagnosis, prevention, and cure of human diseases;
- in the processes of human growth and development;
- in the biological effects of environmental contaminants;
- in the understanding of mental, addictive and physical disorders; and
- in directing programs for the collection, dissemination, and exchange of information in medicine and health, including the development and support of medical libraries and the training of medical librarians and other health information specialists.

Composed of 27 Institutes and Centers, the NIH provides leadership and financial support to researchers in every state and throughout the world.

National Cancer Institute (NCI). http://www.cancer.gov/. Established in 1937 the NCI leads a national effort to eliminate the suffering and death due to cancer. Through basic and clinical biomedical research and training, NCI conducts and supports research that will lead to a future in which we can prevent cancer before it starts, identify cancers that do develop at the earliest stage, eliminate cancers through innovative treatment interventions, and biologically control those cancers that we cannot eliminate so they become manageable, chronic diseases. See also http://www.nih.gov/about/almanac/organization/NCI.htm.

National Eye Institute (NEI). http://www.nei.nih.gov/. Established in 1968 the NEI conducts and supports research that helps prevent and treat eye diseases and other disorders of vision. This research leads to sight-saving treatments, reduces visual impairment and blindness, and improves the quality of life for people of all ages. NEI-supported research has advanced our knowledge of how the eye functions in health and disease. See also http://www.nih.gov/about/almanac/organization/NEI.htm.

National Heart, Lung, and Blood Institute (NHLBI). http://www.nhlbi.nih.gov/index.htm. Established in 1948 the NHLBI provides leadership for a national program in diseases of the heart, blood vessels, lung, and blood; blood resources; and sleep disorders. Since October 1997, the NHLBI has also had administrative responsibility for the NIH Woman's Health Initiative. The Institute plans, conducts, fosters, and supports an integrated and coordinated program of basic research, clinical investigations and trials, observational studies, and demonstration and education projects. See also http://www.nih.gov/about/almanac/organization/NHLBI.htm.

National Human Genome Research Institute (NHGRI). http://www.genome.gov/. Established in 1989 the NHGRI supports the NIH component of the Human Genome Project, a worldwide research effort designed to analyze the structure of human DNA and determine the location of the estimated 30,000 to 40,000 human genes. The NHGRI Intramural Research Program develops and implements technology for understanding, diagnosing, and treating genetic diseases. See also Human Genome Project (The); and http://www.nih.gov/about/almanac/organization/NHGRI.htm.

National Institute on Aging (NIA). http://www.nia.nih.gov/. Established in 1974 the NIA leads a national program of research on the biomedical, social, and behavioral aspects of the aging process; the prevention of age-related diseases and disabilities; and the promotion of a better quality of life for all older Americans. See also Gerontology; and http://www.nih.gov/about/almanac/organization/NIA.htm.

National Institute on Alcohol Abuse and Alcoholism (NIAAA). http://www.niaaa.nih.gov/. Established in 1970 the NIAAA conducts and supports approximately 90 percent of the U.S. scientific investigation of alcohol's effects on health and the causes, consequences, prevention, and treatment of alcohol use disorders. NIAAA's overall mission is to support and promote the best science on alcohol and health for the benefit of all. In this pursuit, NIAAA uses multidisciplinary and transdisciplinary approaches to increase understanding of normal and abnormal biology and behavior related to alcohol use; to improve diagnosis, prevention, and treatment of alcohol use disorders; and to enhance quality health care. See also http://www.nih.gov/about/almanac/organization/NIAAA.htm.

National Institute of Allergy and Infectious Diseases (NIAID). http://www3.niaid.nih.gov/. Established in 1948 the NIAID conducts and supports research to study the causes of allergic, immunologic, and infectious diseases, and to develop better means of preventing, diagnosing, and treating these illnesses. See also http://www.nih.gov/about/almanac/organization/NIAID.htm.

National Institute of Arthritis and Musculoskeletal and Skin Diseases (NIAMS). http://www.niams.nih.gov/. Established in 1986 the mission of NIAMS is to support research into the causes, treatment, and prevention of arthritis and musculoskeletal and skin diseases; the training of basic and clinical scientists to carry out this research; and the dissemination of information on research progress in these diseases.

The Institute also conducts and supports basic research on the normal structure and function of joints, muscles, bones, and skin.

Basic research involves a wide variety of scientific disciplines, including immunology, genetics, molecular biology, structural biology, biochemistry, physiology, virology, and pharmacology. Clinical research includes rheumatology, orthopaedics, dermatology, metabolic bone diseases, heritable disorders of bone and cartilage, inherited and inflammatory muscle diseases, and sports and rehabilitation medicine. See also http://www.nih.gov/about/almanac/organization/NIAMS.htm.

National Institute of Biomedical Imaging and Bioengineering (NIBIB). http://www.nibib.nih.gov/. Established in 2000 the mission of the NIBIB is to improve health by leading the development and accelerating the application of biomedical technologies. The Institute is committed to integrating the physical and engineering sciences with the life sciences to advance basic research and medical care. This is achieved through: research and development of new biomedical imaging and bioengineering techniques and devices to fundamentally improve the detection, treatment, and prevention of disease; enhancing existing imaging and bioengineering modalities; supporting related research in the physical and mathematical sciences; encouraging research and development in multidisciplinary areas; supporting studies to assess the effectiveness and outcomes of new biologics, materials, processes, devices, and procedures; developing technologies for early disease detection and assessment of health status; and developing advanced imaging and engineering techniques for conducting biomedical research at multiple scales. See also http://www.nih.gov/about/almanac/organization/NIBIB.htm.

National Institute of Child Health and Human Development (NICHD). http://www.nichd.nih.gov/. Established in 1962 the mission of the NICHD is to ensure that every person is born healthy and wanted, that women suffer no harmful effects from the reproductive process, and that all children have the chance to fulfill their potential to live healthy and productive lives, free from disease or disability, and to ensure the health, productivity, independence, and well-being of all people through optimal rehabilitation.

In pursuit of this mission, the NICHD conducts and supports laboratory research, clinical trials, and epidemiological studies that explore health processes; examines the impact of disabilities, diseases, and defects on the lives of individuals; and sponsors training programs for scientists, doctors, and researchers to ensure that NICHD research can continue. See also http://www.nih.gov/about/almanac/organization/NICHD.htm.

National Institute on Deafness and Other Communication Disorders (NIDCD). http://www.nidcd.nih.gov/. Established in 1988 the NIDCD conducts and supports research and research training on disorders of hearing and other communication processes, including diseases affecting hearing, balance, smell, taste, voice, speech, and language through:

- Research performed in its own laboratories and clinics
- A program of research grants, individual and institutional research training awards, career development awards, center grants, and contracts to public and private research institutions and organizations
- Cooperation and collaboration with professional, academic, commercial, voluntary, and philanthropic organizations concerned with research and training that is related to deafness and other communication disorders, disease prevention and health promotion, and the special biomedical and behavioral problems associated with people having communication impairments or disorders
- The support of efforts to create devices which substitute for lost and impaired sensory and communication functions
- Ongoing collection and dissemination of information to health professionals, patients, industry, and the public on research findings in these areas.

See also http://www.nih.gov/about/almanac/organization/NIDCD.htm.

National Institute of Dental and Craniofacial Research (NIDCR). http://www.nidcr.nih.gov/. Established in 1948 the mission of the NIDCR is to improve oral, dental, and craniofacial health through research, research training, and the dissemination of health information. We accomplish our mission by:

- Performing and supporting basic and clinical research;
- Conducting and funding research training and career development programs to ensure an adequate number of talented, well-prepared and diverse investigators;

- Coordinating and assisting relevant research and research-related activities among all sectors of the research community;
- Promoting the timely transfer of knowledge gained from research and its implications for health to the public, health professionals, researchers, and policy-makers.

See also http://www.nih.gov/about/almanac/organization/NIDCR.htm.

National Institute of Diabetes and Digestive and Kidney Diseases (NIDDK). http://www.niddk.nih.gov/. Established in 1948 the NIDDK conducts and supports research on many of the most serious diseases affecting public health. The institute supports much of the clinical research on the diseases of internal medicine and related subspecialty fields as well as many basic science disciplines.

The Institute's Division of Intramural Research encompasses the broad spectrum of metabolic diseases such as diabetes, inborn errors of metabolism, endocrine disorders, mineral metabolism, digestive and liver diseases, nutrition, urology and renal disease, and hematology. Basic research studies include biochemistry, nutrition, pathology, histochemistry, chemistry, physical, chemical, and molecular biology, pharmacology and toxicology.

The NIDDK extramural research is organized into four divisions: Diabetes, Endocrinology and Metabolic Diseases; Digestive Diseases and Nutrition; Kidney, Urologic and Hematologic Diseases; and Extramural Activities.

The Institute supports basic and clinical research through investigator-initiated grants, program project and center grants, and career development and training awards. The Institute also supports research and development projects and large-scale clinical trials through contracts. See also http://www.nih.gov/about/almanac/organization/NIDDK.htm.

National Institute on Drug Abuse (NIDA). http://www.nida.nih.gov/. Established in 1973 the NIDA provides national leadership for research on drug abuse and addiction by supporting a comprehensive research portfolio that focuses on the biological, social, behavioral and neuroscientific bases of drug abuse on the body and brain as well as its causes, prevention, and treatment. NIDA also supports research training, career development, public education and research dissemination efforts. Through its Intramural Research Program and grants and contracts to investigators at research institutions around the country and overseas, NIDA supports research and research training on:

- The neurobiological, behavioral, and social mechanisms underlying drug abuse and addiction;
- The causes and consequences of drug abuse, including impact on society and morbidity and mortality in selected populations, e.g., ethnic minorities, youth, women;
- The relationship of drug use to problem behaviors and psychosocial outcomes such as mental illness, unemployment, low socioeconomic status, violence;
- Effective prevention and treatment approaches, including a broad research program designed to develop new treatment medications and behavioral therapies for drug addiction;
- The mechanisms of pain and the search for non-addictive analgesics;
- The relationship of drug abuse to cultural and ethical issues such as health disparities; and
- The relationship of drug abuse to the acquisition, transmission, and clinical course of HIV/AIDS, tuberculosis, and other diseases and the development of effective prevention/intervention strategies.

See also http://www.nih.gov/about/almanac/organization/NIDA.htm.

National Institute of Environmental Health Sciences (NIEHS). http://www.niehs.nih.gov/. Established in 1969 the mission of the NIEHS is to reduce the burden of human illness and disability by understanding how the environment influences the development and progression of human disease. To have the greatest impact on preventing disease and improving human health, the NIEHS focuses on basic science, disease-oriented research, global environmental health, and multidisciplinary training for researchers. The NIEHS achieves its mission through:

- Extramural research and training, funded by grants and contracts, to scientists, environmental health professionals, and other groups worldwide,

- Intramural research conducted by scientists at the NIEHS facility and in partnership with scientists at universities and hospitals,
- Toxicological testing and test validation by the National Toxicology Program, and
- Outreach and communications programs that provide reliable health information to the public and scientific resources to researchers.

See also http://www.nih.gov/about/almanac/organization/NIEHS.htm.

National Institute of General Medical Sciences (NIGMS). http://www.nigms.nih.gov/. Established in 1962 the NIGMS supports basic biomedical research that is not targeted to specific diseases. NIGMS funds studies on genes, proteins, and cells, as well as on fundamental processes like communication within and between cells, how our bodies use energy, and how we respond to medicines. The results of this research increase our understanding of life and lay the foundation for advances in disease diagnosis, treatment, and prevention. NIGMS also supports research training programs that produce the next generation of biomedical scientists, and it has special programs to encourage under-represented minorities to pursue biomedical research careers. See also http://www.nih.gov/about/almanac/organization/NIGMS.htm.

National Institute of Mental Health (NIMH). http://www.nimh.nih.gov/. Established in 1949 the mission of the NIMH is to reduce the burden of mental illness and behavioral disorders through research on mind, brain, and behavior.

Investments made over the past 50 years in basic brain and behavioral science have positioned NIMH to exploit recent advances in neuroscience, molecular genetics, behavioral science and brain imaging; to translate new knowledge about fundamental processes into research-able clinical questions; and to initiate innovative clinical trials of new pharmacological and psychosocial interventions, with emphasis on testing their effectiveness in the diagnostically complex, diverse group of patients typically encountered in front-line service delivery systems. NIMH-funded investigators also seek new ways to translate results from basic behavioral science into research relevant to public health, including the epidemiology of mental disorders, prevention and early intervention research, and mental health service research. See also http://www.nih.gov/about/almanac/organization/NIMH.htm.

National Institute of Neurological Disorders and Stroke (NINDS). http://www.ninds.nih.gov/. Established in 1950 the mission of NINDS is to reduce the burden of neurological diseases—a burden borne by every age group, every segment of society, and people all over the world. To accomplish this goal the NINDS supports and conducts research, both basic and clinical, on the normal and diseased nervous system, fosters the training of investigators in the basic and clinical neurosciences, and seeks better understanding, diagnosis, treatment, and prevention of neurological disorders. See also http://www.nih.gov/about/almanac/organization/NINDS.htm.

National Institute of Nursing Research (NINR). http://ninr.nih.gov/ninr/. Established in 1986 the NINR supports basic and clinical research to establish a scientific basis for the care of individuals across the life span–from the management of patients during illness and recovery to the reduction of risks for disease and disability and the promotion of healthy lifestyles. According to its broad mandate, the NINR implements programs of research to understand and ease the symptoms of acute and chronic illness, to prevent or delay the onset of disease or slow its progression, to find effective approaches to achieving and sustaining good health, and to improve the clinical settings in which care is provided. This research extends to problems encountered by patients' families and caregivers. It also emphasizes the special needs of at-risk and under-served populations. These efforts are crucial in translating scientific advances into cost- effective health care that does not compromise quality. See also http://www.nih.gov/about/almanac/organization/NINR.htm.

National Library of Medicine (NLM). http://www.nlm.nih.gov/. Established in 1956 the Library has a statutory mandate from Congress to apply its resources broadly to the advancement of medical and health-related sciences. It collects, organizes, and makes available biomedical information to investigators, educators, and practitioners, and carries out programs designed to strengthen existing and develop new medical library services in the United States. It conducts research in health communications, supports medical informatics, and provides information services and sophisticated tools in the areas of molecular biology and toxicology/environmental health. The Library creates Web-based services for the general public containing information from the NIH and other reliable sources. See also http://www.nih.gov/about/almanac/organization/NLM.htm.

Center for Information Technology (CIT formerly DCRT, OIRM, TCB). http://www.cit.nih.gov/home.asp. Established in 1964 the CIT incorporates the power of modern computers into the biomedical programs and administrative procedures of the NIH by focusing on three primary activities: conducting-computational biosciences research, developing computer systems, and providing computer facilities. See also http://www.nih.gov/about/almanac/organization/CIT.htm.

Center for Scientific Review (CSR). http://cms.csr.nih.gov/. Established in 1946 the CSR is the focal point at NIH for the conduct of initial peer review, the foundation of the NIH grant and award process. The Center carries out peer review of the majority of research and research training applications submitted to the NIH. In addition, the Center serves as the central receipt point for all such Public Health Service (PHS) applications and makes referrals to scientific review groups for scientific and technical merit review of applications and to funding components for potential award. To this end, the Center develops and implements innovative, flexible ways to conduct referral and review for all aspects of science. See also http://www.nih.gov/about/almanac/organization/CSR.htm.

John E. Fogarty International Center (FIC). http://www.fic.nih.gov/. Established in 1968 the John E. Fogarty International Center (FIC) for Advanced Study in the Health Sciences, the international component of the NIH, addresses global health challenges through innovative and collaborative research and training programs and supports and advances the NIH mission through international partnerships. See also http://www.nih.gov/about/almanac/organization/FIC.htm.

National Center for Complementary and Alternative Medicine (NCCAM). http://nccam.nih.gov/. Established in 1999 the NCCAM s dedicated to exploring complementary and alternative healing practices in the context of rigorous science; training complementary and alternative medicine (CAM) researchers; and disseminating authoritative information to the public and professionals.

To fulfill its mission, the NCCAM supports a broad-based portfolio of research, research training, and educational grants and contracts, as well as various outreach mechanisms to disseminate information. See also http://www.nih.gov/about/almanac/organization/NCCAM.htm.

National Center on Minority Health and Health Disparities (NCMHD). http://ncmhd.nih.gov/. Established in 1993 the mission of NCMHD is to promote minority health and to lead, coordinate, support, and assess the NIH effort to reduce and ultimately eliminate health disparities. In this effort NCMHD will conduct and support basic, clinical, social, and behavioral research, promote research infrastructure and training, foster emerging programs, disseminate information, and reach out to minority and other health disparity communities. See also http://www.nih.gov/about/almanac/organization/NCMHD.htm.

National Center for Research Resources (NCRR). http://www.ncrr.nih.gov/. Established in 1962 the NCRR provides laboratory scientists and clinical researchers with environments and tools that they can use to prevent, detect, and treat a wide range of diseases. This support enables discoveries that begin at the molecular and cellular level, move to animal-based studies, and then are translated to patient-oriented clinical research, resulting in cures and treatments for both common and rare diseases. NCRR connects researchers with each other, as well as with patients and communities across the nation, to bring the power of shared resources and research to improve human health. NCRR's extramural grant support is concentrated in four programmatic Divisions: Division for Biomedical Technology Research and Research Resources, Division for Clinical Research Resources, Division of Comparative Medicine, and Division of Research Infrastructure. See also http://www.nih.gov/about/almanac/organization/NCRR.htm.

NIH Clinical Center (CC). http://clinicalcenter.nih.gov/. Established in 1953 the CC is the clinical research hospital for the National Institutes of Health. Through clinical research, physician-investigators translate laboratory discoveries into better treatments, therapies and interventions to improve the nation's health.

Late in 1997, Vice President Al Gore and Senator Mark O. Hatfield broke ground for the new Mark O. Hatfield Clinical Research Center. The center, completed in 2004, houses 240 inpatient beds, 90 day-hospital stations and research labs. Together, the Magnuson and Hatfield centers provide the environment today's researchers need to spark new medical discovery. See also http://www.nih.gov/about/almanac/organization/CC.htm.

STAFF, National Institutes of Health, Bethesda, MD

NATIONAL RESEARCH COUNCIL (NRC). The Council was established by the National Academy of Sciences in the United States in 1916 to associate the broad community of science and technology with the Academy's purpose of furthering knowledge and of advising the federal government. The Council operates in accordance with general policies determined by the Academy under the authority of its congressional charter of 1863, which establishes the Academy as a private, nonprofit, self-governing membership corporation. The Council has become the principal operating agency of both the National Academy of Sciences and the National Academy of Engineering in the conduct of their services to the government, the public, and the scientific and engineering communities. It is administered jointly by both academies and the Institute of Medicine. The National Academy of Engineering and the Institute of Medicine were established in 1964 and 1970, respectively, under the charter of the National Academy of Sciences.

NATROLITE. The mineral natrolite, one of the zeolites, is a sodium aluminum silicate corresponding to the formula $Na_2Al_2Si_3O_{10} \cdot 2H_2O$. It is orthorhombic, crystallizing in slender prisms of nearly square cross-section which are terminated by relatively flat pyramids. There are also fibrous to compact varieties. Natrolite is a brittle mineral; hardness, 5–5.5; specific gravity, 2.2; luster, vitreous; color, red, yellow, white, or colorless; transparent to opaque. Natrolite is found with other zeolites in fissures and cavities in basaltic and related rocks. Czechoslovakia, France, Italy, Norway, Scotland, Ireland, Iceland, Greenland, and South Africa contain well-known localities for natrolite. In the United States it is found in the Triassic traps of New Jersey; also from Oregon, Washington, Montana, Colorado, and as exceptional crystals from San Benito County, California. Superb crystals occur at Mt. St. Hilaire, Quebec, Canada, and from an asbestos mine in Quebec, crystals up to 3 feet (0.9 meters) long and 4 inches (10 centimeters) in diameter have been found. The name natrolite refers to its soda content.

NATTA, GIULIO (1903–1979). An Italian chemist along with Karl Ziegler won the Nobel Prize for chemistry in 1963 for his fundamental work on catalytic polymerization. In 1954 he developed isotactic polypropylene at his laboratory at the Polytechnic Institute of Milan, which led to wide application of various stereospecific polymers with organometallic catalysts such as triethylaluminum. He was for many years consultant for the Montecatini chemical firm. The researchers of Natta, together with those of Karl Ziegler, made possible the chemical manipulation of monomers to form specifically ordered 3-dimensional polymers having predetermined properties, to which the term *tailor-made* is often applied.

NATURAL COORDINATES. An orthogonal, or mutually perpendicular, system of curvilinear coordinates for the description of fluid motion, consisting of an axis T tangent to the instantaneous velocity vector and an axis N normal to this velocity vector to the left in the horizontal plane, to which a vertically directed axis Z may be added for the description of three-dimensional flow.

NATURAL FREQUENCY (Mathematics). A term broadly applied to any system whose transfer function approximates a second-order differential equation of the form $s^2 + 2\zeta\omega_n s + \omega_n^2$, where ω_n is the natural frequency, ζ is the damping ratio, and s is the Laplace transform operator. In control and feedback systems, for example, continuous oscillation or hunting may occur. The frequency of this oscillation is the natural frequency.

NATURAL GAS. A major source of energy for industrial, commercial, and domestic needs, natural gas is consumed by numerous countries worldwide. Because natural gas is comparatively easy to transport over long distances, usage is not confined to regions that produce it. In addition to energy, natural gas also is a critically important source of industrial chemicals, including numerous hydrocarbon-based organics that find ultimate usage in plastics, films, fibers, solvents, and coatings.

In terms of interest as a fossil fuel for generating electrical power and other energy-conversion processes, natural gas is gaining favor. Compared with coal, natural gas frequently is termed the "clean-burning" fuel. As compared with coal as an energy raw material, the "add-on" costs for treating combustion effluents to satisfy environmental requirements are less for natural gas than for coal, and, consequently, the lower cost benefits of coal are eroding. Further, improvements in natural gas production technology and a brighter outlook for natural gas reserves are contributing to the expansion of natural gas consumption. In addition, large advances have been made in the combustion efficiency of natural gas. For example, the efficiency of some domestic heating appliances has increased to about 95% as compared with 60% or 70% efficiency a relatively few years ago. The cogeneration of heat and electricity in industrial utilities is tending to favor natural gas as the raw fuel. In terms of local natural gas distribution, utilities are finding the use of polyethylene pipe an important cost-saving factor.

In a summary of the Gas Research Institute (GRI)[1], the following observation is made: "The United States, the gas industry, and the gas consumer have entered a dynamic new decade filled with change, challenge, and opportunity. These include the reemergence of the environmental movement, the continuing deregulation of the U.S. natural gas market, the expansion of global trade in the former Soviet Bloc, the emergence of more competitive international and national energy markets, and the rapid expansion of technology options. Three strategic needs are likely to dominate the 1990s for the U.S. gas industry and the gas consumer: (1) ensuring gas deliverability while controlling costs to the consumer; (2) responding to increased concern for the environment; and (3) satisfying a demand for higher quality of energy service."

See also **Electric Power Production and Distribution**.

Composition of Natural Gas

The composition of natural gas varies with the source, but essentially is made up of methane, ethane, propane, and other paraffinic hydrocarbons, along with small amounts of hydrogen sulfide, carbon dioxide, nitrogen, and, in some deposits, helium. Natural gas is found underground at various depths and pressures, as well as in solution with crude-oil deposits. Principal gas deposits are found in the United States, Canada, the former Soviet Bloc, and the Middle East. The analysis of a gas sample taken from the Panhandle natural gas field in Texas is given in Table 1. Because numerous parts of the earth do not have natural gas at all, or where supply is less than demand, much natural gas is transported, notably by pipeline in the gaseous or liquid phase and across the seas in specially-designed LNG (liquefied natural gas) carriers.

Origin and Geology of Natural Gas

The most commonly accepted theory concerning the formation of natural gas is the organic theory. Methane is a product of decaying vegetable matter and in areas of stagnant water is found as *marsh gas* or *swamp gas*. It is theorized that over millions of years, the remains of plants and animals were washed down into lakes, the accumulations covered with layers of mud and stone. The latter became stone while the organic matter decayed through the action of heat and pressure and perhaps from effects of bacteria and radioactivity, forming various hydrocarbons. The hydrocarbons were held in tiny spaces between the particles of sand and porous rock and formed natural gas and petroleum. Often the natural gas so formed made its way through the rock to the surface and escaped. In some areas, however, the layers of sand and porous rock were covered by impermeable rock to form huge reservoirs of natural gas at various levels of pressure. See also **Petroleum**.

[1] "1993–1997 Research & Development Plan," Gas Research Institute, Chicago, Illinois.

TABLE 1. ANALYSIS OF NATURAL GAS FROM NATURAL GAS FIELD IN TEXAS PANHANDLE

Component	Mole Percent
Methane	76.2
Ethane	6.4
Propane	3.8
Normal butane	1.3
Isobutane	0.8
Normal pentane	0.3
Isopentane	0.3
Cyclopentane	0.1
Hexane plus other hydrocarbons	0.35
Nitrogen	9.8
Oxygen	Trace
Argon	Trace
Hydrogen	0.0
Hydrogen sulfide	0.0
Carbon dioxide	0.2
Helium	0.45

Note: Heating value of various natural gases averages between 975 and 1180 Btu/cubic foot (8678-10,502 Calories/cubic meter) at 60°F (15.6°C) and 30 inches (76.2 centimeters) mercury pressure.

The organic theory as usually presented is rather general and vague in many respects. As observed by Ourisson (Université Louis Pasteur, Strasbourg) and colleagues, natural gas, as well as coal and petroleum, are fossil fuels, but fossils of what? Fossil fuels form only if the organic matter is buried before it can become completely oxidized to carbon dioxide by microorganisms. According to the microbial origin concept, as the carbon compounds sink deeper into the Earth under accumulating sediments, they are subjected to high temperatures and undergo chemical reaction, during which oxygen and most other elements are eliminated. This yields a mixture composed in the case of gas and petroleum almost entirely of hydrocarbons (carbon in the case of coal). Since the beginnings of photosynthesis on Earth, it is estimated that 10 quadrillion (10^{16}) tons of carbonaceous material has been stored in sediments. Most of this

material is stored in very dilute form and only under exceptional geologic conditions, is it concentrated to become a viable fuel source. In a twenty-year study, which might be called molecular paleontology, Ourisson and coworkers have been studying the detailed genesis of fossil fuels. Thus far, chemical analysis of the most varied organic sediments reveals a surprising commonality — all appear to derive much of their organic matter from once unknown microbial lipids. This topic is presented in more detail in entry on **Petroleum**.

A few scientists, notably Thomas Gold (Cornell University), have proposed that, in contrast with the organic sediments theory, the prime source of natural gas is primordial, abiotic methane rising from deep within the Earth's mantle. This is sometimes referred to as the "deep-earth gas" hypothesis. In this view, methane flows up around the edges of the shield and is responsible for the oil and gas fields in the North Sea and the southern Baltic. Admittedly, this reservoir of natural gas is presently out of the reach of any foreseeable drilling technology. In some areas, it is suggested that the granite crust may have been fractured and subsequently became porous to the extent that methane may have risen into the crust and have become trapped at accessible depths. Other scientists point out that no evidence supports the concept that a large amount of methane was incorporated in the Earth when it was formed. Available geochemical evidence suggests that the early atmosphere, produced by outgassing of the planetary interior, could not have been rich in hydrogen. Further, if the Earth did at one time contain primordial methane or other hydrocarbons, most of that volatile material would have long since escaped by way of volcanism and diffusion. It is also suggested that the analysis of volcanic basalts shows that the rock in the upper mantle is highly oxidizing, in which case any methane present would have been converted to carbon dioxide.

Some experimental drilling programs underway in Sweden, including a well some 5000 meters in granite bedrock, may shed further light on Gold's hypothesis.

In searching for new fields during the 1960s and 1970s, drillers seeking gas and oil in traditional suspect source reservoirs, would find gas and/or oil in only about 9 of 100 wells drilled. Usually, only 2 or 3 of these wells produced sufficient gas and/or oil to be of commercial value. Whereas the average depth of gas well drilled during this period ranged between 5000 and 6000 feet (1524–1829 meters), some drillers are now aiming at the

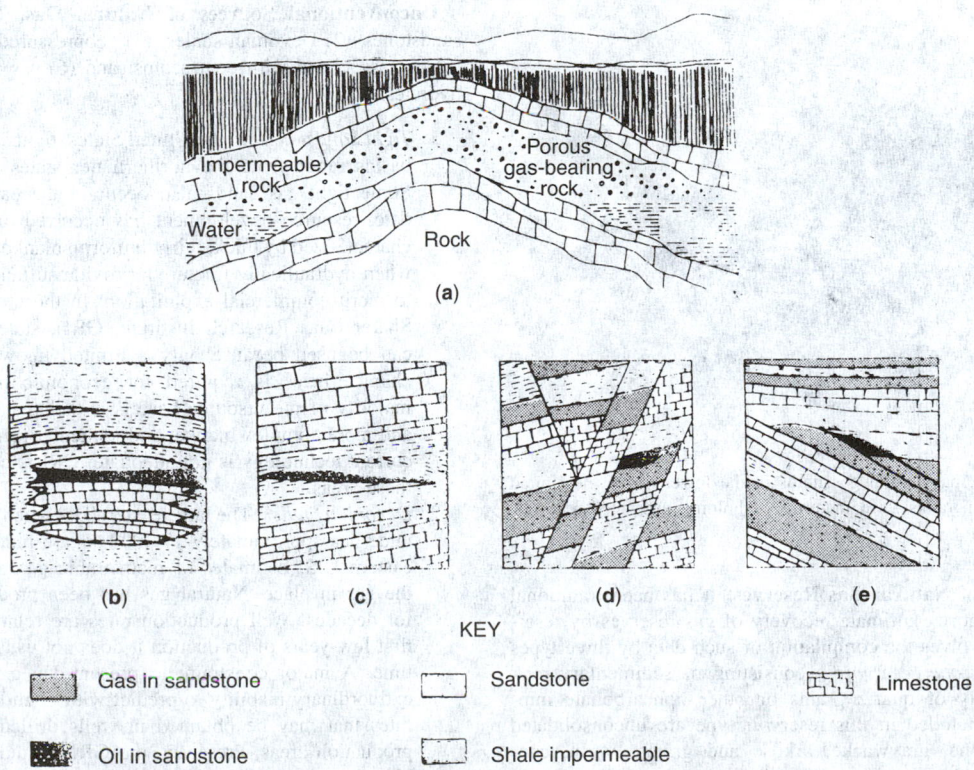

KEY

Gas in sandstone   Sandstone   Limestone

Oil in sandstone   Shale impermeable

Fig. 1.   Types of natural gas reservoirs and entrapments: (**a**) anticlinal trap; (**b**) coral reef trap; (**c**) stratigraphic trap; (**d**) fault trap; and (**e**) unconformity.

30,000-foot (9144-meter) depth. For many years, the deepest gas well in Texas was 28,600 feet (8717 meters). In the early days of offshore drilling, operations were conducted in water only 20–25 feet (6–7.5 meters) deep. There are now many platforms in waters that are deeper by a factor of 20–30 times.

In terms of the conventional or traditional sources, natural gas is found in areas close to exposed or buried mountain ranges. Major deposits of natural gas are found in inclined strata where the rock formations dip away from the crest of a buried hill or the ridge of a buried mountain. Some of the common types of formations in which natural gas is found are shown in Fig. 1. Although natural gas and crude oil are frequently found together, the largest natural gas reserves (about 70% of the estimated reserves) are in deposits neither in contact with, nor dissolved in, oil.

Forecasting Natural Gas Reserves

Since the 1920s, natural gas reserves have been based upon the amount of gas that most likely will be found with oil. Although numerous experts in the field have claimed over the years that this methodology overlooks large amounts of "unassociated" gas, the professionals have been slow to revise their procedures. The concept that much natural gas occurs quite apart from oil is now taking hold. The U.S. Geological Survey, which has proclaimed a gas resource base at about 400 trillion cubic feet, undertook a major study to be completed in the late 1990s. Survey techniques have undergone several dramatic improvements, including the use of computer techniques that will project three-dimensional survey information, as contrasted with past reliance on two-dimensional mapping. A number of experts now believe that the reserves are well over 1,200 trillion cubic feet (34 trillion cubic meters). One independent gas producer has estimated the figure at about 1,500 trillion cubic feet (42.5 trillion cubic meters).

Traditional estimates of oil-associated natural gas reserves historically have rated the former Soviet Bloc as holding nearly 40% of the reserves, Iran about 14%, and the United States nearly 6%, with additional fairly high reserves in Qatar, Algeria, and Saudi Arabia. At one time, North America was attributed to have a 60-year supply, but with revisions in estimating procedures, coupled with increased efficiency in natural gas production and gas combustion efficiency, the future is now believed to be in terms of at least a few centuries. Interest continues, however, in upgrading low-Btu natural gases and developing so-called *substitute natural gas.*

The principal gas fields within the continental United States are indicated in Fig. 2.

Fig. 2. Location of principal natural gas fields in the lower 48 of the United States. Alaska, not shown, ranks third in terms of holding estimated reserves. (*Batelle Memorial Institute.*)

Ultimate Recovery of Natural Gas Reserves. It has been traditional for many years to categorize ultimate recovery of gas reserves by reservoir lithology which involves the compilation of such data by three types of reservoirs: (1) *Sandstone reservoir*—consisting of sedimentary rock composed predominantly of quartz grains or other noncarbonate mineral or rock detritus. Included in this reservoir type are unconsolidated sand, sandstone, siltstone, graywacke, arkose and granite wash, conglomerate, and breccia. (2) *Carbonate reservoir*—composed of sedimentary rock made up predominately of calcite (limestone) and/or dolomite.

(3) *Other reservoirs*—including igneous and metamorphic rocks and some sedimentary rocks, such as fractured shale.

Estimated ultimate recovery is also reported by type of entrapment, of which there are two major types: (1) *Structural trap*—an entrapment in which migration of hydrocarbons in the reservoir rock has terminated primarily because of closure induced by structural deformation, such as folding or faulting. Within this category should also be included entrapments attributed to hydrodynamic forces. (2) *Stratigraphic trap*—an entrapment in which migration of hydrocarbons has terminated because of the pinchout of reservoir rock due either to truncation or to nondeposition or to a facies change in the form of diminished permeability of reservoir rock. Also included in this category are entrapments in which a pinchout of facies change provides part of the barrier to migration of hydrocarbons, with structural elements providing the remaining closure for the entrapment. In these cases, it is recognized that the dominant cause of the accumulation is the lenticularity of truncation of the reservoir rock.

Some estimates indicate that about 65.8% of ultimately recoverable natural gas in the United States will be found in structural traps; the remaining 34.2% in stratigraphic traps.

The estimated ultimate recovery is also reported by the geologic age of the reservoir. It is recognized that problems may arise where the geologic age of a reservoir cannot be determined specifically, such as Permo-Pennsylvanian and Cambro-Ordovician, or where production from reservoirs of different geologic age is combined.

Natural gas liquids occur in either the gaseous phase or in solution with crude oil in the reservoir. They are recovered at the surface as liquids by separation from produced natural gas, by such processes as condensation and absorption in field separators, gasoline plants, and other surface facilities. In this processing, valuable by-products are recovered, such as light oils, natural gasoline, and other petroleum gases such as ethane, propane, and butane. Natural gasoline is blended with gasoline from petroleum refineries to improve starting properties, especially desirable in cold weather. Ethane is a major petrochemical raw material. Propane and butane are made available as LPG (liquefied petroleum gas). Processing of natural gas also removes unwanted material, such as nitrogen, sulfur compounds, carbon dioxide, and water vapor. Some gas fields produce helium, which is extracted cryogenically.

Unconventional Sources of Natural Gas. These include: (1) tight sandstones, (2) Devonian shales, (3) geopressured zones, (4) deep basins, (5) gas associated with coal seams, and (6) gas in the form of methane hydrates.

1. *Tight Sandstones.* In the United States, tight sandstones of the western basins range from the northern tier states to the Mexican border. Some tight gas sands also occur in the eastern United States. To date, resource development has occurred only in the limited areas characterized by thick, fairly uniform, blanket-type formations which, when hydraulically fractured, provide sufficient gas production rates to merit commercial exploitation. In these areas, as pointed out by Sharer (Gas Research Institute, GRI), state-of-the-art technologies can be used because only a limited knowledge of the formation characteristics is required for economic production. However, a majority of the resource base is associated with lower permeability and more complex blanket and lenticular sand formations, for which current technology is not adequate. GRI is concentrating research in these areas.

2. *Devonian Shales.* The large eastern Devonian gas shales resource base underlies approximately 174,000 square miles (453,000 km^2) of the eastern U.S. Estimates of recoverable gas range from 2 to 15% of the gas in place. Natural gas has been produced from these shales for decades. Well production rates are relatively low, but after the first few years of production it does not usually decline rapidly with time. A major constraint to present-day exploitation has been the extraordinary inability to predict with confidence the gas production rates that may be obtained in wells drilled outside the traditional production areas. Presently, the GRI is studying the systematics of historically successful fields, including the Appalachian, Illinois, and Michigan Basins.

3. *Geopressured Zones.* A test well in a geopressure zone was drilled some years ago in Tigre Lagoon in the coastal marshes of southern Louisiana. Known as Edna Delcambre #1, this well produced at a rate of up to 10,000 barrels of water per day from a sandstone aquifer some 12,600 feet (3840 meters) below the surface. Pressure at that depth is nearly 11,000 pounds per square inch (748 atmospheres) and the temperature is 116 °C. Quite an elaborate manifold system is required to collect the gas. The water is disposed by forcing it by its own pressure into another well bore, which penetrates to a depth of 2500 feet (762 meters). Scientists associated with this project had expected about 20 cubic feet/barrel (42 gallon); about 0.6 cubic meter/barrel (159 liters). In actuality, reports indicate that the yield of gas was about 2.5 times that amount.

As explained by specialists in geopressure technology, at great depths (in terms of present technology), the solubility of natural gas in water may be as much as 1000 cubic feet/barrel (28.3 cubic meters/159 liters) at depths of 30,000 feet (9144 meters), whereas that solubility will be reduced by a factor of ten at a depth of 20,000 feet (6096 meters). Under the right combination of geologic and hydrologic conditions, this gas-laden water will move toward the surface, during which process some of the gas will be released from the water in the form of very small bubbles. Ultimately, this gas collects beneath a geologic trap, where conventional free-gas reservoirs are formed. Some authorities now believe that the very deep aquifers are much more extensive than the free-gas reservoirs. It is this gas-saturated water that some scientists believe will be a great source of future natural gas.

The GRI has been investigating the coproduction of gas and water for a number of years. Natural gas from watered-out reservoirs, geopressured aquifers, and high-water-saturated gas-bearing reservoir strata are prime targets. This natural gas is trapped by water such that special production techniques must be used to move the water and remobilize the gas. Although some gas is also dissolved in the water, it is of less significance than the free gas trapped as dispersed bubbles or in pockets or stringers of various sizes in the reservoir rock matrix.

4. *Deep Basins.* These are found at depths between 15,000 and 30,000 feet (4572–9144 meters) and are estimated to contain significant quantities of gas, but generally await the development of advanced production technology and economic incentive.
5. *Gas Associated with Coal Seams.* Methane, the principal constituent of natural gas, is generated during the geologic process of coal formation. A significant portion of this gas is trapped by impermeable strata, and it is present within the fractures and micropores of the coal. (The presence of methane is an ever-present hazard in coal mining.) Major variations in resource estimates are due to uncertainties in the gas content and size of the deeper, not minable coal deposits in the western states that form the major portion of the resource base. Seeking such gas may involve depths as great as 6000 feet (1829 meters) underground. Except for reasons of safety, little effort has been made to recover any of this resource due to high recovery costs, potential uncertainties in production, and deficiencies in state-of-the-art equipment, particularly for the deeper coals. The GRI is concentrating its research efforts on unminable coal because of its large potential as a resource base. While the gas resource associated with mining amounts to about 10% of the energy value of the coal, producers rarely apply new gas recovery technology except where safety is a requirement. Targets of the GRI program are deep coal seams, multiple seams interbedded with shales and sandstones, and deep multiple beds that are too thin to mine.
6. *Methane Hydrates.* Within a certain range of pressures and temperatures, methane and water form hydrates. Described as icelike substances, these hydrates are believed to occur in very substantial quantities, particularly beneath permafrost and in deep-ocean bottoms. Although slush has occurred in gas pipelines under certain conditions for many years, the existence of hydrates in nature was not made known until the mid-1960s. Geologists and hydrologists had previously assumed that gas of this type would have dissipated during earlier geologic ages. This is another area of natural gas resource research awaiting economic incentives.

Exploratory Methods. The principal exploratory methods used are: (1) *Airborne magnetometers,* which seek out anomalies in the magnetic field. Experienced geologists relate these irregularities to the probability of gas reservoirs below the surface. See also **Magnetometer**. (2) *Satellite imagery,* from which surface structures and patterns can be related to previous pattern recognition studies made of surfaces below which gas reservoirs exist. (3) *Gravitometers* are used to detect subtle variations in gravitational pull inasmuch as this is less for a gas reservoir than for continuous dense rock formations. (4) *Seismic methods,* which constitute the most widely-used of exploration methods. See also **Earth Tectonics and Earthquakes**. (5) *Data logging methods,* wherein an instrument is lowered into the borehole and which telemeters back to the surface readings of sonic absorption in an effort to determine the nature and thickness of rock formations. Data loggers operate on the basis of several physical phenomena. (6) *Fossil inspection.* The careful examination of microfossils can assist in fixing the age of rocks that are being penetrated. The condition of the fossils also can be related to probable temperatures to which they have been exposed over geologic periods and these, in turn, can be advantageous in locating possible gas deposits. Usually a combination of two or more exploratory techniques is used.

Liquefied Natural Gas (LNG)

The liquefaction of natural gas for storage and transportation and regasification for final distribution dates back several decades. A few major accidents in the handling of LNG thwarted the progress of the field for a while, but in the early 1970s, LNG was again considered in a major way because of energy-short nations. One of the more serious LNG accidents occurred in Cleveland, Ohio on October 20, 1944, when a storage tank developed a leak with spillage and subsequent fires in the surrounding neighborhood in which 135 persons lost their lives. While liquefaction offers marked storage space savings and convenience, the predominant advantage occurs in connection with both pipeline and ship transportation. Energy-short nations, such as Japan, and some of the European nations, have turned in recent years to the concept of shipping LNG by ship. For example, a large LNG plant at Lumut, Brunei, Borneo went onstream in mid-1974 essentially to furnish LNG to Japan.

Oil- and gas-rich nations, which at one time flared to the atmosphere much of the natural gas that accompanied the production of crude oil, have turned toward conservation—either through reinjection of much of the natural gas underground or through constructing LNG production facilities for shipment of the product overseas. Concurrent with such planning was reevaluation by a number of nations of their own valuable resources and a growing reluctance toward exporting inordinate quantities of gas and oil strictly for money. As of the early 1990s, the shipment of LNG overseas competes with other ways and means for alleviating energy shortages, including coal conversion and gasification, nuclear energy, solar energy, etc.

Three types of liquefaction processes may be used for production of LNG. The standard cascade process, which uses three refrigerants—methane, ethylene, and propane—all circulating in closed cycles, is shown in Fig. 3. There is a separate compressor for each of these refrigerants. The methane and propane are available from the feed gas (natural gas). The ethylene must be furnished separately. Ethane may be used in place of ethylene at a subatmospheric suction pressure. The cascade process has the highest rank in terms of thermal efficiency. As a possible improvement over the cascade process, the mixed refrigerant process was developed in the early 1960s. A single-pressure mixed refrigerant cascade (MRC) system is shown in Fig. 4. In one plant using this process, a hydrocarbon-plus-nitrogen mixture of relatively wide boiling range (N2 through C5) is used as the refrigerant. All of these components can be recovered from natural gas in separate apparatus. In still another system, shown in Fig. 5 a propane and mixture-refrigerant cycle is used. In this process, the cooling load is divided horizontally at about −34.4 °C into an upper portion absorbed by propane and a lower portion absorbed by the mixed refrigerant. In essence, the system is a dual refrigerant cascade in which the lower boiling fluid is a mixture refrigerant. The cascade combination with propane makes it possible to reduce the boiling range of the mixture refrigerant substantially, which improves the thermodynamic efficiency over that of the straight MRC process.

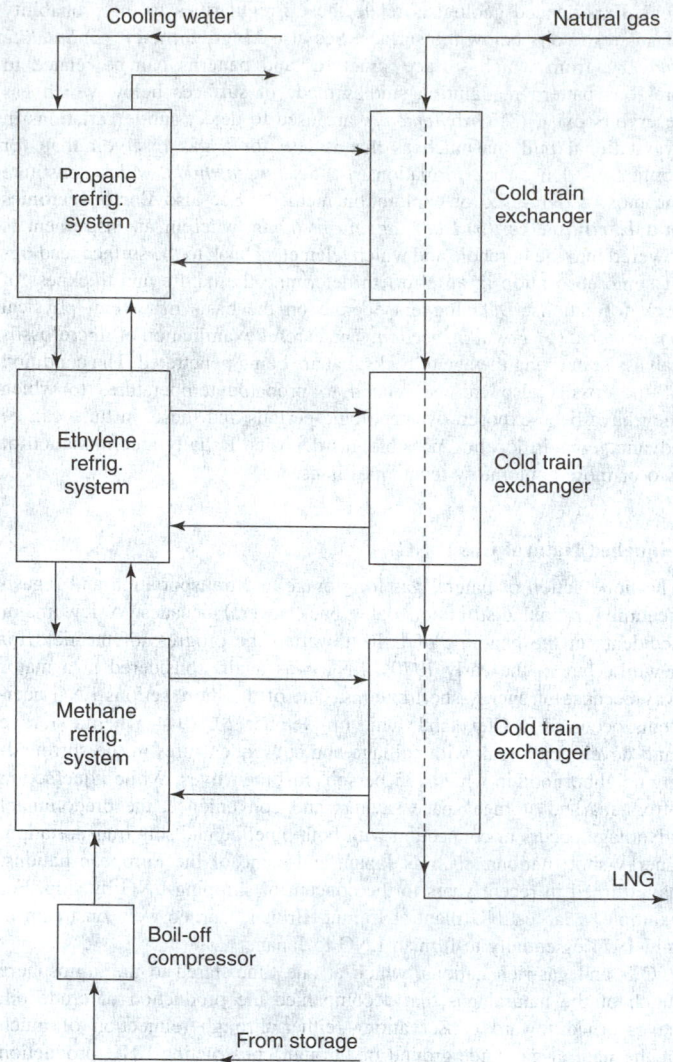

Fig. 3. Conventional or standard cascade system for producing liquefied natural gas (LNG).

Cryogenic Upgrading of Low-Btu Natural Gases

Worldwide, there are substantial reserves of natural gas in which the reservoir formation hydrocarbons are contaminated with nonburning components. The presence of components, such as helium, nitrogen, or carbon dioxide, reduces the heating value of the gas mixture. This can result in the gas being unsuitable for existing transmission and distribution systems. Such contaminated mixtures are termed low-Btu gases if their heating values fall below the minimum standards, regulations, or contract heating value requirements.

Cryogenic processing can be used to upgrade some of these low-Btu gases so as to produce an acceptable high-Btu product. Cryogenic upgrading is a physical process in which subambient temperatures are employed to bring about a separation between the hydrocarbons and nonhydrocarbons in the mixture. The reduction of temperature occurring during cryogenic processing produces a two phase (gas-liquid) mixture. The relative volatilities between the components in the mixture result in selective mass transfer between the two phases. One phase becomes enriched with hydrocarbons and then has a heating value higher than the original gas. The second phase becomes denuded of hydrocarbons and has a heating value below that of the original gas mixture. Frequently, the mass transfer operation requires several theoretical stages in order to achieve the desired product heating value and high hydrocarbon recoveries. While cryogenic upgrading can be applied to gas mixtures containing carbon dioxide or hydrogen sulfide, it has so far only been applied commercially to those hydrocarbon mixtures contaminated with nitrogen and helium.

One of the main considerations in the design and operation of cryogenic upgrading plants is to identify and remove any component from the gas that could adversely affect the operation of the cold sections of the plant. Such components are carbon dioxide, water vapor, and heavy hydrocarbons that have high solidification temperatures and low solubilities. In general, if these components are allowed to remain in the gas to the cryogenic unit, they will form solids during the cooling process that will be deposited on the heat exchanger surfaces. This will lead to fall-off in performance and, possibly, to blockages and plant shutdown.

Particular attention should be paid to identifying any high freezing point components in the low-Btu gas. There exists a range of absorption and adsorption processes to pretreat the low-Btu gas to remove these undesirable components.

A simplified flowsheet of cryogenic upgrading plant is given in Fig. 6. The plant, consisting of two identical trains, is capable of processing 260 million standard cubic feet (7.3 million cubic meters) per day of low-Btu gas (580 Btu/standard cubic foot) (5162 Calories/cubic meter) and upgrading the gas into 143 million standard cubic feet (4 million cubic meters) of high-Btu gas (980 Btu/standard cubic foot; 8722 Calories/cubic meter). The plant stream parameters are indicated in Table 2.

The low-Btu gas is available at 800 psig (54 atmospheres) and is mainly a nitrogen-methane mixture. In addition to a small quantity of helium, the gas also contains small quantities of carbon dioxide, water vapor, and heavy hydrocarbons. The carbon dioxide is removed by washing with monoethanolamine (MEA), the water is taken out on molecular sieve, and the heavy hydrocarbons by adsorption on activated carbon. The gas is then cooled in aluminum plate-fin exchangers against the returning high-Btu product gas and vent gas. The gas is then expanded to 380 psig (26 atmospheres) and a vapor-liquid mixture passes into the H.P. (high pressure) fractionator. The purpose of this fractionator is to bring about an initial separation of the nitrogen-methane and to produce a liquid reflux for the L.P. (low pressure) fractionator.

A nitrogen-enriched vapor flows up the H.P. fractionator, while methane is returned to the sump of this column by a nitrogen reflux stream produced in the tubes of the overhead condenser. The refrigeration required to produce this nitrogen reflux is provided by evaporating some of the liquid methane, from the L.P. column, in the shell of the overhead condenser. Two liquid streams are taken from the H.P. fractionator, and these become the feed and reflux for the L.P. fractionator. The L.P. feed is an enriched methane stream taken from the base of the H.P. fractionator. The L.P. reflux is a high purity nitrogen liquid taken off the H.P. fractionator just below the condenser. The upgrading is completed in the L.P. fractionator. The feed stream is stripped to produce a high-Btu liquid containing 4% nitrogen and having a heating value of 980 Btu/standard cubic foot (8722 Calories per cubic meter). The liquid is pumped from the column sump, evaporated, and superheated against the incoming low-Btu gas. The gas from the top of the L.P. column is mainly nitrogen and is also heated to ambient temperature against the incoming low-Btu gas. By using this arrangement of two distillation columns, the separation of nitrogen and methane can be achieved using only the pressure energy available in the low-Btu gas.

Transportation of Natural Gas

The mode selected for gas transportation depends mainly on: (a) the distance over which the gas must be moved; (b) the geographical and geological characteristics of the terrain (considering both overland and overseas [underseas]) across which the gas must be moved; (c) environmental factors directly associated with the gas transportation mode; (d) the physical characteristics of the gas to be transported, notably, the phase — whether gaseous or liquid; and (e) the construction and projected operating costs of the transportation system, based upon trading off the advantages and limitations over which some flexibility of selection may be present. Aside from economic factors, a system can be engineered to transport either the gaseous or liquid phase, thus giving rise to considerable flexibility in certain situations.

Overland Pipelines. Detailed maps of gas pipelines in the United States and other parts of the world can be found in several references, particularly among the periodicals serving the pipeline industry. Notable among these references is the international petroleum encyclopedia and atlas issued periodically by Petroleum Publishing Co., Tulsa, Oklahoma. Numerous

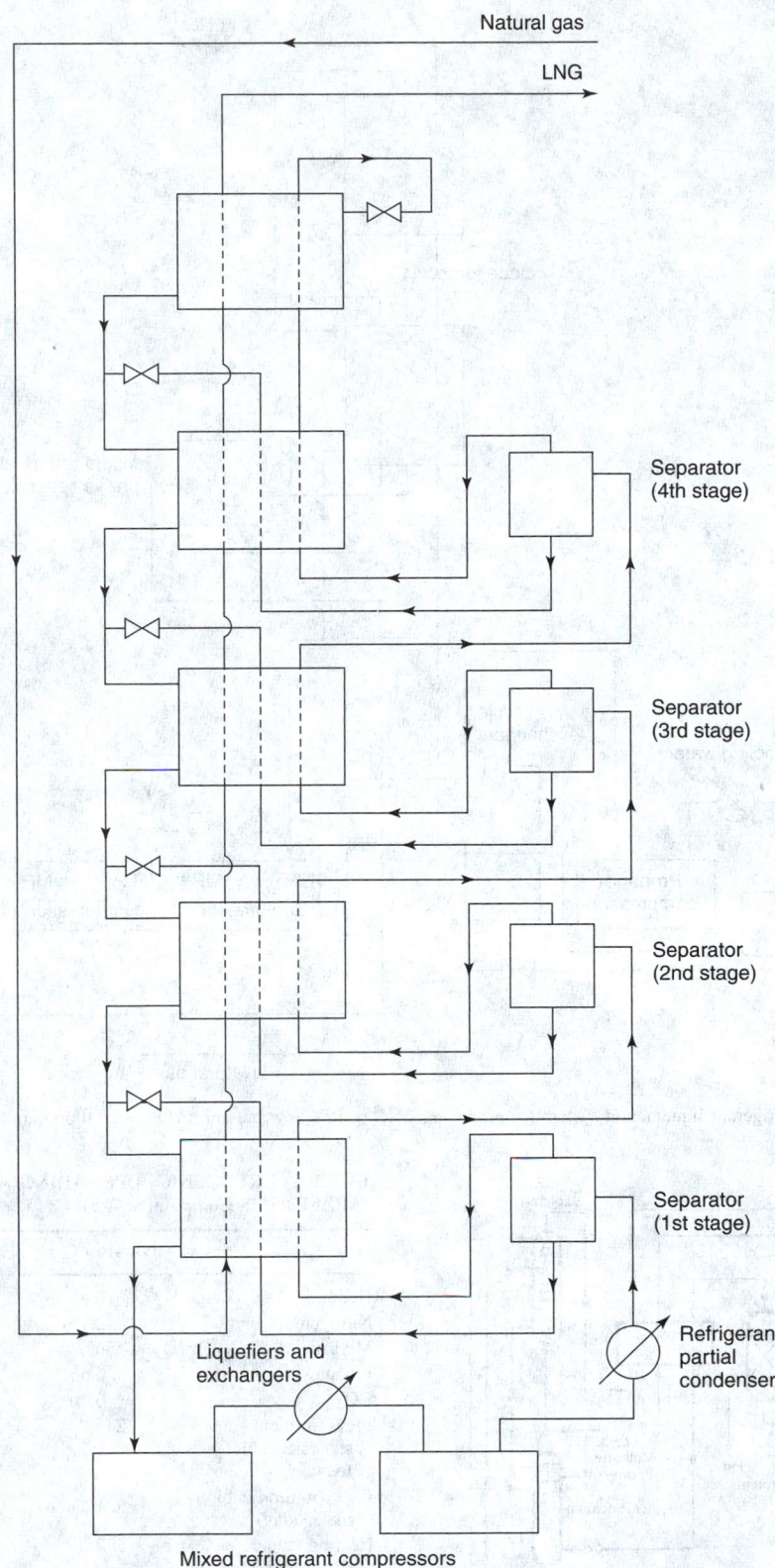

Natural gas

LNG

Separator
(4th stage)

Separator
(3rd stage)

Separator
(2nd stage)

Separator
(1st stage)

Refrigerant
partial
condenser

Liquefiers and
exchangers

Mixed refrigerant compressors

Fig. 4.   Single-pressure, mixed refrigerant cascade system for producing LNG.

trade associations serving the pipeline industry are also excellent sources on pipeline statistics. There are so many pipelines that presentation of this type of information is beyond the scope of this encyclopedia.

Historically, Texas, Louisiana, Oklahoma, and New Mexico have been large producers of natural gas, as well as some significant fields in the West Virginia-Ohio-Pennsylvania area. New developments in Alaska are and will continue to influence the gas transportation and distribution pattern.

Much of the installed gas pipeline ranges from 14 to 30 inches (36 to 76 centimeters) in diameter, the most common ranging from 20 to 35 inches (51 to 89 centimeters) but there is a strong trend toward larger-diameter lines, from 42 inches (107 centimeters) upward. Line pipe is made from high-strength plates, 3/38 inch to 1 inch (1 to 2.5 centimeters) in thickness. Sections of pipe are usually 40 feet (12 meters) long, minimum, ranging up to 60 or 80 feet (18 or 24 meters). Lengths of pipe

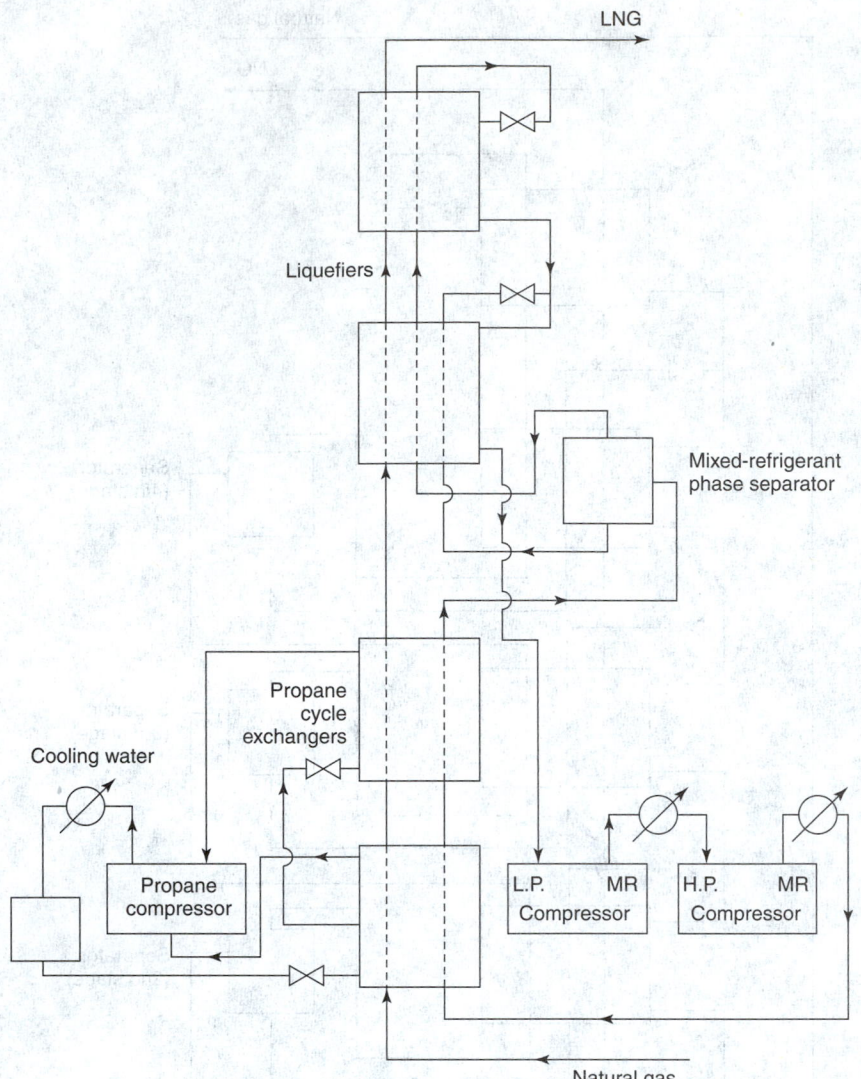

LNG

Liquefiers

Mixed-refrigerant phase separator

Propane cycle exchangers

Cooling water

Propane compressor

L.P.   MR
Compressor

H.P.   MR
Compressor

Natural gas

Fig. 5. Propane-mixed-refrigerant liquefaction system for producing LNG. L.P. = low pressure; H.P. = high pressure; MR = mixed refrigerant.

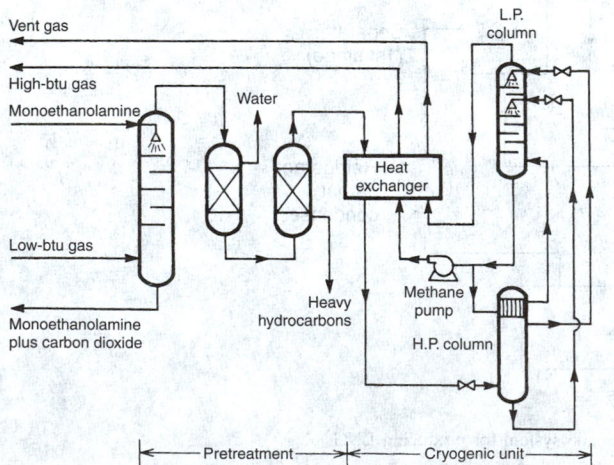

Vent gas

High-btu gas

Monoethanolamine

Water

L.P. column

Low-btu gas

Heat exchanger

Methane pump

Monoethanolamine plus carbon dioxide

Heavy hydrocarbons

H.P. column

|← Pretreatment →|← Cryogenic unit →|

Fig. 6. Plant for nitrogen removal from natural gas using cryogenic upgrading. (*Petrocarbon Developments, Ltd.*)

TABLE 2. CRYOGENIC UPGRADING OF NATURAL GAS — STREAM PARAMETERS Composition — Mol.%

	Low-Btu Gas	High-Btu Gas	Vent Gas	Helium
Helium	0.40	—	0.09	100.00
Nitrogen	42.75	4.00	98.95	—
Methane	56.02	95.09	0.96	—
Ethane +	0.53	0.91	—	—
CO_2	0.30	—	—	—
Flow (million standard cubic feet/day)	246	143	100	0.43
Flow (million cubic meters/day)	7	4	2.8	0.012
Heating value (Btu/standard cubic foot)	580	980	—	—
Heating value (Calories/cubic meter)	5162	8722	—	—

arrive at the scene most often by truck and are strung out by special pipe carriers along the right-of-way so that the construction crews will find them near the place where they are to be installed. Helicopter delivery of pipe is sometimes used where it is impossible for trucks to do the job.

The total weight of steel going into a long-distance pipeline is impressive. For example, a pipe with a wall thickness of 1/2 inch (13 millimeters) and a diameter of 30 inches (76 centimeters) will weigh more than 400 tons (360 metric tons) per mile.

In building very long pipelines, the pipeline company usually employs several construction contractors. The total length of line is divided into a number of sections with separate equipment and crews. Usually, each crew works on not more than 100 miles (161 kilometers). By partitioning the construction task, the entire operation can be speeded up, particularly important in areas where freezing temperatures or rain and mud may interfere with the work.

The numerous machines needed to dig the trench, weld the sections of pipe, apply protective coating to prevent corrosion, lower the pipe into the trench, and cover the trench with earth are known collectively as a *main line spread*. The trench is usually 3 or more feet (1 meter or more) in depth, sufficiently deep to prevent damage by plowing and earth-moving equipment. Depending on the size of the pipe, the trench will range from 2 to 4 feet (0.6 to 1.2 meters) or more in width.

Teams of welders join the pipe sections into a continuous tube. The most modern welding techniques involve automatic welding machines. X-ray equipment is used to inspect welds. When several sections of pipe have been welded together, the continuous tube is lowered gently into the trench by *sideboom tractors*. These machines have cranes or derricks slanted over to one side so that they can pick up the pipe and lower it several feet away from the tractor itself. Pipe purchased from steel mills may come with a coating and wrapping already applied. The thick coating may be of coal tar or asphaltic material, which is then covered with heavy paper or fiberglass. This protective coat-and-wrap is needed to prevent rusting. If bare pipe is used, there are special machines that coat and wrap right on the job just before the pipe is lowered into the trench.

A special piece of equipment, known as the *holiday detector*, is a hoop of metal placed around the pipe after it is coated and wrapped. A small electrical current flows through the hoop. If there is a "holiday," i.e., a spot where there is no coating, the detector alerts the operator. This is brought to the attention of a special crew that coats and wraps bare spots in the pipe.

Since pipelines do not follow an absolutely straight line, bending machines are used to curve the pipe in the vertical, horizontal, or both directions. When a pipeline must cross a river, the contractor will dig or dredge a deep trench in the river bed. The pipe is then surrounded by heavy weights and encased in concrete so that it will not be carried away by the current. If there is a suitable bridge across the river, the pipeline may be hung from the underside of the steel girders of the bridge. In some cases, a special bridge is constructed to carry the pipe across the stream. In crossing a highway or railroad, the pipeline must be put through a tunnel under the structure. A giant auger will be used to bore under the road to accommodate a section of somewhat larger-diameter pipe, forming the tunnel through which the main pipeline passes.

Gas pressures in long-distance pipelines may range from 500 to 5,000 pounds per square inch (34 to 340 atmospheres) with 1,000 psi (68 atmospheres) being quite common. Pressure is boosted to make up for frictional losses by use of compressor stations located every 50 to 100 miles (80 to 161 kilometers) along the pipeline. In terms of lineal velocity, natural gas may travel at a rate of about 15 miles (24 kilometers) per hour; thus, about three days are required to move a molecule of gas over a distance of 1,000 miles (1609 kilometers).

All along the pipeline, there are valves and regulators that may be opened or shut to control the internal pressure, or to cut off the flow entirely if an unexpected break in the line is caused by a flood, earthquake, or other disaster. The valves and regulators can be operated by microwave radio long before any crew could reach them. Stations for reducing the pressure, located near points of consumption, frequently are called *city gates*. These stations measure the amount of gas leaving the main pipeline at this point as well as reducing the pressure.

Marine Pipelines for Gas

With some alterations, the techniques that apply to construction and laying of marine pipelines for gas also apply to fluids, such as oil. Marine pipelines can be underwater in a river, marsh, or ocean, but the predominant industry effort in recent years is the construction of pipelines in the open ocean at increasingly deeper levels. The trend toward deepwater pipelining and construction in harsher environments naturally follows the expansion of the search for offshore gas and oil. This search began in earnest after World War II and is expanding at an ever-increasing rate; even if slowed to some

extent by some environmental concerns in the United States, the rate is rapid in other parts of the world. Worldwide energy needs have caused oil companies to move into areas that only a few years ago would have been too expensive to develop on a practical basis. Lines are now being laid in water depths of several hundred feet (meters) and cover distances of 200 miles (320 kilometers) or more from field to shore. These longer lines are major trunk lines bringing gas and oil to land terminals. Other lines are necessary out at the field to connect platforms to each other; or possibly to connect platforms to sea berths.

The sizing of the pipeline, the design of the pumping and compression systems needed to move the products, the design of the automation systems, and many of the corrosion control procedures are the same regardless of whether the pipeline is on land or at sea. The two major areas of design difference between land and marine pipelines are (1) the stresses incurred in getting the pipeline to the sea bottom, and (2) the necessity of keeping the line stable and in place while it is exposed to forces induced by current and wave.

The stability problem is theoretically simple but is complicated somewhat by the uncertainty of precise values for some of the coefficients used in the calculations. Basically, it is a matter of providing enough weight in the pipe and pipe coating system to provide a net downward force when balanced against the buoyance and the lift force caused by the seawater moving by the pipe. This net downward force, in conjunction with the coefficient or friction for the particular pipe-soil combination under examination, can then mobilize a horizontal resisting force. This should be somewhat larger than the drag force exerted on the pipe by the water motion in order to give the desired safety factor.

Different safety factors or horizontal water velocities may be utilized depending on the operating conditions that will be encountered during the life of the pipeline. For example, many pipelines will be buried beneath the sea bottom at some time interval, ranging from a few weeks to a year or two, after their construction. The exposure of this line to maximum horizontal water velocities caused by storm current and waves is obviously much less than that of a line that will remain on the surface of the sea bottom. It is also obviously necessary to consider whether the line will contain gas, oil, or other substance at the time the design loads may occur.

It is important to carefully consider the foregoing points in the design of the weight coating since the ability of a contractor to safely construct the line relates very closely to the negative buoyancy of the pipe and coating.

The most common method of marine pipeline construction utilizes a floating vessel on which the pipe is assembled in a horizontal position. As additional joints or sections of pipe are added to the already-completed segment, the barge is moved forward, actually moving out from under the completed pipeline. This is sometimes called the "stovepipe" method, named after the manner the pipe sections are added, one after another. This pipeline extends off the stern of the vessel and spans down to the sea bottom. It is supported part of the way down by a construction aid called a "pontoon" or "stinger." This is basically a slender structure pinned to the vessel on one end and with built-in buoyancy that can be controlled so that it floats at the proper angle to the water surface to provide support to the pipeline.

In shallow water, the pipeline is then allowed to span from the end of the pontoon to the sea bottom as a simple beam. As water depths increase, it becomes necessary to add tension to the pipe on the barge. This, of course, changes the analytical problem from one of a simple beam to one of a beam under tension. This analysis must take into account the weight of the pipe, the wall thickness, the type of steel in the pipe, the tension on the pipe, the support of the pontoon, the geometrical configuration of the tension on the pipe, the geometrical configuration of the pipe-pontoon-barge system, and the pipe end condition at the sea bottom.

There are three basic configurations of pipelay vessels in common usage: (1) the barge-type hull; (2) the ship-shape hull; and (3) the semi-submersible vessel. The barge-type hull (Fig. 7) is the most common because of its economy and simplicity, its ability to provide the space and stability for heavy lifts and deck cargo, including pipe, and its shallow draft, permitting work close inshore. The primary disadvantage is its relative sensitivity to sea conditions. In particular, roll and heave motions will shut down pipelay operations in 6-foot to 14-foot (1.8- to 4.2-meter) waves, depending on wave direction and period.

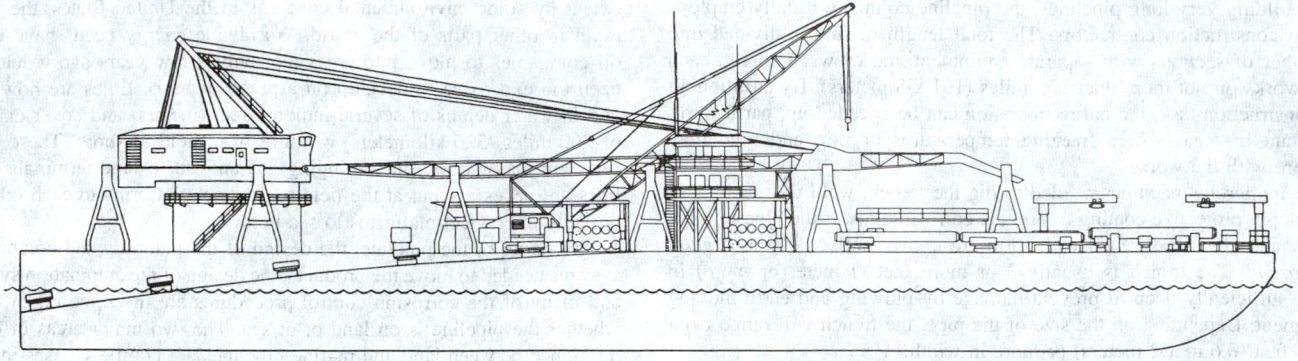

Fig. 7. Pipelay vessel with barge-type hull.

Overseas Shipping of LNG. A key feature of most LNG carriers in operation is the insulation system, which maintains the cargo at −162 °C. In one type of ship, the cargo is carried in five tanks constructed of a thin welded membrane of special steel. Each tank is separated from the inner hull by insulating material. The small fraction of the cargo that boils off because of heat leakage is used as boiler fuel for the propulsion of the ship. On a loaded voyage, this may provide about 90% of the fuel needed. The ships are ballasted for return voyage with seawater carried in separate wing tanks. Some LNG is left in the cargo tanks to ensure a nonexplosive gaseous atmosphere and to keep the tanks cool for the next voyage. Again, boil-off gas provides part of the propulsion fuel. One configuration of an LNG ship-loading system is shown in Fig. 8.

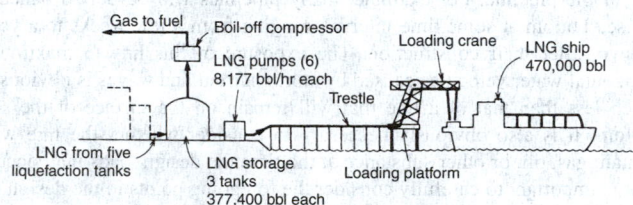

Fig. 8.   LNG ship-loading system. Trestle is 2.6 miles (4.3 km) long.

Safety in handling LNG in ships and at loading and unloading terminals of large scale has been a matter of constant concern. The observation has been made that the LNG gas carried would bury a football field under 125 feet (38 meters) of liquefied gas, or, after conversion to the gaseous phase, 600 football fields to the same depth. One factor that has not been routinely considered in the past is a phenomenon called a *flameless vapor explosion*. It is well known that, if water, for example, could be heated without nucleation occurring on the sides of the vessel, the water temperature could be raised well above the boiling point of 100 °C. If this could be done, and with the continued application of heat, the liquid would suddenly explode in its transition from the liquid to the vapor phase. Although not probable, it is possible that conditions favoring flameless vapor explosion could occur if liquefied natural gas were permitted to escape over a water surface. One scientist has observed that an explosion of this kind is possible when a liquid is 4 to 6% (no more, no less) above its normal temperature of vaporization. It is further observed that an explosion of this nature would not occur when LNG first spreads across a volume of water, but with time and the warming of the LNG such a hazard could occur. While an explosion of this type is not comparable to that from a chemical reaction, the explosion could greatly disperse the LNG over a greater area, thus spreading the zone of risk.

Underground Storage

The largest additional supply of natural gas for peak demands comes from underground storage reservoirs located, for example, close to the northern cities, as compared with the producing wells which may be located in the southwestern area of the country. Some of the storage pools are operated by pipeline companies, but most of the gas in underground storage is owned by the local gas companies that serve metropolitan areas.

The underground reservoirs are filled with gas from the pipelines during the summer months, when all of the fuel that the lines can deliver is not consumed. This method allows the producing wells and the pipelines to operate at fairly steady rates at all times of the year. Also, it is established that a gas field will produce more gas over a longer period if the gas is withdrawn at a steady rate.

Four states — Michigan, Pennsylvania, Illinois, and Ohio — have half of the total underground gas storage pools in the United States, with a total capacity of over 5 trillion cubic feet (142 billion cubic meters). In a typical year, about one-fourth of this volume will be used during cold waves to furnish the additional gas needed to supply homes and apartments.

The most common type of underground reservoir now storing gas is a previously producing gas or oil field. The supplies remaining in these pools are too small, and at too low a pressure, to justify continued production. But, the reservoir rock can hold gas pumped down through the same wells that once took gas out of the ground.

About 90% of the storage pools being used once produced gas or oil. In Pennsylvania there are over 60 such pools close to the large industries and centers of population. There are over 30 such pools in West Virginia, Michigan, Ohio, Kansas, Indiana, New York, and Kentucky, as well as smaller numbers in 13 other states. The gas to be stored is pumped into the old wells by compressors similar to those used to move gas in pipelines. The gas is stored under about the same pressure as originally existed in the field. In developing a gas storage reservoir, a company obtains a lease from the landowners in much the same manner that gas producers do.

The gas industry has been developing underground storage reservoirs for more than 60 years. The first known experiment in storing gas underground was conducted in 1915 in Welland County, Ontario, Canada by the National Fuel Gas Company. The success of this effort prompted the Iroquois Gas Corporation, a subsidiary of National Fuel, to develop, in 1916, the Zoar field south of Buffalo, New York. It was the first storage operation in the United States and is the oldest continuously used reservoir.

During the past 60 years, over 80 companies have invested several billions of dollars in underground storage facilities.

Another kind of underground storage reservoir is called an aquifer. An aquifer is an underground rock structure holding large quantities of water. The underground rock is porous and permeable. The pore spaces are filled with water, and impermeable rock covers the porous rock. Wells are drilled into such formations, and gas is forced into the pores under pressure. As the gas pressure increases, the gas pushes the water farther down into the porous rock, making room for the gas.

There are over 40 aquifers in the United States, located in Illinois, Indiana, Iowa, Kentucky, Minnesota, Missouri, Utah, and Washington. Three unusual reservoirs have been developed: an abandoned coal mine in Colorado; and salt domes in Michigan and Mississippi.

History of Natural Gas as an Energy Resource

It is reported that, perhaps 2,000 years ago, the Chinese piped natural gas from shallow wells through bamboo poles, for burning under large pans to evaporate seawater for salt. The first commercial use of natural gas in the western world was for lighting the streets of Genoa, Italy, circa 1802. The first evidence of natural gas deposits in the United States is found in reports of "burning springs" in various parts of New York, Pennsylvania,

Ohio, and West Virginia. As early as 1626, French missionaries visiting the Indians in northwestern New York recorded that they could ignite gases rising from shallow waters. Many early reports were given of the presence of natural gas along the shores of Lake Erie and in the streams flowing into it. There also are references to "burning springs" in the Ohio River valley and along the Pacific shores of California. It is reported that General George Washington was fascinated by a "burning spring" in the Kanawha Valley, near Charleston, West Virginia, in 1775. Early settlers who drilled wells for water often reported the presence of traces of natural gas. The generally accepted birthplace of the natural gas industry in the United States is Fredonia, New York. Fredonia is located on Canadaway Creek, which empties into Lake Erie in the northwest corner of New York State. William A. Hart is reported to have dug a well in 1825 and obtained sufficient natural gas to light two stores, two shops, and a grist mill. Hollow logs were used for piping. Sufficient gas would accumulate in the well riser during the day to supply the gas lights at dusk. Hart was also instrumental in building the first natural gas lighthouse in 1829 along Lake Erie. The lighthouse, consisting of 13 gas lamps and reflectors in two tiers, served until 1859. In 1858, the first natural gas company in the United States was formed, The Fredonia Gas Light Company.

The consumption of natural gas gradually increased prior to World Wars I and II as more and more small pipelines brought communities within reach of natural gas fields accompanied by the retirement of previous manufactured or town gas facilities (the early forerunners of the substitute natural gas).

Natural Gas–powered Vehicles. The concept of natural gas–powered vehicles has become a *limited* reality in terms of the millions of gasoline- and diesel-fluid-power vehicles. In 1993, Mack Trucks (Allentown, Pennsylvania) and the Gas Research Institute have teamed to research and develop a natural-gas version of the Mack E7™ heavy-duty engine. A prototype of the design was scheduled for testing on a refuse vehicle in the Boston area. If successful, the engine also could be applied to a variety of heavy-duty vehicles, including long-haul tractor/trailers, construction equipment, and road maintenance trucks. The development was propelled by the needs of truck fleet owners who may be required by legislation to operate alternatively fueled vehicles. The engine will be required to meet applicable U.S. Environmental Protection Agency and California Air Resources Board emissions standards while maintaining the performance and reliability of its diesel-fueled counterpart. A 6-cylinder, 12-liter engine will be developed. The vehicle's onboard gas storage will hold the energy equivalent of about 45 gallons (170 liters) of diesel fuel.

Substitute Natural Gas. The oil crisis of the 1970s spawned a number of attempts to create synthetic natural gas. Some of these processes reached pilot and demonstration plant stages and beyond in their development. For example, substitute natural gas (SNG) from sewage wastes has enjoyed impressive success. The anaerobic digestion of a solid waste and water or sewage sludge slurry will produce a methane-rich gas.

Additional Reading

Abelson, P.H.: "The Gas Research Institute," *Science*, 1715 (December 11, 1992).

Bethke, C.M. et al.: "Supercomputer Analysis of Sedimentary Basins," *Science*, 261 (January 15, 1988).

Burnett, W.M. and S.D. Ban: "Changing Prospects for Natural Gas in the United States," *Science*, 305 (April 21, 1989).

Castaneda, C.J.: *A History of the Natural Gas Industry*, Macmillan Library Reference, New York, NY, 1999.

Caton, J. (Editor): *Alternative Fuels and Natural Gas, Volume 3*, American Society of Mechanical Engineers, New York, NY, 1995.

Considine, D.M.: *Energy Technology Handbook*, The McGraw-Hill, Companies, Inc., New York, NY, 1977.

Fischetti, M.: "There's Gas in Them Thar Hills!" *Technology Review (MIT)*, 17 (January 1993).

Fulkerson, W., R.R. Judkins, and M.K. Sanghvi: "Energy from Fossil Fuels," *Sci. Amer.*, 136 (September 1990).

Holtberg, P.: *1993 Policy Implications of the GRI Baseline Projection of U.S. Energy Supply and Demand to 2010*, Gas Research Institute (Washington Operations), Washington, DC, 1993.

Jensen, B.A.: "Improve Control of Cryogenic Gas Plants," *Hydrocarbon Processing*, 109 (May 1991).

Lyons, W.C.: *Standard Handbook of Petroleum and Natural Gas Engineering*, Vol. 1, Butterworth-Heinemann, Inc., Woburn, MA, 2001.

Lyons, W.C.: *Standard Handbook of Petroleum and Natural Gas Engineering*, Vol. 2, Butterworth-Heinemann, Inc., Woburn, MA, 2001.

McCabe, K.A., S.J., Rassenti, and V.L. Smith: "Natural Gas Pipeline Networks," *Science*, 534 (October 25, 1991).

Melvin, A.: *Natural Gas: Basic Science and Technology*, Adam Hilger (London), Taylor & Francis (Philadelphia), 1988.

Ourisson, G., P. Albrecht, and M. Rohmer: "The Hopanoids: Paleochemistry and Biochemistry," *Pure and Applied Chemistry*, **51**(4), 709–729 (April 1979).

Ourisson, G., P. Albrecht, and M. Rohmer: "Predictive Microbial Biochemistry: From Molecular Fossils in Procaryotic Membranes," *Trends in Biochemical Sciences*, **7**, 236–238 (1982).

Ourisson, G., P. Albrecht, and M. Rohmer: "The Microbial Origin of Fossil Fuels," *Sci. Amer.*, 44–51 (August 1984).

Sharer, J.C. and P. O'Shea: "Gas Research Institute's Research Program on Unconventional Natural Gas," *Chem. Eng. Progress*, (February 1986).

Sweetser, R.: *The Fundamentals of Natural Gas Cooling*, Prentice-Hall, Inc., Upper Saddle River, NJ, 1997.

Willett, R. (Editor): *1996 Natural Gas Yearbook*, John Wiley & Sons, Inc., New York, NY, 1995.

Woods, T.J.: *The Long-Term Trends in U.S. Gas Supply and Prices: 1992 Edition of the GRI Baseline Projection of U.S. Energy Supply and Demand to 2010*, Gas Research Institute, Chicago, Illinois, December 1991.

Web References

Gas Research Institute: http://www.gri.org/

The American Petroleum Institute: http://www.api.org/

NATURAL HYDROCOLLOIDS. See **Food Additives**.

NATURAL RUBBER. See **Rubber (Natural)**.

NAUTICAL MILE. The fundamental unit of distance used in navigation, once defined, for purposes of convenience, as 6,080 feet (1,853 meters). Rigorously, the nautical mile was defined as the length of 1 minute of arc on a great circle drawn on the surface of a sphere with the same area as the earth; thus as 6,080.27 feet (1,853.27 meters).

But, owing to the fact that the earth is an oblate spheroid and flattened at the poles, the length of 1 minute of arc measured along a meridian varies in different latitudes. It is shortest at the poles and longest at the equator, having an average length of 6,076.82 feet (1,852.21 meters). To resolve the confusion caused by this variation of the nautical mile with latitude, various countries established standard figures. At last an international agreement was reached whereby the nautical mile was defined, and is presently accepted, as 1,852 meters. For navigation purposes, a good approximation for the number of nautical miles is the number of minutes of arc along a great-circle route.

The nautical mile is frequently confused with the *geographical mile*, which is defined as the length of 1 minute of arc on the earth's equator and has a length of 6,087.15 feet (1,855.36 meters). See also **Units and Standards**.

NAUTICAL SYSTEM. See **Meteorology**.

NAUTICAL TWILIGHT. That period when the upper limb of the sun is below the visible horizon and the center of the sun is not more than 12 degrees below the celestial horizon.

NAVIER-STOKES EQUATIONS. The equations of motion for a viscous fluid that may be written

$$\frac{d\mathbf{u}}{dt} = -\frac{1}{\rho}\nabla p + \mathbf{F} + v\nabla^2\mathbf{u} + \frac{1}{3}v\nabla(\nabla \cdot \mathbf{u}),$$

where p is the pressure, the density, $\mathbf{F}$ the total external force, $\mathbf{u}$ the fluid velocity, and v the kinematic viscosity.

For an incompressible fluid, the term in $\nabla \cdot \mathbf{u}$ (divergence) vanishes and the effects of viscosity then play a role analogous to that of temperature in thermal conduction and to that of density in simple diffusion. Solutions of the Navier–Stokes equations have been obtained only in a limited number of special cases; in atmospheric motion, the effects of molecular viscosity are usually overshadowed by the action of turbulent processes and the Navier–Stokes equations have been of little direct application. The use of the concept of eddy viscosity has overcome this limitation in certain problems. The equations are derived on the basis of certain simplifying assumptions concerning the stress tensor of the fluid; in one dimension they represent the assumption referred to as the Newtonian friction law.

See also **Ekman Spiral**; **Fluid and Fluid Flow**; and **Newtonian Friction Law**.

NAVIGATION. Since development of the early navigation satellites (INMARSAT, et al.) just a few decades ago, the approach to the science and equipment of navigation has undergone a major anatomical change that is geared to ultrasimplification for the end user. The operator no longer requires sophisticated support instrumentation or an understanding of and appreciation for the geometry and mathematics that have typified pre-satellite navigation systems. Refinements of navigation satellite systems, as exemplified by the Global Positioning System (GPS), first applied by the military during the Persian Gulf War (1990–1991), are occurring apace and ultimately to a significant extent will obsolete many of the early and contemporary navigation systems. Because of the installed investment in contemporary systems, possible security restrictions that may be imposed on GPS during wartime, and the ability of some users throughout the world operating on limited budgets, the time of phaseout for contemporary non-satellite–based systems is unknown. Further, there is much interest in the lore of navigation methodologies. This article incorporates, as in prior editions, a chronology of navigation system developments. The article commences, however, with a condensed overview of the GPS system.

Space-Age Triangulation

In the Global Positioning System, satellites with precisely synchronized atomic clocks continuously broadcast their time and location to suitable earthbound computerized receivers. In a fully self-contained manner, the receiver calculates and displays its Earth position. Data received from at least three satellites is required to yield the *latitude*- and *longitude* of the receiver. Data from a fourth satellite is required to yield *altitude*. The receiver is synchronized to the satellites' clocks. Distance from the receiver to the satellite is calculated by measuring the time interval required for the signal to travel from satellite to receiver. When three satellites are in "view" simultaneously, a computer in the receiver quickly works out the precise position by triangulation because only one point on a plane can yield a particular combination of distances from the three satellites. See Fig. 1. Receivers are small and may be handheld.

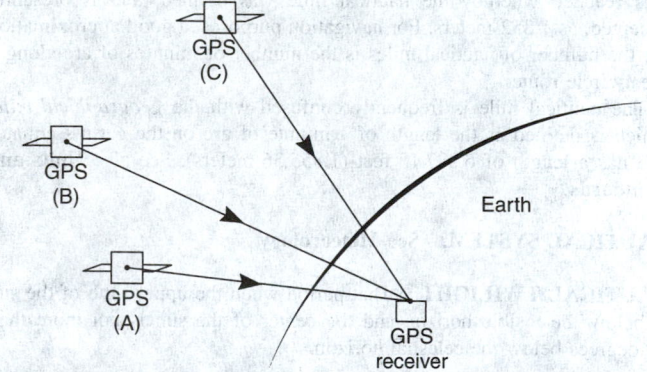

Fig. 1. Global Positioning Satellites (A, B, C), each equipped with a precisely synchronized atomic clock, continuously broadcast their time and location to earthbound receivers. Computers in the receiver calculate distances of receiver from each satellite, this information derived from the time required for transmission of signal from satellite to receiver. By employing the methodology of *triangulation*, the receiver determines the only *acceptable location* that will geometrically satisfy the signal time differences. Only three GPS readings are required to determine *latitude* and *longitude*. Additional data from a fourth GPS will yield *altitude* information. The GPS thus combines advanced satellite technology with simple geometry. See **Satellites (Communications and Navigation)**.

Quite a large number of satellites in orbit are needed to ensure that at least three satellites are in sight at any particular Earthpoint and four in view if altitude information is required as well. As of 1994, the U.S.

military intended to have 24 satellites in place, including three spares. Each satellite orbits Earth once every 12 hours at an altitude of 10,900 miles (17,538 km). The ultimate total cost of the GPS system is estimated at $10 billion upward.

Because of the advantages of the GPS to air, sea, and ground operations during the U.S. military venture Desert Storm, equipment was released for use in the field at least 2 years prior to an established schedule. The GPS markedly increased the accuracy of bombing runs, accurately guided sea-launched missiles to targets, steered ships and personnel safely through mine fields, and advised tank and land-based commanders with reliable information on their exact location even in featureless desert scenarios.

Had the enemy been aware of more detail receiver design and had developed suitable receivers, they could have taken advantage of the GPS. Subsequent analyses of the various encounters did not reveal, however, any enemy use of the system. But this possibility in terms of future military actions has prompted the U.S. military to offer GPS information on two levels by way of an arrangement known as *differential GPS*. For most military uses of GPS, the maximum available accuracy is needed. Reports on early design of the system indicated that accuracy could be expressed in terms of a few millimeters rather than meters. An early satellite-based system had provided scientists in various fields, notably in connection with movement of Earth's crust, with a *fine signal* in the millimeter range.

Although subject to future alteration, the military developed the concept of *selective availability*—that is, (1) a fine and very accurate position reading, and (2) a course location fix, the accuracy of which would be in the range of 100 meters. In this system, a slight distortion of the satellite signal is introduced but still permits a finer, undistorted signal for specially designed receivers for military applications. This approach basically denies location information of the best achievable accuracy to scientists and for a host of demanding applications of the system for commercial use. One scientist has observed, "The military has come up with a system that is useful to mankind. Now they don't want mankind to have full use of it." This scientist is engaged in studies for predicting seismic risks.

For example, commercial uses that require information of high accuracy are found in the monitoring of railway cars on tracks that are quite close to each other and for navigating ocean-going vessels through rock-infested straits where, at a minimum, an accuracy of a few meters is required.

Some scientists have developed electronic means to circumvent the signal distortion problem by using schemes that are identified as *differential GPS*. In one version of this technique, GPS information from numerous satellites and using several receivers will cancel out minor errors. One scientist has observed, "We've learned to live with selective availability, but the solution is costly, by perhaps an order of magnitude increase in the GPS readings required. This in turn is reflected by increases in the costs of data storage, transmission, and analysis."

Although GPS is a major breakthrough for navigation and position-related analyses, the system, like communications satellites, can be plagued with satellite instability problems, faulty antenna pointing, and failure of solar panels for power, among other factors.

The early work required to develop navigation satellites, dating back to the 1970s and ultimately leading to the GPS, is described in considerable detail in the Seventh Edition of this encyclopedia.

Classification of Navigation Sensors

The term *navigation* stems from the Latin word *navigare*, -meaning "to conduct a ship." Traditionally, navigation has been treated under three main classifications: (1) the *sailings*—to find the direction or directions in which to head a ship in order that it may proceed on the most practical course from one place to another (see also **Sailings (The)**), (2) *dead reckoning*—to find the position of a ship in any particular instant, provided that its position at some prior time is given and that, since leaving this position, the headings, distances run on each heading, and the effects of environment and motions of the medium supporting the ship are known (see also **Dead Reckoning**), and (3) *celestial navigation*—to find the position of a ship at any instant by means of observations, not necessarily visual, of one or more objects on the surface of the earth, or on the celestial sphere. Although still a helpful classification, the foregoing obviously is dominated by consideration of ships at sea, not taking into full account air and space navigation. Also, within the last few decades, navigational instrumentation has been put onto a much more sophisticated footing so that a classification of the type shown in Fig. 1 is much more useful.

Further, there are two broad categories of navigation: (1) *absolute navigation*, wherein knowledge of present position is known in relation to an overall earth-coordinate system (latitude and longitude, for example); and (2) *relative navigation*, wherein position is known relative to some special local coordinate or grid system. The accuracy attainable in a relative navigation situation usually is significantly better than that which can be obtained on an absolute basis.

Navigation Sensors

The primary navigation sensors and their functions are listed in Table 1. The more commonly used sensors for both marine and aircraft navigation are included. Frequently, a navigation system will be made up of various combinations of these sensors.

TABLE 1. NAVIGATION SENSORS BY PRIMARY FUNCTION

Meters

Velocity Meters	Altitude or Depth Meters
Pilot log	Pressure altimeter
Electromagnetic log	Pressure-depth meter
RPM tachometer	Radar altimeter
Airspeed (pilot) meter plus wind	Sonar fathometer
Doppler radar	Inertial systems
Laser radar	Ground or other vehicle
Doppler sonar	Trackers and data link
Inertial systems	

Time Meters

Chronometer
Radio-synchronous signals
Time standards (crystal oscillators)

Heading Reference

Magnetic compass	Radiometric celestial tracker
Gyro compass	Inertial systems
Optical celestial tracker	

Position-fix Devices

Sextant	Navigation satellite	Radio transponders
Radio aids	Landmarks	Ground or "mother" ship
LORAN	Optical sighting	Trackers and data link
OMEGA	Radar sighting	Gravity anomaly, map matching
SHORAN	Seamarks	Magnetic anomaly, map matching
RAYDIST	Sonar bottom fixing	
DECCA	Optical celestial tracker	
VOR/DME		
TACAN	Radiometric celestial tracker	

As will be noted from Fig. 2, there are self-contained navigation systems (self-processed information) and externally-controlled systems wherein a data link is required between the vehicle (aircraft, ships, etc.) being navigated and some reference or data point. An abridged list of examples of these two general classes of systems is given in Table 2.

Many factors are involved in selecting the most appropriate navigation system, including the usual parameters of accuracy, cost, size, weight, power requirements, reliability, operational simplicity/complexity, degree of automaticity desired, ruggedness in environmental extremes—factors which have given rise to numerous standards of equipment. Other very important factors for many situations include, for example, worldwide availability, all-weather operability, and continuous versus periodic availability. Certain radio navigation systems, for example, may not embrace

TABLE 2. SELF-CONTAINED AND EXTERNALLY CONTROLLED SYSTEMS

SELF-CONTAINED SYSTEMS
— Completely self-contained
— Self-processed information; no radiation required.
   All measurements internal or in local vicinity of vehicle, such as inertial navigation.
— Self-processed information
— Vehicle is passive; natural earth or natural sky references.
   Natural can include constructed landmarks (buildings, etc.) not purposely placed as navigation aids.
   Optical star trackers.

SELF-CONTROLLED SYSTEMS
— Self-processed information; vehicle passive; earth passive or artificial
— Purposeful navigation landmarks.
   Navigational buoys.
— Self-processed information; vehicle passive; active earth or active sky

LORAN
— Self-processed information; vehicle active; earth passive or artificial; or artificial celestial
   Ranging to lead ship.
— Self-processed information; vehicle active; earth active
   Radar transponders.

EXTERNALLY CONTROLLED SYSTEMS
— Ground or other vehicle processed information and data link; vehicle passive; external tracker passive
   Theodolite optical tracker.
   Skin tracking radar.
— Ground or other vehicle processed information and data link; vehicle active; external tracker passive
   Optical tracking of light on vehicle.
— Ground or other vehicle processed information and data link; vehicle active; external tracker active
   Beacon tracking radar.

Foregoing are examples only; not all-inclusive.

transmitters that cover the entire world. All-weather, night-and-day operation rules out, for example, such systems as optical star trackers (as sole sensors); clouds below an aircraft, for example, rule out laser doppler systems. In some navigational satellite systems, there may not always be a satellite within range, and a fix may have to await radio view when one comes into position. Thus, it is rare when a single navigation sensor can satisfy all desired requirements. Even magnetic compasses and normal-mode gyrocompasses are essentially unusable in the polar regions of the Earth.

Celestial Navigation

In this system, navigation is achieved by means of observing celestial objects. In the year 1837, Captain Thomas Sumner discovered what has since been known as the Sumner Line, and modern celestial navigation may be said to date from that discovery. The methods for determining terrestrial latitude and longitude from sextant observations of altitude of celestial objects are briefly described in **Sumner Line**. During intervening years, much research has been carried out on the theory of celestial line of position, and numerous methods for calculating the data necessary to plot that line have been developed.

It is in order to briefly describe what is meant by a celestial line of position. At any instant, any celestial object is directly at the zenith for some particular spot on the surface of the earth. This point in years past was known as the subsolar, sublunar, or substellar point, depending upon whether the object observed was the sun, the moon, or a planet or star. The term *Ground Position* (GP) is now used no matter what celestial object is used. The terrestrial coordinates of GP may be expressed in terms of celestial coordinates of the object, the latitude being equal to the declination of the object, and the longitude equal to its Greenwich hour angle. Both of these quantities are tabulated for the sun, moon, planets, and navigator's stars in various readily available almanacs. The tabulations are given in terms of Greenwich Civil Time, and this time is used by navigators for recording the times of sextant observations for altitude.

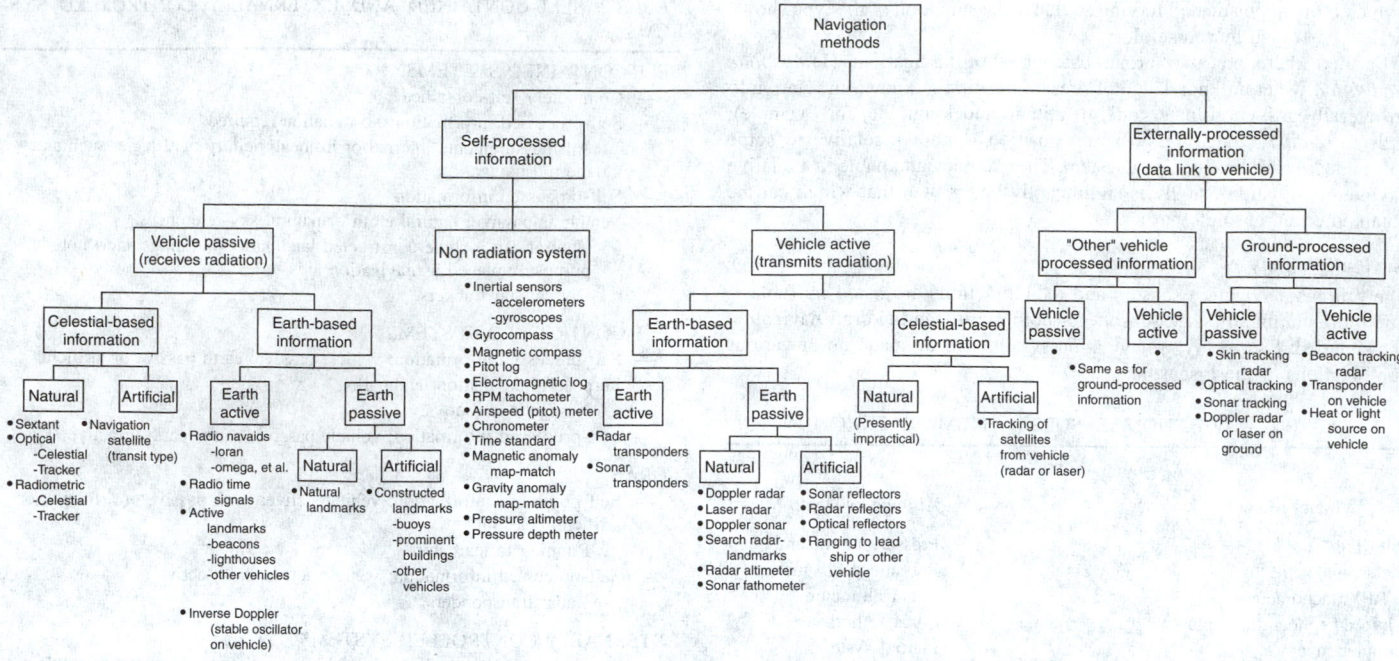

Fig. 2.   Classification of sensors used in navigation systems.

In practice, the line of position is drawn on a Mercator chart, a plotting sheet, or a small-area plotting sheet by employing the geometric proposition that a radius of a circle is always perpendicular to an arc. At some particular Greenwich Civil Time (GCT), the altitude (h_s) of a celestial object is obtained with the sextant or bubble octant. Then, using the dead-reckoning position, or some position close to it, which leads to simplified computations, the values of the altitude (h_c) and the bearing (Z_n) that the object would have at the assumed position and GCT of observation are computed or taken from suitable tables. The dead-reckoning, or assumed, position is now set down on the plotting sheet and a line drawn through it in the direction Z_n. This line is a section of the radius of a circle drawn about the GP, which, in reality, is usually off the plotting sheet. Next, the difference between the computed and observed altitudes ($h_c - h_t$) is taken; this is called the "intercept." If the intercept is zero, the ship must be on a line of position perpendicular to the bearing line and passing through the plotted position. If the intercept is plus (+), the line of position must pass through a point ($h_c - h_t$) minutes of arc, or nautical miles, away from the GP along the bearing line; and if the intercept is minus (−), the line must be between the plotted position and the GP. In either case, the line of position must be perpendicular to the bearing line.

In spite of the fact that the line of position is actually a circle with radius ($90° - h_t$) miles, the line may be drawn as straight in practically all cases. If we assume that the altitude of the object is 80°, the value of ($90° - h_t$) is 10° or 600 miles (965.6 kilometers). In this case, a straight line 60 miles long (96.56 kilometers), perpendicular to the radius, will differ from the actual circle by less than a mile at its extremities. Accordingly, the assumption of a straight line will not lead to appreciable errors, if the altitude is less than 80° and the drawn line is less than 60 miles (96.56 kilometers) long.

Simple statistical analysis shows that the point on the line of position closest to the dead-reckoning point is the most probable position that can be obtained for the ship from a single observation of altitude. In air navigation, this position is referred to as the estimated position (EP). Care must be taken not to confuse this EP with the EP obtained by dead-reckoning navigation in marine navigation. This most probable position, or EP, can be obtained without plotting the line by any of the above methods if the dead-reckoning (DR) position instead of the assumed position is used, since the intercept gives the shortest distance from the DR position to the line.

An example of the use of celestial navigation at sea is given in the following practical case:

During the night, the navigating officer of a ship on passage from England to the United States wishes to check the dead-reckoning position of the ship. The two stars Alpheratz and Altair are well placed for observation. When the navigating officer's watch reads 23 h 40 m 10.0 s, the sextant altitude of Alpheratz is 50° 34'.3. For the purpose of checking the deviation of the steering compass, the bearing of the star is taken by this compass and found to be 121°. At watch reading 23 h 46 m 15.4 s, the sextant altitude of Altair is 50° 20'.7. The watch times must be corrected to obtain Greenwich Civil Time (GCT), and the sextant altitudes corrected for instrumental errors, dip, and refraction to obtain the geocentric altitude (h_t). These corrected results are:

Star	GCT	h_t
Alpheratz	02h 39m 34.0s	52° 27'.4
Altair	02h 45m 39.4s	50° 13'.8

Using the average of the watch times, the dead-reckoning position of the ship is found as latitude 43° 24'.6N and longitude 48° 27'.4W. Since the ship is proceeding at only 16 knots, and since the DR position is probably somewhat in error, this position is used for computing the altitudes and bearings that the stars should have at the GCT observation. These values are found to be:

Star	Bearing	h_c
Alpheratz	098°.4	52° 32'.5
Altair	216°.2	50° 07'.4

The intercepts ($h_c - h_t$) are found: for Alpheratz +5'.1 and Altair −6'.4. These yield two "most probable" positions of the ship, one 5.1 miles from the DR position in the direction 278°.4 (i.e., away from the GP of Alpheratz), and the other 6.4 miles in the direction 216°.2 (i.e., toward the GP of Altair). To determine the fix, the DR position is plotted, the bearing lines to the two GP's drawn through it, the intercepts measured off in the proper direction, and the two lines of positions drawn through the intercepts perpendicular to the lines to the GP points. The point of intersection of the lines of position is the fix at 2345. The actual plotting, on small area plotting sheet, is shown in Fig. 3. From the figure, the fix is found to be in latitude 43° 20'.7N and longitude 48° 35'.2W. To determine the compass deviation, the difference between the observed compass bearing of Alpheratz and the computed bearing is found to be 121° − 098° = 23°. Since the variation in this region is 26 °W, the deviation must be 3 °E.

Proper selection of stars to be observed will yield data of extreme importance to the pilot and navigator. For example, if the object is nearly

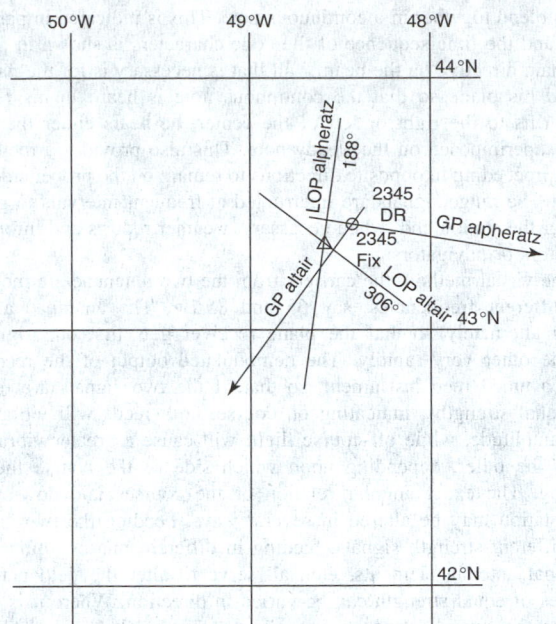

Fig. 3.   Scale drawing for celestial navigation example.

ahead or astern of the plane, the line of position will cross the course nearly at right angles, and the length of the intercept will provide a check on the ground speed being made good. On the other hand, if the object is in a direction approximately perpendicular to the course, the value of the intercept will indicate the accuracy of the wind correction angle.

In many cases, altitudes and bearings of celestial objects may be computed in advance of the actual observing. These predetermined altitudes have many uses in air navigation. If an aircraft is to depart at a definite time and follow a specified course, the altitudes and bearings at indicated times may be computed before the plane leaves the ground. The course to be followed is plotted on a chart, the predetermined DR positions for indicated times are marked, and the bearings of the GP of the object are indicated by lines drawn through the DR positions. The precomputed altitudes are geocentric, but they may be transformed into those expected to be read to the octant at the specified times by applying the various corrections with reversed signs. The navigator measures the altitude at an indicated time, obtains the difference between his value and that predicted, and lays off this distance along the drawn bearing line either toward or away from the GP. In this way, an EP is determined within a few seconds after the observation is completed, and no computing is required during the flight. The pilot is notified to alter heading and air speed to bring the plane back to schedule. If, due to unforeseen conditions, the plane gets so far off scheduled position that the intercepts are more than 150 miles, the predetermined altitudes must be abandoned and regular celestial navigation adopted.

Under some conditions, it may be necessary for a plane to make an accurate landfall (e.g., locate a small island, life raft, etc.) under conditions where celestial navigation must be relied upon. In such cases, the use of precomputed altitudes gives great assistance. First, an estimated time of arrival (ETA) is obtained. Then, using the latitude and longitude of the landfall, a series of altitudes and bearings of a celestial object is computed. The interval of time between computed values depends somewhat on the rapidity with which the values are changing, but is usually about 10 minutes. The series begins at least half an hour before ETA and extends beyond that value. Two curves are then drawn on graph paper, showing altitude and bearing as a function of time. The plotted altitudes are those expected with the octant, i.e., with corrections applied to computed geocentric values. If possible, an object that is approximately ahead or astern of the plane should be selected. About half an hour before the predicted ETA, the pilot alters heading 10° or 15° to the right or left of that predicted for the true course, so that there will be no question as to which side of the landfall he is approaching, and the navigator begins taking altitudes of the object. The navigator plots his observed values, as

a function of time of observation, on the same graph as that showing the predetermined values, and obtains a curve of observed values. At the instant that the observed curve intersects the curve obtained from precomputation, the plane must be on a line of position running through the landfall. The bearing of the celestial object at this instant is read off the plotted bearing curve, the line is drawn at right angles to this bearing through the destination, and the pilot is instructed to alter heading to run down the line.

Radio Navigation

The use of radio aids in navigation for checking the dead-reckoning position of a ship is known as radio navigation. Radio direction-finders were used very early in the development of radio technology to avoid the difficulties of celestial navigation from a ship or aircraft and for emergencies in bad weather. The simplest system uses a directional antenna to locate the direction to several radio stations. A simple triangulation then locates the ship or aircraft with respect to the location of the stations, usually well known and in the map of reference with which the pilot is familiar. The method is complicated by the aircraft velocity, but not by accelerations, weather, or the availability of tables. The pilot can tune in a station near his destination and simply follow the signal to it. Two deficiencies are present: (1) the location of the stations; and (2) the errors inherent in a directional antenna. These problems led to improved radio systems.

The use of radio bearings as lines of positions is best explained by an example from ship navigation. A ship is proceeding on heading 330° at 12 knots. Three radio beacons, A, B, and C, are located in the following positions:

Station	A	B	C
Latitude	29° 30′N	30° 00′N	28° 40′N
Longitude	83° 20′W	81° 40′W	81° 52′W

At 0812 the dead-reckoning position of the ship is $L = 28°\,32'N$ & $Lo = 82°\,42'W$ and at that time, radio bearings, corrected for deviation of the radio compass, are $A = 000°$, $B = 063°$, and $C = 117°$. These must be changed to true bearings by adding to each the heading of the ship, obtaining: $A = 330°$, $B = 033°$, and $C = 087°$. Since these are great-circle bearing, they must be reduced to rhumb-line bearings by applying the correction factors for $A - 0.05$, $B + 0°.2$, and $C + 0°.1$. Then, working either on a mercator chart, mercator plotting sheet, or small-area plotting sheet, the corrected rhumb-line bearings are plotted, and the fix determined as the center of the triangle of intersection of the three lines of position. The position of the fix is $L = 28°\,37'N$ & $Lo = 82°\,44'W$, and, since the sides of the triangle are less than 3 miles, we can assume that the fix is probably correct to within 1 mile. The complete solution is illustrated in Fig. 4 which is drawn on a small-area plotting sheet and labeled in accordance with standard procedure.

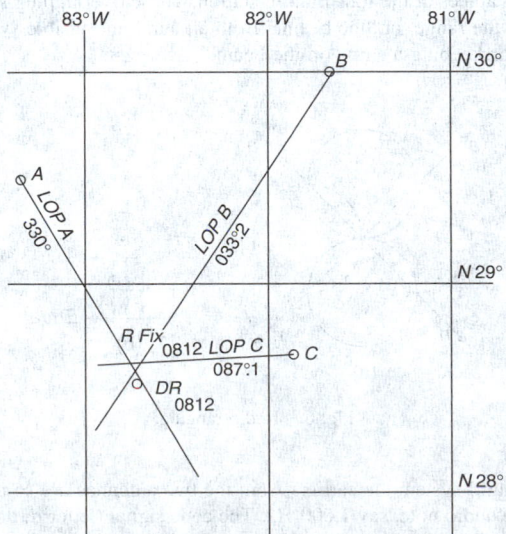

Fig. 4.   Radio navigation scale diagram.

Three corrections must be applied to a radio bearing before it can be used as a line of position on a mercator chart or small-area plotting sheet. Loop antennae and radio direction-finders have deviation corrections, due to the magnetic field of the ship. These must be determined in advance for different headings of the ship and applied to radio bearings as obtained. Then the radio bearing, which is relative, must be changed to true bearing by adding the true heading of the ship. Finally, since radio follows great circles, the bearing must be converted from great-circle to mercator, or rhumb-line, bearing. Figure 5 shows two points, X and Y, plotted on a mercator chart, with the rhumb line, XMY, and the great circle, XGY, connecting the two points. The great circle will always be convex toward the nearest pole, and we have drawn the figure for the Northern Hemisphere with true north indicated both at X and Y. The lines gX and $g'Y$ are tangents to the great circle at X and Y, respectively, and are the directions in which the signal from Y will arrive at X, and that from X will arrive at Y. Let us consider X to be the receiving station. Then the angle R (NXg) represents the great-circle bearing of Y, and the angle B (NXM) the rhumb-line bearing. In this case, it is noted that a correction must be added to the great-circle bearing to obtain the rhumb line. Reversing stations and considering Y the receiver, we see that at this point the correction must be subtracted to obtain rhumb line from great circle.

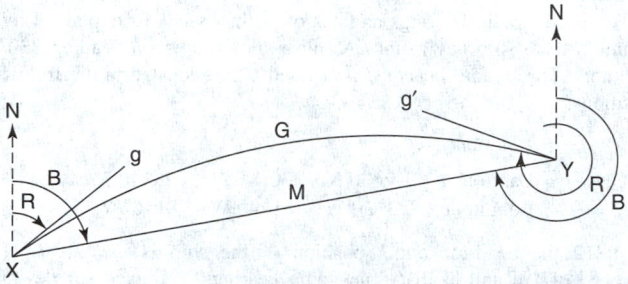

Fig. 5.   Plotting of X and Y on a mercator chart.

In case a navigator is working on a Lambert chart, as is frequently the case in air navigation, the great-circle bearing is close enough to a Lambert line to be plotted without correction other than for deviation and heading.

A radio range is a system of radio signals designed for the purpose of guiding a ship or plane along a designated track toward or away from a specified location. Relatively low frequency (200–400 kHz) has been used. Although there are various modifications, the track is indicated by the intersection of two field patterns from the range antenna system. The usual antenna arrangement is two pairs of cross antennae set 90° in space from one another. This gives two figure-eight field patterns as shown in Fig. 6(a). The patterns overlap in narrow wedge-shaped regions, which have their apices at the transmitting station. These overlapping sectors are known as the range, or "the beam." Both an aural and visible system may be used for keeping a ship "on the beam."

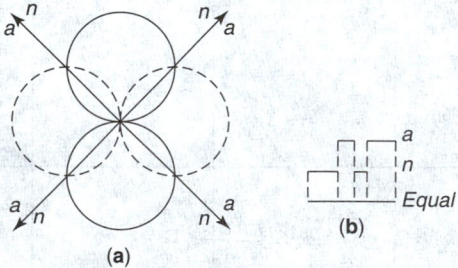

Fig. 6.   Radio ranging.

In the aural system the carriers from the two antennae are so modulated with some audio note, say 1,000 Hz. The code signal (letter a, dot dash) is transmitted from one antenna, and the code signal (letter n, dash dot) from the other, so timed that on the center of the overlapping region the two

signals blend together in a continuous note. This is indicated in part (a) of Fig. 6 and the time sequence of the code characters is shown in part (b). To remain directly "on the beam," all that is necessary is for the navigator to head his plane so that the continuous note is heard in his receiver. If he drifts to the right or left of the center, he hears either the a or n signal superimposed on the steady note. This also provides a method for planes proceeding in opposite directions to remain on the proper side of the airway. The range signals are interrupted at frequent intervals to give the name of the station and, when necessary, weather reports and information to aviators or navigators.

In the visual method, the carriers from the two antennae are modulated with different frequencies, say 65 and 85 Hz. The antennae are then excited alternately so that the plane receiver gets first one signal and then the other very rapidly. The demodulated output of the receiver is fed to a tuned reed instrument, so that, if the two signals are received with equal strengths, indicating on course, both reeds will vibrate with equal amplitude, while off-course flight will cause a greater vibration of one or the other, depending upon which side of the course the plane is flying. The exact angular relation of the courses laid down by the range station may be altered in several ways. Feeding the two antennae with different strength signals, feeding in different phases, utilization of additional antenna elements, etc., all serve to alter the field pattern so the lines of equal strength can be varied in direction. Where it is desired to rotate the courses after their angular relation has been fixed, a double goniometer may be used to feed the antennae. This gives a continuous 360° control of the direction of the beams.

The term *radio compass* has been used loosely over the years. When the loop antenna was first applied to the determination of radio bearings, the term *radio-compass station* was applied to shore installations that would forward, upon request, the bearing of a ship from the station. Next, the term was applied to a group of shore installations, each equipped with a loop antenna, from which the navigating officer of a ship within range could obtain the latitude and longitude of his ship. After the loop antenna and receiving sets had been developed to a state where they could be carried by the ships themselves, the term radio compass was applied to the loop. As new and improved radio equipment became available, the term radio compass was successively applied to any radio device that could be used to determine bearing. A glance through any textbook on navigation, particularly those dealing with air navigation, will yield at least two, and sometimes as many as five, different instruments for radio compass.

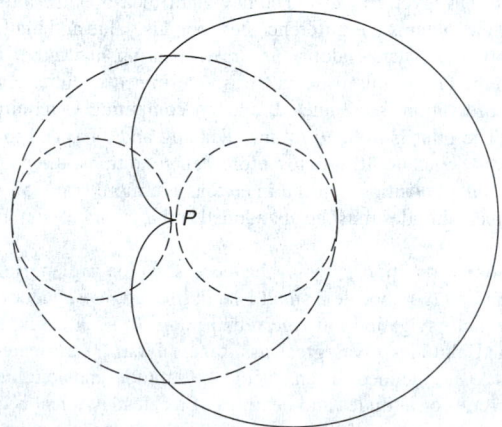

Fig. 7.   Radio compass antenna patterns.

Radio compass is also applied to a direction-finding instrument which has a dial and looks in many respects like an ordinary compass and which is used for heading the ship in much the same manner as a compass. Two radio antennae are used with the instrument—a loop and a nondirectional antenna. The volume controls for signal intensity are so adjusted that the signal strength from the loop and the nondirectional antenna are the same when the station is in the plane of the loop. The two antennae are then fed into a single receiver, and the signal intensity is illustrated in Fig. 7.

The resultant signal intensity from any station is shown on the instrument panel.

With reference to the figure, the length of a line from P to any one of the three curves is proportional to the strength of the signal from a station in the direction toward which the line is drawn. The dotted (figure eight) curve represents the relative intensities in various directions for the loop alone; the dashed (circle) curve represents the intensity for the nondirectional antenna alone; and the full curve that for the combined loop and nondirectional antenna, i.e., for the radio compass.

Hyperbolic Navigation. Hyperbolic navigation is a general method for determining a line of position by measuring the difference in the distance from the navigator of two stations of known position. The difference in distance is found by measuring the difference in time between arrival of signals transmitted from the two stations. A great variety of signaling methods is theoretically possible. Some of these systems are described shortly. Since electromagnetic waves travel with a speed of about ∼186,000 miles (299,274 kilometers)/second, the difference between arrival times of the signals will be very small. The unit of time used in these systems is the microsecond (0.000001 second); a difference between arrivals of one of these units indicates a distance difference of 0.186 miles (0.299 kilometers) from the two transmitters.

Points of constant difference of time between arrival of the two signals will fall on spherical hyperbolas, with the transmitters at the foci. For navigational purposes, one need only consider the lines of intersection of these surfaces with the surface of the earth. The total number of distinguishable lines in any system is equal to the time required for the signal to travel from the master to the slave and back again, divided by the smallest time interval that can be measured by the receiving equipment.

In some systems, two or more slave stations may be used with a single master. The cycle of transmission always begins at the master station, and the signal travels out in all directions. The arrival of the master signal at the slave "triggers off" the slave. Operators continuously monitor both stations, each monitoring the signals for the other, to detect the slightest variations of frequency carrier waves, intervals between pulses, and characteristics of the signals. At the slave station, adjustable delay circuits are available, so that, once the cycle has been started, the slave station may transmit simultaneously with the master or be delayed by any desired amount.

As shown in Fig. 8, certain fundamental lines, representing integral multiples of distance or time difference, are superimposed on regular navigational charts. Tables are published that contain the data for determining the fundamental lines. By graphical interpolation on the chart, or by mathematical interpolation from tables and stored data, the navigator can determine a hyperbolic line of position, using the observed difference in time of arrival of the signals, and the particular stations being used. The accuracy of the line varies from about 200 yards (183 meters) up to about 2 miles (3.2 kilometers), depending upon the distance of the observer from the base line between stations, and the type of equipment in use.

A fix as determined by a hyperbolic navigation system employing one master station, M, and two slaves, S_1 and S_2 (as in GEE navigation) is shown in Fig. 9. The diagram also indicates the value of hyperbolic navigation for "homing" on point A. The navigator obtains a fix at P and then sets his indicating equipment so that the pips from M and S_1 are in coincidence on the display instrument. Then the navigator heads the plane or ship so that, when these pips remain in coincidence, the ship will be following the hyperbola PA. By taking observations on the MS_2 pair at

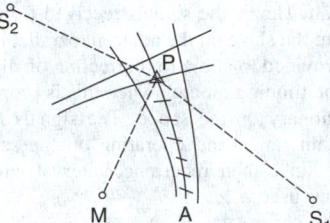

Fig. 9. Fix (hyperbolic navigation).

intervals, the navigator can determine the rate at which the objective is being approached.

DECCA Navigation. This is a system of hyperbolic navigation that employs low-frequency continuous-wave radiation.

The master and any slave station radiate continuous waves, whose frequencies are related by a simple fraction, say one-fourth. The radiations from the two will be in phase when the distances from the stations differ by even multiples of a specified unit. This unit is a function of the wavelengths radiated.

In practice, a master and two slave stations are used. The receiving set reduces all three to a common frequency, and phase meters indicate the relative phase of each slave to the master. The accuracy of the setting of the phase meter is such that differences in distance may be determined with an accuracy of the order of magnitude of 100 feet (30.5 meters), and this is independent of distance from the base line. However, there is complete ambiguity of position, since there is no positive method of determining the number of complete phase changes between the observer and either station.

In spite of the ambiguity mentioned above, the system is of great accuracy and value when used in proceeding to some specified objective. A line of position, involving the master and one of the slave stations, is selected, which passes through the desired objective. The pilot must get the ship onto this line and then set his phase meter. If the pilot then proceeds so that the phase meter setting remains constant, the craft must be following the hyperbola directly to the objective. The hyperbolic lines from the master and the other slave will intersect the hyperbolic track along which the ship is proceeding. The pilot computes the number of complete phase changes that are to be expected, between the point of departure and the objective, along the line that the craft is following. When this number, plus any remaining fractional phase change, has been completed, the pilot must be directly over the objective.

LORAN Navigation. This is a long-distance radio-navigation system for aircraft and ships, utilizing synchronized pulses transmitted simultaneously by widely spaced transmitting stations. Hyperbolic lines of position are determined by measuring the difference in the time of arrival of these pulses. The intersection of two of these lines of position, obtained from either three or four stations, gives a position fix. Standard LORAN operates on frequencies between 1,800 and 2,000 kHz. LORAN C is a widely used version of LORAN that uses pulse signals for more precise time-delay measurement and operates at a frequency of 100 kHz. The range is 2,000 nautical miles (3706 kilometers). LORAN D is a tactical LORAN system that uses the coordinate converter of low-frequency LORAN C.

In standard LORAN, the time systems of the master and slave stations are such that the signal from the master always reaches a ship during the first half of the recurrence cycle, and that from the slave during the second half. This is accomplished by including a delay circuit in the slave timing system that delays the retransmission of the signal received from the master until half the recurrence period has elapsed.

Standard receiving equipment has been designed for ships and planes in which both the receiving and timing units are present, with selector switches permitting the operator to set on the frequency and recurrence rate assigned to any LORAN pair he wishes to use. Differential amplifiers, synchronized by the timing circuit to the recurrence rate of the station, act on both the master and slave signals to deliver them at equal strength to the indicator unit.

The slow sweep of the oscilloscope ("viewing scope") appears as two parallel lines, one covering the first half of the recurrence cycle and the

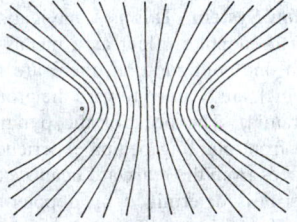

Fig. 8. Fundamental curves for hyperbolic navigation.

other the second half. Hence the signals received from the master appear on one line and from the slave on the adjacent parallel line (see Fig. 10(c)). An adjustment is provided to allow for correction of slight variations in the crystal control of the timing unit and, when this is properly set, the desired signals remain stationary on the scope. The signals from other stations, which may be within range and operating on the same frequency, will drift along the line since their recurrence interval will be different from that of the pair being used.

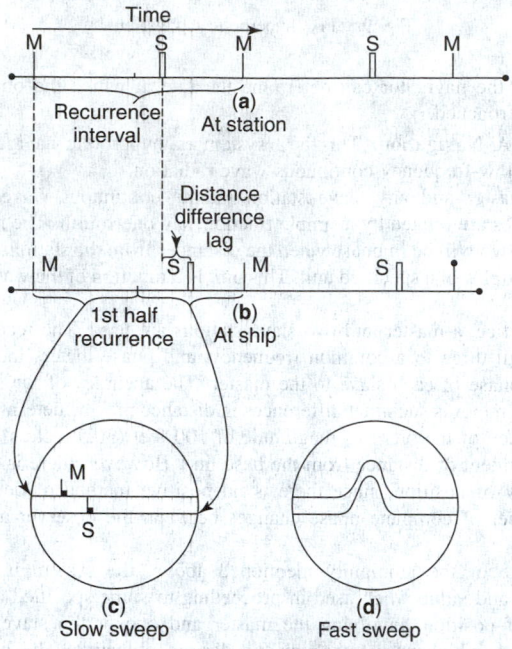

Fig. 10. Characteristics of LORAN signals.

When the signals are properly "set up" on the scope, a set of time markers is thrown on the screen, and a determination is made of the time interval between reception of signals from the master and slave. This will include the delay interval at the slave station, but, since this is standard for each recurrence rate, it may be allowed for. A delay circuit is now introduced, and the signals brought into approximate coincidence. With this adjustment made, a fast sweep spreads out the signals so that close coincidence may be established (see Fig. 10(d)), and the time interval measured to within one microsecond.

Using this measured difference in time of arrival of the two signals, the navigator then uses either tables or LORAN charts to obtain one line of position. The selector switches are then set to the characteristics of another LORAN pair, a second line of position determined, and a fix obtained (see Fig. 11). The accuracy of the determined fix is of the same order of magnitude as that obtained from good celestial navigation and, of course, is independent of the state of visibility.

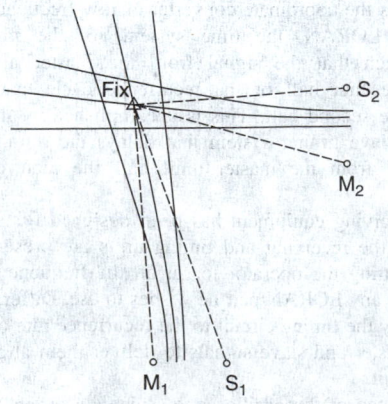

Fig. 11. Obtaining a fix with LORAN.

LODAR. This is a direction finder with which the direction of arrival of LORAN signals is determined free of night effect by observing the separately distinguishable ground and skywave LORAN signals on a cathode ray oscilloscope and positioning a loop antenna so as to obtain a null indication of the component selected to be most suitable.

OMEGA Navigation. This is a long-range system that can provide worldwide coverage with only eight stations. The operating frequency is 10 kHz, and the estimated accuracy is 5 nautical miles (9.3 kilometers) with typical receivers. The system was developed for totally submerged submarines. The operating principle is similar to *delrac*, a British radio-navigation system designed to provide worldwide coverage by using 21 pairs of master-slave stations, with a 3,000-mile (4827 kilometers) range for each pair of stations. Frequencies used in *delrac* are in the band from 10 to 14 kHz. DECCA indicating equipment can be used with *delrac*.

GEE Navigation. This is a vhf (very high frequency) radio navigation system developed in Great Britain and is similar to LORAN. For the transmission of the signals, one master and two or more slave stations are used. The distance between stations is about 75 miles (121 kilometers) and the stations are located approximately on a circle, with the master station between the slaves. The frequencies used are between 20 and 88 MHz, and the length of the pulses are of the order of magnitude of 6 microseconds. The accuracy of the lines of position varies with the square of the distance of the ship from the base lines between stations. On this line, the accuracy is of the order of magnitude of 200 yards (183 meters) when the navigator is on the base line, and about one mile at a distance of 400 miles (644 kilometers) from the base. Since time differences can be read simultaneously from the master and two slave stations, the fix can be determined by simultaneous observations without the necessity of using the running fix method. The system is excellent for "homing" on a particular objective.

TACAN System. An air-navigation system in which a single uhf (ultra-high frequency) transmitter sends out signals that actuate airborne equipment to provide range and bearing indications with respect to the transmitter location when interrogated by a transmitter in the aircraft. Each TACAN station broadcasts a location-identifying Morse-code signal at regular intervals. Also termed tactical air navigation.

SHORAN System. A precision short-range position-fixing system using a pulse transmitter and receiver in an aircraft or other vehicle and two transponder beacons at fixed points. A receiver in the aircraft measures the round-trip times of the signals and converts these into distances to the fixed ground stations. Ordinary triangulation on a map then gives position.

OMNI System. This is a radio system that includes the directional information by modulating its radio signal with a simple dot-and-dash code, one code for a position to the left of the beam and another for a position to the right of the beam. The stations are located on airways and at the approaches to airways and to airports. With the system, the pilot selects his OMNI way point or destination on the radio and listens for the code that tells when the aircraft is to the left or right of the path. The system has been highly refined with onboard computers and displays. The system also is supplemented with distance-measuring equipment (DME). With the latter, the pilot interrogates the station with a transmitter and receives a distance indication from the station on a receiver. The measurement is made by determining the travel time of the radio wave. DME complicated the airborne equipment considerably but opened the way for easy, continuous navigation by using only a single set of equipment. Use of two stations can provide coverage for all locations within their range and thus free the pilot from flying the designated lines toward the OMNI stations. Very-high-frequency omnirange operates in the band from 112 to 118 MHz.

Doppler Navigation System. This is a navigation system for aircraft which makes use of the doppler effect as a means for determining drift and ground speed. In one configuration, there are four beams of pulsed microwave energy, which are beamed toward the ground (along the corners of an imaginary pyramid). The peak of the pyramid is at the aircraft. The echoes from the front-pointing beams experience an upward doppler shift, whereas the echoes from the rearward beams experience a downward doppler shift. Any drift is determined by doppler shift of echoes from beams on either side of the aircraft. The doppler shifts are compared in a computerized system, enabling all necessary navigation information under adverse weather conditions, various altitudes, and with need for reference to ground stations.

In a navigation satellite system, the satellite transmits accurate time signals and position data to a receiver on board a ship or aircraft. A central ground tracking station transmits correction signals to the satellite many times each day to sustain high accuracy of the system. The objective of such satellites, from which radio doppler shift measurements can be made under all weather conditions, is to provide the position of a ship or aircraft anywhere on earth with an accuracy of about 0.5 nautical mile (0.93 kilometer) or better.

NOTE: Numerous other articles in this encyclopedia relate directly or indirectly in navigation. Check alphabetical index.

Additional Reading

Baker, D.J.: "Toward a Global Ocean Observing System," *Oceanus*, 76 (Spring 1991).

Beardsley, T.: "Messages from on High," *Sci. Amer.*, 112 (July 1988).

Bjerklie, D.: "The Electronic Transformation of Maps," *Technology Review (MIT)*, 54 (April 1989).

Carron, M.J. and K.A. Countryman: "Developing Oceanographic Products to Support Navy Operations," *Oceanus*, 67 (Winter 1990/91).

Frye, D.W., W.B. Owens, and J.R. Vales: "Ocean Data Telemetry," *Oceanus*, 46 (Spring 1991).

Garver, J.G., Jr.: "A Love Affair with Maps," *Nat'l. Geographic*, 130 (November 1990).

Grewal, M.S., A.P. Andrews, and L.R. Weill:*Global Positioning Systems, Inertial Navigation, and Integration*, John Wiley & Sons, Inc., New York, NY, 2001.

Grosvenor, G.M.: "New Atlas Explores a Changing World," *Nat'l. Geographic*, 126 (November 1990).

Kiernan, V.: "Guidance from Above in the Gulf War," *Science*, 1012 (March 1, 1992).

Koehr, J.E.: "The United States Navy's Role in Navigation and Charting," *Oceanus*, 82 (Winter 1990/91).

Lawrence, A.: *Modern Inertial Technology: Navigation, Guidance, and Control*, Springer-Verlag, Inc., New York, NY, 1998.

Marden, L.: "Tracking Columbus Across the Atlantic," *Nat'l. Geographic*, 572 (November 1986).

McVey, V.: *The Sierra Club Wayfinding Book*, Little, Brown and Co., Boston, MA, 1989.

Monastersky, R.: "Satellite Secrecy Doesn't Sink Scientists," *Science News*, 358 (December 8, 1990).

Richardson, P.L. and R.A. Goldsmith: "The Columbus Landfall: Voyage Track Corrected for Winds and Currents," *Oceanus*, 2 (Fall 1987).

Tetley, L. and D.M. Calcutt: *Electronic Navigation Systems*, 3rd Edition, Butterworth-Heinemann, Inc., Woburn, MA, 2001.

Turbank, G.: "A New Way to Find Your Way," *Amer. Forests*, 10 (August 1989).

Waldrop, M.M.: "Flying the Electric Skies," *Science*, 1532 (June 30, 1989).

Wolper, J.S.: *Understanding Mathematics for Aircraft Navigation*, McGraw-Hill Professional Book Group," New York, NY, 2001.

Wood, D.: "The Power of Maps," *Sci. Amer.*, 88 (May 1993).

NAVIGATORS' STARS. A list of 55 stars has been designated as the "navigators' stars." The list was selected to cover the entire celestial sphere in such a manner that a navigator, no matter at what season or in what part of the earth he may be operating, will have two or three navigators' stars available for observation. The names and positions of these stars are listed in the Air Almanac, published by the U.S. Naval Observatory, and in a number of other publications. See also **Celestial Sphere and Astronomical Triangle**; and **Navigation**.

NAZCA PLATE. See **Ocean**.

N-DISPLAY. In radar, a display similar to the K-display in which the target appears as a pair of vertical deflections or blips from the horizontal time base. Direction is indicated by the relative amplitude of the vertical deflections; target distance is determined by moving an adjustable signal along the baseline until it coincides with the horizontal position of the vertical deflections. The horizontal control is calibrated in distance. Also called *N-scan, N-scope*, or *N-indicator*.

NEAR-EARTH ASTEROID RENDEZVOUS (NEAR). See **Discovery Program**; and **Space Science Missions: Solar System**.

NEAR SHOEMAKER MISSION. See **Near-Earth Asteroid Rendezvous Mission**.

NEARSIGHTEDNESS. See **Vision and the Eye**; and **Myopia**.

NEBULA. The term nebula (*stella nebulosa*) was originally used by astronomers to describe any luminous spot that remained fixed relative to the stars. Before the application of the telescope, probably the only objects to which the term applied where those that are now referred to as star clusters, although reference was made in the tenth century to the great spiral in Andromeda. Following the application of the telescope, many more nebulous objects were discovered. Originally, these were grouped into three classes: the diffuse nebulae (Fig. 1); the planetary nebulae; and the spiral nebulae. Further research indicated that the spiral nebulae were very different from the other two. Application of the large telescopes proved that, in reality, they are groups of stars. Thus, for this class, the term nebula was dropped and the term galaxy or spiral galaxy used.

Fig. 1. The Great Nebula in Orion (NGC 1976), with NGC 1977 below. (*Lick Observatory*.)

Some nebulae shine only by reflecting the light of stars contained in them ("reflection nebulae," which are dust in the environment of cooler, young stars). Others contain extremely hot stars whose radiation knocks electrons into high energy levels in the nebula gas, exciting gas atoms and allowing them to radiate as the electrons return to their ground states, producing soft glows also called HI regions. Some nebulae are stellar birthplaces; inside them, fresh stars are being produced out of the nebular gas and dust. Other nebulae are debris of explosions marking disruptions of unstable stars, supernova remnants like the Crab Nebula.

With the tendency for a mixture of gas and dust to collect in clouds and condensations, obscuring clouds of interstellar material, known as dark nebulae, may collect. When one of these clouds is in the vicinity of a bright star, the intense radiation from the star will illuminate the cloud, and a bright diffuse nebula is observed. Studies of the spectra of these objects have shown that the light is made up of both reflected starlight and of radiation from the interstellar material. The character of the reflected light is similar to that from the nearby stars. The radiation from the nebulous material itself, however, is quite different in character from starlight. This nebular spectrum is the bright-line type, and is probably produced by the absorption of radiation from the star by the gas atoms, and then reradiation of this energy in frequencies characteristic of the gas atoms and their states of excitation and ionization. This hypothesis has been considered sound because the character of the nebular spectrum depends upon the spectral type of the nearby stars. If the star is hot (B-type), the nebular spectrum is rich with bright lines, but in the vicinity of a relatively cool star (A-type or later), the nebular spectrum is almost entirely that of reflected starlight.

When the spectra of the bright nebulae were first studied, two bright lines were observed that could not be identified with any known terrestrial or solar element. At one time, it was believed that the lines were due to some material that existed only in the nebulae and the name of an element Nebulium was coined. These lines turned out to be due to forbidden transitions of doubly ionized oxygen, [OIII], which, at that time, had not been produced in the laboratory because of the very low density required to prevent collisions from de-exciting the atoms.

In addition to the diffuse nebulae, there are the planetary nebulae (see photograph accompanying entry on **Lyra**), so-called because they have quite definite shapes and look more or less like planets when observed through a telescope. The general appearance of these objects, on detailed photographs, is that of shells of gaseous material. They are generally elliptical in form, and may appear to be made up of several elliptical shells having their axes at various angles to each other or as helices. Frequently, a very blue, and hence, a very hot star is observed at the center of the shell, which is a hot white dwarf, the core of the star whose envelope the nebula represents. A number of these stars prove to be very short (less than one hour) period binaries. The spectra of the planetaries is, in general, the same as that of the diffuse nebulae found in the vicinity of B-type stars. In some cases, where the central star can be studied, variations in its light have been found, and the nebular radiation is found to vary with that of the star. It is evident that the planetary nebulae are actually stars with very extended and attenuated atmospheres. Careful studies of the spectral lines from the planetaries indicate that the shell may be expanding. Expanding shells of gas have been observed around some novae, but the rate of expansion is far greater than that for the planetaries.

Since the advent of radio astronomy, a significant number of different kinds of molecules have been found in the interstellar medium, mostly in dark clouds of dust where light from stars cannot penetrate. However, the first interstellar molecule to be discovered was the radical CH, found accidentally in a star spectrum in 1937 with conventional earth-based optical instrumentation. The radicals CN and CH^+ were found with the Mount Wilson Observatory 100-inch (254-centimeter) optical telescope in 1939. The discovery of other interstellar molecules had to await the development of radio telescopes. The OH radical was the first to be added to the list by radio techniques in 1963. Since that time, and particularly since the early 1970s, a number of new interstellar molecules have been added to the list. These include NH_3, H_2O, H_2CO (formaldehyde), CO, HCN, HC_3N (cyanoacetylene) CH_3OH (methanol), CH_2O_2 (formic acid), CS, CH_3CN (methyl cyanide), SiO, HNC, CH_3C_2H (methyl acetylene), NH_2CHO (formamide), CH_3CHO (acetaldehyde), HNCO (isocyanic acid), H_2CS (thioformaldehyde), H_2S, H_2CNH (methylene amine), and SO (sulfur monoxide), the latter found in 1973.

At one time, some investigators felt that the most important process in interstellar molecule formation was the combination of neutral molecules by radiative association, that is, reaction with photon emission. However, because the chemical activation energy required for neutral molecule reactions acts as a barrier to formation of more complex molecules, it was agreed that the reaction rates would be too slow for interstellar molecule formation. In contrast, the majority of reactions between ions and neutral molecules do not have such a barrier. Heavier ions can be built from lighter ones. Exemplary reactions include: $H_2^+ + H_2 \rightarrow H_3^+ + H$; or $H_3^+ + CO \rightarrow HCO^+ + H_2$. Ions can recombine with electrons to form neutral molecules, thus completing the reaction pathway. Electron density in dense clouds is estimated at 10^{-7} cm^{-3}. Commencing with the foregoing reactions, the building of complex molecules in dense clouds may require the assumption that the source of ionization is the flux of

cosmic rays at energies of 100 MeV and upwards. The reactions also require solid surfaces, like dust grains, on which to occur. Densities greater than 10^4 cm^{-3} also appear necessary. In addition to a chain of reactions commencing with hydrogen, other possible chains could commence from ionized helium, carbon, and oxygen. However, some investigators feel that the hydrogen reactions are probably basic and that the ion-molecule scheme probably would not take place unless HCO^+ is present in the dense cloud.

Theories of interstellar molecule formation include the theory of formation on dust grains and the ion-molecule theory. Estimates indicate that many of the simple diatomic molecules may be formed by collisions in space; the more complex ones, such as H_2CO_2, CH_3CN, and NH_3CO, may be formed on grains of interstellar dust in clouds where the concentration of hydrogen molecules (H_2) is $10^6/cm^3$ or higher. See also list of entries given in **Astronomy**. See also **Hubble Space Telescope (HST)**.

STEVEN N. SHORE, New Mexico Institute of Mining and Technology, Socorro, NM

NEBULE. A unit of atmospheric opacity. One nebule is the opacity of a screen having such transmissivity T_n that

$$T_n^{100} = 0.001,$$

that is, such that 100 of the screens placed in optical series would transmit only one-thousandth of the incident light. One speaks of a given atmosphere or atmospheric stratum as possessing an opacity of a certain number of nebules per kilometer (mile). The transmissivity T of an optical path r km long through an atmosphere of opacity n nebules per km is given by

$$T = T_n^{nr},$$

where T_n is defined above. An opacity of one nebule per km is equivalent to an extinction coefficient of 0.069 per km.

Additional Reading

Middleton, W.E.K.: *Vision through the Atmosphere*, University of Toronto Press, Toronto, ON, Canada, 1980.

AMS

NECROSIS. The local death of cells results in changes in the tissue known as necrosis. These consist of disintegration of the cellular structure with destruction of the nucleus and coagulation or liquefaction of the cytoplasm. The causes of necrosis include interference with the blood supply of a tissue physical injury, and deleterious actions by bacteria or their toxins.

NEEDHAM, JOHN TURBERVILLE (1713–1781). Born in London of recusants, Needham was educated in French Flanders after the death of his parents. After being ordained a Catholic priest, he provided for himself as a teacher in Flanders, England and Portugal. His first essay (1745) describes the microscopical milt-vessels of the squid, and some eels of the blighted wheat, work for which he was elected a Fellow of the Royal Society in 1747, the first Catholic priest to receive this honor. In 1748 Georges Buffon invited him to Paris, where they experimented on microorganisms appearing in organic infusions. Needham sealed various infusions in jars, before and after they had been heated, and reported having seen microscopical organisms emerging from vegetable filaments. He interpreted it as an effect of a vital force able to transform vegetable into animal substances. Needham then started to travel in Europe, as a tutor of young Catholic noblemen. He personally met and corresponded with many naturalists of his time, and performed microscopical observations with them, notwithstanding the rejection (Charles Bonnet 1764) or refutation of his experiments (Lazzaro Spallanzani 1765 and 1776). He was called to Brussels in 1768 for the reorganization of the Royal Academy of Belgium, of which he became the director. See also **Bonnet, Charles (1720–1793)**; **Buffon, Georges Louis (1701–1788)**; and **Spallanzani, Lazzaro (1729–1799)**.

The renewal of the idea of spontaneous generation based on research made by such a skilled experimentalist and microscopist as Needham had two major consequences. On the one hand, it supplied an impulse to the microscopical research made between 1760 and 1790 in Germany, where many authors defended the doctrine. On the other hand, Needham was considered a "loser" in the spontaneous generation quarrel, which tended to bring discredit on the microscopical researches of the eighteenth century. Needham also participated in the fight against materialism, which earned him the sarcasms of Voltaire.

M. J. RATCLIFF, Wellcome Institute for the History of Medicine, London, UK

NÉEL TEMPERATURE. The transition temperature for an antiferromagnetic material. Maximal values of magnetic susceptibility, specific heat, and thermal expansion coefficient occur at the Néel temperature. See also **Antiferromagnetism**.

NEGATOL. A condensation product of *m*-cresolsulfonic acid with formaldehyde. A polymerized dihydroxydimethyldiphenylmethanedisulfonic acid. It is dispersible in water, forming very acidic colloidal solutions. The pH of a 5% dispersion is approximately 1.0.

NEGATRON. A term sometimes applied to the normally occurring negatively charged electron when it must be distinguished from a positron. In many parts of the world the name *negaton* is used instead of negatron. The word negatron is used in this encyclopedia wherever distinction is made between positively and negatively charged electrons.

NEGRO BUG (*Insecta, Hemiptera*). Small shining black bugs with a smooth convex upper surface. They resemble beetles superficially. Most of the abdomen is covered by a greatly enlarged sclerite of the thorax, which also conceals most of the wings.

NEIGHBORHOOD OF A POINT. The interior of some bounded geometric figure (such as a square or circle in the plane) which contains the point. A neighborhood or a point on a line, plane, or surface is usually taken as the set of points within a stated distance of the point (e.g., an open interval on the line or the interior of a circle in the plane, with the point as center). One speaks of a property as holding *in the neighborhood of a point* if there exists a neighborhood of the point in which the property holds, or of a numerical quantity (e.g., curvature) depending on the nature of a curve or surface in the neighborhood of a point if the value of the quantity can be determined from knowledge of the portion of the curve or surface in an arbitrarily small neighborhood of the point. See also **Point**.

NEKTON. The portion of a population made up of animals capable of directive locomotion through a fluid medium. Usually applied only to aquatic animals, including the fishes, although flying creatures constitute a similar part of the terrestrial fauna and may be called an aerial nekton. See also **Ocean**.

NEMATIC LIQUID CRYSTALS. See **Liquid Crystals**.

NEMATODES. Of the phylum *Nemata* or *Nematoda*, these are roundworms or threadworms. They are abundant in fresh and salt water and in the soil; many are internal parasites of animals and plants. Some are parasitic in humans and the domestic animals and are important in relation to human welfare. The body of these worms lacks a spiny proboscis and is marked by slender longitudinal lines along the sides. These lateral lines follow the excretory tubes. There is a wide range of variations in the life cycle. To place *Nemata* in their proper perspective, it is in order to mention the other important phyla of worms: Flatworms (*Platyhelminthes*); ribbon worms (*Nemertea*); spiny-headed worms (*Acanthocephala*); hairworms (*Nematomorpha*); and segmented worms (*Annelida*). Reference to the entry on **Intestinal Nematodes** is suggested.

Nematode Damage to Food Crops

Nematodes are very important economic pests on food crops. Very few crops are immune to attacks of these creatures, which inhabit the soil about the roots of plants. Nematode populations number into the millions and billions, in field crops, orchard operations, greenhouse (glasshouse) facilities, and truck and home gardens. Actually, nematodes have been one of the last of the major crop pests to be well understood, and aggressive research in the field only dates back some 50 to 60 years. Research progress was impeded mainly by difficulties in isolating and preparing the nematodes for detailed examination. Some authorities

place nematology just about one-half century behind entomology, but much progress has been made during the past decade or two and thus the technological gap is narrowing. Among the challenges facing the nematologist today are: (1) development of nematode-resistant varieties of crop plants; (2) cooperation with agricultural engineers in development of more effective means for applying nematicides; (3) education of farmers and large food producers on cultural methods for controlling nematodes, including fallowing and rotation of crops; (4) development of improved systemic nematocides as well as synthetic plant diffusates which stimulate early emergence of nematodes from dormancy; and (5) a continuing program of identifying yet undiscovered species. Continuing work on the classification and nomenclature of nematology also is important.

Although the importance of nematodes to food growth economics was relatively late in being appreciated, a knowledge of the existence of nematodes dates back to ancient times. It is believed that the Guinea worms mentioned in the Old Testament (*Numbers* 21:6–9) as "fiery serpents" were nematodes. Parasitic nematodes were alluded to by Aristotle as early as about 350 B.C. Free-living nematodes were observed in vinegar, and referred to as vinegar eels, as early as the mid-15th century. In the late 1700s, Linnaeus, Scopoli, Steinbuch, and Needham showed a causal relationship between the nematode *Anguina tritici* and the disease known as "cockles" of wheat and other cereal plants. In the late 1800s and early 1900s, Julius Kühn and associates in Germany intensively researched the sugar beet nematode and some authorities credit Kühn with the first use of soil fumigation. He used carbon disulfide on infested sugar beet fields. The life cycle of this pest (*Heterodera schachtii*) was ascertained, providing knowledge upon which effective cultural practices could be established.

Galls on the roots of cucumbers were noted by Berkeley in England as early as 1855. It is believed that the term root-knot nematode was first used in 1879 by Cornu. In 1887, root galls on coffee were described by Goeldi. Bastian, who wrote a monograph on the *Anguillulidae* in 1886, is considered the father of nematology. One of the first full texts on the subject, "Nematodes That Are Important for Agriculture," was authored by the Russian I.N. Filipjev in 1934. Outstanding early work was done by N.A. Cobb at the U.S. Department of Agriculture in the early 1900s. An excellent summary of the history of nematology from its beginnings to the 1960s is given in *Principles of Nematology*, by Gerald Thorne, McGraw-Hill, New York, 1961. Additional and more recent references are listed at the end of this article.

Nature of Nematodes

These economic pests are found essentially wherever soil is found—from deserts and tropical areas to cold, high-altitude mountainous terrains. For example, in 1929 Thorne found several hundred specimens in soil samplings taken from Colorado mountain soils at levels of over 14,000 feet. There appears to be an almost infinite variety of nematodes, the greatest variations probably occurring in marine waters near shallow coastal beaches.

The typical nematode may be described as a slender, quite active animal that ranges from 0.2 to 10 millimeters in length, although the majority are less than 2 millimeters long. The body is usually cylindrical in shape, although several other forms are known, including pear- and lemon-shaped forms. The nematode body is covered with cuticle, a tough, flexible layer of material. In some species, the coating is marked or texturized, which helps the nematologist greatly in identification. However, many are not so conveniently marked, thus requiring detailed microscopic examination to yield identity. On the average, a nematode will undergo four moults in developing from egg to adult. During these stages, the body increases in diameter and length.

Control chemicals for use against nematodes are various fumigants, such as carbon bisulfide, chloropicrin, D-D, EDB, formaldehyde, Fumazone, hydrogen cyanide, methyl bromide, and Nemagon. These and other chemical materials either are banned or are subject to rigid control in some countries.

Nematodes are both *endo-* (inhabit and consume internal organs of host) and *ectoparasitic* (live on surface of host). They can be classified roughly in terms of the portions of the host plant they prefer for habitation. Some of these include:

Bud and Leaf Nematodes (genus *Aphelenchoides*). These nematodes exhibit both endo- and ectoparasitism, a factor determined by Franklin in

1950. The endoparasites are found in leaves; the ectoparasites are found in plant crowns, leaf axils, or inflorescences, where these parts are protected by other folding tissues of the host plant. Nematodes of this type were discovered by Ormerod in England in 1889.

Bulb and Stem Nematodes (genus *Ditylenchus dispaci* (Kühn) Filipjev). There are over 30 species. The symptoms of their presence were first observed by Schwertz in 1855 on clover, oats, and rye, but the nematode causative agent was not revealed until further studies by Kühn in 1857 and others at later dates.

Burrowing Nematodes (genus *Radopholus* (Thorne)). These nematodes were named and classified in 1949. They are endoparasitic, attacking plant roots. They probably are found in tropical and subtropical regions, but have been found in cooler regions as well. *Radopholus similis* (Cobb, 1893) causes banana plant disease (*Musa sapientum*), first noted in Fiji (1890). Also noted on diseased coffee roots in Java (1898) and on diseased sugarcane roots in Hawaii (1907).

Cyst-Forming Nematodes (genus *Heterodera* (Schmidt, 1871)). These include the first of the important nematodes to be associated with plant diseases, namely, the sugar beet nematode (*Heterodera schachtii*) first observed by Schmidt. The female cuticle of this pest transforms into a light-to-dark brown, cyst-like sac. This sac protects the eggs. These cysts are oval or spheroidal in shape and range from 0.4 to 0.8 millimeter in length. Inasmuch as the cyst material does not decompose readily, there are often great accumulations of these bodies in the soil. While older cysts will not have eggs, those of more recent years may contain as many as 600 eggs.

Root-Knot Nematodes (genus *Meloidogyne* (Goeldi)). At one time these were considered to be a single species. Five or more species were established by 1949. These pests are among the most economically important of the nematodes and include the coffee root-knot nematode *Meloidogyne exigua* (Goeldi 1892); the Japanese root-knot nematode, *M. japonica* (Treub, 1885); the northern root-knot nematode *M. hapla* (Chitwood, 1949a); the peanut (groundnut) root-knot nematode *M. arenaria* (Neal, 1889); the Thames root-knot nematode *M. arenaria* thamesi (Chitwood, 1952); the southern root-knot nematode *M. incognita* (Chitwood, 1949a); and the cotton root-knot nematode *M. incognita var. acrita* (Chitwood, 1949a), among others. The root-knot nematodes are often associated with various fungus diseases. Damage is caused by formation of galls on plant roots, causing stunting and wilting and, frequently, expiration of the plant if not controlled.

Root-Lesion Nematodes (genus *Pratylenchinae* (Thorne, 1949)). These have been described since the late 1920s. Because of openings in plant roots caused by the pests, bacteria and fungi may enter and thus these nematodes are often associated with serious diseases from these causes. The root-lesion nematodes cause openings or lesions rather than knots or galls as in the case of the genus *Meloidogyne*. There are about 10 major species and they infest a variety of very important crops, such as coffee, citrus, pineapple, potato, rice, and sugar beet, among many others.

Nematode Damage

Important diseases caused by nematodes affect many crops. Nematodes are found almost universally in the soil and, at one time or another, they contribute to the damage, minor or major, to nearly all plants. There are however, numerous situations where nematode damage can be severe and even catastrophic to some crops if effective control measures are not taken. Only three examples are given here. Much more detail will be found in the Considine (1981) reference listed.

Citrus. A number of nematode species damage various citrus crops. Possibly the most serious situation occurs on citrus in Florida and is a condition known as the *spreading decline of citrus.* See Figs. 1 and 2. Caused by a burrowing nematode, the damage has ranged into the many millions of dollars. Experiments in treating diseased trees, however, point to a factor, still unknown, that also is active in causing spreading decline disease. In some experiments, destruction of the nematodes did not fully prevent spreading of the disease. Symptoms of the disease include stunted trees with subnormal foliage, small fruit, and retarded terminal growth. Wilting is excessive during dry, hot periods. There is also a reduction of young feeder roots. The term *spreading* derives from the fact that the nematodes spread out or migrate from one tree to the next. A study made by Suit and Ford in 1950 indicates that the advance is at the rate of 1.6 trees per year. The nematodes have been known to migrate under highways and railroad rights of way. Rather aggressive methods have had to be used in attempts to eradicate the disease in Florida, including systematic and frequent soil inspections and the destruction by burning of infested trees and planted areas; and the extensive use of very strong chemicals to an average depth of 12 inches (0.3 meter). The persistence of the nematodes is exemplified by the finding of live nematodes within a depth of 12 feet (4 meters) of a very heavily treated area. However, inasmuch as most of the nematodes are found within the top 5 feet (1.5 meters) of the soil, diffusion of the treatment to a depth of 6 to 8 feet (2 to 2.5 meters) is usually effective.

(a)   (b)

Fig. 1.   Young Navel orange trees on Troyer citrange rootstock 2 years after planting: (**a**) trees planted in soil in which citrus nematodes had been killed by preplant soil fumigation; (**b**) tree infected with the citrus nematode soon after planting in nonfumigated, nematode-infested soil in same field as tree at left. (*Agricultural Extension Service, University of California.*)

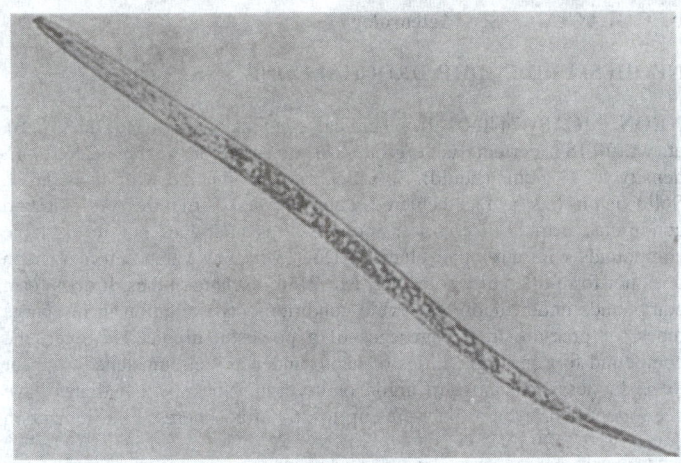

Fig. 2.   Citrus nematode larva. Actual length is $\frac{1}{70}$ inch (0.4 mm). Stylet of feeding mechanism is at head (lower right-hand corner of view). (*Agricultural Extension Service, University of California.*)

Some authorities have observed a close parallel between the spreading decline of citrus disease in Florida and the yellows disease of pepper on Bangka.

The nematode species *Hemicycliophora arenaria* has been found on rough lemon root-stock in California. This pest causes an enlargement of terminal and lateral root-tips, which appear as small knobs. Damage may be the result of secretion of enzymes by the nematodes.

Potato. Nematodes of the genus *Ditylenchus* (Filipjev, 1934) injure potato by producing a progressive dry rot of the tuber. While this damage is proceeding, there is no evidence to be seen from observation of stems and foliage. There may be several strains that cause this condition. Known since 1888, *Ditylenchus destructor* has been a major cause of injury to potato in Europe for a long time. For many years, United States officials intercepted numerous shipments containing the strain.

Presence of the pest can be determined only by cutting into the tuber. In some instances, much of the tuber can be infested. Crop rotation provides no effective control. Fumigation is the main control, and of the fumigants, ethylene dibromide is perhaps most effective. With sufficient fumigant applied, virtual eradication can be accomplished. Resistant varieties have not been successful. Where potatoes are used for silage, the fermentation of the silage kills the pests.

The golden nematode of potatoes, *Heterodera rostochiensis* (Wollenweber, 1923), is very damaging to potato. This nematode was discovered by Kühn in 1881 when he was doing research on the sugar beet nematode. Somewhat later reports were made in Germany and Scotland. The presence of the nematode in much of Europe was confirmed and, in the British Isles, the pest was called the potato root eelworm. During the interim, the nematode has been reported by Israel and, in 1941, the pest was first found in the United States on Long Island, New York. The nematode forms pear-shaped cysts on the potato, thus differing from the lemon-shaped cysts of the sugar beet nematode. This difference removed any doubt that the two pests were different. Since this was a new find in nematology and one affecting a huge market crop, research on the pest was intensive and many countries instituted crop regulatory and quarantine measures.

Rice. The rice stem nematode (*Ditylenchus angustus* (Filipjev, 1936)) is the cause of ufra disease in rice and was observed for the first time by Butler in 1913. This pest and resulting disease poses the greatest threat to success of the rice crop in India. It is found most commonly in an area north of the Bay of Bengal and east of the Ganges River. Even though many hundreds of varieties of rice are grown in these areas, all appear to be susceptible to this pest. Massive infestations in India have been known since 1916. The pest climbs the stems and interferes with the growing process of the plant, after which it consumes leaves and stems — in essence devastating the plant.

Additional Reading

Anderson, R.C.: *Nematode Parasites of Vertebrates: Their Development and Transmission*, 2nd Edition, Oxford University Press, Inc., New York, NY, 2000.

Atkinson, H.J.: *The Physiology of Nematodes*, 2nd Edition, Columbia University Press, New York, NY, 1977.

Considine, D.M., Editor: *Foods and Food Production Encyclopedia*, Van Nostrand Reinhold, New York, NY, 1982.

Croll, N.A.: *The Organization of Nematodes*, Academic Press, Inc., Boca Raton, FL, 1976.

Dropkin, V.H.: *Introduction to Plant Nematology*, 2nd Edition, John Wiley & Sons, Inc., New York, NY, 1980.

Hollis, J.P.: "Action of Plant-Parasitic Nematodes on Their Hosts," *Nematologica*, **9**, 475–494 (1963).

Kontaxis, D.G.: "Nematicides Improve Sugar Beet Yields," *California Agriculture*, **31**(4), 10–11 (1977).

Lee, D.L.: *Biology of Nematodes*, Gordon & Breach Publishing Group, Newark, NJ, 2001.

Radewald, J.D. et al.: "Citrus Nematode Disease and Its Control," *Agricultural Extension Service*, Univ. of California, Berkeley, CA, Bul. AXT-211 (revised periodically).

Rhode, R.A.: "The Nature of Resistance in Plants to Nematodes," *Phytopathology*, **55**, 1159–1162 (1965).

Stone, A.R. et al., Editors: *Concepts in Nematode Systematics*, Academic Press, Inc., Boca Raton, FL, 1983.

Thorne, Gerald: *Principles of Nematology*, The McGraw-Hill Companies, Inc., New York, NY, 1961.

Zuckerman, B.M. et al.: *Plant Parasitic Nematodes: Morphology, Anatomy, Taxonomy, and Ecology*, Academic Press, Inc., Boca Raton, FL, 1981.

NEMERTEA.   Marine worms with a flattened body, often long and ribbon-like in form. They are unsegmented and have no body cavity, hence they are sometimes included with the flatworms as a class of the phylum *Platyhelminthes*. More often they are made a separate phylum because the alimentary tract is a tube opening with an anterior mouth and a posterior anus. Like the free-living flatworms, they have ciliated (cilia) integument. They are also provided with an eversible proboscis enclosed in a dorsal tubular cavity, the rhynchocoel, and associated with but not derived from the alimentary tract.

These worms live among seaweed or at the bottom of the ocean and prey on living animals or eat dead ones. They are not economically important. A few freshwater species and a few parasitic forms are known.

NENCKI REACTION.   The ring acylation of phenols with acids in the presence of zinc chloride, or the modification of the Friedel-Crafts alkylation-acylation procedure by substitution of ferric chloride for aluminum chloride.

NENITZESCU INDOLE SYNTHESIS.   Hydrogenative acylation of cycloolefins with acid chlorides in the presence of aluminum chloride; with five- and six-membered rings, no change in ring size occurs, but with seven-membered rings, rearrangement takes place with formation of a cyclohexane derivative.

NEODYMIUM.   [CAS: 7440-00-8]. Chemical element symbol Nd, at. no. 60, at. wt. 144.24, third in the Lanthanide Series in the periodic table, mp 1,016°C, bp 3,068°C, density 7.004 g/cm^3 (20°C). Elemental neodymium has a close-packed hexagonal crystal structure at 25°C. The pure metallic neodymium is silver-gray in color, the luster becoming dull upon exposure to moist air at room temperatures. When pure, the metal is soft and malleable and may be worked with ordinary equipment. Because the metal is pyrophoric, it must be stored in an inert atmosphere or vacuum. There are seven natural isotopes, ^{142}Nd through ^{146}Nd, ^{148}Nd, and ^{150}Nd. ^{144}Nd is mildly radioactive with a half-life of 10^{10}–10^{15} years. Seven artificial isotopes have been produced. Of the light (or cerium-group) rare-earth metals, neodymium is the third most plentiful and ranks 60th in abundance of elements in the earth's crust, exceeding tantalum, mercury, bismuth, and the precious metals, excepting silver. The element was first identified by C.A. von Welsbach in 1885. Electronic configuration $1s^2 2s^2 2p^6 3s^2 3p^6 3d^{10} 4s^2 4p^6 4d^{10} 4f^3 5s^2 5p^6 5d^1 6s^2$. Ionic radius Nd^{3+} 0.995 Å. Metallic radius 1.821 Å. First ionization potential 5.49 eV; second 10.72 eV.

Other important physical properties of neodymium are given under **Rare-Earth Elements and Metals**.

Primary sources of the element are bastnasite and monazite, which contain from 15 to 25% neodymium. Plant capacity involving liquid-liquid or solid-liquid organic ion-exchange processes for recovering the element is in excess of 200,000 pounds (90,720 kilograms) Nd_2O_3 annually. Metallic neodymium is obtained by electrolysis of fused anhydrous $NdCl_3$ or by the electrolytic reduction of the oxide in molten NdF_3.

Use of elemental neodymium as a colorant for glass was one of the early applications. The color ranges from pure violet to purple and finds use in sunglasses, protective glasses for industry, art objects of glass, tableware, and decorative fiber optics. Use of neodymium in amounts of 3–5% by weight imparts dichroic properties to glass. Neodymium-doped single-crystal yttrium-aluminum oxide garnets (Nd:YAG) have been used in lasers. Research has shown the Nd ion to exhibit laser characteristics in a wide range of compounds and glasses. A formulation of 75% neodymium and 25% praseodymium, frequently called didymium, is used as a metallurgical additive. Within the last several years, it has been found that the use of Nd_2O_3 in barium titanate capacitors increases the dielectric strength of these electronic components over a wider temperature range. Neodymium also has been used as an ingredient of phosphate-type phosphors. Investigations continue into further electronic and optical uses of the element and its compounds.

EDITOR'S NOTE: Extensive research during the early 1980s led to the development of a new and powerful magnet material with the probable composition, $R_2Fe_{14}B$ (where R=a light rare earth). The rare earth predominantly used thus far is neodymium. The recent neodymium-iron-boron material exhibits extremely powerful magnetic qualities as compared with traditional magnet materials. More detail is given under **Rare-Earth Elements and Metals**. Also see **Magnetism**.

Scientists (California Institute of Technology) reported that the isotopic composition of Drake Passage (Antarctica) seawater had been determined. The Antarctic Circumpolar Current, which controls interocean mixing, flows through the Drake Passage. The ratio, $^{143}Nd/^{144}Nd$, was found to be uniform with depth at two experiment stations—with an intermediate value between those of the Atlantic and Pacific Oceans. Further, Piepgras and Wasserburg determined that the Antarctic Circumpolar Current is made up of approximately 70% Atlantic water. It was further reported that cold bottom water from a site in the south-central Pacific has the Nd isotopic signature of the water in Drake Passage. The investigators used a box model to emulate the exchange of water between the Southern Ocean and ocean basis to the north with the isotopic results. An upper limit of about 33 million cubic meters/second was calculated for the rate of exchange between the Pacific and the Southern Ocean. Further determinations of samarium and neodymium were made and found to increase approximately linearly with depth. In essence, the findings suggest that Nd may be a valuable tracer in oceanography and possibly useful in paleo-oceanographic studies. See also **Ocean**; and **Polar Research**.

Additional Reading

Anderson, D.L.: "Composition of the Earth," *Science*, **367** (January 20, 1989).

Cherfas, J.: "Proton Microbeam Probes the Elements," *Science*, **11500** (September 28, 1990).

DePaolo, D.: *Neodymium Isotope Geochemistry*, Springer-Verlag New York, Inc., New York, NY, 1988.

Greenwood, N.N. and A. Earnshaw: *Chemistry of the Elements*, 2nd Edition, Butterworth-Heinemann, Inc., Woburn, MA, 1997.

Letokhov, V.S.: "Detecting Individual Atoms and Molecules with Lasers," *Sci. Amer.*, **54** (September 1988).

Lewis, R.J., Sr.: *Hawley's Condensed Chemical Dictionary*, 13th Edition, John Wiley & Sons, Inc., New York, NY, 1999.

Lide, D. (Editor): *CRC Handbook of Chemistry and 86th Edition—A Ready- Book of Chemical Reference and Physical Data*, CRC Press, LLC., Boca Raton, FL, 2005.

Lugmair, G.W. et al.: "Samarium-146 in the Early Solar System: Evidence from Neodymium in the Allende Meteorite," *Science*, **222**, 1015–1017 (1983).

Piepgras, D.J. and G.J. Wasserburg: "Isotopic Composition of Neodymium in Waters from the Drake Passage," *Science*, **217**, 207–214 (1982).

Robinson, A.L.: "Powerful New Magnet Material Found," *Science*, **223**, 920–922 (1984).

Staff: *ASM Handbook—Properties and Selection: Nonferrous Alloys and Pure Metals*, ASM International, Materials Park, OH, 1990.

White, R.M.: "Opportunities in Magnetic Materials," *Science*, **229**, 11–15 (1985).

K.A. Gschneidner, Jr., and B. Evans, *Rare-Earth Information Center, Institute of Physical Research and Technology*, Iowa State University, Ames, IA.

NEOGLACIAL.   See **Meteorology**.

NEOHESPERIDIN DIHYDROCHALCONE.   See **Sweeteners**.

NEON.   [CAS: 7440-01-9]. Chemical element, symbol Ne, at. no. 10, at. wt. 20.183, periodic table group 18, mp −248.68 °C, bp −246.01 °C, density 1.204 g/cm^3 (liquid). Specific gravity compared with air is 0.674. Solid neon has a face-centered cubic crystal structure. At standard conditions, neon is a colorless, odorless gas and does not form stable compounds with any other element. Due to its low valence forces, neon does not form diatomic molecules, except in discharge tubes. It does form compounds under highly favorable conditions, as excitation in discharge tubes, or pressure in the presence of a powerful dipole. However, the compound-forming capabilities of neon, under any circumstances, appear to be far less than those of argon or krypton. No known hydrates have been identified, even at pressures up to 260 atmospheres. First ionization potential, 21.599 eV.

Neon occurs in the atmosphere to the extent of approximately 0.00182%. In terms of abundance, neon does not appear on lists of elements in the earth's crust because it does not exist in stable compounds. However, because of its limited solubility in H_2O, neon is found in seawater to the extent of approximately 1.5 tons per cubic mile (324 kilograms per cubic kilometer). Commercial neon is derived from air by liquefaction and fractional distillation. For most applications, the gas need not be in a highly pure form, but may be supplied along with small quantities of the other rare gases, such as argon and krypton. The gas finds principal applications in various electronic devices and lamps, but the most familiar application is the neon tubes used mainly in signs. The use of neon signs for identification and advertising signs reached the Iron Curtain countries at a date much later than in the Western countries. Neon emits the familiar orange light. Neon also has been used in certain lasers.

In the 1983 Luberoff reference, the author observes that neon signs, once considered vulgar symbols of a consumer society, are fast becoming icons of a bygone era. However, in recent years a group of preservationists, people who formerly decried the impact of neon advertising, now often defend it. Luberoff points out how a blue-lettered sign (5878 neon-filled glass tubes) became an integral part of the Boston skyline. A study of Boston's signs and lights by the Boston Redevelopment Authority showed that this sign (Citgo) was the only commercial sign that the public thought should remain.

There are three natural isotopes, ^{20}Ne through ^{22}Ne, and four radioactive isotopes, ^{18}Ne, ^{19}Ne, ^{23}Ne, and ^{24}Ne, all with half-lives of less than 5 minutes. Ramsay and Travers first found the element when investigating the properties of liquid air in 1898. The element is easily identified spectroscopically. Neon emits characteristic red and green lines in its spectrum.

Neon in Meteorites. As pointed out by Lewis and Anders, the noble gases are unique among the elements found in meteorites. They are highly volatile and unreactive and they did not condense in even the most primitive meteorites and thus are present at only a minute fraction of their proportion in the sun, ranging from about 10^{-5} for xenon to 10^{-9} for neon and helium. However, very small quantities of these gases are tightly bound in the meteorite and are freed when the host mineral begins to melt or decompose at high temperatures.

Scientists have found three types of neon in meteorites: (1) Primordial or planetary neon (called neon A); (2) solar neon (neon B), which consists of solar-wind neon ions implanted in meteorites that happen to have been at the surface of their parent body; and (3) cosmogenic neon (neon S), formed when cosmic rays passing through the meteorite spall, or shatter, atomic nuclei in their path. Each type has different proportions of the three isotopes of neon. Although the procedure is too detailed for inclusion here, Lewis and Anders explain how, through the use of stepped heating of meteorite materials, the types of neon can be measured. Their ratios to each other provide clues as to what type of star may have been the source of a given meteorite.

Additional Reading

Anderson, D.L.: "Composition of the Earth," *Science*, 367 (January 20, 1989).

Cherfas, J.: "Proton Microbeam Probes the Elements," *Science*, 11500 (September 28, 1990).

Greenwood, N.N. and A. Earnshaw: *Chemistry of the Elements*, 2nd Edition, Butterworth-Heinemann, Inc., Woburn, MA, 1997.

Letokhov, V.S.: "Detecting Individual Atoms and Molecules with Lasers," *Sci. Amer.*, 54 (September 1988).

Lewis, R.S. and E. Anders: "Interstellar Matter in Meteorites," *Sci. American*, **249**(2), 66–77 (1983).

Lewis, R.J., Sr.: *Hawley's Condensed Chemical Dictionary*, 13th Edition, John Wiley & Sons, Inc., New York, NY, 1999.

Lide, D.R.: *CRC Handbook of Chemistry and Physics* 86th Edition, CRC Press, LLC., Boca Raton, FL, 2005.

Luberoff, D.: "But Is It Art? (Neon Signs)," *Technology Review (MIT)*, **86**(5), 76–77 (July 1983).

NEONATAL. Related to or affecting the newborn human child, particularly during the first month after birth.

NEOPLASM. Any new or abnormal overgrowth of cellular tissue. A neoplasm is a cellular tumor and may be either benign or malignant.

NEOPRENE. See **Elastomers**.

NEOTAME. See **Sweeteners**.

NEPER (NP). See **Units and Standards**.

NEPHELINE. Nepheline, of hexagonal crystallization, is a sodium-potassium aluminum silicate $(Na, K)(AlSiO_4)$. It is found in silica-poor geological environments, where there had been insufficient silica to form feldspar. Nepheline rocks are characterized by the absence of quartz within them. They constitute a mineral family group known as the *feldspathoids*. Crystals are extremely rare; usually occurs massive to compact. Luster, is greasy in the massive varieties; vitreous in crystals. Color grades from yellowish to colorless in crystals; gray, green and reddish in massive material. It ranges from transparent to translucent. Hardness is of 5.5–6, specific gravity of 2.55–2.65.

Immense masses of nepheline-rich rocks occur on the Kola Peninsula, the former U.S.S.R., in Norway and in the Republic of South Africa; also in the Bancroft, Ontario, Canada region. Smaller deposits are found in Maine and Arkansas in the United States. Fine crystals are found in lavas on Mt. Vesuvius, Italy.

Nepheline is used extensively in the manufacture of glass.

ELMER B. ROWLEY, Union College, Schenectady, NY

NEPHELOMETRY. Sir John Tyndall noted that particles that are invisible when directly in the path of a strong light become discernible when viewed from the side. Now known as the Tyndall effect, the phenomenon derives from reflection of part of the incident light by the particles. The reflected light is directly proportional to the number of particles in suspension. An instrument for measuring the intensity of reflected light so produced in a nephelometer and may be used for the quantitative determination of small amounts of diverse materials that have the ability to reflect light when in liquid suspension. Examples include the measurement of traces of silver wherein the chloride ion is added to a solution of material containing silver to produce insoluble silver chloride in suspension form. Small amounts of calcium in titanium alloys may be determined by measuring suspensions of the stearate formed in a suitable medium. Nephelometry also finds application in the measurement of bacterial growth rates; for the analysis of cholesterol, glycogen, and enzymes; for controlling the clarity of beverages, water, and wastewater; for solution control in tanning operations; and for any measurement situation where an unknown composition may be transformed into, or related to, a form of suspension.

Nephelometric methods are similar to fluorometric methods in that both involve measurement of scattered light. However, the scattering is inelastic in nephelometry and elastic in fluorometry. Thus, the scattered light measured in fluorometry is of a longer wavelength than the incident light, and both incident and scattered light are of the same wavelength in a nephelometric determination. In fact, the two functions sometimes are combined into one instrument, which may be termed a nefluoro-photometer. When the instrument operates as a nephelometer, it utilizes two Tyndall windows, located opposite each other in a cylindrical sample cell and with their common axis perpendicular to the path of the entering light. The concentration of suspended particles is determined by summing the photocurrents of the two cells. When used as a fluorometer, the instrument measures light emitted by a sample that is excited by incident radiation in the appropriate spectral band. Further, the same instrument can be set up for use as a photometer to measure light transmitted by the sample. Three light sources may be used—an incandescent source for colorimetric or nephelometric applications; a mercury-arc source for fluorometry; and a sodium-arc source with principal emission at 320 and 590 nanometers, when a sharp peak at either of these wavelengths is required, as in the instance of vitamin A determinations.

See also **Analysis (Chemical)**; **Fluorometers**; **Photometers**; and **Turbidimetry**.

NEPHOLOGY. See **Meteorology**.

NEPHOMETER. See **Precipitation and Hydrometeors**.

NEPHOSCOPE. An instrument for determining the direction of cloud motion. There are two basic designs of nephoscope: the direct-vision nephoscope and the mirror nephoscope. Also called *nepheloscope*.

NEPHRITIS. See **Kidney and Urinary Tract**.

NEPHRON. See **Kidney and Urinary Tract**.

NEPTUNE (Planet). Eighth planet from the sun (~3 billion miles; 4.5 billion km), Neptune is about four times the size (diameter) of Earth, with a mass slightly greater than 17 times that of Earth. Its mean density is less than that of Earth. This low density, coupled with the high value of the planet's albedo (0.52), is indicative of the planet's thick atmospheric layer. Neptune has seven confirmed satellites, the best understood of which is Triton.

Neptune is invisible to the naked eye, having a stellar magnitude of only about 7.7. The planet can be observed with a telescope having an aperture greater than 2.5 cm (1 in), but it can be distinguished with such a small instrument only by its change in position against the starry background from night to night.

The planet was discovered in 1846 and will not complete one trip around the sun from its position at that time until the year 2010, a total of 164 years later. Viewed by a large telescope, Neptune appears as a small, circular, somewhat greenish disk without distinctive markings. Thus, for many years, it was difficult to estimate the planet's rotation period. However, in 1928, Moore and Menzel (Lick Observatory) found, from spectroscopic observations employing the Doppler principle, that the planet rotates in the same directional sense as most other members of the solar system and with a rotation period of about 16 hours.

Voyager 2 Encounter with Neptune

The *Voyager 2* spacecraft, initially conceived to visit Jupiter (July 9, 1981) and Saturn (August 26, 1981), was launched from Cape Canaveral, Florida on August 20, 1977, and performed so well that its mission was extended to include later encounters with Uranus (January 24, 1986) and Neptune (August 25, 1987). The dates given are for closest approaches.

Upgrading of System Instrumentation. With the pending investigation of Neptune in mind, after completion of the encounter with Uranus, project managers took advantage of the available time (approximately 3 years) to improve ground system instrumentation as much as possible to upgrade the data return for the Neptune encounter. These efforts included:[1]

- *Attitude control* was altered to reduce angular rates by approximately 25% below those experienced at Uranus, in an attempt to improve compensation for impulses from tape recorder starts and stops, to permit an additional scan platform rate useful for motion compensation near the closest approaches to Neptune and Triton, and to provide for "nodding" image motion compensation (NIMC). This permitted acquisition of motion-compensated images without disrupting the communication link

[1] As reported by E.C. Stone (California Institute of Technology) and E.D. Miner (Jet Propulsion Laboratory).

with Earth or utilizing limited tape recorder resources. Instrument control for the imaging system also was changed to permit exposure durations between 15 and 96 seconds and real-time images with exposure durations in multiples of 48 seconds—capabilities that were used extensively during the Neptune encounter, where light levels were only 40% of those at Uranus.

- *Signal strength* is notably less from Neptune than from Uranus because of the great transmission distance between the spacecraft and Earth-based tracking systems. Each of the three 64-meter–diameter tracking antennas of the National Aeronautics and Space Administration (NASA) Deep Space Network (DSN) were enlarged to 70-meter diameter and improved in shape. A high-efficiency, 34-meter tracking station was added to the Madrid DSN complex. Extensive arraying of antennas was used to further increase the effective collector area. The 64-meter Parkes Radio Telescope in Australia again was made available to enhance the capability of the Canberra DSN tracking antennas. Similarly, Voyager signals collected by the National Radio Astronomy Observatory's Very Large Array (VLA) near Socorro, New Mexico, were combined with signals collected at the Goldstone, California DSN complex. The 27 25-meter antennas of the VLA provided a collecting area equivalent to 2 70-meter antennas. During closest-approach operations on August 25, 1989, the Japanese Institute of Astronautical Science utilized its 64-meter tracking antenna at Usuda, Japan, to augment Voyager science data collection. The spacecraft and all of the ground systems worked flawlessly during the Neptune encounter, testifying to the high level of expertise and teamwork within the Voyager project and its supporting organizations.

- *Trajectory design* was engineered to maximize the information returned during the Neptune-Triton encounters. The team had maximum freedom in this regard because no further missions for Voyager 2 were contemplated. Three primary objectives were sought: (1) a close approach to Triton, including both sun and Earth occultations as viewed from the spacecraft; (2) a close polar passage of Neptune, including both sun and Earth occultations; and (3) timing of the closest approach so that both of the Neptune-Triton occultations occurred at relatively high elevation angles over the Canberra DSN complex.

Voyager 2 **Instrument Systems and Management.** The investigatory disciplines and their management for the Neptune encounter are outlined in Table 1.

TABLE 1. *VOYAGER 2* INVESTIGATIONS AND MANAGERS

Investigation	Principal investigator and affiliation
Imaging (ISS)	B.A. Smith, University of Arizona, Tucson, AZ
Photopolarimetry (PPS)	A.L. Lane, Jet Propulsion Laboratory, California Institute of Technology, Pasadena, CA
Infrared spectroscopy (IRIS)	B.J. Conrath, Goddard Space Flight Center, Greenbelt, MD
Ultraviolet spectroscopy (UVS)	A.L. Broadfoot, University of Arizona, Tucson, AZ
Radio science (RSS)	G.L. Tyler, Stanford University, Stanford, CA
Magnetometry (MAG)	N.F. Ness, Bartol Research Institute, University of Delaware, Newark, DE
Plasma (PLS)	J.W. Belcher, Massachusetts Institute of Technology, Cambridge, MA
Low-energy charged particles (LECP)	S.M. Krimigis, Applied Physics Laboratory, The Johns Hopkins University, Laurel, MD
Cosmic rays (CRS)	E.C. Stone, California Institute of Technology, Pasadena, CA
Plasma waves (PWS)	D.A. Gurnett, University of Iowa, Iowa City, IA
Planetary radio astronomy (PRA)	J.W. Warwick, Radiophysics, Inc., Boulder, CO

Chronology of the Neptune Encounter. Activities of the *Voyager 2* encounter commenced on June 5, 1989, when the spacecraft was 117×10^6 km from the center of the planet. The closest approach (center of Neptune) was a distance of 29,240 km and occurred on August 25, 1989. Radio signals from this position required a transit time to Earth of 4 hours,

6 minutes. The closest approach to Triton occurred on September 10, 1989, when the spacecraft was 39,800 km from the center of the satellite. This encounter period extended to October 2, 1989.

Design of the Neptune science sequences relied mainly on prior telescopic observations from Earth and on Voyager findings at Uranus, although some early *Voyager* data were used to make revisions to later observations. See Fig. 1. Provision was made for late retargeting of imaging frames to newly discovered satellites and rings. Timing of close Neptune and Triton observations also was adjusted at the last possible instant to take advantage of the most recent estimates of geometric event times made by the navigation team.

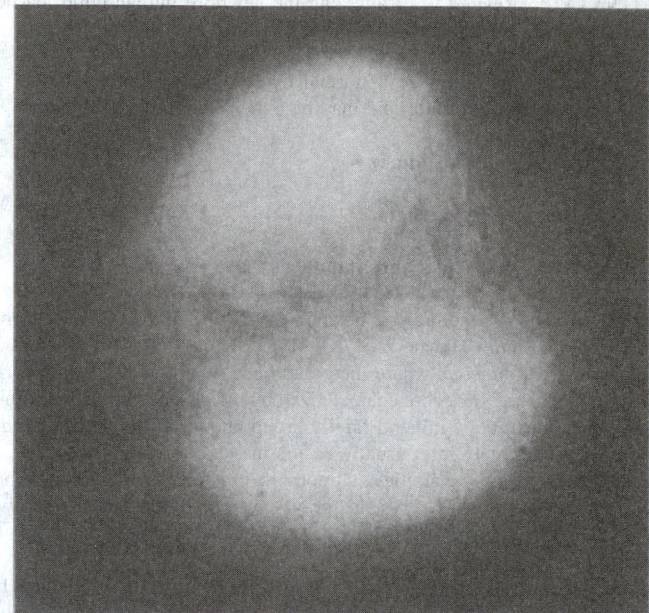

Fig. 1. Features of Neptune in which camera incorporating a charge-coupled device (CCD) was used with a 1.54-meter Catalina telescope. (*University of Arizona*, 1979.)

An initial, comprehensive report of the *Voyager 2* encounter was released in December 1989. Detailed information is available from the Jet Propulsion Laboratory, Pasadena, California. Some of these reports were included in *Science* magazine, issue of December 15, 1989. Analysis of these data, including the formulation of explanations and postulations pertaining to the overall Neptunian system, will be forthcoming for several years into the future. A number of post-encounter papers issued during the early 1990s are listed at the end of this article.

The spacecraft's plutonium power sources may hold out until approximately 2015. Somewhat before that time, *Voyager 2* is expected to encounter the heliopause (very edge of the solar system where the solar wind collides with interstellar media). See also **Voyager Missions to Jupiter and Saturn**.

Atmosphere of Neptune

Images of Neptune were obtained by the narrow-angle camera of Voyager 2 and indicated large-scale cloud features that persist for several months or longer.[2] The periods of rotation of these features about the planetary axis range from 15.8 to 18.4 hours. The atmosphere equatorward of −53° rotates with periods longer than the 16.05-hour period deduced from *Voyager's* planetary radio astronomy experiment. This is presumably the planet's internal rotation period. The wind speeds computed with respect to this radio period range from 20 meters per second eastward to 325 meters per second westward. Thus, it was found that the cloud-top wind speeds are approximately the same for all the planets ranging from Venus to Neptune, even though the solar energy inputs to the atmospheres vary by a factor of 1000.

---

[2] As reported by H.B. Hammel (Jet Propulsion Laboratory) and associated team members (see papers listed).

Neptune has an effective temperature of about 59.3 K. Derivation of Neptune's Bond albedo continues to require a more thorough study of *Voyager 2* instrument data. Neptune, however, appears to emit about 2.7 times as much energy as it absorbs from the sun. This greater contribution of internal heat may be the cause of the greater activity in the Neptunian atmosphere relative to that of Uranus. The horizontal temperature structures of the two atmospheres are quite similar, with the poles and equator at very nearly equal temperatures, while mid-latitudes are several degrees cooler. Temperature in the extreme upper atmosphere is nearly 750 K, but, because of Neptune's larger mass, colder atmosphere, and greater ring distances, the effects of gas drag on ring material are less at Neptune than at Uranus.

A number of prominent cloud features are apparent in images of Neptune's atmosphere, including an Earth-size "Great Dark Spot" (GDS) which occurs near −10° latitude. The GDS is located and is of approximately the same size as Jupiter's Great Red Spot. See Figs. 2 and 3. The GDS rolls in a counterclockwise direction, with a 16-day period. Another, but smaller, dark spot with a bright central core is located at −55° latitude. Bands of lower reflectivity extend from +6° to +25° latitude and from −45° to −70° latitude. The GDS is flanked by cirruslike cloud features. Other similar features occupy relatively narrow latitude ranges near latitudes of −27° and −71°. These are believed to be optically thick upward extensions of the methane (CH_4) cloud deck. The details of these features change with time scales that are smaller as compared with the 16-hour rotation

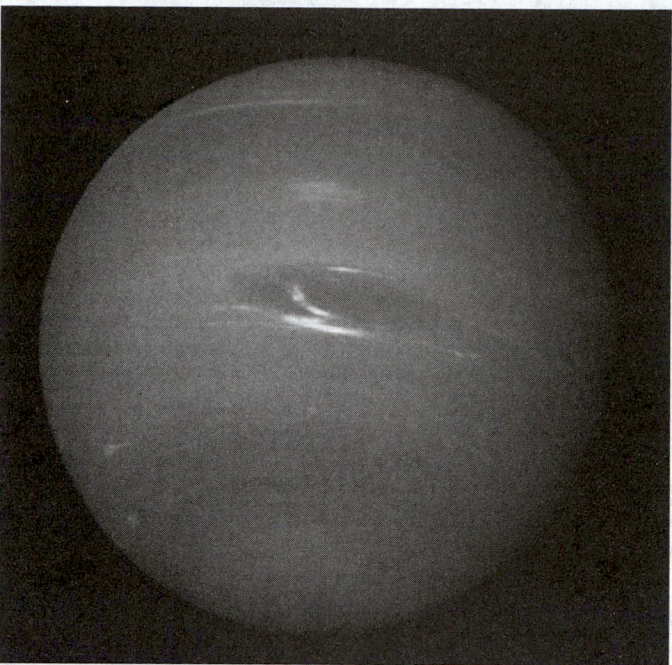

Fig. 3.   This picture of Neptune was produced from the last whole planet images taken through the green and orange filters on the *Voyager 2* narrow angle camera. The images were taken at a range of 4.4 million miles from the planet, 4 days and 20 hours before closest approach. The picture shows the Great Dark Spot and its companion bright smudge; on the west limb the fast moving bright feature called Scooter and the little dark spot are visible. These clouds were seen to persist for as long as Voyager's cameras could resolve them. North of these, a bright cloud band similar to the south polar streak may be seen. (*Jet Propulsion Laboratory, Pasadena, California.*)

period. One of these features, referred to as the "Scooter," is bright and found near the −42° latitude. It is deeper within the atmosphere than the aforementioned cirrus clouds and is postulated to be an upward extension of the deeper cloud deck. Velocities measured with respect to the internal rotation of Neptune indicate wind speeds ranging from about +20 m s^{-1} (prograde) at −54° latitude to −325 m s^{-1} (retrograde) at −22°. The GDS resides in a region with strong wind shears and is believed to lie at a lower level than most of the brighter cloud features, virtually independent of the higher-altitude winds. The cirruslike clouds are found at altitudes of 50 to 100 km above the lower cloud layer. Optically thin layers of haze, believed to be produced photochemically from CH_4, are found at still higher altitudes.

As with the other giant planets, hydrogen (H_2) predominates the Neptunian atmosphere. Although subject to further calculations from measurements taken, the mole fraction of atmospheric helium [He]/[H_2] is estimated to be less than 0.25. Methane is more abundant in Neptune's upper atmosphere than in the atmosphere of Uranus. The absorption of red light by CH_4 gives Neptune its characteristic blue color. Deeper in the atmosphere, acetylene was detected. The signature of an optically thin cloud deck of methane ice was observed in radio occultation data. Strong absorption of radio waves may indicate the presence of small amounts of ammonia (NH_3).

Some further analyses of wind speed data are in order. Some scientists believe that the winds on Neptune are faster than found on any other planet, a surprising situation because of the small amount of energy received from the sun or from the interior of the planet. Past conceptual models of the general circulation of the giant planets do not readily provide a simple explanation as to why the highest winds appear to occur on Neptune, with its assumed low-energy sources. Neptune receives an estimated $\frac{1}{900}$ of the Earth's input of solar energy, but may have wind speeds of nearly 600 meters per second. How the near-supersonic winds can be maintained has puzzled some investigators. Scientists at the Space Science and Engineering Center of the University of Wisconsin–Madison have offered the following hypothesis: "Based on principles of angular

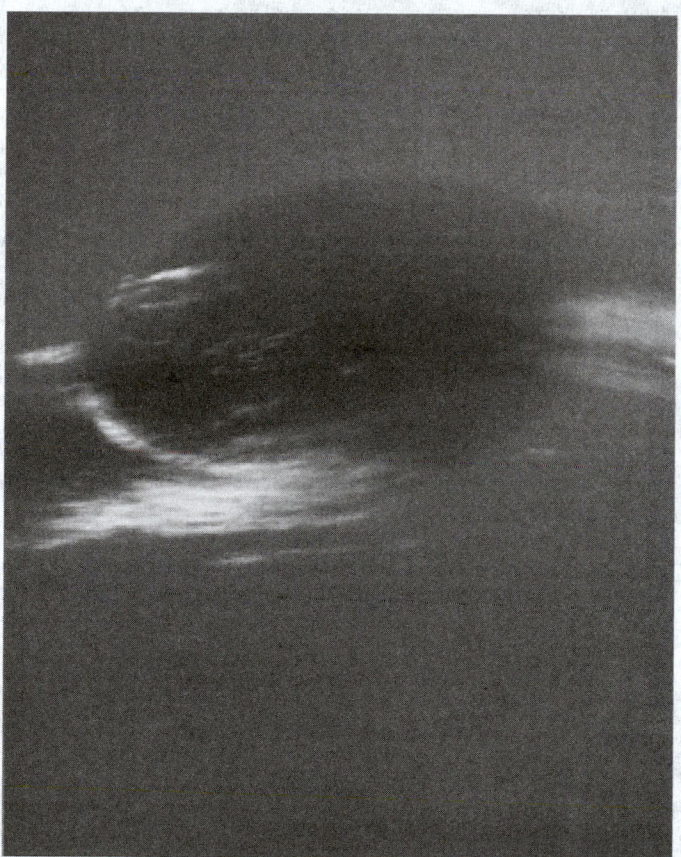

Fig. 2.   This photograph shows the last face-on view of the "Great Dark Spot" that *Voyager* will make with the narrow-angle camera. The image was shuttered 45 hours before closest approach at a distance of 2.8 million kilometers (1.7 million miles). The smallest structures that can be seen are of an order of 50 kilometers (31 miles). The image shows feathery white clouds that overlie the boundary of the dark and light blue regions. The pinwheel (spiral) structure of both the dark boundary and the white cirrus suggest a storm system rotating counterclockwise. Periodic small- scale patterns in the white cloud, possibly waves, are short-lived and do not persist from one Neptunian rotation to the next. This color composite was made from the clear and green filters of the narrow-angle camera. (*National Aeronautics and Space Administration, Jet Propulsion Laboratory, Pasadena, California.*)

TABLE 2. PROPERTIES OF NEPTUNE'S RINGS

Feature	Distance (10^3 km)	Distance (R_N)	Width (km)	Optical Depth	Comments
1-bar atmosphere	24.76	1.000			Equatorial radius of Neptune
	38.	1.5		<0.0001	Inner extent of 1989N3R?
1989N3R	41.9	1.69	1700*	0.0001	High dust content
	49.	2.0		<0.0001	Outer extent of 1989N3R?
1989N2R	53.2	2.15	†	0.01	High dust content
1989N4R (inner)	53.2	2.15		0.0001	Inner edge of "plateau"
1989N4R (outer)	59.	2.4		0.0001	Outer edge of "plateau"
1989N1R	62.9	2.54	15	0.01–0	Contains three bright dusty arcs

*Tabulated width of 1989N3R is full width at half maximum. †1989N2R is narrow and unresolved in Voyager images.
Source: California Institute of Technology and Jet Propulsion Laboratory.

momentum and energy conservation in conjunction with deep convection, leads to a regime of uniform angular momentum at low latitudes. In this model, the rapid retrograde winds observed are a manifestation of deep convection, and the high efficiency of the planet's heat engine is intrinsic from the room allowed at low latitudes for reversible processes, the high temperature at which heat is added to the atmosphere, and the low temperatures at which heat is extracted." (See Suomi/Limaye/ Johnson reference listed.)

A scientist (California Institute of Technology) observes, "Neptune's supersonic winds are not a certainty. The altitudes of the different cloud features, for example, are difficult to confirm, making it hard to compute the clouds' speeds. Moreover, it is difficult to tell from the photos whether the movements represent actual fluid motion of atmospheric masses or merely a wave moving through the atmosphere." (See Eberhart reference listed.)

Rings of Neptune

Earth-based stellar occultation measurements made during the early and middle 1980s alerted investigators to the probable existence of rings or, at least, partial rings (ring arcs) at a number of radial distances from the center of the planet. *Voyager 2* imaging confirmed the presence of a system of at least six rings of prograde, equatorial, and circular rings. The outermost ring occurs at a distance of 62,900 km from the center of the planet. It is described as being composed of three bright, dusty areas. Data on Neptune's rings are given in Table 2.

Narrow rings are believed to be confined by the actions of relatively nearby satellites (shepherds). These rings may serve to prevent material from spiraling inward toward the planet. No ring shepherds have been noted thus far in the Neptunian system. The *Voyager 2* instrumentation, however, was limited to observing satellites of a diameter of 12 km or greater. Thus, tiny satellites may have escaped attention. Thus, it is not known whether or not additional shepherding satellites exist. Such material, if azimuthally unrestrained within the ring, should spread relatively uniformly around the ring within a time span of a few years. Additional encounters with largely enhanced resolution may be required at some future date to fully explain the dynamics of the planet's ring system.

Particles within the rings appear to be smaller than those found in the rings of Uranus. The dust content of one ring (1989N3R) is nearly double that of the other rings and thus compares better with that of the rings of Saturn and Uranus.

From analysis of *Voyager 2* data, C. Porco (Department of Planetary Sciences, University of Arizona) proposes an interesting explanation for Neptune's ring arcs, "A radial distortion with an amplitude of approximately 30 km is traveling through the ring arcs, a perturbation attributable to the nearby satellite Galatea. Moreover, the arcs appear to be azimuthally confined by a resonant interaction with the same satellite, yielding a maximum spread in ring particle semimajor aces of 0.5 km and spread in forced eccentricities large enough to explain the arcs' 15-km radial widths." Additional ring arcs were discovered during the course of the study and provide further support to this model. (See Poroco reference listed.)

TABLE 3. PROPERTIES OF NEPTUNE'S SATELLITES

Satellite Name	Distance (10^3 km)	Distance (R_N)	Period (Hours)	Diameter (km)	Resolution (km per Line Pair)	Normal Albedo
1989N6	48.0	1.94	7.1	54 ± 16	47.2	0.06?
1989N5	50.0	2.02	7.5	80 ± 16	34.8	0.06?
1989N3	52.5	2.12	8.0	180 ± 20	36.8	0.06?
1989N4	62.0	2.50	10.3	150 ± 30	33.8	0.054
1989N2	73.6	2.97	13.3	190 ± 20	8.2	0.056
1989N1	117.6	4.75	26.9	400 ± 20	2.6	0.060
Triton	354.8	14.33	141.0	2705 ± 6	0.8	0.6–0.9
Nereid	5513.4	222.65	8643.1	340 ± 50	86.6	0.14

Source: California Institute of Technology and Jet Propulsion Laboratory.

Triton and Other Satellites[3]

During its approach to Neptune, *Voyager 2* images revealed six new satellites. All satellites orbit the planet in prograde, circular orbits of low inclination. Characteristics of the satellites are summarized in Table 3. Five of the six satellites orbit within 1° of Neptune's equatorial plane. The 1989N6 satellite has an inclination of nearly 6°. Data from Earth-based observations and *Voyager 2* (Figs. 4–10). Fig. 11 shows Triton's inclination to be 157°. Nereid's inclination is 29°. The respective orbital eccentricities of Triton and Nereid are 0.00 and 0.75. Nereid's distance from the center of the planet ranges from 1.39×10^6 to 9.64×10^6 km. Nereid's highly elliptical orbit makes it theoretically unlikely that its rotation and orbital periods are equal. However, *Voyager 2* detected no rotational brightness variations in excess of 10%.

Triton. Much scientific interest has been directed toward this, by far the largest of Neptune's satellites and the existence of which has been known by earthbound observations for several years. Of all known natural bodies in the solar system, Triton has the lowest surface temperature (38 ± 4 K). Triton's atmosphere is predominantly nitrogen (N_2), with the presence of CH_4 in the lower atmosphere. The surface pressure, as measured by the radio science instrumentation aboard *Voyager 2*, is 16 ± 3 microbars. It is believed that a thermal inversion may exist in the lower 5 km of the atmosphere, and thus a tropopause altitude of 25 to 50 km is inferred. It is uncertain whether the clouds and haze layers observed in this region of the atmosphere result from simple condensation or from surface eruptions. The temperature at altitudes above 400 km is 95 ± 5 K. Atmospheric nitrogen is transported from the illuminated polar regions to the unilluminated polar regions.

The surface of Triton is that of a geologically young body and is devoid of heavily cratered terrain. The polar regions (south of latitude −15°) are covered with seasonal ice, believed to be N_2. Spring in Triton's southern region extends for several years—for example, in the present time frame, from about 1960 to the year 2000. Seasonal ice presents a slightly reddish tint believed to result from organic compounds photochemically produced by interaction between methane and

[3] Principal information source: Initial report of Jet Propulsion Laboratory, Pasadena, California.

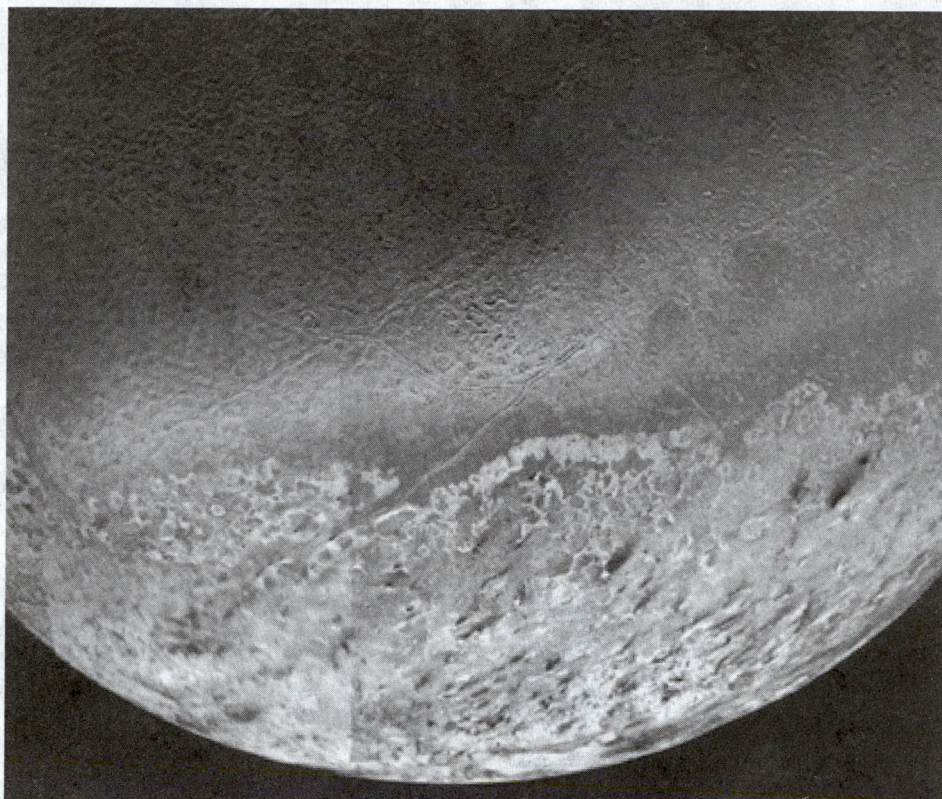

Fig. 4. *Voyager 2* obtained this high-resolution color image of Neptune's large satellite Triton during its close flyby on Aug. 25, 1989. Approximately a dozen individual images were combined to produce this comprehensive view of the Neptune-facing hemisphere of Triton. Fine detail is provided by high-resolution, clear-filter images, with color information added from lower-resolution frames. The large south polar cap at the bottom of the image is highly reflective and slightly pink in color; it may consist of a slowly evaporating layer of nitrogen ice deposited during the previous winter. From the ragged edge of the polar cap northward the satellite's face is generally darker and redder in color. This coloring may be produced by the action of ultraviolet light and magnetospheric radiation upon methane in the atmosphere and surface. Running across this darker region, approximately parallel to the edge of the polar cap, is a and of brighter white material that is almost bluish in color. The underlying topography in this bright band is similar, however to that in the darker, redder regions surrounding it. (*Jet Propulsion Laboratory, Pasadena, California.*)

Fig. 5.   View of about 300 mi (483 km) across Triton's surface. (*National Aeronautics and Space Administration, Jet Propulsion Laboratory, Pasadena, California.*)

nitrogen. Energetic particle bombardment also may assist in producing these reactions. The equatorial regions of Triton at most latitudes contain a thin layer of nitrogen frost. The layer appears as a bright, slightly blue coloration, a layer that does not obscure underlying topography. Observations indicate that northward of the equator there is a variety of terrains. Some scientists have referred to this topology as reminiscent of the "skin of a cantaloupe."

This terrain dominates the western (trailing) hemisphere of Triton and is believed to consist of a dense concentration of pits (dimples) that are crisscrossed by ridges. Few impact craters are discernible in these areas. The terrain of the eastern (leading) hemisphere of Triton is made up of a series of much smoother units, including caldera-like structures of water ice. Frozen lakes are surrounded by successive terraces, possibly due to a series of flooding actions.

Within the polar regions of Triton are numerous wind streaks with albedos that are 10–20% lower than the polar ices. The streaks, which overlie deeper ice deposits, appear to be young, possibly less than 1000 years of age. Two active geyserlike plumes were discovered near the subsolar latitude ($-55°$). As determined by stereoscopic viewing, these plumes rise to an altitude of about 8 km. Above a plume, dense clouds form and serve as a source for a westward wind-driven trail of material more than 100 km long. It has been suggested that the plumes may result from the explosive release of N_2 gas, which carries ice-entrained dark material in the exit nozzle to high altitudes.

In a report by L.A. Soderblom (U.S. Geological Survey, Flagstaff, Arizona) and a team of investigators, they explicitly describe the plume phenomenon: "The radii of the rising columns appear to be in the range of several tens of meters to a kilometer. One model for the mechanism to drive the plumes involves heating of nitrogen ice in a subsurface greenhouse environment; nitrogen gas pressurized by the solar heating explosively vents to the surface carrying clouds of ice and dark particles into the atmosphere. A temperature increase of less than 4 kelvins above the ambient surface value of 38 ± 3 K is more than adequate to drive the plumes to an 8-km altitude. The mass flux in the trailing clouds is estimated to consist of up to 10 kg of fine dark particles per second, or twice as much nitrogen ice and perhaps several hundred or more kilograms of nitrogen gas per second. Each eruption may last a year or more, during which on the order of a tenth of a cubic kilometer of ice is sublimed."

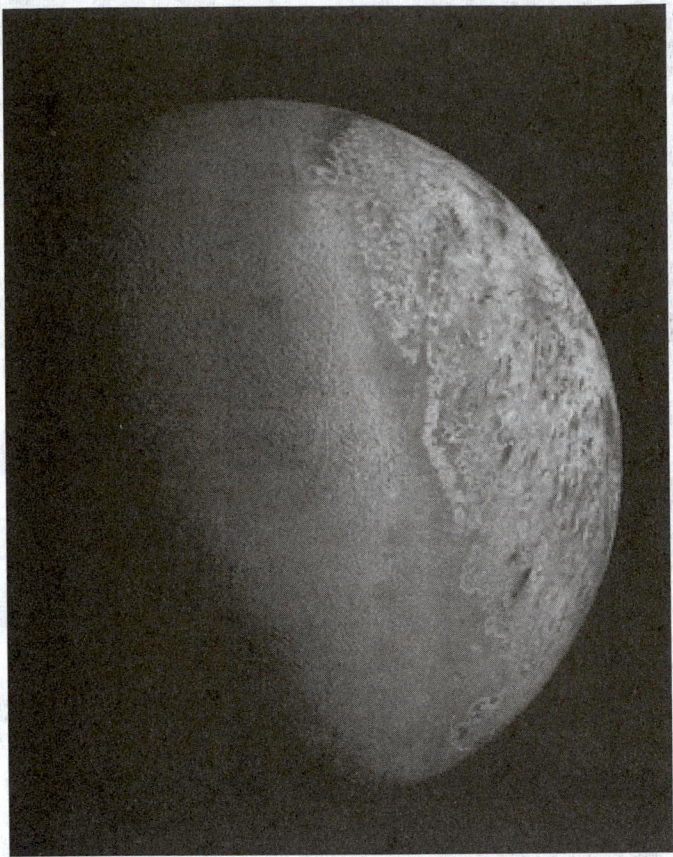

Fig. 6.   This photo-mosaic of Neptune's major satellite Triton was constructed from high-resolution images obtained by the *Voyager 2* spacecraft during its close flyby August 25, 1989. Images taken through Voyager's orange, violet, and ultraviolet filters are displayed as red, green, and blue respectively. The large, pinkish, highly reflective south polar cap is in the lower part of the picture; it may consist of a slowly evaporating layer of nitrogen ice deposited during the previous winter. North of the polar cap's ragged edge, the surface is darker and redder, perhaps because of the action of ultraviolet and magnetospheric radiation upon atmospheric and surface methane. A bluish band, running just north of the edge of the cap, may be freshly deposited nitrogen frost. The western (left) part of the region is dotted with small dimples with raised rims and shallow central depressions; long fractures, which have allowed icy material to ooze up, cross each other and extend into the polar region. Eastward, in the bluish region, an area of smooth plains and low hills, shows dense cratering, extensive resurfacing, and two areas resembling lunar maria, which may have been formed by large scale volcanic flooding. Triton's volcanic and tectonic activities involved icy materials such as frozen methane, nitrogen and water. The dark streaks prominent in the south polar region probably result from active venting, discovered in Voyager 2 images. (*U.S. Geological Survey, Flagstaff, Arizona. Jet Propulsion Laboratory, Pasadena, California.*)

In another approach to the plume phenomenon, A.P. Ingersol and K.A. Tryka (Division of Geological and Planetary Sciences, California Institute of Technology) observe, "Their structure suggests that the plumes are an atmospheric rather than a surface phenomenon. The closest terrestrial analogs may be dust devils, which are atmospheric vortices originating in the unstable layer close to the ground. Since Triton has such a low surface pressure, extremely unstable layers could develop during the day. Patches of unfrosted ground near the subsolar point could act as sites for dust devil formation because they heat up relative to the temperature of 48 K or higher, as observed by the *Voyager* radio science team. Assuming that velocity scales as the square root of temperature difference times the height of the mixed layers, a velocity of 20 m per second is derived from the strongest dust devils on Triton. Winds of this speed could raise particles provided they are a factor of 10^3 to 10^4 less cohesive than those on Earth."

Impact craters are rare on Triton — that is, those that are observable at the 3 to 1.8 km resolution acquired during the mapping sequence of

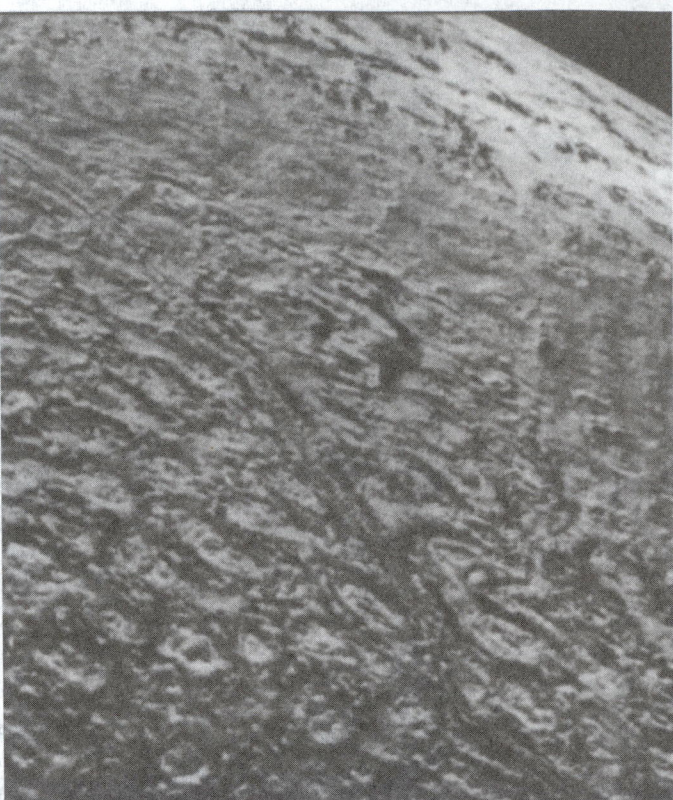

Fig. 7.   Triton from 25,000 mi (40,225 km). Depressions may be caused by melting and collapsing of icy surface. (*National Aeronautics and Space Administration, Jet Propulsion Laboratory, Pasadena, California.*)

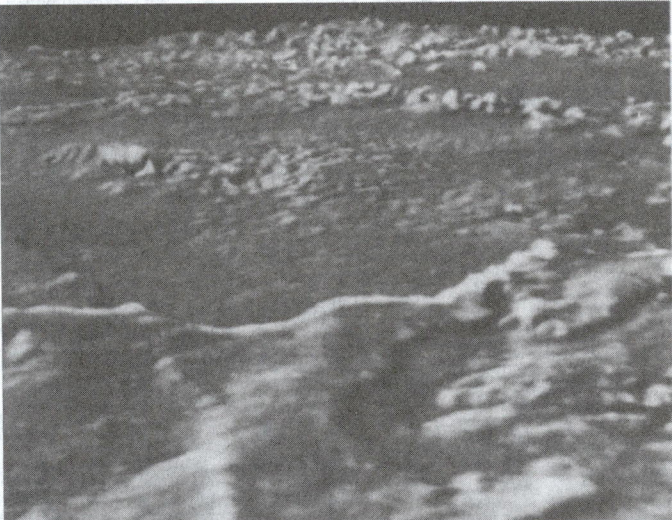

Fig. 8.   Computer-generated perspective view of one of Triton's calderalike depressions. (*National Aeronautics and Space Administration, Jet Propulsion Laboratory, Pasadena, California.*)

Voyager 2. The highest-resolution images obtained show various degrees of smear, making analysis difficult. Thus, it is difficult to compare Neptunian cratering with other observed planets. The largest and uncontested impact crater viewed by *Voyager 2* on Triton is only about 27 km in diameter. Several large quasi-circular features exist, but are believed to be of internal origin. Fresh impact craters on Triton have morphologies similar to those on other icy satellites seen at comparable resolutions. These features include simple sharp-rimmed and bowl-shaped interiors and a few craters with flat floors and central peaks. Based mainly on the similarity of size distribution on Triton

Fig. 9. Satellite 1989N1, discovered by *Voyager 2*. (*National Aeronautics and Space Administration, Jet Propulsion Laboratory, Pasadena, California.*)

Neptune's Magnetosphere

Eight days prior to *Voyager 2's* closest approach to Neptune, a distance of about 470 Neptune radii (R_N), the first indication of a Neptune magnetic field was obtained from radio emissions. Subsequently, the spacecraft crossed a well-defined, detached bow shock at 34.9 R_N. The inbound magnetopause was not as well defined because *Voyager* entered a highly tilted magnetic field at very high magnetic latitude, permitting the first observation of a "pole-on" magnetosphere in which the solar wind is incident on the magnetic polar region rather than the equatorial.

It has been determined that, as the magnetic field rotates with the planet each 16.11 hours, satellites and ring particles sweep through large ranges of magnetic latitude. The incident solar wind deforms the magnetic field, resulting in a well-developed magnetic tail behind the planet. However, as the planet rotates, the magnetosphere configuration changes from pole-on with a cylindrically shaped magnetotail plasma sheet to a more normal planar plasma sheet. Because of this unique geometry and the timing of the flyby, the spacecraft did not cross the plasma sheet.

No evidence exists for Neptunian electrostatic discharges of the kind observed on Saturn and Uranus by planetary radio astronomy (PRS). Many typical plasma waves were detected by the plasma wave instrumentation during the encounter. These included electron plasma oscillations in the solar wind upstream of the bow shock, and chorus, hiss, electron cyclotron waves, and upper hybrid resonance waves in the inner magnetosphere. There was no indication of lightning-generated whistlers.

See also **Voyager Missions to Jupiter and Saturn**; and **Hubble Space Telescope (HST)**.

and Miranda (satellite of Uranus) and the relatively young surface of Triton, comets are believed to be the primary source of cratering. On the other hand, the peculiar size distribution of sharp craters on the "cantaloupe" terrain and other evidence suggests that they are of volcanic origin.

Fig. 10. These two 591-second exposures of the rings of Neptune were taken with the clear filter by the *Voyager 2* wide-angle camera on Aug. 26, 1989 from a distance of 280,000 kilometers (175,000 miles). The two main rings are clearly visible and appear complete over the region imaged. The time between exposures was one hour and 27 minutes. [During this period the bright ring arcs in the outer bright ring were not visible in either picture (they were unfortunately on the opposite side of the planet for each exposure).] Also visible in this image is the inner faint ring at about 42,000 kilometers (25,000 miles) from the center of Neptune, and the faint band which extends smoothly from the 53,000 kilometer (33,000 miles) ring to roughly halfway between the two bright rings. Both of these newly discovered rings are broad and much fainter than the two narrow rings. These long exposure images were taken while the rings were back-lighted by the sun at a phase angle of 135 degrees. This viewing geometry enhances the visibility of dust and allows fainter, dusty parts of the ring to be seen. The bright glare in the center is due to over-exposure of the crescent of Neptune. The two gaps in the upper part of the outer ring in the image on the left are due to blemish removal in the computer processing. Numerous bright stars are evident in the background. Both bright rings have material throughout their entire orbit, and are therefore continuous. (*Jet Propulsion Laboratory, Pasadena, California.*)

Fig. 11.   Neptune and Triton 3 days after flyby. Triton is smaller crescent and is closer to viewer. (*National Aeronautics and Space Administration, Jet Propulsion Laboratory, Pasadena, California.*)

Additional Reading

Belcher, J.W. et al.: "Plasma Observations Near Neptune: Initial Results from Voyager 2," *Science*, 1478 (December 15, 1989).

Broadfoot, A.I.: "Ultraviolet Spectrometer Observations of Neptune and Triton," *Science*, 1459 (December 15, 1989).

Brown, R.H. et al.: "Energy Sources for Triton's Geyser-Like Plumes," *Science*, 431 (October 19, 1990).

Brown, R.H. et al.: "Triton's Global Heat Budget," *Science*, 1465 (March 22, 1991).

Conrath, B. et al.: "Infrared Observations of the Neptunian System," *Science*, 1454 (December 15, 1989).

Cruickshank, D.P. et al.: "Triton: Do We See to the Surface?" *Science*, 283 (July 21, 1989).

Eberhart, J.: "Neptune Marvels Emerge from Data Deluge," *Science News*, 391 (December 16, 1989).

Goldreich, P. et al.: "Neptune's Story," *Science*, 500 (August 4, 1989).

Gore, R.: "Neptune — Voyager's Last Picture Show," *Nat'l. Geographic*, 34 (August 1990).

Gurnett, D.A. et al.: "First Plasma Wave Observations at Neptune," *Science*, 1494 (December 15, 1989).

Hammel, H.B.: "Neptune Cloud Structure at Visible Wavelengths," *Science*, 1165 (June 9, 1989).

Hammel, H.B. et al.: "Neptune's Wind Speeds Obtained by Tracking Clouds in Voyager Images," *Science*, 1307 (September 22, 1989).

Hansen, C.J. et al.: "Surface and Airborne Evidence for Plumes and Winds on Triton," *Science*, 421 (October 19, 1990).

Helfenstein, P. et al.: "Large Quasi-Circular Features Beneath Frost on Triton," *Science*, 824 (February 14, 1992).

Hillier, J. et al.: "Voyager Disk-Integrated Photometry of Triton," *Science*, 419 (October 19, 1990).

Hubbard, W.B. et al.: "Interior Structure of Neptune: Comparison with Uranus," *Science*, 648 (August 9, 1991).

Hunt, A.E. and P. Moore: *Atlas of Neptune*, Cambridge University Press, New York, NY, 1994.

Ingersoll, A.P. and K.A. Tryka: "Triton's Plumes: The Dust Devil Hypothesis," *Science*, 435 (October 19, 1990).

Kerr, R.A.: "Triton Steals Voyager's Last Show," *Science*, 928 (September 1, 1989).

Kerr, R.A.: "The Neptune System in Voyager's Afterglow," *Science*, 1450 (September 29, 1989).

Kerr, R.A.: "Neptune's Triton Spews a Plume," *Science*, 213 (October 13, 1989).

Kerr, R.A.: "A Passion for the Little Things Among the Planets," *Science*, 998 (November 24, 1989).

Kerr, R.A.: "A Geologically Young Triton After All?" *Science*, 1563 (December 22, 1989).

Kerr, R.A.: "Geysers or Dust Devils on Triton?" *Science*, 377 (October 19, 1990).

Kinoshita, J.: "Neptune," *Sci. Amer.*, 82 (November 1989).

Kirk, R.L., R.H. Brown, and L.A. Soderblom: "Subsurface Energy Storage and Transport for Solar-Powered Geysers on Triton," *Science*, 424 (October 19, 1990).

Krimigis, S.M. et al.: "Hot Plasma and Energetic Particles in Neptune's Atmosphere," *Science*, 1483 (December 15, 1989).

Lane, A.L. et al.: "Photometry from Voyager 2: Initial Results from the Neptunian Atmosphere, Satellites, and Rings," *Science*, 1450 (December 15, 1989).

Lunine, J.I.: "Voyager at Triton," *Science*, 386 (October 19, 1990).

Lyons, J.R., Y.L. Yung, and M. Allen: "Solar Control of the Upper Atmosphere of Triton," *Science*, 204 (April 10, 1992).

McElheny, V.: "Neptune's Magnetic Looks," *Technology Review (MIT)*, 12 (January 1990).

Nelson, R.M. et al.: "Temperature and Thermal Emissivity of the Surface of Neptune's Satellite Triton," *Science*, 429 (October 19, 1990).

Ness, N.F. et al.: "Magnetic Fields at Neptune," *Science*, 1473 (December 15, 1989).

Pollack, J.B., J.M. Schwartz, and K. Rages: "Scattering in Triton's Atmosphere: Implications for the Seasonal Volatile Cycle," *Science*, 440 (October 19, 1990).

Polvani, L.M. et al.: "Simple Dynamical Models of Neptune's Great Dark Spot," *Science*, 1393 (September 21, 1990).

Poroco, C.C.: "An Explanation for Neptune's Ring Arcs," *Science*, 995 (August 30, 1991).

Smith, B.A. et al.: "Voyager 2 at Neptune: Imaging Science Results," *Science*, 1433 (December 15, 1989).

Smith, B.A.: "Voyage of the Century," *Nat'l. Geographic*, 34 (August 1990).

Soderblom, L.A. et al.: "Triton's Geyser-Like Plumes: Discovery and Basic Characteristics," *Science*, 410 (October 19, 1990).

Stone, E.C. et al.: "Energetic Charged Particles in the Magnetosphere of Neptune," *Science*, 1489 (December 15, 1989).

Stone, E.C. and E.D. Miner: "The Voyager 2 Encounter with the Neptunian System," *Science*, 1417 (December 15, 1989).

Strom, R.G., S.K. Croft, and J.M. Boyce: "The Impact Cratering Record on Triton," *Science*, 437 (October 19, 1990).

Stromovsky, L.A.: "Latitudinal and Longitudinal Oscillations of Cloud Features on Neptune," *Science*, 684 (November 1, 1991).

Suomi, V.E., S.S. Limaye, and D.R. Johnson: "High Winds of Neptune: A Possible Mechanism," *Science*, 929 (February 22, 1991).

Thompson, W.R. and C. Sagan: "Color and Chemistry on Triton," *Science*, 415 (October 19, 1990).

Tyler, G.L. et al.: "Voyager Radio Science Observations of Neptune and Triton," *Science*, 1466 (December 15, 1989).

Warwick, J.W. et al.: "Voyager Planetary Radio Astronomy at Neptune," *Science*, 1498 (December 15, 1989).

Yelle, R.V.: "The Effect of Surface Roughness on Triton's Volatile Distribution," *Science*, 1553 (March 20, 1992).

Web References

Jet Propulsion Laboratory: http://www.jpl.nasa.gov/solar_system/solar_system_index.html#

NASA: http://www.nasa.gov/projects.html

NEPTUNIAN SERIES. See **Radioactivity**.

NEPTUNIUM. [CAS: 7439-99-8]. Chemical element, symbol Np, at. no. 93, at. wt. 237.0482 (predominant isotope), radioactive metal of the Actinide series, also one of the Transuranium elements. Neptunium was the first of the Transuranium elements to be discovered and was first produced by McMillan and Abelson (1940) at the University of California at Berkeley. This was accomplished by bombarding uranium with neutrons. Neptunium is produced as a by-product from nuclear reactors. ^{237}Np is the most stable isotope, with a half-life of 2.20×10^6 years. The only other very long-lived isotope is that of mass number 236, with a half-life of 5×10^3 years.

^{237}Np is parent of the neptunium $(2n + 1)$ alpha decay series. Other isotopes include those of mass numbers 229–235 and 238–241; metastable forms of ^{236}Np, ^{240}Np and two of ^{237}Np are known. Electronic configuration $1s^2 2s^2 2p^6 3s^2 3p^6 3d^{10} 4s^2 4p^6 4d^{10} 4f^{14} 5s^2 5p^5 5d^{10} 5f^5 6s^2 6p^6 - 6d^1 7s^2$. Ionic radii Np^{4+} 0.88 Å; Np^{3+} 1.02 Å (Zachariasen). Oxidation potential $Np \rightarrow Np^{3+} + 3e^-$, 1.85 V; $Np^{3+} \rightarrow Np^{4+} + e^-$, -0.155 V; $Np^{4+} + 2H_2O \rightarrow NpO_{2+} + 4H^+ + e^-$, -0.739 V; $NpO_2^+ \rightarrow NpO_2^{2+} + e^-$, -1.137 V. See also **Chemical Elements**.

Neptunium has the oxidation states (VI), (V), (IV), and (III) with a general shift in stability toward the lower oxidation states as compared to uranium. The compounds which are formed are very similar to the corresponding compounds of uranium.

The ionic species corresponding to the oxidation states vary with the acidity of the solution; in acid solution of moderate strength the species are Np^{3+}, Np^{4+}, NpO_2^+, and NpO_2^{2+} as in the case of uranium and plutonium. The potential scheme in 1-M HCl is as follows:

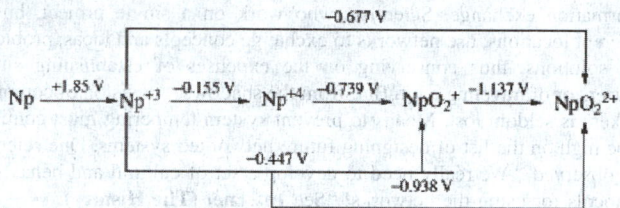

It will be seen that the metal is highly electropositive, in common with the other actinide elements. The $Np^3 \rightarrow Np^{4+}$ couple is reversible and this oxidation can be accomplished by the oxygen of the air. The (IV) state is stable, not oxidized by air, and only slowly oxidized to NpO_2^+ by nitric acid. The $Np^{4+} \rightarrow NpO_2^+$ couple is not readily reversible, whereas the $NpO_2^+ \rightarrow NpO_2^{2+}$ couple is reversible; this is reasonable on the basis that the former involves making or breaking the neptunium-oxygen bonds, whereas the latter does not. The oxidation of NpO_2^+ to NpO_2^{2+} requires moderately strong oxidizing agents. Neptunium differs from uranium and plutonium in that its potential relations are such as to render NpO_2^+ moderately stable with respect to disproportionating, even in solutions containing moderate concentrations of hydrogen ion.

The potentials are altered extensively by change in the hydrogen ion concentration and by the presence of any of a number of anions capable of forming complex ions.

Neptunium ions in aqueous solution possess characteristic colors: pale purple for Np^{3+}, pale yellow-green for Np^{4+} green-blue for Np^{5+}, while NpO_2^{2+} varies from colorless to pink or yellow-green depending on the acid present.

The precipitation reactions of Np^{3+} are similar to those of the tripositive rare earths, those of Np^{4+}, to the other tetrapositive actinides and to Ce^{4+}, and those of NpO_2^{2+} to the corresponding ions of uranium and plutonium. All of the simple salts of NpO^{2+} appear to be soluble.

The neptunium oxide system exhibits complexity similar to that found in the uranium oxide system. Thus, the important oxide is NpO_2 and there exists a range of compositions, depending upon conditions, up to Np_3O_8.

As a metal, Np has a relatively low melting point ($\sim 640\,°C$), is very dense (20.45 g/cm³), and is ductile. The alpha form reacts with hydrogen, carbon, oxygen, sulfur, the halogens, and phosphorus to yield a number of binary compounds.

The important halides of neptunium are the trifluoride, NpF_3, purple or black and hexagonal, the hexafluoride, NpF_6, brown and orthorhombic, the trichloride, $NpCl_3$, white and hexagonal, the tetrachloride, $NpCl_4$, red-brown and tetragonal, and the tribromide, $NpBr_3$, α-form green and hexagonal, β-form green and orthorhombic.

In research at the Institute of Radiochemistry, Karlsruhe, West Germany, during the early 1970s, investigators prepared alloys of neptunium with iridium, palladium, platinum, and rhodium. These alloys were prepared by hydrogen reduction of the neptunium oxide in the presence of finely divided noble metals. The reaction is called a *coupled reaction* because the reduction of the metal oxide can be done only in the presence of noble metals. The hydrogen must be extremely pure, with an oxygen content of less than 10^{-25} torr.

Industrial utilization of neptunium has been very limited. The isotope ^{237}Np has been used as a component in neutron detection instruments. Neptunium is present in significant quantities in spent nuclear reactor fuel and poses a threat to the environment. A group of scientists at the U.S. Geological Survey (Denver, Colorado) has studied the chemical speciation of neptunium (and americium) in ground waters associated with rock types

that have been proposed as possible hosts for nuclear waste repositories. See Cleveland reference.

Additional Reading

Cleveland, J.M., K.L. Nash, and T.F. Rees: "Neptunium and Americium Speciation in Selected Basalt, Granite, Shale, and Tuff Ground Waters," *Science*, **221**, 271–273 (1983).

Fuger, J. and L.R. Morss: *Transuranium Elements: A Half Century*, American Chemical Society, Washington, DC, 1992.

Keller, C. and B. Erdmann: Preparation and Properties of Transuranium Element–Noble Metal Alloy Phases, Proc. 1972 Moscow Symp. Chem. Transuranium Elements, 1976.

Krot, N.N. and A.D. Gel'man: "Preparation of Neptunium and Plutonium in the Heptavalent State," *Dokl. Chem.*, **177**, 1–3, 987–989 (1967).

Lewis, R.J., Sr.: *Hawley's Condensed Chemical Dictionary*, 13th Edition, John Wiley & Sons, Inc., New York, NY, 1999.

Lide, D.R.: *CRC Handbook of Chemistry and Physics* 86th Edition, CRC Press, LLC., Boca Raton, FL, 2005.

Magnusson, L.B. and T.J. LaChapelle: "The First Isolation of Element 93 in Pure Compounds and a Determination of the Half-life of 93Np237," *Amer. Chem. Soc. J.*, **70**, 3534–3538 (1948).

Marks, T.J.: "Actinide Organometallic Chemistry," *Science*, **217**, 989–997 (1982).

McMillan, E. and P.H. Abelson: "Radioactive Element 93," *Phys. Rev.*, **57**, 1185–1186 (1940).

Seaborg, G.T.: "The Chemical and Radioactive Properties of the Heavy Elements," *Chem. Engng. News*, **23**, 2190–2193 (1945).

Seaborg, G.T. and W.D. Loveland: *The Elements beyond Uranium*, John Wiley & Sons, Inc., New York, NY, 1990.

Thayer, J.S. and F.E. Brinckman: *Environmental Chemistry of the Heavy Elements: Hydrido and Organo Compounds*, John Wiley & Sons, Inc., New York, NY, 1995.

NERNST EFFECT. If heat is flowing through a strip of metal and the strip is placed in a magnetic field perpendicular to its plane, a difference of electric potential develops between the opposite edges. This phenomenon, discovered by Nernst in 1886, is analogous to the Hall effect, but with a longitudinal flow of heat replacing the longitudinal electric current. See also **Hall Effect**.

NERNST HEAT THEOREM. For a homogeneous system, the rate of change of the free energy with temperature, as well as the rate of change of heat content with temperature, approaches zero as the temperature approaches absolute zero.

NERNST, HERMANN WALTHER (1864–1941). Nernst was a German Chemist and Physicist. In 1894 he received invitations to the Physics Chairs in Munich and in Berlin, as well as to the Physical Chemistry Chair in Göttingen. He accepted this latter invitation. At Göttingen Nernst founded the Institute for Physical Chemistry and Electrochemistry and became its Director. In 1905 he was appointed Professor of Chemistry, later of Physics, in the University of Berlin, becoming Director of the newly-founded "Physikalisch-Chemisches Institut" in 1924. He remained in this position until his retirement in 1933.

He made major contributions to electrochemistry, thermodynamics, and photochemistry. Nernst's early studies in electrochemistry were inspired by Arrhenius' dissociation theory which first recognized the importance of ions in solution. His heat theorem, known as the Third Law of Thermodynamics, was developed in 1906. In 1918 his studies of photochemistry led him to his atom chain reaction theory. In later years, he occupied himself with astrophysical theories, a field in which the heat theorem had important applications.

He is remembered best for the Nernst Effect, named after him. For his work in thermochemistry Nernst won the Nobel Prize in Chemistry in 1920.

See also **Nernst Effect; Nernst Heat Theorem; Nernst-Thompson Rule**; and **Thermodynamics**.

J. M. I.

NERNST-THOMPSON RULE. A solvent of high dielectric constant favors dissociation by reducing the electrostatic attraction between positive and negative ions, and conversely a solvent of low dielectric constant has small dissociating influence on an electrolyte.

NERVOUS SYSTEM AND THE BRAIN. See **Central and Peripheral Nervous Systems**.

NERVOUS SYSTEM (Fishes). See **Fishes**.

NET. A set of intervals such that every point of a closed linear interval [*a*, *b*] is contained in at least one interval of the set, each interval being called a *mesh* of the net.

NETTLE HAIR. A form of hair-like scale, found on some caterpillars, which causes an irritation of human skin resembling that of nettles. Caterpillars of the io and the brown-tail moths have larger poisonous spines of similar properties. The irritation caused by some species is very mild but others are much more severe.

NETWORK. In elementary terms, a network consists of three or more points that are connected physically by some medium for the purpose of transferring information, energy, or materials from one point to another. Networks may occur naturally, as in the case of the brain and nervous system of living creatures, or networks may be contrived to serve human needs. *Material-transfer* networks are exemplified by gas, oil, and water pipelines. An *energy-transferring* network is typified by the distribution over wires and cables of centrally generated electricity. An *information-transfer* network relies on various segments of the electromagnetic spectrum to convey intelligence from transmitters to two or more receivers. A simple connection between two points usually is referred to as a *link* rather than a network. As of the early 1990s, the word *network* most frequently refers to a *communication network*, the subject addressed by this article.

The points on a network may be referred to as stations or nodes at which information is received/transmitted instrumentally or humanly for storage, action, or sometimes retransmittal. Ultimately, unless wasted, the information will become part of the human need for that information. Information transmitted and received may be of analog or digital form, but the latter predominates by far. Thus, the term *digital network* is very common.

Networking equipment is costly, and the total cost of some nationwide and global networks is tremendous. Networks must be planned for expansion because of the need for more and more information exchange. The concept of information networks as they are known today essentially commenced with telegraphy and telephony. For decades, these were person-to-person communication. See also **Telephony (Telecommunications)**.

Chronology of Computer Networks. The concept of a computer network in the United States and worldwide stemmed from research work at the Rand Corporation in the early 1960s. Late in the 1960s, a network sponsored by the U.S. Defense Advanced Research Projects Agency (DARPA) and known as ARPANET was established. According to D. Van Honweling (University of Michigan), "The notion when ARPANET was established was that it was primarily to share computing resources. As things turned out, that wasn't the way it got used. It got used by human beings who wanted to work with other human beings." In the mid-1970s, a variety of networks joined ARPANET to offer connectivity. These subsidiary networks had the specific objective of linking government and academic institutions together. Concurrently, NASA and the U.S. Department of Energy established their own networks.

Taking note of the lack of a national framework for networking as of the early 1980s, the National Science Foundation (NSF) established the NSFNET designed around the existing six NSF-supported supercomputer centers. The concept was that of including not only the supercomputer centers, but also to link together the several other important networks that had been developed during the mid-1980s. Operation of the network was assigned to private contractors when this "backbone" operated at 56 kilobits per second. A rate of 1.5 million bits per second had been achieved by early 1990, with goals as of 1993 set in excess of 50 million bits per second.

A *packet* is defined as bits that contain address and some fraction of the particular message being sent. On this standard, over 100 million packets per month had been sent in 1988. By 1990, this figure had increased to 2.5 billion packets. By February 1990, 10% of all information ever sent on the network had been sent during 1 month. A survey of NSFNET use indicates: (1) networked mail, 30%, (2) file exchange, 29%, (3) interactive applications, 20%, (4) name lookup, 15%, and (5) other services, 6%.

Of growing value to users of computer networks is the transmission of *motion graphics*. Most scientists agree that moving pictures are more eloquent than words. One scientist has observed, "Most people's comprehension works around visualization. Networks are making that comprehension tremendously more transferable."

Several years ago, the concept of a "Global Village" was envisioned, wherein communications, including graphics — essentially as an extension to traditional radio and television — would permit persons living almost anywhere to participate globally in science, politics, the arts, and other information exchange. Scientists who work on a single project, but at different locations, use networks to exchange concepts and ideas, problems and solutions, thus conserving on the expenses of establishing single quarters or of traveling a lot. But, from the shadows, the vision of computer hackers is seldom lost. Means to prevent system tampering must continue to be high on the list of designing future networked systems. One scientist has observed, "We really need to develop a set of cultural and behavioral protocols for using the networks!" See **Internet (The History)**.

Data Transmission Needs Are Virtually Unlimited. Considering the information worthy of recording that has been generated in centuries past and the macro amounts of data generated each second of every day in present times that are important to some segment of civilized society, the word *overwhelming* seems inadequate. Thousands of needs to retrieve data can be cited, of which just one case — that is, the needs of the National Aeronautics and Space Administration (NASA) — are exemplary. Considering just the time frame that embraced NASA data gathering from spacecraft (*Pioneer* to the *Ultraviolet Explorer*), 6 trillion bytes (1 byte = 8 bits or binary units) were collected. Spacecraft launched during the last few years have been estimated to more than quadruple that amount. The *Hubble Space Telescope* will generate several trillion bytes each year. Introduction of the *Earth Observing System* (EOS) is estimated to generate a trillion bytes of information every few days. (The battery of sensors planned for EOS have been so designed.) Because of data storage and retrieval costs, these are a prerequisite cost accounting factor in budgeting for total mission coverage and, in fact, for selecting only those missions for which adequate data handling support will be available the years that the mission may be productive.

By way of comparison, it has been estimated that the U.S. Library of Congress, with over 19 million books, represents the storage of 6 trillion bites of information.

The data storage and retrieval components of the human genome project also are exemplary of how failure to adequately budget for electronic storage and networking of monstrous quantities of information can threaten ambitious research projects per se. An accounting of the *Bionet* project (National Institutes of Health) is given in the Wallich reference listed.

Examples of network databases are those of the Scientific & Technical Information Network, including such topics as: (1) BEILSTEIN, a comprehensive database in organic chemistry offering a wide range of information for over 1,700,000 substances reported over the past 150 years; (2) JANAF, a joint U.S. Army-Navy-Air Force file produced by the National Institute for Standards and Technology (NIST), containing the chemical thermodynamic properties for over 1000 substances; (3) NBSFLUIDS, which is an NIST file that contains calculation and data generation software useful for creating tables of temperature-dependent thermodynamic and thermophysical properties for 12 cryogenic fluids; and (4) NBSTHERMO, an NIST file of ambient temperatures and chemical thermodynamic properties of 8000 inorganic and numerous two-carbon organic substances.

Expanding the Points in a Network. Even though most of the time communications and transportation networks usually are initially designed to accommodate future as well as present needs, in recent years the satiation of network users is seldom accommodated. The frequent requirement to expand networks today is regarded by many network engineers as virtually inevitable. What initial steps can be taken to assure a minimum cost for network expansion? As reported by B. Cipra (see reference listed), F. Hwang (AT&T Bell Laboratories) and Ding Zhu Du (Princeton University) have proposed the method indicated in Fig. 1. This materializes a conjecture with which mathematicians have been attempting to prove for

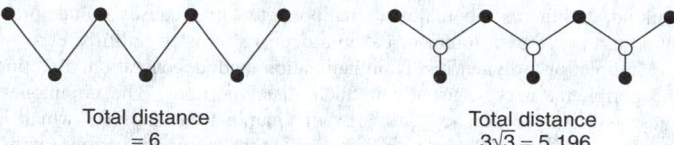

Total distance = 6   Total distance $3\sqrt{3}$ = 5.196

Fig. 1. Designing the "shortest" network. As shown here, a network can be shortened by adding additional points. The example shows a savings of about 13%. Whether or not savings of over 13% can be achieved has not been settled among mathematicians. As indicated by this figure, the greatest savings result when the points are at the vertices of an equilateral triangle. Using the center of the triangle as a hub reduces the length of the network by $\frac{\sqrt{3}}{2}$ (~13%).

many years. In formulating their approach, the engineers have borrowed from the "minimax" problem of game theory. See also **Game Theory**.

The "shortest network" problem has been analyzed in scholarly detail by M. Bern (Xerox Palo Alto Research Center) and R. Graham (AT&T Bell Laboratories). In their paper (see reference listed), the researchers conclude: "Although knowledge about algorithms (including Melzak's) has progressed greatly in recent years, the shortest-network problem remains tantalizingly difficult. The problem can be stated in simple terms, and yet solutions defy analysis. A tiny variation in the geometry of a problem may appear to be insignificant, and yet it can radically alter the shortest network for the problem. The shortest-network problem will continue to frustrate and fascinate us for years to come."

Classification of Network Uses. To ask the question, "Where are communication networks used?" is tantamount to asking, "Where is information used?" An appreciation for the phenomenal needs and uses of networking may be piqued by considering the following:

- Statistical information for business and government. One of the early pioneering users of networks was the travel industry for determining space availability and making reservations. Computerized banking and trading also are early examples of early network use.
- Educational institutions, where most universities today have extensive private networks and the ability to tap into national and global networks. This is expanding rapidly to include public and vocational schools.
- Scientific research, notably for efforts conducted by teams of researchers who represent several disciplines and who do not have their offices and laboratories at a centralized location.
- Transportation and shipping, where goods movements must be closely time related. See also **Navigation**.
- Medical practice, particularly pertaining to crises that must tap distant sources of special expertise and in order to expedite the location and ultimate receipt of organs and fluids for transplantation. Numerous hospitals operate their own networks, which can readily be tapped into national and global networks.
- Earth and space research, as previously mentioned.
- Entertainment program retrieval.
- Manufacturing production, described later in this article.
- Distribution systems, to which much attention has been devoted by economists recently.
- And this list could go on and on.

Distribution and Networking. Information networking has and will continue to impact upon the very fundamental life-styles of business, government, scientific, and other professional people. Quite recently, particular attention has been given to the manner in which networking ultimately may affect materials distribution, selling, and trading. Depending upon a nation's form of government, prices may be fixed or tightly regulated by a central organization or allowed to float in accordance with supply and demand. Even in free-trade economies, some commodity prices (notably of utilities, such as the gas and electricity supply) are regulated by national or state and provincial authorities. As observed by K. McCabe, S. Rassenti, and V.L. Smith (University of Arizona), "Domestically, we have witnessed in the last decade (1980s) uncommon political and economic forces that have resulted in increased reliance on markets to discipline prices, output, and the entry and exit of firms in industries traditionally regulated by state

and federal agencies. This has been part of a worldwide move toward privatization in the socialist and command economies of Great Britain, New Zealand, Eastern Europe, and the (former) Soviet Union. In the United States the extent of deregulation has been less than complete in all of them."

The deregulation movement motivated an experimental study of *auction markets* designed for interdependent network industries, such as natural gas pipelines or electric power systems by the aforementioned researchers. Through the use of an information-transfer network and a computerized dispatch center, decentralized agents would submit bids to buy commodities and offers to sell transportation and commodities. Computer algorithms would determine prices and allocations that maximize the gains from exchange in the system relative to the submitted bids and offers. Details are described in the McCabe reference listed.

Networked Communications for Manufacturing

Although information-exchange networks for manufacturing and processing facilities frequently include wide-area networks (WANs), which exchange information between plants at different locations and with corporate headquarters, the principal need for information exchange occurs at the factory floor level through the use of *local-area networks* (LANs), which provide the coordination and govern the actions of specific machines and processes. See also article on **Local Area Networks**.

As described in the article **Control System**, industrial manufacturing is based largely on the measurement and control of numerous variables (temperature, pressure, dimension, count, material flow, to mention a few). In a typical manufacturing scene, hundreds to thousands of feedback control loops may be present. Information coordination and balance are achieved by way of local area networks.

Because of the multitude of connections required and the great variety of sensors used, severe standards for component design are imposed. Development of these standards is described in the article on Control System Architecture.

Other Networks

The word *network* is used to designate systems other than information-exchange networks. For example, *electric network* stems back to Kirchhoff (1850) and Maxwell (1870).

An electric network may be described as a combination of elements, either as (1) a combination of interconnected devices, such as inductors, capacitors, resistors, and generators, or (2) the abstraction of interconnected branches having the properties of inductance, resistance, capacitance, etc.

An *active network* is one whose output waves are dependent upon sources of power, apart from that supplied by any of the actuating waves, which power is controlled by one or more of these waves.

An *all-pass network* is designed to introduce phase shift or delay without introducing appreciable attenuation at any frequency.

A *differentiating network* is one whose output is the time derivative of its input waveform. Such a network preceding a frequency modulator makes the combination a phase modulator; or, following a phase detector, it makes the combination a frequency detector. Its ratio of output amplitude to input amplitude is proportional to frequency, and its output phase leads its input by 90 degrees.

An *equivalent network* is one in which, under certain conditions of use, it may replace another network. The networks need not be of the same form. For example, one may be electrical, the other mechanical. If one network can replace another network in any system whatsoever without altering in any way the operation of that portion of the system external to the networks, the networks are said to be of *general equivalence*.

A *linear-passive network* is a network such that (1) if currents of any waveform are fed to the terminals of the network, the total energy delivered to the network is non-negative; (2) no voltages appear between any pair of terminals before a current is fed to the network.

Neural networks are described in the articles **Central and Peripheral Nervous Systems**; and **Artificial Intelligence: Neural Networks**.

Additional Reading

Bern, M.B. and R.L. Graham: "The Shortest-Network Problem," *Sci. Amer.*, 84 (January 1989).
Cerf, V.G.: "Networks," *Sci. Amer.*, 72 (September 1991).

Cipra, B.: "In Math, Less is More—Up to a Point," *Science*, 1081 (November 23, 1990).

Dertuozos, M.L.: "Building the Information Marketplace," *Technology Review(MIT)*, 29 (January 1991).

Haavind, R.: "The Smart Tool for Information Overload—Hypertext," *Technology Review (MIT)*, 42 (November–December 1990).

Horgan, J.: "Jukeboxes for Scientists," *Sci. Amer.*, 24 (July 1989).

Kauffels, J.: *Practical LANs Analyzed*, Halstead Press, New York, NY, 1989.

Kay, A.C.: "Computers, Networks and Education," *Sci. Amer.*, 138 (September 1991).

King, J.P.: "Distributed Control Systems," *3.5 in Process/Industrial Instruments and Controls Handbook*, 4th Edition (D.M. Considine, Editor), The McGraw-Hill, Companies, Inc., New York, NY, 1993.

Malone, T.W. and J.F. Rockart: "Computers, Networks and the Corporation," *Sci. Amer.*, 128 (September 1991).

McCabe, K.A., J. Rassenti, and V.I. Smith: "Smart Computer-Assisted Markets," *Science*, 534 (October 25, 1991).

Naugle, M.G.: *The Illustrated Network Book: A Graphic Guide to Understanding Computer Networks*, Thomson Publishing Company, New York, NY, 1997.

Negroponte, N.P.: "Products and Services for Computer Networks," *Sci. Amer.*, 106 (September 1991).

Palca, J.: "Getting Together Bit by Bit," *Science*, 160 (April 13, 1990).

Peterson, I.: "The Electronic Grapevine," *Science News*, 90 (August 11, 1990).

Steen, M.V. and H. Sips: *Computer and Network Organization: An Introduction*, Prentice-Hall, Inc., Upper Saddle River, NJ, 1995.

Tesler, L.G.: "Networked Computing in the 1990s," *Sci. Amer.*, 86 (September 1991).

Wallich, P.: "Who's Minding the Store," *Sci. Amer.*, 20 (October 1989).

White, G.W., U.W. Pooch, and E.A. Risch: *Computer System and Network Security*, DIANE Publishing Company, Collingdale, PA, 1999.

NETWORK SYNTHESIS. Network synthesis is that branch of the theory of electric networks that deals with the systematic determination of the structure and element values of an electric network possessing preassigned characteristics.

NETWORK (Telecommunications). See **Telephony (Telecommunications)**.

NEURALGIA. Pain occurring along the course of a cranial or peripheral nerve or posterior root. Neuralgia is difficult to differentiate sharply from neuritis. The primary distinction is that neuritis best describes the acute and chronic inflammation of the peripheral nerves, while neuralgia is due to acute and chronic inflammation of the pathway stations (ganglia) lying along the course of nervous pathways. The pain in neuralgia is usually of a sharp, shooting, intermittent type and accompanied by increased sensitivity of the skin supplied by the nerve. Many varieties of neuralgia are differentiated according to the body part affected. Common forms include: (1) trigeminal neuralgia, a very severe form marked by agonizing pain over branches of the trigeminal nerve in the face; (2) intercostal neuralgia, which can be mistaken for pleurisy because the intercostal nerves, after leaving the spinal cord, run around each side of the chest between the ribs; (3) Morton's neuralgia—pain in the joint of the third and fourth toe caused by pinching of a nerve in this region; and (4) sciatic neuralgia.

Trigeminal neuralgia also is known as tic douloureux and involves one or more branches of the fifth cranial (trigeminal) nerve. A closely associated neuralgia involves the ninth cranial nerve, often causing severely sharp pains in the throat, spreading to the ear on the same side of the face. Palliative relief in some cases can be obtained from injections of alcohol into the ganglion of the nerve. In more severe cases, surgical procedures, as cutting the sensory root of the nerve, may be required to provide permanent relief. Drugs such as *Dilantin*® Sodium slow the rate of peripheral nerve conduction and have been used in the treatment of trigeminal neuralgia. The successful use of injections of vitamin B_{12} also has been reported. Intercostal neuralgia is sometimes accompanied by a skin eruption in herpes zoster or shingles.

NEURITIS. An inflammation of a nerve, either chronic or acute. Mononeuritis or localized neuritis is the term used when one nerve is involved. When several or many nerves are involved the term multiple neuritis or polyneuritis is used. Localized neuritis develops from injury, infection, chronic intoxication, or metal poisoning. Neuritis, either localized or multiple, complicates many of the infectious diseases, such as

typhoid, diphtheria, tuberculosis, smallpox, etc. Other causes include pressure on a nerve by a tumor or calcium deposits in osteo-arthritis, etc.

Multiple or polyneuritis is inflammation and degeneration affecting the peripheral nerves, usually in their distal portions. The commonest cause is dietary deficiency, primarily of vitamin B_1 (thiamine), which is usually associated with chronic alcoholism; other causes are various toxins, either chemical agents such as arsenical drugs, lead, nitrobenzol, coal-tar products, or toxins produced by bacteria.

The clinical picture is similar no matter what the cause: there is a gradual onset of numbness and tingling of the hands and feet, weakness of the limbs, altered sensation to touch, and pain, loss of deep reflexes, and eventually paralysis and atrophy of the extremities. Treatment is directed first toward removal of the cause.

NEUROCHEMISTRY (THE HISTORY). Neurochemistry can be defined as the biochemistry of the nervous system, and neurochemists are scientists who employ chemical, biochemical and molecular biological concepts and techniques to further our understanding of its structure and function. Although the term "neurochemistry" has been in common use for less than 50 years, efforts to understand the physical basis of the human mind can be traced to ancient Greek philosophers, who concluded that the human brain is the site of origin of body movement and of cognition. They also observed that the brain was composed of fatty material. This remained the state of our scientific knowledge of the brain for nearly 2000 years.

Beginnings: Neurochemical Concepts and Discoveries through the Eighteenth Century

Alchemists of the seventeenth century confirmed the fatty nature of the brain and found further that brain contained phosphorus, a mysterious substance that, once isolated, glowed in the dark and could even burst into flame. Towards the end of the eighteenth century, as alchemy gave way to chemistry, the stage was set for quantitative chemical analyses of biological tissues. The brain, like other animal organs, was found to contain cholesterol, but not triglyceride, the "depot fat" commonly found in other body organs. Brain lipids contained the novel compound sphingosine in combination with fatty acids in the form of sphingolipids such as cerebroside and sphingomyelin. The brain was also found to be enriched in phosphorus-containing lipids, mainly the phosphoglycerides, which accounted to a large degree for the brain's high phosphorus content. See also **Biochemistry (The History)**; and **Lipids**.

The Chemical Composition of the Brain: Discoveries of the Nineteenth Century

Neurochemistry of the nineteenth century consisted largely of successively better separations, analyses and identification of brain lipids and their constituent components. There was relatively little neurochemical research activity of a dynamic nature, such as investigation of brain metabolism, other than the finding that brain function required a constant supply of oxygenated blood, and that a number of potent natural poisons were neurotoxic. In this sense, neurochemistry lagged behind the field of neurophysiology, which had long since documented the electrical nature of nerve conduction and of the activation of muscular contraction. It was perhaps the desire to make brain chemistry more dynamic, together with the old mystique of phosphorus, which led the nineteenth century Dutch biologist Jakob Moleschott to claim "without phosphorus there is no thought!" This marked an early attempt to draw the concepts of mind and brain a bit closer. See also **Physiology (The History)**.

The Neurochemical Uniqueness of Nerve Cells: The First Half of the Twentieth Century

During this period, brain biochemistry came into its own. A major advance was the demonstration that the mechanism of slowing of the frog heartbeat following electrical stimulation of the vagus nerve was chemical in nature. The substance released by vagal stimulation was identified as acetylcholine. From this beginning eventually came the knowledge that impulses are conducted along nerve fibres electrically, while neurons communicate with one another by a chemical process, termed synaptic transmission. With this advance, the playing field was leveled between the neurophysiologists (the "sparks" as they were referred to, because they measured electrical activity) and the biochemists and pharmacologists of the day (called the "soups," because they ground up tissues in order to analyze them). Today, we would refer to the "soups" as neurochemists, and to both groups as neuroscientists, as is discussed below. See also **Central and Peripheral Nervous Systems**; **Neurotransmitters**; and **Synapse**.

During this period, considerable insight was gained into the nature of brain metabolism from experiments on cerebral blood flow. Although the brain of a resting human adult accounts for only 2% of body weight, it was found to account for over 20% of the body's oxygen consumption. This high level of chemical activity was supported via an efficient extraction by the brain of glucose from the blood, followed by its nearly complete oxidation to carbon dioxide and water. The enormous chemical energy consumed by the brain was expended largely for the generation of ion gradients to support nerve impulse conduction and for the complex steps involved in synaptic transmission.

Cellular and Molecular Neurochemistry: The Second Half of the Twentieth Century

Progress in brain biochemistry began to accelerate sharply, and the term "neurochemistry" now became widely recognized. In 1954 the first of four biennial International Neurochemical Symposia was convened at Oxford University by a group of prominent scientists from a variety of backgrounds, including lipid biochemistry, pharmacology, microchemistry, histochemistry, endocrinology and neuropathology. They shared a common interest in the biochemical basis of brain structure and function. From the attendees of these meetings arose the *Journal of Neurochemistry* in 1956 and the inception of the International Society for Neurochemistry in 1965. The *Journal* became the official publication of the Society soon thereafter. Today there are a number of other neurochemical societies (Japanese, American, European, Argentine and Asian Pacific).

Many of the neurochemical advances of the middle and late twentieth century find their antecedents in the discoveries of the first half-century. A number of genetically transmitted neurological diseases were traced to specific metabolic defects. A form of mental retardation associated with urinary excretion of phenylpyruvic acid was traced to a defective liver enzyme. By placing infants on a diet low in the amino acid phenylalanine during the period of rapid brain growth, the damage to brain could be largely avoided–a neurochemical triumph. Improved understanding of brain lipid formation and breakdown continued. Several forms of mental retardation were found each to be associated with accumulation of a specific sphingolipid in neurons, and could be traced to the loss of a normally present lipid-degrading enzyme specific to that disease. Pathological accumulations of additionally discovered sphingolipids, the gangliosides, were described.

The nature of chemical transmission at the interneuronal interface–the synapse–involves many sequential steps, all of which can be completed in a few milliseconds. There are now dozens of known chemical messengers, including neurotransmitters, neurohormones and growth factor molecules, and for each there may be multiple receptor protein types and subtypes, typically embedded in postsynaptic neuronal membranes. This area of research is of great interest because most neuroactive drugs produce their effects at the level of neurotransmitter synthesis and release and postsynaptic signal transduction, via intracellular chemical messengers. Prominent among these "second messengers" are cyclic nucleotides and the cleavage products of the membrane lipid PIP_2.

While most of the brain's neuronal connections are believed to be genetically "hard-wired," some synapses can be modified during an individual's lifetime. Adaptive changes between neurons seen following a trauma, such as a partial brain ablation, or behaviorally in the form of memory of a learning experience, are collectively referred to as neuroplasticity and are believed to be mediated by altered synaptic structure and function. Long-term memory formation requires protein synthesis, a finding that implicates the neuronal cell nucleus in prolonging the altered state. Non-invasive techniques for measuring regional brain metabolism and blood flow have made it feasible to localize regional brain metabolism related to higher brain function, including learning and memory in humans.

The Future of Neurochemistry

The twenty-first century begins with increasing evidence of a new form of brain disease transmission that involves misfolding of prions, constitutive brain proteins, apparently independent of DNA or RNA vectors. Progress in understanding the pathogenesis of the dementias of ageing has revealed

genetic and nongenetic factors that may eventually point to preventative and therapeutic agents.

During the period in the mid-twentieth century that the term "neurochemistry" took root and the various neurochemical societies were established, "neuroscience" and neuroscience societies also appeared. The distinction between the terms "neurochemistry" and "neuroscience" has become progressively blurred. For one thing, most neuroscientists, regardless of their orientation (behavioral, neurophysiological, pharmacological, developmental, etc.), use neurochemical techniques. "Neurochemistry" has thus become generic and, in a sense, a victim of its success. In addition, traditional neurochemists have increasingly turned to interdisciplinary approaches that easily fit within the rubric of neuroscience. How the term "neurochemistry" will be distinguished from the term "neuroscience" in the future is difficult to predict, but it is a certainty that the pace of the neurochemical approach to brain research will continue to accelerate.

Additional Reading

Finger, S.: "Otto Loewi and Henry Dale: the Discovery of Neurotransmitters," In: Finger, S.: *Minds Behind the Brain*, Oxford University Press, New York, NY, 2000.

Finger, S.: *Minds Behind the Brain: A History of the Pioneers and Their Discoveries*, Oxford University Press, New York, NY, 2004.

McIlwain, H.: "Neurochemistry and Related Terms: Their Introduction and Acceptance," *Neurochemistry International*, **12**, 431–438 (1988).

McIlwain, H.: "Biochemistry and Neurochemistry in the 1800s: Their Origins in Comparative Animal Chemistry," *Essays in Biochemistry*, **25**, 197–224 (1990).

Prusiner, S.B.: "Prions," *Proceedings of the National Academy of Sciences of the USA*, **95**, 13363–13383 (1998).

Siegel, G.J., B.R.W. Albers, D.L. Price; S.T. Brady: *Basic Neurochemistry*, 7th Edition, Elsevier Science & Technology Books, New York, NY, 2005.

Sourkes, T.L.: "The Origins of Neurochemistry: The Chemical Study of the Brain in France at the End of the Eighteenth Century," *Journal of the History of Medicine and Allied Sciences*, **47**, 322–339 (1992).

Tower, D.B.: "Origins and Development of Neurochemistry," *Neurology*, **8** (supplement 1), 3–31 (1958).

BERNARD W. AGRANOFF, University of Michigan, Ann Arbor, MI

NEUROFIBROMATOSIS. A condition characterized by multiple tumors in the skin or along the course of peripheral nerves. These tumors occasionally become malignant. Also called von Recklinghausen's disease.

NEUROLEPTIC MALIGNANT SYNDROME. First described and named by Delay and Deniker in 1968, neuroleptic malignant syndrome (NMS) was thought to be a variant of drug fever in which hyperpyrexia and autonomic and other neurological abnormalities developed during phenothiazine therapy. The principal characteristics of NMS are hyperthermia, hypertonicity of skeletal muscles, and fluctuating consciousness, along with instability of the autonomic nervous system. As described by Guzé & Baxter (see reference), common autonomic dysfunctions include pallor, diaphoresis, blood-pressure instability, tachycardia, and cardiac dysrhythmias. The muscular hypertonia consists of a generalized "lead-pipe" increase in tone, which increased muscle tone may result in decreased chest-wall compliance and, as a consequence, breathing problems severe enough to require respiratory support. Disturbances of consciousness may range from agitation or alert mutism to stupor and coma. Estimates of mortality range from 20 to 30%, this depending upon the type of neuroleptic drugs responsible and upon duration of exposure to offending drug. Deaths usually occur between 3 and 30 days after presentation of symptoms. Death may result from respiratory failure, cardiovascular collapse, renal failure, arrhythmias, and thrombo-embolism. Respiratory failure may result from aspiration pneumonia or tachypneic hypoventilation. A primary factor in the causation of NMS may be a decrease in the availability of dopamine in the brain.

Neuroleptic drugs, also referred to as antipsychotic agents or major tranquilizers have been commonly used in the treatment of psychotic illnesses and also used in general medicine as antiemetics; in connection with dissociative anesthesia; and for the treatment of diseases, such as Tourette's syndrome. Although neuroleptic agents have been known for side-effects, NMS, frequently fatal, has not been well recognized.

In 1985, the seriousness of the side effects of neuroleptic or antipsychotic drugs, such as phenothiazines, butyrophenones, and thioxanthenes, among others, prompted the American Psychiatric Association to issue a warning to psychiatrists and physicians, advising them to carefully weigh the benefits of the drugs against the potential for developing *tardive dyskinesia* (or NMS).

Additional Reading

Caroff, S.N.: "The Neuroleptic Malignant Syndrome," *J. Clin. Psychiatry*, **41**, 79–83 (1980).

Culliton, B.J.: "Antipsychotic Drugs (and Tardive Dyskinesia)," *Science*, **229**, 1248 (1985).

Diamond, J.M. and A.B. Santos: "Unusual Complications of Antipsychotic Drugs," *Am. Fam. Physician*, **26**, 153–157 (1982).

Guzé, B.H. and L.R. Baxter, Jr.: "Neuroleptic Malignant Syndrome," *N. Eng. J. Med.*, **313**(3), 163–165 (July 18, 1985).

Hopkins, P.M. and F.R. Ellis: *Hyperthermic and Hypermetabolic Disorders: Exertional Heat Stroke, Malignant Hyperthermia and Related Syndromes*, Cambridge University Press, New York, NY, 1996.

Tollefson, G.I.: "The Neuroleptic Syndrome and Central Dopamine Metabolites," *J. Clin. Psychopharmacology*, **4**, 150–153 (1984).

Yassa, R., D.V. Jeste, and N.P.V. Nair: *Neuroleptic-Induced Movement Disorders*, Cambridge University Press, New York, NY, 1996.

NEURON.   See **Central and Peripheral Nervous Systems**.

NEUROPTERA.   Insects of varied form and habits, including the dobson fly, the golden eyes or lacewings, the alder flies, and the ant lions. The order is characterized by the four membranous wings, usually with a large number of branching veins that are united in some species to form a network. The mouth is formed for biting but in some larvae the mandibles and maxillae fit together to form a piercing and sucking organ. See Fig. 1. The insects are predacious.

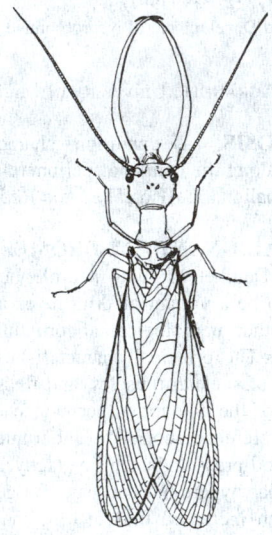

Fig. 1.   Dobson fly, valued as a food for fishes. (*USDA.*)

Neuroptera are of minor economic importance as food for fishes. The larvae of lacewings prey on aphids, hence they are of some assistance in holding these pests in check.

The order contains about 1700 species. The main families of Neuroptera include:

Chrysopidae	Lacewings, aphid-lions, golden-eyed flies
Hemerobiidae	Brown lacewings
Mantispidae	Mantid flies
Myrmeleonidae	Ant lions or doodle bugs
Rhaphidiidae	Snake flies
Sialidae	Dobson flies, alder flies, fish flies

NEUROREGULATORS.   Neuroregulators represent a diverse group of compounds that include both neurotransmitters and neuromodulators. Receptors for neuroregulators on both the cell surface and within the cell represent the molecular targets of the majority of drugs in clinical use. In order to classify an endogenous agent as a neurotransmitter, it must possess a number of general characteristics, which are described. Neurotransmitter receptors are grouped according to sequence homology and structures are given. In many cases human receptors have been cloned.

NEUROSYPHILIS.   See **Syphilis**.

NEUROTENSION.   See **Central and Peripheral Nervous Systems**.

NEUROTRANSMITTERS.   Signaling is a neuron's chief function. Within the central nervous system (CNS), neurons communicate with one another; in the periphery, nerve cells also signal to glands and muscles. Communication between the various cellular components of the organism is essential for complex metazoan animals. In simpler animals and during early stages in the development of higher forms, communication is mediated by hormones and growth factors that diffuse relatively long distances from secretory cells to target tissues. In the nervous system, chemical messengers are secreted from specialized parts of neurons, called *nerve terminals* or *nerve endings*, to act on receptors in the membrane of neighboring target cells. The targets are close by, not much further than the distance between two adjacent cells. These chemical messengers are called *neurotransmitters*.

Until the beginning of the twentieth century, many physiologists thought that neurons in the CNS were all connected together, forming one great continuous cell or syncytium similar to cardiac muscle. If neurons were structured in this way, there would be no need for chemical transmission between nerve cells, because they would all be connected electrically. Other physiologists thought that neurons were discrete, but made close contacts with other neurons at nerve terminals. Supported by Ramón y Cajal's magisterial histological studies of both the central and peripheral nervous systems, Charles Sherrington named these close contacts synapses (from the Greek, to clasp). The debate was not settled until mid-century when the fine structure of nervous tissue could be examined. Electron microscopy revealed both the many varieties of sherringtonian synapses where chemical messengers are released, as well as cytoplasmic bridges between some neurons, called electrical or electrotonic synapses, where neurotransmitters are not required for communication between the cells. See also **Central and Peripheral Nervous Systems**; **Cajal, Santiago y Ramón (1852–1934)**; and **Sherrington, Charles Scott (1857–1952)**.

Otto Loewi, Henry Dale and the History of Chemical Neurotransmission

As might be expected, knowledge about neurotransmitters advanced with the study of synaptic transmission. In the first quarter of the twentieth century, physiological and pharmacological examination of synaptic transmission made use of experimentally advantageous parts of the body, for example the neuromuscular junction–the synapse between the motor neuron and muscle–and nerves that innervate body organs, the heart or adrenal gland, where the effects of nerve stimulation and drugs could be readily assessed. Thus, when the nerve is stimulated, muscles contract, the heart beat slows or increases, and adrenaline (epinephrine) is released from the adrenal medulla. From his pharmacological studies with nerves of the autonomic nervous system, Henry Dale showed that adrenaline simulates the effects of stimulating sympathetic neurons with considerable precision, and correctly inferred that adrenaline is a neurotransmitter at synapses that he called adrenergic. See also **Adrenaline and Noradrenaline**; and **Dale, Henry Hallett (1875–1968)**.

It was not until 1921 that Otto Loewi showed that acetylcholine is released upon nerve stimulation and is the transmitter that slows the heart. His experiment was simple in design. Loewi perfused two frog hearts with Ringer solution. Both beat autonomously at a normal rate. He then stimulated the vagus nerve to one of the hearts. As anticipated, the heart with the stimulated vagus slowed. But so did the other heart. A substance which Loewi first called *Vagusstoffe* was released from the terminals of the stimulated nerve into the perfusate and had affected the other heart. This neurotransmitter was isolated and shown to be acetylcholine. The elegant design of this experiment came to Loewi in a dream that he had on two successive nights. On the first, he awoke from the dream, made some notes, and fell asleep. The next morning he had forgotten the dream and could not decipher his notes. But when he awoke from the dream on the second night, he raced to his laboratory to do the experiment. See also **Acetylcholine**; and **Loewi, Otto (1873–1961)**.

By 1910, both adrenaline and acetylcholine had been isolated as body constituents, and compounds similar in structure had been studied as drugs that mimic the action of the two transmitters. Since many substances, both natural and synthetic, can mimic the effects of nerve stimulation, what are the criteria for identifying a true neurotransmitter? Loewi's experiment suggested some of the criteria that are used today for accepting a substance as a neurotransmitter:

1. The substance must be synthesized in the neuron.
2. The substance must be present in the neuron's presynaptic terminal, and be released on stimulation in sufficient amounts to exert its expected effects on its target cell. (Nerve stimulation, which produces membrane depolarization, leads to the influx of Ca^{2+} ions into nerve terminals. Although there are some exceptions, physiological release of neurotransmitters is dependent on increased intracellular Ca^{2+} concentration.)
3. When applied as a drug to the target cell in reasonable concentrations, the substance must mimic the action of the transmitter exactly. (It is understood that the target cell must have appropriate receptors for the neurotransmitter: even if a genuine transmitter, for example acetylcholine, were to be applied to a cell without acetylcholine receptors, it could not act as a transmitter at that synapse.)
4. A mechanism for removing the neurotransmitter once it has been released is a critical physiological feature of synaptic transmission. As in speech, punctuation is necessary for transmitting a meaningful message. Most transmitters are removed by reuptake into neurons and glial cells, and by diffusion. With acetylcholine, removal is enzymatic, by acetylcholinesterase, which hydrolyses acetylcholine and rapidly removes it from the synaptic cleft.

There are Many Different Neurotransmitters

Substances used as neurotransmitters belong to three distinct families: (1) amines, (2) purines and (3) peptides (Tables 1 and 2). Often called *classical* or *small molecule transmitters*, the members of the amine family include acetylcholine, amino acids and other substances derived from common amino acids through a few enzymatic steps. All of these transmitters are synthesized in the cytosol of nerve endings and then taken up into vesicles for release when the neuron is stimulated.

TABLE 1. AMINE NEUROTRANSMITTERS

Neurotransmitter	Type of compound	Precursor
Ester		
Acetylcholine	Ester	Choline
Amino acids		
Glutamate	α Amino acid	α-Ketoglutarate (glutamine)
Glycine	α Amino acid	L-Serine
γ-Aminobutyric acid GABA)	γ Amino acid	L-Glutamate
Biogenic amines		
Adrenaline (epinephrine)	Catecholamine	Noradrenaline
Noradrenaline (norepinephrine)	Catecholamine	Dopamine
Dopamine	Catecholamine	L-Tyrosine
Octopamine	Phenylamine	L-Tyrosine
5-Hydroxytryptamine (serotonin)	Indoleamine	L-Tryptophan
Histamine	Imidazoleamine	L-Histamine

The vesicles in nerve terminals all contain adenosine triphosphate (ATP). When vesicles release neurotransmitters, they also release ATP. At some

TABLE 2. SOME FAMILIES OF NEUROPEPTIDE TRANSMITTERS[a]

Family	Transmitters
Opioid	Opiocortins, enkephalins, dynorphin, FMRFamide
Neurohypophyseal	Vasopressin, oxytocin, neurophysins
Tachykinins	Substance P, substance K (neurokinin A), neurokinin B
Secretins	Secretin, glucagon, vasoactive intestinal peptide, gastric inhibitory peptide, growth hormone-releasing factor, peptide histidine isoleucineamide
Insulins	Insulin, insulin-like growth factors I and II
Pancreatic polypeptide related	Pancreatic polypeptide, NPY
Gastrins	Gastrin, cholecystokinin

[a]FMRFamide, Phe-Met-Arg-Phe amide; NPY, neuropeptide Y.

synapses there are purinergic receptors that are activated by ATP or its catabolites, adenosine monophosphate (AMP) and adenosine.

Peptides function as neurotransmitters. Although the number of active neuropeptides is not yet known, there are more than 100 but probably less than 500. Most of these peptides can be grouped into a few distinct families because they share similar amino acid sequences. The genes that encode the peptides that belong to a family are likely to be related through evolution. All of these peptide messengers are synthesized in the cell body as large precursor polyproteins and are then processed proteolytically within vesicles. As there is little protein synthesis in nerve endings, peptides must be transported considerable distances from the neuron's cell body to the synapse for release. Like amine transmitters and purines, peptides are also released at nerve terminals in vesicles.

Small Molecule Neurotransmitters Synthesized from Common Cellular Metabolites

Amine Transmitters. The biosynthetic enzyme, choline acetyltransferase, catalyses the esterification of acetyl coenzyme A and choline to acetylcholine (Fig. 1). This reaction occurs only in cholinergic neurons because the transferase is a differentiated property of these neurons and is not expressed in other cells. The enzyme can therefore be used to identify cholinergic neurons, for example by immunohistochemistry with specific antibodies to the enzyme protein. While acetyl coenzyme A is a metabolite that is abundant in all cells, the synthesis of the transmitter acetylcholine is limited by the supply of dietary choline, which must be taken up from the bloodstream or recaptured by the neuron from transmitter that has been released and hydrolyzed by cholinesterase. Nicotinic receptors are activated by the plant alkaloid, nicotine; muscarinic receptors, by the mushroom poison, muscarine. Acetylcholine is the excitatory neurotransmitter at the neuromuscular junction, where it binds to postsynaptic nicotinic acetylcholine receptors, and is the inhibitory (parasympathetic) transmitter to the heart through muscarinic acetylcholine receptors. In the periphery, acetylcholine is also the transmitter for all preganglionic neurons of the autonomic nervous system. There are many cholinergic synapses in the brain, for example of neurons in the nucleus basalis which has widespread projections to the cerebral cortex.

There are over 40 different nicotinic acetylcholine receptors. These receptors are ionotropic, meaning that, upon binding acetylcholine, they open to allow ions from the extracellular space to flow into the postsynaptic neuron. There also are a large number of muscarinic acetylcholine receptors, which are metabotropic. These receptors produce a biochemical change mediated by a G protein within the postsynaptic neuron. It is important to note here that, like other neurotransmitters, acetylcholine can be excitatory at some synapses and inhibitory at others. Moreover,

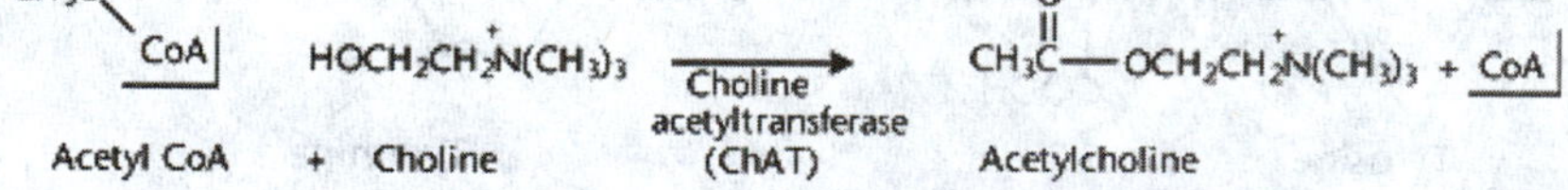

Fig. 1.   Esterification of acetyl coenzyme A to acetylcholine.

depending on the properties of the postsynaptic receptor, acetylcholine can create ion channels in the postsynaptic membrane or it can alter the biochemical properties of the postsynaptic cell.

L-Glutamate is synthesized from abundant precursors that are intermediates in carbohydrate metabolism. The two important routes are from α-ketoglutarate by transamination and from glutamine by hydrolysis. The pathway from glutamine is instructive because it illustrates how most transmitter substances are removed from the synaptic cleft. Once released, this amino acid activates glutamate receptors in the postsynaptic target cell. The concentration of glutamate in the synaptic cleft soon falls because the transmitter is taken up both into the neurons that released it and into nearby glial cells. Uptake is mediated by specific transporter molecules. In glial cells, the glutamate is converted to glutamine by the enzyme, glutamine synthase. Glutamine is then transported into neurons where it is hydrolyzed back to glutamate by glutaminase.

More than half the synapses in the CNS are glutaminergic, as almost all excitatory neurons use glutamate as neurotransmitter. There are at least five excitatory amino acid receptors; these are ionotropic and classified by their distinctive sensitivity to selective agonists. Noteworthy is the N-methyl-d-aspartate (NMDA) receptor, which has been implicated in memory, development and epilepsy. Although active when applied, the amino acid aspartate is not thought to be released as a neurotransmitter. By activating these receptors, the large amount of glutamate released into the brain during a stroke or trauma can cause tissue damage, a process called *excitotoxicity*. There are also at least six metabotropic receptors for glutamate.

Glycine, synthesized from serine by the enzyme, serine hydroxymethylase, is an abundant constituent of body tissues. It is the chief inhibitory neurotransmitter in the spinal cord, rostral part of the brain, and retina.

Γ-Aminobutyric acid (GABA) is the major inhibitory transmitter in the brain. Glutamic acid decarboxylase (GAD) removes a carboxyl group from glutamate using pyridoxal phosphate as a cofactor (Fig. 2).

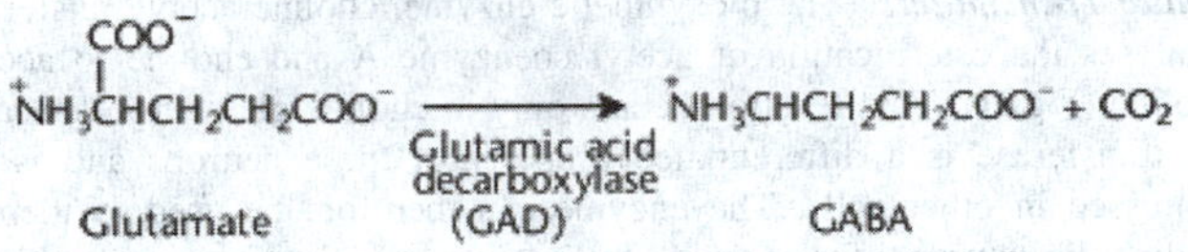

Fig. 2.   Decarboxylation of glutamate to γ-aminobutyric acid (GABA).

GAD immunoreactivity can be used as a specific marker for neurons that release GABA as transmitter. Both glycine and GABA inhibit by activating receptors to allow Cl^- ions to flow into the postsynaptic neuron, leading to hyperpolarization. GABA$_A$ receptors are ionotropic, and are controlled directly; GABAM$_B$ receptors are metabotropic, operating through a Cl^- conductance indirectly.

Biogenic Amines. The biogenic amine transmitters include the catecholamines (dopamine, noradrenaline (norepinephrine) and adrenaline), the phenolamine octopamine, the indoleamine serotonin (5-hydroxytryptamine; 5-HT) and the imidazoleamine histamine. The catecholamines and serotonin are synthesized in pathways with enzymes that are quite similar in amino acid sequence and in mechanism. The first enzymes in both pathways are oxidases that require tetrahydrobiopterin (THB) as cofactor. These oxidases, which are rate limiting in the formation of the transmitters, are specific to transmitter type. In aminergic neurons, the amino acid l-tyrosine is converted to l-dihydroxyphenylalanine (l-DOPA) (Fig. 3).

Precursor tyrosine, an essential amino acid which is supplied in the diet, must be taken up by neurons from the bloodstream. L-DOPA is then decarboxylated by decarboxylase to dopamine (Fig. 4).

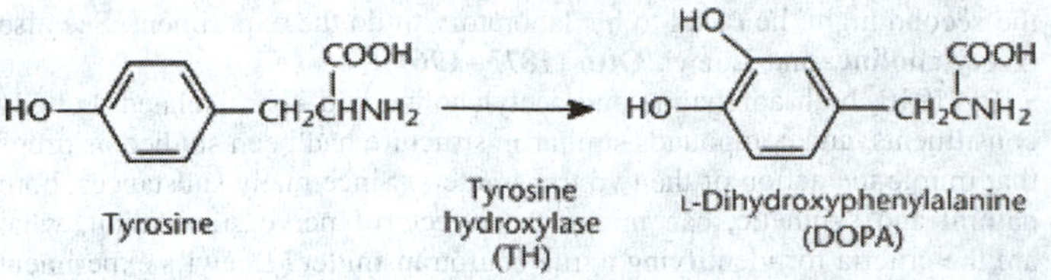

Fig. 3.   Hydroxylation of L-tyrosine to L-dihydroxyphenylalanine (L-DOPA).

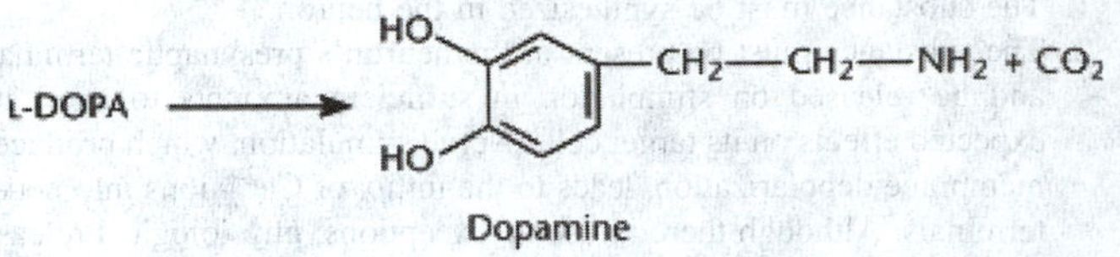

Fig. 4.   Decarboxylation of L-DOPA to dopamine.

Dopaminergic neurons are prominent in the substantia nigra and the ventral tegmental area; they project to the basal ganglia, the hypothalamus and the limbic system, where they play important roles in the regulation of voluntary movement, thinking and planning, and in the addiction and reward system. There are five types of dopamine receptors, all metabotropic, operating through a G protein.

Noradrenaline is formed from dopamine in a reaction catalyzed by dopamine β-hydroxylase (Fig. 5). Dopamine neurons lack the hydroxylase. Noradrenaline is used as a transmitter by cells in the locus ceruleus, which project throughout the cerebral cortex, cerebellum and spinal cord. As proposed by Dale, noradrenaline is the transmitter used by the postganglionic neurons of the sympathetic nervous system.

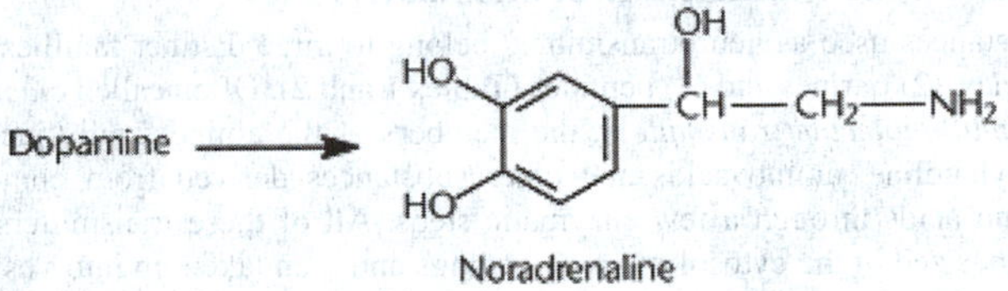

Fig. 5.   Dopamine is hydroxylated to form noradrenaline (norepinephrine).

Noradrenaline can be converted to adrenaline by methylation, a reaction catalyzed by phentolamine-N-methyltransferase and requiring S-adenosylmethionine as methyl donor (Fig. 6).

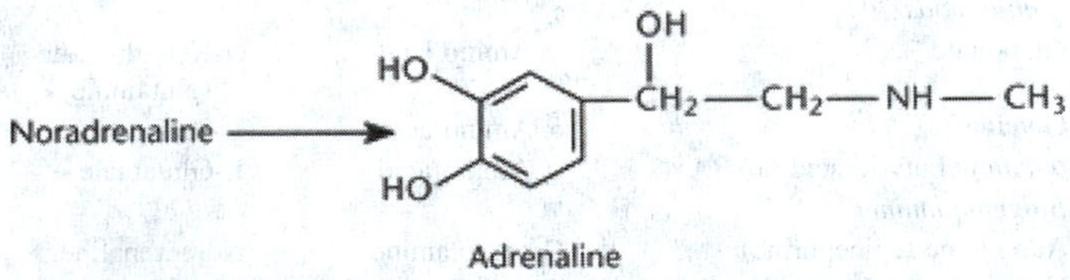

Fig. 6.   Noradrenaline (norepinephrine) is methylated to form adrenaline (epinephrine).

The few neurons that use adrenaline as transmitter are intermingled with noradrenergic cells in the medulla oblongata, solitary tract nucleus and medial longitudinal fascicle. In the adrenal medulla and sympathetic nervous system, some adrenergic cells release adrenaline instead of noradrenaline.

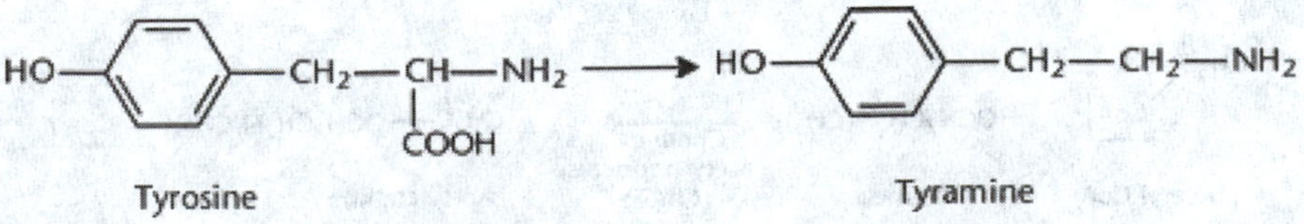

Fig. 7.   Decarboxylation of tyrosine to tyramine.

L-Tyrosine can also be converted to octopamine, a neurotransmitter in invertebrates (crustacea, insects and mollusks). The carboxyl group is first removed by DOPA decarboxylase to produce tyramine (Fig. 7). Octopamine is then formed by dopamine β-decarboxylase (Fig. 8).

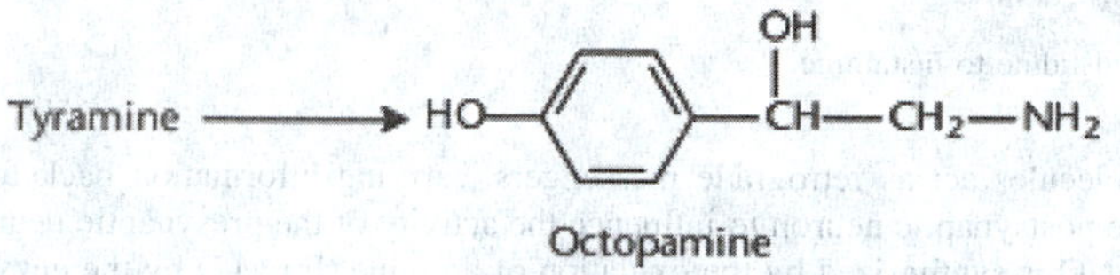

Fig. 8.   Decarboxylation of tyramine to octopamine.

L-Tryptophan is the precursor of serotonin. This essential amino acid is first oxidized to 5-hydroxytryptophan (5-HTP) in a reaction homologous to the formation of L-DOPA from tyrosine (Fig. 9). 5-HTP is then converted to serotonin (5-HT) by the same decarboxylase used in the catecholamine pathway (Fig. 10). See also **Serotonin**.

There are at least 13 human 5-HT receptors, only one of which is ionotropic. Serotonergic cell bodies are situated predominantly in the midline or raphe region of the upper brainstem and pons, with extensive projections through the cortex and spinal cord. Serotonergic transmission is important in sleep and arousal, sexual behavior, aggression and mood.

Histamine has long been recognized as a local hormone or autocoid that is active in inflammation and in the control of smooth muscle, especially in blood vessels and in exocrine glands, notably the secretion of gastric juice. It is synthesized from L-histidine by the enzyme, histidine decarboxylase (Fig. 11). Histamine is inactivated by methylation catalyzed by histamine methyltransferase.

Three types of neuronal histamine receptors have thus far been identified, all of which are metabotropic. The relatively few (approximately 10,000) cells that use histamine as neurotransmitter are located in the posterior hypothalamus with diffuse projections throughout the brain. The physiological role of histamine is uncertain, but it is thought to be involved in the control of appetite and to regulate the secretion of pituitary hormones.

Purine Transmitters. ATP, synthesized in mitochondria, is present in high concentrations in cholinergic and adrenergic vesicles. The evidence for purinergic neurotransmission in the periphery is extremely strong, especially for sympathetic neurons to the vas deferens, smooth muscle of the gut, and dorsal root ganglion cells that form synapses in the dorsal horn of the spinal cord. Purines typically modulate synaptic transmission, for example by inhibiting, through the activation of autoreceptors, further transmitter release from the very neuron from which the purines were originally released. Purine receptors belong to two classes, P_1 and P_2. The P_1 type is sensitive to adenosine and AMP, and the P_2 to ATP and adenosine diphosphate (ADP). Some of the P_2 receptors are ionotropic, but all the rest are metabotropic.

Peptide Neurotransmitters Cleaved from Large Peptide or Protein Precursors

Neuropeptide transmitters are all derived from the class of proteins that pass through the major membrane systems of the cell—the endoplasmic reticulum, Golgi apparatus and the elements of the lysosomal system. Thus, they follow the cellular pathway of other secretory molecules. Neuropeptides are synthesized exclusively in the cell body of the neuron in precursor form and subsequently are processed proteolytically. Unlike the small molecule transmitters that were considered above, peptide

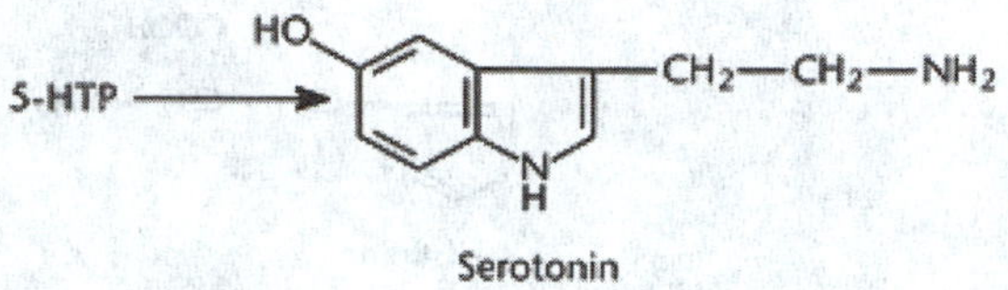

Serotonin

Fig. 10.   Decarboxylation of 5-hydroxytryptophan (5-HTP) to serotonin (5-hydroxytryptamine).

messengers cannot be recycled or resynthesized at nerve terminals, and must be transported from the cell body by axonal transport.

Neuroactive peptides are quite short, usually from 3 to 15 amino acids long, compared with an average precursor protein of about 200 amino acids in length. There are several advantages to this apparently wasteful process. One is amplification. In some polyproteins, more than one copy of a short amino acid sequence is present. By proteolytic processing, several identical neuroactive peptides can be cut out from a single precursor molecule. Another advantage is diversity. Amino acid sequences for more than one type of neuroactive peptide are present in other polyproteins. Processing of these precursors can result in the production of several different neuropeptides for release from the same vesicle. These different transmitters usually have synergistic actions on postsynaptic targets. Another advantage is regional variety: two neurons can express the same gene resulting in the identical polyprotein, but different neuropeptides are produced by the two cells either because their processing peptidases differ or because posttranslational modification differs between the two cells. For example, bulky glycosylation can prevent proteolytic cleavage at particular sites in the molecule. Differential processing occurs with the opiate precursor, proopiomelanocortin (POMC). In the anterior pituitary, POMC is processed to corticotrophin; in the intermediate lobe and arcuate nucleus of the hypothalamus, β-endorphin and α-melanocyte-stimulating hormone are formed.

The opiate family of neuropeptides illustrates some of the general principles of peptide transmitters. Opioid peptides are encoded by three distinct genes, opiocortin, enkephalin and dynorphin. Each produces a polyprotein that is processed to yield peptides with narcotic properties. These peptides all contain the amino acid sequence Tyr-Gly-Gly-Phe. The three opioid genes may have arisen in evolution from a single ancestral gene by duplication and subsequent divergence. Similarly, there are two tachykinin genes; one (neurokinin A) encodes substance P and neurokinin A and the other (neurokinin B) encodes neurokinin B.

Three types of opioid receptors (δ, k, μ) have been cloned. As with essentially all neuropeptides, they are coupled with G proteins and are therefore metabotropic. They are activated by morphine, the alkaloid from the poppy, to produce relief of pain (narcosis) and many other effects (e.g., sleepiness, euphoria and constipation) that are mediated by the different receptor subtypes distributed on postsynaptic target cells throughout the body.

Peptides differ in several important ways from small molecule transmitters. These differences cause neurotransmission by peptides to be slow in onset, to affect targets over a large area, and to last longer than transmission by small molecules. Peptide transmitters are relatively large (even dipeptides or tripeptides are bigger than any small molecule transmitter) and the largest (e.g. insulin) have molecular weights of over 5000 Da. As a result, peptides diffuse slowly. Peptides are not taken back up into neurons or glial cells, nor are they degraded by specific enzymes in the synaptic cleft. Diffusion is the chief mechanism for removing peptide transmitters.

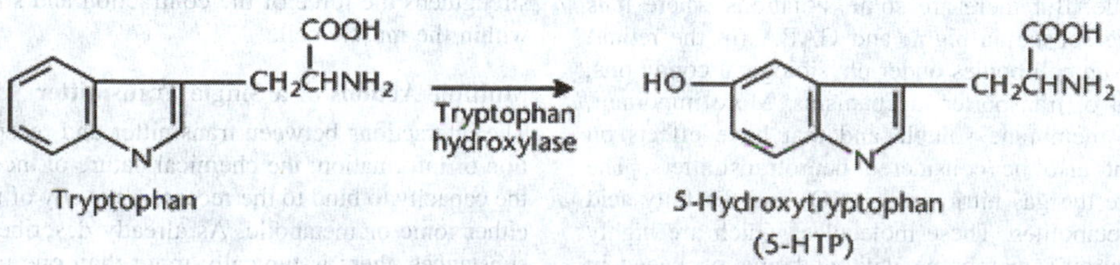

Fig. 9.   Hydroxylation of tryptophan to 5-hydroxytryptophan.

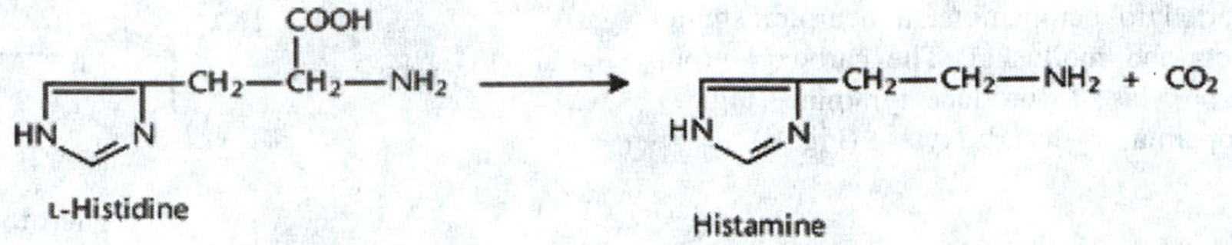

Fig. 11.   Decarboxylation of L-histidine to histamine.

Storage of Neurotransmitters in Vesicles

Perhaps the most important distinction between small molecule transmitters and neuropeptides is the type of vesicle in which they are stored. Two types of vesicle are prominent in nerve endings: small vesicles of about 50 nm in diameter which appear clear or electron-lucent in electron micrographs, and large vesicles of 60–175 nm in diameter, which are electron-dense. The small vesicles are therefore called *lucent*, and the larger vesicles *dense-cored*. The lucent vesicles contain small molecule transmitters and ATP; they lack any core proteins or peptides. The membranes of these vesicles contain special proteins (synaptotagmin, synaptophysin and synaptobrevin) which enable them to undergo rapid fusion with specializations of the nerve terminal's synaptic membrane (docking sites) when the local concentration of Ca^{2+} rises. Ca^{2+} levels rise abruptly near these docking sites through special Ca^{2+} channels, which open when the nerve impulse depolarizes the terminal. As a result of membrane fusion, the contents of the vesicles empty into the synaptic cleft.

Dense-cored vesicles contain peptides and ATP. They also can contain small molecule transmitters, typically biogenic amines. The membranes of dense-cored vesicles do not contain all of the proteins used to facilitate membrane fusion, and they do not dock at specialized synaptic sites. Rather, release occurs by fusion with nerve-terminal membrane at some distance from the Ca^{2+} channels. Nevertheless, the fusion of dense-cored vesicles also requires raised levels of intracellular Ca^{2+} produced by nerve impulses. In contrast to release from small lucent vesicles, however, which occurs rapidly and in response to a sudden and transient burst of Ca^{2+}, the release from the large vesicles occurs only when the Ca^{2+} concentration in the entire nerve terminal becomes increased, usually as a result of repeated firing. So, release of peptides from the larger vesicles tends to be slower than the release of transmitters from small clear vesicles.

Although they both are derived from the Golgi–endosomal system, the two types of vesicle are quite different. The large dense-cored vesicles, which assume their mature form when they leave the transGolgi network for their trip down the axon to the nerve ending, are homologous to secretory granules in nonneuronal secretory cells. In contrast, the membrane destined to become small synaptic vesicles is thought to leave the transGolgi network in larger organelles (endosomes) which also reach the nerve terminal by fast axonal transport. Old membrane from synaptic vesicles that have already undergone cycles of transmitter release is taken back into the terminal by endocytosis. This membrane from the old vesicles fuses with the new membrane transported in endosomes to form new lucent vesicles. Thus, the small vesicle membrane does not carry any cargo from the cell body, whereas the dense-cored vesicle is the only source of neuropeptides. This difference between the two types of vesicles is perhaps the most crucial: both types must actively take up small molecule transmitter substances synthesized in the cytosol of the nerve ending. For this purpose, they have a variety of transmitter-specific pumps or transporters. However, there is no mechanism for transporting peptides across membranes: neuropeptide transmitters must be synthesized in the cell body of the neuron.

It might be presumed that, in order to be a transmitter, a substance must be released from synaptic vesicles. For the substances thus far discussed, this condition is usually true. But there are some situations where it is not. Thus, dopamine (in the substantia nigra) and GABA (in the retina) are known to be released from cell bodies under physiological conditions, probably through a reversal of transporter mechanisms. More important, several molecules that are membrane soluble and that have effects on postsynaptic neurons might also be considered neurotransmitters. The most prominent of these are the gas nitric oxide (NO) and the fatty acid arachidonic acid and its metabolites. These molecules, which are highly reactive, slip through a neuron's membrane without being packaged in vesicles. In some instances, it is thought that these membrane-soluble

molecules act as retrograde messengers, carrying information back from the postsynaptic neuron to influence the activity of the presynaptic neuron.

NO is synthesized by the oxidation of arginine (Fig. 12) by the enzyme nitric oxide synthase together with an electron donor (flavin–adenine dinucleotide (FAD), reduced nicotinamide–adenine dinucleotide phosphate (NADPH) and tetrahydrobiopterin (THB) as cofactor. Arachidonic acid is released from membrane phospholipid by the receptor-mediated activation of a phospholipase A_2.

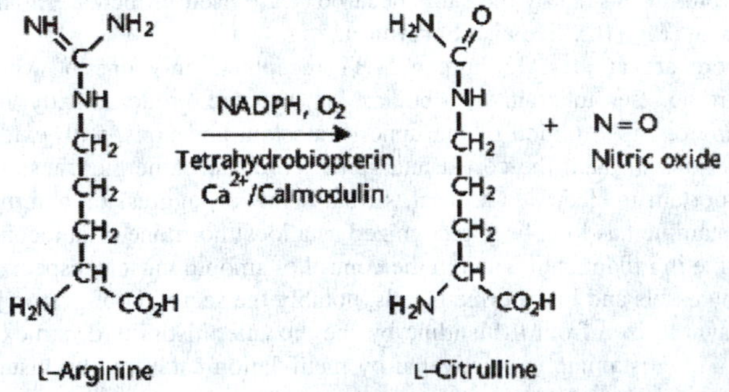

Fig. 12.   Synthesis of nitric oxide (NO) via the oxidation of L-arginine.

Colocalization and Differential Release of Multiple Neurotransmitters

In 1935, Henry Dale proposed that the transmitter biochemistry of a neuron's cell body is the same as that of its terminals. Further elaboration of this idea led to the suggestion that all the terminals of a given neuron would release the same transmitter(s). This complex of ideas is called *Dale's Principle*, and, in its most extreme formulation, reduces to one neuron, one transmitter substance. As with most principles in biology, Dale's has been shown to be invalid in all of its versions, but is none the less useful. Thus, in most instances, neurons do release the same transmitter substances at all of their terminals. But many neurons contain several neurotransmitters. If a neuron contains more than one transmitter, they are said to coexist in that neuron. Usually, a small molecule transmitter exists in small synaptic vesicles and the same small molecule transmitter coexists with a peptide or peptides in large vesicles. These neurotransmitters can be released differentially depending on the pattern and intensity of the stimulus to the neuron. Thus, an amine transmitter (acetylcholine, GABA, serotonin) can be released through small vesicles by brief nerve stimulation, and a neuropeptide together with the same amine transmitter from large vesicles, if the stimulation delivered to the neuron is sustained. See also **Dale, Henry Hallett (1875–1968)**.

The physiological significance of corelease has not yet been determined for every neuron in which it occurs. In some instances, the two transmitters operate synergistically on the postsynaptic target. Thus, motor neurons release acetylcholine and calcitonin gene-related peptide (CGRP). Acetylcholine causes muscle fibres to contract. CGRP activates adenylyl cyclase to synthesize cyclic AMP, which initiates a series of phosphorylations that strengthens the force of the contraction and stimulates energy metabolism within the muscle cell.

Multiple Actions of a Single Transmitter

The interactions between transmitter and receptor represent a transformation of information: the chemical nature of the transmitter determines only the capacity to bind to the receptor, not any of the effects on the target cell, either ionic or metabolic. As already described for individual transmitter substances, there is typically more than one receptor type for each transmitter; indeed, for several substances, there are 10 or more. Because of the variety of receptor types, a transmitter can have more than one action. Most obvious are actions at different synapses. Thus, acetylcholine activates an ionotropic receptor on skeletal muscle fibers, where it is excitatory; when released from vagus nerve terminals, it activates a metabotropic receptor to slow the heart, and therefore is inhibitory. Dopamine and noradrenaline can stimulate the postsynaptic target cell and inhibit further release of the biogenic amine transmitter presynaptically due to the interaction of the transmitter with postsynaptic excitatory receptors and presynaptic autoreceptors.

There are many examples where a transmitter reacts with different postsynaptic receptors. An especially interesting example is glutaminergic transmission at synapses in the hippocampus. Three classes of glutamate receptors regulate excitatory synaptic action in the postsynaptic cell: two receptors are ionotropic, the NMDA receptor and the nonNMDA (kainate, quisqualate and α-amino-3-hydroxy-5-methyl-4-isoxazoleproprionate (AMPA)) receptors. The third type is metabotropic. At normal resting potentials, a pulse of glutamate depolarizes the cell briefly through action on nonNMDA receptors; under these conditions, NMDA receptors contribute very little to the excitatory postsynaptic potential. When the postsynaptic neuron has been depolarized (for example, by a rapid train (tetanus) of impulses), however, a pulse of glutamate causes a much greater depolarization, produced largely by the NMDA receptor. The differences in response, which are dependent upon the previous experience of the synapse and the pattern and intensity of transmitter release, underlie the production of long-term potentiation, a plastic feature of synapses thought to give rise to memory and learning.

A final example of the diversity in transmitter action has been most clearly analyzed in large molluscan *Aplysia* neurons. Acetylcholine can have opposite actions on the same postsynaptic (follower) cell. Acetylcholine first causes the follower cell to depolarize, and then to hyperpolarize. A similar effect occurs with histamine. This kind of response is called *conjoint action*, and is caused by two distinct receptors in the follower cells, each with their own time constants.

Summary

In general, synaptic transmission through neurotransmitters can be viewed as a specialized form of secretion. Transmission through neuropeptides is most like nonneuronal secretion. Release of neuropeptide transmitters is mediated by large vesicles that fuse with the neuron's membrane. The process depends upon the influx of Ca^{2+}. Neurons have developed a more specialized mechanism for releasing amine and purine neurotransmitters. These small molecule transmitters are released when small lucent vesicles dock at specific sites in the synaptic membrane. Docking and membrane fusion is facilitated by special proteins in the vesicle membrane. Once they are released, the neurotransmitter acts on the postsynaptic target by binding to receptors in the membrane. As a rule, each transmitter has more than one receptor, and therefore a single transmitter can have more than one action.

See also **Central and Peripheral Nervous Systems**.

Additional Reading

Burnstock, G.: "Purines as Cotransmitters in the Adrenergic and Cholinergic Neurones," *Progress in Brain Research*, **68**, 193–203 (1986).

Cooper, J.R., F.E. Bloom, and R.H. Roth: *The Biochemical Basis of Neuropharmacology*, 8th Edition, Oxford University Press, New York, NY, 2003.

Dale, H.H.: "Acetylcholine as a Chemical Transmitter of the Effects of Nerve Impulses," *Journal of the Mount Sinai Hospital*, **IV**, 401–429 (1938).

Kandel, E.R., J.H. Schwartz, and T. Jessell: *Principles of Neural Science*, 4th Edition, Appleton & Lange, Stamford, CT, 2000.

Kupfermann, I.: "Functional Studies of Cotransmission," *Physiological Reviews*, **71**, 683–732 (1991).

Loewi, O.: "An Autobiographic Sketch," *Perspectives in Biology and Medicine*, **4**, 3–25 (1960).

Sanes, D.H., W.A. Harris, and T.A. Reh: *Development of the Nervous System*, Elsevier Science & Technology Books, New York, NY, 2005.

Siegel, G.J., B.W. Agranoff, R.W. Albers, and R. Fisher: *Basic Neurochemistry: Molecular, Cellular, and Medical Aspects*, 6th Edition, Lippincott Williams & Wilkins, Philadelphia, PA, 1998.

Sitte, H.H., and M. Freissmuth: *Neurotransmitter Transporters*, Springer-Verlag New York, LLC, New York, NY, 2006.

Sossin, W.S., J.M. Fisher, and R.H. Scheller: "Cellular and Molecular Biology of Neuropeptide Processing and Packaging," *Neuron*, **2**, 1407–1417 (1989).

von Bohlen und Halbach, O., and R. Dermietzel: *Neurotransmitters and Neuromodulators: Handbook of Receptors and Biological Effects*, 2nd Edition, John Wiley & Sons, Inc., Hoboken, NJ, 2006.

Webster, R.A.: *Neurotransmitters, Drugs and Brain Function*, John Wiley & Sons, Inc., New York, NY, 2001.

JAMES H. SCHWARTZ, Columbia University College of Physicians and Surgeons, New York, NY

NEUTRAL. 1. Having no electric charge, or no net electric charge. (Thus, an atom, in which the total negative charge of the electrons is equal to the positive charge of the nucleus which they surround, is neutral.)

2. According to the ionization hypothesis, a concentration of hydrogen ions equal to 1×10^{-7}. (The figure varies a little according to the temperature and the method of determining the degree of ionization of water.) Hydrogen-ion concentrations greater than this figure confer acid properties; lower concentrations occur in alkaline systems.

NEUTRALIZATION. A chemical reaction in which water is formed by mutual interaction of the ions that characterize acids and bases when both are present in an aqueous solution, i.e., $H^+ + OH^- \rightarrow H_2O$, the remaining product being a salt. R. T. Sanderson states: "An aqueous solution containing an excess of hydronium ions is called acidic. It readily releases protons to electron-donating substances. An acqueous solution containing an excess of hydroxyl ions is called basic. It readily accepts protons from substances that can release then, and is in general an excellent donor. No aqueous solution can contain an excess of both hydronium and hydroxyl ions, because when these ions collide, a proton is immediately transferred from the hydronium to the hydroxyl ion, and both become water molecules."

Neutralization occurs with both (1) $Ca(OH)_2 + H_2SO_4 \rightarrow CaSO_4 + 2H_2O$; (2) $HCOOH + NaHCO_3 \rightarrow HCOONa + CO_2 + H_2O$. It should be noted that neutralization can occur without formation of water, as in the reaction $CaO + CO_2 \rightarrow CaCO_3$. Neutralization does not mean the attaining of pH 7.0; rather it means the equivalence point for and acid-base reaction. When a strong acid reacts with a weak base, the pH will be less than 7.0, and when a strong base reacts with a weak acid, the pH will be greater than 7.0.

See also **Acids and Bases**.

NEUTRAL POINT. 1. In the clear sky, one of several points in the principal plane (or very near it) where the degree of linear polarization P = 0 (i.e., polarization is neutral). Stated another way, a neutral point occurs where Stokes parameter $Q = U = 0$. In fact, P often differs indistinguishably from zero within a small area ($\sim 3°$ radius) around the nominal principal point, and thus neutral polarization routinely occurs slightly outside the principal plane. For a single-scattering molecular atmosphere, the neutral points would coincide with the solar and antisolar points. In the real atmosphere, scattering by large particles (e.g., haze) and multiple scattering displace the observed neutral points. The Arago neutral point occurs $\sim 10° - 30°$ above the antisolar point. The Brewster neutral point occurs below the sun and the Brewster point occurs above it. When the sun is low in the sky, the Brewster and Babinet points can be as much as $25° - 35°$ from it, although their angular distances need not be equal. In principle, the Brewster and Babinet points coincide with the zenith sun.

2. Same as hyperbolic point. See also **Hyperbolic Point**.

AMS

NEUTRINO. A neutral particle of very small (presumed zero) rest mass and of spin quantum number $\frac{1}{2}$. This particle was initially postulated to account for the continuous energy distribution of beta particles and to conserve angular momentum in the beta-decay process. Experimental evidence indicates that, for the linear momentum to be conserved in the beta process, there must be a contribution from a departing neutrino. Presumably, a neutrino (or antineutrino) is emitted in every beta transition. The energy of a neutrino emitted in a beta disintegration is assumed equal to the difference between the energy of the particular beta particle and the energy corresponding to the upper limit of the continuous spectrum for that beta transition. The neutrino has also been postulated as one of the particles in pion (π) decay and as two of the particles in muon (μ) decay. These processes, however, lead to two types of neutrinos: an electron-associated neutrino ν_e and a muon-associated neutrino ν_μ. For example, $\pi^+ \rightarrow \mu^+ + \overline{\nu_\mu} + e^+ + \nu_e$ or $\pi^+ \rightarrow e^+ + \nu_e$, whereas neutron decay obeys only the process $n \rightarrow p^+ + e^- + \overline{\nu_e}$, where the bar over ν indicates an anti-particle. The difference between neutrinos was established at Brookhaven in 1962 when it was shown that a beam of neutrinos from the process $\pi^+ \rightarrow \mu^+ + \nu_\mu$ gave rise to the process $\nu_\mu + n \rightarrow p + \mu^-$ but not to $\nu_\mu + n \rightarrow p + e^-$. There is also a neutrino associated with the tau particle. Because of its properties, the neutrino has negligible interactions with matter and has proved difficult to detect. It was first positively identified experimentally in 1956 by Reines and Cowan, Jr. See also **Particles (Subatomic)**.

The term antineutrino usually denotes an antiparticle whose emission is postulated to accompany radioactive decay by negatron emission, such as, for example, in neutron decay into a proton p^+, negatron e^- and antineutrino $\overline{\nu_e}$, expressed by the equation $n \rightarrow p^+ + e^- + \overline{\nu_e}$. Capture of a neutrino by the neutron, $\nu_e + n \rightarrow p^+ + e^-$ would be an equally good description of the process. Positron emission is accompanied by a neutrino, as in the decay $^{64}Cu \rightarrow ^{64}Ni + e^+ + \nu_e$. Orbital electron capture also involves a neutrino, as for example, $e^- + {}^{64}Cu \rightarrow {}^{64}Ni + \nu_e$. Since there is no possibility of charge differentiation between the antineutrino and the neutrino, differentiation between these two particles can be made only on the basis of such properties as the sign of the ratio of magnetic moment to angular momentum.

In the past the terms neutrino and antineutrino were sometimes used in reverse sense to that stated above, i.e., the neutrino is said to accompany negatron emission and the antineutrino, positron emission. The preferred usage has been accepted in order to provide conservation of leptons in the conservation laws.

Neutrino Astronomy. For several years, there has been a marked trend in astronomy to expand the use of the electromagnetic spectrum beyond the visual range in investigating the universe. It now appears that, in addition to infrared, ultraviolet, gamma ray, x-ray, and radio astronomy, among others, increasing attention will be given to neutrino astronomy, i.e., the detection and measurement of neutrinos emanating from such celestial bodies as supernovas and x-ray double stars. Traditionally, neutrinos have been created in large accelerators and used for investigating the characteristics of other elementary particles. It is now reasoned that neutrinos entering the earth's atmosphere from celestial distances may furnish new, heretofore unavailable information.

Search for Neutrinos

Certain mysteries continue to surround the neutrino, emphasizing how far physicists still may be from a full comprehension of the complete array of subatomic particles. Numerous detectors in different locations have been constructed for detecting neutrinos, particularly as they emanate from the sun.

The first recorded attempt to construct a neutrino telescope was undertaken by Davis (Brookhaven National Laboratory) in the 1960s. Davis' principal objective was detection of low-energy neutrinos emitted by the sun. These neutrinos are generated deep within the sun as the result of thermonuclear reactions. It has been estimated that nearly 10% of the energy released by the transmutation of hydrogen in the sun is carried away by neutrinos, which have energies ranging from $\frac{1}{2}$ million eV to about 14 million eV. The solar flux of neutrinos is tremendous. It is estimated that 10^{14} solar neutrinos pass through the human body every second. Davis' detector consisted of a large tank containing 610 tons of tetrachloroethylene, C_2Cl_4. Of the chlorine atoms present, 25% are of the isotope chlorine 37. When this atom captures a neutrino, it is transformed to an atom of argon 37.

An early detector was established in the shaft of the Homestake Mine (South Dakota) in 1968. The data gathered did not correspond with the scientists' expectation. The neutron flux measured was less than one-third that predicted. This discrepancy challenged researchers to question perhaps the "established" concepts of particle physics and indeed the manner in which the sun functions.

Some years later, a Japanese-built detector (*Kamiokande II*), designed to detect the more energetic solar-emitted neutrinos (~ 5 mil electron volts), came up about 50% short of the expected counts, thus reconfirming a shortage of solar neutrons.

Neither of the aforementioned detectors was designed to sense comparatively low-energy (proton-proton) neutrinos, which result from the fusion of two protons.

Subsequently, two additional detectors were built with the objective of seeking lower-energy neutrinos. One of the detectors was located in the Caucasus of the former Soviet Bloc and was named the Soviet-American Gallium Experiment (SAGE). The detector used a gallium metal detector believed to be sensitive to neutrinos, with energy as low as 0.23 mil electron volts. Another similar detector (*Gallox*) was constructed in Italy. At both sites, difficulties with calibration were expected and did occur. Some tenuous observations later were made to suggest a correlation of neutrinos detected with the occurrence of sun spots and with solar acoustic oscillations.

Based upon the foregoing experiences, some researchers observed that the same reluctance to interact with matter is responsible for the neutrino's long range and ability to resist detection. Thus, it was reasoned that an apparatus for detecting neutrinos should be massive and shielded from the interference of other particles and radiation. As a solution to these problems, some researchers proposed a deep underwater muon and neutrino detector (acronym DUMAND).

Fiber optic data cable will stretch from the shore for 30 km to a connector box some 4800 meters below the ocean surface. Strings of nine separate cables will rise vertically about 280 meters above the ocean floor. Each cable, held up by a float, will contain 24 detectors. The apparatus, referred to as a neutrino telescope, will have to pick up at least ten muon events per year from any given 1° patch of sky for the DUMAND scientists to be confident that they have a significant neutrino source, not just a few background pulses from non-neutrino cosmic rays.

Construction of DUMAND, with Japanese and German collaborators, got underway off the west coast of Kona (Hawaii) in 1990 and was expected to be operational within a few years. V. Stenger (University of Hawaii) observes, "The idea of using the ocean floor as a detector actually goes back to the 1960s and people started taking it seriously back in the mid 1970s."

A number of scientists are hopeful that the neutrino telescope will ultimately yield much additional information on neutron stars, supernovas, quasars, etc., which are presumed to be large emitters of neutrinos. Some astronomers estimate that SS 433 puts out 1000 times more energy per second than the brightest stellar object known in the galaxy. Why should such a powerful object have been discovered so recently? The fact is that SS 433 is a comparatively weak source of photons. The accreting matter that gives SS 433 its great power also serves to screen its bright central region from view. Much remains to be learned concerning SS 433 and perhaps neutrino astronomy may supply some answers at a future date.

Laboratory Experimentation on Neutrinos

During the 1960s, L.M. Lederman, M. Schwartz, and J. Steinberger conducted the well-known two-neutrino experiment, which established a relationship between particles, muon and muon neutrinos, electron and electron neutrino. This later evolved into the standard model of particle physics. The Nobel prize in physics was shared by these researchers in 1988.

Massive Neutrino Proposed. J. Simpson (University of Guelph, Ontario) in the late 1980s presented evidence of the existence of a *heavy neutrino* having a mass of 17,000 electron volts (keV, the units of energy that are interchangeable with mass). A renowned neutrino physicist for several years prior, in 1985 Simpson conducted a series of experiments in his laboratory. The objective of his initial experiment was that of measuring the energy of electrons emitted from heavy hydrogen (tritium) in the radioactive process of beta decay. In this process, the energy of the emitted electrons should appear as a spectrum (smooth curve) from zero to a maximum endpoint. However, Simpson noted "occasional" aberrations in the plotted data. This so-called "kink" corresponded with an energy of 17 KeV of the normal plot. This indicated the probable presence of an unknown massive force. In addition to repeated experiments with tritium, Simpson also used Sulfur-35. Simpson then suggested that the aberration, when it occurred, could be attributed to a heavy 17 KeV neutrino.

Simpson's carefully prepared data, when published, attracted wide attention and drew a wide spectrum of professional reactions. Meanwhile, experiments by other laboratories have confirmed Simpson's results. Researchers at the Lawrence Berkeley Laboratory, using Carbon-14, have reported evidence for a 17.2 KeV neutrino in 1.4% of their experiments. Researchers at the Ruder Boskovic Institute in Zagreb (formerly Czechoslovakia), using Iron-55 and Germanium-71, also have found the evidence appearing in 1.5% of their experiments. Thus, confidence in Simpson's claim appeared to be gaining momentum in the early 1990s.

As pointed out by M. Turner (University of Chicago), a massive neutrino would "violate every theoretical prejudice we have in particle physics, astrophysics, and cosmology." J. Bahcall (Institute of Advanced Study at Princeton) observes, "It's a surprise. If it's true, then it's pointing us in a different direction than previous physics suggested."

See also **Particles (Subatomic)**.

Additional Reading

Bahcall, J.N.: *Neutrino Astrophysics*, Cambridge University Press, New York, NY, 1990.

Breuker, H. et al.: "Tracking and Imaging Elementary Particles," *Sci. Amer.*, **58** (August 1991).

Brown, L.M., M. Dresden, and L. Hoddeson: *Pions to Quarks: Particle Physics in the 1950s*, Cambridge University Press, New York, NY, 1989.

Caldwell, D.O.: *Current Aspects of Neutrino Physics*, Springer-Verlag New York, Inc., New York, NY, 2001.

Cence, R.J. et al.: *Neutrino 81*, University of Hawaii, Honolulu, HI, 1981.

Dar, A.: "Astrophysics and Cosmology Closing in on Neutrino Masses," *Science*, **1529** (December 14, 1990).

Dehmelt, H.: "Experiments on the Structure of an Individual Elementary Particle," *Science*, **539** (February 2, 1990).

Florini, E.: *Neutrino Physics and Astrophysics*, Plenum, New York, NY, 1982.

Gutbrod, H. and H. Stöcker: "The Nuclear Equation of State," *Sci. Amer.*, **58** (November 1991).

Horgan, J.: "Three Americans Honored for 1960's Neutrino Experiment," *Sci. Amer.*, **31** (December 1988).

Kim, C.W. and A. Pevsn: *Neutrinos in Physics and Astrophysics*, Gordon & Breach Publishing Group, Newark, NJ, 1993.

Lederman, L.M.: "Observations in Particle Physics from Two Neutrinos to the Standard Model," *Science*, **664** (May 12, 1989).

Mohapatra, R.N. and P.B. Pal: *Massive Neutrinos in Physics and Astrophysics*, World Scientific Publishing Company, Inc., River Edge, NJ, 1997.

Powell, C.S.: "Looking for Nothing: The Taciturn Neutrino Keeps Physicists Guessing," *Sci. Amer.*, **22** (April 1991).

Schwartz, M.: "The First High-Energy Neutrino Experiment," *Science*, **1445** (March 17, 1989).

Selvin, P.: "Is There a Massive Neutrino?" *Science*, **1426** (March 22, 1991).

Stenger, V.J. and J.G. Learned: *High Energy Neutrino Astrophysics: Proceedings of the Workshop*, World Scientific Publishing Company, Inc., Riveredge, NJ, 1992.

Traweek, S.: *Beamtimes and Lifetimes. The World of High Energy Physics*, Harvard University Press, Cambridge, MA, 1992.

Waldrop, M.M.: "A Nobel Prize for the Two-Neutrino Experiment," *Science*, **669** (November 4, 1988).

Waldrop, M.M.: "Solar Neutrino Deficit Confirmed?" *Science*, **1607** (June 29, 1990).

Waldrop, M.M.: "Astrophysics in the Abyss," *Science*, **208** (October 12, 1990).

Winter, K., and G. Altarelli: *Neutrino Mass*, Springer-Verlag New York, Inc., New York, NY, 2003.

Web Reference

The Sudbury Neutrino Observatory: http://www.sno.phy.queensu.ca/

NEUTRON. The discovery of the neutron by Chadwick in 1932 represented a great step forward in the investigation of nuclei of atoms. Chadwick found that a radiation emitted when α-rays from polonium reacted with beryllium could project protons from a thin sheet of paraffin wax. Although the radiation itself produced no observable ionization when passing through a gas, the protons released from the paraffin were detected in an ionization chamber. Inability to produce ionization was interpreted as a lack of electric charge. From measurements of the ionization from the protons, Chadwick deduced that the so-called beryllium radiation must consist of neutral particles with a mass very nearly equal to that of the proton. He announced the discovery of the neutron, a previously unknown particle. It has been confirmed that the neutron has no charge and a mass of 1.088665 atomic mass units. Thus, it is heavier than the proton by 0.00139 mass unit. The introduction of the neutron into nuclear structure produced a sharp change in previously held concepts. Lacking knowledge of the neutron, masses of atomic nuclei had been attributed solely to protons. The number of protons required on this basis for most nuclei greatly exceeded the known charge number. In an attempt to solve this dilemma, a number of electrons were assigned to each nucleus to adjust the charge number to the proper value. This compromise created an even greater problem, that of accommodating so many electrons in the small space occupied by a nucleus. Bringing the neutron into the picture meant that a nucleus contains only enough protons to equal the charge number, with the rest of the mass contributed by neutrons. No additional electrons were required.

Decay. The neutron in the free state undergoes radioactive decay. Elaborate experiments by Robson were required to identify the products of the decay and to measure the half-life of the neutron. He showed that the neutron emits a β-particle and becomes a proton. The half-life was found to be 12.8 minutes. In stable nuclei, neutrons are stable. In radioactive nuclei, decaying by β-emission, the neutrons decay with a half-life characteristic of the nuclei of which they are a part. See also **Radioactivity**.

Detection. Because it is a neutral particle the neutron is detected by means of a secondary charged particle which it releases in passing through matter or by means of the radioactivity which the neutron can induce in stable atoms. Protons may be projected by collisions with neutrons in hydrogenous material and the ionization from the protons can be measured in an ionization chamber, as in the original experiment with neutrons. Secondary charged particles may be the direct result of nuclear disintegration produced by neutrons, as in the case of the reaction $^{10}B + ^1n \rightarrow ^7Li + \alpha$. Commonly, the radioactivity induced in a stable element by neutron capture serves to detect neutrons, and this technique is known as the *activated foil method*. Also, fission may be utilized for detection of neutrons by placing fissionable material inside an ionization chamber and observing the ionization generated by the fission fragments.

Energies. The kinetic energy of neutrons has an important bearing on their behavior when interacting with nuclei. These kinetic energies may range from near zero to as much as 50 MeV. It is therefore natural to classify neutrons in terms of energy according to their properties in each range of energy. For example, energies from zero to about 1000 eV are usually called *slow neutrons*. Because they are more readily captured by nuclei than faster neutrons, slow neutrons are responsible for a large number of nuclear transformations. When slow neutrons have velocities in equilibrium with the velocities of thermal agitation of the molecules of the medium in which they are situated, they are called *thermal neutrons*. The distribution of these velocities approaches the Maxwell distribution

$$dn(v) = Av^2 e^{-(Mv^2/2kT)} dv$$

where v is the neutron velocity, M its mass, k is Boltzmann's constant, and T is the absolute temperature. In the slow neutron range of energies, various atomic nuclei show strong absorption (capture) of neutrons at fairly well-defined energies. Neutrons having energies corresponding to those of the absorption bands are called *resonance neutrons*. Frequently, neutrons with energies greater than 1000 eV and less than 0.5 MeV are termed *intermediate neutrons*. In more general terms, all neutrons with energies greater than 0.5 MeV are called *fast neutrons*. The practical upper limit of neutron energy is set by the devices thus far developed for accelerating charged particles to extremely high energies.

Magnetic Moment and Spin. Alvarez and Bloch succeeded in measuring the moment of the magnetic dipole associated with the known spin of $\frac{1}{2}$ possessed by the neutron. More refined measurements by Cohen, Corngold, and Ramsey of the magnetic moment μn yielded a value of

$$\mu n = 1.913148 \text{ nuclear magnetons}$$

Interactions with Nuclei. Neutrons may be scattered or captured by heavy nuclei. Scattering may be elastic, resulting only in the change of direction of the neutrons, or inelastic, in which the neutron loses part of its energy to the scattering nucleus. Collisions with light nuclei, in the absence of capture, result in communicating considerable fractions of the neutron energy to the target nucleus. A neutron colliding head-on with a proton will give practically all its kinetic energy to the proton. As the mass of the target nucleus increases, the transfer of energy decreases, in accordance with the laws of conservation of energy and momentum. The loss of energy by mechanical impact is utilized in slowing down fast neutrons, a process known as *moderation*. Slow neutrons are most useful, for example, in the production of radioelements from stable elements by neutron capture. A good moderator should have low mass and a small capture cross section.

The rate r of capture of neutrons from a neutron flux F (neutrons cm^{-2} sec^{-1}) incident on a layer of matter having N nuclei per square centimeter is given by

$$r = F\sigma N$$

where σ is the complete probability of capture. Replacing r by dN/dt and writing the flux as nv, where n is the number and v is the velocity of the neutrons, we have

$$dN/dt = nv\sigma N$$

which integrated gives

$$N = N_0 e^{-nv\sigma t}$$

where N is the number of unchanged nuclei in the target area at time t and N_0 is the number at time $t = 0$. The cross section σ is so named because it has the dimensions of an area. The unit for the cross section is the *barn*, equal to $10^{24} cm^2$. When, as is often the case, σ is proportional to $1/v$, the advantage of slow neutrons in capture interactions becomes apparent. When the value of s departs sharply from that predicted by the $1/v$ law, it usually increases over a narrow range of energies, and we have what is called a *resonance*. Slow neutron cross sections are customarily quoted for thermal neutrons at $20\,°C$, corresponding to a value of v of 2200 m/sec. Under these conditions, representative thermal neutron capture cross sections are: boron, 759 barns; cobalt, 38 barns; cadmium, 2450 barns; gadolinium, 46,000 barns; gold, 99.8 barns; helium, 0; lead, 0.170 barn; and oxygen <0.0002 barn.

Additional interactions of neutrons with nuclei include the release of charged particles by neutron-induced nuclear disintegration. Commonly known reactions are $n - p$, $n - d$, and $n - \alpha$. In these cases, the incident neutrons may contribute part of their kinetic energy to the target nucleus to effect the disintegration. Hence, more than mere neutron capture is involved. Then, there is usually a lower threshold for the neutron energy below which the reaction fails to occur. Another important reaction involving neutrons is fission, which may occur under different conditions for either slow or fast neutrons with appropriate fissionable material.

Sources of Neutrons. Any nuclear reaction in which neutrons are released might serve as a source of neutrons. In the initial experiments with neutrons, an α-n reaction was used. Because of the charge on the α-particle, it must have a high kinetic energy to penetrate a nucleus. Thus, polonium α-particles could release neutrons from beryllium. Such a natural source produces relatively few neutrons. The yield of neutrons from charged particle reactions can be increased manyfold by the use of particle accelerators. Here large numbers of charged particles of high energy can be used in the bombardment of the target to release numerous neutrons. Frequently deuterons or protons are used for the bombardment. A far more prolific source is the nuclear reactor. Fission of uranium is usually the source of the neutrons in this case. A nuclear reactor as usually constructed generates neutrons of different energies in various parts of its structure. Neutrons of suitable energy for a given experiment may be brought outside the reactor through channels into appropriate sections of the reactor. See also **Nuclear Power Technology**.

Traditionally, the neutron is regarded as a particle which is a component of nuclei and which exists only briefly in the free state. For many purposes, this view is sufficient. However, it became obvious some years ago from various experiments, for example, in very high energy accelerators, that the neutron must have a complex structure. This view was reinforced by the nature of the decay of the neutron. A ς particle is ejected from the neutron on decay, but it is quite certain that the electron did not exist within the neutron prior to the decay. Rearrangements of an internal structure of the neutron must provide the energy for the formation and ejection of the ς particle. An early theory would have the neutron consist of a proton and a π^- meson bound together so that they oscillate between a completely bound state and a more loosely bound state. This concept might explain the feeble interaction that has also been observed between electrons and neutrons at very short range.

Fermi Age Model. This is a model for the study of the slowing down of neutrons by elastic collisions. It is assumed that the slowing down takes place by a very large number of very small energy changes. Phenomena due to the finite size of the individual losses are ignored. In this model, the word age is somewhat of a misnomer, since its units are those of area rather than time. The name arises because the variable τ, the Fermi age, appears in the Fermi age equation in the same way that time appears in the standard heat-diffusion equation. The equation for the Fermi age in a unit volume of nuclear reactor is $\tau = D\phi/q$, in which D is the diffusion coefficient for fast neutrons, $\phi = nv$ is the fast neutron fluence, and q is the number of neutrons thermalized per second per cubic centimeter. For this purpose fast neutrons are inclusively all neutrons with energies between those acquired at fission and that energy at which they are thermalized.

Ultracold Neutrons. As pointed out by King (Massachusetts Institute of Technology), there is probably as much to be learned between the lowest energy yet reached and zero energy as there is between the highest energy attained and infinite energy. Ultracold neutrons may provide an avenue to very-low-energy research. It should be recalled that a neutron with an energy of 10^7 electron volt is at the low end of the energy scale. A neutron has the energy that would be imparted to an electron by a potential difference of one ten-millionth of a volt (0.1 microvolt). As pointed out by Golub et al. in 1979, this is the amount of energy of a particle in a gas whose temperature is one millidegree K. Unlike high-energy particles, ultracold neutrons move at a rate measured in a few meters per second. Golub et al. have proposed that inasmuch as ultracold neutrons cannot penetrate a solid surface, they can be confined in a metal bottle and by storing over long periods, it may be possible to measure the fundamental properties of the neutron.

Outside the nucleus, the neutron is an unstable particle. Free neutrons are rare in nature. By the process of beta decay, the neutron breaks down into a proton, an electron, and a neutrino (a massive particle). For probing atomic and molecular structures, thermal neutrons have traditionally been used. As explained by Golub et al., ultracold neutrons can be employed in a similar way, but their low energy and long wavelength adapt them to the examination of materials on a somewhat larger scale. At the Technical University (Munich), Steyerl and associates have developed a neutron spectrometer. In conventional neutron spectrometers, particles are analyzed by a magnet that bends their trajectories. In an ultracold-neutron spectrometer, the earth's gravitational field is used. In the device, the neutrons enter the spectrometer in a horizontal movement and are accelerated as they fall a fixed distance to a specimen. Those neutrons that rebound from a target are collected by an exit slit of the instrument. An exchange of energy with the specimen is reflected by the maximum height to which the neutrons rebound.

During the past decade, since their detection, much research has been directed toward methods of extracting, storing, and manipulating ultra-cold neutrons. It now appears that the next period will be one of investigating the neutron per se and possibly of using this new knowledge for the study of other systems of particles.

See **Particles (Subatomic)** for more recent views. See also **Proton**.

Glossary of Neutron Terms

Neutrons are designated according to their energies, including the following:

Thermal neutrons, or neutrons in thermal equilibrium with the substance in which they exist; most commonly, neutrons of kinetic energy about 0.025 eV, which is about $\frac{2}{3}$ of the mean kinetic energy of a molecule at $15\,°C$.

Epithermal neutrons, or neutrons having energies just above those of thermal neutrons; the epithermal neutrons energy range is between a few hundredths eV and about 100 eV.

Slow neutrons (a less definite classification), which may mean either neutrons having energies up to about 100 eV, or thermal neutrons.

Intermediate neutrons, which are neutrons having energies in a range that extends roughly from 100 to 100,000 eV. This range is above that of epithermal neutrons and below that of fast neutrons.

Fast neutrons, which are neutrons with energies exceeding 10^5 eV, although sometimes a lower limit is given.

Resonance neutrons may be either of the following: (1) for a specified nuclide or element, neutrons that have energies in the region where the cross section of the nuclide or element is particularly large because of the occurrence of a resonance. For example, cadmium resonance neutrons have energies between 0.05 and about 0.3 eV. (2) Neutrons having kinetic energies in the region of values for which prominent resonances are encountered in many nuclides; loosely, epithermal neutrons.

Prompt neutrons are those neutrons released coincident with a fission process.

Delayed neutrons are those neutrons released subsequently in a fission process, or, more generally, neutrons emitted by excited nuclei formed in any radioactive process (beta disintegration, in all cases so far known). The neutron emission itself is prompt, so that the observed half life is that of the preceding beta emitter. The situation is similar to that involving gamma-ray emission, which is a competing process. Delayed neutron emission is possible only if the excitation energy of the product nucleus exceeds the neutron binding energy for that nucleus. The chemistry of the delayed neutron emitter is that of the beta activity; thus ^{87}Br, ^{137}I, and ^{17}N are delayed neutron precursors, although the neutron emission actually takes place from excited nuclei of the products ^{87}Kr, ^{137}Xe, and ^{17}O.

Neutron cycle is the average life history of a neutron in a nuclear reactor. The gain in the number of neutrons in a reactor during any individual neutron cycle is given by $n(k-1)$, where n is the number of neutrons in the reactor of the beginning of the cycle and k is the multiplication factor.

Neutron excess is the difference between the number of neutrons and the number of protons in an atomic nucleus. This is found by subtracting the atomic number of that nuclide from the mass number; or by subtracting twice the atomic number from the mass number.

Neutron flux density is the number of neutrons that enter a sphere of unit cross-sectional area per unit of time. This quantity is sometimes defined in terms of a unidirectional beam of neutrons incident perpendicularly upon a unit area, but this definition is less general. It is also sometimes called neutron current density.

Neutron number is the number of neutrons in a nucleus. Its symbol is N. The neutron number for a given nuclide is equal to the difference between the mass number and the atomic number for that nuclide.

Additional Reading

Breuker, H. et al.: "Tracking and Imaging Elementary Particles," *Sci. Amer.*, **58** (August 1991).

Brown, A.: *The Neutron and the Bomb: A Biography of Sir James Chadwick*, Oxford University Press, Inc., 1997.

Gutbrod, H. and H. Stöcker: "The Nuclear Equation of State," *Sci. Amer.*, **58** (November 1991).

Hamilton, J.H.: *Fission and Properties of Neutron-Rich Nuclei*, World Scientific Publishing Company, Inc., River Edge, NJ, 1998.

Harding, A.K.: "Physics in Strong Magnetic Fields Near Neutron Stars," *Science*, **1033** (March 1, 1991).

Ruthen, R.: "Out of Its Field: How the Neutron Responds to A Field That Exerts No Force," *Sci. Amer.*, **26** (October 1989).

Schopper, H.: *Low Energy Neutrons and Their Interaction with Nuclei and Matter: Low Energy Neutron Physics*, Springer-Verlag, Inc., New York, NY, 2000.

Shepherd, G.M.: *Foundations of the Neutron Doctrine*, Oxford University Press, Inc., New York, NY, 1998.

Soloway, A.H., R.F. Barth, and D.E. Carpenter: *Advances in Neutron Capture Therapy*, Kluwer Academic Publishers, Norwell, MA, 1993.

Traweak, S.: *Beamtimes and Lifetimes: The World of High Energy Physics*, Harvard University Press, Cambridge, MA, 1992.

Web Reference

Neutron Scattering Web: http://www.neutron.anl.gov/

NEUTRON ACTIVATION ANALYSIS. This is a method of elemental analysis based upon the quantitative detection of radioactive species produced in samples via nuclear reactions resulting from neutron bombardment of samples. The neutron-induced reactions are of two main types: (1) those induced by very slow (thermal) neutrons, having energies of about 0.025 eV; and (2) those induced by fast neutrons having energies in the range of MeV. The method is used in two different forms. The purely instrumental form is fast and nondestructive and is based upon the quantitative detection of induced gamma-ray emitters by means of multichannel gamma-ray spectrometry. The amount of the element present is usually computed from the photopeak (total absorption peak) height or area of its gamma ray, or one of its principal gamma rays, compared with that of the standard. Where interferences from other induced activities are serious and cannot be removed by decay, spectrum subtraction, or computer solution, one must turn to the radiochemical separation method. Here the activated sample is put into solution and equilibrated chemically with measured amounts (typically 10 milligrams) of added carrier of each of the elements of interest, before chemical separations are carried out. The element to be detected needs then to be recovered in chemically and radiochemically pure form, but it need not be quantitatively recovered, since the carrier recovery is measured and the counting data are then normalized to 100% recovery. This form is slower, but it applies to pure beta emitters, as well as to gamma emitters, and it does eliminate interfering activities.

NEUTRON INTERFEROMETER. See **Gravitation**.

NEUTRON STARS. One of the end points in stellar evolution, in which a star of 1.4 to approximately 3 times the mass of the sun is supported by internal pressure generated by compressed "degenerate" neutrons.

Normal stars result from a balance of forces: the force of gravity pulling inward and a force generated by the energy of the nuclear reactions inside pushing outward. When the star exhausts its core hydrogen, it leaves the "main sequence" of normal stars and contracts under the force of gravity. The increased fusion that follows provides energy for the outer layers to swell, and the star becomes a "red giant." Later, after the helium is exhausted, stars still contain more than about 4 solar masses expand again to become "red supergiants."

When the supergiants have built all their nuclear fuel into iron, they explode as *supernovae* (q.v.). Those that have more than about 1.44 solar masses left after this stage are not stopped by the force of electron degeneracy that supports *white dwarfs* (q.v.). 1.44 solar masses is a theoretically derived value known as the "Chandrasekhar limit," after S. Chandrasekhar of the University of Chicago. Those stars that have more than 1.44 but less than about 3 solar masses (with the value of this upper limit being less accurately known than the value of the lower limit) will contract until the neutrons in the star resist being pushed together any further. This generates a pressure known as "neutron degeneracy pressure." The resulting object is a *neutron star*.

This situation was predicted by J. Robert Oppenheimer and colleagues in the 1930s. It comes from the application of quantum mechanical rules, including the Pauli exclusion principle and the Heisenberg uncertainty principle, to the neutrons. A neutron star contains approximately 2 solar masses in a volume only 20 km or so across, giving a density of billions of tons per cubic centimeter.

Neutron stars may have solid, crystalline crusts about 100 m thick. The neutron stars' atmospheres may be only a few centimeters thick. The concentration of a normal stellar magnetic field by gravitational collapse would correspond to a neutron star magnetic field of billions of gauss.

Though no neutron stars were known until 1968, neutron stars are now detected in various ways. Most of the known neutron stars are pulsars (q.v.), of which over 400 are now known. In these cases, we detect a beam of radio radiation, possibly emitted along the magnetic axis of the star, as it sweeps across space. Periods of rotation of the neutron stars in pulsars range from 1.4 ms to about 3 s. A very few of these pulsars have been detected in the optical or x-ray part of the spectrum. The best-studied example is the pulsar in the Crab Nebula, the remnant of the supernova explosion of 1054. This object is detected across the entire spectrum. In addition to its radio pulsations, its pulsations have been recorded in the x-ray region with the Einstein Observatory spacecraft and in the optical region with telescopes on earth. Another unusual pulsar, in a binary system, has been used as a test of the general theory of relativity. See also **Cosmology**.

Other neutron stars appear in binary x-ray sources. Hundreds were mapped by NASA's High Energy Astronomy Observatory 1 in the late 1970s, and many were then studied individually by the Einstein Observatory, High Energy Astronomy Observatory 2. Perhaps the best studied had been discovered by an early x-ray spacecraft, UHURU. This source, Hercules X-1, shows rapid pulsations every 1.24 seconds, but the pulses turn on and off with a 1.7-day period. This longer period is presumably caused by eclipses by the primary object in the system, the variable star HZ Herculis, while the shorter period is presumably caused by rotation of the neutron star. A third period of about 35 days also exists in Hercules X-1.

The most unusual astronomical object now known is SS433, whose spectral lines vary with an 164-day period over a Doppler range corresponding to velocities of about 25% of the speed of light. SS433 is in our galaxy, and such a velocity in our galaxy is unprecedented. A 13-day periodicity has also been discovered in SS433, indicating that it is a binary system of that period. In leading models, an accretion disk of material has formed around the neutron star, and jets of gas are emitted above and below the plane of the disk, which is inclined to us. As the disk precesses, we alternately see each of the beams approaching and receding with the longer period of the system. It is the material in the beams that has the high velocities.

Discovery of neutron stars has spurred work in stellar evolution of binary systems. Each object in a binary system has a Roche lobe around it, within which the gravitational force of the object is strong enough to resist the tidal forces from outside objects. The Roche lobes define a figure-of-8 around the two members of a binary system. In some binaries, one or both objects swell to fill their Roche lobes, at which time material can flow from one object to the next. The exchange of mass can severely alter the evolution of the objects, and the relative masses of the objects can interchange. When one of the objects is a compact object like a neutron star, mass from the companion flowing through the neck of the figure-of-8 and falling on the neutron star can lead to a surface nuclear explosion that is a powerful emitter of x-rays.

Additional Reading

Buccheri, R., J. Van Paradijs, and M.A. Alpar: *The Many Faces of Neutron Stars*, Kluwer Academic Publishers, Norwell, MA, 1998.

Kawaler, S.D., G. Srinivasan, I.D. Novikov, G. Meynet, and D. Schaerer: *Stellar Remnants*, Springer-Verlag, Inc., New York, NY, 2000.

Meszaros, P.: *High-Energy Radiation from Magnetized Neutron Stars*, University of Chicago Press, Chicago, IL, 1992.

Rothschild, R.E. and R.E. Lingenfelter: *High Velocity Neutron Stars and Gamma-Ray Bursts*, Springer-Verlag, Inc., New York, NY, 1996.

JAY M. PASACHOFF, Williams College, Williamstown, MA

NEUTROSPHERE. The atmospheric shell from the Earth's surface upward in which the atmospheric constituents are for the most part not ionized, i.e., it is electrically neutral. The region of transition between the neutrosphere and the ionosphere is somewhere between 70 and 90 kilometers, depending on latitude and season. See also **Earth**.

NEWCASTLE DISEASE. Also sometimes called *fowl pest, pseudo-fowl pest, Ranikhet disease*, and *avian pneumoencephalitis*, the disease is of viral origin. The disease was first noted in the Dutch East Indies. A series of outbreaks of the disease occurred in England near Newcastle-on-Tyne, from which the name of the disease was derived.

In his excellent reference, R.F. Gordon devotes several pages to the history and characteristics of Newcastle disease and summarizes the situation as of the mid- to late 1970s as follows: "By 1966, a fairly stable situation seemed to have developed in which fully virulent virus had become endemic in the tropics, milder disease in North America and western Europe and an intermediate form in Iran and Arab countries to the West. In 1968, an upsurge of the disease was reported in Iraq. Reports of disease, difficult to control then followed from Lebanon, Israel, Greece and in 1970 from England and Holland and in 1971 to other countries in western Europe. Workers have given the strain the designation Essex '70. In 1971 the United States reported cases of fully virulent Newcastle disease occurring along its southern border and designated such strains as velogenic viscotropic Newcastle disease (VVND). A comparison of the two types of isolate has been made. The Essex '70 type of virus has been observed to be highly pneumotropic and to spread rapidly while the VVND type of isolate seems on average, more lethal, more likely to give rise to visceral lesions and to spread more erratically. Although no distinction can accurately be made by laboratory tests, there is considerable epidemiological evidence to suggest their mode of spread is significantly different. By 1974, the world use of vaccine had increased greatly and in countries such as the UK where slaughter on infected farms was practiced until 1962, there had been heavy expenditure on control. Currently losses from the disease are not high in countries where extensive vaccination is practiced although virulent forms of the virus are now much more widespread than ever before."

NEW GUINEA COSTAL CURRENT. See **Ocean Currents**.

NEW MILLENNIUM PROGRAM (NMP). NASA space science missions have ventured to the moon, explored other planets, traveled to the edges of our solar system, and peered back in time. They have also done what is sometimes even more difficult — studied our own planet, Earth.

These missions have provided astounding views of the universe and new knowledge of our solar system, but there is still so much more to "see" and learn. And, as missions become progressively more daring, and thus more difficult, more advanced capabilities are needed. However, before new, untried technologies are used for the first time on complex exploration missions, engineers and scientists want to make sure they will operate well, and safely, in the hazardous environment of space.

To accomplish this, NASA's Office of Space Science (OSS) and Office of Earth Science jointly established the New Millennium Program (NMP) in 1995 — an ambitious, exciting vision to speed up space exploration through the development and testing of leading-edge technologies. A unique program, managed by the Jet Propulsion Laboratory/California Institute of Technology, NMP provides a critical bridge from initial concept to exploration-mission use. Through NMP, selected technologies are demonstrated in the "laboratory" of space that can't be replicated on Earth.

Since its inception over a decade ago, NMP has validated many innovative technologies for both Earth science and space science missions. Now funded and managed solely out of NASA's newly formed Science Mission Directorate (SMD), the Program continues to demonstrate advanced technologies that will enable space science missions of the 21st century with significant (a several-generation leap) technical capabilities.

Highly advanced technologies are key to more capable, powerful, and efficient spacecraft and science instruments. They are also key to gathering new and exciting scientific knowledge of our solar system and of our universe.

While an emerging technology may seem promising and likely to provide the technical capabilities NASA requires, it may also present an unacceptable risk to any exploration mission using it for the first time in space. The goal of NMP is to reduce the risks to, as well as the costs of, future NASA space science missions.

To meet its goals, the NMP identifies and selects leading-edge technologies that will increase the capability of future Science Mission Directorate missions. To identify the crucial technologies required, technologists are guided by the roadmaps of NASA's three mission areas: Sun-Earth System, Solar System, and Universe. The technical requirements outlined in these roadmaps are matched with technologies emerging from the national "pipeline" of current technology-development efforts. Once selected, these untried technologies are demonstrated on NMP in-space validation missions.

Ion propulsion, for instance, is a technology that was in the development stage for several decades. It was considered too risky to use, but promised a revolution in space propulsion if it did work. So, in 1998 NMP conducted its first technology validation mission — Deep Space 1 (DS1), with a xenon ion drive engine and several other advanced, high-risk technologies: solar electric propulsion; autonomous optical navigation; beacon monitor operations; solar concentrator array; telecommunications devices; microelectronics and spacecraft structure; autonomous operations systems; miniature camera and imaging spectrometer; and miniature ion and electron spectrometer.

Following its launch, three of DS1's technologies had to work within a few minutes of when the spacecraft separated from the rocket that took it into space. Unlike most interplanetary missions, which have many months of coasting with minimal activity before reaching their destinations, DS1 immediately began a very intensive period of demanding experiments to characterize the 12 technologies on board.

And, while the primary focus of NMP's projects is to test advanced instruments, spacecraft systems and subsystems, and mission concepts in spaceflight, they may also return science data as a byproduct of their testing. DS1 is a good example. After the testing period, a bonus mission was conducted during which the revolutionary xenon ion engine propelled the DS1 spacecraft through space to a successful encounter with Comet Borrelly — returning the best images and other science data ever from a comet!

Through NMP, advanced technologies are developed at a lower cost and in less time than has been done in the past. The Program's space-flight validation advances the readiness levels of these emerging technologies, so that they may be applied to operational science missions sooner. In this way, scientific information through space exploration is gained sooner.

Innovative Technologies

To achieve its future science goals, NASA needs smarter, faster, more powerful spacecraft carrying "intelligent," more autonomous instruments. Moreover, these revolutionary advanced spacecraft and instruments will need to communicate with Earth-based receiving stations and with other space-based communication networks at much higher data rates. And, for some missions, both spacecraft and instruments must be smaller in size and operate more efficiently. Significantly advanced technologies will enable higher science return, while others will substantially reduce operational costs.

While developing innovative emerging technologies is an exciting enterprise, it is much more difficult to design specialized equipment and components to withstand the harsh environment of space — zero pressure, extreme temperatures, and deadly high-energy particles — than to design "gadgets" for everyday applications here on Earth. Plus, using untried technology for the first time in complex space exploration missions is highly risky. The New Millennium Program (NMP) identifies and space-flight validates breakthrough technologies that will significantly reduce risks and costs and ultimately benefit future NASA science missions.

NMP's selected technologies encompass small (Space Technology 9) to medium class (Deep Space 1) system-level technologies, which are flight validated by NMP-funded teams. The Program also identifies subsystem technologies (such as those on Space Technology 6 and Space Technology 8). Both classes of technology experiments help enable NASA's **Discovery Program** and **Explorers Program** (medium-class, small, and university-class) space science missions. See also **Discovery Program**; and **Explorers Program**.

Some NMP projects (Deep Space 1, Earth Observing 1, Space Technology 5, and Space Technology 7) incorporate several spacecraft systems and instrument technologies on a single NMP flight validation mission. Other NMP projects (Deep Space 2, Earth Observing 3, and Space Technology 6) fly new instrument concepts for technological advances as secondary payloads (experiments) on spacecraft that have other primary objectives. Under either scenario, the technologies flown are proven under strict cost and schedule constraints.

The New Millennium Program (NMP) taps into the current technology "pipeline" by engaging teams of engineers and scientists from industry, academia, NASA, and other governmental agencies that are developing promising technologies for more capable spacecraft and science instruments. During the NMP technology selection process, these teams are guided by the roadmaps of NASA's three mission areas: Sun-Earth System, Solar System, and Universe. These roadmaps are also used to conceive and design the Program's space-flight validation missions.

Example of technologies selected for improved-capability spacecraft are: ion propulsion; autonomous navigation; more efficient power production and use; and faster, more efficient communications.

Example of technologies selected for improved-capability science instruments are: An advanced imaging camera that makes extremely fine wavelength discriminations; an advanced imaging spectrometer for atmospheric measurements

Other promising technologies include: solar sails; commercial off-the-shelf products for space application; precision formation flying capability; large telescopes; lower cost avionics; on-board decision making; higher date rate communications; lighter spacecraft that require smaller, less expensive launch vehicles; improved thermal management in small spacecraft; and aerocapture system technology.

The launch vehicle is a significant portion of any mission's budget. The larger a spacecraft, the larger and more costly a launch vehicle is required to boost it into space. Another significant portion is allocated to the operations team throughout the mission. Smaller spacecraft and fewer (or no) operations team members translates into smaller budgets — without compromising the science payback.

While NMP's primary focus is to test advanced instruments, spacecraft systems, and concepts in flight, scientific data may, in addition, be returned as a "byproduct." The capabilities now being developed and tested will one day enable NASA's solar system exploratory spacecraft and Earth satellites to answer even more challenging scientific questions.

Completed NMP Missions

Deep Space 1 (DS1). DS1 tested 12 advanced technologies, including the xenon ion propulsion engine. DS1 flew a very successful extended mission to study Comet Borrelly, returning the best comet5 images and data ever to that date. See also **Deep Space 1 (DS1)**.

Earth Observing 1 (EO1)

EO-1 tested two advanced imaging technologies of unprecedented clarity: Hyperion (the world's only hyperspectral satellite sensor utilizing 220-bands of the spectrum at approximately 30-meter or 98.4 ft. spatial resolution) and the Advanced Land Imager (a lightweight, high performance, multi-spectral sensor). The EO-1 spacecraft was also used as a testbed to validate the NMP Space Technology 6 Autonomous Sciencecraft software. EO1 is now (2006) in an extended mission, expanding public access to its data and transforming the mission into an advanced development sensor web and testbed activity. See also **Earth Observing System (EOS)**.

Space Technology 5 (ST5). ST5 flew three miniaturized satellites called small-sats or nanosats, in Earth's magnetosphere. The primary mission of ST5 was to validate several technologies in miniatured form, as well as the methods for manufacturing multiple small satellites. As a secondary science data gathering mission, ST5 mapped the intensity and direction of magnetic fields. This data allows scientists to infer the presence of electrical currents carried by energetic charged particles. Studying this region may yield important information about the space weather that disrupts our communication, navigation, and power systems. The mission was completed in June 2006, with successful validation of all tested technologies. See also **Space Technology 5 (ST5) Mission**.

Space Technology 6 (ST6). The ST6 mission is to validate two revolutionary technologies: the Inertial Stellar Compass (Compass) and the Autonomous Sciencecraft Experiment (Sciencecraft). Compass will provide the capability of low-power, low-mass, and high-precision attitude determination for long duration space science missions. This technology will be particularly beneficial for future small-sat missions. Sciencecraft technology (validation completed on the EO1 testbed in 2004) will equip future missions with significant onboard decision-making, allowing them to take advantage of novel science opportunities. For example, Sciencecraft may be used to capture science phenomena and identify regions where change — new flooding, ice melt, or lava flows — have occurred. Sciencecraft software has already been used on the Mars Exploration Rovers Spirit and Opportunity to detect and image Martian dust devils.

See also **Earth Observing System (EOS) Mission**; and **Space Technology 6 (ST6) Mission**.

In-development NMP Missions

Space Technology 7 (ST7). ST7 will test the Disturbance Reduction System (DRS). Working in conjunction with enhanced sensor technology provided by the European Space Agency (ESA), the enhanced micronewton thrusters of ST7 complete a position-measurement technology that can be used to test theories of gravity, map the gravity fields of the Earth or other planets to study properties of planetary crust and ocean currents, and detect cosmic gravitational waves. ST7's DRS technology, along with ESA's LISA Technology Package Intertial Sensor, will be used by the Laser Interferometer Space Antenna (LISA), the first dedicated in-space gravitational-wave observatory. There is also a preliminary plan to use the DRS in the future EX-5 mission, which will map the Earth's variable gravity field. See also **Space Technology 7 (ST7) Mission**.

Space Technology 8 (ST8). ST8 will space validate four new subsystem-level technologies never before tried in space. These include (1) a Dependable Multiprocessor — a commercial-off-the-shelf (COTS)-based computing architecture that can automatically adapt in real time to the error-causing radiation environment of space; (2) UltraFlex 175 — an ultra-lightweight, deployable solar array system that is a highly efficient power producer for the entire spacecraft; (3) SAILMAST — a technology for strong, ultra-lightweight structures that can be used to support huge, deployable solar sails and other large structures; and (4) the Thermal

Loop—a miniature thermal management system for small spacecraft that tightly controls the operating temperatures of the instruments while using little power. See also **Space Technology 8 (ST8) Mission**.

Near Future NMP Missions

Space Technology 9 (ST9). ST9 will test system-level technology selected from the concept areas of consideration within solar sail system technology, formation flying system technology, system technology for large space telescopes, descent and terminal guidance for pinpoint landing and hazard avoidance, and aerocapturesystem technology for planetary missions.

Infusion of NMP-validated Technologies into Subsequent Science Missions

Technologies that New Millennium Program missions have helped to develop and have validated in actual space mission applications can subsequently be used with confidence by science mission planners. The following are some of the uses these technologies have found so far. Ion Engine; Autonomous Navigation; Beacon Monitor Technology; Small Deep Space Transponder; Autonomous Sciencecraft; and UltraFlex Solar Arrays.

Deep Space 1 Ion Engine. The Dawn mission, scheduled to launch in June 2007, will use an ion engine. This technology was validated well beyond anyone's expectations on the NMP mission Deep Space 1. Dawn, a Discovery Program project, could not carry out its mission with conventional propulsion.

The solar electric (ion) engine technology on Deep Space 1 allowed the spacecraft to precisely shape its trajectory to meet up with asteroid Braille in July 1999 and Comet Borrelly in September 2001. The ion engine operates on a different principle from conventional chemical propulsion. It uses xenon gas as fuel, producing a force as gentle as the weight of a single sheet of paper. However, unlike conventional chemical thrusters, which are used only for orbit maneuvers, the ion engine thrusts continuously. Because in space there is no resistance from friction or drag, and only relatively small gravitational effects, the spacecraft eventually attains speeds much greater than possible with conventional engines. In addition, it is very fuel efficient, so the spacecraft can be small, with a very small "gas tank." See also **Ion Propulsion (Rockets)**.

Dawn will visit the two large asteroids, Vesta and Ceres, orbiting each for several months. The ion engine will allow Dawn to be captured into orbit by Vesta, then six months later, Dawn will be able to spiral away from Vesta and thrust on to Ceres, again spiraling into orbit around that asteroid.

Deep Space 1 Autonomous Navigation. Deep Impact, a NASA Discovery-class mission, was launched in January 2005 and spent the next seven months chasing down Comet Tempel 1. Why? Specifically for the purpose of placing itself—or, actually, a part of itself called the "smart impactor"—right in the path of the comet to create a collision that would blast a crater in the comet's surface. In this way, scientists could get their first look inside a comet.

Scientists knew that the surface of a comet's nucleus is blacker than coal, but theoretically, comets were "dirty snowballs." How did the surface get so black? The concept of a mission to impact a comet had been proposed decades ago, but NASA was not convinced that the impactor could actually hit a comet, especially an active one. But the Deep Impact mission was different.

On July 4, 2005, then, the Deep Impact spacecraft released its "smart impactor," to complete its first and last mission. The Deep Impact flyby spacecraft (as well as many other space- and ground-based telescopes) remained at a safe distance, taking images and spectrogaphic data to learn the composition of the plume of material blasted into space by the impactor.

So how did the smart impactor hit its target? After all, the path of an active comet is not precisely predictable. The nucleus is not spherical, so instead of rotating neatly, it may tumble like a badly-thrown football. Also, the nucleus contains pockets of gas that shoot out in jets like thrusters, pushing the nucleus this way and that.

The smart impactor succeeded because it was equipped with Autonomous Navigation, "AutoNav."

Along with the ion engine, AutoNav was one of 12 technologies validated on the NMP Deep Space 1 mission. AutoNav is basically a computer program full of artificial intelligence. It uses a camera to see and

thrusters to steer. AutoNav guided the Deep Impact "bullet" directly to its target. Using a series of impact correction maneuvers (ICMs) in the last few minutes before impact, the impactor homed in and made a bullseye. Without those last-minute maneuvers, it would have completely missed the nucleus. Humans were 14 light-minutes away (round trip time for a signal to travel from the spacecraft to Earth and back again), so would not have been able to send correction maneuver instructions in time.

Another mission using AutoNav was Stardust. Stardust is the first U.S. space mission dedicated solely to exploring a comet, and the first robotic mission designed to return extraterrestrial material from outside the orbit of the Moon. Launched February 7, 1999, Stardust's primary mission was to collect dust and carbon-based samples during its closest encounter with Comet Wild 2. The rendezvous took place in January 2004. The capsule containing the samples of cometary and interstellar particles successfully returned to Earth January 15, 2006, making a soft landing in the Utah desert. Scientific analysis of the samples is under way. See also **Stardust Mission**.

A version of the AutoNav software, as modified for the Comet Borrelly encounter during Deep Space 1's extended mission, had been uplinked to the Stardust spacecraft to enable it to precisely locate the nucleus of Comet Wild 2.

Deep Space 1 Beacon Monitor Technology:. The New Horizons mission launched January 19, 2006, is the first mission to Pluto and its moon Charon. It has a very long cruise period ahead of it, with a planned Pluto-Charon encounter in July 2015. New Horizons is the first science mission to use the Beacon Monitor technology, one of 12 technologies validated by the Deep Space 1 mission. See also **New Horizons Mission**.

In beacon monitor operations, an on-board data summarization system determines the overall spacecraft health. Then it selects one of 4 radio tones to send to Earth to indicate how urgently it needs contact with the large antennas of the Deep Space Network, NASA's worldwide network of stations used to communicate with probes in deep space. These tones are easily detected with low cost receivers and small antennas, so monitoring a spacecraft that uses this technology will free up the precious resources of the Deep Space Network, thus allowing more spacecraft to explore the solar system without having to expand the network. Each tone is like a single note on a musical instrument. One tone might mean that the spacecraft is fine, and it does not need contact with human operators. Another might mean that contact is needed sometime within a month, while a third could mean that contact should be established within a week. The last is a virtual red alert, indicating the spacecraft and, therefore, the mission is in jeopardy.

Deep Space 1 Small Deep Space Transponder. Deep Space 1 validated a small deep-space transponder (SDST), built by General Dynamics, that combines the receiver, command detector, telemetry modulator, exciters, beacon tone generator, and control functions into one 3-kilogram (6.6-pound) package. The SDST allows X-band uplink and X-band and Ka-band downlink. To achieve the SDST's functionality without a new technology development would require over twice the mass and 4 or 5 individual subassemblies.

The SDST generates the tones needed for beacon monitor operations, conceived to reduce the large demand that would be expected on the Deep Space Network if many missions were in flight simultaneously, as envisioned by NASA.

Since Deep Space 1, the Small Deep Space Transponder has been used on the Dawn, Deep Impact, Mars Odyssey, Mars Exploration Rovers, Mars Reconnaissance Orbiter, Messenger, Stereo, and Spitzer Space Telescope missions.

Space Technology 6 Autonomous Sciencecraft: The Autonomous Sciencecraft software uses artificial intelligence to "decide" what images or data are worth taking and downlinking (transmitting back) to Earth. In this case, the sciencecraft software was uplinked to the Mars Exploration Rovers, Spirit and Opportunity, in June 2006, long after their primary mission was complete. The software enables the rovers to spot dust devils passing across the Martian surface and "know" that they are of interest so should be photographed and the images downlinked to Earth. If no dust devil is passing, there is certainly no sense in making image after image of the same, unchanging landscape, and clogging the data transmission with scientifically insignificant images.

The Autonomous Sciencecraft software was also adapted for and uploaded to the Mars Odyssey spacecraft to study the carbon dioxide ice caps, dust storms, and thermal anomalies observed from its orbit at Mars.

As part of the NMP Space Technology 6 mission, the autonomous sciencecraft technology was uplinked to and validated by the Earth Observing 1 (EO-1) spacecraft, also long after its primary technology validation mission was complete. It worked so well, that it was tried on the Mars rovers. On EO-1, a primary test target was an erupting volcano, Mount Erebus, in Antartica. EO1's new sciencecraft software could detect whether the volcano was active by comparing a new image with the previous. If the software detected any slight difference, the spacecraft could be instructed to record images and other data long before humans ever knew about it. The software on EO1 was also used for flood and ice tracking, as well as imaging clouds.

EO-1's baseline mission was to fly and validate three advanced technology land imaging instruments. The three instruments are the Advanced Land Imager (ALI), the Hyperion hyperspectral imager, and the Linear Etalon Imaging Spectrometer Array (LEISA) Atmospheric Corrector (LAC). These instruments incorporate revolutionary land imaging technologies that will enable future Landsat and Earth observing missions to more accurately classify and map land utilization globally.

EO-1 continues its extended mission with expanded public access to EO-1 image data and the transformation of the mission into an advanced development sensor web and testbed activity. Autonomous Sciencecraft is currently beinhg used to autonmate observations planning and execution so as to significantly reduce the cost of EO-1 extended mission operations.

Space Technology 8 UltraFlex Solar Arrays:. NASA's Exploration Systems Mission Directorate has selected a Lockheed Martin Corp. design for the new Orion crew vehicle to take humans back to the Moon and on to Mars. The baseline design uses UltraFlex solar arrays to supply electrical power to the vehicle. The UltraFlex (175 model) is one of the technologies that will be tested as part of the Space Technology 8 mission. However, NASA (and Lockheed Martin) will be watching the development of the UltraFlex for ST8 in order to benefit from the ground validation testing and then the in-space validation results.

See also **Orion Crew Vehicle**; and **Space Technology 8**.

Education

The New Millennium Program (NMP) integrates mission content into existing educational delivery systems and focuses on inspiring underrepresented minorities to get involved and stay involved in science. NMP's education and public outreach (EPO) plan upholds NASA's guidelines and standards for providing materials and resources to educators, students, and the general public. Education and outreach is accomplished through publicly-accessible websites for each NMP flight-validation mission (linkable above) and through the Program's EPO-dedicated website, **The Space Place** (available in both English and Spanish). http://spaceplace.nasa.gov/en/kids/.

Each NMP mission-specific website explains flight-validation concepts, describes the complement of advanced technologies to be tested, and, if applicable, outlines science goals. And, NMP-sponsored missions leverage The Space Place within their individual EPO plans.

To this end, The Space Place creates unique educational products that demonstrate the principles behind technology and science in space exploration. These products are reviewed by scientists and engineers for technical accuracy and by four teacher-advisors for grade-level appropriateness.

NANCY J. LEON, Jet Propulsion Laboratory, Pasadena, CA

NEWT. *(Amphibia, Urodela)*. Air-breathing salamanders that are at least partially aquatic in habits. The many species are found chiefly in the Northern hemisphere. See also **Caudata(Salamanders)**.

NEWTON-COTES FORMULA. A method of numerical integration. Assume that the integral

$$\int_a^b f(x)\,dx$$

may be approximated by

$$\int_a^b \phi(x)\,dx = A_0 y_0 + A_1 y_1 + \cdots + A_n y_n$$

where the quantities A_i are independent of y_i. By proper choice of these quantities, which may be found by the method of undetermined coefficients, the numerical result may be made very close to the true value of the integral. Special cases of the formula are the trapezoidal rule, the Simpson rule, and the Weddle rule.

NEWTONIAN FLUID.   See **Colloid System; Fluid**.

NEWTONIAN FRICTION LAW.   ((Also called *Newton's formula for the stress*.) The statement that the tangential force (i.e., the force in the direction of the flow) per unit area acting at an arbitrary level within a fluid contained between two rigid horizontal plates, one of which is motionless and the other which is in steady motion, is proportional to the shear of the fluid motion at that level.

Mathematically, the law is given by

$$\tau = \mu \frac{\partial u}{\partial z},$$

where τ is the tangential force per unit area, usually called the shearing stress; μ a constant of proportionality called the dynamic viscosity; and $\partial u / \partial z$ the shear of the fluid flow normal to the resting plate. In deriving this expression Newton assumed that either the speed u of the moving plate or the distance between the plates was so small that, once a steady state was reached, the speed of the fluid increased linearly from zero at the resting plate to the speed u at the moving plate. In this case both the shear of the motion and the shearing stress are constant throughout the fluid.

AMS

NEWTONIAN LIQUIDS.   See **Viscosity**.

NEWTONIAN MECHANICS.   (Also called *classical mechanics*.) A system of mechanics based on Newton's laws of motion. The salient characteristics of Newtonian mechanics are that mass and energy are separately conserved, all physical variables can take on a continuous set of values, the state of a system at any instant uniquely determines its state at any later instant (determinism), and interactions at a distance are instantaneous.

NEWTONIAN TELESCOPE.   A reflecting-type telescope having a 45° mirror located just inside the focus, so that the primary image is observed through a hole in the side of the tube. See also **Telescope (Astronomical-Optical)**.

NEWTON (N).   See **Units and Standards**.

NEWTON RINGS.   An interference phenomenon, easily observed by laying a slightly convex lens upon a flat glass plate. When the lens and plate are arranged so that monochromatic light is reflected at a suitable angle to the observer's eye, the point of contact is seen to be surrounded by a series of concentric, alternately bright and dark rings, which become closer together with increasing radius. The rings are due to the interference of light at the film of air between the glass surfaces, which film increases in thickness with increasing distance from the contact point. If the radius of curvature of the convex surface is R, and if, counting the central contact-spot as the zero ring, we number the rings in order, both bright and dark, from the center out, the radius of the Nth ring in monochromatic light of wavelength λ is approximately

$$a = \sqrt{\frac{NR\lambda}{2}}$$

With white light, the bright rings become colored spectra, the overlapping of which at larger values of N causes the system to become indistinct and disappear.

NEWTON'S FORMULA FOR INTERPOLATION. Let a difference table be given with numerical values of $y_0, y_1, y_2, \ldots$; equally spaced values of the argument, $x_0, x_1, x_2, \ldots$; $h = (x_n - x_0)/n$ and the finite differences $\Delta^n y_k$. Then a value of y for $x = x_k + hu$, not contained in the table, may be found by Newton's formula for forward interpolation

$$y = y_k + u\Delta y_k + \binom{u}{2}\Delta^2 y_k + \cdots + \binom{u}{2}\Delta^n y_k$$

As its name implies, this equation is used for calculation near the beginning of a difference table. Near the end of such a table, Newton's formula for backward interpolation is appropriate

$$y = y_k - v\Delta y_{k-1} + \binom{v}{2}\Delta^2 y_{k-2} - \cdots \binom{v}{k}\Delta^k y_0$$

where $x = x_k - hv$. These equations are also known as the Gregory-Newton formula.

See also **Interpolation**.

NEWTON, SIR ISAAC (1642–1727).

Sir Isaac Newton was one of the greatest scientific thinkers of all ages. Newton's Laws of Dynamics, set forth as three "laws" of force and motion changed the way people viewed the world.

Newton was born in Woolsthrope, Lincolnshire, England. His birth was premature and he was not expected to live. In his early elementary school years, he seemed like an average student. It was not until he was in Trinity College at Cambridge University that Newton began to display his genius for mathematics. It was perhaps, however, the outbreak of the bubonic plague that gave Newton the time to do his greatest thinking. When Cambridge closed due to the plague, Newton went home and during 1665–1666 he was at the height of his creative power. While "thinking" he discovered the calculus, the law of gravity, his three laws of motion, and various properties of optics including measurable, mathematical patterns in the phenomenon of color. During this time, he also invented a reflecting telescope.

In 1684, astronomer Edmond Halley encouraged Newton to put his research on gravity into a book. Newton wrote *Principia*, a work that is now considered one of the greatest of all scientific books.

He was elected a Fellow of the Royal Society of London in 1671, and in 1703, Newton was elected president of the Royal Society of London. He was annually re-elected for the rest of his life. In 1704, his major work, *Opticks* was published and also works on the quadrature of curves and the classification of the cubic curves. In 1705, he was knighted in Cambridge by Queen Anne and was the first scientist with such an honor.

From about 1714 on, Newton became a highly esteemed philosopher and scientist throughout Europe. Today Newton's name is like an historic marker in man's history. Almost all modern physical science and technological advances since the seventeenth century can be traced back to the great thinking of Sir Isaac Newton. See also **Acceleration (Due to Gravity)**; **Calculus**; **Gravitation**; **Newton-Cotes Formula**; **Newton Rings**; **Newton's Formula for Interpolation**; **Newton's Laws of Dynamics**; **Relativity and Relativity Theory**; **Rheology**; and **Telescope (Astronomical-Optical)**.

J. M. I.

NEWTON'S LAWS OF DYNAMICS.

The classical or Newtonian dynamics rests upon certain propositions first enunciated in systematic form by Sir Isaac Newton, which he set forth as three "Laws" of force and motion.

The first Law states, in effect, that bodies of matter do not alter their motions in any way except as the result of forces applied to them. A body at rest remains at rest, or if in motion it continues to move in the same direction with the same speed, unless a force is impressed upon it. It is quite conceivable that Newton's interpretation of "force" was the primitive concept we all have, based on muscular effort, and that he regarded this statement as the expression of a natural law connecting force with motion. On the other hand, he may have recognized in this first Law, as we now do, an objective definition of force, namely, that which is capable of altering bodily motions in the face of an opposition called inertia whose nature is even now not fully understood.

The second Law is made up of two distinct parts. (1) When different forces are allowed to act upon free bodies, the rates at which the momentum changes are proportional to the forces applied. (2) The direction of the change in momentum caused by a force is that of the line of action of the force. Part 1 may now be regarded, from the more rigorous viewpoint of dimensional analysis, as a definition of the standard measure of force. Two forces are judged equal if they produce change of momentum at equal rates; one force is twice as great as another, if it changes the momentum at twice the rate; etc. Moreover, two bodies have equal mass if equal forces produce change of momentum at equal rates; and one body has twice the mass of another if an equal force changes its momentum at half the rate. This is the inertial concept of mass mentioned above. Part 2 emphasizes the vector character of force, and points out that no single force, acting alone, can cause a change of motion in any direction save that of its own line of action. If the effect is apparently in some other direction, as when a string operates over a pulley, the force is always combined with one or more auxiliary forces, the resultant of all of them being in the direction of the observed change of motion.

The third Law asserts the equality of "action and reaction." In the case of forces acting on bodies at rest, the principle is easily illustrated. When a steel truss rests on a pier and presses downward upon it with a force of 100,000 units, the pier exerts an upward thrust or "reaction" against the truss, also of 100,000 units, and this thrust, tending to bend the truss upward, is a most important factor in computing the stresses in the truss members. The Law also applies to forces acting upon bodies free to yield and to receive acceleration; a fact not explicitly stated by Newton, and discussed more fully elsewhere as d'Alembert's principle.

These propositions constituted the unquestioned foundation of dynamics until about the beginning of the twentieth century. So far as practical operations with bodies of ordinary size are concerned, they still answer every purpose. It is only when we consider motions with velocities comparable to that of light, or attempt to analyze the mechanics of bodies of the atomic and electronic order of magnitude, that the Newtonian dynamics breaks down and must be replaced by a system founded upon the postulates of relativity and the concepts of the quantum theory.

NEXT GENERATION SPACE TELESCOPE (NGST).

See **James Webb Space Telescope (JWST)**.

NEYMAN-PEARSON THEORY.

In statistical inference, the theory of tests of hypotheses developed by J. Neyman and E.S. Pearson. It is based on the delimitation of two types of error, the rejection of a true hypothesis and the acceptance of a false hypothesis. Errors of the first kind are controlled at assigned probability levels. Errors of the second kind cannot simultaneously be so controlled, but can be explored for different values of parameters alternative to the true one. Tests that have a smaller probability of the second kind of error are said to be more powerful and part of the theory is devoted to seeking for most powerful tests. The probability of rejection of a false value of a parameter, graphed as ordinate against the value of the parameter as abscissa, is the Power Function of the test.

NIACIN.

[CAS: 59-67-6]. Sometimes referred to as nicotinic acid or nicotinamide and earlier called the P-P factor, antipellagra factor, antiblacktongue factor, and vitamin B_4, niacin is available in several forms (niacin, niacinamide, niacinamide ascorbate, etc.) for use as a nutrient and dietary supplement. Niacin is frequently identified with the B complex vitamin grouping. Early in the research on niacin, a nutritional niacin deficiency was identified as the cause of pellagra in humans, blacktongue in dogs, and certain forms of dermatosis in humans. Niacin deficiency is also associated with perosis in chickens as well as poor feathering of the birds.

Varying in degree in relationship to the length and severity of diet deficiency of niacin, pellagra is clinically manifested by skin, nervous system, and mental conditions. The disease occurs most frequently among the economically deprived and particularly in areas where the diet may be high in maize (corn) intake. The disease was first described by

Gaspar Casal in 1735 and was common in many areas, including Europe, Egypt, Central America, and the southern portion of the United States for many years. The largest outbreak occurred in the United States during the period 1905–1915 and resulted in a high mortality. The medical awareness and understanding of vitamins and dietary deficiencies, coupled with the availability of dietary supplements in staple foods, have resulted in a great lessening in the occurrence of pellagra. Where the disease is found, niacin is a specific for the treatment of acute pellagra. Those afflicted are accustomed to a diet low in protein and made up largely of carbohydrates. Predisposing causes are found idiosyncrasies, chronic alcoholism, and diseases that interfere with the assimilation of a proper diet.

Huber first synthesized nicotinic acid in 1867. In 1914, Funk isolated nicotinic acid from rice polishings. Goldberger, in 1915, demonstrated that pellagra is a nutritional deficiency. In 1917, Chittenden and Underhill demonstrated that canine blacktongue is similar to pellagra. In 1935, Warburg and Christian showed that niacinamide is essential in hydrogen transport as diphosphopyridine nucleotide (DPN). In the following year, Euler et al. isolated DPN and determined its structure. In 1937, Elvhehjem et al. cured blacktongue by administration of niacinamide derived from liver. In the same year, Fouts et al. cured pellagra with niacinamide. In 1947, Handley and Bond established conversion of tryptophan to niacin by animal tissues.

In the physiological system, niacin and related substances maintain nicotinamide adenine dinucleotide (NAD) and nicotinamide adinucleotide phosphate (NADP). Niacin also acts as a hydrogen and electron transfer agent in carbohydrate metabolism; and furnishes coenzymes for dehydrogenase systems. A niacin coenzyme participates in lipid catabolism, oxidative deamination, and photo synthesis.

Nicotinic acid can be converted to nicotinamide in the animal body and, in this form, is found as a component of two oxidation-reduction coenzymes, NAD and NADP, as previously mentioned. Structurally, these are:

Nicotinic acid   Nicotinamide

Nicotinamide adenine dinucleotide (NAD) R* = H.

Nicotinamide adenine dinucleotide phosphate (NADP).

The nicotinamide portion of the coenzyme transfers hydrogens by alternating between an oxidized quaternary nitrogen and a reduced tertiary nitrogen as shown by:

NAD (oxidized)   NADH + H⁺ (reduced)

Enzymes that contain NAD or NADP are usually called dehydrogenases. They participate in many biochemical reactions of lipid, carbohydrate, and protein metabolism. An example of an NAD-requiring system is lactic dehydrogenase which catalyzes the conversion of lactic acid to pyruvic acid. Numerous NAD-dependent enzyme systems are known.

Lactic acid   Pyruvic acid

Distribution and Sources. In plants, niacin production sites occur in leaves, germinating seeds, and shoots. In humans, niacin is not available from intestinal bacteria, but some conversion is made from tryptophan which occurs in tissues.

High niacin content (10–100 milligrams/100 grams) Chicken (white meat), groundnut (peanut), halibut, heart (calf), kidney (beef, pork), liver (beef, calf, chicken, pork, sheep), meat extracts, rabbit (white meat), swordfish, tuna, turkey (white meat), yeast

Medium niacin content (1–10 milligrams/100 grams) Almond (dry), asparagus, avocado, barley, bean (kidney, lima, snap, wax), beef, broccoli, cashew, cheeses (camembert, roquefort, Swiss), chestnut, chicken (dark meat), clam, date (dry), duck, fig (dry), fishes (except those listed under "High"), kale, lamb, lentil (dry), maize (corn), molasses, mushroom, oats, oyster, parsley, pea, potato, prune (dry), rice (brown), rye, soybean (dry), shrimp, walnut, wheat, wheat germ

Low niacin content (0.1–1.0 milligram/100 grams) Apple, apricot, banana, beet, beet greens, berries (black-, blue-, cran-, rasp-, straw-), Brussels sprouts, cabbage, carrot, cauliflower, celery, cherry, chicory, coconut, cucumber, currant, dandelion greens, eggs, eggplant, endive, fig, grape, kohlrabi, lettuce, lemon, melons, milk, onion, parsnip, peach, pear, pecan, pepper, pineapple, plum, pumpkin, radish, raisin (dry), rhubarb, spinach, sweet potato, tangerine, tomato, turnip, watercress

Precursors in the biosynthesis of niacin: In animals and bacteria, tryptophan; and in plants, glycerol and succinic acid. Intermediates in the synthesis include kynurenine, hydroxyanthranilic acid, and quinolinic acid. In animals, the niacin storage sites are liver, heart, and muscle. Niacin supplements are prepared commercially by: (1) Hydrolysis of 3-cyanopyridine; or (2) oxidation of nicotine, quinoline, or collidine.

Bioavailability of Niacin. Factors which cause a decrease in niacin availability include: (1) Cooking losses; (2) bound form in corn (maize), greens, and seeds is only partially available; (3) presence of oral antibiotics; (4) diseases which may cause decreased absorption; (5) decrease in tryptophan conversion as in a vitamin B_6 deficiency. Factors that increase availability include: (1) alkali treatment of cereals; (2) storage in liver and possibly in muscle and kidney tissue; and (3) increased intestinal synthesis.

Antagonists of niacin include pyridine-3-sulfonic acid (in bacteria); 3-acetylpyridine, 6-aminonicotinamide, and 5-thiazole carboxamide. Synergists include vitamins B$_1$, B$_2$, B$_6$, B$_{12}$, and D, pantothenic acid, folic acid, and somatotrophin (growth hormone).

In humans, overdosage of niacin causes a limited toxicity (1 to 4 grams/kilogram) with individual variations in sensitivity.

Additional Reading

FAO United Nations: *Requirements of Vitamin A, Thiamine, Riboflavin and Niacin: Report of a Joint FAO-Who Expert Group*, FAO United Nations, Geneva, Switzerland, 1967. http://www.fda.gov/ *(United States Food and Drug Association)*.

Institute of Medicine, Food Nutrition Board Staff: *Dietary Reference Intakes: Thiamin, Riboflavin, Niacin, Vitamin B6, Folate, Vitamin B12, Pantothenic Acid, Biotin, and Choline*, National Academy Press, Washington, DC, 1998.

Web Reference

United States Food and Drug Association. http://www.fda.gov/

NIACINAMIDE. See **Pyridine and Derivatives**.

NICHE. See **Ecology**.

NICKEL. [CAS: 7440-02-0]. Chemical element, symbol Ni, at. no. 28, at. wt. 58.69, periodic table group 10, mp 1453 °C, bp 2732 °C, density 8.9 g/cm^3 (solid, 20 °C), 9.04 g/cm^3 (single crystal). Elemental nickel has a face-centered cubic crystal structure. Nickel is a silver-white metal, harder than iron, capable of taking a brilliant polish, malleable and ductile, magnetic below approximately 360 °C. When compact, nickel is not oxidized on exposure to air at ordinary temperatures. The metal is soluble in HNO$_3$ (dilute), but becomes passive in concentrated HNO$_3$. The metal does not react with alkalis. Finely divided nickel dissolves 17× its own volume of hydrogen at standard conditions. There are five naturally occurring stable isotopes, ^{58}Ni, ^{60}Ni through ^{62}Ni, and ^{64}Ni. Six radioactive isotopes have been identified, ^{56}Ni, ^{57}Ni, ^{59}Ni, ^{65}Ni, and ^{66}Ni. ^{59}Ni has a half-life of 8×10^4 years, and ^{63}Ni has a half-life of 80 years. The half-lives of the remaining radioactive isotopes are relatively short, expressed in hours and days. The element ranks 21st among the elements in terms of abundance in the earth's crust, the estimated average content of igneous rocks being about 0.02%. In terms of cosmic abundance, nickel ranks 28th among the elements. Nickel ranks 40th in terms of concentration in seawater, the estimated content being about 2.5 tons of nickel per cubic mile (540 kilograms per cubic kilometer) of seawater. Awareness of nickel probably dates back to antiquity, but the element was not firmly identified until 1751 when Axel Fredric Cronstedt isolated the metal from the sulfide ore NiAsS.

First ionization potential 7.33 eV, second, 18.13 eV. Oxidation potentials Ni → Ni^{2+} + 2e$^-$, 0.230 V; Ni^{2+} + 2H$_2$O → NiO$_2$ + 4H$^+$ + 2e$^-$, -1.75 V; Ni + 2OH$^-$ → Ni(OH)$_2$ + 2e$^-$, 0.66 V; Ni(OH)$_2$ + 2OH$^-$ → NiO$_2$ + 2H$_2$O + 2e$^-$, -0.49 V.

Other physical properties of nickel are given under **Chemical Elements**.

In the early 1800s, the principal sources of nickel were in Germany and Scandinavia. Very large deposits of lateritic (oxide or silicate) nickel ore were discovered in New Caledonia in 1865. The sulfide ore deposits were discovered in Sudbury, Ontario in 1883 and, since 1905, have been the major source of the element. The most common ore is pentlandite, (FeNi)$_9$S$_8$, which contains about 34% nickel. Pentlandite usually occurs with pyrrhotite, an iron-sulfide ore, and chalcopyrite, CuFeS$_2$. See also **Chalcopyrite**; **Pentlandite**; and **Pyrrhotite**. The greatest known reserves of nickel are in Canada and Russia, although significant reserves also occur in Australia, Finland, the Republic of South Africa, and Zimbabwe.

Principal producers and/or exporters of nickel include, in diminishing order, Canada, Russia, the United Kingdom, Norway, and Indonesia. Main consumers are the United States, Japan, the United Kingdom, Norway, Germany, Canada, and France.

After beneficiation of the raw ore to form a sulfide concentrate, the latter is roasted to achieve partial oxidation of iron and partial removal of sulfur. The roasted material then is smelted with a flux to eliminate the rock content. At this point, part of the iron goes into the slag. The remaining material is a copper-bearing nickel-iron matte, made up mainly of the sulfides of these metals. The matte is then treated in a Bessemer converter to achieve further removal of iron and sulfur. After controlled cooling, which assists separation, the Bessemer product is finely ground and subjected to magnetic separation and differential flotation. The separated product is an impure nickel sulfide. The sulfide then is sintered to nickel oxide. This product may be marketed for some applications, but the majority of the oxide is cast into anodes for refining into nickel metal by one of two major processes.

In (1) the electrolytic process, a nickel of 99.9% purity is produced, along with slimes which may contain gold, silver, platinum, palladium, rhodium, iridium, ruthenium, and cobalt, which are subject to further refining and recovery. In (2) the Mond process, the nickel oxide is combined with carbon monoxide to form nickel carbonyl gas, Ni(CO)$_4$. The impurities, including cobalt, are left as a solid residue. Upon further heating of the gas to about 180 °C, the nickel carbonyl is decomposed, the freed nickel condensing on nickel shot and the carbon monoxide recycled. The Mond process also makes a nickel of 99.9% purity.

Uses

The three main commercial forms of primary nickel are: (1) electrolytic sheets, (2) pellets resulting from the decomposition of nickel carbonyl, and (3) ferronickel. Traditionally, pellets are favored in Europe, whereas electrolytic nickel is favored in North America. Additional forms of commercial nickel are powder, ingots, shot, and briquettes. Ferronickel, containing 24–48% nickel with the remainder iron, is used mainly in the production of stainless steel. More than half of the nickel produced is used in stainless steels and high-nickel alloys. Additional uses include nickel plating, iron and steel castings, coinage, and copper and brass products.

The main consumer of nickel is austenitic stainless steel, which contains from 3.5 to 22% nickel and 16 to 26% chromium. In these steels, nickel stabilizes the austenite and enhances the ductility of the steel. Nickel, along with chromium, contributes to corrosion resistance. Up to amounts of about 9%, nickel adds strength, hardness, and toughness to many alloy steels. Alloys in the 9% nickel range remain stable at low temperatures and are capable of handling liquefied gases. The lower-nickel steels (0.5 to 0.7%) are ductile, strong, and tough, and find use for many automobile parts, in power machinery, and construction equipment. There are hundreds of nickel-containing alloys, running the gamut from hardenable silver alloy (0.02% Ni) up to malleable nickel (99% Ni).

Wrought Nickel and High-Nickel Alloys

Some of the major nickel alloys, along with wrought nickel, are described in Table 1.

Commercially pure wrought nickel in the form of sheets, wire, and tubing has many uses because of its corrosion resistance. These uses include utensils, food-processing equipment, marine hardware, coinage, and chemical equipment. Electroplated nickel also is used as a protective coating on steel. *Nimonic* alloys, not shown on the table, are based on an 80% Ni, 20% Cr composition. They are high-strength, heat-resistance metals that are age-hardened to increase strength at elevated temperatures — with a useful range of 700–825 °C. *Monel* metal (several types) is a high-strength corrosion-resistant alloy available in many wrought and cast forms for use in processing equipment, marine construction, and household appliances. *K Monel* can be heat treated by precipitation hardening to about 2× the strength of annealed *Monel*. *Hastelloy*-type alloys are well known for their excellent resistance to HCl, H$_2$SO$_4$, and other acids. The *Incoloy*-type alloys (35% Ni approximately) are heat-resistant alloys used mainly as castings for furnace work. The lower-nickel/higher-chromium alloys generally are classified as stainless steels. See also **Iron Metals, Alloys, and Steels**.

Although not of high-tonnage production, several nickel metals serve important uses, such as:

Permalloy, 78.5% Ni, 21.5% Fe; *Hipernik*, 50% Ni, 50% Fe; and *Perminvar*, 45% Ni, 30% Fe, 25% Co — are representative of a group of high-nickel magnetic alloys.

Constantan, 45% Ni, 55% Cu, has high electrical resistivity and a very low temperature coefficient of resistivity. It is extensively used with copper as a thermocouple element.

Nichrome, 80% Ni, 20% Cr (several types with variations of these percentages and additions of other elements, such as silicon in small amounts), is used as resistance wire for heating elements.

Calorite, 65% Ni, 8% Mn, 12% Cr, 15% Fe, also is used in electric heating elements.

TABLE 1. WROUGHT NICKEL AND REPRESENTATIVE NICKEL ALLOYS

Melting Range °C	Poisson's Ratio
Wrought nickel	1,435–0.31
99% Ni, 0.25% Cu, 0.15% C	1,445
Duranickel 301	1,400–0.31
93.9% Ni, 0.05% Cu, 0.15% C, 0.15% Fe, 0.5% Ti, 4.5% Al	1,440
Monel 400	1,300–0.32
66.0% Ni, 31.5% Cu, 0.12% C, 1.35% Fe	1,350
Hastelloy B	1,320 —
63.5% Ni, 0.05% C, 5.0% Fe, 2.5% Co, 1.0% Cr, 28.0% Mo, 0.3% V	1,460
Hastelloy F	1,290–0.305
45.5% Ni, 0.05% C, 20.5% Fe, 2.5% Co, 22.0% Cr, 6.5% Mo, 1% W, 2% (Nb + Ta)	1,295
Inconel 600	1,370–0.29
72% Ni, 0.5% Cu, 0.15% C, 8.0% Fe, 15.5% Cr	1,425
Incoloy 800	1,355–0.30
32.5% Ni, 0.75% Cu, 0.10% C, 45.6% Fe, 21.0% Cr	1,390
Illium G	1,255–0.29
56.0% Ni, 6.5% Cu, 22.5% Cr, 6.5% Mo q	1,340

Note: Recently introduced new or improved nickel alloys include:

Inconel alloy 625 — Low-cycle fatigue resistance has been increased from 70–80,000 to 110–120,000 psi at 10 cycles. This has been achieved through grain size control and improved product cleanliness. Major applications are bellows and expansion joints.

Inconel alloy 725 — An age-hardenable alloy for deep sour gas well service, combining height strength with the attributes of Inconel alloy 625, such as pitting resistance and stress corrosion cracking resistance to salt, hydrogen sulfide, and sulfur at temperatures up to about 230 °C (450 °F) and to sulfide stress corrosion cracking.

Inconel alloy 622 — Modified composition and special thermal mechanical processing give this alloy superior thermal stability and resistance to intergranular attack and localized corrosion. The alloy is particularly suited to acidified halide environments, especially those containing oxidizing acids.

Inconel alloy 925 — An age-hardenable nickel-iron-chromium alloy providing high strength up to 540 °C (1000 °F). Developed for use in gas production applications, such as tubular products, tool joints, and equipment for surface and downhole hardware in gas industry.

Inco alloy 25–6MO — Used for its corrosion resistance in many environments, this is an austenitic nickel-iron-chromium alloy with a substantial (6%) addition of molybdenum. Especially useful for resisting pitting and crevice corrosion in media containing chlorides or other halides. Applications include equipment for handling sulfuric and phosphoric acids, offshore platforms and other marine equipment, and for bleaching circuits in pulp and paper plants.

Alumel, 94% Ni, 2.5% Mn, 0.5% Fe plus small amounts of other elements, is used in thermocouples

Chromel, 35–60% Ni, 16–19% Cr, generally with the balance Fe, also is used as resistance wire and for thermocouples.

Invar, 36% Ni, 64% Fe, has a very low temperature coefficient of expansion and is used for measuring tapes, instruments, and bimetallic thermostats.

Elinvar, 34% Ni, 57% Fe, 4% Cr, 2% W, has a very low temperature coefficient of elasticity which makes it useful for springs in watches and precision instruments.

There are hundreds of special nickel-bearing alloys of proprietary formulations and tradenames.

Alloy with Memory

In seeking a way to reduce the brittleness of titanium, U.S. Navy researchers serendipitously discovered a nickel-titanium alloy having an amazing memory. Previously cooled clamps made of the alloy (*nitinol*) are flexible and can be placed easily in position. When warmed to a given temperature, the alloy hardware then exerts tremendous pressure. Use of conventional clamps for holding bundles of wires or cables in a ship or aircraft structure requires special tools. For this and other applications in industry and medicine, nitinol has been in demand. The alloy, however, is not easy to produce because only minor variations in composition can affect the "snap back" temperature by several degrees of temperature.

Nickel Powders. The use of nickel in powdered form has increased markedly during the last few years. As shown by Fig. 1, nickel powders are available in several types and are used in a variety of products.

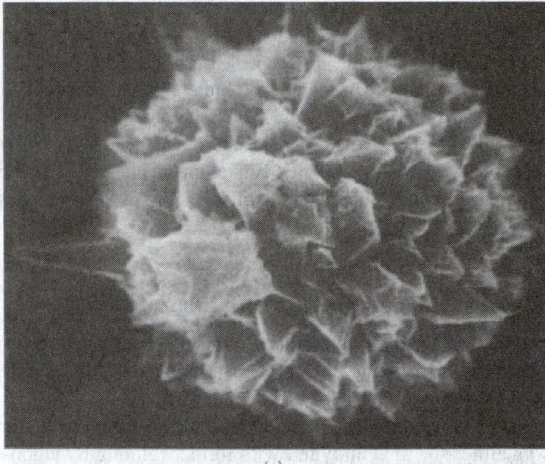

(a)

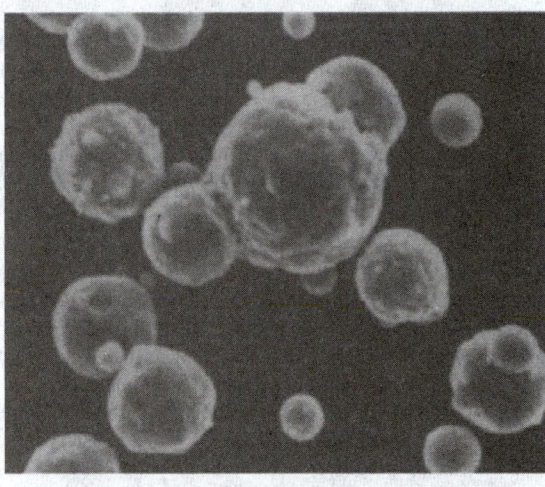

(b)

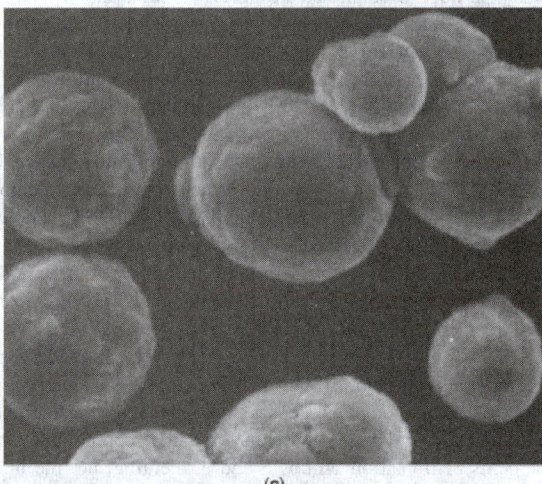

(c)

Fig. 1.   Types of available nickel powders. (**a**) With a surface area of 0.4m²/g. this is a spiked nickel powder of single particles 3–7 microns in diameter, with a bulk density of 2 g/cc. The powder is used in both metal and chemical systems for powder metal parts, getters, magnets, electronic strip, flake, and organo-nickel compounds and nickel salts and soaps. (**b**) High-density nickel powder consisting of 8–12 micron semismooth particles, offering mixability with both metallic and nonmetallic powder systems. Applications include welding rods, nickel aluminide, nickel-columbium additives, abradable seals, powder metal parts, carbide binders, and conductive plastics. (**c**) Spherically shaped, high- purity nickel powder, with a surface area of 0.15m²/g and a Fisher particle size of 8–9 microns. Applications include friction materials, plasma spraying, metal injection molding, welding electrodes, magnets, cemented carbides, and powder metal nickel steels. (*Source: INCO Specialty Powder Products, Saddle Brook, New Jersey.*)

Nickel in Nanometer Materials. Coating a metal with an ultrathin layer of another metal creates properties not found separately in either of the materials. Considerable recent research has been directed toward improving the mechanical properties of bimetallic laminates, sometimes called *composition modulated films*, which have interlayers only a few nanometers thick. Attractive properties also have been found for similar systems, called *nanometer materials*. Nickel has been used in combination with copper, ruthenium, and other metals for producing these new materials.

Production of High-Performance Nickel Alloys

In the production of high-performance alloys, the critical first step of alloying requires sophisticated equipment, stringent controls, and expertise. Several production methods are used.

Air melting in electric-arc or induction furnaces is used for many alloys, sometimes for final alloying, with further refining by argon-oxygen decarburization. Melting in air can result in impurities in some alloys, a problem eliminated by vacuum induction melting, used to produce ingots for direct rolling or for remelting. Remelting is accomplished by two methods, both with precise, computerized control. *Electroslag* remelting uses electrical resistance heating to remelt an ingot (electrode) under molten slag containing fluxes that remove impurities. *Vacuum arc* melting refines the structure of cast electrodes in a contaminant-free chamber. Remelting yields alloys of the highest level of refinement.

Nickel Chemistry and Compounds

With its $3d^8 4s^2$ electron configuration, nickel forms Ni^{2+} ions. Having a nearly complete $3d$ subshell, nickel does not yield a $3d$ electron as readily as iron and cobalt, and trivalent and tetravalent forms are known only in the hydrated oxides, Ni_2O_3 and NiO_2, and a few complexes.

Nickel(II) oxide, NiO, produced by heating the carbonate, is thermally stable. Higher oxides of nickel, including Ni_2O_3 and NiO_2, are known only as hydrates, being prepared by vigorous oxidation of NiO in alkaline solution.

Nickel(II) sulfide, precipitated from Ni^{2+} solutions by ammonium sulfide, may show quite a little departure from stoichiometric composition. Like iron(II) and cobalt(II) FeS and CoS, it has in crystal form an electrical conductivity and other properties similar to a metal or alloy. There is no conclusive evidence that Ni_2S_3 can be prepared, but NiS_2 is known and believed to be, like FeS_2, a compound of Ni^{2+} and the S_2^{2-} ion.

All four dihalides of nickel with the common halogens are known: NiF_2, formed by reaction of hydrofluoric acid or nickel(II) chloride or by thermal decomposition of $[Ni(NH_3)_6][BF_4]_2$, is greenish yellow, while the other three dihalides, formed directly from the elements, are green for the chloride, yellow for the bromide, and black for the iodide. In general, anhydrous Ni^{2+} salts are yellow and the ion $Ni(H_2O)_6^{2+}$ in aqueous solution is green.

Other elements with which nickel forms binary compounds, especially at higher temperature, are boron, carbon, nitrogen, silicon, and phosphorus. Like NiO, these compounds may depart slightly or even considerably from daltonide composition, frequently being interstitial compounds, and with higher elements of transition groups 5 and 6, merging into the interstitial compound-solid solution picture which nickel exhibits with the other transition metals.

Divalent nickel forms two main types of complexes. The first consists of complexes of the spin-free ("ionic" or outer orbital) octahedral type (see also **Ligand** for their discussion) in which the ligands are principally H_2O, NH_3, and various amines such as ethylenediamine and its derivates, e.g., $Ni(H_2O)_6^{2+}$, $Ni(NH_3)_6^{2+}$, $Ni(en)_6^{2+}$. These complexes usually have colors toward the high-frequency side of the spectrum, i.e., violet, blue, and green. The other class consists of tetracovalent square complexes with ligands such as CN^-, the dioximes and their derivatives, and other chelates, which usually have colors on the low frequency side of the spectrum, i.e., red, orange, and yellow. The structure of the nickel-dimethylglyoxime complex is

This compound is of interest not only in analysis, but because by limited oxidation with the halogens it yields a unipositive ion containing trivalent nickel and also because the hydrogen bonds formed to the oxygen atoms are among the shortest known. Similarly, the tetracyanide complex of nickel, $Ni(CN)_4^{2-}$, may be reduced by sodium amalgam to give an ion of composition $Ni(CN)_4^{3-}$, or $(NC)_3Ni-Ni(CN)_3^{4-}$ containing Ni(I). This latter ion forms a potassium salt of nickel(I) of the formula $K_4Ni_2(CN)_6$ which is reduced in liquid NH_3 by metallic potassium to give the compound $K_4Ni(CN)_4$ in which the nickel has an effective valence of zero. Of course, this zero valence also exists in the carbonyls of nickel (and other elements) which, however, are covalent. $Ni(CO)_4$ is prepared by reaction of carbon monoxide with freshly reduced nickel, which occurs at ordinary temperatures and pressures. As with the carbonyls of other metals, the CO groups may be directly or indirectly, partially or completely, replaced by other groups. Derivatives of trivalent phosphorus form many such compounds of general formula $Ni(CO) \times_{4-x}(PR_3)_x$, where R may be one or more of such groups as F, Cl, Br, I, alkyl, aryl, alkoxy, aryloxy, etc.

See also **Nickel (In Biological Systems)**.

Health and Safety Factors

Eye and Skin Contact. Some nickel salts and aqueous solutions of these salts, eg, the sulfate and chloride, may cause a primary irritant reaction of the eye and skin. The most common effect of dermal exposure to nickel is allergic contact dermatitis.

Protective equipment and clothing such as face shields and gloves should be worn and safety showers should be available wherever there is a possibility of being splashed or otherwise contacted by nickel containing solutions. If dermatitis should occur, the possibility that it is nickel related should be brought to the attention of a physician.

Inhalation. Nickel carbonyl is an extremely toxic gas. The permissible exposure limit (PEL) in the United States is 1 part per billion (ppb) in air. Nickel carbonyl may form wherever carbon monoxide and finely divided nickel are brought together. Nickel carbonyl should be used in totally enclosed systems or under good local exhaust.

The potential chronic toxicity is of concern. Based on epidemiological and experimental results, the International Agency for Research on Cancer (IARC) http://www.iarc.fr/, has concluded that all nickel compounds are Category 1, ie, known human carcinogens, and has classified metallic nickel as a Category 2B carcinogen, ie, possibly carcinogenic to humans.

It is good practice to keep concentrations of airborne nickel in any chemical form as low as possible and certainly below the relevant standard. Local exhaust ventilation is the preferred method, particularly for powders, but personal respirator protection may be employed.

Additional Reading

Carter, G.F. and D.E. Paul: *Materials Science and Engineering*, ASM International, Materials Park, OH, 1991.

Greenwood, N.N. and A. Earnshaw: *Chemistry of the Elements*, 2nd Edition, Butterworth-Heinemann, Inc., Woburn, MA, 1997.

Hanson, A.: "The Metal That Remembers," *Technology Review(MIT)*, **26** (May–June 1991).

Houston, J. and P. Feibelman: "Ultrathin Metal Coatings Yield Unique Properties," *Advanced Materials & Processes*, **31** (March 1991).

Lancaster, J.R., Jr.: *The Bioinorganic Chemistry of Nickel*, John Wiley & Sons, Inc., New York, NY, 1988.

Lide, D.R.: *CRC Handbook of Chemistry and Physics*, 88th Edition, CRC Press, LLC., Boca Raton, FL, 2007.

Sax, N.R. and R.J. Lewis, Sr.: *Sax'x Dangerous Properties of Industrial Materials*, 10th Edition, John Wiley & Sons, Inc., New York, NY, 1999.

Staff: "A Quick Reference Guide to Nickel and High-Nickel Alloys," *Advanced Materials and Processes*, **54** (October 1991).

Staff: *ASM Handbook—Properties and Selection: Nonferrous Alloys and Pure Metals*, ASM International, Materials Park, OH, 1991.

Staff: "Metals Forecast," *Advanced Materials & Processes*, **17** (January 1991); **24** (January 1992); **18** (January 1993).

Web References

Nickel Institute: http://www.nidi.org/

Nickel Producers Environmental Research Association (NiPERA): http://www.nipera.org/

NICKEL (In Biological Systems).   Despite its many pharmacological and in vitro actions, convincing evidence showing that nickel is an essential element for some animal species did not appear until the early 1960s. There has been considerable further and more convincing research during the 1970s and early 1980s.

Like most trace elements, nickel can activate various enzymes in vitro, but no enzyme has been shown to require nickel, specifically, to be activated. However, urease has been shown to be a nickel metalloenzyme and has been found to contain 6 to 8 atoms of nickel per mole of enzyme (Fishbein et al., 1976). RNA (ribonucleic acid) preparations from diverse sources consistently contain nickel in concentrations many times higher than those found in native materials from which the RNA is isolated (Wacker-Vallee, 1959; Sunderman, 1965). Nickel may serve to stabilize the ordered structure of RNA. Nickel may have a role in maintaining ribosomal structure (Tal, 1968, 1969). These studies and other information have led to the suggestion that nickel may play a role in nucleic acid and/or protein metabolism.

Nickel also may act to stimulate or inhibit the release of various hormones (Nielsen, 1971, 1972; Dormer et al., 1973; Clay, 1975; Horak-Sunderman, 1975). Nickel has been found to inhibit insulin release from the pancreas (Dormer et al., 1973; Clay, 1975), and stimulates glucagon secretion (Horak-Sunderman, 1975).

Nickel as an essential element in ruminant nutrition has not been proved conclusive as of the early 1980s. However, with nonruminants, some evidence indicates that certain species fed low-nickel diets have a greater infant mortality rate and a general degradation of the reproductive process (Nielsen, 1975; Anke et al., 1973).

Zinc and nickel appear to behave similarly at certain sites in the biological system. Both elements are capable of activating certain enzymes; for example, arginase is an enzyme which can be activated by either element (Parisi-Vallee, 1969). Stimulation of enzyme activity is at a site at which trace element substitutions or interactions may occur. However, some sacrifice of activity usually results when normally occurring metal is replaced by a trace metal. Nucleic acids as well as the ribosomes are likely sites of interaction between nickel and zinc. Both metals are consistently found in high concentrations firmly bound to RNA. It has been suggested that they function in maintaining the structure of RNA, thus preventing conformational changes. Nickel appears to be as effective as zinc at equal concentrations in this respect. Nickel and zinc are also found in ribosomal ash and studies have indicated that both can contribute to ribosomal conformation. The white blood cell is another possible site at which nickel and zinc may interact. Leukocytes are high in zinc and total leukocyte counts as well as differential white cell counts change drastically during a zinc deficiency. The interrelationship between nickel and zinc has been studied in vitro primarily in swine and rats. Their relationship has been studied largely from a substitution standpoint. Nickel appears to substitute for zinc to a certain extent in both species.

Similarly, the relationship between nickel and copper has been under study. One of the major functions of copper is in hemoglobin formation. Hemoglobin and hematocrit values decline rapidly during a copper deficiency. Copper is currently believed to exert its effect on hemoglobin metabolism through ceruloplasmin. Early work also indicated that nickel might be involved in hematopoiesis. Investigators in 1974 found a decreased concentration of copper in the lung and spleen of rats receiving 5 parts per million of nickel in drinking water. High levels of dietary nickel in rats and mice have been reported to decrease the activity of cytochrome oxidase, a copper-containing enzyme.

As pointed out by Eskew, Welch, and Cary in 1983, in contrast with the situation in animals, for which four new essential trace elements were identified in recent years, no new generally essential micronutrient for higher plants has been found since the mid-1950s. When it was found that urease is a nickel-metalloenzyme, this suggested that Ni may play a role in higher plants. Nickel has evidenced a stimulation of growth when urea is the sole source of nitrogen, but has slight or no effect when other nitrogen enrichment sources are used. The aforementioned investigators claim that Ni is essential for nitrogen metabolism in soybeans (*Glycine max* (L.) Merr.), either when nitrogen is furnished as NO_3^- and NH_4^+ or when plants depend upon nitrogen fixation. In experiments, soybean plants deprived of Ni accumulated toxic concentrations of urea (2.5%) in necrotic lesions on their leaflet tips. This occurred regardless of whether the plants were furnished with inorganic N or were dependent on N fixation. Nickel deprivation resulted in delayed nodulation and in a reduction of early growth. The addition of Ni (1 microgram/liter) to the nutrient media prevented urea accumulation, necrosis, and growth reductions. Extrapolating these findings, it is suggested that Ni may be essential for other higher plants.

Toxicity. Nickel contact dermatitis can occur among wearers of nickel-containing jewelry, more common among females than males. This is particularly true of nickel sulfate present in some jewelry. Localization of sites unexpectedly involves the ear lobes, neck, fingers, and wrists. Nickel is a major offender in connection with AECD (allergic eczematous contact dermatitis).

As mentioned earlier, nickel carbonyl is a volatile intermediate in the Mond process for nickel refining. This compound also is used for vapor plating of nickel in the semiconductor industry, and as a catalyst in the chemical and petrochemical industries. The toxicity of the compound has been known for many years. Exposure of laboratory animals to the compound has induced a number of ocular anomalies, including anophthalmia and microphthalmia, and has been shown to be a carcinogenic for rats.

Additional Reading

Anke, M. et al.: "Low Rations for Growthe and Reproduction in Pigs," in *Trace Element Metabolism in Animals*, (W.G. Hockstra et al., editors), University Park Press, Baltimore, MD, 1073.

Clay, J.J.: "Nickel Chloride-induced Metabolic Changes in the Rat and Guinea Pig," *Toxicol. Appl. Pharmacol.*, **31**, 55 (1975).

Considine, D.M. and G.D. Considine, Eds.: "Foods and Food Production Encyclopedia," in *Van Nostrand Reinhold*, New York, NY, 1982.

Dormer, R.L. et al.: "The Effect on Nickel on Secretory Systems," *Biochem.J.*, **140**, 135 (1973).

Eskew, D.L., R.M. Welch, and E.E. Cary: "Nickel: An Essential Micronument for Legumes and Possibly All Higher Plants," *Science*, **222**, 621–623 (1983).

Fishbein, W.N. et al.: "The First Natural Nickel Metalloenzyme: Urease," *Fed. Proc.*, **35**, 1680 (1976).

Horak, E. and F.W. Sunderman, Jr.: "Effects on Ni(II) upon Plasma Glucagon and Glucose in Rats," *Toxicol. Appl. Pharmacol.*, **33**, 388 (1975).

Neilsen, F.H.: "Studies on the Essentiality of Nickel," in *Newer Trace Elements in Nutrition*, (W. Mertz and W.E. Cornatzer, editors), Marcel Dekker, New York, NY, 1971.

Parisi, A.F. and B.L. Vallee: "Zinc Metalloenzymes: Characteristics and Significance in Biology and Medicine," *Amer. J. Clin. Nutr.*, **22**, 1222 (1969).

Spears, J.W and E.E. Hatfield: "Role of Nickel in Animal Nutrition," *Feedstuffs*, 24–28 (June 13, 1977).

Sunderman, F.W., Jr.: "Measurements of Nickel in Biological Materials by Atomic Absorption Spectrometry," *Amer. J. Clin. Path.*, **44**, 182 (1965).

Sunderman, F.W., Jr. et al.: "Eye Malformations in Rats:Induction by Prenatal Exposure To Nickel Carbonyl," *Science*, **203**, 550–552 (1979).

Tal, M.: "On the Role of Zn^{2+} and Ni^{2+} in Ribosome Structure," *Biochem. Biophys. Acta*, **169**, 564 (1968).

Wacker, E.E.C. and B.L. Vallee: "Nucleic Acids and Metals. I. Chromium, Manganese, Nickel, Iron and Other Metals in Ribonucleic Acid from Diverse Biological Sources," *J. Biol. Chem.*, **234**, 3257 (1959).

NICKELINE.   A nickel arsenide mineral, NiAs, crystallizes in the hexagonal system but is usually found massive. Color, light copper; hardness, 5.0–5.5; specific gravity, 7.784; luster, metallic; opaque. Found in several European localities and in the Province of Ontario, Canada; in the United States at Franklin, New Jersey, and Silver Cliff, Colorado. It is an ore of nickel.

NICOTINAMIDE.   See **Pyridine and Derivatives**; and **Vitamins**.

NICOTINE.   See **Alkaloids**.

NICOTINIC ACID.   See **Pyridine and Derivatives**; and **Vitamins**.

NIGHT BLINDNESS.   See **Vision and the Eye**.

NIGHTINGALE *(Aves, Passeriformes)*.   A warbler, *Luscinia megar-hyncha*, of western Europe, noted for its song. Farther east two other species, the eastern, *L. pheilomella*, and Persian, *L. hafizi*, nightingales, are found.

The nightingale is a trim bird, about 6 to 7 inches (15 to 18 centimeters) in length. The coloring is brown with lighter brown underneath. The female does the nesting and brooding, although the male helps with feeding the young. Incubation period is from 13 to 14 days. The eggs are blue or off-white with some markings.

NIGHTJARS AND NIGHTHAWKS *(Aves, Caprimulgiforme)*.   These birds make up the majority of the *Caprimulgiformes*, an order of nocturnal birds with very wide mouths. Nightjars have mottled plumage and a short beak. They are insect-eaters, flying chiefly at night. All continents have some of the numerous species with the exception of Australia. In North America, the nighthawk and whippoorwill are the most widely known representatives of the group, with the poorwill, chuck-will's widow (*Antrostomus carolinensis*), and the Merrill parauque (*Nyctidromus albicollus*) as less widely distributed species. Nightjars have been known since the time of Aristotle and are mentioned in the Bible. The poorwill (*Phalaenoplilus nuttallii*) is about 8 to 10 inches (20 to 25 centimeters) long and makes no formal nest. Eggs are laid in grasslands. This species uses the tactic of displaying a "broken" wing to distract the attention of predators. The legs are short and weak. The species is known for squatting lengthwise on limbs of trees and also for going into hibernation. During such periods, respiration is barely detectable and the temperature drops from a normal figure of about 100 °F to 66 °F (38 °C to 18.9 °C).

The whippoorwill, is a nocturnal bird (*Caprimulgus* (*Antrostomus*) *vociferus*), with a short beak and wide mouth, adapted for taking insects in flight. Its call has been likened to the words used in its name. Often the three syllables are repeated over and over scores of times without cessation.

The common nighthawk (*Chordeiles minor* (*virginianus*)) is widely distributed in North America and winters far into South America. A second species, the Texan nighthawk (*C. acutipennis*) enters the southwestern United States. Nighthawks are characterized by very short beaks and by wide mouths. They are well adapted for catching insects in flight. They commence their flights late in the day. Their flight is easy and powerful and their long dives, terminating in a peculiar hollow boom, are a memorable exhibition.

Goatsuckers were given their odd name in the mistaken belief during Aristotle's time that the birds took milk from domesticated goats. They, like the other nightjars, are characterized by weak legs, a short weak beak, a very wide mouth, and crepuscular habits. The *Uropsalis lyra* is a small bird about 4 inches in length (10 centimeters), but with a lyre-type tail some 27 inches (69 centimeters) in length. This species inhabits the environs of Colombia in South America. The habits are nocturnal. The call is penetrating. Another species is the parauque, a large bird of Mexico and Texas which resembles the poorwill. Also related to the goatsuckers is the frog mouth (*Podargus*) found in the Oriental and Australian regions. See also *Caprimulgiformes*.

NIMBOSTRATUS.   See **Clouds and Cloud Formation**.

NIMBUS.   See **Clouds and Cloud Formation**.

NIMBUS E MICROWAVE SPECTROMETER (NEMS).   See **Nimbus Satellite Program**.

NIMBUS SATELLITE PROGRAM.   Nimbus, a second-generation meteorological satellite which got its name from a cloud formation, is larger and more complex than TIROS (Television Infrared Observation Satellite). See also **Polar-Orbiting Environmental Satellite (POES)**. The Nimbus Technology satellite program was initiated by NASA in the early 1960's to develop an observational system capable of meeting the research and development needs of Earth scientists. The objectives of the program

were to: develop advanced passive radiometric and spectrometric sensors for surveillance of the atmosphere and oceans; develop and evaluate new active and passive sensors for sounding the atmosphere and for mapping surface characteristics; develop advanced space technology and ground data processing techniques for meteorological and scientific research; and participate in global observation programs such as the World Weather Watch (WWW). The Nimbus satellites were launched into near-polar, sun-synchronous orbits and acquired global data twice every 24 hours.

History

On August 28, 2004, NASA celebrated the 40th anniversary of the launch of the Nimbus-1 Earth-observation satellite. See Fig. 1. Starting in 1964 and for the next twenty years, the Nimbus series of missions was the United States' primary research and development platform for satellite remote-sensing of the Earth. The seven Nimbus satellites, launched over a fourteen-year period, shared their space-based observations of our planet for thirty years. Each mission taught scientists not only something new about the Earth system, but also taught them something new about how to create, operate, and improve the technology for observing the Earth from space.

Fig. 1.   The image at left is an artist's drawing of the general design of the Nimbus series of satellites. The solar-panel "wings" move throughout the day to track the Sun during the daylight part of the satellite's orbit. The 10-foot-tall satellite has the attitude control system on top, separated from a 5-foot-diameter "sensory ring" (center) with scaffolding. The sensory ring holds the batteries and electronics for each of the sensors that are mounted underneath the ring (bottom). Image taken from Madrid, C.R., ed. (1978) *The Nimbus 7 Users' Guide*. Goddard Space Flight Center: National Aeronautics and Space Administration.

Over the Nimbus Project's thirty-year history, the satellites increased the scientific community's knowledge of the Earth's atmosphere, land surface and ecosystems, weather, and oceanography. When Nimbus 1 launched in 1964, it gave meteorologists their first global images of clouds and large weather systems. By the time the second Nimbus mission went into orbit in 1966, the sensors onboard were sophisticated enough to measure the temperature of the ocean. Nimbus 3 in 1969 added atmospheric temperature observations and the ability to measure solar radiation above the atmosphere. In 1970, Nimbus 4 began collecting our first global observations of the ozone layer, and by 1972, scientists and engineers had incorporated into Nimbus 5 the capability to measure rainfall over the world's oceans and to map and monitor sea ice. Taking the atmosphere's temperature became more sophisticated with Nimbus 6 in 1975, which gave scientists the first satellite-based measurements of atmospheric temperature at different altitudes. The final mission, Nimbus 7, which was launched in 1978, collected data on ozone, the stratosphere, ocean conditions, and global weather until 1994.

Experimental users quickly came to depend on Nimbus observations for everything from hurricane forecasting to search and rescue. While providing these important operational data, each Nimbus spacecraft also carried instruments that demonstrated and tested new techniques for measuring different parts of the Earth system. In that capacity, the Nimbus missions became a driving force behind remote-sensing innovation.

NASA transferred the technology tested and refined by the Nimbus missions to the National Oceanic and Atmospheric Administration (NOAA) for its operational satellite instruments. The technology and lessons learned from the Nimbus missions are the heritage of most of the Earth-observing satellites NASA and NOAA have launched over the past three decades. Even today, scientists use data from the Nimbus missions to study the Earth system and climate change.

The Earth's Radiation Budget

When it comes to climate and climate change, the Earth's radiation budget is what makes it all happen. Swathed in its protective blanket of atmospheric gases against the boiling Sun and frigid space, the Earth maintains its life-friendly temperature by reflecting, absorbing, and re-emitting just the right amount of solar radiation. To maintain a certain average global temperature, the Earth must emit as much radiation as it absorbs. If, for example, increasing concentrations of greenhouse gases like carbon dioxide cause Earth to absorb more than it re-radiates, the planet will warm up.

One of the most important scientific contributions of the Nimbus missions was their measurements of the Earth's radiation budget. For the first time, scientists had global, direct observations of the amount of solar radiation entering and exiting the Earth system. The observations helped to verify and refine the earliest climate models, and are still making important contributions to the study of climate change. As scientists consider the causes and effects of global warming, Nimbus radiation budget data provide a base for long-term analyses and make change-detection studies possible. The Nimbus technology gave rise to current radiation-budget sensors, such as the CERES instruments on NASA's Terra and Aqua satellites.

This summary was adapted from the presentation of Tom Vonder Harr, of the Department of Atmospheric Science at Colorado State University, on the contribution of Nimbus satellites to the study of Earth's radiation budget.

A Weather Forecasting Revolution

When it comes to weather satellites, it's not a stretch to say that nearly everything that sensors are capable of today has its roots in the pioneering technology tested during the first Nimbus missions. Today, anyone with an internet connection and even the slightest interest can pull up the latest satellite image showing the weather over his or her hometown. But 40 years ago, the idea that we could observe something as intangible as air pressure using a satellite orbiting hundreds of miles above the Earth was revolutionary. With each Nimbus mission, scientists broadened their ability to collect atmospheric characteristics that improved weather forecasting, including ocean and air temperatures, air pressure, and cloudiness. The global coverage provided by Nimbus satellites made accurate 3-5 day forecasts possible for the first time.

The ability of the Nimbus satellites to detect electromagnetic energy in multiple wavelengths (multi-spectral data), in particular the microwave region of the electromagnetic spectrum, made it possible for scientists to look into the atmosphere and tell the difference between water vapor and liquid water in clouds. In addition, they were able to measure atmospheric temperature even in the presence of clouds, a capability that allowed scientists to take the temperature in the "warm core" of hurricanes.

This summary was adapted from the presentations of Bill Smith, of Hampton University and Dave Staelin, of the Department of Electrical

Engineering and Computer Science, at Massachusetts Institute of Technology, on how Nimbus satellites revolutionized the study and prediction of Earth's weather and climate.

The Ozone Layer

Even before the Nimbus satellites began collecting their observations of Earth's ozone layer, scientists had some understanding of the processes that maintained or destroyed it. They were pretty sure they understood how the layer formed: solar radiation breaks apart the stable, double-atom form of oxygen (O_2) into two unstable singles that quickly latch on to whatever is around, sometimes re-forming as O_2, but also occasionally glomming on to an existing O_2 molecule to make ozone (O_3). They knew from laboratory experiments that halogens (chlorine, bromine, etc) could destroy ozone. Finally, weather balloons had revealed that the concentration of ozone in the atmosphere changed over time, and scientists suspected weather phenomena or seasonal change were responsible. But how did all of these pieces of information work together on a global scale?

Scientists conducted experiments from NASA experimental aircraft and proved that atmospheric chemicals such as the chlorofluorocarbons (CFCs) released from refrigerants and aerosol sprays did destroy ozone. As Nimbus 7 satellite observations accumulated between 1978 and 1994, it became increasingly clear that CFCs were creating a hole in the ozone layer each winter season over Antarctica. See Fig. 2. Not only that, but despite some year-to-year variations, it appeared the hole was becoming larger.

Public concern gave rise to the Montreal Protocol http://hq.unep.org/ozone/Montreal-Protocol/Montreal-Protocol2000.shtml, which bound the countries that signed the treaty to phase out the use of ozone-depleting chemicals. Without the Nimbus measurements, we would probably not have been aware of how severe the ozone hole problem was until many years later—perhaps until we began to see alarming increases in the rate of skin cancer and other negative effects of ozone loss.

This summary was adapted from the presentation of Arlin Krueger, of the Joint Center for Earth Systems Technology at University of Maryland, Baltimore County, on how long-term collection of observations of stratospheric ozone led to our awareness of the ozone hole and the dangers it posed to life on Earth.

The Color of the Ocean

Anyone who has lived or stayed near the ocean for a long enough time can tell you how the sea seems to change color from day to day, from deep sparkling blue on a bright sunny day to slate gray beneath a thick layer of clouds. What fewer people know is that the color of the ocean changes as concentrations of sediment, organic matter, and ocean plant life change. These changes in ocean color signal biological processes that affect marine life as well as public health, particularly in coastal areas.

When the Nimbus 7 satellite launched in 1978, it carried on board the first sensor engineered to observe the ocean in visible wavelengths of light. Originally intended to be only a one-year technology demonstration, the Coastal Zone Color Scanner (nicknamed "CZCS") ended up delivering science data over selected test sites for the next 8 years!

With CZCS, NASA gave ocean biologists their first global-scale pictures of ocean plant growth, changing scientists' views of the marine biosphere. Scientists discovered that ocean plant life matched land-based plant life in terms of its rates of photosynthesis and seasonal changes. CZCS data also began to reveal the effects of land-based pollution on coastal ecosystems. The success of the mission paved the way for Sea-viewing Wide Field-of-view Sensor (SeaWiFS) and the ocean science sensors of NASA's Earth Observing System series of satellites in orbit today.

This summary was adapted from a presentation by Wayne Esaias, of NASA's Goddard Space Flight Center, on the legacy of Nimbus 7's Coastal Zone Color Scanner.

A Sea of Change

When the Nimbus 5 spacecraft launched in 1972, scientists planned for its Electrically Scanning Microwave Radiometer to collect global observations of where and how much it rained across the world. However, a new priority for the sensor evolved in the months following its launch: mapping global sea ice concentrations. When Nimbus 7 launched in 1978, technology had improved enough for scientists to distinguish newly formed (i.e., "first

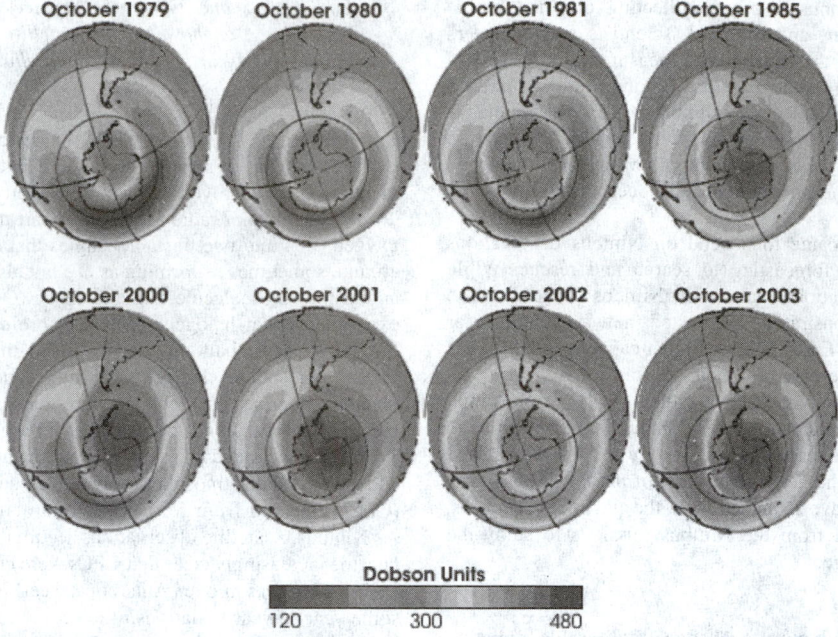

Fig. 2. Nimbus 7 measurements collected over the South Pole beginning in 1978 identified a previously unknown hazard: the large-scale destruction of ultraviolet-blocking ozone by chemicals released into the atmosphere through refrigeration devices and aerosol sprays. These globes show ozone concentrations over Antarctica in selected Octobers from 1979–85 and 2000–2003. Nimbus observations began to point to a drop in ozone (blue areas) as early as 1980, with more extreme decreases developing in 1985. (*courtesy of Paul Newman, Richard Stolarski, Mark Shoeberl, Arlin Krueger*).

year") sea ice from older ice. The data it collected during its 9-year lifespan provide a significant chunk of the long-term record of Earth's sea ice concentration that today's scientists use for studies of climate change.

Among the most serendipitous discoveries that the Nimbus missions made possible was that of a gaping hole in the sea ice around Antarctica in the Southern Hemisphere winters of 1974–76. See Fig. 3. In a phenomenon that has not been observed since, an enormous, ice-free patch of water, called a polynya, developed three years in a row in the seasonal ice that encases Antarctica each winter. Located in the Weddell Sea, each year the polynya vanished with the summer melt, but returned the following year. The open patch of water may have influenced ocean temperatures as far down as 2,500 meters and influenced ocean circulation over a wide area. The Weddell Sea Polynya has not been observed since the event witnessed by the Nimbus satellites in the mid-70s. Without those images, we might never have known that an event like that did — or even could — occur.

> This summary was adapted from a presentation by Per Gloersen of NASA's Goddard Space Flight Center on the contribution the Nimbus missions have made to the collection of long-term records of Earth's "vital signs" that we must have to study the causes and effects of climate change.

Satellite Search and Rescue and Data Collection

Today, it is not too big a challenge for people to figure out exactly where they are on the Earth. Between cell phones and pocket-sized Global Positioning System (GPS) devices, getting your bearings is only as complicated as deciphering your users manual. Thirty years ago, however, locating and tracking the position of something or someone on the Earth's surface was a tougher task.

NASA's Nimbus satellites (beginning with Nimbus 3 in 1969) blazed the trail into the modern GPS era with operational search and rescue and data collection systems. See also **Global Positioning System (GPS)**. The satellites tested the first technology that allowed satellites to locate weather-observation stations set up in remote locations and to command the stations to transmit their data back to the satellite. The most famous demonstration of the new technology was through the record-breaking flight of British aviator Sheila Scott, who tested the Nimbus navigation and locator communication system when she made the first-ever solo flight over the North Pole in 1971.

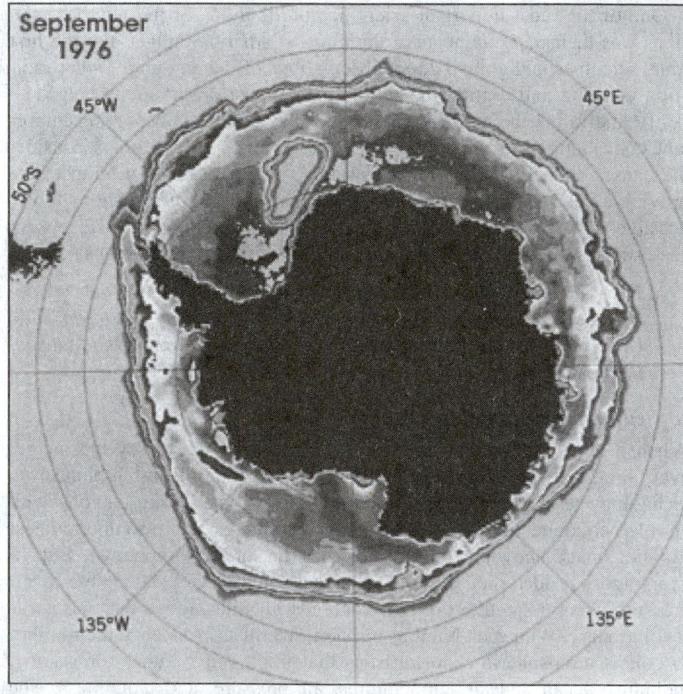

Fig. 3. In the mid-1970s, Nimbus sensors were recording sea ice concentrations in the Southern Ocean surrounding Antarctica when they detected a large ice-free opening in the ice pack. This patch of open water, called a *polynya*, recurred each winter between 1974–76, and has not retuned since. In the image at left, Antarctica is colored black and sea ice concentrations appear in shades of red and orange. The Weddell Sea polynya is the light blue (open water) area in the upper left quadrant of the image. (*Claire Parkinson (NASA GSFC)*).

In a series of "first-ever" experiments, Nimbus satellites tracked the movements of free-floating buoys in the Arctic Ocean, the erratic travels of weather balloons, and the movements of animals from sea turtles to puffins. Scientists set up ground-based observation stations in remote or even dangerous environments—such as right outside a black bear's den in Yellowstone National Park—and used Nimbus sensors to retrieve the data on environmental conditions that the ground station was recording.

The Nimbus ground-to-satellite-to-ground communication system demonstrated the first satellite-based search and rescue system. Among the earliest successes were the rescue of two hot air balloonists who went down in the North Atlantic in 1977 and, later that year, tracking a Japanese adventurer on his first attempt to be the first person to dogsled solo to the North Pole through Greenland. Tens of thousands of people over the past three decades have been rescued through the Search and Rescue Satellite-aided Tracking (SARSAT) operational system on NOAA satellites.

This summary was adapted from a presentation by Charles Cote of NASA's Goddard Space Flight Center on how Nimbus missions became the proving ground for satellite-based search and rescue technology.

Operational History of the Nimbus Satellites

Nimbus-1 (Nimbus-A). Launch Date/Time: August 28, 1964 at 08:52:00 UTC, from Vandenberg AFB, CA, aboard a Thor-Agena B rocket. The primary objectives were to test the Nimbus spacecraft configuration and provide improved cloud photographs using the APT system deployed on TIROS-8. The Advanced cameras and high-resolution infrared radiometers were also to be tested for improved daylight as well as night cloud-cover conditions. The polar-orbiting spacecraft consisted of three major elements: (1) a sensory ring, (2) solar paddles, and (3) the control system housing. The solar paddles and the control system housing were connected to the sensory ring by a truss structure, giving the satellite the appearance of an ocean buoy. Nimbus, depending on the experiments carried, was 3.04 to 3.7 m tall (10 to 12 ft), 1.52 m (5 ft) in diameter at the base, and 3 to 3.96 m (9.8 to 13 ft) across with solar paddles extended. The sensory ring, which formed the satellite base, housed the electronics equipment and battery modules. The lower surface of the torus-shaped sensory ring provided mounting space for sensors and telemetry antennas. An H-frame structure mounted within the centre of the torus provided support for the larger experiments and tape recorders. Mounted on the control system housing which was located on top of the spacecraft, were sun sensors, horizon scanners, gas nozzles for attitude control, and a command antenna. See Fig. 4. Use of a stabilization and control system allowed the spacecraft's orientation to be controlled to within plus or minus 1 degree for all three axes (pitch, roll, and yaw).

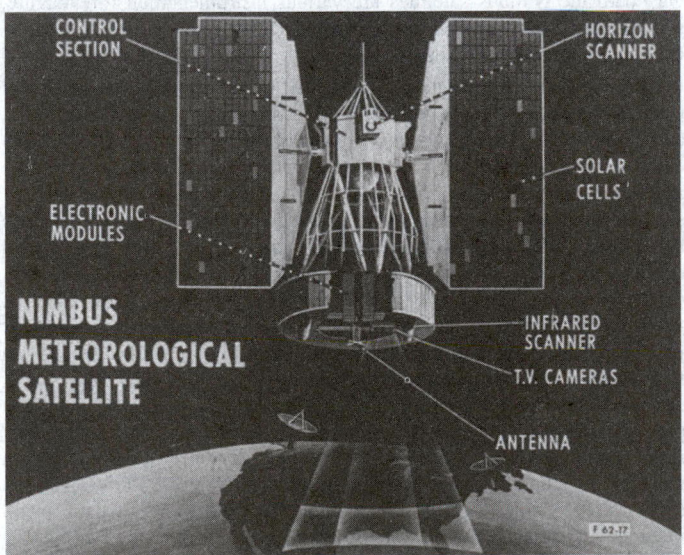

Fig. 4.   Nimbus. (NASA).

The spacecraft carried (1) an advanced vidicon camera system (AVCS) for recording and storing remote cloud-cover pictures, (2) an automatic

picture transmission (APT) camera for providing real-time cloud cover pictures, and (3) a high-resolution infrared radiometer (HRIR) to complement the daytime TV coverage and to measure nighttime radiative temperatures of cloud tops and surface terrain. A short second-stage burn resulted in an unplanned eccentric orbit. Otherwise, the spacecraft and its experiments operated successfully until September 22, 1964. The solar paddles became locked in position, resulting in inadequate electrical power to continue operations.

Nimbus-B. Launch Date: May 18, 1968 from Vandenberg AFB, CA, aboard a Thorad (Long-Tank Thor) rocket. Primary experiment subsystems on Nimbus-B consisted of (1) a satellite infrared spectrometer (SIRS) for determining the vertical temperature profiles of the atmosphere, (2) an infrared interferometer spectrometer (IRIS) for measuring the emission spectra of the earth-atmosphere system, (3) both high- and medium-resolution infrared radiometers (HRIR and MRIR) for yielding information on the distribution and intensity of infrared radiation emitted and reflected by the earth and its atmosphere, (4) a monitor of ultraviolet solar energy (MUSE) for detecting solar UV radiation, (5) an image dissector camera system (IDCS) for providing daytime cloud-cover pictures in both real-time mode, using the real-time transmission system (RTTS), and tape recorder mode, using the high data rate storage system (DHRSS), (6) a radioisotope thermoelectric generator (RTG), SNAP-19, to assess the operational capability of radioisotope power for space applications, and (7) an interrogation, recording, and location system (IRLS) designed to locate, interrogate, record, and retransmit meteorological data from remote collection stations. Two minutes after launch, a booster malfunction forced a command destruct of the rocket and satellite. The rocket and spacecraft debris fell into the Pacific Ocean between Jalama Beach and San Miguel Island.

Nimbus-2 (Nimbus-C). Launch Date/Time: May 15, 1966 at 07:55:00 UTC, from Vandenberg AFB, CA, aboard a Thor-Agena D rocket. The spacecraft carried (1) an advanced vidicon camera system (AVCS) for recording and storing remote cloud-cover pictures, (2) an automatic picture transmission (APT) camera for providing real-time cloud-cover pictures, and (3) both high- and medium-resolution infrared radiometers (HRIR and MRIR) for measuring the intensity and distribution of electromagnetic radiation emitted by and reflected from the earth and its atmosphere. The spacecraft and experiments performed normally after launch until July 26, 1966, when the spacecraft tape recorder failed. Its function was taken over by the HRIR tape recorder until November 15, 1966, when it also failed. Some real-time data were collected until January 17, 1969, when the spacecraft mission was terminated owing to deterioration of the horizon scanner used for earth reference.

Nimbus 3 (Nimbus-B2). Launch Date/Time: April 14, 1969 at 07:54:00 UTC, from Vandenberg AFB, CA, aboard a Thorad-Agena D rocket. Primary experiments consisted of (1) a satellite infrared spectrometer (SIRS) for determining the vertical temperature profiles of the atmosphere, (2) an infrared interferometer spectrometer (IRIS) for measuring the emission spectra of the earth-atmosphere system, (3) both high- and medium-resolution infrared radiometers (HRIR and MRIR) for yielding information on the distribution and intensity of infrared radiation emitted and reflected by the earth and its atmosphere, (4) a monitor of ultraviolet solar energy (MUSE) for detecting solar UV radiation, (5) an image dissector camera system (IDCS) for providing daytime cloud-cover pictures in both real-time mode, using the real time transmission system (RTTS), and tape recorder mode, using the high data rate storage system, (6) a radioisotope thermoelectric generator (RTG), SNAP-19, to assess the operational capability of radioisotope power for space applications, and (7) an interrogation, recording and location system (IRLS) experiment designed to locate, interrogate, record, and retransmit meteorological and geophysical data from remote collection stations. Nimbus 3 was successful and performed normally until July 22, 1969, when the IRIS experiment failed. The HRIR and SIRS experiments were terminated on January 25, 1970, and June 21, 1970, respectively. The remaining experiments continued operation until September 25, 1970, when the rear horizon scanner failed. Without this horizon scanner, it was impossible to maintain proper spacecraft attitude, thus making most experimental observations useless. All spacecraft operations were terminated on January 22, 1972.

Nimbus-4 (Nimbus-D). Launch Date/Time: April 8, 1970 at 08:17:00 UTC, from Vandenberg AFB, CA, aboard a Thorad-Agena D rocket.

Primary experiments consisted of (1) an image dissector camera system (IDCS) for providing daytime cloud-cover pictures, both in real-time and recorded modes, (2) a temperature-humidity infrared radiometer (THIR) for measuring daytime and nighttime surface and cloud-top temperatures as well as the water vapor content of the upper atmosphere, (3) an infrared interferometer spectrometer (IRIS) for measuring the emission spectra of the earth/atmosphere system, (4) a satellite infrared spectrometer (SIRS) for determining the vertical profiles of temperature and water vapor in the atmosphere, (5) a monitor of ultraviolet solar energy (MUSE) for detecting solar UV radiation, (6) a backscatter ultraviolet (BUV) detector for monitoring the vertical distribution and total amount of atmospheric ozone on a global scale, (7) a filter wedge spectrometer (FWS) for accurate measurement of IR radiance as a function of wavelength from the earth/atmosphere system, (8) a selective chopper radiometer (SCR) for determining the temperatures of six successive 10-km layers in the atmosphere from absorption measurements in the 15-micrometer CO_2 band, and (9) an interrogation, recording, and location system (IRLS) for locating, interrogating, recording, and retransmitting meteorological and geophysical data from remote collection stations. The spacecraft performed over 10 years until it was deactivated on September 30, 1980.

Nimbus-5 (Nimbus-E). Launch Date/Time: December 11, 1972 at 07:56:00 UTC, from Vandenberg AFB, CA, aboard a Delta 900 rocket. See Fig. 5.

Fig. 5.   Nimbus E, the fifth spacecraft in the Nimbus series, is shown preparing for launch on December 11, 1972 from the Western Test Range (WTR), Space Launch Complex SLC-2, West, by the Thrust- Augmented Delta vehicle. The satellite was placed in an 1,100-kilometer (683.5 mile) run-synchronous nearly circular polar orbit. The spacecraft was designated Nimbus-5 upon confirmation that it had achieved successful orbit. (*courtesy of NASA*).

Primary experiments included (1) a temperature-humidity infrared radiometer (THIR) for measuring day and night surface and cloud-top temperatures, as well as the water vapor content of the upper atmosphere, (2) an electrically scanning microwave radiometer (ESMR) for mapping the microwave radiation from the earth's surface and atmosphere, (3) an infrared temperature profile radiometer (ITPR) for

obtaining vertical profiles of temperature and moisture, (4) a Nimbus E microwave spectrometer (NEMS) for determining tropospheric temperature profiles, atmospheric water vapor abundances, and cloud liquid water contents, (5) a selective chopper radiometer (SCR) for observing the global temperature structure of the atmosphere, and (6) a surface composition mapping radiometer (SCMR) for measuring the differences in the thermal emission characteristics of the earth's surface. The spacecraft performed over 10 years until it was deactivated on March 29, 1983.

Nimbus-6 (Nimbus-F). Launch Date/Time: June 12, 1975 at 08:12:00 UTC, from Vandenberg AFB, CA, aboard a Two-stage Delta 2910 rocket. The nine experiments selected for Nimbus 6 were (1) earth radiation budget (ERB), (2) electrically scanning microwave radiometer (ESMR), (3) high-resolution infrared radiation sounder (HIRS), (4) limb radiance inversion radiometer (LRIR), (5) pressure modulated radiometer (PMR), (6) scanning microwave spectrometer (SCAMS), (7) temperature-humidity infrared radiometer (THIR), (8) tracking and data relay experiment (T+DRE), and (9) tropical wind energy conversion and reference level experiment (TWERLE). This complement of advanced sensors was capable of (1) mapping tropospheric temperature, water vapor abundance, and cloud water content; (2) providing vertical profiles of temperature, ozone, and water vapor; (3) transmitting real-time data to a geostationary spacecraft (ATS 6); and (4) yielding data on the earth's radiation budget. The spacecraft performed over 7 years until it was deactivated on March 29, 1983 along with Nimbus-5.

Nimbus-7 (Nimbus-G). Launch Date/Time: October 24, 1978 at 08:14:00 UTC, from Vandenberg AFB, CA, aboard a Two-stage Delta 2910 rocket. Eight experiments were selected: (1) limb infrared monitoring of the stratosphere (LIMS), (2) stratospheric and mesospheric sounder (SAMS), (3) coastal-zone color scanner (CZCS), (4) stratospheric aerosol measurement II (SAM II), (5) earth radiation budget (ERB), (6) scanning multichannel microwave radiometer (SMMR), (7) solar backscatter UV and total ozone mapping spectrometer (SBUV/TOMS), and (8) temperature-humidity infrared radiometer (THIR). These sensors were capable of observing several parameters at and below the mesospheric levels.

The craft was placed in Sun-synchronous orbit and transmission of data from all of the experiments was completed as scheduled. For the first time NASA and ESA (European Space Agency) were able to receive data concerning the global atmosphere in real time. The Nimbus-7 spacecraft was turned off in 1994 after16 years of service. The TOMS instrument failed in May of 1993.

Nimbus Instruments

On board the Nimbus satellites are various instrumentation for imaging, sounding, and other studies in different spectral regions. Table 1 summarizes the Nimbus instruments.

Advanced Vidicon Camera System (AVCS). The Advanced Vidicon Camera System (AVCS), which consisted of three cameras, a tape recorder, and an S-band transmitter, recorded and stored a series of remote daytime cloud-cover pictures for subsequent playback to selected ground data acquisition stations. The AVCS cameras were mounted on the satellite sensory ring, facing earthward and deployed in a fan-like array to produce a three-segment composite picture. Each camera covered a 37-deg field of view with the center camera pointing straight down. The optical axes of the other two cameras were directed 35 deg to either side. Each of the cameras employed an f/4 lens with a focal length of 16.5 mm. A potentiometer attached to the solar array controlled the lens opening from f/16 when the spacecraft was over the equator to f/4 when it was near the poles. The 800-scan-line, 2.54-cm-diameter (1-in-diameter) vidicon pickup tubes yielded a linear resolution of better than 1 km (0.6 mi) at nadir from an altitude of 800 km (497 miles). At this altitude, the camera array produced a composite picture covering an area of 830 by 2,700 km (516 by 1,678 miles). Up to 192 pictures (two full orbits of data) or 64 pictures per camera could be stored on tape for subsequent playback to an acquisition station. Using a transmission frequency of 1707.5 MHz, the two orbits of pictures could be telemetered to a ground station in 4 min. The AVCS experiment was highly successful. It provided the first near-global, high-resolution cloud-cover pictures ever assembled and confirmed the decision to use this particular camera assembly as a basis for the first operational satellite system TOS/ESSA (TIROS Operational

TABLE 1. EXPERIMENTAL INSTRUMENTATION ON-BOARD NIMBUS SATELLITES

Acronyms	Instruments	Satellites
Visible Imagery		
AVCS	Advanced Vidicon Camera System	Nimbus-1 and 2
APT	Automatic Picture Transmission System	Nimbus-1 and 2
IDCS	Image Dissector Camera System	Nimbus-3 and 4
Infrared Imaging		
HRIR	High Resolution Infrared Radiometer	Nimbus-1, 2, and 3
MRIR	Medium Resolution Infrared Radiometer	Nimbus-2 and 3
THIR	Temperature-Humidity Infrared Radiometer	Nimbus-4 through 7
Microwave Imaging		
ESMR	Electrically Scanning Microwave Radiometer	Nimbus-5 and 6
SMMR	Scanning Multispectral Microwave Radiometer	Nimbus-7
Sounding		
HIRS	High Resolution Infrared Sounder	Nimbus-6
FWS	Filter Wedge Spectrometer	Nimbus-4
IRIS	Infrared Interferometer Spectrometer	Nimbus-3 and 4
ITPR	Infrared Temperature Profile Radiometer	Nimbus-5
LIMS	Limb Infrared Monitor of the Stratosphere	Nimbus-7
LRIR	Limb Radiance Inversion Radiometer	Nimbus-6
NEMS	Nimbus E Microwave Spectrometer	Nimbus-5
PMR	Pressure Modulated Radiometer	Nimbus-6
SAMS	Stratospheric and Mesospheric Sounder	Nimbus-7
SCAMS	Scanning Microwave Spectrometer	Nimbus-6
SCR	Selective Chopper Radiometer	Nimbus-4 and 5
SIRS	Satellite Infrared Spectrometer	Nimbus-3 and 4
Others		
IRLS	Interrogation, Recording, and Location System	Nimbus-4
BUV	Backscatter Ultraviolet Spectrometer	Nimbus-4
CZCS	Costal Zone Color Scanner	Nimbus-7
ERB	Earth Radiation Budget	Nimbus-6 and 7
MUSE	Monitor of Ultraviolet Solar Energy (Solar UV Monitor)	Nimbus 3, and 4
SAM II	Stratospheric Aerosol Measurement-II	Nimbus-7
SCMR	Surface Composition Mapping Radiometer	Nimbus-5
SBUV/TOMS	Solar Backscatter Ultraviolet/Total Ozone Mapping Spectrometer	Nimbus-7
TWERLE	Tropical Wind Energy Conversion and Reference Level Experiment	Nimbus-6

System/Environmental Science Services Administration). Data from this experiment can be obtained through SDSD.

Automatic Picture Transmission (APT) System. The Automatic Picture Transmission (APT) system was a camera and transmitter combination designed to transmit local daytime, slow-scan television pictures of cloud-cover conditions to properly equipped ground receiving stations on a real-time basis. The camera used a 108-deg wide-angle f/1.8 objective lens with a focal length of 5.7 mm. The camera was mounted facing earthward on the H-frame inside the sensory ring, with its optical axis parallel to the spacecraft spin axis. The actual picture taking required 8 s and the transmission 200 s. Earth-cloud images retained on the photosensitive surface of the 2.54-cm-diameter (1-in-diameter) vidicon were read out at four lines per second to produce an 800-line picture. A 5-W TV transmitter (136.95 MHz) relayed the pictures to local APT stations within communication range See Fig. 6. The faceplate of the vidicon had reticle marks that appeared on the picture format to aid in relating the picture to its geographical position on the earth's surface. At the nominal satellite altitude, a picture covered approximately a 1,660- by 1,660-km square with a horizontal resolution of around 3 km at nadir. The experiment supplied over 1,600 high-quality cloud-cover pictures to participating APT stations during the spacecraft's 3.5-week lifetime. It proved the capability

of weather satellites to provide high-quality daytime local cloud-cover data to operational meteorologists on an essentially real-time basis. Its success bolstered the decision to include such instrumentation in the TIROS Operational System (TOS).

Fig. 6. Automatic Picture Transmission (APT) enabled meteorologists to obtain immediate local area cloud pattern photographs when the Nimbus satellite was within a 1,700-mile range of a receiving station. The APT subsystem pioneered on Nimbus 1, which was launched August 28, 1964, provided direct readout of nighttime and daytime cloud coverage. It transmitted photographic data of synoptic meteorological conditions in areas 1,200nmi square to over 300 ground stations in more than 43 countries. APT was part of three experiments performed on Nimbus-1, with the other experiments being Advanced Vidicon Camera Subsystem (AVCS) and High Resolution Infrared Radiometer (HRIR). This subsystem facsimile receiver required relatively simple ground station equipment at the price of about $30,000 per set.

Backscatter Ultraviolet (BUV) Spectrometer. The Backscatter Ultraviolet (BUV) spectrometer experiment was designed to monitor the vertical distribution and total amount of atmospheric ozone on a global scale by measuring the intensity of UV radiation backscattered by the earth/atmosphere system during day and night in the 2500- to 3400-A spectral band. The primary instrumentation consisted of a double monochromator containing all reflective optics and a photomultiplier detector. The double monochromator was composed of two Ebert-Fastie-type monochromators in tandem. Each monochromator had a 52- by 52-mm grating with 2,400 lines per mm. Light from a 0.05-sr solid angle (subtending approximately a 222-sq-km (138-sq-mile) area on the earth's surface from a satellite height of approximately 1,100 km (684 miles) entered the nadir-pointing instrument through a depolarizing filter. A motor-driven cam step rotated the gratings to monitor the intensity of 12 ozone absorption wavelengths. The detector was a photomultiplier tube. For background readings, a filter photometer measured the reflected UV radiation in an ozone-free absorption area near 3800 A. Signals from both units were read by separate range-switching electrometers with seven ranges. A BUV experiment cycle required 6144 s. Each cycle, in turn, was divided into 192 BUV frames of 32-s duration. Calibration by onboard light sources was performed in 26 of the 192 frames. The other frames were used for experimental data. During each of these data frames, the monochromator measured the intensity of the UV radiation in each of the 12 wavelength bands, while the photometer measured the UV intensity in a single wavelength band. The dwell time at each wavelength was 1.8 s, and, during this interval, four analog

UV intensity measurements were taken at 400-ms intervals in addition to an integrated pulse count measurement of the UV intensity and energetic particle flux. Once each orbit, the field of view was changed to monitor the sun or moon directly. The measurement range of the signal current was from 0.2 to 3000 microamps. The vertical distribution of ozone was obtained by mathematical inversion techniques.

Coastal Zone Color Scanner (CZCS). The Coastal Zone Color Scanner Experiment (CZCS) was designed to map chlorophyll concentration in water, sediment distribution, gelbstoffe concentrations as a salinity indicator, and temperature of coastal waters and ocean currents. Reflected solar energy was measured in six channels to sense color caused by absorption due to chlorophyll, sediments, and gelbstoffe in coastal waters. The CZCS was a multi-channel scanning radiometer which used a rotating plane mirror at a 45 degree angle to the optic axis of a Cassegran telescope. The mirror scanned 360 degrees but only the 80 degrees of data centered on nadir were collected for ocean color measurements. The instrument viewed deep space and calibration sources during the remainder of the scan. The incoming radiation was collected by the telescope and divided into two streams by a dichroic beam splitter. One stream was transmitted to a field stop that was also the entrance aperature of a small polychromator. The radiance that entered the polychromator was separated and re-imaged in five wavelengths on five silicon detectors in the focal plane of the polychromator. The other stream was directed to a cooled mercury cadmium telluride detector in the thermal region (10.5–12.5 micrometer). A radiative cooler was used to cool the thermal detector. To avoid sun glint, the scanner mirror was tilted about the sensor pitch axis on command so that the line of sight of the sensor was moved in 2-deg increments up to 20 deg with respect to the nadir. Spectral bands at 0.443 and 0.670 micrometers centered on the most intense absorption bands of chlorophyll, while the band at 0.550 micrometers centered on the "hinge point," the wavelength of minimum absorption. Ratios of measured energies in these channels were shown to closely parallel surface chlorophyll concentrations. Data from the scanning radiometer were processed, with algorithms developed from the field experiment data, to produce maps of chlorophyll absorption. The temperatures of coastal waters and ocean currents were measured in a spectral band centered at 11.5 micrometers. Observations were made also in two other spectral bands, 0.520 micrometers for chlorophyll correlation and 0.750 micrometers for surface vegetation. The scan width was 1,556 km (967 miles) centered on nadir and the ground resolution was 0.825 km (0.5 mi) at nadir. See Fig. 7.

Earth Radiation Budget (ERB). The ***Nimbus-6 Earth Radiation Budget (ERB)*** experiment measured reflected and emitted terrestrial radiation fluxes in conjunction with solar radiation. The results were used (1) to determine the earth radiation budget, (2) to determine the angular distribution of terrestrial radiation for various meteorological and geographic regimes, and (3) to correlate measurements made using identical but independent channels calibrated to the same standard. Incoming solar radiation from 0.2 to 50 micrometers was normally monitored in 10 spectral intervals as the satellite orbited over the Antarctic, just before it started its northward trip on the daylight side of the earth. Terrestrial radiation measurements were taken continuously in the 0.2- to 4-micrometer, 0.7- to 3-micrometer, and 4- to 50-micrometer intervals. The measurements were taken in two ways. Four channels, using fixed wide-angle optics (133.3-deg field of view), measured the total outgoing radiation integrated over the entire disk of the earth. The second set of measurements was obtained from eight high-resolution narrow-angle scanning channels that measured the terrestrial radiation emanating from a relatively small area over a range of zenith and azimuth angles. The multichannel radiometer employed a biaxial scanning mechanism which enabled measurements to be obtained from the forward horizon to the aft horizon in a 64-s interval. Each axis of the scanning mechanism contained four shortwave channels (0.2 to 4.0 micrometers) and four longwave channels (4.0 to 50 micrometers) with a 0.25- by 5.14-deg field of view. The channels were oriented in a directional fan to cover 20 deg to each side of the orbital plane. The 64-s scan period allowed an area to be measured from up to 17 different angles as the spacecraft passed overhead.

The objective of the ***Nimbus 7 Earth Radiation Budget (ERB)*** experiment, a follow-on to the Nimbus 6 ERB, was to determine the earth radiation budget on both synoptic and planetary scales by simultaneous measurements of incoming solar radiation and outgoing earth-reflected

Fig. 7.   This first image of the global biosphere which was produced by combining data from two different satellite sensors show for the first time the productive potential of the Earth's vegetative biomass. The ocean image is a composite of all data collected during the 20-month period from November 1978 through June 1980 by the Coastal Zone Color Scanner (CZCS) flown on NASA's Nimbus-7 satellite, managed by the Goddard Space Flight Center, Greenbelt, Maryland. The CZCS data show concentrations of marine phytoplankton pigment. Phytoplanktons, the microscopic plants that grow in the sunlight regions of the ocean, form the base of the marine food web. Red and orange colors indicate areas of high plankton concentrations. Yellow and green represent areas of moderate concentration. One of the most notable features in this image is the clear delineation of the equator through increased plant abundance, and the differences between the equatorial Atlantic, Indian, and Pacific Oceans. Plankton concentrations tend to be high nutrients to the sunlit upper ocean layers. Major areas of the South Pacific are blank (black areas=no data) because the CZCS could operate only intermittently. The land-vegetation image is a composite of three years of data from the Advanced Very High Resolution Radiometer (AVHRR) on the National Oceanographic and Atmospheric Agency (NOAA-7) satellite, which measured land-surface radiation in the visible and near-infrared bands to estimate chlorophyll and leaf potential for chlorophyll production. The lighter shades of green highlight tropical and subtropical forests, temperate forests and farmlands, and some drier regions such as savannas and pampas. The yellow shades in the United States Midwest show lower potential, while the darker yellow shades of Northern Hemisphere forests and the dry Australian Outback rank lower. Desert, high mountains, and arctic regions reflect barren conditions, intermittently. The land-vegetation image is a composite of three years of data from the Advanced Very High Resolution Radiometer (AVHRR) on the NOAA-7 satellite, which measures land-surface radiation in the visible and near-infrared bands to estimate chlorophyll and leaf mass. The dark green areas (rain forests) show the highest potential for chlorophyll production. The lighter shades of green highlight tropical and subtropical forests, temperate forests and farmlands and some drier regions such as savannas and pampas. The yellow shades in the United States Midwest show lower potential, while the darker yellow shade of Northern Hemisphere forests and the dry Australian Outback rank lower. Desserts, high mountains, and arctic regions reflect barren conditions. (*courtesy of NASA*).

(shortwave) and emitted (longwave) radiation. Both (1) fixed wide-angle sampling of terrestrial fluxes at the satellite altitude and (2) scanned narrow-angle sampling of the angular radiance components, were used to determine outgoing radiation (reflected and emitted). The ERB subsystem consisted of a 22-channel radiometer containing separate subassemblies to perform the required solar, earth-flux (wide angle), and scanned earth radiance (narrow angle) measurements. The systems used optical filters for spectral discriminations, and for uncooled thermal detectors, thermopile detectors in the solar and fixed-earth-flux channels, and pyroelectric detectors in the scanning channels. The 10 solar channels observed the sun in front of the observatory in the X-Y plane. The solar channels obtained usable solar data only during a period of about 3 min in each orbit when the spacecraft was over the Antarctic region. Their full response field of view (FOV) was 0.18 rad. The solar channel subassembly was pivoted plus or minus 0.35 rad in the X-Y plane to compensate for sun-angle deviation when required. The channel 10c solar channel was a model H-F self-calibrating cavity thermopile used for monitoring the total solar irradiance (0.2 to 50 micrometers). The four fixed earth-flux channels (numbered 11–14) were mounted so that they could continuously view the total earth disk, and record data at 0.25-s intervals. The eight narrow FOV scanning channels were mounted in the scanning head. The NFOV channels consisted of four shortwave (0.2 to 4.8 micrometers) and four longwave (4.5 to 50+ micrometers) channels. The scanning head was gimbal-mounted in the radiometer unit main frame. The FOVs of the telescopes were asymmetric (4.4 by 89.4 mrad) and those of the shortwave and longwave channels were coincident. The 89.4 mrad FOVs of the four pairs of channels were not contiguous, but covered only alternate 89.4 mrad angular intervals along the horizon.

Electrically Scanning Microwave Radiometer (ESMR). The primary objectives of the **Nimbus-5 Electrically Scanning Microwave Radiometer (ESMR)** were (1) to derive the liquid water content of clouds from brightness temperatures over oceans, (2) to observe differences between sea ice and the open sea over the polar caps, and (3) to test the feasibility of inferring surface composition and soil moisture. To accomplish these objectives, the ESMR was capable of continuous global mapping of the 1.55-cm (19.36 GHz) microwave radiation emitted by the earth/atmosphere system, and could function even in the presence of cloud conditions that block conventional satellite infrared sensors. An 83.3- by 85.5-cm (33- by 34-in) radiometer antenna system, deployed after launch, scanned the earth successively at various angles in a plane perpendicular to the spacecraft orbital track, producing a brightness-temperature map of the surface of the earth and its atmosphere. The scanning process was controlled by a computer on board, and it consisted of 78 symmetrically distributed independent scan spots extending 50 deg to either side of nadir. Angular separation of the scan spots allowed for an 8.5% overlap between view positions. From a mean orbital height of 1,100 km (684 miles), the radiometer had an accuracy of about plus or minus 1 °C (33.8 °F) with a spatial resolution of about 25 km (15.5 miles) at nadir. The ESMR data were stored on magnetic tape for transmission to ground acquisition stations.

The **Nimbus-6 Electrically Scanning Microwave Radiometer (ESMR)** measured the earth's microwave emission to provide the liquid water content of clouds, the distribution and variation of sea ice cover, and gross characteristics of land surfaces (vegetation, soil moisture, and snow cover). The two-channel scanning radiometer operated in a 250-MHz band centered at 37 GHz. One channel was used to measure the vertical polarization and the other measured the horizontal polarization. The antenna beam array, a 90- by 20- by 12-cm (35- by 8- by 5 in) box-like structure, was mounted on top of the spacecraft sensory ring and was pointed in the direction of the spacecraft's forward motion and tilted down 45 deg from the satellite antenna axis. The antenna beam scanned the earth in 71 discrete steps for various angles extending up to 35 deg on either side of the orbital plane. The deduced brightness temperatures were expected to be accurate to within 3–5 deg K. Spatial resolution was 20 km (12 miles) in the cross-track direction and 45 km (28 miles) in the direction parallel to the subpoint track.

Filter Wedge Spectrometer (FWS). The **Nimbus 4 Filter Wedge Spectrometer (FWS)** experiment was designed to accurately determine the radiance from the earth-atmosphere system as a function of wavelength by measuring the emitted and reflected infrared radiation in the 1.2- to 2.4- and 3.2- to 6.4-micrometer bands. The instrumentation consisted of (1) a telescope, (2) a rotating disk chopper, (3) a rotating (3.75 rpm) circular interference filter wheel, and (4) a lead selenide detector. The filter wheel was a two-180-deg-segment (one per passband) 100-layer interference filter with the layer thickness linearly increasing as a function of angular position, causing the bandpass to shift toward longer wavelengths. Incoming radiation was reflected off a surface mirror and was collected by a telescope oriented normal to the earth's surface. The telescope had a 3-deg field of view directly below the satellite, and a pole-to-pole strip approximately 57 km (35 miles) wide was viewed on each satellite pass with a 2,461-km (1,529 mi) separation between successive strips at the equator. The telescope focused the collected radiation onto the edge of the multi-toothed chopper wheel that chopped the energy at 333 Hz. After passing through the chopper, the energy was refocused onto the edge of

the circular variable filter at an aperture that acted as both spectrometer slit and a system field stop. The energy was then re-imaged on a lead selenide detector radiatively cooled to 175°K (−145 °F). The incident radiation was sampled 20 times per second, resulting in a spectral intensity plot of 158 points for each passband per revolution. Onboard calibration was accomplished by alternate viewing of the earth and calibration standards by the detector. Spectral plots were analyzed by applying an inversion technique to the radiative transfer equations to obtain the water vapor content. At activation of this experiment on orbit 5, the data output was degraded, exhibiting ice absorption patterns in both channels. On June 8, 1970, the FWS suffered mechanical failure when the drive motor on the chopper wheel failed. No useful data were collected from this experiment.

High Resolution Infrared Radiation Sounder (HIRS). The **Nimbus 6 High Resolution Infrared Radiation Sounder (HIRS)** supported the GARP data test set by providing vertical temperature profiles twice daily on a global basis, extending up to approximately 40 km (25 miles), and information on the water vapor distribution in the troposphere. The HIRS measured radiances primarily in five spectral regions: (1) seven channels near the 15-micrometer CO_2 absorption band, (2) two channels (11.1 and 3.7 micrometers) in the IR window, (3) two channels (8.2 and 6.7 micrometers) in the water vapor absorption band, (4) five channels in the 4.3-micrometer band, and (5) one channel in the visible 0.69-micrometer region. The sounder consisted of a Cassegrain telescope, scanning mirror, dichromatic beam splitter, filter wheel, chopper, and associated electronics. The HIRS scanned the earth's surface in a plane normal to the spacecraft's orbital path with a maximum scan angle of 30 deg to either side of nadir to provide data with a spatial resolution of 25 km (15.5 miles). The instrument was turned off as a precautionary move on May 27, 1976, when the filter chopper motor failed.

High-Resolution Infrared Radiometer (HRIR). The **Nimbus-1, 2, and 3, High-Resolution Infrared Radiometer (HRIR)** was designed (1) to map the earth's nighttime cloud-cover and thus to complement the daytime television (AVCS) coverage and (2) to measure the radiative temperatures of cloud tops and surface terrain. Mounted on the earth-oriented sensory ring, the radiometer measured thermal radiation in the 3.5- to 4.1-micrometer "window" region. The HRIR subsystem consisted of (1) an optical system, (2) an infrared detector (lead selenide photoconductive material), (3) electronics, (4) a magnetic tape recorder, and (5) a filter to minimize attenuation effects of water vapor and carbon dioxide. In contrast to the AVCS camera, no image was formed within the radiometer. The HRIR sensor merely transformed the received radiation into an electrical voltage, which was recorded on the tape recorder for subsequent playback when the satellite came within range of an acquisition station. The radiometer had an instantaneous field of view of about 1.5 deg, which at a nominal spacecraft altitude corresponded to a ground resolution of approximately 8 km at nadir. The radiometer was capable of measuring radiance temperatures from 210 to 330 K. Since the radiometer operated in the 3.5- to 4.1-micrometer region, the daytime pictures include reflected solar radiation in addition to the emitted surface IR radiation. However, the reflected solar radiation did not saturate the instrument, and a usable output was still obtained. In spite of a short operational lifetime (3.5 weeks), the HRIR system successfully demonstrated the feasibility of complete surveillance of surface and cloud features on a global scale during nighttime. With its improved spatial resolution, the radiometer yielded more detailed visual data on the structure of the Intertropical Convergence Zone (ITCZ) and on the formation of tropical storms and frontal systems than had previously been possible.

Image Dissector Camera System (IDCS). The **Nimbus-3 and 4 Image Dissector Camera System (IDCS)** was designed to take daytime cloud-cover photographs. The pictures could be transmitted to APT stations using the real-time transmission system (RTTS) or stored on magnetic tape for subsequent playback to ground acquisition stations. The camera was mounted on the bottom of the satellite sensory ring and pointed vertically down toward the earth at all times. The image dissector was a shutterless electronic scan and step tube mounted behind a wide-angle (108 deg), 5.7-mm focal length lens. Scanning and stepping functions occurred continuously while the satellite progressed along its orbital path. The field of view of the optics was 73.6 deg in the direction of flight and 98.2 deg in a plane normal to the direction of flight. The image was focused by the optics on a photosensitive surface of the image dissector tube. A line-scanning beam scanned the photosensitive surface at 4 Hz with a frame period of 200 s. At the nominal spacecraft altitude of 1,100 km (683.5 miles), each resulting picture was approximately 1,400 km (870 miles) on a side, with a ground resolution of 3 km (1.9 mi) at nadir.

Infrared Interferometer Spectrometer (IRIS). The **Nimbus-3 Infrared Interferometer Spectrometer (IRIS)** experiment was designed to provide information on the vertical structure of the atmosphere and the emissive properties of the earth's surface by measuring the surface and atmospheric radiation in the 5.0- to 20-micrometer band using a modified Michelson interferometer. Incoming radiation was reflected into the instrument from a plane mirror. The radiation was split into two beams that recombined and interfered after reflection on a fixed mirror and a moving Michelson mirror. The recombined beam was then focused on a bolometer detector. Interference effects resulted from the optical path difference between the two beams as the mirror moved. The moving mirror traveled about 2 mm in 11 s to give an interferogram, which was recorded on magnetic tape. The interferograms were transmitted to an acquisition station, where a Fourier transform was performed to produce a thermal emission spectrum of the earth. From these spectra, vertical profiles of temperature, water vapor, and ozone, as well as other parameters of meteorological interest, could be derived. The instrument had a field of view equivalent to a 144-km (89 mi) diameter circle on the surface of the earth at a planned orbital height of 1,100 km (683.5 miles). The experiment was successful, and good data were obtained until the instrument failed on July 22, 1969.

The **Nimbus-4 Infrared Interferometer Spectrometer (IRIS)** experiment was designed to provide information on the vertical structure of the atmosphere and on the emissive properties of the earth's surface by measuring the surface and atmospheric radiation in the 6.25- to 25-micrometer range using a modified Michelson interferometer. Radiation from a cone of the atmosphere, whose base on the surface of the earth was a circle about 94 km (58 miles) in diameter for a nominal satellite altitude of approximately 1,100 km (683.5 miles), was received and reflected by a mirror. The reflected radiation was split into two approximately equal beams by a beamsplitter. After reflection on a fixed and moving mirror, respectively, the two beams interfered with each other with a phase difference proportional to the optical path difference between both beams. The moving mirror traveled about 3.6 mm in 13 s to give an output signal from the bolometer. This signal, an interferogram, was recorded on tape. The interferograms were transmitted to a ground receiving station, where a Fourier transform was performed to produce a thermal emission spectrum of the earth. From these spectra, vertical profiles of temperature, water vapor, and ozone were derived, as well as other parameters of meteorological interest. The instrument had a field of view of 5 deg and a spectral resolution of less than 0.4 micrometer (nominally 1.4 reciprocal centimeters). The IRIS experiment was successful in spite of a transmission conflict with the Real-Time Transmission System (RTTS) that resulted in some periods of lost data after November 28, 1970. The IRIS experiment was turned off on January 25, 1972 to conserve spacecraft power.

Infrared Temperature Profile Radiometer (ITPR). The **Nimbus-5 Infrared Temperature Profile Radiometer (ITPR)** experiment was designed to measure the three-dimensional temperature field in the earth's atmosphere with a spatial resolution of 32 km (20 miles). The radiometer sensed four intervals in the 15-micrometer CO_2 band, one interval in the water vapor rotational band near 20 micrometers and two spectral intervals in the atmospheric window regions near 3.7 and 11 micrometers. The ITPR viewed the earth successively at various angles distributed symmetrically about nadir in a plane normal to the orbital track. Forty-two geographically independent scan spots were taken along a single strip. As the satellite progressed along its orbital path, the radiometer observed 10 such 42-spot strips to form a matrix of independent scan spots. Each matrix was produced in 222 s with the whole scanning sequence repeated every 240 s. The matrix data were recorded on magnetic tape for subsequent playback to a ground acquisition station. Matrix measurements taken in the CO_2 and water vapor absorption bands were used to calculate temperature profiles and total water vapor content in the troposphere and lower stratosphere. The two window measurements helped to detect and eliminate cloud contamination of the radiances, thus permitting actual determination of profiles down to the earth's surface in all but completely overcast areas. Because of the erratic behavior of the scan mechanism which developed shortly after launch, the instrument operated only in the nadir mode except for brief periods.

Interrogation, Recording, and Location System (IRLS). The **Nimbus-4 interrogation, recording, and location system (IRLS)** experiment was designed to collect and retransmit meteorological, geophysical, and other experimental data from remote unmanned data collection stations (platforms) deployed on a global scale. The IRLS could also determine the location and track the movement of such platforms as balloons, ocean buoys, and ships to within an accuracy of 2 km (1.2 miles). The IRLS consisted of (1) a 466-MHz receiver, (2) a 401.5-MHz transmitter, (3) decoding and coding circuits, (4) a range detector, and (5) a 100-kb satellite data memory capable of storing data obtained during each orbit for up to 370 different interrogations. On each orbit pass, when the satellite was within range of an acquisition and command station, the satellite command memory was programmed to communicate with selected platforms during the coming orbit. The satellite stored both the address (number) of each platform and the desired time that each should be contacted. At the appropriate time in orbit, the satellite interrogated each platform, measured the satellite to platform distance by determining the round trip propagation time of the rf signal, received the analog data from the platform, converted it to digital form, and stored it. Upon return to the locale of the ground station, the station commanded the satellite to transmit the stored data and to accept new commands for the next orbit.

Limb Infrared Monitor of the Stratosphere (LIMS). The objective of the Limb Infrared Monitor of the Stratosphere (LIMS) experiment on Nimbus-7 was to map the vertical profiles of temperature and the concentration of ozone, water vapor, nitrogen dioxide, and nitric acid in the lower to middle stratosphere range, with extension to the stratopause for water vapor and into the lower mesosphere for temperature and ozone. This experiment was a follow-on to the Limb Radiance Inversion Radiometer (LRIR) flown on Nimbus 6. The instrument had a six-channel infrared (IR) radiometer that incorporated Hg-Cd-Te detectors cooled by a two-stage solid cryogen cooler. The LIMS instrument consisted of two electronic boxes and the radiometer unit. The radiometer unit consisted of the solid cryogen package (SCP) and the optical-mechanical package (OMP). The OMP contained the optics for the folded telescope, while the SCP contained the detector assembly and cryogen. Radiance from the earth's limb entered the OMP aperature, reflected off the scan mirror to the 18-cm (7 in) diameter off-axis parabolic primary mirror where the radiation was focused and chopped at 945 Hz. The radiation was re-collimated by the secondary mirror and directed through a Lyot stop to a folding mirror and into the detector capsule assembly (DCA). The radiation was then focused through a cadmium telluride lens and through interference filters, which defined the FOVs, and onto an array of discrete mercury cadmium telluride detectors. The detectors were maintained at about 63 K temperature by the cryogen. The LIMS began a scan near 153 km (95 miles) altitude, taking about 12 sec to move near 38 km (24 miles) below the solid limb, then retraced its motion upward. After every second scan pair, the scan mirror was placed in a position to observe radiation from a small cavity blackbody for inflight warm calibration after which the instrument viewed space to obtain a cold calibration point. Calibration data was included in the LIMS output data stream. The LIMS mapped vertical profiles of thermal IR emission coming from the horizon in six bands (6.2, 6.3, 9.6, 11.3, and two at 15 micrometers) of CO_2N, CO_2, O_3, HNO_3, H_2O, and NO_2. Two of the channels were used to determine radiance profiles of emission by CO_2. The profiles were mathematically inverted to obtain temperature versus pressure. The infrared temperature profile, together with radiance profiles in the other spectral bands, were then used to infer the vertical distribution of trace constituents. The temperature was determined to an accuracy of about $1.5\,°K$ ($-457\,°F$). Constituent concentrations were determined with an accuracy of about 20%, with the exception of NO_2 which was determined to within about 50%. Instantaneous vertical field of view at the horizon was 2 km (1.2 mi) for the temperature, ozone, and nitric acid channels, and 4 km (2.5 mi) for the NO_2 and water vapor channels.

Limb Radiance Inversion Radiometer (LRIR). The **Nimbus-6 Limb Radiance Inversion Radiometer (LRIR)** provided calibrated radiance versus altitude profiles by intercepting radiation emanating from an atmospheric path which is tangential to a particular geocentric height. The LRIR sensed radiation in four spectral intervals: (1) the 14.6- to 15.9-micrometer CO_2 band, (2) the 14.2- to 17.3-micrometer CO_2 band, (3) the

8.8- to 10.1-micrometer ozone band, and (4) the 20- to 25-micrometer water vapor rotational band. Measurements taken in the two CO_2 channels and the water vapor channel were used to calculate global temperature and water vapor profiles in the stratosphere and lower mesosphere. In addition, values of the geostrophic wind up to 1 mb (approximately 48 km (30 miles)) were derived analytically from the deduced temperature profiles. The radiometer included an optical system, a scanning mirror, choppers, and associated electronics and employed an ammonia-methane cooler system for three of the four detector channels. While the deduced temperature profiles had an rms accuracy of 3 deg at heights above 15 km (9 miles), the values for ozone were accurate to within 20% at 1 mb. Water vapor values at the same height were within 50%. The instrument functioned successfully until January 7, 1976, when the detector temperature began to rise rapidly, and the instrument was turned off.

Medium-Resolution Infrared Radiometer (MRIR). The *Nimbus-2 and 3 Medium-Resolution Infrared Radiometer (MRIR)* experiment measured the intensity and distribution of electromagnetic radiation emitted by and reflected from the earth and its atmosphere in five selected wavelength intervals from 0.2 to 30 micrometers. Data for heat balance of the earth-atmosphere system were obtained, as well as measurements of water vapor distribution, surface or near-surface temperatures, and seasonal changes of stratospheric temperature distribution. The five wavelength regions were (1) the 6.4- to 6.9-micrometer channel, which covered the 6.7-micrometer water vapor absorption band, (2) the 10- to 11-micrometer band, which operated in the "atmospheric window," (3) the 14- to 16-micrometer band, which covered the 15-micrometer carbon dioxide absorption band, (4) the 5- to 30-micrometer band, which measured the emitted long-wavelength infrared energy for heat budget purposes, and (5) the 0.2- to 4.0-micrometer channel, which yielded information on the intensity of reflected solar energy (albedo). Radiant energy from the earth was collected by a flat scanning mirror inclined at 45 deg to the optical axis. The mirror rotated at 8 rpm and scanned in a plane perpendicular to the direction of motion of the satellite. Each of the four channels contained a 4.33-cm-diameter folded telescope with a 2.8-deg field of view and a thermistor-bolometer. The collected energy was modulated by a mechanical chopper to produce an ac signal. The signal was then amplified and recorded on magnetic tape for subsequent playback to a ground acquisition station. At a satellite altitude of 1,100 km (683.5 miles), a horizontal resolution of 55 km (34 miles) could be obtained.

Monitor of Ultraviolet Solar Energy (MUSE) (Solar UV Monitor. The Nimbus-3 Monitor of Ultraviolet Solar Energy (MUSE) experiment was designed (1) to look for temporal variations in the solar UV flux in five broad bands in the interval 1,150 to 3,000 Å, (2) to measure the solar flux in these regions, and (3) to measure the atmospheric attenuation at these wavelengths as the sensors on board viewed the setting sun after the spacecraft had crossed the terminator in the Northern Hemisphere. The sensors had their maximum response at 1,216 Å, 1,600, 1,800, 2,000, and 2,600 Å. The MUSE instrumentation, which consisted of five vacuum photodiodes housed in an electronics package and a sensor package, was mounted in the rear of the Nimbus spacecraft. All sensors except the 1.216-Å sensor had semi-transparent photocathodes that were deposited on the windows. The 1,800-, 2,000-, and 2,600-Å sensors had aluminum oxide windows, while the 1,216-Å and 1,600-Å sensors had MgF2 and CaF2 windows, respectively. The five spectral regions were determined by the transmittance of the filter or window materials on the short wavelength side, while the long wavelength cutoffs were produced by the varying degrees of opacity of the different photocathode materials. The appropriate bands of UV flux entered the photodiodes and produced a current that was measured by an electrometer and digitized by the Nimbus pulse code modulation (PCM) system. Simultaneously, the solar aspect system measured the angle of incidence of the solar rays and transmitted its digital information to the PCM system. The PCM data were stored on magnetic tape and transmitted on playback to the data acquisition facility. The instrument had a basic 48-s cycle and a one sample per second data rate. The field of view of the sensors was about 90 deg. Solar acquisition began, therefore, at 45 deg prior to the earth day/night terminator and ceased completely at the satellite day/night transition. The instrument had only an inflight electrical calibration sequence because there were no known suitable UV sources that could provide an inflight optical calibration. A similar experiment was flown on Nimbus-4.

Nimbus E Microwave Spectrometer (NEMS). The *Nimbus E (Nimbus-5) Microwave Spectrometer (NEMS)* was designed primarily to demonstrate the capabilities and limitations of microwave sensors for measuring tropospheric temperature profiles, water vapor abundances, cloud liquid water content, and earth surface temperatures. The NEMS could continuously monitor emitted microwave radiation at frequencies of 22.235, 31.4, 53.65, 54.9 and 58.8 GHz. The three channels near the 5-mm oxygen absorption band were used primarily to determine the atmospheric temperature profiles. NEMS provided measurements even in cloud-cover conditions that normally restrict the usefulness of conventional IR data in such situations. The two water vapor channels near 10 mm permitted the water vapor and cloud liquid water content over oceans to be estimated and also to yield an estimated temperature once the surface emissivity had been calibrated by comparison with direct measurements. The three oxygen channels shared a common signal and reference antenna. Both water vapor channels had their own signal and reference antennas. From an average satellite height of 1,100 km (683.5 miles), the NEMS viewed a 180-km (112 mi) diameter circle on the earth's surface. NEMS data were recorded on magnetic tape for subsequent playback to a ground acquisition station. An advancement of this instrument, the Scanning Microwave Spectrometer (SCAMS), was flown on Nimbus-6 later.

Pressure Modulated Radiometer (PMR). The *Nimbus-6 Pressure Modulator Radiometer (PMR)* experiment took radiometric measurements in the 15-micrometer CO_2 band at altitudes between 45 and 70 km (28 and 43.5 miles) on a global scale. By appropriate mathematical retrieval methods, the temperature structures of the upper stratosphere and lower mesosphere were deduced. The pressure-modulation technique permitted the extension of selective chopping techniques to higher altitudes where the pressure-broadened emission lines in the 15-micrometer CO_2 band became so narrow that conventional spectrometers and interferometers had insufficient spectral resolution. In addition to pressure scanning (in discrete steps), the radiometer also employed Doppler scanning along the direction of flight. The PMR comprised two similar radiometer channels, each consisting of a plane scanning mirror, reference blackbody, pressure-modulator cell, and detector assembly. The plane mirror was gold coated and mounted at 45 deg on a 90-deg stepping motor so that the field of view of the channel could be directed to space or to the internal reference blackbody for inflight range and zero calibration. The motor was mounted on a pair of flexible pivots so that the mirror could be rotated through plus or minus 7-1/2 deg from its rest position to give the required Doppler scan. Major components in the pressure-modulator cell were a movable piston, a diaphragm, and a magnetic drive coil. The detector assembly consisted of a field lens, a condensing light pipe, and a pyroelectric flake bolometer. Each radiometer had a field of view that was 20 deg whole-angle across the spacecraft's line of flight and 40 deg whole-angle parallel to the line of flight. The derived temperature values were within 2 deg K at 65 km and about 0.2 deg K near 50 km (31 miles) with a vertical resolution of 10 km (6 miles).

Satellite Infrared Spectrometer (SIRS). The *Nimbus-3 Satellite Infrared Spectrometer (SIRS)* experiment was designed to indirectly determine the vertical temperature profiles of the atmosphere by measuring the infrared radiation emitted from the earth and its atmosphere in seven spectral intervals in the carbon dioxide band (11 to 15 micrometers) and one interval in the atmospheric window centered at 11.1 micrometers. The main components of the Fastie-Ebert fixed-grating spectrometer consisted of (1) a plane, light-collecting mirror to provide a single earth-viewing beam fixed in the vertical, (2) a rotating chopper mirror, (3) a spherical mirror, (4) a 12.7-cm (5-in) diffraction grating with 1250 lines per inch, (5) a set of eight exit slits with a single interference filter, (6) eight wedge-immersed thermistor bolometers, (7) a blackbody radiation source for calibration, and (8) eight preamplifiers and eight operational amplifiers. The incoming radiation was chopped, spectrally dispersed by the diffraction grating, focused on the exit slits as a spectrum by the spherical mirror, and converted to electrical signals. The signals were then amplified and stored on magnetic tape for subsequent playback to a ground acquisition station. The instrument field of view was 11.5 by 11.5 deg centered on nadir. This provided data over an area roughly 220 km (137 miles) on a side at a satellite height of 1,100 km (683.5 miles). Data from the 11.1-micrometer channel yielded surface and/or cloudtop temperatures. Data from the carbon dioxide band could be used to generate temperature-pressure profiles by a mathematical inversion technique. The resulting temperatures had rms errors slightly less than $1\,°C$ ($33.8\,°F$). The SIRS experiment was successful and good data were obtained. On June 21, 1970, the experiment was turned off and all data acquisition effort was transferred to the SIRS experiment on Nimbus-4.

The *Nimbus-4 Satellite Infrared Spectrometer (SIRS)* experiment was designed to determine the vertical temperature and water vapor profiles of the atmosphere by using a Fastie-Ebert fixed-grating spectrometer. The instrument measured the infrared radiation (11 to 36 micrometers) emitted from the earth and its atmosphere in 13 selected spectral intervals in the carbon dioxide and water vapor bands plus one channel in the 11-micrometer atmospheric window. The main components of the spectrometer consisted of (1) a plane, light-collecting mirror to provide one fixed and two variable earth-viewing angles, (2) a rotating chopping mirror that served alternately to collect space radiation and earth radiation, (3) a 2.5-in. diffraction grating with 1250 lines per inch, (4) 14 slits with associated interference filters, (5) 14 thermistor bolometers, and (6) a blackbody source for calibration purposes. The SIRS used a scan mirror to observe 12.5 deg to either side of the subsatellite track. The field of view directly below the SIRS was approximately 215 sq km (134 sq miles). The carbon dioxide band radiation data were transformed to a temperature profile by a mathematical inversion technique. By a similar technique, this information could then be combined with the water vapor band data to obtain a water vapor profile. The 11-micrometer atmospheric window data yielded surface and/or cloudtop temperatures. The SIRS experiment performed normally for several months after launch but began to deteriorate in early 1971. Problems in the SIRS instrument calibration after April 1971, in addition to spacecraft yaw problems, significantly reduced the number of useful soundings obtained. The archival data were produced through April 8, 1971. The experiment operated on a limited time basis until March 6, 1973, when it was placed operationally off.

Scanning-Microwave Spectrometer (SCAMS). The *Nimbus-6 Scanning Microwave Spectrometer (SCAMS)* was designed to map tropospheric temperature profiles, water vapor abundance, and cloud water content to be used for weather prediction even in the presence of clouds, which block conventional satellite infrared sensors. The instrument was an advancement of the Nimbus E microwave spectrometer (NEMS) on Nimbus-5. The SCAMS continuously monitored emitted microwave radiation at frequencies of 22.235, 31.65, 52.85, 53.85 and 55.45 GHz. The three channels near the 5.0-mm oxygen absorption band were used primarily to deduce atmospheric temperature profiles. The two channels near 10 mm permitted water vapor and cloud water content over calm oceans to be estimated separately. The instrument, a Dicke-superheterodyne type, scanned plus or minus 45 deg normal to the orbital plane with a 10-deg field of view. The three oxygen channels shared common signal and reference antennas. Both water vapor channels had their own signals and reference antennas. The absolute rms accuracy of the oxygen channels was better than 2 deg K and that of the water vapor channels better than 1 deg K. The dynamic range for all channels was 0-400 deg K. The ground resolution was approximately 145 km (90 miles) near nadir and 330 km (205 miles) at the scan limit. The instrument ceased functioning on May 31, 1976, due to jamming of the scan mechanism.

Scanning-Multispectral Microwave Radiometer (SMMR). The primary purpose of the Scanning Multichannel Microwave Radiometer (SMMR) on Nimbus-7 was to obtain sea surface temperature and near-surface winds under all-weather conditions for developing and testing global ocean circulation models and other aspects of ocean dynamics. Winds, water vapor, liquid-water content, mean cloud droplet size, rainfall rate and sea ice parameters were also determined. Microwave brightness temperatures were observed with a 10-channel (five-frequency dual polarized) scanning radiometer operating at frequencies of 37, 21, 18, 10.69, and 6.6 GHz. Six Dicke-type radiometers were utilized. Those operating at the four longest wavelengths measured alternate polarizations during successive scans of the antenna; the others operated continuously for each polarization. The antenna was a parabolic reflector offset from the nadir by 42 deg. Motion of the antenna reflector provided observations from within a conical volume along the ground track of the spacecraft. The same instrument was flown on SEASAT 1.

Selective Chopper Radiometer (SCR). The *Nimbus-4 Selective Chopper Radiometer (SCR)* observed the emitted infrared radiation in the 15-micrometer absorption band of carbon dioxide. From these measurements the temperatures of six successive 10-km layers of the atmosphere were determined from earth or cloudtop level to 60-km height. Height resolution was obtained by a combination of optical multi-layer filters and selective absorption of radiation using carbon-dioxide-filled cells within the experiment. The SCR had six channels, which were arranged in three units of two. The four lower channels were called single-cell channels. The optics of each channel consisted of a cantilever-mounted blade shutter that oscillated at 10 Hz and successively chopped the field of view (FOV) between earth and space. The chopped radiation was then passed through a 10-cm path length of carbon dioxide, the pressure being set for each channel to define the viewing depth of the atmosphere. Behind the carbon dioxide path was a narrow-band filter, the centers of which were different for each channel, and a light pipe which focused the radiation on a thermistor-bolometer detector. To obtain adequate height resolution in the upper layers of the atmosphere, the upper two channels operated on a slightly different principle and were known as double-cell channels. The technique consisted of switching the radiation between two half-cells, which were semicircular in shape and of 1-cm path length, and which contained different pressures of carbon dioxide. A movable 45-deg mirror replaced the oscillating shutter used in the lower four channels. During one half-period, earth radiation passed through one half-cell and space radiation through the other. The situation was reversed during the other half-period. The radiation then passed through a light pipe onto a thermistor-bolometer detector. Inflight calibration was carried out by viewing of an internal reference blackbody of known temperature prior to the view of space. The output of each channel was sampled once every second. The upper two channels had a circular FOV approximately 160 km (99 miles) in diameter, and the lower four had a rectangular FOV about 112 km (70 miles) square.

The *Nimbus-5 Selective Chopper Radiometer (SCR)* was designed to (1) observe the global temperature structure of the atmosphere up to 50 km in altitude, (2) make supporting observations of water vapor distribution, and (3) determine the density of ice particles in cirrus clouds. To accomplish these objectives, the SCR measured emitted radiation in 16 spectral intervals separated into the following four groups: (1) four CO_2 channels between 13.8 and 14.8 micrometers, (2) four channels at 15.0 micrometers, (3) an IR window channel at 11.1 micrometers, a water vapor channel at 18.6 micrometers, two channels at 49.5 and 133.3 micrometers, and (4) four channels at 2.08, 2.59, 2.65, and 3.5 micrometers. From an average satellite altitude of 1,100 km (683.5 miles), the radiometer viewed a 48-km (30-mile) circle on the earth's surface with a ground resolution of about 25 km (15.5 miles). A similar experiment was flown on Nimbus 4.

Solar Backscatter Ultraviolet/Total Ozone Mapping Spectrometer (SBUV/TOMS). The objectives of the Solar Backscatter Ultraviolet and Total Ozone Mapping Spectrometer (SBUV/TOMS) on Nimbus-7 were to determine the vertical distribution of ozone, map the total ozone content, and monitor the incident solar ultraviolet (UV) irradiance and ultraviolet radiation backscattered from the earth. The SBUV consisted of a double Ebert-Fastie spectrometer and a filter photometer similar to the BUV on Nimbus-4. The SBUV spectrometer measured solar UV backscattered by the earth's atmosphere at 12 wavelengths between 0.25 and 0.34 micrometer, with a spectral bandpass of 0.001 micrometer. The SBUV used three detectors: a photomultiplier tube (PMT) and a photodiode for the monochromator, and one photodiode for the photometer. Both the monochromator and the photometer have chopper wheels operating at 25 Hz. The SBUV used a depolarizer to eliminate the sensitivity of the grating monochromator to polarization of the backscattered radiation. The instrument's field of view (FOV) at nadir was 0.20 rad. A roughened aluminum diffuser plate viewed the sun for solar-spectral irradiance measurements and for calibration by viewing a mercury-argon lamp. The diffuser plate was driven by a stepper motor to three postions on command: SBUV, TOMS, and STOW. The contribution functions for the eight shortest wavelengths were centered at levels ranging from 55 to 28 km (34 to 17 miles) and were used to infer the vertical ozone profile. The four longest wavelengths had contribution functions in the troposphere which were used to compute the total ozone amount. The SBUV spectrometer had a second mode of operation that allowed a continuous solar-spectral scan from 0.16 to 0.4 micrometer for detailed examination of the extraterrestrial solar spectrum and its temporal variations. A parallel photometer channel at 0.343 micrometer measured the reflectivity of the atmosphere's lower boundary in the same 0.21-rad FOV. The TOMS was a single Ebert-Fastie spectrometer with a fixed grating and an array of exit slits. The TOMS step-scanned across the orbital track 51 deg from the nadir in 3-deg steps with an FOV of approximately 0.052 rad. At each scan position, the earth radiance was monitored at six wavelengths between 0.31 and 0.38 micrometer to infer the total ozone amount. The TOMS completed a cross scan in eight seconds, with one second for retrace, to record 35 scenes per scan. At each scene, a chopper sequentially sampled all six wavelengths four times. The TOMS used the same type of PMT as SBUV, and had a separate mercury-argon lamp for wavelength calibration and a separate depolarizer. The TOMS shared the diffuser plate with SBUV. Both SBUV and TOMS had five scanner modes and a shared electronics module.

Stratospheric Aerosal Measurement II (SAM-II). The objective of the Stratospheric Aerosal Measurement (SAM II) experiment on Nimbus-7 was to provide vertical distribution of stratospheric aerosols in the polar regions of both hemispheres. When no clouds were present in the instantaneous field of view (IFOV), the tropospheric aerosols could also be mapped. The instrument, basically a sun photometer, measured the extinction of solar radiation at 1.0 micrometer wavelength during spacecraft sunrise and sunset. The SAM II instrument package consisted of optics and electronics subassemblies. The optical assembly consisted of gimbals, a flat entrance window (which filters out UV radiation), Cassegrain optics, a flat scanning mirror, Sun acquistion sensors, and a sun-photometer detector package. Solar radiation was reflected from the scan mirror into the Cassegrain telescope forming a solar image at the slit plate, which contained two solar edge sensors for monitoring solar limb crossings on either side of the detector aperature. Solar radiation passed through the aperature, was collected by a field lens, passed through an interference filter for wavelength discrimination, and finally measured by a silicon photodiode detector. The optics assembly was gimbaled in azimuth. After acquisition in azimuth, the mirror servo scanned in elevation until the Sun was acquired. The Sun was then scanned back and forth. The photometer viewed a portion of the solar disk with a 0.145 mrad IFOV and a sampling rate of 50 samples per second. As the spacecraft first viewed the sunrise, the photometer-pointing axis was depressed approximately 0.52 rad with respect to the spacecraft horizontal. The photometer continued looking at the sun until its depression angle was on the order of 0.44 rad (approximately 1.4 minutes observing time). Before sunset, the photometer head rotated 3.14 rad in azimuth and viewed the sun from a depression of approximately 0.44–0.52 rad as the spacecraft orbited to the dark side of the earth. The extinction measurements were inverted for the number density times the aerosol scattering cross section by using the Lambert-Beer Law and assuming the atmosphere to be composed of layers. To determine the stratospheric aerosol optical properties, ground-truth and in situ balloon-borne aerosol measurements were also made.

Stratospheric and Mesospheric Sounder (SAMS). The objective of the Stratospheric and Mesospheric Sounder (SAMS), a pressure broadening spectral radiometer, was to observe emission from the limb of the atmosphere through seven pressure-modulated cells and six detectors in order to determine temperature and vertical concentrations of H_2O, N_2O, CH_4, CO, and NO in the stratosphere and mesosphere. Measurements of zonal wind in this region were attempted by observing the Doppler shift of atmospheric emission lines. Radiation from the limb of the atmosphere was incident on a scan mirror in front of a 15-cm (6-in) aperture telescope. The scan mirror scanned the limb, viewed space for calibration, and viewed the atmosphere obliquely to obtain vertical profiles. There were three adjacent fields of view, each 28 by 2.8 mrad (corresponding to 100 km by 10 km at the limb). The FOVs were focused onto a field-splitting mirror by the telescope which directed radiation to six detectors. Seperation into channels was accomplished through dichroic beam splitters. There were seven pressure modulator cells (PMC), two containing CO_2, the remainder N_2O, NO, CH_4, CO, H_2O. Pressure in the cells could be varied on command by changing the temperature of a small container of molecular sieve material attached to each PMC. The spectral parameters for the H_2O channel were 2.7 micrometers and 25 to 100 micrometers. All other channels lay within the range 4.1 to 15 micrometers. A chopper operating at 250 Hz within the telescope, allowed the measurement of two separate signals from all

detectors, one at 250 Hz and one at the PMC frequency. Comparison of these signals permitted the elimination of emission from interfering gases within a particular spectral interval. In front of the chopper, a small black body at known temperature was introduced for calibration. Accurate measurement of the atmospheric pressure at the level being viewed was obtained from the two signals from one CO_2 channel.

Surface Composition Mapping Radiometer (SCMR). The Surface Composition Mapping Radiometer (SCMR) on Nimbus-5 measured (1) terrestrial radiation in the 8.3- to 9.3-micrometer and 10.2- to 11.2-micrometer intervals and (2) reflected solar radiation in the 0.8- to 1.1-micrometer range. Surface composition and sea surface temperatures could be obtained from these measurements. The SCMR had an instantaneous field of view (FOV) of 0.6 mrad, equivalent to a ground resolution of 660 m at nadir. The scan mirror rotated at 10 rps to provide scan lines 800-km (497-miles) wide across the spacecraft track. The instrument began malfunctioning soon after launch. The last usable data were transmitted on January 4, 1973. A modified instrument, heat capacity mapping radiometer, was flown on the Heat Capacity Mapping Mission (HCMM) later.

Temperature-Humidity Infrared Radiometer (THIR). The Nimbus-4 Temperature-Humidity Infrared Radiometer (THIR) was designed to detect emitted thermal radiation in both the 10.5- to 12.5-micrometer region (IR window) and the 6.5- to 7.0-micrometer region (water vapor). The window channel measured cloudtop temperatures day and night. The other channel operated primarily at night to map the water vapor distribution in the upper troposphere and stratosphere. The instrument consisted of a 12.7-cm Cassegrain system, a scanning mirror common to both channels, a beam splitter, filters, and two germanium-immersed thermistor bolometers. In contrast to TV, no image was formed within the radiometer. Incoming radiant energy was collected by a flat scanning mirror inclined at 45 deg to the optical axis. The mirror rotated through 360 deg at 48 rpm and scanned in a plane normal to the spacecraft velocity vector. The energy was then focused into a dichromatic beam splitter, which divided the energy spectrally and spatially into two channels. Both channels of the THIR sensor transformed the received radiation into an electrical (voltage) output with an information bandwidth of 0.5 to 360 Hz for the 10.5 to 12.5 micrometer channel and 0.5 to 120 Hz for the water vapor channel. The THIR sensor data were normally recorded on tape for subsequent playback to a ground acquisition station. However, direct readout infrared radiometer (DRIR) data could be transmitted to APT ground stations for both day and night portions of the orbit using the Nimbus 4 real-time transmission system (RTTS). At a nominal spacecraft altitude, the window channel had a ground resolution of about 7 km (4.4 miles) and the water vapor channel about 22 km (14 miles) at nadir. The THIR was initially successful but failed on January 11, 1971 (orbit 3,731). It was restarted several times thereafter for very short periods of time before it finally ceased all operations in August 1971. A similar experiment was flown on Nimbus-5, 6 and 7.

Except for data being digitized on board, the Nimbus-7 THIR was of the same design and operation as the THIR flown on Nimbus-4, 5, and 6. The instrument was turned off in 1985 to conserve power.

Tracking and Data Relay (T+DRE). This experiment provided the Nimbus portion of a communication link from Nimbus-6 to ATS to a ground station. The purpose of the experiment was to gain information on the use of such a link for range and rate communications (for satellite geodetic purposes) and for data communication from a low-orbiting spacecraft through a synchronous spacecraft to a ground telemetry station. The instrumentation included an S-band transponder, a command detector/decoder, an antenna programmer, a digital evaluation module, an S-band antenna, and an antenna gimbal assembly. Initial experiment operation was nominal.

Tropical Wind Energy Conversion and Reference Level (TWERLE). The goals of the Nimbus 6 Tropical Wind Energy Conversion And Reference Level Experiment (TWERLE) were closely associated with the objectives of the Global Atmospheric Research Program (GARP) and included (1) measuring upper atmospheric winds in the tropics, (2) studying the relative air motion along isobaric surfaces to determine the rate of conversion of atmospheric potential energy into kinetic energy, and (3) providing direct measurements of various meteorological parameters that served as reference points in adjusting indirect temperature soundings made from satellites. The experiment consisted of two basic components:

(1) approximately 300 constant-level meteorological balloons to yield measurements of winds, temperature, and pressure in the tropics and at southern hemisphere midlatitudes at 150 mb (about 13.6-km altitude), and (2) the Nimbus-6 random access measurements system (RAMS) to provide data collection and location determinations from the balloons. The 3.5-m-diameter (11.5 ft diameter) polyester-mylar balloons were equipped with a transmitter-oscillator, solar power supply, digitizer/modulator, and sensors. The sensors consisted of a radio altimeter having an accuracy of better than plus or minus 20 m, a bead thermistor monitoring the ambient air temperature to an accuracy of 0.5 deg C, and a pressure sensor measuring the 150-mb flight altitude to an accuracy of 0.5 mb. A magnetic cutdown device was used to eliminate any accidental overflights into regions of the Northern Hemisphere north of 20 deg N latitude. The RAMS merely detected each balloon signal (401.2 MHz) and extracted the carrier frequency, balloon identification, and sensor data. This information, along with time references, was stored in digital form for subsequent relay to a ground acquisition station. The balloon's position and velocity were derived from the relative motion between the platform and the satellite by measuring Doppler shifts in the carrier signal received from the balloon. TWERLE was capable of a location accuracy of 5 km (3 miles) and a platform velocity accuracy of 1 m/s. In addition to the TWERLE balloon experiment, many other experiments used RAMS. These experiments used ocean buoys to measure oceanographic and atmospheric parameters.

Web References

Goddard Space Flight Center, Nimbus-7 Images: http://svs.gsfc.nasa.gov/search/Instrument/Nimbus-7.html

Goddard Space Flight Center, TOMS Ozone: http://svs.gsfc.nasa.gov/search/Series/TOMSOzone.html

Goddard Space Flight Center, TOMS Aerosols: http://svs.gsfc.nasa.gov/search/Series/TOMSAerosols.html

REBECCA LINDSEY, NASA's Goddard Space Flight Center

NIOBIC ACID. Any hydrated form of Nb_2O_5. It forms as a white, insoluble precipitate when a potassium hydrogen sulfate fusion of a niobium compound is leached with hot water or when niobium fluoride solutions are treated with ammonium hydroxide. Soluble in concentrated sulfuric acid, concentrated hydrochloric acid, hydrogen fluoride, and bases. Important in analytical determination of niobium. See also **Niobium**.

NIOBIUM. [CAS: 7440-03-1]. Chemical element, symbol Nb, at. no. 41, at. wt. 92.906, periodic table group 5, mp 2,458–2,468 °C, bp 4,742 °C, density 8.6g/cm^3 (20 °C). Elemental niobium has a body-centered cubic crystal structure. The metal has a slightly bluish tinge, is ductile and malleable, and when polished resembles platinum. The metal burns upon being heated in air. There is one natural isotope ^{93}Nb. Seven radioactive isotopes have been identified ^{90}Nb through ^{92}Nb and ^{94}Nb through ^{97}Nb, with a wide range of half-lives. ^{94}Nb has the longest half-life (2×10^4 years). The element was first identified by C. Hatchett in 1801 and was originally called columbium which name persisted for many years. The name still appears widely in the literature, particularly in connection with alloys bearing the element, such as columbium steels.

First ionization potential 6.77 eV; second 13.895 eV; third 24.2 eV. Oxidation potential $Nb \rightarrow Nb^{3+} + 3e^-$, ca. 1.1 V; $2Nb + 5H_2O \rightarrow Nb_2O_5 + 10H^+ + 10e^-$, 0.62 V.

Other important physical properties of niobium are given in Table 1 and under **Chemical Elements**.

Niobium occurs, usually with tantalum, in columbite $Fe(NbO_3)_2$, (80% Nb_2O_5), pyrochlore (50% Nb_2O_5), samarskite (50% Nb_2O_5), chiefly found in western Australia, and South Dakota. Recovered along with tantalum by fusion with potassium bisulfate, and obtained in the residue after subsequent extraction with H_2O. Niobium and tantalum are separated by fractional crystallization of the potassium fluorides, niobium concentrating in the mother liquid and tantalum in the crystals.

TABLE 1. REPRESENTATIVE PROPERTIES OF REFRACTORY METALS

Property	Tungsten	Tantalum	Molybdenum	Niobium
Density, g/cm^3	19.3	16.6	10.2	8.7
Melting point, °C	3,390–3,420	2,996	2,617	2,458–2,468
Boiling point, °C	5,660	5,325–5,525	4,612	4,742
Linear coefficient of expansion per °C	4.3×10^{-6}	6.5×10^{-6}	4.9×10^{-6}	7.2×10^{-6}
Thermal conductivity, 20 °C (cal/cm^2/cm/ °C/s)	0.40	0.13	0.35	0.13
Specific heat, 20 °C (cal/g/ °C)	0.032	0.036	0.061	0.065
Working temperature, °C	1,700	ambient	1,600	ambient
Electrical conductivity	31	13	30	12
Nuclear cross section (thermal neutrons, Barns/atom)	19.2	21.3	2.4	1.1
Tensile strength, 1000 psi				
20 °C	100–500	100–150	120–200	75–150
500 °C	175–200	35–45	35–65	35
1000 °C	50–75	15–20	20–30	13–17
Young's Modulus of Elasticity, psi				
20 °C	59×10^6	27×10^6	46×10^6	14×10^6
500 °C	55×10^6	25×10^6	41×10^6	7×10^6
1000 °C	50×106	22×106	39×10^6	—
Poisson's Ratio	0.284	0.35	0.32	0.38
Corrosion resistance, 100 °C				
Dilute HNO$_3$	See Tungsten	N	R	N
Dilute H$_2$SO$_4$	See Tungsten	N	S	VS
Concentrated H$_2$SO$_4$	See Tungsten	N	S	R
Dilute HCl	See Tungsten	N	S	—
Concentrated HCl	See Tungsten	N	SL	SL
Concentrated Hydrofluoric acid		R	SL	R
Phosphoric acid, 85	N	SL	VS	
Concentrated NaOH		R	N	R

N = no appreciable corrosion.
VS = < 0.0005 inch (0.013 millimeter) per year.
SL = 0.0005–0.005 inch (0.013–0.13 millimeter) per year.
S = 0.005–.01 inch (0.13 0.25 millimeter) per year.
R = >0.01 inch (0.25 millimeter) per year.

The principal uses for the element are in alloys. Niobium also has gained prominence in research as a superconducting material. At the temperatures of liquid helium, niobium becomes a superconductor and, in the form of a fine wire, has been incorporated in a superconducting cell. The element has both size and cost advantages over electronic materials. The alloy Nb_3Sn becomes superconducting at a somewhat higher temperature. Niobium-titanium and niobium-zirconium alloys also have potential as superconductors.

Alloys. Niobium is used in steel, notably stainless steels, to stabilize the carbon present (as carbide) and for preparing niobium carbide, used for dies and cutting tools. Ferroniobium is a strong carbide-forming material and, when added to 18-8 stainless steel, stabilizes areas that may be heat-affected during welding and thus cause subsequent intergranular corrosion. Niobium steels are used for rotors in gas turbines where temperatures up to 700°C must be withstood. Niobium-base alloys find application in fast reactors. Superalloys for very demanding use, as in military applications contain niobium with cobalt and zirconium. When alloyed with titanium, molybdenum, and tungsten, the elevated-temperature hardness of niobium in enhanced, whereas when alloyed with vanadium and zirconium, the strength of niobium up to temperatures of 500°C is increased. Metallurgically, niobium is attractive because of its density, good workability, retention of tensile strength at high temperatures, and its high melting point. In the temperature range 920–1,200°C, niobium has been found superior to most other metals on a strength-to-weight basis for aerospace applications. In multicomponent alloys, zirconium and hafnium when added with niobium add effectively to strength, even more so than molybdenum or tungsten, but there is some sacrifice in ductility.

In metallurgy, niobium is classified as a refractory metal, along with tungsten, tantalum, and molybdenum. A comparison of the four metals is given in the accompanying table.

Niobium in Tool Steels. In the matrix method of tool-steel development, the composition of the heat-treated matrix determines the steel's initial composition. Carbide volume-fraction requirements then are calculated, based upon historical data, and the carbon content is adjusted accordingly. This approach has been used to design new steels in which niobium is substituted for all or part of the vanadium present as carbides in the heat-treated material. Niobium provides dispersion hardening and grain refinement, and forms carbides that are as hard as vanadium, tungsten, and molybdenum carbides.

Chemistry and Compounds. Elemental niobium is insoluble in HCl or HNO_3, but soluble in hydrofluoric acid or a mixture of hydrofluoric and HNO_3.

As might be expected from its $4d^45s^1$ electron configuration, niobium forms pentavalent compounds. However, the stability of its compounds of lower valence is greater than that of the corresponding tantalum compounds, in keeping with the group 5 position of niobium and tantalum. Nevertheless the similarity of the properties of the compounds of the two metals is so great that special methods are required for their separation, such as solvent extraction of the pentachlorides or chromatographic removal of adsorbed TaF_5 with an ethylmethyl ketone-water system. In addition, divalent and tetravalent compounds are known, and an interstitial, nonstoichiometric hydride.

Niobium forms a divalent oxide, NbO, insoluble in water, but readily soluble in acids or NH_4OH. It also gives by direct combination of the metal on heating with oxygen, the pentoxide, Nb_2O_5, which can be reduced by hydrogen at high temperature to NbO_2, and on heating with magnesium to Nb_2O_3.

Niobium(III) halides are known, notably the chloride, $NiCl_3$, which is of particular interest because its solution has been shown to contain Nb^{3+} ions (in equilibrium with $NbCl_6^{3-}$ complex ions).

Tetravalent niobium is believed to occur in the form of $NbOCl_4^{2-}$ ions in a solution obtained, with color change, by reduction of HCl solution of $NbCl_5$, and by inference in similarly reduced solutions of the other pentahalides. Tetravalent niobium also is found in the dioxide (see above) and the carbide, NbC.

Four pentahalides of niobium, NbF_5, $NbCl_5$, $NbBr_5$, and NbI_5 have been prepared by heating the pentoxide with carbon in a current of the halogen. They are hydrolyzed in H_2O, and even in concentrated aqueous solution of the respective halogen acids; the Nb^{5+} ion is apparently not present, but rather complex ions such as $[NbOCl_4]^-$ or $[NbOCl_5]^{2-}$.

The products of partial hydrolysis of the pentahalides are oxyhalides, such as $NbOF_3$, $NbOCl_3$, and $NbOBr_3$. They are designated in the older literature as columbyl or columboxy compounds. The more stable oxyhalogen compounds of niobium are complexes, such as $NbOF_3 \cdot 3NaF$, $NbOF_3 \cdot ZnO \cdot 6H_2O$, and $NbOF_3 \cdot 2KF \cdot H_2O$.

Further complexes of Nb(V) are formed with oxygen-function compounds, such as *o*-dihydroxybenzene and acetylacetone.

The so-called niobic acid is the hydrated pentoxide, $Nb_2O_5 \cdot xH_2O$, insoluble in H_2O.

The metaniobates of the alkali metals, $MNbO_3$, the orthoniobates M_3NbO_4 and the pyroniobates, $M_4Nb_2O_7$, where M is an alkali metal, can be prepared by various alkali carbonate or hydroxide fusion processes.

Niobium forms a nitride, NbN, and a carbide, NbC.

Niobium forms a diamino compound, $(NH_2)_2NbCl_3$, and an ammine complex, $NbCl_5 \cdot 9NH_3$. It forms two cyclopentadienyl compounds, $(C_5H_5)_2NbBr_3$ and $(C_5H_5)Nb(OH)Br_2$. Its other organometallic compounds are essentially oxygen-functional ones, such as $Nb(OCH_3)_5$, $Nb(OC_2H_5)_5$, $Nb(O)(OC_5H_{11})_3$, and $Nb(OC_5H_{11})_5$. These compounds are named as substituted niobanes (thus, the last is pentabutoxy niobane) or as alkyl niobate esters.

Health and Safety Factors

Toxicity data on niobium and its compounds are sparse. The most common materials, eg, niobium concentrates, ferroniobium, niobium metal and niobium alloys, appear to be relatively inert biologically. Limited animal experiments show high toxicity for some salts, which are related to disturbance of enzyme action. Niobium hydride has moderate fibrogenic and general toxic action. Recommended maximum allowable concentrations are 6 mg/m³. Recommended maximum permissible concentration of Nb in reservoir water is 0.01 mg/L. The threshold for affecting clarity and biological oxygen demand (BOD) is 0.1 mg/L.

Unstable niobium isotopes that are produced in nuclear reactors or similar fission reactions have typical radiation hazards. See also **Radioactivity**.

Fire fighting procedures for niobium and niobium hydride powder, suggest letting the fire burn itself out. Small fires can be controlled by smothering with dry table salt or using Type D dry powder fire-extinguishing material. Under no circumstances should water be used, as a violent explosion may result.

Additional Reading

Carter, G.F. and D.E. Paul: *Materials Science and Engineering*, ASM International, Materials Park, OH, 1991.
Gupta, C.K. and A.K. Suri: *Extractive Metallurgy of Niobium*, CRC Press, LLC., Boca Raton, FL, 1994.
Lide, D.R.: *CRC Handbook of Chemistry and Physics*, 88th Edition, CRC Press, LLC., Boca Raton, FL, 2007.
Staff: *ASM Handbook—Properties and Selection: Nonferrous Alloys and Pure Metals*, ASM International, Materials Park, OH, 1990.
Staff: "Tool-Steel Developers Take Note of Niobium," *Advanced Materials 7 Processes*, **15** (June 1991).
Titran, R.H.: "Niobium and Its Alloys," *Advanced Materials & Processes*, **34** (November 1992).

NIRENBERG, MARSHALL WARREN (1927–Present).

Marshall Nirenberg is an American scientist who deciphered the genetic code in living cells and in 1968 shared the Nobel Prize for Physiology or Medicine.

The son of a land developer, Nirenberg was born in New York City but grew up in Orlando. Nirenberg pursued his early interest in biology and chemistry at the University of Florida, and received his PhD in 1957 at the University of Michigan for his work on the enzyme transport mechanism for the sugar hexose in ascites tumor cells. Nirenberg then moved to the National Institutes of Health (NIH) in Bethesda as a postdoctoral fellow to work with DeWitt Stetten, Jr. During his collaboration with William Jakoby on the genetic control of enzymatic induction, Nirenberg became interested in protein synthesis in a cell-free system. He was appointed a staff scientist at NIH in 1960. In a series of experiments conducted with Johann Heinrich Matthaei, a postdoctoral fellow from Germany, Nirenberg discovered in 1961 that poly-U, a synthetic RNA polymer of polyurdylic acid, functions as a template for producing a protein composed of the single amino acid phenylalanine. UUU became the first word of the genetic code deciphered.

It took five years to solve the entire code for twenty amino acids. In this phase of the work Nirenberg faced fierce competition from the eminent scientist Severo Ochoa, and NIH scientists teamed up with him in a remarkable *esprit de corps*. Among them, Philip Leder, Maxine Singer and Leon Heppel assisted Nirenberg by devising enzymatic methods for synthesizing trinucleotides of known sequence. Nirenberg shared the 1968 Nobel Prize for Physiology or Medicine with Har Gobind Khorana and Robert W. Holley for their investigations on the genetic code. Since then, Nirenberg has been exploring the new scientific frontier of neurobiology at NIH. See also **Genetics and Gene Science (Classical)**; **Holley, Robert William (1922–1993)**; and **Ochoa, Severo (1905–1993)**.

B. S. Park
V. A. Harden
National Institutes of Health, Bethesda, MD

NISIN. See **Food Additives**.

NIST. The National Institute of Standards and Technology, the headquarters of which is located in Gaithersburg, Maryland, 20899. NIST replaces the former National Bureau of Standards (NBS), which was established by the U.S. Congress in 1901, with the objectives of: (1) serving as the basis for the nation's physical measurement system; (2) providing scientific and technological services for industry and government; (3) establishing a technical basis for equity in trade, and (4) providing technical services to promote public safety.

As of 2000, NIST is comprised of several divisions and departments, including:

- Technology services
- Manufacturing technology centers
- Standards services
- Standards Code and information
- Standards management
- Weights and measures
- Laboratory accreditation
- Measurement services
- Standard reference materials
- Physical measurement services
- Research and technology applications
- Technology development and small business
- National technology workshop
- Information services
- Research resources development
- Research information services

NITER. This potassium nitrate mineral KNO_3 of orthorhombic crystallization usually occurs as thin crusts, or as silky acicular crystals. It has a hardness of 2, and specific gravity of 2.09–2.14, is of white color, translucent with vitreous luster. It occurs as a surface efflorescence, or in soils rich in organic material in arid regions. World occurrences include Spain, Italy, Egypt, Arabia, India, Russia, the western United States, the Republic of South Africa, and Bolivia, South America. Large quantities were recovered from limestone caves in Tennessee, Kentucky, Alabama and Ohio during the Civil War for use in the manufacture of gunpowder. It is used as a source of nitrogen compounds, for explosives and fertilizers.

NITRATION. Nitration is defined in this article as the reaction between a nitration agent and an organic compound that results in one or more nitro ($-NO_2$) groups becoming chemically bonded to an atom in this compound. Nitric acid is used as the nitrating agent to represent C-, O-, and N-nitrations. O-Nitrations result in esters. N-nitrations are often used as a first step for production of nitramines.

For example, a nitro group is substituted for a hydrogen atom, and water is a by-product. Nitro groups may, however, be substituted for other atoms or groups of atoms. Nitro compounds can also be produced by addition reactions, e.g., the reaction of nitric acid or nitrogen dioxide with unsaturated compounds such as olefins or acetylenes.

Nitrations are highly exothermic, i.e., ca 126 kJ/mol (30 kcal/mol) However, the heat of reaction varies with the hydrocarbon that is nitrated. The mechanism of a nitration depends on the reactants and the operating conditions. The reactions usually are either ionic or free-radical. Ionic nitrations are commonly used for aromatics; many heterocyclics; hydroxyl compounds, e.g., simple alcohols, glycols, glycerol, and cellulose; and amines. Nitration of paraffins, cycloparaffins, and olefins frequently involves a free-radical reaction.

Ionic Nitration Reactions

Acid mixtures containing nitric acid and a strong acid, e.g., sulfuric acid, perchloric acid, selenic acid, hydrofluoric acid, boron trifluoride, or an ion-exchange resin containing sulfonic acid groups, can be used as the nitrating feedstock for ionic nitrations. These strong acids are catalysts that result in the formation of nitronium ions, NO_2^+. Sulfuric acid is almost always used industrially since it is both effective and relatively inexpensive.

The NO_2^+ mechanism has been accepted since about 1950 for the nitration of most aromatic hydrocarbons, glycerol, glycols, and numerous other hydrocarbons in which mixed acids or highly concentrated nitric acid are used. The mechanism has been discussed in detail and critically analyzed. NO_2^+ attacks an aromatic compound (ArH) as follows:

$$ArH + NO_2^+ \longrightarrow \left[Ar \underset{NO_2}{\overset{H}{\diagdown}} \right]^+ \longrightarrow ArNO_2 + H^+ \qquad (1)$$

Nitrosonium ions, NO^+, are, however, the ions employed to start the nitration sequence for easily nitratable aromatic compounds such as phenol.

The kinetics of aromatic nitrations are functions of temperature, which affects the kinetic rate constant, and of the compositions of both the acid and hydrocarbon phase. In addition, a larger interfacial area between the two phases increases the rates of nitration since the main reactions occur at or near the interface. Larger interfacial areas are obtained by increased agitation and by the proper choice of the volumetric % acid in the liquid–liquid dispersion. The viscosities and densities of the two phases and the interfacial tension between the phases are important physical properties affecting the interfacial area.

Increased agitation of a given acid–hydrocarbon dispersion results in an increase in interfacial areas owing to a decrease in the average diameter of the dispersed droplets. As the droplets decrease in size, the ease of separation of the two phases, following completion of nitration, also decreases.

Significant process changes have occurred in many nitration plants. Continuous-flow units are now widely used in the 1990s, replacing batch nitrations. A well-designed continuous-flow plant often offers all of the following advantages, per unit weight of product, as compared to batch units; increased safety, decreased energy requirements, reduced amounts of undesired by-products, fewer environmental problems, reduced labor requirements, and lower operating expenses.

Many nitrated products are explosives, including DNT, TNT, and nitroglycerine (NG). To minimize the potential for run away reactions and explosions, the compositions of the feed acids and reaction conditions are currently better controlled than formerly. In some processes, 99% or more of the feed HNO_3 reacts. Dispersions (or mixtures) of such a waste acid and the nitration product are relatively safe to handle. Also, centrifugal separators are used in many modern processes to rapidly separate the hydrocarbon and used acid phases. Rapid separation greatly reduces the amounts of nitrated materials in the plant at any given time and reduces undesired reactions of the nitrated products.

Considerable effort has been made to minimize energy requirements in the nitrations plants too.

A significant concern in all nitration plants using mixed acid centers on the disposal method or use for the waste acids. They are sometimes employed for production of superphosphate fertilizers. Processes have also been developed to reconcentrate and recycle the acid.

Considerable worldwide interest has occurred in the late 1980s and the early and mid-1990s for nitrations using N_2O_5. Production of nitramines (or N-nitrations) is particularly promising, since these compounds are more stable in the presence of N_2O_5–HNO_3 solutions as compared to mixed acids containing H_2SO_4. Good results have been obtained for the production of the high explosives, cyclotetramethylenetetranitramine or HMX and DADN. Another high explosive, polynitrofluorene, has been produced via C-nitrations, for the first time. The overall exothermicities of the reactions are less when N_2O_5 is used, as compared to mixed acids.

Solutions of CH_2Cl_2 and N_2O_5 have only mild nitrating power. Yet some nitrations are rapid: the N_2O_5 reacts on an almost stoichiometric basis, and only minimal residual nitric acid is present upon completion of the nitration. Some nitrations having unique characteristics can be accomplished.

Free-Radical Nitrations of Paraffins

Both vapor-phase and liquid-phase processes are employed to nitrate paraffins, using either HNO_3 or NO_2. The nitrations occur by means of free-radical steps, and sufficiently high temperatures are required to produce free radicals to initiate the reactions steps.

Free-radical nitrations consist of rather complicated nitration and oxidation reactions. When nitric acid is used in vapor-phase nitrations, the main initiating reaction (eq. 2) produces either $\cdot NO_2$ or $\cdot ONO$. Temperatures of > ca350 °C are required to obtain a significant amount of initiation, and equation is the rate-controlling step for the overall reaction. Reactions 3 and 4 are chain-propagating steps.

$$HNO_3 \rightarrow \cdot OH + NO_2 \qquad (2)$$

$$RH + \cdot OH \rightarrow R \cdot + H_2O \qquad (3)$$

$$R \cdot + HNO_3 \rightarrow RNO_2 + \cdot OH \qquad (4)$$

When nitrogen dioxide is used, the main reaction steps are as in equations 5 and 6.

$$RH + NO_2 \rightarrow R \cdot + HNO_2 \rightarrow R \cdot + \cdot OH + NO \qquad (5)$$

$$R \cdot + \cdot NO_2 \rightarrow RNO_2 \qquad (6)$$

An important side reaction in all free-radical nitrations is production, of unstable alkyl nitrites (eq. 7). They decompose to form nitric oxide and alkoxy radicals (eq. 8) which form oxygenated compounds and lower molecular weight alkyl radicals which can form lower molecular weight nitroparaffins by reactions 4 or 6. The oxygenated hydrocarbons often react further to produce carbon oxides and water.

$$R \cdot + \cdot ONO \rightarrow RONO \qquad (7)$$

$$RONO \rightarrow NO + RO\cdot \qquad (8)$$

Propane is thought to be the only paraffin that is commercially nitrated by vapor-phase processes. Temperature control is a primary factor in designing the reactor, and several approaches have been investigated. A spray nitrator in which liquid nitric acid is sprayed into hot propane is used industrially. Relatively small-diameter tubular reactors, fluidized-bed reactors, and molten salt reactors have all been successfully used in laboratory units.

Health and Safety Factors

The danger of an explosion of a nitrated product generally increases as the degree of nitration increases. Nitroaromatics and some polynitrated paraffins are highly toxic when inhaled or when contacted with the skin. All nitrated compounds tend to be highly flammable.

Additional Reading

Albright, L.F., R.V. Carr, and R.J. Schmitt: *Nitration: Recent Laboratory and Industrial Developments*, American Chemical Society (ACS), Washington, DC, 1996.

Fischer, J.F.: in H. Feuer and A.T. Nielsen, eds., *Nitro Compounds: Recent Advances in Synthesis and Chemistry*, VCH Publishers, New York, NY, 1990, Chapt. 3.

Gilbert, E.: in S.M. Kaye, ed., *Encyclopedia of Explosives and Related Items*, Vol. 9, U.S. Army Armament Research and Development Command, Dover, NJ, 1980, T235–286.

Hill, M.E. and co-workers: in ACS Symposium Series No. 22, American Chemical Society, Washington, DC, 1976, Chapt. 17, pp. 253–271.

Urbanski, T.: *Chemistry and Technology of Explosives*, Vols. 1–3, The Macmillan Co., New York, NY, 1964, 1965; Vol. 4, Permagon Press, Elmsford, NY, 1983.

NITRIC ACID.

[CAS: 7697-37-2]. HNO3, also known as *aqua fortis*, azotic acid, hydrogen nitrate, or nitryl hydroxide, is a chemical of major industrial importance. Because of its properties as a very strong acid and a powerful oxidizing agent, as well as its ability to nitrate organics, nitric acid is essential in the production of many chemicals (eg, pharmaceuticals, dyes, synthetic fibers, insecticides, and fungicides), but is used mostly in the production of ammonium nitrate for the fertilizer industry. By the end of the nineteenth century, its industrial importance had already become established in the production of explosives and dyestuffs. After World War II nitric acid production grew rapidly with the expanding use of synthetic fertilizers. Because of the increased popularity of urea as a fertilizer, production has leveled off in the 1990s. Most growth in demand has come from the production of polyurethanes, fibers, and ammonium nitrate-based explosives. Other uses for nitric acid are in the manufacture of explosives (trinitrotoluene, nitroglycerin, etc), metal nitrates, nitrocellulose, and nitrochlorobenzene, the treatment of metals (eg, the pickling of stainless steels and metal etching), as a rocket propellant, and for nuclear fuel processing.

The first reports of nitric acid have been credited to Arab alchemists of the eighth century. By the Middle Ages it was referred to as *aqua fortis* (strong water) or *aqua valens* (powerful water). From that time onward, nitric acid was produced primarily from saltpeter [CAS: 7757-79-1] (potassium nitrate) and sulfuric acid. In the nineteenth century, Chilean saltpeter [CAS: 7631-99-4] (sodium nitrate) from South America largely replaced potassium nitrate. However, at the beginning of the twentieth century newer manufacturing technologies were introduced. In Norway, where electricity was inexpensive, electric arc furnaces were used to make nitrogen oxides, and subsequently nitric acid, directly from air. The commercial life of these furnaces was relatively brief and most were shut down by 1930. At about the same time, a different production method was being developed. In 1908, at Bochum, Germany, Ostwald piloted a 3-t per day nitric acid process based on the catalytic oxidation of ammonia with air. In 1913 the synthesis of ammonia from coal, air, and water was successfully demonstrated using the Haber-Bosch process. With a secure and economical supply of ammonia, ammonia oxidation became firmly established as an industrial route to nitric acid manufacture. Process developments continued and plant scale increased to commercial quantities in both Europe and the United States. The first full-size plant to be built in the United States was installed in 1917 by Chemical Construction Company (Muscle Shoals, Alabama). The process operated at atmospheric pressure and used multiple ammonia oxidation converters. Since those early days, ammonia oxidation has become the basis of all commercial nitric acid production. There have been many advances in plant design leading to improved process performance and higher production capacities at increased operating pressures. More details on the history of nitric acid and development of the manufacturing process are available.

In the modern ammonia oxidation process, most nitric acid is produced as a weak acid (50–65 wt %). High monopressure processes minimize capital investment, whereas split- or dual-pressure processes optimize ammonia conversion efficiency and catalyst use. Weak acid is suitable for use in the production of fertilizers, but stronger acid (up to 99 wt % acid) is required for many organic reactions of industrial importance. Direct strong nitric (DSN) processes make the acid directly from nitrogen oxides obtained by the oxidation of ammonia. Nitric acid concentration (NAC) processes use extractive distillation to concentrate the weak acid. A dehydrating agent such as sulfuric acid or magnesium nitrate is used to enhance the volatility of HNO3 so that distillation methods can surpass the maximum boiling azeotrope of nitric acid. Several DSN processes can produce weak and strong acids simultaneously.

Nitric acid is a colorless liquid, sp. gr. 1.503 (25 °C), freezing point −41.6 °C, and boiling point 86 °C. The 100% acid is not entirely stable and must be prepared from its azeotrope (constant-boiling mixture) by distillation with concentrated sulfuric acid. Reagent grade HNO_3 is a water solution containing about 68% HNO_3 (weight). This strength corresponds to the constant-boiling mixture of the acid with water, which is 68.4% HNO_3 (weight) at atmospheric pressure and boils at 121.9 °C. Nitric acid is completely miscible with water. It forms two solid hydrates, $HNO_3 \cdot H_2O$ and $HNO_3 \cdot 2H_2O$, with corresponding melting points of approximately −38 and −18.5 °C. Nitric acid is a strong acid and a powerful oxidizer. In dilute solutions, it is almost completely ionized to H^+ and NO_3^- ions and behaves like a strong acid.

With organic compounds, HNO_3 may act as a nitrating agent, as an oxidizing agent, or simply as an acid. The classic example of nitration is its reaction with benzene or toluene in the presence of concentrated H_2SO_4 to form nitrobenzene or nitrotoluene (TNT). An example of oxidation properties is in the oxidation of cyclohexanol by HNO_3 to produce adipic acid, an intermediate of nylon. Behaving like an acid, it forms nitroglycerin by esterification of glycerol in the presence of concentrated sulfuric acid.

An interesting property of HNO_3 is its ability to passivate some metals, such as iron and aluminum. This property is of significant industrial importance, since modern processes for producing the acid depend on it. Modern suitability formulated stainless steel alloys are usefully resistant to nitric acid through a wide range of conditions. The acid's passivity or the metal's resistance to attack is attributed to the formation of a protective oxide layer on the surface of the metal.

Nitric acid is a high tonnage industrial chemical. Much of the production is used in the manufacture of agricultural fertilizers, largely in the form of ammonium nitrate, NH_4NO_3. See also **Fertilizer**. About 15% of the nitric acid produced is used in explosives (nitrates and nitro compounds), and about 10% is consumed by the chemical industry. As the red fuming acid or as nitrogen tetroxide, HNO_3 is used extensively as the oxidizer in propellants for space rockets and missiles.

Production of Nitric Acid. Three commercial methods have been developed for nitric acid production: (1) the reaction between sulfuric acid and sodium nitrate, (2) the thermal combination of oxygen and nitrogen in air, and (3) the catalytic oxidation of ammonia and absorption of the gaseous products in waters. There are numerous variations of these fundamentals processes. The principal process used today is based on the catalytic oxidation of ammonia and absorption of the gaseous products in water. This process was developed by Ostwald (Germany) and based on earlier work of Kuhlmann (France). In the Ostwald process, HNO_3 is produced in a 3-stage operation: (1) Ammonia is oxidized to nitric oxide, (2) the nitric oxide is further oxidized to nitrogen dioxide, and (3) the gases are absorbed in water to yield HNO_3 according to

$$4NH_3 + 5O_2 \rightarrow 4NO + 6H_2O$$

$$2NO + O_2 \rightarrow NO_2$$

$$3NO_2 + H_2O \rightarrow 2HNO_3 + NO$$

The nitric oxide formed in the last equation returns to the gas phase, is reoxidized to nitrogen dioxide, and reabsorbed. These reactions are highly exothermic. In actuality, numerous complex reactions occur in addition to the main reactions just outlined.

In a manufacturing plant, air is preheated, mixed with superheated ammonia vapor, and reacted catalytically over a gauze composed of 90% platinum and 10% rhodium at a temperature of 800–960 °C and operating pressures between 1 and 8.2 atmospheres. The reaction produces nitrogen dioxide, NO_2 and nitric oxide, NO. The latter is oxidized to NO_2 in the reaction train. The NO_2 actually exists in equilibrium with its dimer, N_2O_4. This equilibrium mixture, sometimes referred to as nitrogen peroxide, is absorbed in water in a cooled absorber tower to form HNO_3 at a strength of 55–60% HNO_3.

Health and Safety Factors

Nitric acid and the oxides of nitrogen found in its fumes are highly toxic and capable of causing severe injury and death. It is corrosive, and can destroy human tissue. Nitric acid is regulated by OSHA, which lists it as a Process Safety Hazardous Chemical and Air Contaminant. Under SARSH, the EPA lists it as an Extremely Hazardous Substance and Toxic Chemical.

First-aid practices for the treatment of exposure to nitric acid should be obtained from a current version of the Material Safety Data Sheet or other appropriate safety literature.

Additional Reading

Web Reference

Chemical and Other Safety Information: http://physchem.ox.ac.uk/MSDS/

NITRIDES. At elevated temperatures and pressures, nitrogen combines with most elements to form nitrogen compounds. In the presence of metals and semimetals, it forms nitrides where nitrogen has a nominal valence of −3. Atomic nitrogen, which reacts much more readily with the elements than does molecular nitrogen, forms nitrides with elements that do not react with molecular nitrogen even at very high pressures. The binary compounds of nitrogen may be classified, according to their chemical and physical properties, into four groups: saltlike, metallic, nonmetallic or diamondlike, and volatile nitrides.

An alphabetical list of nitrides is given in Table 1.

TABLE 1. ALPHABETICAL LIST OF NITRIDES

Compound	CAS Registry Number	Formula
Aluminum nitride	[24304-00-5]	AlN
Americium nitride	[12296-96-1]	AnN
Ammonia	[7664-41-7]	NH_3
Antimony nitride	[12333-57-2]	SbN
Arsenic nitride	[26754-98-3]	AsN
Barium nitride	[12047-79-9]	Ba_3N_2
Berkelium nitride	[56509-31-0]	BkN
Berylium nitride	[1304-54-7]	Be_3N_2
Bismuth nitride	[12232-97-2]	BiN
Boron nitride	[10043-11-5]	BN
Bromine nitride	[15162-90-0]	Br_3N
Cadmium nitride	[12380-95-9]	Cd_3N_2
Californium nitride	[70420-43-8]	CfN
Calcium nitride	[12013-82-0]	Ca_3N_2
Carbon nitride	[12069-92-0]	CN_2
Cerium nitride	[25764-08-3]	CeN
Chlorine nitride	[10025-85-1]	Cl_3N
Chromium nitride	[24094-93-7]	CrN
Chromium nitride (2:1)	[12053-27-9]	Cr_2N
Cobalt nitride (2:1)	[12259-10-8]	Co_2N
Cobalt nitride (3:1)	[12432-98-3]	Co_3N
Copper nitride	[1308-80-1]	Cu_3N
Curium nitride	[56509-28-5]	CmN
Cyanogen	[2074-87-5]	$(CN)_2$
Dinitrogen tetraoxide	[10544-72-6]	N_2O_4
Dysprosium nitride	[12019-88-4]	DyN
Erbium nitride	[12020-21-2]	ErN
Europium nitride	[12020-58-5]	EuN
Fluorine nitride	[13967-06-1]	F_3N
Gadolinium nitride	[25764-15-2]	GdN
Gallium nitride	[25617-97-4]	GaN
Germanium nitride	[12065-36-0]	Ge_3N_4
Gold nitride	[13783-74-9]	Au_3N
Hafnium nitride (1:1)	[25817-87-2]	HfN
Hafnium nitride (3:2)	[12508-69-9]	Hf_3N_2
Holmium nitride	[12029-81-1]	HoN
Indium nitride	[25617-98-5]	InN
Iodine nitride	[21297-03-1]	I_3N
Iron nitride (2:1)	[12023-20-0]	Fe_2N
Iron nitride (4:1)	[12023-64-2]	Fe_4N
Lanthanum nitride	[25764-10-7]	LaN
Lead nitride (3:2)	[58572-21-7]	Pb_3N_2
Lead nitride (3:4)	[75790-62-4]	Pb_3N_4
Lithium nitride	[26134-62-4]	Li_3N
Lutetium nitride	[12125-25-6]	LuN
Magnesium nitride	[12057-71-5]	Mg_3N_2
Manganese nitride (2:1)	[12163-53-0]	Mn_2N
Manganese nitride (3:2)	[12033-03-3]	Mn_3N_2
Manganese nitride (4:1)	[12033-07-7]	Mn_4N
Mercury nitride	[12136-15-1]	Hg_3N_2
Molybdenum nitride	[12033-19-1]	MoN
Molybdenum nitride (2:1)	[12033-31-7]	Mo_2N
Neodymium nitride	[25764-11-8]	NdN
Neptunium nitride	[12058-90-1]	NpN
Nickel nitride	[12033-45-3]	Ni_3N
Niobium nitride	[11092-17-4]	NbN
Niobium nitride (2:1)	[12033-43-5]	Nb_2N
Niobium nitride (4:3)	[12163-98-3]	Nb_4N_3
Nitrogen	[7727-37-9]	N_2
Nitrous oxide	[10024-97-2]	N_2O
Phosphorus nitride	[17739-47-8]	PN
Plutonium nitride	[12033-54-4]	PuN
Potassium nitride	[29285-24-3]	K_3N

TABLE 1. (Continued)

Compound	CAS Registry Number	Formula
Praseodymium nitride	[25764-09-4]	PrN
Protactinium nitride	[75733-54-9]	PaN$_2$
Rhenium nitride	[12033-55-5]	Re$_2$N
Rubidium nitride	[12136-85-5]	Rb$_3$N
Samarium nitride	[25764-14-1]	SmN
Scandium nitride	[25764-12-9]	ScN
Selenium nitride	[12033-59-9]	SeN
Silver nitride	[20737-02-4]	Ag$_3$N
α-Silicon nitride	[12033-89-5]	Si$_3$N$_4$
Sodium nitride	[12136-83-3]	Na$_3$N
Strontium nitride	[12033-82-8]	Sr$_3$N$_2$
Sulfur nitride	[28950-34-7]	SN
Tantalum nitride	[12033-62-4]	TaN
Tantalum nitride (2:1)	[12033-63-5]	Ta$_2$N
Tantalum nitride (3:5)	[12033-94-2]	Ta$_3$N$_5$
Tellurium nitride	[59641-84-8]	TeN
Terbium nitride	[12033-64-6]	TbN
Thallium nitride	[12033-67-9]	TlN
Thorium nitride	[12033-65-7]	ThN
Thorium nitride (3:4)	[12033-90-8]	Th$_3$N$_4$
Thulium nitride	[12033-68-0]	TmN
Tin nitride (3:2)	[75790-61-3]	Sn$_3$N$_2$
Tin nitride (3:4)	[75790-62-4]	Sn$_3$N$_4$
Titanium nitride (1:1)	[25583-20-4]	TiN
Titanium nitride (2:1)	[12169-08-3]	Ti$_2$N
Tungsten nitride	[12058-38-7]	WN
Tungsten nitride (2:1)	[12033-72-6]	W$_2$N
Uranium nitride	[25658-43-9]	UN
Vanadium nitride	[24646-85-3]	VN
Vanadium nitride (2:1)	[12209-81-3]	V$_2$N
Ytterbium nitride	[24600-77-9]	YbN
Yttrium nitride	[25764-13-0]	YN
Zinc nitride	[1313-49-1]	Zn$_3$N$_2$
Zirconium nitride	[25658-42-8]	ZrN

Properties

Saltlike Nitrides. The nitrides of the electropositive metals of Group 1 (IA), 2 (IIA), AND 3 (IIIB) form saltlike nitrides having predominantly heteropolar (ionic) bonding and are regarded as derivatives of ammonia. The composition of these nitrides is determined by the valency of the metal, eg, Li3N, Ca3N2, and ScN. The thermodynamic stability of the saltlike nitrides increases with increasing group number. For example, the nitrides of the alkali metals are only marginally or not at all stable, whereas the rare-earth metals are effective nitrogen scavengers in metals and alloys. The saltlike nitrides are generally electrical insulators or ionic conductors. The nitrides of the Group 3 (IIIB) metals are metallic conductors or at least semiconductors, and thus, represent a transition to the metallic nitrides. The saltlike nitrides are characterized by sensitivity to hydrolysis. These compounds react readily with water or moisture to give ammonia and the metal oxides or hydroxides.

Metallic Nitrides. The nitrides of the transition metals of Groups 6 and 7 (IVB–VIIB) are generally termed metallic nitrides because of metallic conductivity, luster, and general metallic behavior. These compounds, characterized by a wide range of homogeneity, high hardness, high melting points, and good corrosion resistance, are grouped with the carbides, borides and silicides as refractory hard metals. Metallic nitrides can be alloyed with other nitrides and carbides of the transition metals to give solid solutions.

Although there are several hundred binary nitrides, only a relative few ternary bimetallic nitrides are known. A group of ternaries of the composition $M_xM'_yN_z$, where M is an alkali, alkaline-earth, or a rare-earth metal and M is a transition or post-transition metal, have been synthesized.

Metallic nitrides are wetted and dissolved by many liquid metals and can be precipitated from metal baths.

Nonmetallic (Diamondlike) Nitrides. The nitrides of some elements of Groups 13 (IIIA) and 14 (IVA) e.g., BN, Si$_3$N$_4$, AlN, GaN, and InN, are characterized by predominantly covalent bonding. These are stable chemically, have high degrees of hardness (e.g., cubic BN) and high melting points, and are nonconductive or semiconductive. The structural elements of diamondlike nitrides are tetrahedral, M$_4$N, which are structurally related to diamond. Although the most common graphite-like form of BN does not contain these structural elements, boron nitride is considered a diamond-like nitride for two reasons: the existence of a diamond-like form at high pressures and the chemical and physical behavior of BN. Diamond-like nitrides have stoichiometric compositions having no homogeneity range and, as a rule, do not form solid solutions with each other.

Volatile Nitrides. The nitrogen compounds of the nonmetallic elements are generally not very stable. These nitrides decompose at elevated temperatures. Some are explosive and decompose upon shock. They form distinct molecules similar to organic compounds, and at low temperatures are gaseous, liquid, or easily volatilized solids. Exceptions are $(SN)_x$, which is polymeric, chemically stable, and has semimetallic properties; and $(PNCl_2)_x$, which has attracted some scientific interest as inorganic rubber. None of the volatile nitrides has obtained any substantial industrial application except ammonia (hydrogen nitride) and nitrogen oxide (oxygen nitride). Gaseous nitrogen fluorides are explosive; Cl$_3$N, a dark-yellow liquid, evaporates somewhat on heating and explodes. I$_3$NNH$_3$ [CAS: 15823-38-8] detonates at the slightest touch.

Preparation

Metals or metal hydrides may be nitrided using nitrogen or ammonia. Pure metal powders or pure metal hydride powders yield nitride products that are nearly as pure as the precursors.

A process based on the reaction of metal oxides rather than more expensive metal powders and nitrogen or ammonia in the presence of carbon is economical and has possibilities for large-scale production. However, the products, which contain oxygen and carbon, are not very pure.

Many nitrides, e.g., BN, AlN, TiN, ZrN, HfN, CrN, Re$_2$N, Fe$_2$N, Fe$_4$N, and Cu$_3$N, may be prepared by the reaction of the corresponding metal halide and ammonia. Nitrides may also be obtained by the reaction of ammonia and oxygen-containing compounds, ammonium-oxo complexes, or oxides and ferrous metal oxides. These nitrides, however, are not very pure and may contain residual oxygen and halogen.

The van Arkel gas decomposition process gives especially pure nitrides and nitride films, which under certain conditions may precipitate as single crystals. The nitrides include TiN, ZrN, HfN, VN, NbN, BN, and AlN.

The nitrides, Si$_3$N$_4$, Ge$_3$N$_4$, Zn$_3$N$_2$, Cd$_3$N$_2$, and Ni$_3$N, may also be produced by thermal decomposition of the corresponding metal amide of imide. Rb$_3$N and Cs$_3$N are obtained by azide decomposition. AlN and Si$_3$N$_4$ can be produced by the carbothermal reduction of intercalation compounds, magadiite- and montmorillonite-polyacrylonitrile. Nitrides low in nitrogen can be synthesized from nitrides having a higher nitrogen content by decomposition in a vacuum or by reduction with hydrogen.

The formation of nitrides from gaseous halides, ammonia, and nitrogen (atomic and molecular) in a plasma processing torch is possible by means of a specific type of plasma processing called cathodic arc plasma deposition (CAPD). Plasma nitriding offers several advantages: It is nonpolluting and energy efficiency, provides flexible deposition conditions without sacrifice of quality, minimizes distortion, and is easily applicable to compound film deposition. Ion implantation directly inserts nitrogen into metal surfaces.

The hardening, i.e., increase in nitrogen content, achieved by nitriding special alloy steels is technologically significant in the heat treatment of high quality parts, such as gears. Hardness properties are imparted by the resulting coatings of needle-shaped precipitates of the nitrides and carbonitrides of iron, aluminum, chromium, molybdenum, etc. The hardness of these coatings exceeds that of the precipitation-hardened parts by ca 30%.

In nitriding or carbonitriding of condensed materials, molten cyanides are used at ca 570 °C. This method produces fairly thick coatings of nitrides or carbonitrides after ca 1 hr without the risk of distortion during surface hardening. See also **Nitriding**.

Wear-resistant layers can be deposited on the surface of nearly every kind of material (e.g., steel, cast iron, and cemented carbides) by a chemical vapor deposition (CVD) process. See also **Thin Films**.

Health and Safety Factors

As a chemical group, toxicity of nitrides generally stems from the possible reactions with water to form toxic fumes (especially ammonia) rather than from the nitride. There are, of course, exceptions. The salt-like nitrides decompose when in contact with water or moisture to form ammonia, which can irritate the respiratory organs and mucous membranes. The metallic nitrides are very stable chemically, but fine powder or dust of the nitrides of the transition metals can be pyrophoric; nitrides of the actinide metals are carcinogenic.

The diamondlike nitrides, especially as dust, can irritate the lungs or cause scratching of the eyes owing to mechanical means. Nitrides of the 11(IB) and 12(IIB) metals and especially the volatile nitrides have to be handled with extreme care because of their instability and high degree of toxicity.

Uses

Nitrides are used for their high strength and hardness, in nuclear applications, solid electrolytes, refractories, abrasives, coatings and lubrication, catalysis, and electronic and optoelectronic applications.

Additional Reading

Albrecht, M., and J. Neugebauer: *Nitride Semiconductors: Handbook on Materials and Devices*, John Wiley & Sons, Inc., New York, NY, 2003.

Bechstedt, F., B.K. Meyer, and M. Stutzman: *Group III-Nitrides and Their Heterostructures: Growth, Characterization and Applications*, John Wiley & Sons, Inc., New York, NY, 2003.

Freer, R. ed.,: NATO ASI Series: *The Physics and Chemistry of Carbides, Nitrides, and Borides*, Vol. 185, Kluwer Academic Publishers, Boston, MA, 1990.

Gil, B.: *Group III Nitride Semiconductor Compounds: Physics and Applications*, Oxford University Press, New York, NY, 1998.

Goldschmidt, H.: *Interstitial Alloys*, Butterworths, London, 1967.

Gubanov, V.A., V.P. Zhukov, and A.L. Ivanovsky: *Electronic Structure of Refractory Carbides and Nitrides*, Cambridge University Press, New York, NY, 2005.

Moustakas, T.D., and J.I. Pankove: *Gallium-Nitride (GaN) II*, Vol. 5, Elsevier Science & Technology Books, New York, NY, 1998.

Rabenau, A.: *Solid State Ionics*, **6**, 277 (1982).

Samsonov, G.V.: *Nitridij*, Naukova Dumka, Kiev, USSR, 1969.

Toth, L.E.: *Transition Metal Carbides and Nitrides*, Academic Press, Inc., New York, NY, 1971.

NITRIDING. Surface hardening of alloy steels by heating the metals to a temperature of 490–650 °C in an atmosphere of partially dissociated NH_3 (ammonia). As in cyaniding, hardening results from the formation of nitrides of iron and of certain alloying elements that may be present in the steel. Much longer heating time is required than in carburizing practice, and while the depth of penetration is generally less, the maximum hardness at the surface is higher, 900–1,100 D.P.H. (Vickers Brinell) compared to 800–900 D.P.H. for an average carburized case. Nitriding also differs from carburizing in that the parts are fully heat-treated to develop the required core properties before the nitriding treatment. Because of the comparatively low temperature of the process, distortion and dimensional changes are at a minimum. Nitrided steels have good corrosion-resistance when used for valves, pump parts, shafting, and bearing surfaces operating in steam, crude oil, gasolines, and gaseous products of combustion. The fatigue strength is also improved by nitriding.

Other typical applications are piston pins, crankshafts, cylinder liners, timing gears, gauges, and ball and roller bearing parts.

NITRIFICATION. Nitrification is the biologically facilitated oxidation of ammonium (NH_4^+) to nitrate (NO_3^-). The process of nitrification is performed by ammonium and nitrite (NO_2^-) oxidizing bacteria found in soil and waters (Table 1). In soil and water, ammonium is supplied through decomposition of organic matter by microbes and/or anthropogenic or natural additions of inorganic nitrogen sources containing ammonia (NH_3), urea, or ammonium. Nitrification consists of two independent microbially facilitated reactions. In the first reaction, *ammonium* oxidizers oxidize ammonium to nitrite, and in the second reaction, *nitrite* oxidizers oxidize nitrite to nitrate.

The purpose of nitrification (microbial facilitated conversion of ammonium to nitrate) is to provide nitrifying bacteria the energy they require for metabolism. The nitrification reactions are represented in Figure 1, and the

$$2\,NH_4^+ + 3O_2 \xrightarrow{\text{Ammonium oxidizers}} 2\,NO_2^- + 2\,H_2O + 4\,H^+$$

$$2\,NO_2^- + O_2 \xrightarrow{\text{Nitrite oxidizers}} 2\,NO_3^-$$

Fig. 1. Nitrification reactions.

TABLE 1. COMMON SPECIES OF NITRIFIERS AND THEIR HABITATS

Bacterial Species	Habitat
Ammonium oxidizers	
Nitrosomonas europea	Soil, water, sewage
Nitrosospira briensis	Soil
Nitrosococcus oceanus	Marine
Nitrosococcus mobilis	Marine
Nitrosococcus nitrosus	Marine
Nitrosolobus multiformis	Soil
Nitrosovibrio tenuis	Soil
Nitrite oxidizers	
Nitrococcus mobilis	Marine
Nitrobacter gracilis	Marine
Nitrobacter winogradski	Soil
Nitrobacter agilis	Soil, water

bacteria involved are listed in Table 1. There are two different microbially mediated nitrification reactions.

Chemoautotrophic nitrification is the use of carbon dioxide and carbonates as a carbon source for cell construction and oxidation of inorganic molecules such as NO_3^- or NO_2^- as an energy source. Chemoautotrophic nitrification is carried out only by a select group of bacteria. Heterotrophic nitrification is the use of organic compounds for both a carbon and energy source. Heterotrophic nitrification is executed by a wide variety of heterotrophic bacteria and fungi. Chemoautotrophic nitrification is believed to be up to 1000 times faster and is better understood than heterotrophic nitrification. Chemoautotrophic nitrification dominates in neutral to alkaline agricultural soils, whereas heterotrophic nitrification dominates in acid forest soils.

Relationship between Soils and Waters

Nitrification is a key component of the nitrogen cycle (Fig. 2). The nitrogen cycle represents the fundamental transformations of nitrogen in the environment. In agricultural soil systems, we strive to manage this cycle better to increase crop production and minimize the negative impacts of nitrogen on the environment. Implementing management practices that will help manage nitrogen in agricultural systems to protect surface and groundwaters is an example. Nitrate, the end product of the nitrification process, is highly soluble in water and is not readily adsorbed to soil particles; therefore, if nitrate is not used by plants or microbes as a nutrient, it can move vertically through soils to groundwater and in runoff to surface waters resulting in problems from nutrient enrichment. Because nitrification is a microbially mediated process, many management practices aimed at reducing nitrate leaching focus on the nitrification process.

Factors Influencing Nitrification

Assuming an adequate supply of ammonium and a sufficient population of nitrifying bacteria, there are several factors that affect nitrification in natural systems. These factors include pH, moisture, temperature, and oxygen concentration.

pH. Soil and water pH influence the availability of many essential nutrients such as calcium and phosphorus. Microbes require these nutrients for growth and development; therefore, pH ranges that promote adequate availability of these nutrients are required for nitrification. Nitrification in soils is optimized at pH 8.5 but can proceed in a pH range of 4.5 to 10. Figure 3 illustrates the relationship between soil pH and nitrification. At low pH levels, many essential nutrients are less available for use by plants and microbes.

Moisture. Nitrifying bacteria require water to live and function. In soils, the optimum water content for nitrification is at field capacity

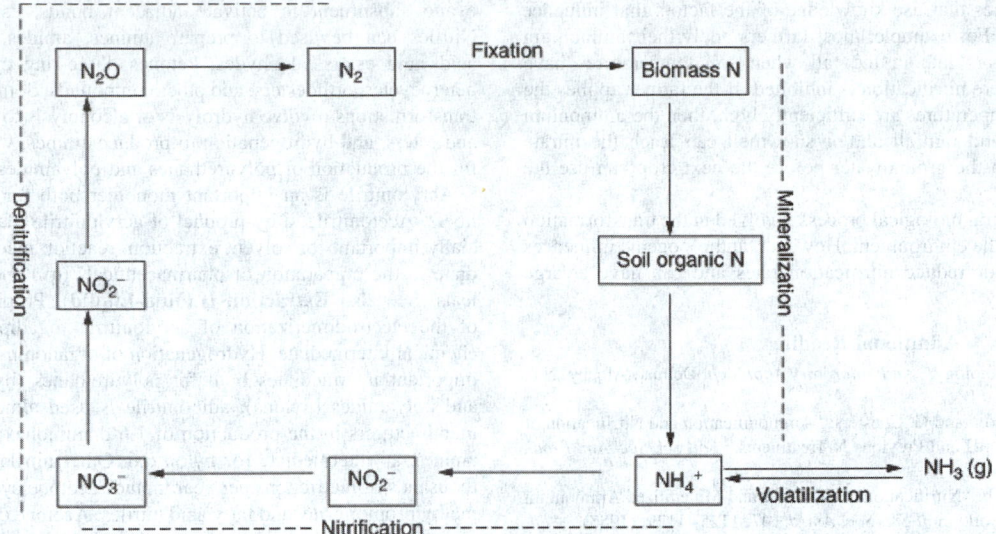

Fig. 2.   Nitrogen cycle.

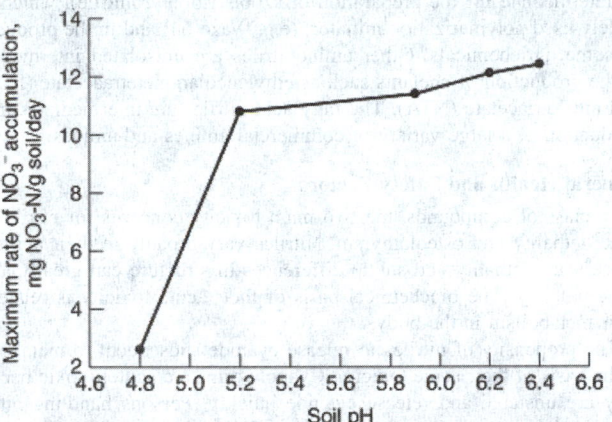

Fig. 3.   Effect of soil pH prior to incubation on nitrate accumulation in soil treated with 169 kg N as NH_4NO_3/ha and incubated at 23 °C for 30 days (*Tisdale, et. al. Ref.*).

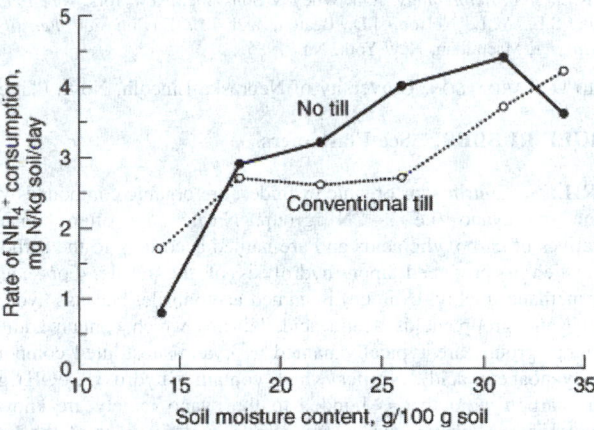

Fig. 4.   Nitrification rate of two tillage systems of a Maury silt loam soil column related to soil moisture content. (*Rice, and Smith Ref.*).

(water remaining in soil after gravity has removed excess water after a period of saturation). Research shows that, as water content increases in soil from dry to an optimum content, nitrification rates increase (Fig. 4). However, waterlogged conditions drastically decrease oxygen content and thus decrease the rate of nitrification. The water content at field capacity varies, depending on the soil physical properties such as texture and organic matter content.

Temperature. Generally, as temperature increases to a point so does the rate of nitrification (Fig. 5). Studies have shown that ammonium is converted to nitrite at temperatures from 0 to 65 °C (32 °F to 149 °F) and nitrite is converted to nitrate from 0 to 40 °C (32 °F to 104 °F). Rates of nitrification dramatically decrease above and below these ranges. Optimum temperatures for nitrification are between 30 and 35 °C (86 °F and 95 °F).

Oxygen Concentration. Figure 1 illustrates the need for oxygen in the nitrification process. Diffusion rates and concentrations of oxygen in soils and water will directly influence the rates of nitrification.

Management of Nitrification in Soil to Reduce Nitrate Losses to Surface and Groundwater

Reducing nitrate losses in agricultural soils that receive large nitrogen (ammonia and ammonium) input from fertilizers is important to maintain high quality water resources. Chemicals called nitrification inhibitors have been developed that inhibit the nitrogen oxidative processes of nitrifying bacteria when added to agricultural soils concurrently with ammonium based fertilizers. In other words, nitrogen in the form of ammonium persists in the soil for longer periods of time that allow for increased plant use of the

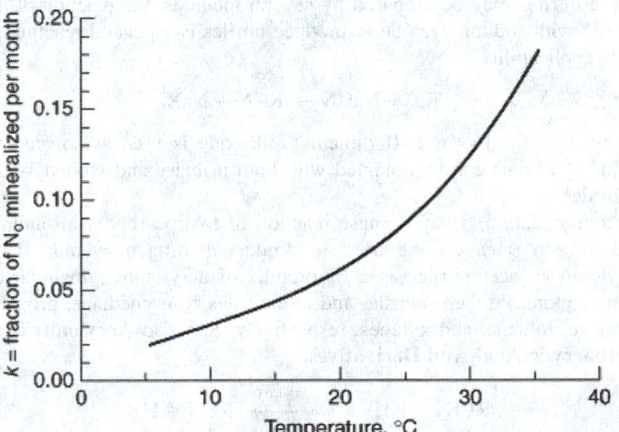

Fig. 5.   Fraction of N mineralized per month, k, in relation to temperature (k was estimated graphically for observed average monthly air temperatures).

ammonium. Ammonium is adsorbed to soil particles; therefore, it does not readily move with water like nitrate. Nitrification inhibitors have a finite effective functionality in soil because soil microbes degrade them over time. Studies have shown that nitrification suppression by the inhibitor dicyandiamide can last 3 months, depending on the amount added with fertilizer.

Management practices that use knowledge of the factors that influence nitrifiers are common. For example, most farmers apply their ammonium based nitrogen fertilizers late in the fall when soil temperatures have declined to a point where nitrification is inhibited. If the farmer applies the fertilizer when soil temperatures are sufficiently high, then the ammonium is nitrified to nitrate and rainfall and/or snowmelt can leach the nitrate out of the root zone to the groundwater before the next crop can use the nitrogen.

Nitrification is a natural biological process involved in the transformation and fate of nitrogen in the environment. However, anthropogenic influences can either accentuate or reduce nitrification rates and can have a large impact on water quality.

Additional Reading

Coyne, M.S.: "Soil Microbiology," *An Exploratory Approach*, Delmar, Albany, NY, 1999.

Dancer, W.S., L.A. Peterson, and G. Chesters: "Ammonification and Nitrification of N as Influenced by soil pH and Previous N Treatments," *Soil Sci. Soc. Am. Proc.*, **37**, 67–69. (1973).

Rice, C.W. and M.S. Smith: "Nitrification of Fertilizer and Mineralized Ammonium in No-till and Plowed Soil," *Soil Sci. Soc. Am. J.*, **47**, 1125–1129. (1983).

Stanford, G., J.N. Carter, D.T. Westermann, and J.J. Meisinge: "Residual Nitrate and Mineralizable Soil Nitrogen in Relation to Nitrogen Uptake by Irrigated Sugarbeets," *Agron. J.*, **69**, 303–308. (1977).

Tate, R.L.: *Soil Microbiology*, John Wiley & Sons, Inc., New York, NY, 1995.

Tisdale, S.L., W.L. Nelson, J.D. Beaton, and J.L. Havlin: *Soil Fertility and Fertilizers*, Macmillan, New York, NY, 1985.

DAVID D. TARDALSON, University of Nebraska-Lincoln, North Platte, NE

NITRILE RUBBER.   See Elastomers.

NITRILES.   Nitriles, or organic cyanides, are organic compounds which contain the cyano (i.e., −CN) group. Nitriles are often considered derivatives of carboxylic acids and are named according to the carboxylic acid which is produced upon hydrolysis of the nitrile. For example, cyanomethane (methyl cyanide) is named acetonitrile, because hydrolysis of its cyano group yields acetic acid. Nitriles which contain additional functional groups are typically named as cyano-substituted compounds, (e.g., cyanoacetic acid). Nitriles which contain a hydroxy (−OH) group on the carbon atom that is bonded to the cyano moiety are known as cyanohydrins. Aliphatic nitriles are named as derivatives of the longest carbon chain and the carbon of the nitrile is included.

General Preparations and Chemical Properties

While nitriles may be prepared by several methods, the reaction of alkyl halides with sodium cyanide to produce nitriles is a general reaction with wide applicability:

$$RX + NaCN \rightarrow RCN + NaX$$

where $X = Cl$, Br, or I. If dimethyl sulfoxide is used as solvent, high yields of nitriles can be obtained with both primary and secondary alkyl chlorides.

Ammoxidation, a vapor-phase reaction of hydrocarbon with ammonia and oxygen (air), can be used to produce hydrogen cyanide (HCN), acrylonitrile, acetonitrile (as a by-product of acrylonitrile manufacture), methacrylonitrile, benzonitrile, and toluinitriles from methane, propylene, butylene, toluene, and xylenes, respectively. See also **Acrylonitrile**; and **Methacrylic Acid And Derivatives**.

$$RCH_3 + NH_3 + O_2 \xrightarrow{\text{catalyst}} RCN + H_2O$$

Addition of HCN to unsaturated compounds is often the easiest and most economical method of making organonitriles. However, the addition of HCN to unactivated olefins and the regioselective addition to dienes is best accomplished with a transition metal catalyst.

Chemistry and Uses of Nitriles

As a class of compounds, nitriles have broad commercial utility that includes their use as solvents, feedstocks, pharmaceuticals, catalysts, and pesticides. The versatile reactivity of organonitriles arises both from the reactivity of the C≡N bond, and from the ability of the cyano substituent to activate adjacent bonds, especially C−H bonds. Nitriles can be used to prepare amines, amides, amidines, carboxylic acids and esters, aldehydes, ketones, large-ring cyclic ketones, imines, heterocycles, orthoesters, and other compounds. Some of the more common transformations involve hydrolysis or alcoholysis to produce amides, acids and esters, and hydrogenation to produce amines, which are intermediates for the production of polyurethanes and polyamides.

Acrylonitrile is an important monomer both for plastics and synthetic fibers. Acetonitrile, a by-product of acrylonitrile manufacture, is commercially important for solvent extraction, reaction media, and as an intermediate in the preparation of pharmaceuticals (qv) and other organic chemicals. See also **Extraction (Liquid-Liquid)**. Propionitrile, a by-product of the electrodimerization of acrylonitrile to adiponitrile, is used as a chemical intermediate. Hydrogenation of organonitriles to amines provides important intermediates both for polyurethanes (by way of isocyanates) and polyamides (nylons); adiponitrile is used almost exclusively by the manufacturers in the production of 1,6-diaminohexane (hexamethylenediamine), an intermediate for nylon 6,6. Other nitriles that are produced in thousands of metric tons per year include acetone cyanohydrin, 2-amino-2-methylpropionitrile, and fatty acid nitriles. Acetone cyanohydrin is an intermediate for the preparation of methyl methacrylate and acrylic resins, (e.g., lucite and plexiglas) and for 5,5-dimethylhydantoin, which is used to make commercial water treatment chemicals. 2-Amino-2-methylpropionitrile is an intermediate for the preparation of azobis(isobutyronitrile), which is a widely used polymerization initiator, (e.g., Vazo 64) and in the production of some agrichemicals. Other aminonitriles are unisolated intermediates in the production of chelants such as ethylenediaminetetraacetate (EDTA) and nitrilotriacetate (NTA). The fatty acid nitriles are intermediates in the production of a large variety of commercial amines and amides.

General Health and Safety Factors

As a class of compounds, the two main toxicity concerns for nitriles are acute lethality and osteolathyrsm. Nitriles vary broadly in their ability to cause acute lethality and subtle differences in structure can greatly affect toxic potency. The biochemical basis of their acute toxicity is related to their metabolism in the body.

The propensity of nitriles to release cyanide subsequent to metabolism is the basis of their acute toxicity. Cyanohydrins are acutely toxic because they are unstable and release cyanide quickly. Persons handling nitriles should take precautions to prevent inhalation of fumes or skin contact.

Acetonitrile

[CAS: 75-05-8]. Acetonitrile (methyl cyanide), CH_3CN, is a colorless liquid with a sweet, ethereal odor. It is completely miscible with water and its high dielectric strength and dipole moment make it an excellent solvent for both inorganic and organic compounds including polymers. Many gases also are highly soluble in acetonitrile. It forms low boiling azeotropes with many organics and high boiling azeotropes with BF_3, $SiCl_4$, and $(CH_3)_4Pb$.

Although acetonitrile is one of the more stable nitriles, it undergoes typical nitrile reactions and is used to produce many types of nitrogen-containing compounds.

Because of its good solvency and relatively low boiling point, acetonitrile is used widely as a recoverable reaction medium, particularly for the preparation of pharmaceuticals. Its largest use is for the separation of butadiene from C_4 hydrocarbons by extractive distillation.

Acetonitrile also is used as a catalyst and as an ingredient in transition-metal complex catalysts. There are many uses for it in the photographic industry and for the extraction and refining of copper. It also is used as a reagent for the preparation of a wide variety of compounds.

Adiponitrile

Adiponitrile (hexanedinitrile, dicyanobutane, ADN), $NC(CH_2)_4CN$, is manufactured mainly for use as an intermediate for hexamethylenediamine (1,6-diaminohexane), which is a principal ingredient for nylon-6,6. BASF has announced the development of a process to make caprolactam from adiponitrile. Caprolactam is used to produce nylon-6.

Pure adiponitrile is a colorless liquid and has no distinctive odor. It is soluble in methanol, ethanol, chloroalkanes, and aromatics but has low solubility in carbon disulfide, ethyl ether, and aliphatic hydrocarbons. At

20 °C, the solubility of adiponitrile in water is ca 8 wt %; the solubility increases to 35 wt % at 100 °C.

Adiponitrile undergoes the typical nitrile reactions.

Adiponitrile is made commercially by several different processes utilizing different feedstocks.

The principal use of adiponitrile is for hydrogenation to hexamethylene diamine leading to nylon-6,6. Adipoquanamine, prepared by the reaction of adiponitrile with dicyandiamide (cyanoguanidine), has typical liquid nitrile properties that suggest its use as an extractant for aromatic hydrocarbons.

α-Aminonitriles

α-Aminonitriles are compounds containing both cyano and amine substituents attached to the same carbon atom. They are versatile synthetic intermediates that are used to make amino acids, agrichemicals, chelants, radical initiators, and water-treatment chemicals. In some cases, aminonitriles produced as intermediates are not isolated, but immediately further reacted, for example by hydrolysis, as is the case in producing ethylene-diaminetetraacetate (EDTA) or nitrilotriacetate (NTA). Isolated and commercially available aminonitriles include 2-amino-2-methylpropanenitrile (aminoisobutyronitrile, AN-64), 2-amino-2-methylbutanenitrile (AN-67), 2-amino-2,4-dimethylpentanenitrile (AN-52), and 1-aminocyclohexane carbonitrile (AN-88). The designations in parentheses arise from their identity as intermediates in the production of azo radical initiators.

In 1990, DuPont began practicing a one-step process in which a ketone is treated simultaneously with both HCN and ammonia at 40–60 °C. This process is both faster and more selective than previous two-step processes.

α-Aminonitriles are stable at modest temperatures (<70 °C) in the absence of water; in the presence of water, they can degrade to their original constituents, i.e., ketone (aldehyde), ammonia and hydrogen cyanide if insufficient ammonia is present. The aminonitriles based on ketones are clear colorless liquids, but sometimes appear yellow to brown depending on the synthetic procedure and the amount of decomposition. They are soluble in polar organic solvents and in aromatic solvents.

α-Aminonitriles may be hydrolyzed to amino acids, such as is done in producing ethylenediaminetetracetate (EDTA) or nitrilotriacetate (NTA). In these cases, formaldehyde is utilized in place of a ketone in the synthesis. The principal use of the ketone-based aminonitriles is in the production of azobisnitrile radical initiators.

Azobisnitriles

Azobisnitriles are efficient sources of free radicals for vinyl polymerizations and chain reactions, e.g., chlorinations. See also **Initiators (Free-Radical)**; **Initiators (Anionic)**; and **Initiators (Cationic)**. These compounds decompose in a variety of solvents at nearly first-order rates to give free radicals with no evidence of induced chain decomposition. They can be used in bulk, solution, and suspension polymerizations; and because no oxygenated residues are produced, they are suitable for use in pigmented or dyed systems that may be susceptible to oxidative degradation.

These compounds are essentially insoluble in water, sparingly soluble in aliphatic hydrocarbons, and soluble in functional compounds and aromatic hydrocarbons.

The azobisnitriles have been used for bulk, solution, emulsion, and suspension polymerization of all of the common vinyl monomers, including ethylene, styrene vinyl chloride, vinyl acetate, acylonitrile, and methyl methacrylate. The polymerizations of unsaturated polyesters and copolymerizations of vinyl compounds also have been initiated by these compounds.

Benzonitrile

[CAS: 100-47-0]. Benzonitrile, C_6H_5CN, is a colorless liquid with a characteristic almondlike odor. It is miscible with acetone, benzene, chloroform, ethyl acetate, ethylene chloride, and other common organic solvents but is immiscible with water at ambient temperatures and soluble to ca 1 wt% at 100 °C. It distills at atmospheric pressure without decomposition, but slowly discolors in the presence of light.

Like acetonitrile, benzonitrile is a powerful solvent for many inorganic and organic materials including some polymers. It can be converted to a large number and variety of derivatives by simple syntheses; e.g., by hydrolysis, it can be converted to either benzoic acid or benzamide. The most important reaction is with dicyandiamide to produce 2,4-diamino-6-phenyl-1,3,5-triazine (benzoguanamine).

Benzonitrile can be produced in high yield by the vapor-phase catalytic ammoxidation of toluene.

The most important commercial use for benzonitrile is the synthesis of benzoguanamine, which is a derivative of melamine and is used in protective coatings and molding resins. See also **Amino Resins**; and **Cyanamides**.

Cyanoacetic Acid and Esters

Cyanoacetic acid, $CNCH_2COOH$, is a strong organic acid with a dissociation constant at 25 °C of 3.36×10^3. It is prepared by the reaction of chloroacetic acid with sodium cyanide. It is hygroscopic and highly soluble in alcohols and diethyl ether but insoluble in both aromatic and aliphatic hydrocarbons. It undergoes typical nitrile and acid reactions but the presence of the nitrile and the carboxylic acid on the same carbon cause the hydrogens on C-2 to be readily replaced. The resulting malonic acid derivative decarboxylates to a substituted acrylonitrile:

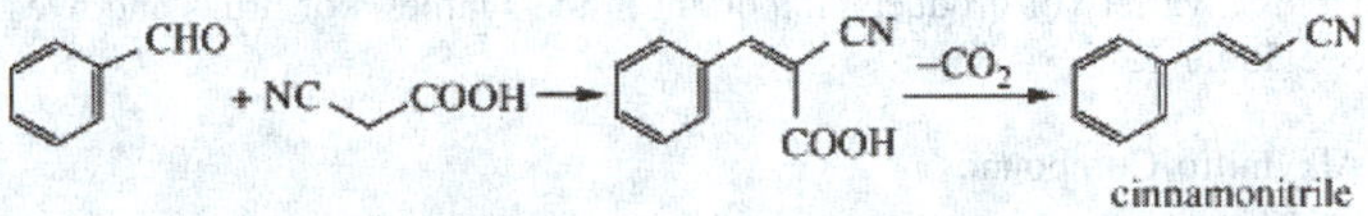

The methyl and ethyl esters of cyanoacetic acid are slightly soluble in water but are completely miscible in most common organic solvents including aromatic hydrocarbons. The esters, like the parent acid, are highly reactive, particularly in reactions involving the central carbon atom. They are prepared by esterification of cyanoacetic acid and are used principally as chemical intermediates.

Although cyanoacetic acid can be used in applications requiring strong organic acids, its principal use is in the preparation of malonic esters and other reagents used in the manufacture of pharmaceuticals. See also **Alkaloids**; and **Vitamin**.

Isophthalonitrile

Isophthalonitrile (1,3-dicyanobenzene, IPN), is a white solid which melts at 161 °C and sublimes at 265 °C. It is slightly soluble in water but readily dissolves in dimethylformamide, N-methylpyrrolidinone and hot aromatic solvents. IPN undergoes the reactions expected of an aromatic nitrile. It is prepared by vapor-phase ammoxidation of meta-xylene. Its principal use is as an intermediate to amines. As a reagent, IPN can be used to convert aromatic acids to nitriles in near quantitative yields.

2-Methylglutaronitrile

Methylglutaronitrile is readily hydrogenated to give 2-methyl-1,5-pentanediamine (DYTEK A, MPMD), used as a comonomer in polyamide fibers and resins, as a curing agent for epoxy coatings, and as its isocyanate in specialty urethanes. A co-product of the DYTEK A process is 3-methylpiperidine, which can be used to produce vulcanization accelerators for rubber curing.

Pentenenitriles

Pentenenitriles are produced as intermediates and by-products in DuPont's adiponitrile process. 3-Pentenenitrile is the principal product isolated from the isomerization of 2-methyl-3-butenenitrile.

3-Pentenenitrile (3PN) is used entirely by the manufacturers to make adiponitrile. cis-2-Pentenenitrile (2PN) can be cyclized catalytically at high temperature to produce pyridine, a solvent and agricultural chemical intermediate. 2PN is also used in the manufacture of pentachloropyridine, an intermediate in the insecticide Dursban, and 1,3-pentadiamine, which is used as a curing agent for epoxy coatings and as a chain modifier in polyurethanes.

Fatty Acid Nitriles

Fatty acid nitriles are produced as intermediates for a large variety of amines and amides. Fatty acid nitriles are produced from the corresponding acids by a catalytic reaction with ammonia in the liquid phase. They have little use other than as intermediates.

Additional Reading

Barton, D.H.R. and W.D. Ollis: *Comprehensive Organic Chemistry*, Vol. 2, Pregamon Press, Oxford, UK, 1979, pp. 528–562.

Hoffman, R.V.: *Organic Chemistry: An Intermediate Text*, 2nd Edition, John Wiley & Sons, Inc., Hoboken, NJ, 2004.

Lide, D.R.: *CRC Handbook of Chemistry and Physics*, 88th Edition, CRC Press, Boca Raton, FL, 2007.

Moss, R.A., M.S. Platz, and M. Jones: *Reactive Intermediate Chemistry*, John Wiley & Sons, Inc., Hoboken, NJ, 2004.

Moury, D.T. *Chem. Rev.*, **42**, 192 (1948).

Solomons, T.W. Graham, and C.B. Fryhle: *Organic Chemistry*, 8th Edition, John Wiley & Sons, Inc., New York, NY, 2003.

Torssell, K.: *Nitrile Oxides, Nitrones and Nitronates in Organic Synthesis: Novel Strategies in Synthesis*, Vol. 20, John Wiley & Sons, Inc., New York, NY, 1988.

NITRO- AND NITROSO-COMPOUNDS. Nitro-compounds contain the nitro-group ($-NO_2$) attached directly to a carbon atom; nitroso-compounds contain the nitroso-group ($-NO$) similarly attached. A very important member of this group is nitrobenzene, which upon reduction yields a variety of products, important in the synthesis of drugs and dyes. See Table 1.

Alkylnitro-Compounds:

Primary	Secondary	Tertiary
$CH_3CH_2 \cdot NO_2$	$(CH_3)_2CH \cdot NO_2$	$(CH_3)_3C \cdot NO_2$
Nitroethane	Nitrodimethylmethane	Nitrotrimethylmethane
	(2 - nitropropane)	

Isomeric Nitrites:

$CH_3CH_2 \cdot ONO$	$(CH_3)_2CH \cdot ONO$	$(CH_3)_3 \cdot ONO$
Ethylnitrite	Isopropylnitrite	1,1 - dimethylethyl nitrite

Alkylnitroso-Compounds:

$(CH_3)_3C \cdot NO$
Nitrosotrimethylmethane

Nitrates:

$CH_3CH_2 \cdot ONO_2$	$(CH_3)_2CH \cdot ONO_2$	$(CH_3)_3C \cdot ONO_2$
Ethylnitrate	Isopropylnitrate	1,1 - dimethylethylnitrate

Nitrosamine:

$(C_2H_5)_2N{:}NO$
Diethylnitrosamine

Benzenoid Nitro- and Nitroso-Compounds:

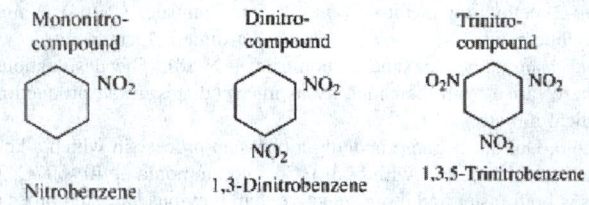

Mononitro-compound	Dinitro-compound	Trinitro-compound
Nitrobenzene	1,3-Dinitrobenzene	1,3,5-Trinitrobenzene

TABLE 1. REPRESENTATIVE NITRO-AND NITROSO COMPOUNDS

Compound	Formula	Melting Point, °C	Boiling Point, °C
	REPRESENTATIVE NITRO COMPOUNDS		
Nitrobenzene [CAS: 98-95-3]	$C_6H_5 \cdot NO_2$	6	211
1,3-Dinitrobenzene [CAS: 99-65-0]	$C_6H_4(NO_2)_2$ (1,3)	90	302
2-Nitrotoluene [CAS: 88-72-2]	$CH_3C_6H_4(NO_2)$ (2)	−11	222
2,4-Dinitrotoluene [CAS: 121-14-2]	$CH_3C_6H_3(NO_2)_2$ (2,4)	70	300
Trinitrotoluene (TNT) [CAS: 118-96-7]	$CH_3C_6H_2(NO_2)_3$ (2,4,6)	81	240 expl.
3-Nitrophenol [CAS: 554-84-7]	$HOC_6H_4 \cdot NO_2$ (3)	96	194 (70 torr)
2,4,6-Trinitrophenol [CAS: 29663-1104] (picric acid) [CAS: 88-89-1]	$HOC_6H_2(NO_2)_3$ (2,4,6)	122 expl.	>300
4-Nitrobenzaldehyde [CAS: 555-16-8]	$C_6H_4(COH)(NO_2)$ (1,4)	58	164 (23 torr)
4-Nitrobenzoic acid [CAS: 62-23-7]	$C_6H_4(COOH)(NO_2)$ (1,4)	240	subl.
4-Nitrobenzyl alcohol [CAS: 619-73-8]	$C_6H_4(CH_2OH)(NO_2)$ (1,4)	93	185 (12 torr)
2-Nitronaphthalene [CAS: 581-89-5]	$C_{10}H_7(NO_2)$ (2)	79	165 (15 torr)
1-Nitroanthraquinone [CAS: 82-34-8]	$C_6H_4(CO)_2C_6H_3(NO_2)$ (1)	230	subl.
2-Nitro-propane [CAS: 79-46-9]	$(CH_3)_2CHNO_2$	−93	120
Nitroethyl alcohol [CAS: 625-48-9]	$CH_2OHCH_2NO_2$	< −80	194
Nitrobromoform (bromopicrin) [CAS: 464-10-8]	NO_2CBr_3	10	expl.
Nitrochloroform (chloropicrin) [CAS: 76006-2]	NO_2CCl_3	−64	112
Nitrofurane	$C_4H_3O \cdot NO_2$	28	
Nitrourea [CAS: 556-89-8]	$\begin{array}{c} NH_2 \\ OC \\ NHNO_2 \end{array}$	155 dec.	
Nitroguanidine [CAS: 556-88-7]	$\begin{array}{c} NH_2 \\ HNC \\ NHNO_2 \end{array}$	246	
1,3-Nitroaniline	$C_6H_4(NO_2)(NH_2)$ (1,3)	114	>285
	REPRESENTATIVE NITROSO COMPOUNDS		
Nitrosobenzene [CAS: 586-96-9]	C_6H_4NO	68	58 (18 torr)
4-Nitrosophenol (4-quinoneoxime) [CAS: 104-91-6]	$C_6H_4(OH)(NO)$ (1,4)	125	144 dec.
4-Nitrosonaphthol-1 (4-naphthaquinoneoxime)	$C_{10}H_6(OH)(NO)$ (1,4) or $C_{10}H_6(O)(NOH)$ (1,4)	193	
2-Nitrosonaphthol-1	$C_{10}H_6(OH)(NO)$ (1,2)	163 dec.	
N-Nitrosomethylaniline	$\begin{array}{c} CH_3 \\ C_6H_5N \\ NO \end{array}$	13	128 (20 torr)
4-Nitrosophenylaniline	$C_6H_5NH \cdot C_6H_4NO$	145	
1-Nitrosonaphthylamine-2	$C_{10}H_6(NH_2)(NO)$ (2,1)	151	
Diphenylnitrosamine	$(C_6H_5)_2N \cdot NO$	66	

dec., decomposes; expl., explodes; sub., sublimes

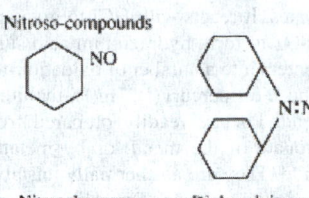

Nitroso-compounds

NO

Nitrosobenzene   Diphenylnitrosar

N:N

Under the proper conditions of concentration of HNO_3 and of temperature, benzene forms mainly nitrobenzene, nitrobenzene forms mainly 1,3-dinitrobenzene, and 1,3-dinitrobenzene, mainly 1,3,5-trinitrobenzene.

When nitrobenzene is treated (1) with zinc and calcium chloride or ammonium chloride solution, beta-phenylhydroxylamine, C_6H_5NHOH, is formed, and from this by treatment with chromic acid or ferric chloride nitrosobenzene is formed, (2) with tin or iron and HCl, aniline, $C_6H_5NH_2$, is formed and from this by treatment with nitrous acid followed by treatment with stannous chloride plus HCl phenylhydrazine, $C_6H_5NH \cdot NH_2$, is formed.

Mono- or poly-substituted nitro-compounds are changed in whole or in part to the corresponding amino-compounds by proper choice of reducing agent and temperature, e.g., in acid medium 1,3-dinitrobenzene yields 1,3-phenylenediamine, $C_6H_4(NH_2)_2(1, 3)$, and with ammonium sulfide yields 3-nitroaniline $(1)H_2NC_6H_4NO_2(3)$. When diphenylnitrosamine is reduced, 1,1-diphenylhydrazine, $(C_6H_5)_2N \cdot NH_2$, is formed.

See also **Nitration**.

Additional Reading

Nishimura, S.: *Handbook of Heterogeneous Catalytic Hydrogenation for Organic Synthesis*, John Wiley & Sons, Inc., New York, NY, 2001.

Patai, S.: *Chemistry of Amino, Nitroso and Nitro Compounds*, Vol. 1, John Wiley & Sons, Inc., New York, NY, 1997.

NITROCELLULOSE. See **Cellulose**.

NITROGEN. [CAS: 7727-37-9]. Chemical element, symbol N, at. no. 7, at. wt. 14.0067, periodic table group 15, mp $-209.86 °C$, bp $-195.8 °C$, critical temperature $-147.1 °C$, critical pressure 33.5 atmospheres, density $1.14 g/cm^3$ (solid), 1.25057 g/L (0°C, 760 torr), 0.9675 (air = 1.0000). Solid nitrogen has a hexagonal crystal structure. Nitrogen at standard conditions is a colorless, odorless, tasteless gas. The gas is slightly soluble in H_2O (2.35 parts nitrogen in 100 parts H_2O at 0°C), the solubility decreasing with increasing temperature (1.55 parts nitrogen in 100 parts H_2O at 20°C). Nitrogen is slightly soluble in alcohol and is essentially insoluble in most other known liquids. There are two naturally occurring isotopes, ^{14}N and ^{15}N, with ^{14}N by far the most abundant (99.635%). Four radioactive isotopes have been identified, ^{12}N, ^{13}N, ^{16}N, and ^{17}N, all with extremely short half-lives measured in seconds or minutes. In terms of abundance in igneous rocks in the Earth's crust, nitrogen does not appear among the first 37 most abundant elements. In terms of abundance in seawater, nitrogen ranks 16th, with an estimated 2,300 tons of nitrogen per cubic mile of seawater. In terms of cosmic abundance, nitrogen ranks 7th. For comparison, assigning a value of 10,000 to silicon, the figure for nitrogen is 160,000 and that for hydrogen, estimated the most abundant, a figure of 3.5×10^8. Of dry air in the earth's atmosphere, disregarding pollutants, 78.09% is nitrogen by volume and 75.54% by weight. In the atmosphere, the nitrogen is mixed with oxygen, argon, the rare gases, CO_2, and H_2O vapor. Nitrogen was first identified as an element by Daniel Rutherford in 1772, and about that same time, Scheele, Priestly, and Cavendish were also working with this "burnt" or "dephlogisticated" air. The name nitrogen from the Latin *nitrum*, or the Greek combination, *nitron* "native soda" and *gene* "forming" was suggested by Chaptal in 1790 when it was learned that niter and nitric acid were formed from this constituent. In the 1780s nitrogen oxides were produced by combining nitrogen and oxygen using an electrical discharge. Nitrogen was first liquified by Cailletet in 1877. In the early 1900s atmospheric nitrogen was first used for large-scale industrial purposes. Calcium cyanamide was first produced in 1895 by the Frank-Caro process. In 1900, Birkeland-Eyde developed the first industrial oxidation of nitrogen. Ostwald was awarded the 1909 Nobel Prize for work that led to the industrial-scale catalytic oxidation of NH_3 to HNO_3. Two more Nobel Prizes were granted for the Haber-Bosch process for the catalytic synthesis of Ammonia from N_2 and H_2. This process reached industrial scale by 1913. In the 1990s, five of the 15 largest volume industrial chemicals produced in the United States contain nitrogen: ammonia, nitrogen (gaseous and liquified), ammonium nitrate, Nitric Acid, and Urea.

Like oxygen, nitrogen is essential to practically all forms of life, making some of the compounds of this element extremely important as foods and fertilizers. Nitrogen serves the important function of diluent in the earth's atmosphere, controlling natural burning and respiration rates that otherwise would proceed much faster with higher concentrations of oxygen. Nitrogen is an important ingredient of numerous inorganic and organic compounds, including alkaloids, amides, amines, cyanides, cyanogens, diazo compounds, hydrazines, imides, nitrates, nitrides, nitrites, nitriles, oximes, purines, pyridines, and ureas. In terms of high-tonnage production, the nitrogen compound NH_3 (ammonia) ranks first with worldwide production exceeding 50 million tons annually.

First ionization potential 14.84 eV; second, 29.47 eV; third, 47.17 eV; fourth, 73.5 eV; fifth, 97.4 eV. Oxidation potentials $H_2N_2O_2 + 2H_2O \rightarrow 2HNO_2 + 4H^+ + 4e^-$, -0.80 V; $N_2O_4 + 2H_2O \rightarrow 2NO_3^- + 4H^+ + 2e^-$, -0.81 V; $HNO_2 + H_2O \rightarrow NO_3^- + 3H^+ + 2e^-$, -0.94 V; $NO + 2H_2O \rightarrow NO_3^- + 4H^+ + 3e^-$, -0.96 V; $NO + H_2O \rightarrow HNO_2 + H^+ + e^-$, -0.99 V; $2NO + 2H_2O \rightarrow N_2O_4 + 4H^+ + 4e^-$, -1.03 V; $2HNO_2 \rightarrow N_2O_4 + 2H^+ + 2e^-$, -1.07 V. $N_2O + 3H_2O \rightarrow 2HNO_2 + 4H^+ + 4e^-$, -1.29 V; $N_2O + H_2O \rightarrow 2NO + 2H^+ + 2e^-$, -1.59 V; $N_2 + H_2O \rightarrow N_2O + 2H^+ + 2e^-$, -1.77 V; $N_2O_4 + 4OH^- \rightarrow 2NO_3^- + 2H_2O + 2e^-$, 0.85 V; $NO + 2OH^- \rightarrow NO_2^- + H_2O + e^-$, 0.46 V; $N_2O_2^{2-} + 4OH^- \rightarrow 2NO_2^- + 2H_2O + 4e^-$, 0.18 V; $NO_2^- + 2OH^- \rightarrow NO_3^- + H_2O + 2e^-$, -0.01 V; $N_2O_2^{2-} \rightarrow 2NO + 2e^-$ $-.10$ V; $N_2O + 6OH^- \rightarrow 2NO_2^- + 3H_2O + 4e^-$ $-.15$ V; $N_2O + 2OH^- \rightarrow 2NO + H_2O + 2e^-$ $-.76$ V. $2NO_2^- \rightarrow N_2O_4 + 2e^-$ $-.88$ V.

Other physical properties of nitrogen are given under **Chemical Elements**.

Industrial Nitrogen

Like many of the elements, the compounds of nitrogen by far exceed the use of elemental nitrogen (discounting its important role as diluent in the atmosphere). Industrially, nitrogen gas is produced as a by-product in the liquefaction of air to produce pure oxygen. For some applications, nitrogen provides an excellent inert atmosphere for electric furnace operations and for the gaseous insulation of transformers. An inert atmosphere is required where air must be excluded. Nitrogen is one of the three main gases used for such atmospheres, the other two being carbon monoxide and hydrogen. In providing an inert atmosphere, nitrogen reduces the velocities of reactions, lowers the partial pressure and reduces the flammability of any active gases that may be present. Since commercial nitrogen usually contains traces of oxygen, H_2O vapor, and CO_2, sufficient to cause some oxidation at high temperatures, methane may be added to make the gas fully inert.

Nitrogen gas also is required for nitriding certain alloy steels, but pure gas is not required. The nitrogen is provided by dissociating ammonia at the process temperatures ranging from 475–650°C. Metals treated in this manner are hardened by the formation of nitrides on their surface (casehardening). In cyaniding, iron-base alloys simultaneously absorb carbon and nitrogen by heating the metals in a cyanide salt. Again, the nitrogen is not required in initial gaseous form. See also **Nitriding**. Several powder metallurgy techniques also utilize dissociated NH_3 atmospheres.

Environmental Aspects of Nitrogen

The oxides of nitrogen are among the most critical of air pollutants—both in their effects and in their abatement. These aspects of nitrogen are discussed under **Pollution (Air)**.

Chemistry and Compounds

Most of the high-tonnage nitrogen-bearing compounds are described elsewhere in this volume. See also **Ammonia**; **Ammonium Chloride**; **Ammonium Hydroxide**; **Ammonium Nitrate**; **Ammonium Phosphates**; **Ammonium Sulfate**; and **Fertilizer**.

In the laboratory, nitrogen, mixed with argon, neon, krypton, and xenon, is obtained from the air by passing it over heated copper to remove the oxygen, or pure by fractional distillation of liquid air whereby the nitrogen distills off before the oxygen. Pure nitrogen may also be obtained by heating such compounds as ammonium nitrite and ammonium dichromate, and collecting the gas. Mixed with carbon monoxide in producer gas, nitrogen may be utilized without separation by first making methyl alcohol from carbon monoxide and hydrogen and then using hydrogen and nitrogen for ammonia. When nitrogen at low pressure is subjected to the silent electric discharge, activated nitrogen is produced. Activated nitrogen displays a golden yellow afterglow upon cessation of the current, increased by cooling and decreased by heating. This form of nitrogen is very active with phosphorus, with alkali metals (forming azides), with the vapor of zinc, mercury, cadmium, arsenic (forming nitrides), with many metallic chlorides (forming a green fluorescence), and with hydrocarbons (forming hydrocyanic acid and cyanides). The transformation of nitrogen to activated nitrogen is partial, and its return to ordinary nitrogen takes place rapidly, in about one minute.

The metal amides and imides are important in the nitrogen system. The amides of the active metals are produced by (1) reaction of the metal with NH_3, (2) reaction of the metal hydride with NH_3, (3) reaction of the metal nitride with ammonia, (4) reaction with another amide, as $KNH_2 + NaI \rightarrow NaNH_2 + KI$ (in liquid NH_3). This last method is generally useful for the preparation of the heavy metal amides and imides from halides and binary halogenoids of the heavy metals. Cadmium amide, $Cd(NH_2)_2$ and lead imide, PbNH, for example, are readily prepared in this way. In some cases neither the amide nor the imide is stable, and the reaction proceeds to the nitride.

$$3HgBr_2 + 6KNH_2NH_3 liq.Hg_3N_2 + 6KBr + 4NH_3$$

The metal amides and imides are very reactive with oxygen, and are often unstable or even explosive. Some nitrides (e.g., of silver, gold, and mercury) are explosive, but others are stable. The latter may be obtained, (1) by reaction with the metal with nitrogen or ammonia at higher temperatures, e.g., aluminum nitride and magnesium nitride, AlN and Mg_3N_2, (2) by deamination of the metal amide or azide on heating, e.g., Ba_3N_2. The great thermal stability of certain nitrides, e.g., those of boron, silicon and phosphorus, BN, Si_3N_4 and P_3N_5, is attributed to polymerization. Many of the transition metal nitrides are interstitial compounds and are hard and metal-like in their properties.

In the nitrogen system, hydrazine is analogous to hydrogen peroxide in the oxygen system, its structure being

$$\begin{array}{c} H \\ H:N:N:H \\ H \end{array}$$

It is readily oxidized, even undergoing auto-oxidation under many conditions, and it is a powerful reducing agent. Like hydrogen peroxide it readily disproportionates (e.g., with a platinum catalyst), giving nitrogen and NH_3. Its reactivity (and other properties) makes it, and its derivative, unsymmetrical dimethylhydrazine, important rocket fuels. It forms addition compounds with many substances, including a monohydrate with H_2O. Hydrazine ($pK_{B1} = 6.04$, $pK_{B2} = 14.88$) forms hydrazinium(1+) compounds, containing the $N_2H_5^+$ ion, analogous to ammonium, and hydrazinium(2+) compounds containing the $N_2H_6^{2+}$ ion.

Hydroxylamine is related in its structure both to hydrazine (see formula above) and to hydrogen peroxide.

$$\begin{array}{cc} H & H \\ H:N:O: & :O:O: \\ H & H \end{array}$$

The chemical properties of hydroxylamine also suggest a compound intermediate between hydrazine and hydrogen peroxide. Its bond lengths are, N−, 1.46 Å, N−H, 1.01 Å, O−H, 0.96 Å, and its angles are H−O−N, 103°, H−N−O, 105°, and H−N−H, 107°. It is a base ($pK_B = 9.02$), forming salts containing the hydroxylammonium ion $HONH_3^+$.

Hydrazoic acid, HN_3, $pK_A = 4.72$, and most of its covalent compounds (including its heavy metal salts) are explosive. It is formed (1) in 90% yield by reaction of sodium amide with nitrous oxide, (2) by reaction of hydrazinium ion with nitrous acid, (3) by oxidation of hydrazinium salts, (4) by reaction of hydrazinium hydrate with nitrogen trichloride (in benzene solution). Hydrazoic acid forms metal azides with the corresponding

hydroxides and carbonates. It reacts with HCl to give ammonium chloride and nitrogen, with H_2SO_4 to form hydrazinium acid sulfate, with benzene to form aniline, and it enters into a number of oxidation-reduction reactions.

The azides, except those of mercury(I), Hg(I), thallium(I), Tl(I), copper, Cu, silver, Ag, and lead, Pb, are readily prepared from hydrazoic acid and the oxide or carbonate of the metal, or by metathesis of the metal sulfate with barium azide. They are all thermally unstable, giving nitrogen and free metal or occasionally nitride. The azide ion appears to resonate between four structures:

$$^-\!:\ddot{N}=\overset{+}{N}=\ddot{N}:^-\,,\ \ ^-\!:\underset{\cdot\cdot}{N}=\overset{+}{N}=\ddot{N}:^-$$

$$^-\!:\underset{\cdot\cdot}{\ddot{N}}-\overset{+}{N}\equiv N:^-\,,\ \ ^-\!:N\equiv\overset{+}{N}-\underset{\cdot\cdot}{\ddot{N}}:^=$$

These structures are in accord with a spacing of 1.15 Åand electronic charges of −0.83, 0.66, and 0.83 on the three nitrogen atoms.

N(I) Compounds. Hydration of nitrogen(I) oxide, N_2O to hyponitrous acid, $H_2N_2O_2$, is not possible. However, the latter decomposes (in three steps) to yield the former, which is thus its anhydride. Spectroscopic studies indicate a linear structure for N_2O, resonating between

$$^-\!:\ddot{N}=\overset{+}{N}=\ddot{O}:\ \text{and}\ \ddot{N}\equiv\overset{+}{N}-\ddot{O}:^-$$

However, heat capacity measurements give a higher entropy at low temperatures than spectroscopic studies do, which is explained by a partial randomness of the structure at low temperatures.

Hyponitrous acid ($pK_{A1} = 7.05$, $pK_{A2} = 11.0$) and its salts are obtained by: (1) reduction of sodium nitrite with (a) sodium amalgam, (b) by electrolysis, (c) by stannous or ferrous salts; (2) by reduction of alkyl nitrates; (3) by reduction of hydroxylamine by noble metal oxides; and (4) by reduction of sodium hydroxylamine monosulfonate in alkaline solution.

Explosive salts such as NaNO can be prepared by the reaction of NO and liquid ammonia solutions of alkali metals. The unstable free acid, HNO, is thought to be an intermediate in many redox reactions of nitrogen compounds.

Nitramide, NO_2NH_2, a weak acid ($pK_A = 6.59$), is relatively more stable than its isomer hyponitrous acid.

N(II) Compounds. Nitrogen(II) oxide is formed in many reductions of nitrous acid, but is best prepared pure by reduction with ferrous ions, Fe^{2+}, or iodide ions, I^- It undergoes many types of addition reactions, but its very slight tendency to dimerize and its low reactivity under ordinary conditions suggest that its odd electron lies in an antibonding orbital of very low energy; and the molecular orbital formulation is

$$NO[KK(z\sigma)^2(y\sigma *)^2(x\sigma)^2(w\pi)^4(v\pi *)]$$

The nitrosyl compounds can be readily classified on the basis of three modes of reaction of the NO molecule in accordance with the above formulation.

1. It can lose (or partly lose) the odd electron to form an ion of the formula

$$:N\equiv O:^+$$

This formula gives rise to ONF, ONCl and ONBr by direct reaction of NO and the halogen. These are covalent compounds. Such salts as $NOBF_4$, $NOPF_6$, $NOAuF_4$, $NOSO_3F$, and $NOHSO_4$, on the other hand, are ionic. These may be considered the salts of nitrous acid acting as a base, $ONOH \rightleftharpoons NO^+ + OH^-$, $pK_B = 18.2$.

2. It can gain an electron to form a negative ion of the formula

$$N\equiv\ddot{O}:^-$$

Thus dry NO reacts with sodium in liquid ammonia to form sodium nitrosyl, NaNO (empirical formula).

3. It can share a pair of electrons to form a coordinate link, as it does in coordination compounds. In most of these, it appears to coordinate as the positive ion, by transfer of an electron to an acceptor metal, which is thereby reduced by 1 unit in oxidation state. This causes, in

some cases, the need for placing a negative charge on the metal. To avoid this, Pauling assumed the presence of four bonding electrons, involving structures of the type

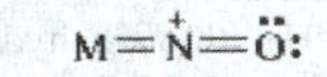

$$M = \overset{+}{N} = \ddot{O}:$$

Nitrogen(III) Compounds. Nitrogen(II) oxide, NO, readily enters into equilibrium with NO_2 to form N_2O_3, nitrogen sesquioxide. The latter is unstable even at room temperature and consists of an equilibrium mixture of the three compounds. Its structure appears to be $ON = N - NO_2$. If an equimolar mixture of NO and NO_2 is cooled and condensed, a blue liquid, bp $3.5\,^{\circ}C$, largely N_2O_3, is obtained. The latter readily combines with H_2O to form nitrous acid, HNO_2 ($pK_A = 3.29$). Nitrous acid is unstable, forming the equilibrium mixture, $3HNO_2 \rightleftharpoons NO_3^- + 2NO + H_3O^+$, which in concentrated solution or on warming is largely displaced to the right ($K = 39.6$ at $30\,^{\circ}C$). Moreover, the NO undergoes further reactions, so that the actual system is complex. One of these reactions is: $NO + OH \rightarrow NO^+ + OH^- \rightleftharpoons NO \cdot OH \rightleftharpoons HNO_2$.

The existence of NO^+ and NO^- helps to explain the kinetics of nitrous acid as an oxidizing agent. It oxidizes I^-, Sn^{2+}, Fe^{2+}, Ti^{3+}, $S_2O_3^{2-}$, SO_2, and H_2S. It reacts with NH_3, urea, sulfonates and some other nitrogen compounds to produce nitrogen. With aromatic amines in the cold, it gives diazo compounds, while with secondary amines it gives nitroso compounds. Nitrous acid also functions as a reducing agent, as in the reactions with permanganate and hydrogen peroxide, in which nitrate ion is formed.

The nitrites vary widely in solubility, those of the alkalies and alkaline earths being very soluble, while those of the heavy metals are only slightly so. Moreover, the latter are relatively unstable, some decomposing at room temperature. The nitrites, like nitrous acid, function either as oxidizing or reducing agents. X-ray and spectroscopic studies give a triangular structure for the nitrite ion, with the $N-O$ bond length 1.13 Å and the $O-N-O$ angle $120-130°$. Values of 1.23 Å and 116° have also been reported. Complex ions containing the NO_2 group may be either nitrito complexes (e.g., $Co(NH_3)ONO^{2+}$) or nitro complexes (e.g., $Co(NH_3)NO_2^{2+}$). The former of these two examples readily isomerizes to the latter.

Nitrosyl fluoride, NOF, and nitrosyl chloride, NOCl, are quite stable, but the bromide decomposes at room temperature. They are prepared by direct union of NO and the halogen, among other methods. Three trihalides, NF_3, NCl_3, and NI_3, are known. The first is a colorless stable gas; NCl_3 is a yellow liquid and NI_3, a brown solid; both are explosive. The contrast in stability is attributed to the large amount of ionic resonance energy of the $N-F$ bond, which gives NF_3 a negative heat of formation.

Nitrogen(IV) Compounds. Nitrogen dioxide, NO_2, readily associates to form the tetroxide, N_2O_4, so that at ordinary temperatures and pressures both forms are present in equilibrium. Since nitrogen dioxide has an unpaired electron, it is paramagnetic and colored (red). N_2O_4 is diamagnetic and colorless. As with NO, the odd electron is in an antibonding orbital but of higher energy so that NO_2 is more reactive and more readily undergoes dimerization. The $N-O$ bond length is 1.20 Å and the angle is 132° (electron diffraction). The structure of N_2O_4 is, on the basis of spectral and entropy considerations,

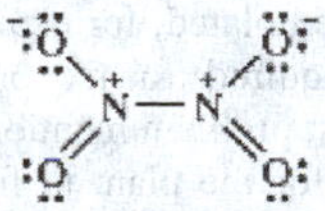

This formula is at variance with Pauling's stability argument, but is supported by Ingold's evidence (Nature, **159**, 743, 1947). Longuet-Higgins has proposed the structures

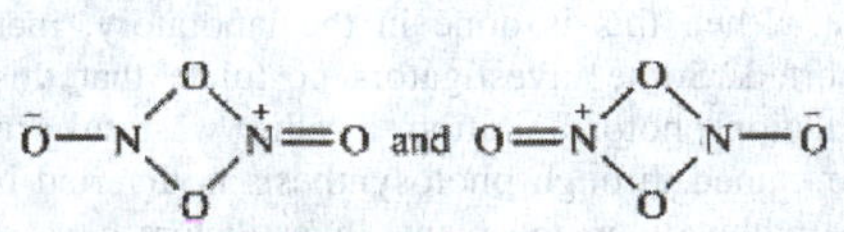

Nitrogen dioxide molecules react with NO to form N_2O_3, in an equilibrium mixture. The equilibrium mixture of NO_2 and N_2O_4 also reacts with water in a series of reactions

$$2NO_2 + H_2O \rightleftharpoons H^+ + NO_3^- + HNO_2$$

$$3HNO_2 \rightleftharpoons H^+ + NO_3^- + NO + H_2O$$

In warm solution, at high acidity, the second reaction is very rapid. In basic solutions the simple disproportionation $N_2O_4 + 2OH^- \rightarrow NO_2^- + H_2O$ takes place.

Nitrogen(V) Compounds. Nitrogen(V) oxide, N_2O_5, the anhydride of nitric acid, is a white solid subliming at $32.4\,^{\circ}C$ and 760 mm. It hydrates readily to HNO_3, is a strong oxidizing agent, and decomposes at $20\,^{\circ}C$ slowly into NO_2 and O_2. Its structure in the gas state consists of the molecules

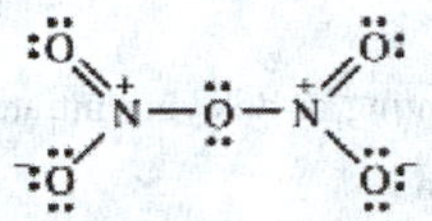

However, x-ray, Raman, and infrared spectra show the crystalline solid to consist of NO_2^+ and NO_3^- ions.

Pure nitric acid, HNO_3, is a colorless liquid boiling with decomposition at $86\,^{\circ}C$ and 760 torr. Upon continued heating it decomposes into NO_2, O_2 and H_2O. It is a fairly strong acid ($K_A = 22$), showing dissociation in concentrated solutions, and the presence of nitryl cation, NO_2^+ (nitronium ion). Solutions of HNO_3 in H_2SO_4 owe many of their properties to ions such as NO_2^+ and NO^+, as well, of course, as to HSO_4^- and oxonium ions.

The properties of HNO_3 are in accordance with resonance between the three electronic structures:

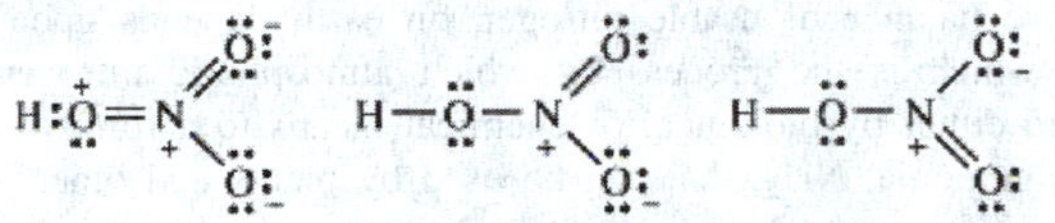

in which the last formula contributes a relatively small proportion to the overall structure. The two $N-O$ bond lengths are 1.22 Å, and $N-O-H$ bond lengths 1.41 Å and 0.96 Å. The $N-O-H$ angle is 90° and the $O-N-O$ angle 130°.

The reactions of nitric acid are of three types: (a) acid-base reactions which are typical of a strong acid; (b) oxidation reactions, such as those with metals and organic materials, the latter often involving carbonization; (c) substitution reactions such as the replacement of $-H$ by $-NO_2$ in aromatic hydrocarbons, to form nitro compounds, or of hydroxyl hydrogen by $-NO_2$ to produce esters of HNO_3.

These esters of nitric acid form one of the two groups of nitrates, the covalent group, which are also exemplified by nitryl hypofluorite and hypochlorite ($FONO_2$ and $ClONO_2$), often called fluorine and chlorine nitrate. Most nitrates, however, are ionic, e.g., salts of HNO_3. All metal nitrate are soluble in H_2O. Anhydrous metal nitrates, such as $Cu(NO_3)_2$. $Ti(NO_3)_4$, $VO(NO_3)_3$, $CrO_2(NO_3)_2$, $Si(NO_3)_4$, can be made by the action of liquid N_2O_4 on the metal (e.g., Cu) or of $ClONO_2$ on the corresponding chloride (e.g., the other examples given above).

The nitrate ion is considered to resonate between three equivalent structures of the form:

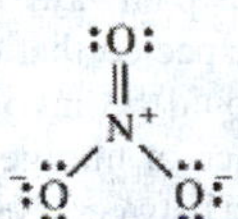

Two nitryl halides, NO_2F and NO_2Cl, are known, as well as nitryl salts, such as NO_2AsF_6, NO_2SbF_6, $(NO_2)_2SiF_6$, NO_2ClO_4, etc. Nitrogen also forms higher oxides, such as NO_3, and possibly NO_4, under action of the electric discharge.

Nitrate Losses from Disturbed Forest Ecosystems

Nutrient losses occur following a forest harvest or other disturbance, whether natural or anthropogenic. Studies have shown a variety of patterns of such losses. Vitousek et al. (1979) report on a systematic examination of nitrogen cycling in disturbed forest ecosystems and show that at least 8 processes, operating in 3 stages in the nitrogen cycle, can delay or prevent solution losses of nitrate from disturbed forests. The study involved 19 forest sites in the United States, including Pack Forest, Findley Lake, and Cascade Head in the northwest; Tesuque Watersheds in the southwest; Lake Monroe in southern Indiana; Coweeta in southwestern North Carolina; and Harvard Forest, Mount Mossilauke, and Cape Cod in the northeastern United States.

The 3 stages and 8 operative processes identified are:

Stage 1. Processes preventing or delaying ammonium accumulation.

- (a) Nitrogen immobilization
- (b) Ammonium fixation
- (c) Ammonia volatilization
- (d) Plant nitrogen uptake

Stage 2. Processes preventing or delaying nitrate accumulation.

- (e) Lag in nitrification
- (f) Denitrification to: $-N_2$, N_2O, or NO_x, $-NH_4$

Stage 3. Processes preventing or delaying nitrate mobility.

- (g) Lack of water
- (h) Nitrate sorption
- (i) Denitrification at depth

The researchers stress that the net effect of all of these processes, except uptake by regrowing vegetation, is insufficient to prevent or delay losses from relatively fertile sites and thus such sites have the potential for very high nitrate losses following disturbance.

Nitrogen Fixation

A positive balance of usable nitrogen on earth depends upon nitrogen fixation which is the process by which atmospheric nitrogen, N_2, is converted either by biological or chemical means to a form of nitrogen, such as ammonia, NH_3, that can be used by plants and other biological agents. Insofar as the total amount of N_2 fixed, the biological processes for converting from N_2 to NH_3 are the most significant. In biological nitrogen fixation, microorganisms, either free-living or in symbiosis with plants (mainly in root nodules), reduce N_2 to NH_3 at atmospheric pressure and within the temperature range of $20-37\,°C$. This natural process is to be contrasted with industrial chemical conversion processes, which may require up to 300 atmospheres of pressure and a reaction temperature range of $200-300\,°C$.

Biological Nitrogen Fixation. The occurrence and importance to soil fertility of biological nitrogen fixation have been known since the early 1800s. The first major finding did not occur until 1960, however, when it was shown that cell-free extracts of the anaerobic bacterium *Clostridium pasteurianum* could be made to fix nitrogen if molecular oxygen, O_2, were rigorously excluded—and also if pyruvic acid, a source of energy and electrons, was supplied. This finding demonstrated that studies no longer were restricted to whole cells, as previously indicated, but that it should be possible to isolate and chemically identify the components of the nitrogen-fixing system.

The first demonstrable product of cell-free N_2 fixation is NH_3, as had been strongly suggested by previous whole-cell studies. Since the reduction of N_2 to $2NH_3$ requires six electrons and since most electron transfer systems known in biochemical pathways involve either a one-or a two-electron transfer, it could be expected that either six one-electron or three two-electron transfer steps would be involved in nitrogen fixation. This would also suggest the existence of nitrogen compounds of valence states (reduction states) intermediate between N_2 and NH_3. However, no such intermediates have been found even in systems using cell-free extracts.

Because of failure to detect intermediates, attention was focused on the mechanism in extracts of *Clostridium pasteurianum* through which electrons were transferred from pyruvic acid to the nitrogen-fixing system. These investigations led to the discovery and isolation of the new electron carrier ferredoxin (Fd) which functioned by accepting electrons released during pyruvate oxidation by enzymes present in the clostridial extracts. The electrons from reduced Fd were transferred to a variety of different acceptors as directed by the cell. For example, some of the electrons from reduced Fd were transferred to hydrogenase, an enzyme which combined the electrons with protons (H^+) to produce molecular hydrogen, H_2, a major by-product of this anaerobe. Other electrons from reduced Fd were transferred via a flavoprotein carrier to nicotinamide adenine dinucleotide

phosphate ($NADP^+$) to yield NADPH, a reduced electron carrier shown to be important in the metabolism of all biological agents. It was also found that electrons from Fd were required for nitrogen fixation when pyruvate was present as supporting substrate.

A major finding was that H_2, through hydrogenase, would act as an electron source for reducing ferredoxin. Thus, in these extracts, H_2 could be used to reduce $NADP^+$ to NADPH and NO_{2-} to NH_3, and Fd was necessary as an intermediary electron carrier. Since Fd is required for pyruvate-supported N_2 fixation, it may be expected that H_2 would support nitrogen fixation, since reduced Fd is readily produced from H_2 in these extracts. Molecular H_2 alone, however, did no support N_2 fixation. This suggested either that a component other than reduced Fd was required, or that H_2, although capable of reducing Fd, was inhibitory to N_2 fixation as prior whole-cell studies had indicated. If an additional component were required, it appeared that it was produced from pyruvic acid, since pyruvic acid supported active N_2 fixation.

Several unsuccessful attempts were made to obtain N_2 fixation in extracts to which H_2, N_2, and one of the other products of pyruvate metabolism, ATP, were added. Active N_2 fixation did occur, however, when another product of pyruvate metabolism, acetyl phosphate, was added in addition to H_2 and N_2. When compounds such as ADP were removed from cell extracts by dialysis, no N_2 fixation occurred unless ADP was added together with phosphate, H_2, and N_2. Acetyl phosphate then was acting as a source of ATP. The reason ATP did not work directly was that a continuous supply of ATP was required, and a high concentration of ATP, if added directly to a cell-free extract, was highly inhibitory to N_2 fixation. In whole cells that are fixing N_2, a continuous supply of ATP is made available during sugar metabolism.

Genetic Manipulation. High on the list of many researcher's agendas for projects using the practical application of recombinant DNA research has been the possible development of a living organism that will produce ammonia—in an effort to lessen dependence upon costly and highly energy-consuming synthetic ammonia fertilizers. However, at symposia held on this topic, these achievements are considered by most researchers as quite long-range. There are fundamental problems difficult to overcome, including: (1) the possibility that increasing biological nitrogen fixation, for which the plant furnishes the energy, can cause a net decrease in crop yields by depriving the plant of nitrogen for the production of certain critical growth elements; and (2) the very rapid-acting inactivation by oxygen of nitrogen-fixation mechanisms. Cloning techniques may be a path toward introducing nitrogen-fixation genes into certain bacteria. One objective is that of developing new forms of bacteria that will enter into symbiotic relationships with crop plants, such as corn (maize) and wheat, that do not possess their own nitrogen fixation symbionts.

In addition to recombinant DNA and molecular cloning techniques, some scientists have combined their research with more conventional genetic techniques. An *E. coli* plasmid capable of carrying nitrogen-fixation genes of *K. pneumoniae* has been developed. Some researchers also believe that nitrogen-fixation genes may be introduced directly into plant cells to result in a plant that requires no nitrogen fertilizer.

In research activities such as these, much knowledge has been gained concerning the energy needs for biological nitrogen fixation. More energy is used than originally contemplated; for example, 20 moles of adenosine triphosphate (ATP) are required to fix one mole of nitrogen. This contributes largely to the first problem mentioned earlier, namely, the great amount of energy required for the plant to fix its own nitrogen, possibly leading to yield reduction.

The well-known nitrogen fixation by rhizobia depends on photosynthesis by the plant. Although the method is essentially impractical, photosynthesis can be increased by blanketing the plant with an atmosphere enriched in carbon dioxide. When this is done in the laboratory, increased legume yields are reported. Some investigators postulate that this effect is the result of a reduction photorespiration, a rather wasteful process in which carbon dioxide gained through photosynthesis is diverted into a series of less productive pathways in the plant. Investigators have also found that 30% of the energy used by the nitrogenase of most rhizobial species goes to producing hydrogen rather than ammonia. Research has also shown that the organisms that perform the nitrogen fixation function in plants are indeed quite diverse in themselves. Thus, new combinations of plants and organisms may increase efficiency in some cases.

As pointed out by Evans-Barber (1977), nitrogen is fixed by a variety of microorganisms in addition to those associated with legumes. Some of these include bacteria located in soils, in decaying wood, and on the surfaces of plant roots. They also include free-living blue-green algae, with fungi, ferns, mosses, liverworts, and higher plants (Hardy-Havelka, 1975). Reviews of numerous nitrogen-fixing organisms are given by Silvester (1976), Dalton (1974), Bond (1974), and Stewart (1974).

Role of Molybdenum in Nitrogen Fixation. Traditionally, molybdenum has been considered a key to the reduction of molecular nitrogen to ammonia by soil- and water-dwelling microorganisms. The metal is believed to be a part of the catalytically active site of nitrogenase, which is the enzyme that accomplishes the reduction. As early as 1980, researchers (North Carolina State University) suggested that the bacterium *Azobacter vinelandii* may have an alternative system for nitrogen fixation, a mechanism that may not require molybdenum. The proposal was regarded with some skepticism until the findings by researchers (Agriculture and Food Research Council Unit of Nitrogen Fixation, University of Sussex, England) in 1985 that confirmed the fact that *A. vinelandii* does have a second fixation system. Mutants of *A. vinelandii* were studied. Genes coding for the nitrogenase proteins were specifically deleted or inactivated. It was found that deletion of all three nitrogenase structural genes did not interfere with the fixation process. However, the process was effective only when molybdenum was not present. It was also found that the "wild" type of bacterium must have Mo in order to reduce nitrogen. Thus, it appears that the alternative mechanism is activated only when the system is subjected to molybdenum starvation. It has been suggested that the alternative system represents an adaptation to molybdenum-poor soils. The extension of these findings to other nitrogen-fixing microorganisms remains to be accomplished. Some further details pertaining to the Sussex investigation are given by Marx (1985).

Madigan (1979) and associates found that photosynthetic purple bacteria can grow with dinitrogen gas as the only source of nitrogen under anaerobic conditions, with light as the energy source. They also found that *Rhodopseudomonas capsula* can fix nitrogen in darkness with alternative energy conversion systems.

See also **Fertilizer.**

Health and Safety Factors

Gaseous nitrogen is nontoxic and nonflammable, but does not support life. Nitrogen should be stored and used only in well-ventilated areas. Special care must be taken entering an enclosed area, which may be enriched in nitrogen.

Liquid nitrogen and its vapor are extremely cold and can rapidly freeze human tissue. Liquid nitrogen spills should be flushed with water to accelerate evaporation. When exposed to liquid nitrogen, carbon steel, rubber, and plastic become embrittled and may fracture under stress. Copper, brass, bronze, Monel, aluminum, and 300 series austenitic stainless steels remain ductile and are acceptable for cryogenic service. Liquid nitrogen in poorly insulated containers can concentrate and condense atmospheric oxygen on the exterior surfaces, which may cause a serious fire hazard. Storage vessels or handling equipment should be provided with multiple pressure relief devices to prevent the buildup of high pressure. A pressure relief valve for primary protection and a frangible disk for secondary protection are commonly provided for on commercial liquid storage vessels.

Uses

Applications for nitrogen are widespread in both its gaseous and liquid phases. Gaseous nitrogen is usually used as an inert blanketing or carrier gas. Liquid nitrogen is used as an expendable nonreactive, nontoxic refrigerant. Very few applications, excepting the large-scale synthesis of ammonia from atmospheric nitrogen, use nitrogen as a reactant.

Chemical Process Industry. Gaseous nitrogen is used extensively in the chemical process industry as a blanketing, inerting, or purging agent to prevent oxidation of sensitive materials or to prevent the formation of an explosive mixture. Chemical storage tanks are frequently maintained under a slight positive pressure of nitrogen. During withdrawal of product from the storage tank, nitrogen is injected into the tank to maintain

headspace pressure; during filling of the storage tank, nitrogen is vented from the tank to atmosphere, a flare system, or a solvent recovery system. Process vessels, piping, and storage vessels are purged with nitrogen to remove oxygen or control reactant levels. Purging is performed by one of three methods, ie, displacement, pressurization, or dilution. Displacement purging is often used to purge piping systems. The volume of nitrogen required corresponds to the volume of the pipe. Pressurization purging involves the repeated pressurization with nitrogen and venting of a vessel to remove the contaminant. Dilution purging assumes complete mixing of the vessel contents and the injected nitrogen, and the contaminant concentration is slowly reduced through time.

Nitrogen is utilized during polymerization operations where oxygen is a polymerization inhibitor. Nitrogen is often used for the pressure transfer of hazardous or flammable liquids and as a drying agent for oxidation-sensitive materials such as nylon resins and fibers. Nitrogen is used to agitate and deoxygenate liquids and solid–liquid mixtures. It is also used as a carrier gas for other reactants such as oxygen in the regeneration of spent reforming catalyst in petroleum refining. Nitrogen is used as a blowing agent in foamed plastic production and in blow molding of plastic containers. It can be used as a stripping agent to remove volatile organic compounds from process wastewater.

Liquid nitrogen is used in cold traps to remove and recover solvents or volatile organic compounds from gas streams to reduce atmospheric emissions. Liquid nitrogen can be used to accelerate the cooldown time for process reactors.

Food Industry. Large volumes of liquid nitrogen are used in the food industry for cryogenic freezing, where it competes with liquid carbon dioxide and mechanical freezing technologies. Cryogenic freezing of frozen foods reduces cell damage, which results in reduced thawed product quality. By freezing the food very rapidly, inter- and intracellular ice crystal growth is minimized, avoiding rupture and damage of the cell walls of the food product. Concentration of water-soluble components between ice crystals during the freezing process is also reduced. Both effects reduce syneresis or drip loss due to release of cellular fluid, an effect that adversely impacts the texture, flavor, aroma, and nutritional value of the thawed product. Cryogenic freezing also reduces dehydration during the freezing process.

Liquid nitrogen also finds application in pressurizing aluminum and plastic food and noncarbonated beverage cans and bottles. Several drops of liquid nitrogen are injected into the can or bottle prior to sealing. After sealing, the nitrogen vaporizes to provide internal pressurization which provides sidewall strength and allows the use of very thin sidewall materials.

Other uses for liquid nitrogen include batch chilling of meat products and cookie dough, and cooling flour and sugar during warm season processing.

Gaseous nitrogen is used in controlled and modified atmosphere food storage and packaging to maintain the quality of fresh fruits and vegetables by retarding oxygen-dependent ripening and decay processes. Controlled atmospheres are extensively used in the storage of apples and pears to allow extended storage prior to sale without significant quality degradation.

Primary Metallurgy and Heat Treatment. In ferrous metallurgy, nitrogen is typically provided by pipeline as a co-product with oxygen to steel producing facilities for use as a scavenging, stirring, and inert blanketing gas during steel processing and casting operations. Nitrogen can be used to replace more expensive argon in the argon–oxygen decarburization (AOD) process during the manufacture of austenitic stainless steels. In nonferrous metallurgy, nitrogen finds application in degassing of aluminum melts to remove hydrogen and reduce oxide formation. Nitrogen is used to shroud and cool aluminum extrusion dies to reduce surface oxidation and increase die life.

In most heat treatment processes for metal fabricated parts, nitrogen serves as a nonreactive, passive constituent of the furnace atmosphere. It can be used alone for the annealing of aluminum, copper, and some low carbon steels. Most often it is used in combination with other gases such as H_2 and CO to produce reactive atmospheres for sintering, carburizing, and carbonitriding ferrous parts. Nitrogen reacts with some stainless steels and cannot be used in their heat treatment. At high temperatures, atomic nitrogen combines with iron to form finely divided nitrides, producing a hardened nitrided or carbonitrided surface layer.

Electronics Manufacturing. In the manufacture of semiconductors, gaseous nitrogen is used in the largest quantity of any gas as an inert carrier in epitaxy and diffusion processes, ion implantation, chemical vapor deposition, annealing, and plasma etching. Nitrogen purity requirements are extremely rigorous, reaching 1 ppb contaminant levels. On-site cryogenic nitrogen plants are often used as the supply mode. If delivered liquid nitrogen is used, careful purging of transfer lines is required to maintain purity. Facility piping is usually electropolished stainless steel.

In electronic component assembly operations, standard purity nitrogen finds uses in atmospheres for wave soldering and infrared reflow soldering machines. This application has grown because manufacturers are converting soldering operations from older technologies that required post-cleaning with chlorofluorocarbons, to no-clean technologies, that require nitrogen atmospheres. Nitrogen is also used extensively in various utility applications such as dry box storage of work-in-process components, blanketing of process chemicals, burn-in oven atmospheres, and nitrogen-driven pneumatic tools.

Oil and Natural Gas Production. Nitrogen is used in oil and natural gas well completion and stimulation applications as well as oil and natural gas enhanced recovery. Applications in well completion and stimulation include use as a foaming agent for fracturing, acidizing, cementing, and gravel packing fluids. Nitrogen up to 95% by volume is used in these fluids to reduce density and hydrostatic gradient in the wells, thereby reducing the amount of energy required to return the fluid to the surface after the operation. Nitrogen for these applications historically has been delivered in liquid form by tanker truck and vaporized on-site. In the 1990s, portable on-site membrane generators had begun to find application.

Nitrogen is used for pressure maintenance in oil and gas reservoirs for enhanced recovery. It is sometimes used as a miscible agent to reduce oil viscosity and increase recovery in deep reservoirs. Other applications include recovery of oil in attic formations, gas cap displacement, and a sweep gas for miscible CO_2 slugs. Nitrogen competes with CO_2, a more miscible gas with Hydrocarbons, in most of these applications. The production mode is typically by on-site cryogenic separation plants.

Two applications well suited to noncryogenic nitrogen production technologies include the replacement of cushion gas with nitrogen in underground natural gas storage reservoirs, pioneered by Gaz de France, and the use of nitrogen to displace natural gas in underground coal seams.

Other Applications. Other applications for gaseous nitrogen include use as a blanketing agent for float glass manufacture, pressurizing agent in aerosols, inflation agent in aircraft tires and landing struts, purging agent for electrical cabinets, and pressurization agent for autoclaves used in the production of composite materials.

A significant application for liquid nitrogen is the cryogenic grinding of plastic or heat-sensitive materials. Plastic materials become brittle at cryogenic temperatures enabling easier grinding or deflashing operations. Cryogenic grinding is performed on both thermoplastic and thermosetting resins, old rubber tires for recycling, spices, coffee, coloring agents and pigments, and wax.

Other liquid nitrogen applications include freezing biological specimens such as livestock semen and whole blood, cryosurgery, shrink fitting metal parts, and paint removal. Liquid nitrogen can be used for ground freezing to allow excavation in unstable wet soils, freeze plugging pipe sections to allow repairs while the rest of the pipeline remains pressurized, and cooling concrete in hot weather.

Additional Reading

Bingham, E., B. Cohrssen, and C. Powell: *Patty's Toxicology, Hydrocarbons - Organic Nitrogen Compounds, Vol. 4*, 5th Edition, John Wiley & Sons, Inc., New York, NY, 2000.

Bond, G.: in *The Biology of Nitrogen Fixation*, (A. Quispel, editor), North-Holland, Amsterdam, 1974.

Cheung, H. and J.H. Royal: Efficiently Produce Ultra-High-Purity Nitrogen On-Site, *Chem. Eng. Progress*, **64** (October 1991).

Clark, J. S.: *Nitrogen, Oxygen and Sulfur Ylide Chemistry (The Practical Approach in Chemistry Series 2002)*, Oxford University Press, New York, NY, 2002.

Dalton, H.: *Crit. Rev. Microbiol*, **3**, 183 (1974).

Evans, H.J. and L.E. Barber: Biological Nitrogen Fixation for Food and Fiber Production, *Science*, **197**, 332–339 (1977).

Graham, P.H. and S.C. Harris: *Biological Nitrogen Fixation*, Unipub, New York, NY, 1984.

Greenwood, N.N. and A. Earnshaw: *Chemistry of the Elements*, 2nd Edition, Butterworth-Heinemann, Inc., Woburn, MA, 1997.

Golterman, H.L.: *Chemistry of Phosphate and Nitrogen Compounds in Sediments*, Kluwer Academic Publishers, Norwell, MA, 2004.

Hardy, R.W.F. and M.D. Havelka: *Science*, **188**, 633 (1973).

Knowles, R. and T.H. Blackburn: *Nitrogen Isotope Techniques*, Academic Press, Inc., San Diego, CA, 1997.

Legocki, A., H. Bothe, and A. Puhler: *Biological Fixation of Nitrogen for Ecology and Sustainable Agriculture*, Springer-Verlag, Inc., New York, NY, 1997.

Leigh, G. J.: *Nitrogen Fixation at the Millennium*, Elsevier Science, New York, NY, 2002.

Lide, D.R.: *CRC Handbook of Chemistry and Physics* 86th Edition, CRC Press, LLC., Boca Raton, FL, 2005.

Madigan, M.T., J.D. Wall, and H. Gest: Dinitrogen Fixation by Photosynthetic Microorganisms, *Science*, **204**, 1429–1430 (1979).

Marx, J.L.: Fixing Nitrogen without Molybdenum, *Science*, **229**, 956–957 (1985).

Metz, C.B.: *Biology of Fertilization*, Academic Press, Inc., San Diego, CA, 1985.

Meyers, R.A.: *Handbook of Chemicals Production*, The McGraw-Hill Companies, Inc., New York, NY, 1986.

Postgate, J.R.: *Nitrogen Fixation*, Cambridge University Press, New York, NY, 1998.

Silvester, W.B.: in *Proceedings of the 1st International Symposium on Nitrogen Fixation*, (W.E. Newton and C.J. Nyman, editors), Washington State University Press, Pullman, WA, 1976.

Staff: Nitrogen Plant Opens in South Carolina, *Chem. Eng. Progress*, 10 (February 1990).

Staff: Can Catalytic Combustion in Jet Engines Zap NOx? *Chem. Eng. Progress*, 19 (March 1992).

Staff: Advanced Catalyst Zaps Nitrogen, *Chem. Eng. Progress*, 21 (July 1992).

Stevenson, F.J. and M.A. Cole: *Cycles of Soils: Carbon, Nitrogen, Phosphorus, Sulfur, Micronutrients*, 2nd Edition, John Wiley & Sons, Inc., New York, NY, 1999.

Stewart, W.D.P.: in *The Biology of Nitrogen Fixation*, (A. Quespel, editor), North-Holland, Amsterdam, 1974.

Vitousek, P.M. et al.: Nitrae Losses from Disturbed Ecosystems, *Science*, **204**, 469–474 (1979).

Web References

Nitrogen: http://www.praxair.com/nitrogen
Nitrogen Fixation: http://academic.reed.edu/biology/Nitrogen/
The Nitrogen Cycle: http://www.physicalgeography.net/fundamentals/9s.html

NITROGEN (Fertilizer). See **Fertilizer**.

NITROGEN GROUP (The). The elements of group 15 of the periodic classification sometimes are referred to as the Nitrogen Group. In order of increasing atomic number, they are nitrogen, phosphorus, arsenic, antimony, and bismuth. The elements of this group are characterized by the presence of five electrons in an outer shell. The similarities of chemical behavior among the elements of this group are less striking than hold for some of the other groups, e.g., the close parallels of the alkali metals or alkaline earths. Although all of the elements of this group have valences in addition to 5+, all do have the 5+ valence in common. Unlike the alkali metals or alkaline earths, for example, the elements of the nitrogen group are not so similar chemically that they comprise a separate group in classical qualitative chemical analysis separations. Three of the five, however, antimony, arsenic, and bismuth are members of the second group in terms of qualitative chemical analysis.

NITROGLYCERIN. See **Coronary Artery Disease (CAD)**; and **Explosive**.

NMR SPECTROSCOPE. See **Nuclear Magnetic Resonance**.

NOBEL, ALFRED BERNHARD (1833–1896). Nobel was a Swedish industrialist and European munitions maker. He was born in Stockholm in 1833. Most of his early education was from tutors and what he learned as he traveled during his teenage years through much of North America and Europe. He learned to speak several languages fluently and studied mechanical engineering.

After his travels, he joined his father's business and worked developing mines, torpedoes, explosives, and other war materials for the Russian czar.

The Nobel factory was financially successful until the end of the Crimean War when it fell into bankruptcy. Alfred then went into business with his brother manufacturing drilling tools for oilfields.

In the 1860's Alfred revolutionized the explosives industry by developing the Nobel detonator, a new fuse for nitroglycerin followed by a safe way to handle nitroglycerin by using an organic packing material to reduce its volatility and producing dynamite. In the late 1880's, Nobel produced ballistite, a smokeless powder.

Nobel's inventions brought him a monetary fortune as he controlled most of the explosive manufacturing factories throughout the world. He was, however, disillusioned that his inventions had mostly applications for war. Nobel personally desired world peace. Throughout his life, he had acted in a humanitarian manner. In his will, Nobel directed that the great majority of his estate be used for the purpose of giving yearly prizes to persons whose personal efforts made outstanding contributions to the advancement of physics, chemistry, medicine and physiology, literature, and world peace. The first Nobel Prizes were awarded in 1901 to Wilhelm K. Roentgen (discovery of X-rays in 1895), J.H. van't Hoff (chemical thermodynamics and osmotic pressure), and E.A. von Behring (diphtheria antitoxin). See also **Nobel Prizes**.

Web Reference

Nobel Foundation: http://www.nobel.se/nobel/nobel-foundation/index.html

J. M. I.

NOBELIUM. [CAS: 10028-14-5]. Synthetic radioactive chemical element, symbol No, at. no. 102, at. wt. 254 (mass number of ^{254}No), radioactive metal of the Actinide series, also one of the Transuranium elements. Nobelium has valences of 2^+ and 3^+. In 1957, a group of American, English, and Swedish scientists bombarded a target of several curium isotopes (largely ^{244}Cm) with a beam of ^{13}C ions from the cyclotron at the Nobel Institute for Physics. They obtained a few alpha particles of 8.5 MeV energy and half-life of 10 minutes. This was considered to indicate the presence of element 102 with a probable mass number of 251 or 253. At that time, the element was named nobelium with assignment of the symbol, No. Further experiments at the University of California, however, failed to confirm this discovery. In April 1958, Ghiorso, Sikkeland, Walton, and Seaborg, working with the heavy ion linear accelerator (HILAC) at Berkeley, showed the isotope 102^{254} to be a product of the bombardment of ^{246}Cm with ^{12}C ions. Confirming experiments at Berkeley in 1966 showed the existence of ^{254}No with a 55-second half-life; ^{252}No with a 2.3 second half-life; and ^{257}No with a 23-second half-life. Four other isotopes are now recognized, including ^{255}No with a half-life of 3 minutes.

In 1973, scientists at Oak Ridge National Laboratory and Lawrence Berkeley Laboratory, produced a relatively long-lived isotope of nobelium through the bombardment of ^{248}Cm with ^{18}O ions. A total half-title of 58 ± 5 minutes was computed from the combined data of both laboratories. See also **Chemical Elements**.

Additional Reading

Ditmer, P.F. et al.: "Identification of the Atomic Number of Nobelium by an X-ray Technique," *Phys. Rev. Lett.*, **26**, 17, 1037–1040 (1971).

Fields, P.R. et al.: "Production of the New Element 102," *Phys. Rev.*, **107**, 5, 1460–1462 (1957).

Flerov, G.N. et al.: "Experiments to Produce Element 102," *Sov. Phys. Dokl.*, **3**, 3, 546–548 (1958).

Ghiorso, A., T. Sikkeland, J.R. Walton, and G.T. Seaborg: "Attempts to Confirm the Existence of the 10-minute Isotope of 102," *Phys. Rev. Lett.*, **1**, 1, 17–18 (1958).

Ghiorso, A., T. Sikkeland, J.R. Walton, and G.T. Seaborg: "Element No. 102," *Phys. Rev. Lett.*, **1**, 18–20 (1958).

Hammond, C.R.: "The Elements," in *Handbook of Chemistry and Physics*, 67th Edition, CRC Press, Boca Raton, Florida, 1986–1987.

Maly, J., T. Sikkeland, R. Silva, and A. Ghiorso: "Nobelium: Tracer Chemistry of the Divalent and Trivalent Ions," *Science*, **160**, 1114–1115 (1968).

Marks, T.J.: "Actinide Organometallic Chemistry," *Science*, **217**, 989–997 (1982).

Mikheev, V.L. et al.: "Synthesis of Isotopes of Element 102 with Mass Numbers 254, 253, and 252," *Sov. At. Energy*, **22**, 93–100 (1967).

Seaborg, G.T. (editor): *Transuranium Elements*, Dowden, Hutchinson & Ross, Stroudsburg, PA, 1978.

Silva, R.J. et al.: "The New Nuclide Nobelium-259," *Nucl. Phys.*, **A216**, 97–108 (1973).

NOBEL PRIZES. In 1895, the will of Alfred Bernhard Nobel, a successful Swedish industrialist and European munitions maker, directed that the great majority of his estate be invested for the purpose of yielding annual prize money to be awarded to persons who, as the result of their personal efforts, made outstanding contributions to the advancement of science, literature, and peace. Initially, the awards were confined to five domains — Physics, Chemistry, Physiology or Medicine, Literature (of an idealistic tendency), and Peace (to promote the fraternity of nations and the abolition or diminution of standing armies and the formation and increase of peace congresses). In later years, the governors of the fund added Mathematics as a qualifying discipline and, in 1968, a Nobel Prize for the Economic Sciences was established. Although a prize for each of the foregoing categories is awarded each year, it is not mandatory that an award be made for each category every year, a factor determined by the governors of the fund. The will became effective when Nobel died on December 10, 1886, but the first prizes were not awarded until 1901 because of the need for legal clarification of Nobel's wishes as demanded by family members who also were mentioned in the will.

In studying the will, attorneys recognized the possible requirement for splitting a given award among two or even three individuals, but with a maximum of three persons per award. Nobel also mentioned certain criteria for selection of award recipients in each field. In connection with the prize for Physics, Nobel mentioned "discovery" or "invention," whereas the words "discovery" or "improvement" were stipulated concerning the prize for Chemistry. In terms of the prize for Physiology and Medicine, the key word was "discovery." These were guiding factors for application by the governors of the fund, especially during the early years.

If adjusted for inflation over the years, the original fund of 27,716,243 Swedish kroners ($7,427,953) was quite significant, especially when allowing for capital gains through investments realized over subsequent years. Money available for the annual prizes is essentially determined by the annual capital generated by the fund, but with the stipulation that at least 10% of that gain be reinvested each year. It is interesting to note that the honor associated with the prize is never split — that is, a Nobel Laureate is so designated even though a prize may be shared by two or three persons. The original will specified that the Royal Academy of Science (Sweden) select winners in Physics and Chemistry, that the Karolinska Institute of Medicine (Stockholm) make the selections in Physiology and Medicine, and that the Swedish Academy (of Letters) select winners in Literature. A committee of the Norwegian Parliament was specified to select winners of the Peace prize, noting "no consideration whatever be paid to the nationality of the candidates."

Over the years, with the addition of new prize categories, the governors of the Nobel Foundation necessarily have amended certain procedures. The Foundation is governed by a five-person board of control made up of one appointment by the King of Sweden and one each as appointed by the aforementioned organizations (now referred to as Nobel Institutes). Members of the selection board are appointed for a period of $4\frac{1}{2}$ years. The science awards are presented on December 10 (anniversary of Nobel's death) of each year at the Stockholm Concert Hall, with personal felicitation of the King of Sweden. The Peace prize is presented in a formal ceremony in Oslo.

The first Nobel prizes were awarded in 1901 to Wilhelm K. Roentgen (discovery of X-rays in 1895), J.H. van't Hoff (chemical thermodynamics and osmotic pressure), and E.A. von Behring (diphtheria antitoxin). Listings of scores of Nobel prizes awarded over a century of progress can be found in a number of references, such as "The Information Please Almanac," 45th Edition, 701–709, Houghton Mifflin Company, Boston, Massachusetts. Specific 1992 winners are described in "U.S. Researchers Gather a Bumper Crop of Laurels," Science, 542 (October 23, 1992). An excellent review of the first half-century of the Nobel Foundation is given in an article by George W. Gray, "The Nobel Prizes," *Sci Amer.*, 1 (September 1949).

Although a chronological listing of Nobel prizes in the sciences provides a good source of tracing the progress of science over the years, not all major discoveries and inventions have been so honored. There are several outstanding scientists who have not been included. Traditionally, Nobel prizes are given for achievements that date back a few to several years rather than for discoveries of the immediate past.

Web Reference

The Nobel Foundation: http://www.nobel.se/nobel/nobel-foundation/index.html

NOBLE FIR.   See **Fir Trees**.

NOBLE GASES.   The noble gases are helium [CAS: 7440-59-7], He; neon [CAS: 7440-01-9], Ne; argon [CAS: 7440-37-1], Ar; krypton [CAS: 7439-90-9], Kr; xenon [CAS: 7440-63-3], Xe; and radon [CAS: 14859-67-7], Rn. These are all members of Group 18 (VIIIA) of the Periodic Table and are characterized by completely filled valence electron shells. Historically, they have been called the rare or inert gases. But although comparatively rare, krypton, xenon, and radon are not completely inert and exhibit reactivity under certain conditions; all three form stable molecules with highly electronegative elements such as F, Cl, and O. Although inert enough, helium and argon are not rare in terms of available quantities; both are bulk items of commerce. Since commercial quantities of helium are produced from helium-bearing natural gas reserves, at some future time, if those reserves become depleted, helium could once again return to being a truly rare gas.

The noble gases were not isolated until the last decade of the nineteenth century making them the most recently discovered group of stable. While seeking more accurate density values for certain gases during the period 1882–1894, Rayleigh noted that the density of nitrogen isolated by removing oxygen from air was consistently about one-half percent greater than that of nitrogen obtained from chemical reactions. Working with Rayleigh, Ramsay treated atmospheric nitrogen with hot magnesium and obtained a residual gas, ~1.25% of the nitrogen, that was completely unreactive. The gas had a relative density of 19.075 (O = 16) and exhibited spectral lines not seen before. In 1894, the discovery of an inert gas was announced by Ramsey and Rayleigh and they called it argon from the Greek *argos*, inactive.

In 1868, within a decade of the development of the spectroscope, an orange-yellow line was observed in the sun's chromosphere that did not exactly coincide with the D-lines of sodium. This line was attributed to a new element which was named helium, from the Greek *helios*, the sun. In 1891, an inert gas isolated from the mineral uranite showed unusual spectral lines. In 1895, a similar gas was found in cleveite, another uranium mineral. This prominent yellow spectral line was then identified as that of helium, which to that time had been thought to exist only on the sun. In 1905, it was found that natural gas from a well near Dexter, Kansas, contained nearly 2% helium. See also **Natural Gas**.

The existence of neon (Greek *neos*, new) was predicted, as was the existence of heavier members of the group. In 1898 krypton (Greek *kryptos*, hidden) was discovered by spectroscopic examination of the residue from a sample of liquid air. Neon was discovered in the same year. A month later, xenon (Greek *xenos*, strange) was isolated from the residue left after distillation of krypton.

Radon-220 [CAS: 22481-48-7], ^{220}Rn, a decay product of thorium, was discovered by Owens and Rutherford in 1900. The more common radon-222, a decay product of radium, was discovered later in the same year and was isolated in 1902.

Helium-3 [CAS: 14762-55-1], ^{3}He, has been known as a stable isotope since the middle-1930s and it was suspected that its properties were markedly different from the common isotope, helium-4. The development of nuclear fusion devices in the 1950s yielded workable quantities of pure helium-3 as a decay product from the large tritium inventory implicit in maintaining an arsenal of fusion weapons. Helium-3 is one of the very few stable materials where the only practical source is nuclear transmutation.

See also **Argon**; **Helium**; **Krypton**; **Neon**; **Radon**; and **Xenon**.

Additional Reading

Massey, A.G.: *Main Group Chemistry*, 2nd Edition, John Wiley & Sons, Inc., New York, NY, 2000.

NOCTILUCA   (*Protozoa, Mastigophora*).   A genus of protozoans with one phosphorescent species. This minute form is sometimes so abundant in the ocean that the water appears luminous at night.

NOCTILUCENT CLOUDS.   See **Clouds and Cloud Formation**.

NODAL LINES.   On a vibrating diaphragm, lines along which no vibration takes place. If the diaphragm is circular they consist of two kinds, concentric nodal circles and nodal diameters.

NODAL POINT.   Of all the rays that pass through a lens from an off-axis object point to its corresponding image point, there will always be one for which the direction of the ray in the image space is the same as that in the object space. See Fig. 1. The two points at which these segments, if projected, intersect the axis are called the nodal points, and the transverse planes through them are called the nodal planes. Only if n and n'', the refractive indices in the object and image spaces, are identical are the nodal planes also the principal planes. C is the optical center of the lens.

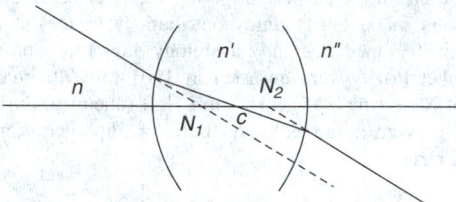

Fig. 1.   Nodal points of lens.

NODE.   The point, line, or surface in a standing wave system where some characteristic of the wave field has essentially zero amplitude. A node is also a terminal of any branch of a network, or a terminal common to two or more branches of a network. The terms junction point, branch point, and vertex are synonymous.

NODES (Line of).   See **Line of Nodes**.

NOGUCHI, HIDEYO (1876–1928).   Noguchi was a Japanese microbiologist who as an original staff member at the Rockefeller Institute in New York worked on a range of infectious diseases including syphilis and yellow fever. Born Seisaku Noguchi in Sanjogata, Honshu, Japan to peasant farmers, he adopted the name Hideyo after graduating from the Tokyo Medical College in 1897. Noguchi had been inspired to study medicine after a surgeon restored some function to his badly burnt left hand, the result of a childhood accident.

After meeting Simon Flexner, who was visiting the Institute for Infectious Diseases, where Noguchi worked as an assistant to Shiasaburo Kitasato, he decided to leave Japan for America. His early work in America, and with a Carnegie grant in Denmark, was on snake venoms.

When the Rockefeller Institute opened in 1904, Flexner invited Noguchi to join the staff. In 1905 together they confirmed *Treponema pallidum* as the cause of syphilis. Noguchi continued to work with *T. pallidum*. In 1913 his observations proved conclusively that paresis and tabes dorsalis represent the final stages of syphilis in the brain and spinal cord.

In 1918 Noguchi began a series of field trips to the western states of America and South America in order to establish the aetiology of bartonellosis (Carrion disease) and trachoma, making important discoveries. He was less successful with yellow fever, incorrectly identifying a spiral organism–*Leptospira icteroides*–as the causal organism. In 1927, in Africa, Adrian Stokes correctly determined that a filterable virus was responsible. With his characteristic energy Noguchi sailed for Africa, frantically conducting fieldwork only to die of the disease himself in 1928.

Despite his significant contributions to medical microbiology Noguchi had a tendency to haste and carelessness and the inappropriate application of bacteriological methods to viruses. He was married in 1912 to Mary Dardis; there were no children. See also **Microbiology**; and **Virology (The History)**.

H. J. POWER, Wellcome Trust Centre for the History of Medicine at UCL, London, UK

NOISE.   Any undesired sound. By extension, noise is any unwanted disturbance within a useful frequency band, such as undesired electric waves in any transmission channel or device. Such disturbances, when produced by other services, are called interference. Noise is also accidental or random fluctuation in electric circuits due to motion of the current carriers. From this concept of noise, the term is used as an adjective to denote unwanted fluctuations in quantities that are desired to remain constant, or to vary in a specified manner. For example, *noise voltage* is a term applied to spontaneous fluctuations in voltage in a component, device or system. The root-mean-square value of these fluctuations then gives a quantitative measure of the noise voltage. With this figure, such other measures as noise power and noise temperature are defined with the necessary stipulation of standard conditions for the acoustic, electric or electroacoustic system.

Background noise is (1) noise due to audible disturbances of periodic and/or random occurrence; (2) in receivers, the noise in absence of signal modulation on the carrier; (3) in recording and reproducing, the total system noise independent of whether or not a signal is present. The signal should not be included as part of the noise.

Thermal noise or Johnson noise is the noise produced by thermal agitation of charges in a conductor. The available thermal noise power produced in a resistance is independent of the resistance value, and is proportional to the absolute temperature and the frequency bandwidth over which the noise is measured, as indicated by the formula

$$N_t = 1.37 \times 10^{-23} T \Delta F$$

in which N_t is the available thermal noise power, T is the temperature of the resistance in degrees Kelvin, and Δf is the bandwidth in cycles per second.

Shot noise is the fluctuation in the current of charge carriers passing through a surface at statistically independent times. It has a uniform spectral density W_i given by

$$W_i = \frac{e I_o}{2\pi}$$

Random noise, exemplified by thermal noise and shot noise, has a uniform energy versus frequency distribution. *White noise* (or *Gaussian noise*) is a random noise having a constant energy per unit bandwidth that is independent of the central frequency of the band.

Over a number of years, society has been plague with ever-increasing noise from traffic, machinery, overamplified and discordant music, rocket and jet engine blasts, and sonic booms. Excessive and constant exposure of the ear to noise can and does cause impairment or loss of hearing. In addition other physical and frequently psychological changes can also occur as the result of noise disturbing sleep, impairing efficiency, and otherwise causing emotional disturbances. During wartime, military personnel often receive partial or total hearing loss after exposure to blasts and gunfire. Nearly 60,000 veterans of World War II, for example, have received compensation for ear damage considered to be a service-connected disability. Occupational deafness (earlier referred to as boiler maker's deafness) is a hearing loss resulting from prolonged exposure to industrial noise. Such damage is permanent, and may be partial or total. Overamplified music, particularly favored by teenagers, also is a cause of hearing impairment although the problem has not been subject to the same degree of analysis as in the case of industrial noises. See also **Hearing and the Ear**; and **Supersonic Aerodynamics**.

One of the frustrating results of noise is the masking effect it produces in reducing the intelligibility of speech. For example, if the speaker and listener are separated by 5 feet (1.5 meters), the levels of noise that will barely permit reliable word intelligibility are 50 decibels (dB) for normal conversation; 57 dB for raised speech; 63 dB for very loud speech; and 69 dB for shouting. As shown in Table 1, these levels are approached or exceeded in several day-to-day industrial and commercial activities. See also **Acoustics**.

Additional Reading

Beranek, L. and I.L. Ver: *Noise and Vibration Control Engineering: Principles and Applications*, John Wiley & Sons, Inc., New York, NY, 1992.
Fahy, F.J. and J.G. Walker: *Fundamentals of Noise and Vibration*, Routledge, New York, NY, 1998.
Kryter, K.D.: *The Effects of Noise on Man*, 2nd Edition, Academic Press, Inc., San Diego, CA, 1985.
Lord, H. and H.A. Evensen: *Noise Control for Engineers*, Krieger Publishing Company, Melbourne, FL, 1987.
Norton, M.P.: *Fundamentals of Noise and Vibration Analysis for Engineers*, Cambridge University Press, New York, NY, 1990.
Ott, H.W.: *Noise Reduction Techniques in Electronic Systems*, 2nd Edition, John Wiley & Sons, Inc., New York, NY, 1991.
Sankar, B.V.: *Vibration and Noise Control*, American Society of Mechanical Engineers, New York, NY, 1998.

NOISE GENERATOR. Electronic noise, by definition, is an unwanted disturbance and its reduction in communications circuits is a constant aim of the electronics engineer. When supplied by a properly controlled generator, however, noise becomes a very useful test signal. Random noise contains no periodic components and its future value is completely unpredictable. It is described by its amplitude distribution and its spectrum. In its most common form, it has a gaussian distribution of amplitudes. Two spectral varieties are useful: (1) "white noise" has a uniform spectral level over a specified frequency range, i.e., equal energy in each hertz of frequency; (2) "pink noise" has equal energy in each octave and, therefore, the energy in each hertz decreases at a rate of 3 decibels per octave. The random-noise signal, embracing a wide range of frequencies and having a randomly varying instantaneous amplitude, closely approximates the signals normally encountered in many electronic circuits and, particularly, in busy communication systems.

Specific applications for electronic noise generators include, as a broadband signal source: (1) intermodulation and cross-talk tests, (2) simulation of telephone line noise, (3) measurements on servo amplifiers, (4) noise interference tests on radar, (5) determining meter response characteristics, (6) setting transmission levels in communication circuits, (7) frequency response measurements; and, as a signal source, for (8) the measurement of reverberation, (9) sound attenuation of ducts, walls, panels, or floors, (10) acoustical properties of materials, (11) room acoustics, and (12) classroom or laboratory demonstrations. With a suitable power amplifier, such devices (13) can be used to drive a loudspeaker to produce high level acoustic noise for fatigue testing of structures and components, and (14) can drive a vibration shaker for structural tests of components and assemblies.

Two noise sources for the audio-frequency range are in common use. One is the semiconductor diode, operated at low current in the reverse breakdown mode. In the instrument illustrated in Fig. 1, noise from a semiconductor diode is amplified, filtered to establish the spectral shape, and further amplified to produce an output level of 3 volts. Internal circuits provide clipping of the peaks of the noise, if desired, or change the spectrum from white to pink noise as required.

TABLE 1. NOISE LEVELS FOR VARIOUS SOURCES AND LOCATIONS

Description of Noise	Noise Level (dB)
Threshold of hearing	0
Rustle of leaves in gentle breeze	10
Quiet whisper (distance of 5 feet)	10
Average whisper (distance of 4 feet)	20
House in country (average situation)	30
House in city (average situation)	40
Apartment (average situation)	40
Hotel	42
Theater (between performances)	42
Small retail establishment	52
Commercial garage	55
Medium-size office	58
Residential street	58
Restaurant	60+
Medium-size retail establishment	62
Factory or warehouse office	63
Large retail establishment	63
Ordinary conversation (distance of 3 feet)	65
Large office	65
Traffic on busy street	68
Factory (light-to-medium work)	78
Riveter (distance of 35 feet)	97
Hammer blows on steel plate (distance of 2 feet)	114
Threshold of pain	130

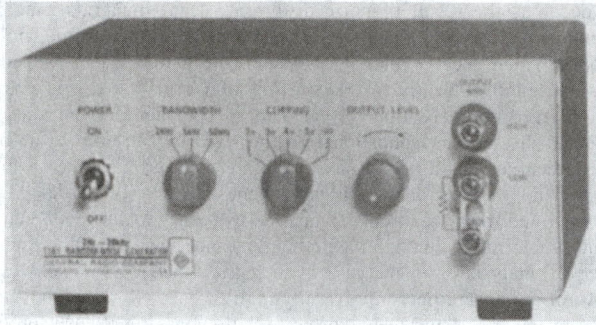

Fig. 1.   Random-noise generator. (*GenRad, Inc.*)

The other noise source is the pseudo-random noise generator, constructed of digital circuit elements connected as a shift register with feedback connections. When the feedback and initialization are such as to produce the maximum length sequence when the shift register is clocked, the rectangular wave output can be taken from the shift register and filtered with a low-pass filter to produce a waveform with good approximation to Gaussian or normal amplitude distribution. Such noise is called pseudo-random noise, because the pattern is precisely repeated each time the shift register sequence is repeated.

The most common noise source for the radio-frequency range is the temperature-limited thermionic diode. Its shot noise, although very low in level, is used because of its excellent spectral flatness and true Gaussian amplitude distribution.

JAMES J. FARAN, Jr., Lincoln, MA

NOISE (Statistics).   Disturbance terms, analogous to noise in the engineering sense, which appear in time-series. They are usually regarded as random elements superposed on the systematic components of the series and have a tendency to obscure its nature.

NOISE TEMPERATURE.   A measure of the noise power of a device or circuit. The noise temperature is the temperature of a resistor that has noise power equal to that of the device or circuit. Specifically, the noise temperature is defined by $T = N/kB$, where N is the noise power within bandwidth B, and $k = 1.38 \times 10^{-23}$ J K^{-1} is Boltzmann's constant. A radar system is characterized by several noise temperatures: the antenna temperature T_a, the receiver temperature T_r, and the transmission line temperature T_l. The transmission line temperature is a measure of the noise power within the receiver bandwidth generated by the resistive losses in the transmission line or waveguide between the antenna and the receiver. The transmission line temperature is frequently combined with either the antenna or the receiver temperature, depending on the reference point for the measurement. The total system temperature is $T = T_a + T_l + T_r$.

AMS

NOISE THRESHOLD.   The power level of a signal below which noise is likely to obscure the signal and above which the signal is discernible.

NOMENCLATURE (CHEMICAL).   Chemical nomenclature embraces several subcategories; names for chemical elements and compounds; names for classes of compounds and substances, such as mixtures and composites; names for particles, processes and transformations, properties, effects, units of measurements, techniques, instruments and apparatus, and even for theories and concepts. Only the first three are considered to be the heart of chemical nomenclature; the others are regarded as terminology, although it is often not at all easy to draw a sharp boundary between the two.

The largest part of the subject is the nomenclature of organic compounds, simply because there are so many of them, and of such diverse nature. The types of compounds and their structures differ considerably among organic, inorganic, and biochemical substances, and each of their respective nomenclatures has developed somewhat differently, although not entirely independently. Macromolecular nomenclature and pharmaceutical nomenclature have practical requirements of their own. It is therefore appropriate to treat each of these several areas separately.

Concern with chemical nomenclature has grown on a broad international scale as the importance of consistent, uniform nomenclature is increasingly recognized. When one compound is known by more than one name or, when one name may refer to more than one compound, serious confusion can result. The effect of such confusion is especially acute in indexes and compilations of names, and it is therefore natural that much of the work in systematizing names of chemical substances has been done in connection with such works. For example, both the Beilstein Institute and the Chemical Abstracts Service (CAS) use International Union of Pure and Applied Chemistry (IUPAC) principles as the basis for their index nomenclature with only slight modification due primarily to indexing considerations. Beginning in 1972, CAS introduced changes in its system designed mainly to reduce the number of variations allowed by the IUPAC rules and recommendations. In the past, the nomenclature of the Royal Society of Chemistry (London) represented a further variant. Various committees, both national and international, are working toward a consistent, systematic nomenclature. Among the areas in which nomenclature plays a key role are patent law, trade and customs regulations, identification of controlled substances, pharmaceutical and health information, and studies of the environment and pollution.

In the United States, the Committee on Nomenclature, Terminology, and Symbols of the American Chemical Society is the clearinghouse for nomenclature recommendations developed by various divisional nomenclature committees of the Society and international nomenclature bodies. Progress is measured not only by improved nomenclature, but also by extension of nomenclature principles to newly developing areas of chemistry and by development of new principles to cope with new types or structures.

Important as names are, they cannot serve all purposes. There are other, complementary means of identifying chemical compounds, e.g., structural formulas, notation systems, and registry numbers. None of these is truly nomenclature, however, which is strictly language based. Although structural formulas are sometimes easier to recognize than names, the reverse may also be true. Structural formulas can be space-consuming, troublesome to reproduce, and difficult to arrange in an order useful for retrieval of information. Systems for representing chemical structures by fragmentation coding, topological coding, and linear notation have been developed that have particular advantages for electronic manipulation. Notation systems are more compact than language-based nomenclature, but are less familiar to most chemists and many suffer from the serious disadvantage of being difficult to recognize. Of the various systems, the one originated by Dyson was once adopted by IUPAC, but has been largely abandoned. The system originated by Wiswesser is still widely applied and has catalyzed the development of significant innovations in computerized handling of information about chemical structures. Registry Numbers generated by Chemical Abstracts Service also provide an unambiguous identification for chemical compounds, but they are assigned arbitrarily and therefore have no intrinsic structural information.

Element Names and Symbols

Although symbols of the elements are not a part of chemical nomenclature, the two are closely related, and symbols have played an extremely important role in chemistry. Ancient Egyptian inscriptions include hieroglyphics for gold, silver, copper, iron, and lead. In the time of alchemy, the known metals were associated with the seven classical planets and the same symbols (which were not letters) were used for both a metal and the corresponding planet. Dalton devised a somewhat similar system based on the circle as a conventional representation of the atom. He and others used symbols to represent composition. Berzelius inroduced the modern, alphabetic symbols, which have become completely international. Interesting accounts of the evolution of chemical symbols have been given. Because of the difficulty of establishing priority of discovery for many of the elements of atomic number >100, especially elements 104 and 105, and because of the need to refer to hypothetical elements with higher atomic numbers, IUPAC developed interim symbols and names for such elements or example, element 104 would have the interim symbol Unq, and the name unnilquadium. Guidelines now exist for determing priority of discovery of a new element and for its name to be suggested by the discoverer. Accordingly, the names and symbols for the tranfermium elements 104–111 have been established as follows:

Atomic number	Name	Symbol
101	mendelevium	Md
102	nobelium	No
103	lawrencium	Lr
104	rutherfordium	Rf
105	dubnium	Db
106	seaborgium	Sg
107	bohrium	Bh
108	hassium	Hs
109	meitnerium	Mt
110	darmstadtium	Ds
111	roentgenium	Rg

See also **Chemical Elements**.

Inorganic Nomenclature

Perhaps no subject in chemistry has undergone less change over the twentieth century than inorganic nomenclature. This longevity attests to the fundamental soundness of the original proposals of Guyton de Morveau that established inorganic chemical nomenclature, and those of Werner that extended and broadened its base; it also suggests why inconsistencies and confusions have remained as well, which have continued to disconcert chemists. In the past quarter century, however, development of inorganic nomenclature has again accelerated and inconsistencies are being eliminated.

In the days of alchemy and the phlogiston theory, no system of nomenclature that would be considered logical in the 1990s was possible. Names were not based on composition, but on historical association, eg, Glauber's salt for sodium sulfate decahydrate and Epsom salt for magnesium sulfate; physical characteristics, eg, spirit of wine for ethanol, oil of vitriol for sulfuric acid, butter of antimony for antimony trichloride, liver of sulfur for potassium sulfide, and cream of tartar for potassium hydrogen tartrate; or physiological behavior, eg, caustic soda for sodium hydroxide. Some of these common or trivial names persist, especially in the nonchemical literature. Such names were a necessity at the time; they were introduced because the concept of molecular structure had not been developed, and even elemental composition was incomplete or indeterminate for many substances.

The System of Guyton de Morveau, Lavoisier, and Co-Workers. The first attempt toward a convenient nomenclature belongs to Guyton de Morveau. His pioneer work led to publication in 1787 of *Methode de Nomenclature Chimique*, written in collaboration with Lavoisier, Berthollet, and Fourcroy, which proved to be a landmark in the development of chemistry.

The fundamental principle of the new nomenclature was that the name of a compound should exhibit the elements involved and their relative proportions, if known. The combinations of oxygen with other elements played a dominant role. Thus, the product of the union of a simple nonmetallic substance with oxygen was called an acid, whereas that of the union of a metal with oxygen was called an oxide. The union of an acid and an oxide produced salt. The acids or oxides were given names in which the generic part was the word "acid" or "oxide" and the specific part was an adjective derived from the name of the other element. The same principle supplied names for sulfides and phosphides.

The names adopted for salts consisted of a generic part derived from the acid and a specific part from the metallic base: The names for salts of acids containing an element in different degrees of oxidation were given different terminations.

Berzelius divided the elements into metalloids (nonmetals) and metals according to their electrochemical character, and the compounds of oxygen with positive elements (metals) into suboxides, oxides and peroxides. His division of the acids according to degree of oxidation has been little altered. He introduced the terms anhydride and amphoteric and designated the chlorides in a manner similar to that used for the oxides.

Established Practice in the English Language. The nearly literal translation of the French terms in English, Russian and other languages resulted in the system whose use has become standard practice in English-speaking as well as other countries. The system has been molded by the fact that elemental composition and valence (or oxidation number) are the principal variables for most inorganic compounds other than the most complex, whereas connectivity and the possibility of isomers have been of little concern.

Modified Forms in Common Use. There are numerous situations in which the foregoing system does not meet all requirements. In the formation of binary compounds, several elements exhibit more than two states of oxidation. One method, recommended by the IUPAC, of handling these situations is the use of prefixes derived from Greek to indicate stoichiometric composition, e.g., titanium dichloride, $TiCl_2$; and dinitrogen oxide (nitrous oxide) N_2O. Other accepted methods of indicating proportions of constituents are the Stock system (oxidation number) and the Ewens-Bassett (charge number) system.

Some elements form acids with more than four oxidation states, requiring other combinations of prefixes and suffixes: $H_4P_2O_6$, intermediate between H_3PO_3 and H_3PO_4, is known as hypophosphoric acid. Here again, the oxidation-number and charge-number systems offer advantages. Ortho-, meta-, and pyro- prefixes or numerical prefixes to denote stages of hydroxylation of acids also find use. In many instances, special names have been created to deal with unusual situations.

Systems of Compounds. The nomenclature system of Guyton de Morveau and co-workers was designed specifically for oxygen compounds. As early as 1826, it became evident that the halogens could play much the same role in many other compounds as oxygen does in the familiar oxygen salts. By 1840, Hare was writing of chloro acids and chloro bases, and recognized classes of salts: oxy-, sulfo- (now called thio-), seleni-, telluri-, chloro-, fluoro-, cyano-, etc. Remsen was a proponent of this system of nomenclature, but it received its fullest treatment from Franklin in connection with his concept of systems of compounds.

The analogies are shown by the following reactions:

$$K_2O + B_2O_3 \longrightarrow K_2O \cdot B_2O_3 \text{ or } KBO_2$$

$$K_2S + B_2S_3 \longrightarrow K_2S \cdot B_2S_3 \text{ or } KBS_2$$

$$KF + BF_3 \longrightarrow KF \cdot BF_3 \text{ or } KBF_4$$

$$K_3N + BN \longrightarrow K_3N \cdot BN \text{ or } K_3BN_2$$

The products resulting from such reactions should, therefore, have analogous names. If KBO_2 is a borate, then KBS_2 is a thioborate, and KBF_4 is a fluoroborate. Similarly, the replacement of an oxygen atom by a sulfur atom or two fluorine atoms is understandable. However, the relationship of K_3BN_2 is less obvious, until one considers the dehydration and deammoniation schemes:

$$B(OH)_3 \xrightarrow{-H_2O} OBOH$$

$$B(NH_2)_3 \xrightarrow{-NH_3} HNBNH_2$$

His scheme of nomenclature for nitrogen compounds, and the names thio-, chloro-, etc, did become widespread, especially for sulfur and halogen compounds.

Although the foregoing pattern of nomenclature is useful, it does lead to some difficulties. Many quaternary compounds contain oxygen and another electronegative element. In the series M_2CO_3, M_2CO_2S, M_2COS_2, and M_2CS_3, the names are carbonates, (mono)thiocarbonates, dithiocarbonates, and trithiocarbonates, respectively. However, in practice both the prefixes mon- and tri- are often omitted, and its is uncertain whether the omission signifies the mono- or the completely substituted compound. The situation is somewhat more complicated when oxygen and fluorine are present in the same compound, because one is bivalent and the other univalent, and the coordination number toward fluorine is different from that toward oxygen: H_3PO_4, H_2PO_3F, HPO_2F_2, and HPF_6. Furthermore, investigators have not always been consistent in choosing the same reference state for the names of the oxygen salts and the halogen salts.

Coordination Compounds. The approach of Werner to the problem of naming ternary and higher order compounds is based on an entirely different point of view. By considering all such substances as complex or coordination compounds, he succeeded in making a wide variety of them according to a single general pattern. To designate the oxidation state of the element serving as the center of coordination, Werner chose the characteristic endings suggested by Brauner, but these have been totally superseded by the oxidation-number and charge-number systems.

The Stock Oxidation-Number System. Stock sought to correct many nomenclature difficulties by introducing Roman numerals in parentheses to indicate the state(s) of oxidation.

The oxidation-number system is easily extended to include other coordination compounds. Even substances represented by the formulas $Na_4Ni(CN)_4$ and $K_4Pd(CN)_4$ create no nomenclature problem; they become sodium tetracyanonickelate(0) and potassium tetracyanopalladate(0), respectively.

The Charge-Number (Ewens-Bassett) System. The oxidation state of an atom as expressed by the oxidation number is a formal concept for partitioning the electric charge between atoms in a molecule or chemical structure. For many chemical structures, this formal procedure may lead to representations of charge distribution that are inconsistent with experiment. Therefore, Ewens and Bassett proposed to express only the total charge on an ion without representing valence and its associated arbitrariness of assigning electronic distribution within a given structure, e.g., titanium(2+) chloride for $TiCl_2$; titanium(3+) for $TiCl_3$; potassium tetrachloroplatinate(2−) for K_2PtCl_4; and sodium tetracyanonickelate(4−) for $Na_4Ni(CN)_4$.

International Agreement. The first report of the Commission for the Reform of the Nomenclature of Inorganic Chemistry was written in 1926 by Délépine. Subsequent rules (1940, 1959) were expanded and improved in 1990 to provide the basis for naming inorganic compounds. They retain most of the well established names for binary and pseudobinary compounds and for the oxoacids of the nonmetals and derivatives.

The IUPAC Commission on Nomenclature of Inorganic Chemistry continues its work, which is effectively open-ended. Guidance in the use of the IUPAC rules as well as explanations of their formulation are available. A second volume on nomenclature of inorganic chemistry is in preparation; it will be devoted to specialized areas. Some of the contents have had preliminary publication in the journal, *Pure and Applied Chemistry*, e.g., "Names and Symbols of Transfermium Elements" in 1944.

Organic Nomenclature

Modern organic nomenclature is such that it can be better understood by first tracing how it developed. Organic substances played a minor role in the *Methode de Nomenclature Chimique*. Eighteen organic acids were given their present names (succinic, malic, etc), and several other substances were mentioned, such as alcohol, ether (including esters as well as true ethers), starch, gluten, and camphor. Gaseous hydrocarbons, the only ones included, were lumped together as carbonated hydrogen gas. Thus a few common names were incorporated into the new method, but no systematic organic names were possible because of lack of knowledge. Little else could be done, for the basis for determining elemental composition, as in empirical formulas, did not yet exist.

The practice of assigning ad hoc names to organic compounds was neither avoidable, nor burdensome when only a small number of compounds were recognized. Such ad hoc names were termed "trivial" or "traditional," to indicate that they contain no encoded structural information. They are useful for common compounds, and many of them are retained to this day, but they are not helpful in understanding chemical relationships. As they proliferated, the number and variety of them became unmanageable. The development of systematic nomenclature was driven by this circumstance, and was made possible by advances in understanding and determining the structure of molecules.

Systematic nomenclature is in essence a scheme for encoding structural information in a name. For organic chemistry, it probably began in 1832, when Justus Liebig's journal, *Annalen der Chemie*, was born and when Liebig and Woehler published their memorable article on the radical of benzoic acid. This radical (C_6H_5CO in the modern formula) they termed "benzoyl," thus coining -yl (from the Greek *hyle*, meaning stuff or material), one of the most useful suffixes in chemistry. By radical or compound radical, they meant a group of atoms that remains unaltered in chemical transformations. The word group is used for almost any portion of a molecule considered as a unit for convenience in naming or otherwise.

The name ethyl soon followed. These two names, one of an acid group (radical) and the other of a hydrocarbon group (radical), may be regarded as the progenitors of the host of group names used in the 1990s. From them it was an easy step to the combinations of benzoyl chloride, ethyl iodide, ethyl oxide, etc, many of which still survive. These binary names are analogous to the binary inorganic names introduced in 1787.

It was many years before organic nomenclature shook off the influence of electrochemical theory and its binary names. Gradually, as facts accumulated, it became clear that this theory must give way to a unitary conception of the molecule. At the same time, the phenomenon of substitution, or replacement of one atom or group of atoms by another, was recognized to be of central importance. Some binary names are still used, either as a true expression, as for salts, or for convenience, as in ethyl sulfide or acetyl chloride, but for the most part the principle of substitution is used without regard to whether such replacement can actually be effected experimentally. Usually the atom replaced is hydrogen, and the replacement may be indicated by either a prefix or a suffix. Thus, in naming CH_3Cl chloromethane rather than methyl chloride, the replacement of one atom of hydrogen in methane, CH_4, by chlorine is indicated. A third group of names is formed by combining a class name with a specific word, as in ethyl alcohol or benzophenone oxime. Whatever the method or combination of methods used, there must be a name for a parent compound to form a basis for it.

By 1866, it was possible for Hofmann to arrange hydrocarbons in series by their empirical formulas, i.e., methane, CH_4, and methene, CH_2; ethane, C_2H_6, ethene, C_2H_4, and ethine, C_2H_2; propane, C_3H_8, propene, C_3H_6, and propine, C_3H_4; and quartane, C_4H_{10}, quartene, C_4H_8, and quartine, C_4H_6. These are known as the Hofmann-Gerhardt names.

Hofmann's scheme has been modified by replacing quartane with butane and continuing the homologous series with the Greek forms pentane, hexane, etc, which are still used. For the C_nH_{2n} series, i.e., the olefins, the names methylene, ethylene, propylene, etc, came into use instead of Hofmann's terms, but the names propene, butene, pentene etc were revived in the Geneva system and are the preferred terms. For C_2H_2, ethine has never replaced the older term acetylene, but propine, butine, etc, reappeared in the Geneva system. The ending -yne is used, as in propyne, etc, to avoid confusion with the ending -ine of organic bases such as aniline.

The Hofmann-Gerhardt names did not distinguish between isomers. Different methods of distinguishing isomers arose: $CH_3CH_2CH_2CH_3$ became normal butane (abbreviated to *n*-butane), and $CH_3CH(CH_3)_2$ isobutane or trimethylmethane. Of olefins, CH_2=$CHCH_2CH_3$ became α-butylene or ethylethylene; CH_3CH=$CHCH_3$, β-butylene or symmetrical dimethylethylene; and CH_2=$C(CH_3)_2$, isobutylene or unsymmetrical dimethylethylene. It thus becomes evident that as the number of carbon atoms and therefore the number of isomers increases, the coining of such names meets with insuperable difficulties. The situation with regard to hydrocarbons had its parallel in the nomenclature of alcohols and other types of compounds.

The Geneva System. The Geneva Conference was strongly influenced by the need for names that would be suitable for systematic indexing of organic compounds. The groundwork was laid by a French subcommission. One of the chief principles of the system was the selection of the longest straight chain of carbon atoms in the molecule as a parent structure. Thus, the names butane, pentane, etc, would refer to the normal (unbranched) isomers only. The parent hydrocarbon could then be modified by attaching to its name one or more prefixes or suffixes to specify chemically characteristic features commonly termed functional groups. A representative selection is given in Lance 1 when two or more different positions of attachment of a prefixed or suffix exist, a position designator, called a locant, is necessary. These are arabic numerals, set off by hyphens, starting with 1 at an end of the chain. Accordingly, CH_3CH=$CHCH_3$ became 2-butene, and $CH_3CH_2CHOHCH_2CH_3$ became 3-pentanol. The position of the locants at the beginning or at the end was considered equally acceptable. In the 1990s however, the official IUPAC recommendation is to place the locant immediately before the feature that it locates, as in but-2-ene and pentan-3-ol for the foregoing examples.

The International Union and the Definitive Report. The next important step was the *Definitive Report of the Commission on the Reform of the Nomenclature of Organic Chemistry* in 1930 at a meeting in Liège. This

TABLE 1. PREFIXES AND SUFFIXES FOR SOME PRINCIPAL FUNCTIONAL GROUPS[A]

Formula	Class name	Prefix	Suffix	Radicofunctional form
$-COOH$	Carboxylic acids	Carboxy	-carboxylic acid[b] or -oic acid	
$-SO_3H$	Sulfonic acids	Sulfo	-sulfonic acid	
$-COOR$	Esters	Alkoxy carbonyl	alkyl -oate or carboxylate[b]	
$-COX$	Acid halides	Halocarbonyl	-oyl halide or carbonyl halide[b]	
$-CONH_2$	Amides	Carbamoyl cyano	-amide or carboxamide[b]	
$-CN$	Nitriles		-nitrile or -carbonitrile[b]	Alkyl[c] cyanide
$-CH=O$	Aldehydes	Formyl	-al or -carbaldehyde[b]	
$-C=O$	Ketones[d]	Oxo	-one	Dialkyl[c] ketone
$-OH$	Alcohols	Hydroxy	-ol	Alkyl[c] alcohol
$-SH$	Thiols (or mercaptans[e])	Sulfanyl (Mercapto[e])	-thiol	Alkyl mercaptan
$-NH_2$	Amines	Amino	-amine	
$=NH$	Imines	Imino	-imine	
$-OR$	Ethers	Alkoxy or aryloxy		(Di)alkyl ether
$-Cl$	Chlorides	Chloro		Alkyl[c] chloride
$-NO_2$	Nitro compounds	Nitro		
$-SO_2$	Sulfones	Alkyl[c] sulfonyl		(Di)alkyl[c] sulfone

[a] In order of precedence.

[b] The shorter form implies no additional carbon atoms, and is used when the group is part of a chain. The long form implies one more carbon atom than the parent structure, and is used when the group is attached to a ring, or for other reasons is not conveniently named as part of a carbon skeleton.

[c] Or aryl.

[d] Both bonds from the carbonyl group must be to a carbon atom.

[e] Has been widely used, but is no longer officially recommended.

report used the Geneva rules as a basis for modification, and many of the 68 Liège rules deal with topics not touched in the original Geneva report.

An important modification of the Geneva system is that the fundamental chain used as a basis in an aliphatic compound is not necessarily the longest chain in the molecule, but must be the longest chain of those containing the maximum number of occurrences of the principal functional group (Rule 18). This shifts the importance for naming from side chains such as methyl and ethyl to functional groups such as $-COOH$ and $-OH$.

The concept of the principal function raises the question of how priority is determined when two or more different functional groups are present. No arbitrary rule can be entirely satisfactory, but an order has been codified in IUPAC recommendations, and an essentially similar order is used by Chemical Abstracts Service. In general, a higher state of oxidation takes precedence over a lower one (Table 1).

The use of prefixes and suffixes for distinguishing the various radicals, groups, and functions has caused some problems, because some groups, e.g., HS−, have borne more than one name, and some names, e.g., anisyl, have had more than one meaning. The *Definitive Report* included only a limited number of prefixes and suffixes. Chemical Abstracts Service publishes its own lists the most recent version can be found in the comprehensive *Guide to the Use of IUPAC Nomenclature of Organic Compounds*. An important departure from earlier recommendations is that the systematic names of acyl groups derived from carboxylic acids must end in -oyl, common traditional or trivial names, such as acetyl and oxalyl, excepted. The purpose of this rule is to distinguish unambiguously between hydrocarbyl groups and acyl groups. Thus anisyl can only mean methoxyphenyl, whereas anisoyl refers only to methoxybenzoyl.

The ending -yl (or -oyl) is standard for univalent groups (with certain traditional exceptions, such as succinoyl). It may be combined with a sign for unsaturation, as in propenyl, $CH_3CH=CH$ thynyl, $CH\equiv C-$. The ending -ylene is one device for denoting a bivalent group in which the two free valences are on different atoms, but, with the exception of methylene, $-CH_2-$, and ethylene, $-CH_2CH_2-$, the ending -diyl, with locants as appropriate, is preferred, as in propane-1,3-diyl, $-CH_2CH_2CH_2-$. When the two free valences of a bivalent group form a double bond, the ending is -ylidene, as in ethylidene, $CH_3CH=$. For a trivalent group forming a triple bond, the ending is -ylidyne, as in ethylidyne, $CH_3C\equiv$.

For indicating the number of groups of the same kind, the prefixes di-, tri-, tetra-, etc are used when the expressions are simple, and bis-, tris-, tetrakis-, etc when they are complex; for example, "dichloro," but "bis(dimethylamino)." The prefix bi- is used to denote the joining of two groups of the same kind together, as in biphenyl, $C_6H_5-C_6H_5$, or the

doubling of a compound with loss of two hydrogen atoms, as in biarsine, H_2AsAsH_2.

A historical account of the development of organic nomenclature from the time preceding the Geneva Conference to fairly recent times is available.

The IUPAC Commission on Nomenclature of Organic Chemistry has continuing responsibility for revising and expanding the rules that appeared in the *Definitive Report*.

A considerable number of trivial or semitrivial (traditional) names have been retained by IUPAC for compelling practical reasons; the approved ones are available. A very brief selection is shown in Table 2. Other contributions to organic nomenclature are available.

TABLE 2. COMPOUNDS USED AS PARENT STRUCTURES WITH TRIVIAL NAMES

Chemical Formula	Name	Systematic Equivalents
C_6H_6	Benzene	
$C_{10}H_6$	Naphthalene	
C_5H_5N	Pyridine	
C_4H_4S	Thiophene	
C_4H_4O	Furan	$-COOH$
C_6H_5OH	Phenol	Benzenol
$C_6H_5NH_2$	Aniline	Benzenamine
$H_2N=H_2$	Hydrazine	Diazane
NH_3	Ammonia or amine	Azane
$HO=H$	Hydrogen peroxide	Dioxidane
CH_3COCH_3	acetone	propan-2-one
H_2NCONH_2	urea	carbamide

Biochemical Nomenclature

The IUPAC Commission of Nomenclature of Biological Chemistry was established in 1921, along with the organic and inorganic commissions. It worked actively and closely with the organic commission. Early subjects of concern were carbohydrates, proteins, enzymes, and fats. More recently, this Commission shared its work with a corresponding Commission of the International Union of Biochemistry, which is not the International Union of Biochemistry and Molecular Biology (IUBMB); this led to the establishment of the Joint Commission on Biochemical Nomenclature (JCBN) in 1964.

The Joint IUPAC/IUBMB Commission has published many recommendations dealing with the nomenclature of natural products. The IUBMB Commission on Nomenclature has also issued a number of recommendations dealing with areas of a more biochemical nature and for naming enzymes.

The presence of many chiral centers in compounds of biochemical significance or natural-product interest has led to the use of stereoparents. These are parent structures having trivial names that imply (without explicitly expressing) a particular steric configuration. Common examples are the names of simple sugars, exemplified by glucose.

Although it is not strictly within the subject of biochemical nomenclature, it is appropriate to mention the existence of standardized generic names for pharmaceutical drugs. Such names are essentially coined, or trivial, names, but often include syllables from the systematic organic names, and endings that reflect a structural class, e.g., -cillin (from penicillin), or an important area of medical application, e.g., -vir (antiviral). Glossaries of these generic names are published periodically and a glossary of United States Approved Names (USAN) is published annually.

Macromolecular Nomenclature (Polymers)

The nomenclature of macromolecules (polymers) can be complicated because there is little or no regularity in the molecules; for such molecules the structural details may also be uncertain. In cases where the macromolecule is a polymeric chain with some uncertainties about regularity in its structure, a simple expedient is to name the polymer based on the monomer(s) that gave rise to it. Thus, there are source-based names such as poly(vinyl chloride).

The first attempt to formulate a systematic nomenclature for polymers was based on the smallest repeating structural unit; it was published in 1952 by a Subcommission on Nomenclature of the IUPAC Commission on Macromolecules. The report covered not only the naming of polymers, but also symbology and definitions of terms. However, these nomenclature recommendations did not receive widespread acceptance. Further progress was slow, with a preliminary report on steric regularity in high polymers published in 1962 and updated in 1966.

In 1967, the Polymer Nomenclature Committee of the American Chemical Society published proposals for naming linear polymers on the basis of their chemical structure, which were then introduced into *Chemical Abstracts (CA) Indexes* and published in their final form in 1968.

A Macromolecular Division of IUPAC was created in 1967, and it created a permanent Commission on Macromolecular Nomenclature, parallel to the other nomenclature commissions. The Commission over the years has issued recommendations on basic terms of polymer science, stereochemical definitions and notations, structure-based nomenclature for regular single-strand organic polymers source-based nomenclature for polymers and copolymers, nomenclature for double-strand organic polymers, irregular single-strand organic polymers, abbreviations for polymers, and, in cooperation with the Commission on Nomenclature of Inorganic Chemistry, quasi-single-strand inorganic and coordination polymers. These or their precursors are collected in a compendium referred to as the IUPAC Purple Book.

Recommendations on additional aspects of macromolecular nomenclature such as that of regular double-strand (ladder and spiro) and irregular single-strand organic polymers continue to be published in *Pure and Applied Chemistry*. Recommendations on naming nonlinear polymers and polymer assemblies (networks, blends, complexes, etc) are expected to be issued in the near future.

Examples of the two macromolecular nomenclature systems are as follows. For source-based names for homopolymers and copolymers: polyacrylonitrile, poly(methyl methacrylate), poly(acrylamide-*co*-vinylpyrrolidinone), polybutadiene-*block*-polystyrene, and poly(propyl methacrylate)-*graft*-poly(1-vinylnaphthalene). Structure-based examples are as follows: poly(oxy-1,4-phenylene) (1), poly(oxyethyl-eneoxyterephthaloy) (2) and poly[imino(1-oxo-1,6-hexanediyl)] (3).

Nomenclature in Other Areas of Chemistry

A number of glossaries of terms and symbols used in the several branches of chemistry have been published. They include physical chemistry, physical-organic chemistry, and chemical terminology (other than nomenclature) treated in its entirety. The IUPAC has also issued recommendations in the fields of analytical chemistry, clinical chemistry, photochemistry, colloid and surface chemistry, ion exchange, spectroscopy, atmospheric chemistry, toxicology, clinical chemistry, pesticides, chemical kinetics, bioinorganic chemistry, computational drug design, medicinal chemistry, combinatorial chemistry, and theorectical organic chemistry.

Additional Reading

American Chemical Society, *Macromolecules*, **1**, 193 (1968).

Beilstein Institut, *Beilstein's Handbuch der Organischen Chemie*, 4th ed., Springer-Verlag, Berlin and Heidelberg, 1972, 4th and 5th Suppls., 1973–1995.

Block, B.P., W.H. Powell, and W.C. Fernelius: *Inorganic Chemical Nomenclature: Principles and Practices*, American Chemical Society, Washington, DC, 1990.

Brown, S.S.: *History of IUPAC, 1989–1999. Supplement to History of IUPAC, 1919–1987*, IUPAC, 2001.

Chemical Abstracts Service, *Chemical Substance Index Names*, Appendix IV in Chemical Abstracts Index Guide, American Chemical Society, Columbus, Ohio, 2002, pp. 176I–327I.

Fresenius, P.: *Organic Chemical Nomenclature: Introduction to the Basic Principles*, John Wiley & Sons, Inc., New York, NY, 1989 (trans. from German ed., 1983).

Godly, E.W.: *Naming Organic Compounds: A Systematic Instruction Manual*, Ellis Horwood Ltd., Chichester, U.K., 1989.

International Union of Biochemistry and Molecular Biology, *Biochemical Nomenclature and Related Documents*, 2nd Edition, Portland Press, London, UK, 1992.

International Union of Pure and Applied Chemistry, Physcial Chemistry Division, Commission on Physicochemical Symbols, Terminology, and Units, I. Mills, T. Cvitas, K. Homann, N. Kallay, and K. Kuchitsu, eds., *Quantities, Unites, and Symbols in Physical Chemistry*, 2nd ed., Blackwell Scientific Publications, Oxford, U.K., 1993.

International Union of Pure and Applied Chemistry, Organic Chemistry Division, Commission on Physical Organic Chemistry, *Pure Appl. Chem.*, **66**, 1077 (1995).

International Union of Pure and Applied Chemistry, *Compendium of Chemical Terminology*, A.D. McNaught, and A. Wilkinson, compilers, Blackwell Science, Oxford, U.K., 1997.

International Union of Pure and Applied Chemistry, Analytical Chemistry Division, J. Inczédy, T. Lengyel, and A. Ure, eds., *Compendium of Analytical Nomenclature*, 3rd ed., Blackwell Science, Ltd., Oxford, U. K., 1998.

International Union of Pure and Applied Chemistry, Clinical Chemistry Division, Commission on Quantities and Units in Clinical Chemistry and International Federation of Chemical Chemistry, Scientific Division, Committee on Quantities and Uses, J.C. Rigg, S.S. Brown, R. Dybkaer, and H. Olsen, eds., *Compendium of Terminology and Nomenclature of Properties in Clinical Laboratory Sciences*, Recomendations 1995, in Blackwell Science Ltd., U. K., 1995.

International Union of Pure and Applied Chemistry, Organic Chemistry Division, Commission on Photochemistry, *Pure Appl. Chem.*, **68** (12), 2223 (1996).

International Union of Pure and Applied Chemistry, Division of Physical Chemistry, *Pure Appl. Chem.*, **31**, 577 (1972); *Pure Appl. Chem.*, **46**, 71 (1976).

International Union of Pure and Applied Chemistry, Analytical Chemistry Division, Commission on Analytical Nomenclature, *Pure Appl. Chem.*, **29**, 619 (1972).

International Union of Pure and Applied Chemistry, Analytical Chemistry Division, *Pure Appl. Chem.*, **30**, 651 (1972); *Pure Appl. Chem.*, **45**, 99 (1976); *Pure Appl. Chem.*, **45**, 104 (1976).

International Union of Pure and Applied Chemistry, Chemistry and the Environment Division, Commission on Atmospheric Chemistry, *Pure Appl. Chem.*, **62**, 11, 2167 (1990).

International Union of Pure and Applied Chemistry, Chemistry and Human Health Division, Commission on Toxicology, *Pure Appl. Chem.*, **65** (9), 2003 (1993).

International Union of Pure and Applied Chemistry, Chemistry and Human Health Division, Commission on Nomenclature, Properties, and Units, *Pure Appl. Chem.*, **68** (4), 957 (1996).

International Union of Pure and Applied Chemistry, Chemistry and the Environment Division, Commission Commission on Agrochemicals and the Environment, *Pure Appl. Chem.*, **68** (5) 1167 (1996).

International Union of Pure and Applied Chemistry, Physical Chemistry Division, Commission on Chemical Kinetics, *Pure Appl. Chem.*, **68** (1), 149 (1996).

International Union of Pure and Applied Chemistry, Inorganic Chemistry Division, Commission on Nomenclature of Inorganic Chemistry, *Pure Appl. Chem.*, **69** (6), 1251 (1997).

International Union of Pure and Applied Chemistry, Chemistry and Human Health Division, Medicinal Chemistry Section, *Pure Appl. Chem.*, **69** (5), 1137 (1997).

International Union of Pure and Applied Chemistry, Chemistry and Human Health Division, Medicinal Chemistry Section, *Pure Appl. Chem.*, **70** (5), 1129 (1998).

International Union of Pure and Applied Chemistry, Chemistry and Human Health Division, Medicinal Chemistry Section, *Pure Appl. Chem.*, **71** (12), 2349 (1999).

International Union of Pure and Applied Chemistry, Organic Chemistry Division, Commission on Physical Organic Chemistry, *Pure Appl. Chem.*, **71** (10), 1919 (1999).

Richer, J.-C., R. Panico, and W.H. Powell: *Guide to the Use of IUPAC Nomenclature of Organic Compounds*, Blackwell, Oxford and London, UK, 1994.

Verkade, P.E.: *A History of the Nomenclature of Organic Chemistry*, F.C. Alderweireldt, and co-workers eds., D. Reidell, Dordrecht, The Netherlands, 1985.

Web References

International Union of Biochemistry and Molecular Biology (IUBMB): http://www.chem.qmul.ac.uk/iubmb/

International Union of Pure and Applied Chemistry (IUPAC): http://www.chem.qmul.ac.uk/iupac/

A Nomenclature of Junctions and Branchpoints in **Nucleic Acids**: http://www.chem.qmul.ac.uk/iubmb/misc/bran.html

Abbreviations and Symbols for **Nucleic Acids, Polynucleotides and their Constituents**: http://www.chem.qmul.ac.uk/iupac/misc/naabb.html

Abbreviations and Symbols for the Description of the Conformation of **Polypeptide Chains**: http://www.chem.qmul.ac.uk/iupac/misc/ppep1.html

Abbreviations and Symbols for the Description of Conformations of **Polynucleotide Chains**: http://www.chem.qmul.ac.uk/iupac/misc/pnuc1.html

Basic Terminology of **Stereochemistry**: http://www.chem.qmul.ac.uk/iupac/stereo/

Enzyme Nomenclature: http://www.chem.qmul.ac.uk/iubmb/enzyme/

Glossary of Terms Used in **Bioinorganic Chemistry**: http://www.chem.qmul.ac.uk/iupac/bioinorg/

Glossary of Terms Used in **Physical Organic Chemistry**: http://www.chem.qmul.ac.uk/iupac/gtpoc/

Glossary of Terms Used in **Medicinal Chemistry**: http://www.chem.qmul.ac.uk/iupac/medchem/

Nomenclature and Symbolism for **Amino Acids and Peptides**: http://www.chem.qmul.ac.uk/iupac/AminoAcid/

Nomenclature of **Carbohydrates**: http://www.chem.qmul.ac.uk/iupac/2carb/

Nomenclature of **Carotenoids**: http://www.chem.qmul.ac.uk/iupac/carot/

Nomenclature and Symbols for **Folic Acid and Related Compounds**: http://www.chem.qmul.ac.uk/iupac/misc/folic.html

Nomenclature of **Glycolipids**: http://www.chem.qmul.ac.uk/iupac/misc/glylp.html

Nomenclature of **Glycoproteins**, **Glycopeptides** and **Peptidoglycans**: http://www.chem.qmul.ac.uk/iupac/misc/glycp.html

Nomenclature of **Lignans** and **Neolignans**: http://www.chem.qmul.ac.uk/iupac/lignan/

Nomenclature of **Lipids**: http://www.chem.qmul.ac.uk/iupac/lipid/

Nomenclature for **Multienzymes**: http://www.chem.qmul.ac.uk/iubmb/misc/menz.html

Nomenclature of **Multiple Forms of Enzymes**: http://www.chem.qmul.ac.uk/iubmb/misc/isoen.html

Nomenclature for Incompletely Specified Bases in **Nucleic Acid Sequences**: http://www.chem.qmul.ac.uk/iubmb/misc/naseq.html

Nomenclature of **Peptide Hormones**: http://www.chem.qmul.ac.uk/iubmb/misc/phorm.html

Nomenclature of **Phosphorus-Containing Compounds** of Biochemical Importance: http://www.chem.qmul.ac.uk/iupac/misc/phospho.html

Numbering of atoms in **Myo-inositol**: http://www.chem.qmul.ac.uk/iupac/cyclitol/myo.html

Recommendations for Nomenclature and Tables in Biochemical Thermodynamics: http://www.chem.qmul.ac.uk/iubmb/thermod/

Nomenclature of **Cyclitols**: http://www.chem.qmul.ac.uk/iupac/cyclitol/

Nomenclature of **Electron-Transfer Proteins**: http://www.chem.qmul.ac.uk/iubmb/etp/

Nomenclature for **Vitamins B-6 and Related Compounds**: http://www.chem.qmul.ac.uk/iupac/misc/B6.html

Nomenclature of **Quinones with Isoprenoid Side-Chains**: http://www.chem.qmul.ac.uk/iupac/misc/quinone.html

Nomenclature of **Retinoids**: http://www.chem.qmul.ac.uk/iupac/misc/ret.html

Nomenclature of Steroids: http://www.chem.qmul.ac.uk/iupac/steroid/

Nomenclature of **Tetrapyrroles**: http://www.chem.qmul.ac.uk/iupac/tetrapyrrole/

Nomenclature of **Tocopherols and Related Compounds**: http://www.chem.qmul.ac.uk/iupac/misc/toc.html

Prenol Nomenclature: http://www.chem.qmul.ac.uk/iupac/misc/prenol.html

Prokaryotic and Eukaryotic translation factors: http://www.chem.qmul.ac.uk/iubmb/misc/trans.html

Revised Section F: **Natural Products and Related Compounds**: http://www.chem.qmul.ac.uk/iupac/sectionF/

Section H: **Isotopically Modified Compounds**: http://www.chem.qmul.ac.uk/iupac/sectionH/

Symbols for Specifying the Conformation of **Polysaccharide Chains**: http://www.chem.qmul.ac.uk/iupac/misc/psac.html

Symbolism and Terminology in **Enzyme Kinetics**: http://www.chem.qmul.ac.uk/iubmb/kinetics/

The Nomenclature of **Corrinoids**: http://www.chem.qmul.ac.uk/iupac/misc/B12.html

NOMENCLATURE (Organic Chemistry). See **Organic Chemistry**.

NOMENCLATURE (PETROLEUM INDUSTRY). Crude oils, complex mixtures of naturally occurring organic liquids, are difficult to characterize in detail. Thus, many of the definitions used by the exploration, production, and refining sectors of the petroleum industry to describe petroleum and its products often lack precision. Even the term petroleum is poorly defined. Although often used synonymously with crude oil, petroleum is also frequently used to include natural gas and even solid hydrocarbons. See also **Natural Gas**. Definitions of materials are commonly given in terms of the processes used to obtain them. Gasoline, for example, is the fraction of crude oil that distills between 15 and 200 °C (60 and 392 °F). Further complications arise because different parts of the petroleum industry use terms in differing ways. For example, wax may refer to material made up predominantly of long-chain alkanes, or it may refer to esters of long-chain alcohols and acids. See also **Waxes**. Even the term hydrocarbons is used loosely indicating all the compounds in crude oils, whether or not these include compounds of nitrogen, sulfur, and oxygen. See also **Hydrocarbons**.

In nature petroleum occurs in subsurface accumulations, or reservoirs, called fields that may be made up of one or more pools. Petroleum compositions vary widely and range from hydrocarbon-rich gases called natural gas, through crude oil liquids, to high molecular weight solids known as reservoir bitumen, residual oil, or tar. Petroleum is generated from kerogen, the high molecular weight, insoluble organic material in source rocks. High subsurface temperatures convert the kerogen to a petroleum-like range of compounds called bitumen. Part of this bitumen moves out of the source rock, in the process of expulsion or primary migration, and moves through permeable rocks to accumulate in a reservoir (secondary migration). The petroleum engineering procedures for bringing this petroleum to the surface are called production.

Traditionally the unit of crude oil production has been the barrel (bbl), equal to 42 U.S. gallons, 5.61 ft^3, 158.8 L, or 0.159 m^3. Increasingly petroleum reserves are given in metric tons, but because one unit is a volume and the other a weight, there can be no unique conversion factor for a material having a range of densities. Fields of $>500 \times 10^6$ bbl (79.5×10^6 m^3) of recoverable oil ($>100 \times 10^6$ bbl (15.9×10^6 m3) in the U.S.) are called giants. Oil density may be reported in any appropriate units, and although metric units are used it is more common to report densities as degrees API (°API) or API gravity, where API stands for American Petroleum Institute. The relationship between density and API gravity is an inverse one defined by the following relationship:

$$°\text{API} = [141.5/\text{specific gravity at } 60°\text{F}] - 131.5$$

Water corresponds to an API gravity of 10; crude oils fall between 10 and 60°API. The most common crude oil values are in the 35–40° range.

Other terms relating to physical properties include viscosity; refractive index; pour point, ie, the lowest temperature at which the oil flows; flash point, i.e., the temperature at which the oil ignites; and aniline point, ie, the minimum temperature at which equal volumes of oil and aniline are completely miscible. These are determined under defined conditions established by ASTM.

Natural gas production is generally given in cubic feet or cubic meters (1000 ft^3 = Mcf = 289.3 m^3). Reserves of a trillion cubic feet (Tcf) (28.3 × 10^{12} m^3) or more form a giant gas field. Natural gas is called dry when methane is the dominant hydrocarbon, and wet if it contains more than 4 L/100 m^3 of natural gas liquids (>0.3 gal/100 ft^3). When gas (or oil) has a bad odor owing to high concentrations of hydrogen sulfide and volatile sulfur compounds it is called sour. Sweet gas has no noticeable odor. For statistical purposes gas is commonly reported as an equivalent amount of oil based on an equivalent heating capacity. The conversion is normally made using 170 m^3 (6000ft^3 = 1 bbl) and leads to a barrel of oil equivalent (boe). See also **Natural Gas**.

Crude oils contain a wide range of hydrocarbons including straight and branched chains, ring compounds, and aromatics, as well as more complex compounds that incorporate nitrogen, sulfur, and oxygen (often called the NSOs), and some nickel and vanadium. The straight-chain, normal alkanes, range from 1 to > 100 carbon atoms. These are often called paraffins in the petroleum industry because of the useful adjective paraffinic. Branched hydrocarbon chains that are nominally built up from repeated isoprene units (2-methyl butane structure) are called isoprenoids or terpenoids, and the 19- and 20-carbon compounds, named pristane and phytane, respectively, are frequently present in high concentrations. Isoprenoids also lead formally to saturated multiring structures. Petroleum chemists use the obsolete word naphthenes for the compounds that organic chemists call alicyclics. A better term, cycloparaffin, is used herein, leading to the adjective cycloparaffinic rather than naphthenic. Some of the characteristic structures in this group can be directly related to molecules synthesized by organisms. Whereas these have been called chemical fossils, it is more usual to call them biological markers or biomarkers. Common examples include the steranes and hopanes.

Aromatic hydrocarbons form a minor but important group of compounds in crude oils and range from single-ring to multiring compounds. The latter are called polycyclic aromatics (PAHs). Small aromatic molecules are environmentally significant and BTEX is commonly used as an abbreviation for benzene–toluene–ethyl benzene–xylenes. Multiringed compounds containing both aromatic and saturated rings may be referred to in the older literature as naphtheno-aromatics. The highest molecular weight fraction of crude oils commonly contains ashphaltenes that are dark in color, NSO-rich, and very aromatic.

Most crude oil is refined to provide useful products and the dominant process is distillation. Petroleum products produced by simple distillation without the use of pressure, cracking, or catalysts are called straight run. Residual material that has too high a molecular weight to distill forms a residuum, often called by such names as asphalt. Naphtha (unrelated to naphthenes) is a distillate of petroleum having a boiling range lower than about 200 or 260 °C (even occasionally up to 350 °C). As a process intermediate, naphtha includes the components used to formulate gasoline and the lighter grades of fuel oils such as kerosene and diesel fuel oil. As a finished product, naphtha usually denotes a more specific type of narrow boiling range material. The terms naphtha and solvent may be used interchangeably. For example, Varnish Makers' & Painters' (VM&P) naphtha has a range of 95–150 °C. The majority of streams within a refinery designated as naphthas are straight-run materials, however the term can also be used for some cracked distillates.

A number of other words that have traditionally been used in the petroleum industry are difficult to define precisely. These refer partly to specific boiling ranges, but also to certain intended uses. Thus, gasoline boils lower than naphtha, and kerosenes generally higher, but these terms are applied to products that are intended as fuels, rather than as solvents.

Gas oil is a product boiling slightly higher (235–425 °C, or sometimes wider) than kerosene. The main feedstock to the catalytic cracking units, it received its name from use as an enriching agent in the production of city or manufactured gas. It is often used as diesel fuel. See also **Feedstocks**.

Cylinder oil is a viscous oil used for lubricating the cylinders and valves of steam engines. It is prepared from cylinder stock. The product from cylinder stock, when filtered and processed, is bright stock. See also **Lubricant**; and **Lubricating Agents**.

Cycle stock (recycle stock) denotes any product that is recycled, that is, taken back to an earlier stage in the process. The term cycle stock is also used for the gas oil-like product of catalytic cracking.

The word distillate is occasionally used by petroleum chemists with a specialized meaning. Although anything that has been distilled is, of course, a distillate, the term distillate is sometimes used to denote distillate fuel oil as opposed to residual fuel oil.

In the petroleum industry the International Union of Pure and Applied Chemistry (IUPAC) system is in widespread use for naming organic compounds. Two points, however, regarding group names and the prefix, iso, call for comment.

The Prefix Iso

In names such as isobutane, isopentane, isobutyl alcohol, and isoamyl alcohol, the prefix iso has a precise meaning, i.e., one methyl group attached to the next-to-terminal carbon atom and no other branch. This notation is also frequently used by petroleum chemists to have a much wider meaning, denoting nothing more than branched-chain. If both meanings persist, any individual use of the prefix becomes ambiguous. Herein, an effort is being made to use branched-chain or just branched consistently for the looser meaning of iso, so that this prefix can be kept for denoting concisely what otherwise would require some circumlocution. An exception is made for the well-established name isooctane, which is 2,2,4-trimethylpentane [CAS: 540-84-1].

See also **Petroleum**; and **Petroleum Refining**.

NONCERTIFIED COLORS. See **Colorants (Foods)**; and **Food Additives**.

NONDESTRUCTIVE TESTING (NDT). The examination of materials and objects for the purpose of detecting defects without in any way harming the test object. NDT contrasts vividly with destructive testing methods, which chemically consume or physically damage the test object, rendering it unfit for use. Whereas destructive testing must be confined to statistical sampling procedures, NDT enables 100% on-line inspection if desired. The trend in recent years has been in this direction, with emphasis on automating and increasing the speed of NDT operations. Another significant trend has been that of *testing work in progress*, as contrasted with earlier procedures which concentrated on testing raw materials and final products. In this way, very helpful information for step-by-step quality control can be provided. There remain, of course, numerous examples of where statistical destructive testing is needed — for example, in determining the ultimate compressive and tensile strengths of materials and parts or checking the corrosion resistance of materials. In recent years, it has proven possible to combine the results of NDT with computerized simulation in some instances to predict failure of test objects under certain conditions. For obvious economic reasons, NDT is preferred by manufacturers over destructive testing this accounts for high acceptance and many advancements which have occurred in NDT methods. For research and development applications, nondestructive methods are sometimes referred to as NDE (nondestructive evaluation).

Traditionally, NDT has been associated with metals and materials of construction for finding potentially unsafe conditions, such as cracks, voids, holes, inclusions, and other inconsistencies, as may be found in metal sheets, plates, bars, tubes, castings, forgings, and weldments. Such defects may arise from faulty manufacturing, or from later use, as the result of corrosion, abrasion, vibration, mishandling, and inattention to required maintenance procedures. In recent years, the applications for NDT have broadened to include all manner of materials — films, coatings, polymers, composites, and ceramics as encountered in a wide variety of industries, including numerous uses in the electronics manufacturing industry. Also, NDT is widely used for on-site inspection of large and heavy equipment, which cannot be detached for testing, after installation, but where periodic checks are required. Examples include the inspection of weldments in pipelines, aircraft engines and structural components, military equipment, bridge structures, etc.

During the last half of the 1980s and well into the 1990s, NTD enjoyed the benefits of measurement and computer technologies that contributed immensely to the speed, accuracy, and reliability of NTD, even though instrument costs have risen markedly as a result. However, the ability to make more measurements within shorter periods of time probably has not increased the unit costs proportionately. A number of measurement techniques entirely new to the NDT field have been added to increase the variety of choices.

Radiographic Methods

Radiographic (X- and gamma-ray technology) method using film was one of the earliest NTD schemes used. Although early systems are undergoing modernization, this comparatively simple method still enjoys acceptance for certain applications. This basic technique has taken on a number of new formats.

Film Images. Images made by the traditional film technique are shown in Fig. 1. To reduce costs and meet environmental restrictions, a dry-silver system was introduced in 1991. The system produces a silver-based image without the use of wet chemistry, using photothermographic technology. The image is developed on exposed film by thermal energy rather than by the traditional method of immersing film in a liquid developer and fixer. Three elements required for dry processing are a specially coated film, fluorescent exposing screen, and a thermal processor. The film has a translucent polyester base similar to that of conventional film. Its ultrafine grain produces detailed images of archival quality.

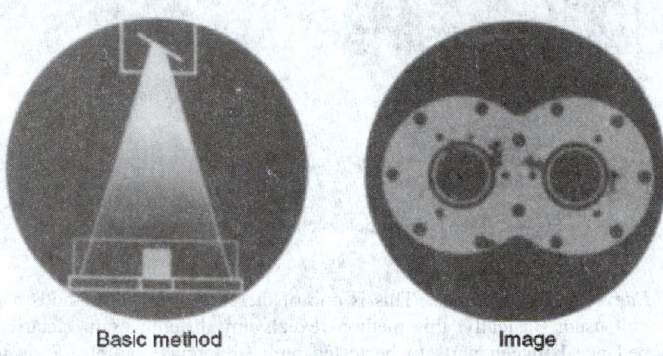

Basic method   Image

Fig. 1. X-rays or gamma rays are used to create a shadow image of light and dark that reveals any flaws or inclusions in a test part.

Traditional radiographic methods use two-dimensional film to record the attenuation of X-rays passing through a three-dimensional object. The result is a shadowgraph in which all object features are superposed. To improve the totality of information obtained, backscattering methods were introduced several years ago.

Principal applications for radiography include the inspection of castings, electrical assemblies, weldments, small, thin, and complex wrought products, some nonmetallics, solid propellant rocket motors, cans or containers, composites, and nuclear reactor fuel rods, among many others.

Chronology of Radiographic Methods. In 1985, researchers at John Hopkins University described a flash X-ray system that uses increased-power X-ray sources to generate very intense short pulses. High-gain X-ray intensifier detectors are used. Exposure times as short as 30 ns are possible and thus microstructural changes due to explosions, heat pulses, and shock waves can be detected. An indirect and direct method are used. In the indirect method, the X-ray diffraction image is converted into a visible light image by a fluorescent screen. The researchers have found that for the indirect method, a multiple-stage image-intensifier system coupled to an external fluorescent screen is the most sensitive and almost instantaneous system. Multiple stages of amplification allow individual X-ray photons to be detected. In more advanced systems, there is inclusion of a microchannel plate where electrons strike the output phosphor and are converted into a strong, visible image.

In the direct method, an X-ray-sensitive vidicon TV camera directly converts the X-ray image into an electronic charge pattern on a photoconductive target, which is read out by a scanning electron beam and displayed visually on a TV monitor.

In addition to testing uses per se, flash-X-ray techniques have been used to study the orientation of single crystals, to study lattice rotation accompanying plastic deformation, to measure the grain boundary migration during recrystallization annealing, and to determine the physical state of exploding materials.

In another technique known as X-ray transmission asymmetric crystal topography, changes in defect structure during polymerization of single crystals have been studied.

Digital Radiography. In this technique, the traditional film is replaced by a linear array of detectors and the X-ray beam is collimated into a fan beam. The object is moved perpendicularly to the detector array, and the attenuated radiation is sampled digitally by the detectors. Data are processed by stored information in the computer's memory to yield a two-dimensional image of the part being inspected.

X-Ray Computed Tomography (CT). In computed tomography, penetrating radiation from many angles is used to reconstruct cross-sectional images of an object. The advantages of CT are exemplified by the inspection of aircraft/aerospace castings for internal defects. Advantages of CT include increased reliability, elimination of unnecessary rejects, and wider use of castings instead of forgings and parts machined from wrought stock. CT has been found to have greater sensitivity (dependent on part size and geometry) than conventional film. CT can spatially define flaw distribution. Aerospace test engineers claim that castings can be measured with an accuracy of better than 0.05 mm (0.002 in), but is adversely affected by the amount of image noise and the edge-detection method used. Computed tomography systems are costly. The general principles of CT are described in article on **X-Ray Scan and Other Medical Imagery**.

Ultrasonic Methods

Typically, ultrasonic images are produced by mechanically scanning an ultrasonic transducer in a raster pattern over an area of a structure and then displaying the reflected or transmitted energy in a suitable format. Usually, the scan is performed in a tank of water or with some form of squirter nozzle. The liquid medium serves to transmit the ultrasonic energy from the transducer into the test material. Conventionally, the data are displayed as C-scans (a plan view image where a color scale is used to display signal amplitude or depth information) or as B-scans (image of a cross section at one particular location of interest, typically with a color indicating signal amplitude).

As indicated by Fig. 2, there are several testing modes: (1) pulse-echo mode, (2) through-transmission mode, (3) reflector-plate mode, and (4) angle-beam mode.

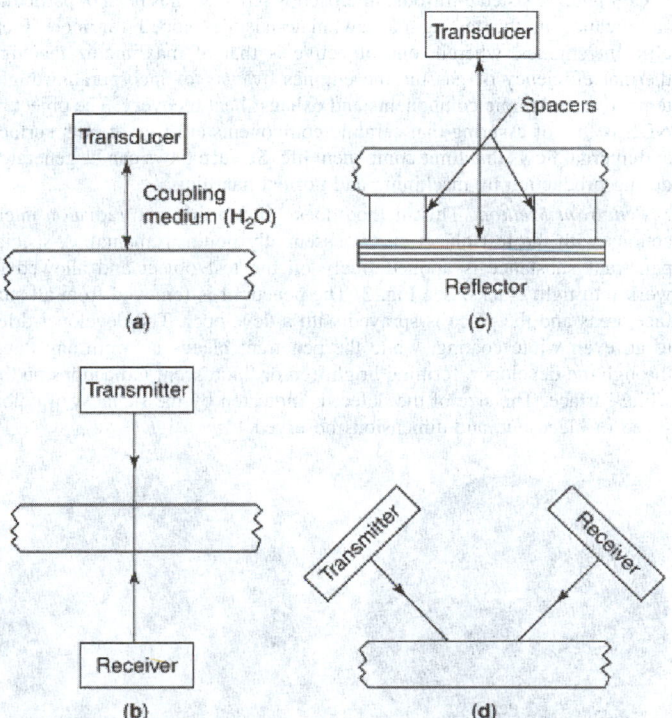

Fig. 2. Ultrasonic NDT methods: (**a**) pulse-echo; (**b**) through-transmission; (**c**) reflector-plate (double-through transmission); and (**d**) angle-beam.

Sonic (<0.1 MHz) and ultrasonic (0.1 to 25 MHz) radiation have been used for many years in NDT. In a simple testing scheme, sonic or ultrasonic vibrations are generated and sent by way of a pulse beam through the part to be tested. The beam travels unimpeded through large parts, may be angled for testing sheet stock, and can impact materials immersed in a liquid. Any flaw reflects vibrations back to the instrument, which indicates the location and size of the discontinuity on a CRT (cathode-ray tube). Access is required to only one side of the material being tested. Although energy can be lost from the ultrasonic beam due to geometrical effects, these can be controlled to increase the sensitivity of attenuation measurements. Hence, microstructural alterations, such as microcracks, foreign particles, precipitates, grain boundaries, interphase boundaries, and dislocation defects, can be detected. Research has shown that attenuation measurements have detected microstructural change during fatigue testing, therefore giving early warning to fatigue-induced failure, as well as measuring oxygen content in titanium welds.

Acoustic methods are applicable to numerous kinds of materials. The method can be used, for example, to reveal fiber/matrix bond strength in polymer-matrix composites. In one method (Wan-li Wu, National Institute of Standards and Technology), a continuous wave argon-ion laser is used to heat a very small area of the composite. The resulting thermal expansion between fiber and resin produces a measurable change due to debinding. Conventional methods of evaluating bond strength are time consuming and tedious. Instead of measuring the thermal stress, the laser power level at which debonding occurs is used as the index of debonding stress. Although sonic scanning techniques can be used to detect voids and cracks at interfaces in polymer-matrix composites, they do not measure the strength of interfacial bonds.

As pointed out by D. Sturges (General Electric Aircraft), "Modern ceramic materials offer many attractive physical and mechanical properties for use in a rapidly growing variety of industrial applications. The critical nature of many applications, however, imposes technical challenges in manufacturing and inspection. One nondestructive evaluation (NDE) technique of major relevance to inspecting ceramics is ultrasonic microscopy (also termed acoustic microscopy), that is, the use of tightly-focused, high-frequency sound beams to form images of the point-to-point reaction of a material to periodic stress waves. This technique offers high sensitivity for the detection of small defects, and often is a complementary technique to X-ray inspection."

Computer-assisted ultrasonic microscopy (CAUM) has been of particular significance in the testing of new materials developed for more fuel-efficient engines, wherein one objective is that of maximizing the high thermal efficiency of gas turbine engines by way of incorporating high-temperature ceramic components and exhaust-heat recovery. The object of NDE is that of assuring that ceramic components are free of both surface and internal flows that limit component life. Surface flaws can be generated during production by machining and normal handling.

Penetrant Method. This method does not depend upon radiation interactions with the test object and is essentially noninstrumental. A special penetrant substance is applied freely on the test object and allowed to work into tight cracks. See Fig. 3. The penetrant is removed from all surface areas and the piece is sprayed with a developer. The developer dries to an even white coating, while the penetrant bleeds up from any flaws through the developer, forming bright-red or fluorescent indications on the white surface. The size of the defect is indicated by the richness of color, speed of bleed-out, and dimensions observed.

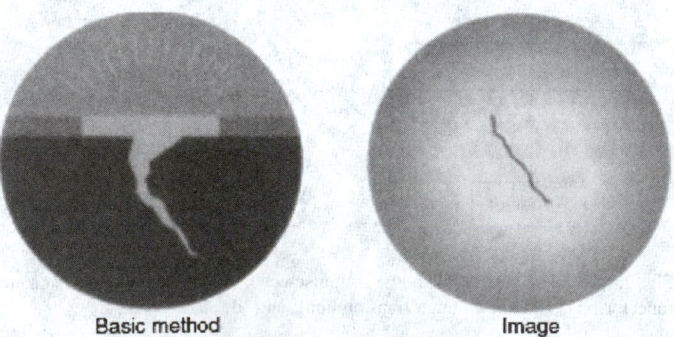

Basic method   Image

Fig. 3.   Penetrant method for detecting flaws.

Because of environmental concerns, a new generation of biodegradable penetrants having sensitivity levels ranging from 1 to 4 has been developed.

The new penetrants are water washable and, in most instances, can be directly discharged into sewers. They are free of petroleum-based solutions.

Magnetic-Particle Method. This method makes use of iron powder to reveal the leakage magnetic field created at a flaw or break when any part is magnetized. The familiar horseshoe magnet best illustrates this principle. (1) If a horseshoe magnet is bent into a circle, the field between the ends attracts and holds magnetic iron powder. (2) If a magnet is made completely closed, the field will be contained entirely within the ring and no iron powder will be attracted. (3) However, if the round magnet is cracked, poles are created at the break, and iron powder is instantly attracted to the cracked area to pinpoint the defect. See Fig. 4.

Basic method   Image

Fig. 4.   Magnetic particle method for detecting flaws.

Eddy-Current Methods. This is one of the earliest NDT methods and is still used. Basically, this method reveals any differences in electrical impedance between parts to be tested and a reference sample. Parts to be examined are passed through a coil or explored with a probe, and a trace appears on a CRT. Since magnetic and electrical characteristics are closely related to metallurgical quantities, a trace position or pattern or a meter reading clearly shows variations in metal hardness and composition, as well as defects. Both ferrous and nonferrous parts can be tested, and various coils, probes, and detector tips are available.

Aside from more sophisticated electronics, a major contribution to improve eddy-current instrumentation has come from the development of the eddy current resonance digitizing (ECRD) method. With this method, eddy-current instrumentation can separate nonferrous alloys based upon characteristics other than simply their conductivity. See Fig. 5.

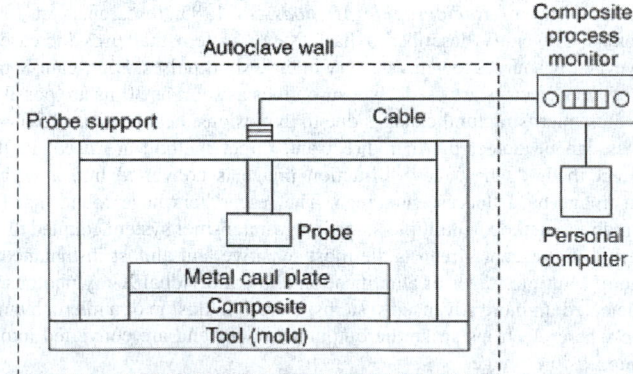

Fig. 5.   Use of eddy-current testing for monitoring a composite cure. Typically, eddy-current testing uses an electronic instrument having a small probe on the end of a flexible electrical cable. The probe is placed either against or close to the target. The target must be electrically conductive to allow the generation of eddy currents. The time-variant nature of the probe's magnetic field causes electric currents to flow in the target material. Higher field frequencies or a more electrically conductive material increase the depth of the eddy-current penetration into the material. The concentration of eddy currents near the surface of the material is referred to as the "skin effect." Eddy currents generate their own magnetic field that opposes the probe's magnetic field. Detection circuits in the instrument sense the impedance changes in terms of phase/amplitude changes in probe-coil voltage. (*Suggested by Bar-Cohen and Nguyen.*)

NDT Outlook

Improvements in current, established technologies and the introduction of new ways to test materials, nondestructively are expected to continue apac. One promising method is *positron annihilation*. The positron is the antiparticle of the electron; thus a positron/electron pair is unstable and will annihilate. In this process, two gamma rays at approximately 180° to one another are emitted from the center of the mass of the pair. A very slight departure from 180° is directly proportional to the transverse component of the momentum of the pair. The momenta of the electrons involved in such collisions can be calculated from the geometry and intensity of the gamma rays. The dynamics of the electron/positron system underlie the use of the technique for the study of defects in materials.

Additional Reading

Adams, T.E. and A.C. Wey: "Nondestructive Sectioning: Alternative to Physical Sectioning," *Advanced Materials & Processes*, **54** (February 1992).

Akuezue, H.C. and S.K. Verma: "Positron Annihilation: NDE at the Atomic Level," *Advanced Materials & Processes*, **26** (March 1992).

Altshuler, T.L.: "Atomic-Scale Materials Characterization," *Advanced Materials & Processes*, **18** (September 1991).

Bar-Dohen, K.H. Nguyen, and R. Botsco: "Eddy Currents Monitor Composites Cure," *Advanced Materials & Processes*, **41** (April 1991).

Bindell, J.B.: "Elements of Scanning Electron Microscopy," *Advanced Materials & Processes*, **20** (March 1993).

Blitz, J.: *Electrical & Magnetic Methods of Nondestructive Testing*, Institute of Physics Publishing, London, UK, 1991.

Bray, D.E. and D. McBride: *Nondestructive Testing Techniques*, John Wiley & Sons, Inc., New York, NY, 1992.

Carter, G.F. and D.E. Paul: *Materials Science and Engineering*, ASM International, Materials Park, OH, 1991.

Cartz, L.: *Nondestructive Testing: Radiography, Ultrasonics, Liquid Penetrant, Magnetic Particle, Eddy Current*, ASM International, Materials Park, OH, 1995.

Cormia, R.D.: "Problem-Solving Surface Analysis Techniques," *Advanced Materials & Processes*, **16** (December 1992).

Dulski, T.R.: "Residual-Element Analysis: Measuring the Minuscule," *Advanced Materials & Processes*, **20** (February 1992).

Engl, H.W. and W. Rundell: *Inverse Problems in Medical Imaging and Nondestructive Testing*, Springer-Verlag, Inc., New York, NY, 1997.

Evans, N.J.: "Impedance Spectroscopy Reveals Materials Characteristics," *Advanced Materials & Processes*, **41** (November 1991).

Hauk, V. and H. Behnken: *Structural and Residual Stress Analysis by Nondestructive Methods: Evaluation, Application, Assessment*, Elsevier Science, New York, NY, 1997.

Hellier, C.J.: *Handbook of Nondestructive Evaluation*, McGraw-Hill Professional Book Group, New York, NY, 2000.

Malhotra, V. and N. Carino: *CRC Handbook on Nondestructive Testing of Concrete*, CRC Press, LLC, Boca Raton, FL, 1990.

McGonnagle, W.: *International Advances in Nondestructive Testing*, Vol. 16, Gordon & Breach Publishing Group, Newark, NJ, 1991.

Michaels, T.E. and B.D. Davidson: "Ultrasonic Inspection Detects Hidden Damage in Composites," *Advanced Materials & Processes*, **34** (March 1993).

Prask, H.J.: "Neutron Probes Tackle Industrial Problems," *Advanced Materials & Processes*, **26** (September 1991).

Staff: "Testing for Materials Selection," *Advanced Materials & Processes*, **5** (June 1990).

Staff: "Computed Tomography Details Casting Defects," *Advanced Materials & Processes*, **54** (November 1990).

Staff: "Nondestructive Examination," *Advanced Materials & Processes*, **63** (January 1992).

Staff: *Nondestructive Testing*, American Society for Testing & Materials, West Conshohocken, PA, 1999.

Sturges, D.: "Sounding Out Ceramic Quality," *Advanced Materials & Processes*, **35** (April 1991).

Webb, S.C.: "PCs Help Optimize Materials Testing," *Advanced Materials & Processes*, **21** (November 1991).

Wu, Wen-li: "Acoustic Emissions Reveal Fiber/Matrix Bond Strength," *Advanced Materials and Processes*, **39** (August 1991).

Xavier Maldague, P.V.: *Theory and Practice of Infrared Technology for Nondestructive Testing*, John Wiley & Sons, Inc., New York, NY, 2001.

Web References

Nondestructive Testing Information Analysis Center: http://www.ntiac.com/
The American Society for Nondestructive Testing (NDT): http://www.asnt.org/
The online Journal of Nondestructive Testing: http://www.ndt.net/

NONEUCLIDEAN GEOMETRY.   See **Geometry**.

NONHYDROSTATIC MODE.   An atmospheric model in which the hydrostatic approximation is not made, so that the vertical momentum equation is solved. This allows nonhydrostatic models to be used successfully for horizontal scales of the order of 100 meters (328 feet), resolving small-scale mesoscale circulations such as cumulus convection and sea-breeze circulations. In recent years, computer power has made mesoscale weather prediction with nonhydrostatic models feasible, and several such models are in routine use by major meteorological modeling groups and operational centers. See also **Hydrostatic Model**.

NONNEWTONIAN LIQUIDS.   See **Viscosity**.

NONNUTRITIVE SWEETENERS.   See **Sweeteners**.

NONSINUSOIDAL WAVE.   Any periodic wave that is not a pure sine wave. Such waves may, however, be analyzed into numerous sine components, utilizing Fourier series, and often circuits may be analyzed by considering these one at a time.

NOR (Circuit).   A computer logical decision element that provides a binary 1 output if all the input signals are a binary 0. This is the overall NOT of the logical OR operation. Output F is positive only when both transistors are cut off. This occurs when both inputs A and B are negative. See Fig. 1.

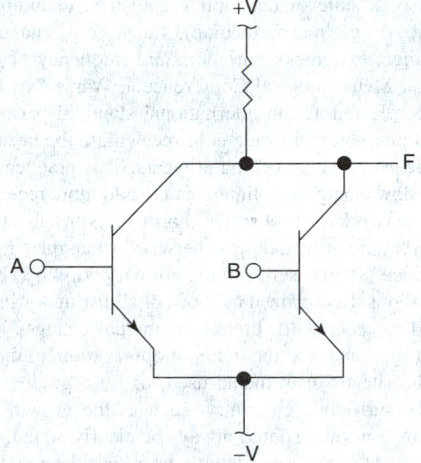

Fig. 1.   Transistor-type NOR circuit.

NOR'EASTER.   Nor'easter is a colloquial term for a macro scale storm whose winds come from the northeast, especially in the coastal areas of the Northeastern United States and Atlantic Canada.

More specifically, it describes a low pressure area whose center of rotation is just off the Carolina coast and whose leading winds in the left forward quadrant rotate onto land from the northeast. The precipitation pattern is similar to other extra-tropical storms. They also can cause coastal flooding, coastal erosion and gale force winds. The storm taps the Atlantic's moisture supply and dumps heavy snow over a densely populated region. The snow and wind may combine into blizzard conditions and form deep drifts paralyzing the region.

Nor'easters are usually formed by an area of vorticity associated with an upper level disturbance or from a kink in a frontal surface that causes a surface low pressure area to develop. Such storms often move slowly in their latter mature stage, frequently intense.

The northeastern United States and Atlantic Canada, particularly New England (coastal cities) and Nova Scotia, are usually hit with several nor'easters each year, most often in the winter and early spring, but also sometimes during the autumn. These storms can leave inches of rain or several feet of snow on the region, and sometimes last for several days. Until the nor'easter passes, thick dark clouds often block out the sun. During a single storm the precipitation can range from a torrential downpour to a fine mist, but it does not stop. Low temperatures and wind gusts of up to 50 miles per hour are also associated with a nor'easter. On very rare occasions, such as the North American blizzard of 2006, and a nor'easter in 1979, the center of the storm can even take on the circular shape more typical of a hurricane and have a small eye.

See also **Blizzard**.

NORFOLK ISLAND PINE.   See **Araucarias**.

NORFTHRUP, JOHN H. (1891–1987).   An American chemist who won a Nobel Prize in chemistry in 1946 along with James B. Sumner and Wendell M. Stanley. His work was primarily concerned with isolation and crystallization of enzymes. Many first included the production of the enzyme trypsin in the laboratory and isolation of the first bacterial virus. He was also responsible for producing diptheria antitoxin in crystalline form. His education was at eastern schools including Harvard, Yale, and Princeton.

NORMAL.   1. Referring to a normal distribution. 2. Regular or typical in the sense of lying within the limits of common occurrence, but sometimes denoting a unique value, as a measure of central tendency. Either sense presupposes a stable probability distribution. 3. As usually used in meteorology, the average value of a meteorological element over any fixed period of years that is recognized as a standard for the country and element concerned.

Often erroneously interpreted by the general public as meaning the weather patterns that one should expect. In the broadest sense, "normals" should consist of a suite of descriptive statistics, including measures of central tendency (e.g., mean, median), range (e.g., standard deviation, interquartile range, extremes), variation, and frequency of occurrence. At the International Meteorological Conference at Warsaw in 1935, the years 1901–30 were selected as the international standard period for normals. Recommended international usage is to recalculate the normals at the end of every decade using the preceding 30 years. This practice is used to take account of the slow changes in climate and to add more recently established stations to the network with observed normals. Normals should be based on actual observations if available; otherwise a recognized method should be used to "reduce" shorter series to the normal period by comparison with neighboring stations. Recognized methods of adjusting for inhomogeneities should be used to account for breaks or gradual changes introduced into the data record by changes in the hours of observation, in the observational practices, in the site or instruments used, or by a gradual change in the character of the surrounding country, such as the growth of a city. The years covered by a normal should always be clearly stated, since averages for different periods of the same length are rarely the same.

AMS

NORMAL CONCENTRATION.   A one normal solution (often abbreviated 1N) contains one gram-equivalent weight of a particular substance dissolved in 1 liter of *solution*. The equivalent weight of a substance may be defined as that weight of the substance that will involve, in a chemical reaction, one atomic weight of hydrogen, or that weight of any other element or portion of a substance, which, in turn, would involve in reaction one atomic weight of hydrogen.

As an example, the chlorine atom of potassium chloride (KCl) also is found in hydrochloric acid (HCl) in combination with one hydrogen atom. Thus, the gram-equivalent weight of KCl is 74.555, which is the same as its gram-molecular weight. A one normal solution of KCl will contain 74.555 grams of the salt per liter of solution.

For a particular solution, the molar and normal concentration are the same only when the gram-molecular and gram-equivalent weights are the same. Sulfuric acid H_2SO_4 represents a case where these values are not the same. This acid contains two active hydrogen ions and, therefore, its gram-equivalent weight is one-half of its gram-molecular weight. Phosphoric acid H_3PO_4 contains three active hydrogen ions. Consequently, the gram-equivalent weight for this acid is one-third that of the gram-molecular weight. Calcium hydroxide $Ca(OH)_2$ contains two active hydroxyl ions, each being equivalent to a hydrogen ion. Therefore, the gram-equivalent weight of $Ca(OH)_2$ is one-half of its gram-molecular weight.

NORMAL EQUIVALENT DEVIATE.   The normal equivalent deviate of a proportion p is the deviate in a normal distribution with unit variance that exceeds a proportion p of the total frequency. Thus N.E.D. $(p) = x$ where

$$p = (2\pi)^{-1/2} \int_{-\infty}^{x} \exp(-\tfrac{1}{2}t^2)\, dt$$

For ease in computation, x is often replaced by y, the probit of p, where $y = x + 5$.

NORMAL (Gaussian) DISTRIBUTION.   One of the standard distributions of statistical theory. The mathematical form may be written

$$f(x) = \frac{1}{\sigma(2\pi)} \exp\left\{ -\frac{1}{2\sigma^2}(x - \mu)^2 \right\}$$

This has mean μ and variance σ^2 and the form of the frequency curve is shown in Fig. 1.

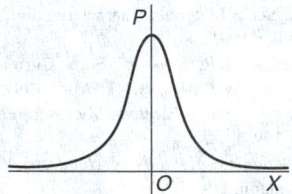

Fig. 1.   Typical Gaussian distribution curve.

The distribution was considered by Laplace and Gauss and is known by various names, that of "normal" being conferred by Karl Pearson, although he admitted that the practical occurrence of data of exactly normal type was somewhat abnormal. However, a number of distributions occurring in practice are approximately of the normal form, such as that of heights of men or women, or errors of measurement in reported observations of a magnitude.

The distribution owes its theoretical importance to a number of features, notably that under the **Central Limit Theorem** (q.v.) a great many of the statistics in common use tend to have a normal distribution in large samples, so that asymptotically the standard error can be used to set probability limits to estimates of parent parameters even in data from non-normal populations.

The distribution is a simple transfer of the "error curve." See also **Error**.

The distribution can be generalized to the case of p variables, being then of the form

$$f\infty \exp\left\{ -\frac{1}{2} \sum_{j,k=1}^{p} \alpha_{jk}(x_j - \mu_j)(x_k - \mu_k) \right\}, \quad -\infty \leq x_j \leq \alpha$$

That is to say, the quantity in the exponential is a general quadratic form, limited only by the fact that the form must be positive definite in order to ensure the convergence to unity of the total frequency.

SIR MAURICE KENDALL, International Statistical Institute, London

NORMAL (Geometry).   Perpendicular. At right angles to a given line or plane.

NORMALIZE (Mathematics).   1. To change in scale so that the sum of squares, or the integral of the squares of the transformed quantity is unit. (See also **Orthogonal Function**.)

2. To transform a random variable so that the resulting random variable has a normal distribution.

3. In computer operations, to adjust the exponent and coefficient of a floating-point result, so that the coefficient is in the prescribed normal range. Also called standardize.

NORMALIZING.   See **Iron Metals, Alloys, and Steels**.

NORMAL (Principal, to a Curve at a Point P).   The normal to the curve at P which lies in the osculating plane at P. A unit vector in the direction of the principal normal is called the *unit (principal) normal*. It is usually taken as directed from *P* to the concave side of the curve.

NORRISH, RONALD G. W. (1897–1978). An English physical chemist who was recipient of the Nobel Prize in 1967 with Manfred Eigen and George Porter. His analysis of reactions of one ten-billionth of a second were made possible by disturbing the chemical equilibrium with short energy pulses. After receiving a doctorate, he went on the Sorbonne before returning to Cambridge to teach. His career was long and distinguished by many awards.

NORTH. For many centuries, north has been the fundamental direction used by navigators and surveyors. During this long period, a loose usage of the term has become prevalent. It seems desirable to set down certain standard meanings accepted by the majority of modern navigators and astronomers.

True north (unless a qualifying adjective is used with north, true north is to be assumed) is the direction along the geographical meridian of the observer, in the plane of the observer's horizon, toward the North Pole of rotation of the earth. When the observer is facing the setting sun, the North Pole is to his right. *Compass north* is the direction of the plane of the horizon toward which the north-seeking end of the compass points. Unless otherwise stated, compass north refers to north defined by the magnetic compass; if another type of compass is used, it should be clearly indicated, e.g., gyrocompass north, etc. *Magnetic north* is the direction in the plane of the observer's horizon toward the north magnetic pole of terrestrial magnetism. For methods of conversion from any one of these three "norths" to any other, see **Compass (Navigation)**. See also **Navigation**.

NORTH AMERICAN HIGH. See **Atmosphere (Earth)**.

NORTH ATLANTIC CURRENT. See **Ocean Currents**.

NORTH ATLANTIC FLOUNDER. See **Flatfishes**.

NORTH ATLANTIC WATERS. In studies of oceanography, numerous bodies of ocean water are designated. The principal bodies in the North Atlantic are:

North Atlantic Central Water. A shallow oceanic water mass extending roughly from the southern parts of Greenland and Iceland to a region described by a line drawn from the northern end of South America to Africa. Temperature range is 8–19 °C (46.4–66.2 °F), salinity ranges from 35.1 to 36.7%.

North Atlantic Deep and Bottom Water. This dense ocean current arises in the Atlantic Ocean near the southeastern tip of Greenland where it meets the warmer water of the Gulf Stream below which it sinks to a depth of from 7,000 to 13,000 feet (2,100 to 3,900 meters) as it creeps southward. This water has been traced as far south as 60 degrees where the colder and heavier Antarctic water forces it to the surface.

North Atlantic Intermediate Water. An oceanic water mass lying at depths between the North Atlantic Deep and bottom water and the North Atlantic central water. Temperature range is from 2.5 to 4.0 °C (36.5 to 39.2 °F), salinity range is from 34.7 to 34.9%. The area of this water is more limited than those of the water masses above and below it. It has a high oxygen content.

NORTH BRAZIL CURRENT. See **Ocean Currents**.

NORTH EQUATORIAL COUNTERCURRENT (NECC). See **Ocean Currents**.

NORTH EQUATORIAL CURRENT (NEC). See **Ocean Currents**.

NORTHER. See **Winds and Air Movement**.

NORTHERN LIGHTS. See **Aurora and Airglow**.

NORTHERN PIKE. See **Pike**.

NORTH KOREA COLD CURRENT. See **Ocean Currents**.

NORTH PACIFIC CURRENT. See **Ocean Currents**.

NORTH PACIFIC WATERS. In studies of oceanography, numerous bodies of ocean water are designated. The principal bodies in the North Pacific are:

North Pacific Central Water. Due to the great size of the Pacific Ocean, it contains more well-developed oceanic water masses than the Atlantic Ocean. Thus, there are both eastern and western north Pacific central surface water masses. They are relatively shallow, extending from the subarctic Pacific water on the north to the Pacific equatorial water on the south, and covering the width of the ocean except for a transition zone on the eastern side. Their temperature ranges are 8 to 18 °C (46.4 to 64.4 °F) and their salinity ranges from 33.8 to 34.9%, the lower values of each existing at lower depths (400 to 700 meters) (1,320 to 2,310 feet) where they meet the north Pacific intermediate water.

North Pacific Deep Water. Measurements at great depths show the existence of an oceanic water mass of practically constant salinity and low temperature below 2,500 to 3,000 meters (8,250 to 9,900 feet) in the south Pacific Ocean, and still deeper in the north Pacific Ocean. Since the Bering Straits are so narrow and shallow, this mass cannot be produced by currents from the Arctic Ocean, as in the north Atlantic deep and bottom water. Since the oxygen content of the Pacific deep water is less in the north Pacific than the south Pacific Ocean, this water mass is believed to be supplied from the antarctic bottom water or the Atlantic deep and bottom water.

North Pacific Intermediate Water. An oceanic water mass lying below the north Pacific central and equatorial waters at depths ranging from 600 to 800 meters (1,980 to 2,640 feet) in the north and to about 200 meters (660 feet) and 1,000 meters (3,300 feet) (two layers) in the south. It is characterized by low salinity and especially low oxygen content.

NORTHROP, JOHN HOWARD (1891–1987). Northrop was born at Yonkers, New York on 5 July 1891, the son of a zoologist. He studied chemistry at Columbia University, New York, where he gained his BSc in 1912, an MA in 1913, and a PhD on enzyme chemistry in 1915. Until 1949 his research career was spent at the Rockefeller Institution for Medical Research at Princeton, New Jersey. From 1949 until his retirement in 1970 he was Professor of Bacteriology and Biophysics at the University of California, Berkeley.

During World War II, Northrop devised a method of fermenting potatoes to produce acetone and this probably awakened his interest in enzymes. Between 1916 and 1924 Northrop also worked on a variety of other subjects with the physiologist Jacques Loeb, including studies of osmosis and the response of organisms to light. Gradually he began to focus on enzymes, their molecular weights and solubilities. When James Sumner announced the successful crystallization of urease in 1926, it was met with hostility and criticism. It was Northrop's support in the early 1930s, by the crystallization of pepsin and trypsin, and several other enzymes, that firmly established that enzymes were high molecular weight proteins. Here Northrop's wide experience of physical chemistry proved beneficial since he was able to use the constant solubilities of proteins, and not just their crystallization, as a test of purity. Using such a test, Northrop was able to establish the existence of enzyme precursors such as pepsinogen which acted as autocatalysers in the increase of enzyme concentration. When Wendall Stanley crystallized the *Tobacco mosaic virus* (TMV) in 1935, Northrop decided to apply his experimental skills to the study of bacteriophages. He was convinced that, like enzymes, phages would turn out to be proteins – the first bacterial enzymes. In 1938, however, when he isolated and crystallized the first staphylococcus phage, it proved to be a nucleoprotein. Nevertheless, right up to the emergence of molecular biology in the 1950s, Northrop remained convinced that phages, like enzymes, were self-reproducing systems. Honor was satisfied in 1951 when he described viral nucleic acid as the autocatalytic agent that promoted phage replication. Before then, in 1946, Northrop, Stanley and Sumner shared the Nobel Prize for Chemistry for "their preparation of enzyme and virus proteins in pure form." Northrop's postwar work consisted of the isolation of diphtheria antitoxin and the pneumococcal antibody. During the final decades of his long and fruitful career he continued to work on phage kinetics and the origins of bacterial viruses. See also **Biochemistry (The History)**; **Loeb, Jacques (1859–1924)**; **Stanley, Wendell M. (1904–1971)**; and **Sumner, James Batcheller (1887–1955)**.

Additional Reading

Northrop, J.H.: *Crystalline Enzymes*, 2nd Edition, Columbia University Press, New York, NY, 1948.

Northrop, J.H.: *The Chemistry and Physiology of Growth*, Columbia University Press, New York, NY, 1949.

Sherby, L.S., and W. Odelberg: *The Who's Who of Nobel Prize Winners 1901–2000*, 4th Edition, Greenwood Publishing Group, Inc., Westport, CT, 2001.

W. H. BROCK, University of Leicester, Leicester, UK

NORWALK VIRUS.   An epidemic of winter vomiting disease in 1968 involved many residents of Norwalk, Ohio. At that time, the causative agent was unknown. However, the illness resembled a vomiting and diarrhea syndrome observed in some regions as early as 1929. The syndrome occurred during the winter and affected mainly teenagers and adults. Similar episodes among close-living groups of people were reported from boarding schools in England during the 1950s. A number of enteroviruses and rotaviruses were identified in earlier years, such as the coxsackieviruses and echoviruses, among others, but not all epidemics had been satisfactorily explained. The Norwalk agent did not exactly fit any of the prior patterns.

Field investigators from the Center for Disease Control (Atlanta, Georgia) were despatched to Norwalk to collect stool specimens and early and convalescent blood. These were frozen for future examination. The National Institute of Health (United States) established a program involving volunteers and electron microscopic examinations, directed at identifying the Norwalk agent. A specific virus was identified. It resembles paroviruses, measures about 27 nanometers (0.027 micrometer) and has since been officially called the Norwalk agent. Three serologically different forms have been identified. One of these is now called the Hawaiian agent. More recently, additional varying agents have been identified in England. A widespread outbreak of Norwalk virus gastroenteritis in Australia was attributed to contaminated shellfish.

Persons infected with the Norwalk agent display (after 12 hours incubation) fever, vomiting, diarrhea, cramps, malaise, and leukopenia over a period of 2 to 3 days. Biopsy of the intestine during the acute phase of the disease has demonstrated that the villi of jejunum become flattened and infiltrated with mononuclear cells. Absorption of xylose and fat is reduced during and after the illness. Treatment is supportive with rehydration therapy for any major loss of fluids. Experience with volunteers indicates that natural immunity as the result of infection may persist only for a matter of weeks or months in many cases. Several volunteers who were rechallenged after 27 and 42 months developed symptoms of the disease. Because of wide variation in the results of the volunteer program, some authorities believe that there may be genetic differences in susceptibility.

Additional Reading

Blacklow, N.R., R. Dolin, and D.S. Fedson: "Acute Infectious Nonbacterial Gastroenteritis: Etiology and Pathogenesis," *Ann. Intern. Med.*, **76**, 993 (1972).

Melnick, J.L.: "Enteroviruses. Viral Infections of Humans: Epidemiology and Control," (A.S. Evans, editor), Plenum Medical Book Co., New York, NY, 1976.

Web Reference

Center for Disease Control and Prevention (CDC): http://www.cdc.gov/health/diseases.htm

NORWAY SPRUCE.   See **Spruce Trees**.

NORWEGIAN CURRENT.   See **Ocean Currents**.

NORWEGIAN TOPKNOT.   See **Flatfishes**.

NOSE FLY (*Insecta, Diptera*).   Also known as the *sheep bot*, this species, *Oestrus ovis* (Linne), acts against sheep, goat, and wild deer and is found throughout North America; it is particularly abundant in Idaho, Montana, New Mexico, and Texas, and generally in the areas lying between these states. The nose fly strikes at the nose of the animal and deposits eggs in the nostrils. The maggots lodged in the nostrils and head sinuses create an inflamed condition and accompanying catarrhal discharge. The insect is known, on occasion, to attack humans.

Where the insect is present, the nostrils of the animal should be treated with pine tar. An early application is required in mid-April in most areas. Specially treated salt logs are obtainable which permit the animals to smear their noses with pine tar in an "automatic" fashion. In times of abundant presence of the flies, some producers herd the animals into darkened retreats where the flies are not active.

NOSE (Odor Receptor).   See **Flavorings**; and **Olfactory System**.

NOT (Circuit).   Also known as an inverter circuit, this is a circuit which provides a logical NOT of the input signal. If the input signal is a binary 1, the output is a binary 0. If the input signal is in the 0 state, the output is in the 1 state. Referring to Fig. 1, if A is positive, the output F is at 0 V inasmuch as the transistor is biased into conduction. If A is at 0 V, the output is at $+V$ because the transistor is cut off. Expressed in Boolean algebra, $F = A'$, where the prime denote the NOT function.

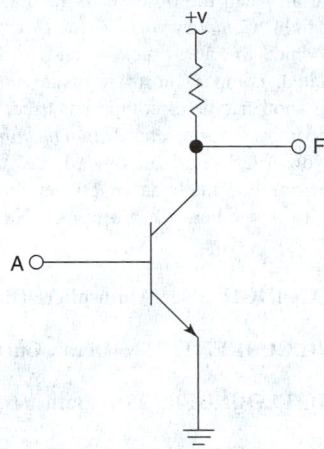

Fig. 1.   Inverter or NOT circuit.

See also **Diode Transistor Logic**.

NOTOCHORD.   A longitudinal stiffening rod found in all embryonic chordates and in the adults of some of the lower members of this phylum (*Chordata*). It lies between the central nervous system and the alimentary tract (digestive system) and is the axis around which the spinal column develops. As the bony structure forms, the notochord is almost crowded out of existence. In the human body small remnants of it form the nuclei pulposi in the intervertebral disks.

NOVA AND SUPERNOVA.   Traditionally, a nova (an early term meaning "new star") has been described as a star that suddenly displays an increasing brilliance and then over a period of time grows fainter. Because of the tremendous distances between the earth and such objects, it is interesting that while some of these events of a cataclysmic nature can be seen by the naked eye from earth as though occurring now, they actually extend far back in time. Most novae that do not attain naked-eye brilliancy are discovered and studied instrumentally.

Supernova 1987A.   On the night of 23/24 February 1987, Ian Shelton, working at the University of Toronto Las Campanas Station in northern Chile, discovered the brightest supernova seen since 1604. A separate article on this historic discovery will be found in this encyclopedia: **Supernova 1987A**.

Brightness of Supernova.   There is no known procedure by which an estimate of the total number of novae appearing each year can be predicted. It has been estimated that ten or more novae reach a brightness of the ninth stellar magnitude or greater each year, based upon historical records. Between 1900 and 1935, only five novae reached conspicuous brightness. More recent bright novae included V1500 Cyg (Nova 1975 Cyg) and the recurrent nova RS Ophiuci.

In Fig. 1, the light curves of three bright novae of the present century are represented. The ordinate scale of brightness is expressed in stellar magnitude. Since magnitude 6 is the limit of naked-eye visibility, the length of time that each was visible to the naked eye may be determined from the time scale given at the bottom. These curves are characteristic of most novae, with the very rapid rise to maximum and then the relatively slow and irregular decline. Examination of photographic records indicates that novae are not actually new stars at all, but rather are faint stars that suddenly increase in intensity. An increase of ten magnitudes is by no means uncommon, and this represents an increase of light intensity amounting to 10,000-fold. The total emitted energy in one outburst is about 10^{45} ergs.

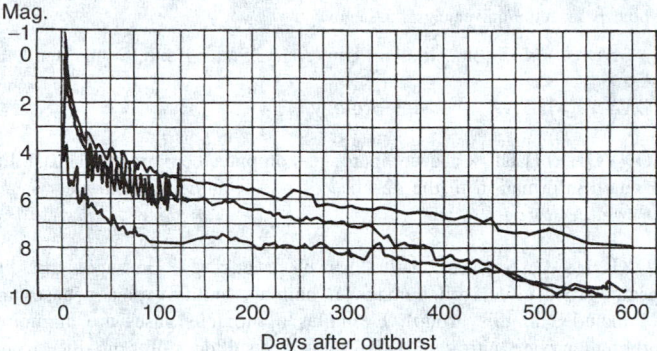

Fig. 1. Light curves of Nova Aquilae, 1918; Nova Persei, 1901; and Nova Geminorum, 1912. They are designated in order of decreasing height. (*Harvard College Observatory Annals.*)

Coupled with the increase in light intensity of a nova is a correspondingly remarkable change in spectral characteristics. Although the spectral changes in different novae vary to a considerable extent, there are certain stages of development that are more or less characteristic of them all. During the period of rise, in the few cases where increasing novae have been detected in time for observation, the stars is of the hot, blue, A-type, with the absorption lines displaced very strongly to the violet. As the star starts to decline, the color changes from white to yellow, and bright lines, particularly of hydrogen and ionized iron, appear. The bright lines then broaden out to bands of irregular structure, which soon completely mask the continuous spectrum of the star. A few days later, dark lines again make their appearance, and these are displaced far to the violet. Soon, bright lines again appear, frequently of the type characteristic of the gaseous nebulae, except that they are broad. Often multiple components, corresponding to separate shells or blobs, are observed. As the brightness of the star further decreases, it eventually settles down to a peculiar O-type spectrum, with bright lines superimposed on a continuous and dark-line absorption spectrum.

Recent infrared observations have revealed that considerable amounts of dust can also be present in the nova ejecta. Radio observations argue for the presence of extensive circumstellar shells pre-existing in the nova environment. Ultraviolet observations show that the period immediately following the nova ejection event is dominated by a strong stellar wind emanating from the still-hot compact star.

Novae have been observed telescopically in some of the exterior galaxies such as the great spiral in Andromeda. Since the distances of some of these objects are at least very approximately known, it is possible to get an approximation to the absolute magnitudes at maxima of the novae observed in them. For these extragalactic novae, we find absolute magnitudes of the order of −4, a value that compares favorably with those determined for the few cases where the distance of a galactic nova is known.

Supernovae fall into two broad categories, mainly on the basis of their light curves. These are:

Type I. These supernovae are around absolute photographic magnitude −18.6 maximum, or more than 200 million times as luminous as the sun. If one of them were placed at the distance of 10 parsecs from Earth, at which distance the sun would be barely visible, the nova would appear 14 times as bright as the full moon does to Earth. The spectra show extremely broad emission bands. The light variations show a rapid rise to maximum, followed at first by a rapid, and later by a slower, decline, usually with a characteristic time scale of 50 to 70 days. See Figs. 2 and 3.

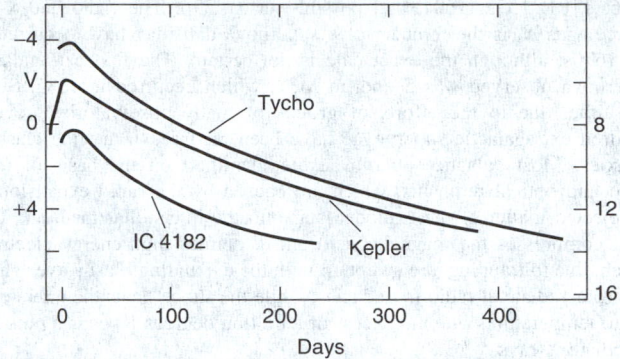

Fig. 2. Light curves of Kepler's and of Tycho's supernovae, as reconstructed by Baade (1945). Scale on right is for the supernova IC 4182. The light curves show that the supernovae of 1572 and 1604 were of Type I. (*After van den Bergh.*)

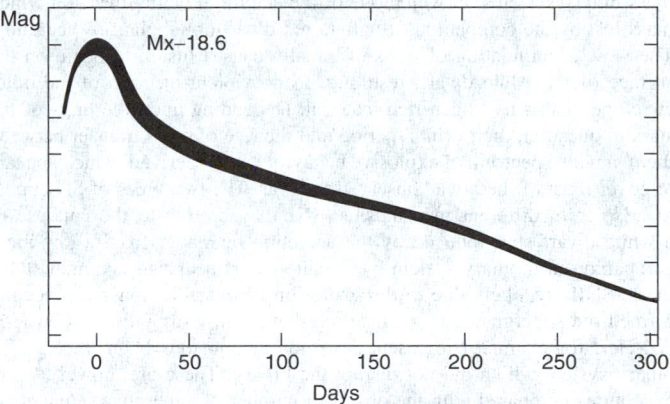

Fig. 3. Composite blue light curve obtained by fitting observations of 38 Type I supernovae. One-magnitude intervals are marked on the ordinates. (*After van den Bergh.*)

Type II. The members of this group reach a maximum luminosity equal to about 20 million suns. After maximum, they fade more slowly at first than do the members of the other group, followed by a more rapid decline after about 100 days. These supernovae (SN) show greater diversity than Type I. See Fig. 4.

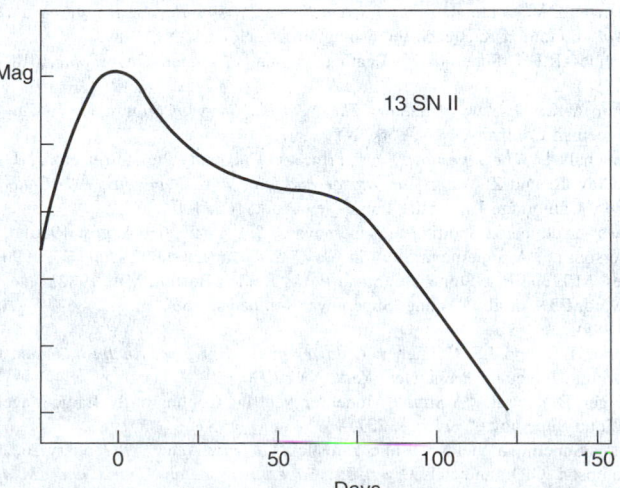

Fig. 4. Composite light curve obtained from 13 Type II supernovae. (*After van den Bergh.*)

Type V. Two extragalactic examples of this class are known. They are more massive than Type II and appear to be associated with very massive (of order 100 $M_{\odot}$) progenitors.

Historical records indicate that supernovae were observed in AD 185, 1006, 1181, 1572, 1604, and possibly in AD 396. The radio and x-ray source, Cas A, is the remnant of a supernova that may have occurred in the 1670s, although the exact date is not certain. The first extragalactic supernova observed was S and, in M31, which occurred in 1885. Since that time, due to the efforts of groups of many observatories, several hundred extragalactic supernovae have been recorded in a wide class of galaxies. Most galactic supernovae are identified on the basis of their radio and optical remnants, which are caused by the rapid expansion of the ejected matter of the exploded star into the interstellar medium. This shell compresses the magnetic field and generates high-energy electrons which, due to trapping, are swept up with the expanding blast wave. These electrons radiate at radio frequencies, while the shock heats the interstellar gas to temperatures often in excess of a million degrees K, causing the gas to radiate X-rays.

The Crab and Vela supernova remnants have pulsars associated with them, while W 50, a weak optical nebula, has been associated with the peculiar x-ray and optical source SS 433. Few other supernova remnants can, however, be unambiguously linked to pulsars at this writing.

Observations indicate that all novae are members of short-period, low-mass binary systems, in which one of the stars in a white dwarf on which mass lost by the companion (usually a red dwarf or subgiant) is accreting. The slow accumulation of mass eventually causes fusion to begin on the surface of the white dwarf, resulting in the violent ejection of the outer envelope of this hydrogen-rich material. Depending upon the mass of the stars in question, their orbital period and the rate of mass transfer between them, a wide spectrum of explosive behavior can be derived, which appears to cover most of the novae observed thus far. The two types of SN appear to arise from different mechanisms. SNI may be due to the collapse of a white dwarf star, induced by the accretion of matter from a low mass companion in a binary system by a white dwarf near its maximum stable mass. SNII are likely due to the explosion of unstable, massive, recently formed red supergiants. It is suggested that some x-ray pulsars (Cen X-3, HZ Her, for example) represent supernova explosions by members of a binary system which did not disrupt the binary. These are, however, rare exceptions compared with the observed number of supernova remnants.

Additional Reading

Asimov, I.: *The Exploding Suns: The Secrets of the Supernovas*, St. Martin's Press, Inc., New York, NY, 1985.

Bethe, H.A. and G. Brown: "How a Supernova Explodes," *Sci. Amer.*, 60–68 (May 1985).

Boss, A.P.: "Collapse and Formation of Stars," *Sci. Amer.*, 40–45 (January 1985).

Chandrasekhar, S.: "On Stars, Their Evolution and Their Stability," *Science*, 226, 497–505 (1984).

Harris, M.J.: "Short-Lived s-Process Gamma-Ray Lines in Type II Supernovae," *Science*, 60 (April 1, 1988).

Imshennik, V. et al.: *Astrophysics and Space Physics Reviews: Supernova 1987a*, Vol. 8, Gordon & Breach Publishing Group, Newark, NJ, 1989.

Kirshner, R.F.: "Supernova — Death of a Star," *National Geographic*, 618 (May 1988).

Mann, A.K.: *Shadow of a Star: The Neutrino Story of Supernova 1987a*, W.H. Freeman Company, New York, NY, 1997.

Marschall, L.: *The Supernova Story*, Princeton University Press, Princeton, NJ, 1994.

McCray, R. and Z. Wang: *Supernovae and Supernova Remnants: IAU Colloquium 145*, Cambridge University Press, New York, NY, 1994.

News: "Automated Spotting of Supernovae," *Sci. Amer.*, 65 (August 1986).

Peterson, I.: "A Supernova Story in Clay," *Science News*, 397 (June 23, 1990).

Rees, M.J. and R.J. Stoneham: *Supernovae*, Reidel, Boston, MA, 1982.

Seward, F.D. et al.: "Young Supernova Remnants," *Sci. Amer.*, 88–96 (August 1985).

Sparke, L.S. and J.S. Gallagher: *Galaxies in the Universe: An Introduction*, Cambridge University Press, New York, NY, 2000.

Spergel, D.N. et al.: "A Simple Model for Neutrino Cooling of the Large Magellanic Cloud Supernova," *Science*, 237, 1471–1473 (1987).

Staff: "Supernova Yields Cosmic Yardstick," *Science News*, 59 (January 26, 1991).

Thompson, G.D. and J.T. Bryan, Jr.: *Supernova Search Charts and Handbook*, Cambridge University Press, New York, NY, 1989.

Waldrop, M.M.: "Supernova 1987A: A Mysterious Stranger," *Science*, 237, 25–26 (1987).

Waldrop, M.M.: "Sighting of a Supernova," *Science*, 235, 1143 (1987).

Waldrop, M.M.: "The Supernova 1987A," *Science*, 235, 1322–1323 (1987).

Waldrop, M.M.: "Feeding the Monster in the Middle," *Science*, 478 (January 27, 1989).

Waldrop, M.M.: "And Now for a Real Crab Nebula," *Science*, 1140 (March 3, 1989).

Weiss, P.L.: "Seeing Supernovas in Galactic Chimneys," *Science News*, 133 (September 1, 1990).

Wheeler, J.G. and R.P. Harkness: "Helium-rich Supernovas," *Sci. Amer.*, 50–58 (November 1987).

Williams, R.E.: "The Shells of Novas," *Sci. Amer.*, 120–131 (April 1981).

Wilson, O.C. et al.: "The Activity Cycle of Stars," *Sci. Amer.*, 104–121 (February 1981).

Web Reference

United States National Aeronautics and Space Administration, Space Science homepage. http://universe.gsfc.nasa.gov/

STEVEN N. SHORE, Indiana University, South Bend, South Bend, IN

NOVA ZEMLYA. See **Meteorology**.

NONVOLATILE. Of a computer or computer component. The ability to retain information in the absence of power as *nonvolatile memory*, or *nonvolatile storage*.

NOXIOUS GAS. Any natural or by-product gas or vapor that has specific toxic effects on humans or animals (military poison gases are not included in this group). Examples of noxious gases are ammonia, carbon monoxide, nitrogen oxides, hydrogen sulfide, sulfur dioxide, ozone, fluorine, and vapors evolved by benzene, carbon tetrachloride, and a number of chlorinated hydrocarbons. Gases that act as simple asphyxiants are not classified as noxious. See also **Ammonia**; **Benzene**; **Carbon Monoxide**; **Carbon Tetrachloride**; **Fluorine**; **Hydrocarbons**; **Nitrogen**; and **Pollution (Air)**.

NPOESS PREPARATORY PROJECT (NPP). See **Earth Observing System (EOS)**.

NRM WIND SCALE. See **Winds and Air Movement**.

NUCLEAR CHEMISTRY. The division of chemistry dealing with changes in or transformations of the atomic nucleus. It includes spontaneous and induced radioactivity, the fission or splitting of nuclei, and their fusion or union; also the properties and behavior of the reaction products and their separation and analysis. The reactions involving nuclei are usually accompanied by large energy changes, far greater than those of chemical reactions; they are carried out in nuclear reactor for electric power production and manufacture of radioactive isotopes for medical use, also (in research work) in cyclotrons. See also **Nuclear Fission**; **Nuclear Fusion; Radiochemistry;** and **Nucleus**.

NUCLEAR FISSION. A type of nuclear reaction in which the compound nucleus splits into two nearly equal parts, rather than ejecting one or a few small nuclear particles, as in most nuclear reactions. Our knowledge of nuclear fission dates back to the mid-1930s when Fermi and his coworkers showed that the number of distinctly different radioactive nuclides that could be induced by neutron bombardment of uranium far exceeded the number expected, unless some previously unknown pattern of isomerism could be found. Furthermore, the radiochemical properties of many of these radio-elements different quite markedly from expectations. For example, both Hahn and Strassman in Germany and Curie and Savitch in France found that certain unknown activities, thought to be radioactive radium, always followed the chemically separated barium fraction rather than the radium fraction. Hahn and Strassman found several other similar examples and were able to show that uranium, when bombarded by neutrons, undergoes what then appeared to be a very unusual nuclear reaction in that the products are radio-elements with about half the atomic number of uranium. These findings were interpreted by Meitner and Frisch as the division of an excited nucleus into nuclei of medium mass, a process that was given the name *nuclear fission*.

The first such process to be extensively studied was fission induced in ^{235}U by thermal neutrons (neutrons with energies of about 0.03 eV). This

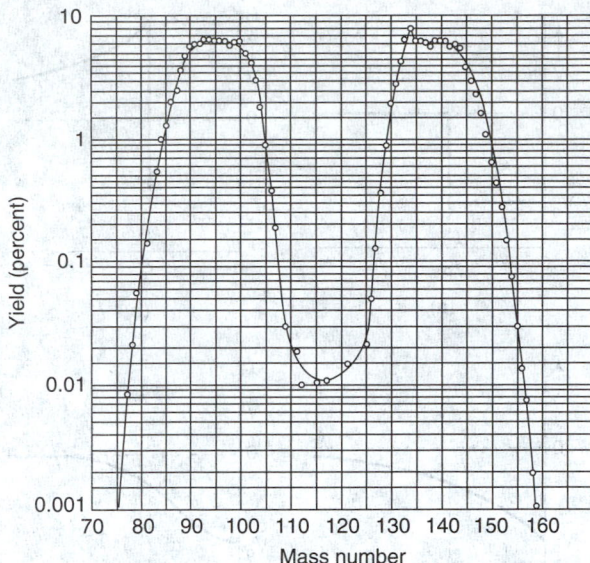

Fig. 1. Mass fission yield curve for ^{235}U + n (thermal).

nucleus, a multiplication factor of one or greater. Maintenance of a chain reaction is essential to the proper functioning of both nuclear weapons and nuclear reactors.

The probability that fission can occur (generally called the cross section for fission) varies widely among different nuclides. Only a few nuclides, such as ^{235}U, have a high probability of undergoing fission when they capture a neutron. In other nuclides, the probability of fission is generally much smaller. As an example, the cross section as a function of incident neutron energy is shown in Fig. 2 for fission of ^{235}U and of ^{238}U. Although fission can be induced in ^{238}U, such a process is possible only if the incident neutron has an energy greater than 1 MeV, whereas neutrons of any energy can induce fission in ^{235}U. The characteristic double hump

reaction, symbolically represented by the equation

$$^{235}\text{U} + n \longrightarrow {}^{236}\text{U} \longrightarrow \text{fission},$$

produces an unstable system which achieves stability by splitting into two large fragments, not by ejecting one or a few small particles.

An individual fission does not produce a unique pair of fragments, but in a large number of such processes, the mass distribution of the fragments can be predicted with reasonable certainty, leading to predictable fission yields. A fission yield, usually expressed as a percentage, describes that fraction of nuclear fission processes that give rise to a specified nuclide or group of isobars. The yields of single nuclides are known as independent yields and those of a set of isobars as mass yields or chain yields. Since two fragments are produced by each fission, the total of all fission yields for a given fission process is 200%. The fission yield curve is different for each mode of induced fission, the most commonly known one being that for thermal neutron induced fission of ^{235}U, shown in Fig. 1. The chemical characteristics of the two fragments vary within limits, so that many elements are formed. Analysis of the fission products shows that most of them are in two mass groups, a "light" group consisting of elements having mass numbers between 85 and 104, and a "heavy" group consisting of elements having mass numbers between 130 and 149. Fragment mass numbers that have been detected range from around A = 70 to around A = 160. The determination of independent yields is made more difficult by the fact that many of the products are highly radioactive and undergo extensive secondary changes, sometimes in extremely short times, a very small fraction of a second.

A most significant aspect of nuclear fission is its great release of energy. The source of this energy is the loss of mass between the initial and final products of the reaction. The total mass of all atoms and nuclear particles produced in a single fission process in less than the original mass of the ^{235}U atom and the neutron that combined with the ^{235}U to induce fission. During fission of ^{235}U, the total energy released because of loss of mass is about 200 MeV. In practical units, the fissioning of 1 gram of ^{235}U yields 24,000 kilowatt-hours of energy.

Another important feature of fission is the presence of neutrons among the reaction products, slightly more than two for each fission of a ^{235}U atom. These neutrons are not an immediate consequence of fission, but are boiled off the original fission products, their release being possible because of the very large amount of available energy. If all these neutrons were captured by other ^{235}U nuclides, the number of available neutrons would multiply by factors of two for every generation of fission processes, a very rapid increase. However, some neutrons escape from the region containing the ^{235}U and others are absorbed in nonfission capture processes. The minimum conditions for a self-sustaining chain reaction is that at least one neutron from each nucleus undergoing fission must cause fission of another

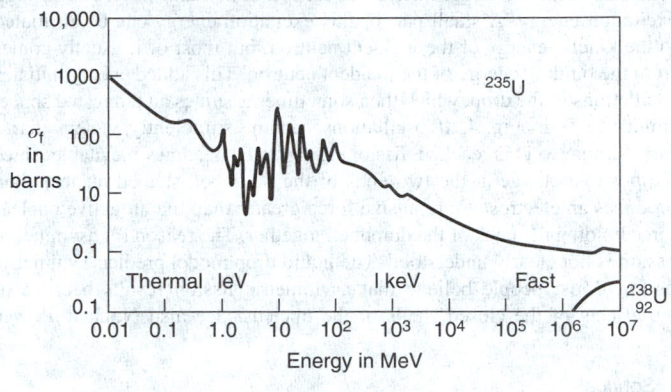

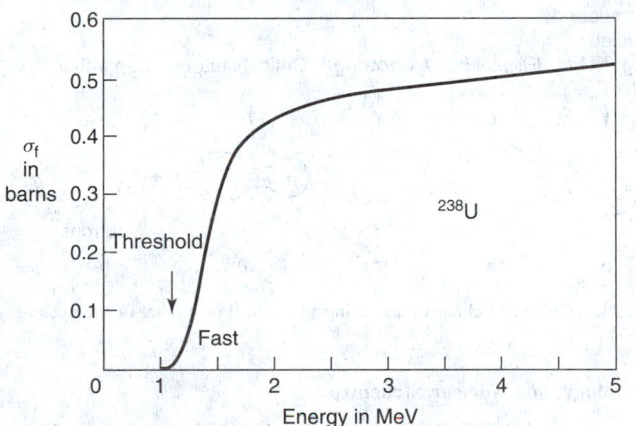

Fig. 2. Fission cross section as function of energy for ^{235}U and ^{238}U.

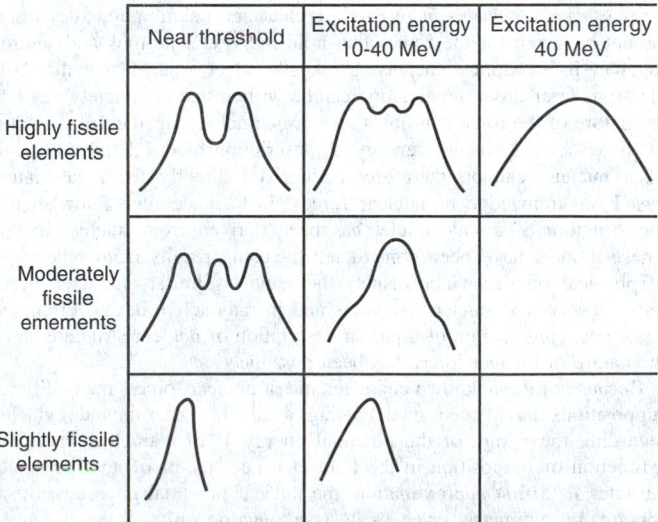

Fig. 3. Mass fission yield curves as function of excitation energy and degree of fission probability.

yield curve of Fig. 1 (asymmetric fission) is common only for low neutron excitation energy and targets consisting of highly fissile elements. For either higher excitation energies or less fissile elements, such as actinium or radium, symmetric fission becomes much more important, creating a triple humped fission-yield curve, shown in Fig. 3. Slightly fissile elements, such as lead and bismuth, or very high excitation energies further emphasize the symmetric mode of fission, also illustrated in Fig. 3. Nuclear fission may be induced by particles other than neutrons, such as alpha particles and photons. In some nuclides, it also occurs spontaneously, although the probability of such occurrence is so low that it has almost no effect on the radioactive decay characteristic of the nuclide.

Nuclear fission has generally been explained theoretically in terms of the liquid-drop model of the nucleus. In this model, the incident neutron combines with the target nucleus to form a compound nucleus at a high excitation energy. A small part of this excitation energy can be attributed to the kinetic energy of the incident neutron, but most of it usually comes from the binding energy of the incident neutron. This added energy initiates oscillations in the drop, which then sometimes assumes an elongated shape, similar to B in Fig. 4. If oscillations become sufficiently violent that a form similar to D is reached, fissioning (form E) becomes inevitable, since the positive charge at the two ends of the dumbbell-shaped nucleus then produces an electrostatic repulsive force greater than the attractive nuclear force holding the neck of the dumbbell together. The reason for asymmetric fission is not clearly understood. The liquid drop model predicts symmetric fission. Most people believe that asymmetric fission results because of the effects of the closed shells of the nucleus. See also **Nuclear Power**

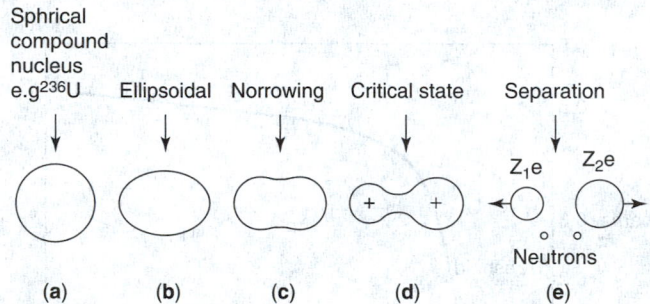

Fig. 4.   Fission mechanism according to liquid-drop model of the nucleus.

Technology; and **Nuclear Structure**.

C. SHARP COOK, The University of Texas at El Paso

NUCLEAR FORCES.   Strong, short-range, attractive forces that interact between the individual nucleons of an atomic nucleus. Unfortunately, despite several decades of research, a clear and unambiguous description cannot be given for the forces that hold individual protons and neutrons together in an atomic nucleus. Unlike the electrostatic force that holds electrons in an atom, no equation can be written that completely describes the nature of the force that holds an atomic nucleus together, or the nature of its associated potential energy. A description of the detailed structure of a nucleus cannot, therefore, be derived directly from calculations based on knowledge of nuclear forces. Instead, detailed knowledge of the structure of atomic nuclei has been derived from nuclear models. These models have been constructed by using results from other fields of physical science which display the same or similar characteristics as those observed in nuclear reactions and in radioactive decay. From such analogies, construction of a partial description of nuclear structure and of the nature of nuclear forces has been possible.

Because of the unknown characteristic of nuclear forces, many different suppositions have been made, using available experimental evidence, regarding the nature of the potential energy V of a nuclear particle as a function of its position in the field of a nucleus, or of another nuclear particle. To a first approximation, the nuclear potential is assumed to be spherically symmetric, such as V is a function only of the distance r from the center of the field, thus being the same in all directions, and is representable by a curve as in Fig. 1 curves (a) to (f).

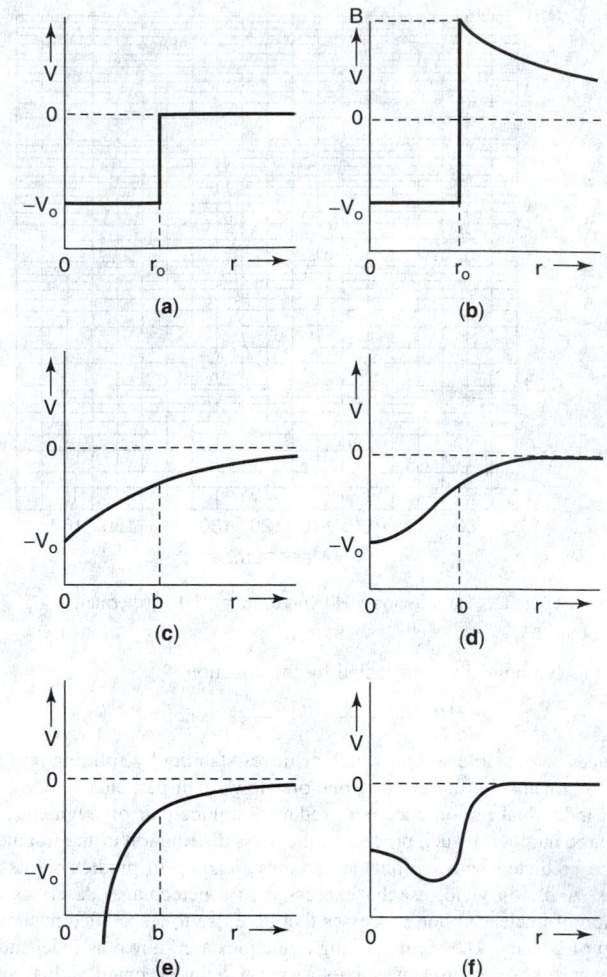

Fig. 1.   Potential energy of a nuclear particle versus distance from the center of the field.

A *potential well* is the name given to a region in which a minimum in the potential is formed; it results from attractive forces. A *potential barrier* is the name given to a region in which there is a maximum in the potential; it results from repulsive forces, either alone or in combination with attractive forces. Some central potentials commonly used as approximations to nuclear potentials are illustrated in the curves. Curve (a) shows a square well potential, which has a constant negative value $-V_0$ for $r \leq r_0$ and zero value for $r \geq r_0$. When this curve represents the potential between two nucleons, r_0 is called the *range of nuclear forces*; when it represents the potential of a nucleus, as this nucleus interacts with an individual nucleon, r_0 is called the *nuclear radius*. Curve (b) shows a square well potential for $r \leq r_0$ with a Coulomb potential resulting from repulsive electrostatic forces, for $r > r_0$. The resulting barrier is called a *Coulomb barrier*, and the maximum energy b is called the barrier height. Such a potential approximates that of a positively charged particle in the field of a nucleus, and is often used in the theory of alpha particle distintegration and nuclear reactions. Curve (c) shows an exponential well, $V = V_0 e^{-r/b}$; curve (d) shows a Gaussian well, $V = V_0 e^{-r/b2}$. Curve (e) shows a Yukawa potential, $V = -(V_0/r)e^{-r/b}$ used in the meson theory of nuclear forces for the interaction between two nucleons; and (f) shows a wine-bottle potential, characterized by a low central elevation. If a high central elevation is present, the resulting barrier is called a *central barrier*, or *repulsive core*.

Although the predominant part of the nuclear potential is the part described above that is derived from the central force produced by the average effects of all other nucleons in the system on the individual nucleon under observation, evidence exists that nonsymmetric tensor and spin-orbit coupling terms must be included in the description of the nuclear potential. These are derived from a tensor force resulting from a coupling between individual pairs of nucleons and from the coupling between spin and orbital angular moments of the individual nucleus, as described by the shell model of the nucleus.

A considerable amount of evidence indicates that nuclear forces are charge-independent, i.e., the neutron-neutron, neutron-proton, and proton-proton forces are identical. The meson theory of nuclear forces, originated by Yukawa, postulates the atomic nucleus being held together by an exchange force in which particles, now called mesons, are exchanged between individual nucleons within the nucleus.

C. SHARP COOK, The University of Texas at El Paso

NUCLEAR FUSION. The character of the atomic nucleus is such that the individual nuclear particles are most tightly bound in elements of intermediate atomic number. When energy is sought, attention is focused on the more loosely assembled elements, releasing energy by splitting (*fissioning*) the heavy isotopes, or by joining (*fusing*) the lighter ones. There is less energy release per fusion reaction than there is per fission reaction, but the reactants are more plentiful and, in many respects, easier to handle. A particular fusion reaction is of interest if the power produced can be sufficiently large to offset the power consumed in generating and maintaining the reacting medium, and if the relevant rates can be large enough so that economically interesting regimes are accessible to modern technology. There are over thirty such reactions possible. The most appealing of the fusion reactions as possible routes to fusion energy are (1) those which involve the heavy hydrogen isotopes, deuterium, $_1^2$H or D; and (2) those which involve tritium $_1^3$H or T. These tend to have the largest fusion reaction probability (cross section) at the lowest energies. Deuterium is abundant, naturally occurring and in wide use now as D_2O in heavy-water-moderated reactors. Tritium is a radioactive isotope with a 12.3-year half-life and does not occur in nature. Tritium emits an electron and decays to stable helium-3.

The deuterium (D-D) reaction chain may be represented by:

$$D + D \longrightarrow {}^3He + n + 3.2 \text{ MeV}$$

$$D + D \longrightarrow T + p + 4.0 \text{ MeV}$$

$$D + T \longrightarrow {}^4He + n + 17.6 \text{ MeV}$$

$$D + {}^3He \longrightarrow {}^4He + p + 18.3 \text{ MeV}$$

$$6D \longrightarrow 2{}^4He + 2p + 2n + 43.1 \text{ MeV}$$

The first two equations represent the fact that the D-D reaction can follow either of two paths, producing tritium and one proton; or helium-3 and one neutron, with equal probability. The products of the first two reactions form the fuel for the third and fourth reactions and are burned with additional deuterium. The net reaction consists of the conversion of six deuterium nuclei into two helium nuclei, two hydrogen nuclei, and two neutrons along with a net energy release of 43.1 MeV. The reaction products — helium, hydrogen, and neutrons — are harmless as contrasted with the myriad fission products obtained in a fission reactor. The neutrons produced may be absorbed in sodium to produce an additional 0.25 MeV per cycle. Therefore, the D-D reaction produces at least 7 MeV per deuterium atom (deuteron) and, with absorption in sodium, more than 10 MeV per fuel atom.

The peak reaction rate coefficient of the D-D reaction is considerably less than that of the deuterium-tritium (D-T) reaction occurring within the (D-D) cycle. Thus, attention tends to focus on the latter. Because tritium does not occur naturally, the reaction must be supplemented by one using lithium to reproduce the tritium fuel:

$$D + T \longrightarrow {}^4He + n + 17.6 \text{ MeV}$$

$$n + {}^6Li \longrightarrow {}^4He + T + 4.8 \text{ MeV}$$

$$D + {}^6Li \longrightarrow 2{}^4He + 22.4 \text{ MeV}$$

This reaction is tritium-regenerating and produces only helium as a reaction product.

The D-T reactor is technologically more complex than the D-D reactor because of the need to facilitate the second reaction (which takes place outside the plasma) and because very energetic neutrons must be slowed down to allow the reaction with lithium to take place. However, the conditions needed to achieve net power output are less demanding than

for the D-D fuel reactor. The D-T reaction will probably be exploited first, but its ultimate, very long term use may be limited by the availability of lithium. See also **Lithium (For Thermonuclear Fusion Reactors).**

Fusion reactions can take place only when the nuclei of the fuel atoms are brought into close enough conjunction. The nuclei are positively charged and so repel each other. This repulsion is equivalent to an energy barrier which can be penetrated with reasonable efficiency only if the reacting nuclei have kinetic energy comparable to the barrier height. The level of kinetic energy required depends upon the particular reaction and the desired reaction rate, but in general, plasmas of interest have average energy per particle in excess of 5 keV. A collection of particles with average energy 5 keV has an effective temperature of at least 10^8 degrees Kelvin. At these temperatures, the gas is completely dissociated into its constituent positively charged nuclei and free electrons. The density ranges between 10^{13} to 10^{14} cm^{-3}. The electrical charge density is such that the behavior of the collection of particles is completely dominated by electrostatic and electromagnetic phenomena. Such a charge-dominated collection of ionized matter is known as *plasma*. This plasma at such extremely high temperatures cannot be confined by walls made of materials, known or imagined. But confinement, even for a nanosecond or less, is required if fusion reactions are to occur. Ways to confine the plasma have been researched for 20–30 years by scientists in a number of countries.

After nearly three decades of effort, fusion ignition, that is, the efficient burnup of deuterium and tritium has been accomplished only in one way, namely, the thermonuclear or hydrogen bomb. In this instance, obviously tremendously greater amounts of energy are released than are required to trigger the fusion reactions. In the case of the hydrogen bomb, an atomic bomb was used to generate the extremely high temperature and the degree of confinement required for hydrogen nuclei to fuse. The methods researched to date for confining the plasma include containment within magnetic fields; inertial confinement methods in which the fuel is pelletized in a special way and fusion reactions are initiated either by laser beams or beams of particles; and by heating the plasma with high-power microwave radiation.

NUCLEAR HEART SCAN. See **Coronary Artery Disease (CAD).**

NUCLEAR MAGNETIC MOMENT. An electrically charged particle of finite size that possesses angular momentum, acts like a small magnet and thus possesses a magnetic moment. For atomic nuclei, the magnitude of this moment is within a range of values between zero and a few nuclear magnetons.

NUCLEAR MAGNETIC RESONANCE (NMR) AND MAGNETIC RESONANCE IMAGING (MRI). Since the discovery of electron spin resonance (ESR) by Zavoiski in 1945 and the codiscovery of nuclear magnetic resonance (NMR) in 1946 by Purcell, Pound, and Torrey (Harvard University) and Bloch, Hansen, and Packard (Stanford University), the field of magnetic resonance has reached a high degree of sophistication and versatility.

Soon after its discovery, the principle was applied by physicists to a variety of research programs. Chemists also became interested in NMR because they recognized its potential for elucidating structure and dynamics in pure and applied chemistry. NMR has gained exceptional acceptance over the intervening years as a tool for yielding details of molecular structure, such interests coinciding with other developments in molecular biology. The use of NMR in medical research dates back about 40 years, commencing with proton NMR measurements on cells and organs of both humans and laboratory animals. During the past 20 years, the applications of NMR to clinical medicine have expanded rapidly. NMR permits the analysis of protein structures, for which the former X-ray methods were inadequate. By the late 1980s, the structures of some 50 important proteins had been determined. NMR allows the investigation of proteins and peptides and other macromolecules in an aqueous medium. Consequently, the effects of pH (hydrogen ion concentration) can be observed and the binding of water at the interior and surface of proteins can be determined.

The molecular weights of proteins studied thus far range from about 5000 to 15,000. The use of three-dimensional Fourier transform NMR may ultimately permit the study of macromolecules up to a molecular weight of 40,000.

In analytical chemistry in the early 1990s, NMR was routinely used to study: (1) polymers, polymer networks, and copolymers, and (2) asphalt, bitumens, tars, and pitches, among numerous examples that could be cited.

Commencing with clinical medicine (circa 1981), which led to the development of NMR and magnetic resonance imaging (MRI) for medical research and diagnostics, the medical applications have captured the "headlines," particularly since the early 1990s. It is interesting to note that in 1981 there were only two MRI devices in the United States. By 1987, the installations had reached about 600 in the United States and about 100 MRI facilities in Europe. As of 1994, these Figures expanded into several hundred installations, representing a tremendous investment by the health care field.

Fundamentals of the Technology

Discovery of the principles of ESR and NMR led to the development of several subtechnologies.

Electron Spin Resonance (ESR). Fundamentally, ESR spectrometers consist of three main parts: (1) a magnet with corresponding power supply to provide a steady dc magnetic field at about 3300 G; (2) a microwave bridge capable of producing an oscillating electromagnetic field at a frequency of about 0.5 GHz (*X* band), which is coupled via waveguide to a high-*Q* microwave reflection cavity; and (3) the associated signal detection including dc field modulation, amplification, and display systems. See Fig. 1.

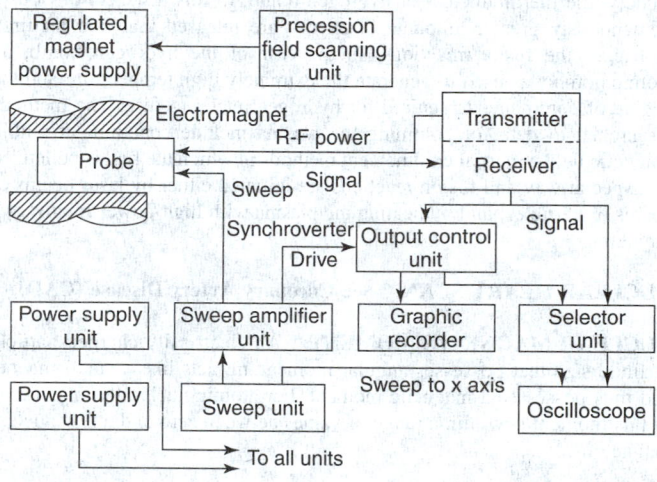

Fig. 1.   Schematic of ESR spectrometer.

Observation of ESR from a particular sample is contingent upon the presence of a macroscopic spin magnetic moment $\overline{\mu}$; i.e., the sample under investigation must contain some minimum number of unpaired electron spins. Upon insertion into the cavity, the sample is subjected to the dc magnetic field H_0, and the unpaired electrons align themselves both parallel (high energy) and antiparallel (low energy) to H_0. The ratio of the spin populations of the high to low energy states is given by the Botzman distribution, $e^{-h\nu/kT}$, where $h\nu$ is the energy difference of the two states. Because of the torque on $\overline{\mu}$ produced by H_0, the spin magnetic moment will precess about H_0, at the Larmor frequency $\omega = \gamma_e H_0$, where γ_e is the gyromagnetic ratio of the electron.

Analysis is accomplished by sweeping the magnetic field until the Larmor frequency of the spin system is identical to the fixed frequency of the oscillating microwave field emanating from the bridge. When the two frequencies are coincident, a net microwave energy will be absorbed by the spin system from the oscillating field because of the excess spin population in the lower energy state. If energy continues to be absorbed, the spin populations will equalize and saturation will occur—no net microwave energy will be absorbed and the signal will disappear. To avoid this situation the spin system interacts with its surroundings (lattice) and transfers the absorbed microwave energy to the lattice in some interval called the spin-lattice relaxation time.

This phenomenon, known as relaxation, acts to restore the spin system to its original Botzman distribution of populations. At equilibrium, microwave energy is being absorbed by the spin system and then transferred to the lattice. This process is monitored via microwave rectification of the reflected cavity signal by a diode detector, preamplification, and lock-in phase detection at the dc field modulation frequency. The signal which appears on the recorder may be a single line from an unpaired electron or a group of lines. The latter are caused by neighboring nuclei with nonzero nuclear spin (hyperfine structure) and by surrounding electric field gradients (quadrupole interactions).

In this manner the unpaired electron spins may be used as a probe for analysis of their immediate microscopic surroundings. Typical instrument sensitivity is such that approximately 10^{-11} mole of a paramagnetic species can be detected, but generally concentrations of 10^{-4} mole give optimum ESR spectra.

Nuclear Magnetic Resonance (NMR). This technique is essentially based on the same principle as ESR, but NMR is capable of detecting nuclei (MHz) instead of electrons (GHz). (Lack of a standardized nomenclature has resulted in numerous modifiers in connection with magnetic resonance instrumentation—electron, proton, nuclear, etc., plus application-related terms, such as silicon-29, oxygen-17, ^{13}C, ^{31}P NMR, etc.)

In nuclear magnetic resonance spectroscopy, a nucleus possessing a magnetic moment when placed in a homogeneous magnetic field will precess about the field axis at a rate which is dependent upon the strength of the field. See Fig. 2. If these nuclei are then brought into contact with an oscillating electromagnetic field of the same frequency as the Larmor precessional frequency of the nuclei, energy will be absorbed by the nuclei spin system from the oscillating ratio frequency field. As stated above, the net energy absorption is proportional to the Boltzman distribution of spin populations and the effective nuclear relaxation times.

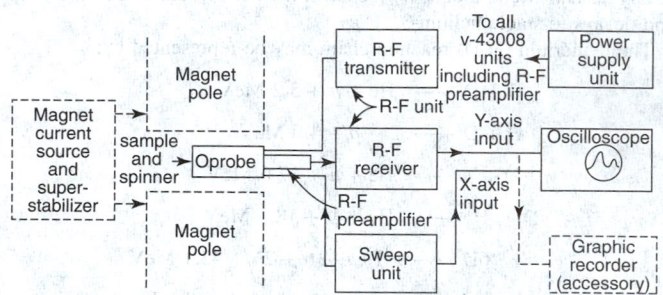

Fig. 2.   Schematic of NMR spectrometer.

Different nuclei will have different precessional frequencies and therefore at a particular field will absorb energy at certain characteristic radio frequencies. Also, nuclei of the same nuclear species (such as hydrogen) will absorb energy at slightly different frequencies, dependent upon their molecular environment. The latter observation makes possible an entire subfield of NMR spectroscopy, termed high-resolution NMR, which is an appropriate method for determining chemical structure and identifying and measuring similar nuclei in two or more different compounds (mixtures).

The major components of a magnetic resonance spectrometer are: (1) a magnet capable of producing a very strong homogeneous field which may be continuously varied over a very small range; (2) a low-power radio frequency oscillator which supplies rf power to a small transmitter coil surrounding the sample; (3) a small receiver coil which also surrounds the sample (but is orthogonal with respect to the transmitter), and feels (4) a sensitive radio receiver (tuned to the same frequency as the transmitter) capable of amplifying any signal which might be induced in the receiver; and (5) a recorder of oscilloscope which can display the resulting spectra.

Various decoupling techniques are used to simplify complex NMR spectra. For example, in a simple two-spin system (homonuclear) giving rise to four NMR lines it is possible to saturate a nucleus at a particular rf frequency and collapse the remaining doublet to a single NMR line. This occurs because the second nucleus sees an averaged interaction from the first rather than two distinct interactions from spins in the high and low energy states. Noise decoupling is predominantly used to decouple different nuclei—for example, hydrogen nuclei C^{13} spectra (heteronuclear). Here the entire hydrogen spectrum is saturated over a range of frequencies (noise) leaving behind the much simplified C^{13} spectrum.

In internuclear double resonance one nuclear line at a given frequency is observed while a second frequency is swept through the remainder of the NMR spectrum. All nuclei which are in some way coupled to the line being observed will enhance or de-enhance the latter as the second frequency passes through the resonance condition. Also, the Fourier transform technique has nicely complimented NMR. Any complex waveform can be converted to a spectrum of frequencies by Fourier transformation.

In NMR, the waveform is a superposition of a set of nuclear precession frequencies with amplitudes decaying due to relaxation and field inhomogeneity. The transformation may be carried out by analog means (spectrum analyzer) or on a small dedicated laboratory computer. The latter appears to be the most convenient solution. Since the free induction decay signal decays with time, whereas the instrumental noise remains constant, the noise content is higher in the tail of the transient signal, and it is possible to improve the overall signal-to-noise ratio by weighting the transient signal with an exponentially decaying function of time. The shorter the time constant of this exponential, the greater the improvement in sensitivity, but this increases the linewidths of the transformed spectrum. The reversal of this procedure can enhance resolution at the cost of sensitivity.

High-Resolution Nuclear Magnetic Resonance of Solids. Important developments in solid-sample NMR techniques of the early 1980s have made NMR of significant interest as a tool for characterizing solid samples—as it has been in the past for the study of liquids. As observed by Maciel, the development of line-narrowing techniques, such as magic-angle spinning (MAS) and high-power decoupling, has led to powerful high-resolution NMR for studying solids. In favorable cases (for example, where high abundances of protons are present) cross polarization (CP) provides a means of circumventing the time hurdle caused by inefficient spin-lattice relaxation in many solids. Combining the CP and MAS approaches for carbon-13 with proton decoupling has become a popular and routine experiment for organic solids. For many nuclides with spin quantum number $1 > \frac{1}{2}$, the central nuclear magnetic resonance transition can be used in high-resolution experiments that involve rapid sample spinning. A continuing stream of other advances in NMR technology bodes well for the characterization of solids by a wide range of nuclides. The complexities of this topic unfortunately are beyond the scope of this encyclopedia.

It is interesting to note that several of the concepts for improving NMR technology, as listed by Levy and Craik, in 1988, already have been partially or fully achieved: (1) two-dimensional Fourier transform (FT NMR); (2) high-resolution NMR in solids; (3) new types of pulse sequences; (4) chemically induced dynamic nuclear polarization; (5) multiple quantum NMR; and (6) NMR imaging (MRI).

Two-Dimensional NMR. Bax and Lerner report on how two-dimensional Fourier transform pulse NMR (2-D FT NMR) has extended the range of applications of NMR spectroscopy into the area of large, complex molecules, such as DNA and proteins. Great spectral simplification can be obtained by spreading the conventional one-dimensional NMR spectrum in two independent frequency dimensions, thus removing spectral overlap, facilitating spectral assignment, and providing additional information. Conformational information related to interproton distances is available from resonance intensities in certain types of two-dimensional experiments. Two-dimensional NMR spectroscopy also has been applied to the study of ^{13}C and ^{15}N to provide connectivity information and greatly improving the sensitivity of these determinations. A traditional NMR spectrum of a sugar and a 2D spectrum are contrasted in Fig. 3.

Correlative spectroscopy (COSY) is an approach that involves correlating groups believed to be coupled to each other and to prove that this coupling does exist. *Spin-echo correlation spectroscopy* (SECSY) is a variation of the COSY technique. Sometimes called *J-resolved spectroscopy*, it is a technique that allows a separation of the chemical shift of a nucleus from the coupling to other nuclei. This simplifies the spectrum and permits one to assign each resonance to a specific nucleus. Sometimes referred to as *nuclear Overhauser effect* (NOESY), this is a technique that makes it possible to determine distances between nonadjacent residues in a peptide chain. Another application, *multiple quantum transitions*, takes advantage of the fact that molecules in a sample are forced to absorb or emit several quanta of energy at one time. For example, the technique can be used to determine which carbon atoms in a molecule are connected to other specific atoms in the molecule.

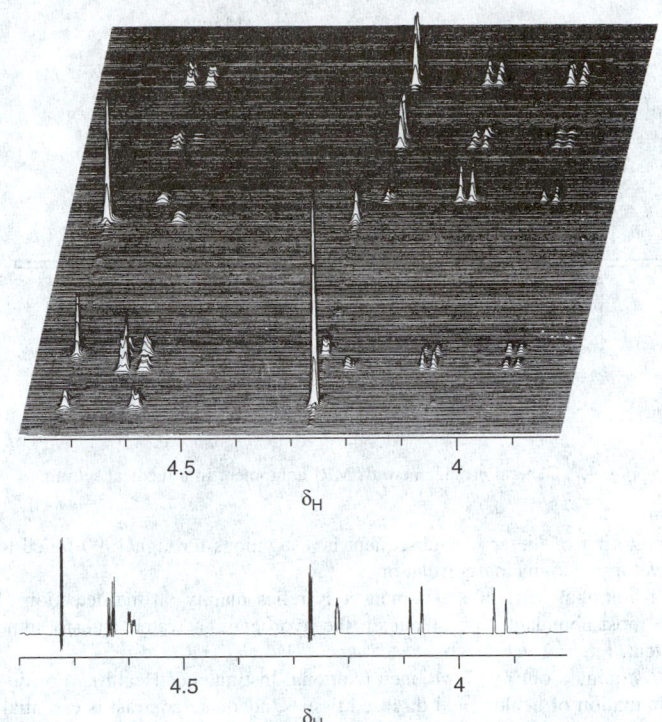

Fig. 3. (*Top*) A two-dimensional nuclear magnetic resonance spectrum of a sugar; (*Bottom*) conventional NMR spectrum of same sugar. The two-dimensional spectrum also can be plotted as a contour map with intensities denoted by color. (*JEOL Inc.*)

Magnetic Resonance Imaging (MRI). In 1973, Lauterbur added a new aspect to NMR basics, namely that of image formation based upon NMR principles. This led to NMR imaging, now commonly referred to as magnetic resonance imaging (MRI).

The source of the imaging photons is not a Roentgen tube, nor an unstable isotope, nor an electron storage ring, but rather it is a radio transmitter. The commonly referred *radio photon* radiation allows photons to pass into the tissues under study, where they are absorbed by nuclear particles rather than by electrons. The nucleons are momentarily excited by the process and then return to their resting state by emitting photons of the same or nearly the same energy that they had absorbed. Because the nuclei of the different chemical elements absorb radio photons of different frequencies, it is possible to use the method to detect the presence of a single element in a sample. When protons are irradiated by radio photons of a frequency that precisely matches their own precessional frequency, resonance occurs. The resonant frequency (Larmor frequency) is determined by the natural rotational velocity (gyromagnetic ratio) characteristic of each species of atomic nucleus and by the strength of the applied static magnetic field, as expressed by: Larmor frequency = Magnetic field × Gyromagnetic ratio. See Fig. 4.

Magnetic resonance imaging is well suited for the imaging of soft tissues. It has been found particularly effective in (1) studying skeletal musculature, notably in the male and female pelvic regions, (2) delineating tumor development and assisting in planning surgery or therapy, and (3) precisely locating tumors, particularly in the brain where, with time, the evolution of hematomas can be studied. For clinical work, MRI installations may be used in two staff shifts per day, holidays, and weekends.

The magnetic fields used are usually 1.5 to 2 teslas (15,000 to 20,000 gauss). Barriers to using higher magnetic fields are magnet cost, additional expense for building the site, and notation by some persons (volunteers) to discomfort when over 4 teslas are used, which also causes the patient to move.

In 1986, Fossel and McDonough (Beth Israel Hospital, Boston) proposed that proton NMR spectroscopy of human plasma possibly could be a "potentially valuable approach to the detection of cancer and the

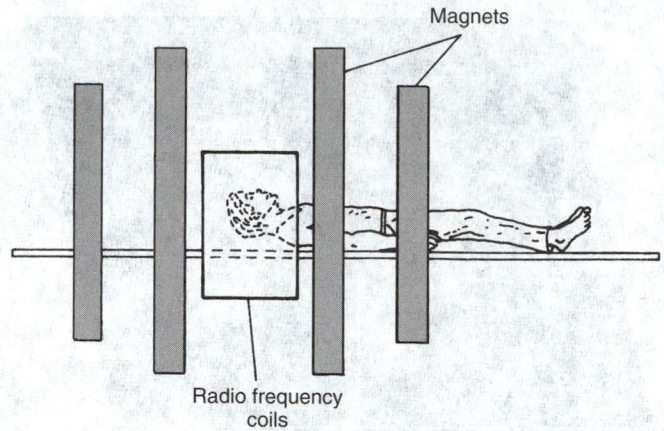

Fig. 4.   General arrangement of MRI equipment in a medical setting.

monitoring of therapy." Subsequent investigations through 1990 failed to develop a convincing correlation.

Traditional MRI of the human body relies mainly on the detection of the most abundant type of nuclei, the hydrogens in water, and, to some extent, fat.

As pointed out by C. Moonen (National Institutes of Health), "For discrimination of healthy and diseased tissues, adequate contrast is essential. Such contrast depends not only on differences in water concentration, but also on the NMR relaxations, which, in turn, are related to local mobilities and interactions."

Magnetic resonance imaging, in addition to providing detailed information about the macroscopic structure and anatomy, also permits the noninvasive spatial evaluation of various biophysical and biochemical processes in living systems. These include the motion of water in processes, such as vascular flow, capillary flow, diffusion, and exchange. Further, the concentrations of various metabolites can be determined for the assessment of regional regulation of metabolism. In the scholarly Moonen paper, examples are given of flow imaging, diffusion imaging, and the imaging of tissue perfusion, of exchange, and of metabolites. These aspects of MRI imaging sometimes are referred to as *functional MRI*.

Imaging of the central nervous system by MRI technology is of great interest. Advantages of MRI include its sensitivity to soft tissue, the contrast between gray and white matter, the paucity of signals from the skull, and the availability of coronal, sagittal, and transverse sections. The procedure has been used to detect posterior fossa tumors, such as acoustic neuromas, and pituitary and parasellar tumors, orbital tumors, multiple sclerosis, and a number of other lesions in the craniovertebral junction, and of the cord and spine. MRI technology, according to a number of authorities, has large potential for use in *noninvasive* cardiological studies, particularly of blood flow imaging. Contrast agents are not required because nuclei in rapidly flowing blood move out of the volume of interest during the interval required for application of the rf pulse and of gradient magnetic fields. The parameters that influence MRI signals at various flow rates are being studied.

In addition to hydrogen atoms, MRI can create, for example, ^{31}P images which are excellent indicators of energy metabolism. ^{23}Na images reflect extracellular and intracellular fluid fluxes.

Echo-Planar Imaging. A major problem with MRI in the past has been the long data acquisition times (up to several minutes). Consequently, MR images are subject to so-called *motional artifacts*, caused by physiological motions (heartbeat, blood flow, bowel peristalsis, breathing) as well as by voluntary movements in severely ill and uncooperative patients, including children. Echo-planar imaging (EPI) permits faster scan times, thus effectively reducing imaging to a fraction of a second as compared with minutes. Although a technical description of how EPI is implemented is too complex for coverage here, echo-planar imaging uses only one nuclear spin excitation per image.

EPI has broadened the use of MRI to include the evaluation of cardiac function in real time, mapping of organ blood pool and perfusion, functional imaging of the central nervous system, depiction of blood and cerebrospinal fluid flow dynamics, and motion picture imaging of the mobile fetus in utero. EPI also has the practical advantages of increasing patient throughput at a lower cost per MRI examination. With these advantages, it is expected that EPI will become an established tool for early diagnosis of some common and potentially treatable diseases, such as ischemic heart disease and stroke.

Comparison with Ultrasonography. The health care community is aware of the high costs of MRI and continues to seek other techniques that may be effective at lower cost. One such comparison was made in connection with prostate cancer. The approach to treatment varies and depends on the extent of cancer at the time of diagnosis.

In a specific comparative study over a period of 15 months, 230 patients were evaluated with identical imaging techniques. It was found that MRI correctly staged 77% of cases of advanced disease and 57% of cases of localized disease. The corresponding Figures for ultrasonography were 66% and 46%. MRI identified only 60% of all malignant tumors measuring more than 5 mm on pathological analysis, while ultrasonography identified only 59%. The study concluded, "The MRI and ultrasonography equipment is not highly accurate in staging early prostate cancer, mainly because neither technique has the ability to identify microscopic spread of disease."

Superconducting Quantum Interference Device (SQUID). These devices are sensitive detectors of magnetic fields. Low-temperature superconductors have been used in the past to sense weak magnetic signals from the brain for medical diagnosis. Such devices, however, required cooling with liquid helium and thus have resulted in costly, unwieldy apparatus. Research is now underway toward using higher-temperature superconductors, such as $YBa_2Cu_3O_7$. Although the higher-temperature semiconductor still requires refrigeration, this can be accomplished with liquid nitrogen, thus resulting in a less costly, simpler, and more portable detecting device.

Additional Reading

Abelson, P.H.: "New Horizons in Medicine," *Science*, 1109 (November 25, 1988).

Ancreasen, N.C.: "Brain Imaging: Applications in Psychiatry," *Science*, 1381 (March 18, 1988).

Bain, L.: "MRI—Safety Issues Stimulate Concern," *Science*, 1245 (May 31, 1991).

Bax, A. and L. Lerner: "Two-Dimensional Nuclear Magnetic Resonance Spectroscopy, "*Science*, **232**, 960–967 (1986).

Edelman, R. et al.: *Clinical Magnetic Resonance Imaging*, 2nd Edition, W.B. Saunders, Philadelphia, PA, 1996.

Hari, R. and O.V. Lounasmaa: "Recording and Interpreting Cerebral Magnetic Fields," *Science*, 432 (April 28, 1989).

Kirkwood, J.R.: *Essentials of Neuroimaging*, 2nd Edition, Churchill-Livingstone, Inc., New York, NY, 1995.

Krestel, E. (Editor): *Imaging Systems for Medical Diagnosis: Fundamentals and Technical Solutions—X-Ray Diagnostics—Computed Tomography—Nuclear Medical Diagnostics—Magnetic Resonance Imaging—Ultrasound Technology*, John Wiley & Sons, Inc., New York, NY, 1990.

Levy, G.C. and D.J. Craik: "Developments in Nuclear Magnetic Resonance Spectroscopy, "*Science*, **214**, 291–299 (1981).

Magin, R. et al. (Editors): *Biological Effects and Safety Aspects of Nuclear Magnetic Resonance Imaging and Spectroscopy*, Vol. 649, New York Academy of Sciences, New York, NY, 1992.

Moonen, C.T. et al.: "Functional Magnetic Resonance Imaging in Medicine and Physiology," *Science*, 53 (October 6, 1990).

Pool, R.: "Putting SQUIDs to Work," *Science*, 862 (August 24, 1990).

Randal, J.: "NMR: The Best Thing Since X-Rays," *Technology Review (MIT)*, 59 (January 1988).

Rifkin, M.D. et al.: "Comparison of Magnetic Resonance Imaging and Ultrasonography in Staging Early Prostate Cancer," *New Eng. J. Med.*, 623 (September 6, 1990).

Shulman, R.: "NMR—Another Cancer-Test Disappointment," *New Eng. J. Med.*, 1002 (April 5, 1990).

Sochurek, H. and P. Miller: "Medicine's New Vision," *National Geographic*, 2 (January 1987).

Stehling, M.K., R. Turner, and P. Mansfield: "Echo-Planar Imaging in a Fraction of a Second," *Science*, 43 (October 4, 1991).

Sutton, D. and J.W.R. Young: *A Short Textbook of Clinical Imaging*, Springer-Verlag, Inc., New York, NY, 1991.

Tamraz, J. and Y. Comair: *Atlas of Regional Anatomy of the Brain Using MRI: With Functional Correlations*, Springer-Verlag New York, Inc., New York, NY, 2000.

Wilson, M. and F. Ruzicka (Editor): *Modern Imaging of the Liver: Applications of Computerized Tomography Ultrasound, Nuclear Medicine and Magnetic Resonance Imaging*, Marcel Dekker, Inc., New York, NY, 1989.

NUCLEAR MAGNETIC RESONANCE SPECTROSCOPY. See **Magnetic Resonance Spectroscopy**.

NUCLEAR MAGNETON. A unit of magnetic moment used in atomic and nuclear physics, defined as

$$\mu_N = eh/4\pi M = 5.050 \times 10^{-27} \text{ ampere-meters}^2$$

in which e is the electronic charge $(1.602 \times 10^{-19} c)$, h is Planck's constant, and M is the rest mass of the proton. This unit is derived for nuclear particles by analogy with the Bohr magneton, which is applicable to the magnetic moment associated with electron orbital angular momentum in an atom. No real nuclear particle, however, has a magnetic moment of exactly one nuclear magneton. The magnetic moment of the proton is $+2.793$ nuclear magnetons, and of the neutron -1.913 nuclear magnetons.

NUCLEAR POTENTIAL. The potential energy V of a nuclear particle as a function of its position in the field of a nucleus or of another nuclear particle. A central potential is one that is spherically symmetric, so that V is a function only of the distance r of the particle from the center of force. A noncentral potential, on the other hand, is one that is not spherically symmetrical, or one that depends upon the relative directions of the angular momenta associated with the particle and the center of force, as well as upon the distance r. A negative potential corresponds to an attractive force, while a positive potential corresponds to a repulsive force.

Although the expression can certainly be applied to the problem of nuclear forces, the usual meaning of a nuclear potential refers to the interaction of a nucleon (neutron or proton) with a complex nucleus. Although the potential energy of a single nucleon inside a nucleus is clearly a rapidly varying function of position and time (since it represents the interaction with a large number of closely packed, fast-moving particles), one may nevertheless speak of the average potential energy, and one may regard this as a smoothly varying function. For a neutron, the nuclear potential is essentially negative inside the nucleus, rising rapidly to zero outside the nuclear radius R. For a proton, the long-range electrostatic repulsion must, of course, be added. Owing to the Pauli exclusion principle, and to the exchange nature of nuclear forces, however, such a potential cannot in general be regarded simply as a function of position, $V = V(r)$; it depends in addition upon the momentum of the particle, which in quantum mechanics does not commute with the position. Hence, the potential must be regarded as a nondiagonal matrix operator $V = \langle r|V|r'\rangle$ in configuration space, or a similar operator in momentum space.

Although the concept of a nuclear potential in this latter sense cannot be defined in a precise way, it has nevertheless been useful, both qualitatively and quantitatively, in the investigations of nuclear structure and nuclear reactions. It has been of particular usefulness in the optical model of nuclear reactions.

NUCLEAR POWER TECHNOLOGY. After a terse review of the physics and chemistry of nuclear fission and the nature of these reactions, some of the design features and operating parameters of the four families of reactors currently installed are described. These units, some of which are approaching their retirement are, of course, the starting basis for the design of next-generation reactors, which appeared prior to the end of the 20th century. Design improvements that constitute these forthcoming systems are described in terms of their status as of the mid-1990s. The topic of radioactive waste handling is summarized. Based upon available information, nuclear power technology in the United States as well as in Canada, France, Japan, the U.K., and other leading industrialized nations is covered.

The first nuclear power plant was installed at Shippingport, Pennsylvania in 1957 and, after serving as a test facility for several years, was dismantled because of "old age." The plant had a capacity of 60,000 kilowatts. Indeed, the plant was extremely small by comparison with hundreds of units installed today in the United States and other major countries of the world.

NATURE OF NUCLEAR FISSION REACTIONS

The energy of a nuclear fission reaction can be computed from the change in mass between reactants and products according to Einstein's law:

$$\Delta E = \Delta mc^2$$

where E is the energy in ergs, m is mass in grams, and c is the velocity of light in centimeters per second. For example, the mass difference in this equation is $\Delta m = 0.2058$ amu (atomic mass units). Therefore, $\Delta E = 931$ MeV/amu $\times 0.2058$ amu $= 191.6$ MeV. The average amount of energy released in the various fission reactions is about 200 MeV (million electron volts). This energy is distributed in the fission process as:

	MeV
Kinetic energy of fission fragments	165
Radioactive-decay energy	23
Kinetic energy of neutrons	5
Prompt gamma-ray energy	7

The energy of a chemical reaction, approximately 3–4 eV, is dramatically lower than that of a nuclear reaction. Hence, the fission of ^{235}U yields 2.5 million times as much energy as the combustion of the same weight of carbon.

The importance of fission in energy (power) production lies in two facts: (1) an exceedingly large amount of energy is released in the fission reaction; and (2) the production of excess neutrons permits a chain reaction. These two circumstances make it possible to design nuclear reactors in which self-sustaining reactions occur with the continuous release of energy. Although described later, it may be pointed out here that nuclear fission is not the only energy-releasing nuclear reaction. The fusion of light nuclides, like hydrogen, into heavier elements is also an energy-producing process.

The heat generated in nuclear power plants is transferred to a working fluid and from this point on the nuclear power plant and the conventional fossil-fueled power plant are essentially similar.

Fission Reaction. In nuclear fission, the nucleus of a heavy atom is split into two or more fragments. The reaction is initiated by the absorption of a neutron. A typical reaction is

$$^{235}_{92}\text{U} + ^{1}_{0}n \longrightarrow ^{137}_{56}\text{Ba} + ^{97}_{36}\text{Kr} + 2^{1}_{0}n + \Delta E$$

In this reaction, a ^{235}U atom absorbs one neutron, becomes unstable, and subsequently fissions into two fission fragments plus two neutrons. This is just one of the many ways in which ^{235}U might fission. The number of neutrons produced in a fission reaction is usually 2 or 3. The excess neutrons produced by the fission reaction provide the means of self-sustaining the chain reaction. Nuclides including ^{233}U, ^{235}U, and ^{239}Pu, which are fissionable by neutrons of all energies, are termed fissile nuclides.

Nuclear Fuels. There are two broad categories of nuclear fuels: (1) the fissile nuclides previously mentioned; and (2) the fissionable nuclides, ^{232}Th (thorium) and ^{238}U. Thermal reactors use fissile nuclides as fuel, while fast reactors are designed to burn fissionable materials. In fast reactors, only a small portion of the ^{232}Th and ^{238}U are fissioned directly. A larger portion of these materials is converted into ^{235}U and ^{239}Pu, respectively, through neutron absorption. Thus, this type of reactor not only consumes fuel, but also produces (breeds) new fuel material. Hence, the term *breeder reactor* is used for reactors designed to take advantage of this phenomenon. Breeding is possible in thermal reactors also, but to a lesser extent. The fuel material in a fast reactor must contain a significant amount (about 10%) of one of the fissile materials. The remainder of the fuel must have a high mass number in order to avoid slowing down the neutrons. The natural reserves of fissionable materials are more than 100 times greater than the reserves of fissile materials. Consequently, from the viewpoint of utilization of available energy resources, fast reactors are of great importance. Breeder reactors are described later.

Moderators. The most important slow-down mechanism is elastic scattering on elements of low mass number. Materials like light and heavy water, beryllium oxide, and graphite are used to slow down, or *thermalize* the neutrons to an energy of about 0.025 eV. As neutrons collide with the nuclei of these atoms, their kinetic energy and speed are gradually reduced until thermal equilibrium is achieved with the reactor structure. The fewer such collisions before deceleration is complete, the less chance of ^{238}U atoms absorbing neutrons.

Critical Mass. Thermal neutrons, which move like atoms in a low-pressure gas, diffuse throughout the reactor. They must be absorbed by a nucleus of the reactor structure, in which case they merely make that nucleus radioactive. Or, they may strike a fissionable atom of ^{235}U, causing fission and, in turn, releasing more neutrons to maintain the reaction. Should the number of neutrons absorbed by the moderator and ^{238}U be greater than about 1.5 excess neutrons emitted from each fission, the chain reaction will not be maintained. Therefore, the reactor core must be designed so that the mass of fuel will be just sufficient to ensure one neutron from each fission causing fission in another atom. A mass and configuration of fissionable material in which this occurs is termed the *critical mass* — or a reactor in which this condition is achieved is said to have "gone critical."

To measure a chain reaction, a multiplication factor k is used to indicate the ratio of neutrons in one generation to those in the preceding generation. Thus, in a constant chain reaction where the total number of neutrons neither increases nor decreases, the heat output is constant and $k = 1$. Should k rise above unity, the rate of fission, and hence the rate of heat productivity, steadily rises. This is so even if k is held constant at its new value. Here lies one *major difference between nuclear reactors and conventional steam generators*. In the latter, heat output is proportional to firing rate. If the firing rate is increased, the steam output is increased; but it remains constant at its new level. In a nuclear reactor, an increase in k results in continuously rising heat output. Only by returning the rate of neutron production to its original ratio can heat output be maintained at its new level.

Reactivity Control. Absorption of excess neutrons, above those needed to maintain a constant reactivity level, provides close control over the degree of reactivity. This is accomplished by inserting materials having a high neutron-capture rate into the core. Control rods of special alloy metals are moved into and out of the cores as required. To start the reactor from shutdown (black start), control rods are partially withdrawn until *k* becomes greater than one. Neutron flux and heat output grow until the desired level is reached. At this point, control-rod movement is quickly reversed to keep *k* at unity. The reactor is shut down by inserting the rods to their full extent. In this position, the rods absorb more than 1.5 excess neutrons per fission and the chain reaction quickly stops. Heat production continues for a time, but is usually dissipated by an auxiliary cooling system.

It is interesting to contrast a nuclear power reactor and a nuclear bomb. The designer of a nuclear fission bomb seeks to release as much fission energy as possible within the shortest possible time (milliseconds). Thus, a bomb is designed to favor *prompt neutrons*. By contrast, the normal operating mode of a nuclear power reactor is one in which prompt neutrons alone cannot sustain a chain reaction, but prompt neutrons together with delayed neutrons can. Only the delayed neutrons are controllable. A power reactor is designed to release fission energy *slowly and smoothly* and in just the right amounts to convert water into steam. Whereas the "fuel" in a bomb is used up essentially in an instant, in a power reactor the energy release is spread over months and years. It has been agreed by physicists for many years that it is physically impossible for a power reactor to explode in the manner of an atomic bomb.

TYPES AND MAJOR CHARACTERISTICS OF NUCLEAR POWER REACTORS

In order, the following types of nuclear fission reactors are described in this section: (1) light water reactors, (a) pressurized water reactors, (b) boiling-water reactors; (2) high-temperature gas-cooled reactors; (3) heavy water reactors; and (4) fast breeder reactors. Military reactors are not described.

Contemporary Light-Water Reactors (LWRs)[1]

These reactors are of two principal designs: (1) *pressurized water reactors* (PWR), and (2) *boiling-water reactors* (BWR). In a PWR, heat generated in the nuclear core is removed by water (reactor coolant) circulating at high pressure through the primary circuit. The water in the primary circuit both cools and moderates the reactor. Heat is transferred from the primary to the secondary system in a heat exchanger, or boiler, thereby generating steam in the secondary system. The BWR differs from the PWR primarily in that boiling takes place in the reactor itself. Comparable steam temperatures are possible at pressures of about 1000 pounds per square inch (6.9 mPa) as contrasted with 2000 psi (13.8 mPa) for pressurized reactors.

Contemporary Boiling Water Reactor (BWR)

Aside from its heat source, the boiling water reactor (BWR) generation cycle is substantially similar to that found in fossil-fueled power plants. One of the first BWRs was the Vallecitos BWR, a 1000 psi (6.9 mPa) reactor which powered a 5 MW electric generator and provided power to the Pacific Gas & Electric Company grid through 1963. Power output capabilities have increased many times during the intervening years as shown by tabular summaries given later in this entry.

---

[1] In nuclear power technology, ordinary water, in contrast with *heavy water*, is termed *light water*.

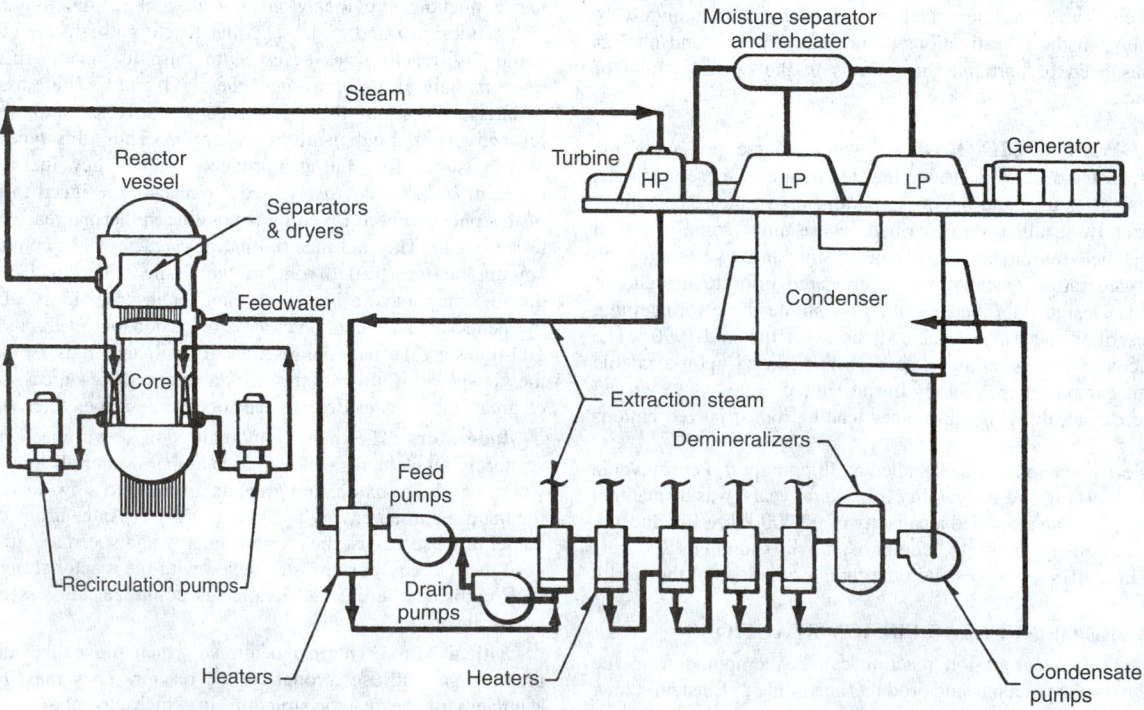

Fig. 1.   Contemporary direct-cycle reactor system used in boiling water reactor. (*General Electric.*)

The direct-cycle boiling water reactor nuclear system (Fig. 1) is a steam generating system consisting of a nuclear core and an internal structure assembled within a pressure vessel, auxiliary systems to accommodate the operational and safeguard requirements of the nuclear reactor, and necessary controls and instrumentation. Water is circulated through the reactor core producing saturated steam which is separated from the recirculation water, dried in the top of the vessel, and directed to the steam turbine-generator. The turbine employs a conventional regenerative cycle with condenser de-aeration and condensate demineralization. The direct-cycle system is used because of its inherently simple design, contributing to reliability and availability.

The steam from a BWR is, of course, radioactive. The radioactivity is primarily ^{16}N, a very short-lived nitrogen isotope (7 seconds half-life) so that the radioactivity of the steam system exists only during power generation. Extensive generating experience has demonstrated that shutdown maintenance on a BWR turbine, condensate, and feedwater components can be performed essentially as a fossil-fuel plant.

The reactor core, the source of nuclear heat, consists of fuel assemblies and control rods contained within the reactor vessel and cooled by the recirculating water system. A 1,220-MWe BWR/6 core consists of 732 fuel assemblies and 177 control rods, forming a core array 16 feet (4.8 meters) in diameter and 14 feet (4.2 meters) high. The power level is maintained or adjusted by positioning control rods up and down within the core. The BWR core power level is further adjustable by changing the recirculation flow rate without changing control rod position, a feature that contributes to excellent load-following capability.

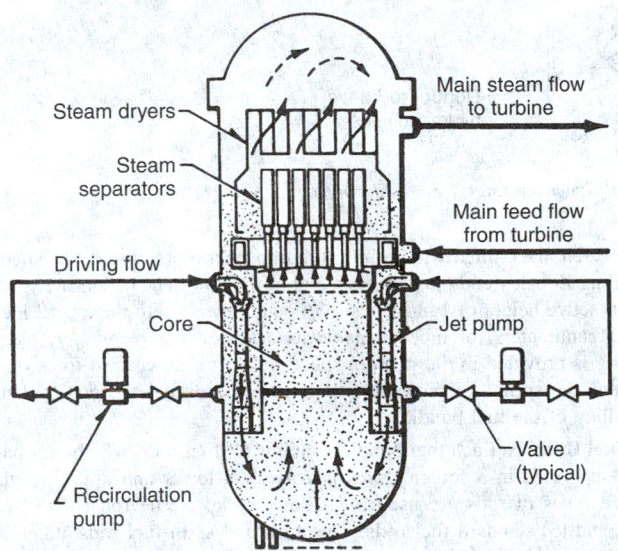

Fig. 2. Steam and recirculation water flow paths of contemporary boiling water reactor. (*General Electric.*)

The BWR is the only light water reactor system that employs bottom-entry control rods. From the very first BWRs, bottom-entry control rods have been used because reactivity and moderator density is highest in the lower part of the core. They provide optimum power shaping characteristics for the type of core where moderator density is varied as a function of power level. Bottom-entry and bottom-mounted control rod drives also allow refueling without removal of rods and drives, and allow drive testing with an open vessel prior to initial fuel loading, or at each refueling operation. The hydraulic system, using reactor system pressure, provides rod insertion forces that are greater than gravity or mechanical systems.

The BWR requires substantially lower primary coolant flow through the core than pressurized water reactors. The core flow of a BWR is the sum of the feedwater flow and the recirculation flow, which is typical of any boiler. Unique to the BWR is the application of jet pumps inside the reactor vessel. See Fig. 2. The jet pumps deliver their driving force from the external recirculation pumps and generate about two-thirds of the recirculation flow within the reactor vessel. The jet pumps also contribute to the inherent safety of the BWR design under loss-of-coolant emergency

conditions because they continue to provide internal circulation with one or both external recirculation loops out of service. The BWR can deliver about one-third power through this natural jet pump circulation mode, a vital capability in effecting a "black start" (a fully fresh start-up of a reactor) of the plant without external power.

The BWR operates at constant pressure and maintains constant steam pressure similar to most fossil-fueled boilers. The BWR primary system operates at pressure about one-half that of a pressurized water reactor primary system, while producing steam of equal pressure and quality.

The integration of the turbine pressure regulator and control system with the reactor water recirculation flow control system permits automated changes in steam flow to accommodate varying load demands on the turbine. Power changes of up to 25% can be accomplished automatically by recirculation flow control alone, at rate of 15% per minute increasing and 60% per minute decreasing. This provides a load-following capability that can track rapid changes in power demand.

Nuclear Boiler Assembly. This assembly consists of the equipment and instrumentation necessary to produce, contain, and control the steam required by the turbine-generator. The principal components of the nuclear boiler are: (1) reactor vessel and internals—reactor pressure vessel, jet pumps for reactor water circulation, steam separators and dryers, and core support structure; (2) reactor water recirculation system—pumps, valves, and piping used in providing and controlling core flow; (3) main steam lines—main steam safety and relief valves, piping, and pipe supports from reactor pressure vessel up to and including the isolation valves outside of the primary containment barrier; (4) control rod drive system—control rods, control rod drive mechanisms and hydraulic system for insertion and withdrawal of the control rods; and (5) nuclear fuel and in-core instrumentation.

Reactor Assembly. This assembly (Fig. 3) consists of the reactor vessel, its internal components of the core, shroud, top guide assembly, core plate assembly, steam separator and dryer assemblies and jet pumps. Also included in the reactor assembly are the control rods, control rod drive housings and the control rod drives.

Each fuel assembly that makes up the core rests on an orificed fuel support mounted on top of the control rod guide tubes. Each guide tube, with its fuel support piece, bears the weight of four assemblies and is supported by a control rod drive penetration nozzle in the bottom head of the reactor vessel. The core plate provides lateral guidance at the top of each control rod guide tube. The top guide provides lateral support for the top of each fuel assembly.

Control rods occupy alternate spaces between fuel assemblies and may be withdrawn into the guide tubes below the core during plant operation. The rods are coupled to control rod drives mounted within housings welded to the bottom head of the reactor vessel. The bottom-entry drives do not interfere with refueling operations. A flanged joint is provided at the bottom of each housing for ease of removal and maintenance of the rod drive assembly.

Except for the Zircaloy in the reactor core, the reactor internals are stainless steel or other common corrosion-resistant alloys. The reactor vessel is a pressure vessel with a single full-diameter removable head. The base material of the vessel is low alloy steel, which is clad on the interior, except for nozzles with stainless steel weld overlay to provide the necessary resistance to corrosion.

The shroud is a cylindrical, stainless steel structure that surrounds the core and provides a barrier to separate the upward flow through the core from the downward flow in the annulus. Two ring spargers, one for low pressure core sprays and the other for high pressure core spray are mounted inside the core shroud in the space between the top of the core and steam separator base. The core spray ring spargers are provided with spray nozzles for the injection of cooling water under emergency conditions. A nozzle for the emergency injection of neutron absorber (sodium pentaborate) solution is mounted below the core in the region of the recirculation inlet plenum.

The steam separator assembly consists of a domed base, on top of which is welded an array of standpipes with a 3-stage separator located at the top of each standpipe. The steam separator assembly rests on the top flange of the core shroud and forms the cover of the core discharge plenum region. In each separator, the steam-water mixture rising through the standpipe impinges on vanes which give the mixture a spin to establish a vortex

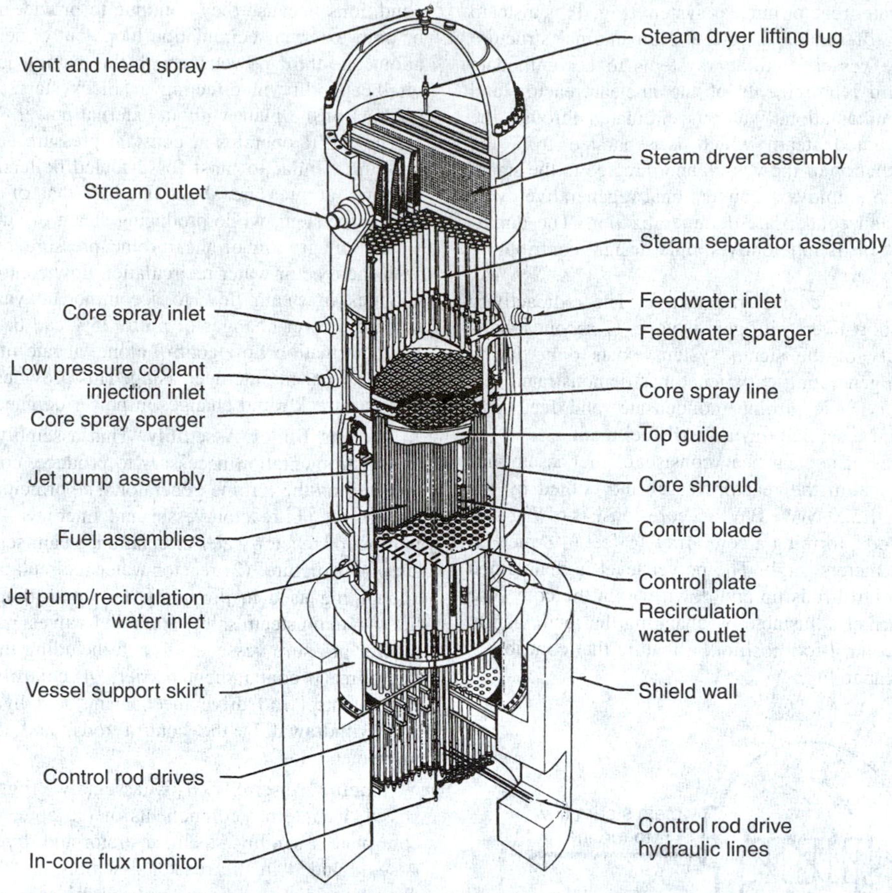
Vent and head spray
Steam dryer lifting lug
Steam dryer assembly
Stream outlet
Steam separator assembly
Core spray inlet
Feedwater inlet
Feedwater sparger
Low pressure coolant injection inlet
Core spray line
Core spray sparger
Top guide
Jet pump assembly
Core shroud
Fuel assemblies
Control blade
Jet pump/recirculation water inlet
Control plate
Recirculation water outlet
Vessel support skirt
Shield wall
Control rod drives
Control rod drive hydraulic lines
In-core flux monitor

Fig. 3. Reactor assembly of contemporary boiling water reactor. (*General Electric.*)

wherein the centrifugal forces separate the water from the steam in each of three stages. Steam leaves the separator at the top and passes into the wet steam plenum below the dryer. The separated water exits from the lower end of each stage of the separator and enters the pool that surrounds the standpipes to join the downcomer annulus flow.

The steam dryer assembly is mounted in the reactor vessel above the separator assembly and forms the top and sides of the wet steam plenum. Vertical guides on the inside of the vessel provide alignment for the dryer assembly during installation. The dryer assembly is supported by pads extending inward from the vessel wall and is held down in position during operation by the vessel head. These vanes are attached to a top- and bottom-supporting member forming a rigid, integral unit. Moisture is removed and carried by a system of troughs and drains to the pool surrounding the separators and then into the recirculation downcomer annulus between the core shroud and reactor vessel wall.

Control Rod Drive System. Positive core reactivity control is maintained by the use of movable control rods interspersed throughout the core. These control rods thus control the overall reactor power level and provide the principal means of quickly and safely shutting down the reactor. The rods are vertically moved by hydraulically actuated, locking piston type drive mechanisms. The drive mechanisms perform both a positioning and latching function, and a scram function with the latter overriding any other signal (scram signifies prompt shutdown).

Core Configuration. The reactor core of the BWR is arranged as an upright cylinder containing a large number of fuel assemblies and located within the reactor vessel. The coolant flows upward through the core. The plan of a typical core arrangement of a large BWR is shown in Fig. 4. The lattice configuration is shown in Fig. 5.

Fuel Rod. A fuel rod consists of uranium dioxide (UO_2) pellets and a Zircaloy-2 cladding tube. The UO_2 pellets are manufactured by compacting and sintering UO_2 powder into cylindrical pellets and grinding to size. The immersion density of the pellets is approximately 95% of theoretical UO_2 density. A fuel rod is made by stacking pellets into a cladding tube that is evacuated, back-filled with helium to atmospheric pressure, and sealed by welding Zircaloy end plugs in each end of the tube. The pellets are stacked to an active height of 148 inches (376 centimeters) with the top 12 inches (30.5 centimeters) of tube available as a fission gas plenum. A plenum spring is provided in the plenum space to exert a downward force on the pellets, the spring keeping the pellets in place during the pre-irradiation handling of the fuel bundle.

Fuel Bundle. Each fuel bundle contains 63 fuel rods, which are spaced and supported in a square (8×8) array by a lower and upper tie plate. Three types of rods are used in a fuel bundle: (1) tie rods; (2) a water rod; and (3) standard fuel rods. The third and sixth fuel rods along each outer edge of a bundle are tie rods. The eight tie rods in each bundle have threaded-end plugs, which screw into the lower tie plate casting. A stainless steel hexagonal nut and locking tab is installed on the upper end plug to hold the assembly together. The water rod not only serves as a spacer support rod, but also provides a source of moderator material near the center of the fuel bundle. This flattens the neutron flux across the bundle, and leads to lower local peaking factors and better utilization of uranium in the interior rods of the fuel assembly.

The initial core will contain fuel assemblies having a common average enrichment ranging from approximately 1.6% (weight) of ^{235}U to 2.2%, depending upon initial cycle requirements. Each assembly will contain different enrichment rods. Selected rods in each assembly will, in addition, be blended with gadolinium burnable poison. The reload fuel will also contain four different enrichment rods with an average enrichment in the range of 2.4 to 2.8%. Different ^{235}U enrichments are used in fuel assemblies to reduce the local power peaking. Low enrichment rods are used in the corner rods and in the rods nearer the water gaps; higher enrichment uranium is used in the central part of the fuel bundle.

Fuel Channel. A fuel channel encloses the fuel bundle. The combination of a fuel bundle and a fuel channel is called a fuel assembly. See Fig. 6. The channel is a square-shaped tube fabricated from Zircaloy 4. The outer dimensions are 5.518 inches (14 centimeters) by 5.518 inches

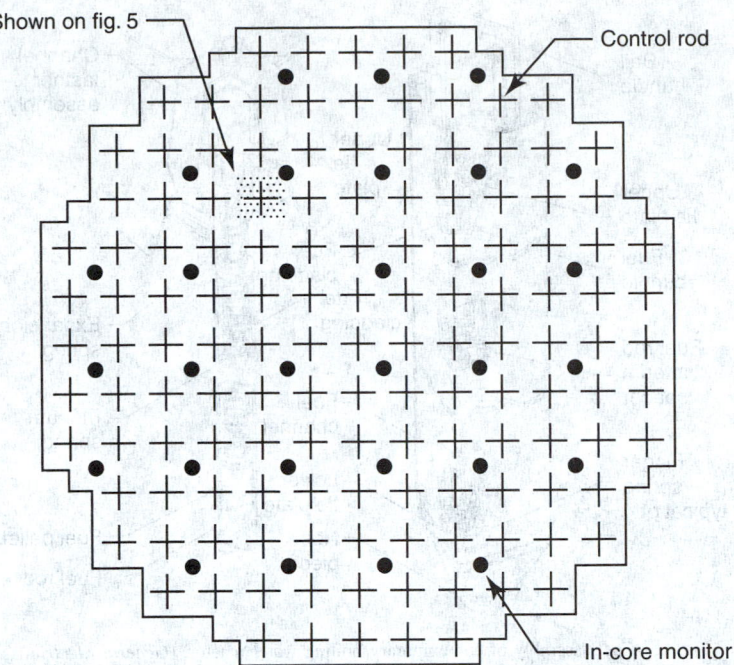

Fig. 4.   Typical core arrangement in contemporary boiling water reactor. (*General Electric.*)

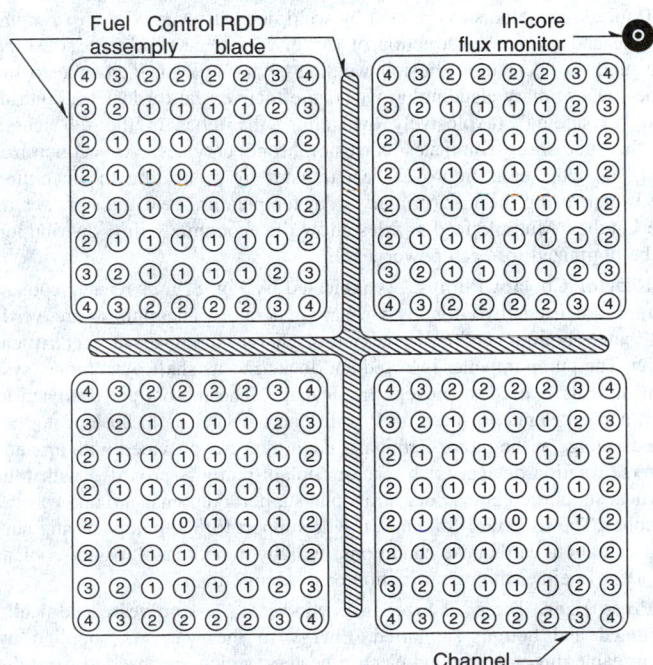

Fig. 5.   Core lattice arrangement in contemporary boiling water reactor. (*General Electric.*)

Neutron Sources. Several antimony-beryllium start-up sources are located within the core. They are positioned vertically in the reactor by "fit up" in a slot (or pin) in the upper grid and a hole in the lower core support plate. The active portion of each source consists of a beryllium sleeve enclosing two antimony-gamma sources. The resulting neutron emission strength is sufficient to provide indication on the source range neutron detectors for all reactivity conditions equivalent to the condition of all rods inserted prior to initial operation. The active source material is entirely enclosed in a stainless steel cladding with an outside diameter of approximately 0.7 inch (1.8 centimeter). The source is cooled by natural circulation of the core leakage flow in the annulus between the beryllium sleeve and the antimony-gamma sources.

Core Design Margins. The reactor core is designed to operate at rated power with sufficient design margin to accommodate changes in reactor operations and reactor transients without damage to the core. In order to accomplish this objective, the core is designed, under the most limiting operating conditions and at 100% rated power, to meet the following bases. (1) The maximum linear heat generation rate, in any part of the core, is always less than 13.4 kW/foot (43.97 kW/meter). (2) The minimum ratio between critical heat flux and fuel operating heat flux, in any part of the core, is always greater than 1.9.

Power Distribution. The design power distribution is divided for convenience into several components: (1) relative assembly power; (2) local; and (3) axial. The relative assembly power peaking factor is the maximum fuel assembly average power divided by the reactor core average assembly power. The local power peaking factor is the maximum fuel rod average heat flux in an assembly divided by the assembly average fuel rod heat influx. The axial power peaking factor is the maximum heat flux of a fuel rod divided by the average heat flux in that rod. Peaking factors vary throughout an operating cycle, even at steady-state full-power operation, since they are affected by withdrawal of control rods to compensate for fuel burnup. The design peaking factors represent the values of the most limiting power distribution that will exist in the core throughout its life.

Because of the presence of steam voids in the upper part of the core, there is a natural characteristic for a BWR to have the axial power peak in the lower part of the core. During the early part of an operating cycle, bottom-entry control rods permit a partial reduction of this axial peaking by locating a larger fraction of the control rods in the lower part of the core. At the end of an operating cycle, the higher accumulated exposure and greater depletion of the fuel in the lower part of the core reduces the axial peaking. The operating procedure is to locate control rods so that the reactor operates with approximately the same axial power shape throughout an operating cycle.

Reactivity Control. The movable boron-carbide control rods are sufficient to provide reactivity control from the cold shutdown condition to the

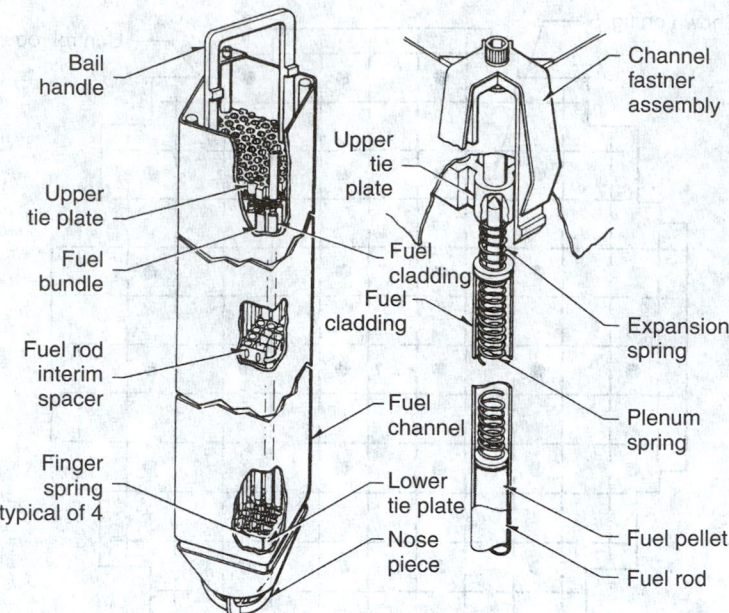

Fig. 6.   Fuel assembly of contemporary boiling water reactor. (*General Electric.*)

full-load condition. Supplementary reactivity control in the form of solid burnable poison is used only to provide reactivity compensation for fuel burnup or depletion effects. The movable control rod system is capable of bringing the reactor to the subcritical when the reactor is an ambient temperature (cold), zero power, zero xenon, and with the strongest control rod fully withdrawn from the core. In order to provide greater assurance that this condition can be met in the operating reactor, the core is designed to obtain a reactivity of less than 0.99, or a 1% margin on the "stuck rod" condition. See Fig. 7.

Reactor Auxiliary Systems include: (1) a reactor water cleanup system for maintaining high reactor water quality by removing fission products, corrosion products, and other soluble and insoluble impurities; (2) a fuel and containment pool cooling and cleanup system—a system which accommodates the beta and gamma radiation heating from the fission products that remain in the spent fuel, as well as drywell heat transferred to the upper containment pool; (3) a closed cooling water system for reactor service consisting of a separate, force circulation loop; (4) emergency equipment cooling system; (5) standby liquid control system; (6) reactor core isolation cooling system; (7) emergency core cooling system; (8) high-pressure core spray system; and (9) residual heat removal system.

Contemporary Pressurized Water Reactor (PWR)

In a typical pressurized water reactor (PWR), heat generated in the nuclear core is removed by water (reactor coolant) circulating at high pressure through the primary circuit. The water in the primary circuit cools and moderates the reactor. The heat is transferred from the primary to the secondary system in a heat exchanger, or boiler, thereby generating steam in the secondary system. The steam produced in the steam generator, a tube-and-shell heat exchanger, is at a lower pressure and temperature than the primary coolant. Therefore, the secondary portion of the cycle is similar to that of the moderate-pressure fossil-fueled plant. In contrast, in boiling-water or direct-cycle systems, steam is generated in the core and is delivered directly to the steam turbine.

The similarities of basic pressurized water reactor design from one manufacturer to the next are more striking than the differences. Therefore, the description of one particular configuration (Combustion Engineering, Inc.) can suffice to convey the general operating principles. The major components of a PWR are: (1) the reactor vessel which contains the oxide fuel core, core intervals, control element assemblies, and in-core instruments; (2) the electrically-heated pressurizer; (3) the electric-motor-driven primary coolant pumps; and (4) the U-tube type steam generators. See Fig. 8. The primary coolant system layout can be fitted into a variety

of containment types and concepts. A prestressed cylindrical containment is common. Figure 9 shows the arrangement in a spherical containment. This type of building lends itself to separation of safeguards equipment, steam lines, and emergency power supplies.

Steam Generators. The basic geometry is shown in Fig. 10. With the nuclear steam supply system operating at 3,817 MW, two steam generators produce a total of 17.18×10^6 pounds (7.89×10^6 kilograms) of steam per hour at 1,070 psia (72.8 atmospheres). The steam generators are constructed, using carbon steel pressure-containing members and Inconel-600 tubes. The tube-sheet is clad by weld deposit for maximum strength; tongue and groove construction of the divider plate places no stress on the tube-sheet cladding. Fusion welding of the end of each tube to the tube-sheet primary cladding provides an effective seal for leakage control, and "expanding" (explosively expanding) the tubes in the full length of the tube-sheet eliminates corrosion-prone crevices. An economizer section on the units improves heat transfer by preheating the incoming feedwater, using the low (primary side) temperature heat transfer area of the U-tubes. Multiple feed nozzles allow the economizer flow distribution to be optimized for each power level.

Reactor Coolant Pumps. As indicated by Fig. 8, four reactor coolant pumps are used, two for each steam generator. The pumps are vertical, single-bottom-suction, horizontal-discharge, motor-driven centrifugal units. The pump impeller is keyed and locked to its shaft. A complex system of seals is used to prevent any leakage. The motors are designed to start and accelerate to speed under full load with a drop to 80% of normal rated voltage at the motor terminals. Each motor is provided with an anti-reverse rotation device. Each reactor coolant pump is provided with four vertical support columns, four horizontal support columns, and one vertical snubber. The structural columns provide support for the pumps during normal operation, earthquake conditions, and any hypothetical loss-of-coolant accident in either the pump suction or discharge line.

Pressurizer. The pressurizer is a cylindrical pressure vessel, vertically mounted and bottom supported. Energy to the water is supplied by replaceable direct-immersion electric heaters, which are inserted from the bottom head of the pressurizer. Nozzles are provided for spray, surge, relief, and instrumentation connections. The pressurizer maintains reactor coolant system operating pressure and, in conjunction with the chemical and volume control system, compensates for changes in reactor coolant volume during load changes, heat-up, and cool-down. During full-power operation, the pressurizer is about $\frac{1}{3}$ full of saturated steam.

Reactor Vessel. This vessel is designed to contain the fuel bundles, the control element assembles, and the internal structures necessary for support of the core. The reactor is a stainless clad, thick-walled, carbon

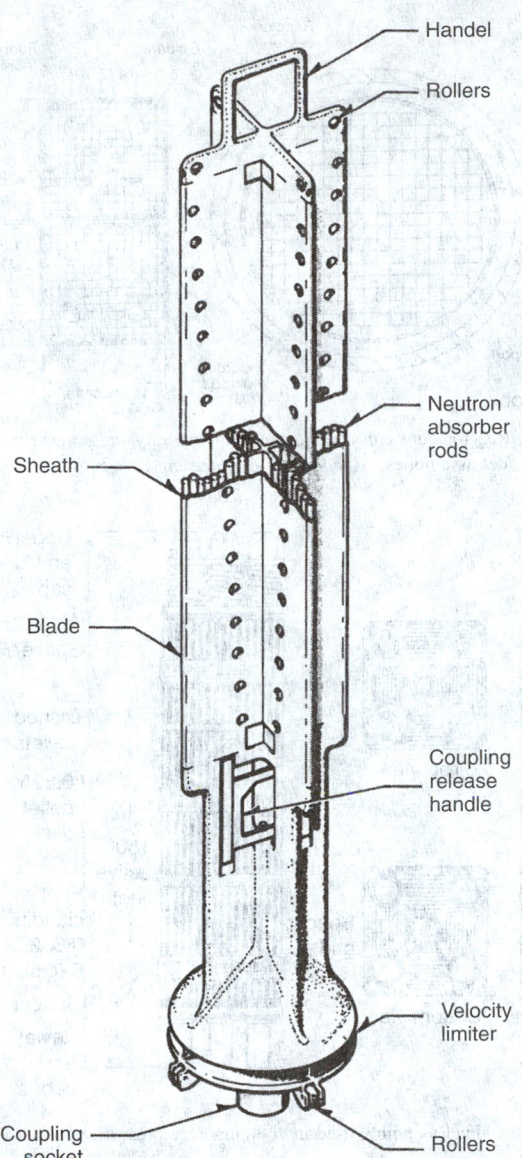

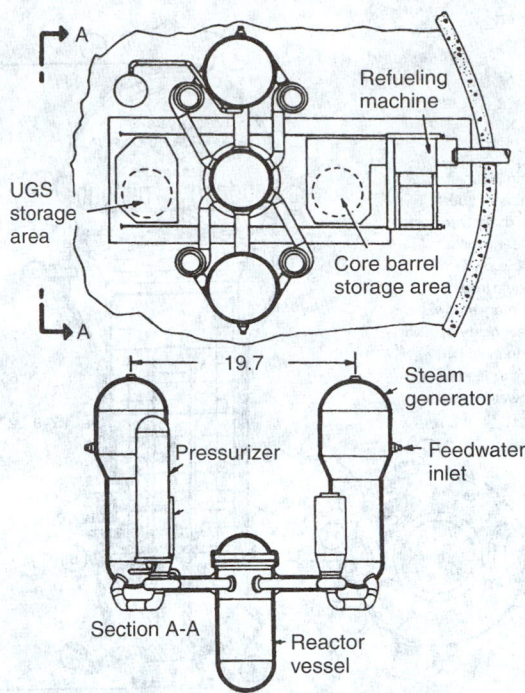

Fig. 8.   Nuclear steam supply system for contemporary pressurized water reactor. (*Combustion Engineering.*)

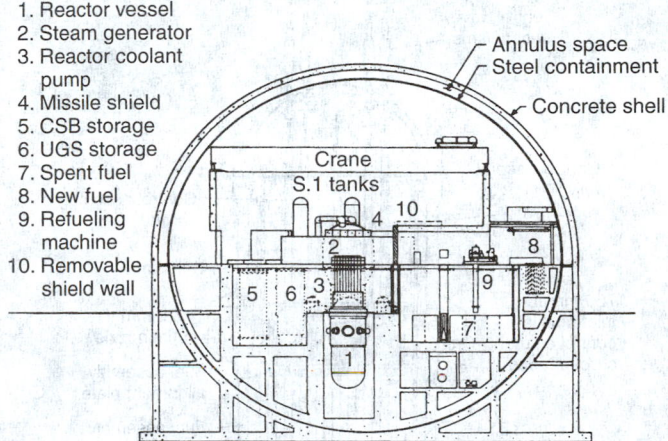

1. Reactor vessel
2. Steam generator
3. Reactor coolant pump
4. Missile shield
5. CSB storage
6. UGS storage
7. Spent fuel
8. New fuel
9. Refueling machine
10. Removable shield wall

Fig. 9.   Spherical containment for contemporary pressurized water reactor. (*Combustion Engineering.*)

Fig. 7.   Control rod used in contemporary boiling water reactor. The cruciform control rods contain 76 stainless steel tubes (19 tubes in each wing of the cross). These tubes are filled with boron carbide powder compacted to approximately 65% of theoretical density. The tubes are seal-welded with end plugs on either end. The individual tubes act as pressure vessels to contain the helium gas released by the boron-neutron capture reaction. The control rods have an active length of 144 inches (365.8 centimeters) of boron carbide, a span of 9.75 inches (24.8 centimeters), and an overall length of 173.75 inches (441.3 centimeters). The control rods can be positioned at 6-inch (15-centimeter) steps and have a nominal withdrawal and insertion speed of 3 inches (7.5 centimeters) per second. Control rods are cooled by the core leakage (bypass) flow. In addition to satisfying initial control effectiveness requirements, it is expected that the control rods will have an average lifetime of approximately 15 full-power years. (*General Electric.*)

steel pressure vessel comprised of a cylinder with two hemispherical heads. The lower head is integrally welded to the vessel shell and contains in-core instrumentation nozzles. The upper closure head, containing the control element drive mechanism nozzles, is attached to the vessel by means of a bolted flange, thus permitting the head to be removed to provide access to the reactor internals. The head flange is drilled to match the vessel flange stud bolt locations.

The vessel flange is a forged ring with a machined ledge on the inside surface to support the core support barrel. The flange is drilled and tapped to receive the closure studs and is machined to provide a mating surface for the reactor vessel closure seals. Sealing is accomplished by using two silver-plated, NiCrFe-alloy, self-energizing O-rings. The space between the

two rings is monitored to detect any inner-ring coolant leakage. The inlet and outlet nozzles are located radially on a common plane below the vessel flange. Extra thickness in the vessel course provides the reinforcement required for the nozzles. Snubbers built into the lower portion of the vessel shell limit the amplitude of any displacement of the core support barrel. Core stops are also built into the reactor vessel to limit the downward displacement of the core support barrel.

Cladding for the reactor vessel is a continuous integral surface of corrosion-resistant material, having $\frac{7}{32}$-inch (0.56 centimeter) nominal thickness, and a $\frac{1}{8}$-inch minimum thickness. The reactor vessel is supported by four vertical columns located under the vessel inlet nozzles. These columns are designed to flex in the direction of horizontal thermal expansion and thus allow unrestrained heat-up and cool-down. The columns also act as a hold-down device for the vessel. The supports are designed to accept normal loads and seismic and pipe rupture accident loads.

The reactor arrangement is shown in Fig. 11. The barrel-calandria guide structure is a rugged (3-inches thick, barrel section) unit that can withstand and protect all control element fingers from the combined effects of seismic

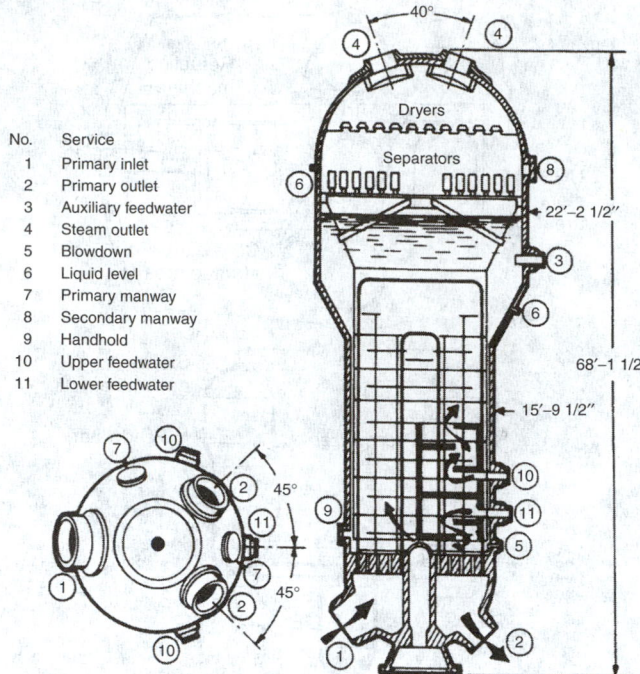

Fig. 10. Steam generator for contemporary pressurized water reactor. (*Combustion Engineering*.)

No.	Service
1	Primary inlet
2	Primary outlet
3	Auxiliary feedwater
4	Steam outlet
5	Blowdown
6	Liquid level
7	Primary manway
8	Secondary manway
9	Handhold
10	Upper feedwater
11	Lower feedwater

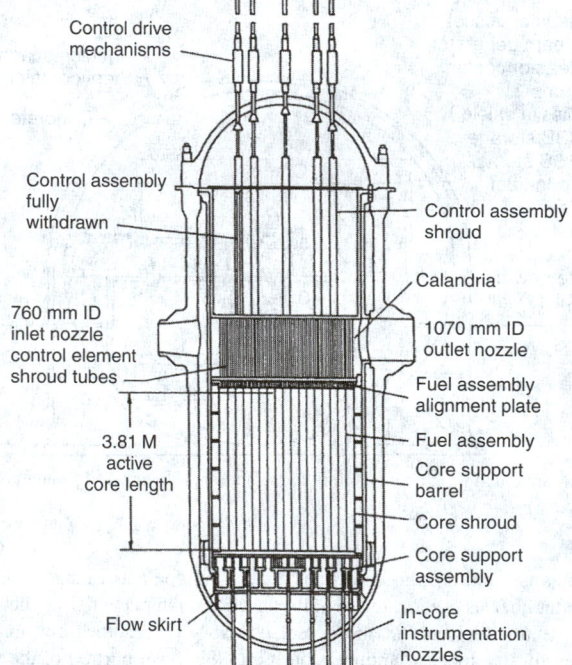

Fig. 11. Reactor arrangement in contemporary pressurized water reactor.

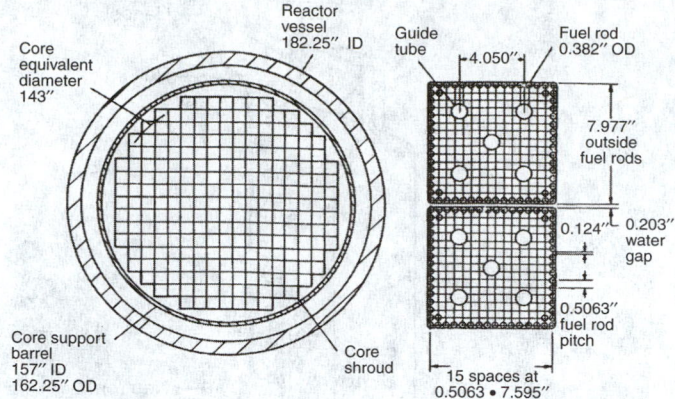

Fig. 12. Reactor core cross section of contemporary pressurized water reactor with 241 fuel assemblies. (*Combustion Engineering*.)

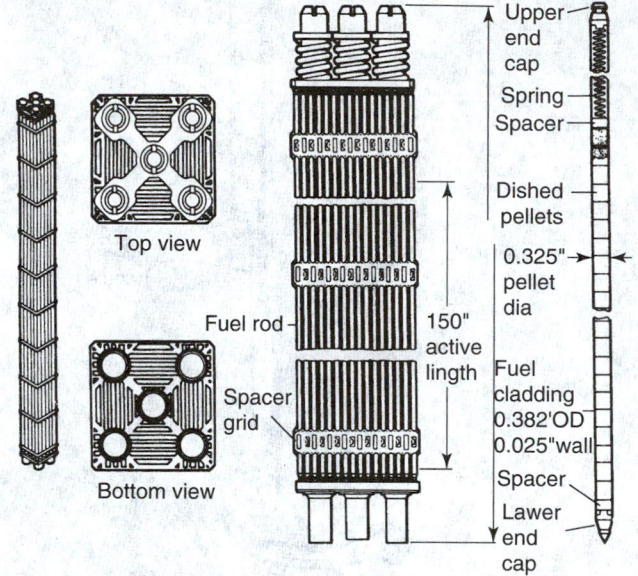

Fig. 13. Fuel assembly used in contemporary pressurized water reactor.

and blowdown loads that may result from a loss-of-cooling accident. The calandria structure fits over the control element guide tubes of the fuel assemblies, aligning all fuel assemblies, and laterally restraining the top ends of the fuel assemblies. With the upper guide structure in place, a continuous guide tube for each control finger is formed, extending from the top of the tube-sheet to the bottom of the fuel assembly. Because of this feature, which isolates every control finger from the coolant crossflow, flexibility is obtained in the number of control fingers that can be attached to one control assembly, i.e., one control element assembly can serve more than one fuel assembly.

Severe emergency core cooling system criteria require that the builders of water reactors increase the linear feet of fuel in the reactor core for the same power in order to reduce LOCA (loss-of-cooling accident) fuel temperatures. In the unit described here, an assembly with a 16×16 fuel rod array of smaller diameter rods is used in the same assembly envelope that was occupied by a 14×14 assembly in earlier designs. This results in a maximum linear heat rate decrease in the assembly of about 25%.

As shown in Fig. 12, the active core is made up of 241 fuel assemblies, all of which are mechanically identical. As indicated by Fig. 13, each fuel assembly contains 236 Zircaloy-clad, UO_2 fuel rods retained in a structure consisting of Zircaloy spacer grids welded at about 15-inch (38.1-centimeter) intervals to five Zircaloy control element assembly guide tubes which, in turn, are mechanically fastened at each end to stainless steel end fittings. The overall length of the fuel assembly is about 177 inches (450 centimeters) and the cross section is about 8 inches (20.3 centimeters) by 8 inches (20.3 centimeters). Each fuel assembly weighs about 1,450 pounds (657.7 kilograms). With reference to Fig. 13, fuel rods, consisting of uranium dioxide (UO_2) pellets of low enrichment canned in thin-walled Zircaloy-4 tubing, are designed to achieve average burnups of about 33,000 MWD/MTU (thermal megawatt days/metric tons of uranium) and peak burnups of about 50,000 MWD/MTU. The design factors limiting burnup of the fuel are the effects on the clad of volumetric changes of the fuel pellet and fission gas release.

As indicated in Fig. 13, the fuel rod consists essentially of 0.325-inch (0.82-centimeter) diameter, 0.390-inch long UO_2 pellets canned in a 0.382-inch (0.97-centimeter) outside diameter Zircaloy-4 tube. The high density fuel pellets are dished at both ends to allow for axial differential thermal expansion and fuel volumetric growth with burnup.

The control element assemblies consist of an assembly of 4, 8, or 12 fingers approximately 0.8-inch (2-centimeter) outside diameter and arranged as shown in Fig. 14. The use of cruciform control rods, as in boiling water and early pressurized water reactors, necessitates large water gaps between the fuel assemblies to ensure that the control rods will scram (prompt shutdown) satisfactorily. These gaps cause peaking of the power in fuel rods adjacent to the water channel compared to fuel rods some distance from the channel.

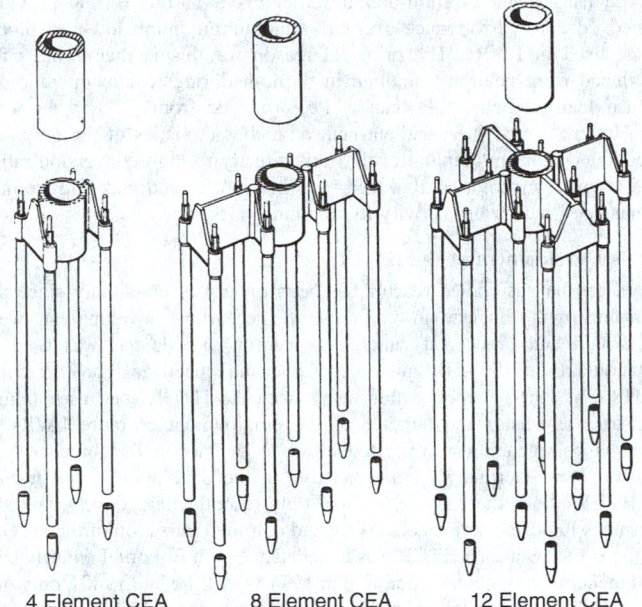

4 Element CEA   8 Element CEA   12 Element CEA

Fig. 14. Four 8- and 12-element control element assemblies. (*Combustion Engineering.*)

A five-hole assembly design was evolved from consideration of the lower peaking effect of smaller (removal of one fuel rod) water holes versus the mechanical advantages of larger (four fuel rods removed) water holes. The larger water holes allowed the use of rugged, 0.9-inch (2.3-centimeter) outside diameter by 0.035-inch (0.89-centimeter) thick Zircaloy guide tubes for the fuel assembly structure. These water holes are distributed relatively uniformly in the reactor core when placed in the 16 × 16 fuel rod lattice. The particular arrangement of water holes was selected in consideration of the water gap between fuel assemblies; the effect of the central water hole in the fuel assembly is balanced by the water gap between fuel assemblies. The mechanical simplicity and ruggedness outweigh the advantage of obtaining a small decrease in local peaking by using very small fingers. The slightly higher peaking associated with the design can be compensated, to a large extent, by varying the enrichment of the fuel in the rods adjacent to the control channel and/or by using water displacers in local hot spots.

The control element assembly, shown in Fig. 15, consists of 0.8-inch (2-centimeter) outside diameter Inconel tubes containing boron carbide pellets as the neutron absorbing material. A gas plenum is provided in order to limit the maximum stress due to generation of internal gas pressure. The individual control fingers are attached mechanically and locked to the various spider assemblies. This allows for simplifications in manufacture, shipping, and assembly of the control element assembly. Because all fingers are removable and replaceable, servicing and disposal problems are decreased. It is intended that the spider assembly and its extension shaft be reused whenever possible.

Design of the upper guide structure permits flexibility in the number of control fingers that can be attached to one control assembly. The standard pattern of control assemblies is shown in Fig. 16. In the standard design, power changes at close to full power, shaping of the radial power distribution, and control of the axial power distribution are best handled by the low worth 4-finger control element assembly entering

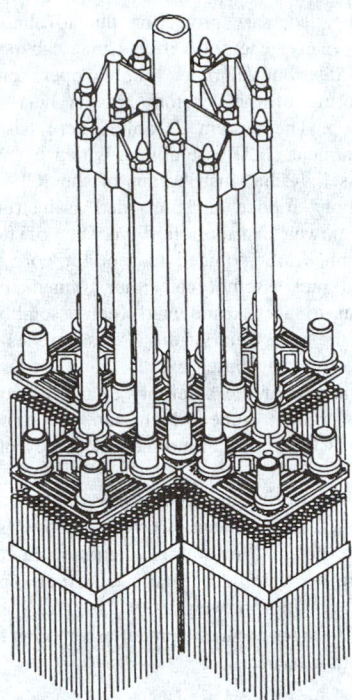

Fig. 15. Control element assembly and fuel for contemporary pressurized water reactor.

a single fuel assembly. Shutdown reactivity control in the peripheral region of the core is handled by the 8-finger control element assembly and in the central region of the core it is handled by the 12-finger control element assembly. The need for the two types of shutdown control element assemblies is to obtain "stuck rod worths" in the high reactivity fuel on the periphery of the core which are about equal to the control element assembly control worth in the lower reactivity central zone of the core.

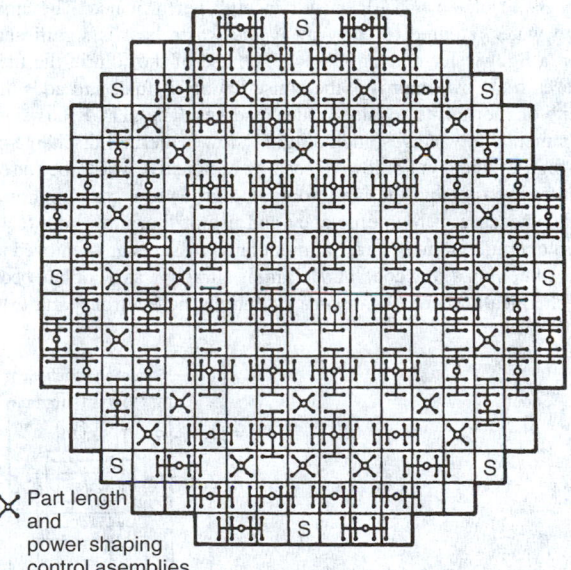

Fig. 16. Standard pattern of control assemblies in contemporary pressurized water reactor core. The pattern provides more-than-sufficient control for self-generated plutonium recycle. For complete open-market plutonium recycle, 4-element control assemblies are added in positions marked S.

Instrumentation. The large size of present water reactors and the nuclear effects which can occur, such as xenon redistribution, stuck rods, and reactivity anomalies require that emphasis be placed on instrumentation and control systems if high plant availability is to be maintained, while providing the necessary protection due to abnormal occurrences. Because there are many reactivity effects that can produce changes in the reactor power distribution, more reliable operation can be obtained by on-line monitoring of the reactor. This is best achieved with in-core instrumentation. The system described here has provision for up to 61 in-core instrument (ICI) assemblies, which enter from the bottom of the reactor vessel. Radial distribution of the ICI is such that every type of fuel assembly, rodded and unrodded, is instrumented, assuming symmetrical core power distribution. Five sets of four symmetrically located ICI assemblies are included to monitor core power tilts. Also, every instrumented fuel assembly is either immediately adjacent to or diagonally adjacent to an instrumented fuel assembly to obtain good radial coverage of the core. Each of the ICI assemblies contains five self-powered fixed detectors distributed axially along the length of the core, a thermocouple at the end of the assembly to monitor outlet temperature, and a dry-well instrument tube which can accommodate a movable detector. This allows for high measurement accuracy of the on-line fixed detector.

The continuous monitoring and processing of the data from over 300 fixed detectors by the core monitoring computer provides the operator with information on core power distribution, maximum linear heat rate in the fuel, departure from nucleate boiling ratio, and fuel exposure. These data can then be used to obtain improved maneuvering of core power using the relative low worth 4-finger control element assemblies for power changes and power distribution control.

High-Temperature Gas-cooled Reactor (HTGR)

Although there have been comparatively few gas-cooled reactors installed for generating commercial nuclear electric power, the concept has a number of operating advantages over light-water reactors and could play an important role in the reactor designs for the next century.

The high-temperature gas-cooled reactor (HTGR) is a thermal reactor that produces desired steam conditions. Helium is used as the coolant. Graphite, with its superior high-temperature properties, is used as the moderator and structural material. The fuel is a mixture of enriched uranium and thorium in the form of carbide particles clad with ceramic coatings.

The high-temperature conditions and high thermal efficiency (approximately 39%) of the (HTGR) result in high performance. The amount of cooling water required to carry away the waste heat is significantly less than in a light-water reactor (LWR). The use of thorium in the fuel cycle decreases fuel cost, improves the conservation of fuel, and adds the large deposits of thorium to available fuel reserves. The HTGR has significant environmental advantages, including: (1) lower thermal discharge because of its high efficiency; (2) low release of radioactive waste because of the high-integrity fuel and the inert coolant; and (3) low consumption of raw materials because of high efficiency and use of thorium in the fuel cycle.

High operating temperatures at moderate pressures are achieved through the use of helium as the coolant. Helium is attractive as a coolant because it: (1) is chemically inert; (2) absorbs essentially no neutrons; and (3) makes no contributions to the reactivity of the system. Carbon dioxide also has been used as a coolant.

Graphite is used as the moderator and core structural material because of (1) excellent mechanical strength at high temperatures; (2) very low neutron-capture cross section; (3) good thermal conductivity; and (4) high specific heat. Graphite has a long history of use in thermal reactors. Because of low neutron-capture cross section, no neutrons are lost within the core through absorption in metallic fuel cladding or structural supports. Graphite also is well suited to high-temperature operations, increasing in strength with temperature up to a point (2,482 °C) well beyond the operating range of the HTGR.

The use of the thorium-uranium fuel cycle in the HTGR provides improved core performance over the plutonium/uranium low-enrichment cycle used in LWRs. The principal reason for this is that fissile ^{233}U produced from neutrons captured in thorium during reactor operation is neutronically a better fuel than ^{239}Pu, produced from ^{238}U in the low-enrichment cycle. The excellent neutronic characteristics of the graphite-moderated thorium/uranium cycle leads directly to high conversion ratios and low fuel inventories. Reduced ^{235}U inventories and make-up requirements spell reduced sensitivity to uranium prices.

Early Development of the HTGR

Work on the gas-cooled reactor has been underway essentially since the dawning of the nuclear power industry. The earliest developments were in Britain and France, at which time carbon dioxide gas was used as the coolant. In 1965, Britain opted for an advanced gas-cooled reactor (AGR). In 1969, France swung away from the HTGR (because of high construction costs) and targeted to the employment of more LWRs as well as commencing a concerted effort to develop a fast breeder type reactor. West Germany has been active in the development and testing of HTGRs since the early 1960s, but only recently (late 1980s) have the Germans indicated serious efforts toward commercialization of the HTGR. In the United States, a HTGR was installed at Peach Bottom, Pennsylvania, commencing commercial operation in 1974. As of the late 1980s, only one other HTGR was installed in the United States, the Fort St. Vrain plant near Denver, Colorado.

A simplified flow diagram of this station, which generates 842 MW (thermal) to achieve a net output of 330 MW (electrical), is given in Fig. 17. The helium coolant, at a pressure of about 700 psi (47.6 atmospheres), flows downward through the reactor core, where it is heated to 777 °C. The coolant flow can be trimmed by the use of orifice valves located at the top of the core that are integral with the control rod drive mechanisms. From the reactor core, the coolant flows through the steam generators. After passing through the steam generators, the helium is returned to the core at a temperature of about 404 °C by four steam-turbine-driven helium circulators. Two identical loops are used, each including a six-module steam generator and two helium circulators. Each loop contributes half of the total output of the nuclear steam supply system, which produces steam at 2,400 psig (163.3 atmospheres) and 538 °C with single reheat to 538 °C. The helium circulators are driven by the exhaust steam from the high-pressure turbine. This steam is then reheated and

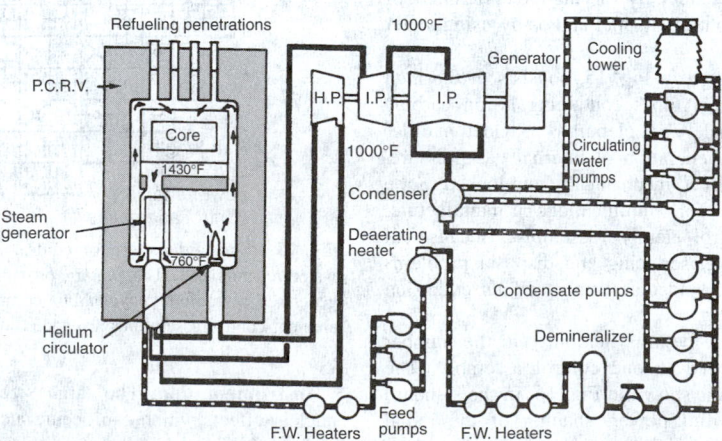

Fig. 17.   Simplified flow diagram of Fort St. Vrain Nuclear Generating Station. (*GA Technologies*.)

returned to the intermediate-pressure turbine. The circulators are also equipped with a Pelton water wheel drive so that they may be driven using the boiler feed pumps for emergency conditions.

The general reactor arrangement is shown in Fig. 18. The prestressed concrete reactor vessel (PCRV) is 31 feet in internal diameter with a 75-foot (23 meters) internal height. The upper and lower heads are nominally 15 feet (4.5 meters) thick, and the walls have a nominal thickness of 9 feet (2.7 meters). Thus, the PCRV provides the dual function of containing the coolant at operating pressure and also providing radiological shielding.

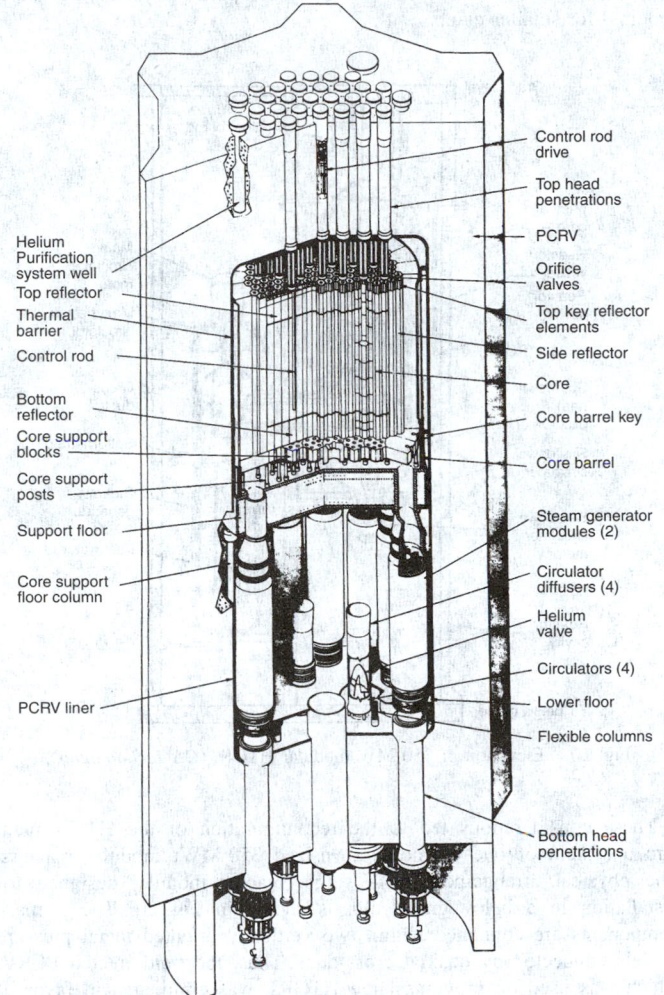

Fig. 18. General reactor arrangement of Fort St. Vrain high-temperature gas-cooled reactor. (*GA Technologies.*)

Reactor Core. The HTGR fuel element is a graphite block, hexagonal in cross section and having a grid of longitudinal fuel holes and coolant channels. The fuel element blocks are stacked in columns of eight blocks each and grouped into fuel regions consisting of a central column surrounded by six columns. Each region rests on a large core support block which, in turn, rests on graphite posts standing on the liner of the central cavity. Hexagonal graphite reflector elements are located above, below, and around the active core. These elements are surrounded by permanent side-reflector blocks to give the entire assembly a circular configuration. The fuel holes contain a rod consisting of ceramic-coated fuel particles in a graphite matrix. The coatings, applied by pyrolitic techniques, are multilayered to ensure a high degree of fission-product confinement. A porous interlayer, or buffer zone, accommodates the expansion of the irradiated fuel and provides storage space for gaseous fission products. The outer layer acts as a fission-product retention barrier and provides structural strength. In effect, the particle coating functions as a miniature spherical pressure vessel.

To achieve a fuel management scheme with the lowest fuel cycle cost consistent with the current thermal and material performance limits, the following parameters are selected: (1) a fuel cycle incorporating uranium/thorium; (2) a fuel lifetime of four years; (3) an average power density of $8.4 \ W/cm^3$; and (4) a refueling frequency of once a year.

The reactor is controlled by two control rods located in each refueling region. All control rod pairs have scram (quick shutdown) capability and are driven by gravity. A backup reserve shutdown system is included. This consists of boronated graphite pellets that can be introduced from hoppers located in each refueling penetration into the core via the cylindrical channels in the central fuel element of each refueling region.

Safeguard Systems. The design of HTGR incorporates many inherent safety features and a number of engineered safeguards. The inherent safety characteristics include negative power and temperature coefficients, assured by the thorium content of the fuel. In addition, the high heat capacity of the large mass of graphite ensures that any core temperature transient resulting from reactivity insertions or interruptions in cooling will be slow and readily controllable. This important safety feature eliminates the need for an emergency core cooling system. Only a residual heat removal system is required for the long-term decay heat, and control of the HTGR is inherently easier than in reactors in which the coolant functions as the moderator. The uranium/thorium fuel contained in the ceramic-coated particles is not susceptible to sudden release of the stored-up fission products as a result of melting. Since the entire primary coolant system is contained within the PCRV, external piping, which might be subject to sudden rupture, is eliminated. Structural strength and integrity of the PCRV is enhanced by the redundant reinforcing steel and prestressed wire tendons. At the maximum credible pressure, the prestressing elements are not stressed above levels experienced during their initial tensioning. As a result, sudden loss of coolant due to prestress failure is not credible.

Second-Generation HTGRs

In addition to upgrading the HTGR at the Fort St. Vrain nuclear power station, efforts have been underway for several years to make both larger and smaller gas-cooled high-temperature reactors. Smaller, modular units could provide the flexibility needed by the public utilities as they plan their expansions for projected increases in electricity requirements. Inherent safety, already a feature of the HTGR, would be enhanced because of the smaller size and low power density of modular units. For example, it is estimated that the power density of a modular gas-cooled reactor would be only 3 kW/liter, as compared with 6 kW/liter for a large reactor and 100 kW/liter for a conventional pressurized-water reactor (PWR) as previously described.

In the new designs, if coolant were lost, the nuclear chain reaction would be terminated by the reactor's negative temperature coefficient after a modest temperature rise. Core diameter of the modular units would be limited so that decay heat could be conducted and radiated to the environment without overheating the fuel to the point where fission products might escape. Thus, inherent safety would be realized without operator or mechanical device intervention.

Large Commercial HTGRs[2]

Following construction of the Fort St. Vrain facility, the HTGR was marketed commercially in direct competition with large pressurized water reactors (PWRs) and boiling water reactors (BWRs). Between 1971 and 1975, ten such reactors were ordered by U.S. utilities. The designs of the commercial HTGR were similar to Fort St. Vrain in that they used the graphite based core structure, helium coolant, prestressed concrete reactor vessel (PCRV) and superheated steam cycle. However, the designs differed in that power outputs were significantly larger and the reactor system was rearranged to accommodate the larger-size components.

[2] This portion of article on HTGRs contributed by R.A. Dean, Sr. Vice President, GA Technologies Inc., San Diego, California.

The large HTGRs had power ratings of 2000 and 3000 MWt which corresponded to net electrical outputs of 770 and 1160 MWe, respectively. An example of the rearranged reactor system is shown in Fig. 19. A multi-cavity PCRV was used to enclose the reactor system instead of the single-cavity PCRV used in Fort St. Vrain. This was a major advancement in PCRV technology and necessitated the development of a circumferential wire-wrap prestressing system instead of circumferential tendons, although the longitudinal tendons were retained. The sizes of the multi-cavity PCRV were approximately 100 feet (30 m) high by 120 feet (36 m) in diameter for the 3000 MWt plant and 100 feet (30 m) high by 105 feet (32 m) in diameter for the 2000 MWt plant.

the HTGR to be entirely inherently safe by virtue of the high temperature structural integrity of the graphite core and ceramic coated fuel. This means that a small HTGR would not require any active safety equipment or any action by the operator in order to prevent release of radioactivity for any accident condition.

A major concern with reducing plant size was the economic impact from reversing the economy of scale. However, it was learned that economy-of-scale effects could be offset by several beneficial factors that apply to smaller nuclear plants. This includes the shift of major portions of the work from the site to the factory; the learning effects appreciated by replication of a larger number of smaller units in a factory environment and the elimination or simplification of many components/systems no longer required for smaller plants.

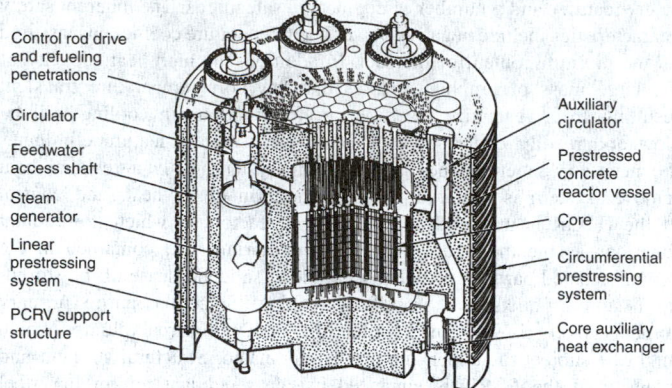

Fig. 19.   Integrated HTGR nuclear steam system (1170 to 3360 MW thermal). (*GA Technologies.*)

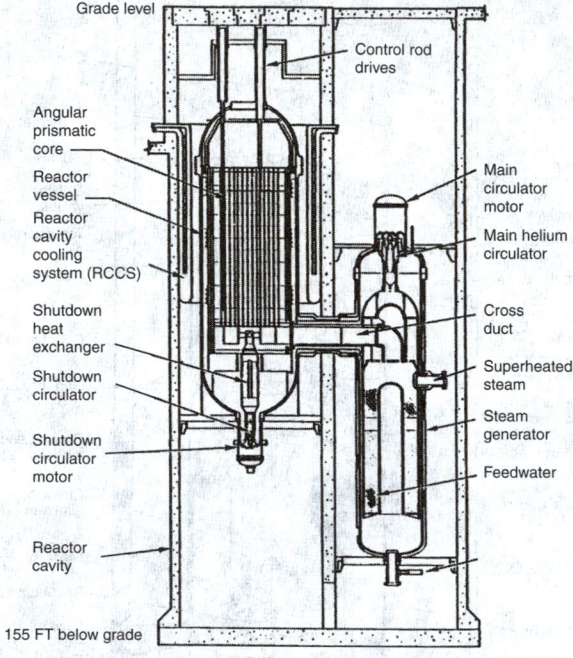

Fig. 20.   Elevation of 350 MW modular HTGR. (*GA Technologies.*)

The graphite reactor core was located in the central cavity of the PCRV. The steam generators and steam-driven main helium circulators were located in vertical cavities arranged around the periphery of the core. The 2000 MWt system had four steam generator-circulator side cavities while the 3000 MWt unit had six such cavities. The hot primary coolant helium (1366 °F; 741 °C) exiting from the bottom of the core collected in the lower core plenum from which it was distributed to the steam generators through the lower cross-ducts. The circulators, located above the steam generators, returned the cool helium (710 °F; 377 °C) through the upper cross-ducts to the upper core plenum. The helium in the upper plenum then flowed down through the core where it was heated.

Another feature of the large HTGRs was the core auxiliary cooling system which provided an independent means of core afterheat removal in the event that the main coolant loops (i.e., steam generators and steam-driven circulators) were shut down. The core auxiliary cooling system consisted of two redundant cooling loops, each capable of removing 100% of the afterheat, for the 2000 MWt plant and three redundant cooling loops, each capable of removing 50% of the decay heat, for the 3000 MWt plant. Each loop contained a motor-driven circulator and water-cooled heat exchanger and circulated flow through the core just as the main loops. Shutoff valves were located in both auxiliary and main loops to assure that helium would not bypass the core through the one system while the other system was in operation.

All ten large commercial HTGRs were ordered during the early 1970s as an indirect consequence of the energy crisis brought about by the oil embargo; all were cancelled by 1976. The combination of the recession plus new emphasis on conservation brought about a rapid reduction in electric energy demand which, in turn, resulted in cancellation of over 100 nuclear power plant orders, including the large HTGRs.

Small Modular HTGRs

In the early 1980s, the major influence on new designs was the renewed emphasis on safety brought about by the accident at the Three Mile Island nuclear plant. The experience from licensing and operation of nuclear plants during the 1970s indicated a need to reduce design complexity and develop passive approaches to reactor safety rather than rely on complex emergency safety systems. HTGR designers in the United States and Europe determined that a substantial reduction in plant size could enable

These considerations led to the reconfiguration of the HTGR plant into a system of one or more downsized 350 MWt modular reactors. The physical arrangement of a single reactor module, designed for installation in a below-grade silo, is shown in Fig. 20. The primary components are contained within two vertically oriented metal pressure vessels connected by coaxial cross-duct. Thus, the field-erected PCRV, which was used on previous large HTGRs, was eliminated in favor of shop-fabricated metal pressure vessels. The use of metallic pressure vessels also facilitates installation in underground silos, which enhances the safety of the plant.

The reactor vessel, which is approximately 72 feet (22 m) high by 22.5 feet (6.9 m) in diameter, contains the graphite core, reflector, and shutdown heat removal system (non-safety). The other vessel contains a single helical coiled steam generator and motor-driven circulator with magnetic bearing instead of water-lubricated bearings as in previous concepts. The size of both vessels is within allowable limits for barge, rail, and overland transportation.

During normal operation, the main circulator transports hot helium at 1266 °F (686 °C) from the bottom of the core to the steam generator which, in turn, produces superheated steam at 1005 °F (541 °C) and 2500 psia. The cold helium at 496 °F (258 °C) is returned to the top of the reactor core. During normal shutdown and refueling, the non-safety auxiliary shutdown heat removal system removes core afterheat if the main heat transport system is not operational.

A principal feature of the modular HTGR is its capacity for safely rejecting core afterheat in a completely passive manner (i.e., without the need for any active core cooling systems) such that any release of fission products from the fuel is prevented during severe accident conditions.

This feature is a result of both the reactor system configuration and the high temperature capability of the fuel. In the event of a loss of forced circulation cooling of the core via either the main circulator/steam generator or auxiliary shutdown circulator/heat exchanger, core afterheat will continue to be safely removed by direct conduction through the core and reflector to the reactor vessel wall. The heat is then dissipated from the reactor vessel surface by radiation and natural convection to cooling panels surrounding the interior surface of the reactor cavity. See Fig. 20, previously mentioned. These panels are part of the Reactor Cavity Cooling System (RCCS) which consists of natural convection air ducts that ultimately transport the core afterheat directly to the atmosphere.

In order to achieve this passive core cooling capability in a reactor with a power level and power density that are economically attractive, the annular core arrangement shown in Fig. 21 was adopted. The active core consists of an annular region of hexagonal graphite fuel blocks containing standard HTGR fuel. Unfueled graphite reflector blocks make up the region inside the active core annulus and the region surrounding the outside of the annulus. This arrangement results in a higher radial heat conductance for fuel at the innermost radius than for a solid cylindrical active core. Thus, the annular core arrangement permits operation at a higher power for a given volume than a solid active core.

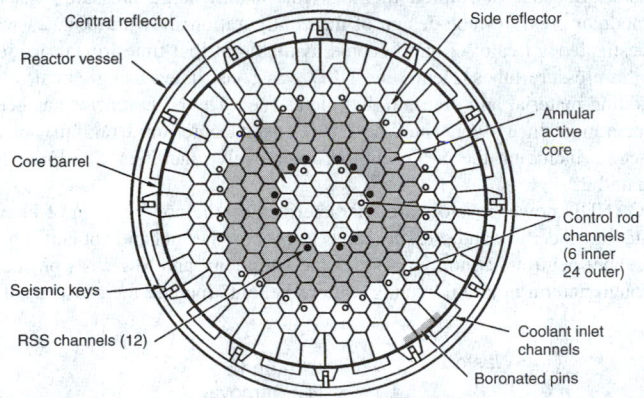

Fig. 21. Reactor core cross section of 350 MW modular HTGR. (*GA Technologies.*)

The modular HTGR uses the same form of fuel as the large HTGR and Fort St. Vrain installation except for the important difference that the ^{235}U enrichment was reduced to about 20% from the previous value of 93%. The fuel is in the form of coated fuel and fertile particles which are bonded into graphite rods and inserted into the hexagonal graphite fuel blocks. See Fig. 22. The fuel and fertile particles consist of uranium oxycarbide and thorium oxide kernels (about 350 micrometers in diameter), respectively, first coated with a porous graphite buffer, followed by three successive layers of pyrolytic carbon, silicon carbide, and pyrolytic carbon. The outer diameter of the coated particles is about 800 micrometers for the uranium particles and slightly larger for the thorium particles. The coatings essentially form a high-temperature refractory-based pressure vessel around each fuel/fertile kernel for the purpose of retaining fission products. Extensive operation and test data on these particles confirm that essentially no failure of the refractory coating occurs if the fuel is maintained below 3272 °F (1800 °C). As previously mentioned, the reactor design parameters were selected such that passive core afterheat removal will prevent this temperature from being reached during any credible accident condition. Thus, the modular HTGR is inherently, passively safe.

The reference plant arrangement features four 350 MWt HTGR modules supplying steam to two turbine generators that produce a net electrical output of 558 MWe at a new plant efficiency of 39.9%. Each reactor module is housed in an independent, vertical, cylindrical, concrete confinement, which is fully embedded in the earth. The four reactor modules share common systems for fuel handling, helium processing, and other essential services. A common control room is used to operate all four reactors

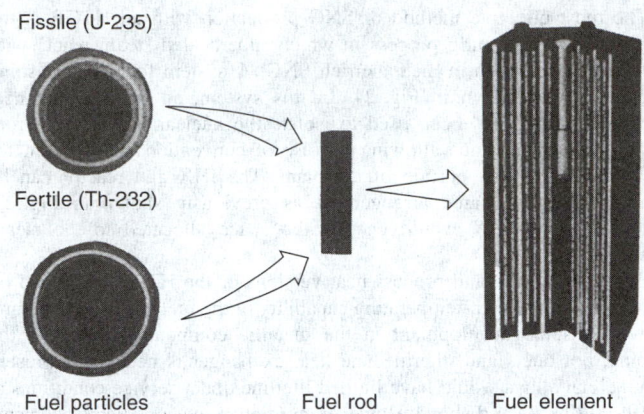

Fig. 22. Fuel components of HTGR. (*GA Technologies.*)

and the turbine plant. Operation of the entire complex is completely automated. Human operator actions are not required for control during power production or to assure safe shutdown during hazardous conditions.

Potential for HTGRs

The HTGR's use of ceramic-coated fuel and graphite moderator enables operation of the reactor core at much higher temperatures than are required for electric power production via a steam Rankine cycle. Core outlet helium temperatures in excess of 1800 °F (982 °C) are achievable without impacting the integrity of the HTGR fuel or core structures. This very high temperature capability opens up the possibility for more efficient methods of power production or direct use of high-temperature thermal energy for process heat applications. See also **Cogeneration (Electricity and Thermal Energy)**.

An attractive electric power producing concept for the 21st century is the HTGR gas-turbine (HTGR-GT) which has the potential for thermal efficiencies over 50% by taking advantage of the high HTGR core outlet temperature. Although several variations in system configuration are possible, the most straightforward HTGR-GT concept is the direct Brayton cycle, illustrated in Fig. 23. This cycle is closed and the helium primary coolant is also the working fluid for the power conversion system. The entire heat source and power conversion system of an HTGR-GT, which is capable of a net electric output of 170 MWe, can be enclosed within the two pressure vessels and cross-duct arrangement similar to the modular steam cycle HTGR previously shown in Fig. 20. The recuperator and precooler would occupy the same space as the steam generator and the turbo-compressor would replace the circulator.

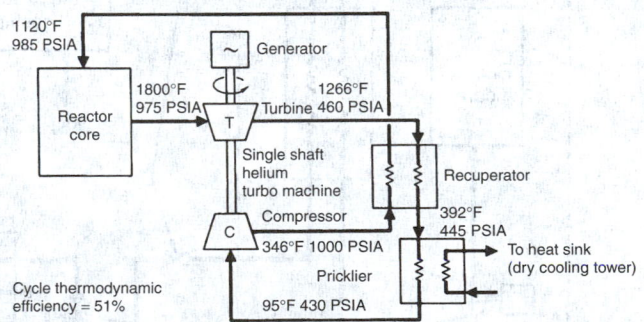

Fig. 23. HTGR gas turbine system with exceptional cycle thermodynamic efficiency. (*GA Technologies.*)

Perhaps the most significant potential use of the HTGR's high temperature capability is the production of synthetic fuels from coal. The HTGR is an important option for supplanting the current consumption of oil and natural gas with synthetic natural gas (SNG). The HTGR can supply the necessary energy for this endothermic process and, therefore, increase the recoverable energy in the SNG product by at least 60% over traditional coal combustion processes.

The most effective method of SNG production with an HTGR is the steam-carbon reforming process in which superheated steam reacts with pulverized coal to form methane-rich SNG. A system for accomplishing this process is shown in Fig. 24. In this system, an intermediate heat exchanger (IHX) has been used to isolate the nuclear heat source from the process steam, thus allowing the use of conventional equipment for the SNG production portion of the plant. The IHX and reactor can be configured in the same arrangement as previously shown in Fig. 23, except that the IHX would occupy the space allocated to the steam generator.

Both gas turbine and process heat versions of the HTGR are based on the demonstrated high-temperature capability of the fuel and core structure. However, some development in the metallic components, such as the turbine, hot ducts and intermediate heat exchanger is necessary. Present commercial alloys would have limited lifetime under service conditions at 1650 °F (899 °C) and above. However, currently envisioned advancements in ceramics and carbon-carbon composites indicate that high-temperature nonmetallic substitutes for metallic alloys will soon be available. These materials advances are the key to making future application of the HTGR a reality.

Heavy Water Reactor (HWR)

During the atomic energy developments in the World War II years and for a period thereafter, the United States, the United Kingdom, and Canada cooperated closely and many of the nuclear scientists of these countries appreciated the merits of heavy water as a moderator. Each of these countries pursued some development of HWRs for commercial power generation, but at different paces and dedication. Only Canada took to the HWR for commercial power generation. See Figs. 25 and 26.

One of the first high-priority nuclear applications of the United States was for naval propulsion. Because of a very tight minimal physical size criterion, LWRs offered advantages over the HWR. The United Kingdom placed emphasis on the production of plutonium for weapons programs. Gas graphite reactors were a reasonable early choice. When commercial nuclear power was recognized as a needed source of energy, because of the accumulated operating experience it was reasonable to adapt the reactors which had already been developed in the United States and the United Kingdom for military purposes. Long-term savings at that time was not a major criterion.

In the postwar years, hydroelectric power amply met a large portion of Canada's power needs and its abundance made nuclear power quite noncompetitive. Canadian utility operators were used to capital-intensive plants combined with low operating costs. In analyzing the prospects for nuclear power in Canada, utility planners and engineers placed a significant value on low fueling costs, and thus neutron economy was paramount. Therefore, when commercial nuclear power studies commenced in Canada in the mid-1950s, the choice was the HWR. This choice was bolstered by experience with, and knowledge about, heavy water production plants gained when Canadian scientists were trading experience from the heavy water-moderated NRX research reactor when the United States was developing the Savannah River production facility, which was dismantled in the early 1990s.

Principal advantages of heavy-water reactors are: (1) more efficient absorption of the energy released in the reactor, (2) greater fuel "burn-up" and, therefore, fuel economy, and (3) refueling can take place while the reactor is in service.

The first Canadian nuclear power demonstration (NDP) reactor was of 20-MWe capacity and was configured similarly to a light water reactor. Because of limited facilities for making large pressure vessels, a modular pressure-tube design of the configuration shown in Fig. 24 was investigated. Zircaloy-2 had become available at that time for fabrication of the pressure tubes. Hence the NPD was constructed using Zircaloy as cladding material and uranium dioxide as fuel. The NPD reactor has been in operation since 1962. The CANDU (Canada Deuterium Uranium) power reactors, including the NPD, number over twelve facilities. See Figs. 27, 28, and 29.

CANDU power reactors are characterized by the combination of heavy water as moderator and pressure tubes to contain the fuel and coolant. Their excellent neutron economy provides the simplicity and low costs of once-through natural uranium cycling. Future benefits include the prospect of a

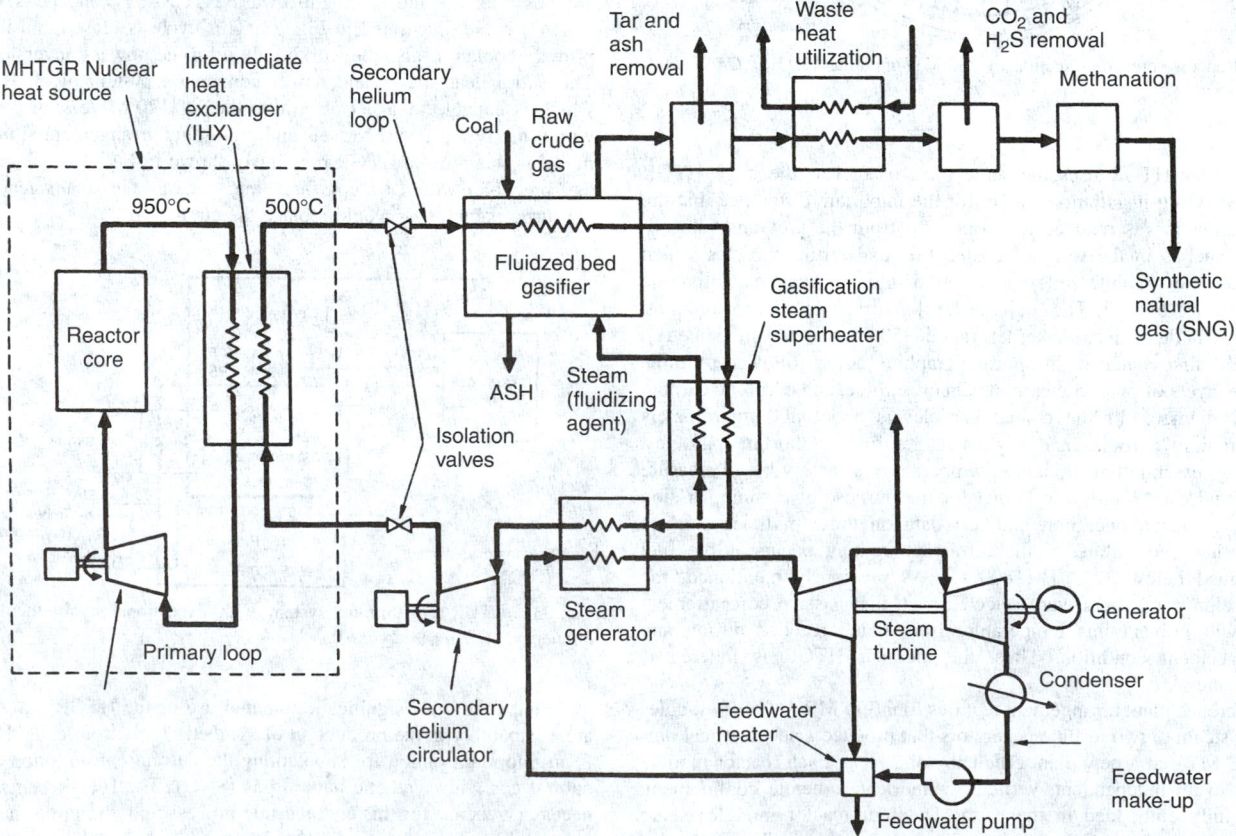

Fig. 24. Advanced process heat HTGR with intermediate loop for producing synthetic natural gas (SNG) by steam gasification of coal. (*GA Technologies.*)

Fig. 25. Series of towers comprising part of the heavy water production plant at Ontario Hydro's Bruce nuclear power complex near Tiverton on the shores of Lake Huron. Heavy water is a clear, colorless liquid that looks and tastes like ordinary water. It occurs naturally in ordinary water in the proportion of approximately one part heavy water to 7000 parts of ordinary water. While ordinary water is a combination of hydrogen and oxygen (H_2O), heavy water (D_2O) is made of up of deuterium — a form, or isotope, of hydrogen — and oxygen. Deuterium is heavier than hydrogen in that it has an extra neutron in its atomic nucleus, so heavy water weighs about 10% more than ordinary water. It also has different freezing and boiling points. It is the extra neutron that makes heavy water more suitable than ordinary water for use in CANDU nuclear reactors as both a moderator and a heat transport medium. (*Ontario Hydro, Toronto, Ontario, Canada.*)

near-breeder thorium fuel cycle to provide security of fuel supply without the need to develop a new reactor, such as the fast breeder. The CANDU system is appropriate for countries of intermediate economic and industrial capacity, such as Canada. Producing heavy water is fundamentally simpler than enriching uranium and commercial heavy water plants have been built in smaller sizes than would be possible for uranium enrichment plants. Although Canada has rather generous supplies and reserves of uranium, there is increasing pressure on Canada to export uranium, a pressure that will probably intensify further if the introduction of fast breeder reactors in other countries is delayed. The current simplest possible fuel cycle for the CANDUs, which is not dependent upon fuel reprocessing, will probably be retained in Canada so long as uranium remains plentiful and comparatively economical. However, for future planning, research to date has indicated that a "self-sufficient thorium cycle" may be practicable in the CANDUs with minimal modification. It has been observed that, at equilibrium, the thorium cycle would require no further uranium. Only small quantities of thorium, which is more abundant than uranium, would be required. Also of interest for the future is *electronuclear breeding*, i.e., the use of electric power to convert fertile to fissile material for neutron economy.

Fast Breeder Reactors

The fast breeder reactor derives its name from its ability to breed, that is, to create more fissionable material than it consumes. This ability stems from the fact that neutrons travel faster than they do in a thermal reactor. The breeding process depends, in part, upon the neutrons maintaining a high speed, or high energy. If their speed or energy is allowed to degrade as occurs in thermal reactors, the number of neutrons produced per absorption in uranium or plutonium decreases. Furthermore, at lower velocities, neutrons tend to be captured in various structural materials of the reactor, and this further reduces the breeding potential. It is important, therefore, in fast reactors to keep the velocity of the neutrons high. Water, which is used as a coolant in some thermal reactors, tends to slow the neutrons down and thus prevent efficient breeding. Therefore, it is necessary to use a coolant that does not slow the neutrons or capture them as they travel through the coolant. Liquid sodium and gaseous helium under pressure are the two principal coolants used to date.

Fuel Cycle Considerations. Approximately 99.3% of uranium as it is found in nature is the isotope ^{238}U and 0.7% is ^{235}U. Uranium-235 is a fissile isotope, that is, if it is struck by a neutron it will split; this fission yields on the average approximately two neutrons and 200 MeV of energy. This amount of energy corresponds to approximately 78 million Btu for every gram of uranium which fissions (3.5×10^{10} Btu/pound) (1.95 Calories/kilogram). Most reactors today are largely dependent upon ^{235}U for their energy. However, some of the neutrons released in fission of ^{235}U also are absorbed in nonfissionable ^{238}U. As the ^{238}U absorbs a neutron, it is transformed into fissionable ^{239}Pu (plutonium). Thus, while the reactor is sustaining the fission process and thereby creating energy, it is

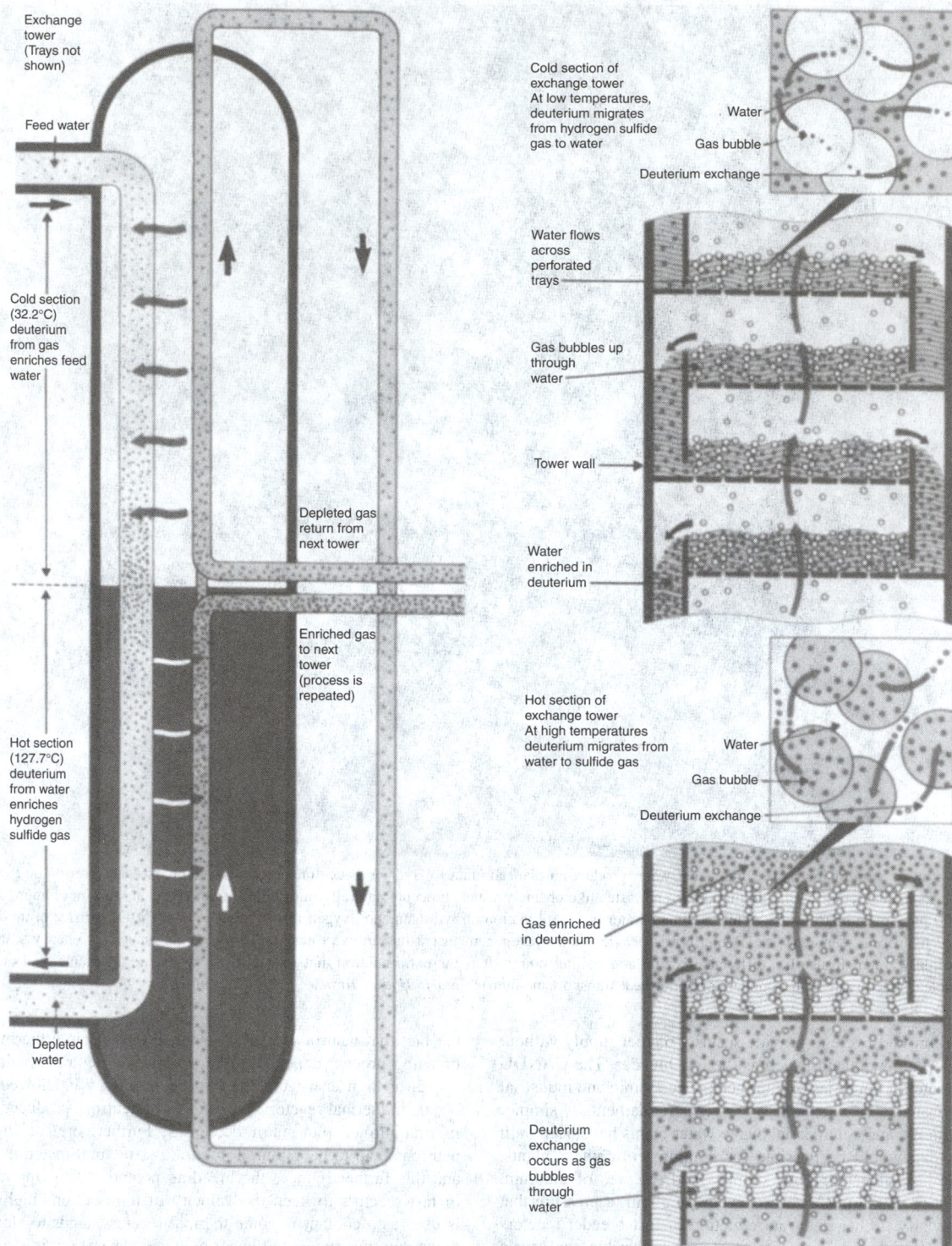

Fig. 26. The production of heavy water is based upon the behavior of deuterium in a mixture of water and hydrogen sulfide. When liquid H_2O and gaseous H_2S are thoroughly mixed, the deuterium atoms exchange freely between the gas and the liquid. At high temperatures, the deuterium atoms tend to migrate toward the gas, while they concentrate in the liquid at lower temperatures. In the first and second stages of production, the towers of a heavy water plant are operated with the top section cold and the lower section hot. Hydrogen sulfide gas is circulated from bottom to top and water is circulated from top to bottom through the tower. In the cold section, the deuterium atoms move toward the water and are carried downward, while in the hot section, they move toward the gas and are carried upward. The result is that both gas and liquid are enriched in deuterium at the middle of the tower. A series of perforated trays are used to promote mixing between the gas and water in the towers. A portion of the H_2S gas, enriched in deuterium, is removed from the tower at the juncture of the hot and cold sections and is fed to a similar tower for the second stage of enrichment.

The first stage of the process enriches the gas from 0.015% deuterium to 0.07%. A second stage further enriches it to about 0.35%. Again, the enriched gas is fed forward to a third stage. The product from this third stage, now in the range of 10 to 30% heavy water, is sent to a distillation unit for finishing to 99.75% purity "reactor-grade" heavy water. Because the production of heavy water uses a toxic gas, H_2S, safety is a top priority at heavy water plants. H_2S is a colorless gas, slightly heavier than air. To safely expel H_2S from the system, it is directed to a flare tower where it is burned off. Initially, each of Ontario Hydro's reactors requires about 800 megagrams, or one year's production from a heavy water plant. After that, less than 1% of the heavy water is lost and has to be replenished each year. (*Ontario Hydro, Toronto, Ontario, Canada.*)

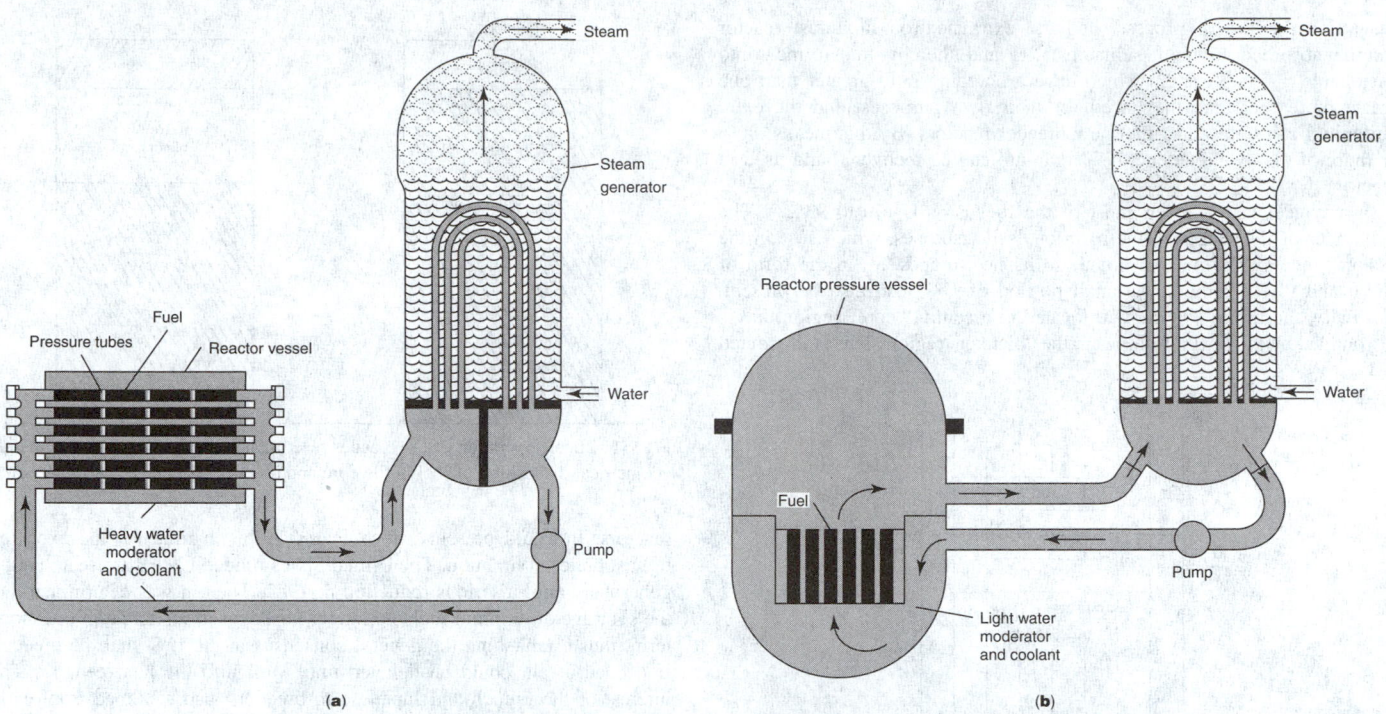

Fig. 27.   Comparison of heavy water reactor (**a**) with light water reactor (**b**). (*After Robertson, Atomic Energy of Canada Limited.*)

Fig. 28.   Pickering (Ontario) "A" generating station's initial performance was outstanding and the station was hailed as a major Canadian technological achievement at the time of its commissioning in February 1971. It reached full power in 3 months, well ahead of schedule. The final Unit 4 went on line in May 1973. At full output (2,160,000 kW), Pickering "A" generates enough power to supply more than 1.5 million homes. In 1974, construction was begun on a twin station, Pickering "B," also shown in this view. The "B" station has the same capacity as its forerunner and all four units became operational in 1986. (*Ontario Hydro, Toronto, Ontario, Canada.*)

Fig. 29.   Darlington, Ontario, nuclear generating station, one of the largest energy projects undertaken in North America. The plant is shown in late stages of construction in 1986. It is located on the shores of Lake Ontario, 5 km southwest of Bowmanville in the Town of Newcastle. Currently nuclear power plants provide one-third of Ontario's electricity needs (the other two-thirds comes from hydro installations). Selected because of its proximity to the residential, industrial, and commercial energy markets of Ontario, the 1200-acre (485-hectare) site has good transportation access, an abundant supply of cooling water from Lake Ontario, relative isolation, and excellent bedrock for station foundations. When the four-unit station is completed it will provide 3,524,000 KW of electricity, enough to serve a city of 3 million people. Electricity from the four units (each 881,000 KW net) will be fed into the Ontario Hydro electricity grid system through 500,000 V transmission lines already crossing the site. The Darlington generating station was approved by the provincial government in July 1977. Site preparation began in 1978 and by 1981, the first concrete was poured for the station's foundations. Construction activity peaked in 1986–1987 with approximately 6800 workers on site. (*Ontario Hydro, Toronto, Ontario, Canada.*)

also generating fresh fuel, which can later be used to create more energy. Unfortunately, this is an inefficient process in present thermal reactors where the neutron velocity is established by the temperature, or thermal energy, so only limited amounts of additional energy are made available by transformation of ^{238}U into ^{239}Pu.

The fast breeder reactor makes possible the recovery of most of the available energy in uranium. This occurs because during fission in the fast breeder nearly three neutrons are released for every neutron absorbed as compared with only approximately two neutrons in a thermal reactor. On the average, between one and two neutrons are necessary for sustaining the fission process, and the extra neutron in a fast reactor can be absorbed in nonfissionable ^{238}U and thereby transformed into fissionable ^{239}Pu. Reactors which have a breeding ratio greater than one create more fuel than they need for their own purposes, and the extra plutonium can be used to fuel new breeder reactors. By this means, 80% or more of the available energy in uranium can be recovered and used in reactors.

In a typical fast breeder, most of the fuel is ^{238}U (90 to 93%). The remainder of the fuel is in the form of fissile isotopes, which sustain the fission process. The majority of these fissile isotopes are in the form of ^{239}Pu and ^{241}Pu, although a small portion of ^{235}U can also be present. Normally, the fissile isotopes are located in a central "core" region that is surrounded by the fertile isotopes in the "blanket" region. This is illustrated in Fig. 30.

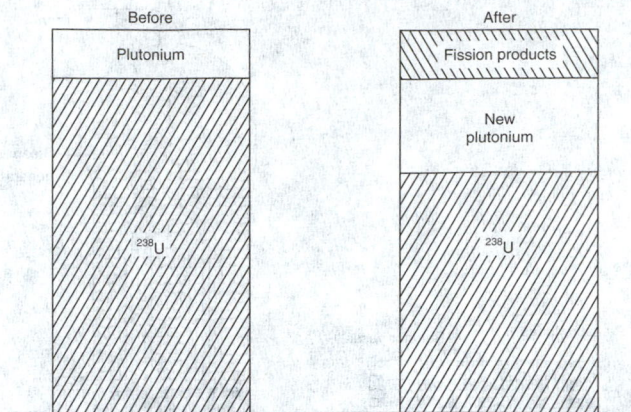

Fig. 31.   Basic operation of the breeder reactor. The illustration does not include geometrical disposition of fuel in the core and blanket system. (*General Electric.*)

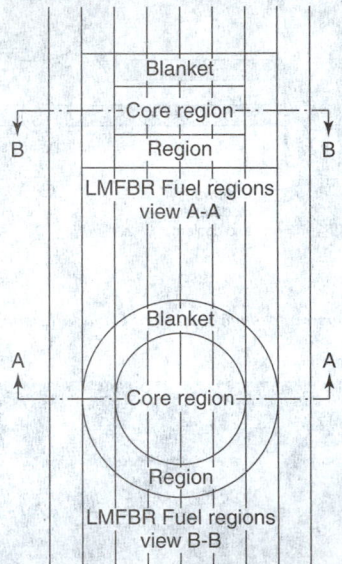

Fig. 30.   Liquid-metal fast breeder reactor core and blanket arrangement.

When the fuel is initially loaded into the reactor, the core region will typically contain from 10 to 15% fissile isotopes with the remainder being ^{238}U. Essentially all of the blanket will be ^{238}U. As energy is extracted from the fissile isotopes, they become depleted (the initial plutonium is gradually used up). However, in a breeder reactor, new plutonium will be formed in the core and blanket regions faster than it is consumed. Additionally, undesirable fission products are formed which must ultimately be removed. This process is schematically illustrated in Fig. 31. The "before" chart represents the new fuel condition and the "after" chart corresponds to the situation when the fuel is removed for reprocessing. Typically, the fuel removed for reprocessing will contain from 1 to 3% new plutonium. It is in this manner that the fast breeder can recover from 80 to 90% of the available energy in uranium resources. Most present reactors require some enrichment of the ^{235}U isotope used to fuel them. This enrichment process requires a plant, which, in turn, uses large amounts of electrical energy. Because the fast breeder converts the fertile isotope ^{238}U into the fissile isotope ^{239}Pu, no enrichment plant is necessary. The fast breeder serves as its own enrichment plant. The need for electricity for supplemental uses in the fuel cycle process is thus reduced.

Fast Breeder Reactors in Perspective

Of the several fundamental ways to use nuclear fission reactions to generate electric power, the fast breeder reactor (notably the liquid-metal-cooled fast breeder reactor, LMFBR) probably has the most checkered history. The fast breeder reactor received its early impetus when there was serious concern over what appeared to be a limited supply of uranium and consequent increasing prices of uranium fuels. With the fast breeder concept offering up to a 100-fold increase in the utilization efficiency of uranium, it appeared to be the logical replacement (second generation) for light-water

reactors. It would present a technical solution in time for the expected tight supplies of uranium. The United States funded LMFBR research quite generously until a serious reduction in 1982. Then, it was determined that the shortage of uranium fuels no longer posed a serious threat in the short-term, thus establishing the general consensus in the U.S. that the breeder, if needed at all, could be delayed until well into the next century. The interest of the French and Japanese in breeders also stemmed from early concerns with uranium shortages and prices, but was much more serious because these countries are extremely uranium poor and, further, these countries have fewer energy options. For example, the cost of coal ranges from 1.6 to 2.5 times the cost of coal in the United States. Continued progress in fast breeder development, particularly in France, also has been accelerated by a very heavy past investment in the technology coupled with a desire to be fully self-contained as regards the generation of electric power.

Particularly, in the United States, because of its several energy fuel options and its current "rethinking of nuclear power," the principal nuclear power research targets no longer include uranium supplies. Rather, the targets are the lowering of capital costs for existing light-water reactor technology, shortening the plant construction and licensing lead times, achieving higher plant availability (essentially eliminating long power outages), and increasing plant safety. From this "rethinking" process, the U.S. Congress suspended funding for the Clinch River Breeder Reactor.

It is interesting to note that, as of the early 1990s, most authorities were not quite willing to "forget" the fast breeder altogether. The differences in views essentially reside in timing. The fast breeder, even though costly to build today, does offer several temptations. For example, experience gained from the Experimental Breeder Reactor (Idaho Falls, ID) which commenced operation in 1964 is reported to operate better now than when it was first built, showing no evidence of corrosion. It is suggested that if such experience were extrapolated the useful life of an LMFBR could be between 100 and 150 years! Weinberg (1986 reference listed) recalls, "that one of Newcomen's original steam engines, built in the mid-18th century, continued to operate until 1918." It does appear that the economy of the LMFBR is somewhat analogous to large hydroelectric projects that require very large investments of capital, but coupled with low operating costs, provide really cheap electric power once they are fully amortized. However, one usually must wait a generation before this situation occurs. Weinberg also observes, "Because the breeder requires little, if any, mining of uranium, its environmental impact is much smaller, at least at the front end of the fuel cycle, than is the impact of the LWR. The roughly 300,000 tons of depleted uranium stored outside the diffusion plants, if used in breeders, could fuel our (U.S.) entire electric system for centuries!"

Davis (1984 reference listed) observes that there is sufficient know-how today to build and operate fast breeder reactors with confidence of their safety and reliability, as exemplified with large units in France. Davis further suggests that the principal problems remaining in fast breeder technology include: (1) a reduction in overall costs to make the fast breeder competitive with coal and the light-water reactors — an estimated reduction factor of 1.5 to 2; (2) assurance of adequate safeguard on the plutonium fuel cycle; (3) more demonstrations of commercial-scale operations, such as the engineering scale-up of important system components—pumps and steam generators; (4) implementing a large, overall system which must include parallel facilities for fuel processing and refabrication; and (5) setting in place the reprocessing of light-water reactor fuel to provide the "start up" plutonium for the fast breeders.

Some technical successes with the fast breeder have occurred in France: (1) demonstrations of a positive breeding gain of 0.15 ± 0.04 in a complete breeder fuel cycle, and (2) demonstration at several laboratories of uranium and plutonium oxide fuel elements that can sustain more than 10% burnup of the original mixture of $^{238}U/^{239}Pu$ before the fuel has to be reconstituted, representing a tenfold improvement (Weinberg, 1986). It has also been observed that, compared with a number of so-called alternative fuels, the new Super-Phoenix (France) plant can produce electricity at costs that are markedly less than electrical energy from solar-powered photovoltaics, for which funding remains significant. Another proposed inexhaustible power source, nuclear fusion, still remains in an early stage of development. See **Fusion Power**.

LMFBR Design Principles. There are many design differences among the reactor designs, including: (1) primary coolant system arrangement; (2) refueling mechanism design and arrangement; (3) steam generator type and arrangement; (4) core support method; (5) structural material choices; and (6) safety features. Perhaps the most noticeable difference is that of the primary system arrangement. This difference is schematically illustrated in Figs. 32 and 33. The system of Fig. 32 corresponds to a "loop" or "piped" arrangement where the reactor, pumps, and intermediate heat exchangers are located separate from each other and piping carries the sodium from one point to the other. The "pool" or "tank" arrangement of Fig. 33 includes the reactor, intermediate heat exchangers and pumps in one large pool of sodium which is contained in a separate tank. Each concept has advantages and disadvantages. The pool concept is somewhat easier to design for certain hypothetical accident situations. The loop concept is easier to construct and to maintain.

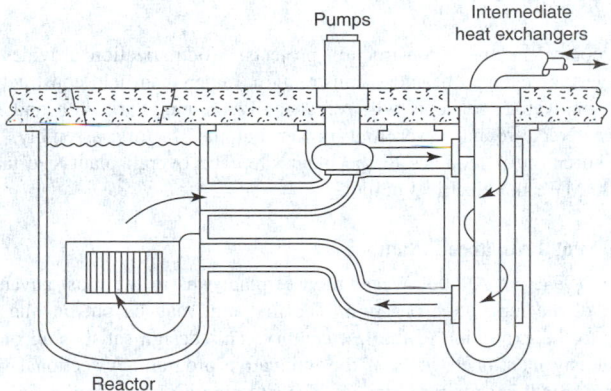

Fig. 32. Loop arrangement in the liquid-metal fast breeder reactor. (*General Electric.*)

The flow circuit for an LMFBR where two sodium circuits are included is shown schematically in Fig. 34. The reactor is cooled by the primary sodium, which becomes radioactive as it picks up heat in passing through the core or fueled region. In this particular arrangement, the sodium is heated to 560 °C and flows through pipes (schematically shown as a single line in the figure) to the intermediate heat exchangers. In the heat exchangers, the primary sodium transfers heat to the nonradioactive sodium. After being cooled to 393 °C in the heat exchangers, the primary sodium is pumped back into the reactor where it again repeats the circuit. The nonradioactive secondary sodium is circulated from the intermediate heat exchangers through steam generators where the heat from the sodium is transferred to water, which becomes superheated steam for use in the turbine. The cooled secondary sodium is pumped back through the intermediate heat exchangers where the process is repeated. Steam from the steam generators is used to turn the rotor of the turbine generator to

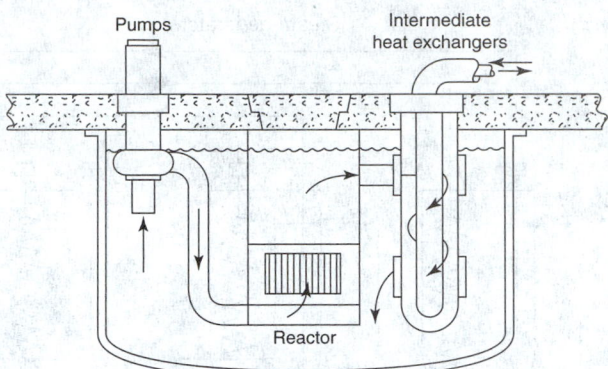

Fig. 33. Pool arrangement in the liquid-metal fast breeder reactor. (*General Electric.*)

generate electricity. In the arrangement shown, 1,200 MW of electricity are generated at a net overall efficiency of 39%. This relatively high efficiency is possible because of the excellent thermal characteristics of sodium.

Nuclear Power Innovations for 1995 and Beyond

Increasing concerns over the impact of fossil fuel-burning electric power generating plants on the environment, the accelerating demands for electric power, and a growing dependence on foreign oil supplies brought about a resurgence of interest in nuclear power reactor research during the mid-1980s. Advanced programs were established to create new plant designs that would reflect past experience to achieve an extremely high degree of operating safety, competitive construction, and operating costs, including the streamlining of the licensing process and consumer confidence in nuclear power technology. Thus, during the past decade, much private and government-sponsored research has been directed toward nuclear reactor research in three areas, namely, light water reactors (LWRs), high-temperature gas-cooled reactors (HTGRs), and liquid metal-cooled reactors (LMRs).

The innovative program commenced with an impressive nuclear power base—some 107 nuclear plants with full-power licenses operating in the United States and producing 18% of the nation's electricity—worldwide, 414 plants in 26 countries generate 298,000 megawatts of electricity (MWe), accounting for 16% of the world's generating capacity.

A survey of reactor developments of the three aforementioned types reveals a number of common generic technical features. These include passive stability, simplification, ruggedness, ease of operation, and modularity.

Overall goals for the innovative program can be summarized as follows:

1. Assured safety with features that minimize the negative consequences of human error, especially a reduction in the chance of occurrence of severe core damage by at least a factor of 10 less than former, contemporary designs.
2. Significantly simpler designs, with increased safety and performance margins in key operational parameters.
3. High reliability throughout a lifetime on the order of 60 years and an increase in plant availability to 85% or greater than the contemporary average of less than 70%.
4. Reduction in capital, operating, maintenance, and fuel costs to meet the economic competition with coal-burning generators. A reduction in construction time to the range of 3 to 5 years as compared with more than 10 years, which has been the experience with some of the later contemporary reactors.
5. A modular design that is standardized at a high quality level and thus predictably licensable.

Passive Stability. Passive design characteristics ensure core stability by eliminating the potential for a runaway chain reaction. In the innovative program, this has been a hallmark from the outset of the program. Passive characteristics are internal governors—that is, physical laws ensure that the reaction rate decreases instantaneously as the temperature of the coolant or fuel or the power of the reactor increases, without the need for external control devices.

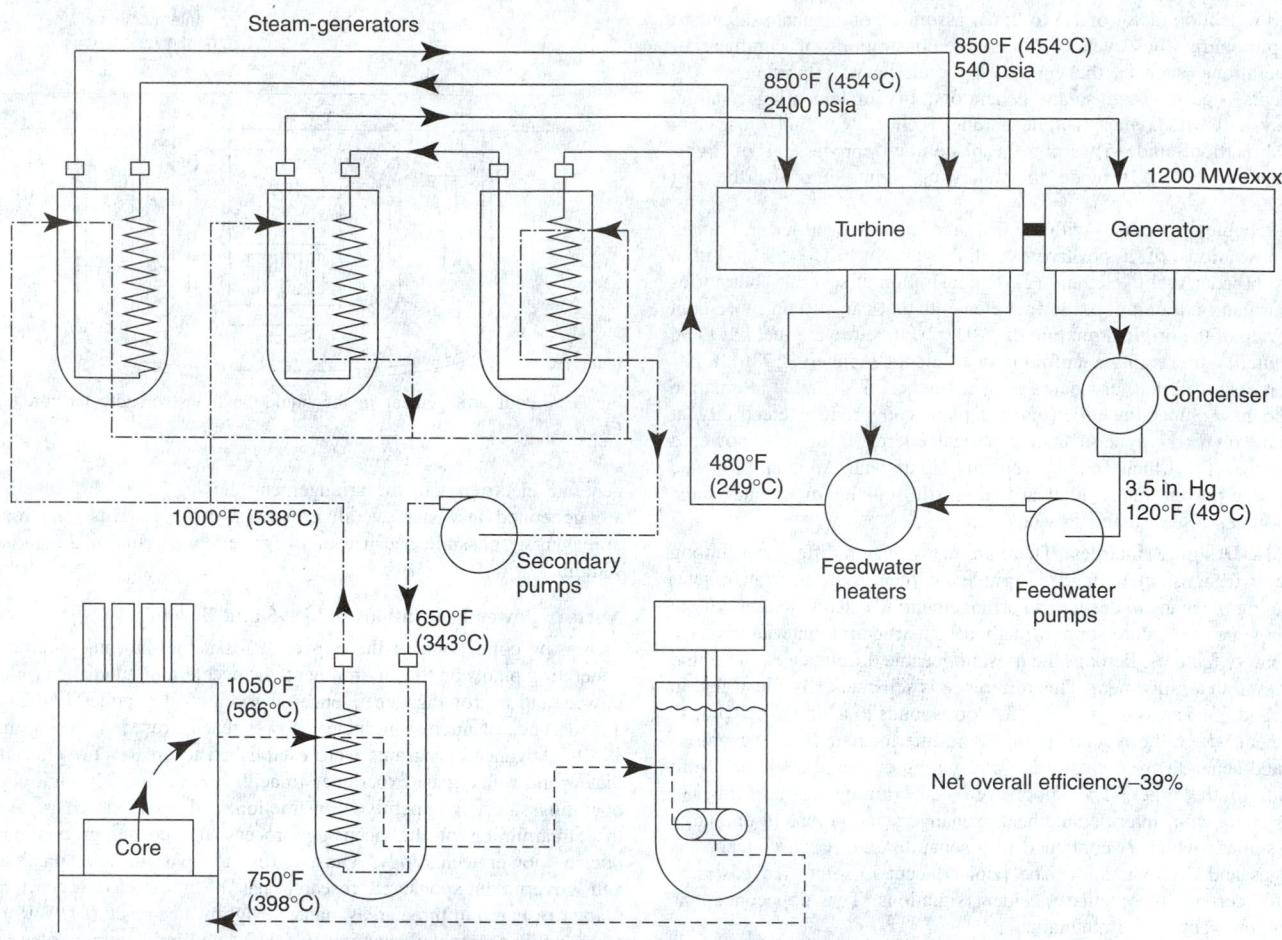

Fig. 34.   Liquid-metal fast breeder reactor flow circuit. (*General Electric*.)

Ruggedness. In some past designs, long-term reliability has been impaired by attempts to achieve the highest in efficiency and economic performance. In response to this negative experience, the margin in certain key performance parameters is being increased in order to lessen the burden on the equipment. By reducing power densities and coolant temperatures, higher reliability will be achieved over a longer lifetime. Past field experience has identified more effective methods of coolant chemistry control and materials selection, factors that will contribute to the long-lived reliability of the components of future systems. Greater emphasis is being given to the selection of proven, high-quality materials and components and on improved methods and quality control over assembly and construction.

Ease of Operation. Thorough investigation of the Three Mile Island incident some years ago showed that a lack of attention had been given to the human factor. The innovation program is addressing this problem in several ways. The computer and telecommunication revolution has made it more practical to use improved technology and human engineering methodologies to revamp the control room and the reactor instrumentation system. These improvements will make the plant easier to operate and provide the operator with a greatly increased amount and quality of information on plant conditions. Graphic displays, diagnostic aids, and expert systems are being developed for such advanced control rooms.

The other design goals complement the new technology to make the operator's task even easier. The passive safety features substantially extend the response time required of the operators in an emergency condition. The margin being built into the systems provides broader normal operating regimes and longer response times for operator action. Greater emphasis is being placed on simplification of operating procedures.

Modularity. Economic competitiveness requires that the construction time be shortened dramatically. Modular construction techniques are a key contributor to achieving this goal and are a proven approach to

cost control in major construction projects. Modularization provides for a larger percentage of factory construction, rather than field construction. New innovative concepts will rely heavily on modularization and will be centered around lower unit power outputs, factory assembly, and transportation of modules to the plant site. The overall plant size target is 600 MWe net electrical output.

The AP600 Advanced Plant

As of 1994, the AP600 (Westinghouse) plant was in the most advanced stage of the innovative program. The first unit may be put on line just prior to the expiration of the last century. This design satisfies all of the previously mentioned goals of the innovative program. A sectional view of the AP600 is shown in Fig. 35. The site plan is shown in Fig. 36.

AP600 Passive Safety System Details. These features reduce operator responsibilities and add an extra margin of safety over contemporary PWR designs. See Fig. 37 on p. 2501. Large volumes of water stored in the containment eliminate the need for operator action to assure make-up water, either for small leaks that may occur during normal operation or for a major loss of coolant accident (LOCA). A passive plant is a system that assures public safety even if the operators fail to act.

The passive residual heat removal heat exchangers remove core decay heat if steam generator heat removal is not available. Passive residual heat removal heat exchangers in the in-containment refueling water storage tanks are connected to the reactor coolant system (RCS) piping, forming a full-pressure, closed, natural circulation cooling loop.

Two core make-up tanks provide borated make-up water whenever the normal make-up system is unavailable. The tanks are located above the reactor coolant system loop piping and kept at system pressure by steam lines from the pressurizer. These tanks function at any system pressure, using only gravity as a motive force. If the reactor protection system detects a need for make-up water, core make-up tanks discharge and isolation valves open automatically, allowing the tanks to drain into the reactor vessel.

Two accumulators provide the high make-up flows initially required by a large LOCA. These tanks contain 1700 cubic feet (about 48 cubic meters) of borated water pressurized with 300 cubic feet (about 8.5 cubic meters) of nitrogen at 700 psi (4.8×10^6 Pa). The accumulators are isolated from the RCS by check valves. Each accumulator is paired with a core make-up tank, the pair sharing an injection line to the

Fig. 35. Sectional view of the 600 MWe pressurized water reactor (PWR) expected to be operational by 1995. Considered the PWR of the future, the plant is designed for a minimum useful life span of 60 years and features numerous economic and safety features, including passive systems for ultimate protection. (*Joint project of Westinghouse, the Electric Power Research Institute, and the U.S. Department of Energy.*)

Legend:

1. Fuel Handling Area
2. Concrete Shield Building
3. Steel Containment
4. Passive Containment Cooling Water Tank
5. Passive Containment Cooling Air Baffles
6. Passive Containment Cooling Air Inlets
7. Equipment Hatches (2)
8. Personnel Hatches (2)
9. Core Make-up Tanks (2)
10. Steam Generators (2)
11. Reactor Coolant Pumps (4)
12. Integrated Head Package
13. Reactor Vessel
14. Pressurizer
15. Depressurization Valve Module Location
16. Passive Residual Heat Removal Heat Exchangers
17. Refueling Water Storage Tank
18. Technical Support Center
19. Main Control Room
20. Integrated Protection Cabinets

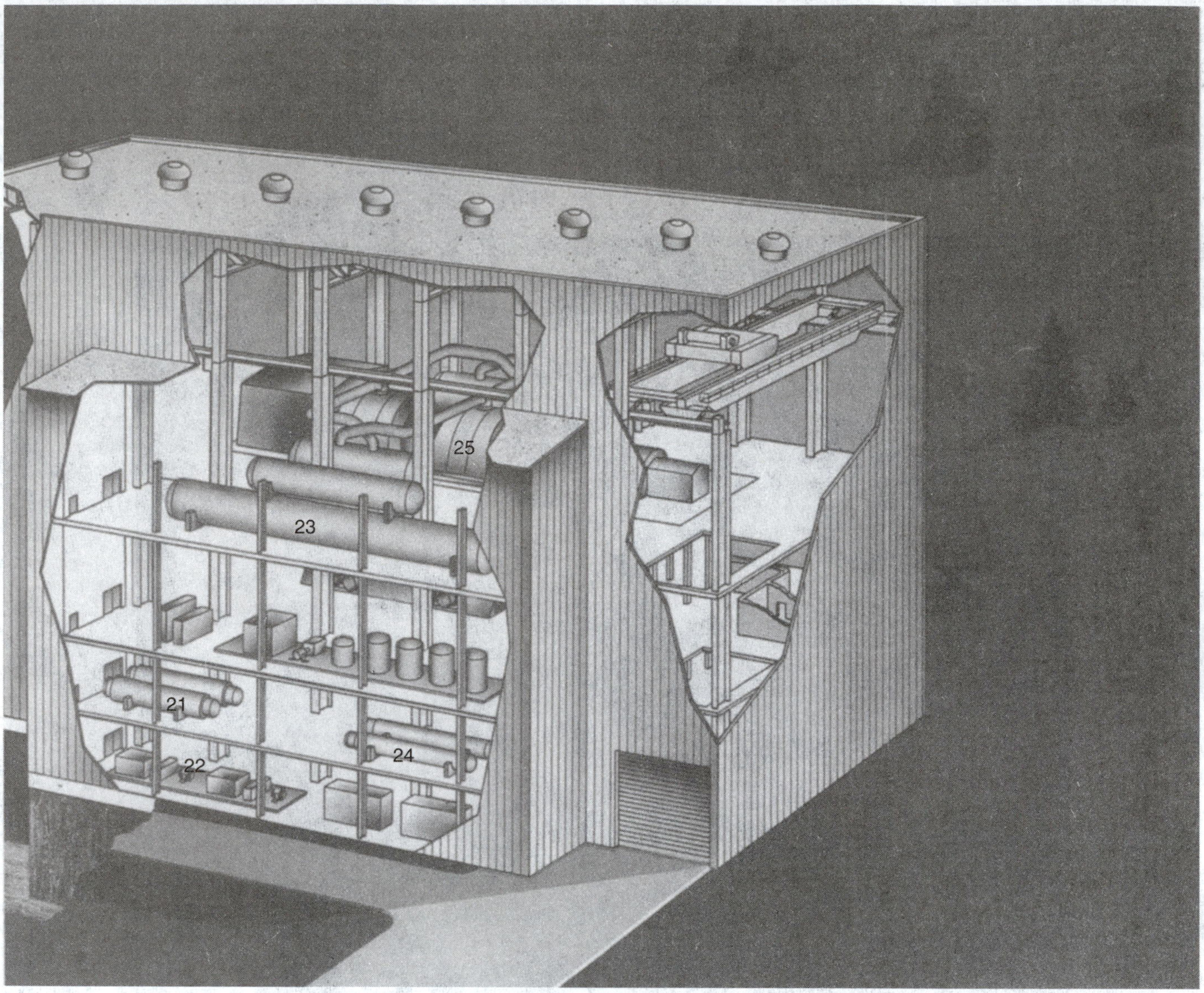

Fig. 35. (*Continued*)

21. High Pressure Feedwater Heaters
22. Feedwater Pumps
23. Deaerator
24. Low Pressure Feedwater Heaters

25. Turbine/Generator
26. Spargers (2)
27. Accumulators (2)
28. Main Steam Line

29. Feedwater Line
30. Passive Containment Cooling Air Flow

reactor vessel downcomer. This ensures that at least one accumulator/core make-up tank pair would be available following an injection line LOCA.

The in-containment refueling water storage tank provides 500,000 gallons (about 1900 cubic meters) of water with a gravity head above the core. This water inventory is sufficient to flood the containment above the level of the reactor core and provide decay heat removal by natural circulation.

The automatic depressurization system depressurizes the RCS if core make-up tank level is low. Depressurization allows gravity injection from the in-containment refueling water storage tank, which is at atmospheric pressure. To ensure that the automatic depressurization system works when needed, while minimizing the consequences of spurious valve operation, the system provides phased depressurization with two redundant sets of valves connected to the pressurizer. The discharge is sparged into the in-containment refueling water storage tank. The automatic depressurization system valves are arranged in three stages to reduce peak flow rates. A fourth depressurization stage is provided directly on the RCS hot leg.

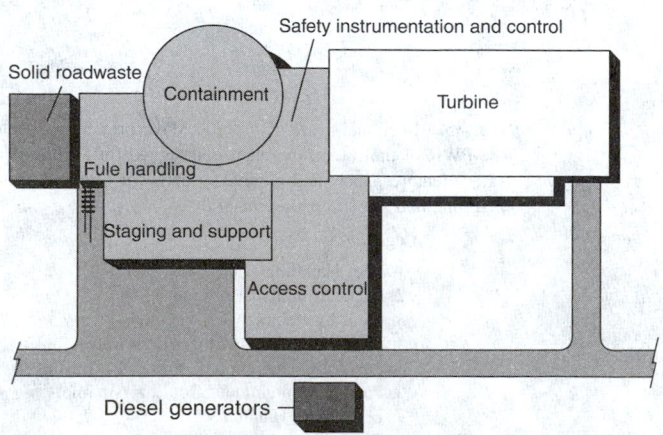

Fig. 36. Site plan for the advanced pressurized water reactor (PWR). (*Westinghouse Electric Corporation, Energy Systems.*)

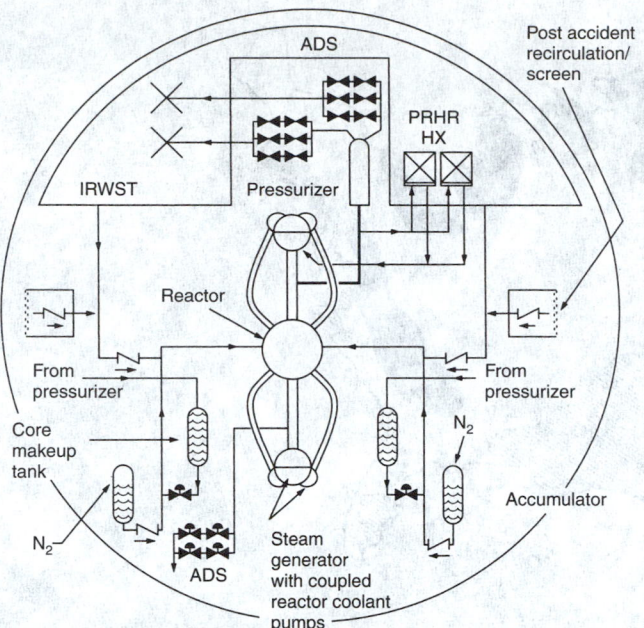

Fig. 37. Schematic representation of the in-containment passive safety injection system (PSIS). IRWST = in-containment refueling water storage tank. PRHR-HX = passive residual heat removal heat exchanger. ADS = automatic depressurization system (four stages). (*Westinghouse.*)

The passive containment cooling system provides the safety grade ultimate heat sink that prevents the containment shell from exceeding its design pressure of 45 psig [3.1 × 10^5 Pa (g)]. The system uses natural air circulation between the steel containment shell and the concrete shield building. During postulated accidents, air cooling is enhanced by draining water onto the steel containment shell. The water is provided by gravity from a 350,000-gallon (about 1300 cubic meters) annular tank in the roof of the shield building. This tank has sufficient water to provide three days of cooling.

RADIOACTIVE WASTES

Decisions pertaining to the location of radioactive waste sites mainly derive from political and sociological sources. A condensed overview of the technical aspects is given here.

The radioactive wastes associated with nuclear reactors fall into two categories: (1) *commercial wastes*—the result of operating nuclear-powered electric generating facilities; and (2) *military wastes*—the result of reactor operations associated with weapons manufacture. Because the fuel in plutonium production reactors, as required by weapons, is irradiated less than the fuel in commercial power reactors, the military wastes contain fewer fission products and thus are not as active radiologically or thermally. They are nevertheless hazardous and require careful disposal.

Nuclear power plants use fuel rods with a life span of about three years. Each year, roughly one-third of spent fuel rods are removed and stored in cooling basins, either at the reactor site or elsewhere. Typical modern nuclear power plants discharge about 30 tons of the spent fuel per reactor per year. Comparatively little of the radioactive wastes, as is currently reliably known worldwide, has been processed for return to the fuel cycle. Actually, fuel reprocessing causes a net increase in the volume of radioactive wastes, but, as in the case of military wastes, they are less hazardous in the long term. Nevertheless, the wastes from reprocessing also must be disposed of with great care.

Spent fuel from a reactor contains unused uranium as well as plutonium-239 which has been created by bombardment of neutrons during the fission process. Mixed with these useful materials are other highly radioactive and hazardous fission products, such as cesium-137 and strontium-90. Since reprocessed fuels contain plutonium, well suited for making nuclear weapons, concern has been expressed over the possible capture of some of this material by agents or terrorists operating on behalf of unfriendly governments that do not have a nuclear weapons capability.

Categories of Wastes by Content

In addition to the two source categories previously mentioned, radioactive wastes are classified in accordance with their content:

High-level wastes contain 99.9% of the nonvolatile fission products, 0.5% of the uranium and plutonium, and all the actinides formed by transmutation of the uranium and plutonium in the reactors. Among the actinides are neptunium and americium. High-level wastes are either the aqueous wastes resulting from reprocessing; or the spent-fuel rods to be disposed of in the absence of reprocessing.

Cladding wastes are comprised of solid fragments of Zircaloy and stainless steel cladding (tube in which the fuel is placed) and other structural elements of the fuel assemblies remaining after the final cores have been dissolved.

Low-level transuranic wastes are solid or solidified materials which contain plutonium or other long-lived alpha-particle emitters in known or suspected concentrations higher than 10 nanoCuries per gram and external radiation levels after packaging sufficiently low to allow direct handling.

Intermediate level transuranic wastes are solids or solidified materials that contain long-lived alpha-particle emitters at concentrations greater than 10 nanoCuries per gram and which have, after packaging, typical surface dose rates between 10 and 1000 mrems/hour due to fission product contamination.

Nontransuranic low-level wastes are diverse materials which are contaminated with low levels of beta- and gamma-emitting isotopes, but which contain less than 10 nanoCuries of long-lived alpha activity per gram.

Permanent Disposal Methodologies

Because of many doubts among various authorities pertaining to the permanent geologic depositories, as previously mentioned, considerable effort continues as regards semipermanent and permanent types of depositories. Many methodologies involve the so-called "sequence of barriers" approach. The first barrier is the form in which radioactive materials are embedded—vitrification, calcination, etc. The requirements for the first barrier are that it not be corrosive and possess excellent thermal stability and mechanical integrity. Wastes generate much heat during their initial decade of confinement. This affects decisions as regards the wasteform and the second barrier, the frequently mentioned canister which encapsulates the wasteform. The principal function served by the canister is protection of the material during the collection and transportation (to geologic site) phases. The canisters also should provide excellent protection of their contents for a minimum of 50 years, just in case it is desired to retrieve the wastes at some future date. Canisters must resist corrosive chemicals, they must withstand extremely high radiation fluxes caused by fission-product decay and the heat generated by the decaying wastes. It is interesting to note that an unprotected stainless steel canister will not resist structural deterioration arising from salt brines for that long a period. Provision must be made for cooling the canisters, either by air or water. The canisters should be designed to permit maximum heat transfer and, currently, the cylinder and annulus configurations are preferred. A third barrier would be the geologic site itself, obviously impervious to water penetration and in a seismically stable location. To fulfill all the foregoing requirements (and more), consideration is being given to phasing the waste storage procedure, possibly storing the canisters for water cooling during the first few years, after which air cooling would suffice.

The predominant preference appears to be the underground depository option. Along these lines, it is interesting to review the rock formations in the United States, as shown in Fig. 38.

Commenting briefly on other proposed methodologies, as shown in Fig. 39: (a) In *solution-mined cavities*, it is proposed that chemical solutions would be used to mine cavities in appropriate media, such as rock salt; (b) in the *drilled-hole matrix*. A series of large-diameter holes would be drilled into the geologic media to depths up to 2 kilometers to form a grid of holes. The solid wastes would be packed into these holes, then sealed. (c) In the *rock-melting concept*, liquid wastes (no solidification) are poured into a subterranean cavity, which would be created by an underground explosion. (d) In the *hydrofracture concept*, liquid radioactive wastes are converted into a type of grout (cement or cementlike materials used). This grout is pumped under high pressure into

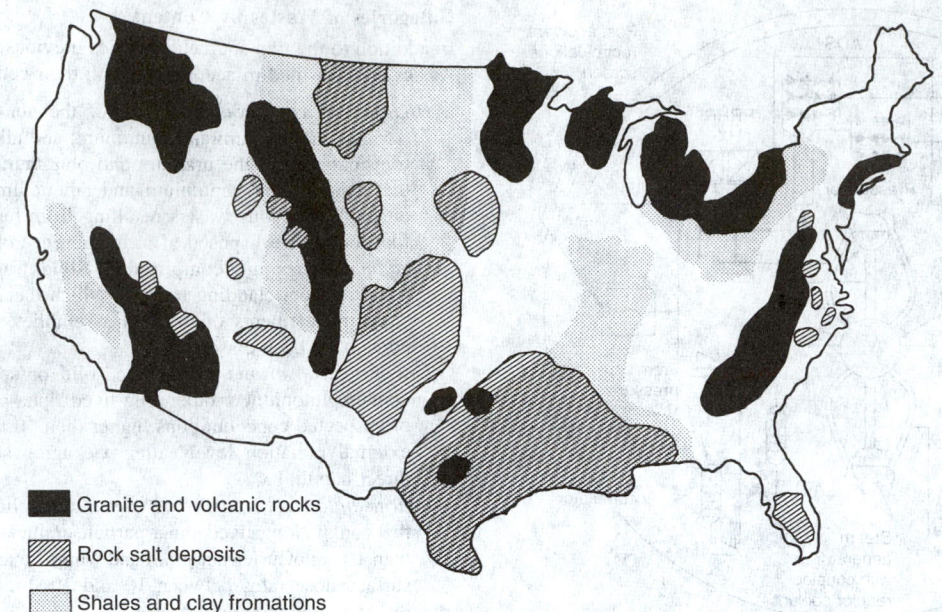

Granite and volcanic rocks

Rock salt deposits

Shales and clay fromations

Fig. 38. Some authorities prefer rock salt deposits for the permanent disposal of nuclear wastes on the assumption that the heat generated by radioactive decay would fuse salt and wastes into an impermeable mass. Other experts question the integrity of salt formations. Increasing attention has turned to hard media, such as the granitic and basaltic rocks, and to shale and clay formation, with the hope that the extensive occurrence of such formations would minimize the need to transport wastes over long distances.

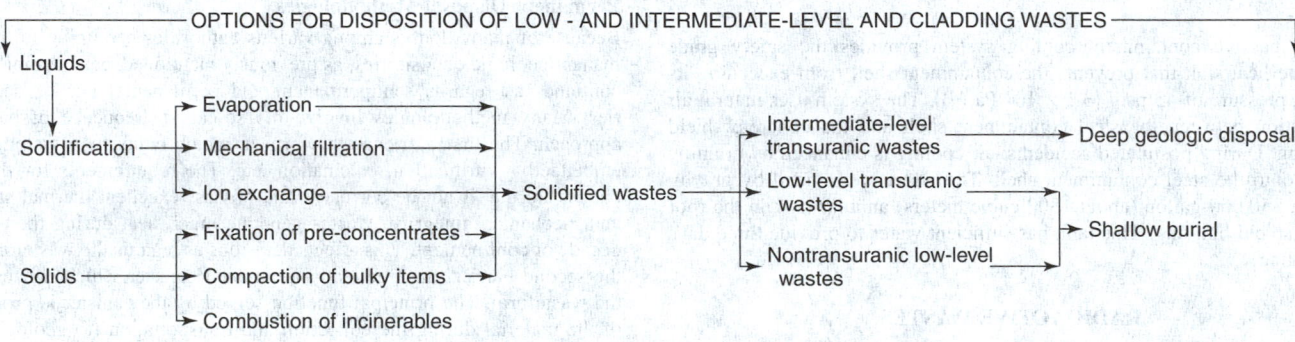

OPTIONS FOR DISPOSITION OF LOW - AND INTERMEDIATE-LEVEL AND CLADDING WASTES

Liquids

Solidification → Evaporation → Mechanical filtration → Ion exchange → Solidified wastes

Solids → Fixation of pre-concentrates → Compaction of bulky items → Combustion of incinerables

Solidified wastes → Intermediate-level transuranic wastes → Deep geologic disposal

Low-level transuranic wastes

Nontransuranic low-level wastes → Shallow burial

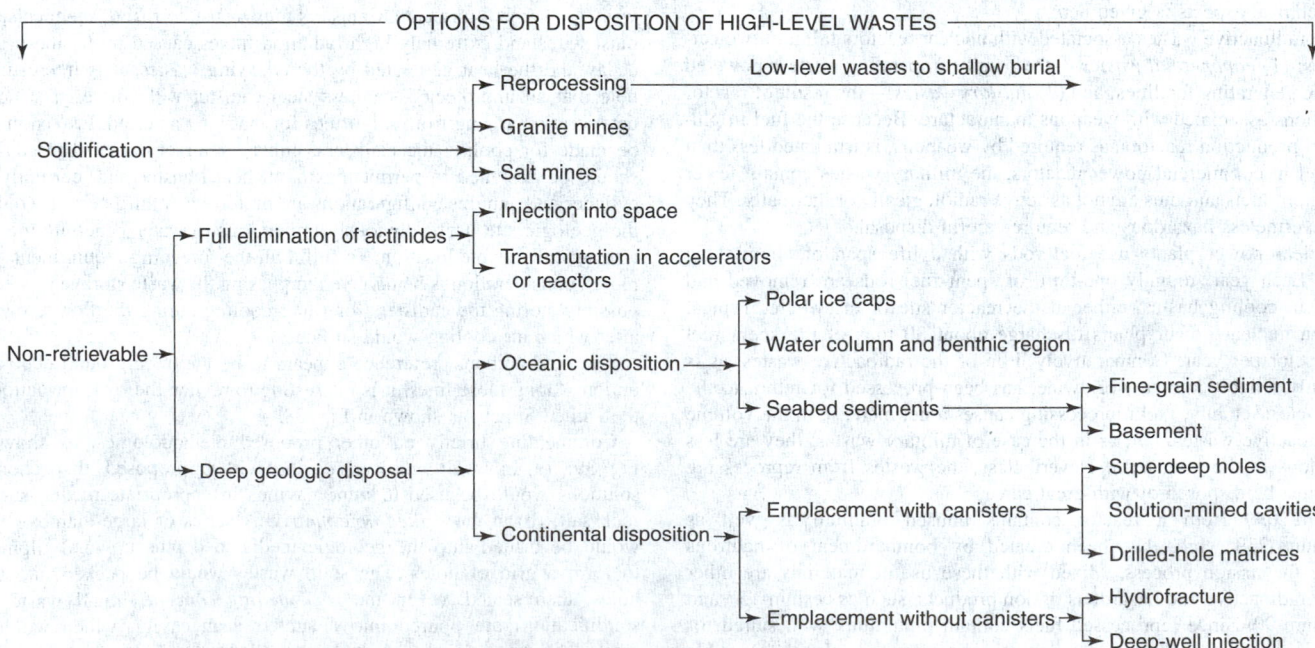

OPTIONS FOR DISPOSITION OF HIGH-LEVEL WASTES

Solidification → Reprocessing → Low-level wastes to shallow burial

→ Granite mines

→ Salt mines

Full elimination of actinides → Injection into space

→ Transmutation in accelerators or reactors

Non-retrievable

Deep geologic disposal → Oceanic disposition → Polar ice caps

→ Water column and benthic region

→ Seabed sediments → Fine-grain sediment

→ Basement

→ Continental disposition → Emplacement with canisters → Superdeep holes

→ Solution-mined cavities

→ Drilled-hole matrices

→ Emplacement without canisters → Hydrofracture

→ Deep-well injection

Fig. 39. Panorama of options for consideration in the disposition of radioactive wastes. (*Top portion of chart*: Options for disposition of low- and intermediate-level and cladding wastes. *Bottom portion of chart*: Options for disposition of high-level wastes.)

shale as deep as 1 kilometer. The pressure of the operation causes the underlying shale to fracture and the wastes fill up the cracks so formed. This procedure has been used for years in the petroleum field. See also **Petroleum**. (e) In the *polar ice concept*, the wastes would melt through the ice (although this approach would require considerable new technology); or the wastes would be placed on the surface of the ice or anchored within the ice. Advantages include long distances from populations and excellent thermal cooling. Disadvantages include extensive transport and poor retrievability. This method is not high on the list of choices mainly because of too many unknown factors that will require considerable research and experimentation. (f) *Oceanic disposition*, in addition to the polar ice cap concept, are subduction zones and other deep sea trenches and rapid sedimentation areas. A "Seabed Disposal Program Annual Report" states, "Placing high-level wastes on the seafloor, i.e., in the water column, effectively puts the waste contained directly into the biosphere. Since it is difficult to conceive of a practical man-made wasteform/container system that would survive without releasing radionuclides for hundreds of thousands of years in a marine environment, one must assume the radioactive material would eventually enter the ecosystem." The subseabed sediments in the central North Pacific are loosely packed, fine-grained, deepsea "red clays" and have not been fully dismissed from continuing investigation.

Additional Reading

Apostolakis, G.: "The Concept of Probability in Safety Assessments of Technological Systems," *Science*, 1359 (December 7, 1990).

Bodansky, D.: *Nuclear Energy: Principles, Practices, and Prospects*, American Institute of Physics, College Park, MD, 1996.

Bothwell, R.: *Nucleus. The History of Atomic Energy Limited*, Univ. of Toronto, Toronto, Canada, 1988.

Bruschi, H. and T. Andersen: "Turning the Key," *Nuclear Engineering International*, 1 (November 1991).

Cobb, C.E., Jr. and K. Kasmauski: "Living with Radiation," *National Geographic*, 403 (April 1989).

Cottrell, A.: "Safe Energy," *Review (University of Wales)*, 38 (Autumn 1987).

Davis, W.K.: "Problems and Prospects for Nuclear Power," *Chem. Eng. Progress*, **80**(6), 11–16 (June 1984).

Golay, M.W. and N.E. Todreas: "Advanced Light-Water Reactors," *Sci. Amer.*, 82 (April 1990).

Golay, M.W.: "Longer Life for Nuclear Plants," *Technology Review (MIT.)*, 25 (May/June 1990).

Goldschmidt, B.: *Atomic Rivals*, Rutgers University Press, New Brunswick, NJ, 1990.

Green, S.J.: "Solving Chemical and Mechanical Problems of PWR Steam Generators," *Chem. Eng. Progress*, 31 (July 1987).

Hansen, K. et al.: "Making Nuclear Power Work: Lessons from Around the World," *Technology Rev. (MIT)*, 30 (February 1989).

Hodgson, P.E.: *Nuclear Power, Energy and the Environment*, World Scientific Publishing Company, Inc., River Edge, NJ, 2000.

Jerome, F.: "Yo-Yo Journalism and Nuclear Power," *Technology Review (MIT)*, 73 (April 1989).

Kairi, S.P.: "Outage Risk Management," *EPRI J.*, 34 (April/May 1992).

Kurdsunoaeglu, B., A. Perlmutter, and S.L. Mintz: *The Challenges to Nuclear Power in the Twenty-First Century*, Kluwer Academic Publishers, Norwell, MA, 2000.

Lester, R.K.: "Rethinking Nuclear Power," *Sci. Amer.*, 31 (March 1986).

Marshall, E.: "Counting on New Nukes," *Science*, 1024 (March 2, 1990).

Miller, P. and R.H. Ressmeyer: "A Comeback for Nuclear Power? Our Electric Future," *National Geographic*, 60 (August 1991).

Roberts, L.: "British Radiation Study Throws Experts into Tizzy," *Science*, 24 (April 6, 1990).

Shoup, R.L.: "International Waste Management Symposium," *Nuclear Safety*, **18**(4) (1977).

Skerret, P.J.: "Will the Public Say Yes to Nukes?" *Technology Review MIT*, 8 (April 1991).

Slovic, P., J.H., Flynn and M. Layman: "Perceived Risk, Trust, and the Politics of Nuclear Waste," *Science*, 1603 (December 13, 1991).

Spinard, B.I.: "U.S. Nuclear Power in the Next Twenty Years," *Science*, 707 (December 12, 1988).

Staff: "AP600—A Cost Competitive Power Source," *Energy Digest*, 4 (Pittsburgh, Pennsylvania (Fall 1991).

Staff: International Atomic Energy Agency, *Choosing the Nuclear Power Option: Factors to Be Considered*, Bernan Associates, Lanham, MD, 1998.

Staff: University Press Cambridge, *Resistance to New Technology: Nuclear Power, Information Technology and Biotechnology*, Cambridge University Press, New York, NY, 1997.

Staff: IAEA, *Nuclear Power Reactors in the World, April 1997*, Bernan Associates, Lanham, MD, 1997.

Suzuki, T.: "Japan's Nuclear Dilemma," *Technology Review (MIT)*, 41 (October 1991).

Taylor, J.J.: "Improved and Safer Nuclear Power," *Science*, 318 (April 21, 1989).

Weinberg, A.M. and I. Spiewak: "Inherently Safe Reactors and a Second Nuclear Era," *Science*, **224**, 1398–1402 (1984).

Web References

History of Nuclear Power Plant Safety: http://users.owt.com/smsrpm/nksafe/
International Atomic Energy Agency: http://www.iaea.org/worldatom/
NRC Short History: http://www.nrc.gov/SECY/smj/shorthis.htm
Nuclear History Site: http://geocities.com/RainForest/Andes/6180/
U.S. Nuclear Regulatory Commission: http://www.nrc.gov/

NUCLEAR SPIN. The intrinsic angular momentum of the atomic nucleus due to rotation about its own axis. It is usually designated I and has the magnitude, $\sqrt{I(I+1)}h/2\pi \approx I(h/2\pi)$, where I is the nuclear spin quantum number which has different (integral or half-integral) values (including zero) for different nuclei. In spectroscopy, the nuclear spin is of importance for the explanation of the hyperfine structure, and of the intensity alternation in band spectra.

NUCLEAR STRUCTURE. The nucleus of an atom of atomic number Z and mass number A contains Z protons and $A-Z$ neutrons, bound together under the influence of shortrange nuclear forces much as molecules are bound together in a drop of liquid. The strength of binding may be determined by subtracting the actual mass of the atom from the mass of its constituent particles considered as free particles. The binding energy E_B is then related to this mass defect ΔM by Einstein's relation, $E_B = \Delta Mc^2$, where c is the velocity of light. The precise value of E_B depends upon the nucleus concerned, and upon how many neutrons and protons it contains but it is of the order of 8 MeV per nucleon in most nuclei.

The binding energy determines whether the nucleus is stable or unstable. Among the lighter nuclei, the ones which are stable are those in which the number of protons is approximately equal to the number of neutrons, so that $A \approx 2Z$. In heavier stable nuclei, there is an excess of neutrons over protons owing to the repulsive electrostatic forces between the protons. Thus, the most stable oxygen nucleus is ^{16}O, containing 8 protons and 8 neutrons, while the most nearly stable uranium nucleus is ^{238}U, containing 92 protons and 146 neutrons. Nuclei containing a disproportionate number of neutrons tend to be unstable and decay radioactively by emission of electrons whereby neutrons are converted into protons; those containing an excess of protons similarly tend to decay by emission of positrons or by capture of orbital electrons.

It has been shown that the nucleus is approximately spherical in shape and of volume proportional approximately to its mass. It is, however, capable of executing oscillations about the spherical form, and in certain circumstances may even acquire a permanent deformation. The heaviest nuclei are unstable under deformation, as a result of which they undergo spontaneous fission. These properties may be described qualitatively by regarding the nucleus as an electrically charged drop of liquid possessing volume energy and surface tension.

Although the nucleus is normally found in its lowest energy state, it may be produced as the result of a nuclear reaction, or through radioactivity in a number of excited states whose detailed properties may differ quite markedly from the lowest state. If formed in an excited state, it will decay, normally by the emission of electromagnetic radiation (gamma rays) to the lowest state, or by the emission of particles to another nucleus.

NUCLEAR TRANSMUTATION. The transformation of one nuclide into another, which differs from it in nuclear charge, mass, or stability, i.e., in a nuclear reaction or a process of radioactivity. Such changes occur in natural radioactive processes, but the general need for a systematic notation for expressing them came only with the great number of transmutations discovered after particle accelerators provided high-energy ion beams capable of penetrating the Coulomb barrier of all stable atomic nuclei.

Two representative transmutation equations are

$$^{27}Al + n \longrightarrow {}^{27}Mg + {}^{1}H$$

which shows the transmutation of aluminum atoms of mass number 27, by bombardment with neutrons, to magnesium atoms of mass number 27, with the emission of a proton, and

$$^9Be + \gamma \longrightarrow {}^8Be + n$$

which shows the transmutation of beryllium atoms of mass number 9, under gamma-ray bombardment, to beryllium atoms of mass number 8, with the emission of a neutron.

The two reactions above may also be expressed in condensed from as

$$^{27}Al(n, p)^{27}Mg$$

and

$$^9Be(\gamma, n)^8Be.$$

NUCLEATE BOILING.   See **Boiler (Steam Generator)**; and **Boiling**.

NUCLEATION.   The process of initiation of a new phase in a super-cooled (for liquid) or supersaturated (for solution or vapor) environment; the initiation of a phase change of a substance to a lower thermodynamic energy state (vapor to liquid condensation, vapor to solid deposition, liquid to solid freezing). In nature, heterogeneous nucleation is the more common where such a change takes place on small particles of different composition and structure. Homogeneous nucleation occurs when the change of state centers upon embryos that exist in the same initial state as the changing substance. In this case, the nucleation system contains only one component, and it is termed homogeneous nucleation. In meteorology, particularly in cloud physics, a number of types of nucleation are of interest. The process by which cloud condensation nuclei initiate the phase change from vapor to liquid is important in all cloud formation problems. The physical nature of freezing nuclei that may be responsible for the conversion of drops of supercooled water into ice crystals is critically important in precipitation theory, as is the clarification of the role of homogeneous nucleation near $-40\,^\circ$C ($-40\,^\circ$F). Thermodynamically, all nucleation processes involve free energy decrease associated with the bulk phase change and the free energy increase associated with the creation of new interfaces between phases.

NUCLEI (Cloud Formation).   See **Atmosphere-Ocean Interface**.

NUCLEI (Sublimation).   See **Precipitation and Hydrometeors**.

NUCLEIC ACIDS.   Nucleic acids are polymeric materials formed from nucleotides and essential to all organisms. Deoxyribonucleic acid (DNA), most often a double-helical biopolymer, encodes the genetic information contained in each cell. Ribonucleic acid (RNA) constitutes a more diverse class of biopolymers that are able to adopt both helical and other more complex tertiary structures. The structural diversity of RNAs enables these molecules to carry out a variety of intracellular functions, including transmitting the genetic message to the site of protein synthesis. See also **Proteins**. Both DNA and RNA interact with a host of other molecules, e.g., proteins, drugs, as well as other RNAs and DNAs. The specificity of these interactions, key to all biological processes, is related to recognition of specific structural elements including sites for hydrogen-bond formation and regions where sequence-dependent structural variation complements the binding molecule.

Development of techniques to synthesize oligonucleotides, i.e., short, well-defined sequences of DNA or RNA, has provided the opportunity to study nucleic acid structure in detail. In addition, oligonucleotides have proved invaluable in analytical procedures used in genetic engineering, protein engineering, affinity chromatography, and forensics, as well as in medicine. See also **Chromatography**; and **Forensic Chemistry**. The unique ability of nucleic acids to bind to self-complementary sequences has been exploited in the design of oligonucleotide probes, new drug development and as a framework for constructing molecular devices (nanotechnology). The completion of the Human Genome Project has ushered in a new era where techniques and products based on oligonucleotides will become increasing important. See also **Human Genome Project (The)**.

DNA Structure

The structure of DNA is characterized by its primary sequence, secondary helical structure, and higher order structure or topology. The primary sequence of DNA refers to the atomic connectivities required to construct the polynucleotide chain. The helical conformation of these polynucleotide chains constitutes the secondary structure of DNA. Sequence-dependent structural diversity and flexibility are important DNA characteristics and play a crucial role in biological processes. The organization of helical DNA in topologically distinct three-dimensional conformations represents the higher order structure. Higher order structural features, in particular the supercoiling of DNA, are thought to have a profound influence on the dynamic processes and biology of nucleic acids within living cells.

The DNA double helix was first identified by Watson and Crick in 1953. Not only was the Watson-Crick model consistent with the known physical and chemical properties of DNA, but it also suggested how genetic information could be organized and replicated, thus providing a foundation for modern molecular biology.

The primary structure of DNA is based on repeating nucleotide units, where each nucleotide is made up of the sugar, i.e., 2'-deoxyribose, a phosphate, and a heterocyclic base, N. The most common DNA bases are the purines, adenine (A) and guanine (G), and the pyrimidines, thymine (T) and cytosine (C) (see Fig. 1). The base, N, is bound at the 1'-position of the ribose unit through a heterocyclic nitrogen.

The nucleotides are linked together via the phosphate groups, which connect the 5'-hydroxyl group of one nucleotide and the 3'-hydroxyl group of the next to form a polynucleotide chain (Fig. 1a). DNA is not a rigid or static molecule; rather, it can adopt a variety of helical motifs.

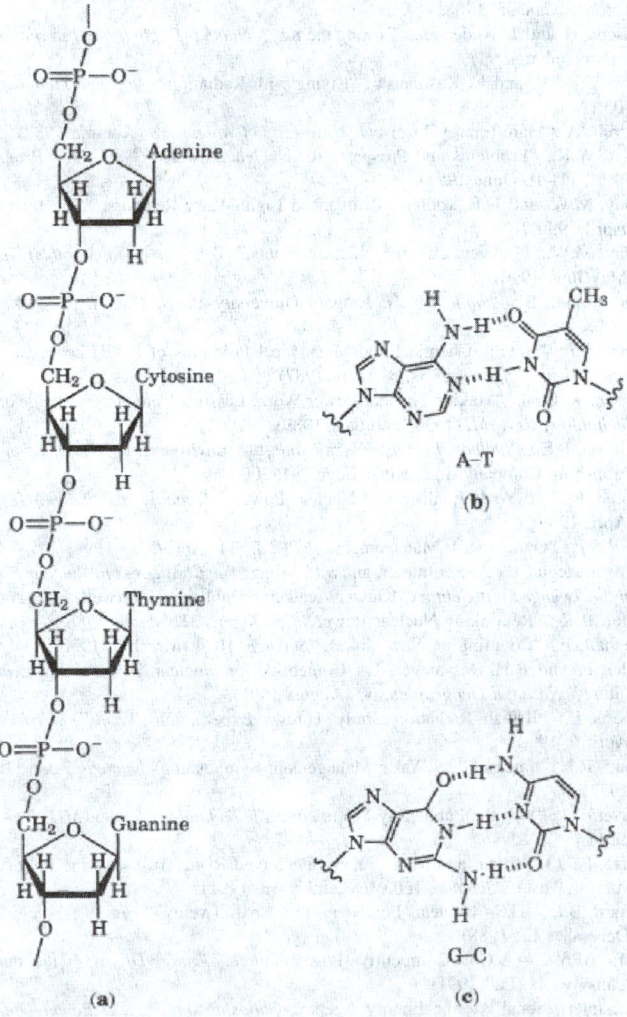

Fig. 1.   Elements of DNA structure: (**a**) a deoxypolynucleotide chain, which reads d(ACTG) from $3' \rightarrow 5'$ or d(GTCA) from $3' \rightarrow 5'$; and (**b**) and (**c**) the Watson-Crick purine–pyrimidine base pairs, A–T and G–C, respectively, where — represents attachment to the deoxyribose.

In A- or B-form DNA, two self-complementary polynucleotide strands associate with one another to form a right-handed double helix. The two polynucleotide chains are antiparallel.

In addition to A- and B-form DNA, several other helical conformations have been identified. Among these, the most well-studied is Z-DNA, a left-handed helix first characterized by x-ray crystallographic analysis of the oligonucleotides d(CGCGCG) and d(CGCG). Other alternating purine—pyrimidine sequences, in particular alternating CG sequences, have been shown to adopt the Z-conformation at high ionic strength.

RNA Structure

RNA has a variety of functions within a cell; for each function, a specific type of RNA is required. Messenger RNA (mRNA) serves as intermediaries for carrying genetic messages from the DNA to the ribosomes where protein synthesis takes place. Ribosomal RNA (rRNA) serves both structural and functional roles in the ribosome; it is diverse, both in terms of its size and structure. Transfer RNAs (tRNAs) are small molecules that have a central role in protein synthesis. Other RNA molecules, called ribozymes, function as enzymes to catalyze chemical transformations. Although ribozymes most often catalyze cleavage of the RNA phosphodiester backbone, they have also been shown to participate in cleavage of DNA, replication of RNA, and reactions with phosphate monoesters. Other RNAs are associated with enzymes to form riboprotein complexes involved in many biological processes. The multifunctional character of RNA, particularly the involvement of RNA in enzymatic processes, has led to the hypothesis that life on earth evolved from RNA, and that RNA had both the genetic and catalytic functions commonly associated with DNA and proteins, respectively.

The primary structure of RNA is similar to that of DNA, but with a few notable exceptions. First, in RNA, instead of thymine, the pyrimidine base uracil (U) occurs, forming a complementary base pair with adenine in regions of double-stranded RNA. Also, a wide variety of ribonucleotides having modified or minor bases are found in naturally occurring RNA, one of the most common of which is pseudouridine. In human tRNAs, as many as 25% of the bases are nonstandard. Over 80 modified bases have been characterized in naturally occurring tRNA; although the role of base modification is not clear, it may be important for biological recognition.

The other important feature of the primary structure of RNA is the presence of the 2′-hydroxyl group in ribose. Although this hydroxyl group is never involved in phosphodiester linkages, it does impose restrictions on the helical conformations accessible to double-stranded RNA.

RNAs are single-stranded molecules that fold, allowing different regions of the ribonucleotide to form distinct secondary structural elements. When self-complementary regions of the RNA strand are aligned, duplex regions, which may have Watson-Crick base pairs, are formed. In contrast to DNA, double-stranded regions in RNA are much more likely to have unusual base-pairing between noncomplementary bases and to incorporate non-Watson-Crick base-pairing. Owing to the steric requirements of the 2′-hydroxyl group on the ribose sugar, these duplex regions are constrained to an A-form helix, i.e., a 3′-endo sugar conformation. Although double-stranded RNA has the general features of an A-form helix, actual duplex characteristics, such as rise per base pair, groove dimensions, and base pair displacement from the helical axis, may vary.

The functional diversity of RNA is directly related to its structural diversity. In contrast to DNA, RNA molecules are synthesized as single-stranded polynucleotides that fold to give complex tertiary structures. These structures, which incorporate hairpins, loops, bulges, and junctions between single-stranded and double-stranded regions, exhibit long-range interactions within the folded tertiary structure. Long-range intramolecular interactions serve to stabilize the three well-characterized RNA structures.

Oligonucleotide Synthesis

Synthetic oligonucleotides are widely used in scientific investigations. Most synthetic oligonucleotides are produced for use as primers in the polymerase chain reaction (PCR), a widely used analytical technique having commercial applications in diagnostic medicine, genetic engineering (qv), and forensics. See also **Forensic Chemistry**. A large volume of oligonucleotides are also synthesized for use as primers in DNA-sequencing. Demands for sequencing primers have increased rapidly to support large-scale DNA mapping and sequencing efforts such as the human genome

project. Although the quantities of oligonucleotides used are relatively small, these materials are essential in many areas of basic research.

In molecular biology, synthetic oligonucleotides are used as linkers in gene-cloning and to introduce site-directed mutations in genes. Synthetic oligonucleotides are required for structural, biochemical, and biophysical studies of DNA and RNA. Oligonucleotides also are important for examining the association of proteins and small molecules, e.g., for intercalating drugs with nucleic acids on a molecular level.

The first procedures for oligonucleotide synthesis, typically carried out in solution, made use of H-phosphonate and phosphotriester chemistry. These approaches are useful in some large-scale syntheses and in syntheses of various oligonucleotide analogues. Most modern procedures, however, are based on solid-phase phosphoramidite chemistry. Automated oligonucleotide synthesizers are commercially available, as are the required reagents and phosphoramidites. Together these permit the rapid production of custom oligonucleotides and oligonucleotide analogues.

Modified Oligonucleotides. Much of the interest in modified oligonucleotides is related to use as antisense agents. Antisense agents are typically short (15–30 base pairs in length) oligonucleotides having sequences that are complementary to coding or regulatory regions within mRNA, although some antisense oligonucleotides have also been designed to target DNA. The antisense sequence recognizes and binds to a complementary sequence via the formation of a double-stranded duplex that have normal Watson-Crick base-pairing. Antisense oligonucleotides can inhibit gene expression at the translational level. The potential to design oligonucleotides having the ability to recognize and inhibit specific genes makes the antisense approach promising in the development of new therapeutic agents. In addition, antisense oligonucleotides can be used in research to elucidate gene function by providing a mechanism for regulating a gene artificially.

Although all natural antisense oligonucleotides are short RNA sequences, most of the synthetic antisense oligonucleotides are deoxyoligonucleotides. In the design of an effective antisense oligonucleotide, several factors must be considered. First, the oligonucleotide must be specific, binding with high affinity to a single sequence within the target RNA. A second consideration is stability within the cellular environment. Thus all unmodified oligonucleotides are degraded too rapidly to be used effectively as therapeutic agents. A significant research effort has been directed toward discovering chemical modifications that can increase the nuclease resistance of the oligonucleotide backbone.

An effective therapeutic agent must also have the ability to reach its target sequence *in vivo*. In order to enhance membrane transport, antisense oligonucleotides are frequently modified by covalent attachment of carrier molecules or lipophilic groups.

Antisense oligonucleotides are usually designed to inhibit gene expression by interfering with the translation of mRNA. One mechanism for this type of inhibition involves binding the oligonucleotide to the translation-initiation sequences of the mRNA, which prevents ribosome association and protein synthesis. Another potential mechanism involves hybrid formation at some other sequence within the mRNA, thus impeding translocation of the ribosome along the mRNA strand by steric blocking. These two mechanisms are based on blocking a sequence of RNA or DNA by double-stranded duplex formation using a specific antisense oligonucleotide.

A less specific mechanism based on the action of Rnase H, an enzyme catalyzing single-strand cleavage of RNA, may be predominate for unmodified, thioate and dithioate oligonucleotides. In the Rnase H mechanism, the duplex formed by the antisense oligonucleotide and the target RNA is a substrate for Rnase H. The enzyme cleaves the RNA at the complexes site rendering the RNA vulnerable to further degradation and inactivation by cellular exonucleases. The oligonucleotide, which is probably not a substrate for Rnase H, can target multiple copies of complementary RNA. Where applicable, antisense oligonucleotide action mediated by the Rnase H mechanism has been shown to be a potent inhibitor of gene expression.

Modified oligonucleotides can also be designed for binding to double-stranded DNA by forming a triple helix.

Oligonucleotides can also inhibit gene expression at the transcriptional level by binding to a single-stranded or open sequence of DNA. In this mechanism, the antigene oligonucleotide is designed to be complementary to a regulatory sequence preceding a gene. For example, an oligonucleotide complementary to the *lac* operator sequence (repressor protein binding site) has been found to inhibit specifically β-galactosidase synthesis in *E. coli*. Normally, expression of the lactose-metabolizing enzyme, β-galactosidase, is blocked at the transcriptional level by the repressor binding to the operator sequence. In the presence of a lactase metabolite, the repressor is converted to a nonbinding form and dissociates from the operator, which results in the transcription of the gene and β-galactosidase production. However, in the presence of the antisense oligonucleotide, the synthesis of β-galactosidase is inhibited. The antisense oligonucleotide can then act as a repressor by binding to an open or single-stranded region within the operator.

Although development of modified oligonucleotides as antisense and antigene agents is a principal focus of research in the 1990s, there are many other interesting applications that have greater immediate commercial significance. Included among these are applications using nucleic acid probes, which are oligonucleotides that have been modified by the attachment of a detectable chemical group. Probes can be designed to recognize RNA or DNA sequences characteristic of specific eukaryotic genes, viruses, or bacteria. Several analytical and diagnostic procedures have been developed based on the hybridization of the probe with its target sequence and the subsequent detection of the hybrid by the group attached to the oligonucleotide. Probes, particularly useful in automated sequencing protocols, may contain fluorescent groups, phosphors, radioactive tracers, etc. In addition, probes can be designed to help elucidate the structure of biological molecules. DNA-binding molecules, including intercalators, alkylating agents, and photosensitive molecules, have also been linked to oligonucleotides as a way of directing a drug to a specific DNA sequence. These modifications often enhance binding as well.

Oligonucleotide Bioconjugates

Although many molecules, including proteins and small intercalating and groove-binding ligands, bind to RNA and DNA, only nucleic acids are able to bind with the high specificity required to recognize a single sequence within the 3×10^9 base-pairs of the human genome. The unique specificity of oligonucleotides can be exploited to direct a multitude of other chemical agents to a sequence of interest by attaching these agents to oligonucleotides through molecular linkers. Oligonucleotides labeled with fluorescent or other detectable groups provide nucleic acid probes that can be used to screen a large pool of DNA for a specific sequence. Cationic or lipophilic groups can be attached to improve the binding and bioavailability of antisense oligonucleotides. Bioconjugates have also been widely used in research because they enable scientists to learn more about the structure and function of nucleic acids and ligands that bind to them.

Several strategies have been devised to attach various chemical groups to oligonucleotides. Groups can be attached to the 3'- or 5'-terminus of the oligonucleotide, along the backbone through the phosphate or the 2'-hydroxy group of ribose, or to modified purines or pyrimidines.

See also **Adenosine**; **Adenosine Phosphates**; **Amino Acids**; and **Protein**.

Additional Reading

Caruthers, M.H. and co-workers: *Methods in Enzymology*, Vol. 154, Academic Press, Inc., New York, NY, 1987, pp. 287–313; Vol. 211, 1992, pp. 3–20.

Clayton, J., and C. Dennis: *50 Years of DNA*, Palgrave Macmillan, New York, NY, 2003.

Gesteland, R.F., T.R. Cech, and J.F. Atkins: *The RNA World*, 2nd Edition, Cold Spring Harbor Laboratory Press, Cold Spring Harbor, New York, NY, 1999.

Goni, J.R., X. de la Cruz, and M. Orozco: *Nucleic Acids Res.*, **32**, 354 (2004).

Herdewijn, P.: *Oligonucleotide Synthesis: Methods and Applications*, Humana Press, Totowa, NJ, 2004.

Khachigian, L.M.: *Synthetic Nucleic Acids as Inhibitors of Gene Expression*, CRC Press, LLC, Boca Raton, FL, 2004.

Lodish, H., P. Matsudaira, and A. Berk: *Molecular Cell Biology*, 5th Edition, W. H. Freeman and Company, New York, NY, 2003.

Mahato, R.I.: *Biomaterials for Delivery and Targeting of Proteins and Nucleic Acids*, CRC Press, LLC, Boca Raton, FL, 2004.

Moldave, K.: *Progress in Nucleic Acid Research and Molecular Biology*, Elsevier Science & Technology Books, New York, NY, 2005.

Neidle, S.: *Nucleic Acid Structure and Recognition*, Oxford University Press, New York, NY, 2002.

Nellen, W.: *Small RNAs: Analysis and Regulatory Functions*, Springer-Verlag New York, LLC, New York, NY, 2005.

Ronot, X. and Y. Usson: *Imaging of Nucleic Acids and Quantification in Photitonic Microscopy*, CRC Press LLC, Boca Raton, FL, 2001.

Rosemeyer, H. and M. Volkan Kisakurek: *Perspectives in Nucleoside and Nucleic Acid Chemistry*, John Wiley & Sons, Inc., New York, NY, 2000.

Saenger, W.: *Principles of Nucleic Acid Structure*, Springer-Verlag, New York, NY, 1984.

Web References

Biochemistry of Nucleic Acids, http://web.indstate.edu/thcme/mwking/nucleic-acids.htm

Nucleic Acid Symbols, http://www.chem.qmul.ac.uk/iupac/misc/naabb.html

Nucleic Acids, http://www.visionlearning.com/library/science/biology-1/BIO1.1-nucleic_acids.htm

Nucleic Acids Research, http://nar.oxfordjournals.org/

Nucleic Acid Database (NDB), http://ndbserver.rutgers.edu/atlas/

NUCLEONICS. The applications of nuclear science in physics, chemistry, biology, and other sciences, including military science, and in industry, and the techniques associated with these applications.

NUCLEONS. Two nuclear particles, the proton and the neutron, and their antiparticles are known as *nucleons*. The rest mass of the proton is 1.0076 amu; that of the neutron, 1.0089 amu. The antiproton bears the same relation to the proton that the electron does to the positron, i.e., its charge is equal and opposite and its mass is the same, the charge being equal in magnitude to the electronic charge. Protons and antiprotons also annihilate each other when they collide, the reaction of a single pair producing positive and negative pions or kaons. If the proton and antiproton do not collide, but experience a "near miss," then an exchange of charge can occur, resulting in the formation of a neutron-anti-neutron pair. The four nucleon particles are fermions and have a spin angular momentum quantum number of $\frac{1}{2}$. See also **Neutron**; **Particles (Subatomic)**; and **Proton**.

NUCLEOPHILE. An ion or molecule that donates a pair of electrons to an atomic nucleus to form a covalent bond. The nucleus that accepts the electrons is called an *electrophile*. This occurs, for example, in the formation of acids and bases according to the Lewis concept, as well as in covalent bonding in organic compounds.

NUCLEOPHILIC REACTION. A reaction in which a nucleophilic reagent attacks an electrophilic compound. The reagent is taken to be the inorganic substance (in the case of reactions of inorganic and organic substances) or the simpler of two reacting organic compounds. The electron pair for the bond formed is furnished by the nucleophilic reagent.

NUCLEOPROTEINS AND NUCLEIC ACIDS. Nucleic acids are compounds in which phosphoric acid is combined with carbohydrates and with bases derived from purine and pyrimidine. Nucleoproteins are conjugated proteins consisting of a protein moiety and a nucleic acid. Originally, nucleoproteins were thought to occur only in the nuclei of cells, but it was later established that they are far more widely distributed, being found in cells of all types, animal and plant. They are found in the chromosomes, in the genes, in viruses, and bacteriophages.

The protein portion of the nucleoproteins is basic in nature and being complex in structure may form several types of linkage, depending upon the type of nucleic acid. In gastric digestion or hydrolysis with weak acid, nucleoproteins yield protein and nuclein. The latter in pancreatic digestion or hydrolysis with weak alkali yields additional protein and nucleic acid. See also **Nucleic Acids**.

NUCLEOSIDE. A compound of importance in physiological and medical research, obtained during partial decomposition (hydrolysis) of nucleic acids and containing a purine or pyrimidine base linked to either *d*-ribose, forming ribosides, or *d*-deoxyribose, forming deoxyribosides. They are nucleotides minus the phosphorus group. See also **Adenosine**; and **Nucleic Acids**.

NUCLEUS. 1. The nucleus of an atom is the positively charged core, with which is associated practically the entire mass of the atom, but only a minute part of its volume.

2.   The nucleus of a molecule is a group of atoms connected by valence bonds so that the atoms and their bonds form a ring or closed structure, which persists as a unit through a series of chemical changes.

3.   A group of cell bodies in the central nervous system of vertebrates. Examples of such groups are the red nucleus in the midbrain, through which impulses are routed for the control of subconscious muscular movements, and Deiter's nucleus, lying at the junction of the medulla with the hindbrain. Through this center impulses pass for muscular action involved in the maintenance of equilibrium.

4.   A somewhat spherical or oblong body in most living cells. This nucleus contains the chromosomes, which, in turn, bear the genes of heredity. The nucleus also contains a nucleolus, or sometimes two or more nucleoli and a basic ground substance, the nucleoplasm. A nuclear membrane surrounds it on the outside, but this membrane is very porous, allowing materials to pass through rather freely.

5.   Condensation nuclei take part in phase changes, as in the formation of clouds; in seeding concentrated solutions to bring about crystallization, precipitation, etc.

NUCLIDES.   See **Chemical Elements.**

NULL.   In direction-finding systems wherein the output amplitude is a function of the direction of arrival of the signal, the minimum output amplitude (ideally zero). The null is frequently employed as a means of determining bearing. The term *minimum* is often used to indicate an imperfect null.

NULL VECTOR.   A vector $\mathbf{A}_\mu$ of zero length ($\mathbf{A}_\mu \mathbf{A}_\mu = 0$). In special relativity theory the displacement between two events on the path of a photon is a null vector.

NUMBER THEORY.   The theory of numbers is concerned primarily with the properties of the integers, or whole numbers, $0, \pm1, \pm2, \ldots$. The integers form a ring, i.e., when one or more of the operations addition, subtraction, and multiplication are applied to any two integers, the result is always an integer.

Numbers can be classified in many ways. For example, odd and even numbers, prime numbers, square numbers, and perfect numbers. These classes of numbers were mentioned by Euclid (circa 300 B.C.). The Chinese knew in 500 B.C. that $2^p - 2$ is divisible by p, for prime numbers p. Euclid proved that there exist infinitely many primes. The proof is by contradiction. Assume p is the largest prime. Let M be the product $2 \cdot 3 \cdot 5 \cdots p$ of all primes up to p. Then $M + 1$ is either a prime number itself, or is divisible only by primes greater than p.

While most results of number theory are easy to understand, it has taken brilliant mathematicians many years to construct proofs of many and some of the most interesting conjectures remain unproved.

Number theory has been extended to the study of the specific properties of other classes of numbers, such as rational, algebraic, and transcendental numbers. Discussed here are elementary number theory, algebraic number theory, analytic number theory, and Diophantine approximations.

Elementary Number Theory. The main concern of this branch of number theory is divisibility. If a and b are integers, and $a = bc$ for some c, then b is a factor or divisor of a, written symbolically as $b|a$. Thus, $1|a$ and $a|a$ for every integer a. A prime number $p > 1$ is an integer that has only 1 and p as divisors. Other integers greater than 1 are called composite.

A concept known as the fundamental theorem of arithmetic which was proved by Euclid states that every integer n greater than 1 can be expressed as a product of primes in only one way when the order of prime factors is disregarded.

A perfect number is an integer that is equal to the sum of its proper divisors less than itself. The five least are: 6; 28; 496; 8,128; 33,550,336. Euclid showed that $(2^n - 1)2^{n-1}$ is a perfect number if and only if 2^{n-1} is a prime number. Numbers of the form $2^p - 1$ are known as Mersenne numbers. All even perfect numbers are of Euclid's type. No odd perfect numbers have been found.

Let m denote a positive integer. Every integer a can be represented in the form $a = qm + r$, where q is an integer and r is an integer which has one of the m values $0, 1, \ldots, m - 1$. Two integers a and b are congruent modulo m if the value of r is the same for a and b when expressed

in the above form. Symbolically, $a \equiv b (\mathrm{mod}\ m)$. If $a \equiv b (\mathrm{mod}\ m)$, then $m|(a - b)$. The congruence relationship is an equivalence, i.e., it is reflexive, symmetric, and transitive. The relation defines equivalence classes of numbers congruent to each other, which are called residue classes. There are m residue classes modulo m.

The symbol (a, b) is frequently used to denote the greatest common divisor of integers a and b, assumed at least one of which is nonzero. If $(a, b) = 1$, a and b are coprime.

If $a \equiv b (\mathrm{mod}\ m)$, $(a, m) = (b, m)$. If $(a, m) = 1$, the residue class of a modulo m is said to be prime to m. Those residue classes which are prime to m form a reduced system of residue classes mod m. Euler introduced the function $\Phi(m) = m\Pi_{p/m}(1 - 1/p)$ to denote the number of reduced residue classes modulo m. Some of the properties of Φ are:

$$\Phi(mn) = \Phi(m)\Phi(n) \text{ if } (m, n) = 1$$

and

$$\Phi(p^r) = p^r(1 - 1/p)$$

where p is any prime and r is a positive integer.

Two important congruences are Wilson's Theorem

$$(p - 1)! \equiv -1 (\mathrm{mod}\ p)$$

and Fermat's theorem

$$a^{p-1} \equiv 1 (\mathrm{mod}\ p), \text{ where } p \text{ is a prime.}$$

Euler generalized the latter to

$$a^{\Phi(m)} \equiv 1 (\mathrm{mod}\ m), \text{ where } (a, m) = 1$$

The linear congruence $ax \equiv b (\mathrm{mod}\ m)$ is solvable for x if and only if $d|b$, where $d = (a, m)$. There are exactly d solutions (incongruent to each other).

If $m = p$ is a prime number then $ax \equiv 1 (\mathrm{mod}\ p)$ has exactly one solution for each a which is coprime with p. This solution is the reciprocal of a modulo p which shows that residue classes modulo p form a finite field.

The congruence $a_0 x^n + a_1 x^{n-1} + \cdots + a_n \equiv 0 (\mathrm{mod}\ m)$ is of degree n, if m does not divide a_0. If $m = p$, the number of solutions cannot exceed n but a solution may not exist.

If $(a, m) = 1$, the least positive e such that $a^e \equiv F (\mathrm{mod}\ m)$ must divide $\Phi(m)$. If $e = \Phi(m)$, a is called a primitive root of m and the powers of $a, a^2, \ldots, a^{e-1}$ form a reduced set of residues modulo m. A modulus m has primitive roots only when $m = 2, 4, p^k, 2p^k$, where p is an odd prime. The period of the periodic decimal into which the reduced fraction a/m can be expanded has the length $\Phi(m)$ or a divisor thereof. It has the length $\Phi(m)$ only when the base (10 for decimal expansion) is a primitive root modulo m.

Fibonacci numbers were discovered by Leonardo of Pisa in 1202 A.D. The first two Fibonacci numbers are 1; subsequent Fibonacci numbers are obtained by adding the previous two Fibonacci numbers. The first seven are 1, 1, 2, 3, 5, 8, and 13. The sequence grows exponentially with the nth number in the sequence being approximately equal to $(1/\sqrt{5})((1 + \sqrt{5})/2)^n$.

A theorem known to the Chinese in ancient times and known as the Chinese remainder theorem can be stated as:

If $(m_i, m_j) = 1$ for $1 \leq i < j \leq n$, then the system of linear congruences

$$x \equiv a_1 (\mathrm{mod}\ m_1)$$

$$x \equiv a_2 (\mathrm{mod}\ m_2)$$

$$\cdots$$

$$x \equiv a_n (\mathrm{mod}\ m_n)$$

has a unique solution modulo m, where $m = \Pi_{i=1}^n m_i$.

Gödel showed how to use this theorem as a coding trick, in which an arbitrary finite sequence of numbers can be encoded as a single number.

In 1970, Yuri Matyasevich completed the proof that no procedure can be devised to determine the existence of a solution to a Diophantine equation. This proof involves the Chinese remainder theorem, Fibonacci numbers, and Diophantine equations. In 1900, David Hilbert posed 23 major problems for mathematicians. Matyasevich's results complete the solution of Hilbert's tenth problem.

If $x^2 \equiv a \pmod{m}$ is solvable, a is called a quadratic residue modulo m, otherwise a quadratic nonresidue. Let $m = p$ be an odd prime. The Legendre symbol (a/p) is defined to be $+1$, if a is a quadratic residue modulo p and as -1 if a is a quadratic nonresidue. It is assumed that $a \not\equiv 0 \pmod{p}$.

The Legendre symbol is a character of the multiplicative group of residue classes modulo p:

$$\left(\frac{a}{p}\right) \cdot \left(\frac{b}{p}\right) = \left(\frac{ab}{p}\right)$$

If q is another odd prime

$$\left(\frac{p}{q}\right)\left(\frac{q}{p}\right) = (-1)^{(p-1)/2 \cdot (q-1)/2}$$

Which is known as the reciprocity law of quadratic residues. Also,

$$\left(\frac{-1}{p}\right) = (-1)^{(p-1)/2}$$

Jacobi generalized the Legendre symbol by defining

$$\left(\frac{a}{p_1 p_2 \cdots p_n}\right) = \left(\frac{a}{p_1}\right)\left(\frac{a}{p_2}\right) \cdots \left(\frac{a}{p_n}\right), \text{ and } \left(\frac{a}{-k}\right) = (ak),$$

if a is positive.

The reciprocity law still holds for any odd numbers p and q not both negative.

An expression, $f(x, y) = ax^2 + bxy + cy^2$, is a binary quadratic form. A number m is represented by the form f if there exists a pair of integers u and v such that $f(u, v) = m$. The representation is called primitive if $(u, v) = 1$. The determinant of the form f is defined as $d = ac - (\frac{1}{4})b^2$. The discriminant $D = -4d$.

The forms, $f = ax^2 + bxy + cy^2$ and $F + AX^2 + BXY + CY^2$, are equivalent if there exists a transformation $x = \alpha x + \beta y, Y = \gamma x + \delta y$ with $\alpha\delta - \beta\gamma = \pm 1, \alpha, \beta, \delta$ all integers so that $F(X, Y) = f(x, y)$.

The determinants of equivalent forms are equal.

Since $af = (ax + \frac{1}{2}by)^2 + dy^2$, if $d > 0$, $a \neq 0$ then af is positive for any x and y not both zero. Therefore, f represents numbers of only one sign, that of a. If $d < 0$, f represents both positive and negative numbers.

For binary forms with a positive determinant, the number of classes of integral, positive-definite, binary quadratic forms with a given determinant is finite.

An automorph of f is a unimodular transformation carrying f into itself. For example, $x = -x_1$, $y = -y_1$.

In the above, a, b, c may denote any real numbers. Now consider f integral; i.e., a, b, c, integers. The greatest common divisor (g.c.d.) of a, b, c is called the divisor of f. If the g.c.d. is 1, f is called primitive.

All the automorphs of the primitive form $ax^2 + bxy + cy^2$ of discriminant D are given by $x = \frac{1}{2}(t - bu)X - cuY$, $y = auX + \frac{1}{2}(t + bu)Y$ where t, u range over all the integral solutions of the so-called Pell's equation, $t^2 - Du^2 = 4$.

It can be shown that if D is a positive nonsquare integer, all integral solutions of Pell's equation are given by $\frac{1}{2}(t + u\sqrt{D}) = \pm[\frac{1}{2}(T \pm U\sqrt{D})]^k$, $k = 0, 1, 2, \ldots$, and T and U are the least solution in positive integers.

Fermat stated and Euler proved that every prime that is congruent to 1 mod 4 is the sum of two squares. For such a prime number, -1 is a quadratic residue. Lagrange found that every natural number is the sum of at most 4 squares.

Algebraic Number Theory. The notions of primality, integers, divisibility, etc., which originated in arithmetic, have been extended to systems other than ordinary integers. The ordinary integers are referred to as rational integers to distinguish them from integers in fields other than the real number field. Gauss defined a complex number as $a = a_0 + a_1 i$, where a_0 and a_1 are rational and $i^2 = -1$. The conjugate of a denoted by a^* is given by $a^* = a_0 - a_1 i$.

The product aa^{**} is rational and is called the norm of a denoted by $N(a)$ which is equal to $a_0^2 + a_1^2$. If $a = bc$, then $a^* = b^* c^*$ and $aa^{**} = (bb^*) \cdot (cc^*)$ which shows that the norm of b is a factor of the norm of a. As with ordinary integers $b|a$ if $a = bc$ in complex integers.

A complex integer is a number of the form $a = a_0 + a_1 i$ where a_0 and a_1 are rational integers.

The only complex numbers that divide all complex integers are those of norm 1, ± 1 and $\pm i$. If $b|a$ the associates ub of b, where u is any unit, are also factors of a. If the complex integer p has no factors other than associates and units, p is called a prime.

The complex number $p = p_0 + p_1 i$ is a complex prime if $N(p)$ is a rational prime such that $N(p) \equiv 1 \pmod{4}$, $p = \pm(1 \pm i)$, or if p is the associate of a rational prime which is congruent to 3 modulo 4. One can prove that $p|ab$ only if $p|a$ or $p|b$ and thus deduce the fundamental theorem of arithmetic for complex integers. Virtually all theorems in the theory of ordinary numbers have analogues for complex numbers.

An algebraic number field $R(\theta)$ of degree n is generated by the root θ of an algebraic equation $a_0 x^u + a_1 x^{u-1} + \cdots a_n = 0$ where all a_i are rational and $a_0 \neq 0$, is an algebraic integer if all a_i are rational integers and $a_0 = 1$. The algebraic integers in an algebraic number field form an integral domain. Consider numbers of the form $a = a_0 + a_1\theta$, where $\theta^2 = -5$. The conjugate a^* and the norm of a are defined in a manner analogous to the complex numbers. The number 21 can be expressed as $21 = (1 + 2\sqrt{-5})(1 - 2\sqrt{-5})$ or as $21 = (4 + \sqrt{-5})(4 - \sqrt{-5})$, where the factors on the right hand side of each equation are prime in this system. This example demonstrates that the fundamental theorem of arithmetic (uniqueness of factorization) does not always hold true for such algebraic number fields.

Uniqueness of factorization can be restored by the introduction of ideals. An ideal a in the algebraic number field $K = R(\theta)$ is a set of integers of K such that: (1) if α and β are in a, then $(\alpha + \beta)$ is in a; and (2) if α is in a and ξ is any algebraic integer in K than $\alpha\xi$ is in a. A number α is replaced by its principal ideal (α) which consists of all numbers $\alpha\xi$, where ξ runs through all integers of K.

Ideals, which were invented for use in number theory, are now used extensively in higher algebra. Other generalizations of numbers occur in the theory of linear algebras. The algebra of quaternions is an example. See also **Quaternion**.

Analytic Number Theory. This involves the use of the methods of calculus and function theory to study the properties of the integers. The most famous problem is to determine the number $\pi(x)$ of prime numbers up to X. Gauss and Legendre suggested asymptotic formulas for $\pi(x)$. The simplest formula is $\pi(x) \sim (x/\log x)$. The symbology $f(x) \sim g(x)$ indicates that $f(x)$ is asymptotic to $g(x)$. Two functions, $f(x)$ and $g(x)$, are said to be asymptotic when the ratio $f(x)/g(x)$ tends to one as a limit as x tends to infinity.

This formula was not proved until 1896. The methods involved the use of the Riemann-Zeta Function $\zeta(s)$ defined by the series $\zeta(x) = \Sigma^\infty 1/n^s$. Where s is a complex variable $s = \sigma + it$. The series is convergent only for $\sigma > 1$ but $\zeta(s)$ can be defined for $\sigma \leq 1$ by using analytic continuation and is found to be a function with only a simple pole of residue 1 at $s = 1$. By using the theorem of unique prime number factorization, it can be shown that $\zeta(x) = \Pi_p 1/(1 - p^{-s})$, where p runs over all prime numbers, $\zeta(s)$ has no zeros for $\sigma = 1$. Riemann knew that $\zeta(s) = 0$ for s equal any negative even integer and that all the remaining zeros satisfied $0 < \sigma < 1$. Riemann conjectured that all the nontrivial zeros of $\zeta(s)$ lie on the line $\sigma = \frac{1}{2}$. This conjecture was contained in a memoir of 1859 and remains unproved.

P.G.L. Dirichlet proved that there exist infinitely many prime numbers which satisfy $p \equiv a \pmod{m}$ where a and m are given coprime numbers. Dirichlet defined a more general set of functions $L(s, x) = \Sigma_{n=1}^\infty X(n)/n^s$ which are known as Dirichlet L-functions. Where $s = \sigma + it$ as before and $X(n)$ is a character of the multiplicative group of residue classes modulo m and coprime to m. Therefore, X satisfies the equations $X(a_1) = X(a_2)$ for $a_1 \equiv a_2 \pmod{m}$ and $X(a)X(b) = X(ab)$. $X(n)$ is zero if n and m contain a common factor greater than 1. $X(n)$ can have values ± 1, 0, or roots of unity.

The conjecture that there are no zeros of the L-functions with real part $\sigma > \frac{1}{2}$ is known as the Generalized Riemann Hypothesis.

Additive Number Theory. Srinivasa Ramanujan was a protegé of G.H. Hardy. Hardy once visited Ramanujan in a hospital in England. Hardy remarked that the taxi in which he came had the rather uninteresting number 1729 for its license number. Ramanujan immediately remarked that 1729 was interesting because it was the smallest positive integer that could be expressed as the sum of two cubes in two different ways, namely $1729 = 12^3 + 1^3 = 10^3 + 9^3$. This is an example of an additive number theory result. Partitions and polygonal numbers, which occur in additive number theory, are discussed below.

A partition of n is a decomposition of the number n into additive parts. Repetitions in the additive parts are allowed and the order is irrelevant. Let $p(n)$ denote the number of partitions of n. Euler gave

$$1 + \sum_{n=1}^{\infty} p(n)x^n = \prod_{k=1}^{\infty} (1 - x^k)^{-1}$$

as a generating function for $p(n)$. Generating functions can be constructed for partitions that are subject to various restrictions. Euler observed that if $E(n)$ is the number of partitions of n into an even number of unequal parts and $U(n)$ is the number of partitions into an odd number of unequal parts, then $E(n) - U(n) = 0$ unless n is of the form $\left(\frac{1}{2}\right)m(3m \pm 1)$, when $E(n) - U(n) = (-1)^m$.

An asymptotic form for $p(n)$ was developed by Hardy and Ramanujan, the first term of which is

$$p(n) \sim \frac{1}{4n\sqrt{3}} \exp\left(\pi \sqrt{\frac{2n}{3}}\right).$$

Rademacher gave an exact expression for $p(n)$ as the sum of a convergent infinite series. Ramanujan observed that $p(n)$ possesses various congruence properties such as

$$p(5n + 4) \equiv 0 \pmod{5}$$

and

$$p(7n + 5) \equiv 0 \pmod{7}.$$

A polygonal number of order m is given by $x + \frac{1}{2}(m - 2)(x^2 - x)$, where $x = 0, 1, 2, \ldots$. The value of m is equal to the number of sides of a polygon; i.e., $m = 3$ gives triangular numbers, $m = 4$ gives square numbers, etc. Fermat stated that every positive integer is the sum of m polygonal numbers of order m. This theorem was proved by Legendre for $m = 3$, Lagrange for $m = 4$, and by Cauchy for the remaining values of m.

Polygonal numbers of order 4 are squares of integers. Fermat's theorem says that every positive integer can be expressed as the sum of one, two, three, or four nonzero squares. Waring stated an extension of this theorem to higher powers. Every positive integer can be expressed as the sum of at most $g(k)$ of kth powers. Another function $G(k)$ which gives the least number of kth powers required to represent all but a finite number of positive integers is of even greater interest. Clearly, $G(k) \leq g(k)$. Lagrange's result indicates that $G(2) = g(2) = 4$. Other results include $g(3) = 9$, $G(3) \leq 7$, $19 \leq g(4) \leq 35$, $G(4) = 16$, $37 \leq g(5) \leq 54$, and $G(5) \leq 23$. Vinogradow proved that $G(k) \leq k(3\log k + 10)$.

Vinogradow also proved that every sufficiently large odd number is the sum of three odd primes.

Diophantine Approximations. Every number processed by a computer or written in decimal form with a finite number of digits is a rational number. Therefore, there is some practical interest in the errors introduced by approximating irrational numbers by rational fractions. By using continued fractions, it can be shown that if θ is any irrational number, then there are infinitely many fractions p/q such that $|\theta - p/q| < 1/q^2$. A continued fraction is an expression of the form

$$a_1 + \cfrac{b_1}{a_2 + \cfrac{b_2}{a_3 + \cdots}}$$

which is abbreviated $a_1 + (b_1/a_2) + (b_2/a_3) + \cdots$. If all b_i's are equal to 1, the continued fraction is called a simple continued fraction. The simple continued fraction

$$a_1 + \cfrac{1}{a_2 + \cfrac{1}{a_3 + \cdots}}$$

is sometimes abbreviated as $[a_1, a_2, a_3, \ldots]$. The finite simple continued fractions

$$c_1 = [a_1] = a_1$$

$$c_2 = [a_1, a_2] = a_1 + \frac{1}{a_2}$$

$$c_3 = [a_1, a_2, a_3] = a_1 + \cfrac{1}{a_2 + \cfrac{1}{a_3}}$$

are called the convergents of the simple continued fraction $[a_1, a_2, \ldots a_k]$.

In general, it can be shown that $c_1 < c_3 < c_5 \cdots < x < \cdots c_3 > c_4 > c_2$, where x is any irrational number. If x is rational the last convergent is equal to x (in this case the continued fraction is finite).

A periodic continued fraction is a continued fraction of the form $[a_1, a_2, \ldots, a_n, \overline{b_1, b_2, \ldots b_m}]$ where n is a nonnegative integer and m is a positive integer. The period of the continued fraction is the sequence of repeating terms $b_1, b_2, \ldots b_m$. The length of the period is m.

A quadratic irrational is an irrational number that is a solution of $ax^2 + bx + c = 0$ where a, b, and c are integers. Every periodic continued fraction represents a quadratic irrational and every quadratic irrational can be expressed as a periodic continued fraction. It can be shown that if k is not a square then $\sqrt{k}$ can be expressed as

$$\sqrt{k} = [a_1, \overline{a_2, a_3, a_4, \ldots, a_4, a_3, a_2, a_1}].$$

Liouville proved that if ξ is any real algebraic number of degree n and k is any constant, then there exist at most a finite number of rational fractions p/q such that $|p/q - \xi| < (k/q^{n+1})$. This result can be used to construct transcendental numbers. Choose a number λ with a sufficiently rapid sequence of rational approximations such that there are an infinite number of rational fractions p/q that satisfy $|p/q - \lambda| < (k/q^{n+1})$.

DONALD R. HODGE, The BDM Corporation, Vienna, VA

NUMERICAL CONTROL. Since the Industrial Revolution (circa 1760), there has been a continuing search for more effective manufacturing methods. As part of this unrelenting process, numerically controlled (NC) machines were developed in the early 1960s and have been and continue to be widely used. These developments occurred before the widespread use of electronic computers, and, of course, the personal computer was unknown at that time. The simplest NC machines of the 1960s were *open-loop* — that is, the two or three axes of a machine tool were directed to manipulate the machining of a part in two or three dimensions by controlling the axes of the machine. There was no feedback to the controller. See Fig. 1. However, *closed-loop* systems with feedback to the controller were introduced shortly thereafter. See Fig. 2. Numerically coded coordinate data were fed to the early machines by means of punched paper (later mylar) tape. The tapes proved to be cumbersome to prepare and very difficult to store and retrieve over long periods of time. As a consequence of these inconveniences, but limited to rather large installations, data were transmitted to the controller by means of digitized electric signals. Unfortunately, these modernized early systems were limited to sharing mainframe computers because smaller computers with the required capacity were still unavailable. Later, less expensive computers with the required data-handling capacity were introduced. These led to the tying of NC systems with CAD/CAM data centers. In what is now termed *distributed numerical control* (DNC), one computer can feed part programs to *multiple* machine tools. The advent of mini- and microcomputers, plus the ability to address multiple serial ports on a real-time basis, gained acceptance of this technology by manufacturers. Contemporary NC systems continue to cut down job lot sizes, reduce setup time, and trim machining time and the skill level of the direct labor force. It should be emphasized further that NC systems are now tied together at the factory floor level by *local area networks* and that these, in turn, can receive and feed information to plant-wide and corporate-wide networks.

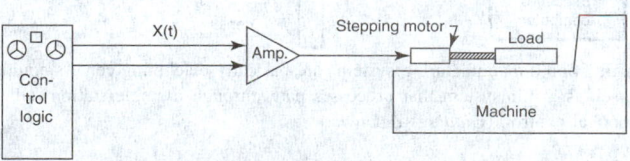

Fig. 1.   Open-loop numerical control system.

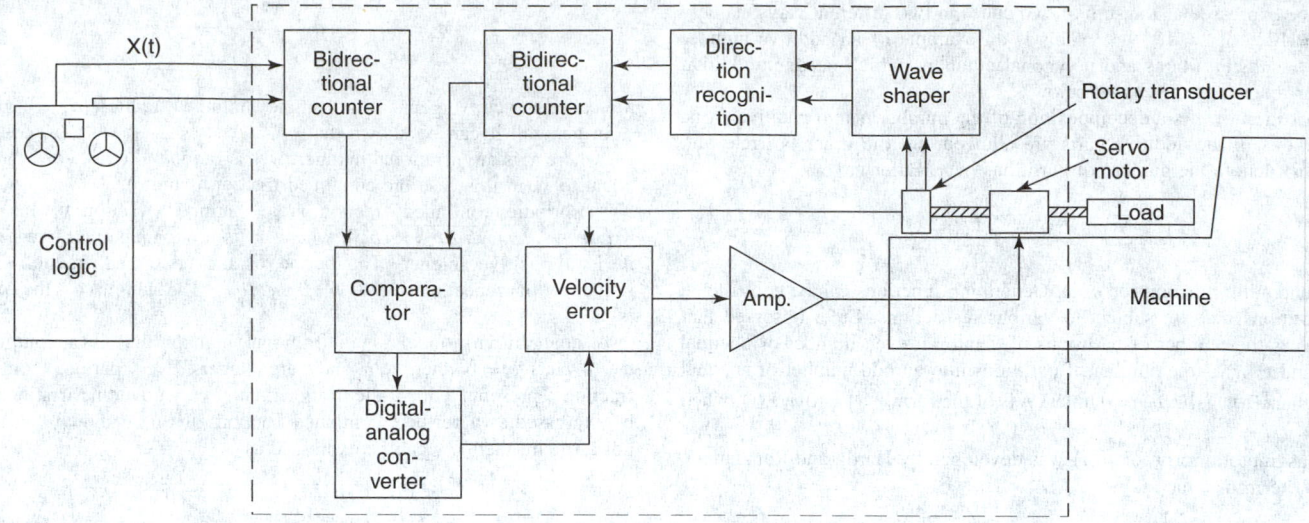

Fig. 2.   Closed-loop numerical control system.

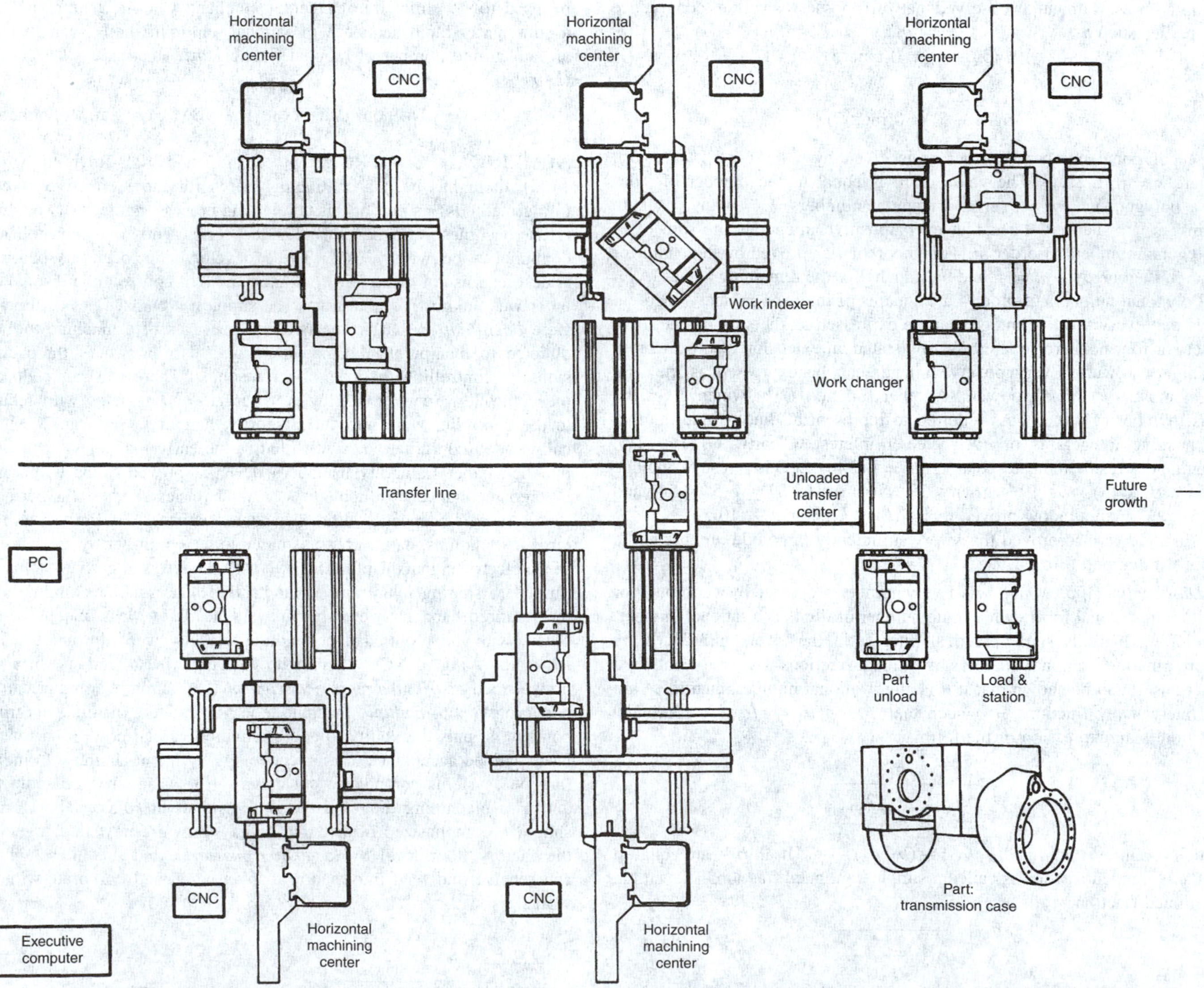

Fig. 3.   Some manufacturing systems are basically machining centers equipped with work handling options and joined by transfer equipment. Each machining center in itself is a minisystem that processes parts through a single station rather than through sequential stations. PC = programmable controller; CNC = computerized numerical control. (*Giddings & Lewis.*)

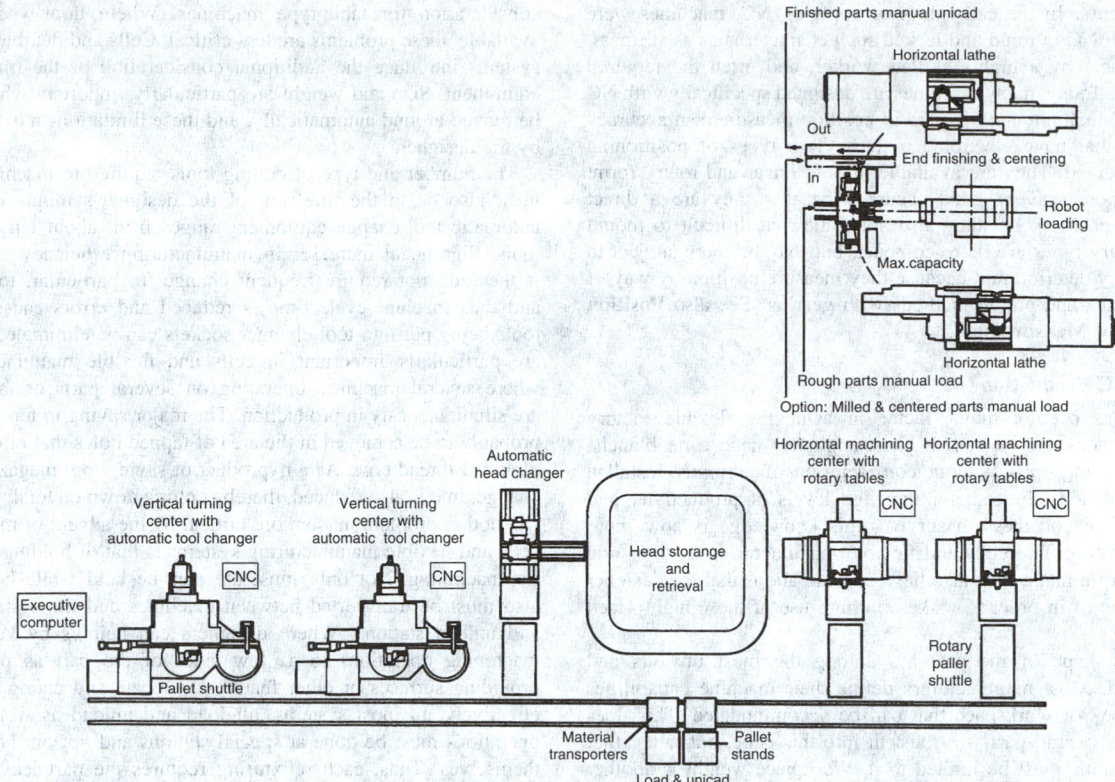

Fig. 4. Mid-volume manufacturing is frequently best served with general-purpose equipment and controls arranged as production modules. Thus, familiar and proven parts and components can be programmed by a master computer that has been preprogrammed for any number of parts. (*Giddings & Lewis.*)

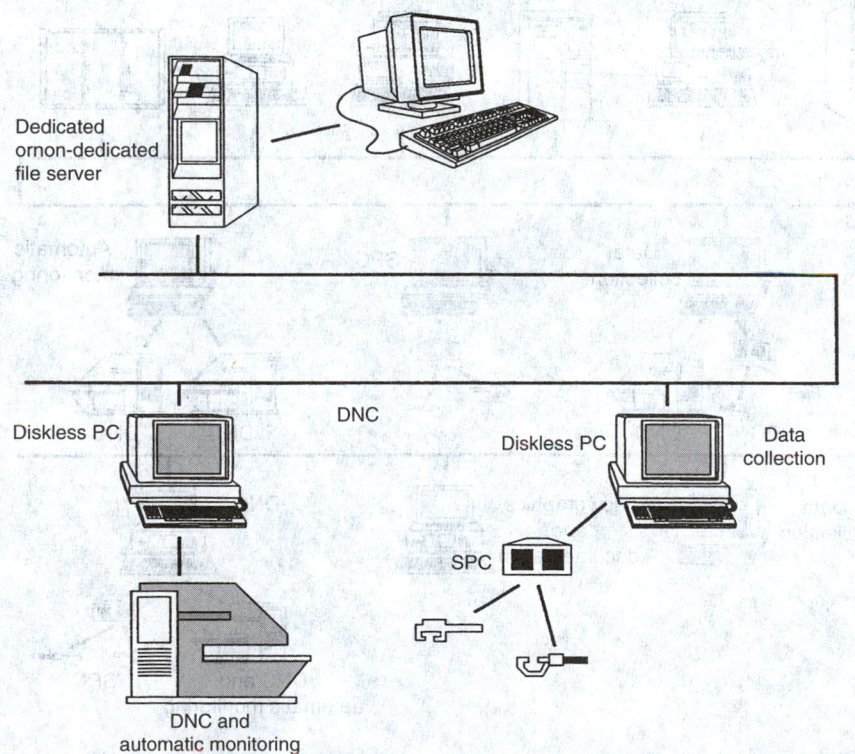

Fig. 5. Entry-level network for direct numerical control (DNC).

Basic system: (1) Entry-level standard network, (2) dedicated or nondedicated file server, (3) based on Intel 286 microprocessor, (4) supports up to five active users, (5) controls application information over network, (6) applications can be integrated in seamless environment through factory network control system.

Features: (1) High performance — files are transferred at 10 Mbs, (2) easy-to-use menu-driven utilities put the network supervisor in control, (3) security — allows supervisor to restrict network access, (4) single-source database for complete control of all files, (5) cost-effective network uses diskless PCs, (6) virtually unlimited expansion capabilities.

Applications: Statistical process control (SPC), (2) direct numerical control (DNC), (3) data collection, (4) view graphics and documentation, (5) automatic monitoring. (*CAD/CAM Integration, Inc.*)

NC Components. In the early days of applying NC, machines were not designed with NC in mind and lacked such characteristics as stiffness, accuracy obtainable by a highly skilled worker, and often the required production speed. Today, many machines are designed specifically with NC as an objective. Much attention is paid to position measurement accuracy and by devices that have electronic outputs. Two types of positioning devices have emerged. They are available in both linear and rotary form. Linear devices possess very good accuracy because they are a direct measure of axis position, but they also are somewhat difficult to mount and protect. Rotary types are more compact, but may be more subject to wear and accuracy deterioration because they measure position by way of lead screws or rack-and-pinion methods with gearing. See also **Position and Displacement Measurement**.

Designing for NC Production

Although NC has been a major factor in achieving flexible factory automation, this does not mean that plant engineers have carte blanche in terms of designing parts without consideration of currently installed manufacturing methods. Indeed, for the high levels of productivity and quality needed for a business to survive, this knowledge is now more important than ever. While cells and flexible manufacturing systems add to the flexibility of a plant, they also have considerations that the designer needs to be aware of in order to make optimum use of these high-priced installations.

The size and shape of the part are among the most obvious and important criteria. Most manufacturers define their machine capabilities in terms of "cube" of workspace that can be accommodated. This does not mean that the overall part size must fit into this cube, but rather that machining operations must be limited to it. Workpiece weight is another

consideration for table-type machines. Where floor-type machines are available, these problems are less critical. Cells and flexible manufacturing systems introduce the additional consideration of the material handling equipment. Size and weight are particularly important where parts are to be moved around automatically, and these limitations must also be known by the designer.

The number and type of cutting tools required to machine a workpiece must also be in the forefront of the designer's mind. The capacity of automatic tool-change equipment ranges from about ten to one-hundred tools. Significant increases in manufacturing efficiency can be achieved if these do not require frequent change. In particular, tool-change time and thus machine-cycle time is reduced and errors caused by incorrect tools being put into tool-changer sockets can be eliminated. These factors are particularly important in cells and flexible manufacturing systems where several machines operating on several parts or families of parts are simultaneously in production. The major saving in tool quantity is still probably to be achieved in the area of tapped holes that often require three tools per thread type. As a byproduct of saving tool magazine space, tool-change time is also reduced, thereby cutting down on total machining time.

Another problem made more critical by the advent of multiple machine cells and flexible manufacturing systems is that of holding the workpieces for machining. Not only must the part be held while being cut, but it also must be transported between machines and the system or cell load and unload stations. Where designers can help is by keeping required machining operations to as few sides of the part as possible and by providing surfaces or other features to locate and clamp on quickly and effectively. In most systems, all load and unload as well as refixturing operations must be done at special stations and not on the machine tools themselves. Thus, each refixturing requires the part leaving a machine,

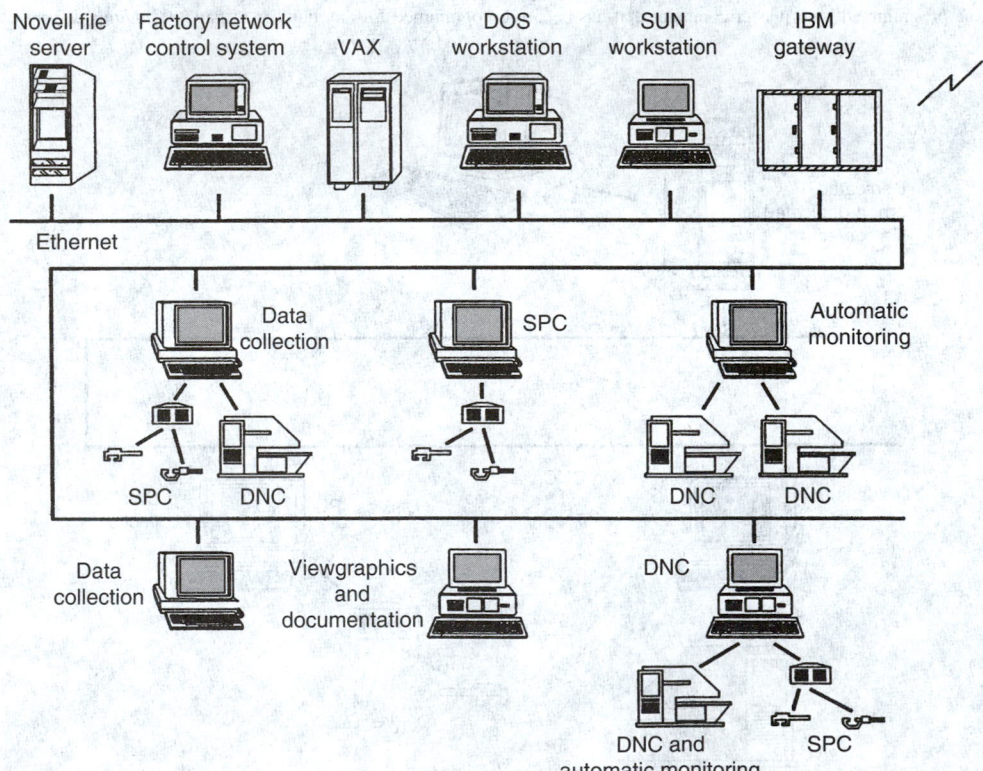

Fig. 6. Advanced NC network.

Basic system: (1) Controls application information of network, (2) applications can be integrated in seamless environment through factory network control system, (3) designed for medium to large business, (4) dedicated 386-based file server technology, (5) supports up to 100 active users on network, (6) high-performance network operating system.

Features: (1) Provides high functionality while maintaining high performance levels, (2) multiuser, multi-tasking architecture allows user to perform many operations simultaneously, (3) enhanced features for security, system reliability, and network management, (4) cell configurations decrease exposure to manufacturing down-time, (5) single-source database for complete control of files, (6) cost-effective network — uses diskless PCs, (7) simple-to-use menu-driven utilities make learning and adding applications easy.

Applications: (1) Statistical process control (SPC), (2) direct numerical control (DNC), (3) data collection, (4) view graphics and documentation, (5) automatic monitoring. (*CAD/CAM Integration, Inc.*)

traveling to a special station, and then traveling back again to continue being machined. This involves significant time as well as making the material transporter unavailable for other operations and can have an important detrimental effect on the efficiency of the system as a whole.

The designer must also know the rotary axis capability of the manufacturing plant. Many rotary Tables can only position in 15-degree steps, and any subdivisions beyond this angle can result in severe difficulties for the shop floor. The cost increment between Tables with indexing as opposed to full positioning capability to any angle can be very large, and thus designs that call for this requirement in order to be manufactured may be very expensive.

The designer must also keep in mind the part programmer and the capabilities available to simplify the task of producing effective NC tapes. Features such as origin shift, subroutines, probing, etc., previously mentioned are targeted at making programming easier in certain commonly found manufacturing situations. By designing the part to take advantage of these features, the designer can enhance the success of the manufacturing process.

Numerical control can be an effective production tool, not only for the very large manufacturing complexes, but for smaller shops as well. A panorama of contemporary systems is shown by Figures 3 through 6.

NUMERICAL DATA FILES.   See **File Types**.

NUMERICAL MODELING.   In oceanography, the prediction of flow evolution via numerical construction of approximate solutions to the governing equations. Solutions are obtained by assigning discrete values to temporal and spatial derivatives in order to convert the governing differential equations into algebraic equations that can be solved by using computational methods. Because computational resources are finite, no one technique is ideal for all applications. Some models define the equations on very fine spatial intervals. This approach furnishes solutions that are very accurate, but that span only small spatial regions (spatial scales of a few meters, at present). At the other extreme, some models span entire ocean basins by using large spatial intervals (hundreds of kilometers). Here, approximation of unresolved motions is a crucial and difficult issue. Similar trade-offs must be made with respect to temporal solutions. Numerical models also differ in the equations and boundary conditions that are employed. The most general model commonly used in oceanography includes momentum conservation via the incompressible Navier–Stokes equations with the Boussinesq approximation, mass conservation via the incompressibility condition, and equations expressing conservation of heat energy and salt (e.g., Gill 1982). For large-scale applications, the hydrostatic approximation is usually made. The vertical coordinate may be the geometric height, or a convenient substitute such as density, pressure, logarithm of pressure, or potential temperature. Surface boundary conditions generally express fluxes of momentum, heat, and freshwater from the atmosphere. Basin-scale models use boundary conditions that approximate the effects of bottom topography. Smaller-scale models typically specify periodic conditions at the side boundaries and an energy radiation condition at the bottom.

Additional Reading

Gill, A.E.: *Atmosphere–Ocean Dynamics*, Elsevier Science & Technology Books, New York, NY, 1982.
Haidvogel, D.B., and A. Beckmann: *Numerical Ocean Circulation Modeling*, Imperial College Press, London, UK, 1998.
Norbury, J., I. Roulstone: *Large-Scale Atmosphere-Ocean Dynamics: Analytical Methods and Numerical Models*, Vol. 1, Cambridge University Press, New York, NY, 2002.

NUSSELT NUMBER.   The nondimensional parameter, defined as

$$N_u = \frac{Q}{\Delta T}\frac{d}{k}$$

where Q is the heat loss or heat transfer from a solid body, ΔT is the difference of temperature between the body and its surroundings, d is the scale size of the body, k is the thermal conductivity of the surrounding fluid. The Nusselt number is useful in the reduction of measurements of free and forced convective loss of heat either from the same body in different conditions or from different bodies of geometrically similar shapes. See also **Heat Transfer**.

NUTATION.   In the case of a spinning object (e.g., a top or gyroscope, a particle, or an astronomical body), the inclination of the axis to the vertical will vary periodically between certain limiting angles. This motion is called nutation. It is a variation in precession. In astronomy, nutation is caused by the attracting force of the sun and the moon tending to pull the equatorial bulge of the earth into the plane of the ecliptic. The amount of this force is changing throughout the year as the declinations of the sun and the moon change. For example, twice during each year, both the sun and the moon are on the equator, and at those times, their precessional forces are zero. The principal nutation is due to the periodic change in the plane of the moon's orbit, and has a period of about 19 years. Most of the nutation effects are periodic in character.

NUTHATCH *(Aves, Passeriformes)*.   Small climbing birds which cling in any position to the bark of trees as they search for food. They eat both insects and seeds. The most common North American species is the white-breasted nuthatch, *Sitta carolinensis*, found throughout the United States east of the Rockies. The red-breasted nuthatch, *S. canadensis*, is more common in Canada and the mountains but migrates in winter as far as the southern states. The pigmy nuthatch, *S. pygmaea*, is a western species. Nuthatches are found on all other continents except South America, although in Africa they are confined to the north.

NUTMEG TREE.   Of the family *Myristicaceae*, the nutmeg tree grows to about 60 feet (18 meters) in height and has pointed lanceolate leaves. Nutmegs are the seeds of this tree which is native to the Molucca Islands. The trees are dioecious, pistillate and staminate flowers being borne on separate trees. Since the trees are frequently grown in cultivation, especially in favorable localities, it is necessary to plant some of both sexes to ensure cross-pollination and seed formation. The flowers are pale yellow, the fruit a dark orange-colored berry containing a single large brown seed. Surrounding the seed is a branched deep-red aril, which upon drying becomes pale brown. The seed is the nutmeg of commerce; the aril is mace. Both the seed and the aril contain an aromatic oil, but only the poor-quality seeds are used for extraction of oil. The tree begins to bear at about 6 years of age and may produce for nearly a century. The tree produces best when planted at an altitude of from 700 to 1,800 feet (213 to 549 meters) in a tropical climate. The average tree will produce about 20 pounds (9 kilograms) of nutmegs per year. An acre produces about 1,200 pounds (544 kilograms) of green nutmegs annually. From this quantity, about 150 pounds (68 kilograms) of mace are produced; and 720 pounds (327 kilograms) of dried nutmegs. See also **Flavorings**.

NUTRIENTS (Soil).   See **Fertilizer**.

NUTRITIONAL SCIENCE (THE HISTORY).   The development of nutritional science followed two largely separate lines. One studied the food requirements of humans and animals, at first considering only protein and energy; this developed from knowledge of physiology and chemistry. The second rose from the study of diseases that were gradually realized to result from deficiency of a micronutrient. We will consider each line in a separate historical sequence, following developments up to about 1950. More recent work, including study of the biochemical functions of nutrients, will be considered in other articles. See also **Biochemistry (The History)**; and **Physiology (The History)**.

Protein

The term "protein" was not coined until 1839, but there had already been the concept of "animal substance." Blood serum, minced muscle, egg white etc., all were gluey, set hard when heated, and when decaying in warm, moist conditions turned alkaline with a foul smell.

It was a problem for early scholars to understand how a diet of leaves and grains could turn into "animal substance" in a growing animal when they had quite different properties and turned acidic when left to rot. Jacopo Beccari, of the University of Bologna, showed in 1728 that by adding water to white wheat flour and kneading it under a stream of water, so that the starch washed out, a residue, "gluten," remained. This had the properties of "animal substance" and he believed it to be the hidden, true nutrient in wheat.

By the 1780s, French chemists had identified the main alkaline vapor from decayed "animal substance" as ammonia, which contained the element nitrogen, whereas starch, sugars and fats contained only carbon, hydrogen and oxygen.

Nitrogen was the largest component of the atmosphere, but could animals use it to synthesize their own animal substance, or did they have to rely on the smaller amounts present in plants? François Magendie, a French physiologist, reported in 1816 that dogs fed on either sugar and water, or fat and water, after a few days lost appetite and eventually died. He concluded that dogs needed nitrogenous foods. See also **Magendie, François (1783–1855)**.

Herbivorous animals, which subsisted on foods of lower nitrogen content, were also tested. Jean Baptiste Boussingault (1802–1887) reported results with a cow fed for several weeks on a ration of hay and potatoes that kept her at constant weight. Analysis showed that she received 202 g nitrogen (N) per day. The milk secreted each day contained 46 g N and her combined excreta another 109 g N. Since input exceeded output there was no reason to believe that a cow obtained additional N from the air. See also **Boussingault, Jean Baptiste (1802–1887)**.

Chemists compared the animal substances found in various animal tissues and in plants. They all contained approximately 16% N, but their solubility differed greatly. In 1838 Gerrit Mulder, a Dutch physician, claimed that they had a common radical, protein, containing 40 atoms of carbon, 62 of hydrogen, 10 of nitrogen and 12 of oxygen; the differences coming from the radical being combined with different proportions of phosphorus and/or sulfur.

Justus von Liebig in Germany adopted the idea, and added that they all had a ratio of exactly 4 carbon atoms to 1 of nitrogen. He generalized that the powers of animals to synthesize compounds were very limited; they normally obtained from plants the range of nitrogenous substances that they needed, but were able to make some slight changes such as removing phosphorus or sulfur atoms. Starch and sugars could only either be oxidized or converted to fat. In other words, substantial chemical synthesis occurred in plants, and animals could only oxidize and degrade, to complete the balance of nature between the two kingdoms.

The French chemist Jean Dumas had been working on similar lines with Boussingault. They believed that Liebig had stolen their ideas without really understanding them. They denied that animals converted carbohydrates to fat, which required the removal of oxygen rather than its addition. Liebig replied that pigs fed on starchy foods such as potatoes obviously laid down fat. The French argued that pigs' diet also contained sufficient fat. Boussingault carried out a careful balance study, and finally admitted defeat.

Liebig's reputation as an authority on nutrition (albeit an "armchair" one) soared. He next argued that, since muscle tissue contained nothing but protein, the energy for muscular contraction must come from its breakdown, with the nitrogenous fragment being converted to urea that appeared in urine. This meant that human capacity for physical work was proportional to the consumption of protein, the only true nutrient. Carbohydrate served only to combine with oxygen that would otherwise be dangerous to body systems. Carbon dioxide production increased with physical exercise because heavier breathing needed to cool the body increased the introduction of oxygen. There were protests that inactive rich people ate most protein, whilst laborers ate mainly bread, but Liebig's idea was widely accepted.

Results appearing to contradict Liebig's idea were obtained by Edward Smith in a London prison. Convicts carried out heavy labor on only 3 days per week and their daily excretion of urea was no greater after exercise than on rest days. In 1865 in Switzerland, Adolf Fick and Johannes Wislicenus carried out a critical trial, climbing a mountain while eating a protein-free diet, and showed, from the small amount of urea they excreted, that the energy needed for the climb could not have come solely from the breakdown of protein.

The proof depended on the concept of conservation of energy. James Joule and others had shown that the heat needed to raise 1 kg of water by 1 °C was equivalent to the energy needed to raise 423 kg against the force of gravity by 1 m. The men had ascended nearly 2000 m, and had done, on average, an absolute minimum of 139 000 kg-m of work, equivalent to 323 kilocalories. From their excretion of urea during and for 7 hours after the climb, it was calculated that each had metabolized 37 g protein. Edward Frankland, in London, had found that complete combustion of muscle protein (with a small correction for combustible energy remaining

in urea) yielded 4.37 kcal g^{-1}. The maximum energy yield from 37 g was therefore only 162 kcal. Even if muscles were 100% efficient in converting latent energy into work, and with no allowance for the work of the heart and lungs during the climb, protein could not have been their only muscle fuel. It was concluded that a muscle did not consume itself as a fuel, but relied mainly on carbohydrates or fat.

The German School was not converted, and argued that a high-protein diet also contributed nervous energy. Carl Voit recommended a daily intake of 120 g for a working man, the choice of people able to make their own selection. Wilbur Atwater, the authority in the USA, believed that the standard should be higher still. This was investigated by Russell Chittenden, Professor of Physiological Chemistry at Yale. After working with academic colleagues, Yale athletes and volunteer soldiers from the Medical Corps, his final conclusion in 1904 was that 64 g protein was sufficient for a man of average weight to function, physically and mentally, at full potential. This was bitterly debated but was confirmed in later studies. "Look after the calories and the protein will look after itself" came to be an axiom for adults. See also **Atwater, Wilbur Olin (1844–1844)**.

Energy

Antoine Lavoisier had carried out respiration studies with Armand Séguin in the 1780s. They found increased expiration of "fixed air" (carbon dioxide) during physical work, and concluded that it came from the oxidation, or combustion of body fat, which was the source of animal heat. From trials with guinea-pigs they concluded that the heat produced by a living system was the same as that produced by external combustion of the same amount of carbon. See also **Lavoisier, Antoine Laurent (1743–1794)**.

By 1842 French chemists had concluded that a man needed to "combust" enough material to yield 2500–3000 kilocalories of heat per day, to maintain body temperature. By 1866 it was realized that dietary fats and carbohydrates were serving as fuels for production of physical work. For the next 40 years work was concentrated on measuring the energy value of foods by burning them in a "bomb calorimeter" containing oxygen at high pressure, with the vessel immersed in water whose rise in temperature could be measured. The values obtained then required a correction for their digestibility determined in actual feeding trials.

Wilbur Atwater led such studies in the USA, and also the construction of a calorimeter that a subject could live in for several days, the heat output being trapped by circulating water pipes. Carbon dioxide output and oxygen consumption could also be measured. By having a subject do known amounts of physical work on a stationary bicycle, muscular efficiency was found to be approximately 25%, whether it was mainly fat or carbohydrate supplying the energy; that is to say, of the energy liberated, 25% appeared as physical work and 75% as heat.

By 1900 Atwater's group could rank the economic value of foods in terms of the relative costs of providing equivalent amounts of protein and energy. One general conclusion was that fruits and leafy vegetables were uneconomical and, by implication, unnecessary—a conclusion reversed after discovery of the vitamins.

Deficiency Diseases

Scurvy. This condition, often known as "the sailor's disease" became a problem once voyages of ten or more weeks at sea began in the late fifteenth century. Typically, crew members became weak and stiff, with loose teeth and big, flat blood blisters over their body. The Latin name for the condition was *Scorbutus*, first anglicized to *scorby* and then *scurvy*. See also **Ascorbic Acid (Vitamin C)**.

It was soon found that eating citrus, other fresh fruit or fresh green food resulted in a rapid recovery. But on long voyages fresh fruit would go mouldy, and search was made for substitutes. One theory, dating from 1600, was that scurvy came from bodies being out of balance in the alkaline direction, since acid fruit was a corrective. The College of Physicians in London recommended supplying sulfuric acid to naval surgeons, which they could give diluted with water.

This was finally tested in what has been called "the first controlled clinical trial." In 1747, James Lind, a young surgeon with no college education, took 12 sailors with similar cases of scurvy and assigned two to each of six treatments in current use. After only 6 days, the pair receiving two oranges and one lemon per day had recovered. The other groups, which included treatments with sulfuric acid, vinegar or fermented cider, showed no improvement. See also **Lind, James (1716–1794)**.

Lind later received professional training as a physician and published a treatise where he wrote that: "although acids agree in certain properties, they differ widely in others, and especially in their effects upon the human body." Little notice was taken of Lind's work and it was not until 1795 that lemons or limes became a standard issue in the British Navy. Lind did not believe that the active factor in lemon juice or other "anti-scrobutics" was needed by people ashore because, in his native Scotland, they remained healthy throughout the winter without any such "anti-scrobutics." He had forgotten their use of turnips, another source of the active factor.

Goitre. One element discovered in the early 1800s was iodine, isolated from the ash of seaweed. By 1820 in Switzerland, salts of iodine had proved an effective treatment for goitre, characterized by swelling of the thyroid gland in the neck. There were also several demonstrations that goitre and cretinism occurred where food and water were particularly low in iodine. Boussingault, who had begun his professional work in South America, recommended iodination of salt in low-iodine areas there in 1831. Unfortunately, iodine, for which only 0.15 mg per day are needed, was sometimes given in unnecessarily large doses with toxic effects, and its use became discredited. August Hirsch, in his *Handbook of Geographical and Historical Pathology* (1883), wrote that the iodine-deficiency theory was: "a short-lived opinion … endemic goitre and cretinism have to be reckoned amongst the infective diseases." See also **Thyroid Gland**.

Even after the discovery by 1900 that iodine was concentrated in the thyroid gland, it took 20 years for studies to begin again on the value of routine iodine supplements. But by 1923, David Marine had shown their value in reducing the incidence and severity of goitre among schoolgirls in Ohio, and iodized salt was finally reintroduced.

Beriberi. Beriberi, unknown in Europe, was called "the national disease of Japan" by Western physicians working there in the 1880s. It was characterized first by weakness and loss of feeling in the legs, then (in many cases) by a dropsical swelling of the body, and finally by difficulty in breathing and heart failure.

It had already been seen by European physicians in their Asian colonies. Thomas Christie reported from Ceylon (now Sri Lanka) in 1804 that it came from "want of a stimulating and nourishing diet" but, unlike scurvy, "did not respond to acid fruits." Dutch naval surgeons believed that the disease came from native diets being low in both protein and fat. But, as the germ theory of disease became dominant, it was assumed that a microorganism would be found as the cause.

While engaged in such work, Christiaan Eijkman, a young Dutch physician in Batavia (now Jakarta), Indonesia, made a chance observation. In 1889, while trying to infect chickens with blood from beriberi patients, he noticed birds, whether injected or not, that had developed a leg weakness reminiscent of beriberi. This did not recur and he discovered that when the condition was developing, the birds had been fed left-over cooked rice from the adjoining hospital, and at other times on raw, feed-grade rice. In further trials he found that chickens fed on fully milled "white rice" would develop "polyneuritis" while those given the less highly milled brown rice would not. See also **Eijkman, Christiaan (1858–1930)**.

A study in the many small prisons in Java showed that beriberi was a serious problem only where white rice was the staple food, but not in those using brown rice. This showed the value of animal models for studying a human nutritional disease. The work to identify the active factor in the skin and germ of rice grains was to take another 40 years. See also

Nutritional Anemia. In the Victorian period, anemia was common among young women. The concentration of hemoglobin (the oxygen-carrying molecule) in their blood was abnormally low. It was known also that most of the body's iron was present as hemoglobin and, in the 1830s, the French physician Pierre Blaud had recommended ferrous sulfate pills as a certain remedy. But this was not generally followed. See also **Anemias**.

There were two problems. First was Liebig's generalization that the animal kingdom could not synthesize. One German physician wrote: "it is as unlikely that humans could make hemoglobin from inorganic iron as that they could make protein from potassium nitrate and starch." Second, the method used to analyze the iron content of foods was nonspecific, and bread was considered as good a source as meat. Anemia was explained as the result of autointoxication.

In 1893 Ralph Stockman, in Edinburgh, improved the analytical procedure and demonstrated the low iron intake of anemic women. He also confirmed that iron salts given either by mouth or injection produced responses in blood haemoglobin levels.

The Twentieth Century

Amino Acids. Liebig had envisaged food proteins being transferred intact to animal tissues. However, in normal digestion, proteins were found to be broken down into the same small amino acid units as were obtained when proteins were boiled in strong acids. Clearly, animals could re-synthesize proteins from these breakdown products. Some vegetable proteins were also found to have very different proportions of individual amino acids from the animal proteins, so that the need for re-synthesis became understandable. But, "Were all proteins equally nutritious?" Thomas Osborne, of the Connecticut Agricultural Experiment Station, had spent many years isolating plant proteins and in 1909 began collaboration with Lafayette Mendel, a nutritionist at Yale. They fed young rats' purified diets with Osborne's preparations as their sole protein source. See also **Amino Acids**.

One experiment used zein, the main storage protein in maize grains and known to lack two amino acids – tryptophan and lysine. On this diet the rats lost 20 g per head in 2 weeks. With added tryptophan and lysine, they gained 22 g. Further work showed that both tryptophan and lysine were "essential" amino acids for rats, and also for humans, meaning that we cannot synthesize them even though they are essential components of our tissue proteins. In contrast, some other amino acids such as glycine can be synthesized.

Most plant proteins had some of each amino acid found in animal proteins, though often at lower levels. As expected, they supported some growth but had a lower "biological value" in feeding trials — more being required for a given synthesis of animal tissue than with protein having a better balance of amino acids.

The value of the proteins in grain was generally limited by their relatively low lysine content and that of legumes (beans and peas) by methionine. Mixtures of the two classes of foods partly supplemented each other's deficiencies. Protein nutrition was now seen as a matter of "amino acid nutrition" and methods of analysis were developed so that the amino acid contributions of any particular dietary mixture could be calculated.

Vitamins. From 1900 to 1950 a major interest was the discovery and synthesis of what came to be called vitamins. Several lines of work converged in this direction. See also **Chick, Harriette (1875–1977)**

By 1905, the findings of the Dutch in Java were confirmed by British scientists in Malaysia, and workers in several countries began to concentrate the anti-beriberi factor from rice bran, and to test known compounds that they guessed might be active, using either chickens or pigeons as their model species.

In 1905, two Norwegian medical scientists, Axel Holst and Theodor Frölich, investigating the problem of beriberi in Norwegian sailing ships, tried using guinea-pigs to produce a model of the disease. To their surprise, the disease that developed when the animals were fed on milled grains was recognizable as scurvy, which was confirmed by their responding to lemon juice.

Between 1905 and 1912, pellagra had become an important, disabling disease in the southern states of the USA. Known previously only in southern Europe, it was characterized by a repulsive, flaky dermatitis on parts of the skin exposed to the sun, diarrhea and mental disorder. It appeared that Italian immigrants to the region might have brought the "infection" with them. Joseph Goldberger, a senior US public health officer, and some of his colleagues put this idea to the test by eating skin or excreta from sufferers but none developed the disease. However, sufferers were characterized by low milk consumption, and in orphanages Goldberger was able to correct the problem by giving supplements of milk and eggs. They then found that dogs fed a diet similar to that of pellagra sufferers developed a condition with some similarities to pellagra. From testing many possible supplements, brewer's yeast proved most potent. Its value was confirmed in a mental hospital, and in the late 1920s the Red Cross distributed it as a preventive.

In 1912, Casimir Funk, a young Polish scientist working in London, had reported the isolation of crystals that cured polyneuritis in pigeons fed on white rice. He believed that they contained an amine ($-NH_2$) group.

He then suggested that beriberi, scurvy and pellagra were all caused by dietary deficiencies of different "vital amines," a term later abbreviated to vitamins. In one sense, Funk was wrong: he had not isolated the anti-beriberi factor though, when finally isolated, it was found to contain an amine group. But the factors later found to prevent scurvy and pellagra was chemically very different. Nevertheless, Funk had devised "a captivating word that focused the attention of scientists." See also **Funk, Casimir (1884–1948)**.

So far the work described sprang from investigation of human diseases. Further studies came from curiosity as to whether the animal kingdom's nutritional requirements were limited to protein, carbohydrates, fat and minerals, as assumed by Liebig and his successors.

One important study was begun by Professor Edwin Hart at the University of Wisconsin, Madison. Dairy cattle were fed rations made up from single grains–wheat, oats or maize (corn). They were balanced for the known nutrients, and for fiber, by varying the proportions of grain, hay and straw. In short, only the "corn" rations kept the animals in good health. Cows on the other rations had calves born dead or weakly.

Elmer McCollum had been hired to do the analytical work for these studies. He decided that he could only make rapid progress by using a smaller and rapidly breeding species. He showed how rats needed a fat-soluble material, named vitamin A. See also **McCollum, Elmer Verner (1879–1967)**. and **Vitamin A**.

Two scientists in Europe had already started to work with either mice or rats, feeding them on what were called "purified" diets, that is to say a mixture of purified protein, carbohydrate, fat and either a mix of pure salts or the ash prepared by incinerating milk powder. Cornelis Pekelharing in Utrecht and Frederick Gowland Hopkins in Cambridge (UK) both wrote, in 1905 and 1906, respectively, that they had found that mice deteriorated on purified diets, but that small additions of milk restored them. See also **Hopkins, Frederick Gowland (1861–1947)**.

During the 1930s the water-soluble "vitamins" postulated by Funk were all isolated, identified and synthesized: ascorbic acid (or vitamin C) against scurvy, thiamin (or vitamin B_1) against beriberi, and nicotinic acid (or niacin) and riboflavin against pellagra. Of these only nicotinic acid was a previously known compound.

Further work with rats and chickens revealed the existence of pantothenic acid, pyridoxine and biotin as water-soluble factors, vitamins D, K and E as fat-soluble ones in addition to vitamin A, and lastly the need for more trace minerals including copper and zinc. See also **Vitamin E**.

Fats. The French chemist Michel Chevreul had shown in the 1820s that fats (including vegetable oils) were compounds of glycerol and fatty acids. The proportions of individual fatty acids influenced their properties. Some, the unsaturated acids, reacted with oxygen to give an oil its "drying" properties. However, in the nineteenth century, dietary fat was thought of merely as a concentrated source of energy (9 kcal g^{-1}) compared with carbohydrates (4 kcal g^{-1}), and not to be essential because it could be synthesized by both humans and animals from carbohydrate. See also **Fatty Acids**.

In 1930, George and Mildred Burr in Minnesota demonstrated that rats fed on a fat-free diet developed scaly skin and abnormal kidneys, and died early. The condition could be completely reversed by giving linseed oil and cod liver oil. Giving pure linoleic acid helped to almost the same extent and this was described as 'an essential fatty acid'.

The finding was contested but finally accepted. Soon after, a class of hormones, the prostaglandins, were discovered that had structures related to linoleic acid and other highly unsaturated fatty acids.

More Diseases

Rickets. By 1900 it was realized that many mineral elements must be essential nutrients since they were always present in animal bodies. They included sodium, potassium, and the mineral constituents of bone, calcium, magnesium, phosphate and fluoride.

Rickets had been a serious problem in the big cities of northern Europe and North America. Infants in their first two years of life were at risk from developing weak bones that would bend on pressure, leading to the "bandy legs" associated with life in the slums. But, although rickety bones had a lower-than-normal mineral content (mainly calcium phosphate), giving mineral supplements did not help.

Cod liver oil was a traditional treatment in some areas, and others had learned to expose an infant's skin to the sun. By 1920, with the help of X-rays, it seemed clear that both were effective, and that it was the ultraviolet portion of sunlight that was active. Unfortunately, this was the portion most absorbed by the smoke over industrial cities. It was discovered that vitamin D, the active fraction in cod liver oil, was also synthesized from sterols in skin irradiated with ultraviolet light. See also **Vitamin D**.

Macrocytic (big cell) Anemia. In 1928 Lucy Wills, an English physician, began to investigate the anemia common amongst pregnant women factory workers in Bombay. Unlike iron deficiency anemia, the red cells were abnormally large rather than small. The only effective supplement was yeast extract. A similar condition was produced in monkeys. When an unknown growth factor for bacteria was identified, it was also found active for monkeys, and patients in India, and given the name folic acid, or folacin. See also **Folic Acid**.

Another condition recognized since 1850 was pernicious anemia, another progressive condition characterized by bigger-than-normal cells. It did not respond to iron or to folacin. Pernicious anaemia patients also showed degeneration of the gastric mucosa. In the late 1920s, William Castle and colleagues in Boston showed that the normal stomach lining secretes an "intrinsic factor" which, when mixed with either meat or liver and fed to a sufferer from the disease, produced improvement. The active "extrinsic factor" was isolated from liver in 1948 and named vitamin B_{12} (or cobalamin because the molecule included a cobalt atom). It is a large molecule and poorly absorbed unless combined with "intrinsic factor." Cobalamin seems to be synthesized only by certain microorganisms. Humans have traditionally obtained it from the meat or milk of ruminants who, in turn, obtained it from bacteria in their gut. See also **Vitamin B_{12} (Cobalamin)**.

Kwashiorkor. Kwashiorkor, a disease of infants, was first described in 1933 by an English physician, Cicely Williams, working in West Africa. It was characterized by swelling of the hands and feet, skin peeling off in a "crazy pavement" pattern, and an impression of misery. There was a high mortality and on autopsy the liver was seen to be fatty. She considered it to be a nutritional deficiency of some kind. See also **Malnutrition**.

The same disease, under different names, was later realized to exist in other areas of Africa, South America, Southeast Asia and the West Indies. Workers found that it could be treated with high-protein milk powders, though recovery was slow. It was probably the result of using weaning foods that were low in protein and also too bulky for consumption sufficient to meet energy requirements.

Recent Studies

Since 1950 it has become clear that diets supplying all the known nutrients in adequate amounts can still be suboptimal over the long term.

Coronary Heart Disease. Heart failure caused by the blockage of an artery in the muscular wall of the heart kills many men between the ages of 30 and 50 in more affluent countries. The plaques that block the passage of blood through the arteries contain a high proportion of cholesterol, and this condition of atherosclerosis can be produced in rabbits by adding cholesterol to their feed. See also **Cerebrovascular Diseases**; **Cholesterol**; **Congestive Heart Failure**; **Coronary Artery Disease (CAD)**; and **Ischemic Heart Disease**.

Further studies on humans demonstrate that, although dietary cholesterol has some effect in raising the level of blood serum cholesterol, feeding saturated fat has a larger effect, whilst polyunsaturated fatty acids have the opposite influence. From 1957 on, recommendations have been made that people should consume less of saturated fat sources, including eggs and butter.

Cancer. Observations of the differing incidences of various types of cancer in different countries have led to suspicions that diet influences risk. By 1940, several studies had shown that underfed rats or mice developed fewer tumors than those eating freely. Epidemiological studies then showed that the 5-fold range in age-adjusted death rates from breast cancer in different countries correlated strongly with national fat consumption. See also **Cancer and Oncology**.

Many measures of affluence tend to move together—fat, sugar, obesity, physical inactivity and so on. But this kind of observation has stimulated much ongoing work. There have also been suggestions from both animal experiments and epidemiological studies that vegetables contain factors having a partial protective effect. See also **Obesity**.

General Conclusions

Nutritional science began with breaking down nutritional requirements to individual chemicals and identifying them. By 1950, that work seemed complete. Since then studies have turned to the further problem of understanding how the many compounds present in foods may influence resistance to the development of cancer and other diseases, for good or for ill.

Additional Reading

Carpenter, K.J.: *The History of Scurvy and Vitamin C*, Cambridge University Press, New York, NY, 1986.

Carpenter, K.J.: *Protein and Energy: a Study of Changing Ideas in Nutrition*, Cambridge University Press, New York, NY, 1994.

Carpenter, K.J.: *Beriberi, White Rice and Vitamin B*, University of California Press, Berkeley, CA, 2000.

Darby, W.J., and T.H. Jukes: *Founders of Nutrition Science*, 2 Vols. American Institute of Nutrition, Bethesda, MD, 1992.

Galdston, I.: *Human Nutrition Historic and Scientific*, International Universities Press, New York, NY, 1960.

Guggenheim, K.Y.: *Nutrition and Nutritional Diseases: the Evolution of Concepts*, DC Heath, Lexington, MA, 1981.

Hallgren, B., Ö. Levin, S. Rossner, and B. Vessby: *Diet and Prevention of Coronary Heart Disease and Cancer*, Raven Press, New York, NY, 1986.

McCollum, E.V.: *A History of Nutrition*, Houghton Mifflin, Boston, MA, 1957.

O'Hara, M.J.: *Elizabethan Dietary of Health*, Coronado Press, Lawrence, KA, 1997.

Williams, R.R., and T.D. Spies: *Vitamin B₁ (Thiamin) and its use in Medicine*, Macmillan, New York, NY, 1938.

KENNETH J. CARPENTER, University of California, Berkeley, CA

NUX VOMICA TREE. Of the family *Loganiaceae*, the nux vomica tree is a small-to-moderate size tree found in India, Sri Lanka, Burma, Thailand, and Australia. The trunk is often crooked, short and thick. The leaf is smooth with three to five veins and ovate. The flower is light green, small and of a tubular shape. The fruit is hard with a gelatinous pulp. The tangerine-size fruit contains from one to five disk-shaped seeds. The seeds contain powerful alkaloids, including strychnine and brucine.

NYLON. [CAS: 63428-83-1]. $(C_6H_{11}NO)_n$. Generic name for a family of polyamide polymers characterized by the presence of the amide group $-CONH$. By far the most important are nylon 66 (75% of U.S. consumption) and nylon 6 (25% of U.S. consumption). Except for slight difference in melting points, the properties of the two forms are almost identical, though their chemical derivations are quite different. Other types are nylons 4, 9, 11, and 12.

The first nylon developed (type 6/6) was discovered in 1938 by W. H. Carothers. Since that time, nylons have filled an important role for industry and the consumer in various formulations, shapes, and forms, e.g., oriented fibers, which are subsequently processed into fabrics, fishing line, and other monofilament uses; injection-molded nylons, used as bearings, gears, and other parts subjected to wear and impact; extruded nylon tubing and hose, used in large quantity because of its chemical inertness, high strength, and flexibility; oriented nylon strip used as strapping for packaging, displacing traditional steel strapping; and heavy cast-nylon parts, frequently used in the textile, paper-making, and bottle-handling fields.

Most nylons exhibit a combination of high melting point, high strength, impact resistance, wear resistance, chemical inertness, and a low coefficient of friction.

Types of Nylon. Type 6/6 and type 6 nylons are widely used, dominating the field of textile fibers. Nylon 6/10, a lower-strength material produced in less volume, is used for industrial applications requiring improved moisture stability and high dielectric strength. It also has a lower melting point, lower specific gravity, and higher cost than types 6/6 and 6. Nylons 11 and 12 appeared considerably later than the other formulations. Generally, these nylons have a lower order of moisture absorption and are thus preferred where consistent properties are required in the presence of moisture.

They are also more chemically inert, flexible, and in certain cases, are transparent.

Formulations. Types 6/6 and 6/10 are formed by the condensation of diamines with dibasic organic acids into linear chains containing amide groups. Types 6, 11, and 12 are self-condensed amino acids.

Type 6/6:

$$NH_2(CH_2)_6NH_2 + HOOC(CH_2)_4COOH \rightarrow$$
Hexamethylenediamine   Adipic acid

$$[NH(CH_2)_6NHCO(CH_2)_2CO]_n + H_2O$$
Polyhexamethyleneadipamide

Type 6/10:

$$NH_2(CH_2)_6NH_2 + HOOC(CH_2)_8COOH \rightarrow$$
Hexamethylenediamine   Sebacicacid

$$[NH(CH_2)_6NHCO(CH_2)_8CO]_n + H_2O$$
Polyhexamethylenesebamide

Type 6:

$$NH(CH_2)_5CO \rightarrow [NH(CH_2)_5CO]_n$$
ε-Caprolactam   Polycaprolactam

Type 11:

$$NH_2(CH_2)_{10}COOH \rightarrow [NH(CH_2)_{10}CO]_n$$
Aminoundecanoicacid   Polyaminoundecanamide

Type 12:

$$NH(CH_2)_{11}CO \rightarrow [NH(CH_2)_{11}CO]_n$$
Laurolactam   Polydodecanolactam

In addition to the basic nylons, a variety of copolymers can be manufactured, some of which are commercially available. Nylons and nylon copolymers can be blended to form alloys with specific customized properties.

Nylons can be modified by the addition of certain plasticizers, fillers, reinforcements, and stabilizers. Ordinarily, nylons used for injection molding, such as type 6/6, have relatively low molecular weights (on the order of 15,000 to 20,000). High molecular weights are available to provide higher melt viscosity for nylon resins which are to be extruded into tubing or shapes. The molecular weight of nylon generally is determined by the ASTM relative-viscosity test.

Nylon resins usually are supplied in the form of cylindrical or rectangular diced pellets. Most commercial nylon molding resins are nontoxic. If a large amount of residual monomer is retained in the resin, as can occur with certain unextracted formulas, the material should not be in prolonged contact with food because of the possibility of monomer leaching.

Nylons require modification or stabilization to improve their resistance to certain environmental effects. Unstabilized nylon is degraded by ultraviolet light. The most widely used stabilizer has been approximately 2% well-dispersed carbon black, which has proved effective in the absorption of ultraviolet light. The nylons are considered adequate for outdoor applications if they are not exposed to direct sunlight.

See also **Caprolactam**; **Fibers**; and **Polyamide Resins**.

NYMPH. If the wing pads of an insect are developed on the outside of its body, the insect is said to have a *simple* or *gradual metamorphosis*. The insect during this growing stage is called a *nymph*. This situation is contrasted with the case where the wing pads are developed internally during the growing stage. In this case, the growing stage is called a *larva*, and the insect is said to have a *complete* or *complex metamorphosis*. Nymphs generally appear something like miniature adults and have the same general life style. In most cases of larva, the size, shape, locomotion, and eating habits of the larva contrast sharply with the adult form. See also **Larva**.

NYQUIST FREQUENCY. For data defined at equal time-intervals *t*, the frequency of a sine or cosine term with a period double the interval

+ NaOH →

EVP (ethylidene vinylpyrrolidinone)

EBVP (ethyldiene-bis-vinylpyrrolidinone)

t. Frequencies higher than this will not be directly detectable by spectral analysis.

NYQUIST, HARRY (1889–1976). Nyquist, born in Sweden, was an American engineer, mathematician and scientist. His work on frequency response has been used in analysis and design of electronic amplifiers and servomechanisms.

See also **Frequency Response**; **Nyquist Frequency**; **Nyquist Rate (Signaling)**; and **Stability (System)**.

J. M. I.

NYQUIST RATE (Signaling). In transmission, if the essential frequency range is limited to B cycles per second, $2B$ is the maximum number of code elements per second that can be unambiguously resolved, assuming the peak interference is less than half a quantum step. This rate is generally referred to as signaling at the Nyquist rate, and $\frac{1}{2}B$ is called the Nyquist interval.

NYSTAGMUS. See **Vision and the Eye**.

NYTRIL. See **Fibers**.

O

OAK TREES. Of the *Fagaceae* family (beech family), the oaks are trees and shrubs (*Quercus*) essentially of the north temperate region. All of the more northern species are deciduous plants. Many of those in the southern part of the range have evergreen leaves and often are called live oaks. In Asia and the Pacific coast of North America, oaks are found in regions approaching tropical conditions. The oaks make up the greater part of the beech family.

The oak sends down a root system as deep as 15 feet (4.5 meters) into the ground. The over 300 species of oaks have simple alternate leaves. The flowers are of two kinds, borne on the same tree. The pistillate flowers are borne singly and are surrounded by an involucre of many scales beyond which the stigmas protrude. The staminate flowers are borne in long slender pendant catkins. Pollination is by wind. The fruit is an acorn, a nut of characteristically cylindrical shape, capped by the small persistent style-base, and seated in the scaly involucre, which forms a cup partially or almost wholly surrounding the nut.

Many of the oaks are valuable trees, yielding woods that have a variety of uses. In early times, before the day of the sawmill, oaks were much used in the construction of buildings. Often the oaks used for this purpose were split into thin planks, a method of preparation that served well to bring out the attractive grain of the wood. This grain is due partly to the numerous large vessels which are formed periodically every spring and appear as very evident dark lines or streaks in the wood, and partly to the large vascular rays which appear as irregular flakes, especially when the wood is split in a radial plane. In modern construction, oak is often used as paneling or flooring. To obtain the best grain, the wood is quarter-sawed, that is, cut in such a way that the flat surfaces shall be as nearly radial as is possible. Because of its beauty and also its durability, oak wood is also much used in furniture making. In America the principal species used for wood is white oak, *Quercus alba*. In Europe several species are used, among them the English oak, *Quercus robur*. Often these European oaks are trees of remarkable size, and are preserved because of their rugged beauty. The wood is very strong and durable, and finds considerable use in ship construction. Formerly much more was used for this purpose.

Some engineering properties of oak wood are given in Table 1.

Accidents sometimes cause the formation of oak wood of special properties and value. The trunks of fallen trees may lie buried for long periods of time in bogs or elsewhere. Sometimes, when removed, these logs are found to be perfectly sound and to have developed a rich dark brown or nearly black color, which makes them especially sought after for furniture making. Such oak is known as bog oak. Living trees frequently develop large irregular growths, known as burls, in which a very irregular much-contorted grain is found. The custom of cutting back the top of the tree, causing the development of numerous adventitious buds, a practice known as pollarding, causes a similar irregular grain. These burls are used for making veneers. Another species of oak, *Quercus Suber*, yields cork.

Oaks are valuable sources of tannin. In many species the bark is the source of the tannin, but in *Quercus aegilops*, a native of Eastern Europe and Asia, a tannin known as valonia is obtained from the cup and the young acorns.

African oak, a strong, heavy wood, comes from African trees of other genera than *Quercus*. This wood is rarely used, due to the difficulty of removing the heavy wood from its native forest.

Oak wilt, a serious disease caused by the fungus *Endoconidiophora fagacearum*, has become widespread in the Central and Eastern States. Great damage to the oak forests there is threatened unless some means of controlling the disease can be found.

Periodically, American Forests revises its National Register of Big Trees. The numerous kinds of oaks holding records in the United States (1986)

are given in Table 2. The record holding blue oak is shown in Fig. 1 on p. 2517; the swamp white oak in Fig. 2 on p. 2517.

White Oak. Considered by many foresters as the outstanding tree of the many oaks, the white oak (*Quercus alba*) receives its name from its light-colored bark. The white oak occurs widely throughout the eastern United States, ranging from central Maine to northern Florida, and west from southern Quebec through southern Ontario and southern Michigan, through Wisconsin, southeastern Minnesota, much of Iowa, eastern Kansas, eastern Oklahoma to eastern Texas, excluding a narrow belt along the Gulf of Mexico. The white oak prefers well-drained soil. The largest white oaks are found in the valleys of the western slopes of the Allegheny Mountains as well as the bottom lands of the lower Ohio Basin. Some white oaks live for several centuries, a few are known to date back over 800 years.

Black Oak. This is another common and large oak which is found in the eastern United States, ranging from southern Maine and northern Vermont westward through southern Ontario, southern Michigan to southeastern Minnesota, and south to northern Florida and eastern Texas, eastern Oklahoma, eastern Kansas, southeastern Nebraska and Iowa. The black (*Quercus velutina*) oak prefers rich, well-drained, gravelly soils. However, it is not usually found in great numbers in areas with exceptionally rich soils because of the black oak's intolerance for shade caused by competitive trees that thrive in such soils. The ash and tuliptree are frequently found associated with the black oak.

Pin Oak. The natural range of the pin oak (*Quercus palustris*) extends from southwestern New England to northern North Carolina, and from Ohio to Kentucky and western Tennessee. Its distribution also includes southeastern Iowa, eastern Kansas, northeastern Oklahoma and northern Arkansas. The pin oak is not considered an important source of lumber, but rather it is preferred for urban and ornamental plantings. If encountered in normal logging operations, however, it may be marketed as red oak.

Northern Red Oak. The range of this tree (*Quercus rubra*) extends from Maine south and westward, including New England, New York, Pennsylvania, Maryland, Delaware, Virginia and West Virginia, Ohio, Michigan, Indiana, Wisconsin, all but extreme western Minnesota, all of Iowa except the northwest corner, all of Missouri, Illinois, Kentucky, Tennessee, all but the southern portion of Alabama, the northwestern portions of Georgia and South Carolina, and much of North Carolina except the coastal and southeastern portions of that state. Timber from the tree is widely used for general construction, flooring, furniture, railroad ties, posts, and poles, and interior finish. It is estimated that West Virginia, North Carolina, and Tennessee probably have the largest stands.

Bur or Mossycup Oak. This is generally considered second to the white oak in terms of size and grandeur. The tree (*Quercus macrocarpa*) is predominant in the midwestern United States and southern Quebec, Ontario, Manitoba, and parts of Nova Scotia in Canada. Although the bur oak is found in New England, its principal range includes New York, Michigan Ohio, Indiana, Illinois, much of Kentucky and a small part of Tennessee, as well as North and South Dakota, Minnesota, Wisconsin, Iowa, and Missouri. The tree also is found in eastern Nebraska, eastern Kansas, much of Oklahoma except the extreme western portion of that state, and in a north-south corridor of Texas, essentially in the central portion of that state. The bur oak grows slowly and some authorities do not consider it mature prior to an age of 200 to 300 years. The tree is popular for urban plantings and has demonstrated an unusual ability to withstand the smoke and pollution of cities.

NOTE: A concise and convenient reference that details the major oaks (and over a score of other principal trees found in the United States)

TABLE 1. SELECTED ENGINEERING PARAMETERS OF OAK WOODS

Parameter	Red Oak		White Oak	
	Green	Oven-Dried	Green	Oven-Dried
Moisture content, %	80	12	70	12
Specific Gravity	—	0.66	—	0.70
Modulus of rupture, psi	8500	14,400	8100	13,900
MPa	58.7	99.4	55.9	95.9
Modulus of elasticity, psi	1360	1810	1200	1620
MPa	9.4	12.5	8.3	11.2
Crushing strength (compression parallel to grain), psi	3520	6920	3520	7040
MPa	24.3	47.7	24.3	48.6
Shear, psi	1220	1830	1270	1890
MPa	8.4	12.6	8.8	13.0
Compressive strength (compression perpendicular to grain), psi	800	1260	850	1410
MPa	5.5	8.7	5.9	9.7
Tensil strength (perpendicular to grain), psi	740	760	820	770
MPa	5.1	5.2	5.7	5.3
End hardness, pounds	1050	1490	1110	1300
kilograms	476	676	503	590
Side hardness, pounds	1030	1300	1070	1330
kilograms	467	590	485	603

Source: U.S. Forest Products Laboratory.

TABLE 2. RECORD OAK TREES IN THE UNITED STATES[1]

Specimen	Circumference[2]		Height		Spread		Location
	Inches	Centimeters	Feet	Meters	Feet	Meters	
Ajo oak (1998) (*Quercus turbinella* var. *ajoensis*)	82	208	32	9.8	40	12.2	Arizona
Arizona white oak (1999) (*Quercus arizonica*)	133	338	56	17.1	47	14.3	Arizona
Arkansas oak (1996) (*Quercus arkansana*)	140	356	88	26.8	100	30.5	Mississippi
Bear oak (1992) (*Quercus ilicifolia*)	34	86	41	12.5	30	9.1	West Virginia
Bigelow oak (1999) (*Quercus durandii* var. *breviloba*)	111	282	54	16.5	38	11.6	Texas
Black oak (1999) (*Quercus velutina*)	322	818	86	26.2	105	32	Connecticut
Blackjack oak (1999) (*Quercus marilandica*)	144	366	94	28.7	65	19.8	Georgia
Blue oak (1974) (*Quercus douglasii*)	243	617	94	28.7	48	14.6	California
Bluejack oak (1992) (*Quercus incana*)	119	302	54	16.5	54	16.5	Florida
Bur oak (1995) (*Quercus macrocarpa*)	322	818	96	29.3	103	31.4	Kentucky
California black oak (1972) (*Quercus kelloggii*)	338	859	124	37.8	115	35.1	Oregon
Canyon live oak (1998) (*Quercus chrysolepis*)	422	1072	95	29	126	38.4	California
Chapman oak (1989) (*Quercus chapmanii*)	81	206	45	13.7	50	15.2	Florida
Cherrybark oak (1991) (*Quercus falcata* var. *pagodaefolia*)	324	823	124	37.8	136	41.5	Virginia
Cherrybark oak (1993) (*Quercus falcata* var. *pagodifolia*)	342	869	110	33.5	108	32.9	Virginia
Chestnut oak (1997) (*Quercus prinus*)	222	564	144	43.9	70	21.3	Tennessee
Chinkapin oak (1995) (*Quercus muehlenbergii*)	258	655	110	33.5	92	28	Kentucky
Chisos oak (1982) (*Quercus graciliformis*)	65	165	66	20.1	36	11	Texas

TABLE 2. (*Continued*)

Specimen	Circumference[2]		Height		Spread		Location
	Inches	Centimeters	Feet	Meters	Feet	Meters	
Coast live oak (1999) (*Quercus agrifolia*)	338	859	58	17.7	75	22.9	California
Darlington oak (1992) (*Quercus hemisphaerica*)	234	594	96	29.3	95	29	Georgia
Delta post oak (1988) (*Quercus stellata* var. *paludosa*)	118	300	108	32.9	56	17.1	Texas
Dunn oak (1995) (*Quercus dunnii*)	83	211	37	11.3	36	11	Arizona
Dunn oak (1999) (*Quercus dunnii*)	85	216	35	10.7	40	12.2	California
Durand oak (typ.) (1997) (*Quercus durandii* var. *durandii*)	189	480	95	29	106	32.3	Georgia
Emory oak (1998) (*Quercus emaryi*)	192	488	54	16.5	86	26.2	Arizona
Emory oak (1993) (*Quercus emaryi*)	186	472	56	17.1	92	28	Arizona
Engelmann oak (1968) (*Quercus engelmanni*)	129	328	78	23.8	100	30.5	California
English oak (1997) (*Quercus robur*)	187	475	88	26.8	91	27.7	Ohio
English oak (1993) (*Quercus robur*)	178	452	102	31.1	89	27.1	Washington
Gambel oak (1981) (*Quercus gambelii*)	216	549	47	14.3	85	25.9	New Mexico
Georgia oak (1999) (*Quercus georgiana*)	73	185	75	22.9	63	19.2	Georgia
Graves oak (1982) (*Quercus gravesii*)	154	391	42	12.8	40	12.2	Texas
Graves oak (1976) (*Quercus gravesii*)	145	368	51	15.5	41	12.5	Texas
Gray oak (1993) (*Quercus grisea*)	216	549	45	13.7	73	22.3	New Mexico
Havard oak (1986) (*Quercus havardii*)	40	102	30	9.1	23	7	Texas
Interior live oak (1982) (*Quercus wislizeni*)	268	681	90	27.4	69	21	California
Lacey oak (1989) (*Quercus glaucoides*)	107	272	58	17.7	96	29.3	Texas
Laurel oak (1993) (*Quercus laurifolia*)	267	678	93	28.3	122	37.2	Alabama
Live oak (typ.) (1976) (*Quercus virginiana* var. *virginiana*)	439	1115	55	16.8	132	40.2	Louisiana
Mexican blue oak (1999) (*Quercus oblongifolia*)	120	305	65	19.8	69	21	New Mexico
Myrtle oak (1986) (*Quercus myrtifolia*)	69	175	36	11	35	10.7	Florida
Netleaf oak (1998) (*Quercus rugosa*)	88	224	47	14.3	36	11	Arizona
Northern pin oak (1999) (*Quercus ellipsaidalis*)	184	467	128	39	92	28	Ohio
Northern red oak (1999) (*Quercus rubra*)	294	747	98	29.9	97	29.6	Massachusetts
Northern red oak (1997) (*Quercus rubra*)	257	653	134	40.8	81	24.7	North Carolina
Nuttall oak (1991) (*Quercus nuttallii*)	280	711	118	36	85	25.9	Louisiana
Oglethorpe oak (1999) (*Quercus oglethorpensis*)	117	297	69	21	69	21	Georgia
Overcup oak (1987) (*Quercus lyrata*)	258	655	156	47.5	120	36.6	North Carolina

(*continued*)

TABLE 2. (*Continued*)

Specimen	Circumference[2]		Height		Spread		Location
	Inches	Centimeters	Feet	Meters	Feet	Meters	
Pin oak (1991) (*Quercus palustris*)	240	610	110	33.5	112	34.1	Tennessee
Post oak (typ.) (1987) (*Quercus stellata* var. *stellata*)	236	599	85	25.9	88	26.8	Virginia
Post oak (typ.) (1996) (*Quercus stellata* var. *stellata*)	237	602	84	25.6	88	26.8	Georgia
Sand live oak (1995) (*Quercus virginiana* var. *geminata*)	189	480	81	24.7	106	32.3	Florida
Sand live oak (1995) (*Quercus virginiana* var. *geminata*)	181	460	94	28.7	100	30.5	Florida
Sand post oak (1995) (*Quercus stellata* var. *margaretta*)	157	399	87	26.5	92	28	Florida
Scarlet oak (1995) (*Quercus caccinea*)	248	630	120	36.6	93	28.3	Kentucky
Shingle oak (1997) (*Quercus imbricaria*)	208	528	105	32	62	18.9	Ohio
Shumard oak (typ.) (1994) (*Quercus shumardii* var. *shumardii*)	249	632	190	57.9	88	26.8	Tennessee
Silverlear oak (1994) (*Quercus hypoleucoides*)	123	312	69	21	52	15.8	Arizona
Southern red (typ.) (1999) (*Quercus falcata* var, *falcata*)	312	792	150	45.7	156	47.5	Georgia
Swamp chestnut oak (1998) (*Quercus michauxii*)	276	701	105	32	216	65.8	Tennessee
Swamp chestnut oak (1989) (*Quercus michauxii*)	197	500	200	61	148	45.1	Alabama
Swamp white oak (1992) (*Quercus bicolor*)	228	579	120	36.6	92	28	Maryland
Texas oak (1999) (*Quercus shumardii* var. *texana1*)	108	274	60	18.3	59	18	Texas
Texas live oak (1999) (*Quercus virginiana* var. *fusiformis*)	295	749	42	12.8	98	29.9	Texas
Toumey oak (1994) (*Quercus toumeyi*)	68	173	27	8.2	33	10.1	Arizona
Turbinella (typ.) (1993) (*Quercus turbinella* var. *turbinella*)	160	406	43	13.1	49	14.9	Nevada
Turkey oak (1994) (*Quercus laevis*)	127	323	72	21.9	75	22.9	Florida
Valley oak (1984) (*Quercus lobata*)	348	884	163	49.7	99	30.2	California
Vasey oak (1982) (*Quercus pungens* var. *vaseyana*)	45	114	48	14.6	40	12.2	Texas
Vasey oak (1996) (*Quercus pungens* var. *vaseyana*)	61	155	39	11.9	32	9.8	Texas
Water oak (1996) (*Quercus nigra*)	278	706	120	36.6	111	33.8	Louisiana
White oak (1996) (*Quercus alba*)	382	970	96	29.3	119	36.3	Maryland
Willow oak (1986) (*Quercus phellos*)	318	808	73	22.3	132	40.2	Mississippi

[1]From the "National Register of Big Trees," American Forests (by permission).
[2]At 4.5 feet (1.4 meters).

Fig. 1.   Record blue oak tree. See Table 2. (*E. Logel.*)

Fig. 2.   Record swamp white oak tree. See Table 2. (*D.J. Preston.*)

is "Knowing Your Trees," edited by G.H. Collingwood, Warren D. Brush and Devereux Butcher, published by American Forests, Washington, DC, (revised periodically).

Cork Oaks. The cork oak (*Quercus suber*) is found in Europe and Africa. Cork cells are found in the outer bark of most woody-stemmed plants, but in amounts too small and with too-brittle walls to be of any commercial value. But in the *Quercus Suber*, the cork cells become a very large part of the tissue of the bark, and have been used for centuries. The cork oak tree is a medium-size tree seldom much over 50 feet (15 meters) in height, growing particularly in those countries bordering the Mediterranean Sea. The evergreen leaves are small, ranging from $1\frac{1}{2}$

to 3 inches (3.8 to 7.6 centimeters) long, and about an inch wide, with slightly toothed margins. The bark of the tree soon becomes rough and deeply furrowed, but is of little value except as ground cork or as a source of tannin. When the tree is about 20 years old, this first-formed bark is removed, care being taken not to injure the phloem and cambium layers. Within 10 years a new cork layer has formed. This layer is the first of many layers which are removed once every 10 years or so throughout the life of the tree. Removal is generally done in the early summer at a time when hot dry winds will not cause injury to the unprotected phloem and cambium.

After removal, the cork is air-dried for a time, then boiled to soften it and to remove some of the tannin. The outer part of the bark is scraped off, and the rest pressed out flat and dried. It is then ready to ship.

The physical properties of cork account for its many uses. It is very light and buoyant, more than 50% of its volume being air, and hence is used in the manufacture of floats, life-preservers, and so forth. The living protoplasm of the cork cells dries up early in their development, leaving hollow cells, each containing a small mass of air which expands after compression. Therefore, cork is very resilient, and is frequently used as a core on which to wind yarn or string in the manufacture of baseballs. In the early stages of their formation, the walls of cork cells are cellulose, but this is soon impregnated by a waterproof and nonabsorbent lipoid substance, suberin. Therefore cork is used in making handles for fishing rods, shoe-soles, and cork stoppers. Since the hollow cork cells are poor conductors both of heat and sound, cork is much used as insulating material. For this use cork is ground up and then pressed into sheets with various binding materials, giving much larger sheets than can be obtained from the tree. Ground cork is also an important constituent of linoleum, gaskets, and other products.

Cork is traversed by lenticels, loose masses of porous tissue, which appear as dark spots or holes in stoppers. Usually in making stoppers the bark is cut so that these will be transverse in the stopper. In making stoppers, the forms are first punched out as cylinders, and then trimmed down by machine to the required tapering shape. New packaging materials and techniques have displaced cork from a number of formerly exclusive applications for it.

English Oak. This is a comparatively small oak, the height ranging from 20 to 40 feet (6 to 12 meters). The spread is about 20 feet (6 meters). The tree is cone-shaped with a life expectancy of about 50 years. It is easily cultivated and is often used as a windbreak or as an ornamental tree between sidewalks and curbs in large cities. Careful grooming produces beautiful specimens. The bark is light-gray.

OARFISH *(Regalecus glesne).*   The longest bony (rather than cartilaginous) fish known to exist in the ocean. The fish may reach a length of more than 17 meters (56 feet) and weigh up to 300 kilograms (~660 pounds). The fish, seldom observed, is called the "king of the palace under the sea" by the Japanese and a sea serpent by others. The oarfish prefers warm, temperate waters worldwide and lives at depths ranging from 20 to 200 meters (66 to 660 feet). The lifespan of the fish is unknown. The fish has a long red dorsal fin that rises to a manelike crest atop its head.

The oarfish is described as very elusive and secretive, with few sightings reported over the centuries. Morton Brunnich, a Dutch naturalist, first described the fish in the scientific literature. In 1771, Brunnich found a specimen washed up on a beach in Norway. Fewer than 25 sightings subsequently have been reported. Another was found on the shore off Orkney (Scotland) in 1808. In 1906, possibly the closest encounter with a live oarfish occurred off the shore of the Island of Sumbawa (Indonesia). Although the crew of a nearby ship was not successful in enticing the fish, a crew member described it, "A long and very beautiful fish came to the surface at the ship's bow. Baited rigs were thrown to it, but it took no notice of them. With its vivid red crest and dorsal fin, scarlet streamers on its sides, and blue of its head and intense shine of silver on its body, it was probably the most beautiful creature I've ever seen."

The last sighting of an oarfish (well-preserved remains) was found on the beach at Malibu, California, in 1963. The specimen is displayed at the Los Angeles County Museum of Natural History. More detail can be found in article by Cheryl Lyn Dybaa (*Oceanus*, 98, Spring 1993).

OASIS EFFECT.   See **Meteorology**.

OATS *(Avena sativa; Gramineae).* Oats are annual cereal grasses native in temperate regions of the Old World. The several species of oat plants are characterized by their closed leaf-sheath, a wide-branched panicle, and by a special type of inflorescence. In the florets of wild oat plants the lemma has the midrib prolonged as a prominent awn, the basal portion of which is spirally twisted. In many cultivated forms this awn has been eliminated. The grain of oats is not easily separated from the surrounding husk, composed of the lemma and palea, which are neither palatable nor digestible. See also **Grasses.**

Oats are principally adapted to growing in a climate having cool summers and abundant moisture. However, the plants are very hardy and tolerant of adverse conditions.

Oats are an important livestock feed, with approximately 95% of the United States crop used for this purpose. Rolled oats and oatmeal are important breakfast cereals for human consumption in some regions of the world. Use of oats in bakery products is quite limited because oat flour contains no gluten. However, oat flour can be mixed in relatively small portions with other flours and thus impart a distinct flavor. Also, because of a natural antioxidant contained in oat flour, the substance finds use for preserving certain foods that contain fats, such as peanut (groundnut) butter, oleomargarine, and lard. Oat flour also has been used for dusting potato chips and salted nuts, as well as for a coating on papers used to contain fatty products, such as bacon and coffee.

Oat hulls are the primary source of furfuraldehyde, which is used industrially as a solvent and chemical intermediate. Over the years, limited amounts of oatmeal have been used in soaps.

Botany. The several species of oat plants are characterized by their closed leaf-sheath, a wide-branched panicle, and by a special type of inflorescence. See Fig. 1. In the florets of wild oat plants, the lemma has the midrib prolonged as a prominent awn, the basal portion of which is spirally twisted. In many cultivated forms, this awn has been eliminated, mainly in the interest of processing for livestock feed. The grain of oats is not easily separated from the surrounding husk, composed of the lemma and palea, neither of which is palatable or digestible.

The principal parts of the common oat (*A. sativa*) are shown in Fig. 2. Oats are naturally self-pollinated.

As shown by Fig. 3, the oat kernel or caryopsis is spindle-shaped and furrowed on one side. The color is a light buff and the hairs are fine and silky. The kernel usually is from $\frac{5}{16}$- to $\frac{7}{16}$-inch (7.5 to 10.5 millimeters) long and from $\frac{1}{16}$- to $\frac{1}{8}$-inch (1.5 to 3 millimeters) wide. Of the total grain, the kernel makes up between 65 and 75% of the weight. The two main parts of the kernel are the endosperm and embryo. The term *groat* is frequently used to designate the oat caryopsis.

Fig. 2. Oat inflorescence and details of components. (1) Panicle of Avena sativa. (2) Distal or top part of panicle, bearing four spikelets. (3) Lateral view of spikelet (one-flowered) in anthesis, showing separated lemma and pales with one branch of plumose stigma and three stamens (item 7) protruding. (4) Lateral view of lemma, showing dorsal attachment of awn. (5) Ventral view of palea. (6) Lodicules. (7) Stamens. (8 to 13) Lateral views of a floret before, during, and after anthesis. (14) Diagrammatic longitudinal dorsal-ventral section of floret, showing lemma (X) palea, androecium, and gynoecium. (15) Diagrammatic cross section of spikelets of three-flowered spikelet before anthesis: (a) lower or outer glume; (b) upper or inner glume; (c) lemma or flowering glume; (d) palea; (e) anthers; (f) stigma; (g) lemma or secondary floret with enclosed pales, stamens, and stigma; (h) rudimentary tertiary floret. (16 to 18) Pistil before, during, and after anthesis. (19) Apical portion of stigma, greatly enlarged, showing adhering pollen grains. (20) Cross section of anther showing four lobes. (21) Pollen grains (enormously enlarged). (22) Floret, ventral, and dorsal view of kernel and caryopses (smaller than natural size). (23) Oat kernel. (24 to 26) Caryopsis or groat (enlarged). (27) Cross section of caryopsis. (*USDA diagram from original sketch by Boettcher and Hughes.*)

Fig. 1. Oat panicles: (left) unilateral, "side," or "horse-mane" oat with 7 whorls or branches; (right) equilateral, or spreading, or "tree" panicle with 5 whorls or branches. (*USDA photo.*)

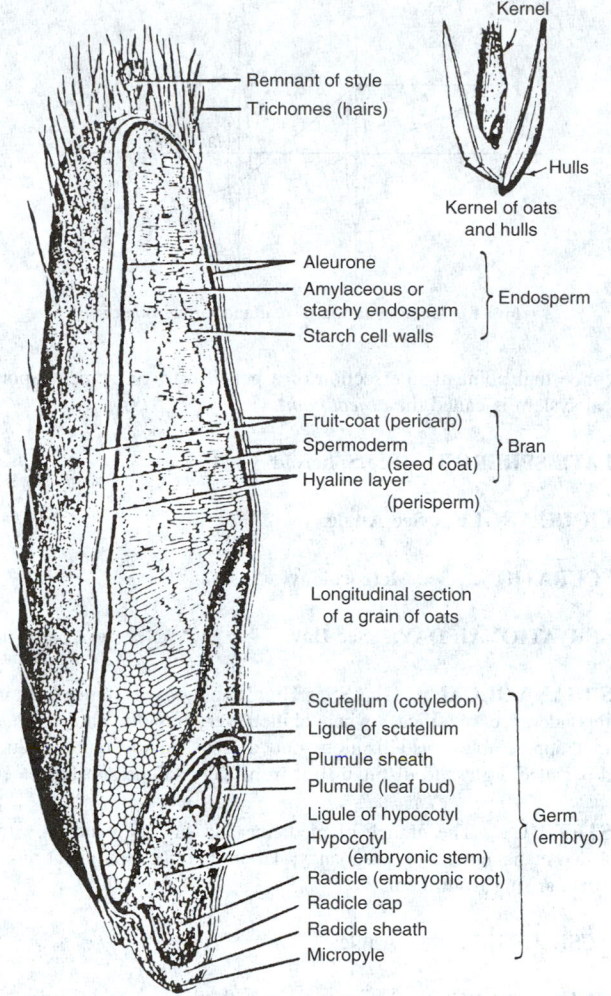

Kernel

Remnant of style
Trichomes (hairs)

Hulls

Kernel of oats
and hulls

Aleurone
Amylaceous or
starchy endosperm   } Endosperm
Starch cell walls

Fruit-coat (pericarp)
Spermoderm
          (seed coat)   } Bran
Hyaline layer
          (perisperm)

Longitudinal section
of a grain of oats

Scutellum (cotyledon)
Ligule of scutellum
Plumule sheath
Plumule (leaf bud)
Ligule of hypocotyl
Hypocotyl   } Germ
          (embryonic stem)   (embryo)
Radicle (embryonic root)
Radicle cap
Radicle sheath
Micropyle

Fig. 3.   Parts of the oat caryopsis. (*The Quaker Oats Company.*)

Species and Varieties. The manner of designating and identifying the various species and subspecies of oats is technically complex and mainly of importance to plant breeders.

Characteristics that are desirable in oat varieties include short straw, strong straw that resists lodging, high test weight, high yield, and resistance to crown rust, stem rust, yellow dwarf disease, and smut.

In planting oats, tillage of soil that is too wet causes severe compaction and reduced yields. When oats follow maize (corn) in rotation, the stalks should be chopped and either diced or plowed. One tandem dicing just ahead of planting provides an excellent seedbed when oats follow soybeans or other beans. Oats usually are seeded at the rate of 2 to 3 bushels per acre (71 to 107 kilograms per hectare) and at a depth of 1 to 2 inches (2.5 to 5 centimeters) to achieve 20 to 30 plants per square foot (215 to 323 plants per square meter).

Oats may be seeded alone strictly as an oats crop, or they may be used as a companion crop when seeding legumes and/or grasses. The seeding of the two crops together represents a compromise between obtaining the highest yield of oats and a good stand of the legumes or grasses. A vigorous, late-maturing oat crop may result in a poor, weak stand of forage as the two crops compete for water and soil nutrients.

Oats have the capacity to respond to fertilizer, especially nitrogen. Phosphorus and potassium are needed to balance the nitrogen. The cool, moist soils that are common early in the spring are not conducive to bacterial action that will release nitrogen, or to the high availability of phosphorus.

When oats and a companion crop of legumes and grasses are seeded, the fertility program must be adequate to feed both crops. Greatest response will be obtained when the fertilizer is drilled with the seed oats and the legume seed is placed in a band directly over the fertilizer band. One hundred bushels (1440 kilograms) of oats with 2 tons (1814 kilograms) of straw will require 90 pounds (40.5 kilograms) of nitrogen; 60 pounds (27 kilograms) of P_2O_5; and 115 pounds (51.8 kilograms) of K_2O. Various soils will release these nutrients at different rates due to the organic matter content variations, rotation program used, soil temperature, and the parent material of the soil. A typical fertilizer application will be from 40 to 60 pounds per acre (45 to 67 kilograms per hectare) each of nitrogen, phosphate, and potash. Oats are a short-season crop. Every growing day is important, and a crop that is deficient in any nutrient even when only 1 to 4 inches (2.5 to 6 centimeters) tall will probably not recover from any significant span of inadequate nutrition.

Additional Reading

Dendy, D.A. and B.J. Dobraszczyk: *Cereals and Cereal Products: Chemistry and Technology*, Aspen Publishers, Inc. Gaithersburg, MD, 2000.

Forsberg, R.A.: *Breeding Oat Cultivars Suitable for Production in Developing Countries: 1996 Report*, DIANE Publishing Company, Collilngdale, PA, 1998.

Marshall, H.G. and M.E. Sorrells: *Oat Science and Technology*, American Society of Agronomy, Madison, WI, 1992.

Seibold, R.: *Cereal Grass: Nature's Greatest Health Gift*, Keaets Publishing, Inc., Chicago, IL, 1994.

Simons, M.: *Crown Rust of Oats and Grasses*, The American Phytopathological Society, St. Paul, MN, 1970.

Webster, F.H.: *Oats Chemistry and Technology*, American Association of Cereal Chemists, St. Paul, MN, 1986.

Wood, P.J.: *Oat Bran*, American Association of Cereal Chemists, St. Paul, MN, 1993.

Web References

American Association of Cereal Chemists: http://www.scisoc.org/aacc/
The American Phytopathological Society: http://www.apsnet.org/

OBESITY. The accumulation and storage of body fat in excess of that required by a statistical normal individual of same age, frame size, sex, and height. When caloric intake exceeds expenditure, excess fat accumulates—either existing fat cells are swelled with additional fat, or new fat cells are formed, or both actions may occur together. Scientists are not certain. There is the general opinion among specialists that the number of fat cells within the human body increases in the fetus and newborn and later in childhood and adolescence, while during other periods, particularly in later life, the number of cells does not increase, but rather the amount of fat in each cell may increase. However, in persons who are over 50% overweight, there is an increase in both cell numbers and fat crowding. Researchers have estimated that an average adult body contains between 3 and 5×10^{10} fat cells. It has further been estimated that each cell contains about 0.5 microgram of triglycerides. Thus, the fat content ranges between 26 and 44 pounds (12 and 20 kilograms). Obesity in individuals ranges from being just a little bit fat (plump, paunchy, etc.) to an extreme overweight condition where the person's life is threatened and where living is uncomfortable and clumsy. Treatment of the severely obese includes total starvation over a period in a hospital environment. This approach has proved successful, but many specialists now prefer to use a so-called protein-sparing diet in such cases. See also **Starvation.** In more recent years, an intestinal bypass may be considered. A majority of physicians consider this a solution of last resort because of the morbidity and mortality presently associated with the procedure. Consequently, the approach is generally reserved for situations where serious complications, such as congestive heart failure, debilitating arthritis, anoxia, among others, are present as well as the failure of prior weight reduction procedures.

The mental and emotional aspects of obesity are emphasized by the fact that only 10% or fewer of the individuals who commence a dieting regimen and who achieve early success will maintain their reduced weight beyond one, two, or three years, depending upon the individual. Thus, calorie control through dieting and the possible use of other procedures, such as jaw wiring, acupuncture, transcendental meditation, yoga, hypnosis, and gonadotropin injections, are, unfortunately, of a transient nature. Such procedures require consistent enforcement of sensible eating habits, essentially over a lifetime, once the initial objectives are achieved. Those persons who do ultimately succeed in controlling or eliminating obesity have, through a combination of physical and mental conditioning, managed to alter their fundamental eating and exercise habits.

There is no shortage of references on special diets for losing weight or counseling to participate in more vigorous exercise. Dieting, other than the simplistic, common sense approach of voluntarily reducing caloric intake, should be supervised by a physician who is well acquainted with the health of the individual. Some weight-reduction diets can seriously affect the health of some individuals. Most of the millions of words written to describe the benefits of special diets essentially are wasted because of the transient nature of responding to them. The benefits of a sensible exercise program for a majority of individuals are undeniable and exercise, of course, can contribute to a weight-loss program. In the interest of realism, however, it should be pointed out that from the standpoint of calorie burning, exercise is consistently exaggerated. For example, 10 to 15 minutes of jogging will only consume 200–300 calories. Nevertheless, regular exercise contributes to the successful maintenance of weight loss.

For many years, drug therapy for obesity has been a topic of controversy and remains so. In tests using placebos, it has been demonstrated that amphetamines, including their analogues (fenfluramine, etc.) can result in increased weight loss over a period of several months. These drugs appear to be mainly effective where emotional stress is a prime component of obesity. But such drugs are contraindicated for some individuals with other problems and thus require the judgment of a physician in their use.

The fact that obesity has been experienced by the human race since antiquity attests to the stubborn nature of the problem. The connection between obesity and inheritance is obvious, but not understood. Obviously, much additional research is required to understand the chemistry of obesity as well as of its psychiatric and heritable nature. Quite recently, some scientists have been taking bold and new approaches to developing this understanding. For example, some scientists have been investigating the systemic changes that occur during certain periods in the lives of various mammals — these initial studies indicating that probably through hormonal control, there are preprogrammed periods of fasting. These are not confined simply to hibernation, but to periods when anorexia (loss of appetite or distaste for food) functions at times for survival of individuals and preservation of species. Bull seals, for example, go without feeding for many weeks while minding their duties of defending territory and harem. Preprogrammed anorexia appears to be part of such processes as migrating and molting, i.e., not eating even when food is immediately available. Researchers are also looking to anorexia nervosa, principally a disorder of teenagers, who literally starve themselves because of a fear of fatness. This is a complex disease and still poorly understood. See also **Anorexia**; and **Diet**.

OBJECTIVE PRISM. In the ordinary laboratory spectrograph a collimator is necessary in order that the light from the narrow slit may be sent through the dispersing agent in a parallel beam. At the principal focus of the camera lens the images of the slit in the different radiations are commonly known as the spectral lines.

The stars are at such tremendous distances that they subtend infinitesimal angles and the light from them reaches the earth in parallel beams. Hence the slit and collimating lens of the laboratory spectrograph are unnecessary and the dispersing agent, usually a prism, may be placed directly before the objective lens of the telescope or astrographic camera. With such an instrument, commonly known as an objective prism, instead of single-point images of the stars, there appear on the photographic plate a series of dots, each of them in some particular radiation from the stars, or, in other words, the spectra of the stars. In order that the spectra may have appreciable width or that the spectral lines may have length instead of appearing as mere dots, the refracting edge of the prism is set parallel to the spherical coordinate of right ascension and the telescope driven a bit too fast or too slow, slightly "trailing" the images.

OBJECT (Real). In geometrical optics, an object (or image) is termed *real* if from each point of it, light diverges toward the optical system. The first object (O) in Fig. 1 is real. An example of a *virtual* image is I_1, which would have been formed by lens L_1 had not lens L_2 been interposed. I_1 acts as a virtual object for lens L_2. See also **Mirrors and Lenses**. The

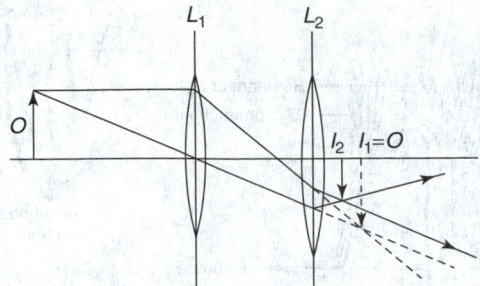

Fig. 1.   Demonstration of real and virtual objects.

real or virtual point of intersection of a pencil of rays incident upon an optical system is called the *object point*.

OBLATE SPHEROID. See **Spheroid**.

OBLIQUE ANGLE. See **Angle**.

OBSCURATION. See **Meteorology**.

OBSERVATIONAL DAY. See **Day**.

OBSIDIAN VOLCANIC GLASS. Highly acidic lavas (those containing a preponderance of silica), when chilled very rapidly and congealed, without appreciable crystallization, into a rigid liquid solution. Such a solid is called a glass to distinguish it from a crystalline substance.

OBSTETRICS. The branch of medicine and surgery which has to do with the management of pregnancy, labor and the complications and disorders arising from them.

OBTUSE ANGLE. See **Angle**.

OBUKHOV LENGTH. A parameter with dimension of length that gives a relation between parameters characterizing dynamic, thermal, and buoyant processes. At altitudes below this length scale, shear production of turbulence kinetic energy dominates over buoyant production of turbulence. It is defined by

$$L = \frac{-u_*^3 T_v}{kg Q_{v0}}, \tag{1}$$

where k is von Kármán's constant, u_* is the friction velocity (a measure of turbulent surface stress), g is gravitational acceleration, T_v is virtual temperature, and Q_{v0} is a kinematic virtual temperature flux at the surface. The parameter was first described by Obukhov in 1946, and therefore should not be called the Monin–Obukhov length, even though there is a Monin–Obukhov similarity theory that uses it. The Obukhov length, of order one to tens of meters, is the characteristic height scale of the dynamic sublayer. The Obukhov length is zero for neutral stratification, and positive (negative) for stable (unstable) stratifications. The dimensionless Obukhov length z/L (where z is height above the surface) is used as a stability parameter, with $z/L = 0$ for statically neutral stability, and is positive (negative) in a typical range of 1 to 5 (−5 to −1) for stable (unstable) stratification.

See also **Buckingham Pi Theory**; and **Dimensional Analysis**.

OCCIPITAL LOBE. See **Central and Peripheral Nervous Systems**.

OCCULTATION. When a large body passes in front of a smaller body, an occultation occurs. Thus, when the moon passes between a star and the earth, the light of the star is cut off, and the star is said to be occulted. The term eclipse, which is technically correct for this phenomenon, is reserved for circumstances involving the sun, earth, and moon, and for such things as eclipsing binaries, satellites of Jupiter, etc. As the moon passes between the earth and a star in its revolution about the earth, the star disappears behind the eastern edge of the moon and reappears again at the western edge. The disappearance and reappearance are practically instantaneous, proving both that the moon has no sensible atmosphere and that the star appears sensibly as a point of light. The interval between disappearance and reappearance depends fundamentally upon how closely the center of the moon appears to pass across the star.

Since the disappearance and reappearance are practically instantaneous, the times of the phenomena can be determined with great precision. Observations of the instants of occultation of stars by the moon may be used to determine the position of the moon extremely accurately. A large number of such occultations are observed both by professional and amateur astronomers, and the results are used to verify and correct the theory regarding complicated motions of the moon. The differences in the time of occultation of a given star, as observed by widely separated observers, may be used to determine both the distance of the moon and the difference in terrestrial longitude of the two observers. This method of determination of difference in longitude was the most accurate available before the development of modern methods for distribution of Greenwich time.

If the diffraction pattern differs from that of a point source, it is possible to obtain stellar diameters by observing occultations photoelectrically with high time resolution equipment. This same technique was used at radio frequencies to measure the sizes of the two components of the radio source 3C273.

OCEAN.

The surface area of the earth is approximately 510.1×10^6 square kilometers (196.9×10^6 square miles). Of this total area, the oceans and adjacent seas, forming a series of interconnected saline bodies, cover 375.55×10^6 square kilometers (145 million square miles), or over 70% of the earth. The oceans and seas are not evenly distributed. Approximately 81% of the surface of the Southern Hemisphere is covered by ocean waters; about 61% of the Northern Hemisphere is covered. Because the oceans are continuous and interconnected, the names assigned are somewhat arbitrary or depend upon natural land boundaries, such as the continental land masses or various island chains.

Traditionally, there are five oceans: the Atlantic, the Pacific, the Indian, the Arctic, and the Antarctic. The Antarctic Ocean, lacking any precise natural boundaries, is sometimes considered an extension of the Atlantic, the Pacific, and the Indian oceans. The adjacent bodies of salt water and various subdivisions of the oceans are generally known as seas, but local usage may also sanction such terms as gulfs, bays, channels, and straits, terms that are sometimes used interchangeably. The names *epicontinental, epeiric,* and *Mediterranean* seas are less frequently used. An epicontinental or epeiric sea is a sea on the continental shelf or within a continent. A Mediterranean sea is a type of epicontinental sea that is deep and that connects with an ocean by a narrow opening.

Among the large and important seas or designated regions of the principal oceans are the South China Sea, the Caribbean Sea, the Mediterranean Sea, the Bering Sea, the Gulf of Mexico, the Sea of Okhostk, the Sea of Japan, Hudson Bay, the East China Sea, the Andaman Sea, the Black Sea, the Red Sea, the North Sea, the Baltic Sea, the Yellow Sea, the Gulf of California, and the Persian Gulf. It should be noted that some landlocked bodies of water, usually called lakes, are historically called seas — thus, the Caspian Sea, the Dead Sea, and the Sea of Galilee, among others.

From a geochemical standpoint, the oceans and seas may be classified and named in still another way, i.e., in terms of ocean water masses. Determinations of the physical and chemical properties of water samples from the oceans have shown that they may be divided into regions in which the properties of samples are relatively constant. These regions are characterized not only by their geographical locations, but also by their depths. Thus, there is not only a water mass described as *arctic surface water,* but also *arctic deep water, antarctic surface water, antarctic intermediate water, antarctic bottom water,* and *antarctic circumpolar water,* among others.

The properties most widely measured to characterize the water masses are: density, expressed in terms of the unit sigma; temperature, expressed in degrees Celsius; and salinity. In actual practice, the last is not determined directly, but is computed from various measurements, such as electrical conductivity, refractive index, or some other property whose relation to salinity is well established. Chlorinity is especially useful for this purpose, since the ratio of chloride content to the other elements in seawater is virtually constant despite the variation in total salinity from one water mass to another. Since in the same water mass the salinity varies with the temperature, the most widely used function for characterizing a given water mass is its T-S (temperature-salinity) graph. For that reason, temperature and salinity ranges are generally given in the various entries in this encyclopedia pertaining to the major oceanic water masses. These entries include: **Antarctic Waters** (bottom, circumpolar, intermediate, and surface); **Arctic Waters** (deep and surface); **Indian Ocean Water**; **Mediterranean Water**; **North Atlantic Waters** (deep and bottom, central, and intermediate); **North Pacific Waters** (central and intermediate); **Pacific Equatorial Water**; **South Pacific Central Water**; **South Pacific Current**; and **Subarctic Pacific Water**.

The sizes and depths of the three major oceans are:

Pacific Ocean 166.24×10^6 square kilometers (64,186,300 square miles). Average depth: 4188 meters (13,740 feet).
Atlantic Ocean 86.56×10^6 square kilometers (33,420,000 square miles). Average depth: 3735 meters (12,254 feet).
Indian Ocean 73.43×10^6 square kilometers (28,350,500 square miles). Average depth: 3872 meters (12,704 feet).

Later in this entry other physical and chemical characteristics of the oceans and seas are described in more detail.

Ocean Volume and Depth

The volume of the oceans and their seas is nearly 1.5×10^9 cubic kilometers (350 million cubic miles). The average depth is approximately 4 kilometers (2.5 miles). An average depth figure is useful for some calculations, but it should be emphasized that the depths of the oceans vary over a wide range. The oceans are not the deepest in their middle portions, as sometimes believed, but rather the greatest depths occur in trenches found along the continental margins. As indicated in Table 1, the greatest known depth is the Mariana Trench of the Pacific Ocean, which is over 11,000 meters (6.9 miles) below sea level. The continental shelves are about 200 meters (656 feet) below sea level.

Sea Level

Sea level changes occur principally as the result of glaciers forming or melting. In an ice age, the sea level drops; in an age of ice melting, the sea level rises. At present, the earth is in an interglacial period. A schematic sea level curve for the eastern United States is shown in Fig. 1. The curve indicates a stabilization of sea level (at a low point) some 15,000 B.P. (before present), followed by a relatively rapid rise in level (in terms of geological time), which tapered off about 5000 years ago. At the latter time, it is estimated that the sea level was about 5 meters (15 feet) below its present level. On the average, since that time, the sea level has risen about one millimeter per year and thus is gradually encroaching on the continents. Such worldwide changes over long periods of time are termed *eustatic.*

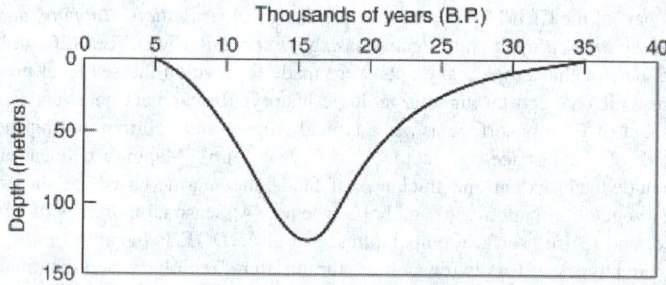

Fig. 1.   Sea-level curve for last 40,000 years. (*After Milliman and Emery.*)

There are also noneustatic sea level changes as the result of tectonism, which is prevalent along the Pacific Coast. These changes are also caused as the result of glacial rebound (local adjustments to prior glaciation), as well as seasonal changes that result from freshwater inflow and heating and cooling cycles in the ocean (*steric effects*). Records of tide gages, which reflect both eustatic and noneustatic changes, along the New England

TABLE 1. PRINCIPAL DEEP TRENCHES OF THE OCEAN

Name	Location		Depth Meters	Depth Feet
PACIFIC OCEAN				
Mariana Trench	11°21′N	142°12′E	11,033	36,200
Tonga Trench	23°15.3′S	174°44.7′W	10,882	35,702
Kuril Trench	44°15.2′N	150°34.2′E	10,542	34,587
Philippine Trench	10°24′N	126°40′E	10,539	34,578
Izu Trench	30°32′N	142°31′E	10,374	34,033
Kermadec Trench	31°52.8′S	177°20.6′W	10,047	32,964
Bonin Trench	24°30′N	143°24′E	9,156	30,041
New Britain Trench	06°34′S	153°55′E	9,140	29,988
Yap Trench	08°33′N	138°02′E	8,527	27,976
Japan Trench	36°08′N	142°43′E	8,412	27,600
Palau Trench	07°40′N	135°04′E	8,138	26,700
Aleutian Trench	50°53′N	176°23′E	8,100	26,574
Peru-Chile Trench	23°18′S	71°41′W	8,064	26,454
New Hebrides Trench	20°36′S	168°37′E	7,570	24,837
Ryukyu Trench	25°15′N	128°32′E	7,507	24,629
Mid-America Trench	14°02′N	93°39′W	6,669	21,880
ATLANTIC OCEAN				
Puerto Rico Trench	19°35′N	66°17′W	8,648	28,374
South Sandwich Trench	55°14′S	26°29′W	8,252	27,072
Romanche Gap	00°16′S	18°35′W	7,864	25,800
Cayman Trench	19°12′N	80°00′W	7,535	24,720
Brazil Basin	09°10′S	23°02′W	6,119	20,076
INDIAN OCEAN				
Java Trench	10°15′S	109° (approx) E	7,725	25,344
Ob Trench	(not accurately fixed)		6,874	22,553
Vema Trench	(not accurately fixed)		6,402	21,004
Agulhas Basin	(not accurately fixed)		6,195	20,325
Diamantina Trench	35°00′S	105°35′E	6,062	19,889
ARCTIC OCEAN				
Eurasia Basin	82°23′N	19°31′E	5,450	17,880
MEDITERRANEAN SEA				
Ionian Basin	36°32′N	21°06′E	5,150	16,896

Source: U.S. Mapping Agency, Hydrographic/Topographic Center

coast since about 1940 show a submergence of the coast at a rate of about 3 millimeters per year, whereas along the Gulf coast, the rate is about 15 millimeters (continental submergence) per year. Although erosion by wind and wave actions associated with hurricanes and winter storms, coupled with ecological disturbances, are principally responsible for beach and coastal area deterioration, the slow changes in sea level also participate in these processes. Coastlines are discussed later in this entry.

Much of the research on sea level changes was conducted during the International Decade of Ocean Exploration (IDOE) during the 1970s http://unesdoc.unesco.org/images/0001/000138/013802eo.pdf, particularly as part of the CLIMAP (*Climate: Long-Range Investigation, Mapping, and Prediction*) project http://gcmd.nasa.gov/records/GCMD_FE00126.html. As part of that project, attempts were made to develop the sea level profile as it changed throughout geologic history. Researchers constructed a model of Earth's surface as per a typical August and February during the peak of the last ice age (some 18,000 years ago). Mapping delineation included: (1) extent and thickness of land- and marine-based ice sheets, (2) vegetation patterns, (3) global sea level, (4) seasonal extremes of sea ice, and (5) sea-surface temperatures. See also **IDOE Program**.

Four types of fossils (coccoliths, foraminifera, radiolaria, and diatoms), sedimentation rates, and oxygen isotope data were analyzed and compared with conditions of the present time. Fortunately, extensive data were already available from the deep-sea core library maintained by the Lamont-Doherty Geological Observatory of Columbia University http://www.ldeo.columbia.edu/. With these data, it was possible to reconstruct many of the ocean's characteristics at the time of the last ice age. Samples documented all major temperature gradients, surficial water masses, and circulation patterns of glacial world oceans. By determining ice volumes, global sea level changes, which were caused by transfer of

water from oceans to ice caps, could be calculated. Maximum and minimum ice sheet models recorded sea level changes of 150 to 100 meters (492 to 328 feet). These models were further refined by using oxygen isotope data and yielded an overall value of a drop in sea level of 150 meters (492 feet). The sea-surface temperatures were based upon quantitative counts of microfossils.

Scientists observed that the area covered by permanent ice was substantially different during the glacial maximum as compared with the situation today. It is believed that, in the Northern Hemisphere, the huge land-based ice sheets attained a thickness of about 3 kilometers (1.9 miles) and the extent of pack ice and marine-based ice sheets significantly increased. As regards the Southern Hemisphere, the most striking contrast was the greater extent of sea ice. Thus, it is postulated that, combined with a sea level lower by 150 meters (492 feet), these developments caused significant changes in the surface characteristics of the earth. It has been reasoned that, on land, grasslands, steppes, sandy outwash plains, and deserts spread at the expense of forests; whereas the extent of snow-covered land increased significantly. The reconstruction of land vegetation was based upon pollen distribution and types. The aforementioned changes, coupled with an increase in glaciation, are assumed to have caused an increase in surface albedo over modern values.

In more recent studies, Gornitz, et al. (Goddard Space Flight Center), have analyzed large amounts of data derived from tide-gauge stations throughout the world. These data indicate that the mean sea level rose by about 12 centimeters (4.7 inches) over the past century. The sea level change has a high correlation with the trend of global surface air temperature. A large part of the sea level rise, according to these investigators, can be accounted for in terms of the thermal expansion of the upper layers of the ocean. The data also indicate weak indirect evidence for a net melting of the continental ice sheets.

The researchers suggest that global warming due to increasing atmospheric carbon dioxide could melt the marine West Antarctic ice sheet, raising the global sea level some 5 to 6 meters (16 to 20 feet). They point out that a sea level rise of as little as 15 centimeters (6 inches) could double the probability of damaging storm surges on the coast of Britain, causing substantial beach erosion and the intrusion of seawater into low-lying areas that are now freshwater regions. Data from over 700 stations were obtained from the Institute for Oceanographic Science, Birkenhead, England http://www.marine.gov.uk/bodc.htm. Individual station records were reduced to a common reference point by fitting a least-squares regression line to sea level as a function of time and by defining the zero point to be the value of the regression curve for 1940. The annual mean sea level curves for stations within a geographical region were then averaged to yield a mean sea level curve for each region. The researchers also attempted to remove the long-term (usually 6000 years) sea level trends from the station data in order to obtain short-term sea level fluctuations, which are perhaps more appropriate for correlation with global climate variations in the past century. The cause of the long-term trend is uncertain. It has been argued that as much as 90% of it is residual isostatic uplift of continents due to the removal of the Wisconsin ice sheets. However, the long-term trend may contain a eustatic component, for example, due to a change in volume of the Antarctic or Greenland ice sheets.

As observed by Gornitz, et al., the estimates for long-term sea level change are based on ^{14}C dating of measured positions of shoreline indicators in the geologic records, for example, mollusks, corals, and brackish-water peats. Further adjustments in the data were made and are described in the Gornitz et al. reference (listed).

In another study, M.F. Meier (U.S. Geological Survey, Tacoma, Washington) http://geology.wr.usgs.gov/docs/stateinfo/WA.html, investigated the contribution of small glaciers to global sea level. Meier observed long-term changes in glacier volume and hydrometeorological mass balance models, which yield information on the transfer of water from glaciers, excluding those in Greenland and Antarctica, to the oceans. The average observed volume change for the period 1900–1961 was scaled to a global average by the use of the seasonal amplitude of the mass balance. The data are used to calibrate models for estimating the changing contribution of glaciers to sea level for the period 1884 to 1975. Although the error band is large, these glaciers appear to account for a third to half of observed rise in sea level, approximately that fraction not yet explained by thermal expansion of the ocean. More details are given in the Meier reference (listed).

Although too detailed to describe here, it should be noted that in the mid-1970s, geologists (Peter Vail, et al., Exxon Production Corp.) announced a new approach to estimating and forecasting sea level. In deep drilling activities, mainly in connection with petroleum exploration, numerous unconformities (missing sections of sediment laid down during a given geologic time period and later eroded away) have been found with considerable consistency worldwide. The same unconformities are found on such widely dispersed, tectonically unrelated margins that it would appear only a global sea level change could link them, due mainly to the waxing and waning of major ice sheets. A part of the complexity is that while sea level changes occur, so do the levels of the continental margins, which have been estimated to be sinking at a rate of 1 to 2 centimeters (0.4 to 0.8 inches) per thousand years. Thus, it is concluded that even if global sea level remains unchanged, the sea still will creep up the shelf as it sinks — and even if global sea level drops fast enough to overtake subsidence and cause the sea's edge to retreat down the margin, sediments washing off the continent can stop short of the sea and fill in space created by subsidence above the waterline. For a further introduction to this topic, reference to Kerr is suggested. Also, see the Watts references.

Seawater Temperature

The mean temperature of the oceans is about 3.9 °C (~39 °F). This figure is useful in certain calculations, but obviously specific waters vary widely as affected by geographic location, season of the year, and depth. In torrid zones, the temperature of the surface waters may be between 24 and 27 °C (75 and 80 °F). Temperature decreases markedly toward the Polar Regions, where the temperature is about −2 °C (28 °F), except where ocean currents of higher or lower temperature exert thermal influence. With exception of

these currents, it was once believed that at any given depth the ocean waters within any given region were quite uniform.

In connection with the IDOE program, previously mentioned, the NOR-PAX project http://scilib.ucsd.edu/sio/archives/siohstry/onr.html (North Pacific Experiment ca. 1998) science team found the existence of enormous pools of surface water abnormally warmer or cooler than the mean by 1 to 1.5 °C (33.8 to 34.7 °F). These pools were found to be up to 300 meters (984 feet) deep, some 1,500 kilometers (932 miles) in lateral extent, and with time periods of some 2.5 years in duration. See also http://gale.ssec.wisc.edu/willy/caljet/, http://www.ssec.wisc.edu/experiment/, http://scilib.ucsd.edu/sio/archives/siohstry/onr.html.

It has long been established that, when compared with the atmosphere, the ocean has a great capacity to store heat and energy. The ocean also has large inertia that creates a potential for delayed feedback on climatic time scales. Thus, the time duration of the anomalous pools of water was not surprising. By using computer models and numerical techniques, the investigators were able to forecast both the shapes and intensities of anomaly patterns — up to a season in advance. These proved useful in shedding light on the upper-ocean processes as well as for predicting long-term ocean effects on the marine atmosphere. Essentially, in a continuation of the 1970s NORPAX project, scientists have more recently improved their computer modeling techniques. However, the interface between ocean and atmosphere is severely complex, particularly as the interface affects inland weather. This problem is explored in some detail in the article on **Atmosphere-Ocean Interface**.

Seawater temperature also varies with the ocean's currents. See also **Ocean Currents**.

Because solar energy is the most important source of heat for the oceans, there exists (except in Polar Regions) a very shallow layer of relatively warm water at the surface. Under this is a boundary layer called the *thermocline* in which the temperature drops quite rapidly with increasing depth until the large mass of water at its lower boundary is encountered, below which the temperature falls only very slightly with increasing depth. At the Polar Regions, the temperature of the water at or near the surface does not show significant differences with the waters at a greater depth. The effect of depth on seawater temperatures varies with locale. For example, in the Caribbean area, within a short distance of land, surface water temperatures may be in the 27–29 °C (~81–84 °F) range, yet the temperature at the 600–900-meter (2,000–3,000-foot) level may be as low as 4–7 °C (~39–45 °F). Use of seawater temperature differentials is the basis of one means for capturing solar energy for useful work. See also **Solar Energy**.

Sea-Surface Temperature (SST). As described by Cane (see reference), El Niño conditions occur when there are very marked alterations in SST. The El Niño of 1982–1983 has been very well documented in the literature and is described in some detail in this encyclopedia in the article on **El Niño**. Briefly, in July 1982, conditions in the eastern equatorial Pacific seemed normal. By October, however, the SST was nearly 5 °C (41 °F) above normal and the sea level at the Galápagos Islands had risen by 22 centimeters. The anomalies at depth were even greater. A huge influx of warm water had increased the heat content of the upper ocean at a rate that exceeded the climatological surface heat flux by a factor of more than 3, and the thickness of the warm layer was now greater than all previously observed values. Temperatures at the South American coast were near normal, but within a month they too would rise sharply. It became obvious that what had been labeled as a "warm event" had become a major El Niño. The severe nature and rather unusual developmental behavior of the 1982–1983 El Niño provided a great incentive for oceanographers and meteorologists to probe the extremely complex relationship between the oceans and climate.

Density

The density of seawater is determined by temperature and by the concentration of salts (salinity). The density increases as the temperature falls to the freezing point and also with increasing salinity. Accordingly, because the denser water tends to sink, and temperature is the most important determinant, the waters with the greatest density are found in the cold polar and arctic seas. While density of pure water at 4 °C (39.2 °F) is equal to 1, the density of seawater ranges over somewhat higher values, which vary with proximity to shores, rivers, etc., as well as with geographic location and depth. Representative average values are 1.026–1.028. However, seawater, like fresh water, has its maximum density about 4 °C (39.2 °F) above its freezing point, and as it becomes colder than 3–4 °C (37.4–39.2 °F), it becomes less dense, and rises. This fact explains why water cannot freeze at the bottom of the sea.

The high density and fluidity of seawater make it difficult to lower people and equipment below the ocean surface. The immense hydrostatic pressures at great depths will crush all but the strongest of vessels. Materials selection is difficult because very high stresses can result when equipment is constructed of materials having differing compressibilities. Thus, ocean technology, like so many areas of technology is confronted with corrosion, strength, and other material-related problems.

The combination of density and low viscosity provides excellent conditions for surface travel on the oceans. There is a high lift-drag ratio for ships, making it possible to move heavy loads with reasonable speed and relatively little power. However, this same combination of characteristics contributes to major hazards of shipping, notably the creation of large wind-waves. Obviously, if the water were more viscous, as in the case of a heavy oil or syrup, the wind would not have sufficient force to build up high waves. Conversely, if the water were considerably less viscous, the force and hence hazard present in waves would not be present. It is also of interest to observe that the high surface tension of seawater contributes to its not sticking readily to surfaces, allowing removal (except salt residues) by evaporation within short periods.

Seawater Composition

Because the composition of seawater is affected mainly by the addition of dissolved salts brought to it by the great rivers, the composition differs from one region to the next. On the average, the salt content of seawater is about 3.5% (weight). Although there are many compounds and chemical elements in seawater, the principal dissolved salts are:

	% (Weight) of all Dissolved Salts
Sodium chloride (NaCl)	77.76
Magnesium chloride (MgCl$_2$)	10.88
Magnesium sulfate (MgSO$_4$)	4.74
Calcium sulfate (CaSO$_4$)	3.60
Potassium sulfate (K$_2$SO$_4$)	2.46
Magnesium bromide (MgBr$_2$)	0.22
Calcium carbonate (CaCO$_3$)	0.34

Aside from mineral values, exploitation of which may increase in the future, salinity is probably the least desirable characteristic of seawater. See also **Ocean Water**. The water obviously cannot be consumed in any quantity by most land forms without adverse effects. Although saline waters (mainly brackish) can be used with limited success agriculturally in modern drip irrigation systems, generally they are not acceptable. The electrolytic chemical activity resulting from ionized salts, together with frequent presence of dissolved oxygen, exerts very adverse corrosive attacks on metals and other substances that come in contact with seawater. Salt water is a relatively good conductor of electricity and thus the ocean can be penetrated only slightly by radio waves. In contrast, seawater is an excellent conductor of sound waves of low frequency. For example, an explosion that may be detected only a distance of a half-mile or so on land can be heard a thousand or more miles under water. See also **Sonar**. The U.S. Navy has had a program under consideration for several years that would bring long-wave radio communication with submarines on a worldwide basis. See also **Antenna (Communications)**.

In separate articles on the many *chemical elements*, the content of each element in seawater is given. See also the table on chemical abundance in the summary article on the **Chemical Elements**.

Chlorinity is a measure of the chloride content, by mass, of seawater (grams per kilogram of seawater, or per cubic mile). Initially, chlorinity was defined as the weight of chlorine in grams per kilogram of seawater after the bromides and iodides had been replaced by chlorides. To make the definition independent of atomic weights, chlorinity is now defined as 0.3285233 times the weight of silver equivalent to all the halides present. See also **Chlorinity**.

Pressure/Depth of Ocean

At the average depth of the oceans (4,000 meters; 13,124 feet), the pressure is about 388 atmospheres. At the Mariana Trench, previously mentioned, the pressure is on the order of 1070 atmospheres. Pressure increases at the rate of one atmosphere per each 10.3 meters (33.9 feet) of depth, not considering corrections for temperature and varying composition.

Ocean Currents

Currents in the oceans represent the movement (flow) of water, usually in large amounts, as the result of tides, differences in water densities, the stress of various internal and external pressures, and of the wind. Ocean currents can flow for thousands of kilometers (thousands of miles). They are very important in determining the climates of the continents, especially those regions bordering on the ocean. Perhaps the most striking example is the Gulf Stream, which makes northwest Europe much more temperate than any other region at the same latitude. Another example is the Hawaiian Islands, where the climate is somewhat cooler (subtropical) than the tropical latitudes in which they are located because of the California Current.

Surface ocean currents are generally wind driven and develop their typical clockwise spirals in the northern hemisphere and counter-clockwise rotation in the southern hemisphere due to the imposed wind stresses. In wind driven currents the Ekman spiral effect, results in the currents flowing at an angle to the driving winds. The areas of surface ocean currents move somewhat with the seasons, this is most notable in equatorial currents.

Deep ocean currents are driven by density and temperature gradients. Thermohaline circulation, also known as the ocean's conveyor belt, refers to the deep ocean density-driven ocean basin currents. These currents that flow under the surface of the ocean, and are thus hidden from immediate detection, are called submarine rivers. These are currently being researched by floating devices, which maintain their depth according to slightly differing densities of waters. Upwelling and downwelling areas in the oceans are areas, where significant vertical movement of ocean water is observed.

Knowledge of surface ocean currents is essential in reducing costs of shipping, since they reduce fuel costs. In the sail-ship era knowledge was even more essential. A good example of this is the Agulhas current, which long prevented Portuguese sailors from reaching India. Even today, the round-the-world sailing competitors employ surface currents to their benefit. See also **Ocean Currents**.

Ocean-Continent Boundaries and Margins

About a century ago, Suess (German geologist) proposed that at one time in the development of the earth's surface there was no South Atlantic Ocean, but rather Africa and South America comprised one very large continent, which Suess chose to call *Gondwanaland*. Suess based his proposals largely upon observing the excellent fit (jigsaw puzzle analogy) between the eastern coast of South America and the west coast of Africa, rather vividly apparent from a cursory examination of a map of the world, particularly if constructed with polar coordinates. Then, in 1912, another German scientist, Wegener, extended the concept of the South America–Africa "connection" and proposed the prior existence of a supercontinent, which he proposed to call *Pangea*. See also **Wegener, Alfred Lothar (1880–1930)**. It is interesting to note that, during the interim, researchers have demonstrated a number of similarities in rock, fossil, and land forms that provide a rather striking correspondence between eastern South America and western Africa. These include certain similarities of land forms, such as mountains, deserts, cliffs, and flat-topped peaks. See diagram in entry on **Earth Tectonics and Earthquakes**.

As stressed by McCoy/Rabinowitz, "The South Atlantic has intrigued scientists for decades—from Suess and Wegener to Bullard to many researchers today. This archetypal example of the development of an ocean basin by continental drift is particularly well displayed by Bullard's interpretation, where the small irregularities in refitting continental margins seem remarkably slight considering the millions of years of erosion and deposition along these margins. This, however, emphasizes an important point; such continental margins preserve ancient events with little modification. Technically, they define an ocean/continent boundary. In addition to the remarkable mesh of these margins is the geological link of older rocks, fossils, structural alignments, and even some ancient physiographic features that are found on the two opposing land masses, which are today separated by a vast ocean."

Although various postulates had been developed, a concerted program to obtain a better understanding of the sea floor did not commence until the mid-1940s. Important to the early efforts were maps developed by Heezen and Tharp which indicate a worldwide ridge and rift network on the ocean floor. This proved to be a submarine chain of mountains some 65,000 kilometers (40,389 miles) long and extending into all oceans. These mountains follow a global zone of shallow earthquakes as well as heat flows.

The greater depths of the oceans are found in trenches at the margins of the oceans. Excepting the Indonesia, Antilles, and Scotia deeps, the major trenches are found in the Pacific Ocean floor. However, as pointed out by *Uyeda*, in a broad sense, even these three exceptions may be regarded as part of the Pacific margin because they comprise the belts of active volcanoes, sometimes called the Ring of Fire around the Pacific. These are also belts of high seismic activity. Margins with trenches are Pacific-type or active margins.

Continental Displacement. Sometimes less appropriately called **continental drift**, continental displacement is a general term that can be used in connection with many aspects of the theory originally propounded at length by Wegener. In a pioneering way, Wegener postulated the displacement of large plates of continental (sialic) crust, moving freely across a substratum of oceanic (simatic) crust, but the mechanisms involved were so implausible to most geologists at that time that the concept was generally discredited for many decades. In more recent years, however, new evidence has been found and more acceptable alternative mechanisms have been proposed — so that the original theory has gained a greater degree of credence: (1) the continents have remained relatively fixed as the earth has expanded, leaving progressively wider gaps of oceanic areas between; (2) the continents have moved away from each other by sea-floor spreading along a median ridge or rift, producing new oceanic areas between the continents; or (3) the masses propelled away from the ridges consist of thick plates, composed of both continental and oceanic crust, which have moved in various directions independently of each other. Some contemporary geologists and oceanologists suggest that a true explanation of world tectonics will combine some or all of the foregoing explanations, to which will be added concepts yet to be created.

In deference to perspective, it is in order to review very briefly the content of the aforementioned concepts. The *expanding earth concept* suggests that the diameter of the earth has grown progressively larger through time, perhaps by a third or more during recorded geologic time, as a result of changes in atomic and molecular structure in the core and lower mantle, without change in actual mass. The theory has been linked with continental displacement and sea-floor spreading, although these have also been otherwise explained. Arguments and evidence for an expanding Earth were eloquently presented by Holmes. But, as pointed out by Burke, "For a variety of reasons, most earth scientists found and still find the idea of substantial expansion of the earth hard to accept (for one thing, the amount of work required against gravity is mind-boggling)."

In another concept, a *world rift system* is proposed as a major tectonic element of the earth consisting of mid-oceanic ridges and their associated trenches, such as those found along the Mid-Atlantic Ridge. The rift system is believed to be the locus of tensional splitting and upwelling of magma that has resulted in sea-floor spreading.

Another more recent and well-accepted concept is that of *plate tectonics*. The essence of this concept is that global tectonics is based on an Earth model characterized by a relatively small number (10 to 25) of large, broad, thick *plates* (composed of areas of both continental and oceanic crust and mantle), each of which "floats" on some viscous underlayer in the mantle and moves more or less independently of the others and grinds against them like ice floes in a river, with much of the dynamic activity concentrated along the periphery of the plates, which are propelled from the rear by *sea-floor spreading*. It has been suggested that the continents form a part of the plates and move with them, much as logs frozen in the ice floes. In 1965, Wilson, reporting in *Nature*, suggested that the rigid plates of the lithosphere moved along tectonically active boundaries of *three kinds*. These observations lead to an increased validity of the general continental displacement or drift theory as applied to an interpretation

of the Atlantic margins. As observed by Burke, "These margins formed at divergent plate boundaries and, as the ocean grew and the divergent boundary continued to spread, the two continental fragments were carried symmetrically away from each other. At the same time, their edges were draped in young sediments. The contrasting character of Pacific coasts became understandable once it was recognized that they are places where the lithosphere is returning to the mantle and that most Pacific margins mark boundaries across which plates converge."

The third type of margin, known as a transfer margin, from present knowledge is believed to occur less frequently. In this type of margin, plates slide past each other.

The classification of margins into three well definable categories has provided for a convenient framework for investigating and reporting on ocean-continent boundaries. In the recent literature, margins thus are frequently described as: (1) *Atlantic-type* margins; (2) *Pacific-type* or active margins; and (3) *transfer* margins.

An understanding of continental margins is no longer solely a matter of scientific interest, but is economically very important today because exploration for oil and gas sometimes occurs along these margins and associated shelves. It should be pointed out that these exploratory activities have significantly contributed to an understanding of the margins and of ocean formation. See also **Petroleum**.

A margin is made up of three depth zones, proceeding in the following order from the continent toward the ocean basin: (1) *Continental shelf*, generally ranging from 30 to 300 +kilometers (19 to 186 miles) in width and usually with a gentle slope to a depth of less than 200 meters (656 feet). The seaward end of the shelf is called the *shelf break*. (2) *Continental slope*, a narrow, plunging face, extending from the shelf break to the sea floor to a depth of some 3,000–4,000 meters (9,843–13,123 feet). (3) At a point where the sea-floor gradient drops below 1 in 40, the *continental rise* extends further to a depth of some 4,000–6,000 meters (13,123–19,685 feet) in a much more gentle slope that may be from 80 to 500 +kilometers (50 to 311 miles) long. See also entries on **Continental Rise; Continental Shelves**; and **Continental Slopes**. See also Fig. 2.

In 1953, an International Commission defined the continental shelf, shelf edge, and continental borderland as: "The zone around the continent extending from the low-water line to the depth at which there is a marked increase of slope to greater depth. Where this increase occurs the term shelf edge is appropriate. Conventionally the edge is taken at 100 fathoms (600 feet), but instances are known where the increase of slope occurs at more than 200 fathoms (1,200 feet) or less than 65 fathoms (390 feet). When the zone below the low-water line is highly irregular and includes depths well in excess of those typical of continental shelves, the term continental borderland is appropriate."

Continental shelves comprise about 18% of the earth's total land area. Depending upon the advance or retreat of glaciers, the shelves are alternately exposed and drowned. The shelves underlie about 7.5% of the total oceans. In total area, the shelves comprise about 25.9 million square kilometers (10 million square miles).

Exploration is far from complete in terms of determining the specific parameters of the numerous continental shelves. Data are most detailed in terms of the shelves around Alaska, the southern California coast and Baha California, the northern shore of the Gulf of Mexico and the coast around Florida and northward to Nova Scotia, the coasts around Ireland and the British Isles and the northern coasts of France and Germany, and parts of the coasts of the Sea of Japan, the Yellow Sea, and the East China Sea. Continental shelf data are particularly scant pertaining to Greenland, the north central and north eastern coast of Russia (Kara Sea, Laptev Sea, etc.), most of the Canadian Arctic, Labrador, Newfoundland, Iceland, most of the coastline of South America (excepting Venezuela and Argentina), and most of Africa, Australia, and New Zealand. The gathering of continental shelf information on hand has been assisted importantly as the result of off-shore oil exploration activities and continuing studies which followed the IDOE projects.

The most accessible rocks, of course, provide information only on the most recent history of a shelf. Particular attention has to be given to properly identifying whether a rock sample is actually from the underlying bedrock of the shelf, or possibly from an adjacent projecting hill. There is also the possibility that rocks may have been laid down by ancient glaciers or streams and thus not native to the shelf under study. Unfortunately,

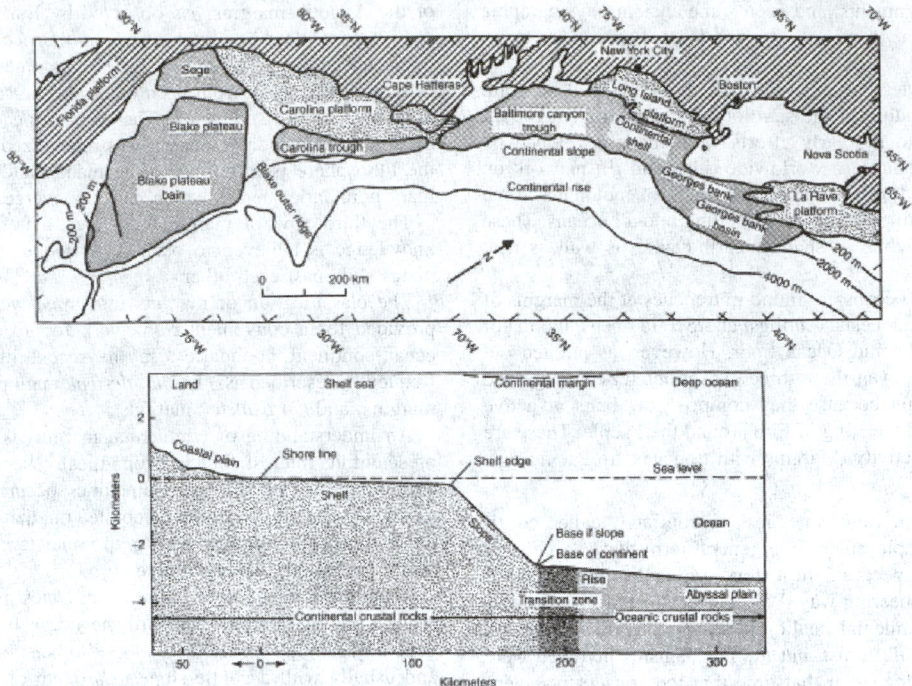

Fig. 2.   (Top) Continental margin along the eastern coast of the United States. Major sediment-filled basins and platforms are shown. *SEGE* = Southeast Georgia Embayment. (*After Schlee* et al.) (Bottom) Geomorphic features of the continental margin. (*National Petroleum Council.*)

from the standpoint of geological study, many of the drillings into the shelves have penetrated salt domes and folds rather than normal structural features. Consequently most information pertaining to the shelves has come from geophysical methods, notably seismic reflection and refraction and measurements of geomagnetism and gravity.

Investigations to date indicate two principal types of shelves: (1) Under lying sedimentary strata; and (2) underlying igneous and metamorphic rocks. Many of the continental shelves comprise the top surface of very long, essentially pyramidically-shaped sedimentary strata. Their position against the continents is maintained by long and narrow fault blocks. Along the perimeter of the Pacific Ocean, noted for its tectonic activity, there are also deep trenches, which parallel the base of the continental slope. It is estimated that a geologic dam extends for thousands of kilometers along the coastline of the western United States. Rocks found on the Farallon Islands (granitic) off San Francisco are estimated at about 100 million years of age, but it was not until about 25 million years ago that they were pushed up to form the dam of which they are a part. Several such fault dams also are estimated to have risen during the last 500 million years in the Yellow Sea. A similar dam existed along the eastern coast of the United States. With passage of time the trench was filled with sediments. These spilled over the former dam, resulting in the continental slope. The slope is maintained by the angle of slope of the sediments and, as might be expected, some instability has been evidenced by landslides and features of erosion as detected by seismic probing.

Prior to investigations of the last few decades, it was believed that sediments deposited on the shelves gradually became finer as the edge of the shelf was approached—from gravel and sand at the shore to very fine sand and ultimately silt and clay to form the so-called mud line at the edge. Samplings have not indicated this to be true. The current conclusion is that the size of sediment grains is not related to distance from shore. The former concept only holds for wave-controlled areas between the shore and perhaps depths up to 20 meters (66 feet).

It is estimated that a majority of the sediment found on the continental shelves was deposited during the last 15,000 years, that is, after the last lowering of the sea level by glaciers. The finding of fossilized plant and animal life in the sediments attests to the time when the shelf areas were dry.

Atlantic-Type Margins. It is estimated that the Atlantic Ocean started to open about 200 million years ago (Triassic Period), commencing with the separation of South America from Africa and, later, the separation of North America from Europe. As pointed out by Ross, "The breakup probably started with a broad upwelling of the crust, followed by thinning, and then gradual splitting or rifting apart. The two new continents then continued spreading apart with new ocean crust forming by volcanic activity in the resulting central rift. In the early stages of the split, the new sea was long and narrow, probably with isolated depressions. Its connection with the ocean was restricted, and evaporitic conditions often developed, which led to extensive salt deposits."

The accumulation of sediment in the Atlantic-type margins is far greater than that which occurred on Pacific margins. This is as may be expected because the majority of sediment-carrying rivers (Amazon, Congo, Mississippi, Niger, Rhine, etc.) drain into the Atlantic Ocean. It is with these wedges of sediment that oil and gas are associated.

Most of the sediment introduced into the Atlantic Ocean comes from just ten rivers. A number of new concepts have been added to the original concept of displacement (drift) in an effort to further refine a model of the Atlantic-type margin. One of these refinements deals with *subsidence*, which apparently occurs when the margins cool as they age. Marginal subsidence of a mature nature, relatively speaking, is found along the eastern United States. This subsidence has been estimated at about 3 kilometers (1.9 miles), occurring over a 150-million year period. In contrast, higher ground is found along the margins of younger oceans, such as the Red Sea, associated with which are the hills found in Egypt and the Sudan.

Many contemporary geologists and oceanologists suggest that possibly the Red Sea is a relatively new body of seawater slowly expanding to form a larger ocean. Research has indicated that the Red Sea is opening to the northeast at a rate of one to a few centimeters per year, causing a slow compression of the Persian Gulf against the Asian continent. See Fig. 3. In this process, it is postulated that the Persian Gulf region is being thrusted (subducted), along with the related coastal area of Iran, under the adjacent continental region. Ross (1979) projects that it will require some 25 million years for the Persian Gulf to be completely subducted, at which time Saudi Arabia and Iran will collide.

It is believed by some scientists that continent-to-continent collision already is taking place along parts of the Gulf of Oman. Ross summarizes this by stating, "Translation movements, where plate boundaries slide by parallel to each other, occur in the Gulf of Aqaba (Elat), the Dead Sea, and the Jordan Rift Valley up to northern Syria. The entire length of this

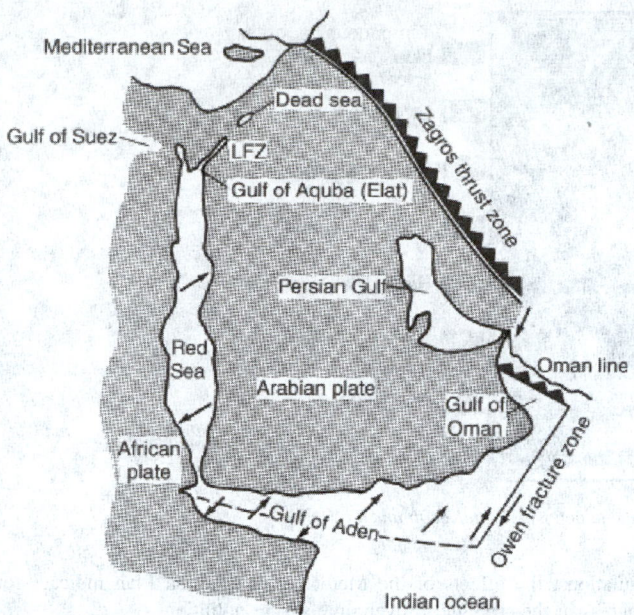

Fig. 3. Features of the Arabian Plate showing relationship with African Plate. The spreading of the sea floor is represented by full arrows (Red Sea and Gulf of Aden). Translation is indicated by half-arrows. Subduction is indicated by wedged zone. *LFZ* = Levant fracture zone. (*After Ross.*)

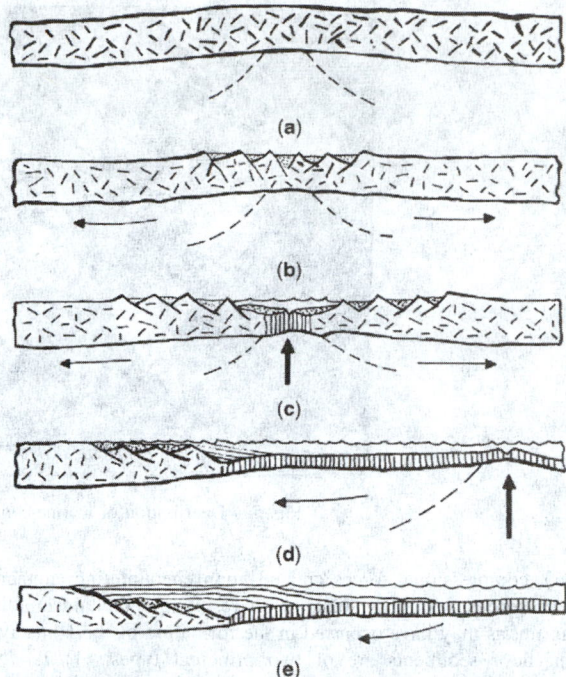

Fig. 4. Formation and evolution of a passive continental margin: (**a**) doming, (**b**) rifting, (**c**) youthful drifting, (**d**) intermediate drifting, and (**e**) mature drifting. (*After Oceanus.*)

feature is often called the Levant Fracture Zone. A similar type of motion also occurs in the Indian Ocean along the Owen Fracture Zone and the Oman Line, which offsets the subduction along the Zagros Thrust Zone and Gulf of Oman."

Although the application of plate tectonics to a description of the Red Sea as being in an early stage of sea-floor spreading in current times is convincing to some scientists, not all investigators accept this explanation of the origin of the Red Sea. Questions include: To what extent is the current width of the Red Sea ascribable to sea-floor spreading? To what degree is this attributable to thinning and stretching of the continental crust that took place prior to and during the early spreading? These questions are explored by Francheteau and LePichon (reference listed).

An important geologic aspect of the Red Sea is the hot brine areas which occur along the bottom of the central rift. The economic value of these sediments (underlying the brines) has been emphasized by many investigators from a number of countries.

The sediments which form the thick wedges at Atlantic margins undergo changes with the passage of time and thus are good indicators for determining the age of existing oceans. See Fig. 4. For example, it is estimated that the Red Sea commenced to take form as recently as 25 million years ago, whereas it is estimated that the South Atlantic Ocean is some 130 million years old, and the Central Atlantic Ocean some 160 million years old. Some scientists believe that the type of sediments being collected today in the East African Rift typify the phenomenon of early sediment accumulation. It has been further postulated that after initial sediment accumulation, formation of limestone structures occurs. Evidence indicates that widespread limestone accumulation in the Central and South Atlantic occurred over a period of some 20 million years, and this was followed by accumulation of marine sands and muds. Of course, as an ocean matures, the distribution of sediment along the margins becomes less uniform. Generally, past alterations in sea level have been attributed to cyclic melting and freezing of the polar ice caps. In more recent years, many geologists and oceanologists now believe that another major cause of sea level changes has been the changes in the rate of sedimentation, i.e., the rate at which the ocean floors are made.

It is postulated that because the young ocean floor is hot, it requires a larger volume—thus displacing water and causing a rise in sea level. Relative to the sea floor, the sea level was perhaps twice as high as the present sea level during the Cretaceous Period (136–65 million years ago). It has been postulated that this striking rise in sea level was the result of a combination of large new accumulations of sediment coupled with the

greater volume required for the hot, young materials forming new oceanic crust. At much later times, the crust cooled and subsided, adding another factor to the complex series of events that affects sea level.

Ocean Floor and Crust — Active Margins

At one time, geologists believed that the earth's crust was a reasonably stable layer forming an envelope around the fluid mantle and core. It was envisioned that crustal blocks floated on a plastic mantle. This concept was not supported by evidence, particularly of the kind assembled during the past few decades. Rather, it is now envisioned that at least three concepts are at work, i.e., sea-floor spreading, continental drift, and plate tectonics. Crustal plates apparently move apart at a rate of 1 to 10 centimeters (0.4 to 4 inches) annually. As they spread apart, there is an upwelling of basaltic material that causes a ridge. Upon cooling, the molten rock adheres to the crust that is moving away on each side of the fissure. When the movement is slow (about 3 centimeters (1.2 inches) per year), a rather steep sloping of the sea floor on either side of the ridge occurs. Steep escarpments are produced because the sinking dominates the process informing the slope. On the other hand, when movement of the plates is at a faster rate, the horizontal spreading is rapid in contrast with the sinking of the crust. The steepness of the slope is determined by the balance between these two actions.

The Atlantic Ocean is known for slow spreading. The Pacific Ocean is known for fast spreading. Thus, in a continuous process running over millions of years, there is a slow but steady spreading and renewal of the sea floor, the spreading ranging from 1 to 10 centimeters (0.4 to 4 inches) per year. It has been estimated that the crust sinks about 9 centimeters (3.5 inches) per year for the first ten million years after it forms, dropping to somewhat over 3 centimeters (1.2 inches) per 1000 years for the next 30 million years, and to about 2 centimeters (0.8 inches) per 1000 years after that. However, not all crust sinks, a notable example being that of the southern Mid-Atlantic Ridge where it is estimated that the sea floor level has not changed for the last 20 million years. Regions where spreading occurs in multiple directions are termed spreading centers. These centers also move.

When a crustal plate grows, its leading edge is destroyed at an equal rate. In some instances, the edge may slide under the oncoming edge of another plate and return to the asthenosphere. Such action causes a deep trench. The Mariana Trench and Tonga Trench in the Pacific are examples. If plate movement is relatively slow (5 to 6 centimeters (2 to 2.4 inches)

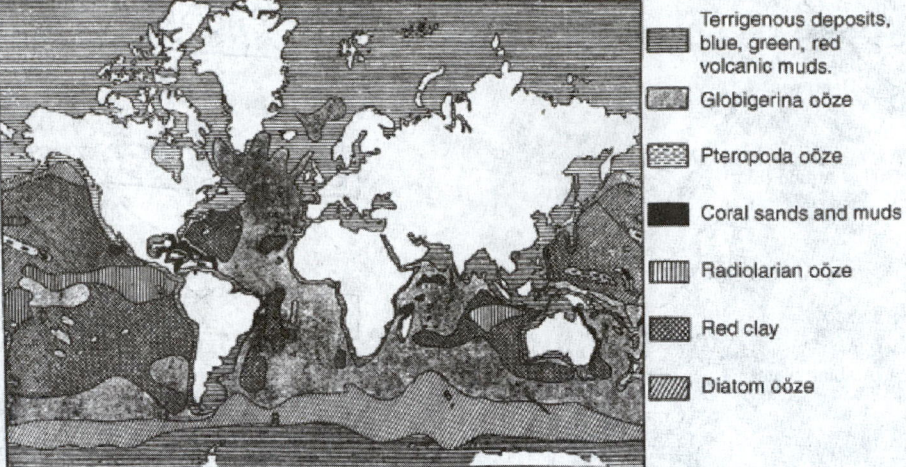

Fig. 5.   Distribution of marine-continental and oceanic deposits. (*After Scott and Collet.*)

per year), compressional forces created upon encountering another plate may be absorbed within the plate, with resulting mountain building by buckling up, as may have occurred in the formation of the Himalayas.

Ocean floor sediments are of two principal types: (1) *Terrigenous deposits* of terrestrial origin; and (2) *pelagic deposits*, consisting of matter produced within the sea itself. Close to the edges of the great land masses a variety of sands, silts, clays, and marls will be found on the continental shelves and epicontinental seas and, to a lesser extent, on the continental slopes. This terrigenous matter is deposited by rivers or secured by the effects of erosion on the coastlines and subsequently deposited by the action of currents. Such accumulations may total many hundreds of feet in depth. Other deposits have been formed by icebergs carrying materials that are released and settle to the ocean floor as the berg melts. In certain deep parts of the ocean floor there are deposits of red clay, believed to be of terrestrial origin. These are of extremely fine texture and are carried in suspension in the ocean currents until they finally settle to the ocean floor. The maximum accumulation of this red clay is believed to be at the rate of $\frac{1}{25}$ of an inch (1 millimeter) per 1000 years. See Fig. 5.

The deep sea oozes of pelagic origin are of two kinds. The *calcareous ooze* consists of the shells of the single-celled *Globigerina*, of a small mollusk, the pteropod, or of certain plants. These calcium carbonate shells are found only in depths less than approximately 14,500 feet (4,420 meters); at greater depths, the available carbon dixoide is able to dissolve such shells. The second type of *pelagic* ooze is that of s*iliceous* origin, consisting of the silicon skeletons produced by organisms known as radiolaria and diatoms. Because it is not subject to dissolution, this siliceous ooze is found at very great depths. The rough average of these sediments over the entire ocean floor is estimated to be about 2,000 feet (610 meters).

For many years, the mysteries of the Mediterranean Sea have been probed by oceanographers who have found the Mediterranean an excellent laboratory for geological and geophysical studies and for physical oceanography. Scientists continue to probe the geological history of the two complex deep-sea basins that comprise the Mediterranean. As some researchers have observed, "If rocks could tell tales, the Rock of Gibraltar would be a master storyteller."

In the late 1880s, a number of European geologists suggested a number of theories pertaining to the tectonic origin of the Mediterranean, but any substantiation had to await the development of modern deep-sea drilling technology and oceanographic instrumentation. For a number of years, some researchers proposed that the Mediterranean was a huge dry desert, situated some 3,000 meters (9,843 feet) below present sea level. It was likened to a "death valley" scenario. Proof of this concept has been lacking, and the theory no longer has strong support among the professionals. In 1990, D. Stanley (National Museum of Natural History) remarked, "I believe the seafloor remained almost continually covered by very saline waters, perhaps one hundred to several hundred meters deep."

Almost an entire issue of the prestigious *Oceanus* magazine (Spring 1990) is devoted to the various scientific aspects of this interesting sea. It is also noteworthy to report that, after more than a decade of research and

regulations, the effects of the Mediterranean Action Plan indicate some progress in the struggle to overcome severe pollution.

Subduction. As early as the late 1920s, Holmes (University of Edinburgh) and others suspected that the tectonic features at active margins were the result of the workings of some common process. It was further suggested that the process may be a down-thrusting mantle convection current. These concepts were later refined in the light of an emerging theory of plate tectonics and it was proposed that the leading edge of a rigid oceanic plate may be the specific material being thrust downward into the mantle. Subduction, a concept originally proposed by Alpine geologists, may be defined as the process of one crustal block descending beneath another, by folding or faulting or both. A subduction zone may be defined as an elongate region along which a crustal block descends relative to another crustal block, e.g., the descent of the Pacific plate beneath the Andean plate along the Andean trench.

Because evidence indicates that mid-oceanic ridges are continually being generated, it is logical to assume that they are being consumed elsewhere in order to keep the surface area constant.[1] Thus, the subduction or consumption of oceanic plates may be the logical consequence of sea-floor spreading. As observed by Uyeda, active margins, with deep trenches and seismicity, were the most obvious localities. Uyeda asks, "Can the process of subduction really explain the tectonic features at active margins?" In studying a number of active regions, including the Benioff-Wadati zone underlying Japan, the seismicity of an area generally fits the concept of subduction. As of the early 1990s, the subduction theory does not offer a full explanation, particularly concerning the *thermal regime* of the subduction zones. High heat flow in the back-arc basins, the low velocity and high attenuation anomalies of the mantle wedge, and the active arc volcanism all require further explanation. Perhaps the answers lie with the formulation of two or more modes of subduction. Two such modes are described by Uyeda.

Mid-Ocean Ridges Research. Investigations accelerated markedly during the 1980s and early 1990s toward understanding the mid-ocean ridges. These ridges wrap around Earth for over 70,000 kilometers (44,500 miles) and have been likened to the "seam of a baseball." These ridges are considered to be the most volcanically active mountain chain in the entire solar system.

---

[1] It is of historical interest to introduce the theory of the *contracting earth*, widely believed in the 19th and the first part of the 20th centuries—to the effect that orogenic and other structures of the earth were produced by compression of the crust during its gradual contraction on the surface of a cooling, but originally molten globe (a familiar textbook illustration of the time was a dried apple). The theory has since been discredited, as the evidence shows that the earth is not cooling and contracting in the manner then believed.

It should also be mentioned that the expanding earth concept is described earlier in this entry.

Contemporary researchers tend to support the concept of a hierarchy in the segmentation of mid-ocean ridges. As described by K. Macdonald (University of California, Santa Monica), "*First-order* segments are generally hundreds of kilometers long, persist for millions to tens of millions of years, and are bound by relatively rigid, plate-transform faults, called *first-order discontinuities*. A first-order segment usually is divided into several *second-order* segments." The latter are shorter, are not as long lived, and are bounded by non-rigid, second-order segments of smaller magnitude, third- and fourth-order discontinuities. It has been established that these discontinuities can migrate along the length of a ridge.

It is interesting to note that the chain of active volcanoes that comprise the mid-ocean ridges expel, during an average year, ten times more lava than that which flowed from Mt. St. Helens in 1980.

Oceanographers in the past generally have directed their research of the mid-ocean ridges on a segment-by-segment basis. In their studies, the theory of plate tectonics served as a foundation. The principal plates and ridges are indicated on a map contained in the entry on **Earth Tectonics and Earthquakes**. With improvements in drilling techniques and exploratory instrumentation, scientists now can concentrate on the macrostructure of the mid-ocean ridge system. Pertaining to the *systematics* that have emerged, Macdonald comments, "Is the architecture of the global mid-range ridge system really so orderly, or is this concept of a 'segmentation hierarchy' merely a human construct?"

As of 1992, D. Blackman and T. Stroh (InterRidge, an international project) http://www.interridge.org/, summarize their key goals as including: (1) characterizing the global ridge structure; (2) understanding crustal accretion and upper-mantle dynamics; (3) charting the variability over time of volcanic and hydrothermal systems; (4) mapping biological colonization and evolution at ridge crests; (5) determining the properties of multiphase

materials at ridge crests; and (6) developing technology for ridge-crest experimentation.

Examples of new technological advances include chemical sensors that detect minute changes in trace elements and compounds (such as hydrogen sulfide, methane, iron, manganese, and oxygen), geodetic instruments to measure uplift and tilt of volcano flanks, broadband ocean-bottom seismometers, and deep-water temperature and chemical profiling systems. Systems that can deploy and manipulate these sensitive instruments will also be required and may take the form of remotely operated seafloor vehicles or manned submersibles.

P. Lonsdale and C. Small (Scripps Institution of Oceanography) reported in 1992 of studies on seafloor spreading, a process that creates new material to fill in gaps between Earth's separating crustal plates, which results in broad elevations with spreading centers along their crests. Examples of how mid-ocean ridges and ocean basins are created are given. See Fig. 6.

In 1992, W. Bryan (Woods Hole Oceanographic Institution) reported on studies of the evolution of deep-sea volcanology. It is interesting to note that the first volcanic rocks from a mid-ocean ridge were accidentally sampled during cable-laying operations in the North Atlantic in 1874. As reported by Bryan, "Throughout the first half of the 20th century the seafloor was widely assumed to be basaltic, but evidence for this assumption was still sketchy and indirect. A 'basaltic' and therefore 'volcanic' seafloor was consistent with the arguments based on isostasy and bathymetry that remain valid today. The continents must stand high, because they are composed of relatively thick, light granitic rock that literally floats higher on the underlying mantle than does the thinner, heavier rock comprising the oceanic crust. Also, petrologists generally assumed that basalts of volcanic islands such as Hawaii or Iceland were representative of the rocks to be found on the deep seafloor. Although

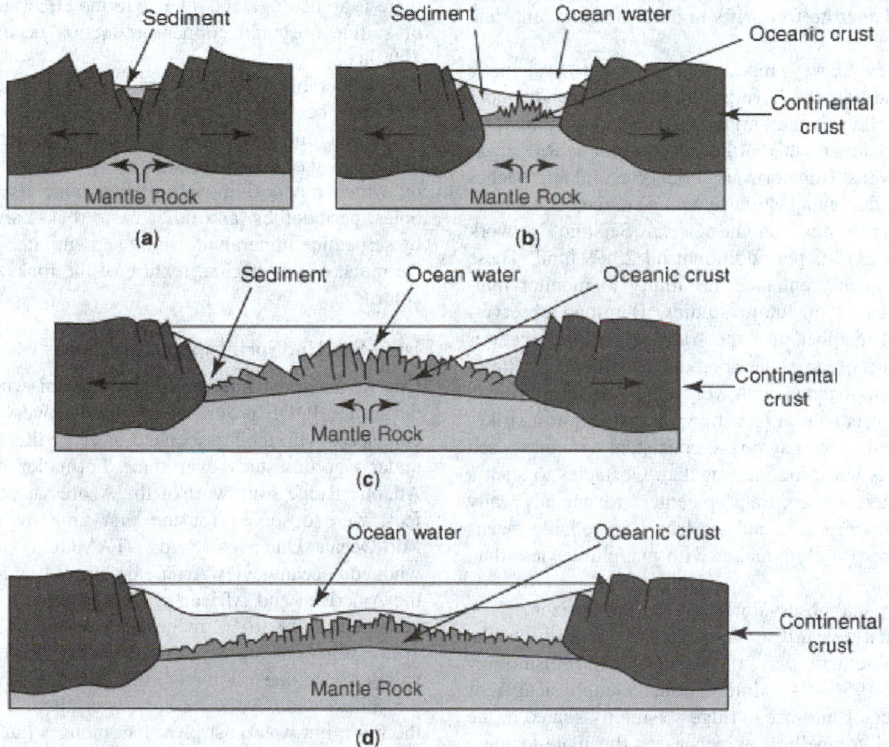

Fig. 6.   Development and demise of a mid-ocean ridge: (**a**) Rift valley—the deep central cleft with a mountainous floor in the crest of a mid-ocean ridge. The valley results from plate separations, as fast-spreading ridges and upwelling magma fill the rift and smooth the topography, while at slow-spreading ridges the upwelling magma does not fill the rift, but adheres to the trailing edge of the spreading plates. Example of a rift valley is the East African Rift. (**b**) Continued crustal separation produces a gap that is partly filled by sediment washed off the continents and partly by melting of the mantle to produce oceanic crust. Example is the Gulf of California. (**c**) As the gap between separating continents continues, the gap increases and oceanic crust formation by seafloor spreading at the crest of a rifted mid-ocean ridge develops. Example is North Atlantic Ridge. (**d**) When continental separation ceases, seafloor spreading stops and the mid-ocean ridge subsides as it cools and gradually becomes covered with sediment. (*After Lonsdale and Small.*)

there are often striking differences between continental volcanic rocks and the deeper crustal rocks on which they have erupted, the shaky logic of this analogy as applied to the seafloor does not ever appear to have been challenged."

When the reality of seafloor spreading and plate tectonics became generally accepted in the mid-1960s, mid-ocean ridge volcanoes proved to be the answer needed to explain the process for creating new seafloor. Deep-sea drilling programs proved invaluable toward reaching conclusions drawn today. One of the greatest challenges remaining is that of long-term observation of several of the numerous volcanoes located along the Mid-Atlantic Ridge. Also, models will be required to prove if hot spots are the locations of upwelling "mantle plumes" that carry fresh, hot, and previously undepleted mantle from a deep, previously untapped source. The principal hot spots on the Mid-Atlantic Ridge now recognized include the Azores and Tristan de Cunha in the south Atlantic.

J. Karson http://www.nicholas.duke.edu/people/faculty/karson2.html#pubs (Duke University) is investigating the tectonics of slow-spreading ridges. Karson observes, "There is a growing awareness that fast- and slow-spreading ridges function in very different ways. The sputtering magma supply of slow-spreading ridges results in substantial periods of plate separation that involve stretching and faulting of relatively cool oceanic lithosphere with little or no magmatism. The fault patterns of the median valley appear to mimic those of continental rifts; however, at least locally, very highly stretched and thinned masses of crust and upper mantle occur. The median-valley geology and fault structure documented by near-bottom studies predicts a very heterogeneous geological structure in slow-speed crust. This result is yet to be clearly defined or reconciled with the geophysical expression of the crust away from spreading centers. Future studies of the geometry and kinematics of faulting on slow-spreading ridges will determine the nature of faulting over much larger areas than have been studied to date, and will help contribute to the overall understanding of how the lithosphere is pulled apart to form rifts in both the continents and the seafloor."

E. Bergman (U.S. Geological Survey) reported in 1992 that "Systematic studies of mid-ocean ridge earthquakes have produced many insights concerning the tectonics of accreting plate boundaries. The largest transform earthquake in the last three decades on the northern Mid-Atlantic Ridge was a magnitude 7 event on the Vena Transform in 1962." Several approaches are being investigated for deploying high-dynamic range, broadband seismometers in deep ocean basins for the Ocean Seismic Network http://msg.whoi.edu/osn/EOS/EOS_paper_additional_jul_23_99.html. These types of instruments would greatly enhance the ability to monitor mid-ocean ridge seismicity. In terms of future studies, Bergman observes, "Transform faults are an end member of a spectrum of geologic features associated with offsets of mid-ocean ridge spreading segments. Little is known about the seismicity associated with very small offsets. From a seismological point of view, it is natural to define a transform as a strike-slip focal mechanism. This definition may not be consistent with one based on morphology. The issue has yet to be investigated. Obstacles to such a study include obtaining sufficiently accurate epicenters to unequivocally place earthquakes on small ridge offsets and the lack of a reliable means to determine focal mechanisms for earthquakes with magnitudes less than about 5."

Permanent seafloor geophysical observatories are on the horizon.

It is interesting to note that *seaquakes* were reported as early as the 19th century, but not in a systematic way. In the late 1950s, seismology was investigated during the 1956–1957 International Geophysical Year (IGY) project. The importance of mid-ocean ridge seismicity soared in the late 1960s, when it provided compelling evidence for the plate-tectonic hypothesis.

Ophiolites. Ophiolites are sections of the oceanic crust and the subjacent upper mantle that have been uplifted or emplaced to be exposed within continental crustal rocks. The concept of crust formation at mid-oceanic ridges has been largely postulated on the basis of measurements of magnetic and seismic patterns and upon the examination of sediments recovered by exploration drilling vessels—all well-established methods, but techniques which are undergoing continuous improvement. In more recent years, the study of ophiolites has made a major contribution to the understanding of crust formation.

An *ophiolite* may be described as a *group* of mafic and ultramafic igneous rocks ranging from spilite and basalt to gabbro and peridotite, including rocks rich in serpentine, chlorite, epidote, and albite derived from them by later metamorphism, whose origin is associated with an early phase of the development of a geosyncline. The term was originated by Steinman in 1905. Ophiolites are found on continents and for many years were not directly associated with oceanic crust. The existence of ancient oceans over present continental material was mainly suspected as the result of finding the fossilized remains of marine plants (diatoms) and marine animals (shells, etc.). For example, in the middle of the 15th century, the Swiss naturalist Hemerli found some fossils in Alpine rocks, well removed from the present oceans and thousands of meters above sea level. Considerably later, marine fossils from the Triassic period (200–230 million years ago) were found in the high peaks of the Himalayas. An association of mountains with the remains of marine activity became established and, during more recent periods, much additional evidence of the mountain-ocean relationship has been gathered.

An association of ophiolites with ocean crust in recent years has been established as the result of oceanographic technology, including seismic probing, drilling, dredging, and observations from submersibles. As pointed out by O'Connell, the resemblance between ophiolites and what is known about ocean crust is so strong that ophiolites are thought to be huge slabs of oceanic crust that broke off subducting ocean crust and were pushed onto continents and island arcs. O'Connell further observed that, as fragments of ocean crust, ophiolites hold intriguing scientific information about the location of ancient oceans and the formation of ocean crust, and they also can be the site of important mineral deposits. In some ophiolite complexes, such as at Hare Bay in northwestern Newfoundland, no economically recoverable minerals have been found, but in others, such as Troodos on Cyprus, there are extensive deposits. Sulfide minerals may occur as large deposits in the basalts of the ophiolites. It is interesting to note that there is archeological evidence to the effect that possibly the first smelting of sulfide ores and copper production occurred on Cyprus as early as 4000 B.C.

Considerable thought and research has gone into describing how ophiolites may be formed and become oceanic crust. See Fig. 7. Ultramafics frequently undergo a process known as serpentinization. This may be defined as the process or state of hydrothermal alteration (metasomatism) by which magnesium-rich silicate minerals (olivine, pyroxenes, amphiboles, periodotites, and other basic rocks) are converted into or replaced by serpentine minerals, forming serpentinite. This process tends to obliterate much of the original texture of the rocks. Serpentinite is a dark green color.

Deep-Sea Hot Springs and Cold Seeps

One of the most interesting findings, of which there were many, made during the IDOE program was the discovery of deep-sea hot springs and cold seeps with their associated oases of life. These phenomena have been under vigorous study ever since. Formation of oceanic crust on the Mid-Atlantic Ridge southwest of the Azores at depths of 2,700 meters (8,858 feet) were observed for the first time by the IDOE French-American Mid-Ocean Undersea Study (FAMOUS) team in 1974, http://www.whoi.edu/oceanus/viewArticle.do?id=2512&archives=true. Along this line, the American and African crustal plates are separating at a rate of 2 to 3 centimeters (0.8 to 1.2 inches) per year.

Investigations showed that the ridge crest has a fault-bounded central valley about one kilometer (0.6 mi) deep and 2 to 4 kilometers (1.2 to 2.5 miles) wide. Volcanic rock was found extending across the width of the floor, but was most prevalent along a line of central valley hills. Mid-Atlantic Ridge lava formations are shown in Figs. 8, 9, and 10. As observed by Davin and Gross (1980), systematic compositional variation in the lavas across the valley floor apparently reflects a zoning or evolution in the underlying magma chamber. However, it is believed that several lava flows with discrete geochemical characteristics may result from mantle-derived magma moving into the chamber. This compositional zoning is one of the most important discoveries made by the scientists of the FAMOUS project and thus far has not been documented at any other location. Two fracture zones were studied extensively through photography, dredging, and submersible dives and little evidence of recent faulting was observed, although microearthquakes are frequent along the faults. The FAMOUS

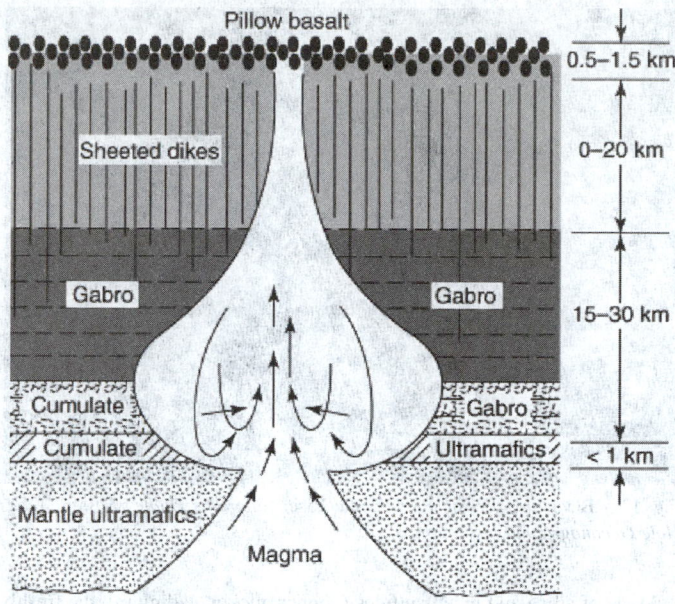

Fig. 9.   Bulbous pillow lavas with cracked crust. Note sponge growing on pillow in left center of view. To the right is elongate or "Cousteau" pillow. Submersible shown is the Alvin. (*Woods Hole Oceanographic Institution.*)

Fig. 7.   Schematic diagram of process that leads to production of cumulate rocks (ophiolites) that in some ways resembles the formation of sedimentary rocks. Magma protrudes through an opening in the mantle ultramafics (rocks of the mantle that have been partially melted, deformed, and recrystallized) and rises to form a "magma chamber," visualized as having the form of a thin-necked, upright vase. It is envisioned that convection results as the magma at the top and adjacent to the edges of the chamber cool and thus migrate downward, while the new, warmer material rises. It is theorized that this may be the manner in which oceanic crust (or some of it) is formed. (*After O'Connell,* 1979; *and Peterson et al.,* 1974.)

Fig. 8.   Lava formation known as collapsed blister pillow is caused by tectonic activity on rift valley floor of Mid-Atlantic Ridge. (*Woods Hole Oceanographic Institution.*)

Fig. 10.   Lava formation found on Mid-Atlantic Range. (*Woods Hole Oceanographic Institution.*)

project demonstrated that detailed geological mapping of rough, deep-sea terrain can be done by submersibles, and it assisted an understanding of the principal components in the seafloor spreading process, namely, a narrow rifted valley with an axial volcanic ridge fed by an underlying magma chamber.

The Nazca Plate off the west coast of South America also was studied during IDOE for its tectonic plate cycle. This includes the generation of new crust along the East Pacific Rise and processes at the zone of continental plate collision where oceanic plate is partly subducted along the Peru-Chile Trench and assimilated beneath South America. See Fig. 11. Researchers found submarine hot springs on the seafloor, around which

new mineral deposits were forming. As observed by Davin and Gross, "These features were first observed at the Galápagos Rift in 1977, and then more extensively in 1978 and 1979 in a larger-scale cooperative program among French, Mexican, and American scientists at the Rivera Fracture Zone off the west coast of Mexico near the mouth of the Gulf of California. Seawater circulates through newly formed oceanic crust, removing heat and reacting chemically with the rocks. Recently formed volcanic rock (erupted at temperatures of about 1200 °C (2192 °F) causes very hot waters to be discharged at temperatures of about 350 °C (662 °F) through vents on the ocean floor. These vents have been observed on the East Pacific Rise project off Mexico. Vents occur on fresh basalt in clusters and narrow bands about 250 meters (820 feet) across and several kilometers (miles) long. Individual vents form irregular chimneys nearly 10 meters (33 feet)

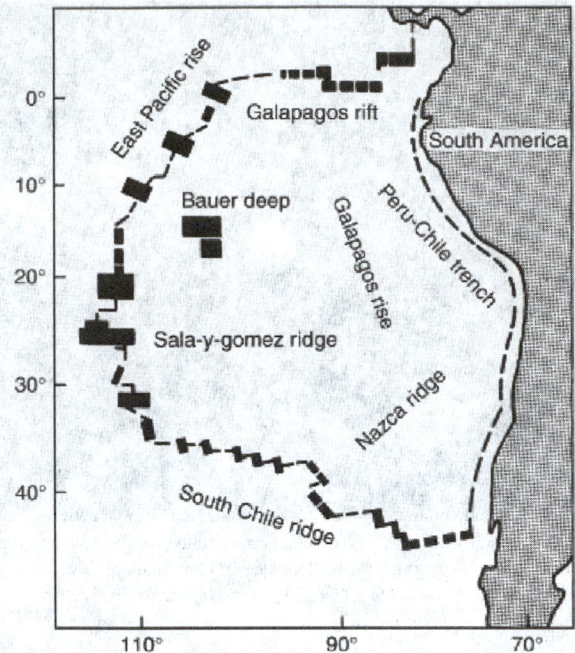

Fig. 11.   Nazca Plate off west coast of South America. New crust is being formed along East Pacific Rise. (*After Davin and Gross, 1980.*)

Fig. 13.   Black smokers on East Pacific Rise. (*Photo by Dudley Foster, Woods Hole Oceanographic Institution.*)

Made of silica and metal sulfides (copper, nickel, cadmium), the freshly precipitated metal sulfides are carried upward as plumes. At 300 °C (572 °F), seawater density is only 0.7 grams/cc and therefore a buoyant plume is formed. Ultimately, the sulfide particles settle and enrich the ridge crust sediment. It is interesting to note that filter-feeding organisms, which live off bacteria that grow in hydrogen sulfide, are found at these hot springs.

In an effort to explain metalliferous deposits on the Nazca Plate, a so-called geo-still concept was developed. This is well explained by Davin and Gross: "A major portion of the sediments on the moving plate descends into the subduction zone. As the plate reaches mantle depths, materials are heated and ore-forming solutions move into the overlying rocks while refractive materials remain in the mantle. Molten rock (magma) rises in large batholiths to within a few kilometers (mile) of the surface. As the magma cools, copper is concentrated in deposits near the top of the formation. Erosion subsequently exposes these deposits for exploitation."

Based upon IDOE data and additional explorations, a highly schematic representation of the "black smoker" geochemical process is shown in Fig. 14. As explained by Edmond (Massachusetts Institute of Technology), black smokers form by the precipitation of $CaSO_4$ (anhydrite) and Fe, Zn, and Cu sulfides. The hot solutions exiting the smoker are buoyant relative to the surrounding cold water. Thus, they rise and disperse into the water column. Edmond also observes that generally void spaces in rocks are filled with water. When molten material from the mantle intrudes into the crust, this water is raised to high temperatures. If there is sufficient permeability, the water will convect to the seafloor, where it forms *hot springs*. If this condition occurs on land (for example, the geysers and hot springs at Yellowstone Park, U.S.A.), much of the hot water recirculates back into the rock, thus making it much more difficult to understand the chemistry of such springs. The situation is much simpler at mid-ocean spreading centers. Seawater enters the highly permeable crust through tectonically induced faults and through contraction cracks caused by rapid cooling. The heated seawater exits through the undersea vents, rises through the water column above the vent orifices, and dissipates. The comparatively simple chemistry of the two reactants—basalt and seawater—facilitate experimentation in the laboratory and the construction of models.

In addition to the earlier studies at the Galápagos Spreading Center, previously mentioned, hydrothermal activity at the ocean bottom in other regions has been observed and studied. These include hot vents and hydrocarbon seeps found in the Sea of Cortez (Gulf of California). In this area, a whole system of spreading centers at which new oceanic crust is formed by cooling of molten rock has been found. As described by Lonsdale (University of California, San Diego) http://sio.ucsd.edu/rab/act_detail.cfm?state=%26(N%2F.U%5C%2F_%0A, where the system comes ashore (beneath the Colorado River delta in the Salton Trough), there are high-temperature geothermal fields (for possible later exploitation for electricity generation). See also **Geothermal Energy**. Other spreading centers have been found in the Guaymas Basin

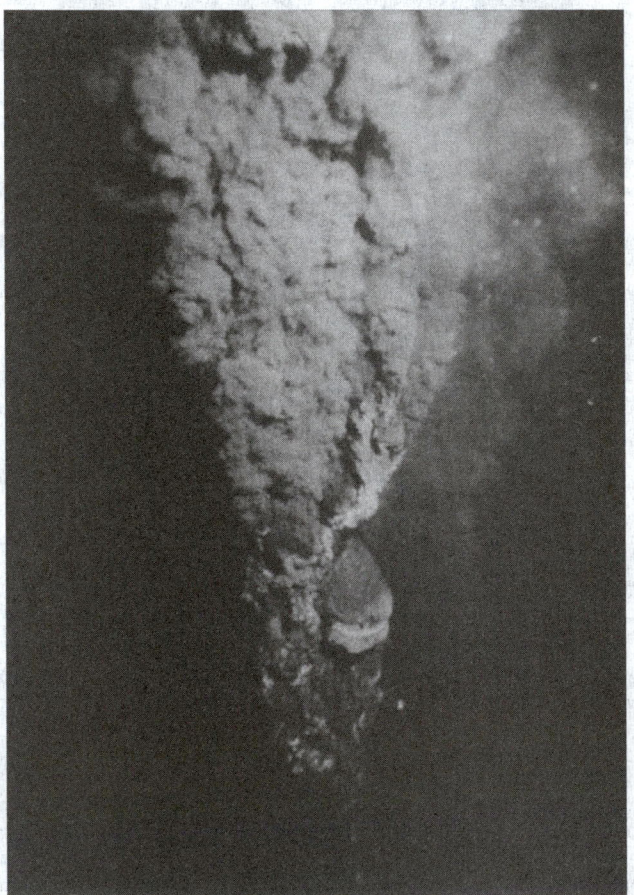

Fig. 12.   Black smoker on East Pacific Rise (21° N). (*Photo by Dr. R.D. Ballard, Woods Hole Oceanographic Institution.*)

high and about 4 meters (13 feet) across. The discharges, 1 to 2 meters (3.3 to 6.6 feet) per second, form dark, smokelike plumes in the overlying waters. The scientists who first saw the vents in 1979 called them "black smokers." See Figs. 12 and 13.

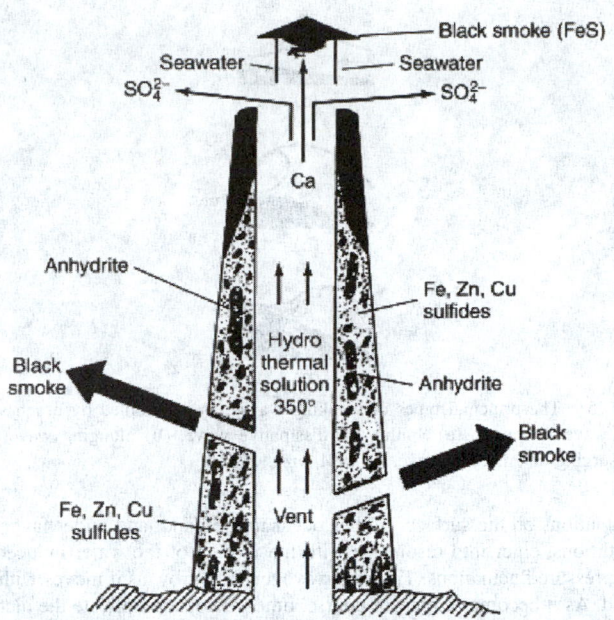

Fig. 14. Highly schematic diagram of a black smoker found in certain hydrothermal areas of the sea floor. These smokers are formed by the precipitation of anhydrite (CaSO₄) and iron, zinc, and copper sulfides. The hot solutions exiting the smoker are buoyant relative to the surrounding cold seawater and consequently rise and disperse into the water column. (*After Oceanus.*)

in the central gulf. Marine polymetallic sulfide deposits have also been found on the Juan de Fuca Ridge and also are presumed to exist on the Gorda Ridge nearby (in the U.S. Exclusive Economic Zone). Broadus and Bowen (Woods Hole Oceanographic Institution) explored the feasibility and economics of mining such sulfide deposits.

When exploring the processes of formation and subsequent erosion of the passive continental margin off the west coast of Florida, a group of researchers serendipitously found exotic communities with abundant organisms at a depth of 3,266 meters (10,715 feet) in the abyssal Gulf of Mexico. The Atlantic gulf community was found to contain the same types of organisms that characterize the Pacific vent communities — white bacterial mats; large, dense beds of mussels; numerous small gastropods; the shells of live mussels; thick patches of 1-meter long tube worms; red-fleshed vesicomyid clams; galatheid crabs; and eel-like zoracid fishes. The Pacific vent communities are found immediately adjacent to hydrothermal vents associated with mid-ocean ridge crest magma sources. As pointed out by the researchers, until the Florida discovery, these were the only known large deep-sea biological communities that receive their primary energy from chemical sources rather than from solar radiation via photosynthesis. It is proposed that the chemical energy is released by bacterially mediated oxidation of reduced inorganic compounds, such as hydrogen sulfide (H₂S) coming from the hot vents. This is a process of *chemosynthesis* and is described in some detail by Jannasch (Woods Hole Oceanographic Institution). See also **Chemosynthesis**.

In 1994, research of hydrothermal vent systems is continuing. As pointed out by M. Tivey http://www.whoi.edu/science/MCG/dept/personnel/scientist_tivey.html (Woods Holes Oceanographic Institution), "Hydrothermal systems transfer large amounts of heat and mass from Earth's interior to the oceans. Fluids exiting the chimneys are metal-rich, hot, and acidic, and vent at velocities on the order of meters per second. A striking feature of black smoker chimneys is how remarkably thin their walls are. They range in thickness [5 in (12.7 cm) to ¼ in (6.3 mm)]. Across this thin layer is a temperature difference of 300 °C or more." See also **Black Smoker**.

Principal questions for which answers are being sought include: (1) Where is all the fluid coming from? (2) How does it circulate? (3) How does it get so hot? (4) How do the fluids become metal-rich? (5) Where do the particulates come from? (6) How do chimneys form?

As posed by researcher Tivey, more advanced questions would include: (7) What is the extent of hydrothermal venting at mid-ocean spreading centers and back-arc basins? (see also **Volcano**) (8) What is the significance

of variation in fluid composition, temperature, flow rate, and composition of solid precipitates among hydrothermal sites? (9) How long are vent sites active? (10) Does fluid composition change with time and, if so, on what time scale? (11) What proportion of minerals is deposited at the vent site versus dispersed into the water columns as black smoke?

As part of an ongoing National Science Foundation–funded project to study the genetics and dispersal mechanisms of organisms inhabiting vent environments, several biological expeditions to the East Pacific and the Gulf of Mexico were staged in 1991. One exploration was that of the *Rose Garden* hydrothermal vent site (Galápagos Rift). Previously, it had been visited in 1979, 1985, and 1988. No major changes were noted from prior visitations. Other known vents also were visited. Numerous species of life were found and cataloged. The research vessel used was DSV *Alvin*. An informative chart listing the various vent and seep regions and their known resident fauna is available from researcher R.A. Lutz (Woods Hole Oceanographic Institution).

The research party also investigated seeps at several locations. A seep is a place of contact between deep-sea sediments and limestone walls where hypersaline waters seep onto the seafloor and feed sulfide-dependent biological communities.

In 1992, D. Toomey (University of Oregon) http://www.uoregon.edu/~drt/cv.html, reported on progress being made in the tomographic imaging of spreading centers. The researcher observes, "In recent years, working models of oceanic spreading centers have evolved from two-dimensional, steady-state idealizations to more realistic three-dimensional, time-dependent systems. The new dimension added to the working models is the pervasive along-strike variability of mid-ocean ridge processes, notably in the production of melt beneath the spreading center. Current hypotheses suggest that ascending melt within the mantle is focused into magmatic centers separated on the order of tens to a hundred kilometers. Each magmatic center supplies the greater portion of melt and heat to a single ridge segment. Within an individual ridge segment, processes such as faulting, hydrothermal circulation, and magmatic accretion vary systematically as a function of distance from the magmatic center. The hypothetical structural unit, consisting of a local maximum of magmatism bounded by along-axis minima, became known as a spreading-center segment or cell. This simple model of cellular segmentation provides an improved, but controversial, working hypothesis for mid-ocean ridge studies."

Tomographic studies of seismic velocity structure beneath local segments of the East Pacific Rise, the Mid-Atlantic Ridge, and the Icelandic rift represent a new and powerful approach to the seismological study of divergent plate boundaries.

Wave Action in the Ocean

Wind blowing across the surface of water exerts a force on the surface of the water in the direction of the wind. This interface phenomenon between the ocean and the atmosphere (see also **Atmosphere-Ocean Interface** for description from meteorological viewpoint) is far from understood in any degree of detail. The rotation of the earth is a fundamental contributing force. Fluid friction is also an important factor and poorly understood. There is a certain depth below which both current and frictional forces associated with it become very small. In the layer above that level, friction is important. This upper layer is called the *Ekman layer*. See also **Ekman Spiral (Oceanography)**; and **Winds and Air Movement**. Using l to denote the length of a wave, C, the speed of the waveform and T, the period (the time it takes to move one wavelength) for waves in general is

$$C = \frac{1}{T} \tag{1}$$

which is obvious if we substitute units in some system, such as the length of one wave in feet, the period in seconds, and the speed in feet per second.

The three principal types of ocean waves are due to (1) the forces that produce the tides, (2) the wind, and (3) earthquakes (seismic waves or tsunami).

The waves produced by the tidal forces are progressive waves because their length (one-half the circumference of the earth) is so great in comparison with the depth of the oceans. For such waves the wave velocity, C, varies with the square root of the depth, being given by the expression

$$C = \sqrt{gd} \tag{2}$$

where d is depth and g is the acceleration of gravity 32.16 feet (9.8 meters) per second. Substituting this value, we find that the wave velocity at a depth of 500 feet (152 meters) would be 127 feet (38.7 meters) per second, while at 5,000 feet (1524 meters) it would increase to 399 feet (122 meters) per second. Note that this speed is that of the waveform; that of the water is far smaller, being only about 2.3 feet (0.7 meter) per second (at crests and troughs, where it is greatest) for a wave having an amplitude of 10 feet (3 meters).

The assumption of a channel of infinite length, on which Equation (2) for a progressive wave was based, is only strictly true in the oceans surrounding Antarctica. It fails completely in tidal basins and estuaries, where the types of waves due to the tides are stationary waves rather than progressive waves. Here we are concerned with stationary waves. The periods of these stationary waves are determined by the dimensions of the basin. Assuming that these are constant, which would obviously not be true of any actual basin, a simple relationship can be derived for the period of the stationary wave:

$$T = \frac{2l_B}{\sqrt{g d_B}} \qquad (3)$$

where l_B is the (constant) length of the basin and d_B its (constant) depth. (The width is not significant for such an ideal basin.) When the period of the stationary wave so computed is the same, or very close to, that of the tides, then a condition of resonance is obtained which produces the abnormally high tides that occur in certain basins. (The tidal range in the Bay of Fundy may exceed 50 feet (15.2 meters) in spring at times when the moon is at perigee, i.e., the point in its orbit closest to the earth.)

Another tidal phenomenon, which differs from the foregoing in that it involves a traveling wave, is the *tidal* bore, a relatively massive wave that moves up a river. The height of such a wave depends upon the slope of the river bed (as well as the depth and the current). Favorable combinations of these factors, resulting in a rapid change of surface elevation, produce a large bore or a series of small ones, especially during periods of abnormal high tides.

The common waves of the ocean, as well as the greater ones occurring during storms, are produced by the wind. For such waves, a general equation is

$$C = \sqrt{\frac{gl}{2\pi}\left(\tanh\frac{2\pi d}{l}\right)} \qquad (4)$$

where C is the velocity of the waveform as before, l is the wavelength, tanh means the hyperbolic tangent, and d is the depth. This equation clarifies the difference between waves in deep water and those in shallow water. If d is large relative to l so that $2\pi d/l$ is large, then the tanh term is close to 1 (since $\tanh x$ approaches 1 rapidly as x increases, being 1 to four places of decimals when x is 6.5). In that case, Equation (4) simplifies to

$$C = \sqrt{\frac{gl}{2\pi}} \qquad (5)$$

which is the equation for all wind-produced waves when in deep water. When the ratio d/l is very small, as it is for the waves produced by the tidal forces, then the value of $\tanh x$ approaches x closely, and $\tanh 2\pi\, d/l$ approaches $2\pi\, d/l$, so that cancellation in Equation (4) gives $C = \sqrt{gd}$, as given in Equation (2) for those semidiurnal waves of the tides.

The velocity of waves in shallow water, however, cannot be represented by either Equation (2) or Equation (5) since for such values of $\tanh 2\pi$ d/l the value of the tanh term must be found and used in Equation (4). Of the three properties of water waves, period T, wave velocity (C), and wavelength (l), the period is the least changed by movement into shallow water, but the wavelength and amplitude are the quantities that change. These changes are not uniformly related; as a wave moves from deep water into a depth such that $d/l = 0.5$, the amplitude (height) begins slowly to decrease, reaching a value of about 90% of the original amplitude when $d/l = 0.06$, after which the amplitude increases until the wave breaks. The theoretical breaking point occurs when $l/d = 7$, but few waves attain such heights, most of them being far lower in relation to their lengths. See Fig. 15.

There are a number of theories of wave generation by the wind. They usually include the effect of wind turbulence, with the resulting pressure

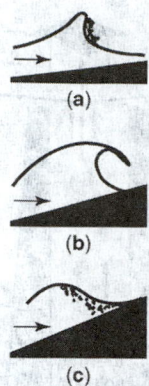

Fig. 15.   The principal types of breaking waves, as determined by beach slope and wave steepness: (**a**) Spilling or dissipative wave; (**b**) plunging wave; and (**c**) surging or reflective wave. (*After Aubrey, 1981.*)

fluctuations on the surface. Eddies occur in the wind, and under favorable conditions, enter into resonance with the motion of the water induced by the pressure fluctuations. Thus the wave tends to grow as it moves with the wind. As it becomes larger, of course, other factors complicate the picture. In fact, analysis of the waves usually is a complicated operation, although the methods of harmonic analysis have been widely applied.

The wave analyzers that have been developed analyze the wave trace into the periods that compose it. The figure shows the considerable number and range of periods in the wave pattern of a typical storm sea. The energy in the wave system is the important consideration for many purposes; this is represented by the peak period, that is, the period of maximum energy. Even this, however, is an oversimplification in that it does not take into consideration the interaction of the components and other variables. This fact explains the difficulties in wave forecasting from meteorological data.

The various formulas developed for this purpose must take into account the speeds of the surface wind and the gradient wind, their duration and the distance of the observer from the region of wave generation, since the wind-produced waves are attenuated with distance. These are the major factors; others include the motion of the storm area, the effects of tidal currents, the temperatures of air and sea, and the effects of any shallow depths between storm and observer.

In recent years wave prediction has become important for another reason, i.e., for dealing with seismic waves of tsunamis, which are waves produced by earthquakes. Incorrectly called "tidal waves," tsunamis have no relation to the tide. Since such waves, because of their great size, often cause great damage in coastal areas, the prediction of their time of arrival, and their height, is of great importance. Their height depends upon the depth of water, for essentially the same reasons as those given for wind-generated waves. Thus waves only a foot or more high in the open sea may break on the shore at heights of 30 feet (9 meters) or more. This great amplification is due to their length and speed. Their length is great, being on the order of tens of miles, or even over 100 miles (161 kilometers), and thus as explained earlier in this entry, their speed depends on the depth of the sea over which they travel. As stated there, their speed over a depth of 5,000 feet (1524 meters) would be 399 feet per (122 meters) second, or about 270 miles (434 kilometers) per hour. Since the average depth of the Pacific Ocean is much greater than 5,000 feet (1524 meters) there is less than a day after the earthquake occurs to prepare for the tsunami at distances as great as 8,000–10,000 miles (12,875–16,093 kilometers).

Coastlines and Coastal Waters

The shape and characteristics assumed by a stretch of oceanic coastline reflect a myriad of geological factors and continuous chemical and biochemical processes. A major geological factor controlling shoreline features is plate tectonics, which influence the width and bathymetric detail of the continental shelf as well as the local rise and fall of sea level. Principal factors of an immediate nature that contribute to coastline and beach modification include winds, waves, tides, storms (notably hurricanes), and human influences on the ecology of coastal waters. Coastlines sometimes are referred to as Atlantic-type or Pacific-type, as shown by Fig. 16.

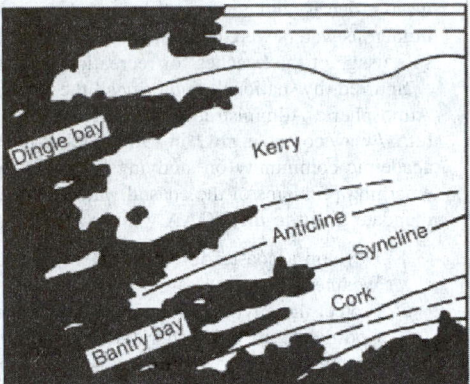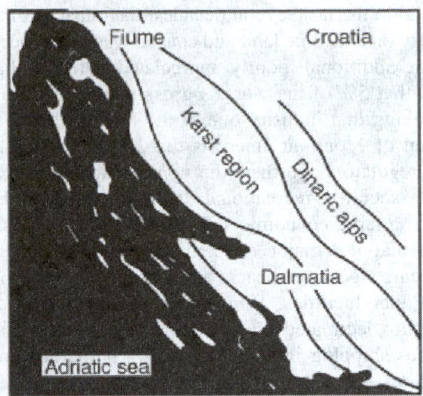

Fig. 16.   Types of coastlines. (Left) Atlantic-type in which geologic structure trends at a high angle to the coast; (Right) Pacific-type, where structure trends parallel to the coast. Suess came up with this type of classification over 100 years ago. (*After Holmes, 1965.*)

In terms of geological time, a rising sea level results in beaches migrating landward and hence contracting. Changes of river drainage patterns, faulting, slumping, and biological changes (particularly as regards coral and mangrove beaches) are very short term geologically.

In many areas of the world, there is concern with island safety, where the coastline is a predominant factor. Pilkey and Neal (see reference) established a set of guidelines for evaluating island safety, showing high, moderate, and low hazard potential.

Factors constituting a high hazard include: (1) an island elevation of less than 5 feet (1.5 meters); (2) the absence of dunes; (3) an island of narrow width; (4) an island located near an existing inlet and where there is no salt marsh backup; (5) a known shoreline erosion rate in excess of 3 feet (1.5 meters) per year; (6) a location where, in the past, dunes have been destroyed and new inlets created; (7) a history of overwash; (8) little or very little vegetation; (9) poor footing materials, such as compactible layers of peat and clay; (10) poor drainage during the rainy season; (11) an inadequate or contaminated water supply; (12) unsuitability for sanitary installation because of improper sediment and soil that may intersect the water table; (13) the presence of many shells in the soil; and (14) the presence of finger canals.

Coastal waters (sometimes called *coastal oceans*) may be defined as the region extending from the beaches out across the *continental shelf*, slope, and rise. As pointed out by K. Brink http://www.whoi.edu/dept/profile. go?id=1341 (Woods Hole Oceanographic Institute), "Using this definition, the offshore edge would often be near that of the purely politically defined Exclusive Economic Zone or EEZ. There is, however, a true scientific cohesiveness to the essentially geological definition above. For example, current patterns over the continental shelf and slope tend to be distinctly different from those of the open ocean, and the consequent shelf physical processes make this region the most biologically productive area of the world's ocean .... The coastal ocean absorbs most of the impact that land-based activities have on the ocean, including river outflow and wind transport of particles and chemicals into the sea. These effects make the coastal ocean important scientifically and economically."

Although the year 1980 was dedicated in the United States as the Year of the Coast in an effort to refocus attention of scientists and the lay public on the 128,772 km (80,000 mi) of the nation's coastline, an impressive degree of interest in this topic was not exhibited until 1988, as explained in the following pages.

Degradation of Coastlines and Coastal Waters

Progressive deterioration of many of the world's shorelines and beaches has been known by the scientific community and by the educated and concerned politicians and lay public for many years. But as so often is the case, some form of crisis must occur before corrective measures are taken. This crisis, which received wide attention by the news media and politicians, occurred in 1988 when several of the ocean beaches in the northeastern United States (New Jersey and New York) had to be closed to recreation. The basic cause of this crisis was water pollution of the most revulsive and frightening nature, such as the appearance of medical wastes

(including syringes and other hospital wastes), not to mention ordinary garbage. Polluting wastes, of course, are *only one* of several problems that are affecting coastal waters. Some solutions to coastal problems reside in science and technology, but, to a much greater extent, answers must be provided by people and their governments. There are no technical cures of a breakthrough nature that can be called upon as a "quick fix" for coastal water problems. Rather, the role of science is to understand the infrastructure of coastal technology and, with increasingly accurate information on hand, to enhance the wisdom of political regulators and of the great numbers of people who dwell in the coastal and nearby regions, or who frequent the coastal areas as tourists and recreation seekers.

Scope of Coastal Water Problems. Worldwide, there are comparatively few countries that do not share some coastline with the oceans. Total global coastlines are estimated to extend about 440,000 km (273,400 mi). An impressive portion of the total coastline length has been affected adversely over scores of years, mainly as the result of human activities. Devastation, of course, is more obvious in some areas than in others. Degradation, for example, is quite apparent along the eastern and western United States, including the Gulf of Mexico to the south. The southern shorelines of the European countries, particularly lying along the Mediterranean, Adriatic, and Aegean seas; the islands of the Caribbean sea; the Arabian sea, including the Persian Gulf; the Baltic sea; and the North sea have been seriously affected. Some of the coastline problems are essentially universal; others are specific to certain lengths of the coastline.

This article is concerned mainly with the ocean coastlines, but degradation of freshwater coastlines, notably the Great Lakes of North America, the shores of massive rivers worldwide, and, for example, the virtual disappearance of the mineral-laden Aral Sea, which typifies the environmental neglect of the former Soviet bloc, are part of the total coastline problem.

Classes of Coastal Water Problems. The degradation of coastal waters is exemplary of what can occur when there are severely conflicting interests at work that can lead to strong pro and con positions in terms of solving problems.

Population Pressures. Demographics reveal considerable evidence for supporting population pressure as one of the root causes of present coastline problems. A poorly appreciated statistic is that 45% of the population of the United States lives in coastal counties, such as Nantucket, Massachusetts, although coastal regions constitute only 10% of the land area in the country (excluding Alaska). It is estimated that if current trends continue, by the year 2015, this coastal population will have grown by 15 million, to 127 million persons.

In recent years, Italy has experienced a tremendous immigration of people from other countries (18,000 Albanian refugees in 1991, for example), prompting future Italian restrictions. It is interesting to note that a large percentage of immigrants to countries on the ocean coasts tend to remain near the coast. People pressure exacerbates coastal water problems because of increased production of local wastes and the quest for housing along or near the shorelines. (It is interesting to note that population pressure is one of the root causes of nearly all environmental concerns facing the world today.)

Economic Pressures. Within the last several decades, the impact of the tourist and recreation industry on coastline land and coastal waters has been tremendous. There are some additional, poorly appreciated statistics that are relevant. It is estimated that 5% of the world's gross national product is expended on tourism, making that industry one of the largest of global enterprises and thus a group of economic interests that has considerable clout with politicians and regulators. Further, the construction industry, which builds hotels and associated recreational facilities (restaurants, nightclubs, casinos, etc.) has a large economic stake in coastal locations.

It is interesting to note that the major concern with coastal waters of a comparatively few years ago was beach and property protection in connection with recreational facilities. Primary concern was that of building sea walls and restoring decimated beaches. Commercial interest in coastline areas also embraces shipping interests, including ports, lagoons for pleasure boating and fishing, and simply the overall economic pressures generated by so many large cities and towns that are located immediately adjacent to a coastline. The population of such places is constantly growing. Boston's polluted harbor, considered one of the worst environmentally disturbed bodies of water in the world, dramatizes the effects of pollution on coastal waters. Hong Kong harbor is another example. See also **Water Pollution**.

Another economic pressure exerted on coastal waters management is the fisheries industry. It is estimated that roughly 75% of the world's total fish catch comes from coastal waters worldwide. The outlets for waterborne inland pollution are the estuaries and deltas of rivers and direct dumping into the ocean. While pollution may be greatest in an estuary, ocean currents mix and transport pollutants over long distances along the coastlines. These carry significant concentrations of agricultural chemicals and fertilizers along the coastline for significant distances. The dwindling fish catch in many areas is attributed to: (1) poisoning of numerous fish and crustaceans with chemicals that kill or adversely affect the reproductive processes of these creatures; and (2) nutrient overenrichment, which nurtures undesirable plants and other organisms, thus creating a condition known as *hypoxia* (depletion of oxygen). Thus, the fisheries industry has a large stake in what occurs biologically along the world's coastlines. Particularly, this impact has adversely affected fishing along the eastern seaboard of the Atlantic ocean in recent years.

For obvious reasons, polluted beaches also affect coastline tourism and vacationing.

Positive Pressures of Coastal Science. Scientists have recognized comparatively recently that the problems of coastal waters involve a networking of actions and reactions, both scientific and geographic and that, in addition to understanding the purely physical forces to which the oceans are subject (that is, currents, tides, seiches, tropical storms and hurricanes, the geometry and tectonics of the ocean shelves), equal if not greater attention must be given to biological and biochemical forces that persist in coastal waters. By providing greater understanding of these ecological and biological factors, coastal scientists can assist effectively in developing long-term protocols for guiding future coastline development.

Coastal Ocean Processes Program (CoOP)

It is along the foregoing lines that the CoOP program http://www.skio. peachnet.edu/coop/, was established in early 1990 and defined as a program — *to obtain a new level of quantitative understanding of the processes that dominate the transports, transformations, and fates of biologically, geologically, and chemically important matter on the continental margins.*
Specific objectives include:

- Coastal air-sea fluxes and couplings, such as how carbon dioxide finds its way from the ocean to the atmosphere or vice versa.
- Fluxes of matter through the seabed, such as sediment deposition or the release of chemicals from the bottom.
- Land-derived effects, such as the fate of river-borne nutrients.
- Chemical and biological transformations within the water column, such as how plants grow in response to a chemical change.

National Oceanic and Atmospheric Administration (NOAA) Program

The New Jersey and New York beach crises of 1988 may, in historic perspective, be regarded as a fortuitous omen of the future and a call for immediate action. In that year, the beaches were cluttered with hospital wastes, debris, and garbage; dead and dying dolphins were washed ashore; numerous waters were contaminated by toxics and nutrients — all affecting safe usage of the beaches for recreation and damaging fisheries.

Spurred by national authorities, the existing National Oceanic and Atmospheric Administration created the *Coastal Ocean Program* (COP) http://www.cop.noaa.gov/, in 1989 to focus not only on the NOAA but also the academic community on studying intensely the longstanding and rapidly emerging problems of the coastal waters of the United States. The general mandate given to the NOAA was:

1. Develop a coastal forecast system in cooperation with the National Weather Service;
2. Protect the environmental quality of the coastlines; and
3. Coordinate all U.S. federal science efforts as they pertain to coastal science.

Four of several specific NOAA subprograms are described here, and since 1989 considerable progress has been made in all areas:

- *CoastWatch* is a survey and communication program. It is a system of regional data-access sites supported by a central data processing and distribution center. The program allows managers and researchers rapid access to satellite and in-situ data. Recently developed remote sensors provide effective tools for coastal water researchers long before they see descriptions of them in the literature.
- *Watershed and Habitat-Change Analysis.* Coastal wetlands are among Earth's most productive ecosystems. It is estimated that nearly half of the U.S. wetlands have been lost through draining, filling, and other forms of degradation, largely in the interest of land recovery and development.

Not only are breeding grounds for waterfowl and other wildlife lost, but wetlands also support commercial species of crustaceans, such as shrimp. Loss of wetlands also contributes to the pollution of coastal waters because wetlands assist in protecting the coasts from considerable water runoff. Part of the NOAA program is that of developing a protocol for use by federal and state agencies as well as for academic researchers in mapping coastal habitats and adjacent uplands. Satellite mapping, already in place by the NOAA, is an important element of this program.

This program was applied first to the Chesapeake Bay watershed to analyze changes in land cover and habitats for the world's largest estuary. Completed in 1992, this study shows the effects of population increases in the Washington, DC, and Baltimore, Maryland areas and reveals a 1% loss in wetland habitats over the time frame 1984–1989.

An important point is that the study differentiated between pollution from rivers and tributaries and atmosphere-derived pollutants.

- **Nutrient Overenrichment Studies** Agricultural and suburban runoff, industrial waste, and raw sewage contain nutrients that enter the coastal waters in excessive amounts and can kill some organisms of value and cause some undesirable species to proliferate. A condition referred to as hypoxia (depletion of oxygen) occurs commonly, especially in the Gulf of Mexico. To address this problem, the NOAA has created the Nutrient Enhanced Coastal Ocean Productivity (NECOP) program. The Mississippi River basin, which drains one-third of the continental United States, carries huge quantities of nutrient-rich water to the Gulf.
- **Coastal Fisheries Ecosystem (CFE) Program** This study focuses on the ecological processes that affect commercial fish populations. As pointed out by L. Wenzel (Sea Grant Fellow) and D. Scavia (NOAA), "CFE grows out of a new direction in fisheries science, an integrated approach to understanding fisheries within the context of their ecosystems. The need for such an approach is clear. In the Bering Sea pollock fishery, for example, the largest single-species fishery in the world, the Russian and the U.S. 200-mile (322 kilometers) exclusive economic zones enclose a high-seas 'doughnut hole' open to foreign fleets. Since the mid-1980s, this 'hole,' the central basin of the Bering Sea, has been heavily fished. The Northwest Fisheries Management Council needs to know how this fishery affects stock sizes in the U.S. exclusive economic zone. Recruitment (the number of adult fish added to the population each year) varies dramatically in pollock populations, creating enormous uncertainty in stock assessments. To improve these predictions, COP-funded scientists conduct genetic analyses of pollock stocks, to see how much stocks from foreign and U.S. waters mix, and they mount field studies to determine what physical oceanographic and ecological factors

affect pollock survival in the critical egg and larval periods. Once these factors have been identified, monitoring programs can be established to improve the scientific basis for pollock management."

In summary, future ventures of the NOAA's COP program will be directed to:

- Developing a coastal forecast system in cooperation with the National Weather Service.
- Protecting environmental quality.
- Coordinating U.S. federal science efforts as they pertain to coastal ocean science.

Additional Reading

Aksenov, V.V. and A.B. Karasev: "Satellite Oceanography," *Oceanus*, 69 (Summer 1991).

Aubrey, D.G.: "Perspectives from a Shrinking Globe," *Oceanus*, 9 (Spring 1993).

Baker, D.J.: "Toward a Global Ocean Observing System," *Oceanus*, 76 (Spring 1991).

Baker, E.T.: "Megaplumes," *Oceanus*, 84 (Winter 1991/1992).

Bartholomew, C. and C. Mullen: "The Ocean Versus Deep-Water Salvage," *Oceanus*, 73 (Winter 1990/1991).

Bergman, E.A.: "Mid-Ocean Rise Seismicity," *Oceanus*, 60 (Winter 1991/1992).

Bischof, J.: *Ice Drift, Ocean Circulation and Climate Change*, Springer-Verlag, Inc., New York, NY, 2001.

Bjerklie, D.: "Getting Heated Up Over Climate Research," *Technology Review (MIT)*, 10 (November/December 1991).

Blackman, D. and T. Stroh: "RIDGE and InterRidge," *Oceanus*, 21 (Winter 1991/1992).

Bonatti, E.: "Not So Hot 'Hot Spots' in the Oceanic Mantle," *Science*, 107 (October 5, 1990).

Bowen, M.F.: "Jason's Mediterranean Adventure," *Oceanus*, 61 (Spring 1990).

Bradley, R.S.: *Principles of Ocean Physics*, 2nd Edition, Morgan Kaufmann Publishers, Orlando, FL, 2000.

Brekhovskikh, L.M. and V.G. Neiman: "The History of Soviet (Russian) Oceanology," *Oceanus*, 20 (Summer 1991).

Brink, K.H.: "The Coastal Ocean Processes (CoOP) Effort," *Oceanus*, 47 (Spring 1993).

Britton, J.C. and B. Morton: *Shore Ecology of the Gulf of Mexico*, University of Texas Press, Austin, TX, 1989.

Broadus, J.M. and R.V. Vartanov: "The Oceans and Environmental Safety," *Oceanus*, 14 (Summer 1991).

Brown, N.: "The History of Salinometers and CTD (Conductivity, Temperature, Depth) Sensor Systems," *Oceanus*, 61 (Spring 1991).

Bryan, W.B.: "Exploring Pacific Seafloor Ashore: Magadan Province, USSR (Russia)," *Oceanus*, 48 (Summer 1991).

Bryan, W.B.: "From Pillow Lava to Sheet Flow: Evolution of Deep-Sea Volcanology," *Oceanus*, 42 (Winter 1991/1992).

Bulloch, D.K and G. Reiger: *Wasted Ocean*, Lyons & Burford Publishers, Inc., New York, NY, 1989.

Burke, K.: "The Edges of the Ocean," *Oceanus*, 22(3), 2–9 (1979).

Cane, M.A.: "Oceanographic Events during El Niño," *Science*, 222, 1189–1195 (1983).

Choi, H.: *Advances in Geosciences: Volume 5—Oceans and Atmospheres (OA)*, World Scientific Publishing Company, Inc., River Edge, NJ, 2006.

Davin, E.M. and M.G. Gross: "Assessing the Seabed," *Oceanus*, 23, 1, 20–32 (1980).

Deacon, M.B., Rice, T., and C. Summerhayes: *Understanding the Oceans: Marine Science in the Wake of HMS Challenger*, Taylor & Francis, Inc., Philadelphia, PA, 2001.

Duedall, I.W. and M.A. Champ: "Artificial Reefs: Emerging Science and Technology," *Oceanus*, 94 (Spring 1991).

Edmond, J.M. and K. Von Damm: "Hot Springs on the Ocean Floor," *Sci. Amer.* 78–93 (April 1983).

Edmond, J.M.: "The Geochemistry of Ridge Crest Hot Springs," *Oceanus*, 27(3), 15–19 (1984).

Elderfield, H.: *The Oceans and Marine Geochemistry: Treatise on Geochemistry*, Volume 6, Pergamon Press, New York, NY, 2006.

Emelyanov, E.M.: *Barrier Zones in the Ocean*, Springer-Verlag New York, LLC, New York, NY, 2005.

Falnes, J.: *Ocean Waves and Oscillating Systems: Linear Interactions Including Wave-Energy Extraction*, Cambridge University Press, New York, NY, 2005.

Francheteau, J.: "The Oceanic Crust," *Sci. Amer.*, 114–129 (September 1983).

Fredj, G., et al.: "Mediterranean Biology," *Oceanus*, 43 (Spring 1990).

Frisbee, K.S.: "Deep Water Over Complex Tectonics," *Oceanus*, 56 (Spring 1990).

Garrett, C. and L.R.M. Maas: "Tides and Their Effects," *Oceanus*, 27 (Spring 1993).

Given, D.: "Underwater Technology in the USSR (Russia)," *Oceanus*, 67 (Spring 1991).

Goldberg, E.D.: "Competitors for Coastal Open Space," *Oceanus*, 12 (Spring 1993).

Golden, F.: "A Quarter-Century Under the Sea (*Alvin*)," *Oceanus*, 2 (Winter 1988/1989).

Gornitz, V., Lebedeff, S., and J. Hansen: "Global Sea Lever Trend in the Past Century," *Science*, 215, 1611–1614 (1982).

Gurvich, E.: *Metalliferous Sediments of the World Ocean: Fundamental Theory of Deep-Sea Hydrothermal Sedimentation*, Springer-Verlag New York, LLC, New York, NY, 2005.

Hartwig, E.O.: "Trends in Ocean Science," *Oceanus*, 96–100 (Winter 1990/1991).

Hass, P.M. and J. Zuckman: "The Mediterranean Is Cleaner," *Oceanus*, 38 (Spring 1990).

Holmes, A.: *Principles of Physical Geology*, 2nd Edition, Chapman & Hall, New York, NY, 1992.

Ivanov, Y.A.: "Physical Oceanography: A Review of Recent Soviet (Russian) Research," *Oceanus*, 81 (Summer 1991).

Jensen, J.J. and J. Hovermale: "Numerical Air/Sea Environmental Protection," *Oceanus*, 40 (Winter 1990/1991).

Julian, M. and P.R. Ryan: "Introduction: The Med," *Oceanus*, 4 (Spring 1990).

Kagan, B.A., J.T. Houghton, and A.J. Dessler: *Ocean Atmosphere Interaction and Climate Modeling*, Cambridge University Press, New York, NY, 2006.

Kantha, L.H., and C.A. Clayson: *Numerical Models of Oceans and Oceanic Processes*, Elsevier Science & Technology Books, New York, NY, 2000.

Karson, J.A.: "Tectonics of Slow-Spreading Ridges," *Oceanus*, 51 (Winter 1991/1992).

Kerr, R.A.: "Vail's Sea-Level Curves Aren't Going Away," *Science*, 226, 677 (1984).

Komen, G.J., K. Hasselmann, L. Cavaleri, M. Donelan, and S. Hasselmann: *Dynamics and Modelling of Ocean Waves*, Cambridge University, Press, New York, NY, 2006.

Knox, G.: *Biology of the Southern Ocean*, CRC Press, LLC, Boca Raton, FL, 2006.

Lacombe, H.: "Water, Salt, Heat, and Wind in the Mediterranean," *Oceanus*, 26 (Spring 1990).

LePichon, X. and J. Francheteau: "A Plate-Tectonic Analysis of the Red Sea-Gulf of Aden Area," *Tectonophysics*, 46, 369–406 (1978).

Lindau, R.: *Climate Atlas of the Atlantic Ocean: Derived from the Comprehensive Ocean Atmosphere Data Set (Coads)*, Springer-Verlag, Inc., New York, NY, 2001.

Liss, P.S., and R.A. Duce: *Sea Surface and Global Change*, Cambridge University Press, New York, NY, 2005.

Liu, J.: "The Segmented Mid-Atlantic Range," *Oceanus*, 11 (Winter 1991/1992).

Lonsdale, P. and C. Small: "Ridges and Rises: A Global View," *Oceanus*, 26 (Winter 1991/1992).

Lutjeharms, J.R.E., F.A. Shillington, and C.M. Duncombe Ray: "Observations of Extreme Upwelling Filaments in the Southeast Atlantic Ocean," *Science*, 774 (August 16, 1991).

Lutz, R.A.: "The Biology of Deep-Sea Vents and Seeps," *Oceanus*, 75 (Winter 1991/1992).

Macdonald, K.C. and P.J. Fox: "The Mid-Ocean Range," *Sci. Amer.*, 72 (June 1990).

Macdonald, K.C., Scheirer, D.S., and S.M. Carbotte: "Mid-Ocean Ridges: Discontinuities, Segments, and Giant Cracks," *Science*, 986 (August 30, 1991).

Macdonald, K.C.: "Introduction: Mid-Ocean Ridges, The Quest for Order," *Oceanus*, 9 (Winter 1991/1992).

Mann, K.H., and J.R. Lazier: "Dynamics of Marine Ecosystems," *Biological-Physical Interactions in the Oceans*, 3rd Edition, Blackwell Publishing Professional, Malden, MA, 2005.

McCoy, F.W. and P.D. Rabinowitz: "The Evolution of the South Atlantic," *Oceanus*, 22, 3 (1979).

Meier, M.F.: "Contributions of Small Glaciers to Global Sea Level," *Science*, 226, 1418 (1984).

Milliman, J.D.: "Sea Levels: Past, Present, and Future," *Oceanus*, 40 (Summer 1989).

Monastersky, R.: "Predictions Drop for Future Sea-Level Rise," *Science News*, 397 (December 18, 1989).

Mooers, C.N.K. and A.R. Robinson: "Turbulent Jets and Eddies in the California Current and Inferred Cross-Shored Transports," *Science*, 223, 51–53 (1984).

O'Connell, S.: "Ophiolites," *Oceanus*, 22(3), 33 (1979).

Ostenso, N.A., Metalnikov, A.P., and B.I. Imerekov: "A History of USSR (Russia)–US Cooperation in Ocean Research," *Oceanus*, 87 (Summer 1991).

Peltier, W.R. and A.M. Tushinham: "Global Sea Level Rise and the Greenhouse Effect," *Science*, 806 (May 19, 1989).

Pilkey, O.H. and W.J. Neal: "Barrier Island Hazard Mapping," *Oceanus*, 38 (Winter 1980–1981).

Potter, T.D. and B.R. Colema: *Handbook of Weather, Climate, and Oceans*, The McGraw-Hill Companies, Inc., New York, NY, 2001.

Prager, E. and S. Earle: *The Oceans*, The McGraw-Hill Companies, Inc., New York, NY, 2001.

Ross, D.A.: "The Red Sea: A New Ocean," *Oceanus*, **22**(3), 33–39 (1979).

Roughgarden, J., Gaines, S., and H. Possingham: "Recruitment Dynamics in Complex Life Cycles," *Science*, 1460 (1988).

Ryan, P.R.: "A Challenge to Alvin from the USSR (Russia)," *Oceanus*, 67 (Winter 1988/1989).

Schouten, H. and J. Whitehead: "Ridge Segmentation: A Possible Mechanism," *Oceanus*, 19 (Winter 1991/1992).

Siedler, G., J. Gould, and J. Church: *Ocean Circulation and Climate*, Academic Press, Inc., San Diego, CA, 2001.

Smith, H.D.: *Oceans: Key Issues in Marine Affairs*, Kluwer Academic Publishers, Norwell, MA. 2004.

Smith, W.O., Jr.: *Polar Oceanography*, Academic Press, Inc., San Diego, CA, 1990.

Spindel, R.C. and P.F. Worcester: "Ocean Acoustic Tomography," *Sci. Amer.*, 94 (October 1990).

Staff: NOAA, *Central and Western Pacific Ocean and Indian Ocean*, The McGraw-Hill Companies, Inc., New York, NY, 2005.

Stanley, D.J.: "In Search of the Origins of the Mediterranean," *Oceanus*, 16 (Spring 1990).

Stow, D.: *Oceans: An Illustrated Reference*, University of Chicago Press, Chicago, IL, 2006.

Suess, E. and J. Thiede: *Coastal Upwelling*, Plenum Publishing Corporation, New York, NY, 1983.

Sverdrup, K.A., A.B. Duxbury, and A.C. Duxbury: *Introduction to the World's Oceans*, 8th Edition, The McGraw-Hill Companies, Inc., New York, NY. 2003.

Tippie, V.K. and J.H. Cawley: "Modernizing NOAA's Ocean Services," *Oceanus*, 84 (Spring 1991).

Tivey, M.K.: "Hydrothermal Vent Systems," *Oceanus*, 68 (Winter 1991/1992).

Toomey, D.R.: "Tomographic Imaging of Spreading Centers," *Oceanus*, 92 (Winter 1991/1992).

Uyeda, S.: "Subduction Zones," *Oceanus*, **22**(3), 52–62 (1979).

Van Dover, C.L.: "Diving the Soviet (Russian) Mir Submersibles," *Oceanus*, 8 (Summer 1991).

Vartanov, R.V.: "Dynamics of Ocean Ecosystems: A National Program in Soviet (Russian) Biooceanology," *Oceanus*, 66 (Summer 1991).

Vine, A.C.: "The Birth of Alvin," *Oceanus*, 10 (Winter 1988/1989).

Watts, A.B. and J. Thorne: *Nature (London)*, **311**, 365 (1984).

Weller, R.A., et al.: "Three-Dimensional Flow in the Upper Ocean," *Science*, **227**, 1552 (1985).

Wenzel, L. and D. Scavia: "NOAA's Coastal Ocean Program: Science for Solutions," *Oceanus*, 85 (Spring 1993).

Whitehead, J.A.: "Giant Ocean Cataracts," *Sci. Amer.*, 50 (February 1989).

Wilson, J.T.: "A New Class of Faults and Their Bearing on Continental Drift," *Nature*, **207**, 343–347 (1965).

Wright, L.: *Morphodynamics of Inner Continental Shelves*, CRC Pres, LLC, Boca Raton, FL, 1995.

Zilanov, V.K.: "Living Marine Resources," *Oceanus*, 29 (Summer 1991).

Web References

Costal Ocean Processes (CoOP): http://www.skio.peachnet.edu/coop/

Scripps Institution of Oceanography: http://www.sio.ucsd.edu/

Scripps Research Oceanography: http://www.sio.ucsd.edu/research/oceano-graphy.html

The French-American Mid-Ocean Undersea Study (FAMOUS): http://www.historychannel.com/deepseadetectives/?page=history

The National Oceanic & Atmospheric Administration (NOAA): http://www.noaa.gov/

Woods Hole Department of Applied Ocean Physics and Engineering: http://www.whoi.edu/science/AOPE/dept/

Woods Hole Department of Biology: http://www.whoi.edu/science/B/dept/

Woods Hole Department of Geology & Geophysics: http://www.whoi.edu/science/GG/dept/

Woods Hole Department of Marine Chemistry and Geochemistry: http://www.whoi.edu/science/MCG/dept/

Woods Hole Department of Physical Oceanography: http://www.whoi.edu/science/PO/dept/

Woods Hole Oceanographic Institution (WHOI): http://www.whoi.edu/science/science.html

OCEAN-ATMOSPHERE INTERFACE. See **Atmosphere-Ocean Interface**.

OCEAN CONVEYOR BELT. The global recirculation of water masses that determines today's climate. The conveyor belt is driven by the sinking of North Atlantic Deep Water (NADW) through cooling of the surface water in the Greenland and Labrador Seas. NADW flows southward through the Atlantic below a depth of 3000 m. When it reaches the Antarctic Circumpolar Current (ACC), some of it continues into the Indian and Pacific Oceans at depth, enters the Atlantic through the Drake Passage and returns to the North Atlantic. Most NADW, however, rises very close to the surface in the ACC, where it freshens considerably through contact with surface water and enters all three oceans as Antarctic Intermediate Water at depths of 700–1000 meters (2,296–3,280 feet). Antarctic Intermediate Water penetrates into the Northern Hemisphere, being slowly entrained by central water, the water mass above it. Pacific central water enters the Indian Ocean through the Indonesian Seas. It then joins Indian central water to flow eastward and then southward in the subtropical gyre. Agulhas Current eddies carry it into the Atlantic, where it moves northward with the Benguela and Brazil Currents and in the Gulf Stream system toward the Greenland and Labrador Seas to cool and sink again, thus completing the conveyor belt circulation.

OCEAN CURRENTS. Currents in the oceans represent the movement (flow) of water, usually in large amounts, as the result of tides, differences in water densities, the stress of various internal and external pressures, and of the wind. Forces of less importance include frictional and Coriolis effects. The movement may be either horizontal or vertical and includes the currents which are a part of the general pattern (*climatic circulation*) as well as the more transitory movements (*synoptic*).

Significant new information has been collected on ocean currents during the past decade, notably by programs in connection with the IDOE Program (The International Decade of Ocean Exploration (1971–1980). The IDOE Program was established by the Marine Sciences Act of 1966 and motivated both by anticipated discoveries of useful and important marine resources and by scientific curiosity. As with other areas of physical oceanography, data from the IDOE venture continues to be analyzed and scientists have not squared completely in their framing of new concepts as the result of new data. For this reason, in this edition of this encyclopedia, traditional concepts based upon many years of study and thought prior to IDOE are included. A major observation of the MODE (Mid-Ocean Dynamics Study) and the POLYMODE (Polygon Mid-Ocean Dynamics Experiment) projects, for example, was the discovery that ocean currents, such as the Gulf Stream and the Circumpolar Current, are not relatively quiescent streams, as previously assumed, but rather they spawn numerous rings and eddies.

Although ocean currents move very slowly when compared with the winds or the rivers, they are capable of moving large masses of water; indeed, it is estimated that the Gulf Stream opposite Chesapeake Bay moves between 70 and 90 million cubic meters (3,178,319,959 cubic feet) of water per second. Our earliest knowledge of such current arose from ocean navigation in which a ship sighted on a fixed point, A, and was headed in the direction of that point. In the presence of a current, the ship will be displaced toward A' and from the difference between A and A' it is possible to determine the direction and average surface velocity of the current. Current measurement techniques improved over the years, including platforms, floating markers, and radioactive tracers. In studies of many years ago, almost exclusive dependence was placed on the drift bottle, a bottle that is partially loaded with sand (so that it is almost completely submerged to minimize the effects of the surface winds) and set free at a given time with a card to the finder enclosed requesting that it be returned, showing place and date where found. In earlier years, the Ekman current meter was used. This is a meter in which a propeller turns at a speed proportional to the water velocity and a tail fin keeps the propeller pointed into the current. See also **Ekman Spiral (Oceanography)**. The meter can be lowered to the desired depth on a steel wire and by various ingenious mechanisms the number of turns of the propeller in a given time can be determined as can the direction of the current, hence the speed of the current. Others, of the drift type, such as the Swallow floats, utilize neutral density floats that can be set to float at a predetermined depth from which they emit electronically sounds that can be detected by appropriate devices at a distance up to several miles. In every instance, however, the accuracy and utility of these and other devices depends on the accuracy with which the position of the meter or ship can be determined. If either of these is determined within as little as 300 feet (91 meters) by use of Loran or Decca, its velocity readings are accurate within 1–2%. It is also possible to use ships and buoys anchored in very deep water. Ocean currents and circulation are also studied by satellite altimetry. Major ocean currents are the *surface currents* and *density currents*.

Classification of Ocean Currents

Surface Currents. These are currents that move along the surface and are restricted to mainly the upper 600 feet (183 meters) or so of the sea and to a maximum depth of perhaps 1000 feet (302 meters). These currents vary somewhat with latitude, being shallower in the lower latitudes. These surface currents arise mainly from wind stress directly or indirectly various internal pressure forces. Due to friction, the wind does not merely skim the surface of the sea, but introduces a wind effect, or stress, that carries the surface water along with it. This layer, in turn, sets up a turbulence in the deeper levels. However, due to the Coriolis effect, the surface water is deflected to the right of the wind direction in the Northern Hemisphere due to the Earth's rotation, and each lower layer is dragged still further to the right until the direction of the drift is opposite to that at the surface. The velocity of the drift, however, decreases rapidly downward and is only $\frac{1}{23}$ that of the surface velocity at the point at which the drift is in opposite direction to that of the wind. In addition, of course, the average current of water will be at an even greater angle to the wind than the surface current. The Coriolis effect is slight near the equator and increases as the distance from that line increases.

Density Currents. Also called *subsurface currents*, these currents consist of those in which the water begins to flow downward when it is more dense than the water next to it. These may arise from several causes; evaporation, from polar waters, and from turbidity. An illustration of the first is the Mediterranean Sea in which the hot, dry climate evaporates far more water from the sea than it receives from rainfall or from rivers feeding into it. Accordingly, the Mediterranean has a greater concentration of salts than the Atlantic Ocean with which it is connected by the Strait of Gibraltar, the floor of which is only 900 feet (274 meters) below sea level. As a result, the much less dense Atlantic water flows in at the surface and the much more dense water from the Mediterranean moves into the Atlantic under the inflow.

Polar Creep. Polar waters are also responsible for density currents. In such cases—known collectively as *polar creep*, cold water from the polar regions gradually sinks down and moves toward the Equator and beyond until it encounters still colder water and is gradually forced to the surface. Due to the great depth at which such currents are found, it is difficult to estimate their rate of travel, but it is believed to be no more than a mile (1.6 kilometers (1 mile) per day—taking no less than 20 years to pass the Equator. It is believed by oceanographers that these currents are of great importance to the animal life in the seas; first, by carrying dissolved oxygen to the great depths at which such life is found; second, by *upwelling*—that is, by carrying minerals from the bottom of the sea to the top where they are used for growth by the microscopic plant and animal life (plankton) which, in turn, provide sustenance for much larger animals of the sea, including the fish.

Upwelling is encountered along the western coasts of continents, such as the coast of southern California. The displaced surface water is transported away from the coast by the action of winds parallel to it or by diverging currents. Upwelling also may occur in the open ocean, where cyclonic circulation is relatively permanent or where southern trade winds cross the equator. In 1991, J.R. Lutjeharms and associate researchers (University of Cape Town) http://www.uct.ac.za/, reported on the observation of extreme upwelling filaments in the southeast Atlantic Ocean. The researchers note, "Oceanic upwelling regimes play a principal role in the ecology of the eastern boundary areas of most ocean basins, in particular that of the northeast Pacific, the southeast Pacific, and southeast Atlantic oceans. It has been suggested that in many instances biological productivity of upwelling areas is concentrated at the fronts that separate cold upwelled water from the adjacent ocean surface water. Frontal behavior may therefore be an important element in the potential primary productivity of upwelling areas as a whole. Moreover, by raising deeper water to the sea surface where it is warmed by insolation and atmosphere-to-ocean heat transfer, upwelling regimes modify the temperature and salinity of substantial volumes of water. The full areal extent of upwelling regimes, as delineated by their upwelling fronts, is therefore a factor in global water mass modification and thus by implication in climate."

Surveillance by satellite has been an effective tool for gathering data on upwelling.

So-called *ocean cataracts* were noticed initially by oceanographers in the 1960s. Inasmuch as the ocean is in thermal equilibrium, it follows that heat flowing upward in the process must equal the heat flowing downward. As observed by J. Whitehead (Woods Hole Oceanographic Institute) http://www.whoi.edu/scitechdir/staff.go?id=235, "Because convection is an extremely efficient mechanism for transferring heat, the downward convective channels do not have to be very large in cross-sectional area to balance the heat transferred by the oceanwide warming of the deep water. The narrow currents of sinking cold water are in fact the precursors to the ocean cataracts."

These water-in-water flows can be enormous. The largest waterfalls on land pale by comparison with the relatively few ocean cataracts that have been located. In studying ocean cataracts, oceanographers use radioactive isotopes in addition to temperature measurements. Tritium, left over from nuclear bomb tests, has been used. Eight ocean cataracts have been studied—Denmark Strait and the Iceland-Faroe Passage in the north Atlantic; Discovery Gap and Strait of Gibraltar in the middle Atlantic; Ceara Abyssal Plain and Romansch Fracture in the mid-south Atlantic; South Shetland Islands and Filchner Shelf off the tip of South America and the coast of Antarctica. Exemplary of the tremendous flows is the rate of about 5 million cubic meters per second through the Denmark Strait cataract. It should be pointed out that the Shetland Islands cataract is powered primarily by salinity differences rather than by thermal differences.

Turbidity Currents. The third type of density current is the *turbidity current*—a current made more dense than the surrounding water by the addition of mud or silt. Their origin is the subject of much dispute, they have not ever been actually observed in the ocean but they are believed to be responsible for the deposits of sand and shells at great distances out on the continental slope. The regional currents include the Agulhas current. Antarctic bottom water, the Antarctic circumpolar current, Antarctic immediate water, the Antilles current, the Caribbean current, the Cromwell current, the East Greenland current, the equatorial countercurrent, the Florida current, the Guiana current, the Gulf Stream, the Gulf Stream countercurrent, the Humboldt current, the Irminger current, the Labrador current. North Atlantic deep water, the North Atlantic current, the South Atlantic current, the South Equatorial current, the South Pacific current, and the West-Wind drift. Many of these are described in separate alphabetical entries in this encyclopedia and below.

Equatorial Current System. The system of ocean currents found in the upper Atlantic and Pacific Oceans between 20°S and 20°N. Its major components are the westward flowing North and South Equatorial Currents (NEC and SEC), which occupy most of the region. The Equatorial Intermediate Current (EIC) is a subsurface band of intensified westward movement between 2°S and 2°N at depths between 300 and 1,000 meters (984 and 3,281 feet). All other components of the system are narrow (200–400 kilometer (124–249 mile) bands of eastward flow: the North and South Equatorial Countercurrents (NECC and SECC), found at depths between 0 and 200 m and located between 5° and 10°N and 5° and 10°S, respectively; the Equatorial Undercurrent (EUC), found on the equator at a depth of 200 meters (656 feet); and the North and South Subsurface Countercurrents (NSCC and SSCC), found on either side of the EIC at depths between 400 and 700 meters (1,312 and 2,297 feet). The major elements of the equatorial current system, particularly the NEC, SEC, NECC, and EUC, are also seen in the Indian Ocean during the northeast monsoon season (December–April) but are significantly modified during the southwest monsoon season. In addition, the Indian equatorial jet is a feature not seen in the other oceans.

Types of Ocean Currents

Agulhas Current. Agulhas Current also called *Agulhas stream*. The Agulhas current is the major western semi of the subtropical gyre in the southern Indian Ocean and one of the swiftest ocean currents with mean speeds of 1.6 m s^{-1} and peak speeds exceeding 2.5 m s^{-1}. Its total transport of 70 Sv (70×10^6 m^3 s^{-1}) near 31°S and up to 135 Sv (135×10^6 m^3 s^{-1}) near 35°S is also among the largest of all ocean currents. The Agulhas Current is fed mainly from the East Madagascar Current and to a smaller degree from the Mozambique Current. When passing the Agulhas Bank, the current produces significant upwelling.

To the south of the Cape of Good Hope, the current flows west to southwestward first but turns sharply eastward when reaching the Agulhas Current retroflexion region near 40°S, 20°E. Eddies spawned in this region continue to move westward and turn northward to join the Benguela Current. The transport of water from the Indian into the Atlantic Ocean through the eddies is an important part of the global ocean conveyor belt.

Alaska Current. The Alaska current is an eastern semi of the North Pacific subpolar gyre. It is a shallow current carrying relatively warm water northward and thus has a climate influence similar to that exercised by the North Atlantic and Norwegian Currents on the climates of northwestern Europe, though on a smaller scale. It flows cyclonically around the Gulf of Alaska, feeding into the Alaskan Stream. Freshwater from the many rivers of Canada and Alaska reduces the water density near the coast; the result is a pressure gradient normal to the coast that constrains the current geostrophically to the coastal region and increases its speed to 0.3 m s^{-1}.

Alaskan Stream. The Alaskan Stream is the continuation of the Alaska Current along the southern side of the Aleutian Islands. The distinction between the Alaskan Stream and the Alaska Current is gradual, and the two currents are sometimes regarded as one. They are, however, of different character, the Alaska Current being shallow and variable but the Alaskan Stream reaching to the ocean floor. Despite its modest speed of 0.3 m s^{-1}, it is a western boundary current. Most of the water of the Alaskan Stream feeds directly into the Oyashio. Some of its flow enters the Bering Sea between the Aleutian Islands (most of it between 168° and 172°W) and follows a cyclonic path before feeding into the Kamchatka Current, thus eventually also contributing to the Oyashio.

Aleutian Current. Aleutian Current also called the *Subarctic Current*. The Aleutian current is the southern, eastward flowing current of the subpolar gyre in the North Pacific. It is fed by the outflow from the Oyashio and lies north of the North Pacific Current, with which it establishes the polar front in the west and experiences much water exchange as it proceeds eastward. As it approaches the coast of North America, it divides to form the northward flowing Alaska Current and the southward flowing California Current.

Algerian Current. A narrow intense current along the Algerian coast in which the transport of water from the Atlantic Ocean into the Mediterranean Sea is concentrated. The Algerian Current originates from the Almeria–Oran front and for the first 300 kilometers (186 miles) is less than 30 kilometers (19 miles) wide, with average speeds of 0.4 m s^{-1} and maximum speeds of 0.8 m s^{-1}. See also **Meteorology**.

Angola Current. As a relatively strong southward current along the Angolan coast. The current forms the eastern side of a cyclonic gyre centered on 13°S, 4°E and is driven by the extension of the South Equatorial Countercurrent. It reaches to depths of at least 300 meters (984 feet) and attains its maximum speed of 0.5 m s^{-1} just below the surface.

Antarctic Circumpolar Current (ACC). The ACC is an eastward flowing current, also known as the West Wind Drift that circles Antarctica and extends from the surface to the ocean floor. With a volume transport of 130 Sv (130×10^6 m^3 s^{-1}) it is the largest of all ocean currents. Current speed in the ACC is comparatively modest (0.1 m s^{-1}, but larger in fronts), the large transport being achieved by the current's great depth. Seventy-five percent of the transport occurs in the polar and subantarctic fronts that make up only 20% of the ACC area. Interannual variability is about 15% of the mean but can reach 40% on occasions. The ACC is influenced by bottom topography, which causes deflections from its general westward path and eddy formation, particularly at the Scotia Ridge, the Kerguelen Plateau, and the Macquarie Ridge. The eddies are instrumental for the poleward transport of heat across the current, which would otherwise block meridional heat transfer. See also **Antarctic Waters**.

Antilles Current. The Antilles Current is one of the currents of the North Atlantic subtropical gyre flowing along the northern side of the Greater Antilles. Its water is supplied from the northern branch of the North Equatorial Current, drawn mainly from the Gulf Stream recirculation and therefore identical to water of the Sargasso Sea. East of the Straits of Florida, the Antilles Current merges with the Florida Current. The continuity of the Antilles Current as a steadily westward flowing current is somewhat in doubt, but westward transport does exist in the mean.

Arctic Polar Front. The Arctic polar front is the frontal zone between the subtropical and subpolar gyres of the Northern Hemisphere. In the Atlantic Ocean it is established by the meeting of the warm and saline Gulf Stream and the cold and fresh Labrador Current and extends as a temperature and salinity front, sometimes also known as the cold wall, from south of Newfoundland and the Grand Banks northeastward to the central North Atlantic. In the Pacific Ocean it consists of two parts, separated by the Japanese islands. The larger fresh is formed by the confluence of the warm and saline Kuroshio and the cold and fresh Oyashio and seen as a temperature and salinity front extending eastward from Japan near 35°N. The smaller fresh extends across the Sea of Japan in the west, where it separates the warm and saline Tsushima Current from the cold and fresh Mid-Japan Sea Cold Current.

Azores Current. These currents are one of the North Atlantic subtropical gyre. It receives some 15 Sv (15×10^6 m^3 s^{-1}) from the North Atlantic Current near 45°W and continues as an eastward current along 35°N to eventually feed into the Canary Current and close the gyre.

Baffin Current. The Baffin Current is a southward flowing current on the eastern side of Baffin Bay with speeds of 0.2–0.4 m s^{-1}. It is fed by low-salinity water from the Arctic Ocean, thus contributing to the freshwater budget between the Pacific and Atlantic, and by the West Greenland Current. It feeds the Labrador Current.

Benguela Current. The Benguela Current is the eastern boundary current of the South Atlantic subtropical gyre. It transports 20–25 Sv ($20–25 \times 10^6$ m^3 s^{-1}) northward along the coast of Namibia with speeds of about 0.2 m s^{-1}, gradually leaving the coast between 30° and 25°S. As an eastern boundary current, the Benguela Current is associated with strong coastal upwelling that reaches as far north as 18°S. The upwelling is strongest in spring and summer (October–February) when the trade winds are steady; during winter (July–September) it extends northward but becomes intermittent because the trades are interrupted by passing low pressure systems.

Bering Slope Current. These currents are located in the eastern part of the subpolar gyre in the deep (western) part of the Bering Sea. Currents in this gyre are weak (0.1–0.2 m s^{-1}) but reach to the ocean floor and are associated with eddy formation at oceanic ridges. A southward countercurrent with maximum speeds of 0.25 m s^{-1} near a depth of 150 meters (492 feet) exists between the shelf and the northward-flowing Bering Slope Current, indicating its character as the eastern boundary current of a subpolar gyre.

Boundary Currents. These are ocean currents with dynamics determined by the presence of a coastline. They fall into two categories: 1) western boundary currents, which are narrow, deep-reaching, and fast-flowing currents, not unlike jet streams, associated with current instability and eddy shedding; and 2) eastern boundary currents, which are shallow, cover a wider region, are of moderate strength, and are often associated with coastal upwelling and a subsurface countercurrent along the continental slope. Both are integral parts of the circulation in oceanic gyres. The rotation of the earth causes an accumulation of energy on the western side, which has to be dissipated in boundary currents; this gives the western boundary currents typical widths of 100 kilometers (62 miles) and typical speeds of 2 m s^{-1} and causes them to shed eddies frequently to increase the dissipation of energy. No similar requirement of energy dissipation exists on the eastern side, so eastern boundary currents can be broad and slow. Their special character as a boundary current results from coastal upwelling, which brings the thermocline to the surface and as a result produces a temperature front and an associated geostrophic maximum in the current speed, known as the coastal jet. Because of the upwelling, eastern boundary currents are atmospheric heat sinks. Western boundary currents are atmospheric heat sinks if they move cold water toward the equator, which occurs in the subpolar gyres, and atmospheric heat sources where they move tropical water into temperate regions, as in the subtropical gyres.

Brazil Current. The Brazil Current is the western boundary current of the South Atlantic subtropical gyre. Fed by the South Equatorial Current, it flows as a narrow, swift current from 10°S southward along the South American coast. The Brazil Current is one of the weaker western boundary currents with a total transport of less than 30 Sv (30×10^6 m^3 s^{-1}), half of this occurring over the shelf. Current speed is below or near 1 m

s^{-1}. The current separates from the coast somewhere between 33° and 38°S, forming a front with the northward flowing Malvinas (Falkland) Current and continuing southward on the eastern side of the front. The separation point is more northward in summer (December–February). The southernmost extent of the Brazil Current varies between 38° and 46°S on a two-month timescale associated with eddy shedding. The current continues eastward as the South Atlantic Current.

California Current. The California Current is the eastern boundary current of the North Pacific subtropical gyre. It flows southward along the coast of Washington and California and is associated with coastal upwelling during spring and summer (April–September). The associated low sea surface temperatures (about 15 °C (59 °F) produce a coastal strip of sea fog along the otherwise extremely hot coastline. In autumn and winter (October–March) the upwelling is replaced by northward flow known as the Davidson Current, which reaches its peak speed of 0.2–0.3 m s^{-1} in January.

Canary Current. The Canary Current is the eastern boundary current of the North Atlantic subtropical gyre. It flows southward along the coast of Africa from Morocco to Senegal and is associated with coastal upwelling. The upwelling reaches its southernmost extent in winter (December–March) when the trade winds are strongest; it then extends past Cape Blanc (21°N), where the Canary Current separates from the African coast.

Caribbean Current. The Caribbean Current is a strong, swift current passing from east to west through the Caribbean Sea. The current is the major pathway for water from the Southern into the Northern Hemisphere in the global ocean conveyor belt. It is also an element of the western boundary current system of the North Atlantic subtropical gyre and thus associated with high speeds and eddy shedding: 0.2 m s^{-1} in the Grenada Basin; 0.5 m s^{-1} in the Venezuela, Columbia, and Cayman Basins; and 0.8 m s^{-1} near Yucatan Strait. Eddies can produce occasional current reversal from westward to eastward in all basins. Most of the water from the Caribbean Current leaves through Yucatan Strait, but a small amount returns eastward from the Caribbean Sea into the Atlantic as the Caribbean Countercurrent.

China Costal Current. A surface current in the East China Sea and Yellow Sea flowing southward along the Chinese coast from Bohai Gulf to Taiwan. It is driven by the northeast monsoon in winter and continues against the southwest monsoon through the summer, strengthened by river runoff from monsoonal rainfall. Taking in most of the waters from the Yangtze River, it contributes greatly to the increased summer transport of the Tsushima Current.

Cromwell Current. See **Equatorial Undercurrent (EUC)**.

Davidson Current. The Davidson Current is the countercurrent of the Pacific Ocean running north along the west coast of the United States (from northern California to Washington to at least latitude 48°N) during the winter months.

East Auckland Current. The continuation of the East Australian Current along the eastern coast of New Zealand's North Island; the current is thus part of the western boundary current system of the South Pacific subtropical gyre. During summer (December–March) it continues southward as the East Cape Current to reach the Chatham Rise. During winter (June–August) some of it separates from the shelf and flows eastward into the Pacific, forming a temperature front along 29°S.

East Australian Current. The EAC is the western boundary current of the South Pacific subtropical gyre. It originates in the Coral Sea near 1°S from the South Equatorial Current and flows southward along the east Australian coast. Although it is the weakest of all western boundary currents with a mean transport of little more than 15 Sv (15 × 10^6 m^3 s^{-1}), its speed is rarely less than 1.5 m s^{-1}. The current is stronger in summer (December–March). It separates from the Australian coast between 31° and 34°S to flow toward the northern tip of North Island, New Zealand, shedding about three eddies per year in the process. Its eastward passage from Australia to New Zealand is known as the Tasman Front, which separates warm tropical water in the Coral Sea from the subtropical water of the Tasman Sea.

East Greenland Current. A southward flowing current along Greenland's east coast that forms part of the North Atlantic subpolar gyre and at the same time constitutes the major outflow route of Arctic water into the Atlantic. This water has a salinity of 30–33 psu and a temperature below −1 °C (30.2 °F). Some of it is diverted just north of Denmark Strait and northeast of Iceland into the East Iceland Current, which carries it toward the Norwegian Sea as part of the formation process of Arctic Bottom Water. The remainder is joined south of Denmark Strait and southwest of Iceland by the northwestward flowing Irminger Current, which brings the water of the subpolar gyre. Transport estimates are 5 Sv (5 × 10^6 m^3 s^{-1}) for the East Greenland Current and 8–11 Sv (8–11 × 10^6 m^3 s^{-1}) for the Irminger Current. The combined flow continues around the southern tip of Greenland into the West Greenland Current.

East Korea Warm Current. This current branches off from the Tsushima Current as it enters the Japan Sea through the Korea Strait. It follows the Korean coast northward to 36°–38°N where it meets the North Korea Cold Current to establish the polar front of the Japan Sea. It continues northeastward along the southern side of the polar front, shifting its path every few months and shedding eddies along the way. It rejoins the Tsushima Current before reaching 40°N.

East Madagascar Current. The East Madagascar Current is one of the western boundary currents of the subtropical gyre in the southern Indian Ocean. It originates near 20°S where the South Equatorial Current splits east of Madagascar into a northward and a southward flowing branch. The current carries the 20 Sv (20 × 10^6 m^3 s^{-1}) of the southern branch southward along Madagascar as a swift, deep, and narrow boundary current. South of Madagascar some of this water moves westward to join the Agulhas Current, but most of it flows northward along the west coast of Madagascar to at least 15°S before also turning westward to join the Mozambique Current.

Equatorial Current System. The system of ocean currents found in the upper Atlantic and Pacific Oceans between 20°S and 20°N. Its major components are the westward flowing North and South Equatorial Currents (NEC and SEC), which occupy most of the region. The Equatorial Intermediate Current (EIC) is a subsurface band of intensified westward movement between 2°S and 2°N at depths between 300 and 1,000 meters (984 and 3,281 feet). All other components of the system are narrow (200–400 kilometers (124–249 miles)) bands of eastward flow: the North and South Equatorial Countercurrents (NECC and SECC), found at depths between 0 and 200 meters (0 and 656 feet) and located between 5° and 10°N and 5° and 10°S, respectively; the Equatorial Undercurrent (EUC), found on the equator at a depth of 200 meters (656 feet); and the North and South Subsurface Countercurrents (NSCC and SSCC), found on either side of the EIC at depths between 400 and 700 meters (1,312 and 2,297 feet). The major elements of the equatorial current system, particularly the NEC, SEC, NECC, and EUC, are also seen in the Indian Ocean during the northeast monsoon season (December–April) but are significantly modified during the southwest monsoon season. In addition, the Indian equatorial jet is a feature not seen in the other oceans.

Equatorial Deep Jets. A vertical stack of alternately eastward and westward ocean currents below the thermocline within a degree of the equator. First discovered in the Indian Ocean, equatorial deep jets have been most extensively observed and described in the Pacific. There they have a dominant vertical wavelength of 300–400 m, and are most clearly seen at depths between 500 and 2000 m. The eastward and westward relative current extreme values may be superimposed on a larger vertical scale flow that may be either eastward or westward. For example, an eastward jet may appear in a given measurement as a relative minimum in a larger-scale westward flow. On a given longitude the depths of the jets vary interannually, but there is no clear evidence for steady vertical propagation.

Equatorial Undercurrent (EUC). The EUC is a subsurface current flowing eastward along the equator. The ECU is a narrow, swift-flowing ribbon with a thickness of 200 meters (656 feet) and a width of at most 400 kilometers (249 miles), it displays the largest current speeds of the equatorial current system. In the Pacific, where it is also known as the Cromwell Current, it flows with a speed of 1.5 m s^{-1} at a depth of 200 meters (656 feet) in the west, rising to a depth of 40 meters (132 feet) in the east. In the Atlantic its core is at a depth of 100 meters (328 feet) and its speed exceeds 1.2 m s^{-1}. In the Indian Ocean it exists as a flow ribbon centered on a depth of 200 meters (656 feet) during the northeast monsoon season (December–April); during the remainder of the year this flow gets incorporated into the eastward flowing southwest monsoon current. In all oceans the EUC swings back and forth between two extreme positions 90–150 kilometers (56–93 miles) either side of the equator with a two- to three-week period.

Florida Current. The Florida Current is one of the western boundary currents of the North Atlantic subtropical gyre. A deep, narrow, and swift current, it originates from the Loop Current of the Gulf of Mexico. It enters the Atlantic Ocean through the Florida Straits with a transport of 30 Sv (30×10^6 m^3 s^{-1}) and speeds of 1.8 m s^{-1} and follows the coast northward for 1,200 kilometers (746 miles) to Cape Hatteras (35°N). Input from the Antilles Current and entrainment from the Gulf Stream recirculation in the Sargasso Sea increase its transport to 70–100 Sv ($70–100 \times 10^6$ m^3 s^{-1}) before it leaves the coast at Cape Hatteras to continue as the Gulf Stream.

Guiana Current. The Guiana Current flows northwestward along the northern coast of South America (the Guianas). The Guiana Current is an extension of the South Equatorial Current (flowing west across the ocean between the equator and 20°S), which crosses the equator and approaches the coast of South America. Eventually, it is joined by part of the North Equatorial Current and becomes, successively, the Caribbean Current and the Florida Current.

Guinea Current. An ocean current flowing along the south coast of northwest Africa into the Gulf of Guinea.

Gulf Stream. The Gulf Stream is one of the western boundary currents of the North Atlantic subtropical gyre and one of the swiftest ocean currents with one of the largest transports. A deep, narrow, and swift current, it continues from the Florida Current for 2,500 kilometers (1,553 miles) in a northeastward direction, penetrating into the Atlantic as a free jet. It reaches its maximum transport of 90–150 Sv ($90–150 \times 10^6$ m^3 s^{-1}) near 65°W before beginning to lose water to the Sargasso Sea; this water rejoins the Florida Current as the Gulf Stream recirculation. Some 50–90 Sv ($50–90 \times 10^6$ m^3 s^{-1}) continue northeastward past the Grand Banks (50°W), where the current is also known as the Gulf Stream Extension and the North Atlantic Current (or North Atlantic Drift). It forms a marked temperature and salinity front with the Labrador Current, which meets the Gulf Stream Extension from the north and then flows parallel to it. As a free jet, the Gulf Stream develops instabilities in the form of meanders that eventually break off as eddies, also known as rings. Cyclonic (cold core) rings contain cold Labrador Current water and drift slowly southwestward into the Sargasso Sea. Anticyclonic (warm core) rings contain warm Sargasso Sea water and drift southwestward in the Slope Water found between the Gulf Stream and the continental shelf. See also **Gulf Stream Countercurrent**.

Guyana Current. One of the western boundary currents of the Atlantic Ocean and part of the pathway for water from the southern into the Northern Hemisphere in the global ocean conveyor belt. Flowing northward along the northern coast of South America, it originates from the North Brazil Current, receives a contribution from the South Equatorial Current, and continues as the Caribbean Current. The continuity of the Guyana Current as a permanent northward current is in doubt, but northward flow does exist in the mean. Eddies related to flow instability typical for western boundary currents have been reported.

Humboldt Current. See **Peru/Chile Current**.

Kamchatka Current. (Also known as the East Kamchatka Current.) The Kamchatka Current is the western part of the subpolar gyre in the deep (western) part of the Bering Sea. It flows southward along the Kamchatka Peninsula with speeds of 0.2–0.3 m s^{-1}. Most of its water leaves the Bering Sea and forms one of the two sources of the Oyashio.

Kuroshio. Kuroshio also called *Japan Current*. The Japan Current is one of the western boundary currents of the North Pacific subtropical gyre. The Japan current is a deep, narrow, and swift current, it continues from the Philippines Current in a northeastward direction from Taiwan along the continental rise of the East China Sea, through Tokara Strait, and close to

the eastern coast of Japan. At 35°N it separates from the coast and flows eastward into the Pacific as a free jet known as the Kuroshio Extension. It forms a marked temperature and salinity front with the Oyashio, which meets the Kuroshio Extension from the north and then flows parallel to it. Like all other western boundary currents, the Kuroshio develops instabilities and sheds eddies. Its unique characteristic is that south of Honshu it switches between three quasi-stable paths across the Izu Ridge at irregular intervals of 18 months to several years. Volume transport in the Kuroshio increases downstream and reaches 57 Sv (57×10^6 m^3 s^{-1}) in the Kuroshio Extension, increasing seasonally by 15% during summer. The current's path in the extension is characterized by large meridional excursions in the so-called First and Second Crest at 145° and 152°E. On approaching the Shatsky Rise at 157°E, the Kuroshio Extension divides into several paths that tend to recombine before the Emperor Seamounts near 170°E cause the current to split again and disintegrate. The flow then continues as the North Pacific Current.

Labrador Current. The Labrador Current is the western boundary current of the North Atlantic subpolar gyre. The Labrador Current receives considerable input of Arctic Surface Water originating from the East Greenland Current and supplied through the West Greenland and Baffin Currents. Its mean transport is close to 35 Sv (35×10^6 m^3 s^{-1}). The current is strongest in February when it carries 6 Sv more water than in August. It is also more variable in winter, with a standard deviation of 9 Sv in February but only 1 Sv in August. Near the Grand Banks it forms the polar front, also known as the cold wall, with the northward flowing Gulf Stream, with which it shares the shedding of, eddies. The cold water of the Labrador Current allows icebergs from western Greenland to travel as far south as 40°N, the latitude of southern Italy.

Leeuwin Current. The Leeuwin Current is the eastern boundary current of the south Indian Ocean. It occupies the latitude band of the subtropical gyre but is not part of it; rather, it is found as a narrow and swift southward flowing current along the west Australian shelf opposing the broad northward flow of the subtropical gyre (the West Australian Current) farther offshore. The Leeuwin Current runs against the prevailing wind; it is driven by the alongshore pressure gradient caused by the connection between the Pacific and Indian Oceans north of Australia. Its water, which is of tropical origin, cools as it proceeds southward, producing convection and a continuous deepening of the surface mixed layer along the current's path. The associated heat loss is a significant heat gain for the atmosphere.

Liman Current. The Liman Current is a cold southward flowing current between Sakhalin and the Asian mainland carrying water from the Sea of Okhotsk into the Sea of Japan.

Loop Current. The Loop Current is the passage of water through the Gulf of Mexico from Yucatan Strait to the Straits of Florida and the connection between the Caribbean and Florida Currents. The Loop Current is part of the western boundary current system of the North Atlantic subtropical gyre and as such is swift flowing, extending to great depth, and prone to instabilities. Its path includes a large northward excursion into the gulf beyond 27°N but retreats to 25°N when shedding an eddy. Eddies drift slowly westward into the central and western Gulf of Mexico.

Malvinas Current. The Malvinas Current is a jetlike northward looping excursion of the Antarctic Circumpolar Current east of southern Argentina; also known as the Falkland Current. Somewhere between 33° and 38°S it meets the Brazil Current and turns eastward, forming an intense temperature and salinity front.

Mid-Japan Sea Cold Current. The slow southward movement of cold water in the central Japan Sea that, when meeting the northward flowing Tsushima Current, establishes the Arctic Polar Front that stretches from southern Korea to Hokkaido.

Mindanao Current. One of the western boundary currents of the Pacific Ocean. It flows southward with great speed along the island of Mindanao and carries a transport of 25–35 Sv ($25–35 \times 10^6$ m^3 s^{-1}) from the North Equatorial Current into the North Equatorial Countercurrent. It is part of the Mindanao Eddy.

Mozambique Current. The Mozambique Current is one of the western boundary currents of the subtropical gyre in the southern Indian Ocean. It flows southward along the east coast of Mozambique through the

Mozambique Channel and contributes about 30 Sv (30×10^6 m^3 s^{-1}) to the Agulhas Current. Only about 6 Sv of this transport enter the Mozambique Channel from the north, fed from the northern branch of the South Equatorial Current; the remainder comes from the East Madagascar Current, which makes a northward loop south of Madagascar before it joins the Mozambique Current north of 15°S.

New Guinea Costal Current. A seasonal western boundary current through Vitiaz Strait and along the northern coast of New Guinea. The current flows northwestward from April to November and is then a link between the northern branch of the South Equatorial Current and the Halmahera Eddy. It reverses during December to March, opposing the New Guinea Coastal Undercurrent underneath, when the Philippines experience monsoon winds blowing from the northeast in the Northern Hemisphere and from the northwest south of the equator.

North Atlantic Current. Also known as North Atlantic Drift, West Wind Drift. This eastward flowing current originates from the Gulf Stream Extension east of the Grand Banks (about 40°N, 50°W). It initially forms part of the Atlantic subtropical gyre but separates from it after less than 500 kilometers (311 miles) near 45°W, turning northeastward and following the Arctic Polar Front, also known as the North Wall, with a transport of some 30 Sv (30×10^6 m^3 s^{-1}). Some of this water enters the subpolar gyre through mixing across the polar front and feeds the Irminger Current, but most of it is delivered to the Norwegian Current. The North Atlantic Current carries warm subtropical water much farther north than any other current of the Northern Hemisphere. As a result the climate of northern Europe is much milder than the climate of Alaska or northern Siberia, both of which are located at comparable latitude.

North Brazil Current. One of the western boundary currents of the Atlantic Ocean and part of the pathway for water from the Southern into the Northern Hemisphere in the global ocean conveyor belt. Flowing northward along the coast of northern Brazil with speeds up to 0.8 m s^{-1}, it originates from the South Equatorial Current and continues as the Guyana Current.

North Equatorial Current (NEC). The NEC is a broad region of uniform westward flow that forms the southern part of the Northern Hemisphere subtropical gyres driven by the trade winds. Being directly wind driven, the NEC responds quickly to variations in the wind field and is therefore strongest in winter (February). In the Atlantic Ocean it is found between 8° and 30°N with speeds of 0.1–0.3 m s^{-1}. In the Pacific Ocean it has similar speed but is limited to 8°–20°N. In the Indian Ocean it exists only during December–April when the northeast monsoon produces the same wind forcing as the Northern Hemisphere trade winds. It then runs as a narrow current of 0.3 m s^{-1} from Malacca Strait to Sri Lanka where it bends southward and accelerates in the region 60°–75°E to 0.5–0.8 m s^{-1} between 2°S and 5°N and continues along the equator.

North Equatorial Countercurrent (NECC). The NECC are a band of eastward flow between the westward flowing North and South Equatorial Currents. The location and strength of the NECC is determined by the intertropical convergence zone (ITCZ) of the atmosphere. In the Pacific Ocean it is strongest in May–January when it flows between 5° and 10°N with 0.4–0.6 m s^{-1}; in February–April it is restricted to 4°–6°N with speeds below 0.2 m s^{-1} and disappears east of 110°W. In the Atlantic Ocean the NECC is observed between 5° and 10°N with speeds of 0.1–0.3 m s^{-1}; it is strongest during August when it flows from South America into the Gulf of Guinea and weakest in February when it is restricted to the region east of 20°W. In the Indian Ocean the NECC exists during the northeast monsoon season only and is then the only countercurrent; it is therefore mostly called the Equatorial Countercurrent. It is centered on 5°S, again the location of the ITCZ.

North Korea Cold Current. The North Korean Cold Current is a southward flowing current along the northern Korean coast fed from the Liman Current. At 37°–38°N it meets the East Korean Warm Current to establish the Arctic Polar Front of the Japan Sea. It is the source of the Mid-Japan Sea Cold Current, which withdraws water from it along its way to reinforce the polar front across the entire length of the Japan Sea.

North Pacific Current. also known as the West Wind Drift. This eastward current forms the northern part of the North Pacific subtropical gyre. It originates from the Kuroshio Extension east of the Emperor Seamounts (170°W) and maintains the Arctic Polar Front with the Aleutian or Pacific Subarctic Current, with which it experiences much mixing as it proceeds eastward. A broad band of flow some 2,000 kilometers (1,243 miles) wide, it feeds its water into the California Current on approaching the North American coast.

Norwegian Current. This type of current flows northeastward along the Norwegian coast. It is the continuation of the North Atlantic Current and discharges about 10 Sv (10×10^6 m^3 s^{-1}) into the Arctic Ocean. Because its waters are of subtropical origin it has a significant impact on the climate of northern Europe; ports in northern Norway located at 70°N are ice-free throughout the year.

Oyashio. The Oyashio is a western boundary current of the North Pacific subpolar gyre. Fed by the Alaskan Stream and Kamchatka Current, the Oyashio flows southward along the Kuril Islands and the east coast of northern Japan to 39°–40°N and occasionally as far south as 36°N. Near southern Hokkaido (43°N) the current splits into two branches known as the First and Second Oyashio Intrusion. In the region east of Tsugaru Strait (41°N) it is dominated by eddies; one or two cyclonic (warm core) eddies are formed each year. Every six years or so one of them grows into a giant eddy that then dominates the region for a year. Farther downstream the Oyashio meets the northward flowing Kuroshio to form the Arctic Polar Front. Both currents maintain their own frontal systems, the Oyashio Front being characterized by temperatures of 2°–8°C (35.6–46.6°F) and feeding into the Aleutian Current, while the Kuroshio Front has temperatures of 10°–15°C (50–59°F) and feeds into the North Pacific Current.

Pacific Subarctic Current. See **Aleutian Current**.

Peru/Chile Current. The Peru/Chile Current is the eastern boundary current of the south Pacific subtropical gyre, also known as the Humboldt Current. The Peru/Chile current originates where part of the water that flows toward the east across the subantarctic Pacific Ocean is deflected toward the north as it approaches South America. It flows northward along the coast of Chile, Peru, and Ecuador, and is associated with the economically most important coastal upwelling of the World Ocean. The upwelling region extends from 40°S into the equatorial region, where the current separates from the coast and turns toward the west, joining the South Equatorial Current, and the coastal upwelling blends into the equatorial upwelling belt.

Philippines Current. One of the western boundary currents of the North Pacific subtropical gyre. It flows northward from just north of Mindanao (10°N) to Taiwan where it continues as the Kuroshio. It is fed from the North Equatorial Current, from which it continues to entrain water along its way.

Portugal Current. One of the eastern boundary currents of the North Atlantic subtropical gyre. Flowing southward along the Portuguese coast, it is the continuation of the Azores Current and continues as the Canary Current. Because of its location outside the trade-wind belt, it is not associated with significant upwelling.

RAFOS Technology. A Lagrangian method for the acoustic observation of ocean currents in the interior of the ocean. The main components of a RAFOS system are a minimum of three fixed sound sources moored at intermediate depths and free-drifting floats labeling a certain water mass. The term RAFOS (SOFAR spelled backwards) was selected for this technology to indicate that the sound transmission direction is the opposite of the SOFAR technology. In both cases, the minimum in vertical profiles of sound velocity in the ocean is utilized for communication between the sources and receivers. RAFOS floats are expendable receivers that listen for coded signals from the moored RAFOS sound sources. After their underwater mission is terminated (duration up to two years), RAFOS floats drop a ballast weight, ascend to the surface, and transmit their internally stored data via satellite link to a shore-based receiving station. The RAFOS technology enables acoustic observations of ocean currents to within the millimeter per second range.

Somali Current. A prominent western boundary current in the northern Indian Ocean. During the northeast monsoon season the Somali Current flows southward from 5° to 1°N in December, expanding to 10°N–4°S in January–February and contracting again to 4°N–1°S in March. It is then fed from the North Equatorial Current and discharges into the Equatorial Countercurrent. During all these months its speed is 0.7–1.0 m s^{-1}. During the southwest monsoon the current develops into an intense northward jet with extreme surface speeds; 2 m s^{-1} have been reported for May and 3.5 m s^{-1} for June. The jet is fed from the South Equatorial Current and flows along the eastern coast of the Horn of Africa; part of it continues along the Arabian Peninsula as the East Arabian Current. South of 5°N the jet is shallow; southward flow continues below a depth of 150 m. North of 5°N the jet deepens and embraces the permanent thermocline. During its northward phase the Somali Current is associated with strong upwelling between 2° and 10°N. The upwelled cold water turns offshore near Ras Hafun (11°N), forming a large anticyclonic eddy with a diameter of about 500 kilometers (311 miles) known as the Great Whirl. Eventually the water from the Somali Current enters the Southwest Monsoon Current.

South Atlantic Current. The SAC is an eastward current that forms the southern part of the South Atlantic subtropical gyre. It is fed by the Brazil Current and follows the subtropical front, gradually losing water to the subtropical gyre and Brazil Current recirculation. About 20–25 Sv (20–25 × 10^6 m^3 s^{-1}) reach the African coast and continue as the Benguela Current. Cooling of the warm Brazil Current water along its path makes the South Atlantic Current a heat source for the atmosphere.

South Equatorial Current (SEC). The SEC is the broad region of uniform westward flow driven by the trade winds that forms the northern part of the Southern Hemisphere subtropical gyres. Being directly wind driven, the SEC responds quickly to variations in the wind field and is therefore strongest in winter (August). In the Atlantic Ocean it is found between 3°N and 25°S with speeds of 0.1–0.3 m s^{-1}. In the Pacific Ocean it covers the same latitude band but attains 0.6 m s^{-1} and a transport of about 27 Sv (27 × 10^6 m^3 s^{-1}) in August; this decreases to 7 Sv in February. In the Indian Ocean it occupies the latitude band 8°–30°S during the northeast monsoon (December–April) and expands northward to 6°S in September during the southwest monsoon, with speeds close to 0.3 m s^{-1} throughout the year.

South Equatorial Countercurrent (SECC). The SECC are a band of weak eastward flow in the Atlantic and Pacific Oceans embedded in the South Equatorial Current near 8°S, caused by a wind stress minimum in the Southern Hemisphere trade winds. In the Pacific Ocean the SECC is controlled by the Asian–Australian monsoon and is strongest during the northwest monsoon (December–April), with speeds approaching 0.3 m s^{-1}; it is barely seen during the remainder of the year. East of the date line it decreases rapidly in strength and is absent from the eastern Pacific during most of the year. In the Atlantic Ocean it is weak, narrow, and variable and has its largest speed of little more than 0.1 m s^{-1} often below the surface at a depth of 100 meters (328 feet).

South Indian Ocean Current. This eastward current forms the southern part of the subtropical gyre in the Indian Ocean. It is fed by the Agulhas Current and follows the subtropical front, gradually losing water to the subtropical gyre and Agulhas Current recirculation. East of Africa it begins with 60 Sv (60 × 10^6 m^3 s^{-1}) and arrives off the coast of western Australia with 10 Sv, which continues as the West Australian Current. Cooling of the warm Agulhas Current water along its path makes the South Indian Ocean Current a heat source for the atmosphere.

South Pacific Current. The SPC is the eastward current that forms the southern part of the South Pacific subtropical gyre. It is fed by the East Australian Current and its continuation, the East Auckland Current and East Cape Current, and follows the subtropical front. It is much weaker than the South Atlantic and South Indian Ocean Currents, carrying little more than 5 Sv (5 × 10^6 m^3 s^{-1}). Nevertheless, cooling of the warm water from the western boundary currents along its path makes the South Pacific Current a heat source for the atmosphere. It feeds its water into the Peru/Chile Current.

Subtropical Countercurrent. An eastward flowing narrow current in the center of a subtropical gyre where the general water movement is weakly westward. The three subtropical countercurrents found in the North Pacific at 20°–26°N between Hawaii and Asia extend to depths of at least 800 meters (2,625 feet) and have speeds of 0.15 m s^{-1}. Other subtropical countercurrents have been reported north of the Hawaiian Ridge and in the Coral Sea.

Surface Current. The current at the sea surface, which is partly due to the effects of wind and waves. Empirically, it is found that the drift current at the surface is 2.5%–3.0% of the wind speed, and it decreases rapidly in the uppermost meter.

Tsushima Current. The Trushima Current is a warm current flowing northward through the Japan Sea along the west coast of Kyushu and Honshu. The Tsushima Current is a branch of the western boundary current of the North Pacific subtropical gyre, which is split by the Japanese islands. It branches off the Kuroshio near 30°N to enter the Japan Sea through Korea Strait, where it carries 1.3 Sv (1.3 × 10^6 m^3 s^{-1}, about 2% of the total Kuroshio transport) with speeds near 0.4 m s^{-1} in summer (August) but only 0.2 Sv with less than 0.1 m s^{-1} in winter (January). Most of the summer transport is fed into the East Korea Warm Current but rejoins the Tsushima Current after the East Korea Warm Current separates from the coast at 36°–38°N. The seasonal variability of the Tsushima Current effects the hydrography of the southern Japan Sea greatly, reducing surface salinity from 35 psu in winter to below 32.5 psu in summer when the current carries low salinity water from the Yellow Sea. Most of the Tsushima Current rejoins the Kuroshio through the eastward flowing Tsugaru Warm Current, which passes through Tsugaru Strait (the passage between Honshu and Hokkaido). This current runs into the Oyashio near 42°N, which forces it to flow southward on the shelf along the east coast of Honshu to meet and join the northward flowing Kuroshio near 35°N. Another part of the Tsushima Current continues farther north, pushing the polar front to its most northern position in the Pacific, to enter the Sea of Okhotsk between Hokkaido and Sakhalin. This water traverses the Sea of Okhotsk as the Soya Warm Current, a rapid current with speeds reaching 1 m s^{-1} that stays close to the coast. Current shear between the fast-flowing coastal water and the offshore region persistently produces eddies of between 10 and 50 kilometers (6 and 31 miles) in diameter. The water leaves the Sea of Okhotsk near 46°–47°N to flow south between the Oyashio and the east coast of Hokkaido.

West Australian Current. The WAC is the eastern boundary current of the subtropical gyre in the Indian Ocean. It covers the eastern half of the southern Indian Ocean in a broad northward movement but does not reach to the Australian coast, where the Leeuwin Current flows southward instead.

West Greenland Current. The West Greenland Current is a current of the North Atlantic subpolar gyre flowing northward along the west coast of Greenland. It is fed from the East Greenland Current and from the cyclonic circulation of the Labrador Sea and achieves a transport of over 30 Sv (30 × 10^6 m^3 s^{-1}). Some of its water turns westward at 64°N to join the Labrador Current; the remainder continues northward as a relatively warm current through Davis Strait and into Baffin Bay, where it can be seen as a temperature maximum of greater than 1°C (33.8°F) at a depth of 500 meters (1,640 feet). This water eventually turns westward as well to feed the Baffin Current and through it the Labrador Current.

Zanzibar Current. A western boundary current in the Indian Ocean that flows permanently northward from 10°S along the east African coast. Also known as the East African Coastal Current.

During the northeast monsoon the Zanzibar Current flows against the wind, meeting the southward flowing Somali Current at 1°N in December, 4°S in February (when the Zanzibar Current is weakest), and at the equator in April. Throughout this period it continues across the equator as an undercurrent that feeds into the Equatorial Countercurrent. During the southwest monsoon season the current strengthens considerably, attaining speeds of 2 m s^{-1} and a transport of 15 Sv (15 × 10^6 m^3 s^{-1}) and feeding into the Somali Current, which flows northward during that season.

Continuing Research on Ocean Currents

Since the IDOE Program, considerable scientific research has been directed, during the past decade, to better understanding the nature of ocean currents. For example, the instantaneous California Current, as investigated by Mooers and Robinson (see reference), is seen to consist of intense meandering current filaments (jets) intermingled with synoptic-mesoscale eddies. These quasi-geostrophic jets entrain cold, upwelled coastal waters and rapidly advect them far offshore. This behavior accounts for the elongated, cool surface features that are seen extending across the California Current region in satellite infrared imagery. The associated advective mechanism should provide significant cross-shore transports of heat, nutrients, biota, and pollutants. The California Current is the major eastern boundary current of the North Pacific. Its flow regime is important for fisheries and climate-related processes; for oil and gas recovery operations and waste disposal; for biological, chemical and geological investigations; and for physical oceanographic studies. The source of these eddies and their role in the local internal dynamics of the California Current have yet to be determined, but are under serious study.

In a pioneering research effort, Weller and colleagues (Woods Hole Oceanographic Institution) http://www.whoi.edu/scitechdir/staff.go?id =234, have made measurements from the research platform, FLIP, which provide some of the first direct observations of three-dimensional flow within the surface mixed layer of the ocean. Relatively narrow regions of downwelling flow were found within the mixed layer, in coincidence with bands of convergent surface flow. At mid-depth in the mixed layer, the downwelling flow had magnitudes of up to 0.2 meter (0.66 feet) per second and was accompanied by a downwind, horizontal jet of comparable magnitude. There is some evidence that these motions transport heat and phytoplankton within the mixed layer. The researchers observed that during the day incoming solar radiation heats the upper ocean, but the amount of radiation that penetrates the ocean decays exponentially. As a result, a shallow, warm layer may tend to form at the surface during midday. At night the surface of the ocean loses heat to the atmosphere and fluid at the surface may be cooler than at the interior. This three-level flow may play an important role in the biology of certain oceanic life forms.

Since the beginning of the U.S. Global Change Research Program (USGCRP) http://www.usgcrp.gov/, the World Ocean Circulation Experiment (WOCE) http://woce.nodc.noaa.gov/wdiu/, has been considered one of the most important efforts to be undertaken by USGCRP. See entry on **Global Change**; and **Global Change Research Program (GCRP)**.

Additional Reading

Bischof, J.: *Ice Drift, Ocean Circulation and Climate Change*, Springer-Verlag, Inc., New York, NY, 2001.
Bradley, R.S.: *Principles of Ocean Physics*, 2nd Edition, Morgan Kaufmann Publishers, Orlando, FL, 2000.
Dijkstra, H.A.: *Nonlinear Physical Oceanography: A Dynamical Systems Approach to the Large Scale Ocean Circulation and El Nino*, 2nd Edition, Springer-Verlag New York, LLC, New York, NY, 2005.
Earle, S.A., and T. Cahill: *Atlas of the Ocean: The Deep Frontier*, national Geographic Society, Washington, D.C., 2001.
Kagan, B.A., J.T. Houghton, and A.J. Dessler; *Ocean Atmosphere Interaction and Climate Modeling*, Cambridge University Press, New York, NY, 2006.
Liss, P.S., and R.A. Duce: *Sea Surface and Global Change*, Cambridge University Press, New York, NY, 2005.
Lutjeharms, J.R.E., F.A. Shillington, and C.M. Duncombe Ray: "Observations of Extreme Upwelling Filaments in the Southeast Atlantic Ocean," *Science*, 774 (August 16, 1991).
Mooers, C.N.K. and A.R. Robinson: "Turbulent Jets and Eddies in the California Current and Inferred Cross-Shored Transports," *Science*, **223**, 51–53 (1984).
Potter, T.D. and B.R. Colema: *Handbook of Weather, Climate, and Oceans*, The McGraw-Hill Companies, Inc., New York, NY, 2001.
Prager, E. and S. Earle: *The Oceans*, The McGraw-Hill Companies, Inc., New York, NY, 2001.
Siedler, G., J. Gould, and J. Church: *Ocean Circulation and Climate*, Academic Press, Inc., San Diego, CA, 2001.
Staff: NOAA, *Central and Western Pacific Ocean and Indian Ocean*, The McGraw-Hill Companies, Inc., New York, NY, 2005.
Stow, D.: *Oceans: An Illustrated Reference*, University of Chicago Press, Chicago, IL, 2006.
Sverdrup, K.A., A.B. Duxbury, and A.C. Duxbury: *Introduction to the World's Oceans*, 8th Edition, The McGraw-Hill Companies, Inc., New York, NY, 2003.

OCEAN (Hydrology). See **Hydrology**.

OCEANICITY. See **Climate**.

OCEAN MIXING. Any process or series of processes by which parcels of ocean water with different properties are brought into intimate small-scale contact, so that molecular diffusion erases the differences between them. There is a distinction between stirring, which moves the water parcels into intimate contact, and mixing, the final process of molecular diffusion that blends the water parcels together. The term "mixing" is currently used to describe all of the processes, including molecular diffusion.

OCEAN PERCH. See **Redfish**.

OCEAN RESEARCH VESSELS. Principal types of oceanographic research vessels are surface craft and manned or unmanned submersibles. Some craft are quite specialized, as in the case of a drilling vessel. In recent years, scientific satellites also have contributed to oceanographic research.

Chronology of Ocean Research Vessels

The era of the early mapping of the seas and the use of vessels for other kinds of scientific studies is somewhat obscure. It is known that, as early as 1838, the U.S. Navy designed two "exploring vessels" for use in the U.S. Exploring Expedition of 1838–1842. In 1838, the Coast Survey schooner *Nautilus* was built specifically for surveying, as was the steamer *Blake* in 1874. Alexander Agassiz carried out some oceanographic studies with the *Blake*. In 1879, the U.S. Fish Commission ordered a coal-burning steamer, the *Fish Hawk*, for fisheries research. The vessel was complete with a floating fish hatchery. The *Alexander Agassiz*, designed and built in San Diego in 1907 and first assigned to the West Coast marine station, was a sailing vessel equipped with two gasoline engines. The ship was used in an area of the Pacific Ocean from Point Conception to the Mexican border and seaward to about 120 miles (193 kilometers). The region included depths to 1100 fathoms. (1 fathom = 6 feet = ∼1.8 meters) The *Agassiz* was equipped with dredges, trawls, closing nets, current meters, and other oceanographic instruments of that era. The ship served the Scripps Institution of Oceanography (La Jolla, California), the successor of the West Coast marine station, for about ten years. The *Atlantis*, probably the first American ship especially designed for oceanographic research in the modern sense, entered operations in 1931 as the research vessel for the newly formed Woods Hole Oceanographic Institution (Woods Hole, Massachusetts), originally funded by the Rockefeller Foundation. At a length of 142 feet (43 meters) the *Atlantis* was quite large for a sailing ketch, but a fine size for a research vessel, striking a good balance between seaworthiness, range, personnel requirements, and research capability. The *Atlantis* was a pioneering ship and her scientific accomplishments were impressive. Work with the *Atlantis* included studies of the Gulf Stream and its meanders, investigations of the geophysical properties of the sea floor; studies of the ocean's sound properties, studies of deep midwater fauna, and investigations of submarine canyons. This vessel was replaced by the *Atlantis II* in 1963. Other earlier vessels included the R.V. *E.W. Scripps* and the *R.V. Vema*. Earlier vessels charged principally with making ocean depth surveys included the *H. M.S. Challenger*, which operated over the period 1872–1876, and the U.S. Navy surveying ship *Nero*, which made a survey for a proposed telegraphic cable between Honolulu and Manila by way of Guam and Yokohama. The greatest depths accurately measured during that survey were in the range of 5100 and 5270 fathoms.

Submersible research craft applied much of the technology developed for military submarines. Interest in submarines dates back to the early 1600s. Van Dribel invented a submarine rowboat in 1624; Le Son built the *Rotterdam* in 1652; Bushnell built a submersible boat (the *Turtle*) in 1776; Fulton built the *Nautilus* for the French government in 1800; and the partially submersible torpedo boat *David* was constructed by the Confederacy during the War Between the States, and probably sparked the potential for submersible craft in military engagements. The military submarine did not gain significant prominence until World War I. Charles Beebe pioneered the use of a bathysphere (nonnavigable) in 1934. Lowered by a steel cable from a surface ship, Beebe could attain depths of a little over 923 meters (3028 feet). The practical use of navigable submersible vessels for research work dates back only a few decades.

Contemporary Ocean Research Vessels

Robertson Dinsmore (Woods Hole Oceanographic Institution) classifies ocean research surface craft as follows:

1. Large Ships — Length, 200 feet (61 meters) or longer. These craft make expeditions of long duration and carry 20 to 25 scientists with a like number of crew. Capable of cruising 250 to 280 days at sea,

or up to a year on extended voyages. These ships play a major role in research.

2. Intermediate-Size Ships — Length, 150 to 200 feet (37 to 61 meters) for carrying 12 to 16 scientists and a crew of about 12 persons. Because of their lower operating costs, they are used for most of the shorter oceanographic cruises — 2 to 3 weeks' duration. They are limited by the sea state they can operate in and have limited laboratory and storage space.

3. Small Vessels — Length, 80 to 150 feet (24 to 37 meters), generally considered coastal vessels. They carry from 9 to 12 scientists on short exploratory cruises of 1 to 2 weeks' duration. They are used mainly for small projects close to shore. There are also a number of vessels of less than 80 feet (24 meters) in length, usually located near the main oceanographic research centers. Some are also located in the Great Lakes.

R.V. Atlantis II. This vessel was named for the Woods Hole Oceanographic Institution's first research vessel, previously mentioned. Under a grant from the National Science Foundation (NSF), the vessel was built in 1961–1962 and commissioned in 1963. The vessel has worked in all disciplines and has traveled worldwide. In recent years, the ship has been engaged in intensive geological and geophysical studies in the Atlantic. The vessel participated extensively in some of the IDOE (International Decade of Ocean Exploration) programs. See also Ocean. In 1979, the ship underwent a major mid-life refit and was converted from steam to diesel power to reduce operating costs and increase her range and selection of ports. Programs for improving the effectiveness of the vessel continue as funding permits.

A diagram of the *Atlantis II* is shown in Fig. 1. There are four laboratories aboard the ship. These encompass 404 square meters (4350 square feet) and are fitted with outlets for fresh water, seawater, oxygen, and other gases. There is a controlled-temperature aquarium aboard. The ship was designed for efficiency and versatility. She carries a general purpose shipboard computer for data analysis at sea and a precision graphic recorder to electronically record depth measurements. Gyro compass repeaters, speed indicators, and winch line pull indicators are located around the ship for the use of scientists. Hydrophones are attached to the underwater hull to receive sound, and large electric patch panels are installed in all major laboratories for rapid communication and recording of information. The ship has a fully equipped machine shop and an explosives magazine for safe storage of small depth charges used in seismic refraction profiling. A bulbous underwater observation chamber in the bow is equipped with six viewing ports. The ship is air conditioned and is equipped with underwater lights for night work, a large stern ramp, large uncluttered deck space, and an enclosed crow's nest for daytime lookout.

R.V. Knorr and Other Surface Vessels. The *Knorr* participated in a major way in the Titanic Expeditions and numerous other research projects over the past several years. The *Knorr* is 74.6 meters (245 feet) long, with a displacement of 2075 long tons, a range of 16,900 kilometers (10,000 miles) at 11 knots, and accommodations for a crew of 25 and scientific party of 24. The ship was built in 1969. There is also the *Oceanus*, which is 54 meters (177 feet) long, with a displacement of 962 long tons, a range of 12,067 kilometers (7,500 miles) at 14 knots, and accommodations for a crew of 12 and scientific party of 12. The ship was built in 1975. The *Knorr* is owned by the U.S. Navy; the *Oceanus* by the NSF.

The University-National Oceanographic Laboratory System (UNOLS). The UNOLS was established in 1971 with the responsibility of coordinating and scheduling oceanographic research ships and to provide opportunities to scientists who normally do not have access to research vessels. During the IDOE, previously mentioned, this arrangement proved extremely worthwhile. Generally, the membership in UNOLS is defined as "those academic institutions that operate significant federally funded oceanographic facilities." See Table 1. The membership also includes some 40 smaller laboratories that hold associate memberships and participate in the use of seagoing science facilities.

Surface vessels of a nature similar to *Atlantis II, Knorr,* and *Oceanus* are used by other institutions and countries for oceanographic research, but their number is quite limited in the light of research required. Much of the information gathered for oceanographic research is obtained through the use of these special research vessels which are equipped with a large variety of instruments for measuring depth, temperature, salinity, magnetism, variations in gravitational force, seismic probes, and for taking and analyzing samples. Excellent underwater photographic equipment is usually aboard, as well as excellent navigational aids. Numerous improvements have been made in recent years. An example is the sound energy used in connection with acoustic and seismic equipment. At one time dynamite was the major source of sound energy. More recent energy sources include the electric spark, compressed air and propane gas. Hydrophones that trail behind the vessel receive reflected energy from the bottom. These signals are duly processed and tape recorded, thus providing a continuous seismic-reflection profile. Information from depths of several kilometers below the sea floor can be obtained. This makes possible the construction of geological cross sections. Together with samples from dredging or drilling, much information has been gained pertaining to the structure of the continental shelves. The larger research vessels include their own drilling equipment.

Internationally Operated Ocean Research Surface Vessels. Those countries in addition to the United States that own and operate ten or more

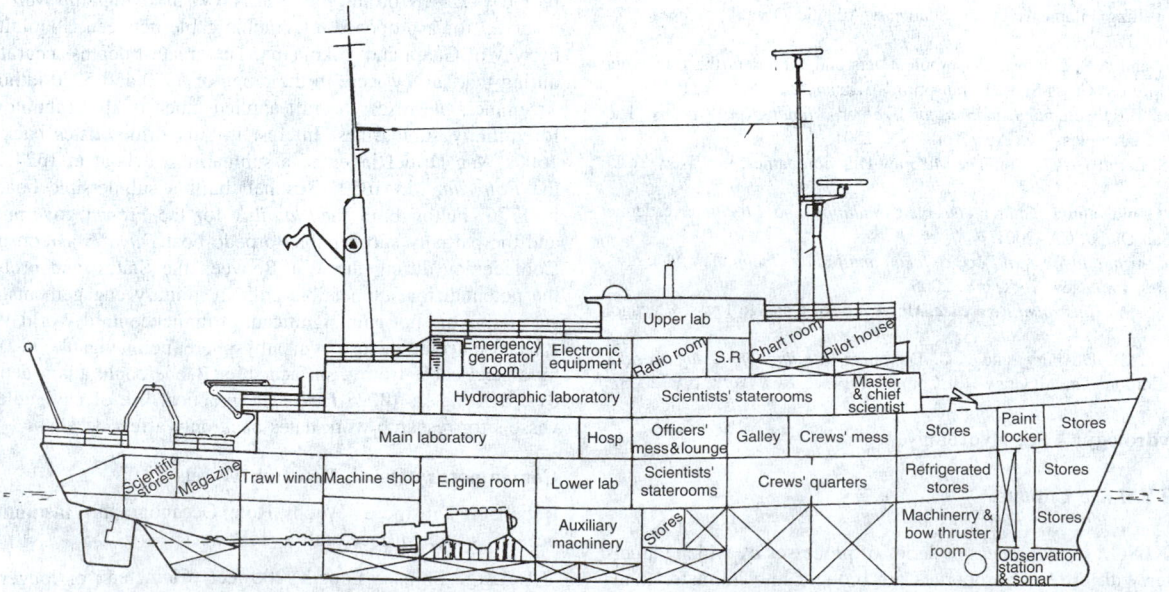

Fig. 1.   Inboard profile of the *R.V. Atlantis II*, considered a large ocean research surface vessel. Ships of this type are periodically updated as technology advances and as funding permits. (*Woods Hole Oceanographic Institution*.)

TABLE 1. UNOLS FLEET OF SURFACE VESSELS FOR OCEAN RESEARCH

Ship's Name	Length Ft.	Length M	Built/ Converted	Crew/ Scientists	Owner	Operating Laboratory
LARGE SHIPS						
Melville	245	75	1970	22/26	U.S. Navy	Scripps
Knorr	245	75	1969	24/25	U.S. Navy	W.H.O.I.
Atlantis II	210	64	1963	24/25	W.H.O.I.*	W.H.O.I.
T.G. Thompson	208	63	1965	22/19	U.S. Navy	U. Washington
T. Washington	208	63	1965	19/23	U.S. Navy	Scripps
Conrad	208	63	1963	25/18	U.S. Navy (Columbia U.)	Lamont-Doherty
INTERMEDIATE SHIPS						
Oceanus	177	54	1975	12/12	N.S.F.	W.H.O.I.
Wecoma	177	54	1975	12/16	N.S.F.	Oregon State U.
Endeavor	177	54	1976	12/16	N.S.F.	U. Rhode Island
Gyre	174	53	1973	11/18	U.S. Navy	Texas A. M.
Columbus Iselin	170	52	1972	12/13	U. Miami*	U. Miami
New Horizon	170	52	1978	12/13	S.I.O.	Scripps
Fred H. Moore	165	50	1967/1978	9/20	U. Texas	U. Texas
Kana Keoki	156	48	1967	12/16	U. Hawaii	U. Hawaii
SMALL SHIPS						
Cape Florida	135	41	1981	9/10	N.S.F.	U. Miami
Cape Hatteras	135	41	1981	9/12	N.S.F.	Duke U
Alpha Helix	133	40.5	1965	12/12	N.S.F.	U. Alaska
Ida Green	130	40	1965/1972	7/12	U. Texas	U. Texas
Cape Henlopen	122	37	1975	6/12	U. Delaware	U. Delaware
Velero IV	110	34	1948	11/12	U.S.C.	U. Southern California
Ridgely Warfield	106	32	1967	8/10	J.H.U.*	Johns Hopkins U.
E.B. Scripps	95	29	1965	5/8	S.I.O.	Scripps
Cayuse	80	24	1968	7/8	N.S.F.	Moss Landing Marine Lab
Longhorn	80	24	1971	5/10	U. Texas	U. Texas
Laurentian	80	24	1974	5/8	U. Michigan	U. Michigan

*Funded by National Science Foundation.
UNOLS = University-National Oceanographic Laboratory System.
W.H.O.I = Woods Hole Oceanographic Institution.
S.I.O. = Scripps Institution of Oceanography.
N.S.F. = National Science Foundation.
O.S.U. = Oregon State University.

surface vessels for ocean research include: Russia[1], 194 vessels; Japan, 94; United Kingdom, 39; France, 27; Canada, 25; former West Germany, 15; Brazil, 12; Sweden, 11; and Argentina, Australia, and Italy, 10 each.

Special mention should be made of the French vessel *Le Suroit*, operated by the Institute Français de Recherche sur l'Exploitation des Mers (Toulon, France), which played an important role in the discovery of the *Titanic*.

The efficacy and sophistication of modern underwater research was widely acclaimed by the scientific community and by the public worldwide when the discovery of the resting site of the *Titanic* was announced in early September 1985. Culminating years of research and planning by an American and French team of researchers, the moment of discovery was given as very early on the morning of 1 September. The location: approximately 360 miles (580 km) off Grand Banks, Newfoundland at a depth of 13,000 feet (396 m) of water. On 9 September 1985, Robert D. Ballard (Woods Hole Oceanographic Institution) and leader of the American team observed that "the *Titanic* lies on a gently sloping alpine-like countryside overlooking a small canyon below. [Her] bow faces north and the ship sits upright on the bottom, [two of her] mighty stacks still pointing upward. There is no light at this great depth. It is quiet and peaceful, a fitting place for the remains of this greatest of [peacetime] sea tragedies to rest. May it forever remain that way." It was learned later that the great ship had broken into two parts, with the stern section lying some 1830 feet (558 m) behind the bow and facing in an opposite direction. In addition to the great skills used by the leaders and crew of this special expedition, the finding and later close-up investigations of the ship are a testament to the oceanographic technology available throughout the 1980s and notably the research ships used—the surface ships *Knorr* (United States) and the *Le Suroit* (France) and the underwater vehicles

Argo (U.S.), the SAR (France), and for the 1986 re-expedition to the *Titanic*, the robotic submersible, the *Jason Jr.* (*J.J.*). Other very important elements of oceanographic technology used included customized sonar, photography, and imagery techniques. Because the *Titanic* expedition has been described in almost limitless detail in both the scientific and lay literature, the fascinating story is not repeated here.

Submersibles

In commenting on advances in ocean technology over the past few decades, scientists have cited several developments which have made major contributions: (1) the deep-diving submarine (submersibles); (2) the ability to drill in deep water; (3) the ability to navigate precisely—improving knowledge of mid-ocean position from a range of 1 to 5 miles (1.6 to 8 kilometers) down to a position fix within 0.1 to 0.01 mile (0.16 to 0.016 kilometer) if within 500 miles (805 kilometers) of land and to 10 feet (3 meters) if within 10 miles (16 kilometers) of land; (4) the availability of television and side-looking sonar for inspection of the sea floor; and (5) among the most important of advances, the vastly improved communications links between remotely located instruments and receiving stations and the efficacy of communication between submersibles and the mother research vessel.

Improved television tubes can amplify light by a factor of 30,000, thus eliminating the need for artificial lighting and the consequent problems of backscatter. Also included with the foregoing advances would be greatly improved materials of construction, such as better steels, high-strength aluminum alloys designed for marine uses, titanium, glass, fiberglass, and plastics.

As of 1994, it is estimated that well over one hundred manned and unmanned submersible craft are in use as oceanographic exploratory tools. These craft are used not only for fundamental oceanographic research, but also by petroleum and gas producers in exploratory operations, notably operating off the continental shelves of the United States and in the

[1] As of 1994, the status of oceanographic research in Russia and other former Soviet bloc countries is uncertain.

North Sea. The use of submersibles in connection with military operations remains obscure.

In early submersibles, personnel and instrumentation were housed in a metallic sphere about 2 meters (6.6 feet) in diameter that incorporated from one to several viewports. The inside of the sphere was maintained at atmospheric conditions and thus no special suits or decompression were required. High-pressure technology for constructing submersibles dates back at least to the 1960s. In 1960, the *Trieste* went to a depth of more than 10 kilometers (6.2 miles). Over the years, the earlier technology has been refined. In an average operating dive, a vessel may remain submerged for 6 to 10 hours and may be on the bottom for 3 to 7 hours in waters some 3 kilometers (nearly 2 miles) deep. Rate of ascent is about 2 kilometers (1.25 miles) per hour. Movement over the bottom, depending upon local circumstances, will range between 1 and 2 knots. A submersible may range over a number of kilometers if it moves steadily.

For certain kinds of surveys, in deep-sea exploration as in exploring outer space, there is no substitute for human observers. However, there are many measurements where submersible instrument packages perform well and frequently at lower cost. A number of advanced surface instruments have been adapted to submersible usage. Examples include the narrow-beam multichannel echo sounding system (SONARRAY) developed by the U.S. Navy and the surface side-scan system (GLORIA) developed in the United Kingdom. Some instruments for deep-sea research can be dropped and can be commanded remotely to become operational. Among these are seismometers and current meters. Submersible instrument packages also may be deep-towed. The Scripps Deep-Tow, for example, is equipped with multiple sensors and accurate navigation. In contrast with manned submersibles, which sometimes yield only photographic or subjective information, submersible instruments commonly will yield graphic and quantitative data. Small transponders may be attached to nearly any instrument package to provide information on its position.

In exploration of the Galápagos Rift in 1979, scientists of the Woods Hole Oceanographic Institution employed a system known as *Angus* for the first time. An acronym for acoustically navigated underwater system, *Angus* is an unmanned 2-ton sled that is towed across the bottom terrain. It can take some 3000 colored pictures of the bottom in a 15–16-hour-period. Previously it had been towed at about 3.6 meters (11.8 feet) from the bottom, giving pictures of a relatively small area. With an advanced lighting system, *Angus* can provide pictures covering one-half acre (0.2 hectare or 1765 square meters) in one frame.

DSV Alvin. Probably the world's most active manned submersible is the *Alvin*, operated by the Woods Hole Oceanographic Institution, which has made well over 1700 dives of exploration to depths up to 13,120 feet (~4000 meters). The craft undergoes regular updating to incorporate the latest technological developments. The diagram given in Fig. 2 indicates the general layout of this submersible. The term *propeller* and *thruster* with reference to a submersible essentially are synonymous. The *Alvin* was a key vehicle in the exploration of the *Titanic* remains.

Funds for the construction of *Alvin* were provided by the Office of Naval Research. The Bureau of Ships of the U.S. Navy assisted in the preparation of performance specifications for its design and construction, and the Applied Sciences Division of Litton Industries designed and built the original vehicle. *Alvin's* original 2.1-meter (7-foot) diameter pressure sphere of high-strength steel 3.39 centimeters (1.33 inches) in thickness for operational depth to 1829 meters (6000 feet) was replaced in 1973 with a 4.90-centimeter (1.93-inch) thick titanium alloy hull provided by the Naval Ship Systems Command. The titanium hull doubled the vehicle's capability without increasing the weight. The pressure sphere accommodates a pilot and two scientific observers as well as instrumentation and life support equipment for endurance up to 72 hours.

In 1978, *Alvin's* 7-meter (23-foot) aluminum frame was replaced with a 7.6-meter (25-foot) titanium frame and an optional second arm installed. The new frame allows for increased instrumentation and can accommodate a fourth battery pack for additional endurance. The submersible was designed to require minimum assistance from large vessels. A front view is given in Fig. 3.

Navigation equipment and other instruments for *Alvin* include gyro compass and gyro repeater, magnetic compass; nose-mounted horizontal-scanning sonar system; indicators for depth, speed, list, trim and variable ballast; echo sounder, battery voltmeters, ammeters and ground detector; and five viewports. Electric power is provided by banks of lead-acid batteries, 60- and 30-volt dc systems, 40.5 kWh total. Communications equipment includes closed circuit television, sonar telephone (voice or code), and marine band (VHF) radio telephone.

In the event of accident or malfunction, occupants of *Alvin* can be returned safely to the surface. Each of the batteries can be dropped to reduce the vehicle's weight, and the two mechanical arms can be dropped should they become hopelessly entangled. As a last resort, the pressure sphere itself, carrying the personnel, can be mechanically disconnected from the rest of the vehicle. The sphere is buoyant and will float to the surface. A closed circuit rebreather with a 6-hour capacity is provided for each occupant in case a fire in the sphere produces noxious fumes. Chemical fire extinguishers are carried.

Comparison of *Alvin* with submersibles capable of operating at exceptional depths can be noted from Fig. 4.

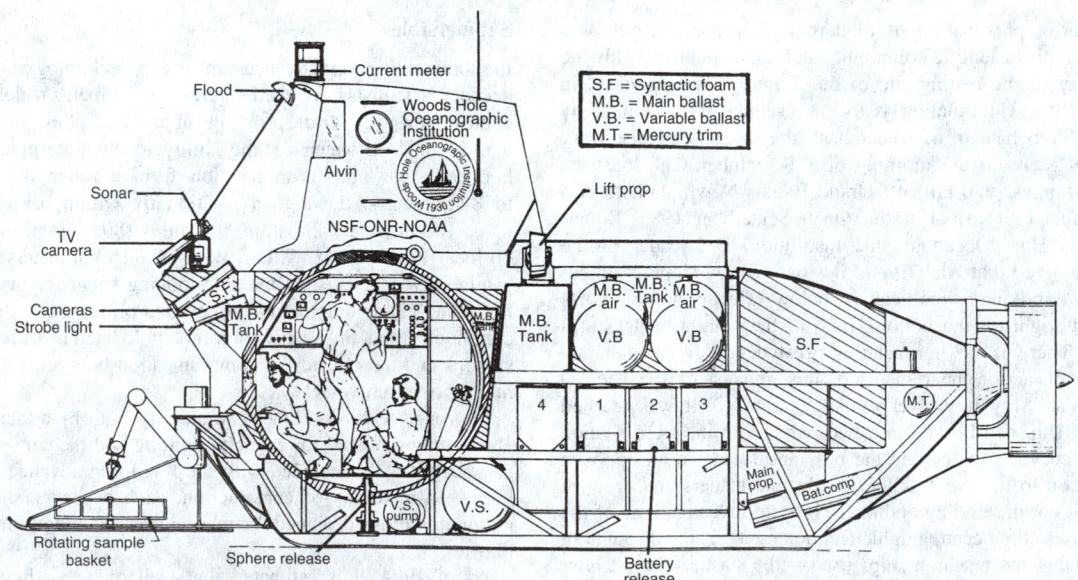

Fig. 2.   Inboard profile of the *R.V. Alvin*, one of the most active of manned submersibles. Design details are periodically updated. The vehicle is designed to carry a pilot and two scientific observers and, in recent years (for and since the *Titanic* Expedition), is linked to a robotic submersible by a 61-meter (200-foot) long tether. The submersible's images are transmitted to *Alvin*, thus enabling *Alvin's* occupants to maneuver it precisely in confined spaces. (*Woods Hole Oceanographic Institution.*)

Fig. 3. Front view of *Alvin*, showing manipulating arms, cameras, lights, and numerous other accessories. (*Woods Hole Oceanographic Institution.*)

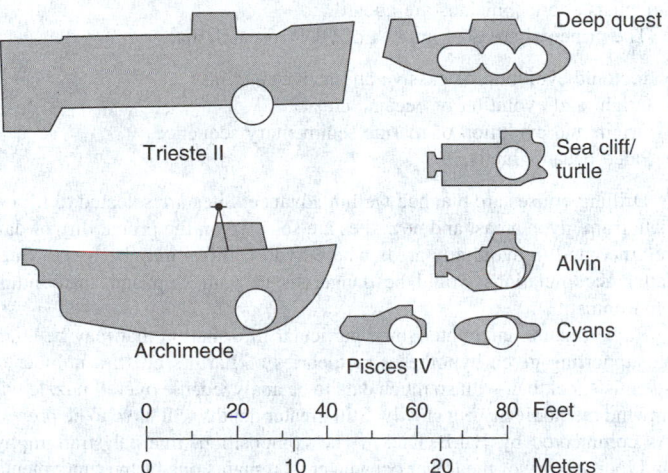

Fig. 4. Profiles of submersibles capable of reaching exceptional ocean depths. (*After Hertzler and Grassie.*)

Projects in which *Alvin* is continually engaged include studies of deep-sea geology, geophysics, and biology, as well as establishment of and periodic visits to deep ocean stations. For several years the vessel has played a primary role in investigations of plate tectonic spreading centers. Seventeen dives in 1974 to the rift valley of the Mid-Atlantic Ridge gave scientists their first close-up view of the tectonic landscape now known to be common to spreading centers—as part of project FAMOUS (French-American Mid-Ocean Undersea Study). In 1976, *Alvin* continued these studies with dives to 3,658 meters (12,000 feet) in the Cayman Trough. In

1977, *Alvin* made the first of a continuous series of dives to warm water vents on the Galápagos Rift, where diving scientists discovered clusters of unusual animal life.

Robotic Underseas Vehicles

Although frequently called underwater *robots*, contemporary unmanned submersibles are not true robots, but are designated more accurately as *teleoperators*—that is, they are remotely operated vehicles under the control of human operators. Teleoperators are becoming more self-sufficient and automatic, capable of making local decisions, and are approaching the status of true robots. However, some form of master programming will be required by even the most advanced designs. At some early date in the future, it is envisioned that preprogrammed robots will be able to survey large stretches of the seafloor for long periods without assistance from surface mother vessels. In such cases, the power supply will have to be self-contained within the submersible, and information collected by the craft will be transmitted or stored until the robot is retrieved. Such devices will be designed to take greater risks in some cases without endangering the well-being of any human operators.

In 1988, a Finnish firm constructed two identical submersibles capable of accommodating three people for the Russian oceanographic research program. The two craft, named *Mir 1* and *Mir 2*, were carried by a mother research vessel, *Akademik Mstislav Keldysh*, a ship that could support 130 people and incorporate 18 laboratories. The blimp-shaped submersibles were designed to dive to 6000 meters (19,686 feet) and thus can explore the seafloor at greater depths than the *Alvin* previously described. In their first few years of use, the *Mirs* explored parts of the seafloor of the Atlantic and Pacific oceans, and plans call for their exploration of the essentially unknown seafloor of the Indian ocean. The *Mir* submersibles also completed a survey of the *Titanic* resting site in 1991.

Teleoperated craft for underwater studies and in use for well over a decade stemmed from the development of teleoperators for the handling of radioactive and other dangerous materials.

Jason Jr. This is a small robotic (unmanned) type of tiny submersible that is designed to operate in conjunction with a manned submersible, such as *Alvin*. *J.J.* is a mere 28 inches (71 cm) in length and, on land, weighs 250 pounds (113 kg). The hull of the small craft is made of syntactic foam (incorporating billions of microscopic air-filled glass spheres bonded by epoxy). The result is that it is essentially weightless in water. *J.J.* is controlled by a console held by a pilot onboard the *Alvin*. Buttons in the handgrip activate the robot's vertical thrusters and control the tether to *Alvin*. A photo trigger activates the still camera aboard *J.J.* and a companion button adjusts the tilt of the video camera on *J.J.* A video monitor in *Alvin* displays *J.J.'s* field of vision. Switches in *Alvin* remotely control the motors and lights on *J.J.* and a joystick controls horizontal movements of the robot. A larger version of *J.J.* is now being designed. It will be considerably more sophisticated than *J.J.* and will include robot arms for retrieving samples. The versatility and maneuverability (in and out of very tight spaces) were proved during the meticulous examination of parts of the *Titanic* wreckage.

As observed by D. Yoerger (Woods Hole Oceanographic Institution), "One view of teleoperated submersibles is that they are replacements for manned submersibles, with advantages in terms of endurance, safety, and cost. While these are strong points for remotely operated vehicles (ROVs), *Jason* was not intended strictly to replace manned submersibles. It was designed to perform a variety of functions, many of which complement rather than duplicate the strong points of a manned vehicle. For example, manned submersibles are rarely used for sonar survey, due in part to problems of storing or processing data in small spaces with limited manpower. *Jason's* high-bandwidth telemetry system and precise control capabilities make it a very effective sonar platform."

ROVs are capable of a variety of interesting tasks. For example, within the last few years, *Jason* has participated in several projects of an archeological nature. One of these ventures was the exploration of two War of 1812 vessels, the *Hamilton* and *Scourage*, which lay on the bottom of Lake Ontario at a depth of 90 meters (255 feet). The craft was able to move over the decks of these historic vessels for electronic still camera surveys. In an expedition conducted in 1989, *Jason's* manipulator excavated the remains of an ancient Roman shipwreck at a depth of

760 meters (2500 feet) in the Mediterranean Sea. The computer-controlled compliance of *Jason's* arm enabled recovery of delicate ceramic jars, dishes, and oil lamps without damage. An earlier ROV (*Jason Jr. (J.J.)*) received worldwide acclaim when it was deployed in the second expedition to the *Titanic* in 1986.

ROV *Jason* carries a wide variety of sensors and manipulators for operation at depths to about 7000 meters (20,000 feet). ROV *Jason* can maintain heading and depth, follow precise track lines, and move automatically under the pilot's supervision. The manipulator arm automatically reacts to contact force to grasp objects firmly yet without damage. Although the human pilot is always in charge, the control task is apportioned between the pilot and the automatic systems in the craft.

Although ROVs have participated in a number of newsworthy events, they are designed primarily for gathering scientific knowledge. Currently there is much interest in *untethered* vehicles that can move without the restriction of cable. As of the early 1990s, a number of such craft are in the design or construction and testing phase.

Operational as of 1991, the French oceanographic agency developed the *Epulard*. The craft operates in a telerobotic mode, with its activity supervised by shipboard operators through an acoustic communications link. A similar craft, *EAVE*, is under development at the University of New Hampshire. The craft will have completely autonomous capabilities for search and survey tasks.

The Woods Hole Oceanographic Institute is building a robotic vehicle called the *ABE* (Autonomous Benthic Explorer). The *ABE* is designed to perform surveys in deep-sea hydrothermal vent areas independently of a surface vessel. These vents, located at several thousand meters, are extremely dynamic. As pointed out by D. Yoerger, "Flows of hot, chemical-laden water and volcanic activity occur unpredictably. Traditional submersible or ROV expeditions cannot remain on station long enough to observe many of these changes. *ABE* will remain on the bottom for extended periods of time, mostly in a low-power "sleep" mode, eventually as long as one year. Periodically *ABE* will repeat video surveys and measure oceanographic parameters such as water temperature, salinity, and optical properties. *ABE* will complement other vent-studying techniques. While *ABE's* observations will not be as high in quality as *Jason's* or *Alvin's*, *ABE* is designed to make observations when these traditional vehicles cannot be on station. Likewise, through its mobility *ABE* will complement fixed instrumentation.

Initially the vehicle will execute preplanned tasks. With experience, it is believed that more instrumentation and measurement quality and flexibility can be added in future redesigns. For example, *ABE* may be programmed to react to data from other seafloor sensors, such as an ocean bottom seismometer, and could report on seismic activity by way of an acoustic link.

In the United Kingdom, autonomous vehicles are being planned. One will be used for geological surveys, and the other to survey the water column. Japan has plans underway to construct a diesel-powered, full-ocean-depth autonomous submersible to survey the Mid-Ocean Ridge.

Deep-Ocean Drilling

As may be gleaned from the prior article on Ocean, knowledge of ocean processes and phenomena has increased many times over the past few decades and appears to be proceeding at an accelerating rate into the mid-1990s. The combined achievements of the numerous subsciences that comprise ocean knowledge are coalescing to form a unification of past principles. There is a major unknown, however—that being how little scientists know about the composition and structure of the two-thirds of Earth's crust that underlies the oceans. As observed by H.J.B. Dick (Woods Hole Oceanographic Institution), "While it has often been said that the surface of Mars is better known than the seafloor of Earth, the situation is far worse for the oceanic crust." This is explained, of course, by the severe difficulties in retrieving evidence. Conventional geological techniques simply have not worked to solve this problem. Among these techniques, one would include the use of surface mapping with geophysical sensing at depth to interpolate Earth's deep structure. Researcher Dick further points out, "It is increasingly apparent that the only way to obtain direct and precise knowledge of the composition and structure of the oceanic crust is to drill into it."

Deep-sea drilling dates back to the 1960s and the Mohole project. The project, almost constantly plagued by budget deficiencies, produced a few successes, but insufficient data to prove old theories or propose new concepts. The project, however, can be credited for sharpening the awareness, on an international scale, of the need for information and for the establishment of the Deep Sea Drilling Project, later followed by the Ocean Drilling Program (ODP). This program represents an international partnership of scientists and governments to explore Earth's origin and evolution beneath the seafloor.

The drill ship *JOIDES Resolution* operates a continuous series of cruises for the purpose of drilling and retrieving long cylindrical samples of sediment and rock samples or *cores*. Additional information is obtained from measurements made in the drilled holes. ODP is funded by the U.S. National Science Foundation and contributions from 18 other countries. Ten leading oceanographic institutions provide planning and management. These include Texas A&M University, the Lamont-Doherty Geological Observatory, and the Scripps Institution of Oceanography (University of California, San Diego). During the time frame (1985 to 1992), 77,500 meters (254,275 feet) of cores from 683 holes at 279 sites have been recovered. This represents nearly 50 cruises to sites in the Atlantic, the Pacific, the northern and southern polar seas, and the Indian Ocean. These cores are archived at three marine institutions in the United States, from which thousands of core samples have been passed to researchers in more than 38 countries throughout the world.

As described by V. Cullen (Woods Hole Oceanographic Institution), "Many lengths of 9.5-meter (31-foot) drill pipe are attached together to lower a large drill bit to the seafloor. This takes about 12 hours in 5,500 meters (18,045 feet) of water. Core barrels are then lowered through the drill pipe to receive and contain the core material. When a length of about 9.5 meters (31 feet) has been drilled, the core barrel is raised to the ship, where technicians recover the long cylinder of sediment or rock, cut it into 1.4 meter (5-foot) sections, and begin documenting and describing its origin, appearance, and contents."

The *JOIDES Resolution* is equipped with seven different laboratories, representing expertise in geochemistry, geophysics, paleontology, petrology, paleomagnetics, and sedimentology. After photographing and preparing meticulously written descriptions, the cores are packed and stored under refrigeration. The ship is operated by a crew of 68. Approximately 50 scientists and technicians are aboard.

The principal areas of interest of ODP research are:

- Tectonic evolution of passive and active margins.
- Origin and evolution of oceanic crust.
- Origin and evolution of marine sedimentary sequences.
- Paleo-oceanography.

Drilling cruises are planned well in advance. Sites are selected to fill out data from given areas, and new sites are selected on the probability of data returned and to avoid any areas where hydrocarbons may be found. Such latter accumulations would be dangerous to both ship and surrounding environment.

Rather than to encounter some particular information that may be a key to supporting given hypotheses or theories of Earth's crust, a number of scientists feel that, with so much data to be analyzed, the overall puzzle will unwind methodically but slowly. Still greater depths will have to be probed. As commented by H.J.B. Dick, "The composition, internal stratigraphy, and rock history of the lower oceanic crust remains one of the fundamental unanswered questions of Earth science."

For many years, the so-called "layer-cake" or "infinite onion model" was accepted as describing the processes that occur between the mantle and the surface of the oceans. The model is shown in Fig. 5 in a highly schematic fashion.

After drilling in 1987, a tectonically exposed section of the lower oceanic crust on the southwestern Indian Ridge, some of the plans for future drilling operations were changed. It was found that drilling lower-oceanic crust was much easier and accompanied by nearly total rock recovery, and required no new technology. It was also found that the compositional diversity of the rocks closely resembled those found in fossil magma chambers exposed on land, but that the processes for rock formation differed markedly from the dynamic and physical processes that appear on land. In 1989, leaders

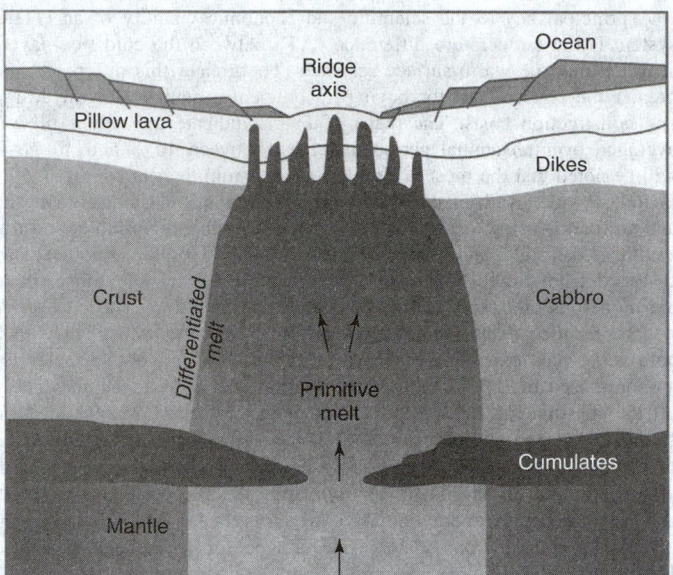

Fig. 5. Highly schematic sketch of "layer cake" or "infinite onion" model. This model has been accepted for several years, but now is in question. The model as shown portrays a large steady-state magma chamber that underlies the ridge system. The chamber is disrupted only by very large fracture zones. Here oceanic crust grows continuously to eventually make up the lower two-thirds of oceanic crust. (*After Oberlander.*)

of the ODP proposed "that while drilling a single deep hole through the oceanic crust should be a long-term goal, a major program of drilling directly into the lower oceanic crust and mantle should be done over the next decade." Offset drilling, as used by sedimentologists, is the strategy proposed.

See also **Earth Tectonics and Earthquakes**.

Additional Reading

Ballard, R.D.: "A Long Last Look at TITANIC," *National Geographic*, 698–727 (December 1986).

Cullen, V.: "Ocean Drilling Program," *Oceanus*, 31 (Winter 1992/1993).

Detrick, R.S. and J.C. Mutter: "New Seismic Images of the Ocean Crust," *Oceanus*, 54 (Winter 1992/1993).

Dick, H.J.B.: "A New Mandate for Deep-Ocean Drilling," *Oceanus*, 26 (Winter 1992/1993).

Dinsmore, R.: "The University Fleet," *Oceanus*, 5–14 (Spring 1982).

Foster, D.: "DSV Alvin: Some Dangers and Many Delights," *Oceanus*, 17 (Winter 1988/1989).

Given, D.: "Underwater Technology in the USSR (Russia)," *Oceanus*, 67 (Spring 1991).

Golden, G.: "A Quarter-Century Under the Sea," *Oceanus*, 2 (Winter 1988/1989).

Hanson, L.C. and S.A. Earle: "Submersibles for Scientists," *Oceanus*, 31 (Fall 1987).

Herbst, D.: "In the Wake of a Modern Jason," *Oceanus*, 84 (Summer 1989).

Honjo, S.: "From the Gobi to the Bottom of the North Pacific," *Oceanus*, 45 (Winter 1992/1993).

Hotta, H.: "Deep Sea Research Around the Japanese Islands," *Oceanus*, 32 (Spring 1987).

Kaharl, V.A.: *Water Baby: The Story of Alvin*, Oxford University Press, Inc., New York, NY, 1990.

Kirn, T.F.: "Tuna Sub," *Technology Review (MIT)*, 10 (October 1991).

Odani, Y.: "Kalro, A Unique Research Vessel," *Oceanus*, 35 (Spring 1987).

Ryan, W.B.F.: "An Introduction — Down to the Sea in a Ship," *Oceanus*, 10 (Winter 1991/1993).

Smith, D.K.: "Illuminating the Seafloor," *Oceanus*, 74 (Winter 1992/1993).

Staff: *Japanese Deepsea Research*, 28 (January 3, 1992).

Tagawa, S.: "Deep Submersible Project," *Oceanus*, 29 (Spring 1987).

Van Dover, C.L.: "Diving the Soviet Mir Submersibles," *Oceanus*, 8 (Summer 1991).

Vine, A.C.: "The Birth of Alvin," *Oceanus*, 10 (Winter 1988/1989).

Williams, A.J., 3rd: "Ocean Engineering," *Oceanus*, 4 (Spring 1991).

Yoerger, D.R.: "Robotic Undersea Technology," *Oceanus*, 32 (January 1881).

Web References

Scripps Institution of Oceanography: http://www.sio.ucsd.edu/
The National Oceanic & Atmospheric Administration (NOAA): http://www.noaa.gov/
Woods Hole Marine Operations: http://www.marine.whoi.edu/ships/ships_vehicles.htm

OCEAN RESOURCES (Energy).   In one way or another, it is highly likely that over the next several decades large amounts of energy will be derived from the oceans and seas of the world. This observation refers to means in addition to the production of oil and natural gas from the sedimentary deposits beneath the oceans.

Admittedly, during the past decade, both scientific and economic interests have dwindled considerably for two reasons: (1) much greater optimism regarding natural gas supplies, particularly under the continents, hence slowing undersea exploration; and (2) a glut in the petroleum supplies of the world which has persisted for approximately a decade. However, all authorities do not agree that the pursuit of oceanic energy should be permitted to atrophy. There are some schools of thought that forecast a negative turnaround in our energy supply fairly early in the 21st century, at which time aggressive interest in oceanic energy will reemerge.

Numerous sources of energy from the oceans have been proposed and investigated over a period of many decades. These include: (1) *Ocean thermal energy conversion* (OTEC), wherein advantage is taken of thermal differences in layers of seawater, a technology in which sizeable investments have been made during the past decade or so and a concept that dates back to the 1880s. (2) *Ocean wave energy conversion* — in actuality, a manifestation of wind power, because waves on the ocean surface are generated by atmospheric winds acting over large areas and long periods of time. This concept also dates back a number of years, but has received only marginal scientific and engineering attention as compared with OTEC technology. Much remains to be learned pertaining to the basic science of ocean wave action; much of the attention to date has been focused on the interaction of waves with coastlines, where coastline conservation rather than waves as a source of energy has been the major concern. (3) *Ocean current energy conversion*, also a concept dating back many years. It was in 1835 when Gaspard Gustave de Coriolis published a paper in which he analyzed the distortions of fluid motions resulting from Earth's rotation — the Coriolis effect. See also **Coriolis Effect**. This effect combines with thermal and climatic forces as components of a mighty global heat engine to create current streams and gyres, as associated with the Gulf Stream and other major ocean currents. The wind plays an important, but not exclusive role in the creation of gyres.* Much has been learned concerning ocean currents during the past decade (as, for example, from the IDOE programs described in entry on **Ocean**). From a practical application standpoint, energy from ocean currents is in a very early stage of investigation. (4) *Tidal energy* has been exploited on comparatively small scale for several centuries, dating back to the 11th Century when tidal mills for grinding grain were used in Gaul, Andalusia, and England. It has furnished regionally significant electrical energy for northern France since 1966, when the world's largest tidal generating station installed in the maritime estuary of the Rance River (Saint Malo-Dinard, Brittany) was dedicated. Tidal energy is unlikely to be a truly major factor in the world's future energy supplies simply because, without the expenditure of tremendous funds, there are relatively few economically viable locations to exploit. However, as explained in a separate entry on **Tidal Energy**, the science and technology required for energy from the source have been well established over a number of years. (5) *Energy from saline gradients* is a relatively new concept, even though the forces of osmosis have been understood for many years. The phenomenon of osmosis was first observed by Abbé Nollet in 1748. An osmotic pressure difference equal to a theoretical waterfall of 240 meters (787 feet) exists at the mouth of rivers where fresh water enters the sea. Assuming that energy extraction

---

* A gyre is a great, closed, circular motion of water in each of the major ocean basins, centered on a subtropical high-pressure region; its movement is generated by convective flow of warm surface water poleward, by the deflective effect of the Earth's rotation, and by the effects of prevailing winds. The water within each gyre turns clockwise in the Northern Hemisphere; counterclockwise in the Southern Hemisphere. See also **Coriolis Effect**.

technology can be developed, a large potential would exist at the Amazon River (Brazil), the La Plata-Parana River (Argentina), the Congo, Yangtze, Ganges, and Mississippi rivers, as well as the Great Salt Lake (United States) and the Dead Sea (Israel/Jordan).

Further ocean-related energy sources include (6) *submarine geothermal springs* and *geothermal aquifers*; (7) *oceanic biomass* (it is estimated that the oceans of the world fix 10^{10} tons or more of carbon per year into organic material, mainly through photosynthesis by microscopic plants, the phytoplankton — with the possible future conversion of biomass to liquid fuels); (8) *resources for nuclear power processes*, including uranium, thorium, lithium, and deuterium (9) *direct conversion of ocean winds* by surface-located wind machines and wind farms (see also **Wind Power**); and (10) *ice from icebergs*, not only as a source of fresh water, but also used as a heat sink for nuclear, fossil, and even OTEC-type power generation. When two phases of a substance are present in a nearly isothermal environment, rather high temperature gradients and potential power can be achieved with small energy inputs, by a change in phase of one form.

Ocean Energy Thermal Conversion (OTEC)

It was d'Arsonval who in 1881 drew the attention of the scientific community to the enormous amounts of energy that may be available from thermal differences in layers of seawater. Not until 1930, however, did the French physicist Georges Claude operate the first crude ocean thermal difference power plant at the edge of Mantanzas Bay, Cuba. Besieged with fabrication, logistics, and weather problems, Claude finally succeeded in laying a 1.6-meter (diameter) by 1.75 kilometer (length) cold water pipe from an on-shore plant out into the bay, reaching a depth of about 700 meters. From a seawater temperature difference of $14\,°C$, Claude's turbine generated 22 kilowatts (electric) of power. The plant operated on an open Rankine cycle system, i.e., steam generated from the seawater was used directly as the working fluid for the turbine. Whereas a conventional steam-turbine system operates with high pressures and large temperature differences, the Claude installation required a very low pressure (vacuum) to evaporate. The steam had pressures of approximately 0.02 atmosphere in the condensers, where it was condensed by mixing with cold water falling as rain. At these pressures, the specific volume (cubic meters/kilogram or cubic feet/ pound) of the steam was very high — so that the diameter of the turbine required for a given power level was some 30 to 50 times larger than that required for a conventional plant. Thus, with only a 1-meter (diameter) turbine available, Claude had deliberately oversized the cold water pipe by a factor of ten in cross-sectional area to avoid too much warming of the cold water as it was drawn up to the condenser. As a result, Claude put more power into the vacuum pump than he obtained from the turbine. Nevertheless, the principle of conversion of low-quality heat to power had been demonstrated. A diagram of Claude's design is given in Fig. 1.

A principal key to the scientific and economic viability of an OTEC system is the temperature difference (ΔT) between the cold (low-layer) seawater and the warm surface seawater. The greater this difference, the greater the chances are for having an efficient system at lowest design and construction costs. The maps of Fig. 2 indicate the values of ΔT averaged over an annual period that exist between $40\,°S$ and $40\,°N$. It will be noted that the most favorable regions (from the standpoint of ΔT) include a wide swath of the Pacific Ocean eastward of China, Indonesia, and northern Australia. There is also a large favorable region in the central Pacific Ocean south and west of Mexico. Another favorable region extends eastward from South America to west central Africa. Of course, there are numerous other factors that enter into consideration of a location. It will be recalled that Claude operated from a shore-located plant with a cold water pipe extending out into the ocean. An unfavorable factor for a shore-based facility is the relative shallowness of waters immediately off the coastlines in most places. Another factor that determines location selection, among other considerations, is the method selected for utilization of the power generated. There are two principal options in this regard: (1) transmission of the power by submarine electric cable to the shore; and (2) manufacture energy-intensive products aboard the OTEC platform at sea. Most likely candidate products would be hydrogen (see also **Hydrogen (Fuel)**), ammonia, or chlorine. More complex operations, such as electrowinning of magnesium and aluminum and fertilizer production, have been suggested. Precise station keeping would be mandatory for the electricity transmission scheme, whereas the product option would permit periodic location changes. However, even with the product option, mid-ocean locations, which would involve long distances, are not attractive from the standpoint of transportation economics, not only of the products, but of crews and maintenance material.

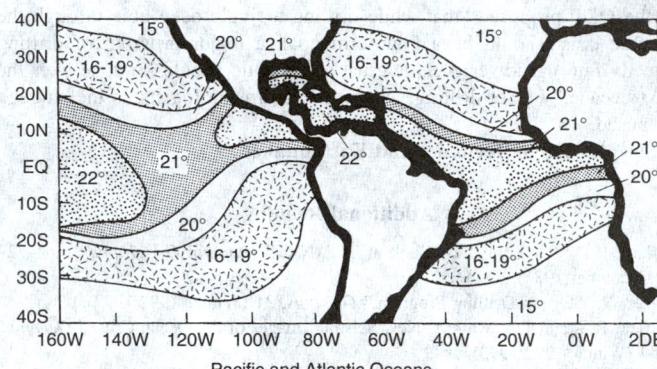

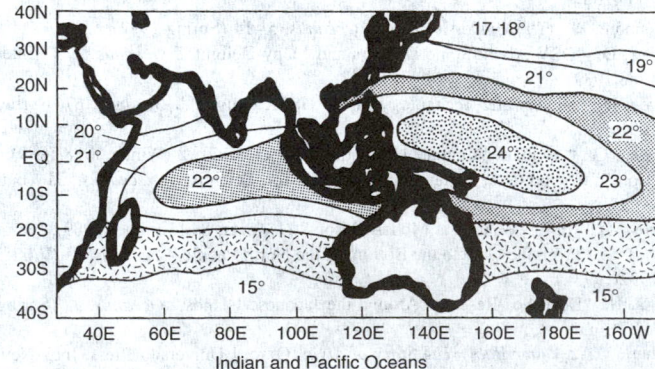

Fig. 2. Average annual temperature difference (ΔT) between ocean surface water temperature and temperature of water at a depth of 1000 meters (3281 feet). ΔT is in degrees Celsius. Solid black areas are waters of less than 1000 meters in depth. (Suggested by R. Cohen, 1980.)

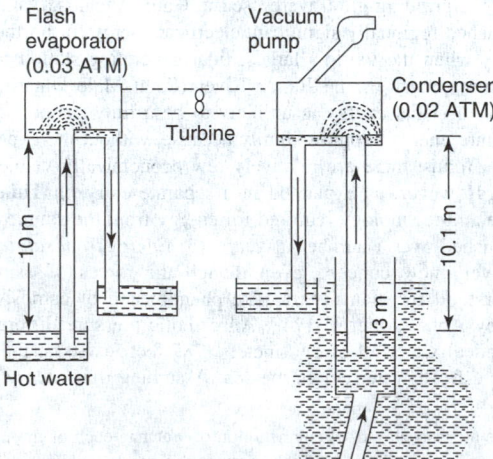

Fig. 1. Open Rankine cycle system used by Claude in 1930 experiment. Flash-evaporated seawater drove a turbine and then was recondensed by cold water falling in the condenser. Details are available in "Power from Tropical Seas," by G. Claude, *Mech. Eng.*, 52, 1039–1044 (December 1930).

Open-Cycle System. A modern version of Claude's open-cycle system is shown in Fig. 3. The state of open-cycle system technology is somewhat behind that of the closed system approach, at least in the United States. Considerably larger and more complex equipment is required for the open-cycle system. Very large turbines, comparable to wind turbines, and

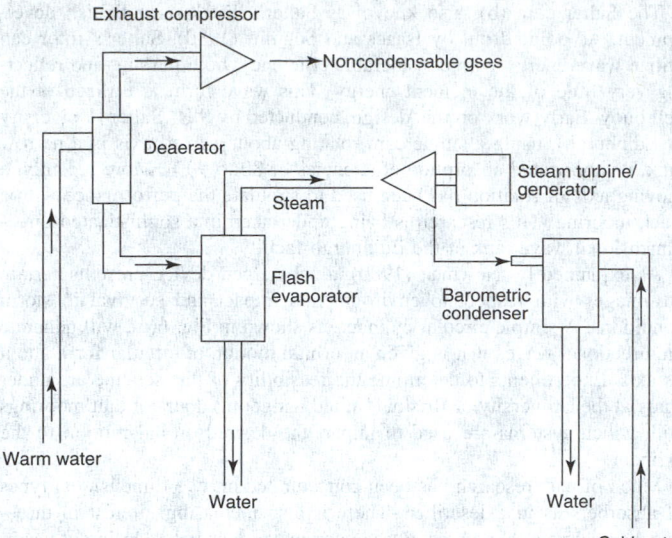

Fig. 3. Open-cycle OTEC system. In an open system, the ocean surface temperature (warm water) is adequate to flash evaporate seawater. The escaping steam causes a turbine wheel (connected to an electric generator) to turn. Spent vapor is then cooled in a condenser by cold water pumped from depth. Unlike the closed-cycle system, the condensate need not be returned to the evaporator.

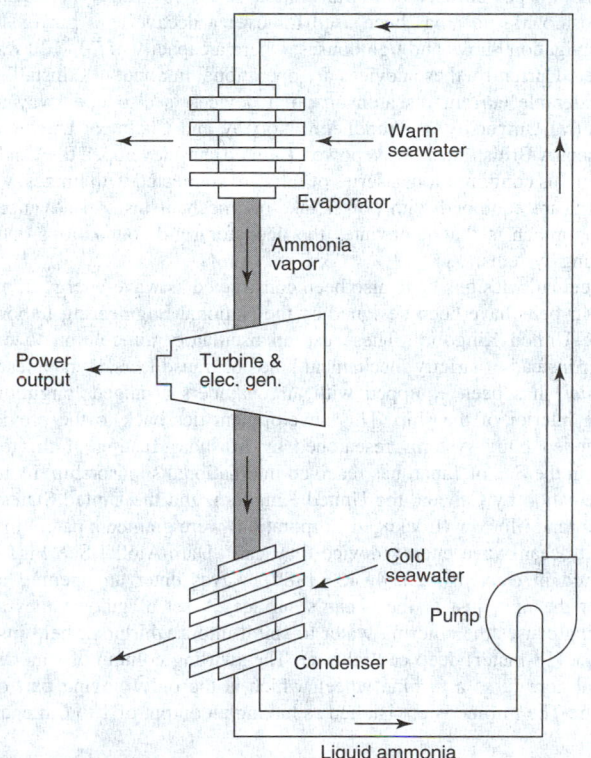

Fig. 4. A closed-cycle OTEC system. Cold fluid, such as ammonia, is heated and vaporized in an evaporator. Vapor at high pressure moves blades of a turbine which drives a generator, as it is expanding toward the cooling system (condenser) in which the vapor is condensed back into liquid. The ammonia (fluid chosen for system shown) is vaporized by the narrow temperature differential between the warm and cold seawater.

de-aerators for removing dissolved gases from seawater are required. More recent open-cycle system studies have been pointing toward cost-effective solutions of the turbine and degasification problems.

Closed-Cycle System. As shown by Fig. 4, the closed-cycle system requires a refrigeration-type working fluid, such as ammonia or propane,

which is evaporated and recondensed continuously in a closed loop to drive a turbine. Because of the small ΔT, currently envisioned OTEC plants must by their very nature be of very low efficiency — in the neighborhood of 2.5%. Under normal circumstances (requiring some costs for fuel), such a plant would be considered hopelessly inefficient for serious consideration. But since this inefficiency is reflected only against initial costs of heat exchangers and working fluid, the approach holds promise in an energy-conscious world. Most investigators to date have found ammonia to be the best working fluid from the point of view of heat transfer — about twice as effective as propane, the next best candidate. The technology for handling ammonia has been established for many years. In these plants, usable power can be generated when there is a ΔT of as little as 14.7 Celsius degrees (27 Fahrenheit degrees), but locations providing a ΔT of a minimum of 20 Celsius degrees (36 Fahrenheit degrees) are the most desirable. So-called grazing OTEC platforms, readily movable from one location to another, could feasibly take advantage of seasonal changes in ΔT.

The state of OTEC technology is probably pretty well summarized by Isaacs and Schmitt (1980): "Problems with ocean thermal energy conversion include the design and stability of the intake pipe; biofouling of pipes, heat-exchange surfaces, and structure; construction, control, and maintenance of heat exchangers of unprecedented dimensions and required efficiency; corrosion; power transmission; and environmental effects. The best approach to these problems will probably involve testing modules of the final assembly both ashore and afloat where conditions are appropriate."

One estimate of the thermal-gradient flux available on an annual basis over the global surface is 1028 ergs. See the figure showing the spectrum of various energy quantities in the entry on **Energy**.

Ocean Wave Energy Conversion

The surfaces of the oceans and seas are rarely smooth. The forces of wind power exerted over vast expanses of the sea tend to be integrated in the form of waves that reflect wind energy from long distances. It has been estimated that wave-energy fluxes in the open sea or against the coasts may vary from a few watts to a megawatt per meter. The total wave power incident upon the coastlines of the world has been estimated by various researchers to be $2-3 \times 10^{12}$ watts. Wave energy impinging upon the coastlines varies considerably from one location to the next, as well as with seasons.

Wave energy is distributed in a thin layer of the ocean (less than 100 meters; 328 feet in depth). Wave-energy fluxes are greatest in summer and in the zones of the westerlies and trade winds. As pointed out by Newman (1980), the energy per unit of horizontal area is proportional to the wave period and to the square of the wave height. This energy is carried along at a reduced speed, known as the group velocity, which in deep water is half the velocity of the individual wave crests. The product of the energy per unit area and the group velocity is the rate of energy flux per unit width of wave front.

One of the earliest approaches to retrieving energy from wave action is the most obvious, i.e., the conversion of vertical displacement into a useful form of energy. A simplistic device is illustrated in Fig. 5. Vertical displacement has been the basis of the use of wave energy in whistling and lighted navigational buoys for many years. It has been reported that a moderate-scale power plant is being constructed in the Orient on the basis of a sophisticated version of this principle.

Wave motors brought out from time to time have depended for their operation largely on the lifting power of the waves. One installed at Atlantic City, New Jersey as early as 1911 consisted of six 4-foot (1.2-meter) cylindrical floats 4 feet (1.2 meters) high. These, each weighing about 3100 pounds (1406 kilograms), were lifted 2 feet (0.6 meter) by the waves about 11 times per minute. They drove a horizontal shaft by means of chains and ratchets, developing only 12 horsepower (12.2 metric horsepower). Steadiness was obtained through the use of heavy flywheels.

A wave motor employing a hydraulic ram for raising a portion of the water to a high level was proposed by Smith in 1927 (*Mech. Eng.*, September 1927, page 995). The waves were envisioned as entering a scoop, which would be connected to the ram by a long drive pipe. It was envisioned that the apparatus would be automatically adjusted for vertical level with tide changes.

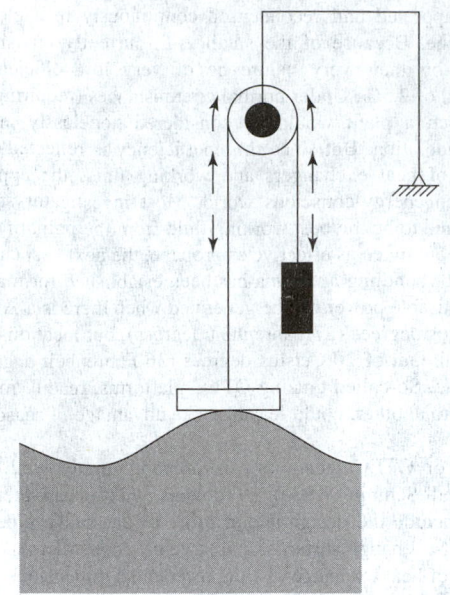

Fig. 5. Schematic diagram of simplistic vertical-type wave-energy pick-up or absorber.

A wave-operated pump was developed at the Fountain for Ocean Research (San Diego, California) in 1976. This design is based on the inertial interaction of two integrated systems—a buoy connected to a long vertical pipe flooded with water. As explained by the researchers (Isaacs, Castel, and Wick, 1976), the water column responds to such low frequencies that it cannot follow the sea surface when uncoupled from the pipe and float. Controlled by a simple check valve, the water in the pipe is forced to accelerate upward by the motion of the buoy, and it continues to flow upward as the buoy and pipe drop. The hydraulic pressure in a reservoir then increases and is limited only by the maximum vertical acceleration and the pipe length. For a 100-meter pipe and a maximum estimated acceleration of 0.5 g, the pressure will increase to about 5 atmospheres. Before this pressure is reached, the water is allowed to escape continuously under pressure and to interact with a turbine or Pelton wheel, and thus power is generated. Tests of this system at sea have indicated a 25% efficiency. Advantages of the system as claimed by the inventors include simplicity, relative invulnerability, response to a broad band of wave frequencies, and multiplication of wave pressure.

In studying energy pick-up or absorbing devices that would interface in some way with energy-carrying waves, some researchers have turned to studying wave-making devices which have been used to simulate waves, required for various studies as, for example, coastline erosion research. Three such devices are shown in Fig. 6. Commenting briefly on these devices, Newman (1980) observes that if the wedge device (a) is driven in an oscillatory manner (in the direction parallel to the back side), the fluid disturbance will be confined essentially to the front. If the apex is sufficiently deep, the resulting waves will be trapped on the front side of the wedge, radiating away from this side in one direction.

The Salter cam (b), also known as Salter's Duck, is a British development. As pointed out by Isaacs and Schmitt (1980), Salter's rotor can utilize wave energy with considerable efficiency, both passing and reflecting very little of the incident energy. This wave form is utilized in the bell buoy. Early work on this design, conducted by S.H. Salter (University of Edinburgh), using a single cam rotating about a fixed axis in a narrow tank, demonstrated absorption efficiencies of 80 to 90%. More recently, a moving axis of rotation has been used to simulate the performance with a slack mooring. This research is being undertaken in a sophisticated three-dimensional wave tank at the Edinburgh facility.

As explained by Newman (1980), a submerged device presents certain advantages with respect to environmental impact and survival in storm conditions. A simple circular cylinder, as shown in Fig. 6(c), will generate unidirectional waves if it is given an orbital motion of circular form about its axis. Experiments to determine the feasibility of this scheme are under study at the University of Bristol (United Kingdom). Pairs of taut moorings with winch systems are used to impart the desired orbital motion to the cylinder.

Much of past research has been concentrated on two-dimensional types of absorbers as just described. There is also increasing interest in three-dimensional point absorbers. Based upon experience in the design of wave-power sources for buoys and lighthouses, it is well established that such devices respond equally to waves from all directions. In certain situations, wave-energy transducers sensitive to multidirectional waves are attractive. Particular attention in recent years has been given to small-scale point absorbers. An analogy can be made between a wave-energy point absorber and a simple radio antenna, the wire diameter of which is not important in terms of the power received or transmitted. A device of this type, developed by Yoshio Masuda, has been used for over a decade as a power source for navigation buoys and lighthouses, where a capacity of 70–120 watts is needed. Currently, these devices are operational in about 400 installations. Considerable current research on point devices is also underway at the Technical University of Trondheim, Norway and Chalmers University in Sweden. A British firm (Wavepower, Limited) has developed the Cockerell raft. In this configuration, a series of rafts are connected with hinges, which, in turn, are equipped with power take-off mechanisms. An advantage of this approach is that it obviates the need for rigid foundations and taut mooring systems.

Pneumatic designs have also been constructed as wave-energy transducers. Air bells have been designed by the National Engineering Laboratory in the United Kingdom. These use an oscillatory air column to extract power instead of strictly mechanical kinds of transducers. A Japanese ship (Kaimei) has been equipped with air chambers arranged longitudinally in the interior of the ship. This development ties back to the previously mentioned buoy systems researched by Masuda. Testing of this equipment in the Sea of Japan has received international sponsorship, including participation by Canada, the United Kingdom, and the United States.

Wirt and Morrow (Lockheed Corporation) were granted a patent in 1979 for an ocean wave energy device they call "Dam-Atoll." See Fig. 7. As reported in Oceanus, 22, 4, 43 (1980), waves enter an opening at the top of the unit, just at the ocean's surface. A set of guide vanes at the opening causes the entering water to spiral into a whirlpool, held inside a 60-foot (18-meter) deep central core. The swirling column of water in the central core turns a turbine wheel, which is the only moving part of the system. The turbine is envisioned as having an output of 1 to 6 megawatts

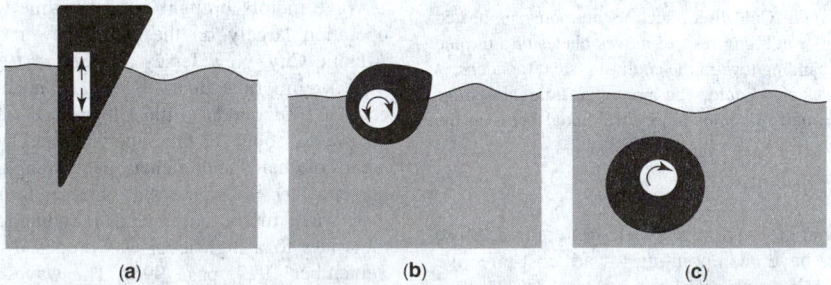

Fig. 6. Types of wavemakers, also considered as wave-energy absorbers: (a) unidirectional wedge-shaped device; (b) the Salter cam; and (c) submerged cylinder for generating unidirectional waves through orbital motion. (After Newman.)

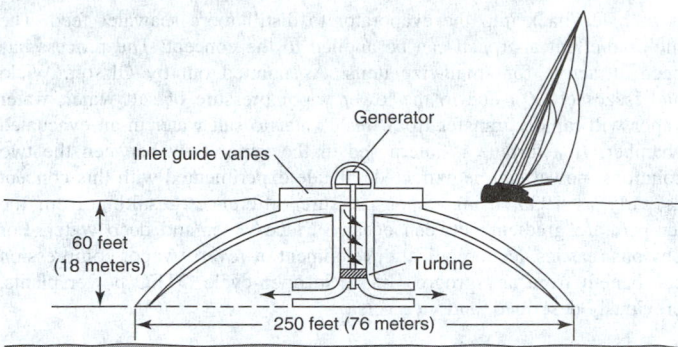

Fig. 7. Ocean wave energy device (*Dam Atoll*) for converting wind-powered wave energy into electricity. Arrows in the central cylinder indicate the whirlpool action of the water in the central cylinder, which acts like a giant flywheel to keep the turbine spinning even though wave action may be intermittent.

(electric). The inventors suggest that such devices could be anchored off the windy beaches of the world, where there is about 40 megawatts of power available per kilometer (64.4 megawatts/mile) of beach. This is estimated as sufficient to furnish the domestic power requirements for about 40,000 people. It has been envisioned that it would be possible to anchor 500 to 1000 such units along the coast of the Pacific Northwest, for example, that would furnish an electric generating capacity comparable to that of the Hoover Dam. In addition to electricity generation, other uses for the device suggested include a means for cleaning up and recovering oil spills, protecting beaches from wave erosion, forming calm harbors in the open sea, and desalinating seawater through the reverse osmosis process.

The design of systems to capture wave power has intrigued scientists for many years, and many other schemes and concepts have been proposed, well beyond the scope of this encyclopedia. In particular, reference is suggested to the proceedings listed (University of Kent; Gothenburg Symposium; and University of Delaware), as well as the Newman (1980), Ross (1979), and Isaacs and Schmitt (1980) reference listed.

Ocean Current Energy Conversion

The ocean currents or gyres are a major storage medium for solar energy. The major ocean current energy flux has been estimated at 1025 ergs (annually) as compared with the ocean-wave energy flux of 1027 ergs (annually). Even though the energy densities are relatively low, a number of proposals to exploit this energy source have been made over the years. A central theme of several proposals has been the application of the venturi-tube principle, wherein currents are directed to flow into throatlike configurations, thus increasing their velocity. As mentioned by Isaacs and Schmitt (1980), if the potential for using currents of flowing water has been so attractive, one would expect the technology to be in practice along the great rivers. Instead, the practice in the past has been one of building dams to achieve higher current velocities over small spatial intervals.

A major step toward advancing the ocean current technology of recent years is the Coriolis concept first proposed in 1973 by W.J. Mouton (Tulane University).

Lissaman (1980) observes that energy calculations for an array of 242 large turbines (each rated at 83 megawatts) about 170 meters (558 feet) in diameter and moored in the Gulf Stream, in a relatively compact array of about 30 kilometers (18.6 miles) cross-stream dimension and 60 kilometers (37 miles) streamwise extent, could produce about 10,000 megawatts of electricity, a significant portion of Florida's electrical needs. This is an energy equivalent of about 130 million barrels of oil per year. Cost effectiveness studies thus far have been encouraging. Late in 1978, work funded by the U.S. Department of Energy involved the design, analysis, and water tests of a 1-meter (3.2-foot) diameter model at the David Taylor Model Basin in the U.S. Naval Ship Research and Development Center, Bethesda, Maryland. Additional studies are scheduled, with the sea test of a full-scale prototype scheduled for completion in 1984. For more details, see Lissaman (1979, 1980), Richardson et al. (1969), and Stommel (1965) references.

Energy from Saline Gradients

The phenomenon of osmotic pressure is explained in the entry on **Osmotic Pressure**. The diagram of Fig. 8 portrays the osmotic pressure difference between fresh water and salt water, this amounting to the energy equivalent of a waterfall some 240 meters (787 feet) in height, which theoretically occurs at the mouth of every river that exits fresh water into the oceans. Since one foot of water is equivalent to 0.0295 atmosphere, this pressure difference also can be expressed as 23.2 atmospheres. Although scientists agree that the energy, as the result of salient gradients, is in place, there is no consensus at this early phase of a young technology as to how this energy can be extracted and converted to a practical, usable form. Certain concepts are immediately suggested as the result of past related experiences, particularly in connection with various processes, such as electrodialysis and reverse osmosis, which have become part of desalination and permeable membrane technology. See also **Desalination**; and **Dialysis**.

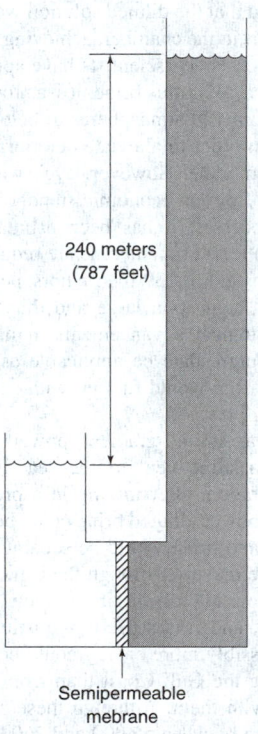

240 meters (787 feet)

Semipermeable mebrane

Fig. 8. Magnitude of osmotic pressure difference between fresh water and seawater.

In 1976, Leitz (U.S. Bureau of Reclamation) described a "dialytic battery" in this regard. This concept is essentially the reverse of the electrodialysis process used for desalination. Briefly, in the desalination procedure, electrodes impress an electric field across a cell through which saline water flows. Cations migrate to the cathode; anions migrate to the anode. The positive ions pass through cation-permeable membranes; negative ions pass through anion-permeable membranes. Between alternate membranes, the water becomes enriched with or depleted of salts. The energy requirements are in proportion to saline content, but increase rapidly with attainment of high purity. For the latter reason, electrodialysis has proved best suited for the purification of brackish waters rather than normal seawater.

As explained by Salisbury (1978), in the dialytic battery, fresh water is poured into half of the cells; salt water into the other half. As they mix, they generate an electric current. The cells are adjacent and separated by selective membranes. Half allow only positive ions to pass through, while the other half allow only passage of negative ions. The membranes are arranged so that the positive ions flow in one direction and negative ions travel in the other, thus generating an electric current. In a system of this type, it is estimated that the membrane would be the most costly component of the basic system. However, the costs of arranging a steady supply of fresh water, and an equal volumetric supply of seawater, coupled with the means to exit the mixed water in such a way as not to pollute the fresh water supply pose construction and engineering problems.

McCormick (U.S. Naval Academy) has suggested a "pressure retarded osmosis" process. Again, this is essentially the reverse of another desalination process, i.e., reverse osmosis. Briefly, in the desalination procedure, the saline water is pressurized above its osmotic pressure — then flows over a semipermeable membrane. The membrane is permeable to water, but not to dissolved solids. The fresh water is transported through the membranes. Energy requirements increase less rapidly with increasing saline content than with the previously described electrodialysis process. Tailoring of membranes to different saline concentrations is technologically feasible. High-pressure equipment is required for this process.

In the proposal of McCormick, brine would be pumped at about 120 atmospheres of pressure into a large chamber. A large pipe made of a semipermeable membrane (to permit passage of water, but no dissolved solids) would rest inside the chamber. The membrane pipe would be filled with fresh water, which would flow through the pipe in an attempt to dilute the brine. It is suggested that the diluted solution would then flow through a turbine from an exit port in the chamber, achieving a force twice that of the initial hydraulic pressure. Israeli scientists have applied this method, using brine from the Dead Sea. With this brine, it was found necessary to reduce the applied pressure from 120 atmospheres to below 100 atmospheres and to use a diluted brine in order to alleviate deterioration and compaction of the semipermeable membrane. However, even with these alterations, the process appears viable from an economic standpoint.

With membrane processes, it has been estimated that 70,000 square meters (753,200 square feet) of membrane are required to generate approximately 10,000 horsepower (10,139 metric horsepower). With current technology, obviously this suggests a large and thus land-based installation. However, with technological advancements in membrane materials, the system may at some future date be applicable as a source of power for ships, in which case, a ship would fuel up with salt rather than petroleum or coal prior to leaving port.

It is obvious that a salinity-gradient power plant would not have to be located on a coastline next to the sea. Rather, as proposed by Intertechnology Corporation, after mixing in a pressure retarded osmosis facility and generating power, diluted brine could be pumped to a solar still, where the fresh water and brine would be separated and thus, as renewed raw materials, would be recycled through the osmotic generator. Scientists recently have made some interesting observations as regards the salinity-gradient concept. Wick and Isaacs (Scripps Institution of Oceanography) have suggested that possibly more energy could be produced from the salt in the salt domes along the Gulf Coast than from the oil and natural gas which are associated with them. Although these salt domes already have produced on the average between 55,000 and 260,000,000 barrels of oil, it has been suggested that only in the highest yielding of these domes is the energy potential of the oil greater than that of the salt. Along these lines, some scientists have been disturbed by the suggestion that these brines be pumped out and dumped into the sea so as to prepare a place to store parts of the nation's strategic oil reserve. However, because salinity-gradient technology is only in a very early phase, it is not likely that such appeals will be listened to seriously.

Wick and Isaacs (1978) have reported estimates that, in the case of the Thompson salt dome (Ft. Bend, Texas) the oil energy (in MW-year) is 3.1 times that of the salt energy (in MW-years), but that in many other domes, the reverse situation holds. For example, it is estimated that the East Tyler dome (Smith, Texas) has a salt energy potential that is over 16,500 times that of its oil potential; that the Bethel dome (Anderson, Texas), the salt energy potential is over 1600 times that of its oil energy potential.

Reverse vapor compression as a process to utilize salient-gradient energy also has been proposed. In this concept, the vapor pressure difference between high-salinity and low-salinity waters is exploited. Membranes are not required. As in the cases of the other two major approaches just described, the reverse vapor compression has its reverse analog in the vapor compression distillation process used for desalination. It has been shown that, in principle, most desalination schemes that are reversible are capable of producing power when directed to *salination*. In vapor compression distillation, the vapor from boiling brine is compressed mechanically, thus increasing the vapor temperature and pressure. The compressed vapor

is then fed back into the evaporator to distill more seawater feed. The multiple-effect approach can be applied to the concept. The process has been attractive for small-size units. As pointed out by Olsson, Wick, and Isaacs (1979), due to the lower vapor pressure of salt water, water vapor will rapidly transfer from fresh water to salt water in an evacuated chamber. If a turbine is interposed in the vapor flow between the two solutions, power can be extracted. Claude experimented with this concept as early as 1930, using vapor pressure differences resulting from the temperature gradient between ocean surface water and deep water. For obvious reasons, technological development of reverse vapor compression can benefit from any progress made in open-cycle OTEC power plants, previously described, and vice versa.

Submarine Geothermal Springs and Geothermal Aquifiers

During the past decade, the discovery of a number of submarine geothermal springs and geothermal aquifiers has been reported, some of these in connection with the IDOE projects of the 1970s. Discoveries have been made in the eastern Mediterranean Sea and in the Pacific Ocean off Hawaii, among other locations. These areas are not easy to find because their thermal effects usually do not penetrate upper layers of the ocean, but rather they are trapped below the thermocline. As may be expected, the most concentrated geothermal heat is associated with areas of tectonic activity, such as volcanic islands. Some of these sources are capable of releasing in excess of 106 MW-years of energy in a single eruption. It has been reported that a suspected underwater explosive event destroyed all sea-level culture in the eastern Mediterranean (1500 B.C.). Possible exploitation of submarine geothermal energy sources is in a very early stage of research.

Oceanic Biomass

Use of dried sea plants as a source of energy by simply burning them dates back many centuries in coastal areas where seaweeds were more available than wood. Modern planners are now directing their attention to the utilization of wet biomass, using such processes as anaerobic digestion to produce low-grade gas (50–60% methane). The basic chemical building blocks are also available in these materials to produce alcohol and other organic fuels. Seaweed farms as a source of energy have been envisioned. Seaweed as a source of foods and food additives is described briefly in the entry on **Seaweeds**. The giant kelp *Macrocystis pyrifera* is also a potential energy source. In the mid-1970s, an Ocean Food and Energy Farm (OFEF) program was commenced, jointly funded by the Energy Research and Development Administration, the American Gas Association, and the U.S. Navy. An open-ocean farm to cover some 100,000 acres (40,469 hectares), some 12.5 miles (21 kilometers) on a side and located about 100 miles (160 kilometers) off southern California was proposed. The farm substrate would be maintained at a depth of approximately 100 feet (30 meters). The area would be made up of flexible triangular modules, about 1000 feet (304 meters) on a side and covering about 10 acres (4 hectares). The modules would be held in position by diesel-powered propulsors. Through the use of wave-powered pumps, nutrient-rich water would be upwelled from a depth of about 300 feet (91 meters). Kelp plants would be attached to the substrate, with one plant per each 363 square feet (33.7 square meters). It is estimated that about 4 years would be required for a crop to mature, after which the standing crop would be harvested by ship at the rate of six times per year. It is estimated that the yield would approximate 15 dry tons per acre (33.3 metric tons per hectare) per year. About 53% of this would be organic biomass.

Prior to proceeding with such a large commercial operation, a research program was commenced. The first test farm was made operative in late 1978, but the initial plantings did not survive. A second test is currently underway. As pointed out by Rhyther (1980), open-ocean energy farming of seaweeds must be regarded as a long-term prospect that cannot be expected to be realized in a time frame of less than tens of years. Much remains to be learned about the basic biology of the plants, particularly their nutrition and growth, and factors that control their organic productivity.

As reported in the entry on **Ocean**, explorations of the Galápagos Rift in 1979 excited the scientific community concerning the concept of chemosynthesis as the primary source of organic nutrients for life found in the vicinity of submarine thermal vents. It has been hypothesized that this particular food chain begins with the production of bacterial biomass, which then leads to the massive but highly localized animal communities found in the vent areas. This has opened up an entirely new concept of the chemosynthetic production of biomass. Instead of using light for the reduction of carbon dioxide, some bacteria use the energy liberated by the oxidation of certain electron sources with free oxygen. Thus, as observed by Jannasch (1980), these bacteria produce organic matter chemosynthetically rather than photosynthetically. For energy, they use hydrogen sulfide and other sulfur compounds.

Additional Reading

Blevins, R.W., J.T. Stadter, and R.D. Weiss: "At-Sea Test of Large Diameter Steel Cold Water Pipe," *IEEE Symp. Ocean Engng.*, New York, NY (February 3, 1980).

Chang, P.Y. and R.A. Barr: "A Frequency-Domain Approach to Analyzing the Dynamics of OTEC Plant Platforms and Cold Water Pipes," *IEEE Symp. Ocean Eng.*, New York, NY (February 3, 1980).

Charlier, R.H. and J.R. Justus: *Ocean Energies: Resources for the Future*, Elesiver Science, New York, NY, 1992.

Cohen, R.: "Energy from Ocean Thermal Gradients," *Oceanus*, 22, 4, 12–22 (1980).

Considine, D.M.: "Tidal Energy," in *Energy Technology Handbook*, (D.M. Considine, editor), The McGraw-Hill Companies, Inc., New York, NY, 1977.

d'Arsonval, A.: "Utilisation des forces naturelles: Avenir de l'electricité," *La Revue Scientifique*, 370–372 (1881).

Duff, G.F.D.: "Tidal Power in the Bay of Fundy," *Technology Review (MIT)*, 81, 2, 34–42 (1978).

Dugger, G.L.: "Ocean Thermal Energy Conversion," in *Energy Technology Handbook*, (D.M. Considine, editor), The McGraw-Hill Companies, Inc., New York, NY, 1977.

Dugger, G. (editor): *Proceedings, Sixth OTEC Conference (June 19–22, 1979)*, U.S. Govt. Printing Office, Washington, DC, 1979.

Guenther, D.A., Jones, D., and W. Chiou: "Power Extraction from Deep Ocean Waves," *IEEE Symp. Ocean Engng.*, New York, NY (February 3, 1980).

Hartline, B.K.: "Tapping Sun-Warmed Ocean Water for Power," *Science*, 209, 794–796 (1980).

Hunt, J.: *Petroleum Geochemistry and Geology*, W.H. Freeman Company, San Francisco, CA, 1995.

Isaacs, J.D. and W.R. Schmitt: "Ocean Energy: Forms and Prospects," *Science*, 207, 265–273 (1980).

Jannasch, H.W. and C.O. Wirsen: "Chemosynthetic Primary Production at East Pacific Sea Floor Spreading Centers," *BioScience*, 29, 592–598 (1979).

Kash, D., et al.: *Energy under the Ocean: A Technology Assessment*, University of Oklahoma Press, Norman, OK, 1973.

Lissaman, P.B.S.: "Energy Available from Arrays of Ocean Turbine Systems Moored in the Florida Current," Rept. AUR 7038, AeroVironment, Inc., 1977.

Lissaman, P.B.S.: "The Coriolis Program," *Oceanus*, 22, 4, 23–28 (1980).

McCormick, M. and Y. Kim (Editors): *Utilization of Ocean Waves—Wave to Energy Conversion*, American Society of Civil Engineers, Reston, VA, 1987.

Mei, C.C.: *The Applied Dynamics of Ocean Surface Waves*, World Scientific Publishing, Inc., Riveredge, NJ, 1989.

Metz, W.D.: "Ocean Thermal Energy," *Science*, 198, 178–180 (1977).

Newman, J.N.: "Power from Ocean Waves," *Oceanus*, 22, 4, 38–45 (1980).

Olsson, M., Wick, G.L., and J.D. Isaacs: "Salinity Gradient Power: Utilizing Vapor Pressure Differences," *Science*, 206, 452–454 (1979).

Pauling, J.R.: "An Equivalent Linear Representation of the Forces Exerted on the OTEC Cold Water Pipe by Combined Effects of Waves and Currents," *IEEE Symp. Ocean Engng.*, New York, NY (February 3, 1980).

Pierre, A.: "Tides in the Competition between Energy Sources," *Rev. Fr. Energ.* (September–October 1966).

Rau, G.H. and J.I. Hedges: "Carbon-13 Depletion in a Hydrothermal Vent Mussel: Suggestion of a Chemosynthetic Food Source," *Science*, 203, 648–649 (1979).

Richardson, W.S., Scmitz, W.J., and P.P. Niiler: "The Velocity Structure of the Florida Current from the Straits of Florida to Cape Fear," *Deep, Sea Res.*, 16, 225–234 (1969).

Ross, D.: *Energy from the Waves*, Pergamon, New York, NY, 1979.

Ryan, P.R.: "Harnessing Power from Tides," *Oceanus*, 22, 24–65 (1980).

Ryther, J.H.: "Fuels from Marine Biomass," *Oceanus*, 22, 45–63 (1980).

Salisbury, D.F.: "Plugging into Salt," *Technology Review (MIT)*, 80, 7, 8–9 (1978).

Scotti, R. and T. McGuinnes: "Design and Analysis of the OTEC Cold Water Pipe," *IEEE Symp. Ocean Engng.*, New York, NY (February 3, 1980).

Staff: "Offshore Wind Systems and a New Wave Energy Device," *Oceanus*, 22, 4, 46–47 (1980).

Stambaugh, K. and W. Jawish: "Structural Analysis and Design of a Cold Water Pipe for a 40 MWe Span OTEC Platform," *IEEE Symp. Ocean Engng.*, New York, NY (February 3, 1980).

Stommel, H.: *The Gulf Stream—A Physical and Dynamical Description*, Univ. California Press, Berkeley, California, 1965.

Whitmore, W.F.: "OTEC: Electricity from the Ocean," *Technol. Rev. (MIT)*, 81, 1, 58–63 (1978).

Wick, G.L.: "Power from Salinity-Gradients," *Energy*, 3, 95–100 (1978).

Wick, G.S.: "Salt Power," *Oceanus*, 22, 4, 29–37 (1980).

Web References

Scripps Institution of Oceanography: http://www.sio.ucsd.edu/

The National Oceanic & Atmospheric Administration (NOAA): http://www.noaa.gov/

Woods Hole Oceanographic Institution (WHOI): http://www.whoi.edu/science/science.html

Woods Hole Department of Applied Ocean Physics and Engineering: http://www.whoi.edu/science/AOPE/dept/

Woods Hole Department of Geology & Geophysics: http://www.whoi.edu/science/GG/dept/

OCEAN RESOURCES (Living). The oceans and seas contribute in a major way to the sustenance and wealth of Earth's human inhabitants as well as to many other life forms. Several of these resources are described in separate entries in this encyclopedia. See also **Crustaceans (Edible)**; **Fishes**; **Fish Meals, Oils, and Protein Concentrates**; **Mollusks**; **Seaweeds**; and **Turtles**. Numerous specific fishes are described separately. See list of entries in entry on **Fishes**.

Ocean Life Forms and Balances. Marine biology and ecology are concerned with the study of the plants and the animals in the sea and of their relationships with one another and with the environments. Such studies pertain to the manner in which these organisms are adapted to the various chemical and physical properties of the seawater, including various pollutants, to the motions and currents of the sea, to the availability of light at various depths, and to the solid surfaces that constitute the sea floor.

Marine life is classed in three groups: *benthos*, *nekton*, and *plankton*. The first are the plants and animals living on the sea bottom such as the permanently fixed or immobile forms (sponges and corals), the various creeping forms (crabs, snails), and others that burrow. Barnacles, the larger seaweeds, and sea squirts are also members of this group. *Nekton* are swimming animals that can move freely and that are capable of migration from one place to another. *Plankton* are the floating and drifting small animals (*zooplankton*) and plants (*phytoplankton*) capable of only limited locomotion, if at all.

Animal life of one kind or another exists at all depths of the oceans and in great abundance; indeed, nearly half of all classes of animals are marine. Plant life, however, is much less abundant and consists chiefly of the phytoplankton and the large seaweeds and algae. Nevertheless, the animal life is almost wholly dependent on the phytoplankton for its existence and these minute forms, in turn, have the same requirements for growth as other green plants. The most important of such requirements is sunlight without which photosynthesis cannot take place. Accordingly, these minute forms, which constitute the primary source of food in the oceans, are restricted to a rather shallow layer of water (generally no more than 450 feet (137 meters) deep under even the most favorable condition) called the *photic zone*, the zone in which there is sufficient light for plant growth. In addition, such growth depends on the availability of certain minerals that may be in short supply (thus limiting further growth), notably the phosphates and nitrates. Other salts and carbon dioxide are present in such quantities that they do not normally constitute any impediment to such growth. The shallow zone to which the plankton are restricted as well as the various limitations on several necessary nutrients has important consequences for animal life throughout the oceans.

A thin uppermost layer, of which the photic zone is a part, contains most of the animal life within the oceans. Called the *epipelagic zone*, it is no more than 700 feet (213 meters) in depth. Beneath it is the twilight zone (*mesopelagic zone*) in which there is sufficient light for animals to use, but less than required for plant photosynthesis. Below this is the *bathypelagic zone* into which small amounts of light penetrate; this is succeeded, finally, by the abyssal pelagic zone.

Within the photic zone, the phytoplankton would soon consume several of the available salts were these not replenished from the abundant supplies at lower depths. In the temperature oceans, the warmer waters rising to the surface as the colder waters sink during winter bring an ample replenishment of such nutrients and the phytoplankton grows rapidly in spring, consuming much of the replenished supply of nutrients. Growth slows during the summer and finally ceases in the absence of one or more necessary salts until the cycle is repeated the following winter and spring. In tropical regions, where light penetrates the oceans to much greater depths and no winter mixing occurs of surface waters with deeper waters richer in nutrients, the phytoplankton is much more sparsely distributed and to a much greater depth than occurs elsewhere. The balance of the oceans and, by far the greater part, is abundant in such salts since there is not sufficient light to permit utilization of such salts by plant life.

At the greater depths there is no food production, and animal life in such zones is dependent on the remains of surface organisms falling from above or brought down, together with the oxygen required for animal life, by the mixing of waters of different densities. Thus limited, animal life is also much more sparse than in the surface layer.

The phytoplankton are consumed chiefly by their animal counterparts, the zooplankton—the copepod Crustaceans and the somewhat larger euphausid crustaceans—as well as the larvae of most of the larger forms that occupy the seas. These in turn, constitute the food source for still larger animals and, in general, each form tends to feed on smaller organisms and is itself the food supply for somewhat larger organisms, the whole resting on the existence in great amounts of the simple plant forms the phytoplankton.

The central fact in the ecology of the oceans is that their plant life consists essentially of the phytoplankton, which are minute in size and can live only in those top levels of the ocean to which sufficient sunlight can penetrate to permit photosynthesis. This depth varies roughly from 350 feet (107 meters) in the clearer waters of the oceans to as little as 50 feet (15 meters) in the more turbid coastal waters.

With the exception of such negligible amounts of plant material as may be carried into the sea by rivers or as provided by algae, the phytoplankton is the sole source of the plant food upon which all the animals in the sea depend, since marine animals, like those on land, cannot synthesize living matter from the chemical materials in seawater. Of course, the supply of the food for the animals of the sea is not limited to the upper levels where the phytoplankton can live; both this plant food and food derived from upper-level animals can reach the lower levels by the sinking of the bodies of dead organisms. However, all the animals of the sea do not live directly on phytoplankton; many of them are carnivores, whose food is entirely animal. An important intermediary is the zooplankton, which preys upon the phytoplankton and in turn forms the food supply of larger organisms. The zooplankton is a general name for a large number of groups of organisms, which include small jellyfish, arrow worms and crustacea; in the latter group the genus *Calanus* is particularly important, forming as it does the principal food of the herring, which in turn is eaten by larger animals.

One of the most striking facts about marine ecology is the stratification of organisms by depth. While the fish are broadly classified into pelagic fish, which spend much of their time near the surface, and the demersal fish, which tend to stay near the bottom, this classification has been developed chiefly in fishing waters, and there is considerable evidence that in the parts of the ocean of greatest depth there are more than two zones in which the characteristic fish and other animals remain.

The general term for the organisms living on the bottom of the sea is the Benthos. Some of them live entirely on fine particles suspended in the water, others on the materials that lie on or in sea beds, while many others are carnivorous. Thus, the first of these groups includes the animals that form colonies (polyzoa), such as the corals, as well as many of the mollusca including mussels, oysters, scallops and calms. Some of these animals bury themselves in the sediment, and being provided with siphons, draw in the seawater. Many of the second class of animals, i.e., those which feed on materials in the sea bottom, also burrow into it. They include various species of worms, as well as hearturchins, brittle stars and other animals. The carnivorous group includes many invertebrates, some of which feed on animals in the seawater, as the starfish feed on mollusks, and the bristle worms feed on many of the smaller animals. There are also carnivores, like the sea mouse, which burrow into the sea bed to eat the worms. In many parts of the sea the invertebrate predators consume so much of the available animal food that the food supply remaining for the fish seriously limits their population. This is especially true on the sea bottom, because so many important commercial fish are demersal, including cod, haddock,

hake and plaice. In fact, the pelagic fish, i.e., those which obtain their food near the surface, such as the mackerel, herring and tuna, especially the first two of these, are not limited by the food consumption of the predators because they feed on plankton directly. These fish are, however, limited by themselves being prey, especially during their earlier stages of growth.

Plankton. Phytoplankton require nutrients for successful growth similar to green plants on land. Some of these are almost always present in excess of requirements, but others, e.g., phosphorus and nitrogen, are greatly depleted during the periods of active algal growth, which commonly occur in spring and autumn in northern open coastal waters. The plant population then declines and remains minimal until the nutrient supply is restored from deeper levels by vertical mixing and by decomposition of vegetation and animal remains. Light is a limiting factor to aquatic plant growth in high latitudes at certain seasons. Temperature exerts a selective influence on both plants and animals, but does not in itself appear to limit the total quantity of plankton developed.

Phytoplankton is for the most part composed of a variety of unicellular yellow-green algae dominated by diatoms, armored dinoflagellates, naked flagellates, and coccolithophorides,* which together according to some authorities account for more than half of the organic production of the earth. See Fig 1. The relative abundance of component groups varies regionally and seasonally. Diversity of species is greatest in tropical waters and least in boreo-arctic regions. Diatoms, distinguishable by their siliceous shells, occur singly and in chains, have one or more chromatophores, and during flowering season, or "blooms," flourish in great abundance in coastal waters. The dinoflagellates are more mobile, possess flagella, and like diatoms dominate at times in the phytoplankton during bloom periods, particularly in warm neritic (near surface and/or coast) waters. At these times luminous species often cause brilliant bioluminescence. Some, possessing chlorophyll, are truly phytoplankton, but others are either parasitic, or feed like animals and are classed as zooplankton. See also **Diatom**; and **Dinoflagellata**. Little is understood about the naked flagellates because of their minute size and the difficulty in preserving them. However, they are now considered the principal food of most planktonic filter feeders and may play an equally important, if not more

Fig. 1.   Diatoms (*magnification 1000×: Bausch & Lomb*).

important, role than diatoms, particularly in neritic waters. In the open ocean, the naked flagellates tend to be augmented, if not replaced, by small calcareous coccolithophorides.

Other pelagic yellow-brown algal types at times present include the brownish colonial flagellate *Phaeocystis*, which forms gelatinous clusters,

---

* Coccolithophore is any of numerous, minute, mostly marine, planktonic biflagellate protists having brown pigment-bearing cells that at some phase of their life cycles are encased in a sheath of coccoliths to form a complex calcareous shell.

particularly in northern coastal waters, and the globular green *Halosphaera*, widespread over the open ocean. A blue-green alga *Trichodesium* appears at times as yellowish-brown patches on the surface in warm climates.

The principal research tool in plankton biology is the towed net. It has provided most of the basic information pertaining to the distribution, diversity, productivity, and behavior of zooplankton. During the 1970s, an alternative method of study was developed—direct underwater observation and collection of animals by divers. Studies have shown that much essential information about the lives of planktonic animals cannot be derived simply by collecting the animals with a net. As pointed out by Harbison and Madin (1979), this new perspective is essential to understanding the place of pelagic animals in the open ocean—one of the largest and least known of all environments on Earth.

Coccolithophore. Like any other type of phytoplankton, coccolithophores are one-celled marine plants that live in large numbers throughout the upper layers of the ocean. Unlike any other plant in the ocean, coccolithophores surround themselves with a microscopic plating made of limestone (calcite). These scales, known as *coccoliths*, are shaped like hubcaps and are only three one-thousandths of a millimeter in diameter.

What coccoliths lack in size they make up in volume. At any one time a single coccolithophore is attached to or surrounded by at least 30 scales. Additional coccoliths are dumped into the water when the coccolithophores multiply asexually, die or simply make too many scales. In areas with trillions of coccolithophores, the waters will turn an opaque turquoise from the dense cloud of coccoliths. Scientists estimate that the organisms dump more than 1.5 million tons (1.4 billion kilograms) of calcite a year, making them the leading calcite producers in the ocean.

Most phytoplankton need both sunlight and nutrients from deep in the ocean. The ideal place for them is on the surface of the ocean in an area where plenty of cooler, nutrient-carrying water is upwelling from below. In contrast, the coccolithophores prefer to live on the surface in still, nutrient-poor water in mild temperatures.

Coccolithophores do not compete well with other phytoplankton. Yet unlike their cousins, coccolithophores do not need a constant influx of fresh food to live. They often thrive in areas where their competitors are starving. Typically, once they are in a region, they dominate and become more than 90 percent of the phytoplankton in the area.

Coccolithophores live mostly in subpolar regions. Some other places where blooms occur regularly are the northern coast of Australia and the waters surrounding Iceland. In the past two years, large blooms of coccolithophores have covered areas of the Bering Sea. This surprises many scientists since the Bering Sea is normally a nutrient-rich body of water.

Coccolithophores are not normally harmful to other marine life in the ocean. The nutrient-poor conditions that allow the coccolithophores to exist will often kill off much of the larger phytoplankton. Many of the smaller fish and zooplankton that eat normal phytoplankton also feast on the coccolithophores. In nutrient-poor areas where other phytoplankton are scarce, the coccolithophores are a welcome source of nutrition.

In the long term, the plants seem to be good for the environment. Coccolithophores make their coccoliths out of one part carbon, one part calcium and three parts oxygen ($CaCO_3$). So each time a molecule of coccolith is made, one less carbon atom is allowed to roam freely in the world to form greenhouse gases and contribute to global warming. Three hundred twenty pounds of carbon go into every ton of coccoliths produced. All of this material sinks harmlessly to the bottom of the ocean to form sediment.

The coccolithophores' short-term effect on the environment is somewhat more complex. This effect again has to do with the formation of their coccoliths and the chemical reaction involved in the process. The chemical reaction that makes the coccolith also generates a carbon dioxide molecule, a potent greenhouse gas, from the oxygen and carbon already in the ocean. While much of the gas is sucked back in by the coccoliths (all plants take in carbon dioxide for food) some of it escapes into the atmosphere and immediately becomes part of the greenhouse gas problem. Scientists are concerned in the short term that greenhouse gases will cause the upper layers of the ocean to become more temperate and stagnant. This would increase the number of coccoliths in the world, which would produce more greenhouse gas.

The coccolithophores also affect the global climate in the short term by increasing the oceans' albedo. Albedo is the fraction of sunlight an object reflects—higher albedo values indicate more reflected light. Coccolithophore blooms reflect nearly all the visible light that hits them. Since most of this light is being reflected, less of it is being absorbed by the ocean and stored as heat.

Ctenophora. The comb jellies or sea walnuts, constituting a small phylum of marine pelagic animals related to the coelenterates, are sometimes included in this phylum. The phylum includes the peculiar Venus' girdle, a transparent, ribbonlike animal whose longitudinal axis is across the middle of the slender body.

The ctenophores differ from the coelenterates in the following structures: (1) The alimentary tract opens to the exterior at the end of the body opposite to the mouth. (2) The body bears rows of ciliated plates, the combs, which give the animals one of their common names. These are organs of locomotion. (3) Colloblasts or adhesive cells take the place of nematocysts.

The classification is as follows:

Class *Tentaculata*. A pair of long tentacles.
   Order *Cydippida*. Body spherical to cylindrical. Tentacles long.
   Order *Lobata*. Tentacles replaced in adult by fringe of short tentacles around mouth.
   Order *Cestida*. Body ribbon-like. Venus' girdle.
Class *Nuda*. Tentacles absent. Body conical to ovoid.
Order *Beroida*. Wide mouth and pharynx; numerous side branches to meridional gastrovascular canal.

Ctenophores are among the most beautiful and voracious of the marine plankton. They consume enormous quantities of copepods, fish larvae, and other planktonic forms and thus are an important part of the ocean food chains. Ctenophores are highly dependent upon their ciliary paddles, called comb plates. As pointed out by Tamm (1980), each comb plate, or ctene, consists of a transverse band of thousands of long cilia, up to 2 millimeters in length, which beat together as a unit. Ctenophores are the largest known animals that use cilia for locomotion, and the longest cilia are found in the comb plates of ctenophores. Thus, ctenophores offer an excellent medium for studying cilia and flagella, as reported in considerable detail by Tamm (1980).

Mass Extinctions in the Ocean. Considerable attention has been given during the past few years to evidence that indicates the occurrence of mass extinctions of certain forms of life in the ocean. It appears that during brief intervals over the past 700 million to one billion years, a number of marine animals and plants died out. Various hypotheses have been offered to explain such mass extinctions, including loss of some species due to the cooling of the sea. Such extinctions not only have affected lower sea life forms, such as single-celled algae and plankton, but also large swimming reptiles and whales. Probably the best known of these extinctions is that which occurred about 65 million years ago at the end of the Cretaceous period. This event apparently eliminated most marine species at about the same time it is estimated that dinosaurs etc. became extinct on land. In a mass extinction, some species can recover, or new but related species may evolve. In other instances, evidence indicates the complete elimination of certain species. Changes in sea level also are implicated as causes of mass extinctions. The ratio of various chemical isotopes in skeletons, oozes, and other geologic sediments and materials in the ocean provide clues as to the timing and effects of mass extinctions. In addition to isotopic carbon dating, which has been well established for many years, other isotope ratios are now widely used, including, for example, the isotopic composition of neodymium in ocean waters and remnants. For further detail, see the separate article of **Mass Extinctions**.

International Decade of Ocean Exploration

The decade of the 1970s represented an accelerated program of research into nearly all aspects of ocean science. See also **Ocean**. Programs directed specifically to ocean biology included an intensive study of the Coastal Upwelling Ecosystems Analysis (CUEA), the initial goal of which was to predict phytoplankton distribution and growth in upwelling ecosystems on the basis of mesoscale observations of the critical forcing processes, mainly wind and circulation, but also including biological processes of grazing, predation, and nutrient regeneration of zooplankton, fish, and benthos.

On the basis of the published work, it is possible to cite a series of advances resulting from the CUEA project. This research tested the hypothesis that upwelling results from the tight coupling of a set of physical and biological processes, that this coupling is understandable, and that it can therefore be the basis for long-term management and use of the biological resources of upwelling ecosystems. The research also established the relationship between local winds and productivity and, less precisely, the relationships between very large-scale variations and productivity and nature of the ecosystem. Costlow and Barber (1980) suggest that given any coastal upwelling regime, with knowledge of its shelf width and latitude, it is possible to predict how variations in the local winds will affect primary productivity. The researchers give as an example the northwest African region at the latitude of Cape Verde or Daker, where increased storm frequency will enhance primary productivity inasmuch as in those regions there are frequent periods of nutrient depletion in the surface coastal waters.

Another IDOE biology program was the Controlled Ecosystem Pollution Experiment (CEPEX), which had three objectives: (1) to determine the effects of various pollutants on the microbial, phytoplankton, and zooplankton components of a large, field-based experimental ecosystem; (2) to evaluate changes in nutrient uptake kinetics related to pollutant stress; and (3) to identify the chemical variations that may occur in experimental ecosystems subjected to pollutant stress over specific periods of time. The effort involved scientists from nine American institutions as well as Canadian and British scientists.

Additional Reading

Attrill, M. (Editor): *Rehabilitated Estuarine Ecosystems: The Thames Estuary, Its Environment and Ecology*, Chapman and Hall, New York, NY, 1998.

Betzer, P.R., et al.: "The Oceanic Carbonate System: A Reassessment of Biogenic Controls," *Science*, **226**, 1074–1077 (1984).

Capone, D.G. and E.J. Carpenter: "Nitrogen Fixation in the Marine Environment," *Science*, **217**, 1140–1142 (1982).

Corliss, B.H., et al.: "The Eocene/Oligocene Boundary Event in the Deep Sea," *Science*, **226**, 806–810 (1984).

Costlow, J.D. and R. Barber: "IDOE Biology Programs," *Oceanus*, **23**, 1, 52–61 (1980).

Dawes, C.J.: *Marine Botany*, John Wiley and Sons, Inc., New York, NY, 1998.

Druffel, E.M.: "Banded Corals: Changes in Oceanic Carbon-14 During the Little Ice Age," *Science*, **218**, 13–19 (1982).

Fenical, W.: "Natural Products Chemistry in the Marine Environment," *Science*, **215**, 923–928 (1982).

Grassle, J.F.: "Hydrothermal Vent Animals: Distribution and Biology," *Science*, **229**, 713–717 (1985).

Harbison, G.R. and L.P.M. Nadin: "Diving — A New View fo Plankton Biology," *Oceanus*, **22**, **2**, 18–27 (1979).

Haymon, R.M., Koski, R.A., and C. Sinclair: "Fossils of Hydrothermal Vent Worms from Cretaceous Sulfide Ores of the Samali Ophiolite, Oman," *Science*, **223**, 1407–1409 (1984).

Hsü, K.J., et al.: "Mass Mortality and Its Environmental and Evolutionary Consequences," *Science*, **216**, 249–256 (1982).

Jannasch, H.W. and M.J. Mottl: "Geomicrobiology of Deep-Sea Hydrothermal Vents," *Science*, **229**, 717–725 (1985).

Kerr, R.A.: "The Ocean's Deserts are Blooming," *Science*, **232**, 1345 (1986).

Koehl, M.A.R.: "The Interaction of Moving Water and Sessile Organisms," *Sci. Amer.*, **247**(6), 124–134 (1982).

Levinton, J.: *Marine Biology: Function, Biodiversity, Ecology*, Oxford University Press, Inc., New York, NY, 1995.

Lewin, R.: "Life Thrives Under Breaking Ocean Waves," *Science*, **235**, 1465–1466 (1987).

Littler, M.M., et al.: "Deepest Known Plant Life Discovered on an Uncharted Seamount," *Science*, **227**, 57–59 (1985).

Martinez, L., Silver, M.W., King, J.M., and A.L. Alldredge: "Nitrogen Fixation by Floating Diatom Mats: A Source of New Nitrogen to Oligotrophic Ocean Waters," *Science*, **221**, 152–154 (1983).

Piepgras, D.J. and G.J. Wasserburg: "Isotopic Composition of Neodymium in Waters from the Drake Passage," *Science*, **217**, 207–214 (1982).

Rau, G.H.: "Hydrothermal Vent Clam and Tube Worm 13C/12C: Further Evidence of Nonphotosynthetic Food Sources," *Science*, **213**, 338–339 (1981).

Raup, D.M. and J.J. Sepkoski, Jr.: "Mass Extinction in the Marine Fossil Record," *Science*, **215**, 1501–1503 (1982).

Stanley, S.M.: "Mass Extinctions in the Ocean," *Sci. Amer.*, **250**(6), 64–72 (1984).

Takahashi, K., Hurd, D.C., and S. Honjo: "Phaeodarian Skeletons: Their Role in Silica Transport to the Deep Sea," *Science*, **222**, 616–618 (1983).

Tamm, S.: "Cilia and Ctenophores," *Oceanus*, **23**, **2**, 50–59 (1980).

Thiede, J.: "Reworked Neritic Fossils in Upper Mesozoic and Cenozoic Central Pacific Deep-Sea Sediments Monitor Sea-Level Changes," *Science*, **211**, 1422–1424 (1981).

Valiela, I.: *Marine Ecological Processes*, Springer-Verlag, New York, Inc., New York, NY, 1995.

Vernick, E.I., et al.: "Climate and Chlorophyll — Long-Term Trends in the Central Pacific Ocean," *Science*, **238**, 70–73 (1987).

Other References

Alexander, L., et al.: *Large Marine Ecosystems: Patterns, Processes, and Yields*, AAAS Books, Waldorf, MD, 1992.

Beauchamp, B., et al.: "Cretaceous Cold-Seep Communities and Methane-Derived Carbonates in the Canadian Arctic," *Science*, 53 (April 7, 1989).

Blaxter, J.H.S. and A.J. Southwardz: *Advances in Marine Biology*, Vol. 30, Academic Press, Inc., San Diego, CA, 1995.

Duedall, I.W. and M.A. Champ: "Artificial Reefs: Emerging Science and Technology," *Oceanus*, 94 (Spring 1991).

Grassle, J.F.: "A Plethora of Unexpected Life," *Oceanus*, 41 (Winter 1988/1989).

Grigg, R.W.: "Paleoceanography of Coral Reefs in the Hawaiian-Emperor Chain," *Science*, 1737 (June 24, 1988).

Holden, C.: "Ocean-Slick Yardstick," *Science*, 1484 (September 27, 1991).

Holden, C.: "Picture-Perfect Plankton," *Science*, 681 (February 7, 1992).

Horgan, J.: "A Grave Tale — Do Whale Remains Help Life Spread on the Deepest Ocean Floor?" *Sci. Amer.*, 18 (January 1990).

Kasteleijn, H.W.: "Marine Biological Research in the Galapagos: Past, Present, and Future," *Oceanus*, 33 (Summer 1987).

Kerr, R.A.: "An About-Face Found in the Ancient Ocean," *Science*, 1359 (September 20, 1991).

Steele, J.H.: "The Message from the Oceans," *Oceanus*, 4 (Summer 1989).

Van Dover, C.L.: "Do 'Eyeless' Shrimp See the Light of Glowing Deep-Sea Vents?" *Oceanus*, 47 (Winter 1988/1989).

Vinogradov, M.E.: "Dynamics of Ocean Ecosystems: A National Program in Russian Biooceanology," *Oceanus*, 66 (Summer 1991).

Walbran, P.D., et al.: "Evidence from Sediments of Long-Term Acanthaster planci Predation on Corals of the Great Barrier Reef," *Science*, 847 (August 25, 1989).

Ward, F. and J. Greenberg: "Florida's Coral Reefs are Imperiled," *Nat'l. Geographic*, 114 (July 1990).

Zilanov, V.K.: "Living Marine Resources," *Oceanus*, 29 (Summer 1991).

Web References

Scripps Institution of Oceanography: http://www.sio.ucsd.edu/

Scripps Research Oceanography: http://www.sio.ucsd.edu/research/oceanography.html

The National Oceanic & Atmospheric Administration (NOAA): http://www.noaa.gov/

Woods Hole Oceanographic Institution (WHOI): http://www.whoi.edu/science/science.html

Woods Hole Department of Biology: http://www.whoi.edu/science/B/dept/

Woods Hole Department of Marine Chemistry and Geochemistry: http://www.whoi.edu/science/MCG/dept/

OCEAN RESOURCES (Mineral). Since antiquity, the oceans and seas have been a major source of salt (sodium chloride) and continue to be so. Today, solar sea salt is produced in about 60 countries. The People's Republic of China, Australia, Mexico, India, Brazil, the Bahamas, Spain, and France are among the leading producers. At present, about 38% of the sodium chloride produced is evaporated from seawater. The value is estimated at over $400 million per year. Solar salt is very important to many countries, such as Japan, where there are few or no salt deposits. For several decades, seawater has been a significant source for bromine (production commenced by DuPont in 1931); iodine (from kelp, once very important, but no longer an economic source); magnesium (production started by Dow in 1941); potassium; sulfur; and several other elements and their compounds. Today, over 13% of the requirements for bromine come from seawater, as do over 70% of magnesium metal and 33% of magnesium compounds required by industry. Sulfur, associated with the cap rock of salt domes, has been produced from two salt domes just off the coast of Louisiana for many years. For a few decades, the continental shelves under the oceans have been producing large volumes of natural gas and petroleum. And also, for a few decades, the oceans have provided fresh water for many regions through various desalination processes. There are over 500 desalination plants in operation or under construction, with plants in arid and semiarid locations, such as the Middle East, but also in some highly urbanized regions and cities where fresh water is in short supply, such as in Italy and the Netherlands.

Many of the ore deposits found on the continents are the result of ancient oceans. Tin is found in offshore deposits, such as in Indonesia, in Cornwall (Saint Ives Bay), and Phuket Island off the west coast of the Malay Peninsula.

Within the past 20–30 years, much interest has been shown in manganese nodules on the seafloor in various locations. More recently, the discovery of hot brines in the Red Sea, "black smokers" on the East Pacific Rise and suspected in many other locations, and ophiolites has excited the scientific community and attracted industrialists because these phenomena are associated with metals, such as cadmium, copper, nickel, and zinc. These findings have largely resulted from the funding provided for geological and oceanographic research as part of the Deep Sea Drilling Project (DSDP) and the International Decade of Ocean Exploration (IDOE), projects which have been in place since the late 1960s.

More details pertaining to most of these ocean raw materials will be found in a number of specific entries in this encyclopedia. See also **Bromine**; **Chemical Elements**; **Desalination**; **Magnesium**; **Manganese**; **Natural Gas**; **Ocean**; **Ocean Research Vessels**; **Petroleum**; **Sodium Chloride**; and **Sulfur**.

The resource potential of the oceans awaits further technological development. It is interesting to note that the famous German chemist, Fritz Haber, spent more than eight years after World War I in attempts to recover gold from seawater in order to pay the German war debt. The results were disappointing, but large quantities of gold are in very large quantities of seawater. Currently, there is considerable interest in attempts to recover uranium from seawater, particularly by nations with no assured supply. Should fusion power come to fruition, after a few years, the ocean may be looked to as a source of lithium. Beach sands also have received considerable attention in recent years as sources of metals and other materials. Marine beaches may contain gold, silver, platinum, and diamonds in addition to magnetite, cassiterite, chromite, columbite, ilmenite, rutile, scheelite, zircon, monazite, and wolframite. Heavy-mineral beach sands are usually commercially worked for the titanium content of the rutile and the ilmenite. The same sands may also be processed to recover thorium from monazite and zircon for use in foundry sands. Currently, marine beaches are mined for heavy mineral production in Australia, Brazil, India, Madagascar, Mozambique, Sierra Leone, South Africa, and Sri Lanka, among other countries.

Diamonds are found in the seafloor sediments on the coast of the Kalahari Desert in southwest Africa.* The origin of the diamonds is obscure, but it is generally believed that basaltic and kimberlite pipes exist on the ocean floor as on the nearby land. There is a relative abundance of gemstones in the marine deposits and a few large stones have been recovered. Dredging began in 1961, using suction dredges capable of operating in waters to depths of 50 meters. Because of rough seas on this exposed coast, a number of barges were lost and the operation was concluded. However, in recent years a subsidiary of DeBeers is using a dredge protected by a seawall, thus permitting mining offshore about 120 meters at depths of 90 meters.

Calcium carbonate often precipitates from tropical or subtropical waters when the water becomes supersaturated due to enrichment of the carbonate content by intense biological photosynthesis and by solar heating of carbon dioxide-rich cooler waters. The aragonite precipitates as single needles in the shallow waters at a rate of about one millimeter of wet sediment per year. Continuing deposition leads to cementation and the formation of successive concentric sheaths known as oöids. The most extensively studied oölithic aragonite deposit is that distributed over the 250,000-square-kilometer (96,525-square-mile) Great Bahama Bank on the continental shelf near islands of the Bahamas. Most of the areas are less than 5 meters deep and are composed of quite pure calcium carbonate containing higher levels of strontium and uranium than are found in limestones of biological origin. Similar deposits occur in the Gulf of Batabanó (Cuba) and in the Mediterranean Sea off Egypt and Tunisia, as well as on the Trucial coast of the Persian Gulf.

* Acknowledgement of assistance obtained from W.F. McIlhenny, The Dow Chemical Company, Freeport, Texas in preparation of several of the following paragraphs is hereby made.

Iron is a common constituent of marine sediments. Magnetite is found in beach sands and iron is common in glauconitic marine silicates. Iron oxides and sulfides occur where anaerobic conditions and elevated temperatures are found, as in the hot, salty brines found near rifts. Iron is a major constituent of the ferromanganese nodules.

Magnetite-rich iron sands have been dredged from the ocean floor just off Kagoshima Bay (Japan) in water averaging from 15 to 40 meters in depth. Iron sand concentrates were produced in Japan as recently as 1976, although a major marine iron sand operation in Kyushu ceased operation in 1966.

Marine sand and gravel for fill and for aggregate have been produced on all coasts of the United States, particularly from San Francisco and San Pedro Bays in California and from Long Island Sound. Marine sand and gravel are found in significant quantities in the United Kingdom.

Phosphorites (marine apatites) are dense, light-brown-to-black concretions, ranging in size from sands to nodules and irregular masses. Phosphorites have been found off Argentina, Chile, Japan, Mexico, Peru, South Africa, and Spain, and several islands in the Indian Ocean. Some also have been found off the west coast of North America and on the eastern North American continental shelf. These deposits occur where water upwelling transports phosphorus and where the rate of sedimentation is slow. The nodules are usually found as a monolayer on the surface. The mineralogy of the marine phosphorites is similar to western U.S. land deposits, which were almost certainly marine in origin. Phosphorites are quite constant in composition, containing 45–47% calcium oxide and 29–30% phosphorus trioxide. Seawater is generally saturated with tricalcium phosphate so that, under the oxidative conditions normally present, the phosphates precipitate in colloidal form and accrete to existing surfaces, rather than forming a phosphorite suspension. Although most of the phosphorite is believed to have formed during the Miocene epoch, it is believed that precipitation is currently taking place. The largest known seafloor phosphorite deposit is off the coast of California from Point Reyes to the Gulf of California along the inner edge of the continental shelf. Additional deposits have been found on the edges of the Blake Plateau east of Florida. A recovery project was commenced in 1962–1963, but failed to materialize.

Glauconite or green sands (a hydrated silicate with potassium, iron, and aluminum as cations) is widely distributed on the ocean floor in both ancient and more recent marine sediments. Glauconite is often found with phosphorite and occurs on the tops of banks, submerged hillcrests, and on slopes in water from 50 to 2000 meters in depth. Glauconites are known off the coasts of Africa, Australia, China, Japan, Portugal, South America, the United Kingdom (Scotland), the United States (California and the Atlantic shelf), and New Zealand. A 130-square-kilometer (50-square-mile) deposit has been identified on the Santa Monica shelf off California.

Submarine Hydrothermal Deposits

Discovery of the East Pacific Rise hot springs has created extensive interest and plans are underway to commence a four-year, multi-institutional project to explore the East Pacific Rise for additional areas of hot spring activity and ore deposition. The major objective of the program will be to examine the nature of hydrothermal processes along the mid-ocean ridge system from the slow-spreading to the very fast-spreading segments, such as at 10–30° South. The project will involve the use of surface ships, deeply towed instrument packages, new high-precision multibeam echo sounding for making highly accurate topographic maps of the seafloor, and ultimately manned submersibles, such as *Alvin*. See also **Ocean Research Vessels**.

The knowledge of submarine hydrothermal deposits was advanced by a large measure in 1979 when the hot springs on the East Pacific Rise at 21° north were discovered. Unlike the warm springs discovered on the Galápagos Spreading Center a few years ago, the springs on the East Pacific Rise are hot, with water venting at temperatures as high as 350 °C and at velocities of several meters per second. These formations are precipitating large quantities of sulfide ore and minerals rich in copper, zinc, and iron. See also **Ocean**. The precipitates form chimneys around the individual vents that spout black or white smoke composed of precipitated crystals of sulfides and other minerals. The discovery is the most exciting and significant in this field since the discovery of the Red Sea hot brines and metal deposits (Mottl, 1980).

By *hydrothermal* is meant hot water. When deposits are formed by chemical precipitation from hot solutions, they are termed hydrothermal.

Hydrothermal deposits on land represent a very important class of economically retrievable ore deposits and provide a significant percentage of various metals, such as copper, zinc, lead, silver, gold, tin, molybdenum, among others. Mottl (1980) suggests that five factors are involved in forming hydrothermal ore deposits: (1) a source of the ore metals; (2) a source of water that dissolves and later precipitates the metals, concentrating them during the total cycle; (3) a source of heat; (4) a pathway between the site where metals are dissolved and precipitated and the site where they are finally deposited which is permeable and permits solution flow; and (5) the ultimate collection or deposition site. For preservation, it is also important that ores be deposited in places where they will not be eroded away, as by weathering. Because so many factors are involved, there is a wide variety of hydrothermal ore deposits.

In terms of submarine hydrothermal ore deposits, the source of heat is the thermal energy associated with the formation of new oceanic lithosphere along the mid-ocean ridge system, where the seafloor is spreading apart and basaltic magma wells up. Because of tensional forces present, the newly formed crust becomes fractured, allowing seawater to percolate down through the fractures. During this percolation, the seawater is heated by contact with hot rock and commences to react with the rock, leaching metals that may be present. Because of the lesser density of the seawater (due to temperature), it rises and ascends to the seafloor and exists at submarine hot springs. Because there are several factors involved, there is, as on land, a wide variety of submarine hydrothermal ore deposits. Much remains to be understood and to confirm some of the early postulates, as given above, pertaining to the actual formation of submarine deposits.

For example, the concentrations of ore metals in most natural waters are quite low, particularly so in "normal" seawater. Measuring these low concentrations has been a problem of marine chemistry for many years. It is interesting to note that when artificial seawater, made up from pure reagent chemicals, is exposed to metallic elements, the ultimate solutions produced will contain from 100 to 1000 times the concentrations of these metals as compared with natural seawater.

The first submarine hot springs discovered along a mid-ocean ridge, those at the Galápagos Spreading Center, were emitting water at only 20°C, but the chemistry of this water indicated that it had reacted with basalt at 350–400°C. Then came the discovery of the 350° springs on the East Pacific Rise. Currently, the chemistry of this water is being studied at the Massachusetts Institute of Technology. To date, no submarine hydrothermal deposit has been sufficiently studied that all components contributing to its formation are known. Nevertheless, data at hand as of the early 1980s suggest some intriguing relationships among known deposits along mid-ocean ridges and point out the importance of special situations in producing and preserving large deposits.

Offshore Oil and Gas Resources

Although oil and gas exploration and production activities which occur offshore involve an extension of continents (the continental margins), they are nevertheless considered more in the general terms of oceanological rather than continental resources (the latter generally considered land or above-sea-level resources). Various aspects of offshore oil and gas production are described briefly in the entries on **Natural Gas**; and **Petroleum**.

As we go into the next century, because of an apparent glut of petroleum on world markets and because of much greater optimism pertaining to the ultimate natural gas reserves in the world, the emphasis on the exploration for petroleum and natural gas in underwater locations has markedly diminished.

Manganese Nodules

Deep-sea nodules, comprised mainly of manganese and iron oxides, have been found in abundance over large areas of the deep ocean floor that have been examined to date. In some locations, the nodules have been found to contain generous proportions of nickel, copper, cobalt, molybdenum, and vanadium, as well as manganese and iron. From the standpoint of potential commercial exploitation, a deposit is not considered promising unless the nickel-copper content is 1.8% (weight) or greater. On this basis, one of the most promising areas is the Clarion and Clipperton fracture zone, an immense area some 4400 kilometers (2730 miles) long and 900 kilometers

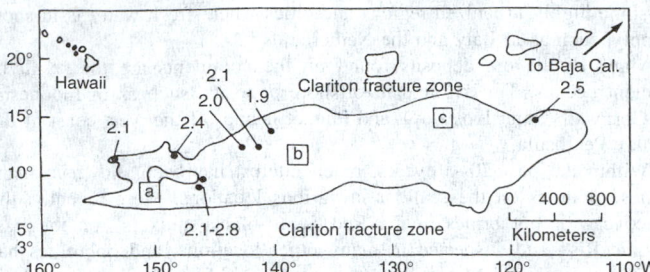

Fig. 1.   Regions where manganese nodules containing more than 1.8% nickel-copper occur in the northeastern equatorial Pacific Ocean. Numbers indicate average percent of nickel-copper in one-degree squares. Areas a, b, and c indicate locations of activity carried out as part of Deep Ocean Mining Environmental Studies Program. (*After McKelvey, U.S. Geological Survey.*)

(560 miles) wide at its widest point. This zone is located southeast of Hawaii and southwest of Baja California. See Fig. 1.

An investigation of the origin and distribution of manganese nodules and the processes by which they selectively concentrate copper, nickel, and other metals was one of the first major projects under the sea beds Assessment project of the IDOE program. At a workshop attended by over a hundred scientists from various countries, the most likely locations for the exploration and study of manganese nodules were selected. The north central Pacific was identified as the zone where the nodules have the highest metal content. A team of American scientific investigators proposed that a comprehensive field and laboratory program be initiated to relate the high metal content to the local geological conditions. Data gathering was concerned along a transect that both academic and industrial scientists agreed could serve as a potential mining site. In addition to dredge sampling and piston cores, bathymetric measurements, sidescan sonar, and high-resolution television pictures were obtained. See Fig. 2. The results provided a broad-scale picture of the conditions under which nodules form, but the mechanisms for concentrating specific metals are still not well defined.

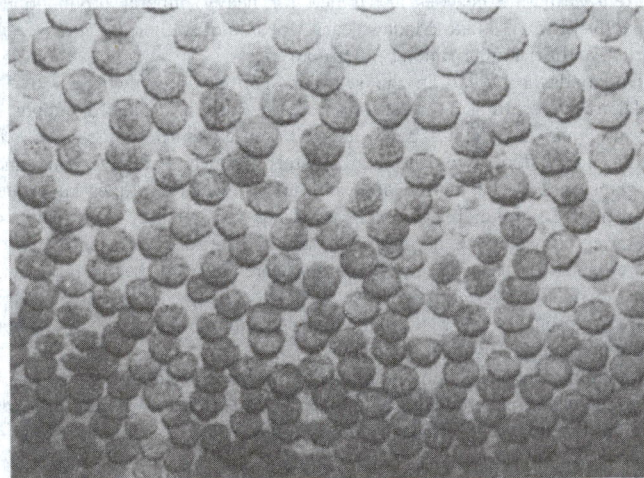

Fig. 2.   Type of manganese nodules found in north central Pacific Ocean zone. (*Woods Hole Oceanographic Institution.*)

Although the nodules vary widely in their composition over the world oceans, metals are concentrated in three distinct types. One type comprises the nickel-copper-rich nodules of the Clarion-Clipperton variety, which is mainly formed in the equatorial regions. Another type, high in cobalt (1% or more) and low in nickel and copper, appears to be most commonly formed on sea mounts. The third type is high in manganese (35% or more), but low in other metals; it is known mainly on the eastern side of the Pacific Basin. As of the early 1980s, the most economically attractive were the cobalt-rich nodules.

The nodules form in a layered structure around a nucleus, which may be almost any material on the ocean floor. Most deep-sea nodules tend to be spherical or oblate in form. Nodules may occur up to 25 centimeters in diameter, but they average about 5 centimeters. The deposits may occur as slabs or agglomerates, or as incrustations on rocks or as pavement in some areas. The nodules are disorderly crystalline materials with layers of MnO_2 (mixed Mn^{2+}-Mn^{4+} oxides) alternating with $Mn(OH)_2$ and $Fe(OH)_3$. Excess iron appears as a mixture of goethite ($Fe_2O_3 \cdot H_2O$) and lepidocrocite.

The nodules are formed by the oxidation and precipitation of iron and manganese. The oxidation of Mn^{2+} is catalyzed by a reaction surface to a tetravalent state that absorbs additional Fe^{2+} or Mn^{2+} which, in turn, becomes oxidized. A surface is required and the initial deposition may be of iron oxide, possibly from volcanic or geothermal sources. Proper conditions of pH, redox potential, and metal ion concentration are found in deep ocean waters. The rate of accumulation appears to be very slow. The growth also may be discontinuous, and is estimated at a faster rater rate near the continental margins.

Precious Coral

Only a few species of coral have a combination of beauty, hardness, and luster, such the black coral species of Hawaii (*Antipathes dichotoma* and *Antipathes grandis*). These are highly valued by the jewelry trade. There are also a few red, pink, gold, and bamboo varieties that are in demand. Black coral also occurs in the Gulf of California and in the Pacific Ocean off Baja California, plus a few scattered locations in the Pacific Ocean east of Australia and north of New Zealand. Traditional sources of red and pink corals have been the Mediterranean Sea, various locations in the western Pacific Ocean, ranging from the Philippines, the Ryuku and Bonin islands and south of Japan. There is also a string of precious red and pink coral beds northwest of Hawaii and off the Cape Verde islands in the Atlantic Ocean off west Africa. A submersible vessel, the *Star II*, operated by Maui Divers of Hawaii, Ltd., is used to harvest pink coral (*Corallium secundum*) from the Makapuu bed. State regulations permit the collection of only 4400 pounds (1996 kilograms) within a 2-year period.

Additional Reading

Amsbaugh, J.K. and J.L. Van der Voort: "The Ocean Mining Industry: A Benefit for Every Risk?" *Oceanus*, 22–27 (Fall 1982).
Andreae, M.O. and H. Raemdonck: "Dimethyl Sulfide in the Surface Ocean and the Marine Atmosphere: A Global View," *Science*, **221**, 744–747 (1983).
Burroughs, T.: "Ocean Mining," *Technology Review (MIT)*, **87**(3), 54–60 (1984).
Clark, J.P.: "The Rebuttal: The Nodules are Not Essential," *Oceanus*, 18–21 (Fall 1982).
Cooke, R.: "Metals in the Sea," *Technology Review (MIT)*, **87**(3), 61–65 (1984).
Cronan, D.S.: *Underwater Minerals*, Academic Press, Inc., San Diego, CA, 1980.
Curtis, C.: "The Environmental Aspects of Deep Ocean Mining," *Oceanus*, 31–36 (Fall 1982).
Dordrecht, Y.: *Transfer of Technology for Deep Sea-Bed Mining: The 1982 Law of the Sea Convention and Beyond*, Kluwer Academic Publishers, Norwell, MI, 1995.
Fitzgerald, W.F., Gill, G.A., and J.P. Kim: "An Equatorial Pacific Ocean Source of Atmospheric Mercury," *Science*, **224**, 597–599 (1984).
Heath, G.R.: "Manganese Nodules: Unanswered Questions," *Oceanus*, 37–41 (Fall 1982).
Johnson, K.S.: "In Situ Measurements of Chemical Distributions in A Deep-Sea Hydrothermal Vent Field," *Science*, **231**, 1139–1141 (1986).
Knecht, R.W.: "Introduction: Deep Ocean Mining," *Oceanus*, 3–11 (Fall 1982). ·
Koski, R.A., et al.: "Metal Sulfide Deposits on the Juan de Fuca Ridge," *Oceanus*, 42–46 (Fall 1982).
MacLeish, W.H.: *The Struggle for Georges Bank*, Atlantic Monthly Press, New York, NY, 1985.
Manheim, F.T.: "Marine Cobalt Resources," *Science*, **232**, 600–608 (1986).
Moore, J.G. and G.W. Moore: "Deposit from a Giant Wave on the Island of Lanai, Hawaii," *Science*, **226**, 1312–1315 (1984).
Mortlock, R.A. and P.N. Froelich: "Hydrothermal Germanium Over the Southern East Pacific Rise," *Science*, **231**, 43–45 (1986).
Mottl, M., Holland, H.D., and R.F. Corr: "Chemical Exchange During Hydrothermal Alteration of Basalt by Seawater," *Geochim. Cosmochim. Acta*, **43**, 869–884 (1980).
Mottl, M.J.: "Submarine Hydrothermal Ore Deposits," *Oceanus*, **23**, **2**, 18–27 (1980).
Pendley, W.P.: "The Argument: The U.S. Will Need Seabed Minerals," *Oceanus*, 12–17 (Fall 1982).
Post, A.: *Deepsea Mining and the Law of the Sea*, Kluwer Academic Publishers, Norwell, MI, 1983.
Riggs, S.R.: "Paleoceanographic Model of Neogene Phosphorite Deposition, U.S. Atlantic Continental Margin," *Science*, **223**, 123–131 (1984).
Rona, P.A.: "Mineral Deposits from Sea-Floor Hot Springs," *Sci. Amer.*, 84–92 (January 1986).
Siegel, M.C. and S. Turner: "Crystalline Todorokite Associated with Biogenic Debris in Manganese Nodules," *Science*, **219**, 172–174 (1983).
Vogt, P. and B. Tucholke: *The Western North Atlantic Region*, Geological Society of America, Inc., Boulder, CO, 1986.

Web References

Scripps Institution of Oceanography: http://www.sio.ucsd.edu/
Scripps Research Oceanography: http://www.sio.ucsd.edu/research/oceanography.html
The National Oceanic & Atmospheric Administration (NOAA): http://www.noaa.gov/
Woods Hole Oceanographic Institution (WHOI): http://www.whoi.edu/science/science.html
Woods Hole Department of Marine Chemistry and Geochemistry: http://www.whoi.edu/science/MCG/dept/
Woods Hole Department of Geology & Geophysics: http://www.whoi.edu/science/GG/dept/

OCEAN SURFACE TOPOGRAPHY MISSION (OSTM/JASON-2). See **Earth Observing System (EOS)**; and **Space Science Missions: Earth**.

OCEAN THERMAL ENERGY CONVERSION (OTEC). Utilization of ocean temperature differentials between solar-heated surface water and cold deep water as a source of electric power. In tropical areas such differences amount to 35–40 °F. A pilot installation now operating near Hawaii utilizes a closed ammonia cycle as a working fluid, highly efficient titanium heat exchangers, and a polyethylene pipe 2000 feet long and 22 inches inside diameter to handle the huge volume of cold water required. Alternative uses for such a system, such as electrolysis of water, ammonia production, and desalination, are envisaged. There has been active interest in the possibilities of this energy source in France from the time of d'Arsonval (1881) that continues, especially in Japan and Hawaii. Ongoing research indicates that OTEC may be harder to commercialize than once projected.

Web References

Hawaii OTEC: http://www.hawaii.gov/dbedt/ert/otec_hi.html
NREL: Ocean Thermal Energy Conversion: http://www.nrel.gov/otec/what.html
The Australian Renewable Energy Site: http://acre.murdoch.edu.au/ago/ocean/ocean.html

OCEAN (Tides). See **Tides**.

OCEAN WATER. An electrolyte solution containing minor amounts of nonelectrolytes and composed predominantly of dissolved chemical species of fourteen elements O, H, Cl, Na, Mg, S, Ca, K, Br, C, Sr, B, Si, and F (Table 1). The minor elements, those that occur in concentrations of less than 1 ppm by weight, although unimportant quantitatively in determining the physical properties of seawater, are reactive and are important in organic and biochemical reactions in the oceans.

Dissolved Species

The form in which chemical analyses of seawater are given records the history of our thought concerning the nature of salt solutions. Early analytical data were reported in terms of individual salts NaCl, $CaSO_4$, and so forth. After development of the concept of complete dissociation of strong electrolytes, chemical analyses of seawater were given in terms of individual ions Na^+, Ca^{++}, Cl^-, and so forth, or in terms of *known* undissociated and partly dissociated species, e.g.,HCO_3^-. In recent years there has been an attempt to determine the thermodynamically stable dissolved species in seawater and to evaluate the relative distribution of these species at specified conditions. Table 1 lists the principal dissolved species in seawater deduced from a model of seawater that assumes the dissolved constituents are in homogeneous equilibrium, and (or) in equilibrium, or nearly so, with solid phases.

TABLE 1. ABUNDANCES OF THE ELEMENTS AND PRINCIPAL DISSOLVED CHEMICAL SPECIES OF SEAWATER, RESIDENCE TIMES OF THE ELEMENTS

Element	Abundance (mg/L)	Principal species	Residence time (years)
O	857,000	H_2O; $O_2(g)$; SO_4^{2-} and other anions	
H	108,000	H_2O	
Cl	19,000	Cl^-	
Na	10,500	Na^+	2.6×10^8
Mg	1,350	Mg^{2+}; $MgSO_4$	4.5×10^7
S	885	SO_4^{2-}	
Ca	400	Ca^{2+}; $CaSO_4$	8.0×10^6
K	380	K^+	1.1×10^7
Br	65	Br^-	
C	28	HCO_3^-; H_2CO_3; CO_3^{2-}; organic compounds	
Sr	8	Sr^{2+}; $SrSO_4$	1.9×10^7
B	4.6	$B(OH)_3$; $B(OH)_2O^-$	
Si	3	$Si(OH)_4$; $Si(OH)_3O^-$	8.0×10^3
F	1.3	F^-; MgF^+	
A	0.6	$A(g)$	
N	0.5	NO_3^-; NO_2^-; NH_4^+; $N_2(g)$; organic compounds	
Li	0.17	Li^+	2.0×10^7
Rb	0.12	Rb^+	2.7×10^5
P	0.07	HPO_4^{2-}; $H_2PO_4^-$; PO_4^{3-}; H_3PO_4	
I	0.06	IO_3^-; I^-	
Ba	0.03	Ba^{2+}; $BaSO_4$	8.4×10^4
In	<0.02		
Al	0.01	$Al(OH)_4^-$	1.0×10^2
Fe	0.01	$Fe(OH)_3(s)$	1.4×10^2
Zn	0.01	Zn^{2+}; $ZnSO_4$	1.8×10^5
Mo	0.01	MoO_4^{2-}	5.0×10^5
Se	0.004	SeO_4^{2-}	
Cu	0.003	Cu^{2+}; $CuSO_4$	5.0×10^4
Sn	0.003	$(OH)?$	5.0×10^5
U	0.003	$UO_2(CO_3)_3^{4-}$	5.0×10^5
As	0.003	$HAsO_4^{2-}$; $H_2AsO_4^-$; H_3AsO_4; H_3AsO_3	
Ni	0.002	Ni^{2+}; $NiSO_4$	1.8×10^4
Mn	0.002	Mn^{2+}; $MnSO_4$	1.4×10^3
V	0.002	$VO_2(OH)_3^{2-}$	1.0×10^4
Ti	0.001	$Ti(OH)_4?$	1.6×10^2
Sb	0.0005	$Sb(OH)_6^-?$	3.5×10^5
Co	0.0005	Co^{2+}; $CoSO_4$	1.8×10^4
Cs	0.0005	Cs^+	4.0×10^4
Ce	0.0004	Ce^{3+}	6.1×10^3
Kr	0.0003	$Kr(g)$	
Y	0.0003	$(OH)?$	7.5×10^3
Ag	0.0003	$AgCl_2^-$; $AgCl_3^{2-}$	2.1×10^6
La	0.0003	La^{3+}; $La(OH)^{2+}?$	1.1×10^4
Cd	0.00011	Cd^{2+}; $CdSO_4$	5.0×10^5
Ne	0.0001	$Ne(g)$	
Xe	0.0001	$Xe(g)$	
W	0.0001	WO_4^{2-}	1.0×10^3
Ge	0.00007	$Ge(OH)_4$; $Ge(OH)_3 O^-$	7.0×10^3
Cr	0.00005	$(OH)?$	3.5×10^2
Th	0.00005	$(OH)?$	3.5×10^2
Sc	0.00004	$(OH)?$	5.6×10^3
Ga	0.00003	$(OH)?$	1.4×10^3
Hg	0.00003	$HgCl_3^-$; $HgCl_4^{2-}$	4.2×10^4
Pb	0.00003	Pb^{2+}; $PbSO_4$	2.0×10^3
Bi	0.00002		4.5×10^5
Nb	0.00001		3.0×10^2
Tl	<0.00001	Tl^+	
He	0.000005	$He(g)$	
Au	0.000004	$AuCl_2^-$	5.6×10^5
Be	0.0000006	$(OH)?$	1.5×10^2
Pa	2.0×10^{-9}		
Ra	1.0×10^{-10}	Ra^{2+}; $RaSO_4$	
Rn	0.6×10^{-15}	$Rn(g)$	

Adapted from Goldberg, E. D., "Minor Elements in Seawater," in J. P. Riley and G. Skirrow, Eds., "Chemical Oceanography," vol. 1, Academic Press, New York, 1965, pp. 164–165.

Both associated and nonassociated electrolytes exist in seawater, the latter (typified by the alkali metal ions Li^+, Na^+, K^+, Rb^+, and Cs^+) predominantly as solvated free cations. The major anions, Cl^- and Br^-, exist as free anions, whereas as much as 20% of the F in seawater may be associated as the ion-pair MgF^+, and IO_3^- may be a more important species of I than I^-. Based on dissociation constants and individual ion activity coefficients the distribution of the major cations in seawater as sulfate, bicarbonate, or carbonate ion-pairs has been evaluated at specified conditions by Garrels and Thompson (1962).

About 10% each of Mg and Ca is tied up as the sulfate ion-pair. It is likely that the other alkaline earth metals, Sr, Ba, and Ra, also exist in seawater partly as undissociated sulfates; about 60% and 21%, respectively, of the total $SO_4^=$ and HCO_3^- are complexed with cations, and two-thirds of the $CO_3^=$ is present as the ion-pair $MgCO_3^0$.

The activities of Mg^{++} and Ca^{++} obtained from the model of seawater proposed by Garrels and Thompson have recently been confirmed by use of specific Ca^{++} and Mg^{++} ion electrodes, and for Mg^{++} by solubility techniques and ultrasonic absorption studies of synthetic and natural seawater. The importance of ion activities to the chemistry of seawater is amply demonstrated by consideration of $CaCO_3$ (calcite) in seawater. The total molality of Ca^{++} in surface seawater is about 10^{-2} and that of $CO_3^=$ is 3.7×10^{-4}; therefore the ion product is 3.7×10^{-6}. This value is nearly 600 times greater than the equilibrium ion activity product of $CaCO_3$ of 4.6×10^{-9} at 25 °C and one atmosphere total pressure. However, the activities of the free ions Ca^{++} and $CO_3^=$ in surface seawater are about 2.3×10^{-3} and 7.4×10^{-6}, respectively; thus the ion activity product is 17×10^{-9} which is only 3.7 times greater than the equilibrium ion activity product of calcite. Thus, by considering activities of seawater constituents rather than concentrations, one is better able to evaluate chemical equilibria in seawater; an obvious restatement of simple chemical theory but an often neglected concept in seawater chemistry.

Constancy and Equilibrium

The concept of constancy of the chemical composition of seawater, i.e., that the ratios of the major dissolved constituents of seawater do not vary geographically or vertically in the oceans except in regions of runoff from the land or in semienclosed basins, was first proposed indirectly in 1819 by Marcet and expanded later by Forchammer and Dittmar. The concept was established on a purely empirical basis whereas in actual fact there is a theoretical basis for the concept.

Barth (1952) proposed the concept of residence (passage) time of an element in the oceanic environment and formalized this concept by the equation

$$\lambda = \frac{A}{dA/dt},$$

where λ is the residence time of the element, A is the total amount of the element in the oceans, and dA/dt is the amount of the element introduced or removed per unit time. Seawater is assumed to be a steady-state solution in which the number of moles of each element in any volume of seawater does not change; the net flow into the volume exactly balances the processes that remove the element from it. Complete mixing of the element in the ocean is assumed to take place in a time interval that is short compared to its residence time. Table 1 shows the residence times of the elements, and Table 2 compares the residence times of some elements on the basis of river input and removal by sedimentation. For the major elements the results are strikingly similar and suggest that at least as a first approximation seawater is a steady-state solution with a composition fixed by reaction rates involving the removal of elements from the ocean approximately equalling rates of element inflow into the ocean. Thus, as a first approximation the steady-state oceanic model implies a fixed and constant seawater composition and provides a theoretical basis for the concept of the constancy of the chemical composition of seawater. However, it is possible that at any time, t_0, for example, the present, the ratios of the major dissolved constituents in the open ocean may be nearly invariant simply because the amounts of new materials introduced by streams and other agents to the ocean are small compared to the amounts in the ocean, and these new materials are mixed into the oceanic system relatively rapidly. But over time periods of 1000 to 2000 years or more

TABLE 2. THE RESIDENCE TIMES OF ELEMENTS IN SEA-WATER CALCULATED BY RIVER INPUT AND SEDIMENTATION

Element	Amount in ocean (in units of 10^{20} g)	Residence time in millions of years	
		River input	Sedimentation
Na	147.8	210	260
Mg	17.8	22	45
Ca	5.6	1	8
K	5.3	10	11
Sr	0.11	10	19
Si	0.052	0.035	0.01
Li	0.0023	12	19
Rb	0.00165	6.1	0.27
Ba	0.00041	0.05	0.084
Al	0.00014	0.0031	0.0001
Mo	0.00014	2.15	0.5
Cu	0.000041	0.043	0.05
Ni	0.000027	0.015	0.018
Ag	0.0000041	0.25	2.1
Pb	0.00000041	0.00056	0.002

After Goldberg, E. D., "Minor Elements in Seawater," in J. P. Riley and G. Skirrow, Eds., *Chemical Oceanography*, vol. 1, Academic Press, New York, 1965, p. 73

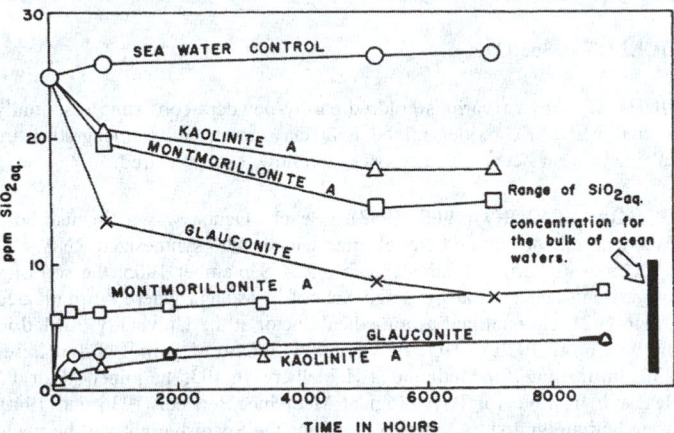

Fig. 1. Concentration of dissolved silica as a function of time for suspensions of silicate minerals in seawater. Curves are for 1-g (<62 μm) mineral samples in silica-deficient (SiO_2 in water was initially 0.03 ppm) and silica-enriched (SiO_2 was initially 25 ppm) seawater at room temperature. Notice that the minerals react rapidly and that the dissolved silica concentration for individual minerals becomes nearly constant at values within or close to the range of silica concentration in the oceans. From Mackenzie, F. T., Garrels, R. M., Bricker, O. P., and Bickley, F., "Silica in Seawater: Control by Silica Minerals" *Science*, **155**, 1404 (1967)

the major ionic ratios can only remain constant if the ocean is a steady-state solution whose composition is controlled by mechanism(s) other than simple mixing.

Further insight into the constancy concept can be gained by exploring possible mechanisms governing the steady-state composition of seawater. The steady-state solution could be simply a result of the rates of major element inflow into the oceans being equal to rates of outflow by biologic removal, flux through the atmosphere, absorption on sediment particles, and removal in the interstitial waters of marine sediments. For example, Ca^{++} carried to the oceans by streams is certainly removed, in part, in sea aerosol generated at the atmosphere-ocean interface and transported into the atmosphere, later to fall as rain or dry fallout on the continents. However, recent theoretical and experimental work suggests that seawater may be modeled as a steady-state solution in equilibrium with the solids that are in contact with it. Sillén has modeled the oceanic system as a near-equilibrium of many solid phases and seawater. Experimental work has shown that aluminosilicate minerals typical of those in the suspended load of streams and in marine sediments react rapidly with seawater containing an excess or deficiency of dissolved silica. Reactions involving

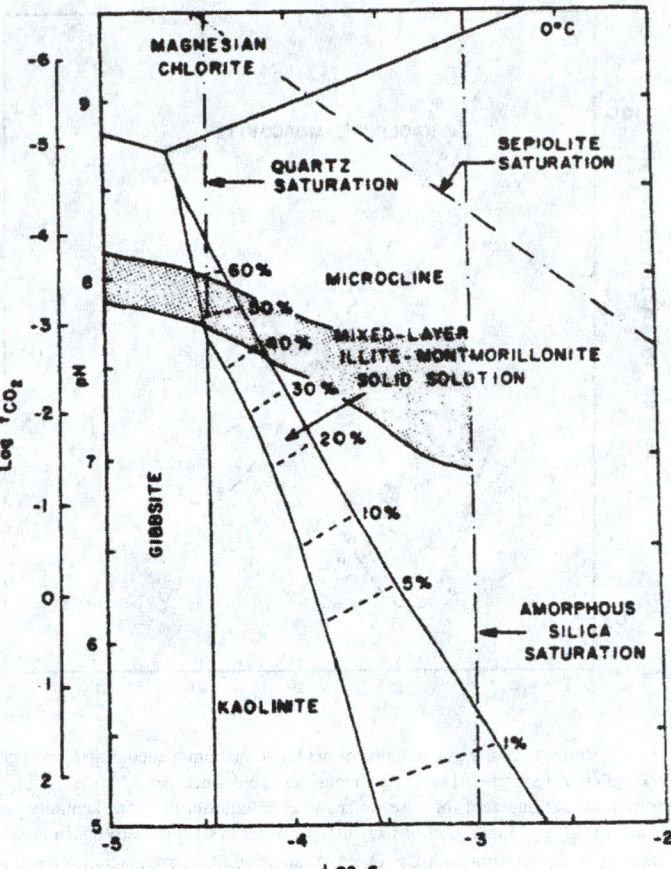

Fig. 2. Logarithmic activity diagram depicting equilibrium phase relations among aluminosilicates and seawater in an idealized nine-component model of the ocean system at the noted temperatures, one atmosphere total pressure, and unit activity of H_2O. The shaded area represents the composition range of seawater at the specified temperature, and the dot-dash lines indicate the composition of seawater saturated with quartz, amorphous silica, and sepiolite, respectively. The scale to the left of the diagram refers to calcite saturation for different fugacities of CO_2. The dashed contours designate the composition (in % illite) of a mixed-layer illitemontmorillonite solid solution phase in equilibrium with seawater. From Helgeson, H. C. and Mackenzie, F. T., 1970, Silicate-seawater equilibria in the ocean system: Deep Sea Res.)

these aluminosilicates may control on a long-term basis the activities of H_4SiO_4 and other constituents in seawater. Thus it has begun to emerge that the composition of the oceans represents an approximation of dynamic equilibrium between the water and the solids that are carried into it in suspension or are precipitated from it by the continuous evaporation and renewal by streams. Therefore, if seawater is a solution in equilibrium with solid phases, or even closely approaches such a system, then the *ion activity ratios* of the major dissolved species would be fixed and the chemical composition of the ocean would be "constant." Consequently, the activity of Ca^{++} in the ocean is not simply a result of removal processes involving sea aerosol, adsorption and so forth but is controlled by solid-solution equilibria. A model leading to nearly invariant ion activity ratios geographically and vertically at any time, t_0, in the oceans based on mixing rates alone may be sufficient to explain the constancy of seawater composition but is somewhat misleading and uninformative when considered in light of the recent advances in treating the oceans as an equilibrium system.

Some limitations of the equilibrium model of seawater do exist. Sillén has pointed out that based on equilibrium calculations all the nitrogen in the ocean-atmosphere system should be present as NO_3^- in seawater; however, most of the nitrogen is present as N_2 gas in the atmosphere. Also, the concentrations of the major alkaline earth elements, Mg, Ca, and Sr, in seawater may vary slightly with depth or geographic location.

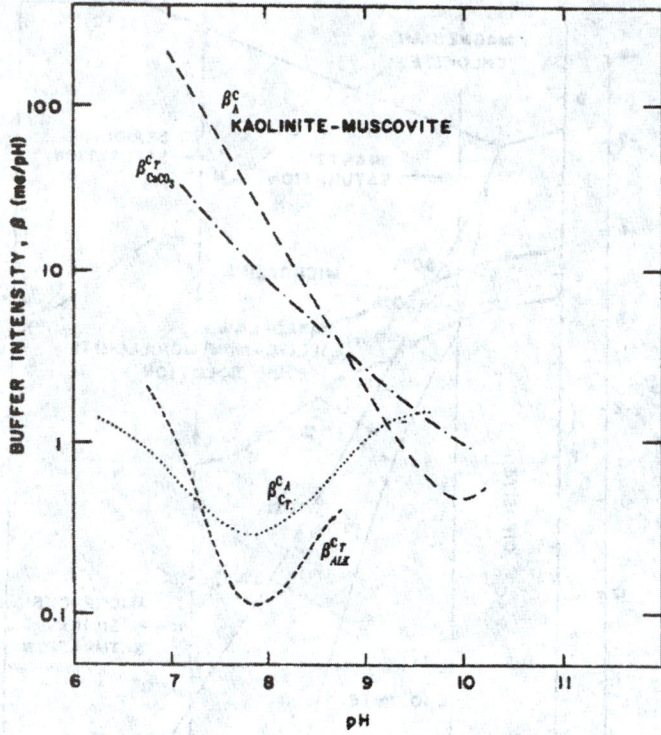

Fig. 3.   Buffer intensity as a function of pH for some homogeneous and heterogeneous chemical systems. The buffer intensities are defined for $\beta^{C_A}_{\text{Kaolinite-muscovite}}$, addition of a strong acid (or base) to seawater in equilibrium with kaolinite and muscovite; $\beta^{C_T}_{\text{CaCO}}$ 3, addition of CO_2 in a seawater system of zero noncarbonate alkalinity in equilibrium with $CaCO_3$; $\beta^{C_A}_{C_T}$, addition of a strong acid (or base) to a seawater solution of constant total dissolved carbonate; and $\beta^{C_T}_{Alk}$, addition of total CO_2 to a seawater solution of constant alkalinity. Data from Morgan, J. J., preprint, "Applications and Limitations of Chemical Thermodynamics in Natural Water Systems")

Buffering and Buffer Intensity of Seawater

The view has long been held that hydrogen-ion buffering in the oceans is due to the $CO_2-HCO_3^- -CO_3^=$ equilibrium. Within recent years this view has been challenged, and the importance of aluminosilicate equilibria in maintaining the pH of seawater emphasized. The buffer intensity of a system is of thermodynamic nature and is defined as

$$\beta^{c_i}_{c_j} = \frac{dC_i}{dpH},$$

where $\beta^{c_i}_{c_j}$ is the pH buffer intensity for incremental addition of C_i to a closed system of constant C_j at equilibrium. Homogeneous buffer intensities are defined for systems without solid phases, e.g., the addition of a strong acid to a carbonate solution, whereas heterogeneous buffer intensities are defined for systems with solid phases, e.g., the addition of a strong acid to a solution in equilibrium with calcite, $CaCO_3$, or with kaolinite and muscovite. The homogeneous buffer intensities for the range of seawater and interstitial marine water pH values (7.0 to 8.3) are about 10- to 100-fold less than the heterogeneous intensities involving equilibrium between calcite and seawater or kaolinite, muscovite, and seawater. Both of these heterogeneous equilibria represent large capacities for resistance to seawater pH changes. Unfortunately, the kinetic aspects of these buffer systems have not been investigated quantitatively. However, it is apparent that aluminosilicate equilibria have buffer intensities equal to and perhaps greater than (the buffer intensities of most aluminosilicate equilibria in natural waters have only been qualitatively evaluated) the $CO_2-CaCO_{3(s)}$ equilibria in seawater. Small additions of acid or base to the oceans could be buffered by the homogeneous equilibrium $CO_2-HCO_3^- -CO_3^=$. However, large incremental additions of acid or base or additions over a duration of time would involve the heterogeneous carbonate and aluminosilicate equilibria; the relative importance of each

would depend on the buffer intensities of the various equilibria and the relative rates of aluminosilicate and carbonate reactions.

For geologically short-term processes on the order of a few thousands of years, it is likely that the carbon dioxide-carbonate system regulates oceanic pH. The long-term pH is controlled by an interplay of various near-equilibria involving carbonates and silicates.

See also **Ocean**.

Additional Reading

Crompton, T.R.: *Analysis of Seawater*, Springer-Verlag New York, LLC, New York, NY, 2005.
Grasshoff, K., K. Kremling, and M. Ehrhardt: *Methods of Seawater Analysis*, 3rd Edition John Wiley & Sons, Inc., New York, NY, 1999.

FRED T. MACKENZIE, Northwestern University, Evanston, IL

OCEAN WAVES.   Waves, on the surface of the ocean, generated by wind action. Their dynamics are governed by the influence of gravity and, for short waves of wavelength less than about 20 centimeters (8 inches), surface tension. The waves are dispersive, with angular frequency, ω, being given by

$$\omega = (gk + Tk^3)^{1/2},$$

where g is the acceleration due to gravity, $k = 2\pi/\lambda$ is the wavenumber, λ is the wavelength, and T is the surface tension divided by the water density.

See also **Ocean**.

OCELOT.   See **Cats**.

OCHER.   Any of various colored earthy powders consisting essentially of hydrated ferric oxides mixed with clay, sand, etc. Some grades are calcined (burnt ocher). The colors are yellow, brown, or red.

OCHOA, SEVERO (1905–1993).   Severo Ochoa was a Spanish-born American biochemist and Nobel prize winner who synthesized RNA.

Ochoa was born at Luarca, Spain on 24 September 1905, the son of a lawyer, and educated at the High School in Malaga where he graduated BA in 1921. He qualified as a medical doctor at the University of Madrid in 1929. From 1929 to 1931 he worked with Otto Meyerhof at the Kaiser Wilhelm Institute for Medicine at Heidelberg. In 1935 he joined Madrid's Research Institute for Physiological Medicine. Between 1936 and 1940, when he immigrated to America to escape the Spanish civil war, he made further studies at Heidelberg, Plymouth and Oxford. In 1940 he joined the medical school at Washington University, St Louis, and worked with C. F. Cori. In 1945 he was elected to the New York College of Medicine's chair of biochemistry. In 1974 he joined the Roche Institution of Molecular Biology in Nutley, New Jersey. He became an American citizen in 1956, but on retirement in 1985 he returned to Spain as honorary professor of Biology at the University of Autonoma, Madrid, where he died on 1 November 1993. He was married but had no children. He was elected Fellow of the Royal Society in 1985. Although he lived abroad most of his life, he is recognized in Spain as one of its most illustrious scientists. See also **Cori,Carl Ferdinand (1896–1984)** and **Meyerhof, Otto Fritz (1884–1951)**.

Long before his work on RNA Ochoa had acquired fame as a biochemist and pharmacologist for his work on glyoxalase and the effects of adrenal secretions on muscular contraction (1930s); and for his isolation of the Krebs cycle intermediate, triphosphopyridine nucleotide (TPN), in the 1950s. It was, however, Crick and Watson's work on the structure of DNA that inspired him to search for the enzyme that assembles nucleotides into DNA. In 1955 he identified a bacterial enzyme in sewage that would create RNA *in vitro*. His student Arthur Kornberg followed this up with a complete synthesis of RNA. In 1959 Ochoa and Kornberg shared the Nobel prize in physiology or medicine for their discovery of the enzyme that synthesizes RNA from nucleoside diphosphates in the presence of magnesium ions. See also **Biochemistry (The History)**; and **Kornberg Arthur (1918–1997)**.

Additional Reading

Kornberg, A., and B.L. Horecker: *Reflections on Biochemistry in Honour of Severo Ochoa*, Pergamon Press, Oxford, UK, 1976.

Ochoa, S.: "Enzymatic Synthesis of Nucleic Acid like Polynucleotide's," *Science*, **122**, 907–910 (1955).

Ochoa, S.: *Enzymatic Synthesis of RNA: Nobel lectures Physiology or Medicine 1942–1962*, Elsevier, Amsterdam, Netherlands, 1964. pp. 639–664.

Ochoa, S.: *Macromolecules: Biosynthesis and Function*, Academic Press, New York, NY, 1964.

Ochoa, S.: "The Pursuit of a Hobby," *Annual Review of Biochemistry*, **48**, 1–30 (1980).

Ochoa, S.: *Viruses, Oncogenes, and Cancer*, Karger, New York, NY. 1985.

Rudavsky, S.: "Severo Ochoa," In: Krapp, K.M.: *Notable Twentieth-Century Scientists*, Gale Research, Detroit, MI, 1998, pp. 337–339.

NOEL G. COLEY, The Open University, Milton Keynes, UK

OCTAHEDRITE. See **Anatase**.

OCTAL. A number system used on binary digital computers. The system is based on a radix of 8. One octal digit corresponds to three binary digits, thus allowing easy conversion from one system to the next. The correspondence between numbers in decimal, binary, and octal notation is given in Table 1.

TABLE 1. COMPARISON OF DECIMAL, OCTAL, AND BINARY NOTATION

Decimal	Octal	Binary
0	0	0
1	1	1
2	2	10
3	3	11
4	4	100
5	5	101
6	6	110
7	7	111
8	10	1000
9	11	1001
10	12	1010
11	13	1011
12	14	1100
13	15	1101
14	16	1110
15	17	1111
16	20	10000

OCTANE NUMBER. A number indicating the antiknock properties of an automotive fuel mixture under standard test conditions. Pure normal heptane (a very high-knocking fuel) is arbitrarily assigned an octane number of zero, while 2,2,4-trimethylpentane, or isooctane (a branched-chain paraffin), is assigned 100. Thus, a rating of 80 for a given fuel indicates that its degree of knocking in a standard test engine is that of a mixture of 80 parts isooctane and 20 parts *n*-heptane.

Octane ratings as high as 115 have been obtained by addition of tetraethyllead to isooctane. Premium leaded gasolines have a research octane rating of about 100, but this value drops to 85–90 for unleaded gasolines, though it may be improved by addition of methyl-*tert*-butyl ether. The octane rating scale ends at 125, and any higher figure is meaningless.

Research octane numbers are those obtained under test or laboratory conditions; they generally run about ten points higher than the so-called motor octane numbers, which represent actual road operating conditions.

See also **Petroleum**.

Web References

ASTM International: http://www.astm.org/DATABASE.CART/REDLINE_PAGES/D2700.htm

Exxon Company, U.S.A.: http://www.prod.exxon.com/exxon_productdata/lube_encyclopedia/octane_number.html

OCTAVE. The interval between two sounds having a basic frequency ratio of two. By extension, the octave is the interval between any two frequencies having the ratio 2:1. The interval, in octaves, between any two frequencies is the logarithm to the base two (3.322 times the logarithm to the base 10 of the frequency ratio).

OCTOPUS. See **Mollusks**.

OCUBA WAX. See **Bayberry Shrubs and Trees**.

OCULAR HYPERTENSION. Ocular hypertension occurs when the intraocular pressure in the eyes is above the normal range, but it has not yet affected vision or damaged the structure of the eyes. Normal eye pressure usually ranges between 10 and 21 mm of mercury. Pressure consistently above 21 indicates ocular hypertension. The condition can develop into glaucoma, a serious disease that causes damage to the optic nerve.

Most at risk for developing ocular hypertension are African Americans and those with a family history of the condition. It is also more common in those who are nearsighted, have high blood pressure, or are diabetic. Because ocular hypertension has no outward symptoms, people over the age of 40 and those in a high-risk category for glaucoma should have their pressure checked every year. A pressure check is a painless procedure that is part of any comprehensive eye examination.

In simple terms, ocular hypertension is caused by an excessive buildup of fluid inside the eye. This fluid, or aqueous humor, nourishes the cornea, iris, and lens, and maintains intraocular pressure. A typical eye produces about 4 cc of fluid a day, which circulates and then drains out of the eye. If the drainage system becomes clogged or if too much fluid is produced, pressure inside the eye can build up. The reasons for this are not fully understood.

There are normally no symptoms of ocular hypertension, which is one of the reasons why regular eye examinations are so important. Although ocular hypertension in itself is not sight threatening, if pressure within the eye builds to the point where it damages the optic nerve (glaucoma), eyesight can be permanently damaged.

An instrument called a tonometer is used to check eye pressure. There are two types of tonometers. One is called an applanation tonometer and uses an instrument that looks somewhat like a pen. After numbing eye drops are administered, the instrument is applied gently to the front surface of the eye and provides a pressure reading. The other type is a noncontact tonometer, which directs a warm puff of air toward the eye without touching it.

Neither ocular hypertension nor glaucoma can be prevented or cured, and ocular hypertension does not usually require treatment unless it progresses to glaucoma. Some doctors may, however, treat the condition with eye drops or other medicines as a precautionary measure. After you are diagnosed with ocular hypertension, your eye health must be monitored closely. See also **Glaucoma**; and **Tonometer**.

Vision Rx, Inc., Elmsford, NY

ODD-EVEN RULE OF NUCLEAR STABILITY. The rule, based on the number of stable nuclides, that nuclides with even numbers of both protons and neutrons are most stable; those with an even number of protons and an odd number of neutrons, or vice versa, are somewhat less stable; and those with odd numbers of both protons and neutrons are least stable.

ODONATA. The dragonflies and damsel flies, an order of insects containing moderate to large species with slender bodies and four narrow net-veined wings. Their flight is powerful and they are well adapted to take their insect prey on the wing. The mouth is formed for biting. In the immature stages these insects are aquatic. Commonly called devil's darning needles. The order contains about 2700 species and is represented in all parts of the world.

ODOR. An important property of many substances, manifested by a physiological sensation caused by contact of their molecules with the olfactory nervous system. Odor and flavor are closely related, and both are profoundly affected by submicrogram amounts of volatile compounds. Attempts to correlate odor with chemical structure have produced no definitive results. Objective measurement techniques involving chromatography are under development. Even potent odors must be present in a concentration of 1.7×10^7 molecules/cc to be detected. It has been authentically stated that the nose is 100 times as sensitive in detection of threshold odor values as the best analytical apparatus.

Many compounds have a characteristic odor that is an effective means of identification. Toxic and noxious gases have distinctive odors often utilized for warning purposes. An important exception is carbon monoxide, which is almost odorless. The penetrating, banana-like odor of amyl acetate has been used in mine rescue work. Among the most powerful unpleasant odors are those of organic sulfur compounds, especially ethyl mercaptan (skunk). Organic substances having a pleasant odor are broadly designated as aromatic, regardless of chemical nature. The cyclic aromatic (benzene) series of essential oils have a pleasant odor and are the basis of perfumes and fragrances. Odor research, including evaluation by test panels, is conducted at the Olfactronics and Odor Sciences Center at Illinois Institute of Technology, Chicago, IL.

See also **Odor Modification**.

Web Reference

Illinois Institute of Technology: http://www.iit.edu/

ODORANT. A substance having a distinctive, sometimes unpleasant odor that is deliberately added to essentially odorless materials to provide warning of their presence. For example, mercaptan derivatives may be added to natural gas for this purpose. In a broad sense, perfumes are odorants that are added to cosmetics, toilet goods, etc, largely for consumer appeal.

See also **Odor**; and **Odor Modification**.

ODOR DETECTION (Physiology of). See **Flavors and Essences**; **Olfactory System**; and **Pollination**.

ODOR MASKING. Addition of a substance with strong odor to obtain a less-offensive effect without changing the composition of the original odorous substance.

See also **Odor Modification**.

ODOR MODIFICATION.

Olfaction

Olfaction begins when an odorant stimulates the olfactory receptor cells, triggering the opening or closing of the ion channels, which in turn convert this stimulus to an electrical response to the olfactory bulb and ultimately to other parts of the brain. The olfactory neurons send messages to the olfactory bulbs, structures about the size and shape of peach pits, located on the underside of the large overhanging frontal lobes of the cerebrum. The fact that there is a gene responsible for odor receptor proteins was discovered in 1991.

Perception of odor is therefore a physical mechanism by which information is processed in the brain. Day by day new odors are appearing, all of which are immediately accepted and sorted within the seemingly unlimited categories of the olfactory brain. The brain not only recognizes this information, but evaluates it, sorts it, and associates it with experiences, events, likes, and dislikes.

Some substances are odorous, others are not. Humans can smell at a distance; if one smells the roses in a garden, it is not ordinarily considered that part of the rose is in contact with the nose. Substances of different chemical constitution may have similar odors.

The sense of smell is rapidly fatigued. Fatigue for one odor does not affect the perception of other dissimilar odors, but will interfere with the perception of similar odors. Two or more odorous substances may cancel each other out; this compensation means that two odorous substances smelled together may be inodorous.

Odor travels downwind. Many animals have a keener sense of olfaction than humans. Insects have such extraordinary keenness of smell that it may be a different modality of the chemical sense from that known to humans.

Odors

Odors have been classified according to Carolus Linnaeus, the eighteenth century Swedish botanist who proposed seven odoriferous qualities: aromatic, fragrant, musky, garlicky, goaty, repulsive, and nauseous. Later

in the twentieth century, ethereal (fruity) and empyreumatic (burnt organic matter), together with subdivisions of Linnaeus' classification, were added. In the 1990s, researchers concentrate less on categorizing odors, and more on how people detect and interpret them. Although the average person can name only a handful of common odors, this limitation results from memory retrieval failure, rather than a failure to detect the differences.

Odors are measured by their intensity. The threshold value of one odor to another, however, can vary greatly. Detection threshold is the minimum physical intensity necessary for detection by a subject where the person is not required to identify the stimulus, but just detect the existence of the stimulus. Accordingly, threshold determinations are used to evaluate the effectiveness of different treatments and to establish the level of odor control necessary to make a product acceptable. Concentration can also produce different odors for the same material.

Evaluation Methodologies

Industry has standardized procedures for the quantitative sensory assessment of the perceived olfactory intensity of indoor malodors and their relationship to the deodorant efficacy of air freshener products. Synthetic malodors are used for these evaluation purposes. These malodors should be hedonically associated to the "real" malodor, and must be readily available and of consistent odor quality. These malodors should be tested in various concentrations and be representative of intensities experienced under normal domestic conditions. Panelists are trained to evaluate malodor intensity and the degree of modification.

Modification

Masking can be defined as the reduction of olfactory perception of a defined odor stimulus by means of presentation of another odorous substance without the physical or removal chemical alteration of the defined stimulus from the environment. Masking is therefore hyperadditive; it raises the total odor level, possibly creating an overpowering sensation, and may be defined as a reodorant, rather than a deodorant. Its end result can be explained by the simple equation of $1 + 1 => 2$.

Odor masking does little or nothing to control malodors; it merely covers them up. Many materials used in masking odors are aldehydes, which are very chemically reactive and usually comprise the top note of a fragrance. Odor masking is used in many areas of household, industrial, and institutional use via products that mask such malodors as pet smells, smoke, cooking, and numerous other odors. The forms by which masking is executed vary, and can be solid, liquid, and aerosol.

Counteraction, sometimes referred to as neutralization, occurs when two odorous substances are mixed in a given ratio and the resulting odor of the mixture is less intense than that of the separate components. The acceptable term to describe this occurrence is compensation. Materials that can accomplish this are basically organic odors which are highly polarized, have a strong affinity for each other, and may also have a low vapor pressure. Some of these molecules have the ability to compensate physiologically for certain malodor materials; others to react chemically with them. Counteraction occurs when the compensating substrate is able to form a coordinate bond with osmophoric sites unique to malodor molecules, such as amino- and thio- moieties. The result is overall reduction in odor; the malodor is transformed into an acceptable state, often with some residual freshening odor. This result lowers the total odor perception and can be exemplified by $1 + 1 = < 2$.

Commercial Aspects

Translating odor modifiers into consumer products results in forms, such as solids, liquids, and aerosols, for a market defined as products "for the nose." This includes products that cover up or eliminate odors, perfume the home, or cleanse the air. The categories of this market can be broken out as traditional air fresheners, cat litter products, aroma care, air purification, and disinfectant in both consumer and industrial applications.

Behavior Modification by Odor

Although odorous materials no doubt impact each other, much discussion centers around the ability of odorous materials to influence human behavior.

Odors play a much greater role in human behavior than previously thought. The sense of smell provides a direct link with the function of the brain; therefore, the further study of olfaction can only advance the learning of causes and effects of stimuli to the brain.

The future in research will certainly lead to a better understanding of how odors are recognized, sorted, and classified. Studies promise, among other things, to determine whether perceptually similar, but structurally different, odors share the same class of receptor proteins, whether responses to odors can be modified, and possibly why olfactory neurons regenerate but other neurons do not.

Additional Reading

Lawless, H.T.: *Chem. Senses Flavor*, **14**(3), 349–360 (1989).
Lord, S.: *Vogue*, 171 (Dec. 1991).
McCord, C.P. and W.N. Witheridge: *Odors Physiology and Control*, McGraw-Hill Book Co., Inc., New York, NY, 1949.
Zwaardemaker, H.C.: *Arch. Anat. Physical*, 423–432 (1900).

OFF-AXIS PARABOLIC MIRROR. A mirror in the shape of a paraboloid of revolution will reflect a parallel beam to a single focus within the incident beam, if the beam is parallel to the axis of the paraboloid. In order to reflect a beam to a point off of the original beam, a part only of a paraboloidal mirror is used. Such a part is therefore known as an off-axis parabolic mirror. It is commonly manufactured by cutting it from the larger total paraboloid, because of the difficulty of making small, paraboloidal sections.

OFFLINE. In terms of instruments and computers, offline describes situations where such devices are not directly in the dynamic control or data system loop, but rather they are auxiliary entities. Information collected by offline equipment may be used to adjust dynamic conditions within a loop, but not by the usual instantaneous feedback method. Rather, offline equipment is used in a supervisory manner for making periodic adjustments.

OFFSET. See **Proportional Control**.

OFFSHORE DRILLING. See **Petroleum**.

OGG. See **Data Compression**.

OGIVE. A body of revolution formed by rotating a circular arc about an axis that intersects the arc; the shape of this body; also, a nose of a projectile or the like so shaped.

OHM, GEORGE SIMON (1789–1854). Ohm was a German physicist whose early interest in mathematics was inspired by his father who educated him at home. He was a mathematics and physics teacher at the Jesuit Gymnasium of Cologne when he began his own experimental work after he learned of Oersted's discovery of electromagnetism in 1920. He wanted to teach at a university so he began to work towards the publication of his experimental results. He is remembered for finding and stating in 1827, the relationship between voltage, resistance, and current in an electric circuit referred to as Ohm's law. Ohm's law appears in his famous book *Die galvanische Kette, mathematisch bearbeitet* in which related his theory of electricity. His work was not immediately accepted in Germany. Recognition finally came in the early 1840s from British Scientists. The Royal Society of London awarded him its Copley Medal in 1841. In 1845, he became a full member of the Bavarian Academy. In 1852, Ohm achieved the position of chair of physics at the University of Munich. The unit in which resistances are measured has been named the ohm after him.

See also **Ohm's Law**.

J. M. I.

OHMIC CONTACT. A contact between two materials, possessing the property that the potential difference across it is proportional to the current passing through.

OHMMETER. See **Electrical Instruments**.

OHM'S LAW. This very familiar law of electric conduction, stated by George Simon Ohm in 1827, is expressible in various forms, of which the following is typical: The steady electric current in a metallic circuit is proportional to the constant total electromotive force operating in the circuit: $I = KE$. The constant K, known as the "conductance" of the circuit, is the reciprocal of the resistance R; so that the equation may be written in the more usual form.

$$I = \frac{E}{R}$$

OIL. See **Petroleum**.

OIL GAS. A gas made by the reaction of steam at high temperature on gas oil or similar fractions of petroleum, or by high-temperature cracking of gas oil. One typical analysis is heating value 554 Btu/ft^3, illuminants 4.2%, carbon monoxide 10.4%, hydrogen 47.6%, methane 27.0%, carbon dioxide 4.6%, oxygen 0.4%, nitrogen 5.8%, autoign temp 637F (336°C).

OIL SANDS. See **Tar Sands**.

OILS, ESSENTIAL. The volatile etherial fraction obtained from a plant or plant part by a physical separation method is called an essential oil. The physical method involves either distillation (including water, steam, water and steam, or dry) or expression (pressing). For the most part, essential oils represent the odorous part of the plant material, and therefore these oils have traditionally been associated with the fragrance and flavor industry. See also **Perfume**. Since essential oils frequently occur as a very small percentage by weight of the original plant material, the processing of large quantities is often required to obtain usable amounts of oil. As a result, expression of an essential oil is only employed in those cases where both the form of the natural plant material, such as a citrus peel, and the quantity of oil present make the process feasible.

It has frequently been observed that the aroma of an essential oil is substantially different from that of the plant before processing. Because this phenomenon is largely the result of the treatment of the plant material with heat or hot water, various other methods have evolved over the years in an attempt to obtain a concentrate of the volatiles which more truly represents the aroma of the original. With the exception of the method of expression, almost all of these involve treatment of the plant material with one or more organic solvents (or mixtures thereof) followed by concentration of the extracted solute. Solvent extraction frequently yields, in addition to the volatile oil, various quantities of semi- or nonvolatile organic material such as waxes, fats, fixed oils, high molecular weight acids, pigments, and even alkaloidal material. However, because solvent extraction often results in a product with superior and more representative odor properties to that of a distilled oil, many natural products critically important to the flavor and fragrance industry are available as various extracts in addition to an essential oil.

Some of the commonly used botanical extracts include the following.

Absolute. This is concentrated extract obtained by treatment of a concrete or other hydrocarbon-type extract of a plant or plant part with ethanol.

Absolute Oil. This is the steam distillable portion of an absolute. Frequently, the absolute oil possesses superior odor properties to that of the corresponding essential oil.

Aroma Distillate. Used by the flavor industry, aroma distillates are the product of continuous extraction of the plant material with alcohol at temperatures between ambient and 50°C followed by steam distillation, and, lastly, concentration of the combined hydro-alcoholic mixture.

Concrete. Hydrocarbon extracts of plant tissue, concretes are usually solid to semisolid waxy masses often containing higher fatty acids such as lauric, myristic, palmitic, and stearic as well as many of the nonvolatiles present in absolutes.

Infusion. Infusion botanical extracts are tinctures that have been concentrated by either total or partial removal of the alcohol by distillation.

Oleoresin. Natural oleoresins are exudates from plants, whereas prepared oleoresins are solvent extracts of botanicals, which contain oil (both volatile and, sometimes, fixed), and the resinous matter of the plant. Natural oleoresins are usually clear, viscous, and light-colored liquids, whereas prepared oleoresins are heterogeneous masses of dark color.

Pommade. These are botanical extracts prepared by the enfleurage method wherein flower petals are placed on a layer of fat which extracts the essential oil.

Resin and Resinoid. Natural resins are plant exudates formed by the oxidation of terpenes. Many are acids or acid anhydrides. Prepared resins are made from oleoresins from which the essential oil has been removed. A resinoid is prepared by hydrocarbon extraction of a natural resin.

Tincture. This is prepared by aqueous alcoholic extraction of the raw plant material. Since the extract is not further concentrated, the plant extract is not exposed to heat.

Essential oils are isolated from various plant parts, such as leaves (patchouli), fruit (mandarin), bark (cinnamon), root (ginger), grass (citronella), wood (amyris), heartwood (cedar), gum (myrrh oil), balsam (tolu balsam oil), berries (pimento), seeds (dill), flowers (rose), twigs and leaves (thuja oil), and buds (cloves).

Exceptions to the simple definition of an essential oil are, for example, garlic oil, onion oil, mustard oil, or sweet birch oils, each of which requires enzymatic release of the volatile components before steam distillation. In addition, the physical process of expression, applied mostly to citrus fruits such as orange, lemon, and lime, yields oils that contain from 2–15% nonvolatile material. Some flowers or resinoids obtained by solvent extraction often contain only a small portion of volatile oil, but nevertheless are called essential oils. Several oils are dry-distilled and also contain a limited amount of volatiles; nonetheless they also are labeled essential oils, eg, labdanum oil and balsam oil Peru. The yield of essential oils from plants varies widely. For example, nutmegs yield 10–12 wt % of oil, whereas onions yield less than 0.1% after enzymatic development.

The function of the essential oil in the plant is not fully understood. Microscopic examination of plant parts that contain the oil sacs readily shows their presence. The odors of flowers are said to act as attractants for insects involved in pollination and thus may aid in preservation and natural selection. Essential oils are almost always bacteriostats and often bacteriocides. Many components of essential oils are chemically active and thus could participate readily in metabolic reactions. They are sources of plant metabolic energy, although some chemists have referred to them as waste products of plant metabolism. Exudates, which contain essential oils, eg, balsams and resins, act as protective seals against disease or parasites, prevent loss of sap, and are formed readily when the tree trunks are damaged.

The cultivation of essential oil-bearing plants has kept pace with modern agricultural methods. Hybrids are grown to yield oils of specific odor, flavor, or other properties. New plants have been developed, eg, lavandin, and new essential oils are isolated and evaluated every year. Very few are commercialized. New growing areas for specific oils are opened that offer quality or economic advantages, such as soil conditions, irrigation, and the availability of labor. Most oils are prepared close to their source so as to provide access to the freshly harvested plant material and to keep processing costs low. Exceptions include the absolutes and other extracts of natural oleoresins, such as myrrh, olibanum, and labdanum, which tend to maintain their odor quality over a longer period of time. A few oils are still produced under very primitive conditions.

Concentrated or folded oils are processed by various physical means to remove wholly or partly undesirable or nonflavor components, such as terpenes or sesquiterpenes, which have poor alcohol and water solubility, very low flavor value, and poor stability. Although this group, for the most part, comprises citrus oils with high terpene contents which cause clouding in drink applications, other oils such as spearmint are included.

Aroma chemicals are isolates, or chemically treated oils or components of oils. Some components are removed physically, others chemically. In most cases, they are further purified by distillation.

A number of other valuable aroma chemicals can be isolated from essential oils, eg, eugenol from clove leaf oil, which can also, on treatment with strong caustic, be isomerized to isoeugenol, which on further chemical treatment can be converted to vanillin. Sometimes the naturally occurring component does not require prior isolation or concentration, as in the case of cinnamaldehyde in cassia oil which, on direct treatment of the oil by a retro-aldol reaction, yields natural benzaldehyde. This product is purified by physical means.

Commercial Essential Oils

Commercial essential oils include rose, jasmin, orange flower (neroli) oil, lavender and lavandin, geranium oil, citronella oil, bergamot oil, lime oil, orange oil, grapefruit oil, sandalwood oil (East Indian), patchouli oil, vetiver oil, galbanum oil, myrrh, oakmoss, tonquin musk, ambergris, tobacco, osmanthus, olibanium, amyris oil, anise oil, anise oil (star), sweet basil oil, bay oil, bitter orange oil, black pepper, bois de rose oil, cannaga oil, caraway oil, cardamom oil, cassia oil, cedarleaf oil, cedarwood oil, Roman chamomile oil, cinnamon bark oil, citronella oil, clove bud oil, coriander oil, cornmint oil, eucalyptus oil, ginger oil, juniper oil, labdanum oil, lemon oil, nutmeg oil, oregano oil (Spanish), orris, palmarosa oil, peppermint oil, petitgrain bigarade oil, pimento berry oil, pine oil, rosemary oil, sage oil (Dalmatian), sage (clary) oil, spearmint oil (native), tagetes oil, thyme oil, turpentine oil, wintergreen oil, and ylang ylang.

Safety and Regulatory Aspects

Essential oils possess a variety of biological properties which may result in varying responses by humans on exposure. An important factor in these effects is the dose to which one is exposed. Thus, essential oils may have both beneficial and toxic effects, depending on their dose. The potential for biological effects from essential oils is not surprising; many botanical species are known to contain substances that possess biological properties, and their identification has contributed significantly to knowledge of biochemistry and physiology as well as the development of therapeutic agents, e.g., quinine and digitalis.

The toxicities of many essential oils have been reported in monographs. Most essential oils used by the flavor and fragrance industries are relatively nontoxic or slightly toxic on acute oral or dermal exposure, and are considered safe when used at levels present in consumer products. In general, the levels of fragrances and flavors in consumer products, and thus the levels of any essential oil ingredients, are relatively low.

Because essential oils are used predominantly by the flavor and fragrance industries, these commercial oils must undergo the same scientific scrutiny as all other flavor and fragrance substances and must be in compliance with all applicable health, safety, and environmental regulations. Guidelines and regulations on the use of essential oils in fragrances differ from those applying to essential oils used in flavors.

Many essential oils have been designated by the FDA or by the expert panel of FEMA as Generally Recognized As Safe (GRAS) for their intended use in foods and flavors. The use and safety of these GRAS substances are continuously being reviewed and the list of GRAS substances updated.

Many countries have adopted chemical substance inventories in order to monitor use and evaluate exposure potential and consequences. In the case of essential oils used in many fragrance applications, these oils must be on many of these lists. New essential oils used in fragrances are subject to premanufacturing or premarketing notification (PMN). PMN requirements vary by country and predicted volume of production. They require assessment of environmental and human health-related properties, and reporting results to designated governmental authorities.

Essential oils are also influenced by legislation that regulates specific products that may contain these oils, e.g., the U.S. Food, Drug, and Cosmetic Act and the European Community Cosmetic Directive. Essential oils would not be anticipated to be of environmental concern, considering that they originate from botanical sources. Thus, natural processes exist to degrade essential oils and recycle their components effectively in the environment.

Additional Reading

Arctander, S.: *Perfume and Flavor Materials of Natural Origin*, 1960.

Bauer, K., D. Garbe, and H. Surburg: *Common Fragrance and Flavor Materials: Preparation, Properties and Uses*, 4th Edition, John Wiley & Sons, Inc., New York, NY, 2001.

Kubeczka, K.H. and V. Formaecek: *Essential Oils Analysis by Capillary Gas Chromatography and Carbon-13 NMR Spectroscopy*. 2nd Edition, John Wiley & Sons, Inc., New York, NY, 2002.

Mookherjee, B.D. and C.J. Mussinan, eds.: *Essential Oils*, Allured Publishing Corp., Wheaton, IL., 1981.

Mookherjee, B.D. and R.A. Wilson: in *On Essential Oils*, Synthite Industrial Chemicals Private Ltd., Synthite Valley, Kolenchery, India, 1986, pp. 281–329.

Srinivas, S.R.: *Atlas of Essential Oils*, the Bronx, NY, 1986.

OILS (Fixed).   See **Fixed Oils**.

OIL SHALE.   This term refers to a carbonaceous rock that can produce oil when heated to pyrolysis temperatures of 427 to 538 °C (800–1,000 °F). The oil cannot be extracted with organic solvents at room and moderate temperatures with known technology. The oil precursor in the rock is a high-molecular-weight organic polymer, *kerogen*. An elemental analysis of kerogen derived from the upper zones of Colorado and Utah oil shales is: Carbon, 80.5% (weight); hydrogen, 10.3%; nitrogen, 2.4%; sulfur, 1.0%; and oxygen 5.8%. Usually, the host rock is comprised of dolomite, calcite, quartz, and clays.

Status of Oil Shale Technology

As with most other alternative sources of energy and notably hydrocarbons, the so-called oil glut of the 1980s essentially brought to a halt the further development of oil shale technology. With the fading of federal support in connection with the demise of the Synthetic Fuels Corporation (SFC) in the mid-1980s and the abandonment by most petroleum companies of their earlier interest in alternative energy sources, the shale oil program is essentially a "dead issue" as of the late 1980s. It is estimated that few researchers are now engaged in advancing the technology. One of the last major investments made in oil shale retorting was sponsored by Union Oil Company in connection with its plant located near Parachute Creek, Colorado.

Resources

Within a radius of about 200 kilometers (124 miles) of a center point where the borders of Colorado, Utah, and Wyoming intersect, an area comprised of four great basins—the Green River, Piceance, Uinta, and Washakie Basins as shown on the map in Fig. 1—some authorities estimate there is the equivalent of 2 trillion barrels of oil to be found—more than 50 times the total reserves for petroleum in the United States as presently known. These shales can be expected to yield from 25 to 30 gallons of oil per ton of shale. Of the explored basins, it is estimated that about 83% of these shales are in Colorado; 8.8% in Utah; and 8.2% in Wyoming. These deposits are generally referred to as occurring in the Green River formation. These oil shale resources occur in beds at least 30 meters (98 feet) in thickness. A comparatively small percentage may be mined by surface techniques. For the deeper veins, essentially underground mining technology or in-situ recovery technology will be required. Geologists consider that these deposits date from the Eocene era.

There are also Eastern oil shales, which represent petroleum locked in older shales dating back to the Devonian and Mississippian eras. It is estimated that these shales contain some 2 trillion barrels of oil and underlie an area of over a million square kilometers (400,000 square miles) in Michigan, western Pennsylvania, eastern Ohio, southwestern Indiana, eastern and southern Illinois, most of Kentucky and Tennessee, much of Oklahoma, and northern Alabama. They are sometimes broken down into what are known as the Eastern and Midwestern oil shale deposits. In Michigan, the shale is approximately 60 meters (197 feet) thick and is in a basin at depths ranging from about 0.8 kilometer (0.5 mile) to outcroppings in three of the northern counties. In general, the Eastern oil shales become thicker and deeper toward the east.

Processing Oil Shale

Work on the production of petroleum-like materials from oil shale in the Green River formation dates back many years. One of the first major efforts was that undertaken by the U.S. Bureau of Mines in 1944. This program involved two 40-ton capacity retorts that operated between 1947 and 1951, with a production of some 920 runs and a total consumption of 37,500 tons of raw shale. This was a batch process and much was learned from this experience.

In a status report (1983), one of the largest (12,800 tons/day) and the last of numerous oil shale retorting processes (Union Oil Company) is described in considerable detail by Duir, Griswold, and Christolini in the February 1983 issue of *Chemical Engineering Progress* (pp. 45–50). This paper includes several diagrams and provides an excellent starting point for the reader who may be interested in this topic.

As of 1995, there were only a few commercial oil shale facilities operating in the world. These facilities are located in countries where

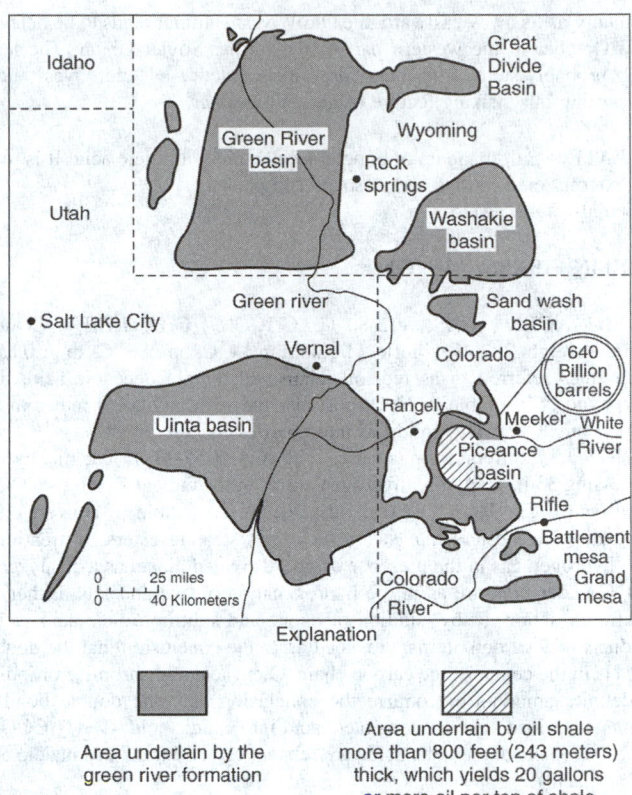

Fig. 1.   Distribution of oil shale in the Green River formation. (*U.S. Geological Survery.*)

the economic, political, and environmental requirements for commercial oil shale development are met. There are commercial oil shale facilities in Brazil, China, Estonia, and Israel. No commercial oil shale facilities have existed in the United States because the costs of shale oil processing exceed those associated with conventional petroleum crude processing.

Additional Reading

Berkowitz, N.: *Fossil Hydrocarbons: Chemistry and Technology*, Elsevier Science, New York, NY, 1997.
Lee, S.: *Oil Shale Technology*, CRC Press LLC., Boca Raton, FL, 1990.
Selly, R.C.: *Elements of Petroleum Geology*, 2nd Edition, Elsevier Science, New York, NY, 1997.

OKAPI.   See **Giraffe and Okapi**.

OKRA.   See **Malvaceae**.

OLAH, GEORGE A. (1927–).   Born in Hungary, now an American citizen, he won the Nobel Prize for chemistry in 1994 for his work with carbocations. These are positively charged hydrocarbons with lifetimes on the order of microseconds. Olah developed methods of studying carbocations with different physical techniques, changing the direction of this field. He received a Ph.D. from the Technical University of Budapest in 1949.

OLD WIVES' SUMMER.   A period of calm, clear weather, with cold nights and misty mornings but fine warm days, which sets in over central Europe toward the end of September; comparable to Indian summer. See also **Indian Summer**

It has been explained as a transition between the summer and winter pressure types. In summer, central Europe is dominated by the Azores high, from which a wedge of high pressure extends to southwestern Germany. In winter, the dominant feature is the Siberian high, from which a ridge extends across Switzerland. Between these two stages there is often a period, on the average occurring between 18 and 22 September, during which an independent anticyclone forms over Germany. As this gradually drifts away eastward, the Old Wives' summer tends to be delayed until October in the western part of the former Soviet Union. The term itself probably stems from the widespread existence of "old wives' tales" concerning this striking feature of autumn weather.

OLEATE.   Salt made up of a metal or alkaloid with oleic acid. It is used for external medications and in soaps and paints.

See also **Oleic Acid**.

OLEFIN FIBERS.   See **Fibers**.

OLEIC ACID.   [CAS: 112-80-1]. $CH_3(CH_2)_7CH:CH(CH_2)_7 \cdot COOH$, formula weight 282.45, colorless liquid, mp 14 °C, bp 286 °C, sp gr 0.854. Sometimes referred to as red, oil, elaine oil, or octadecenoic acid, this compound is insoluble in H_2O, but miscible with alcohol or ether in all proportions. Oleic acid solidifies into colorless needle crystals.

Oleic acid differs from stearic acid [CAS: 57-11-4] chemically by possessing 33 instead of 35 hydrogen atoms in the radical $C_{17}H_{33} \cdot COOH$ (oleic acid), $C_{17}H_{35}COOH$ (stearic acid). It is possible to convert oleic acid and oleate esters into stearic acid and stearate esters by treatment with hydrogen gas in the presence of finely divided nickel as a catalyzer at 250 °C under pressure as in the hydrogenation of oils and fats. Either by careful oxidation, or by addition of ozone and splitting, oleic acid yields products of 9 carbon atoms, thus leading to the conclusion that the double bond is in the center of the carbon chain. Oleic acid adds bromine or iodine in definite amounts to confirm the conclusion that one double bond is contained. Nitric acid converts oleic acid into elaidic acid, $C_{17}H_{33}COOH$, mp 51 °C (oleic and elaidic acids are related, cis- and trans-, as maleic and fumaric acids).

Oleic acid may be obtained from glycerol trioleate, present in many liquid vegetable and animal nondrying oils, such as olive, cottonseed, lard, by hydrolysis. The crude oleic acid after separation of the water solution of glycerol is cooled to fractionally crystallize the stearic and palmitic acids [CAS: 57-10-3], which are then separated by filtration, and fractional distillation under diminished pressure. Oleic acid reacts with lead oxide to form lead oleate, which is soluble in ether, whereas lead stearate or palmitate is insoluble. From lead oleate oleic acid may be obtained by treatment with H_2S (lead sulfide, insoluble solid, formed). With sodium oleate, a soap is formed. Most soaps are mixtures of sodium stearate, palmitate, and oleate.

Representative esters of oleic acid are: methyl oleate, $C_{17}H_{33}COOCH_3$, bp 190 °C at 10 millimeters pressure; ethyl oleate, $C_{17}H_{33}COOC_2H_5$, bp 205 °C at 10 millimeters pressure; glyceryl trioleate (triolein), $C_3H_5(COOC_{17}H_{33})_3$, bp 240 °C at 18 millimeters pressure.

Oleic acid is used in the preparation of metallic oleates, such as aluminum oleate for thickening lubricating oils, for water-proofing materials, and for varnish dryers. The glyceryl ester of oleic acid is one of the constituents of many vegetable and animal oils and fats.

See also **Vegetable Oils (Edible)**.

OLEORESIN.   Any of a number of mixtures of essential oils and resins characteristic of the tree or plant from which they are derived. Most types are semi-solid and tacky at room temperature, becoming soft and sticky at high temperatures. They have various distinctive odors. See also **Balsam**; **Oils, Essential; Resins (Natural)**; and **Rosin**.

OLEUM.   The Latin word for oil, applied to fuming sulfuric acid. (Sulfuric acid was originally called oil of vitriol.)

OLFACTORY SYSTEM.   In humans, this system provides the sense of smell and also contributes to taste. Most sensations described as taste or flavor actually are aromas. When chewing or swallowing, odor-bearing molecules progress through the back of the mouth and into the olfactory cells, which are linked by nerve cells to the brain. Obviously, smell and taste are extremely important to the marketers of foods, beverages, cosmetics, deodorants, soaps, cleaners, and a variety of household products. See also **Flavors and Essences**.

As pointed out by A.R. Newman (see reference listed), most odorants are a select group of small, hydrophobic molecules with masses up to 300

Da, which is typical of volatile molecules. Some non-volatiles, such as oils, can be detected as aerosols. Complex structure-function relationships have been developed to classify odorants. Generally, many odor-producing molecules include one or two polar functional groups, often containing an oxygen or sulfur atom. Classification of odors is made difficult by the fact that they can be described only in subjective terms, such as *minty, fishy, pungent*, and so on. Moreover, the vast majority of natural odors are complex mixtures of odorants in which subtle variations in the relative ratios may, for instance, distinguish the bouquet of one wine from another.

As observed by A.R. Hirsch (Smell and Taste Treatment and Research Foundation, Chicago, Illinois), "Ability to smell and taste varies widely among individuals and even the same individual varies in sensitivity under different conditions...In general, women have greater olfactory sensitivity than men, and their sensitivity is keenest at ovulation...Young people are more sensitive than old; nonsmokers are more sensitive than smokers; hungry people are more sensitive than those who are satiated; heavy eaters especially are much more sensitive to smell and taste impressions when they are hungry."

A phenomenon sometimes referred to as "smell blindness" occurs in nearly half of the human population, particularly in the lack of sensitivity (*anosmia*) to the odor of androsterone, a volatile steroid found in human perspiration. Nearly every individual has a few deficiencies in sensing the full spectrum of odors. Aside from the odor of combustible materials and certain animal odors, such as skunks, few odors announce imminent dangers, and hence they are not regarded as serious handicaps.

Researchers at the Monell Chemical Senses Center (Philadelphia, Pennsylvania) have discovered that anosmias can be reversed. This is accomplished by continued repetitious exposure to certain "blind spots." Investigators suspect that repeated exposure to odorous molecules, such as androsterone, stimulates certain olfactory receptors to multiply. Neurons of the olfactory system have two distinguishing factors—an ability to detect specific molecules and the ability to reproduce themselves. Thus, the correction of an anosmia may be the result of a multiplication of the neurons, or more molecular receptors may be increased on each neuron.

It is important to note that these are the only neurons in the body known to regenerate and repopulate themselves. One researcher has observed, "If we could learn what causes the turnover in olfactory neurons, we might learn how to stimulate other nerves to regenerate. Some researchers suggest a possible connection with the histocomplex, which is important to immune response. Thus, olfactory-directed research may provide further understanding of the human immune system.

Newman suggests further that three properties define olfactory response: (1) *threshold value* (minimum amount detected), (2) *intensity* (response), and (3) *type of odor* (physical and/or chemical properties). See reference listed.

In 1986, the staff of *National Geographic* magazine conducted a survey of the "Human Sense of Smell" by sending out thousands of forms containing six sealed patches representing different aromas and including a questionnaire for respondents to complete. The study was designed by the Monell Chemical Senses Center. The survey questionnaire stated, "Of all our senses, smell is the least understood. You have probably seen all the colors you will ever see and heard all the tones you will ever hear. But smell (and flavor, which is mostly smell) seems to have no such limits, and how do you measure it?"

As pointed out by the Editor of *National Geographic*, "Smell and emotion are so entwined with experience that each of us may perceive the same odor with far different feelings. Depending on one's early exposure to horses, the aroma of a stable might delight one person, frighten another, and sadden a third."

The results of the survey were published in the October 1987 issue of the magazine. The highlights included the findings pertaining to six fundamental odors:

- Androsterone (*sweat*)—difficult to identify by many people; women are more sensitive to this odor than men are.
- Isoamyl acetate (*banana*)—readily detectable, but difficult to identify; discerned equally by men and women.
- Galaxolide (*musk*)—difficult to detect and identify; recognized by women more easily than by men.

- Eugenol (*cloves*) — easily detected and identified by both sexes.
- Mercaptans (*gas*) — easy to detect, but less obnoxious to older than younger people.
- Rose — easy to detect and identify by women and only slightly less so by men.

G. Lynch and R. Granger (University of California, Irvine) have combined their respective skills in neurobiology and computer science in constructing a computer model based on the olfactory receptors of rats. Basically, individual odors are simulated on bar codes. This has led the researchers to note that the neuronal cells closest to the entrance of the nose make a rough "sort" of incoming molecules, and other cells located further within the nose sort out the more subtle differences.

P. Bartlett (University of Warwick, U.K.) demonstrated an "electronic nose" at the Pittsburgh Conference (analytical chemistry) in 1991. The device is comprised of an array of twelve stannic oxide (SnO_2) sensors for discriminating the aromas of foods, beverages, and perfumes. Although having possible use in product manufacture, the device is strictly an electronic analogue and does not duplicate mammalian processes.

Olfactory Systems in Lower Animals. Sense organs of many lower animals that live in water are more logically interpreted as organs of taste or as more primitive chemoreceptors, since materials must reach them in solution. Some aquatic insects and fishes, however, have olfactory organs enclosed in cavities which open to the exterior. Whether reached by water or not, they are classed as olfactory organs from their resemblance to such organs in related terrestrial forms.

The olfactory organs of insects are most abundant on the antennae. They consist of blunt processes or flat plates, associated with sensory nerve endings. In some cases they are grouped in the lining of depressions. As many as 39,000 have been reported on a single antenna and five or six thousand are frequently present.

In the vertebrates the olfactory cells lie in the epithelium lining a pair of olfactory pits which form in the embryo as depressions at the anterior end of the head. These cells are connected with the brain by the fibers of the olfactory nerves, the most anterior pair of cranial nerves of known function. In the air-breathing vertebrates the olfactory epithelium becomes part of the lining of the nasal passages.

Additional Reading

Amato, I.: "Evolving an Electronic Schnozz," *Science,* 1431 (March 22, 1991).

Billmeyer, B.A. and G. Wyman: "Computerized Sensory Evaluation System," *Food Technology,* 100 (July 1991).

Brennan, P., Kaba, H., and E.B. Keverne: "Olfactory Recognition: A Simple Memory System," *Science,* 1223 (November 30, 1990).

Gardner, J., et al.: *NATO ASI Series*, Springer-Verlag, Berlin, 1990.

Gibbons, R.: "The Intimate Sense of Smell," *National Geographic,* 324 (September 1986).

Gilbert, A.N. and C.J. Wysocki: "The Smell Survey Results," *National Geographic,* 514 (October 1987).

Hirsch, A.R.: "Smell and Taste (Foods)," *Food Technology,* 96 (September 1990).

Margolis, F.L. and T.V. Getchell: *Molecular Neurobiology of the Olfactory System: Molecular, Membranous and Cytological Studies*, Kluwer Academic Publishers, Norwell, MA, 1988.

Marshall, E.: "Don't Underestimate the Nose," *Science,* 1021 (March 1, 1991).

Newman, A.R.: "Electronic Noses," *Analytical Chemistry,* 585A (May 15, 1991).

Ross, P.E.: "Smelling Better: Smell-Blindness 'Cure' May Point to Olfactory Mechanism," *Sci. Amer.,* 32 (March 1990).

Shulman, S.: "Banana Neurons," *Technology Review (MIT)*, 18 (August/September 1988).

Stone, H., McDermott, B.J., and J.L. Sidel: "The Importance of Sensory Analysis for the Evaluation of Quality," *Food Technology,* 88 (June 1991).

Swanson, L.W.W., Hokfelt, T., and A. Bjorklund: *Integrated Systems of the CNS: Cerebellum, Basal Ganglia, Olfactory System*, Vol. 12, Elsevier Science, New York, NY, 1996.

OLIGOCENE. A geologic period of the Tertiary, of the Lower Cenozoic era of the geologic time-scale. The term was proposed by Beyrich in 1854. Type locality near Paris, France. Maximum thickness of strata in Italy. This period began approximately 36,000,000 years ago and lasted for about 16,000,000 years. In the United States marine sediments overlap the Cretaceous and earlier Tertiary sediments of the Atlantic border of South Carolina and the Gulf of Mexico. Marine sediments also occur on the Pacific Coast. Terrestrial sediments are well developed in the easterly Great Plains and Oregon (John Day Basin). The "Bad Lands" of South Dakota, eastern Wyoming, and North Dakota (Black Hills) are important collecting localities for the fossil mammals of this period. In the Paris Basin occur fresh and brackish water deposits, which contain numerous fossil vertebrates, invertebrates, and plants (see also **Paleobotany**). The Oligocene formations are also well developed in Germany, and in the Alps. The marine invertebrates and fishes of the Oligocene are similar to those in the Eocene. Among the mammals the true carnivores have replaced the creodonts. The principal types of mammals are the Archaetherium (giant pig). Poebrotherium (ancestor of the camels), Mesohippus (early horse), Hyracodon (cursorial rhinoceros), and Hoplophoneus (progenitor of the saber-toothed cats). For mineral resources of this period see also **Tertiary**.

OLIGOCLASE. See **Feldspar**.

OLIGURIA. See **Kidney and Urinary Tract**.

OLINGOS. See **Raccoons**.

OLIVE TREE. Of the family *Oleaceae*, the olive is a small tree (*Olea europaea*) indigenous to the eastern Mediterranean region. It has lanceolate evergreen leaves, small inconspicuous flowers, and a purplish drupe, the flesh of which is very bitter in the natural state. Cultivation of the tree has continued through many centuries, gradually spreading not only to all Mediterranean countries, but abroad to suitable regions both in the Old and the New Worlds. In the United States, olive growing is largely restricted to California. The wood of the tree is used to a limited extent.

The olive fruit contains from 60 to 85% oleic acid, 7 to 14% palmitic acid, and 4 to 12% linoleic acid. It also contains stearic and myristic acids. Its iodine value is 85%.

The principal product is the oil, which is expressed from the flesh of the fruit. To obtain this oil the fruit is picked when ripe and usually allowed to dry a bit to remove some of the water contained in the flesh. Pressing freshly picked fruit yields a much higher grade of oil. Pressing is often done in a rather primitive machine. The fruit is first crushed and then firmly pressed. The best grade of oil is known as virgin oil.

Olive oil is widely used as a food or in the preparation of food, owing to its characteristic color, odor, and flavor. Cheaper grades are used in soap-making. Sardine packers require large quantities for packing their product.

OLIVINE. The mineral olivine is a silicate of magnesium and ferrous iron corresponding to the formula $(Mg,Fe)_2(SiO_4)$. Olivine is the group name for the isomorphous series forsterite, Mg_2SiO_4, and fayalite, Fe_2SiO_4. The ratio of magnesium to iron varies considerably but the more common olivines are richer in Mg than in Fe. Olivine crystallizes in the orthorhombic system, usually in flattened prismatic forms, also granular and massive. It has a conchoidal fracture and is rather brittle; hardness, 6.5–7; specific gravity, 3.22–4.39; luster, vitreous; color, olive to gray-green; may be yellowish-brown from the oxidation of the iron. It is transparent to translucent. Olivine occurs both in igneous rocks as a primary mineral as well as in certain rocks of metamorphic origin. It has also been discovered in meteorites.

Olivine crystallizes from magmas that are rich in magnesia and low in silica and which form such rocks as gabbros, norites, peridotites and basalts. The metamorphism of impure dolomites or other sediments in which the magnesia content is high and silica low seems to produce olivine.

Transparent olivines of good color are sometimes used as a gem, often called *peridot*, the French word for olivine; it is also called chrysolite from the Greek meaning gold. Olivine occurs in the lavas of Vesuvius and Monte Somma and in the Eifel district of Germany. Gem material comes from St. John's Island in the Red Sea, Upper Burma, and from Minas Geraes, Brazil. In the United States olivine localities are Orange County, Vermont; Webster and Jackson counties, North Carolina. Arizona and New Mexico have also furnished some gem material.

See also **Peridotite**.

OLLAND CYCLE. A modulating system used in some forms of chronometric radiosonde. The meteorological transducing elements are so designed that each of them controls an electrical contact arm that sweeps a sector of a rotating wire helix or spiral. During a rotation of the helix or spiral each electrical contact momentarily touches the wire at a point in the cycle determined by the value of the meteorological parameter being measured. The chronometric pulses are translated into their meteorological equivalents at the receiving station by a chronograph operating in synchronism with the rotation of the helix or spiral.

OMBROSCOPE. An instrument that indicates the presence of precipitation. The ombroscope consists of a heated, water-sensitive surface that indicates by mechanical or electrical techniques the occurrence of precipitation. The output of the instrument may be arranged to trip an alarm, to record on a time chart, to raise the top on a convertible automobile, etc.

OMEGA SUN. A two-image inferior mirage of the setting or rising sun in which the erect image of the solar disc, which forms the top of the omega, touches the inverted segment of the lower image, which forms the base of the omega. When the sun is even lower, all that is seen are two images of the upper segment of the sun floating back to back above the optical horizon and having an outline similar to that of an American football. The stage at which only two images of the upper rim are seen is one way in which the green flash is formed. See also **Inferior Mirage**.

ONAGER.   See **Horses, Asses, and Zebras**.

ONCOGENES.   As study of the molecular and cellular basis of cancer progresses, it is becoming clear that no two cancers are identical: tumor development is a complex process, and there are many paths to malignancy. Nevertheless, certain tenets persist: that cancer arises as the result of genetic change; that this leads to loss of control over cellular proliferation, and that usually several genetic errors are required to reach the full neoplastic phenotype. Deregulated cellular proliferation may arise in two main ways: through the loss of genes that normally check cell growth (the tumor suppressors) or by the gain of function of genes that either promote cellular proliferation or prevent cell death (the oncogenes) (from Greek *onkos*, tumor). Some cancer-associated genes, for example those involved in DNA repair, do not fall easily into either category; others such as *P53* are tumor suppressors in wild-type form, but can act as dominant oncogenes in certain mutant forms. In many instances of adult human cancer, both features–oncogenic activation and loss of tumor suppressor activity–may be identified.

A typical oncogene has dominant activity, and requires only one allele to be activated in order to be oncogenic. An oncogene may be viral or cellular in origin; a viral oncogene may be unique to the virus, or a homologue of a cellular 'proto-oncogene'. An oncogene can promote transformation *in vitro*, and tumor formation in transgenic animals; further, its deregulation is usually recurrently associated with malignancy. Of over a hundred oncogenes that have been described, many are not entirely typical, but the model remains useful.

Viral-induced Tumors

The concept of cancer-causing genes arose from early observations that viruses can induce tumor formation in animals. Both retroviruses and DNA viruses can carry oncogenes, which often resemble cellular genes. Viruses may also induce tumors through integration into the host genome, bringing a cellular gene under control of a viral promoter, or a virus may promote tumor formation through its suppressive effect on the host immune system. Despite these varied mechanisms, the contribution viruses make to cancer in human and animal populations is relatively small, although the discovery of virally induced tumors has had far-reaching implications for our understanding of oncogenes. See also **Oncogenic Viruses**.

Retroviruses. Pioneering work on retroviral tumors was performed in the early part of the twentieth century, particularly by Peyton Rous who described the Rous sarcoma virus (RSV), which induced tumor formation in chickens. RSV is a typical acute retrovirus, an RNA virus that copies its RNA to DNA by reverse transcription after infection of a cell. The DNA is inserted into the host genome, where it can persist and be inherited by subsequent cell generations. Work on RSV and other acute retroviruses demonstrated not only tumor formation by the viruses, but their ability to transform fibroblasts [Martin, 1970]. Transformation is an *in vitro* phenomenon which has proved a useful assay for oncogenic viruses–and oncogenes themselves. Transformed cells do not respond to growth inhibitory signals such as cell–cell contact, and they have fewer requirements for growth factors than their untransformed counterparts. They are 'immortal', in that they can proliferate indefinitely, and have an unusually rounded morphology. They thus appear as fast-growing colonies of rounded cells which can be quantified on a culture plate. Studies on a form of mutant RSV that failed to induce transformation led to the identification of the first viral oncogene, *src*. See also **Rous, Francis Peyton (1879–1970)**.

Oncogenes like *src* are thought to have been captured from the host genome during viral integration, although they have often undergone mutation during their viral passage. While retroviral oncogenes thus have cellular counterparts, the context in which they are carried renders them oncogenic. They are driven by the promoter within the retroviral long terminal repeat (LTR), which produces unregulated, often higher, expression than the normal cellular genes (Figure 1). Furthermore, retroviral oncogenes may encode proteins with properties different to those of the cellular proteins, as a result of mutation. Importantly, retroviral oncogenes are unnecessary for replication of the virus. Indeed, they often carry a survival disadvantage as a result of tumor induction, and in all likelihood are maintained artificially within the experimental environment, with low natural prevalence or significance.

Chronic retroviruses generally induce tumor formation by the alternative mechanism of gene insertion, whereby random insertion of the viral genome into the host genome can cause activation of cellular genes, causing them to become oncogenic (Figure 1). Because the process of gene activation is random, and because the resulting gene activation is often lethal for the cell, it can take many years for a tumor to be induced by chronic retroviral infection.

Viral Tumors in Humans

There are a small number of human malignancies in which a viral aetiology has been demonstrated, or strongly suspected. However, no acute retrovirus has been associated with any human tumor, and only two chronic retroviruses are linked to cancer in humans. Human immunodeficiency virus (HIV) is associated with a number of malignancies through its immunosuppressive effects, which prevent an adequate T-cell response to malignant cells. This is particularly marked when the malignant cells are expressing foreign antigens, for example from other viruses, which would normally provoke a host response. Moreover, HIV can have a more direct effect on tumor formation by inducing capillary growth, contributing to the pathogenesis of Kaposi sarcoma and perhaps facilitating spread of other tumors. Human T cell leukemia virus type I (HTLV-I), another chronic retrovirus, has been identified as the aetiological agent of adult T-cell leukemia, endemic in certain areas of the world including central Africa and the Caribbean basin. Only 1–4% of HTLV-I carriers develop T-cell leukaemia, and this only after a latency of 20–30 years. See also **Human Imunodeficiency Viruses (HIV)**.

DNA viruses are also not commonly oncogenic in humans–surprisingly, given their frequency of infection in higher animals, and their tendency to promote cell proliferation. One reason is that malignant transformation confers no survival benefit to the virus. Second, DNA virus replication is accompanied by expression of immunogenic proteins, leading to destruction of the host cell. Where such virus-associated tumors are common, therefore, is in the context of impaired T-cell immunity. For example, tumors associated with immunosuppression due to HIV infection include Epstein–Barr virus (EBV)-associated lymphomas, Human herpes virus 8 (HHV-8)-positive Kaposi sarcoma, and anogenital carcinoma associated with Human papillomavirus (HPV) infection. Third, there are nonimmune mechanisms thought to operate within and between cells that suppress the function of viruses and their oncogene products. See also **Herpesviruses (Human)**; and **Papillomaviruses**.

EBV is a lymphotropic herpesvirus, endemic in all human populations but only rarely associated with malignant change. Acute EBV infection can give rise to fatal lymphoproliferation of B cells, and indeed T cells, but only in the context of severe immunosuppression such as exists after

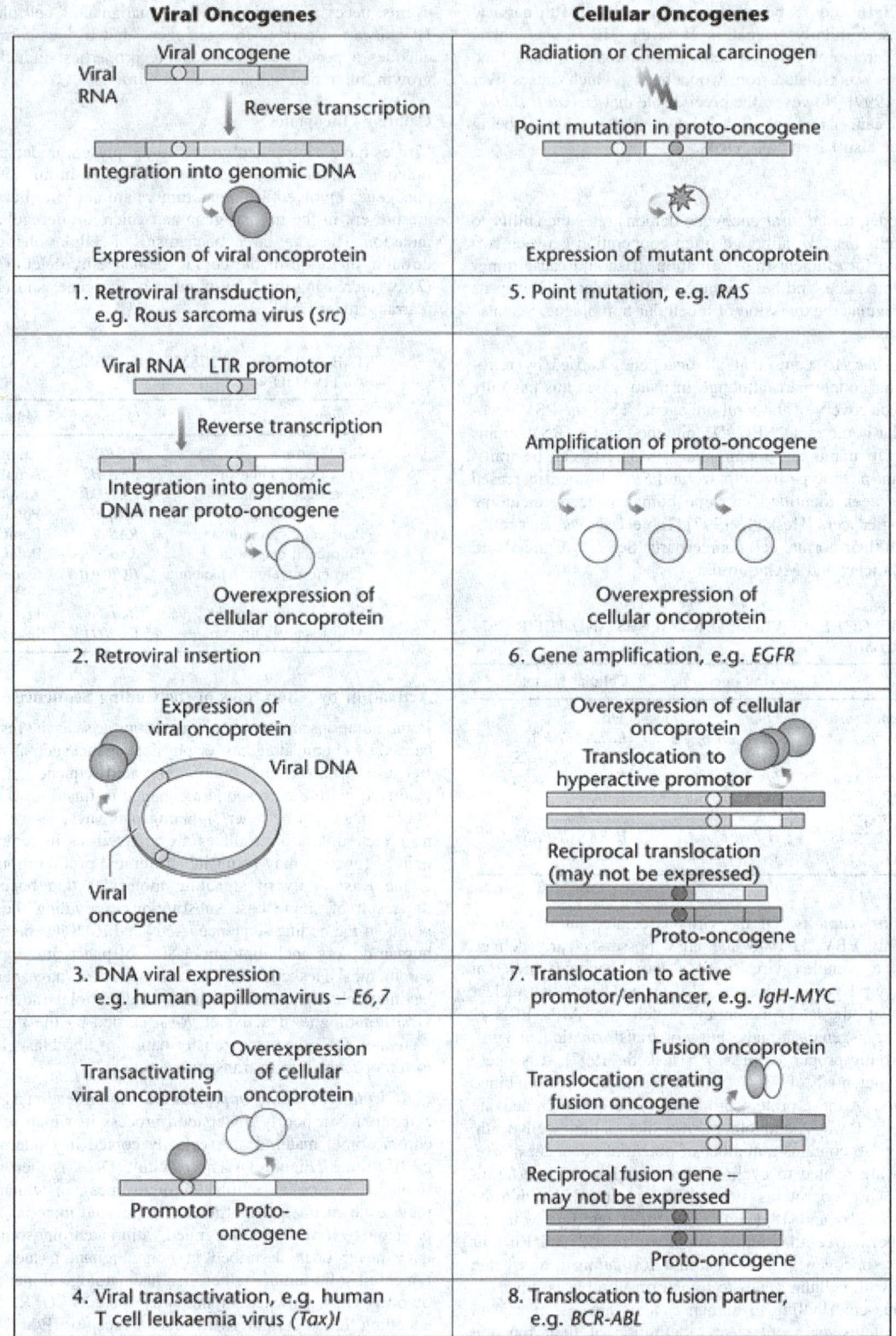

Viral Oncogenes

Cellular Oncogenes

Viral oncogene

Viral RNA

Reverse transcription

Integration into genomic DNA

Expression of viral oncoprotein

1. Retroviral transduction, e.g. Rous sarcoma virus (*src*)

Radiation or chemical carcinogen

Point mutation in proto-oncogene

Expression of mutant oncoprotein

5. Point mutation, e.g. *RAS*

Viral RNA   LTR promotor

Reverse transcription

Integration into genomic DNA near proto-oncogene

Overexpression of cellular oncoprotein

2. Retroviral insertion

Amplification of proto-oncogene

Overexpression of cellular oncoprotein

6. Gene amplification, e.g. *EGFR*

Expression of viral oncoprotein

Viral DNA

Viral oncogene

3. DNA viral expression e.g. human papillomavirus – *E6,7*

Overexpression of cellular oncoprotein

Translocation to hyperactive promotor

Reciprocal translocation (may not be expressed)

Proto-oncogene

7. Translocation to active promotor/enhancer, e.g. *IgH-MYC*

Transactivating viral oncoprotein

Overexpression of cellular oncoprotein

Promotor   Proto-oncogene

4. Viral transactivation, e.g. human T cell leukaemia virus *(Tax)l*

Fusion oncoprotein

Translocation creating fusion oncogene

Reciprocal fusion gene – may not be expressed

Proto-oncogene

8. Translocation to fusion partner, e.g. *BCR-ABL*

Fig. 1.   Mechanisms of oncogene activation. LTR, long terminal repeat

bone marrow transplantation, or through rare genetic defects. Similarly, reactivation of latent EBV, leading to lymphoproliferative tumors, may occur with chronic immunosuppression, either iatrogenic (posttransplant) or virus-induced (AIDS-associated). In nonimmunosuppressed patients, Burkitt lymphoma and nasopharyngeal carcinoma have an established association with EBV. Burkitt lymphoma was long suspected to have an infectious cause, due to its geographically limited occurrence, and the link was initially demonstrated by electron microscopic observation of EBV

particles in tumor cultures [Epstein, et al.: 1964]. Some types of Hodgkin disease and some T-cell lymphomas also have an association with the virus. See also **Leukemias**.

Other DNA viruses involved in human malignancy include HHV8, another herpesvirus identified from Kaposi sarcoma in acquired immune deficiency syndrome (AIDS) patients, now with an established role in the aetiology of this tumor [Moore and Chang, 1998]. HPV, a DNA virus, has over 60 genotypes, 11 of which are associated with human cancers, HPV-16 and 18 being strongly associated with cervical and anal carcinomas. Similarly, Hepatitis B virus (HBV) was linked with hepatocellular carcinoma through seroepidemiological studies, and a closely related virus was isolated from woodchucks, which causes liver cancer [zur Hausen, 1999]. However, the precise role that *Hepatitis B virus* plays in human liver cancer remains ill-defined, with no oncogenes being clearly identified. See also **Liver**; and **Virus**.

Viral Oncogenes

According to the model, a true viral oncogene demonstrates the ability to transform primary cells *in vitro*, although often cooperation between two oncogenes is required for efficient transformation. It should cause tumor formation in transgenic mice, and be associated with cancers, either in its viral form, or as deregulated expression of a cellular homologue. See also **Tumorigenly Assays**.

Retroviral Oncogenes. It seems that all oncogenes carried by retroviruses have a cellular counterpart, although in many cases this has only been identified after discovery of the viral oncogene. Thus the RSV oncogene *src* has a cellular homologue, SRC. Despite the fact that RSV strains are not tumorigenic in mammals, mammalian fibroblasts can be transformed to a malignant phenotype by both *src* and *SRC*; further, increased *SRC* expression has been identified in some human cancers, including colon, skin and breast cancers [Hesketh, 1997]. Table 1 shows other retroviral oncogenes, with their normal cell counterparts. See also **Cancer and Oncology**; **Colon Cancer**; and **Melanoma**.

TABLE 1. EXAMPLES OF RETROVIRAL ONCOGENES AND THEIR CELLULAR HOMOLOGUES

Retrovirus	Viral oncogene	Cellular homologue
Abelson murine leukaemia virus	*abl*	ABL
Avian erythroblastosis virus	*erbA, erbB, ets*	ERBA, ERBB (EGFR), ETS
FBJ murine osteosarcoma virus	*fos*	FOS
Avian sarcoma virus-17	*jun*	JUN
Avian myelocytomatosis virus	*myc*	MYC
Murine sarcoma virus	*H-ras, K-ras*	H-RAS, K-RAS
Rous sarcoma virus	*src*	SRC

Viral Oncogenes in Humans. Of the virus-associated human tumors mentioned above, only EBV, HHV-8 and HPV possess clearly defined oncogenes. EBV is a complex virus with a large genome; in latent infections, about 11 viral genes are expressed, grouped into the nuclear antigens or EBNAs, and the latent membrane proteins or LMPs. EBNA1 induces tumours in transgenic animals, but not transformation *in vitro*; EBNA2 has transforming properties *in vitro* [Hesketh, 1997]. It is likely that the oncogenic potential of EBV is the result of the combined contribution of several of these protein products, some of which activate cellular proliferative pathways, and others of which inhibit cell death. The HHV-8 genome also contains a number of potential oncogenes: here, the protein products are related to cyclin D, interleukins and interferon-responsive factors. Two oncoproteins, E6 and E7, are encoded within the genome of HPV types 16 and 18 (Figure 1); these interfere with the function of two important cellular tumor suppressors, p53 and Rb [zur Hausen, 1999]. These oncogenes result in tumor formation in transgenic mice, and cooperate with cellular genes to transform fibroblasts *in vitro*.

Other viral proteins contributing to human cancers are less typical in their role as oncoproteins, or await clearer definition of their function in tumor formation. HBV frequently shows integration during chronic infection, although this is not part of its life cycle. *Cis*-activation of adjacent cellular genes may result from such integration, while transactivating viral proteins such as pX may contribute to aberrant expression of cellular genes. HTLV-I similarly possesses the viral gene, tax, which transactivates the viral LTR through protein–protein interaction (Figure 1). In addition it can transactivate a number of cellular genes, including those encoding cytokines, cytokine receptors and transcription factors, and it is this function that is thought to contribute to malignant transformation. The *tax* gene cannot, however, induce transformation alone, and there may be other HTLV-I genes, and probably nonviral cellular

events, necessary for inducing the malignant T-cell phenotype. Similarly, HIV does not carry a true oncogene, but it does carry the gene *Tat* that encodes a protein with angiogenic properties, inducing endothelial cell growth, migration and invasion *in vitro*.

Cellular Oncogenes

Viruses have contributed greatly to the present understanding of oncogene function, but have had a less significant role in human oncogenesis. Most oncogenes involved in human tumors are not viral but cellular–genes that are present in the normal genome, which are deregulated or activated by mutation. There are three mechanisms of cellular oncogenic activation: (1) through alteration of the coding sequence by deletion or point mutation; (2) by increasing the copy number of the gene; and (3) by chromosomal rearrangement (Table 2).

TABLE 2. MECHANISMS OF CELLULAR ONCOGENE ACTIVATION

Malignancy	Oncogene	Mutation
Glioblastoma	*EGFR*	Amplification
Breast carcinoma	*ERB-B2*	Amplification
Breast carcinoma	*CCND1*	Amplification
Neuroblastoma	*ERB-B2*	Point mutation
Pancreatic carcinoma	*RAS*	Point mutation
Colorectal carcinoma	*RAS*	Point mutation
Chronic myeloid leukaemia	*BCR-ABL*	Gene fusion
Burkitt lymphoma	*MYC*	IgH juxtaposition
Follicular lymphoma	*BCL2*	IgH juxtaposition
Mantle cell lymphoma	*CCND1*	IgH juxtaposition

Activation by Alterations of the Coding Sequence

Point mutations and intragenic mutations arise as the result of DNA damage caused by chemical agents, or physical agents such as radiation (Figure 1). Because of base changes, the amino acid sequence of an encoded protein is altered, with a corresponding change in function. If the protein is a key player in growth or other signaling pathways, loss of function will have no effect (unless both alleles are affected–as in tumor suppressors), but gain of function may result in accelerated proliferation. A prime example is the Ras family of signaling molecules that become hyperactive as the result of single base substitutions, providing these occur at crucial points in the coding sequence. Activated RAS has been identified in many human tumors, including about 80% of pancreatic and 40% of colorectal carcinomas [Hesketh, 1997]. Such mutations have been demonstrated to be inducible by chemical toxins or ultraviolet radiation, and there is a viral homologue of activated *RAS*, carried by the murine sarcoma virus. Activated *RAS* mediates transformation of fibroblasts in culture, and leads to tumor formation in transgenic animals.

Activation by Gene Amplification. Amplification as a route to oncogene activation is a poorly understood process in which several megabases of chromosomal material are typically copied in tandem, up to perhaps 50 or 100 times (Figure 1). The resulting DNA is then retained episomally (outside the normal cellular chromosomes), in which case cell division may result in unequal portions of replicated material being passed to the progeny, or it may be reintegrated in a chromosome. Such phenomena have never been described in nonmalignant tissues, and appear to be typical of solid tumors rather than haematological malignancies. Examples of oncogenes activated in this way include EGFR, NEU and CCND1, encoding the epidermal growth factor receptor (EGFR), the related erbB2 receptor and cyclin D1, respectively.

Activation by Chromosomal Rearrangement. Oncogenes may also be activated through chromosomal rearrangement, i.e. translocations, interstitial deletions or inversions. These tend to result in the gene being brought under new transcriptional control, leading to over expression, or in the gene being truncated, making it hyperactive, or in the gene being fused to another, leading to a hybrid oncoprotein with functions combining those of both gene partners. Haematological malignancies are particularly associated with chromosomal rearrangements: a well-known example of a fusion gene results from the translocation of ABL on chromosome 9 to the BCR locus on 22. The result, *BCR-ABL*, produces a new Bcr-Abl fusion protein. This molecular change is very strongly associated with chronic myeloid leukaemia (CML) in which it is frequently the only abnormality.

A typical translocation in lymphomas results in a variety of genes, often called *B-cell leukaemia/lymphoma genes*, (*BCL-*), being translocated near to one of the immunoglobulin loci, where they are brought under control of the immunoglobulin promoter and/or enhancer. Since the normal lymphocyte expresses high levels of immunoglobulin, any gene translocated to the same locality, such as *MYC* in Burkitt lymphoma; *BCL2* in follicular lymphoma, or *CCND1* in mantle cell lymphoma, will also be expressed at high levels. See also **Leukemias**.

Virus-associated Chromosomal Rearrangements. The *IgH-MYC* translocation, together with similar translocations of *MYC* to the immunoglobulin light chain loci, are universal findings in Burkitt lymphoma, even in the sporadic cases where EBV is not involved. The close association between a chromosomal translocation and a viral oncogenic effect, however, remains something of a puzzle–are they separate phenomena, both important in the development of lymphoma? If so, non-EBV-related cases are more difficult to explain. Another speculative explanation has been that EBV increases the probability of specific gene translocation, either directly at the gene level, or indirectly by selecting for cells containing that translocation. Interestingly, it has recently been demonstrated that an adenovirus oncogene, E1A, specifically induces a fusion oncogene identical to that found in the human tumor Ewing sarcoma [Kirn and Hermiston, 1999].

Normal Functions of Oncogenes

Both cellular oncogenes and most viral oncogenes derive from normal cellular genes or proto-oncogenes. These genes encode proteins that function in essential cellular pathways: in general, a proto-oncogene is likely to be pro-proliferative, and/or pro-survival, and/or an inhibitor of differentiation. These functions of a cell are governed by factors in the extra cellular environment such as cell–cell contact and growth factors, and signals are relayed from the cell surface to the cell nucleus by signaling cascades. Proto-oncogenes may be divided into broad groups according to their position in such cascades: they may be growth factors; cell surface receptors; membrane transducers; intracellular signaling proteins or transcription factors. The pathways are complex, converging and diverging, and the endpoints–proliferation, survival and state of differentiation–are often interdependent. Nevertheless, even an incomplete understanding of normal proto-oncogene function provides insights into the role of oncogenes in tumorigenesis.

Growth Factors and Receptors. Cell surface receptors that function in growth regulation fall into several super families. One of these includes the platelet-derived growth factor (PDGF) receptor, the ligand for which has an oncogenic homologue, *SIS*, which when over expressed causes excessive signaling through the PDGF pathway. Another of the receptor tyrosine kinase families is the epidermal growth factor (EGF) receptor family. In this case, the gene encoding the EGF receptor itself is amplified in some human tumors [Hanahan and Weinberg, 2000], causing it to function as an oncogene. A related gene, *NEU*, is also over expressed in tumors and has a viral homologue *erbB*, which encodes a truncated EGF receptor. In this form, the receptor is constitutively active, and does not require ligand binding to function. The EGF receptor, like most growth factor receptors, is a tyrosine kinase, signaling to downstream molecules through membrane-associated binding and phosphorylation (Figure 2).

Membrane Transducers. Two important categories of protein that transmit signals from receptors to cytoplasmic molecules are the 'non-receptor tyrosine kinases' such as the protein encoded by *SRC*, and GTP-binding proteins or 'G-proteins', such as members of the Ras family. This family is comprised of closely related genes *H-RAS*, *K-RAS* and *N-RAS*, encoding proteins of 21 kDa (hence, $p21^{ras}$). Each $p21^{ras}$ protein binds one molecule of guanosine triphosphate (GTP) or guanosine diphosphate (GDP); phosphorylation of bound GDP converts $p21^{ras}$ from 'off' mode to 'on' mode, mediating binding and activation of downstream regulators like Raf and MEK. *RAS* becomes oncogenic usually as the result of point mutations, which prevent the hydrolysis of $p21^{ras}$-bound GTP, maintaining it permanently in the 'on' position.

Serine/Threonine Kinases. Further downstream, many molecular mediators function as serine/threonine kinases, each kinase phosphorylating

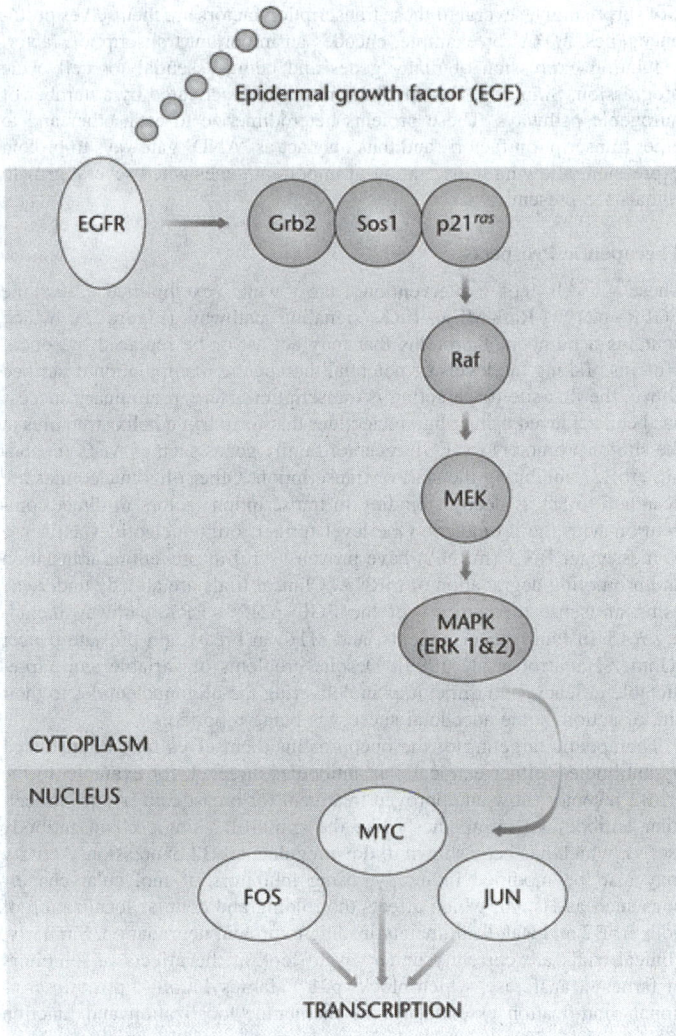

Fig. 2.   The epidermal growth factor receptor (EGFR) signaling pathway

other kinases in cascade fashion. These include the protein product of *AKT*, the cellular homologue of an avian viral oncogene. The Akt protein is an important cytoplasmic kinase that lies downstream of several growth factor and cytokine receptors, and upstream of cell cycle and apoptotic regulators.

Cell Cycle Regulators. The proliferation state of the cell is determined by its position in, and passage through, the cell cycle. Passage through the cycle is in turn governed by a number of proteins that take their cues from signaling cascades, allowing the cell to move from one stage to the next. Some of these proteins can function as oncogenes–such as cyclin D1, encoded by *CCND1*, which normally governs entry of the cell into S phase from G_1. When over expressed as a result of gene amplification or immunoglobulin H (IgH) translocation, the increased cyclin D1 promotes cycling and proliferation.

Inhibitors of Apoptosis. In parallel with the mitogenic pathways are the pathways leading to cell death, which must be inhibited in order for malignant proliferation to proceed in exponential fashion. Inhibitors of apoptosis can thus function as oncogenes: the paradigm here is Bcl2, an antiapoptotic molecule bound to the mitochondrial membrane, serving to prevent caspase activation. Bcl2 is overexpressed in many lymphomas [Hesketh, 1997].

Transcription Factors. The endpoints of mitogenic, cell cycle and apoptotic signaling usually lie within the nucleus, where expression of proteins is altered at the level of transcription. Many pathways converge on transcription factors that promote or inhibit transcription of important genes.

Not surprisingly, several of these transcription factors are themselves proto-oncogenes. *MYC*, for example, encodes an important transcription factor, regulating expression of many genes and being essential for cell cycle progression. Similarly, Fos and Jun proteins are activated by a number of mitogenic pathways. These proteins heterodimerize to each other and to other transcription factors, and thus may act as 'AND' gateways to protein expression, allowing transcription of important genes when several growth signals are present.

Therapeutic Prospects

These levels of intervention are well exemplified by the EGFR–p21^{ras}–Raf–MEK–ERK signaling pathway (Figure 2), which contains a number of proteins that may act as, or be replaced by, oncoproteins. Taking the levels of potential therapeutic manipulation described above, the first site for targeting is transcription. Here, preliminary success has been achieved using oligonucleotides that form triple helix structures at the site of promoters for EGF receptor family genes such as *NEU* (encoding erbB2*)*, inhibiting the start of transcription. Other oligonucleotides are designed to act as decoys, binding to transcription factors in direct competition with the promoters. One level further, oligonucleotides antisense to messenger RNA (mRNA) have proven useful in preventing translation and promoting degradation of mRNA. Clinical trials are already underway using antisense to members of the EGF–p21^{ras}–ERK pathway including *RAS* in lung cancer patients, and *MYC* in breast and prostate cancer [Gomez-Navarro, et al.: 1999]. Despite problems of variable and unpredictable efficacy, and difficulties in delivering the oligonucleotides to their site of action, some anecdotal success is being reported.

Therapeutic targeting of the oncoproteins themselves may be mediated by antibodies–either extracellular antibodies directed, for example, to the erbB2 receptor (now an approved treatment for breast cancer), or intracellular antibodies ('intrabodies') like the anti-erbB2 single chain antibody (scFv), which has been shown to downregulate erbB2 expression. Activity may also be modified indirectly, using inhibitors of molecular chaperones such as Hsp90, which affects the folding and cellular localization of both erbB2 and Raf-1, ultimately leading to protein degradation. Similarly, clinical trials are currently underway to look at the effects of inhibitors of farnesyltransferase, which blocks p21^{ras} farnesylation–a posttranscriptional modification essential for its membrane localization and function [Seckl, 2000]. Finally, progress is being made using immunotherapeutic approaches to counter oncogenic members of the EGFR signaling pathways. Thus, T-cell responses have been generated to erbB2 and mutant p21^{ras}, both *in vitro* and *in vivo*.

Bcr-Abl as a Target for Intervention. Oncogenes are genes that are directly involved in the development of malignancy, and may well be essential for this process. They are often derived from normal cellular genes, or proto-oncogenes, whose functions in cell proliferation, differentiation and survival have become deregulated. Oncogenes may be carried by viruses, or represent cellular genes that have been activated by viruses or by mutation. They usually operate in key cell signaling pathways, and have become important molecular targets for the development of novel anticancer agents, with some preliminary success.

Additional Reading

Epstein, M.A., B.G. Achong, and Y.M. Barr: "Virus Particles in Cultured Lymphoblasts from Burkitt's Lymphoma," *Lancet*, **1**, 702–703 (1964).

Gomez-Navarro, J., D.T. Curiel, and J.T. Douglas: "Gene Therapy for Cancer," *European Journal of Cancer*, **35**, 2039–2057 (1999).

Hanahan, D., and R.A. Weinberg: "The Hallmarks of Cancer," *Cell*, **100**, 57–70 (2000).

Hesketh, R.: *The Oncogene and Tumour Suppressor FactsBook*, 2nd Edition, Academic Press, London, UK, 1997.

Kirn, D., and T. Hermiston: "Induction of an Oncogenic Fusion Protein by a Viral Gene–A New Chapter in an Old Story," *Nature Medicine*, **5**, 991–992 (1999).

Martin, G.S.: "Rous sarcoma Virus: A Function Required for the Maintenance of the Transformed State," *Nature*, **227**, 1021–1023 (1970).

Moore, P.S., and Y. Chang: "Kaposi's Sarcoma-associated Herpesvirus-encoded Oncogenes and Oncogenesis," *Journal of the National Cancer Institute Monographs*, **23**, 65–71 (1998).

Peters, G., and K.H. Vousden: *Oncogenes and Tumour Suppressors: Frontiers in Molecular Biology*, Vol. 19, Oxford University Press, Oxford, UK, 1997.

Seckl, M.J.: "Growth Factor and Cell Signaling Inhibitors as Novel Anticancer Agents," *Cancer Topics*, **11**, 1–4 (2000).

Verfaillie, C.M., R.S. McIvor, and R.C.H. Zhao: "Gene Therapy for Chronic Myelogenous Leukemia," *Molecular Medicine Today*, **5**, 359–366 (1999).

zur, Hausen H.: "Viruses in Human Cancers," *European Journal of Cancer*, **35**, 1878–1885 (1999).

AMANDA R. PERRY, Institute of Cancer Research, Sutton, Surrey, UK

ONCOGENIC VIRUSES. Since the isolation of *Epstein–Barr virus* (EBV) from Burkitt lymphoma (BL) cells in 1964, there have been a number of other viruses that have been demonstrated to be associated with the aetiology of human cancers. While there have been some false alarms along the way, the viruses described in this article have been shown to be the aetiological agents of various cancers. One of the problems with confirming that a virus is the aetiological agent of a particular cancer is that infection with the viruses themselves are common in populations throughout the world, allowing for some geographical variations, yet only a small fraction will present with cancer. The one exception may be *Human herpesvirus-8* (HHV-8) or *Kaposi sarcoma-associated herpes virus* (KSAHV), which, on present serological data, suggests that infection is primarily in individuals with Kaposi sarcoma.

Human cancers are caused by both deoxyribonucleic acid (DNA)- and ribonucleic acid (RNA)-containing viruses, and the article will describe the viruses and the associated cancers and present information on the mechanisms of oncogenesis.

DNA Viruses

There are three DNA virus families that contain amongst their number cancer causing viruses. They are the *Herpesviridae* (HHV-8, EBV), *Papovaviridae* (certain human papillomavirus types) and the *Hepadnaviridae* (*Hepatitis B virus*; HBV).

Herpesviridae

Human Herpesvirus 8/Kaposi Sarcoma-associated Herpesvirus.

Genomic Organization. HHV-8 was first discovered in Kaposi lesions in 1994 when part of the genome was isolated by the polymerase chain reaction (PCR) [Chang, *et al.*: 1994]. The whole genome was subsequently isolated from cell lines, which were derived from body cavity-based primary effusion lymphomas (PELs) also caused by this virus. PEL cell lines infected with HHV-8 have been established and are the only cell culture system at present that will maintain the virus *in vitro*, although the virus can be transmitted to 293 cells (human embryonic kidney cell line transformed by adenovirus 5), but without significant amplification. The genome of HHV-8 has been sequenced and is approximately 165 kbp in length, with variable length terminal repeats at the ends and in between a single long unique region of approximately 140.5 kbp, which contains all the coding sequence of the virus. The genomic organization places the virus among the gammaherpesviruses, a group which includes EBV and *Herpesvirus saimiri*. This genomic organization of HHV-8 is the same, whether isolated from PEL cells or from Kaposi sarcoma tumors, indicating that the virus is truly involved in the development of two different tumors. See also **Herpesviruses (Human)**.

Integration. Some of the DNA tumor viruses integrate their genome into that of the host cell chromosomes, but it is unclear at present if this process is essential for the tumorigenesis or for the various transformed phenotypes seen *in vitro*. So far, the genome of HHV-8 does not appear to integrate into the chromosomes of tumor cells, and the virus can be rescued and transmitted to 293 cells, although replication is very low and detection by PCR is necessary. The genome is circularized in lytic infection and replication is thought to occur by the 'rolling circle' model.

Activation of Cell Growth. The HHV-8 genome sequence has 81 open reading frames (ORFs) and they have been named according to their homology to those of *Herpesvirus saimiri*. Although HHV-8 does not appear to have genes homologous to those in EBV associated with transformation of cells in culture, the virus does have a number of genes with sequence homologies to mammalian genes. Many of these cellular genes function in cell growth regulation and cell cycle control, and so the activity of the viral counterparts may deregulate normal cell growth control. Preliminary studies on the function of these genes suggest that their function is consistent with this idea. The pirated genes include viral homologues of cyclin D, interleukin 6 (IL-6), homologues of complement-binding proteins, three macrophage inflammatory proteins, Bcl-2, interferon regulatory factor, IL-8 receptor-like G-protein coupled receptor and, finally, an adhesive molecule related to neural cell adhesion molecule (NCAM).

The virus life cycle, like other gammaherpesviruses, has two phases, a latent and a lytic stage. In the former, the DNA is maintained in cells and only replicates in synchrony with the replication cycle of the cell, with little or no infectious virus produced. In the lytic phase, the viral DNA is amplified with the production of late messenger RNA (mRNA) and viral particle production. In latency, there are only a limited number of transcripts expressed, two of which code for viral cyclin D and IL-6.

Viral Cyclin D (v-cyclin D). Cyclin D is a family of cyclins that act to allow progression from G_1 into the S phase of the cell cycle and are therefore at a pivotal point in cell cycle regulation. Cyclin D regulates progression through G_1 and into S by binding to cyclin-dependent kinases 4 and 6 (CDK4 and CDK6), which phosphorylate the retinoblastoma protein (RB), resulting in the inactivation of RB and release of transcription factors, whose activity is necessary for the synthesis of many of the proteins and enzymes needed for DNA synthesis. Cyclin D binds to CDK4 and 6. Inhibition of the kinase activity by cyclin-dependent kinase inhibitors, such as p16, p21 and p27, results in inhibition of RB phosphorylation resulting in cell cycle arrest. The v-cyclin D binds to CDK6 and phosphorylates RB in a similar fashion to the cellular cyclin/CDK6 complex. However, unlike the cellular complex, v-cyclin D/CDK is not inhibited by the cyclin-dependent kinase inhibitors [Swanton *et al.*, 1997]. Most tumors require inhibition of the RB checkpoint control in G_1 for successful proliferation, and so the virus may help to achieve this by producing a cyclin, which is not regulated by the normal cell-regulating proteins, such as the cyclin-dependent kinases. Therefore normal messages, which would cause cell cycle arrest, may not function efficiently in HHV-8-infected cells.

Interleukin 6. The cellular form of the cytokine IL-6 has been shown to stimulate growth of a number of human malignancies, especially of B-cell origin. The viral IL-6 is 25% homologous to cellular IL-6 and the viral form has been shown to act on signaling pathways involving JAK/STAT (Janus kinase/signal transducer and activator of transcription) receptor/transcription factors. Therefore, IL-6 may be important for the growth and transformation of B cells but it is unclear what role this cytokine has on Kaposi sarcoma, as these cells do not express the viral IL-6, indicating some cell specificity of expression.

Viral Bcl-2. This is a homologue of the cellular Bcl-2 protein which is involved in rescuing cells from apoptosis, and is a member of a larger set of cellular proteins that can either prevent or stimulate apoptosis. This viral protein is expressed during lytic infection and may function to inhibit apoptosis induced by the lytic phase, thereby allowing more time for replication of the virus. The role in any oncogenic pathway is unclear at present.

Other Potential Oncoproteins Expressed by HHV-8. There are other proteins coded for by HHV-8, which are compatible with an involvement in stimulating cell growth, an essential property of an oncoprotein. The G-protein coupled receptor has homology with the IL-8 receptor and is constitutively active in infected cells. However, since this viral gene product is not expressed in latently infected cells, its role is not clear, although over expression in NIH3 T3 cells causes' transformation. There is a latency-expressed protein called LANA, which has some features of a transcription factor, but the function is unknown. An HHV-8-encoded transforming protein called K1 is structurally similar to lymphocyte receptors and has a tyrosine-based activation motif involved in signal-transducing activity [Lee, et al.: 1998]. See also **Oncogenes**.

Significance for Human Cancers. HHV-8 is associated with malignant transformation of two cell types, endothelial cells and B lymphocytes. Both the endemic form of Kaposi sarcoma in African and Mediterranean countries and the most recently observed form in human immunodeficiency virus (HIV)-infected males are caused by the same virus. The virus is found in the majority of lesions by PCR technology and serological studies have found >80% of patients with Kaposi sarcoma are HHV-8-antibody positive, whereas the level of positivity in the UK and USA in blood donors is 1–5%. See also **Humman Immunodeficiency Viruses (HIV)**.

HHV-8 has also been linked to rare lymphoproliferative diseases or body cavity-based primary effusion lymphomas (BCBL/PELs) and to other rare B-cell lymphomas such as Castleman disease and immunoblastic lymphadenopathy. One problem early on was that many of the BCBL/PEL cells also contained the genome of EBV, so it was difficult to differentiate aetiology. Now that EBV-negative tumors have been observed, the evidence of guilt has shifted to HHV-8. There are other malignancies that have been associated with HHV-8 infection, such as squamous cell carcinomas (SCCs) and angiosarcomas, but these findings have not been reproduced by others and are therefore unlikely candidates.

Epstein-Barr Virus. This was the first human virus to be associated with cancer. Unlike *Herpes simplex virus*, EBV has very limited host range and infects and replicates in B cells and epithelial cells. However, replication in the former is low, and only under certain circumstances is lytic infection observed. The receptor for EBV, CD21 (immunoglobulin super family member), is found on B cells and a related receptor has been observed on epithelial cells.

Genomic Organization. EBV is a double-stranded DNA virus of approximately 172 kbp in length encoding for more than 85 gene products and is a gammaherpesvirus. Like HHV-8, the genome has terminal repeats at the ends of the genome separated by a unique region coding for the viral proteins. The genome circularizes after infection of the target B cells via the terminal repeat sequences.

Integration. While integration has been observed in some cells that have been grown out from tumor tissue, this process is probably not a factor in tumorigenesis *in vivo*.

Activation of Cell Growth. There are nine EBV proteins that have lymphocyte immortalization functions *in vitro*: the Epstein–Barr nuclear antigen (EBNA) family (EBNA-1, -2, -3A, -3B, -3 C, EBNA-LP) and the latent membrane proteins 1, 2A and 2B (LMP-1, -2A, -2B). After infection, EBV exhibits both a latency and lytic phase of persistence in the infected host. There are three classes of latency depending on viral gene expression and, in the simplest form, Burkitt lymphoma (BL) cells express only the EBNA-1 protein. EBNA-1 functions to maintain the viral genome as an episome, while the other EBNA proteins regulate viral transcription. While these EBNA factors activate transcription through binding to certain transcription factors, such as PU.1 and CBF-1, the mechanism by which this activity causes immortalization of B cells is not clear. LMP-1 has six hydrophobic transmembrane domains and its expression mimics changes in the B cell associated with activation. LMP-1 mimics constitutively activated tumor necrosis factor receptor (TNFR) family members [Mosialos *et al.*, 1995], which include TNFR, CD40 receptor and the CD30 receptor resulting in the activation of nuclear factor $_kB$ (NF $_kB$), an important transcription factor in B cells. LMP-2A and -2B proteins have 12 transmembrane domains and they appear to act together to increase signal transduction in cells through Src and Syk pathways.

Significance for Human Cancer. EBV is associated with a number of different cancers and some of these have only been observed recently, since the advent of immunosuppressive therapies or infection with HIV resulting in the acquired immune deficiency syndrome (AIDS). EBV has been known for some time to be associated with BL and nasopharyngeal carcinoma. Both these cancers have geographically restricted distributions, with BL found in young children in the central part of Africa, 10° north and south of the equator, while nasopharyngeal carcinoma is found predominantly in the southern coastal region of the People's Republic of China. See also **Acquired Immune Deficiency Syndrome (AIDS)**.

Posttransplant lymphoproliferative disease is a well-recognized risk of immunosuppressive therapy or infection with HIV and the development of AIDS. Individuals who are seronegative for EBV before transplant show the highest risk for the development of post transplant lymphomas. The majority of cases are associated with EBV and the cancers can be poly- or monoclonal. They include, poly- and monomorphic diffuse, large cell immunoblastic lymphomas, small noncleaved cell lymphomas, and diffuse large cell lymphomas. In AIDS, primary central nervous system, small noncleaved cell lymphomas are associated with EBV, as are non-Hodgkin lymphomas. Hodgkin disease has been associated with EBV infection for a number of years without really conclusive proof of an aetiological role. Recent evidence suggests that EBV DNA and

RNA is found in the Reed–Sternberg cells (presence of these cells is pathognomonic of Hodgkin disease). Thus, EBV is now associated with a number of lymphoproliferative diseases, which are mainly of B-cell origin. See also **Hodgkin's Disease**.

Papoviridae. Certain members of the subfamily *Papillomavirinae* have been shown to cause lower genital tract cancers; the most common of these is cervical cancer. Human papillomaviruses (HPVs) replicate in stratified epithelial cells and so are the aetiological agents of SCCs at various sites. See also **Papillomaviruses**.

Genomic Organization. This genus has a small circular genome of approximately 8 kbp DNA. HPV mRNA is transcribed off one strand of the DNA by host cell RNA polymerase II. There are five early genes coding for proteins, which are involved in DNA replication (E1 and E2) or activation of cell growth (E6, E7 and E5), and three late genes, two of which code for the major and minor capsid proteins (L1 and L2, respectively) and one of unknown function (E4).

Integration. During the malignant phase of the disease, when the lesion is confined to the epithelium, the viral DNA is in an episomal form, and amplification of the genome and viral particle production occurs in the upper parts of the epithelium. In malignant cells, the genome is integrated in over 70% of cases, while in the rest there is evidence of only episomal copies. While integration is not specific to any chromosomal site or specific sequence in the circular viral genome, all of the malignant cells investigated continue to express E6 and E7 proteins, suggesting that these proteins are essential for maintenance of the malignant phenotype. Integration often takes place in the region of the E1 and E2 ORFs, and it is thought that this may relieve the transcriptional inhibition of the E2 protein, allowing increased expression of E6 and E7, as seen *in vitro*. It is not clear if the episomal copies observed in a minority of cancers have mutations in the genome which may exhibit the properties of integrated viral DNA, such as mutations in E2 that may derepress its activity.

Activation of Cell Growth. Like most of the tumor viruses, the cell/growth-activating proteins affect either the signal transduction pathways or cell cycle control.

E6. E6 binds to the E6-associated protein (E6-AP) and p53 in a heterotrimeric complex [Huibregtse *et al.*, 1991]. The result of this binding is the premature degradation of p53 through the ubiquitin pathway. E6-AP is in fact a ubiquitin protein ligase. Since one of the functions of p53 is to control the passage of cells through the G_1 phase of the cell cycle, any abrogation of this activity could lead to uncontrolled cell cycle progression. E6 has also been shown to bind to a protein called *E6-binding protein* (E6-BP), which has sequence identity to ERC-55, a calcium-binding protein of unknown function.

E7. E7 binds to the retinoblastoma family of proteins, RB, p107 and p130. Most of the studies have been on the activity of E7 and RB. Normally, late in G_1 RB is phosphorylated, and this releases transcription factors such as E2F, which are necessary for the transcription of genes, whose products are important for DNA synthesis. E7 can cause this same release of factors in the absence of phosphorylation of RB and so can drive cells into S phase in an uncontrolled manner. E7 has also been shown to bind the AP-1 family of transcription factors and modulate their activity. Very recently RB has been shown to bind a histone deacetylase (HDAC) activity [Brehm, et al.: 1998], which may explain the ability of this protein to repress transcription. E7 can release this repression but it can only do it using two domains of the E7 protein, the RB domain and the zinc-finger domain. Therefore it appears that E7 can bind or compete with the deacetylase enzyme for binding to RB.

E1 and E2. Both these proteins are involved in the replication of the viral genome, and they bind to specific sequences in the origin region. E1 has ATPase and helicase activity, while E2 has been shown to be a repressor of transcription from the major early promoter at p97, which controls transcription of E6 and E7 genes. The repressive activity of E2 is removed when the HPV-16 genome is integrated, leading to up regulation of E6 and E7, which may be important in malignant conversion.

Significance for Human Cancer. Certain HPVs cause epithelial cancers, especially of the genital tract. Cervical cancer is the most common and, in fact, worldwide it is the leading cancer-related cause of death in women.

The genital types causing cancer are 16, 18, 31–35, 51, 52, 56, 58, 59, 61, 67–70 and 73; and others involved in benign genital infections are types 6, 11, 42–44, 53–55 and 66.

Like all cancers of viral aetiology, the progression from infection to cancer is a long process in normal immunocompetent individuals. The highest incidence of HPV-16 infection is in the 18–30 year age group; however, the highest incidence of cancer is in the individuals >35 years of age. While 20% of young people may be infected with the oncogenic viruses, only a small number (1 in 10 000) will develop malignant disease. HPV has also been shown to be associated with SCC in immunosuppressed patients, although the types involved are different from the genital ones and are not well defined. In a rare autorecessive disease, epidermodysplasia verruciformis, in patients who are partially immunosuppressed, there is a high incidence of SCC, and the viruses most commonly isolated are HPV-5 and -8.

Hepandnaviridae. HBV is the only human member of this family, which is one of the aetiological agents of hepatitis and hepatocellular carcinoma (HCC). The other viral cause of liver cancer (HCV) is discussed in the section on RNA viruses.

Genomic Organization. HBV has a double-stranded, circular DNA genome. However, in the virion the double strand is incomplete and is completed only on infection of liver cells. The replication of this DNA virus is unusual because the viral polymerase has a reverse transcriptase activity for synthesis of new DNA genomes from viral RNA templates. The genome is approximately 3 kbp in size and it has four ORFs, which are coded off the same strand. The largest ORF codes for the viral polymerase, which has RNA- and DNA-dependent activity. The second largest ORF codes for the surface proteins, which are important for diagnosis of infection, while the third ORF codes for the core protein, which encapsidates the viral genome. The fourth ORF codes for a protein called the X protein, because initially no function was attributable to this protein. However, the X protein is now known to have a transcriptional activation function.

Integration. Integration of the viral genome into chromosomal DNA occurs late in the chronic phase of the disease, but it is unclear if integration is really necessary for malignant conversion. One result of integration is reduced viral load observed late in the disease process. Integration also occurs in HCC caused by woodchuck and ground squirrel hepatitis viruses (also members of the *Hepadnaviridae* family) adjacent to the *myc* family of genes. However, in woodchucks, integration usually results in upregulation of N-*myc2*, while in ground squirrels there is an upregulation of c-*myc* [Transy, et al.: 1992]. Therefore, induction of HCC in these two closely related species does not appear to occur through the same activated cellular oncogenes. No single gene appears to be upregulated in HCC in humans caused by HBV.

Activation of Cell Growth. HBV, unlike the other human viruses so far discussed, does not code for an obvious viral oncoprotein. Apart from the structural proteins and the polymerase, HBV codes for the X protein, which has been shown to be an activator of cellular genes. However, the role of this protein in the tumorigenic process is unclear, especially since the virus causing HCC in ducks and other fowl do not appear to code for such a protein. There is evidence that the X protein is necessary for viral replication *in vivo*, and as such would be important in the persistence of the virus. HCC results from long-term persistence of the virus and it may be due to the constant destruction of infected cells by anti-HBV cytotoxic T lymphocytes and the subsequent proliferation of new liver cells infected with the virus. The continued proliferation may result in the outgrowth of altered cells, which are prone to malignant conversion.

Significance for Human Cancer. HBV is a major cause of HCC, and there is a higher incidence in parts of the world where HBV infection is contracted early in life (neonatally) and persistence is common (20–30%). In Western countries such as the USA and northern Europe, infection usually occurs in adulthood and persistence occurs in approximately 1–10% of infected individuals. There is now a successful vaccine consisting of the surface antigen (HBsAg) of the virus, which protects against infection of HBV. Since the vaccine protects against infection and subsequent complications, it is hoped that the incidence of HCC caused by HBV will decrease in coming decades.

RNA Viruses

There are two families of RNA viruses with a member that causes cancer. One, *Hepatitis C virus* (HCV) belongs to the *Flaviviridae* family, and is in the genus *Hepacivirus*, while the other is part of the *Retroviridae* family and is called *Human T-cell lymphotropic virus* (HLTV), types I and II, and belongs to the genus *Oncovirinae*. See also **Flaviviruses**.

Hepacivirus.

Genomic Organization. HCV is a positive-sense, single-stranded RNA virus with a genome of approximately 10 kb. The genome codes for 10 proteins, including three structural proteins, an RNA-dependent RNA polymerase and various proteases. There are certain hypervariable regions of the genome, which have necessitated the division into four genotypes, and there is some geographical variation in the distribution of the types. For instance, types 1, 2 and 3 have a worldwide distribution, with types 1a and 1b predominating in North America, while type 4 predominates in Africa. See also **Hepatitis Viruses**.

Integration. Integration does not occur with this virus as it is strictly an RNA virus and has no capacity to be reverse transcribed into DNA.

Activation of Cell Growth. Although HCV does not code for a classical oncoprotein, or even any proteins that are obviously involved in cell proliferation, there is now evidence that the core protein (encapsidates the genomic RNA) is able to alter host cell regulation and has been reported to activate cellular protooncogenes and repress p53 transcription [Ray, et al.: 1997]. However, the mechanism of malignant conversion is unclear and the role of the immune response has been difficult to determine. As with HBV; HCC does not develop for a number of years or decades after initial infection, so long-term immunological destruction of hepatocytes followed by cell proliferation may have a role, as it is thought to do with HBV-induced HCC.

Significance for Human Cancer. HCV infection leads to persistent infection in 50–80% of individuals, a much higher rate than seen for HBV in adults. HCV is increasingly recognized as a cause of HCC, and this will presumably continue as HBV immunization increases. Approximately 10–20% of persistent carriers will develop cirrhosis of the liver, and, of these, 1.5% will go on to develop HCC. At present, no other cancer has been associated with HCV infection.

Retroviridae. While many avian and mammalian retroviruses cause a number of different cancers in animals, only one retrovirus has so far been demonstrated to cause cancer in humans. This virus is HTLV-I, with a related virus, HTLV-II, not known to be associated with malignancies and other pathogenic processes.

Genomic Organization. This is a positive-sense, single-stranded RNA virus, which, like all retroviruses, carries two copies of the genome in the virion. It has the familiar retrovirus genome of 9 kb and is organized with terminal repeats at either end. The genome codes for viral structural proteins (Gag-core protein, Pol-polymerase and Env-envelope proteins) and the organization of these is similar to that of other retroviruses. In addition, located at the 3′ end of the genome are at least four ORFs coding for regulatory proteins.

Integration. As part of their natural replication process, retroviruses integrate proviral DNA sequences into the host cell chromosome, as they code for an RNA-dependent DNA polymerase which will produce a double-stranded complementary DNA molecule with long terminal repeat sequences which allow efficient integration in a sequence-independent manner. This keeps all the coding capacity of the genome intact and so progeny RNA is produced from the integrated sequences.

Activation of Cell Growth. HTLV-I codes for a number of proteins with sequences near the 3′ end of the genome, often called the pX region. Two of the best-characterized proteins are Tax and Rex. Tax can transactivate the HTLV-I promoter through binding to the transcription factor CREB (cAMP response element-binding protein) and the coactivator CBP (CREB-binding protein), mediating the interaction of the latter with the basal transcription machinery [Kwok, et al.: 1996]. In fact, Tax will activate a number of different cellular promoters through CREB binding and also binding NF$_k$B transcription factors, resulting in the expression of a number of cellular genes, including IL-2, TGF-β, GM-CSF, IL-2α receptor and c-*myc*. IL-2 is an important growth factor for T cells, the host cells for HTLV-I.

Rex is a nuclear protein and regulates the balance between spliced and nonspliced mRNA, thus favoring the transcription and translation of the *gag*, *pol* and *env* genes over the cell regulatory genes expressed through multiply-spliced transcripts. Rex also appears to affect cellular processes, and is thought to act with Tax to increase expression of IL-2α receptor, by an as yet unknown mechanism.

Significance for Human Cancer. HTLV-I causes adult T-cell leukemia/lymphoma, which is more common in Japan, West and Central Africa and in the Caribbean population of African descent. Infection typically precedes cancer by several decades, and there is a very tight association between the development of cancer and HTLV-I seropositivity. However, the lifetime risk of an HTLV-I-seropositive carrier is estimated to be 2–4%, and most have acquired the infection as children. The virus is also the cause of a nonmalignant disease known as tropical spastic paresis or HTLV-I-associated myelopathy, a chronic progressive demyelinating disease. See also **Leukemias**.

Additional Reading

Brehm, A., E.A. Miska, D.J. McCance, et al.: "Retinoblastoma Recruits Histone Deacetylase Activity to Repress Transcription," *Nature*, **391**, 597–601 (1998).

Chang, Y., E. Cesarman, M.S. Pessin, et al.: "Identification of Herpes-like DNA Sequences in AIDS-associated Kaposi's Sarcoma," *Science*, **265**, 1865–1869 (1994).

Huibregtse, J., M. Schneffner, and P.M. Howley: "A Cellular Protein Mediates Association of p53 with the E6 Oncoprotein of Human Papilloma Virus Types 16 and 18," *EMBO Journal*, **10**, 4129–4135 (1991).

Kwok, R.P., M.E. Laurance, J.R. Lunblad, et al.: "Control of cAMP-regulated Enchancers by the Viral Transactivator Tax through CREB and Co-activator CBP," *Nature*, **380**, 642–646 (1996).

Lee. H., J. Guo, M. Li, et al.: "Identification of an Immunoreceptor Tyrosine-based Activation Motif of K1 Transforming Protein of Kaposi's Sarcoma-associated Herpes Virus," *Molecular and Cellular Biology*, **18**, 5219–5228 (1998).

McCance, D.J.: *Human Tumor Viruses*, ASM Press, Washington, DC, 1998.

Mosialos, G., M. Birkenbach, R. Yalamanchill, et al.: "The Epstein–Barr Virus Transforming Protein LMP1 Engages Signaling Proteins for the Tumor Necrosis Factor Receptor Family," *Cell*, **80**, 389–399 (1995).

Ray, R.B., R. Steele, K. Meyer, and R. Ray: "Transcriptional Repression of p53 Promoter by Hepatitis C Virus Core Protein," *Journal of Biological Chemistry*, **272**, 10983–10986 (1997).

Straight, S.W., B. Herman, and D.J. McCance: "The E5 Oncoprotein of Human Papillomavirus Type 16 Inhibits the Acidification of Endosomes in Human Keratinocytes," *Journal of Virology*, **69**, 3185–3192 (1995).

Swanton, C., D.J. Mann, B. Fleckstein, et al.: "Herpes-cyclin.cdk6 Complexes Evade Inhibition by cdk Inhibitor Proteins," *Nature*, **390**, 184–187 (1997).

Transy, C., G. Fourel, W.S. Robinson, et al.: "Frequent Amplification of c-*myc* in Ground Squirrel Liver Tumors Associated with Past or Ongoing Infection with Hepadnavirus," *Proceedings of the National Academy of Sciences of the USA*, **89**, 3874–3878 (1992).

Tunely, E.I.: *New Research on Oncogenic Viruses*, Nova Science Publishers, Inc., Huntington, NY, 2007.

Umar, C.S.: *New Developments in Epstein-Barr Virus Research*, Nova Science Publishers, Inc., Huntington, NY, 2006.

DENNIS J. McCANCE University of Rochester, Rochester, NY

ONCOLOGY. See **Cancer and Oncology**.

ONION. Of the family *Amaryllidaceae* (amaryllis family), the common dry onion of commerce (*Allium cepa*) is related to a great number of other species of the genus and of similar odor and taste. Closely related species are chive, garlic, leek, shallot, and Welsh onion. The great majority of the over 500 species of *Allium* are wild plants of no commercial significance, but that possess the characteristic odor and taste. The onion is consumed raw and is cooked by almost all means known. The onion is in countless thousands of cooking recipes, used by peoples of various cultures throughout the world.

Some authorities regard the general region of northwestern India, Afghanistan, western Tien Shan, and the former Soviet bloc republics of Tajik and Uzbek, as the most likely area of origin of *Allium cepa*. But the vegetable also may be native to Turkey, the Near East, and the Mediterranean region. The Welsh or Japanese onion (*Allium fistulosum*), widely grown in the Orient, is regarded as being native to central and western China. The onion is mentioned in the Old Testament, Numbers 11:5, "We remember the fish, which we did eat in Egypt freely; the cucumbers, and the melons, and the leeks, and the onions, and the garlic." A number of species of onion were described by Theophrastus as early as 322 B.C. Pliny offered directions for cultivation of the onion in 210 A.D. Chaucer mentioned the onion in 1340 and European botanists in the 1500s regarded the onion as one of "the commonest vegetables."

In the United States, there are two crops of onions per year. Spring crops in Arizona, California, and Texas account for about 21% of total annual production. The summer crop, accounting for the bulk of onion production, is spread over numerous states. The total dry onion production in the United States approaches 2 million tons. There are numerous varieties of commercially grown onions. See Fig. 1.

Fig. 1.   Various shapes of onions: (1) flattened globe; (2) globe; (3) high globe; (4) spindle; (5) Spanish; (6) flat; (7) thick-flat; (8) Granex; (9) top. (*By permission from "Onions" by Dr. Henry A. Jones, Desert Seed Company, El Centro, California.*)

The two major insect pests of onion are the onion maggot and the onion thrip. Onions also are attacked by the bulb and stem nematode, causing a condition known as onion bloat. Soft rot is the most prevalent causes of onion loss in storage. It is a bacterial infection. Black mold may be destructive during storage and transit of onions grown in Texas, California, Arizona, and New Mexico. To control this condition, bulbs should be protected from moisture in the field during and after pulling and during transit. During storage, onions require adequate vertical air channels between bags to permit good ventilation and refrigeration.

ONLINE. In terms of instruments and computers, online describes situations where such devices participate directly in the dynamic control or data system loop. Information collected by online equipment is immediately processed and fed into a closed-loop system with no delay or human intervention.

ONSAGER, LARS (1903–1976). A Norwegian chemist who won the Nobel Prize for chemistry in 1968. He studied and wrote on the theory of electrolytic conduction and theory of dielectrics. He also worked with superfluids and crystal statistics and reciprocal relations in irreversible processes. After receiving his doctorate in Norway, he came to the U.S. and became a citizen. See also **Dielectric Theory**

ONSHORE WIND.   See Winds and Air Movement.

ONYCHOMYCOSIS (OM).   Mycosis is any disease caused by a fungus. Onychomycosis (OM) refers to a fungal infection that affects the toenails or the fingernails. It may involve any component of the nail unit, including the nail matrix, the nail bed, or the nail plate. OM is not life threatening, but it can cause pain, discomfort, and disfigurement and may produce serious physical and occupational limitations. Psychosocial and emotional effects resulting from OM are widespread and may have a significant impact on quality of life.

Half of all nail disorders are caused by OM, and it is the most common nail disease in adults. Toenails are much more likely to be infected than fingernails. The incidence of OM has been increasing and is related to diabetes, a suppressed immune system, and increasing age. Adults are 30 times more likely to have OM than children. In fact, only 2.6% of children younger than 18 years are reported to have onychomycosis, but as many as 90% of elderly people have onychomycosis.

When the nails are exposed to a warm moist environment, a fungus can develop on the nail or under its outer edge. This is called onychomycosis. Depending on the type of fungus, the nail may turn yellow, gray, brown, or black. The nail may become brittle and crack. It may separate from its bed. The surrounding skin may be red, itchy or swollen.

The main subtypes of OM are distal lateral subungual OM (DLSO), white superficial OM (WSO), proximal subungual OM (PSO), endonyx OM (EO), and candidal OM. Patients may have a combination of these subtypes. Total dystrophic OM refers to the most advanced form of any subtype.

Pathophysiology

The pathogenesis of OM depends on the clinical subtype.

DLSO (Distal Lateral Subungual OM). In DLSO, the most common form of OM, the fungus, usually a dermatophyte, invades the space between the tip of the toe and the nail tip or the skinfolds at the sides of the nails. The toenails turn yellow and separate from the nail bed beneath them.

WSO (White Superficial OM). Less common than DLSO, this affects only the surface of the nail turning it white and crumbly in spots or all over the nail surface. This is the most easily treated infection in that it can be simply scraped off the nail and a topical antifungal medication applied.

PSO (Proximal Subungual OM). This is the least common of nail fungi and may first appear as a white or yellowish spot on the nail close to the cuticle. From there, it can progress into a plaque that collects on the underside of the nail. Then, debris may collect under the nail and it may lift off its bed and even be shed entirely. This infection is usually treated with a systemic drug.

EO (Endonyx OM). EO is a variant of DLSO whereby the fungi infect the nail via the skin and directly invade the nail plate.

Candidal OM. Candidal nail infection may manifest in 3 ways: onycholysis, paronychia, or chronic mucocutaneous disease. Onycholysis may be caused primarily by yeast, or the organism may secondarily colonize onycholytic nails. Candidal paronychia is usually secondary to trauma of the nail fold. Chronic mucocutaneous candidiasis affects the nail plate and eventually infects the proximal and lateral nail folds. Total dystrophic OM involves the entire nail unit and may include permanent scarring of the nail matrix. The infection can turn nails yellow or green or the nail may look opaque. This infection can be treated topically.

Treatment

Medical Care. Several years ago the medical management of OM was limited to topical therapy and 2 unreliable systemic drugs: griseofulvin and ketoconazole. Topical therapy is beneficial only for mild cases involving the very distal nail plate. The use of griseofulvin and ketoconazole is plagued by high relapse rates (70–85%), prolonged treatment regimens (10–18 mo for toenails), constant laboratory monitoring, and numerous adverse effects. The introduction of newer oral agents has revolutionized the medical treatment of OM and reduced potential adverse effects and drug interactions. As the rate of recurrence remains high, even with newer agents, the decision to treat should be made with a clear understanding of the cost and risks involved as well as the risk of recurrence.

Topical antifungals. The use of topical agents should be limited to cases involving less than half of the distal nail plate or for patients unable to tolerate systemic treatment. Agents include amorolfine (approved in other countries), ciclopirox olamine 8% nail lacquer solution, sodium pyrithione, bifonazole/urea (available outside the United States), propylene glycol-urea-lactic acid, the imidazoles, and the allylamines.

Topical treatments alone are generally unable to cure OM because of insufficient nail plate penetration. Ciclopirox solution has been reported to penetrate through all nail layers but has low efficacy when used as monotherapy. It may be useful as adjunctive therapy in combination with oral therapy or as prophylaxis to prevent recurrence in patients cured with systemic agents.

Oral therapy. The newer generation of oral antifungal agents (itraconazole and terbinafine) has replaced older therapies in the treatment of OM. They offer shorter treatment regimens, higher cure rates, and fewer adverse effects. Fluconazole (not approved by the Food and Drug Administration [FDA] for treatment of OM) offers an alternative to itraconazole and terbinafine. Derivatives of fluconazole may also be available soon. The efficacy of the newer antifungal agents lies in their ability to penetrate the nail plate within days of starting therapy. Recent evidence shows better efficacy with terbinafine than other oral agents.

To decrease the adverse effects and duration of oral therapy, topical and surgical treatments may be combined with oral antifungal management.

Surgical Care. Surgical approaches to OM treatment include surgical nail avulsion and matrixectomy by chemical or mechanical means.

- Chemical removal by using a 40–50% urea compound should be reserved for patients with very thick nails or for those who may not tolerate mechanical avulsion.
- Removal of the nail plate should be considered an adjunctive treatment in patients undergoing oral therapy.
- A combination of oral, topical, and surgical therapy can increase efficacy and reduce cost.

Additional Reading

Daniel, C.R.: *Diagnosis of Onychomycosis and Other Nail Disorders,* Springer-Verlag New York, LLC, New York, NY, 1996.

Staff: *Onychomycosis: A Medical Dictionary, Bibliography, and Annotated Research Guide to Internet References,* ICON Health Publications, San Diego, CA, 2004.

ONYCHOPHORA. Small, soft-bodied, creeping animals, slightly like caterpillars in appearance. They are of limited distribution in warm countries and have no common name. The name of one genus, *Peripatus,* is sometimes applied indiscriminately to all members of the group.

Onychophora are regarded as the most primitive of the terrestrial arthropods. Their structure suggests the ancestral form of the insects.

ONYX. See **Agate**.

ONYX MARBLE. See **Travertine**.

OÖLITE. The term is from the Greek meaning egg and stone. Oölites are well-rounded sand-like particles, originally formed of calcite but sometimes subsequently altered to either dolomite, or entirely silicified. The structure is typically concentric about a nucleus, and often with radial lines. Oölites are relatively common constituents of limestones, often forming distinct beds. Oölites are now forming on the shores of Great Salt Lake, but no authentic cause is known of marine oölites being formed at the present time. Coarse-grained oölites, in which the particles are about the size of peas, are called pisolites, from the Greek, meaning pea and stone.

OOLOGY. The study of eggs and egg-producing processes in various egg-bearing animals. A good reference on this topic is "Egg Incubation in Birds and Reptiles," by D. Charles Deeming and Mark W.J.K. Ferguson (Editors), Cambridge University Press, New York (1992).

OOZE. Ooze is a general term used to designate the mud found on the ocean bottoms at abyssal depths and composed largely of the calcareous and silicified shells of minute surface living marine organisms, called plankton.

OPACITY. Imperviousness to radiation, especially to light; the property of stopping the passage of light rays numerically expressed as the reciprocal of the transmittance. Density (photographic) or optical density is given by

$$d = \log O = \log 1/T$$

where d is density, O is opacity, and T is transmittance. This usage should now be discarded in favor of the term absorbance.

OPACUS. See **Clouds and Cloud Formation**.

OPAH (*Osteichthyes*). Of the order *Allotriognathi,* family *Lampridae,* the opah (*Lampris regius*) is the only member of this family. See Fig. 1. It may be described as a laterally compressed fish with an oval shape. The fish is noted for its spectacular coloration and patterning with a blue-to-gray upper surface, rose red undersurface, body covered with white spots, jaws, and fins vermilion, and the eyes set within a gold-colored area. The opah is a large fish, measuring up to 6 feet (1.8 meters) in length and weighing up to 600 pounds (272 kilograms). Because the flavor is excellent, the opah would make an excellent commercial item were it abundant. It is well distributed throughout the seas. The depth at which the opah may be most abundant has yet to be determined. During a number of years, fewer than 50 specimens have been taken from the waters of southern California and the Pacific northwest to Alaska, where it is known to range.

Fig. 1.   Opah (*Lampris regius.*)

OPAL. The mineral opal, long classified as an amorphous mineral gel, has been found by X-ray analysis to consist of a microcrystalline aggregate of crystallites of cristobalite. On this basis, opal may be considered as a variety of cristobalite bearing the same relationship to that mineral as chalcedony does to quartz. Opal is hydrous silica, $SiO_2 \cdot nH_2O$, with variable water content. It never occurs in crystal form; usually as irregular veins or masses, or as pseudomorphous replacements after wood or fossilized material such as bones and shells. Opaline silica occurs in many forms: geyserite from geyser deposits, siliceous sinter (fiorite) form siliceous waters of hot springs, and diatomite (diatomaceous earth) from siliceous shells of diatoms and comparable microscopic species. It has a conchoidal fracture; hardness 5.5-6.5; specific gravity 2.1-2.3; luster, vitreous or greasy to dull; color very variable, colorless, white, milky-blue, gray, red, yellow, green, brown, and black. Often a beautiful play of colors may be observed in the gem varieties. The color play in opals is attributed to three different mechanisms: finely divided pigmentation of foreign material; light interference by open-spaced grid of cristobalite crystallization; and reflected light. It may well be that two or all three causes may contribute to the color effect in any given opal specimen. Before a more complete understanding of opal color is established these phases seem to be of prime significance.

Besides the gem varieties, which show the delicate play of colors, there are other kinds of common opal, such as: the milk opal, a milky bluish to greenish kind; resin opal, which is honey-yellow with a resinous luster; wood opal, resulting from the replacement of the organic matter of wood by opal, and hyalite, a colorless glass-clear opal sometimes called Muller's Glass. Opal is deposited at relatively low temperatures and may occur in the fissures of almost any type of rock. Hungary, Australia, Honduras, Mexico and in the United States Nevada and Idaho, have been sources of gem opals. Hyalite comes from Czechoslovakia, Mexico, Japan and British Columbia. Other common varieties of opal are widespread in their occurrence. The word opal is derived from the Latin *opallus*.

OPAQUE PLASMA.   A plasma through which an electromagnetic wave cannot propagate and is either absorbed or reflected. In general, a plasma is opaque for frequencies below the plasma frequency. The fact that a plasma is opaque over a certain frequency range will change the radiation properties within that frequency range. Any radiation emitted within the volume of the plasma is quickly absorbed. In this opaque region, therefore, the plasma can only radiate from its surface.

OPAQUE SKY COVER.   See **Meteorology**.

OPEN-CHANNEL FLOWMETERS.   See **Flow Measurement (Liquids and Gases)**.

OPEN DELTA.   This is a three-phase transformer connection using two single-phase transformers connected to form a V or open delta across the three lines. Such a connection has about 58% of the capacity of full delta using transformers of the same rating. It is often used for temporary work anticipating a later completion of the delta or for emergency service when one transformer of the complete delta requires servicing.

OPEN-LOOP CONTROL.   See **Feedback Control**.

OPEN UNIVERSE.   See **Cosmology**.

OPERAND.   An entity to which an operation is applied. The operand, for example, may be a portion of a computer instruction, or it may be identified by the address part of the instruction.

OPERATING CHARACTERISTIC.   In quality control and decision theory generally, a measure of the probabilities of accepting a false hypothesis for varying values of the parameter specified by that hypothesis. For example, if a batch of items is to be accepted or rejected on the basis of the proportion of defective items in a sample from the batch, the OC curve would graph the proportion of defectives as abscissa against the probability of acceptance as ordinate. A good acceptance rule would then have a curve falling rapidly to zero as the proportion of defectives increased. Considered upside down (i.e., graphing the probability of *rejection* as ordinate) the OC curve is equivalent to the graph of the Power Function. See also **Neyman-Pearson Theory**.

OPERATING SYSTEM (Computer).   An integrated collection of service routines for supervising the sequencing of programs by a computer and may provide debugging, input/output control, accounting, compilation, storage assignment, data management, and related services. Essentially synonymous with monitor system and executive system. See also **Program (Computer)**.

OPERATIONAL AMPLIFIER.   See **Amplifier**.

OPERATIONAL WEATHER LIMITS.   See **Meteorology**.

OPERATIVE TEMPERATURE.   In the study of human bioclimatology, one of several parameters devised to measure the cooling effect of the air upon a human body. It is equal to the temperature at which a specified hypothetical environment would support the same heat loss from an unclothed, reclining human body as the actual environment.

In the hypothetical environment, the wall and air temperatures are equal and the air movement is 7.6 cm s^{-1}. From experiments it has been found that

$$\text{operative temperature} = 0.48T_r + 0.19[v^{1/2}T_a - (v^{1/2} - 2.76)T_s],$$

where T_r is the mean radiant temperature, T_a the mean air temperature, T_s the mean skin temperature (all in degrees Celsius), and v the air speed in centimeters per second.

Additional Reading

Buettner, K.J.K.: in *Compendium of Meteorology*, American Meteorological Society, Boston, MA, 1951.
Newburgh, L.H.: *Physiology of Heat Regulation and the Science of Clothing*, W. B. Saunders, Philadelphia, PA, 1968.

AMS

OPERATOR (Mathematics).   The symbolic direction to perform an operation such as addition, multiplication, differentiation, extraction of roots, etc., or some combination of these operations. A linear operator is the most common case. It obeys the distributive law, $\mathbf{A}[f(x) + g(x)] = \mathbf{A}f(x) + \mathbf{A}g(x)$ and $\mathbf{A} \cdot cf(x) = c\mathbf{A}f(x)$, where c is any constant. If the order of applying operators to a function is immaterial, the operators are commutative. Suppose $\mathbf{A} = a +$ and $\mathbf{B} = b+$, with a, b constant, are applied to a function of x, then $\mathbf{AB}f(x) = a + b + f(x) = \mathbf{BA}f(x)$ and $\mathbf{A}$, $\mathbf{B}$ commute. However, if $\mathbf{P} = \partial/\partial x$ and $\mathbf{Q} = x$, then $\mathbf{P}$ and $\mathbf{Q}$ are not commutative for $\mathbf{PQ}f(x) = f(x) + \mathbf{QP}f(x)$.

If $\mathbf{A}$ and $\mathbf{B}$ are noncommutative, their commutator is $(\mathbf{A}, \mathbf{B}) = \mathbf{AB} - \mathbf{BA}$. According to quantum theory, if the commutator vanishes for two operators that represent dynamical variables, then the measurement of one of these variables does not interfere with that of the other.

A differential operator involves one or more differentiations. Examples are $D = d/dx$, $D^2 = d^2/dx^2$, $D^{(n)} = d^n/dx^n$; $F(u) = f(x)u'' + g(x)u'' + (x)u$. In the more general case of an nth-order operator

$$L(u) = \sum_{i=0}^{n} [f_i(x)u^{(r_i)}]^{(S_i)}$$

the adjoint operator is

$$L(u) = \sum_{i=0}^{n} (-1)^{r_i + S_i} [f_i(x)u^{(S_i)}]^{(r_i)}$$

If $L(u) = \overline{L}(u)$, the operators are self-adjoint. Any second-order differential operator can be made self-adjoint with an integrating factor exp $\int ((g - f')/f)dx$.

See also **Del**; and **Laplacian**.

OPERCULUM.   See **Fishes**.

OPHIOLITES.   See **Ocean**.

OPHITIC TEXTURE.   A term proposed by Michel-Levy, in 1877, for the characteristic texture of dolerites, in which the pyroxene crystals are penetrated by laths of plagioclase feldspar. This type of texture differs from poikilitic in that in the latter type of texture the pyroxene crystals entirely enclose a number of laths of plagioclase.

OPHIUROIDEA.   The brittle stars, a class of echinoderms resembling starfishes with a well-marked disk and slender arms. The class is distinguished chiefly by this sharp demarcation of disk and arms, which accompanies the restriction of visceral organs to the disk. In addition the tube feet are without suckers and the madreporite lies on the oral surface.

OPHTHALMOLOGY.   That branch of medical science that deals with the structure, functions, and diseases of the eye and of the visual system.

OPOSSUM. See **Marsupialia**.

OPPENHEIMER, J. ROBERT (1904–1967). Oppenheimer was an American scientist whose areas of achievement include, invention, physics, and technology. His interest in science, is believed to have been sparked by a German grandfather who gave him a mineral collection. He wrote letters to famous geologists and was invited at age twelve to give a lecture at the New York Mineralogical Society. Oppenheimer went to Harvard University and excelled at theoretical physics. After graduation he studied at the famous Cavendish Laboratory at Cambridge, England under Ernest Rutherford. In 1929, he took positions at Berkeley and Cal Tech in the United States. He was always regarded as an exceptional teacher and excellent theoretician. He devoted early research to the study of subatomic particles such as electrons and positrons and published 16 papers on quantum physics. His work lead to many later finds including neutron stars.

Oppenheimer's name is almost synonymous with the atomic bomb. In 1941, after learning the Germans had split the atom, President Roosevelt established the Manhattan Project. In June 1942, Oppenheimer was appointed its scientific director. Oppenheimer recruited and coordinated the effort of hundreds of scientists at a research station at Los Alamos, New Mexico for the Manhattan Project. The scientists were working to produce the first atomic bomb. At 5:30 A.M., Monday, July 16, 1945, Oppenheimer witnessed the first explosion of an atomic bomb in the New Mexico dessert.

After the bombing of Hiroshima and Nagasaki, Oppenheimer received much publicity including Time magazine referring to him as "The Father of the Atomic Bomb". On January 12, 1946, he was given the Presidential Medal of Merit. Americans were full of gratitude for his work.

After the war, Oppenheimer was always concerned about atomic weapon usage. In 1947, he chaired the U.S. Atomic Energy Commission. He wrote government reports about atomic energy and gave more than 200 speeches to the public. In 1949, Oppenheimer became the director of the Institute for Advanced Study in Princeton, New Jersey. In 1953, however, at the height of U.S. anti-communist feelings, Oppenheimer lost his security clearance because of his associations with friends who were sympathetic to communism. With the loss of his security clearance, Oppenheimer lost his influence on America's science policy. However, in 1963 Oppenheimer was presented with the Enrico Fermi award (the highest award a physicist can receive) by President Lyndon B. Johnson.

Oppenheimer is considered one of the greatest scientists of the twentieth century.

See also **Manhattan Project (The)**; and **Neutron Stars**.

J. M. I.

OPTICAL AIR MASS (symbol m). A measure of the length of the path through the atmosphere to seal level traversed by light rays from a celestial body, expressed as a multiple of the path length for a light source at the zenith. Originally called, simply, *air mass*. Also called *airpath*.

OPTICAL ANOMALY. The behavior of certain organic compounds, such as those whose molecules contain conjugated double bonds, in which the observed values of the molar refraction are not in accord with the values calculated from the known equivalents.

OPTICAL ANTIPODES. Two compounds composed of the same atoms and atomic linkages, which differ in their structural formulas only in that one is the mirror image of the other. The term is commonly applied to substances containing an asymmetric atom, or bond, in which the plane of polarized light is rotated to the right by one of the optical antipodes, and to the left by the other. See also **Amino Acids**.

OPTICAL BRIGHTENER. Also referred to as optical bleach; colorless dye; fluorescent brightener. A colorless fluorescent, organic compound that absorbs UV light and emits it as visible blue light. The blue light masks the undesirable yellow of textiles, paper, detergents, and plastics. Some examples are derivatives of 4,4′- diaminostilbene-2,2′-disulfonic acid, coumarin derivatives such as 4-methyl-7-diethylaminocoumarin.

OPTICAL CENTER (of a Lens). A point so located on the axis of a lens that any ray, which in its passage through the lens passes through this point, has its incident and emergent parts parallel. See also **Mirrors and Lenses**.

OPTICAL CHARACTER RECOGNITION (OCR). A system for automatically identifying handwritten or printed characters by one of several types of photoelectric devices for the purpose of providing electronic identification input to data processing systems. OCR systems are applicable where there are voluminous amounts of printed input data, as encountered, for example, by banks, insurance companies, retail credit firms, brokerage houses, warehouse accounting, mail and postal systems, etc. In addition to the several types of OCR systems, there are other character identification approaches, notably magnetic ink character recognition systems. See also **Magnetic Ink Character Recognition (MICR)**. Each approach has its advantages and disadvantages and sometimes selection of the most effective approach is quite difficult.

As with MICR, to take advantage of full electronic differentiation in character reading, some modification of the general appearance and shape of letters and numbers is required. The numerals as modified for MICR (and illustrated in that alphabetical entry in this volume) also are useful in OCR systems with the exception, of course, that magnetic ink is not required. In this approach (sometimes termed the one-dimensional approach), the signal obtained is the amount of material that is sensed through a slit—a single function of time corresponding in duration to the time required for the character to pass by the slit. One-dimensional optical approaches are limited, however, to small numbers of accurately printed characters. Two-dimensional systems, although considerably more complex, make much more sensing data available.

A number of sensing approaches have been conceived, and some have been quite successful. In the optical masking approach, an image of the character is projected on a set of masks. The total amount of light passing through the masks is collected by a photodetector. It is necessary to relate the mask designs to the expected character shapes in such a way that the mask that permits the largest signal will be the mask that identifies with a given character. Obviously, in such a system the characters being measured must be reasonably uniform both in terms of size and optical characteristics.

In spot scanning, one small character segment is covered at a time. In one spot scanner, there is a rotating disk between the light source and the character, with a pattern of slots that breaks the character into distinguishable elements. The device may be limited to from 300 to 500 characters per second and is designed to one highly stylized type font. In one type of electronically generated spot scanner, a vidicon tube similar to that used in telecasting picks up the character image. The surface of the tube is scanned by an electron beam that breaks the character down into digital components. In another method (flying-spot scanner), a cathode-ray tube generates a beam of light that moves across the character in a scan pattern. A lens system projects the reflected light to photomultipliers that translate black-and-white areas into electrical signals. Speeds up to 2,000 or more characters per second are obtainable. The systems are costly and require high printing quality of specific types of fonts.

In another system (retinal sensing), a two-dimensional matrix of photosensors is used. These sense an entire moving character rather than a segment of it. Early configurations of this system had a character resolution approaching that of the human eye and could read up to 2,400 characters per second. Essentially, the device is a mosaic-image sensor, or a mosaic of photocells onto which each character to be read by the system is focused. The photocells are physically constrained and thus their dimensions with respect to each other can be held constant to avoid character distortion. The mosaic, like the scanner, is one character width wide and three character heights high. Behind each photocell in the retina, there is a silicon chip that is sensitive to black, white, and shades of gray between. The photocells are interconnected so that each single cell "sees" not only its own portion of the character being read, but also the portions covered by other cells around it. This arrangement enables the device to judge relationships, dismiss smudges as not being part of the characters, and accept even the light portion of the character because the area next to the character is even lighter. The recognition logic establishes relative values between each cell and those surrounding it.

In recent years, particularly because of the urgent needs for automation in postal and package handling and manufacturing situations, OCR systems have been undergoing constant change and improvement. See also **Pattern Recognition**.

OPTICAL CRYSTAL.

A comparatively large crystal, either natural or synthetic, used for infrared and ultraviolet optics, piezoelectric effects, and shortwave radiation detection. Examples are sodium chloride, potassium iodide, silver chloride, calcium fluoride, and (for scintillation counters) such organic materials as anthracene, naphthalene, stilbene, and terphenyl.

OPTICAL EMISSION SPECTROCHEMICAL ANALYSIS.

In this analytical technique, an optical device is used to analyze radiation from electrically excited sample atoms. The analyzing device provides monochromatic images whose intensities are measured and related to the concentration of the elements within the sample that produces the specific radiation measured. The technique is precise and rapid, and adaptable to solid, powder, or liquid samples.

More than seventy elements may be detected by standard procedures. Atomic gases, such as O, N, H, He, Ar, Ne, Kr, Xe, and Rn and the halogens are excluded. Nonmetallic substances, such as C, S, and Se, require vacuum path spectrometers for optimum detection and measurement. Analytical ranges may extend from fractional parts per million to about 40% concentration. Computer-controlled photoelectric optical emission spectrometers will output printed percent concentrations for 30 to 50 elements per sample in just a few minutes. This form of analytical instrumentation is used widely in production and quality control, as well as for research studies.

A schematic diagram of an optical emission spectrometer is given in Fig. 1. Various means are used to introduce the sample, whether solid or liquid, into an excitation stand where energy is imparted to it by some form of excitation source. The atoms composing the sample are excited and therefore emit their characteristic radiations, which are then separated by a grating in the spectrometer into line spectra. The light of selected element lines is isolated by slits and focused on phototubes. The sensitivity is adjusted by attenuating the high voltage from the high voltage supply. The intensity of a spectrum line can be correlated with the concentration of the element producing it. It is therefore necessary to measure intensities with very high precision.

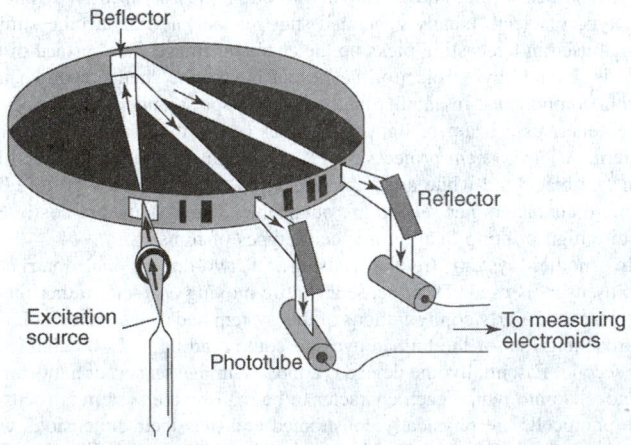

Fig. 1.   Operating principle of optical emission spectrometer

Sample atoms may be excited by absorbing specific energies from an electric discharge. These atoms, raised to higher-than-usual energy levels, are unstable, and revert to their stable states by emitting the absorbed radiation according to the relation:

$$E_2 - E_1 = h\nu = hc/l \tag{1}$$

where

$E =$ energy, eV
$h =$ Planck's constant (6.624×10^{-27} erg-second)
$\nu =$ frequency, Hz
$\lambda = c/\nu =$ wavelength (in Å$= 10^{-8}$ centimeter)
$c =$ velocity of light (3×10^{10} centimeters/second).

Atomic transitions may be expressed in terms of wavelength, and qualitative analysis may be performed by wavelength determination and identification.

The commonly used dispersive device is the diffraction grating, which produces spectra by light interference according to the relation:

$$N\lambda = d(\sin\alpha \pm \sin B) \tag{2}$$

where

$N =$ an integer
$\lambda =$ wavelength
$d =$ grating constant (width of single groove)
$\alpha =$ angle of incident light
$B =$ angle of diffracted light.

For constant a and the same sin B, integer values of N produce spectra of $\frac{1}{2}\lambda$, $\frac{1}{3}\lambda$, etc., called *spectral orders*.

A grating ruled on a spherical surface combines the properties of the diffraction grating with the focusing ability of the optical surface. Such a device, with radius of curvature R, focuses spectra as images of the entrance (primary) slit on the circumference of a circle of diameter R, when the entrance slit is also located on the circumference of the circle.

The usual measure of how well a grating separates individual wavelengths is given by the reciprocal linear dispersion, in angstroms per millimeter, as follows:

$$\text{Å/mm} = \frac{d\cos B}{Nf} \tag{3}$$

Thus, dispersion is governed by the fineness of the grating ruling d and the focal length f of the focusing element.

The concentration C of an irradiating element is related to the intensity I of the emitted spectral line, according to the relationship:

$$I = kC^n \tag{4}$$

where k and n depend on the excitation conditions employed. Accuracy and precision are improved by use of an internal standard reference line of another element of constant concentration. The relationship becomes:

$$\frac{I_x}{I_r} = k_1 C_x^{n_1} \tag{5}$$

where I_x and I_r are the intensities of spectral lines emitted by elements x and r; C_x is the concentration of element x; and k_1 and n_1 are constants depending on the line pair and on the excitation conditions. The relative intensities of lines having different excitation energies depend on the temperature of the spark discharge column.

The source unit must vaporize and excite a portion of the sample, which is generally used as one of the electrodes between which the electric discharge takes place. No single excitation source is ideally suited for all applications of emission spectrochemistry. Trace impurities in metals, alloying constituents in high concentrations, biological substances, ceramics, slags, oils, nonconductors, refractories—all may require different excitation techniques and sample preparation procedures. Table 1 summarizes the important characteristics of the commonly used spectrochemical source units.

Photographic radiation detection may be used, but film emulsion response is not linear. Film calibrations are required to relate measured densities with the intensities producing these densities before *intensity* versus *concentration* working curves can be formulated. Although quite general in application at one time, the photographic technique is slower than photoelectric radiation detection wherein each beam whose intensity is to be measured is directed onto a photomultiplier detector through a suitably sized exit (secondary) slit. The output of the detectors is transmitted to the measuring console, where it is translated into the readout format of the system.

TABLE 1. SPECTROCHEMICAL EXCITATION SOURCE UNITS

Type	Voltage	Current, A	Characteristics
Dc arc	220	3–30	Most sensitive, least reproducible, quantitative analysis; trace element quantitative analysis.
Ac arc	2500	5	Good sensitivity, more reproducible, best use for self-electrode metal analysis.
High-voltage spark interrupted auxiliary gap	5000 15–40,000	2–5 3–20 RF	Least sensitive, most precise, ±1% or better. Excites higher energy lines. Parameter selection allows variations between arc-like and spark-like in spectral excitation.
Multisource	1000	Peak discharge currents from 5 to in excess of 500 A at time constants ranging from 8 to less than 1 ms.	Sensitive and precise. Parameter selection allows wide variety of controlled unidirectional and oscillatory charges variable from arc-like to spark-like in spectral excitation.

Many dramatic changes in the development of readout electronics have occurred over the last 15 to 20 years. Modern systems use integrated circuit and digital computing devices. The engineer is no longer required to design complex circuitry to perform the basic tasks of control. Software now becomes the tool by which timing, sequencing, and logic control are accomplished.

Generally, spectrometer systems fall into two major functional categories—system control and data handling. The digital controller with its controlling and computing capabilities is ideal for handling both tasks with a minimum of effort required by the circuit design engineer.

Particularly during the last decades, several new analytic techniques have been developed that, when appropriate, have a tendency to displace former traditional methodologies.

Additional Reading

Crouch S.R.: *Spectrochemical Analysis*, Pearson Custom Publishing, Boston, MA, 1988.

Thorne, A.P., S. Johansson, and U. Litzen: *Spectrophysics: Principles and Applications*, Springer-Verlag New York, Inc., New York, NY, 2001.

OPTICAL FIBER SYSTEMS.   Optical fibers are hair-thin structures (usually cylindrical in shape) capable of transmitting light signals with extremely low signal loss and at very high digital pulse rates. Fibers are available in a variety of sizes and material compositions and with a wide range of optical performance. Although commercially available fibers are solid structures, they function as "light pipes" that guide rays of light and are therefore sometimes called *lightguides*. When used to connect a light source to a light receiver (photodetector) to form a communication system, the fiber carries *photons* instead of the *electrons* used in traditional metal-conductor communication links. Although a laser light source containing a narrow range of optical wavelengths is preferred for carrying signals the farthest and fastest, light-emitting diodes (LEDs) are also used, especially for short distance communication links within buildings. Thus, the three key elements of a lightwave communication system are the light source, the optical fiber, and the photodetector.

Fiber-Optic Systems in Perspective

As early as 1841, D. Colladon demonstrated light guiding by a jet of water, and in the following year J. Babinet showed the phenomenon in a bent glass rod. However, these experiments did not receive wide publicity until John Tyndall duplicated and popularized the same effect in 1854. In 1880, shortly after the telephone was invented, Alexander Graham Bell proposed telecommunications using lightwaves. See Fig. 1. Patented as the "Photophone," the concept depended on the free propagation of light through the atmosphere. Of course, a century ago Bell had no powerful steady light source such as the laser, and even on very clear days atmospheric disturbances severely limited the practical distance over which undisturbed light could travel.

These experiments were followed by the development of light pipes to illuminate homes (W. Wheeler), bent glass rods to illuminate body cavities for dentistry and surgery (Roth and Reuss), and surgical lamps (D. Smith).

As early as 1910, the possibility of guiding electromagnetic waves by internal reflection within long cylinders of dielectric material was

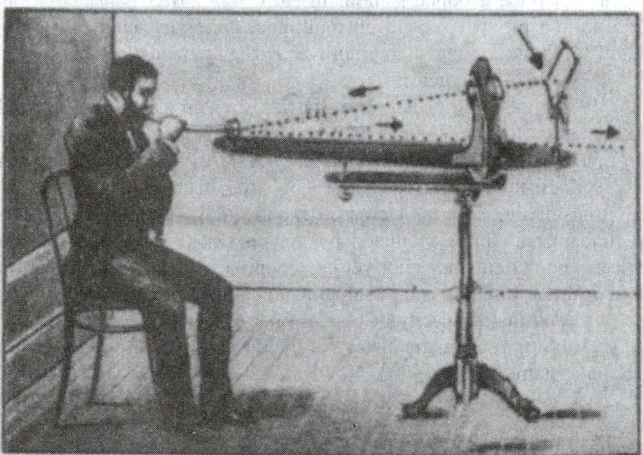

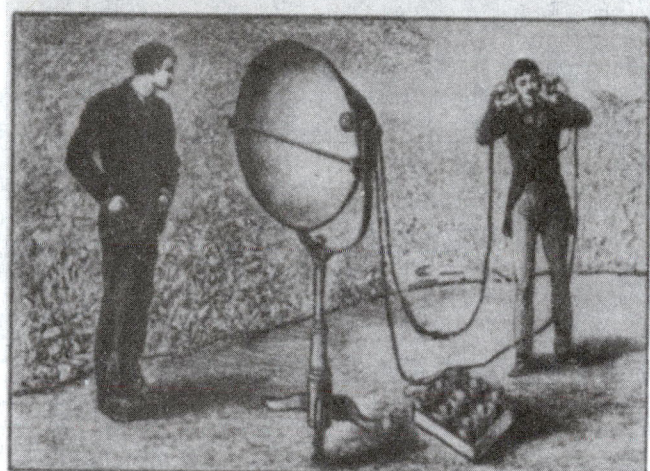

Fig. 1. Old woodcuts showing photophone patented by Alexander Graham Bell in 1880, representing the first attempt to utilize light for the transmission of sound. (*Top*) Sunlight was reflected and focused by a lens onto a mechanism that was vibrated by sound waves (speech), thus modulating the intensity of the exiting light beam. (*Bottom*) At the receiving end of the system, variations of intensity of the light changed the resistance of a selenium photocell, thus controlling an electric current input to the receiving telephone.

investigated on a theoretical basis (D. Hondros and P. Debye). The first quantitative experimental investigations took place in 1920. Although some interest was shown in conducting light by glass rods, a serious rekindling of interest in lightwave communication had to await the first laboratory demonstration of a laser in 1960 (T.H. Maiman, Hughes Aircraft Company). Earnest efforts to devise shielded waveguide structures were made shortly thereafter. Because the glasses available in the early 1960s possessed prohibitively large absorption and scattering losses, some early experimental lightguides consisted of gas-filled underground conduits (some 20 centimeters (7.87 inches) in diameter) that incorporated lenses at various intervals to refocus the light and change its direction when required. These systems were unsatisfactory on several counts: bulk, expense, and the extreme sensitivity to temperature and alignment of the components.

In the mid-1960s, Charles K. Kao of STL Laboratories, determined that the fundamental limit on glass transparency was less than 20 decibels (dB) per kilometer (km), which is low enough to make glass fibers practical for communications. (A loss of 20 dB/km means that the optical power after 1 km is 1% of the amount at the beginning.) This spurred investigators to explore methods for making purer glass. In 1970, fibers with attenuation less than 20 dB/km were made and demonstrated in the laboratory (Maurer, Keck and Schultz, Corning), and this was reduced to 4 dB/km by 1972.

Combining these fibers with emerging semiconductor lasers with lifetimes reaching 1000 hours (Bell Labs, 1973) and photodetectors, enabled lightwave communications to become practical in the mid-1970s. The first nonexperimental fiber-optic link became operational in Dorset, England (1975) to service a police station. In early 1976, Bell Labs started tests at 45 million bits per second on multimode fibers installed in Norcross, Georgia. Fibers continued to improve as their attenuation decreased to 0.47 dB/km (M. Horiguchi).

In 1980, video signals were carried by optical fibers $2\frac{1}{2}$ miles (4 kilometers) for the Winter Olympic Games in Lake Placid, New York. The first long-haul intercity installations (AT&T, Washington-New York; New York-Boston) were made in 1983. After that, the capacity of fiber optic transmission systems increased exponentially. Despite this progress, the fundamental limits predicted by the physics of photonics materials, devices, and systems have not yet been approached (Kogelnik). The challenge of future research and development continues to be a fuller exploitation of the ultimate capacity of optical fibers.

Fig. 2 shows a schematic history of the development of optical communication systems.

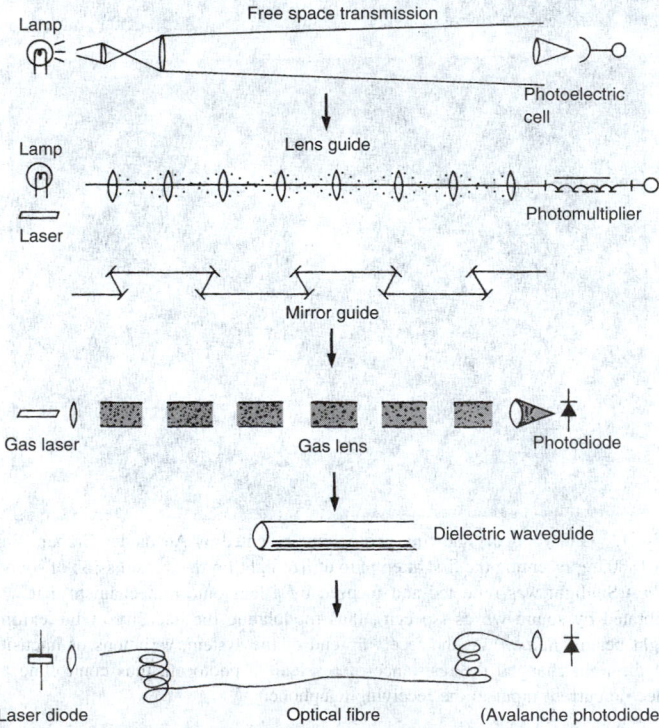

Fig. 2. Schematic representation of the development of optical communication systems. (*Adapted from Suematsu and Iga.*)

Seldom does a new technology have as many practical advantages as found in optical fiber systems and become available in only a decade of concentrated research and development. As solid state physics and semiconductors created the electronics industry, fiber-optic systems have revolutionized the telecommunications industry.

Fiber optic systems are more economical than their alternatives—copper wire, radio relay, and satellite. The regeneration of signals sent on copper cables is necessitated at several mile intervals, whereas the distance on optical fibers can be over a thousand miles by using optical amplifiers approximately every 50 miles.

The first fiber optic systems used light in the 850 to 875 nanometer (nm) wavelength region (the so-called "first window"), and were followed with systems operating near 1300 nm (the "second window") where fiber loss is smaller, and then near 1550 nm (the "third window") where fiber loss is even smaller. In the year 2000, fibers were carrying digital signals at 10 billion bits per second on one or more (as many as 40) wavelengths in the 1550 nm window.

Because optical fibers are nonconducting, fiber optic systems provide excellent electrical isolation and immunity from electrical interference. Signal losses are much lower in fibers (as low as 0.20 dB/km) compared to other guided transmission media, such as twisted copper pairs, coaxial cable, and metallic waveguides. In addition, the bandwidth or information carrying capacity of fibers is far greater. When one or more optical fibers are packaged into cables, the cables are smaller and more flexible than their metallic counterparts.

In 1984, British Telecom laid the first submarine fiber optic cable to carry regular traffic, to the Isle of Wight. In 1988, service began on the first transatlantic fiber-optic cable (TAT-8) from Tuckerton, New Jersey to France and England. Similar undersea communication links between Japan and Guam and other Pacific locations began in 1989. With the development of optical fiber technology proceeding at such a rapid rate, decisions as to whether or not to delay undersea cables for further advancements sometimes are difficult to make.

Earlier "fantastic" claims that 24,000 simultaneous telephone calls on a single pair of fibers at rates of up to 1.7 billion bits per second could be made, that full-length motion pictures such as *Gone with the Wind* could be fed to a home memory unit in one second, or that major symphonies, such as Beethoven's Fifth, could be transmitted in less than 1/50 of a second have already been surpassed. For example, here is listing of some "hero" experiments illustrating an optical fiber's capability as of the year 2000.

- Using 160 wavelengths of light with each wavelength carrying 40 billion bits per second of information, NEC researchers transmitted 6.4 trillion bits of information per second over a special fiber 186 kilometers long.
- Using a single wavelength of light on an experimental TrueWave® fiber, Bell Labs scientists transmitted 320 billion bits of information per second over a distance of 200 kilometers.
- Bell Labs demonstrated 3.28 trillion bits per second over a 300 km of experimental TrueWave fiber. This experiment used 82 wavelengths of light with each operating at 40 billion bits per second.

Types of Optical Fibers

Optical telecommunications fibers fall into two main categories: *multimode fiber*, and *single-mode* (also called *monomode*) *fiber*. Multimode fibers receive their name because they can propagate many (hundreds) of light modes, which can be thought of as paths taken by the light as it travels in the fiber. Single-mode fibers, on the other hand, propagate only one mode. Multimode and single-mode fibers can be further divided into subset categories. For example, multimode fibers can be either *step-index* or *graded-index*, and single-mode fibers can be *dispersion-unshifted*, *dispersion-shifted*, or *nonzero-dispersion shifted*. Each of these, in turn, can be further differentiated. For example, two types of graded-index multimode fibers have either 50 micron or 62.5 micron core diameters. Two types of dispersion-unshifted fiber have either depressed clad or matched clad refractive index profiles. Or they may have a typical high loss at the "water peak" wavelength near 1385 nm or low loss at this wavelength, such as with Lucent's AllWave™ fiber.

Single-mode fibers have many advantages over multimode fibers. See Fig. 3. Compared to the 50 or 62.5 micron core diameter of typical graded-index multimode fibers, some single-mode fibers have a core diameter near 8 micrometers. Single-mode fibers also have a core-cladding refractive index difference of a few tenths of a percent compared to 1 or 2% for multimode fibers. Their smaller core diameter and refractive index difference allow single-mode fibers to propagate light in only one clearly defined path (or mode). Because this reduces multipath effects that spread the arrival times of a short input pulse, single-mode fibers have high bandwidth, meaning that they can carry signals at higher bit rates than multimode fibers.

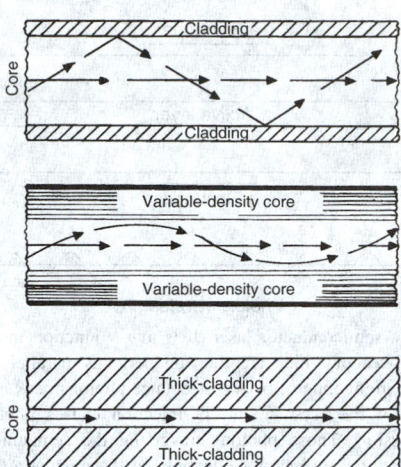

Fig. 3. Fundamental classes of optical fibers: (**a**) Step-index fiber made up of glasses of two different densities, the core and the cladding. Lightwaves travel in a zig-zag fashion down the core, bouncing off one side and then the other side of the core-cladding interface; (**b**) Graded-index fiber where the glass in the core varies in density — hence the light travels in a smooth, curving path, causing less distortion of transmitted information; and (**c**) single- or monomode fiber where the core is very small relative to the wavelength of the transmitted light causing the light to move down the fiber in a straight line — hence resulting in very low distortion. In modern designs, the distortion has been reduced to almost the theoretical low limit.

Various types of optical fibers are used for specific applications. For example, multimode fibers are used primarily in enterprise systems: buildings, offices, campuses. Special single-mode transmission fibers exist for submarine applications, and for metropolitan and long-haul terrestrial applications. And in addition to these transmission fibers, there are various "specialty" fibers for performing dispersion compensation (dispersion compensating fiber), optical amplification (erbium-doped fiber), and other special functions.

Characteristics Affecting Optical Fiber Performance

Phase Index of Refraction. Phase index of refraction is the ratio of the phase velocity of light in a vacuum to its velocity in another medium, such as glass. The value of this parameter depends on wavelength, and the composition, temperature and pressure of the medium. The higher the refractive index of a material, the lower the phase velocity of light in the material, and the more a light ray is bent as it enters the material from air.

Numerical Aperture (*NA*). Numerical aperture describes an angle just outside the fiber's end face that determines the largest angle that a light ray can have to the fiber axis and still be captured and propagate within the fiber. The formula from Snell's law governing the numerical aperture number of a fiber is

$$NA = \sqrt{n_1^2 - n_2^2}$$

where n_1 and n_2 are the phase refractive indices of the fiber's core and cladding, respectively.

Most optical fibers have numerical apertures between 0.15 and 0.4, and these correspond to light acceptance half-angles of about 8 and 23 degrees. Typically, fibers having high *NAs* exhibit greater loss and lower bandwidth.

Light Loss or Attenuation through a Fiber. The amount of light at the output of a fiber is smaller than at the input. This attenuation is expressed in decibels per kilometer (dB/km), which is a relative power unit according to the formula

$$\alpha(dB) = -10 \log \frac{P_o}{P_i}$$

where P_o/P_i is the ratio of optical power at the output to the optical power launched into the fiber at the input. This ratio is smaller than 1, and the logarithm of a number smaller than 1 is negative. Consequently, the minus sign in the equation makes the attenuation value a positive number. A comparison of the ratio of output power to input power in percent to

the quantity in dB is as follows:

$$80\% \text{ transmission} = \text{a loss of } \sim1 \text{ dB}$$
$$50\% \text{ transmission} = \text{a loss of } 3 \text{ dB}$$
$$10\% \text{ transmission} = \text{a loss of } \sim10 \text{ dB}$$
$$1\% \text{ transmission} = \text{a loss of } \sim20 \text{ dB}$$

Bandwidth. Bandwidth is a measure of information-carrying capacity. An optical fiber's bandwidth can be expressed either in the time domain as pulse dispersion in nanoseconds per kilometer (ns/km), or in the frequency domain as frequency passband in megaHertz-kilometers (MHz-km). Light pulses spread or broaden as they pass through a fiber depending on the material used and its design. A fiber's bandwidth limits the rate at which optical pulses can be transmitted and decoded without error at the terminal end of the optical fiber. In general, optical fibers with small core diameter and low numerical aperture have higher bandwidth and lower loss.

Glass Processing

Depending on a fiber's application, a number of glass compositions may be used. However, for low-loss applications, the options become increasingly limited. Multicomponent glasses containing a number of oxides are not suited to making very low-loss fibers. Multicomponent glasses are prepared by essentially standard optical melting procedures, but with special attention given to details for increasing transmission and controlling defects from later fiber drawing steps. In contrast, low-loss fibers are usually made from pure fused silica doped with minor constituents. These require special manufacturing techniques.

Several methods have been used for manufacturing low-loss optical fibers, including the rod-in-tube method, the double-crucible method, and the more recent and widely used chemical vapor deposition (CVD) methods. CVD methods can be divided into two main categories: inside processes and outside processes. With the inside processes, such as modified chemical vapor deposition (MCVD) and plasma-activated chemical vapor deposition (PCVD), the chemical reactions that form the glass occur inside a glass starting tube (sometimes called a "substrate" tube). This starting tube serves as a containment vessel during deposition and eventually becomes part of the fiber. With outside processes, such as outside vapor deposition (OVD) and vapor axial deposition (VAD), vapor deposition occurs on the outside surface of a starting rod, and the reaction that forms the glass occurs near a burner.

With all CVD methods, silica and other glass-forming oxides and dopants are deposited at high temperatures on an object. In the inside processes, the object is collapsed and becomes part of the fiber, whereas in the outside processes, the object is removed and the resultant *soot blank* is dried and sintered. This produces a thick cylindrical *preform*. A long, thin fiber can then be drawn from this preform at high temperature, or the preform can first be made larger by either depositing more glass on its outer surface or by placing the preform in a glass overclad tube.

There are numerous variations on these processes and these are considered proprietary by most manufacturers. See Fig. 4.

Improvements in Lasers

Just a few years ago, the so-called C^3 laser (cleaved-coupled-cavity) appeared, wherein the alignment of two conventional semiconductor lasers yields a beam of exceptional purity that enables communication systems to send signals at rates as great as billions of bits, or binary digits per second. Just as recently as the late 1980s, commercial lightwave systems were limited to somewhat less than 2 million bits per second, but nevertheless a rate that permits the transmission of 24,000 simultaneous telephone calls on a single pair of fibers.

The cleaved-coupled-cavity laser developed by W.T. Tsang and colleagues (AT&T Bell Laboratories) was designed especially for use in fiber optics telecommunication systems and probably as much as the optical fibers themselves will contribute to the gross claims, as mentioned earlier in the article, made for faster, higher-capacity optical links.

Traditionally, semiconductor lasers have been used in optical communication systems. They are about the size of a grain of salt and produce pulses of light from pulses of electric current. Their electrical requirements

Fig. 4.   Experimental optical fiber designs are tested by creating glass preforms, heating them in a special furnace located above the research fiber drawing tower (as shown) and drawing the fiber from the preform onto a collection drum shown in the foreground. The fiber is then used in tests conducted with light generators (lasers and LEDs) and photodetectors. The single-mode fiber is rated for high-capacity transmission, while lower-capacity multimode fibers are well matched to various economical sources and detectors. (*AT&T Bell Laboratories.*)

are minimal (generally a few mA at 1 or 2 V). These lasers can generate infrared (IR) light where optical fibers are most nearly transparent. Unlike a gas laser, for example, the semiconductor laser is mechanically stable and reliable.

For comparison, a simple semiconductor laser is illustrated and briefly described in Figs. 5 and 6. A buried-heterostructure laser is shown in Fig. 7. The alignment of two lasers to form the C^3 laser is shown in Fig. 8. The turnability of a C^3 laser is illustrated in Fig. 9.

As observed in Tsang, the C^3 laser configuration offers at least three advantages in optical communication systems: (1) The exceedingly monochromatic output of the laser eliminates the problem of chromatic dispersion in optical fibers. This facilitates the transmission of digital information in a single-mode fiber at a wavelength of 1.55 micrometers, that is, the wavelength at which a silica fiber is most nearly transparent to electromagnetic radiation. In a test, using the C^3 laser, scientists at AT&T Bell Laboratories transmitted digital information in an optical fiber more than 120 kilometers long at a rate of one gigabit (10^9 bits) per second without reamplification along the path. The frequency of error was less than two bits in 10^{10}. (2) By coupling several C^3 lasers, each tuned to a different wavelength, to a single optical fiber, wavelength-division multiplexing can make it possible to carry several independent messages on the fiber. (3) As proposed by Tsang, at rates on the order of a billion switchings per second, it is possible to shunt the output wavelength of a C^3 laser among as many as 15 modes spaced about 2 nm apart. Thus,

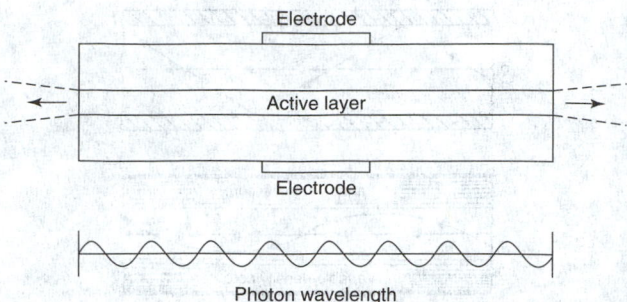

Fig. 5.   A simple semiconductor laser. It is a *p-n* junction in a semiconductor crystal the end faces of which are flat and perfectly parallel. Thus, the faces form a pair of semireflecting mirrors that bounce photons back and forth through the *active layer* of the crystal. Current injection causes photons to arise by spontaneous emission. Those photons traversing the semiconductor cause an avalanche of *stimulated* emission. Reflections at the mirrors are self-reinforcing provided that the wavelength of the photon fits evenly into the *length* of the laser. See also Fig. 6.

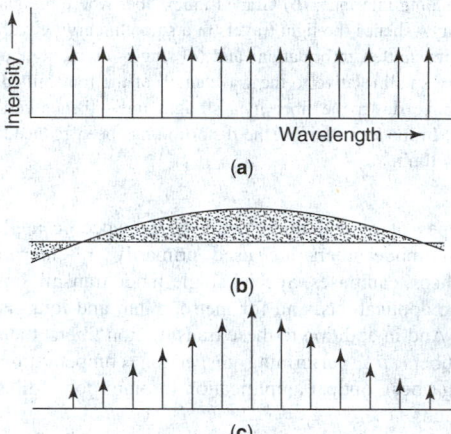

Fig. 6.   Simple semiconductor laser energy diagram. This schematic diagram illustrates why the output beam of the laser jumps randomly among several wavelengths. Fundamentally, the laser resonates at the *infinite number* of wavelengths that fit evenly into the length of the laser as indicated in (**a**). On the other hand, the *p-n* junction produces photons in only a narrow range of wavelengths — the *gain profile*, as indicated in (**b**). Therefore, the beam emitted by the laser includes only the resonant wavelengths positioned within the profile, as shown in (**c**). (*After Tsang.*)

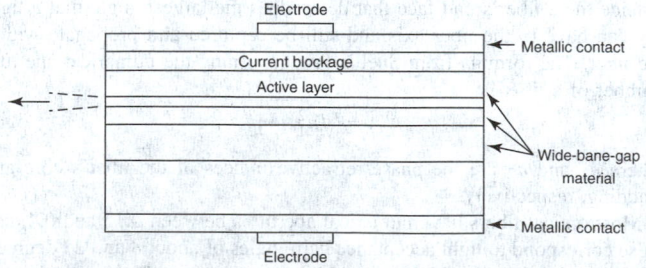

Fig. 7.   In an effort to improve the performance of the basic semiconductor laser, researchers developed the buried-heterostructure laser. In this configuration, the *p-n* junction is reduced to a "tube" that runs the length of the semiconductor crystal. This tube is surrounded by layers of semiconductor whose wide band gap raises the electrical barrier confining charge carriers within the tube. The wide-band-gap material also confines the photons produced at the junction. The laser beam spreads because of diffraction occurring where the beam emerges from the face of the device.

the single-wavelength transmission of data, with high-power and low-power pulses representing the binary digits 1 and 0, respectively, yields to multiple-wavelength transmission.

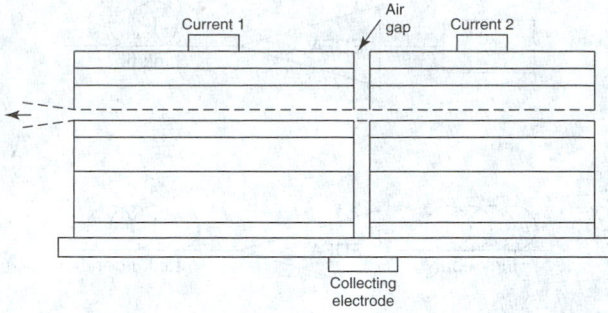

Fig. 8.   The C³ laser consisting of two aligned lasers. The half-lasers have different lengths and thus their resonant wavelengths are differently spaced. Only a few of them match. The mismatches are suppressed. Among the matches, only one is near the peak gain. Thus, the C³ laser beam is made up almost exclusively of that wavelength. Tests have shown that the probability of the beam "jumping" to another wavelength is less than 1 in 10 billion beam samplings.

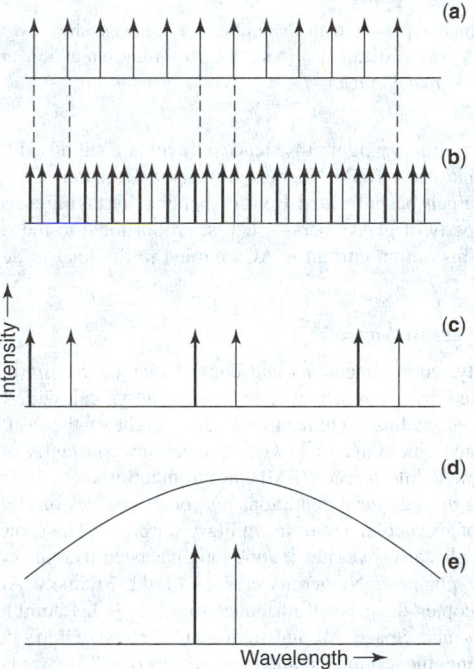

Fig. 9.   The energy diagram of the C³ laser illustrates its tunability by current injection. As explained by Tsang, the current to one of the half-lasers elevates it above its *lasing* threshold. Thus, its resonant modes are fixed as indicated in (a). The other (second) half-laser is maintained below its lasing threshold. Thus, it still has resonant modes, as shown by the thin black lines in (b). But, some of the resonant modes [thin black lines in (c)] match the modes of the light-emitting half-laser. The match at the peak of the laser's gain profile (d) determines the wavelength of the C³ laser's output (e). Now the current applied to the second half-laser is changed. The change alters its resonant modes [heavy black lines in (b)]. Thus, a new set of matching resonances is established [heavy black lines in (c)] and with it a new output wavelength [heavy black line in (e)]. (*After Tsang.*)

In the future, it is expected that the C³ laser may become part of optical logic circuitry. This will be feasible because of the laser's tunability and from the electrical isolation of the two half-lasers. See Fig. 10.

One problem of the past results from the fact that much energy is wasted at the beginning of an optical link. H.M. Presby (Bell Laboratories) asserts in connection with the loss of about half the laser light, "It seemed like an awful waste, like throwing away half a tank of gas." Through redesigning the tips of optical fibers into carefully sculpted microlenses, a surprising amount of light was captured. Optical problems could be solved by designing an aspheric, hyperbolic shape. In a process that is akin to shaping a piece of metal in a lathe, the researchers rotate silicon fibers in and around a carbon dioxide laser beam.

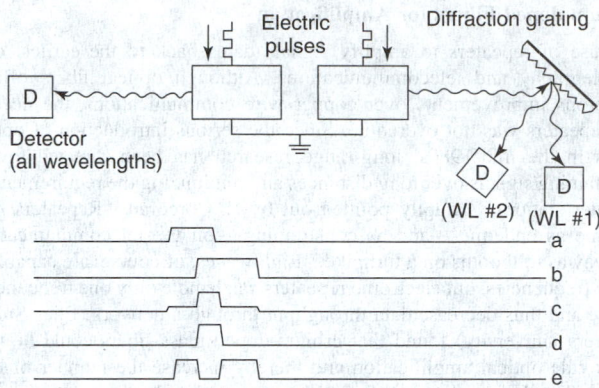

Fig. 10.   Use of the C³ laser in an optical logic circuit. The laser's tunability and the electrical isolation of the two half-lasers make this application possible. Visualize the application of independent trains of electric pulses applied to the half-lasers, as indicated in (**a**) and (**b**). Simultaneous pulses cause the emission of light at wavelength #1. A pulse to one of the half-lasers causes the emission of light at wavelength #2. Detection of light at both wavelengths is equivalent to the logic operation OR (**c**). Detection of wavelength #1 is equivalent to the logic operation AND (**d**). Detection of wavelength #2 is equivalent to the logic operation EXCLUSIVE OR (**e**). (*After Tsang.*)

Researchers also have been investigating microlasers. These lasers range from 1 to 5 microns in diameter and are "carved" from a multilayered semiconductor substrate. It is estimated that a million such lasers occupy but a square centimeter on a chip. See Fig. 11.

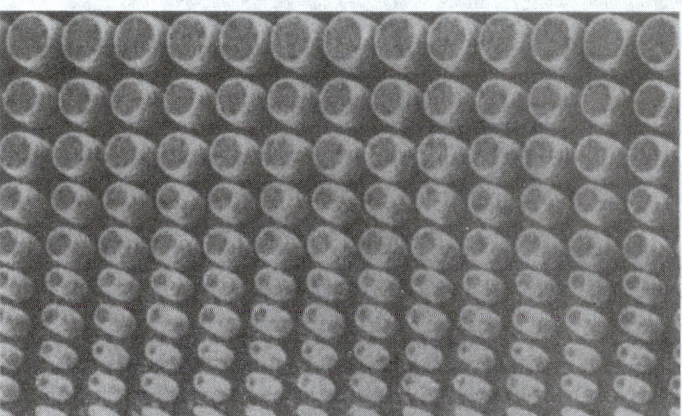

Fig. 11.   An array of microscopic lasers sculpted from a multi-layered semiconductor substrate. The lasers range from 1 to 5 microns in diameter, and a million of them occupy a square centimeter on a chip. (*AT&T Bell Laboratories.*)

Light-Emitting Diode for Telecommunications

In telecommunications, a light-emitting diode (LED) operates similarly to a laser in that it sends pulses of light signals representing speech or data through an optical fiber. Lasers are powerful and are preferred for sending light signals over great distances, such as between cities. Lasers require a measurable amount of power and require temperature regulation. An LED system developed by GTE Laboratories (LOC-LED System) operates on less than one-tenth the power required for a typical laser transmitter and requires no temperature control or means to adjust light signals, thus greatly simplifying the circuitry. The LOC-LED system generates sufficient optical power to fulfill most needs of fiber optic systems over a range of about 6 to 8 miles (9.7–12.9 km), which is a typical span for a local service loop. The system has been tested at temperatures from −4 to +185 °F (−20 to +85 °C). No prior commercially available LED device generates as much power while using so little energy as the LOC-LED.

Fiber optics in telecommunications is discussed further in the article on **Telephony (Telecommunications)**.

Erbium-doped Fibers for Amplification

The use of repeaters to amplify signals dates back to the earliest days of telegraphy and telecommunications. Although optical fibers offered numerous improvements over copper wire communications, the need to use repeaters was not overcome. Since the serious introduction of optical fibers in the mid-1980s, long-range research has been directed toward maintaining signals over long distances and minimizing the requirement for repeater stations. As aptly pointed out by E. Corcoran, "Repeaters have become as endemic—and as constraining—on the telecommunications freeway as tollbooths on a turnpike." Lightwaves, of course, are capable of many frequencies, but electronic repeaters can handle only one frequency at a time and thus decrease data throughput through a network. Elias Snitzer (Rutgers University) found that erbium-doped glass fibers could be used to provide optical amplification and thereby increase the number of light frequencies that can be amplified. When pumped with a light source, more erbium ions in these fibers are forced to a higher energy level (known as *population inversion*) than remain at lower levels. As a transmission signal carrying information enters the erbium fiber, some of the excited erbium ions give up their energy to the information signal—thereby amplifying them. The amplification is purely optical. The information signal does not first need to be converted to an electrical signal.

Researchers (AT&T Bell Laboratories) have demonstrated that erbium-doped devices (Fig. 12) can "boost signals traveling at any bit rate, transmission networks can be upgraded by simply changing the transmitters (and receivers). Virtually any video data or voice signal can be dumpled 'like marbles' into one end of the 'transparent light pipes' and roll out intact at the other end."

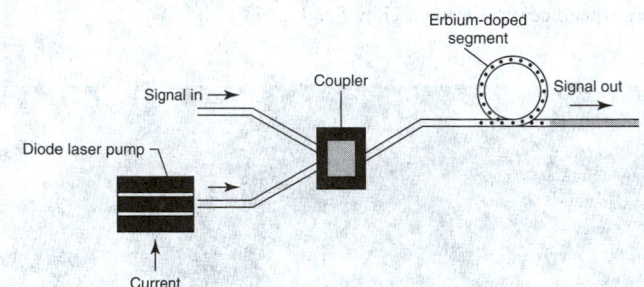

Fig. 12.   Erbium-doped light amplifier. Light from a diode laser pump excites erbium ions in segment of an optical fiber. These ions then emit light and thus boost a passing optical signal. (*After AT&T Bell Laboratories.*)

Earlier research on the erbium-doped principle had been conducted by University of Southampton, U.K., and Japanese communications engineers. Currently, scientists envision that the new devices will be applied first to cable television and transoceanic telecommunications network. As pointed out by E. Corcoran, "Fiber amplifiers could enable a video broadcaster to lay one cable from the transmission center to a neighborhood, then split up the signal, amplify it and route the fibers directly into the homes." The Japanese already are at work on an all-optical network under the Pacific Ocean to be operable by 1996. For local communications loops, L.J. Andrews (GTE) observes, "Cost is the critical issue." John Mellis (BTD, Ipswich, U.K.) says, "The challenges are all engineering now ... Because optical amplifiers are being seriously considered in transoceanic submarine systems, they will certainly happen, if not in this decade then the next." Another professional in the field observes, "There's absolutely no doubt that this is the thing to work on. It's changed how we think about fiber-optic systems."

There are three different locations in which optical amplifiers can be located in a network. See Fig. 13. Used immediately after a laser, booster amplifiers deliver output powers higher than 100 mW, and can therefore increase the output power of a laser by more than an order of magnitude. Preamplifiers increase the strength of an optical signal before it enters a conventional receiver.

Solitons. A *soliton* may be defined as a wave that retains its shape indefinitely. Bright-pulse solitons are pulses of light that travel over long distances without dispersing or broadening. Each of the wavelengths that

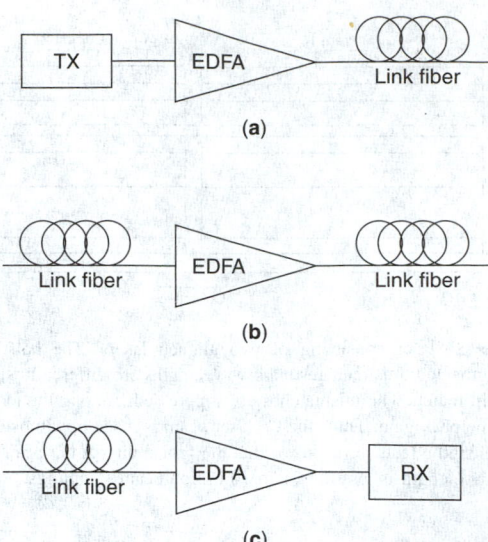

Fig. 13.   Three types of optical amplifier as distinguished by their location in a network: (**a**) Postamplifier (booster), (**b**) in-line amplifier, and (**c**) optical preamplifier. (*From J. Augé.*)

comprise an *ordinary* light pulse tend to travel at a slightly different speed. D.R. Grischkowsky (IBM Thomas J. Watson Research Center) observes that a pulse can be prevented from dispersing by taking advantage of an optical property of glass fibers—that is, proportional to the received light intensity. This output current is AC coupled to the load to deliver the RF signal.

Optical Fibers in Aircraft

Traditionally, communications and control in modern aircraft have been accomplished by as much as 240 km (in a typical wide-body jet) of electric wires, adding significantly to the weight of the craft. Even with insulation and shielding, such wire connections constantly are subject to electromagnetic interference (EMI), not to mention increased vulnerability to lightning during storm conditions and other sources of electromagnetic radiation of particular note to military aircraft. The potential of an "electronic blizzard" over a war zone long has been recognized as a hazard by military planners. Numerous crashes of the Sikorsky Aircraft Black Hawk helicopter have been attributed to EMI. R.J. Baumbick (National Aeronautics and Space Administration, Cleveland, Ohio) notes, "Wires are becoming the dominant antennae in aircraft." By replacing copper lines with optical fibers, the weight of cabling required can be reduced by an estimated 50%. See Fig. 14.

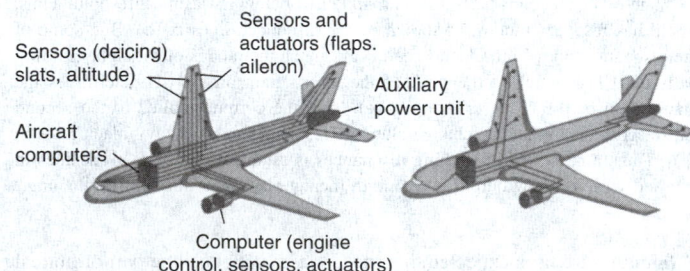

Fig. 14.   Contrast of (left) hard-wire aircraft communications and controls and (right) use of fiber optics. (*After United Technologies Research Center.*)

For the use of light systems in aircraft, researchers recognize the need for developing improved light sources for transmitters. Lasers do not serve well because of the high temperatures encountered in supersonic flight and of intense heat from aircraft engines. Powerful light-emitting diodes are required. A difficult problem is that of using optical components in place of traditional electric control of hydraulic actuators (control of flaps, rudders, and other flight-control surfaces). Sensors that heat a small amount of hydraulic fluid and thus build up pressure that can be amplified by the actuator's hydraulic system are being developed and tested.

Fiber Optics in Biomedicine

Sensors incorporating glass or plastic optical fibers have demonstrated several advantages over electrosensors for biomedical applications. These sensors involve no electrical connections and hence are safe from that standpoint; the leads are quite small and flexible: they can be incorporated in catheters for multiple sensing; where required, they can be implanted for relatively long periods. The fibers are considerably less than 1 millimeter in diameter. Where designed for simplicity, they often can be considered disposable.

As reported by Peterson (National Institutes of Health) and Vurek (Sorenson Research Corp.), there are three principal types of *in vivo* fiber-optic sensing configurations—*photometric* (or bare-ended fiber), *physical parameter sensors* in which a transducer at the end of a fiber alters the light signal in accordance with the values of the parameter measured, and *chemical probes*. In the latter, a suitable reversible reagent fixed at the end of a fiber provides spectrophotometric or fluorometric analysis. The earliest use of fiber optic sensors was in connection with reflectometry, spectrophotometry, and fluorometry where no transducer was used. In oximetry, the hemoglobin content of blood is measured spectrophotometrically. Where dyes are injected into the blood, fiber optic sensors can be used for measuring blood flow, cardiac output, and perfusion. When a microtransducer is attached at the end of a fiber optic conductor, temperature and pressure measurements can be made. A notable application is the use of a temperature sensor in connection with the hyperthermal treatment of cancer. Accuracy of $\pm 0.1\,°C$ can be obtained. Sometimes such devices are used in multiple locations. For example, a layer of liquid crystals at the end of optical fibers will produce changes in light scattering due to temperature change. Fiber optic sensors have been designed for monitoring intracranial and intracardiac pressure. Chemical sensors have been developed to measure pH, PO_2, PCO_2, and glucose, among other chemical variables. The details of these biomedical applications for fiber optic sensors are well developed in the Peterson-Vurek paper (reference listed).

Fiber Optic Communications in Botany

It has been known for centuries that the position of a plant relative to its source of light will affect the manner in which the plant grows and develops, including its shape. Thus, there arose expressions such as "a plant seeks or reaches for the light." Horticulturists and farmers carefully lay out their greenhouses and fields so that maximum advantage can be taken of available sunlight or, in some cases, artificial light. Only in recent years, however, has it been suspected that plants may possess means for further utilizing the radiation which they receive. Researchers Mandoli (Stanford University) and Briggs (Carnegie Institution) have observed that, in addition to depending upon light as a source of energy (photosynthesis), light is also used as a means of communication. For example, the tissues of plant seedlings can guide light through distances measured in centimeters (a distance of 4.5 centimeters has been demonstrated in the laboratory). The exact method of light transmission within plants has not been fully established, but researchers suspect that natural fibers are involved—so called "light pipes."

In the initiation and control of several physiological processes in plants, lightwave communications may serve in a way comparable to the nervous system of animals. Some of the factors influenced by light receptors include the time for a seed to germinate, the angle a shoot should take to counter gravitational forces, the rate with which a leaf should develop, and the time when a plant should bloom. Laboratory findings show that the amount of energy needed for light signaling is several orders of magnitude less than the light used for photosynthesis.

Light-sensitive detectors are pigment cells, of which the molecules making up a substance known as *phytochrome* are among the most important. These photoreceptors are sensitive to various parts of the light spectrum and thus play distinctive roles in managing different parts of a plant's physiology. In addition to the spectral distribution of light as received by the photosensitive cells, other important factors include the amount of light received, the direction from which the light is received, and the duration of the light signal.

This relatively recent area of botany and biology admittedly is in an early stage of development. Also see the article on **Etiolation**.

Fiber Optics in the Private Network

Private Networks, new and existing, must be well planned and carefully structured. Their network cabling solution may require a balance of both copper and fiber to cost effectively meet today's needs and support the high-bandwidth applications of the future, such as multimedia and full motion digital video-conferencing. Lucent Technologies, SYSTIMAX® SCS product families, supports these new and existing networks.

Rapidly evolving applications and technologies will drastically increase the speed and volume of traffic on LAN/WAN networks. Ensuring that your structured cabling solution is designed to accommodate the higher transmission rates associated with these evolving bandwidth intensive applications will be critical. Some examples include:

- Multimedia workstation
- Networked Scientific Modeling
- Imaging, Radiography, Computer Aided Design/Computer Aided Manufacturing (CAD/CAM)
- Asynchronous Transfer Mode (ATM)
- High-Definition Television (HDTV)
- Array processor workstations
- Mass memory database transfer
- Videophone
- Photonic (lightwave) switches and processors.

Local Area Networks (LANs). A LAN is a data communications system that enables users to access common data processing (PCs, minicomputers, and mainframe computers) and peripheral equipment (printers and fax machines). LANs, are created by using workstations with adapter cards and connecting them to file servers (where the operating system/software resides) and printers. Gateways are used to connect LANs to other LANs or operating systems like large mainframes where there is a need to share departmental or corporate computing systems. A LAN can be as simple as a few workstations working off a file server or as complex as putting hundreds of workstations on a network that runs between floors of a building or between a number of buildings in a campus environment. LANs, which were originally designed so that users could share and access a few expensive printers or controllers, have expanded into essential telecommunications networks. Today, LANs are used for file and printer sharing, electronic mail, shared databases, point-of-sale, and order entry systems.

LANs deployed on different floors or buildings are typically connected with multimode fiber. However, newer high-speed LAN topologies like full motion video do utilize single-mode fiber in some long-distance route applications where the excellent transmission characteristics of single-mode fiber are required.

LAN Topologies. SYSTIMAX SCS offers cabling architecture options for Fiber-to-the-Desktop installations: the traditional **Hierarchical Star architecture** and the new **Single Point Administration architecture**.

The traditional **Hierarchical Star architecture** is designed for maximum flexibility. Cross-connect facilities are provided in both the telecommunications closets and the main equipment room. The riser backbone cables can be sized with low counts which allow only distributed active equipment, or for greatest flexibility, with high counts which permit both distributed and centralized active equipment. The horizontal cross-connect facility helps ensure the greatest life span for the system by allowing the active equipment to be located closer to the work areas. The short horizontal runs can support applications at higher speeds than the longer combined horizontal/riser runs used with centralized active equipment. Also, this architecture is standards-compliant with both riser design approaches.

Single Point Administration architecture is designed for simplicity and cost-effectiveness for centralized equipment. This approach provides direct connections from all work areas (offices) to the cross-connect in the main equipment room, forming a single point of administration optimized for centralized active equipment. The single point of administration provides the simplest circuit management possible by eliminating the need to cross-connect circuits in multiple places. It also provides the ability to connect users in different areas of a building directly to the same LAN segment, reducing traffic on bottleneck-prone bridges and routers. Single point administration also provides three cost benefits. First, it eliminates the need for horizontal cross-connects, saving passive hardware costs. Second, it consolidates active equipment, reducing the number of idle ports in the system, thereby saving active equipment costs. Finally, this type of architecture eases network administration and maintenance, reducing technical support staff effort.

The LAN topology is the physical layout of a LAN—how the controllers, workstations (primarily PCs), and other equipment are connected by the cable. The three basic LAN topologies are: Bus; Star; and Ring.

The bus topology has all the workstations on the network attached (via the information outlet) to a single cable that carries the signal in both directions through the network. The bus network, can be expanded by adding several segments together with bridges, routers, and repeaters. The Institute of Electrical and Electronics Engineers Inc. (IEEE) Standard 802.3 is an example of a bus topology. In a star topology, all of the nodes or workstations are connected with unshielded twisted pair (UTP) or fiber optic cable to a centrally located common controller or concentrator. The central control point permits centralized network administration, management, and troubleshooting. StarLAN is an example of a star topology, as is IEEE 802.3. In a ring topology, the network forms a ring. A token carries data through the network. Workstations are connected as in a star topology to a central administration point, such as a multistation access unit (MAU). Token Ring (IEEE 802.5) is an example of a ring topology.

The LANs, which were introduced back in the 1980s, were based on copper cable, with very few fiber applications being used. Most of the more recently introduced LANs (FDDI, 10BASE-F, and DFON) primarily use fiber optics but offer interfaces (bridges and routers) with the older copper-based networks. The more recent trend is in the use of "smart hubs" or concentrators, which allow both bus and ring (Ethernet and Token Ring) topologies to be mixed in the same electronics within the communications closet. In addition, the backbone network, which links the hubs/ concentrators together, could be on a higher speed LAN topology such as FDDI running at 100 Mbps.

The advent of new lower cost optoelectronics for LAN applications has spurred a growing interest in using fiber all the way to the workstation. These total fiber networks offer easy migration to even higher speed applications like ATM. See also **Local Area Networks**.

Fiber Optics in Industrial Instrumentation and Networks

The combination of light-transmitting optical cable and miniature silicon sensors has resulted in the development of a new measurement technology for various industrial processes. Three variables may be measured with this technique—temperature, pressure, and refractive index. The new systems are immune to electromagnetic and radio-frequency interference, they provide more accuracy in electrically noisy environments, and their miniature size improves response and causes minimal process disturbances.

The fiber optic sensors utilize an extrinsic Fabry-Perot interferometer to spectrally modulate light in proportion to pressure, temperature, or refractive index variations. Because they are based on spectral modulation instead of amplitude modulation, they are not affected by such common problems as fiber bending, connector losses, and aging.

Pressure Sensing. As shown by Fig. 15, a cavity resonator is constructed using a float-bottom pocket-etched into a refractory glass substrate. The pocket is 2–3 wavelengths deep and covered with a diaphragm. The two reflecting surfaces (bottom of pocket and the diaphragm) form an interferometer. The cavity is evacuated, thus permitting the diaphragm to deflect. This deflection is a function of absolute pressure. In effect, the path length changes. A parallel can be made with the action of a soap bubble. As the bubble changes size and hence film thickness, interference effects occur. These reinforce reflected light at certain wavelengths, enhancing transmission of light through the film at other wavelengths. Color (frequency) changes occur as a result.

The cavity resonator (sensor) is 0.015 in (0.4 mm) in all three dimensions. The resonator is joined to a multimode fiber by a glass capillary 0.032 in (0.8 mm) in diameter. The light source's spectrum is centered at 850 nm. As the effective cavity depth changes as a function of the measured variable, the cavity's reflectance spectrum shifts. This spectral shift modulates (skews) the incoming light spectrum. A dichroic filter and

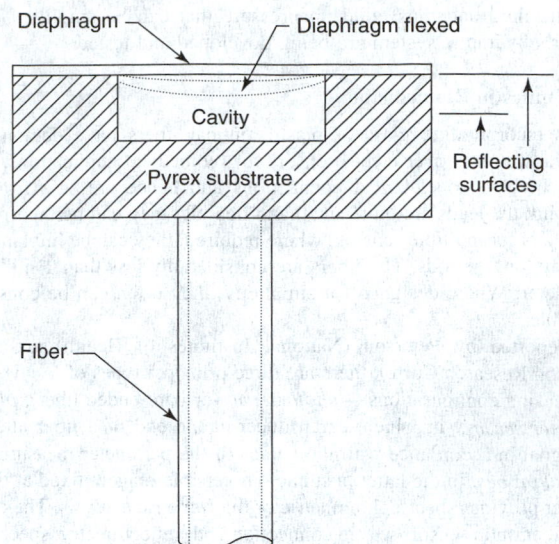

Fig. 15.   Fiber-optic pressure sensor. As light passes into a cavity resonator formed by a glass substrate and a flexible diaphragm, it is reflected from both surfaces, forming an interference pattern that changes as the diaphragm flexes with pressure changes. (*Yazbak, Foxboro, Massachusetts.*)

photodiodes are then used to discern differences in the returning light's spectrum strength within two wavebands.

Temperature Sensing. As shown in Fig. 16, a layer of silicon (whose refractive index changes with temperature) is placed in the optical path in place of the evacuated cavity, as previously described. The second reflector (glass) is rigid. The effective path length thus changes with temperature.

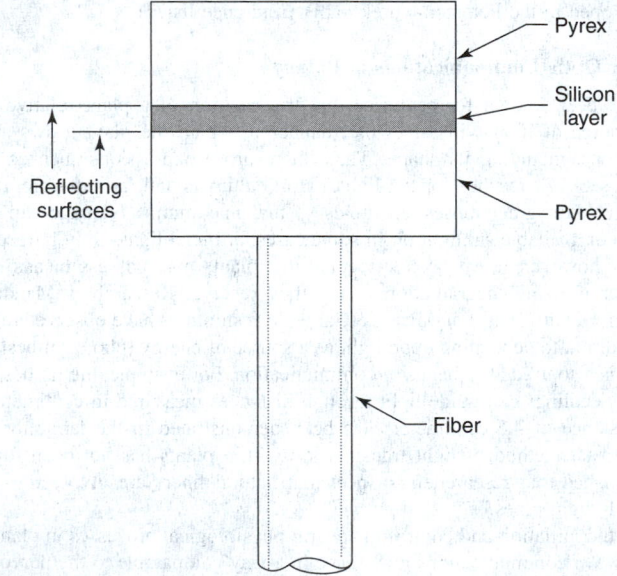

Fig. 16.   Fiber-optic temperature sensor. A thin layer of silicon placed in the optical path exhibits a large change in refractive index with temperature, changing the effective path length. (*Yazbak, Foxboro, Massachusetts.*)

Refractive Index. This property is the ratio of the velocity of light in a vacuum to the velocity of light in a transparent material. The property is used extensively in the processing industries for measuring component concentrations or for tracking changes in molecular makeup, such as materials that are reacting. The hydrogenation of food oils is an example of such a process.

The extent of hydrogenation (the degree of saturation of a double bond in the ester chain of an edible oil) may be indicated by conventional laboratory titrations of the iodine number. This is a time-consuming, grab-sampling method. Continuous measurement and control via refractive index measurement is far more efficient. As shown by Fig. 17, the fluid of interest is drawn by capillary action into a duct through a glass substrate, and the effective path length varies in proportion to the refractive index of the fluid. Excellent correlation of refractive index with the iodine number of hydrogenated oil is shown in Fig. 18.

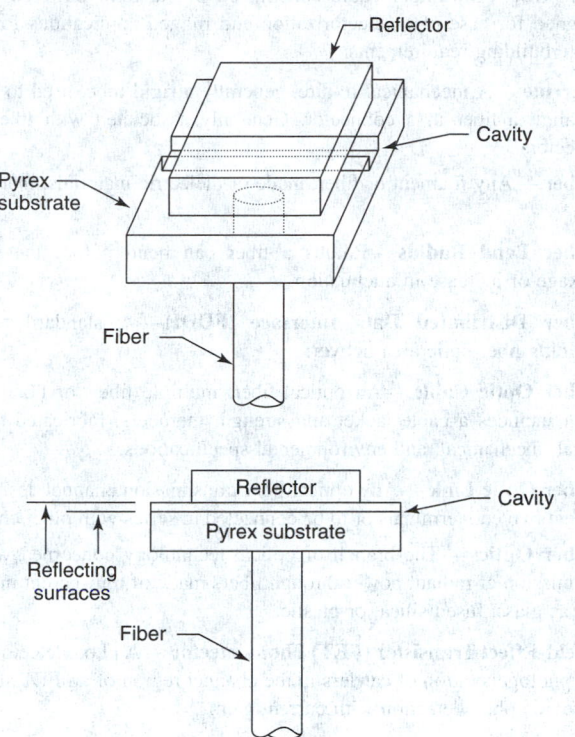

Fig. 17. Fiber-optic refractive index sensor. Fluid of interest is drawn by capillary action into a duct through a glass substrate. The effective path length varies in proportion with the refractive index. (*Yazbak, Foxboro, Massachusetts.*)

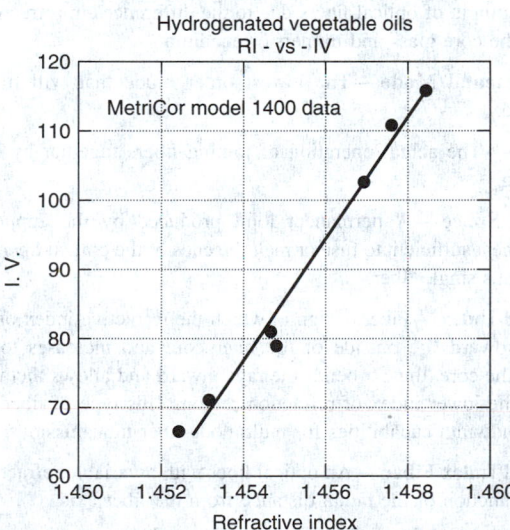

Fig. 18. Correlation of refractive index with iodine number of hydrogenated oil. (*Yazbak, Foxboro, Massachusetts.*)

Note: The foregoing information on fiber optic sensors was furnished by Gene Yazbak, Foxboro, MA.

Optical Fiber Terminology

An abridged glossary of terms used to describe optical fiber products and processes would include:

Absorption — A physical mechanism in fibers that attenuates light by converting it into heat-thereby raising the fiber's temperature. In practice the temperature increase is slight and difficult to measure. Absorption arises from tails of the ultraviolet and infrared absorption bands, from impurities such as the OH–ion, and from defects in the glass structure.

Adapter — A mechanical media termination device designed to align and join fiber optic connectors. Often referred to as a coupling, bulkhead, or interconnect sleeve.

Adapter Efficiency — The efficiency of optical power transfer between two components.

Adapter Loss — The power loss suffered when coupling light from one optical device to another.

Aramid Yarn — Strength elements that provide tensile strength and provide support and additional protection of the fiber bundles. Kevlar is a particular brand of aramid yarn.

Armor — Additional protective element beneath outer jacket to provide protection against severe outdoor environments. Usually made of plastic-coated steel, it may be corrugated for flexibility.

Attenuation — The decrease in magnitude of power, of a signal in transmission between points. A term used for expressing the total loss of an optical system, normally measured in decibels (dB) at a specific wavelength.

Attenuation Coefficient — The rate of optical power loss with respect to distance along the fiber, usually measured in decibels per kilometer (dB/km) at specific wavelength. The lower the number, the better the fiber's attenuation. Typical multimode wavelengths are 850 and 1300 manometers (nm); single-mode wavelengths are 1310 and 1550 nm. Note: When specifying attenuation, it is important to note whether the value is average or nominal.

Avalanche Photodiode (APD) — A photodiode designed to take advantage of avalanche multiplication of photocurrent. As the reverse-bias voltage approaches the breakdown voltage, hole-electron pairs created by absorbed photons acquire sufficient energy to create additional hole electron pairs when they collide with ions; thus a multiplication or signal gain is achieved.

Axial Ray — A light ray that travels along the axis of an optical fiber.

Backbone Cabling — The portion of premises telecommunications cabling that provides connections between telecommunications closets, equipment rooms, and entrance facilities. The backbone cabling consists of the transmission media (optical fiber cable), main and intermediate cross-connects, and terminations for the horizontal cross-connect, equipment rooms, and entrance facilities. The backbone cabling can further be classified as interbuilding backbone (cabling between buildings), or intrabuilding backbone (cabling within a building).

Bandwidth-Distance Product — The information-carrying capacity of a transmission medium is normally referred to in units of MHz-km. This is called the bandwidth-distance product or more commonly bandwidth. The amount of information that can be transmitted over any medium changes according to distance. The relationship is not linear, however. A 500 MHz-km fiber does not translate to 250 MHz for a 2 kilometer length or 1000 MHz for a 0.5 kilometer length. It is important, therefore, when comparing media, to ensure that the same units of distance are being used.

Bandwidth Limited Operation — The condition prevailing when the system bandwidth, rather than the amplitude of the signal, limits performance. The condition is reached when modal dispersion distorts the shape of the waveform beyond specified limits.

Beamsplitter — A device used to divide an optical beam into two or more separate beams.

BER (Bit Error Rate) — In digital applications, the ratio of bits received in error to bits sent. BERs of one errored bit per billion (I × IO-9) sent are typical.

Buffer — Material used to protect optical fiber from physical damage, providing mechanical isolation and/or protection. Fabrication techniques include tight or loose tube buffering as well as multiple buffer layers.

Buffering — (1) A protective material extruded directly on the fiber coating to protect it from the environment (tight buffered); (2) extruding a tube around the coated fiber to allow isolation of the fiber from stresses in the cable (buffer tubes).

Buffer Tubes — Extruded cylindrical tubes covering optical fibers(s) used for protection and isolation.

Bundle — Many individual fibers contained within a single jacket or buffer tube. Also a group of buffered fibers distinguished in some fashion from another group in the same cable core.

Cable — An assembly of optical fibers and other material providing mechanical and environmental protection.

Cable Assembly — Optical fiber cable that has connectors installed on one or both ends. General use of these cable assemblies includes the interconnection of optical fiber cable systems and opto-electronic equipment. If connectors are attached to only one end of a cable, it is known as a pigtail. If connectors are attached to both ends, it is known as a jumper or patch cord.

Cable Bend Radius — Cable bend radius during installation implies that the cable is experiencing a tensile load. Free bend infers a smaller allowable bend radius since it is at a condition of no load.

Central Member — The center component of a cable. It serves as an antibuckling element to resist temperature induced stresses. Sometimes serves as a strength element. The central member material is either steel, fiberglass, or glass-reinforced plastic.

Centralized Cabling — A cabling topology used with centralized electronics connecting the optical horizontal cabling with intra-building backbone cabling passively in the telecommunications closet.

Chromatic Dispersion — Spreading of a light pulse caused by the difference in refractive indices at different wavelengths.

Cladding — The dielectric material surrounding the core of an optical fiber.

Coating — A material put on a fiber during the drawing process to protect if from the environment and handling.

Composite Cable — A cable containing both fiber and copper media per article 770 of the National Electric Code (NEC).

Connecting Hardware — A device used to terminate an optical fiber cable with connectors and adapters that provide an administration point for cross-connecting between cabling segments or interconnecting to electronic equipment.

Connector Panel — A panel designed for use with patch panels; it contains either 6, 8, or 12 adapters pre-installed for use when field-connectorizing fibers.

Connector Panel Module — A module designed for use with patch panels, it contains either 6 or 12 connectorized fibers that are spliced to backbone cable fibers.

Core — The central region of an optical fiber through which light is transmitted.

Core Eccentricity — A measure of the displacement of the center of the core relative to the cladding center.

Core Ellipticity (non-circularity) — A measure of the departure of the core from roundness.

Critical Angle — The smallest angle from the fiber axis at which a ray may be totally reflected at the core/cladding interface.

Cutoff Wavelength — The shortest wavelength at which only the fundamental mode of an optical waveguide is capable of propagation.

Data Rate — The maximum number of bits of information, which can be transmitted per second, as in a data transmission link. Typically expressed as megabits per second (Mbps).

Dielectric — Nonmetallic and, therefore, nonconductive. Glass fibers are considered dielectric. A dielectric cable contains no metallic components.

Entrance Facility — An entrance to a building for both public and private network service cables including the entrance point at the building wall and continuing to the entrance room or space.

Fan-Out — Multifiber cable constructed in the tight buffered design. Designed for ease of connectorization and rugged applications for intra- or interbuilding requirements.

Ferrule — A mechanical fixture, generally a rigid tube, used to protect and align a fiber in a connector. Generally associated with fiber optic connectors.

Fiber — Any filament or fiber, made of dielectric materials, that guides light.

Fiber Bend Radius — Radius a fiber can bend before the risk of breakage or increase in attenuation.

Fiber Distributed Data Interface (FDDI) — A standard for a I 00 Mbit/s fiber optic area network

Fiber Optic Cable — An optical fiber, multiple fiber, or fiber bundle which includes a cable jacket and strength members, fabricated to meet optical, mechanical, and environmental specifications.

Fiber Optic Link — Any optical fiber transmission channel designed to connect two end terminals or to be connected in series with other channels.

Fiber Optics — The branch of optical technology concerned with the transmission of radiant power through fibers made of transparent materials such as glass, fused silica, or plastic.

Field-Effect Transistor (FET) Photodetector — A photodetector employing photogeneration of carriers in the channel region of an FET structure to provide photodetection with current gain.

Fresnel Reflection — The reflection of a portion of the light incident between two homogeneous media having different refractive indices. Fresnel reflection occurs at the air/glass interfaces at entrance and exit ends of an optical fiber.

Fresnel Reflection Losses — Reflection losses that are incurred at the input and output of optical fibers due to the differences in refraction index between the core glass and immersion medium.

Fundamental Mode — The lowest order mode that will travel in a waveguide.

Fusing — The actual operation of joining fibers together by fusion or by melting.

Fusion Splice — A permanent joint produced by the application of localized heat sufficient to fuse or melt the ends of the optical fiber, forming a continuous single fiber.

Graded-Index — Fiber design in which the refractive index of the core is lower toward **the** outside of the fiber core and increases toward the center of the core, thus, it bends the rays inward and allows them to travel faster in the lower index of refraction region. This type of fiber provides higher bandwidth capabilities for multimode fiber transmissions.

Graded Index Fiber — An optical fiber with a variable refractive index that is a function of the radial distance from the fiber axis.

Horizontal Cabling — That portion of the telecommunications cabling that provides connectivity between the horizontal cross-connect and the work-area telecommunications outlet. The horizontal cabling consists of transmission media, the outlet, the terminations of the horizontal cables, and horizontal cross-connect.

Horizontal Cross-Connect (HC) — A cross-connect of horizontal cabling to other cabling, e.g., horizontal, backbone, equipment.

Hybrid Cable — A fiber optic cable containing two or more different types of fiber, such as 62.5 μm multimode and single-mode.

Index Matching Fluid — A fluid with an index of refraction close to that of glass that reduces reflections caused by refractive-index differences.

Index Matching Material — A material, often a liquid or cement whose refractive index is nearly equal to the core index. Used to reduce Fresnel reflections from a fiber end face.

Insertion Loss — The attenuation caused by the insertion of an optical component; in other words, a connector or coupler in an optical transmission system.

Interbuilding Backbone — The portion of the backbone cabling between buildings. (See Backbone Cabling.)

Intermediate Cross-Connect (IC) — A secondary crossconnect in the backbone cabling used to mechanically terminate and administer backbone cabling between the main cross-connect and horizontal cross-connect.

Intrabuilding Backbone — The portion of the backbone cabling within a building. (See Backbone Cabling.)

Irradiance — Power density at a surface through which radiation passes at the radiating surface of a light source or at the cross section of an optical waveguide. The normal unit is Watts per centimeters squared, or W/cmu2d.

Jumper — Optical fiber cable that has connectors installed on both ends. (See Cable Assembly.)

Laser Diode (LD) — Light Amplification by Stimulated Emission of Radiation. An electro-optic device that produces coherent light with a narrow range of wavelengths, typically centered around 780 nm, 1320 nm, or 1550 nm. Lasers with wavelengths centered around 780 nm are commonly referred to as CD Lasers.

Lasing Threshold — The lowest excitation level at which a laser's output is dominated by stimulated emission rather than spontaneous emission.

Launching Fiber — A fiber used in conjunction with a source to excite the modes of another. fiber in a particular way. Launching fibers are most often used in test systems to improve the precision of measurements.

Leaky Modes — In the boundary region between the guided modes of an optical waveguide and the lightwaves which are not capable of propagation, there are so-called leaky modes which are not guided but are capable of limited propagation with increased attenuation. Leaky modes are a possible source of errors in the measurement of fiber loss, but their effect can be reduced by mode strippers.

Light — In the laser and optical communication fields, the portion of the electromagnetic spectrum that can be handled by the basic optical techniques used for the visible spectrum extending from the near ultraviolet region of approximately 0.3 micron, through the visible region and into the mid infrared region of about 30 microns.

Lightwaves — Electromagnetic waves in the region of optical frequencies. The term "light" was originally restricted to radiation visible to the human eye, with wavelengths between 400 and 700 manometers (nm). However, it has become customary to refer to radiation in the spectral regions adjacent to visible light (in the near infrared from 700 to about 2000 nm) as "light" to emphasize the physical and technical characteristics they have in common with visible light.

LXE — Fiber Optic Express Entry

MDPE — Abbreviation used to denote medium density polyethylene. A type of plastic material used to make cable jacketing.

Macrobending — Macroscopic axial deviations of a fiber from a straight line, in contrast to microbending.

Main Cross-Connect (MC) — The centralized portion of the backbone cabling used to mechanically terminate and administer the backbone cabling, providing connectivity between equipment rooms, entrance facilities, horizontal cross-connects, and intermediate cross-connects.

Material Dispersion — The dispersion associated with a non-monochromatic light source due to the wavelength dependence of the refractive index of a material or of the light velocity in this material.

Mechanical Splicing — Joining two fibers together by permanent or temporary mechanical means (vs. fusion splicing or connectors) to enable a continuous signal. The CamSplice is a good example of a mechanical splice.

Megahertz (MHz) — A Unit of frequency that is equal to one million cycles per second.

Microbending — Curvatures of the fiber which involve axial displacements of a few micrometers and spatial wavelengths of a few millimeters. Microbends cause loss of light and consequently increase the attenuation of the fiber.

Micrometer (gm) — One millionth of a meter; 10-6 meter, Typically used to express the geometric dimension of fibers, for example, 62.5 μm.

Mini Bundle Cable — Loose tube cable in which the buffer tube contains two or more fibers, typically 6 or 12 fibers.

Modal Dispersion — Pulse spreading due to multiple light rays traveling different distances and speeds through an optical fiber.

Modal Noise — Disturbance in multimode fibers fed by laser diodes. It occurs when the fibers contain elements with mode-dependent attenuation, such as imperfect splices, and is more severe the better the coherence of the laser light.

Mode — A term used to describe an independent light path through a fiber, as in multimode or single-mode.

Mode Field Diameter — The diameter of the one mode of light propagating in a single-mode fiber. The mode field diameter replaces core diameter as the practical parameter in single-mode fiber.

Mode Mixing — The numerous modes of a multimode fiber differ in their propagation velocities. As long as they propagate independently of each other, the fiber bandwidth varies inversely with the fiber length due to multimode distortion. As a result of inhomogeneities of the fiber geometry and of the index profile: a gradual energy exchange occurs between modes with differing velocities. Due to this mode mixing, the bandwidth of long multimode fibers is greater than the value obtained by linear extrapolation from measurements on short fibers.

Modes — Discrete optical waves that can propagate in optical waveguides. They are eigenvalue solutions to the differential equations which characterize the waveguide. In a single-mode fiber, only one mode, the fundamental mode, can propagate. There are several hundred modes in a multimode fiber which differ in field pattern and propagation velocity. The upper limit to the number of modes is determined by the core diameter and the numerical aperture of the waveguide.

Mode Scrambler — A device composed of one or more optical fibers in which strong mode coupling occurs. Frequently used to provide a mode distribution that is independent of source characteristics.

Modified Chemical Vapor Deposition (MCVD) Technique — A process in which deposits are produced by heterogeneous gas/solid and gas/liquid chemical reactions at the surface of a substrate. The MCVD method is often used in fabricating optical waveguide preforms by causing gaseous material to react and deposit glass oxides. Typical starting chemicals include volatile compounds of silicon, germanium, phosphorus, and boron, which form corresponding oxides after heating with oxygen or other gases. Depending on its type, the preform may be processed further in preparation for pulling into an optical fiber.

Modulation — Coding of information onto the carrier frequency. This includes amplitude, frequency, or phase modulation techniques.

Monochromatic — Consisting of a single wavelength. In practice, radiation is never perfectly monochromatic but, at best, displays a narrow band of wavelengths.

Multimode Distortion — The signal distortion in an optical waveguide resulting from the superposition of modes with differing delays.

MuItimode Fiber — An optical waveguide in which light travels in multiple modes. Typical core/cladding size (measured in micrometers) is 62.5/125.

Multiplex — Combining two or more signals into a single bit stream that can be individually recovered.

Multi-User Outlet — A telecommunications outlet used to serve more that one work area, typically in open-systems furniture applications.

Nanometer (nm) — A unit of measurement equal to one billionth of a meter; 10-9 meters. Typically used to express the savelength of light, for example, 1300 nm.

Near Field Radiation Pattern — Distribution of the irradiance over an emitting surface; in other words, over the cross section of an optical waveguide.

Numerical Aperture — A measure of the range of angles of incident light transmitted through a fiber. Depends on the differences in index of refraction between the core and the cladding. (The number that expresses the light gathering ability of a fiber. Related to acceptance angle.)

Optical Time Domain Reflectometer (OTDR) — A method for characterizing a fiber wherein an optical pulse is transmitted through the fiber and the resulting backscatter and reflections to the input are measured as a function of time. Useful in estimating attenuation coefficient as a function of distance and identifying defects and other localized losses.

Optical Waveguide — Dielectric waveguide with a core consisting of optically transparent material of low attenuation (usually silica glass) and with cladding consisting of optically transparent material of lower refractive index than that of the core. It is used for the transmission of signals with lightwaves and is frequently referred to as fiber. In addition, there are planar dielectric waveguide structures in some optical components, such as laser diodes, which are also referred to as optical waveguides.

Optoelectronic — Pertaining to a device that responds to optical power, emits or modifies optical radiation, or utilizes optical radiation for its internal operation. Any device that functions as an electrical-to-optical or optical-to-electrical transducer.

PE — Abbreviation used to denote polyethylene. A type of plastic material used for outside plant cable jackets.

PVC — Abbreviation used to denote polyvinyl chloride. A type of plastic material used for cable jacketing. Typically used in flame-retardant cables.

PVDF — Abbreviation used to denote polyvinyl difluoride. A type of material used for cable jacketing. Often used in plenum-rated cables.

Photocurrent — The current that flows through a photosensitive device, such as a photodiode, as the result of exposure to radiant power.

Photodiode — A diode designed to produce photocurrent by absorbing light. Photodiodes are used for the detection of optical power and for the conversion of optical power into electrical power.

Pigtail — A short length of optical fiber for coupling optical components. It is usually permanently fixed to the components.

PIN Diode — A semiconductor device used to convert optical signals to electrical signals in a receiver.

PIN-FET Receiver — Optical receiver with a PIN photodiode and low noise amplifier with a high impedance input, whose first stage incorporates a Field-Effect Transistor (FET).

PIN Photodiode — A diode with a large intrinsic region sandwiched between p-doped and n-doped semiconducting regions. Photons in this region create electron hole pairs that are separated by an electric field, thus generating an electric current in the load circuit.

Preform — A glass structure from which an optical fiber waveguide may be drawn.

Prefusing — Fusing with a low current to clean the fiber end. Precedes fusion splicing.

Primary Coating — The plastic coating applied directly to the cladding surface of the fiber during manufacture to preserve the integrity of the surface.

Rayleigh Scattering — Scattering by refractive index fluctuations (inhomogeneities in material density or composition) that are small with respect to wavelength.

Receiver — A detector and electronic circuitry to change optical signals into electrical signals.

Receiver Sensitivity — The optical power required by a receiver for low error signal transmission. In the case of digital signal transmission, the mean optical power is usually quoted in Watts or dBm (decibels referred to I milliwatt).

Reflection — The abrupt change in direction of a light beam at an interface between two dissimilar media so that the light beam returns into the media from which it originated.

Refraction — The bending of a beam of light at an interface between two dissimilar media or in a medium whose refractive index is a continuous function of position (graded index medium).

Refractive Index — The ratio of the velocity of light in vacuum to that in an optically dense medium.

Repeater — In a lightwave system, an optoelectronic device or module that receives an optical signal, converts it to electrical form, amplifies or reconstructs it, and retransmits it in optical form.

Riser — Pathways for indoor cables that pass between floors. It is normally a vertical shaft or space. Also a firecode rating for indoor cable.

Scattering — A property of glass that causes light to deflect from the fiber and contributes to optical attenuation.

Single-Mode Fiber — Optical fiber with a small core diameter (typically 9 μm) in which only a single-mode, the fundamental mode, is capable of propagation. This type of fiber is particularly suitable for wideband transmission over large distances, since its bandwidth is limited only by chromatic dispersion.

Spontaneous Emission — This occurs when there are too many electrons in the conduction band of a semiconductor. These electrons drop spontaneously into vacant locations in the valence band, a photon being emitted for each electron. The emitted light is incoherent.

Step Index Fiber — A fiber having a uniform refractive index within the core and a sharp decrease in refractive index at the core/cladding interface.

Stimulated Emission — This occurs when photons in a semiconductor stimulate available excess charge carriers to the emission of photons. The emitted light is identical in wavelength and phase with the incident coherent light.

Telecommunications Closet (TC) — An enclosed space for housing telecommunications equipment, cable terminations, and cross-connects. The closet is the recognized cross-connect between the backbone and horizontal cabling.

Threshold Current — The driving current above which the amplification of the lightwave in a laser diode becomes greater than the optical losses, so that stimulated emission commences. The threshold current is strongly temperature dependent.

Tight-Buffered Cable — Type of cable construction whereby each glass fiber is tightly buffered by a protective thermoplastic coating to a diameter of 900 micrometers. Increased buffering provides ease of handling and connectorization.

Total Internal Reflection — The total reflection that occurs when light strikes an interface at angles of incidence greater than the critical angle.

Transmitter — A driver and a source used to change electrical signals into optical signals.

Wavelength Division Multiplexing (WDM) — Simultaneous transmission of several signals in an optical waveguide at differing wavelengths.

Zero-Dispersion Wavelength—Wavelength at which the chromatic dispersion of an optical fiber is zero. Occurs when waveguide dispersion cancels out material dispersion.

Additional Reading

Adrian, P.: "Technical Advances in Fiber-Optic Sensors: Theory and Applications," *Sensors*, 23 (September 1991).

Agrawal, G.: *Fiber-Optic Communication Systems*, John Wiley & Sons, Inc., New York, NY, 1997.

Amato, I.: "The Natural Roots of Fiber Optics," *Science News*, 414 (December 23–30, 1989).

Augé, J., et al.: "Progress in Optical Amplification," *Microwave J.*, 62 (June 1993).

Baumbick, R.J. and J. Alexander: "Fiber Optics Sense Process Variables," *Control Eng.*, **27**, 3, 75–77 (1980).

Bobb, L.C. and P.M. Shankar: "Tapered Optical Fiber Components and Sensors," *Microwave J.*, 219 (May 1992).

Corcoran, E.: "Light Talk: U.S. and Japanese Compete to Put Optical Fibers in the Home," *Sci. Amer.*, 74 (October 1989).

Corcoran, E.: "Light Traffic: Optical Amplifiers Promise to Unclog Lightwave Communication," *Sci. Amer.*, 106 (March 1991).

Corcoran, E.: "Avoiding the Potholes on Optical Highways," *Sci. Amer.*, 143 (April 1992).

Desurvire, E.: "Lightwave Communications: The Fifth Generation," *Sci. Amer.*, 114 (January 1992).

Dutton, H.: *Understanding Optical Communications*, Prentice-Hall, Inc., Upper Saddle River, NJ, 1999.

Furse, C. and R. Haupt: "Down to the Wire," *IEEE Spectrum* (February 2001).

Gabel, D.: "Fiber Optics on the Rise," *Electronic Buyers' News*, 36 (January 28, 1991).

Grimes, G.: "Microwave Fiber-Optic Delay Lines: Coming of Age in 1992," *Microwave J.*, 61 (August 1992).

Hamilton, K.J.: "Fiber Optic Sensors Grow Into Networks," *InTech*, 20 (February 1991).

Hecht, J.: *City of Light*, Oxford University Press, New York, NY, 1999.

Henkel, S.: "Single Optical Fiber Does It All for Smart Transmitters," *Sensors*, 8 (January 1992).

Holden, C.: "Plugging Into the Pacific Ocean," *Science*, 599 (August 11, 1989).

Horgan, J.: "Dark Solutions: Physicists Generate Durable Pulses of Darkness," *Sci. Amer.*, 24 (May 1988).

Ito, T., K. Fukuchi, K. Sekiya, D. Ogasahara, R. Ohhira and T. Ono: "6.4 Tb/s (160 × 40 Gb/s) WDM Transmission Experiment with 0.8 bits/Hz Spectral Efficiency," European Conference on Optical Communication, post-deadline paper, September 2000.

Jones, W.B. Jr.: *Introduction to Optical Fiber Communication Systems*, Oxford University Press, Inc., New York, NY, 1995.

Kazovsky, L., et al.: *Optical Fiber Communication Systems*, Artech House, Inc., Norwood, MA, 1996.

Kogelnik, H.: "High-Speed Lightwave Transmission in Optical Fibers," *Science*, **228**, 1043–1048 (1985).

Ledwith, A.: "Glasses for Fibre Optic Communications," *Review (University of Wales)*, 15 (Spring 1988).

Mandoli, D.F. and W.R. Briggs: "Fiber Optics in Plants," *Sci. Amer.*, 90–98 (August 1984).

McHugh, P.: "Fiber Optics Extend the Reach of Photoelectric Sensors," *Instruments & Control Systems*, 57 (August 1989).

Nicholson, P.J.: "An Introduction to Fiber Optics," *Microwave J.*, 26 (June 1991).

Nicholson, P.J.: "An Overview of the Synchronous Optical Network," *Microwave J.*, 24 (December 1991).

Nielsen, T.N., et al.: "3.28 Tb/s (82 × 40 Gb/s) Transmission Over 3 × 100 km of Nonzero-dispersion Fiber Using Dual C- and L-band Hybrid Raman/Erbium-doped Inline Amplifiers," Optical Fiber Communication Conference, post-deadline paper 29, March 2000.

Papannareddy, R.: *Introduction to Lightwave Communication Systems*, Artech House, Inc., Norwood, MA, 1997.

Peterson, J.I. and G.G. Vurek: "Fiber-Optic Sensors for Biomedical Applications," *Science*, **224**, 123–127 (1984).

Pratsinis, S.E. and S.V.R. Mastrangelo: "Material Synthesis In Aerosol Reactors (Optical Fiber Manufacture)," *Chem. Eng. Progress*, 65 (May 1989).

Raybon, G., et al.: "320 Gbit/s Single-channel Pseudo-linear Transmission over 200 km of Non-zero Dispersion Fiber," Optical Fiber Communication Conference, post-deadline paper 29, March 2000.

Refi, J., *Fiber Optic Cable—a LightGuide*, abc TeleTraining, Inc., Geneva, IL, 1991.

Stix, G.: "Light Flight: Optical Fibers May be the Nerves of New Aircraft," *Sci. Amer.*, 120 (May 1991).

Suematsu, Y. and K.I. Iga: *Introduction to Optical Fiber Communications*, John Wiley & Sons, Inc., New York, NY, 1982.

Tsang, W.T., N.A. Olsson and R.A. Logan: "High-Speed Direct Single-Frequency Modulation with Large Tuning Rate and Frequency Excursion in Cleaved-Coupled-Cavity Semiconductor Lasers," *Applied Physics Letters*, **42**(8), 650–652 (April 15, 1983).

Woracek, D.: "Fiber Optic Sensors Endure Microwaves," *InTech*, 24 (February 1991).

Yazbak, G.: "Fiberoptic Sensors Solve Measurement Problems," *Food Technology*, 76 (July 1991).

Web References

CoreTek Inc: http://www.coretekinc.com/
General Cable: http://www.generalcable.com/
Lucent Technologies: http://www.lucent.com/ofs/
Nanoptics, Inc: http://www.nanoptics.com/

Lucent Technologies, Optical Fiber Solutions, Norcross, GA

OPTICAL GLASS. Glass to be useful for lenses, prisms and other optical parts through which light passes, as distinguished for mirrors, must be completely homogeneous. This includes freedom from bubbles, striae, seeds, strains, etc. In order to reduce aberrations, the optical designer needs many different kinds of glass. A few typical types are described in Table 1. The v-number is the reciprocal of the dispersive power of the glass.

TABLE 1. VARIOUS GLASSES

	Type	n_D	v-Number
Borosilicate	Crown	1.5170	64.5
Barium	Crown	1.5411	59.5
Spectacle	Crown	1.5230	58.4
Light	Flint	1.5880	53.4
Ordinary	Flint	1.6170	38.5
Dense	Flint	1.6660	32.4
Extra dense	Flint	1.7200	29.3

OPTICAL IMAGES (Graphical Construction). The image of a object point may be located to first order accuracy by drawing any two of three easily located lines.

Given a lens L, its optical axis $x - y$, its foci F_1 and F_2 and an object point O (Fig. 1.):

1. Draw a line OA parallel to $x - y$ and then the line AF_2.
2. Draw a line OF_1, extend it to the lens at B, and then extend it from B parallel to the optical axis.
3. Draw the line OC, and continue it without deviation through the lens.

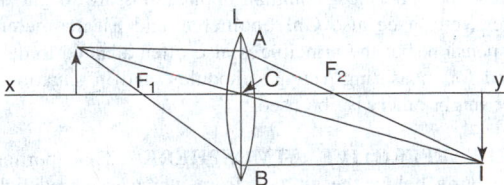

Fig. 1. Principal pathways of simple lens system.

The three lines should meet at the point I, the image of O. If, instead of converging to a point, the three lines are diverging, trace each of them back, and they should meet at a point to the left of the lens indicating a virtual image. See also **Geometrical Optics**. If the three lines are parallel the image is at infinity. This same method of construction may be applied to any lens or curved mirror.

These are the three easily located lines. If it becomes desirable to trace some other ray (tracing a single ray through more than one lens) the following construction holds. (Fig. 2.)

EA is the ray to be traced through the lens. Draw mm through F_2 perpendicular to the optical axis. Draw CD parallel to EA. The ray will follow the path AD after leaving the lens.

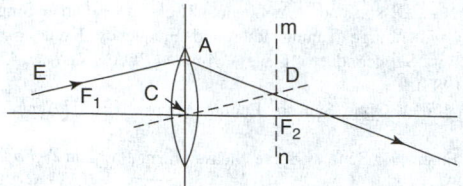

Fig. 2.   Method for tracing specific rays through lens system.

OPTICAL IMAGING PROBE.   An optical particle probe that records the size and shape of the shadow of each particle that intercepts and attenuates the illumination by a laser beam.

Of these, shadowing probes differentiate particle shadows from the light of the unobstructed beam with linear arrays of optically activated diodes; for example, the one-dimensional cloud probe normally for sizing 10-μm to either 300- or 600-μm cloud hydrometeors, the two-dimensional cloud probe normally for sizing and imaging 25–800-μm cloud hydrometeors, and the two-dimensional precipitation probe normally for sizing and imaging 200-μm to 6.4-mm precipitation hydrometeors. The electronics record either the maximum one-dimensional width of the shadow or two dimensions and shape of the shadow, to estimate particle size and type (raindrop, crystal growth habit, etc.). Another type, the cloud particle imager, illuminates each particle with a pulsed laser and records size, shape, and detailed structure of each hydrometeor with a high-resolution solid-state digital imaging camera, normally to size 5-μm–2.3-mm hydrometeors. Also called an *optical array probe*.

AMS

OPTICAL ISOMER.   Either of two kinds of optically active three-dimensional isomers (stereoisomers). One kind is represented by mirror-image presence of one or more asymmetric carbon atoms in the compound (glyceraldehyde, lactic acid, sugars, tartaric acid, amino acids). The other kind is exemplified by diastereoisomers, which are not mirror images. These occur in compounds having two or more asymmetric carbon atoms; thus, such compounds have 2_n optical isomers, where n is the number of asymmetric carbon atoms.

See also **Optical Rotation**.

OPTICAL ISOMERISM.   See **Isomerism**.

OPTICAL LEVER.   A common device for amplifying and measuring small rotations. The object rotated carries a small mirror, which, reflecting a beam of light, deflects it through twice the angle of rotation to be measured. Light from a lamp is thus reflected as a bright spot moving along a scale, or the image of a fixed scale is viewed in the mirror by means of a reading telescope. The most common applications are to galvanometers, electrometers, etc., (see also **Galvanometer**; and **Electrometer**) using a torsion suspension; but the principle is also often adapted to devices such as that used for measuring Young's modulus, and in situations where a micrometer might otherwise be used.

OPTICALLY EFFECTIVE ATMOSPHERE.   That portion of the atmosphere lying below the altitude from which scattered light at twilight still reaches the observer with sufficient intensity to be discerned. Also called *effective atmosphere*. The top of this region lies between 50 and 60 kilometers (31 to 37 miles).

OPTICALLY HOMOGENEOUS.   Homogeneous on the scale of the wavelength of the electromagnetic radiation of interest.

Pure liquid water is optically homogeneous over the visible spectrum because one cubic wavelength of water contains many molecules, whereas a cloud of water droplets is not optically homogeneous. Optical homogeneity is more general than transparency, usually restricted to visible wavelengths. A body is said to be transparent if it transmits images. Optical homogeneity is necessary for transparency but not sufficient. A sufficiently thick sample of an absorbing optically homogeneous material would be described as opaque rather than transparent.

AMS

OPTICALLY SMOOTH.   Smooth on the scale of the wavelength of the electromagnetic radiation of interest. No surface is absolutely smooth, if for no other reason than matter is composed of molecules in motion. An approximate criterion for smoothness is the Rayleigh criterion. A surface is reckoned to be optically smooth if $d < \lambda/(8\cos\theta)$, where d is the surface roughness (e.g., root-mean-square roughness height measured from a reference plane), λ is the wavelength of the incident illumination, and θ is the angle of incidence of this illumination. Thus, a surface that is smooth at some wavelengths is rough at others, or that is rough at some angles of incidence and smooth at others (e.g., near-grazing angles).

AMS

OPTICAL MICROSCOPE.   A magnifying lens system that utilizes light in the visible wavelength range of the electromagnetic spectrum (5000 Å). A convex glass lens bends or focuses light waves because of the difference in density between glass and air. Invented in 1590 by the Janssen brothers and later improved by van Leeuwenhoek, the compound microscope has three lenses: a condenser lens, which concentrates the incident light; an objective lens, which gives an enlarged reverse image of the specimen; and a projector lens, which further enlarges the image and return it to normal position. Its maximum resolving power is 0.5 micron, compared with 100 microns for the human eye. The compound microscope is particularly useful in studying bacterial and other microorganisms in their natural state without interfering with their behavior. It has been of untold benefit to biologists and bacteriologists and also has innumerable uses in chemical and metallurgical research, as well as in forensic chemistry.

OPTICAL MODE.   A type of thermal vibration of a crystal lattice whose frequency is nearly independent of wave number. The optical modes may be thought of as internal vibrations of the molecules or unit cells of the lattice, loosely coupled from cell to cell. In ionic crystals, this leads to strong absorption in the infrared because of the fluctuating dipole moment as the ions of opposite sign move relative to one another. The optical modes contribute to the specific heat.

OPTICAL PARTICLE PROBE.   Any of a class of instruments for sizing and counting large numbers of individual aerosols or hydrometeors to characterize populations, or also for imaging and classifying the shapes of individual hydrometeors by measuring the optical illumination, scattering, or attenuation of a laser beam by each particle and recording the result electronically. Also called *electro-optical particle probe*.

Included are optical scattering probes and optical imaging probes. Primarily airborne instruments, they can be adapted for stationary use at the surface for precipitation or fog monitoring, in some cases with forced ventilation to create a particle flux.

AMS

OPTICAL PATH.   In a medium of refractive index n, the product of the geometrical distance d and the refractive index. When there are several segments $d_1, d_2 \ldots$ of the light path in substances having different indices $n_1, n_2, \ldots$, the optical path is found from the relationship:

$$\text{optical path} = n_1 d_1 + n_2 d_2 + \cdots = \sum_i n_i d_i$$

and in a medium in which n varies continuously:

$$\text{optical path} = \int n\, ds$$

where ds is an element of length along the path. According to the Fermat principle, the optical path connecting two points has an extreme value.

OPTICAL PATHLENGTH.   (Sometimes called *optical path* or *optical length*.) The line integral of the refractive index (real part) along a ray connecting two points in an optically homogeneous medium. Because refractive index depends on frequency, so does optical pathlength. Although optical pathlength may be shortened to optical path, there is a difference between a path and its length.

See also **Optical Thickness**.

OPTICAL PUMPING. The process of "pumping" atoms from one hyperfine quantum state to another by a process of resonant fluorescent scattering of light. The original purpose was to facilitate the detection and measurement of the radio-frequency fine and hyperfine structure, which otherwise is observable by magnetic resonance and atomic beam methods, yet which cannot be directly observed by optical spectroscopy, primarily because of the Doppler width of spectral lines. Optical pumping also embraces any experimental work in which fine structure, radio-frequency spectroscopic measurements, polarization of electrons or nuclei, atomic cross sections, and oscillator strengths are measured or produced by means of polarized, filtered, or modulated light, and perhaps detected by the light as well. The distinction between measurements made in this way and those of conventional spectroscopy arises from the fact that the wavelength of the light is not used to measure energy splittings.

Several interesting light modulation effects may be observed in pumping experiments. Light beat and modulation effects occur when an atom scatters light while it simultaneously is undergoing hyperfine transitions. If a sample is oriented in the Z direction, and a radio-frequency field excites a Zeeman resonance, the atoms will precess coherently. An additional polarized light beam passing through the sample will be modulated at the Larmor frequency. As an inverse effect, if the pumping light itself is modulated at the appropriate radio frequencies, transitions will be induced.

Atomic gyroscopes may be constructed by using the optically pumped angular momentum and the light to detect precession. An important application also has been that of producing population inversion required in the operation of masers and lasers, particularly potassium and rubidium vapor, and chromium in ruby.

See also **Laser**.

OPTICAL PYROMETER. A device for measuring the temperature of an incandescent radiating body by comparing its brightness for a selected wavelength interval within the visible spectrum with that of a standard source; a monochromatic radiation pyrometer. Temperatures measured by optical pyrometers are known as brightness temperatures and except for black bodies are less than the true temperatures.

OPTICAL ROTATION. The change of direction of the plane of polarized light to either the right of the left as it passes through a molecule containing one or more asymmetric carbon atoms, e.g., sugars. The direction of rotation, it to the right, is indicated by either a plus sign (+) or a $d-$; if to the left, by a minus sign (−) or an $l-$. Molecules having a right landed configuration (D) usually are dextrorotatory, D(+), though they may be levorotatory, D(−); those having a left-handed configuration (L) are usually levorotatory, L(−), but may be dextrorotatory d(+). Compounds having this property are said to be optically active and are isomeric. The amount of rotation varies with the compound but is the same for any two isomers, though in opposite directions.

See also **Optical Isomer**.

OPTICAL SCATTERING PROBE. A type of optical particle probe that measures the forward scattering caused by each particle or ensemble of particles intercepting a laser beam. The scattering is converted to an equivalent size via calibration and scattering theory for spheres. Examples are the passive cavity aerosol spectrometer probe, normally for sizing 0.1–3.0-μm diameter aerosols, the forward scattering spectrometer probe, normally for sizing either 0.3–20-μm or 0.5–47-μm aerosols and cloud droplets, and the optical cloud drop spectrometer, normally for sizing 2–200-μm cloud droplets.

AMS

OPTICAL THICKNESS.

1. Also absorption optical thickness, scattering optical thickness, total optical thickness, optical depth.) The (dimensionless) line integral of the absorption coefficient, or of the scattering coefficient, or of the sum of the two, along any path in a scattering and absorbing medium. More often than not, the qualifiers "absorption," "scattering," and "total" are omitted (context sometimes being sufficient to determine which is meant). Optical thickness also depends on the wavelength of the radiation of interest. In a uniform medium, optical thickness

has a simple physical interpretation as the length of a path in units of mean free path. Optical thickness and optical depth are used more or less synonymously; if there is a distinction between them it is that optical thickness is applied to an entire path through a medium. Normal optical thickness is optical thickness along a vertical path. See also **Optical Pathlength**.

2. The degree to which a cloud modifies the light passing through it. Optical thickness depends on the physical constitution (crystals, drops, droplets), and the form; the overall effect depends on the scatter parameter and the phase function for the particles as well as their concentration and the vertical extent of the cloud.

AMS

OPTICAL TURBULENCE. Irregular and fluctuating gradients of optical refractive index in the atmosphere. Optical turbulence is caused mainly by mixing of air of different temperatures, and particularly by thermal gradients which are sufficient to reverse the normal decrease in density with altitude, so that convection occurs.

OPTIC NEURITIS. An inflammation of the optic nerve, which is the bundle of nerve fibers that starts at the back of the eye and carries light impulses from the retina to the brain. The retina receives light signals like the film in a camera, and the optic nerve transmits these signals to the brain where they are processed and turned into images, giving us the ability to see. If some or all of these nerve fibers become inflamed, the optic nerve becomes swollen and the fibers do not work properly, causing blurry vision. Depending on the number of inflamed nerve fibers, vision can range from near normal to extremely poor.

Optic neuritis is a relatively rare condition that can affect both adults and children. In adults, the condition usually affects one eye; in children it often affects both eyes at the same time. The most common age for developing the condition is in the 30s, but it can affect people of all ages.

Optic neuritis is usually associated with other medical problems, particularly viral infections and multiple sclerosis (MS). The disorder is often the first symptom of MS, and about 40% of people who experience optic neuritis eventually develop multiple sclerosis. The condition can also develop from an abuse of tobacco or alcohol, or from exposure to toxic substances such as lead or wood alcohol. Often the actual cause of optic neuritis cannot be determined.

Symptoms of optic neuritis usually come on suddenly and may include blurry or dim vision (as though the lights are being turned down) and a fading of colors. There can also be pain in the eye socket, impaired depth perception, and blind spots in the field of vision. Loss of vision usually occurs over a period of 2 to 5 days and generally improves within 4 to 12 weeks.

The condition can be difficult to diagnose because the symptoms are similar to those caused by several other eye problems. One of the first steps in diagnosing optic neuritis is an examination of the optic nerve with an instrument called an ophthalmoscope. The optic nerve enters the back of the eye and is visible to the doctor as a small disc. Any swelling of this inside part of the nerve can be detected, but if the swelling occurs behind the eye, it is not visible and other diagnostic procedures are necessary. These procedures may include tests such as an ultrasound, CT scans, or visual brain wave recordings. Other standard tests used in diagnosing optic neuritis are color vision, peripheral vision, and the reaction of the pupil to light.

Unfortunately, there is no consistently reliable treatment for optic neuritis. Steroids such as cortisone and prednisone are sometimes prescribed, but their effectiveness in treating the problem is questionable. Most people who develop the condition, however, recover normal vision without treatment. Recovery may take from a few weeks to a few months, depending on the severity of the condition.

Vision Rx, Inc., Elmsford, NY

OPTICS. Originally that branch of physical science which treats of the phenomena of light and of vision. Today, because of the constantly increasing importance of ultraviolet and infrared radiation, optics has come to include all phenomena associated in any way with electromagnetic waves with wavelengths greater than x-rays and shorter than microwaves. Numerical limits of this wavelength region are not definitely defined.

Since the advent of devices such as the electron microscope and the cathode ray tube, in which beams of particles are focused to form images, the study of the behavior of such instruments is also called optics, usually with an appropriate modifier.

Scores of articles in this encyclopedia are devoted to specific aspects of the optical- and light-related sciences. Check the alphabetical index for subjects related to lenses and lens systems, light, mirrors and reflectors, optical fiber technology, optical instruments and materials, photography, prisms, and vision and the eye. See also **Geometrical Optics**; **Optical Fiber Systems**; **Photography and Imagery**; **Telephony**; and **Thin Films**.

OPTICS (Fiber). See **Optical Fiber Systems**.

OPTIMUM MAGNIFICATION. The maximum value of the numerical aperture of a dry lens is 1.0. Oil-immersion objectives with numerical apertures up to 1.65 have been constructed. The minimum distance between points which are just resolved is thus 2.7×10^{-5} centimeters for a dry lens, and 1.6×10^{-5} centimeters for an oil-immersion lens, using oblique illumination and a wavelength of 5,500 Å. The maximum useful magnification is thus about 800 for dry objectives, and 1,200 for oil-immersion objectives.

OPTOMETRY. A branch of optics dealing with the optical performance of the individual eye, and with measurements upon it. See also **Vision and the Eye**.

ORAL CAVITY. The cavity usually called the mouth. It is formed in the vertebrates of an embryonic depression, the stomodaeum, which forms in the ectoderm of the under side of the head and unites with the embryonic gut just behind its anterior end. The depression is deepened by the growth of processes from the body wall at the level of the pharynx, which forms the upper and lower jaws. Later the olfactory pits break through to join it and in this stage, which persists in the amphibians, the cavity is common to the respiratory and digestive systems. In a more advanced stage shelf-like partitions grow out from the lateral walls and join to form the palate, which divides the cavity into respiratory and oral portions, as in humans. Also called the buccal cavity.

ORANGE TREE. See **Citrus Trees**.

ORANGUTAN. See **Anthropoids**.

ORBIS. See **Antelope**.

ORBITALS. This article embraces both atomic and molecular orbitals.

Atomic

From spectroscopic studies, it is known that when an electron is bound to a positively charged nucleus only certain fixed energy levels are accessible to the electron. Before 1926, the old quantum theory considered that the motion of the electrons could be described by classical Newtonian mechanics in which the electrons move in well defined circular or elliptical orbits around the nucleus. However, the theory encountered numerous difficulties and in many instances there arose serious discrepancies between its predictions and experimental fact.

A new quantum theory called wave mechanics (as formulated by Schrödinger) or quantum mechanics (as formulated by Heisenberg, Born and Dirac) was developed in 1926. This was immediately successful in accounting for a wide variety of experimental observations, and there is little doubt that, in principle, the theory is capable of describing any physical system. A strange feature of the new mechanics, however, is that nowhere does the path or velocity of the electron enter the description. In fact it is often impossible to visualize any classical motion that could be consistent with the quantum mechanical picture of the atom.

In this theory the electron is viewed as a three-dimensional standing wave. The pattern of the wave is described by a wave function ϕ (analogous to the amplitude of a water wave). This one-electron wave function is called

an atomic orbital. Since the wave function can be positive or negative (and real or complex) it does not describe an observable property of the electron. However, the square of the wave function (ϕ^2 or ϕ times its complex conjugate) is always positive and real, and can be identified with the probability of finding the electron at any point. This was first suggested by Born, but has now received ample experimental support. Hence, when the wave function is calculated for any electron we can determine the regions in space where the electron is most likely to be found, though we cannot say what type of motion results in that particular probability pattern.

Atomic orbitals are usually labeled by a set of designating numbers called quantum numbers. The one that determines the energy of the resulting state (for hydrogen) is given the symbol n and called the "principal quantum number." It assumes the values 1, 2, 3, 4, 5, ... to infinity, with increasing electron energies. The second quantum number is given the symbol l. It can be identified with the angular momentum of the electron due to its orbital motion, and assumes values of 0, 1, 2, 3, ... to $(n - 1)$. For historical reasons the orbitals with these values are referred to as $s, p, d, f, \ldots$ orbitals respectively. Hence a $3d$ orbital is one for which $n = 3$ and $l = 2$. The third quantum number is usually given the symbol m and is difficult to define in the absence of an external field. However, under all conditions it can assume $2l + 1$ values.

From the rules given in the previous paragraph it can easily be shown that for $n = 1$, 2 and 3 there are a total of 14 allowed orbitals. For an isolated hydrogen atom all orbitals must be spherically symmetrical and have the shapes shown in series (**a**) of Fig. 1. These are plots of the probability function (ϕ^2) for the orbitals and hence the darkest areas represent regions where the electron is most likely to be found. Though there are actually three $2p$ orbitals, three $3p$ orbitals and five $3d$ orbitals, in the absence of an external field the orbitals within a given set are identical in shape and energy, and are said to be degenerate. If a direction is defined by the presence of a magnetic field or the approach of another atom, the degeneracy is removed. The shapes of the allowed orbitals under these conditions are shown in series (**b**) of Fig. 1. The quantum number m is well defined for these orbitals and its value is given beside each orbital. For the discussion of bonding in polyatomic molecules another set of orbitals is useful. These are shown in series (**c**) of Fig. 1. The quantum number m is not well-defined for these orbitals, and they are usually labeled according to the axis along which they lie.

In addition to the orbitals shown in Fig. 1 there are "hybrid" orbitals that are not stationary states for the electron in an isolated atom. They can be obtained by taking a linear combination of the standard orbitals in Fig. 1. Since the electron distribution is "off center" they are useful only for atoms that are perturbed by an electric field (Stark-effect) or by the approach of other atoms as occurs in chemical-bond formation.

In addition to the three quantum numbers discussed above, experimental evidence requires an additional quantum number m_s, which by analogy to classical mechanics is attributed to an intrinsic (i.e., position-independent) property of the electron called "spin." Unlike the other quantum numbers, however, it can assume only two values ($\pm\frac{1}{2}$). As we shall see, this fact determines the orbital population of the many-electron atom.

It is an unfortunate consequence of the mathematical complexity of the quantum mechanical equations that the hydrogenic atom (i.e., one-electron atom) is the only system for which an exact probability distribution (ϕ^2) can be obtained. Approximate methods must be used to calculate the wave functions for the many electron atoms. We begin with the natural assumption that the wave function for such a more complex atom (ψ) can be obtained by taking a product of the appropriate one electron functions (ϕ). However, even if we ignore for the moment the coulombic repulsion between electrons, there are two fundamental postulates of quantum mechanics that complicate the picture. One is that electrons are indistinguishable. Hence, when their positions are interchanged, the probability function (ψ^2) must remain unchanged. This means that the wave function (ψ) must either remain the same or only change signs when electron positions are interchanged. A second postulate (called the Pauli principle) is that the total wave function must change signs when electron positions are interchanged. From these requirements it can be seen that when two electrons are placed in the same orbital (i.e., assigned the same orbital wave function ϕ) the total wave function will change signs (as required by the Pauli principle) only if the electrons have different spin

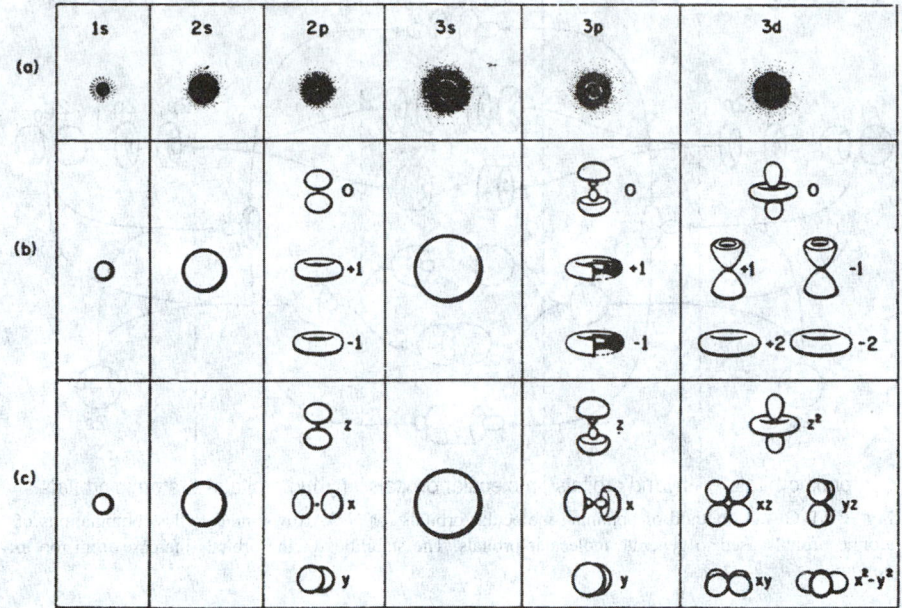

Fig. 1.   Hydrogen atomic orbitals: (**a**) for isolated atoms; (**b**) with one direction defined; (**c**) with three directions defined.

quantum numbers ($+\frac{1}{2}$ and $-\frac{1}{2}$). It is hence a general rule that hydrogenic orbitals can only be occupied by a maximum of two electrons and these must have opposite spin.

If we begin with the most tightly bound orbital and add two electrons to each orbital the so called "ground state configuration" of the atom is obtained. For example the electronic configuration of the silicon atom is written $1s^2$, $2s^2$, $2p^2$; $3s^2$, $3p^2$. The orbital shapes and energies are, however, considerably altered by electron-electron repulsion, especially between electrons whose orbitals overlap appreciably. This has several marked effects. For a given value of n, all the orbitals no longer have the same energy. The binding energy now decreases with increasing values of the quantum number l. Secondly, because of the coulombic repulsion between electrons, they will tend to occupy separate orbitals whenever feasible (for example the $3p$ orbitals configuration in the silicon atom is actually $3p^1$, $3p^1$). Furthermore, when electrons are forced together into the region of one hydrogenic orbital it is quite likely that electron-electron repulsion (always greater than 25 kcal) leads, in effect, to slightly different orbitals for each electron.

For chemical purposes we are most interested in the shapes of orbitals in which the valence (outermost) electrons reside. By assuming that the inner electrons act only to screen some of the positive charge on the nucleus the valence electrons can be shown to assume the shape of the appropriate hydrogenic orbital, with the insignificant difference that the inner nodes in the orbitals shown in Fig. 1. are drawn in closer to the nucleus because the shielding is poorer in this region. It is worth noting that for the many-electron atom, the indistinguishability of electrons has the effect of making only the total probability function physically meaningful. For this reason and because of the repulsion between electrons, the individual one-electron wave functions are so correlated that, though the "independent orbital concept" remains a very useful approximation, it is not fundamental to the problem.

Molecular

By analog with atomic orbitals, the wave function (ψ) for one electron in a molecule is called a molecular orbital, and the probability of finding the electron at any point is similarly given by the value of ψ^2 at that point. Just as in the case of atomic orbitals, an exact solution of the equations is possible only for the one-electron molecule (H_2^+) and it is only for this species that accurate molecular orbitals can be obtained.

The shapes of 10 of the most tightly bound orbitals of H_2^+ are shown in Fig. 2. Three quantum numbers can be used to label these orbitals. The one that is always well defined is given the symbol λ and can be identified with the component of the orbital angular momentum along

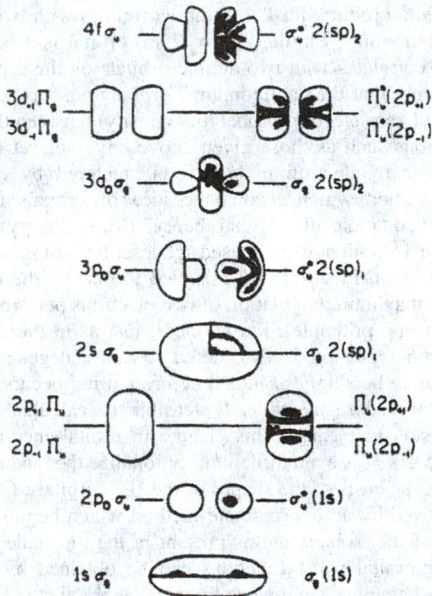

Fig. 2.   Molecular orbitals of H_2^+. The "united atom" designation is given on the left-hand side and the "separate atom" designation on the right-hand side.

the internuclear axis. It can take on values 0, 1, 2, 3, ... for which the orbitals are called σ, π, δ, ... respectively. The other two quantum numbers are defined differently depending on the nuclear separation. When the internuclear distance is short, as the case of H_2^+, it is convenient to consider the atomic orbital that would result from a given molecular orbital if the two nuclei were made to coalesce (in our imaginations). In the case of homonuclear diatomic molecules, for example, the molecular orbital designated $3d\ \sigma$ has $\lambda = 0$ and correlates to a $3d$ atomic orbital when the nuclei coalesce (i.e., the atomic quantum numbers n and l become well-defined at short internuclear distance). On the other hand, when the internuclear distance is large, a more useful and significant label is one which identifies the atomic orbitals with which a given molecular orbital correlates when the distance between the two nuclei approaches infinity. For this "separate atom" designation an additional symbol must be used to distinguish between the two molecular states that can arise from a given pair of atomic states. Chemists find it most useful to use the superscript * to indicate the higher energy (antibonding) orbital and the absence of

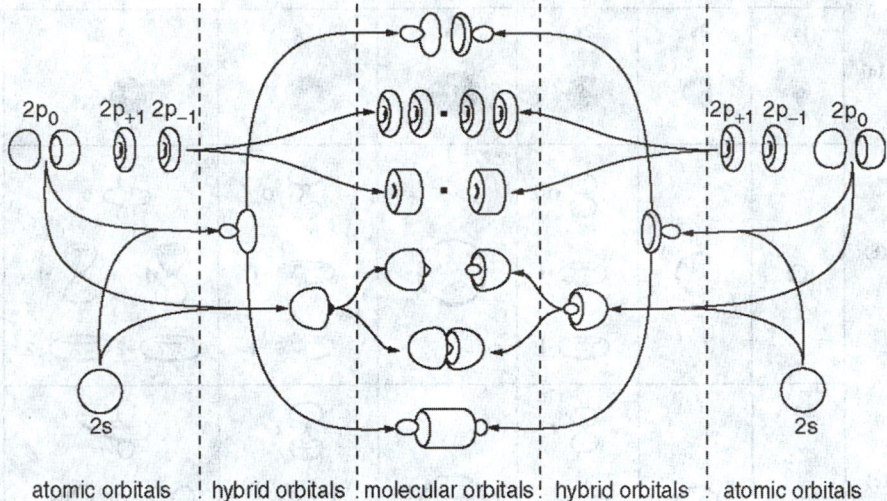

atomic orbitals | hybrid orbitals | molecular orbitals | hybrid orbitals | atomic orbitals

Fig. 3. L.C.A.O. method of obtaining molecular orbitals for N_2. Arrows indicate the combinations of atomic orbitals used to generate molecular orbitals. The stabilities of the orbitals increase from top to bottom.

* to indicate the lower energy (bonding) orbital. When the difference in symmetry between the two states is important the symbols g (gerade-symmetric) and u (ungerade-symmetric) are used. Thus, the orbital $\sigma^*(2p_x)$ is an antibonding orbital with $\lambda = 1$ that correlates with two $2p_x$ orbitals on the separated atoms. Similarly, a $\sigma_g 2(sp)$ orbital is a bonding orbital with $\lambda=0$ that correlates with two atomic orbitals on the separated atoms. It is worth noting that the "antibonding" orbitals possess a nodal surface (i.e., a region of zero electron-probability density) between the two nuclei.

Exact solutions such as those given above have not yet been obtained for the usual many-electron molecules encountered by chemists. The approximate method which retains the idea of orbitals for individual electrons is called "molecular-orbital theory" (M. O. theory). Its approach to the problem is similar to that used to describe atomic orbitals in the many-electron atom. Electrons are assumed to occupy the lowest energy orbitals with a maximum population of two electrons per orbital (to satisfy the Pauli exclusion principle). Furthermore, just as in the case of atoms, electron-electron repulsion is considered to cause degenerate (of equal energy) orbitals to be singly occupied before pairing occurs.

It has not proved mathematically feasible to calculate the electron-electron repulsion that causes this change in orbital-energies for many-electron molecules. It is even difficult to rationalize the qualitative changes in sequence on the basis of the shapes of the H_2^+ orbitals. Greater success has been achieved by an approximate method which begins with orbitals characteristic of the isolated atoms present in the molecule, and assumes that molecular orbital wave functions can be obtained by taking linear combinations of atomic orbital wave functions (abbreviated L.C.A.O.). For homonuclear diatomic molecules, the atomic orbitals used are normally those with which a given molecular orbital correlates. An example of how molecular orbitals are manufactured in the L.C.A.O. approximation is shown in Fig. 3. According to this theory, the relative molecular-orbital energies will depend on (a) the spread of orbital energies on the individual atoms (since this determines the extent of hybridization) and (b) on the internuclear distance which is largely determined by the relative number of bonding and antibonding orbitals that are filled.

The L.C.A.O. approximation has also proved useful for the description of heteronuclear diatomic molecules, although valence-bond theory has been somewhat more successful in its quantitative calculations of bond energies. For these molecules the selection of appropriate atomic orbitals to be used in linear combination is governed by considerations of symmetry, energetics and overlap. These considerations also apply to the formation of molecular orbitals in polyatomic molecules by the L.C.A.O. approximation. In the case of polyatomic molecules it is always possible to develop either (a) molecular orbitals that extend between only two atoms (localized M.O.s) or (b) molecular orbitals that extend over the entire molecule (delocalized M.O.s). In the case of saturated molecules, the two descriptions are practically equivalent. However, for molecules with conjugated double bonds and especially for aromatic compounds, delocalized molecular orbitals which extend over several atoms must be used. When the L.C.A.O. method is used to develop molecular orbitals in complexes between metal ions (possessing d orbitals) and ligands such as CO, OH⁻, NH_3, CN⁻ etc., the procedure is called Ligand Field Theory.

A thorough test of the L.C.A.O. molecular orbitals has been possible only for H_2^+ where both the wave functions and orbital energies are known accurately. Such a comparison shows that though the shapes of the wave functions are reasonably represented by the approximation, their energies can be appreciably in error, especially for excited state. Nevertheless, the L.C.A.O. approximation has proved the most fruitful method of obtaining molecular orbitals that has been developed to date.

Additional Reading

Clark, T., and R. Koch: *Chemist's Electronic Book of Orbitals*, Springer-Verlag New York, LLC., New York, NY, 1998.

Dias, J.R.: *Molecular Orbital Calculations Using Chemical Graph Theory*, Springer-Verlag New York, LLC., New York, NY, 1997.

Gil, V.: *Orbitals in Chemistry: A Modern Guide for Students*, Cambridge University Press, New York, NY, 2000.

Rauk, A: *Orbital Interaction Theory of Organic Chemistry*, 2nd Edition, John Wiley & Sons, Inc., New York, NY, 2000.

E. A. OGRYZLO, University of British Columbia, Vancouver, British Columbia, Canada

ORBITAL VELOCITY. As an ocean wave propagates, the water itself moves in an approximately circular motion (in deep water), with the amplitude of the motion decaying with depth below the sea surface. The velocity associated with the fluid motion is known as the orbital (or particle) velocity.

ORBIT (ASTRONOMY). The path that a celestial object follows in its motions through space, relative to some selected point, is known as the orbit of the object. The solution of the two-body problem indicates that, in the case of two objects moving under the influence of their mutual gravitational attractions, the relative orbit of one to the other will be a conic section. The character of the conic will depend upon initial conditions. In the case of members of the solar system, because of the very large mass, relatively, of the sun in comparison with any of the other members, the orbits of the members may be conveniently represented as ellipses with the sun at one focus. In the case of the satellites of the various members, the orbits of the satellites may be represented as ellipses with the primary object at one focus. Any deviations from the two-body problem may be treated as departures or perturbations from the simple conic.

To define completely the orbit of an object at any particular instant, and to permit the determination of the position of the object in this orbit at any future time, six quantities must be calculated, known as the elements

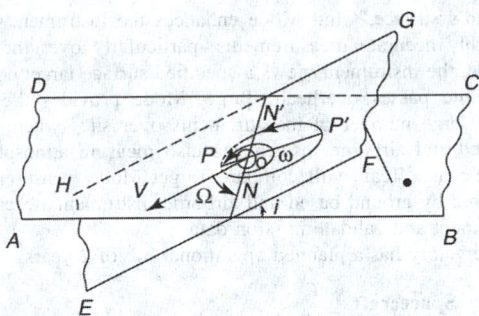

Fig. 1.   Diagram of orbit of member of solar system.

of the orbit. In Fig. 1, we have a diagram of a planetary orbit about the sun. The plane $ABCD$ represents the plane of the ecliptic, and the plane $EFGH$ represents the plane containing the orbit. These two planes must intersect in a line, NN', which must pass through the sun, since both the earth (the plane of the ecliptic being the plane of the earth's orbit) and the planet are revolving about the sun. The elliptical orbit of the planet is represented by $PNP'N'$, with the direction of motion of the planet as indicated, PP' representing the major axis of the ellipse, and P the point where the object is closest to the sun (the perihelion point). The line SV represents, in the plane of the ecliptic, the direction to the vernal equinox. We will assume that we are looking down on the plane of the ecliptic from the direction of its north pole. The points N and N', where the planet is on the ecliptic, are known as the nodes; the point N, where the planet is passing from south to north of the ecliptic, is the ascending node; and the point N' is the descending node. To define the orbit plane, the orbital element, i, the inclination, gives the angle between the orbit plane and the plane of the ecliptic; while the element, Ω (the angle VSN), gives the celestial longitude of the ascending node. The size and shape of the orbit in its plane are given by the orbital elements a, semimajor axis (usually expressed in astronomical units) and e, the eccentricity of the conic. To locate the position of the conic in the orbit plane, we use either ω (the angle NSP measured in the direction of motion of the planet), or, more commonly, $\overline{\omega}$, the longitude of perihelion, which is the sum of the two angles, Ω and ω. It should be noted that Ω is measured in the plane of the ecliptic, whereas Ω is measured in the orbit plane, so that is not strictly a longitude. To locate the position of the object in its orbit, the element T, which is some epoch or date when the object is at perihelion, and the rate at which the object is moving in the orbit are necessary. Occasionally, we find the rate of motion of the object given as an orbital element, but it is strictly not an independent element, since it may be derived from a. Listing the six elements as defined above, we have i, Ω, a, e, ω, and T. Unfortunately, several different systems of symbolic notation have been used by different authors, and great care must be exercised in interpreting any symbolic description of an orbit.

In describing the orbits of satellites, the plane frequently used is the plane of the planet's orbit about the sun, and the inclination, i, of the satellite orbit is referred to it. Also, instead of locating the position of the orbit in the plane by perihelion, the point where the satellite is closest to the planet is used. In the case of satellites of Jupiter, the point is known as perijove. In the case of binary star orbits, i gives the inclination of the orbit plane to the plane perpendicular to the line of sight, and the point where the smaller star is closest to the primary is called periastion, for obvious reasons.

The problem of orbit computation is far too complex to be included in a work of this character. In general, three observations of an object, giving the spherical coordinates of the object and the times of observation, are sufficient to compute a set of preliminary elements. From these preliminary elements, subsequent positions of the object may be computed and compared with observed positions. From the differences between observed and computed positions, the preliminary elements are corrected until a satisfactory representation of all observations is obtained. Before a definitive orbit can be obtained (i.e., an orbit that will give accurate positions for a long period of time), the perturbations due to other objects must be computed.

History

Orbits were first analysed mathematically by Johannes Kepler who formulated his results in his three laws of planetary motion. First, he found that the orbits of the planets in our solar system are elliptical, not circular (or epicyclic), as had previously been believed, and that the sun is not located at the center of the orbits, but rather at one focus. Second, he found that the orbital speed of each planet is not constant, as had previously been thought, but rather that the speed of the planet depends on the planet's distance from the sun. And third, Kepler found a universal relationship between the orbital properties of all the planets orbiting the sun. For each planet, the cube of the planet's distance from the sun, measured in astronomical units (AU), is equal to the square of the planet's orbital period, measured in Earth years. Jupiter, for example, is approximately 5.2 AU from the sun and its orbital period is 11.86 Earth years. So 5.2 cubed equals 11.86 squared, as predicted.

Isaac Newton demonstrated that Kepler's laws were derivable from his theory of gravitation and that, in general, the orbits of bodies responding to the force of gravity were conic sections. Newton showed that a pair of bodies follow orbits of dimensions that are in inverse proportion to their masses about their common center of mass. Where one body is much more massive than the other, it is a convenient approximation to take the center of mass as coinciding with the center of the more massive body.

See also **Earth Orbiting Satellite Theory**; **Kepler's Laws of Planetary Motion**; and **Peturbation**.

ORBITING CARBON OBSERVATORY (OCO).   The Orbiting Carbon Observatory (OCO) is a NASA Earth System Science Pathfinder (ESSP) Program mission designed to make precise, time-dependent global measurements of atmospheric carbon dioxide (CO_2) from an Earth orbiting satellite. See also **Earth System Science Pathfinder (ESSP) Program**.

CO_2 is a critical component of the Earth's atmosphere. Since the beginning of the industrial age, the concentration of CO_2 has increased by about 25%, from about 280 parts per million to over 370 parts per million. Scientific studies indicate that CO_2 is one of several gases that trap heat near the surface of the Earth. These gases are known as greenhouse gases. Many scientists have concluded that substantial increases in the abundance of CO_2 will generate an increase in the Earth's surface temperature. Historical records provide evidence of this trend, which is often called *global warming*. Current research indicates that continuing increases in atmospheric CO_2 may modify the environment in a variety of ways. These changes may impact ocean currents, the jet stream and rain patterns. Some parts of the Earth might actually cool while the average temperature increases. Thus, a more correct term for this phenomenon is climate change. See also **Carbon Dioxide**; **Global Warming**; and **Greenhouse Gases**.

CO_2 can enter the atmosphere from a variety of sources. Some sources are natural, such as rotting plants, forest fires and ordinary breathing. Human activities augment the emission of CO_2 into the atmosphere. Automobiles, factories and home heating units burn fossil fuels such as oil, coal and natural gas. Burning these fossil fuels releases CO_2 into the atmosphere.

Other natural processes remove CO_2 from the atmosphere. Plants use sunlight to photosynthesize CO_2 and water into sugar and other carbohydrates. The oceans also absorb atmospheric CO_2. Sea creatures incorporate the CO_2 dissolved in sea water into their shells. After these creatures die, their shells fall to the bottom of the ocean. Over time, these sediments form carbonate rocks. Processes that absorb CO_2 from the atmosphere are often referred to as sinks. The complete process of CO_2 exchange is known as the carbon cycle.

To better understand the carbon cycle, the Carbon Dioxide Information Analysis Center of the U. S. Department of Energy tracks and monitors CO_2 emissions from a global network of ground-based sites. This network provides a tremendous amount of insight into the global abundance of CO_2 and its variability over changes in seasons. Unfortunately, the global network does not include enough stations to resolve the spatial distribution of CO_2 sources and sinks at the scale of continents or ocean basins. Thus, even with these extensive measurements, the processes that regulate the exchange of CO_2 between the atmosphere, the oceans, and the biosphere are not completely understood.

Using a space-based platform, OCO will collect a far greater number of high resolution measurements which in turn will provide the distribution of CO_2 over the entire globe. These measurements will be combined with data from the ground-based network to provide scientists with the information that they will need to better understand the processes that regulate atmospheric CO_2 and its role in the carbon cycle. This enhanced understanding is essential to improve predictions of future atmospheric CO_2 increases and their impact on the climate. This information could help policy makers and business leaders make better decisions to ensure climate stability and, at the same time, retain our quality of life.

The Jet Propulsion Laboratory, http://www.jpl.nasa.gov/, will lead the OCO effort. Orbital Sciences Corporation, http://www.orbital.com/ and Hamilton Sundstrand Sensor Systems will partner with JPL to realize this vital mission.

Observatory Overview

The Orbiting Carbon Observatory (OCO) is a dedicated spacecraft that carries a single instrument comprised of three high resolution grating spectrometers. The instrument, developed by Hamilton Sundstrand Sensor Systems, will acquire the most precise measurements of atmospheric CO_2 ever made from space. The spacecraft, developed by Orbital Sciences Corporation, is based upon the LeoStar-2 architecture. The same LeoStar-2 design was used on the successful Earth orbiting SORCE and GALEX missions. The Observatory will be launched from the Vandenberg Air Force Base in California on a dedicated Taurus XL rocket in September, 2008. The Observatory will fly in a near polar orbit that enables the instrument to observe most of the Earth's surface at least once every sixteen days.

The abundance of CO_2 in the atmosphere varies both with time of day and with season. OCO measurements must record changes in CO_2 abundance over annual seasonal cycles. To remove the effect of changes in CO_2 abundance over each day, and discriminate between seasonal variations and long term changes, OCO will always acquire measurements at the same time of day. Thus, the spacecraft will fly in Sun synchronous orbit so that all observations take place at about 1:18 PM.

The Observatory will fly in loose formation with a series of other Earth orbiting satellites known as the Earth Observing System Afternoon Constellation, or the A-train. This coordinated flight formation will enable researchers to correlate OCO data with data acquired by other instruments on Earth observing spacecraft. In particular, Earth scientists will compare OCO data with nearly simultaneous measurements acquired by the Atmospheric Infrared Sounder (AIRS) instrument. The AIRS instrument files on the Earth Observing System Aqua platform.

To provide the mission with additional flexibility, the Observatory will acquire data in three different measurement modes. In Nadir Mode, the instrument views the ground directly below the spacecraft. In Glint Mode, the instrument tracks near the location where sunlight is directly reflected

on the Earth's surface. Glint Mode enhances the instrument's ability to acquire highly accurate measurements, particularly over the ocean. In Target Mode, the instrument views a specified surface target continuously as the satellite passes overhead. Target Mode provides the capability to collect a large number of measurements over sites where alternative ground based and airborne instruments also measure atmospheric CO_2. The OCO Science Team will compare Target Mode measurements with those acquired by ground based and airborne instruments to calibrate the OCO instrument and validate mission data.

The Observatory has a planned operational life of 2 years.

Observatory Spacecraft

The design and architecture of the OCO spacecraft bus is based on the successful Solar Radiation and Climate Experiment (SORCE), http://lasp.colorado.edu/sorce/ and (Galaxy Explorer) GALEX missions, http://www.galex.caltech.edu/. The spacecraft structure is made of honeycomb panels that form a hexagonal shape. This structure houses the instrument and the spacecraft bus components. See Fig. 1. The total weight of the Observatory is about 530 kg (1,170 lb). Panels with solar cells are attached and stowed such that the whole structure fits inside the small fairing of the Taurus launch vehicle. A metal ring, mounted to the bottom of the structure, attaches the Observatory to the launch vehicle and separates the two after launch.

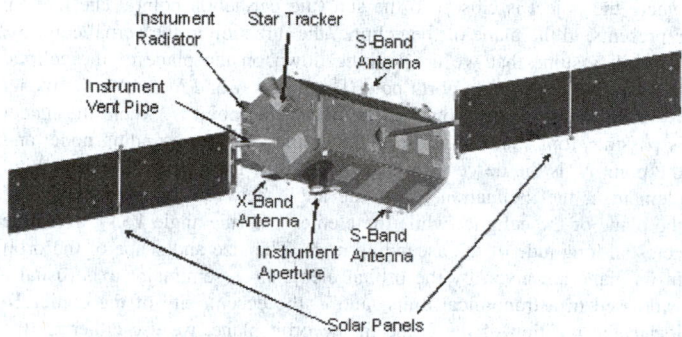

Fig. 1.   Illustrated drawing of the Orbiting Carbon Observatory (OCO).

The on-board computer, which is designed to fly in the harsh space environment, controls the spacecraft bus components. This computer hosts software, which receives commands from an Earth station through an S-band antenna and returns telemetry and science data back to Earth using a high data rate X-band transmitter.

The spacecraft computer manages the pointing of the spacecraft. Ground commands tell the computer where to point the instrument. The computer

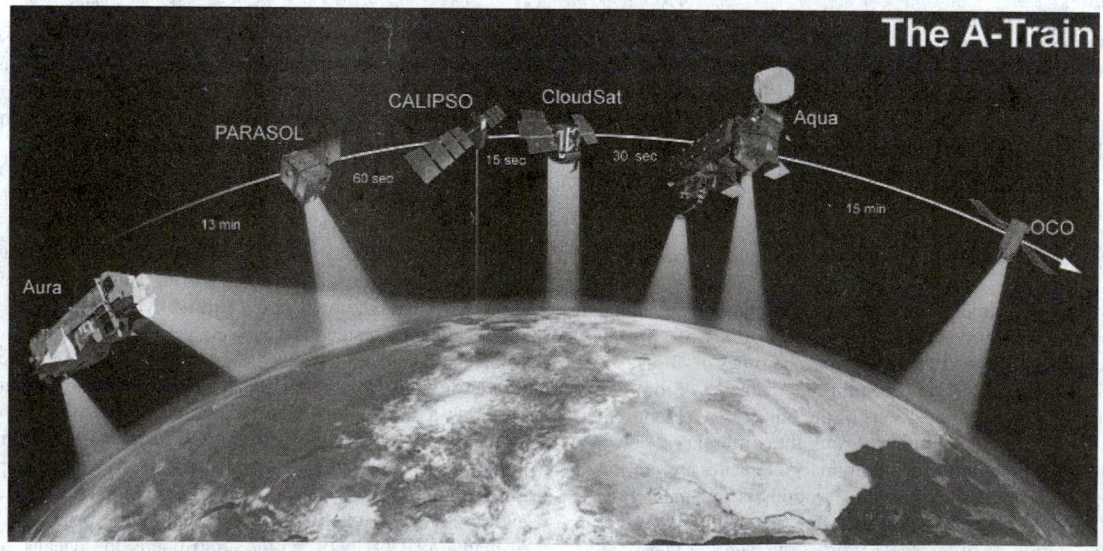

Fig. 2.   The Earth Observing System Afternoon Constellation, or the A-train.

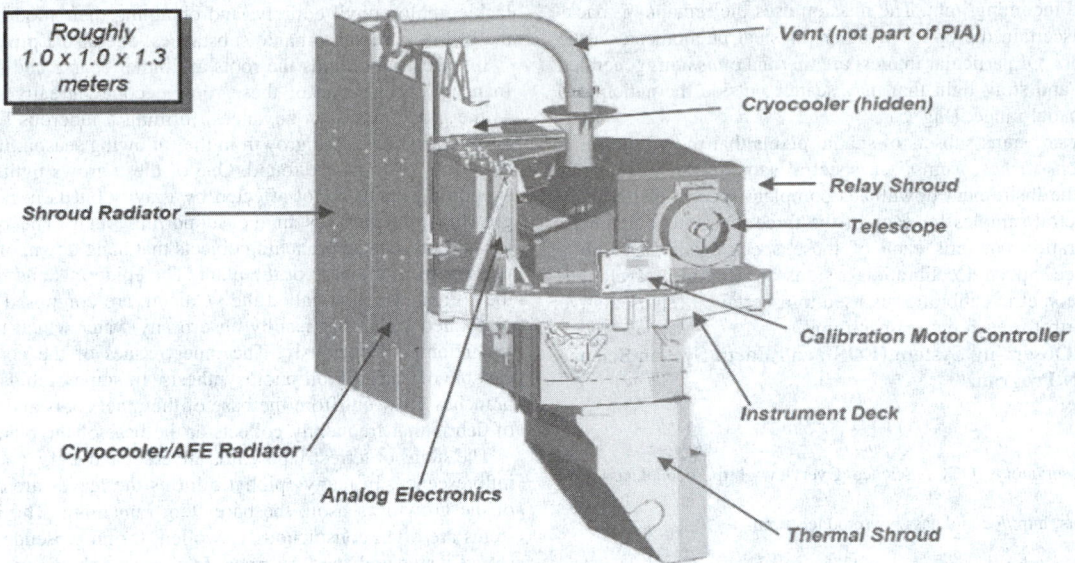

Roughly
1.0 x 1.0 x 1.3
meters

Vent (not part of PIA)

Cryocooler (hidden)

Relay Shroud

Telescope

Calibration Motor Controller

Instrument Deck

Thermal Shroud

Shroud Radiator

Cryocooler/AFE Radiator

Analog Electronics

Fig. 3.   Orbiting Carbon Observatory (OCO) spectrometers.

uses four wheels to move the spacecraft. A star tracker verifies that the spacecraft has reached the correct orientation. In addition to pointing the instrument, the spacecraft must know where on Earth the footprint of the instrument is located. An on-board Global Positioning System (GPS) receiver provides that information.

Spacecraft software ensures that the solar arrays face the sun so that adequate power is always available to charge the battery and run all the components and the instrument. The power required to run the entire observatory is equivalent to the power needed for nine common household light bulbs.

The OCO spacecraft flies in formation with several other spacecraft in what is called the *Aqua train* or *A-Train*. See Fig. 2. OCO uses a hydrazine based propulsion system and four thrusters, located below the spherical tank, to maintain the spacecraft's location within this formation.

Instrument Design

The Observatory carries a single instrument that incorporates three classical grating spectrometers. Each spectrometer detects the intensity of radiation within a very specific narrow band at Near Infrared (NIR) wavelengths (a range of the electromagnetic spectrum with slightly higher wavelength than that of visible light.) The three spectrometers share a common structure, a cryogenic cooler, and an input telescope. See Fig. 3.

The telescope consists of an 11 centimeter (4 inch) aperture, as well as a primary and a secondary mirror. The relay optics assembly includes a collimating mirror, fold mirrors, dichroic beam splitters, band isolation filters and re-imaging mirrors. Each spectrometer consists of a slit, a two-lens collimator, a grating, and a two-lens camera. Each of the three spectrometers has an essentially identical layout. Minor differences among the spectrometers, such as the coatings, the lenses and the gratings, account for the different bandpasses that are characteristic of each channel. The focal ratios of the instrument optics range from f/1.6 to f/1.9.

To implement an optically fast, high-spectral-resolution measurement system, the OCO instrument combines refractive and reflective optical techniques. Since the light in the common telescope and relay optics assembly has not yet been separated into the three distinct wavelength bands, these instrument subsystems primarily use reflective optics. On the other hand, the extremely narrow channel bandpasses make potential chromatic aberrations in the spectrometers negligible, which enables the use of refractive optics. See Fig. 4.

The instrument focal plane is a two dimensional array of light detectors. Each dimension contains 1024 detectors. Along one dimension, the detectors progressively vary in their sensitivity to spectral wavelength or color. Detectors along the other dimension acquire data that are representative of different spatial location. OCO employs all 1024 of the pixels in the spectral dimension but only 220 of the pixels in the spatial dimension. Of the 220 pixels that are used in the spatial dimension, only

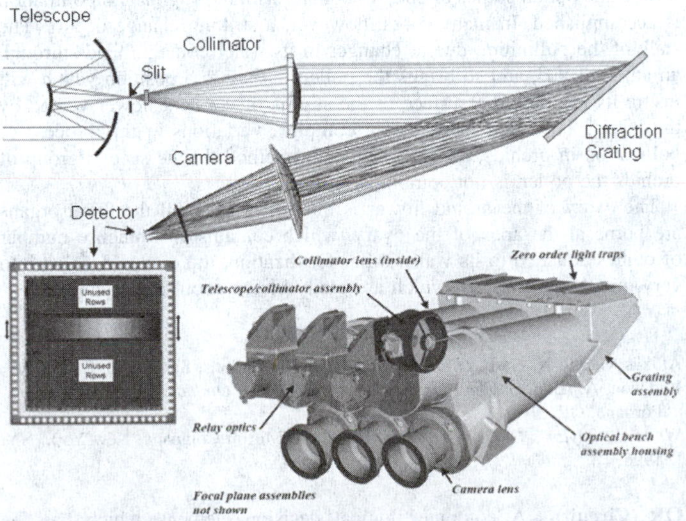

Telescope

Collimator

Slit

Diffraction
Grating

Camera

Detector

Collimator lens (inside)

Zero order light traps

Telescope/collimator assembly

Grating
assembly

Relay optics

Optical bench
assembly housing

Focal plane assemblies
not shown

Camera lens

Fig. 4.   The Orbiting Carbon Observatory (OCO) Spectrometer and assembly.

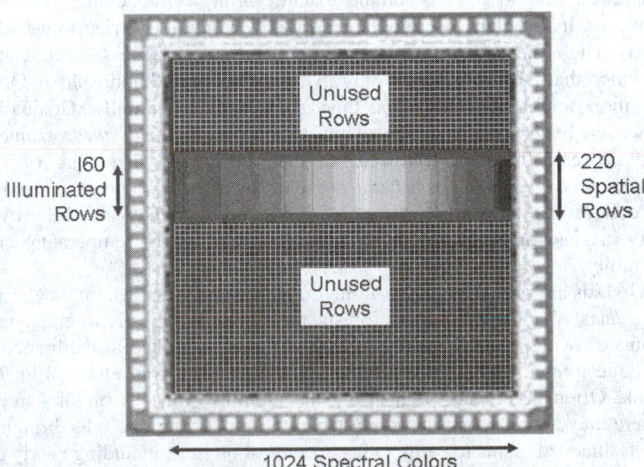

Unused
Rows

160
Illuminated
Rows

220
Spatial
Rows

Unused
Rows

1024 Spectral Colors

Fig. 5.   The Orbiting Carbon Observatory (OCO) calibration overview.

160 are exposed to incoming light. The mission uses the remaining 'back-of-slit pixels' to ascertain the effect of other physical phenomena on the focal plane detectors. Of particular interest are thermal emissions generated by the instrument and stray light that may scatter outside the anticipated wavelength and spatial range. (Fig. 5)

Instrument software sums subsets of spatial pixels that record the same spectral color. Each of these sums is a spectral samples. In glint, nadir and target mode, the instrument downloads complete spectra that contain 1024 summed spectral samples for each of the three instrument channels. Radiometric calibration converts each of these spectral samples into a meaningful radiance. Spectral calibration assigns an accurate wavelength to each sample. Geometric calibration is used to ascertain a representative location on the Earth for each entire spectrum.

See also **Earth Observing System (EOS)**; and **Earth System Science Pathfinder (ESSP) Program**.

Web Reference

Orbiting Carbon Observatory (OCO) Science Overview: http://oco.jpl.nasa.gov/science.html
Science Data Products: http://oco.jpl.nasa.gov/products.html.

ORBITING RETRIEVABLE FAR AND EXTREME ULTRAVIOLET SPECTROMETER (ORFEUS) MISSION.   See **Space Science Missions: Universe**.

ORCHARD GRASS.   See **Grasses**.

ORCHID (*Orchidaceae*).   The Orchids, which form the second largest family of plants, are the highest development of the Monocotyledons, just as the Composite Family marks the highest point of evolution reached by the Dicotyledons. But between the two families a most striking contrast exists. The Composites are most beautifully formed to enjoy a more abundant life; the numerous small flowers are massed in compact heads, so pollination is almost certain of accomplishment. The reduction of the fruit to a single ovule and the presence of various barbs or scales or bristles, called the pappus, greatly favor the probability of successful continuance of the species. In the 9,000–10,000 species of Orchids, the individual flowers are usually very conspicuous objects often of bizarre form and rare beauty. But they depend entirely on insects for pollination and often exhibit elaborate modifications to insure successful insect pollination. The seeds are minute and borne in tremendous numbers, few of which germinate and grow to maturity. As a result, orchid plants are relatively rare, a fact which has contributed not a little to the zeal with which collectors have sought these plants.

Originally, orchids were known only from the reports brought back by travelers from the tropics, who spoke of the brilliant colors, the curious forms and the delightful fragrance, and also of the mystery and folklore that often attached to orchids. Later, botanists gained a wider knowledge of the family from dried specimens, in which color, fragrance, and to a considerable extent, form, were lacking. In time, however, living plants were obtained by collectors and carried to western lands. There they were grown usually without much success, since it was assumed that they could grow only in a very hot humid atmosphere. Only the most tolerant species stood this, and they only partially. Gradually, however, better understanding of the plants' requirements was obtained, and successful culture followed, so that during the first third of the nineteenth century orchids became popular. Necessarily, they are expensive to collect, for they are not easy to transport, and they can be grown only in glass houses under fairly uniform conditions of temperature and humidity.

Orchids are primarily plants of the tropical rain forests, where they grow in greatest abundance. In these forests, they are found mostly as epiphytes, plants growing on other plants. Such plants grow attached to the branches of large trees, or massed in a crotch of the tree, or even attached to the trunk. Often they occur high up on the topmost branches of lofty trees, where they are inaccessible except when the supporting tree falls, bringing all its attached plants to earth. Other species of orchids, including nearly all those found in extratropical regions, are terrestrial. All orchids, wherever they grow, are herbaceous perennials. A few species are saprophytes,

lacking chlorophyll entirely, and obtaining their food by absorption from the soil of complex organic substances. See also **Epiphytes**.

In terrestrial orchids the roots are rather coarse and sparsely branching. In many species one of these roots becomes greatly swollen with food, as the growing season advances, forming a tuberous body which will be used to promote rapid growth in the following season. In epiphytic orchids two kinds of roots are found. One of these grows tightly appressed to the supporting plant, is not affected by gravity but is negatively phototropic, growing into crevices in the supporting plant. The other roots are the aerial roots, coarse branching objects that hang down, often in conspicuous masses, from the base of the plant. The epidermis and the outer portion of the cortex, which is called the velamen, are composed of dead cells with perforated walls that readily absorb any water which may come to them and retain it tenaciously. The inner tissues of the cortex are green and capable of carrying on photosynthesis. In some orchids slender absorbing branches grow out from the base of the other roots and penetrate the mass of debris that frequently collects at the base of the plant.

The stems of terrestrial orchids are erect and leafy, and terminated by the inflorescence. In many epiphytic forms the leaves are dropped at the end of the growing season, the bare stem remaining. The internodes of such stems are often conspicuously swollen, forming pseudobulbs in which are stored water and food reserves. In other epiphytic species the leaves are fleshy and serve for storage.

The most characteristic and in many species the most conspicuous part of the plant is the flower. In all orchids the flowers are irregular but formed on a very uniform pattern. In all there is a six-parted perianth composed of two groups of three members each. One of the inner three differs from all the others and is designated the lip or labellum. This becomes variously fringed in some orchids, broadly expanded in others, and even saccate (slipper-like), as in the Lady's Slipper. It is commonly much more brilliantly colored than the other five parts and gives to the flower its showiness. It serves as a landing place for insects, and is often a definite factor in bringing about pollination. Often various outgrowths such as spurs add further complexity to the flower. The essential organs of the flower, the stamens and pistil, are united into a single body, the column, which is a characteristic feature of the orchid flower. Its structure varies somewhat in the different groups of the family. In orchids there are only one or in some species, two, anthers present. The greater number of species have a single anther, which has two lobes, each filled with a mass of pollen grains held together by fine elastic threads. Such a mass is called a pollinium. The threads all unit to form a slender stalk which ends in a sticky mass resulting from the breaking down of certain cells in the rostellum. The latter is a special organ, which represents one of the stigmas of the flower.

An insect comes to the orchid flower seeking the nectar it contains. In getting this nectar the insect comes in contact with the sticky mass of cells which become firmly cemented to some part of the body, often the eyes, or the antennae. When the insect leaves the flower, it drags the pollinium out. The latter may be in a position so that when the insect enters the next flower, the pollen comes directly in contact with the stigma and pollination is accomplished. In many cases, however, a striking change occurs. The stalk of the pollinium, due to changes in its water content, bends through an angle of $90°$, and so brings the pollen mass into a position which will insure its reaching the surface of the stigma of the next flower visited by the insect. There are many, often complex, variations in this process of pollination in orchids, all on this general method. In the second group of orchids the pollen is not combined into masses.

The ovary of the orchid flower is interior; that is, all the floral organs are borne at the apex of the ovary, which contains an immense number of ovules attached to its walls. After fertilization, the ovules develop into very minute, light seeds, which are easily blown about by air currents.

Additional Reading

Allikas, G. and N. Nash: *Orchids*, Advantage Publications, San Diego, CA, 2000.
Pridgeon, A. (Editor): *The Illustrated Encyclopedia of Orchids*, Timber Press, Inc., Portland, OR, 1992.
White, J.: *Taylor's Guide to Orchids*, Houghton Mifflin Company, New York, NY, 1996.

OR (Circuit).   A computer logical decision element, which has the characteristics of providing a binary 1 output if any of the input signals

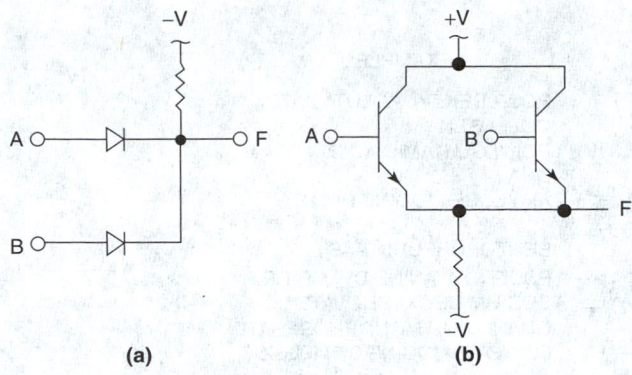

Fig. 1.   OR circuits. (**a**) Diode or gate-type; (**b**) transistor-type.

are in a binary 1 state. This is expressed in terms of Boolean algebra by $F = A + B$. A diode and a transistor representation of this circuit is shown in Fig. 1. In the diode-type OR circuit, if either (or both) input signal A or B is positive, the respective diode is forward-biased and the output F is positive. The number of allowable input signals to the diode Or gate is a function of the back resistance of the diodes. The input transistors of the transistor-type OR circuit are forced into higher conductivity when the respective input signal becomes positive. Thus, the output signal becomes positive when either or both inputs are positive.

See also **Exclusive OR Circuit**.

THOMAS J. HARRISON, IBM Corporation, Boca Raton, FL

ORDER-DISORDER THEORY AND APPLICATIONS. Phase transitions in binary liquid solutions, gas condensations, order-disorder transitions in alloys, ferromagnetism, antiferromagnetism, ferroelectricity, antiferroelectricity, localized absorptions, helix-coil transitions in biological polymers and the one-dimensional growth of linear colloidal aggregates are all examples of transitions between an ordered and a disordered state.

The two quantities which apparently must be used to explain or describe these phenomena are the presence of a potential between the particles or spins and a small volume which must be assigned to each particle so that two particles or spins cannot occupy the same space. All the above phenomena may be semi-quantitatively treated by a single statistical model called the Ising model which consists of a lattice in space, each site of which possesses either a 0 or a 1. Thus two rows of a two dimensional lattice might look like

$$\ldots 011000011011110011 \ldots$$

$$\ldots 110110010100110111 \ldots$$

The 0's and 1's represent different particles or spins depending upon the system being studied and their definition will be given below for different systems. The disordered state corresponds to a random array of 0's and 1's. The ordered state is described by an ordered arrangement of the 0's and 1's. The total number of lattice sites equals N. The number of 0's and 1's on the lattice are respectively n_0 and n_1 and $n_1 + n_0 = N$.

The dimensionality of the lattice depends on the physical nature of the phase. For transitions in linear biopolymers and associative colloids, a one-dimensional lattice is used because these materials become ordered in a one-dimensional manner even though they actually exist in three dimensions. For absorption of gas onto a surface, a two-dimensional lattice is sufficient. For bulk phase changes, however, a three-dimensional lattice must be used.

The great advantage of this model is that it gives essentially the same explanation for a large number of seemingly completely diverse physical and chemical phenomena, often with quantitative success. This permits a deeper insight into the statistical thermodynamic behavior of different types of matter.

The calculation of the thermodynamic properties begins by selecting values for w_{00}, w_{11}, w_{10}, which are the potential energies between like and unlike particles or spins occupying nearest neighbor sites. Since two particles cannot occupy one lattice point, the energy of repulsion of two particles on the same site is considered infinite. The energy of a given configuration is then given by

$$E_{\text{Conf}} = n_{11}w_{11} + n_{10}w_{10} + n_{00}w_{00} \tag{1}$$

where n_{11}, n_{00} and n_{10} are the number of nearest neighbor pairs. The probability of a given configuration is given by Boltzmann's theorem as

$$P(E_{\text{Conf}}) = \frac{e^{(-E_{\text{Conf}}/kT)}}{\sum\limits_{\text{Conf}} e^{(-E_{\text{Conf}}/kT)}} \tag{2}$$

where the summation occurs over all possible configurations, keeping the number of 0's and 1's constant. The restriction of a constant number of 0's and 1's may be removed by letting the lattice interact with its surroundings. The probability that the lattice possesses a given number of 1's and a configurational energy, E_{Conf}, is given by

$$P(E_{\text{Conf}}, n_1) = \frac{e^{-Xn_1/kT} e^{-E_{\text{Conf}}/kT}}{\sum\limits_{n_1=0}^{N} e^{-Xn_1/kT} \sum\limits_{\text{Conf}} e^{-E_{\text{Conf}}/kT}} \tag{3}$$

where X is a suitable thermodynamic quantity such as the chemical potential or magnetic field, etc.

For a lattice gas, the 0 and 1 stand for an empty and filled site respectively. Consequently, w_{11} is the attractive potential energy between two gas molecules when they occupy adjacent sites on the lattice, and $w_{00} = w_{10} = 0$.

For binary solutions including linear associative colloids, the 0 and 1 represent solvent and solute respectively; w_{11} and w_{00} are the potential energies between like molecules and w_{10} is the potential energy between unlike molecules.

For a ferromagnet the 0 and 1 represent spins of $-\frac{1}{2}$ and $+\frac{1}{2}$ respectively. It is generally assumed that the potential energy between like spins is always the same, i.e., $w_{11} = w_{00}$. To show how the quantity, X of Eq. (3), is obtained, consider a ferromagnet interaction with an external magnetic field. The total energy is

$$E = (n_0 - n_1)H \cdot d + E_{\text{Conf}}$$

where H is the magnetic field strength, and d is the magnetic dipole moment of a spin. Thus for a ferromagnet $X = 2H \cdot d$ since $N - 2n_1 = n_0 - n_1$.

The above described order-disorder transitions are all three-dimensional phase transitions and occur with essentially infinite sharpness unless the condensed phase exists in a colloidally dispersed state. Recently, it has been shown that certain polymers and associative colloids, particularly those of biological interest, have one-dimensional order-disorder transitions which may be explained in exactly the same terms that describe the three-dimensional phase changes discussed above. However, because the condensed state is colloidally dispersed and the ordering occurs only in one dimension, it may be shown that such transitions cannot be infinitely sharp.

Figure 1 illustrates the transition from an ordered helical to a disordered state which occurs in a large number of colloidal systems. For the helix-coil transition in polymers, the 1 and 0 represent, respectively, a hydrogen bonded turn of the helix and an unhydrogen bonded section of randomly fluctuating polymer. In an associative colloid, the 0 or 1 represent respectively a cell containing a solvent or colloid monomer. In either case, there exists an energy of attraction between 1's which leads to the formation of a one dimensional ordered helix. The fact that the transition tends to be rather sharp comes about because the 01 and 10 configurations which always occur at the beginning and end of a helical sequence are energetically unfavorable. This means that configurations with a lot of ends containing many short segments are suppressed since the configurational energy is large when n_{10} is large and this leads to a small Boltzmann factor for this configuration.

The fact that so many diverse phenomena can be correlated and explained by such a simple model makes it possible to develop tables of equivalent thermodynamic properties. This was first done by Yang and Lee who showed the equivalence between the properties of the lattice gas and Ising ferromagnet. Hill extended their analogies to the case of binary liquid mixtures. After the development of the helix-coil transition theory

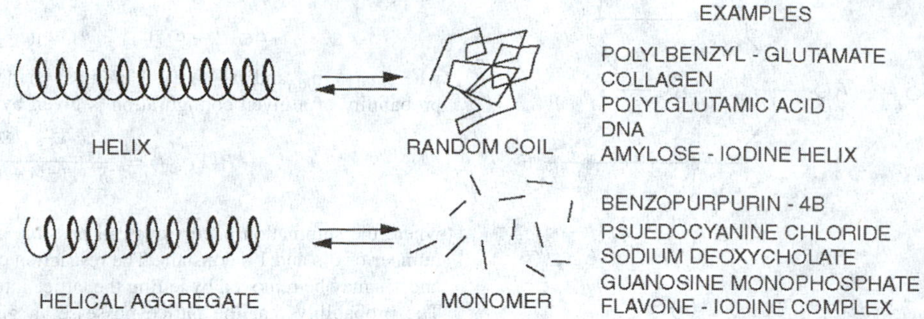

Fig. 1.   Illustration of the helix-coil transitions in biopolymers (*top*). Helix-monomer transition in associative biocolloids (*bottom*).

by Zimm and Bragg and others, Peticolas gave a corresponding table of equivalent thermodynamic properties between polymers and associative colloids. Thus the force-length curve for a helical polymer is the one dimensional analogue of the three dimensional pressure-volume curve for gas condensation and is equivalent to the chemical potential-mole-fraction curve for the associative linear colloid.

W. L. PETICOLAS, University of Oregon, Eugene, OR

ORDER (Of).   If a function $f(x)$ becomes $|x^k f(x)| < K$ as x approaches 0 *or* ∞, where K is a positive number independent of x and not zero, then $f(x)$ is said to be of order $1/x^k$. In symbols, $f(x) = \mathbf{O}(x^{-k})$, where this notation is called the *ordor-symbol*. The limiting process 0 or ∞ is not always stated explicitly but inferred from the context. In the special case where $\lim x^k f(x) = 0$, one writes $f(x) = \mathbf{o}(x^{-k})$. If $f(x)$ is bounded, one writes $f(x) = \mathbf{O}(1)$. In any case, $\mathbf{O}(x^k) \times \mathbf{O}(x^n) = \mathbf{O}(x^{k+n})$.

ORDER OF MAGNITUDE.   For two functions u and v to be of the same order of magnitude near t_o means there are positive numbers ε, A, and B such that

$$A < \left|\frac{u(t)}{v(t)}\right| < B \text{ if } 0 < |t - t_0| < \varepsilon.$$

If this is true for $v(t)$ replaced by $[v(t)]^r$, then u is of the rth order with respect to v. If $\lim_{t \to t_0} u(t)/v(t) = 0$, u is of lower order than v and one writes $u = \mathbf{o}(v)$; if $\lim_{t \to t_0} |u(t)/v(t)| = \infty$, u is of higher order than v. Also, one says that u is of the order of v on a set S (usually a deleted neighborhood of some t_0) and writes $u = \mathbf{O}(v)$ if there is a positive number B such that $|u(t)/v(t)| < B$ if $t \varepsilon S$. The statements $u = \mathbf{o}(1)$ and $u \to 0$ are equivalent; $u = \mathbf{O}(1)$ on S is equivalent to "u is bounded on S." If $\lim_{t \to t_0} u(t)/v(t) = 1$, then u and v are asymptotically equal at t_0. These concepts also are used when $t_0 = \pm\infty$ with appropriate changes. For example, if $\{u_n\}$ and $\{v_n\}$ are sequences and there are numbers B and N such that

$$\left|\frac{u_n}{v_n}\right| < B \text{ if } n > N,$$

then u_n is of the order of v_n and $u_n = \mathbf{O}(v_n)$.

ORDINARY RAY.   That magnetoionic wave component deviating the least, in most of its propagation characteristics, relative to those expected for a wave in the absence of the Earth's magnetic field. More exactly, if at fixed electron density, the direction of the Earth's magnetic field were rotated until its direction is transverse to the direction of phase propagation, the wave component whose propagation is then independent of the magnitude of the Earth's magnetic field. Also called *ordinary-wave component*. See **Magnetic Double Refraction**; and **Magnetoionic Theory**.

ORDOVICIAN.   A period of the Paleozoic Era. Type locality, border of Wales and Shropshire, England.

The formations of this system were first studied and described by Charles Lapworth in 1879. The Ordovician period began 440 million years ago

and lasted for 60 million years. In 1874 J.D. Dana had already proposed a post-Cambrian system, the Canadian. This system has only been recently recognized in Britain (Northwest Highlands of Scotland) where it is still considered as upper Cambrian by some British geologists (1937). In 1911, E.O. Ulrich proposed a new system (period) called the Ozarkian, to include some of the upper Cambrian and some of the lower Canadian. There is still considerable doubt as to the systemic importance of the Ozarkian, especially as it appears to be absent in Great Britain. Ordovician formations are well exposed in North America due to the fact that a large part of the continent was submerged during this period. The marine sediments are of two principal types, limestones and shales, the former contain "shelly fossils" and the latter, principally graptolites. Thus the American Ordovician covers an enormous area and has an endless variety of local developments, for during the vast lapse of time included in the period, the shallow epeiric seas were continually shifting their positions, outlines and depths. The United States Ordovician contains no igneous rocks with the exception of volcanic ashes, called bentonites. The formations of this system are well exposed in Portugal, Switzerland, Bohemia, Austria, Hungary, Ireland, eastern Asia, China, Manchuria, Siberia, the Himalayas, Burma, Morocco, Australia, New Zealand, northern Argentina, Bolivia, and eastern Peru. The maximum thickness of Ordovician strata, 40,000 feet, occurs in Australia. All classes of marine invertebrates occur as fossils, a number of which classes are now extinct. The principal types are graptolites, corals, echinoderms, bryozoans, pelecypods, nautiloids, trilolites, and ostracods. (See also **Invertebrate Paleontology**.) The strata of eastern North America are an important source of petroleum. Because of the wide surficial distribution of the Ordovician limestones they are much quarried for foundation structures, road metal, flux for the reduction of iron ores, and especially for the lime used in mortar, whitewash, fertilizers and Portland cement. In Vermont and Tennessee occur valuable deposits of marble. Lead and zinc ores in rocks of Ordovician age are mined in Iowa, southern Wisconsin and northern Illinois. Phosphates of the same age are mined in central Tennessee. In eastern North America the Ordovician closed with a period of mountain building named the Taconic Revolution by J.D. Dana, in 1895.

ORE BENEFICIATION.   See **Iron**.

ORE BLOCK.   A section of an orebody, usually rectangular, that is used for estimates of tonnage and quality of the orebody.

OREBODY.   A continuous, well-defined mass of material of sufficient ore content to make extraction economically feasible.

OREGANO.   A savory herb which has become quite popular in the past several years. It was introduced into the United States and England from the Mediterranean area. Probably the original herb is a member of the Mint family (see also **Mint Family**), *Origanum vulgare*, and most of the oregano imported from Europe comes from this plant. Mexican oregano, of indistinguishable flavor and aroma, however, comes from species of *Lippia* of the family *Verbenaceae*. Thus the name oregano is a general term applying to a particular herb flavor, rather than to a particular species of plant.

ORES. Mineral aggregates in which the valuable metalliferous minerals are sufficiently abundant to make the aggregates worth mining. Types and origins of ore deposits are illustrated in Fig. 1. See also **Copper**; and **Iron**.

ORGAN. A multicellular structure made up of various tissues for the performance of some complex function. The stomach, for example, contains in its walls tissues of all of the five principal divisions. Its function is the digestion of food and this end is gained by the cooperative exercise of the simpler functions of all of the tissues composing it.

ORGANELLE. A portion of a cell having specific functions, distinctive chemical constituents, and characteristic morphology; it is a unit subsystem of a cell. Examples are mitochondria and chromosomes. Organelles are often closely associated with enzymes. The lysosome (an enzyme-bearing organelle) has been synthesized.

ORGANIC CHEMISTRY. The term *organic*, which means *pertaining to plant or animal organisms*, was introduced into chemical terminology as a convenient classification of substances derived from plant or animal sources. Early it was believed that organic compounds could arise only through the operation of a vital force inherent in the living cell. However, Wöhler discovered in 1828 that the organic compound urea (see also Wöhler, Friedrich), identified in urine by Rouelle in 1773, could be produced by heating the inorganic salt ammonium cyanate:

$$NH_4^+OCN^- \longrightarrow H_2NCONH_2$$
$$\text{Ammonium cyanate} \qquad\qquad \text{Urea}$$

Subsequently, the association of organic compounds only with living organisms was discontinued.

The term *organic* has persisted, but the modern definition of *organic chemistry* has changed to mean the *chemistry of carbon compounds*. Sometimes a few carbon compounds are excluded from this category, such as carbon dioxide, CO_2; metal carbonates, e.g., Na_2CO_3; carbonyls, e.g., $Ni(CO)_4$; cyanides, e.g., KCN; carbides, e.g., CaC_2; and a few others, but this exclusion is somewhat arbitrary. The designation *organic* is still pertinent because the chemistry of carbon compounds is more important to everyday life than that of any other element.

The uniqueness of carbon stems from its ability to form strong carbon-carbon bonds that remain strong when the carbon atoms are simultaneously bonded to other elements. Whereas both the carbon-hydrogen and carbon-fluorine compounds CH_3CH_3 and CF_3CF_3 are highly stable and relatively unreactive, the corresponding compounds in which the carbon atoms are replaced by boron, silicon, phosphorus, and others either are thermodynamically unstable or highly reactive.

Theoretically, an infinite number of different carbon compounds can exist. Carbon atoms alone or in combination with other atoms, such as oxygen, nitrogen, etc., can join to form linear, branched, and cyclic chains of nearly any length. One, two, or three bonds may be shared between two carbon atoms. In most stable organic compounds, the total number of bonds to each carbon atom is four.

Classification of Organic Compounds

A subject with the wide scope that characterizes organic chemistry requires a logical approach to its organization so that knowledge can be gathered and applied manageably. Molecular structure has become the key method for classifying this subject. Scientists use two-dimensional diagrams or three-dimensional models to depict molecular structures. Although these analogies are sometimes crude representations of actual molecules, they are useful for communicating information about the molecules. Structural diagrams characteristic of some basic types of organic molecules are shown in Table 1. A more detailed discussion on naming organic compounds is given later.

The compounds shown in Table 1 contain only carbon and hydrogen and are called hydrocarbons. Most organic compounds that contain other kinds of atoms in addition to carbon and hydrogen are considered as formally derived from hydrocarbons in which a hydrogen atom has been replaced

TABLE 1. EXAMPLES OF STRUCTURAL DIAGRAMS OF HYDRO-CARBONS

Type of Compound	Formula	Name
Linear alkane	Or $CH_3CH_2CH_2CH_2CH_2CH_3$ or $CH_3(CH_2)_4CH_3$ or n-C_6H_{14} or	n-Hexane
Branched alkane	CH_3 | $CH_3CH_2CH_2CH_2CH_3$ Or	2-Methylpentane
Monocyclic alkane	or	Cyclohexane
Bicyclic alkane		Bicyclo[2.2.1] heptane or Norbornane
Polycyclic alkane		Pentacyclo [5.3.0.0.^{2,5}−0^{3,9} 0^{4,8}] decane or 1,3-Bishomocu-bane
Alkene	$CH_3C=CCH_2CH_3$ or	*trans*-2-Pentene
Alkyne	$CH_3CC≡CH_2CH_3$ or	2-Pentyne
Aromatic	or	Benzene
Polymer	$X(CH_2)_n$ Y ($n \geq 1000$, X and Y vary according to how polymer was prepared)	Polymethylene or Polyethylene

by another atom or collection of atoms. However, these derivatives usually are not formed directly from the hydrocarbons. The other atoms usually are referred to as functional groups. A group of compounds having the same functional group is referred to as a family. Table 2 shows a number of such families. The R and R′ in the formulas of Table 2 represent any hydrocarbon in which a hydrogen atom has been removed from the position to which the functional group is attached.

Molecules with common features also are grouped into more specialized families. The compounds in one such family, carbohydrates, contain only carbon, hydrogen, and oxygen. The hydrogen and oxygen atoms are in the

TABLE 2. EXAMPLES OF SOME MAJOR FAMILIES OF ORGANIC COMPOUNDS

Family	General Structure	Family	General Structure
Alcohol, phenol	ROH	Isocyanate	RN=C=O
Ether	ROR′	Thiol	RSH
Aldehyde	$\overset{O}{\overset{\|}{RCH}}$	Sulfide	RsR′
Ketone	$\overset{O}{\overset{\|}{RCR'}}$	Sulfoxide	$\overset{O}{\overset{\|}{RSR'}}$
Carboxylic acid	$\overset{O}{\overset{\|}{RCOH}}$	Sulfone	$\overset{O}{\overset{\|}{\underset{\underset{O}{\|}}{RSR'}}}$
Ester	$\overset{O}{\overset{\|}{RCOR'}}$	Sulfonic acid	$\overset{O}{\overset{\|}{\underset{\underset{O}{\|}}{RSOH}}}$
Amine	RNH$_2$	Chloride	RCl
Amide	$\overset{O}{\overset{\|}{RCNH_2}}$	Bromide	RBr
Nitrile	RC≡N	Iodide	RI
Isonitrile	$R\overset{+}{N}≡\bar{\bar{C}}$	Organolithium	RLi
Nitro compound	$R\overset{+}{N}\overset{\nearrow O}{\underset{\searrow O^-}{}}$	Heterocycle	R⌒Y
Nitroso compound	RN=O		(Y is an atom othan than carbon such as N, O, Si, P, S, etc.: Y is bonded to R at two or more positions.) Examples:
Imine	$\overset{NH}{\overset{\|}{RCR'}}$		, pyridine;
Azo compound	RN=NR′		
Diazo compound	$RR'C=\overset{+}{N}=\bar{N}$		, thiophene.
Diazonium salt	RN≡N$^+$X$^-$ (anion)		

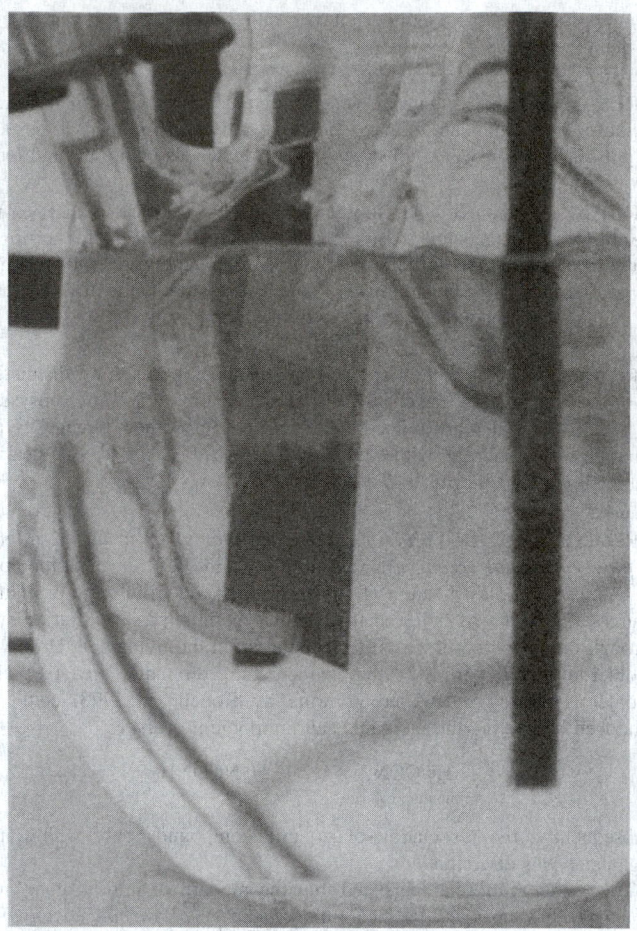

Fig. 1. Apparatus for testing corrosion of metal coated with a plastic. (*The Dow Chemical Company.*)

same ratio as in water (2H:1O), hence the suffix *-hydrate*. Other specialized families include terpenes, alkaloids, steroids, lipids, proteins, enzymes, vitamins, and organometallic compounds.

Although these classifications are important for education and documentation in organic chemistry, research usually is rather specialized.

It is often oriented in a practical way, not to the structure of compounds or to the kinds of atoms they contain, but to the manner in which the compounds are used. A partial list of such uses includes plastics, pharmaceuticals, insecticides, fungicides, herbicides, paints, petroleum, fuels, dyes, photography, and adhesives. See Fig. 1.

An important area of organic chemistry is that which deals with life and living substances. Organized under the title biochemistry, this is a subfield of organic chemistry, since most of the compounds involved contain carbon. Numerous advances have been made recently in biochemistry, and of all the areas of organic chemistry it probably will produce the greatest progress in the next decade. Noteworthy advances can be expected in the areas of biochemistry relating to medicine and human health. Some of the categories along which the study of biochemistry is organized

are proteins, peptides, amino acids, nucleoproteins, enzymes, nucleotides, carbohydrates, lipids, steroids, carotenoids, porphyrins, nucleic acids, vitamins, and hormones. These topics are described in detail elsewhere in this volume.

Theoretical organic chemistry is another field that has progressed rapidly in recent years. Chemists have derived molecular orbital symmetry rules that allow understanding and predicting the stereochemistry and relative rates of organic reactions in electronic ground and excited states. In a ground state molecule, all electrons are in their lowest energy levels, whereas, in an excited state molecule, at least one electron is in a higher energy level. For example, the Woodward-Hoffman orbital symmetry rules for concerted reactions predict that ground state (thermal) cycloaddition reactions involving $4n + 2$ (where n is an integer) π-electrons, and excited state (photochemical) cycloaddition reactions that involve $4n$ π-electrons, may occur via a concerted process (Scheme 1).

A concerted process or reaction occurs without the involvement of an intermediate. The stereochemistry of the reactants is retained in the product, and the reaction is usually more facile than a comparable nonconcerted reaction. The other combinations of ground state, excited state, $4n + 2$, $4n$ reactions cannot be concerted reactions.

Another area that has received increased attention is environmental organic chemistry. Reactions that organic compounds undergo when they are released to the environment are becoming as significant as the reactions by which the compounds are prepared or the reactions that take place in the use of the compounds. Some environmentally important types of reactions are hydrolysis, oxidation, sunlight-initiated photochemical decomposition, and biodegradation by microbes.

Only a limited discussion of the large field of organic chemistry can be given in the space allotted for this article. Consult alphabetical index for further information on specific topics. The reader is also directed to the references at the end of this article for examples of more detailed treatments of organic chemistry. Refs. Schmerling; Richey; and Morrison

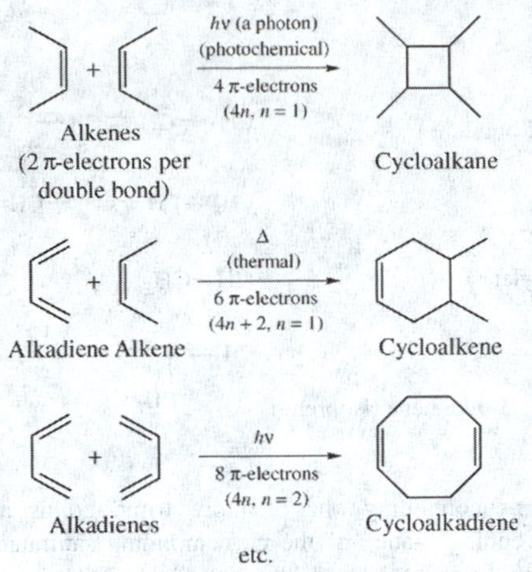

Scheme 1

et al. represent texts that treat organic chemistry at elementary, intermediate, and advanced levels, respectively.

Nomenclature of Organic Compounds

With the foregoing review in mind, the reader will appreciate the difficulties associated with naming organic molecules. Originally, chemical names were indicative of the sources of compounds. For example, *catechol* was the name given to a compound isolated from the natural product *gum catechu*. Chemists have coined other nonsystematic names such as *cubane* or *basketene* to pictorially describe molecules. These nonsystematic names are called common or trivial names. The names *phenol, acetic acid*, and *styrene* are also nonsystematic but are widely understood in chemistry.

As the number of known organic molecules increased, a systematic approach to nomenclature was required. To minimize confusion in communicating chemical information, a name should be consistent with other systems in use and should clearly define the structure of a molecule. Specialists in organic chemistry have developed nomenclatures that are logical for their disciplines, thus devising systems for naming alcohols, antibiotics, carboxylic acids, etc.

Systematic nomenclature on a worldwide scale began in 1892 when a committee of the International Chemical Congress established a set of standards known as the Geneva Rules for naming organic compounds. The International Union of Pure and Applied Chemistry (IUPAC) http://www.iupac.org/dhtml_home.html was formed in 1919 and further developed this nomenclature system. In 1886 in the United States, the American Chemical Society (ACS) established a Committee on Nomenclature. The ACS and IUPAC have developed parallel rules for naming organic compounds.

Alternative rules within the latter system may allow assigning more than one unambiguous name to a compound. The controlled alphabetic listing in the *Chemical Substances Index* of Chemical Abstracts Service (CAS), a part of the ACS, requires that each compound have a unique name. This convention ensures that all information for a single compound such as $H_{2n}-CH_2-CH_2-OH$, which can be unambiguously named 2-aminoethanol, 2-aminoethyl alcohol, 2-hydroxyethylamine, etc., will appear in the index under one name: ethanol, 2-amino-. Because universal nomenclature systems are complex and sometimes inconvenient, some chemists have retained the older methods of naming molecules. These older names are used in the parts of this section that do not deal specifically with nomenclature.

The following paragraphs introduce some basic areas of organic chemical nomenclature. Further details can be found in *Chemical Substances Index Names* and *Nomenclature of Organe Chemistry* listed and end of this entry.

In addition to trivial names, some of the other categories of organic compound names are:

Generic name: one that indicates a class of compounds; e.g., *alkanes, esters*.

Parent name: a base from which other names are derived; e.g., *ethanol*, from *ethane*; *butanoic acid*, from *butane*.

Systematic name: a name composed of syllables defining the structure of a compound; e.g., *chlorobenzene*, 2-methylhexane.

Substituitive name: one describing replacement of hydrogen by a group or element; e.g., 1-methylnaphthalene, 2-chloropropane.

Replacement name: a name describing compounds that have carbon replaced by a hetero atom; also called "a" nomenclature; e.g., 2-azaphenanthrene.

Subtractive name: one that indicates removal of specified atoms; e.g., in the aliphatic series names ending in -*ene* or -*yne*, such as *ethene* or *ethyne*, and names involving anhydro-, dehydro-, deoxy-, nor-, etc.

Additive name: one that signifies addition between molecules and/or atoms without replacement of atoms; e.g., *styrene oxide*.

Conjunctive name: a combination of two names, one of which represents a cyclic structure and the other an acyclic chain, with one hydrogen atom removed from each; e.g., *benzenemethanol*.

Fusion name: a combination that results from linking with an "o" two names of cyclic systems fused by two or more common atoms; e.g., *benzofuran*.

Multiplicative name: nomenclature describing the symmetrical repetition of radicals about a central unit; e.g., 2, 2′-oxybis (*ethanol*).

Hantzsch-Widman name: a name devised by Hantzsch and Widman for describing heterocyclic systems, in which the prefix denotes a hetero atom(s) and the suffix denotes the ring size and degree of saturation; e.g., *oxirene, aziridine*.

Von Baeyer name: a name that describes alicyclic bridged systems; e.g., *bicyclo* [2.2.1] *heptane*.

The procedure for naming a compound involves some or all of the following steps, depending on the structure of the molecule under study: (1) the type of nomenclature to be used (conjunctive, multiplicative, etc.) is chosen; (2) the parent structure is named; (3) the prefixes, suffixes, and names of functional and substituent groups that were not included in (2) are attached; (4) the numbering is completed.

Hydrocarbons

Acyclic Hydrocarbons. A knowledge of the structural features of hydrocarbon skeletons is basic to the understanding of organic chemical nomenclature. The generic name of saturated acyclic hydrocarbons, branched or unbranched, is *alkane*. The term *saturated* is applied to hydrocarbons containing no double or triple bonds.

The simplest saturated acyclic hydrocarbon is called methane. Names of the higher, straight-chain (*normal*) homologs of this series contain the termination "-ane," as shown in Table 3. The structures of the first four members of the series in Table 3 are: CH_4, CH_3-CH_3, $CH_3-CH_2-CH_3$, and $CH_3-CH_2-CH_3$. The structures of subsequent members of this series are formed by inserting additional CH_2 units.

TABLE 3. EXAMPLES OF SATURATED ACYCLIC HYDROCARBONS

Molecular Formula	Name	Molecular Formula	Name
CH_4	methane	$C_{18}H_{38}$	octadecane
C_2H_6	ethane	$C_{19}H_{40}$	nonadecane
C_3H_8	propane	$C_{20}H_{42}$	icosane
C_4H_{10}	butane	$C_{21}H_{44}$	henicosane
C_5H_{12}	pentane	$C_{22}H_{46}$	docosane
C_6H_{14}	hexane	$C_{23}H_{48}$	tricosane
C_7H_{16}	heptane	$C_{30}H_{62}$	triacontane
C_8H_{18}	octane	$C_{31}H_{64}$	hentriacontane
C_9H_{20}	nonane	$C_{32}H_{66}$	dotriacontane
$C_{10}H_{22}$	decane	$C_{40}H_{82}$	tetracontane
$C_{11}H_{24}$	undecane	$C_{50}H_{102}$	pentacontane
$C_{12}H_{26}$	dodecane	$C_{100}H_{202}$	hectane
$C_{13}H_{28}$	tridecane	$C_{101}H_{204}$	henhectane
$C_{14}H_{30}$	tetradecane	$C_{102}H_{206}$	dohectane
$C_{15}H_{32}$	pentadecane	$C_{110}H_{222}$	decahectane
$C_{16}H_{34}$	hexadecane	$C_{120}H_{242}$	icosahectane
$C_{17}H_{36}$	heptadecane	$C_{132}H_{266}$	dotriacontahectane
		$C_{200}H_{402}$	dictane

Univalent groups derived from the preceding acyclic hydrocarbons by removal of one hydrogen atom from a terminal carbon atom are named by replacing the ending "-ane" with "-yl."

EXAMPLES:

$$\text{ethyl} \quad CH_3-CH_2-$$
$$\text{butyl} \quad CH_3-CH_2-CH_2-CH_2-$$

A saturated branched acyclic hydrocarbon is named by numbering the longest chain from one end to the other, and the positions of the side chains are indicated by the lowest possible numbers. The numbers precede the group, and are separated from them by a hyphen.

EXAMPLES:

$$\overset{1}{C}H_3-\overset{2}{C}H-\overset{3}{C}H_2-\overset{4}{C}H_2-\overset{5}{C}H_3$$
$$\underset{CH_3}{|}$$

2-methylpentane (not 4-methylpentane)

$$\overset{6}{C}H_3-\overset{5}{C}H-\overset{4}{C}H_2-\overset{3}{C}H-\overset{2}{C}H-\overset{1}{C}H_3$$
$$\;\;\;\;|\;\;\;\;\;\;\;\;\;\;\;\;\;\;|\;\;\;\;\;|$$
$$\;\;\;CH_3\;\;\;\;\;\;\;CH_3\;\;CH_3$$

2,3,5-trimethylhexane (not 2,4,5-trimethylhexane)

If two groups are attached to the same carbon atom, the number is repeated.

EXAMPLES:

$$\overset{}{\underset{}{CH_3}}$$
$$\overset{1}{C}H_3-\overset{2}{C}H_2-\overset{3}{C}-\overset{4}{C}H_2-\overset{5}{C}H_3$$
$$\underset{CH_3}{|}$$

3,3-dimethylpentane (not 3-dimethylpentane)

For some purposes, such as alphabetical listing of basic skeleton names, inverted word order is used.

EXAMPLES:

pentane, 2-methyl-
benzene, chloro-

The two propyl groups are distinguished by calling them *normal*-propyl or *n*-propyl, $CH_3-CH_2-CH_2-$, and isopropyl or i-propyl, $[(CH_3)_2-CH-]$, in common usage. The latter is called 1-methylethyl in systematic nomenclature. The butyl groups are named as follows:

Structure	Systematic Name	Trivial name
$CH_3-CH_2-CH_2-CH_2-$	butyl	*n*-butyl
$CH_3-CH-CH_2-$ $\quad\quad\vert$ $\quad\;\;CH_3$	2-methylpropyl	*i*-butyl
CH_3-CH_2-CH- $\quad\quad\quad\quad\vert$ $\quad\quad\quad\;CH_3$	1-methylpropyl	*s*-butyl (*s = secondary*)
CH_3 $\vert$ CH_3-C- $\vert$ CH_3	1,1-dimethylethyl	*t*-butyl (*t = tertiary*)

Hydrocarbons that contain one or more double bonds are called "unsaturated," and are named by replacing the ending "-ane" of the corresponding saturated hydrocarbon with the ending "-ene," "-adiene," or "-atriene," etc. The generic names of unsaturated hydrocarbons are *alkene, alkadiene, alkatriene*, etc. The double bonds receive the lowest possible numbers.

In the following examples, the names listed in the *Chemical Substances Index* of the CAS system are given first. Other names are given in parentheses.

EXAMPLES:

2-butene (2-butylene)
$$\overset{1}{C}H_3-\overset{2}{C}H=\overset{3}{C}H-\overset{4}{C}H_3$$

1,4-hexadiene
$$\overset{1}{C}H_2=\overset{2}{C}H-\overset{3}{C}H_2-\overset{4}{C}H=\overset{5}{C}H-\overset{6}{C}H_3$$

ethene (ethylene)
$$CH_2=CH_2$$

2-methyl1-1,3-butadiene (isoprene)
$$\overset{1}{C}H_2=\overset{2}{C}-\overset{3}{C}H=\overset{4}{C}H_2$$
$$\underset{CH_3}{|}$$

Hydrocarbons containing one or more triple bonds are named by replacing the ending "-ane" of the corresponding saturated hydrocarbon with the ending "-yne," "-adiyne," "-atriyne" etc. The triple bonds receive the lowest possible numbers. Double bonds take precedence over triple bonds when there is a choice in numbering.

EXAMPLES:

1-butyne
$$\overset{1}{C}H\equiv\overset{2}{C}-\overset{3}{C}H_2-\overset{4}{C}H_3$$

1-hexane-3,5-diyne
$$\overset{1}{C}H_2=\overset{2}{C}H-\overset{3}{C}\equiv\overset{4}{C}-\overset{5}{C}\equiv\overset{6}{C}H$$

Unsaturated branched acyclic hydrocarbons are numbered in the same manner as alkanes. The longest chain is chosen as the parent. If the alkene or alkyne contains two or more chains of equal length, the chain containing the maximum number of double bonds is chosen as the parent.

Univalent or multivalent groups derived from alkenes and alkynes are named as follows:

EXAMPLES:

ethenyl (vinyl)	$\overset{2}{C}H_2=\overset{1}{C}H-$
ethynyl	$\overset{2}{C}H\equiv\overset{1}{C}-$
2-propynyl	$\overset{3}{C}H\equiv\overset{2}{C}-\overset{1}{C}H_2-$
methylidyne	$CH\equiv$
ethylidene	$CH_3-CH=$
ethylidyne	$CH_3-C\equiv$

Alicyclic Hydrocarbons. Saturated monocyclic hydrocarbons, or "cycloalkanes," are named by attaching the prefix "cyclo" to the name of the acyclic unbranched alkane.

EXAMPLES:

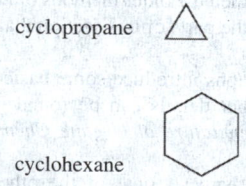

cyclopropane

cyclohexane

Univalent groups derived from unsubstituted cycloalkanes are named *cyclopropyl, cyclohexyl*, etc., in a manner analogous to that used for naming acyclic alkanes. The carbon atom with the free valence is numbered as 1.

EXAMPLES:

cyclopropyl

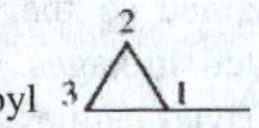

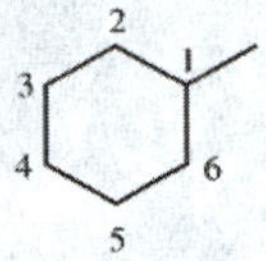

cyclohexyl

Unsaturated monocyclic hydrocarbons are named by substituting "-ene," "-adiene," "-atriene," "-yne," "-adiyne," etc., in the name of the corresponding cycloalkane. Double and triple bonds are given numbers as low as possible.

EXAMPLES:

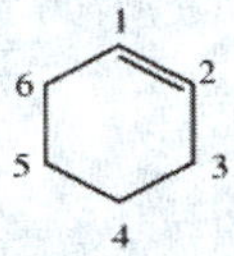

cyclohexene

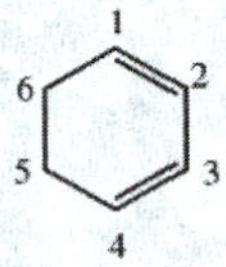

1,3-cyclohexadiene

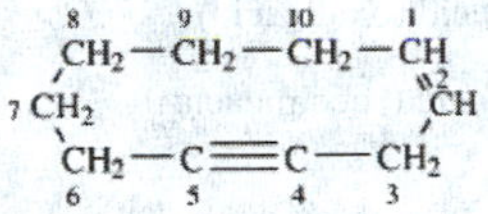

$CH_2-CH_2-CH_2-CH$
$CH_2 \qquad CH$
$CH_2-C\equiv C-CH_2$

1-cyclodecen-4-yne

Aromatic Hydrocarbons. Aromatic hydrocarbons generally are considered those which have the characteristic chemical properties of benzene. Many such compounds are known more commonly by their trivial names than by their systematic names.

EXAMPLES:

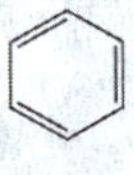

benzene

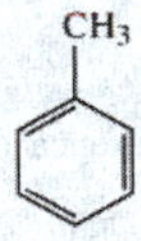

CH_3

methylbenzene (toluene)

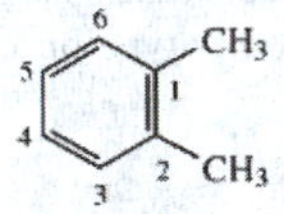

CH_3
CH_3

1,2-dimethylbenzene (*o*-xylene)

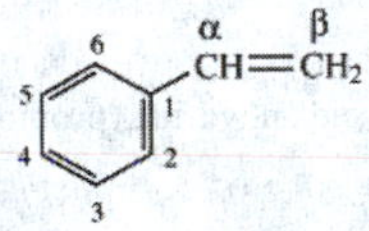

$\alpha \quad \beta$
$CH=CH_2$

ethenylbenzene (styrene)

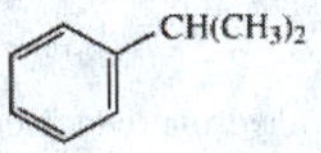

$CH(CH_3)_2$

(1-methylethyl)benzene (cumene)

The terms *ortho*, *meta*, and *para* (abbreviated *o*, *m*, and *p*) refer to the location of substituents on the benzene ring and are equivalent to 1,2-, 1,3-, and 1,4-substitution in systematic nomenclature, respectively. The lowest numbers possible are given to substituents.

Fused aromatic systems are named by prefixing the largest parent trivial names with combining forms such as benz(o)- and naphth(o)-. Hydrocarbons that contain five or more fused benzene rings in a linear arrangement are named from a numerical prefix followed by "-acene."

EXAMPLES:

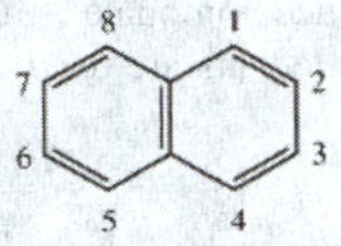

naphthalene

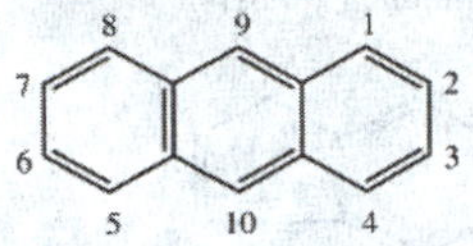

anthracene

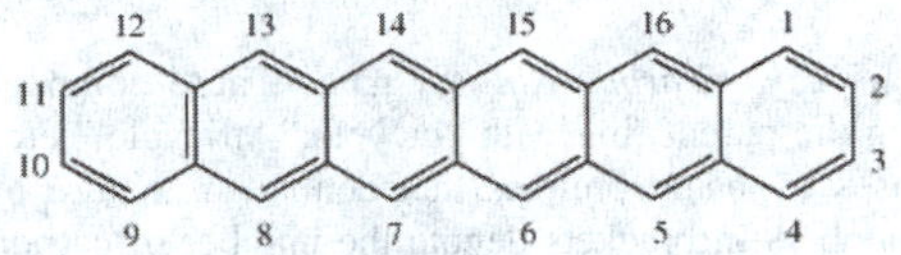

hexacene

Fusion prefixes are used to designate to which side of the parent hydrocarbon a substituent ring is attached.

EXAMPLES:

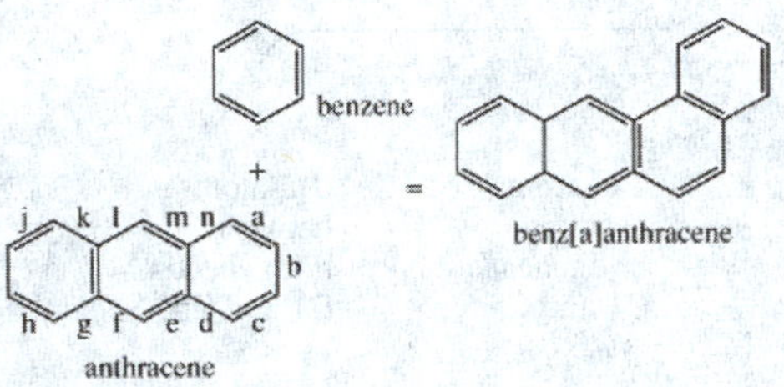

benzene

+

anthracene

=

benz[a]anthracene

Hydrogenation products of complex aromatic ring systems that are not treated as alicyclic hydrocarbons are named by prefixing "dihydro," "tetrahydro," etc., to the parent name. The lowest locants are used. "Perhydro" is used in trivial nomenclature to indicate a fully hydrogenated compound.

EXAMPLES:

1,2-dihydronaphthacene

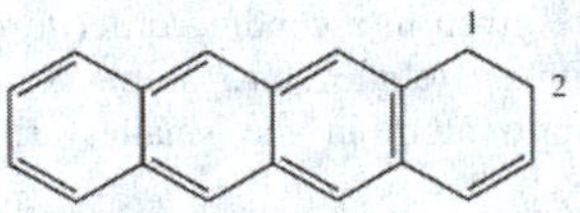

docosahydropentacene (perhydropentacene)

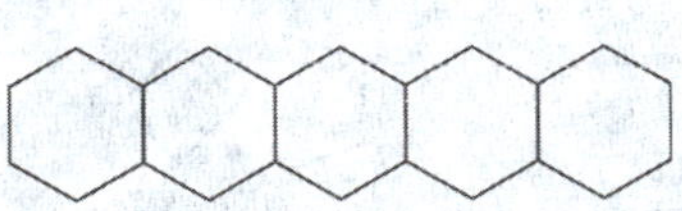

Multiple unsubstituted assemblies of benzene rings are named by using the appropriate prefix with the radical name "phenyl."

EXAMPLES:

1, 1'-biphenyl

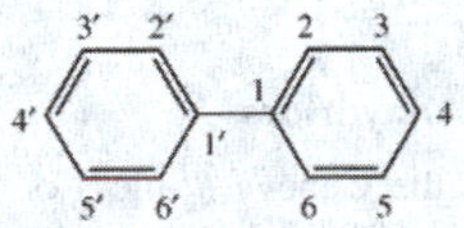

1, 1':4', 1''-terphenyl (*p*-terphenyl)

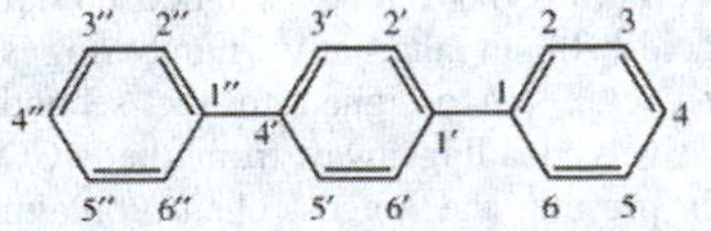

Indicated hydrogen in aromatic systems is assigned to angular or non-angular positions when needed to accommodate structural features in systematic nomenclature. The lowest locants are used.

EXAMPLE:

H-indene (not 3*H*-)

Bridged Hydrocarbon Ring Systems. These compounds are named by prefixing the parent ring with "bicyclo," "tricyclo," etc., or in some complex cases by using prefixes that denote the nature of the bridges. The three numbers in brackets denote the number of carbon atoms in each of the three bridges in descending order.

EXAMPLES:

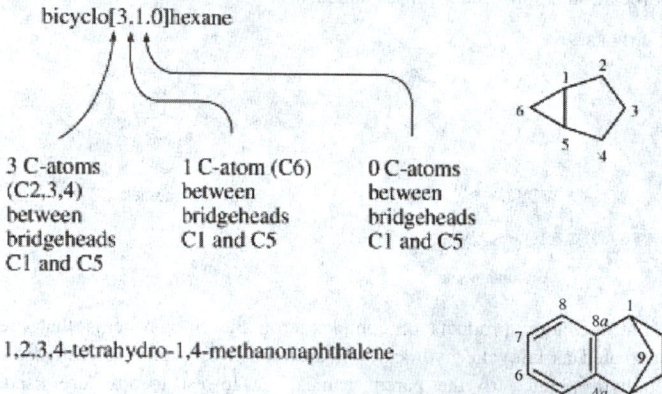

bicyclo[3.1.0]hexane

| 3 C-atoms (C2,3,4) between bridgeheads C1 and C5 | 1 C-atom (C6) between bridgeheads C1 and C5 | 0 C-atoms between bridgeheads C1 and C5 |

1,2,3,4-tetrahydro-1,4-methanonaphthalene

Spiro Hydrocarbon Ring Systems. Spiro systems contain pairs of rings or ring systems that have only one atom (a "spiro atom") in common. The name of the simplest monospiro system is formed by prefixing the acyclic hydrocarbon name with "spiro" and numerals separated by periods. The numerals are given in ascending order to define the number of atoms in each ring linked to the spiro atom. Numbering begins at the atom next to the spiro atom in the smallest ring for monospiro systems.

EXAMPLES:

spiro[3,4]octane

dispiro[5.1.7.2]heptadecane

Carboxylic Acids and Their Anhydrides

Acids are named according to the Geneva ("-oic") or "-carboxylic" system. They are regarded as derived from parent hydrocarbons having the same number of carbon atoms so that CH_3 is replaced by COOH. The carbon atom of the carboxyl group is assigned number one in aliphatic monocarboxylic acids. In an alternative numbering system, the Greek letters *alpha, beta,* etc., are assigned to the second, third, etc., carbon atoms, respectively, leading away from the −COOH group. When chain branching is present, the longest chain containing the carboxylic

acid group at one end is chosen for naming the molecule. Unsaturated aliphatic acids are named so the longest chain includes the maximum number of unsaturated linkages. Double bonds are given preference over triple bonds. Trivial names are retained for some common molecules.

EXAMPLES:

formic acid	HCOOH
acetic acid	$\overset{2}{C}H_3\overset{1}{C}OOH$
hexanoic acid (caproic acid)	$CH_3(CH_2)_4COOH$
octadecanoic acid (stearic acid)	$CH_3(CH_2)_{16}COOH$
2-propenoic acid (acrylic acid)	$\overset{3}{C}H_2{=}\overset{2}{C}H\overset{1}{C}OOH$
2-methylbutanoic acid	$\overset{4}{C}H_3\overset{3}{C}H_2\overset{2}{C}H\overset{1}{C}OOH$ with CH_3
cyclobutanecarboxylic acid	—COOH (cyclobutane)
benzoic acid	—COOH (benzene)
ethanedioic acid (oxalic acid)	HOOC–COOH
butanedioic acid (succinic acid)	HOOC–CH₂CH₂–COOH
1,2-benzenedicarboxylic acid (phthalic acid)	benzene with two COOH

Conjunctive nomenclature may be used for naming cyclic acids. It is applied to any ring system attached by a single bond to one or more acyclic hydrocarbon chains, each of which bears only one principal functional group.

EXAMPLE:

cyclohexaneacetic acid — CH₂COOH

Acid anhydride names are formed from systematic, Geneva, trivial, conjunctive, or other type of acid names.

EXAMPLES:

propanoic acid anhydride (propionic anhydride)   CH₃CH₂CO–O–CH₃CH₂CO

benzoic acid anhydride (benzoic anhydride)

See also **Carboxylic Acids**.

Alcohols

Monohydric alcohols are named by adding "-ol" to a molecular skeleton name. Carbon chains, unsaturation, etc., are numbered in a manner analogous to that used for carboxylic acids. (See preceding section.)

EQUATIONS:

methanol (methyl alcohol) CH_3OH

cyclohexanol (cyclohexyl alcohol)

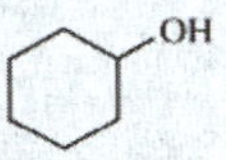

2-propen-1-ol (allyl alcohol) $\overset{3}{C}H_2=\overset{2}{C}H\overset{1}{C}H_2OH$

See also **Alcohols**.
The simplest aromatic hydroxy compound is called phenol:

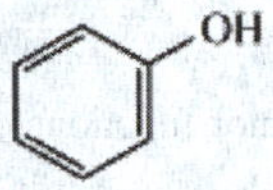

Esters

Simple esters are named on the basis of their alcohol and acid functions. Carbon chains, unsaturation, etc., are numbered in a manner analogous to that used for carboxylic acids.

EXAMPLES:

ethyl acetate (inverted name: acetic acid, ethyl ester) $CH_3COOC_2H_5$

propyl 2-butenoate $\overset{4}{C}H_3\overset{3}{C}H=\overset{2}{C}H\overset{1}{C}OOCH_2CH_2CH_3$

methyl benzoate

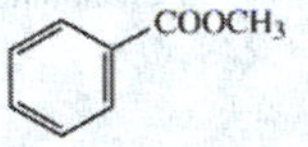

See also **Esters**.

Ethers

Alkoxy compounds are commonly called ethers. In current CAS nomenclature they are named as derivatives of functional parent compounds, hydrocarbons, etc., by use of "oxy" radicals.

EXAMPLES:

1, 1'-oxybis(ethane) (ethyl ether or diethyl ether) $C_2H_5OC_2H_5$

(hexyloxy)cyclopropane (cyclopropyl hexyl ether) $O(CH_2)_5CH_3$

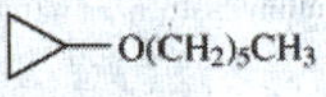

methoxybenzene (anisole)

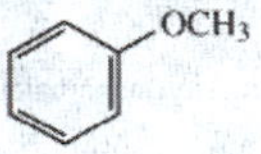

See also **Ethers**.
Aldehydes are named from the corresponding acids by use of "-carboxaldehyde" and "-al" suffixes.

EXAMPLES:

acetaldehyde CH_3CHO

2-butenal $\overset{4}{C}H_3\overset{3}{C}H=\overset{2}{C}H\overset{1}{C}HO$

3-methylbenzaldehyde

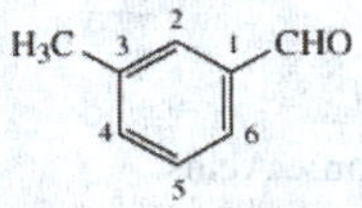

See also **Aldehydes**.

Ketones

Ketones are named by use of the characteristic suffix "-one."

EXAMPLES:

2-propane (acetone) $\overset{1}{C}H_3\overset{2}{C}O\overset{3}{C}H_3$

1-hexen-1-one (butylketene) $CH_3(CH_2)_3\overset{2}{C}H=\overset{1}{C}O$

4-cyclopentyl-2-butanone $\overset{4}{C}H_2\overset{3}{C}H_2\overset{2}{C}O\overset{1}{C}H_3$

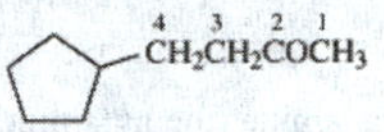

diphenylmenthanone (benzophenone)

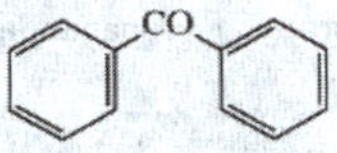

See also **Ketones**.

Peroxides

Simple peroxides are named as follows:

EXAMPLES:

ethyl methyl peroxide $C_2H_5OOCH_3$

benzoyl peroxide

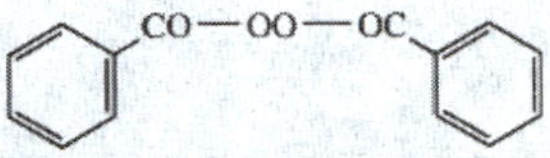

Halogenated Compounds

Hydrocarbons, esters, etc., which have one or more hydrogen atoms replaced by a halogen atom are named so that the substituents have the lowest possible numbers. When multiple functions are present, they are numbered according to precedences established by IUPAC or CAS nomenclature rules. Both trivial and systematic nomenclatures are used.

EXAMPLES:

chloromethane (methyl chloride) CH_3Cl

2-iodobutane (*sec*- butyl iodide) $\overset{1}{C}H_3\overset{2}{C}H\overset{3}{C}H_2\overset{4}{C}H_3$ with I

1-bromo-2-fluorobenzene

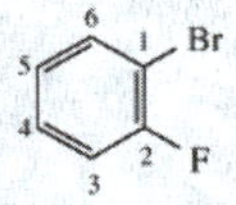

3-chloropentanoic acid(β-chlorovaleric acid) $\overset{5}{C}H_3\overset{4}{C}H_2\overset{3}{C}H\overset{2}{C}H_2\overset{1}{C}OOH$ with Cl

chloroethane (vinyl chloride) $CH_2=CHCl$

1,1-dichloroethene (vinylidenechloride) $CH_2=CCl_2$

tetrabromomethane (carbon tetrabromide) CBr_4

Acid halides are named as follows:

EXAMPLES:

acetyl chloride CH_3COCl

6-heptenoyl chloride $\overset{7}{C}H_2=\overset{6}{C}H(CH_2)_4\overset{1}{C}OCl$

cyclohexanecarbonyl bromide

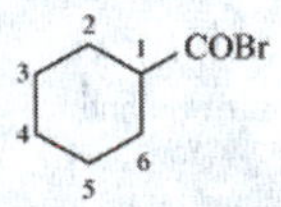

benzoyl fluoride

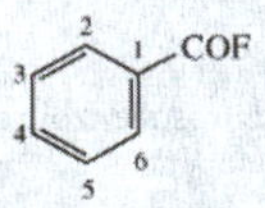

See also **Chlorinated Organics**.

Nonheterocyclic Nitrogen Compounds

Amines are named by adding the suffix "-amine" either to the name of the hydrocarbon or to the hydrocarbon radical. A second system names all amines as derivatives of primary amines.

EXAMPLES:

methanamine (methylamine) CH_3NH_2

N,*N*-dipropyl-1-propanamine (tripropylamine) $(CH_3CH_2CH_2)_3N$

benzenamine (aniline)

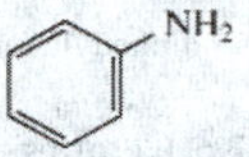

N-ethylcyclohexanamine (*N*-ethylcyclohexylamine)

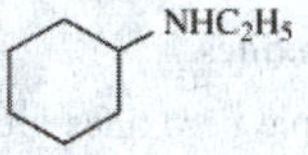

Imines are named from the hydrocarbon by the addition of the suffix "-imine."

EXAMPLES:

ethanimine $\overset{2}{C}H_3\overset{1}{C}H{=}NH$

2,4-cyclopentadien-1-imine

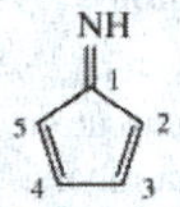

See also **Amines**.

Names of amides are based on the corresponding acids. Thus, "-oic acid" becomes "-amide," and "-carboxylic acid" becomes "-carboxamide."

EXAMPLES:

acetamide $\overset{2}{C}H_3\overset{1}{C}ONH_2$

benzamide

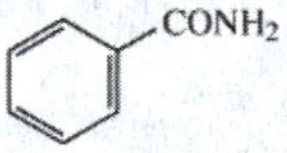

cyclohexanecarboxamide

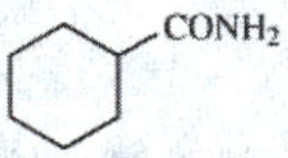

N-methylpentadecanamide $\overset{15}{C}H_3(CH_2)_{13}\overset{1}{C}ON\overset{N}{H}CH_3$

N,*N*-dimethyl-2,4-pentadienamide $\overset{5}{C}H_2{=}\overset{4}{C}H\overset{3}{C}H{=}\overset{2}{C}H\overset{1}{C}O\overset{N}{N}(CH_3)_2$

See also **Amides**.

Nitro and nitroso compounds are named so that the substituents have the lowest possible numbers.

EXAMPLES:

2-nitrobutane $\overset{1}{C}H_3\overset{2}{C}H(NO_2)\overset{3}{C}H_2\overset{4}{C}H_3$

4-nitrosobenzoic acid

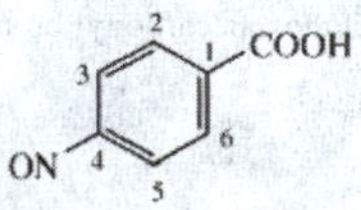

See also **Nitro- and Nitroso-Compounds**

Nitrile names are formed from common names of carboxylic acids.

EXAMPLES:

acetonitrile CH_3CN

benzoinitrile (phenyl cyanide)

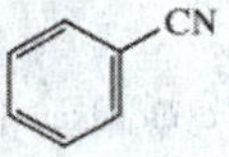

3-butenenitrile $\overset{4}{C}H_2{=}\overset{3}{C}H\overset{2}{C}H_2\overset{1}{C}N$

ethenetetracarbonitrile (tetracyanoethylene)

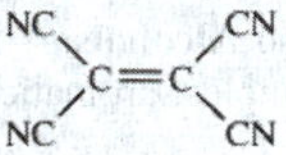

In the presence of more senior functional groups, the nitrile function is expressed by the prefix "cyano."

EXAMPLES:

4-cyanobenzamide

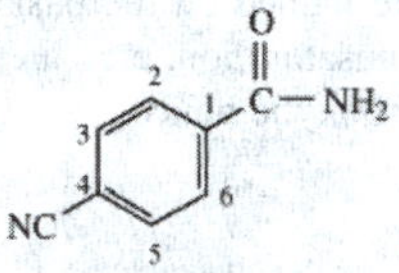

3-cyanobutanoic acid $\overset{4}{C}H_3\overset{3}{C}H(CN)\overset{2}{C}H_2\overset{1}{C}OOH$

Nonheterocyclic Sulfur Compounds

Sulfur compounds are named similarly to oxygen compounds.

EXAMPLES:

(methylthio)benzene (methyl phenyl sulfide)

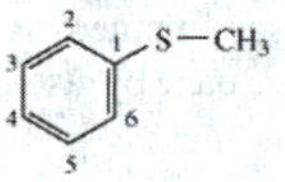

1,1′-sulfinylbis(benzene) (diphenyl sulfoxide)

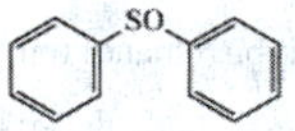

2-(propylsulfonyl)naphthalene (2-naphthyl propyl sulfone)

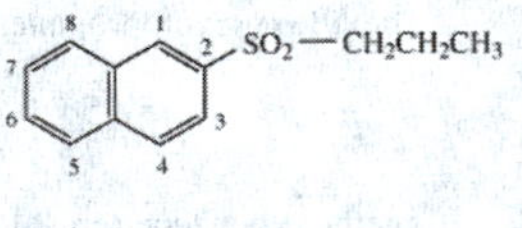

2-butanethiol (2-mercaptobutane) $\overset{1}{C}H_3\overset{2}{C}H(SH)\overset{3}{C}H_2\overset{4}{C}H_3$

2-propanethione (thioacetone) $CH_3C({=}S)CH_3$

N-methylbenzenesulfonamide

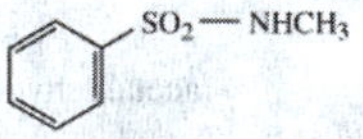

benzenesulfonic acid

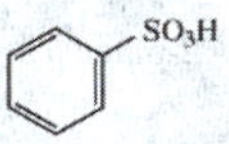

See also **Sulfonic Acids**.

Heterocyclic Compounds

Some of the common heterocyclic nitrogen, oxygen, and sulfur compounds and their numbering systems are shown:

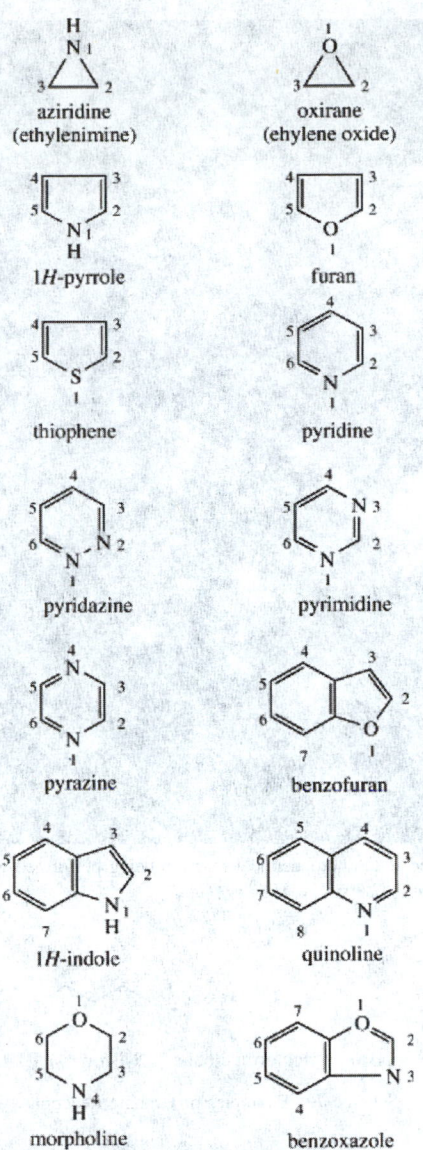

aziridine
(ethylenimine)

oxirane
(ehylene oxide)

1H-pyrrole

furan

thiophene

pyridine

pyridazine

pyrimidine

pyrazine

benzofuran

1H-indole

quinoline

morpholine

benzoxazole

Organic Reaction Mechanisms and Processes

Many reactions occur in which one organic compound is converted into another. The molecular details of the intermediate steps by which compounds are converted into new products are called reaction mechanisms. The four broad classes of reaction mechanisms are: cationic, anionic, free radical, and multicenter processes in which neither charged species nor odd electron species is involved. Examples of each type will be given, but many variations can exist within each type. Also, varying degrees of sophistication exist in our knowledge of the exact reaction pathways that organic compounds follow. The examples discussed show only the major steps involved.

Cationic and anionic mechanisms involve species that have either positive or negative charges, respectively (heterolytic reactions). An example of a reaction that proceeds via a cationic mechanism is the hydrolysis of t-butyl bromide (Scheme 2).

$$(CH_3)_3CBr + H_2O \longrightarrow (CH_3)_3COH$$
t–Butyl bromide   Water   t–Butyl alcohol

$$+ (CH_3)_2C=CH_2 + HBr$$
Isobutylene   Hydrogen bromide

Scheme 2

Mechanism:

$$(CH_3)_3CBr \rightarrow (CH_3)_3C^+ + Br^-$$

$$(CH_3)_3C^+ + H_2O \rightarrow (CH_3)_3COH + H^+$$

$$(CH_3)_3C^+ \rightarrow (CH_3)_2C=CH_2 + H^+$$

The addition of methanol to methyl acrylate in the presence of sodium methoxide is an example of a reaction that proceeds by an anionic mechanism (Scheme 3).

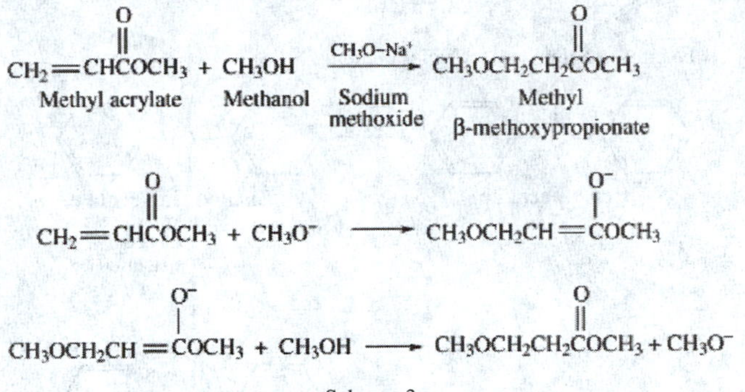

$$CH_2=CHCOCH_3 + CH_3OH \xrightarrow{CH_3O-Na^+} CH_3OCH_2CH_2COCH_3$$
Methyl acrylate   Methanol   Sodium methoxide   Methyl β-methoxypropionate

$$CH_2=CHCOCH_3 + CH_3O^- \longrightarrow CH_3OCH_2CH=COCH_3$$

$$CH_3OCH_2CH=COCH_3 + CH_3OH \longrightarrow CH_3OCH_2CH_2COCH_3 + CH_3O^-$$

Scheme 3

Mechanism:
Free radical (homolytic) reactions involve species with an unpaired electron. The ultraviolet-light-initiated reaction of methane with chlorine is an example (Scheme 4).

$$CH_4 + Cl_2 \xrightarrow{h\nu} CH_3Cl + HCl$$
Methane Chlorine   Methylchloride Hydrogen chloride

$$Cl_2 \xrightarrow{h\nu} 2Cl\cdot$$

$$Cl\cdot + CH_4 \longrightarrow CH_3\cdot + HCl$$

$$CH_3\cdot + Cl_2 \longrightarrow CH_3Cl + Cl\cdot$$

Scheme 4

A number of reactions do not seem to belong to any of the above mechanistic types. Such processes are referred to as multicenter reactions. The Diels-Alder cycloaddition reaction of 1,3-butadiene with maleic anhydride is an example (Scheme 5). No charged or odd electron intermediates seemingly are involved in this reaction.

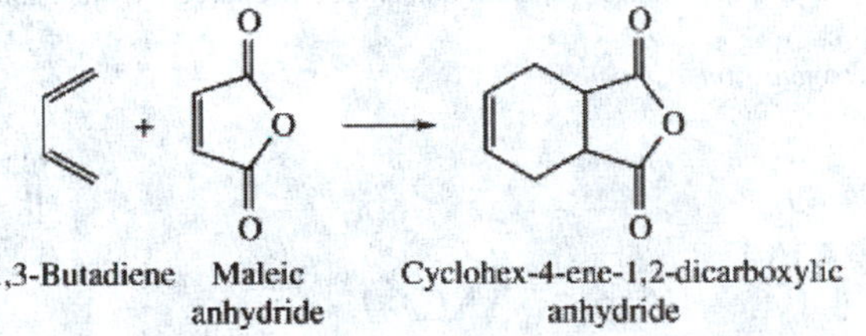

1,3-Butadiene   Maleic anhydride   Cyclohex-4-ene-1,2-dicarboxylic anhydride

Scheme 5

The reactions shown in Schemes 2–5, with the exception of the photodissociation of Cl_2 to Cl atoms (Scheme 4), occur in molecules that are in electronic ground states. Reactions also can occur in molecules existing in excited electronic states. Commonly, these excited states are produced by irradiating the reactants with ultraviolet or visible light, hence the term *photochemistry*. When a molecule is in an excited state, its reactions are often different from those it normally exhibits in its ground state. An organic compound often can exist in more than one excited state, as shown in Scheme 6.

Excited state molecules usually differ from their ground state counterparts by having dissimilar electronic and geometric configurations, and shorter lifetimes, e.g., 10^{-2} to 10^{-10} second. See Fig. 2.

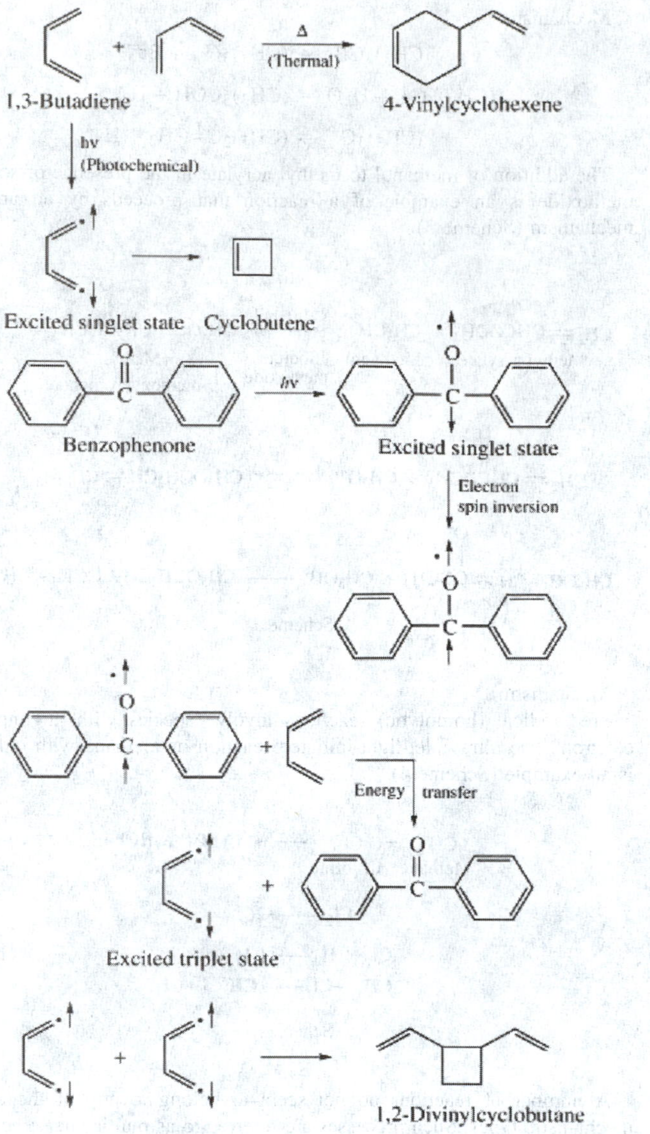

1,3-Butadiene + (diene) →Δ (Thermal)→ 4-Vinylcyclohexene

↓ hv (Photochemical)

Excited singlet state   Cyclobutene

Benzophenone →hv→ Excited singlet state

↓ Electron spin inversion

↓ (triplet benzophenone)

Benzophenone + (diene) →Energy transfer→

Excited triplet state +   Benzophenone

+ → 1,2-Divinylcyclobutane

Scheme 6

A few additional examples of organic reactions are shown in Scheme 7. Many novel and complex compounds can be prepared by these and similar reactions.

Hydrogenation

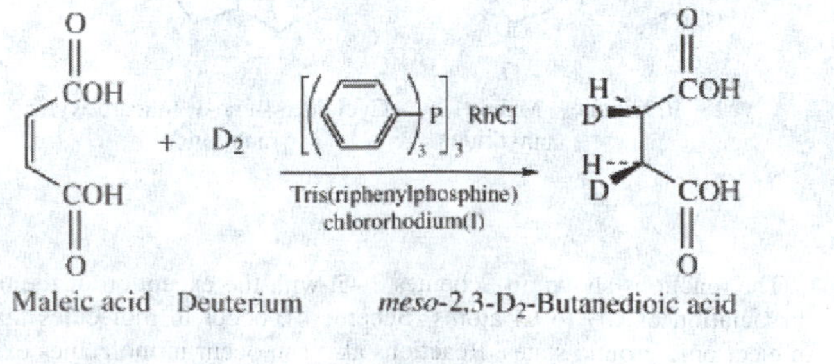

Maleic acid + Deuterium + [(C₆H₅)₃P]₃RhCl Tris(triphenylphosphine) chlororhodium(I) → *meso*-2,3-D₂-Butanedioic acid

Carbene Addition

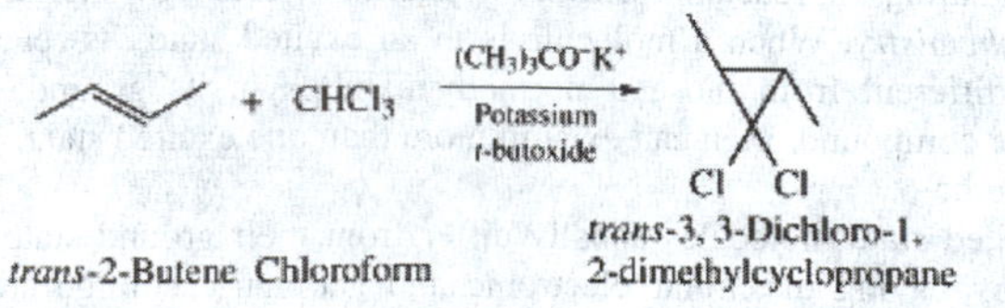

trans-2-Butene + Chloroform →(CH₃)₃CO⁻K⁺ Potassium *t*-butoxide→ *trans*-3,3-Dichloro-1, 2-dimethylcyclopropane

Fig. 2.   Excited-state photochemical reaction being performed with a laser light source. (*The Dow Chemical Company.*)

Free Radical Addition

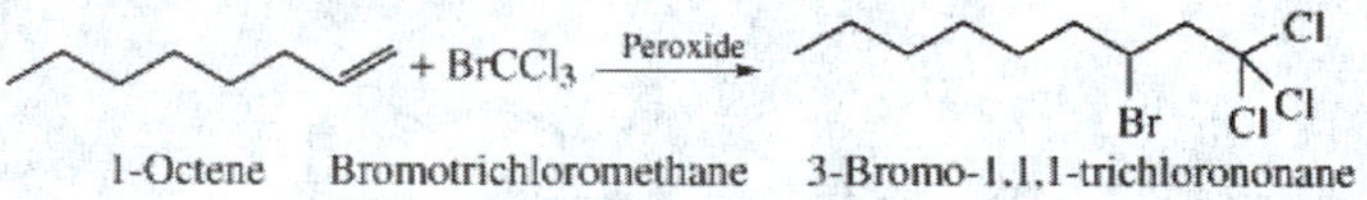

+ BrCCl₃ →Peroxide→

1-Octene   Bromotrichloromethane   3-Bromo-1,1,1-trichlorononane

Scheme 7.   Examples of Organic Reactions.

Oxidative Cleavage

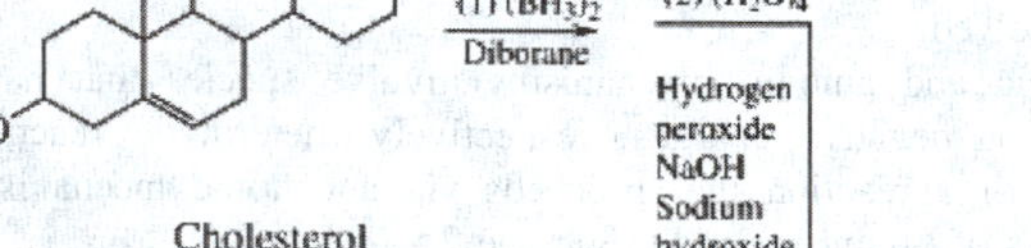

→(1) O₃ Ozone→ →(2) Zn Zinc H₂O→

trans-3-Hexene   Propionaldehyde

Hydroboration-Oxidation

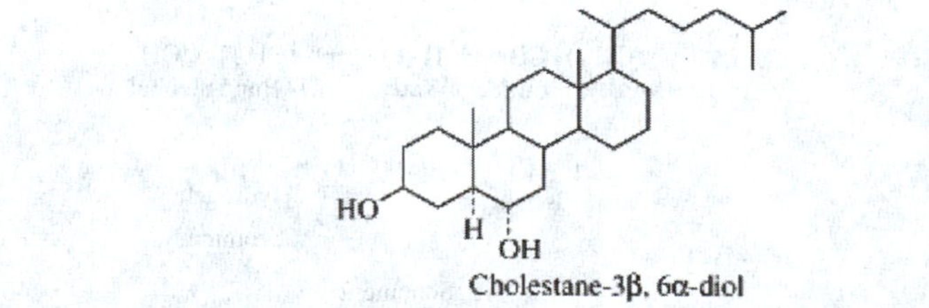

Cholesterol →(1) (BH₃)₂ Diborane→ →(2) (H₂O)₄ Hydrogen peroxide NaOH Sodium hydroxide→ Cholestane-3β, 6α-diol

Oxidation

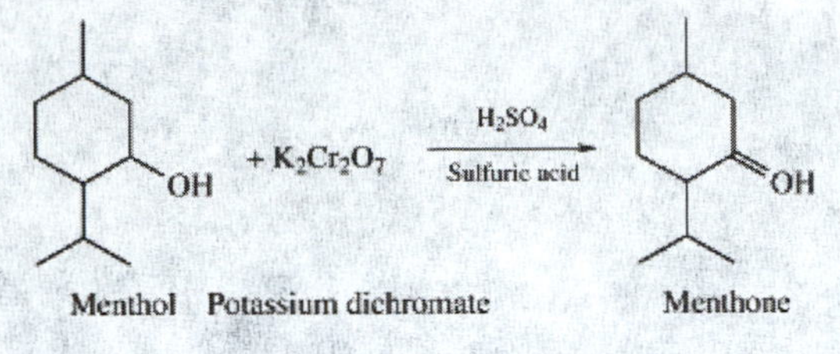

OH
+ K2Cr2O7
H2SO4
Sulfuric acid
OH
Menthol
Potassium dichromate
Menthone

Reduction

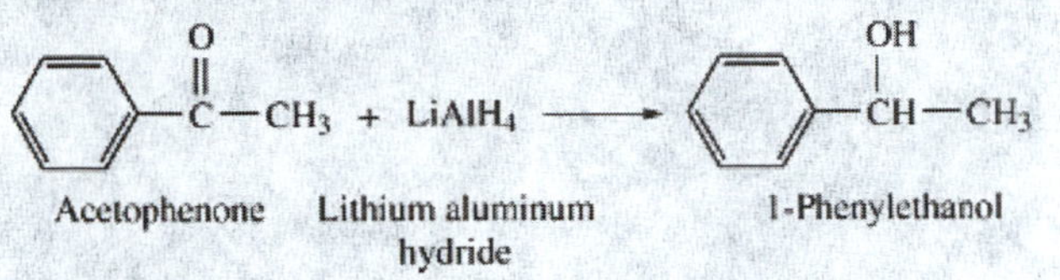

O
C—CH3
+ LiAlH4
OH
CH—CH3
Acetophenone
Lithium aluminum hydride
1-Phenylethanol

Chlorination-Amination

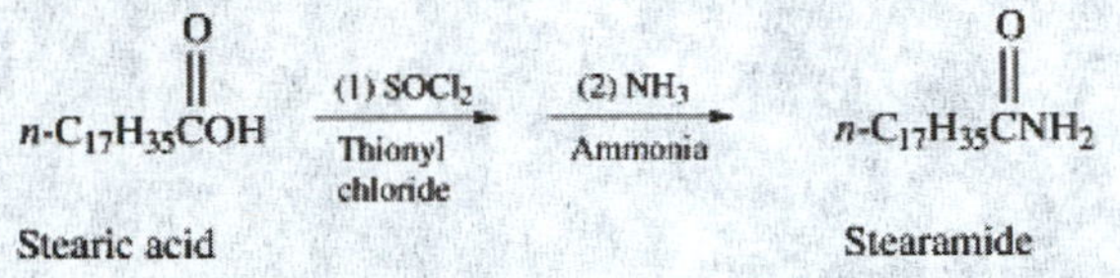

O
n-C17H35COH
(1) SOCl2
Thionyl chloride
(2) NH3
Ammonia
O
n-C17H35CNH2
Stearic acid
Stearamide

Substitution

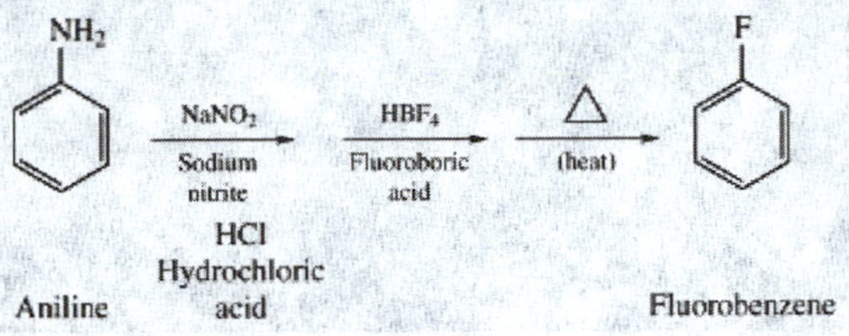

NH2
NaNO2
Sodium nitrite
HBF4
Fluoroboric acid
(heat)
F
HCl
Hydrochloric acid
Aniline
Fluorobenzene

Benzyne Cycloaddition

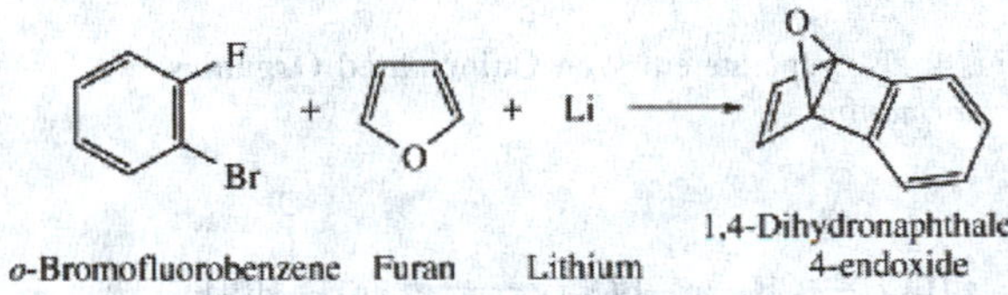

F
Br
+
O
+ Li
O
o-Bromofluorobenzene
Furan
Lithium
1,4-Dihydronaphthalene-1, 4-endoxide

Photoisomerization

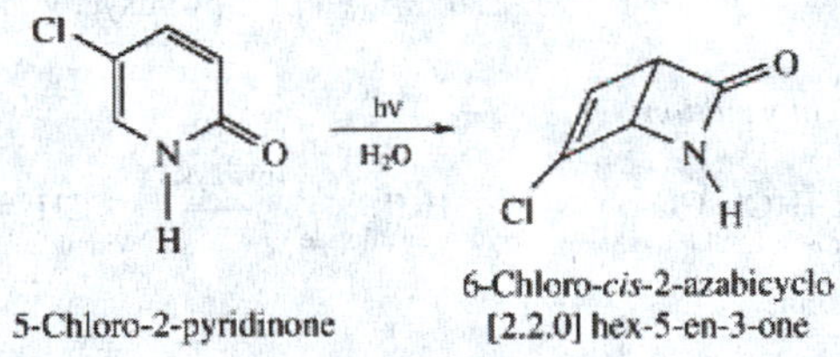

Cl
N
O
H
hv
H2O
O
N
Cl
H
5-Chloro-2-pyridinone
6-Chloro-cis-2-azabicyclo [2.2.0] hex-5-en-3-one

Many organic reactions are referred to by the inventor's or discoverer's name. A few name reactions are shown in Scheme 8.

Friedel-Crafts Alkylation

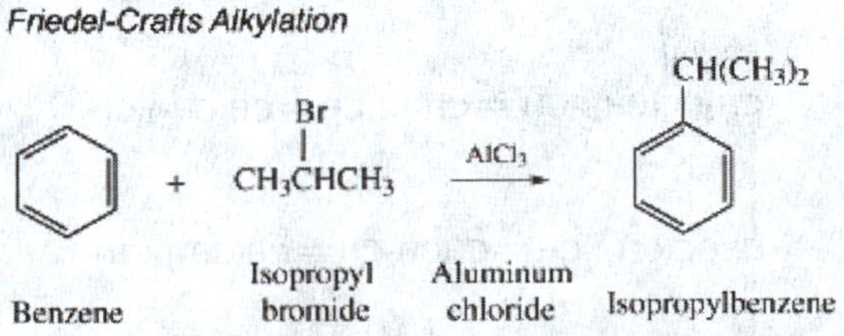

+
Br
CH3CHCH3
AlCl3
CH(CH3)2
Benzene
Isopropyl bromide
Aluminum chloride
Isopropylbenzene

Scheme 8. Examples of Organic Name Reactions.

See also **Friedel-Crafts Reaction**.

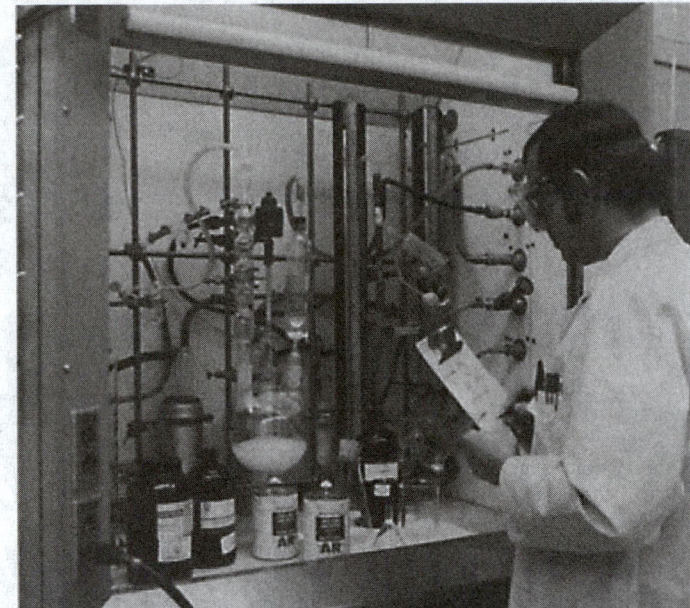

Fig. 3. Typical laboratory apparatus showing preparation of 2-phenyl-2-decanol by a Grignard reaction. (*The Dow Chemical Company*.)

Grignard Reaction (Fig. 3).

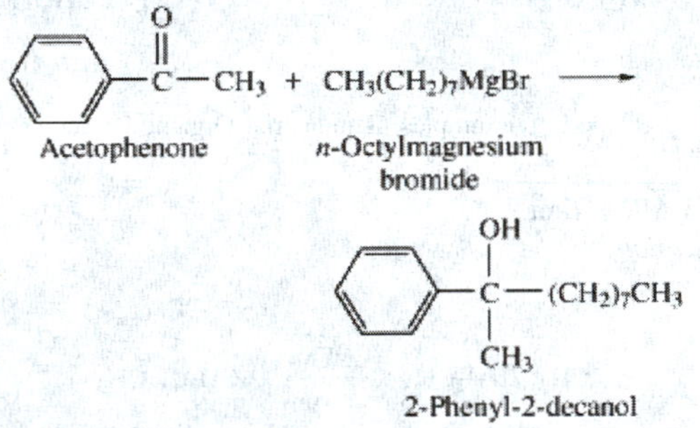

O
C—CH3
+ CH3(CH2)7MgBr
Acetophenone
n-Octylmagnesium bromide
OH
C—(CH2)7CH3
CH3
2-Phenyl-2-decanol

See also **Grignard Reactions**.

Baeyer-Villiger Oxidation

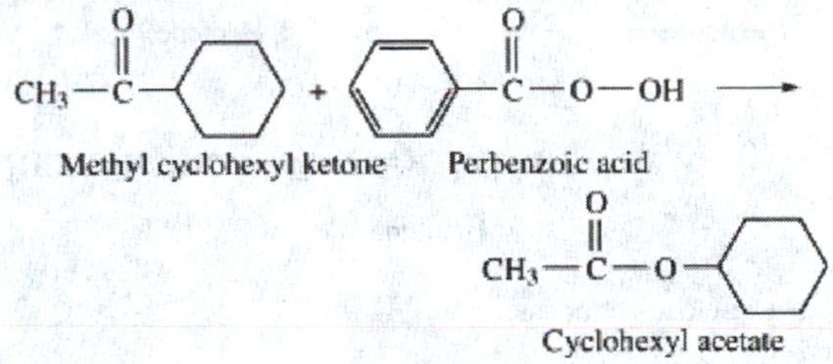

O
CH3—C
+
O
C—O—OH
Methyl cyclohexyl ketone
Perbenzoic acid
O
CH3—C—O
Cyclohexyl acetate

Meerwein-Ponndorf-Verley Reduction

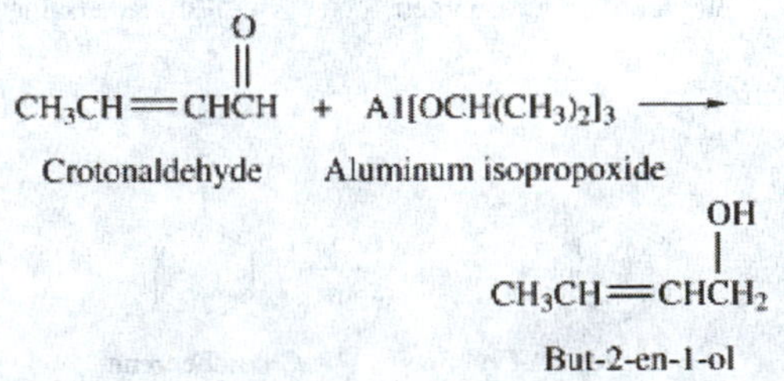

O
CH3CH=CHCH
+ Al[OCH(CH3)2]3
Crotonaldehyde
Aluminum isopropoxide
OH
CH3CH=CHCH2
But-2-en-1-ol

Skraup Quinoline Synthesis

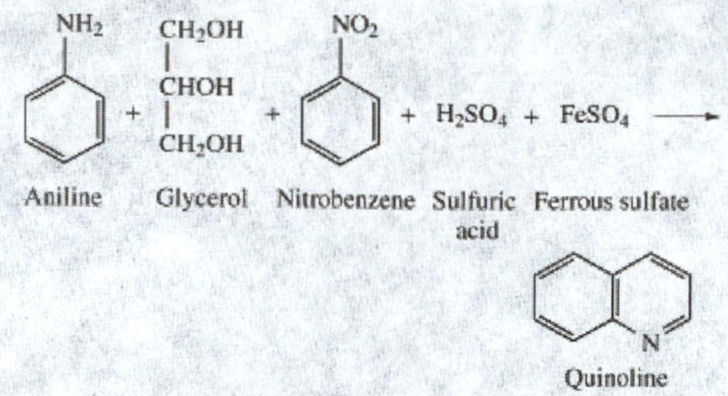

Aniline + Glycerol + Nitrobenzene + Sulfuric acid + Ferrous sulfate $\longrightarrow$

Quinoline

A very active area of organic chemistry is the synthesis of complex natural products. In these syntheses, numerous reactions, of which those in Schemes 7 and 8 are examples, are often employed serially to convert a starting compound into a final product that occurs in nature.

These syntheses are useful because they serve to verify structure that has been assigned, provide an alternate source of the compound if a larger supply is needed, or provide a route to derivatives or analogs of the natural material. The derivatives may possess enhanced properties, such as biological activity, that are not found in the natural product.

Some important industrial organic processes are shown in Scheme 9. Although most of these reactions involve mixtures of isomers or homologs as reactants and products, for simplicity only the major components are shown.

Alkylation

$$(CH_3)_3CH + CH_2=CHCH_3 \xrightarrow[\text{catalyst}]{HF} CH_3CH=CHCH_2CH_3$$

Isobutane   Propylene   2,3-Dimethylpentane

Scheme 9.   Examples of Industrial Organic Processes.

See also **Alkylation**.

Isomerization

$$CH_3(CH_2)_2CH_3 \xrightarrow[\text{catalyst}]{AlCl_3} CH_3CHCH_3$$

n-Butane   *i*-Butane

Cracking

$$CH_3(CH_2)_{14}CH_3 \xrightarrow[\text{catalyst}]{\text{Silica alumina}} CH_3(CH_2)_4CH=CH_2$$

n-Hexadecane   1-Heptene

$$+ CH_3CH_2CH=CH_2 + CH_3CH_2CHCH_3$$

1-Butene   Isopentane

See also **Cracking Process**.

Oxidation

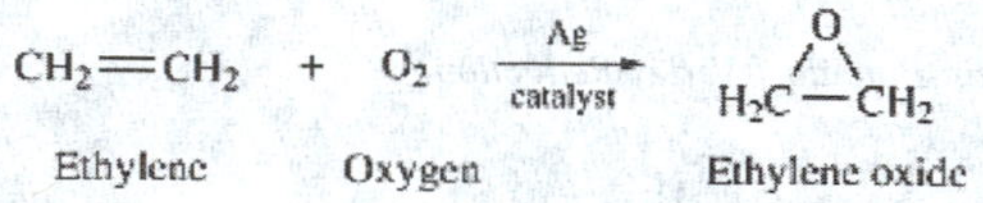

$$CH_2=CH_2 + O_2 \xrightarrow[\text{catalyst}]{Ag} H_2C-CH_2$$

Ethylene   Oxygen   Ethylene oxide

Chlorination

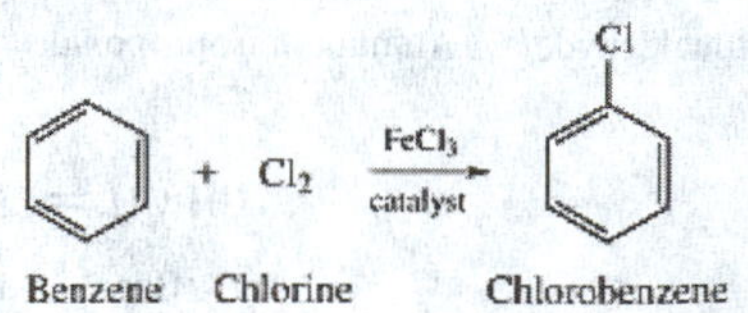

Benzene + Cl$_2$ $\xrightarrow[\text{catalyst}]{FeCl_3}$ Chlorobenzene

Benzene   Chlorine   Chlorobenzene

Fig. 4.   Typical industrial reactor for the preparation of chlorobenzene by the chlorination of benzene. (*The Dow Chemical Company.*)

See Fig. 4; also separate entry on **Chlorinated Organics**.

Hypochlorination

$$CH_3CH=CH_2 + HOCl \longrightarrow CH_3CHCH_2Cl$$

Propylene   Hypochlorous acid   1-Chloro-2-hydroxypropane

Dehydrochlorination

$$CH_3CF_2CL + HCL \xrightarrow{\text{pyrolysis}} CH_2=F_2$$

1-chloro-1,1-difluroethane   Hydrogenchloride   Vinylideneflouride

Bromination

$$CH_2=CH_2 + Br_2 \rightarrow CH_2BrCH_2Br$$

Ethylene   Bromine   Ethylenedibromide

Hydrogenation

$$CH_2OC(CH_2)_7CH=CHCH_2CH=CH(CH_2)_4CH_3$$
$$CHOC(CH_2)_7CH=CHCH_2CH=CH(CH_2)_4CH_3$$
$$CH_2OC(CH_2)_7CH=CHCH_2CH=CH(CH_2)_4CH_3$$

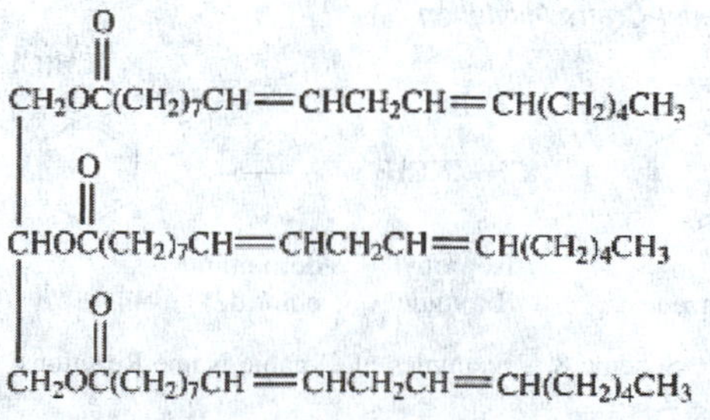

Trilinolein

Fermentation

Sucrose →(*Aspergillus niger* (a microorganism)) Citric acid

Coupling

p-Sulfobenzenediazonium chloride + Sodium 2-naphtholate →

Orange II (a dye)

Sulfonation

Naphthalene + H$_2$SO$_4$ → α-Naphthalenesulfonic acid

Sulfuric acid

Hydrolysis

CH$_2$=CHCH$_2$Cl$^+$ NaOH → CH$_2$=CHCH$_2$OH
Allyl chloride   Sodium hydroxide   Allyl alcohol

Ammonolysis

p-Chloronitrobenzene + NH$_3$ → *p*-Nitroaniline

Ammonia

See also **Amination**.

Hydration

CH$_2$=CH$_2$ + H$_2$SO$_4$ + H$_2$O → CH$_3$CH$_2$OH
Ethylene   Sulfuric acid   Water   Ethanol

+H$_2$ →(Ni catalyst) Tristearin

Hydrogen

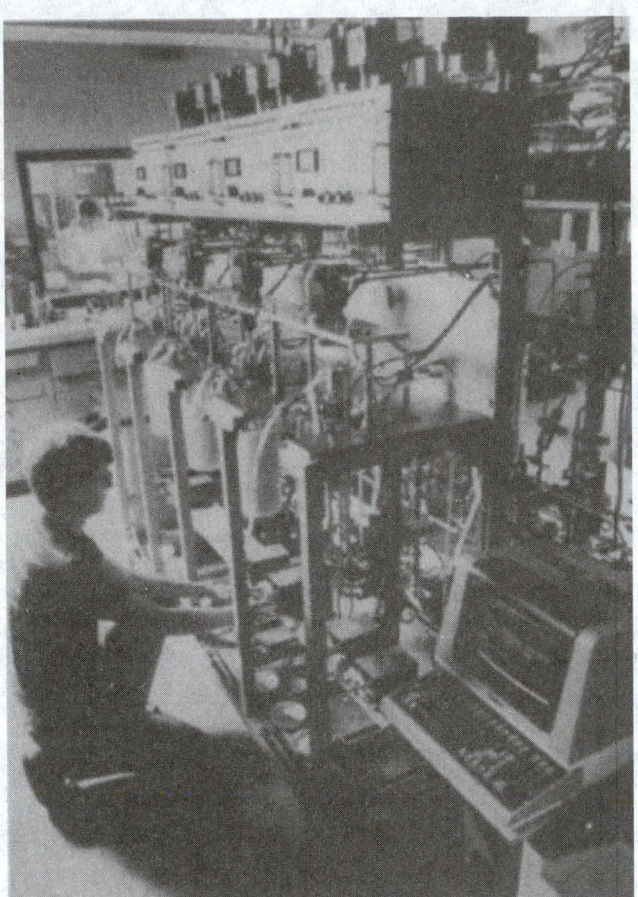

Fig. 5.   Laboratory fermentation process equipment. (*The Dow Chemical Company.*)

See Fig. 5 and also the separate entry on **Fermentation**.

Hydrogenolysis

CH$_3$(CH$_2$)$_{10}$COCH$_3$ + H$_2$ →(Copper-chromic oxide catalyst) CH$_3$(CH$_2$)$_{11}$OH
Methyl laurate   Hydrogen   Lauryl alcohol

Dehydrogenation

⬡—CH$_2$CH$_3$ →(Iron-chromic oxide catalyst) ⬡—CH=CH$_2$
Ethylbenzene   Styrene

Nitration

Toluene + HNO$_3$ + H$_2$SO$_4$ → 2,4,6-trinitrotoluene (TNT)

Nitric acid Sulfuric acid

See also **Nitration**.

Hydroformylation

$$CH_3CH=CH_2 + CO + H_2 \longrightarrow CH_3(CH_2)_2CH$$

Propylene   Carbon Hydrogen   *n*-Butyraldehyde
             monoxide

Esterification

Phthalic anhydride Octanol   Dioctyl phthalate

$$+ C_8H_{17}OH \xrightarrow[\text{catalyst}]{H_2SO_4}$$

Vinyl Polymerization

$$CH_2=CHCl \xrightarrow[\text{initiator}]{K_2S_2O_8} (CH_2CHCl)_{\sim 1,000-10,000}$$

Vinyl chloride   Polyvinyl chloride

Condensation Polymerization

$$HOC(CH_2)_4COH + H_2N(CH_2)_6NH_2$$

Adipic acid   Hexamethylenediamine

$$\downarrow \Delta \text{ (heat)}$$

$$HO\left[CH(CH_2)_4CNH(CH_2)_6NH\right]_{\sim 50-90} H$$

Poly(hexamethyleneadipamide)
[Nylon 66]

Additional Reading

Bickford, M., J.I. Kroschwitz, M. Howe-Grant: *Kirk-Othmer Encyclopedia of Chemical Technology, Concise*, 4th Edition, John Wiley & Sons, Inc., New York, NY, 2003.

Chemical Substance Index Names: Section IV from the Chemical Abstracts Index Guide, American Chemical Society, Columbus, Ohio, 1985 — and references cited therein.

Green, M.M., and H.A. Wittcoff: *Organic Chemistry Principles and Industrial Practice*, John Wiley & Sons, Inc., New York, NY, 2003.

Gokel, G.W.: *Dean's Handbook of Organic Chemistry*, 2nd Edition, The McGraw-Hill Companies, Inc., New York, NY, 2003.

Graham Solomons, T.W.: *Fundamentals of Organic Chemistry*, John Wiley and Sons, Inc., New York, NY, 1999.

Graham, Solomons T.W., and C.B. Fryhle: *Organic Chemistry, Study Guide*, 8th Edition, John Wiley & Sons, Inc., New York, NY, 2003.

Hellwinkel, D.: *Systematic Nomenclature in Organic Chemistry: A Directory to Comprehension and Application on Its Basic Principles*, Springer-Verlag, Inc., New York, NY, 2001.

Hoffman, R.V.: *Organic Chemistry: An Intermediate Text*, 2nd Edition, John Wiley & Sons, Inc., Hoboken, NJ, 2004.

McMurry, J.: *Fundamentals of Organic Chemistry*, Thomson Learning Publications, Fresno, CA, 1997.

McMurry, J. and M. Castellion: *Fundamentals of Organic and Biological Chemistry*, 2nd edition, Prentice-Hall, Inc., Upper Saddle River, NJ, 1998.

Morrison, R.T. and R.N. Boyd: *Organic Chemistry*, 6th Edition, Prentice-Hall, Inc., Upper Saddle River, NJ, 1992.

Nomenclature of Organic Chemistry; Sections A, B, C, D, E, F and H combined, Pergamon Press, Oxford, 1979. (IUPAC nomenclature.)

Richey, H.G., Jr.: *Fundamentals of Organic Chemistry*, Pentrice-Hall, Inc., Upper Saddle River, NJ, 1983.

Schmerling, L.: *Organic and Petroleum Chemistry for Nonchemists*, Penn-Well Publishing, Tulsa, OK, 1981.

Smith, M.B. and J. March: *March's Advanced Organic Chemistry: Reactions, Mechanisms, and Structure*, 5th Edition, John Wiley & Sons, Inc., New York, NY, 2001.

Wade, L.G. Jr.: *Organic Chemistry*, 4th Edition, Prentice-Hall, Inc., Upper Saddle River, NJ, 1998.

Weissermel, K., and Hans-Jurgen Arpe: *Industrial Organic Chemistry*, 4th Edition, John Wiley & Sons, Inc., New York, NY, 2003.

ORGANIC LIGHT-EMITTING DIODES (OLEDs). The general phenomenon of *electroluminescence* (EL) encompasses many processes, including that of common inorganic light-emitting diodes (LED) devices, their organic counterparts, and even cathode ray tubes (CRTs), vacuum fluorescent displays (VFDs), and field emission displays (FEDs) (In the last three cases, the EL process is called *cathodoluminescence*). The term organic light-emitting diode (OLED) refers specifically to light-emitting diode devices (with p-n junctions) made from organic materials. See also **Inorganic Light-Emitting Diode Displays**; and **Cathode-Ray Tube**.

The term *organic electroluminescence* EL is often used incorrectly to refer to an OLED device; organic EL in the true sense of the term does not operate like a diode, because there is no p-n junction. (Instead, light is emitted from within an active layer, wherein the applied electric field has raised various transition metal dopants to excited energy levels, which emit light as they relax. For example the inorganic EL device uses manganese as a dopant in a zinc sulfide host, while organic EL devices have an organic base or even a transition metal-organic compound as the active layer.) The OLED devices currently being pursued, and which are the main subject here, are true diodes.

The ubiquitous device known as the LED (light-emitting diode), based on inorganic semiconducting materials, was introduced commercially in 1969. In LED devices, electrons and holes are electrically injected into a p-n junction device, where they recombine and emit light; this same general principle applies in an OLED device, except that organic materials are used in place of semiconductors. The emission of light from organic materials subjected to an electric field has been known since at least the early 1960s. OLED technology dates back to work on conducting polymers in England in the early 1970s and at the University of Pennsylvania in 1977 on "synthetic metals." Concerted efforts in industry did not occur until the end of the 1980s, fueled primarily by a convergence of positive developments. New materials, processes, and electronics have been developed; more researchers have entered the field; and new prototypes have shown promising performance. These factors clearly point to the likelihood of viable commercial OLED products in the near future. Such products have the potential to offer very attractive features: monolithic light emission (no backlight required) over a large area, a low voltage, and full color.

Unlike inorganic LED devices, OLED devices are created with transparent organic materials using large-area, thick-film deposition processes. There is no need to grow a crystal, saw it, polish it, or dope it. It is an inherently large-area process, not a chip-oriented process. And, unlike inorganic LED devices, which must be mounted as individual lamps to create a large-area, high information content display, OLED devices can be made on a single, solid substrate. Most OLED devices are fabricated on a transparent anode (indium tin oxide, or ITO) layer, which is itself deposited on a glass substrate. One type of device (the earliest) uses small molecules that are vacuum-deposited to form a hole transport layer, on top of which a layer of emissive material and metallic cathode are deposited. An alternative approach uses fluorescent polymers such as poly (*p*-phenylenevinylene) (PPV) and polyfluorene, prepared in solution and spin-casted.

Ink jet printing techniques have been used to pattern polymeric OLED devices.

Steady technical progress has been made in both the small-molecule and polymer approaches. A major advance in small-molecule OLED devices was announced in a 1985 U.S. patent from researchers at the Eastman Kodak Company[1]. Since that time, researchers at Kodak have amassed a very significant patent portfolio so that Kodak has become the dominant intellectual property holder in small-molecule OLED devices. Many companies have licensed these patents and have gone on to make further technical breakthroughs.

Equally dramatic progress in polymer-based OLED devices was announced by researchers at Cambridge University[2] in 1990; in 1993 they[3] patented this technology. Devices based on this technology consist of a series of polymer layers deposited by spin- or knife-coating. Ownership of this and subsequent patents was transferred to Cambridge Display Technology (CDT), which is now the major intellectual property player in polymer OLED devices.

In 1997 a 256×64 pixel, monochrome car radio display, using technology licensed from Kodak was made. In 1997 and 1998 Color passive matrix OLED devices in a 320×240 pixel format was demonstrated Monochrome, active matrix, 320×240 pixel prototypes were produced. In total, more than 80 companies are investigating OLED devices.

Laboratory devices based on small-molecule and polymer technologies have excellent technical performance today. Neither technology seems to have a major performance edge over the other. However, the early commercial products perform rather modestly compared to what has been accomplished in the laboratory, where, for example, small full-color televisions have been demonstrated. Many major companies have moved rapidly to start or expand research and development programs and prototype product development. This spreading interest in OLED technology stems from the perceived potential advantages that it offers:

Flexible substrates (displays shaped to product design; continuous coating); emissive devices (back lights and color filters not needed); ease of fabrication; design latitude; design simplicity; fast switching speed (video display capability); low weight; high brightness (useful in bright environments); wide color gamut; low voltage (compatible with TFT active matrix drivers); monolithic light emission with 140 to 160° viewing angle (vertical and horizontal); and rugged.

Although the results so far have been achieved quite rapidly, at least from the perspective of historical flat panel display development, there is still much work to be done. Lifetime, particularly packaging for long life, remains a key issue with these technologies. The techniques for making large, full-color video displays are just in the concept stage. Also, there are a number of business challenges relating to displacing the cathode ray tube (CRT) and liquid crystal display (LCD) with such a different technology as the OLED device. These areas of future concern are as follows: Operating lifetimes are too short; operation in extreme conditions has not been proved; improvements in color and video capability needed; people are unfamiliar with the technology; manufacturing infrastructure not yet established; and claimed that it will be inexpensive, but this is unproven.

Many researchers feel confident that when enough resources are available, these problems will be solved. Manufacturing any radically new type of display is sure to be a much longer process than originally planned, but it is rare that so many companies with similar technologies show positive preliminary results so quickly.

As noted previously, OLED devices emit light through the recombination of holes and electrons in a junction region. This physical phenomenon lends itself to a number of basic device designs, which are discussed in this section.

Monochromatic OLED Devices

The simplest organic LED device that one might construct is shown schematically in Figure 1. This structure contains one type of hole transport layer and one type of electron transport layer, so it is monochromatic. Three layers are shown as constituting the light-emitting heart of the device: hole and electron transport layers and a recombination zone, shown in black. All of these three elements need not be present. The recombination region need not be a separate material but could lie in either of the transport layers. Further, the mobility of one of the injected charges could be significantly higher than the mobility of the other. This would allow the use of a single charge transport layer, with the recombination occurring near one of the electrodes. Devices on the today make use of these possibilities.

Although the specific materials and fabrication processes for this type of device will be discussed later, it is worth noting a few features at this time. The cathode is normally an opaque metal, so the light is emitted through the transparent indium tin oxide (ITO) anode and substrate. In Figure 1 the cathode and anode are drawn such that they both differ in size from the transport layers in order to illustrate one of the marked advantages of OLED devices. Because the in-sheet electrical conductivity of the transport layers is very low (organic materials typically conduct far more poorly than inorganic semiconductors), light will be emitted only from the region under the cathode in Figure 1. A patterned cathode gives patterned light emission, with each pixel being electrically addressable

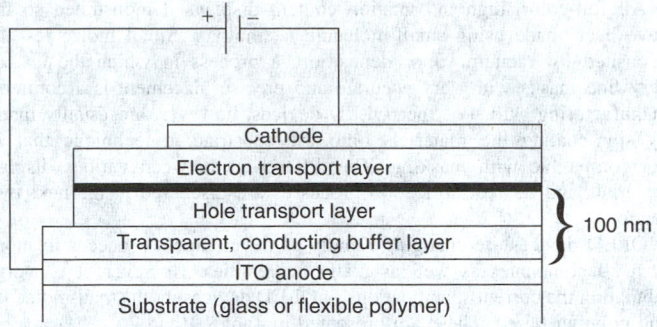

Fig. 1.   Monochromatic OLED Device Structure.

independently. Further, the techniques for depositing the various layers in the device can yield uniform film properties over significant area. These two facts taken together mean that it is surprisingly easy to fabricate extended OLED devices, which can function as monochrome displays (with centimeter-scale diagonal dimensions in today's products, and tens of centimeters diagonal in the laboratory). Patterned cathodes can be deposited through a mask in a single step; one need not tile a large number of point LEDs together to obtain the display.

White OLED devices can be considered monochromatic because they emit "one" color of light — white (which is, of course, composed of all colors). They can be made in a variety of ways, of which the easiest is doping a hole or electron transporting polymer layer with red, green, and blue fluorescent dyes chosen in concentrations so that the emission is white. Very bright white OLED devices have been made in this fashion.

Full-color OLED devices, which can produce a broad gamut of colors by combining different proportions of red, green, and blue light, present much larger technical challenges. Thus, the first product marketed by Pioneer was a monochrome green display. Newer generations of Pioneer's products are displays with separate red, green, and blue areas. They are not full-color; they are tri-color. The issues can be appreciated by examining Figure 2, which is a schematic representation of a single full-color pixel. To span the entire color spectrum, it is necessary to use separate blue, green, and red subpixels, for which the voltage must be controlled individually. The materials in each subpixel differ from each other, at least in doping and probably in the small molecule or polymer itself. This means that the depositions of the three light-emitting layers cannot be done simultaneously; they must be done sequentially through masks. For large-area displays (such as computer monitors and televisions), manufacturing problems can arise with the registration of individual masks to high positional accuracy. Several companies have demonstrated full-color, 5-inch-diagonal passive matrix displays, and Sanyo has demonstrated a full-color active matrix display, but commercial products are not yet available.

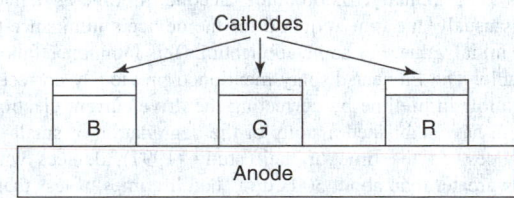

Fig. 2.   Full-Color OLED Pixel.

An approach proposed by Universal Display Corp., Princeton, New Jersey, (*http://www.universaldisplay.com/*) is to locate three primary color devices in a vertical stack in order to produce full color. The SOLED, or Stacked OLED, device could improve pixel density (resolution) when it is eventually implemented in a manufacturing line. The potential drawback is that light from the bottommost device must pass through two other devices on its way to the viewer's eye. An alternative structure used in some color TFEL systems is to pair up red and green on one level and to stack the blue on top in a two-layer device.

All full-color, high information content displays demonstrated so far have been made using small-molecule technology. Small molecules are deposited by vacuum vapor deposition, a process in which the use of very fine masks with very accurate and precise placement is a common manufacturing skill. Polymer OLED devices, however, are usually made by spin-coating the materials onto the substrate, a technique that is not compatible with masking. The differences between various display materials and device fabrication processes are discussed in the next two sections.

OLED devices are changing rapidly as development proceeds in more than 80 companies as well as many universities. However, it is worth tabulating the current specifications of OLED devices, both small-molecule and polymer based. These are presented in Tables 1 and 2.

TABLE 1. PERFORMANCE PARAMETERS FOR THREE SMALL-MOLECULE OLED DEVICE

	Blue	Green	Red
Operating voltage (V)	10	8	9
Luminous efficacy (lm/W)	0.56	3.9	1.3
Luminance (cd/m^2)	355	1,980	77

TABLE 2. PERFORMANCE PARAMETERS FOR THREE POLYMER OLED DEVICE

	Blue	Green	Red
Operating voltage (V)	4.13	4.07	3.5
Luminous efficacy (lm/W)	1.29	7.3	1.3
Luminance (cd/m^2)	340	1,900	300

One set of materials does not perform significantly better than the other. The major distinction is in operating voltage, where small-molecule voltages are about twice as great as polymer voltages. While smaller voltages are clearly desirable, it is possible that further research will diminish this gap.

The luminances in both tables are certainly adequate for OLED devices to be used as computer monitors or televisions. Much higher luminances have been observed in the laboratory, with a few reports of 100,000 cd/m^2, which is adequate for some outdoor lighting in daytime. However, lifetimes at such luminances are measured in seconds; such bright devices with long operating lifetimes are not yet on the development horizon.

The lifetime of an OLED device is an important concern, and state-of-the-art parameters are changing rapidly. At the luminances listed in Tables 1 and 2, it is common today for experienced laboratories to observe lifetimes in excess of 10,000 hours. These are generally acceptable lifetimes in a practical sense, but there are two further issues. First, in the particular case of monitors, even 10,000 hours may be too short; a computer monitor operating 10 hours per day would require only a bit over three years to accumulate 10,000 hours. Second, lifetime as defined in the literature is usually the time required for the device's luminance to fall to 50% of its initial value. For some applications this luminance falloff would be intolerable. The Pioneer display mentioned previously corrects for the temporal falloff in lifetime by correcting the drive current electronically.

Sato et al. have published a study of the degradation of small-molecule OLED devices. At the time of their study (1997), devices with initial luminances greater than about 300 cd/m^2 had lifetimes of less than 10,000 hours. For initial luminances of 500 cd/m^2, the lifetime was only about 4,000 hours, and the lifetime dropped to about 300 hours for an initial luminance of about 1,500 cd/m^2. A similar compilation of data obtained in 1999 would likely show improved lifetimes at all luminance levels.

The color performance of OLED devices matches well with the established NTSC and PAL standards for high definition television.

Organic Light-Emitting Molecules

Small Molecules. The selection of organic molecules suitable for OLED devices involves considering several variables. High electroluminescence efficiency is correlated with high fluorescence efficiency, although this is not a very restrictive requirement because many organic molecules have high fluorescence efficiency. More important is the fact that practical devices must operate at a low voltage (well below 10 volts). To obtain applied electric fields large enough for useful electroluminescence, the organic films must be very thin (a few tens of nanometers). They must also be very uniform in thickness and composition to assure uniform light intensity over the entire device. And finally, ease of manufacturing and cost considerations come into play.

Also, the emitted light must be bright. The brightness is determined by the rate of electron-hole recombination and is proportional to the current, after the threshold has been reached. This implies a high electrical conductivity for the transport layers so that, for a reasonable voltage of less than 10 V, emission is not limited by the number of charges available for recombination. For inorganic semiconducting materials, it is easy to find materials with useful electrical conductivities that also satisfy all other device variables. Organic materials are significantly less conductive, however, and present a serious challenge.

Effective conduction also relies on the electrodes' ability to efficiently inject electrons at one end and holes at the other end of the device. Thus, the anode work function should match the energy of the highest occupied molecular orbital (HOMO, roughly equivalent to the valence band) of the hole conduction layer. Similarly, the cathode work function should match the energy of the lowest unoccupied molecular orbital (LUMO, roughly equivalent to the conduction band) of the electron conduction layer.

Full-color light-emitting devices require the availability of blue, green, and red emitters, each of which satisfies all of the conditions discussed previously, and each of which can be deposited into the device without affecting any of the previously deposited materials. A fairly complete color palette can be constructed from small-molecule emitters.

Finally, a practical device must emit bright light for tens of thousands of hours of operation. Small-molecule OLED devices usually contain amorphous electron-transporting and hole-transporting layers, which must remain unchanged (that is, must not crystallize) through the normal exposure to high electric fields, high photon fluxes inside the device, and operating temperatures significantly higher than room temperature.

A large number of conjugated organic molecules that satisfy all of the criteria above well enough to qualify for commercial device applications have been found.[4,5] Figure 3 shows examples of the molecular structure of the electron- and hole-transporting layers for a prototype Kodak device. This device produces a fairly pure green light.

The performance of small-molecule OLED devices can be improved by doping the electron transport layer, the hole transport layer, or both with photoluminescent dyes. This has been shown to widen the color gamut, increase efficiencies, and reduce line widths of the emissive bands.[5] In the case of Alq (or one of its derivatives), successful dopants include perylene (blue), coumarin-6 (green), or DCM2 (red).

Although it is not strictly a p-n device, a new small-molecule technology from Opsys warrants mentioning here. Opsys has developed a class of materials called organolanthanide phosphors (OLPs) that rely on light emission from a lanthanide ion embedded in an organic shell (thus, they are organic EL devices, not OLED devices). The reason this development is significant is that it offers the potential for very stable devices through the separation of the light-emitting components from the more fragile organic components. Stability is a primary concern, as short lifetime continues to be a drawback of organic EL and OLED devices.

Polymers. Organic polymers are composed of finite-length chains of atoms called *segments*, which are attached to a carbon-carbon backbone. The bulk polymer is made up of a three-dimensional, irregular array of polymer chains. As with the organic molecules described above, most commercial polymers are poor electrical conductors. However, those that do conduct have electrical conductivity far superior to that of small-molecule materials, rivaling that of semiconductors for current flow along the polymer backbone. Polymer materials are generally applied to the OLED device as monomers by a spin-on process. The monomers are then polymerized through light (usually ultraviolet), heat, or both. See also **Colloid System**.

The selection of a suitable polymer for OLED applications involves issues similar to those discussed for small molecules, but there are some differences. First, it is the polymer film, rather than the monomer molecules from which it was made, that must have the desired electroluminescence qualities. Second, unlike amorphous films of organic molecules, polymer

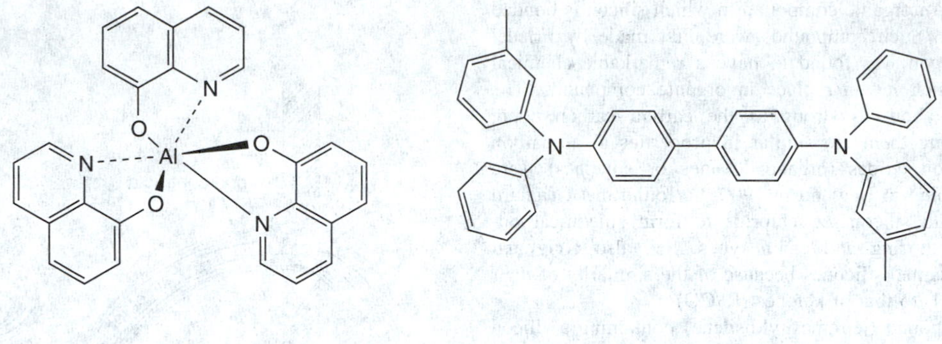

tris(8-hydroxyquinolinato)aluminum (Alq)
Electron transport material

TPD
Hole transport material

Fig. 3. Electron- and hole-transporting molecules.

films are unlikely to crystallize with exposure to electric fields, large photon fluxes, or high temperatures.

However, there are two common difficulties with polymer films that do not occur with small-molecule films. The first is exposure to temperatures higher than the glass transition temperature (Tg), the temperature at which the polymer chains have enough energy to make significant changes in their position or orientation. Upon cooling back below Tg, the polymer film will have a different microscopic structure, which will almost certainly degrade or destroy performance. Tg is typically $50-100\,°C(122-212\,°F)$, but has been pushed above $200\,°C(392\,°F)$ in some materials. Second, as polymer films age, they often lose free volume, which is called *shrinkage*. The speed and magnitude of this effect depend on the specific polymer and its state of crosslinking, but, overall, shrinkage causes polymer OLED devices to have a limited shelf life.

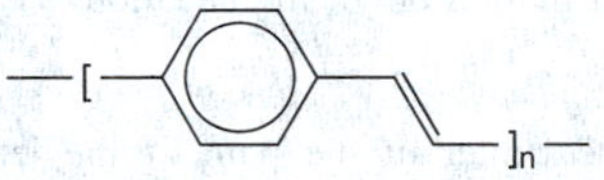

Fig. 4. Structure of *p*-phenylenevinylene (PPV).

The structure of one commonly used polymer, *p*-phenylenevinylene (PPV), is shown in Figure 4. PPV or a comparable material will be used in the orange-emitting polymer devices being released by the PolyLED business unit of Philips and has also been employed successfully by CDT. See also family of articles catalogued under **Flat Panel Display Technology**, including **(Inorganic) Light-Emitting Diode Displays**; and **Semiconductor**.

For additional reading, refer to Flat Panel Display Technology entry.

Stanford Resources, Inc., San Jose, CA

ORGANISMS (Dioecious). Organisms that produce only one type of reproductive gamete; they are usually classified as either male or female. Diecious organisms are the opposite of monecious organisms, which bear both male and female gametes. In botanical usage, however, the word diecious may be applied to trees that bear distinct male and female flowers or cones even though both may be on the same tree. Most higher animals, including all of the arthropods and chordates, are diecious, but most of the higher plants are monecious. Most of the flowers produced by the angiosperms bear both male and female organs.

ORGANOBORANE. A compound composed of an unsaturated organic group and a borane obtained by the hydroboration reaction. Such compounds are useful catalytic reagents in organic syntheses of some complexity, e.g., *cis-* or *trans*-olefins, optically pure alcohols, alkanes, and ketones. Prostaglandins and insect pheromones have been synthesized by this means. A particularly versatile example is triphenylboron $B(C_6H_5)_3$.

See also **Borane; Carborane**; and **Hydroboration**.

ORGANOCLAY. A clay such as kaoliln or montmorillonite, to which organic structures have been chemically bonded; since the surfaces of the clay particles, which have a latticelike arrangement, are negatively charges, they are capable of binding organic radicals. When this type of structure is in turn reacted with a monomer such as styrene, a complex known as a polyorganosilicate graft polymer results.

ORGANOGENY. The formation of organs in the embryo. After the formation of the germ layers and their initial stages of differentiation, each layer or its subordinate parts gives rise to certain of the organs characteristic of the adult.

ORGANOLEPTIC. A term widely used to describe consumer testing procedures for food products, perfumes, wines, and the like in which samples of various products, flavors, etc. are submitted to groups or panels. Such tests are a valuable aid in determining the acceptance of the products and thus may be viewed as a marketing technique. They also serve psychological purposes and are an important means of evaluating the subjective aspects of taste, odor, color, and related factors. The physical and chemical characteristics of foods are stimuli for the eye, ear, skin, nose, and mouth, whose receptors initiate impulses that travel to the brain, where perception occurs.

ORGANOMETALLIC COMPOUNDS. An organic compound composed of a metal attached directly to carbon (RM); such compounds have been prepared of practically all the metals, as well as with such non-metals as silicon and phosphorus. Metallic salts (soaps) of organic acids are excluded. Examples are diethylzinc (the first known organometallic). Grignard compounds such as methy magnesium iodide (CH3MgI), and metallic alkyls such as butyllithium (C4H9Li), tetraethyllead, triethyl aluminum, tetrabutyl titanate, sodium methylate, copper phthalocyanine, and metallocenes. Some are highly toxic or flammable; others are coordination compounds.

Reactive and moderately reactive organometallic compounds will react with all functional groups; two major types of reaction in which they are involved are oxidation and cleavage by acids. Probably the most important organometallic reactions are those involving addition to an unsaturated linkage. Many of them are powerful catalysts and form useful coordination complexes.

See also **Catalysis; Coordination Compounds**; and **Metallocenes**.

ORGANOPHOSPHORUS COMPOUND. Any organic compound containing phosphorus as a constituent. These fall into several groups, chief of which are the following: (1) phospholipids, or phosphatides, which are widely distributed in nature in the form of lecithin, certain proteins, and nucleic acids; (2) plasticizers, insecticides, resin modifiers, and flame-retardants; (3) pyrophosphates, e.g., tetraethyl pyrophosphate, which are the basis for a broad group of cholinesterase inhibitors used as insecticides; (4) phosphoric esters of glycerol, glycol, sorbitol, etc., which are components of fertilizers. While many of these compounds play an important part in animal metabolism, those in group (3) are toxic and should be handled with extreme care. See also **Flame Retardants**.

ORGANOSILICON. An organic compound in which silicon is bonded to carbon (organosilane). Such compounds were first made by Friedel and Crafts in 1863. Silicon was found to have a remarkable chemical similarity to carbon, which it can replace in organic compounds. The silicon–carbon bond is about as strong as the carbon–carbon bond, and compounds containing them are similar in properties to all-carbon compounds. Organosilicon oxides (organosiloxanes or silicones) were discovered by F. S. Kipping in England in 1900; he found that Grignard reagents would react with silicon tetrachloride to form silicon-carbon-bonded polymers or both ring and chain types. See also **Grignard Reactions**. These were named silicones because of the similarity of their empirical formula (R_2SiO) to that of ketones (R_2CO).

An organosilicon compound (tetramesityldisilene) containing a silicon to silicon double bond has been synthesized. It is a crystalline solid, mp 176 °C, and has reactive properties similar to olefins. Compounds of the type are silylenes.

ORGANOSOL. Colloidal dispersion of any insoluble material in an organic liquid; specifically the finely divided or colloidal dispersion of a synthetic resin in plasticizer in which dispersion the volatile content exceeds 5% of the total.

See also **Plastisol**.

ORGANOTIN COMPOUNDS. A family of alkyl tin compounds widely used as stabilizers for plastics, especially rigid vinyl polymers used as piping, construction aids, and cellular structures. Some have catalytic properties. They include butyl tim trichloride, dibutyltin oxide, etc., and various methyltin compounds. They are both liquids and solids.

ORIENTAL FRUIT MOTH (*Insecta, Lepidoptera*). An inconspicuous gray moth with a wingspan of about $\frac{1}{2}$-inch (12 millimeters). The larva of this species, *Grapholitha* or *Laspeyresia molesta* (Busck), is a pink worm with a brown head and up to $\frac{1}{2}$-inch (12 millimeters) long. The larva bores into twigs and new shoots of apple, apricot, peach, pear, plum, and quince. The twigs and shoots are destroyed in short order. The larva also bores into the stem end of fruit and eats the pulp. This damage also increases the susceptibility of the tree to disease. The insect was probably imported from the Orient about 1915 on nursery stock. It is well established in the eastern United States and has been found on peach in some of the western states.

In some regions, early maturing varieties of peach and apricot may be planted, in which case the fruit often can be picked before the pest attacks the fruit. Damage can be prevented in this manner even in heavily infested areas. The soil around an infested tree should be cultivated to a depth of 1 to 4 inches (2.5 to 10 centimeters) about 1 to 3 weeks before blooming time. This cultivation will kill many of the overwintering larvae in the soil.

When the fruit moth becomes a serious pest, a dust impregnated with a light-grade mineral oil can be applied. An effective formulation is: 60% sulfur, 35% 300-mesh talc; and 5% light-grade mineral oil, by weight. The dust should be applied at 5-day intervals, starting about 20 days before the peaches are picked. The oil dusts act as irritants and not as poisons.

ORIFICE. An orifice is an opening having a closed perimeter through which a fluid may discharge. The orifice may be open to the atmosphere, which is the case of free discharge, or it may be partially or entirely submerged in the discharged fluid. The standard orifice is the sharp-edged orifice shown in Fig. 1, but other types, such as the well-rounded orifice, the partially rounded orifice, the Borda mouthpiece, and the short tube orifice, have their special uses. An orifice may be very small, as in the case of those used for leak ports or for calibration, or large, as illustrated by sluice gates in a dam. The head of water on the orifice is measured from the water level surface to the center line of the orifice. Should the head above the orifice be so small as to be less than approximately the vertical dimension of the orifice, the following remarks will not apply, as this would come under the special case of large orifices under low heads.

The streamlines in water approaching a sharp-edged orifice converge on the orifice from all directions, and so continue to converge for approximately one-half of the orifice diameter downstream. The jet contracts to a section somewhat smaller in diameter than the orifice, after which it increases in size. The contracted section is known as the *vena contracta*. The ratio of the cross section of the jet at the vena contracta

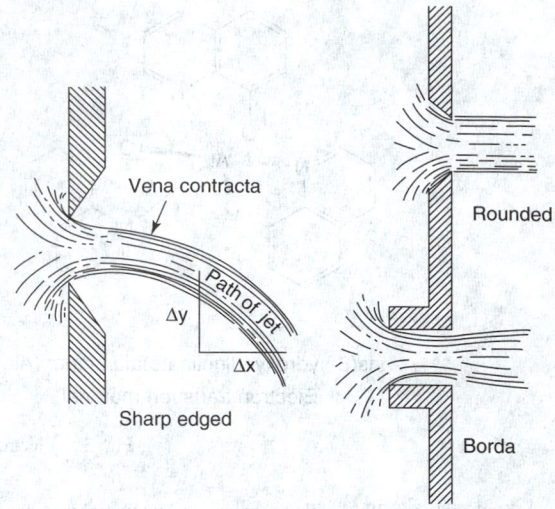

Fig. 1. Orifices.

to the area of the orifice is known as the contraction coefficient. Friction in the orifice slows the velocity to a somewhat lower value than the ideal free spouting velocity, which is $\sqrt{2gh}$. The ratio of actual to spouting velocity is the velocity coefficient. Since the discharge is the product of velocity and area, the discharge coefficient is the product of velocity and contraction coefficients. It has a numerical value of 0.61 for the average sharp-edged orifice. The discharge from an orifice of area a is

$$Q = .61a\sqrt{2gh}$$

The path taken by a jet discharging freely horizontally under head of h is parabolic in shape due to the pull of gravity acting on a particle having, originally, horizontal motion only. The equation

$$x^2 = 4C_v^2 hy$$

gives a curve of the center of the path of the jet, C_v is the velocity coefficient which averages 98% for sharp-edged orifices. Suppression of the contraction of a jet increases the discharge from an orifice. An orifice on the side wall of a tank near the bottom has a higher coefficient of contraction than one which is located farther away from the bottom. Similarly an orifice with the upstream edges rounded has a higher coefficient of contraction than one with sharp edges. The discharge may be as much as 30% greater for well-rounded orifices. Orifices which are submerged, orifices which are squared instead of circular, and orifices in which the water approaches with a high velocity, cannot be treated by the equation given above without corrections being made for these special conditions.

The foregoing discussion relates to the flow of water through an orifice. Orifices are much in use for measuring flows of vapors and gases. The method employed is to place an orifice of some type in the pipe or duct carrying the fluid. By means of a manometer, or pressure gauge, the upstream and downstream pressures are measured, and the discharge can be determined from those readings coupled with the known area of the orifice. The flow of a gas through an orifice depends on the area of the orifice, the upstream pressure, the temperature, and a factor which involves gas constants, such as the ratio of the specific heats at constant pressure and constant volume, and the ratio of the upstream and downstream pressures. The formula for the weight of gas flowing is $C_1 C_2 aP/\sqrt{T}$ (pounds per second). C_1 is the constant just mentioned, C_2 the velocity of approach correction, P the upstream pressure, a the area of the orifice, and T the absolute temperature. Many steam flow meters are based on the principle of the orifice. A sharp-edged, or thin-plate, orifice is clamped between the flanges at some joint of a flanged steam line. Pressure leads are taken from upstream and downstream sections to an instrument which is a pressure-measuring device, but which may be calibrated to read steam flow. See also **Flow Measurement (Liquids and Gases)**.

ORIGINS PROGRAM. Have you ever looked up at the night sky, marveling at the vastness of the Universe and your own connection to it?

It is hard to communicate the full sense of wonder that floods through us at such a moment, but we all understand. At least once, the dimly glittering night sky has stopped us in our tracks, bringing quiet contemplation of how the Universe came to be and what our relationship is to everything within it.

NASA's Origins Program seeks to answer two enduring human questions:

Where do we come from? To answer this, we need to understand the astronomical, physical, planetological, and biological processes necessary to generate and sustain life on earth. Knowing "where we come from" means understanding how the great chain of events unleashed after the Big Bang culminated in us and in everything we observe today. It is the story of our cosmic roots, told in terms of all that precedes us; the origin and development of galaxies, stars, planets, and the chemical conditions necessary to support life. See also **Cosmology**.

Are we alone? To answer this we need to understand the building blocks of life and the conditions necessary for life to arise. We need to search our solar neighborhood to see if such conditions exist elsewhere. We need to search for signatures of life. Knowing our uniqueness—"whether we're alone" in the cosmos—depends on our search for life-sustaining planets and on our understanding of its glorious diversity here on Earth. Only by seeing the innumerable possibilities on our home planet can we be sure that we'll recognize life if and when we find it somewhere else.

Ever since humans became capable of thought and reason, we have pondered these questions. Our ancestors, huddled around their ancient campfires, must have wondered about such mysteries. The questions are abstract, and profound in their implications; yet seem so natural that the youngest children gathered in modern classrooms ask them today.

Our generation is privileged to live in an era in which advances in science and technology allow us to investigate these intriguing questions. While the questions can be simply stated, the scientific and technical foundations needed to answer them are challenging.

Over the course of the next two decades, the Origins Program will develop the sophisticated telescopes and technologies that will bring us the information we seek. Although the questions are challenging, our generation is privileged to have the technological ability to reveal the possibilities for the first time. Just as the Greeks were known for democracy, the Egyptians for pyramids, and the Romans for roads, our civilization may well be remembered for discovering life beyond our own planet, forever changing our perception of the Universe and our place within it.

Science

Approximately 15 billion years ago in cosmic history, the first galaxies took shape from vast clouds of early chemical elements.

In the furnace of stars, life-sustaining chemicals such as carbon and oxygen came into being. Then, in awe-inspiring blasts from dying stars, life's chemicals blew out into space, only to condense anew into stars like our sun and planets like Earth.

Through the mixing of these vital chemicals and energy, the living Universe blossomed with the earliest self-replicating organisms and the profusion of life on our planet. Seeing similar chemical conditions wherever we look in the cosmos, the hope of finding life somewhere else rises inevitably within us.

To seek answers to the two defining questions of the Origins Program, scientists have outlined four goals that will speed us on our way to discovery:

Goal One. To understand how galaxies formed in the early universe. One of the biggest mysteries in science is how the Universe went from a uniform, relatively smooth structure to the clumpy, galaxy-strewn expanse we observe today. Our telescopes are beginning to look farther and farther back in cosmic time, but we have almost no information on the span stretching from 100 million years to 1 billion years after the Big Bang. In many senses, we can think of this age as "the cosmic dark ages." For Origins, it is also one of the most interesting times of all, because that is when the first galaxies began to form from a vast sea of tiny particles. Consider it a time for the seeds of our cosmic roots.

Gravity's Role in Galaxy Formation. To understand how galaxies formed, the Origins Program will be taking a look at gravity in great detail. In the early Universe, uniformly distributed matter began to gather as gravity weakly acted upon it. Small variations, or ripples, began to appear as matter accumulated in different regions and began to grow. The result? The first galaxies were born. By accurately mapping the amount, distribution, and chemical content of gaseous matter in the early Universe, Origins missions will tell us more about how this process took place. Detecting light from the Universe's very first generation of stars will also tell us when the first structures began to shine, generating the chemicals necessary for life.

How Galaxies Produce Chemicals for Stars, Planets, and Living Organisms. Life as we know it depends on the complex chemistry of organic matter, and yet the very early universe was made only of hydrogen, helium, and deuterium. None of the atoms necessary for life–carbon, for instance–had yet been formed. One of the triumphs of 20th-century science is understanding that the elements needed for building planets and for supporting life are forged in the fiery furnace of stars. The chemical elements produced by these stars are gravitationally bound to the galaxies in which the stars "live." By looking at the chemical composition of galaxies over cosmic time, we can see how their store of heavy elements grows throughout the ages. We want to understand how such chemical enrichment takes place and how new stars and planets can form the increasingly available materials. The answer will tell us if the development of a giant galaxy like our own Milky Way is essential to the eventual emergence of life. See also **Galaxy**.

Goal Two. To understand how stars and planetary systems form and evolve. Great clouds of gas and dust give birth to the stars, as neighboring atoms pull together in a gravitational bond. This bond grows stronger and stronger as more material packs in. See also **star**. So massive and dense does the central core become that nuclear fusion bursts on the scene, and a young star begins to shine. The energy is so intense that it blows most of the remaining gas away, leaving a disk of dusty material from which planets form. These "leftover" particles orbit the new star, knocking and gravitating together to form new worlds. Our observations have shown that this birthing process is going on all the time. We're lucky, in fact, to have so many examples of star and planetary systems, all at various stages of formation. The Origins Program wants to study these transformations in much greater detail, hoping to answer perhaps the most intriguing question of all: whether our solar system (with its life-sustaining Earth) is a common outcome of planet formation or very rare.

Young Planetary Systems in Formation. Most stars have a twin nearby–or even multiple companion stars. As we know quite well, our own star (the sun) seems to be an exception. That's good news for us, of course, because our planet Earth does quite well with its stable, single sun for life-giving energy. But why are we so different? The answer probably stems from initial conditions in the gas cloud from which our solar system emerged. That's why we want to study the life histories of young stars and planetary systems all the way back to their origins. Measuring the temperature, density, and chemical conditions of the planetary disks will help us understand how, and how often, solar systems like our own emerge. Greater analysis of dust disks at later stages will also tell us more about the role of gravity in planet formation. As small dust particles coagulate into larger and larger grains to form planets, they increasingly draw in nearby materials as they orbit around their parent stars. This depletion of surrounding material produces detectable gaps in the disks, giving us a "high sign" that faraway planets are in formation.

Mature Planetary Systems. What really drives the Origins quest forward is our desire to know if, somewhere out there, another Earth exists in a mature planetary system. The best way to find out is to take a census of nearby stars and any planetary systems around them. Origins will therefore survey about a thousand of the closest stars to us, as well as a significant sample of more distant ones. In our inventory of the galactic "neighborhood," we want to take a look at the orbital characteristics and physical properties of planets to find out what they're like. Mass, temperature, and atmospheric composition will be particularly important. We will also try to study the parent stars in greater detail to find out how their characteristics might influence the kinds of planets that end up forming. We might even be able to say when the earliest planets began to take shape in our own home galaxy, the Milky Way. By far the most fascinating search, however, will be our survey of planets that are about the same mass as Earth. Analyzing their characteristics will help us determine how frequently habitable worlds like our own are born. With that knowledge, we should begin to know whether life is exceedingly rare... or a cosmic imperative instead.

Goal Three. To determine whether habitable or life-bearing planets exist around nearby stars.

What does it mean to look for an Earth-like planet? The question is actually more complex than you might think. Our own home planet has been around for about 4.7 billion years, and it has not always been the world we know today. Think of the ice age... or the dinosaur age... or much further back—some 2 billion years ago—when living material on Earth had only just begun to pump oxygen into the atmosphere. While we would not have survived back then, Earth was abundant in life forms that could. To find a habitable planet, then, we have to expand our mental horizons to encompass our planet's past as well as its future—and all of the possibilities in between. We have to think about what would happen if Earth were slightly different—larger or smaller, warmer or colder, with different gravitational or chemical conditions—and what effect that would have on the possibility for life.

Worlds Where Life Can Thrive. Based on our only example (Earth!), life seems to need a couple of key ingredients: liquid water, key chemicals such as carbon, and a source of energy for the complex chemical reactions that take place in all living organisms. Of the 100 billion stars in our galaxy alone, there are certainly enough stable stars out there to provide energy for life's chemical transformations. We also find plenty of carbon and other necessary chemicals just about everywhere. Based on those two factors, the odds look good that life is possible somewhere else, but that still leaves the question of liquid water. One of the most important steps is to figure out what kinds of planets are likely to have water in a flowing, life-enabling form, and then to find out how commonly they form. For a habitable planet, location is everything. You do not want to be too close to the sun, or water would boil away. Too far away, and water would freeze. A stable, more circular orbit is also important, because wild swings toward and away from the sun would not support a stable supply of liquid water over time. Those simple limits help us narrow down our search to the "Goldilock's Zone"—that is, a "geographical" band around stars where planets would neither be too hot nor too cold, but rather "just right." The Origins census of nearby stars will look for planets in this comfort zone, giving us a sense of how many other life-supporting worlds might exist.

Identifying Life-Bearing Worlds. No matter how different conditions on other world might be, we do know one thing: life and its environment are inextricably linked. Life changes a planet's condition as it takes in food and energy and releases waste products. A changing planetary environment, due to biological, geophysical, or climatic activity, in turn causes life to adapt, resulting in the rich diversity in plants and animals that we encounter all around the world. Exploring this continuous cause-and-effect dance on Earth will help us comprehend the intertwining relationship between life and host planets everywhere. Nowhere is this relationship more apparent than in the observed characteristics of an atmosphere, especially combined with the planet's temperature and orbital location. Origins will particularly look for the presence of carbon dioxide, ozone, water, and chemical combinations that would not tend to occur in nature without biological activity. Of course, it is not enough to look for chemicals that only indicate the presence of life as we find it throughout Earth's history. See also **Earth**. The Origins Program will therefore be developing a catalog of all the possible chemical signatures for habitable planets and for life on them. In this research, it will be very important to identify how atmospheric gases produced by geological activity differ from those produced by life. After all, we would not want to mistake a barren planet, bursting with gaseous chemicals, for one abundant in life. Reconstructing the environmental and biological history of Earth, while identifying the extreme environments in which life has flourished here, will give us a good sense of the widest environmental limits in which life is possible.

Goal Four. To understand how life forms and evolves. Life on Earth covers the gamut, from the most primitive life forms to the most complex. With the help of microscopes, we have peered into everything from the world of single-celled organisms to the 100-trillion cells that make up the human body. No matter which living system we study, we find that each cell is largely made of proteins, and that each protein itself is a complex molecule made of millions and millions of atoms, arranged just so. It turns out that six atoms are particularly important to life: hydrogen, carbon, nitrogen, oxygen, phosphorus, and sulfur. Ninety-eight percent of the material in all living organisms is made of these atoms. Among them, carbon is the most important of all, since it has a special chemical property that likes to bind with other atoms. Carbon is therefore the "glue" that holds life's large and complex molecules together. That is why we tend to talk about "carbon-based life"—it is the only life we know. See also **Carbon; Hydrogen(Fuel); Nitrogen; Oxygen; Phosphorus**; and **Sulphur**. While we can generally say that life is based on a few select atoms and the presence of moisture and energy, which hardly accounts for life in all its glory. How living systems first emerged from these basic conditions remains a fundamental mystery that the Origins Program will explore.

How Matter is Organized Into Living Systems. The first thing we want to understand is how organic (carbon-based) molecules formed on our planet. Some scientists propose that comets, meteorites, and cosmic dust may have brought them to Earth. Others propose that they formed in the early atmosphere, in hydrothermal vents in the ocean, or in geothermal environments on land. The geological record of Earth, as well as the composition of bodies in our solar system, may provide key insights into the source of organics. The next critical step is trying to understand the all-important leap from basic organic materials to organized systems that could process energy and nutrients and transform them for life's biological processes. Origins laboratory research will study chemical reactions under conditions similar to Earth's early environment to understand how early biological structures may have emerged. Of particular interest is how the ancient counterparts of modern cells developed from simple structures, and came to process energy, metabolize, and transfer information to succeeding generations (genetics). Of course, Earth's biological processes may not be the same as those on another world, as chemical conditions on other planets might have produced substantially different organisms. Having no other life-sustaining world for comparison, we do not know which of the properties we observe are necessary for life, and which are specific to life on our own planet. Therefore, the Origins Program will create laboratory models that exhibit "life-like" properties and use a variety of chemical "ingredients" for their make-up. In that way, we will be more confident about detecting life, however different, on another world.

Limits to Life in Different Planetary Environments. The abundance of life in the universe also depends on the ability of living systems to adapt to their environments. We know life on Earth thrives in a number of different places, and can find evidence of past life in fossil records when conditions on our planet were vastly different than today. By studying the Earth's history and micro-environments, we will better understand not only the conditions in which life can operate, but also the environments in which it can not. In order to establish the limits for life, Origins research will look at self-replicating molecules and other systems, seeking clues for the way in which living systems adapt and pass on genetic information that enables future generations to survive. The one thing we know is that life is hardy, and finds a niche in wide range of environments, even those that seem impossibly extreme and hostile. Identifying the environmental limits to life will improve our ability to analyze the potential for life on other worlds. Because the diversification and survival of early life on Earth depended on the ability of microbial communities to live in harmony with their environment, Origins researchers will also take a look at how these communities cooperate and compete to harvest energy and nutrients. These studies are important because microbial life accounts for most of the living material on Earth. It is likely that another life-sustaining world would be filled with these simple forms of life too. Understanding the biosignatures (the life signs, or markers) left by these microbial communities in planetary rocks and atmospheres will help us identify the chemical signature for life that we hope to observe on a distant world someday. Astrobiology is the scientific discipline that will make all of these studies possible. See also **Astrobiology**.

Missions

For the first time in history, humanity is on the verge of having the technological capability to explore age-old questions about our cosmic origins and the possibility of life beyond Earth.

Even with the best talents and technologies available, our plans are ambitious and daring. To collect faint light from the first-ever galaxies and from Earth-sized planets around distant stars, we would essentially need telescopes the size of Texas. We have not achieved that capability here on Earth, let alone in space, and even if we could do it, the effort would triple the national debt.

That is why the Origins Program has embarked on a series of closely linked missions that build on prior achievements. As each Origins mission makes radical advances in technology, innovations will be fed forward, from one generation of missions to the next. By the end of the decade, we will have combined the very best imaging, formation flying, and other visionary technologies, giving us the power of enormous telescopes at a fraction of the cost.

Support for these missions is provided by the Interferometry Science Center (ISC), http://isc.caltech.edu/ a science operations and analysis service sponsored by the Origins theme and operated by the California Institute of Technology. The ISC facilitates timely and successful execution of projects that use interferometry, a key technology in the Origins Program.

While the majority of Origins efforts are focused on developing space-based observatories above Earth's atmosphere, investigations on the ground pave the way for future achievements in orbit.

Space-Based Observatories

The Origins missions form a family, in which each generation passes on a rich technological heritage to those that come after. Much like in human families, each mission has something unique to contribute, yet is closely tied to the others to form a supportive web. Origins currently has four chronological generations that move technology and knowledge forward.

The Precursor Missions

Hubble Space Telescope (HST). Our first-ever, long-term space observatory, HST reveals stunning views of the Universe with 10-times better resolution than any ground-based telescope.

Since 1990, NASA's Hubble Space Telescope (HST) has given us over 14,000 images of the Universe, allowing us to see stars exploding, galaxies colliding, planets forming, and other spectacular wonders that occur throughout our dynamic, evolving, ever mysterious Universe.

A cooperative program between NASA and the European Space Agency (ESA), the HST is the world's first long-lived, space-based observatory. See also **Hubble Space Telescope**.

Far Ultraviolet Spectroscopic Explorer (FUSE). FUSE looks at the Universe in a whole new light by studying objects in the ultraviolet portion of the spectrum, which is unobservable with other telescopes.

For hundreds of years, astronomers were only able to explain the Universe through the visible light that their eyes could see. NASA's Far Ultraviolet Spectroscopic Explorer (FUSE) gives astronomers a new tool for their exploration: a space telescope that studies the far ultraviolet light that is both invisible to us and is largely filtered out by Earth's atmosphere.

Far ultraviolet light in the Universe can help us understand more about conditions right after the Big Bang, the dispersion of chemical elements in galaxies, and the composition of interstellar gas clouds from which stars and planets form. A complement to other Origins missions.

FUSE's focus on the far ultraviolet permits astronomers to study the many important atoms, ions, and molecules that cannot be investigated otherwise.

FUSE was developed for NASA by the Johns Hopkins University, which has the primary responsibility for all aspects of the mission. Collaboration also comes from the Canadian and French space agencies, which share in observing time.

FUSE was launched on June 24, 1999, with a projected operational life of three years. For further information on the FUSE Mission see: http://fuse.pha.jhu.edu/

Space Infrared Telescope Facility (SIRTF). Able to see infrared (heat) radiation, SIRTF will peer through the veil of gas and dust that obscures most of the Universe from view.

Giant clouds of gas and dust block most of the Universe from view. NASA's Space Infrared Telescope Facility (SIRTF) will lift "the cosmic veil, "looking through these clouds to reveal stars forming in the heart of dusty galaxies, brown dwarfs, and even galaxies that existed near the beginning of time. It will also be able to characterize the disks of gas and dust around stars from which planets eventually form.

Because infrared radiation measures heat, astronomers have to cool the telescope to near absolute zero ($-460°$ Fahrenheit) so that the telescope can observe distant places in the Universe without interference from the heat of the near-Earth environment. With proper shielding from the sun, SIRTF will be launched into an Earth-trailing solar orbit that allows the telescope to cool rapidly. Rather than carrying large amounts of onboard cryogen (coolant), SIRTF also pioneered an innovative "warm launch" architecture that allows the telescope to cool in the frigid vacuum of space. This innovative design significantly reduced mass and launch costs.

SIRTF is the final element of NASA's Great Observatories Program, a series of four space-borne observatories designed to study the universe over many different wavelengths. The observatory is also a bridge to the Origins Program, providing information that will help us understand the formation and development of galaxies, stars, and planets. See also **Great Observatory Program**.

The SIRTF Telescope was launched into space by a Delta rocket from Cape Canaveral (Fig. 1), Florida on 25 August 2003. During its 2.5-year mission, Spitzer will obtain images and spectra by detecting the infrared energy, or heat, radiated by objects in space between wavelengths of 3 and 180 microns (1 micron is one-millionth of a meter). Most of this infrared radiation is blocked by the Earth's atmosphere and cannot be observed from the ground.

Fig. 1. ASA's freshly painted Stratospheric Observatory for Infrared Astronomy (SOFIA) 747SP aircraft sits outside a hangar at L-3 Communications Integrated Systems' facility in Waco, Texas. (*NASA*)

In December 2003, four months after its launch, NASA formally gave the Spitzer Telescope its new name finally retiring the old **SIRTF, the Space Infrared Telescope Facility**. See also **Chandra X-Ray Observatory**; and **Spitzer Space Telescope (SST)**.

Stratospheric Observatory for Far Infrared Astronomy (SOFIA). The world's largest airborne telescope, SOFIA will make observations that are impossible for even the largest and highest of ground-based telescopes. Another advantage SOFIA offers is the opportunity for teachers and the media to experience science in action, on-board as guests.

The Stratospheric Observatory for Infrared Astronomy (SOFIA) is a specially outfitted Boeing 747-SP aircraft. See Figs 1 & 2. The observatory, which features a German-built 100-inch (2.5 meter) diameter infrared telescope weighing 20 tons, is approaching the flight test phase (October, 2006) as part of a joint program by NASA and DLR (the German Aerospace Center, Deutsches Zentrum fuer Luft- und Raumfahrt). SOFIA's science and mission operations are being planned jointly by Universities Space Research Association (USRA) and the Deutsches SOFIA Institut (DSI). Once operational, SOFIA will be the world's primary infrared observatory during a mission lasting up to 20 years, as well as an

Fig. 2. Stratospheric Observatory for Infrared Astronomy (SOFIA) 747SP aircraft. (*NASA*)

outstanding laboratory for developing and testing instrumentation and detector technology, and a national education & public outreach facility putting educators from across the U.S. and Germany in contact with frontier scientific research. SOFIA will be based at NASA's Ames Research Center at Moffett Federal Airfield near Mountain View, California.

One of SOFIA's primary goals will be to study the properties of the interstellar medium—clouds of gas and dust that lie between stars in a galaxy. These clouds are important, because new stars and planets will eventually form from them. SOFIA will measure the infrared (heat) emissions from the clouds, seeking to understand their chemical makeup. Particularly interesting is the presence and abundance of carbon, which cools the interstellar medium and alters its subsequent chemical evolution. Carbon chemistry is important to study, because it is the basis of life as we know it. SOFIA will also study star and planet formation, along with other important Origins questions.

SOFIA is scheduled for its first flight test in 2007, and the first scientific observational flights are planned for 2008, with an operational life of 20 years or more. For further information on the SOFIA Mission see: http://sofia.arc.nasa.gov/; and http://www.sofia.usra.edu/Sofia/sofia.html

The First Generation Missions

Our seeking continues with missions that all serve as technological parents of second-generation Terrestrial Planet Finder.

StarLight. StarLight's technologies will enable future spacecraft to detect Earth-sized planets around other stars.

StarLight's two small telescopes can achieve the resolution of a telescope mirror 125 meters (137 yards) in diameter—wider than a football field is long!

StarLight's formation flying technologies will control the distance between the two spacecraft to within less than 1 centimeter (0.4 inch) and the angle (bearing) to within 3 arc minutes (873 μ radians).

StarLight is scheduled for launch in 2005, with an operational life of 12 months or more. For further information on the StarLight Mission see: http://starlight.jpl.nasa.gov/

Space Interferometry Mission (SIM). With a pinpoint accuracy several hundred times greater than any previous mission, SIM will begin identifying stars that have planetary systems around them.

Out of all the stars in the night sky, we wonder which ones might have planets swirling around them. NASA's Space Interferometry Mission (SIM) will continue the search by identifying stars that "wobble," that is, stars that are pulled back and forth as orbiting planets move from one side of the star to the other. If SIM sees such a gravitational tug, it infers the presence of planets. While this method of finding planets is indirect, it paves the way for Terrestrial Planet Finder and other missions that will eventually image the distant worlds first discovered by SIM.

While ground-based observatories have identified large, gaseous planets around other stars, only SIM's incredible precision will allow us to begin detecting the extremely tiny star wobble caused by an Earth-sized planet.

In its efforts to find Earthlike planets that lie closer to their parent stars, SIM will also pioneer a technique to block out the star's light so that the tiny, faint, orbiting planets can be seen. Its ability to measure the position of stars several hundred times more accurately than any previous program will also allow to astronomers to determine the size and age of the Universe.

In an Earth-trailing solar orbit, SIM will receive continuous solar illumination, avoiding the occultations that would occur in an Earth orbit. (While this orbit is similar to the Spitzer Space Telescope, SIM is an optical telescope and thus does not need to block out heat from the sun.)

SIM is scheduled for launch in 2014, with an operational life of five years or more. For further information on the SIM Mission see: http://planetquest.jpl.nasa.gov/SIM/sim_index.cfm

James Webb Space Telescope (JWST). Nearly four times the size of Hubble's mirror yet ultra-lightweight, NGST will study the very first stars and galaxies to emerge in the Universe.

The JWST will look back to an extremely important period in the early history of the Universe when the first stars and galaxies began to form. While we have a fairly good understanding of the Universe in other periods, we have no observations during this time when the Universe was between one million and a few billion years old. JWST's studies will help us understand the shape and chemical composition of the universe, the evolution of galaxies, and the nature of unseen "dark matter."

JWST will study infrared (heat) emissions from this early time, seeing objects 400 times fainter than those currently studied with large ground-based telescopes or the current generation of space-based infrared telescopes. At 6.5 meters (21.3 feet), JWST's primary mirror is more than two-and-a-half times as large as Hubble's, giving it much more light gathering capability. Its tennis-court sized sunshade will help eliminate the heat from sun, which is necessary for reducing heat "pollution" from the surrounding environment.

JWST is an international collaboration among NASA, the European Space Agency (ESA), and the Canadian Space Agency (CSA). NASA has the overall responsibility for the JWST mission; ESA provides the Near Infrared Spectrograph, Mid-Infrared Instrument Optics Assembly, and the Ariane Launch Vehicle; and the CSA provides the Fine Guidance Sensor/Tunable Filter Instrument. Goddard Space Flight Center (GSFC).

http://www.nasa.gov/centers/goddard/home/index.html, manages the JWST project and provides ISIM components; the Jet Propulsion Laboratory (JPL) http://www.nasa.gov/centers/goddard/home/index.html, manages the Mid-Infrared Instrument; Ames Research Center (Ames).

http://www.nasa.gov/centers/ames/home/index.html, is providing the Detector Technology Development; Johnson Space Center (JSC).

http://www.nasa.gov/centers/johnson/home/index.html, provides the Observatory Test Facilities; Marshall Space Flight Center (MSFC).

http://www.nasa.gov/centers/marshall/home/index.html, provides the Mirror Technology Development and Environmental Research; and Glenn Research Center (GRC) http://www.nasa.gov/centers/glenn/home/index.html, provides the Cryogenic Component Development. Academic and Industry Partners include: Northrop Grumman Space Technologies http://www.st.northropgrumman.com/; The Space Telescope Science Institute http://www.stsci.edu/resources/; Ball Aerospace http://www.ballaerospace.com/; ITT http://www.itt.com/; ATK http://www.atk.com/; University of Arizona http://www.arizona.edu/; and Lockheed Martin.

http://www.lockheedmartin.com/wms/findPage.do?dsp=fnec&ti=100.

JWST is scheduled for launch in 2013 (approx.), with an operational life of 5 to 10 years or more. For further information on the JWST Mission see: http://www.jwst.nasa.gov/. See also **James Telescope (JWST)**.

The Second Generation Mission

The culmination of a decade's work, this mission will combine preceding technologies to begin revealing whether life is a cosmic imperative.

Terrestrial Planet Finder (TPF). Flying four advanced telescopes in formation, TPF will give us the first "family portraits" of other planetary systems, and maybe even a picture of a planet where life might exist. See Figs. 3 & 4.

NASA's Terrestrial Planet Finder (TPF) will study all aspects of planets, from their formation to their final characteristics. In addition to measuring the size, temperature, and placement of Earth-sized and other planets,

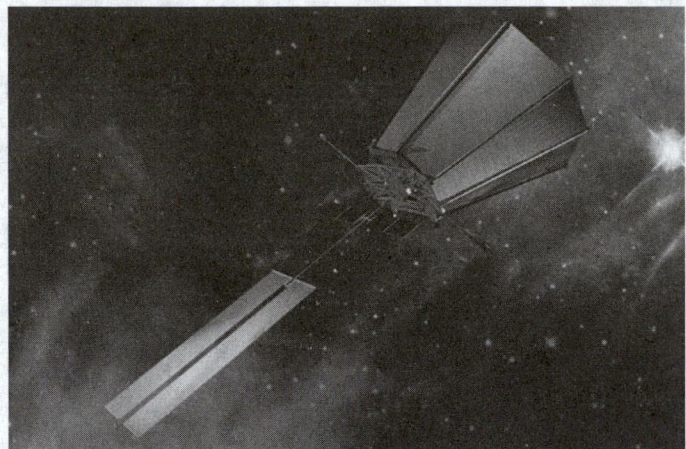

Fig. 3.   Terrestrial Planet Finder comprises two complementary observatories: a visible-light coronagraph (above), to launch around 2014, and a formation-flying infrared interferometer, to launch before 2020 (Fig. 4). (*NASA*)

Fig. 4.   Terrestrial Planet Finder comprises two complementary observatories: a visible-light coronagraph, to launch around 2014 (Fig. 3), and a formation-flying infrared interferometer (above), to launch before 2020.

TPF will look for gases such as carbon dioxide, water vapor, ozone, and methane that would indicate that a far-away planet could, or even does, support life.

TPF will find the tiny, faint planets around distant stars by reducing the glare of their parent stars a hundred-thousand times, taking pictures of planetary systems as far away as 50 light years. With pictures a hundred times more detailed than those of the Hubble Space Telescope, TPF will also allow us to study the black hole at the center of the Milky Way and other exciting phenomena in the universe.

TPF is scheduled for launch in 2014 (approx.), with an operational life of six years or more. For further information on the TPF Mission see: http://planetquest.jpl.nasa.gov/TPF/tpf_index.cfm.

The Third Generation Missions

For now, these missions remain just a vision because the required technology is not on the immediate horizon. Today's missions, however, put us on the path toward such monumental achievements.

Life Finder (LF). Once we identify any habitable planets, Life Finder would fly telescopes at even larger distances to detect chemicals that actually reveal biological activities, that is, the presence of life.

If earlier Origins missions lead us to discover another world with life-sustaining conditions, it does not necessarily mean that life has actually emerged there. Life Finder (LF) will seek to determine if a distant planet with the right living conditions actually has an abundance of living creatures!

Life Finder will be even more sensitive than Terrestrial Planet Finder, but the principle of characterizing a planet's conditions is the same. If a planet harbors life, biological activity on the planet will impact the atmosphere, just as it does on our own home planet. When we analyze the radiation coming from the planet, we can look for much finer dips in the energy. These dips indicate the presence of methane and other chemicals we do not expect to find in nature unless biological activity is pumping it into the atmosphere.

We will also have to keep in mind the history of Earth in relation to a new world. The simplest life forms existed on Earth well before an abundance of oxygen appeared in the atmosphere, which in turn allowed multicelled organisms to flourish. Origins astrobiology research will help us expand our knowledge of "life signs" that would appear at different stages in a planet's history, as well as signs that would appear given a planetary chemistry that is not exactly the same as our own. With these insights, we will give ourselves the best possible chance of recognizing life if and when we find it somewhere else.

Planet Imager (PI). If we found a planet with life, we would not leave it to our imagination! To create a picture would require a number of telescopes flying in formation to achieve the power of a telescope 360 kilometers (225 feet) wide.

Just imagine knowing that another Earthlike planet is out there... and that for the first time we know exactly where it is! That could never be the end of the story, because our thirst for knowing about that distant planet would be enormous. Would it have continents like ours? Oceans as large? Clouds? Or would it be foreign enough to rival the creations of our very best science-fiction writers?

This new earth, this Terra Nova, would be at the forefront of our imaginations. We'd never be content just knowing it exists, we would want to see it for ourselves! The Planet Imager (PI) mission will produce pictures of single planets at much higher resolution than any preceding mission. Instead of seeing a planet as a single dot, we would strive for a larger image, consisting of as many pixels as possible. The greater the number of pixels, however, the more complex the mission.

We do have an idea of what we would need to accomplish our goal: an array of interferometers that each carried NGST-sized telescopes (about eight meters (26.2 feet). They would have to fly in exquisitely precise formation... over distances of 6,000 kilometers (3730 miles) or more! Right now, we're not even close to having the technology to accomplish the task, but the generations of Origins missions leading up to Planet Imager will pave the way.

Ground-Based Observatories

Reaching toward incredibly capable space observatories must begin by stretching here at home. The following ground-based efforts provide the technological bedrock for future Origins missions.

Two Micron All-Sky Survey (2MASS). The Two Micron All-Sky Survey (2MASS) used two 1.3 meter (51-inch) telescopes at Mount Hopkins, Arizona, and Cerro Tololo, Chile to conduct the most thorough census ever made of our Milky Way galaxy and the nearby universe. The work was completed in February 2001, but data processing will continue through 2002. Catalogues produced from 2MASS data contain more than 300 million stars and galaxies, including previously undetected star-forming regions, as well as galaxies behind the disk of the Milky Way. The 2MASS data will help scientists prepare for future infrared space missions, including the Spitzer Space Telescope.

For further information on the Two Micron All-Sky Survey (2MASS) project: See http://www.ipac.caltech.edu/2mass/

Keck Interferometer. By connecting the twin Keck telescopes and combining incoming light, Keck will function as a single, much larger and more powerful telescope.

To expand the capabilities of the world's largest telescopes even farther, the Origins Program is in the process of connecting the twin Keck telescopes. By combining the light paths from each twin, astronomers will gain the capability of a single telescope the size of the distance between them. That's about the equivalent of an 85-meter (279-foot) mirror; almost the size of a football field.

With the addition of four proposed 1.8-meter (5.9-foot) "outrigger" telescopes that are located nearby, Origins astronomers eventually hope to have the ability to simulate a telescope with mirrors anywhere between 25 and 140 meters (82 and 459 feet).

For further information on the Keck Interferometer Array project: See http://planetquest.jpl.nasa.gov/Keck/keck_intro.cfm.

Palomar Testbed Interferometer (PTI). By combining light from telescopes at the Palomar Observatory, PTI takes the first crack at developing technologies that combine light coming in from distant objects in the cosmos.

Located near San Diego, the Palomar Testbed Interferometer (PTI) is developing some of the technologies needed for the Keck Interferometer and the Space Interferometry Mission (SIM). PTI uses multiple telescopes to measure interference fringes created when light gathered by the telescopes is combined and processed. This technique enables scientists to measure the positions and distances between stars with great accuracy.

PTI's dual-star tracking system, the first of its kind, will help provide measurements that detect a star's "wobble," allowing us to infer the presence of planets orbiting around the star. PTI has also been used to measure the sizes of dwarf, giant, and supergiant stars; the sizes of emissive regions around young stellar objects; and binary star orbits. It is the first interferometer to have directly measured the diameter changes of a Cepheid variable star, and directly measured the rotational oblateness of a rapidly rotating star.

For further information on the Palomar Testbed Interferometer (PTI) project: See http://www.astro.caltech.edu/palomar/pti.html.

The Large Binocular Telescope Interferometer (LBTI). Two 8.4-meter (27.5-foot) class telescopes on Mount Graham, Arizona, will be linked to create an infrared interferometer capable of imaging distant galaxies and other faint objects over a wide field-of-view.

Because of its unique geometry and relatively direct optical path, the LBTI will offer science capabilities that are different from other interferometers. It will be cable of providing high-resolution images of many faint objects over a wide field-of-view, including galaxies in the Hubble Deep Field with 10 times the Hubble resolution.

Nulling techniques will enable the LBTI to study emissions from faint dust clouds around other stars. These dust clouds reflect light and give off heat, and so interfere with the search for planets. By helping the characterize these emissions, the LBTI will provide critically needed data for the design of the Terrestrial Planet Finder, a later mission that will study planets orbiting nearby stars.

For further information on The Large Binocular Telescope Interferometer (LBTI) project: See http://medusa.as.arizona.edu/lbto/index.htm. For more information on the Origins Program: See http://origins.jpl.nasa.gov/index1.html.

See also **Astrobiology**; **Space Science Mission: Universe**; and **Universe (The)**.

Web References

Astronomical Society of the Pacific: http://www.astrosociety.org/
German Aerospace Center: http://www.dlr.de/en/desktopdefault.aspx/tabid-1/86_read-7261/
SOFIA Links: http://www.sofia.usra.edu/Links/links.html
Universities Space Research Association: http://www.usra.edu/

NASA/Jet Propulsion Laboratory, Pasadena, CA

ORIOLE *(Aves, Passeriformes).* Brightly colored birds of the family *Oriolidae*, found in all parts of the Old World. The North American birds to which this name is applied belong to the family *Icteridae* and are more closely related to the blackbirds than to the true orioles. The orchard oriole, *Icterus spurius*, and the Baltimore oriole, *I. galbula*, are the most widely known species of the latter group.

ORION CREW EXPLORATION VEHICLE. Lessons from the past are guiding NASA's next step into the future, as the space agency prepares to replace the space shuttle with an Apollo-style vehicle for human explorers. See Fig. 1. America will send a new generation of explorers to the moon aboard NASA's Orion crew exploration vehicle.

The Orion crew exploration vehicle is named for one of the brightest and most recognizable star formations in the sky. It will be a multi-purpose capsule—the central member of a family of spacecraft and shuttle-derived

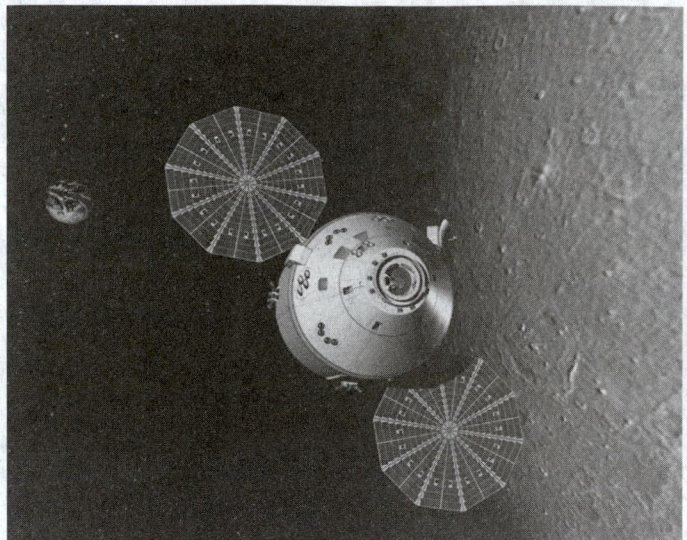

Fig. 1. Orion in lunar orbit. (Image courtesy of Lockheed Martin Corp.).

launchers that NASA's Constellation Program is developing to carry astronauts back to the moon and then onward to Mars and other destinations in the solar system. The first flight with astronauts aboard is planned for no later than 2014. Orion's first flight to the moon is planned for no later than 2020. See also **Constellation Program**.

A component of the Vision for Space Exploration, Orion's development is taking place in parallel with missions to complete the International Space Station using the space shuttle before the shuttle is retired in 2010.

In what amounts to one of the most significant NASA procurements in more than 30 years, two industry teams, Northrop Grumman/Boeing and Lockheed Martin, spent 13 months refining concepts, analyzing requirements and sketching designs for Orion. On Thursday, Aug. 31, 2006, managers of NASA's Exploration Systems Mission Directorate revealed that Lockheed Martin Corp. of Bethesda, MD., was awarded the contract to build Orion.

The contract with Lockheed Martin has a seven-year base valued at about $3.9 billion for design, development, testing and evaluation of the new spacecraft. Production and sustaining engineering activities are contract options worth more than $4 billion through 2019.

Versatility will be Orion's trademark. It is being designed to fly to the moon, but could also be used to service the International Space Station in low-Earth orbit. See Fig. 2. "Our intent is to keep the destination focusing the design but we are not excluding the possibility of using Orion for other things, such as de-orbiting the Hubble Space Telescope in the 2020s or making a trek to an asteroid," said Jeff Hanley, who manages the Constellation Program from the Johnson Space Center in Houston.

Orion improves on the best features of Project Apollo and the Space Shuttle Program, increasing the likelihood of success. Going with known technology and known solutions lowers the risk. Although Orion borrows its shape and aerodynamic performance from Apollo, the new capsule's updated computers, electronics, life support, propulsion and heat protection systems represent a marked improvement over legacy systems.

Unlike the winged space shuttle orbiter, which is mounted beside its external fuel tank and boosters for liftoff, Orion will be placed on top of its booster to protect it from ice, foam, and other launch system debris during ascent. See Fig. 3. Placing the spacecraft on top of the launch vehicle also allows the addition of an abort system that can separate capsule and crew from the booster in an emergency.

Orion will be similar in shape to the Apollo spacecraft. The Apollo-style heat shield is the best understood shape for re-entering Earth's atmosphere, especially when returning directly from the moon. Among the most obvious improvements is the command module's size. Orion will be 5 meters (16.5 feet) in diameter and have a mass of about 22.7 metric tons (25 tons). Orion will have more than 2.5 times the interior volume of the three-seat Apollo capsules that carried astronaut crews to the moon for missions lasting only several hours to several days in the late 1960s and early 1970s. See Fig. 4 and 5. Orion will be crucial for developing a sustained human presence on

Fig. 2.   Artist's concept of Orion Crew Exploration Vehicle rendezvous with the International Space Station. (Image courtesy of NASA/John Fassanito and Associates.)

Fig. 4.   Exploded view of the Crew Exploration Vehicle. (Image courtesy of Lockheed Martin, Corporation.)

Fig. 3.   Artist's concept of NASA's Ares I crew launch vehicle. Launched by a shuttle-derived solid rocket booster and upper stage engine similar to those used during Apollo, the crew of four would rendezvous with the ship carrying the lunar module in earth orbit before heading to the moon. (Image courtesy of NASA).

Fig. 5.   Artist's concept of Orion docked with a lunar lander in orbit around the moon. (Image courtesy of Lockheed Martin, Corporation.)

the moon. It will be able to carry four astronauts to the moon and support missions of up to six months.

A launch abort system atop the Orion capsule will be capable of pulling the spacecraft and its crew to safety in the event of an emergency on the launch pad or at any time during ascent. See Fig. 6.

NASA is leveraging the talent and resources of the entire agency in the design and development of Orion. While Constellation Program management resides at Johnson, all 10 of the agency's field centers are making important contributions. See also **Constellation Program**.

Journey to the Moon

For missions to the moon, NASA will use two separate launch vehicles, each derived from a mixture of systems with heritage rooted in Apollo, space shuttle and commercial launch vehicle technology.

An **Ares V** cargo launch vehicle will precede the launch of the crew vehicle **Ares 1**, delivering to low-Earth orbit the Earth departure stage and the lunar module that will carry explorers on the last leg of the journey to the moon's surface. Orion will dock with the lunar module in Earth orbit, and the Earth departure stage will propel both on their journey to the moon. Once in lunar orbit, all four astronauts will use the lunar landing craft to travel to the moon's surface, while the Orion spacecraft stays in

lunar orbit. Once the astronauts' lunar mission is complete, they will return to the orbiting Orion vehicle using a lunar ascent module. The crew will use the service module main engine to break out of lunar orbit and head to Earth.

Orion and its crew will reenter Earth's atmosphere using a newly developed thermal protection system. Parachutes will further slow Orion's descent through the atmosphere.

Ares Rockets

On June 30, 2006, NASA announced the names of the next generation of launch vehicles that will return humans to the moon and later take them to Mars and other destinations. The crew launch vehicle will be called Ares I, and the cargo launch vehicle will be known as Ares V. See Fig. 7.

"It's appropriate that we named these vehicles Ares, which is a pseudonym for Mars," said Scott Horowitz, associate administrator for NASA's Exploration Systems Mission Directorate, Washington. "We honor the past with the number designations and salute the future with a name that resonates with NASA's exploration mission."

The "I and V" designations pay homage to the Apollo program's Saturn I and Saturn V rockets, the first large U.S. space vehicles conceived and developed specifically for human spaceflight.

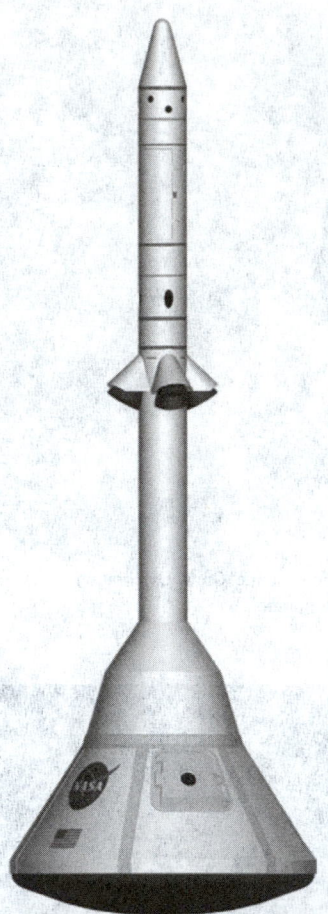

Fig. 6. Launch Abort System of the Crew Exploration Vehicle. (Image courtesy of Lockheed Martin, Corporation.)

Fig. 7. Artist concept of the Ares V, left, and Ares 1. (*Image courtesy of NASA.*)

The crew exploration vehicle, which will succeed the space shuttle as NASA's spacecraft for human space exploration, will be named later. This vehicle will be carried into space by Ares I, which uses a single five-segment solid rocket booster, a derivative of the space shuttle's solid rocket booster, for the first stage. A liquid oxygen/liquid hydrogen J-2X engine derived from the J-2 engine used on Apollo's second stage will power the crew exploration vehicle's second stage. The Ares I can lift more than 24,948 kilograms (55,000 pounds) to low Earth orbit.

Ares V, a heavy lift launch vehicle, will use five RS-68 liquid oxygen/liquid hydrogen engines mounted below a larger version of the space shuttle's external tank, and two five-segment solid propellant rocket boosters for the first stage. The upper stage will use the same J-2X engine as the Ares I. The Ares V can lift more than 129,727 kilograms (286,000 pounds) to low Earth orbit and stands approximately 110 meters (360 feet) tall. This versatile system will be used to carry cargo and the components into orbit needed to go to the moon and later to Mars.

Ares 1 Rocket. NASA is already at work developing hardware and systems for the Ares I rocket that will send future astronauts into orbit. Built on cutting-edge launch technologies, evolved powerful Apollo and space shuttle propulsion elements, and decades of NASA spaceflight experience, Ares I is the essential core of a safe, reliable, cost-effective space transportation system–one that will carry crewed missions back to the moon, on to Mars and out into the solar system.

Ares I is an in-line, two-stage rocket configuration topped by the Orion crew vehicle and its launch abort system. See Fig. 8. In addition to the vehicle's primary mission–carrying crews of four to six astronauts to Earth orbit–Ares I may also use its 25,401 kilogram (25-ton) payload capacity to deliver resources and supplies to the International Space Station, or to "park" payloads in orbit for retrieval by other spacecraft bound for the moon or other destinations.

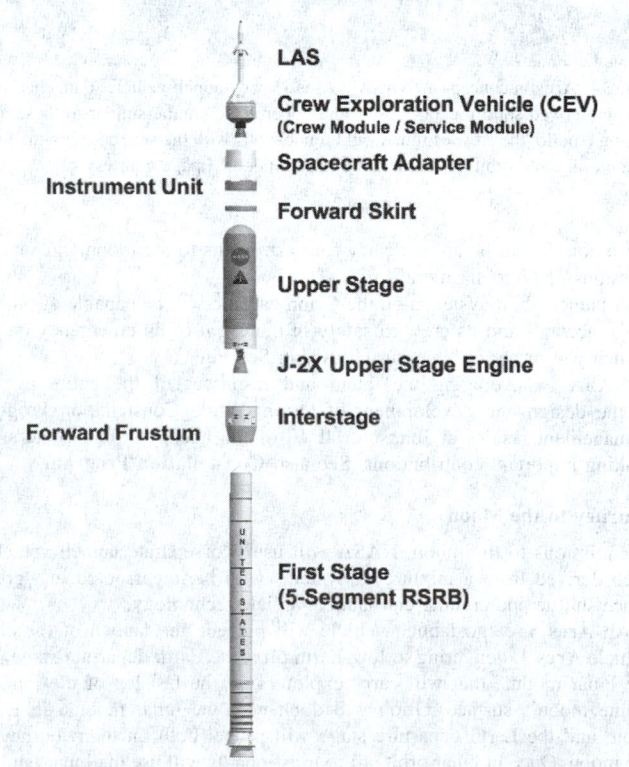

Fig. 8. Expanded view of the Ares 1. (*Image courtesy of NASA.*)

During launch, the first-stage booster powers the vehicle toward low Earth orbit. In mid-flight, the reusable booster separates and the upper stage's J-2X engine ignites, putting the vehicle into a circular orbit.

Ares 1 First Stage. The Ares I first stage is a single, five-segment reusable solid rocket booster derived from the Space Shuttle Program's reusable solid rocket motor, which burns a specially formulated and shaped solid propellant. See Fig. 9.

Fig. 10.   A J-2 engine undergoes static firing. (*Image courtesy of NASA.*)

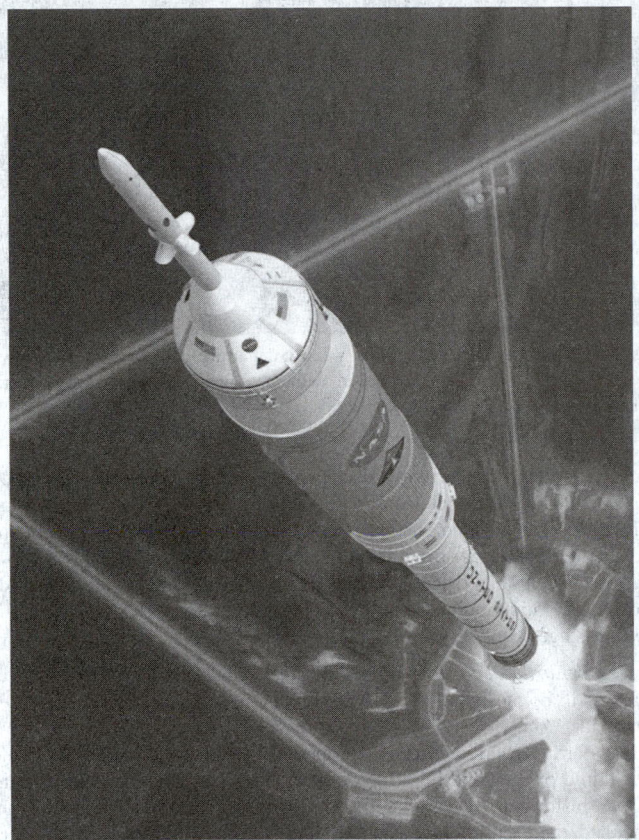

Fig. 9.   Artist concept of Ares 1 during launch. (*Image courtesy of NASA.*)

A newly designed forward adapter will mate the vehicle's first stage to the upper stage, and will be equipped with booster separation motors to disconnect the stages during ascent.

Ares I Upper Stage / Upper Stage Engine. The Ares I second, or upper, stage is propelled by a J-2X main engine fueled with liquid oxygen and liquid hydrogen. See Fig. 10.

The J-2X is an evolved variation of two historic predecessors: the powerful J-2 engine that propelled the Apollo-era Saturn IB and Saturn V rockets, and the J-2 S, a simplified version of the J-2 developed and tested in the early 1970s but never flown.

Ares V Rocket. NASA is planning and designing hardware and propulsion systems for the Ares V cargo launch vehicle–the "heavy lifter" of America's next-generation space fleet.

During launch, the Ares V first stage and core propulsion stage power it upward toward Earth orbit. After separation from the spent core stage, the upper stage–also known as the Earth Departure Stage–takes over, and by a J-2X engine puts the vehicle into a circular orbit. See Fig. 11. The cargo vehicle's propulsion system can lift heavy structures and hardware to orbit or fire its engines for trans-lunar injection, a trajectory designed to intersect with the moon. Such lift capabilities will enable NASA to carry a variety of robust science and exploration payloads to space and could possibly take future crews to Mars and beyond.

Ares V First Stage. The first stage of the Ares V vehicle relies on two, five-segment reusable solid rocket boosters for lift-off, illustrated in Fig. 12.

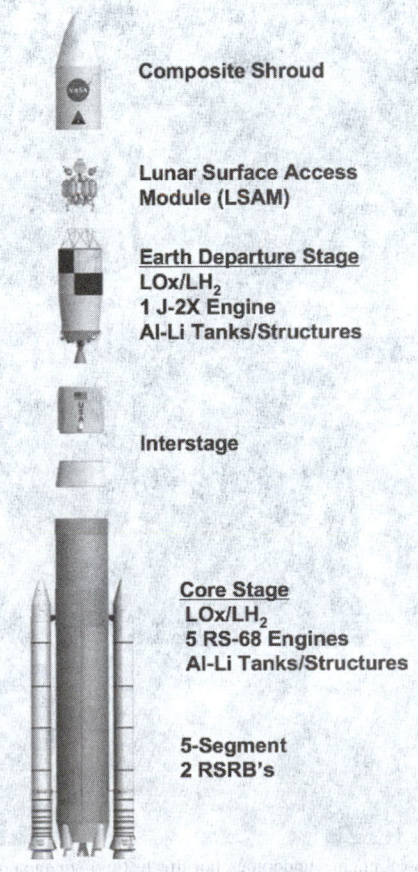

Composite Shroud

Lunar Surface Access Module (LSAM)

Earth Departure Stage
LOx/LH$_2$
1 J-2X Engine
Al-Li Tanks/Structures

Interstage

Core Stage
LOx/LH$_2$
5 RS-68 Engines
Al-Li Tanks/Structures

**5-Segment
2 RSRB's**

Fig. 11.   Expanded view of the Ares V. (*Image courtesy of NASA.*)

Fig. 12.   Artist concept of Ares 5 five-segment reusable solid rockets. (*Image courtesy of NASA.*)

Fig. 14.   Concept image of the Ares V Earth departure stage in orbit, shown with the Crew Exploration Vehicle docked with the Lunar Surface Access Module. (*Image courtesy of NASA*).

Derived from the space shuttle solid rocket boosters, they are similar to the single booster that serves as the first stage for the cargo vehicle's sister craft, Ares I.

Ares V Core Stage / Core Stage Engine. The twin solid rocket boosters of the first stage flank a single, liquid-fueled central booster element. Derived from the space shuttle external tank, the central booster tank delivers liquid oxygen/liquid hydrogen fuel to five RS-68 rocket engines — a modified

version of the ones currently used in the Delta IV launcher developed in the 1990s by the U.S. Air Force for its Evolved Expendable Launch Vehicle program and commercial launch applications. The RS-68 engines serve as the core stage propulsion for Ares V. See Fig. 13.

Atop the central booster element is an interstage cylinder, which includes booster separation motors and a newly designed forward adapter that mates the first stage with the Earth Departure Stage.

Ares V Earth Departure Stage / Engine. The Ares V Earth Departure Stage will be designed by NASA's Marshall Space Flight Center in Huntsville, Ala.

The Earth Departure Stage is propelled by a J-2X main engine fueled with liquid oxygen and liquid hydrogen. The J-2X is an evolved variation of two historic predecessors: the powerful J-2 upper-stage engine that propelled the Apollo-era Saturn IB and Saturn V upper stages and the J-2 S, a simplified version of the J-2 developed and tested in the early 1970s but never flown.

The Earth Departure Stage separates from the core stage and its J-2X engine ignites mid-flight. See Fig. 14.

Once in orbit, the Orion crew capsule — the astronaut module delivered to orbit by Ares I — docks with the orbiting Earth Departure Stage carrying the Lunar Surface Access Module, which will ferry astronauts to and from the moon's surface. Once mated with the crew module, the departure stage fires its engine to achieve "escape velocity," the speed necessary to break

Fig. 13.   An RS-68 engine undergoes hot-fire testing. (*Image courtesy of Pratt and Whitney Rocketdyne.*)

Fig. 15.   Artist concept of crew vehicle and lander in lunar orbit. (*Image courtesy of NASA.*)

free of Earth's gravity, and the new lunar vessel begins its journey to the moon.

The Earth Departure Stage is then jettisoned, leaving the crew module and Lunar Surface Access Module mated. Once the four astronauts arrive in lunar orbit, they transfer to the lunar module and descend to the moon's surface. The crew module remains in lunar orbit until the astronauts depart from the moon in the lunar vessel, rendezvous with the crew module in orbit and return to Earth.

Lunar Surface Access Module. Anchored atop the Earth Departure Stage is a composite shroud protecting the Lunar Surface Access Module, or LSAM. See Fig. 15.

This module includes the descent stage, developed by the Marshall Space Flight Center, that will carry explorers to the moon's surface; and the ascent stage, developed by the Johnson Space Center, that will return them to lunar orbit to rendezvous with the crew exploration module, their ride home to Earth.

Web References

Ares Launch Vehicles: http://www.nasa.gov/mission_pages/constellation/ares/rocket_science.html.

Constellation Program: http://www.nasa.gov/mission_pages/constellation/main/index.html

AMIKO NEVILLS, National Aeronautics and Space Administration (NASA)

ORION (The hunter). This constellation is on the whole, the richest and most impressive of all of the constellations. The "belt and sword" of Orion are frequently referred to in both ancient and modern literature. Despite the wide area covered, the physical characteristics of many of the stars are so similar that there is considerable evidence in support of the theory that they have a common origin.

The star Betelgeuse must be considered an exception to this class, for it is quite different from the other bright members of the constellation. Its color is a distinct yellow-orange as contrasted with the blue-white tint of the typical "Orion star."

The middle "star" of the sword is not a star at all, but is a huge emission nebula. This is one of the very few nebulae that can be seen to any degree of satisfaction through a small instrument. Of course, up to a certain point, the larger the instrument, the finer the view.

Several of the bright stars in the constellation are multiple objects, and the possessor of a 4-inch telescope will find the stars of Orion worth more than a passing glance. (See map accompanying entry on **Constellations**.)

ORNITHOLOGY. The scientific study of birds. Oölogy is a special division dealing with the eggs of birds. See also **Birds**.

OROGENY. Literally, the process of formation of mountains. The term came into use in the middle of the nineteenth century, when the process was thought to include both the deformation of rocks within the mountains, and the creation of the mountainous topography. Only much later was it realized that the two processes were mostly not closely related, either in origin or time. Today, many geomorphologists and a few geologists use orogeny for the formation of mountainous topography; most geologists regard this process as postorogenic and epeirogenic. By present geological usage, orogeny is the process by which structures within mountain areas were formed, including thrusting, folding, and faulting in the outer and higher layers, and plastic folding metamorphism, and plutonism in the inner and deeper layers. This usage has practical advantages; only in the very youngest, late Cenozoic mountains is there any evident casual relation between rock structure and surface landscapes. Little such evidence is available for the early Cenozoic, still less for the Mesozoic and Paleozoic, and virtually none for the Precambrian — yet all the deformational structures are much alike, whatever their age, and are appropriately considered as products of orogeny. *Tectogenesis*, a synonym of orogenesis, in its present meaning, is the process of by which mountainous areas are formed, e.g., folding or thrusting, without implication of the formation of mountainous topography. (*American Geological Institute, by permission.*)

OROGRAPHIC. Relating to mountains and mountain effects. Often refers to influences of mountains or mountain ranges on airflow, but also used to describe effects on other meteorological quantities such as temperature, humidity, or precipitation distribution. A major effect is orographic lifting. See also **Orographic Lifting**.

AMS

OROGRAPHIC CLOUD. See **Clouds and Cloud Formation**.

OROGRAPHIC LIFTING. Ascending air flow caused by mountains.

Mechanisms that produce the lifting fall into two broad categories: 1) the upward deflection of horizontal larger-scale flow by the orography acting as an obstacle or barrier; or 2) the daytime heating of mountain surfaces to produce anabatic flow along the slopes and updrafts in the vicinity of the peaks. The first category includes both direct effects, such as forced lifting and vertically propagating waves, and indirect effects, such as upstream blocking and lee waves. Even though this term strictly refers only to lifting by mountains, it is sometimes extended to include effects of hills or long sloping topography. When sufficient moisture is present in the rising air, orographic fog or clouds may form.

AMS

OROGRAPHIC PRECIPITATION. Precipitation caused or enhanced by one of the mechanisms of orographic lifting of moist air. Examples of precipitation caused by mountains include rainfall from orographic stratus produced by forced lifting and precipitation from orographic cumuli caused by daytime heating of mountain slopes. Many of the classic examples of locations having excessive annual precipitation are located on the windward slopes of mountains facing a steady wind from a warm ocean. As another example, wintertime orographic stratus (cap clouds) often produce the major water supply for populated semiarid regions such as the mountainous western United States, and as a result these cloud systems have been a target of precipitation enhancement, cloud-seeding projects intended to produce snowpack augmentation. Orographic precipitation is not always limited to the ascending ground, but may extend for some distance windward of the base of the barrier (upwind effect), and for a short distance to the lee of the barrier (spillover). The lee side with respect to prevailing moist flow is often characterized as the dry rain shadow.

AMS

OROGRAPHIC VORTEX. Atmospheric vortex or whirlwind produced by flow over or past mountains and other obstacles. Orographic vortices exist over a wide range of scales and orientations, from eddies of a few tens to a few hundreds of meters across, oriented in any direction, and shed by individual peaks or other topographic obstacles, to synoptic-scale cyclones, vertical vortices that form or intensify in the lee trough downwind of mountain-range scale barriers. Eddies of several hundred meters to a few tens of kilometers across (the larger scale representing approximately the scale of the mountain producing them) contribute to aircraft turbulence and enhance damage during downslope windstorm conditions. They often are the result of periodic shedding from the obstacle that produced them. Especially strong vertical vortices have been called "mountainadoes," indicating a resemblance to mountain tornadoes. Under strong convective heating conditions vortices spawned in the mountains sometimes continue downwind over the heated plains and participate in the initiation of dust devils. See also **Winds and Air Movement**.

AMS

OROGRAPHY. 1. The nature of a region with respect to its elevated terrain. 2. That branch of geomorphology that deals with the disposition and character of hills and mountains.

AMS

ORPIMENT. This mineral, like realgar, its frequent associate, is an arsenic sulfide. Oripment, however, is the trisulfide corresponding to the formula As_2S_3. It is monoclinic, usually in foliated, granular, or powdery aggregates; hardness, 1.502; specific gravity, 3.49; with a resinous to somewhat pearly luster; color, various shades of lemon yellow; translucent to nearly opaque. Oripment is found in association with realgar, although a somewhat rarer mineral. It is believed to be formed from the alteration of other arsenic-bearing minerals. It occurs in Czechoslovakia, Romania, Macedonia, Japan, and in the United States in Utah, Nevada, and

Wyoming. The name oripment is derived from a corruption of the Latin *auripigmentum*, meaning golden paint, because of its color as well as the belief that it contained gold.

ORTHOCLASE.   See **Feldspar**.

ORTHOGONAL ANTENNAS.   In radar, a pair of transmitting and receiving antennas, or a single transmitting-receiving antenna, designed for the detection of a difference in polarization between the transmitted energy and the energy returned from the target.

ORTHOGONAL FUNCTION.   The word *orthogonal* means, in general, *perpendicular*, thus the three axes of a rectangular Cartesian coordinate system are orthogonal in pairs, hence mutually orthogonal. Conditions for orthogonality are readily expressible in vector notation, for if the scalar product of two vectors in two dimensions vanishes, $\mathbf{A} \cdot \mathbf{B} = 0$, the two vectors are perpendicular to each other or orthogonal. The concept is easily generalized to n-dimensional space by assuming two quantities with components $A_i, B_i (i = 1, 2, \ldots, n)$ and they are then orthogonal if

$$\sum_{i=0}^{n} A_i B_i = 0$$

If the vector space involved has an infinite number of dimensions and if the components A_i, B_i are continuously distributed so that the index i becomes a continuous variable, the two functions are orthogonal if

$$\int_a^b \mathbf{A}(x) \cdot \mathbf{B}(x) \, dx = 0$$

The limits of integration are needed to specify the range of x for which the functions are defined. They may be finite or infinite.

Arbitrary functions may be made orthogonal by the Schmidt process. Suppose such a set $f_1(x), f_2(x), \ldots$ is defined over the range $a \leq x \leq b$, so that

$$\int_a^b f_i(x) f_j(x) \, dx = 0; i \neq j$$

Then presumably

$$\int_a^b f_i^2(x) \, dx = c_i^2, \text{a constant}$$

It is frequently convenient to redefine the functions so that they are *normalized*. Thus, if $c_i \phi_i(x) = f_i(x)$, then

$$\int_a^b \phi_i(x) \phi_j(x) \, dx = \delta_{ij}$$

and the functions $\phi_i(x)$ are said to form an *orthonormal* set. If x is complex, generalization of this procedure is possible.

A set of functions $f_i(x)$ is *complete* if an arbitrary function $F(x)$ satisfying the same boundary conditions as the functions of the set can be expanded as

$$F(x) \sum_{i=1}^{\infty} A_i f_i(x)$$

the A_i being constant coefficients. If the set is an orthonormal one, the expansion coefficients are readily found by integration

$$A_i = \int F(t) f_i(t) \, dt$$

with appropriate limits to the definite integral. Similar, but more complicated integrals result when the functions are not normalized. These procedures are generalizations of those used in a Fourier series expansion.

ORTHOGONAL GROUP.   See **Lie Group**.

ORTHOGONAL POLYNOMIALS.   If it is desired to fit a polynomial function

$$y = \alpha_o + \alpha_1 x + \alpha_2 x^2 \ldots$$

to observed data by the method of least squares, ordinary multiple regression techniques can be used, but it is often more convenient to use orthogonal polynomials, especially if the degree of the required polynomial is unknown. These polynomials are such that, if the polynomial is of degree r, then

$$P_0(x) = 1 \text{ and } \sum_i P_r(x_i) P_s(x_i) = 0$$

for all $r \neq s$. For equally spaced points, the values of the polynomials have been fairly extensively tabulated; rewriting the equation as $y = \beta_0 + \beta_1 P_1(x) + \beta_2 P_2(x) + \ldots$, the estimates b of the β's are given by $b_r = \sum y_i P_1(x_i) / \sum [P_r(x_i)]^2$, and the sum of the squares of the residuals by $\sum y^2 - \sum \{b_r \sum [P_r(x_i)]^2\}$. This technique has the advantage that the estimate of any one of the β's is not affected by the fitting of other terms; this does not apply to direct estimates of the α's.

ORTHOGONAL SQUARES.   It may be possible to superimpose two $n \times n$ Latin squares in such a way that any letter of the first occurs just once with every letter of the second. The resulting array is called a Graeco-Latin square, and the two squares are said to be orthogonal. In certain cases further squares may be superimposed in an orthogonal manner, up to a maximum of $(n - 1)$ squares. The letters of these squares, together with the rows and columns, define $(n + 1)$ sets of $(n - 1)$ contrasts which together account for all the contrasts between the n^2 cells of the square. A complete set of $(n - 1)$ orthogonal squares has been shown to exist for $n = p^s$ where p is prime; no Graeco-Latin square exists for $n = 6$, but contrary to an old conjecture, a square exists for $n = 10$. See also **Latin Square**.

ORTHOGRAPHIC PROJECTION.   A method of representing solid objects on a plane surface, using parallel rays or projectors, in contrast to a cone or rays as used in perspective. The rays are perpendicular to the plane of representation, in contrast to oblique representation or projection, in which the rays are parallel, but at an angle to the plane of representation. In auxiliary projection, the plane of representation is inclined with respect to the principal axis of the object, although the rays are perpendicular to the plane of representation, and usually perpendicular to an inclined face of the object.

(a) Orthographic projection

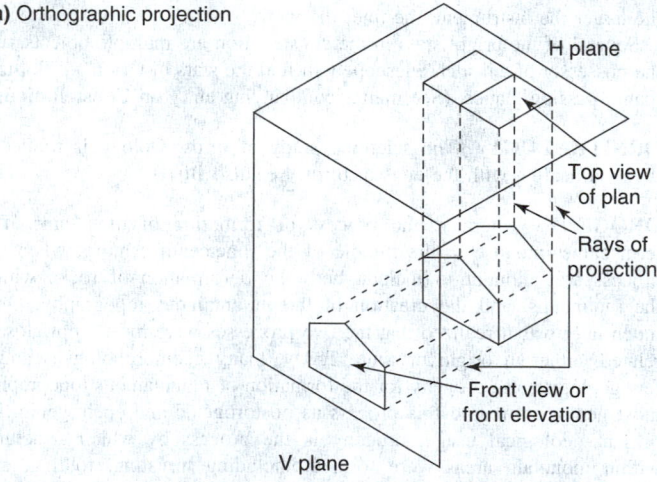

(b) Orthographic views

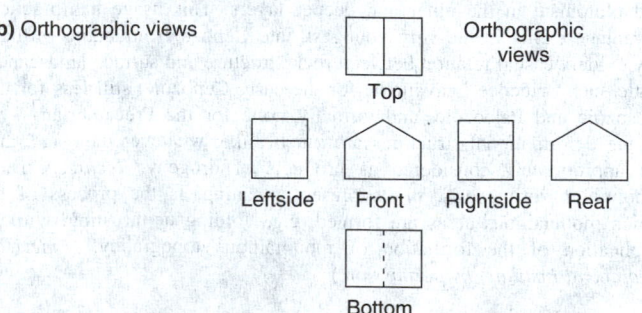

Fig. 1.   **(a)** Orthographic projection. **(b)** Orthographic views.

Six principal orthographic views are possible, although usually only four: the top view or plan, the front view or front elevation, and the left-side and right-side views or side elevations are employed. The usual arrangement of views in American practice is shown in Fig. 1; in Europe, a view arrangement in which the top view is below the front view, and the so-called right-side view is to the left of the front or top view, referred to as first angle projection, is extensively used.

See also **Perspective**.

ORTHOKERATOLOGY.

Orthokeratology (ortho-K) is a nonsurgical system of treating myopia (nearsightedness) and astigmatism with a sequential series of specialized contact lenses that gradually flatten the cornea. The procedure uses rigid gas permeable contact lenses with very high oxygen permeability to gradually change the curvature of the front surface of the eye. These lenses are worn for a specific period of time then changed for new lenses with a slightly different shape. This repetitious method of fitting and changing the curves allows the anatomical shape of the cornea to slowly change, reducing the dependence on contact lenses or eyeglasses. Orthokeratology produces no permanent change in the structure of the cornea.

Although the procedure has been in existence for many years, recent improvement in lens technology and the availability of the corneal topographer, used for mapping the surface of the eye, have made the process more precise and dependable. The procedure works best for up to four diopters of myopia and two diopters of astigmatism. Refractive errors above this level may be reduced but not totally corrected.

To understand how orthokeratology works, it is first necessary to understand the visual function of the eye. See also **Visual Function (Eye)**.

If the cornea is too steep or if the eye is too long from front to back, light rays are focused in front of the retina, resulting in myopia. If the front of the cornea is unevenly curved, it causes the light rays to fall on different spots on the retina, resulting in astigmatism.

The objective of ortho-K is to correct nearsightedness and/or astigmatism by reshaping the cornea in order to correct for imperfections.

Ortho-K contact lenses, called corneal molds, are fitted in progressive stages to gradually reshape the cornea towards less curvature and a more spherical shape. Similar to a contact lens, the mold works as you wear it, whether awake or asleep. The program length usually varies between three and six months, depending on the degree of visual error and the rates of change. Over this period of time, you gradually decrease the time you wear the mold to establish the minimum wear time to maintain the corneal shape and good functional vision.

During the first phase of the program, lasting several days to several weeks, the visual changes occur quite rapidly, requiring frequent examinations and progressive lens changes. The second phase of the process, called the holding phase, lasts from three to six months and is designed to train the corneal tissue to retain its shape over a 24-hour period. In the final phase, lenses called retainer lenses are prescribed to stabilize the new corneal shape. The schedule for wear of these lenses is determined on an individual basis, but usually begins with full-time and then is reduced to short periods of the day or overnight. The cornea will return to its original shape if the retainer is not work on an ongoing basis.

Orthokeratology is a controversial procedure, according to some eye doctors. See also **Astigmatism**; **Cornea**; **Myopia (Nearsightedness)**; and **Retina**.

Vision Rx, Inc., Elmsford, NY

ORTHOPTERA.

Insects of many forms with many common names, constituting one of the larger orders. The principal members of the group are the grasshoppers, crickets, mantises, stick insects, and cockroaches. Locusts are grasshoppers. Some authorities place these major forms in separate orders, but the general tendency is to group them together.

The order is characterized by biting mouthparts and gradual metamorphosis (Paurometabola). When wings are present there are two pairs, the front wings somewhat thickened as tegmina and the hind pair broad and membranous, folding beneath the tegmina when at rest.

These insects have a wide range of habits and occupy many terrestrial habitats. The locusts are important crop pests in some areas and cockroaches are common household pests.

The main families of Orthoptera include:

Blattidae	Roaches and cockroaches
Gryllidae	Crickets, tree crickets, mole crickets
Locustidae (formerly *Acrididae*)	Grasshoppers or locusts
Mantidae	Praying mantis
Phasmidae	Walking-sticks, walking-leaves
Tettigoniidae	Long-horned grasshopper, green meadow grasshoppers, katydids, cave and camel crickets

ORTHOSCOPIC SYSTEM.

An optical system corrected for both distortion and spherical aberration. Also called rectilinear system.

ORTHO-STATE.

1. In diatomic molecules, such as hydrogen molecules, the ortho-state exists when the spin vectors of the two atomic nuclei are in the same direction (i.e., parallel), whereas the para-state is the one in which the nuclei are spinning in opposite directions.

2. In helium the ortho-state is characterized by a particular mode of coupling of the electron spins. See also **Helium**.

ORTHOTOMIC SYSTEM.

An optical system that contains only rays which may be cut at right angles by a properly constructed surface.

ORYX.

See **Antelope**.

OSBORN, HENRY FAIRFIELD (1857–1935).

Henry Osborn was an American evolutionary biologist and powerful, long-time head of the American Museum of Natural History, New York.

Born into a wealthy old stock New England family, Osborn did not become interested in natural history until his junior year at Princeton College (later University). Under the tutelage of the Swiss geologist Arnold Guyot and Scottish philosopher James McCosh, he absorbed a belief that evolution was a purposeful, divinely ordered process. Later studying abroad with Francis Balfour and T. H. Huxley, Osborn added a strict empiricism to his metaphysical beliefs. Finally, working with the eccentric Edward Drinker Cope of Philadelphia, Osborn developed a human evolutionary system that sought to combine predestination with free will, which he called *aristogenesis*. See also **Cope, Edward Drinker (1840–1897)**; **Huxley, Thomas Henry (1825–1895)**.

As head of the American Museum of Natural History, Osborn helped transform that institution from a dusty Victorian cabinet of curiosities into a major research centre. Osborn is best known for being the sponsor of numerous large-scale fossil-hunting expeditions led by Barnum Brown, Walter Granger, William Diller Matthew, William Beebe, and others. The most spectacular of these was the Central Asiatic Expedition to the Gobi Desert of Mongolia in the 1920s. Led by Roy Chapman Andrews, the expedition was an attempt to prove that Central Asia was the cradle of humankind, but instead brought back a large cache of dinosaur fossils including the first known dinosaur eggs. Osborn named both *Tyrannosaurus rex* (found by Brown) and *Veloceraptor* (found by Andrews). See also **Dinosaurs (Dinosauria)**; **Paleontology (The History)**.

BRIAN REGAL, Boston, MA

OSCILLATION.

An oscillation is one complete period of vibratory or periodic motion, for example, the whole succession of states that takes place before the motion begins to repeat itself. For example, one oscillation of a pendulum bob is a complete excursion from where it started back to its original position with the *same* velocity (magnitude and direction). The time of one oscillation is called one period and the number of oscillations per second (the reciprocal of the period) is called the frequency. The definition cannot be applied strictly to nonperiodic motions. In such motions the period is usually taken as the time between successive zeros of the displacement.

There are three types of oscillations: (1) Oscillations that continue in a circuit or system after the applied force has been removed, the frequency of the oscillations being determined by the parameters in the system or circuit, commonly referred to as shock-excited oscillations. (2) The oscillation of some physical quantity of a body or system when the externally applied forces consist either of those which do no work, or of those which are derivable from a potential that is invariant during the time under consideration, or both. (3) That type of oscillatory motion into which a suitable system not subject to external driving forces is capable of being excited by a displacement from an equilibrium position.

The frequency of oscillation is the number of complete oscillations of a given system per unit time, commonly symbolized by v or f. The frequency is the reciprocal of the period, the time for one complete oscillation. Sometimes the angular frequency, symbolized by ω, is used for greater convenience in manipulating trigonometric functions. The angular frequency has the unit radian per unit time and is equal to $2 \times$ (frequency).

Modes of Oscillation. In the case of standing waves, the boundary conditions restrict the possible frequencies of oscillation to a discrete set of values. The whole set constitutes the modes of oscillation of the system. The oscillations of a string clamped at both ends, and the oscillations of sound waves in a closed or open pipe are examples of cases where boundary conditions impose modes.

A degenerate oscillating system is a vibrating system with several degrees of freedom in which the frequencies associated with two or more degrees of freedom may be equal in magnitude.

OSCILLATOR.   As applied to electronic equipment, a circuit that generates a periodic signal. The waveform may be sinusoidal or nonsinusoidal. As the term is most often used, however, it refers to a source signal which has a waveform closely approximating a sine wave. The oscillator is the heart of radio transmitters used for communication, navigation, radar, and similar functions requiring a source of radio-frequency energy, since it generates the high-frequency carrier signal essential for achieving the desired function. The early oscillators used in radio consisted of inductance-capacitance circuits to which a surge of electrical energy was applied by the breakdown of a spark gap or an arc. This energy then surged back and forth between the inductance and capacitance until it was all dissipated as radiation and as circuit losses. This type of oscillator produced damped oscillations and is no longer used. Modern oscillators use vacuum tubes or transistors in various circuit arrangements. Since both of these electronic elements provide amplification, they can be used as oscillators by feeding back some of the output energy to the input circuit so that the device effectively drives itself, i.e., furnishes its own input signal. The various oscillator circuits in common use effect this function by different means. Shown in the figures are various circuits employing transistors and vacuum tubes which find use as practical oscillators.

For extremely stable frequency characteristics, crystal-controlled oscillators, one form of which is shown in Fig. 1, are most often used. In these use is made of the frequency selective properties obtainable from a quartz crystal by means of the piezoelectric effect. An extremely sharp resonance characteristic is obtained thereby which permits feedback of signal from output to input inadequate to sustain oscillation only over a very narrow range of frequencies. This results is extremely good frequency stability; a variation of 1 part in 20,000,000 in representative of this type of oscillator.

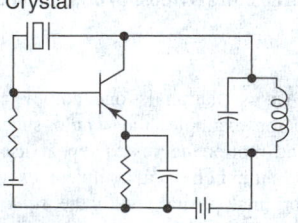

Fig. 1.   Crystal oscillator.

Where continuously adjustable frequency output is needed some type of self-controlled oscillator is needed. The Hartley oscillator, shown in Fig. 2 is one of the simplest. The energy is fed back from output to input circuits through the inductive coupling of the two sections of the coil. The

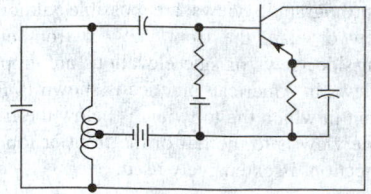

Fig. 2.   Hartley oscillator.

frequency is determined by the inductance and capacitance values in the tuned circuit.

A very similar circuit is the Colpitts oscillator shown in Fig. 3. It differs from the Hartley circuit only in the manner in which energy is fed back from the output to the input, the coupling being accomplished by a capacitance voltage divider rather than a tapped coil.

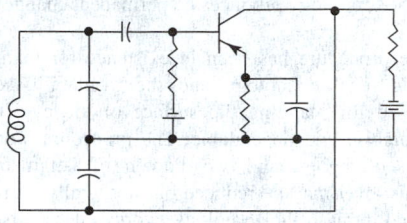

Fig. 3.   Colpitts oscillator.

A transistor blocking oscillator circuit is shown in Fig. 4.

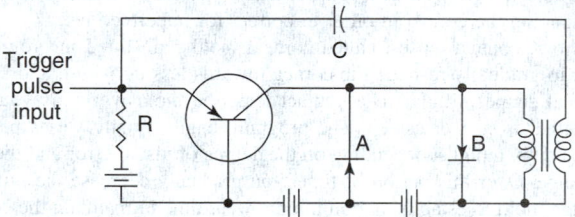

Fig. 4.   Transistor blocking oscillator circuit.

OSCILLOMETRY.   The measurement of capacitance produced by the dielectric constant and conductivity of a liquid sample in a cell. The cell is all glass, constructed as two concentric cylinders with bonded inner and outer metal surfaces to form the plates of a capacitor, thus the sample contacts only the glass walls. See Fig. 1.

A frequency of 5 MHz is used for small capacitive reactances of the cell walls.

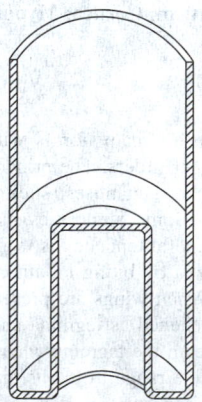

Fig. 1.   Sectional view of cell used in oscillometry.

A dry cell (air filled) is connected to an oscillator circuit which is tuned to a resonant frequency of 5 MHz, by adding parallel capacitance inside the measuring instrument. The sample is then inserted into the cell with a resulting increase in the cell capacitance and calibrated parallel capacitance in the instrument removed until a return to the resonant condition is indicated. Thus the amount of capacitance removed is equal to the capacitance change of the cell.

Chemical titrations may be performed as point plotted capacitance change or by observing frequency change indicated on a meter or recording device. Dielectric constant measurements are determined by calibration using known standards or by mathematical equations derived using simulated cell constants. Solution conductivity effects a change in capacitance in the cell which cannot be distinguished from changes due to the dielectric constant. Generally conductivity measurements are made using known standards and the assumption that the dielectric constant is unchanging.

OSCILLOSCOPE. An instrument primarily for making visible the instantaneous value of one or more rapidly varying electrical quantities as a function of time or of another electrical or mechanical quantity (IEEE). An *oscillograph* is an oscilloscope capable of recording as well as displaying the aforementioned measurements and is a form of $X - Y$ recorder. For decades the oscilloscope has been a workhorse in the development, testing, and troubleshooting of electrical and electronic equipment. A number of refinements (improving the performance, making the instrument more convenient to use, etc.) have been made, but the basic principles remain. History records that the first known use of a cathode-ray tube, the primary element of the oscilloscope, to display changing phenomena occurred as early as 1897, which was the same year that the tube was used to prove that cathode rays were electrified and would respond to a magnetic field.

An oscilloscope consists of several basic elements: (1) cathode-ray tube (CRT), (2) time-base generator, (3) vertical deflection channel, and (4) power supply. Oscilloscope construction over the years has taken full advantage of modern solid state, semiconductor electronics, as well as improvements in phosphors which are used in CRTs for other applications. Some CRT-based terminals and workstations are, in essence, highly sophisticated oscilloscopes.

See also **Spectrum Analysis**.

OSCULATING ORBIT. The ellipse that a satellite would follow after a specific time t (the epoch of osculation) if all forces other than central inverse-square forces ceased to act from time t on. An osculating orbit is tangent to the real, perturbed, orbit and has the same velocity at the point of tangency.

OSCULATING PLANE. The plane, if it exists, through a point P of a space curve which has contact of higher order with the curve than any other plane through the point; the limiting position of the plane through P, P', P'' as the points P', P'' approach P along the curve. See also **Curve (Space)**.

OSMIUM. [CAS: 7440-04-2]. Chemical element, symbol Os, at. no. 76, at. wt. 190.2, periodic table group 8, mp 3,015° to 3,075°C, bp 4,927° to 5,127°C, density 22.6g/cm³ (solid), 22.8g/cm³ (single crystal) (20°C). Elemental osmium has a close-packed hexagonal crystal structure. Compact osmium is a bluish-white metal and is not attacked by acids. Discovered by Tennant in 1804. The seven stable isotopes of osmium are ^{184}Os, ^{186}Os through ^{190}Os, and ^{192}Os. Electronic configuration $1s^2 2s^2 2p^6 3s^2 3p^6 3d^{10} 4s^2 4p^6 4d^{10} 4f^{14} 5s^2 5p^6 5d^6 6s^2$. Ionic radius Os$_4^+$ 0.65 Å. Metallic radius 1.3377 Å. Oxidation potential Os + $4H_2O \rightarrow OsO_4 + 8H^+ + 8e^-$, −0.85V.

Chemical Properties

Finely divided Os oxidizes in air, producing the poisonous and volatile tetroxide. The compact metal is not attacked by nonoxidizing acids. The finely divided metal dissolves in fuming HNO_3, aqua regia, and alkaline hypochlorite solutions. When fused with Na_2O_2 or KNO_3 and KOH, the metal is converted to the corresponding water-soluble osmate, K_2OsO_4.

The brown or black insoluble osmium(IV) oxide, OsO_2, can be made by heating Os with a limited amount of O_2 or with osmium(VIII) oxide. This compound forms a brown to black-blue dihydrate that can be prepared by reducing a solution of the tetroxide or by hydrolyzing a solution of sodium hexachloroosmate, Na_2OsCl_6.

Osmium(VIII) oxide, the most important compound, is formed in one of the reactions unique to the platinum metals. Its ease of formation and volatility make it useful in a purification step for the refining or analysis of Os. The tetroxide is readily formed by heating the metal in air or distilling an osmium-containing solution from HNO_3. Although an aqueous solution of osmium(VIII) oxide is neutral to litmus, it is a weak acid with first dissociation constant of about 8×10^{-13}. Osmium(VIII) oxide is soluble in water, alcohol, and ether. The compound is widely used as a stain for tissues. When an alkaline solution of osmium(VIII) oxide is reduced with alcohol or KNO_2, an osmate(VI) is formed. Potassium osmate(VI) is formed by adding an excess of KOH to such a solution, resulting in the precipitation of violet crystals of $K_2OsO_4 \times 2H_2O$. The osmate(VI) ion is probably better written as $OsO_2(OH)_2$.

Osmium(II) chloride can be prepared by heating osmium(III) chloride in vacuum at 500°C. This dark-brown compound is insoluble in HCl or H_2SO_4. NHO_3 or aqua regia oxidizes it to the tetroxide. Osmium(III) chloride is best made by decomposing ammonium hexachloroosmate(IV), $(NH_4)_2OsCl_6$, in a Cl_2 stream at 350°C. The brown hygroscopic powder sublimes above 350°C, and at about 560°C it disproportionates into the tetrachloride and dichloride. Osmium(IV) chloride is formed from the elements at 650–700°C. The black compound slowly dissolves in water, eventually forming the dioxide. The free acid, H_2OsCl_6, is stable in solution and can be made by refluxing osmium(VIII) oxide with HCl and alcohol. The ammonium salt can be precipitated by adding NH_4Cl to such a solution. This salt is reduced to the metal when heated in H_2. The potassium salt is well known. Both are brownish-red solids yielding orange solutions in water.

Recent studies have established the reaction product of Os metal and F_2 at 300°C to be the hexafluoride, OsF_6. This yellow volatile solid had previously been described as an octafluoride. Osmium(VI) fluoride melts at 33.4°C and boils at 47.5°C. OsF_6 can be reduced to a pentafluoride and a tetrafluoride. The pentafluoride is a blue-gray crystalline solid that melts at 70°C to a green viscous liquid and boils at 226°C. The tetrafluoride distills at about 290°C. Potassium hexafluoroosmate(V), $KOsF_6$, can be made by reacting KBr, osmium(IV) bromide, and bromine trifluoride. The white powder dissolves in water to form a colorless solution that hydrolyzes to yield some osmium(VIII) oxide. On addition of 1 equiv of KOH to a fresh solution, an orange color develops, O_2 is evolved, and yellow crystals of potassium hexafluoroosmate(IV), K_2OsF_6, form.

Os forms many complexes with nitrite, oxalate, carbon monoxide, amines, and thio ureas. The latter are important analytically. Osmium forms the interesting aromatic "sandwich" compound, osmocene. A *metallocene* is described under **Ruthenium**. See also **Chemical Elements**; and **Platinum and Platinum Group**.

Proton nuclear magnetic resonance (NMR) is a widely used tool for researching biomolecules. Although much too detailed to report here, Zai-Wei Li and Henry Taube (Stanford University) reported in 1992 that they have had success in analyzing for certain molecules by using a dihydrogen osmium complex on a versatile 1H NMR recognition probe.

Osmium Isotopic Ratios in Paleogeology

As pointed out in 1983 by J.M. Luck and K.K. Turekian (Yale University), one of the most creative concepts regarding the cause of the many paleontologic extinctions at the Cretaceous-Tertiary boundary is the one put forward by Alvarez et al. in 1980 involving the impact of a large asteroid or comet with the earth at that geologic time period. The Alvarez hypothesis stemmed from the finding of an exceptional chemical signature (high iridium concentration) at the Cretaceous-Tertiary boundary at Gubbio, Italy, a signature that was later found in other marine as well as continental sections which bracket the Cretaceous-Tertiary boundary. Because of the radioactive decay of rhenium-187 (4.6×10^{10} years), the osmium-187/osmium-186 ratio changes in planetary systems as a function of time and the rhenium-187/osmium-186 ratio. For a value of the ^{187}Re/^{186}Os ratio of about 3.2, typical of meteorites and the earth's mantle, the present ^{187}Os/^{186}Os ratio is about one. The earth's continental crust has an estimated $^{187}Re/^{186}Os$ ratio of about 400. Thus for a mean age of the continent of 2×10^9 years, a present $^{187}Os/^{186}Os$ ratio of about 10 is expected. Marine manganese nodules show values (6 to 8.4), which are compatible with this expectation if an allowance for a 25% mantle osmium supply to the oceans is made. The Cretaceous-Tertiary boundary iridium-rich layer in the marine section at Stevns Klint, Denmark, yields a $^{187}Os/^{186}Os$ ratio of 1.65 and the one in a continental section in the Raton Basin, Colorado, is 1.29. The investigators conclude that the simplest explanation is that these represent osmium imprints of predominantly meteoritic origin.

As reported in 1992 by M.F. Horan, J.W. Morgan and J.N. Grossman (U.S. Geological Survey), and R.J. Walker (University of Maryland), "Rhenium and osmium concentrations and the osmium isotopic compositions of iron meteorites were determined by negative thermal ionization mass spectrometry. Data for the IIA iron meteorites define an isochron with an uncertainty of approximately ±31 million years for meteorites ~4500 million years old. Although an absolute rhenium-osmium closure age for this iron group cannot be as precisely constrained because of uncertainty in the decay constant of ^{187}Re, an age of 4460 million years ago is the minimum permitted by combined uncertainties. These age constraints imply that the parent body of the IIAB magmatic irons melted and subsequently cooled within 100 million years after the formation of the oldest portions of chondrites. Other iron meteorites plot above the IIA isochron, indicating that the planetary bodies represented by these iron groups may have cooled significantly later than the parent body of the IIA irons."

Additional Reading

Anderson, D.L.: "Composition of the Earth," *Science*, 367 (January 20, 1989).

Cherfas, J.: "Proton Microbeam Probes the Elements," *Science*, 1150 (September 28, 1990).

Davis, J.R.: *Metals Handbook*, 2nd Edition, ASM International, Materials Park, OH, 1998.

Greenwood, N.N. and A. Earnshaw: *Chemistry of the Elements*, 2nd Edition, Butterworth-Heinemann, Inc., Woburn, MA, 1997.

Horan, M.F., et al.: "Rhenium-Osmium Isotope Constraints on the Age of Iron Meteorites," *Science*, 1118 (February 28, 1992).

Krebs, R.E.: *The History and Use of Our Earth's Chemical Elements: A Reference Guide*, Greenwood Publishing Group, Inc., Westport, CT, 1998.

Lewis, R.J., Sr.: *Hawley's Condensed Chemical Dictionary*, 13th Edition, John Wiley & Sons, Inc., New York, NY, 1997.

Lide, D.R.: *CRC Handbook of Chemistry and Physics* 86th Edition, CRC Press, LLC., Boca Raton, FL, 2005.

Luck, J.M. and K.K. Turekian: "Osmium-187/Osmium-186 in Manganese Nodules and the Cretaceous-Tertiary Boundary," *Science*, **222**, 613–615 (1983).

Sinfelt, J.H.: "Bimetallic Catalysts," *Sci. American*, **253**(3), 90–98 (1985).

Staff: *ASM Handbook—Properties and Selection: Nonferrous Alloys and Pure Metals*, ASM International, Materials Park, OH, 1990.

Zai-Wei, Li and H. Taube: "Use of a Dihydrogen Osmium Complex as a Versatile 1 H NMR Recognition Probe," *Science*, 210 (April 10, 1992).

OSMOSIS.   Passage of a pure liquid (usually water) into a solution (e.g., of sugar and water) through a membrane that is permeable to the pure water but not to the sugar in the solution. This passage can also occur when the two phases consist of solutions of different concentration. The membrane is called semipermeable when the molecules of the solvent, but not those of the solute, can penetrate it. This pushing of water through a membrane into a solution results from the greater tendency of water molecules to escape from water than from a solution. The term *osmosis* us usually restricted to movement through a solid or liquid barrier that prevents the phases from mixing rapidly. In test apparatuses parchment or collodion membranes are used; in plants and animals the cell wall acts as a diffusion barrier. The pressure exerted by osmosis is substantial and accounts for the elevation of sap from root systems to the tops of trees. Osmosis is considered an essential characteristic of growth.

Reverse osmosis is used as a method of desalting seawater, recovering wastewater from paper mill operations, pollution control, industrial water treatment, chemical separations, and food processing. This method involves application of pressure to the surface of a saline solution, thus forcing pure water to pass from the solution through a membrane that is too dense to permit passage of sodium and chlorine ions. Hollow fibers of cellulose acetate or nylon are used as membranes, since their large surface area offers more efficient separation.

See also **Desalination**; **Diffusion**; and **Membrane Separations Technology**.

OSMOTIC COEFFICIENT.   A factor introduced into equations for nonideal solutions to correct for their departure from ideal behavior, as in the equation:

$$\mu = \mu_x^0 + gRT \ln x_1$$

in which μ is the chemical potential, μ_x^0 is a constant, representing a standard value of the chemical potential, R is the gas constant, T the absolute temperature, x_1 is the mole fraction of solvent, and g is the osmotic coefficient.

OSMOTIC PRESSURE.   Pressure that develops when a pure solvent is separated from a solution by a semipermeable membrane which allows only the solvent molecules to pass through it. The osmotic pressure of the solution is then the excess pressure which must be applied to the solution so as to prevent the passage into it of the solvent through the semipermeable membrane.

Because of the similarity in the relations for osmotic pressure in dilute solutions and the equation for an ideal gas, van't Hoff proposed his bombardment theory in which osmotic pressure is considered in terms of collisions of solute molecules on the semipermeable membrane. This theory has a number of objections and has now been discarded. Other theories have also been put forward involving solvent bombardment on the semipermeable membrane, and vapor pressure effects. For example, osmotic pressure has been considered as the negative pressure which must be applied to the solvent to reduce its vapor pressure to that of the solution. It is, however, more profitable to interpret osmotic pressures using thermodynamic relations, such as the entropy of dilution.

A number of methods have been developed for measurement of osmotic pressure.

In the Berkeley and Hartley method, a porous tube with a semipermeable membrane such as copper ferrocyanide deposited near the outer wall, and a capillary tube attached to one end contains the pure solvent. The solution surrounds the tube and is enclosed in a metal vessel to which a pressure may be applied which is just sufficient to prevent the flow of solvent into the solution. Berkeley and Hartley also developed a dynamic method for measuring osmotic pressure.

Simple osmometers have also been developed by Adair particularly for aqueous colloidal solutions. A thimble-type collodion membrane is attached to a capillary tube and contains the solution. When equilibrium is established the difference in level inside and outside the capillary is measured. Capillary corrections are made. For organic solvents a dynamic type osmometer may be used. A membrane of large surface area is clamped between two half cells and attached to each half cell is a fine capillary observation tube. With such an apparatus, equilibrium is rapidly established between solution and solvent contained in the half cells. The volume of the half cell may be small (about 20 cubic centimeters). The level of the solvent is usually arranged to be a little below the equilibrium position, and the height of the solvent in the capillary as a function of time is measured. This procedure is repeated with the level of the solvent just above the equilibrium position. A plot is then made of the half sum of these readings.

Since the osmotic pressure is related to the concentration of dissolved solute particles, it is related to the lowering of the freezing point and elevation of the boiling point.

The relation between osmotic pressure and lowering of the freezing point and the elevation of the boiling point may be expressed by the relation:

$$\Pi = \frac{LT}{\overline{v} \, T_0^2} \Delta T$$

where L is the molar heat of fusion or of vaporization of the solvent, T the temperature at which the osmotic pressure is measured, T_0 the freezing point or the boiling point of the solvent, $\overline{v}$, the partial molar volume of the solvent, and Π, the osmotic pressure.

Moreover, the relation to lowering of the vapor pressure is

$$\Pi = \frac{RT}{\overline{v}} \ln \frac{p_0}{p}$$

or

$$\Pi = -\frac{RT}{\overline{v}} \ln x_0 = -\frac{RT}{v} \ln(1-x)$$

For dilute solutions,

$$\Pi = cRT$$

where Π is the osmotic pressure, $\overline{v}$ *the partial molar volume of the solvent in the solution*, p_0 the vapor pressure of pure solvent, p the partial vapor pressure of the solvent in equilibrium with the solution, x_0 the mole fraction of the solvent, x the mole fraction of the solute, c the concentration of the solution in moles per liter, R, the gas constant, and T, the absolute temperature.

OSMOTIC PRESSURE (Cell).   See Cell (Biology).

OSTEICHTHYES (Bony Fishes).

The *Osteichthyes*, or bony fishes, are the most abundant group of fishes at the present day, as well as throughout the recent fossil history of the fishes. They can be divided into two major groups—the *Sarcopterygii*, or lobe-finned fishes, and the *Actinopterygii*, or ray-finned fishes. The *Sarcopterygii* includes the Tetrapoda as its most derived clade, because when lobe-finned fish colonized the land, they evolved into tetrapod land vertebrates, thus making the 'fish' part of the group paraphyletic. Each group has a rich fossil record, and each has representatives alive today. The *sarcopterygians* are represented by the coelacanth and six species of lungfish, while the *actinopterygians* are represented by more than 22 000 species of teleosts, as well as two types of birchir, 36 species of sturgeons and paddlefishes, one species of bowfin and seven species of gar. See also **Fishes**; **Tetrapod Limbless Locomotion**; **Tetrapod Walking and Running**; and **Tetrapod Climbing and Swinging**.

Osteichthyan fishes are characterized by the presence of a well-ossified bony internal skeleton, bony scales and fins that are supported by bony rods, or lepidotrichia. These scales and fin elements provide a high degree of rigidity and improved swimming power.

The internal anatomy of osteichthyans is unique among fishes in the presence of a gas-filled organ in the middle of the body. The sarcopterygians have this as a lung, enabling many taxa to effectively breathe air, while the actinopterygians have this as a swim-bladder, enabling them to regulate their position vertically in the water column.

The *Sarcopterygii* and the *Actinopterygii* are readily distinguishable by their fin morphology: actinopterygians have 'ray-fins' supported by a series of narrow cartilaginous or bony rods called radials, while sarcopterygians have fleshy 'lobe-fins' supported by a single basal radial and with muscles that can modify the posture of the fin.

Within this basic design, the typical view of a fish as being a slippery, grey aquatic animal that possesses fins and breathes through gills does apply to some species, but many more species deviate from this 'standard' pattern. In some fishes, the body is elongated, in others it is flattened, compressed or shortened, while the fins may be elaborately extended, reduced or absent. Size may vary from 10 mm (1/4 in) to over 20 m (66 ft), and weight may vary from 1.5 g (0.05 ounce) to thousands of kilograms (pounds). The limits of this diversity become even more apparent when we look through the fossil record of the osteichthyans.

Sarcopterygians—"Lobe-Finned" Fishes

Sarcopterygians are an extremely ancient group, with fossils being found back to the Lochkovian, in the Lower Devonian. The majority of sarcopterygians, commonly known as *rhipidistians* are extinct, but the group survives in the tetrapods and the extant sarcopterygians–the coelacanths and the lungfish. These continue to survive, though few in number and diversity. See also **Coelacanths (Osteishthyes)**; and **Lungfishes (Osteishthyes)**.

The early members of all sarcopterygian groups were very similar in shape to each other, and also to early actinopterygians, while the extant forms are morphologically extremely derived.

Coelacanths originated in freshwater, but moved to a mainly marine habitat in the Permian and Triassic. They have a fossil record reaching to the Cretaceous, and then no examples were known until 1938, when one was caught off the coast of Southern Africa. Typical coelacanths have short bodies with large dorsal, anal and paired fins. The tail is divided into three sections. They also possess many unique skull features, such as absence of the maxilla, and presence of numerous paired snout bones containing large sensory pores. Coelacanths, even the recent *Latimeria chalumnae*, have maintained a relatively stable gross anatomy through time.

From their origin, lungfishes (or dipnoans) were a highly characteristic group, easily recognized by a number of features in their skull roof and dentition. They were built to allow a strong bite, with numerous interconnected skull bones relating to powerful jaw muscle insertions on the inside of the skull roof. The massive size of the lower jaws, and the specialized dental arrays, also reflects a skull capable of exerting great power in the bite. The bodies have two equally-sized dorsal fins and long feathery pectoral and pelvic fins. The extant forms show a varying degree of differentiation from the primitive dipnoan form. The Australian *Neoceratodus* is by far the most primitive, having changed little in shape in more than 100 million years. The South American and African forms (*Lepidosiren* and *Protopterus*), are long, slender fishes, with greatly reduced pectoral and pelvic fins, mainly used for a sensory function.

Actinopterygians—"Ray-Finned" Fishes

The *Actinopterygii* is the largest group of fishes at the present day. They have a rich and diverse fossil record back to the Ludlow, in the Mid-Silurian.

A primitive actinopterygian can be exemplified by *Cheirolepis*. The body is slender and is covered by many microscopic scales. The tail is strongly heterocercal and there are large, triangular, dorsal and anal fins, and paired pectoral and pelvic fins. A major feature of the skull is a long gape with many sharp teeth set along the jaws. The eye is small and placed anteriorly. Several primitive fish characters are retained: a pineal opening between the frontals, and an open spiracular slit.

The most primitive extant actinopterygians are *Polypterus*, the birchir, and *Calamoichthys* (= *Erpetoichthys*), the reedfish. These are elongate fishes with numerous spines forming the dorsal fin. The next most primitive representatives are the sturgeons and paddlefishes. Sturgeons (*Huso, Acipenser* and *Scaphirhynchus*) are characterized by five series of bony scutes running across their flanks and dorsal midline, while paddlefishes (*Polyodon* and *Psephurus*) have an extremely elongate rostrum, used for sensory purposes, and filter feeding apparatuses. See also **Paddlefishes (Osteichthyes)** and **Sturgeons (Osteichthyes)**. Gars (*Lepisosteus* and *Atractosteus*), the first extant members of the *Neopterygii* (a clade composed of the teleosts, gars, bowfins, and several extinct groups of advanced actinopterygians), again are slender, with an elongate mouth full of very sharp, pointed teeth. See also **Gars (osteichthyes)**; and **Bowfin (Osteichthyes)**. *Amia*, the bowfin, shows many of the more advanced actinopterygian characters, such as a tail with the body lobe not reaching the end, an interopercular bone and the maxilla freed from the cheek series. All these small anatomical changes meant significant benefits for locomotion and feeding mechanisms.

The culmination of actinopterygian evolution was the teleosts, a group containing most of the fishes alive today. One particular key innovation gave rise to the initial teleost radiation, and that was the evolution of certain bones in the tail skeleton, called uroneurals, which stiffened the dorsal lobe of the tail and supported a series of dorsal fin rays. This simple change gave the actinopterygians greater swimming power and allowed a greater variety of body shapes to evolve. One other major change, occurring slightly later in the history of the group, was modification of the jaw structure. The premaxillae became able to move independently of the main upper jaw bone, opening up a whole new landscape of possible innovation and feeding mechanisms, which teleosts have exploited to the full. The more derived teleosts, or Acanthopterygii, have taken this versatility one step further. They have generally lost their jaw teeth, moving the dental array back onto the pharyngeal series. This has led to the acanthopterygians being able to protrude their jaws in a number of efficient ways, turning the mouth into a versatile prehensile device.

Phylogeny

There is still much to be learnt about the phylogeny of osteichthyans. Not only is there little consensus about the placing of the major sarcopterygian groups, there is also lack of resolution of the living actinopterygian clades.

Within the sarcopterygians, it is agreed that the sister group of tetrapods is a group called the panderichthyids, elongated, tetrapod-like fishes, followed by the paraphyletic osteolepiform taxa. One view puts the porolepiforms, another group of extinct sarcopterygians, as sister group to that clade (the tetrapodomorphs), followed by the coelacanths, then the lungfishes. The contrary view suggests that the porolepiforms and lungfishes form a clade, which is then sister group to the tetrapodomorph clade. Then the sister group to all of those is the coelacanths.

The living sister group of actinopterygian teleosts is also still debated. Some researchers suggest *Amia*, while others think it is *Lepisosteus*. Whichever turns out to be correct, all three groups are contained within the neopterygian clade. The sister groups to this clade are a series of ever more primitive fossil stem-group neopterygian and palaeoniscoid taxa, including the sturgeons and paddlefishes close by the neopterygian clade, and polypterids near the base of the actinopterygian clade.

There is also lack of consensus within teleost phylogeny. The largest order, the *perciforms*, is still not well characterized, and the positions of various teleost families is also widely debated. However, it is clear that, within the teleosts, the most primitive living groups are the osteoglossids (arapaimas (*Arapaima gigas*), arawanas (*Osteoglossum* spp.), elopomorphs (e.g. tarpons (Family *Megalopidae*), eels (Order *Anguilliformes*)) and clupeomorphs (e.g. herrings and sardines (Family *Clupeidae*) and anchovies (Family *Engraulidae*)). The rest of the teleosts are contained within an advanced clade, the Euteleostei. This has the salmoniforms (e.g. salmon and trout (both Family *Salmonidae*)) and the ostariophysans (huge diversity including milkfish (Family *Chanidae*), carps (Family *Cyprinidae*), catfish (Order *Siluriformes*), characins (Family *Characidae*) and electric eels (Family *Electrophoridae*)) as the basal groups, with the remaining, more advanced, teleosts designated as the Neoteleostei. The most derived forms of this group are the Percomorpha (huge diversity, e.g. opah (Family *Lampridae*), oarfish (Family *Regalecidae*), dories (Family *Zeidae*), sand eels (Family *Hypoptychidae*), trumpetfish (Family *Aulostomidae*), goatfishes (Family *Mullidae*), butterflyfish (Family *Chaetodontidae*), angelfish (Family *Pomacanthidae*), cichlids (Family *Cichlidae*), damselfish (Family *Pomacentridae*), wrasse (Family *Labridae*)) and the Atherinomorpha (flying fishes (Family *Exocoetidae*), needlefish (Family *Belonidae*), killifish (Family *Fundulidae*)), with successive sister groups of the Paracanthopterygii (e.g. cod (Family *Gadidae*), hake (Family *Merlucciidae*), frogfish (Family *Antennaridae*), anglerfish (Order *Lophiiformes*)), *Myctophiformes* (lanternfishes (Family *Myctophidae*)), *Aulopiformes* (e.g. deep sea lizardfish (Family *Synodontidae*), lancetfish (Family *Alepisauridae*)) and *Stomiiformes* (e.g. bristlemouths (Family *Gonostomatidae*), hatchetfishes (Family *Sternoptychidae*)).

See also **Anchovy and Anchoveta (Osteichthyes)**; **Angelfish (Osteichthyes)**; **Angelfishes (Osteichthyes)**; **Carp (Osteichthyes)**; **Catfishes (Osteichthyes)**; **Cichlids (Osteichthyes)**; **Codfishes (Osteichthyes)**; **Eels (Osteichthyes)**; **Hake (Osteichthyes)**; **Hatchet Fishes (Osteichthyes)**; **Herring (Osteichthyes)**; **Iniomous Fishes (Osteichthyes)**; **Milkfish (Osteichthyes)**; **Oarfish (Regalecus glesne)**; **Salmon (Osteichthyes)**; **Tarpon (Osteichthyes)**; **Trout (Osteichthyes)**; and **Wrasses (Osteichthyes)**.

Diversity and the Fossil Record

Sarcopterygians. The oldest lobefin fossils are found in the Early Devonian of Arctic Canada, Spitzbergen and China. These consist of skulls, some belonging to a small clade called the *porolepiforms*, and some that appear to be intermediate between lungfishes and other sarcopterygians. See also **Fossil Record**; and **Fossils in Phylogenetic Reconstruction**.

Generic diversity rises to around 60 taxa in the Upper Devonian, before slowly falling to the present day low. A large proportion of the early sarcopterygian radiation is composed of the *osteolepiforms*, an extinct, paraphyletic group of *rhipidistians*. Some of the most prolific collections of these come from the Old Red Sandstone beds of North Scotland, deposits which were laid down in a large subtropical lake. In the lake, smaller lobefins such as *Dipterus* (a lungfish) and *Osteolepis* were scavengers or low-level predators, while a top predator was the porolepiform *Glyptolepis*. At the end of the Devonian, most of the osteolepiforms became extinct, leaving lungfishes and coelacanths to become the dominant sarcopterygian groups.

During the later Paleozoic and the Mesozoic, coelacanths became fairly conservative in shape and size, although some managed to reach sizes of 1.5 m (5 ft). Lungfishes reached the same sizes, but quickly diverged from the ancestral bodyplan. Family and generic diversity of both groups remained low through the Mesozoic and the Tertiary. See also **Mesozoic**.

Actinopterygians. The fossil record of the actinopterygians shows a completely different story that starts with a slow increase in numbers through the Paleozoic and Mesozoic, with a huge radiation occurring later, in the Cretaceous and early Tertiary. The oldest actinopterygian fossils are scales found in the Late Silurian of China and Russia.

In general, the Paleozoic actinopterygians are a homogeneous group, but some variations on the primitive actinopterygian pattern start emerging by the Carboniferous. Over 60 genera of these primitive 'paleoniscoid' fishes are known from the late Paleozoic and early Mesozoic.

The first neopterygians are known from the Permian, though their initial radiation did not occur until the Triassic, recorded in such sites as the Northern Italian Zorzino Limestone. These fishes show the first major actinopterygian ecological radiation. Many have crushing teeth for feeding on hard-shelled invertebrates, while the variety of body plans, with some taxa fusiform and others deep-bodied, is unmatched in earlier fossil sites. Approximate 50 genera of these early neopterygians are known from the early Mesozoic.

The enormous diversity of teleosts can easily be demonstrated from living fish. Teleosts encompass marine food fishes such as cod (Family *Gadidae*), anchovy (Family *Engraulidae*) and salmon (Family *Salmonidae*), brightly coloured tropical fishes such as butterfly fishes (Family *Chaetodontidae*) and angel fishes (Family *Pomacanthidae*), freshwater fishes such as trout (Family *Salmonidae*) and perch (Family *Percidae*), and more bizarre looking fishes such as the seahorse (Family *Ephippidae*), sunfish (Family *Molidae*) and oarfish (Family *Regalecidae*). See also **Anchovy and Anchoveta (Osteichthyes)**; **Angelfishes (Osteichthyes)**; **Codfishes (Osteichthyes)**; **Oarfish (Regalecus glesne)**; **Perches and Darters (Osteichthyes)**; **Salmon (Osteichthyes)**; **Seahorses (Osteichthyes)**; **Sunfishes (Osteichthyes)**; and **Trout (Osteichthyes)**.

Habitats and Abundance

Almost all natural bodies of water bear fish life, from Arctic lakes, through mountain streams, large rivers, tropical lagoons, muddy estuaries, to the deep ocean. All these environments can be broken down into many more sub-environments, all of which will have their own individual fish faunas.

Sarcopterygians. The Australian lungfish (*Neoceratodus*) is relatively uncommon, confined to four rivers in Queensland. In contrast, the African (*Protopterus*) and the South American (*Lepidosiren*) lungfishes are more common, and live in swamps that are likely to dry up. In consequence, they can actively breathe air. As the waters recede, they dig burrows in the mud, in which they aestivate. A small tube connects the burrows to the outside, and water entering this when the rains come makes the lungfish cough, waking it up. See also **Lungfishes (Osteichthyes)**.

Latimeria, the coelacanth, is a slow-moving fish that generally drifts in Indian Ocean currents. They live at depths of 100–700 m, and feed on fishes and the occasional cuttlefish. See also **Coelacanths (Osteichthyes)**; and **Cuttlefish (Mollusca, Cephalopoda)**.

Actinopterygians

Freshwater–Temperate Streams and Rivers. It is difficult to describe a 'standard' freshwater fish fauna. Warmer water fish faunas will differ from colder water faunas, Asian assemblages will differ from African ones, which will differ from American ones.

In North America, warm, slow-moving, streams are characterized by such fish as smallmouth bass (*Micropterus dolomeiui*), catfish and a wide variety of small, or deep-bodied, fishes, such as darters (Family *Etheostomatidae*). Cold water, generally quick-moving, streams are characterized by trout (Family *Salmonidae*) and sculpins (Family *Cottidae*). In small streams, the dominant fishes usually feed on insects that drop into the water, or on detritus. In larger rivers, more trophic levels and a greater diversity, including endemic forms, are found. See also **Bass (Osteichthyes)**; **Catfishes (Osteichthyes)**; and **Sculpins (Osteichthyes)**. The Mississippi River is an important centre for fish diversity containing sturgeons (Family *Acipenseridae*), paddlefish (Family *Polyodontidae*), bowfins (Family

Amiadae) and gars (Family *Lepisosteidae*), trout-perches (Family *Percopsidae*) and pygmy sunfishes (Family *Elassomatidae*), and more modern teleosts such as perches (Family *Percidae*), pikes (Family *Esocidae*), catfishes (Order *Siluriformes*, generally Family *Ictaluridae*) and bass (Family *Moronidae*). See also **Bowfin (Osteichthyes)**; **Gars (Osteichthyes)**; and **Pike (Osteichthyes)**.

The Amazon, in South America, is home to a large and rapidly growing fish fauna. This is characterized by a large number of 'relict' fishes, such as the osteoglossomorph arawanas (*Osteoglossus* species) and arapaimas (*Arapaima gigas*), as well as characins (Family *Characidae*) such as the piranha (*Serrasalmus* sp.), tetra (various genera) and pacu (*Colossoma macropomum*), cichlids (Family *Cichlidae*) and catfish (Order *Siluriformes*). See also **Characids (Osteichthyes)**; and **Cichlids (Osteichthyes)**.

Freshwater — Tropical Lakes. The tropics contain an enormous variety of fish species, often with extraordinary specializations for feeding and reproduction. An excellent example of this is Lake Malawi which now contains over 500 species of endemic fishes, most of which have evolved in the last 200–300 years. Such rapid evolution is possible as the fishes involved are cichlids. Their well-developed territorial habits make the species subject to isolation, and thus rapid speciation. Furthermore, the highly developed pharyngeal jaws have permitted extraordinary feeding specializations to evolve. Among the basic types of feeding seen in the African cichlids are epilithic algae feeders, periphyton collectors, leaf choppers, mollusk feeders, invertebrate pickers, zooplankton feeders, scale eaters, fin choppers, piscivory, and egg and embryo feeders.

Marine Environments. Most marine fish live on or near the edges of the continents, from the intertidal regions to the edge of the continental shelf. Within this region, there are a wide variety of habitat types, each inhabited by a distinctive set of fishes. These include rocky intertidal areas, exposed beaches, mudflats, salt marshes, mangrove swamps, seagrass flats, kelp beds, rocky seafloors, soft bottom areas and hypersaline pools. A typical near-shore community would contain bottom-associated fishes (e.g. surfperches (Family *Embioticidae*), wrasses (Family *Labridae*), porgies (Family *Sparidae*) and blennies (Family *Blenniidae*)) and open-water fishes (e.g. anchovies (Family *Engraulidae*), herrings (Family *Clupeidae*), jacks (Family *Carangidae*) and flatfish (Order *Pleuronectiformes*)). A very specialized near-shore habitat is the coral reef, home of a huge diversity of colorful teleosts (e.g. wrasses (Family *Labridae*), damselfish (Family *Pomacentridae*), angelfish (Family *Pomacanthidae*), surgeonfish (Family *Acanthuridae*), puffers (Family *Tetraodontidae*) and eels (various families)). See also **Blennies (Osteichthyes)**; **Flatfishes (Osteichthyes)**; **Herring (Osteichthyes)**; **Porgies (Osteichthyes)**; **Puffers (Osteichthyes)**; **Surgeonfishes (Osteichthyes)**; and **Wrasses (Osteichthyes)**.

The surface waters of the deeper oceans provide a rather featureless habitat for only 2% of teleost species. However, these fishes are of great importance to humans, as they provide most of the large fisheries. These fishes include herrings (Family *Clupeidae*), sardines (Family *Clupeidae*), cod (Family *Gadidae*), salmon (family *Salmonidae*) and tuna (Family *Scombridae*). The distribution of these fishes is strongly influenced by physical conditions, with fish shoals being associated with areas of upwelling and oceanographic fronts. See also **Tuna (Osteichthyes)**.

A specialized open-water habitat is the abyssal ocean. Most of the fishes living there are from three families–angler fishes (Order *Lophiiformes*), *aulopiforms* and eels (various families). They live in an extreme environment, and thus are very specialized in both feeding strategy and body form. Light organs may be present, and eyes may be absent. Many attract mates by producing sounds with the swimbladder. Many eat anything they come into contact with.

A second extreme open-water environment is the polar regions. The Antarctic has a diverse fish fauna comprising around 270 species, dominated by a mainly endemic group, the Nototheniodei (containing the cod icefish, crocodile icefish, Antarctic dragonfish and plunder fish), containing approximately 100 species. Over 80% of Antarctic fish species are found solely in that region. There is much less diversity in the Arctic, with only 110 species recorded from the arctic America, with the total for the whole Arctic not presumed to be much greater. The assemblage is dominated by sluggish, bottom-dwelling groups such as the lumpfishes (Family *Cyclopteridae*), sculpins (Family *Cottidae*), wolffishes (Family *Anarhichadidae*) and flounders (Order *Pleuronectiformes*, various

families). Because of the action of sea ice around the Arctic shores, most fishes live in deeper waters offshore. The polar fishes show various adaptations to the cold, icy environment, including the production of peptide or glycopeptide antifreezes and the modification of feeding apparatuses to pick microscopic food from the underside of the ice.

Behavior (Habits and Lifestyles)

Specialized behavior in fishes is primarily concerned with three important activities–feeding, reproduction and escape from enemies. See also **Fish: Ingestion**.

Schooling. A school of fish is a group that swims at about the same speed in roughly parallel orientation, while maintaining constant nearest-neighbor distance. Traveling schools range in shape from long, thin lines, to wedge shapes, ovals or squares. Grouping is protective behavior as there is safety in numbers. Furthermore, the twisting, fast-moving mass of silvery fish makes it difficult for a predator to pick an individual out of the school. See also **Fish: Swimming**.

Vertical Migration. Many fish undergo daily journeys through the water column, in search of phytoplankton that rises and sinks. Herbivores generally are at depth during the day, and at the surface during the night. Day is a risky time for fishes to be in the surface waters because of avian and piscine predators, and to add to this, there are energetic advantages to digesting food in colder, deeper waters.

Certain migrations are limited by the range of pressures through which the fish can swim. Cod (*Gadus* spp.), for example, has a rigorous decompression schedule that makes it possible for them only to move tens of meters, while swordfish (*Xiphius gladius*), in contrast, can easily migrate 600 m (1,969 ft) vertically.

Horizontal Migration. Large seasonal migrations are generally related to feeding and spawning. Maturing Atlantic cod (*Gadus morhua*) make a yearly migration from offshore, over-wintering feeding grounds to the Norwegian coast to spawn in the spring. Some oceanic tunas are even more impressive, making huge trans-oceanic migrations. Long-fin tunas move from the mid-Pacific to the west coast of America or Japan each year, and another tuna, the albacore, has been shown to migrate around the North Pacific, following the 14 °C (57 °F)isotherm. One of the most remarkable migrations is undertaken by the salmon, for example *Oncorhynchus* sp., the Pacific salmon. These have been shown to return with an amazing degree of accuracy to the stream in which they were spawned, after wandering as adults for several years and covering thousands of kilometers (miles) through the Pacific Ocean.

Symbiosis. One example is cleaning symbiosis, in which cleaner species remove parasites from host fishes. This is particularly important in reef communities, where cleaner wrasses set up cleaning stations on pieces of coral to service larger fishes. The exchange of food for protection is also relatively common, with a good example being the wholly beneficial association of anemones with damselfish (Family *Pomacentridae*) such as the clown fish. Juvenile stromateid teleosts (Family *Stromateoidei*), on the other hand, have a somewhat ambiguous relationship with the Portuguese man-o-war, sometimes living unscathed among the tentacles, but at other times being found only as a partially digested corpse.

Life Histories. Fish display a fascinating array of reproductive behavior patterns. Three major categories can be recognized: nonguarders, guarders and bearers. Nonguarders do not protect their eggs once spawning has been completed. These fishes distribute gametes into the water column, and the resultant larvae and young are left to float in the plankton. Examples of these are sturgeons, tuna and sardines. This type of pelagic spawning is common on coral reefs, where some of the most complicated mating systems of fishes can be seen, including harems and sex changing. See also **Fish: Reproduction**.

Guarders guard the embryos until they hatch, and frequently tend the larval stages as well. Territorial and courtship rituals are commonly seen, and some guarders, for example sunfishes, construct nests for their embryos.

Bearers are fish that carry their embryos, and sometimes their young as well, around with them, either externally or internally. Many forms carry spawned young in the mouth (catfishes and cichlids), while others carry the young in pouches (seahorses and pipefish). Others carry eggs internally, giving birth to large young (surfperches and some poeciliids).

While most fishes simply transform from embryos, through a juvenile stage, to an adult life-form, some fishes show spectacular metamorphoses through their life cycle. An example of this is the eel. The eel has leptocephalus larvae, which are thin, transparent and leaflike, living by drifting passively in the plankton. The leaflike shape apparently allows the larvae to obtain a large proportion of their nutrition by direct absorption of dissolved organic matter. The leptocephalus larvae become more eel-shaped after at least one year of passive living.

Perciforms. The *Perciforms*, with over 9,200 species, is the largest order of vertebrates. From humble beginnings in the Cretaceous, they flourished in the Tertiary, and diversified in shape, size and number of species to the present day. Most *perciforms* are adapted for life as predators in shallow ocean water or in lakes. Though the interrelationships of the families is constantly being debated, the actual membership of the group is clearer, and is based on, among several other characteristics, the possession of spines in front of the fins, and having a dorsal fin split into two segments.

The membership of the Perciforms is vast, and includes: ocean basses (Family *Acropomatidae*), dottybacks (Family *Pseudochromidae*), sunfish (Family *Centrarchidae*), perches (Family *Percidae*), cardinalfish (Family *Apogonidae*), remoras (Family *Echeneidae*), jacks (Family *Carangidae*), snappers (Family *Lutjanidae*), goatfishes (Family *Mullidae*), butterfly-fish (Family *Chaetodontidae*), angelfish (Family *Pomacanthidae*), cichlids (Family *Cichlidae*), damselfish (Family *Pomacentridae*), wrasse (Family *Labridae*), parrotfish (Family *Scaridae*), cod icefish (Family *Nototheniidae*), sandperches (Family *Pinguipedidae*), stargazers (Family *Uranoscopidae*), blennies (Family *Blennidae*), dragonets (Family *Callionymidae*), gobies (Family *Gobiidae*), rabbitfish (Family *Siganidae*), mullets (Family *Mugilidae*), barracudas (Family *Sphyraenidae*) and gouramis (Family *Belontiidae*).

Additional Reading

Ahlberg, P.E.: "A Re-examination of Sarcopterygian Interrelationships, with Special Reference to the Porolepiformes," *Zoological Journal of the Linnean Society*, **103**, 241–287 (1991).

Ahlberg, P.E.: "*Elginerpeton Pancheni* and the Earliest Tetrapod Clade," *Nature*, **373**, 420–425 (1995).

Benton, M.J.: *The Fossil Record 2*, Chapman & Hall, London, UK, 1993.

Benton, M.J.: *Vertebrate Paleontology*, 2nd Edition, Chapman & Hall, London, UK, 1997.

Lauder, G.V., and K.F. Liem: "The Evolution and Interrelationships of the Actinopterygian Fishes," *Bulletin of the Museum of Comparative Zoology*, **159**, 95–197 [Review of Actinopterygian phylogeny] (1983).

Long, J.A.: *The Rise of the Fishes*, The Johns Hopkins University Press, Baltimore, MD, 1995.

Maisey, J.G.: *Discovering Fossil Fishes*, Henry Holt, New York, NY, 1996.

Moyle, P.B., and J.J. Cech Jr.: *Fishes: An Introduction to Ichthyology*, 3rd Edition, Prentice-Hall, Inc., Upper Saddle River, NJ, 1996.

Nelson, J.S.: *Fishes of the World*, 3rd Edition, John Wiley and Sons, Inc., New York, NY.

Patterson, C.: "Bony Fishes," In: Prothero, D.R., and R.M. Schoch eds., *Major Features of Vertebrate Evolution, Short Courses in Paleontology*, Vol. 7, Paleontological Society, Knoxville, TN, 1994, pp. 57–84.

Paxton, J.R., and W.N. Eschmeyer: *Encyclopedia of Fishes*, 2nd Edition, Academic Press, Inc., San Diego, CA, 1998.

Stiassny, M.L.J., L.R. Parenti, and G.D. Johnson: *The Interrelationships of Fishes 2*, Academic Press, San Diego, CA, 1996.

REBECCA HITCHIN, University of Bristol, Bristol, UK

OSTEOARTHRITIS. This is the most common joint disease. As much as 80% of the population has radiographic evidence of osteoarthritis by the age of 65. Although only about 60% of patients with radiographically detectable osteoarthritis have symptoms, 15 to 30% of all visits to general practitioners may be attributed to difficulty with ambulation, largely due to osteoarthritis.

Osteoarthritis is a degenerative joint disease and, among people over age 50 (especially women), is the most common musculoskeletal problem. Under age 40, the disease usually is the result of an injury, from sports or occupational strain or of congenital abnormalities. Most often, the disease develops in the weight-bearing joints, such as the hips, knees, and spine and, to some extent, in joints that are used repetitively, such as the fingers.

Cartilage is a firm, rather rubbery material that covers the end of each bone. The normal purpose of cartilage is to act as a cushion and thus provide a smooth surface between the bones. In osteoarthritis, however, the cartilage breaks down, its surface becoming thin and uneven. Eventually, the ends of the bones may be left unprotected. Then the bones grind against each other, causing pain. With the development of pain, many patients tend to use affected joints less, and this leads to muscle weakness and stiffness of joints. A spur also may form in the surrounding bone tissue.

The three major contributing causes of osteoarthritis are injury, time, and weight. Overuse of a joint with time increases wear; obesity contributes to the stress placed on the joints and damages them prematurely. Overweight persons are more apt to develop osteoarthritis of the hips and knees earlier than their thinner neighbors.

The most common symptom is aching pain when the affected joint is in use. Except in advanced cases, this pain will subside when the affected joint is rested. Osteoarthritis does not always become worse if the aforementioned precautionary measures are taken. Although the pain may become rather constant, the common symptoms of *rheumatoid arthritis*, such as prolonged morning stiffness, swollen lymph nodes, fatigue, and fever usually are not present in osteoarthritis. Physical examination of a patient with osteoarthritis usually reveals joint tenderness and sometimes swelling. When moved, the joints sometimes cause *crepitus* (popping, clicking, or grating sounds). Enlargement of bone ends also may be present, such as the finger ends (Heberden's nodes) or the middle joints (Bouchard's nodes). A radiographic examination is required to yield complete information for the physician.

Degenerative joint disease of the spine can take several forms, but usually affects the vertebral bodies or disks, or both. These deformities are most severe when they occur in the middle to lower cervical spine and in the lower lumbar region. In contrast with ankylosing spondylitis, spinal fusion is uncommon in osteoarthritis.

Osteoarthritis and Exercise. Todd Schmidt (Hughston Sports Medicine Foundation, Columbus, Georgia), reports, "Forty years after they had begun training, one group of former champion runners had a lower incidence of osteoarthritis than a nonrunning group of the same age. Comparisons of the spines, hips, knees, and feet of a group of runners averaging 25 miles (40 km) per week with a group of nonrunners showed no difference in cartilage thickness, spur formation, grinding, and joint stability. Another study found that former college runners were no more likely than anyone else to have symptomatic osteoarthritis." The study dealt with normal uninjured joints. Joints that have meniscus and cartilage damage or ligament instability are clearly at an increased risk for osteoarthritis. Because joints are designed for motion and shock absorption, running on an appropriate surface appears to be physiologic and nontraumatic. Further research along these lines is continuing.

Heritable Factors. There is substantial qualitative evidence that osteoarthritis occurs with greater frequency in some families. Research along these lines has been underway for several years. No discovery of a breakthrough nature, however, has been made as of the early 1990s. For those readers of scholarly interest in this subject, reference to the Knowlton paper (reference listed) is suggested.

More is known about the rare Marfan syndrome, which is considered to be an inherited connective tissue disorder. There are some 40 to 60 cases per million population. The most prominent clinical manifestations occur in the skeletal, ocular, and cardiovascular systems. Kainulatinen (Kuopio University Central Hospital, Finland), with reference to a recent study, observes: "The chromosomal localization of the mutation in Marfan syndrome is a first step toward the isolation and characterization of the defective gene and serves as a diagnostic test in families in which cosegregation of these markers with the disease has been confirmed."

Additional Reading

Bradley, J.D., et al.: "Comparison of An Antiinflammatory Dose of Ibuprofen, An Analgesic Dose of Ibuprofen, and Acetaminophen in the Treatment of Patients with Osteoarthritis of the Knee," *New Eng. J. Med.*, 87 (July 11, 1991).

Hamanishi, C.: *Advances in Osteoarthritis*, Springer-Verlag, Inc., New York, NY, 1999.

Hosie, G. and J. Dickson: *Managing Osteoarthritis in Primary Care*, Blackwell Science, Inc., Malden, MA, 2000.

Kainulainen, K., et al.: "Location of Chromosome 15 of the Gene Defect Causing Marfan Syndrome," *New Eng. J. Med.*, 935 (October 4, 1990).

Knowlton, R.G., et al.: "Genetic Linkage of a Polymorphism in the Type II Procollagen Gene (COL2A1) to Primary Osteoarthritis Associated with Mild Chondrodysplasia," *New Eng. J. Med.*, 326 (February 22, 1990).

Koopman, W.J.: *Arthritis and Allied Conditions: A Textbook of Rheumatology*, 14th Edition, Lippincott Williams & Wilkins, Philadelphia, PA, 2000.

Liang, M.H. and P. Fortin: "Management of Osteoarthritis of the Hip and Knee," *New Eng. J. Med.*, 125 (July 11, 1991).

Moskowitz, R.W., J.A. Buckwalter, D.S. Howell, et al.: *Osteoarthritis: Diagnosis and Medical/Surgical Management*, W.B. Saunders Company, Philadelphia, PA, 2001.

Shipman, P., et al.: *The Human Skeleton*, Harvard University Press, Cambridge, MA, 1985.

OSTEOLOGY. A division of vertebrate anatomy which deals with the skeletal system. The study of bones. Comparative osteology is the study and comparison of the bones of different races and different species of organisms.

OSTEOMYELITIS. See **Bone**.

OSTRICH *(Aves, Struthioniformes; Struthio Camelus).* The only living representative of the suborder *Struthiones*, family *Struthionidae*. Its relatives inhabited wide areas in Asia, Europe, and Africa since the Eocene about 55 million years ago. Eight extinct species all belonged to the same genus as the surviving species. See Fig. 1.

Fig. 1. Ostriches.

Attaining 3 meters ($9\frac{1}{2}$ feet) in height and weighing more than 150 kilograms (330 pounds), the male ostrich is the largest living bird. The head and about $\frac{2}{3}$ of the neck are sparsely covered with short, hair-like, degenerated feathers, which make it appear naked. The skin is variably colored, depending on the subspecies. The legs are particularly strong and long. The foot has two toes: a large strongly clawed third toe and weaker, generally clawless fourth (outside) toe. The first and second toes are absent. The feathers have no secondary or aftershaft. There are 50 to 60 tail feathers. The wing has 16 primary, 4 alular, and 20–23 secondary feathers. The wing feathers and rectrices have changed to decorative plumes.

Ostriches live on the open savannah, in the dry South African bushland or on the wide sand plains of deserts which have hardly any plants, and also dense bush. See Fig. 2. This adaptable grazer is at home even in steep rocky mountain country. Depending upon habitat and seasons, they eat various grasses, bushes, and forage on trees. Plants that store water help them through dry seasons but cannot meet all water requirements for a longer period of time. Without open water, they eventually die of thirst. They supplement their plant diet as much as possible with animal food, such as invertebrates and small vertebrates which they chase by rather ungracefully zig-zagging after them. Long periods of abundant food influence the reproductive readiness of the ostrich. Even in the most unfavorable weather, during long droughts, or during occasional local rains, the ostrich is a very adaptable and opportunistic breeder which can rear a few young in any season.

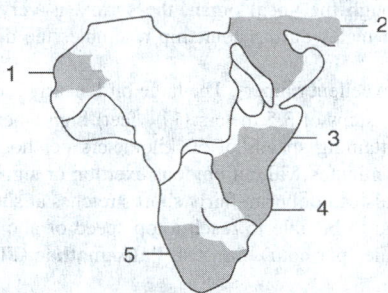

Fig. 2. The ostrich (*Struthio camelus*) was distributed in these areas of Africa and Asia Minor until a few decades ago. Since then it has been exterminated in many areas. Subspecies: (1) North African ostrich, *Struthio camelus camelus*; (2) Arabian ostrich, *Struthio camelus syriacus*; exterminated; (3) Somali ostrich, *Struthio camelus molybdophanes*; (4) Maasai ostrich, *Struthio camelus massaicus*; (5) South African ostrich (*Struthio camelus australis*.)

The social life of ostriches is one of the most complex in the animal world. In rainless periods, when wandering, and in the common grazing grounds at watering places, they often form peaceful aggregations of up to 680 birds, but the individual flocks remain recognizable. Social contacts between birds of different groups are initiated by approaching the other in a submissive posture, with lowered head and tail down. Often a family of a herd adopts the chicks or young of another. Single males may join together and form "schools" of half-grown ostriches, with whom they wander about for days or weeks. For the communal sandbath, each flock seeks out a sandy depression.

Depending upon local circumstances and the composition of the population, ostriches live in monogamy or polygamy. Pair formation generally takes place within the large flocks. The "goading" of the old females and the animosities between the males often lead to displays and "dances" of the entire flock. The most common mating pattern is polygyny, in which a male generally lives with a "head female" and two "auxiliary" females. The head female tolerates the others and all lay their eggs into a common nest. In most cases, the experienced head female drives the auxiliary females away from the nest as soon as they have finished laying.

The male ostrich has far more to do during breeding than merely fertilizing the eggs. He is a genuine father. He scratches out the nest depression in a sandy spot, often in a dried-out river bed. Even before the eggs are going to be laid he guards the nest. Finally he sits down on it, the female lays her eggs before his breast, and he pushes them under his body with his neck and beak. The male ostrich incubates from the late afternoon until early morning; the female, therefore, does not have to sit for long since she only incubates in the hottest hours of the day.

When holding an ostrich egg, one may wonder how the chick could possibly get out of the shell without help. The shell is as thick as china. It weighs about $1\frac{1}{2}$ kilograms (3 pounds), about as much as 25–30 chicken eggs. Ostrich eggs are good to eat; they taste almost like chicken eggs. When refrigerated, an ostrich egg keeps up to a year and remains edible. It takes about 2 hours to hard-boil one.

Behavioral interactions between chicks and parents begin a few days before hatching, with pleasant-sounding contact calls from the chicks within the eggs. Actual hatching may take hours or even days. Soon after hatching the nestlings swallow the first small stones for grinding food, and already on the first day most chicks are ready to explore the vicinity of the nest. When in danger from a terrestrial enemy, such as a jackal, the adults try to decoy with conspicuous zigzag running, wing beating, and moot calls; at the appropriate time one parent calls the young, which meanwhile have cowered motionless, and leads them away.

One to three months after hatching, and sometimes much later in areas thinly populated by ostriches, the families rejoin the large flocks. It is believed that ostriches live thirty to seventy years.

Excellent eyesight and acute hearing are their most important senses, they often make the ostrich an unintentional but reliable sentinel for many African grazing animals, such as zebras, antelopes, and gazelles. They are capable of producing pleasant-sounding calls, moot and harsh guttural sounds, hissing and snorting, and a far reaching territory and courtship song. The volume and quality of these calls are not inferior to those of song birds, although the vocal organ, the syrinx, is very primitive. The calling of ostrich males during courtship resembles the distant roaring of lions.

Ostriches are excellent runners. The large bird, nearly 3 meters ($9\frac{1}{2}$ feet) high, easily takes steps of 3.5 meters ($11\frac{1}{2}$ feet) high when running. They are capable of attaining speeds of 50 kilometers per hour (31 miles per hour) for 15–30 minutes without obvious exertion or signs of exhaustion. Other wild animals can only run fairly short stretches at such a high speed. Ostriches are said to be able to reach a top speed of about 70 kilometers per hour ($43\frac{1}{2}$ miles per hour). See also **Paleognathae (Flightless Birds)**; and **Ratites**.

OSTWALD, WILHELM (1853–1932). A German chemist who won the Nobel Prize for chemistry in 1909. He was considered to be a founder of modern physical chemistry. His work involved research in catalysis, the rates of chemical reactions, equilibrium, and conductivity of organic acids. He was an admirer of Mach and did not readily accept the atomic theory. He received honorary doctorates from several universities in Germany, Great Britain and the USA, and was made an honorary member of learned societies in Germany, Sweden, Norway, the Netherlands, Russia, Great Britain and the USA.

OTITIS MEDIA.   See Hearing and the Ear.

OUTCROP.   Everywhere beneath the soil at greater or lesser depths there exist the solid continuous masses of rock that make up the earth's crust. Wherever rock appears protruding through the soil cover it is spoken of as an outcrop.

OUTFLOW JET.   Nocturnal cold-air jet flowing out of the mouth of a valley or canyon as it opens onto a plain. When the flow is fully developed, its depth near the exit is about the same as the height of the valley or canyon sidewalls there. It represents a continuation or extension of the mountain (downvalley or downcanyon) breeze generated in the valley by surface cooling. The jet often achieves peak wind speeds outside the valley. This acceleration is probably due to the conversion of potential to kinetic energy, as the column of cold air flowing out of the valley, freed of the confining sidewalls, fans out in the horizontal and compresses vertically. Another factor is the release of the flow from surface friction along the sidewalls. Maximum speeds in the jet begin near the surface early in the evening, but for well-developed jets at exits from deep valleys the highest speeds may be at 300–500 meters (984–1,640 feet) or more above the ground. In deep, extensive valley drainage systems, peak speeds may exceed 10 m s^{-1} on clear nights.

AMS

OUTGASSING.   The removal of gas from a metal by heating at a temperature somewhat below melting, while maintaining a vacuum in the space around the metal.

OUTPUT UNIT.   In computer terminology, a unit which delivers information from the computer to an external device or from internal storage to external storage.

OUZEL (or Ousel).   Aves, Passeriformes. 1. The ring ouzel, *Turdus torquatus*, a mountain bird (Aves) of central and northern Europe. A thrush. 2. The blackbird of Europe and eastern Asia, also a thrush. 3. The water ouzel, or dipper. The name water ouzel applies to several species of the Northern Hemisphere which haunt rapid streams. Although not a swimming bird it wades into deep water, where it is said to make its way by using both feet and wings. The North American species, *Cinclus mexicanus*, is characteristically a western mountain bird. Its dull gray color is more than offset by its interesting habits and by a glorious song, heard all too rarely. See also **Thrush**.

OVAL OF CASSINI.   A higher plane curve, also known as a cassinoid or cassinian, defined as the locus of a point which moves so that the product of its distance from two fixed foci ($\pm a$, 0) equals k^2, a constant. Its equation in Cartesian coordinates is $(x^2 + y^2 + a^2)^2 - 4a^2x^2 = k^2$. The curve cuts the X-axis at the two real points $\pm c$, where $k = c^2 - a^2$ and at two other points, real if $\sqrt{2a} > c$ but imaginary if $\sqrt{2a} < c$. Under the same conditions, it cuts the Y-axis in two real points or in four imaginary points. Thus, when $k < a$, there are two separated ovals each enclosing a focus and the curve is said to be *bipartite*. If $k > a$, the two ovals merge into one. When $k = a$, the curve is known as the lemniscate of Bernoulli.

Cassini ovals are useful in the study of optics. Another curve of similar shape is the Cartesian oval.

OVARIES.   See Gonads; Hormones; and Pituitary Gland.

OVERCAST

1. Descriptive of a sky cover of 1.0 (95% or more) when at least a portion of this amount is attributable to clouds or obscuring phenomena aloft; that is, when the total sky cover is not due entirely to surface-based obscuring phenomena. In aviation weather observations, an overcast sky cover is denoted by the symbol "symbol"; it may be explicitly identified as thin (predominantly transparent); otherwise a predominantly opaque status is implicit. An opaque overcast sky cover always constitutes a ceiling. *See* obscuration.

2. Popularly, the cloud layer that covers most or all of the sky. It generally suggests a widespread layer of clouds such as that considered typical of a warm front.

See also **Sky Cover**.

OVERFOLD.   A term used by structural geologists to define an overturned anticline resulting in a recumbent fold often simulating a normal stratigraphic pile of strata for which it may be mistaken.

OVERLAP.   1. In some magnetic amplifier and rectifier circuits, the undesirable flow of current for a period after the supply voltage has gone to zero (and perhaps reversed) caused by energy stored in some circuit inductance.

2. In a magnetic amplifier, the simultaneous conduction of the half wave elements during certain intervals of supply voltage cycle.

3. In geology, the gradual burial of the land mass or mountain slopes from which the sediments were derived.

OVERRANGE.   Of a system or element, any excess value of the input signal above its upper range-value or below its lower range-value. See also **Range and Span (Instrument)**.

OVERSEEDING.   A cloud seeding operation in which an excess of nucleating material is released over that required to produce the desired effect. As the term is normally used, the excess is relative to that amount of nucleating material expected to produce the maximum amount of precipitation received at the ground. For example, in seeding a supercooled water cloud with dry ice or silver iodide in order to increase precipitation, the addition of too much seeding material will create too many ice crystals insofar as they compete for the available water vapor, and few of them will grow large enough to fall as precipitation. There are unresolved questions about what composes the optimal amount of seeding material to produce the desired effect. These questions include how to estimate the right amount of material and when and where to deliver it to the cloud; it is difficult to assign quantitative values that will result in overseeding. See also **Cloud Seeding**.

OVERTHRUST. A low-angle fault of the thrust or compressional type frequently resulting in the translocation of a mass of rocks several miles. Because of the low angle of the thrust as well as the possible inversion of great piles of formations, this type of faulting or failure of the lithosphere has led, and still leads, to considerable technical difficulties in determining the stratigraphic history in regions where low angle faulting predominates. See Fig. 1.

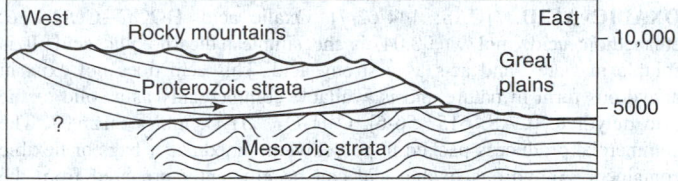

Fig. 1. Diagrammatic structure section showing how the great body of Proterozoid strata has been thrust-faulted from the west for long distances over upon Late Mesozoic strata in Glacier National Park, Montana. The Mesozoic beds were folded by action of the overriding block. Length of section about 25 miles (40 kilometers). Vertical scale, much exaggerated.

OVERVOLTAGE. The excess of observed decomposition voltage of an aqueous electrolyte over the theoretical reversible decomposition voltage.

OVINAE. See **Goats and Sheep**.

OVIPAROUS REPRODUCTION. The production of eggs that hatch after expulsion from the body. Frogs, birds, and many forms of reptiles and fish, as well as many insects and other invertebrate animals, have oviparous reproduction. See also **Fishes**.

OVIPOSITOR. An organ used by some of the arthropods to deposit their eggs. It consists of a maximum of three pairs of appendages formed to transmit the egg, to prepare a place for it, and to place it properly. In some of the insects the organ is used merely to attach the egg to some surface, but in many parasitic species (see also **Hymenoptera**) it is a piercing organ as well. It is used by the grasshoppers to force a burrow in the earth to receive the eggs and by cicadas to pierce the wood of twigs for a similar purpose. Both long-horned grasshoppers and sawflies cut the tissues of plants by means of the ovipositor. None of these examples is quite as remarkable as the ichneumon flies (parasitic *Hymenoptera*) which have a slender ovipositor several inches long, used to drill into the wood of tree trunks. These species are parasitic in the larval stage on the larvae of woodboring insects, hence the egg must be deposited in the burrow of the host.

The sting of wasps and bees is also an ovipositor, in this case highly modified and associated with poison glands. Some roach-like fish have an ovipositor as a tubular extension of the genial orifice in the breeding season for depositing eggs in the mantle cavity of the pond mussel.

OVOVIVIPAROUS REPRODUCTION. The condition where the young are born alive, but they hatch from eggs within the body of the mother just before being born. There is no placental connection between the young and the mother to furnish food before the birth. Sharks, rattlesnakes, certain snails, some minnows, and flesh flies are examples of the great variety of animals which may have this type of reproduction. It should not be confused with viviparous reproduction, where there is a placental connection between mother and embryo before birth, nor oviparous reproduction, where the eggs are laid before hatching. See also **Fishes**.

OWEN, RICHARD (1804–1892). Richard Owen was an English comparative anatomist and palaeontologist who founded the British Museum (Natural History). Born in Lancaster, Richard Owen was the second of six

children and the younger of two sons. His father, Richard Owen Sr., was a West India merchant who died when Richard was five years old. After attending the local grammar school (1810–1820), where he was reportedly 'lazy and impudent', he became apprenticed to local surgeons/apothecaries (1820–1824). In 1824, he matriculated at Edinburgh University, but stayed only about half a year, moving in 1825 to St Bartholomew's Hospital, London, to work as prosector to the lectures by John Abernethy. In 1827, at the age of 22, on Abernethy's recommendation, Owen started out on his life-long career as a museum curator, being made assistant curator to William Clift at the Hunterian Museum of the Royal College of Surgeons. A major part of his duties was the preparation of a Descriptive and Illustrated Catalogue of the collections. This work formed the basis of his skill as the leading British comparative anatomist of his generation–a skill he demonstrated in a series of classic memoirs and reports. At this early stage in his career, being crucially dependent on the patronage of Oxbridge Anglicans, brought up on the natural theology of William Paley, Owen perfected the functionalist approach in comparative anatomy. Examples of this work are his *Memoir on the Pearly Nautilus* (1832) as well as his work of a decade later on the extinct moas from New Zealand. In 1837, at the age of 32, Owen became the first permanent Hunterian Professor of Comparative Anatomy and Physiology at the College of Surgeons.

Like the great hero of his youth Georges Cuvier, Owen specialized in the study of vertebrate osteology. He perfected the use of dental characteristics for classificatory purposes. His *Odontography* (1840–1845) has proved a gold mine of information to anatomists till the present day. The odontographical expertise proved particularly useful in the field of vertebrate paleontology, and in 1842 he published a 'Report on British fossil reptiles', which had been delivered in 1839 to the British Association for the Advancement of Science (BAAS); in this publication he introduced the order *Dinosauria*. In a further report to the BAAS, of 1846, Owen significantly diverged from his earlier functionalist line and, borrowing from German idealist morphology, he presented his classic work' 'On the archetype and homologies of the vertebrate skeleton', published as a book in 1848.

In 1842 Owen succeeded Clift as Conservator of the Hunterian Museum. From the outset of his museum career, Owen had been concerned with the expansion of both collections and buildings. His goal was to turn the Hunterian Museum into a national museum of natural history. This proved impossible and in 1856 he resigned from his employment at the College of Surgeons to become the first superintendent of the natural history collections at the British Museum (1856–1883). From that time on, Owen fought hard to have a separate museum for his collections constructed, developing his views on museum organization in a rare treatise of its kind, *On the Extent and Aims of a National Museum of Natural History* (1862). After a series of major Parliamentary debates, his plans were realized and the Natural History Museum in South Kensington was erected (1873–1881). Two years later, Owen retired from public service to the grace-and-favour Sheen Lodge in Richmond Park. During his tenure at the British Museum (Natural History), Owen's publications mainly concerned vertebrate fossils. Among the many fine contributions of this period was his description of the *Archaeopteryx* (1863).

Owen's work on the vertebrate archetype led him to believe in an evolutionary, natural origin of species. To this he gave cautious, some would say cryptic expression in a variety of publications, for example in his *On the Nature of Limbs* (1849). Strong criticism from his Anglican, Paleyite patrons turned Owen into a closet evolutionist, but he enormously resented it when later he was portrayed by Charles Darwin, with whom Owen had enjoyed a good working relationship, as an arch-creationist. Part of his resentment, too, was the fact that Darwin, by coming out so explicitly and brilliantly in support of organic evolution, outshone Owen. Further tension between the two men resulted from the fact that Darwin and his immediate circle of adherents tried to sabotage Owen's plan for a national museum of natural history. These tensions were reflected in Owen's notoriously hostile essay for the *Edinburgh Review* (1860) on 'Darwin on the origin of species', but an excessive preoccupation with this review by Darwinians ever since has hindered a balanced appreciation of Owen's considerable accomplishments. See also **Paleontology (The History)**.

Additional Reading

Desmond, A.: *The Politics of Evolution, Morphology, Medicine, and Reform in Radical London*, University of Chicago Press, Chicago, IL, 1989.

Gruber, J.W., and J.C. Thackray: *Richard Owen Commemoration*, Natural History Museum Publications, London, UK, 1992.

Owen, R.: *Odontography; or, a Treatise on the Comparative Anatomy of the Teeth; their Physiological Relations, Mode of Development, and Microscopic Structure, in the Vertebrate Animals*, 2 vols, Baillière, London, UK, 1840–1845.

Owen, R.: *On the Archetype and Homologies of the Vertebrate Skeleton*, Van Voorst, London, UK, 1848.

Owen, R.: *On the Extent and Aims of a National Museum of Natural History*, Saunders Otley, London, UK, 1862.

Richards, E.: "A Question of Property Rights: Richard Owen's Evolutionism Reassessed," *British Journal for the History of Science*, **20**, 129–171 (1987).

Rupke, N.A.: *Richard Owen, Victorian Naturalist*, Yale University Press, New Haven, CT, 1994.

Sloan, P.R.: *Richard Owen. The Hunterian Lectures in Comparative Anatomy May–June, 1837*, University of Chicago Press, Chicago, IL, 1992.

NICOLAAS A. RUPKE, Göttingen University, Göttingen, Germany

OWENS DUST RECORDER. An instrument for rapidly obtaining samples of airborne dust; a type of dust counter. The dust particles pass through a cylindrical chamber and are drawn at high velocity through a narrow slit in front of a microscope cover glass. The fall in pressure due to expansion of the air passing through the slit causes the formation of a moisture film on the glass to which the dust adheres. An analysis for quantity and size of the particles on the cover glass is made with the aid of a microscope. The vacuum required to operate the instrument is developed by an attached hand pump.

OWLS (*Aves, Strigiformes*). The *Strigiformes* comprise an order of birds with hooked beaks and talons of the birds of prey, but differing from the hawks and related forms in having the eyes directed forward and in their nocturnal habits. The common name for this order is owl. There are about 130 or more species. The head and neck of the owl is modified so that it can see nearly 270 degrees by turning its head. The ears and eyes are highly specialized. Although the eye cannot be turned in the socket, the owl can see small darting objects with very little light, substituting head movement for eye movement. It is reported that an owl can catch an insect with no more light than a candle would produce at 1,170 feet (351 meters). The eye is surrounded by a small disk of radiating feathers. The ear openings are unusually large and well adapted to detecting the slightest of sounds. The legs are strong and the claws are sharp. When landing from flight, the owl spreads its tail to serve as a brake.

Owls of various species utter different kinds of sounds, ranging from whistles, to hoots, and some sounds described as laughing. These birds nest in trees, on rocks, or on the ground, depending upon species. The egg is shaped like a sphere. The young are covered with white down. The adults have soft thick plumage that contributes to silencing their flight. A barn owl is shown in Fig. 1.

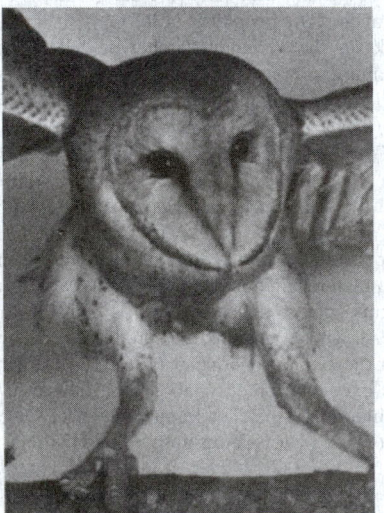

Fig. 1.   Barn owl. (*A.M. Winchester.*)

North American species range in size from the 6-inch (15 centimeters) Acadian owl to the 2-foot (0.6 meter) great horned owl (*Bubo virginanus*). There is a similar range of sizes in the other parts of the world—from the pigmy owls of the Old World to the great hawk-owl of Australia. The owl is represented on all continents.

Although owls are characteristically predacious, they do not disdain insects as food. The smaller species catch mice and other small animals; the larger owls prey upon animals as large as rabbits. The great horned owl will sometimes maraud a poultry yard. See also **Strigiformes**.

For several years the most famous of North American owls has been the spotted owl, whose habitat is found in the forests of the Northwest. This owl has been the subject of extensive environmental disputes.

OXALIC ACID. [CAS: 144-62-7]. Oxalic acid, $HOOC-COOH$, or ethanedioic acid, mol wt 90.04, is the simplest dicarboxylic acid. It is soluble in water, and acts as a strong acid. This acid does not exist in anhydrous form in nature and is available commercially as a solid oxalic acid dihydrate [CAS: 6153-56-6], $C_2H_2O_4 \cdot 2H_2O$, mol wt 126.07. The commercial product is packed in polyethylene-lined paper bags or flexible containers. Anhydrous oxalic acid can be efficiently prepared from the dihydrate by azeotropic distillation in a low boiling solvent that can form a water azeotrope, such as benzene and toluene.

Oxalic acid was synthesized for the first time in 1776 by Scheele through the oxidation of sugar with nitric acid. Then, Wöhler synthesized it by the hydrolysis of cyanogen [CAS: 460-19-5] in 1824.

The potassium or calcium salt form of oxalic acid is distributed widely in the plant kingdom. Its name is derived from the Greek *oxys*, meaning sharp or acidic, referring to the acidity common in the foliage of certain plants (notably *Oxalis* and *Rumex*) from which it was first isolated. Oxalic acid is found in spinach, rhubarb, etc. Oxalic acid is a product of metabolism of fungi or bacteria and also occurs in human and animal urine; the calcium salt is a principal constituent of kidney stones.

Oxalic acid is used in various industrial areas, such as textile manufacture and processing, metal surface treatments leather tanning, cobalt production, and separation and recovery of rare-earth elements. Substantial quantities of oxalic acid are also consumed in the production of agrochemicals, pharmaceuticals, and other chemical derivatives.

Physical Properties

The physical and thermochemical constants of anhydrous oxalic acid and oxalic acid dihydrate are summarized in Table 1.

Reactions

The reactions of oxalic acid, including the formation of normal and acid salts and esters, are typical of the dicarboxylic acids class. Oxalic acid, however, does not form an anhydride.

On rapid heating, oxalic acid decomposes to formic acid, carbon monoxide, carbon dioxide, and water. In aqueous solution, it is decomposed by uv, x-ray, or γ-radiation with the liberation of carbon dioxide. Photodecomposition also occurs in the presence of uranyl salts.

Oxalic acid is a mild reducing agent, and is oxidized by potassium permanganate in acid solution to give carbon dioxide and water. Oxalic acid is catalytically reduced by hydrogen in the presence of ruthenium catalyst to ethylene glycol, and electronically reduced to glyoxylic acid.

Oxalic acid reacts with various metals to form metal salts, which are quite important as the derivatives of oxalic acid. It also reacts easily with alcohols to give esters.

Manufacture

Many industrial processes have been employed for the manufacture of oxalic acid since it was first synthesized. The following processes are in use worldwide: oxidation of carbohydrates, the ethylene glycol process, the propylene process, the dialkyl oxalate process, and the sodium formate process. Sodium formate process is no longer economical in the leading industrial countries, except for China.

Nitric acid oxidation is used where carbohydrates, ethylene glycol, and propylene are the starting materials. The dialkyl oxalate process is the newest, where dialkyl oxalate is synthesized from carbon monoxide and alcohol, then hydrolyzed to oxalic acid. This process has been developed by UBE Industries in Japan.

TABLE 1. PHYSICAL AND THERMOCHEMICAL PROPERTIES OF OXALIC ACID AND ITS DIHYDRATE

Property	Value
OXALIC ACID, ANHYDROUS, $C_2H_2O_4$	
Melting point, °C	
α	189.5
β	182
Density d_4^{17}, g/mL	
α	1.900
β	1.895
Refractive index, β, n_4^{20}	1.540
Vapor pressure (solid, 57–107 °C), kPa[a]	$\log_{10} P = -(4726.95/T) + 11.3478$
Specific heat (solid, −200 to 50 °C), J/g	$C_p^b = 1.084 + 0.0318t$
Heat of combustion, ΔE_c (at 25 °C), kJ/mol[c]	−245.61
Standard heat of formation, ΔH_f (at 25 °C), kJ/mol[c]	−826.78
Standard free energy of formation, ΔG_f (at 25 °C), kJ/mol[c]	−697.91
Heat of solution (in water), kJ/mol[c]	−9.58
Heat of sublimation, kJ/mol[c]	90.58
Heat of decomposition, kJ/mol[c]	826.78
Specific entropy, S (at 25 °C), J/(mol · K)[c]	120.08
Logarithm of equilibrium constant, $\log_{10} K_f$	122.28
Thermal conductivity (at 0 °C), W/(m · K)[d]	0.9
Ionization constant	
K_1	6.5×10^{-2}
K_2	6.0×10^{-5}
Coefficient of expansion (at 25 °C), nL/(g · K)	178.4
OXALIC ACID DIHYDRATE, $C_2H_2O_4 2H_2O$	
Mp, °C	101.5
Density d_4^{20}, g/mL	1.653
Refractive index, n_4^{20}	1.475
Standard heat of formation, ΔH_f (at 18 °C), kJ/mol[c]	−1422
Heat of solution (in water), kJ/mol[c]	−35.5
pH (0.1 M soln)	1.3

[a] To convert $\log_{10} P_{kPa}$ to $\log_{10} P_{mm\,Hg}$, add 0.875097 to the constant, $T = K$.
[b] To convert C_p, J/g, to C_p cal/g, divide both terms of the equation by 4.184.
[c] To convert J to cal, divide by 4.184.
[d] To convert W/(m · K) to (Btu · in.)/(h · ft^2 · °F), divide by 0.1441.

Many attempts have been made to synthesize oxalic acid by electrochemical reduction of carbon dioxide in either aqueous or nonaqueous electrolytes.

Health and Safety Factors

Oxalic acid is caustic and corrosive to humans. The severity of symptoms associated with oxalic acid poisoning is related to the concentration and quantity ingested. Oxalic acid removes calcium in the blood, forming calcium oxalate, and severe damage to the kidney may occur because of the insoluble calcium oxalate.

Uses

Because rare-earth oxalates have low solubility in acidic solutions, oxalic acid is used for the separation and recovery of rare-earth elements. The oxalic acid process for anodizing aluminum was developed in Japan. In addition to oxalic acid, inorganic oxalate salts are also used in coloring anodic coatings. Oxalic acid is a constituent of cleaners that are used for automotive radiators, boilers, and steel plates before phosphating. As a chelating agent, oxalic acid forms water-soluble complexes on metal surfaces during cleaning and rinsing.

In pulp bleaching, oxalic acid serves as a bleaching agent, but is often used together with other bleaching agents because of its relatively high cost. Oxalic acid is also used for the bleaching of cork, wood (particularly veneered wood), straw, cane, and natural waxes.

Oxalic acid has various uses in fabric cleaning, application of dyestuff, and modifying properties of cellulose fabrics. Oxalic acid is used as a pH modifier in leather tanning by tannin and basic chromium sulfate. It also functions as a bleaching agent for leather. It is used for marble polishing especially in Italy. It not only removes iron veins by forming water-soluble iron oxalate, but also serves as a polishing auxiliary. Starch powder is heated together with oxalic acid and hydrolyzed to produce millet jelly. Oxalic acid functions as a hydrolysis catalyst, and is removed from the product as calcium oxalate. This application is carried out in Japan. Oxalic acid is also used for the production of cobalt, as a raw material of various agrochemicals, and pharmaceuticals, for the manufacture of electronic materials, for the extraction of tungsten from ore, for the production of metal catalysts, as a polymerization initiator, and for the manufacture of zirconium and beryllium oxide.

Derivatives

Oxalic acid forms neutral and acid salts, as well as complex salts.

Ammonium Oxalate. This salt [CAS: 1113-38-8], $(NH_4)_2C_2O_4$, mol wt 124.10, exists as a monohydrate [CAS: 6009-70-7] or in anhydrous form. Anhydrous ammonium oxalate is obtained when the monohydrate is dehydrated at 65 °C. It is used for textiles, leather tanning, and precipitation of rare-earth elements.

Ammonium Iron(III) Oxalate. This mixed salt [CAS: 29696-35-3], $(NH_4)_3[Fe(C_2O_4)_3]$, mol wt 374.04, is produced as an emerald-green crystalline trihydrate [CAS: 13268-42-3]. The compound is not stable to light. It was once used extensively in the manufacture of blueprinting papers.

Potassium Hydrogen Oxalate. Potassium acid oxalate [CAS: 127-95-7], KHC_2O_4, mol wt 146.15, exists as a monohydrate. It is of historical interest because it is the salt of sorrel found in vegetation and the first oxalate isolated.

Potassium Oxalate. The monohydrate [CAS: 6487-48-5], $K_2C_2O_4 \cdot H_2O$, mol wt 184.24, is produced as a colorless crystalline material or a white powder. The anhydrous salt [CAS: 583-52-8], mol wt 166.22, is obtained when the monohydrate is dehydrated at 160 °C. The monohydrate is preferred as a reagent in analytical chemistry and in miscellaneous uses principally because of its high solubility as compared with other simple neutral oxalates; the saturated solution, at 0 °C, contains about 20 wt %, and at 20 °C, about 25 wt % $K_2C_2O_4$.

Sodium Oxalate. This salt [CAS: 62-76-0], $Na_2C_2O_4$, mol wt 134.01, is obtained in such high purity and is so stable that it is used as a titrimetric standard.

Calcium Oxalate. The monohydrate [CAS: 5794-28-5], $CaC_2O_4 \cdot H_2O$, mol wt 128.10, is of importance principally as an intermediate in oxalic acid manufacture and in analytical chemistry; it is the form in which calcium is frequently quantitatively isolated.

Nickel Oxalate. This salt NiC_2O_4, mol wt 146.7, is produced as a greenish white crystalline dihydrate [CAS: 6018-94-6]. Nickel oxalate is used for the production of nickel catalysts and magnetic materials.

Yttrium Oxalate. This compound [CAS: 126476-37-7], $Y_2(C_2O_4)_3$, mol wt 441.91, exists as a trihydrate, nonahydrate [CAS: 7100-75-6], or heptadecahydrate. The compound is used for the production of a red fluorescent material for color television.

Dialkyl Oxalates. Oxalic acid gives various esters. Dialkyl esters ROOC—COOR, are industrially useful, but monoalkyl esters, ROOC—COOH, are not. The dialkyl esters are characterized by good solvent properties and serve as starting materials in the synthesis of many organic compounds, such as pharmaceuticals, agrochemicals, and fine chemicals. Among the diesters, dimethyl, diethyl, and di-n-butyloxalates are industrially important.

Oxamide. This diamide [CAS: 471-46-5], $H_2NCOCONH_2$, mol wt 80.07, is sparingly soluble in water and insoluble in various organic solvents. It melts at about 350 °C, with accompanying decomposition. Because of the low solubility in water, the compound is granulated and used as a slow-release nitrogen fertilizer. Conventional nitrogen fertilizers such as ammonium sulfate, urea, ammonium nitrate, and ammonium phosphate, are soluble in water, and thus are easily lost as run-off when it rains. On the contrary, oxamide stays in the soil longer. Therefore, it is gradually decomposed by microorganisms in the soil and utilized by plants for longer periods.

Oxalyl Chloride. This diacid chloride [CAS: 79-37-8], ClCOCOCl, mol wt 126.9, is produced by the reaction of anhydrous oxalic acid and phosphorus pentachloride. The compound vigorously reacts with water, alcohols, and amines, and is employed for the synthesis of agrochemicals, pharmaceuticals, and fine chemicals.

Additional Reading

Jpn. Pat. 6,126,977-B (1986), S. Tahara and co-workers (to UBE Industries).
Sarver, L.A. and P.H.M.P. Briton: *J. Am. Chem. Soc.*, **49**, 943 (1927).
U.S.Pat. 2,057,119 (1936), G.S. Simpson (to General Chemical (Allied Chemical).
Werneck S. and R. Pinner: *The Surface Treatment and Finishing of Aluminum and Its Alloys*, Robert Draper, Ltd., Teddington, UK, 1972.

OX BOW LAKE. The type of lake developed on the flood plains, especially in the delta regions, of large rivers. Ox bow lakes represent the detached or truncated former meanders of the main stream, which, from time to time, straighten their courses by forming cutoffs or shortcuts.

OXIDATION AND OXIDIZING AGENTS. Many years ago, the term *oxidation* signified a reaction in which oxygen combines chemically with another substance. The term now has a much broader meaning and includes any reactions in which electrons are transferred. Oxidation and reduction always occur simultaneously (redox reactions), and the substance which gains electrons is termed the *oxidizing agent*. For example, cupric ion is the oxidizing agent in the reaction: Fe (metal) $+ Cu^{++} \rightarrow Fe^{++} + Cu$ (metal). Here, two electrons (negative charges) are transferred from the Fe atom to the Cu atom. Thus the Fe becomes positively charged (is oxidized) by loss of two electrons, while the Cu receives the two electrons and becomes neutral (is reduced). Electrons may also be displaced within the molecule without being completely transferred away from it. Such partial loss of electrons likewise constitutes oxidation in its broader sense and leads to the application of the term to a large number of processes which at first inspection might not be considered to be oxidations. Reactions of a hydrocarbon with a halogen, for example: $CH_4 + Cl_2 \rightarrow CH_3Cl + HCl$, involves partial oxidation of the methane. Also, when a halogen addition to a double bond is made, this is regarded as an oxidation.

Dehydrogenation is also a form of oxidation, when two hydrogen atoms, each having one electron, are removed from a hydrogen-containing organic compound by a catalytic reaction with air or oxygen, as in oxidation of alcohols to aldehydes. See also **Dehydrogenation**.

Oxidizing agents are widely used throughout the chemical and petrochemical industries. It is also interesting to note that, while a primary thrust in food processing is to prevent oxidation (associated with rancidity and spoilage in foods), some powerful oxidizing agents are required by some processes to perform the function of bleaching. See also **Bleaching Agents**.

As one example, after a crude fat or oil is refined to remove its impurities, it must be further treated by bleaching to remove coloring materials that are typically present. Sulfuric and metaphosphoric acids and hydrogen peroxide have been used for this purpose. Calcium hypochlorite is used in bleaching sugar syrup prior to crystallization. Not all effective bleaching agents can be used because of their toxicity and inability to remove them completely. The aim is to remove all traces of the bleaching agent during subsequent processing so that they do not occur in the final product. Calcium peroxide, acetone peroxide, and benzoyl peroxides are other bleaching agents sometimes used in the food industry.

The use of oxidizers, such as potassium bromate, potassium iodate, and calcium peroxide as dough modifiers in the baking industry dates back many years. However, their mechanism has not been fully explained. Among authorities, there are at least two major viewpoints. It has been proposed that oxidizers inhibit proteolytic enzymes present in flour. It has also been proposed that the number of $-S-S-$ bonds between protein chains is increased, forming a tenacious network of molecules. This action leads to a tougher, drier, and more extensible dough. The need for oxidizers is less with well aged flours and in connection with supplemented flours that have been effectively *brominated* at the mills. When used properly,

oxidizers contribute to improve appearance, brighter crumb, and better texture of breads. Oxidizers do not appear to interfere with the generation of gas by yeast or leavening chemicals. However, oxidizers used to excess can destroy the desirable properties of the dough and end-products.

Additional Reading

Afanas'ev, I.B.: *Oxidation and Antioxidants in Organic Chemistry and Biology*, Marcel Dekker, Inc., New York, NY, 2004.
Gitler, C., and A. Danon: *Cellular Implications of Redox Signaling*, Imperial College Press World Scientific, London, UK, 2003.
Grabke, H.J.: *Oxidation of Intermetallics*, John Wiley & Sons, Inc., New York, NY, 1998.
Hodnett, B.K.: *Heterogeneous Catalytic Oxidation: Fundamental and Technological Aspects of the Selective and Total Oxidation of Organic Compounds*, John Wiley & Sons, Inc., New York, NY, 2000.
St. Angelo, A.J.: *Lipid Oxidation in Food*, American Chemical Society (ACS), Washington, DC, 1992.

OXIDATION NUMBER. In its original and restrictive sense, the number of electrons which must be added to a cation to neutralize the charge. The concept has been extended to anions by assignments of negative oxidation numbers. Moreover, it has been further extended, first to all atoms or radicals joined by electrovalent bonds, and then to covalent compounds in which the shared electrons are distributed equally. For the broadest use of the concept, the expression "oxidation state" is often used.

OXIDATION POTENTIAL. The potential drop involved in the oxidation (i.e., ionization) of a neutral atom to a cation, of an anion to a neutral atom, or of an ion to a more highly charged state (e.g., ferrous to ferric).

OXIDATION-REDUCTION INDICATOR. A substance that has a color in the oxidized form different from that of the reduced form and can be reversibly oxidized and reduced. Thus, if diphenylamine is present in a ferrous sulfate solution to which potassium dichromate is being added, a violet color appears with the first drop of excess dichromate.

See also **Indicator**.

OXIDATIVE COUPLING. A polymerization technique for certain types of linear high polymers. Oxidation of 2,6-dimethylphenol with an amine complex of copper salt as catalyst forms a polyether, with splitting off of water. The product is soluble in aromatic and chlorinated hydrocarbons; insoluble in alcohols, ketones, and aliphatics. It is thermoplastic and unaffected by acids, bases, and detergents. It has a very broad useful temperature range (from -170 to $+190\,°C$). It is also dimensionally stable and has good electrical resistance. Oxidative coupling of diacetylenes and dithiols also yields promising polymers.

OXIDE. A mineral in which metallic atoms are bonded to oxygen atoms.

OXIDIZING MATERIAL. Any compound that spontaneously evolves oxygen either at room temperature or under slight heating. The term includes such chemicals as peroxides, chlorates, perchlorates, nitrates, and permanganates. These can react vigorously at ambient temperatures when stored near of in contact with reducing materials such as cellulosic and other organic compounds. Storage areas should be well ventilated and kept as cool as possible.

OXIMES. One of a number of compounds that result from the interaction of aldehydes, ketones, and other carbonyl-containing substances with hydroxylamine, e.g., acetone yields acetoxime

$$\begin{array}{c} CH_3 \\ \diagdown \\ C{=}O + H_2NOH \rightarrow \\ \diagup \\ CH_3 \end{array} \qquad \begin{array}{c} CH_3 \\ \diagdown \\ C{=}NOH + H_2O \\ \diagup \\ CH_3 \end{array}$$

OXONIUM COMPOUNDS. Coordination compounds, commonly of certain oxygen-containing organic substances, with mineral acids, of the general type $[R_2O]HCl$. These compounds bear a strong resemblance to the oxonium (hydronium) ion, which is a proton in combination with a water molecule, and is the form in which protons commonly exist in aqueous solutions.

OXO PROCESS. The oxo process, also known as hydroformylation, is the reaction of carbon monoxide and hydrogen with an olefinic substrate to form isometric aldehydes as shown in Equation (1). The ratio of isomeric aldehydes depends on the olefin, the catalyst, and the reaction conditions.

$$RCH{=}CH_2 + CO + H_2 \xrightarrow{\text{catalyst}} RCH_2CH_2CHO \qquad (1)$$
$$+ R(CH_3)CHCHO$$

If a double-bond shift occurs, the number of aldehyde isomers is increased.

Synthesis gas, a mixture of CO and H_2, also known as syngas, is produced for the oxo process by partial oxidation (eq. 2) or steam reforming (eq. 3) of a carbonaceous feedstock, typically methane or naphtha. The ratio of CO to H_2 may be adjusted by cofeeding carbon dioxide, CO_2, as illustrated in equation 4, the water gas shift reaction.

$$2CH_4 + O_2 \rightarrow 2CO + 4H_2 \qquad (2)$$

$$CH_4 + H_2O \rightarrow CO + 3H_2 \qquad (3)$$

$$CO_2 + H_2 \rightleftharpoons CO + H_2O \qquad (4)$$

$$2CH_4 + CO_2 + O_2 \rightarrow 3CO + 3H_2 + H_2O \qquad (5)$$

The overall process for producing a 1:1 CO to H_2 ratio by partial methane oxidation and the water gas shift reaction is represented by equation 5.

History

The oxo reaction proceeds most frequently in the presence of a Group 8–10 (VIII) metal catalyst in the liquid phase, most particularly with members of Group 9, the Co-Rh-Ir triad. The earliest catalyst, hydrocobalt tetracarbonyl, [CAS: 16842-03-8] $HCo(CO)_4$ was an outgrowth of Fischer-Tropsch investigations carried out prior to World War II on the effect of olefins on hydrocarbon synthesis. The hydroformylation reaction, as practiced in the early days using cobalt catalysis, presented formidable requirements of high pressure, containment of the hydrogen, containment of carbon monoxide, and handling of the toxic and unstable metal carbonyls.

The search for catalyst systems which could effect the oxo reaction under milder conditions and produce higher yields of the desired aldehyde resulted in processes utilizing rhodium. Oxo capacity built since the mid-1970s, both in the United States and elsewhere, has largely employed tertiary phosphine-modified rhodium catalysts.

Propylene [CAS: 115-07-1] is the predominant oxo process olefin feedstock. Ethylene [CAS: 74-85-1], as well as a wide variety of terminal, internal, and mixed olefin streams, are also hydroformylated commercially. Branched-chain olefins include octenes, nonenes, and dodecenes from fractionation of oligomers of C_3-C_4 olefins as well as octenes from dimerization and codimerization of isobutylene and 1- and 2-butenes.

Linear terminal olefins are the most reactive in conventional cobalt hydroformylation.

Oxo aldehyde products range from C_3 to C_{15}, ie, detergent range, and are employed principally as intermediates to alcohols, acids, polyols, and esters formed by the appropriate reduction, oxidation, or condensation chemistry.

The classic challenges in oxo technology are simultaneously to achieve high reaction rate, high selectivity to the desired aldehyde, and to utilize a highly stable catalyst.

Catalysts

Catalysts used are unmodified cobalt, ligand-modified cobalt, ligand-modified rhodium and rhodium modified with ionic phosphine ligands.

Future Trends. In addition to the commercialization of newer extraction/decantation product/catalyst separations technology, there have been advances in the development of high reactivity oxo catalysts for the conversion of low reactivity feedstocks such as internal and α-alkyl substituted α-olefins. These catalysts contain (as ligands) ortho-*t*-butyl or similarly substituted arylphosphites, which combine high reactivity, vastly improved hydrolytic stability, and resistance to degradation by product aldehyde, which were deficiencies of earlier, unsubstituted phosphites.

Uses

n-Propanol and *n*-propyl acetate account for about 70% of the U.S. propionaldehyde derivative market. These compounds are used principally in flexographic and gravure inks which require volatile solvents to prevent smearing and ink accumulation on the printing presses. Some propanol is also converted into *n*-propylamines which are important pesticide intermediates. *n*-Propanol is also employed as a precursor for glycol ethers.

The highest volume oxo chemical in the United States, *n*-butyraldehyde, is converted mainly into *n*-butanol, employed chiefly to produce butyl acrylate and methacrylate. In contrast, the principal *n*-butyraldehyde derivative in Europe and Japan is 2-ethylhexanol, the precursor to the poly(vinyl chloride) (PVC) plasticizer, DOP.

1,4-Butanediol [CAS: 110-63-4] (BDO) goes primarily into tetrahydrofuran [CAS: 109-99-9] (THF) for production of polytetramethylene ether glycol (PTMEG), used in the manufacture of polyurethane fibers, eg, Du Pont's Spandex. THF is also used as a solvent for PVC and in the production of pharmaceuticals.

The principal C_5 valeraldehyde derivatives, *n*-amyl and 2-methylbutyl alcohols, are used predominantly to make zinc diamyldithiophosphate lube oil additives, which are employed primarily in automotive antiwear applications. Similarly, the *n*-valerate and 2-methylbutyrate esters of pentaerythritol and trimethylolpropane are used in aeromotive synlube formulations and as refrigerant lubricants.

C_7-C_9 oxo-derived acids are the principal derivatives of the C_7-C_9 oxo aldehydes, and in analogy to C_5 oxo aldehyde market applications, are used chiefly to make neopolyol esters which are employed almost entirely in aeromotive applications.

Several alcohols in the C_6-C_{13} range are produced by oxo reactions and are used in both plasticizer and detergent applications. Linear $C_{12}-C_{15}$ alcohols are employed primarily in detergent applications.

Safety, Health, and Environmental Concerns

Oxo plants employ mixtures of highly toxic, flammable gases under pressure at high temperatures and require strict adherence to established operating safety codes and emergency reporting procedures to local, state, and federal authorities. In the United States, carbon monoxide is classified as both an acute, fire, and sudden release hazard.

The carbon monoxide component of the oxo reactant gases presents the most immediate human health hazard.

Additional Reading

Brown, C.K. and G. Wilkinson: *J. Chem. Soc. (A)*, 1392 (1970).
Chemical Economics Handbook, Oxo Chemicals Report, SRI International, Menlo Park, CA., Jan. 1991 and preliminary 1994 draft.
Frohning, C.D. and C.W. Kohlpaintner: in B. Cornils and W.A. Herrmann, eds., *Applied Homogeneous Catalysis with Organometallic Compounds*, Vol. 1, VCH Publishers, Weinheim, Germany, 1996, pps. 29–90.
Pruett, R.L.: *Adv. Organometal. Chem.*, **17**, 1 (1979).

OXY. The radical $-O-$ in organic compounds, performing in a manner similar to the oxo radical ($O=$) and the epoxy radical ($-O-$).

OXYACID. An acid that contains oxygen, such as chloric acid ($HClO_3$).

OXYAZO COMPOUNDS. Compounds of the type $RN{=}NC_6H_4OH$, containing both the azo group $-N{=}N-$, and a hydroxyl group $-OH$, both attached to carbon atoms in the same ring. These compounds are commonly produced by the action of diazo compounds upon phenols in alkaline solution. They constitute a class of dyes. See also **Dyes**.

OXYGEN. [CAS: 7782-44-7]. Chemical element, symbol O, at. no. 8, at. wt. 15.9994, periodic table group 16, mp $-218.4\,°C$, bp $182.96\,°C$, critical temperature $118.8\,°C$, critical pressure 49.7 atmospheres, density 1.568g/cm^3 (solid), 1.429 g/L ($0\,°C$). Solid oxygen has a cubic crystal structure. Oxygen at standard conditions is a colorless, odorless, tasteless gas. Oxygen is slightly soluble in H_2O (4.89 parts oxygen in 100 parts H_2O at $0\,°C$), the solubility decreasing with increasing temperature (2.6 parts oxygen in 100 parts H_2O at $30\,°C$; 1.7 parts oxygen in 100 parts H_2O at $100\,°C$). Oxygen is slightly soluble in alcohol. Molten silver dissolves up to $10\times$ its volume of oxygen, but easily gives up the gas upon cooling. There are three stable isotopes, ^{16}O through ^{18}O. Three radioactive isotopes have been identified, ^{14}O, ^{15}O, and ^{19}O, with short half-lives measured in seconds and minutes. In terms of abundance in igneous rocks in the earth's crust, oxygen ranks first, with an average composition by weight of 46.6%. In terms of abundance in seawater, oxygen also ranks first, with an estimated 4 billion tons of oxygen per cubic mile of seawater. In terms of cosmic abundance, oxygen ranks eighth. For comparisons, assigning a value of 10,000 to silicon, the figure for oxygen is 220,000 and that for hydrogen, estimated the most abundant, is 3.5×10^8. Of dry air in the earth's atmosphere, 23.15% is oxygen by weight; 20.98% by volume. In the atmosphere, the oxygen is mixed with nitrogen, argon, the rare gases, CO_2, and H_2O vapor. Oxygen first was identified by Priestly in 1774 when he was experimenting with mercuric oxide. In the same year, Scheele also identified the element. Oxygen is required for burning and combustion, although the conditions of combustion vary widely. For example, phosphorus burns in air at the low temperature of $34\,°C$ when ignited. The temperature if ignition for ether in air is $340\,°C$, for ethyl alcohol in air, $560\,°C$, kerosene in air, about $300\,°C$, and hydrogen in air, about $600\,°C$. The oxidation process may occur with the rapidity and violence of an explosion, or may be as slow as the rusting of iron. Nearly all known species of living things require oxygen in some form, either free or chemically bound. First ionization potential 13.614 eV; second, 34.93 eV; third, 54.87 eV. Oxidation potentials $H_2O_2 \rightarrow O_2 + 2H^+ + 2e^-$, -0.68 V; $3H_2O \rightarrow \frac{1}{2}O_2 + 2H_3O^+ \ (10^{-7}M) + 2e^-$, -0.815 V; $3H_2O \rightarrow \frac{1}{2}O_2 + 2H^+ + 2e^-$, 1.229 V; $4H_2O \rightarrow H_2O_2 + 2H_3O^+ + 2e^-$, -1.77 V; $3H_2O \rightarrow O(g) + 2H^+ + 2e^-$, -2.42 V; $HO_2^- + OH^- \rightarrow O_2 + 2H_2O + 2e^-$, 0.075 V; $4OH^- \rightarrow O_2 + 2H_2O + 4e^-$, -0.401 V; $3OH^- \rightarrow HO_2^- + H_2O + 2e^-$, -0.87 V; $OH^- \rightarrow OH + e^-$, -1.4 V. Other physical properties of oxygen are given under **Chemical Elements**

Allotropic Forms

The three known allotropic forms of oxygen are (1) the ordinary oxygen in the air, with two atoms per molecule, O_2, (2) ozone, O_3, with three atoms per molecule, and (3) the rare, very unstable, nonmagnetic, pale-blue O_4. The latter breaks down readily into two molecules of O_2.

When oxygen is subjected to the silent electric discharge, activated atomic oxygen is produced. Atomic oxygen displays an afterglow upon cessation of the current, and the oxygen is notably active with hydrogen bromide, forming bromine; with H_2S, forming sulfur, SO_2, sulfur trioxide, and H_2SO_4; with CS_2, forming carbon monoxide, CO_2, and SO_2, and, strangely, reduced molybdenum trioxide to a white oxide not reducible with hydrogen. The concentration of atomic oxygen obtainable by the silent electric discharge through oxygen is estimated at 20%.

The normal electron distribution of the electrons of the oxygen atom is $1s^2 2s^2 2p_x{}^2 2p_y{}^1 2p_z{}^1$, with 2 unpaired electrons in the $2p$ orbitals. The covalent or partly covalent compounds of oxygen would be expected to have $90°$ bonding angles. But in many cases they have values significantly greater (ca. $104°$ for R_2O and $105°$ for H_2O). This suggests the promotion of a $2s$ electron to a $2p$ orbital (i.e., $2p_y$ orbital), still leaving two unpaired electrons (a $2s$ and a $2p_z$ electron), and permitting partial sp^3 hybridization (which is incomplete because sufficient energy is not available) but producing bond angles between $90°$ and $109°$ for the sp^3 tetrahedral structure, with covalent-polar bonds.

The oxygen molecule is paramagnetic with a moment in accord with two unpaired electrons. In molecular orbital terms, the configuration is written

$$O_2[KK(z\sigma)^2(y\sigma^*)^2(x\sigma)^2(w\pi^*)^4(v\pi^*)^2]$$

in which KK designates the complete $1s$ shells of the two atoms, which are nonbonding, the term $(z\sigma)^2$ denotes the bonding effect of one pair of $2s$ electrons, one from each of the O atoms, $(y\sigma^*)^2$ denotes the antibonding effect of the second pair, the $(x\sigma)^2$ term represents the s-bond formed by one pair of p-electrons, $(w\pi)^4$ represents the 2 π-bonds formed by the other two pairs of p-electrons, while the $(v\pi^*)^2$ term denotes the last

pair of p-electrons, which go into the next π subshell (two orbitals) with unpaired spins, and are antibonding.

Ozone

Ozone [CAS: 10028-15-6]. O_3, obtained by electrical discharge through oxygen or high-current electrolysis of sulfuric acid, is considered on the basis of electron diffraction studies to have an O–O–O bond angle of $127 \pm 3°$ and O–O bond length of 1.26 ± 0.02Å. Its structure is considered to resonate among several forms, chiefly

Ozone is an unstable blue gas, of characteristic odor, liquefiable at $-12\,°C$, formed when ordinary oxygen is subjected to electrostatic discharge, density 1.5 times that of oxygen gas, mp $-251.4\,°C$, bp $-111.5\,°C$. It is explosive by percussion or under variations of pressure. Ozone reacts (1) with potassium iodide, to liberate iodine, (2) with colored organic materials, e.g., litmus, indigo, to destroy the color, (3) with mercury, to form a thin skin of mercurous oxide causing the mercury to cling to the containing vessel, (4) with silver film, to form silver peroxide, Ag_2O_2, black, produced most readily at about $250\,°C$, (5) with tetramethyldiaminodiphenylmethane $(CH_3)_2N \cdot C_6H_4 \cdot CH_2 \cdot C_6H_4 \cdot N(CH_3)_2$, in alcohol solution with a trace of acetic acid to form violet color (hydrogen peroxide, colorless; chlorine or bromine, blue; nitrogen tetroxide, yellow). In contrast to hydrogen peroxide, ozone does not react with dichromate, permanganate, or titanic salt solutions. Ozone reacts with olefin compounds to form ozonide addition compounds. Ozonides are readily split at the olefinin-ozone position upon warming alone, or upon warming their solutions in glacial acetic acid, with the formation of aldehyde and acid compounds which can be readily identified, thus serving to locate the olefin position in oleic acid, $C_{17}H_{33} \cdot COOH$, as midway in the chain $(CH_3(CH_2)_7CH:CH(CH_2)_7COOH$. Ozone is used (1) as a bleaching agent, e.g., for fatty oils, (2) as a disinfectant for air and H_2O, (3) as an oxidizing agent. See also **Aerosol**; and **Ozone**.

The protective effects of an ozone layer in the stratosphere of the earth have been known for many years. Ozone prohibits full penetration of ultraviolet radiation from the sun to the surface of the earth. Much research has been conducted and is still underway to determine the extent to which certain chemical pollutants may be destroying the ozone layer gradually and, among other factors, causing marked warming of the earth. **Role of Oxygen in Water.** The solvent properties of H_2O are due in great part to the dipole moment of its molecules (1.8 debye units) and its high dielectric constant (ca. 78). Its hydrogen atoms form hydrogen bonds with electronegative atoms such as fluorine, nitrogen, or oxygen. In fact, the H_2O molecules associate in H_2O by this mechanism. Also, the oxygen atoms of H_2O because of their residual negative charges are electrically attracted by cations, so that the H_2O molecules arrange themselves around cations, facilitating solution and ionization. In the same way, H_2O molecules surround anions by attraction of the positive ends of the dipoles. By these two processes, as well as the dissociation of water into oxonium and hydroxide ions, it forms hydrates with many compounds. Moreover, H_2O readily reacts with large numbers of compounds because of these properties. Thus the hydrolysis of covalent halides that have at least one lone pair of electrons is initiated by the donation of a proton by the H_2O, followed by splitting off of hydrogen chloride.

Oxides

Oxygen forms oxides with all the elements except some inert gases. Oxides are said to be normal when they contain no oxygen atoms that are bonded to each other, as in the peroxides. The normal oxides may be divided into three groups, basic, acidic, and neutral. The basic oxides, which react with or dissolve in H_2O to produce alkaline solutions, are formed by the alkali and alkaline earth elements (except beryllium) by the lighter Lanthanides and actinium, by silver(I), thallium(I) and lead(II). The oxides of the nonmetals and of the transition metals in their higher oxidation states are in general acidic. The oxides lying in the positions between the two groups exhibit both basic and acidic properties (amphiprotic or amphoteric) such as those of aluminum, tin(II) and iron(III), Al_2O_3, SnO, and Fe_2O_3.

The known facts about the structure of hydrogen peroxide, H_2O_2, are that the O—H distances are 0.97Å, the O—O distances 1.47Å, the HOO angles 94°, and the dihedral angle between the planes of the two O—H radicals 97°. The O—O—O is essentially a single one. In the liquid, H_2O_2 is somewhat more self-ionized than water. In water $pK_A = 11.75$, $pK_B = 17$. Its reactions may be oxidizing or reducing. Thus, it oxidizes Fe(II) to Fe(III), Ti(III) to Ti(IV) and SO_3^{2-} to SO_4^{2-}; but it reduces MnO_4^- (acid solution) to Mn^{2+}. Peroxides are known for the alkali and alkaline earth metals, as well as zinc, cadmium, mercury, thorium, uranium, plutonium, etc. However, not all compounds of formula MO_2 (where M is a metal atom) are peroxides; some are merely dioxides, as MnO_2, PbO_2, etc., others are superoxides, such as NaO_2, KO_2, RbO_2, CsO_2, CaO_4, SrO_4, and BaO_4. These last compounds contain the group O_2^-, as evident from their paramagnetism and crystal structure. Perhydroxyl, the free acid corresponding to the superoxides, is unstable ($H_2O_2 \rightarrow HO_2 + H^+ + e^-$, $E° = 1.5V$; $HO_2 \rightarrow O_2 + H^+ + e^-$, $E° = +0.13V$). It is a moderately strong acid, $pK_A = 2.2$.

The peroxyacids containing —O—OH groups, are formed with all the transition elements in groups 4, 5, 6 of the periodic table, with main group elements 4 and 5 as well as elements of atomic numbers from boron to sulfur, inclusive. Representative peroxyacids are peroxymonosulfuric acid,

$$H:\overset{\displaystyle :\overset{\cdot\cdot}{O}:}{\underset{\displaystyle :\overset{\cdot\cdot}{O}:}{O}:S:O:O:}H$$

and peroxychromic acid,

$$H:\overset{\displaystyle :\overset{\cdot\cdot}{O}:}{\underset{\displaystyle :\overset{\cdot\cdot}{O}:}{O}:O:Cr:O:O:}H$$

The only peroxydiacids are formed by sulfur, phosphorus, carbon and boron, of which the most important is peroxy disulfuric acid

$$H:\overset{\displaystyle :\overset{\cdot\cdot}{O}:\quad :\overset{\cdot\cdot}{O}:}{\underset{\displaystyle :\overset{\cdot\cdot}{O}:\quad :\overset{\cdot\cdot}{O}:}{O}:S:O:O:S:O:}H$$

although peroxy bridge compounds are also formed by certain transition element complexes, e.g., $[Co(NH_3)_5OOCo(NH_3)_5]^{4+}$ and $[Co(NH_3)_5 OOCo(NH_3)_5]^{5+}$.

Industrial Oxygen

As with hydrogen, the electrolysis of water offers one approach to the production of pure oxygen. However, the economics are as unfavorable for oxygen production in this manner as for hydrogen. See also **Hydrogen**. For industrial oxygen production, air is the raw material. Using air, processes are of two major types: (1) liquid-oxygen processes wherein the oxygen is fractionally distilled from liquid air, and (2) gaseous-oxygen processes. See also **Cryogenics**.

Because of the relatively high energy costs of compressing and refrigerating involved in oxygen production, many processes have been developed and tested over the years, a high percentage of these later abandoned. An idea of the alternatives which face the process designer can be gathered from scanning the methods available specifically in the area of producing refrigeration for these processes: (1) Joule-Thomson effect only; (2) Joule-Thomson effect plus auxiliary refrigeration with an ordinary liquid-vapor cycle at moderate- or high-temperature levels, i.e., relative to liquid-air temperature; (3) Joule-Thomson effect plus approximately reversible expansion of the air or products in an expander; (4) refrigeration essentially due only to approximately reversible expansions of auxiliary

fluid or fluids operating in liquid-vapor cycles, i.e., the cascade process; and (5) processes using an auxiliary nitrogen-liquefaction cycle.

Designers also face the choice of capacity of an oxygen plant. Costs per unit weight of oxygen made are lowered as the capacity of the plant goes up. For example, a plant with a capacity of 2000 tons (1800 metric tons) per day will produce oxygen at approximately 50% of the cost per unit weight as a plant with a 200-ton (180-metric ton) capacity per day.

The demand for industrial oxygen has created the need for several new plants during the past 20 years. Capacities for most recent plants range from about 1100 tons (990 metric tons) per day to 2500 tons (2250 metric tons) per day.

An oxygen pipeline system was established in western Europe in the late-1970s that is 592 miles (956 kilometers) long. The Eastern Network of this system serves 30 consumers in France, Luxembourg, and West Germany; the Northern Network serves some 40 additional users in France, Belgium, and the Netherlands.

Uses

In addition to the requirements by the chemical industry for oxygen as a reactant, either directly from the air or in purer, more concentrated form as from a separation plant, significant quantities of purified oxygen are used for welding and cutting metals. Oxygen of a purity of 99.5% is required for oxyacetylene and oxyhydrogen torches. When combined in proper proportions, acetylene and oxygen yield a flame with a temperature of about 3,480 °C. Oxyhydrogen flames are somewhat lower in temperature, but they are particularly useful for welding light-gage aluminum and magnesium allows and for underwater cutting. In welding applications, a reduction in purity of oxygen used from 99.5% to 99.0% will cut welding efficiency by over 10%. During the past several years, basic oxygen steelmaking has increased requirements for pure oxygen. In this process, nearly pure oxygen is introduced by means of a lance into molten iron and scrap. The oxygen combines with carbon and other unwanted elements and refines raw steel in much less time than the older open-hearth furnaces. The basic oxygen process exceeded the open-hearth process in terms of output to the United States in 1970 for the first time. On the total scale of consumption, relatively limited amounts of oxygen go into medical and life-support applications, as required for emergency situations in aircraft at high altitudes.

Role of Oxygen in Corrosion

Oxygen and oxidizing agents exert both a positive and negative influence on corrosion of metals. On the one hand, an oxidizing agent may form a protective oxide film on the surface of certain metals, aluminum being an excellent example, which essentially arrests corrosion by many external agents. On the other hand, the presence of oxidants may increase the rate of corrosion by supporting cathode reactions. As an example, Monel metal fully resists attack by oxygen-free 5% H_2SO_4 at room temperature. The corrosion rate rises, however, in almost direct proportion to oxygen content. A 20% oxygen content will cause a corrosion rate of about 150 mdd (milligrams of metal corroded per square decimeter per day). A concentration of 40% will increase the rate to about 250 mdd; a concentration of 80% to about 450 mdd. The oxygen need not be present in all of the acid contained in the metal vessel, but simply present in that concentration at the interface of metal, acid, and surrounding atmosphere. The effect of oxidizing salts on corrosion can be dramatic. Several factors, in addition to oxygen, affect corrosion, including the presence of other metals (electromotive-force displacements of one metal by another), temperature, acidity, and velocity. These factors are discussed further under **Corrosion**.

Oxygen Toxicity of Plants

As early as 1801, Huber and Senebier observed that grains develop more satisfactorily in an atmosphere containing a mixture of 3 parts nitrogen and 1 part oxygen than in an atmosphere containing 3 parts of oxygen and 1 part nitrogen. Considerably later, in 1878, Bert noted that the earlier observations also apply to the development of many plant species and are not peculiar to grains. Bert further suggested that excessive oxygen may slow down various reactions involving fermentation. It was not until much later, in the mid-1940s, that scientists (Dickens, Haugaard, and Stadie) further confirmed that enzymes are inactivated by oxygen excesses. They particularly stressed this fact in connection with enzymes that contain a sulfhydryl group in the active site. Machaelis (1946), Barron (1946), and Gilbert (1963) later pointed out that molecular oxygen alone acts in a rather sluggish manner in this regard and that, therefore, a special process or phenomenon must be involved. Molecular oxygen can be reduced only by accepting one electron at a time.

A number of scientists in the late 1960s through the mid-1970s pointed out that many sources in biological systems produce oxygen *free radicals*. For example, some oxidative enzymes which contain flavin as a prosthetic group proceed by a radical mechanism. When illuminated, chloroplasts produce superoxide ions and singlet oxygen. Because of its singlet configuration, the latter is not hindered in its interactions with biological materials. As pointed out by Griffiths and Hawkins, singlet oxygen can be formed from the ground state when energy, usually in the form of light, is supplied n the presence of a photosensitizer. The compounds that are photosensitized include many dyes and pigments, such as chlorophyll, flavins, and hematoporphyrins. The interaction between the sensitizer and oxygen results in the transfer of electrons, with the formation of superoxide ion. McCord and Fridovich 1969 discovered the enzyme superoxide dismutase. Their later findings show that aerobic organisms contain it, giving further credence to the proposal that all oxygen-metabolizing organisms from superoxide free radicals as a result of a univalent reduction of oxygen. As pointed out by Kon (1978), those free radicals that are toxic to the organism, by themselves or through interaction with other active forms of oxygen, are dismutated by the action of this enzyme.

There is a close relationship between oxygen toxicity and radiation on enzymes, DNA, and fats. Gerschman et al. (1954) showed that the same substances that afford protection against oxygen poisoning also increase resistance to radiation. Their results were further strengthened by experiments that demonstrated that additive nature of the two effects. Work on the effects that free radicals have on some of the polysaccharides used in food processing was commenced by Kon and Schwimmer and reported in 1977.

Environmental Aspects of Oxygen

Gaseous oxides, notably those of carbon, nitrogen, and sulfur which result from the combustion of fossil fuels and numerous industrial processes, comprise a large portion of the air pollution problem. These compounds are discussed under the specific elements and, in particular, are described under **Pollution (Air)**. In connection with the pollution of water in streams, lakes, ponds, rivers, etc., the content of dissolved oxygen in water is of prime concern. Dissolved oxygen must be available to support fish and other desirable living species in natural waters, and sufficient additional oxygen must be available in the water to effect biological degradation of both natural and manufactured materials which reach the water. The overuse of streams for disposal purposes in many instances has almost fully depleted the dissolved oxygen available for life support and hence has given rise to the term "dead" lakes or streams. Two terms are widely used: (1) BOD (biological oxygen demand) which is the requirement for dissolved oxygen in water to degrade or decompose organic matter within a measured time period at a given temperature, and (2) COD (chemical oxygen demand) which is the requirement for dissolved oxygen in water to combine with chemicals, essentially of an inorganic nature, which are introduced into a stream as the result of disposal operations. These aspects of oxygen are discussed under **Water Pollution**.

Earth's Oxygen Supply

The manner in which the earth's present oxygen system and reserves were formed has been the subject of much postulation for many years. Many of the details remain unclear and unconfirmed. In a theory proposed by Berkner-Marshall (1964, 1965), as the earth's atmosphere evolved, there was a slow buildup of the concentration of oxygen—proceeding from a trace to the present content of 23.15% (weight). This theory also proposes that the oxygen content of the atmosphere fluctuated from time to time in a major and relatively rapid manner. There is speculation that these major alterations may have accounted for the extinctions of animal and life forms that took place at the ends of the Paleozoic and Mesozoic eras. For example, there was a great reduction in life in the latter part of the

Permian period (Paleozoic era) when many kinds of strange reptiles and trilobites disappeared and seem to have left no descendents. Plant life declined greatly too during the late Paleozoic. From thousands of species in the Pennsylvanian period, there remained only a few hundred during the late Permian. Numerous explanations, particularly of a climatic nature, have been offered for these periods of reduction in life.

As pointed out by Van Valen (1971), photosynthesis does not produce a net change in oxidation. Except in bacterial photosynthesis, oxygen production is accompanied by a stoichiometrically equal quantity of reduced carbon. Thus, almost all of the oxygen is eventually used to oxidize reduced carbon. Predominantly, this oxidation occurs as the result of respiration in animals and plants. Further oxidation occurs as the result of forest fires. As observed by Borchert (1951), the only net gain in oxygen equals the amount of reduced carbon buried, as in the form of peat, black mud, and similar sediments. It has been estimated that most individual molecules of carbon remain reduced only for relatively short periods (months or years) because animals and plants have geologically very short lives. Plants respire and so oxidize some reduced carbon almost immediately. Other net sources of oxygen include nitrogen fixation and the photolysis of water in the upper atmosphere. Some investigators have considered these sources quantitatively unimportant, although Brinkmann (1969) suggests that this process would produce, over the earth's history $(4.5 \times 10^9$ years), about seven times the present mass of oxygen in the atmosphere.

Numerous ways have been proposed to explain a net loss of molecular oxygen. Oxidation of volcanic gases, ferrous iron, sulfur, sulfide, and manganese, and the accretion of hydrogen from the solar wind are among these. Such processes are sometimes referred to as *oxygen sinks*. Estimates by Holland (1962) indicate that the net gain and net loss over geologic time are essentially in balance.

Van Valen has posed the question, "What can happen if photosynthesis is suddenly and drastically reduced?" Under such conditions, at a new steady state, production of oxygen and its consumption in the oxidation of carbon would be equal. But, before the new steady state occurs, would animals and decomposers use up much of the previously stored carbon in plants, thus creating a new loss of oxygen? Several investigators have observed that even if all the carbon in all organisms now alive were oxidized, this would decrease the atmospheric concentration of oxygen by less than 0.1% of its present value. And, further, still less than 1% of the present oxygen concentration would be used if all the reduced carbon available in soils and the like were reduced.

Much more detailed explanation of the stability of atmospheric oxygen is contained in the excellent review by Van Valen(1971).

As pointed out by Broecker (1970), the earth's oxygen supply is frequently included in lists of concerns over alterations in the environment, particularly as brought about by anthropogenic activities. Several investigators have made a number of observations which tend to invalidate any claims that oxygen is in danger of serious depletion. Broecker observes that each square meter of the earth's surface is covered by 60,000 moles of oxygen gas. Further, plants living in the ocean and on land produce about 8 moles of oxygen per square meter of surface each year. It is also observed that animals and bacteria destroy nearly all of the products of this photosynthetic activity—thus they use an amount of oxygen nearly equal to that generated by plants. Using the rate at which organic carbon enters the sediments of the ocean as a measure of the amount of photosynthetic product preserved each year, Broecker estimates this to be about 3×10^{-3} mole of carbon per square meter per year. This corresponds to approximately 1 part in 15 million of the oxygen present in the atmosphere. It is estimated, however, that this small amount of oxygen is probably being destroyed by a number of processes, including oxidation of reduced carbon, iron, and sulfur (weathering mechanisms). Broecker points out that the oxygen content of the atmosphere is thus well buffered, particularly in terms of relatively short time spans (100 to 1000 years).

Over a period of time, people have recovered about 10^{16} moles of fossil carbon and the fuels containing this carbon have been oxidized as sources of energy. Byproduct carbon dioxide from this combustion represent about 18% of the carbon dioxide content of the atmosphere. Two moles of atmospheric oxygen are used to liberate each mole of carbon dioxide from fossil fuel sources. Broecker points out that this process uses up only 7 out of every 10,000 available oxygen molecules. It is estimated that if these fuels are burned at an accelerating rate (5% per year), by the end of this century, only about 0.2% of available oxygen (20 molecules in every 10,000) will be used. It is estimated that if all known fossil fuels were ultimately burned, only 3% of available oxygen would be consumed. In terms of urban oxygen needs, particularly for automotive combustion needs, it is estimated that carbon monoxide levels in the atmosphere (in terms of physiological damage) would reach intolerable levels before the oxygen content of the atmosphere would have decreased by 2%.

The case of anthropogenic alterations of photosynthetic rates and its possible effects on oxygen supply has been covered previously by the observation that stoppage of all photosynthetic activity would require less than 1% of the present oxygen concentration.

Sverdrup et al.(1942) estimated that the oxygen content of deep sea water averages about 2.5 cubic centimeters at standard temperature and pressure per liter (0.1 mole per cubic meter). Thus, there are about 250 moles of oxygen gas in the deep sea for each square meter of earth surface. The oxygen content of the deep-sea waters is renewed about every 1000 years. The magnitude of this oxygen reservoir is tremendous. Broeker emphasizes this by observing that if the entire terrestrial photosynthetic product were dumped each year into the deep sea, the supply of deep-sea oxygen would last 50 years. But, if the waste products of 1 billion people were limited to 100 kilograms of dry organic waste per year, this would consume 0.01 mole of oxygen per square meter of earth surface and the deep-sea oxygen supply would last some 25,000 years.

In the summary of this report, Van Valen (1971) states, "There are three processes weakly concentration-dependent that keep changes in concentration of atmospheric pressure from being a random walk—inhibition of net photosynthesis by oxygen, the passage of hydrogen through the oxidizing part of the atmosphere before it escapes from the earth, and burial of reduced carbon in anaerobic water. A stronger regulator seems desirable but remains to be found. The cause of the initial rise in oxygen concentration presents a serious and unresolved quantitative problem."

And, in summary of his report, Broeker (1970) states, in part, "It can be stated with some confidence that the molecular oxygen supply in the atmosphere and in the broad expanse of open ocean are not threatened by human activities in the foreseeable future. Molecular oxygen is one resource that is virtually unlimited."

Additional Reading

Baukal, C.: *Oxygen Enhanced Combustion*, CRC Press, LLC, Boca Raton, FL, 1998.

Berkner, L.V. and L.C. Marshall: in *The Origin and Evolution of Atmospheres and Oceans*, (P.J. Brancazio and A.G.W. Cameron, editors), pages 102–126, Wiley, New York, NY, 1964.

Berkner, L.V. and L.C. Marshall: *J. Atmos Sci.*, **22**, 225 (1965).

Borchert, H.: *Geochim. Cosmochim. Acta*, **2**, 62 (1951).

Brinkmann, R.T.: *J. Geophys. Res.*, **74m** 5355 (1969).

Brocker, W.S.: "Man's Oxygen Reserves," *Science*, **168**, 1537–1538 (1970).

Dickens, R.: "The Toxic Effects of Oxygen on Brain Metabolism and on Tissue Enzymes," *Biochem. J.*, **40**, 145, 170 (1946).

Gerschman, R., et al.: "Oxygen Poisoning and X-irradiation: A Mechanism in Common," *Science*, **119**, 623 (1954).

Gilbert, D.L.: "The Role of Pro-Oxidants and Anti-Oxidants in Oxygen Toxicity," *Radiation Res. Suppl.*, **3**, 44 (1963).

Griffiths, J. and C. Hawkins: "Mechanistic Aspects of the Photochemistry of Dyes and Their Immediates," *J. Soc. Dyers Colorists*, **89**, 173 (1973).

Holland, H.D.: in *Petrologic Studies: A Volume in Honor of A.F. Buddington*, (A.E.J. Engel, H.L. James, and B.F. Leonard, editors), pages 447–477, Geological Society of America, Washington, DC, 1962.

Kon, S. and S. Schwimmer: "Depolymerization of Polysaccharides by Active Oxygen Species Derived from Xanthine Oxidase Systems," *Food Biochem.*, **1**, 141 (1977).

Kon, S.: "Effects of Oxygen Fee Radicals on Plant Polysaccharides," *Food Technol.*, **32**, 5, 84–94 (1978).

Kruk, I.: *Environmental Toxicology and Chemistry of Oxygen Species: The Handbook Of Environmental Chemistry*, Springer-Verlag, New York, Inc., New York, NY, 1997.

Lide, D.R.: *CRC Handbook of Chemistry and Physics*, 88th Edition, CRC Press, LLC, Boca Raton, FL, 2007.

McCord, J.M. and I. Fridovich: "Superoxide Dismutase: An Enzymatic Function for Erythrocuprein," *J. Biol. Chem.*, **244**, 6046 (1969).

Michaelis, L.: "Fundamentals of Oxidation and Reduction," in *Currents in Biochemical Research* (D.E. Green, editor), pages 207, Wiley, New York, NY, 1946.

Sawyer, D.: *Oxygen Chemistry*, Oxford University Press, New York, NY, 1999.

Sundquist E.T. and W.S. Broecker, Eds.: *The Carbon Cycle and Atmospheric CO2*, American Geophysical Union, Washington, DC, 1985.

Sverdrup, H.U., Johnson, M.W., and R.H. Fleming: *The Oceans, Their Physics, Chemistry and General Biology*, Prentice-Hall, Englewood Cliffs, New Jersey, 1942.

Van Valen, L.: "The History and Stability of Atmospheric Oxygen," *Science*, **171**, 439–443 (1971).

OXYGEN BALANCE. Oxygen content relative to the total oxygen required for oxidation of all carbon, hydrogen, and other easily oxidizable elements to carbon dioxide, water, etc.

OXYGEN CELL. An electrolytic cell whose emf is due to a difference in oxygen concentration at one electrode compared with that at another electrode of the same material.

OXYGEN CONSUMED (OC; COD; DOC). A measure of the quanity of oxidizable components present in water. Since the carbon and hydrogen, but not the nitrogen, in organic matter are oxidized by chemical oxidants, the oxygen consumed is a measure only of the chemically oxidizable components and is dependent on the oxidant, structure of the organic compound, and manipulative procedure. Since this value does not differentiate stable from unstable organic matter, it does not necessarily correlate with the biochemical oxygen demand value. It is also known as chemical oxygen demand (COD) and dichromate oxygen consumed (DOC).

See also **Biochemical Oxygen Demand(BOD)** and **dissolved Oxygen(DO)**.

OXYGEN DEBT. A term used to refer to the buildup of a need for oxygen through anaerobic respiration of muscle cells in a higher vertebrate animal during violent exercise. When the energy demands are too great to be satisfied by the aerobic respiration, the cells turn to anaerobic respiration. Lactic acid is an end product of such respiration; this acid tends to accumulate in the muscles and some of it diffuses out into the blood and accumulates in the liver. When the activity ceases, deep breathing continues and the extra oxygen is used to reconvert the lactic acid back to pyruvic acid and to carry the pyruvic acid on through the tricarbocyclic acid cycle. It may also be reconverted back to glucose and glycogen. We say that the muscles have built up an oxygen debt during the very active exercise and this is repaid in the continued deep breathing during rest following the exercise.

OXYGEN GROUP. The elements of group 16 of the periodic classification sometimes are referred to as the Oxygen Group. In order of increasing atomic number, they are oxygen, sulfur, selenium, tellurium, and polonium. The elements of this group are characterized by the presence of six electrons in an outer shell. The similarities of chemical behavior among the elements of this group are less striking than hold for some of the other groups, e.g., the close parallels of the alkali metals or alkaline earths. With exception of oxygen, all elements of the group have a valence of 4+, in addition to other valences. All of the elements with the exception of polonium also have a valence of 2−. Unlike the alkali metals or alkaline earths, for example, the elements of the oxygen group are not so similar chemically that they comprise a separate group in classical qualitative chemical analysis separations. Tellurium and selenium do appear together among the rarer metals of the second group in terms of qualitative chemical analysis.

OXYGEN SINK. A reservoir consisting of a chemical element or compound that combines readily with oxygen and thus removes it from the atmosphere. During the early part of Precambrian time, sulfur iron, and other elements and compounds served as important oxygen sinks, preventing oxygen from accumulating in the atmosphere.

OXYL PROCESS. A method for directly producing higher alcohols by catalytically reducing carbon monoxide with hydrogen.

OXYTOCIN. A polypeptide hormone which is secreted by the posterior lobe of the pituitary gland of mammals and other vertebrates. Oxytocin exerts a stimulating effect upon the muscles of the breast (milk-ejection) and those of the uterus of mammals. It is sometimes used medically to stimulate labor in cases of difficult childbirth and to time the onset of labor. See also **Central and Peripheral Nervous Systems**.

OYASHIO. See **Ocean Currents**.

OZOCERITE. Sometimes spelled ozokerite, this is a natural, brown to jet black mineral (paraffin) wax comprised mainly of hydrocarbons. The melting point is variable. The material is soluble in chloroform. When heated with sulfuric acid (20–30%) from 120–200 °C, ozocerite yields ceresine. Sometimes called earth wax, fossil wax, mineral wax, and native paraffin.

OZONE. [CAS: 10028-15-6]. Ozone, O_3, is an allotropic form of oxygen first recognized as a unique substance in 1840 by the German-Swiss chemist Christian Friedrich Schönbein. He noted ozone appeared during thunderstorms and named the gas ozone for its peculiar smell. Ozone is derived from a Greek word *ozein*, meaning to smell. Its pungent odor is detectable at ~0.01 ppm. It is thermally unstable and explosive in the gas, liquid, and solid phases. In addition to being an excellent disinfectant, ozone is a powerful oxidant not only thermodynamically, but also kinetically, and has many useful synthetic applications in research and industry. Its strong oxidizing and disinfecting properties and its innocuous by-product, oxygen, make it ideal for the treatment of water. Indeed, the most important application of ozone is in the treatment of drinking water, which began in Europe in 1903. The treatment of swimming pool water was also developed in Europe. Another important ozone application is for odor control in industrial processes and municipal wastewater-treatment plants. Ozone also is used on a large scale for the treatment of municipal secondary effluents. Industrial high quality water supplies are also treated with ozone. In addition, ozone has applications in the treatment of cooling-tower water and in pulp bleaching. Advanced oxidation processes employing ozone in combination with uv, H_2O_2 greatly improves the reactivity of ozone toward organic contaminants via generation of hydroxyl radicals.

Ozone, which occurs in the stratosphere 15–50 km (9–31 miles) in concentrations of 1–10 ppmv, is formed by the action of solar radiation on molecular oxygen. It absorbs biologically damaging ultraviolet radiation (230–320 nm), prevents the radiation from reaching the surface of the earth, and contributes to thermal equilibrium on earth. The concentration of ozone in the stratosphere is being depleted by release of chlorine and bromine containing substances (eg, bromo- and chlorofluorocarbons and methyl bromide), which are photolyzed forming chlorine and bromine atoms that catalytically destroy ozone. The production and use of such substances are being phased out according to recommendations of the Montreal protocol. Ozone also is present in the troposphere. Although concentrations in remote areas are low, much higher levels occur in urban areas due to increased concentrations of carbon monoxide, hydrocarbons, and nitrogen oxides released primarily by internal combustion engines.

Properties

At ordinary temperatures, pure ozone is a pale blue gas ($d = 2.1415$ g/L at 0 °C (32 °F) and 101.3 kPa (1 atm)) that can be condensed to an indigo blue liquid, which freezes to a deep blue-violet solid, mp−192.5 °C (−314.5 °F), bp −111.9 °C (−169.4 °F). The solubility of gaseous ozone at atmospheric pressure and 0 °C is 1.1 g/L H_2O. Gaseous ozone can be adsorbed by porous solid substrates such as silica gel and is often used in this form in organic synthesis.

Ozone is endothermic, thus it can burn or detonate by itself and represents the simplest combustible and explosive system. The concentration threshold for spark-initiated explosion of liquid ozone in oxygen at −183 °C (−297 °F) is 18.6 mol% O_3; the concentration limit for shock wave-initiated detonation of gaseous ozone-oxygen at 25 °C (77 °F) is 9.2 mol % O_3. Explosions of gaseous ozone can be initiated by shock wave, electrical spark, heat, or sufficiently intense light flash. Explosion of pure liquid ozone and concentrated solutions in oxygen can be initiated by impurities, sudden change in temperature or pressure, heat, electrical spark, or mechanical shock.

Ozone is a triangular molecule; its bond angle (116.8°) was established by microwave spectroscopy. The bond length of the ozone molecule

(0.1278 nm) is intermediate to that of a single and double oxygen bond, corresponding to a bond order of 1.7. Ozone is diamagnetic with C_{2v} symmetry and has a low dipole moment of 1.77×10^{-30} C•m (0.53 D). Based on Pauling resonance concepts, the structure of ozone is a hybrid, principally of form (**1**), with a small contribution from (**2**).

The bonding in (**1**) consists of two σ-bonds and one three-center–four-electron π-bond. Molecular orbital and valence bond calculations indicate that the 1,3-diradical (**3**), having two σ-bonds and weak bonding between the singly occupied π-atomic orbitals on the terminal O atoms, may contribute significantly to the ground-state structure of ozone. Other studies suggest that ozone may have a hypervalent structure, with bonding similar to (**1**), and negligible diradical character.

Thermochemical Decomposition

The decomposition of gaseous ozone is sensitive not only to homogeneous catalysis by light, trace organic matter, nitrogen oxides, mercury vapor, and peroxides, but also to heterogeneous catalysis by metals and metal oxides.

In pure water, the decomposition of ozone at 20 °C (68 °F) involves a complex radical chain mechanism, initiated by OH^- and propagated by O_2^- radical ions and HO radicals.

Hydrogen peroxide greatly accelerates the decomposition of ozone in alkaline solutions because of formation of HO_2^-, which reacts rapidly with ozone to form the radical ion O_2^-.

Photochemical Decomposition

Gaseous ozone is decomposed to oxygen atoms and molecules by absorbing radiation in the visible and uv spectrum: $O_3 + h\nu \rightarrow O_2 + O$.

In contrast to photolysis of ozone in moist air, photolysis in the aqueous phase can produce hydrogen peroxide initially because the hydroxyl radicals do not escape the solvent cage in which they are formed $O(^1D) + H_2O$ H_2O_2. Hydrogen peroxide is photolyzed slowly to hydroxyl radicals: $H_2O_2 + h\nu \rightarrow 2HO$. In pure water, HO radicals can decompose ozone, while in natural or polluted water, the HO radicals can oxidize undesirable solutes. The photolysis also forms $O(^3P)$ atoms, which can oxidize organic compounds. This is the basis of the O_3/uv process, one of the advanced oxidation processes (AOPs).

Chemistry of Ozone

The inorganic chemistry of ozone is extensive, encompassing virtually every element except most noble metals, fluorine, and the inert gases.

Ozone reacts rapidly with various free radicals and radical ions such as O, O_2^-, H, HO, N, NO, Cl, and Br. Some of these radicals (HO, NO, Cl, and Br) can initiate the catalytic decomposition of ozone.

The strong electrophilicity of ozone is manifested in its reaction with a wide variety of organic and organometallic functional groups, e.g., olefins, acetylenes, aromatics (carbocyclic and heterocyclic), activated C−H bonds (acetals, alcohols, aldehydes, ethers, and glycosides), unactivated C−H bonds (alkanes, cycloalkanes, and alkyl aromatics), deactivated C−H bonds (carboxylic acids and ketones), C−N and N−N bonds, Si−H and Si−C bonds, organometallic bonds (e.g., Grignard reagents), and nucleophiles (e.g., ammonia, amines, amino acids, arsines, disulfides, hydroxylamines, nitriles, phosphites, selenides, sulfides, and thioethers). Ozone also acts as a nucleophile, e.g., in its reaction with carbocations.

Atmospheric Ozone

Stratosphere. In the stratosphere, the region of the Earth's atmosphere from 6 to 31 miles (10 to 50 kilometers) above the surface, the chemical compound ozone plays a vital role in absorbing harmful ultraviolet radiation from the sun. During the past 20 years, concentrations of this important compound have been threatened by human-made gases released into the atmosphere, including those known as CFCs. These chemical compounds as well as meteorological conditions in the stratosphere affect the concentration of stratospheric ozone.

NOAA uses satellite, airborne and ground-based systems to continuously monitor stratospheric ozone as well as the chemical compounds and atmospheric conditions that affect its concentration. NOAA's Aeronomy Laboratory, http://www.al.noaa.gov/, Earth System Research Laboratory, http://www.cmdl.noaa.gov/, Climate Prediction Center, http://www.cpc.ncep.noaa.gov/, and the National Climatic Data Center, http://www.ncdc.noaa.gov/oa/ncdc.html, are actively involved in monitoring and research, which enhances the scientific understanding of ozone and the processes affecting its concentration in the stratosphere. See also **Stratospheric Ozone**.

Troposphere. Ozone in the troposphere is a greenhouse gas, a health hazard and harmful to plants and materials. In contrast to stratospheric ozone, which is necessary for life on earth, increases in tropospheric ozone are a cause for concern. Tropospheric ozone is highly variable. Highest concentrations are typically found in the upper troposphere, where air mixes with ozone-rich stratospheric air, and in polluted urban regions in summer. Before satellites provided a daily picture of global total ozone, scientists assumed the middle troposphere and the remote tropical troposphere were clean since they were usually removed from these two sources. Now troposheric ozone research is rapidly studying and trying to understand the processes responsible for the global tropospheric ozone distribution. See also **Tropospheric Ozone**.

Ozone Generation

The physical mass-transfer rate of ozone into water is affected by the gaseous ozone concentration, temperature, pressure, composition of the solution (ie, pH, ionic strength, and the presence of reactive substances), gas dispersion, turbulence, mixing, which depends on the type of contactor. Mass transfer of ozone into water in a bubble column as a function of pH has been measured.

Ozone Contractors and Dispersion Devices. Efficient transfer of ozone into solution requires the dispersion of gaseous ozone into small bubbles. This is accomplished in various types of ozone contactors, eg, porous diffuser bubble columns (co- or countercurrent flow), mechanically agitated vessels, turbine mixers, tubular reactors, in-line static mixers, as well as negative pressure (venturi) and positive pressure injectors. In turbines and injectors, ozone and water are forced or drawn concurrently through a small opening. Under intense mixing, bubbles are sheared and mixed thoroughly, decreasing the liquid film thickness but increasing both interfacial area and contact time. Plate and packed columns and spray towers also can be employed to increase the gas–liquid contact. In plate and packed columns, ozone and water can flow co- or countercurrently. Single or multiple contact chambers (up to 5) can be used. Contact time varies from 2–20 min, depending on the application. Faster ozone-transfer rates result in faster disinfection rates. Because ozone is a strong oxidant, corrosion-resistant materials of construction such as stainless steel, glass, and Teflon, should be employed.

Off-Gas Treatment. Ozone-transfer efficiencies vary with the number of stages and are typically > 90%. However, since even a 95% ozone absorption efficiency can result in a contactor off-gas containing as much as 740-ppmw ozone (based on a 1.5 wt% O_3 feed gas), treatment is required to reduce the ozone concentration to an acceptable maximum level of 0.2 mg m^{-3}. Ozone in the vent gases from water-treatment ozone contact chambers is destroyed mainly by thermal (300–350°C (572–662°F) for ≤5 s) and/or catalytic means, and sometimes by wet granular-activated carbon (GAC). Another option is recycling the off-gas to points in the water-treatment system having a high ozone demand. Dilution of ozone vent gases with air has been employed whenever practical. When oxygen is used as the feed gas, it can be recycled to the ozone-generation step; however, once-through operation is common in order to avoid redrying costs.

Uses

Ozone is used in the treatment of drinking water and in industries where high purity water is required (e.g., breweries, pharmaceuticals, and electronics). Ozone is also used in industrial wastewater pollution control,

wastewater disinfection, and odor control; in the treatment of process water, such as cooling tower water; in the treatment of swimming pools and spas; in pulp bleaching; and in organic synthesis, as a selective oxidant.

Among other uses, ozone therapy, employing $O_3 - O_2$, is increasingly being employed and studied in dentistry, veterinary and sports medicine, and proctology. Ozone is used as an aquatic oxidant and disinfectant in zoos, large aquariums, as well as fish and shrimp hatcheries. Ozone also is used for food preservation, in cold storage rooms, brewery cellars, hotel and hospital air ducts, and air conditioning systems. Ozone has also been used in textile bleaching and in the bleaching of esters, oils, fats, waxes, starch, flour, ivory, etc. Oxidation of Ag^+ by ozone is employed commercially to produce high purity AgO. The use of ozone as a chemical agent decontaminant has been patented.

Health and Safety Factors

As a constituent of the atmosphere, ozone forms a protective screen by absorbing radiation of wavelengths between 200 and 300 nm, which can damage DNA and be harmful to life. Consequently, a decrease in the stratospheric ozone concentration results in an increase in the UV radiation reaching the earth's surfaces, thus adversely affecting the climate as well as plant and animal life. For example, the incidence of skin cancer is related to the amount of exposure to uv radiation.

Research in 1974 indicated that chlorine radicals from photodegradation of CFCs (chlorofluorocarbons) can destroy ozone. Because of these studies and the fact that the two main varieties CFC-11 ($CFCl_3$) and CFC-12 (CF_2Cl_2) can persist in the atmosphere for 75–100 years, the United States in 1978 banned their use in aerosols, eg, hair sprays and certain deodorants. However, pressure to eliminate CFCs slackened until 1985, when the Antarctic ozone hole was discovered. See also **Ozone Hole**.

Confirmation of the destruction of ozone by reactive chlorine and bromine from halofluorocarbons has led to international efforts (through the United Nations Environment Program) to reduce emissions of ozone-destroying CFCs and other ozone destroying substances into the atmosphere. The 1987 Montreal Protocol on Substances That Deplete the Ozone Layer and its various revisions established schedules for phasing out production and use of various CFCs and other ozone depleting substances based on their "ozone depletion potential (ODP)". The schedules were based on comparing the projected ozone loss for a given release of a CFC compared to that for CFC-11. Substances with an ODP close to that of CFC-11 were scheduled for rapid phase out, whereas those with lower ODPs were scheduled for slower phase out. Replacement substances, eg, HCFCs (hydrochlorofluorocarbons) were scheduled for later phase out in favor of substances with lower ODPs. The recommendations of the Montreal Protocol have been incorporated into the U. S. EPA's Clean Air Act. See also **Montreal Protocol**.

Ozone can be toxic to plants, animals, and fish. The lethal dose, LD_{50}, for albino mice is 3.8 ppmv for a 4-h exposure; the 96-h LC_{50} for striped bass, channel catfish, and rainbow trout is 80, 30, and 9.3 ppb, respectively. Although ozone is toxic to humans, it is even more toxic to plants, eg, exposure to 0.1 ppmv results in a reduction in the photosynthesis rate by a factor of 2. Small, natural, and anthropogenic atmospheric ozone concentrations can increase the weathering and aging of materials, eg, plastics, paint, textiles, and rubber. For example, rubber is degraded by reaction of ozone with carbon–carbon double bonds of the rubber polymer, requiring the addition of aromatic amines as ozone scavengers. See also **Antioxidant**. An ozone decomposing polymer (noXon) has been developed that destroys ozone in air or water.

The toxicity of ozone to humans is largely related to its powerful oxidizing properties. The odor threshold of ozone varies among individuals but most people can detect 0.01 ppm in air, which is well below the limit for general comfort. The symptoms experienced on exposure to 0.1–1 ppm ozone are headache, throat dryness, irritation of the respiratory passages, and burning of the eyes caused by the formation of aldehydes and peroxyacyl nitrates. Exposure to 1–100 ppm ozone can cause asthma-like symptoms such as tiredness and lack of appetite. Short-term exposure to higher concentrations can cause throat irritations, hemorrhaging, and pulmonary edema.

Ozonation of drinking water produces various by-products such as aldehydes, ketones, carboxylic acids, organic peroxides, epoxides, nitrosamines, *N*-oxy compounds, quinones, hydroxylated aromatic compounds, brominated organics, and bromate ion. Although some of these compounds are potentially toxic or carcinogenic, most bioassay-screening studies have shown that ozonated water induces substantially less mutagenicity than chlorinated water.

See also **Ozone Depletion (Science of)**.

Additional Reading

Bailey, P.S.: *Ozonation in Organic Chemistry*, Vols. 1 and 2, Academic Press, Inc., New York, 1978–1982.

Beltran, F.J.: *Ozone Reaction Kinetics for Water and Wastewater Systems*, CRC Press LLC, Boca Raton, FL, 2003.

Bocci, V.: *Oxygen-Ozone Therapy: A Critical Evaluation*, Kluwer Academic Publishers, Norwell, MA, 2002.

Bocci, V.: "Ozone," *A New Medical Drug*, Springer-Verlag New York, LLC, New York, NY, 2005.

Brasseur, G.P.J.J.Orlando, and G.S. Tyndall: *Atmospheric Chemistry and Global Change*, Oxford University Press, New York, NY, 1999.

Dessler, A.: *Chemistry and Physics of Stratospheric Ozone*, Elsevier Science, New York, NY, 2000.

Gottschalk, C., J.A. Libra, and A. Saupe: *Ozonation of Water and Waste Water*, Wiley-VCH, New York, NY, 2000.

Hewitt, C.N., and A.V. Jackson: *Handbook of Atmospheric Science*, Blackwell Publishing, Malden, MA, 2003.

Hoigné, J. and H. Bader: *Water Res.*, **17**, 185 (1983).

Hrubec, J.: *Handbook of Environmental Chemistry*, Springer-Verlag, New York, NY, 1998.

Kogelschatz, U.B.Eliasson, and M. Hirth: *Ozone Sci. Eng.*, **10**, 367 (1988).

Lambright, W.H.: *NASA and the Environment: The Case of Ozone Depletion* (Monographs in Aerospace History Series #38), United States Government Printing Office, Washington, DC, 2005.

Langlais, B., D.A. Reckhow, and D.R. Brink: *Ozone in Water Treatment*, Lewis Publishers, Chelsea, MI, 1991.

Lewis, R.J., and N. Irving Sax: *Sax's Dangerous Properties of Industrial Materials*, 10th Edition John Wiley & Sons, Inc., New York, NY, 1999.

Lide, D.R.: *CRC Handbook of Chemistry and Physics*, 88th Edition, CRC Press LLC, Boca Raton, FL, 2007.

Parker, L. and W.A. Morrissey: *Stratospheric Ozone Depletion*, Nova Science Publishers, Inc., Huntington, NY, 2003.

Rakness, K.L.: *Ozone in Drinking Water Treatment: Process Design, Operation, and Optimization*, American Water Works Association, Denver, CO, 2005.

Roshchina, V.V.: *Ozone and Plant Cell*, Springer-Verlag New York, LLC, New York, NY, 2003.

Seinfeld, J.H. and S.N. Pandis: *Atmospheric Chemistry and Physics: From Air Pollution to Climate Change*, John Wiley & Sons, Inc., New York, NY, 1997.

Staff: *Standard Methods for the Examination of Water and Wastewater*, 18th ed., American Public Health Association, Washington, DC. 1994, pp. 4–105.

Wennberg, P.O. and co-workers: *Science*, **266**, 398 (1994).

OZONE DEPLETION (Science Of).

The Earth's ozone layer protects all life from the sun's harmful radiation, but human activities have damaged this shield. Less protection from ultraviolet light will, over time, lead to higher skin cancer and cataract rates and crop damage. The U.S., in cooperation with over 160 other countries, is phasing out the production of ozone-depleting substances in an effort to safeguard the ozone layer.

The term **ozone depletion** is used to describe two distinct but related observations: a slow, steady decline, of about 3% per decade, in the total amount of ozone in the earth's stratosphere during the past twenty years and a much larger, but seasonal, decrease in stratospheric ozone over the earth's Polar Regions during the same period. (The latter phenomenon is commonly referred to as the "ozone hole.") The detailed mechanism by which the polar ozone holes form is different from that for the mid-latitude thinning, but the proximate cause of both trends is believed to be catalytic destruction of ozone by atomic chlorine and bromine. See also **Ozone Hole**. The primary source of these halogen atoms in the stratosphere is photodissociation of chlorofluorocarbon (CFC) compounds, commonly called freons, and bromofluorocarbon compounds known as halons, which are transported into the stratosphere after being emitted at the surface.

In 1974, chemists Frank Sherwood Rowland and Mario Molina, at the University of California-Irvine, began studying the impacts of CFCs in the earth's atmosphere. They discovered that CFC molecules were stable enough to remain in the atmosphere until they got up into the middle of the stratosphere where they would finally (after an average of 50–100 years

for two common CFCs) be broken down by ultraviolet radiation releasing a chlorine atom. Rowland and Molina then proposed that these chlorine atoms might be expected to cause the breakdown of large amounts of ozone (O_3) in the stratosphere. Their argument was based upon an analogy to contemporary work by Paul J. Crutzen and Harold Johnston, which had shown that nitric oxide (NO) could catalyze the destruction of ozone. (Several other scientists, including Ralph Cicerone, Richard Stolarski, Michael McElroy, and Steven Wofsy had independently proposed that chlorine could catalyze ozone loss, but none had realized that CFCs were a potentially large source of chlorine.) Crutzen, Molina, and Rowland were awarded the 1995 Nobel Prize for Chemistry for their work on this problem.

The environmental consequence of this discovery was that, since stratospheric ozone absorbs most of the ultraviolet-B (UV-B) radiation reaching the surface of the planet, depletion of the ozone layer by CFCs would lead to an in increase in UV-B radiation at the surface, resulting in an increase in skin cancer and other impacts such as damage to crops and to marine phytoplankton.

After publishing their pivotal paper in June 1974, Rowland and Molina testified at a hearing before the U.S. House of Representatives in December, 1974. As a result significant funding was made available to study various aspects of the problem and to confirm the initial findings. In 1976 the U.S. National Academy of Sciences (NAS) released a report that confirmed the scientific credibility of the ozone depletion hypothesis. NAS continued to publish assessments of related science for the next decade.

The provisions of the Protocol include the requirement that the Parties to the Protocol base their future decisions on the current scientific, environmental, technical, and economic information that is assessed through panels drawn from the worldwide expert communities. To provide that input to the decision-making process, advances in understanding on these topics were assessed in 1989, 1991, 1994, 1998 and 2002 in a series of reports entitled Scientific assessment of ozone depletion.

The Ozone Layer

The Earth's atmosphere is divided into several layers. The lowest region, the troposphere, extends from the Earth's surface up to about 10 kilometers (6.2 miles) in altitude. Virtually all human activities occur in the troposphere. Mt. Everest, the tallest mountain on the planet, is only about 9 km (5.6 miles) high. The next layer, the stratosphere, continues from 10 km (6.2 miles) to about 50 km (31 miles). Most commercial airline traffic occurs in the lower part of the stratosphere.

As shown in Fig. 1, most atmospheric ozone is concentrated in a layer in the stratosphere, about 15–30 kilometers (9.3–18.6 miles) above the Earth's surface. Ozone is a molecule containing three oxygen atoms. It is blue in color and has a strong odor. Normal oxygen, which we breathe, has two oxygen atoms and is colorless and odorless. Ozone is much less common than normal oxygen. Out of each 10 million air molecules, about 2 million are normal oxygen, but only 3 are ozone. See also **Ozone**.

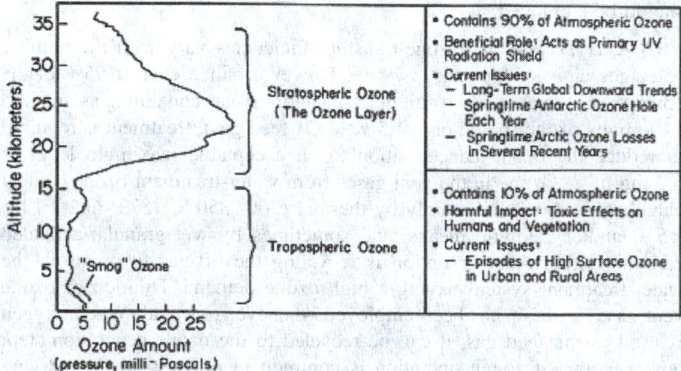

Fig. 1.   Atmospheric Ozone. (graph courtesy of World Meteorological Organization, Scientific Assessment of Ozone Depletion: 1998, WMO Global Ozone Research and Monitoring Project–Report No. 44, Geneva, 1998).

However, even the small amount of ozone plays a key role in the atmosphere. The ozone layer absorbs a portion of the radiation from the sun, preventing it from reaching the planet's surface. Most importantly, it absorbs the portion of ultraviolet light called UVB. UVB has been linked to many harmful effects, including various types of skin cancer, cataracts, and harm to some crops, certain materials, and some forms of marine life.

At any given time, ozone molecules are constantly formed and destroyed in the stratosphere. The total amount, however, remains relatively stable. The concentration of the ozone layer can be thought of as a stream's depth at a particular location. Although water is constantly flowing in and out, the depth remains constant.

While ozone concentrations vary naturally with sunspots, the seasons, and latitude, these processes are well understood and predictable. Scientists have established records spanning several decades that detail normal ozone levels during these natural cycles. Each natural reduction in ozone levels has been followed by a recovery. Recently, however, convincing scientific evidence has shown that the ozone shield is being depleted well beyond changes due to natural processes.

Ozone Depletion

Ozone depletion is the result of a complex set of circumstances and chemistry. For over 50 years, chlorofluorocarbons (CFCs) were thought of as miracle substances. They are stable, nonflammable, low in toxicity, and inexpensive to produce. Over time, CFCs found uses as refrigerants, solvents, foam blowing agents, and in other smaller applications. Other chlorine-containing compounds include methyl chloroform, a solvent, and carbon tetrachloride, an industrial chemical. See also **Carbon Tetrachloride**. Halons, extremely effective fire extinguishing agents, and methyl bromide, an effective produce and soil fumigant, contain bromine. All of these compounds have atmospheric lifetimes long enough to allow them to be transported by winds into the stratosphere. Because they release chlorine or bromine when they break down, they damage the protective ozone layer. The discussion of the ozone depletion process below focuses on CFCs, but the basic concepts apply to all of the ozone-depleting substances (ODS).

In the early 1970s, researchers began to investigate the effects of various chemicals on the ozone layer, particularly CFCs, which contain chlorine. They also examined the potential impacts of other chlorine sources. Chlorine from swimming pools, industrial plants, sea salt, and volcanoes does not reach the stratosphere. Chlorine compounds from these sources readily combine with water and repeated measurements show that they rain out of the troposphere very quickly. In contrast, CFCs are very stable and do not dissolve in rain. Thus, there are no natural processes that remove the CFCs from the lower atmosphere. Over time, winds drive the CFCs into the stratosphere.

The ozone depletion process Fig. 2, begins when CFCs and other ozone-depleting substances (ODS) are emitted into the atmosphere (1). Winds efficiently mix the troposphere and evenly distribute the gases. CFCs are extremely stable, and they do not dissolve in rain. After a period of several years, ODS molecules reach the stratosphere, about 10 kilometers (6.2 miles) above the Earth's surface (2).

Strong UV light breaks apart the ODS molecule. CFCs, HCFCs, carbon tetrachloride, methyl chloroform, and other gases release chlorine atoms, and halons and methyl bromide release bromine atoms (3). It is these atoms that actually destroy ozone, not the intact ODS molecule. It is estimated that one chlorine atom can destroy over 100,000 ozone molecules before it is removed from the stratosphere (4).

Ozone is constantly produced and destroyed in a natural cycle, as shown in Fig. 3. However, the overall amount of ozone is essentially stable. This balance can be thought of as a stream's depth at a particular location. Although individual water molecules are moving past the observer, the total depth remains constant. Similarly, while ozone production and destruction are balanced, ozone levels remain stable. This was the situation until the past several decades.

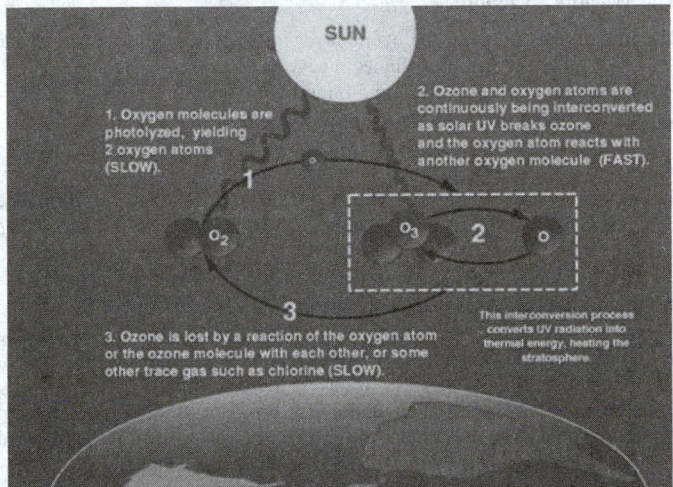

Fig. 3. Ozone cycle. (image courtesy of NASA/GSFC).

Large increases in stratospheric chlorine and bromine, however, have upset that balance. In effect, they have added a siphon downstream, removing ozone faster than natural ozone creation reactions can keep up. Therefore, ozone levels fall.

Since ozone filters out harmful UVB radiation, less ozone means higher UVB levels at the surface. The more the depletion, the larger the increase in incoming UVB. UVB has been linked to skin cancer, cataracts, damage to materials like plastics, and harm to certain crops and marine organisms. Although some UVB reaches the surface even without ozone depletion, its harmful effects will increase as a result of this problem.

Large volcanic eruptions can have an indirect effect on ozone levels. Although Mt. Pinatubo's 1991 eruption did not increase stratospheric chlorine concentrations, it did produce large amounts of tiny particles called aerosols (different from consumer products also known as aerosols). These aerosols increase chlorine's effectiveness at destroying ozone. The aerosols only increased depletion because of the presence of CFC — based chlorine. In effect, the aerosols increased the efficiency of the CFC siphon, lowering ozone levels even more than would have otherwise occurred. Unlike long-term ozone depletion, however, this effect is short-lived. The aerosols from Mt. Pinatubo have already disappeared, but satellite, ground-based, and balloon data still show ozone depletion occurring closer to the historic trend.

One example of ozone depletion is the annual ozone "hole" over Antarctica that has occurred during the Antarctic Spring since the early 1980s. Rather than being a literal hole through the layer, the ozone hole is a large area of the stratosphere with extremely low amounts of ozone. Ozone levels fall by over 60% during the worst years. See also **Polar Research**.

In addition, research has shown that ozone depletion occurs over the latitudes that include North America, Europe, Asia, and much of Africa, Australia, and South America. Over the U.S., ozone levels have fallen 5–10%, depending on the season. Thus, ozone depletion is a global issue and not just a problem at the South Pole.

Reductions in ozone levels will lead to higher levels of UVB reaching the Earth's surface. The sun's output of UVB does not change; rather,

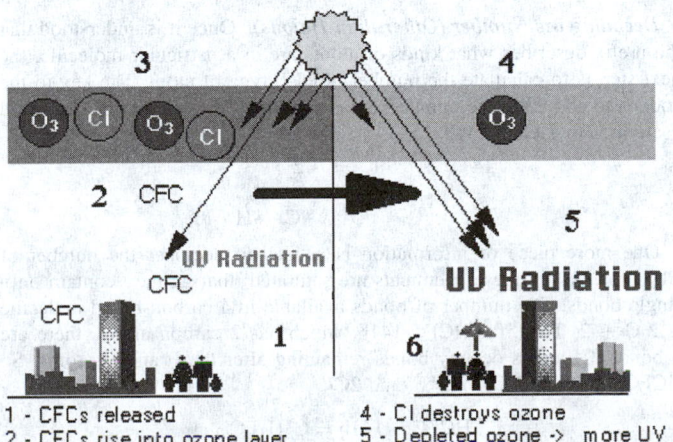

1 - CFCs released
2 - CFCs rise into ozone layer
3 - UV releases Cl from CFCs
4 - Cl destroys ozone
5 - Depleted ozone -> more UV
6 - More UV -> more skin cancer

Fig. 2. The process of ozone depletion. (image courtesy of EPA).

less ozone means less protection, and hence more UVB reaches the Earth. Studies have shown that in the Antarctic, the amount of UVB measured at the surface can double during the annual ozone hole. Another study confirmed the relationship between reduced ozone and increased UVB levels in Canada during the past several years.

Laboratory and epidemiological studies demonstrate that UVB causes nonmelanoma skin cancer and plays a major role in malignant melanoma development. In addition, UVB has been linked to cataracts. All sunlight contains some UVB, even with normal ozone levels. It is always important to limit exposure to the sun. However, ozone depletion will increase the amount of UVB, which will then increase the risk of health effects. Furthermore, UVB harms some crops, plastics and other materials, and certain types of marine life.

The World's Reaction

The initial concern about the ozone layer in the 1970s led to a ban on the use of CFCs as aerosol propellants in several countries, including the U.S. However, production of CFCs and other ozone-depleting substances grew rapidly afterward as new uses were discovered.

Through the 1980s, other uses expanded and the world's nations became increasingly concerned that these chemicals would further harm the ozone layer. In 1985, the Vienna Convention was adopted to formalize international cooperation on this issue. Additional efforts resulted in the signing of the Montreal Protocol in 1987. See also **Montreal Protocol**.

After the original Protocol was signed, new measurements showed worse damage to the ozone layer than was originally expected. In 1992, reacting to the latest scientific assessment of the ozone layer, the Parties decided to completely end production of halons by the beginning of 1994 and of CFCs by the beginning of 1996 in developed countries.

Because of measures taken under the Protocol, emissions of ozone-depleting substances are already falling. Based on measurements of total inorganic chlorine in the atmosphere, which stopped increasing in 1997 and 1998, stratospheric chlorine levels have peaked and are no longer increasing. The good news is that the natural ozone production process will heal the ozone layer in about 50 years.

Ozone-Depleting Substances

Class I Ozone-Depleting Substances. The CFCs are listed in Table 1.

The numbers in the Ozone Depletion Potential "ODP-1" column are from Table 1–5 of *The Scientific Assessment of Ozone Depletion, 2002*, a report of the World Meteorological Association's Global Ozone Research and Monitoring Project. The ODPs in the "ODP-1" column that were not in Table 1–5 of the 2002 report have not been updated since 1998 and are from *The Scientific Assessment of Ozone Depletion, 1998*. The "ODP-2" column numbers are from the Montreal Protocol, and the "ODP-3" column numbers are from 40 CFR Part 82, stratospheric ozone protection regulations required by title VI of the Clean Air Act amendments.

All GWP values represent global warming potential over a 100-year time horizon. The numbers in the "GWP-1" column are from Table 1–6 of *The Scientific Assessment of Ozone Depletion, 2002*, a report of the World Meteorological Association's Global Ozone Research and Monitoring Project. The GWPs in the "GWP-1" column that were not provided Table 1-6 of the 2002 report have not been updated since 1998 and are from *The Scientific Assessment of Ozone Depletion, 1998*. "GWP-2" column numbers are from the Intergovernmental Panel on Climate Change (IPCC) *Second Assessment Report: Climate Change 1995*, "GWP-3" column numbers are from the *IPCC Third Assessment Report: Climate Change 2001*, and "GWP-4" column numbers are from 40 CFR Part 82, stratospheric ozone protection regulations required by title VI of the Clean Air Act amendments.

Under section 602 of the Clean Air Act, EPA was required to list in the Federal Register the GWPs for ozone-depleting substances. That list was published on January 19, 1996. See also http://www.epa.gov/docs/fedrgstr/EPA-AIR/1996/January/Day-19/pr-372.html.

Blanks in the data indicate that the information was not shown in the original source.

Note: Global warming potentials (GWPs) for selected non-ozone-depleting substances are provided in Table 3.

Class II Ozone-Depleting Substances. Class II substances are listed in section 602 of the Clean Air Act, and comprise all HCFCs. See Table 2.

All the class II substances and their isomers are regulated under the accelerated phase out. These compounds are numbered according to the ASHRAE Standard 34 scheme. See numbering scheme for ozone-depleting substances and their substitutes below.

Blanks in the data indicate that the information was not shown in the original source.

Global Warming Potentials (GWPs) of ODS Substitutes. The GWP represents how much a given mass of a chemical contributes to global warming over a given time period compared to the same mass of carbon dioxide See Table 3. Carbon dioxide's GWP is defined as 1.0.

The HFCs are numbered according to the ASHRAE Standard 34 scheme. None of the chemicals listed below depletes the ozone layer. GWP values for ozone-depleting substances are listed in Table 1.

Numbering Scheme for Ozone-Depleting Substances and their Substitutes

Various ozone-depleting substances (CFCs, HCFCs, HBFCs, and halons) and their substitutes (HFCs, HCs, and PFCs) are numbered according to a system devised several decades ago and now used worldwide. Although it may seem confusing, in fact it provides very complex infor-mation about molecular structure and also easily distinguishes among various classess of chemicals. For example, it is not intuitive that 1,1-dichloro-1-fluoroethane (CFCl2-CH3) is an ozone-depleting substance. The designation HCFC-141b, in contrast, immediately conveys its ozone-depleting nature while concisely describing its structure.

Prefixes. The first step, and one that will provide a valuable way to check the results, is to understand the prefixes CFC, HCFC, HFC, PFC, and Halon. In CFCs and HCFCs, the first "C" is for chlorine (atomic symbol: Cl), and in all of them, "F" is for fluorine (atomic symbol: F), "H" is for hydrogen (atomic symbol: H) and the final "C" is for carbon (atomic symbol: C). It is the chlorine that makes a substance ozone-depleting; CFCs and HCFCs are a threat to the ozone layer but HFCs are not. PFC is a special prefix meaning "perfluorocarbon". "Per" means "all," so perfluorocarbons have all bonds occupied by fluorine atoms. Finally, halons are a general term for compounds that contain C, F, Cl, H, and bromine (atomic symbol: Br). Halon numbers are different from the others and will be discussed separately. As you decode the structure, double-check the results against the prefix. For example, an HFC contains no chlorine, so your results should not show any Cl atoms. (see Table 4)

Compounds used as refrigerants may be described using either the appropriate prefix above or with the prefixes "R-" or "Refrigerant." Thus, CFC-12 may also be written as R-12 or Refrigerant 12.

Blends of refrigerants are assigned numbers serially, with the first zeotropic blend numbered R-400 and the first azeotropic blend numbered R-500. Blends that contain the same components but in differing per-centages are distinguished by capital letters. For example, R-401A contains 53% HCFC-22, 13% HFC-152a, and 34% HCFC-124, but R-401B contains 61% HCFC-22, 11% HFC-152a, and 28% HCFC-124. For the composition of refrigerant blends see http://www.epa.gov/ozone/snap/refrigerants/refblend.html.

Decoding the Number (Other than Halons). Once it is understood that the prefix describes what kinds of atoms are in a particular molecule, the next step is to calculate the number of each type of atom. The key to the code is to add 90 to the number; the result shows the number of C, H, and F atoms. For HCFC-141b:

$$141 \quad + \quad 90 \quad = \quad 2 \quad 3 \quad 1$$

$$\text{\#C} \quad \text{\#H} \quad \text{\#F}$$

One more piece of information is needed to decipher the number of Cl atoms. All of these chemicals are saturated; that is, they contain only single bonds. The number of bonds available in a carbon-based molecule is 2 C + 2. Thus, for HCFC-141b, which has 2 carbon atoms, there are 6 bonds. Cl atoms occupy bonds remaining after the F and H atoms. So HCFC-141b has 2 C, 3H, 1F, and 2Cl:

$$\text{HCFC} - 141b = C_2H_3FCl_2$$

Notice that the HCFC designation (hydrochlorofluorocarbon) is a good double-check on the decoding; this molecule does, indeed, contain H, Cl, F,

TABLE 1. CLASS 1 OZONE-DEPLETING SUBSTANCES

Chemical Name	Lifetime, in years	ODP1 (WMO 2002)[a]	ODP2 (Montreal Protocol)	ODP3 (40 CFR)	GWP1 (WMO 2002)	GWP2 (SAR)	GWP3 (TAR)	GWP4 (40 CFR)	CAS Number
Group I (from section 602 of the CAA)									
CFC-11 (CCl_3F) Trichlorofluoromethane	45	1.0	1.0	1.0	4680	3800	4600	4000	75-69-4
CFC-12 (CCl_2F_2) Dichlorodifluoromethane	100	1.0	1.0	1.0	10720	8100	10600	8500	75-71-8
CFC-113 $C_2F_3Cl_3$) 1,1,2-Trichlorotrifluoroethane	85	1.0	0.8	1.0	6030	4800	6000	5000	76-13-1
CFC-114 $C_2F_4Cl_2$) Dichlorotetrafluoroethane	300	0.94	1.0	1.0	9880		9800	9300	76-14-2
CFC-115 C_2F_5Cl) Monochloropentafluoroethane	1700	0.44	0.6	0.6	7250		7200	9300	76-15-3
Group II (from section 602 of the CAA)									
Halon 1211 (CF_2ClBr) Bromochlorodifluoromethane	16	6.0	3.0	3.0	1860		1300		353-59-3
Halon 1301 (CF_3Br) Bromotrifluoromethane	65	12	10.0	10.0	7030		6900		75-63-8
Halon 2402 $C_2F_4Br_2$) Dibromotetrafluoroethane	20	<8.6	6.0	6.0	1620				124-73-2
Group III (from section 602 of the CAA)									
CFC-13 (CF_3Cl) Chlorotrifluoromethane	640	1.0	1.0	1.0	14190		14000	11700	75-72-9
CFC-111 C_2FCl_5) Pentachlorofluoroethane		1.0	1.0	1.0					354-56-3
CFC-112 $C_2F_2Cl_4$) Tetrachlorodifluoroethane		1.0	1.0	1.0					76-12-0
CFC-211 C_3FCl_7) Heptachlorofluoropropane		1.0	1.0	1.0					422-78-6
CFC-212 $C_3F_2Cl_6$) Hexachlorodifluoropropane		1.0	1.0	1.0					3182-26-1
CFC-213 $C_3F_3Cl_5$) Pentachlorotrifluoropropane		1.0	1.0	1.0					2354-06-5
CFC-214 $C_3F_4Cl_4$) Tetrachlorotetrafluoropropane		1.0	1.0	1.0					29255-31-0
CFC-215 $C_3F_5Cl_3$) Trichloropentafluoropropane		1.0	1.0	1.0					4259-43-2
CFC-216 $C_3F_6Cl_2$) Dichlorohexafluoropropane		1.0	1.0	1.0					661-97-2
CFC-217 C_3F_7Cl) Chloroheptafluoropropane		1.0	1.0	1.0					422-86-6
Group IV (from section 602 of the CAA)									
CCl_4 Carbon tetrachloride	26	0.73	1.1	1.1	1380	1400	1800	1400	56-23-5
Group V (from section 602 of the CAA)									
Methyl chloroform $C_2H_3Cl_3$) 1,1,1-Trichloroethane	5.0	0.12	0.1	0.1	144		140	110	71-55-6
Group VI (listed in the Accelerated Phaseout Final Rule)[b]									
Methyl bromide (CH_3Br)	0.7	0.38	0.6		5		5		74-83-9
Group VII (listed in the Accelerated Phaseout Final Rule)[b]									
$CHFBr_2$		1.0	1.0						
HBFC-12B1 (CHF_2Br)		0.74	0.74						
CH_2FBr		0.73	0.73						
C_2HFBr_4		0.3–0.8	0.3–0.8						
$C_2HF_2Br_3$		0.5–1.8	0.5–1.8						
$C_2HF_3Br_2$		0.4–1.6	0.4–1.6						
C_2HF_4Br		0.7–1.2	0.7–1.2						
$C_2H_2FBr_3$		0.1–1.1	0.1–1.1						
$C_2H_2F_2Br_2$		0.2–1.5	0.2–1.5						
$C_2H_2F_3Br$		0.7–1.6	0.7–1.6						
$C_2H_3FBr_2$		0.1–1.7	0.1–1.7						
$C_2H_3F_2Br$		0.2–1.1	0.2–1.1						
C_2H_4FBr		0.07–0.1	0.07–0.1						
C_3HFBr_6		0.3–1.5	0.3–1.5						
$C_3HF_2Br_5$		0.2–1.9	0.2–1.9						
$C_3HF_3Br_4$		0.3–1.8	0.3–1.8						
$C_3HF_4Br_3$		0.5–2.2	0.5–2.2						
$C_3HF_5Br_2$		0.9–2.0	0.9–2.0						
C_3HF_6Br		0.7–3.3	0.7–3.3						
$C_3H_2FBr_5$		0.1–1.9	0.1–1.9						
$C_3H_2F_2Br_4$		0.2–2.1	0.2–2.1						
$C_3H_2F_3Br_3$		0.2–5.6	0.2–5.6						
$C_3H_2F_4Br_2$		0.3–7.5	0.3–7.5						

(Continued)

TABLE 1. *(Continued)*

Chemical Name	Lifetime, in years	ODP1 (WMO 2002[a])	ODP2 (Montreal Protocol)	ODP3 (40 CFR)	GWP1 (WMO 2002)	GWP2 (SAR)	GWP3 (TAR)	GWP4 (40 CFR)	CAS Number
$C_3H_2F_5Br$		0.9–1.4	0.9–1.4						
$C_3H_3FBr_4$		0.08–1.9	0.08–1.9						
$C_3H_3F_2Br_3$		0.1–3.1	0.1–3.1						
$C_3H_3F_3Br_2$		0.1–2.5	0.1–2.5						
$C_3H_3F_4Br$		0.3–4.4	0.3–4.4						
$C_3H_4FBr_3$		0.03–0.3	0.03–0.3						
$C_3H_4F_2Br_2$		0.1–1.0	0.1–1.0						
$C_3H_4F_3Br$		0.07–0.8	0.07–0.8						
$C_3H_5FBr_2$		0.04–0.4	0.04–0.4						
$C_3H_5F_2Br$		0.07–0.8	0.07–0.8						
C_3H_6FBr		0.02–0.7	0.02–0.7						
Group VIII (from the Chlorobromomethane Phaseout Final Rule)									
CH_2BrCl Chlorobromomethane	0.37		0.12	0.12					

[a]*The Scientific Assessment of Ozone Depletion, 2002* updated a limited number of GWPs and ODPs (semiempirical values for all updated ODPs except CFC-114 and CFC-115, which are model-derived). All GWPs and ODPs that were not updated in 2002 are 1998 values that have not changed.

[b]In July 1992 , EPA issued a final rule (57 FR 33754) implementing section 604 of the Clean Air Act Amendments of 1990. That section limits the production and consumption of a set of chemicals known to deplete the stratospheric ozone layer. EPA controls production and consumption by issuing allowances or permits that are expended in the production and importation of these chemicals. These allowances can be traded. See also http://www.epa.gov/ozone/title6/phaseout/accfact.html.

and C. The "b" at the end describes how these atoms are arranged; different "isomers" contain the same atoms, but they are arranged differently. The letter designation for isomers is discussed below.

Another example: HFC-134a.

$$134 + 90 = 2 \quad 2 \quad 4$$

$$\#C \quad \#H \quad \#F$$

Again, there are 6 bonds. But in this case, there are no bonds left over after F and H, so there are no chlorine atoms. Thus:

$$HFC - 134a = C_2H_2F_4$$

In this case, too, the prefix is accurate: this is an HFC (hydrofluorocarbon), so it contains only H, F, and C, but no chlorine.

One final example: PFC-218.

$$218 + 90 = 3 \quad 0 \quad 8$$

$$\#C \quad \#H \quad \#F$$

This time, there are $2 \times 3 + 2 = 8$ bonds. However, there are no bonds left over after F, so there are no chlorine atoms or H atoms. Thus:

$$PFC - 218 = C_3F_8$$

Once again, the prefix is accurate: this is a PFC (perfluorocarbon), so it contains only F, and C.

Note: Any molecule with only 1 C (e.g., CFC-12) will have a 2-digit number, while those with 2 C or 3 C will have a 3-digit number.

Decoding the Number (Halons). Halon numbers directly show the number of C, F, Cl, and Br atoms. The numbering scheme above does not give a direct number for the number of Cl atoms, but that can be calculated. Similarly, Halon numbers do not specify the number of H atoms directly. Note that you don't need to add anything to decode the number:

$$Halon \quad 1 \quad 2 \quad 1 \quad 1$$

$$\#C \quad \#F \quad \#Cl \quad \#Br$$

For this molecule, there are $2 \times 1 + 2 = 4$ bonds, all of which are taken by Cl, F, and Br, leaving no room for any H atoms. Thus:

$$Halon \ 1211 = CF_2ClBr$$

Isomers. Isomers of a given compound contain the same atoms but they are arranged differently. Isomers usually have different properties; only one isomer may be useful. So far, we've deciphered the "HCFC" and the "141" of something like HCFC-141b, and we now know the specific atoms in the molecule. The remaining piece of the puzzle is determine the arrangement for a particular isomer. Since all of the compounds under discussion are based on carbon chains (1-3 carbon atoms attached in a line

of single bonds: e.g.,C−C−C), the naming system is based on how H, F, Cl, and Br atoms are attached to that chain.

A single C atom can only bond with 4 other atoms in one way, so there are no isomers of those compounds. For two-carbon molecules, a single lower-case letter following the number designates the isomer. For three-carbon molecules, a lower-case two-letter code serves this purpose.

2-Carbon (Ethane-Derived) Chains. First, consider two-carbon molecules. For example, HCFC-141, HCFC-141a, and HCFC-141b all have the same atoms (2C, 3H, 1F, and 2Cl), but they are organized differently. To determine the letter, total the atomic weights of the atoms bonded to each of the carbon atoms. The arrangement that most evenly distributes atomic weights has no letter. The next most even distribution is the "a" isomer, the next is "b," etc. until no more isomers are possible.

A common way of writing isomers' structure is to group atoms according to the carbon atom with which they bond. Thus, the isomers of HCFC-141 are:

HCFC-141: $CHFCl-CH_2Cl$ (atomic weights on the 2 carbons $= 3.75$ and 55.5)

HCFC-141a: $CHCl_2-CH_2F$ (atomic weights on the 2 carbons $= 21$ and 72)

HCFC-141b: $CFCl_2-CH_3$ (atomic weights on the 2 carbons $= 3$ and 90)

For HFC-134, the isomers are:

HFC-134: CHF_2-CHF_2
HFC-134a: CF_3-CH_2F

Another way of writing out chemical structures is to specify, for each of the Cl, F, and Br atoms, the ordinal number of the carbon to which they are bonded and to use numerical prefixes (2=di, 3=tri, 4=tetra, 5=penta, etc.) to specify the total number of each kind of atom. The suffix for the molecular name depends on the number of carbons. Molecules with 1 C end in "methane" (since there are no isomers of methane-derived molecules, they have no letter designation), " 2 C end in "ethane," and 3 C end in "propane." It is assumed that any bonds not occupied by Cl, F, or Br are occupied by H, so H atoms are not specified. So, the isomers of HCFC-141 can be written as:

HCFC-141: $CHFCl-CH_2Cl$ 1,2-dichloro-1-fluoroethane
HCFC-141a: $CHCl_2-CH_2F$ 1,1-dichloro-2-fluoroethane
HCFC-141b: $CFCl_2-CH_3F$ 1,1-dichloro-1-fluoroethane

For HFC-134, the isomers are:

HFC-134: CHF_2-CHF_2 1,1,2,2-tetrafluoroethane
HFC-134a: CF_2-CH_2F 1,1,1,2-tetrafluoroethane

CFC-12 does not have any isomers, since it contains only 1 C. In addition, there is no need to number the carbons. Thus, its name is difluorodichloromethane.

TABLE 2. CLASS II OZONE-DEPLETING SUBSTANCES

Chemical Name	Lifetime, in years	ODP1 (WMO 2002[a])	ODP2 (Montreal Protocol)	ODP3 (40 CFR)	GWP1 (WMO 2002)	GWP2 (SAR)	GWP3 (TAR)	GWP4 (40 CFR)	CAS Number
HCFC-21 ($CHFCl_2$) Dichlorofluoromethane	1.7	0.04	0.04		148		210		75-43-4
HCFC-22 (CHF_2Cl) Monochlorodifluoromethane	12.0	0.05	0.055	0.05	1780	1500	1700	1700	75-45-6
HCFC-31 (CH_2FCl) Monochlorofluoromethane		0.02	0.02						593-70-4
HCFC-121 C_2HFCl_4) Tetrachlorofluoroethane		0.01–0.04	0.01–0.04						354-14-3
HCFC-122 $C_2HF_2Cl_3$) Trichlorodifluoroethane		0.02–0.08	0.02–0.08						354-21-2
HCFC-123 $C_2HF_3Cl_2$) Dichlorotrifluoroethane	1.3	0.02	0.02–0.06	0.02	76	90	120	93	306-83-2
HCFC-124 C_2HF_4Cl) Monochlorotetrafluoroethane	5.8	0.02	0.02–0.04	0.02	599	470	620	480	2837-89-0
HCFC-131 $C_2H_2FCl_3$) Trichlorofluoroethane		0.007–0.05	0.007–0.05						359-28-4
HCFC-132b $C_2H_2F_2Cl_2$) Dichlorodifluoroethane		0.008–0.05	0.008–0.05						1649-08-7
HCFC-133a $C_2H_2F_3Cl$) Monochlorotrifluoroethane		0.02–0.06	0.02–0.06						75-88-7
HCFC-141b $C_2H_3FCl_2$) Dichlorofluoroethane	9.3	0.12	0.11	0.1	713		700	630	1717-00-6
HCFC-142b $C_2H_3F_2Cl$) Monochlorodifluoroethane	17.9	0.07	0.065	0.06	2270	1800	2400	2000	75-68-3
HCFC-221 C_3HFCl_6) Hexachlorofluoropropane		0.015–0.07	0.015–0.07						422-26-4
HCFC-222 $C_3HF_2Cl_5$) Pentachlorodifluoropropane		0.01–0.09	0.01–0.09						422-49-1
HCFC-223 $C_3HF_3Cl_4$) Tetrachlorotrifluoropropane		0.01–0.08	0.01–0.08						422-52-6
HCFC-224 $C_3HF_4Cl_3$) Trichlorotetrafluoropropane		0.01–0.09	0.01–0.09						422-54-8
HCFC-225ca $C_3HF_5Cl_2$) Dichloropentafluoropropane	1.9	0.02	0.025		120		180	170	422-56-0
HCFC-225cb $C_3HF_5Cl_2$) Dichloropentafluoropropane	5.8	0.03	0.033		586		620	530	507-55-1
HCFC-226 C_3HF_6Cl) Monochlorohexafluoropropane		0.02–0.1	0.02–0.1						431-87-8
HCFC-231 $C_3H_2FCl_5$) Pentachlorofluoropropane		0.05–0.09	0.05–0.09						421-94-3
HCFC-232 $C_3H_2F_2Cl_4$) Tetrachlorodifluoropropane		0.008–0.1	0.008–0.1						460-89-9
HCFC-233 $C_3H_2F_3Cl_3$) Trichlorotrifluoropropane		0.007–0.23	0.007–0.23						7125-84-0
HCFC-234 $C_3H_2F_4Cl_2$) Dichlorotetrafluoropropane		0.01–0.28	0.01–0.28						425-94-5
HCFC-235 $C_3H_2F_5Cl$) Monochloropentafluoropropane		0.03–0.52	0.03–0.52						460-92-4
HCFC-241 $C_3H_3FCl_4$) Tetrachlorofluoropropane		0.004–0.09	0.004–0.09						666-27-3
HCFC-242 $C_3H_3F_2Cl_3$) Trichlorodifluoropropane		0.005–0.13	0.005–0.13						460-63-9
HCFC-243 $C_3H_3F_3Cl_2$) Dichlorotrifluoropropane		0.007–0.12	0.007–0.12						460-69-5
HCFC-244 $C_3H_3F_4Cl$) Monochlorotetrafluoropropane		0.009–0.14	0.009–0.14						
HCFC-251 $C_3H_4FCl_3$) Trichlorofluoropropane		0.001–0.01	0.001–0.01						421-41-0
HCFC-252 $C_3H_4F_2Cl_2$) Dichlorodifluoropropane		0.005–0.04	0.005–0.04						819-00-1
HCFC-253 $C_3H_4F_3Cl$) Monochlorotrifluoropropane		0.003–0.03	0.003–0.03						460-35-5
HCFC-261 $C_3H_5FCl_2$) Dichlorofluoropropane		0.002–0.02	0.002–0.02						420-97-3
HCFC-262 $C_3H_5F_2Cl$) Monochlorodifluoropropane		0.002–0.02	0.002–0.02						421-02-03
HCFC-271 C_3H_6FCl) Monochlorofluoropropane		0.001–0.03	0.001–0.03						430-55-7

[a] *The Scientific Assessment of Ozone Depletion, 2002* updated a limited number of GWPs and ODPs (semiempirical values for all updated ODPs except CFC-114 and CFC-115, which are model-derived). All GWPs and ODPs that were not updated in 2002 are 1998 values that have not changed.

TABLE 3. GLOBAL WARMING POTENTIALS OF HFCS AND PFCS

Chemical	Atmospheric Lifetime	GWP	Use
HFC-23 (CHF_3)	270	12240	Byproduct of HCFC-22 used in very-low temperature refrigeration blend and component in fire suppression. Also used for plasma etching and cleaning in semiconductor production.
	264	11700	
	260	12000	
HFC-32 (CH_2F_2)	4.9	543	
	5.6	650	Blend component of numerous refrigerants.
	5.0	550	
HFC-41 (CH_3F)	2.4	90	
	3.7	150	Not in use today.
	2.6	97	
HFC-43-10mee $C_5H_2F_{10}$)	15.9	1610	
	17.1	1300	Cleaning solvent
	15	1500	
HFC-125 C_2HF_5)	29	3450	
	32.6	2800	Blend component of numerous refrigerants and a fire suppressant.
	29	3400	
HFC-134 $C_2H_2F_4$)	9.6	1090	
	10.6	1000	Not in use today.
	9.6	1100	
HFC-134a (CH_2FCF_3)	14	1320	One of the most widely used refrigerant blends, component of other refrigerants, foam blowing agent, fire suppressant and propellant in metered-dose inhalers and aerosols.
	14.6	1300	
	13.8	1300	
HFC-143 $C_2H_3F_3$)	3.5	347	
	3.8	300	Not in use today.
	3.4	330	
HFC-143a $C_2H_3F_3$)	52	4400	
	48.3	3800	Blend component of several refrigerant blends.
	52	4300	
HFC-152a $C_2H_4F_2$)	1.4	122	Blend component of several refrigerant blends and foam blowing agent. Also used as an aerosol propellant.
	1.5	140	
	1.4	120	
HFC-227ea C_3HF_7)	34.2	3660	Fire suppressant and propellant for metered-dose inhalers, and refrigerant.
	36.5	2900	
	33.0	3500	
HFC-236fa $C_3H_2F_6$)	240	9650	Refrigerant and fire suppressant.
	209	6300	
	220	9400	
HFC-236ea $C_3H_2F_6$)	10.7	1350	
	—	—	Not in use today.
	10.0	1200	
HFC-245ca $C_3H_3F_5$)	6.2	682	
	6.6	560	Not in use today; possible refrigerant in the future.
	5.9	640	
HFC-245fa $C_3H_3F_5$)	7.6	1020	
	—	—	Foam blowing agent and possible refrigerant in the future.
	7.2	950	
HFC-365mfc $C_4H_5F_5$)	8.6	782	
	—	—	Some use as a foam blowing agent; possible refrigerant in the future.
	9.9	950	
Perfluoromethane (CF_4)	50000	5820	
	50000	6500	Plasma etching and cleaning in semiconductor production and low temperature refrigerant.
	50000	5700	
Perfluoroethane C_2F_6)	10000	12010	
	10000	9200	Plasma etching and cleaning in semiconductor production.
	10000	11900	
Perfluoropropane C_3F_8)	2600	8690	Plasma etching and cleaning in semiconductor production, low temperature refrigerant and fire suppressant.
	2600	7000	
	2600	8600	
Perfluorobutane C_4F_{10})	2600	8710	
	2600	7000	Fire suppressant and refrigerant where no other alternatives are technically feasible.
	2600	8600	
Perfluorocyclobutane (c-C_4F_8)	3200	10090	
	3200	8700	Not used much if any. Refrigerant where no other alternatives are technically feasible.
	3200	10000	
Perfluoropentane C_5F_{12})	4100	9010	Not used much if any. Precision cleaning solvent-low use refrigerant where no other alternatives are technially feasible.

TABLE 3. (Continued)

Chemical	Atmospheric Lifetime	GWP	Use
	4100	7500	
	4100	8900	
Perfluorohexane C$_6$F$_{14}$)	3200	9140	Precision cleaning solvent-low use, refrigerant and fire suppressant where no other alternatives are technially feasible.
	3200	7400	
	3200	9000	
	740	10970	
NF$_3$	—	—	Plasma etching and cleaning in semiconductor production.
	—	—	
Sulfur hexafluoride (SF$_6$)	3200	22450	Cover gas in magnesium production, casting dielectric gas and insulator in electric power equipment fire suppression. Also used as a discharge agent in military systems and formerly an aerosol propellant.
	3200	23900	
	3200	22200	
HFE-7100 C$_4$F$_9$OCH$_3$)	5.0	397	
	—	—	Cleaning solvent and heat transfer fluid.
	5.0	390	
HFE-7200 C$_4$F$_9$OC$_2$H$_5$)	0.77	56	Cleaning solvent and heat transfer fluid.
	—	—	
	0.77	55	

TABLE 4. PREFIXES IN NUMBERING SCHEME

Prefix	Meaning	Atoms in the Molecule
CFC	Chlorofluorocarbon	Cl, F, C
HCFC	Hydrochlorofluorocarbon	H, Cl, F, C
HBFC	Hydrobromofluorocarbon	H, Br, F, C
HFC	Hydrofluorocarbon	H, F, C
HC	Hydrocarbon	H, C
PFC	Perfluorocarbon	F, C
Halon	N/A	Br, Cl (in some but not all), F, H (in some but not all), C

3-Carbon (Propane-Derived) Chains. Molecules with 3 C atoms are more complicated to name. The first letter designates the atoms attached to the middle carbon atom, and the second letter designates decreasing symmetry in atomic weights of atoms attached to the outside carbon atoms. Unlike 2 C chains, however, the most symmetric distribution is the "a" isomer, instead of omitting the letter entirely.

Atoms on Middle Carbon	Code Letter
Cl$_2$	a
Cl, F	b
F$_2$	c
Cl, H	d
H, F	e
H$_2$	f

For example, HCFC-225ca is:

C$_3$HF$_5$Cl$_2$ (3C = 8 bonds),
CF$_3$ − CF$_2$ − CHCl$_2$, and
1,1,1,2,2-pentafluoro-3,-dichloropropane

When no isomers are possible, no letters are used. For example, there is only one way to arrange 3C and 8F, so it is written as PFC-218 and not PFC-218ca.

The EPA Global Programs Division

In addition to regulating the end of production of the ozone-depleting substances, the U.S. Environmental Protection Agency (EPA) implements several other programs to protect the ozone layer under Title VI of the Clean Air Act, http://www.epa.gov/oar/caa/. These programs include refrigerant recycling, http://www.epa.gov/ozone/title6/608/index.html; product labeling, http://www.epa.gov/ozone/title6/labeling/index.html; banning non essential uses of certain compounds, http://www.epa.gov/ozone/title6/

noness/index.html; and reviewing substitutes, http://www.epa.gov/ozone/snap/index.html.

Scientific Assessment of Ozone Depletion: 2002

This is the executive summary of the most recent World Meteorological Organization and United Nations Environmental Programme assessment. It contains the most up-to-date understanding of ozone depletion and reflects the thinking of over 275 international scientific experts who contributed to its preparation and review. See also http://esrl.noaa.gov/csd/assessments/.

Ozone Depletion Glossary

An abridged glossary of terms used to describe ozone depletion would include:

Aerosol. (1) A small droplet or particle suspended in the atmosphere, typically containing sulfur. Aerosols are emitted naturally (e.g., in volcanic eruptions) and as the result of human activities (e.g., by burning fossil fuels). See also **Aerosol.** There is no connection between particulate aerosols and pressurized products also called aerosols. (2) A product that relies on a pressurized gas to propel substances out of a container. Consumer aerosol products in the US have not used ozone-depleting substances (ODS) since the late 1970s because of voluntary switching followed by federal regulation. The Clean Air Act and EPA regulations further restricted the use of ODS for non-consumer products. All consumer products, and most other aerosol products, now use propellants that do not deplete the ozone layer, such as hydrocarbons and compressed gases.

Carbon Tetrachloride (CCl$_4$). A compound consisting of one carbon atom and four chlorine atoms. Carbon tetrachloride was widely used as a raw material in many industrial uses, including the production of CFCs, and as a solvent. Solvent use ended when it was discovered to be carcinogenic. It is also used as a catalyst to deliver chlorine ions to certain processes. Its ozone depletion potential is 1.2.

Chlorofluorocarbon (CFC). A compound consisting of chlorine, fluorine, and carbon. CFCs are very stable in the troposphere. They move to the Stratosphere and are broken down by strong ultraviolet light, where they release chlorine atoms that then deplete the ozone layer. CFCs are commonly used as refrigerants, solvents, and foam blowing agents. The most common CFCs are CFC-11, CFC-12, CFC-113, CFC-114, and CFC-115. The ozone depletion potential (ODP) for each CFC is, respectively, 1, 1, 0.8, 1, and 0.6. See also **Chlorofluorocarbons(CFC).**

Class I Substance. One of several groups of chemicals with an ozone-depletion potential of 0.2 or higher. See Table 1.

Class II Substance. A chemical with an ozone-depletion potential of less than 0.2. See Table 2.

Clean Air Act (CAA). Law amended by Congress in 1990. Title VI of the CAA directs EPA to protect the ozone layer through several regulatory and voluntary programs. Sections within Title VI cover production of ozone-depleting substances (ODS), the recycling and handling of ODS, the evaluation of substitutes, and efforts to educate the public.

Column Ozone. Ozone between the Earth's surface and outer space. Ozone levels can be described in several ways. One of the most common measures is how much ozone is in a vertical column of air. The Dobson unit is a measure of column ozone. Other measures include partial pressure, number density, and concentration of ozone, and can represent either column ozone or the amount of ozone at a particular altitude.

Dobson Unit (DU). A measurement of column ozone levels. If 100 DU of ozone were brought to the Earth's surface, it would form a layer 1 millimeter thick. In the tropics, ozone levels are typically between 250 and 300 DU year-round. In temperate regions, seasonal variations can produce large swings in ozone levels. For instance, measurements in Leningrad have recorded ozone levels as high as 475 DU and as low as 300 DU. These variations occur even in the absence of ozone depletion, but they are well understood. Ozone depletion refers to reductions in ozone below normal levels after accounting for seasonal cycles and other natural effects. For a graphical explanation, see NASA's TOMS site: http://jwocky.gsfc.nasa.gov/teacher/basics/dobson.html. See also **Dobson Unit.**

Federal Register (FR). The daily publication containing all federal government actions. The Federal Register is the formal method of communication for any Notice, Notice of Proposed Rulemaking (NPRM), or Final Rulemaking (FRM) issued by the US government. Once published in the FR, rules are collected in the Code of Federal Regulations. The FR is available at many libraries. FR cites areas similar in form to 11 FR 12345, where 11 is a number corresponding to the year (e.g., 62 is 1997) and 12345 represents the page number (pages are numbered continuously through the year; the first page published in each year is page number 1). Thus, a Notice whose cite is 62 FR 10700 was published beginning at page 10700 in 1997. It is usually helpful to obtain the date as well, since it is difficult to guess the date solely on the page number.

Global Warming Potential (GWP). A number that refers to the amount of global warming caused by a substance. The GWP is the ratio of the warming caused by a substance to the warming caused by a similar mass of carbon dioxide. Thus, the GWP of CO_2 is defined to be 1.0. CFC-12 has a GWP of 8,500, while CFC-11 has a GWP of 5,000. Various HCFCs and HFCs have GWPs ranging from 93 to 12,100. Water, a substitute in numerous end-uses, has a GWP of 0. See Table 1 and 2.

Halon. A compound consisting of bromine, fluorine, and carbon. The halons are used as fire extinguishing agents, both in built-in systems and in handheld portable fire extinguishers. Halon production in the U.S. ended on 12/31/93 because they contribute to ozone depletion. They cause ozone depletion because they contain bromine. Bromine is many times more effective at destroying ozone than chlorine. At the time the current U.S. tax code was adopted, the ozone depletion potentials of halon 1301 and halon 1211 were observed to be 10 and 3, respectively. These values are used for tax calculations. Recent scientific studies, however, indicate that the ODPs are at least 12 and 6, respectively. Note: technically, all compounds containing carbon and fluorine and/or chlorine are halons, but in the context of the Clean Air Act, "halon" means a fire extinguishing agent as described above. See Table 1.

Hydrobromofluorocarbon (HBFC). A compound consisting of hydrogen, bromine, fluorine, and carbon. Although they were not originally regulated under the Clean Air Act, subsequent regulation added HBFCs to the list of class I substances. See Table 1.

Hydrocarbon (HC). A compound consisting of carbon and hydrogen. Hydrocarbons include methane, ethane, propane, cyclopropane, butane, and cyclopentane. Although they are highly flammable, HCs may offer advantages as ODS substitutes because they are inexpensive to produce and they have zero ozone depletion potential, very low global warming potential (GWP), and low toxicity.

Hydrochlorofluorocarbon (HCFC). A compound consisting of hydrogen, chlorine, fluorine, and carbon. The HCFCs are one class of chemicals being used to replace the CFCs. They contain chlorine and thus deplete stratospheric ozone, but to a much lesser extent than CFCs. HCFCs have ozone depletion potentials (ODPs) ranging from 0.01 to 0.1. Production of HCFCs with the highest ODPs will be phased out first, followed by other HCFCs.

Hydrofluorocarbon (HFC). A compound consisting of hydrogen, fluorine, and carbon. The HFCs are a class of replacements for CFCs. Because they do not contain chlorine or bromine, they do not deplete the ozone layer. All HFCs have an ozone depletion potential of 0. Some HFCs have high GWPs.

Methyl Bromide (CH_3Br). A compound consisting of carbon, hydrogen, and bromine. An effective pesticide, this compound is used to fumigate soil and many agricultural products. Because it contains bromine, it depletes stratospheric ozone and has an ozone depletion potential of 0.6.

Methyl Chloroform (CH_3CCl_3). A compound consisting of carbon, hydrogen, and chlorine. Methyl chloroform is used as an industrial solvent. Its ozone depletion potential is 0.11.

Nanometer. A distance of one billionth of a meter. The nanometer, or nm, is a common unit used to describe wavelengths of light or other electromagnetic radiation such as UV. For example, green light has wavelengths of about 500-550 nm, while violet light has wavelengths of about 400-450 nm. One billionth is a tiny number. One foot is about one billionth the distance of 48 round-trips between Los Angeles and Washington, DC.

Ozone-Depleting Substance(s) (ODS). A compound that contributes to stratospheric ozone depletion. ODS include CFCs, HCFCs, halons, methyl bromide, carbon tetrachloride, and methyl chloroform. ODS are generally very stable in the troposphere and only degrade under intense ultraviolet light in the stratosphere. When they break down, they release chlorine or bromine atoms, which then deplete ozone. See Table 1 and 2.

Ozone Depletion Potential (ODP). A number that refers to the amount of ozone depletion caused by a substance. The ODP is the ratio of the impact on ozone of a chemical compared to the impact of a similar mass of CFC-11. Thus, the ODP of CFC-11 is defined to be 1.0. Other CFCs and HCFCs have ODPs that range from 0.01 to 1.0. The halons have ODPs ranging up to 10. Carbon tetrachloride has an ODP of 1.2, and methyl chloroform's ODP is 0.11. HFCs have zero ODP because they do not contain chlorine. See Table 1 and 2.

Perfluorocarbon (PFC). A compound consisting of carbon and fluorine. PFCs have extremely high global warming potentials (GWPs) and very long lifetimes. They do not deplete stratospheric ozone, but EPA is concerned about their impact on global warming.

UV (ultraviolet radiation). Ultraviolet radiation is a portion of the electromagnetic spectrum with wavelengths shorter than visible light. The sun produces UV, which is commonly split into three bands: UVA, UVB, and UVC. UVA is not absorbed by ozone. UVB is mostly absorbed by ozone, although some reaches the Earth. UVC is completely absorbed by ozone and normal oxygen.

UVA. A band of ultraviolet radiation with wavelengths from 320-400 nanometers produced by the Sun. UVA is not absorbed by ozone. This band of radiation has wavelengths just shorter than visible violet light.

UVB. A band of ultraviolet radiation with wavelengths from 280-320 nanometers produced by the Sun. UVB is a kind of ultraviolet light from the sun (and sun lamps) that has several harmful effects. UVB is particularly effective at damaging DNA. It is a cause of melanoma and other types of skin cancer. It has also been linked to damage to some materials, crops, and marine organisms. The ozone layer protects the Earth against most UVB coming from the sun. It is always important to protect oneself against UVB, even in the absence of ozone depletion, by wearing hats, sunglasses, and sunscreen. However, these precautions will become more important as ozone depletion worsens.

UVC. A band of ultraviolet radiation with wavelengths shorter than 280 nanometers. UVC is extremely dangerous, but it is completely absorbed by ozone and normal oxygen (O_2).

Additional Reading

Anderson, S.O., and K.M. Sarma, (Foreword by Kofi Annan): *Protecting the Ozone Layer: The United Nations History*, Earthscan/James & James Publishers, London, UK, 2005.

Gillespie, A.: *Climate Change, Ozone Depletion and Air Pollution: Legal Commentaries with Policy and Science Considerations*, Brill Academic Publishers, Inc., Boston, MA, 2005.

Hoffman, M.J., and J.N. Rosenau: *Ozone Depletion and Climate Change: Constructing a Global Response*, State University of New York Press, Albany, NY, 2005.

Lambright, W.H.: *NASA and the Environment: The Case of Ozone Depletion (Monographs in Aerospace History Series #38)*, United States Government Printing Office, Washington, DC, 2005.

Parker, L., and W.A. Morrissey: *Stratospheric Ozone Depletion*, Nova Science Publishers, Inc., Hauppauge, NY, 2003.

Parson, E.A.: *Protecting the Ozone Layer: Science and Strategy*, Oxford University Press, New York, NY, 2003.

Staff, Intergovernmental Panel on Climate Change: *Safeguarding the Ozone Layer and the Global Climate System*, Cambridge University Press, New York, NY, 2005.

Web References

The Case of Ozone Depletion: http://history.nasa.gov/monograph38.pdf.

The Implementation of Title VI of the Clean Air Act: http://www.hhs.gov/asl/testify/t980402a.html.

Reports to the Nation: Our Ozone Shield. http://www.ogp.noaa.gov/library/rtnf92.htm.

U. S. Environmental Protection Agency: http://www.epa.gov/ozone/science/index.html.

U. S. Environmental Protection Agency (EPA), Washington, DC

OZONE GENERATION. Ozone can be generated by a variety of methods, the most common of which involves the dissociation of molecular oxygen electrically or photochemically (UV). The short-lived oxygen atoms (lifetime $\sim 10^{-5}$ s) react rapidly with oxygen molecules to form ozone. The widely employed technique of electric discharge produces much higher concentrations than the uv technique and is more practical and efficient for production of large quantities. A less common method of ozone formation is electrochemical generation.

Commercial production and utilization of ozone by electric discharge (also called corona or silent discharge) consists of five basic unit operations: gas preparation, electrical power supply, ozone generation, contacting (i.e., ozone dissolution in water), and destruction of ozone in contactor off-gases.

Discharge Characteristics. The energy for chemical reaction is transferred to oxygen molecules by energetic electrons producing atoms, excited molecules, and ions. In an ozone generator, the feed gas (oxygen or air) passes between two closely spaced electrodes (one of which is coated with a dielectric) under an applied nominal potential of ~ 10 kV. A corona or dielectric barrier discharge occurs when the gas becomes partially ionized, resulting in a characteristic violet glow in air. Corona discharge consists of numerous randomly distributed, low current (but high current density) microdischarge pulses. The approximately columnar streamers or filaments (100–200 μm diameter) emanating from the metal electrode discharge at the dielectric and extinguish within 10 ns.

Because ozone formation occurs only within these microdischarge channels, ozone-production efficiency for the most part depends on the strength of the microdischarges, which is influenced by a number of factors, eg, the gap width, pressure, properties of the dielectric and metal electrode, power supply, and moisture. In weak discharges, a significant fraction of the energy is consumed by ions, whereas in stronger discharges, almost all of the discharge energy is transferred to electrons responsible for the formation of ozone. The optimum is a compromise that avoids energy losses to ions but at the same time obtains a reasonable conversion efficiency of oxygen atoms to ozone.

Ozone Generation from Oxygen. Use of oxygen for ozone generation is advantageous because of its purity and absence of by-products.

Oxygen is dissociated into atoms by inelastic collisions with energetic electrons (6–7 eV): $O + e^- \rightarrow 2 O(^3P \text{ and } ^1D) + e^- \rightarrow 2O(3P \text{ and } 1D) + e^-$. Whereas oxygen atoms are formed in nanoseconds, their subsequent reactions occur on a microsecond time scale. The highly reactive oxygen atoms can recombine in the gas phase and on the wall. For the gas phase, the reaction is $2\ O + M \rightarrow O_2 + M + 498.3$ kJ, where $M = O_2$ and $k^{25°C} = \sim 1 \times 10^{11}$ M^{-2} s^{-1}. Recombination of oxygen atoms is unimportant until $O/O_2 > 0.01$. The main reaction is with oxygen molecules in the presence of a third body, M, forming ozone: $O + O_2 + M \rightarrow O_3 + M + 106.5$ kJ. The lower the relative oxygen atom concentration, O/O_2, the greater the fraction of oxygen atoms forming ozone.

Ozone formed in the above reaction is initially in a vibrationally excited state (O_3^*). Although most is quenched by collision with other molecules, a small fraction can react faster with oxygen atoms than ground-state ozone. Vibrationally excited ozone also can be formed by collision with (^1D) oxygen atoms and with vibrationally excited oxygen molecules. Vibrationally excited singlet oxygen molecules ($^1\Delta_g$ and $^1\Sigma_g^+$) can be formed by electron impact or by recombination of oxygen atoms.

Ozone can be destroyed thermally, by electron impact, by reaction with oxygen atoms, and by reaction with electronically and vibrationally excited oxygen molecules. Processes involving ions, e.g, O_2^-, O_2^+, O^-, O^+, and O_3^-, are of minor importance. The reaction $O_3 + O(^3P) \rightarrow 2\ O_2$, is exothermic and can contribute significantly to heat evolution. Efficiently cooled ozone generators with typical short residence times (seconds) can operate near ambient temperature where thermal decomposition is small.

Ozone Concentration. Experimental studies show that the ozone concentration increases with specific energy (eV/O_2) before reaching a steady state. The steady-state ozone concentration varies inversely with temperature but directly with pressure, reaching a maximum at about 101.3 kPa (1 atm). Above atmospheric pressure the steady-state ozone concentration decreases with pressure, apparently due to the pressure dependence of the rate constant ratio $k(O_2)/k(O_3)$ for the reactions of O_2 and O_3 with energetic electrons. The preparation of ozone from oxygen presents fewer operational problems than that from air because significant amounts of moisture and large concentrations of nitrogen are absent. However, small amounts of nitrogen ($\sim 4\%$) actually increase the ozone concentration.

Ozone Generation from Air. Although the use of air for ozone generation has the advantage that air, unlike oxygen, is readily available, the concentration of ozone produced with air is lower than that produced with oxygen. In addition, the presence of moisture in air interferes with discharge formation and reaction kinetics and creates potential for corrosion that can adversely affect the performance of the ozone generator and increase the need for maintenance.

The basic chemistry of ozone generation from oxygen is more complex when air is employed because of formation of nitrogen atoms, vibrationally excited nitrogen molecules, and nitrogen oxides. Nitrogen atoms are formed by the dissociation of nitrogen molecules by electron impact; they can generate oxygen atoms via the following reactions: $N + O_2 \rightarrow NO + O$ and $N + NO \rightarrow N_2 + O$. Oxygen atoms also can be formed by the dissociation of molecular oxygen by vibrationally excited nitrogen molecules. Thus, atomic nitrogen and excited nitrogen molecules enhance the formation of ozone by increasing the atomic oxygen concentration.

Nitric oxide can initiate catalytic decomposition of ozone, as previously discussed, by the reactions $NO + O_3 \rightarrow NO_2 + O_2$ and $NO_2 + O \rightarrow NO + O_2$. Nitrogen dioxide also can destroy ozone: $NO_2 + O_3 \rightarrow NO_3 + O_2$. NO_3 reacts with NO to form NO_2 as indicated by $NO_3 + NO \rightarrow 2\ NO_2$. Nitrogen dioxide and trioxide are in equilibrium with nitrogen pentoxide: $NO_2 + NO_3 \rightarrow N_2O_5$. Decreasing the temperature and increasing the pressure shifts the equilibrium to the right, reducing the effect of lower oxides on the decomposition of ozone. Nitrogen pentoxide can react with oxygen atoms to form nitrogen dioxide. Only N_2O_5 and N_2O are present in the gases exiting the discharge. The nitrous oxide (N_2O) that is formed is inert toward ozone. Of less importance is the destruction of ozone by nitrogen atoms, $N + O_3 \rightarrow NO + O_2$, which is much slower than that by oxygen atoms.

Ozone Concentration. Because of the formation of nitrogen oxides, a steady-state ozone concentration cannot be obtained; instead, due to the buildup of nitrogen oxides, an increase in residence time in the discharge results in a decrease in ozone concentration beyond the maximum value. Thus, there is an optimum residence time for maximum ozone production.

Suppression of Nitrogen Oxides. The concentration of nitrogen oxides during preparation of ozone from air increases linearly with the energy density in the discharge, causing a decrease in the formation rate of ozone.

Most commercial ozone generators produce 0.5 kg (1.1 lbs) of nitrogen oxides for every 100 kg (220 lbs) of ozone generated. The formation of nitrogen oxides at a given energy density is minimized by decreasing the residence time and temperature, increasing the pressure, and reducing the dew point of air.

Feed Gas Preparation. Air and oxygen are the two gas sources for preparation of ozone.

The use of oxygen for industrial ozone generation is significant and increasing. Oxygen provides a higher ozone concentration and more efficient ozone dissolution than air, and does not add nitrogen oxides to the water. It is prepared from dry, filtered air by liquefaction and fractional distillation. Liquid oxygen (LOX) can be prepared on-site or purchased from vendors. Oxygen is sometimes used to enrich air-fed systems. Although oxygen-rich off-gases from ozone contactors can be recycled, more often they are discarded to avoid re-drying costs.

Air is widely used as the feed gas for commercial ozone generators. The air feed gas to the ozone generator should be dry and free of foreign matter. Filtered ambient air is drawn into the plant by vacuum, blower, or compressor. The pressure of the treated air can vary from subatmospheric to >400 kPa (4 atm). Since compression heats the air, cooling is necessary. The air is filtered again to remove oil droplets that can foul the desiccant dryers and interfere with ozone generation. Any hydrocarbons in the air can be removed with activated carbon. Moisture is removed by desiccant-drying or a combination of refrigerant- and desiccant-drying. Desiccant-drying is accomplished by using molecular sieves, silica gel, or activated alumina, all of which are capable of regeneration. Liquid water droplets in refrigerant-dried air should be removed by filtration prior to contacting the desiccant dryers. A final filtration is necessary to remove desiccant dust particles down to 1μm. The efficiency of ozone generation decreases with increasing moisture content in the air. At high dew points, nitrous and nitric acids are deposited within the ozone generator, decreasing performance and substantially increasing the maintenance frequency. The air feed to the ozone generator should have a dew point of at least $-60\,°C$, corresponding to a moisture content of £20 ppmv; some systems, however, operate at a dew point of $-80\,°C$ $(-112\,°F)$. A sensor should be placed in the air stream entering the generator that can shut the system off and sound an alarm if the dew point increases above the desired level. In high pressure systems, the pressure of the compressed air prior to entering the ozone generator is reduced by means of a pressure-reducing valve. The pressure employed depends on the ozone generator type and can vary from 100 to 240 kPa (0–20 psig). The pressure of the ozone generator feed should be maintained at a constant level to avoid affecting power draw and applied voltage.

Cooling Requirements. Since the majority of the electrical energy input to the electric discharge is dissipated as heat, cooling is necessary to minimize decomposition of ozone and extend dielectric life. Double-sided cooling is more effective than single-sided cooling in removing heat from the ozone generator. The gas exiting an efficiently cooled ozone generator normally is near ambient temperatures where the rate of decomposition is low.

Electrical Characteristics. Electrical energy to the ozone generator is provided by a power supply, a frequency converter, and a transformer. Ozone formation is directly proportional to the power consumed in the discharge at constant O_3 concentration, temperature, and pressure. The average discharge power consumption P (W) is given by $P = 4C_d v_s f[v_o - v_s(C_d + C_g)/C_d]$; where C_d and C_g are the dielectric and gap capacitances (F), v_s is the peak gap sparking potential (V), v_o is the peak driving potential (V), and f is the frequency (Hz). The sparking potential is given by $v_s(O_2) = 26.55 pt_g + 1480$ and $v_s(air) = 29.64 pt_g + 1350$, where p is the absolute pressure (kPa) and t_g is the gap thickness (mm). For a given geometry, gap, and pressure, the power consumed by the discharge can be increased by operating at higher driving voltages and frequencies, and employing thinner dielectrics having higher dielectric constants.

The potential and frequency employed in commercial ozone generators varies with the type of design and can range from 5 to 20 kV and 50 to 3000 Hz, respectively. Typically, high frequency ozone generators operate at lower voltages, where the expected lifetime of the high voltage electrode is virtually unlimited. Although lower voltage decreases the ozone production rate, when combined with high frequency it can produce more ozone per unit electrode area. Modern ozone generators operating at 10 kV rms and 600–1000 Hz employ power densities of 3–4 kW m^{-2}, resulting in production densities of 0.2–0.25 kg hm^{-2} in air and 0.35–0.45 kg hm^{-2} in oxygen.

Ozone Concentration and Yield. The output of an ozone generator can be increased by raising the power input at constant temperature and feed gas flow rate, but the increase in output is less than proportional unless the gas flow is increased to maintain a constant ozone concentration. Raising the flow rate at constant power input decreases the ozone concentration, but increases the ozone and energy yields. At low flow rates, although the ozone concentration is high, the yield is low because the specific energy is high. At higher flow rates, the ozone concentration decreases; the yield approaches a limiting value because the specific energy does not change much at low ozone concentrations.

Commercial ozone generators are available with a wide range of production rates. One manufacturer markets ozonators with production rates from oxygen as high as 247, 202, and 171 kg h^{-1} at 7, 10, and 12 wt% O_3 and as high as 134 and 103 kg h^{-1} from air at 2.1 and 3.5 wt% O_3. Higher production rates are obtained by combining multiple units. Lower capacity units also are available with production rates as low 50 g h^{-1}. The required ozone concentration depends on the application; concentrations as high as 16 wt% have been produced commercially from oxygen. Lab-scale ozonators also are manufactured with production rates of ~10 g h^{-1}.

Ozone Generator Design. A better understanding of discharge physics and the chemistry of ozone formation has led to improvements in power density, efficiency, and ozone concentration, initiating a trend toward downsizing.

The basic configuration of an electric discharge cell consists of two closely spaced electrodes (one of which is coated with a dielectric), supplied with high voltage alternating current (ac) and filled with a flowing oxygen-containing gas. The gap width varies from 1 to 3 mm, depending on whether oxygen or air is employed. The purpose of the dielectric, usually made of glass or ceramic, is to limit current flow, resulting in the formation of a relatively cold plasma. A thin dielectric with a high dielectric constant facilitates heat removal and improves ozone-generating efficiency. The higher the applied voltage, the thicker the electrode should be to prevent failure by electrical arcing. The dielectric must be strong enough to withstand mechanical shock and prevent puncturing by the applied voltage. High peak voltages induce dielectric failure, as do high power densities, the latter on account of its dielectric heating.

The electric discharge ozone generator is equivalent to a gas-phase reactor having internal heat generation; its design also bears some similarity to heat exchangers such as shell and tube. The gap width influences both the voltage requirement and the back pressure and should be uniform to avoid hot spots. Small gap widths typically are employed to facilitate heat removal; the smaller the discharge gap, the greater the power efficiency.

Types of Ozone Generator. Since 1906, a number of different ozone generators have been developed, including the plate-type (water- or air-cooled), the horizontal tube (water-cooled), and the vertical tube (water- or oil- and water-cooled) generators. Originally introduced in 1906, plate-type ozone generators have experienced operational problems and have been discontinued for use in some countries, even though many installations remain operational and the technology is still being promoted by some manufacturers.

Other Methods of Ozone Generation

Ultraviolet Light. *Short Wavelength Light.* The mechanism of the practical photochemical production of ozone is similar to that in the stratosphere; ie, oxygen atoms, formed by the photodissociation of oxygen by short-wavelength uv radiation (£240 nm), react with oxygen molecules to form ozone. At low conversions, the limiting quantum yield is ~2. The steady-state ozone concentration (~3.5 mol% max) depends on temperature, pressure, and whether oxygen or air is employed; the time-to-steady-state depends on the light intensity. Efficiencies as high as 9% can be obtained, at low ozone concentration, by using narrow-band uv radiation produced by a xenon excimer laser operating at 172 nm.

Short and Long Wavelength Light. In practice, ozone concentrations obtained by commercial uv devices are low. This is because the low intensity, low pressure mercury lamps employed produce not only the 185-nm radiation responsible for ozone formation, but also the 254-nm radiation that destroys ozone, resulting in a quantum yield of ~0.5 compared to the theoretical yield of 2.0. Furthermore, the low efficiency (~1%) of these lamps results in a low ozone production rate of ~2 g kWh^{-1}.

Typical output of a commercial 40-Wt uv lamp using air is ~0.5 g h^{-1} of ozone and a maximum concentration of 0.25 wt%. However, these maximum ozone yields and concentrations cannot be obtained simultaneously by the uv method. The low concentrations of ozone available from uv generators preclude their use for water treatment because the transfer efficiencies of ozone from air into water is low and large volumes of carrier gas must be handled. More than 44 kWh are required to generate one kg of ozone from dry air by UV radiation under high gas flow rates and low concentrations.

Electrolysis. High current density electrolysis of aqueous phosphate solutions at room temperature produces ozone and oxygen in the anodic gas. Electrolysis of 68 wt% sulfuric acid can produce 18–25 wt% ozone in oxygen when a well-cooled cell is used. Although electrolysis of water can produce high concentrations of ozone, the output is low, and the cost is several times more than that of electric discharge processes.

Radiochemical. Ozone can be prepared radiochemically by irradiation of gaseous or liquid oxygen with β- and γ-rays from radioactive isotopes or a nuclear reactor. Although its energy efficiency is greater than that of ozone produced by electric discharge, this complex process has never been commercialized due to problems associated with recovery of ozone and separation of by-products and radioactive material.

Miscellaneous. In the laboratory, pure liquid ozone can be produced quantitatively by cooling a stream of atomic oxygen in oxygen at liquid nitrogen temperatures ($-196\,°C$ ($-321\,°F$): $O + O_2 + M \rightarrow O_3 + M$, where M is the cold reactor wall. Pure stable gaseous ozone also can be prepared quantitatively by electric discharge (using a Tesla coil) through oxygen cooled with liquid nitrogen followed by warming to room temperature.

See also **Ozone**.

OZONE HOLE.

The ozone hole is not technically a "hole" where no ozone is present, but is actually a region of exceptionally depleted ozone in the stratosphere over the Antarctic that happens at the beginning of Southern Hemisphere spring (August–October). The average concentration of ozone in the atmosphere is about 300 Dobson Units; any area where the concentration drops below 220 Dobson Units is considered part of the ozone hole as shown in Fig. 1.

Chlorofluorocarbons and Ozone

Many people have heard that the ozone hole is caused by chemicals called CFCs, short for chlorofluorocarbons. CFCs escape into the atmosphere from refrigeration and propellant devices and processes. In the lower atmosphere, they are so stable that they persist for years, even decades. This long lifetime allows some of the CFCs to eventually reach the stratosphere. In the stratosphere, ultraviolet light breaks the bond holding chlorine atoms (Cl) to the CFC molecule. A free chlorine atom goes on to participate in a series of chemical reactions that both destroy ozone and return the free chlorine atom to the atmosphere unchanged, where it can destroy more and more ozone molecules. For those who know the story of CFCs and ozone, that is the part of the tale that is probably familiar.

The part of the story that fewer people know is that while the chlorine atoms freed from CFCs do ultimately destroy ozone, the destruction doesn't happen immediately. Most of the roaming chlorine that gets separated from CFCs actually becomes part of two chemicals that—under *normal* atmospheric conditions—are so stable that scientists consider them to be long-term reservoirs for chlorine. So how does the chlorine get out of the reservoir each spring?

Polar Stratospheric Clouds and Ozone

Under normal atmospheric conditions, the two chemicals that store most atmospheric chlorine (hydrochloric acid, and chlorine nitrate) are stable. But in the long months of polar darkness over Antarctica in the winter, atmospheric conditions are unusual. An endlessly circling whirlpool of

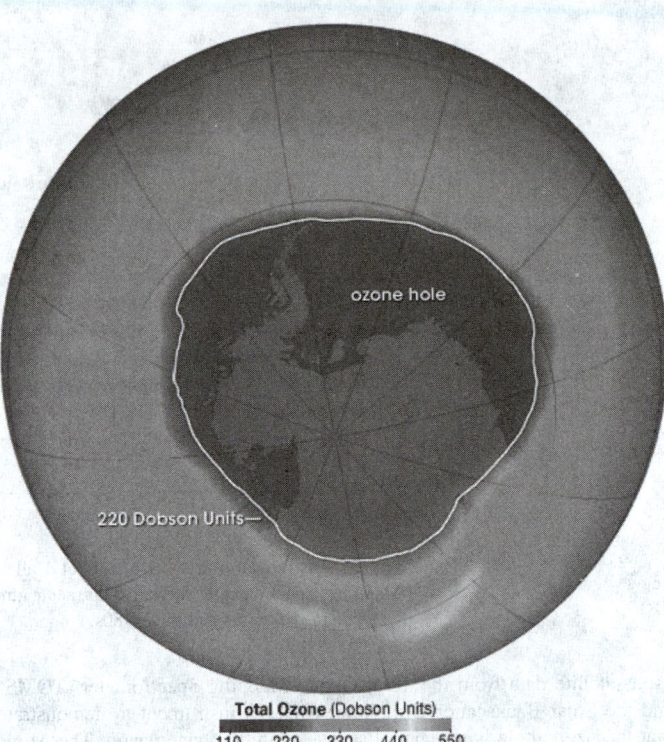

Fig. 1. This map shows the ozone hole on October 4, 2004. The data were acquired by the Ozone Monitoring Instrument on NASA's Aura satellite. See also http://www.knmi.nl/omi/research/news/indexen.html.

stratospheric winds called the polar vortex isolates the air in the center. Because it is completely dark, the air in the vortex gets so cold that clouds form, even though the Antarctic air is extremely thin and dry. Chemical reactions take place that could not take place anywhere else in the atmosphere. These unusual reactions can occur only on the surface of polar stratospheric cloud particles, which may be water, ice, or nitric acid, depending on the temperature.

These reactions convert the inactive chlorine reservoir chemicals into more active forms, especially chlorine gas (Cl_2). When the sunlight returns to the South Pole in October, UV light rapidly breaks the bond between the two chlorine atoms, releasing free chlorine into the stratosphere, where it takes part in reactions that destroy ozone molecules while regenerating the chlorine (known as a catalytic reaction). A catalytic reaction allows a single chlorine atom to destroy thousands of ozone molecules. Bromine is involved in a second catalytic reaction with chlorine that contributes a large fraction of ozone loss. The ozone hole grows throughout the early spring until temperatures warm and the polar vortex weakens, ending the isolation of the air in the polar vortex. As air from the surrounding latitudes mixes into the polar region, the ozone-destroying forms of chlorine disperse. The ozone layer stabilizes until the following spring.

History of the Ozone Hole

Throughout the 20th century, discoveries and observations trickled in that would allow scientists to understand how human-made chemicals like chlorofluorocarbons create a hole in the ozone layer over Antarctica each spring.

As early as 1912, Antarctic explorers recorded observations of unusual veil-type clouds in the polar stratosphere, although they could not have known at the time how significant those clouds would become. In 1956, the British Antarctic Survey set up the Halley Bay Observatory on Antarctica in preparation for the International Geophysical Year (IGY) of 1957. In that year, ozone measurements using a Dobson Spectrophotometer began. See Fig. 2.

These measurements gave the first clues that there was trouble in the ozone layer. There was no ozone hole at all until the late 1970s. In 1985, a group of scientists (Joseph C. Farman, Brian G. Gardiner, and Jonathan D. Shanklin) published in the journal Nature the first paper on observations of springtime losses of ozone over Antarctica. In 1986, NASA scientists

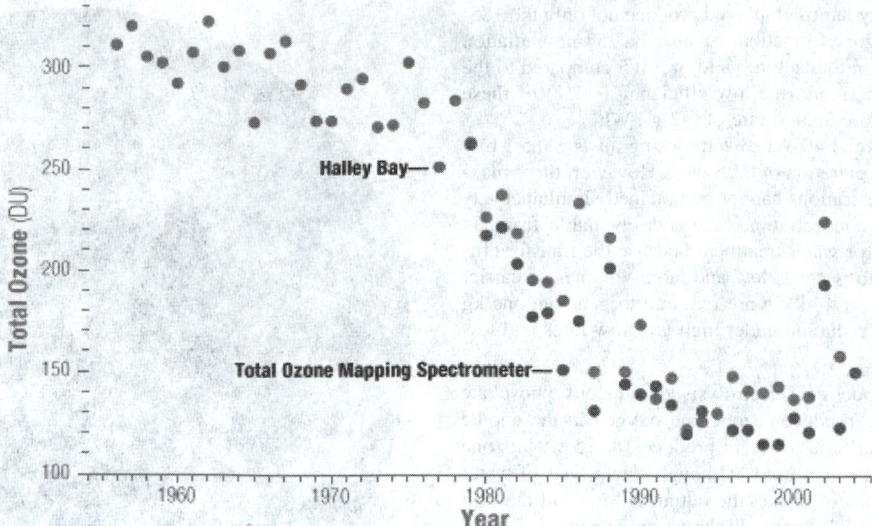

Fig. 2. Instruments on the ground (at Halley Bay) and high above Antarctica (the Total Ozone Mapping Spectrometer) measured an acute drop in total atmospheric ozone during September and October in the early- and mid-1980s.

used satellite data from the Total Ozone Mapping Spectrometer (TOMS) and the Solar Backscatter Ultraviolet (SBUV) instrument to demonstrate that the ozone hole is a regional-scale Antarctic phenomenon. The ozone hole appeared first over the colder Antarctic because the ozone-destroying chemical process works best in cold conditions. The Antarctic continent has colder conditions than the Arctic, which has no land-mass. As the years have gone by the Ozone Hole has increased rapidly and is as large as the Antarctica continent. The hole lasts for only two months, but its timing could not be worse. Just as sunlight awakens activity in dormant plants and animals, it also delivers a dose of harmful ultraviolet radiation. After eight weeks, the hole leaves Antarctica, only to pass over more populated areas, including The Falkland Islands, South Georgia and the tip of South America. This biologically damaging, high-energy radiation can cause skin cancer, injure eyes, harm the immune system, and upset the fragile balance of an entire ecosystem. News about the ozone hole that forms over Antarctica each October has spread around the world. The ozone hole can be as big as 1.5 times larger than the United States. See Fig. 3.

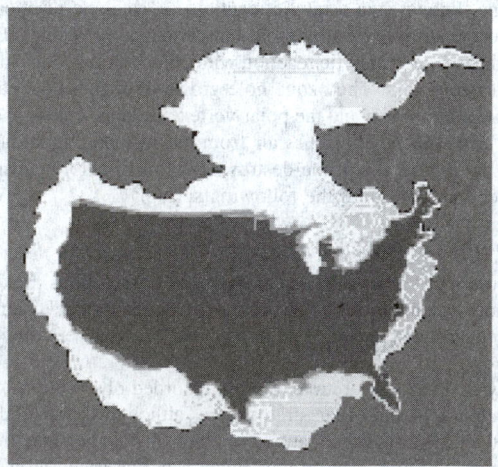

Fig. 3. Size comparison Antarctica-United States.

Table 1 is a list of geographic areas to be used as references in conceptualizing just how large the ozone hole can get. The following are referable areas:

Between 1986–87, several papers suggested possible mechanisms for the ozone hole, including chemical, dynamical (meteorological), and

TABLE 1. GEOGRAPHIC REFERENCE AREAS

Geographic Area	Square Kilometers	Square Miles
Australia	8,923,000	3,445,190
United States	9,363,130	3,615,124
Europe	10,498,000	4,053,300
Antarctica	13,340,000	5,150,602
Russia	17,078,000	6,593,853
North America	25,349,000	9,787,304
Africa	30,335,000	11,712,409
South Pole to 70 S	15,300,000	5,907,363

solar cycle influences. Among the key papers explaining the atmospheric chemistry of CFCs and ozone depletion was one by Susan Solomon and several colleagues (http://remus.jpl.nasa.gov/info.htm). The paper also emphasized the need for polar stratospheric clouds to explain the reaction chemistry. Also in 1986, Michael B. McElroy and colleagues described a role for bromine in ozone-depleting reactions. Paul Crutzen and Frank Arnold proposed that the polar stratospheric clouds could be made of nitric acid trihydrate, which would explain the clouds' presence at an altitude and temperature that should not have been cold enough for the tiny amount of pure water vapor present in the stratosphere to condense.

Observational evidence of the role of chlorine in ozone loss continued to mount during that same period. For example, the National Ozone Expedition (NOZE) measured elevated levels of the chemical chlorine dioxide (OClO) during the springtime ozone hole from McMurdo Research Station. Then in 1987, the Antarctic Airborne Ozone Expedition flew the ER-2 and DC-8 research aircraft from Punta Arenas, Chile, into the Antarctic Vortex. See Fig. 3.

The aircraft observations produced the "smoking gun" linking CFC-derived chlorine to the ozone hole. The flight data showed a negative correlation between chlorine monoxide (ClO) and ozone: the higher the concentration of ClO, the lower the concentration of ozone. In 1988, the husband and wife team Mario and Luisa Molina described the chemical reactions through which ClO catalyzes the extremely rapid destruction of ozone.

NASA has been monitoring the status of the ozone layer through satellite observations since the 1970s, beginning with the TOMS sensors on the Nimbus satellites. The latest-generation ozone-monitoring technology, the Ozone Monitoring Instrument (OMI), is flying onboard NASA's Aura satellite. See also **Aura Satellite**.

The Antarctic ozone hole will reach sizes on the order of 8–10 million square miles (20,719,904 — 25,899,881 square kilometers) nearly every year until about 2018 or so. Around 2018, things should slowly start

Fig. 4. Aircraft measurements in the late 1980s confirmed the link between CFCs, chlorine, and ozone loss. Here a NASA ER-2 high-altitude research aircraft lifts off from Kiruna, Sweden on a mission to study Arctic ozone. (Image courtesy of NASA Dryden Flight Research Center).

improving, and somewhere between 2020 and 2025, scientists we will be able to detect that the ozone hole is actually beginning to decrease in size. Eventually the ozone hole will go back to its normal level around 2068, nearly 20 years later than previously believed.

See also **Montreal Protocol**; **Ozone**; and **Ozone Depletion (Science of)**.

Web References

European Ozone Research Coordinating Unit: http://www.ozone−sec.ch.cam.ac.uk/.
Jonathan Shanklin British Antarctic Survey: http://www.antarctica.ac.uk/met/jds/.
NASA Ozone Hole Watch: http://ozonewatch.gsfc.nasa.gov/.
Ozone Hole History: http://www.theozonehole.com/ ozoneholehistory.htm.
Ozone Hole Tour University of Cambridge, UK: http://www.atm.ch.cam.ac.uk/tour/ index.html.

OZONOLYSIS. (1) Oxidiation of an organic material, i.e., tall oil, oleic acid, safflower oil, cyclic olefins, carbon treatment, peracetic acid production by means of ozone. (2) The use of ozone as a tool in analytical chemistry to locate double bonds in organic compounds and a similar use in synthetic organic chemistry for preparing new compounds. Under proper conditions, ozone attaches itself at the double bond of an unsaturated compound to form an ozonide. Since many ozonides are explosive, it is customary to decompose them in solution and deal with the final product.

See also **Oxygen**.

OZONOSPHERE. A region of the atmosphere from about 15 to 60 km (9 to 37 miles, roughly the extent of the stratosphere), which contains large concentrations of ozone. Also called *ozone layer*. See also **Ozone**.

The ozone concentration peaks at about 10^{13} molecules per cubic centimeter near 20 km (12.4 miles), while the mixing ratio peaks at slightly higher altitude (about 9–10 ppm at 30 km (19 miles)). Ozone in this region is produced from photolysis of O_2 molecules and is destroyed by reactions involving the oxides of nitrogen, chlorine, and hydrogen. Because of the strong UV absorption spectrum of ozone, the ozone layer effectively limits penetration of UV radiation to the earth's surface to wavelengths longer than 290 nm. See also **Atmospheric Shell**.

Additional Reading

Crutzen, P.J.: "The Influence of Nitrogen Oxides on the Atmospheric Ozone Content," *Quart. J. Roy. Meteor. Society*, **96** (408), 320–325 (1971).
Crutzen, P.J.: "Ozone Production Rates in the Oxygen–Hydrogen–Nitrogen Atmosphere," *J. Geophys. Res.*, **76**, 7311–7327 (1971).
Johnston, H.S.: "Reduction of Stratospheric Ozone by Nitrogen Oxide Catalysts from Supersonic Transport Exhaust," *Science*, **173**, 517 (1971).
Molina, M.J., and F.S. Rowland: "Stratospheric Sink for Chlorofluorocarbons: Chlorine Atom Catalyzed Destruction of Ozone," *Nature*, **249**, 890 (1974).
Stolarski, R.S., and R.J. Cicerone: "Stratospheric Chlorine: A Possible Sink for Ozone," *Can. J. Chem.*, **52**, 1610–1615 (1974).

AMS

P

PAAR TURBIDIMETER. A visual-extinction device for measurement of solution turbidity. The length of the column of liquid suspension is adjusted until the light filament can no longer be seen.

See also **Nephelometry**.

PACA. See **Rodentia**.

PACEMAKER. See **Arrhythmias (Cardiac)**; and **Cardiovascular Devices**.

PACIFIC COD. See **Codfishes**.

PACIFIC DEEP WATER. A water mass found between 1,000 and 3,000 meters (3,281 and 9,843 feet) in the Pacific Ocean. It is characterized by very sluggish flow and correspondingly low oxygen content. See also **Ocean**.

PACIFIC EQUATORIAL WATER. An immense surface oceanic water mass, extending from Central America to the East Indies. It is broader on the east, extending from 20°N to 20°S latitude, but it narrows greatly on the western side of the Ocean. Its surface temperature range is 8–15°C (46.4 to 59°F), salinity range 34.6–35.15%, but both properties decrease with depth below the lower values.

PACIFIC HIGH. See **Atmosphere (Earth)**.

PACKING (Absorption Column). See **Absorption (Process)**.

PACKIT. See **Data Compression**.

PHACOEMULSIFICATION. Phacoemulsification is a modified version of extracapsular cataract extraction (ECCE) and is the most common surgical procedure for removing cataracts. As in other forms of ECCE, phacoemulsification involves removing the eye's natural lens while leaving in place the back of the capsule, which holds the lens in place. The difference with phacoemulsification is that the cataract is broken into tiny pieces that are suctioned from the eye through a smaller incision than that required by other forms of cataract surgery. Healing and rehabilitation are faster with this procedure, and there is little, if any discomfort.

To understand how the phacoemulsification technique works, it is important to understand what a cataract is and how it interferes with vision. The eye works like a camera with two lenses. The first lens is the cornea, a clear membrane that covers the front of the eye. The second lens is the eye's natural crystalline lens, which is held in place by a capsule located behind the pupil. The cornea is responsible for about 70% of the eye's focusing power, while the natural lens fine-tunes the image.

When the natural lens becomes cloudy, usually because of the aging process, it keeps light rays from passing through or diffuses the light in such a way that vision becomes fuzzy or hazy. This cloudy lens is called a cataract. The object of cataract surgery is to remove this hazy lens and to replace it with a plastic prescription lens that is permanently implanted in the eye.

In phacoemulsification cataract surgery, the surgeon makes a very small incision, about 1/8th of an inch, in the white of the eye near the outer edge of the cornea. A small ultrasonic probe is inserted through this opening and, oscillating at 40,000 cycles per minute, is used to break up (emulsify) the cataract into tiny pieces. The emulsified material is simultaneously suctioned from the eye by the open tip of the same instrument. The hard central core of the cataract (the nucleus) is removed first, followed by extraction of the softer, peripheral cortical fibers that make up the remainder of the lens. The front (anterior) section of the lens capsule is removed along with the fragments of the natural lens. The back (posterior) portion of the capsule is left in place to hold and maintain the correct position for the implanted intraocular lenses.

After removal of the cataract, a prescription intraocular lens, or IOL, is permanently implanted in the lens capsule to replace the natural crystalline lens of the eye that was removed during the surgery. This lens is rolled inside a tiny hollow tube and inserted through the same incision that was used to remove the cataract. The folded lens is pushed out of the tube by a tiny plunger and, as it unfolds, is positioned by the surgeon in the center of the lens capsule. The new lens is held in place by microscopic, spring-like wires that are attached to the implant.

The tiny incision made during phacoemulsification surgery generally requires no stitches and heals itself in a few days. Antibiotic and steroid eye drops may be given to diminish inflammation, to prevent infection, and to keep the eye moistened for several days following surgery.

Phacoemulsification cataract surgery is one of the most effective surgical procedures performed in the United States today, and a large percentage of patients are very satisfied with the results. See also **Cataract**; and **Extracapsular Cataract Extraction (ECCE)**.

Vision Rx, Inc., Elmsford, NY

PADDLEFISHES (*Osteichthyes*). Of the order *Chondrostei*, family *Polyodontidae*, the paddlefishes resemble to some extent some species of shark. Although classified as *Osteichthyes*, they are a strange combination of cartilage and bone. These fishes are named for the rather tremendous paddle that is attached to the nose, equal in length to about one-third the length of the body of the fish. The small barbels under the paddle are reminiscent of sturgeons.

Paddlefishes are known for very rapid growth and development. When fully grown, a paddlefish may measure up to 6 feet (1.8 meters) in length, with a paddle an additional 2 feet (0.6 meter) in length and about 4 inches (10 centimeters) in width. Biologists have not determined the use for the paddle because the fish obtains its food by swimming with its mouth wide open. The main diet is comprised of crustaceans and planktonic organisms.

Only two kinds of paddlefishes are known: (1) the *Psephurus gladius*, found in the Yangtse River valley; and (2) the *Polyodon spathula*, found in the Mississippi valley. It is considered to be an acceptable food fish. In the United States, the fish is sometimes called the spoon-bill, spoon-beaked sturgeon, and spoon-billed catfish, or simply spooner.

PAGET'S DISEASE. See **Bone**; and **Dermatitis and Dermatosis**.

PAINT AND FINISH REMOVERS. The term finish denotes the final process of manufacturing. Finishing operations include such processes as clear coating (varnishes and lacquers), painting, plating, anodizing, phosphatizing, galvanizing, and blueing, all of which take place at the terminal point of manufacturing. Finishing is defined as the process of coating or treating a surface for the purpose of protecting and/or decorating the product. The useful life of most usable objects is greater than the finish. This results in a periodic need to remove and replace the finish.

The physical properties of finish removers vary considerably due to the diverse uses and requirements of the removers. Finish removers can be grouped by the principal ingredient of the formula, method of application, method of removal, chemical base, viscosity, or hazardous classification. Except for method of application, a paint remover formulation usually has one aspect of each group, by which it can be used for one or more applications.

Finish removers are applied by brushing, spraying, troweling, flowing, or soaking. Removal is by water rinse, wipe and let dry, or solvent rinse. Removers may be neutral, basic, or acidic. The viscosity can vary from water thin, to a thick spray-on, to a paste trowel-on remover. The hazard classification, such as flammable or corrosive, is assigned by the U.S. Department of Transportation (DOT) for the hazardous materials contained in the remover.

Organic Finish Removers

Methylene Chloride Finish Removers. Methylene chloride formulas are the most common organic chemical removers. The low molar volume of methylene chloride allows it to rapidly penetrate the finish by entering the microvoids of the finish. When the solvent reaches the substrate, the remover releases the adhesive bond between the finish and the substrate and causes the finish to swell. The result is a blistering effect and an efficient rapid lifting action. Larger molecule solvents generally cannot cause this lifting action and must dissolve the finish. When methylene chloride is used in amounts of 78% or more, even with flammable cosolvents, the mixture is nonflammable. A typical methylene chloride base remover includes cosolvents, activators, evaporation retarders, corrosion inhibitors, thickeners, and wetting agents.

Typical cosolvents include methanol [CAS: 67-56-1], ethanol [64-17-5], isopropyl alcohol [CAS: 67-63-0], or toluene. The selection of cosolvents depends on the requirement of the formula and their interaction with other ingredients. Methanol is a common cosolvent in methylene chloride formulas since it has good solvency and is needed to swell cellulose-type thickening agents. A typical methylene chloride formula used to strip wood is as follows: methylene chloride 81.1%; 1toluene 2.1%; paraffin wax (ASTM 50−53 °C mp), 1.6%; methycellulose, 1.2%; methanol, 7.8%; and mineral spirits 6.2%.

The rate of stripping or the stripability on catalyzed urethane and epoxy resin finishes can be increased by adding formic acid, acetic acid, and phenol. Sodium hydroxide, potassium hydroxide, and trisodium phosphate may be added to the formula to increase the stripability on enamel and latex paints. Other activators include oleic acid [CAS: 144-62-07], trichloroacetic acid [CAS: 76-83-9], ammonia, triethanolamine [CAS: 102-71-6], and monoethylamine [CAS: 75-04-7]. Methylene chloride-type removers are unique in their ability to accept cosolvents and activators that allow the solution to be neutral, alkaline, or acidic. This ability greatly expands the number of coatings that can be removed with methylene chloride removers.

Paraffin wax vapor barriers are used in water rinse removers that can disperse the wax without coating the substrate. In soak tank applications, water is sometimes floated on top of an all-solvent, neutral pH, non-water rinse remover to prevent evaporation. Flotation devices that cover the exposed surface area may be used with other formulas.

Corrosion inhibitors are used to protect both the container and the metal substrate being stripped. Acid activated removers use inhibitors to block corrosion on active metals. Typical inhibitors are propylene oxide [CAS: 75-56-9], butylene oxide [CAS: 9106-88-7], triethylammonium phosphates, and sodium benzoate [CAS: 532-32-1].

Health and Safety. Remover formulas that are nonflammable may be used in any area that provides adequate ventilation. Representative personal breathing zone air samples of persons working with these removers should be taken according to Occupational Safety and Health Administration (OSHA) recommendations. The current OSHA standard for methylene chloride includes a Permissible Exposure Limits (PELs) of 25 ppm. Expressed as an 8-h time-weighted average (TWA) PEL means that no employee may be exposed to an airborne concentration of methylene chloride in excess of 25 parts of methylene chloride per million parts of air (25 ppm) averaged over an 8-h period. Short-term exposure limit (STEL) requires that no employee is exposed to an airborne concentration of methylene chloride in excess of 125 parts of methylene chloride per million parts of air (125 ppm) as determined over a sampling period of 15 min.

The environment can be monitored with passive detection badges or by active air sampling and charcoal absorption tube analysis. The vapor of methylene chloride produces hydrogen chloride and phosgene gas when burned. Methylene chloride-type removers should not be used in the presence of an open flame or other heat sources such as kerosene heaters.

Persons exposed to methylene chloride removers should wear protective clothing and eye protection.

Environmental Impact. Methylene chloride is nonphotochemically reactive and is not listed as an ozone (qv) depleter. Methylene chloride removers can easily be recovered from paint chips and other residue sedimentation, thus allowing recovery of remover and its continued use. This greatly increases the useful life of the remover and, when mixed with fresh remover, eliminates the need for disposing of the used remover. This process requires no special recovery equipment. The high volatility of methylene chloride allows the waste residue from the stripping process to be easily dried. The resulting waste is normally considered hazardous because of the amounts of heavy metals from old finishes.

Petroleum and Oxygenate Finish Removers. Many older finishes can be removed with single solvents or blends of petroleum solvents and oxygenates. Varnish can be removed with mineral spirits, shellac can be stripped with alcohols, and lacquers can be removed with blends of acetates and alcohols (lacquer thinners). The removal mechanism is one of dissolving the coating, then washing the surface or wiping away the finish. This method is often used to reamalgamate or liquefy old finishes on antique items of furniture.

Petroleum and oxygenate finish removers, may contain ketones, esters, low molecular weight aromatics, and glycol ethers. Cosolvents include methanol, *n*-butanol [CAS: 71-36-3], *sec*-butyl alcohol [CAS: 78-92-2], or xylene [CAS: 1330-20-7]. Sodium hydroxide or amines are used to activate the remover. Paraffin wax is used as an evaporation retarder though its effectiveness is limited because it is highly soluble in the petroleum solvents. Cellulose thickeners are sometimes added to liquid formulas to assist in pulling the paraffin wax from the liquid to form a vapor barrier or to make a thick formula. Corrosion inhibitors are added to stabilize the formula for packaging.

Wetting agents are used to make a water rinse remover. Water rinse removers are normally used for removing paint, where the surfactants help remove paint and remover from the substrate. Solvent rinse removers or wipe and dry formulas may be used for stripping clear finishes. A typical petroleum and oxygenate formula is as follows: toluene 21%; acetone19%; alkyl acetate 31%; methyl ethyl ketone19%, and butyl alcohol10%.

This is a liquid scrape-off remover for brush or soak applications. Clean up is with a solvent that is compatible with finish to be used, or wipe and dry.

Health and Safety. Petroleum and oxygenate formulas are either flammable or combustible. Flammables must be used in facilities that meet OSHA standards for use with flammable liquids. Adequate ventilation that meets the exposure level for the major ingredient must be attained. Extreme caution must be taken to prevent the possibility of fire when using flammable removers.

Natural rubber, neoprene, or other gloves suitable for use with the remover formula must be worn. Canister respirators are available for most petroleum and oxygenate remover solvents. Symptoms of long-term exposure should be identified from the MSDS of the stripper.

Environmental Impact. Most petroleum and oxygenate removers are photochemically reactive and classed as volatile organic compounds (VOCs). Recovery and disposal of this type of remover after use is difficult because the finish is resolubilized by the remover. The dissolved finish cannot be separated from the spent remover and the whole mixture must be disposed as a liquid hazardous waste. Recovery by distillation of the solvents is dangerous because the nitrocellulose from lacquer finish may cause autoignition in the still.

Other Organic Removers. Concerns over the reported toxicity and carcinogenicity of methylene chloride have stimulated research for alternative solvents in remover formulas. N-Methylpyrrolidinone [CAS: 872-50-4] and dibasic esters (dimethyl glutarate [CAS: 1119-40-0] or dimethyl adipate [CAS: 627-93-0]) have been used in removers. They remove single-component finishes but work much more slowly than methylene chloride, petroleum, and oxygenate group removers. They have little success on epoxy and catalyzed finishes.

Health and Safety. Both *N*-methylpyrrolidinone and dibasic esters have very low vapor pressure which limits worker exposure to vapors. Manufacturers recommend that the same safety precautions be taken as with other organic solvents. Hazardous location requirements must be considered if the formula is flammable. Ventilation that reduces vapors to manufacturer's recommended exposure levels should be used. Protective clothing must be worn during use.

Environmental Impact. The volume of waste remover from these products is remarkably increased when compared to methylene chloride, petroleum, and oxygenate removers, since both *N*-methylpyrrolidinone and dibasic esters have low vapor pressures. Recovery of the remover after use is difficult because the finish is resolubilized by the remover.

Inorganic Finish Removers

Liquid Alkaline Removers. This group consists of alkaline materials that are dissolved in water then heated to an appropriate temperature to remove finishes. In a typical application, a hot water bath large enough to submerge an item is used. Various alkaline materials may be used to provide the desired alkalinity. Sodium or potassium compounds are preferred, such as sodium hydroxide [CAS: 1310-73-2], sodium carbonate [CAS: 5968-11-6], sodium silicates, mono-, di-, and trisodium phosphates, tetrasodium pyrophosphate [CAS: 7722-88-5], and sodium tripolyphosphate [CAS: 7758-29-4] and potassium hydroxide. Activators, such as phenol, gluconic acid [CAS: 526-95-4], and alkali metal gluconate are dissolved into a suitable solvent that itself can be dissolved in and added to the caustic solution. Surfactants are added to aid in removal of the solution and increase the penetration of the finish. Various suitable surfactants of the anionic, nonionic, and cationic types may be used, provided they are soluble and effective in the alkaline stripping solution. A typical aqueous alkaline remover formula is sodium hydroxide 87.0%; sodium carbonate 6.0%; sodium hydrogen phosphates 3.0%; sodium gluconate 3.0%; phosphate ester 0.5%; and pine oil 0.5%. This aqueous alkaline remover is used for stripping the finish from wood or ferrous metals at a mix ratio of 30–600 g/L (0.25–5 lbs/gal).

Paste-Type Alkaline Removers. Sodium hydroxide, potassium hydroxide, or other caustic compounds are blended to make these types of removers. Polymer-type thickeners are added to increase the viscosity that allows the remover to be applied with a brush, trowel, or spray. Some of these products use a paper or fabric covering to allow the remover finish mixture to be peeled away. The most common application for this group of removers is the removal of architectural finishes from the interior and exterior of buildings. The long dwell time allows for many layers of finish to be removed with one thick application of remover.

Sodium hydroxide and salt can be heated to a fused state in baths to allow the removal of finishes from ferrous metals. The most common use of this method is the removal of heavy concentrations of paint on conveyer parts and hangers used in production spray systems.

Health and Safety. Protective clothing that is compatible with the remover formula must be worn. Caustic soda baths should be ventilated to remove vapors from the work area. Most caustic removers are corrosive and cause severe burns with minimal contact to the skin.

The liquid from spent caustic soda baths must be disposed of or treated as a hazardous waste. The finish residue may contain heavy metals as well as caustic thus requiring treatment as a hazardous waste.

Manufacturing and Processing

Finish removers are manufactured in open or closed kettles. Closed kettles are preferred because they prevent solvent loss and exposure to personnel. To reduce air emissions from the solvents, condensers are employed on vent stacks.

Standard 0.25 or 0.50 lb (227 g) tin coated cans are used for packaging liquid with neutral and mildly alkaline base formulas; polypropylene is used for acid–base removers. Steel and polypropylene drums are used for industrial removers. Viscous removers are packaged in removable top containers. Dry caustic removers are packaged in bag-lined boxes or fiber drums.

The DOT has established standards for the packaging and labeling of hazardous materials offered for shipment by public transportation. The Consumer Product Safety Commission (CPSC) has set standards for retail labeling and packaging. OSHA and the U.S. EPA have labeling requirements.

See also **Paints and Coatings**.

Additional Reading

Staff: *Industrial Users of Paint and Finish Removers*, Paint Remover Manufacturer's Association, 1992.

Staff: *Solvents Used in Paint Removers*, Paint Remover Manufacturer's Association, Sept. 1991.

Staff: *Chemical Protective Clothing for Furniture Stripping*, Department of Health and Human Services, Washington, D.C., Mar. 11, 1991.

Staff: *Methylene Chloride Consumption By Paint and Coating Removal Groups*, Paint Remover Manufacturer's Association, 1992.

Web References

General Chemical Corp. http://www.removepaints.com/products/ps/Wood_strip_3900.asp

Parks Corporation, Paint/Finish Removers, http://www.newparks.com/parkspaint.html

Parks Corporation, Solvents & Cleaners–Paint/Finish Removers, http://www.newparks.com/solvents.html

The Sherwin-Williams Company, http://www.sherwin-williams.com/pro/

PAINTS AND COATINGS. Traditionally, paints and other coatings have been considered in terms of protecting and decorating buildings, houses, furniture, automobiles, toys, boats, machines, and the like. These products represent a large industry estimated at about $ 15 billion per year worldwide. Nevertheless, these kinds of coatings account for only 10–15% of the total coatings industry.

One researcher[1] has classified coatings in two technically defined categories.

1 Type I — Products manufactured and sold with several coating layers. Examples are the aforementioned applications, but also include color photographic film, graphic arts films for printing, coated papers for printing, coated containers for food packaging, magnetic storage media for computers and audio/visual equipment, optical disks for digital data storage, adhesive tapes, wallpaper, and metallized films.

2 Type II — Coatings that become an integral part or a key intermediary for a device or piece of equipment. Sometimes, the term *core technology* is used for such applications. There is a wide variety of coating products in this classification. They would include photoresists for circuit board manufacture, thick- and thin-film coatings on integrated circuits, adhesives for bonding metals, phosphor coatings on electronic display screens, stain repellents on fibers, dyes on fibers, ceramic glazes, encapsulated time-release drugs, and thin-film photovoltaic cells.

The foregoing categories of coatings immediately dramatize the versatility and specialization of a huge industry that has developed essentially during the past three decades. With the continuing interest in materials composites, continued rapid growth is expected. Coating technology is a complex undertaking fraught with numerous design and manufacturing problems. These are particularly difficult in scaling up laboratory procedures to high production rates on the factory floor. Consequently, a number of successful processes remain proprietary. The technology has demanded much of chemists and chemical and mechanical engineers, with particular emphasis on an understanding of fluid dynamics.

Although the two foregoing classifications may serve a very useful purpose at the scientific level, for the purposes of this article, the first part is devoted to traditional surface coating products that serve protective and decorative purposes. The second part, in less depth, addresses coatings that are used for special purposes as represented, (e.g., by uses in electronic devices).

Protective and Decorative Coatings[2]

Paint, coating, and finish are terms used to describe a wide variety of materials designed to adhere to a substrate and act as a thin, plastic-like layer. Paints are available for decorative, protective, and other purposes. They can be decorative by covering defects (being opaque), by changing color, or by providing a desired gloss or sheen. Protective uses include shielding metals from corrosion, protecting plastic from degradation caused

---

[1] E.D. Cohen, E.I. Du Pont De Nemours & Co., Parlin, New Jersey.

[2] Some of this information was provided by The Sherwin-Williams Company, Cleveland, Ohio.

by ultraviolet light, acting as a moisture barrier and providing mar and scratch resistance for wood or plastic surfaces. A paint can also be used for its special spectral properties (e.g., light absorbing for heating swimming pools; radar absorbing for military vehicles; etc.), or unusual physical properties (e.g., strippable coatings for the interior of paint spray booths, insulating coatings for electrical parts, etc.).

Paints are generally liquids before they are applied to a surface. When applied, they should completely replace the substrate/air interface with a substrate/paint interface (called wetting). The forces of attraction created by this wetting process are responsible for the paint's adhesion. The paint then dries and/or cures to form a hard film. Drying is the physical action of solvent leaving the film, while curing refers to a chemical reaction which connects polymer chains of the paint together.

Composition

Thousands of raw materials are used to manufacture coatings, but they can generally be classified into four categories: (1) binders or resins, (2) pigments, (3) solvents, and (4) additives. Binders and resins are generally organic compounds, usually polymeric or oligomeric in nature, which provide a continuous matrix in the final film and have a major influence on the toughness, flexibility, gloss, chemical resistance and cure/dry properties of the coating.

Pigments. These are finely divided powders (particles between 0.1 and 50 micrometers in diameter) which are dispersed throughout the binder. In addition to reinforcing the final film, much as they do in composite plastics, they influence a coating's resistance to abrasion and corrosion, and they also are the major factor in the gloss, color, and opacity of a coating.

Solvents or Thinners. These substances have a major effect on a paint's application viscosity and also affect a coating's cure dry properties, and often its toxicity. Most solvents evaporate and leave the film during the drying process, although certain "reactive dilutents" are designed to react with the resin and become part of the binder system. Water is a popular solvent because of its low cost, low toxicity, and nonpolluting nature. Almost half of the coatings currently produced use water as a major solvent. Disadvantages of using water include the effect of humidity on the drying characteristics of the paint and the difficulty of making a water-resistant film from materials suspended in water. Major organic solvents include mineral spirits, ketones, acetates, alcohols, and xylene.

Additives. Among the more important classes of additives used in coatings are: (1) *surfactants*, which are used to suspend pigment and binder particles; (2) *thickeners* to obtain proper rheology (especially in latex paints); (3) *plasticizers*, which lower the glass transition temperature of the binder and increase the flexibility of the coating; (4) *antifoam agents* to prevent bubbles in aqueous paints; (5) *antiskin agents*, which prevent the formation of a dry layer on top of the paint while it is still in the can; (6) *preservatives*, such as biocides and mildewcides to protect the binder from microscopic organisms both before and after application; (7) *ultraviolet light absorbers* to protect the binder and/or substrate from degradation due to sunlight; and (8) a variety of surface *conditioners* and *lubricants*, which help the film adhere to the substrate or protect the film by giving it a lubricated surface. Additives will often interact and coating formulators must be careful to watch for synergistic and antagonistic effects.

Ratio of Components. The ratio of the four aforementioned components (binders, pigments, solvents, and additives) greatly influences the properties of a coating. The volume fraction of the solid film (which is pigment) is referred to as the *pigment volume concentration*. The concentration where there is barely sufficient binder to fill in the voids between the pigment particles is referred to as *critical pigment volume concentration* (or *CPVC*). When the composition of a coating is changed from below its CPVC to above it, the properties of the coating begin to dramatically degrade (except for opacity, which increases). The performance changes are primarily due to the pockets of air which form in the final film because of the shortage of binder. With the exception of flat architectural (house) paints, fillers, and certain primers, almost all coatings have a pigment volume concentration less than the CPVC. The CPVC is a function of a pigment combination's oil adsorption and particle size distribution.

The ratio of solvent to nonvolatiles is important since the volatile organic compound (VOC) composition of coatings is increasingly a target of government regulations. Usually the VOC of a coating is expressed in units of *mass per volume* and is calculated by multiplying the density of the coating by that fraction of the coating that is volatile. For coatings using water (or certain chloroalkanes) as a solvent, the effect of these "exempt" solvents is subtracted out before the calculation is made. Current VOC limits in the United States range from 250 grams per liter for some California architectural alkyds to 450 grams per liter for some furniture finishes.

Binder Classifications

Paints are often classified by the type of binder they include. The most common classifications (with percent of total coatings used in 1985) are: Latexes (31%); waterborne (10%); non-aqueous dispersions (2%); solvent-borne (55%); and one hundred percent solids coatings (2%). A small volume of paint is made with silane binders.

Latexes. These are dispersions of high-molecular-weight polymer particles in an aqueous medium. Since the polymer is in a suspended form, the viscosity of the mixture is almost exclusively a function of the viscosity of the continuous phase (i.e., water) and is not affected by the molecular weight of the polymer. This permits the use of higher-molecular-weight material than can be used with a solution-type approach. The film forms (i.e., the paint dries) when the water evaporates and the spherical latex particles are forced together, overcoming the steric and ionic forces which had been stabilizing them. Once the particles touch, the surface tension of the latex causes the individual particles to coalesce (i.e., the polymers in adjoining particles entangle), aided by the capillary forces created by the evaporating water. Latex particles are generally 0.1 to 0.5 micrometer in diameter, although particle sizes as much as an order of magnitude on either side of these values is used for specialized purposes.

The two most commonly used latex systems are acrylic systems (40% of usage), which perform very well, but are relatively expensive, and the vinyl-acrylic copolymer systems (57% of usage and growing), which do not perform as well, but are less expensive.

Latexes are usually considered separate from other waterborne binders because the method employed for synthesizing and suspending them is very different. For latexes, emulsion polymerization is used and no organic solvent is required to obtain or stabilize the emulsion. In contrast, most waterborne binders are synthesized in organic solvent solutions and then "let down" with water. Often, some quantity of organic solvent must remain or the resulting aqueous mixture will not be stable. Most waterborne resins need to be cured. Reactions commonly used include the oxidative polymerization of unsaturated aliphatic chains, the reaction with aminoplast resins, and the reaction of epoxies to form ether or ester bonds. Waterborne resin compositions include acrylics, polyesters (including alkyds), urethanes, phenolics, and epoxies, among others.

Non-Aqueous Dispersions (NAD$_s$). These are the solvent-borne analogues of latexes. They were used as automotive finishes in the 1970s, but their use is now declining. The use of NADs as an auxiliary binder, however, is increasing as it has been found that the addition of a small amount of NAD can improve a coating's drying characteristics and rheology.

Solvent-Borne Coatings. These cover a wide variety of resins including alkyds. Alkyds are polyesters made from soya, linseed, or other oils and are a major factor in architectural, automotive, and industrial maintenance usages, among other applications. Acrylics are known for their exterior durability and are the major binder in automotive coatings. Epoxies are used mostly in automotive, industrial maintenance, metal container, and coil coatings. Polyurethanes are isocyanate-based binders and are used where excellent properties are required, such as in the magnetic media, magnet wire, industrial maintenance, and deck coatings fields. Polyesters, other than alkyds, are used in coil, metal furniture, metal container, appliance, and automotive coatings. Amino crosslinkers primarily are modified melamines and are important in metal container, automotive, coil, and metal furniture applications, among others.

In addition to drying, most solvent-borne coatings undergo some type of cure. Chemical reactions used for the cure of solvent-borne coatings include the oxidative crosslinking of unsaturated carbon bonds (alkyds and polyesters), the reaction of melamine derivatives to form ureas (polyesters, alkyds, acrylics, and amino resins), ether and ester formation by epoxies (epoxies, polyesters and acrylics), and urethan formation by isocyanates. Curing is especially important in low VOC coatings because the viscosity of a resin is a function of the molecular weight of the material. The demand for lower VOC (i.e., higher solids) coatings has resulted in a move toward lower-molecular weight and more reactive solvent-borne coatings. About one-quarter of the solvent-borne coatings currently used are considered "high solids" (i.e., they have a VOC of 350 grams/liter or less).

The thrust for higher solids also has been a factor in the growth of *100% solids* coating technologies. These technologies include polymerization initiated by ultraviolet radiation or an electron beam; the use of powdered coatings which coalesce when sufficient heat is applied; and vapor cure technology, where a reactive resin is crosslinked by exposure to a reactive vapor. While the capital investment required by these technologies is high, their use is expected to continue to grow because of their efficient use of material and the superior properties that can be obtained. Another means of using a *solventless system is hot melt coatings*. These are applied at high temperatures without solvent and dry by cooling them to room temperature.

Silanes. Silanes and silane derivatives dominate the small market for inorganic binders. These materials are used both in combination with organic binders and by themselves. As co-binders, they increase the chemical and moisture resistance of a film. When used by themselves, they form brittle, very chemical-resistant films. Silane coatings are usually more expensive than their organic counterparts and must be kept dry before application.

Pigment Classifications

Pigments used in the paint industry are commonly classified by function: (1) *Hiding* (or *prime*) pigments scatter light and are used to obtain opacity; (2) *extender* (or "inert") pigments are used to reinforce the binder, increase the pigment volume concentration, lower gloss, and lower the cost of a paint; (3) *colored* pigments which are used to tint a paint can be either inorganic or organic; (4) *metallic* pigments are used for corrosion prevention and appearance reasons; and (5) *protective* and other *functional pigments* can be used to add special features to a coating.

Hiding Pigment. The refractive index of hiding pigments must be sufficiently different from the refractive index of the binder (usually about 1.5) if light is to be effectively scattered. Two crystal structures of titanium dioxide (commonly referred to as titanium) are the most widely used hiding pigments in the paint industry with the rutile version being more popular than anatase because the refractive index of rutile (2.76) is higher than that of anatase (2.55). Rutile is also more thermally stable and photostable, although the chalking property of the anatase pigment has been used to advantage in making "self-cleaning" paints. Like most inorganic pigments, titanium dioxide is a naturally occurring compound that must be mined, crushed and processed before it is a suitable raw material for paint. The optimum diameter for light scattering is about 0.2 micrometer.

Other naturally occurring hiding pigments include zinc oxide (refractive index = 2.01) and zinc sulfide (refractive index = 2.37). At one time, lead carbonate (refractive index = 2.0) was a leading hiding pigment. In addition to hiding, zinc oxide is a fungistat (i.e., it inhibits the growth of fungi).

Pockets of air (refractive index = 1.0) encapsulated in plastic are also used as light-scattering pigments. The size of these synthetic pigments range from 0.6 to 20 micrometers, depending upon the number of 0.5-micrometer bubbles per particle.

Extender Pigment. The major classes of extender pigments are as follows:

Calcium carbonate (also called *whiting*). *Calcium carbonate* is inexpensive, has a low binder demand, and is not colored, but is acid sensitive.
Clay (or *kaolin*) covers a wide variety of materials which are inert and inexpensive, but it is more yellow than calcium carbonate.
Talcs are very inexpensive and easily suspended pigments.
Silicas (mostly silicon dioxides) are low cost, inert, and very hard.

Other extender pigments include barium sulfate, feldspar, diatomite, and mica.

Color Pigments. Chrome yellow is the leading inorganic color pigment. It is used primarily in traffic-marking paint for roads and highways. The yellow, red, and brown versions of iron oxide also are important and are used in a variety of industrial and architectural coatings.

There are hundreds of organic color pigments used in the paint industry, the vast majority of which are synthetic. Since many of these are vulnerable to ultraviolet light or chemical degradation, great care must be taken in choosing pigments that are suitable for a given paint and its intended usage.

The most commonly used black pigments are carbon blacks. In addition to being efficient light absorbers, some varieties of these small-particle-size materials impart electrical conductivity and thixotropy to paints. The leading inorganic black pigment is black iron oxide. This material is used primarily because it is easier to disperse than the carbon blacks. See also **Carbon Black**.

Metallic Pigments. The leading metallic pigment is zinc dust, which is used mostly in zinc-rich primers, where it acts as a *passivating agent*. Aluminum flake is used for the silvery metallic appearance that it imparts.

Anticorrosive Pigments. Several pigments are used primarily because of their anticorrosive properties. Chromates (zinc, strontium, and lead, if permitted) are the most effective anticorrosive pigments, but the chronic toxicity danger associated with them is a matter of serious concern. Barium metaborate, red lead, and borosilicates have been the principal nonchromate materials used. Other pigments used for special purposes include iron oxide for magnetic media and copper oxide in marine coatings to prevent barnacle and algae growth.

Tributyltin acetate [CAS: 56-36-0], $(C_4H_9)_3SnOOCCH_3$, gained widespread usage in marine paints. The loss of ship performance and efficiency resulting from the growth of barnacles, seaweeds, tubeworms, and other organisms on boat bottoms has been known since the time of the Phoenicians, when copper strips were fastened to hulls to prevent fouling. Various navies and ocean shippers worldwide have found tributyltin acetate (TBT) effective, particularly in tropical waters. Unfortunately, the ingredient in paint has been found to be toxic. One authority in ocean chemistry has observed that TBT is the most toxic compound man has introduced into the marine environment. In recent years, numerous studies and evaluations of TBT-based paints have been made. Some countries have regulations against its use. Details on the adverse growth of oysters, for example, have been reported. Some of these observations are covered in the Champ reference listed.

Coating Manufacturing Process

There are usually two steps in the paint-making process: (1) dispersing the pigment (called *grinding*) and (2) mixing in the raw materials not used in grinding (called *letting down*). Except for NADs, solvent-borne coatings are made by grinding the pigment into a binder/solvent solution. The polymer serves a dual purpose in the step: (1) It thickens the mixture, increasing the dispersing efficiency of energy put into the system, and (2) it adsorbs onto the surface of the pigment particles, stabilizing them in suspension. Once stabilized, the suspension is let down by stirring in the remaining raw materials. For most waterborne paints (including all latexes), the grinding step consists of dispersing the pigment in water. The binder is not included since it is not stable enough to withstand the grinding process. Instead, surfactants are used to help wet and stabilize the pigment. Paints prepared in this manner add the binder in the letdown.

Mills. Several types of machines (called *mills*) are used for grinding pigment. *Media mills* have a chamber where the pigment, binder and a solid media are all ground together. The grinding action of the media particles on one another provides the shear needed to breakdown the pigment agglomerates. Some media mills have chambers that hold an entire batch at one time, while others have smaller chambers through which a batch is passed in a continuous flow. Batch mills (those falling into the former category) include those using pebbles, ceramic beads, or steel shot as media. Continuous mills (where the batch flows through the chamber) usually use sand or ceramic beads as the media.

Roller mills have large, closely placed rollers capable of grinding very thick pigment suspensions. Adjoining rollers are turned in opposite directions and at high speeds. The point where adjoining roller surface separate is subjected to sufficient shear to pull the pigment agglomerate apart.

High-speed dispersers (HSDs) are a third general type of mill. HSDs use blades attached to rotating shafts to disperse the pigment, much as an egg beater is used to disperse flour. HSDs are currently the most widely used grinding equipment. HSDs or similar stirring devices are generally used for the letdown regardless of the type of mill used in the grinding step. See also **Ball, Pebble, and Rod Mills**.

Application Methods

Coatings can be applied in a variety of ways depending upon the nature of the substrate and the viscosity of the coating. Brush, roller and pressure pads are popular methods of application for architectural and industrial maintenance coatings. Advantages of these methods include lack of capital investment and the ability to apply coatings on site. Disadvantages include their labor intensity and their limitation to use *ambient cure* coatings. Air, and especially airless, spray equipment can apply coatings much faster than a brush or roller, but this method requires more equipment and is not as adaptable. Spraying is used for architectural, industrial maintenance, wood furniture, automotive refinish and other coatings.

Electrostatic Spraying. In this method, a paint with a negative charge is applied to a substrate with a positive charge. This is the method of choice for automotive, metal furniture, appliance, machinery, and metal container coatings. The equipment for electrostatic spray costs more than regular spray equipment, but the transfer efficiency can be much higher. A drawback is that electrostatic spraying can be used only if the object to be painted can hold an electrical charge and does not have deep crevices.

Electrodeposition. Another way of using electricity to paint is electrodeposition (ED). In this method, the object to be coated is dipped into a vat of charged aqueous coating. An opposite electrical charge is then applied to the object and the paint is attracted to the surface of the charged object. Having been painted, the object is removed from the vat, rinsed, and baked. Electrodeposition requires a very large capital investment, does not allow for color changes, works only for conductive (metal) objects, and is only suitable for coatings that are baked. Ambient cure paints do not have the long-term stability needed for ED. Nevertheless it is the greatly preferred application method for automotive primers and many other metal products because of its high transfer efficiency—desirable for both environmental and economic reasons.

Dip Tanks. This method, which does not use electricity, is especially suitable for small objects. The main disadvantage to dipping is the difficulty of getting an even coat without *drips*.

Roller Coaters and Sheet Coaters. These are often used in coil coating where very large, flat surface areas need to be painted quickly. In these methods, the objects to be painted are passed between a doctor blade and applicator rolls. These methods are suitable only for coatings that are to be baked.

Surface Preparation of the Substrate. This is extremely important for all methods of paint and coatings application. The failure of a paint system is often due *not* to the paint itself, but because of a failure in surface preparation. For example, an anticorrosive paint applied to a rusty surface will not be effective if the rust falls off taking the new paint with it. For wood and plastic surfaces, old paint or a weathered surface layer may have to be removed. For older metal objects, the removal of corrosion is often required. Sandblasting is one method to remove both the old paint and any corrosion. For new metal objects, a phosphate or chromate layer is often chemically bonded to the metal to provide a surface to which a coating can easily adhere.

Paints for Specific Functions

Paints are often separated into the types of jobs they perform. *Primers* are meant to be applied to bare substrates and then covered with a topcoat. As such, they must have good adhesion and good recoatability, but color and light stability are usually unimportant. When used over metal, primers are usually expected to provide corrosion protection. *Sealers* are similar to primers except they are used over porous substrates. Sealers eliminate the leaching of material from the substrate and also prevent paint components from migrating to the substrate. *Surfacers* are highly pigmented paints that are applied in a thick layer to mask surface irregularities and to allow good adhesion by a topcoat. *Fillers* are a type of surfacer that is used to fill holes. *Stains* are low-solids coatings applied to wood to accentuate its grain. Most interior stains require a topcoat.

Some finishes do not require a primer. Varnishes are clear, tough, and usually have a glossy finish. Exterior varnishes give wood some protection from sunlight. *Shellacs* are a type of varnish which offers the advantages of a quick dry and easy sanding, but which is sensitive to water and has a limited shelf life.

Topcoats are usually applied over an undercoat (i.e., primer, stain, etc.). A topcoat must protect its undercoat from environmental damage and should provide the appearance characteristics desired for the particular application. Topcoats are available in a variety of colors, textures, and glosses. One specialized topcoat is *lacquer*, which is a solution of resin in organic solvents. Lacquers dry, but do not cure. Because of their high VOCs, the use of lacquers is declining. *Enamel* is a term used to describe a glossy, opaque topcoat.

A unique two-coat, topcoat system is used in the automobile industry. A basecoat is applied to a primed surface to provide opacity, color, and a metallic appearance, while the clearcoat provides gloss and a mirrorlike finish (referred to as *distinctness of image*).

Various regulatory agencies in the United States and other countries have set limits on the use of low-volatile organic coatings. These are widely used in the chemical, petrochemical, and metallurgical industries to resist corrosion. Presently, high-temperature paints and coatings are largely silicone in nature. These are considered the best-performing coatings for smokestacks, boilers, mufflers, furnaces, incinerators, combustion chambers, and jet engines. High solids polyorganosiloxane polymers (silicone resins) are used to protect steel piping and other equipment where high temperatures accelerate deterioration of ferrous substrates. They also are used on storage tanks where appearance is an issue, as in the case of oil refinery tanks.

In addition to their high-temperature properties, high solids polyorganosiloxane polymers are resistant to numerous chemicals. For example, electric power generation frequently creates sulfuric acid as a byproduct. The process often involves extremely high temperatures as well. Silicone-based coatings currently appear to be the only materials capable of withstanding such conditions.

Special-purpose Coatings

These coatings include the use of materials (generic coatings) for purposes other than their contribution to protecting and decorating a structure or product. There is a multitude of such uses, and only a few can be described here because of space limitations.

Electronic Microstructure Fabrication. A microstructure may be defined as a pattern formed on or imbedded in the surface of some substrate material. Microstructure implies that the transverse dimensions of the patterns are in the microscopic and submicroscopic range, factors that determine the scale of electronic circuit integration. With the progress of electronic component miniaturization, such ranges have advanced from small-scale integration (SSI) to medium-scale integration (MSI) to large-scale integration (LSI) and to very-large-scale integration (VLSI). As of the early 1990s, such integration has progressed to millions and billions of transistors (e.g., on one integrated circuit). See also **Microelectronics**.

Patterns on microstructures may be formed in layers or insulators deposited on a surface or may consist of chemical or physical modification of shallow regions of the substrate. Traditionally, the most important use of microstructure fabrication has been the creation of large numbers of transistors, diodes, resistors, and capacitors fabricated with the interconnections that enable them to perform useful electronic functions on a single piece or "chip," usually silicon. Recent trends in miniaturization include two additional classifications (i.e., *application-specific* integrated circuits (ASICs) and *very-high-speed* integrated circuits (VHSICs).

Since the beginning of the concept of integrated circuits several years ago, the "yardstick" of dimension has been the micron (micrometer). A micrometer equals 1/1,000,000 of a meter, or 1/25,000 of an inch, or 1000 nanometers, or 10,000 angstroms. The wavelength of light, by comparison, extends from approximately 4000 to 7000 angstroms. Approximately 150 half-micron-wide lines would fit within the width of a human hair. The "resists" used in the fabrication of microminiaturized components fall into the "coatings" category in the parlance of modern coatings technology.

A *simplified* example will illustrate the process of microstructure fabrication. With reference to Fig. 1, an *n*-type region has been created by diffusion of a donor impurity into a surface of *p*-type silicon, forming a *p-n* junction diode. There is a metal contact to the *n*-region, and the contact line is insulated from the *p*-type surface by a layer of silicon dioxide. The diameter of the diode is on the order of 10 micrometers.

The fabrication begins with the application of a layer of photoresist to the oxidized surface of a silicon wafer. The photoresist is then exposed to light in the region where the diode is to be formed. Photoresist is a polymeric mixture that is deposited as a thin layer, perhaps 1 μm thick, upon an SiO_2 film on a silicon wafer. Irradiation with light in the near UV region of the spectrum modifies the chemical properties of the photoresist, and, in "positive" photoresist, makes it more soluble in certain developers. Thus, one step frequently employed in microstructure fabrication is the projection of the image of a mask onto the photoresist layer. It becomes possible to remove the exposed region of the photoresist by dissolving it with a suitable developer. The SiO_2 layer can then be removed from the areas that were exposed to light by hydrogen fluoride etches. The

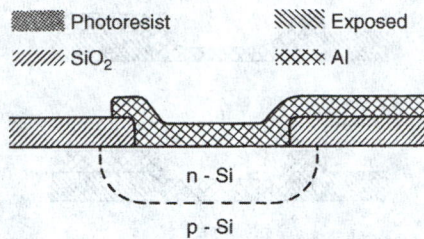

Photoresist   Exposed

SiO₂   Al

Fig. 1. A diode fabricated on the surface of a wafer of silicon. An *n*-type region has been created by diffusing a donor impurity through an opening in a layer of SiO₂ on the silicon. Electrical contact is made to the *n* region by a deposited aluminum conductor. The SiO₂ insulates the silicon from the aluminum.

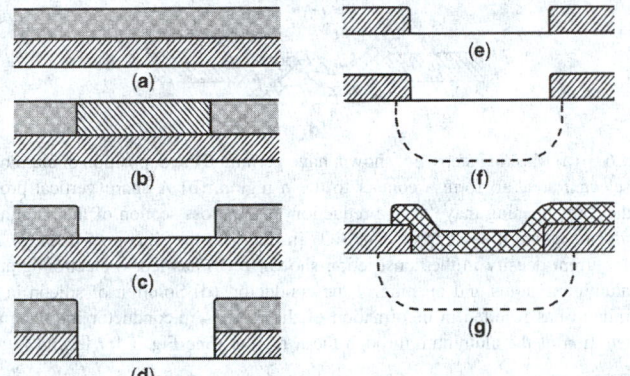

(a)

(b)

(c)

(d)

(e)

(f)

(g)

Fig. 2. Process steps used to produce the structure shown in Fig. 1: (a) A film of SiO₂ has been formed by oxidizing the silicon and a layer of photoresist has been deposited on the SiO₂. (b) Shading shows a region of the photoresist that has been exposed to light and thereby made more soluble. (c) The exposed photoresist has been removed. (d) An etchant that reacts with the SiO₂, but not with the photoresist, has been removed. (e) Another solvent has been used to remove the unexposed photoresist. (f) Donor atoms have diffused into the silicon through the opening in the SiO₂ to produce an *n*-type region. (g) Additional masking steps, not shown, have permitted aluminum to be evaporated onto the diode in a pattern that forms a contact to the n region of the diode. (See Fig. 1 for legend.)

photoresist is resistant to HF etches and the SiO₂ in the unexposed areas is not affected by the etch. After etching, the remaining photoresist can be removed by a solvent, leaving a silicon substrate covered with SiO₂ only in the unexposed areas. The SiO₂ film acts as a barrier to the contact of impurities in a gaseous phase with the silicon. Thus, when the silicon wafer covered by the patterned SiO₂ film is exposed, to, for example, a gas containing phosphorus at high temperatures, the phosphorus, being very soluble in silicon, diffuses into the exposed areas rapidly. An idealized description of this sequence of process steps is shown in Fig. 2. The effect of this doping is very important in electronics, since phosphorus is a donor impurity and a region of n-type or electron conductivity is produced where it is present.

Many physical phenomena, however, obstruct the formation of the ideal structure depicted in Fig. 2. The technical literature is well supplied with papers devoted to each of the steps illustrated in Fig. 2. None is as straightforward as appears at first sight. It is instructive to discuss them further, since much of the essence of microstructure fabrication is revealed by examining them in detail.

Figure 2(a) suggests that the thickness of the photoresist and SiO₂ layers are independent of position. While the layer thickness is not an extremely critical process parameter, its control cannot be entirely neglected, as the time needed for the subsequent developing or etching steps depend on it. The wafers used in modern silicon technology have diameters of three or more inches, and maintaining uniformity of layers and process parameters across a wafer is not a trivial task. Also, very high standards of cleanliness must be maintained, as any particulate contamination will affect the resist adversely.

Figure 2(b) shows a well-defined boundary between the exposed and unexposed areas of the photoresist. In fact, the dimensions of the structures

produced in modern microelectronics are comparable to the wavelength of the exposing light, so that diffraction prevents such sharp contrast from being achieved. Furthermore, high-resolution projection exposure schemes require that the light be monochromatic to avoid the problems of chromatic aberration in the lenses. The photoresist must be reasonably transparent to insure that its full thickness is exposed to the light. The silicon surface is, however, reflective, so that the interference of the incident and reflected light produces standing waves in the photoresist and nonuniform exposure of the photoresist in the vertical direction. Complicated effects of this kind are clearly important to microstructure fabrication. It must also be apparent that, as the amount of exposure received is a continuous function of position, the time of development required to remove a given region of photoresist will also be a continuous function, and that the profile of the developed photoresist will depend on the time of development. In particular, the size of the opening in the photoresist, to which Fig. 2 is oriented, will depend on the time.

Resists can also be exposed with focused electron beams, as used in electron microscopes, instead of light. The electrons, however, pass through the resist layer into the substrate, where they are scattered, and some eventually return from the substrate to the resist, exposing it at a distance from the intended opening. Great care is needed to allow for the backscattering phenomenon in calculating exposures for nearby openings. See **Electron Beam Lithography**.

Development of the photoresist proceeds somewhat as shown in Fig. 3, with simultaneous lateral and vertical removal of material. The tapered edge of the resist film may be a disadvantage, because the exact point at which the film is thick enough to protect the underlying SiO₂ layer during the succeeding etching step, and thus the size of the hole that will be produced in the SiO₂, is not clearly defined. Prolonging the development beyond the point shown in Fig. 3(b) allows continued lateral development and increase in the size of the opening. Also, however, achieving perfect adhesion of the photoresist film to the SiO₂ is difficult, and the developer may invade the interface between the two layers, producing the undesirable result shown in Fig. 3(d).

The developed photoresist, Fig. 2(c), is then used as a mask to etch the SiO₂ layer. Again, perfection is hard to achieve. Etching for too short a time will leave a certain amount of photoresist in the hole. Etching for too long a time can cause undercutting, as shown in Fig. 4(c). After removal of the photoresist, the wafer is exposed to a diffusant, affording additional opportunities for deviations from idealized behavior. Time and temperature of diffusion are important and can produce results resembling Figures 5(a)

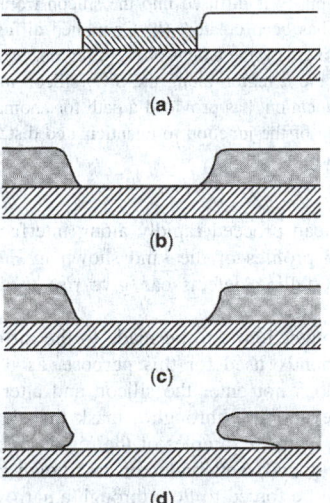

(a)

(b)

(c)

(d)

Fig. 3. Development of the photoresist: (a) Although exposure increases the solubility of photoresist in the developer, there is only a finite ratio of dissolution of the exposed photoresist to that of the unexposed region. In addition, the exposure received is not a perfect step function at the boundary of the exposed areas. Thus, dissolution proceeds laterally as well as vertically. (b) The opening in the photoresist has penetrated to the surface of the SiO₂. (c) With continued development the opening continues to enlarge. (d) Poor adhesion of the photoresist to the SiO₂ has allowed the development to penetrate the interface. (See Fig. 1 for legend.)

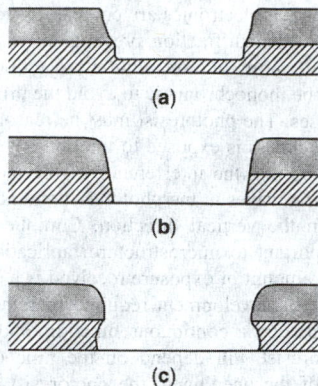

(a)

(b)

(c)

Fig. 4. The opening in the photoresist is used to mask the etching of a hole in the SiO$_2$: (**a**) An early stage of development. (**b**) The opening in the SiO$_2$ has reached the silicon. (**c**) Prolonged etching can lead to removal of SiO$_2$ underneath the photoresist masks. (See Fig. 1 for legend.)

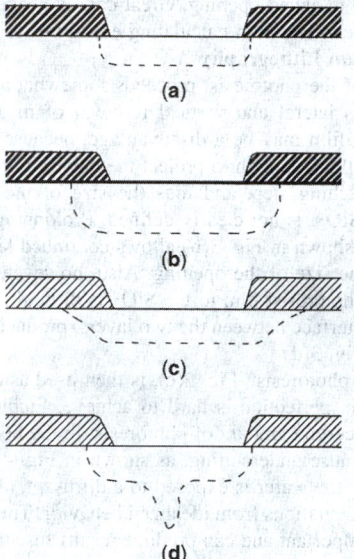

(a)

(b)

(c)

(d)

Fig. 5. A donor impurity is diffused into the silicon from a gaseous phase. (**a**) A shallow n region has been created. (**b**) Continued diffusion, longer times, or higher temperatures increase the extent of the n region. (**c**) Surface diffusion has caused spreading of the n region along the SiO$_2$-silicon interface. (**d**) A crystal defect, such as a dislocation, has provided a path for anomalously high diffusion and led to penetration of the junction to unanticipated distance from the surface. (See Fig. 1 for legend.)

or 5(**b**). Diffusion can proceed rapidly along interfaces in certain cases, leading to junction profiles of the kind shown in Fig. 5(**c**). Preferential diffusion along crystalline defects can give rise to a profile resembling that shown in Fig. 5(**d**).

Next, a metal connection is to be made to the diffusion-doped region. Aluminum is frequently used for this purpose, as it has high electrical conductivity and does not enter the silicon and alter its properties. The aluminum is also evaporated through a mask that defines the shape and location of the conductor. Examples of the region of contact between the aluminum and the doped semiconductor are shown in Fig. 6. It is seen that the current will be forced to flow through a narrow constriction if the profiles are as shown in Fig. 6(**b**). The high current densities may cause electromigration of the aluminum atoms, leading to the open circuit shown in Fig. 6(**c**).

Also, silicon is somewhat soluble in aluminum. One can thus encounter the situation shown in Fig. 6(**d**), where enough silicon has been dissolved to allow the metal to completely penetrate the doped region, shorting the junction.

None of the problems illustrated in Figs. 2, 3, 4, 5, and 6 is insurmountable. A great many ingenious ways to avoid the difficulties

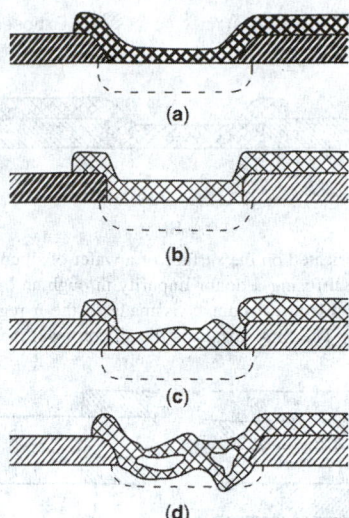

(a)

(b)

(c)

(d)

Fig. 6. (**a**) Masking steps, not shown, have permitted the deposition of aluminum in selected areas to form a contact to the n region. (**b**) A sharp vertical profile in the SiO$_2$ opening may cause a reduction of the cross section of the aluminum conductor where it passes from the SiO$_2$ insulator to the silicon surface. (**c**) The high current density in the constriction shown in (**b**) has led to electromigration of aluminum atoms and opening of the conductor. (**d**) Solution of silicon in the aluminum has resulted in deformation of the metal-semiconductor interface and penetration of the aluminum through the n region. (See Fig. 1 for legend.)

described are known. Sometimes these are guided by physical or chemical knowledge; frequently they are empirical fixes.

The microstructure engineer must also be aware of constraints that have little to do with chemistry and materials science, but tend to be more closely related to mechanical technology. One of the most difficult of these is registration or alignment. A substrate is usually passed through several process steps in order to form a desired structure. For example, fabricating a transistor may involve diffusing a base dopant through a window in SiO$_2$, subsequently diffusing an emitter dopant through a somewhat smaller opening in an SiO$_2$ layer, and finally, using masking layers to form contacts to these transistor elements. It is necessary to ensure that the emitter region is created in the correct position within the previously diffused base region. The mask that is used to define the emitter region lithographically must be precisely located with respect to the geometrical structure already established on the substrate by previous processing steps. This positioning is known as alignment. It may require that separate structures that can be easily located but have no electronic function be provided on the substrate.

Alignment all over a large substrate is made more difficult by dimensional changes that may take place during processing at high temperatures. Materials soften at high temperatures and can deform under the force of gravity or the stresses that accompany temperature gradients and contacts between different materials.

Further, economic factors also constrain the utilization of microstructure fabrication technology. These are the factors that control the cost of production, such as throughput, the rate at which substrates can be processed by the fabrication tools, capital investment required, and demands on operator time and skill. Electron beam exposure, for example, provides high resolution but uses expensive equipment that works slowly. Naturally, all of the elements of cost must be weighed against the value of the product produced.

Chemical Vapor Deposition. Deposition of tungsten, molybdenum, and their silicides by chemical vapor deposition (CVD) is of relatively recent interest in the microelectronics industry. These materials are useful for gates and interconnects in metal oxide semiconductors (MOS) devices. Aluminum, the widely used interconnect material, has a comparatively low melting point (600 °C) and a markedly different coefficient of thermal expansion (compared to silicon), so that over a period of years researchers have been seeking an alternative for aluminum.

In using CVD for microelectronics applications, the deposit thickness must be as uniform as possible. This can best be achieved by conducting the deposition in a surface-controlled, not a diffusion-controlled, regime.

In this way, effects of reactor geometry on deposition rates are minimized and irregular-shaped substrates will tend to be uniformly coated.

An allied process is *metal-organic chemical vapor deposition* (MOCVD). In this process, two or more metal-organic chemicals (example: trimethylgallium) or one or more metal-organic sources and one or more hydride sources (example: arsine, AsH$_3$) are used to form the corresponding intermetallic crystalline solid solution. MOCVD materials technology is a vapor-phase growth process that is used to study the basic physics of novel materials and to grow complex semiconductor device structures, particularly for new optoelectronic and photonic systems. The process is reported in some detail by Dupuis.

Ion Implantation. In this process, alloying elements are introduced into a host material by accelerating the ions to a high energy level and allowing them to strike the surface of the host. The impinging atoms penetrate into the substrate material to a depth of one micrometer or less, depending upon atomic number and energy of the atom. Although it has a number of other applications metallurgically, the process has been of interest in the semiconductor industry primarily in connection with doping substrates with elements in Periodic Table Groups 13 and 15. Use of the process in fabricating a Schottky barrier gate used in a metal-semiconductor field-effect transistor (MESFET) is described in the article on **Semiconductors**. See also **Ion Implantation**.

Multidisciplinary Characteristics of Microstructure Fabrication. Upon observing the practice of microstructure fabrication, one cannot fail to notice a resemblance to certain aspects of modern metallurgy and chemical engineering. For example, the precipitates produced by metallurgical processing have a dimensional scale similar to that of electronic microstructures. Inhomogeneities on a scale of 0.01 ϕm to 10 ϕm control the desirable properties of a structure. This preoccupation with the properties of solids on a microscopic scale produces a common interest in techniques and in interactions with basic science. Thus, both microstructure fabrication and physical metallurgy: (a) rely on phenomena that take place in the solid state; (b) depend on analytical tools that are capable of chemical analyses with the highest possible spatial resolution; (c) involve the motion of atoms through solids, controlled by diffusion, solution, nucleation, and precipitation; (d) involve interface phenomena at the contact between different solids; and (e) are sensitive to crystal defects.

It must further be noted that both metallurgy and microstructure fabrication are practical disciplines, they are oriented toward the economic production of structures that have a useful role in commerce and industry. In this respect both are engineering rather than scientific disciplines. On the other hand, their deep probing of phenomena on an atomic scale and under unusual conditions produce new discoveries and lead to new concepts that enhance basic science.

This is not to say that microstructure fabrication is a branch of metallurgy. The detailed motivation of the two disciplines is rather different, metallurgy concentrating on the mechanical properties of solids, while microstructure fabrication controls the electronic properties of structures made from magnetic and optical materials, semiconductors, metals, and insulators. The basic difference, of course, is that microstructure fabrication involves control of the fabrication process in detail at the dimensional level of the structure, while metallurgical processing exercises control at a much grosser level. The application of lithography, with the attendant use of exposure tools, clean rooms, resists, masks, and etchants is the province of microstructure fabrication. Crystalline defects are usually undesirable in microstructures; the metallurgist can frequently use them to advantage. Metallurgy also encompasses its extractive aspects. There is no doubt that microstructure fabrication is a distinct activity.

The technique of microstructure fabrication has grown up as an art in response to a continuous economic and functional motivation to push to the smaller and smaller. Chemistry, physics, and empiricism are combined into the creation of novel physical structures that have enormous economic impact, that, indeed, form the basis of whole new industries. Progress has been made by adaptation and invention to meet the needs of the moment. By and large, the art has grown rapidly, adapting and improving old methods to new situations and inventing as needed and possible. It is common experience that obtaining reproducible results requires very careful control of all aspects of the fabrication processes, such as temperatures, pressures and time. High standards of cleanliness and reagent purity must be enforced. Seemingly identical apparatuses and starting

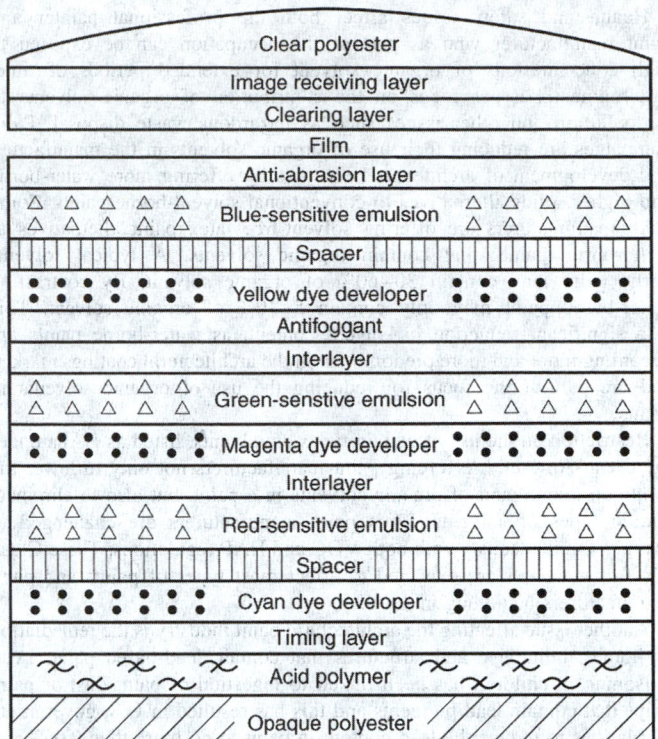

Fig. 7.   Schematic sectional view of layers composing the "thickness" of a color photographic film. (*After Cohen.*)

materials often yield different results. Recipes must be carefully followed, with little understanding of which aspects of a process are critical or what contaminants are important, and why.

The rapid development has outstripped basic understanding of the fundamental mechanisms underlying the techniques. It must be recognized, however, that the optimal exploitation of microstructural technology, the maximization of performance, yields, and utilization of available silicon area will depend on a detailed interpretation of each step in fabrication as a chemical or physical process. Furthermore, unanticipated phenomena are encountered and inject surprises into basic science. A new interdisciplinary field of applied science has emerged.

Photographic Color Film
One of the early examples of coating technology was for building up layers in color film. These films exemplify the use of multiple layers of coatings that are "built in" to the final product. In use, the camera lens also will be coated. See Fig. 7. See also **Photography and Imagery**.

Adhesives. Commonly, adhesives are applied by coating products and parts that must be held together. Some adhesives require application at the time of usage; others are pre-applied in semi-dry or dry form, but with a "sticky" surface. Adhesives range from natural products, such as glue to plastics and polymers. The epoxies are exemplary of modern adhesives. With the continuing replacement of many metal parts with plastics, adhesives are displacing welding and mechanical fasteners to a dramatic degree. In the final essence, most adhesives must form a binding coating on one form or other. See also **Adhesives**.

Environmental, Health, and Safety Factors
The most significant environmental and health issues affecting the paint and coatings industry are regulations to lower the VOC content for virtually all types of paints and to restrict the use of certain solvents known as hazardous air pollutants (HAPs) under the federal Clean Air Act. Except for the water in a latex paint or in other water-based coatings, solvents used in house paints are mostly all VOCs. Several states, along with the U.S. EPA, have implemented environmental regulations to restrict the VOC content of paints, as mandated by the Clean Air Act. These regulations are aimed at minimizing the emission of organic compounds from paints that contribute to the formation of air pollution in the form of smog or ground-level ozone.

Health and safety issues affect both the professional painter and paint manufacturer who as part of the occupation can be exposed to high concentrations of organic solvent for extended periods of time. Environmental issues focus on the contribution of organic solvents to air pollution and other issues such as hazardous waste disposal. Paint companies are reducing their use of organic solvents in the manufacture and development of architectural coatings by offering more water-borne and higher solids alternatives to conventional solvent-borne paints. Some paint manufacturers are offering solvent-free latex paint alternatives to water-borne paints that contain organic solvents. A typical solvent-borne paint may contain 30–60% of organic solvent. By contrast, a water-based paint may only contain 5–10% of organic solvent. This is a significant reduction in solvent content, as water-borne paints are becoming more and more predominant in the architectural coatings market, and are part of the focus for reducing the use of organic solvents in paints.

Restriction on the use of certain types of solvents, listed as HAPs under the Clean Air Act, are forcing paint manufacturers not only to lower the limits on the amount of organic solvents in a paint, but also to eliminate certain types of solvents. Thus paint manufacturers are challenged to comply simultaneously with both VOC and HAP regulations. These Clean Air Act mandates are expected to affect most types of paints and paint manufacturers beginning in 1996.

Another issue affecting the architectural paint industry is the remediation of homes, buildings, and structures that contain lead-based paint. Lead poisoning in children has been linked to ingestion of paint dust or paint chips that contain lead pigments and this has resulted in U.S. government regulations to reduce the lead content in paint to no more than 0.06%.

Additional Reading

Anderson, D.G.: "Coatings (Analysis of)," *Analytical Chemistry*, **87R** (June 15, 1991).

Bieleman, J.: *Additives for Coatings*, John Wiley & Sons, Inc., New York, NY, 2000.

Champ, N.A. and F.L. Lowenstein: "TBT: The Dilemma of High-Technology Antifouling Paints," *Oceanus*, 69 (Fall 1987).

Cohen, E.D.: "Coatings: Going Below the Surface," *Chem. Engineering Progress*, 19 (September 1990).

Cohen, E.D. and E.J. Lightfoot: "A Primer on Forming Coatings," *Chem. Engineering Progress*, 30 (September 1990).

Coyle, D.J., C.W. Macosko, and L.E. Scriven: "Fluid Dynamics of Reverse Roll Coating," *Amer. Inst. of Chem. Eng's J.*, 16 (February 1990).

Dupuis, R.D.: "Metalorganic Chemical Vapor Deposition of III — V Semiconductors," *Science*, **226**, 623–629 (1984).

Deligny, P., P.K.T. Oldring, and N. Tuck: *Resins for Surface Coatings, Alkyds & Polyesters*, Vol. 22, John Wiley & Sons, Inc., New York, NY, 2001.

Elliott, D.: *Microlithography: Process Technology for IC Fabrication*, The McGraw-Hill Companies, Inc., New York, NY, 1986.

Fettis, G.: *Automotive Paints and Coatings*, John Wiley & Sons, Inc., New York, NY, 1995.

Finzel, W.A.: "Use Low-VOC (Volatile Organic Compounds) Coatings," *Chem. Engineering Progress*, 50 (November 1991).

Florio, J., and D.J. Miller: *Handbook of Coatings Additives*, 2nd Edition, Marcel Dekker, Inc., New York, NY, 2004.

Freitag, W. and D. Stoye: *Paints, Coatings and Solvents*, 2nd Edition, John Wiley & Sons, Inc., New York, NY, 1999.

Glass, J.E.: *Technology for Waterborne Coatings*, Vol. 663, American Chemical Society, Washington, DC, 1997.

Herman, H.: "Plasma-Sprayed Coatings," *Sci. Amer.*, 112 (September 1988).

Lambourne, R.: *Paint and Surface Coatings: Theory and Practice*, Prentice-Hall, Inc., Upper Saddle River, NJ, 1993.

Marrs, J.M.: "Ultraviolet Light for Photochemical Deposition," *Chem. Eng. Progress*, 31–34 (January 1986).

Satas, D. and A. Tracton: *Coatings Technology Handbook*, 2nd Edition, Marcel Dekker, Inc., New York, NY, 2000.

Schweizer, P.M.: "Visualization of Coating Flows," *J. Fluid Mechanics*, **193**, 285 (1988).

Scriven, L.E. and W.J. Suszynski: "Take a Closer Look at Coating Problems," *Chem. Engineering Progress*, 24 (September 1990).

Shumay, W.C., Jr.: "High-Performance Adhesives on the Line," *Adv. Material & Processes*, 82 (August 1987).

Staff: *Paints Related Coating and Aromatics*, American Society for Testing & Materials, West Conshohocken, PA, 1998.

Sturge, J.M., V. Walworth, and A. Shepp: *Imaging Processes and Material: Neblette's*, 8th Edition, John Wiley & Sons, Inc., New York, NY, 1997.

Tobiason, T.L.: "Choosing a Finish for Your Electronic Package," *Instruments & Control Systems*, 85 (May 1990).

Various: 6th International Coating Process Science and Technology Symposium Papers, *Amer. Inst. of Chem. Engineers*, New York, NY, 1992.

Weldon, D.G.: *The Failure Analysis of Paints and Coatings*, John Wiley & Sons, Inc., New York, NY, 2001.

Web References

Benjamin Moore & Company, http://www.benjaminmoore.com/

Duron Paints, http://www.duron.com/products.html

Krylon Products Group, http://www.KPG-Industrial.com

PPG Architectural Finishes, Inc. http://www.ppg.com/ppgaf/pittsburgh/index.htm

Rust-Oleum Corp, http://www.rustoleum.com/brand.asp?frm_brand_id=14&SBL=1

The Sherwin-Williams Company, http://www.sherwin-williams.com/pro/

PAIR PRODUCTION. The conversion of a photon into a negatron and a positron when the photon traverses a strong electric field, usually that surrounding a nucleus but occasionally that of an electron. The electric charge is conserved, since the negatron and positron have charges of equal magnitude and opposite sign.

PALEOBOTANY. This is the study of ancient plants known today only through fossil remains. These are of two general types: prints and petrifactions. Much of our knowledge of fossil plants has been derived from the study of petrifactions in which the original structure of the plant has been replaced by mineral material.

Paleobotany yields much interesting information about the plants of prehistoric times, notably of the Carboniferous and Devonian periods. The best known fossil plants are those which grew during the Carboniferous period of the Paleozoic era. The many types of coal are goechemical compounds and mixtures of plants, including their spore cases, oxidized cellulose (mother of coal), lignite, and bacterial byproducts. The nature of the processes by which coal has been formed is such that coal is often not the best material to study to get a picture of the fossil plants composing it. But found associated with the coal are mineral masses called coal balls. Within these, plant remains are often preserved with beautiful detail. Microscopic studies of peels and thin sections of coal balls have made it possible for the paleobotanist to reconstruct many of the Carboniferous plants.

PALEOCENE. The earliest period in the Cenozoic Era of the geologic time-scale. It is also spoken of as basal Tertiary. The term was first proposed by Schimper in 1874. The greatest thickness of the formations of this system occur in Wyoming. The Paleocene began approximately 60 million years ago and lasted for about 10 million years. In the United States, the sediments are mainly of terrestrial origin, occurring as intermontane deposits of sands, gravels, and clays. The Paleocene is chiefly remarkable in that the dinosaurs have been replaced by the archaic or earliest types of mammals.

PALEOCLIMATE (OR GEOLOGICAL CLIMATE). Climate for periods prior to the development of measuring instruments, including historic and geologic time, for which only proxy climate records are available.

PALEOCLIMATIC SEQUENCE. The sequence of climatic changes in geologic time. It shows a succession of oscillations between warm periods and ice ages, but superimposed on this are numerous shorter oscillations. The tendency to regard the whole of a geologic period, lasting for 20 million years and more, as having a single type of climate is a great oversimplification, as is shown by the succession of glacial and interglacial periods in the Quaternary. Even the warm periods are known to have been made up of successions of climates of different degrees of warmth; but until much more information is available, it will not be possible to set out in detail the sequence of changes of the earlier paleoclimates.

PALEOCLIMATOLOGY. Paleoclimatology is the study of climates in the Earth's past. The methods used to study younger (Quaternary) and older (pre-Quaternary) climates are similar, but climates can usually be studied in the Quaternary with much higher resolution. Pre-Quaternary paleoclimatology is more of a challenge, not only because the rocks have been exposed to erosion and tectonics longer but also because the continents have moved. See also **Quaternary Period**; and **Ice Ages**.

The same computer models that have been used to predict weather and climate changes in the future have been adapted to the study of paleoclimates. These have become important tools in palaeoclimatology because they provide hypotheses about climate that can be tested with data collected from the geological record.

Climate Models

Numerical climate models have become increasingly sophisticated in the last 20 years, as has their application to the study of paleoclimates. The models are adaptations of those written originally for weather prediction. The foundations for such models are the equations for conservation of momentum, mass and energy, the hydrostatic equation, and the atmospheric equation of state (gas law). These are calculated from specified starting conditions for cells and their neighbors in a three-dimensional grid, and the more complex the grid, the larger the number of calculations. A typical grid is 4° latitude by 7° longitude and 12 layers deep, resulting in a large enough number of calculations that, only with the advent of supercomputers, was the use of such models practicable. Paleoclimate modeling studies are, in some ways, more difficult than modeling studies of the modern atmosphere because the starting conditions are not always known with great accuracy and assumptions that might be made for the modern atmosphere cannot necessarily be made for the ancient one. See also **Climate**.

The grid that can be handled even with modern supercomputers is much coarser than many atmospheric processes. For example, thunderstorms, which are important for convection (that is, vertical heat transfer), operate on horizontal scales of a few tens of kilometers (miles), much less than 4° latitude or even the 1° latitude of higher-resolution models. Many other processes also operate on subgrid scales. Thus, climate modelers must parameterize these processes, which means that the effects of the processes must be assumed mathematically before the models are run. Differences in the results among different climate models are largely the result of different assumptions, and much of the most active research in climatology is focused on improving our understanding of subgrid-scale processes in order to improve parameterization.

The climate models are modified and tested based on evidence of past climates. This evidence comes in a variety of forms, some quantitative, most qualitative. Even the qualitative evidence is useful, however, because it can still provide information about trends and amplitudes of climate change.

Kinds of Evidence of Paleoclimate

Information about palaeoclimate comes from the record of rocks, fossils and, for the Holocene, ice caps. The types of rocks and fossils that provide the most information on palaeoclimate are listed in Table 1. For both, simply where they are located can tell a lot about what the climate was like. For example, salt deposits—called *evaporites* by geologists—form by evaporation of sea or lake water and are indicative of warm, dry climates or hot climates that have only a short rainy season. Deposits formed by ice action occur only where the climate was cold. Although ice can form on mountains, the chances that such glacial deposits will be preserved are small because mountains are erosional environments. Therefore, glacial deposits tend to occur in higher latitudes, and how far they extend toward the equator can be an indication of how large the ice caps were and, by extension, how cold global climate was.

Although a few individual types of plants and animals are indicative of specific palaeoclimates, more often the information about climate comes from the organisms' biogeography, that is, their geographic distribution. For example, some marine diatom species live only in cold water whereas others live only in warm water, while still others prefer waters of intermediate temperature. By mapping out the distributions of different groups of diatoms through geological time, geologists can get a sense of how the distribution of temperature changed in the oceans through time. Even if the species are extinct, their temperature tolerances can be inferred from the patterns, so that at least relative changes in temperature can be studied. Many groups of organisms exhibit latitudinal diversity gradients, in which the abundance of species decreases with increasing latitude. The

TABLE 1. TYPES OF FOSSILS AND ROCKS USED TO INTERPRET PALEOCLIMATE (THE SPECIFIC SOURCE OF INFORMATION FOR EACH IS SHOWN IN PARENTHESES)

Fossils
Marine:
    Foraminifera (morphology, biogeography, stable isotopes)
    Calcareous nannofossils (biogeography)
    Marine ostracods (biogeography)
    Marine diatoms (biogeography, stable isotopes)
    Dinoflagellates (biogeography)
    Brachiopods (biogeography, stable isotopes)
    Molluscs (morphology, biogeography, stable isotopes)
    Corals (biogeography, stable isotopes)

Terrestrial:
    Pollen (biogeography)
    Nonmarine ostracods (morphology, stable isotopes)
    Nonmarine diatoms (biogeography)
    Plants (biogeography, morphology)
    Freshwater and terrestrial molluscs (biogeography, stable isotopes)
    Vertebrates (biogeography, morphology)

Rocks
Marine:
    Chalk (distribution, chemistry)
    Biogenic siliceous rock (distribution)
    Phosphorites (distribution)
    Organic matter and organic-rich rocks (distribution, chemistry)
    Reefs (distribution)
    Clay minerals (distribution, type)

Terrestrial:
    Wind-blown deposits (distribution)
    Salt deposits (distribution, type)
    Ancient soils (distribution, type)
    Lacustrine deposits (lake level, type)
    Glacial deposits (distribution) and ice (chemistry)

steepness of the diversity gradient can be an indication of the steepness of the latitudinal temperature gradient.

Some organisms also have morphologies that are adaptations for particular climates. Perhaps the most useful of these is the different forms of the leaves of flowering plants, which vary with temperature and rainfall. The proportion of species with entire (smooth) leaf margins to those with non-entire (serrated, toothed, or lobed) margins is closely related to the mean annual temperature where the plants were growing. Size is also related to climate. The leaves of plants in tropical rainforests tend to be large. At the other extreme, leaves of plants in deserts tend to be very small (or absent). Many other morphological adaptations of leaves are also useful. The major margin types of leaves of flowering plants had evolved by the Late Cretaceous, so this method is useful from that time to the present.

Paleoclimatologists who study the Quaternary record have an advantage using fossils in that many species found in the Quaternary record are still living today. This means that paleoclimatologists can look at the climates in which the modern species are living and directly infer the same climate where those species occurred in the past. In contrast, most species of organisms in the older record are extinct. Paleoclimatologists sometimes use what is called the nearest-living-relative method, in which the climatic tolerances of the closest relative are taken as the climatic tolerances of the extinct species. This works well for a few organisms, but is not reliable in most cases. See also **Ice Ages**.

An important tool in paleoclimatology is the use of stable isotopes, particularly of oxygen and carbon. Shells of aquatic animals record the isotopic composition of the waters in which the animals live; in addition, marine limestones record the isotopic composition of sea water. The shells and limestones are composed of calcite ($CaCO_3$), and the carbon in the calcite is mostly ^{12}C, the most common carbon isotope, with some ^{13}C. The oxygen in the calcite is mostly ^{16}O, the most common oxygen isotope, with some ^{18}O.

The ratio of ^{18}O to ^{16}O in sea water is directly related to the volume of the ice caps. The water in the ice caps comes from the oceans, and the water molecules in the ice caps are rich in ^{16}O. This means that the greater the ice-cap volume, the less ^{16}O is left in the rest of the ocean. In addition, the ratio of ^{18}O to ^{16}O in limestone and shells is also related to the temperature of the sea water; if the sea water is warm, the ratio will be lower.

Foraminifera (Table 1) are particularly good recorders of the isotopic composition of sea water, and thus many paleoclimatologic studies use the stable isotopic composition of foraminiferan shells to understand the fluctuations in the ice caps and changes in ocean temperatures. The use of oxygen isotopes in terrestrial animals is more complicated because evaporation, rainfall, and a number of other processes can affect the isotopic composition of terrestrial waters. See also **Foraminifera**.

In the oceans, the stable isotopes of carbon can provide information about the productivity of the overlying waters and large-scale changes in the carbon cycle; on land, carbon isotopes have been useful for understanding fluctuations in the atmospheric concentration of the greenhouse gas carbon dioxide and the ecology of the plants that lived on the land surface. Organic matter is preferentially enriched in ^{12}C compared with the total global carbon reservoir. In the oceans, high productivity in the surface waters will result in the transport of greater amounts of organic matter to the ocean bottom. This means that the remaining carbon in the surface waters becomes more and more enriched in ^{13}C and the shells of organisms living there, such as foraminifera, will also be more enriched in ^{13}C. By examining the change in the carbon isotopic composition of these shells through time, changes in productivity can be determined.

On land, carbon isotopes are studied from limestone nodules that form in some soils. These nodules are formed partially from carbon dioxide respired by plant roots. The types of plants, which can be related to climate, can sometimes be determined from the carbon isotopic composition of the nodules. Where the type of plant is constant, the carbon isotopic composition of the soil nodules can provide information about the atmospheric component of carbon dioxide in the soil, which is related to the atmospheric concentration of the gas.

Case Studies

Pollen studies are very important for understanding continental climate in the Quaternary era. Pollen are commonly taken from cores in lake sediments. In the younger part of the era, it is even possible to study climate change that is expressed along altitudinal gradients because mountain building is rarely significant over such short time scales. Thus pollen from lakes at different altitudes can provide a third dimension—elevation—to the geographic picture of climate change. Plant macrofossils, that is, the remains of fragments or organs such as leaves, are not used as commonly in Quaternary studies as in the older geological record because pollen are more common, easier to sample, and easily related to living taxa. An exception to this rule is the plant macrofossils found in packrat middens. Packrat middens have proved to be a rich source of palaeoclimatic information for the younger record. The importance of packrat middens is that they are more densely distributed than lakes in dry areas such as the Great Basin of the western United States and they occur even at high elevations. See also Pollen: Structure, Development and Function.

Packrat middens are accumulations of crystallized urine, faecal pellets and plant parts brought in by the packrats. They are commonly stratified, such that the oldest material is on the bottom and the youngest on the top, but the patterns within each midden can be determined by radiocarbon dating. Although packrats are North American, small mammals on other continents, such as hyrax in the Middle East and stick-nest rats in Australia, build similar structures.

Packrat middens are common in the Mojave Desert, an extremely arid region of California. Through detailed radiocarbon dating and study of the plant fragments in middens from the area, workers such as G. R. Spaulding have found that juniper woodlands, which today are restricted to high mountain tops, occurred at much lower elevations before about 8000 years ago. For example, in south-central Nevada, *Juniperus osteosperma* is uncommon below 1,700 m (5,577 ft), but occurred at elevations as low as 800 m (2,625 ft) until 13 000 years ago. Similarly, limber pine (*Pinus flexibilis*) occurs above 2,600 m (8,530 ft) today, but as low as 1,600 m (5,249 ft) until 13 000 years ago. The lowland vegetation called *desertscrub*, which today occurs to elevations as high as 1,600 m (5,249 ft), was restricted to elevations below 1,000 m (3,281 ft) until about 13 000 years ago.

A paleoclimatologic study from the much older record, that of the Triassic period shows how using a variety of data can result in a quite clear picture of climate even in the distant past. Mutti and Weissert 1995, studying rocks in the Italian Alps, used oxygen isotope data from carbonate platforms—large deposits of shallow-water marine limestone—along with other information. The deposit contains a series of high-relief surfaces, called karst, that indicate that the limestone was partially dissolved at times before deposition resumed. The sequence is also interrupted by river and salt deposits.

The existence of karst was further evidence of high rainfall because so much limestone had been dissolved; the karst surface shows that the relief on the dissolved surface was as much as 120 m (394 ft). See also **Hydrology**. In addition, some of the limestone showed features that indicate that the carbonate platform dried out. These features alternated with karst, and Mutti and Weissert interpreted this to mean that climate had alternated between wet and dry. The river deposits were of a type, which occur in regions with high but seasonal rainfall, and they contain material that was probably formed in soils under alternating wet and dry climates. Thus all the evidence compiled by Mutti and Weissert led to the same conclusion: a warm climate that was highly seasonal with respect to rainfall.

Major Changes in Climate in Earth History

Caution must be used in making broad statements about global climate in that changes observed in one part of the world may or may not indicate global changes. The major changes that affect global climate are waxing and waning of the ice caps. Major glacial episodes were in the Late Proterozoic, the end of the Ordovician period, the later Carboniferous period to the beginning of the Permian period, and the Pleistocene. Not surprisingly, these were the coolest intervals in Earth history.

Two times of strong monsoonal climates have also characterized palaeoclimate since the end of the Proterozoic. Monsoonal climate is characterized by reversal of circulation between winter and summer and by concentration of precipitation in the summer months. During the Late Carboniferous to the Early Jurassic, when much of the continental area was aggregated into a single, large continent called Pangea, much of the continent was affected by a large-scale monsoonal circulation. About 8 million years ago, the strong Asian monsoon began; this circulation affects climate in southern Asia, most of Africa, and the Indian Ocean. See also **Climate**; and **Geologic Time Scale**.

Additional Reading

Bradley, R.S.: *Paleoclimatology: Reconstructing Climates of the Quaternary*, Academic Press, London, UK, 1999.

Hoefs, J.: *Stable Isotope Geochemistry*, Springer-Verlag New York, LLC, New York, NY, 2004.

Maher, B.A., and R. Thompson: *Quaternary Climates, Environments and Magnetism*, Cambridge University Press, New York, NY, 2000.

Mutti, M., and H. Weissert: "Triassic Monsoonal Climate and its Signature in Ladinian-Carnian Carbonate Platforms (Southern Alps, Italy)," *Journal of Sedimentary Research*, **B65**, 357–367 (1995).

Parrish, J.T.: *Interpreting Pre-Quaternary Paleoclimate from the Geologic Record*, Columbia University Press, New York, NY, 1998.

Spaulding, W.G.: "Vegetational and Climatic Development of the Mojave Desert: The Last Glacial Maximum to the present," In: Betancourt, J.L., T.R. Van Devender, and P.S. Martin eds., *Packrat Middens*, University of Arizona Press, Tucson, AZ, 1990 pp. 166–199.

Wolfe, J.A.: "Paleoclimatic Estimates from Tertiary Leaf Assemblages," *Annual Review of Earth and Planetary Sciences*, **23**, 119–142 (1995).

JUDITH TOTMAN PARRISH,University of Arizona, Tucson, AZ

PALEOGNATHAE (Flightless Birds). Current evolutionary consensus suggests that there are two broad subdivisions within living birds (Neornithes), respectively neognathous birds (Neognathae) and paleognathous birds (Paleognathae). Paleognaths, the subject of this article, are largely characterized by their primitive skull morphology. The group includes a number of well-known flightless birds, such as the ostrich, rhea, emu, cassowaries and kiwis as well as the small, flighted tinamous. See also **Ostrich (Aves, Struthioniformes; Struthio Camelus)**.

Basic Design

For almost 200 years, anatomists have recognized the presence of two broad subdivisions within living birds, the modern flying birds

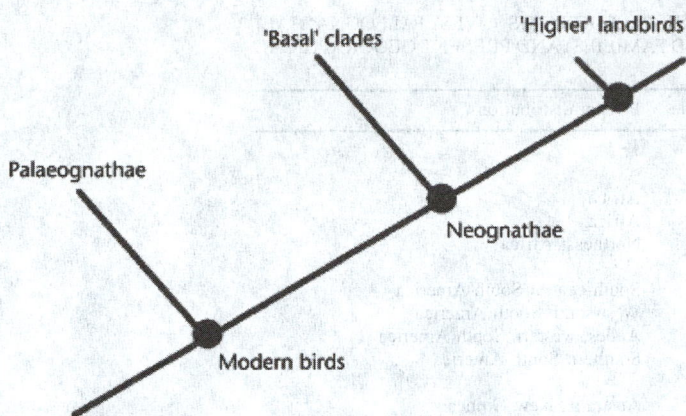

Fig. 1. The broad subdivision within living birds into the Paleognathae and Neognathae (see text for details).

(Neognathae) and their large, flightless relatives (Paleognathae, Figure 1). Paleognaths have a primitive internal skull (palatal) morphology (see below), characteristic of this group. Today, the group comprises a number of well-known flightless birds, the ostrich, rhea, cassowary, emu, kiwi, as well as the weakly flighted tinamou. Excluding tinamous, considered the most primitive of the recent paleognaths, the remaining species are often grouped together and referred to as ratites (Fig. 1). See also **Birds**.

A close evolutionary relationship among the living paleognathous birds was first suggested by the Darwinian biologist T. H. Huxley in 1867; all these birds have very immobile palates (the underside of the skull), including a number of bones firmly sutured together. All other groups of modern birds (Neognathae) possess a lighter and much more mobile skull. Within Paleognathae, a diagnostic feature of the ratites is their breastbone that lacks a keel, the bony process used for the attachment of flight muscles, although this does vary within the group. Ostriches, for example, have an extremely curved sternum compared with the flattened sternum of the kiwi. This point is illustrated still further when the fossil ratites such as the moa (New Zealand) and elephant bird (Madagascar) are examined, they possess wide, flattened sternae. The structure of the pelvis is also often cited as an indication of ratite inter-relationships–these birds all possess a pelvis with all three parts separate, unfused and open-ended. See also **Huxley, Thomas Henry (1825–1895)**.

Diversity

Although the paleognaths warrant their own separate division within Neornithes (Paleognathae), on the whole this is not a very speciose (diverse) group. According to current ornithological classifications, only a single species of living ostrich (family Struthionidae), two species of rhea (family Rheidae), four cassowaries and emus (family Casuariidae) and three kiwis (family Apterygidae) are recognized. By far the most diverse family of recent paleognaths are the tinamous (family Tinamidae) of Central and South America, of which over 40 species have been named Table 1. Regarding the extinct taxa, it was once thought that there were as many as 38 species of moa (family Dinornithidae) in New Zealand, but over the last 25 years, this has been condensed to just 11. Indeed, new research involving nuclear DNA sequences (examining the exact order of base pairs in a segment of DNA) has suggested that the true number of moa species may in fact be far less even than recent estimates. Sexual size dimorphism must also be considered as a factor, thus reducing even further the species numbers—although undetectable from their fossil remains, moa females were consistently larger than males of the same species. DNA comparisons have established that variations in moa skeletons found previously are likely due to different sexes of birds and not due to species differences. The elephant birds (family Aepyornithidae) of Madagascar are far less speciose in comparison with the moa. Indeed, there are only two known genera, *Aepyornis* and *Mullerornis*, comprising of three and four species respectively. See also **Aves (Birds)**.

Habitats

The habitats of living paleognaths vary widely, as is the case in most living birds. Their geographical distribution is confined mainly to the continents and countries of the southern hemisphere, and some tinamous that are found in Mexico and Central America.

Tinamous range in abundance from southern Mexico to Patagonia (Argentina), and inhabit a number of ecologically diverse habitats within the Neotropics. These birds range from the Amazonian jungle to the high altitude slopes of the Andes (up to 4,300 m (14,000 ft) high). See also **Tinamous (Tinamidae)**. Kiwis are nocturnal birds that live in a variety of habitats ranging from the seashore, where they use their long bills to probe the substrate for invertebrates, to scrub habitats and tall forests; these birds are known only from the islands of New Zealand. The moa was also found in New Zealand before its extinction, which coincided with human settlement, and recent studies have suggested that these birds occupied three principal habitats. The first of these, uplands, supported the growth of herbs and clump grasses and was inhabited by genera such as *Megalapteryx* and *Pachyornis*. The second, lowland wet forests, had a high level of rainfall as well as a covering of tall forests with continuous canopy, home to genera such as *Anomalopteryx* and *Dinornis*. Finally, lowland dry climate zones, contained a mix of forest margins, grass and shrub land, and was inhabited by genera such as *Euryapteryx* and *Anomalopteryx*. Cassowaries live in the forests of New Guinea, their adjacent islands and the forested regions of northern Australia, whereas emus are confined to Australia. However, in the relatively recent geological past, these birds were much more widespread in range—only several thousand years ago, at least three species went extinct from Tasmania and smaller islands off the coast of Australia. Rheas, like the other ratites, vary in their habitats as a result of differing food requirements. The Greater Rhea (*Rhea americana*) is quite abundant in tall grass habitats, compared with the open scrub favored by the Lesser Rhea (*Rhea pterocnemia*).

Habits and Life History

In terms of their habits, most patterns are common to all of these birds. Eggs are laid by the female from a single, functional oviduct. However, aspects of nesting, egg laying, and general life histories vary widely within paleognaths.

Tinamous forage entirely on the ground and can run rapidly; they are relatively weak fliers when in the air. These birds are predominantly herbivores, feeding on a variety of seeds and plant material, but their diet can also include insects. They build scruffy looking nests on the ground, the female producing a clutch of up to 12 eggs. Tinamou chicks are highly sociable and are able to run around soon after hatching. Kiwis, on the other hand, are nocturnal, forest dwelling birds that feed by probing undergrowth for invertebrates (mainly worms and the like) with their long bills. These birds spend the daylight hours asleep in burrows; they lay very large single eggs, incubated by the male for 90 days that are the largest known (relative to body size) of any bird.

Like the emus and rheas, ostriches live and forage in small groups that can contain up to 60 individuals, and eat a wide variety of foods, including plants and seeds, insects and even small mammals and reptiles. They are able to swim very well, but are most famous for their speed along the ground. These large, bipedally adapted birds are able to sprint at up to 95 km (60 miles)h^{-1}; unlike the other ratites that have three toes, ostriches have only two, and retain obvious wings and tails.

Since cassowaries are largely nocturnal and solely solitary forest dwellers, they have become well adapted to living in obstructed habitats. On their heads, they have evolved a bony forehead plate (termed a "casque") that is used to deflect obstructions, such as leaves and overhead branches. These birds are also known to be very strong swimmers, an adaptation to crossing fast flowing rivers; because of this, they have few feathers on the extremities of their tails and wings. Cassowaries also have a reputation for aggressiveness; they have been known to attack and kill humans. In general, however, these birds eat mainly fruit and small, soft plant materials, although they have been reported to dispatch both larger plants and, on occasion, small animals. Emus also have poorly developed wings and are able to swim well; in common with the other large ratites, they primarily consume plant material.

Fossil History

The fossil record of paleognathous birds largely consists of specimens that have been classified within one, or other, of the living groups. The fossil history of the tinamous, for example, is known to extend back to deposits of

TABLE 1. THE GENERA AND SPECIES OF RECENT PALEOGNATHS (AVES, PALEOGNATHAE) GIVING HIGHER-LEVEL TAXONOMY (ORDERS AND FAMILIES) AND PRESENT GEOGRAPHICAL DISTRIBUTION

Scientific (Latin) name	Common (English) name	Present distribution
Order Struthioniformes		
Family Struthionidae		
Struthio camelus	Ostrich	Africa
S. camelus camelus	African ostrich	Africa
S. camelus molybdophanes	Somali ostrich	Northeast Africa
Family Rheidae		
Rhea american	Greater rhea	South-central South America
Rhea pennata	Lesser rhea	West-south South America
R. pennata tarapacensis	Puna rhea	Andes, western South America
R. pennata pennata	Darwin's rhea	Southern South America
Family Casuariidae		
Casuarius casuarius	Southern cassowary	Australia, New Guinea
C. bennetti	Dwarf cassowary	New Guinea
C. unappendiculatus	Northern cassowary	New Guinea
C. novaehollandiae	Emu	Australia
Family Apterygidae		
Apteryx australis	Brown kiwi	New Zealand
A. owenii	Little spotted kiwi	Southern New Zealand
A. haastii	Greater spotted kiwi	Southern New Zealand
Order Tinamiformes		
Family Tinamidae		
Tinamus tao	Grey tinamou	Amazon
T. solitarius	Solitary tinamou	Southeast South America
T. osgoodi	Black tinamou	Western South America
T. major	Great tinamou	Mexico and Central America, South America
T. guttatus	White-throated tinamou	Amazon
Nothocercus bonapartei	Highland tinamou	Central America, South America
N. julius	Twany-breasted tinamou	Andes, western South America
N. nigrocapillus	Hooded tinamou	Andes, Peru, Bolivia
Crypturellus berlepschi	Berlepsch's tinamou	Northwestern Couth America
C. cinereus	Cinereous tinamou	Amazon
C. soui	Little tinamou	West, Southwestern South America
C. ptaritepui	Tepui tinamou	Patagonia
C. obsoletus	Brown tinamou	South America
C. obsoletus obsoletus	Brown tinamou	South America
C. obsoletus traylori	Traylor's tinamou	Eastern Peru
C. cinnamoneus	Thicket tinamou	Mexico and Central America
C. undulatus	Undulated tinamou	South America
C. transfasciatus	Pale-browned tinamou	South America
C. strigulosus	Brazilian tinamou	Amazon, southeast Brazil
C. boucardi	Slaty-breasted tinamou	Mexico and Central America
C. kerriae	Choco tinamou	Colombia
C. erythropus	Red-legged tinamou	Amazon
C. erythropus columbianus	Colombian tinamou	Northwest Colombia
C. erythropus idoneus	Santa Marta tinamou	Northwest South America
C. erythropus saltuarius	Magdalena tinamou	Colombia
C. erythropus erythropus	Red-legged tinamou	Amazon
C. duidae	Grey-legged tinamou	Northwest Amazon
C. noctivagus	Yellow-legged tinamou	East Brazil
C. atrocapillus	Black-capped tinamou	Peru, Bolivia
C. atrocapillus atrocapillus	Black-capped tinamou	Peru
C. atrocapillus garleppi	Garlepp's tinamou	Bolivia
C. variegatus	Variegated tinamou	Amazon
C. brevirostris	Rusty tinamou	Amazon
C. bartletti	Barlett's tinamou	West Amazon
C. parvirostris	Small-billed tinamou	South America
C. casiquiare	Barred tinamou	Amazon
C. tataupa	Tataupa tinamou	South South America
Rhynchotus rufescens	Red-winged tinamou	South South America
Nothoprocta taczanowskii	Taczanowski's tinamou	Andes, Peru, Bolivia
N. kalinowskii	Kalinowski's tinamou	Andes, Peru
N. ornata	Ornate tinamou	Andes, west South America
N. pentlandii	Andean tinamou	Andes
N. cinerascens	Brushland tinamou	South-Central South America
N. perdicaria	Chilean tinamou	Andes, Chile
N. curvirostris	Curve-billed tinamou	Andes, western South America
Nothura darwinii	Darwin's nothura	Mountains, South America
N. chacoensis	Chaco nothura	South-central South America
N. maculosa	Spotted nothura	East, southeast Brazil
N. minor	Lesser nothura	Southeast Brazil
N. boraquira	White-bellied nothura	Southeast South America

(continued)

TABLE 1. (*Continued*)

Taoniscus nanus	Dwarf tinamou	Southeast Brazil
Eudromia elegans	Elegant-crested tinamou	Southwestern South America
E. formosa	Quenbracho crested tinamou	South-central South America
Tinamotis pentlandii	Puna tinamou	Andes, southwestern South America
T. ingoufi	Patagonian tinamou	Andes, southern South America

middle Miocene age in Argentina (about 15 million years ago (Ma), as does the known fossil record of ostriches (genus *Struthio*) in Europe. Recent cassowaries and rheas have a much older documented fossil history—the known record of cassowaries can be extended to the late Oligocene (some 26 Ma) but is restricted to Australia; fossils from the Paleocene (at least 50 Ma) of Argentina and Brazil have been assigned to rheas. See also **Fossil Record**.

However, by far the most interesting fossil paleognaths are a group termed the lithornithids. These fossil birds, described from Eocene (55 Ma) rocks in Europe and North America, were first recognized as primitive paleognaths in the 1980s, but cannot be classified within any of the recent families. This fossil material had, for over a 100 years, previously been assigned to a number of other groups of birds. Representatives of these birds are known on the basis of very good fossil material.

Currently, three separate genera of lithornithid paleognaths are recognized on the basis of over 100 fossil specimens: *Lithornis*, *Paracathartes* and *Pseudocrypturus*; these birds are like tinamous when compared with the recent forms in which they had well-developed wings and were clearly capable of flight. Evidence from this group of enigmatic fossil birds constrains the minimum age for the evolution of flightlessness in the recent paleognath lineages to sometime "after" the Eocene (55 Ma).

Phylogeny

The phylogeny, or evolutionary relationships, of the paleognathous birds has been a source of debate among ornithologists for many years. However, with the advent of molecular methods to study evolution, at last a consensus of opinion appears to be emerging. Within Paleognathae, the tinamous are considered to be the most primitive of the living forms, and have been placed basally with respect to the ratites (the ostrich, rhea, cassowary and emu) (Fig. 2). This basal position for the tinamou reflects their classification

within a single family (Tinamidae) within a separate order (Tinamiformes). Because they are fossils, based on morphological evidence alone, the flighted lithornithids have been placed as a sister group to the extant tinamous (Fig. 2). Within ratites, a close evolutionary relationship between the cassowaries and emus is reflected in their placement within the single order Casuariiformes.

As can be seen from Figure 2, these birds form a close relationship with their sister clade the ostrich and rhea.

Additional Reading

Abourachid, A., and S. Renous: "Bipedal Locomotion in Ratites (Palaeognathiform): Examples of Cursorial Birds," *Ibis*, **142**, 538–549 (2000).

Beddard, F.E.: *The Structure and Classification of Birds*, Longmans, Green and Company, London, UK, 1998.

Bertelli, S., N.P. Giannini, and P.A. Goloboff: "A Phylogeny of the Tinamous (Aves: Palaeognathiformes) Based on Integumentary Characters," *Systematic Biology*, **51**(6), 959–979 (2002).

Bledsoe, A.H.: "A Phylogenetic Analysis of Postcranial Skeletal Characters of Ratite Birds," *Annals of the Carnegie Museum*, **57**, 73–90 (1988).

Carroll, R.L.: *Vertebrate Paleontology and Evolution*, W. H. Freeman and Company, New York, NY, 1988.

Chatterjee, S.: *The Rise of Birds*, Johns Hopkins University Press, Baltimore, MD, 1997.

Feduccia, A.: *The Origin and Evolution of Birds*, Yale University Press, New Haven, CT, 1997.

Houde, P.: "Palaeognathous Carinate Birds from the Early Tertiary of North America," *Science*, **214**, 1236–1237 (1981).

Huxley, T.H.: "On the Classification of Birds and on the Taxonomic Value of Certain of the Cranial Bones Observable in that Class," *Proceedings of the Zoological Society of London*, **1867**, 415–472 (1867).

Huynen, L., C. Millar, R.P. Scofield, et al.: "Nuclear DNA Sequences Detect Species Limits in Ancient Moa," *Nature*, **425**, 175–178 (2003).

Lee, K., J. Felsenstein, and J. Crancraft: "The Phylogeny of Ratite Birds: Resolving Conflicts Between Molecular and Morphological Data Sets," In: Mindell, D.P.: *Avian Molecular Systematics and Evolution*, Academic Press, New York, NY, 1997.

Worthy, T.H., and R.N. Holdaway: *The Lost World of the Moa: Prehistoric Life of New Zealand*, Indiana University Press, Bloomington, IN, 2002.

Leona M. Leonard,
Gareth J. Dyke,
University College Dublin, Belfield, Ireland

PALEOMAGNETISM. The study of natural remanent magnetization in order to determine the intensity and direction of the earth's magnetic field in the geologic past.

PALEONTOLOGY. The study of fossils. A fossil is the evidence of the former existence of an organism, either animal or plant. Fossils may be classified according to their method of fossilization as follows: (1) Actual remains, sharks' teeth, ear bones of whales, chitin, etc. (2) Petrifications. Minute replacements in which the original organic matter has been completely or partially replaced by mineral matter. The principal replacing minerals are calcite, quartz, chert, and pyrite. (3) Molds and casts of interiors and exteriors. (4) Prints of leaves, jellyfish, etc., sometimes showing carbonized traces of organic matter, as in the case of some fossil plants. (5) Coprolites, fossil excrement. (6) Tracks, trails, and burrows.

A valuable reference is *"Principles of Paleontology"*, by D.M. Raup and S.M. Stanley, Freeman, San Francisco, 1979.

PALEONTOLOGY (The History). An understanding of fossils is essential to the science of paleontology. However, the true nature of fossils was not generally accepted until the seventeenth century and paleontology as a science in the modern sense has only existed since the

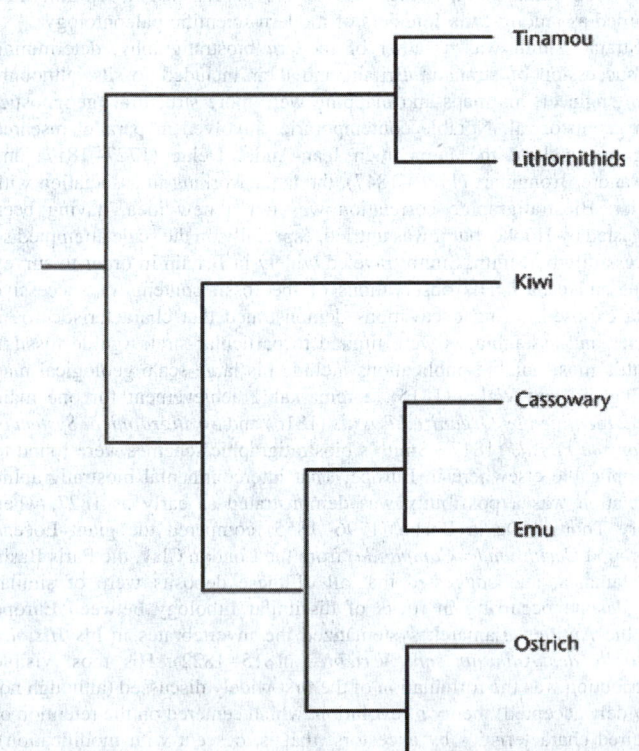

Fig. 2. Consensus regarding the evolutionary relationships of the Palaeognathae (including the fossil lithornithids).

late eighteenth to early nineteenth centuries. Nevertheless, fossils have been recognized as curious objects since man's prehistory. Paleolithic and younger sites of human occupation have yielded fossil shells of benthic mollusks, brachiopods, crinoid columnals, trilobites, corals, sponges and sharks' teeth, all pierced to allow their suspension in necklaces, bracelets and head-dresses. Bronze Age humans have been found buried with, for example, tests of the Chalk echinoids and belemnite guards. Further, there is evidence that fossil collecting was undertaken not just by primitive *Homo sapiens*, but also by the Neanderthals.

Greek philosophers may be regarded as the first true scientists and their interpretation of fossils relied on direct observations. Amongst others, Xenophanes of Colophon (570–470 BC) and Xanthus of Lydia (Turkey) (fifth century BC) accepted that fossil shells found in rocks on mountains indicated a marine origin. Herodotus (*c*.484–420 BC) speculated that the sedimentary sequences of the Nile delta were generated in a series of floods. If a single flood deposited one sedimentary layer, then floods must have occurred over many thousands of years to produce such a thick sedimentary sequence. Unfortunately, such ideas about fossils and time did not persist and were superseded by more mystical explanations. The most persistent was that fossils were the product of some latent plastic character of the Earth itself (the *vis plastica* or *virtus formativa*), probably related to the Aristotelian theory of the spontaneous generation of living organisms, and were thus no more than mineralogical accidents of nature rather than records of past life. Alternatively, fossils were considered to be the product of some influence from the heavenly bodies.

Although even today Chinese folklore considers fossil vertebrates to be the remains of dragons, as early as the sixth century the Chinese philosopher Li Tao-Yuan (died AD 527) referred to fossilized "stone fishes." In medieval Europe, Pleistocene cave bear remains were similarly reputed to be derived from dragons. Further, the legend of a universal deluge — the Biblical flood — was an important influence on later geological and paleontological thought. Thus, fossils were considered to be antediluvian relics transported by the flood and deposited far from the sea. The most "renowned" of these relics was documented by the Swiss Johann Scheucher (1672–1733) who described an articulated fossil skeleton as the remains of a sinner drowned by the flood (*Homo diluvii testis*); this specimen was later re-identified as a Miocene salamander by Cuvier. Although convincing arguments against fossils being sports of the flood or *vis plastica* were formulated (but not published) by such a noted Italian Renaissance philosopher as Leonardo da Vinci (1452–1519), the universal deluge was not truly laid to rest as the unique fossilization event until the emergence of a scientific geology and paleontology in the nineteenth century. See also **Fossil Record**.

Seventeenth and Eighteenth Centuries

The seventeenth and eighteenth centuries saw notable developments of relevant technology, such as the microscope, and also in the organization of both natural history and natural historians with, for example, the founding of the Royal Society in 1660 and the growth of importance of museums. However, ideas founded on observations were not as free to develop. At this time many European thinkers, such as the English physician Martin Lister (1638–1711), continued to support the common view that fossils were inorganic sports of nature, although such ideas were largely discarded during the eighteenth century. That there were other considerations involved in assessing the true nature of fossils was demonstrated by the English theologian and naturalist John Ray (1627–1705), who did not support their organic origin because he did not want to "put a weapon into the Atheists'' hands'. In contrast, the Englishman Robert Hooke (1635–1703), a physicist and naturalist, studied fossils using a compound microscope, and compared the fine structure of fossil and extant organisms; for example, he attempted to identify fossil plants using internal anatomy. His list of paleontological ideas included early discussions of evidence for extinction, evolution, biostratigraphy and paleoecology. Further, Hooke was an early advocate of taphonomic studies, recognizing the fundamentally different states of preservation of moulds and casts. See also **Hooke, Robert (1635–1702); and Ray, John (1627–1705)**.

The Dane Niels Stensen (or Nicolaus Steno) (1638–1687), a physician and naturalist, was another eminent geological and paleontological intellect of the seventeenth century. He drew theoretical geological cross-sections to illustrate his views on the evolution of the crust by dissolution and collapse, and was the first to define the principle of superposition. He argued that tonguestones or *Glossopetrae*, which were then considered to be products of the *vis plastica* that grew in the form of snakes tongues, were in fact the teeth of fossil sharks. This deduction was arrived at following his dissection of a large stranded shark in 1666. He also realized that the included fossil biota could provide indications as to the ambient palaeoenvironment during the formation of a rock unit; that is, fossils were used to define biofacies. However, he failed to influence contemporary geological thought and his achievements remained essentially unrecognized until the nineteenth century. See also **Geological Time Scale**.

One of the saddest figures in the history of paleontology was Johannes Beringer, who made a large collection of specimens from the Würzburg area, many of which were fantastic fakes manufactured by rivals in an effort to discredit him. These were illustrated in Beringer's *Lithographiae Wirceburgensis* (1726). Their subsequent exposure as fakes led to the downfall of Beringer's rivals, which had not been their intention, while he retained his post and even persisted in believing that most of the forgeries were true relics!

The French naturalist George Louis Leclerc, comte de Buffon (1707–1788) published notable speculations on the succession of biotas, evolution, "lost species" (extinction) and the great antiquity for the Earth. Buffon was one of the first naturalists to use the rock record as a source of information regarding the history of the Earth, although he also embraced theoretical approaches that in turn ignored these "archives." This led him to correlate the days of the Book of Genesis with long periods of time during which the Earth slowly evolved, published without supporting data from the rock record and despite opposition from the Church. See also **Buffon, Georges Louis (1707–1788)**.

Nineteenth Century: Some Notable Figures

The late eighteenth and early nineteenth centuries were the time of the "great flowering" or "heroic period" of geology. Paleontology became established as a science at this time, with the institution of research and teaching centers in geology. Amongst a number of notable paleontological endeavors, the efforts of three scientists are worthy of special mention, their work reflecting three major research programs: the Englishman William Smith (1769–1839); and the Frenchmen Jean Baptiste Lamarck (1744–1829) and Baron Georges Cuvier (1769–1832). This trio might be regarded as amongst the founders of modern scientific paleontology.

"Strata" Smith was a father of modern biostratigraphy, determining the succession of strata in Britain and their included fossils, although, as an engineer, his maps and mapping were more structural (geognostic) than geohistorical. Notable contemporaries involved in parallel research programs include the Frenchmen Jean-André Deluc (1727–1817) and Alexandre Brongniart (1770–1847), the latter working in association with Cuvier. Biostratigraphic correlation was not a new idea, having been suggested by Hooke, but it was untried, especially on the scale attempted so successfully by Smith. Smith traveled widely in Britain in order to survey mines and canals. His observations of the fossil contents of successive strata exposed during excavations demonstrated that characteristic fossil species and assemblages were limited to particular strata (guide fossils). Smith's most notable publications include his large-scale geological map of England and Wales (1815), a remarkable achievement for one man, *Strata Identified by Organized Fossils* (1816) and *Stratigraphical System of Organized Fossils* (1817). Smith's biostratigraphic schemes were found to be applicable elsewhere in Europe. That intercontinental biostratigraphic correlation was a possibility was demonstrated as early as 1827, when Henry Thomas De la Beche (1796–1855) compared the giant Eocene gastropod *Cerithium* (= *Campanile*) from the London Clay, the Paris Basin and Jamaica, and suggested that all of these deposits were of similar age, despite occurring in rocks of dissimilar lithology between Europe and the Antilles. Lamarck systematized the invertebrates in his *Histoire Naturelle des Animaux sans Vertébres* (1815–1822). His most visible contribution was the formulation of the first widely discussed (although not as widely accepted) theory of evolution, which centered on the retention of acquired characteristics by ancestors (that is, descent with modification). As stated, this theory lacked a demonstrable mechanism whereby acquired characteristics could be retained. It was also vigorously attacked by Cuvier, who recognized the weaknesses of Lamarck's uniformitarian argument and denigrated it as a rival to his own catastrophist theories that were based on the immutability of species. See also **Lamarck, Jean–Baptiste Pierre Antoine de Monet de (1744–1829)**.

Cuvier and Brongniart, who had learnt something of Smith's methodologies, published a geological map of the Paris "basin" in 1811 (although of much smaller areal extent than Smith's 1815 map), the first to be based on geohistorical, rather than geognostic, principles. However, Cuvier's most enduring achievement was his development and application of the rigorous techniques of comparative anatomy of fossil vertebrates. His law of correlation demonstrated in detail in his *Recherches sur les Ossemens Fossiles* (1812) (Research on the Fossil Ossemens), enabled interpretation of the whole structure of a skeleton even if it was only incompletely known. A notable example of his abilities and methods was the recognition that the fossilized skulls of mammoths, by then identified as derived from elephants by Buffon and others, were derived from ancient and extinct proboscideans rather than extant taxa. The acknowledgement and acceptance of extinction was a notable achievement at a time when many naturalists still believed in the fixity of species. Extinction was an important component of Cuvier's widely accepted catastrophist theory. He postulated that the Earth had been affected by multiple major extinction events, the most recent being the Biblical Deluge, each of which was followed by a new creation of life. This led him to insist that fossil humans could not exist, even though their bones had already been found at high stratigraphic levels in Europe. See also **Cuvier, Georges Leopold Chretien Frederic Dagobert Baron de (1769–1832)**.

Paleontology and the Geological Column

The development of an internationally applicable set of subdivisions of the rock and fossil record was a major achievement of geology and paleontology of the first half of the nineteenth century. The names of and subdivisions between the geological systems are not arbitrary, but relate to particular features recorded by the rock and fossil record. The names of the systems arose in a slightly haphazard way, but all are significant, relating either to the region where the principal features of this part of the rock record were first recognized, or commemorating a distinctive feature of a subdivision. For example, the Carboniferous, first named by William Daniel Conybeare and John Phillips in 1822, literally means "carbon containing," in recognition of the abundant coal-bearing deposits found within this system in North America and Eurasia.

For a system to be recognized in the rock record at a particular locality, it is necessary to utilize broadly applicable tools of correlation, that is, fossils. Until the utility of fossils in correlation was confirmed, it was not possible to extend the limits of any stratigraphic subdivision beyond the local level. This is well demonstrated by reference to the Cambrian–Silurian controversy. The subdivision of the Lower Paleozoic into three distinct systems–Cambrian, Ordovician and Silurian–was due to the work of three British geologists, Adam Sedgwick (1785–1873), Charles Lapworth (1842–1920) and Roderick Impey Murchison (1792–1871), respectively. Their pioneer work on the Lower Paleozoic was undertaken in Wales, hence the Celtic references in the system names:

- Silurian–named after the Silures, an ancient Celtic tribe of the Welsh Borders with England.
- Ordovician–named after the Ordovices, an ancient Celtic tribe of central Wales.
- Cambrian–from the Latin *Cambria*, Wales.

In the early nineteenth century, the Lower Paleozoic rocks of Wales were considered to be a Transition Series between the underlying crystalline basement (Precambrian in modern terms) and the overlying Old Red Sandstone (Devonian of modern usage). The Transition Series was separated from both the underlying and overlying rocks by unconformities. Murchison (south Wales) and Sedgwick (north Wales) began their joint investigations of the Transition Series in 1831. They published their first report in 1835: Murchison recognized the importance of the included fossils when describing the Silurian, but Sedgwick defined the Cambrian System on the basis of lithology and structure, ignoring the paleontology. Murchison's subsequent discovery of fossils similar to those of the Silurian in Sedgwick's Cambrian led to the suggestion that the Cambrian was merely the lowest part of the Silurian. It was not until 1879 that Lapworth showed, with the aid of detailed biostratigraphic evidence,

that the Transition Series had a 3-fold subdivision, the systems being separated by major unconformities. The "middle" system he named the Ordovician.

The Scotsman Charles Lyell (1797–1875) subdivided the Tertiary into three series, the Eocene, Miocene and Pliocene, in 1833. Subsequently, further subdivision resulted from improved knowledge of the fossil and rock record. Lyell's subdivisions were related to the included marine faunas (mainly benthic mollusks) and was the first attempt at a statistical biostratigraphy, a methodology entirely in keeping with his uniformitarian ideas. Lyell recognized that the more ancient the fauna, the less resemblance it would bear to the modern biota. By identifying the elements of a fossil fauna and comparing it to that known at the present day, Lyell deduced that it should be possible to determine the relative antiquity of the fossil mollusks. However, for such a chronometer to work, it was necessary that species were evolving and becoming extinct at a uniform rate, and are not interrupted by "discontinuities" such as major extinction events, which make pre-extinction faunas look relatively more ancient. Thus, this approach is of only limited use, being affected by regional and global extinctions. For example, it cannot be extended back into the Mesozoic due to the effects of the end Cretaceous mass extinction. See also **Lyell, Charles (1797–1875)**; **Mollusca**; and **Mollusks**.

Darwin and Paleontology

The Englishman Charles Darwin (1809–1882) was undisputedly the most notable naturalist of the nineteenth century. Darwin's geological achievements tend to be overshadowed by his contributions to the life sciences, particularly his theory of evolution by natural selection. However, this is in part due to the relatively scant attention paid to the fossil record in *The Origin of Species* (1859), in which Darwin perpetrated a fallacy that persists to plague paleontology, that of "the imperfection of the geological record." Darwin largely ignored paleontology as a source of reliable information to support his own theories, because the fossil record as then known failed to provide the abundant evidence of morphological intermediates and gradual evolutionary changes that Darwin's theory required. However, what was ignored by Darwin was enthusiastically embraced by late nineteenth and twentieth century paleontologists and biologists trying to reconstruct patterns of evolution, recognizing paleontology as the only direct test for their theories. The interweaving of ideas in an interdisciplinary program of paleontology, biogeography and evolutionary morphology enabled these early evolutionary theorists, both Darwinian and non-Darwinian, to explore the history of life. The fossil record has thus become recognized as a primary source of information on evolutionary questions. Darwin was only partly right in recognizing the fossil record to be incomplete; it is a key source of data for evolutionary studies of many groups of organisms. Fossil evidence is not always available, but where it is it can make a unique contribution. See also **Biogeography (The History)**.

Ancient Saurians

Fossils of gigantic fossil saurians were known at least as early as the seventeenth century, although they were not accurately identified as reptiles until the nineteenth century. Notable examples included the original skull of *Mosasaurus* from Maastricht, referred variously to the crocodiles, whales and varanoid lizards, the latter identification finally confirmed by Cuvier. Of the Jurassic marine lizards, notably sold to collectors by the Englishwoman Mary Anning (1799–1847) and her family (but also found elsewhere in Europe), it was only in 1821 that the Englishmen William Daniel Conybeare (1787–1857) and De la Beche correctly identified the affinities of *Ichthyosaurus* and *Plesiosaurus*. The dinosaurs were first named as such by the Englishman Richard Owen (1804–1892), the "English Cuvier," in 1841, although their true morphology was not understood until later, as evidenced by the life size reconstructions dating from the mid-1850s that can still be seen in Crystal Palace in south London. Amongst the early collectors, the Englishman Gideon Algernon Mantell (1790–1852), best known for describing *Iguanodon*, is notable for popularizing the "Age of Reptiles" which preceded the "Age of Mammals." See also **Dinosauria (Dinosaurs)**; **Lizards (Reptilia, Sauria)**; and **Owen, Richard (1804–1892)**.

The center of vertebrate, particularly dinosaur, research moved from Europe to North America in the later nineteenth century, when new and rich deposits of their bones were discovered as the continent was explored.

Amongst the more notorious paleontologists of the past two hundred years are the Americans Edward Drinker Cope (1840–1897) and Othniel Charles Marsh (1831–1899), who waged the celebrated "dinosaur wars" as they fought to obtain and describe these rich collections of Mesozoic and Cenozoic vertebrate remains. Their contribution to vertebrate systematics is undoubted, as they both described hundreds of new tetrapod taxa. However, the rivalry between the two men led to public accusations and defamatory attacks that appeared in the press and led to censorship by the government. See also **Cope, Edward Drinker (1840–1897)**; **Marsh, Othniel Charles (1831–1899)**; and **Systematics: Historical Overview**.

Twentieth Century

Paleontology in the twentieth century has claimed an importance to evolutionary studies that was denied by Darwin. In part this has been driven by improved methodological advances, particularly the use of cladistic techniques in phylogenetic analyses, but also by technological advances, particularly applications of the electronic computer in manipulating large databases. It has also been helped by a progressive development that has seen paleontology become less geological and more biological in outlook. However, the monographic coverage of all fossil groups has been improved throughout the century, continuing a program initiated in the nineteenth century that underpins both the "new evolutionary paleontology" and traditional paleontological areas such as biostratigraphy, despite appearing to be the Cinderella side of the subject. Other technological advances have enabled paleontology to exploit hitherto unsuspected or untapped sources of information for investigating even familiar fossils, most notably the scanning electron microscope (micropaleontology, microstructure), stable isotope geochemistry (paleoenvironmental data, dating) and the search for ancient biomolecules. Paleontology has also permitted the analysis and provided controls for investigations of continental drift and plate tectonics, using the data provided by biogeographic changes through time. Paleontological data provided early support for the ideas on continental drift of the German Alfred Wegener (1880–1930), which were only accepted into the geological mainstream once a driving mechanism had been deduced in the 1960s. See also **Continental Drift**; **Earth Tectonics and Earthquakes**; and **Wegener, Alfred Lothar (1880–1930)**.

Precambrian paleontology became an active field of research only in the latter half of the twentieth century, since which time late Precambrian faunas, termed Ediacara faunas, have been found at over 30+ sites worldwide. The Ediacara fauna takes its name from the type locality 600 kilometers (373 miles) north of Adelaide, Australia. Although the medusoids and related, unmineralized forms found at this locality were originally recorded as early Cambrian, they were subsequently reinterpreted as late Precambrian and described in detail by Austrian Martin F. Glaessner (1906–1989) and co-workers over a number of years. Preservation of so many taxa lacking mineralized hard parts is also exceptional. All of the Ediacaran body fossils were multicellular, a major change in body organization after the pre-Ediacaran, which was dominated by unicellular prokaryotes and eukaryotes. Glaessner considered the fauna to be dominated by cnidarian-like organisms. However, the German Adolf Seilacher has reinterpreted the Ediacaran biota not as the ancestors of Phanerozoic animal phyla, but as an evolutionary dead end composed of a group of organisms, the Vendozoa, with a body structure like an inflated bag (pneu structures).

Other exceptionally well-preserved biotas, comprising organisms that do not normally form part of the rock record, permit the determination of faunal and floral characteristics that usually must be, at best, inferred. Such biotas add to the knowledge of those groups that are not generally preserved in the fossil record, especially those organisms that do not have mineralized skeletons ("soft-bodied"). Such an unusual fauna is given the Germanic name of a *Conservitat Lagerstätten* ("fossil mother lode"). *Lagerstätte* are not particularly common in the fossil record, but still occur about every 20 million years in the Phanerozoic and have formed the subject of intense study in the last third of the twentieth century. Most notable amongst these deposits is the Middle Cambrian Burgess Shale, which has yielded many tens of thousands of specimens. The modern interpretation of this fauna was started by the Englishman Harold B. Whittington and his students, who recognized the peculiar morphology of many of the included taxa. The Burgess Shale is dominated by animals that lacked a mineralized skeleton in life; only about a third of the animal taxa in this fauna had mineralized hard parts. It thus provides evidence that the Cambrian Radiation led to a great diversification of organisms without a mineralized skeleton, as well as those with preservable hard parts. The Burgess Shale probably represents a typical mid Cambrian fauna; what is abnormal is the preservation, which includes those unmineralized organisms not commonly found in coeval deposits. See **Burgess Shale**.

The Darwinian theory that evolution occurs by the gradual accumulation of morphological changes (phyletic gradualism) suggests that most evolution occurs at a phyletic level, that is, within established species. Speciation is regarded as a relatively rare event which involves a gradual divergence in morphologies. However, this view was supplemented by a second pattern recognized from the fossil record. In 1972 the Americans Niles Eldredge and Stephen Jay Gould related most evolutionary change to rapid branching events, with periods of relative stasis between speciations, a pattern they considered to be well-illustrated from the fossil record and which they called *punctuated equilibrium*. The determination of the true pattern of speciation from the paleontological evidence continues to be the subject of energetic debate. Punctuational branching events are geologically fast and result in descendants of divergent morphology, although the question of fidelity and time resolution of the rock record places constraints on such determinations. Indeed, phyletic gradualism and punctuated equilibrium represent extreme views of the evolutionary process, and both probably occur.

Mass extinctions became a prominent area of evolutionary research in 1980 when the American Luis Alvarez and co-workers provided geochemical evidence that suggested the extinction at the Cretaceous–Tertiary (K–T) boundary was driven by the impact of an extraterrestrial bolide, that is, a comet or a meteorite. The evidence for this came from the platinum group elements (particularly iridium) concentrated at the boundary. Alvarez et al. suggested a theory to account for both the anomalously high iridium levels and the mass extinction which depended upon a large bolide, of the order of 10 kilometers (6 miles) in diameter and traveling at 10 to 50 km s^{-1}, striking the Earth at the end of the Cretaceous. Such an impact would eject about 60 times the bolide's weight into the atmosphere as pulverized rock. The K–T boundary iridium anomaly has now been recognized from about 100 different sequences worldwide, both marine and terrestrial in origin, and the impact site may be at Chicxulub Crater off the Yucatan Peninsula in the Caribbean. However, the K–T event is the only major mass extinction for which good evidence of impact is available and the search for iridium anomalies at other extinction horizons has failed to produce conclusive evidence of impact.

Early in 1984 the Americans David M. Raup and Jack J. Sepkoski Jr. (1948–1999) analyzed the last appearance data (extinctions) of all marine families and found strong statistical indications of a periodicity of mass extinctions, with a frequency of about every 26 million years since the end of the Permian. As there is no known terrestrial mechanism that could drive such a pattern of extinctions, extraterrestrial causes were invoked, although evidence for bolide impacts does not exist for most of the events involved. However, assuming that the causes were bolide impacts, three principal potential driving mechanisms were suggested whereby the Earth could be bombarded every 26 million years. The impacting bolides could have been comets, dislodged from the Oort Cloud (a reservoir of comets orbiting the sun far beyond the planets) by some disturbing force, either due to oscillation of the solar system about the galactic plane, disturbance by an unknown tenth solar system planet or disturbance by an unknown sister star to our sun. Despite the excitement caused by this theory, strong doubts have been expressed regarding both the data used and the statistical analysis. See also **Mass Extinctions**.

Future Trends

In 1968, the American Raymond C. Moore (1892–1974) and co-workers predicted future developments in paleontology. From our vantage point, there are certain glaring omissions, and others that have developed subsequently and could not have been predicted. Paleontological research in 1968 was more geological than biological, explaining the absence of a separate subheading "Evolution." Extinctions, biotic radiations and taphonomy were absent from the list. Modern phylogenetic techniques and geochemical investigations only subsequently became issues in paleontology. In 1968 it would have been reckless to predict the microcomputer and how it now underpins all science.

In similar vein, the American David Jablonski recently looked ahead, and proposed an agenda for research that is important to all aspects to paleontology. Four important questions were asked:

- What rules govern biodiversity dynamics, and do they apply at all temporal and spatial scales?
- Why are major evolutionary innovations unevenly distributed in time and space?
- How does the biosphere respond to environmental perturbations at global and regional scales?
- How have biological systems influenced the physical and chemical nature of the Earth's surface, and vice versa?

These questions look "right," just as the predictions of Moore et al. were convincing when originally published. However, it is possible to look at palaeontology from a different direction. The American Arthur J. Boucot estimated in 1999 that we only know something like 10–15% of the genera and species that existed and are fossilized, so there is another 85–90% of the fossil record that remains unknown. Boucot very much feels that we need to go out there and carry on collecting, rather than undertaking more synthesis of an incomplete data set. Probably, we need both. See also **Paleoclimatology**.

Glossary

An abridged glossary of terms used in paleontology would include:

Biofacies;—Rock units distinguished by similar lithological and paleontological criteria were formed in similar environments. A biofacies is a rock unit that contains an assemblage of fossils characteristic of a particular environment.

Conservitat Lagerstätten;—Exceptionally well-preserved biotas consisting of organisms that do not normally form part of the rock record, principally those that have an unmineralized or poorly mineralized skeleton.

Geognosy;—The study of the three-dimensional structure and relationships of rock masses, particularly as applied to mines and mining.

Guide Fossils;—A guide or zone fossil is a fossil species that characterizes a particular interval of the rock record, which is termed the biozone of that species. Guide fossils are essential tools in local and international biostratigraphic correlation.

Law of Correlation;—A methodology, based on the recognition of the conservative nature of the relationships and morphologies of elements of the vertebrate (particularly tetrapod) skeleton, that permits accurate prediction of the overall morphology and function even upon the basis of incomplete specimens.

Mass Extinction;—Mass extinctions are episodic events of global extent, which lead to the demise of numerous, diverse groups of organisms in a variety of environments and are caused by major environmental disturbances.

Phyletic Gradualism;—A theory that suggests that most evolution occurs at a phyletic level that is, within established species. Speciation is regarded as a relatively rare event which involves a gradual divergence in morphologies.

Punctuated Equilibrium;—Unlike phyletic gradualism, relates most evolutionary change to rapid branching events, with periods of relative stasis between speciations.

Taphonomy;—The study of the processes of fossilization, essentially those mechanisms that act on a once-living organism following death. This includes those that act upon a carcass between death and final burial, and the changes in the chemistry of the fossil that mainly occur after burial.

Additional Reading

Benton, M.J.: *Vertebrate Palaeontology*, 3rd Edition, Blackwell Publishers, Malden, MA, 2004.
Bowler, P.J.: *Life's Splendid Drama: Evolutionary Biology and the Reconstruction of Life's Ancestry, 1860–1940*, University of Chicago Press, Chicago, Chicago, IL, 1996.
Carrano, M., R. Blob, T. Gaudin, and J. Wible: *Amniote Paleobiology: Perspectives on the Evolution of Mammals, Birds, and Reptiles*, University of Chicago, Press, Chicago, IL, 2006.
Dean, D.R.: *Gideon Mantell and the Discovery of Dinosaurs*, Cambridge University Press, Cambridge, UK, 1999.
Edwards, W.N.: *The Early History of Paleontology*, British Museum (Natural History), London, UK, 1976.
Everhart, M.J.: "Oceans of Kansas," *A Natural History of the Western Interior Sea*, Indiana University Press, Bloomington, IN.
Foote, M., and A. Miller: *Principles of Paleontology*, 3rd Edition, W. H. Freeman Company, New York, NY, 2006.
Gohau, G.: *A History of Geology*, Rutgers University Press, New Brunswick, Canada, 1990.
Hammer, O., and D.A. Harper: *Paleontological Data Analysis*, Blackwell Publishers, Malden, MA, 2005.
Jones, R.W.: *Applied Palaeontology*, Cambridge University Press, Cambridge, UK, 2006.
Kenrick, P., and P. Davis: *Fossil Plants*, Smithsonian Institution Press, Washington, DC, 2004.
Martin, A., and G. Bertola: *Introduction to the Study of Dinosaurs*, 2nd Edition, Blackwell Publishers, Malden, MA, 2005.
Owen, R.: *Palaeontology: The Evolution Debate, 1813–1870*, Vol. 6, Taylor & Francis, Inc., Philadelphia, PA, 2003.
Palmer, D.: *Prehistoric Past Revealed: The Four Billion Year History of Life on Earth*, University of California Press, Berkeley, CA. 2003.
Rudwick, M.J.S.: "Cuvier and Brongniart, William Smith, and the Reconstruction of Geohistory," *Earth Sciences History*, **15**, 25–36 (1996).
Secord, J.A.: *Controversy in Victorian Geology: The Cambrian–Silurian Dispute*, Princeton University Press, Princeton, NJ, 1986.

STEPHEN K. DONOVAN, The Natural History Museum, London, UK

PALEOPHYTIC. A paleobotanic division of geologic time. The term signifies the time period during which pteridophytes were abundant, occurring between the development of algae and the appearance of the first gymnosperms. See also **Paleobotany**.

PALEOZOIC. The era of "ancient" life or the age of invertebrates. Subdivided from the base up into the following periods: Cambrian, Ozarkian, Canadian, Ordovician (lower Paleozoic), Silurian, Devonian (middle Paleozoic), Mississippian, Pennsylvanian, Permian (upper Paleozoic). The lower Paleozoic was characterized by: (a) the first known marine faunas; (b) dominance of trilobites; (c) rise of animals with hard shells (Cambrian); (d) rise of nautiloids, armored fishes, and corals; and (e) the first evidence of colonial life (Ordovician). The middle Paleozoic was characterized by: (a) the rise of the lung-fishes and scorpions (Silurian); (b) the first known land floras (Devonian), not very different from those of the Pennsylvanian; and (c) earliest evidence of a terrestrial vertebrate. The upper Paleozoic was characterized by: (a) ancient sharks and echinoderms (Mississippian); (b) primitive reptiles and insects (Pennsylvanian); (c) periodic glaciation and extinction of many Paleozoic groups during and after the Permian; and (d) rise of modern insects, land vertebrates, and ammonites (Permian). The Paleozoic Era began about 500 million years ago and lasted for approximately 300 million years.

PALLADIUM. [CAS: 7440-05-3]. Chemical element symbol Pd, at. no. 46, at. wt. 106.4, periodic table group 10, mp 1554°C, bp 2970°C, density 12.02 g/cm^3 (solid), 12.25 g/cm^3 (single crystal) (20°C). Elemental palladium has a face-centered cubic crystal structure. The six stable isotopes of palladium are ^{102}Pd, ^{104}Pd, through ^{106}Pd, ^{108}Pd, and ^{110}Pd. The seven unstable isotopes are ^{100}Pd, ^{101}Pd, ^{103}Pd, ^{107}Pd, ^{109}Pd, ^{111}Pd, and ^{112}Pd. In terms of earthly abundance, palladium is one of the scarce elements. Also, in terms of cosmic abundance, the investigation by Harold C. Urey (1952), using a figure of 10,000 for silicon, estimated the figure for palladium at 0.0091. No notable presence of palladium in seawater has been found.

Palladium was discovered by William Hyde Wollaston in 1803. This element was named by Wollaston in 1804 after the asteroid Pallas, which was discovered two years earlier. Wollaston found element 46 in crude platinum ore from South America. He did this by dissolving the ore in aqua regia, neutralizing the solution with sodium hydroxide, NaOH, precipitating platinum as ammonium chloroplatinate through treatment with ammonium chloride, NH$_4$Cl, and then adding mercuric cyanide to form the compound palladium cyanide. Finally, he heated the resulting compound in order to extract palladium metal.

Electronic configuration $1s^22s^22p^63s^23p^63d^{10}4s^24p^64d^{10}$. Ionic radius Pd^{2+} 0.50 Å (Wyckoff). Metallic radius 1.3755. First ionization potential 8.33 eV; second 19.8 eV. Oxidation potentials Pd→ Pd^{2+}+2e$^-$ 4

M HClO$_4$, -0.83 V, Pd$+2$OH$^-$ $\rightarrow$ Pd(OH)$_2+2$e$^-$, -0.1 V. Further physical properties are given under **Platinum and Platinum Group**.

Pd has some similarities with both Ni and Ag and many with Pt. Pd dissolves more readily in acids than any other member of the platinum group of metals. In aqua regia, the metal dissolves quickly. Even the compact metal dissolves slowly in HCl. In finely divided form, it is quite soluble in all acids. When heated in air at red heat, the monoxide, PdO, is formed. Pd is similarly converted to the dihalides under the same conditions when it is exposed to F$_2$ or CL$_2$. The metal is not affected by H$_2$S.

The black compound, palladium(II) oxide [CAS: 1314-08-5] is formed by fusing palladium(II) chloride [CAS: 7647-10-1] with NaNO$_3$ to 600°C and then leaching out the salts with water. This strong oxidizing agent is easily reduced to the metal by H$_2$. The compound is insoluble in water and acids, including aqua regia. The hydroxide, Pd(OH)$_2$, is made by the hydrolysis of palladium(II) nitrate. The compound is soluble in acids, and water is evolved on heating, but even at 500–600°C some water still remains. At this temperature, the compound starts to lose O$_2$.

Palladium(III) oxide, P$_2$O$_3$, is made as a hydrate by careful oxidation of a solution of palladium(II) nitrate either by anodic oxidation or ozone treatment at -8°C. This unstable brown powder reverts to the monoxide in about 4 days. When heated, the compound loses water and may explode as it changes to the monoxide.

Palladium(II) chloride is formed by direct combination of the elements at 500°C. It is the only stable chloride over 500–1500°C. The red crystals are partly soluble in water and completely soluble in HCl. The fraction insoluble in water is probably a polymer. Palladium(II) chloride also is the product obtained by evaporation of a solution of Pd in HCl. Palladium(II) bromide can also be made from the elements.

When KI is added to a solution of palladium(II) chloride, an insoluble diiodide is precipitated. The dark red-black crystals are soluble in excess iodide with formation of the tetraiodide complex ion. Palladium(II) iodide evolves iodine at 100°C, the decomposition to the elements being complete at 330–360°C. The black compound, palladium(III) fluoride, is made by direct combination of the elements. On reduction, the brown difluoride is formed.

Divalent Pd forms many planar complexes with a coordination number of 4. The tetrachlorides are quite soluble. When a solution of palladium(II) chloride is oxidized with chlorite or chlorate ion, Pd(IV) is formed, which has a coordination number of 8. The addition of NH$_4$Cl to such a solution precipitates ammonium hexachloropalladate(IV) as a red compound. It is somewhat less stable than the platinum analog.

The soluble yellow-brown palladium(II) nitrate is formed by dissolving finely divided Pd in warm HNO$_3$ and then crystallizing the compound from this solution. The analogous sulfate is similarly formed from H$_2$SO$_4$. It crystallizes as a red-brown dihydrate. Both these compounds easily hydrolyze.

Palladium(II) sulfide is precipitated as a brown powder by adding H$_2$S to a solution of palladium(II) ion. When this sulfide is heated with sulfur at 400°C, the insoluble disulfide is formed. The excess sulfur can be extracted with CS$_2$ to yield the gray-black crystalline palladium(IV) sulfide. This compound is not soluble in single acids but is soluble in aqua regia.

Some Pd complexes are important analytically or in the refining of Pd. The yellow dimethylglyoxime compound is quantitatively precipitated from a HCl solution of palladium(II) chloride by the addition of an alcoholic solution of dimethylglyoxime. Palladium(II) has a great affinity for nitrogen-containing ligands. The di- and tetramine find use in refining.

Palladium, as with other members of the platinum group, exhibits catalytic activity for various reactions. One of its best known uses is in conjunction with other platinum metals in the catalytic converters of present-day automobiles.

As reported by Chung-Chiun Liu et al. (*Science*, **207**, 188–189, 1980), a palladium-palladium oxide miniature pH electrode has been developed for pH measurement. The miniature wire-form electrode exhibits a super-Nernstian behavior and gives a mean pH response of 71.4 mV per [pH] (standard deviation, 5.3 mV). The electrode may find applications in biological, medical, and clinical studies.

Uses

Alloys for electrical relays and switching systems in telecommunication equipment, catalyst for reforming cracked petroleum fractions and hydrogenation, metallizing ceramics, "white" gold in jewelry, resistance wires, hydrogen valves (in hydrogen separation equipment), aircraft spark plugs, and protective coatings.

See also **Chemical Elements**.

Additional Reading

Carter, G.F. and D.E. Paul: *Materials Science and Engineering*, ASM International, Materials Park, OH, 1991.

Davis, J.R.: *Metals Handbook*, 2nd Edition, ASM International, Materials Park, OH, 1998.

Greenwood, N.N. and A. Earnshaw: *Chemistry of the Elements*, 2nd Edition, Butterworth-Heinemann, Woburn, MA, 1997.

Krebs, R.E.: *The History and Use of Our Earth's Chemical Elements: A Reference Guide*, Greenwood Publishing Group, Inc., Westport, CT, 1998.

Lide, D.R.: *CRC Handbook of Chemistry and Physics* 86th Edition, CRC Press, Boca Raton, FL, 2005.

Parker, P.: *McGraw-Hill Encyclopedia of Chemistry*, 2nd Edition, The McGraw-Hill Companies, Inc., New York, NY, 1993.

Sinfelt, J.H.: "Bimetallic Catalysts," *Sci. Amer.*, 90–98 (September 1985).

Staff: *ASM Handbook—Properties and Selection: Nonferrous Alloys and Pure Metals*, ASM International, Materials Park, OH, 1990.

Stwertka, A. and E. Stwertka: *A Guide to the Elements*, Oxford University Press, Inc., New York, NY, 1998.

Web References

Stillwater Palladium, http://www.stillwaterpalladium.com/

The Platinum Group Metals Database, http://www.platinummetalsreview.com/jmpgm/index.jsp

WebElements.com Palladium, http://www.webelements.com/webelements/elements/text/Pd/index.html

Linton Libby, Chief Chemist, Simmons Refining Company, Chicago, IL

PALMER DROUGHT SEVERITY INDEX (PDSI). An index formulated by Palmer (1965) that compares the actual amount of precipitation received in an area during a specified period with the normal or average amount expected during that same period. The PDSI is based on a procedure of hydrologic or water balance accounting by which excesses or deficiencies in moisture are determined in relation to average climatic values. Values taken into account in the calculation of the index include precipitation, potential and actual evapotranspiration, infiltration of water into a given soil zone, and runoff. This index builds on Thornthwaite's work (1931, 1948), adding 1) soil depth zones to better represent regional change in soil water-holding capacity; and 2) movement between soil zones and, hence, plant moisture stress, that is, too wet or too dry.

See also **Drought Indices**.

Additional Reading

Palmer, W.C.: *Meteorological Drought Research Paper 45*, U.S. Weather Bureau, Washington, D.C., 1965.

Web Reference

Palmer Drought Severity & Crop Moisture Indices: http://www.cpc.ncep.noaa.gov/products/analysis_monitoring/cdus/palmer_drought/

PALMITIC ACID. [CAS: 57-10-3]. CH$_3$(CH$_2$)$_{14}$COOH, formula weight 256.42, white crystalline powder, mp 64°C, bp 271.5°C, sp gr 0.849. The acid is insoluble in H$_2$O, moderately soluble in alcohol, soluble in ether. About 60% of the content of palm oil is palmitic acid.

Palmitic acid is present as cetyl ester in spermaceti from which, by hydrolysis, the acid may be obtained; it is present in bee's wax as the melissic ester; and in most vegetable and animal oils and fats, in greater or lesser amounts, as glyceryl tripalmitate or as mixed esters, along with stearic and oleic acids. Palmitic acid is separated from stearic and oleic acids by fractional vacuum distillation and by fractional crystallization. With NaOH, palmitic acid forms sodium palmitate, a soap. Most soaps are mixtures of sodium stearate, palmitate, and oleate.

Representative esters of palmitic acid are: methyl palmitate, C$_{15}$H$_{31}$COOCH$_3$, mp 30°C, bp 195°C at 15 mm pressure; ethyl palmitate, C$_{15}$H$_{31}$COOC$_2$H$_5$, mp 24°C, bp 185°C at 10 mm pressure; cetyl palmitate, C$_{15}$H$_{31}$COOC$_{16}$H$_{33}$, mp 54°C; glyceryltripalmitate (tripalmitin), C$_3$H$_5$(COOC$_{15}$H$_{31}$)$_3$, mp 65°C, bp 310°C, approximately.

As the glyceryl ester, palmitic acid is one of the constituents of many vegetable and animal oils and fats.

Palmitic acid finds use in the production of cosmetics, food emulsifiers, pharmaceuticals, plastics, and soaps. One commercial formulation contains 95% palmitic acid, 4% stearic acid, and 1% myristic acid; another preparation contains 50% palmitic acid and 50% stearic acid.

See also **Vegetable Oils (Edible)**.

PALMITOLEIC ACID. [CAS: 373-49-9]. Also called *cis*-9-hexa decanoic acid, formula $CH_3(CH_2)_5CH:CH(CH_2)_7COOH$. This is an unsaturated fatty acid found in nearly every fat, especially in marine oils (15–20%). At room temperature, it is a colorless liquid. Insoluble in water; soluble in alcohol and ether; mp $1.0\,°C$; bp $140–141\,°C$ (5 millimeters pressure). Insoluble in water; soluble in alcohol and ether. Combustible. Palmitoleic acid is used in organic synthesis, and as a standard in chromatographic analysis. See also **Vegetable Oils (Edible)**.

PALM OIL.   See **Vegetable Oils (Edible)**.

PALM TREES (*Palmaceae*). This family of plants contains some 1,100 species, most of which are tropical. Many are of large size. A tall woody stem bearing at its top a crown of large compound leaves characterizes most of them, although some are short and bushy, while a few are vinelike. Only rarely does branching occur, although many do develop numerous basal offshoots. The leaves are either pinnately or palmately compound. The inflorescence is either a simple or a compound spike, surrounded by a spathe, which may become extremely large. The flowers are regular, with their parts in threes or multiples of three. Wind pollination generally occurs in palms. The fruit is a berry, a drupe, or a nut, which in many species has a fibrous mesocarp.

Many species of palms are planted extensively, because of their ornamental habit, in tropical and subtropical countries. Large numbers are also grown as greenhouse or conservatory plants; a few species are of great economic value.

Date Palm (*Phoenix dactylifera*). The mature trunk of this tree often attains a height of 75–100 feet (22.5–30 meters). From the base of this trunk, numerous adventitious roots extend into the soil to a depth of 20 feet (6 meters) or more. From the base of the stem, many basal offshoots develop. The pinnate leaves are from 10 to 20 feet (3 to 6 meters) long, with the individual pinnate sharp-pointed, linear, and from 1 to 3 feet (0.3 to 0.9 meter) long. These leaves remain attached for an indefinite time to the trunk, giving it a very ragged appearance. In cultivation, the old leaves are removed. The inflorescence is a large branched spike enclosed in a large tough spathe. See Fig. 1. In a mature tree, a single flower cluster

Fig. 2.   A bunch of Barhee dates, showing wire spreader that serves to provide ventilation in the center. (*USDA photo.*)

may be composed of several thousands of flowers. The flowers are of two different sexes, borne on different plants, and are small, wax-white, and of firm texture. The pistillate (see also **Flower**) flowers have three carpels, each with a short curved stigma. The carpels are almost surrounded by the perianth, composed of three united sepals and three petals. The staminate flowers, bearing six stamens, are of larger size than the pistillate and more showy. The fruit is a drupe. See Fig. 2. After fertilization, only one of the three carpels develops, the other two being suppressed. At first the carpel is wax-white, but some time after fertilization, it becomes green and remains so during growth. As the drupe becomes mature, the color changes to yellow or red or a blending of these colors, according to the variety of date. When ripe, the color of the fruit varies from pale straw color through deep amber to a deep purple. The seed of the date contains an abundance of endosperm and a rather small embryo.

Many varieties of dates are grown in cultivation. Since cross-pollination is usually necessary to produce seeds, propagation by seeds cannot be used to increase the number of plants. For such plants would not remain true to the seed-bearing parent in type but would be affected by the pollen from the staminate plant; such crossing would cause the formation of new varieties with the possibility that less desirable forms might arise. Furthermore, about half the seedlings would be staminate, producing no fruit. To perpetuate desirable varieties, propagation is largely by means of the basal offshoots. These shoots are cut from the parent and planted, ensuring new plants identical with the parent.

Cultivation of the date has been carried on in Arabia and adjoining countries since prehistoric times. It is probable that the plant is native to this region, even though it is unknown in the wild state. Today the date palm is grown widely in regions having a sufficiently hot climate. The Mesopotamian valley is the chief producing region of the world. In the United States, the date can be grown successfully only in a very limited region in the southern parts of California, Nevada, and Arizona.

The principal product of the plant is the edible fruit. In the Orient, the fruit is often fermented, yielding a very potent alcoholic drink, and also vinegar. Various materials used in building houses, and in making baskets, ropes, and household articles are also obtained from the tree in regions where other sources of such materials may be lacking.

Another useful palm is the oil palm, *Elaeis guineensis*, a native plant occurring in large numbers in the forests of the west coast of Africa. The plant is also extensively cultivated elsewhere in Africa and in the East Indies, particularly in Sumatra. From the fruit of this palm, two valuable oils are obtained. The orange-yellow pericarp yields palm oil, much used in the making of soaps. The endosperm yields a white pleasantly flavored

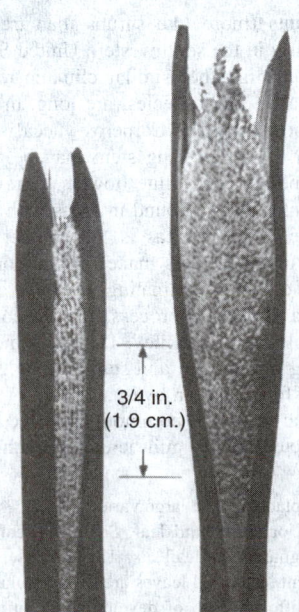

3/4 in.
(1.9 cm.)

Fig. 1.   Flower clusters of the date palm as they appear when the spathe first opens. (Left) female; (right) male. (*USDA diagram.*)

oil known as palm-kernel oil, also used in making soaps. In Africa, the fruits are eaten by the natives.

In India, *Phoenix sylvestris* is widely cultivated as a plant from which sugar is obtained. Many other palms are used as sources for sugar in regions where they grow.

Areca Palm (*Areca catechu*). This tree is native to Malaysia and is extensively cultivated in the southern Asian countries and in the East Indian islands. It is a slender unbranched tree that reaches a height of about 40 feet (12 meters) with a trunk diameter of from 3 to 4 inches (7.6 to 10 centimeters). It bears at its top a crown of 6 to 10 large leaves. The fruit is slightly more than an inch in diameter and has a fibrous rind surrounding the very hard seed. This seed (the *betel nut*) is chewed by many people in the Asiatic countries. To prepare them for chewing, the fruits are gathered just before maturity, boiled, sliced, and dried. A piece of this dried fruit, together with a bit of lime, is wrapped in a betel leaf and the whole placed in the mouth. Chewing this preparation causes an abundant flow of saliva, which is colored brick red and stains the mouth and teeth. The betel leaves are obtained from an entirely different plant, the black pepper (*Piper betle*).

Coconut Palm (*Cocos nucifera*). To the natives of tropical islands, the coconut is of great value. It provides shelter, dishes, food, and drink. However, products of the tree are valuable throughout the world, well over 1 million tons of coconut oil being produced annually. The coconut tree is a striking plant having a columnar trunk some 60 to 80 feet (18 to 24 meters) in height, bearing at its top a crown of bright green pinnate leaves, each leaf ranging from 15 to 20 inches (38 to 50.8 centimeters) in length. The fruit is a large nut composed of three united carpels. The shell of the coconut has three distinct layers; the outer, or epicarp, is thin, smooth, and brown; the middle, or mesocarp, is thick and fibrous; it is the source of the fiber coir. Usually these two layers are removed before the coconuts are shipped, but not infrequently the entire fruit is exhibited as a curiosity by dealers in fruits. Within is the hard brown shell or endocarp. This contains the seed, which consists of an embryo, located under one of the germ pores at the end of the endocarp, and the endosperm, which is of two parts, one the familiar white meat of the coconut, the other the milk which partially fills the cavity of the seed. It is within the meat that the embryo is embedded. The thin brown skin immediately investing the meat and sticking to it when it is removed from the shell is the inner seed-coat. The hard shell of the coconut is much used as a dipper or as a vessel for storage of various substances.

The leaves of the tree yield a fibrous material used by the native islanders and are also used at times as a thatch for shelters. The trunk of the tree is sometimes used by cabinet makers for ornamental work.

Commercially, the most valuable product of the coconut is copra, the dried meat of the nut. This is obtained by splitting the nut and drying the meat, preferably by the sun. Artificial drying is done in regions of great humidity. From the meat is obtained coconut oil, which forms about 63% of the meat. The natives obtain this oil by various means. The hot sun of the tropics may cause it to dry out of the pounded meat. Crude presses are often used to squeeze the oil from the dried meats. Or again the crushed dried meats may be placed in hot water, the melted oil rising to the top and being skimmed off. Any one of these methods is sufficient to supply the moderate needs of the natives, but utterly inadequate to meet the requirements of the civilized nations.

For modern treatment, copra is shipped to large factories. Here it is cleaned and ground, then heated and pressed. The meat is then ground up once more, cooked in water, and pressed by powerful hydraulic presses. By this method, nearly all the oil contained in the copra is extracted.

The principal use of the oils is in the making of soap, especially in soaps that will produce a copious lather. Considerable quantities are also used in making shortenings. Some oil finds use in salad oils and in confectionery.

The copra cake remaining after the oil is expressed may be used as a stock feed.

Seychelles Palm. This tree is found on Seychelles Island in the Indian Ocean northeast of Madagascar. It is of interest because the tree does not bear fruit until it is about 100 years of age. The tree is straight with a bulbous trunk, marked by numerous holes. Roots penetrate into these holes, providing stabilization of the tree against gales. The leaves are large and fan-shaped, ranging from 10 to 12 feet (3 to 3.6 meters) in length. The fan portion covers an area of about 7 by 15 feet (2.1 to 4.5 meters). A single tree will have from 20 to 30 leaves. The leaves are so constructed that rain drains down to the base of the tree. Once the fruit commences, it takes from 10 to 12 months to ripen. The fruit contains the largest seed known in the plant world, the nuts weighing from 30 to 40 pounds (13.6 to 18.1 kilograms) at maturity. The nut is brown, smooth, with a thorny shell. The shell is easily torn away. The fruit is called *durian* and is egg-shaped with a very thick rind. The edible part is soft and creamy, rather custard-like in texture.

The tree was discovered on the aforementioned island in 1742. The nuts can be found on the market throughout Southeast Asia.

Raffia Palm (*Raphia pedunculata*). This African palm tree has leaves of great length, sometimes as much as 25 feet (7.5 meters). The leaf epidermis may be removed in strips. These strips, known commercially as raffia, are used in basketwork and for tying. The leaves curl so that the raffia must be obtained before they become dry. The tree is extensively found on Madagascar.

For record palm trees in the United States, see Table 1.

Development of Palm Leaves. In a most unusual and interesting article, D.R. Kaplan (*Sci. Amer.*, **249** (1), 98–107, July 1983) describes in detail well beyond the scope of this encyclopedia the development of palm leaves. The compound leaves of most plants usually arise either from differential growth or from selective cell death. Not so with palm leaves. Their development pathway combines both of these processes. Representative of the content of the Kaplan article is the following: "Palm-leaf blades typically show one or the other of two major configurations that reflect differences in the distribution of growth during the course of dissection.[1] One of the two is characterized as *pinnate* because of its feathery appearance. The other is characterized as palmate because it is fanlike, rather like a hand with its fingers spread. Pinnate fronds tend to have short petioles; palmate fronds have long ones. What makes the developmental mode of palm leaves distinctive is that at first their blade surfaces are thrown into a series of pleats known as plications. A process of tissue separation along certain of these pleats then cleaves the pleated surface into a series of leaflets. It is easy to see, in looking at a fully developed leaf of the palmate kind, that the segmented nature of the leaf originates from the pleating of the leaf blade. Indeed, as with a Japanese fan, the leaf can be closed by pressing the pleats together. Since each leaflet is V-shaped in cross section, it appears to have been cut out of a pleated surface."

It has been observed that some molecular biologists tend to extrapolate their findings of developmental controls that operate in lower plants to the higher plants, an approach which may lead to oversimplification of the growth or developmental processes. Studies of palm leaf development tend to indicate that many more and different molecular control systems may exist than previously assumed.

Yuccas. These plants (more like shrubs than trees) are abundant in desert areas, particularly in the southwestern United States and in Mexico, although they are found in other similar climatic regions of the world. Yucca leaves, all basal in most species, are long, thick, and shiny green with dagger-like points. Thus, the Carneros yucca is sometimes referred to as Spanish dagger. The flowering stem may be several feet tall and bears many pendant bell-shaped white flowers. A few species have been urbanized and frequently will be found in areas with a mild climate.

An unusual feature of the yuccas is the manner by which they are fertilized. The shape of the flowers makes pollination extremely difficult and consequently the common pollinating insects avoid them. Pollination is accomplished by a symbiotic process, first discovered in 1872 by a Missouri state entomologist, C.V. Riley. The yucca moth (or Navajo yucca borer) deposits an egg in the pistil and stuffs the hole in which the egg is laid with pollen, thus fertilizing an ovary.

A yucca of particular interest is commonly referred to as the Joshua tree, which enjoys a protected area ($\frac{1}{2}$ mil acres) in southern California at the

---

[1] Leaves of flowering plants have a large variety of shapes and sizes. A common variant is the dissected or compound leaf. See also **Leaf**. The blades of these leaves are cut into segments or leafleats. Kaplan observes that, in terms of comparative development, dissected leaves are of particular interest because they clearly illustrate how different path of development can lead to leaves that are closely similar in appearance. The giant fronds of palm trees are the largest and most complex of all dissected leaves.

TABLE 1. RECORD PALM TREES IN THE UNITED STATES[1]

Specimen	Circumference[2]		Height		Spread		Location
	Inches	Centimeters	Feet	Meters	Feet	Meters	
Buccaneer palm (*Pseudophoenix sargentii*)	26	66	25	7.6	8	2.4	Florida
Coconut palm (1979) (*Cocos nucifera*)	60	152	93	28.3	27	8.2	Hawaii
Royalpalm (1995) (*Roystonea elata*)	50	127	99	30.2	18	5.5	Florida
Florida Silverpalm (1979) (*Coccothrinax argentata*)	19	48	27	8.2	8	2.4	Florida
Florida Silverpalm (1994) (*Coccothrinax argentata*)	21	53	25	7.6	7	2.1	Florida
California washingtonia (fanpalm) (1991) (*Washingtonia filifera*)	120	305	83	25.3	21	6.4	California
California washingtonia (fanpalm) (1991) (*Washingtonia filifera*)	100	254	101	30.8	22	6.7	California
California washingtonia (fanpalm) (1997) (*Washingtonia filifera*)	141	358	66	20.1	18	5.5	California
PALMETTOS							
Cabbage palmetto (1994) (*Sabal palmetto*)	69	175	60	18.3	14	4.3	Florida
Mexican palmetto (1995) (*Sabal mexicana*)	61	155	50	15.2	15	4.6	Texas
Mexican palmetto (1995) (*Sabal mexicana*)	61	155	45	13.7	20	6.1	Texas
Saw palmetto (1994) (*Serenoa repens*)	22	56	20	6.1	13	4	Florida
Saw palmetto (1987) (*Serenoa repens*)	27	69	21	6.4	8	2.4	Florida
YUCCAS							
Beaked yucca (1994) (*Yucca rostrata*)	48	122	16	4.9	9	2.7	Texas
Carneros (Spanish-dagger) yucca (1977) (*Yucca carnerosana*)	51	130	25	7.6	10	3	Texas
Faxon yucca (1991) (*Yucca faxoniana*)	91	231	18	5.5	9	2.7	Texas
Joshua tree yucca (1999) (*Yucca brevifolia*)	155	394	46	14	38	11.6	California
Mojave yucca (1987) (*Yucca schidigera*)	66	168	24	7.3	7	2.1	California
Moundlilly yucca (L1998) (*Yucca gloriosa*)	106	269	33	10.1	31	9.4	California
Schott yucca (1997) (*Yucca schottii*)	43	109	15	4.6	12	3.7	Arizona
Sooptree yucca (1996) (*Yucca elata*)	62	157	30	9.1	11	3.4	Arizona
Torrey yucca (1987) (*Yucca torreyi*)	86	218	23	7	6	1.8	New Mexico
Trecul yucca (1991) (*Yucca treculeana*)	24	61	30	9.1	9	2.7	Texas

[1] From the "National Register of Big Trees," American Forests (by permission).
[2] At 4.5 feet (1.4 meters).

junction of the Mojave and Colorado deserts. The Joshua Tree National Monument is located 140 miles (225 km) east of Los Angeles. The record specimen (see accompanying table) is located in the San Bernardino National Forest (California). Most experts consider calling this plant a tree is uncongenial scientifically — because it does not produce annual growth rings, the trunk is fibrous, the bark is soft and corklike, and its branching is erratic, among other factors. It has been reported that early pioneers likened the appearance of the Joshua "tree" to the biblical Joshua waving his arms to encourage the pioneers to push forward. Further lore of this plant is reported by S.L. Keith in an article, "A Tree Named Joshua." *American Forests*, 38–41, July 1982.

Additional Reading

Heinerman, J.: *Aloe Vera, Jojoba, and Yucca*, Keats Publishing, Inc., Chicago, IL, 1990.

Henderson, A., G. Galeano, and R. Bernal: *Field Guide to the Palms of the Americas*, Princeton University Press, Princeton, NJ, 1997.

Howard, F.W., D. Moore, R. Giblin-Davis, and R. Abad: *Insects on Palms*, Oxford University Press, Inc., New York, NY, 1999.

Irish, M., et al.: *Agaves, Yuccas, and Related Plants: A Gardener's Guide*, Timber Press, Portland, OR, 2000.

Jones, D.L.: *Palms throughout the World*, Smithsonian Institution Press, Washington, DC, 1995.

Web Reference

www.usda.gov (Search: Palm Trees)

PALYNOLOGY. The study of pollen of seed plants and spores of other embryophytic plants, whether living or fossil, including their dispersal and applications in stratigraphy and paleoecology.

PAMPAS DEER. See **Deer**.

PAMPERO. See **Winds and Air Movement**.

PANCAKE ICE. Roughly circular accumulations of frazil ice, usually less than about 3 meters (10 feet) in diameter, with raised rims caused by collisions. This form of ice is common in the Antarctic. Pancake ice may develop from grease ice or shuga.

See also **Frazil Ice**; **Grease Ice**; **Ice**; and **Shuga**.

AMS

PANCREAS.   An elongated glandular organ located in the midportion of the abdominal cavity. The organ secretes juices that aid in digestion, and certain of its cells produce the endocrine hormone *insulin*, which aids in the regulation of blood sugar levels. The pancreas is a long, soft, yellowish-gray gland which lies transversely on the posterior abdominal wall, its right end enclosed by the curve of the duodenum and its left end touching the spleen. The gland lies, for the most part, behind the stomach. It is about 6 inches (15 centimeters) long and weighs about 3 ounces (85 grams). The gland secretes a clear, watery, alkaline fluid (the *pancreatic juice*), which passes to the duodenum through the pancreatic duct. Pancreatic juice is one of the chief chemical agents in digestion, for it contains enzymes that break down starch into sugar, fats into glycerine and fatty acids, and proteins into peptones and amino acids. Additionally, the pancreas also secretes directly into the bloodstream the antidiabetic hormone, *insulin*. See also **Diabetes Mellitus**. Patients who must have the pancreas removed are able to live only so long as they receive injections of insulin. It is the hormone function of the pancreas that makes it a part of the endocrine system. See also **Endocrine System**; and **Hormones**.

Pancreatic juice flows into the intestine when acid material comes into contact with the duodenal mucosa. A hormone, *secretin*, is liberated from the mucous coat of the duodenum when an acid substance is in contact with the coat. This hormone enters the bloodstream and is conveyed to the pancreas in a few seconds. There, it stimulates the glandular cells to produce pancreatic juice. There is also nervous control of pancreatic secretion, so that even the thought of food may stimulate its secretion.

Pancreatitis remains a disease of considerable mystery. In the United States, it has been reasonably well established that cholelithiasis (gallstones) and chronic alcoholism together account for about 80% of cases. Other diverse causes include hyperlipidemia, ductal obstruction, viral infection, some drugs, and impaired pancreatic perfusion. In an excellent article (reference listed), W.H. Steinberg, (George Washington University) reviews the numerous hypotheses that attempt to explain at least some of the probable causes of acute pancreatitis. Numerous cases of acute pancreatitis remain *unclassified* — that is, *idiopathic* (cause unknown).

The symptoms of acute pancreatitis are excruciating pain in the region of the organ and radiating toward the back. Unlike the undulating pain of gallbladder disease, the pain is essentially continuous and may persist for several hours or days, but because of its severity, the patient usually seeks medical attention promptly. When the pain reaches a peak plateau, it is usually accompanied by nausea and vomiting and, in about half of the cases, fever of 100–101 °F (37.8–38.3 °C) will be present even though infection may not be obvious. If the condition is allowed to persist without medical attention for several hours, shock and obtundation may occur in about half of the cases. Through differential diagnosis, the physician will check out the symptoms and laboratory findings to make certain that the symptoms are not caused by other conditions associated with abdominal pain, such as acute cholecystitis, perforated duodenal ulcer, myocardial infarction, and sometimes pneumonia, which can produce severe pain in the epigastrium as well as fever.

In view of the severity of pancreatitis, the mortality rate of an acute attack is comparatively low (about 5% of cases). However, where acute suppurative or hemorrhagic pancreatitis is present, the mortality ranges from 50 up to 90%. Where acute pancreatitis is related to alcoholic intake, the patient is prone to developing chronic disease. Statistics indicate that between 10 and 20% of chronic alcoholics will be found to have chronic pancreatitis when examined during surgery or autopsy. Chronic pancreatitis is also evident in patients, notably males approaching 40 years of age, who consume inordinate quantities of protein and fat.

Post-Cardiac Surgery. In 1990, more than one-half million cardiac surgical operations were performed in the United States. Pancreatitis is an infrequent but well-recognized complication of cardiac surgery, including cardiac transplantations and pediatric procedures. It has been reported that ischemia and drugs, such as phenylephrine, norpinephrine, narcotics, steroids, and calcium chloride contribute to "post-pump" pancreatitis. The cause of these complications is poorly understood. In 1991,

C. Fernández-Del Castillo (Harvard Medical School and Massachusetts General Hospital) studied 300 consecutive patients undergoing cardiac surgery with cardiopulmonary bypass. The results of that study: "Evidence of pancreatic cellular injury was detected in 80 patients (27%), of whom 23 had associated abdominal signs or symptoms and 3 had severe pancreatitis." Conclusions: "Pancreatic cellular injury, as indicated by hyperamylasemia of pancreatic origin, is common after cardiac surgery. The administration of large doses of calcium chloride is an independent predictor of pancreatic cellular injury and may be a cause of it."

Cancer of the Pancreas. Middle-aged and elderly men are the most common sufferers from cancer of the pancreas. The disease is only one-third as frequent in women. When the tumor originates in the head of the gland and blocks the bile duct, a steadily increasing jaundice appears, and increasing itching of the skin develops. When the cancer begins in the body of the organ, pain is the most common symptom. The pain is usually constant and of a deep, penetrating character that radiates through to the back, between the shoulder blades. If the cancer occurs in the tail of the gland, there may be no symptoms until the growth has spread to the liver, with resulting loss of weight and general impairment of bodily function. Treatment is removal of the whole gland, a surgical procedure made possible because of means available for controlling shock and the availability of substances which can substitute for the secretions of the gland.

As observed by B.A. Chabner and M.A. Friedman (National Cancer Institute), "Islet-cell carcinoma of the pancreas is a rare, slow-growing neoplasm with a broad range of endocrinologic manifestations". C.G. Mortel (Mayo Clinic, Rochester, Minnesota) tested the efficacy of several drugs in 105 randomly assigned patients with advanced islet-cell carcinoma. Conclusions of the study: "The combination of streptozocin and doxorubicin is superior to the current standard regiment of streptozocin plus fluorouracil in the treatment of advanced islet-cell carcinoma."

Sometimes agglomerations of glandular tissue that simulate cancer occur in the pancreas. These are called islet cell tumors and may not be malignant. Because of the possibility of their becoming malignant, the tumors should be removed. Sometimes this mass of glandular tissue secretes so much insulin that weakness and faintness occur. The patient may even lapse into coma because of the extent to which the blood is depleted of its sugar. In recent years, cases have been reported in which islet cell tumors are associated with peptic ulcerations of the jejunum.

For the preoperative localization of endocrine tumors, such as insulinomas (99% of cases), endoscopic ultrasonography is reported as a highly sensitive and specific procedure to be used after clinical and laboratory diagnosis has been established.

Transplantation. H. Katz (University of Minnesota) and a group of researchers studied a small number of patients with insulin-dependent diabetes mellitus after they had a pancreas-kidney transplantation and determined that carbohydrate metabolism was similar to that in nondiabetic subjects who received the same immunosuppressive agents after kidney transplantation. Pancreas transplantation provides a means of normalizing plasma glucose concentrations, thus offering a way to delay or prevent the long-term complications of diabetes. Whether long-term systemic delivery of insulin has deleterious effects on carbohydrate metabolism in humans remains unknown.

K. Pyseoqaki (University of Minnesota) and a group of researchers reported in 1992 that the transplantation of pancreatic islets rather than the whole pancreas has been introduced as a treatment to diabetes mellitus. The group studied five patients ranging in age from 12 to 37 years. If successful, islet transplantation would have advantages over transplantation of the entire pancreas. However, there are only a few reports of transient successful transplantation of cadaveric islets in humans. The study conclusions: "Intrahepatic transplantation of as few as 265,000 islets can result in the release of insulin and glucagon at appropriate times and in prolonged periods of insulin independence."

Additional Reading

Becker, K.L., W. Hung, C.R. Kahn, et al.: *Principles and Practice of Endocrinology and Metabolism*, 3rd Edition, Lippincott Williams & Wilkins, Philadelphia, PA, 2001.

Castillo, C. Fernández-Del, et al.: "Risk Factors for Pancreatic Cellular Injury After Cardiopulmonary Bypass," *N. Eng. J. Med.*, 382 (August 8, 1991).

Chabner, B.A. and M.A. Friedman: "Progress Against Rare and Not-so-Rare Cancers," *N. Eng. J. Med.*, 563 (February 20, 1992).

PANDAS 3919

Cruickshank, A.H. and E.W. Benbow: *Pathology of the Pancreas*, 2nd Edition, Springer-Verlag Inc., New York, NY, 1995.

Doppman, J.L.: *Pancreatic Endocrine Tumors–The Search Goes On*, N. Eng. J. Med., 1770 (June 23, 1992).

Howard, J.M., R. Prinz, Y. Idezuki, and I. Ihse: *Surgical Diseases of the Pancreas*, Lippincott Williams & Wilkins, Philadelphia, PA, 1997.

Katz, H., et al.: "Effects of Pancreas Transplantation on Postprandial Glucose Metabolism," N. Eng. J. Med., 1278 (October 31, 1991).

Laufer, H. and R.G.H. Downer: *Endocrinology of Selected Invertebrate Types*, John Wiley & Sons, Inc., New York, NY, 1988.

Lee, S.P., J.F. Nicholls, and H.Z. Park: Billiary Sludge As a Cause of Acute Pancreatitis, N. Eng. J. Med., 589 (February 27, 1992).

Lloyd, R.V.: *Endocrine Pathology*, Springer-Verlag, Inc., New York, NY, 1990.

Lott, J.A.: *Clinical Pathology of Pancreatic Disorders*, Vol. 2, Humana Press, Totowa, NJ, 1997.

Moertel, C.B., et al.: "Streptozocin-Doxorubicin, Streptozocin-Fluorouracil, or Chlorozotocin in the Treatment of Advanced Islet-Cell Carcinoma," N. Eng. J. Med., 510 (February 20, 1992).

Moore, W.T. and R.C. Eastman: *Diagnostic Endocrinology*, 2nd Edition, Mosby-Year Book, Inc., St. Louis, MO, 1996.

Owen, D.A. and J.K. Kelly: *Pathology of the Gallbladder, Biliary Tract and Pancreas*, W. B. Saunders Company, Philadelphia, PA, 2000.

Pyzdrowski, K.L., et al.: "Preserved Insulin Secretion and Insulin Independence in Recipients of Islet Autografts," N. Eng. J. Med., 220 (July 23, 1992).

Reber, H.A.: "Acute Pancreatitis–Another Piece of the Puzzle," N. Eng. J. Med., 423 (August 8, 1991).

Rösch, T., et al.: "Localization of Pancreatic Endocrine Tumors by Endoscopic Ultrasonography," N. Eng. J. Med., 1721 (June 25, 1992).

Russell, R.C.G., D. Carr-Locke, M.G. Sarr, and H.G. Beger: *The Pancreas*, Vol. 2, Blackwell Science Inc., Malden, MA, 1998.

Skerrett, P.J.: "New Transplant Method Evades Immune Attack," Science, 1248 (September 14, 1990).

Steinberg, W.M.: "Acute Pancreatitis—Never Leave a Stone Unturned," N. Eng. J. Med., 635 (February 27, 1992).

Trey, C. and C.C. Compton: "Fever Three Weeks after an Operation for Pancreatic Cancer," N. Eng. J. Med., 318 (February 1, 1990).

Waters, D.L., et al.: "Pancreatic Function in Infants Identified as Having Cystic Fibrosis in a Neonatal Screening Program," N. Eng. J. Med., 303 (February 1, 1990).

PANDAS *(Mammalia, Carnivora, Ailuridae)*. These animals are of the family *Ailuridae*; they were once classified together with the raccoons. The Lesser Panda (*Ailurus fulgens*) (also called Red Panda), reaches a length of 79 to 112 centimeters (31 to 44 inches), has a tail length of 28 to 48.5 centimeters (11 to 19 inches), and weighs from 3 to 4.5 kilograms (6.5 to 10 pounds). Two subspecies are distinguished. The head is unusually large, and the snout is short and pointed. The ears are medium-sized and are tapered. The eyes are small and brown, and the nose is black. There are 38 teeth, with three premolars in the upper jaw and four lower premolars. The trunk is supported by short, bear-like legs. The toes have short, partially retractile claws. The soles of the hands and feet are covered with woolly hair, and the lesser panda walks in a plantigrade manner. The fur is soft and long, and the underfur is very thick. The chin and inside of the ears are white. A rust-red stripe beneath each eye separates the white snout from the white cheeks. The upper side of the body is rust-brown and yellowish, while the forehead tends toward a rust-yellow hue. A white spot is located above each eye. The tail is long and bushy and is reddish-brown with distinctly separate light reddish rings. The underside, and the rear of the ears and limbs, are gleaming black. See Fig. 1.

The habitats of the lesser panda are mountain forests and bamboo thickets from 1,800 to 4,000 meters (5,906 to 13,124 feet) along the southeastern slopes of the Himalayas. Its distribution extends from Nepal across Bhutan, Sikkim, northern Assam, northern Burma, and as far east as the western Chinese provinces Yunnan and Szechuan. The western lesser panda (*Ailurus fulgens fulgens*) inhabits the western part of the distribution, while Styan's lesser panda (*Ailurus fulgens styani*) inhabits Szechuan, Yunnan, and northeastern Burma. Styan's lesser panda has a larger body, higher forehead region, and a more intense rust-yellow coloration than the western lesser panda.

Lesser pandas feed chiefly on bamboo shoots, juicy grasses, roots, berries, and fruits, occasionally taking young birds, eggs, small rodents, and insects. They are crepuscular and nocturnal animals, which may be found equally often in trees or on the ground. On the ground the lesser

Fig. 1. Pandas.

panda runs about at a gallop with its back arched and tail raised; it places its hands inward with each step. The hairy soles reduce the danger of slipping on wet, smooth branches in the foggy, high-altitude forests; they also reduce heat loss on snow-covered and ice-encrusted rocky mountain ground. Similar hair covering on the soles is found in the giant panda and the polar bear.

Although lesser pandas can leap up to 1.5 meters (5 feet), they avoid doing so if they can climb to wherever they are going. The sharp claws are put to good use when the panda is climbing. During the day this very heat-sensitive species sleeps in shady branches or in tree hollows, lying coiled up on its side. The tail is used as a pad or to cover the face. The lesser panda often stretches out on a tree limb and sleeps there. It washes itself in a catlike manner; first it licks the soles of the hands for some period of time, and then it rubs the hands across the forehead and the ears. Food is skillfully seized with the hands and is eaten bit by bit, often with the panda sitting upright.

Lesser pandas are rarely seen in their native habitat. They are asocial and typically live alone, but sometimes in pairs. Calls to conspecifics include shrill cries, a whistle which is often uttered repeatedly, a peeping, and a birdlike chirping. If the lesser panda is endangered it withdraws into a rock crevice or a tree. When it is greatly excited, the anal glands secrete a penetrating, musky liquid; generally males use the anal glands to mark tree branches as they forage for food or to mark their territory on the ground.

In spring the female bears 1 to 4 young after a 130-day gestation period. The newborn young is about 20 centimeters (7.8 inches) long, 5 centimeters (2 inches) of which is the tail. The upper side is pale rust-red, with a gray-white head; the characteristic markings of the face are not yet visible, but the black spot on the rear side of the ears can already be clearly seen. The prominent black hair of the legs and belly of the adult exist as a dark gray color in the young panda, and the ring pattern on the tail cannot yet be seen. After 2 to 4 weeks the eyes open. After the young are six weeks old they begin to attain adult coloration.

The Giant Panda (*Ailuropoda melanoleuca*) is the closest related species to the lesser panda. It is the only member of its genus, *Ailuropoda*, and it has not changed its appearance since the Pleistocene. The overall length from the nose to the base of the tail reaches 150 centimeters (59 inches), the tail length is 16 centimeters (6.3 inches), and the weight reaches 125 kilograms (275 pounds). The head is short, pug-nosed, and looks quite powerful because of the greatly developed chewing musculature and the fur in front of the ears. The nose is naked; the ears are round and look as if they were simply placed artificially on top of the head or glued there. The rear of the ears is very steep. The fur is short and thick, and in the rear it is felt-like and gleaming. The white areas often have a yellowish tinge. The feet are oriented to the inside, and the front and rear feet have 5 toes each. Much of the soles of the feet is hair-covered; the giant panda walks in a semi-plantigrade manner. It has the most powerful, broad chewing teeth of any carnivore. See Fig. 2. Distribution is restricted to a very small part of the Hsifan mountain region in western Szechuan.

The panda is most prevalent in steep, rocky, moist mountains with a rather mild subtropical climate where there is an ample bamboo supply;

Fig. 2.   Giant panda. (*A.M. Winchester*.)

such areas are very difficult for a man to traverse. It is generally found at altitudes from 1,500 to 4,000 meters (4,921 to 13,124 feet), where there is often a great deal of snow. The Chinese name for the panda is *beshiung-chin* (white bear). Its black/white coloration appears very striking, as does that of the zebra, but in both cases this color effectively breaks up the outline, as it were; the contrast makes it difficult to recognize the panda in the wild. Pandas generally sit on their rear legs in moderately high branches; the dark parts of their fur approximate the color of the tree limbs, while the white portions are against the sky. Thus the panda is quite effectively camouflaged.

The panda moves through its home range using the same paths and tunnels again and again, sometimes using people-made paths as well. These paths connect feeding sites with rock or tree dens and sleeping spots. The sleeping den is padded with twigs and leaves. Pandas move in a bear-like walk, or they trot and make short leaps. See Fig. 2. They ford streams and rivers. Food and other objects can be grabbed with the clenched hands. Karl Max Schneider reports that vision is the most highly developed sensory modality of the panda, but its hearing, smelling, and taste are also developed. Its dampened roar is rarely heard, only when attacking and during courtship.

Additional Reading

Angel, H.: *Pandas*, Voyageur Press, Stillwater, MN, 1998.

Entwistle, A. and N. Dunstone: *Priorities for the Conservation of Mammalian Diversity: Has the Panda Had Its Day?* Cambridge University Press, New York, NY, 2000.

Lumpkin, S. and J. Seidensticker: *Smithsonian Book of Giant Pandas*, Smithsonian Institution Press, Washington, DC, 2002.

Maple, T.L.: *Saving the Giant Panda*, Longstreet Press, Inc., Atlanta, GA, 2000.

Schaller, G., et al.: *The Giant Pandas of Wolong*, University of Chicago Press, Chicago, IL, 1985.

Web Reference

www.nationalgeographic.com (Search: Panda from National Geographic Society homepage)

PANDEMIC.   See **Contagion.**

PANDEMIC INFLUENZA.   An influenza pandemic is a global outbreak of disease that occurs when a new influenza A virus appears or "emerges" in the human population, causes serious illness, and then spreads easily from person to person worldwide. Pandemics are different from seasonal outbreaks or "epidemics" of influenza. Seasonal outbreaks are caused by subtypes of influenza viruses that already circulate among people, whereas pandemic outbreaks are caused by new subtypes, by subtypes that have never circulated among people, or by subtypes that have not circulated among people for a long time. See Table 1. Past influenza pandemics have led to high levels of illness, death, social disruption, and economic loss.

For additional information on seasonal flu: http://www.hhs.gov/flu/.

For additional information on pandemic flu: http://www.pandemicflu.gov.

Appearance (Emergence) of Pandemic Influenza Viruses

There are many different subtypes of influenza or "flu" viruses. The subtypes differ based upon certain proteins on the surface of the virus (the hemagglutinin or "HA" protein and the neuraminidase or "NA" protein.

TABLE 1. DIFFERENCES BETWEEN SEASONAL (WINTER) FLU AND PANDEMIC FLU

Seasonal Flu	Pandemic Flu
Outbreaks follow predictable seasonal patterns; occurs annually, usually in winter, in temperate climates	Occurs rarely (three times in 20th century - last in 1968)
Usually some immunity built up from previous exposure	No previous exposure; little or no pre-existing immunity
Healthy adults usually not at risk for serious complications; the very young, the elderly and those with certain underlying health conditions at increased risk for serious complications	Healthy people may be at increased risk or serious complications
Health systems can usually meet public and patient needs	Health systems may be overwhelmed
Vaccine developed based on known flu strains and available for annual flu season	Vaccine probably would not be available in the early stages of a pandemic
Adequate supplies of antivirals are usually available	Effective antivirals may be in limited supply
Average U.S. deaths approximately 36,000/yr	Number of deaths could be quite high (e.g., U.S. 1918 death toll approximately 500,000)
Symptoms: fever, cough, runny nose, muscle pain. Deaths often caused by complications, such as pneumonia	Symptoms may be more severe and complications more frequent
Generally causes modest impact on society (e.g., some school closing, encouragement of people who are sick to stay home)	May cause major impact on society (e.g. widespread restrictions on travel, closings of schools and businesses, cancellation of large public gatherings)
Manageable impact on domestic and world economy	Potential for severe impact on domestic and world economy

Pandemic viruses emerge as a result of a process called "antigenic shift," which causes an abrupt or sudden, major change in influenza A viruses. These changes are caused by new combinations of the HA and/or NA proteins on the surface of the virus. Such changes result in a new influenza A virus subtype. The appearance of a new influenza A virus subtype is the first step toward a pandemic; however, to cause a pandemic, the new virus subtype also must have the capacity to spread easily from person to person. Once a new pandemic influenza virus emerges and spreads, it usually becomes established among people and moves around or "circulates" for many years as seasonal epidemics of influenza. The U.S. Centers for Disease Control and Prevention (CDC) and the World Health Organization (WHO) have large surveillance programs to monitor and detect influenza activity around the world, including the emergence of possible pandemic strains of influenza virus.

Influenza Pandemics during the 20th Century

During the 20th century, the emergence of several new influenza A virus subtypes caused three pandemics, all of which spread around the world within a year of being detected.

- 1918-19, "Spanish flu," [A (H1 N1)], caused the highest number of known influenza deaths. (However, the actual influenza virus subtype was not detected in the 1918-19 pandemic). More than 500,000 people died in the United States, and up to 50 million people may have died worldwide. Many people died within the first few days after infection, and others died of secondary complications. Nearly half of those who died were young, healthy adults. Influenza A (H1 N1) viruses still circulate today after being introduced again into the human population in 1977.
- 1957-58, "Asian flu," [A (H2 N2)], caused about 70,000 deaths in the United States. First identified in China in late February 1957, the Asian flu spread to the United States by June 1957.
- 1968-69, "Hong Kong flu," [A (H3 N2)], caused about 34,000 deaths in the United States. This virus was first detected in Hong Kong in early 1968 and spread to the United States later that year. Influenza A (H3 N2) viruses still circulate today.

Both the 1957-58 and 1968-69 pandemics were caused by viruses containing a combination of genes from a human influenza virus and an avian influenza virus. The 1918-19 pandemic virus appears to have an avian origin.

Vaccines to Protect Against Pandemic Influenza Viruses

A vaccine probably would not be available in the early stages of a pandemic. Scientists around the world work together when developing a new vaccine against influenza to select the virus strain that will offer the best protection against that virus. Manufacturers then use the selected strain to develop a vaccine. Once a potential pandemic strain of influenza virus is identified, it takes several months before a vaccine will be widely available. If a pandemic occurs, the U.S. government will work with many partner groups to make recommendations guiding the early use of available vaccine.

Antiviral Medications to Prevent and Treat Pandemic Influenza

Four different influenza antiviral medications (amantadine, rimantadine, oseltamivir, and zanamivir) are approved by the U.S. Food and Drug Administration (FDA) for the treatment and/or prevention of influenza. All four usually work against influenza A viruses. However, the drugs may not always work, because influenza virus strains can become resistant to one or more of these medications. For example, analyses have shown that some of the 2004 H5 N1 viruses isolated from poultry and humans in Southeast Asia are resistant to two of the medications for influenza (amantadine and rimantadine).

More recently, testing of seasonal influenza A (H3 N2) isolates from people in the United States during the current influenza season (2005-06) has shown that a high percentage of circulating viruses are resistant to amantadine and rimantadine. As a result, on January 14, 2006 CDC issued a Health Alert Notice (HAN), http://www.cdc.gov/flu/han011406.htm, recommending that neither amantadine nor rimantadine be used for the treatment or prevention (prophylaxis) of influenza A in the United States for the remainder of the 2005-06 influenza season. The CDC and other public health agencies will continue to monitor both seasonal and avian influenza viruses for resistance to influenza antiviral medications.

Preparing for the Next Pandemic

Many scientists believe it is only a matter of time until the next influenza pandemic occurs. The severity of the next pandemic cannot be predicted, but modeling studies suggest that the impact of a pandemic on the United States could be substantial. In the absence of any control measures (vaccination or drugs), it has been estimated that in the United States a "medium–level" pandemic could cause 89,000 to 207,000 deaths, 314,000 and 734,000 hospitalizations, 18 to 42 million outpatient visits, and another 20 to 47 million people being sick. Between 15% and 35% of the U.S. population could be affected by an influenza pandemic, and the economic impact could range between $71.3 and $166.5 billion.

Influenza pandemics are different from many of the threats for which public health and health-care systems are currently planning:

- A pandemic will last much longer than most public health emergencies and may include "waves" of influenza activity separated by months (in 20th century pandemics, a second wave of influenza activity occurred 3 to 12 months after the first wave)
- The numbers of health-care workers and first responders available to work can be expected to be reduced. They will be at high risk of illness through exposure in the community and in health-care settings, and some may have to miss work to care for ill family members.
- Resources in many locations could be limited, depending on the severity and spread of an influenza pandemic.

Because of these differences and the expected size of an influenza pandemic, it is important to plan preparedness activities that will permit a prompt and effective public health response. The U.S. Department of Health and Human Services (HHS) supports pandemic influenza activities in the areas of surveillance (detection), vaccine development and production, strategic stockpiling of antiviral medications, research, and risk communications. In May 2005, the U.S. Secretary of HHS created a multi-agency National Influenza Pandemic Preparedness and Response Task Group. This unified initiative involves CDC and many other agencies (international, national, state, local and private) in planning for a potential pandemic. HHS has worked with organizations and professional associations at international, federal, state,

and local levels to develop a comprehensive Pandemic Influenza Plan, http://www.hhs.gov/pandemicflu/plan/, in conjunction with the President's National Strategy for Pandemic Influenza, http://www.whitehouse.gov/break homeland/pandemic-influenza.html.

See also **Avian Influenza**; and **Influenza**.

Web References

H5 N1 Outbreaks and Enzootic Influenza: http://www.cdc.gov/ncidod/EID/vol12no01/05-1024.htm

Influenza Pandemics of the 20th Century: http://www.cdc.gov/ncidod/EID/vol12no01/05-1254.htm

Pandemic Influenza Threat and Preparedness: http://www.cdc.gov/ncidod/EID/vol12no01/05-0983.htm

Vaccines and Antiviral Drugs in Pandemic Preparedness: http://www.cdc.gov/ncidod/EID/vol12no01/05-1068.htm

Vaccines for Pandemic Influenza: http://www.cdc.gov/ncidod/EID/vol12no01/05-1147.htm

1918 Influenza: the Mother of All Pandemics: http://www.cdc.gov/ncidod/EID/vol12no01/05-0979.htm

PANETH TECHNIQUE. Method demonstrating the existence of free radicals (e.g., methyl) or atoms, which is based on the removal of metallic "mirror" by a stream of gas containing the radicals. The reaction products can be collected and assayed.

PANNUS. See **Clouds and Cloud Formation**

PANTHER. See **Cats**.

PANTOGRAPH. A type of trolley frequently used on electric locomotives to connect with the overhead trolley wire. The pantograph is a parallel mechanism and is also used in parallel rulers and drafting machines. The principle is best explained by the simple example of the parallel ruler. In this instrument, the straight edges are always parallel, but the perpendicular distance between them may be varied up to the length of the connecting links. Two straight edges are joined with pin joints by two connecting links having points of connection so spaced that the rulers and the connecting links form a parallelogram.

PANTOTHENIC ACID. [CAS: 137-08-6] Pantothenic acid (PA), also known as vitamin B_5, is essential to all forms of life. Its name is derived from the Greek word *pantos* that means "everywhere", which is appropriate for this widely, distributed vitamin. Pantothenic acid is a constituent of coenzyme A, which participates in numerous enzyme reactions. CoA was discovered as an essential cofactor for the acetylation of sulfanilamide in the liver and of choline in the brain. CoA is particularly important in the initial reaction of the TCA cycle (citric acid cycle) of carbohydrate metabolism and energy production. These factors are described in greater detail in the entry on **Coenzymes**. Pantothenic acid is unique among the vitamin group, in that it was one of the first to be isolated, using as a basis a microbiological assay method. Even more unique is the fact that its structure was largely determined, using a highly quantitative biological yeast test, long before it was isolated or obtained in concentrated form. R.J. Williams and coworkers described it as an acid with an ionization constant lower than that of an alpha-hydroxy acid, but about right for a hydroxy acid in which the hydroxyl group was farther removed from the carboxyl group.

In 1901, Wildiers described Bios, an essential for yeast growth. In 1933, Williams isolated crystalline Bios from yeast and named it pantothenic acid. In 1938, Williams isolated pantothenic acid from liver; and, in 1939, Jukes determined liver antidermatitis factor (chick) to be identical with yeast factor. Also, in 1939, Woolley et al. demonstrated beta-alanine as a vital part of pantothenic acid:

(Pantoic acid)   (Beta-alanine)

d (+) Pantothenic acid ($C_6H_{17}O_5N$)

In 1940, Harris, Folkers, et al. reported structure determination and synthesis and crystallization of pantothenic acid. In 1950, Lipmann et al. discovered coenzyme A; and, in 1951, Lynen characterized the coenzyme A structure.

Pantothenic acid participates as part of coenzyme A in carbohydrate metabolism (2-carbon transfer-acetate, or pyruvate), lipid metabolism (biosynthesis and catabolism of fatty acids, sterols, +phospholipids), protein metabolism (acetylations of amines and amino acids), porphyrin metabolism, acetylcholine production, isoprene production.

Distribution and Sources. Particularly high in pantothenic acid content are yeasts, animal glands and organs. Fruits have a low content.

High pantothenic acid content (2.0–10.0 milligrams/100 grams)
Beef (brain, heart, kidney, liver), chicken (liver), cod ovary, groundnut (peanut), herring, lamb (kidney), pea (dry), pork (kidney, liver), royal jelly, sheep (liver), wheat bran and germ, yeast.
Medium pantothenic acid content (0.5–2.0 milligrams/100 grams)
Avocado, bean (lima), beef, broccoli, carrot, cauliflower, cheese, chicken, clam, kale, lamb, lentil (dry), mackerel, mushroom, oats, pea, pork (bacon, ham), rice, salmon, soybean, spinach, walnut, wheat.
Low pantothenic acid content (0.1–0.5 milligram/100 grams)
Almond, apple, banana, bean (kidney), cabbage, grape, grapefruit, honey, lemon, lettuce, lobster, milk, molasses, onion, orange, oyster, peach, pear, pepper (white and sweet), pineapple, plum, potato, shrimp, tomato, turnip, veal, watercress.

Pantothenic acid is produced commercially by synthesis involving the condensation of *d*-pantolactone with salt of *β*-alanine. Some of the dietary supplement forms include calcium pantothenate, dexpanthenol, and panthenol.

Precursors in the biosynthesis of pantothenic acid include α-ketoiso valeric acid (pantoic acid), uracil (*β*-alanine), and aspartic acid. Intermediates in the synthesis include ketopantoic acid, pantoic acid, and *β*-alanine.

Some of the unusual features of pantothenic acid noted by investigators include: (1) it promotes amino acid uptake; (2) it is potentiated by zinc in preventing graying of hair in rats; (3) it promotes resistance to stress of cold immersion; (4) there is a deficiency of pantothenic acid in tumors; (5) it is required for chick hatchability; (6) it is useful in treating vertigo, postoperative shock, and poisoning with isoniazid and curare; (7) it is useful in accelerating wound healing; and (8) it is useful in treating Addison's disease, liver cirrhosis, and diabetes.

See also **Vitamin**.

Additional Reading

Williams, R.J.: "Pantothenic Acid," in *The Encyclopedia of Biochemistry*: (R.J. Williams and E.M. Lansford, Jr., editors), Van Nostrand Reinhold, New York, NY, 1967.

Web References

Pantothenic acid, Linus Pauling Institute's Micronutrient, http://lpi.oregonstate.edu/infocenter/vitamins/pa/
Pantothenic acid, http://www.eagle-min.com/faq/faq102.htm
Vitamin B₅ (Pantothenic Acid), http://www.umm.edu/altmed/ConsSupplements/VitaminB5PantothenicAcidcs.html

PAPAGAYO. See **Winds and Air Movement**.

PAPAYA TREE. Of the family *Caricaceae*, the papaya or papaw tree (*Carica papaya*) is a tropical American tree with a straight barely branching trunk from 6 to 25 feet (1.8 to 7.5 meters) tall. On the upper portion of the stem is borne a crown of large compound, long-petioled leaves. The pale yellow, fragrant flowers are of two kinds, pistillate and staminate, borne on different plants. Papayas are, therefore, dioecious plants. The smooth-skinned fruits vary considerably in shape and size, those of wild plants being not much larger than eggs, while cultivated fruits are much larger. The orange-colored flesh is sweet and juicy and surrounds a central cavity, which contains the numerous seeds. Usually the fruit is eaten raw but may be used in salads or cooked. See Fig. 1.

From the fruit and sap of the plant is obtained an enzyme, papain, which is used as a digestive aid, its action being much like that of pepsin. The leaves, cooked with meat, are said to tenderize meat.

Fig. 1.   Close-up of cluster of papayas (*Carica papaya*). (*Papaya Administrative Committee, Honolulu, Hawaii*.)

The plant is widely introduced in many tropical lands. It is extensively grown in the Hawaiian Islands and to some extent in Florida and California. Propagation is mainly by seed.

A record tree, determined by the American Forests, is located in Homestead, Florida with a circumference of 20 inches (50.8 centimeters), a height of 25 feet (7.62 meters), and a spread of 4 feet (1.22 meters).

The name papaw, sometimes given to this fruit, is also applied to a North American tree, *Asimina triloba*, with which it should not be confused.

PAPER. Paper is a sheet material comprised of individual fibers that are brought together by the removal of water on a screen fabric or a between a pair of such fabrics. The first recorded account of paper that meets the definition just given was from ancient China. It is recorded that in the year 105 AD, Cai Lun, a courtier serving the emperor of the Han dynasty, made significant advances in the strength and quality of paper sheets.

Early accounts of paper describe its preparation from a variety of materials, including fibers from the inner bark of the paper mulberry tree. Ancient paper also was prepared from cast-off textile fibers, making paper the oldest known product prepared from recycled materials. Handcraft papermakers continue to use fibers from such plants as flax, hemp, cotton, and a variety of grasses. A common feature of all such fibers is that they contain cellulose, a natural polysaccharide. See also **Cellulose**. The polysaccharide molecules provide typical papermaking fibers with a water-loving character. Once the fibers have been liberated from each other, they are readily dispersible in aqueous solution.

Though the word paper is derived from the word papyrus, the two products have different origins. Papyrus was developed in ancient Egypt. Unlike paper, the process of forming papyrus sheets does not involve individualized fibers. Rather, flat reeds of the Cyperus papyrus species are pounded repeatedly to make them flexible. The softened reeds then are pressed together into crisscross layers and dried.

During the middle ages in Western lands the use of papyrus was generally abandoned in favor of parchment, derived from leather. Meanwhile, the ancient technology of papermaking, long kept as a trade secret of the ancient Chinese, gradually began to spread. The manufacture of paper from bark and bamboo spread from China to Japan, where its manufacture began in about the year 610. Papermaking from flax and hemp spread through Central Asia, the Middle East, and eventually into Europe. The first European paper was made in Spain in 1150, in France by 1189, in Germany by 1320, and in England by 1494. Papermaking was introduced in the United States in ~1700 by William Rittenhouse in Philadelphia.

In light of the fact that the vast majority of modern paper is prepared from wood fibers, it is perhaps surprising that such technology did not arise until the mid-1800s. The relative lateness of wood's emergence as a major fiber source for paper probably can best be explained by the very strong and chemically resistant lignin that joins fibers together in natural wood. Lignin is a natural, cross-linked phenolic resin. Papermakers have used basically two kinds of approaches to separating the fibers, either applying drastic mechanical force or applying harsh chemical treatments to dissolve the lignin.

Chemical and Material Composition

The words mixture and composite are useful when describing the chemical composition of paper. With a few exceptions, which will be noted, paper's components do not become covalently bonded to each other during manufacture. Rather, the ingredients are mainly held together by reversible hydrogen bonds, in addition to London dispersion forces. The reversible nature of the bonding can be demonstrated by soaking paper in water, followed by agitation, as in the case of a kitchen blender. Though paper's composition varies widely, depending largely on its intended use, key ingredients of typical paper products are cellulosic fibers, mineral products, and bonding agents, including starch and latex products. Other additives, which seldom individually make up >1% of paper's mass, include polyelectrolytes, inorganic coagulants, processing aids such as defoamers and biocides, as well as a variety of additives designed to impart functional characteristics to the paper.

Chemical Composition of Fibers. The fiber in typical paper can be called "cellulosic", meaning that cellulose [CAS: 9004-34-6] is its most prominent component. Cellulose is present not only in wood, but also in various non-woody plants, such as straw, sugarcane (bagasse), reeds, and hemp. Cellulose is a polysaccharide, ie, a polymer comprised of sugar subunits. The stereochemistry of the six-carbon sucrose units that comprise cellulose, as well as the nature of the β-1,4 glycosidic linkages between those units, give rise to a unique tendency to form linear fibrous structures. Localized crystalline regions within these structures tend to be interposed by noncrystalline regions within the cell walls of the fibers.

Wood, the major source of cellulosic papermaking fibers, can likewise be described as a composite. Its four main solid ingredients are held together mainly by physical chemical forces, rather than by a high degree of covalent bonding between them. Three of the ingredients, cellulose, hemicellulose, and lignin, are polymeric. The fourth ingredient, the extractives portion of wood, is actually a class of monomeric resinous materials.

The proportion of different types of fiber in paper depends on the intended use. Softwood fibers, having typical lengths of 2.5–4 mm, are desirable for products requiring high levels of tear, tensile, and fold strength, as well as resistance to edgewise compression forces. Products taking advantage of such attributes of softwood fibers include paper bags, as well as linerboard for corrugated containers. Premium towel and tissue papers are often comprised of softwood fibers. By contrast, hardwood fibers, having fiber lengths closer to 1 mm, are prized for applications demanding smoothness, as in the case of printing papers. To simultaneously achieve the strength, smoothness, and other specified properties of many grades of paper, it can be advantageous to employ blends of the two main categories of wood fibers, as mentioned.

Hemicellulose [CAS: 9034-32-6] is a polysaccharide, like cellulose; however, its structure is more complicated and more varied. From the papermaker's perspective, hemicellulose can be considered as a natural "dry-strength agent". Because of their various side groups, the hemicellulose molecules generally are unable to form the type of crystalline structure for which cellulose is noted. The dominant form of hemicellulose present in hardwood species is 4-o-methyl-D-glucuronoxylan [CAS: 9062-57-1]. By contrast, the hemicellulose in softwood is mainly glucomannan [CAS: 76081-94-2]. The presence of carboxylic acid functions on glucose side-groups of the glucuronoxylans helps to explain a generally higher negative surface charge density of hardwood fibers, compared to softwood fibers prepared by similar pulping methods. The acid groups also contribute to the swelling of fibers in water. High correlations have been observed between the ability of fibers to swell in water, the compliance of their surfaces, and the development of bonds between them when paper is dried. See also **Hemicellulose**.

The third major component of papermaking fibers, especially prominent in natural wood and mechanical pulp fibers, is lignin [CAS: 8068-05-1], which consists of naturally cross-linked phenol–propane units having a variety of ether, ester, and carbon–carbon bond connections between them. Lignin poses significant challenges for papermakers. Not only is lignin very difficult to chemically remove from fibers, but it also has a dark, unstable color. See also **Lignin**.

The extractives component of wood includes a variety of fatty acid, triglyceride, and unsaponifiable organic compounds. In addition, mechanical fibers obtained from conifers, such as pine, can contain ~1% by mass of resin acids, including levopimeric acid, abietic acid, and other terpenoids.

Common Grades of Paper and Paperboard

It can be useful to distinguish between paper products, ie, products having basis weights below ~100–150 g/m^2 versus paperboard products, which is generally thicker and stiffer.

Paper Grades. Newsprint paper usually is comprised mainly of mechanical pulps, such as thermomechanical pulp (TMP) or groundwood. The fiber portion also may contain a minor portion of bleached kraft pulp. Newsprint also can be made from deinked old newspaper (ONP), often supplemented by de-inked old magazine paper (OMP) or mixed office waste (MOW). Though newsprint traditionally was not a filled grade, there is a trend toward increased use of clay or calcium carbonate, especially in publications that use colored images.

The base-stock for light-weight coated (LWC) paper is often quite similar to that of newsprint. However, the coating operation places high requirements on sheet uniformity, including the absence of holes, as well as cross-directional uniformity in weight or density. The coatings, which can comprise up to about one-half of the mass of the final product, are usually applied right on the paper machine, using blade coaters. The main uses of LWC are in magazines, as well as for advertising inserts in newspapers and direct mail.

Uncoated papers for offset lithography and xerographic printing, including use in personal printers, are usually mainly comprised of bleached kraft fibers. Precipitated calcium carbonate (PCC) is the most widely used filler in such grades, and filler levels are most often in the range 5–30%, based on dry mass. Most uncoated fine papers contain ~0.5–1.5% of cationic starch that was added to the fiber slurry. Additional starch, added at a size press, is often in the range of 2–5% of product mass. The most common basis weights are within 40–100 g/m^2. Surface strength is especially important in products to be printed by offset lithography. Controlled stiffness, lack of curl tendencies, and controlled electrical conductivity are critical in papers used for xerography.

Bleached kraft pulp is also the main component of products intended for high end graphic uses, including fine rotogravure and offset images for posters, calendars, brochures, and fancy reports. A characteristic feature of such grades is that multiple layers of coatings are often applied. Both on-machine, off-machine, and combinations of coating strategies are used.

Paper for grocery bags, fast-food bags, and similar applications must have a high resistance to tearing. The fiber of choice is softwood kraft, which may be unbleached or bleached. To maximize the paper strength, the filler content may be low or zero. Wet-strength (see the section Permanent Wet-Strength Resins) and internal sizing treatments (see the section Internal Sizing) are sometimes used to resist rupture if the bag comes into contact with wet food, rain, etc.

Requirements of wrapping papers are very diverse, but they often involve resistance to various types of fluids. Some other wrapping papers need to be heavily treated with internal sizing agents or barrier coatings to protect various contents from aqueous fluids. Fluorochemical treatments, silicones, and laminated layers may be used to hold out aggressive fluids, such as meat juices, in the case of butcher wrap.

The term décor paper denotes the uppermost cellulosic layer of saturated laminate products used for furniture and countertops. For example, décor paper carries the imitation wood image for many furniture products that actually are made from pressed, resin-filled fiberboard. High opacity, uniformity, and a highly porous nature are key requirements of the grade. The high opacity requirements usually imply a relative high usage of titanium dioxide as a filler. A uniform and porous sheet is needed so that resins, such as melamine-formaldehyde resin can penetrate thoroughly into the sheet during a saturating operation.

Filter papers are another type of product that requires uniform and relatively high porosity. A highly porous sheet can be achieved by use of kraft fibers having a relatively low degree of refining. Because many filter papers become wet during use, it is common to employ wet-strength agents in their preparation. Filter papers also can be prepared with a wide range of non-wood fibers, including synthetics.

The everyday use of tissue paper products is in contrast to the rather complex and varied methods involved in its production. Bleached softwood kraft pulp has been a major component of tissue grades, though there is increased use of hardwood fibers (especially for softness), as well as use of recycled fibers from mixed office waste. The traditional method for preparation of tissue paper involve use of a Yankee dryer. Conditions of adhesion between the web and the polished cylinder surface are optimized, mainly by chemical addition, in order to achieve a controlled degree of crepe at a doctor blade. To increase adhesion to the Yankee surface, papermakers may add starch, wet-strength resins, or polyamines. To reduce adhesion, they may use emulsified paraffin [CAS: 8012-95-1], silicone oil, or polyethylene glycol. Alternative methods of tissue preparation include use of low wet-press conditions, through-air drying, and multiple creping. In addition, tissue products often have two or more ply construction, as well as perforations.

In addition to usually having a higher basis weight, towel grades of paper usually differ from tissue grades with respect to wet-strength requirements. Whereas many tissue paper products must be compatible with the flushing requirements of septic systems, most towel grades need sufficient wet-strength to withstand the pressures of rubbing, while in the wet state. Depending on the application, the wet-strength requirements may be met either with temporary wet-strength agents, such as glyoxylated, or permanent wet-strength agents, such as polyaminoamide-epicolorohydrin.

Paperboard Grades. The top and bottom layers of typical corrugated boxes are ideally formed from unbleached softwood kraft fibers of intermediate yield or lignin content. An increasing proportion of linerboard products are composed partly or completely of old corrugated container (OCC) pulp. Up to ~10% content of filler can be used in some linerboard grades, especially in cases where the boxes will be used for point-of-sale advertising. There is a trend towards greater use of multiply forming, in which the top layer may be comprised of bleached fiber, often with a significant level of calcium carbonate filler. Linerboard products are pressed and calendered to relatively high density. Most linerboard products are internally sized to resist water, though it is important not to have such high resistance as to defeat the starch adhesive used in corrugating operations (see below). The most important physical attribute of linerboard is its resistance to edgewise crushing.

Corrugating medium is the type of paperboard that will be compressed into a repeating sinusoidal pattern to act as a spacer within a combined-board corrugated container. Stiffness and compression resistance are key attributes. High yield kraft of neutral sulfite semichemical pulps are preferred, though recycled fibers are also used. Because the resins in the high yield fibers can cause excessive development of water resistance, which can interfere with gluing of the board construction, rewetting agents are often applied at the time of manufacture.

Fine bleached board products include posterboard, file-folder, index card, and similar grades. Some of the desired properties can include brightness, uniform surface characteristics, and stiffness. Bleached kraft fibers are usually the main ingredient, and fillers are often used. Smoothness, gloss, and various printing characteristics can be improved by coatings, as already discussed.

Base-stock for paper cups, milk carbons, juice containers, and similar products is generally made from unfilled bleached kraft pulp. Key attributes include stiffness, resistance to water penetration, and compatibility with plastic laminates. Fluid resistance is generally achieved by the wet-end addition of alkylketene dimer (AKD) size, which is often added together with cationic starch, which contributes to strength. In the case of board made for aseptic packages, rosin size emulsion is often used in combination with the AKD size in order to withstand the effects of peroxide, which is used as a disinfectant. Liquid packaging board may be treated with additional starch at a size press to develop stiffness and better adhesion to an extruded plastic layer. Extruded polyethylene is used for milk cartons. Citrus juices generally require the use of multi-layered plastic laminates, having the ability to resist a variety of penetrants, including oxygen from the air, as well as water vapor.

See also **Papermaking and Finishing**; and **Pulp (Wood) Production and Processing**.

Additional Reading

Bristow, J.A., and P. Kolseth: *Paper Structure and Properties*, Marcel Dekker, Inc., New York, NY, 1986.
Casey, J.P.: *Pulp and Paper Chemistry and Chemical Technology*, 3rd ed., Vol. I, John Wiley & Sons, Inc., New York, NY, 1980.
Mark, R.E., J. Borch, C.C. Habeger, and M.B. Lyne: *Handbook of Physical Testing of Paper*, **2** Volumes, 2nd Edition, Marcel Dekker, Inc., New York, NY, 2001.
Smith, D.C.: *History of Papermaking in the United States (1691–1969)*, Lockwood Publishing Co., New York, 1970.

Web References

American Forest & Paper Association, http://www.afandpa.org.
Magic Material Paper, http://www.gutenberg-museum.de/index.php?id=75&language=e
Paper Online, http://www.paperonline.org/
United States Government Printing Office: Government Paper Specification Standards, http://www.access.gpo.gov/qualitycontrol/paperspecs/txtindex.html

PAPERMAKING AND FINISHING. One of the most important factors in the progress of civilization has been paper, a thin flat tissue composed of closely matted fibers obtained almost entirely from plant sources. In modern life paper finds a variety of uses, for writing, for containers, wrappers, wall covering, and — perhaps most important — in all forms of printing: newspapers, magazines, books.

The art of making paper seems to have been discovered first by the Chinese, who were making paper as early as the beginning of the Christian era. From China the process was carried to Arabia and thence to Europe. Paper was not an important article at first and, since it is not a very durable substance under ordinary conditions, could not compete with parchment or vellum as a medium for the written word. In the fifteenth century writing became more general and the demand for a cheaper material increased. Paper became an important product. At this time paper was made largely from vegetable fibers reclaimed from cloth (especially linen), as had been done since the invention of paper in China. This paper was made entirely by hand, as is done even today in the manufacture of certain expensive types of paper. In making handmade paper, a pulp is formed by soaking the vegetable fibers in water in a vat. From this vat the pulp is dipped out in a mold, the bottom of which is a fine screen. By a deft motion of this mold, the soft pulp is spread over the screen in a thin layer of matted fibers. The water in the pulp drains off, leaving a rather firm mass, which is turned out on a piece of felt. More pieces of half-dried pulp spread on felt are added. The whole pile is then pressed to squeeze out more of the water, press the fibers closer together and form a firm sheet. These are then removed from between the felts, pressed again, and dried. During the final treatment surface sizing is added to render a surface more suitable to receive ink. Sheets of hand-made paper are naturally of limited size and expensive.

To meet the great demand for paper, machine methods were developed. This increased demand for paper also led to the utilization of material that could be obtained in quantities much greater than rags. Out of this developed the vast pulp industry, which today converts vegetable material, mostly soft woods such as spruce and fir, as well as poplar, into a white felt-like mass of fibrous substance, known as pulp. See also **Pulp (Wood) Production and Processing**.

Wet End. A modern paper machine begins with a flow spreader or distributor, conveying a dilute fiber suspension (0.1–1% fibers) to a

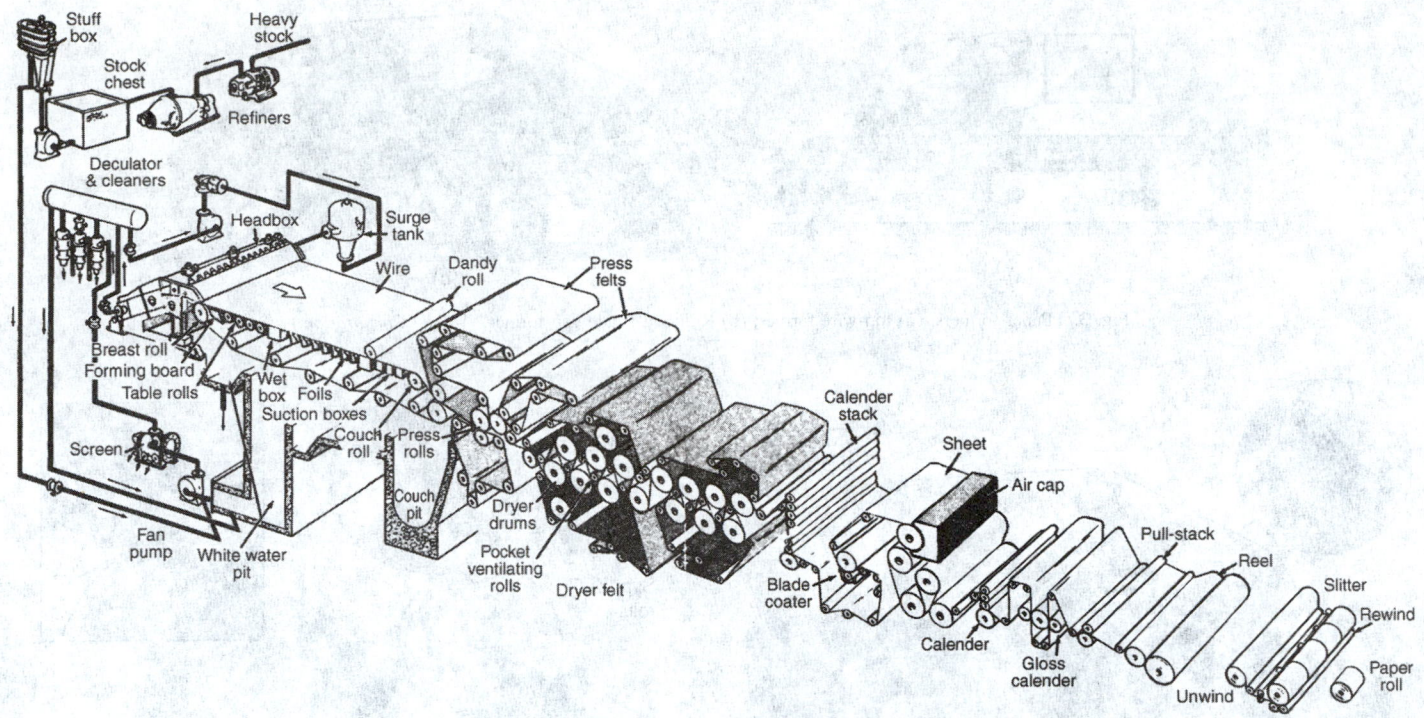

Fig. 1. Fourdrinier machine for producing printing-grade paper. (*Beloit Corporation.*)

headbox which delivers a jet of the suspension or slurry through a slice (sluice) across the full width of the machine, almost 400 inches (~10.2 meters) in some large machines. In the headbox, the fibers are dispersed, and the flow is rectified as well as possible so that the jet is delivered onto a moving endless fine-mesh wire screen with uniform composition, flow rate, and velocity. The pressure in the headbox and its slice opening are adjusted so that the jet velocity matches the speed of the wire screen, which may be up to 4000 feet (~1220 meters) per minute for newsprint. The proper stock flow per unit width corresponds to the desired *basis weight* of the paper. (Basis weight is weight per unit area and varies with grades and sizes of papers.)

The dispersion of fibers in the headbox is brought about by subjecting the slurry or suspension to shear stresses, usually with turbulence. Various designs have been developed to accomplish this.

As shown in Fig. 1, the most common type of paper machine is the Fourdrinier, in which the moving wire screen is in the form of an endless conveyor belt stretched between two large rolls. The roll situated under the headbox slice is called the *breast roll*. The roll located generally at the end of the straight wire run is the *couch roll*. Drainage of the slurry through the wire screen is induced by several types of driving forces. In the early, slow-speed machines, the principal force was gravity. Later, the hydrodynamic action of table rolls, which support the wire and rotate with it, began to play an important part in drainage as speed increased. More recently, foils came into use, i.e., rigid stationary, hydrodynamically shaped elements which support the wire and exert a pumping action through the wire screen. Other means are perforated or slotted boxes with vacuum over which the wire runs. When only water is drained, they are called *wet boxes*. When applied toward the dry end of the wire screen, they also draw air through the wet paper mat and are called *suction boxes*. Other equipment configurations have been developed to meet these objectives. On all modern Fourdriniers (Fig. 2), a forming board located close to the breast roll is used to scrape off the water, drained initially by gravity, from the bottom of the wire.

An important development in paper forming is the use of a top wire dewatering unit placed above the wire of the Fourdrinier. The top wire units use various dewatering elements such as rolls, foils, and vacuum boxes. The top wire units add drainage capacity to an existing Fourdrinier plus improved symmetry through the thickness of the sheet similar to the twin wire formers. An example of this type of former is shown in Fig. 3(**a**) (*Valmet, SymFormer MB*). See also Fig. 4.

Fig. 2. Fourdrinier machine located at Blandin Paper Company, Grand Rapids, Michigan. (*Beloit Corporation.*)

A more recent development is the *roll and blade gap former* (*Valmet SpeedFormer*) shown in Fig. 3(**b**). In this type of machine, the fiber suspension is confined between two wire screens, and water is removed through both wires either simultaneously or alternately. This two-sided drainage leads to greater symmetry of distribution of fines and other nonfibrous particles through the thickness of the sheet. A significant feature of twin-wire forming is the elimination of the free surface of the fiber-water suspension while the sheet is being formed. This greatly reduces the larger-scale disturbances (waves, streaks, and jumps) which occur at higher speeds on Fourdrinier wires.

Not all the fiber and other solid materials are retained by the forming wire. For this reason and because so much water is used in the papermaking process, the *white water* removed in the sheet-forming process is recirculated in the overall system. A large part of it is added directly to the high-consistency stock and fed back to the headbox, while a small portion goes into a *save-all* device, which recovers much of the solids from the white water. These extracted fibers and other solids are returned and added to the suspension. The clarified water is used in showers for cleaning wires and felts and other purposes so that only a small amount of the reused water eventually is discharged.

Press Section. At the end of the forming system, the *paper web* is transferred from the wire to a *press felt*, a fine-textured, usually synthetic fabric. At this point, the web contains about 4 or 5 parts water to 1 part

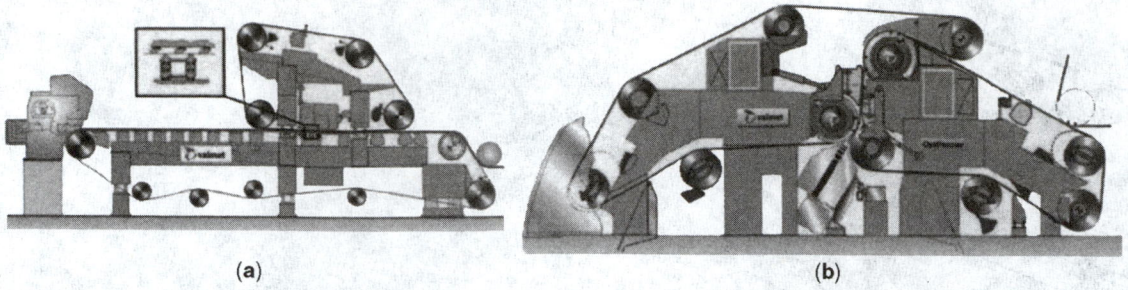

(a)   **(b)**

Fig. 3.   Paper formers: (**a**) top-wire former; (**b**) Roll and Blade gap former. (*Valmet Corporation.*)

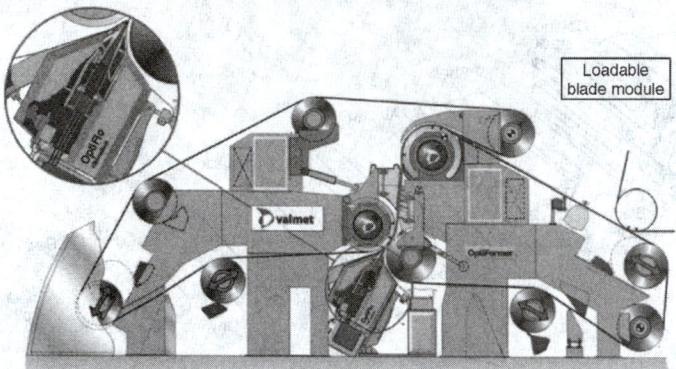

Fig. 4.   Roll and blade gap. (*Valmet Corporation.*)

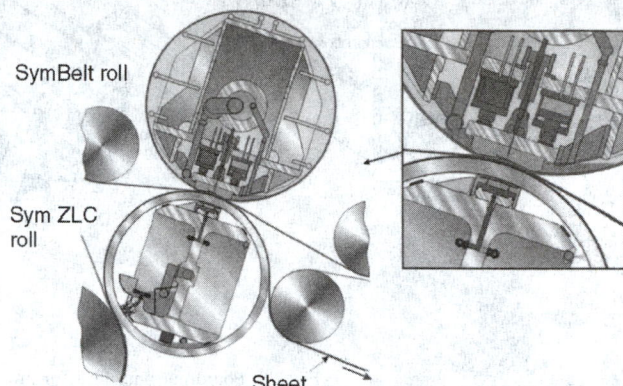

Fig. 5.   The basic belt nip press. (*Valmet Corporation.*)

solids. The wet paper web and one or more press felts pass through two or more press-roll nips, where water is squeezed out. Pressing also compacts the paper mat. This increases the potential interfiber contact areas where bonds will be formed.

The early *plain press* used a pair of metal and rubber-covered solid rolls. The expressed water had to flow out of the nip in the upstream direction, parallel to the paper web, as in an old washing-machine wringer. Nip pressures were then limited by the damage to the wet web (crushing) caused by this lateral flow. Although the plain press was improved in many ways, later development work led to the *fabric press* in which the felt contacting rolls are wrapped with a relatively coarse and incompressible mesh fabric. In another development (Beloit Ventanip press), the felt contacting rolls have narrow, closely spaced circumferential grooves. In both types, the lateral flow is virtually eliminated.

While the development of the modern presses has achieved high performance with simple constructions, the remaining problems of flow resistance and web rewetting leave room for improvement. It is generally recognized that mechanical removal of water is much less costly than drying. This has led to the development of presses with very long nips (*Beloit Extended Nip Press*) as shown in Fig. 5. These presses use a shoe typically of 10 in. (25 cm) width to replace one roll in the roll press configuration and use oil lubrication between the stationary shoe and a moving impervious belt. An additional 20–40% of the remaining water in the sheet can be mechanically removed by these presses with their longer residence time under high pressure.

As the machine speed has continuously has increased the dynamic forces to the web have increased and it has forced to support the web with felts to avoid web breaks. Before 30's typically the web had open draw between press and wire section which was soon replaced with pick-up suction roll arrangement. In this arrangement web is picked up with a suction roll from wire and supported by press felt to the first nip. Most recent fast paper machines are using the same principle of suction roll and supporting fabric not only through press but also through dryer section.

Dryers. After water removal by pressing has been done to the extent practical with present technology, the paper web leaves the press section with 1–2 parts of water to 1 part fibers. Most of this remaining water, down to 5–10%, must be removed by evaporative drying. In the most common method, the paper web is passed over a series of staggered cast-iron drums internally heated by condensing steam at pressures ranging up to approximately 10.2 atmospheres. The paper web is held in contact with the rotating drums by means of dryer felts or fabrics under tension. The diameter of the dryer drums is typically 5–6 feet (1.5–1.8 meters). There may be as many as 100 of them in heavyweight paperboard machines. These dryer drums are shown in the panoramic view. See Fig. 1.

Ventilating devices that blow air of controlled temperature and humidity through the dryer felts into the spaces between adjacent dryers are used. Here the air is confined by the sheet and felt runs. These pocket ventilating systems, together with greater control of the flow patterns within the dryer hood (which usually encloses the entire dryer section) have led to significant improvements in cross-machine uniformity of paper drying. This results in paper and board of improved suitability for modern high-speed converting and printing operations.

Other types of dryers, including radiant heating, dielectric and microwave heating, and high-velocity, hot air impingement, have been developed. These devices are generally applied to drying coated paper where sheet contact to a solid surface may be detrimental during drying. Wider application has been limited because of low thermal efficiencies and high capital costs.

Size Press and Coaters. Many printing grades of paper and paperboard are coated with an aqueous suspension of pigments (such as clay) in adhesives (such as starch) to provide a smoother surface, control the penetration of inks, and improve the pick resistance, appearance, brightness, and opacity. These and other materials are also applied, such as *functional coatings*, to provide such features as water resistance, pressure sensitivity for carbonless copying, and a wide variety of other properties. The appropriate materials may be added to the papermaking furnish during some stage of stock preparation (called *internal sizing*). Application of sizing or coating to one or both surfaces of the formed and dried sheet, rather than as internal sizing, simplifies the sheet-forming process and provides better control of surface properties.

The principal methods of surface coating may be classified as roll, blade, and air-knife coating, according to the method used to apply and control the final coating-layer thickness and smoothness. One version of coater (Beloit Billblade) simultaneously coats and smooths both surfaces of the paper web by running it down through the nip between a blade and a roll

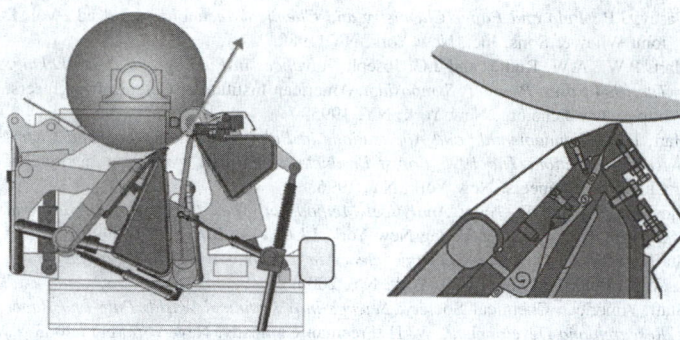

Fig. 6. Jet coater head. (*Valmet Corporation.*)

Fig. 7. Multinip calender as part of paper machine. (*Valmet Corporation.*)

while maintaining two puddles, one between the web and the roll and the other between the flexible blade and the web, thus eliminating the necessity for two coating stations.

A more recent development is the short dwell coater (*Beloit Short Dwell*). The coater consists of a captive pond just before the blade that limits the contact time between sheet and coating material as shown in Fig. 6. The back flow assists in removing the boundary layer of air coming in with the sheet. The shorter contact time results in less coating penetration. Superior coating quality and improved runnability, due to fewer web breaks, have been achieved.

After sizing or coating, the solvent, usually water, must be removed from the coating by evaporative drying. With some coating formulations and paper grades, drying can be done on ordinary steam-heated drums without damage to the coated surface, particularly if the surface of the first drum is smooth (sometimes chrome-plated). However, it is often desirable to do the initial drying with air impingement or radiant heating. Surface coating can be done on the machine as a step in the paper-machine operation, as shown in Fig. 1.

Calenders and Winders. Nearly all paper grades are calendered after they have been dried to the desired final moisture content. Ordinary calendering involves passing the paper web through one or more nips between metal rolls with high linear pressures. The calendering process flattens out the paper structure by virtue of the high pressure and "irons" the sheet. Calendering causes bulk reduction, which often is not desired, and surface smoothing, which is desired. The results strongly depend upon moisture content, calender-roll temperature, roll pressure, and speed.

In *supercalendering*, an off-machine operation, the calender rolls consist of alternating chilled-steel and paper-filled rolls, i.e., paper disks clamped on a steel shaft. These roll fillers have to be replaced periodically. Very high pressures are used. The increased pressure and shear forces associated with deformation of the relatively soft paper roll and the very high roll pressures impart a smoother, glossier surface to the web than ordinary calendering with all-metal rolls. This type of calendering is frequently used on coated sheets to provide a glossy coated surface. Recently polymer covered rolls have been replacing paper-filled rolls and due to better durability have given an opportunity to locate this calender as part of papermachine. Fig. 7.

There are other process configurations for the various coating effects and specifications desired.

Other Types of Machines. Although the Fourdrinier machine is used for making almost all grades of paper and board, other designs are sometimes more advantageous. The *cylinder machine*, invented at about the same time as the Fourdrinier, consists of a rotating cylindrical mold covered with a wire screen and partially submerged in a vat. The stock flows into the vat, and a mat is formed on the cylinder under a hydraulic head difference between the stock level in the vat and the white-water level inside the cylinder. The wet mat is picked up by a felt running through the nip between a couch roll and the cylinder. The cylinder machine is used for making multiply board, employing several vats in series. Because of slow speed and other limitations, the cylinder machine is becoming obsolete. In recent years, several new types of machines have emerged.

The most recent multiply machines use headboxes with simultaneous delivery of two different stocks from the headbox slice (*Beloit Strataflo Headbox*), followed by top wire dewatering units (*Beloit Bel Bond*).

A secondary headbox with additional top-wire dewatering follows the first unit on the forming wire. Other versions of multiply formers use mini-fourdiniers on top of the primary forming wire in various configurations.

Environmental Concerns and Production Efficiency

Reducing Environmental Impacts. The papermaking process, by its nature, has potential to be environmentally friendly and sustainable. Paper is composed mainly of natural, renewable raw materials. Cellulosic fiber used in papermaking is derived from photosynthesis, which takes carbon dioxide from the atmosphere and converts it into O_2 and biomass. Most of the material in paper can be recycled, reducing the amount that gets disposed in other ways. Though relatively inexpensive, in comparison with other manufactured products, paper and paperboard can maintain most of their strength and other properties for many years. Responsible use of forestlands by the paper industry promotes tree planting and growth on lands that might otherwise be used for urban sprawl and other uses that do not take up CO_2 from the atmosphere.

Two particular environmental concerns for papermaking are energy use and the handling of liquid and solid wastes. As was noted in the discussion of the wet-pressing and drying of paper, the evaporation of water is the biggest single energy use in a typical paper mill. Fortunately, in the case of integrated pulp and paper facilities, much of that energy can come from steam generated by combustion of lignin byproducts and other organic chemicals in the spent pulping liquors. The incineration of such liquors in the kraft process results in regeneration of the pulping chemicals, in addition to providing high pressure steam to generate electricity. It can be said that the paper machine gets much of its the energy for free, since the drying operation can use relatively low pressure steam that is no longer efficient for use in generators. As discussed earlier, evaporative energy requirements can be reduced by installing extended-nip wet-presses, which can increase the solids content of the web entering the drying operation. Attention to dryer-felt tensions, condensate handling within dryer cylinders, and ventilation of the air-pockets surrounding dryer cylinders all have the potential to improve the efficiency of drying.

Year by year, less fresh water is used, on average, to produce a unit mass of paper in various production facilities. This trend is motivated by such factors as the cost of fresh water, limits on the concentrations or amounts of substances in discharged water, and the costs associated with wastewater treatment. Water-reduction strategies usually involve replacing fresh water with recirculated process water. Recirculation occurs at various levels within a typical paper machine. First, with few exceptions, the primary circuit of flow involves dilution of incoming fiber slurry (the thick stock) by process water that has been removed during the forming process (the white water). White water also is used for regulating the filterable solids level (consistency) of fiber slurry before such operations as refining and metering of stock flows. Also, fresh water may be replaced by strained and filtered white water in various showers, such as those used to control build-up of foam or to wash roll or fabric surfaces in the forming section of the machine.

Water recirculation has a side-benefit of generally reducing losses of fine cellulosic materials, mineral, and other colloidal material to the wastewater treatment system. The three main principles by which save-all fiber recovery systems operate are filtration, flotation, and sedimentation. If these are operated efficiently, the net loss of cellulosic materials can be <1%. Effective use of a retention aid during formation or at a save-all can further reduce losses of solids from the paper machine system.

Pulp and paper mills have installed primary and secondary treatment systems to control liquid effluents. Primary treatment involves settling basins and/or tanks, ie, clarifiers. These remove ~85–100 wt % of solids, eg, fiber fines, mineral particles, and colloidal matter. Primary sludges, which are removed by the primary clarifiers, generally cannot be reused by the mill to make the same product. However, many mills use the sludges in lower grade products. Secondary wastewater treatment involves the controlled growth of bacterial sludge. Nutrients, such as nitrogen or phosphorous, may be added as needed to support growth of the microorganisms. Both aerobic and anaerobic treatment systems are used. Equipment for various biological waste treatments includes lagoons, activated sludge with air and oxygen, trickling filters, clarifiers, modified biological systems containing activated carbon, and combinations of these. Secondary treatments generally remove 90–95% of the biological oxygen demand (BOD), most solids, and most of the toxicity. Color is more difficult to remove, sometimes requiring so-called tertiary treatment steps. Sludges from secondary treatment tend to be extremely hydrophilic and are difficult to dewater.

As more process water is recycled, more nonpathogenic microbial growth, ie, slime, tends to develop in the mill system. Entrained slime can produce blemishes and holes in the paper web. Some strategies to reduce the impact of slime include treatment of the system with organic biocides, or with oxidizing agents such as chlorine dioxide or hydrogen peroxide. To reduce the toxicity associated with biocide use, progress has been achieved with certain surfactants that interfere with the buildup of biofilms on the surfaces of processing equipment. The efficiency of any form of slime control is greatly increased by appropriate mill cleaning procedures. Thus, it is common practice to periodically stop operations and clean the paper machine system with highly alkaline conditions, usually with the addition of detergents. Acidic treatments also can be used, depending on the nature of the materials that tend to deposit in a given paper machine system. More recently, enzymes have become widely used for periodic cleanings, avoiding the need for high temperatures or strong base or acid solutions.

Sludge Handling and Disposal. The two kinds of sludges generated by pulp and paper mills, primary and secondary, can differ considerably in character. Primary sludges mainly comprise cellulosic and mineral matter. Secondary sludges are largely biological, and they and harder to handle and dewater. In addition, wastes resulting from several sources within the mill can include, bark, sawdust, dirt, knots, pulpwood rejects, fly-ash, cinders, and slag.

Solids content of wet sludges can be 1–40 wt.%. The operating plan for a land-disposal site depends on the fluid state of the sludge. Sludges of 30–35 wt.% solids from a paper recycling or deinking operation may be in a highly fluid state, whereas a low ash, high fiber pulp mill sludge at 15–20 wt.% solids may be quite dry and stable. Sludge handling processes can include thickening, stabilization, conditioning, dewatering, and incineration. Though most sludges are landfilled, some are used in lower grade products such as landscaping bricks, or used as soil conditioners for agriculture. Some primary sludges from papermaking or coating operations have been incinerated to recover their mineral content, which has then been used for products not requiring a high brightness.

Water Quality Assessment. Starting in the mid-1970s, there has been increasing emphasis on the assessment and control of effluent water quality. Key indicators of effluent water quality are the BOD, the total suspended solids (TSS), the absorbable organic halides (AOX), chloroform content, and the water's toxicity towards various organisms. A single organism as an indicator of stream quality has been replaced by community compositional and structural analysis. Thus, total effluent effects on a broad scale can be realized. Other measured parameters are algal assays, fish surveys, sediment mapping, effluent plume mapping, sediment oxygen demand, and socioeconomic impacts.

Additional Reading

Bajpai, P., and R. Kondo: *Biotechnology for Environmental Protection in the Pulp and Paper Industry*, Springer-Verlag New York, LLC, New York, NY, 1999.

Casey, J.P.: *Pulp and Paper Chemistry and Chemical Technology*, 3rd ed., Vol. I, John Wiley & Sons, Inc., New York, NY, 1980.

Hart, P.W., A.W. Rudie, and J.C. Joseph: *Advances in Pulping and Papermaking: The 1994 Forest Products Symposium*, American Institute of Chemical Engineers, http://www.aiche.org, New York, NY, 1995.

Hart, P.W.: *Fundamentals and Applications in Pulping, Papermaking, and Chemical Preparation: The 1995 Forest Products Symposium*, American Institute of Chemical Engineers, New York, NY, 1996.

Sjostrom, E., and R. Alen: *Analytical Methods in Wood Chemistry, Pulping and Papermaking*, Springer-Verlag New York, LLC, New York, NY, 1999.

Staff: American Chemical Society: *Science in a Technical World: Pulp and Paper*, Worth Publishers, Inc., New York, NY, 2000.

Staff: American Chemical Society: *Science in a Technical World: Pulp and Paper Research and Development*, W. H. Freeman Company, New York, NY, 2003.

Web References

American Forest and Paper Association, http://www.afandpa.org/
Institute of Paper Science and Technology, http://www.ipst.edu/
International Paper, http://www.internationalpaper.com/
Paper Online, http://www.paperonline.org/
TAPPI, http://www.tappi.org/
Valmet Corporation, http://www.valmet.com/
Weyerhaeuser, http://www.weyerhaeuser.com/

JIPI JAAKKOLA, Valmet Corporation, Paper Machine Product Manager, Charlotte, NC

PAPER WASP (*Insecta, Hymenoptera; Family Vespidae*).   The hornets and yellow-jackets. Wasps that build their nests of coarse paper made by chewing fragments from weathered or partially decayed wood. Some build subterranean nests and others suspend the nest from the eaves of buildings or from the boughs of trees.

PAPILLOMAVIRUSES.   Papillomaviruses are a diverse and intriguing group of small deoxyribonucleic acid (DNA) viruses that cause hyperproliferative lesions (better known as warts) of cutaneous and mucosal epithelium. These viruses are ubiquitous in the animal kingdom and infect reptiles, birds and mammals. One remarkable feature of papillomaviruses is their strong species and tissue restriction. Numerous different papillomaviruses may also infect a single species, and more than 150 human papillomavirus (HPV) genotypes have been identified. Each type is characterized by a unique DNA sequence and is defined numerically. HPVs are important human pathogens and are associated with significant morbidity as a consequence of genital infections. These viruses are a common sexually transmitted disease (STD), and epidemiological studies have shown that 20–40% of young women in various countries have been diagnosed with genital HPV infections. The most severe impact of human papillomavirus infection in humans is that a small proportion of these lesions progress to anogenital, oral and skin cancers. Almost all cervical cancers and their precursor lesions (93%) contain HPV DNA. Anogenital cancers are associated with infections by a subset of about 20 different HPV types (predominantly HPV-16 and -18) that are defined as 'high-risk' or 'intermediate-risk' HPV types. Many other HPV infections of genital and skin tissues lead to benign but unsightly warts. Some benign lesions, however, can cause life-threatening disease, such as laryngeal papillomatosis. A majority of HPV infections are naturally eliminated, most probably by host cell-mediated immunity. A supporting observation is that HPV infections are especially common in immunocompromised patients. Current treatment of persistent lesions consists mainly of ablative procedures, and recurrences may be as high as 50%. See also **Sexually Transmitted Diseases**.

Classification

Papillomaviruses are small DNA viruses that belong to the *Papovaviridae* [Howley, 1996]. Included in this family are the polyomaviruses and papillomaviruses. All papovaviruses are nonenveloped, contain double-stranded circular DNA, and assemble in the nucleus of the infected cell. The papillomavirus DNA is associated with host cell histones or histone-like proteins and is encapsidated by 72 pentameric capsomeres arranged on a skewed icosahedral lattice. The viral genome consists of between 7.3 and 8.0 kb, and contains 8–10 open reading frames (ORFs) which

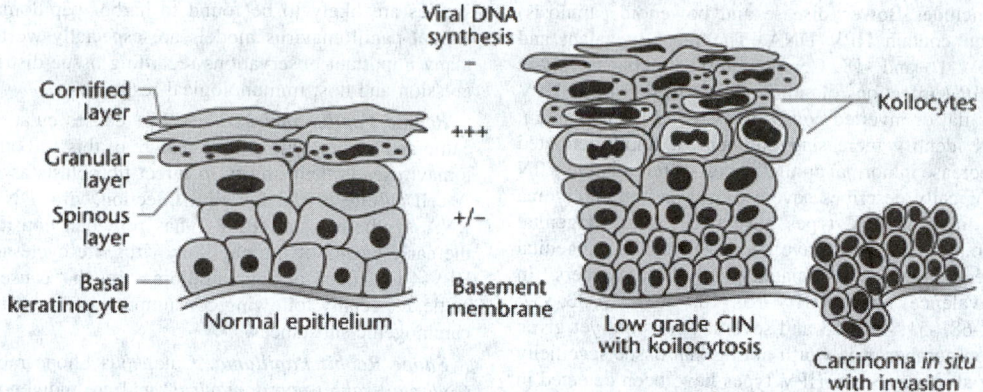

Fig. 1. Left: normal epithelium showing cell layers (basal, spinous, granular and cornified). Center: epithelium representing a low-grade cervical intraepithelial neoplasia (CIN) lesion with koilocytosis and showing that viral DNA synthesis is maximal in the granular layer. Viral DNA is predominantly episomal, and virions are present only in the granular and cornified layers. Right: carcinoma *in situ* with local invasion is shown on the bottom left. HPV DNA in these cells is integrated into host chromosomes.

code for early (E) and late (L) viral proteins. Viral transcription initiates from one DNA strand only and extensive splicing of viral ribonucleic acid (RNA) occurs. The viral life cycle is confined to, and completed within, the differentiating epithelial keratinocyte, and infectious virions are produced in the fully differentiated upper layers (Figure 1). Three major viral oncogenes, E5, E6 and E7, are involved in the stimulation of epithelial cell hyperproliferation leading to the classic raised appearance of most warts. Two early gene products, E1 and E2, regulate viral DNA replication. Another early gene product, E4, disrupts keratin filament formation in differentiating keratinocytes. These events may contribute to virion release from the fully differentiated, virus-laden squames. Two late genes, L1 and L2, make up the coat proteins of the papillomavirus capsid. A small noncoding region (400–800 bp) between the late gene L1 and the early gene E6 contains numerous regulatory elements, including the replication origin.

Papillomaviruses have been classified into a number of subgroups based upon several criteria, including viral DNA/protein sequence homologies, tissue specificity and an association with benign versus malignant lesions [Myers, et al., 1997]. The complete DNA sequence has been determined for about 65 human and about 15 animal papillomaviruses. Partial DNA sequences are available for an additional 60 HPV types, and new HPVs are being isolated and characterized each year. All papillomavirus genomes have a very similar organization of viral genes and regulatory elements. Fragments of the major coat protein (L1) and an early gene (E6) have been used to establish phylogenetic trees by DNA sequence comparisons. An HPV type has been operationally defined as having a nucleotide sequence that differs from the homologous nucleotide sequence of every other HPV type by at least 10%. Isolates of the same type are referred to as 'subtypes' or 'variants', and differ from each other usually by less than 5% at the nucleotide level. HPVs have been subdivided into several major super groups based upon phenotypic and clinical observations. The major supergroup (A) is represented by the 'mucosal/genital' papillomaviruses, which predominantly infect mucosal epithelium of the ororespiratory and anogenital tracts. The next largest supergroup (B) contains mostly papillomaviruses found in cutaneous infections of patients with a rare genetic disorder known as epidermodysplasia verruciformis (EV). This supergroup of HPVs, therefore, is usually described as the 'cutaneous/EV' papillomaviruses. A smaller supergroup (E) contains the HPV types associated with benign cutaneous papillomas common in the general population.

The complete genome sequence has been determined for about 15 animal papillomaviruses, and partial sequence information is available for another 20 animal types. The first complete papillomavirus sequence was established for Bovine papillomavirus type 1 (BPV-1) in 1982 [Chen, et al., 1982]. Complete genomic sequences are available for papillomaviruses that infect cattle, horses, rabbits, dogs, rodents, sheep, nonhuman primates and several ungulates. Partial sequence information is available for several additional rodent papillomaviruses and more than 10 primate papillomaviruses. Sequence comparisons between human and animal papillomaviruses indicate that diversity between papillomavirus types is increased with increasing evolutionary distances between hosts. These observations suggest a hypothesis of an ancient papillomavirus–host association.

Types of Human Papillomas

Papillomas associated with HPV infections have been detected in numerous sites of the body but are confined to cutaneous and internal mucosal regions. In addition, HPV DNA has been detected in many human cancers, including cervical, vaginal, penile and anal cancers, various cutaneous cancers including nonmelanoma skin cancer, and in a proportion of cancers of the sinuses, oropharynx, oesophagus, trachea, larynx and lung.

Cutaneous Warts (Verrucae). Perhaps the most familiar of all HPV infections are the benign common warts of the hands and feet. Typical common hand warts (*verrucae vulgaris*) are often found on the backs of the hands and fingers of small children, and are dome-shaped with multiple conical projections described as papillomatosis. This particular morphological feature provides the descriptive name for the wart viruses. Histological features of *verruca vulgaris* include hyperkeratosis (thickening of the corneal layer) and papillary epidermal hyperplasia (acanthosis). The upper epidermis contains koilocytes–enlarged keratinocytes with pyknotic nuclei surrounded by a clear halo. Deep plantar warts (*verrucae plantaris*) on the soles of the feet can be found in adults as well as children. These lesions are characterized additionally by the presence of abundant intracytoplasmic inclusion bodies in the outer layers. Flat warts of the hands, feet and face (*verrucae plana*) are also grouped with the common warts. Almost all of these infections are eliminated by host cell-mediated immunity, leading to wart regression and clearance of the papillomas. Some plantar warts, however, can be resistant to treatment. The majority of common warts, harbor HPV-1, -2, -3 and -4 (HPV-7 for meat-, poultry- and fish-handlers). Less common cutaneous papillomas contain DNA from HPV-41, -63 and -65.

Genital Warts (Condyloma Acuminata, Buschke-Löwenstein Tumor, Brown Disease, Bowenoid Papulosis). The most important group of HPV infections are the genital or venereal warts. These lesions have been subcategorized into flat, inverted and papillary condylomas. Disease associated with genital HPV infections also includes carcinoma *in situ* of the uterine cervix (see below). The most recognizable of the genital warts are the benign condyloma acuminata. These infections can be large and appear on the external genitalia of both sexes, as well as internally in the vaginal wall, vagina and perianal canal. The predominant HPV types of condyloma acuminatum are the so-called 'low-risk' HPV types 6 and 11. Additional HPV types associated with less frequent benign condylomatous lesions include HPV-40, -42, -43, -44, -54, -55, -83 and several others. Giant condylomas (Buschke–Löwenstein tumour) also involve penile and anal sites and predominantly contain HPV-6 and -11. These latter lesions are locally invasive, but rarely metastasize.

Carcinoma *in situ* (includes Bowen disease and bowenoid papulosis) of anogenital sites often contain HPV DNA. The most prevalent and well recognized are HPV-16 and -18. Cervical intraepithelial neoplasia (CIN) associated with HPV infections of 'intermediate to high-risk' HPV types typically begin as flat or inverted condylomas (condyloma planum; CIN I). Grades of CIN identify increasing numbers of undifferentiated malignant cells and a decrease in normal epithelial cell differentiation. CIN III, therefore, morphologically describes severe dysplasia and carcinoma *in situ*. More than 30 different HPV types have been found in genital infections, and approximately 20 types have been detected in anogenital cancers. The more frequent types associated with these cancers (in decreasing order of prevalence) include HPV-16, -18, -45, -31, -33, -52, -58, -35, -56, -59, -39, -68, -51, -26, -55 and several others not yet given a numerical identification number. It is worth noting that tissue specificity of all HPV types is not absolute. Some HPV types have been detected in both cutaneous and mucosal tissues. In addition, some genital condylomas of benign phenotype may contain high-risk HPV types, and DNA from some low-risk HPV types has been found in squamous cell carcinomas. See also **Sexually Transmitted Diseases**; and //Wart (Verrucae).

Oral and Laryngeal Papillomas. A small but significant number of patients present with papillomas of the mucosa of the ororespiratory tract. Especially difficult to cure are papillomas of the larynx. Although laryngeal papillomas are associated with low-risk HPV-6 and -11, these usually benign infections can cause life-threatening disease by obstruction of airways, and from the debilitating effects of frequent treatments for recurrences. Squamous cell carcinoma of some oral and head and neck cancers contain high-risk HPV-16, -18 and -31, and, occasionally, low-risk HPV-6 and -11 [Schwartz et al., 1998].

Macular Lesions. Cutaneous flat papillomas known as macular lesions appear on patients with a rare autosomal recessive disease, epidermodysplasia verruciformis (EV), and have yielded many different HPV types. Some of these lesions progress to malignant squamous cell carcinomas [Jablonska, et al.: 1972], and the predominant HPV types in such cancers are HPV-5 and -8. Patients with EV tend to have depressed cell-mediated immunity. Recent studies have indicated that certain other skin diseases such as psoriasis may also contain HPV DNA (e.g. HPV-5). In addition, numerous new HPV types have been detected in nonmelanoma skin cancers of immunosuppressed patients [de Villiers, et al.: 1997). These new types have been grouped together with the cutaneous/EV HPV types. Other HPV types found in macular lesions include HPV-9, -12, -14, -15, -17, -19, -20, -21, -47 and -49.

Subclinical or Latent Infections. An important category of HPV-associated disease is subclinical or latent infection. It is becoming increasingly clear that not all HPV infections permanently manifest as macroscopic disease. Subclinical infections represent a significant reservoir of infection for (1) host-to-host transmissions, (2) future sites for macroscopic disease following reactivation by a variety of environmental cofactors (immune suppression, carcinogens, stress-related effects, hormonal changes, exposure to ultraviolet light, wounding and coincidental interactions with other STD agents, to name a few), and (3) future sites for malignant progression following reactivation. Subclinical or latent infections represent the bulk of all human genital HPV infections and are detectable only by highly sensitive screening systems such as polymerase chain reaction (PCR) and reverse transcriptase PCR (RT-PCR), which amplify low levels of viral DNA and messenger RNA (mRNA), respectively.

Types of Animal Papillomas

Animal papillomavirus infections parallel HPV infections in cutaneous and mucosal tissue distribution. Several broad subcategories of animal papillomas can be defined: ungulate fibropapillomas, cutaneous papillomas and oral and genital papillomas. Cutaneous and mucosal animal papillomaviruses display features in common with human counterparts, including tissue distribution, genomic organization of viral genes, viral gene functions and ultrastructural identity of capsids. The earliest animal models of papillomavirus infections include bovine fibropapillomas, the Shope rabbit papilloma and canine oral papillomatosis. Cutaneous papillomas of cattle and rabbits can be massive in size. Papillomaviruses have been detected in most mammals and several birds and reptiles. These observations suggest that, with rigorous screening and sensitive detection systems, all animal species are likely to be found to harbor papillomavirus infections. Four animal papillomavirus models are especially worthy of note because of many important observations regarding tissue distribution, malignant progression and host immunological responses.

Bovine Papillomaviruses. BPV-1 causes cutaneous fibropapillomas in cattle and horses. One unique feature of this and other ungulate fibropapillomaviruses is their ability to infect fibroblasts as well as epithelial cells. A cell culture model for viral infection, viral DNA replication and viral RNA synthesis using BPV-1 has provided considerable information on the early events of the papillomavirus life cycle and viral gene function. BPV-4 is a mucosotropic papillomavirus that causes alimentary cancers in cattle, especially following consumption of bracken fern shown to contain carcinogenic substances.

Shope Rabbit Papilloma. Cutaneous Shope rabbit papillomas contain *Cottontail rabbit papillomavirus*) and are indigenous to cottontail rabbits of the midwestern plains states of the United States.

Canine Oral Papillomatosis. Dogs infected with *Canine oral papillomavirus* provide a model to examine papillomavirus infections in mucosal tissues of intact hosts. Natural transmission is common, and vaccines can be assessed for protection against both experimental and natural infection.

Rodent Papillomaviruses. Rodent papillomavirus infections include the skin lesions of the multimammate rat and the European harvest mouse, and mucosal lesions of the hamster. The interesting features of the first two models are the development of papillomas as the animal's age, and the presence of the viral genome in most tissues of the body. High levels of viral mRNA, DNA and virus capsids are detectable only within the skin lesions.

Mechanism of Oncogenesis

A role for HPVs in human anogenital cancers was suggested in the 1970s [zur Hausen, 1977]. Cloning and sequencing of papillomavirus genomes has provided opportunities to test transforming functions of papillomavirus genomes as well as individual viral genes. Interestingly, high-risk HPV types transform cultures of human keratinocytes, whereas low-risk HPV types do not. Because papillomaviruses complete their life cycle in differentiating epithelium, these viruses have developed strategies to both regulate and disrupt normal epithelial cell differentiation and host cell DNA replication. Three early viral genes have been shown to play a major role in papillomavirus-mediated transformation leading to oncogenesis.

E5. BPV-1 was shown to transform fibroblast cell lines in the early 1960s. Molecular cloning, sequencing and mutational analysis of the BPV-1 genome first identified a small, hydrophobic, 44-amino acid protein (E5) as the major transforming protein of BPV-1. The strong transforming activity of BPV-1 E5 is shared by other ungulate fibropapillomaviruses. In contrast, E5 proteins of HPVs show only weak transforming activity. Further research has shown that papillomavirus E5 proteins interact with several growth factor receptors (e.g. platelet-derived growth factor receptor (PDGFR) and epidermal growth factor receptor (EGFR)), the 16-kDa subunit of the vacuolar H^+-ATPase (adenosine triphosphatase), and the α-adaptin-like protein. The predicted role of E5 in cellular transformation involves E5 dimerization and transmembrane interaction with vacuolar H^+-ATPase subunits and growth factor receptors, leading to receptor dimerization in the absence of ligand. The consequences of these interactions are inhibition of the acidification of host cell endosomes, with retention of nondegraded receptors and upregulation of receptor density.

E6. E6 proteins are 16–19 kDa in size and are present at very low levels in both the nucleus and cytoplasm. The oncogenic nature of papillomavirus E6 proteins can be demonstrated by *in vitro* cell transformation assays. The first potential mechanism of E6 transforming ability was the demonstration of an association of E6 with the tumor suppressor protein p53. Thus, E6-associated p53 complexes are targeted for degradation via the ubiquitin pathway. Loss of p53 is hypothesized to relax DNA repair and thus allows for an accumulation of mutations. Other functional domains of E6 proteins are involved in transcriptional activation, apoptotic pathways and interactions with a variety of proteins whose role in tumor progression are still under investigation. Proteins that have been shown to interact with the multifunctional E6 protein include: p53, E6-AP (E6-associated protein), E6-BP (E6-binding protein), paxillin (regulator of cell adhesion to the extracellular matrix), hDLG (human homologue of the *Drosophila* discs large tumour suppressor protein), IRF-3 (interferon regulatory factor 3), zyxin (a focal adhesion contact protein), CBP (CREB-binding protein, which is involved in transcriptional control) and several members of a family of apoptosis-promoting proteins such as BAK (*Bcl-2* homologous antagonist/killer). Current research is focusing on the interactions between E6 and these proteins at the molecular and cellular level to determine more precisely the role of E6 in oncogenesis. Potential discriminating activity between E6 proteins from high- and low-risk HPV types have been found for p53 and CBP interactions. HPV E6 proteins of high-risk HPV types strongly associate with p53 and CBP, whereas E6 proteins of low-risk HPV types do not.

Additional characteristics of E6 proteins include: four CX_2C zinc-finger motifs involved in DNA binding, inhibition of HPV E7-induced apoptosis, resistance to TNF (tumor necrosis factor)-mediated lysis in cells sensitive to TNF, and activation of the telomerase enzyme [Klingelhutz et al., 1996]. Telomerase adds repeat sequences to chromosome ends and is believed to play a role in cellular immortalization. There is a general consensus that E6 proteins mediate their transforming activity via both p53-dependent and p53-independent mechanisms.

E7. Papillomavirus E7 proteins also disrupt the functions of host cell-cycle regulatory proteins, including interactions with the tumor suppressor protein pRB and other 'pocket proteins', p107 and p130. E7 proteins are approximately 100 amino acids in size, and contain two CX_2C zinc-finger DNA-binding motifs, a pRB-binding domain, a casein kinase II (CKII) phosphorylation site, and regions that show homology with other viral oncogenes. Other activities include cell transformation, immortalization and transactivation, and tumor formation in HPV transgenic mice. HPV E7 proteins interact with the cyclin/cyclin-dependent kinase system, leading to reactivation of host DNA replication machinery in postmitotic, differentiated epithelial cells, and to increased viral DNA synthesis. Low-risk HPV E7 proteins do not demonstrate the transforming potential of high-risk HPV E7 proteins in a manner analogous to HPV E6 proteins of high- versus low-risk HPV types.

Role in Cervical Cancer. Uterine cervical cancer is the second leading cause of cancer-related deaths in women worldwide. Approximately 500 000 deaths per year are attributed to cervical cancer. Age-adjusted incidence rates vary from 10 to 40 per 100 000 per year. There is universal agreement now that HPVs are the central aetiological factor in this cancer. In fact, associations between cervical cancer, sexual intercourse and an infectious agent(s) have been recognized for many years. With the advent of improved molecular biological techniques for HPV detection, almost all cervical cancers and their precursor lesions (93%) have been shown to contain HPV DNA [Bosch, et al.: 1995]. In fact, a recent reassessment of the 'HPV-negative' samples by these investigators has revealed that most of the latter lesions actually contain HPV DNA. These new studies have led the authors to conclude that 'the occurrence of HPV negative cervical carcinomas is extremely rare or perhaps non-existent'. Although the connection between cervical cancer and certain HPV types is the most widely recognized, these HPV types are involved in cancers of other mucosal sites, as discussed above. The mechanism(s) by which certain HPV types trigger events leading to cervical cancer have come from a variety of experimental and epidemiological studies. A summary of the most important observations is listed below:

- DNA of certain HPV types (approximately 20 types) has been found in almost all cervical cancers.
- Precursor lesions containing high-risk HPV DNA precede carcinoma *in situ* of the cervix, which precedes malignant metastatic cervical cancer.
- Viral oncogenes (E6 and E7) of high-risk HPV types transform cultures of human keratinocytes and various animal cell lines.
- Viral DNA of high-risk HPV types in cervical and anogenital cancer cells is often integrated into the host chromosomes, whereas DNA of low-risk HPV types is usually extrachromosomal or episomal. Integration of the viral DNA leads to upregulation of E6 and E7 gene expression. One mechanism for this upregulation includes the loss of function by the viral E2 protein (gene disrupted by integration), which can negatively regulate viral E6 and E7 expression.
- Viral DNA integration has been found to occur at chromosomal break points or fragile sites. Such integration may lead to potential

disregulation of certain cellular oncogenes or tumour suppressor genes. Surveys of viral/host DNA integration sites have revealed the presence of hybrid viral/host mRNA and proteins.

- Particular molecular variants of HPV-16 are associated with anogenital neoplasms more frequently than are other HPV-16 variants [Xi, et al.: 1998].

One noticeable and intriguing observation is that cervical cancer is by far the most prevalent of the HPV-associated anogenital cancers in immunocompetent patients, despite HPV infection of mucosal tissues of other genital and ororespiratory sites in both genders. A region of the cervix in which columnar epithelium changes to squamous epithelium (called the *transition zone*) is a particularly vulnerable site for HPV infections and subsequent malignant progression.

There is a general consensus that HPVs are necessary but not sufficient for the development of cervical cancer. HPVs play a role in the development of precancerous lesions by stimulating epithelial hyperproliferation and disrupting cell cycle regulatory events, thereby setting the stage for malignant progression. Various cofactors have been suggested by epidemiological and other studies. The end result often includes chromosomal abnormalities in cervical cancer cells, which is a feature that is common to many other human tumors. See also **Cancer and Oncology**; and **Oncogenes**.

Epidemiology and Controls

Incidence. Most attention has focused on genital HPV infections, whose incidence varies widely from study to study, and with the methods used for HPV detection. Genital HPV infections are infrequent in young children, reach a maximum level in sexually active young adults, and decrease with further increase in age. Several recent developments in HPV research have triggered a reassessment of HPV incidence in various patient populations. These include: (1) the increased awareness of the abundant number of different HPV viral types, and methods for their individual detection; (2) the predominance of subclinical or latent infections versus macroscopic or clinically evident disease; (3) the predominant role of host cell-mediated immunity in the control of HPV disease; and (4) the development of more sensitive viral DNA detection methods (PCR) for measuring subclinical HPV infections.

In a cohort of college women studied over a 3-year period, the cumulative incidence of HPV infection was 43%, with 60% of the women having infections at some time during the study period [Ho, et al.: 1998]. The median duration of HPV infection was 8 months, indicating that the bulk of genital HPV disease in these patients was transient. Persistent infections occurred in 9% of the cohort. Risk factors for HPV persistence included older age, infection with multiple HPV types, high viral load within the lesions and infection with a high-risk HPV type at the previous visit. Several important conclusions are: (1) men represent an important source of viral transmission; (2) the transient nature of most of the HPV infections in young adults strongly suggest an immunologically-mediated response leading to regression; (3) a variety of risk factors contribute to the outcome of persistence and progression of genital HPV infections; and (4) transmission must also occur from subclinical infections. Since these studies confirm HPV infection by PCR of viral DNA, it is not clear what proportion of these 'infections' represent transcriptionally-active infections, or latent infections that may be activated at a later time by various cofactors, or merely represent the presence of viral DNA in an inactive, noninfectious state. Further support for a role of host immunity in the control of HPV disease stems from data obtained from immunocompromised patients. Transplant recipients on immunosuppressive therapy and patients infected with Human immunodeficiency virus (HIV) have an increased incidence of HPV infections and HPV-associated anogenital cancers. Invasive cervical carcinoma was added to the Centers for Disease Control and Prevention (CDC) case definition of acquired immune deficiency syndrome (AIDS) in 1993. See also **Acquired Immune Deficiency Syndrome (AIDS)**; and **Human Immunodeficiency Viruses (HIV)**.

Transmission. The cellular and molecular events of natural transmission of HPV infections in humans are poorly understood. Common hand and foot warts in children are believed to spread by person-to-person contact with infected sites. Sexual intercourse is clearly indicated for genital HPV spread. Juvenile onset recurrent respiratory papillomatosis is believed to be acquired during vaginal delivery from mothers with genital HPV infections. Shed virus from the upper layers of productive lesions that is rubbed into sites of microabrasion in the uninfected epithelium during intercourse is the accepted explanation for natural transmission of genital HPV infections. Target cells for new infections are believed to be basal epithelial cells exposed to virions following microtrauma to the epithelium. Studies with animal models of papillomavirus infection have used experimentally-induced wounding and papillomas which contain abundant levels of infectious virions. However, the high incidence of human genital infections in sexually active young adults, coupled with less frequent macroscopic disease and the difficulty in detecting abundant levels of virions in many human lesions, suggest that natural transmissions may also occur from subclinical infections.

Controls

Prevention. Prevention of the spread of genital HPV infections is a topic that is seldom discussed. Condoms are ineffective protection against the spread of exophytic condylomata. Males are clearly the vector for heterosexual transmission to females, but male penile infections often are insignificant, and seldom develop into cancerous lesions. Post infection preventive measures in developed countries include early detection by the Papanicolaou (PAP) test, which looks for abnormal cells from cervical scrapings. Preinfection protective measures, such as antiviral microbicidal agents, are unavailable. Future prevention is aimed at protective vaccines using HPV virus-like particles (VLPs) and early viral protein- and/or DNA-based reagents for activation of protective immunity. Clinical trials to determine the efficacy of some of these vaccine strategies are currently underway.

Treatment. Treatment of HPV infections is largely ablative. Methods include topical or intralesional treatments with caustic agents such as podophyllin, podophylotoxin, trichloroacetic acid, acyclic nucleotides such as 1-[(S)-3-hydroxy-2-phosphonylmethoxypropyl]cytosine (HPMPC) and 5-fluorouracil. Electrodessication, electrocautery, carbon dioxide laser and conventional surgery are used for the management of extensive or refractory disease. Immunomodulators (interferons, interleukins and immunopotentiating agents such as Imiquimod) and photoablative therapy have been used with some success. New treatments include phytochemicals such as indole-3-carbinol, a compound naturally present in cruciferous vegetables. Overall efficacies for these treatments range from 22 to 95% for clearance of exophytic genital warts. Recurrence rates are often high, and range from 25 to 50% [Auborn and Steinberg, 1990]. A significant component of successful treatment includes active specific cell-mediated immunity. Thus, recalcitrant lesions in some patients may be attributable to potential genetic susceptibilities, immune suppression and coincidence of certain genetic disorders that result in impaired immunity. Additional factors include correlations between cervical cancer and certain major histocompatibility complex (MHC) class II polymorphism, and between anogenital cancer and the presence of 'variants' of high-risk HPV types [Xi, et al., 1998]. See also **Antiviral Drugs**.

Additional Reading

Auborn, K.J., and B.M. Steinberg: "Therapy of Papillomavirus-induced Lesion," In: Pfister, H.: *Papillomaviruses and Human Cancer*, CRC Press, LLC, Boca Raton, FL, 1990, pp. 203–224.

Bosch, F.X., M.M. Manos, N. Munoz, et al.: "Prevalence of Human Papillomavirus in Cervical Cancer: A Worldwide Perspective," *Journal of the National Cancer Institute*, **87**, 796–802 (1995).

Campo, M.S.: *Papillomavirus Research: From Natural History to Vaccines and Beyond*, Canister Academic Press, Norwich, UK, 2006.

Chen, E.Y., P.M. Howley, A.D. Levinson, and P.H. Seeburg: "The Primary Structure and Genetic Organization of the Bovine Papillomavirus Type 1 Genome," *Nature*, **299**, 529–534 (1982).

Davy, Clare, and J. Doorbar: *Human Papilloma Viruses: Methods and Protocols*, Springer-Verlag New York, LLC, New York, NY, 2005.

de Villiers, E.M., D. Lavergne, K. McLaren, and E.C. Benton: "Prevailing Papillomavirus Types in Non-melanoma Carcinomas of the Skin in Renal Allograft Recipients," *International Journal of Cancer*, **73**, 356–361 (1997).

Garcea, R., and D. DiMaio: *Papillomaviruses*, Springer-Verlag New York, LLC, New York, NY, 2007.

Ho, G.Y.F., R. Bierman, L. Beardsley, C.J. Chang, and R.D. Burk: "Natural History of Cervicovaginal Papillomavirus Infection in Young Women," *New England Journal of Medicine*, **338**, 423–428 (1998).

Howley, P.M.: "*Papillomavirinae*: the Viruses and Their Replication," In: Fields, B.N., D.M. Knipe, and P.M. Howley, et al.: *Fields Virology*, 5th Edition, pp. 2045–2076, Lippincott Williams & Wilkins, Philadelphia, PA, 2006.

Jablonska, S., J. Dabrowski, and K. Jakubowicz: "Epidermodysplasia Verruciformis as a Model in Studies on the Role of Papovaviruses in Oncogenesis," *Cancer Research*, **32**, 583–589 (1972).

Klingelhutz, A.J., S.A. Foster, and J.K. McDougall: "Telomerase Activation by the E6 Gene Product of Human Papillomavirus Type 16," *Nature*, **380**, 79–82 (1996).

Knipe, D.M., P.M. Howley, M.A. Martin, and D.E. Griffin: *Fields Virology, Vol. 2*, 5th Edition, Lippincott Williams & Wilkins, Philadelphia, PA, 2006.

McCance, D.J.: *Human Papilloma Viruses: Perspectives in Medical Virology*, Vol. 8, Elsevier Science & Technology Books, New York, NY, 2002.

Mindel, A.: *Genital Warts: Human Papillomavirus Infection*, Hodder Arnold Publication, New York, NY, 1995.

Myers, G., H. Baker, K. Munger, et al.: *Human Papillomaviruses 1997*, Theoretical Biology and Biophysics Group T-10, Los Alamos, NM, 1997.

Rous, P., and J.W. Beard: "The Progression to Carcinoma of Virus-induced Rabbit Papilloma (Shope)," *Journal of Experimental Medicine*, **62**, 523–548 (1935).

Schwartz, S.M., J.R. Daling, D.R. Doody, et al.: "Oral Cancer Risk in Relation to Sexual History and Evidence of Human Papillomavirus Infection," *Journal of the National Cancer Institute*, **90**, 1626–1636 (1998).

Tommasino, M.: *Papillomaviruses in Human Cancer: The Role of E6 and E7 Oncoproteins*, Landes Bioscience, Austin, TX, 1997.

Xi, L.F., C.W. Critchlow, C.M. Wheeler, et al.: "Risk of Anal Carcinoma *in situ* in Relation to Human Papillomavirus Type 16 Variants," *Cancer Research*, **58**, 3839–3844 (1998).

zur Hausen, H.: "Human Papillomaviruses and their Possible Role in Squamous Cell Carcinomas," *Current Topics in Microbiology and Immunology*, **78**, 1–30 (1977).

NEIL D. CHRISTENSEN, Pennsylvania State University, Hershey, PA

PAPRIKA.   See **Spices**.

PARABENS.   See **Antimicrobial Agents (Foods)**; and **Food Additives**.

PARABOLA.   A conic section obtained by a cutting plane parallel to an element of a right circular conical surface. It is the locus of a point that moves so that its distance from the *directrix* equals its distance from the focus; thus, the eccentricity is unity.

The standard equation in rectangular Cartesian coordinates is $y^2 = 2px$. The coordinates of its focus are $x = p/2$, $y = 0$ and its directrix is parallel to the Y-axis at $x = -p/2$. The straight line through the focus and perpendicular to the directrix is the *axis* of the parabola. The point where the parabola crosses the axis is the *vertex*. When the curve is placed in its standard position, the axis is the X-axis, and the vertex is the coordinate origin.

The *latus rectum* of the parabola has length $2p$. Its center is at infinity; hence, it is a noncentral conic.

The parabola has no asymptote. Its equation in polar coordinates is $r = 2a/(1 - \cos\theta) = \sec^2 \theta/2$. If the tangents at the extremities of the latus rectum are taken as coordinate axes, the parabola is tangent to the new coordinate system, and its equation is $x^{1/2} + y^{1/2} = a^{1/2}$. The evolute of a parabola is a semicubical parabola.

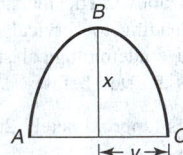

Fig. 1.   Points and dimensions of parabola.

With reference to Fig. 1,

$$\text{the length of arc } ABC = \sqrt{4x^2 + y^2} \frac{y^2}{2x} \ln \frac{2x + \sqrt{4x^2 + y^2}}{y};$$

and the area of section $ABC = \frac{4}{3}xy$.

See also **Conical Surface**; and **Conic Section**.

PARABOLA (Cubical).   A higher plane curve represented by the equation $y = ax^3$. There is a point of inflection at the origin, and the X-axis is tangent to the curve at that point. See Fig. 1.

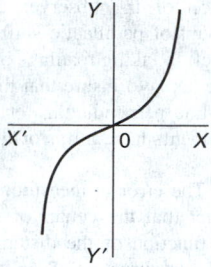

Fig. 1.   Cubical parabola.

See also **Curve (Higher Plane)**.

PARABOLA (Semicubical).   A higher plane curve represented by the equation $y^2 = ax^3$. It is also know as Neil's parabola. There is a cusp of the first kind at the origin, where the X-axis is a double tangent. It is the evolute of an ordinary parabola and a special case of a binomial curve, $y = C^{1-n}x^n$. See Fig. 1.

See also **Curve (Higher Plane)**.

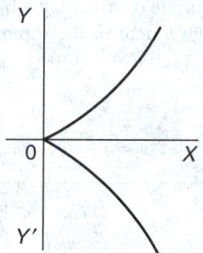

Fig. 1.   Semicubical parabola.

PARABOLIC COORDINATE.   In a limiting case of confocal paraboloidal coordinates, only two surfaces result. There are two families of parabolas with common foci at the origin of a rectangular coordinate system, one extending toward the positive and the other toward the negative Z-direction. Rotate the families about this axis to obtain paraboloids of revolution (ξ, η = constant) and add planes from the Z-axis (ϕ = constant). The position of a point is then given by

$$x = \xi\eta\cos\phi$$
$$y = \xi\eta\sin\phi$$
$$z = \tfrac{1}{2}(\eta^2 - \xi^2)$$

See also **Coordinate System**.

PARABOLIC CYLINDRICAL COORDINATE.   A curvilinear coordinate system similar to parabolic coordinates, with coordinate surfaces consisting of two families which are parabolic cylindrical (ξ, η = constant) and a family of planes (z = constant). A point in this system is given by

$$x = \xi\eta$$
$$y = \tfrac{1}{2}(\eta^2 - \xi^2)$$
$$z = z$$

See also **Coordinate System**.

PARABOLIC MIRROR.   See **Spherical Aberration**.

PARABOLIC SPIRAL.   A special kind of spiral, the equation of which in polar coordinates is $r^2 = a\theta$. It is also called Fermat's spiral. The windings of it get closer together as it approaches the pole. It has no asymptote. See Fig. 1.

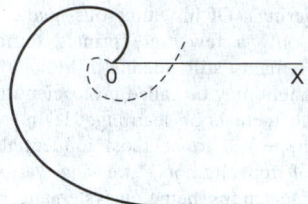

Fig. 1.   Parabolic spiral.

PARABOLOID.   A noncentral quadric surface described by the equation

$$\frac{x^2}{a^2} \pm \frac{y^2}{b^2} = \frac{z}{c}$$

With the upper sign, the surface is an elliptic paraboloid. Sections parallel to the XZ- and YZ-planes are parabolas; sections parallel to the XY-plane at $z = k$ are ellipses if k/c is positive, but the locus is imaginary if k/c is negative. When $a = b$, the elliptic sections become circles, and the quadric is a surface of revolution which could be obtained by rotating a parabola about the Z-axis.

A hyperbolic paraboloid results when a minus sign is taken in the equation for the surface. Sections parallel to the XZ- and YZ-planes are parabolas; sections parallel to the XY-plane are hyperbolas. When placed in its standard position, as shown by the equation, the origin of the coordinate system is a saddle point for the hyperbolic paraboloid. No surface of revolution is obtained in this case. If one member of a confocal quadric system becomes a paraboloid, all three of the surfaces are paraboloids. The equation for this case may be taken as

$$\frac{x^2}{A-q} + \frac{y^2}{B-q} = \frac{2z}{C} - \frac{q}{C^2}$$

where $A > B$. When $q > A$ or $q < B$, the surfaces are elliptic paraboloids with vertex on the negative and positive Z-axis, respectively; when $A > q > B$, the surfaces are hyperbolic paraboloids.

See also **Quadric Surface**.

PARABOLOIDAL COORDINATE.   A curvilinear coordinate system of confocal paraboloids. If λ, μ, ν are the real roots of a cubic equation in a parameter describing such surfaces, they are elliptic paraboloids (λ, μ = const.); $-\infty < \lambda < b^2$; $a^2 < \nu < \infty$ and hyperbolic paraboloids ((μ = const.), $b^2 < \mu < a^2$ where $a > b > c$ are also constants).

As in the case of ellipsoidal coordinates, a convention is needed for the sign of the coordinates. They are related to rectangular coordinates by the equations

$$x^2 = \frac{(a^2 - \lambda)(a^2 - \mu)(a^2 - \nu)}{(b^2 - a^2)}$$

$$y^2 = \frac{(b^2 - \lambda)(b^2 - \mu)(b^2 - \nu)}{(a^2 - b^2)}$$

$$z = \frac{1}{2}(a^2 + b^2 - \lambda - \mu - \nu)$$

See also **Coordinate System**.

PARACELSUS (1493/94–1541).   Paracelsus was a Swiss physician and philosopher of Swabian origin who propelled the development of medicine, chemistry and pharmacology in the Renaissance.

Paracelsus being a later nickname, Theophrastus of Hohenheim was born near the Benedictine abbey at Einsiedeln, Switzerland, where his mother was a bondsman. His father, William, an illegitimate offspring of a noble Swabian family, the Bombasts of Hohenheim, moved in 1502 to Villach, Carinthia, to be their town physician. Paracelsus received his early training from his father and from several clergymen and took, in about 1515, a doctorate in medicine and surgery at Ferrara, Italy. After years of travel, service as an army surgeon and attempts to settle in Salzburg and Strasbourg, he reached the climax of his career in 1527, when he was appointed town physician and lecturer to the city and university of Basel.

Being of quarrelsome temperament, everywhere his endeavors to reform medieval medicine soon led to conflicts with local authorities, causing further endless wanderings. Of his numerous medical, philosophical and theological writings only a few were printed during his lifetime and a considerable part remains still unedited. Although passing tribute to natural magic, his thought may be called protoscientific, stressing personal experience rather than mere book learning. He introduced alchemy into pharmacology, which proved to be most influential, and discarded the prevailing concept of four humors, stressing various other causes of disease instead. His teachings being unsystematic, rather unintelligible and somewhat dark, Paracelsus soon became the subject of legendary tales and remains controversial to this day. The present research focuses on correcting biographical inaccuracies and on scientific editions and interpretations of his works. See also **Physiology (The History)**.

Additional Reading

Gillispie, C.C.: 1970–1980 *Dictionary of Scientific Biography*, Charles Scribner's Sons, New York, NY.

Pagel, W.: *Paracelsus. An Introduction to Philosophical Medicine in the Era of the Renaissance*, Karger, Basel, Switzerland, 1982.

Web Reference

The Zurich Paracelsus Project: http://www.paracelsus.unizh.ch/

U. L. GANTENBEIN, Institute for History of Medicine, Zurich, Switzerland

PARAFFIN.   (1) Also called alkane. A class of aliphatic hydrocarbons characterized by a straight or branched carbon chain; generic formula C_nH_{2n+2}. Their physical form varies with increasing molecular weight from gases (methane) to waxy solids. They occur principally in Pennsylvania and mid-continent petroleum. (2) Paraffin Wax. See **Waxes**.

See also **Organic Chemistry**.

PARAFFINS.   See **Organic Chemistry**.

PARAFOVEAL VISION.   Vision in which the eye is so oriented toward the pertinent light source as to have the light fall upon some portion of the retina surrounding the fovea. Also called *scotopic vision*. See **Foveal Vision**.

The portion of the retina used in this type of vision contains receptors known as rods. Although these rods do not permit the sort of color-sensing vision possible with the cones in the central or foveal region of the retina, they have the useful property of responding to very low illuminance, particularly after dark adaptation is complete. Nighttime vision is performed primarily with the rods.

PARAKEETS.   See **Parrots and Cockatoos**.

PARALLAX (Astronomy).   Parallax effects are very important in astronomy. In transferring from one system of coordinates to another, parallax effects must always be applied when the location of the origin changes. For example, observations are taken from the surface of the earth, and geocentric parallax must be applied when transferring the observations to the center of the earth. Distances of many celestial objects are expressed in terms of parallactic angles. The angle subtended by an equatorial radius of the earth at the distance of the sun is known as solar parallax. The angle subtended by the radius of the earth at the distance of any other member of the solar system is known as the *horizontal parallax* of that object. In measuring the distances of the stars, the parallactic shift due to the revolution of the earth about the sun is employed; the angle subtended by one astronomical unit at the distance of the star is known as stellar parallax of the object. As the characteristics of solar motion become more completely known, parallactic shifts, called secular parallax, due to this motion will undoubtedly be used for the determination of distances.

Stellar distances are given by

$$D = \frac{1}{\pi}$$

where π is the parallactic shift in seconds of arc and D is in parsecs.

See also **Dynamical Parallax**; **Secular Parallax**; **Solar Parallax**; **Spectroscopic Parallax**; and **Stellar Parallax**.

PARALLAX (Instrument).   As an observer moves about, the relative positions of distant objects seem to change. This apparent change, actually due to change of position of the observer, is technically known as parallactic shift. The amount of parallactic shift is inversely proportional to the distance of the object. In taking readings of instruments using scales and pointers, care must be taken to insure that the eye of the observer and the pointer are both in a line perpendicular to the plane of the scale. To facilitate this, some instruments have a mirror in the plane of the scale.

PARALLAX ERROR.   The error in measurement between two pairs of antenna caused by the fact that the center of the two baselines do not coincide. This error is a function of the distance of the target from the baseline, as well as its relative direction.

PARALLEL A/D CONVERTER.   This is the simplest design among the electronic analog-to-digital converters. A comparator is provided for each quantization level in an *n*-bit converter. See also **Quantization**. A digital representation of the input signal is obtained by appropriately decoding the output of the multiple comparators. Reference is made to the example and schematic diagram in Fig. 1. A 2-bit converter is shown. Comparators 1, 2, and 3 are biased with reference voltages of 0.75 V_r, 0.5 V_r, and 0.25 V_r. V_r is the full-scale reference voltage and is equal to the full-scale input range of the A/D converter. All the comparators that are biased at a level less than the level of the input signal provide a binary 1 output when the input signal is applied. In the example shown, an input signal of 0.6 Vr will result in a 1 output from comparators 2 and 3. The output of comparator 1 will be a 0. Through appropriate decoding, the binary digital representation of 1 0 is yielded. The binary state at the input and output of each logic block is included in the diagram.

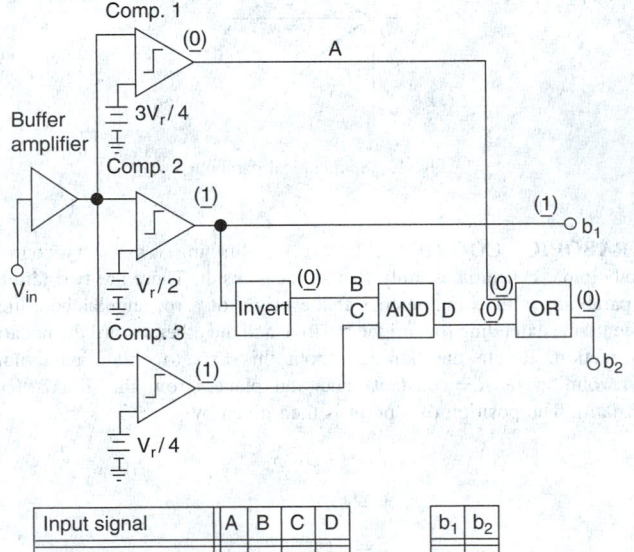

Input signal	A	B	C	D		b_1	b_2
$3V_r/4 < V_{in}$	1	0	1	0		1	1
$V_r/2 < V_{ir} < 3V_r/4$	0	0	1	0		1	0
$V_r/4 < V_{in} < V_r/2$	0	1	1	1		0	1
$V_i < V_r/4$	0	1	0	0		0	0

Fig. 1.   Parallel analog-to-digital converter.

Even though the example provides only 2-bit resolution, the addition of more comparators and decoding logic can increase the resolution. However, each additional bit of resolution requires a doubling of the necessary hardware. Consequently, the parallel A/D converter normally is used only for converters of fewer than 5 or 6 bits resolution. Not only must the hardware be increased, but increasing the resolution makes marked demands on the threshold stability of the comparators in the interest of avoiding output ambiguity and excessive errors.

Simplicity for low-resolution requirements and high speed are the principal advantages of the parallel A/D converter. The settling times of the comparators and logic delays are the main factors that determine the speed.

THOMAS J. HARRISON, IBM Corporation, Boca Raton, FL

PARALLELEPIPED.

A polyhedron with parallelograms as faces, thus, a six-sided prism with parallelograms as bases. The general case is an oblique parallelepiped where the lateral edges are not perpendicular to the bases. If the edges are perpendicular, it is called right; if the six faces are rectangles, rectangular; if the six faces are all squares, a cube. Sometimes the figure is also called a parallelepipedon and, incorrectly, parallelopiped or parallelopipedon.

The volume of a parallelepiped is conveniently written in vector analysis as the scalar triple product **[ABC]** where **A**, **B**, **C** are three vectors describing the edges of the parallelepiped. Its lateral area and volume can also be given by the same equations used for a prism.

PARALLELOGRAM.

A quadrilateral with opposite sides parallel and equal in length. Its consecutive angles are complementary. If a, b are the lengths of its sides and C is the acute angle between them, the area of a parallelogram is $A = ab \sin C = bh$, where h is the distance between the bases, b. Its area is also given by $2A = d_1 d_2 \sin D$, where d_1, d_2 are the lengths of the two diagonals and D is the acute included angle. A diagonal divides the parallelogram into congruent triangles, the two diagonals bisect each other, and their point of intersection is the center of gravity of the figure.

If two vectors are drawn from the same origin and the same vectors are used to construct a parallelogram, the sum of the vectors equals the length of the diagonal drawn from the origin. This is the well-known parallelogram law of elementary physics, which describes the combination of forces, acceleration, velocity, etc.

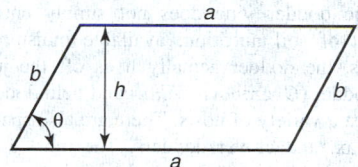

Fig. 1.   Principal parameters of parallelogram.

With reference to Fig. 1, area $A = ah = ab \sin \theta$, where a and b are the lengths of the sides of the parallelogram, h is the height, and θ is the angle between the sides. In a rhombus, which is an equilateral parallelogram, $A = 1ab$, where a and b are the lengths of the diagonals.

See also **Vector**.

PARALLEL SAILING.

A term used to designate the problem of converting the distance traversed by a ship along a parallel of latitude into difference of longitude. This problem is solved under **Departure (Navigation)**.

PARALLEL-SERIAL A/D CONVERTER.

This type of analog-to-digital converter is somewhat similar to the successive-approximation A/D converter because the conversion is accomplished in the form of a series of distinct steps. The speeds of parallel-serial A/D converters are not as high as those of parallel A/D converters, but practical attainment of resolution is considerably better, that is, 12- to 14-bit resolution at speeds over 100,000 samples/second.

A parallel-serial A/D converter is shown schematically in Fig. 1. Here the A/D converter is comprised of four 3-bit D/A converters (digital-to-analog) and four amplifiers. Each has a gain of 8 which corresponds to the 3-bit resolution of each section of the A/D converter, that is, $2^3 = 8$. A 3-bit conversion is accomplished by the seven comparators shown and in a fashion similar to that employed in parallel A/D converters. See also **Parallel A/D Converter**. The voltage that corresponds to the 3-bit conversion is subtracted from the input signal by digital-to-analog converter #1 (DAC 1). This difference signal then is amplified by a factor of 8. The next three bits are then determined in a similar manner in the next stage of the A/D converter. Inasmuch as three bits are determined simultaneously (nearly), five steps yield a 15-bit quantization of the input signal. To accomplish this in a successive-approximation A/D converter requires a 15-step procedure.

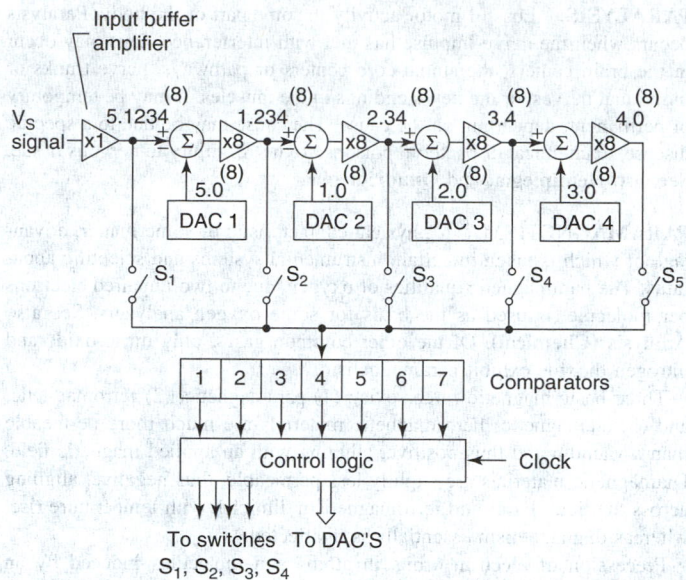

Fig. 1.   Parallel-serial analog-to-digital converter.

With reference to Fig. 1, assume an input voltage of 5.163086 V. Also assume a full-scale range of the A/D converter of 8.0 V. The operation of this A/D converter configuration is best explained in terms of octal numbers. Thus, the input voltage is 5.1234^8. The superscript 8 shows that the number is expressed in octal code. The conversion process commences with the closing of switch S_1 and applying the input signal to the parallel comparators. The output of the comparators shows that the input is in excess of 5^8 but is less than 6^8. Hence, DAC 1 is set to 5^8 and a voltage of 0.1234^8 is applied to the first gain of 8 amplifier. By opening S_1 and closing S_2, the output of the amplifier, 1.234^8 V is applied to the parallel comparators. This parallel conversion yields the determination that the amplified difference voltage is in excess of 1^8 but less than 2^8 V. Hence, DAC 2 is set to 1.0^8 V to furnish a difference signal of 0.234^8 V. Other steps follow which involve the closing of switches S_3, S_4, and S_5, and these actions result in the octal result of 5.1234^8, or the 15-bit binary result 101001010011100.

In the device shown in Fig. 1, it is required that each stage of the converter settle completely before the subsequent 3-bit conversion is commenced. Further circuitry is added for error correction so that if a particular stage has not completely settled, with a resultant incorrect 3-bit conversion, the conversion in the following stage will provide a signal to correct the prior 3-bit representation.

Although the parallel-serial A/D converter has relatively high-speed conversion, the design requires a greater hardware complexity than that of a successive-approximation A/D converter with equal resolution. The parallel-serial A/D converter is not as fast as the parallel A/D converter, but it requires considerably less hardware for resolutions of more than 6 bits. Further, precise comparator adjustments, required in a high-resolution parallel A/D converter, are avoided with the parallel-serial A/D converter.

See also **Analog-to-Digital Converter**.

Thomas J. Harrison, IBM Corporation, Boca Raton, FL

PARALLEL SURFACES.

The surface S' is said to be parallel to the surface S if the distance measured from S to S' along the normal to S at a point P is independent of the position of P on S. If P' is the point at which the normal to S at P intersects S', then the normals to S and S' at P and P', respectively, have the same directions. If S' is parallel to S, then S is parallel to S'.

PARALLEL TRANSMISSION.

The system of information transmission in which the characters of a word are transmitted (usually simultaneously) over separate lines, as contrasted to serial transmission.

PARALYSIS.   Loss of motor activity in some part of the body. Paralysis occurs when the nerve impulse has met with interference. This may occur in the brain centers, the spinal-cord centers or pathways, nerve trunks or individual nerves, or the nerve endings in the muscles. It may be temporary or permanent, depending on the cause. The cause may be due to a specific disease, to chemical or bacterial poisons, toxins, or injury to nervous tissue. See also **Hemiplegia**; and **Quadriplegia**.

PARAMAGNETISM.   A physical characteristic of some matter, advantage of which is taken in certain instrumental systems and scientific apparatus. The paramagnetic qualities of oxygen, due to two unpaired electrons per molecule, is used as the basis for some oxygen analyzers. See also **Analysis (Chemical)**. Of the other common gases, only nitric oxide and nitrogen dioxide exhibit paramagnetism.

Three basic magnetic forces exist: (1) paramagnetic, (2) ferromagnetic, and (3) diamagnetic. Ferromagnetic materials are much more permeable than a vacuum and thus positive, aligning with an applied magnetic field. Diamagnetic materials are slightly less permeable, thus negative, aligning across the field. Para- and ferromagnetism diminish with temperature rise, whereas diamagnetism essentially is unaffected.

Precession of electron orbits in atoms and molecules induced by an applied field causes diamagnetism in all matter even when paramagnetism dominates. Unpaired electrons in atoms or molecules cause para- and ferromagnetism. Normally, electrons of opposite spin pair off, netting zero magnetism per pair. Unbalanced spin moment of unpaired electrons yield both para- and ferromagnetism. Paramagnetic matter has unpaired electrons in outer electron shells. Thermal agitation retards atomic or molecular alignment, and thus the net moment is weak. Ferromagnetic matter, such as iron, cobalt, and nickel, has unpaired electrons in the next-to-outer shell. Interatomic forces cause molecular alignment and permanent magnetism results. See also **Magnetism**.

PARAMETER.   An arbitrary constant, as distinguished from a fixed or absolute constant. Any desired numerical value, subject in some cases to certain restrictions. Thus, in the set of equations $(x - a)^2 + (y - b)^2 = r^2$ for circles in the plane, there are three parameters, two describing the position of the center and one giving the radius, and for each choice of a, b, and r a corresponding circle is obtained. Similarly, the equations of a curve $x = \phi(t), y = \psi(t)$ is said to be in parametric form and each choice for the parameter t produces a corresponding point on the curve.

Parameterization is the representation, in a dynamic model, of physical effects in terms of admittedly oversimplified *parameters*, rather than realistically requiring such effects to be consequences of the dynamics of the system.

PARAMETERIZATION.   The representation, in a dynamic model, of physical effects in terms of admittedly oversimplified parameters, rather than realistically requiring such effects to be consequences of the dynamics of the system.

See also **Convective Adjustment**; **Mellor-Yamada Parameterization**; and **Subgrid-Scale Process**.

PARAMETER (Statistics).   An unknown constant appearing in the specification of a probability distribution, or in the specification of a model, e.g., the constants in a regression equation. The domain of permissible variation of the parameters defines the class of distribution or models of given form under consideration.

PARAMETRIC EQUATIONS.   A set of equations in which the independent variables or coordinates are each expressed in terms of a parameter. For example, instead of investigating $y = f(x)$, or $F(x, y) = 0$, it is often advantageous to express both x and y in terms of a parameter u: $x = g(u); y = G(u)$. The parameter may or may not have a useful geometric or physical interpretation.

PARANA PINE.   See **Araucarias**.

PARAQUE.   See **Nightjars and Nighthawks**.

PARASITE.   An organism living in close association with another organism, deriving its nourishment from the host, and harming the host organism in the process.

PARASITIC PLANTS.   Normally, when one considers injuries to plants and trees by parasites, the causal factors that immediately come to mind are fungi, viruses, and certain kinds of insects and other animal pests. Some plants, known as *phanerogamic parasites*, also can be economically important in some regions on some valuable trees and crop plants. These are plants, such as the mistletoes, Spanish moss, and dodder, among others, which attach themselves to other plants and grow on them depriving the host plant of nutrients, light, and moisture — in essence inhibiting normal growth of the host plants through processes of deprivation and strangulation.

Parasitic plants that strangle the host include a number of climbing vines, of which bittersweet (*Celastrus scandens L.*) is an example. Parasitic plants which in essence suck or consume vital juices of a host plant include the mistletoes, whose sticky seed becomes attached to the bark of the host, germinating and thus utilizing the host as its growth medium and depriving the host of space. Most tree branches so "infected" ultimately expire. The dwarf mistletoes (*Arceuthobium*) operate in a similar fashion against the western yellow pine tree. An interesting description of the mistletoe is given by J.M. Haller (*American Forests*, 11–15, December 1978).

The damage wrought by the foregoing parasitic plants is generally confined to large trees used for shade or timber. Food crops are most seriously affected by the dodders (*Cuscuta* spp.), which can seriously injure such field crops as alfalfa, clover, lespedeza, onion, and sugar beet. The parasitic dodder vine does not simply entwine an affected host to deprive it of soil nutrition, available moisture, and light, but like the mistletoes, the dodder actually lives off the juices of the host plant. Common dodder (*C. gronovii Willd.*) and field dodder (*C. pentagona Engelm.*) will infest a variety of hosts. There are also specializing dodders, such as alfalfa, flax, and clover dodders. The most effective control is that of eliminating dodder seed from crop seed through use of special processing machinery.

PARASITISM.   See **Symbiosis (Ecology)**.

PARASITOLOGY (The History).   Traditionally parasitology has been concerned with the harmful effects of organisms living and feeding inside or on another living organism. Although parasites may belong to any of the following groups — bacteria, yeast, fungi, algae, viruses, protozoa, helminths and arthropods — in effect parasitologists have focused on the internal zooparasites. Bacteria, viruses and fungi of plant origin are the provenance of microbiologists. Before the discovery of unicellular parasites c. 1880s onwards, worms or helminths were the major focus of attention.

Introduction

The large internal parasites of man have been recognized for centuries. Increasing interest in the natural history of these organisms is evident from the seventeenth century, particularly in Europe. Improvements in microscopy ultimately assisted in identifying parasites and studying their morphology, although debates raged about their origin, fuelling disputes on spontaneous generation. Parasites of animals, particularly those of domestic livestock, are also responsible for disease in humans, forming an important part of veterinary medicine, for example the tapeworms of beef and pork (*Taenia saginata* and *T. solium*). See also **Platyhelminthes**; and **Protozoa**.

In the latter part of the nineteenth century, the intensification of European imperialism in the tropics stimulated a greater interest in the study of tropical diseases. Studies of aetiology, informed by the tenets of the germ theory, determined that unicellular parasites or protozoa caused many diseases. The unraveling of complex parasite lifecycles involving intermediary hosts, mainly arthropods and some mollusks followed and fostered enthusiastic attempts to control and eradicate diseases by attacking these vectors. The histories of parasitology and entomology, particularly medical entomology, are closely allied. Control and/or eradication proved more difficult than anticipated for a variety of reasons: epidemiological, economic and political. In the latter half of the twentieth century, the concepts and tools of molecular biology have led to an increase in fundamental knowledge. Continued basic research in parasite biochemistry, physiology, nutrition, immunology and greater understanding of the nature of host/parasite relationships indicate new directions for future control efforts, particularly by the use of vaccines. Currently problems of parasite drug resistance are threatening to create fresh difficulties.

Early History to Late Nineteenth Century

Accurate understanding of the nature of parasitism and the relationship between parasites and specific diseases is a recent phenomenon, related to developments in natural history and theories of disease causation. However, there is plentiful evidence of the awareness of the presence of human and animal parasites in antiquity. Before the nineteenth century, besides observable parasites, credible doctors and their patients associated much ill health with real and imaginary creatures. There were suggestions of a parasitic origin for venereal disease, cancer and a belief that worms could mutate into other animals, e.g. frogs, lizards, salamanders, snakes, leeches and lice within the body. In the pre-Linnaean era, worms, as segmented animals, formed part of the insect group of which frogs and lizards were also members. In other cultures, parasites played different roles. Far from holding them detrimental, Chinese physicians believed that the presence of at least three kinds of worms was essential to maintain health. See also **Annelida**.

The Ebers papyrus of ca. 1550BC describes macroparasites visible to the naked eye, i.e. worms and ectoparasites such as flies and lice. Recognition of schistosomiasis, the most significant parasitic disease in the Nile valley, is extant but its aetiology remained unknown. Aristotle's researches in comparative anatomy revealed the presence of worms in a range of animals. He followed the Hippocratic writings, as did Galen after him and referred to tapeworms (*Taenia* species), round worms (*Ascaris lumbricoides*) and threadworms or pin worms (*Enterobius vermicularis*)—all inhabiting portions of the gastrointestinal tract in man. The presence of guinea-worm (*Dracunculus medinensis*) was well known along the shores of the Red Sea and the biblical story of Moses and the "fiery serpents" (Numbers 21:6) reputedly refers to an epidemic of guinea-worm during the Exodus. This interpretation is open to doubt. However, guinea worm is one of the success stories of modern medical science's attempts to deal with parasitic diseases by breaking the route of transmission. Dnyaneshvar A. Turkhud's (1868–1926) research in India showed that cyclopoids, the intermediate host, can be killed by treating sources of water with potassium permanganate and easily removed from infected water by passing it through a piece of fine cloth. The World Health Organization's safe drinking water campaign, launched in 1981, and national dracunculiasis eradication programs established in endemic countries, have resulted in an annual fifty percent reduction in the global incidence with eradication projected in the near future.

There were moderate increases in the knowledge during the Roman and medieval periods. By the seventeenth century parasitic helminths featured as part of the burgeoning interest in natural history. With the aid of his unique microscopes, Antony van Leeuwenhoek (1632–1723) observed in 1681 specimens of *Giardia duodenalis*. Not until 1954 was *G. duodenalis* identified conclusively as the pathogen responsible for giardiasis: a form of diarrhea with mild abdominal pain most serious when it occurs in combination with immune deficiency syndromes. Leeuwenhoek's microscopy remained unequalled until the nineteenth century and naturalists, in addition to new discoveries and classificatory schemes, focused on the form and life history of the larger known parasites. Their work included identifying different sexes of the same parasite and uniting purportedly different species, which were but the mature and cystic forms of the same organism. For instance, Johann A. E. Göze (1731–1793) in 1784 described the cysts long known in measly pork as helminthic. He commented on the morphological similarities between the head of this worm and those of tapeworms found in intestines but did not unite these two forms as a single species. In 1845, Félix Dujardin (1801–1860), basing his argument on experimental demonstrations of cystic worms metamorphosing into adult worms, suggested that intestinal tapeworms (*Taenia solium*) and cystic forms (*Cysticercus cellulosae*) of pork were one species. Against this idea was the doctrine of spontaneous generation — the idea that living forms can arise directly from non-living matter by

chemical or physical forces. Belief in this theory in general waxed and waned but it did provide a useful rationale for the origin of parasites and the presence of large organisms in human intestines in particular: they obviously had not been ingested as such! The feeding experiments of Gottlob F. H. Küchenmeister (1821–1890) in 1855 confirmed at autopsy Dujardin's supposition. Contemporaries debated the ethics of deliberately giving cystic worms in food to condemned prisoners. See also **Dujardin, Félix (1801–1860)**; **Giardiasis**; and **Leeuwenhoek, Antoni van (1632–1723)**.

The early enthusiasts of parasitology came from a variety of backgrounds. Many were amateurs combining parasitology research with other interests in the natural world and indeed other careers. As comparative anatomists and keen dissectors, initial observations of parasites were often by chance. Discovery and classification dominated their research and output, although theories of the origin of parasites were also important. Francesco Redi (1626–1697), clinician, naturalist, poet and linguist, discovered several new species of helminths through dissection and provided the first illustrations of others in his *Osservazioni intorno agli animali viventi che si trovano negli animali viventi* (1684). Nicolas Andry (1658–1742), known also for coining the word orthopedics, offered a classificatory system based on intestinal worms and those living outside the intestine in his magnum opus *De la génération des vers dans le corps de l'homme* (1700). He repeated this division when describing the ill effects caused by the parasites. Andry opposed theories of spontaneous generation and suggested instead that all animals develop from seeds entering the human body via air or food. Peter Simon Pallas (1741–1811), surgeon and explorer of the Russian Empire, published *Miscellanea zoologica* in 1766, two years after his election to the Royal Society of London. Although not devoted entirely to parasitology, Pallas convincingly differentiated mollusks from worms. Further, in a large section on hydatids — cystic worms — he differentiated these from serous cysts by classifying them as non-adherent or adherent. Göze, a German Protestant pastor and amateur naturalist and microscopist, dedicated his revised classification to Pallas. In *Versuch einer Naturgeschichte der Eingeweidewürmer thierischer Körper*, published in 1782, Göze divided parasites into eleven genera: *Ascaris*, *Trichocephalus*, *Gordius*, *Cucullanus*, *Strongylus*, *Pseudoechinorynchus*, *Planaria*, *Fasciola*, *Taenia* and *Chaos*. The last group formed a catchall for the increasing numbers of known microscopic living organisms some of which he regarded as parasitic. The growth of knowledge justified the separation of helminthology from general zoology, but the discipline remained descriptive and reliant upon the theory of spontaneous generation.

The number and complexity of texts devoted to parasitology increased in the nineteenth century. Carl A. Rudolphi (1771–1832), trained as a doctor and a veterinary surgeon, published an immense two-volume (1808 and 1810) text, *Entozoorum, sive vermium intestinalium historia naturalis* and followed this in 1819 with *Entozoorum synopsis cui accedunt mantissima duplex et indices locupletissima*. These publications effectively trebled the known number of species. J. S. Olombel and Johann G. Bremser (1767–1827) provided current syntheses of the clinical aspects of parasitism in their respective volumes *Remarques sur les maladies vermineuses* (1816) and *Ueber lebende Würmer im lebenden Menschen* (1819). Casimir Joseph Davaine (1812–1882) published in 1860 his *Traité des entozaires et des maladies vermineuses de l'homme et des animaux domestiques*. This impressive work included a classification dividing the genera into six groups, the first of which, *Protozoaires* was a refinement of the Linnaean order *Chaos*. The division into three groups: *Vibrions*, *Trichomonas* and *Paramecium* reflected improved observation at the microscopic level. Davaine was also the first to describe the human parasites — *Pentatrichomonas hominis*, an intestinal flagellate and *Inermicapsifer madagascariensis*, a rare cestode. He advocated the microscopic examination of faeces to assist in diagnosis of worm infections and showed that the eggs of *Ascaris* remain infective for long periods in a damp environment.

In addition to improved classification, there were also significant conceptual advances. In 1842, the Dane Johannes J. S. Steenstrup (1813–1897) published a short text on the phenomena of alteration of generations — *Om fortplantning og udvikling gjennem vexlende generations raekker, en saere-gen form for opfostringen i de lavere dyreklasser*. Steenstrup synthesized accounts of animals that appeared to develop from parents unlike themselves with his own observations on the cercariae found in the freshwater habitats of mollusks *Planorbis cornea* and *Limnaeus stagnalis*. He identified the cercariae as cystic forms of the liver fluke *Distoma* and elucidated the life cycle by analogy with the ova and embryos of another *Trematode*, the *Monostroma*. Dujardin worked on *Trematodes* and *Cestodes* in a similar way and was probably the first to suggest that these organisms spend part of their lives in an intermediate host although he failed to publish these observations. See also **Steenstrup, J Japetus S (1813–1897)**.

Continental Europe led the discipline in the early years of the nineteenth century. In Britain, Matthew Baillie (1761–1823) reported that tapeworm was uncommon in England when briefly reviewing parasites in his seminal *Morbid Anatomy of Some of the Most Important Parts of the Human Body* (1793). In the mid-nineteenth century, translations of European articles appeared under the auspices of the Ray Society (the work of George Busk (1807–1886)) and the Sydenham Society. English editions of Küchenmeister's, *Manual of Animal and Vegetable Parasites* and von Siebold's, *Tape and Cystic Worms*, provided the first systematic texts on parasites. In 1845, three years after its publication in Danish, Steenstrup's *On Alternation of Generation; of the Propagation and Development of Animals through Alternate Generations* appeared in English. It was not until the 1860s and the work of Thomas S. Cobbold (1828–1886) that Britain made either significant original observations or contributions to the newer literature. Cobbold graduated from Edinburgh in 1851 with a medical degree but devoted his early career to research in biology, quickly focusing on the parasites he observed during dissection. In 1864, he published *Entozoa, an Introduction to the Study of Helminthology, More Particularly to the Internal Parasites of Man*. Cobbold established himself as a consultant physician, specializing in parasitic diseases, on the strength of the warm reception of his work at home and on the continent. He benefited, as did others, from techniques of embedding, sectioning and staining for enhanced microscopy introduced by K. G. F. Rudolf Leuckart (1822–1898). Leuckart's work appeared as *The Parasites of Man and the Diseases which proceed from them: A Textbook for Students and Practitioners* in English translation in 1886. See also **Leuckart, Karl Georg Friedrich Rudolf (1822–1898)**.

Cobbold published a new, more ambitious textbook in 1879. *Parasites; a Treatise on the Entozoa of Man and Animals Including some Account of the Ectozoa* aimed to help the growing number of workers in this field to identify their own specimens. A growing interest in parasitism and the expansion of the European presence in the tropics brought a steady flow of new material to London for examination, much of which found its way to Cobbold's laboratory. In his capacity as the home expert, he wrote, in 1877, to *The Lancet* announcing Joseph Bancroft's (1836–1894) discovery of the adult *Filaria bancrofti* and indeed named the worm thus. As further testimony to Cobbold's prominence at least in the English speaking world, he presented Patrick Manson's discovery of the development of microfilaria in mosquitoes to the Linnean Society in March 1878. A year earlier Manson had confirmed that this parasite was responsible for elephantiasis arabum or filariasis while working in Amoy, China. See also **Manson, Patrick (1844–1922)**.

In addition to the traditional helminths of parasitology, interest in the microparasites found fertile ground among doctors in Europe and the tropics. In the 1860s the work of Louis Pasteur (1822–1895) on fermentation demonstrated that microorganisms were responsible for this and apparently similar processes, for example putrefaction. He and others such as Robert Koch (1843–1910) proceeded to correlate the presence of specific organisms — germs — with particular diseases and ascribe their cause to these germs. Their work led to the discipline of bacteriology. There was some debate about according parasites, particularly the macroparasites, germ status. Following Charles L. A. Laveran's (1845–1922) observations of plasmodia in 1880 and Camillo Golgi's (1843–1926) association of the blood phase of the parasite with the characteristic pattern of fever in malaria (1889), the significance of the unicellular organisms was gradually accepted. See also **Bacteriology; Bacteriology (The History); Golgi, Camillo (1843–1926); Koch, Heinrich Hermann Robert (1843–1910)**; and **Pasteur, Louis (1822–1895)**.

The announcement of Manson's work on microfilaria provided a stimulus to the elucidation of complex lifecycles involving parasites, vectors and human disease. Although much of this effort concerned tropical parasites in European colonies, interest in aspects of parasitology continued elsewhere. Support for the protozoal origin of malignant tumors succeeded

interest in a bacterial aetiology in the 1890s. Theobald Smith (1859–1934) and Fred Lucius Kilborne (1858–1935) demonstrated for the first time the arthropod (*Boöphilus annulatus*) transmission of a protozoan (*Babesia bigemina*) responsible for Texas cattle fever in a series of papers published in 1891 and 1892. Attempts to control this veterinary problem targeted the vector with the development of tickicides, applied by dipping, to eliminate the tick. In malarious areas of Italy, discovery of the role of the mosquito in 1898 accelerated systematic government sponsored bonification (improvement) programs. More generally while parasitology via tropical medicine looked outwards, in nineteenth and twentieth-century Western Europe, the growth of a state sanitary infrastructure including meat inspection, a general rise in living standards, the promotion of personal hygiene and better standards of veterinary education, reduced the incidence of many helminth diseases. Correspondingly their significance in medical terms diminished as their prevention became part of routine public health work. Exceptionally in the USA, the Health Board of the Rockefeller Foundation launched a campaign in 1909 to eradicate infection with hookworm (*Necator americanus*; *Ancylostoma duodenale*). Although the program used anti-helminthic drugs, their primary means of achieving eradication was to initiate latrine building and teach sanitary habits to rural populations. See also **Babesiosis; Smith, Theobald (1859–1934)**.

In contrast, in what were largely the tropical colonial empires of European countries, parasitic infections requiring a high temperature or level of humidity for development and "cosmopolitan" diseases of poverty, poor hygiene and overcrowding remained prevalent. In some instances, attempts to improve the economy and/or levels of food production increased the incidence of parasitic diseases. The irrigation of the Punjab in India in the nineteenth century and parts of the Sudan, around Geizra, in the inter-war period increased rates of malaria and schistosomiasis respectively.

Tropical Parasitology in the Twentieth Century

After widespread acceptance of the parasitic aetiology and vector transmission of malaria and the foundation of the discipline of tropical medicine, parasitology became increasing interconnected with medical entomology. Tropical medicine provided basic post-graduate training for doctors seeking employment in the colonies. Such a scientific and cultural environment provided the framework for many of the discoveries and subsequent strategies for control of parasitic diseases in the early years of the twentieth century. Europeans tended to dominate this field either working for colonial medical services or serving as members of expeditions or commissions sent to the tropics. There were important exceptions. Carlos Chagas (1879–1934) in Brazil described the parasite and vector of the American form of trypanosomiasis in 1909. Several Japanese scientists worked on schistosomiasis in Japan determining the identity of *Schistosoma japonicum* and its intermediate snail hosts such as *Oncomelania nosophora*. The race was on to discover parasites, correlate the presence with existing clinical descriptions and determine means of transmission. The results dominated the first three decades of the twentieth century. Of the major parasitic infections of man: malaria, amoebiasis, trypanosomiasis, leishmaniasis, toxoplasmosis, ascariasis, hookworm, schistosomiasis, filariasis, onchocerciasis and taeniasis, only the vectors of leishmaniasis remained to be identified after 1930. The discovery that species of Phlebotomus sandflies carry both cutaneous and visceral leishmaniasis (Kala-Azar) followed each other in 1941 and 1942. See also **Leishmaniasis; Malaria**; and **Trypanosomiasis**.

The new knowledge of aetiology and transmission promised attempts at control or eradication of various parasitic diseases. Vectors were more often the targets and early results often predicted more than they subsequently delivered. In some instances, the reliance on vector control rebounded when failure made it apparent that the parasite element had received insufficient attention. Studies of the subtleties of parasite behavior and indeed identity were required in order to develop alternate strategies, even if direct antiparasitic methods would still be used to support re-styled vector control operations. Attempts to develop drugs effective against parasites enjoyed some limited success. Quinine from the bark of the chinchona tree had been used since the seventeenth century in the treatment of malaria, and arsenic compounds were introduced in the 1900s to treat sleeping sickness: both had unpleasant side effects, the latter drugs sometimes causing blindness and death. However, anti-parasitic drugs were at the forefront of inter-war pharmacology with the German drug houses taking the lead. Despite advances in tropical chemotherapy, the cost, necessity for easy administration and threats caused by itinerant drug sellers remain problematic.

In the post-war period following on from the work of the Rockefeller Foundation and the League of Nations Committees, the World Health Organization launched "vertical" disease-specific eradication programs, many of which were aimed at parasitic diseases. These global programs managed and run at the national level have met with mixed success. In the case of malaria, the technical difficulties of insecticide resistant mosquitoes and drug resistant protozoa have been further complicated by economic and administrative problems. The onchocerciasis (river blindness) control program in West and Central Africa has been successful in preventing blindness in many of the original participant countries by aerial spraying of the breeding sites of the black fly (*Simulium damnosum*) until the fifteen year life-expectancy of the worm (*Oncherocerca volvulus*) is expended. The recent introduction of Ivermectin as a anti-helminthic drug has helped but maintaining freedom from the parasite and preventing re-infection from non-participating areas remain as threats. Such programs provide classic demonstrations of the need for a close relationship between social and political stability, socio-economic development and disease containment and the need for long term guaranteed funding. The importance of education at the community level and indeed community involvement are acknowledged as key elements in combating parasitic diseases.

Since the 1980s and the emergence of AIDS, the effects of parasitic disease have increased. In the developing world, malaria, trypanosomiasis, Kala Azar and the ubiquitous infections of intestinal worms and protozoa such as *Entamoeba histolytica* increase the burden on the immune system in HIV+ cases, intensifying the experience of this new virus. In the developed world, diseases such as crytosporidiosis (*Cryptosporidium parvum*) have received a new prominence in the last twenty years because AIDS patients are susceptible to the chronic diarrhea it causes. It is difficult to treat and the cysts resist current techniques for water purification.

Locating parasitology within the discipline of tropical medicine, particularly schools of tropical medicine, may have hampered its development as a fundamental science. A disease-centered focus, dominated by the biology of particular host and parasites, left less time for understanding parasitism, as a phenomenon. For instance understanding of the immunology of parasitism lagged behind the understanding of bacterial and viral immunology. In the 1960s and 1970s, the World Health Organization funded several centers in the developing and developed world in an attempt to promote such research. The dominance of molecular biology in understanding the form and function of living organisms has broadened the scope for understanding the complex and intimate inner world of hosts and parasites.

Additional Reading

Baker, J., R. Muller, and D. Rollinson: *Advances in Parasitology*, Elsevier Science & Technology Books, New York, NY, 2006.

Cox, F.E.G.: *Illustrated History of Tropical Diseases*, The Wellcome Trust, London, UK, 1996.

Farley, J.: *The Spontaneous Generation Controversy from Descartes to Oparin*, Johns Hopkins University Press, Baltimore, MA, 1982.

Farley, J.: "Parasites and the Germ Theory of Disease," In: Rosenberg, C.E., and L. Golden: *Framing Disease; Studies in Cultural History*, Rutgers University Press, New Brunswick, NJ, 1992, pp. 33–49.

Foster, W.D.: *A History of Parasitology*, E & S Livingstone, Edinburgh. Scotland, UK, 1965.

Gosling, P.J., and D.W. Hendricks: *Dictionary of Parasitology*, CRC Press, LLC, Boca Raton, FL, 2005.

Grove, D.I.: *A History of Human Helminthology*, CAB International, New York, NY, 1990.

Kean, B.H., K.E. Mott, and A.J. Russell: *Tropical Medicine and Parasitology: Classic Investigations*, Cornell University Press, Ithaca, NY, 1978.

Power, H.J.: *Tropical Medicine in the Twentieth Century: A History of the Liverpool School of Tropical Medicine 1898–1990*, Kegan Paul, London, UK, 1999.

Roberts, L.S., J. Janovy, and P. Schmidt: *Foundations of Parasitology*, 7th Edition, The McGraw-Hill Companies, Inc., New York, NY, 2004.

Scott, H.H.: *A History of Tropical Medicine*, Edward Arnold, London, UK, 1939.

Worboys, M.: "The Origins and Early History of Parasitology," In: Warren, K.S., and J.Z. Bowers: *Parasitology: a Global Perspective*, Springer- Verlag New York, LLC, New York, NY, 1983, pp. 1–18.

HELEN J. POWER, Wellcome Trust Centre for the History of Medicine at UCL, London, UK

PARA-STATE. 1. In diatomic molecules, such as hydrogen molecules, the para-state exists when the spin vectors of the two atomic nuclei are in opposite (i.e., antiparallel) directions, forming a singlet state, $S = 0$, whereas the ortho-state is the one in which the nuclei are spinning in the same direction.

2. In helium, the para-state is one group or system of terms in the spectrum of helium that is due to atoms in which the spin of the two electrons are opposing each other. Another group of spectral terms, the orthohelium terms, is given by those helium atoms whose two electrons have parallel spins. Because of the Pauli Exclusion Principle, a helium atom in its ground state must be in a para-state.

PARATHYROID GLANDS. Four glands with a combined mass scarcely greater than that of a large pea produce parathyroid hormone (PTH) which is one of three main substances required to control the amount of calcium in the extracellular fluid and in the bone marrow. The parathyroid glands are adherent to and sometimes embedded within the back part of the thyroid gland. All of these structures lie in the lower front portion of the neck.

The influence of the parathyroid glands on the regulation of calcium concentrations in the blood of mammals was first recognized by MacCullum and Voegtlin, who reported in 1909 that removal of the glands produced a marked drop in serum calcium levels—to the point of causing tetany, which is described later. That the glands also affect phosphate was demonstrated in 1911 by Greenwald, who produced a rapid fall in urinary phosphate by parathyroidectomizing dogs. In 1924 and 1925, Hanson and Collip independently prepared physiologically active extracts of beef parathyroid. Similarly prepared hormone substances are available today for treating acute hypoparathyroidism, but because patients soon become refractory to injections of the substance, long-continued use of the extract is no longer indicated. During the intervening 75 years, a great deal has been learned concerning calcium and phosphorus metabolism and of the role of the parathyroid glands. Current, quite successful therapies for parathyroid disorders were made possible only through an improved understanding of the endocrinological processes involved.

The second of the three main substances involved in calcium metabolism is *vitamin D*. This major component of the endocrine system regulates the metabolism of bone minerals. The synthesis of active vitamin D commences with ultraviolet radiation of 7-dehydrocholesterol in the skin, producing previtamin D_3. This is converted to vitamin D_3 (cholecalciferol). See also **Vitamin D**. The last of the three substances is *calcitonin*, a polypeptide which at one time was believed to be secreted by the parathyroid glands, but now known to be produced by the parafollicular cells of the thyroid gland.

As with the other endocrine glands (see also **Endocrine System**), disorders mainly develop as the result of over- or underproduction of hormones secreted.

Primary Hyperparathyroidism. In the great majority of patients, excessive PTH is produced as the result of a single adenoma (tumor) of the gland. Primary hyperparathyroidism also may originate from hyperplasia (abnormal increase in volume caused by growth of new cells), or carcinoma. Cancer of the parathyroid is a rare cause and only occurs in 3–5% of cases. Because primary hyperparathyroidism is increasingly diagnosed in an early stage, the more severe symptoms, such as renal stones, osteitis fibrosa cystica, anemia, constipation, and renal failure are not usually presented to the physician. Rather, the patient complains of generalized weakness, which may be accompanied by weight loss, pallor, and keratopathy (deterioration of the cornea). Definite diagnosis requires laboratory measurement of the serum calcium level. Significant bone changes may be revealed by radiologic examination. Diagnosis is extremely important because, if this is positive, correction of a serious disorder almost always involves surgery. Where there is only one adenoma, this will be removed and the remainder of the glands will be left intact. Some surgeons prefer that all four glands be exposed prior to making a decision. Preoperatively, the patient will be given large dosages of 1α-hydroxyvitamin D_3, during which time serum calcium levels are watched carefully. Postoperatively, the patient must be monitored carefully for any appearance of hypoparathyroid tetany. Usually with those patients who have a single adenoma, with a normal second gland, there will be a successful return to normal calcium levels.

In *ectopic hyperparathyroidism*, the symptoms may parallel those of primary hyperparathyroidism, but the source of PTH may arise from a malignancy. Some researchers have reported abnormal PTH in many cancer patients, regardless of the type of tumor or the presence or absence of metastases.

Hypercalcemia. The serum calcium level is regulated by gastrointestinal absorption of calcium, oseteoclastic resorption of mineral from bone stores (see also **Bone**), and tubular reabsorption of calcium from the glomerular filtrate (see also **Kidney and Urinary Tract**). The treatment of hypercalcemia is determined by the underlying cause. Drugs used include prednisone, mithramycin, and calcitonin.

Hypoparathyroidism. This condition may arise from a deficiency in the effective production of PTH by the parathyroid glands; or by a failure of tissues to react to it. Rarely, the condition may occur in newborn infants as a result of hemorrhage that partially destroys the parathyroid gland. It is most common in adults who have undergone surgical removal of the parathyroid tissues as an occasional consequence of removal of diseased thyroid tissue. The most common symptom is a tightening and a spasm of the muscles, most evident in the position assumed by the fingers and toes. An inability to straighten the fingers and toes is referred to as *carpopedal spasm*. There is usually considerable apprehension in the form of a sense of impending doom. There may be difficulty in inhalation, spasm of muscles in the larynx, vomiting, and abdominal pain. This is *tetany*. In persons who develop the disease more gradually, the symptoms are less severe. Treatment of the hypocalcemia resulting from hypoparathyroidism is by a combination of calcium and vitamin D therapy.

The reader who has a scholarly interest in more details concerning the biochemistry of parathyroid hormone and its possible relation to the hypercalcemia of cancer is referred to the references listed.

Additional Reading

Bilezikian, J.P.: "Parathyroid Hormone-Related Peptide in Sickness and in Health," *N. Eng. J. Med.*, 1151 (April 19, 1990).

Bilezikian, J.P., R. Marcus, and M. Levine: *The Parathyroids*, 2nd Edition, Academic Press, Inc., San Diego, CA, 2001.

Eisenberg, B.: *Imaging of the Thyroid and Parathyroid Glands: A Practical Guide*, Churchill Livingstone, Inc., Philadelphia, PA, 1997.

Epstein, K. et al.: "Transient Hypoparathyroidism During Acute Alcohol Intoxication," *N. Eng. J. Med.*, 721 (March 14, 1991).

Livolsi, V.A. and R.A. DeLellis: *Pathology of the Parathyroid and Thyroid Glands*, Lippincott Williams & Wilkins, Philadelphia, PA, 1993.

Marcus, R., M.A. Levine, and J.P. Bilezikian: *The Parathyroids: Basic and Clinical Concepts*, Lippincott-Raven, Philadelphia, PA, 1994.

Nussbaum, S.R., R.D. Gaz, and A. Arnold: "Hypercalcemia and Ectopic Secretion of Parathyroid Hormone By An Ovarian Carcinoma with Rearrangement of the Gene for Parathyroid Hormone," *N. Eng. J. Med.*, 1324 (November 8, 1990).

PARCEL. An imaginary volume of fluid to which may be assigned various thermodynamic and kinematic quantities. The size of a parcel is arbitrary but is generally much smaller than the characteristic scale of variability of its environment.

PARCEL METHOD. A method of testing for instability in which a displacement is made from a steady state under the assumption that only the parcel or parcels displaced are affected, the environment remaining unchanged. Also called *path method*.

Although this method has been applied to various problems (e.g., inertial instability), its most familiar context is with vertical displacements from hydrostatic equilibrium, in which the parcel displaced is assumed to undergo adiabatic temperature changes, and the buoyant force resulting from its contrast with the unchanged environment leads to the criterion for stability, $\gamma < \Gamma$, where γ is the lapse rate of virtual temperature with height and Γ is the dry- or saturation-adiabatic lapse rate, according to the condition of the parcel.

See also **Slice Method**.

AMS

PARENCHYMA. In plants, the term parenchyma refers to a tissue composed of living cells, usually having only thin, primary cell walls and varying widely by morphology and metabolism. These cells are the oldest type of plant cells.

The classification of cells, tissues and tissue systems in plants has been a subject of long-standing controversy among botanists. Separating cells and tissues into distinct categories is problematic because different cell types may overlap in their characteristics of structure and function. A traditional classification system, based on the topographic continuity of tissues, views the vascular plant body as being composed of three tissue systems: dermal (protective covering of the plant body), vascular (primary conductive tissues—xylem and phloem), and fundamental (ground-tissue system—all other tissues) [Esau, 1965]. Parenchyma is one of the most common ground tissues. A second, more fundamental classification of plant cells is based on the nature of their cell walls. Thus, the three categories of cell types are the parenchyma (with thin primary cell walls), the collenchyma (with thickened primary cell walls), and the sclerenchyma (with thick primary and secondary cell walls) [Mauseth, 1988]. Within each cell type, it is possible to subdivide the cells into classes based on morphology, function, or some other feature of interest. For example, parenchyma can be divided into at least five functional classes: synthetic, structural, boundary, transport and storage. This article will utilize the latter classification system because it is relevant to parenchyma cell form and function. See also **Collenchyma**; **Epidermis (Plant)**; **Sclerenchyma**; and **Xylem**.

The word parenchyma stems from the Greek words *para* (beside) and *en-chein* (to pour), and the combination of the two yields the original meaning of parenchyma, as a semi liquid substance "poured beside" more solid, preexisting tissues [Esau, 1965]. In plants, parenchyma cells were probably the phylogenetic precursor of more specialized tissues. Evidence suggests that early land plants arose most directly from freshwater algae (classified with modern charophycean algae), which consisted entirely of branched filaments, or perhaps an early form of parenchyma [Graham, 1993]. Primitive parenchyma cells probably had only one plastid (monoplastidic), accompanied by well-defined pyrenoids, where photosynthetic systems were concentrated. Research has shown that in charophycean algae, freshwater environments with shallow and turbulent waters favor the development of cells and structures that reduce water loss and promote the capacity for the land—plant reproductive cycle. Thus, many of the modifications on the parenchyma cell prototype may have evolved as strategies for the successful invasion of land by aquatic plants. See also **Algae**.

Structure

Cell Walls. As mentioned previously, the structural component that parenchyma cells most often have in common are thin primary cell walls. However, there are exceptions to this classification. Thick cell walls develop often in storage parenchyma, such as the endosperm of date palm (*Phoenix dactylifera*) that eventually becomes thinner during germination [Esau, 1965], and also in sclereids that develop from parenchyma in the secondary phloem of dicots. Additionally, parenchyma cells occurring in the secondary xylem are relatively thick and frequently lignified.

Cell Shape. Because parenchyma cell walls are very thin, past researchers considered cell shape to be a function of surface tension and/or pressure from surrounding cells. Cells were considered to maximize volume and minimize surface area. Thus, under total isolation, cells should assume a spherical shape; and under mutual contact and pressure on all sides, cells should be a geometrically perfect, 14-sided, polyhedron with 8 hexagonal and 6 quadrilateral faces [Esau, 1965; Mauseth, 1988]. Two artificial systems were created to infer whether the primary influence of cell shape involved (1) pressure, subjecting lead shot placed in metal cylinders to pressure, or (2) surface tension, allowing soap foam bubbles placed in a container to self-adjust [Matzke, 1939, 1946]. The polyhedron shape found in some parenchyma cells (e.g. pith of various plants or callus cells in culture) suggested an intermediate form between the resultant shape of the lead shot and the soap bubbles, leading to the conclusion that both pressure and surface tension had roles in shaping the cells. However, other studies suggested that to achieve the perfect polyhedron shape, cells would require only uniform stress and could not be formed via cell division—two conditions unlikely for developing cells. Additionally, there are many other parenchyma cells abundant in plants that do not adhere to influences of pressure or surface tension, such as those with folds or internal ridges, and those with flattened, elongate, curved or stellate shapes. Thus, it is unlikely that the physical forces of pressure and surface tension are the sole controllers of cell shape, and equally likely that other controls from the cells or organisms themselves may be involved.

Cell Contents. The contents of parenchyma cells vary according to their specific functions. Further, a specific type of cell may vary its contents diurnally, seasonally or developmentally [Sauter and van Cleve, 1994]. Cells specialized for photosynthesis (chlorenchyma) contain chloroplasts that fluctuate in their starch contents over the course of a single day. Other nonphotosynthetic cells may have storage plastids, such as chromoplasts (pigmented cells in flowers and fruits) or amyloplasts (storage cells for large quantities of starch). Cells designed for water storage often contain large vacuoles filled with watery or mucilaginous contents, possibly enhancing the capacity of cells to absorb water. Plasmodesmata (connections between cells) are abundant in cells specialized for transport. The presence or absence, abundance, and size of other various cell components (e.g., nucleus, ribosome's, mitochondria, ergastic substances, microtubules, etc.) are also related to cell function and are discussed further in the section on Cell Types and Function.

Distribution

Parenchyma cells occur throughout plants, often as extensive continuous masses (parenchyma tissue), or associated with other cell types in morphologically heterogeneous tissues [Esau, 1965]. Some tissues composed primarily of parenchyma cells are the pith and cortex of stems and roots, the photosynthetic cells of leaves, tubers and fruits, and seed endosperms. In contrast, mixtures of cell types occur in the vascular tissues, where ray parenchyma cells and secretory parenchyma cells occur alongside the conducting sclerenchyma cells of the xylem (tracheids and vessel elements). The way parenchyma cells are arranged within tissues is dependent on function. For example, the arrangement of spongy mesophyll within a leaf creates large, interconnected intercellular spaces that are continuous with the atmosphere via open stoma on the lower epidermis (Figure 1), thus exposing ample cell surfaces for carbon dioxide penetration and photosynthetic fixation. In contrast, the leaf's uppermost cell layers (the epidermis) and the next two layers (palisade parenchyma) have no intercellular spaces, thus deterring water evaporation. The lower epidermal cells of the leaf are also tightly packed, except for the stoma, which are bordered by guard cells that close under conditions of high water stress.

Cell Type and Function

Synthetic Parenchyma. One type of synthetic parenchyma is the chlorophyll-rich, photosynthetic chlorenchyma cell. The main function of these cells is to conduct photosynthesis, that is, capture and convert light energy into chemical energy stored in the bonds of carbohydrates [Mauseth, 1988]. The chlorenchyma cell maximizes light interception by using its large vacuole to press chloroplasts (centers for photosynthesis) uniformly along the cell walls. Two examples of such cells in leaves/needles are palisade parenchyma and spongy mesophyll.

A second type is the meristematic cell, which synthesizes new cells by readily absorbing sugars, water and inorganic nutrients. These cells are usually small with few numbers of each organelle, allowing for rapid duplication, division and enlargement to the required size and shape for its functional role. An exception is the long, meristematic cell that produces vascular cells of the woody xylem, a design strategy that is thought to be favored over requiring daughter cells to undergo extensive elongation.

A third type is the secretory cell, which secretes substances that are eventually transported to exterior plant surfaces (e.g. cutin, mucilages), or to a cavity or duct within the plant (e.g. resin, gums, oils, latex), or stored in the cell vacuole (e.g. myrosinase). Secretory cells have a thin, permeable cell wall. In these cells, vesicles typically envelop substances, transport them to the plasmalemma, fuse to it and void contents to the exterior, or simply break down the cell wall. Secretory substances may provide plants with defense against insect or pathogen attack, protection from desiccation, or compartmentalization after wounding. See also **Gums and Mucilages**.

Structural Parenchyma. Structural parenchyma includes aerenchyma, whose primary function is to provide ample intercellular spaces in the tissue to promote gas exchange. Examples are lobe-shaped spongy mesophyll cells in various leaves and the star-shaped cells in banana (*Musa*) leaves that promote flexibility, or aerenchyma cells in submerged or bulky organs

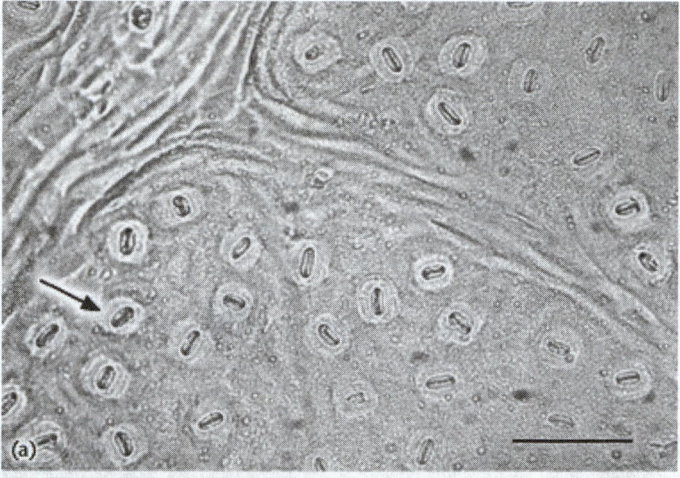

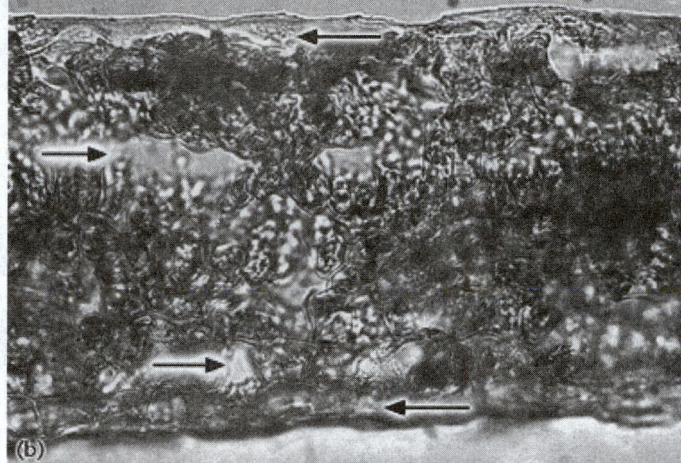

Fig. 1.   Leaf parenchyma cells from *Quercus* spp. (a) Surface replica of lower epidermis, showing guard cells of stoma (arrow) around leaf veins (20 × objective, bar = 50 μm). (b) Cross-section showing upper and lower epidermis (blue arrows) and large intercellular spaces (black arrows) surrounded by spongy mesophyll, filled with chloroplasts (40 × objective, bar = 25 μm). (c) Lobe-shaped spongy mesophyll cell (arrow) (100 × objective, bar = 10 μm).

(roots, stems or leaves), which are arranged to create a prominent airspace system to promote oxygen diffusion. The provision of both synthetic and structural functions by spongy mesophyll cells is an example of the overlapping roles served by many parenchyma cells.

Boundary Parenchyma. This type of parenchyma usually separates two regions of a plant (e.g., the endodermis in roots or some stems, separating the conducting tissue from the cortex, Figure 2a), or a plant from its environment (e.g., the epidermis on many plant surfaces, Figure 2b). These

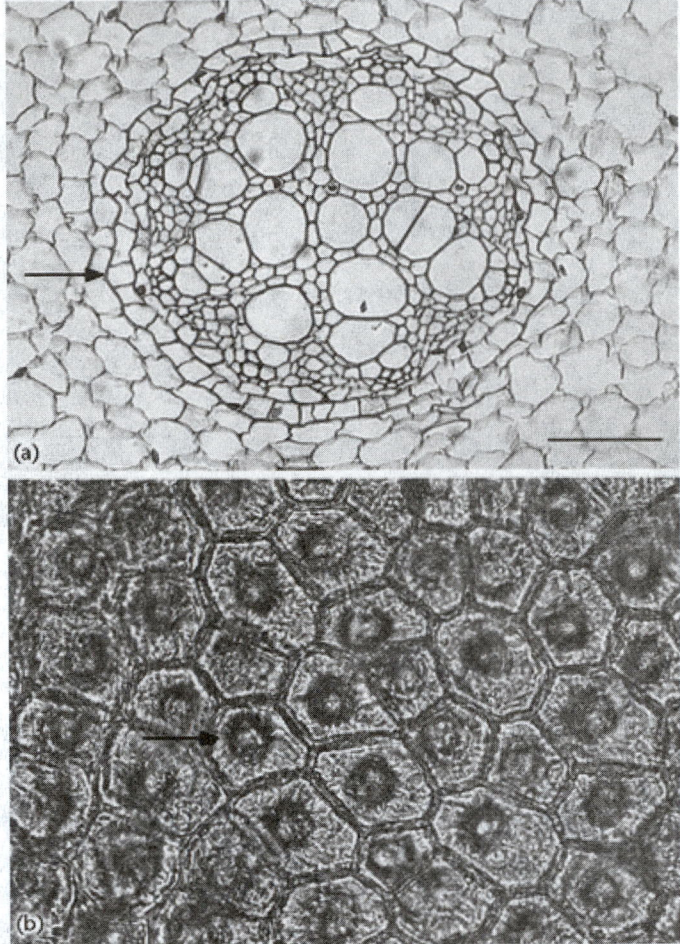

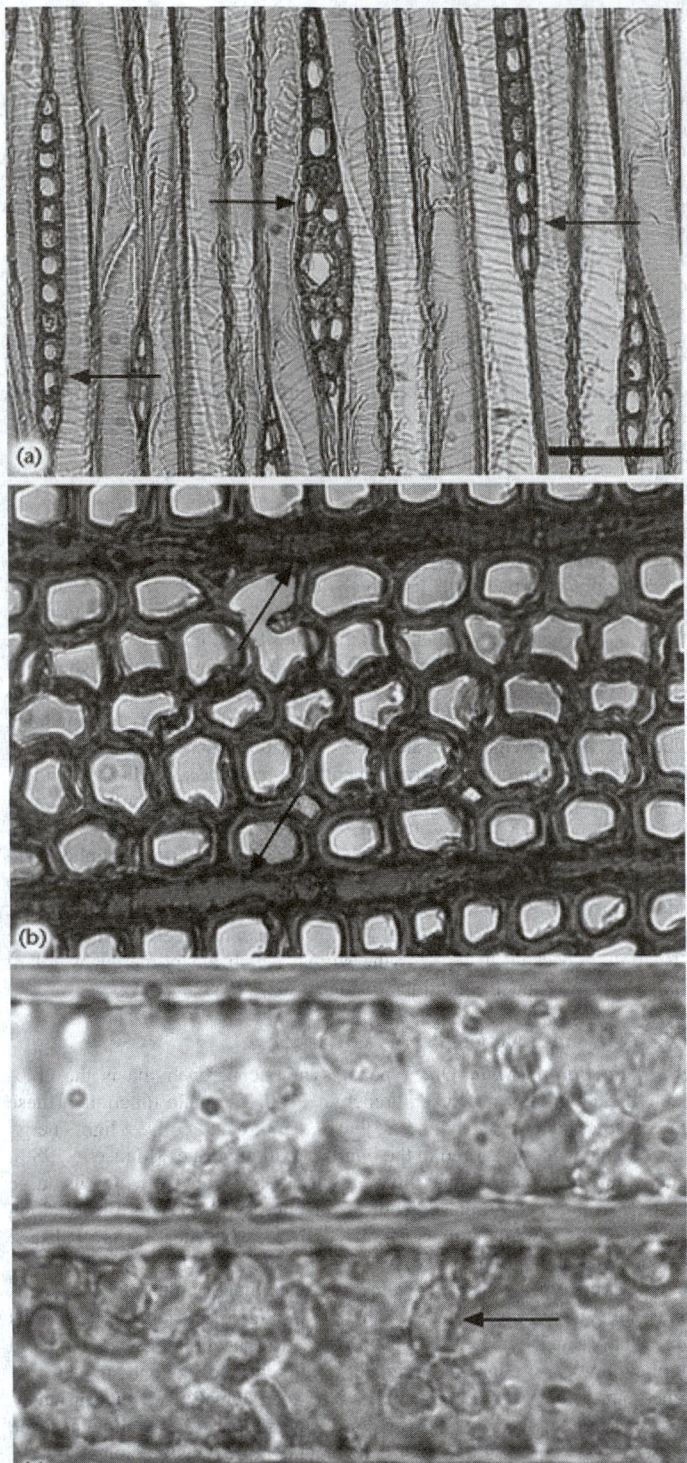

Fig. 2. Boundary parenchyma. (**a**) Cross-section from a root of *Lilium*; the single-cell layer ring of endodermis cells (arrow) encircles the vascular tissue, and the cortex, consisting entirely of thin-walled storage parenchyma, surrounds the endodermis (10 × objective, bar = 100 μm). (**b**) Epidermal peel from a petal of *Rosa*, showing the tightly interconnected parenchyma cells with large nuclei (arrow), probably involved in the synthesis of the light pink pigment within each cell (40 × objective, bar = 25 μmm).

cells are usually attached to one another firmly to prevent entry or loss of substances and resist tearing and puncture. Secretory cells are often associated with boundary parenchyma, providing additional protection. In contrast to the impenetrable endodermis with no intercellular spaces, the epidermis of some tissue is often semipermeable, for example, to allow for diffusion of water and nutrients into roots, or gases (via stoma) into leaves.

Transport and Storage Parenchyma. Some cells are specifically designed for transport, others for storage, while still others are specialized for both. The form of transport parenchyma is dependent on the substance being transported. Transfer cells transport large quantities of material over short distances, sieve elements conduct sugars and other materials over long distances, and light transmission cells channel light for subterranean plants from the soil surface to below ground. Storage parenchyma is abundant in fruits, seed and tubers, where sugars, starches, proteins or oils are often stored. Because these cells are metabolically active in the accumulation and release of products, they usually are of intermediate size to retain an optimal nuclear proportion of the cell [Mauseth, 1988]. Other cells, which are specialized for water storage (e.g. in cacti), are larger with a greatly expanded vacuole and a smaller nuclear proportion. See also **Fruit**; **Plant Reproduction**; and **Seed**.

Longitudinal and radial xylem parenchyma cells, which are specialized for transport and storage of various materials, contrast with the surrounding xylem cells (historically referred to as prosenchyma), which are specialized for mechanical support and conduction of water. As their names suggest,

Fig. 3.   Xylem parenchyma cells from a stem of *Pseudotsuga menziesii* (Mirb.). (a) Tangential section showing ray parenchyma (blue arrows) and epithelial parenchyma around a resin duct (black arrow) interspersed among the hollow tracheids with spiral thickenings on their cell walls (20 × objective, bar = 50 μm). (b) Cross-section of ray parenchyma (arrows) (40 × objective, bar = 25 μm). (c) Radial section of ray parenchyma containing starch granules (arrow) (100 × objective, bar = 10 μm).

these parenchyma cells run through the length and width of the xylem and are aligned end-to-end in long, interconnected chains. Ray cells often occur in bands three to over twenty cells tall, extending radially from inner bark to pith, with individual cells usually stopping in different annual growth rings or parts of the plant (Figure 3). Although not directly continuous, the ray system is probably connected by the longitudinal parenchyma cells which often abut the ray cells, running in the perpendicular direction. Thus, photosynthetic products, ions, nutrients and other substances directed toward the xylem are readily transported throughout the plant and/or stored in parenchyma [Lev-Yadun and Aloni, 1995]. Xylem parenchyma is also involved in water storage [Holbrook, 1991], compartmentalization after wounding [Shigo, 1984], heartwood formation [Magel, 2007], and mechanical padding in woody lianas [Putz and Holbrook, 1991].

Ontogenetic Development

Because all meristematic cells (even the zygote itself) are parenchyma, parenchyma cells arise from preexisting parenchyma cells. Under the traditional classification system (above), parenchyma tissue of the main plant body (i.e., cortex, pith, leaves, flowers) differentiates from the ground meristem [Esau, 1965]. Parenchyma of the primary vascular tissue derives from the procambium and epidermal parenchyma from the protoderm. Secondary parenchyma, such as that in secondary vascular tissue, or in the periderm (bark) of woody stems, develops from the vascular cambium and cork cambium, respectively. After formation, parenchyma cells mature and develop morphologic and metabolic traits according to their specialized roles. The specifics of this process are beyond the scope of this discussion, but can be obtained by further reading of Mauseth [1988], or other texts listed in the Further Reading section.

Note: The author thanks Dr. Everett Hansen for providing the prepared slide of *Lilium* from the collection of plant specimens of the Department of Botany at Oregon State University.

Glossary

An abridged glossary of terms used in the discussion of parenchyma would include:

Cambium — A sheet like fundamental type of meristem. The vascular cambium produces the secondary xylem and phloem; the cork cambium (phellogen) produces the cork tissue of the bark. Both cambia are considered to produce secondary tissues.

Cortex — In the primary plant body, the cells interior of the epidermis and exterior of/surrounding the vascular tissues.

Endodermis — A sheath of cells surrounding the vascular tissues of all roots and of some stems and leaves.

Endosperm — In angiosperms, a specialized storage tissue that is produced (in addition to the zygote) upon the pollen's fertilization of the ovule, which provides nutrients to the growing embryo. Endosperm tissue also provides the bulk of the world's food supply in the form of cereal grains.

Epidermis — In the primary body, the outermost layer of cells of the stem, leaf, or root —but not root cap. See also **Epidermis (Plant)**.

Ergastic Substance — Any material that is present in a cell as a storage product or crystal.

Microtubule — A proteinaceous tubule, constructed of tubulin, which acts as a cyctoskeleton, giving the cell a shape that is later maintained by the cell wall and being responsible for pulling the organelles to specific sites within the cell.

Mitochondrion (pl.: mitochondria) — An organelle that is universally present in plant cells and that is responsible for aerobic respiration.

Myrosinase — An enzyme produced in the central vacuoles of specialized secretory cells called myrosin cells and used in plant defense against insect or animal attack.

Nucleus (pl.: nuclei) — An organelle bounded by two membranes and containing the DNA responsible for most hereditary information.

Palisade parenchyma — In the mesophyll of a leaf, a region in which the chlorenchyma cells are elongate and arranged parallel to one another and perpendicular to the epidermis.

Periderm — The technical term for bark, consisting of the cork (phellem) that is produced by the cork cambium (phellogen) and any epidermis, cortex, and primary or secondary phloem that may be exterior to the cork cambium.

Phloem — A complex tissue that may consist of solely parenchyma or of both parenchyma and sclerenchyma, responsible for the translocation of organic nutrients, especially photosynthetic sugars, downward from the leaves to the lower plant parts. The conducting elements of the phloem are the sieve elements.

Phylogenetic — A term referring to the natural evolutionary relationships of an organism.

Plasmalemma (pl.: plasmalemmae or plasmalemmas) — Also known as plasma membrane, the membrane surrounding the entire protoplast.

Plasmodesma (pl.: plasmodesmata) — A small hole in a primary wall, through which the protoplasm of one cell is in contact with that of adjacent cells. Plasmodesmata may occur individually or in clusters called primary pit fields.

Procambium (pl.: procambia) — A portion of the subapical meristem that produces the cells that mature into the primary xylem and primary phloem. In woody plants, a portion of the procambium is converted into the fascicular cambium (forms the vascular bundle).

Protoderm — Any set of cells that develops into an epidermis.

Pyrenoid — A proteinaceous structure, involved in starch synthesis, found in the chloroplast of hornworts and numerous algae.

Ribosome — A particle consisting of ribosomal RNA and proteins and responsible for translating mRNA into protein.

Spongy mesophyll — A portion of the leaf mesophyll characterized by the cells having little contact among one another and frequent intercellular spaces.

Stoma (pl.: stomata) — A term used loosely to refer to any of the following: the pore and guard cells of a stomatal complex, just the pore, or the entire stomatal complex, located in the plant epidermal layer.

Xylem — A complex tissue in plants that has many functions, the three most obvious being: the conduction of water and solutes, mechanical support of the whole plant and its parts, and storage of water and nutrients. The conducting, sclerenchyma cells of the xylem are called tracheids and vessel elements.

Additional Reading

Beck, C.B.: *Introduction to Plant Structure and Development*, Cambridge, University Press, New York, NY, 2005.

Esau, K.: *Plant Anatomy*, 2nd Edition, John Wiley & Sons, Inc., New York, NY, 1965.

Evert, R., and S.E. Eichhorn: *Esau's Plant Anatomy: Meristems, Cells, and Tissues of the Plant Body: Their Structure, Function, and Development*, 3rd Edition, John Wiley & Sons, Inc., Hoboken, NJ, 2006.

Graham, L.E.: *Origin of Land Plants*, John Wiley & Sons, Inc., New York, NY, 1993.

Holbrook, N.M.: "Stem Water Storage," In: Gartner, B.L.: *Plant Stems, Physiology and Functional Morphology*, Academic Press, San Diego, CA, 1995, pp. 151–169.

Lev-Yadun, S., and R. Aloni: "Differentiation of the Ray System in Woody Plants," *Botanical Review*, **6**, 45–84 (1995)

Magel, E.A.: "Biochemistry and Physiology of Heartwood Formation," In: Savidge, R, J. Barnett, and R. Napier: *Cell and Molecular Biology of Wood Formation*, pp. 363–376, BIOS Scientific Publishers, Oxford, UK, 2000.

Matzke, E.B.: "Volume — shape Relationships in Lead Shot and their Bearing on Cell Shapes," *American Journal of Botany*, **26**, 288–295 (1939).

Matzke, E.B.: "The Three-dimensional Shape of Bubbles in Foam — an Analysis of the Role of Surface Forces in Three-dimensional Cell Shape Determination," *American Journal of Botany*, **22**, 58–80 (1946)

Mauseth, J.D.: "Parenchyma," *Plant Anatomy*, pp. 43–50, Benjamin/Cummings, Menlo Park, CA, 1988,

Putz, F.E., and N.M. Holbrook: "Biomechanical Studies of Vines," In: Putz, F.E. and H.A. Mooney: *The Biology of Vines*, Cambridge University Press, Cambridge, UK, 1992, pp. 73–97.

Rudall, P.: *Anatomy of Flowering Plants: An Introduction to Structure and Development*, 3rd Edition, Cambridge University Press, New York, NY, 2007.

Sauter, J.J., and B. van Cleve: "Storage Mobilization and Interrelations of Starch, Sugars, Protein and Fat in the Ray Storage Tissue of Poplar Trees," *Trees*, **8**, 297–304 (1994).

Shigo, A.L.: "Compartmentalization: a Conceptual Framework for Understanding how Trees Grow and Defend Themselves," *Annual Review of Phytopathology*, **22**, 189–214 (1984).

MICHELE L. PRUYN, Oregon State University, Corvallis, OR

PARESTHESIA.   Any abnormal sensation on the surface of the body. It may be described by the patient as burning, tickling, itching, pricking, etc. Paresthesias occur in nervous-system diseases involving the spinal cord, in pernicious anemia, and polyneuritis.

PARHELION.   See Sundogs.

PARKES PROCESS. A standard process for the separation of silver from lead. From 1 to 2% molten zinc is added to the lead-silver mixture, heated to above the melting point of zinc. A scum containing most of the silver and zinc forms on the surface; this is separated and the silver recovered. The separation of silver is not complete, and the process is repeated several times.

PARKINSON'S DISEASE. Parkinson's disease (PD) is a prototypal neurodegenerative disease — having a rich history of description, etiology, treatments, pathology, and neurochemistry. However, despite its classical clinical picture, its meaning and perception have varied for more than one hundred and eighty years.

Parkinson's disease, was recognized as a syndrome of tremor, stooped posture and shuffling gait by Dr. James Parkinson in 1817. The most prominent pathological characteristic, loss of neural pigments seen both grossly and by histopathology of the *substantia nigra* (SN) was noted in 1921 following viral (influenza) post-encephalitic Parkinsonism. Progress in assessing degeneration in other pigmented brainstem nuclei, the identity of neurotransmitters, and the loss of dopamine (DA) and serotonin (5-HT) were recognized following the introduction of Rauwolfia alkaloids from India as antihypertensives in the 1950's and subsequent development of parkinsonism due to the neurochemical depletion as a side-effect of such treatment. Early surgical therapeutic attempts were nonspecific and included destruction of the spinal motor pathways, resulting in paralysis, for tremor reduction. Pallidotomy and thalamotomy (vide infra) were performed following the introduction of human stereotactic instruments and brain maps in the 1950's; these early semi-empiric procedures were effective in eliminating hyperkinetic symptoms such as tremor and rigidity, but were rapidly supplanted by the inadvertent introduction of dopa. Cotzias in 1967 reported three of sixteen patients who benefited from dopa supplementation. This however, was in an effort to reconstitute neuromelanins by its precursors of phenylalanine and then dopa, not in order to replace DA. This first trial resulted in the known side effects of dopa therapy such as induction of hallucinosis, choreo-athetotic "dyskinesias", and sleep disturbances and the effects of impure preparations such as fever, seizures and rashes in thirteen of the sixteen patients. Despite these distressing results, two patients showed dramatic improvement and interestingly had undergone pallidotomy by Cooper prior to the drug trial, presaging the synergism of pallidotomy and medical (dopaminergic, serotonergic and antiglutaminergic) combined treatments, (vide infra).

This new operative definition of Parkinson's disease based on response of early symptoms to dopaminergic medications did much to initiate circular reasoning regarding dopamine treatment and research between medical and scientific investigations of Parkinson's disease for the twenty years of 1967 to 1987.

Several factors were responsible for the resolution of the dopaminergic loss definition of Parkinson's disease during the late 1980's. Of primary importance was the failure of *L-dopa* replacement therapy and novel dopaminergics to affect the progression of the clinical or pathological disease. Fortuitously, a street-drug cogener of meperidine, MPTP was identified as an oxidative toxin, which resulted in rapid onset Parkinsonism in young addicts in San Francisco. The employment of this substance, which becomes an active toxin MPP+ following the action of monoamine oxidase B (found primarily in serotonergic and dopaminergic neurons) in primate models, lead to new understanding of brainstem glutaminergic influences of primary import in the akinetic symptoms of supervening levodopa failure in advanced Parkinson's Disease. Further interest was generated for toxic and oxidative (metabolic) stress etiologies and a broader focus on the underlying disease process. Lackluster mechanistic attempts to "replace dopamine" by structural surgical methods such as adrenal medullary chromaffin cell brain auto- grafts and fetal mesencephalic isolated dopamine neuron allografts (1984–1992) were quickly supplanted by renewed interest in conventional stereotactics directing lesions at targets identified by the new MPTP model such as the disinhibited and putatively hyperactive subthalamic nucleus, (STN). The STN's overamplified target, the posterior ventromedial portion of the globus pallidus interna (PVP/GPi) has since been the target of selective lesioning or disruptive high-frequency electrical stimulation using implanted electrodes (vide infra), with excellent therapeutic results.

Further characterization of metabolic deficits such as impaired activity of complex I — Coenzyme Q_{10} NAD, cytochrome oxidase in mitochondria (with its maternal pattern of inheritance), the role of serotonergic losses in levodopa failure and in Parkinson's dementia, and clinical subtypes, were also undertaken. Current endeavors are aimed at enhancement of (dopaminergic) neuronal survival using trophic factors, e.g. glial-derived neurotropic factor GDNF, and genetic engineering for dopamine DA, serotonin 5HT and GDNF augmentation using viral vectors, transformed cell lines and neural stem cells.

Pathology

The pathological hallmarks of PD are depigmentation of the substantia nigra and other pigmented brainstem nuclei (e.g. locus coeruleus), progressive loss of dopaminergic neurons and subsequent loss of the other monoaminergic cells (i.e. serotonin and norepinephrine). The loss of dopaminergic neurons is variable and is not related to the severity of disease for the post-encephalitic form is associated with 95% DA loss or greater and is a mild clinical syndrome characterized by slow progression to akinesia; contrariwise classical idiopathic PD may demonstrate 50% to 90% DA cell loss with poor correlation to more progressive akinesia, although 80% DA cell loss is often cited as a threshold sufficient to induce the clinical presentation of PD. These DA-independent findings are exemplified by a recent animal model of low dose MPTP with the primate subjects showing up to 99% DA cell loss in the SN with 97% DA unavailability in the striatal targets, yet no signs or symptoms of parkinsonism. The explanation for this is centered on the noted overabundance of serotonin and it's compensatory role in this instance, possibly mimicking the post-encephalitic subtype of PD.

The pathognomic eosinophilic cytoplasmic inclusion called *Lewy body*, located primarily in pigmented neurons, is found in classical types of PD. This characteristic histopathological finding is absent in the post-encephalitic form and the MPTP model, and MPTP exposed human Parkinson patients, indicating the model as a milder or incomplete expression of idiopathic PD. In so-called *Lewy Body Dementia* (LBD) they are found in cortical and subcortical locations, their density possibly related to the aggressiveness or speed of disease progression as they must be considered transient. In Parkinson's Dementia Complex (PDD) it is the density of the Lewy bodies (LBD) found in the brain stem, which is the only known correlation to the severity of the dementia. In amyotrophic lateral sclerosis (ALS- also called motor neuron disease), if associated with PDD, Lewy bodies will be found in the cervical anterior horn cells. The distribution of Lewy bodies in all of these instances is highly consistent with the distribution of another marker, the neurofibrillary tangle (NFT). This product of cytoskeleton degeneration is common in neurodegenerative disease including post-encephalitic P.D., Alzheimer's dementia (ALZ), schizophrenia, and pugilism. Despite their apparent differing etiologies, the distribution of the NFTs which overlaps the LB's in PDD respects identical nuclei and cell fields consisting of phylogenetically old "reticular areas" and their projections, including hypothalamic and limbic areas. Moreover, the distribution of lesions in Wernicke's encephalopathy is an inclusive subset of this distribution.

Neurochemistry

The brainstem areas involved by LB's and NFT's, which share a common denominator in neurodegenerative diseases are marked by the tri-localization of melatonin (MLT), beta-endorphin (B-END), and glial derived neurotropic factor (GDNF) cell surface receptors. This commonality alludes to the protective role of intrinsic opioids (B-END), the importance of tryptophan-serotonin-melatonin metabolism (and its dual supportive physiological role with opioids), and an understandable dependence on trophic protective factors such as GDNF.

Monoamine loss in PD follows the pattern of early DA and 5-HT loss in most idiopathic cases. Prior to the introduction of L-dopa, cerebrospinal fluid (CSF) studies of patients with idiopathic disease showed an overall 60% to 70% DA loss and 40% to 55% reduction in serotonin. In the 15% of PD patients without tremor (Type B, vide infra) with predominant akinetic, gait and postural instability problems, the primary breakdown product of serotonin, 5-hydroxyindoleacetic acid (5-HIAA), is found in half the concentration as the classic tremor type idiopathic PD, Type A.

In Parkinson's disease progressive loss of 5-HIAA is correlated with advanced akinetic symptoms and DA medication failure. Of particular impact is the excitatory glutaminergic activity in advanced states which has an adverse effect on neuronal oxidative stress, possible white-matter degeneration and is ultimately responsible for the akinetic state (vide infra: Pathophysiology).

Relevant to the issue of neurochemistry in PD is a discussion of the biochemistry of metabolic oxidative stress in PD. As part of normal metabolism up to 3% of oxygen can be converted to reactive oxygen species (ROS) such as hydrogen peroxide or the hydroxyl radical. In neurodegenerative diseases, ROS are either raised beyond capacity of protective mechanisms such as superoxide dismutase (SOD), glutathione (GSH), or MLT or these latter systems may fail. In the former case, the first cytochrome oxidase reaction, complex I, is known to be poorly functioning in the mitochondrial respiratory chain of patients with idiopathic PD. The body's inability to mitigate ROS damage is implicated in cancer, aging and neurodegenerative disease.

Epidemiology

It is estimated that one million North Americans are affected by Parkinson's disease, with the incidence increasing with age and it is more common in males. However, with the demographics of aging, pollution, and stress, PD incidence as well as prevalence is felt to be increasing. One population study in Massachusetts (non-patients) found that greater than 15% of those over 75 years old showed two or more symptoms of PD such as shuffling gait, poor facial expression, poverty of movement, stooped posture, etc., but not tremor. The racial distribution of PD is skewed being most common among fair Caucasians, less common in Asians and Latinos, and least common among Blacks, being rare in unmiscegenated Africans.

All links to central nervous system (CNS) damage including head trauma, high fever, encephalitis, nutritional or neurochemical depletion states (e.g. pellagra), toxic exposure (e.g. herbicides, insecticides), MPTP and other street drugs of abuse which damage monoaminergic systems (e.g. methamphetamine), and even emotional or 'silent trauma' (e.g. post traumatic stress disorder or prolonged untreated depression) have been linked as risk factors for PD. Clinically, such stressors as elective operative surgical interventions (e.g. coronary artery bypass grafts), motor vehicle accidents, death of a primary relative, and even career or financial tragedies are commonly noted as a date of onset of symptoms. Onset before age 45 is considered "young onset" or Narabayashi's "juvenile type" and has a stereo-typed presentation, course, and response to treatment, including induction of bilateral uncoordinated spontaneous movements referred to as dyskinesias.

Clinical Expression, Symptoms, Course and Prognosis

The premonitory and earliest symptoms and signs of PD include loss of sense of smell (anosmia), decreased eye blink rate and facial expression, monotone soft speech, stooped posture, a poverty of spontaneous movements, a feeling of chronic fatigue and then often a resting, "pill-rolling" tremor of the thumb and index finger. Sleep disturbance, a passive personality, and small handwriting (micrographia) with progressively decreasing amplitude of the letters and words in a sentence are characteristic. In the akinetic or Type B expression of the disease there is no tremor but antecedent stress (usually unrecognized) and depression are constant.

Most patients show asymmetric appearance and progression of appendicular hyperkinesias (i.e. tremor, cogwheel rigidity of the arms), but as the symptoms and disease progress, akinesia supervenes and the symptomatology becomes more axial evidenced by stooped posture, balance problems (postural instability), smaller, shuffling steps and such problems as being unable to roll over in bed. Many of the signs can be viewed as motor system loss of control over gravity which implicates the gamma motor system of importance in the pathophysiology of the disease. Undermedicated or under treated, the course of the disease can lead to death within 10 years, even in recently reported studies. In classic type, appropriately treated patients currently experience 10 to 15 years of quality functioning, whereas tremor dominant unilateral symptoms imply 20 or more years of functionality. In all cases, stereotactic surgical interventions may mitigate symptoms and disease for an additional 3 to 10 years. Patients without tremor at onset have a more aggressive progression of disabilities due to akinesia, with only 8 years of independent quality of life, on average. Estimates of from 20 to 80% are given for the occurrence of depression in PD, exceeding the later figure in advanced or akinetic patients. Depression is rare in highly asymmetrically affected patients such as tremor dominant type (representing 5% of cases) and in the young-onset type. Other non-motor signs and symptoms such as autonomic instability (e.g. episodes of profuse sweating, fluctuating blood pressure) also occur based upon poor hypothalamic modulation, possibly related to serotonin deficiency.

Dementia commonly supervenes in elderly, akinetic patients and is of the frontal type. Neurologic exam reveals frontal lobe release signs. Anosmia and a glabellar reflex are constant, and their absence challenges the diagnosis of PD. A snout reflex, palmomental reflex, and the inability to alternatively simultaneously open and close opposite hands confirm the clinical impression of frontal dysfunction. Neurocognitive testing (i.e. using the Wisconsin Card Sort or Stroop Test) will confirm the diagnoses as needed.

Frequently, the differential diagnoses of central hypothyroidism, vitamin B_{12} deficiency, and serotonin deficiency, will be noted as a triad, which when aggressively treated will result in dramatic improvements of these reversible forms of dementia often obfuscated by the constellation of co-morbid Parkinson's symptoms. Other differentials in this clinical scenario include so-called normal pressure hydrocephalus (NPH) with its triad of urinary incontinence, dementia, and gait problems, which can exist concurrently and confound the diagnosis whilst contributing to the parkinsonism. Parkinson's Plus Syndromes (PD Plus), which represent gross structural degenerations and atrophy of widespread CNS systems (beyond the limited pigmented, monoaminergic cell losses in PD) may appear to mimic PD in the first 2 to 5 years of recognition. However, the occurrence of intact sense of smell, poor response to pharmacologics, predominant axial symptoms, paucity of tremor, nuchal rigidity, or the inability to walk within less than 5 years of symptoms all weigh against PD and toward so called PD Plus, for which there is no satisfactory treatment at present.

Treatment

Treatment of PD includes pharmacologic replacement or substitution for losses of neurotransmitters (i.e. L-dopa or ergoloid-derived dopaminergic agonists for DA, selective serotonin re-uptake inhibitors (SSRIs) such as Prozac and B vitamin supplementation for 5-HT, and e.g. l-threo DOPS for norepinephrine) are employed. Other tactics are aimed at reducing glutaminergic influences using amantadine and/or magnesium. Alternative rational strategies focus on the use of antioxidants, such as lipoic acid, green tea, selenium, pycnogenol, and/or a vegetarian diet replete in these nutrients and minerals.

The clinical pharmacology of these treatments requires acumen and patient education. Many patients on L-dopa/carbidopa preparations may remain under-medicated as constipation or amino acid competition (especially dairy protein) block intestinal absorption and also active transport across the blood brain barrier. In decade-long treatment programs, DA therapy may be met with failure, partly due to the second hit of monoamine serotonergic loss or co-morbid conditions following a sequential pathophysiology, viz. serotonin loss reduces thyrotropin releasing hormone activity from the hypothalamus; resulting triiodothyronine deficiency blunts monoamine receptor sensitivity and also leads to gastric mucosal/intrinsic factor losses which result in vitamin B_{12} deficiency and increased oxidative stress and symptomatology.

Pathophysiology

Following losses of monoaminergic function in PD, the extrapyramidal motor system of the basal ganglia and brain stem become dysregulated. Loss of DA inhibition to the putamen results in demonstrated neuronal hyperactivity of the globus pallidus (GPi) interna and further amplification from the disinhibited subthalamic nucleus. The inhibitory gabanergic pallidal outflow via the tegmental bundle of the ansa lenticularis restricts the activity of the midbrain locomotor center and the tegmental pedunculopontine nucleus (PPN). The resulting brainstem and reticulospinal inhibition results in axial akinetic symptoms, gait problems, postural instability, stooped posture, and impairment of anti-gravity activities. Projections from the GPi and brainstem to the thalamus and cortex are responsible for appendicular hyperkinesias such as tremor. See Fig. 1. Because of this mechanism of akinesia associated with an imbalanced neurotransmitter and hyperglutaminergic state, (glutamate being a neuronal excitatory transmitter and so-called "excitotoxin") then increased oxidative stress, the progression of the disease may follow the poorly treated or undermedicated clinical state.

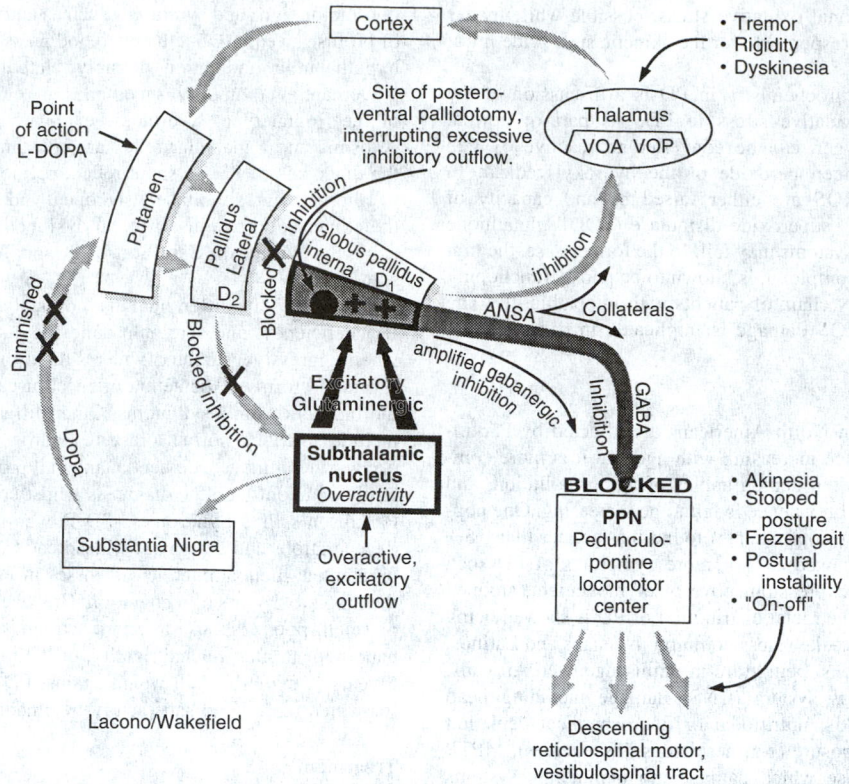

Fig. 1.   Proposed mechanism of the inhibition of the pedunculopontine locomotor nucleus (PPN). In this model, the inhibition is caused by the subthalamic amplification of the pallidal γ-aminobutyric-acid-inhibitory output and results in akinesia, gait freezing, and postural instability. In Type B patients, partial interruption of abnormal pallidal efferents produced by posteroventral pallidotomy (PVP) allows the reversal of brain stem inhibition, reversing akinesia. In Type A patients, the concomitant effects on the motor thalamus (by posteroventral pallidotomy via ansa lenticularis collaterals) accounts for the benefits of posteroventral pallidotomy in decreasing tremor, eliminating rigidity and dyskinesia, and reversing akinesia. Despite its appeal, the model in humans is limited. VOA, ventralis oralis anterior; VOP, ventralis oralis posterior.

Stereotactic Surgical Interventions

Beginning in 1948–55, Spiegel and Wycis, Narabayashi, Nashold, Cooper and others had begun to use precise instruments and brain maps allowing them free access with probes to therapeutically explore deep areas of the human brain. These techniques became widely used in the 1960's following Cooper's serendipitous discovery of the ventral lateral thalamus as an improved target for tremor. These techniques were all but forgotten following the widespread introduction of L-dopa in 1967–70. The advent of computerized CT and MR imaging along with advancing disease in a population of PD patients with so-called *levodopa failure syndrome* spurred new stereotactic therapeutic interventions including fetal brain allografts and adrenal autografts. During this era of increased interest in stereotactics and PD the utility of Leksell's posteroventral pallidotomy technique was rediscovered by Laitinen and Iacono. The ability of a conventional radiofrequency (heat) lesion within the pallidum to eliminate or improve both positive (e.g. tremor, dyskinesia) and negative (akinetic) symptoms was surprising and at first perplexing. Iacono (1994) proposed that partial interruption of abnormal pallidal efferents to the brainstem by posteroventral pallidotomy (PVP) allows reversal of brainstem inhibition of the PPN-locomotor center and reticulo-spinal projections reversing akinesia. See Figs. 1 and 2. This effect was found to be synergistic with medication therapeutic effects. Advances in technique have produced concomitant PVP interruption of subthalamic afferent amplification of the GPi and more direct measures to damage or suppress the subthalamic nucleus. Chronically implanted deep brain electrodes for high-frequency stimulation (DBS) of the refined thalamotomy target for tremor, of the pallidum, and STN for akinesia have been employed to create a reversible lesion effect. The PVP is successful in greater than 80% of operated patients and is able to reverse all symptoms of PD up to 90% of normal with 50% of patients benefiting for more than five years.

Research and Cure

Despite hope of cure beginning with the advent of L-dopa and rekindled before the subsequent failures of fetal grafts, the cure for PD has remained evanescent. However, now on the verge of human genetic treatments utilizing transformed cells or retroviral delivered DNA for DA, 5-HT, or GDNF, the possibility of advancing PD treatment seems imminent. Moreover, the most recent discoveries of neural stem cell capabilities for neuronal repopulation and CNS regeneration add new dimensions to the treatment of PD. Optimism is generated by new research indicating that the adult human brain ordinarily repopulates up to 20% per year of neurons in the hippocampus (an area damaged in ALZ). Parenthetically, serotonin may represent the trophic stimulation for this stem cell derived regeneration. Stem cells in general have broad, flexible ontogenic potential, can migrate and adjust by differentiation to local demands. Utilizing, controlling and then understanding the enigma of embryonic development and aging involving genetic expression is an endearing promise for the cure of neurodegenerative disease. See also **Blood-Brain Barrier**; and **Central and Peripheral Nervous Systems**.

An abridged glossary of terms used to describe Parkinson's disease would include:

Acetylcholine—a chemical which acts as a neurotransmitter. An imbalance between dopamine and acetylcholine results in some Parkinson's disease symptoms.

Action tremor—a tremor that increases when the hand is moving voluntarily.

Agonist—a drug which increases neurotransmitter activity by stimulating the dopamine receptors directly.

Akinesia—no movement.

Amantadine (Symmetrel)—an anti-Parkinson drug.

Anticholinergics—anti-Parkinson drugs that block the action of acetylcholine, thereby rebalancing it in relation to dopamine and reducing rigidity and tremor; e.g., Artane, Cogentin.

Antihistamines—drugs that are often used to relieve cold or allergy symptoms (i.e., Benadryl) but may also be effective in reducing tremor.

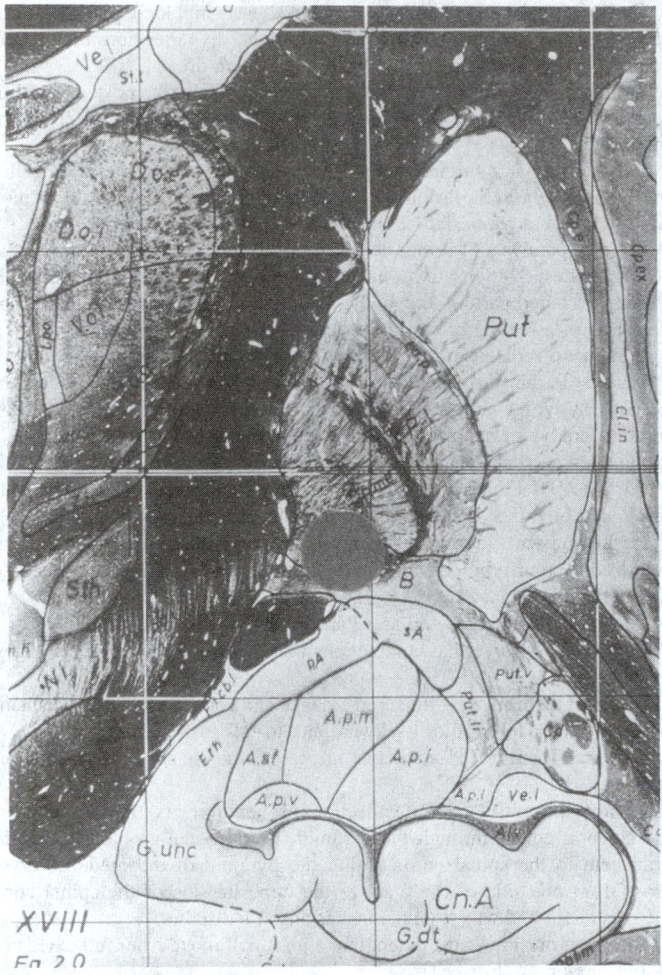

Fig. 2. Coronal brain map showing pallidotomy lesion site and location.

Ataxia — loss of balance.

Athetosis — slow, involuntary movements of the hands and feet.

Atrophy — wasting, shrinkage.

Autonomic Nervous System — that part of the nervous system that is responsible for automatic functions, such as the heartbeat, digestion, salivation.

Axon — the long, hair like extension of a nerve cell that carries a message to the next nerve cell.

Basal Ganglia — several large clusters of nerve cells deep in the brain below the cerebral hemispheres; crucial in coordinating motor commands. Include the striatum and the substantia nigra.

Benign Essential Tremor — A condition characterised by tremor of the hands, head, voice, and sometimes other parts of the body. Essential tremor often runs in families and is sometimes called *familial tremor*. It is sometimes mistaken for a symptom of Parkinson's. However, this is an action tremor and there is no rigidity or bradykinesia.

Beta-Blockers — Drugs which block the action of epinephrine at certain sites. Usually used to treat hypertension and heart disease, they may be effective in the treatment of benign essential tremor.

Bilateral — both sides of the body.

Biofeedback — a behavior modification in which patients are taught to partially control unconscious bodily functions, such as blood pressure or heart rate.

Blink rate — the number of times per minute that the eyelid automatically closes. A normal rate may be 10 to 30 per minute; for the parkinsonian it may be 0 to 5 per minute.

Blepharospasm — forced eyelid closure. See also **Blepharospasm**.

Blood-brain barrier — the protective membrane that separates circulating blood from brain cells.

Body scheme — the ability to identify body parts or to relate body parts to each other; the ability to sense one's position in space.

Bradykinesia — slowness of movement, gradual loss of spontaneous movement.

Bradyphrenia — slowness of thought processes.

Bromocriptine (Parlodel) — a dopamine agonist and anti Parkinson drug.

Bruxism — grinding of teeth and clenching of jaw muscles.

Buccinator — a muscle of the face and cheek.

Central nervous system — the brain and the spinal cord.

Cerebellum — a large structure consisting of two halves (hemispheres) located in the lower part of the brain; responsible for the coordination of movement and balance.

Cerebrum — consists of two parts (lobes), left and right, that form the largest and most developed part of the brain; initiation and coordination of all voluntary movement take place within the cerebrum. The basal ganglia are located immediately below the cerebrum.

Chorea — rapid, jerky, dance like movement of the body. May result from high doses of levodopa and/or long term levodopa therapy.

Cogwheel Rigidity — Stiffness in the muscles, with a jerky quality when arm and leg joints are repeatedly moved.

Constipation — Diminished ability of intestinal muscles to move feces (stool), often resulting in very hard stool. A common problem in Parkinson's.

Corpus striatum — a part of the brain that helps regulate motor activities.

Cortex — the outer layer of the cerebrum, densely packed with nerve cells.

Cryothalamotomy — a surgical procedure in which a supercooled probe is inserted into a part of the brain called the thalamus in order to stop tremors.

Dendrite — a threadlike extension from a nerve cell that serves as an antenna to receive messages from the axons of other nerve cells.

Delusions — a condition in which the patient has lost touch with reality and experiences hallucinations and misperceptions.

Dementia — loss of intellectual abilities.

Deprenyl (Eldepryl, Seleginine, Jumex) — anti-Parkinson drug. A drug that slows the breakdown of chemicals like dopamine by inhibiting the action of certain enzymes. It's increase effects of dopamine in the brain.

Dopa decarboxylase — an enzyme present in the body that converts levodopa to dopamine.

Dopa decarboxylase — inhibitors anti-Parkinson drugs that block the enzyme dopa decarboxylase.

Dopamine — a chemical substance, a neurotransmitter, found in the brain that transmits impulses from one nerve cell to another and regulates movement, balance, and walking. It is the substance that is lost in PD.

Dopamine Agonist — Drugs that mimic the effects of dopamine and stimulate the dopamine receptors

Dopaminergic — a chemical that works like, or has the same effect as, dopamine.

Drug holiday — a 3- to 14-day withdrawal of levodopa after long-term treatment when side effects of levodopa outweigh benefits; rarely done today because of the severe effects of drug withdrawal.

Drug Induced Parkinsonism — Parkinson's symptoms which have been caused by drugs used to treat other conditions, e.g., neuroleptic drugs, and reserpine, use to be used to treat hypertension.

Dyskinesia — an abnormal involuntary movement including athetosis and chorea. Can result from long-term use of high doses of levodopa.

Dysphagia — difficulty in swallowing. Dystonia a slow movement or extended spasm in a group of muscles.

Edema — tissue swelling due to excessive fluid.

Enzyme — a substance that speeds up a specific chemical reaction but that is not itself consumed in the reaction.

Euphoria — a feeling of well-being or elation; may be drug related.

Extensor (muscle) — any muscle that causes the straightening of a limb or other part.

Extrapyramidal system — the system of nerve cells, nerve tracts and pathways that connects the cerebral cortex, basal ganglia, thalamus, cerebellum, reticular formation, and spinal neurons; it is concerned with the regulation of reflex movements such as balance and walking. The extrapyramidal system is damaged in Parkinson's disease.

Festination — short, shuffling steps; involuntary speeding up of the gait. Quick forward steps. A symptom characterized by small, Flexor (muscle) any muscle that causes the bending of a limb or other body part.

Freezing — Temporary, involuntary inability to move.

Ganglion — a cluster of nerve cells.

Globus pallidus — The inner part of the lenticular nucleus. The lenticular nucleus and the caudate nucleus form the Striatum.

Gray matter — the darker-colored tissues of the central nervous system; in the brain, the gray matter includes the cerebral cortex, the thalamus, the basal ganglia, and the outer layers of the cerebellum.

Hormone — a substance secreted by a gland that is transported in the bloodstream to various organs in order to regulate or modify bodily functions.

Hypokinesia — Abnormally diminished motor activity.

Idiopathic — An adjective meaning "of unknown cause". The usual form of Parkinson's is idiopathic Parkinson's.

Incontinence — involuntary voiding of the bladder or bowel.

Intention Tremor — one occurring when the person's attempts voluntary movement.

Lenticular nucleus — This group of cells along with the caudate nucleus form the Striatum or Corpus Striatum.

Levodopa — the single most effective anti-Parkinson drug, which is changed into dopamine in the brain, usually combined with carbidopa (a dopa decarboxylase inhibitor) as Sinemet.

Levodopa-Induced Dyskinesias — A side effect of medication, which may occur with prolonged use. These abnormal, involuntary movements may be alleviated by reducing the amount of medication.

Lewy body — a pink-staining sphere, found in the bodies of dying cells, which is considered to be a marker for Parkinson's disease.

Livido Reticularis — A purplish or bluish mottling of the skin seen usually below the knee and sometimes on the forearm in persons under treatment with the drug amantadine (Symmetrel).

Micrographia — a change in handwriting with the script becoming smaller and more cramped.

Monoamine oxidase (MAO) — an enzyme that breaks down dopamine. There are two types of MAO "A" and "B." In Parkinson's disease, it is beneficial to block the activity of MAO B.

MPTP — a chemical produced during an attempt to make a synthetic narcotic. MPTP destroys the cells of the substantia nigra cells and produces a disease that mimics Parkinson's disease.

Myoclonus — jerking, involuntary movements of the arms and legs. May occur normally during sleep.

Neostriatum — Vital part of the brain comprised of two basal ganglia (caudate and putamen).

Neuroleptic Drugs — (Also called major tranquilizers) A class of drugs which act as dopamine antagonists (by blocking some dopamine receptors). They can aggravate symptoms of Parkinson's. This class includes Haloperidol (Haldol), and the phenothiazines, e.g., Compazine, Stelazine, Chlorpromazine, etc.

Neuron — a cell specialized to conduct and generate electrical impulses and to carry information from one part of the brain to another.

Neurotransmitters — chemical substances that carry impulses from one nerve cell to another; found in the space (synapse) that separates the transmitting neuron's terminal (axon) from the receiving neuron's terminal (dendrite).

Nigral — of or referring to the substantia nigra.

Nigrostriatal Degeneration — Degeneration of the nerve pathways from Substantia Nigra to the striatum. These pathways are normally rich in dopamine and are those affected in PD.

Norepinephrine — a neurotransmitter found mainly in areas of the brain that are involved in governing autonomic nervous system activity, especially blood pressure and heart rate.

On-off phenomena — abrupt changes in performance during the day caused by the taking effect or wearing off of anti-parkinson drugs. A change in the patient's condition, with sometimes rapid fluctuations between uncontrolled movements and normal movement, usually occurring after long-term use of levodopa and probably caused by changes in the ability to respond to this drug.

Orthostatic hypotension — a large decrease in blood pressure upon standing; may result in fainting. Orthostatic hypertension may occur spontaneously in PD or may be related to certain drugs.

Palilalia — A symptom of Parkinsonism, especially the postencephalitic form, in which a word or syllable is repeated and the flow of speech is interrupted.

Pallidotomy — a surgical procedure in which a part of the brain called the *globus pallidus* is lesioned in order to improve symptoms of tremor, rigidity, and bradykinesia.

Palsy — paralysis of a muscle or group of muscles.

Paraesthesia — Sensations, usually unpleasant, arising spontaneously in a limb or other part of the body, variously experienced as Òpins and needlesÓ or a feeling of warmth or coldness (thermal paresthesias).

Parkinson's Disease — That form of Parkinsonism originally described by James Parkinson' as a chronic, slowly progressive disease of the nervous system characterised clinically by the combination of tremor, rigidity, bradykinesia, and stooped posture, and pathologically by loss of the pigmented nerve cells of the Substantia Nigra in the brain.

Parkinson's Facies — A stolid masklike expression of the face, with infrequent blinking; it is characteristic of Parkinson's.

Parkinsonism — a term referring to a group of conditions that are characterized by four typical symptoms — tremor, rigidity, postural instability, and bradykinesia.

Pergolide (Permax) — an anti-Parkinson drug.

Peristalsis — wave like contractions that move food through the digestive tract.

Postural instability — impaired balance and coordination, often causing patients to lean forward or backward and to fall easily.

Postural Tremor — Tremor that increases when hands are stretched out in front.

Pyramidal pathway — a collection of nerve tracts that travel from the cerebral cortex through the pyramid of the medulla oblongata in the brainstem to the spinal cord. Within the pyramid of the medulla, fibers cross from one side of the brain to the opposite side of the spinal cord; the pyramidal pathway is intact in Parkinson's disease.

Range of motion — the extent that a joint will move from full extension to full flexion.

Resting tremor — a tremor of a limb that increases when the limb is at rest.

Retropulsion — the tendency to step backwards if bumped from the front or upon initiating walking, usually seen in patients who tend to lean backwards because of problems with balance.

Rigidity — increased resistance to the passive movement of a limb. A symptom of the disease in which muscles feel stiff and display resistance to movement even when another person tries to move the affected part of the body, such as an arm.

Sialorrhea — drooling.

Sinemet — Trade name for the antiparkinson drug that is a mixture of levodopa and carbidopa. This drug combination contains a ratio of levodopa 4 mg. or 10 mg. to carbidopa 1 mg. (Sinemet 100/25, Sinemet 250/25).

Sinemet CR — Controlled-release Sinemet. 200 mg. Levodopa with 50 mg. Carbidopa in a capsule contained in a matrix (outer layer) releasing the drug more slowly in the body. These capsules are not to be taken all at once, but rather in separate doses over the course of a day.

Spasm — a condition in which a muscle or group of muscles involuntarily contract.

Stereotactic Surgery — Surgical technique, that involves placing a small electrode in an area of the brain to destroy a tiny amount of brain tissue.

Striatonigral Degeneration — This is a degeneration of the nerve pathways travelling from the striatum to the Substantia Nigra. People with this degeneration also appear to have Parkinsonism. However, they respond differently to drug therapy than people with Parkinson's.

Striatum — part of the basal ganglia, it is a large cluster of nerve cells, consisting of the caudate nucleus and the putamen, that controls movement, balance, and walking; the neurons of the striatum require dopamine to function.

Substantia nigra — movement-control center in the brain where loss of dopamine-producing nerve cells triggers the symptoms of Parkinson's disease; substantia nigra means "black substance," so called because the cells in this area are dark. A small area of the brain containing a cluster of black-pigmented nerve cells that produce dopamine which is then transmitted to the striatum.

Sustention (postural) — tremor a tremor of a limb that increases when the limb is stretched.

Synapse — a tiny gap between the ends of nerve fibers across which nerve impulses pass from one neuron to another; at the synapse, an impulse causes the release of a neurotransmitter, which diffuses across the gap and triggers an electrical impulse in the next neuron.

Tardive Dyskinesia — This is a movement disorder associated with long-term use of neuroleptic drugs such as Chlorpromazine, Haloperidol, Loxapine, etc. Movements of a person with tardive dyskinesia are similar in appearance to those of a person with levodopa induced dyskinesias, but the causes of the two conditions are different.

Thalamotomy — Operation in which a small region of the thalamus is destroyed, achieved by stereotactic techniques. Tremor and rigidity in Parkinsonism and other conditions may be relieved by thalamotomy.

Thalamus — Anatomical term designating a mass of gray matter centrally placed deep in the brain near its base and serving as a major relay station for impulses travelling from the spinal cord and cerebellum to the cerebral cortex.

Toxin — A poisonous substance.

Tremor — Rhythmic shaking and involuntary movement of part(s) of the body as a result of sequential muscle contractions.

Tyrosine — the amino acid from which dopamine is made.

Unilateral — Occurring on one side of the body. Parkinson's symptoms usually begin unilaterally.

Vomiting Center — Term referring to an area of the brain where the nausea and vomiting reflex may be triggered by some medications.

"Wearing Off" Phenomenon — Waning of the effect of the last dose of levodopa, associated with abrupt reduction or loss of mobility.

White matter nerve tissue — that is paler in color than gray matter because it contains nerve fibers with large amounts of insulating material (myelin). The white matter does not contain nerve cells. In the brain, the white matter lies within the gray layer of the cerebral cortex. See **Gray matter**.

Glossary with permission from The Iacono Neuroscience Clinic: http://Pallidotomy.com/index.html

Additional Reading

Barbeau, A.: "The Pathogenesis of Parkinson's Disease: A New Hypothesis," *Can. Med. Ass. J.*, **87**, 802–807 (1962).

Cooper, I.S., G. Bravo: "Chemopallidectomy and Chemothalamectomy," Presentation at the meeting of the Harvey Cushing Society, Detroit, Michigan, (April 26, 1957).

Cotzias, G.C. et al.: "Modification of Parkinsonism—Chronic Treatment with L-dopa," *NEJM*, **280**, No 7, 337–345 (1969).

Coyle, J.T., P. Puttfarcken: "Oxidative Stress, Glutamate, and Neurodegenerative Disorders," *Science*, **262**, 689–695 (1993).

Fahn, S. et al.: "Monoamines in the Human Neostriatus: Topographic Distribution in Normals and in Parkinson's Disease and Their Role in Akinesia, Rigidity, Chorea, and Tremor," *J. Neurological Sciences*, **14**, 427–455 (1971).

Halliday, G.M. et al.: "Loss of Brainstem Serotonin and Substance P-containing Neurons in Parkinson's Disease," *Brain Research*, **510** (1990).

Horner, P.J., F.H. Gage: "Regenerating the Damaged Nervous System," *Nature*, **407**, 963–70 (2000).

Hornykiewicz, O.: "Dopamine in the Basal Ganglia: Its Role and Therapeutic Implications (Including the Clinical use of L-dopa)," *Br. Med. Bull.*, **29**, 172–178 (1973).

Iacono, R.P. and B.S.J. Nashold: "Stereotactic Neurosurgery, In Iacono Robert, P., M.D. Iacono, FACS, B.S.J. Nashold, ed. *Textbook of Surgery: The Biological Basis of Modern Surgical Practice*, W.B. Saunders, Philadelphia, PA, (1991).

Iacono, R.P. et al.: "Stereotactic Pallidotomy Results for Parkinson's Exceed Those of Fetal Graft," *The American Surgeon*, **60**, 776–782 (1994).

Iacono, R.P. et al.: "The Results, Indications, and Physiology of Posteroventral Pallidotomy for Patients with Parkinson's Disease," *Neurosurgery*, **36**, 1118–1127 (1995).

Iacono, R.P., et al.: "Chronic Anterior Pallidal Stimulation for Parkinson's Disease," *Acta Neuroch*, **137**, 106–112 (1995).

Iacono, R.P., et al.: "Concentrations of Indoleamine Metabolic Intermediates in the Ventricular Cerebrospinal Fluid of Advanced Parkinson's Patients with Severe Postural Instability and Gait Disorders," *J. Neural. Transm.*, **104**, 451–459 (1997).

Laitinen, L.V. et al.: "Leksell's Posteroventral Pallidotomy in the Treatment of Parkinson's Disease," *J. Neurosurg*, **76**, 53–61 (1992).

Lindner, M.D. et al.: "Implantation of Encapsulated Catecholamine and GDNF-producing Cells in Rats with Unilateral Dopamine Depletions and Parkinsonian Symptoms," *Exp. Neurol.*, **132**, 62–76 (1995).

Liu, H. etal: "A Comparative Study on Neurochemistry of Cerebrospinal Fluid in Advanced Parkinson's Disease," *Neurobiology of Disease*, **6**, 35–42 (1999).

Olanow, C.W.: "An Introduction to the Free Radical Hypothesis in Parkinson's Disease," *Ann. Neurol.*, **32**, S2–S9 (1992).

Parkinson, J.: *An Essay on the Shaking Palsy*, Sherwood, Neely and Jones, London, (1817).

Spiegel, E.A. et al.: "Long Range Effects of Electropallidotomy in Extrapyramidal and Convulsive Disorders," *Neurology*, **8**, 738–743 (1958).

Svennilson, E. et al.: "Treatment of Parkinsonims by Stereotactic Thermolesions in Pallidal Region," *Psychiatr Neurol Scand*, **35**, 358–377 (1960).

Tasker, R.R.: Thalamotomy," *Neurosurg Clin. N. Am.*, **1**, 841–864 (1990).

Wichmann, T. et al.: "The Primate Subthalamic Nucleus. III. Changes in Motor Behavior and Neuronal Activity in the Internal Pallidum Induced by Subthalamic Inactivation in the MPTP Model of Parkinsonism," *J. Neurophysiol*, **72**, 521–530 (1994).

Web References

National Parkinson Foundation, Inc.: http://www.parkinson.org/
Parkinson Alliance.net: http://www.parkinsonalliance.net/
The American Parkinson's Disease Association: http://www.apdaparkinson.com/

R.P. IACONO, M.D., F.A.C.S.

PAROTITIS. Infection of the parotid glands. Parotitis may occur as a complication of any prolonged illness, especially after operations or illnesses in the aged and debilitated. Mumps is a form of parotitis resulting from infection with a virus that has a specific affinity for the parotid gland. See also **Mumps**.

PARROTFISHES (*Osteichthyes*). Of the order *Percomorphi*, family *Scaridae*, the parrotfishes are brightly colored marine fishes with a prominent beak formed by the partial coalescence of the teeth—thus their name. They occur in tropical waters, chiefly about coral reefs. They are herbivorous and are known to erode tropical reefs. They usually remove a bit of coral along with vegetation. Because of their powerful teeth, they are able to grind bits of coral, and hence, this does not interfere with their digestive processes. In 1955, Dr. Howard Winn discovered that some of the parrotfishes construct an envelope made of secreted mucous for covering themselves at night. The exact mechanism required to form the envelope and the actual need for the envelope remain to be fully investigated. The size range of parrotfishes is great, varying from lengths of 18 inches (45 centimeters) and under, up to 6 feet (1.8 meters).

PARROTS AND COCKATOOS (*Aves, Psittaciformes*). The parrots and related species, including cockatoos, macaws, and parakeets (paraquets) are all psittaciformes. These birds are characterized by a strong hooked beak, adapted for opening nuts and seeds. They have two toes directed forward and two toes directed backward.

The parrot is represented on all continents except Europe but is confined chiefly to tropical and subtropical areas. Parrots eat nuts and other fruits and seeds, although the kea parrot (*Nestor notabilis*) of New Zealand is reported to have acquired a taste for mutton. The brilliant colors of many parrots and their ready imitation of human speech have led to their being kept as cage birds for many centuries; hence, they are familiar well beyond their natural range.

There are many subsidiary forms of parrots. The nestor parrots of New Zealand and the neighboring islands are dark-colored birds, including the kea and the kaka. A curious member of the order is the owl-parrot or kakapo of New Zealand, which constitutes a distinct family. It is a flightless bird of owl-like appearance, barring its more brilliant colors, and is largely nocturnal in habits. The cockatiel is a small Australian parrot related to the cockatoos. The conures are small parrots of numerous species found from Mexico into South America. Their prevailing colors are green and yellow. The Carolina paraquet of North America is a member of this group. Broadtails are a group of parrots and parakeets found only in Australia, Norfolk Island, and Tasmania. The Budgerigar is an Australian parakeet, chiefly green with a blue tail and yellow face. In captivity, this species is also known as the Australian lovebird. Other names for the species include zebra, shell, and warbling grass-parakeet.

The cockatoo is a crested parrot whose beak is transversely ridged on the under surface of the hook. The tail is short and broad. Cockatoos occur in Australia and Oriental regions. The bill is sharp, short, and curved. The tongue is worm-like, tiny, and adapted for seizing and grasping various dietary items. Most cockatoos are white but also occur in various colors and with quite a range of sizes. They are frequently cherished by natives as pets. Numerous young birds are also taken for export as cage birds.

The cockatoo's nest is principally in holes of eucalyptus trees in the deep forest. The egg is white and is often deposited on bare wood chips or wood debris. The young are hatched naked and blind. They are fed by regurgitation of the parents for the first 3 months.

The black cockatoo (*Calyptorhynchus magnificus*) is the largest of the species and is a wanderer. With its strong bill, the bird cracks nuts that would require a hammer or vise for a human to break open. The bird's tongue assists much in extracting the kernel from the nut. The white cockatoo (*Cocatua galerita*) is a sulfur crested cockatoo of Australia. The white crested cockatoo (*C. alba*), found in the Molucca Islands, and the rose crested cockatoo (*C. moluccensis*), found on the island of Ceram, are representative of numerous varieties of cockatoos, embracing a wide range of colors and specific characteristics. See also **Psittaciformes**.

PARSEC.

The parsec is a unit of distance used for expressing distances between stars and other members of the sidereal universe. Technically, an object is at a distance of one parsec when it has a stellar parallax of $1''$ (one second of arc); or, in other words, one astronomical unit would subtend an angle of one second at the distance of one parsec. Thus, $D(\text{in parsecs}) = 1/\pi$.

$$1 \text{ parsec} = 3.26 \text{ light-years}$$
$$= 206{,}265 \text{ astronomical units}$$
$$= 1.924 \times 10^{13} \text{ miles}$$
$$= 3.084 \times 10^{13} \text{ kilometers}$$
$$= 3.084 \times 10^{18} \text{ centimeters}$$

Within recent years, in the discussion of distances between extragalactic objects, the parsec is not large enough to be convenient, and the terms kiloparsec (1,000 parsecs), and megaparsec (1,000,000 parsecs), are being used.

See also **Astronomical Unit**; and **Light-Year**.

PARSEVAL'S THEOREM.

A theorem relating the product of two functions to the products of their Fourier series components.

If the functions are $f(x)$ and $F(x)$, and their Fourier series components have respective amplitudes a_n, b_n and A_n, B_n, Parseval's theorem states that under certain general conditions

$$\frac{1}{\pi} \int_{-\pi}^{\pi} f(x)F(x)dx = \frac{1}{2}A_0 a_0 + \sum_{n-1}^{\infty}(A_n a_n + B_n b_n).$$

There is an analogous theorem for Fourier transforms. See also **Fourier Transform**.

AMS

PARSHALL FUME.
See **Flow Measurement (Liquids and Gases)**; and **Flume**.

PARSLEY.
See **Flavorings**.

PARTHENOGENESIS.
The development of eggs without fertilization. In some groups of animals, eggs normally develop in this way, either for a series of generations, interrupted occasionally by a normal fertilization, or as a special part of the reproductive process associated with the development of some young from fertilized eggs.

PARTIAL CORRELATION.
The correlation between the residuals of two random variables (variates) with respect to common regressors. Denoting the regression function of two

variates y and z with respect to a common set of regressors $x_1, x_2, \cdots x_n$ by Y and Z, the coefficient of partial correlation between y and z is defined as the coefficient of simple linear correlation between $(y - Y)$ and $(z - Z)$. To estimate the partial correlation, it is usually necessary to resort to sample approximations Y' and Z' of Y and Z. In that case, the estimate of the partial correlation is the sample value of the coefficient of simple, linear correlation between $(y - Y')$ and $(z - Z')$. In the simplest case in which Y' and Z' are taken as linear functions of a single variable x, the sample estimate $r_{yz} \cdot x$ of the partial correlation coefficient is given by the formula

$$r_{yzx} = \frac{r_{yz} - r_{yx}r_{zx}}{[(1 - r_{yx}^2)(1 - r_{zx}^2)]^{1/2}},$$

where the symbol r_{uv} denotes the sample coefficient of linear correlation between any pair of variates u, v.

See also **Correlation** and **Regression**.

AMS

PARTIAL DIFFERENTIAL EQUATION.
Partial derivatives of the unknown function are involved. The general linear second-order partial differential equation in two variables is

$$A(x, y)\frac{\partial^2 f}{\partial x^2} + B(x, y)\frac{\partial^2 f}{\partial x \partial y} + C(x, y)\frac{\partial^2 f}{\partial y^2}$$
$$= D(x, y)\frac{\partial f}{\partial x} + E(x, y)\frac{\partial f}{\partial y} + G(x, y)f + H(x, y)$$

The two families of curves given by

$$A\,dy = [B \pm (B^2 - AC)]^{1/2}dx$$

having solutions λ, μ = constant, are the families of characteristic curves. When the differential equation is rewritten with λ, μ as independent variables, the equation is in normal form. The three cases arising are *elliptic, hyperbolic*, and *parabolic* partial differential equations. Another special case is Euler's equation, if A, B, C are constants and the right-hand side of the equation is zero. The general second-order equation can be obtained from it by a linear transformation of independent variable.

The boundary conditions appropriate to the various forms of the partial differential equations are different, but they are always needed to fix the functional form of the solution. Problems dealing with the physics of fields usually lead to partial differential equations. See also **Mathematical Physics (Equations of)**.

Some properties of the three special cases are: (1) *Elliptic*, where $A(x, y)C(x, y) > B^2(x, y)$ for all x, y. The characteristic curves become functions of the complex variable. Writing $\lambda = (u + iv)$, the other solution is $\mu = (u - iv)$, and the normal form of the equation is

$$\frac{\partial^2 \phi}{\partial u^2} + \frac{\partial^2 \phi}{\partial v^2} = P(u, v)\frac{\partial \phi}{\partial u} + Q(u, v)\frac{\partial \phi}{\partial v} + R(u, v)\phi$$

The specification of Dirichlet or Neumann conditions along a closed boundary assures a unique solution. The Laplace, Helmholtz, and Poisson equations are of this type.

(2) *Hyperbolic*, where $B^2(x, y) > A(x, y)$ for all x, y. The characteristic curves are all real, and the normal form is

$$\frac{\partial^2 \phi}{\partial \lambda \partial \mu} = P(\lambda, \mu)\frac{\partial \phi}{\partial \lambda} + Q(\lambda, \mu)\frac{\partial \phi}{\partial \mu} + R(\lambda, \mu)\phi$$

Unless the boundary coincides with a characteristic, unique solutions are obtained with specified boundary values and normal derivatives (Cauchy conditions). When the boundary is closed, the Cauchy conditions overdetermine the solution. The wave equation is an example of this type.

(3) *Parabolic*, where $B^2(x, y) = A(x, y)C(x, y)$ for all x, y. There is only one set of characteristic curves given by $Ady = Bdx$, having the solution λ = constant. The normal form is then

$$\frac{\partial^2 \phi}{\partial x^2} = P(x, \lambda)\frac{\partial \phi}{\partial \lambda} + Q(x, \lambda)\frac{\partial \phi}{\partial x} + R(x, \lambda)\phi$$

The *diffusion equation*, $\partial^2\phi/\partial x^2 = a^2\partial x/\partial t$, where t is the time, is an example of this type. Dirichlet conditions on a boundary open at the end toward increasing t result in a unique solution.

Methods of solving partial differential equations must be sought in texts on the subjects.

PARTIALLY BALANCED INCOMPLETE BLOCKS. Partially balanced incomplete blocks form a very general class of experimental design in which not all treatments occur in every block. The designs are such that:

1. Each treatment is replicated r times.
2. Given any treatment, the remainders fall into groups of n_1, $n_2, \ldots, n_m$ such that every treatment of the ith group occurs λ_i times in the same block as the given treatment, the n_i and λ_i being independent of the treatment at the start. Two treatments that occur in the same block λ_i times are called ith associates, and the groups of treatments are called associate classes.
3. Given two treatments which are ith associates, the number of treatments which are simultaneously jth associates of the first and kth associates of the second is denoted by p^i_{jk}. This number must be independent of the particular pair of ith associates at the start.

Under these conditions, estimates of the treatment differences and their standard errors can be obtained without difficulty.

PARTIAL PRESSURE. The pressure exerted by each component in a mixture of gases. In a mixture of perfect gases

$$p_i = \frac{n_i R T}{V}$$

The partial pressure of i is then the same as if component i occupies the same volume at the same temperature in the absence of the other gases. This is the Dalton law, which is treated more fully under that heading.

PARTICLE. Any discrete unit of material structure; the particulate basis of matter is a fundamental concept of science. The size ranges of particles may be summarized as follows: (1) Subatomic: protons, neutrons, electrons, deuterons, etc. These are collectively called fundamental particles. (2) Molecular: includes atoms and molecules with size ranging from a few angstroms to half a micron. (3) Colloidal: includes macromolecules, micelles, and ultrafine particles such as carbon black, resolved via electron microscope, with size ranges from 1 millimicron up to lower limit of the optical microscope (1 micron). (4) Microscopic: units that can be resolved by an optical microscope (includes bacteria). (5) Macroscopic: all particles that can be resolved by the naked eye.

See also **Carbon Black**; and **Particles (Subatomic)**.

PARTICLE ACCELERATOR. A device in which the speed of charged subatomic particles (protons, electrons) and heavier particles (deuterons, alpha particles) can be greatly increased by application of electric fields of varying intensity, often in conjunction with magnetic fields. It is possible to accelerate electrons and protons to speeds approaching the speed of light if sufficiently high voltage is used. Straight-line (linear) accelerators are used for protons, and doughnut-shaped betatrons for electrons; other types are the Van de Graaf electrostatic generator, the synchrotron, and the cyclotron. Before the development of nuclear reactors, the cyclotron was used to accelerate deuterons for use in bombarding stable nuclei to produce neutrons for inducing artificial radioactivity, fission and formation of synthetic (transuranic) elements.

See also **Cyclotron**; and **Particles (Subatomic)**.

PARTICLES. Components of the atmosphere composed of solid or liquid matter. Particles may be both released from the earth's surface, such as dust or smoke, or formed in the atmosphere, as in rain or ice particles or sulfate aerosol. The particles in the atmosphere are usually defined in terms of their size, or diameter. Particles less than 100 μm in diameter are referred to as aerosol particles. These are aerodynamically stable and settle out only slowly (strictly speaking, the term aerosol refers to the gas–particle colloidal system, not just the particulate phase). The aerosol is usually divided into three modes: the Aitken mode (diameter less than 0.5 μm), the accumulation mode (0.5–2.0 μm), and the coarse mode (greater than 2 μm). The Aitken and accumulation modes are collectively referred to as fine particles. The larger particles of the coarse mode compose clouds and hydrometeors such as rain and sleet, which precipitate out.

PARTICLE SIZE. This term refers chiefly to the solid particles of which industrial materials are composed (carbon black, zinc oxide, clays, pigments, and the like). The smaller the particle, the greater will be the total exposed surface area of a given mass. Activity is a direct function of surface area; i.e., the finer a substance is, the more efficiently it will react, both chemically and physically. A colloidal pigment is a more effective colorant than a coarse one because of the greater surface area of its particles. A pound of channel carbon black has a surface area of 18 acres, which largely accounts for its powerful reinforcing effect in rubber. Thus, ultrafine grinding of powders is of utmost importance in such products as paints, cement, plastics, rubber, dyes, pharmaceuticals, printing inks, and numerous others.

See also **Carbon Black**; **Colloid Chemistry**; **Particle**; **Sedimentation**; and **Surface Chemistry**.

PARTICLES (Subatomic). For many years, the atom was traditionally described as having a central positively charged nucleus possessing considerable mass, but of minute dimension—this nucleus surrounded by a number of electrons in orbits at a relatively great distance from the nucleus. The number of electrons and their orbital arrangement determined the chemical properties of the atom, with the atoms of each chemical element possessing their own unique configuration. Recognition of the electron, the first elementary (presumably indivisible) particle, by J.J. Thomson and his associates in the 1890s ushered in an era of interest in *subatomic particles*. Ultimately, this led to the discipline of *high-energy physics*.

Organization of Matter — A Chronology[1]
A better understanding of the building blocks of nature has been a goal for many centuries, extending back to the period in Greek history of Anaxagoras of Ionia (500–428 B.C.), who held that "there was an infinite number of different kinds of elementary atoms, and that these, in themselves motionless and originally existing in a state of chaos, were put in motion by an eternal, immaterial, spiritual, elementary being, from which motion the world was produced." The concept of atoms appeared from time to time in medieval works, although the concepts expressed now seem vague. However, they seem to have been based on the idea that there could be a limit to the divisibility of matter and, consequently, the idea of a final indivisible particle out of which large pieces of matter could be built.

Atoms and Molecules. In the early 1800s, it became clear that chemical reactions could be most simply explained if each chemical element was thought of as composed of very small, identical entities characteristic of the chemical element. Thus there arose a rather well-defined idea of a chemical element composed of identical atoms, as distinguished from a compound composed of groups of different atoms combined into molecules. During the later part of the 1800s, the kinetic theory of gases made use of the idea of atoms and molecules in explaining the behavior of gases. During this period, few scientists still doubted the actual material existence and reality of atoms.

Electrons. It is perhaps rather curious that the idea of atoms became really well established only after it became clear that the atoms were not in any true sense indivisible, but that instead they probably had a complex structure that should be investigated. Since these investigations required equipment and methods that had been developed by physicists rather than chemists, the physicists took the lead and the work became known over a long period as *atomic physics*, or the physics of atomic structure. As mentioned previously, this era was inaugurated by Thomson, who first isolated and established the existence of electrons. He showed that electrons have only about 1/2000 the mass of the lightest known atom, hydrogen. He also showed that these particles, as indicated by their name, carry negative electrical charges. It was later shown by Millikan that all the electronic charges are the same. Thus, the identification of electrons as small electrically charged pieces of matter, and as constituents of all matter, became firmly established.

Electrical Neutrality. Since it was clear that normal matter is electrically neutral, it had to be assumed that each atom contained a positive electrical charge, as well as negative electrons. J.J. Thomson developed the picture of a somewhat spherical, jelly-like mass of positive electricity in which electrons are located at various positions, bound by "quasi-elastic" forces.

[1] Some of the concepts mentioned in this brief historical review have long since been abandoned or altered.

A principal means of investigating the structure of atoms was the examination of light emitted by the material in the gaseous state. This light was found to consist of a number of discrete wavelengths, or colors. Each of these wavelengths was associated, in the early days of the last century, with a mode of vibration of the electrons in the positive jelly. In particular, Lorentz (University of Leiden) was able to show that such electrons, when placed in a magnetic field, would have their modes of vibration changed in a way that explained the findings of Zeeman, who had made early observations of the wavelengths of the light emitted by a radiating gas in a magnetic field.

During 1910–1911, Sir Ernest Rutherford suggested an experiment, carried out by Geiger and Marsden, in which alpha particles from a radioactive source were scattered from thin foils. The angles at which the alpha particles were scattered were found to be such as could best be described by the close approach of a heavy positively charged particle, the alpha particle, to another heavier and more highly positively charged particle, representing the scattering atom.

Nuclear Atom. From the results of the experiments, Rutherford concluded that the mass in the positive charge of an atom, instead of being distributed throughout the volume of a sphere of the order of 10^{-8} centimeter in radius, was concentrated in a very small volume of the order of 10^{-12} centimeter in radius. He thus developed the idea of a nuclear atom. The atom was pictured as a small solar system with the very heavy and highly charged nucleus occupying the position of the sun, and with electrons moving around it, as planets in their respective orbits.

Although this picture of nuclear atoms served to describe the alpha-particle scattering experiments, it still left many questions unsolved. One of these questions referred to the apparent stability of the atoms. An electron moving around the nucleus would tend to emit radiation, to lose its energy, and thereby to spiral into the nucleus. Why did it not do so? Why did the atoms all seem to be quite stable, and all to be of approximately the same size, even though some contain 90 or more electrons, while hydrogen contains only one?

Electron Motion Around the Nucleus. The first approach to a treatment of these problems was made by Niels Bohr in 1913 when he formulated and applied rules for "quantization" of electron motion around the nucleus. Bohr postulated states of motion of the electron, satisfying these quantum rules, as peculiarly stable. In fact, one of them would be really permanently stable and would represent the ground state of the atom. The others would be only approximately stable. Occasionally an atom would leave one such state for another and, in the process, would radiate light of a frequency proportional to the difference in energy between the two states. By this means, Bohr was able to account for the spectrum of atomic hydrogen in a spectacular way. Bohr's paper in 1913 may well be said to have set the course of atomic physics on its latest path.

Correlation with Chemical Properties. Out of the experimental work on the scattering of alpha particles and the theoretical work of Bohr, there grew a fairly definite picture of an atom that could be correlated with its chemical properties. The chemical properties were determined in the first place by the nuclear charge. The nucleus contained most of the atomic mass and carried an electric charge equal to an integral number of positive charges, each of the same magnitude as an electronic charge. This positive nucleus then accumulated around itself a number of electrons just sufficient to neutralize its positive charge and form a neutral atom.

Atomic Number. The number of positive charges or the number of negative electrons around the nucleus was designated as the atomic number of the atom. These showed a close parallelism with the arrangement of atoms in the periodic system. Through the formulation of a number of rules based upon Bohr's picture of quantized orbits, the periodic system of the elements could be understood. Hydrogen was given one electron, and helium two. The two electrons in helium constituted a "closed shell" which exhibited almost perfect spherical symmetry and chemical inactivity.

Thus, during the years after 1913, the feeling grew that the chemical properties of atoms could be pretty well understood. The idea that there were undiscovered elements, as indicated by gaps in the periodic system, was reinforced. These elements and more have since been discovered.

Quantum Mechanics. It was not until 1925 that Bohr's ideas were developed into a mathematical form complete enough and precise enough to permit their general application, under the name *quantum mechanics.*

This development associated with the names of Dirac, Heisenberg, and Schrödinger, provided the basic laws which permit, in principle, the complete and quantitative description of an atom consisting of a heavy positively charged nucleus, and surrounded by enough electrons to make the whole system electrically neutral. See also **Quantum Mechanics**.

Electron Spin. One of the properties of electrons that became evident during the study of optical spectra of atoms was that of *electron spin.* The suggestion was made by Uhlenbeck and Goudsmit in 1925 that one of the features of such spectra could be understood if each electron had associated with it a quantity called spin, which is similar in many ways to angular momentum. Each electron also has a certain magnetic moment which affects the energy in the presence of a magnetic field.[2] This property also has been incorporated into the wave concepts of quantum mechanics.

In order to learn more about the role of spin in colliding protons, in 1973, using the Zero Gradient Synchrotron (ZGS) at the Argonne National Laboratory, Krisch and colleagues (University of Michigan) scattered beams of polarized protons from targets in which the protons were also polarized, i.e., the spinning was all in the same direction. During the series of experiments, it was found that when the beam and the target were polarized in the same direction, violent proton-proton collisions occurred with much greater frequency than where the beam and target were spinning in a like direction. Under the latter circumstance, it appeared that the particles would pass each other, but would not interact.

Some years later (in the late 1970s), this research group further investigated the spin-collision phenomenon of protons, but used a different accelerator, i.e., the Alternating Gradient Synchrotron (AGS) located at the Brookhaven National Laboratory. The latter apparatus made it possible to study the particles at much higher energy levels: (1) an energy level up to 18.5 GeV (beam and target both polarized; and (2) up to 28 GeV (with only the target polarized). Several unexpected findings were yielded by the experiments:

- Effects of spin appear to oscillate with an increase of collision energy of the protons.
- Spin directions of the particles continue to make a difference even at high energy levels. (Normally, one would reason that at high energy levels, the difference in spin directions and the effects of spinning would become smaller simply because the spin of the proton is believed to be constant and this would tend to be overwhelmed by the higher collision energy. This was not the case and provoked suspicion that much less is known about the proton than formerly believed.)

Out of clues gained from experiments with the AGS, one central question, as posed by researcher Krish in his 1987 paper (reference listed), is posed: What does the observed difference between scattering to the left and to the right mean? Krisch observed that perhaps (as some theorists suggest) both the violence and the energy (28 GeV) of the experiments were much too low for a fundamental theory, such as quantum chromodynamics (QCD), to apply. With higher energies, the scattering difference between left and right may soon be measured at the 70- to 800-GeV proton synchrotrons at Seupukhov, CERN, and Fermilab, and even higher levels of the proposed Superconducting Supercollider (SSC).

See also **Quantum Mechanics**; and **Quarks**.

Neutrons and Protons. By 1932, it had been established that atomic nuclei are made of comparatively small numbers of neutrons and protons. Even prior to the use of particle accelerators and the birth of high-energy

---

[2] Even with acceptance of the spin concept, in terms of high-energy experiments, most scientists believed that spin effects would be observed only in low-energy atomic collisions. In the 1950s, C.L. Oxley (University of Rochester) noted large spin effects in high-energy collisions (several hundred million eV). Experimentation in this area, however, was rather limited until the late 1950s, when researchers (University of California Berkeley) proposed constructing polarized proton targets. In a technique involving low temperatures and strong magnetic fields, it was possible to cause electrons to spin in the same direction and, using another technique (microwave radiation) to cause neighboring protons also to spin in one direction. Interesting experiments followed at some laboratories, but by and large many high-energy physicists in the field considered spin as relatively unimportant and it would be even less important as collision energy levels were increased. This assumption has been disproved. Spin direction does seem to be important to collisions even at high energy levels.

physics, other experiments continued to "hint" at the need of additional subatomic particles to satisfy any theory that would unify scientists' understanding of the atom's infrastructure.

A quantum theory of nuclei was made possible by the discovery of the proton and the neutron. The nuclear interaction responsible for holding the nucleus together (against disruptive electrostatic repulsion of the protons) was found to be of an entirely new kind, much stronger than the electric interaction at short distances, but decreasing very much more rapidly with distance. The various complex nuclei differ in the number of protons and neutrons they contain.

By that time, the theory of the interactions between electrons and photons had developed to the point where the electrostatic repulsion or attraction between electrically charged particles could be understood in terms of the exchange of photons between them. In the lowest nontrivial approximation, it gave the Coulomb law for small velocities. The basic interaction was the emission and absorption of "virtual" photons by charged particles.

Pions and Other Particles. A similar mechanism could be invoked to explain the short-range nuclear interaction—i.e., it is due to the exchange of particles, which have nonzero masses which are a fraction of nuclear mass. These theoretical considerations predicted the existence of a set of three particles called pions, which were ultimately discovered.

Another kind of particle and another kind of interaction were discovered from a detailed study of beta radioactivity in which electrons with a continuous spectrum of energies are emitted by an unstable nucleus. The corresponding interactions could be viewed as being due to the virtual transmutation of a neutron into a proton, an electron, and a new neutral particle of vanishing mass called the neutrino. The theory provided such a successful systematization of beta decay rate data for several nuclei that the existence of the neutrino was well established more than 20 years before its experimental discovery. The beta decay interaction was very weak even compared to the electron-photon interaction.

Meanwhile, the electron was found to have a positively charged counterpart called the positron; the electron and positron could annihilate each other, with the emission of light quanta. The theory of the electron did in fact predict the existence of such a particle. It was later found that the existence of such "opposite" particles (antiparticles) was a much more general phenomenon than once surmised.

With intensification of particle physics research, many more particles were discovered and a classification of these particles into five families was proposed—the photon family, electron family, muon family, meson family, and baryon family. Most of these particles are unstable and decay within a time which is often very small by normal standards, but which is many orders of magnitude larger than the time required for any of these particles to traverse a typical nuclear dimension. There is a wide variety of reactions between them, but they could be understood in terms of three basic interactions—the *strong* (or nuclear), *electromagnetic*, and *weak* interactions.

The Nuclear Force. The nuclear forces and the interactions between pions and nucleons are strong; the electron-electron and electron-photon interactions are electromagnetic; the beta decay interactions are weak.

As mentioned previously, by 1932 it was known that nuclei are made of comparatively small numbers of neutrons and protons. A new force was discovered (in addition to the electromagnetic and gravitational forces) that held the positive protons and electrically uncharged neutrons together in the nucleus. This nuclear force was very strong, but of limited range. Its "quantum," the particle analogous to the photon in the electromagnetic field, was of nonzero rest mass. This particle, later called the π-meson or pion, was predicted by Yukawa in 1936 and discovered by Lattes, Occhialini, and Powell in 1947.[3] For a short time, it appeared that physicists had achieved a clear, simple, and correct theory of the

fundamental constitution of matter. However, shortly thereafter, two new and unpredicted particles were reported. The first of these was another meson, somewhat like the pion but more massive. The second was a hyperon, i.e., a strongly interacting particle heavier than the neutron.

With the continuing discovery of more particles, investigators began to suspect that these particles were not in themselves fundamental or elementary, but that they had an internal structure. This paralleled the experiences of the 1800s when the large number of different types of atoms discovered suggested that atoms must have structure. Properties of particles also suggested an internal structure. For example, the neutron's total electric charge is indistinguishable from zero down to very fine limits, yet the neutron has a sizeable magnetic moment.

The discoverers of neptunium (1940), plutonium (1940), americium (1944), berkelium (1949), californium (1950), einsteinium (1952), fermium (1953), mendelevium (1955), and lawrencium (1961) gained much knowledge in the area of high-energy physics.

Research directed toward creating a nuclear bomb also contributed to an improved understanding of high-energy physics.

Exotic atomic nuclei may be described as structures that do not occur in nature, but are produced in collisions. These nuclei have abundances of neurons and protons that are quite different from the natural nuclei. In 1949, M.G. Mayer (Argonne National Laboratory) and J.H.D. Jensen (University of Heidelberg) introduced a spherical-shell model of the nucleus. The model, however, did not meet the requirements and restrains imposed by quantum mechanics and the Pauli exclusion principle. Hamilton (Vanderbilt University) and Maruhn (University of Frankfurt) reported on additional research of exotic atomic nuclei in a paper published in mid-1986 (see reference listed). In addition to the aforementioned spherical model, there are several other fundamental shapes, including other geometric shapes with three mutually perpendicular axes—prolate spheroid (football shape), oblate spheroid (discus shape), and triaxial nucleus (all axes unequal).

In 1964, M. Gell-Mann and G. Zweig (California Institute of Technology) independently pointed out that all the known hadrons (i.e., particles that interact via the strong nuclear force) could be constructed out of simple combinations of three particles (and their antiparticles). These hypothetical particles had to have slightly peculiar properties (the most peculiar being a fractional electric charge). Gell-Mann called these hypothetical particles *quarks* (referring to a sentence in James Joyce's work *Finnegan's Wake*, "Three quarks for Muster Mark"). The theory proposed postulated that three quarks bind together to form a baryon, while a quark and an anti-quark bind together to form a meson. With supposition that the binding is such that the internal motion of the quarks is nonrelativistic (which requires the quarks be massive and sit in a broad potential well), then many quite detailed properties of the hadrons could be explained.

The purpose of the quark model was that of explaining the diversity of the hadrons; not to deal with the internal structure of any particle. But awareness of the model created a natural tendency among investigators to associate newly observed particles (among the poorly understood debris

---

[3] Even though the meson was first predicted by the Japanese scientist Yukawa in 1935, the development of high-energy physics in Japan proceeded slowly from an experimental standpoint. It was not until 1975 that Japan established the National Laboratory for High Energy Physics (KEK). Rather than following traditional research approaches by way of building a proton synchrotron, for example, scientists at the University of Tokyo, in collaboration with physicists in the United States and Canada and later with CERN, decided to construct a meson facility with a powerful superconducting muon channel. In his report on the evolution

of meson science in Japan, Yamazaki (1986 reference cited) lists at least four advantages of opting for a meson facility: (1) It made possible the measurement of $\mu - e$ decay time spectra in a much wider time range (0 to 20 μsec) than previously possible without background, enabling muon-spin relaxation functions (mainly long-time behavior) to be determined precisely; (2) extreme external conditions pulsewise (pulsed RF, laser, high magnetic fields) could be applied; (3) because the time of muon arrival is uniquely defined, any time-dependent transient phenomena could be investigated; and (4) rare events could be selected from continuous backgrounds. Beyond the editorial scope here, Yamazaki describes in considerable detail the numerous accomplishments of KEK during the 1980s, including an improved understanding of nuclear structure from the viewpoint of quark structure. As mentioned, since nucleons and mesons are composed of quarks and since nucleons are densely packed in a nucleus, whether the nucleons in nuclei keep their free identities (mass, size, magnetic moment) is a rewarding problem to investigate. Recently, a new type of hypernuclear spectroscopy has emerged from KEK. Yamazaki defines meson science as an interdisciplinary since it uses "second generation" particles (muons and K mesons) for the creation and detection of exotic states in matter. To study this interesting frontier, scientists strongly sense the need for experimental facilities that will provide meson beams a hundred times as strong as those available today. Plans are underway along these lines. See also **Mesons**.

from particle experiments) with the hypothetical quarks. A number of properties of *partons* (a name given by Feynman, California Institute of Technology) were measured, including intrinsic spin angular momentum, and these were found to be consistent with the predictions of the quark model. Such observations, of course, added credence to the model.

In the 1960s, the quest for a grand unification theory—a theory that would explain all elementary particles and all forces acting between them—grew in intensity among most investigators who had the good fortune of discovering so many new particles, accompanied by the realization that the ultimate structure of matter was more complex than envisioned in the earlier years. The instrumental means for research (accelerators with higher and higher energies) were getting ahead of the theoretical aspects of the topic. Many particles resulting from collisions were found in the debris of experiments—their presence without plausible explanations. Many questions were posed—for instance, why four kinds of force, each with its own characteristic strength, the strengths differing by nearly 40 orders of magnitude (electromagnetism with its infinite range, the weak force for all practical purposes extending out only 10^{-15} centimeter)? For a while, prospects of a unified theory were dim, but a number of theories were proposed and given sufficient serious attention to warrant planning of experimental tests. As pointed out by Glashow (Nobel Prize, Physics, 1979, shared with Salam and Weinberg), in his Nobel Lecture, "In 1956, when I began doing theoretical physics, the study of elementary particles was like a patchwork quilt. Electrodynamics, weak interactions, and strong interactions were clearly separate disciplines, taught and separately studied. There was no coherent theory that described them all. Developments such as the observation of parity violation, the successes of quantum electrodynamics, the discovery of hadron resonances, and the appearance of strangeness were well-defined parts of the picture, but they could not be easily fitted together."

In the early years of investigation, the weak force and the electromagnetic force were regarded as indistinguishable. They were of the same strength and possessed the same infinite range, and they were transmitted by four bosons, all of which were massless. The forces manifested a symmetry, that is, they could be interchanged freely. It was believed that no matter which of these forces was applied, the net effect was the same. These views were later to be altered in the light of the process called *spontaneous symmetry breaking*.[4]

The principle of charge conjugation symmetry states that if each particle in a given system is replaced by its corresponding antiparticle, then it would not be possible to tell the difference. For example, if in a hydrogen atom the proton is replaced by an antiproton and the electron is replaced by a positron, then this antimatter atom will behave exactly like an ordinary atom—if observed by "persons also made of antimatter." In an antimatter universe, the laws of nature could not be distinguished from the laws of an ordinary matter universe.

However, it turns out that there are certain types of reactions where this rule does not hold, and these are just the types of reactions where conservation of parity breaks down. For example, consider a piece of radioactive material emitting electrons by beta decay. The radioactive nuclei are lined up in a magnetic field which is produced by electrons traveling clockwise in a coil of wire, as seen by an observer looking down on the coil. Because of the asymmetry of the radio-active nuclei, most of the emitted electrons travel in the downward direction. If the same experiment were done with similar nuclei composed of antiparticles and the magnetic field were produced by positron current rather than an electron current, then the emitted positrons would be found to travel in the upward, rather than in the downward, direction. Interchanging each particle with its antimatter particle has produced a change in the experiment.

However, the symmetry of the situation can be restored if we interchange the words "right" and "left" in the description of the experiment at the same time that we exchange each particle with its antiparticle. In the

above experiment, this is equivalent to replacing the word "clockwise" with "counterclockwise." When this is done, the positrons are emitted in the downward direction, just as the electrons in the original experiment. The laws of nature are thus found to be invariant to the simultaneous application of charge conjugation and mirror inversion.

Time reversal invariance describes the fact that in reactions between elementary particles, it does not make any difference if the direction of the time coordinate is reversed. Since all reactions are invariant to simultaneous application of mirror inversion, charge conjugation, and time reversal, the combination of all three is called *CPT* symmetry and is considered to be a very fundamental symmetry of nature.

A relatively recent type of space-time symmetry has been introduced to explain the results of certain high-energy scattering experiments. This is *scale symmetry* and it pertains to the rescaling or "dilation" of the space-time coordinates of a system without changing the physics of the system. Other symmetries, such as chirality, are more of an abstract nature, but aid the theorist in an effort to bring order into the vast array of possible elementary particle reactions.

A feature of quantum field theory is that the quanta of the fields are initially massless. Spontaneous symmetry breaking offers a mechanism by which weak field quanta, for example, can acquire masses. Unification of weak and electro-magnetic forces may be viewed thus in the following manner at short distances (high energies), the masses of the weak field quanta become unimportant and thus original symmetry is restored. Symmetry in this context refers to the properties of the equations of motion of particles in the field theories. Spontaneous symmetry breaking occurs when solutions of the equations do not display full symmetry. Some physicists have likened this to a ball moving on a roulette wheel, whose equations of motion are symmetrical about the axis of rotation even though it always stops in an asymmetric position.

The first direct evidence that the proton has not only size, but structure was provided by an experiment at the Stanford Linear Accelerator Center (SLAC) in 1970. Previously it had been established that the proton is not a point-like particle, but has a finite size—a diameter of about 10^{-13} centimeter. Although it is only about 1/100,000 the size of an atom, it is still measurable. This is unlike certain other particles, notably the electron, for which no extension has been noted, so that the electron can be regarded as a mathematical point. In the experiment, electrons were raised to an energy of some 20 billion electron volts and struck protons and neutrons in the atoms of a stationary target. The angular distributions of the scattered electrons and of other particles created in the collisions were carefully monitored. Most of the electrons, as expected, passed through the target with little change in direction. An unexpected excess of widely scattered particles was produced, however—much greater than if the proton were diffuse and homogeneous. The excess of the widely scattered particles was attributed to a mass embedded within the proton, estimated at no more than $\frac{1}{50}$ the diameter of the proton. In later experiments, a target was illuminated by means of muons (like electrons but with a mass 200 times greater); and by a beam of neutrinos (which lack both mass and electric charge). The results of the original and later experiments were consistent and the deep scattering of particles was attributed to collisions between the incident leptons and some "hard" constituent of the proton.

Theories and postulations continue to be developed concerning the symmetry of nature. For example, in a 1986 paper, H.E. Haber (University of California, Santa Cruz) and G.L. Kane (University of Michigan) observe that *supersymmetry* could represent the next step in the quest for a few simple laws that explain the nature of matter. Physicists are seeking evidence to test the theory. As described, in supersymmetry, for every ordinary particle that exists there is a so-called "superpartner" having similar properties, with exception of the quantity referred to as *spin*, previously discussed here. In a 1985 paper, C. Quigg (Fermi National Accelerator Laboratory) further describes current theories and hypotheses pertaining to elementary particles and forces and observes that a coherent view of the fundamental constituents of matter and the forces governing them is beginning to emerge. A present goal of physicists is to merge disparate theories into a single comprehensive description of natural events. It is agreed among most high-energy specialists that to reach that goal, greater and greater energy must be brought to the experiments. The Superconducting Supercollider (SSC) would be one of these. The complex concept of CP invariance is described in a 1988 paper by R. Adair

---

[4] Prior to 1956, it was believed that all reactions in nature obeyed the law of conservation of parity, so that there was no fundamental distinction between left and right in nature. However, Yang and Lee pointed out that in reactions involving the weak interaction between particles, parity was not conserved, and that experiments could be devised that would absolutely distinguish between right and left. This was the first example of a situation where a spatial symmetry was found to be broken by one of the fundamental interactions.

(Brookhaven National Laboratory). In this postulate, the claim is made that without CP invariance there would be no matter in the universe. Adair observes that if the approximate symmetry between matter and antimatter that has been observed were perfect, the universe would be elegantly simple but virtually empty of matter and of creatures made up of that matter who could contemplate that elegance. It is proposed that the existence of the universe as currently known comes from a flaw in a symmetry exhibited by a universal mirror (the CP mirror), i.e., a symmetry that requires that the outcomes of some events in nature should remain the same on changing matter to antimatter and viewing the result in a mirror.

Before discovery of the hadron particle (designated *psi* or *J*), and after much experimental and theoretical effort, physicists had about concluded that three massive, fractionally charged entities (quarks) were the primary building blocks of the universe. However, discovery of the psi particles in 1974 indicated a fourth quark was required. Previously, in the three-quark model, all mesons were made up of one quark and one antiquark; baryons, of three quarks; and all anti-baryons, of three antiquarks. Prior to 1974, all of the known hadrons could be accommodated within this basic scheme. Three of the possible meson combinations of quark-antiquark could have the same quantum numbers as the photon, and hence could be produced abundantly in e^+e^- annihilation. These three predicted states had all been found.

As pointed out by Richter (1977), the first proposal of a theory based on four quarks rather than three was published in 1964 by Amati and others. The motivation at that time was more esthetic than practical, and these models gradually expired for want of an experimental fact that called for more than a three-quark explanation. In 1970, Glashow explained in a paper that a fourth quark was required to explain the nonoccurrence of certain weak decays. The fourth or *c* quark was assumed to have a charge of $+\frac{2}{3}$, like the *u* quark, and also to carry +1 unit of a previously unknown quantum number, called *charm* by Glashow, which was conserved in both the strong and electromagnetic interactions, but not in the weak interactions. Discovery of the psi particles demonstrated a more compelling need for the fourth quark. Richter observes that the four-quark model of hadrons seemed to account, in at least a qualitative fashion, for all of the main experimental information that had been gathered about the psions, and by the early part of 1976, the consensus for charm had become quite strong.

In 1977, the *upsilon particle* was found as the result of energetic collisions between protons and copper nuclei. The upsilon particle has a mass three times greater than any other subatomic entity yet detected (early 1980s). Researchers on this experiment from Columbia University, the State University of New York (Stony Brook), and the Fermi National Accelerator Laboratory reported that, with a mass at its lower energy state equivalent to 9.0 GeV and masses in excited states equivalent to 10 and 10.4 GeV, the upsilon particle has been interpreted as consisting of a massive new quark (the fifth) bound to its antiquark. Confirming experiments were also conducted at the Deutsches Elektronen Synchrotron (DESY) located near Hamburg, Germany. The quantum attribute of the fifth quark was named "bottom." With a fifth quark reported, many physicists felt that finding a sixth quark ("top") was highly probable.

In 1979, the Nobel Prize (Physics) was awarded to Glashow and Weinberg (both of Harvard) and Salam, a Pakistani physicist, in recognition of the significance of the theory which unites the weak force with the electromagnetic force. But most scientists recognize this finding as only a milestone in a series of predictions that include the existence of new particles so massive that they cannot be expected to appear at the energies thus far available to physicists.

The chemists of the nineteenth century once thought that all material substances were comprised of only 36 elements.[5] Over the years, these expanded to over 100 elements. Fifty years ago, it was proclaimed that the elements were made of electrons, protons, and neutrons. Then, commencing in the 1940s, many other particles were found, as described here. Then for a while it seemed that elementary matter could be reduced to three particles — the quarks. But quarks have multiplied in number, with a sixth quark now seriously proposed. Will there be too many quarks? Perhaps hypothetical particles will be proposed from which the quarks are

comprised. Possibly the ultimate answer will lie with the "mathematical groups that order the particles rather than in truly elementary objects."

In the late 1980s and thereafter, the *superstring* theory was drawing the attention of most theoretical physicists. Out of this theory may emerge the long sought concept that will account for all four fundamental forces. Quantum theories have been formulated for three of the four known forces of nature — strong, weak, and electromagnetic interactions. This comprises one of two basic foundations. The other, of course, is Einstein's general theory of relativity, which relates the force of gravity to the structure of space and time. A quantum theory of gravity is missing. Some scientists currently feel that there is much potential for the string theory to bring this unification about. String theory was first proposed by Y. Nambu (University of Chicago) in 1970. Traditional models of elementary particles are based on quantum field theory, which involves dimensionless points and quantum numbers, but with no specified internal structure. In studying one of the alternatives to the foregoing, a dual resonance model, Nambu observed that it was equivalent mathematically to the interaction of bits of string. The dual resonance model had been constructed to show how one hadron should scatter off another. The theory envisions the strings about 10^{-35} meter long (10^{20} times smaller than the diameter of a proton). As observed by M. Green (Queen Mary College, University of London) in a 1986 paper (reference listed), string has extension; it can vibrate like a violin string. The harmonic, or normal, modes of vibrations are determined by the tension of the string. In quantum mechanics, waves and particles are dual aspects of the same phenomenon. Thus, each vibrational mode of a string corresponds to a particle. It is envisioned that two strings interact when they touch their tips together and are fused into one — or one string may split into two parts. Strings can absorb energy in a collision and may ripple and rotate (at the speed of light). The original string theory initially was limited to providing a satisfactory "explanation" for describing bosons (pi meson and rho meson), i.e., particles that have integral numbers of spin angular momentum units. Attempts to describe particles with half-integral spin, such as fermions (proton, neutron), failed. However, in 1976, a suggestion was made by Scherk, a French physicist, to the effect that a string model could represent a fermion, but with the proviso that a fermion was matched by a corresponding boson. Incidentally, this is the type of correspondence required by the principle of supersymmetry. One of several difficulties that lie ahead for the superstring theory is that of matching its mathematical concepts (26 dimensions — 25 space and 1 time) with the real world of 4 dimensions. As one physicist has observed (E. Witten, Princeton University), it is a complete mystery what string theory is at a fundamental level.

State of the Science (Early 1990s)

In terms of comprehending the ultimate nature of matter, most scientists would agree that tremendous progress has been made over the past century and a half, but that a grand unifying theory of particles and their structures and interaction may continue to elude researchers for many years.

Number of Particle Families. How many families of matter may exist? Three, four, or more? An acceptable number among researchers today is three. Three family entities make up matter — the stars, the planets, molecules, and the atoms in the paper upon which this is printed. These fundamental particles are the "up" quark, the "down" quark, and the electron. Some other researchers are not quite so confident. One is reminded of the quotation from Jonathan Swift:

So naturalists, observe, a flea Hath smaller fleas that on him prey: And these have smaller still to bite'em And so proceed ad infinitum.

Within the limitations of contemporary knowledge, there are three families of fundamental particles, the approximate properties of which are given in Table 1. In 1991, G. Feldman (Harvard University) and J. Steinberger (CERN and 1988 Nobelist for discovery of the muon nutrino) observed, "Many questions remain unanswered. Why are there just three families of particles? What law determines the masses of their members, decreeing that they shall span 10 powers of 107? These problems lie at the center of particle physics today. They have been brought one step closer to solution by the numbering of the families of matter."

In their reference listed, Feldman and Steinberger describe how experiments at CERN and SLAC, using electron-positron collisions, showed that there are only three families of fundamental particles in the universe.

---

[5] Line of thought suggested by L.M. Lederman (Columbia University) in a paper on "The Upsilon Particle," *Sci. Amer.*, **239**, 4, 80 (1978).

TABLE 1. THREE FAMILIES OF FUNDAMENTAL PARTICLES

	Charge	Mass in Billions of Electron Volts (GeV)		
		Electron Family	Muon Family	Tau Family
Quarks	2/3	*Up* ~0.01 GeV	*Charm* ~1.5 GeV	*Top* Est. 89GeV*
	−1/3	*Down* ~0.01 GeV	*Strange* ~0.15 GeV	*Bottom* ~5.5 GeV
Leptons	0	*Electron neutrino** <2 × 10⁻⁸GeV	*Muon Neutrino** <2 × 10⁻⁴GeV	*Bottom* <5.5 GeV
	−1	*Electron* 5.11 × 10⁻⁴GeV	*Muon* 0.106 GeV	*Tau* 1.78 GeV

→ = Relative Increase in Mass →
(*Mass unknown)
Data source: G.J. Feldman and J. Steinberger (see reference listed).

By contrast, as D. Cline (University of California) observed in 1988, "Several theorists think a new quark should exist in the vicinity of 246 GeV. One of the notable features of the standard model is its prediction that at high enough energies the various forces begin to unify. In particular, the electromagnetic force and the weak and strong nuclear forces should become a single 'grand unified' force. The forces should be unified at the incredible energy of 10^{15} GeV, considerably beyond what can ever be attained by an accelerator on earth. The extrapolation of measured values of fundamental parameters from low energies to the grand-unified energy scale would require the existence of a new massive quark for consistency."

Nuclear Equation of State. In a late 1991 paper, H. Guthrod (Institute of Heavy-Ion Research, Darmstadt) and H. Stöcker (University of Frankfurt) are developing a nuclear equation of state in order to clarify the "new" states of matter and conditions that may occur inside a supernova and the organization of the universe. The authors observe, "Nuclear matter in its normal phase resembles a liquid. Increasing the temperature or density 'boils' nuclei into the hadron gas phase. Under extreme density but low temperature, nucleons could become 'frozen,' forming condensates, further heating or compression may produce the plasma phase, which would consist of free quarks and gluons. The gas and plasma phases may exist simultaneously over a wide region. Particles that have strange quarks, such as multistrange, metastable objects ('memos') and strangelets, may also form."

This approach parallels our consideration of the equation of state that applies to "ordinary" matter, such as gases, liquids, and solids that exist in macrostructured materials. See also **Equation of State**.

Particle Accelerators

Subatomic particles, such as electrons, positrons, and protons, can be accelerated to high velocities and energies, usually expressed in terms of center-of-mass energy, by machines that impart energy to the particles in small stages or nudges, ultimately achieving in this way very high-energy beams, measured in terms of billions and even trillions of electron volts. Thus, in terms of their scale, particles can be made to perform as powerful missiles for bombarding other particles in a target substance or for colliding with each other as they assume intersecting orbits. Because the particles are empowered with high energy, their smashing encounters are conducive to breaking the particles into their constituents. Instruments or machines used to arrange these particle encounters are known as *particle accelerators* and are very large, their dimensions frequently measured in terms of a few miles or kilometers. Thus, in a sense, it is ironic that the largest tools of science are required to seek knowledge concerning the smallest particles of matter that make up the universe and the earth. Theoretically, if at some future date sufficient energy can be imparted to two particles, their head-on collision will yield a complete fireball of disintegration, such that the absolute, indivisible particles of matter will be lain bare. A further division of matter then would be theoretically impossible. This is a goal of many particle physicists and perhaps achievable within a relatively few decades. Or, on the other hand, would then new theories be formulated to show that the true absolutes have indeed not yet been found?

Electromagnetic forces are used to accelerate particles, requiring that the particles must have an electric charge. Protons (+1) and electrons (−1) are commonly used as the media for particle-physics experiments, although not exclusively. The particles must be accelerated within a vacuum because otherwise they would collide with the molecules of air. The tube-shaped enclosures for the speeding particles are maintained under a vacuum of about 10^{-9} torr. The particles are set in motion by an electric field, which in the simplest configuration is an applied high voltage across a pair of electrodes. The positive electrode attracts electrons; the negative electrode attracts protons. It may be pointed out that an acceleration of this kind occurs in the ordinary television receiver. This basic principle, which applied to the early accelerators, remains the basis of the current operating and planned accelerators of the future. A simple electrode arrangement cannot sustain a potential of over a few million volts because of breakdown of insulation and arcing, and thus is confined to the simplest kind of accelerator where high energies are not required. In a practical sense, particles must be provided with energy in a large number of stages or "nudges." Thus, as in a *linear accelerator* (linac), the stages are strung out along a straight path, each stage requiring a radio-frequency oscillator which sets up an alternating electric field which is connected to each set of electrodes. Many radio-frequency cavities are formed along the line, and to provide the correct parity so that the particles will be accelerated rather than retarded the oscillators for the successive cavities must be synchronized. Effectively, an electromagnetic wave that travels continuously through the evacuated chamber is set up. It has been suggested by one physicist that the particle "rides the electrical wave as a surfer rides a water wave." In a less costly arrangement, known as the *synchrotron*, the particles are made to follow a circular or closed curve rather than a linear course. Groups of particles may circle a ring of this kind several million times while they are increasing their energy and velocity. Only one or a few radio-frequency cavities are required because energy is picked up each time the particle completes a revolution.

In addition to linear accelerators and synchrotrons, the accelerated particles of which ultimately are directed to strike a fixed target, there are *colliding-particle machines* in which particles are made to collide head-on. These machines are similar to synchrotrons except one bunch, or cluster of particles, travels in one direction while another bunch travels in the opposite direction. Where the colliding particles have the same rest mass and, after acceleration, have the same energy, the center-of-mass energy is the sum of the two beam energies. Thus, two beams with energies of 50 GeV (billion electron volts), for example, can provide the colliding force of one beam of 100 GeV against a fixed target. As will be shown later, head-on collision apparatus requires the use of storage rings, and systems are arranged in various configurations. Electron-positron storage rings are particularly efficient for creating new elementary particles from high-energy collisions.

The energy acquired by the particles in an accelerator is expressed in electron volts (eV), the amount of energy gained by any particle bearing a charge equal to that of an electron when it falls through a potential difference of 1 volt. Thus, 10^{3}eV = 1keV (kiloelectron volt); 10^{6}eV = 1MeV (megaelectron volt or one million eV); 10^{9}eV = 1GeV (gigaelectron volt or one billion eV); and 10^{12}eV = 1TeV (teraelectron volt or one trillion eV). The large accelerators have been in the GeV range. Plans call for machines in the TeV range. Inasmuch as the mass and energy can be freely interconverted, the mass of a particle is usually expressed in terms of its energy equivalent in eV. The mass of the proton thus is 938 MeV.

High-energy machines require a supply not only of the type of particle desired, but also particles that have been preliminarily accelerated. Electrons are comparatively easy to generate as inputs to accelerators, as by the Cockcroft-Walton generator. Protons are obtained by ionizing hydrogen atoms.

As early as 1928, E.O. Lawrence constructed one of the first particle accelerators out of laboratory glassware. This was only a few inches in diameter. Principles remain essentially the same even today, but the size of accelerators has increased tremendously. They no longer are parts of laboratories, but rather the detection and laboratory aspects of modern accelerators are appended around the accelerator. Modern accelerators cost many millions of dollars and a major installation requires numerous scientists and scores of support people, aided by a number of digital computers. In the future, costs may become so high that several countries, working together, will have to support particle-physics research facilities, as already exemplified by the European Organization for Nuclear Research (abbreviation CERN for the former name, *Conseil EuropÉen pour la Recherche NuclÉaire*) with facilities located in Geneva, Switzerland, and adjacent land in France.

Particle Generators. Direct voltage for charging particles may be obtained in two basic ways: (1) by means of a cascade process, such as the cascade rectifiers or voltage-multiplying circuits; and (2) by charging up a terminal through actual transportation of the charge.

The Cockcroft-Walton generator is of the first type and consists of several stages of a voltage-doubling circuit together with an ion source and a suitably designed discharge tube. Although it is possible to use electrons, these accelerators are usually used as positive-ion sources and can provide dc currents up to about 10 mA and energies up to about 1.5 MeV without special pressure tanks. With this type of accelerator, Cockcroft and Walton were able, in 1932, to induce the first nuclear reaction using artificially accelerated protons. The reaction was: $^1H + ^7Li \rightarrow ^8Be \rightarrow 2\,^4He$. See Fig. 1.

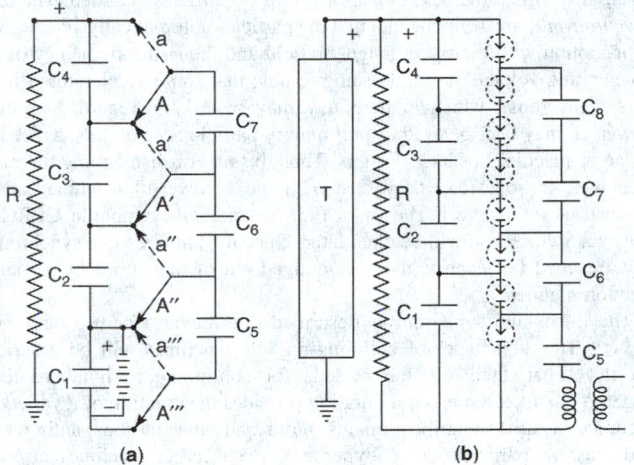

Fig. 1. Cockcroft and Walton scheme in which rectified A.C. voltage is used to charge a series of capacitors to a high D.C. potential, which is then applied to the acceleration of charged nuclear particles. The principle of operation can best be described by the apparatus using mechanical switches shown in A. Condensers of equal capacity are represented by C_1, C_2, C_3, etc. If the switch blades are down, the battery B is connected across condensers C_1 and C_5. On moving the blades up (dashed lines), condenser C_5 transfers half its charge to C_2. When the blades are moved down again, C_2 transfers half its charge to C_6 and C_5 is recharged to full capacitance. A continuous up and down motion of the blades results in a transfer of charge up the condenser bank until every condenser is charged to the voltage of the battery. The total voltage applied to the discharge tube, symbolized here by the resistance R, is the sum of the voltage across the condensers C_1, C_2, C_3 and C_4 in series. In the actual apparatus (part B), an alternating voltage was applied through a transformer and the switching action was accomplished by the use of rectifier tubes.

The second method is employed in the *electrostatic*, or *Van de Graaff*, generator where a row of corona points sprays charge onto a moving belt that carries the charge to the field-free region inside a spherical metal terminal. Currents of about 1 mA for electrons, and up to about 500 µA for positive ions, are obtained in the range 1 to 5 MeV with a precision of about 0.1%. The whole apparatus is enclosed in a pressure tank and operated at about 10 atmospheres. In this form, the maximum energy is close to 10 MeV and is limited by breakdown between the terminal and its surroundings. However, double the energy can be attained by accelerating negative ions to the positively charged terminal; then, through electron-stripping, positive ions are created which can be accelerated again as they pass from the terminal to ground. Such tandem Van de Graaff generators, in two- and three-stage variations, can provide particles in the 10 to 30 MeV range, with high precision. See Fig. 2.

An electric field may also be produced by a time-varying magnetic field. The changing magnetic flux in the central core of a pulsed cylindrical electromagnet induces a transverse electric field that accelerates the particles. These travel in a doughnut-shaped vacuum chamber located

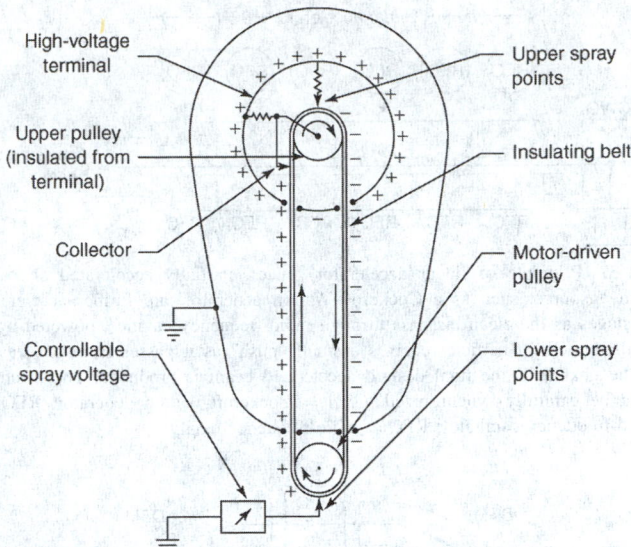

Fig. 2. Diagram of Van de Graff electrostatic belt generator.

between the poles of the magnet surrounding the core. The magnetic field between these poles keeps the particles traveling in a circle, but it must be carefully designed to keep the particle orbits within the vacuum chamber during each pulse. Although betatrons can accelerate positively charged particles, they have been used for electrons. The electrons can be extracted, but they usually bombard an inner target to produce beams of X-rays, which can be as intense as 1400 roentgens/minute at a distance of 1 meter from the target. Pulsing rates vary from 30 to 60 times per second.

Other types of accelerators use various forms of rf electric fields, at relatively low voltage, which are applied many times in a given direction to the particles and are prevented from influencing them when the rf field is reversed.

Linear Accelerators. The linear accelerator (*linac*) has the advantage that the accelerated beam is easily extracted for experimental use. In principle, it is capable of producing well-focused beams of higher intensity than are available from circular machines of the synchrotron type. It does, however, require very high power levels at frequencies where conversion equipment is relatively expensive. For a given final energy, a linear accelerator will usually be materially more expensive than a synchrotron.

The rf fields used for acceleration are set up in a long cylindrical cavity whose axis is to be the axis of the accelerated beam. Hence, for acceleration the field pattern must have a major electric field component parallel to the axis. This requirement is satisfied by the TM_{01} waveguide mode in which a paraxial electric field has its maximum strength at the axis and falls to zero at the cavity wall. Azimuthal magnetic fields lie in planes normal to the axis, have small values near the axis and increase to maximum values at the cavity walls. Usually the field pattern is maintained by coupling to these magnetic fields by loops or apertures excited by external power sources. Corresponding to the high rf magnetic field at the wall, paraxial currents flow in the walls and are responsible for a major function of the power loss in the system. When high electric fields are required on the axis to accelerate to high energy in reasonable distances, the wall currents are correspondingly high.

Both standing wave and traveling wave patterns can be used in linear accelerators. If traveling waves are used, the phase velocity of the waves must be made equal to the velocity of the particles accelerated; as the particle velocity increases, the phase velocity must also increase. But, phase velocities in simple waveguides always are greater than the velocity of light, and loading must be introduced to reduce the phase velocity to the desired value. This can be accomplished by the introduction at intervals of washer-shaped irises.

The operating principles of a linear accelerator are shown in Fig. 3.

Synchrotrons. A particle is made to follow a circular (or other closed curve) orbit by arranging a number of magnets in a ring. The principle is illustrated by Fig. 4. Two kinds of magnets are required. Dipole magnets (two poles) generate a uniform magnetic field. Spaced around the ring,

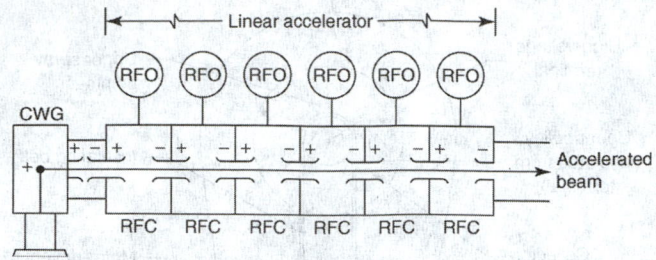

Fig. 3. Principle of linear accelerator (linac). Partially accelerated electrons from a source, such as a Cockcroft-Walton generator, are further accelerated by stages as the electrons pass through radio-frequency cavities, powered by rf oscillators. Each particle receives a small "push" as it passes from one cavity to the next until the final desired accelerated beam is produced. The machine must be carefully synchronized. CSG = Cockcroft-Walton generator; RFO = radio-frequency oscillator; RFC = radio-frequency cavity.

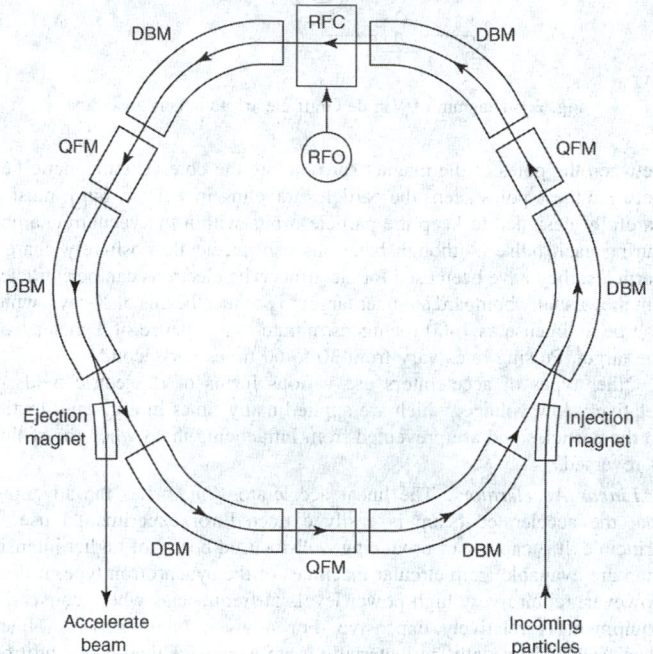

Fig. 4. Principle of synchrotron. One radio-frequency cavity (there may be several) provides a small "push" each time a particle passes through it. Unlike the linear accelerator, the synchrotron requires only one or few rf cavities. Dipole bending magnets keep the particles on their proper course. Focusing magnets keep the particles in a narrow beam, thus preventing undesired scattering. Particles enter the machine through an injection magnet and leave through an ejection magnet. DBM = dipole bending magnet; QFM = quadrupole focusing magnet; RFC = radio-frequency cavity; RFO = quadrupole focusing magnet; RFC = radio-frequency cavity; RFO = radio-frequency oscillator.

these magnets bend the particle trajectory. To keep a concentrated beam of particles, quadrupole magnets (two north poles and two south poles), which have no effect on deflecting the particles, are used to focus them. Acting like lenses, these magnets form the particles into a narrow beam. Depending upon the size and general configuration of a synchrotron, radio-frequency cavities may be variously interspersed among the magnets where the actual acceleration occurs. Special magnets are used for injecting particles into the ring and for extracting the accelerated beam of particles.

The first substantial synchrotron was built as early as 1952. Known as the *Cosmotron*, it was installed at Brookhaven National Laboratory on Long Island, New York. The device achieved energies up to 3 GeV. Two years later, the *Bevatron*, with energies up to 6.2 GeV, was installed at the University of California at Berkeley. A shortcoming of these earlier designs was the magnet system, which provided inadequate focusing of the beam. A system of strong focusing was introduced to later-generation synchrotrons. As pointed out by R.R. Wilson (1980). "The shape of the magnetic field can be described mathematically as being partly uniform (the

dipole component) and partly a gradient in a direction transverse to the orbit of the beam (the quadrupole component). The quadrupole component was made stronger and was alternated in sign, so the oscillations of the particles around the desired orbit were more frequent, but of smaller amplitude. As a result of this alternating gradient, the aperture of the magnets and the bore of the vacuum chamber could be smaller. It is the invention of the synchrotron and of strong focusing that has made the very large accelerators of today economically feasible." Synchrotrons with this design are known as *alternating gradient synchrotrons* (AGS).

The operation of a synchrotron is cyclic, i.e., a bunch of particles will be injected, with the bending magnets precisely adjusted to cause the particles to closely follow the curvature of the evacuated chamber. But, as the particles increase in energy, the field strength in the bending magnets must also be increased. Upon achieving their desired or maximum possible energy level, the particles are extracted—possibly going directly to bombard a target, or to supply an even more powerful synchrotron for further acceleration. When experiments with the first group or bunch of particles have been completed, the magnetic field is reduced to its original level, after which the next bunch or group of particles is added. The term *synchrotron* stems from the fact that the particles automatically synchronize their motion with the rising magnetic field and the rising frequency of the accelerating voltage. It is interesting to note that some accelerators (linacs or synchrotrons), which in their day may have been regarded as most powerful, may later be used as preliminary particle accelerators, as feeders to larger machines of later designs. Thus, instead of dismantling the older machines, in some cases the cost of a more powerful machine can be reduced. An example is shown in Fig. 5. Another example at CERN in Geneva, Switzerland will be cited later. Once the particles in a synchrotron have reached full energy, they are nudged out of their orbit by a special ejection magnet.

The linacs and synchrotrons described thus far are for use with *fixed targets*. The advantages of colliding-particle machines will be described shortly. What, then, are the reasons for continuing to bombard fixed targets? The fixed-target configuration provides the creation of a variety of secondary beams (neutrinos, muons, pions and other mesons, antiprotons, and massive particles called hyperons). Fixed-target machines also are

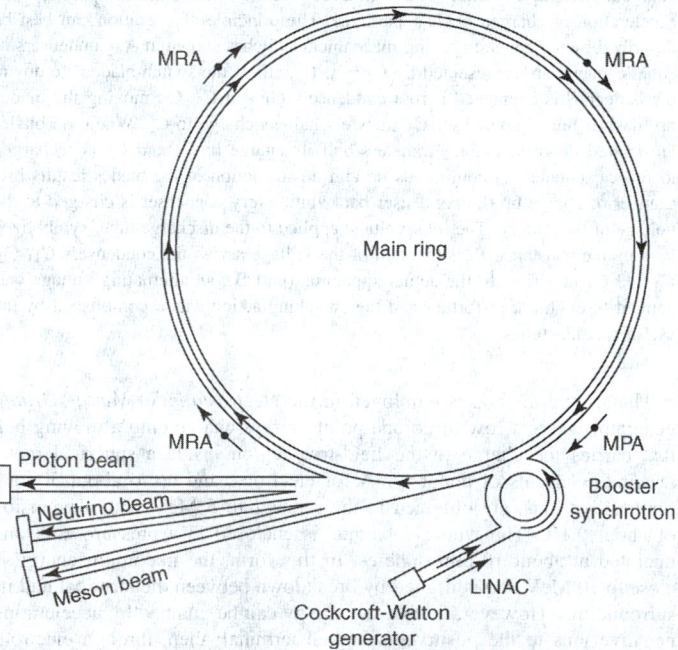

Fig. 5. An early fixed-target accelerator comprised of a large main-ring synchrotron with four stages of acceleration (MRA); a booster synchrotron; a linear accelerator (linac), and a Cockcroft-Walton generator. Protons are accelerated to 0.75 MeV in the Cockcroft-Walton generator; to 300 MeV in the linac; to 8 GeV in the booster synchrotron; and to 400–500 GeV in the main-ring synchrotron. Experiments are not limited to accelerated protons, but also can be conducted with beams of secondary particles (mesons and neutrinos) which are knocked out of the target by impacting protons.

effective in furnishing particles to larger accelerators and storage rings. When particles are beamed at a fixed target, an interaction is assured. In contrast, since in a colliding-particle configuration the great majority of particles will simply pass each other without colliding, the number of events of interest may only occur at the rate of comparatively few per day.

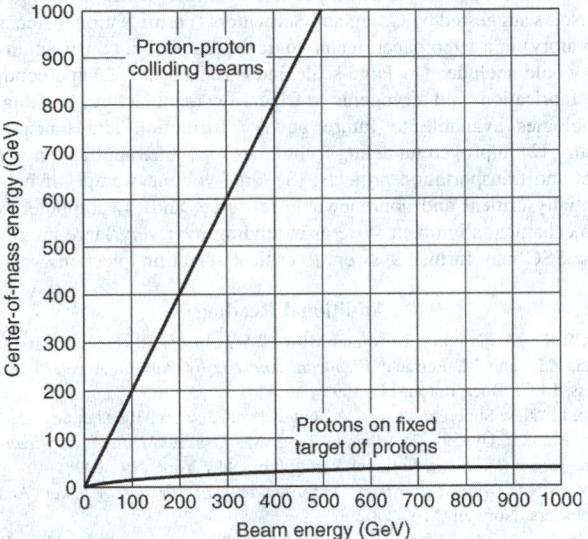

Fig. 6. Comparison of effective collision energy. Colliding-beam devices double the beam energy effectiveness. In fixed-target accelerators, the center-of-mass energy is proportional to the square root of the beam energy (at low energy) and at higher-energy levels, this rises even more slowly because of relativistic effects. (*After R.R. Wilson.*)

Colliding-Particle Machines. As mentioned earlier, the effective beam energy essentially can be doubled when particles can be made to collide head-on with each other. This is shown dramatically by the graph of R.R. Wilson (1980) and given in Fig. 6. Wilson had made an effective comparison of the resolving power of an accelerator with that of a microscope. In a microscope, the ultimate limit to resolution is the wavelength of radiation that illuminates the specimen. Thus, a visual lightwave microscope is limited to distinguishing objects of 10^{-5} centimeter or larger. Since a particle can be described (quantum mechanics) as a wave, the wavelength is inversely proportional to the momentum of the particle. Thus, one objective in accelerator design is improvement of resolution by reduction of particle wavelength. The early large accelerators had an effective resolution of about 10^{-16} centimeter (1/1000 the diameter of the proton).

When a particle and its antiparticle, such as an electron and a positron, or a proton and an antiproton, are used in head-on collision experiments, acceleration of the particles can be accomplished in one ring. This is because electrons and positrons, for example, behave in the same way in terms of their response to magnetic and electric fields. Thus, both particles can be injected into the same ring, one to follow an orbit in a clockwise direction; the other in a counterclockwise direction. Upon injection of a cluster of each type of particle, collisions occur at two points diametrically opposed. This arrangement provides maximum utilization of the equipment.

Other advantages of using a particle and its antiparticle, particularly the electron and the positron, is that when collisions occur, a less confusing splash of debris occurs.

When advantage of the single-ring, head-on collision approach cannot be taken, then *storage rings* are required. These rings are designed much like synchrotrons. Their primary purpose usually is not to accelerate the particles, but rather to maintain or "store" their energy while they continue to circulate in orbits in the ring. Just sufficient energy (usually furnished by a single radio-frequency cavity) is added to overcome losses, mainly due to synchrotron radiation. The storage rings are physically arranged tangentially to the main synchrotron ring so that precision transfer of particles from the storage rings to the main ring can be affected. Another

concept was introduced in the early 1970s at the European CERN facility, where two interlaced rings that store counter-rotating proton beams are used. These rings cross over at eight points around their circumference. There are seven interaction zones that will accommodate detectors.

Scientists at the National Laboratories of the National Committee for Nuclear Energy (C.N.E.N.) at Frascati, Italy pioneered the first electron-positron ring in 1959. Each beam of the machine was 0.25 GeV and yielded a center-of-mass energy of 0.5 GeV. This ring was later moved to the Orsay laboratory outside Paris, France. The number of such rings grew quite rapidly.

Synchrotron Radiation

The electromagnetic radiation emitted as a result of continual acceleration toward the axis of rotation of charged particles moving in a magnetic field is generally known as synchrotron radiation because it was first observed during the operation of electron synchrotrons. The rate of emission of synchrotron radiation varies with the fourth power of the particle energy, inversely with the radius of curvature, and inversely with the fourth power of the rest mass. Because of the rest mass relationship, no significant quantity of this radiation is formed from proton trajectories in magnetic fields, but an easily measurable loss of energy to synchrotron radiation occurs when electrons with high kinetic energies spiral through a magnetic field. Many microwave radiations observed in radio-astronomy measurements are believed to have been formed as synchrotron radiation.

At the beam energies that have been attained to date, synchrotron radiation is not a major factor in designing a proton accelerator, but it is the principal limitation on the energy of electron accelerators. Lessening the curvature of a synchrotron and thus increasing the circumference will reduce synchrotron radiation. For this reason, the radiation is practically negligible in a linear accelerator (linac). Synchrotron radiation energy loss in a storage ring must be made up by providing additional energy from one or more radio-frequency cavities. However, synchrotron radiation is not entirely wasted energy. The effect of the radiation is to damp out small excursions of the electrons away from their main trajectory, making a beam of electrons easier to control than a beam of protons.

Aside from its effects in high-energy experimentation, synchrotron radiation is of interest and value primarily as a source of tunable coherent x-rays. As summarized in a 1985 paper by Atwood, Halbach, and Kim (Lawrence Berkeley Laboratory), a modern 1- to 2-billion-eV synchrotron radiation facility (based on high-brightness electron beams and magnetic undulators) would generate coherent, laser-like, soft x-rays of wavelengths as short as 10 angstroms. The radiation would also be broadly tunable and subject to full polarization control. Radiation with these properties could be used for phase- and element-sensitive microprobing of biological assemblies and material interfaces as well as research on the production of electronic microstructures with features smaller than 1000 angstroms. These short wavelength capabilities, which extend to the K-absorption edges of carbon, nitrogen, and oxygen, are neither available nor projected for laboratory XUV lasers. Higher-energy storage rings (5 to 6 billion eV) would generate significantly less coherent radiation and would be further compromised by additional x-ray thermal loading of optical components.

Much interest and limited application of synchrotron radiation by medical professionals appeared in the late 1980s. Synchrotron radiation is intense and its brilliance extends continuously over a broad bandwidth from infrared to hard x-rays. For example at the Stanford Synchrotron Radiation Laboratory, the output in the x-ray region is more than 100,000 times that of the most powerful x-ray tube. The natural collimation of synchrotron radiation can be monochromatized by different crystals and by adjusting the angle at which the crystal intercepts the radiation, the energy of the monochromatic beam can be selected. The tunability and intensity of the monochromatic radiation are currently making transvenous angiographic applications possible. See also **X-Ray Scan and Other Medical Imagery**.

Third-Generation Synchrotrons. The first run of experiments at the new, $500-million European Synchrotron Radiation Laboratory (ESRF), located in the Dauphiné Alps, was scheduled to commence in the early 1990s. The facility will provide the brightest continuous x-rays in the world, but have been eclipsed by a new Japanese facility in the mid-1990s. The ESRF will generate x-rays two orders of magnitude more brilliant than any prior facility. Stronger x-ray sources are of an advantage to biologists, physicists, chemists, and other researchers because of their shortened exposure time and increased resolution. As pointed out by one American physicist who plans to use the facility, "We'll be able to get a picture in minutes, seconds, or fraction of a second and there will be much less specimen damage when the photons enter and leave a specimen in a millisecond."

A large variety of experiments was scheduled, representing numerous scientific disciplines. Another similar but slightly larger facility was scheduled for the Riso National Laboratory in Roskilde, Denmark, sometime in 1996.

The most powerful synchrotron now is at Harima Science Garden City, approximately 60 mi (96 km) west of Osaka, Japan. It will be known as the SPring-8 (Super Photon Ring-8 GeV). It is envisioned that it may be able to fulfill the x-ray crystallographer's dream—that is, x-ray holography. Hiromichi Kamitsubo, project head, states, "X-ray holography would enable the direct visualization, not just see complex x-ray diffraction patterns of the structure of materials—as if we were seeing them with a microscope." Three-dimensional holograms created in a photosensitive material require high-powered and coherent x-rays. The SPring-8 will have an energy rating of 8 GeV, with a main ring circumference of 4708 ft (1435 m) and 80 beam lines. Estimated cost is $1 billion.

Superconducting Super Collider (SSC)

As mentioned earlier, the ultimate fate of the SSC will be determined by economic and political factors. Generally, the scientific community in the United States and worldwide, in fact, looks to the SSC for important information on the structure of matter. Specific goals include:

1. *The origin of mass.* As explained by Quigg and Schwitters, the current model of the weak interactions, previously mentioned, suggests that a similar situation may be realized in the universe. The field involved is not a magnetic field, but rather what is called a Higgs field. A major objective of the SSC will be to clarify the exact nature of the Higgs field and its interactions with other matter. It is currently assumed that the Higgs field pervades the universe because the total density is minimized in its presence. To change the magnitude of the field significantly (or to manifest one of its particles), sufficient energy density must be supplied to overcome an assumed natural tendency of the field to return to its normal universal background value. The energy levels of the SSC is expected to be sufficient for this.
2. *Families of particles.* Quigg and Schwitters observe that one theoretical approach to understanding particle families postulates new symmetries under which different families are interchanged. A facet of this concept involves combining family symmetry and gage symmetries of the strong, weak, and electromagnetic interactions into one all-encompassing symmetry, but hidden by the presence of suitable background Higgs fields. Implementing such symmetries could lead to a new class of "family interactions" mediated by heavy bosons that may be created at the levels made possible by the SSC.
3. *Chirality* (a peculiar asymmetry in the weak interactions of the observed quarks and leptons) can be investigated.
4. *Supersymmetry* will be investigated, and may lead to unified field theories that include gravity.
5. *Compositeness* is a characteristic of some particles initially investigated at CERN. Better understanding of the cosmos through SSC experiments is also envisioned. The SSC will make it possible to simulate the conditions that prevailed about 10^{-15} second after the primordial Big Bang explosion when the temperature of the universe was estimated about 10^{17} K.

All of the foregoing factors are explained in considerable detail in the Quigg and Schwitters reference listed.

The new accelerator complex will be based on the accelerator principles and technology that were developed for construction of the Fermilab Tevatron, coupled with extensive experience with superconducting magnets gained over the past two decades. The energy of the SSC will be equal to 20 times that of the Tevatron collider, with a total collision energy of 40 TeV. Racetrack in shape with a circumference of 87 km (52 mi), the SSC will utilize 10,000 helium-cooled superconducting magnets. The SSC will utilize an injector system consisting of a linear accelerator followed by two circular accelerators. The diameter of the main ring used in the final acceleration phase will be about 30 km (18.7 mi),

depending upon details yet to be developed pertaining to the magnets that will be used to guide the protons. The superconducting magnet system will require many hundreds of miles of cryogenic plumbing, including several hundred thousand vacuum joints to assure the establishment and maintenance of superconductivity conditions. The linear dimensions of the SSC will be roughly 15 times those of the Tevatron and about 4 times those of the electron-positron collider being constructed by CERN. As suggested by Quigg and Schwitters (Fermi National Accelerator Laboratory) in a 1986 paper, technological fallout from construction of the SSC would include: (1) large-scale industrialization of superconducting wire fabrication and cryogenic refrigerator manufacture, making such technologies available to future power distribution and transportation systems; (2) improved tunneling techniques for future application to public works and transportation projects; (3) large-volume storage of helium, a potentially critical and nonrenewable resource; and (4) computer control and mechanical alignment systems extending over very large areas.

The SSC may furnish answers to critical scientific questions.

Additional Reading

Adair, R.K.: "A Flaw in a Universal Mirror," *Sci. Amer.*, 50 (February 1988).
Alfassi, Z.B. and M. Peisach: *Elemental Analysis by Particle Accelerators*, CRC Press, LLC., Boca Raton, FL, 1991.
Amato, I.: "New Superconductors: A Slow Dawn," *Science*, 306 (January 15, 1993).
Ando, M., C. Uyama, M. Ibaraki, and M. Osaka: *Medical Applications of Synchrotron Radiation*, Springer-Verlag Inc., New York, NY, 1998.
Atutov, S.N.: *Trapped Particles and Fundamental Physics*, Kluwer Academic Publishers, Norwell, MA, 2002.
Bertsch, G.F., L. Frankfurt, and M. Strikman: "Where Are the Nuclear Pions?" *Science*, 773 (February 5, 1993).
Bertschinger, E.: *Uniting Cosmology and Particle Physics*, Freeman, Salt Lake City, UT, 1992.
Branco, G.C., Q. Shafi, and J.I. Silva-Marcos: *Recent Developments in Particle Physics and Cosmology*, Kluwer Academic Publishers, Norwell, MA, 2001.
Breuker, H., et al.: "Tracking and Imaging Elementary Particles," *Sci. Amer.*, 58 (August 1991).
Brown, F.R. and N.H. Christ: "Parallel Supercomputers for Lattice Gauge Theory," *Science*, 1393 (March 18, 1988).
Brown, L.M., M. Dresden, and L. Hoddeson: *From Pions to Quarks: Particle Physics in the 1950s*, Cambridge University Press, New York, NY, 1989.
Chanowitz, M.S.: "The Z Boson," *Science*, 36 (July 6, 1990).
Chupp, E.L.: "Transient Particle Acceleration Associated with Solar Flares," *Science*, 229 (October 12, 1990).
Cline, D.B.: "Beyond Truth and Beauty: A Fourth Family of Particles," *Sci. Amer.*, 50 (August 1988).
Cline, D.B., C. Rubbia, and S. van der Meer: *The Search for Intermediate Vector Bosons*. (March 1982). A classic reference in The Laureates' Anthology, 133, Scientific American, Inc., New York, NY, 1990.
Conte, M. and W.M. MacKay: *An Introduction to the Physics of Particle Accelerators*, World Scientific Publishing Company, Inc., River Edge, NJ, 1991.
Dawson, J.M.: "Plasma Particle Accelerators," *Sci. Amer.*, 54 (March 1989).
Dehmelt, H.: "Experiments on the Structure of an Individual Elementary Particle," *Science*, 539 (February 2, 1990).
Donoghue, J.F., E. Golowich, and B.R. Holstein: *Dynamics of the Standard Model*, Cambridge University Press, New York, NY, 1994.
Dunning, F.B. and R.G. Hulet: *Atomic, Molecular, and Optical Physics: Charged Particles*, Vol. 29, Academic Press, Inc., San Diego, CA, 1995.
Ericson, T. and W. Weise: *Pions and Nuclei*, Oxford University Press, Inc., New York, NY, 1988.
Ezhela, V.V., B. Armstrong, and J.D. Jackson: *Particle Physics: One Hundred Years of Discoveries: An Annotated Chronological Bibliography ANNOTATED*, Springer-Verlag Inc., New York, NY, 1996.
Feldman, G.J. and J. Steinberger: "The Number of Families of Matter," *Sci. Amer.*, 70 (February 1991).
Flam, F.: "CERN's New Detectors Take Shape," *Science*, 180 (April 10, 1992).
Flam, F.: "Neural Nets: "A New Way to Catch Elusive Particles?,"" *Science*, 1282 (May 29, 1992).
Gottfried, K. and V.F. Weisskopf: *Concepts of Particle Physics*, Oxford University Press, Inc., New York, NY, 1997.
Graham, D.: "Testing Physicists' GUTS (Grand Unified Theory)," *Technology Review (MIT)*, 10 (May–June 1988).
Gutbrod, H. and H. Stocker: "The Nuclear Equation of State," *Sci. Amer.*, 58 (November 1991).
Helliwell, J.R.: *Macromolecular Crystallography with Synchrotron Radiation*, Cambridge University Press, New York, NY, 1992.

Hermann, A., et al.: *History of CERN*, North-Holland Elsevier Science, New York, NY, 1990.

Hoddeson, L., M. Riordan, M. Dresden, and L.M. Brown: *Rise of the Standard Model: Particles Physics in the 1960s and 1970s*, Cambridge University Press, New York, NY, 1997.

Lederman, L.M.: "Observations in Particle Physics from Two Neutrinos to the Standard Model," *Science*, 664 (May 12, 1980).

Lederman, L.M.: "The Tevatron," *Sci. Amer.*, 48 (March 1991).

Myers, S. and E. Picasso: "The LEP Collider," *Sci. Amer.*, 54 (July 1990).

Leader, E.: *Spin in Particle Physics*, Cambridge University Press, New York, NY, 2001.

Martin, B.R. and G. Shaw: *Particle Physics*, 2nd Edition, John Wiley & Sons, Inc., New York, NY, 1997.

Month, M. and M. Dienes: *The Physics of Particle Accelerators*, Vol. 2, American Institute of Physics, College Park, MD, 1997.

Olive, K.A.: "The Quark-Hadron Transition in Cosmology and Astrophysics," *Science*, 1194 (March 8, 1991).

Peterson, I.: "Quantum Interference," *Science News*, 363 (December 2, 1989).

Peterson, I.: "Protons and Antiprotons Held in the Balance," *Science News*, 38 (July 21, 1990).

Peterson, R.J. and D.D. Strottman: *Pion-Nucleus Physics*, American Institute of Physics, College Park, MD, 1997.

Peterson, I.: "Beyond the Z," *Science News*, 204 (September 29, 1990).

Polchinski, J.G.: *String Theory: Superstring Theory and Beyond*, Vol. 2, Cambridge University Press, New York, NY, 1998.

Pool, R.: "The Hunting of the Quark—Computer Style," *Science*, 46 (April 3, 1992).

Quigg, C.: *Gauge Theories of the Strong, Weak and Electromagnetic Interactions*, Perseus Publishing, Boulder, CO, 1997.

Rees, J.R.: "The Stanford Linear Collider," *Sci. Amer.*, 58 (October 1989).

Rice, T.M.: "Can Europe Keep up the Pace in Condensed Matter Physics?," *Science*, 482 (April 24, 1992).

Riordan, M.: "The Discovery of Quarks," *Science*, 1287 (May 29, 1992).

Rothman, T.: "Ambidextrous Universe: New Particles Blur Distinction Between Fermions and Bosons," *Sci. Amer.*, 26 (May 1989).

Rubbia, V.: "The European Strategy in Particle Physics," *Science*, 484 (April 24, 1992).

Ruthen, R.: "Quark Quest," *Sci. Amer.*, 32 (March 1993).

Ruthen, R.: "Attractive and Demure," *Sci. Amer.*, 30 (May 1993).

Sarkar, S.: *Big Bang Laboratory for Particle Physics*, Cambridge University Press, New York, NY, 2002.

Schmidt, V.: *Electron Spectrometry of Atoms Using Synchrotron Radiation*, Cambridge University Press, New York, NY, 1997.

Schramm, D.N. and G. Steigman: "Particle Accelerators Test Cosmological Theory," *Sci. Amer.*, 66 (June 1988).

Selvin, P.: "How Do Particles Put on Weight?" *Science*, 173 (January 8, 1993).

Shifman, M., M.A. Shifman, and B.L. Ioffe: *At the Frontier of Particle Physics: Handbook of QCD*, World Scientific Publishing Company, Inc., River Edge, NJ, 2001.

Staff: *Particle Physics Phenomenology*, World Scientific Publishing Company, Inc., River Edge, NJ, 1997.

Sundaresan, M.K.: *Handbook of Particle Physics*, CRC Press, LLC., Boca Raton, FL, 2001.

Taubes, G.: "Are Neutrino Mass Hunters Pursuing a Chimera?" *Science*, 731 (May 8, 1992).

Waldrop, M.M.: "SLAC Feels the Thrill of the Chase," *Science*, 771 (May 10, 1989).

Weinberg, S.: *Discovery of Subatomic Particles*, 2nd Edition, Cambridge University Press, New York, NY, 2003.

Wilcek, F.: "Anyons," *Sci. Amer.*, 58 (May 1991).

Willeke, K.: *Physics of Particle Accelerators: An Introduction*, Oxford University Press, Inc., New York, NY, 2000.

Wilson, E.J.N.: *An Introduction to Particle Accelerators*, Oxford University Press, Inc., New York, NY, 2001.

Yam, P.: "Spin Cycle: Rotating Nucleii Share A Few Moments of Inertia," *Sci. Amer.*, 26 (October 1991).

Yan, Y.T., J.P. Naples, and M.J. Syphers: *Accelerator Physics at the Superconducting Super Collider*, Springer-Verlag Inc., New York, NY, 1995.

Zotter, B.W. and S. Kheifets: *Impedances and Wakes in High Energy Particle Accelerators*, World Scientific Publishing Company, Inc., River Edge, NJ, 1998.

Pre–1988 References

Adair, R.: *The Great Design: Particles, Fields, and Creation*, Oxford University Press, Inc., New York, NY, 1987.

Atwood, D., K. Halbach, and Kwange-Je Kim: "Tunable Coherent X-rays," *Science*, **228**, 1265–1272 (1985).

Brambilla, N., and G. Prosperi: *Quark Confinement and the Hadron Spectrum: Proceedings of the 5th International Conference*, World Scientific Publishing Company, Inc., Riveredge, NJ, 2003.

Barnett, R.M., H.E. Haber, and G.L. Kane: "Supersymmetry—Lost or Found?" *Nuclear Physics*, **B267**(3, 4) 625–678 (April 21, 1986).

Bengtsson, T., et al.: "Nuclear Shapes and Shape Transitions," *Physica Scripta*, **29**(5) 402–430 (May 1984).

Black, J.K., et al.: "Measurement of the CP-Nonconservation Parameter e 1/e," *Physical Review Letters*, **54**(15) 1628–1630 (April 15, 1985).

Broglia, R.A., C.H. Casso: *Frontiers in Nuclear Dynamics*, Plenum, New York, NY, 1985.

Court, G.R., et al.: "Energy Dependence of Spin Effects," *Physical Review Letters*, **57**(5), 507–510 (August 4, 1986).

Crosbie, E.A., et al.: "Energy Dependence of Spin-Spin Effects in p-p Elastic Scattering at 90°," *Physical Review*, **23**(3) 600–603 (February 1, 1981).

de Rujula, A.: "Superstrings and Supersymmetry," *Nature*, **320**(6064), 678 (April 24, 1986).

Eichten, E., et al.: "Supercollider Physics," *Reviews of Modern Physics*, **56**(4), 579–707 (October 1984).

Ellis, J.: "Hope Grows for Supersymmetry," *Nature*, **313**(6004), 626–627 (February 21, 1985).

Glashow, S.L.: "Toward a Unified Theory: Threads in a Tapestry," in *Nobel Lectures*, Elsevier Science, Amsterdam and New York, NY, 1981.

Green, M.B.: "Unification of Forces and Particles in Superstring Theories," *Nature*, **314**(6010), 409–414 (April 4, 1985).

Green, M.B.: "Superstrings," *Sci. Amer.*, 48–60 (September 1986).

Haber, H.E. and G.L. Kane: "The Search for Supersymmetry: Probing Physics Beyond the Standard Model," *Physics Reports*, **117**(2, 3), 75–263 (January 1985).

Haber, H.E. and G.L. Kane: "Is Nature Supersymmetric?" *Sci. Amer.*, 52–60 (June 1986).

Hamilton, J.H., P.G. Hansen, and E.F. Zganjar, *Reports on Progress in Physics*, **48**(5) 631–708 (May 1985).

Hamilton, J.H.: "Magic Numbers, Reinforcing Shell Gaps and Competing Shapes in Nucleii," *Progress in Particle and Nuclear Physics*, **15**, 107–134 (1985).

Krisch, A.D.: "Collisions between Spinning Protons," *Sci. Amer.*, 42–50 (August 1987).

Letessier, J., and J. Rafelski: *Hadrons and Quark Gluon Plasma*, Cambridge University Press, New York, NY, 2002.

Lipkin, H.J.: "Colour Theory in a Spin," *Nature*, **324**(6092), 14–16 (November 6, 1986).

Martin, J.A., W. Greiner: "Potential Energy Surface Model of Collective States," in *High-Angular Momentum Property of Nuclei* (N.R. Johnson, Ed.) Harwood Academic Publishers, New York, NJ, 1983.

Mulvey, J.H.: *The Nature of Matter*, Oxford University Press Inc., New York, NY, 1981.

Nadis, N.: "Anti-Proton Fishing," *Technology Review (MIT)*, 15 (July 1987).

News: "Antiprotons Captured at CERN," *Science*, **233**, 1383–1384 (1986).

News: "Bright Synchrotron Sources Evolve," *Science*, **235**, 841–842 (1987).

News: "CERN Panel Backs New Accelerator," *Science*, **235**, 1567 (1987).

News: "Soviets Plan Huge Linear Collider," *Science*, **238**, 16–17 (1987).

Quigg, C.: "Elementary Particles and Forces," *Sci. Amer.*, 84–95 (April 1985).

Quigg, C. and R.F. Schwitters: "Elementary Particle Physics and the Superconducting Super Collider," *Science*, **231**, 1522–1527 (1986).

Richter, B.: "From the Psi to Charm: The Experiments of 1975 and 1976," *Science*, **196**, 1286–1297 (1977).

Sachs, R.G.: *The Physics of Time Reversal*, University of Chicago Press, Chicago, IL, 1987.

Scherk, J.: "An Introduction to the Theory of Dual Models and Strings," *Reviews of Modern Physics*, **47**(1), 123–164 (January 1975).

Schwartz, J.H., E. Witten: *Superstring Theory*, Cambridge University Press, New York, NY, 1987.

Schwarzschild, B.M.: "Polarized Scattering Data Challenge Quantum Chromodynamics," *Physics Today*, **38**(8), 17–20 (August 1985).

Smith, T.P.: *Hidden Worlds: Hunting for Quarks in Ordinary Matter*, Princeton University Press, Princeton, NJ, 2003.

Sutton, C.: *The Particle Connection*, Simon and Schuster, New York, NY, 1984, van der Meer, S.: "Stochastic Cooling and the Accumulation of Antiprotons," *Science*, **230**, 900–906 (1985).

Waldrop, M.M.: "String as a Theory of Everything," *Science*, **229**, 1251–1253 (1985).

Weinberg, S.: *The Discovery of Subatomic Particles*, W.H. Freeman, New York, NY, 1983.

Wilson, R.R.: "The Next Generation of Particle Accelerators," *Sci. Amer.*, 42–57 (January 1980).

Yamazaki, T.: "Evolution of Meson Science in Japan," *Science*, **233**, 334–338 (1986).

Zichichi, A.: *From Quarks and Gluons to Quantum Gravity: Proceedings of the International School of Subnuclear Physics*, World Scientific Publishing Company, Inc., Riveredge, NJ, 2003.

Zweig, G.: "Quark Catalysis of Exothemal Nuclear Reactions," *Science*, **201**, 973–979 (1978).

PARTICLE VELOCITY. In ocean wave studies, the instantaneous velocity of a water particle undergoing orbital motion. It has the scalar value

$$\frac{\pi}{T} H e^{-2\pi z/L},$$

where T is the wave period, H the wave height, z the depth below still-water level, and L the wave length. At the crest, its direction is the same as the direction of progress of the wave; at the trough it is in the opposite direction.

See also **Orbital Velocity**.

PARTICULATE MATTER. Solid or liquid matter that is dispersed in a gas, or insoluble solid matter dispersed in a liquid, that gives a heterogeneous mixture.

PARTICULATES (PRECIPITATION). The term for solid or liquid particles found in the air. Some particles are large or dark enough to be seen as soot or smoke. Others are so small they can be detected only with an electron microscope. Because particles originate from a variety of mobile and stationary sources, their chemical and physical compositions vary widely. Particulate matter can be directly emitted or can be formed in the atmosphere when gaseous pollutants such as sulfur dioxide (SO_2) and NO_x react to form fine particles. See also **Aerosol**; and **Electrostatic Precipitator**.

PARTITION. A term in mathematics used in a number of ways: 1. A partition of a set S is its division into a number of subsets called *cells*, which must be *exhaustive* (that is, every element of S must belong to one of the subsets) and which must also be *disjoint* (no member of S can belong to more than one of the subsets). For some types of sets, the last requirement is better stated in the form that the intersection of any two sets is zero.

2. The partition of a positive integer is its expression as a sum of positive integers, the number of partitions of a given positive integer being the number of ways in which it can be so expressed. Special types of such partitions result when restrictions are imposed upon the process, such as limiting the number of integers in the sum or requiring that they be different.

3. The partition of a permutation may be described as follows. Consider a permutation

$$\begin{pmatrix} a_1, a_2, \ldots, a_n \\ b_1, b_2, \ldots, b_n \end{pmatrix}$$

of n objects, where the notation indicates that a_i is replaced by b_i. This permutation can also be expressed by cycles $(abc \ldots d)(ef) \ldots$, etc., indicating that a is to be replaced by b, b by $c, \ldots, d$ by a, e by f, f by e, and so forth. Suppose now that there are α cycles of degree (i.e., length) 1, β cycles of degree 2, etc. Then we can conveniently denote this property of the original permutation by the symbol $(1^\alpha 2^\beta 3^\gamma \ldots)$, which is called a partition of the permutation. This concept is useful in group theory, since every finite group can be represented as a group of mutations.

See also **Permutation**; and **Permutation Group**.

PARTITION FUNCTION. An expression giving the distribution of molecules in different energy states in a system

$$Z = \sum q_r e^{-\varepsilon_r/kT}$$

where Z is the partition function, q_r the statistical weight of the rth state of energy ε_r, k is the Boltzmann constant, T, the absolute temperature, and the summation is taken over all the energy states of the system. The energy levels ε_r may be those attributed to rotation, translation, vibration, or electronic energies, etc.

PARTRIDGE (*Aves, Galliformes*). Game birds of numerous species, related to the pheasants and turkeys. The francolins of Asia and Africa are included here. True partridges are similar to the quails. The latter, although more generally known in North America by the name quail, are members of the group. They are found over Europe, Asia, Africa, and North America. Aside from the common quail or bob-white of North America, the names quail and partridge are both applied to the several western species, and in the southern states even the bob-white becomes the partridge. To confuse the term still further the ruffed grouse is often called a partridge in the northern states.

The francolin is a bird of Africa and the Oriental region, related to the partridges. The spur-fowl is a long-tailed Indian and Ceylonese partridge, which resembles pheasants.

See also **Galliformes**; and **Quail**.

PARVOVIRAL ENTERITIS. This is a serious disease of dogs, particularly of younger animals. Morbidity is high and mortality approaches 50%. At present the incubation period is unknown. Affected animals stop eating, may vomit, and become depressed and weak. The disease appears to be most common in kennels. Parvoviruses were first isolated from feces of asymptomatic dogs in 1970. The first report of parvoviruses being related to diarrhea in puppies was in 1977. Other species in which the virus has been associated with enteric disease include cats, rabbits, rodents, and calves. The disease features hemorrhagic enteritis that may involve most of the small intestine and, in some cases, the colon. Major microscopic changes in the intestinal tract have been confined primarily to the small intestine. Considerable research is underway to produce a more effective vaccine against this disease.

PASCAL. The SI derived unit of pressure. One pascal (Pa) is equal to 1 newton m^{-2}. The kilopascal (kPa) is the preferred unit for atmospheric pressure, but the more familiar millibar (mb) is the unit of pressure generally used by meteorologists, by international agreement; 1 mb = 1 hPa (hectopascal). For a typical sea level pressure, 102.345 kPa = 1023.45 hPa = 1023.45 mb.

See also **Units and Standards**.

PASCAL, BLAISE (1623–1662). Pascal was a French mathematician, physicist, and philosopher. He was home-schooled by his father who had unorthodox views on education and told him he was not going to study mathematics until he was fifteen. But Pascal was a prodigy and by the time he was twelve he had already worked on geometry by himself and discovered that the sum of the angles, of a triangle are two right angles. By the age of sixteen, he published a paper on conic sections, which was a groundbreaking theorem. At age nineteen, he invented the first digital calculator, which added and subtracted through use of a series of cogged wheels, in order to help his father with his work on collecting taxes.

Pascal is best known for Pascal's law. This principle states that fluid in vessels transmits pressure equally in all directions. Pascal also performed a series of experiments on atmospheric pressure and proved, air has weight and that air pressure can create a vacuum.

Pascal is also remembered for his theory of probability and Pascal's triangle can be used to calculate probabilities. Pascal is credited with the invention of both the syringe and the hydraulic press.

Pascal's scientific investigations led to valuable contributions for man but he is also remembered for his religious and philosophical writings. His Pensees contains "Pascal's wager" which claims belief in God is rational because, "If God does not exist, one will lose nothing by believing in him, while if he does exist, one will lose everything by not believing."

See also **Digital Computer Systems**; **Pascal Triangle**; and **Pressure**.

J. M. I.

PASCAL'S LAW. See **Pressure**.

PASCAL TRIANGLE. If the coefficients of $(x + y)^k$ in the binomial series are arranged as shown, successive coefficients can be obtained as a sum of two numbers in the preceding line. The second figure in each line is the value of k.

```
         1
        1   1
       1   2   1
      1   3   3   1
     1   4   6   4   1
    1   5  10  10   5   1
   1   6  15  20  15   6   1
```

```
                1                       1
              1     2                 1
            1     3       6     3       1
          1     4       6     4       1
        1     5       10     20     10     5       1
      1     6       15       35     15     6       1
    1     7       21       56       70     56   28   8   1
  1     8       28       84     126     126   84   36   9   1
1    10    45    120    210    252    210    120    45   10   1
```

Other forms of the triangle are often shown, especially that in the shape of an isosceles triangle, with unity at the apex and unities along the sides, as shown below.

The triangle can easily be extended by simple additions; hence the coefficient in a binomial expansion can be determined to any order with a minimum effort.

Pascal (1623–1662) also showed that the triangle could be used to find the number of combinations when selecting k objects from n objects, since this also equals the binomial coefficient,

$$\binom{n}{k}$$

See also **Binomial Series**; and **Probability**.

PASCHEN-BACK EFFECT. In a strong magnetic field, the anomalous Zeeman effect changes into a pattern similar to the normal effect, and this is known as the Paschen-Back effect. Each energy level with a given value of L, the electronic orbital angular momentum splits into $(2L + 1)$ components characterized by the magnetic quantum numbers $M_L = L, L - 1, \ldots, -L$, and each level with a given value of M_L splits into $(2S + 1)$ components with quantum numbers $M_s = S, S - 1, \ldots, -S$, where S is the resultant electron spin. The selection rules are $\Delta M_L = 0, \pm 1$, $\Delta M_s = 0$. Lines with $\Delta M_L = 0$ are plane polarized with electric vector parallel to the direction of the applied magnetic field; those with $\Delta M_L = \pm 1$ are plane polarized with components perpendicular to the field. See also **Atomic Spectra**; and **Hyperfine Structure**.

PASCHEN (Law of). The spark potential between electrodes in a gas depends on the length of the spark gap and the pressure of the gas in such a way that it is directly proportional to the mass of gas between the two electrodes, i.e., the sparking potential is a function of the pressure times the density of the gas. See also **Discharge (Gaseous)**.

PASCHEN SERIES. See **Energy Level**.

PASSERIFORMES *(Aves)*. An exceptionally large order of birds comprising more than a fifth of all living bird species. The length ranges from 7.5 to 110 centimeters (3 to 43 inches), and the weight is from 4.8 to 1350 grams (0.1 to 48 ounces). These birds have 4 toes (the babbling thrush is the only member of this order with a greatly regressed first toe). The toes all originate from the same level on the tarsus, and they are generally free to the base. There is always 1 toe (generally the largest) directed to the rear; this toe cannot be rotated forward. The claw of the rear toe is, with few exceptions, larger than that of the middle anterior toe. Passeriformes have a bony palate with a design only rarely seen in other orders. There is always a distinct sternal keel, but only traces of an appendix. The young hatch with their eyes closed; the inside of the mouth is brightly colored, often with dark spots inside the mouth as well as other juvenile developments which disappear later. The birds have a worldwide distribution with the exception of a very few remote oceanic islands and areas near the poles.

There are 4 suborders: 1. The Broadbills (*Desmodactylae*), in which the flexor tendons of the third toe (the middle front toe) and those of the rear toe are joined together. The front toes are fused at their bases. 2. The Noisemakers (*Clamatores*), in which the flexor tendons of the toes are separate; the lower syrinx has one or two tensor muscle pairs inserted on the half-rings of the trachea either in the middle, through the entirety, or only at its end. 3. The Lyre Birds (*Suboscines*) also have separate flexor muscles in the toes; the lower syrinx has two or three pairs of tensor muscles inserted at both ends of the tracheal half-rings. 4. The Songbirds (*Oscines*), in which the flexor tendons of the toes are separate; the lower syrinx has four to nine pairs of tensor muscles inserted at both ends on the tracheal half-rings.

Passeriformes are land birds even though some of them may get food from water not far from shore. These birds evolved from ground-dwelling forms of tree-dwelling birds, as all have the typical perching foot. Their four toes are suitable for grasping branches, stalks, wires, etc. The grasp remains firm even when the bird is asleep, because the flexor tendon and its sheath rest inside one another, and each must be freed before the toes can extend and the bird can fly, hop off, or fall down. Only a few members of this order never fly at all.

Separate entries are included in this volume on the following passeriformes:

Bird of Paradise	**Finch**	**Raven**
Blackbird	**Gnatcatcher**	**Redstart**
Bluebird	**Grackle**	**Robin**
Bluethroat	**Jay**	**Shrike**
Bobolink	**Junco**	**Sparrow**
Bowerbird	**Kingbird**	**Starling**
Broadbills	**Lark**	**Swallow**
Bulbul	**Lyrebird**	**Tanager**
Bullfinch	**Magpie**	**Thrasher**
Bunting	**Manakin**	**Thrush**
Canary	**Martin**	**Tit**
Cardinal	**Meadowlark**	**Warbler**
Chatterer	**Myna**	**Waxwing**
Chickadee	**Nightingale**	**Weaverbird**
Cowbird	**Nuthatch**	**Wren**
Creeper	**Oriole**	
Crow	**Ouzel**	

Other species of interesting passeriformes are described in alphabetical order as follows:

Babbler — found in the Ethiopian and Indian regions, particularly those of the family *Crateropotidae*.

Beccafico — this is an Italian name translated "fig-eater" or "fig-pecker," said to apply to the European garden warbler, *Sylvia hortensis*. The English call this bird the "pretty chap." This small bird is a favorite among gourmets in Venice and elsewhere on the Tables of Italy, France, and Greece. The term Beccafico is also used for other warblers when used for food in these countries. An annual feast of the beccafico is called the Beccaficata.

Bee-Martin — the common kingbird of North America, *Tyrannus*. The name appears to be undeserved because the bird eats very few, if any, bees.

Bishop-Bird — any bird of several brightly colored species of African weaverbirds which make up the genus *Pyromeland*. The name also has been applied to some of the brightly colored birds of North America, especially by the early settlers of Louisiana.

Broadbill — an Oriental bird with a shallow, but very broad beak.

Brambling — a finch of the Old World, which nests in northern Europe and Asia. *Fringilla montefringilla*.

Calandra — a European lark noted for its song. The name is sometimes applied to other related species of the Old World.

Cassique — a South American bird related to Old World starlings. Several species.

Catbird — a common North American bird, *Damatella carolinensis*, related to the mockingbird and the thrashers. Although quietly colored in slate gray, it is a welcome resident because of its singing. The term catbird is also used to describe an Australian bowerbird.

Chaffinch — name for birds of several species found in Europe and western Asia.

Chat — name applied to several species of birds, usually designated by a compounded word, as the stone-chat and whinchat of Europe, the yellow-breasted chat of North America (*Icteria virens*), and several North African species. The European wheatear and hedge warbler are also called chats.

Chough — a Eurasian bird related to the crows and resembling them in form and color. A few other Asian birds are known as chough-thrushes.

Cock of the Rock — name for birds of several species found in tropical South America. The males are crested and brilliantly colored. See also **Chatterer**.

Cotinga — name for a group of Brazilian chatterers closely related to the bell birds.

Crossbill — a small seed-eating bird of the Northern Hemisphere whose mandibles cross at the tip when the mouth is closed. This adaptation enables the bird to open seeds and fruits, such as cones, very readily.

Dickcissel — small American bird, *Spiza americana*, related to the buntings. It is found in open country and is distinguished by its yellow breast and black throat patch.

Dipper — the water ouzel. See also **Ouzel (or Ousel)**.

Drongo — the king crow of southern Asia and Africa. Several species, mostly black, forming a family not closely related to the true crows.

Dunnock — the European hedge sparrow, *Prunella modularis*.

Fieldfare — a common thrush, *Turdus pilaris*, of northern Europe. See also **Thrush**.

Fire Eye — a common species of ant bird found in Brazil.

Flower Pecker — a brightly colored bird of the Oriental and Australian regions, related to the sun birds and having remarkable nests.

Forktail — a bird found in India and related to the European chats.

Grassquit — Jamaican name for small birds more commonly called buntings.

Hangnest — a group of birds whose nests are woven of vegetable fiber, grass, and hair and are suspended from small branches. The Baltimore oriole is a common North American species. Others occur from the southwestern states to Brazil.

Honey Creeper — species of small birds of tropical South America and the West Indies related to the warblers. These birds visit flowers in a fashion similar to hummingbirds but are incapable of hovering flight. One species is called the banana-quit.

Honey Eater — species of birds of the Australian region. They have long tongues with which they secure nectar from flowers. The group includes the parson bird, stitch bird, and several species called white eyes.

Honey Pecker — species of small brilliantly colored birds of the Oriental and Australian regions, related to the sun birds. One Australian species is called the diamond bird.

Huia — a New Zealand bird related to the starlings. The male has a short, straight beak, whereas that of the female is long and curved.

Manucode — a name applied to a few smaller birds of paradise, found in several islands of the Australian region. Derived from the generic name *Manucodiata*, which is a corruption of a Malay name.

Mavis — the European song thrush, *Turdus philomelus*, a bird that resembles the wood thrush of North America.

Munia — any bird of numerous species of weaver finches constituting the genus *Munia*. They are native to Africa and the Oriental region. The most common species is the rice bird, paddy bird, or Java sparrow, which is regarded as a pest in the rice fields. It has been valued as a cage bird in Europe.

Nutcracker — species of birds related to the crows and jays. The relatively few species are confined to the northern parts of the Northern Hemisphere. The Clarke nutcracker, *Nucifraga columbiana*, lives principally at higher altitudes in the mountains of western North America. The bird is associated with coniferous forests.

Ovenbird — the European willow wren, *Phylloscopus trochilus*, and other birds that build domed nests. Also, a North American warbler, *Seiurus aurocapillus*, sometimes called the golden-crowned thrush.

Also, South American birds of several species, which build mud nests resembling old-fashioned ovens. Genus *Furnarius*.

Ox-Pecker — species of African birds of a group related to the starlings. The common name refers to their habit of climbing about the bodies of domestic cattle in search of ticks and other external parasites. They also visit wild animals for the same purpose.

Piping Crow — Australian birds of several species related to the crows and jays, but not true crows. They are black and white, whence comes their other name, Australian magpie. Unlike the true crows, these birds are quite musical and can be taught to whistle tunes and to speak. They are frequently maintained as cage birds.

Pipit — small quietly colored birds of numerous species related to the wagtails and warblers. They are widely distributed, but most species occur in the Old World. In North America, the common pipit, *Anthus spinoletta*, nests in the north and at high altitudes in the western mountains; the other known species, Sprague's pipit, *A. spraguei*, is a bird of the plains.

Pitta — species of small, brightly colored birds of the Old World, also called the ant thrushes, water thrushes, and ground thrushes. They are only superficially like the true thrushes.

Plant Cutter — several species found only in the temperate zones of South America. They have short, thick beaks with finely serrate edges. They are related to the chatterers.

Redpoll — a small bird related to the finches. It is named for its red crown. One species nests in the northern parts of the Northern Hemisphere and is known in Europe as the mealy redpoll. Europe has another species, the lesser redpoll, and Asia and North America have the related hoary redpoll, *Acanthis hornemanni*, which only occasionally enters the northern United States.

Rice Bird — a term used for the American bobolink. Also for the Java sparrow.

Rifle Bird — a bird of paradise found in Australia and New Guinea. The several species make up the genus *Ptilorhis*.

Rook — a European bird, *Corvus frugilegus*, related to the crows. Its black plumage is glossed with purple, and the face of the adult is usually naked and of a gray color.

Shama — a jungle bird of the Oriental region. The several species are found in the Malayan area, in the Philippines, in India, and on various Pacific islands. They are shy birds. The Indian species is maintained as a cage bird for its beautiful song.

Stonechat — a bird of central and northern Europe, *Saxicola torquata*. The male has a black head and back, a white collar, and reddish underparts. The female is of a brown coloration.

Towhee — a North American bird related to the finches and sparrows. The common eastern species, *Pipilo erythrophthalmus*, is a black and white bird with red-brown sides. It is seen chiefly on the ground and nests chiefly beneath tangled thickets. From its call, the bird is also known as the chewink. Four other species are found in the west, three congeneric with the eastern towhee, and a fourth, the green-tailed towhee. *Oberholseria chlorura*, is placed in a related genus.

Tree Creeper — a small dull-colored bird with a long curved beak. It seeks its prey, consisting of insects and small creatures, in the crevices of bark, moving about the trunks and branches of trees in almost any position. The group is represented in North America by the brown creeper, *Certhia familiaris*, and several subspecies.

Tree Pie — a bird of the Oriental region related to the magpies. The colors of these birds include shades of brown, black, and gray. The beaks are relatively short. In habits, the birds resemble the magpies.

Troupial — a name derived from the French and applied variously to members of the family *Icteridae*, including the orioles, blackbirds, and New World grackles. The name has been used by various writers for the grackles and other birds as mentioned. Also spelled troopial.

Wheatear — a bird, *Oenanthe oenanthe*, related to the thrushes and bluebirds. The bird nests in the northern part of Europe and in Alaska and is widely distributed in the Old World and occasionally in the United States during its southern migrations.

Whinchat — a small European bird related to the bluebirds and thrushes. The bird nests in the far north and winters in Africa.

Woodhewer — a small brown bird of a family found only from Mexico to southern South America. The family includes more than 200 species, mostly limited to the temperature parts of the continent. Among them are the ovenbird previously described.

Yellowbird — the American goldfinch, *Carduelis* (*Astragalinus, Spinus*) *tristis*, and the yellow warbler, *Dendroica aestiva*. Only the male of the former species is yellow, and it has the crown and wings black and the tail marked with black. The yellow warbler is more generally yellow in both sexes. The name yellowbird is not commonly used.

Zosterops — a small bird of the Old World tropics with white rings usually around the eyes. All of the numerous species are birds of small size and quiet colors.

See also **Birds**.

PASSIVE NETWORK (Electronic).   A grouping of resistors, capacitors, or resistors and capacitors required to accomplish the purpose of an electronic circuit. In many different circuit applications, prepackaged units replace clusters of comparably rated low-power resistors and capacitors. Applications include "pull ups" and "push down" transitions among logic circuits, line- and sense-amplifier terminations, decoupling, light-emitting display (LED) drives, ac coupling, supply filtering, line matching, and resistors for current limiting and pulse separating, among other uses. Such networks are available in a variety of packaging formats. See also **Capacitor (Electrical)**; and **Resistor (Circuit)**.

PASSIVITY.   When iron is immersed in concentrated nitric acid, there is no visible reaction (Keir, 1790), although dilute nitric acid results in a marked reaction with iron. Upon removal of the iron from the concentrated nitric acid and immersion in copper sulfate solution, the iron is not plated by copper, although this occurs with ordinary iron. Iron in such a condition is described as passive iron, and the phenomenon is known as passivity. See also **Iron Metals, Alloys, and Steels**.

PASSWORDS (Computer).   A password is a sequence of characters required for access to a computer system. A password is used to identify and authenticate each user in a multi-user system and to prevent users from accessing data that they are not meant to see. Passwords are widely used on the Internet, from logging onto a Web site to accessing a particular application on a server. In fact, passwords are the most commonly used form of authentication on computer systems today because they are easy to set up and administer in automated systems. However, they present several security issues of which system administrators should be cognizant. These include password guessing, sniffing, and cracking. Administrators should implement policies and mechanisms to help prevent against these threats.

The ancient folk tale of Ali Baba and the forty thieves mentions the use of a password. In this story, Ali Baba finds that the phrase "Open Sesame" magically opens the entrance to a cave where the thieves have hidden their treasure. Similarly, modern computer systems use passwords to authenticate users and allow them entrance to system resources and data shares on an automated basis. The use of passwords in computer systems likely can be traced to the earliest timesharing and dial-up networks. Passwords were probably not used before then in purely batch systems.

The security provided by a password system depends on the passwords being kept secret at all times. Thus, a password is vulnerable to compromise whenever it is used, stored, or even known. In a password-based authentication mechanism implemented on a computer system, passwords are vulnerable to compromise due to five essential aspects of the password system:

- Passwords must be initially assigned to users when they are enrolled on the system;
- Users' passwords must be changed periodically;
- The system must maintain a "password database";
- Users must remember their passwords; and
- Users must enter their passwords into the system at authentication time.

Because of these factors, a number of protection schemes have been developed for maintaining password security. These include implementing policies and mechanisms to ensure "strong" passwords, encrypting the password database, and simplifying the sign-on and password synchronization processes. Even so, a number of sophisticated cracking tools are available today that threaten password security. For that reason, it is often advised that passwords be combined with some other form of security to achieve strong authentication.

Types of Identification/Authentication

Access control is the security service that deals with granting or denying permission for subjects (e.g., users or programs) to use objects (e.g., other programs or files) on a given computer system. Access control can be accomplished through either hardware or software features, operating procedures, management procedures, or a combination of these. Access control mechanisms are classified by their ability to verify the authenticity of a user. The three basic verification methods are as follows: What you have (examples: smart card or token); What you are (examples: biometric fingerprint or iris pattern); and What you know (examples: PIN or password).

Of all verification methods, passwords are probably weakest, yet they are still the most widely used method in systems today. In order to guarantee strong authentication, a system ought to combine two or more of these factors. For example, in order to access an ATM, one must have a bank card and know his or her personal identification number (PIN).

History of Passwords in Modern Computing

Conjecture as to which system was the first to incorporate passwords has been bandied about by several computing pioneers on the Cyberspace History List-Server (CYHIST). However, there has not been any concrete evidences as yet to support one system or another as the progenitor. The consensus opinion favors the Compatible Time Sharing System (CTSS) developed at the Massachusetts Institute of Technology (MIT) Computation Center beginning in 1961. As part of Project MAC (Multiple Access Computer) under the direction of Professor Fernando J. "Corby" Corbató, the system was implemented on an IBM 7094 and reportedly began using passwords by 1963. According to researcher Norman Hardy, who worked on the project, the security of passwords immediately became an issue as well: "I can vouch for some version of CTSS having passwords. It was in the second edition of the CTSS manual, I think, that illustrated the login command. It had Corby's user name and password. It worked — and he changed it the same day."

Passwords were widely in use by the early 1970s as the "hacker" culture began to develop, possibly in tacit opposition to the ARPANET. See also **Internet (The History)**. Now, with the explosion of the Internet, the use of passwords and the quantity of confidential data that those passwords protect have grown exponentially. But just as the 40 thieves' password protection system was breached (the cave could not differentiate between AliBaba's voice and those of the thieves), computer password systems have also been plagued by a number of vulnerabilities. Although strong password authentication has remained a "hard" problem in cryptography despite advances in both symmetric (secret-key) and asymmetric (public-key) cryptosystems, the history of password authentication is replete with examples of weak, easily compromised systems. In general, "weak" authentication systems are characterized by protocols that either leak the password directly over the network or leak sufficient information while performing authentication to allow intruders to deduce or guess at the password.

In 1983, the U.S. Department of Defense Computer Security Center (CSC) published the venerable tome *Trusted Computer System Evaluation Criteria,* also known as the Orange Book (http://www.boran.com/security/tcsec.html). This publication defined the assurance requirements for security protection of computer systems that were to be used in processing classified or other sensitive information. One major requirement imposed by the Orange Book was accountability: "Individual accountability is the key to securing and controlling any system that processes information on behalf of individuals or groups of individuals" [Latham, 1985].

The Orange Book clarified accountability as follows:

- Individual user identification: Without this, there is no way to distinguish the actions of one user on a system from those of another.
- Authentication: Without this, user identification has no credibility. And without a credible identity, no security policies can be properly invoked because there is no assurance that proper authorization's can be made.

The CSC went on to publish the *Password Management Guideline* (also known as the Green Book) in 1985 "to assist in providing that much needed credibility of user identity by presenting a set of good practices related to the design, implementation and use of password-based user authentication mechanisms." The Green Book outlined a number of steps that system security administrators should take to ensure password security on the systemand suggests that, whenever possible, they be automated. These include the following 10 rules [Brotzman, 1985]:

- System security administrators should change the passwords for all standard user IDs before allowing the general user population to access the system.
- Each user ID should be assigned to only one person. No two people should ever have the same user ID at the same time, or even at different times. It should be considered a security violation when two or more people knowthe password for a user ID.
- Users need to be aware of their responsibility to keep passwords private and to report changes in their user status, suspected security violations, etc. Users should also be required to sign a statement to acknowledge understanding of these responsibilities.
- Passwords should be changed on a periodic basis to counter the possibility of undetected password compromise.
- Users should memorize their passwords and not write them on anymedium. If passwords must be written, they should be protected in a manner that is consistent with the damage that could be caused by their compromise.
- Stored passwords should be protected by access controls provided by the system, by password encryption, or by both.
- Passwords should be encrypted immediately after entry, and thememorycontaining the plaintext password should be erased immediately after encryption.
- Only the encrypted password should be used in comparisons. There is no need to be able to decrypt passwords. Comparisons can be made by encrypting the password entered at login and comparing the encrypted form with the encrypted password stored in the password database.

The system should not echo passwords that users type in, or at least should mask the entered password (e.g., with asterisks).

Password Security — Background

Information Theory. Cryptography is a powerful mechanism for securing data and keeping them confidential. The idea is that the original message is scrambled via an algorithm (or cipher), and only those with the correct key can unlock the scrambled message and get back the plaintext contents. In general, the strength of a cryptographic algorithm is based on the length and quality of its keys. Passwords are a similar problem. Based on their length and quality, they should be more difficult to attack either by dictionary, by hybrid, or by brute-force attacks. However, the quality of a password, just as the quality of a cryptographic key, is based on entropy. Entropy is a measure of disorder. See also **Encryption**; and **Information Theory**.

Example of Entropy. Say a user is filling out a form on a Web page and the form has a space for "Sex," and leaves six characters for entering either "female" or "male" before encrypting the form entry and sending it to the server. If each character is a byte; (i.e., 8 bits); then $6 \times 8 = 48$ bits will be sent for this response. Is this how much information is actually contained in the field, though?

Clearly, there is only one bit of data represented by the entry — a binary value — either male or female. That means that there is only one bit of entropy (or uncertainty) and there are 47 bits of redundancy in the field.

This redundancy could be used by a cryptanalyst (someone who analyzes cryptosystems) to help crack the key.

Fundamental work by Claude Shannon during the 1940s illustrated this concept, that is, that the amount of information in a message is not necessarily a function of the length of a message (or the number of symbols used in the message) [Sloane & Wyner, 1993]. Instead, the amount of information in a message is determined by how many different possible messages there are and how frequently each message is used.

The same concepts apply to password security. A longer password is not necessarily a better password. Rather, a password that is difficult to guess (i.e., one that has high entropy) is best. This usually comes from a combination of factors (see **Guidelines for Selecting a Good Password**). The probability that any single attempt at guessing a password will be successful is one of the most critical factors in a password system. This probability depends on the size of the password space and the statistical distribution within that space of passwords that are actually used.

Over the past several decades, Moore's Law has made it possible to brute-force password spaces of larger and larger entropy. In addition, there is a limit to the entropy that the average user can remember. A user cannottypically remember a 32-character password, but that is what is required tohave the equivalent strength of a 128-bit key. Recently, password cracking tools have advanced to the point of being able to crack nearly anything a system could reasonably expect a user to memorize (see **Password Length and Human Memory**).

Cryptographic Protection of Passwords. Early on, the most basic and least secure method of authentication was to store passwords in plaintext (i.e., unencrypted) in a database on the server. During authentication, the client would send his or her password tothe server, and the server would compare this against the stored value. Obviously, however, if the password file were accessible to unauthorized users, the security of the system could be easily compromised.

In later systems, developers discovered that a server did not have to store a user's password in plaintext form in order to perform password authentication. Instead, the user's password could be transformed through a one-way function, such as a hashing function, into a random-looking sequence of bytes. Such a function would be difficult to invert. In other words, given a password, it would be easy to compute its hash, but given a hash, it would be computationally infeasible to compute the password from it (see **Hashing**). Authentication would consist merely of performing the hash function over the client's password and comparing it to the stored value. The password database itself could be made accessible to all users without fearof an intruder being able to steal passwords from it.

Hashing. A hash function is an algorithm that takes a variable-length string as the input and produces a fixed-length value (hash) as the output. The challenge for a hashing algorithm is to make this process irreversible; that is, finding a string that produces a given hash value should be very difficult. It should also be difficult to find two arbitrary strings that produce the same hash value. Also called a *message digest* or *fingerprint*, several one-way hash functions are in common use today. Among these are Secure Hashing Algorithm-1 (SHA-1) and Message Digest-5 (MD-5). The latter was invented by Ron Rivest for RSA Security, Inc. and produces a 128-bit hash value. See Table 1 for an example of output generated by MD5. SHA-1 was developed by the U.S. National Institute of Standards and Technology (NIST) and the National Security Agency (NSA) and produces 160-bit hash values. SHA-1 is generally considered more secure than MD5 due to its longer hash value Table 1.

TABLE 1. OUTPUT FROM THE MD5 TEST SUITE

For the Input String	The Output Message Digest is
""(no password)	d41d8cd98f00b204e9800998ecf8427e
"a"	0cc175b9c0f1b6a831c399e269772661
"abc"	900150983cd24fb0d693f7d28e17f72
"message digest"	f96b697d7cb7938d525a2f31aaf161d0
"abcdefghijklmnopqrstuvwxyz"	c3fcd3d76192e4007dfb496cca67e13b
"ABCDEFGHIJKLMNOPQRSTUVWXYZabcdefghijklmnopqrstuvwxyz0123456789"	d174ab98d277d9f5a5611c2c9f419d9f
"123456789012345678901234567890123456789012345678901234567890 57edf4a22be3c955ac49da2e2107b67a 01234567890123456 78901234567890"	

Microsoft Windows NT uses one-way hash functions to store password information in the Security Account Manager (SAM). There are no Windows32 Applications Programming Interface (API) function calls to retrieve user passwords because the system does not store them. It stores only hash values. However, even a hash-encrypted password in a database is not entirely secure. A cracking tool can compile a list of, say, the one million most commonly used passwords and compute hash functions from all of them. Then the tool can obtain the system account database and compare the hashed passwords in the database with its own list to see what matches. This is called a *dictionary attack* (see **Password Cracking Tools**).

To make dictionary attacks more difficult, often a salt is used. A salt is a random string that is concatenated with a password before it is operated on by the hashing function. The salt value is then stored in the userdatabase, together with the result of the hash function. Using a salt makesdictionary attacks more difficult, as a cracker would have to compute the hashes for all possible salt values.

A simple example of a salt would be to add the time of day; for example, if a user logs in at noon using the password "pass," the string that would be encrypted might be "1p2a0s0s." By adding this randomness to the password, the hash will actually be different every time the user logs in (unless it is at noon every day). Whether a salt is used and what the salt actually is depends upon the operating system and the encryption algorithm being used. On a FreeBSD system, for example, there is a function called crypt that uses the DES, MD5, or Blowfish algorithms to hash passwords and can also use three forms of salts.

According to Cambridge University professor of computing Roger Needham, the Cambridge Multiple Access System (CMAS), which was an integrated online–offline terminal or regular input-driven system, may have been among the earliest to implement such one-way functions. It first went online in 1967 and incorporated password protection. According to Needham: "In 1966, we conceived the use of one-way functions to protect the password file, and this was an implemented feature from day one" (R. Needham, personal communication, April 11, 2002).

One-way hashing is still being used today, although it does not address another weakness—in a networked environment, it is difficult to transmit the password securely to the server for verification without its being captured and reused, perhaps in a replay attack. To avoid revealing passwords directly over an untrusted network, computer scientists have developed challenge–response systems. At their simplest, the server sends the user some sort of challenge, which would typically be a random string of characters called a nonce. The user then computes a response, usually some function based on both the challenge and the password. This way, even if theintruder captured a valid challenge–response pair, it would not help him or her gain access to the system, because future challenges would be different and require different responses.

These challenge-and-response systems are referred to as one-time password (OTP) systems. Bellcore's S/KEY is one such system in which a one-timepassword is calculated by combining a seed with a secret password known only to the user and then applying a secure hashing algorithm a number of times equal to the sequence number. Each time the user is authenticated, the sequence number expected by the system is decremented, thus eliminating the possibility of an attacker trying a replay attack using the same password again. One-time passwords were more prevalent before secure shell (SSH) and secure sockets layer (SSL) systems came into widespread use.

Password Cracking Tools

Password-Cracking Approaches. As mentioned earlier, passwords are typically stored as values hashedwith SHA-1 or MD5, which are one-way functions. In other words, this entire encyclopedia could be hashed and represented as eight bytes of gibberish. There would be no way to use these eight bytes of data to obtain the original text. However, password crackers know that people do not use whole encyclopedias as their passwords. The vast majority of passwords are 4 to 12 characters. Passwords are also, in general, not just random strings of symbols. Because users need to remember them, passwords are usually words or phrasesof significance to the user. This is an opportunity for the attacker to reduce the search space.

An attacker might steal a password file–or sniff the wire and capture the user ID/password hash pairs during logon–and then run a password-cracking tool on it. Because it is impossible to decrypt a hash backto

a password, these programs will try a dictionary approach first. The program guesses a password—say, the word "Dilbert." The program then hashes "Dilbert" and compares the hash to one of the hashed entries in thepassword file. If it matches, then that password hash represents the password "Dilbert." If the hash does not match, the program takes another guess. Depending on the tool, a password cracker will try all the words in a dictionary, all the names in a phone book, and so on. Again, the attacker doesnot need to know the original password–just a password that hashes tothe same value.

This is analogous to the "birthday paradox," which basically says, "If you get 25 people together in a room, the odds are better than 50/50 that two of them will have the same birthday." How does this work? Imagine a person meeting another on the street and asking him his birthday. The chances of the two having the same birthday are only 1/365 (0.27%). Evenif one person asks 25 people, the probability is still low. But with 25 people in a room together, each of the 25 is asking the other 24 about their birthdays. Each person only has a small (less than 5%) chance of success, but trying it 25 times increases the probability significantly.

In a room of 25 people, there are 300 possible pairs (25*24/2). Each pair has a probability of success of $1/365 = 0.27\%$, and a probability of failure of $1-0.27\% = 99.726\%$. Calculating the probability of failure: $99.726\%^{300} = 44\%$. The probability of success is then $100\%-44\% = 56\%$. So a birthday match will actually be found five out ofnine times. In a room with 42 people, the odds of finding a birthday match rise to 9 out of 10. Thus, the birthday paradox is that it is much easier to find two values that match than it is to find a match to some particular value.

If a wave of dictionary guesses fails to produce any passwords for the attacker, the cracking program will next try a hybrid approach of different combinations—such as forward and backward spellings of dictionarywords, additional numbers and special characters, or sequences of characters. The goal here again is to reduce the cracker's search space by trying "likely" combinations of known words.

Only after exhausting both of these avenues will the cracking program start in on an exhaustive or brute-force attack on the entire password space. And, of course, it remembers the passwords it has already tried and will not have to recheck these either during the brute-force search.

Approaches to Retrieving Passwords. Most password-cracking programs will first attempt to retrieve password hashes to begin their cracking processes. A sophisticated attacker will not try to guess passwords by entering them through the standard user interface because the time to do so is prohibitive, and most systems can be configured to lock a user out after too many wrong guesses.

On Microsoft Windows systems, it typically requires the "Administrator" privilege to read the password hashes from the database in which they are stored. This is usually somewhere in the system registry. In order to access them, a cracking tool will attempt to dump the password hashes from the Windows registry on the local machine or over the network if the remote machine allows network registry access. The latter requires a target Windows machine name or IP address.

Another method is to access the password hashes directly from the file system. On Microsoft Windows systems, this is the SAM. Because Windows locks the SAM file where the password hashes are stored in the file system with an encryption mechanism known as SYSKEY, it is impossible to read them from this file while the system is running. However, sometimes there is a backup of this file on tape, on an emergency repair disk (ERD), or in the repair directory of the system's hard drive. Alternately, a user may boot from a floppy disk running another operating system such as MS-DOS and be able to read password hashes directly from the file system. This is why securityadministrators should never neglect physical security of systems. If an attacker can physically access a machine, he or she can bypass the built-in file system security mechanisms (see **Recovering Windows NT Passwords**).

Todd Sabin has released a free utility called PWDUMP2 that can dump the password hashes on a local machine if the SAM has been encrypted with the SYSKEY utility that was introduced in Windows NT Service Pack 3. Once a user downloads the utility, he or she can follow the instructions on the Webpage to retrieve the password hashes, load the hashes into a tool such as L0phtCrack, and begin cracking them.

Password Sniffing. Instead of capturing the system user file (SAM on Windows or /etc/passwd or /etc/shadow on Unix/Linux), another way of collecting user IDs and passwords is through sniffing network traffic. Sniffing uses some sort of software or hardware wiretap device to eavesdrop on network communications, usually by capturing and deciphering communications packets. According to Peiter "Mudge" Zatko, who initially wrote L0phtCrack: "Sniffing is slang for placing a network card into promiscuous mode so that it actually looks at all of the traffic coming along the line and not just the packets that are addressed to it. By doing this one can catch passwords, login names, confidential information, etc" [Zatko, 1999b].

L0phtCrack offers an "SMB Packet Capture" function to capture encrypted hashes transmitted over a Windows network segment. On a switched network, a cracker will only be able to sniff sessions originating from the local machine or connecting to that machine. As server message block (SMB) session authentication messages are captured by the tool, they are displayed in the SMB Packet Capture window. The display shows the source and destination IP addresses the user name, the SMB challenge, the encrypted LAN manager hash, and the encrypted NT LAN manager hash, if any. To crack these hashes, the tool saves the session and then works on the captured file.

Recovering Windows NT Passwords. Or, why physical security is still important. Norwegian software developer Petter Nordahl-Hagen has built a resource ("The Offline NT Password Editor") for recovering Windows passwords on workstations. His approach bypasses the NTFS file permissions of Windows NT, 2000, and XP by using a Linux boot disk that allows one to reset the Administrator password on a system by replacing the hash stored in the SAM with a user-selected hash. His program has even been shown to work on Windows 2000 systems with SYSKEY enabled. An MS-DOS version also exists, as does a version that boots from CD-ROM instead of floppy disk.

Thus, physical access to the workstation can mean instant compromise, unless, perhaps the system BIOS settings are also password-protected and do not allow a user to boot from floppy or CD-ROM (however, several attacks against BIOS settings have also been published).

Types of Password-Cracking Tools. Password-cracking tools can be divided into two categories—those that attempt to retrieve system-level login passwords and those that attack the password protection mechanisms of specific applications. The first type includes programs such as L0phtcrack, Cain & Abel, and John the Ripper. Some sites for obtaining password-cracking tools for various platforms, operating systems, and applications are included in the web reference section at the end of this article.

The Russian company ElcomSoft has a developed a range of programs that can crack passwords on Microsoft Office encrypted files, WinZip or PKZip archived files, or Adobe Acrobat (PDF) files. The U.S. federal government charged ElcomSoft with violating the Digital Millennium Copyright Act of 1998 for selling a program that allowed people to disable encryption software from Adobe Systems that is used to protect electronic books. The case drew attention after ElcomSoft programmer Dmitry Sklyarov was arrested at the DefCon 2001 convention in July, 2001 (US. ElcomSoft & Sklyarov FAQ, n.d.).

Password Security Issues and Effective Management

Enforcing Password Guidelines. The FBI and the Systems Administration and Networking Security (SANS) Institute; http://www.sans.org/, released a document summarizing the "Twenty Most Critical Internet Security Vulnerabilities." The majority of successful attacks on computer systems via the Internet can be traced to exploitation of security flaws on this list. One of items on this list is "accounts with no passwords or weak passwords." In general, these accounts should be removed or assigned stronger passwords. In addition, accounts with built-in or default passwords that have never been reconfigured create vulnerability because they usually have the same password across installations of the software. Attackers will look for these accounts, having found the commonly known passwords published on hacking Web sites or some other public forum. Therefore, any default or built-in accounts also need to be identified and removed from the system or else reconfigured with stronger passwords.

The list of common vulnerabilities and exposures (CVE) maintained by the MITRE Corporation (http://www.cve.mitre.org) provides a taxonomy for more than 2000 well-known attacker exploits. Among these, nearly 100

have to do with password insecurities, and another 250 having to do with passwords are "candidates" currently under review for inclusion in the list.

SANS suggests that to determine if one's system is vulnerable to such attacks, one needs to be cognizant of all the user accounts on the system. First, the system security administrator must inventory the accounts on the system and create a master list. This list should include even intermediate systems, such as routers and gateways, as well as any Internet-connected printers and print controllers. Second, the administrator should develop procedures for adding authorized accounts to the list and for removing accounts when they are no longer in use. The master list should be validated on a regular basis. In addition, the administrator should run some password strength-checking tool against the accounts to look for weak or nonexistent passwords.

Many organizations supplement password control programs with procedural or administrative controls that ensure that passwords are changed regularly and that old passwords are not reused. If password aging is used, the system should give users a warning and the opportunity to change their passwords before they expire. In addition, administrators should set account lockout policies, which lock out a user after a number of unsuccessful login attempts, and cause him or her to have his password reset.

Microsoft Windows 2000 and Windows XP include built-in password constraint options in the "Group Policy" settings. An administrator can configure the network so that user passwords must have a minimum length, a minimum and maximum age, and other constraints. It is important to require a minimum age on a password. The following outlines the minimal criteria for selecting "strong" passwords.

Guidelines for Selecting a Good Password
Length.

- Windows systems: seven characters or longer
- Unix, Linux systems: eight characters or longer

Composition.

- Mixture of alphabetic, numeric, and special characters (e.g., #, @, or !)
- Mixture of upper and lower case characters
- No words found in a dictionary
- No personal information about the user (e.g., any part of the user's name, a family member's name, or the user's date of birth, Social Security number, phone number, license plate number, etc.)
- No information that is easily obtained about the user, especially any part of the user ID
- No commonly used proper names such as local sports teams or celebrities
- No patterns such as 12345, sssss, or qwerty
- Try misspelling or abbreviating a word that has some meaning to the user (Example: "How to select a good password?" becomes "H2sagP?")

Password Aging and Reuse. To limit the usefulness of passwords that might have been compromised, it is suggested practice to change them regularly. Many systems force users to change their passwords when they log in for the first time, and again if they have not changed their passwords for an extended period (say, 90 days). In addition, users should not reuse old passwords. Some systems support this by recording the old passwords, ensuring that users cannot change their passwords back to previously used values, and ensuring that the users' new passwords are significantly different from their previous passwords. Such systems usually have a finite memory, say the past 10 passwords, and users can circumvent the password filtering controls by changing a password 10 times in a row until it is the same as the previously used password.

It is recommended that, at a predetermined period of time prior to the expiration of a password's lifetime, the user ID it is associated with be notified by the system as having an "expired" password. A user who logs in with an ID having an expired password should be required to change the password for that user ID before further access to the system is permitted. If a password is not changed before the end of its maximum lifetime, it is recommended that the user ID it is associated with be identified by the system as "locked." No login should be permitted to a locked user ID, but the system administrator should be able to unlock the user ID by changing the password for that user ID. After a password has been changed, the lifetime period for the password should be reset to the maximum value established by the system.

Social Engineering. With all the advances in technology, the oldest way to attack a password-based security system is still the easiest: coercion, bribery, or trickery against the users of the system. Social engineering is an attack against people, rather than machines. It is an outsider's use of psychological tricks on legitimate users of a computer system; usually to gain the information (e.g., user IDs and passwords) needed to access a system. The notorious "hacker" Kevin Mitnick, who was convicted on charges of computer and wire fraud, told a Congressional panel that he rarely used technology to gain information and used social engineering almost exclusively. Mitnick spent 59 months in federal prison. See Federation of American Scientists, www.fas.org/irp/congress/2000_hr/030200_mitnick.htm).

According to a study by British psychologists, people often base their passwords on something obvious and easily guessed by a social engineer. Around 50% of computer users base them on the name of a family member, a partner, or a pet. Another 30% use a pop idol or sporting hero. Another 10% of users pick passwords that reflect some kind of fantasy, often containingsome sexual reference. The study showed that only 10% use cryptic combinations that follow all the rules of "tough" passwords [Brown, ref.].

The best countermeasures to social engineering attacks are education and awareness. Users should be instructed never to tell anyone their passwords. Doing so destroys accountability, and a system administrator should never need to know it either. Also, users should never write down their passwords. A clever social engineer will find it if it is "hidden" under a mouse pad or inside a desk drawer.

Single Sign-On and Password Synchronization. One issue that has irritated users in large secure environments is the burgeoning number of passwords they have to remember to access various applications. A user might need one password to log onto his or her workstation, another to access the network, and yet another for a particular server. Ideally, a user should be able to sign on once, with a single password, andbe able to access all the other systems on which he or she has authorization.

Some have called this notion of single sign-on the *Holy Grail* of computer security. The goal is admirable—to create a common enterprise security infrastructure to replace a heterogeneous one. And it is currently being attempted by several vendors through technologies such as the Open Group's Distributed Computing Environment (DCE), MIT's Kerberos, Microsoft's ActiveDirectory, and Public-Key Infrastructure (PKI)-based systems. However, few, if any, enterprises have actually achieved their goal. Unfortunately, the task of changing all existing applications to use a common security infrastructure is very difficult, and this has further been hampered bya lack of consensus on a common security infrastructure. As a result, the disparate proprietary and standards-based solutions cannot be applied to every system. In addition, there is a risk of a single point of failure. Should one user's password be compromised, it is not just his local system that can be breached but the entire enterprise.

Password synchronization is another means of trying to help users maintain the passwords that they use to log onto disparate systems. In this scheme, when users periodically change their passwords, the new password is applied to every account the user has, rather than just one. The main objective of password synchronization is to help users remember a single, strong password. Password synchronization purports to improve security because synchronized passwords are subjected to a strong password policy, and users who remember their passwords are less likely to write them down.

To mitigate the risk of a single system compromise being leveraged byan intruder into a network-wide attack: Synchronized passwords should be changed regularly, and users should be required to select strong (hard to guess) passwords when synchronization is introduced.

Unix/Linux-Specific Password Issues. Traditionally on Unix and Linux platforms, user information, including passwords, is kept in a system file called*/etc/passwd*. The password for each user is stored as a hash value. Despite the password being encoded with a one-way hash function and a salt as described earlier, a password cracker could still compromise system security if he or she obtained access to the /etc/passwd file and used a successful dictionary attack. This vulnerability can be mitigated by simply moving the passwords in the /etc/passwd file to another file, usually named /etc/shadow, and making this file readable only by those who have administrator or "root" access to the system.

In addition, Unix or Linux administrators should examine the password-file (as well as the shadow password file when applicable) on a regular basis for potential account-level security problems. In particular, it should be examined for the following:

- Accounts without passwords.
- UIDs of 0 for accounts other than root (which are also superuser accounts).
- GIDs of 0 for accounts other than root. Generally, users don'thave group 0 as their primary group.
- Other types of invalid or improperly formatted entries.

User names and group names in Unix and Linux are mapped into numericforms (UIDs and GIDs, respectively). All file ownership and processes use these numerical names for access control and identity determination throughout the operating system kernel and drivers.

Under many Unix and Linux implementations (via a shadow package), the command pwck will perform some simple syntax checking on the password file and can identify some security problems with it. pwck will report invalid usernames, UIDs and GIDs, null or nonexistent home directories, invalid shells, and entries with the wrong number of fields (often indicating extra or missing colons and other typos).

Microsoft-Specific Password Issues. Windows uses two password functions—a stronger one designed forWindows NT, 2000, and XP systems, and a weaker one, the LAN Manager hash, designed for backward compatibility with older Windows 9X networking login protocols. The latter is case-insensitive and does not allow passwords to bemuch stronger than seven characters, even though they may be much longer. These passwords are extremely vulnerable to cracking. On a standard desktop PC, for example, L0phtCrack can try every short alphanumeric password in a few minutes and every possible keyboard password (except for special ALT-characters) within a few days. Some security administrators have dealt with this problem by requiring stronger and stronger passwords; however, this comes at a cost.

In addition to implementing policies that require users to choose strong passwords, the CERT Coordination Center provides guidelines for securing passwords on Windows systems [CERT,2007]:

- Using SYSKEY enables the private password data stored in the registry to be encrypted using a 128-bit cryptographic key. This is a unique key for each system.
- By default, the administrator account is never locked out; so it is generally a target for brute force logon attempts of intruders. It ispossible to rename the account in User Manager, but it may be desirable to lock out the administrator account after a set number of failed attempts over the network. The NT Resource Kit provides an application called passprop .exe that enables Administrator account lockout except for interactive logons on a domain controller.
- Another alternative that avoids all accounts belonging to the Administrator group being locked over the network is to create a local account that belongs to the Administrator group, but is not allowed to log on over the network. This account may then be used at the console to unlock theother accounts.
- The Guest account should be disabled. If this account is enabled, anonymous connections can be made to NT computers.
- The Emergency Repair Disk should be secured, as it contains a copy of the entire SAM database. If a malicious user has access to the disk, he or she may be able to launch a crack attack against it.

Password-Cracking Times. Let us start with a typical password of six characters. When this password is entered into a system's authentication mechanism, the system hashes it and stores the hashed value. The hash, a fixed-sized string derived from some arbitrarily long string of text, is generated by a formula in such a way that it is extremely unlikely that other texts will produce the same hash value—unlikely, but not impossible. Because passwords are not arbitrarily long—they are generally 4 to 12 characters—this reduces the search space for finding a matching hash. In other words, an attacker's password-cracking program does not need to calculate every possible combination of six-character passwords. It only needs to find a hash of a six-character ASCII-printable password that matches the hash stored in the password file or sniffed off the network.

TABLE 2. PASSWORD CRACKING TIMES

Number of Chars in Password	Number of Possible Combinations of 95 Printable ASCII Chars	Time to Crack (in hours)[a]	Number of Possible Combinations of ALL 256 ASCII Chars	Time to Crack (in hours)[a]
0	1	0.0	1	0.0
1	95	0.0	256	0.0
2	9025	0.0	65536	0.0
3	857375	0.0	16777216	0.0
4	81450625	0.0	4294967296	0.0
5	7737809375	0.0	1099511627776	0.3
6	735091890625	0.2	281474976710656	78.2
7	69833729609375	19.4	72057594037927900	20016.0
8	6634204312890620	1842.8	18446744073709600000	5124095.6
9	6.E+17	2.E+05	5.E+21	1.E+09
1	06.E+19	2.E+07	1.E+24	3.E+11
1	16.E+21	2.E+09	3.E+26	9.E+13
1	25.E+23	2.E+11	8.E+28	2.E+16
1	35.E+25	1.E+13	2.E+31	6.E+18
1	45.E+27	1.E+15	5.E+33	1.E+21
1	55.E+29	1.E+17	1.E+36	4.E+23
1	64.E+31	1.E+19	3.E+38	9.E+25

[a] Assume 1 billion hash & check operations/second.

Because an attacker cannot try to guess passwords at a high rate through the standard user interface (as mentioned earlier, the time to enter them is prohibitive, and most systems can be configured to lock the user out after too many wrong attempts), one may assume that the attacker will get them either by capturing the system password file or by sniffing (monitoringcommunications) on a network segment. Each character in a password is a byte. One does not typically need to consider characters with a leading zero in the highest-order bit, because printable ASCII characters are in codes 32 through 126. ASCII codes 0–31 and 127 are unprintable characters, and 128–255 are special ALT-characters that are not generally used for passwords. This leaves 95 printable ASCII characters.

If there are 95 possible choices for each of the six password characters, this makes the password space $95^6 = 735,091,890,625$ combinations. Modern computers are capable of making more than 10 billion calculations per second. It has been conjectured that agencies such as the NSA havepassword-cracking machines (or several machines working in parallel) that could hash and check passwords at a rate of 1 billion per second. How fast could an attacker check every possible combination of six-character passwords? $735,091,890,625/1,000,000,000 =$ about 12 minutes (see Table 2)

What if the system forces everyone to use a seven-character password? Then it would take the attacker 19 hours to brute-force every possible password. Many Windows networks fall under this category. Due to the LAN Manager issue, passwords on these systems cannot be much stronger than seven characters. Thus, it can be assumed that any password sent on a Windows system using LAN Manager can be cracked within a day. What if the system enforces eight-character passwords? Then it would take 77 days to brute-force them all. If a system's standard policy is to require users to change passwords every 90 days, this may not be sufficient.

Password Length and Human Memory

Choosing a longer password does not help much on systems with limitations such as the LAN Manager hash issue. It also does not help if a password is susceptible to a dictionary or hybrid attack. It only works if the password appears to be a random string of symbols, but that can be difficult to remember. A classic study by psychologist George Miller showed that humans work best with the magic number 7 (plus or minus 2). So it stands to reason that once a password exceeds nine characters, the user is going to have a hard time remembering it [Miller, 1956].

Here is one idea for remembering a longer password. Security professionals generally advise people never to write down their passwords. But the user could write down half of it — the part that looks like random letters and numbers — and keep it in a wallet or desk drawer. The other part could be memorized — perhaps it could be a misspelled dictionary wordor the initials for an acquaintance, or something similarly memorable. When concatenated together, the resulting password could be much longer than nine characters, and therefore presumably stronger.

Some researchers have asserted that the brain remembers images more easily than letters or numbers. Thus, some new schemes use sequences of graphical symbols for passwords. For example, a system called PassFace, developed by RealUser, replaces the letters and numbers in passwords with sequences or groups of human faces. It is one of several applications that rely on graphical images for the purpose of authentication. Another company, Passlogix, has a system in which users can mix drinks in a virtual saloon or concoct chemical compounds using an onscreen periodic table of elements as a way to log onto computer networks.

Note: *Portions of this article originally appeared in the Internet Encyclopedia, John Wiley & Sons, Inc., New York, NY.*

Additional Reading

Barbalace, R.J.: *How to Choose a Good Password (and why you should),* Retrieved 2007 from MIT Student Information Processing Board Web site: http://www.mit.edu/afs/sipb/project/doc/passwords/passwords.html

Brotzman, R.L.: *Password Management Guideline (Green Book),* Department of Defense Computer Security Center, Fort George G. Meade, MD, 1985.

Botzum, K.: *Single Sign on—A Contrarian View,* Retrieved 2007 from IBM Software Services for WebSphere Web site: http://www7b.software.ibm.com/wsdd/library/techarticles/0108_botzum/botzum.html

Brown, A.: *U.K. Study: Passwords Often Easy to Crack,* Retrieved 2007 from http://CNN.com Web site: http://www.cnn.com/2002/TECH/ptech/03/13/dangerous.passwords/index.html

Burnett, M., and D. Kleiman: *Perfect Passwords: Selection, Protection, and Authentication,* Syngress Publishing, Rockland, MA, 2005.

CERT Coordination Center: *Windows NT Configuration Guidelines,* Retrieved 2007 from CERT Web site: http://www.cert.org/tech_tips/win_configuration_guidelines.html

Curry, D.A.: *Improving the Security of Your Unix System,* Retrieved 2007 from Information and Telecommunications Sciences and Technology Division, National Institutes of Health Web site: http://www.alw.nih.gov/Security/Docs/unix-security.html Frisch, A., and E. Frisch: *Essential System Administration,* 3rd Edition, O'Reilly Media, Inc., Sebestopol, CA, 2002.

Latham, D.C.: *Trusted Computer System Evaluation Criteria (Orange Book),* Department of Defense National Computer Security Center, Fort George G. Meade, MD, 1985.

Litchfield, D.: *Hackproofing Oracle Application Server (A guide to Securing Oracle 9),* Retrieved 2007 from NGSSoftware Web site: http://www.nextgenss.com/papers/hpoas.pdf

Luby, M. and C. Rackoff: "A Study of Password Security," *Journal of Cryptology,* **1**(3), 151–158 (1989).

Miller, G.A.: "The Magical Number Seven, Plus or Minus Two: Some Limits on our Capacity for Processing Information," *The Psychological Review,* **63**, 81–97 (1956).

Morris, R.T., and K. Thompson: "Password Security: A Case History," *Communications of the ACM,* **22**(11), 594–597 (1979).

Russell, R.: *Hack Proofing Your Network,* Syngress Publishing, Rockland, MA, 2002.

Salkever, A: *Picture This: A Password You Never Forget*, Retrieved 2007 from BusinessWeek.com Web site: http://www.businessweek.com/bwdaily/dnflash/may2001/ nf20010515_060.htm

Sanjour, J., A. Arensburger, and A. Brink: *Choosing a Good Password*, Retrieved 2007 from Computer Science Department, University of Maryland http://www.cs.umd.edu/faq/Passwords.shtml

Schneier, B.: *Secrets & Lies: Digital Security in a Networked World*, John Wiley & Sons, Inc., New York, NY, 2000.

Sloane, N.J.A., and A.D. Wyner: *Claude Elwood Shannon: Collected Papers*, IEEE Press, New York, NY, 1993.

Smith, R.E.: *Authentication: From Passwords to Public Keys*, Addison Wesley Longman, Boston, MA, 2001.

Tippett, P.: *Stronger Passwords Aren't: Information Security*, Retrieved 2007 from TruSecure Corporation Web site: http://www.infosecuritymag.com/articles/june01/columns_executive_view.shtml

Zatko, P. "Mudge," *Vulnerabilities in the S/KEY One Time Password System*, L0pht Heavy Industries, Inc., Web site: http://www.unix.geek.org.uk/~arny/junk/skeyflaws.html

Web Reference

Cain & Abel [computer software]: http://www.oxid.it
Elcomsoft [computer software]: http://www.elcomsoft.com/prs.html
Intertek [computer software]: http://www.intertek.org.uk/downloads
John the Ripper [computer software]: http://www.openwall.com/john
Nordahl-Hagen, P. NET Password Recovery [computer software]: http://home.eunet.no/~pnordahl/ntpasswd/
Passfilt [computer software]: http://support.microsoft.com/kb/q161990/
Passlogix [computer software]: http://www.passlogix.com
RealUser [computer software]: http://www.realuser.com
RSA: http://www.rsa.com/
RSA Sign-On Manager: http://www.rsa.com/products/SOM/technology_back grounders/ SECSOM_TB_1005.pdf#xml=http://www.rsa.com/programs/texis.exe/webinator/search/xml.txt?query=Hashing&pr=rsadotcom&prox=page&rorder=500&rprox=500&rdfreq=500&rwfreq=500&rlead=500&sufs=0&order=r&cq=&id=45cd6fbfa

JEREMY RASMUSSEN, Spyris Electronics, LLC

PASTEURIZATION.   Heat treatment of milk, fruit juices, canned meats, egg products, etc. for the purpose of killing or inactivating disease-causing organisms. For milk, the minimum exposure is $62\,°C$ for 30 min or $72\,°C$ for 15 sec, the latter being called flash pasteurization. Although this treatment kills all pathogenic bacteria and also inactivates enzymes that cause deterioration of the milk, the shelf life is limited. To prolong storage life, temperatures of $80-88\,°C$ for 20-40 sec must be used. Complete sterilization requires ultrahigh pasteurization at from $94\,°C$ for 3 sec to $150\,°C$ for 1 sec. In-can heating at $116\,°C$ for 12 min and $130\,°C$ for 3 min is also employed for maximum stability and long storage life. Some meat products are pasteurized by α-radiation.

See also **Milk And Milk Products**.

PASTEURIZER.   See **Heat Transfer**.

PASTEUR, LOUIS (1822–1895).   Pasteur was a French chemist and microbiologist who made important contributions to biology, medicine, chemistry, and industry.

As a small child, Pasteur showed traits of becoming a scientist. He was fascinated by the local chemist, that made medicine for sick customers. He patiently and carefully observed the things the chemist did and then went home and made drawings of the herbs and roots the man used. Even before finishing high school, Pasteur's study of chemical crystals won attention of the scientific world. Most of what Pasteur is famous for is his work concerning the effects of microbes. He found that living organisms, microbes, cause fermentation. His discovery was important both for theoretical science and for industry. Pasteur's studies showed microbes could be killed by heat. His discovery made winemaking a more scientific process. Pasteur applied the same idea to milk. The process of keeping milk free from bacteria is named pasteurization after him.

During the 1800s the theory of spontaneous generation was raging in the scientific circles. Pasteur's work proved that food and other organic matter does not spontaneously generate microbes and settled the controversy.

Pasteur also discovered a vaccine to prevent rabies and another vaccine to prevent anthrax. Pasteur's greatest achievement was the founding of the science of microbiology.

See also **Fermentation**; **Grapes and Wines**; and **Rabies**.

J. M. I.

PATCH (Computer Program).   A section of coding inserted into a computer routine to correct an error or to alter the routine. Often, it is not inserted into the actual sequence of the routine being corrected, but rather it is placed elsewhere, with an exit to the patch and a return to the routine provided. Also, the act of altering a routine by using a patch. See also **Program (Computer)**.

PATENT LOG.   A term applied to any one of a large group of instruments for recording the speed of a ship through the water and, also, the distance run through the water in a given interval of time.

The screw of the ship itself is, in a sense, a patent log, for the speed of the ship through the water is proportional to the revolutions per minute of the screw, and the distance run is proportional to the total number of revolutions in a given time. However, the distance that the ship will move for a single revolution depends upon a number of variable factors, such as the trim of the vessel, the speed of the vessel, the state of the sea, etc.

The earliest, simplest, and, perhaps, the most reliable of the various types of patent logs is the so-called taffrail log. This instrument consists of a spinner, which is towed astern of the ship, well beyond the turbulence produced by the screw. The revolutions of the spinner are transmitted to a recording mechanism, which was originally at the stern, or taffrail, of the ship. In modern installations, the recording dials may be located on the bridge or wherever they will be of most use to the navigating staff. The dials show the speed of the ship at any instant, and also the distance run after the indicator was set to zero. The instrument must be continually watched to see that it is not fouled by seaweed or debris thrown overboard from the ship. Furthermore, it must be frequently checked by the log chip and line, or some other method, to be certain that the blades have not been bent by objects floating in the water.

The principle of the Pitot tube is used in another type of patent log. The tube itself is below the ship, at the turning center, and operates dials similar to those of the other instruments.

See also **Course**; and **Navigation**.

PATH.   An edge train in which each internal vertex is of degree two and each terminal vertex is of degree one. See also **Vertex**.

PATHFINDER ELEMENT.   An element present in small proportions less than 1% generally metallic in nature, associated with ore deposits at the time of formation. Mapping of the concentration variation of the selected element serves to locate the main ore deposit. Examples are zinc as the pathfinder for lead, copper, and silver ores, and molybdenum associated with copper deposits.

PATHFINDER MISSION TO MARS.   On July 4, 1997, Mars *Pathfinder* landed safely on the surface of Mars. Designed under the new "faster, cheaper, and better" *Discovery* program philosophy, the lander deployed and navigated a small rover named "*Sojourner* " onto the Ares Valles landing site and began collecting data from its onboard scientific instruments. Designed primarily as an entry, descent and landing demonstration, *Pathfinder* returned 2.3 billion bits of new data, including over 17,000 images, 16 chemical analyses of rocks and soil, and 8.5 million individual temperature, pressure and wind measurements. *Sojourner* traversed approximately 100 meters (330 feet) clockwise around the lander exploring about 200 square meters (2,153 square feet) of area. See Fig. 1. The mission captured the imagination of the public, garnered front page headlines during the first week of mission operations, and went on to became one of NASA's most popular missions. A total of about 56.6 million people visited the *Pathfinder* Web Pages during the first month of the mission, with 4.7 million people visiting the Web Pages on July 8, 1997 alone, making the *Pathfinder* landing by far the largest Internet event in history up to that time.

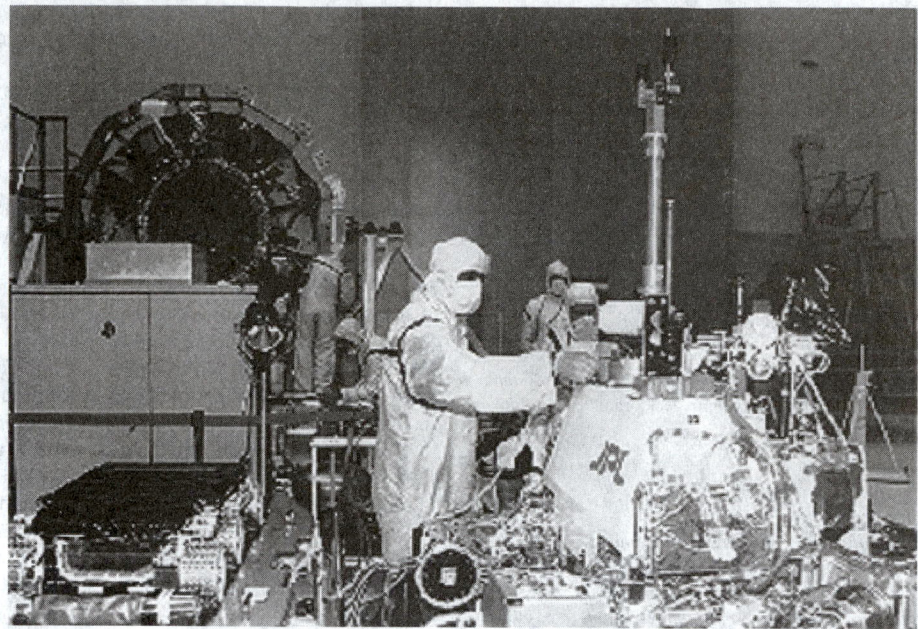

Fig. 1.   Mars *Pathfinder*, rover, and cruise stage being unpacked at the Kennedy Space Center. (*Image courtesy of KSC/NASA.*)

Mission Summary. The Mars *Pathfinder* mission was the second mission launched under the National Aeronautics and Space Administration's (NASA) *Discovery* Program. The *Discovery* missions were developed for small planetary missions with a maximum three-year development cycle and a cost cap of $150 million (Fiscal Year 1992) for development that focused on engineering, science, and technology objectives. Originally conceived as an engineering demonstration of key technologies and concepts for use in future missions to Mars, the primary objective was to demonstrate a low-cost cruise, entry, descent, and landing system that could safely place a variety of science instruments on the surface of Mars. For *Pathfinder*, the cost of the mission was $171 million (Fiscal Year 1996), the *Delta II* launch vehicle was an additional $55 million, the development and operations of the rover cost an additional $25 million, and $14 million was allotted for operations.

Mission and Spacecraft Overview

The *Pathfinder* spacecraft or flight system consisted of three major components: the cruise stage, the entry decent subsystem, and the lander, which consisted of the science instruments and the rover.

Cruise Phase. The cruise phase of the Mars *Pathfinder* mission began with the successful launch atop a *Delta II* rocket from the Kennedy Space Center in Florida on December 4, 1996. See Fig. 2. Once in earth orbit, the spacecraft was given a final boost with the help of a solid-fuel rocket motor called a *Payload Assist Module (PAM-D)*. This 'kick-stage' gave the spacecraft just the right amount of velocity increase it needed to escape Earth's gravity and enter its own orbit around the Sun. Once spent, the third stage was jettisoned.

At separation from the upper stage, the spacecraft was in Earth's shadow and spinning at 20 rpm. An onboard sequence of events was activated once the separation microswitch detected the separation. The Deep Space Network (DSN) initiated spacecraft acquisition and lockup activities using a 34-meter (112-foot) antenna located in the California desert. **See Antenna**. As soon as acquisition occurred, the engineering telemetry broadcast by the spacecraft was received on the ground at a rate of 40 bits per second (b/s). This telemetry consisted of a combination of real time engineering data and stored data from launch, separation, and Earth/Sun acquisition. See Fig. 3.

The spacecraft automatically determines its orientation in space by first determining the location of the Sun with respect to the spin axis of the spacecraft using a Sun sensor located on the top of the cruise stage. This procedure, known as Sun acquisition, was supposed to provide the spacecraft with the information it needed to reduce the spin rate from 20 rpm to a nominal 2 rpm. But due to some difficulties during launch, it

Fig. 2.   Mars *Pathfinder* launch onboard a Delta II on December 4, 1996. (*Image Courtesy of KSC/NASA.*)

was soon discovered that two of the five sensors had been damaged with an unknown, foreign substance. A software patch was developed which corrected the problem and by using the data from the three working sensors, engineers were able to slow the spacecraft down. Once the spacecraft had cleared the moon's orbit and safely spun down to 2 rpm, the star scanner was activated. After star identification had been confirmed, the *Attitude and Information Management (AIM)* computer calculated the spacecraft's orientation and position, and started its seven-month trip to Mars.

Mars *Pathfinder* used an Earth-Mars transfer orbit. The total flight time from Earth to Mars took seven months. See Fig. 4 for a view of the interplanetary trajectory, as it would look from above the Sun. During the seven-month cruise to Mars, a number of activities were performed to maintain the health of the entry vehicle, lander and rover. Navigation was

Fig. 3.   Mars *Pathfinder* in cruise configuration. The red panels are the solar cells that will supply power during the seven month cruise. (*Image courtesy of JPL/NASA.*)

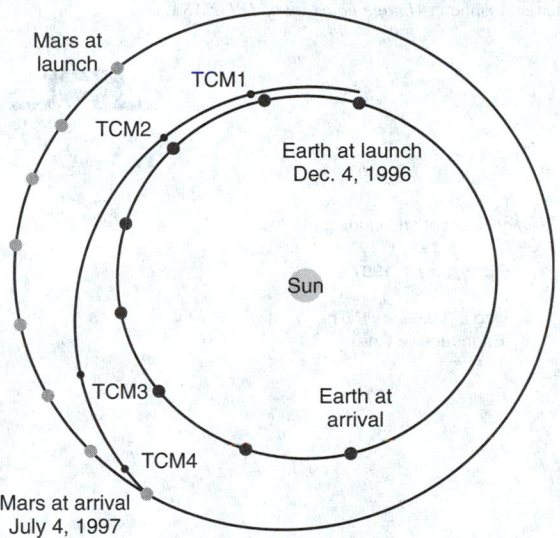

Fig. 4.   Mars *Pathfinder* Cruise Trajectory.

required to maintain the flight path, and the various spacecraft subsystems were monitored and adjusted as needed to keep them operating at peak efficiency.

Cruise activities began once the spacecraft was safely out of Earth orbit. After the attitude was established and the spacecraft was determined to be healthy, the flight team began a two-week initial characterization and calibration period. Systems included the solar array and battery, thermal control, attitude determination and control, and the communication subsystems. The primary spacecraft activities during the first month of cruise were to collect and downlink relevant engineering telemetry and tracking data, initial spacecraft health checks and calibrations, and attitude maneuvers to maintain the correct Earth/Sun geometry. One health check of the Rover and Science Instruments occurred on December 19, 1996.

Measurements of the spacecraft range to Earth and the rate of change of this distance were collected during every *DSN* station contact and sent to the navigation specialists of the flight team for analysis. They used this data to determine the true path the spacecraft was flying, and to determine corrective maneuvers needed to maintain the desired trajectory. The first of four *Trajectory Correction Maneuvers (TCMs)* was scheduled on January 4, 1997 to correct any errors collected from launch. The

magnitude of this maneuver was less than 75 meters (246 feet) per second. Navigation was an ongoing activity that continued until the spacecraft entered the atmosphere of Mars on the 4th of July.

After TCM-1, the flight team transitioned from a "spacecraft checkout mode" to a more routine "spacecraft monitoring mode". DSN tracking coverage was reduced from three contacts a day to three per week to allow other spacecraft like Mars *Global Surveyor* and *Galileo* to use the DSN time. Spacecraft health and performance telemetry was downlinked at 40 b/s or greater during each tracking pass.

A key activity that took place during cruise was the designing and building of command sequences that dictated to the spacecraft how it was to perform each of the activities required. Each cruise command sequence was generated and tested, and then uplinked approximately once every four weeks during one of the regularly scheduled DSN passes. The uplink generation process required 14 days for planning, sequence generation, verification, and commanding.

Two more trajectory correction maneuvers were performed in early February and early May to further reduce any navigation guidance errors. TCM-2 required less than 10 meters (33 feet) per second, and TCM-3 was smaller still, less than 1 meter (3.3 feet) per second. These two maneuvers further reduced any guidance error detected from navigation measurements during cruise.

Starting 45 days prior to entry, tracking was increased to three passes a day and the flight team stepped up its preparation for atmosphere entry and landing. A final health and status check of the instruments and rover was performed on June 4, 1997. A fourth and final trim maneuver was performed on June 24, requiring less than 0.50 meters (1.65 feet) per second due to the accuracy of the previous maneuvers. On June 30, the spacecraft performed a turn to the entry attitude, where it remained until atmosphere entry. The roll thrusters increased the spacecraft spin rate from 2 to 10 rpm for entry. At that time, the cruise phase ended and the flight team transitioned to the entry, landing, and surface operations phases. See Fig. 5.

During the final day of approach, the navigation team produced orbit solutions on a regular basis, and adjustments were made to the computer programs that determine when the parachute should be deployed. At 6 hours out, the final adjustments were made, and the flight team made final preparations for atmosphere entry.

Entry, Decent, and Landing Phase. The fast-paced approach of *Pathfinder* to Mars began with venting of the heat rejection system's cooling fluid about 90 minutes prior to landing. See Fig. 6. This fluid is circulated around the cruise stage perimeter and into the lander to keep the lander and rover cool during the seven month cruise phase of

Fig. 5.   Artist renditions of Mars *Pathfinder* as it enters the Martian atmosphere. (*Image courtesy of JPL/NASA.*)

Cruise stage separation
(8500 km, 6100 m/s)
landing - 34 min

Entry
(125 km, 7600 m/s)
landing - 4 min

Parachute deployment
(6-11 km, 360-450 m/s)
landing - 2 min

Heatshield separation
(5-9 km, 95-130 m/s)
landing - 100 s

Lander separation
bridle deployment
(3-7 km, 65-95 m/s)
landing - 80 s

Radar ground aquisition
(1-5 km, 60-75 m/s)
landing - 32 s

Airbag inflation
(300 m, 52-64 m/s)
landing - 8 s

Rocket ignition
(50-70 m, 52-64 m/s)
landing - 4 s

Bridle cut
(0-30 m, 0-25 m/s)
landing - 2 s

Mars *Pathfinder*

entry, descent and landing

Friday, July 4, 1997

landing at 10:05 am PDT
(Earth receive time)

Deflation /
petal latch firing
landing + 15 min

Airbag retraction /
lander righting
landing + 115 min

Final retraction
landing + 180 min

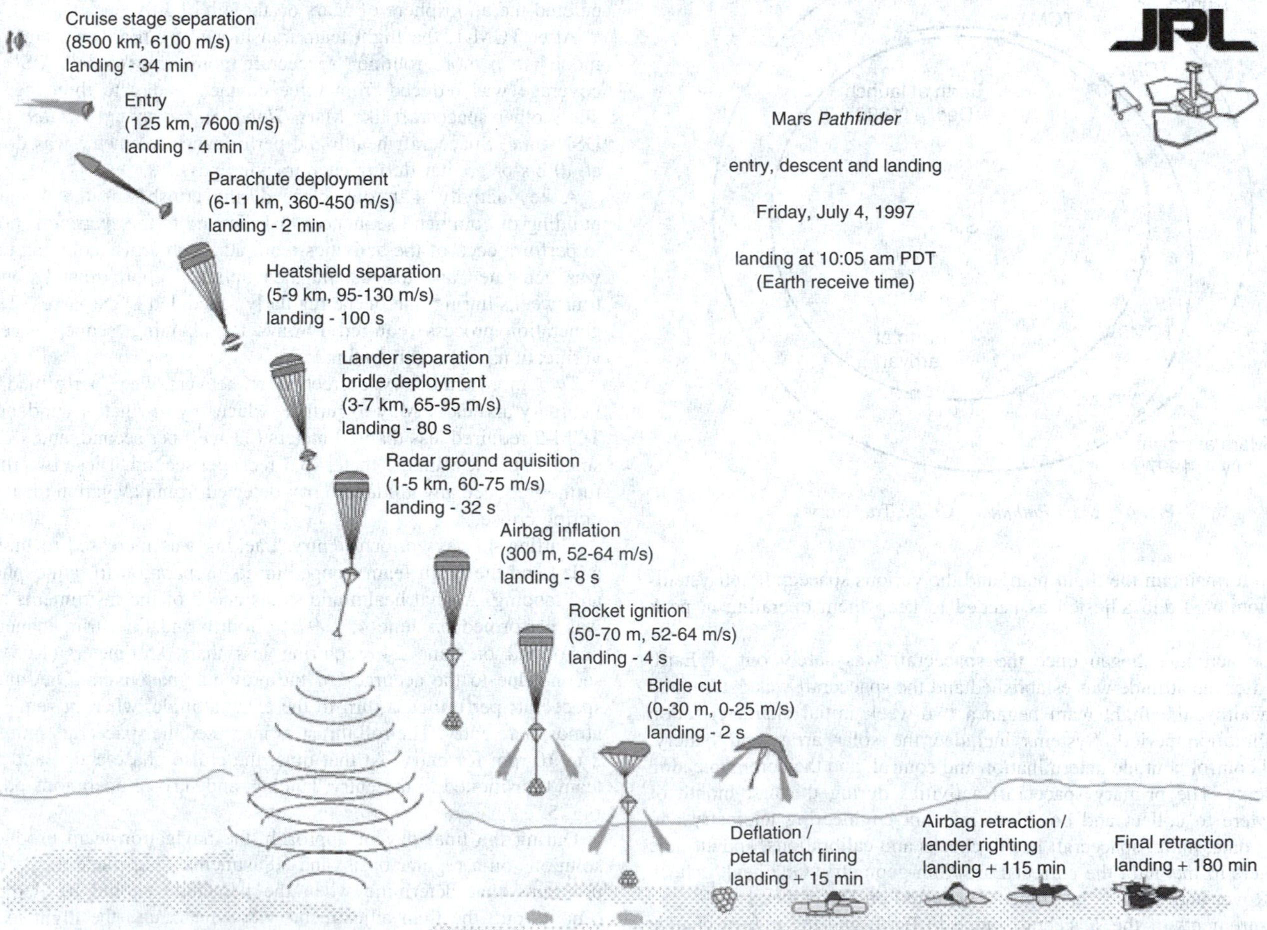

Fig. 6.   Entry, Decent, and Landing schematic for July 4, 1997. (*Image courtesy of JPL/NASA.*)

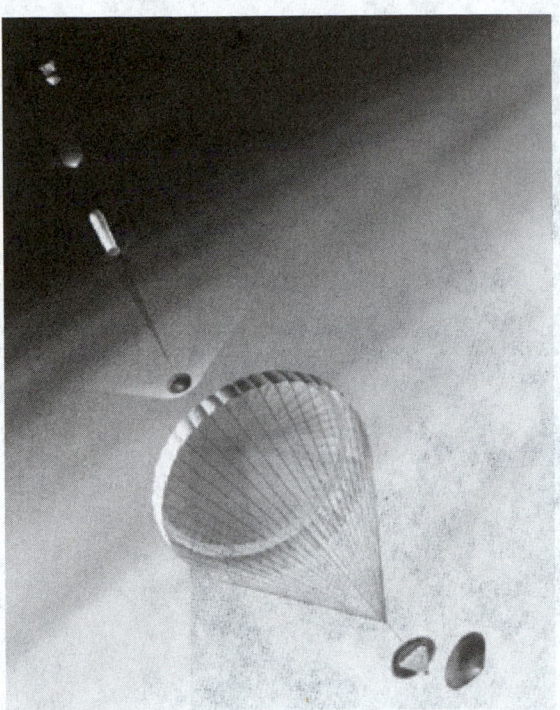

Fig. 7.   Artist rendition of *Pathfinder* entering the Martian atmosphere. (*Image courtesy of JPL/NASA.*)

than 100 megawatts of thermal energy. The Martian atmosphere slowed the vehicle from 7.5 kilometers per second to only 400 meters per second (900 miles per hour).

Entry deceleration of up to 20 gees, detected by on-board accelerometers, set in motion a sequence of preprogrammed events that are completed in relatively quick succession. Deployment of the single, 8-meter (24-foot) diameter parachute occurred 2 minutes and 14 seconds after atmospheric entry at an altitude of 9.4 kilometers (6 miles) above the surface. The parachute was similar in design to those used for the *Viking* program but had a wider band around the perimeter, which helped to minimize swinging.

The heatshield was pyrotechnically separated from the lander 20 seconds later and dropped away. See Fig. 7. The lander soon begins to separate from the backshell and "rappels" down a metal tape on a centrifugal braking system built into one of the lander petals.

The slow descent down the metal tape places the lander into position at the end of a braided Kevlar tether, or bridle, without off-loading the parachute or placing excessive loads on the backshell. The 20-meter (66-foot) bridle provides space for airbag deployment, distance from the solid rocket motor exhaust stream and increased stability. Once the lander was lowered into position at the end of the bridle, the radar altimeter was activated and began a timing sequence for airbag inflation, backshell rocket firing and the cutting of the Kevlar bridle.

The lander's Honeywell radar altimeter acquired the surface about 28 seconds prior to landing at an altitude of about 1.6 kilometers (1 mile). The airbags were inflated 18 seconds later before landing at an altitude of 355 meters (less than 1/4 of a mile) above the surface. See Fig. 8. The airbags had two pyro firings, the first of which cut the tie cords and loosened the bags. The second firing, 0.25 seconds later, and 4 seconds before the rockets fired, ignited three gas generators that inflated the three 5.2-meter (17-foot) diameter bags to a little less than 1 psi. in less than 0.3 seconds.

The conical backshell above the lander contains three solid rocket motors each providing about a ton of force for over 2 seconds. The computer in the lander activates them. Electrical wires that run up the bridle close relays in the backshell which ignite the three rockets at the same instant.

The brief firing of the solid rocket motors at an altitude of 98 meters (323 feet) was intended to essentially bring the downward movement of the lander to a halt some 12 meters, +/−10 meters (40 feet, +/−30 feet) above the surface. In reality, the rockets fired approximately 21 meters (69 feet) above the surface. The bridle separating the lander and heatshield were then cut from the lander, resulting in the backshell driving up and

the mission. Its mission fulfilled, the cruise stage was then jettisoned from the entry vehicle about one-half hour prior to landing at a distance of 8,500 kilometers (5,100 miles) from the surface of Mars. Several minutes before landing, the spacecraft began to enter the outer fringes of the atmosphere about 125 kilometers (75 miles) above the surface. Spin stabilized at 2 rpm, and traveling at 7.5 kilometers (4.5 miles) per second the vehicle entered the atmosphere at a shallow 14.8-degree angle. A shallower entry angle would result in the vehicle skipping off the atmosphere, while a steeper entry would not provide sufficient time to accomplish all of the entry, descent and landing tasks. A *Viking* -derived aeroshell (including the heatshield) protected the lander from the intense heat of entry. At the point of peak heating the heatshield absorbed more

Fig. 8.   One of the many airbag test performed prior to lift-off. Each lobe of the airbags consists of six spheres, with four lobes, one for each of the pedals. The airbags total 16 feet from the ground to the top. (*Image courtesy of JPL/NASA.*)

Fig. 9.   Artist renditions of the landing of Mars *Pathfinder* on the surface of Mars. (*Image courtesy of JPL/NASA.*)

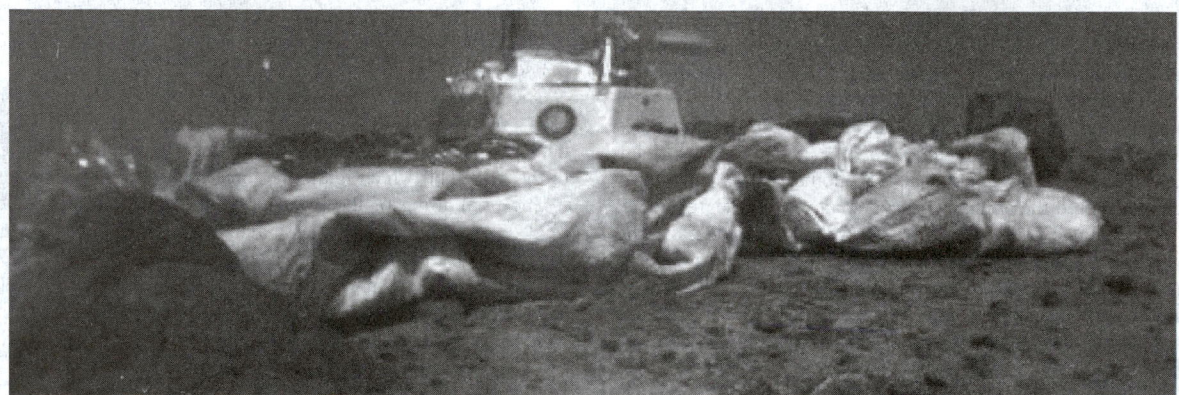

Fig. 10.   Rover view of the lander on the surface of Mars. Notice how far the airbags retracted. (*Image courtesy of JPL/NASA.*)

into the parachute under the residual impulse of the rockets, while the lander, encased in airbags, fell to the surface. See Fig. 9.

Because it was possible that the backshell could be at a small angle at the moment that the rockets fire, the rocket impulse imparted a large lateral velocity to the lander/airbag combination. In fact the impact could have been as high as 25 meters per second (56 mph) at a 30-degree grazing angle with the terrain. It was expected that the lander could have bounced at least 12 meters (40 feet) above the ground and soared 100 to 200 meters (330 to 660 feet) between bounces. Tests of the airbag system verified that it was capable of much higher impacts and longer bounces. In reality, an onboard instrument calculated at least 15 bounces with the first bounce up to 12 meters (40 feet) high without any airbag rupture.

Once the lander had settled on the surface, pyrotechnic devices in the lander petal latches were blown to allow the petals to be opened. The latches locking the sturdy side petals in place were necessary because of the pulling forces exerted on the lander petals by the deployed airbag system. In parallel with the petal latch release, a retraction system began slowly dragging the airbags toward the lander, breaching vent ports on the side of each bag, in the process deflating the bags through a cloth filter. The airbags were drawn toward the petals by internal lines extending between attachments within the airbags and small winches on each of the lander sides. It took about 64 minutes to deflate and fully retract the bags. See Fig. 10.

There is one high-torque motor on each of the three petal hinges. If the lander had come to rest on its side, it would have to be righted to the base petal by opening a side petal with a motor drive to place the lander in an upright position. Once upright, the other two remaining petals would have been opened.

About three hours was allotted to retract the airbags and deploy the lander petals, but on Mars the whole operation only took 87 minutes because *Pathfinder* came to rest on the base petal. At this time, the lander's X-band radio transmitter was turned off for the first time since before it was launched on December 4, 1996. This saved battery power and allowed the transmitter electronics to cool down after being warmed up during entry without the cooling system. It also allowed time for the Earth to rise well above the local horizon so that it would be in a better position for communications with the lander's low-gain antenna.

Science Instruments and Objectives

The Mars *Pathfinder* project landed a single vehicle on the surface of Mars, which included a microrover, *(Sojourner)*, and several science instruments. See Fig. 11. *Sojourner's* mobility provided the capability of "ground truthing" the local landing area and investigating the surface with three additional science instruments: A stereoscopic imager with spectral filters on an extendible mast (IMP), an Alpha Proton X-Ray Spectrometer (APXS), and an Atmospheric Structure Instrument/Meteorology package

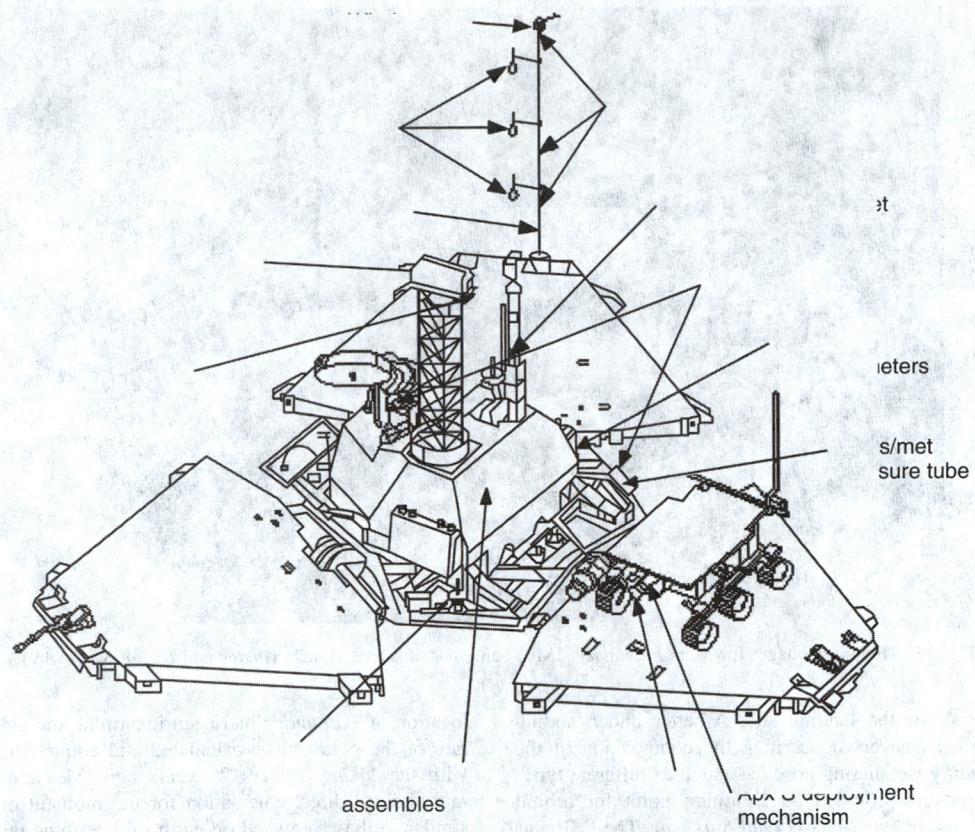

Fig. 11.   Computer drawing of the lander components. (*Image courtesy of JPL/NASA.*)

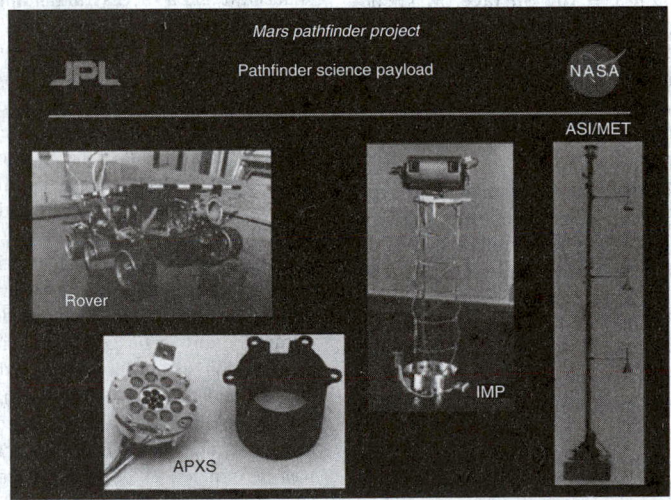

Fig. 12.   Mars *Pathfinder* instrument package. Imager for Mars *Pathfinder* (IMP), Alpha Proton X-Ray Spectrometer (APXS), an Atmospheric/Meteorology. (ASIMET.)

(ASI/MET). See Fig. 12. These instruments allowed for investigations of the geology and surface morphology at submeter to a hundred meters scale, the geochemistry and petrology of soils and rocks, the magnetic and mechanical properties of the soil as well as the magnetic properties of the dust, a variety of atmospheric investigations and rotational and orbital dynamics of Mars.

Landing downstream from the mouth of a giant catastrophic outflow channel (Ares Vales) offered the potential for identifying and analyzing a wide variety of materials in the crust, from the ancient heavily cratered terrain to intermediate-aged ridged plains to reworked channel deposits. Examination of the different surface materials allowed first-order scientific investigations of the early differentiation and evolution of the crust, the development of weathering products and the early environments and conditions that have existed on Mars.

Surface Morphology and Geology at Meter Scale. The Imager for Mars *Pathfinder* (IMP) examined Martian geologic processes and surface-atmosphere interactions similar to what was observed at the *Viking* landing sites. See Fig. 13. Observations of the general landscape, surface slopes and the distribution of rocks were obtained by panoramic stereo images at various times of the day. IMP was also designed to monitor any dust or sand deposition, erosion or other surface-atmosphere interactions. A basic understanding of the surface and near-surface soil properties was obtained by the rover and lander imaging of rover wheel tracks, holes dug by rover wheels, and examining any surface disruptions caused by airbag bounces or retractions.

Petrology and Geochemistry of Surface Materials. The Alpha-Proton X-Ray Spectrometer (APXS) and the visible to near-infrared spectral filters on the IMP determined the dominant elements that made up the rocks

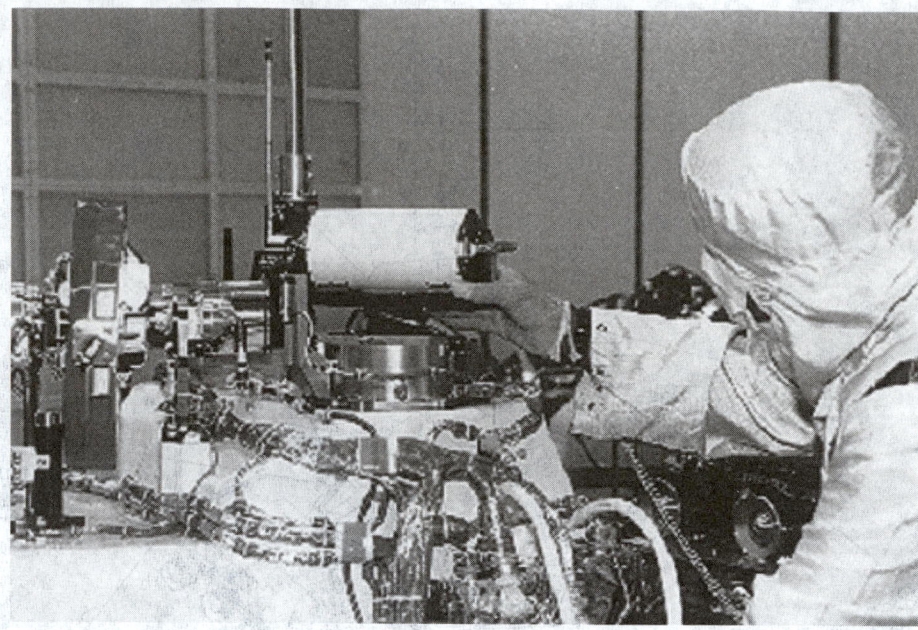

Fig. 13.   Figure 7 Imager for Mars *Pathfinder* (IMP) being tested before launch. (*Image courtesy of KSC/NASA.*)

and other surface materials of the landing site. A better understanding of these materials provided answers concerning the composition of the Martian crust, and secondary weathering products (such as different types of soils). These investigations provided a calibration point for orbital remote sensing observations such as Mars *Global Surveyor*. The IMP was able to obtain full multi-spectral panoramas of the surface and underlying materials exposed by the rover.

Magnetic Properties and Soil Mechanics of the Surface. Magnetic targets were distributed at various points around the spacecraft. Multi-spectral images of these targets identified the magnetic minerals that make up the airborne dust. Using the IMP images, it was possible to identify the mineral composition of the rocks. Detailed examination of the wheel-track images also gave a better understanding of the mechanics of the soil surrounding the landing site.

Atmospheric Structure as Well as Diurnal and Seasonal Meteorological Variations. The Atmospheric Structure Instrument/Meteorology (ASI/MET) experiment was able to monitor the temperature and density of the atmosphere during Entry, Descent and Landing (EDL). In addition, three-axis accelerometers were used to measure atmospheric pressure during entry. Once on the surface, meteorological measurements such as pressure, temperature, wind speed and atmospheric opacity were obtained on a daily basis. Thermocouples mounted on a one meter (3.3 foot) high mast examined temperature profile with height. Wind direction and speed were measured by a wind sensor mounted at the top of the mast, as well as three windsocks interspersed at different heights on the mast. Understanding this data was important for identifying the forces that act on small particles carried by the wind. Regular sky and solar spectral observations using the IMP monitored windborne particle size, particle shape, distribution with altitude and the abundance of water vapor.

Rotational and Orbital Dynamics of Mars. The Deep Space Network (DSN), by using two-way X-Band and Doppler tracking of the Mars *Pathfinder* lander once it was on the surface, was able to address a variety of orbital and rotational dynamics questions. Spacecraft ranging involves sending a code to the lander and measuring the time required for the lander to echo the code back to the Earth-based station. By dividing this time by the speed of light, results can be accurate within 1 to 5 meters (3 to 16 feet) of the distance from the station to the spacecraft. As the lander moves relative to the tracking station, the velocity between the spacecraft and Earth causes a Doppler shift in frequency. Measuring this frequency shift provided an accurate measurement of the distance from the station to the lander. After a few months of observing these features, the Mars *Pathfinder* lander location was determined within a few meters. Once the exact

location of *Pathfinder* had been identified, the orientation and precession rate of the pole can be calculated and compared to measurements made with the *Viking* landers 20 years ago. Measurement of the precession rate allowed direct calculation for the moment of inertia. Measurements similar to these are used on earth to determine the makeup of the earth's interior.

Surface Science Phase. After receiving data indicating the health of the spacecraft and a successful landing, commands were sent to the spacecraft to unlatch the IMP camera and the high gain antenna. The first task of the lander was to determine the location of the Sun. To do this, the IMP scanned the horizon for the brightest spot on the horizon. Once the Sun's location was determined, the high gain antenna was directed towards Earth and the first images were received around 4:30 p.m. PDT. The first received data included the mission success panorama, stereo images of both rover ramps, and spacecraft engineering data which included the health status of the spacecraft and the status of the airbag retraction. See Fig. 14.

After examining the imagery, it was determined that the airbags had not fully retracted from the rover petal and that it would not be safe to deploy the rover petals. Commands were sent up to reclose the rover petal, retract the airbag further, and then redeploy the rover petal. After careful examination of a second set of images, the ramps were determined to be safe and the rover was commanded to stand up. A full panorama of the landing site was also returned on the first day of operation and the rover was driven down the rear ramp the following day (Sol 2). After it was determined to be safe to deploy the IMP camera, the camera mast was deployed to a height of 0.8 meters (2.6 feet) at the end of Sol 2. After some minor communication errors between the rover modem and the lander, the rover deployed the APXS at the surface for its first soil sample. See Fig. 15.

Lander Site Location

When the first images had arrived, five prominent horizon features and two small craters were identified in both lander horizon and *Viking Orbiter* images. This enabled the lander to be located within 100 meters (330 feet) of other surface features at 19.13 °N, 33.22 °W in the U.S. Geological Survey reference frame.

Characteristics of the landing site were determined to be consistent with its prelanding predication of a flat, level flood plain composed primarily of materials left behind by the Tiu and Ares catastrophic floods. The surface is composed of pebbles, cobbles and boulders that closely resemble depositional surfaces found from catastrophic floods on Earth. Two nearby peaks identified as "Twin Peaks," appear to be streamlined hills in IMP images; this is consistent with prelanding predictions of

Fig. 14.   Mission success panorama acquired on July 4, 1997. (*Image courtesy of JPL/NASA.*)

Fig. 15.   Sojourner on the surface of Mars. Rock to the left is Barnacle Bill. (*Image courtesy of JPL/NASA.*)

Fig. 16.   End of day image of the rover. (*Image courtesy of JPL/NASA.*)

Viking Orbiter images of the region. Rocks identified in the Rock Garden are imbricated in the direction of the predicted flow; again agreeing with prelanding predictions Troughs are also visible throughout the scene and have been interpreted to be erosional features produced by the turbulent flood waters. Large boulders can be found perched on top of smaller rocks (i.e. Yogi), consistent with deposition by a flood. Except for later eolian activity, the site appears little altered since it formed up to a few billion years ago.

A variety of soil types have also been identified at the *Pathfinder* landing site. See Fig. 16. These soils appear to be composed of poorly crystalline ferric-bearing materials. Elemental compositions of soil units measured by the APXS are similar in composition to those measured at both of the *Viking* landing sites. Due to the distance between the *Pathfinder* site and the two *Viking* landing sites, the similarities in soil compositions suggest that the compositions are influenced by globally distributed airborne dust.

Rocks that have been identified at the *Pathfinder* site are primarily dark gray and partially covered with coatings of bright dust and/or weathered surfaces. See Fig. 17. From the rock chemistry measured by the APXS they appear to be similar to basalt, basaltic andesites, and andesites found on Earth. Rover close-up and IMP images display rocks with a variety of different morphologies, textures and fabrics. Some of the rocks have been hypothesized to be conglomerates composed of rounded pebbles embedded in a finer matrix. Rocks such as these may be the source of numerous rounded pebbles and cobbles that were identified throughout the site. If

these rocks are conglomerates, their formation suggests that running water was present to smooth and round the pebbles and cobbles over long periods of time. The rounded materials would then be deposited into a finer grained sand and clay matrix and lithified before being carried to the Ares site. This suggests a warmer and wetter past in which liquid water was stable on the surface.

The magnetic properties experiment identified the airborne magnetic dust that was deposited on most of the magnetic targets. The dust is characterized as a light yellowish brown, with clay-sized silicate particles and a small amount of a magnetic mineral (believed to be maghemite). The present interpretation for the maghemite formation is that iron was dissolved out of crystal materials by water and that the maghemite is a freeze-dried secondary precipitate.

Observations from wheel tracks and soil mechanics experiments illustrate that the subsurface consisted of a variety of different materials with different physical properties. See Fig. 18. Rover tracks observed in bright drift material preserved individual cleat marks indicating that they are compressible deposits of very fine-grained dust. Several cloddy deposits found at the landing site appear to be composed of poorly sorted dust, sand-sized particles, lumps of soil, and small rock granules and pebbles.

The atmospheric opacity was determined to be 0.5 and changes slightly higher at night as well as early in the morning due to clouds. The sky is

Fig. 17.   End of day IMP image of the rover in the Rock Garden. (*Image courtesy of JPL/NASA.*)

Fig. 19.   Clouds observed at the Ares Valles landing site. (*Image courtesy of JPL/NASA.*)

Fig. 18.   Rover on Merimaid Dune. (*Image courtesy of JPL/NASA.*)

a light yellowish brown color composed of micron-sized particles and water vapor. See Fig. 19. The upper atmosphere, above 60 kilometers (36 miles) altitude, was determined to be very cold and different from warmer measurements obtained by both *Viking* landers. The differences in the measurements may be attributed to seasonal variations at the time of landing. The multiple peaks in the landed pressure measurements and the entry and descent data are indicative of dust uniformly mixed in a warm lower atmosphere.

The meteorology measurements at the site identified diurnal and higher order temperature fluctuations. The barometric minimum was reached at the site on Sol 20 indicating the maximum extent of the winter south polar cap. Temperatures changed abruptly with time and height in the morning; these observations suggest that the warming of the cold morning air by the Sun created upward moving small eddies. Winds were

fairly consistent, and dust devils were detected repeatedly throughout the mission.

Daily Doppler tracking and less frequent two-way ranging during communication sessions between the spacecraft and Deep Space Network antennas resulted in a solution for the location of the lander and the direction of the Mars rotation axis. Combined with earlier results from the *Viking* landers, the estimated precession constant constrains the core radius of Mars to be between 1,300 and 2,000 kilometers (780 to 1200 miles).

From all of the scientific results that have been completed so far, early Mars appears to have been very similar to an early Earth. Some of the materials that make up the crust may be similar to terrestrial continental crust materials. The rounded pebbles, cobbles suggest a possible conglomerate, which supports water rich early Mars. This would imply that the early environment of Mars was warmer and wetter than today and liquid water may have been in equilibrium. Further Mars missions may be able to answer these questions.

R.C. ANDERSON, JPL. Pasadena, CA

PATHOGENESIS.   The pathway followed in the development of a disorder or disease — from early clinical findings and onset through the many stages that may follow.

PATHOGENIC.   Capable of producing disease. The term most commonly is applied to bacteria, viruses, or fungi possessing this property.

PATHOLOGY.   The study of disease, particularly by laboratory methods, including the bacteriology, virology, parasitology, etc. of pathogenic organisms.

PATHOLOGY (Plants).   See **Botany**.

PATHWAY.   A sequence of reactions, usually of a biochemical nature, in which more-complex substances are converted to simple end products, as in the degradation of the components of foods to carbon dioxide and water. Its course is determined largely by preferential factors involving coenzymes and other catalysts. An example is the TCA cycle, which is the common pathway in the degradation of foodstuffs and cell constituents to carbon dioxide and water.

PATINA.   1. The geochemically altered surface of any discrete object such as a mineral, pebble, or rock. 2. A film, usually green, formed on copper and bronze after long atmospheric exposure. 3. A term used by archeologists to describe the altered surface of artifacts.

PATTERN RECOGNITION. It is difficult to make a case for a general unified theory pertaining to pattern recognition. Rather, it may be described as a field of technical interests comprised of various concepts, tools, and techniques. The general objective of pattern recognition is that of classifying an unknown pattern—to place it into one of several classes of patterns. The need for automatic recognition of patterns rises constantly in terms of information processing systems involved in the scientific, medical, business, military, etc., areas. There are numerous instances, of course, where patterns are easily identified by human visual recognition. In such cases, pattern recognition techniques assist in improving the communication between humans and machines. In many other situations, however, patterns are very difficult to recognize quickly through human perception. Detailed examination of electrocardiograms would be an example, but by no means one of the more difficult areas. During the past decade, one of the most rapidly developing and exciting applications for pattern recognition has been in connection with machine vision in robotic systems. See also **Machine Vision (Recognition and Applications)**; and **Robots and Robotics**.

Pattern recognition technique may be divided into three parts: (1) first, *measurements* must be made of the object or event that is to be recognized; (2) measurements will contain relevant as well as much irrelevant data, and thus, pertinent features must be *extracted* that will better characterize the pattern classes; and (3) once features have been established, the input pattern is *classified* to one of a series of pattern categories.

Hundreds and even thousands of measurements may be required to convert a physical pattern to electric signals suitable for feature extraction. Elimination of noise is a particularly critical problem. In character recognition, locating a character on a line and isolating it from touching characters are not trivial problems. Even more difficult is the segmentation of speech patterns.

The objective of the extraction of features is to provide an intermediate process between measurement and classification that will make the design and implementation of the classifier feasible. The formulation of a set of adequate features is largely empirical and intuitive, based on knowledge of the particular pattern recognition problem and characteristics of the various techniques of classification. Features should emphasize differences between classes and deemphasize differences within classes. It is advantageous for features to be invariant for irrelevant variations in the patterns, which might be translation skew and sizes. Statistically independent features can simplify the design and implementation of the classifier. Generally, a feature cannot be evaluated with any certainty as an individual, but must be tested as a member of a set of features. For example, the two features shown in Fig. 1 taken together result in overlapping classes for the given samples. However, if either one of these features is considered by itself, the classes have significant overlap, as can be seen by projecting the samples on either feature axis. This illustrates the requirement of evaluating combinations of features, which aggravates the difficulty of feature selection. This also indicates the relevancy of context in the general sense to pattern recognition, whether it be the relationships among letters to form words or strokes to form letters.

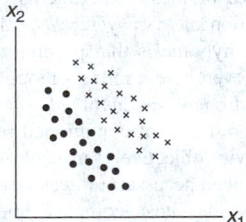

Fig. 1. Statistically dependent features.

Applications. Character recognition provides an automatic means of entering machine-printed and hand-printed alphanumeric data into systems. Machine-printed characters, including special fonts for character recognition, have been recognized without the use of features by template matching or correlation techniques operating on a binary array representation of the character. Features are also used in machine-print application

and are generally required for the recognition of hand-printed characters. A few of the many features that have been investigated are strokes, line endings, line intersections, loops and bays with various sizes, orientations, and locations. See also **Magnetic Ink Character Recognition (MICR);** and **Optical Character Recognition (OCR)**.

Fingerprint classification is normally concerned with identifying fingerprint samples as belonging to various basic types. Generally, some scheme of tracing the ridges in the fingerprints produces directed line segments, line endings, and line intersections as features. From these features, pattern types, such as arches, loops, and whorls, are detected.

The automatic interpretation of aerial photographs and other remote sensor images has application in navigation, earth resources studies, and military reconnaissance. Geometric features, such as dots, straight lines, parallel lines, and x-detectors are of interest in this application. Features providing texture information can be determined from the spatial frequency of the optical density. The analysis of particle tracks in bubble chambers is another form of automatic interpretation of photographs. Events are recognized with the aid of features describing the curvature and end-points of the tracks and intersections of tracks.

Electroencephalogram classification and speech recognition are examples of waveform patterns in which frequency spectrum analysis is important. The energy in selected frequency bands and certain aperiodic waveshapes are features that have been investigated for classifying electroencephalograms. In speech recognition, features such as format frequencies are derived from the energy-frequency-time spectrum called the speech spectrogram. See also **Voice and Sound Production**.

Particularly the approach to classification in many difficult cases requires rather sophisticated mathematical approaches beyond the scope of this article.

Additional Reading

Bezdek, J.C., J. Keller, R. Krisnapuram, and M.R. Pal: *Fuzzy Models and Algorithms for Pattern Recognition and Image Processing*, Kluwer Academic Publishers, Norwell, MA, 1999.

Casacuberta, F. and A. Senfeliv: *Advances in Pattern Recognition and Applications*, World Scientific Publishing Company, Inc., River Edge, NJ, 1994.

Chen, C.H., L.F. Pau, and P.S. Wang: *Handbook of Pattern Recognition and Computer Vision*, 2nd Edition, World Scientific Publishing Company Inc., River Edge, NJ, 2000.

Pal, S.K. and S. Mitra: *Neuro-Fuzzy Pattern Recognition: Methods in Soft Computing*, John Wiley & Sons, Inc., New York, NY, 1999.

Pal, N.R.: *Pattern Recognition in Soft Computing Paradigm*, World Scientific Publishing Company Inc., River Edge, NJ, 2001.

Theodoridis, S. and K. Koutroumbas: *Pattern Recognition*, Academic Press, Inc., San Diego, CA, 1998.

Yu, F.T.S. and S. Jutamulia: *Optical Pattern Recognition*, Cambridge University Press, New York, NY, 1998.

PATTINSON PROCESS. Process for the removal of silver from lead. The silver-lead mixture is melted in one of a series of pots and allowed to cool slowly. The lead, that is free from silver or poorer in silver separates out as crystals, which are removed, leaving the silver-rich lead in the molten state. From a number of such operations in series, a lead rich in silver is obtained, collected, and the silver recovered.

See also **Parkes Process**.

PAULI EXCLUSION PRINCIPLE. The statement that any wave function involving several identical particles must be antisymmetric (must change sign) when the coordinates, including the spin coordinates, of any identical pair are interchanged. If the particles in a system can be considered as occupying definite quantum states, it follows from the principle that no more than one particle of a given kind can occupy a particular state; hence the name, exclusion principle. The principle applies to fermions, but not to bosons. Since electrons, protons, and neutrons are fermions, the Pauli exclusion principle must be used in the assignment of particles to quantum states in theories of atomic and nuclear structure.

PAULING, LINUS CARL (1901–1994). Dr. Linus Pauling was a famous American chemist. He was born in Portland, Oregon and his father was a pharmacist. When he was four, his family moved to Condon, Oregon. On the south edge of town, there was a creek and Pauling sent much time exploring the creek bed and collecting minerals. Eventually Pauling would establish structures for these minerals at the California Institute of Technology.

Pauling enrolled in Oregon Agricultural College (Oregon State University) and was majoring in chemical engineering, but when his mother became ill, he quit school and began working. Because he had show outstanding promise, he was offered a position to be an instructor of quantitative analysis at the college. He taught the class and graduated with a degree in chemical engineering. He attended Cal Tech graduate school. His Ph.D. dissertation was on the crystal structure of different minerals. Pauling received a Guggenheim fellowship to study at the University of Munich where he began applied the concept of quantum mechanics to chemical bonding. Then he took an assistant professorship in chemistry at Cal. Tech and published his paper "The Nature of the Chemical Bond." In 1931, he received the Langmuir Prize of the American Chemical Society for his noteworthy work. In 1933, he was made a member of the National Academy of Sciences. He was 32 years old at that time and he was the youngest appointment ever made.

His serious interest molecular biology began about 1935. He was intrigued by the question of how protein molecules were constructed. As a professor at the California Institute of Technology, he was known for giving "baby toy lectures" because he made models of molecules out of string, rod- and-ball structures, and plastic bubbles in different colors, shapes, and sizes. One day, working with paper, he sketched atoms and chemical bonds and folded them in different ways and discovered the basic structure of the protein molecule.

Pauling, with the help of C.D. Coryell, analyzed the effects of the oxygenation of hemoglobin molecules by measuring their magnetic susceptibility. In 1936, along with A.E. Mirsky, Pauling developed a theory of native, denatured, and coagulated proteins. And in 1950, he and R.B. Corey described the structure of several molecules including muscles, fingernails, hair, and other tissues.

During World War II, Pauling chose not to work on the Manhattan Project for the development of the atomic bomb. He became worried about the radiation the atomic bomb produced and helped organize the Pasadena Federation of Atomic Scientists, a group of scientists working for safe control of nuclear power. He also joined the Emergency Committee of Atomic Scientists, known as the Einstein Committee since it was chaired by Albert Einstein. The aim of the committee was to educate people about atomic weapons. In 1947, he was awarded the presidential Medal of Merit by President Truman for his work on crystal structure, nature of chemical bond, and his efforts to bring about world peace.

In 1954, Pauling received the Nobel Prize in chemistry for his research into the nature of the chemical bond and its application to the elucidation of the structure of complex substances. On October 10, 1962 he was awarded the Nobel Peace Prize for his efforts towards nuclear test ban treaty.

Pauling, in recent years, researched the chemistry of the brain and mental illness, the cause of sickle-cell anemia, and the effects of large doses of Vitamin C on the common cold and on cancer. On August 19, 1994, Pauling, himself, died of cancer at the age of 93.

See also **Electronegativity**.

J. M. I.

PAULI, WOLFGANG ERNST (1900–1958). Pauli was an Austrian theoretical physicist. After WWII, he became an American citizen. When just 20 years of age he wrote "The Theory of Relativity." Later he wrote articles on "Quantum Theory" and "Principles of Wave Mechanics." He is most remembered for formulating the "Pauli exclusion principle". This principle says that two electrons in an atom can never exist in the same state. This is important concept for modern physics. Pauli was awarded the Nobel Prize in physics in 1945 for this discovery.

See also **Pauli Exclusion Principle**; and **Quantum Mechanics**.

J. M. I.

PAULOWNIA. See **Catalpa Trees**.

PAVLOV, IVAN PETROVITCH (1849–1936). Ivan Pavlov was a Russian physiologist, whose classic works in the field of digestion were awarded the Nobel Prize in 1904. He also discovered conditioned reflexes and studied the physiology of higher nervous activity. Pavlov was born in the town of Ryazan on 14 September 1849.

Pavlov was deeply influenced by the ideas of Russian democratic writers (Belinskiy, Pisarev, Tchernishevskiy, Dobrolubov) and by the studies in physiology by G.H. Lewes, L Herman, and especially by I.M. Setchenov. To pursue this new interest he entered the faculty of mathematics and physics in St Petersburg University. This was the choice that defined his future life – that of an atheist and scientist.

Pavlov's scientific studies began at the University under the guidance of Professor F. Ovsiannikov and Professor I. Tsion, and lay in the field of the physiology of digestion.

In 1875, Pavlov received his PhD at the University and entered the Medical and Surgical Academy as a third-year student (in 1881, it became the Military Medical Academy (MMA)). At the Academy he continued his experiments in physiology under the guidance of Professor K. Ustimovitch. Being acknowledged as a masterful experimenter, Pavlov was invited by the famous Russian internist Professor S.P. Botkin (1878–1890) to work at the Laboratory of Physiology, which was a part of the Clinic of Internal Medicine headed by Botkin. Pavlov continued his scientific studies there and also guided the work of a group of students.

In 1890 Pavlov was appointed Professor of Pharmacology at the Military Medical Academy and five years later he was appointed to the then vacant Chair of Physiology, which he held till 1925.

It was at the Institute of Experimental Medicine in the years 1891–1900 that Pavlov did the bulk of his research on the physiology of digestion. It was here that he developed the surgical method of the *chronic* experiment with extensive use of fistulas, which enabled the functions of various organs to be observed continuously under relatively normal conditions. This discovery opened a new era in the development of physiology, for until then the principal method used had been that of *acute* vivisection, and the function of an organism had only been arrived at by a process of analysis. This meant that research into the functioning of any organ necessitated disruption of the normal interrelation between the organ and its environment. Such a method was inadequate as a means of determining how the functions of an organ were regulated or of discovering the laws governing the organism as a whole under normal conditions - problems which had hampered the development of all medical science. With his method of research, Pavlov opened the way for new advances in theoretical and practical medicine. With extreme clarity he showed that the nervous system played the dominant part in regulating the digestive process, and this discovery is in fact the basis of modern physiology of digestion. Pavlov made known the results of his research in this field, which is of great importance in practical medicine, in lectures which he delivered in 1895.

Pavlov's research into the physiology of digestion led him logically to create a science of conditioned reflexes. In his study of the reflex regulation of the activity of the digestive glands, Pavlov paid special attention to the phenomenon of *psychic secretion*, which is caused by food stimuli at a distance from the animal. By employing the method - developed by his colleague D. D. Glinskii in 1895 - of establishing fistulas in the ducts of the salivary glands, Pavlov was able to carry out experiments on the nature of these glands. A series of these experiments caused Pavlov to reject the subjective interpretation of <<psychic>> salivary secretion and, on the basis of Sechenov's hypothesis that psychic activity was of a reflex nature, to conclude that even here a reflex - though not a permanent but a temporary or conditioned one - was involved.

This discovery of the function of conditioned reflexes made it possible to study all psychic activity objectively, instead of resorting to subjective methods as had hitherto been necessary; it was now possible to investigate by experimental means the most complex interrelations between an organism and its external environment.

In 1903, at the 14th International Medical Congress in Madrid, Pavlov read a paper on "The Experimental Psychology and Psychopathology of Animals." In this paper the definition of conditioned and other reflexes was given and it was shown that a conditioned reflex should be regarded as an elementary psychological phenomenon, which at the same time is a physiological one. It followed from this that the conditioned reflex had a clue to the mechanism of the most highly developed forms of reaction in animals and humans to their environment and it made an objective study of their psychic activity possible.

Subsequently, in a systematic program of research, Pavlov transformed Sechenov's theoretical attempt to discover the reflex mechanisms of psychic activity into an experimentally proven theory of conditioned reflexes.

As guiding principles of materialistic teaching on the laws governing the activity of living organisms, Pavlov deduced three principles for the theory of reflexes: the principle of determinism, the principle of analysis and synthesis, and the principle of structure.

The development of these principles by Pavlov and his school helped greatly towards the building-up of a scientific theory of medicine and towards the discovery of laws governing the functioning of the organism as a whole.

Experiments carried out by Pavlov and his pupils showed that conditioned reflexes originate in the cerebral cortex, which acts as the ≪prime distributor and organizer of all activity of the organism≫ and which is responsible for the very delicate equilibrium of an animal with its environment. In 1905 it was established that any external agent could, by coinciding in time with an ordinary reflex, become the conditioned signal for the formation of a new conditioned reflex. In connection with the discovery of this general postulate Pavlov proceeded to investigate *artificial conditioned reflexes*. Research in Pavlov's laboratories over a number of years revealed for the first time the basic laws governing the functioning of the cortex of the great hemispheres. Many physiologists were drawn to the problem of developing Pavlov's basic laws governing the activity of the cerebrum. As a result of all this research there emerged an integrated Pavlovian theory on higher nervous activity. See also **Central and Peripheral Nervous Systems**.

Pavlov lived through the Russian Revolution and the civil war that followed it. Lenin himself recognized his genius and provided financial backing for his research; the new Soviet government built a research complex dedicated exclusively to his experiments.

Additional Reading

Hergenhahn, B.R., and M.H. Olson: *Introduction to the Theories of Learning*, Prentice-Hall, Inc., Upper Saddle River, NJ, 2004.

Pavlov, I.P., and G.V. Anrep: *Conditioned Reflexes: An Investigation of the Physiological Activity of the Cerebral Cortex*, Dover Publications, Mineola, NY, 1927.

Pavlov, I.P.: *I. P. Pavlov: Selected Works*, University Press of the Pacific, Honolulu, HI, 2001.

Todes, D.P.: "Ivan Pavlov," *Exploring the Animal Machine*, Oxford University Press, New York, NY, 2000.

Todes, D.P.: *Pavlov's Physiology Factory: Experiment, Interpretation, Laboratory Enterprise*, Johns Hopkins University Press, Baltimore, MD, 2002.

PAWPAW. See **Laurel Family**.

PEA. See **Bean**.

PEACH-TREE BORER (*Insecta, Lepidoptera; Synanthedon*). Any of three species of moths whose larvae burrow in the sapwood and inner bark of peach trees, sometimes killing younger trees. The moths belong to the family *Aegeriidae*, characterized by the long body and narrow wings, more or less free from scales. All three species attack other fruits as well as peaches.

The larvae of all species are destroyed by digging them out of their burrows and by burning badly infested boughs or entire trees.

PEACH TREES. See **Rose Family**.

PEACH-TWIG BORER (*Insecta, Lepidoptera*). The larva of a small moth, *Anarsia lineatella*, which bores in the young twigs and fruit of peaches and other fruit trees. It is an important pest, chiefly in the western part of the United States, where various spraying programs have been found effective in controlling it.

PEA FAMILY. See **Leguminosae**.

PEAFOWL (*Galliformes, Pavoninae*). Related to the subfamily of turkeys (*Meleagridinae*), peafowl are among the largest gallinaceous birds. They are very long-legged. The extraordinary long tail coverts of the male have strong shafts and magnificent colors and extend far beyond the tail, which has 20 feathers. They form a train with ocellate designs that is erected and spread like a fan to display during courtship. The fan is supported by hidden tail feathers that are much shorter. The feathers have no aftershaft. There are no feathers around the eyes, but rather a crest on the crown. The legs of the male have no spurs. The colors of the female are plain; she has no train and her tail has 18 feathers. There is one genus (*Pavo*) with two species: (1) the *Indian peafowl* (*Pavo cristatus*), short-legged with a marked sex dimorphism, and (2) the green peafowl (*Pavo muticus*), which has a longer and slimmer neck and longer legs, giving it a larger, more pompous appearance. Sex dimorphism is slight. Its crown carries a long bundle of erect, shiny green feathers, resembling a sheaf of ears and having narrow, spiky vanes. The plumage of both sexes is predominantly green. Like the female of the Indian peafowl, the green peafowl female has no train. See Fig. 1.

Fig. 1.   Male green peafowl. (*Pavo muticus*.)

In the wild, the peafowl prefer dense jungle on hilly territory near water. There they live polygamously in small family bands. In the early morning and evening, these large birds seek food in open fields. Wherever natives protect the peacock as the symbol of the god Krishna, it becomes very tame. Then it roams the fields even during the daytime, selecting high trees in the villages for sleeping quarters. The peacock enjoys a special reputation in India as an exterminator of cobras. In fact, it enjoys eating young cobras. As a result, this poisonous snake soon disappears from peacock-inhabited territories.

Peacock calls frequently warn game animals of the presence of tigers and leopards. This is understandable. The bird itself is one of the most common prey of the big cats. Its predilection for snakes and its warning calls have made it a highly valued and popular bird in its home country, probably long before it was introduced as an ornamental bird in parks and gardens of other parts of the world.

In India, the mating season of the peacock essentially coincides with the rainy seasons. The cock surrounds himself with a "harem" of two to five hens. The cock never courts the hen directly, but promptly turns his back on her as soon as she approaches. This, in turn, causes the hen to run around to face him. This curious behavior may be repeated several times. Finally, the hen lies down before the cock, whereupon he folds his fan and treads upon her, in the usual fowl-like manner. Ethologists have various explanations for the peacock's spectacular courtship behavior. One naturalist has observed that the peacock's magnificent fan is used to attract hens from long distances. In the opinion of another naturalist, the spreading fan indicates "feeding" to the hen, even though there is no food. Among many gallinaceous birds, the cock does indeed offer food to the hen and then mates with her.

Peacock chicks practice fan spreading frequently. Their small wings vibrate, and their tiny tail feathers are erected as if a fan were already spread above. Even hens, particularly young hens, sometimes assume this "pompous" attitude. It is mainly during the mating season that the peacock utters its loud, "meowing" call. Both sexes utter this call, but it is more frequently uttered by the males. The Indian natives interpret these calls as "minh-ao" which means "there will be rain." Peacocks do call more frequently before imminent thunderstorms.

The hen hides her nest in the underbrush, but occasionally she uses hollows between the main branches of strong trees, abandoned nests of birds of prey, or even buildings. She lays 3 to 5 creamy-white, thick-shelled eggs, which she incubates for 28 days. After birth, the peacock chicks like to be close under the mother's tail. They grow slowly, and small feather crowns appear after they are 1 month old. But not before they are 3 years old do the young cocks' trains reach their full size. The train can grow for up to 6 years and may reach a length of 160 centimeters (5.2 ft).

The peacock is probably the oldest known ornamental bird. More than 4000 years ago, it was introduced to the cultures of Mesopotamia via trade routes and from there to the Mediterranean nations. Its magnificent coloring, its adherence to its habitat, the ease with which it can be bred, and the way it gets along with other birds have made it the ideal ornamental bird, particularly in large parks. Besides, it destroys very little plant life and is insensitive to fluctuations in the climate.

Many experts consider the green peafowl to be the most beautiful and most impressive of all gallinaceous birds. Green peafowl do not "meow," but instead utter loud trumpet tones that sound like "hah-o-han." They are considered keen-eyed, watchful, and cautious jungle inhabitants. Rarely does an enemy that is sneaking up on them escape their attention. The green peafowl is not resistant to winter weather and requires frost-free shelter during the cold season. Moreover, the green peafowl have an extraordinarily lively, fierce, and courageous temper. The males do not get along well with each other either in "flying cages" or in parks. They become aggressive toward people. When crossbred with the Indian peafowl, a beautiful breed, the Spalding peacock, emerges. They can be raised from hybrids. More adapted to people, these birds are known to live up to 20–30 years.

The subfamily, *Congo peafowl*, is a link between the peafowl and the guinea fowl.

PEAKING CIRCUIT. A circuit used to extend the frequency response of a video amplifier. In a *shunt peaking circuit* (Fig. 1), the impedance of the load circuit of an amplifier stage is raised when a small inductance, included in series with the load resistor, is caused to resonate with the circuit capacitance. In *series peaking*, a small inductance is included in series with the connection to the input of the next stage. At the higher frequencies, this inductance resonates with the input capacitance of the succeeding stage, thus improving the frequency response. Both series and shunt peaking may be included in the same amplifier stage.

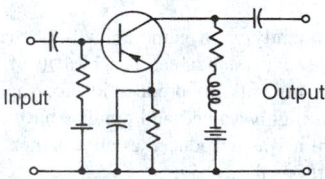

Fig. 1.   Shunt peaking circuit.

PEAKING TRANSFORMER. A transformer which is supplied, through a large series impedance, with a voltage of many times the magnitude necessary to produce normal flux densities in the core. Twice each cycle, therefore, the flux travels rapidly from one saturation region to the other, resulting in a sharp pulse of induced voltage in a secondary winding. The secondary winding is sometimes made of high-permeability material with a smaller cross-sectional area, to increase the saturation effects resulting in a sharper pulse. Used to provide firing pulses for ignitrons and thyratrons, it is sometimes also called an impulse transformer. See also **Transformer**.

PEAK LOAD. See **Gas and Expansion Turbines**; and **Hydroelectric Power**.

PEANUT-GROUNDNUT (*Arachis hypogaea; Leguminosae*). Generally known as the *groundnut* outside the United States, this plant is a

native legume of South America that is now widely cultivated in warm climates throughout the world.

The plant is a bush annual with pinnately compound leaves and rather showy yellow flowers. After fertilization, the flower stalk elongates greatly and bends downward so that the ovary is pushed into the ground. There it develops into the familiar peanut or groundnut with its two or more seeds. See Fig. 1.

Fig. 1.   Mature and immature peanut (groundnut) pods, showing fruiting habit of the plant. (*USDA photo.*)

Large crops are raised annually in the United States. Continuing research on the peanut, because of its economic importance, is conducted at the Tifton, Georgia, experimental station of the U.S. Department of Agriculture.

Large crops are raised annually. Much of the crop (Fig. 2) is roasted in the shell or shelled and salted, and so marketed. Large quantities are ground for peanut butter or crushed for peanut oil. In some areas, boiled peanut are well regarded as snack food. See also **Leguminosae**.

Harvesting practices are among the most critical in peanut culture because of their far-reaching effect on yield, quality, and total value of the crop. In modern harvesting procedures, digging, shaking, and winnowing are accomplished in a single operation. Special machines permit peanuts to be combine harvested directly from the windrow. (A windrow is a row of any crop lined in a row prior to separation.) Whole plants interlaced as a ribbon are brought into the combine, where pods are removed, cleaned, stemmed, and conveyed into a bulk bin on the combine. See Figure 3.

PEANUT OIL. See **Vegetable Oils (Edible)**.

PEARL. A gem formed by bivalve mollusks, particularly by several marine species known as pearl oysters. Pearls are formed as a protection against the irritation caused by foreign objects, either parasites or bits of gravel, which lodge inside the shell. A fold of soft tissue envelops the foreign particle and deposits layer after layer of nacre on it, similar to the mother-of-pearl lining the shell.

PEARSON DISTRIBUTIONS. A family of distributions devised by Karl Pearson in 1895 and subsequently. If f is the frequency and x the

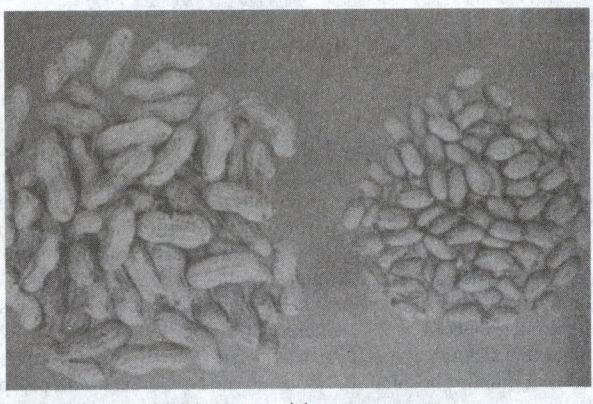

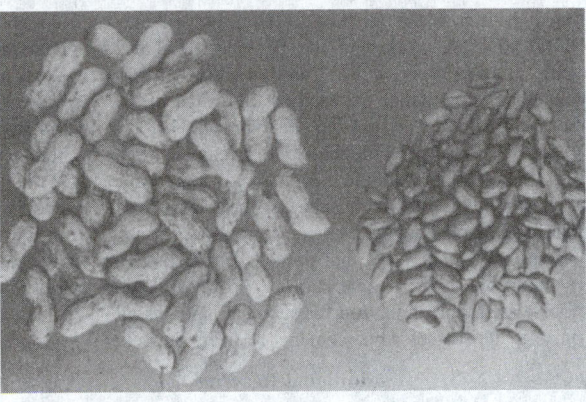

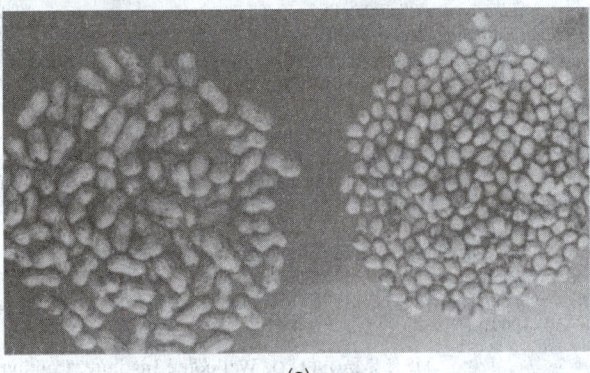

Fig. 2. Varieties of peanuts (groundnuts), showing pods at left and shelled nuts at right: (**a**) Virginia Bunch variety; (**b**) Holland Jumbo variety; (**c**) Dixie Spanish variety. (*USDA photos.*)

Fig. 3. Peanut (groundnut) combine with a capacity ranging from 0.5 to 2 acres (0.2 to 0.8 hectare) per hour. Note parking jack in foreground, which is designed to keep the combine at correct height for attachment to tractor. (*Lilliston Corporation.*)

variate, the family is defined by

$$\frac{1}{f}\frac{df}{dx} = \frac{a-x}{b_0 - b_1 x + b_2 x^2}$$

Pearson distinguished 12 types, according to the values of the constants. Type J is the form

$$f = \frac{x^{p-1}(1-x)^{q-1}}{\beta(p, q)}$$

and is also known as the *Beta distribution*.

Type III is of the form

$$f = \frac{x^{p-1}e^{-x}}{\Gamma(p)}$$

and is also known as the *Gamma distribution*.

The normal distribution is also one of the Pearson system. The other types are less important. Taken as a class, the distributions provide a very flexible set, to one of which many practically occurring distributions may be satisfactorily fitted.

PEARSON, KARL (1857–1936). Karl Pearson wan an English mathematician and statistician who provided the foundation to the modern theory of mathematical statistics.

The second son of William Pearson and Fanny Smith, Karl Pearson was born in London on 27 March 1857, but was of Yorkshire descent. He came from a family of dissenters who were of Quaker stock. By the time Pearson was in his 20s, he had rejected Christianity and become a Freethinker. Though he did not regard himself as an atheist, "he vigorously denied the possibility of a god," Politically, he was a socialist whose outlook was similar to the Fabians, but he never joined the Fabian Society.

In 1875 Pearson received an Open Fellowship from King's College, Cambridge to read mathematics. He graduated with honors, being the Third Wrangler in the Mathematics Tripos, soon after he received a fellowship from King's College, which he held for seven years. In 1903 he was made an Honorary Fellow of King's.

Having started his career as an elastician (that is, someone who derived mathematical equations for elastic properties of matter), Pearson pursued a number of areas before he settled on mathematical statistics. He left for Heidelberg in April 1879 to "improve his German to study physics and metaphysics." His time in Germany was a period of self-discovery. During his time in Heidelberg Pearson read Berkeley, Fichte, Locke, Kant and Spinoza, but he eventually abandoned philosophy because "it made him miserable." Soon afterwards he realized that he "would never be a great mathematician or physicist because he was not a born genius." A year later, he took up rooms back in London at the Inner Temple and read law at Lincoln's Inn. He was called to the Bar at the end of 1881 and practiced the law for a very short time only. By 1882 Pearson had decided that he did not want to pursue the law because it depressed him, and he decided instead to "devote his time to the religious producing of German literature before 1300." From 1882 to 1884, he lectured on German society from the medieval period up to the sixteenth century. In the late spring of 1884 he was offered a post in German at Cambridge.

Pearson, however, found all these pursuits deeply dissatisfying and he "longed to be working with symbols rather than words." He had been writing papers on optics, ether squirts and on the theory of elastic solids and fluids. He deputized mathematics at King's College, London and University College London (UCL) in 1883. Between 1879 and 1884 he applied for more than six mathematical posts and in June 1884 he received the Chair of Mechanism and Applied Mathematics at UCL. During Pearson's first six years at UCL, he taught mathematical physics to engineering students.

Pearson was a founding member of the Men's and Women's Club established in 1885 "for the free and unreserved discussion of all matters in any way connected with the mutual position and relation of men and women." Among the various members was Marie Sharpe whom he married in June 1890. They had three children, Sigrid, Helga and Egon. Six months after his marriage, he took up another teaching post in the Gresham Chair of Geometry, at Gresham College in the City of London, which he held for three years concurrently with his post at UCL. Between February 1891 and November 1893, Pearson delivered 38 lectures. His first eight lectures formed the basis of his book, *The Grammar of Science*, which was published in several languages.

Pearson's last twelve Gresham Lectures signified a turning-point in his career owing to his relationship with W.F.R. Weldon—who was the first biologist Pearson met who was interested in using a statistical approach for problems of Darwinian evolution. Their emphasis on Darwinian population of species not only implied the necessity of systematically measuring variation, but prompted the re-conceptualization of statistical populations. Moreover, it was this mathematization of Darwin which led to a paradigmatic shift for Pearson from the Aristotelian essentialism underpinning the earlier use and development of social and vital statistics. Weldon's questions not only provided the impetus for Pearson's seminal statistical work, but led eventually to the creation of the Biometric School at UCL. See also **Evolution (History)**; and **Weldon, Walter Frank Raphael (1860–1906)**.

Following the success of his Gresham Lectures, Pearson began to teach statistics to students at UCL in October 1894. By 1895, he worked out the mathematical properties of the product-moment correlation coefficient (which measures the relationship between two continuous variables) and simple regression (used for the linear prediction between two continuous variables). In 1896, Pearson introduced matrix algebra into statistical theory and also introduced the coefficient of variation, the standard error of estimate, multiple regression and multiple correlation. From 1896 to 1911, Pearson devised more than 18 methods of correlation. In 1900, he devised the tetrachoric correlation and the phi-coefficient for discrete variables. Nine years later, he devised the biserial correlation when one variable is continuous and the other is discontinuous. Pearson was elected a Fellow of the Royal Society in 1896 and awarded their Darwin Medal in 1898.

At the turn of the century, Pearson reached a fundamental breakthrough in his development of a modern theory of statistics when he found the exact chi-square distribution and devised the chi-square (χ^2, P) goodness of fit test. This achievement was the outcome of the previous eight years of curve fitting for asymmetrical distributions. Four years later, he extended this to the analysis of manifold contingency tables and introduced the chi-square test of independence (which R.A. Fisher termed the chi-square statistic in 1923).

In the twentieth century Pearson established and ran four laboratories. He set up the Drapers' Biometric Laboratory in 1903 following a grant from the Worshipful Drapers' Company (who funded Pearson annually for work in this laboratory until his retirement in 1933). The methodology incorporated in the Drapers' Biometric Laboratory was twofold: the first was mathematical and the second involved the use of numerous instruments. The problems investigated by the biometricians included natural selection, Mendelian genetics and Galton's law of ancestral inheritance, craniometry, physical anthropology and theoretical aspects of mathematical statistics. By 1915, Pearson established the first degree course in mathematical statistics in Britain. A year after Pearson had established the Biometric Laboratory, the Worshipful Drapers' Company gave him a grant so that he could establish an Astronomical Laboratory equipped with a transit circle and a four-inch equatorial refractor. See also **Genetics and Gene Science (Classical)**; and **Mendel, Gregor Johann (1822–1884)**.

In 1907, Francis Galton (who was then 85 years old) wanted to step down as Director from the Eugenics Record Office which he had set up three years earlier, and he asked Pearson if he would take it on. Pearson took on the directorship reluctantly and renamed the office the Galton Eugenics Laboratory. Pearson made very little use of his biometric methods in this Laboratory; instead he developed a completely different methodology for problems relating to eugenics involving actuarial death

rates and a very highly specialized use of family pedigrees assembled in an attempt to discover the inheritance of various diseases. In 1924, Pearson set up the Anthropometric Laboratory, which was made possible by a gift from one of Pearson's students, Ethel Eldeteron. See also **Galton, Sir Francis (1822–1911)**.

When Galton died in January 1911, his estate was bequeathed to UCL and he named Pearson as the first Professor of Eugenics. The Drapers' Biometric and the Galton Eugenics Laboratories, which continued to receive separate funding, became incorporated into the Department of Applied Statistics. Pearson then proceeded to raise funding for a new building for his department. Adequate funding had been raised by 1914 and contracts for the fittings had been made. These developments and further biometric work were shattered by the onset of the First World War. The new laboratory building was taken over by the government to be used as a military hospital. Pearson and his co-workers took on special war duties. Pearson was not able to occupy the building until 4 December 1922.

His wife, Marie Sharpe, died in 1928 and in 1929 he married Margaret Victoria Child, a co-worker in the Biometric Laboratory. Pearson was made Emeritus Professor in 1933 and was given a room in the Zoology Department at UCL, which he used as the office of *Biometrika* from his retirement until his death in 1936. Pearson was offered honors twice by King George V, but he declined on both occasions. Pearson believed that "all medals and honors should be given to young men, they encourage them when they begin to doubt whether their work was of value."

Pearson's statistical work not only provided continuity from the mathematical and statistical work that preceded him, but also engendered the modern theory of mathematical statistics in the twentieth century which, in turn, provided the foundation for such statisticians as R. A. Fisher who went on to make further advancements for a modern theory of statistics.

Additional Reading

Eisenhart, C.: *Karl Pearson: Dictionary of Scientific Biography*, vol. 10, Charles Scribner's Sons, New York, NY, 1974. pp. 447–473.

Mackenzie, D.: *Statistics in Britain 1865–1930: The Social Construction of Scientific Knowledge*, Edinburgh University Press, Edinburgh, Scotland, 1981.

Magnello, M.E.: "Karl Pearson's Gresham Lectures: W.F.R. Weldon, Speciation and the Origins of Pearsonian Statistics," *British Journal for the History of Science*, **29**: 43–64 (1996).

Magnello, M.E.: "Karl Pearson," In: Armitage, P., and T. Colton: *Encyclopedia of Biostatistics*, vol. 4, pp. 3308–3315, Wiley, Chichester, UK, 1998.

Magnello, M.E.: "Karl Pearson's Mathematisation of Inheritance: From Galton's Ancestral Heredity to Mendelian Genetics (1895–1909)," *Annals of Science*, **55**, 35–94 (1998).

Magnello, M.E.: "The Non-correlation of Biometrics and Eugenics: Rival Forms of Laboratory Work in Karl Pearson's Career at University College London." Part 1, *History of Science*, **37** (March), 79–106; Part 2, *History of Science*, **37** (June), 123–150 (1999).

M. E. MAGNELLO, Wellcome Institute, London, UK

PEAR TREES. See **Rose Family**.

PEAT. Semicarbonized residue of plants formed in water-saturated environments (bogs and marshes). It occurs in surface layers 3–10 ft thick and has a water content of 85%. Before peat can be used for chemical or fuel purposes it must be field-dried to a water content of 30–40%. Since the dried product is susceptible to autoignition, storage conditions must minimize this risk. Peat is easily converted to hydrocarbons and is an excellent source of natural gas; when dry it can be used directly as a fuel. The U.S. has peat sources second only to those of the former U.S.S.R., located in Alaska, the north-central states, and Maine, where processing on a large scale is planned. Their total energy content is said to be equivalent to 240 billion barrels of petroleum. The peat can be gasified for production of methanol after mechanical dewatering. Experimental conversion studies have been under way for some time. Substantial quantities of oil, ammonia, and sulfur can be obtained as by-products. See also **Coal**.

PEAT BOG. See **Bog**; and **Swamp**.

PEA WEEVIL (*Insecta, Coleoptera*). A beetle which attacks growing peas in the pod. The most effective control measures are to avoid planting infested seed or to fumigate it with carbon disulfide before planting.

PEBBLE MILL.   A jacketed steel cylinder rotating on a horizontal axis and containing flint or porcelain pebbles as the grinding medium. Its operation is similar to that of a ball mill. It is used for grinding and mixing of dry chemicals, pigments, food products, and the like. Pebble mills are usually lined with alumina, buhrstone, or similar material to protect the walls from wear. See also **Ball, Pebble, and Rod Mills**.

PECAN TREE.   See **Hickory and Wingnut Trees**.

PECCARY.   See **Suines**.

PECLET NUMBER (symbol Np_e).   A nondimensional number arising in problems of heat transfer in fluids. It is the ratio of heat advection to heat diffusion and may be written $Np_e = Ul/k$ where U is a characteristic velocity; l is a characteristic length; and k is the thermometric conductivity. Also, $Np_e = N_{Re}N_{Pr}$ where N_{Re} is the Reynolds number and N_{Pr} is the Prandtl number.

PECTINS.   A *pectic substance* is a group designation for those complex carbohydrate derivatives that occur in or are prepared from plants and contain a large proportion of anhydrogalacturonic acid units, which are thought to exist in a chainlike combination. The carboxyl groups of polyglacturonic acids may be partially esterified by methyl groups and partly or completely neutralized by one or more bases. The general term *pectin* (or *pectins*) designates those water-soluble pectinic acids of varying methyl ester content and degree of neutralization which are capable of forming gels with sugar and acid under suitable conditions. The term *protopectin* is applied to the water-insoluble parent pectic substances which occur in plants and which upon restricted hydrolysis yield pectin or pectinic acids. *Pectic acids* is a term that is applied to pectic substances mostly composed of colloidal polygalacturonic acids and essentially free from methyl ester groups. The salts of pectic acids are either normal or acid *pectates*. The term *pectinic acids* is used for colloidal polyglacturonic acids containing more than a negligible proportion of methyl ester groups. Pectinic acids, under suitable conditions, are capable of forming gels with sugar and acid, or, if suitably low in methosyl content, with certain metallic ions. The salts of pectinic acids are either normal or acid *pectinates*.

Pectins occur commonly in plants, particularly in succulent tissues, and are characterized by the polygalacturonic acids that are fundamental to their structure. The pectins are important emulsifying, gelling, stabilizing, and thickening agents used in the preparation of numerous food products. About 75% of the pectins produced are used in making fruit jams, jellies, marmalades, and similar products. Additional uses include the preparation of mayonnaise, salad dressings, malted milk beverages, frozen dessert mixes, and frozen fruits and berries (to prevent leakage upon thawing), among others. The addition of a dilute pectin solution to milk coagulates the casein. In many food products, the use of pectins as stabilizers is preferred, since they blend better into the flavor complex than do many gums, starches, or a number of carbohydrate derivatives. Pectin jellies do not melt at temperatures below 49 °C, a distinct advantage over gelatin gels that require refrigeration. Pectins also have a number of nonfood uses, including pharmaceuticals and cosmetics.

The location of various pectin substances in plant tissues is well established. Pectins make up most of the middle lamella in unripe fruit and are to be found in the cell walls and in small proportion in all plant tissues. The genesis and fate of pectins in plant tissues have not been fully determined.

Citrus peel, apple pomace from juice manufacture, and beet pulp left over from the manufacture of sucrose are common commercial sources of pectins. After some preliminary purification of the raw material, the extraction is usually performed with hot dilute acid (pH = 1.0–3.5 in a temperature range of 70–90 °C). The pectin is then precipitated from the extract with ethanol or isopropanol, or with metal salts (copper or aluminum). The metal ions have to be subsequently removed by washing with water or acid ethanol. Specific formulas for denatured ethanol for use in pectin manufacture are used. The precipitates are purified, dried, and pulverized to form the yellowish-white powder of commerce.

Pectin substances in solution behave as typical colloids. See also **Colloid System**. Dry, purified pectins are light in color and soluble in hot water to the extent of 2–3%. The pH of pectin solutions is usually 2–3.5.

The proportion of sugar which pectin will form into a firm jelly determines the *jelly grade* of the pectin. In a jelly, jam, or marmalade, the proportions of total solids of sugars, the pH, and the proportion and nature of the pectin used will determine the extent of jellification obtained. The use of pectin in fruit jams and related products is approved by most food regulators because the addition is believed to compensate for an incidental natural deficiency.

In pectic acids, all carboxyl groups are free, or at least not present as the methyl ester. Under suitable conditions, pectins will form jellies with sugar and acid, whereas the low-ester pectins will form *gels* with traces of polyvalent ions. The general structure of pectin is:

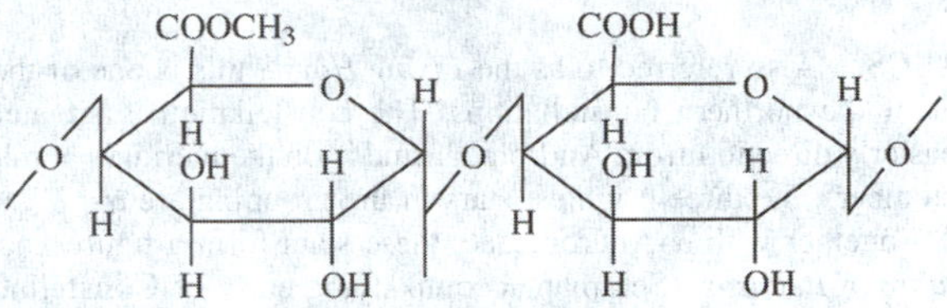

Additional Reading

Additional Reading

Fishman, M.L. and J.J. Jen: *Chemistry and Function of Pectins*, American Chemical Society, Washington, DC, 1986.
Lutz, P.L., H. Schols, R. Visser, and F. Voragen: *Advances in Pectin and Pectinase Research*, Kluwer Academic Publishers, Norwell, MA, 2003.
Quilici-Timmecke, J.: *New Nutrients against Cancer: Modified Citrus Pectin, Soybeans, Lycopene and Other '90's Cancer Fighters*, Keats Publishing, Inc. Chicago, IL, 1998.
Seymour, G.B., and P. Knox: *Pectins and Their Manipulation*, CRC Press LLC., Boca Raton, FL, 2002.
Walter, R.H.: *The Chemistry and Technology of Pectin*, Academic Press, Inc., San Diego, CA, 1997.
Wood, W.A. and S.T. Kellogg: *Biomass: Lignin, Pectin, and Chitin*, Vol. 161, Academic Press, Inc., San Diego, CA, 1988.

PECTORAL FIN.   See **Fishes**.

PEDERSEN, CHARLES JOHN (1904–1989).   An American born in Korea, Pedersen received a M.S. from M.I.T. in 1927. In 1987 he was awarded the Nobel Prize for Chemistry for his work in elucidating mechanisms of molecular recognition, which are fundamental to the enzymic catalysis, regulation and transport. He reported that alkali metal ions could be bound by crown ethers into a more rigid, layered structure, in which the alkali metal ion was bound into the center of the ring. This field of study is called host-guest chemistry.

PEDIATRICS.   That branch of medicine concerned with the prevention, diagnosis, and treatment of the diseases and disorders of children.

PEDICULOSIS.   Infestation of the body with lice. Infestation of the scalp, most prevalent in women, is caused by *Pediculus humanus* var. *capitis* (head lice). The condition is accompanied by severe pruritus (itching) and sometimes is complicated by secondary infections, often arising from contamination of the skin as the result of aggressive scratching. Left unattended, the hair becomes matted with exudation of the lice. Treatment includes gamma benzene hexachloride shampoo or benzyl benzoate (20% solution). Occasionally, a second application will be required within one week. Physicians usually prescribe a systemic antibiotic to allay secondary infection if such is indicated. In body infestation, the cause is *Pediculus humanus* var. *corporis* (body lice) which tend to live within the crevices of clothing, but which attack the skin, forming macules and wheals. Pruritus can be severe. Scratching may initiate secondary infection. The best solution is destruction of infested clothing, but if garments must be retained, they should be boiled once or several times and ironed with a hot iron. Pediculosis pubis (sometimes called *crabs*) results from infestation of the pubic area with *Phthirus pubis*. Uncommonly, the condition also may be found on the eyelashes. Examination reveals lice attached to the skin and lice eggs attached to the hair shafts. Macules (bluish in color) are formed when the lice suck blood from the skin—in the pubic area and sometimes on the thighs.

Gamma benzene hexachloride lotion or cream can be used for treatment. This agent should not be used on the face or eyelids. Physicians usually prescribe 0.5% physotigmine for local application in the case of eyelash infestation.

PEDIGREE METHOD. See **Plant Breeding**.

PEDOLOGY. The study of the origin and classification of soils. The investigation of the regolith as a fundamental natural resource or the basis of terrestrial plant life.

PEEWIT. See **Waders, Shorebirds, and Gulls**.

PEGASUS. Also referred to as the *Flying Horse*, this is one of the most ancient of the northern constellations. The constellation is situated in a southeasterly direction from Andromeda and is on the meridian at midnight in September. The three brightest stars in the grouping are α-, β-, and γ-Pegasi. Together with α-Andromedae, these stars form a figure known as the *square of Pegasus*. (See map accompanying entry of **Constellations**.)

PEGMATITE. The term *pegmatite*, derived from the Greek word meaning "joined together," was first applied by Haüy in 1822 to a peculiar interpenetrating growth of quartz and feldspar sometimes called graphic granite from its resemblance to written characters, particularly those of the Hebrew language. Pegmatite is also used to designate those coarse-grained dikes and sheets, chiefly of granite or syenite, that are apophyses of stocks or batholiths, or of the residual magma, during their congelation. The individual minerals may often reach great size. Granite pegmatites are chiefly composed of alkali feldspar and quartz with some muscovite or biotite but may carry such minerals as tourmaline, topaz, beryl, fluorite, apatite, garnet, lepidolite, etc. See also **Mineralogy**.

PELAGIC. See **Ocean Resources (Living)**.

PELECANIFORMES *(Aves)*. Among the many bird groups whose members live on or at the water, the Pelecaniformes are of especially striking appearance because of the peculiar structure of their feet. Their toes are joined by more or less well-developed webs, but the web in their case includes, in contrast to ducks and geese, the hind toe, which is directed forward and to the inside. This "paddle foot" is found in all members of this order.

Birds of this order are medium sized to very large, and feed exclusively on animal food. Most species obtain their food from the sea. There are a total of 6 well- differentiated families, with 7 genera.

The Tropic-Birds (*Phaethontidae*, genus *Phaethon*) with 3 species; the Pelicans (*Pelecanidae*, genus *Pelecanus*) with 7 species; the Cormorants (*Phalacrocoracidae*, genus *Phalacrocorax*) with 28 species; the Anhingas or Darters (*Anhingidae*, genus *Anhinga*) with 2 species; the Gannets (*Sulidae*, genera *Morus* and *Sula*) with 9 species between them; and the Frigate-Birds (*Fregatidae*, genus *Fregata*) with 5 species.

The Tropic-Birds (family *Phaethontidae*, with only 1 genus, *Phaethon*) are birds of the high seas with a pigeon-like flight. The length is 80–100 centimeters (31–39 inches), of which the body length is 30–45 centimeters (12–18 inches). The wing span measures 92–109 centimeters (36–43 inches) and the weight is 300–750 grams ($10\frac{1}{2}$–$26\frac{1}{2}$ ounces). They are predominantly white with a very long central tail feather. The short legs are far back on the body. Tropic-birds can hardly walk, but they can dig and scrape. There are 3 species: The Red-Billed Tropic-Bird (*Phaethon aethereus*), which reaches a length of 100 centimeters (39 inches); the White-Tailed Tropic-Bird (*Phaethon lepturus*), which reaches a length of 80 centimeters (31 inches); and the Red-Tailed Tropic-Bird (*Phaethon rubricauda*), which is the largest species, with a length of 100 centimeters (39 inches).

The largest birds of this order are generally known as the Pelicans (family *Pelecanidae*). The length is 170–180 centimeters (67–71 inches) and the wingspan is up to almost 300 centimeters (118 inches) in the Dalmatian pelican. The weight ranges from 7–14 kilograms (15–31 pounds). They appear clumsy, but because of the air in the bones and the skin, they are relatively light birds with large bodies, long broad wings, fairly long necks, and gigantic beaks. Between the branches of the lower mandible there is a distensible skin pouch; the upper mandible serves merely as a

flat lid to cover it. The tongue is minute, the legs are short, the feet are large, and the 4 toes are connected by webs.

Pelicans float high on the water and they carry their wings slightly raised; since they have no wing pockets, the beak rests on the slightly curved neck. In flight the head is drawn back onto the shoulders. The flight is light and elegant; gliding often alternates with wing beats. The food consists exclusively of fish, which are scooped up in the bill; only the brown pelican is a plunge diver. They are sociable birds that fly in small groups or larger flocks, mostly in a diagonal line with respect to the direction which they are travelling. They search for food together, and often nest in very large colonies of up to several thousand birds. They breed on a base of reeds and branches, and on shore often on just a few feathers.

The Cormorant family (*Phalacrocoracidae*) has only 1 genus, *Phalacrocorax*. The length is 48–92 centimeters (19–36 inches), and the weight is 0.7–3.4 kilograms ($1\frac{1}{2}$–8 pounds). They are equally well adapted for flying and for swimming. Relatively clumsy on land, they are still not as helpless as some other members of the order. They lie rather deeply in the water when swimming, since the air spaces in their bones are very small. There are no external nares, and the edges of the beak are somewhat toothed. The head and neck have powerful muscles for closing the beak, which in part originate from special long, sesamoid bones behind the back of the head, and which are needed to maintain a grip on fish that have been caught. The plumage usually has crests in the breeding season. They are distributed worldwide, with 29 species, more species than in all the other Pelecaniformes together.

Two species of freshwater Pelecaniformes must be considered as a separate family because of a number of peculiarities of their body structure, although they are in many respects close to the cormorants. These are the Anhingas or Darters (family *Anhingidae*, genus *Anhinga*). The length is 90 centimeters (35 inches). The bill is straight and pointed, and is finely toothed on both cutting edges. The neck has 20 vertebrae and, when the bird is sitting or flying, it is bent into an S or even a G shape. The tail is long, stiff, and rounded at the end. They walk and swim underwater when diving. There are 2 species: The American Anhinga (*Anhinga anhinga*), with no subspecies; the Old World Anhinga (*Anhinga rufa*), which has 3 subspecies.

Anhingas live on freshwater fish and other aquatic animals. Like herons, they have dagger-like beaks and long G-shaped necks. But while herons wade in shallow water and stalk or wait for their prey above the water surface, anhingas hunt for their prey underwater. A special hinge and muscle arrangement on the eight and ninth cervical vertebrae enables them to thrust the head rapidly forward so that the prey is stabbed and stunned.

Like the plumage of the cormorants, the anhinga's is water permeable. This adaptation reduces buoyancy and enables anhingas to submerge silently without attracting the attention of prey animals or enemies. They often swim with only the head and the thin neck projecting above the surface, and when doing so they really resemble a swimming snake; hence they have another name, the snake-bird.

In the air they soar and glide like pelicans, skillfully using thermal air currents. Anhingas are very well adapted to both air and water; they are gliders as well as spear fishers.

The Gannets or Boobies are predominantly black and white seabirds (family *Sulidae*). The length is 70–100 centimeters ($27\frac{1}{2}$–39 inches), and the weight is 1.5–3.5 kilograms (3–8 pounds). The strong beak is pointed and conical, with fine serrations on the cutting edge near the tip. The bare parts of the face, throat, and feet are often colored. The strong feet have well-developed webs. Gannets breathe through a specially constructed palate. There are 2 genera with 9 species.

All gannets obtain their food by diving. Where they are numerous, this method of feeding makes a fascinating spectacle.

The Frigate-Birds (family *Fregatidae*) exceed the members of all the other families in aerial mastery. Almost half of their weight consists of the breast muscles and feathers; their load per unit of wing surface is extraordinarily small.

There is only 1 genus, *Fregata*, with a length of 75–112 centimeters ($29\frac{1}{2}$–44 inches), a wing span of 176–230 centimeters (69–$90\frac{1}{2}$ inches), and a weight of up to 1.5 kilograms (3 pounds). The beak is long and bent into a hook at the tip. The wings are narrow and the lower arm and hand bones are strongly elongated. The tail is deeply forked, often spread and then closed again in flight. The small feet are almost without webs. They are restricted to tropical and subtropical seas; they live mainly where flying fish are common, in water of at least 25 °C. (77 °F.).

There are 5 species. The Magnificent Frigate-Bird (*Fregata magnificens*) is the largest. Its length is 103–112 centimeters ($40\frac{1}{2}$–44 inches), the wing span is 230 centimeters ($90\frac{1}{2}$ inches), the beak is 12 centimeters ($4\frac{1}{2}$ inches), and the weight is 1.5 kilograms (3 pounds). The males have a white breast band and brownish upper. The Ascension Frigate-Bird (*Fregata aquila*) weighs 1.2 kilograms ($2\frac{1}{2}$ pounds). The female has a somewhat greenish gloss, while the males are brownish on the upper breast, nape, and wing band. In the Christmas Frigate-Bird (*Fregata andrewsi*), both the male and female have the same coloring. The Lesser Frigate-Bird (*Fregata ariel*); has a length of 82 centimeters (32 inches). The Great Frigate-Bird (*Fregata minor*); reaches a length of 95 centimeters (37 inches).

Although frigate-birds sometimes fly out over the sea, they tend to breed all year round and, as a rule, stay near their home islands. When a frigate-bird is seen at sea, land is generally not far away. Their ability to find their way is such that these birds are used to send messages in the South Sea islands, as are carrier pigeons elsewhere in the world.

Frigate-birds generally build their nests in low shrubs or trees, and only rarely on bare ground. The nesting colonies are generally close to those of other seabirds, particularly terns or gannets, which frigate-birds rob of their prey or young. See also **Pelicans and Cormorants**.

PELE'S HAIR. A fibrous, basic, natural glass (tachyllite). The congealed liquid lava blown out of volcanoes. Type locality, the Hawaiian Islands.

PELICANS AND CORMORANTS (*Aves, Pelecaniformes*). These birds are widely distributed and of relatively few species. Millions of pelicans, cormorants, and boobies inhabit the islands off the coast of Peru, as evidenced by the views of Fig. 1. The birds almost completely cover the islands when they come in to roost in the evening. They deposit a tremendous amount of droppings (guano), which dries and hardens quickly in the prevailing dry climate. As many as 750 tons of guano per acre per year may be deposited. Food for this large number of birds is provided by abundant fish life in the surrounding cold waters.

The common pelican of North America (*Pelecanus erythrorhynchos*) is chiefly white with orange beak and feet during the breeding season. The bird is large with a long neck, short legs, and webbed toes. See Fig. 2. The beak is very long, and the lower mandible bears a flexible pouch, which can be distended to accommodate a large amount of food. In addition to fish, the pelican consumes other aquatic forms. The pelican lives in colonies. Some species build nests on the ground; others in low bushes or trees. The birds fly at high elevations and often soar in formation. The pelican has practically no voice. The pelican winters along the sea coast, and in summer, it may be found around inland lakes. The white pelican of Africa and Asia is much like the American pelican except the plumage has a delicate pink coloration. Possibly the most attractive pelican is the black and white pelican of Australia (*P. conspicillatus*). Other species display brown and gray plumage. See also Fig. 3.

The cormorant is a large bird of slender build with a moderately long neck and slender beak, slightly hooked at the tip. The feet are webbed, and the birds are strong swimmers and divers. They live entirely on fish. The several species are widely distributed, the common cormorant (*Phalacrocorax*) occurring in eastern North America, Europe, Asia, and northern Africa. In Japan and China, cormorants are kept in captivity to be used for fishing. A ring or strap around the bird's neck keeps it from swallowing the fish that it catches, although some are said to be so well trained that they bring fish to their keepers without this check. The birds are eaten in some parts of the world.

Fig. 2. American white pelican. Bill shows horny growth that appears in breeding season (*New York Zoological Society*).

Fig. 1. Guano birds. (*R.C. and G.E. Murphy*.)

Fig. 3. Formation of pelicans over coastal waters of northern Florida. (*G.D. Considine*.)

The darter (*Anhinga*) is a long-necked diving bird with a long, sharp beak, found in all continents. Resembling the cormorants, the darter is also known as the snake bird or snake neck, and certain species are called wryneck and water turkey. The frigate bird (*Fregata*) is a marine bird related to the cormorant. It is of slender, long-winged build and powerful in flight. The bird lives chiefly on fish sometimes forcibly obtained from other birds. Also related to the cormorant is the gannet (*Sula*), a large fishing bird found on the seacoasts in the higher latitudes of both hemispheres. It breeds in dense colonies on steep rocky cliffs. See also **Pelecaniformes**.

PELLAGRA.   See Niacin.

PELLET. A small unit of a light, bulky material compressed into any of several shapes and sizes, usually either spherical or rectangular. The operation is performed on a pellet mill, which consists essentially of a pair of steel rollers around which rotates a circular perforated metal die. Material is fed into the chambers above and below the inner face of the die. As the die turns, in contact with the rollers, the latter also turns, thus compressing the material and forcing it through the holes in the die at the point of tangency, where the extruded segment is sheared off by knives. Pelletizing is advantageous for fluffy particulates that are difficult to handle in loose form, e.g., carbon black, clays, plastic molding powders, etc. Binding materials called excipients are often used. See also **Ball, Pebble, and Rod Mills**.

PELL'S EQUATION.   See Number Theory.

PELORUS (or Dumb Compass). An instrument once widely used on board ship for taking bearings of external objects. The pelorus consists fundamentally of a circular plate, heavily ballasted and mounted in gimbals. This plate has two pairs of indicators, one pair parallel to the keel of the ship and the other perpendicular to the keel. Concentric with the circular plate, and capable of being rotated independently of each other about a vertical axis, are a graduated dial plate and an alidade or arm for reading angles. The dial plate is graduated in a manner similar to the compass card and may be clamped to the circular plate in any desired position. The alidade carries sighting vanes, and the line through these vanes passes through the axis of rotation of the instrument, carrying indicators at each end, which may be read on the dial plate.

PELTIER EFFECT.   See Thermocouple.

PELTON WHEEL.   See Hydroelectric Power.

PELVIC FIN.   See Fishes.

PELVIS. Literally, a basin. Commonly applied to the bony pelvis of man, which is a compact pelvic girdle comparable to that of vertebrates generally (skeletal system). The human pelvis is properly the basin-like abdominal cavity containing the viscera, supported by the bony pelvis, which is composed of the hip bones on either side, and in front, and the sacrum and coccyx. The pelvis rests upon the lower extremities and supports the spinal column. Within the pelvis are found the rectum, bladder, and generative organs. There are certain sexual differences in the pelvis. The female pelvis is lighter, more slender, with a cavity that is larger, less funnel-shaped, and shorter. When the female pelvis resembles the male type or is deformed by bony disease such as rickets, childbirth is interfered with and either made difficult or impossible by the vaginal route.

The pelvis of the kidney is the principal cavity, which receives the urine from all subordinate divisions and discharges it to the ureter.

PEMPHIGUS VULGARIS. This is classified as a *vesiculobullous* (watery blisters and sacs or cysts containing fluid and affecting mucous membrane) disease. The disease is not commonly seen in patients before the fourth to sixth decade of life. Localized lesions in the mouth may be the only early symptoms. Untreated, the disease progressively involves the mucous membrane surfaces of the mouth, esophagus, vagina, and cervix. When the disease becomes extensive, large areas of the skin may become denuded, leading to the loss of protein and fluid from those areas. Prior to the administration of glucocorticosteroids, serious denudation led to debilitation and even death. The physician will administer glucocorticosteroids with discretion because of the side effects associated with them. It is important that means be taken to prevent infection of denuded areas and thus systemic antibiotics are frequently administered. The etiology of the vesiculobullous diseases is poorly understood.

PENCIL BEAM. Emission, from an antenna, having the form of a narrow conical beam.

PENCIL-BEAM ANTENNA. A unidirectional antenna, so designed that cross section of the major lobe by planes perpendicular to the direction of maximum radiation are approximately circular, and having a very small angular cross section.

PENCIL FISHES.   See Characids.

PENDULUM. In considering a gravity pendulum, let a rigid body of mass M swing on an axis located at distance r above its center of mass, and with respect to which axis the moment of inertia of the body is I. This motion is mathematically not simple, but if the amplitude of swing is small, certain terms in the differential equation of motion may be disregarded and the remaining equation readily solved. The solution gives as the period of the complete oscillation

$$T = 2\pi\sqrt{\frac{I}{Mgr}}$$

in which g is the acceleration of a freely falling body. Huygens found experimentally, what may be proved theoretically, that for any given period of oscillation greater than a certain minimum there are two different distances r for which I/r, and hence T, has the same value. If these two distances are laid off on opposite sides of the center of mass, the two points resulting are "conjugate points" (center of suspension and center of oscillation). Denoting the whole distance between these points by l, the period of swing when the pendulum is suspended at either of them is

$$T = 2\pi\sqrt{\frac{l}{g}}$$

This is the same as the period for an "ideal simple pendulum," i.e., a single particle of mass m suspended by a weightless thread of length l, for which $r = l$, and $I = ml^2$. Kater utilized this principle in his well-known reversible pendulum.

It was Huygens who first adapted the pendulum to regulate a mechanism for keeping time and thereby gave us the common clock.

PENDULUM CLOCK. A clock that uses a pendulum as its frequency standard. Galileo discovered the principle of *isochronism* of the pendulum in 1582, observing experimentally that a pendulum's frequency, or period of swing, seemed independent of its arc. Not until 1641, however, did he begin to adapt the pendulum to clocks. In 1656, the Dutch scientist, Christian Huygens (who 20 years later developed the hairspring to regulate the oscillations of the spring-driven balance wheel) substituted a pendulum for the foliot bar in the then-existing weight-driven clocks. Theoretically, Huygens recognized that the period of the swing of the pendulum, unlike the oscillation rate of the foliot bar, would be practically independent of the clock's drive system. In Huygens' clock, built in 1657 by Salomon Coster, a pendulum was weight-driven through a verge and foliot escapement. However, the pendulum was required to maintain an arc of some 40 degrees.

Huygens recognized that a pendulum with such a large arc is not isochronous and determined that its bob should swing in a cycloidal, or more U-shape, curve. His English contemporary William Clement solved the problem by inventing the anchor escapement that permits the use of a small arc of 3 or 4 degrees. Only within such a small arc is a pendulum practically isochronous and, therefore, most accurate. The small arc also minimizes power requirements and made practical narrower clock cases.

A later development, the so-called deadbeat escapement invented in England by George Graham in 1715, perfected the anchor escapement and was employed in many observatory pendulum clocks for more than 200 years.

The first electrically driven pendulum clock was built in 1843 by Alexander Bain, a Scot. The bob swung between two permanent magnets.

In 1873, George Airy, British Astronomer Royal, designed the first electric pendulum clock to compensate for variations in barometric pressure—which otherwise affect the period of swing. His clock became recognized as the standard time reference throughout the world until 1922, when the Shortt electric pendulum clock replaced it as the primary standard at the Royal Greenwich Observatory outside London.

In 1898, James Rudd, an Englishman, had built the first so-called free pendulum clock that used two pendulums: a master, or "free," pendulum that drove no mechanism; and a slave pendulum that drove an anchor escapement. The Shortt clock was an adaptation of this system. Its slave pendulum always ran slow as compared to the master, with a maximum difference of only $\frac{1}{240}$ of a second reported at the Royal Greenwich Observatory.

The U.S. National Bureau of Standards (NBS) also used the Shortt pendulum clock as its primary standard. Quartz crystal clocks, which in the 1930s and early 1940s could not maintain comparable accuracies for periods longer than three months, replaced the Shortt pendulum as the primary NBS standard during 1946–50. The accuracy of the Shortt clock ranged to 1 part in 4,000,000, equivalent to a variation of 0.020 second per day.

After the replacement of pendulum clocks as primary standards, a new Riefler electric pendulum eliminated the slave pendulum by substituting a diaphragm and pin-hole on the master pendulum. A pencil beam fixed on the diaphragm activates a photo cell that periodically penetrates the pin-hole as the pendulum swings, creating a precise frequency standard without burdening the pendulum with a mechanical escapement. Proposals for a similar system had been impractical previously because of the relatively short life of earlier vacuum tubes. However, electric free pendulum clocks, which must be set on ultrastable mountings and operate in a controlled-atmosphere environment, are too large and delicate for consumer use. See also **Clock**.

W.O. BUENNETT
J.J. CARPENTER
F. DOSTAL
E. VAN HAAFTEN, New York, NY

PENDULUM DAY. See **Day**.

PENDULUM (Foucault). See **Foucault Pendulum**.

PENEPLAIN. Meaning nearly a plain. A physiographic term implying a broad flat erosional surface which has been finally developed regardless of the structure, relative hardness, and solubility of the rocks of the region. In the case of widespread unconformities, the plain of erosion that truncates the subjacent deformed rocks and underlies the superjacent formations is an ancient peneplain. Local topographic features which rise above the peneplain are called monadnocks, after the type, Mt. Monadnock, in New England.

PENETRANT. Any agent used to increase the speed and ease with which a bath or liquid permeates a material being processed by effectively reducing the interfacial tension between the solid and liquid. Penetrants are widely used in the textile, tanning, and paper industries for improving dyeing, finishing, etc., operations. Sulfonated oils, soluble pone oils, and soaps are popular among the older penetrants, and the salts of sulfated higher alcohols are typical of the synthetic organics developed for this purpose.

See also **Wetting Agent**.

PENGUIN (*Aves, Sphenisciformes, Spheniscidae*). Flightless marine birds which inhabit mainly the coasts of the Antarctic, but they are also common in the southern temperate cool zone. Northward, one species is found as far as the Galápagos Islands, which lie on the Equator; they also occur on the subtropical coasts of South America, South Africa, and Australia.

The shape of their bodies makes penguins excellent swimmers. They can remain in the water for a long time without being harmed by the cold, yet they can move quickly and with agility on land. Their adaptation to life in the water is developed to a high degree, similar to that shown by the seals among mammals.

Penguins generally swim on or just below the surface of the sea; under water they reach a speed of 36 kilometers (22 miles) per hour. At higher speeds they dive alternately up and down like dolphins; this enables them to breathe at regular intervals without interrupting their forward movements, and to reduce friction by lubricating the plumage with air bubbles. The heat-conserving air-cushion under their plumage gives their bodies a buoyancy which is not altogether favorable for prolonged diving. Only rarely do they dive for longer than 2 to 3 minutes without surfacing.

On land or on ice, penguins walk upright, although in snow they often move by sliding forward on their bellies. The rock-penguins and their relatives hop in an upright posture with both feet simultaneously. At the edge of the ice, penguins often pick up speed under water so that they shoot out of the waves and land securely on the ice upright on both feet.

Warm-blooded animals that live in cool regions or in cold water must conserve their heat. Since penguins are relatively small, the ratio between body surface and body volume, even in a large penguin, is less advantageous than it is in any other warm-blooded vertebrate that spends much time in the water; it is therefore particularly important for penguins to counteract the heat loss at sea. This is done in part through elevation of the metabolic rate; penguins are much more lively in the water than on land and thus produce more metabolic heat, thereby causing the body temperature to remain constant. A further help is the 2–3 centimeter ($\frac{3}{4}$-1 inch) thick layer of subcutaneous fat, particularly in the polar penguins; further, their thick, waterproof plumage and the air trapped beneath it form a very effective protection.

Such heat insulation is very useful in a cold climate and is essential in enabling penguins to survive in and near the Antarctic. The heat insulation of the Antarctic adelie penguins is so effective that while they are incubating snow can accumulate on them without melting. In warmer areas, and sometimes even in the particularly cold ones, heat insulation becomes a problem; on land penguins are always in danger of overheating, especially when they fight, run, or are very active in some way. Of the two insulating layers the fat is the lesser problem, for it is penetrated by blood vessels. These can dilate so that more blood reaches the outermost skin layer, where it cools off. The protective plumage, however, allows less heat exchange. Penguins can erect the feather tips somewhat, but the downy under layer still holds back much of the heat. The plumage is very resistant to disturbances by wind.

Large penguins lay only one egg; the rock hoppers on Tristan da Cunha often lay three eggs, but most species have a clutch of two eggs and only occasionally one or three. The weight of the egg ranges from $1\frac{1}{2}$ to 4% of the body weight. The incubation period lasts from 33 to 62 days.

The young of the same clutch hatch at the same time or, at most within a day of each other; they are covered with a sparse downy coat and are carefully brooded until, at the age of 6 to 10 days, they begin to regulate their own body temperatures. They keep their thick, almost woolly, down plumage until they are almost fully-grown. They are fed by both parents, who regurgitate food for them; within a short time they almost reach the size of a fully grown penguin. The down is replaced by a new plumage and the young, without parental help or guidance, go into the water.

Most penguins are sociable. They breed in groups or in large noisy colonies; they take to the water in flocks and seek out the feeding grounds in large swarms. One of their largest and most densely populated habitats lies at the edge of Antarctica; on several Antarctic and sub-Antarctic islands, breeding colonies of hundreds of thousands, or even of millions, of penguins have been found. In lower latitudes their numbers are generally fewer.

Penguins feed on floating (planktonic) and swimming animals, particularly small fish, small floating crabs, and squids. Although a few species limit themselves to certain animals, as the black-footed penguin apparently does to fish or the gentoo penguin to small crabs and the largest penguins to fish and squids, nevertheless they usually take whatever is most abundant.

The largest living penguins are the Giant Penguins (genus *Aptenodytes*). There are two species, the Emperor Penguin (*Aptenodytes forsteri*; see Fig. 1), which reaches a length of 115 centimeters (45 inches) and weighs up to 30 kilograms (66 pounds), and the King Penguin (*Aptenodytes patagonica*), which reaches a length of 95 centimeters (37 inches) and weighs up to 15 kilograms (33 pounds).

Although the emperor penguin is only a little larger than the king penguin, it weighs almost twice as much. It is a bird of the high Antarctic

Fig. 1.   Adult Emperor penguin with young. (*New York Zoological Society.*)

and has long, dense plumage and very large fat stores. The king penguin, on the other hand, is a bird of the sub-Antarctic and temperate-cold zone. Its body is more slender and its plumage is less dense. During most of the year it has little subcutaneous fat. The vividly colored patches on the sides of the head of both species are shown off prominently during courtship; if they are colored dark experimentally, the bird fails to attract a partner and hence does not breed. Neither species builds a nest; instead they carry the egg on top of the feet and walk around with it.

The Adelie Penguins (genus *Pygoscelis*) reach a length of 72.5–75 centimeters (28.5–29.5 inches). There are three species: the Adelie Penguin (*Pygoscelis adeliae*), which reaches a length of 70 centimeters (27.5 inches) and a weight of 5 kilograms (11 pounds); the Chinstrap Penguin (*Pygoscelis antarctica*), which reaches a length of 68 centimeters (27 inches) and a weight of 4.5 kilograms (10 pounds); and the Gentoo Penguin (*Pygoscelis papus*), including the subspecies the Great Northern Gentoo Penguin (*Pygoscelis papua papua*), the Southern Gentoo Penguin (*Pygoscelis papua ellsworthii*), and the Macquarie Gentoo Penguin (*Pygoscelis papua taeniata*). All penguins of this genus have long, curved tail feathers, which sweep behind them like brooms as they walk. The adelie penguin lives farthest south on the coasts of the Antarctic continent and the barren islands which surround it. See Fig. 2.

Fig. 2.   Penguins from Adelie, Antarctica

The Crested Penguins (genus *Eudyptes*) reach a length of 70–72 centimeters (27.5–28 inches). There are four species. The Macaroni Penguin (*Eudyptes chrysolophus*) has two subspecies: *Eudyptes chrysolophus chrysolophus*, which reaches a length of 70 centimeters (27.5 inches) and

a weight of 4.2 kilograms (9 pounds), and the Macquarie Island Macaroni Penguin or Royal Penguin, *Eudyptes chrysolophus schlegeli*, which reaches a length of 62 centimeters (24 inches). The Rock Hopper Penguin (*Eudyptes crestatus*) reaches a length of 55 centimeters (22 inches) and a weight of 2.5 kilograms (5.5 pounds). The Erect-Crested Penguin (*Eudyptes atratus*) reaches a length of 67 centimeters (26 inches) and a weight of 3.6 kilograms (8 pounds). The Fiordland Penguin (*Eudyptes pachyrhynchus*); which reaches a length of 55 centimeters (22 inches) and a weight of 3 kilograms (6.5 pounds).

The macaroni penguin is the most southern representative of this genus; the others live in warmer water. Using its sharp claws, the rock hopper can climb on rocks when it allows the waves to throw it on land. It lives on many islands with either temperate-warm or temperate-cold climates, and is also found in unexpectedly large numbers on Heard Island in the sub-Antarctic. Macaroni penguins which breed with adelie penguins on the edge of the Antarctic have feathers that are on longer than those of the erect-crested penguins which live on the much warmer islands around New Zealand or the rock hoppers of Tristan da Cunha and New Amsterdam.

The Black-Footed Penguins (genus *Spheniscus*) are small to medium-sized, reaching a length of 50–71 centimeters (20–28 inches). There are four species: the Black-Footed Penguin (*Spheniscus demersus*), which reaches a length of 70 centimeters (27.5 inches) and a weight of 2.9 kilograms (6 pounds); the Magellan Penguin (*Spheniscus magellanicus*), which reaches a length of 70 centimeters (27.5 inches) and a weight of 4.9 kilograms (11 pounds); the Peruvian or Humboldt Penguin (*Spheniscus humboldti*), which reaches a length of 65 centimeters (25.5 inches) and a weight of 4.2 kilograms (9 pounds); and the Galápagos Penguin (*Spheniscus mendiculus*), which reaches a length of 53 centimeters (21 inches) and a weight of 2.2 kilograms (5 pounds).

In this genus the beak is high and strong at the base with longitudinal grooves; it is used for digging. These penguins are short-tailed and have a smooth plumage. In front of the eyes and on the chin they have red to black bare spots. Their webs are often white spotted. They rub their beaks and necks on one another in a greeting ceremony unlike other penguins.

The Yellow-Eyed (or Yellow-Crowned) Penguin (*Megadyptes antipodes*) is about the size of the gentoo penguin. Its length is 75 centimeters (29.5 inches) and reaches a weight of 5.2 kilograms (11.5 pounds). It has a fairly long beak. There is a black and yellow crown patch, and behind it is a golden-yellow stripe across the upper ear coverts. The nape has slightly elongated feathers. The eyes are pale yellowishgreen. It resides on New Zealand and a few nearby islands.

The Dwarf Penguins (genus *Eudyptula*) are even smaller than the Galápagos penguins. They reach a length of 40–42 centimeters (16–16.5 inches). There are two species with several subspecies. The Little Penguin (*Eudyptula minor*) has three subspecies: a southern subspecies (*Eudyptula minor minor*), reaching a length of 40 centimeters (16 inches) and a weight of 2.5 kilograms (5.5 pounds); a northern subspecies (*Eudyptula minor novaehollandiae*), which reaches a length of 41 centimeters (16 inches), and a weight of 2.2 kilograms (5 pounds); and the Chatham Island Little Penguin (*Eudyptula minor iredalei*), which reaches a length of 39 centimeters (15 inches) and a weight of 2.1 kilograms (4.5 pounds). The White-Flippered Penguin (*Eudyptula albosignata*) reaches a length of 40 centimeters (16 inches), and a weight of 2.4 kilograms (5.2 pounds).

All dwarf penguins have very similar ways of life. Their nests are generally well hidden. They dig holes up to 2 meters (6$\frac{1}{2}$ feet) long or use holes dug out by shearwaters, as well as those found in natural rock or earth caves, under rocks or plants. They come ashore only after sunset. They spend the night on shore, usually in their breeding areas, where they live the whole year. At dawn they return to the sea. See also **Sphenisciformes**.

Web References

Adelie Penguins (Pygosceli adeliae): http://www.siec.k12.in.us/~west/proj/penguins/adelie.html

African Penguins (*Spheniscus demersus*): http://www.siec.k12.in.us/~west/proj/penguins/africanpen.html

Chinstrap Penguins (*Pygosceli Antarctica*): http://www.siec.k12.in.us/~west/proj/penguins/chinstrap.html

Emperor Penguins (*Aptenodytes forsteri*): http://www.siec.k12.in.us/~west/proj/penguins/emperor.html

Erect–Crested Penguins (*Eudyptes sclateri*): http://www.siec.k12.in.us/~west/proj/penguins/erect.html

Fiordland Penguins (*Eudyptes pachyrhynchus*): http://www.siec.k12.in.us/~west/proj/penguins/fiord.html

Galapagos Penguins (*Spheniscus mendiculus*): http://www.siec.k12.in.us/~west/proj/penguins/galap.html

Gentoo Penguins (Pygosceli papua): http://www.siec.k12.in.us/~west/proj/penguins/gentoo2.html

Humbolt Penguins (*Spheniscus humboldti*): http://www.siec.k12.in.us/~west/proj/penguins/humbolt.html

King Penguins (*Aptenodytes patagonicus*): http://www.siec.k12.in.us/~west/proj/penguins/king.html

Little Blue (Fairy) Penguins (*Eudyptula minor*): http://www.siec.k12.in.us/~west/proj/penguins/little.html

Magellanic Penguins (*Spheniscus magellanicus*): http://www.siec.k12.in.us/~west/proj/penguins/magell.html

Macaroni Penguins (*Eudyptes chrysolophus*): http://www.siec.k12.in.us/~west/proj/penguins/mac.html

Rockhopper Penguins (*Eudyptes crestatus*): http://www.siec.k12.in.us/~west/proj/penguins/rock.html

Royal Penguins (*Eudyptes schlrgeli*): http://www.siec.k12.in.us/~west/proj/penguins/royal.html

Snares Island Erect-crested Penguins (Eudyptes robustus): http://www.siec.k12.in.us/~west/proj/penguins/snare.html

Yellow-Eyed Penguins (*Megadyptes antipodes*): http://www.siec.k12.in.us/~west/proj/penguins/yellow.html

PENICILLIN.   See **Antibiotic**.

PENICILLIN (The History). Because it was developed during World War II, and like radar resulted from the Allied collaborative effort, but unlike the atom bomb saved rather than destroyed lives, penicillin acquired a powerful symbolic as well as a practical role in the postwar reconstruction of many countries. This symbolic role was portrayed in the film *The Third Man*, scripted by Graham Greene, in which penicillin, far from being the life-saver it had become, played a more sinister role, as an impure, adulterated substance sold on the black market in the shadowy world of postwar Vienna.

Traditional histories of penicillin are to a large extent the product of this period, when the wartime veil of secrecy was being lifted. In these, penicillin was presented as a miracle drug, discovered by accident by Alexander Fleming in 1928, developed by Howard Florey and his team at the School of Pathology in Oxford during World War II, then produced on a large scale in time for D-day by the American pharmaceutical industry, whose methods were later borrowed, often at high cost, by vanquished and liberated countries alike.

However, alternatives to such traditional views have begun to emerge: the rationales behind Fleming's discovery and Florey's invention of penicillin have been questioned, the American contribution has been subjected to closer scrutiny, and finally the part played by penicillin in the postwar reconstruction of many other countries has begun to be investigated. This article provides an overview of the traditional accounts, while introducing some of the departures taken by more recent histories of penicillin.

Early Ideas About Antibiosis

If penicillin as we know it today, that is to say therapeutic penicillin, is the product of World War II, the mold itself, and the antibacterial substance it produces, which Fleming named 'penicillin', has a much longer history. The use of fungi in medicine goes back to ancient times. However, early ideas about bacterial antagonism, or 'antibiosis', arose towards the end of the nineteenth century, when the germ theory of disease was being formulated. An early observation of the 'struggle for life' between bacteria and the *Penicillium* mould was made by the physicist John Tyndall in 1876. This was followed by the study by Louis Pasteur and Jules Joubert of antagonism between bacteria and the anthrax bacillus. Their study envisaged the use of the phenomenon in therapeutics, but their ideas centered on immunity, rather than on direct antibacterial action.

Nevertheless, the idea of using a harmless bacterium to inhibit a pathogenic one took root in the wake of Pasteur's work. Some clinicians even recommended infecting patients suffering from chronic infections with erysipelas (an inflammatory disease caused by a variety of streptococcus). Studies of the phenomenon *in vivo* as well as *in vitro* multiplied. However, at a time when the main thrust of therapeutics was directed towards

vaccination and serum therapy, the often conflicting evidence made these studies appear unconclusive, and they were not followed by direct therapeutic application. The word 'antibiosis' was coined by Vuillemin in 1889, but was little used until 1928, when it was resurrected, and given more precise meaning by G. Papacostas and J. Gaté in their volume on the use of microbes in therapy: *Les Associations Microbiennes: leurs applications thérapeutiques*.

Although in 1945 S. A. Waksman, who developed streptomycin, defined 'antibiotic' as a 'chemical substance, of microbial origin, that possesses antibiotic powers', the meaning of the word would soon encompass substances derived from other sources, such as plants and animal tissue. See also **Waksman, Selman Abraham (1888–1973)**.

Alexander Fleming and the Penicillium Mold. In 1928, Fleming, who was working at the time on a vaccine against influenza in the inoculation department of St Mary's Hospital, London, made a chance observation of staphylococci that had dissolved as a result of contamination by a mould of the culture plate on which they had been grown. This mold was later identified as *Penicillium notatum*, and Fleming named the antibacterial substance it produced: 'penicillin'. Some of the facts he established in the experiments he devised to study its action were to form the basis of the study, ten years later, by Florey and his team in Oxford. However, at the time, he advocated it mainly as a reagent for the isolation of the suspected causal agent of influenza, *Bacillus influenzae*. See also **Fleming, Sir Alexander (1881–1955)**.

The Sulfonamides and Competitive Antagonism

Throughout the 1930s, penicillin could be found as a laboratory tool in bacteriological laboratories in Britain, and elsewhere: for instance, at the Pasteur Institute in Paris, to which Fleming had donated a sample of penicillin for its mycological collection. In Britain, the biochemist Harold Raistrick made a study of the chemistry of penicillin at the School of Hygiene and Tropical Medicine in London, as part of a wider project on microorganisms. He established some important chemical facts, but as with earlier work done by Fleming himself and carried out under his supervision, this research met a dead end.

In 1935, a discovery was made that would lead, indirectly, to the development of penicillin as a therapeutic agent: the isolation by the chemist Ernest Fourneau and his group at the Pasteur Institute of the active principle of a red dye, Prontosil Rubrum, which had been shown by G. J. P. Domagk of the Bayer laboratories to be effective against streptococci. This was *p*-amino-benzenesulfonamide (otherwise known as sulfanilamide). The discovery led to a new class of chemotherapeutic drugs, the sulfonamides, which were active against a wide range of bacteria, but not the often deadly staphylococci. See also **Sulfonamide Drugs**; and **Sulfonamides (The History)**.

At the Institute of Pathology of the Middlesex Hospital in London, Paul Fildes, D. D. Woods and Selbie studied the mechanism of action of sulfanilamide. They showed that it prevented bacteria from using a substance essential to their growth. By providing a chemical explanation of the therapeutic activity of the drug, their study spurred the search for other substances or organisms that displayed similar competitive antagonism towards bacteria.

In 1939–1940, the Frenchman R. J. Dubos, who had been working under Waksman at Rutgers University in the United States, began his own project to study the bacteria present in soil, and isolated the active principle of the bacteria *Bacillus brevis* which was known as tyrothricin, and which contained two antibiotics: gramicidin and tyrocidine. On account of their toxicity, they were limited in use to local application. Meanwhile, Waksman carried on with his own project, which led him to develop streptomycin, the antibiotic that would revolutionize the treatment of tuberculosis after the war.

Howard Florey and Penicillin as a Chemotherapeutic Drug

Antibiotics were therefore 'in the air' when, in 1938, Florey turned his attention to a number of microorganisms with known antibacterial action, including penicillin. As a leading figure in pathology, he was familiar with the research on penicillin, and with ideas of competitive antagonism between microorganisms. Having obtained funding and gathered together a multidisciplinary research team, which included the biochemists E. B. Chain and Norman Heatley, he began a large project to study microorganisms with known antibacterial action. See also **Florey, Howard Walter (1898–1968)**; and **Chain, Ernst Boris (1906–1979)**.

Far from curtailing their plans, the declaration of war in September 1939 gave impetus to the project, which after a series of preliminary tests focused on penicillin as the most promising vein to follow. Florey designed biological and chemical experiments that were to establish penicillin as a new kind of chemotherapeutic agent, effective against staphylococci, and less toxic than the sulfonamides, at the time the medical profession's favored weapon against wound and other bacterial infections. The results were published in the *Lancet*, in August 1940.

To convince the military, as well as the medical, profession of the value of penicillin, he turned the School of Pathology into a small-scale production plant for the surface-culture of penicillin in order to begin human trials. However, before the results had been published in another *Lancet* article in August 1941, he left with Heatley on a mission to the United States, to investigate the possibility of producing penicillin on a large scale, at a time when the British government was seeking US aid for the war effort.

The British program intersected with work that had already begun in the United States after publication of the first *Lancet* article, and was part of a wider program to develop antibiotic chemotherapy. Nevertheless, Florey's visit had the effect of focusing and coordinating the US effort. By 1942, a two-pronged collaborative program, which provided a model for similar programs after the war, was in place. It aimed to produce penicillin on a large scale by fermentation methods, and to find the route to its synthesis. Although the deep-fermentation methods developed under the aegis of the American Office of Scientific Research and Development were never completely superseded, this research eventually led to the synthetic and semi synthetic penicillin's.

After Florey returned to England, he became involved in large-scale clinical trials in Britain and on the battlefront in North Africa, relying on penicillin produced by both British and American pharmaceutical companies. By D-day, penicillin had acquired the status of a wonder-drug, which would be associated in the public mind with the Liberation of Europe.

Other Wartime Efforts to Develop Penicillin

Despite the embargo that separated Britain from the Continent, scientific publications reached occupied countries via neutral ones, and several instances can be found of independent penicillin research going on under occupation. For example, in France, the chemical group Rhône-Poulenc and researchers at the Pasteur Institute carried out semi-clandestine work under the watchful eye of I. G. Farben and the German authorities. By D-day, they were in a position to produce small amounts of penicillin destined for the Leclerc division, first extracted from the urine of American soldiers, then from cultures produced at a plant set up by the Provisional Government in the centre of Paris. Later, Rhône-Poulenc entered into an agreement with the American company Merck to import deep-fermentation technology. Thus, in France, as in other countries, penicillin soon acquired a mythical status, becoming associated with the resistance and with readmission into the Allied camp, but also with dependence on American know-how and assistance after the war had ended. While experiencing pride at having been the discoverers of this miraculous drug, the British would also begin to feel resentment at having been cheated of the huge profits it brought American industry. In penicillin, they would see the stark illustration of their talent for discovery, and of their apparent inability to develop the fruits of this talent.

Postwar Penicillin

Large-scale clinical trials in Britain and the United States ascertained an increasing range of activity for penicillin, found to be effective against several types of venereal disease, and set a pattern for later clinical trials. This had a profound effect on the national health services set up in the aftermath of war, as penicillin, once it was mass-produced and became more widely available, was one of the best-selling drugs on an ever-growing drugs bill. It was modified for the increasingly complex market for medicines. Since it tended to be inactivated in the presence of acid, it was combined with a buffer so that it could be taken orally, rather than being injected. However, with increasing use, and sometimes abuse, the ineffectiveness of penicillin against certain types of bacteria, which

had evolved a resistance to it, became an acute problem. Although the semi synthetic penicillin's developed since the 1950s have targeted such bacteria, and the search for new and better antibiotics has never ceased, the time when penicillin will no longer be the miracle cure it was during the second half of the twentieth century may be drawing near. See also **Antibiotics**; and **Hodgkin, Dorothy Mary Crowfoot (1919–1994)**.

Additional Reading

Bickel, L.: *Rise up to Life: A Biography of Howard Florey Who Made Penicillin and Gave it to the World*, Angus and Robertson, London, UK, 1970.
Bud, R.: "Penicillin and the New Elizabethans," *British Journal of the History of Science*, **31**, 305–333 (1998).
Bud, R.: "Penicillin," *Triumph and Tragedy*, Oxford University Press, New York, NY, 2007.
Chen, W.: "The Laboratory as Business: Sir Almroth Wright's Vaccine Programme and the Construction of Penicillin," In: Cunningham, A. and Williams, P.: *The Laboratory Revolution in Medicine*, Cambridge University Press, Cambridge, UK, 1992, pp. 245–292.
Lax, E.: *The Mold in Dr. Florey's Coat: The Story of the Penicillin Miracle*, Henry Holt & Company, Inc., New York, NY, 2005.
Williams, T.I.: *Howard Florey: Penicillin and After*, Oxford University Press, Oxford, UK, 1984.

Viviane M. Quirke, Oxford Brookes Business School, Wheatley, UK

PENIS. See **Gonads**.

PENITENT ICE. A spike or pillar of compacted snow, firn, or glacier ice caused by differential melting and evaporation. Necessary for this formation are 1) air temperature near freezing; 2) dewpoint much below freezing; and 3) strong insolation. Consequently, penitent ice is most developed on low-latitude mountains, especially the Chilean Andes, but has been found in polar regions. Penitents are oriented individually toward the noonday sun, and usually occur in east–west lines. The term is derived from the Spanish *nieve penitente* (penitent snow), which is still widely used throughout the literature.

AMS

PENMAN-MONTEITH METHOD. A micrometeorological method to determine the rate of moisture transport (i.e., evaporation rate E or latent heat flux) away from a surface employing the concept of aerodynamic resistance and requiring measurements of net radiation R_n, ground heat flux G, temperature, humidity and wind speed.

The latent heat flux is given by

$$\lambda E = \frac{S(R_n - G) + \rho C_p [e_s(T_z) - e(z)]/r_a}{S + \gamma[(r_a + r_c)/r_a]},$$

where λ is the latent heat of vaporization; r_c and r_a are aerodynamic resistances; C_p and are the specific heat and density of air, respectively; S is the rate of change of saturation vapor pressure with temperature; γ is the psychrometric constant; e_s is the saturation vapor pressure; T_z is the temperature at a height z above the surface; and e is the vapor pressure. The product λE is the latent heat flux.

Additional Reading

Stull, R.B.: *Introduction to Boundary Layer Meteorology*, Springer-Verlag New York, LLC, New York, NY, 1988.

AMS

PENNING DISCHARGE. A direct-current discharge where electrons are forced to oscillate between two opposed cathodes and are restrained from going to the surrounding anode by the presence of a magnetic field. It is sometimes referred to as a pig discharge since the device was originally used as an ionization gage (Penning ionization gage). It is used as a plasma-beam source by permitting the plasma to stream out along the magnetic field through a hole in one of the cathodes.

PENNING EFFECT. An increase in the effective ionization rate of a gas due to the presence of a small number of foreign metastable atoms. For instance, a neon atom has a metastable level at 16.6 volts and if there are a few neon atoms in a gas of argon which has an ionization potential of 15.7 volts, a collision between the neon metastable atom with an argon atom may lead to ionization of the argon. Thus, the energy which is stored in the metastable atom can be used to increase the ionization rate. Other gases where this effect is used are helium, with a metastable level at 19.8 volts, and mercury, with an ionization level at 10.4 volts.

PENNSYLVANIAN PERIOD. A period of the Paleozoic era. A systematic term first proposed by H.S. Williams in 1891. Type locality, Pennsylvania. The period began about 250,000,000 years ago. The term Pennsylvanian is roughly equivalent to the more general term Upper Carboniferous. In Britain this system is referred to as the Coal Measures. During this period there were many oscillations of sea level with relatively rapid alternations of marine and terrestrial sediments and freshwater swamp deposits in which were formed important coal beds. The principal occurrence of the formations of this system are in the Allegheny Plateau region of the eastern United States, westward to the Ohio River.

PENSTOCK. Some considerable surface distance usually separates the intake works and turbines in a medium- or high-head hydroelectric project. Even where an open canal is employed to carry the water from the forebay of a diversion dam to intake works located near the plant, there is still a considerable span to be abridged by a closed water conduit of the pressure type. This water conduit is called the penstock and is always circular in form because that shape is best adapted to withstand internal pressure.

PENTADACTYL APPENDAGE. The form of vertebrate appendage which is regarded as the fundamental terrestrial limb from which all of the specialized appendages of animals above the fishes have been derived.

The two pairs of vertebrate limbs, pectoral and pelvic, are similar in structure and both are attached to the girdles of corresponding name. Each girdle consists, in the primitive state, of three pairs of bones meeting at the articulation of the limbs. On each side of the body, one bone extends toward the back and two toward the middle of the body below. The skeletal structure of the limbs is made up of regions movable to one another and includes a single bone in the segment next to the body, followed by two bones. Then follows a group of small bones, and last five divergent series of moderately long bones, which extend into the digits. Of the more constant bones in the pectoral girdle of vertebrates with limbs, the dorsal bone is the scapula, and the two ventral bones are an anterior clavicle (of dermal origin) and a posterior coracoid. This girdle may contain a procoracoid between the last two bones. The bones of the pectoral appendage (forelimb) are, in the order described above, the humerus, the ulna and radius, the carpals, the five metacarpals of the hand, and the five series of phalanges in the digits. In the pelvic girdle of vertebrates with limbs, the dorsal bone is the ilium, and the two ventral bones are an anterior pubis and a posterior ischium. The bones of the pelvic appendage (hind limb) are the femur, the tibia and fibula, the tarsi, the metatarsals, and the phalanges.

Specialized appendages, such as the wings of birds, flippers of marine mammals, and the legs of the hoofed species, show either a simplification or a slight increase in complexity of the skeletal system and in some cases a consolidation by webbing of the digits or a still more compact fleshy union between them. In all of these cases, however, the basic pentadactyl structure is still present.

PENTLANDITE. The mineral sulfide of iron and nickel corresponding to the formula $(Fe, Ni)_9S_8$. It is isometric, appears in granular masses; hardness, 3.5–4; specific gravity, 5.0; color, bronze-yellow; opaque. Occurs with pyrrhotite, millerite, and nickeline. The best known deposit of pentlandite is at Sudbury, Ontario, Canada, where it is associated with a nickel-bearing pyrrhotite.

PENTOSAN. A complex carbohydrate (hemicellulose) present with the cellulose in many woody plant tissues, particularly cereal straws and brans, characterized by hydrolysis to give five-carbon-atom sugars (pentoses). Thus the pentosan xylan yields the sugar xylose ($HOH_2C \bullet CHOH \bullet CHOH \bullet CHOH \bullet CHO$) that is dehydrated with sulfuric acid to yield furfural ($C_5H_4O_2$).

See also **Carbohydrates**.

PENTOSES. See **Carbohydrates**.

PENUMBRA. See **Eclipse**.

PEPPERIDGE TREE. See **Tupelo Trees**.

PEPPER (*Piper nigrum; Piperaceae*). The pepper plant, *Piper nigrum*, is a woody climbing shrub, which is indigenous in India. It is aided in climbing by the adventitious roots formed at the nodes. The ovate leaves are evergreen. The flowers are minute, without petals, and borne in slender spikes. The fruit is a bright red berry less than $\frac{1}{4}$ inch diameter. Each berry contains a single seed. On drying, the pericarp becomes black and wrinkled.

The berries are gathered before they are ripe and dried, usually by the sun. The dried berries are separated from the stem and ground, producing black pepper. If the pericarp is removed from the berry, leaving the seed and endocarp, the ground product is known as white pepper. The pericarp may be removed by using mature berries and soaking them to soften the pericarp. Or machines may rub off the dried pericarp, a method used in western countries. Pepper is one of the most extensively used of all spices. White pepper is less pungent than black, and hence not so conspicuous when used in cooking. Pepper is grown mostly in India and in the Malaysian regions. Propagation is usually by cuttings, which begin to fruit within four or five years, after which they bear continuously but somewhat irregularly for many years.

Related to *Piper nigrum* is *Piper Betle*, a perennial creeping vine native to Java, but widely grown in tropical Asia. From its leaves is prepared a chew with the nut of the Areca palm. *Piper Cubeba* is another species, also native of Java and the Molucca Islands, which yields a volatile oil, which is used medicinally.

PEPTIZATION. Stabilization of hydrophobic colloidal solutions by addition of electrolytes which provide the necessary electric double layer of ionic charges around each particle. Such electrolytes are known as peptizing agents. The ions of the electrolyte are strongly adsorbed on the particle surfaces. Stable solutions of nonionizing substances acquire a charge in contact with water by preferential adsorption of the hydroxyl ions, which may be considered peptizing agents. The term is also loosely applied to the softening or liquefaction of one substance by trace quantities of another, analogous to the digestion of a protein by an enzyme (pepsin).

PEPTONE. A secondary protein derivative that is water-soluble, not coagulated by heat, and not precipitated on saturation of its solutions with ammonium sulfate.

PERCHED. 1. In hydrology, a term describing the ground water table when it is separated from an underlying layer of ground water by an unsaturated layer of rock.

2. In glaciology, a term describing a glacial boulder, erratic, resting on a prominent topographic position.

3. In a physiography, a term describing a stream whose bed is separated from the top of the ground water table by a dry, unsaturated zone.

PERCHES AND DARTERS (*Osteichthyes*). Of the order *Percomorphi*, suborder *Percoidea*, family *Percidae*, there are several well known species, many of which are considered fine food fishes. The yellow perch (*Perca flavescens*) is a freshwater food fish that lives in lakes and streams in the United States from Iowa to South Carolina and northward into Canada. It comprises a significant part of Great Lakes fishery operations. Yellow perches are relatively small, a large fish weighing about a pound with a length of 14 to 16 inches (35 to 40 centimeters). In the mid-1800s, a yellow perch weighing $4\frac{1}{2}$ pounds (2 kilograms) was recorded. *Perca fluviatilis* (European perch) is quite similar to the American yellow perch, but a bit larger, sometimes weighing up to 6 pounds (2.7 kilograms). Occurring as far east as Siberia in European fresh waters, it is widely distributed but not found in southern Italy, northern Scandinavia, or Spain. Some perches also are found in the Baltic Sea.

Stizostedion vitreum (walleye) is a favorite sporting fish and one of the valuable food fishes in Lake Erie. Most walleyes are in the 1 to 4 pounds (0.45 to 1.8 kilograms) range, but one was recorded at 25 pounds (11.3 kilograms). *Stizostedion canadense* (sauger) appears much like the walleye but is somewhat smaller.

Darters appear to be confined to temperate North American waters east of the Rocky Mountains. They are very quick, small, and prefer a bottom habitat. The majority do not exceed 4 inches (10 centimeters) in length, rarely attaining a length of 8 to 9 inches (20 to 27 centimeters). Some are brilliantly colored.

The *Acerina cernua* is a small species of perch occurring in the slow gravelly streams of England. It is commonly known as the ruffe or pope.

PERCHLORIC ACID AND PERCHLORATES. When in a +7 valence state and combined with oxygen, chlorine forms a family of compounds known as the perchlorates. The perchlorate anion, ClO_4^-, as progenitor is derived from perchloric acid. [CAS: 7601-90-3]. $HClO_4$ is one of the strongest of the mineral acids. The perchlorates are more stable than the other chlorine oxyanions, i.e., chlorates, ClO_3^-; chlorites, ClO_2^-; or hypochlorites, OCl^-. See also **Hypochlorites**. Essentially, all of the commercial perchlorate compounds are prepared either directly or indirectly by electrochemical oxidation of chlorine compounds. The perchlorates of practically all the electropositive metals are known, except for a few cations having low charges.

The most outstanding property of the perchlorates is their oxidizing ability. On heating, these compounds decompose into chlorine, chlorides, and oxygen gas. Aqueous perchlorate solutions exhibit little or no oxidizing power when dilute or cold. However, hot concentrated perchloric acid is a powerful oxidizer and whenever it contacts oxidizable matter extreme caution is required. The acidified concentrated solutions of perchlorate salts must also be handled with caution. Ammonium perchlorate (AP), [CAS: 7790-98-9], is one of the most important perchlorates owing to its high (54.5%) O_2 content and the absence of residue on decomposition. These properties, along with a long shelf life, make it a useful rocket propellant. See also **Rocket Propellants**.

Actual perchlorate production is difficult to determine in any given year, because AP is classified as a strategic material. Future production is expected to depend mostly on space programs.

Properties

Chlorine Heptoxide. The anhydride of perchloric acid is chlorine heptoxide. [CAS: 10294-48-1]. Cl_2O_7 is also known as dichlorine heptoxide. It is obtained as a colorless oily liquid by dehydration of perchloric acid using a strong dehydrating agent such as phosphorus pentoxide, P_2O_5.

$$2HClO_4 + P_2O_5 \rightarrow Cl_2O_7 + 2HPO_3 \qquad (1)$$

The Cl_2O_7 decomposes spontaneously on standing for a few days. The acid dehydration reaction requires a day for completion at $-10\,°C$ and explosions can occur. Upon ozonation of chlorine or gaseous ClO_2 at $30\,°C$, Cl_2O_7 is formed.

Chlorine heptoxide is more stable than either chlorine monoxide or chlorine dioxide; however, the Cl_2O_7 detonates when heated or subjected to shock. It melts at $-91.5\,°C$, boils at $80\,°C$, has a molecular weight of 182.914. It is soluble in benzene, slowly attacking the solvent with water to form perchloric acid; it also reacts with iodine to form iodine pentoxide and explodes on contact with a flame or by percussion. Reaction with olefins yields the impact-sensitive alkyl perchlorates.

Perchloric Acid. Pure anhydrous perchloric acid, $HClO_4$, is quite unstable. In aqueous solution, however, $HClO_4$ is a familiar and useful reagent. The acid is the strongest simple acid and the perchlorate ion the least polarizable negative ion known. Perchloric acid is commonly obtained as an aqueous solution, although the pure anhydrous compound can be prepared by vacuum distillation as a colorless liquid, which freezes at $-112\,°C$ and boils at $16\,°C$ at 2.4 kPa (18 mmHg) without decomposition. The pure acid cannot be distilled at ordinary pressures and explodes at $90\,°C$ after standing at room temperature for 10–30 days. The aqueous solution can be concentrated by boiling at 101 kPa (1 atm) at $203\,°C$, at which point an azeotropic solution is attained which contains 72.4% $HClO_4$.

A number of hydrates of perchloric acid, $HClO_4 \cdot nH_2O$, where $n = 1$, 2, 2.5, 3, and 3.5, are known. These are commonly referred to as the hydronium or oxonium perchlorates, $H_3O^+ClO_4^-$, because of the analogy between the x-ray patterns of these species and ammonium perchlorate.

The combination of oxidizing effect, acidic strength, and high solubility of salts makes perchloric acid a valuable analytical reagent. It is often employed in studies where the absence of complex ions must be ensured.

Ammonium Perchlorate. Ammonium perchlorate is a colorless, crystalline compound having a density of 1.95 g/mL and a molecular weight of 117.5. It is prepared by a double displacement reaction between sodium perchlorate and ammonium chloride, and is crystallized from water as the anhydrous salt. The perchlorates, especially those of the light metals and ammonium ion, are favored as solid oxidizers for rocket propellants.

A newer approach developed for producing commercial quantities of high purity AP involves the electrolytic conversion of chloric acid, [CAS: 7790-93-4] to perchloric acid, which is neutralized by using ammonia gas:

Alkali Metal Perchlorates. The anhydrous salts of the Group 1 (IA) or alkali metal perchlorates are isomorphous with one another as well as with ammonium perchlorate. Crystal structures have been determined by optical and x-ray methods. With the exception of lithium perchlorate, the compounds all exhibit dimorphism when undergoing transitions from rhombic to cubic forms at characteristic temperatures. Potassium perchlorate, [CAS: 7778-74-7]. $KClO_4$ the first such compound discovered, is used in pyrotechnics and has the highest percentage of oxygen (60.1%).

The alkali metal perchlorates are either white or colorless, and have increasing solubility in water in the order of Na > Li > NH_4 > K > Rb > Cs. The high solubility of sodium perchlorate, $NaClO_4$, makes this material useful.

Group 11 (IB) Perchlorates. Copper and silver perchlorates have been studied quite extensively. Copper(I) perchlorate, [CAS: 17031-33-3], $CuClO_4$, and copper(II) perchlorate, [CAS: 13770-18-8], $Cu(ClO_4)_2$, form a number of complexes with ammonia, pyridine, and organic derivatives of these compounds. The copper perchlorate is an effective burn-rate accelerator for solid propellants.

The silver perchlorate [CAS: 7783-93-9] salt, $AgClO_4$, is deliquescent and forms a light-sensitive monohydrate that can be dehydrated at $43\,°C$ and is soluble in a variety of organic solvents. Explosions of silver perchlorate have been reported. Gold forms organic perchlorate, [CAS: 42774-61-8], complexes as well as complexes with silver, e.g., $(C_6H_5)_3AgAu(C_6F_5)_2ClO_4$.

Alkaline-Earth Perchlorates. Anhydrous alkaline-earth metal perchlorates can be prepared by heating ammonium perchlorate in the presence of the corresponding oxides or carbonates. The alkaline-earth perchlorates are unusually soluble in organic solvents.

Group 12 (IIB) Perchlorates. The zinc perchlorate, [CAS: 13637-61-1], cadmium perchlorate, [CAS: 13760-37-7], mercury(I) perchlorate, [CAS: 13932-02-0], and mercury(II) perchlorate, [CAS: 7616-83-3], all exist.

Group 13 (IIIA) Perchlorates. Perchlorate compounds of boron and aluminum are known. Boron perchlorates occur as double salts with alkali metal perchlorates, eg, cesium boron tetraperchlorate [CAS: 33152-95-3], $Cs(B(ClO_4)_4)$. Aluminum perchlorate, [CAS: 14452-95-3], $Al(ClO_4)_3$, forms a series of hydrates.

Group 3 (IIIB) and Inner Transition-Metal Perchlorates. The rare-earth metal perchlorates of yttrium and lanthanum have been reported, as have tetravalent cerium perchlorate, [CAS: 14338-93-3], $Ce(ClO_4)_4$, and uranium perchlorate.

Group 14 (IVA) Perchlorates. Perchlorates containing organic carbon have been reported, as have diazonium perchlorates, oxonium perchlorates, and the perchlorate esters. Extreme caution must be used in working with organic perchlorates; many decompose violently when heated, contacted with other reagents, or subjected to mechanical shock. The diazonium perchlorate of p-phenylenediamine, $ClO_4N_2(C_6H_4)N_2ClO_4$, was reported in 1910 to be the most explosive substance known.

Group 4 (IVB) Perchlorates. Titanium tetraperchlorate, [CAS: 13498-15-2], is known.

Group 15 (VA) Perchlorates. Nitrogen perchlorates have been used as oxidizers in rocket propellants. Hydrazine perchlorate, [CAS: 13762-80-6], $NH_2NH_3ClO_4$, and hydrazine diperchlorate, $ClO_4NH_3NH_3ClO_4$, have been investigated as oxidizers for propellant systems.

Other Group 15 perchlorates include nitronium perchlorate, NO_2ClO_4, also called nitryl or nitroxyl perchlorate; nitrosyl perchlorate, [CAS: 15605-28-4], $NOClO_4$; and phosphonium perchlorate, $P(OH)_4ClO_4$.

Group 5 (VB) Perchlorates. Vanadyl perchlorate, [CAS: 67632-69-3], $VO(ClO_4)_3$, has been prepared.

Group 16 (VIA) Perchlorates. A perchlorate compound perchloryl sulfate, [CAS: 43059-05-8], $SO_4(ClO_4)_2$ was produced by the low temperature electrolysis of a 12-NH_2SO_4 and 3-$NHClO_4$ solution. This compound is a strong oxidizer; reaction with toluene, acetone, benzene, or alcohol at room temperature produces an exothermic and explosive reaction. The $SO_4(ClO_4)_2$ is soluble in Freon and CCl_4 without reaction.

Group 6 (VIB) Perchlorates. Both divalent and trivalent chromium perchlorate compounds, [CAS: 13931-95-8;13527-21-9], have been reported. Chromyl perchlorate has been suggested for a gas-generating system operating at $-45\,°C$.

Group 17 (VIIA) Perchlorates. Fluorine perchlorate, [CAS: 37366-48-6], $FClO_4$, is formed by action of elemental fluorine and 60–70% aqueous perchloric acid solution. The compound is normally a gas. It melts at $-167.5\,°C$ and boils at $-15.9\,°C$. It is extremely reactive and explosive in all states.

The perchloryl fluoride, [CAS: 7616-94-6], $FClO_3$, the acyl fluoride of perchloric acid, is a stable compound. Normally a gas having a melting point of $-147.7\,°C$ and a boiling point of $-46.7\,°C$, it can be prepared by electrolysis of a saturated solution of sodium perchlorate in anhydrous hydrofluoric acid. Some of its uses are as an effective fluorinating agent, as an oxidant in rocket fuels, and as a gaseous dielectric for transformers.

Other Transition Element Perchlorates. Both divalent and trivalent manganese perchlorate compounds, [CAS: 13770-16-6; 13498-03-8] are known. Perchlorates of Fe, Co, Ni, Rh, and Pd have been produced as colored crystals.

Manufacture

Perchloric Acid. Several techniques have been employed in the manufacture of perchloric acid, including thermal decomposition of chloric acid, anodic oxidation of chloric acid, irradiation of chlorine dioxide solutions, electrolysis of hydrochloric acid, oxidation of hypochlorites by ozone, ion exchange, and electrodialysis of perchlorate salts.

Perchlorates. Historically, perchlorates have been produced by a three-step process: (1) electrochemical production of sodium chlorate; (2) electrochemical oxidation of sodium chlorate to sodium perchlorate; and (3) metathesis of sodium perchlorate to other metal perchlorates. The advent of commercially produced pure perchloric acid directly from hypochlorous acid means that several metal perchlorates can be prepared by the reaction of perchloric acid and a corresponding metal oxide, hydroxide, or carbonate.

Shipping and Handling

Perchloric acid and perchlorates are classified as strong oxidizers and emit toxic fumes when decomposed; contact with combustible, flammable, or reducing materials must be avoided. Perchloric acid and perchlorates must be shipped in accordance with the U.S. Department of Transportation hazardous material regulations. Handling these compounds requires the procedures and safety precautions specified by the product supplier. Perchlorates contain a self-sustaining source of oxygen, thus fires involving perchlorates must be extinguished with water. A class of more hazardous compounds is formed by mixing inorganic perchlorates with finely divided metals, sulfur, or organic compounds and must be handled with the same precautions as explosives.

Health and Safety Factors

Perchlorates are unstable materials and are an irritant to the body whenever they come in contact with it. Skin contact must be avoided. They are flammable by chemical reaction and are powerful oxidizers. All perchlorates are potentially hazardous when in contact with reducing materials.

Perchloric acid is a poison by ingestion and subcutaneous routes. It is a severe irritant to eyes, skin, mucous membranes. Ammonium perchlorate and sodium perchlorate are moderately toxic by ingestion and parenteral routes.

Uses

Perchloric acid is used in analytical chemistry for the determination of trace metal constituents in oxidizable substances as well as in the production of high purity metal perchlorates; it has also been introduced as a stable reaction media in the thermocatalytic production of chlorine dioxide. Perchlorates are primarily used in ammonium perchlorate as an oxidizer in the formulations of propellant for solid rocket motors. Perchlorates are used in the production of explosives, pyrotechnics, and in solid, slurried, and gelled blasting formulations. Both magnesium and lithium perchlorates are used in dry batteries. Other perchlorates have found application in oxygen-generation systems, adhesive bonding of steel plates, and the recovery of potassium from brines such as $KClO_4$.

Additional Reading

Keller, C.L.: *Hazardous Materials Regulations*, Tariff No. BOE, U.S. Department of Transportation, Washington, DC., Apr. 23, 1992.

Lide, D.R.: *CRC Handbook of Chemistry and Physics*, 86th Edition, CRC Press LLC., Boca Raton, FL, 2005.

Schumacher, J.C.: *Perchlorates, Their Manufacture & Uses*, ACS Monograph 146, Reinhold Publishing Corp., New York, NY, 1960.

Smith, G.F.: *Perchloric Acid*, 2nd Edition, G.F. Smith Chemical Co., Columbus, OH, 1951.

Woodard, K.E. Jr., D.W. Cawlfield, and S.K. Mendiratta, *Chloric Acid: A New Electrochemical Product*, 183rd Electrochemical Society Meeting, Honolulu, HI, May 1993.

PERCUTANEOUS. Penetrating through the skin as contrasted with entry through breaks and lesions of the skin. For example, some hepatitis viruses invade the body through the skin.

PERFECT FLUID. See **Fluid**.

PERFECT GAS. A perfect gas may be defined by the following two laws: The Joule law: the energy per mole, U, depends only on the temperature; the Boyle law: at constant temperature, the volume V occupied by a given number of moles of gas varies in inverse proportion to the pressure.

By combination of these two laws we obtain the equation of state for perfect gas:

$$pV = nRT \qquad (1)$$

where R is the gas constant, T, the absolute temperature. (It is also called the *perfect gas law*.)

The perfect gas is an abstraction to which any real gas approximates according to the nature of the gas and the conditions. For a given temperature and composition, the perfect gas condition is approached when the density tends to zero. From a molecular point of view, the perfect gas laws correspond to the behavior of a system of molecules whose interactions may be neglected in expressing the thermodynamic equilibrium properties. However, even at a low density, the transport properties depend essentially on the interactions.

The thermodynamic properties of a perfect gas are, of course, especially simple. For example, the difference between the molar heat capacities at constant pressure and constant volume is equal to the gas constant R,

$$C_p = C_v = R \qquad (2)$$

The value of R is 0.08205 liter-atm. degree^{-1} mole^{-1}, which in cgs units is equal to 8.314×10^7 g cm^2 sec^{-2} degree^{-1} mole^{-1}. This relationship, Formula (2), applies only approximately to real gases.

However, the way in which either C_p or C_v depends on the temperature can be calculated only from statistical mechanics.

PERFUMES. A blend of pleasantly odorous substances created for use in a wide variety of applications, ranging from expensive couturier perfumes to cosmetics, personal grooming products, laundry products, household cleaning products, and many others. They are created from a palette of several thousand materials, most of which are manufactured by chemical processing methods. Until late in the nineteenth century, fragrances were derived exclusively from natural sources, which put a limitation on where and how they could be used. The large increase in the use of perfumes since then would not have been possible without chemical progress that allowed the synthesis and commercial production of many new odorants. As a result, fragrances, once a luxury, have been incorporated routinely into a great number of products in daily use.

The use of fragrance, deeply rooted in human experience, predates written history. The earliest evidence of such activity is archaeological, found in the tombs of the First Dynasty Egyptian Kings (5100–3500 BC). Small alabaster vases found along with personal grooming tools are taken as an indication of the use of perfumes and cosmetics. Such vases found in the tomb of King Tutankhamen (1350 BC) in 1926 contained some oily material which had remained elusively fragrant. Analytical methods used at that time were able to indicate only that the material appeared to be made mostly of animal fat and some sort of balsamic substance. Early indications of perfume-making, dating back to 3000 BC, have been found in the former Mesopotamia. Extraction pots for herbal and fragrance preparations found there may have been the world's oldest distillation apparata.

The ancients used fragrances for a variety of reasons. Perfumes were offerings to the gods, probably in the form of incense; and they were used for aesthetic purposes as part of personal grooming in daily life. Their use in Egyptian embalming rituals may well have been an early example of odor masking. It has been suggested that priests were the first perfumers. There are many Old Testament references to perfume ("anointing with oils"), beginning mainly in the book of Exodus. Thus it seems likely that the ancient Hebrews learned to use fragrance from the Egyptians.

Among the earliest writings on compounded perfumes, ie, deliberately created mixtures rather than simple extracts, are those of the Greek, Theophrastus, from 370 BC. He even referred to the use of oil bases to make perfumes long-lasting, something that remains a challenge to the modern perfumer. Although such writings typically focus on Western history, there are ancient writings and artifactual evidence from India and China indicating that the use of perfumes and cosmetics developed independently in those areas, where fragrance use is traditional in the cultures.

By the thirteenth century AD, essential oils were being produced along with medicinal and herbal preparations in pharmacies. Around this time improvements in distillation techniques were made, in particular the development of the alembic apparatus, which would eventually establish the characteristic qualities of such materials. As a result, many of the essential oils in use today are derived from those produced in the sixteenth and seventeenth centuries in terms of odor character, even though production methods have continued to evolve. The current practice of aroma therapy is an indication of this common root of medicinal and fragrance chemistry.

During the middle of the nineteenth century, chemists began to investigate the compositions of natural fragrance and flavor materials. As the science developed, it became possible to identify and then synthesize many specific chemicals. This was followed by the use of chemicals as perfume ingredients, first as materials isolated from essential oils and then as synthetically produced naturals such as vanillin and coumarin [CAS: 91-64-5]. See also **Vanillin**. Late in the nineteenth century, it was discovered that synthetic materials that are unrelated to natural chemicals could have great value as perfume ingredients; nitro musks and the ionones were among the first such materials found. Since that time, synthetic chemicals have grown steadily in numbers and total use compared with materials of natural origin.

Perfume Creation

Perfumes are usually considered in two broad categories, as either fine or functional (household product) fragrances. Fine fragrances include perfumes, colognes, men's colognes, aftershaves, and fragrances for cosmetic products. For the purposes of this article, functional products include all personal and household cleaning products that are perfumed, such as bar soaps, detergents of various types, fabric softeners, bleach formulations and air fresheners. The technical and economic requirements for various kinds of perfumes differ widely and are taken into account during their creation. Creation of a successful fragrance can be a lengthy, painstaking, and costly process. The investment in each fragrance that reaches the marketplace is quite substantial. For these reasons, fragrance formulas are held as trade secrets. Patent protection is generally not suitable for this type of intellectual property because most perfume components are in the public domain; it is the unique combination of ingredients that affords competitive advantage. The formulas are significant assets of the

companies that produce them and are therefore protected by elaborate security arrangements.

The creation of perfumes is a commercial art practiced in the medium of odoriferous substances; it is highly specialized and individualistic. A wide variety of ingredients, numbering in the thousands and varying greatly in chemical composition, is available to perfumers. The choice of ingredients can be based on aesthetic or technical grounds, depending on the intended use of the perfume. Often a fragrance creation begins with a concept inspired by an existing perfume, a newly available ingredient, or a newly appreciated odor facet of an existing material. The perfumer creates an accord based on the inspiring note. The new accord may be used in conjunction with an existing fragrance base, or a new composition may be composed around the original theme. Computer-assisted design methods are utilized by perfumers to help with formulation changes and to keep track of the large material base available to them.

The performance of a fragrance over time in its intended application is an important consideration. A traditional view holds that perfumes can be viewed as a blends with a top note continuing into a middle and on to an end note. This is clearly an oversimplification, since it all components of a fragrance would be expected to reach the nose simultaneously, but in differing amounts depending upon volatility. Thus what is perceived is a blend whose character will depend upon such factors as odor threshhold and impact (dose vs response) for each ingredient. Over time, the more volatile ingredients (traditionally the top note) will be dissipated to the point where they contribute less and less to the overall impression. The rate at which this happens is a function of the overall matrix in which the ingredients sit as well as individual component properties. The perfumer must smooth the odor profile of a formula so that there are no discontinuities in odor impact as the different components evaporate. The amount of an ingredient in a fragrance formulation can vary widely, from 10–20% for some materials to trace levels (parts per million) for others. The cost of an ingredient or its odor strength can be a limitation on its use.

Perfumes and ingredients are evaluated initially by a simple "bioassay": smelling them from paper strips or blotters, and from the various media in which they may be used (eg, soap bars, laundry detergent, etc.). This apparently straightforward exercise is made extremely complex by the ability of human beings to perceive hundreds of thousands of different odor nuances, by individual differences in odor perception, and by differences in the way individuals describe odors. The phenomenon of fatiguing or adaptation to odors is also a complicating factor, which is often not taken into account. The simple criterion of a material smelling good or pleasant is entirely inadequate to determine its value as an odorant. In fact, some materials, both of natural and synthetic origin, smell rather unpleasant by themselves and yet are important in perfumery. Even after years of experience in the hands of many perfumers, new ways to use a particular fragrance ingredient can be discovered and become the basis of a new fragrance type.

Perfume Ingredients

The classical materials of perfumery are natural products. These are mostly of vegetative origin, with some obtained from animal secretions. Just about any part of a plant can be used, including flowers, fruits, leaves, twigs, roots, and wood, depending on the amount and quality of essence it contains. Perfume materials of animal origin include tincture of tonquin musk, civet gum, and beaver castoreum. Such materials have been or are being replaced by synthetic substitutes for environmental, political, or economic reasons. Even though synthetics continue to grow and have displaced naturals in overall usage, the latter are deeply rooted in the art of perfumery and so remain extremely valuable in perfume creation; in addition, naturals provide the odor reference points by which synthetics are often judged or described.

Natural Products. Various methods have been and continue to be employed to obtain useful materials from various parts of plants. Essences from plants are obtained by distillation (often with steam), direct expression (pressing), collection of exudates, enfleurage (extraction with fats or oils), and solvent extraction. Solvents used include typical chemical solvents such as alcohols and hydrocarbons. Liquid (supercritical) carbon dioxide has come into commercial use in the 1990s as an extractant to produce perfume materials. Generally the extracts produced this way are of very good quality, but differ from those traditionally produced. For this reason, they usually cannot replace the older products in existing perfumes, but they may become part of the pallette for new creations.

The principal forms of natural perfume ingredients are defined below; the methods used to prepare them are described in somewhat general terms because they vary for each product and supplier. This is a part of the industry that is governed as much by art as by science.

Concretes are produced by extraction of flowers, leaves, or roots, usually with hydrocarbon solvents. After removal of the solvent by distillation, the concrete is obtained as a thick, waxy residue. Such materials are used in some fine fragrances, but the waxes they contain can give rise to solubility problems. For this reason, concretes are often dissolved in alcohol to make tinctures, or in other low-odor diluents. Production of concretes, especially flower concretes, usually takes place where the botanicals are grown since the odors of such materials deteriorate rapidly after harvesting.

Absolutes are prepared from concretes by further processing to remove materials that can cause solubility problems in perfumes. This is done by dissolution in alcohol, filtering, and removal of the solvent, ultimately at reduced pressures. The resulting products are viscous, oily materials that may be diluted with low-odor substances such as diethyl phthalate.

Concretes and absolutes, both obtained by total extraction of the plant material and not subject to any form of distillation other than solvent removal, are complex mixtures containing many chemical types over a wide molecular weight range. These products have powerful odors and contribute in important ways to the perfumes in which they are used.

Essential oils are produced by distillation of flowers, leaves, stems, wood, herbs, roots, etc. Distillations can be done directly or with steam. The technique used depends mostly on the desired constituents of the starting material. Particular care must be taken in such operations so that undesired odors are not introduced as a result of pyrolytic reactions. This is a unique aspect of distillation processing in the flavor and fragrance industry. In some cases, essential oils are obtained by direct expression of certain fruits, particular of the citrus family. These materials may be used as such or as distillation fractions from them. See also **Oils, Essential**.

Naturally Derived Materials. The following are descriptions of some of the most important naturally-derived materials in use. Importance in this context is defined in terms of total value, which may derive from expensive, low volume materials with great aesthetic value to relatively inexpensive and widely-used products. In some cases, the oils are distilled to provide individual chemicals for use as such or as chemical intermediates. Materials produced in this way from a given natural source are usually not interchangeable with those from other naturals or synthetics. This may be due to optical isomerism, which can have a significant effect on odor, or to trace impurities.

Bergamot oil is produced by cold expression from peels of fruits from the small citrus tree, Citrus bergamia. Bergamot is grown mainly in southern Italy and northern and western Africa.

Bois de rose oil is obtained by steam distillation of wood chips from South American rosewood trees, Aniba rosaeodora. The tree, a wild evergreen, grows mainly in the Amazon basin.

Cedarwood oil is obtained from different parts of the world. They are produced mainly by steam distillation of chipped heartwood, but some are produced by solvent extraction. The oils, which vary significantly in chemical composition, are used in perfumes as such, but the main uses are as distillation fractions and chemical derivatives.

Citronella oil is produced in Ceylon, China, Java, and Brazil by steam distillation of similar, but not identical, grasses. The main constituents of the oil are citronellal [CAS: 106-23-0], geraniol [CAS: 624-15-7], and citronellol [CAS: 106-22-9].

Clove leaf oil is produced mainly in Madagascar and Indonesia. It is obtained by distillation of leaves and twigs of the Eugenia caryophyllata. The material from Madagascar is considered of superior quality to those from the other areas, because its eugenol [CAS: 97-53-0] content is in the ranges 82–92%.

Galbanum gum is an exudate collected from large umbelliferous plants of the Ferula species, which grow wild in the Middle East. The gum is

extracted with alcohol to produce a resinoid or steam-distilled to produce an oil. The odor of galbanum blends excellently with lilac fragrances. In modern perfumery, it is used to give a greenish top note. Galbanum oil and resinoid are complex mixtures from which many materials have been identified. Of these, a group of isomeric 1,3,5-undecatrienes has been identified as key odor contributors, in particular the 3(E),5(Z)-isomer [CAS: 51447-08-6].

Jasmine is one of the most precious florals used in perfumery. The concrete of jasmine is produced by hydrocarbon extraction of flowers from Jasminum officinale (var. Grandiflorum). The concrete is then converted to absolute by alcoholic extraction, filtration and solvent removal. It is produced in many countries, the most important of which is India, followed by Egypt. Jasmine products are rather expensive and are produced in relatively small amounts compared with other materials. However, jasmine is particularly important in perfume creation for its great power and aesthetic qualities. Four of the principal odor contributors to jasmine are cis-jasmone [CAS: 488-10-8], methyl jasmonate [CAS: 91905-97-4], benzyl acetate [CAS: 140-11-4], and indole [CAS: 120-72-9].

Lavandin oil is produced from Lavandula hybrida, is a plant species is of recent origin, unknown until the late 1920s. It is a hybrid of two common lavenders, Lavandula officinalis and Lavandula latifolia. Lavandin is cultivated mainly in southern France and has become one of the most produced and used natural perfumery materials.

Oakmoss, Evernia prunastri, is a lichen that grows mostly on oak and spruce trees. It is collected mostly in the Czech Republic, Croatia, and Morocco. A cheaper quality, which is also called tree moss or mousse d'arbre, grows on spruce and pine trees. Oakmoss is worked into a variety of products, including a concrete, resinoid, and absolute, the last of which is the most used. Materials of this type are typically green in color and are thus limited in some functional perfumes.

Orris is produced from rhizomes of Iris pallida and Iris germanica. The plants are found and cultivated mostly in Italy, but also in Morocco and China. It is used in perfumery as an absolute, a steam-distilled essential oil, and a concrete.

Patchouli oil is produced by steam distillation of the dry leaves of Pogostemon cablin, a shrub-like plant that originated in the Philippines and Indonesia. Most production is done in Indonesia. Patchouli oil has a wonderfully rich odor profile, which is described as warm, sweet, herbaceous, spicy, woody, and balsamic. It is relatively inexpensive for a natural product and is usually available in abundance.

Petitgrain oils are produced by steam distillation of leaves and twigs of the bitter orange tree, Citrus aurantium, the same species used to produce orange flower oil. The so-called biogarde oil is produced from the true bitter orange tree grown in southern France, Italy, Spain, and northern Africa. Petitgrain Paraguay, by far the most used material of this type, is produced from the bitter-sour variety in South America. The odor of these oils is fresh, bitter, and floral, with woody undertones. They are used widely in perfumery, particularly in citrus colognes and floral bouquet perfumes.

Sandalwood is one of the oldest materials in fragrance use. Its oil is produced by steam distillation of coarsely ground wood and roots of Santalum album, a comparatively small, slow-growing tree. Its world production is concentrated in India and Indonesia, the latter being of less preferred quality. In order to obtain a good oil and high yield, only trees that are over 30 years in age are used. This limits the supply of the oil and makes it expensive. Although sandalwood oil is still valuable in some fine fragrance applications, much of the sandalwood odor of perfumes produced since the 1980s is a result of the excellent synthetic materials that have been introduced. The main odor contributors to sandalwood oil are alpha-santalol [CAS: 115-71-9] and beta-santalol [CAS: 77-42-9].

Vetivert oil is steam-distilled from cleaned, dried, and chopped rootlets of Vetiveria zizanoides, a tall perennial grass normally grown for up to 20 months prior to harvesting. Most production of this material is in Haiti, Indonesia, Reunion, and, of a poorer quality, in China. The oil has a heavy woody-earthy odor and an undertone of precious wood.

Violet leaf absolute is produced by the usual extraction methods from Viola odorata (var. *Victoria*). It is grown mainly in the south of France and Egypt. Although this material is not produced in large amounts, it is quite valuable in perfumery for its powerful green leafy and floral character, an odor that belongs to many floral bouquets.

Aroma Chemicals. The use of aroma chemicals in perfumery has been growing since they were first introduced. A number of practical advantages account for this trend. Probably foremost among them is that the growing use of fragrance in the world outstripped the ability to produce enough natural materials, particularly the aesthetically important concretes and absolutes. Increasing world population and the need for food-farming have displaced some of the land area and labor required for production of perfume ingredients. Naturals, especially those produced from flowers, have become more and more expensive due to limited supply and rising labor costs. Some essential oils, particularly those derived from slow-growing trees such as sandalwood, have also been limited in supply for environmental or political reasons. Quality control of synthetics is straightforward compared with naturals because the raw material and product compositions are much less complex. Also, synthetics are not subject to variation in quality and supply due to growing conditions.

Aroma chemicals have grown to be such a major part of the fragrance industry because of their availability at relatively low costs. This has been possible because synthetic fragrance materials are produced from a wide variety of starting materials, from both petrochemical and renewable sources. The most important renewable source is turpentine, followed at some distance by cedarwood oil.

Aroma chemicals are not limited to any particular functional group, but some are more common than others. The majority of materials contains a single oxygenated functional group, although there are also important materials containing two and even three. Most of these are manufactured by syntheses of one to four chemical steps; some of the starting materials are isolated by distillation from abundant natural sources, eg, cedarwood and clove leaf oils. Some hydrocarbons are important fragrance ingredients, as well as several sulfur or nitrogen-containing materials, but the latter two classes are generally used in very small amounts to provide special nuances.

Aroma chemicals are used in perfumes over a wide range of concentrations. The importance of a material to the overall creation should not be judged only by the amount used. Some very minor components impart essential character that may be the key to the commercial success of a perfume. Thus it is characteristic of chemical manufacturing in the fragrance industry that many different materials are produced over a wide range of production volumes. Larger fragrance companies often attempt to simplify their product lines by giving up production of some low-volume products. However, the demands for novelty and richness in perfume creation mitigate against these efforts. It is reasonable to assume that manufacturers of aroma chemicals will continue to produce wide ranges of materials.

Safety, Regulatory, and Environmental Aspects

The fragrance industry has a long record of safety, largely on account of the nature and sources of its ingredients, and how its products are used. The approach to product safety, used successfully for many years, is based on individual ingredients rather than finished perfumes. Far more testing of fragrance formulations would be required than is done for ingredients because the rate of new fragrance creation is high. By examining individual ingredients and setting appropriate limits on their use, it is possible to ensure the safety of fragrances as they are created. This approach is accepted by relevant governmental agencies around the world, such as the U.S. Food and Drug Administration.

The industry supports the International Fragrance Association http://www.ifraorg.org/ (IFRA) that strengthens scientific criteria and develops guidelines for the safe and environmentally sound use of fragrances. IFRA is composed of two arms governed by the IFRA Executive Committee: a scientific arm, the Research Institute for Fragrance Materials (RIFM), and a Communications arm (IFRA). The Research Institute for Fragrance Materials http://www.rifm.org/ (RIFM) is an internationally recognized scientific organization that collects, generates, and disseminates information on the safety of perfume ingredients. This information may originate from published or unpublished sources, or through RIFM's ongoing research program. The findings are reviewed by an independent expert panel of academicians and published in peer-reviewed journals such as Food & Chemical Toxicology. The activities of RIFM are harmonized with those of the International Fragrance Association (IFRA). IFRA, whose members come from various national associations of fragrance manufacturers (eg, from the Netherlands, France, Germany, United States, and Japan), is concerned with all aspects of safety evaluation and regulation in the industry, into which it has introduced self-regulatory discipline. Its primary function is the formulation and continuous updating of the Code of Practice for the Fragrance Industry. The Code provides guidelines on manufacturing practices, as well as toxicological methodology, safety assessments and standards on the safe usage of fragrance materials. These guidelines and standards have been followed by the fragrance industry since the early 1970s.

Fragrances must comply with all applicable regulations and legislation that address occupational and consumer health, safety, and environmental concerns. Many countries have adopted chemical substance inventories in order to monitor use and evaluate exposure potential and consequences. For most fragrance applications, all ingredients must be on these lists. New substances must be subjected to premanufacturing or premarketing notification (PMN). PMN requirements vary by country and by the predicted volumes of production, and/or import; they require assessments of environmental and human health-related properties, and reporting of the results to designated governmental authorities.

Evaluation of the safe use of perfumes is an ongoing process. It is conducted mainly through self-disciplinary efforts of the fragrance industry and involves continuing investigation of the safety of new and existing ingredients. The manufacture and use of perfumes must comply with the growing body of environmental and human health-related regulations worldwide and must meet the rigorous standards set forth by the industry as well as the many legislative requirements throughout the world.

See also **Odorant**.

Additional Reading

Aftel, M.: *Essence and Alchemy: A Natural History of Perfume*, Gibbs Smith, Layton, UT, 2004.

Bauer, K., D. Garbe, and H. Surburg: *Common Fragrance and Flavor Material*, 4th Edition, John Wiley & Sons, Inc., New York, NY, 2001.

Butler, H.: *Poucher's Perfumes, Cosmetics and Soaps*, 10th Edition, Kluwer Academic Publishers, Norwell, MA, 2000.

Calkin, R.R., J.S. Jellinek: *Perfumery: Practice and Principles*, John Wiley & Sons, Inc., New York, NY, 1994.

Curtis, T., and D.G. Williams: *Introduction to Perfumery*, 2nd Edition, Micell Press, Dorset, UK, 2001.

Gimelli, S.P.: *Aroma Science*, Micelle Press, Dorset, UK, 2001.

Kraft, P., and K.A.D. Swift: *Perspectives in Flavor and Fragrance Research*, John Wiley & Sons, Inc., Hoboken, NJ, 2005.

Muller, P.M., and D. Lamparsky: *Perfumes: Art, Science and Technology*, Kluwer Academic Publishers, Norwell, MA, 2002.

Rowe, D.J.: *Chemistry and Technology of Flavors and Fragrances*, Blackwell Publishing, Oxford, UK and CRC Press LLC, Boca Raton, Fla., 2005.

Web References

Fabulous Fragrances, http://www.fabulousfragrances.com/

Research Institute for Fragrance Materials (RIFM), http://www.rifm.org/

PERHUMID CLIMATE.   See **Climate**.

PERIAPSIS.   The orbital point nearest the center of attraction.

PERIASTRON.   That point of the orbit of one member of a binary star system at which the stars are nearest to each other. That point at which they are farthest apart is called *apastron*.

PERICYNTHIAN.   That point in the trajectory of a vehicle that is closest to the moon.

PERIDERM.   A collective term for the tissues composing the outer bark of older stems of woody plants. When the stem increases in diameter as a result of the production of secondary vascular tissues by the cambium, the epidermis is no longer adequate as a protective layer. Cells of the cortex become meristematic and begin to divide, producing a new protective tissue. The dividing cells are called cork cambium or phellogen, while most of the cells produced become cork. These cork cells are small and rectangular; the protoplasm soon disappears leaving only the suberin-impregnated cell walls. Later, even the outer cells of the phloem become cork cambium, so that the bark of old trees contains only phloem and periderm.

Roots form a similar periderm. In this case, it is the pericycle cells that become cork cambium.

PERIDOTITE.   The term peridotite is derived from peridor, the French word for olivine.

It is a coarse-grained igneous rock related to gabbro, which consists of olivine and proxene in varying proportions. Certain peridotites contain spinel, chromite, or mica as accessories.

Rocks consisting essentially of olivine alone are known as dunites, the name coming from the occurrence of this rock in the Dun mountains of New Zealand. In the United States, this mineral is found in North Carolina, South Carolina, and Georgia, where corundum is associated with the dunite in commercial quantities. The olivine of peridotites alters readily to the mineral serpentine, often to such an extent that the rock itself is called a serpentine. As mentioned above, the peridotites may contain chromite or other valuable minerals, often to such an extent that they may be commercially exploited, for nickel, platinum, and precious garnet.

Kimberlite from which diamonds are secured is commonly called a mica peridotite but is more closely related to the lamprophyres. See also **Kimberlite**.

PERIFOCUS.   The point on an orbit nearest the dynamical center (focus). The pericenter is at one end of the major axis of the orbital ellipse.

PERIGEE.   That orbital point nearest the Earth when the Earth is the center of attraction. That orbital point farthest from the Earth is called *apogee*. Perigee and apogee are used by some writers in referring to orbits of satellites, especially artificial satellites, around any planet or satellite, thus avoiding coinage of new terms for each planet and moon.

PERIGLACIAL CLIMATE.   See **Climate**.

PERIHELION.   The point in the orbit of any member of the solar system at which the object is closest to the sun. Since the orbits are all conic sections with the sun at one focus, perihelion must lie on the line of apsides of the conic. The point is diametrically opposite to aphelion.

In orbits of satellites and of other objects related to primaries other than the sun, terms similar to perihelion are used to indicate points in the orbit closest to the primary. For example, the point in the moon's orbit closest to the earth is known as perigee; the corresponding point in an orbit of a satellite of Jupiter is known as perijove; and a similar point in a double star orbit is known as periastron. See also **Orbit (Astronomy)**.

PERIMETRY.   Perimetry is a visual field test of the eye that checks for problems of peripheral (side) vision. Peripheral vision is used primarily for detecting objects and for directing central vision so it is possible to see those objects in detail. A loss of peripheral vision results in a condition called tunnel vision and can lead to legal blindness. Severe glaucoma causes the loss of peripheral vision.

The cells that make up the retina are responsible for the ability to see detail, brightness, and color. There are two types of photoreceptor cells in the cornea: rods and cones. The rods specialize in work at low light levels, and the cones provide sharp vision, color, and contrast discrimination. People with achromatopsia have defective cone cells and must rely on their rod photoreceptors for vision.

In normal eyes, there are about six million cone photoreceptors, located mainly in the macula at the center of the retina. These cells are primarily responsible for sharp, straight-ahead vision and also for the ability to distinguish colors. There are 100 million rod receptors, located mostly at the periphery of the retina. The rods are more sensitive to light than cones are, but rods are not able to differentiate among colors, nor can they perceive shades of gray, black, and white.

There are different variations in the severity of symptoms among individuals with achromatopsia. The rarest and most severe is called complete rod monochromatism, where there is a total lack of cone

function. People with this disorder are extremely sensitive to light, even in normally lit rooms. They also have symptoms of poor visual acuity and nystagmus, which is involuntary movement of the eyes. Other less severe variations of the disorder are known as incomplete rod monochromatism and blue cone monochromatism. The type depends on which cones are affected.

A perimetry test is easy and comfortable. Sometimes a doctor administers the test, but usually a trained technician administers it. To take the test, you sit with your head in a chin rest at the edge of a large, bowl-shaped instrument with a fixed spot in the center, usually a yellow or green light.

There are two types of perimetry tests. In the Goldman kinetic perimeter test, the patient stare directly at the spot as the technician moves objects or lights of different size and brightness from the side. When the object or light is seen, the patient pushs a button.

The threshold static automated perimetry, which is the other type of perimeter test, uses stationary objects of light that blink on and off in various parts of the visual field.

With either method, each eye is tested independently. The maps of visual sensitivity, made by either of these methods, are very important in diagnosing diseases of the visual system.

The perimetry test usually takes no more than 30 minutes, and the results are available immediately. See also **Peripheral Vision**.

Vision Rx, Inc., Elmsford, NY.

PERIOD.   See **Geologic Time Scale**.

PERIODIC FUNCTION.   One for which a constant a exists so that $f(z) = f(z + na)$, where n is an integer. The function repeats itself periodically with the fundamental period a. Typical examples are sin z, cos z, with periods $2n$; e^z, with fundamental period, $2\pi i$. Functions may be doubly periodic such that the periods are of the form $(na + n'a')$, both n and n' integers but not both zero. See also **Elliptic Integral**.

PERIODIC LAW.   Originally stated in recognition of an empirical periodic variation of physical and chemical properties of the elements with atomic *weight*, this law is now understood to be based fundamentally on atomic *number* and atomic *structure*. A modern statement is; the electronic configurations of the atoms of the elements vary periodically with their atomic number. Consequently, all properties of the elements that depend on their atomic structure (electronic configuration) tend also to change with increasing atomic number in a periodic manner.

PERIODIC TABLE OF THE ELEMENTS.   When the chemical elements are arranged in a matrix on the basis of increasing atomic numbers, a pattern of periodicity among the physical and chemical characteristics emerges. See Fig. 1. By no means is the resulting matrix perfect, but the resemblance of characteristics among groups of elements arranged in this manner is indeed both striking and illuminating. Attempts to classify the elements date back to the early work of de Chancourtois (1862) and Newlands (1863), but the discovery of the relationship between atomic-number groupings and characteristics was made by Dmitri Mendeleev in 1869. One year later, Lothar Meyer independently showed the periodicity of the elements in terms of atomic volumes. Meyer defined the latter characteristic as the atomic weight divided by the specific gravity of the element in the solid state.

Although there have been numerous refinements to Mendeleev's early tabulation, fortified by the discovery and isolation of several elements then unknown the fundamental principles of the matrix are the same. The conventional table is shown in Fig. 2. The information also can be presented in polar fashion as shown in Fig. 3. It is interesting to note that as one proceeds clockwise around the circle the atomic numbers appear consecutively and that 18 sectors of the circle become the bases for families or groups of elements.

Notation for designating the grouping of the elements was changed and officially accepted in the mid-1980s. The new notations are used in Fig. 2. They are summarized in Table 1.

Thus, the members of the alkali metals (Group 1), alkaline earths (Group 2), halogens (Group 17), and so on, all bear resemblance, one element to the other, within any given group. There are two significant breakpoints in any representation of periodicity, namely, commencing with

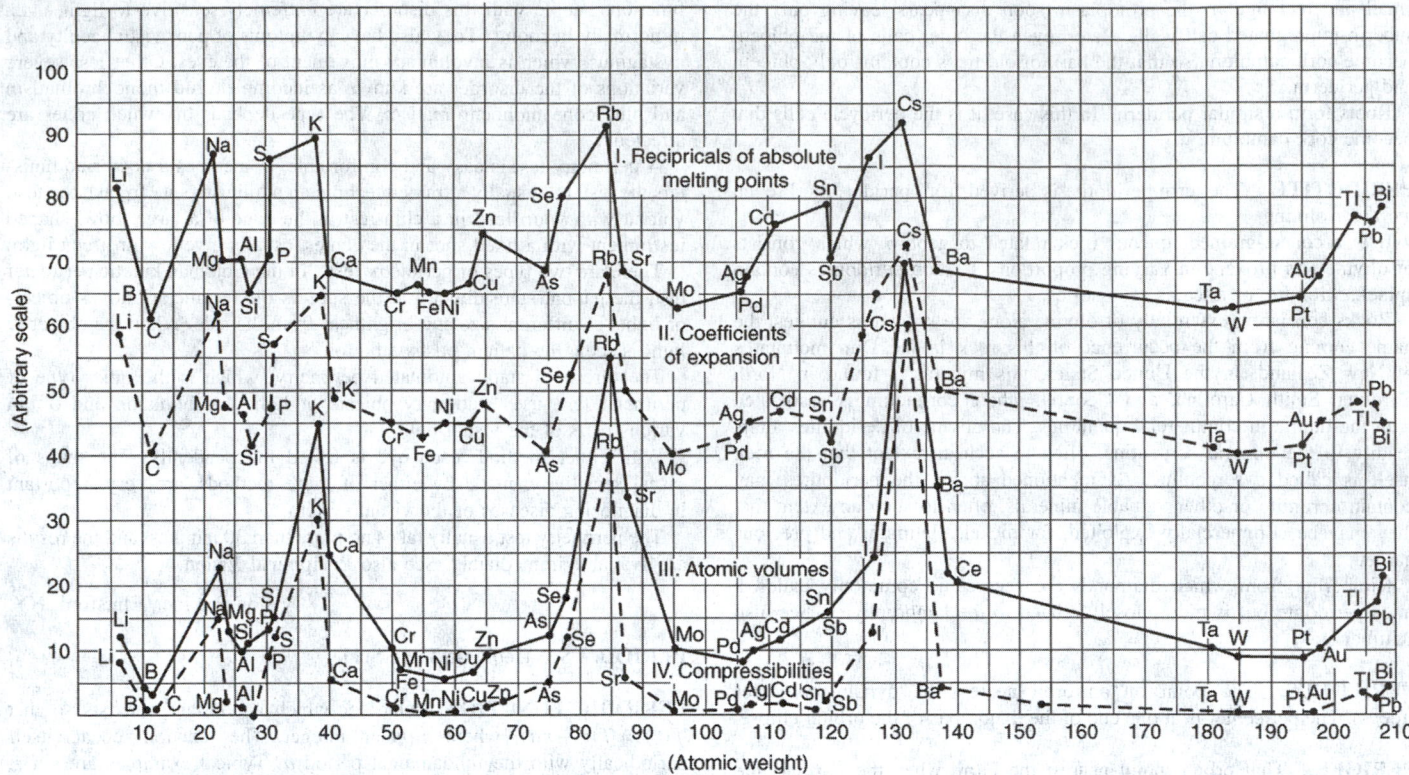

Fig. 1.   Pattern obtained when various parameters are plotted against increasing atomic weight of chemical elements.

TABLE 1. REVISED CHEMICAL ELEMENT GROUP NOTATION VERSUS PRIOR SCHEMES

Revised Notation	Elements in Revised Grouping	Prior IUPAC Form	Prior CAS Version
1	H, Li, Na, K, Rb, Cs, Fr	IA	IA
2	Be, Mg, Ca, Sr, Ba, Ra	IIA	IIA
3	Sc, Y, La, Ac	IIIA	IIIB
4	Ti, Zr, Hf, Rf	IVA	IVB
5	V, Nb, Ta, Db	VA	VB
6	Cr, Mo, W, Sg	VIA	VIB
7	Mn, Tc, Re, Bh	VIIA	VIIB
8	Fe, Ru, Os, Hs	VIIIA	VIII
9	Co, Rh, Ir, Mt	VIIIA	VIII
10	Ni, Pd, Pt, Ds	VIIIA	VIII
11	Cu, Ag, Au	IB	IB
12	Zn, Cd, Hg	IIB	IIB
13	B, Al, Ga, In, Tl	IIIB	IIIA
14	C, Si, Ge, Sn, Pb	IVB	IVA
15	N, P, As, Sb, Bi	VB	VA
16	O, S, Se, Te, Po	VIB	VIA
17	F, Cl, Br, I, At	VIIB	VIIA
18	He, Ne, Ar, Kr, Xe, Rn	VIIIA	VIIIA

Lanthanides and Actinides not included in group numbering.

atomic number 57 (lanthanum) and atomic number 89 (actinium). Attempts to place the elements which follow—in the one case, atomic numbers 59 through 71, and in the other case, atomic numbers 90 through 103, in the underlying geometric matrix (whether tabular or circular) do not succeed. These separate groups are known as the *Lanthanides* (rare earths) and the *Actinides*, respectively. Upon completion of the Lanthanide series (with lutetium, atomic number 71), the orderly geometry resumes with hafnium, atomic number 72 (Group 4) and continues through actinium, atomic number 89. The probable positions of elements 104 through 110 are indicated in Groups 4 through 8, respectively.

An amazing result of Mendeleev's pioneering classification was the prediction of elements yet to be discovered. Mendeleev found that he could maintain geometric logic of his table only if he allowed for some blank spaces in his table. He further reasoned that elements later would be discovered that would occupy these vacant positions and, thus, Mendeleev predicted the existence of gallium, scandium, and germanium. In fact, Mendeleev gave a preliminary name to scandium, calling it eka-boron and predicted the probable properties of the element. The element later was isolated by Lars Fredrik Nilson in 1879. Mendeleev lived to see his prediction confirmed.

In retrospect, with a much fuller understanding of the underlying electronic and particle structure of the elements, most aspects of the periodicity of the elements come as no surprise, but the fact remains that Mendeleev, Meyer, and others made these striking observations without benefit of over 100 years of additional knowledge. The periodicity of the elements is demonstrated by the accompanying graph which plots atomic weights along the abscissa versus an arbitrary ordinate for various observed physical characteristics. See also **Chemical Elements**.

PERIOD LUMINOSITY LAW.   It has been observed that the cepheids of populations I and II form distinct sequences, where their absolute magnitudes are very closely related to their periods. Hence, upon observing the period and apparent magnitude (m) of a cepheid, one can read off the absolute magnitude (M) and then calculate its parallax (π) by means of

$$m - M = 5 \log D - 5$$

where D is in parsecs. The term $m - M$ is often referred to as the *distance modulus*. It is interesting to note that these pulsating stars also follow the relation

$$P \sqrt{\rho} = \text{constant}$$

where P is their period of pulsation, and hence allow us to calculate their mean density (ρ).

See also **Cepheids**.

PERIODONTITIS.   Sometimes called pyorrhea or gingivitis, this condition is caused by degenerative changes in the periodontium and is characterized by purulent discharge or inflammation. Periodontal disease usually begins at the edge of the gums, along sites that frequently are inadequately cleaned when brushing the teeth. The microorganisms that cause periodontitis generally differ from those that produce dental caries. See also **Caries,**

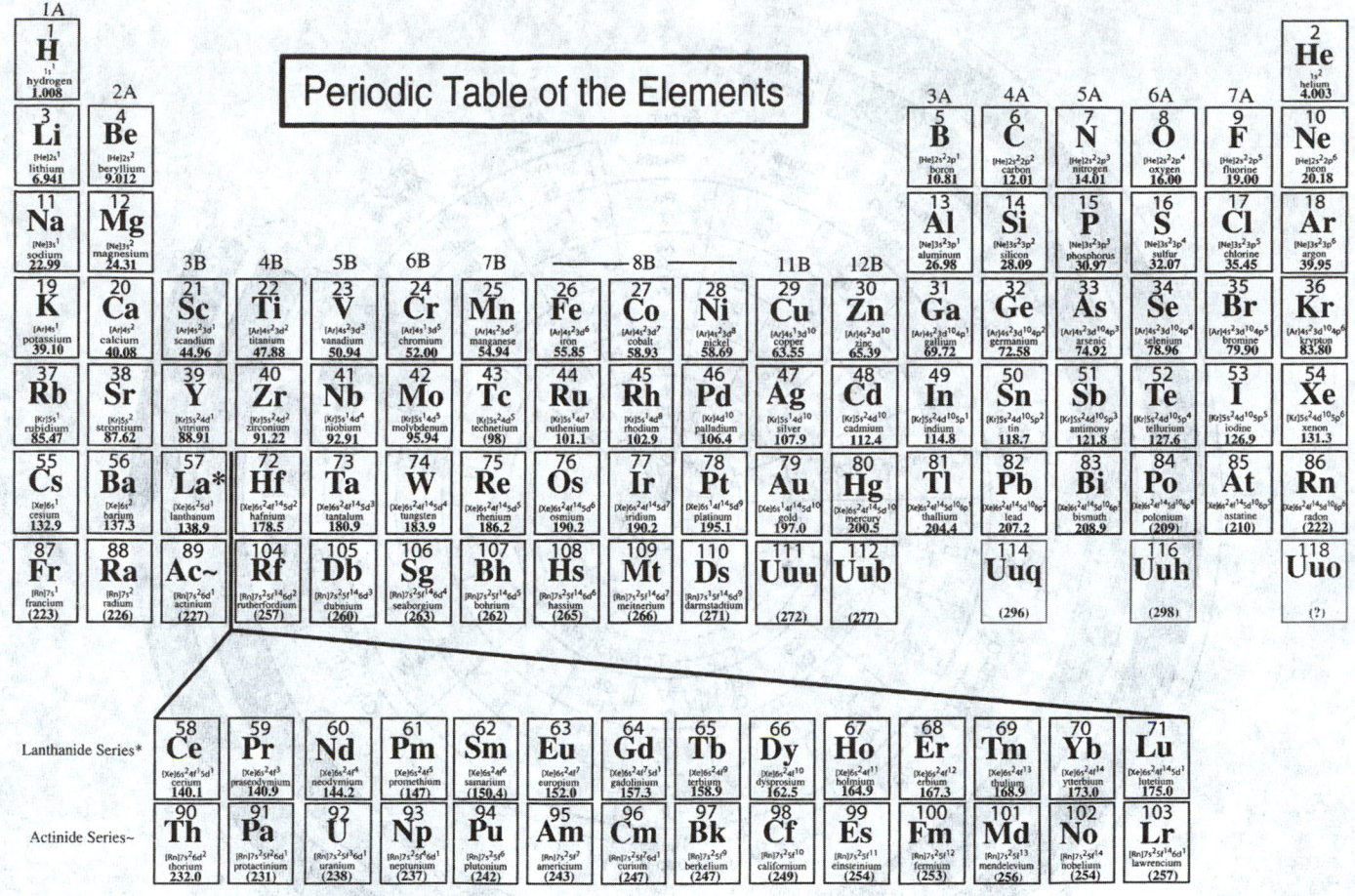

Periodic Table of the Elements

Los Alamos National Laboratory Chemistry Division

Fig. 2.   Conventional representation of periodic table.

Cariology, and Dentistry. Whereas caries results from microorganisms that are nurtured from the food supply of the patient, those organisms that cause periodontitis find their nutrients in the tissue. Thus, a person who consumes little carbohydrate may be free of cavities, but could have periodontal disease. This is demonstrated in some countries where there is little to eat and little dental care available. There may be few cavities among the population, but periodontal disease may be prevalent.

In periodontal disease, microorganisms produce toxic products, which irritate the tissues, causing swelling and redness. In time, underlying connective tissue fibers are destroyed, leaving the gum tissues weak. A space is formed between the edge of the gum and tooth (*periodontal pocket*) which provides an ideal environment for the continued growth of bacteria, the latter ultimately destroying soft tissue and bone. Serum and blood provide an even richer diet at the gingival margin. The process, without attention, continues until teeth are lost.

PERIPHERAL EQUIPMENT (Computer System). The auxiliary machines which may be placed under the control of a computer. Examples include card readers, card punches, magnetic tape units, high-speed printers, scanners, diagnostics, test, and measuring devices. Peripheral equipment may be used on-line or off-line, depending upon the computer design, task requirements, and costs. A device is said to be on-line when it is under the control of the central processing unit, and information reflecting current activity is introduced into the data-processing system as it occurs. A device is off-line when it is not in direct communication with the central processing unit.

With the continuing developments in electronic microminiaturization and the wide acceptance of the personal computer (PC) for industrial as well as business applications, the trend has been toward integrating as many functions as possible into the computer unit per se.

PERIPHERAL THROMBOSIS.   See Thrombosis.

PERIPHERAL VISION.   Peripheral vision, or side vision, is that part of vision that detects objects outside the direct line of vision. For instance, when you read a word on a page, you are using your central vision, but it is your side vision that tells you if the word is at the beginning or end of a sentence, or at the top or bottom of a page. Your peripheral vision also tells you where to look if someone enters the room or if a car is approaching from the side. Like most people, you are probably not aware of the limitations that would exist without peripheral vision, because you are constantly moving your eyes in order to focus with your central vision.

The difference between central and peripheral vision becomes apparent when you understand the visual function of the eye. The eye works like a camera with two lenses: the cornea at the front of the eye and the natural crystalline lens behind the pupil. The cornea is responsible for about 70 percent of the eye's focusing power, while the natural lens fine-tunes the image before it is focused on the retina at the back of the eye. The retina works like the film in a camera, receiving light rays and sending them through the optic nerve to the brain where they are converted into images.

There is a small area in the center of the retina called the macula, which is less than 1/4 inch in diameter, that is responsible for sharp, clear central vision and ability to perceive color. The densely packed photoreceptor (light-sensitive) cells in the macula control the eye's central vision and are responsible for the ability to read, drive a car, watch television, see faces, and distinguish detail. The rest of the retina handles peripheral vision that enables the eyes to see objects off to the side while looking forward.

There are two types of photoreceptor cells in the cornea; rods and cones. The rods specialize in work at low light levels, and the cones provide sharp vision and discrimination. The macula contains a high concentration of cones, which accounts for the sharper focus of straight-ahead vision,

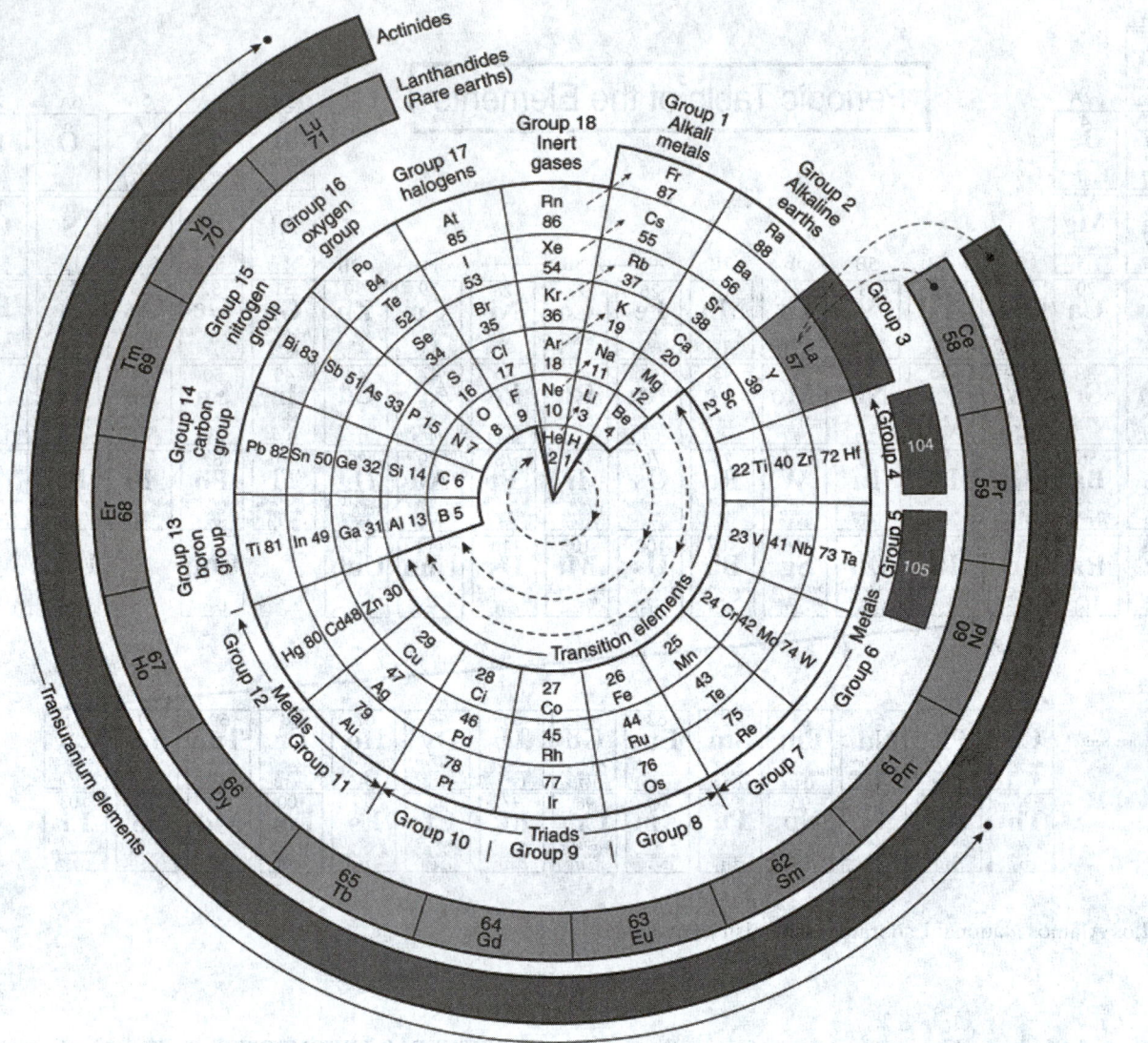

Fig. 3. Polar representation of periodic relationships of the elements. (*Source: Omnibix, U.S.A.*) At upper right is shown the conventional representation.

particularly in bright light. Most of the rods are located in the periphery of the retina. This is why you can often see faint objects more clearly if you do not look directly at them. A dim star, for instance, is best seen when your eyes are not aimed directly at it.

When you see something out of the corner of your eye, its image focuses on the periphery of your retina and you are unable to distinguish the color of what you are seeing. And, because there are fewer rods, the ability to resolve the shapes of objects at the periphery of vision is limited. Normal peripheral vision, called visual field, for one eye is approximately 150 degrees from side to side. For both eyes, it is approximately 180 degrees.

Loss of peripheral vision results in a condition called tunnel vision, which is like looking through a tunnel. Tunnel vision can be caused by several disorders including glaucoma, retinitis pigmentosa, and strokes.

Most eye examinations include a perimetry or visual field test to check peripheral vision. The perimetry test is used to detect and monitor damage from glaucoma and other conditions that may affect the visual pathway from the eye to the brain. In medical terms, perimetry is a systematic measurement of visual field function. An instrument called a perimeter is used to plot the central or peripheral field of vision.

Loss of peripheral vision can actually be more difficult than the loss of visual acuity. However, many low vision aids are available for people who have lost all or a portion of their peripheral vision. Training and optical devices, such as prisms, mirrors, reverse telescopes, and minus lenses, can improve awareness of the environment and independent travel ability.

Vision Rx, Inc.; Elmsford, NY

PERISSODACTYLA *(Mammalia).* Hoofed animals with an odd number of toes, in which the axis of the foot passes through the middle digit. This digit is larger than the others and is symmetrical. Organization of the *Perissodactyla* is shown in Table 1, along with references to specific entries in this volume which describe the various genera in the order of *Perissodactyla.*

TABLE 1. PERISSODACTYLA (Odd-toed Hoofed Mammals)

	In This Volume
EQUINES	See also **Horses, Asses, and Zebras.**
Horses (*Equus caballus* and *przewalskii*)	
Asses (*E. hemionus* and *asinus*)	
Zebras (*E. burchelli. . . .*)	
Grevy's Zebra (*Dolichohippus*)	
TAPIRINES	See also **Tapir.**
Malayan Tapir (*Tapirus indicus*)	
Central American Tapir (*T. bairdii*)	
South American Tapirs (*T. terrestris*)	
Mountain Tapir (*T. roulini*)	
RHINOCEROTINES	See also **Rhinoceros.**
One-horned Rhinoceroses (*Rhinoceros*)	
Asiatic Two-horned Rhinoceroses (*Didimoceros*)	
African Black Rhinoceros (*Diceros*)	
Ceratotheres (*Ceratotherium*)	

PERISTALTIC MOVEMENT.   See **Digestive System (Human)**.

PERITONITIS.   A widespread inflammation of the peritoneum, the membranous lining of the abdominal cavity. The surface area available to infection is large, thus usually producing profound symptoms. The large, moist, warm conditions prevailing in the peritoneum make an ideal environment for the rapid growth and development of bacteria. There are several avenues of access to the peritoneum by bacteria, including a perforation of the gastrointestinal tract, such as may occur with a gastric or duodenal ulcer that may break through, or the rupture of an inflamed and distended appendix. In these instances, the colon bacillus is frequently the causative agent. Other conditions favoring entrance of bacteria into the peritoneum include typhoid fever, obstructions of the bowel, serious dysentery, diverticulitis, and severe gallbladder conditions. Less frequently, bacteria may gain entry via the female genital tract. Pelvic peritonitis results from infection of the Fallopian tubes, most often with the gonococcus. Tuberculous peritonitis occurs in the course of tuberculous infection of the bowel or the genitourinary tract. Pneumococcus peritonitis is seen as a complication of nephrosis. A form of chemical peritonitis may occur if gastric juices, such as bile or pancreatic juices, spill over into the abdominal cavity. In relatively few instances, the peritoneum infection may take the form of local abscesses rather than be of a generalized nature.

Peritonitis may also result from penetrating trauma of the abdominal wall, such as caused by knife or gunshot wounds. Such externally inflicted injury carries with it not only hemorrhage and traumatic shock, but also the possibility of tetanic peritonitis and infection by anaerobic bacteria.

Prior to the availability of sulfonamide drugs and antibiotics, peritonitis was a most dreaded condition. The condition is most serious and requires immediate attention, but because of available treatment properly and timely administered, some of the former grave aspects of the condition have been diminished. The patient with peritonitis is acutely ill, with fever, rapid pulse and respirations, and a distended, tender, rigid abdomen. Treatment varies with cause and will almost always include chemotherapy with or without surgical intervention.

A. C. V.

PERIWINKLE (*Mollusca, Gasteropoda*).   Marine snails with a thick conical spiral shell. They live in shallow water, in the tidal zone, and along the shore. Some of the many species are very widely distributed, and the genus *Littorina* is represented in all parts of the world. An edible species has been introduced into the United States and is now common on the Atlantic coast north of Delaware Bay.

PERKIN, SIR WILLIAM HENRY (1838–1907).   An English chemist who was the first to make a synthetic dyestuff (1856). He studied under Hofman at the Royal College of London. Perkin's first dye was called mauveine, but he proceeded to synthesize alizarin and coumarin, the first synthetic perfume. In 1907 he was awarded the first Perkin Medal, which has ever since been awarded by the American Division of the Society of Chemical Industry for distinguished work in chemistry. Not withstanding the fact that Perkin patented and manufactured mauve dye in England, the center of the synthetic dye industry shifted to Germany, where it remained until 1914.

See also **Hofmann, August Wilhelm**.

PERLITE (or Pearlstone).   An unusual form of siliceous lava composed of small spherules of about the size of bird shot or peas. It is grayish in color with a soft pearly luster. The spherules often show a concentric structure and are believed to be formed as a result of a peculiar spherical cracking developed while cooling. They may be confused with oölites, which are classified as concretions.

PERMAFIL.   A mixture in which the liquid undergoes complete polymerization and hardens without the necessity of any evaporation. Anareobic permafils harden out of contact with air.

PERMANENT FROST.   The last geographic remnant of the last great Ice Age is permafrost, or permanently frozen ground. Permafrost exists today in many places on the earth, including the northern most parts of America, Eurasia, and the Antarctic. In the coldest regions, permafrost has developed and existed for millions of years. Today, it covers about 20% of the earth's surface.

Frozen soil or rock that remains at a temperature of $0\,^\circ$C ($32\,^\circ$F) or below for 2 or more years is permafrost. Permafrost is defined solely by temperature and is not determined by the amount of snow cover, soil moisture content, or its location. Permafrost may contain more than 30% ice or no ice at all, and may be topped by meters of snow cover or no snow at all. The defining characteristic of permafrost is its temperature.

Remains of the Great Ice Ages

What remains from the last Ice Age is nearly one-fifth of the Earth's surface still permanently frozen. During the past 2 million years, the great ice sheets that covered much of the earth's Northern Hemisphere prevented the surface below them from thawing. This freeze seeped deeper and deeper into the earth, forming thick layers of permanently frozen ground, permafrost. As time passed, much of this permafrost thawed due to the end of the Ice Age. As the earth's surface began to warm again, permafrost levels declined. However, across the Northern Hemisphere and in Antarctica, much of the ground remains permanently frozen, even during the summer months. Other permafrost forms when the ground cools enough during the winter to prevent it from thawing in the following summer.

Though the determining characteristic of permafrost is ground temperature, the determining cause is not air temperature but atmospheric climate. Temperature instead affects the distribution and thickness of permafrost.

Geography of Permafrost

Permafrost is located throughout the Northern and Southern Hemispheres. The largest continuous mass of permafrost in the Southern Hemisphere is beneath the thick ice of Antarctica. This colossal mass alone comprises about 9% of the earth's surface. Permafrost is more widespread in the Northern Hemisphere. It is found as far north as Greenland and as far south as the heights of the Himalayas. Most of the Northern Hemisphere's permafrost, which makes up 23% of the hemisphere, occurs in the Eastern Hemisphere in parts of Siberia, eastern Russia, northern parts of Mongolia and China, and in the Tibetan Plateau. Other significant permafrost comprises most of Alaska (85% of the state) and parts of northwestern Canada. More than 50% of the surface of Canada is permafrost. Farther north, permafrost levels decline rapidly as land gives way to the Arctic Ocean; yet, a shelf of permafrost has been discovered even under the sea.

Permafrost is separated into two categories: continuous and discontinuous. Continuous implies permafrost everywhere except underneath bodies of water. Discontinuous permafrost is broken into two subdivisions: widespread and sporadic. In the widespread permafrost zone, 50–90% of the land is permafrost. In the sporadic zone, only 10–50% of the ground is permafrost.

Life in a Permafrost Region

A widely held misconception is that a permafrost zone is barren and inhospitable to any form of plant or animal life. On the contrary, the Arctic permafrost zone is home to a variety of mammals, birds, insects, and evergreen forest plant life that covers much of the Northern Hemisphere's permafrost zone.

In Arctic summers, the terrain is alive with animal life. Insects and other invertebrates inhabit the shallow wetlands. Birds feed from these insects and the plant life; they are provided ideal breeding and nesting grounds during the summer months. Snow geese and tundra swans are among the more common visitors of more than 100 bird species, which also include some hawks, owls, and cranes. Polar bears and caribou are also common in the Arctic's permafrost region. Musk oxen, other species of bears, several rodent species, foxes, squirrels, and hares often accompany these Arctic inhabitants. However, when the brief summer comes to a close and the frigid cold reasserts itself upon the region, most animal life leaves in what is essentially a mass exodus.

The trees and shrubs that grow in permafrost forests are much smaller than those in the south and are far more resistant to freezing conditions. Forests tops are evidenced by species of spruce, birch, and aspen trees, and are carpeted below by shrubs, heath, and moss.

Permafrost Fears

Scientists fear that changes to the permafrost regions pose a serious threat to humans and animals. The Earth's natural warming, or human-influenced global warming, has begun to melt the thick permafrost. Also, removal of vegetation and organic cover, as well as forest fires, can also have temperature-raising effects on permafrost regions, causing further thawing of the Permafrost. If this problem continues, the permafrost could turn into a slush mixture. This would cause permanent damage to buildings, roads, pipelines, and other infrastructure in the Arctic region. In particular, oil pipelines, originally set in solid permafrost, would be disrupted. The results could be disastrous, releasing oil into an already fragile ecosystem. Scientists believe this chain of events has begun, but admit that their investigations still remain at a very early stage. However, in Alaska and other Arctic regions, builders are already taking more precautions before building structures on permafrost.

Web References

Geophysical Institute, University of Alaska, Fairbanks: http://www.gi.alaska.edu/snowice/Permafrost-lab/, and http://www.gi.alaska.edu/snowice/

International Permafrost Association: http://www.geo.uio.no/IPA/

The Geological Survey of Canada: http://gsc.nrcan.gc.ca/index_e.php

ARTHUR M. HOLST, Philadelphia Water Department, Philadelphia, PA

PERMANENT MEMORY.   In computer terminology, storage of information that remains intact when the power is turned off. Also called *nonvolatile storage*.

PERMANENT-PRESS RESIN.   A thermosetting resin used as a textile impregnant or fiber coating to impart crease resistance and permanent hot-creasing to suitings, dress fabrics, etc. Chemicals such as formaldehyde and maleic anhydride are the basis of these products. The resin is "cured" after the fabric has been tailored into a garment. A permanent-press fabric that requires no resin has been developed (a blend of polyester with cotton or rayon).

See also **Maleic Anhydride, Maleic Acid, and Fumaric Acid.**

PERMEABILITY.   As applied in magnetism, *absolute permeability* is the ratio B/H, where B is magnetic flux induction and H is magnetizing force. *Relative or specific permeability* is $\mu_s = B/(\mu_0 H)$, where μ_s is specific permeability and μ_0 is the permeability of free space. See also **Magnetism.**

PERMEABILITY (Soil).   See **Hydrology.**

PERMEAMETER.   An electromagnet arranged for magnetizing a specimen, and for allowing measurement of the flux through the specimen and the magnetizing force at the surface. Various types of permeameter have been designed, aiming at ease of operation, accuracy of results, or use with high-coercivity materials.

PERMEATION.   As applied to gas flow through solids, the passage of gas into, through, and out of a solid barrier having no holes large enough to permit more than a small fraction of the gas to pass through any one hole. The process always involves diffusion through the solid and may involve various surface phenomena, such as sorption, dissociation, migration, and desorption of the gas molecules.

PERMENORM.   Nickel-iron alloy produced by magnetic annealing and drastic cold reduction and used for mechanical rectifiers and low-frequency amplifiers. This alloy has a rectangular hysteresis loop that eliminates arcing at the contacts of mechanical rectifiers, as well as other desirable properties.

PERMIAN PERIOD.   The name of a geologic period. Type locality. Province of Perm, Russia. The formations of this system were first studied and described by R.I. Murchison, in 1841. The Permian period began about 230 million years ago and lasted for about 30 million years. Only the upper Permian appears to be represented in North America. Eastern North America was undergoing uplift and erosion during this period, while the seas invaded the West from the Arctic and Gulf. Increasing uplift of the continents and mountain building culminated in the Appalachian Revolution (U.S.) and the Hercynian Revolution (Europe). Increasing aridity and

widespread equatorial continental glaciation are disclosed by tillites in Australia, Tasmania, New Zealand, India, South America, and Massachusetts.

PERMUTATION.   Given n distinguishable objects or elements, each different arrangement of the elements is a permutation. The number of permutations is $n(n-1)(n-2)\cdots 3\cdot 2\cdot 1 = n!$ Several different symbols, such as $_nP_n$, $P_{n,n}$ or $P(n,n)$ are used to indicate this result. If the n things are taken r at a time $(r < n)$,

$$_nP_r = n(n-1)(n-2)\cdots(n-r+1) = \frac{n!}{(n-r)!}$$

When n_1 of the elements are all alike of the first kind, n_2 of the second kind, etc., so that $n_1 + n_2 + \cdots n_m = n$, the number of permutations is

$$\frac{n!}{n_1!n_2!\cdots n_m!}$$

This result also applies if the elements are separated into m parts with n_i elements in the i th part. If the number of elements in each of the m parts is not specified, but each part must contain at least one element, the number of permutations is

$$\frac{n!(n-1)!}{(n-m)!(m-1)!}$$

This number is increased to

$$\frac{(m+n-1)!}{(m-1)!}$$

if empty parts are permitted.

A *combination* is an arrangement of objects or elements, where the order of arrangement is not distinguished. Thus, given the three letters a, b, c, the possible permutations are (abc), (acb), etc., six in number, but there is only one combination. With symbols as before,

$$_nC_r = \frac{n(n-1)(n-2)\cdots(n-r+1)}{r!} = \frac{n!}{r!(n-r)!}$$

This number is identical with the $\binom{n}{r}$, the coefficient in the binomial series. Moreover, $_nC_r =_n P_{n-r}$; $_nP_r = r!_nC_r$.

The number of combinations of n different elements into m specified parts, with empty parts allowed, is m^n; of n identical elements into m different parts with empty parts is

$$\frac{(n+m-1)!}{n!(m-1)!}$$

but when at least one element is in each part, the number is

$$\frac{(n-1)!}{(m-1)!(n-m)!}$$

Finally, the total number of combinations of n things taken $1, 2, 3, \ldots, n$ at a time is $\Sigma_{i=1}^n {}_nC_i = 2^n - 1$.

PERMUTATION GROUP.   Its elements, $n!$ in number, are the various permutations or rearrangements of a standard arrangement of n symbols of objects. A typical element is

$$S = \begin{pmatrix} s_1 & s_2 & \cdots & s_n \\ 1 & 2 & \cdots & n \end{pmatrix}$$

meaning that the operation S replaces 1 by s_1, 2 by s_2, etc. If another element is indicated by T, then ST, the rearrangement designated by T followed by S, the resulting permutation is also in the group.

A permutation sending s_1 into s_2, s_2 into s_3, etc., and finally s_n into s_1 is called a *cycle* on n letters. It is usually written as $(s_1, s_2, \ldots, s_n)$. The degree of a cycle equals the number of symbols permuted. A cyclic permutation of degree two is a *transposition*. Any permutation may always be written as a product of transpositions, either even or odd in number. The permutation is then said to be even or odd.

The group of all permutations of n letters or objects, of order $n!$, is called the *symmetric group*. The even permutations of n objects form a *subgroup* of the symmetric group. Its order is $n!/2$ and it is called the *alternating group*.

Every group is *isomorphous* to some permutation group. It is easy to find a representation of a permutation group by using a permutation matrix.

Each row and column of such a matrix has but one non-zero element and that is unity. The row and column thus designate the initial and final locations of the object permuted.

See also **Permutation**.

PEROVSKITE. The mineral perovskite is calcium titanate, essentially $CaTiO_3$, with rare earths, principally cerium, proxying for Ca, as do both ferrous iron and sodium, and with columbium substituting for titanium. It crystallizes in the orthorhombic system, but with pseudo-isometric character; fracture subconchoidal to uneven; brittle; hardness, 5.5; specific gravity, 4; luster, adamantine; color, various shades of yellow to reddish-brown or nearly black; transparent to opaque. It is found associated with chlorite or serpentine rocks occurring in the Urals, Baden, Switzerland, and Italy. It was named for Von Perovski.

PEROXIDES AND PEROXIDE COMPOUNDS (Inorganic). A peroxide or peroxo compound contains at least one pair of oxygen atoms, bound by a single covalent bond, in which each oxygen atom has an oxidation number of -1. The peroxide group can be attached to a metal, M, through one (1) or two (2) oxygen atoms, or it can bridge two metals (3):

Peroxides should be distinguished from several other types of compounds having similar names. The higher oxides of lead, manganese, and other elements, although sometimes called peroxides, are not peroxides as defined herein because these contain no oxygen–oxygen bond. Similarly, compounds such as the perchlorates and permanganates are not peroxides. It is preferable for true peroxides to be designated by the prefixes *peroxo* or *peroxy*. In the IUPAC nomenclature, *peroxo* is used for inorganic compounds, *peroxy* for organic compounds.

All the simple peroxides form hydrogen peroxide on contact with water. Many inorganic peroxides tend, as does H_2O_2, to decompose evolving oxygen.

In 1811 Gay-Lussac and Thenard first described the peroxides of barium, sodium, and potassium. Hydrogen peroxide was isolated several years later in 1818 by Thenard. Because of the advantages in stability and handling, the solid peroxides were of greater practical use than hydrogen peroxide solutions at this time. Sodium peroxide gained industrial importance as a bleaching agent in the textile industry and was first produced on a commercial scale in 1899. Sodium peroxide lost its importance when hydrogen peroxide could be more cost-effectively manufactured by the anthraquinone process.

Today the most important industrial inorganic peroxo compound is sodium perborate tetrahydrate, which was discovered in 1898 by Tanatar. It has been produced on a commercial scale since 1907 and is still the preferred bleaching agent in powder detergents. Other industrially important inorganic peroxo compounds are the sodium, potassium, and ammonium salts of peroxodisulfuric acid ($H_2S_2O_8$), the triple salt of potassium peroxomonosulfate ($2\ KHSO_5 KHSO_4 K_2SO_4$) and calcium peroxide (CaO_2). From the hydrogen peroxide addition compounds, sodium carbonate peroxohydrate finds increasing use in laundry detergents and dishwashing formulations, since products with good shelf life properties can be manufactured economically using new coating technologies. Urea peroxide is produced on a several-hundred tonnes scale only and has limited applications in pharmaceutical products.

Group 1 (IA) Peroxides

Peroxides of all the alkali metals having the formula M_2O_2 are known. There are several general methods of preparation: reaction of the metal and oxygen, reaction of the metal monoxide and oxygen, thermal decomposition of the superoxide, and reaction of alkaline solutions of the metal and hydrogen peroxide.

Alkali metal peroxides are stable under ambient conditions in the absence of water. They dissolve vigorously in water, forming hydrogen peroxide and the metal hydroxide. They are strong oxidizing agents and can react violently with organic substances. Only lithium peroxide and sodium peroxide have been commercialized.

Lithium Peroxide. Lithium peroxide [CAS: 12031-80-0], Li_2O_2, Mr 45.9, 2.36 g/cm^3, has an active oxygen content of 34.8%, which is the highest of all metal peroxides. Lithium peroxide is pale yellow solid, stable at ambient temperature, and not hygroscopic. On heating to about 300 °C, it loses oxygen and forms lithium monoxide. The commercial product contains about 96% Li_2O_2. Unlike the other alkali peroxides, the lithium salt cannot be made by direct reaction between the metal and oxygen. It is made commercially by reaction of aqueous lithium hydroxide and hydrogen peroxide, which yields the peroxohydrate trihydrate, and is then dehydrated. Lithium peroxide has not attained industrial importance because of comparatively high cost. Lithium peroxide is a strong oxidizer and can promote combustion when in contact with combustible materials. It is a powerful irritant to skin, eyes, and mucous membranes; Protective clothing should be worn when handling lithium peroxide. The LD_{50} has not been determined, and there is no designated threshold limit value (TLV). However, five grams of many lithium compounds can be fatal.

Sodium Peroxide. Sodium peroxide [CAS: 1313-60-6], Na_2O_2, is a pale yellow solid, stable at ambient temperature, and hygroscopic. On heating, it starts to liberate oxygen at about 300 °C and decomposes rapidly above its melting point of 460 °C.

The commercial product is a powder containing a minimum of 96% Na_2O_2 and approximately 20% active oxygen. It is made commercially by oxidizing the molten metal with either oxygen or air enriched in oxygen. As of the mid-1990s, sodium peroxide has only a few special applications, including chemical analysis and the extraction of platinum from its ores by the Leidie process.

Although neither inflammable nor self-igniting, sodium peroxide is highly inflammable when mixed with oxidizable substances. Such mixtures burn violently, even in the absence of air.

Sodium peroxide is a powerful irritant to skin, eyes, and mucous membranes; protective clothing should be worn when handling it.

Group 2 (IIA) Peroxides

All the elements of Group 2 form peroxides, with the exceptions of beryllium and radium. There are two general methods of preparation: reaction of the metal or monoxide with oxygen, and reaction of the hydroxide with aqueous hydrogen peroxide. These peroxides are more stable in the presence of water than the Group 1 peroxides, primarily because of insolubility in water. Calcium peroxide is used on a large scale; magnesium, strontium, barium, and zinc peroxides have small-scale uses; whereas cadmium and mercury peroxides have no commercial uses at all. A general account of these peroxides is available.

The materials are generally made by triturating the oxides, or hydroxides, with aqueous hydrogen peroxide and drying the solid products. The commercial products are typically mixtures of the peroxides with varying amounts of hydroxides, oxides, carbonates, hydrates, and peroxohydrates.

Magnesium Peroxide. Magnesium peroxide [CAS: 14452-57-4] MgO_2, has not been prepared in the pure state. The hydrates $MgO_2 \cdot 0.5H_2O$ [CAS: 77883-47-7] and $MgO_2 \cdot H_2O$ have been postulated. The product is a white powder containing about 25% MgO_2 and 7% active oxygen. This material is sparingly soluble in water but reacts with water slowly, forming hydrogen peroxide and liberating oxygen gas. It is used in medicine as a stomach antacid and as an antiseptic. There are minor uses for magnesium peroxide in household products, veterinary medicine, and metallurgy. Magnesium peroxide is a strong oxidizer and can cause fire when in contact with combustible materials. It is a powerful irritant to skin, eyes, and mucous membranes.

Calcium Peroxide. Pure calcium peroxide [CAS: 1305-79-9], CaO_2, has been prepared, but the commercial product is a mixture made by reaction of calcium hydroxide and hydrogen peroxide. Commercial material contains either 60 or 75% CaO_2, the remainder is a poorly defined mixture of calcium oxide, hydroxide, and carbonate.

An important application of calcium peroxide is for curing the polysulfide sealants used in double-glazing window units. Calcium peroxide is also used at several gold mines in Australia to increase the recovery of gold and reduce the consumption of cyanide. Solid calcium peroxide can also be used in the heap-leaching of lean gold ores. A proprietary form of calcium peroxide for this purpose is sold by FMC (United States) under the trademark PermeOx.

PermeOx is also used to improve the bioremediation of soils contaminated with creosote or kerosene, to deodorize sewage sludges and wastewater, and to dechlorinate wastewater and effluents. See also **Odor Modification**.

Calcium peroxide has several horticultural and agricultural applications, particularly in Japan. Usually used in the form of granules, it acts by providing extra oxygen for germinating plants and other organisms.

Calcium peroxide has been used for many years as a dough conditioner in the United States, but not in Europe, where this use is not permitted. Another industrial application of calcium peroxide is as an oxidizing agent in the production of certain titanium–aluminum alloys.

Calcium peroxide is among the safest of the inorganic peroxides, presenting no significant hazard with regard to skin contact or absorption, inhalation, and ingestion; but it may be irritating to the skin under humid conditions. Airborne dust is irritating to the eyes, nose, throat, and lungs, but poses no significant long-term inhalation hazard.

Strontium Peroxide. Strontium peroxide [CAS: 1314-18-7], is a light-colored solid of good thermal stability. Commercial strontium peroxide contains about 85% SrO_2 and 10% active oxygen. The only substantial application for this compound is in pyrotechnics. Strontium peroxide produces a red color in flames.

Strontium peroxide is a strong oxidizer and can cause fire when in contact with combustible materials. It is a powerful irritant to skin, eyes, and mucous membranes.

Barium Peroxide. Barium peroxide [CAS: 1304-29-6], BaO_2, was the first-known peroxo compound. It was used until mid-1900 in the manufacture of oxygen by the Brin process and of hydrogen peroxide by the Thenard reaction. It is the only Group 2 peroxide that can be prepared at atmospheric pressure by heating the metal monoxide in air.

The commercial product is a dull yellow powder containing about 90% BaO_2 and about 8.5% active oxygen; the remainder is mainly barium carbonate and barium hydroxide. Today barium peroxide has only limited use in the manufacture of pyrotechnics.

Barium peroxide is a strong oxidizer and can cause fire when in contact with combustible materials. It is a powerful irritant to skin, eyes, and mucous membranes. Consequently, it is also toxic via the subcutaneous route; protective clothing should be worn during handling. The LD_{50} value (mouse, oral) is 50 mg/kg.

Group 12 (IIB) Peroxides

Zinc Peroxide. Zinc peroxide [CAS: 1314-22-3], ZnO_2, is a yellow solid and generally similar in its properties to magnesium peroxide. The commercial product is a pale yellow powder containing about 55% ZnO_2 and 9% active oxygen. It is stable in dry air but loses its oxygen in moist air and on heating. It is insoluble in water but dissolves in dilute acid, liberating hydrogen peroxide.

Zinc peroxide is used as an accelerator in rubber-compounding, as a curing agent for synthetic elastomers, and as a deodorant for wounds and skin diseases. Zinc peroxide is a powerful irritant to skin, eyes, and mucous membranes. The systemic toxicity is similar to that of zinc oxide, for which the LD_{50} (rat, oral) is 7950 mg/kg.

Zinc peroxide is a strong oxidizer and can cause fire when in contact with combustible materials.

Group 13 (IIIB) Peroxides

Boron Compounds.

Nomenclature. The naming of sodium perborate, one of the most important commercial boron compounds, has long been confused.

The stoichiometry of the most important hydrate is $NaBO_3 \cdot 4H_2O$, often called *sodium perborate tetrahydrate*. Only one of the oxygen atoms acts as an oxidant. The formula has often been written as $NaBO_2 \cdot H_2O_2 \cdot 3H_2O$ and the compound referred to as a trihydrate. The crystal structure, established in 1961, contains the cyclic diperoxodiborate anion $[B_2(O_2)_2(OH)_4]^{2-}$ and six associated water molecules. The crystallographically derived names are used to avoid confusion. The prefix *per* usually signifies an element in its highest valence state. Because the boron is always trivalent, the prefix *peroxo*, recommended by IUPAC, is used herein. The commercial name used in the detergent industry is sodium

perborate tetrahydrate. *Sodium Perborate Tetrahydrate.* The compound sodium perborate tetrahydrate (sodium peroxoborate hexahydrate) [CAS: 10486-00-7], $Na_2[B_2(O_2)_2(OH)_4] \cdot 6H_2O$, Mr 153.9, ρ 1.731 g/cm^3, theoretical active oxygen content 10.38%, melts in its own water of crystallization at 65.5°C. The solubility in water is 0.152 g/L at 20°C and 0.388 mol/L at 40°C. The commercial product is white crystalline solid with a bulk density of 0.65–0.90 kg/L.

The traditional electrochemical process of sodium perborate tetrahydrate is no longer of commercial importance. Today it is produced by mixing hydrogen peroxide with an alkaline solution of sodium borate made from a boron mineral, usually technical grade borax pentahydrate or the minerals kernite and tincal. See also **Boron**. The mineral is dissolved in aqueous sodium hydroxide and the solution is clarified; hydrogen peroxide and a stabilizer such as magnesium silicate are then added at a temperature <30°C. The product crystallizes when cooled to ambient temperature and is removed by filtration or centrifuge.

In aqueous solution, all the sodium peroxoborates dissociate for the most part into boric acid, or its anion, and hydrogen peroxide. Peroxoborate species are also present in these solutions, depending on the pH and the concentration for the species type. The nature of these species has been extensively examined by conventional physicochemical methods, by nmr, and by Raman spectroscopy. Both monomeric and polymeric species are usually present. There is some evidence suggesting that these peroxoborates are more reactive than hydrogen peroxide alone under similar conditions.

Sodium perborate tetrahydrate has been an important commercial bleaching agent for more than 90 years. See also **Bleaching Agents**. It is an important ingredient of many household detergents and cleaning products. These applications account for approximately 90% of the total market. The textile industry generally uses hydrogen peroxide for bleaching, but there are a few areas in which sodium perborate tetrahydrate is preferred.

The product is considered nonhazardous for international transport purposes. However, it is an oxidizing agent sensitive to decomposition by water, direct sources of heat, catalysts, and so on. Decomposition is accompanied by the liberation of oxygen and heat, which can support combustion and cause pressure bursts in confined spaces. Decomposition in the presence of organic material is rapid and highly exothermic.

Sodium Perborate Trihydrate. The compound sodium perborate trihydrate (sodium peroxoborate tetrahydrate) [CAS: 28962-65-4], $Na_2B_2(O_2)_2[(OH)_4] \cdot 4H_2O$, Mr 135.9, 1.86 g/cm^3, formerly written as $NaBO_3 \cdot 3H_2O$ theoretical active oxygen content 11.8 wt%, melts at 81.7°C. A number of procedures have been published for producing this hydrate; most involve the crystallization from water at 40–50°C. The crystal structure shows the presence of the same cyclic peroxoborate anion as that in the sodium peroxoborate hexahydrate. Stability and vapor pressure determinations indicate that sodium perborate trihydrate is the thermodynamically stable phase above 15°C. It has, however, never been commercialized because of its slow dissolution in water.

Sodium Perborate Monohydrate. Sodium perborate monohydrate [CAS: 10332-33-9] (sodium peroxoborate), $Na_2[B_2(O_2)_2(OH)_4]$, written as $NaBO_3 \cdot H_2O$, Mr 99.8, 2.12 g/cm^3, has a theoretical active oxygen content of 16.0 wt%. It has no definite melting point. Sodium perborate monohydrate is known only as a microcrystalline powder made by dehydrating the trihydrate. The crystal structure has not been determined, but the vibrational spectrum indicates the presence of the same cyclic peroxodiborate anion (four) as that in the peroxoborate hexahydrate as well as in the peroxoborate trihydrate.

The commercial product has an active oxygen content of at least 15%. This product has replaced the tetrahydrate in some compact household detergents and dishwashing formulations because it dissolves faster and has a greater content of active oxygen per unit volume of granular product. Because of the short washing cycles in North America, sodium perborate monohydrate is the preferred product in this market. Sodium perborate monohydrate is also the preferred product used in denture cleansers.

The toxicity of sodium perborate monohydrate is similar to that of the tetrahydrate. Sodium peroxoborate is an eye irritant, but not a skin irritant. Absorption through large areas of abraded or damaged skin can give systemic boron poisoning.

Anhydrous Sodium Perborate. Anhydrous sodium perborate [CAS: 7632-04-4], $NaBO_3$, is an undefined, powdery material. It should perhaps be regarded more as an amorphous assemblage of radicals than as a defined compound.

Anhydrous sodium perborate effervesces in water. It is used mainly as an ingredient in denture-cleanser tablets to enhance dissolution. Anhydrous sodium perborate is irritating to eyes, skin, and mucous membranes.

Group 14 (IVB) Peroxides

Peroxocarbonates. Peroxocarbonates contain the C−O−O− group and should be distinguished from the carbonate peroxohydrates. Although no crystal structures have been determined, the nature of the peroxocarbonates has been deduced from vibrational spectra. These compounds can be prepared by three general methods: reaction of carbon dioxide and a solution of the metal hydroxide in hydrogen peroxide, anodic oxidation of normal carbonates at low temperatures, and oxidation of aqueous solutions of carbonates with elemental fluorine. Only the peroxocarbonates of the alkali metals are known.

Peroxosilicates. No solid peroxosilicates are known.

Peroxotin Compounds. Older literature records some tin peroxides or peroxohydrates, but these claims have not been substantiated. In contrast, organometallic peroxotin compounds are well established.

Group 15 (VB) Peroxides

Peroxonitrous Acid and Its Salts. Peroxonitrous acid [CAS: 14691-52-2] HOONO, is an isomer of nitric acid, HNO_3, to which it rapidly converts. The half-life of peroxonitrous acid at $0\,°C$ is 10 s; at $27\,°C$, 0.23 s. It has been known since 1904 that the yellow solution made by mixing nitrous acid and hydrogen peroxide at low temperature contains a stronger oxidant than either ingredient alone, but the chemistry involved was not put on a sound basis until 1994. Additional preparatory methods are also available.

Peroxonitrous acid can decompose by two pathways: isomerization to nitric acid, and dissociation into the hydroxyl radical and nitrogen dioxide.

Peroxonitrite is believed to be present in the crystals of nitric trihydrate that form in the stratosphere and in Martian soil. See also **Extraterrestrial Materials**. Peroxonitrous acid may be present in mammalian blood and other biochemical systems.

Peroxophosphoric Acids and Their Salts. The peroxophosphoric acids are structurally similar to the peroxosulfuric acids, and behave similarly. They are hydrolyzed more rapidly, but the oxidation reactions of their anions are kinetically inhibited.

In its usual impure form (H_3PO_4 is the main contaminant), peroxomonophosphoric acid [CAS: 13598-52-2] is a viscous, colorless liquid. Peroxomonophosphoric acid can be prepared by the hydrolysis of peroxodiphosphates in aqueous acid and by the reaction of hydrogen peroxide with phosphorus pentoxide. It is not produced or used commercially and the salts that have been prepared are unstable and impure. Pure peroxodiphosphoric acid [CAS: 13825-81-5], $H_4P_2O_8$, can be obtained by anodic oxidation of phosphoric acid.

Tetrapotassium peroxodiphosphate [CAS: 15593-49-9], $K_4P_2O_8$, is a colorless, crystalline solid, soluble in water to 42.2 wt % at $0\,°C$ and 51.2 wt % at $40\,°C$.

Tetrapotassium peroxodiphosphate has been investigated as desizing agent and ingredient in toothpaste as an anticalculus agent and a bactericide. However, until today the peroxophosphates have no commercial use mainly because of their higher costs compared to the peroxosulfate salts.

Arsenic Peroxides. Arsenic peroxides have not been isolated; however, elemental arsenic, and a great variety of arsenic compounds, have been found to be effective catalysts in the epoxidation of olefins by aqueous hydrogen peroxide. Transient peroxoarsenic compounds are believed to be involved in these systems. Compounds that act as effective epoxidation catalysts include arsenic trioxide, arsenic pentoxide, arsenous acid, arsenic acid, arsenic trichloride, arsenic oxychloride, triphenyl arsine, phenylarsonic acid, and the arsenates of sodium, ammonium, and bismuth. To avoid having to dispose of the toxic residues of these reactions, the arsenic can be immobilized on a polystyrene resin.

Group 16 (VIB) Peroxides

Peroxosulfuric Acids and Their Salts. Two kinds of peroxosulfuric acid are known: peroxomonosulfuric and peroxodisulfuric acids. Neither is available commercially in the pure state. Both had been important intermediates in the electrochemical production of H_2O_2. See also **Hydrogen Peroxide**. Peroxodisulfuric acid $H_2S_2O_8$ is not of commercial importance. However, their salts, the peroxodisulfates or also called persulfates, are widely used as oxidizing agents and radical initiators.

Peroxomonosulfuric acid, H_2SO_5, in the form of Caro's acid has found industrial applications in the mining industry and pulp bleaching. The term *Caro's acid* is commonly used as a synonym for peroxomonosulfuric acid, but is better reserved for the equilibrium mixture with sulfuric acid and hydrogen peroxide. From the salts of peroxomonosulfuric acid only the potassium salt in form of the triple salt $2\,KHSO_5 \cdot KHSO_4 \cdot K_2SO_4$ has gained industrial applications.

Peroxomonosulfuric Acid. Peroxomonosulfuric acid [CAS: 7722-86-3], H_2SO_5, when pure, forms colorless crystals that melt with decomposition at $45\,°C$. One of its protons is strong, as in sulfuric acid, but its other proton, which is on the peroxide group, is weak ($pKa = 9.4$).

Peroxomonosulfuric acid oxidizes cyanide to cyanate, chloride to chlorine, and sulfide to sulfate. It readily oxidizes carboxylic acids, alkenes, ketones, aromatic aldehydes, phenols, and hydroquinone. Under neutral and alkaline conditions peroxomonosulfate oxidizes H_2O_2 yielding oxygen gas. Peroxomonosulfuric acid hydrolyzes rapidly in the presence of water to hydrogen peroxide and sulfuric acid. It is usually made and used in the form of Caro's acid. Peroxomonosulfuric acid exhibits significantly better anti-microbial properties than H_2O_2 and has good overall effectiveness against bacteria and viruses

Caro's Acid. Caro's acid is named after Heinrich Caro (1834–1910), who first described its preparation and oxidizing properties in 1898. Caro's acid is the equilibrium mixture that results from mixing hydrogen peroxide and sulfuric acid. These liquids mix instantly, generating a considerable amount of heat.

The reaction is highly exothermic, requiring removal of the heat of reaction in order to avoid yield losses. There are two manufacturing technologies: the isothermal and the adiabatic process.

Because of novel and cost-effective on-site generation technologies, Caro's acid is finding increasing application in hydrometallurgy, pulp bleaching, effluent treatment, and electronics. There are several applications of Caro's acid in hydrometallurgy. Caro's acid has been used in Australia as an oxidant in the acid leaching of uranium ores. Another hydrometallurgical use of Caro's acid is for detoxifying residual cyanide in effluents from gold leaching.

Caro's acid is effective in delignification and bleaching of wood pulp in (totally chlorine-free (TCF)) bleaching sequences, replacing chlorine and chlorine dioxide. When conditions are carefully controlled, the mechanical properties of the final paper are not impaired. These processes were developed in the 1980s and commercialized in the 1990s.

Caro's acid is highly corrosive and a powerful oxidant. Its acidic properties are similar to those of sulfuric acid of equivalent strength. A strong irritant, it is toxic and should always be handled accordingly. No specific toxicological data are available. Depending on the strength of the product, Caro's acid should be transported in accordance with the relevant regulations pertaining to the most appropriate sulfuric acid solution or to those of liquid oxidizers not otherwise specified (NOS).

Peroxomonosulfates. A number of sodium, potassium, ammonium, cesium, lithium, and rubidium salts of peroxomonosulfuric acid have been described in the literature. Only the potassium salt in form of the triple salt $2\,KHSO_5 \cdot KHSO_4 \cdot K_2SO_4$ [CAS: 3722266-5], Mr 614.8, 2.313 g/cm^3, theoretical active oxygen content of 5.2 wt%, is of commercial importance.

The triple salt is produced by partial neutralization of Caro's acid with aqueous KOH. The water introduced with the raw materials and formed during neutralization reaction is removed by vacuum at ∼25 °C at 1 kPa.

Its main use is in denture cleaners that function without mechanical assistance, in which discolorations are bleached and organic deposits are oxidized. In combination with sodium perborate monohydrate it generates effervescent oxygen. It is also used in dishwashing detergents and hard surface cleaners because of its bleaching and antimicrobial properties. It is used in the metal-fabricating industry as a mild etchant and pickling agent, and in the electroplating industry for detoxifying cyanide solutions. In the latter application, it is mixed with hydrogen peroxide. In water treatment it has found application as a shock treatment agent for swimming-pool water. It quickly oxidizes malodorous chloramines to nitrogen and nitrate.

Another application is its use in wastepaper recycling for the destruction of wet-strength resins. Finally, it is used in the textile industry for rendering wool (qv) shrink-resistant and nonfelting. In general, peroxomonosulfates have fewer uses in organic chemistry than peroxodisulfates. However, the triple salt is used for oxidizing ketones to dioxiranes, which in turn are useful oxidants in organic chemistry.

Peroxodisulfuric Acid. Also called *persulfuric acid*, and *Marshall's acid*, peroxodisulfuric acid [CAS: 13445-49-3], $H_4S_2O_8$, when pure, forms colorless crystals that melt with decomposition at 65°C. Peroxodisulfuric acid is a strong acid but not stable. It is seldom isolated but is synthesized and used in solution. Solutions of peroxodisulfuric acid are reasonably stable when cool but hydrolyze rapidly to hydrogen peroxide and sulfuric acid when heated in strongly acidic solutions.

Peroxodisulfates. The salts of peroxodisulfuric acid are commonly called persulfates, three of which are made on a commercial scale: ammonium peroxodisulfate [CAS: 7727-54-0], $(NH_4)_2S_2O_8$; potassium peroxodisulfate [CAS: 7727-21-1], $K_2S_2O_8$; and sodium peroxodisulfate [CAS: 7775-27-1], $Na_2S_2O_8$. The peroxodisulfates are all colorless, crystalline solids, stable under dry conditions at ambient temperature but unstable above 60°C. All the peroxodisulfates are made commercially by electrolytic processes.

The peroxodisulfate ion in aqueous solution is one of the strongest oxidizing agents known. The principal use of the peroxodisulfate salts is as initiators for olefin polymerization in aqueous systems, particularly for the manufacture of polyacrylonitrile and its copolymers. See also **Acrylonitrile Polymers**. These salts are used in the emulsion polymerization of vinyl chloride, styrene–butadiene, vinyl acetate, neoprene, and acrylic esters. Etching of printed circuit boards and removal of photoresists are also important applications. Desizing of woven textiles and finishing of furs are both long-established applications.

Another application in the textile industry is as a bleaching agent in denim finishing. Sodium persulfate–based formulations are also being sold for shock treatment of swimming-pool water. Other established applications include curing grouts for soil stabilization, initiating polymerization of graphite filament coatings, for thinning of starch, and as the active ingredient in hair bleaching formulations.

Peroxodisulfates can be employed as oxidants in organic chemistry. Heat, light, gamma rays, or transition-metal ions initiate these reaction.

The three peroxodisulfates are all toxic and irritating to skin, eyes, and mucous membranes. Skin sensitization, irritative and contact dermatitis has also been reported.

Persulfate salts should be handled only when wearing suitable protective clothing. The peroxodisulfates assist combustion by releasing oxygen, and must be stored away from combustible materials. Contamination by rust or traces of many metals can cause catalytic decomposition. The peroxodisulfates should be transported in accordance with international transport regulations pertaining to Class 5.1, oxidizing substances.

Other Metal Peroxides

Transition-Metal Peroxides. A 1964 review paid tribute to the significance of transition-metal peroxide and peroxide chemistry for the catalysis of oxidations and the storage and use of oxygen in biological systems. Since that time, many more inorganic peroxo compounds have been isolated and several more transition-metal-catalyzed organic reactions have been. However, transition-metal peroxides, as isolated species, have no place in chemical technology because they are too dangerously explosive.

Transition metals can be divided into two groups according to the characteristics of their peroxides. The first group comprises those metals that, in their highest oxidation states, have no d electrons, e.g., Ti^{4+} and W^{6+}. These metals form peroxides from hydrogen peroxide. The peroxo species act as electrophiles.

The other group of transition metals comprises those metals that retain d electrons in their normal valence states, e.g., Co^{3+} and Pt^{2+}. These metals form peroxides from dioxygen or from hydrogen peroxide. Their colors result from $d–d$ transitions. These peroxo species act as nucleophiles.

Transition-metal-catalyzed oxidations may or may not proceed via peroxo complexes. Twelve important industrial organic oxidation processes catalyzed by transition metals, many of which probably involve peroxo intermediates, have been tabulated. Even when peroxo intermediates can be isolated from such systems, it does not necessarily follow that these are true intermediates in the main reaction.

Actinide Peroxides. Many peroxo compounds of thorium, protactinium, uranium, neptunium, plutonium, and americium are known. The crystal structures of a number of these have been determined. Perhaps the best known are uranium peroxide dihydrate [CAS: 1344-60-1], $UO_4 \cdot 2H_2O$, and, the tetrahydrate [CAS: 15737-4-5], $UO_4 \cdot 4H_2O$, which are formed when hydrogen peroxide is added to an acid solution of a uranyl salt.

Uranium peroxide has found several applications in the nuclear energy industry. It provides a method for precipitating uranium from solution without introducing any extraneous cations or anions. It has been used in the extraction of uranium from its ores, where ammonia and hydrogen peroxide are used to precipitate uranium from leachate. It is also used in some fuel cycles.

Peroxohydrates

Peroxohydrates are crystalline adducts containing molecular hydrogen peroxide. These are commonly called *perhydrates*, but this term is better avoided because per historically implied the maximum oxidation state and hydrate implies the presence of water, neither of which applies to peroxohydrates. They have also been called *hydroperoxidates*.

Peroxohydrates are usually made by simple crystallization from solutions of salts or other compounds in aqueous hydrogen peroxide. They are fairly stable under ambient conditions, but traces of transition metals catalyze the liberation of oxygen from the hydrogen peroxide.

Sodium Carbonate Peroxohydrate. Known commercially as sodium percarbonate, sodium carbonate peroxohydrate [CAS: 15630-89-4] does not contain the C—O—O—C group and is not a peroxocarbonate. Crystallized sodium carbonate peroxohydrate is a white salt with an orthorhombic unit cell. Technical sodium percarbonate obtained by a crystallization process consists of coarse, compact grains with a spherolitic crystal form. The bulk density of commercial sodium percarbonate can range from 800 to 1200 g/L, depending on the synthesis process and conditions. The solubility in water is 131 g/L at 10°C, 154 g/L at 20°C, and 180 g/L at 30°C. The solubility of percarbonate is reduced in the presence of other salts such as Na_2CO_3, Na_2SO_4, and NaCl.

Sodium carbonate peroxohydrate was first prepared by in 1899 from hydrogen peroxide and soda ash. The product was isolated by precipitation with ethanol. It was not correctly identified as a hydrogen peroxide addition compound until 1909 when Riesenfeld and Reinhold elucidated its chemical structure. There are three basic production processes used for the industrial manufacturing of sodium percarbonate, yielding products with different properties with regard to bulk density, active oxygen content, and stability.

Sodium percarbonate is used mainly as an alternative for sodium perborates as bleaching agent in laundry powders automatic dishwashing detergents and color-safe bleaches. Other applications for sodium percarbonate are for industrial textile bleaching, for tripe bleaching, and in denture cleaners. In North America sodium percarbonate is also used in toothpaste for improved cleaning and disinfection.

Small quantities of sodium carbonate peroxohydrate have been used as an oxidizing agent in synthetic organic chemistry. In nonaqueous systems it serves as a source of concentrated hydrogen peroxide and thus can be used to make peroxycarboxylic acids from carboxylic anhydrides, chlorides, or imidazolides. The peroxohydrate does not need high solubility to react. A suspension of sodium carbonate peroxohydrate in tetrahydrofuran, or dichloromethane presaturated with water, can be used. These heterogeneous reactions can be accelerated by ultrasonic radiation.

The LD_{50} (rat, oral) of sodium carbonate peroxohydrate is 1034 mg/kg. The occupational exposure limit is 10 mg/m^3 per 40-hour week. The compound is a skin and eye irritant; inhalation of dust can cause irritation to the mucous membranes and the respiratory system. Decomposition in the presence of organic material can be rapid and highly exothermic. It should be handled by using eye protection, rubber gloves, and industrial footwear.

Other Peroxohydrates. Other peroxohydrates include those of potassium, rubidium, and cesium carbonates, $M_2CO_3 \cdot 3H_2O$; ammonium carbonate peroxohydrate, $(NH_4)_2CO_3 \cdot H_2O_2$; and urea peroxohydrate, $CO(NH_2)_2 \cdot H_2O_2$.

Urea peroxohydrate is an irritant to skin, eyes, and mucous membranes. The U.S. Food and Drug Administration approves it as an over-the-counter drug.

Peroxopolyoxometallates

Polyoxometallates, derived from both isopoly acids and heteropoly acids, are important homogeneous oxidation catalysts. The metals involved are vanadium, niobium, tantalum, molybdenum, and tungsten. The reactions involved are the oxidation of a wide range of organic compounds by hydrogen peroxide or organic hydroperoxide.

Superoxides

The superoxides are ionic solids containing the superoxide, O_2^-. Superoxides of all of the alkali metals have been prepared. Alkaline-earth metals, cadmium, and zinc all form superoxides, but these have been observed only in mixtures with the corresponding peroxides. The tendency to form superoxides in the alkali metal series increases with increasing size of the metal ion.

Metal superoxides are yellow-to-orange solids. Strong oxidizing agents, they react vigorously with most organic materials and reducing agents, and oxidize many metals to their highest oxidation states.

Sodium superoxide [CAS: 12034-12-7], NaO_2, is a yellow solid. No applications are known.

Potassium superoxide [CAS: 12030-8-5], KO_2, is a canary yellow solid that melts at 450–500 °C when pure. Potassium superoxide, a strong oxidizing agent, is similar to the Group 1 metal peroxides. Potassium superoxide is produced commercially by spraying molten potassium into an air stream, which may be enriched with oxygen.

Mine Safety Appliances Co. (MSA) manufactures potassium superoxide in the United States for use in self-contained breathing equipment. There are several published uses for potassium superoxide in organic chemistry, e.g., for oxidizing aromatic compounds and for initiating anionic polymerization.

On contact with skin and mucous membranes, potassium superoxide is converted to potassium hydroxide, which is corrosive and irritating. The reaction with moisture is exothermic and may induce further decomposition with the production of oxygen.

Other superoxides include rubidium superoxide [CAS: 12137-25-6], RbO_2; cesium superoxide [CAS: 1201861-0], CsO_2, are formed by direct reaction of the elements, but are most readily prepared by oxidation of solutions of the metals in liquid ammonia. They are not produced commercially.

Calcium superoxide [CAS: 12133-35-6], $Ca(O_2)_2$; strontium superoxide [CAS: 12169-210], $Sr(O_2)_2$; and barium superoxide [CAS: 55837-89-3], $Ba(O_2)_2$, have all been obtained in low yield and purity by treating the corresponding peroxides with hydrogen peroxide and heating. Heating the metal peroxides or their peroxohydrates often yields products containing some superoxides. These superoxides are not produced commercially.

Ozonides

The ozonides are characterized by the presence of the ozonide ion, O_3-. They are generally produced by the reaction of the inorganic oxide and ozone. Sodium ozonide [CAS: 1205854-7], NaO_3; potassium ozonide [CAS: 12030-89-6], KO_3; rubidium ozonide [CAS: 12060-047], RbO_3; and cesium ozonide [CAS: 12053-67-7], CsO_3, have all been reported. Ammonium ozonide [CAS: 12161-20-5], NH_4O_3, and tetramethylammonium ozonide [CAS: 78657-29-1], $(CH_3)_4 \cdot NO_3$, have been prepared at low temperatures. Whereas the inorganic ozonides are of potential importance as solid-oxygen carriers in breathing apparatus, they are not produced commercially.

Additional Reading

Bertsch-Franck, B. and co-workers: in R. Thompson, ed., *Industrial Inorganic Chemicals: Production and Use*, Royal Society of Chemistry, Cambridge, U.K., 1995, pp. 188–198.

Connor, J.A., and E.A.V. Ebsworth, *Adv. Inorg. Chem. Radiochem.*, **6**, 279, (1964).

Gerhartz, W. ed.: *Ullman's Encyclopedia of Industrial Chemistry*, 5th ed., Vol. A19, VCH, Weinheim, Germany, 1991.

Lide, D.R.: *CRC Handbook of Chemistry and Physics*, 86th Edition, CRC Press LLC., Boca Raton, FL, 2005.

McKillop, A. and W.R. Sanderson: *Tetrahedron*, **51**(22), 6145 (1995).

Morgan, C.A.: *Mellor's Comprehensive Treatise on Inorganic and Theoretical Chemistry*, Suppl., Vol. 5. Longman, London, U.K., 1980.

Waldemar, A.: *Peroxide Chemistry*, John Wiley & Sons, Inc., New York, NY, 2003.

PEROXIDES AND PEROXIDE COMPOUNDS (Organic). Organic peroxides are compounds possessing one or more oxygen-oxygen bonds. They are derivatives of hydrogen peroxide, HOOH, in which one or both hydrogens are replaced by a group containing carbon (R, R′), i.e., ROOH or ROOR′. The ultimate source of the oxygen-oxygen linkage in organic peroxides is oxygen; either from direct air oxidation or from reactions of organic compounds with peroxidic materials derived from oxygen, e.g., hydrogen peroxide, alkali metal peroxides, ozone, or other organic peroxides. Organic peroxides are intermediates or products in air oxidation of many synthetic and natural organic compounds. They are involved in many biological processes including development of rancidity in fats, loss of activity of vitamin products, and firefly bioluminescence. Some biological products contain a peroxide group. Organic peroxides are also involved in gum formation in lubricating oils, prepolymerization of some vinyl monomers, and degradation of olefin polymers.

Almost all organic peroxides are thermally and photolytically sensitive owing to the facile cleavage of the weak oxygen–oxygen bond. This cleavage is a unimolecular (first-order) reaction. The thermal decomposition rates are affected by the structure of the organic peroxide and the decomposition conditions.

Thermal decomposition of peroxides initially forms oxygen-centered free radicals from the oxygen–oxygen bond homolysis. These radicals are reactive intermediates generally having very short lifetimes, i.e., half-life times less than 10^{-3} s. Because they form useful free radicals, they are used commercially as initiators for free-radical reactions.

The first synthesis of an organic peroxide was that of dibenzoyl peroxide [CAS: 94-36-0] (BPO) in 1858. In the early 1900s, BPO was employed as a bleaching agent for edible oils and, subsequently, for grain flours. The use of organic peroxides as free-radical initiators for vinyl monomer polymerization became important during World War II due to increased demand for plastics and synthetic rubber for military tires. Approximately 100 different organic peroxides in well over 300 formulations are commercially produced throughout the world as free-radical initiators for polymerizing vinyl monomers, grafting of monomers onto polymers, curing agents for unsaturated resins, rubber, and elastomers, cross-linking of thermoplastics (e.g., polyethylene), modification/degradation of polypropylene, halogenations, anti-Markovnikov additions to terminal olefins (e.g., formation of primary mercaptans), and telomerizations. Some are used as bleaching agents (i.e., for grain flours and fabrics), olefin epoxidizing agents, and active species in a variety of other applications, e.g., the use of BPO as the active antibacterial component in acne medications.

Organic peroxides can be classified according to peroxide structure. There are seven principal classes: hydroperoxides; dialkyl peroxides; α-oxygen substituted alkyl hydroperoxides and dialkyl peroxides; primary and secondary ozonides; peroxyacids; diacyl peroxides (acyl and organosulfonyl peroxides); and alkyl peroxyesters (peroxycarboxylates, peroxysulfonates, and peroxyphosphates).

Hydroperoxides

There are two main subclasses of hydroperoxides: organic (alkyl) hydroperoxides, i.e., ROOH, and organomineral hydroperoxides, i.e., $R_mQ(OOH)_n$, where Q is silicon, germanium, tin, or antimony. The alkyl group in ROOH can be primary, secondary, or tertiary. Except for ethylbenzene hydroperoxide, only *tert*-alkyl hydroperoxides are commercially important.

Properties. Alkyl hydroperoxides can be liquids or solids. Those having low molecular weight are soluble in water and are explosive in the pure state. Alkyl hydroperoxides are stronger acids than the corresponding alcohols and have acidities similar to those of phenols. *tert*-Alkyl hydroperoxides can be purified through their alkali metal salts.

Hydroperoxides exist as hydrogen-bonded dimers in nonpolar solvents and readily form hydrogen-bonded associations with ethers, alcohols, amines, ketones, sulfoxides, and carboxylic acids. Other physical properties of hydroperoxides have been reported.

Hydroperoxides can react with or without cleavage of the oxygen–oxygen bond. Reactions resulting in scission of the oxygen–oxygen bond involve heterolytic, homolytic, or metal-promoted oxidation–reduction reactions.

Alkyl hydroperoxides are reduced readily to the corresponding alcohols; many such reductions are quantitative and useful for analytical methods. Alkyl hydroperoxides have been used as oxidizing or hydroxylating reagents in organic syntheses.

Bases, such as potassium or sodium hydroxide, piperidine, and pyridine, react with primary and secondary hydroperoxides to form aldehydes or ketones. *tert*-Alkyl hydroperoxides form stable alkali metal salts with caustic; however, when equimolar amounts of the hydroperoxide and its sodium salt are present in aqueous solution, rapid decomposition to *tert*-alcohol and oxygen occurs.

Acids react with alkyl hydroperoxides in two different ways, depending on the hydroperoxide structure and the acid strength.

Hydroperoxides are photo- and thermally sensitive and undergo initial oxygen–oxygen bond homolysis, and they are readily attacked by free radicals undergoing induced decompositions.

Hydroperoxides are decomposed readily by multivalent metal ions, i.e., Cu, Co, Fe, V, Mn, Sn, Pb, etc., by an oxidation-reduction or electron-transfer process. Depending on the metal and its valence state, metallic cations either donate or accept electrons when reacting with hydroperoxides. Either one or two electrons may be transferred depending on the metal. With most transition metals, e.g., Cu, Co, and Mn, both valence states react with hydroperoxides via one electron transfer. Thus, a small amount of transition-metal ion can decompose a large amount of hydroperoxide and, consequently, inadvertent contamination of hydroperoxides with traces of transition-metal impurities should be avoided.

The reactions of *tert*-alkyl hydroperoxides with ferrous ion generate alkoxy radicals. These free-radical initiator systems are used industrially for the emulsion polymerization and copolymerization of vinyl monomers, e.g., butadiene–styrene. Alkyl hydroperoxides are among the most thermally stable organic peroxides. However, hydroperoxides are sensitive to chain decomposition reactions initiated by radicals and/or transition-metal ions. Such decompositions, if not controlled, can be autoaccelerating and sometimes can lead to violent decompositions when neat hydroperoxides or concentrated solutions of hydroperoxides are involved.

Organomineral hydroperoxides undergo thermal and photolytic homolyses.

Synthesis. Hydroperoxides have been prepared from several types of peroxygen compounds including hydrogen peroxide or sodium peroxide, ozone, oxygen, and other organic peroxides. Hydrogen peroxide (H_2O_2) and its anions are powerful nucleophiles and react with reagents RX to form ROOH and HX, where X can be sulfate, acid sulfate, alkane- and arenesulfonate, chloride, bromide, hydroxyl, alkoxide, perchlorate, etc. RX can also be an alkyl orthoformate or *tert*-alkyl carboxylate.

Electron-rich olefins react with hydrogen peroxide under acidic conditions to form hydroperoxides, presumably by means of a carbonium ion intermediate, e.g., *tert*-butyl hydroperoxide from isobutylene.

Organomineral hydroperoxides have been prepared from hydrogen peroxide and organomineral halides, hydroxides, oxides, peroxides, and amines. If HX is an acid, ammonia is used to prevent acidic decomposition.

$$R_mQX_n + nH_2O_2 \rightarrow R_mQ(OOH)_n + nHX$$

Many hydroperoxides have been prepared by autoxidation of suitable substrates with molecular oxygen. These reactions can be free-radical chain or nonchain processes, depending on whether triplet or singlet oxygen is involved.

Many organic peroxides of metals have been hydrolyzed to alkyl hydroperoxides. Saponification of *tert*-alkyl peroxyesters yields alkyl hydroperoxides and carboxylic acids or their alkali metal salts.

Dialkyl Peroxides

Dialkyl peroxides have the structural formula R–OO–R′, where R and R′ are the same or different primary, secondary, or tertiary alkyl, cycloalkyl, and aralkyl hydrocarbon or hetero-substituted hydrocarbon radicals. Organomineral peroxides have the formulas $R_mQ(OOR)_n$ and R_mQOOQR_m, where at least one of the peroxygens is bonded directly to the organo-substituted metal or metalloid, Q. Dialkyl peroxides include cyclic and bicyclic peroxides where the R and R′ groups are linked, e.g., endoperoxides and derivatives of 1,2-dioxane. Also included are polymeric peroxides, which usually are called poly(alkylene peroxides) or alkylene–oxygen copolymers, and poly(organomineral peroxides), where Q = As or Sb.

Properties. Metalloid peroxides behave as covalent organic compounds and most are insensitive to friction and impact but can decompose violently if heated rapidly. Most solid metalloid peroxides have well-defined melting points and the more stable liquid members can be distilled.

Acyclic di-*tert*-alkyl peroxides efficiently generate alkoxy free radicals by thermal or photolytic homolysis. Primary and secondary dialkyl peroxides undergo thermal decompositions more rapidly than expected owing to radical-induced decompositions. Such radical-induced peroxide decompositions result in inefficient generation of free radicals.

The low molecular weight primary dialkyl peroxides are shock-sensitive and explosive, with sensitivity decreasing with increasing molecular weight. Decomposition products from primary and secondary dialkyl peroxides include aldehydes, ketones, alcohols, hydrogen, hydrocarbons, carbon monoxide, and carbon dioxide.

Because di-*tert*-alkyl peroxides are less susceptible to radical-induced decompositions, they are safer and more efficient radical generators than primary or secondary dialkyl peroxides. They are the preferred dialkyl peroxides for generating free radicals for commercial applications.

The susceptibility of dialkyl peroxides to acids and bases depends on peroxide structure and the type and strength of the acid or base. In acidic environments, unsymmetrical acyclic alkyl aralkyl peroxides undergo carbon–oxygen fission, forming acyclic alkyl hydroperoxides and aralkyl carbonium ions. The latter react with nucleophiles, X^-.

Substitution reactions on dialkyl peroxides without concurrent peroxide cleavage are known e.g., the nitration of dicumyl peroxide and the chlorination of di-*tert*-butyl peroxide.

The polymeric peroxides, $(-OOCH_2CXH-)_n$, where X = H, C_6H_5, CH=CH$_2$, etc., are viscous liquids or amorphous solids having as many as 10 repeating units. These compounds usually explode when heated. The products obtained from the thermal or photodecomposition show that cleavage of both oxygen–oxygen and carbon–carbon bonds occurs. The type and amounts of products formed depend on the decomposition conditions and the structure of the peroxide.

Unsaturated aliphatic endoperoxides form bis(epoxides) and/or epoxy aldehydes upon thermolysis. The endoperoxides of polynuclear aromatic compounds are crystalline solids that extrude singlet oxygen when heated, thus forming the parent aromatic hydrocarbon. Endoperoxides undergo carbon–oxygen cleavage in acids and oxygen–oxygen bond cleavage in bases, and they are more easily reduced than dialkyl peroxides.

1,2-Dioxetanes have very low activation enthalpies (ca 109 kJ/mol), therefore, they are unstable at low temperatures and generally cleave thermally or photochemically at the oxygen–oxygen and carbon–carbon bonds. Upon fragmentation, chemiluminescence occurs and two carbonyl compounds are produced in the absence of trapping agents. 1,2-Dioxetanes are reduced to diols, epoxides, or allylic alcohols; the dioxetane structure and the reducing system determine which product forms or predominates.

Dioxiranes are three-membered cyclic ring peroxides that are expected to be very unstable owing to ring strain. They are effective oxygenating agents for epoxidations of olefins, allenes, polycyclic aromatic hydrocarbons, enols, and α, β-unsaturated ketones; for insertions of oxygen into X–H bonds of alkanes, primary and secondary alcohols, aldehydes, and silanes; and for oxidations of sulfides (to sulfoxides and sulfones), imines (to nitrones), and primary amines (to nitro compounds). In these reactions, the dioxirane transfers oxygen to the substrate and generates the ketone from which the dioxirane was derived.

Most organomineral peroxides are hydrolytically unstable and readily hydrolyze to alkyl hydroperoxides or hydrogen peroxide.

Consequently, most organomineral peroxides must be prepared and stored under anhydrous conditions.

Basic hydrolysis of secondary alkyl-substituted silicon and germanium peroxides results in oxygen–oxygen bond cleavage.

The reduction of alkyl-substituted silicon and tin peroxides with sodium sulfite and triphenylphosphine has been reported. Alkyl-substituted aluminum, boron, cadmium, germanium, silicon, and tin peroxides undergo oxygen-to-metal rearrangements, as in the following equations:

$$R_3Si-OO-SiR_3 \rightarrow R_2Si(OR)OSiR_3$$

$$R_2B-OO-R \rightarrow RB(OR)_2$$

Organomineral peroxides also undergo thermal and photo-induced homolysis, yielding free radicals that are effective for initiating polymerization of vinyl monomers.

Synthesis. Dialkyl peroxides are prepared by the reaction of various substrates with hydrogen peroxide, hydroperoxides, or oxygen. They also have been obtained from reactions with other organic peroxides.

α-Oxygen-Substituted Hydroperoxides and Dialkyl Peroxides

Dialkyl peroxides and hydroperoxides which have either a hydroxy, hydroperoxy, alkoxy, or alkylperoxy group on the carbon adjacent to the parent peroxide group are considered separately from the parent compounds due to their unique reactions and properties, but mainly because of their unique syntheses. Their primary preparation from aldehydes and ketones via reaction with hydrogen peroxide, alkyl hydroperoxides and peroxyacids is unique and makes it almost impossible to discuss them without referring to the parent carbonyl compound(s).

The α-oxygen-substituted hydroperoxides and dialkyl peroxides comprise a great variety as shown in Figure 1. When discussing peroxides derived from ketones and hydrogen peroxide, (1) is often referred to as a ketone peroxide monomer and (2) as a ketone peroxide dimer.

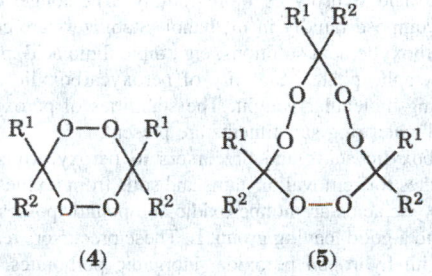

(1)　　　　**(2)**　　　　**(3)**

(4)　　　　　　**(5)**

Fig. 1.   Varieties of α-oxygen-substituted hydroperoxides and dialkyl peroxides. R^1, R^2, R^3 = H or alkyl; X, Y = OH, OOH, OR^4, $OSiR_3$, or OOR^5; R^4, R^5 = alkyl; and R^3 and R^5 may also be acyl, $C(=O)R^6$.

Syntheses and Properties. An example of the complex equilibrium that exists for mixtures of carbonyl compounds and hydrogen peroxide is that from aldehydes and hydrogen peroxide. Hydroxyalkyl hydroperoxides (1, X = OH, R^3 = H) and di(hydroxyalkyl) peroxides (2, X = Y = OH) are formed; cyclic diperoxides (4) are formed in some cases, e.g., from benzaldehyde with concentrated sulfuric acid. Hydroxyalkyl hydroperoxides are the principal products when equimolar amounts of aldehyde and hydrogen peroxide are used at low temperatures. Di(hydroxyalkyl) peroxides are obtained by using excess aldehyde or higher temperatures. These reactions occur without catalysts but occur at much faster rates in the presence of acids.

Starting with ketones and hydrogen peroxide in the presence of a catalytic amount of acid, mixtures of up to eight components have been identified, i.e., (1, X = OH, R^3 = H), (1, X = OOH, R^3 = H), (2, X = Y = OH), (2, X = Y = OOH), (2, Y = OH, Y = OOH), (3), (4), and (5). The ketone structure and reaction conditions, i.e., acid strength, reactant molar ratios, temperature, and time, determine which compounds form and predominate. Mixtures of several peroxide structures usually are present. Individual peroxides have been isolated from several ketones under different conditions.

The pure peroxides should be handled with extreme caution since most, especially those derived from the low molecular weight ketones, are shock- and friction-sensitive and can explode violently. Methyl ethyl ketone peroxide (MEKP) mixtures are produced commercially only as solutions containing <40 wt% MEKPs in solvents, commonly dialkyl phthalates.

Hydroxyalkyl Hydroperoxides. These compounds, represented by (1, X = OH, R^3 = H), may be isolated as discreet compounds only with certain structural restrictions, e.g., that one or both of R^1 and R^2 are hydrogen, i.e., they are derived from aldehydes, or that R^1 or R^2 contain electron-withdrawing substituents, i.e., they are derived from ketones bearing α-halogen substituents. Other hydroxyalkyl hydroperoxides may exist in equilibrium mixtures of ketone and hydrogen peroxide.

Alkoxyalkyl Hydroperoxides. These compounds (1, X = OR^4, R^3 = H) have been prepared by the ozonization of certain unsaturated compounds in alcohol solvents. Alkoxyalkyl hydroperoxides are more commonly called ether hydroperoxides. They form readily by the autoxidation of most ethers containing α-hydrogens, e.g., dioxane, tetrahydrofuran, diethyl ether, diisopropyl ether, di-*n*-butyl ether, and diisoamyl ether. From certain ethers, e.g., diethyl ether, the initially formed ether hydroperoxide can yield alcohol on standing, or with acid treatment form dangerously shock-sensitive and explosive polymeric peroxides.

Hydroxyalkyl Alkyl Peroxides and Hydroxyalkyl Peroxyesters. Hydroxyalkyl alkyl peroxides (1, X = OH, R^3 = alkyl) are reasonably stable and usually can be distilled under a vacuum.

Alkoxyalkyl Alkyl Peroxides. *tert*-Butyl tetrahydropyran-2-yl peroxide (1, where R^3 = *tert*-butyl, H = OR^4, R^1 = H, R^2 and R^4 = 1, 4 -butanediyl, has been isolated. This is one of many examples of alkoxyalkyl alkyl peroxides which may be prepared by reaction of hydroperoxides with vinyl ethers.

1,2,4-Trioxacycloalkanes. 1,2,4-Trioxanes (1), X = OR^4; R^3 and R^4 = alkylene) are generally prepared by the interaction of aldehydes with zwitterionic intermediates made from reaction of singlet oxygen with olefins. They can also be prepared by catalyzed reaction of ketones or aldehydes with 1,2-dioxetanes or endoperoxides, and they can be prepared directly from certain hydroperoxides.

Geminal Dihydroperoxides. These dihydroperoxides as described previously (1, X = OOH, R^3 = H) can be made from many different carbonyl compounds. These peroxides can also be synthesized by perhydrolysis of ketals. Low molecular weight dihydroperoxides are soluble in water and are explosive when pure. They have been reduced to the corresponding ketones with hydriodic acid or zinc and acetic acid. Hydrolysis also gives the corresponding ketones. In the presence of catalytic amounts of acids or on prolonged storage, solutions of dihydroperoxides form equilibrium amounts of hydrogen peroxide and di(hydroperoxyalkyl) peroxides and ultimately equilibrium amounts of cyclic triperoxides.

Diperoxyketals and Diperoxyacetals. Aromatic aldehydes react with alkyl hydroperoxides in the presence of strong acid catalysts such as sulfuric acid to form diperoxyacetals (1, X = OOR^5; R^1 = H, R^2 = Ar, R^3 = R^5 = alkyl). Diperoxyketals (1, X = OOR^5; R^1, R^2, R^3, R^5 = alkyl) are generally prepared by acid-catalyzed reaction of a ketone with two equivalents of an alkyl hydroperoxide.

Diperoxyketals are solids or colorless liquids and are soluble in common organic solvents and insoluble in water. In the pure state, the low molecular weight compounds can decompose violently when heated, and addition of concentrated sulfuric acid can result in flaming decompositions. There are many commercial diperoxyketals, and they are usually diluted with solvents for improved safety.

Tertiary diperoxyketals (1, X = OOR^5, R^1, R^2 = alkyl, R^3, R^5 = tertiary alkyl) are excellent free-radical initiators. Such diperoxyketals are stable, especially those with R^3 = R^5 = *tert* - butyl. Less thermally stable diperoxyketals are those derived from cyclic ketones and those with bulkier *tert*-alkyl groups, e.g., *tert*-amyl, *tert*-octyl, *tert*-cumyl. Commercial members of this group all have R^3 = R^5, and thermally decompose to free radicals by cleavage of only one oxygen–oxygen bond initially, usually followed by β-scission of the resulting alkoxy radicals. For acyclic diperoxyketals, β-scission produces an alkyl radical and a peroxyester. Owing to similarity of thermal stability, the peroxyester decomposes almost simultaneously.

Diperoxyketals, and many other organic peroxides, are acid-sensitive, therefore removal of all traces of the acid catalysts must be accomplished before attempting distillations or kinetic decomposition studies. The low molecular weight diperoxyketals can decompose with explosive force and commercial formulations are available only as mineral spirits or phthalate ester solutions.

Di(hydroxyalkyl) Peroxides. The lowest molecular weight member of this group (2, X = Y = OH),di(hydroxymethyl) peroxide ($R^1 = R^2 = H$) is a dangerously explosive solid. With increasing molecular weight, di(hydroxyalkyl) peroxides become liquids and eventually solids of decreasing explosive nature and water solubility. In solution, these dialkyl peroxides exist in equilibrium with other α-oxygen-substituted peroxides, carbonyl compounds, and hydrogen peroxide.

Formaldehyde reacts with di(hydroxymethyl) peroxide and phosphorus pentoxide to form di(hydroxymethoxymethyl) peroxide (2), where X = Y = OCH_2OH, $R^1 = R^2 = H$.

Reaction of 1,3- and 1,4-diketones ($n = 1$ or 2) with hydrogen peroxide yields cyclic di(hydroxyalkyl) (X = OH) or di(hydroperoxyalkyl) (X = OOH) peroxides (6).

(6)

The di(hydroxyalkyl) peroxide (2) from cyclohexanone is a solid which is produced commercially. The di(hydroxyalkyl) peroxide (2) from 2,4-pentanedione (7, $n = 1$; X = OH) is a water-soluble solid which is also produced commercially (see Table 5). Both these peroxides are used for curing cobalt-promoted unsaturated polyester resins.

Hydroxyalkyl Hydroperoxyalkyl Peroxides. There is evidence that hydroxyalkyl hydroperoxyalkyl peroxides (2, X = OH, Y = OOH) exist in equilibrium with their corresponding carbonyl compounds and other α-oxygen-substituted peroxides. Thermal decomposition of hydroxyalkyl hydroperoxyalkyl peroxides produces mixtures of starting carbonyl compounds, mono and dicarboxylic acids, cyclic diperoxides, carbon dioxide, and water.

Di(hydroperoxyalkyl) Peroxides. Low molecular weight di(hydroperoxyalkyl) peroxides (2, X = Y = OOH) are dangerously prone to explosive decomposition when they are pure. Some have been characterized by acylation to the corresponding diperoxyesters.

Cyclic Peroxides. Cyclic diperoxides (4) and triperoxides (5) are solids and the low molecular weight compounds are shock-sensitive and explosive. The melting points of some characteristic compounds of this type are given in Table 5.

Polymeric α-Oxygen-Substituted Peroxides. Polymeric peroxides (3) are formed from the following reactions: ketone and aldehydes with hydrogen peroxide, ozonization of unsaturated compounds, and dehydration of α-hydroxyalkyl hydroperoxides; consequently, a variety of polymeric peroxides of this type exist. Polymeric peroxides are generally viscous liquids or amorphous solids, are difficult to characterize, and are prone to explosive decomposition.

Ozonides and Ozonization

Unsaturated compounds undergo ozonization to initially produce highly unstable primary ozonides, i.e., 1,2,3-trioxolanes, also known as molozonides, which rapidly split into carbonyl compounds (aldehydes and ketones) and 1,3-zwitterion intermediates. The carbonyl compound-zwitterion pair then recombine to produce a thermally stable secondary ozonide, also known as a 1,2,4-trioxolane.

Most ozonolysis reaction products are postulated to form by the reaction of the 1,3-zwitterion with the extruded carbonyl compound in a 1,3-dipolar cycloaddition reaction to produce stable 1,2,4-trioxanes (ozonides) as shown; with itself (dimerization) to form cyclic diperoxides (4); or with protic solvents, such as alcohols, carboxylic acids, etc., to form α-substituted alkyl hydroperoxides. The latter can form other peroxidic products, depending on reactants, reaction conditions, and solvent.

In the presence of alcohols, the ozonization products are alkoxyalkyl hydroperoxides (1, X = OR^4, R = H_3):

By-products include ozonides. Other peroxidic products including polymeric peroxides and polymeric ozonides can form, depending on reaction conditions, solvent, and olefin used. A variety of cyclic diperoxides (4) have been obtained by ozonolysis of olefins.

Cyclic 1,2,4-trioxanes (7 and 8) have been obtained from the photosensitized oxidation of furans. These compounds are 2,3,7-trioxabicyclo [2.2.1] hept-5-ene (7) and 2,3,7-trioxabicyclo [2.2.1] heptane (8).

(7)   (8)

Peroxyacids

There are two broad classes of organic peroxyacids: peroxycarboxylic acids, $R[C(O)OOH]_n$, where R is an alkyl, aralkyl, cycloalkyl, aryl, or heterocyclic group and $n = 1$ or 2, and organoperoxysulfonic acids, RSO_2-OOH.

Three peroxyacids are produced commercially for the merchant market: peroxyacetic acid as a 40 wt % solution in acetic acid, m-chloroperoxybenzoic acid, and magnesium monoperoxyphthalate hexahydrate. Other peroxyacids are produced for captive use, e.g., peroxyformic acid generated *in situ*, as an epoxidizing agent.

Properties. Physical properties of peroxyacids have been extensively reviewed in the literature. Aliphatic peroxyacids are characterized by sharp unpleasant odors, the intensity of which decreases with increasing chain length. They also are irritating to the skin and mucous membranes.

Organic peroxyacids are not noted for their stability and many lose active oxygen during storage at room temperature. Those that are water soluble hydrolyze slowly to the parent acid and hydrogen peroxide; however, peroxyformic acid hydrolyzes more rapidly. The longer-chain aliphatic members decompose rapidly in methanol. Stabilizers are commonly used for peroxycarboxylic acid solutions, e.g., dipicolinic acid, phytic acid, and pyro- and metaphosphates. Stability of peroxycarboxylic acids increases with increasing molecular weight. The stabilities of peroxybenzoic acids are enhanced when ring substituents are present.

Peroxycarboxylic acids and precursors to peroxycarboxylic acids are used as bleaches for removal of stains and soils from textiles. Precursors to peroxycarboxylic acids are nonperoxidic compounds possessing a reactive acyl group and a good leaving group, L. These precursors react under basic conditions with hydrogen peroxide, inorganic perborates, and inorganic percarbonates to generate peroxycarboxylic acids or salts:

Thermal decompositions of peroxycarboxylic acids and their salts can proceed by free-radical and nonradical paths. Often the decomposition products and the rate are affected by the nature of the solvent. Peroxycarboxylic acids undergo photodecomposition and radical-induced decomposition. They also are decomposed by a variety of metals, metal ions, and complexes.

Peroxycarboxylic acids are among the most powerful organic peroxide oxidizing agents. The main industrial uses of these acids are in the manufacture of epoxides, synthetic glycerol, and epoxy resins. They also have been used as disinfectants, fungicides, and bleaching agents and for shrink-proofing wool.

Synthesis. Many different methods for the preparation of peroxyacids have been described. The most widely used method is the direct, acid-catalyzed equilibrium reaction of 30–98 wt % hydrogen peroxide with carboxylic acids: The equilibrium also can be shifted to the right by removing water azeotropically and/or under a vacuum. Chelating agents may be added during processing to reduce metal-catalyzed decompositions. Sulfuric acid, methanesulfonic acid, and sulfonic acid ion-exchange resins are the most commonly used acid catalysts.

Other methods for preparing peroxycarboxylic acids include: (1) autoxidation of aldehydes, (2) reaction of acid chlorides, anhydrides, or boriccarboxylic anhydrides with hydrogen or sodium peroxide, and (3) basic hydrolysis or perhydrolysis of diacyl peroxides.

Organoperoxysulfonic acids and their salts have been prepared by the reaction of arenesulfonyl chlorides with calcium, silver, or sodium peroxide; treatment of metal salts of organosulfonic acids with hydrogen peroxide; hydrolysis of di(organosulfonyl) peroxides, $R^2S(O)-OO-S(O)R^2$, with hydrogen peroxide; and sulfoxidation of saturated, nonaromatic hydrocarbons, e.g., cyclohexane.

Other Peroxyacids. Benzeneperoxyseleninic acid has been prepared *in situ* from benzeneseleninic acid and hydrogen peroxide and is used to epoxidize terpenic olefins and Baeyer-Villiger oxidation of cyclic ketones.

Acyl Peroxides

The acyl peroxide class is characterized by the following structures:

$$R^1-\overset{O}{\overset{\|}{C}}-OO-\overset{O}{\overset{\|}{C}}-R^2 \qquad R^1O-\overset{O}{\overset{\|}{C}}-OO-\overset{O}{\overset{\|}{C}}-OR^2 \qquad R^1-\overset{O}{\overset{\|}{C}}-OO-\overset{O}{\overset{\|}{C}}-OR^2$$

$$(9) \qquad\qquad\qquad (10) \qquad\qquad\qquad (11)$$

$$R^1-\overset{O}{\overset{\|}{C}}-OO-\overset{O}{\underset{O}{\overset{\|}{S}}}R^2 \qquad R^1\overset{O}{\underset{O}{\overset{\|}{S}}}-OO-\overset{O}{\underset{O}{\overset{\|}{S}}}R^2$$

$$(12) \qquad\qquad\qquad (13)$$

Acyl peroxides of structure (9) are known as diacyl peroxides. In this structure R^1 and R^2 are the same or different and can be alkyl, aryl, heterocyclic, imino, amino, or fluoro. Acyl peroxides of structures (10) to (13) are known as dialkyl peroxydicarbonates, *OO*-acyl *O*-alkyl monoperoxycarbonates, acyl organosulfonyl peroxides, and di(organosulfonyl) peroxides, respectively. R^1 and R^2 in these structures are the same or different and generally are alkyl and aryl. Many diacyl peroxides (9) and dialkyl peroxydicarbonates (10) are produced commercially and used in large volumes.

Properties. Almost all liquid diacyl peroxides (9) and concentrated solutions of the solid compounds are unstable to normal ambient temperature storage; many must be stored well below 0°C. Most of the solid compounds are stable at ca 20°C but many are shock-sensitive. Other physical constants and properties have been reviewed.

Diacyl peroxides (9) decompose when heated or photolyzed (<300 mm). Although photolytic decompositions generally produce free radicals, thermal decompositions can produce nonradical and radical intermediates, depending on diacyl peroxide structure. Symmetrical aliphatic diacyl peroxides of certain structures, i.e., diacyl peroxides (9, $R^1 = R^2 = $ alkyl) without α-branches or with a mono-α-methyl substituent, and diaroyl peroxides (9, $R^1 = R^2 = $ aryl) thermally decompose almost exclusively by homolysis.

Diaroyl peroxides and diacyl peroxides without α-branches are significantly more thermally stable than those with mono- or di-α-substituents. The primary use of most commercial diacyl peroxides (9, $R^1 = R^2 = $ alkyl or aryl) is initiation of free-radical reactions.

Diacyl peroxides (9, $R^1 = R^2 = $ alkyl or aryl) also undergo radical induced decomposition either by direct radical displacement on the oxygen–oxygen bond, or by radical addition to, or abstraction from, the hydrocarbyl group adjacent to the peroxide.

Diacyl peroxide decompositions also are catalyzed by the metal ions of copper, iron, cobalt, and manganese.

This radical-generating reaction has been used in synthetic applications, e.g., aroyloxylation of olefins and aromatics, oxidation of alcohols to aldehydes, etc.

Hydrolysis and perhydrolysis of diacyl peroxides yields peroxycarboxylic acids. Carbanions react by displacement on oxygen.

Amines also react with diacyl peroxides by nucleophilic displacement on the oxygen–oxygen bond forming an ion pair intermediate.

Dialkyl peroxydicarbonates (10) undergo thermolysis to form two alkoxycarbonyloxy radicals that subsequently undergo β-scission to form CO_2 and alkoxy radicals.

These low temperature peroxides are susceptible to radical-induced decompositions. This susceptibility largely accounts for the hazards associated with their production and storage. In contrast to diacyl peroxides (9); the true first-order decomposition rates for dialkyl peroxydicarbonates (10) are not affected by the nature of the R group. In free-radical scavenging solvents, e.g., trichloroethylene, the decomposition rates of di-*n*-propyl, diisopropyl, di-*sec*-butyl, dicyclohexyl, di-2-ethylhexyl, and deheaxadecyl peroxydicarbonates are all essentially the same.

Dialkyl peroxydicarbonates are used primarily as free-radical initiators for vinyl monomer polymerizations. Dialkyl peroxydicarbonate decompositions are accelerated by certain metals, concentrated sulfuric acid, and amines. Violent decompositions can occur with neat or highly concentrated peroxides.

Acyl organosulfonyl peroxides such as acetyl cyclohexanesulfonyl peroxide are efficient radical initiators for vinyl chloride polymerization.

Di(arenesulfonyl) peroxides ($R^2 = R^1 = $ aryl) react with aromatic solvents to form aryl arenesulfonates. These peroxides also form 1:1 adducts with styrene and form hydrobenzoin diarenesulfonates with stilbenes. Di(benzenesulfonyl) peroxide decomposes in water to phenol and sulfuric acid.

Synthesis. Symmetrical diacyl peroxides (9, $R^2 = R^1 = $ alkyl or aryl) are prepared by the reaction of an acyl chloride or anhydride with sodium peroxide or hydrogen peroxide and a base. Unsymmetrical diacyl peroxides (9, $R^2 \neq R^1 = $ alkyl or aryl) are prepared by the reaction of acid chlorides or anhydrides with peroxycarboxylic acids in the presence of a base. Polymeric diacyl peroxides can be prepared from the reaction of dibasic acid chlorides, e.g., succinoyl, fumaryl, sebacoyl, and terephthaloyl chlorides, with sodium or hydrogen peroxide. Cyclic diacyl peroxides can be generated from suitable dibasic acid chlorides and sodium or hydrogen peroxide, especially in dilute solutions. Symmetrical or unsymmetrical diacyl peroxides (9, R^1, $R^2 = $ alkyl or aryl) can be synthesized directly from carboxylic acids and hydrogen peroxide or from peroxycarboxylic acids with dicyclohexylcarbodiimide or *N,N*-dicarbonyldiimidazole as condensing agents.

Diacyl peroxides (9, $R^2 = R^1 = $ alkyl or aryl) have been obtained from the oxidation of carboxylic acid potassium salts by Kolbe electrolysis or by elemental fluorine.

Dialkyl peroxydicarbonates (10) are produced by reaction of alkyl chloroformates with sodium peroxide. *OO*-Acyl *O*-alkyl monoperoxycarbonates (11) are obtained from the reaction of alkyl chloroformates with peroxycarboxylic acids in the presence of a base. Symmetrical di(organosulfonyl) peroxides (20, R = R21) have been prepared by the reaction of organosulfonyl chlorides with sodium peroxide or hydrogen peroxide in the presence of a base. Acyl organosulfonyl peroxides (12) are prepared from the organosulfonyl chlorides and a metal salt of a peroxycarboxylic acid. Acetyl cyclohexanesulfonyl peroxide has been produced commercially by the sulfoxidation of cyclohexane, C_6H_{12}, in the presence of acetic anhydride.

Potassium salts of the peroxides are prepared from the reaction of Caro's acid, H_2SO_5, with acyl chlorides, chloroformates, or organosulfonyl chlorides in the presence of potassium hydroxide.

Alkyl Peroxyesters

Peroxyesters include the alkyl esters of peroxycarboxylic acids; monoperoxydicarboxylic acids; diperoxycarboxylic acids; monoperoxy- and diperoxycarbonic acids; monoperoxy- and diperoxyoxalic acids; peroxycarbamic acids; peroxysulfonic acids; and peroxyphosphoric acids.

Synthesis. Peroxyesters are prepared by the reaction of alkyl hydroperoxides R'OOH, with acylating agents, e.g., acid chlorides, anhydrides, ketenes, organosulfonyl chlorides, phosgene, alkyl chloroformates, oxalyl chloride, alkyl chlorooxalates, isocyanates, carbamoyl chlorides, carboxylic acids, and esters, under appropriate reaction conditions. Reactions with acylating agents that generate hydrogen chloride are carried out in the presence of a base, e.g., pyridine or sodium hydroxide, or by using the sodium or potassium salt of the hydroperoxide.

Properties. Alkyl peroxyesters are hydrolyzed more readily than the analogous nonperoxidic esters and yield the original acids and hydroperoxides from which they were prepared rather than alcohols and peroxyacids. The *tert*-alkyl peroxyesters undergo homolysis, thermally and photochemically, to generate free radicals.

Primary and secondary alkyl peroxyesters thermally decompose by a nonradical process, giving almost quantitative yields of carboxylic acids and carbonyl compounds. *tert*-Alkyl peroxyesters are much less sensitive to radical-induced decompositions than diacyl peroxides. Induced decomposition is only significant in peroxyesters containing nonhindered α-hydrogens or α, β-unsaturation.

Peroxyesters decompose by an electron-transfer process catalyzed by transition metals. This reaction has been used synthetically to bond an acyloxy group to appropriate coreactive substrates.

Criegee rearrangement competes with homolysis in *tert*-alkyl peroxyesters, $RC(O)-OOCR^1R^2R^3$, in which R is strongly electron-withdrawing and the *tert*-alkyl group, i.e., $CR^1R^2R^3$, contains a group with high migratory aptitude and ability to stabilize adjacent carbonium ions. The rearrangement converts the peroxyester to a nonperoxidic ester.

The main industrial use of *tert*-alkyl peroxyesters is in the initiation of free-radical chain reactions, primarily for vinyl monomer polymerizations.

Manufacture and Processing. Owing to the inherent hazards of organic peroxides, they are almost never distilled or confined during manufacture. Generally, open reactors are employed that can be easily vented and deluged with water if an unanticipated exotherm occurs. The preferred materials of reactor construction are 316 stainless steel, plastic, and glass. Significant cooling capacity is required to handle reaction exotherms and to maintain temperature. Because over 100 different organic peroxides are produced commercially, organic peroxide producers manufacture many organic peroxides in the same equipment. Batch processing is generally employed when relatively small production volumes are required, whereas semicontinuous and continuous processing are employed when larger production volumes are required and when safety is a primary issue. Continuous processes are significantly safer to operate than batch processes as smaller amounts of organic peroxides are continuously in process. Besides safer continuous processing, another trend has been the use of reactants of higher purity. These process improvements have resulted in reduced environmental impact as unplanned process decompositions have been decreased and waste streams have been reduced.

Health and Safety Factors

Toxicology. In general, organic peroxides are characterized by a low order of acute toxicity. Most organic peroxides have some oxidizing properties and are irritants. Most of the available toxicity data on commercial organic peroxides are summarized in the literature. There is limited evidence to suggest that organic peroxides are carcinogenic.

Decomposition Hazards. The main causes of unintended decompositions of organic peroxides are heat energy from heating sources and mechanical shock, e.g., impact or friction. In addition, certain contaminants, i.e., metal salts, amines, acids, and bases, initiate or accelerate organic peroxide decompositions at temperatures at which the peroxide is normally stable. These reactions also liberate heat, thus further accelerating the decomposition. Commercial products often contain diluents that desensitize neat peroxides to these hazards. Commercial organic peroxide decompositions are low order deflagrations rather than detonations.

The organic peroxides and peroxide compositions produced commercially are those that can be manufactured, shipped, stored, and used safely. Organic peroxides can be thermally and mechanically desensitized by wetting or by dilution with suitable solvents, inert solid fillers, or insoluble liquids (suspension of solid peroxides in liquid plasticizers or water, and emulsions of liquid peroxides in water).

Recommendations for safe handling and storage of commercial organic peroxides are available from organic peroxide manufactures.

In 1984 the United Nations (UN) Committee on the Transportation of Dangerous Goods, made up of experts from Prins Maurits Laboratory (TNO), Bundesanstalt für Materialprüfung (BAM), and the organic peroxide producers, developed a test procedure for the classification of organic peroxide compositions for transport purposes. The test procedure was accepted by most of the industrial countries of the world. The Department of Transportation (DOT) mandated that the United States peroxide industry would comply with the UN classification system by October 1993. Material Safety Data Sheets (MSDS) and the organic peroxides producers' recommendations should be followed carefully for handling and storage of organic peroxide compositions.

Uses

There are more than 100 commercially available organic peroxides in well over 300 formulations, e.g., neat liquids and solids, and pastes, powders, solutions, dispersions, and emulsions, that have utility in many commercial applications.

Excluding the peroxyacids, which are used primarily as epoxidizing and bleaching agents, approximately 90% of the commercial organic peroxides are consumed by the polymer industry.

They are used in the polymer industry as thermal sources of free radicals. They are used primarily to initiate the polymerization and copolymerization of vinyl and diene monomers.

Organic peroxides also are used as flame-retardant synergists for polystyrene, for preparing block and graft copolymers, for reactive processing, for reducing the molecular weight of polypropylene (i.e., controlled rheology or vis-breaking), for curing adhesives, for drying alkyd resin films, and for initiating cationic polymerization with cyclic ethers and maleic anhydride.

BPO is the preferred bleaching agent for flour and has been used to bleach gums, waxes, fats, and oils. It is the active ingredient in many acne medications. Diacyl peroxides have been used as burnout agents for acetate yarn, drying agents for Chinawood oils, and as free-radical sources in many organic syntheses. Di-*tert*-butyl peroxide is used as an ignition accelerator for diesel fuels and has been used in many organic syntheses either as a source of *tert*-butoxy (photo) or methyl (thermal) radicals.

Additional Reading

Ando, W. ed.: *Organic Peroxides*, John Wiley & Sons, Inc., New York, NY, 1992.
Lide, D.R.: *CRC Handbook of Chemistry and Physics*, 88th Edition, CRC Press LLC., Boca Raton, FL, 2007.
Patai, S. ed.: *The Chemistry of Peroxides*, John Wiley & Sons, Inc., New York, NY, 1983; S. Patai, ed., *Supplement E2: The Chemistry of Hydroxyl, Ether and Peroxide Groups*, John Wiley & Sons, New York, NY, 1993.
Swern, D. ed.: *Organic Peroxides*, Vols. I–III, Wiley-Interscience, New York, NY, 1970–1972.
Waldemar, A.: *Peroxide Chemistry*, John Wiley & Sons, Inc., New York, NY, 2003.

PERSEIDS.   The name given to the most reliable of all meteor showers. Although the Leonids provide some very brilliant displays about three times each century, their appearance in the intervening years is not at all striking. The Perseids make their appearance during August of each year. Because of the fact that the Perseid showers never are as striking as the Leonids, we should not expect to find them referred to as frequently in the ancient writings. Extensive search has been made through the old records, however, and we find mention of the Perseids as far back as 830 A.D. The first determination of the radiant point in the constellation of Perseus was apparently made in 1834.

In appearance, the members of the Perseid shower are as striking as those from any other radiant point. Coming as they do during the month of August, when the nights are warm, they are seen by large numbers of people, who are always impressed by the relatively slow motion, distinctly reddish appearance, and trails, frequently of several seconds' duration, which characterize the members of this swarm. See also **Meteor Shower**.

PERSEUS.   A rich and brilliant constellation of the northern sky. Because Perseus lies right in the Milky Way, it presents many beautiful fields for the opera glass or the small telescope, particularly the field of the bright star α Persei. The star Algol (β Persei) is the famous eclipsing variable star whose striking changes in light intensity caused the Arabs to name it the demon star.

In Perseus is to be found a famous double-star cluster, one of the finest objects in the entire sky for an observer with a small telescope. On a clear moonless night, using a relatively low power on the instrument, an observer will be well rewarded for his search for this wonderful object. See map accompanying entry on **Constellations**; and **Galaxy**.

PERSIMMON TREES.   Of the family *Ebenaceae* (ebony family), there are several species of persimmon trees. The fruits of these trees also are called persimmons. Especially esteemed for their fruits are *Diospyros virginiana*, the American persimmon, and *D. kaki*, the Japanese persimmon. *Diospyros virginiana* is a large American tree, 60–100 feet (18 to 30 meters) high, with rather thick ovate-oblong leaves and pale yellow axillary flowers. The fruit is a large globular berry an inch or more in diameter, orange-yellow in color, and very astringent until fully ripe. See Fig. 1. The astringent quality is due to the presence of much soluble tannin, which is gradually formed into an insoluble compound as the fruit ripens, so that the mouth-puckering quality is nearly lost. Frost action has been considered by many to be the cause of the change in the fruit. The American persimmon is hardy as far north as Rhode Island. The Japanese persimmon is a smaller tree, seldom growing more than 40 feet (12 meters) tall, and is less hardy. Its fruits are larger than those of the American tree, and of reddish color. Both trees have a hard dark wood.

See Table 1.

Also of the ebony family is the ebony tree (*Diospyros ebenum*), a tree native of India and Ceylon. It is a large tree with entire leathery leaves and axillary flowers. The wood of the tree is divided sharply into a soft

Fig. 1.   Persimmon fruit (Diospyros virginiana.) (*USDA photo.*)

white sapwood of little value and a hard very dark heartwood. The latter is much used for inlay work, for black piano keys, for musical instruments, and for handles of various instruments. Many other species of *Diospyros* have dark woods used as a substitute for true ebony. The wood of several other trees, especially that of the pear tree, are frequently stained to imitate ebony.

PERSISTENCE.   1.   The previous value in a time series. Thus, if $x(t)$ denotes the present value, the value of persistence would be $x(t-1)$, whence the latter value is regarded as "persisting." It is used as an objective standard in the verification of weather forecasts. Sometimes it is extended to mean $x(t-h)$ where h is arbitrary.

2.   (Also called *constancy*, or *steadiness*.) With respect to the long-term nature of the wind at a given location, the ratio of the magnitude of the main wind vector to the average speed of the wind without regard to direction.

3.   (Also called *inertial forecast*.) In general, the tendency for the occurrence of a specific event to be more probable, at a given time, if that same event has occurred in the immediately preceding time period.

AMS

PERSISTENCE FORECAST.   In meteorology, a forecast that the future weather condition will be the same as the present condition. The persistence forecast is often used as a standard of comparison in measuring the degree of skill of forecasts prepared by other methods, especially for very short projections.

See also **Persistence**.

AMS

PERSONAL EQUATION.   In making measurements of any character, every observer, no matter how skilled he may be, is bound to make certain errors. These errors are of two kinds: accidental errors which will be small in the case of a good observer and which will be distributed in accordance with the laws of probability; and systematic errors or errors which are always in the same direction and of approximately the same magnitude. As an example of systematic errors we may cite the case of the observation of the transit of a star across the reticle of a meridian circle. In this case a good observer will always press the chronograph key either slightly too early or slightly too late, depending upon the observer.

The value of the systematic error is known as the personal equation of the observer. Personal equation must be determined empirically for each observer under a variety of different observing conditions. For a good observer, the personal equation remains remarkably constant over long periods of time and may be applied directly to any observation.

See also **Error**.

PERSPECTIVE.   A method of representing solid objects on a plane surface, as they would appear to an observer's eye when viewed from a given point. The projectors or visual rays converge from the object to the eye, forming a cone or pyramid of rays. The intersection of these

TABLE 1. RECORD PERSIMMON TREES IN THE UNITED STATES[1]

Specimen	Circumference[2]		Height		Spread		Location
	Inches	Centimeters	Feet	Meters	Feet	Meters	
Common Persimmon (1999) (*Diospyros virginiana*)	88	224	132	40.2	30	9.1	Massachusetts
Common Persimmon (1987) (*Diospyros virginiana*)	136	345	66	20.1	85	25.9	Arkansas
Common Persimmon (1999) (*Diospyros virginiana*)	96	244	121	36.9	42	12.8	Georgia
Common Persimmon (1995) (*Diospyros virginiana*)	95	241	120	36.6	40	12.2	South Carolina
Common Persimmon (1995) (*Diospyros virginiana*)	85	216	132	40.2	37	11.3	South Carolina
Texas persimmon (1965) (*Diospyros texana*)	68	173	26	7.9	32	9.8	Texas

[1]From the "National Register of Big Trees," American Forests (by permission).
[2]At 4.5 feet (1.4 meters).

rays with the picture plane results in a perspective drawing. Two types of perspective representation are in common use — parallel and angular. In the former, all horizontal lines parallel to the picture plane, and all vertical lines, appear horizontal and vertical; all other parallel lines will intersect, if extended, at one or more common points, called vanishing points. In angular perspective, vertical lines appear vertical; all other parallel lines will intersect, if extended, at one or more vanishing points. See Fig. 1.

See also **Orthographic Projection**.

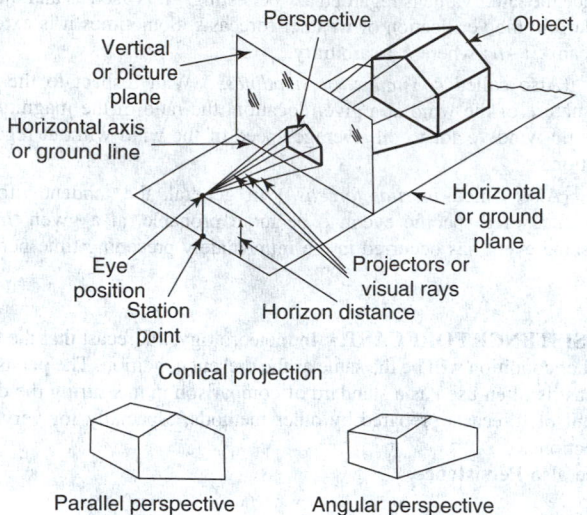

Fig. 1. Types of perspective.

PERSPIRATION. A secretion formed by the sweat glands of the skin of some of the mammals. It consists chiefly of water with small quantities of other materials in solution. The dissolved substances include fatty acids, urea, sodium and potassium chloride, phosphates, lactic acid, and cholesterin. By evaporation at the surface of the skin, the sweat plays an important part in the regulation of body temperature when the surrounding air is too warm for adequate radiation. It is also important as an excretory medium.

PERS SUNSHINE RECORDER. A sunshine recorder of the type in which the timescale is supplied by the motion of the sun. The instrument, which is pointed at the celestial pole, consists of a hemispherical mirror mounted externally on the optical axis of a camera. The lens of the camera forms an image of the sun that is reflected by the hemispherical mirror so that as the sun moves across the sky, the image traces an arc of a circle on the photographic paper.

AMS

PERTHITE. An alkali feldspar comprising parallel or subparallel intergrowths. The potassium-rich phase, usually microcline, seems to be the host from which the sodium-rich phase, usually albite inclusions, exsolved. The exsolved areas typically form blebs, films, lamellae, small strings, or irregular veinlets and usually are visible to the naked eye.

PERTURBATION. 1. A small contribution to a physical quantity, such that the problem into which the quantity enters can be solved exactly or in a far simpler manner than otherwise if the perturbation is neglected. The form in which a perturbation is most frequently used in both classical and quantum mechanics is a small additional energy, called the *perturbation energy*.

2. Any departure introduced into an assumed steady state of a system. The magnitude of the departure is often assumed to be small so that product terms in the dependent variables may be neglected; the term *perturbation* is therefore sometimes used as synonymous with small perturbation. The perturbation may be concentrated at a point or in a finite volume of space; or it may be a wave (sine or cosine function); or, in the case of a rotating system, it may be symmetric about the axis of rotation.

3. In molecular spectra, perturbations cause the displacement of a band from its regular position in the band system (vibrational perturbation) or the displacement (and/or weakening) of corresponding lines in the different branches of a band (rotational perturbation). A perturbation observed in the spectrum is indicative of the presence of a perturbation (shift) of one of the energy levels involved due to interaction with another level of the same, or nearly the same, energy.

4. A much more frequent use is made of perturbation methods in quantum mechanics. The mathematical complexity of many quantum mechanical problems is such that one cannot hope to obtain exact solutions. However, good predictions can sometimes be obtained by means of perturbation theory, if one can assume that the actual system differs only slightly from a simpler system for which the problem can be solved, and the neglected difference can be dealt with as a perturbation of this simpler unperturbed system. The effect of a weak electromagnetic field on an atom, for instance, can be dealt with as a perturbation, and the transition probabilities between the energy states of the unperturbed atom can be calculated by means of perturbation theory. A weak interaction between two particles can be dealt with as a perturbation in the collision process of the two particles. The perturbation methods can be time-independent or time-dependent, according to whether the unperturbed states are described by time-independent wave functions or time-dependent ones. If the strength of a weak interaction between two systems is proportional to a constant parameter, the wave functions and energy values of the wave equation can be expanded in powers of this constant. The zero-order approximation is given by the unperturbed wave functions and energy values which are independent of this parameter. These determine the first-order approximations together with that part of the wave equation that is linear in the coupling parameter. By successive approximations one obtains expressions for second, third, and higher order perturbations in the wave function and the energy. If an unperturbed energy state is degenerate, that is, if two or more states have the same unperturbed energy, the effect of the perturbation has to be taken into account first between these degenerate states.

PERTURBATION (Astronomy). Within the solar system, it fortunately happens that the attraction of one body is dominant, i.e., the attraction of the sun upon any planet is far greater than the attractions of all of the other planets combined. In the case of satellites, the attraction of the primary is preponderant. In such cases, a close approximation to the true orbit may be obtained by neglecting the attractions of other objects and obtaining a preliminary orbit by the methods of solution of the two-body problem.

Using the Keplerian ellipse thus obtained, it is possible to find at any instant, to a high degree of approximation, the distance of the object from other members of the solar system and hence the attractions of these other objects. The effects of these attractions in changing the motion of the object under consideration may be computed, a second approximation to the true position may be obtained, and this can be used to obtain more accurate values to the attracting forces. Usually, the second approximation is sufficiently accurate for all practical purposes.

The influences that the attractions of the other members of the solar system have on the motions of the object under consideration are known as perturbations. In the case of nearly circular orbits, as in the case of the motions of the planets about the sun, it is possible to obtain, in the form of infinite series, an analytic expression for the perturbations. Such a solution is known as general perturbations. If the orbit is highly eccentric, as in the case of many comet orbits, it is not possible to obtain any general analytic expression, and the perturbations are known as special perturbations.

Due to perturbations, the orbits of objects have slow, steady changes in the elements, known as secular perturbations, and also relatively short oscillations about an average value, which are known as periodic perturbations. In actual practice, it is customary to list both the secular and periodic perturbations in Tables from which the accurate position of the planet at any desired instant may be computed. In the case of the perturbations of the moon, one of the most complicated of all perturbation problems, E.W. Brown's Tables of the moon fill three quarto volumes totaling more than 360 pages.

Perturbations have played an important part in astronomical discovery. After the planet Uranus had been discovered and accurately observed, the preliminary orbit, plus perturbations from all known objects, did not accurately represent the observed positions. These deviations could only be explained on the basis of another planet outside the orbit of Uranus; accurate computations were made, and the planet Neptune was discovered. Deviations of observed positions of Neptune from those computed from the orbit stimulated the search that led to the discovery of the planet Pluto. In the case of the planet Mercury, a perturbation in the longitude of perihelion in the orbit could not be explained on the basis of gravitational theory. This led to a fruitless search for a planet between Mercury and the sun. The perturbation was later explained on the basis of the theory of relativity and was one of the early triumphs of this theory.

See also **Kepler's Laws of Planetary Motion**; and **Orbit (Astronomy)**.

PERTURBATION TECHNIQUE.   A mathematical technique to eliminate linear terms in an equation in order to retain the nonlinear (turbulence) terms. Variables such as potential temperature (θ) or velocity (U) can be partitioned into mean (slowly varying) and perturbation (rapidly varying) components. Mean components or averages are often represented with an overbar, while perturbation quantities are indicated with a prime:

$$\theta = \bar{\theta} + \theta' \text{ or } U = \bar{U} + u'.$$

When substituted in the equations of motion or other budget equations, the resulting equations have terms that explicitly describe the mean and turbulence components, and the interaction between these components. Next, the whole perturbation equation can be averaged, which eliminates the linear terms (terms having only one perturbation variable, such as $u'\partial\theta/\partial x$). The remaining nonlinear terms (terms that have products of two or more perturbation quantities, such as $\overline{u'\partial\theta/\partial x}$) represent turbulent fluxes, variances, or correlations.

See also **Turbulent Flux**.

Additional Reading

Star, V.P.: *Physics of Negative Viscosity Phenomena*, The McGraw-Hill Companies, New York, NY, 1968.
Stull, R.B.: *Introduction to Boundary Layer Meteorology*, Springer-Verlag New York, LLC, New York, NY, 1988.

AMS

PERTUSSIS (Whooping Cough).   A disease primarily of young children, caused by a small aerobic, gram-negative coccobacillary *Bordetella pertussis*, which adheres to ciliated epithelial cells in the respiratory tract. This attachment is followed by ciliostasis and subsequent loss of the ciliated cells. Humans are the only known reservoir of the organism. Occasionally asymptomatic infections have been identified, but there is no evidence that these are important in the spread of the disease and there are no chronic carriers.

The disease, still under mandatory reporting to the Centers for Disease Control (Atlanta, Georgia) was prominent prior to 1950, during which year about 800 cases per 100,000 population were reported. With a few intermediate peaks, the incidence declined to a plateau in the early 1960s of less than 10 cases per 100,000. The incidence of the disease has lessened since then and, in 1984, less than one case per 100,000 population was reported—with the majority of reported cases involving children less than one year old. This decrease in incidence appears attributable to the widespread vaccination program introduced in the late 1940s.

The disease is spread by droplets from infected patients. The incubation period is from 12 to 20 days, when minor upper respiratory symptoms are found. These persist for about two weeks, during which period the disease is most contagious. Toward the end of this period, a dry hacking cough appears and becomes progressively worse. After one or two weeks, the cough becomes paroxysmal and prolonged coughing attacks are followed by the characteristic "whoop" which is produced by forced inspiration through a partially closed glottis. Frequently, efforts to expel thickened sputum will induce vomiting.

The appearance of fever suggests a secondary bacterial infection. Otitis media and pneumonia are the most common complications and most noninfectious complications are extremely rare.

Diagnosis is symptomatic and confirmed by isolation of the causative organism in culture.

Treatment of an active case of pertussis is generally supportive. Cough medicines are of no value, nor is passive immunization. The organism is sensitive to a variety of antibiotics, including erythromycin, tetracycline,

chloramphenicol, and probably trimethoprin-sulfamethoxazole. Early treatment during the catarrhal stage will shorten the course of the clinical illness, but if withheld until the paroxysmal stage, antibiotics will have no effect. Even though there is no clinical benefit, however, patients in the coughing stage should receive antibiotics to render them noninfectious to others.

Prevention of *B. pertussis* infection relies upon use of vaccines. Because this is a killed whole bacterial preparation, containing both toxins and antigens, it has various side effects that have been persistently criticized. Attempts to produce a "purer" vaccine have not been wholly successful. An extract vaccine was associated with fewer side effects than the whole cell vaccine, but doubts have been expressed concerning its efficacy. The whole cell vaccine is commonly given together with tetanus and diphtheria toxoids as a series of injections at one or two month intervals, starting at 6 to 12 weeks of age.

Additional Reading

Birkebaek, N.H., M. Kristiansen, T. Seefeldt, et al.: "Bordatella Pertussis and Chronic Cough in Adults," *Clin. Infect. Dis.*, **29**, 1239–1242 (1999).
Decker, M.D., K.M. Edwards, M.C. Steinhoff, et al.: "Comparison of 13 Acellular Pertussis Vaccines: Adverse Events," *Pediatrics*, **96**(suppl), 557–566 (1995).
Evans, A.S. and P.S. Brachman: *Bacterial Infections of Humans: Epidemiology and Control*, 3rd Edition, Plenum Medical Book Company, New York, NY, 1998.
Gangarosa, E.J., A.M. Galazka, L.M. Phillips, et al.: "Impact of Anti-vaccine Movements on Pertussis Control: The Untold Story," *Lancet*, **351**, 356–361 (1998).
Guris, D., P.M. Strebel, B. Bardenheir, et al.: "The Changing Epidemiology of Pertussis in the United States: Increasing Reported Incidence in Adolescents and Adults, 1990–1996," *Clin. Infect. Dis.*, **28**, 1230–1237 (1999).
Halperin, S.A., D. Schiefele, L. Barreto, et al.: "Comparison of a Fifth Dose of a Five-component Acellular or a Whole Cell Pertussis Vaccine in Children Four to Six Years of Age," *Pediatr. Infect. Disease J.*, **18**, 772–779 (1999).
Orenstein, W.A., S. Hadler, and M. Wharton: "Trends in Vaccine-Preventable Diseases," *Semin. Pediatr. Infect. Dis.*, **8**, 23–33 (1997).
Plotkin, S.A. and W.A. Orenstein: *Vaccines*, 3rd Edition, W. B. Saunders Company, Philadelphia, PA, 1999.
Peter G.: *1997 Red Book: Report of the Committee on Infectious Diseases*, 24th Edition, American Academy of Pediatrics, Elk Grove Village, IL, 1997.
Staff CDC: "Pertussis Vaccination: Use of Acellular Pertussis Vaccines Among Infants and Young Children: Recommendations of the Advisory Committee on Immunization Practices (ACIP)," *MMWR*, **46**(RR-7), 1–25 (1997).
Staff CDC: "Pertussis—United States, January 1992–June 1995," *MMWR*, **44**, 525–529 (1995).
Staff CDC: "Transmission of Pertussis From Adult to Infant—Michigan, 1993," *MMWR*, **44**, 74–76 (1995).
Staff Institute of Medicine: Adverse Effects of Pertussis and Rubella Vaccines. National Academy Press, Washington DC, 1991.
Staff Institute of Medicine: *Adverse Events Associated with Childhood Vaccines: Evidence Bearing on Causality*, National Academy Press, Washington, DC, 1994.

Web Reference

Centers for Disease Control and Prevention: http://www.cdc.gov/health/diseases. htm#P and http://www.cdc.gov/nip/publications/pink/pert.pdf

R. C. V.

PERU/CHILE CURRENT.   See **Ocean Currents**.

PERUTZ, MAX FERDINAND (1914–2002).   Max Perutz was an Austrian-born British biochemist who determined the structure of the oxygen-binding protein haemoglobin by X-ray diffraction analysis.

Perutz was born in Vienna, Austria on 19 May 1914. Educated at the University of Vienna, he became interested in biochemistry and wanted to enter Gowland Hopkins's laboratory at Cambridge. Herman Mark, Professor of Physical Chemistry in Vienna, was impressed by J. D. Bernal's X-ray diffraction work and persuaded Perutz to join Bernal's laboratory at Cambridge instead. In September 1936 he began his graduate studies on the structure of minerals, but during the summer vacation in 1937 Perutz met the Austrian biochemist Felix Haurowitz, who suggested that he should investigate the structure of hemoglobin. This was valuable advice as Cambridge has a long tradition of hemoglobin research from Joseph Barcroft's work on its oxygen equilibrium curve in 1907 to David Keilin's work on cell respiration, haem proteins and the cytochromes. G. S. Adair, the Cambridge physiologist, gave Perutz a sample of horse methemoglobin, the oxidized form of hemoglobin, and with this Perutz began his X-ray diffraction studies. See also **Barcroft, Joseph (1872–1947)**.

Hemoglobin crystals produce good X-ray patterns and by determining the dimensions of the unit cell, Perutz concluded that the molecule was a dimer with two identical halves.

Bernal moved to Birkbeck College, London in 1938 and Lawrence Bragg became the new Cavendish Professor of Physics. Bragg was excited by the prospect of applying X-ray techniques to large protein molecules and secured a grant from the Rockefeller Foundation to allow Perutz to continue his work; he received his doctorate in 1940. His career was then interrupted due to the war. He had to obtain a residence permit for his parents who had fled from Nazi persecution and was himself interned as an enemy alien for six months. In 1943 he was required to help in an absurd project to build a gigantic aircraft carrier made of ice.

After the war Perutz directed a Medical Research Council Unit in Cambridge supported by his only student, J. C. Kendrew. The interpretation of his X-ray photographs posed problems because they contained the intensities but not the phases of the diffracted rays without which it was impossible to determine the structure. Perutz solved this problem in 1954, using a method of isomorphous replacement with heavy atoms first applied to the alums in Bragg's laboratory at Manchester. By attaching mercury and gold or silver atoms to the haemoglobin molecule and comparing the X-ray diffraction patterns with those of the pure protein, the phases could be determined. During the next five years Perutz obtained X-ray data for six isomorphous derivatives with heavy metal atoms at different positions and by 1959 he was able to construct a three-dimensional map of the hemoglobin molecule. He showed that it was composed of four separate polypeptide chains forming a tetrahedral structure, with four haem groups enclosed in pockets of the globin near the molecule's surface. The fold of the chains in hemoglobin resembled that of Kendrew's structure for myoglobin from sperm whale muscle, determined two years earlier. Perutz's method opened up protein crystallography and has been extended to several hundred other proteins, enzymes, antibodies and viruses. See also **Kendrew, John Cowdery (1917–1997)**.

In 1962 Perutz shared the Nobel Prize for Chemistry with J. C. Kendrew. Having determined the structure of hemoglobin, he continued to study its properties. The sigmoidal shape of its oxygen equilibrium curve had been attributed to interaction between the haem groups, but their wide separation made it difficult to imagine easy interaction between them. Perutz therefore turned to Haurowitz's 1938 observation that crystals of oxy- and deoxyhemoglobin have different symmetries, suggesting a change in structure during the take-up and release of oxygen. At the Pasteur Institute in Paris in the early 1960s, Jacques Monod, François Jacob and Jean-Pierre Changeux suggested that the sigmoid curve arose from allosteric behavior in which the binding of the first molecule of oxygen makes the take-up of subsequent molecules easier. Monod suggested that complex proteins, made up of several subunits, can exist in two or more reversible arrangements, a relaxed structure (R) in which the subunits are unconstrained and a tense structure (T) in which they are linked by additional bonds.

In 1962 Perutz and Hilary Muirhead found that the change in crystal form was due to a rearrangement of subunits during the binding and release of oxygen. The mechanism was unknown until 1970 when the constraints in the T structure predicted by Monod were found to be salt bridges between the subunits broken on transition to the R structure triggered by movement of the iron atoms relative to the porphyrin, the conjugated protein group. In deoxyhemoglobin the iron is displaced from the plane of the porphyrin group and the molecule becomes dome-shaped; in oxyhemoglobin the iron lies in a plane with the porphyrin. An electronic transition and change of coordination of the iron is also involved. In deoxyhemoglobin the iron is five-coordinated and in high spin, whereas in oxyhemoglobin the iron is six-coordinated and in low spin. Perutz predicted that there should be a reciprocal relationship between the spin-state of the iron and the two structures of haemoglobin and this was later confirmed. The transfer of oxygen to the tissues and return transport of carbon dioxide back to the lungs depends on the take-up of protons to convert carbon dioxide to soluble bicarbonate, a process reversed in the lungs. With J. V. Kilmartin, Perutz showed that the uptake and release of protons is linked to the formation of salt bridges in the T structure of hemoglobin and their rupture in the R structure. Perutz also studied mutant forms of hemoglobin characteristic of certain diseases such as sickle cell anaemia. In 1956, Vernon Ingram, a protein chemist working with Perutz

and Kendrew, found that sickle cell hemoglobin differs from the normal form by the replacement of one pair of valine residues by glutamine acid. Perutz was later asked to develop anti-sickling agents using X-ray analysis to determine their binding sites. This work led him to formulate general rules governing the interaction of drugs with proteins. See also **Anemias**.

Perutz also investigated the formation of glacial ice and the speed of flow of glaciers. He found that the surface of a glacier moves faster than its base. He later became interested in the effects of science on society, expressing his views in several books and articles.

Additional Reading

Hagan, W.J.: In: James, L.K.: *Nobel Laureates in Chemistry 1901–1992*, (History of Modern Chemistry Series), American Chemical Society and the Chemical Heritage Foundation, Washington DC. 1993, pp. 425–441.

Perutz, M.F.: "X-ray Analysis of Hemoglobin (Nobel Address)," *Science*, **140**, 863–869 (1963).

Perutz, M.F.: "Hemoglobin Structure and Respiratory Transport," *Scientific American*, **239**, 92–125 (1978).

Perutz, M.F.: "Origins of Molecular Biology," *New Scientist*, **85**, 326–329 (1980).

Perutz, M.F.: *Wish I'd made you Angry Earlier: Essays on Science, Scientists and Humanity*, Cold Spring Harbor Laboratory Press, Cold Spring Harbor, NY, 1998.

NOEL G. COLEY, The Open University, Milton Keynes, UK

PETALITE. The mineral petalite, lithium aluminum silicate ($LiAlSi_4O_{10}$), is monoclinic, although crystals are rare, this mineral usually occurring in cleavable, foliated masses, whence the name petalite, from the Greek meaning a *leaf*. Its hardness is 6–6.5; specific gravity 2.39–2.46. It is brittle with subconchoidal fracture; perfect basal cleavage; luster, vitreous, colorless to white or gray but may be greenish or reddish; transparent to translucent. Petalite occurs in granite pegmatites with sodium-rich feldspar, quartz, and lepidolite; has been found in Sweden; the former U.S.S.R.; on the Island of Elba; and in the United States at Bolton, Massachusetts, and Peru, Maine. It is interesting to note that lithium was first discovered in this mineral. See also **Lithium**.

PETALS. The petals of many plants are readily recognizable as the expanded floral organs that confer shape, color and pattern to flowers. Morphologically they are formed from the second whorl of organs (the corolla) which lies within the first whorl of organs (the calyx composed of sepals) of a flower. The organs of other whorls may develop an appearance similar to petals. This is the case in many monocotyledonous flowers where the organs of the first two whorls (the perigone) are very similar and are referred to as tepals (iris and tulip, for example). Petals are thought originally to have served a protective function. In primitive plants they appear to have developed from protective leaves or bracts surrounding the sporangia (reproductive cells, Weberling, 1981). Structures similar to petals can still be seen in certain types of fern. Petals have developed in flowering plants that are pollinated by insects, and their specialization as broad, attractive signaling devices is thought to have occurred in conjunction with the evolution of insects as pollinators. The most obvious attributes of petals in serving to attract pollinators are their size, form, color and pattern. In addition, petals may form landing platforms and traps for pollinating insects, adaptations of form which require exquisite cellular specializations. In some flowers they also serve as sites for nectar synthesis and scent production. See also **Flower**.

The Genetic Control of Petal Development

Initiation of Petal Formation. The induction of flower development usually occurs in two stages: induction of the inflorescence meristem and induction of the floral meristem. A number of genes have been identified which control floral meristem induction [Coen and Meyerowitz, 1991]. Some of these genes also serve roles in determining the formation of the floral organ primordia. They may control expression of the floral homeotic genes that regulate organ identity. Homeotic genes are defined by their mutant phenotypes. In the case of those controlling floral organ identity, mutation results in organs of particular whorls adopting the fate of other whorls [Weigel and Meyerowitz, 1994]. From analysis of such mutants in model dicotyledonous species, *Antirrhinum majus* (snapdragon) and *Arabidopsis thaliana* (thailcress), the identity of the four whorls of organs in the flower can be shown to be governed by three functions which control the identity of pairs of whorls (Figure 1). These functions are referred to

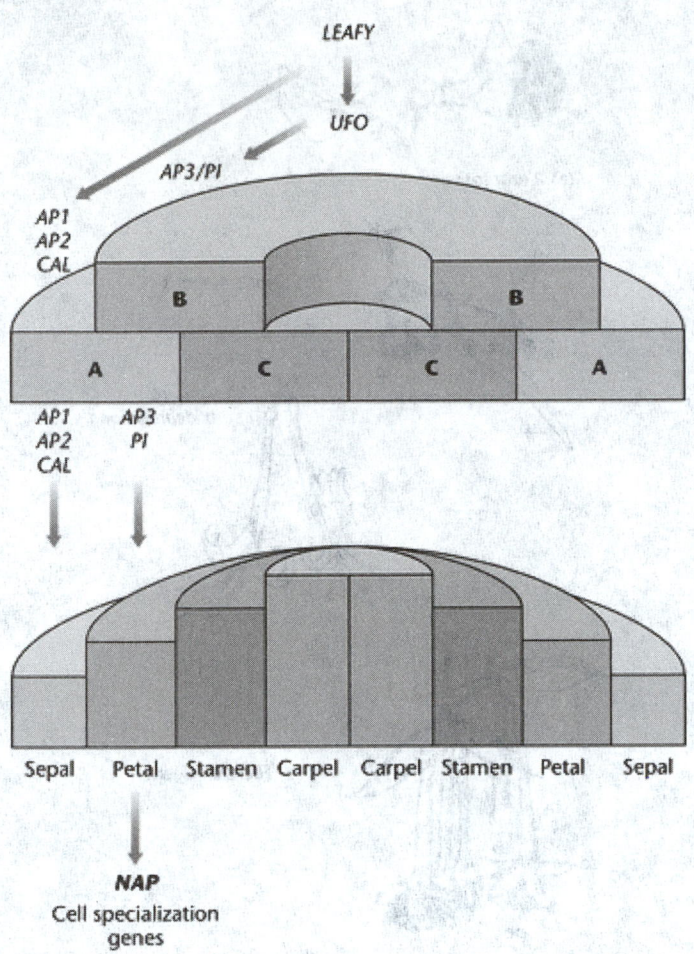

Fig. 1.   Diagram of the ABC model for the determination of floral organ identity.

as the A, B and C functions. The "A" function determines the identity of whorls 1 and 2, "B" of whorls 2 and 3 and "C" of whorls 3 and 4. The activity of functions "A" and "C" is also thought to be mutually antagonistic. Therefore the identity of petals in whorl 2 is determined by the combination of "A" and "B" functions. Some of the genes defining these functions have been identified, including the "B" function genes *APETALA3* (*AP3*) and *PISTILATA* (*PI*) from *A. thaliana* (*DEFICIENS* and *GLOBOSA* from *A. majus*) and the "A" function genes *APETALA1* (*AP1*), *CAULIFLOWER* (*CAL*) and *APETALA2* (*AP2*) from *A. thaliana*. Squamosa from *A. majus* is structurally equivalent to *AP1* but its function in regulating the identity of whorls 1 and 2 has not yet been proven. *AP1*, *CAL*, *AP3* and *PI* (and the equivalent genes from *A. majus*) all encode transcription factors belonging to the MADS-box family. *AP2* encodes a transcription factor of another family, unique to plants. The combinatorial action of these genes determines the identity of the organs in whorl 2 as petals. However the mechanics of this regulatory cascade are by no means clear. The expression of the homeotic genes in particular whorls is thought to be regulated by the floral meristem identity genes (*LEAFY* and *AP1* in *A. thaliana*), modulated by cadastral genes (e.g., *UNUSUAL FLORAL ORGANS* (*UFO*) in *A. thaliana* or *FIMBRIATA* in *A. majus*), which define the boundaries within the floral primordia and control cell proliferation during floral organ growth [Parcy et al., 1998]. Other genes may affect the number of organs that develop in particular whorls. *ETTIN* (*ETT*) from *A. thaliana* affects, among other things, the number of petals in whorl 2. It operates independently of the meristem identity genes, but its activity may be partially redundant with another gene, *PERIANTHA* (*PAN*) which affects organ number in the first three whorls. *ETT* encodes transcription factor, related structurally to other proteins which have been associated with responses to the plant growth regulator auxin. Interestingly *pan* mutants have five sepals, five petals and five stamens (as distinct from the normal 4,4,6 arrangement in Brassicaceae). It has been suggested that *PAN* activity may be a specific function that has developed in this

family to modify the 5,5,5 pattern often observed in other angiosperm families.

Development of Petal Form. Once petal primordia have been initiated, the development of petals involves regulated cell division and expansion. Clonal analysis has shown that, concurrent with primordium formation, cells are delimited to particular organs, and in expanding petals, which are dorsoventrally flattened, epidermal cells rapidly become associated with either the abaxial or adaxial surface. One gene has been defined as a primary target activated by AP3/PI in, *A. thaliana*, *NAP* [Sablowski and Meyerowitz, 1998]. The product of this gene is structurally similar to proteins involved in growth and maintenance of apical meristems. *NAP* is thought to serve functions in promoting cell division and cell expansion leading to the development of petal form. It may encode another transcription factor in the hierarchy of morphological control.

In many flowers, individual petals are fused to form a tube. Fusion may occur at the point of primordial initiation (congenital sympetally) or the individual petals may fuse after primordium formation (false sympetally). Fused or separate petals in dicots are traits often used for classification between plant families, indicating that these features tend to be maintained over long periods of time. However, in monocots the same traits are used for the distinction of genera, indicating them to be more variable in the evolution of new species.

The mechanics of the control of cell division and expansion in petal formation is currently not well understood, although it is thought that the organ identity genes may serve late roles in regulating these processes. The final stages of petal development involve cellular specialization, which usually involves the production and enhancement of color in specific patterns. There is also an inhibition of processes associated with the functioning of leaves, such as the breakdown of chloroplasts or their conversion to leucoplasts. Again, there is evidence for the homeotic genes being involved in switching off conflicting developmental programs [Perbal, et al., 1996].

Cellular Specification in Petals. Color in flowers may be based on different pigment types: anthocyanins, carotenoids and betacyanins. Anthocyanins and carotenoids are most common, accumulating in vacuoles and plastids respectively. Anthocyanins are formed from phenylalanine by a branch of phenylpropanoid metabolism. Their color range is from yellow through red to purple and blue. Particular shades may be created by derivatization, pH, metal chelation or copigmentation with other chemicals in solution. Several genes controlling flower color have been identified. Generally these are transcription factors which regulate expression of the structural genes encoding the enzymes of anthocyanin biosynthesis, and members of the MYB, bHLH and WD40 families have been shown to be important in regulating flower colour, as they are in regulating the color of other plant tissue [Martin and Paz-Ares, 1997]. The differential expression of the genes regulating color may be responsible for creating color patterns. Clearly because many color patterns are integrated over the whole flower, the control of the genes controlling color must be complex, although it is likely to involve the activity of the homeotic genes either directly or indirectly. See also **Anthocyanins**; and **Flavonoids**.

An estimated 80% of angiosperm species also develop specially shaped cells on the inner petal epidermis, the surface that faces prospective pollinators. Generally this specialization involves the formation of conical epidermal cells rather than the normal flat ones and is believed to be an adaptation to enhance the pigment intensity of the petals; the undulating surface of the petal cells serving as tiny lenses to focus light into the pigment-containing epidermal cells, which reduce the amount of white light reflected from the petal surface. The actual morphology of the petal surface may also serve as a tactile recognition signal to insect pollinators. Genes belonging to the family encoding MYB-transcription factors have been shown to direct the formation of conical cells (*MIXTA* in *A. majus* and *PhMYB1* in *Petunia hybrida*). Conical cells are developmentally related to certain types of plant hair (trichome), and have probably evolved from trichomes to serve specific roles in petals to attract pollinators. These genes, as with the genes regulating floral pigmentation, are under developmental control, which is probably, at least in part, a late function of the floral homeotic genes.

Petal Senescence. Petal senescence is a controlled process and involves the reclamation of carbon and nitrogen invested in petal development, for redistribution to other parts of the plant. Senescence may occur very rapidly, as in the day lily where phloem export is initiated as petal autolysis begins, just 32 h after petal opening. Senescence of some flowers is associated with ethylene production, and can be accelerated by exogenously supplied ethylene. Transgenic carnations in which ethylene production has been inhibited show delayed senescence. Ethylene stimulates the activity of enzymes involved in its own synthesis, in an autocatalytic manner. Ethylene also induces the synthesis of senescence-related proteins such as glutathione-*S*-transferases, thiol proteases (the latter being associated with nitrogen mobilization), phospholipases, phosphatidic acid, phosphatases and lipoxygenases, which are associated with membrane turnover. The process of senescence may eventually become autocatalytic as membrane damage leads to free radical production and further membrane damage.

In some flowers petal senescence is induced by pollination and in other flowers similar but distinct changes in floral attractiveness occur in response to pollination [van Doorn, 1997]. Changes in color, flower closure, petal wilting and termination of scent production tend to occur in flowers of subtropical/temperate species following pollination. For these plants pollinator numbers may be limiting and petals from pollinated flowers may be lost to focus the attention of pollinators on unfertilized flowers. Such changes in floral attraction are often, although not always, stimulated by ethylene. It has been shown that short-chain fatty acids, synthesized by the gynoecium upon pollination, are transported to the petals where they increase cellular sensitivity to ethylene, possibly by modifying membrane properties. In some species, floral attraction is maintained, following pollination, for longer than in unpollinated flowers. In these cases it has been argued that the maintenance of floral form and color enhances the overall attractiveness of the whole plant.

The Shape of Petals and Functional Anatomy

The majority of flowering plants are pollinated by animals, and most of these are pollinated by insects. In wind-pollinated plants petals are usually greatly reduced or missing. See also **Pollination**.

In animal-pollinated flowers there are general attractive devices such as petal color, form and scent, and rewards for pollinators including pollen, nectar and feeding hairs. One other constraint on petal development is that flowers must be able to withstand the mechanical stresses associated with routine contact between the pollinators and the stigma/style.

Floral form (which is usually largely dependent on petal form) is often favored by groups of pollinators, whereas other aspects of attraction such as color, scent, pollen, nectar and fruit bodies are more often directed towards specific pollinators. It is thought that pollen was the original attractant for pollinators, other attractants having developed secondarily, and that decoration of the androecium may have originally identified the target for pollinators.

Beetle Flowers. The earliest pollinators were probably beetles, which have vertical mouth parts restricting their pollen searches to open or disc-shaped flowers. Consequently beetle-pollinated flowers are usually open and robust with minimal visual signals, but with scent as an important attractant. In members of the genus *Gutteria* there is a more elaborate specialization of the petals which turn yellow and inflex at anthesis to form a pollination channel. The beetles lay eggs in the pollination chamber and pollinate at the same time. Ovule production is in excess of those consumed by the beetle larvae.

Bee Flowers. Bee pollination requires formation of a landing platform, usually on zygomorphic flowers with tubes. Other attractions include sweet scent and yellow, violet and blue petal colors. Petal form may be elaborated to first supply pollen to visiting bees and then to receive pollen (Figure 2). In labiate flowers two upper (dorsal petals) form a hood and the three lower petals form a landing platform or lower lip (Figure 2b). The anthers and style may lie below the hood, protected and able to donate and receive pollen from the bee's back (nototribal). Alternatively, the style and stamens may lie under the lower lip and contact the pollinator's ventral surface (sternotribal; (Figure 2a). Although the 2:3 arrangement of petals in a zygomorphic flower is most common, the allocation may be 4:1 as in *Lonicera* spp. (Figure 2c) and in the personate flowers of *Nemesia* spp.

The ray florets of Compositae are extreme examples of labiate flowers. The central flowers (disc florets) are radially symmetrical, whereas the outer flowers (ray florets) are zygomorphic. The two dorsal petals may be suppressed in ray florets or all five petals may go to form the lower lip. See also **Composite Family**.

Fig. 2. Illustrations of different types of petal specialization. (a) Zygomorphic flower of pea adapted for sternotribal pollination. The rear petal or standard(s) provides a signal while the two lower petals (the keel) and the two lateral petals or wings (w) form a landing platform. The stamens and style lie beneath the keel and wings. se, sepals. (b) Zygomorphic flower of *Linaria* adapted for nototribal pollination by bees. The two dorsal petals form in the hood (h) while the ventral and two lateral petals form the landing platform (lp). The base of the corolla is extended to form a spur (sp) in which nectar, secreted from nectaries on the gynoecium, accumulates. (c) Zygomorphic flower of *Lonicera hildebrandiana* adapted for butterfly pollination. Four dorsal petals form the hood (h). (d) Flowers of *Fuchsia fulgens*, adapted for pollination by hummingbirds. The petals (pe) and sepals (se) are inserted at the upper margin of an extension of the floral axis, the hypanthium (hy). (e) Flowers of *Aristolochia clematitis* adapted for pollination by trapped flies (pitfall flowers). The peregone (p) forms a tube to trap the insects. One flower is shown in section with the carpels (c) at the base.

Petal tissue may occasionally serve as nectaries, as in *Valeriana* and *Centranthus*. However, the petals more often form a spur or reservoir for the nectar which may be secreted from other organs, but collect in the reservoir for harvest by insect pollinators. Where petals do serve as nectaries they may be considerably modified in form (nectar leaves) as observed in *Nigella* or *Aquilegia*. Nectaries are very diverse but consist of glandular tissue (often subepidermal) that produces sugars. The sugars are often exuded through trichomes or modified stomata.

Some flowers produce oil as an attractant from oil glands (eliaophores). Such oils may be produced by a variety of different plant species to attract specific pollinators such as solitary hairy bees. The positioning of the oil glands may be part of the pollination mechanism. In other species oils are fragrant as, for example, in neotropical orchids and *Gloxinia*, which attract Euglossine bees who use the oils collected during pollination to mark their mating territories.

Scent is often produced from cells of the petals (osmophores). These may be single-layer tissues or multilayered, with a mesophyll production layer and an epidermal emission layer. The production layer may have associated reserve tissue, often storing starch, to supply energy and carbon skeletons for scent production. Scents are derived from a variety of different secondary metabolites including monoterpenoids and phenylpropanoids. Scent production may be specifically localized in the petals to give further spatial guidance to pollinators.

Fly Flowers. Flowers adapted for fly pollination often grow in sunny locations and have nectar readily available since these insects have only short proboscises. Others, which are shade plants, may attract flies with specific odors and colour/patterning; for example dung-fly flowers combine putrid odors (similar to carrion and urine) with dull greenish-purple coloration. Purple/brown coloration over the vascular tissue may further mimic carrion, creating the atmosphere of a place suitable for egg-laying.

The pitfall flowers are highly specialized for fly pollination. In flowers such as birthwort (*Aristolochia clematitis*; (Figure 2c) the first and second whorl organs (perigone) form a flask-shaped tube. Above a widened base the tube narrows and is lined by downward-pointing hairs. The insertion of the hair cells into the general epidermis is such that if pushed up they press against the epidermis and will not bend. They will bend downwards easily, however. A scent produced by the perigone attracts *Diptera* spp., which fall down the tube. If the stigma is pollinated by the visitors the trap hairs wither. The male organs develop and the *Diptera* become covered with pollen. The flower bends over to the horizontal position and the captive pollinators are released to pollinate another flower.

Butterfly Flowers. Flowers adapted to butterfly pollination often have long narrow corolla tubes. Many flowers are open at night and have a horizontal or pendular habit. The attractions are scent and color which may be yellow, red, blue, purple or white. Amongst moth-pollinated flowers white is predominant along with strong scent and evening opening, for example honeysuckle (*Lonicera* spp.; (Figure 2c).

Bird Flowers. Bird-pollinated flowers are usually tubular, bell-shaped or bottlebrush. Landing platforms are absent as pollinators hover or use adjacent twigs, which are more robust than the flowers. Petals are brightly colored, often juxtaposing striking colors (e.g. *Abutilon*, *Fuchsia*; Figure 2d) but smell is relatively unimportant. The reward is usually large amounts of watery nectar. See also **Flower**.

Evolution of Petals

Clearly, evolution of specialized form and attractants has involved coadaptation with insect pollinators [Weberling, 1981]. This is illustrated most vividly in deceptive flowers, where the form and coloration of the flowers actually mimics the pollinating insects. In species such as the bee orchid and fly orchid the mimicking of insect appearance is largely a function of the perigone (first and second whorl organs). It may be reinforced by a scent that mimics the pheromones used by the insects in mating.

Evolution of Bilateral Symmetry. Irregular flowers, in which radial symmetry is lost to bilateral symmetry, provide an adaptation to insect pollination that appears to have evolved independently on numerous occasions [Coen, et al.: 1995]. Several genes associated with the establishment of bilateral symmetry have been identified in *A. majus* (*CYCLOIDEA*, *DICHOTOMA*, *RADIALIS*) [Luo, et al.: 1996]. *A. majus* belongs to a large monophyletic group, the Lomiales, which all have irregular flowers suggesting that a common mechanism involving related genes may underlie

asymmetry in all member species. Other genetic mechanisms may dictate asymmetry in unrelated species. In some regular flowers bilateral symmetry may have been lost, which would mean that irregularity evolved on fewer independent occasions.

Evolution of Petal Loss. Studies of related species of Brassicaceae that show selective petal and/or stamen loss indicate that petal primordia are formed but arrest early [Bowman and Smyth, 1998]. This suggests that modification of homeotic gene function is not the direct cause of petal loss. It is possible that petal loss is associated with changes in the activity of genes defining organ boundaries such as *UFO/FIMBRIATA* or with changes in genes acting downstream of the a + b functions. It is not known if these specific examples of petal loss represent mechanisms generally employed in species with reduced petals.

Evolution of Flower Color. Flower color and pattern also show remarkable diversity within petals. The understanding of color formation is greatest for the anthocyanins, where color is dependent not only on the pigments themselves, but also on the presence of related compounds (copigments) which may modify their color in solution. Clearly the control of synthesis of pigments and copigments along related biochemical pathways gives considerable flexibility to potential color and pattern variation within flowers. Other associated compounds, the flavones and flavonols absorb strongly in the ultraviolet (UV) range, and can be seen as colors by some pollinators (e.g. bees). The regulatory genes controlling anthocyanin production have more specialized target genes in flowers than in other tissues such as seeds. This may allow metabolic flexibility in the synthesis of anthocyanins and flavones/flavonols to give more complex colors and patterns as adaptations serving the role of petals in insect-mediated pollination [Martin and Paz-Ares, 1997].

Additional Reading

Borochov, A., and W.R. Woodson: "Physiology and Biochemistry of Flower Petal Senescence," *Horticultural Review*, **11**, 15–43 (1989).

Bowman, J.L., and D.R. Smyth: "Patterns of Petal and Stamen Reduction in Australian Species of *Lepidium* L. (Brassicaceae)," *International Journal of Plant Science*, **159**, 65–74 (1998).

Coen, E.S., and E.M. Meyerowitz: "The War of the Whorls: Genetic Interactions Controlling Flower Development," *Nature*, **353**, 31–37 (1995).

Kay, Q.O.N., H.S. Daoud, and C.H. Stirton: "Pigment Distribution, Light Reflection and Cell Structure in Petals," *Botanical Journal of the Linnean Society*, **83**, 57–84 (1981).

Luo, D., R. Carpenter, C. Vincent, L, Copsey, and E. Coen: "Origin of Floral Asymmetry in *Antirrhinum*," *Nature*, **383**, 794–799 (1996).

Martin, C., and J. Paz-Ares: "MYB Transcription Factors in Plants," *Trends in Genetics*, **13**, 67–73 (1997).

Parcy, F., O. Nilsson, M.A. Busch, I. Lee, and D. Weigel: "A Genetic Framework for Floral Patterning," *Nature*, **395**, 561–566 (1998).

Perbal, M-C, G. Haughan, H. Saedler, and Z. Schwarz-Sommer: "Non-cell-Autonomous Function of the *Antirrhinum* Floral Homeotic Proteins DEFICIENS and GLOBOSA is Exerted by their Polar Cell-to-cell Trafficking," *Development*, **122**, 3433–3441 (1996).

Sablowski, R., and E.M. Meyerowitz: "A Homolog of *NO APICAL MERISTEM* is an Immediate Target of the Floral Homeotic Genes *APETELA3/PISTILLATA*," *Cell*, **92**, 93–103 (1998).

van Doorn, W.G.: "Effects of Pollination on Floral Attraction and Longevity," *Journal of Experimental Botany*, **48**, 1615–1622 (1997).

Weberling, F.: *Morphology of Flowers and Inflorescences*, Cambridge University Press, Cambridge, UK, (1992).

Weigel, D., and E.M. Meyerowitz: "The ABCs of Floral Homeotic," *Cell*, **78**, 203–209 (1994).

Web Reference

John Innes Centre (JIC): http://www.jic.ac.uk/corporate/index.htm

CATHIE MARTIN, John Innes Centre, Norwich, UK

PETIOLE. See **Leaf**.

PETRELS AND ALBATROSSES (*Aves, Procellariiformes*). These birds essentially comprise the procellariiformes, an order of marine birds known for their powerful flight. The feet are webbed, and the horny sheath of the beak is composed of several parts. Among the petrels are the fulmars and shearwaters, as well as several forms that bear the name petrel. There are many species, and many of them bear one or more names in sailors' vernacular. The Cape pigeon (*Daption capensis*) is a well-known petrel of the Southern Hemisphere, and the little stormy petrel of the North Atlantic (*Hydrobates pelagicus*), under the name Mother Carey's chicken, is probably the most familiar of all to ocean travelers. The name puffin is commonly applied to the Atlantic species of petrel; and the name shearwater to those of the Pacific. Together, they constitute the genus *Puffinus*.

Fig. 1.   Royal albatross. (*Sketch by Glenn D. Considine*.)

The albatross is a large marine bird known from its habit of following ships for many hours without alighting. There are several species, belonging to *Diomedea* and allied genera. See Fig. 1. These birds are commonly called sea gulls although the term albatross is sometimes reserved for the species chiefly found in the South Seas. Of a large number of species, these birds display similar characteristics. One egg is hatched per time in a nest constructed of sticks, usually on the ground. They participate in spectacular courtship dancing, making loud shrieks during the performance.

The Laysan albatross is found in the environs of the mid-Pacific islands and summers in the Aleutians. Unlike most species, this albatross does not follow ships. The blacktailed albatross lives along the Pacific coasts, feeding on squid and fish. The feathers are dark, and the bird often rests on the surface of the water. The range is wide—from Japan and the Pacific islands and along the North American coast from Mexico to Alaska. See also **Procellariiformes**.

PETROCHEMICALS.   Chemicals derived from petroleum and, more specifically, substances or materials manufactured from a component of crude oil or natural gas. See Fig. 1 on p. 2686. In this sense, ammonia and synthetic rubber made from natural gas components are petrochemicals. Many of these chemicals are described in separate articles in this encyclopedia. Check alphabetical index.

Among the most important petrochemicals manufactured are:

Acetic acid	Ethylene dichloride	Phenol
Acetone	Ethylene glycol	Polyethylene
Acrylonitrile	Ethylene oxide	Polypropylene
Benzene	Formaldehyde	Polyvinyl chloride
Cumene	Isopropyl alcohol	Styrene
Cyclohexane	Maleic anhydride	Toluene
Ethylbenzene	Methanol	Vinyl chloride
Ethylene	Phthalic anhydride	Xylenes

The chemical unit operations, such as distillation, extraction, and various separation operations, and the chemical unit processes, such as alkylation, dehydrogenation, hydrogenation, and isomerization, are essentially identical to those operations used in the manufacture of chemicals from other sources.

In order to save materials transportation costs, a petrochemical plant frequently will be located adjacent to a petroleum refinery or gas processing plant. Short pipelines can be used in place of leasing long pipelines or having to depend upon tank car shipments by rail or truck. This also contributes to the overall safety of production. A representative petrochemical plant adjacent to a refinery is shown in Fig. 2 on p. 2687.

An excellent report of the petrochemical industry is prepared annually by the *Oil and Gas Journal* and *Hydrocarbon Processing* magazine.

PETROGENESIS.   That branch of petrology that deals with the origins of rocks. Practically a synonym for petrology unless, as is usually the practice, confined to the igneous (and possibly the metamorphic) rocks.

PETROLATUM.   A semisolid or liquid mixture of hydrocarbons derived by distillation of paraffin-base petroleum fractions. The solid form (mineral jelly) may be either water-white or pale yellow. Its chief uses are in mild ointments, cosmetics, softener in rubber mixtures and food processing (release agent in bakery products, dehydrated fruits and vegetables), defoaming agent (beet sugar, yeast). The liquid form (white mineral oil) is used as a laxative, textile lubricant and dispersing agent. There are three grades of both solid and liquid types with various specifications (USP, NF, and FCC).

PETROLEUM.   A natural oil, ranging in color through black, brown, and green, to a light amber shade. It is often termed crude oil, and consists principally of hydrocarbons, that is, compounds of carbon and hydrogen, but varying amounts of oxygen-, nitrogen-, and sulfur-bearing compounds are almost invariably present. The term *mineral oil*, which was and is often used as a synonym for petroleum, is inadequate, for most geologists believe that it was derived from organic material resulting from reactions of organic materials such as plants and animals buried in sedimentary rocks. The more important of these geologic formations in which petroleum is found are the Tertiary period of the Cenozoic era (50% of the world's oil production comes from these rocks, including regions in California and the Gulf Coast of the United States. Russia, Venezuela, Malaysia, Iran, and Iraq); the Cretaceous period of the Mesozoic era (including the East Texas, Kuwait, and Bahrein fields); the Jurassic period of the Mesozoic era (including the Arkansas and Rocky Mountain regions of the United States, and Saudi Arabia); and the Mississippian period of the Paleozoic era (including the West Texas, Pennsylvania, and Mid-Continent regions of the United States, and the Alberta, Canada, fields).

Petroleum oils vary considerably in composition, even when closely associated geographically. In some areas of the United States, for example, crude oils near the surface may have quite a different chemical composition from those found in deeper strata. Depth alone, however, does not correlate significantly with composition.

Analysis of typical crude oils found in representative areas of the United States are given in Table 1. It may be generalized that crudes found in the eastern and midwestern sections of the United States are predominantly sweet and paraffinic; those found along the Gulf Coast usually are naphthenic; those occurring in the inland southwest are sour and naphthenic; and those found along the west coast are asphaltic. The waxy, sweet paraffinic oils found in Pennsylvania first became prominent because of the high quality of lubricating oils and greases that could be made from them. The severe stresses imposed by the bearings and close-fitting reciprocating surfaces of machinery led to the development of refining processes and the discovery of additive materials whereby many other crude oils also can be transformed into excellent lubricants. Even Pennsylvania oils require special refining and additives to meet present quality specifications.

Analyses of some crude petroleums found outside the United States are given in Table 2, which illustrates the variety of crudes existent, but is not intended to give a full representation of worldwide petroleum source compositions.

API Gravity. This parameter (API stands for American Petroleum Institute), expressed in "degrees," is mathematically related to specific gravity and can be determined with a hydrometer. The specific gravity of water (arbitrarily defined as unity) is 10.00 when expressed as degrees API. API gravity usually, although not infallibly, indicates the gasoline and kerosine contents of the crude. As an example, the Mississippi, Texas, New Mexico, and Louisiana crudes have API gravities between approximately 35 and 40; as do the Arabian, Iranian, and Colombian crudes. The gasoline content (that fraction boiling below about 400°F (204°C) of these crudes ranges from about 25% to over 35% by volume. The kerosene portions of such "light" crudes also are usually high. In contrast, Wyoming sour crude with an API gravity of 17.9 contains but 6% gasoline and about

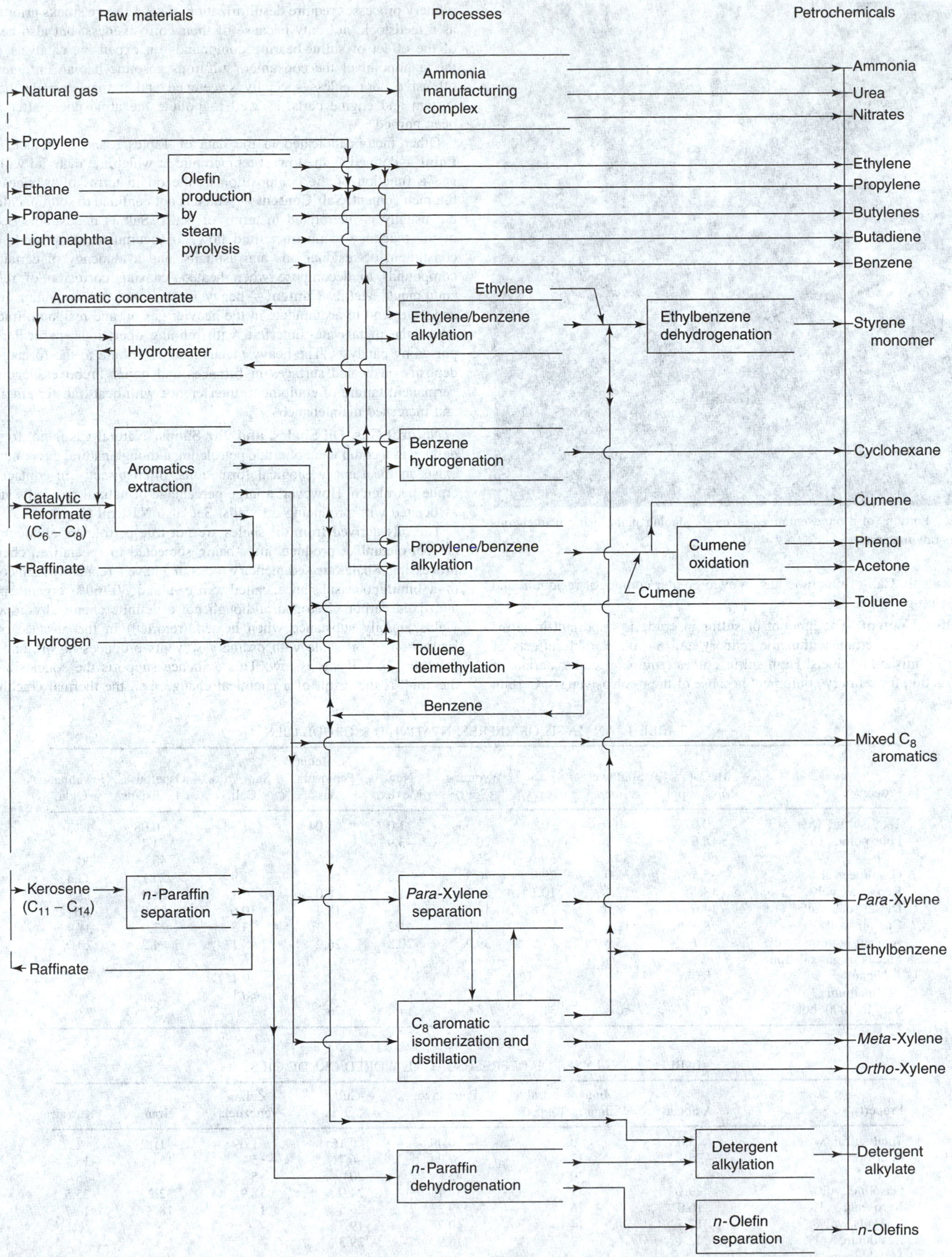

Fig. 1. Interlocking processes and flow of materials in a representative petrochemical complex. (*UOP, Inc.*)

Fig. 2.   Portions of a representative solvent-producing plant, using petrochemicals as raw materials.

40% asphalt. California crude has an even greater content of residuum and almost no gasoline.

Sulfur Content. The amount of sulfur in crude is important in terms of handling the crude within the refinery and the undesirable effects of sulfur in finished products. High-sulfur crudes require special materials of construction for refinery equipment because of their corrosiveness. Certain refinery processes require desulfurization of sour charge stocks prior to use as a feedstock, not only because of their corrosiveness, but also because of the effect of sulfur-bearing compounds on expensive catalysts. From the standpoint of the consumer, sulfurous gasoline has an unforgettably offensive odor unless specially sweetened and it may corrode the fuel system and engine parts, as well as pollute the atmosphere after it has been burned.

Other factors indicated in the data of Tables 1 and 2 include: **Pour Point**—defined as the lowest temperature at which the material will pour and a function of the composition of the oil in terms of waxiness and bitumen content; **Salt Content**—which is not confined to sodium chloride, but usually is interpreted in terms of NaCl. Salt is undesirable because of the tendency to obstruct fluid flow, to accumulate as an undesirable constituent of residual oils and asphalts, and a tendency of certain salt compounds to decompose when heated, causing corrosion of refining equipment; **Metals Content**—heavy metals, such as vanadium, nickel, and iron, tend to accumulate in the heavier gas oil and residuum fractions where the metals may interfere with refining operations, particularly by poisoning catalysts. The heavy metals also contribute to the formation of deposits on heated surfaces in furnaces and boiler fireboxes, leading to permanent failure of equipment, interference with heat-transfer efficiency, and increased maintenance.

Natural Gas, Oil Shales, and Tar Sands. Natural gas is not formally defined as a component of crude petroleum, although natural gas commonly exists in the same geological formations, often directly in contact with crude petroleum. However, a large percentage of natural gas wells are not associated with producing oil wells. See also **Natural Gas**.

The oils derived from oil shales are not true petroleum, although they are petroleumlike products after being subjected to specialized chemical processing. Shales are sedimentary rocks that have a relatively high content of a bituminous substance called *kerogen* and 30–60% organic matter and fixed carbon. Kerogen, although not a definite chemical compound, yields an oily substance when heated (retorted) in the absence of air. Extraction of oil shale with ordinary solvents produces no oil, and their solubility in solvents is low. This evidence supports the conclusion that the "oil" is the result of a chemical change, i.e., the thermal cracking or

TABLE 1. ANALYSIS OF REPRESENTATIVE U.S. CRUDE OILS

Property	McComb, Mississippi	Southwest Texas	East Texas	Wyoming (Sour)	New Mexico	N. Kenia Peninsula, Alaska	San Ardo, Calif.	Ospelousas, Louisiana	Velma, Okla.
Total sulfur, wt%	0.07	0.45	0.2	3.33	1.0	1.04	1.93	0.08	1.13
Pour point, °C	15.6	−1.1	12.8	−20	−3.9			4.4	
°F	60	30	55	−5	25			40	< −30
Gasoline, vol%	35.5	32.0	29.0	6.3	37.8	14.4	1.9	26.1	22.3
Kerosene, vol%	18.1	12.1	10.1	9.1		18.0	16.1	18.9	17.3
Diesel fuel, vol%	14.6	38.0	13.8	14.0		18.4	10.6	22.9	8.5
Gas oil, vol%	28.1	12.6		30.7	41.2	22.3	23.3	27.9	31.9
Asphalt bottoms, vol%	3.7	5.3	47.1	39.9	20.8	26.9	48.1	4.2	20.0
Metals in gas oils, ppm									
Nickel	0.06						0.15		
Vanadium	0.08						<0.1		
Salt, lb/1000 bbl	4	<0.5	31	0.6	14	76		5	78

TABLE 2. ANALYSES OF REPRESENTATIVE WORLD CRUDE OILS

Property	Arabian	Minas, Central Sumatra Topped	Putomayo, Colombia	Gulf Nigeria	Zulia, Venezuela	Iran	Kuwait
Total sulfur, wt%	3.05	0.2	0.49	0.16	1.69	1.12	2.62
Pour point, °C	−36.1	−17.8	7.2	−6.7	< −15	15	< −15
°F	−33	0	45	20	<5	5	<5
Gasoline, vol%	29.1	11	34.1	24.9	18.9	32.2	25.5
Kerosene, vol%	16.0	16	9.3	26.5	14.1	18.3	13.7
Gas oils, vol%	12.5	14	40.7	19.3			
Residuum, vol%	42.4	59	15.9	29.3			
Metals in gas oils, ppm							
Vanadium	0		25	7			
Nickel	0		11	5			
Iron	3						
Salt, lb/1000 lb	12		trace	5			

fragmenting (pyrolysis) of the molecules that make up kerogen. See also **Oil Shale**.

Tar sands is an expression commonly used in the petroleum industry to describe sandstone reservoirs impregnated with a very heavy viscous crude oil which cannot be produced through a well by conventional production techniques. Two other terms, *bituminous sands* and *oil sands*, are gaining favor. The heavy viscous petroleum substances impregnating the "tar sands" are called asphaltic oils. See also **Tar Sands**.

Petroleum processing and petroleum end-products are described in the article on **Petroleum Refining** immediately following this article.

Origin and Geology of Petroleum

Among the general theories for explaining the origin of petroleum, the most widely accepted is the *organic theory*, which can be quickly summarized. Over millions of years, rivers flowed to the seas, carrying large volumes of mud and sand to be spread out by currents and tides over the sea bottoms near the gradually changing shorelines. New deposits were distributed, layer upon layer, over the floors of the seas. Because of the increasing weight of these accumulations, the sea floors slowly sank, building up a thick series of mud and sand layers. High pressure and chemical forces ultimately converted these layers into sedimentary rocks of the type that often contain petroleum—the sandstones, shales, limestones, and dolomites. The organic theory further stipulates most importantly that tiny marine organisms were buried with the silt. In an airless environment and under high pressures and elevated temperatures, these carbon- and hydrogen-containing minuscule life-forms were converted over an extremely long time span into hydrocarbons. This theory, of course, requires acceptance of the concept of drastically altered shorelines, because obviously oil deposits are found in many parts of the world long distances from the present coastlines.[1]

Geologists find it particularly difficult to trace the history of a given hydrocarbon deposit, because the oil and gas may have moved as the result of numerous seismic events, again occurring over a very long time span. A past requisite for commercially exploitable hydrocarbon deposits was prior movement and concentration of large quantities of hydrocarbons in various forms of *traps*. In contrast with oil shales and tar sands, natural gas and petroleum flow relatively easily in permeable underground structures and, consequently, tend to concentrate, greatly assisting the economic exploitation of these materials.

The movement of petroleum from the place of its origin to the traps where accumulations are found is believed to have occurred in an upward direction. This movement took place as the result of the tendency for oil and gas to rise through the ancient seawater with which the pore spaces of the sedimentary formations were filled when originally laid down. An underground porous formation or series of rocks which occur in some

[1] In a more recent, alternate theory, Thomas Gold (Cornell University) suggests that, in contrast with the organic sediments theory, the prime source of natural gas is primordial, abiotic methane rising from deep within the earth's mantle. This is discussed in further detail in the article on **Natural Gas**.

shape favorable to the trapping of oil and gas must also be covered or adjoined by a layer or rock that provides a covering or seal for the trap. A seal of this type, frequently called a *cap rock*, stops further upward movement of petroleum through the pore spaces.

As oil and gas gathered in the upper part of a trap, because of differences in weight of gas, oil, and salt water, these fluids also separated vertically, much in the same manner as if these materials were all present in a bottle. Thus gas, if any is present, is found in the highest parts of the trap, followed by oil (and oil with gas) below the gas, and finally salt water below the oil. Experience has indicated that the salt water seldom was completely displaced by oil or gas from the pore spaces, even within the trap. Even in the midst of oil and gas accumulation, pore spaces within the trap may contain from 10 to 50% or more of salt water. It appears that the remaining water (termed *connate water*) fills the smaller pores and also exists as a coating or film, covering the rock surfaces of the larger pore spaces; thus oil and/or gas are apparently contained in water-jacketed pore spaces. The geological structures called traps are petroleum reservoirs, i.e., they are the oil and gas fields that are explored and produced. All oil fields contain some gas, but the quantity may range widely. See also *Natural Gas*.

Types of Oil Accumulations

A composite diagram showing different types of oil and gas accumulations is given in Fig. 1.

Structural Traps. The attitude of the rocks, whether they are folded, fractured, displaced, or otherwise disturbed, is called their geologic structure. Traps that are due to geologic structure are known as *structural traps*.

A common structural trap, the *anticline*, is an upward bulge in the rock layers which forms an arch capable of holding oil under its apex. The buoyancy of oil and gas carries them upward through porous rock layers into the apex until they are trapped by an impermeable layer. Anticlinal type of folded structure is shown in Fig. 2. Reservoirs formed by folding of the rock layers or strata usually have the shapes shown in Fig. 2(a) and (b). These traps were filled by upward migration of oil and/or gas through the porous strata or beds to the location of the trap. Further movement was arrested by a combination of the forms of the structure and the seal or cap rock provided by the formation covering the structure.

Examples of domal structures are the Conroe Oil Field in Montgomery County, Texas and the Old Ocean Gas Field in Brazoria County, Texas. Another example of a reservoir formed by an anticlinal structure is the Ventura Oil Field in California.

Another type of structural trap is the *fault trap*. A fault is a fracture in the earth's crust along which movement has occurred such that a porous rock layer is offset by a nonporous layer. The oil moving along a porous stratum is dammed or blocked by an impermeable shale or limestone. See Fig. 3. Examples of fields of this type occur along the Mexia fault zone of East-Central Texas.

The *salt dome* is another interesting form of structural trap. This type of trap is found along the Gulf Coasts of Texas and Louisiana and in western Colorado and Utah. This type of structure resulted from the upward thrust

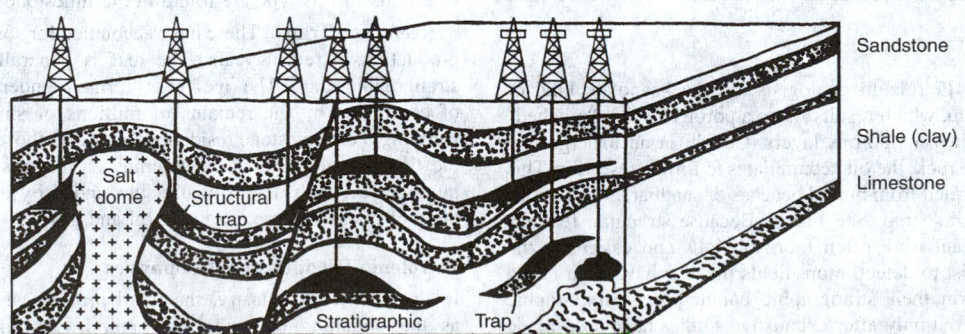

Fig. 1. Composite diagram showing different types of oil and gas accumulations (shown in solid black). *Structural traps* where petroleum deposits may have accumulated are often found along the edges of salt domes or along fault lines. *Stratigraphic traps* may exist where reservoir rock is "pinched off" by denser strata. These accumulations are the "pools" or "reservoirs" which, singly or in groups, compose an oil or gas "field." The pores of the reservoir rock contain oil or gas or both, always accompanied by briny water. The fluids tend to be layered, with gas at the top of the trap, oil in the middle, and water underneath. Most of the petroleum which ever existed has been obliterated, either by attenuation in the earth's crust, or by exposure to heat and pressure high enough to break down its chemical bonds. The accumulations that do exist have endured against long odds. Additional detail on oil and gas reservoirs will be found in Figs. 2 through 7.

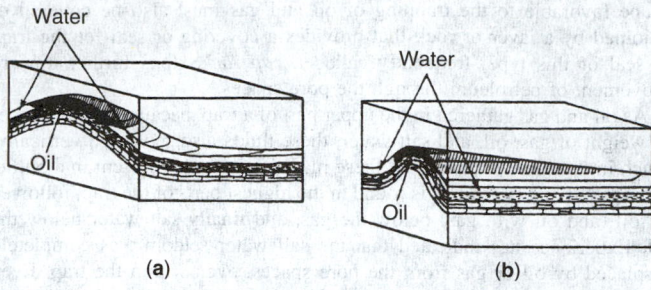

Fig. 2.   Examples of anticline structural traps: (**a**) Oil accumulates in a dome-shaped structure. The dome is circular in outline. (**b**) Anticlinal trap that is long and narrow, differing from the dome configuration.

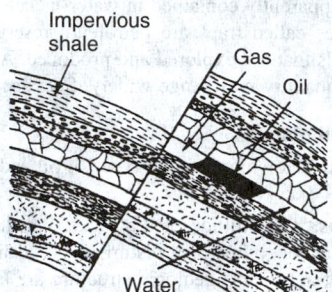

Fig. 3.   Example of a fault structural trap. The oil is confined in traps like this because of the tilt of the rock layers and faulting.

of a great mass of salt far below the earth's surface. When a salt dome rose through a layer of oil-bearing sedimentary rock, oil may have been trapped in anticlines above the dome, or in structures similar to faults along its flanks. One famous example of a salt-dome reservoir is the Spindletop field, near Beaumont, Texas. It was "brought in" in 1901 by a 100,000-barrel-a-day gusher, giving birth to the modern petroleum industry. Another example of a salt-dome field is the Sugarland Oil Field in Fort Bend County, Texas. See Fig. 4.

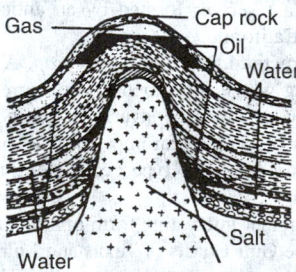

Fig. 4.   Example of a salt-dome structural trap. One of the earliest and greatest reservoirs of this type was Spindletop, brought into production in 1901 and located near Beaumont, Texas.

Stratigraphic Traps. Petroleum geologists also seek another kind of trap, the *stratigraphic trap*, which results when a porous layer is "pinched" or phased out between two nonporous layers. Caught in an underground envelope of impermeable rock, the oil accumulates to form a reservoir. This type of trap may have formed from buried beaches or sandbars. The famous East Texas field is a "strat" trap. See Fig. 5. Because structural features such as anticlines and faults are often more obvious and easier for the geologist and geophysicist to detect, more fields thus far have been found in structural traps than in their stratigraphic counterparts. Stratigraphic traps are usually discovered only after exhaustive studies have been made of rock samples from outcrops and from core samples from wells drilled over large areas.

The serpentine plug, shown in Fig. 6, is an interesting type of trap, an example of which is the Hilbig Field in Bastrop County, Texas. As illustrated, a porous serpentine plug has formed a reservoir within itself by intruding into nonporous surrounding formations.

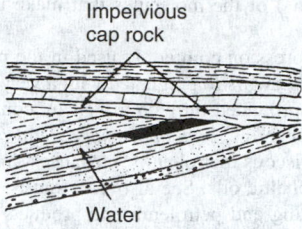

Fig. 5.   Example of a stratigraphic trap. This unconformity represents the condition where upward movement of oil has been halted by the impermeable cap rock laid down across the cutoff (possibly by water or wind erosion) surfaces of the lower beds. This type of reservoir is found in the great East Texas field.

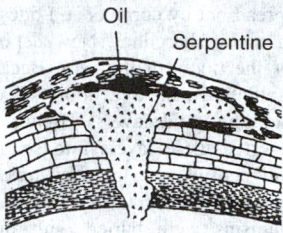

Fig. 6.   Example of a serpentine plug as found in the Hilbig Field in Bastrop County, Texas.

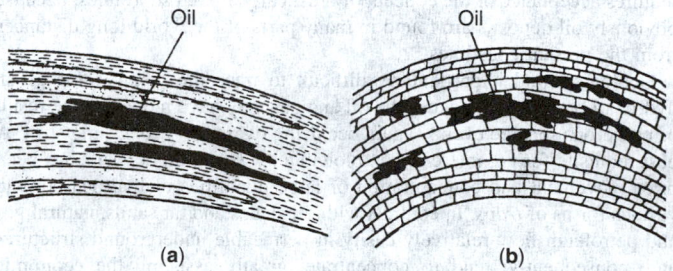

Fig. 7.   Lens-type traps: (**a**) sandstone; (**b**) limestone.

Lens-Type Traps. These form in limestone and sand. In this type of trap the reservoir is sealed in its upper regions by abrupt changes in the amount of connected pore space within a formation. A trap formed in sand is shown in Fig. 7(a). An example is the Burbank Field in Osage County, Oklahoma. This type of trap may occur in sandstones where irregular deposition of sand and shale occurred at the time the formation was laid down. In these cases, oil is confined within the porous parts of the rock by the nonporous parts of rock surrounding it. A lens-type trap formed in limestone is shown in Fig. 7(b). In limestone formations there are frequent areas of high porosity with a tendency to form traps. Examples of limestone reservoirs of this type are found in the limestone fields of West Texas.

Reef-Type Traps. These have accounted for some of the most important production in recent years. The reef is generally considered a type of stratigraphic trap. The reef was formed under the right combination of conditions by the remains of millions of small underwater animals. Building their limestone residences on top of those built by their ancestors, the tiny creatures produced columns or mounds, often reaching several hundred feet high. If eventually surrounded by impermeable rock layers, the reef could become a trap for oil and gas.

Petroleum Production — Geophysics

In the petroleum industry, the word *production* is generally interpreted as the obtaining of crude oil, natural gas, and other related natural hydrocarbons, whether located in reservoirs under the land or under the oceans. Production generally does not relate to the manufacture of final end-products, such as naphtha, gasoline, diesel fuel, lubricating oils, etc. Products obtained by the processing of natural hydrocarbons are commonly referred to as *refined*. In a sense, then, petroleum production is analogous to other natural raw material mining, such as coal and metals mining.

The fundamental phases of petroleum production include: (1) the initial *exploration* required to find heretofore undiscovered oil and gas reservoirs; (2) *primary* and *secondary recovery* methods, which make use of both naturally occurring (or *primary*) reservoir energy and the application of *secondary* energy sources, such as the injection of gas or water; and (3) *enhanced oil recovery* used to increase ultimate oil production beyond that achievable with primary and secondary methods. Enhanced oil recovery (EOR) methods increase the proportion of the reservoir by improving the "sweep" efficiency, reducing the amount of residual oil in the swept zones (increasing the displacement efficiency), and reducing the viscosity of thick oils.

Petroleum Exploration

The exploration required to identify previously undiscovered oil and gas reservoirs is grossly affected by the economics and politics of the worldwide oil markets (supply and demand). The experience of one year cannot easily be extrapolated to that of subsequent years.

Although the petroleum geologist has for several years employed highly sophisticated instrumentation and computer technology, the fact remains that the geologist, in many respects, is still a *detective*, often commencing with the barest of clues and following the search hopefully to a successful conclusion. Decisions based on uncertain information are normal in petroleum exploration. How to capitalize on the creativity of the geological and geophysical scientists further complicates decision making.

Modern petroleum exploration geophysics is based on three important earth properties: (1) density, (2) magnetization; and (3) acoustic response.

Gravity Surveys. Gravitational pull is an effective way to measure the density of the strata lying far below the earth's surface. The *gravimeter* came into oil exploration use in about 1900. If there are dense rocks near the surface, a spring-loaded weight in the instrument will weigh more — obviously since denser rocks exert greater gravitational pull than those which are less dense. Usually, the denser rocks are older and their presence near or far from the surface indicates an underlying shallow or deep basement. Thus, a structural "high," such as an anticline, will often appear as a high reading on the gravimeter. Conversely, a structural "low," such as a syncline, will yield opposite effects. The low density of a salt dome will be reflected on the gravimeter as having a lesser gravity value. When using a gravity meter offshore, it is necessary to place the instrument so that it remains level, regardless of the movement of the vessel. The rise and fall of the ship and other effects of its travel must be measured by other equipment and taken into account when the gravity data are processed.

Magnetic Surveys. The natural magnetic properties of rocks can be useful in the early stages of petroleum exploration. More than three centuries ago, it was found that a freely suspended magnetic needle will move away from a level position in response to the presence of nearby iron. In 1879, this principle was applied in the first *magnetometer*, a simple device combining a magnetized needle with a compass. Both types and depths of rocks far underground affect how much and in which direction the needle will move.

Using a magnetometer, earth scientists can measure the depths at which basement rocks lie. These rocks are usually from the Precambrian period, dating back over 600 million years. They contain large amounts of magnetite, a naturally occurring iron oxide having magnetic properties. Where petroleum-bearing sedimentary rocks overlie these basement rock formations, the magnetometer can be used to indicate variations in the thickness of these sedimentary layers. This information can provide the geologist with a valuable clue as to the probability that those sediments, bent by the underlying basement rocks, may provide the right conditions for the accumulation of recoverable quantities of oil or natural gas.

It is not always necessary to place instruments on the ground to measure magnetic attraction of the subterranean rock layers. Magnetometers can also be used from aircraft flying high above the surface to determine quickly areas of general interest. Use of airborne magnetometers also eliminates the need for and the problems sometimes associated with personnel actually getting on the site.

Seismic Surveys. Gravity and magnetic measurements are considered preliminary. The principal detailing tool used is the seismograph.[2]
See also **Earth Tectonics and Earthquakes**.

The type of experiment used to make a seismic survey in petroleum exploration will be determined by the type of information required, the types of rock strata anticipated, environmental considerations, the type of terrain, and, of course, the economic costs involved.

In petroleum exploration, reflection seismic surveying is the method predominantly used. The seismograph records variations in the way rocks reflect sound waves sent downward from a surface source. The reflected sound waves vary with the type, depth, density, and dip of the rocks encountered. The returning sound waves made from a series of points along the survey path can then be displayed graphically to form a seismic record for interpretation by earth scientists. These principles are diagrammed and explained in Fig. 8.

Four sources of energy may be used to generate the sound waves: (1) controlled explosives; (2) vibrators; (3) weight dropping (where heavy weights are dropped to create the waves); and (4) compressed gases (producing bursts of energy using compressed air or propane). In exploring onshore sources, controlled explosives and vibrators are most often used. In offshore work, compressed air guns are most frequently used. These sources have been shown to cause virtually no damage to marine life.

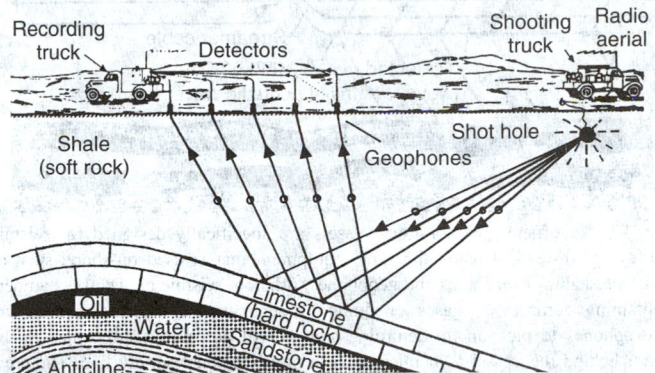

Fig. 8. Onshore seismic field operation, using the shot hole energy method. Energy from a controlled explosion is directed toward underlying rock structures and is reflected to indicate shape of formation below — in this case an anticline. Geophones placed on the ground surface by the surveying crew measure and record the reflected acoustic energy. With the survey truck safely away from the shot hole, the explosives are detonated by an assigned radio frequency from the truck.

When *controlled explosives* are used, the most common method is to place carefully measured charges in shallow "shot holes" a few inches in diameter drilled from a truck-mounted drill. These charges are then detonated to produce the sound waves needed. In less accessible areas, a portable drill may be used and, in certain environmentally sensitive areas, the charges may be mounted on stakes above the ground to minimize plant disturbance.

Truck-mounted vibrators are frequently used. Metal plates, used singly or in groups, are pressed against the ground and vibrated briefly creating a frequency sweep similar to that used in radar, to send a brief acoustic signal into the ground. The vibrators cause little disturbance to the surrounding area and are often used in sensitive locations, in cities and along highways, where dynamite charges would be less acceptable. See Fig. 9.

The principles of seismic surveying are the same regardless of the type of equipment used. As the waves strike the various strata, some are reflected back to the surface, where they are picked up by sensing devices called *geophones* (on land) and *hydrophones* (offshore). See Fig. 10. These

---

[2] The first "seismograph" is recorded to have been an inverted bronze urn with a pendulum inside and used by the Chinese in about 160 A.D. to announce earthquakes. Seismographs were "reinvented," so to speak, during World War I to locate heavy German artillery. In 1920, seismography for use in petroleum exploration was first demonstrated.

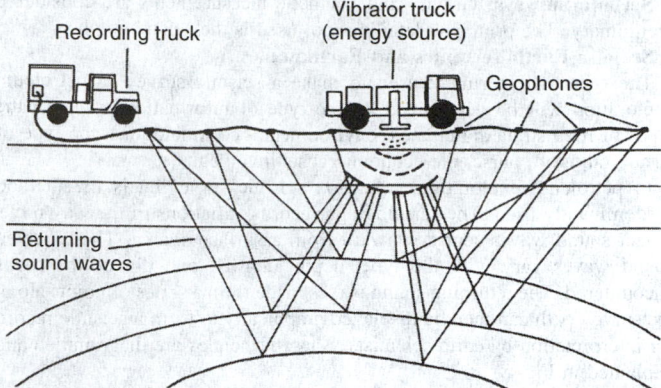

Fig. 9. Seismic survey using a truck-mounted vibrator as the energy source for creating reflected acoustic sounds. This method is used in sensitive areas near cities and along highways because disturbance to the environment is minimal. (*American Petroleum Institute.*)

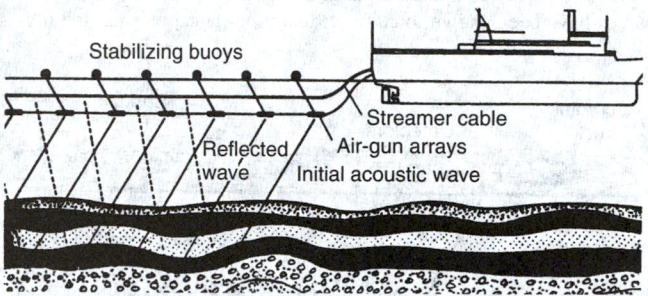

Fig. 10. In offshore exploration, vessels are specifically designed for seismic surveying. Instead of using the same equipment that is used on-shore, such as "surface shaking" machines and geophones, offshore seismic crews use chambers containing compressed gases or fluids to generate the acoustic signals and hydrophones to pick up the returning sounds. Air-gun arrays are trailed in the water behind the vessel as it plies along a predetermined survey line. The crew activates the chambers at set intervals from onboard controls connected by cables to the air guns. Other cables contain arrays of hydrophones to detect sounds that echo off the underlying strata. The vessels are kept on course through the use of radar, loran, and satellite navigation equipment. (*After Exxon.*)

sensors convert acoustic information into electrical signals, whence the data are recorded on magnetic tape for processing at a computer center.

Seismic Data Processing. Continuing advances in technology, such as fiber optics, have enabled the capabilities of seismic surveys to be markedly increased so that, where required, a 3-dimensional picture of the rocks in the subsurface can be obtained. Advanced processing has also enabled discrete anomalies in seismic data from individual rock horizons to be analyzed. Under certain conditions, the presence of hydrocarbons can be directly detected. This analysis, sometimes referred to as "bright spot" technology, has been responsible for numerous discoveries in the Gulf of Mexico in recent years.

Raw data gathered from seismic surveys must be processed to compensate for and to remove a variety of distortions—unwanted noises created by weathered near-surface rocks, normal time delays, and echoing by rebounding acoustic waves—to provide the clearest possible image of the strata below. Computers can restore these distortions in a fraction of the time that was formerly required to adjust the data painstakingly by hand. Advanced techniques not only permit presentations in three dimensions, but also in color, and to create contour maps and models of subterranean features. However, even with the use of sophisticated tools, there remains a large measure of uncertainty. History has shown repeatedly that a prospective area rejected by one petroleum firm has been accepted by another and proved to be successful.

After thorough analysis of seismic and other exploration data, the next steps are management decision making and approval to proceed with exploratory drilling. The ultimate exploratory tool is the *drill bit*.

Exploratory Drilling

Onshore Drilling. When Col. Edwin Drake brought in the first commercial oil well in 1859, he struck oil at a depth of 59 feet, 8 inches (18.2 meters). See Fig. 11. Most early wells were less than 400 feet (122 meters) deep. Shallow oil and gas wells were fully exploited many years ago. Deep producing wells today often exceed depths of 25,000 feet (7,620 meters) and dry holes have been drilled to a depth in excess of 31,000 feet (9,449 meters). In an average year, wildcat wells reach a depth of about 6,000 feet (1,829 meters). Depth, however, is only one of the factors that makes the search for petroleum difficult in modern times. Increasingly, drilling must be done in remote places where it is costly to bring in materials and labor. For example, onshore locations, such as those found on the North Slope of Alaska, can result in drilling costs that are 10 times as high as they would be in the lower 48 states.

Fig. 11. Colonel Edwin Drake (right) and Peter Wilson, a druggist who endorsed a $500 bank loan for Drake, confer in front of the world's first commercial oil well near Titusville, Pennsylvania. Initially, Drake rigged a large wheel powered by steam to raise and lower a cable and iron bit. Later connected to a crude drill pipe and pump, this well produced about 35 barrels a day. (*ca. 1861.*)

The Rotary Drill. This is the most commonly used method and consists of a rotary drill, a power source, a derrick and lifting and lowering devices, and a bit attached to a length ("string") of tubular high-tensile-strength steel. See Fig. 12. The drill string passes through a rotary table that turns it and thus provides the torque needed for the drilling operations. The weight applied to the formation is also a critical factor.

During the drilling operation, a special *drilling mud* (mixture of clay, water, and chemical additives) is pumped down through the hollow drill string and bit into the borehole. The fluid is forced up the borehole and through the area between the drill string and the casing (the "annulus") to the surface. There it is cleaned and recirculated into the well. The fluid helps to cool and lubricate the bit, control the pressures within the well, provide a protective and stabilizing coating to some permeable formations, and brings the rock cuttings up the borehole to the surface. The consistency of the fluid is carefully monitored and adjusted to compensate for pressure changes within the well, as the bit penetrates the various rock strata.

"Spudding" is the actual start of drilling a well and is akin to the first shovel of dirt at groundbreaking. A large bit, frequently from 18 to 38 inches (46–97 cm) in diameter, is used to drill a hole to a depth of from 10 to 100 feet (3–30 meters). The hole is then lined with a conductor pipe ("casing"). The space between the casing and the drilled hole (the "borehole") is filled with cement.

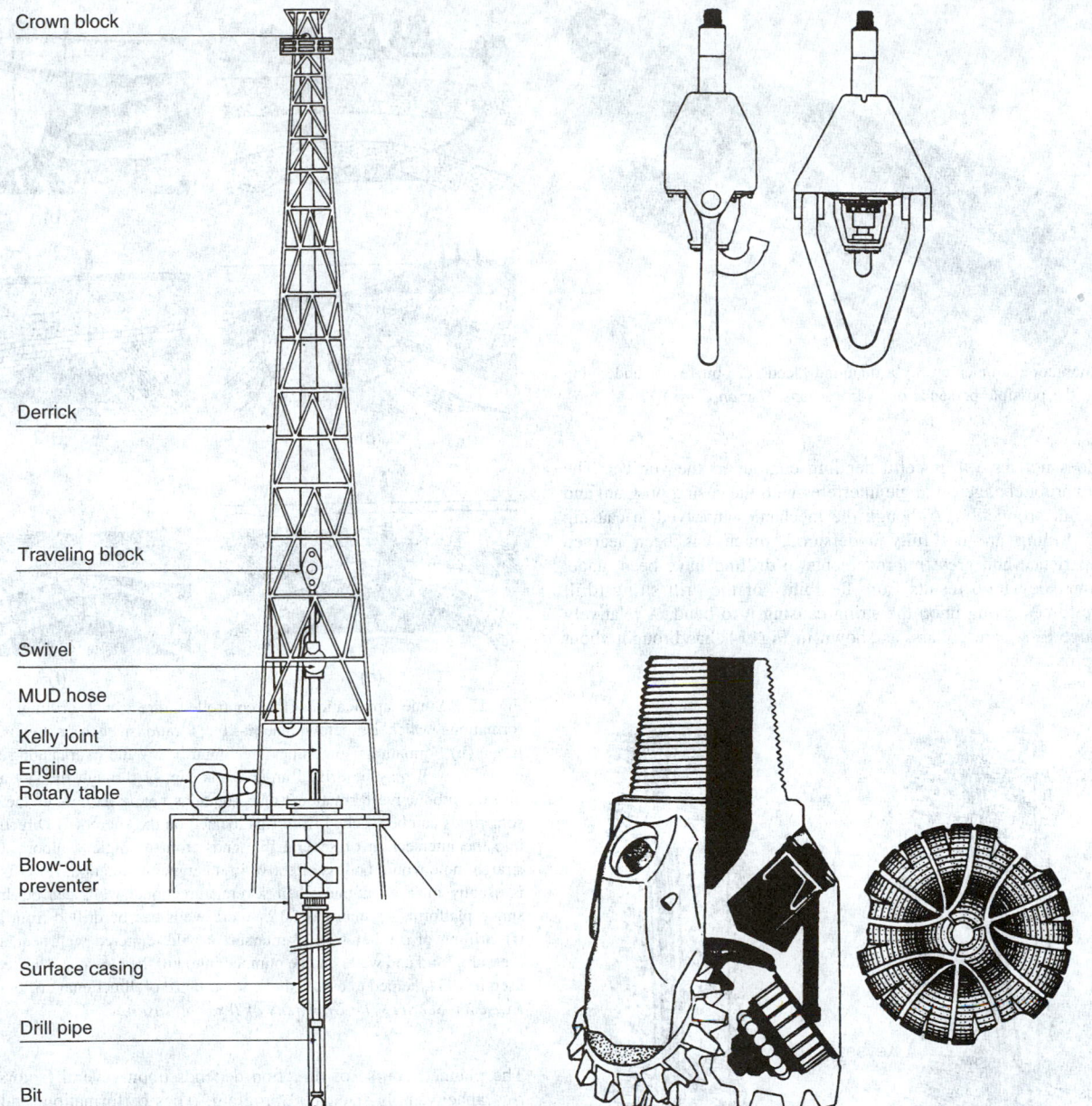

- Crown block
- Derrick
- Traveling block
- Swivel
- MUD hose
- Kelly joint
- Engine
- Rotary table
- Blow-out preventer
- Surface casing
- Drill pipe
- Bit

Fig. 12.   A rotary rig has four systems. The *rotary system* consists of a turntable, a swivel, a square or hexagonal pipe length called a "kelly," which transmits rotary motion from the turntable to the drill pipe, and the drill "string" itself. A *circulating system* of pumps, hoses and other apparatus keeps mud circulating through the well. The *hoisting system* includes the derrick, a drawworks, hoisting blocks, and other equipment needed to lift and lower heavy pipe joints and casing. The *power system* usually consists of diesel engines and generators, set apart from the rig, which provide power for the electric motors that drive the rotary, hoisting, and pumping equipment. The elevated floor allows installation of blowout preventer beneath the platform.

Shown in the upper right of this view is the *swivel* (front and side view) which permits the drill pipe to rotate while mud is pumped down to clean the hole. Shown in the lower right is a three-cone drill bit, with cutaway showing the bearings on which it rotates; and at the extreme lower right, the face of a diamond bit revealing openings through which fluids may pass. (*American Petroleum Institute; Exxon Corp.*)

Drilling is then resumed using a smaller bit and, after the borehole reaches several hundred feet (meters), the bit and drill string are hoisted out of the well and another length of pipe ("surface casing") is lowered into the borehold and cemented in place. Besides preventing the generally unconsolidated surface formation from sloughing into the hole, the casing also protects the freshwater strata ("aquifers") from being contaminated by the drilling mud.

As the drilling proceeds, additional casings of concentrically smaller diameter are lowered into the well and sealed in place until the final depth ("target zone") is reached. During the drilling, the drill bit and string must be removed from the well whenever the bit becomes dull and requires changing or cores are taken from the well. The coring process involves a special cylindrical rock bit, generally with a diamond-encrusted face and a cylinder ("core barrel") into which the core passes and is retained for

recovery at the surface. These cores are analyzed to determine the type of rocks penetrated and their porosity, permeability, chemical analysis and possible hydrocarbon content. See Fig. 13.

Drilling Geometry. Deviation surveying was introduced into oil-well drilling technology in 1929. Before that time, it was generally assumed that a hole properly started as a vertical hole would remain essentially vertical. In many instances, this was not a realistic assumption because many "vertical" holes were found to be quite crooked. Crooked holes not only caused operational problems, but also resulted in false indications of depth. Since the early 1930s, drilling contracts usually have specified a maximum deviation of 3 to 5 degrees. The problem of drilling a straight hole usually is simpler with uniform materials, such as limestone, and more difficult when laminar formations of sandstone and shale are encountered. Often of even greater concern than a crooked hole is an irregular, "jagged"

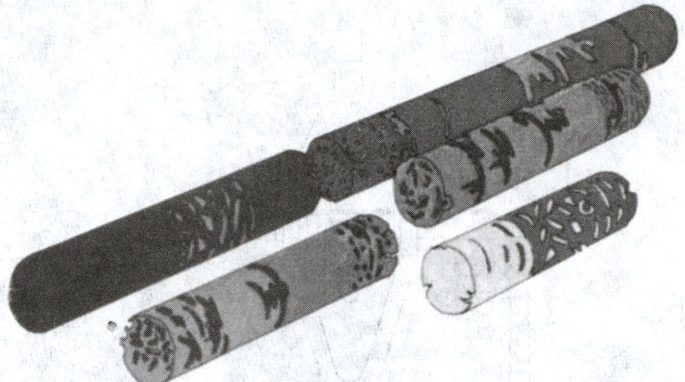

Fig. 13. Rock core samples cut by a diamond-faced core bit reveal underlying structure and the possible presence of hydrocarbons. (*Exxon Corp.*)

hole that does not have a graceful bending contour in the vertical. The presence of abrupt changes in angle interferes with the casing program and ultimately with production. Although the mechanics involved in causing nonvertical drilling are not fully understood, much has been learned through experience and great improvements in drilling have been made. For one thing, deviation results from flexibility of the drill string (drill collars), the forces acting upon the string causing it to bend. A relatively simple change to square collars, as shown in Fig. 14, has brought about marked improvement.

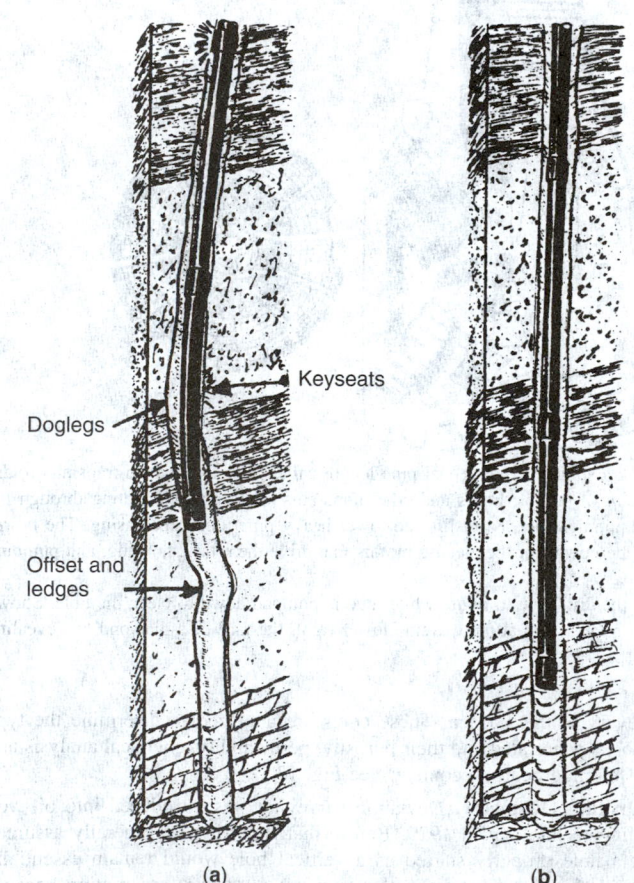

(a)   (b)

Fig. 14. Example of a crooked hole (**a**) drilled without a square collar, and a relatively straight hole (**b**) drilled with a square collar. (*Drilco.*)

Of course, it is frequently desirable to utilize a controlled directional drilling technique. There are several reasons, as indicated by Fig. 15. The three most commonly patterned directional holes are shown in Fig. 16.

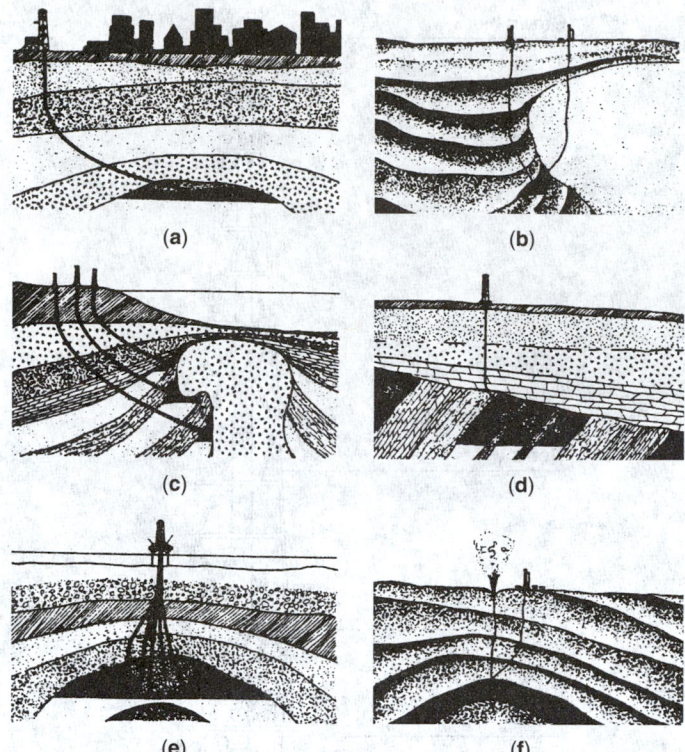

(a)   (b)

(c)   (d)

(e)   (f)

Fig. 15. Some applications for controlled directional drilling: (**a**) Reaching formations which lie below inaccessible locations, such as towns, rivers, and lakes. (**b**) Formations sometimes are found below the overhanging cap of a salt dome. A well may be drilled around this cap, or through the salt and deflected into the productive formation. (**c**) Formations below harbors or the ocean floors sometimes can be reached from rigs located on the shore. (**d**) Directional drilling into the intersection of several oil sands from a single wellbore. Obviously, a straight hole would be less effective in this type of situation. (**e**) Offshore drilling is usually most economic when several directional wells can be drilled from a single platform. As many as 20 or more wells can be drilled from a small area. (**f**) Drilling of a relief well to intersect a wild, cratered well near the source of pressure. Mud and water can be pumped in to kill the blowout. This technique, first used in 1934, helped to establish the importance of directional drilling. (*Petroleum Extension Service, The University of Texas at Austin.*)

The planned course of direction depends upon several factors, including rig capacity, hole size, mud program, types of formation, and the casing program. Meticulous surveying is required to achieve the desired results. Several types of drilling tools may be required.

Offshore Drilling. When exploration moves offshore, standard drilling equipment obviously must be supplemented by some sort of structure that provides a stable platform for operations. The structure also must be movable, given the odds against a single wildcat finding commercial quantities of petroleum.

The first offshore exploration in the 1930s in the Gulf of Mexico was conducted from rigs on barges which could be towed to drilling sites and submerged to rest on the bottom during operations. These were forerunners of the twin-hulled submersible rig, which has an upper hull housing crew quarters and working spaces and a lower hull providing the buoyancy needed to move the unit. See Fig. 17(**a**) on p. 2695.

The jack-up or self-elevated rig, introduced in the 1950s, is a barge with movable legs, which can be lowered to the sea floor and the barge jacked into drilling position above the water. Jack-ups are used in water depths up to about 300 feet (91 meters). See Fig. 17(**b**) on p. 2695.

In deeper waters, exploration is conducted from floating rigs, including submersibles and drill ships. Drill ships with conventionally shaped hulls of seagoing vessels are not so stable as semisubmersibles (semis) in rough waters, but can be moved from location to location much faster.

A critical requirement of all floating rigs is the ability to maintain position over the wellhead while drilling proceeds. Semis and drill ships

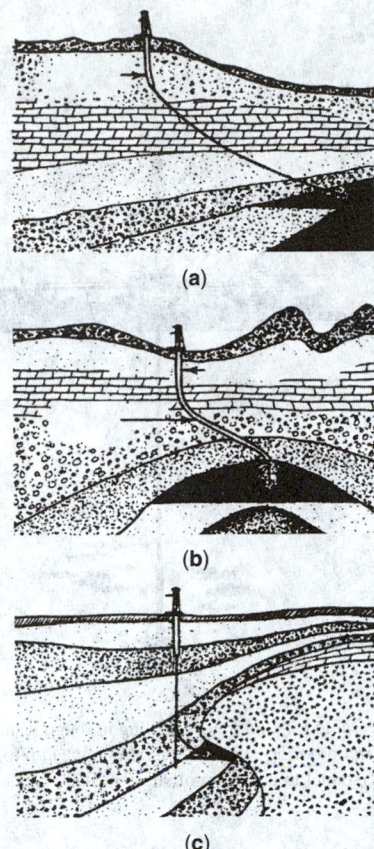

(a)

(b)

(c)

Fig. 16. Principal types of directional drilling patterns: (**a**) The most widely used directional drilling pattern is one in which the initial deflection is obtained at relatively shallow depth. Then, surface casing is set and cemented through the deviated section of hole. From that point, the angle is maintained as a straight line to the target zone. (**b**) This pattern is also initially deflected at shallow depth. Surface casing is set and cemented. Drilling continues on a straight line to a point where the hole is gradually returned to vertical. After intermediate casing is set, drilling is continued to final depth. This type hole is used when undesirable formations must be penetrated and isolated with an intermediate casing string. (**c**) In this pattern, deflection commences at a greater depth. Drift angle is maintained on a straight line to the target. This type hole may be used for exploratory drilling from a dry hole. Normally, the deflected part of this hole is not protected by casing during drilling operations. (*Petroleum Extension Service, The University of Texas at Austin.*)

use either multiple anchors or "dynamic positioning" systems to keep on station. A dynamic positioning system uses thruster engines which, responding to signals from acoustic beacons on the sea floor, automatically make the adjustments required to maintain the rig in position. Hydraulic devices keep a constant tension on the drill string to prevent the up-and-down motion of the sea from being transmitted to the drill bit.

Semis and drill ships find limited use in arctic waters where ice covers the sea most of the year.

In offshore operations, exploration wells are almost always plugged and abandoned even when they strike petroleum. Their sole function is to find oil or gas and to delineate the reservoir. The operator uses this information to pick a location for a permanent production platform from which development wells will be drilled to recover as much petroleum as economically possible. In onshore operations, however, successful exploration wells also become producers.

Measuring Well Characteristics

At selected intervals during the drilling, generally before the casing is run or when formations with hydrocarbon indications are encountered, measurements may be taken of the characteristics of the borehole and surrounding strata. Wire line logging tools are used.

In the early days of the industry, little was known about "downhole" geophysics, that is, the physical characteristics of the subsurface strata, how they might be measured, and what could be learned from such measurement. Since the Drake well, more than 3 million wells have been drilled in the United States. An estimated 27,000 fields have been found. With each new discovery, additional data become available and patterns begin to emerge. It was not until the late 1920s that technological changes were introduced that would have a lasting impact on "logging" (recording) the characteristics within the well during exploratory drilling. The first well-logging device (electrical resistivity log), invented by Conrad Schlumberger (France), was introduced into the United States. In 1934, a second development, the *spontaneous potential* (SP) *curve*, was introduced.

The *resistivity log* was lowered by cable into the borehole, with the drill bit and drill stem removed, thus enabling the recording of the electrical resistivity of the rock layers that the bit had penetrated. The record helped to identify the hydrocarbon content of the reservoir rock (since both oil and gas have different resistivities than does salt water). It was also used to correlate rock horizons between wells, which proved to be an invaluable tool in subsurface mapping.

The *SP Curve* recorded the differences in natural electrical potential between the fluids in the adjacent formations and those within the uncased borehole. This curve was soon accepted as an indicator of the porosity of the rock strata and as a means of locating the boundaries of rock beds. See Fig. 18.

Since then, a number of downhole measurements have been devised that use radioactive, acoustic, and electrical methods.

Presently, televiewers are used to look at rock features in boreholes and computers are programmed to compare, synthesize, and integrate the new range of measurements, thus providing more reliable information and definition of rock properties and formation fluids thousands of feet (meters) below the surface.

Until recently, these measurements could be taken only when the drill bit and string were removed from the borehole. However, in the mid-1980s, a new dimension was added, namely, *measurement while drilling* (MWD) instruments. These devices are mounted above the drill bit and around the drill string to provide a continuing source of data on downhole characteristics. This advancement reduces the drilling downtime previously required when measurements were taken.

Testing. Modern wire line logs will indicate with a good degree of accuracy the potential of a hydrocarbon zone. If the zone is sufficiently promising to warrant further study, a formation test will be undertaken.

Generally, a drill stem test is carried out—either in the open hole or after the hole has been cased. However, the case hole test is the most reliable.

Basically, the drill stem test involves attaching a tubing assembly to the end of the drill pipe, isolating the test zone with rubber packers, and perforating the zone. The tool is then opened so that the fluids or gas in the formation can flow up the drill pipe for metering at the surface. During this process, extensive pressure measurements are taken, which can help to indicate the extent of the reservoir and the rate at which the hydrocarbons could be recovered. Prior to describing how a well is finally completed (if the hole is not dry!), it is in order to describe the forces utilized to transfer the oil from the reservoir to the surface.

Well Drive Systems

It is convenient to classify oil and gas reservoirs in terms of the type of natural energy and forces available to produce the oil and gas. At the time oil was forming and accumulating in reservoirs, pressure and energy in the gas and salt water associated with the oil were also being stored, which would later be available to assist in producing the oil and gas from the underground reservoir to the surface. Obviously, since oil cannot lift itself from reservoirs to the surface, it is largely the energy in the gas or the salt water (or both), occurring under high pressures with the oil, that furnishes the force to drive or displace the oil through and from the pores of the reservoir into the wells.

In nearly all cases, oil in an underground reservoir has dissolved in it varying quantities of gas that emerges and expands as the pressure in the reservoir is reduced. As the gas escapes from the oil and expands, it drives oil through the reservoir toward the wells and assists in lifting it

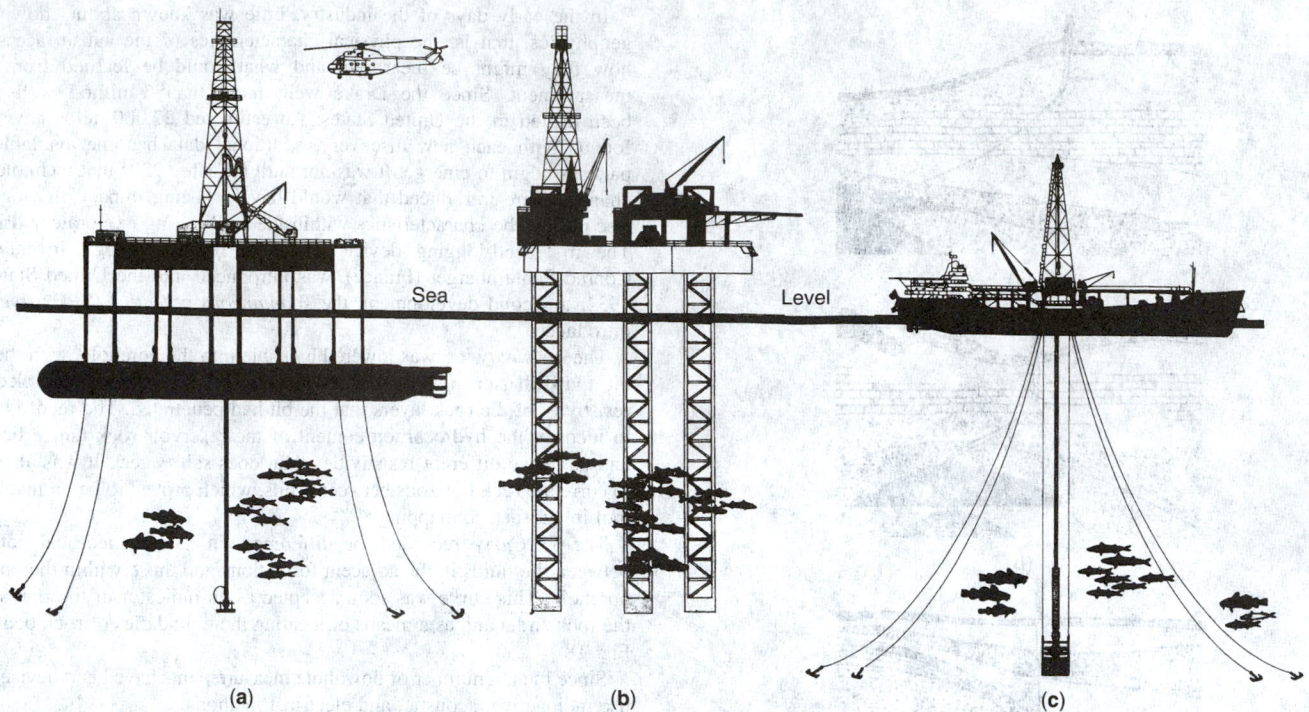

(a)   **(b)**   **(c)**

Fig. 17.   Offshore drilling schemes: (**a**) Big, pontoon-mounted "semisubmersible" rig that is indispensable for exploratory drilling in rough water; (**b**) self-elevating or *jack-up* drilling rig widely used in water depth of less than 300 feet (91 meters); (**c**) turret mooring allows a drill ship to head into prevailing winds and currents while positioned over the well. Helicopters and boats are required to transport personnel to and from offshore well sites. (*Exxon Corp.*)

to the surface. Reservoirs in which the oil is produced by dissolved gas escaping and expanding from within the oil are called *dissolved-gas-drive reservoirs*. See Fig. 19.

Often more gas exists with the oil in a reservoir than the oil can hold dissolved in it under the existing conditions of pressure and temperature in the reservoir. This extra gas, being lighter than the oil, occurs in the form of a cap of gas over the oil. This condition was previously illustrated by Figs. 3 and 4. See also Fig. 20. Such a gas cap is an important additional source of energy because, as production of oil and gas proceeds and as the reservoir pressure is lowered, the gas cap expands to help fill the pore spaces formerly occupied by the oil. Where conditions are favorable, some of the gas coming out of the oil is conserved by moving upward into the gas cap to further enlarge the gas cap. As compared with the dissolved-gas drive, the *gas-cap drive* is more effective, yielding greater recovery of oil. The gas-drive process is typically found with the discontinuous, limited, or essentially closed reservoirs of the types previously shown in Figs. 6 and 7(a). See also Fig. 20.

Where the formation containing an oil reservoir is quite uniformly porous and continuous over a large area, as compared with the size of the soil reservoir per se, very large quantities of salt water exist in surrounding parts of the same formation, often directly in contact with the oil and gas reservoir. This condition is demonstrated by previously shown Figs. 2, 3, 4, and 5. These large quantities of salt water occur under pressure and provide a large additional store of energy to assist in producing oil and gas. A situation like this is termed *water-drive reservoir* and is shown in Fig. 21. The energy supplied by the salt water comes from expansion of the water as pressure in the petroleum reservoir is reduced by production of oil and gas. Water will compress, or expand, to the extent of about one part in 2500 per 100 psi (6.8 atmospheres) change in pressure. Although this effect is slight with reference to small quantities, the phenomenon becomes of importance when changes in reservoir pressure affect large volumes of salt water that are often contained in the same porous formation adjoining or surrounding a petroleum reservoir.

The expanding water moves into the regions of lowered pressure in the oil- and gas-saturated portions of the reservoir caused by production of oil and gas, and retards the decline in pressure. In this way, the expansive energy in the oil and gas is conserved. As shown in Fig. 4, the expanding water also moves and displaces the oil and gas in an upward direction out of the lower parts of the reservoir. By this natural process, the pore spaces

vacated by oil and gas produced are filled with water, and oil and gas are progressively moved toward the wells.

The water drive is generally the most efficient oil-production process. Oil fields in which water drive is effective are capable of yielding recoveries ranging up to 50% of the oil originally in place, if (1) the physical nature of the reservoir rock and of the oil are conducive to the process, (2) care is exercised in completing and producing the wells, and (3) the rate of withdrawal of products is optimal.

When pressures in an oil reservoir have fallen to the point where a well will not produce by natural energy, some method of artificial lift must be used. Oil-well pumps are of three general types: (1) pumps located at the bottom of the hole run by a string of rods, (2) pumps at the bottom of the hole run by high-pressure liquids, and (3) bottom-hole centrifugal pumps. Another method involves the use of high-pressure gas to lift the oil from the reservoir.

Well Completion

Production casing must be set through which the oil and/or gas can be brought safely to the surface. The "pay zone" (productive area) is then sealed off with cement. With the production casing in place, hollow charges are fired through it into the production formation and the drilling mud is gradually displaced, so that the hydrocarbons can flow into the well-bore and up to the surface. There, a "Christmas tree" (an assembly of valves and special connectors) is attached to the top of the production casing. This device controls the flow of oil or gas into the gathering pipelines. See Fig. 22.

Deepwater Production

Two basic types of platforms may be used in deepwater production—*fixed leg* and *compliant*. Each has its advantages and limitations. These facilities can provide all the functions required for drilling, completing, producing, and maintaining conventional wells or a combination of conventional and subsea wells.

Nearly all offshore fields have been developed with fixed-leg platforms. In 1947, for example, a fixed-leg platform weighing 1200 tons was in

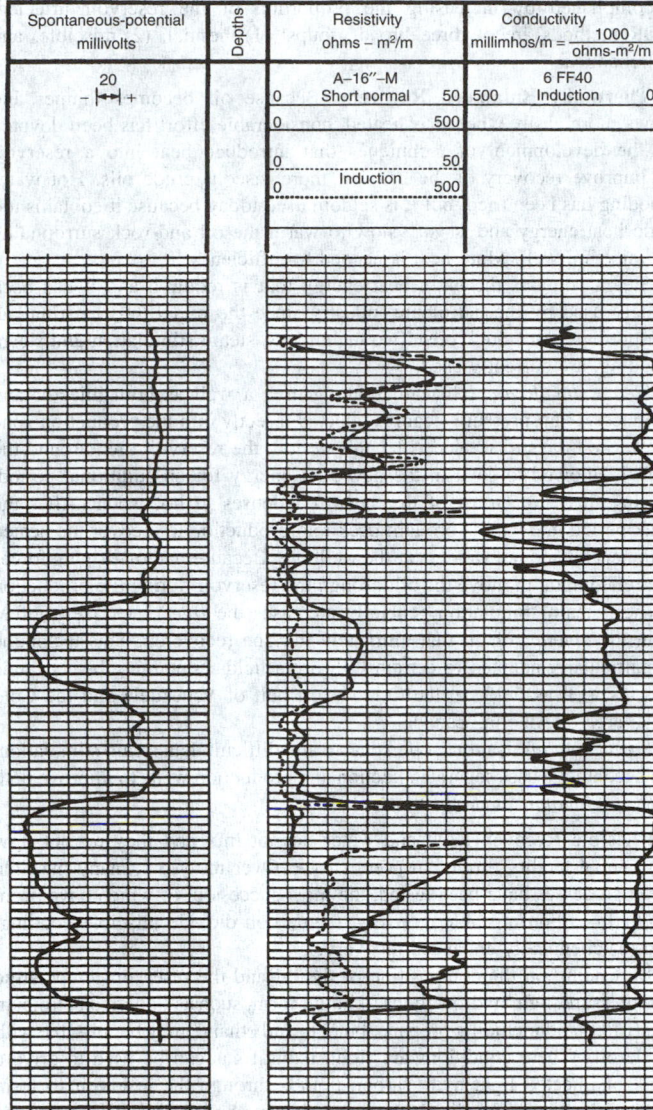

Spontaneous-potential millivolts	Depths	Resistivity ohms – m²/m		Conductivity millimhos/m = $\dfrac{1000}{\text{ohms-m}^2/\text{m}}$	
20 − |↔| +		A–16″–M Short normal		6 FF40	
		0	50	500 Induction 0	
		0	500		
		0 Induction	50		
		0	500		

Fig. 18. Oil well logging provides valuable information on the "down-hole" characteristics of oil and gas wells. Spontaneous potential (SP), electrical resistivity/conductivity logs are frequently recorded simultaneously.

Fig. 19. Dissolved-gas drive. (*Texas Mid-Continent Oil and Gas Association.*)

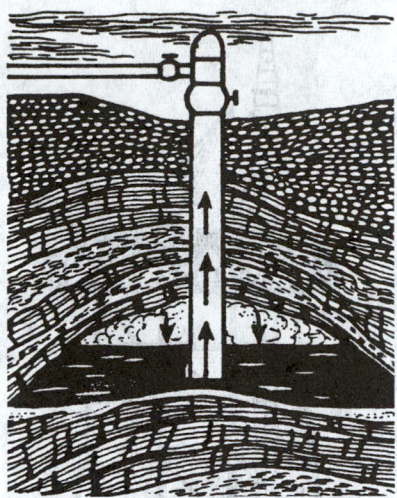

Fig. 20. Gas-cap drive. (*Texas Mid-Continent Oil and Gas Association.*)

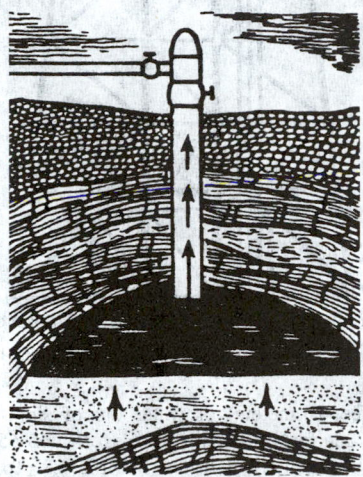

Fig. 21. Water drive. (*Texas Mid-Continent Oil and Gas Association.*)

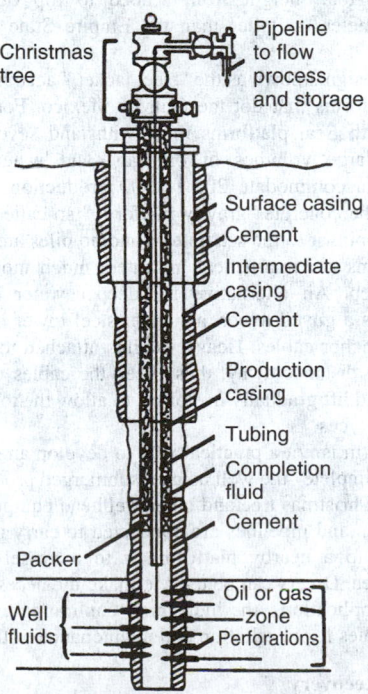

Fig. 22. Completed oil well showing the flow of oil into and up the well to the pipeline connection at the Christmas tree. (*American Petroleum Institute.*)

Fig. 23.   Fixed-leg offshore drilling and production platform of the "steel jacket" variety. (*Exxon Corp.*)

operation in 20 feet (6 meters) of water out of sight of land. Twenty years later, such platforms weighing in excess of 6500 tons were in use in 340 feet (104 meters) of water. See Fig. 23. To date (1986), the tallest fixed-leg platform, weighing 58,000 tons, is located in 1025 feet (312 meters) of water in the Gulf of Mexico. This platform, completed in 1979, has a total height from seabed to top of the derricks of 1265 feet (386 meters)—taller than the Empire State Building in New York City.

This type of design, known as the "steel jacket," accounts for most of the hundreds of platforms that dot the Gulf of Mexico. For larger oil fields, such as the North Sea, platforms must withstand severe environmental forces, handle large volumes of oil, gas, and water, support heavy equipment, and accommodate 200 to 300 production workers. Here, a favored type is the concrete "gravity platform," so called because its own immense weight pins it to the sea bottom and no piles are needed to secure it. Rigid platforms are impractical in waters much more than 1000 feet (305 meters) deep. An alternative for deeper water is a "compliant" structure, such as a guyed tower, a slender steel tower held in place by a radial array of anchor cables. Heavy weights attached to the cables lie on the bottom some distance away; these keep the cables taut under normal sea conditions and lift gradually in storms, to allow the tower to tilt slightly to absorb wave forces.

When a platform is not a practical way to develop an offshore field, the operator may "complete" the well using a submerged production system, in which case, the Christmas tree and other wellhead equipment are installed on the sea bottom and pipelines are connected to carry off the petroleum, either to shore, to a nearby platform, or to a vessel or storage buoy moored in the area. Divers can be used to make the necessary connections. For deepwater applications, the industry is continuing to develop remote, diverless techniques for installation and maintenance of these completions.

Enhanced Oil Recovery

Enhanced oil recovery (EOR) methods increase ultimate oil production beyond that achievable with primary and secondary methods. This is accomplished by increasing the proportion of the reservoir affected. EOR methods are of three broad groups: (1) thermal, (2) miscible, and (3) chemical.

Thermally Enhanced Recovery. Because oil becomes thinner and flows more easily when it is heated, considerable effort has been devoted to the development of techniques that introduce heat into a reservoir to improve recovery of the heavier, more viscous crude oils. Hot water flooding has been tried, but it is seldom used today because it contains too little heat energy and is very slow to warm the oil and rock surrounding an injection well. More heat is needed for efficiency.

Steam contains the extra heat energy that is required and it has been widely used by the petroleum industry since the mid-1960s to stimulate the production of thick oils. Two techniques, steam stimulation and steam flooding, are currently used.

Steam stimulation or steam soaking uses a well as both injector and producer. High-pressure steam is injected directly into the production zone for several days to weeks. After this period, the reservoir area around the well is allowed to soak in the new heat energy for an additional period. During this time, most of the steam condenses to hot water. After the soak period, the well is brought back into production to recover the heated (thinner) oil and hot water near the wellbore. Because natural driving forces are relied upon to move the oil through the reservoir during the production phases, steam stimulation generally increases the *rate* of recovery rather than the *amount* of oil that ultimately may be recovered. This technique is particularly adapted to certain California fields containing heavy crude oils, as well as fields in the Orinoco oil belt of Venezuela and the Cold Lake area of Alberta, Canada.

Steam flooding is more sophisticated and difficult than steam stimulation. This technique uses separate injection and production wells to improve both *rate* and *amount* of production.

Miscible Recovery. Oil and water do not mix and they do not flow with equal facility through a porous rock. Over the years, many miscible flood processes have been tested, the most successful of which have been: (1) hydrocarbon miscible recovery; (2) carbon dioxide miscible flooding; and (3) chemically enhanced recovery.

Depending on the composition of the oil and the reservoir temperature and pressure, light hydrocarbons in liquid form, such as liquefied petroleum gas (LPG) and including propane, butane, and ethane, may be miscible with crude oil. Where conditions are right, natural gas can be used to drive a bank of injected light hydrocarbon liquids through the reservoir to form a miscible flood. The disadvantage of this method is that it involves the prolonged use of valuable hydrocarbon liquids, some of which may never be recovered. The use of natural gas alone has received consideration in special situations where high pressure, combined with low reservoir temperature, may result in miscibility.

Carbon dioxide miscible recovery is a preferred method. Carbon dioxide may not be initially miscible with crude oil, but when it is forced into an oil reservoir, some of the smaller, lighter hydrocarbon molecules in the contacted crude will vaporize and mix with the CO_2, forming a wall of enriched gas (CO_2 plus light hydrocarbons). If the temperature and pressure of the reservoir are suitable, this wall of enriched gas will mix with more of the crude, forming a "bank" of miscible solvents capable of efficiently displacing large volumes of crude oil. Carbon dioxide is found in underground deposits and can be produced through wells similar to gas wells. But its production and transportation to the oil reservoir can add significantly to the cost. A project has been proposed that would involve the construction of a long CO_2 pipeline from a Colorado CO_2 well to a large oil field in Texas. The economic feasibility of the project remains to be proved.

The use of chemicals to coax more oil out of the ground has been investigated for many years. Chemically enhanced methods are of three major types: (1) polymer flooding; (2) surfactant flooding; and (3) alkaline flooding.

Experts estimate that prudent but aggressive application of enhanced recovery technology to known reservoirs in the United States could result in the production of 20 to 30 billion barrels of oil that might otherwise be lost forever. The amount of oil that might be recovered from known oil fields worldwide is estimated to be in the range of 100 to 200 billion barrels. See Fig. 24.

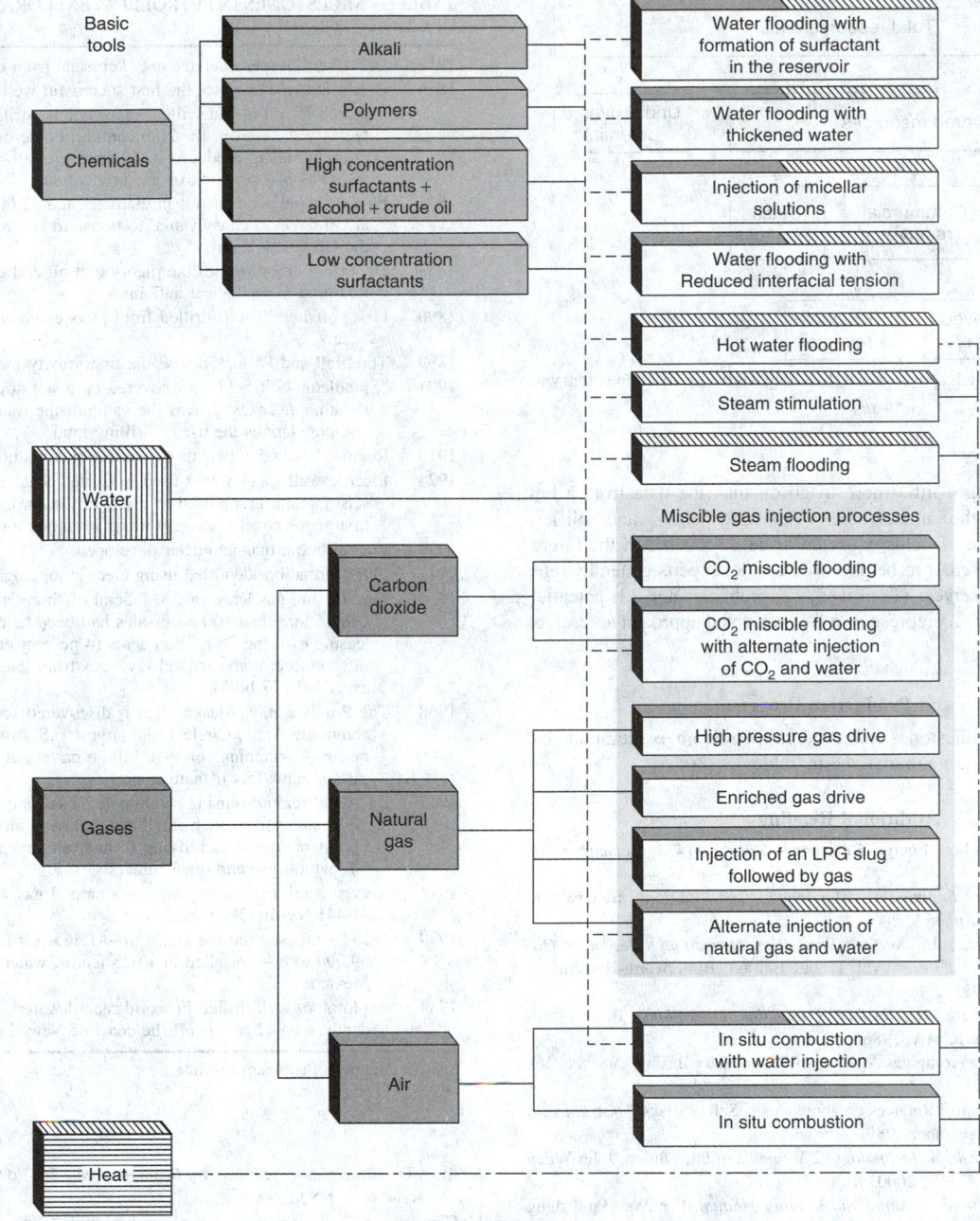

Fig. 24.   Enhanced oil recovery techniques employ heat, gases, chemicals, and water—singly or in combinations—to reduce the factors that inhibit oil recovery and to augment reservoir energy. (*Exxon Corp.*)

Petroleum Reserves

In 1988, the U.S. Geological Survey reduced its prior estimate of the oil and gas remaining to be discovered in the United States by 40%. As pointed out by R.A. Kerr (reference listed), "If the new estimates hold up, it would solidify a new realism in the agency's view of energy resources. In 1972, the agency claimed that there were 450 billion barrels of oil and 2100 trillion cubic feet of gas left to be found—Figures that the USGS itself soon characterized as four times too high." A revised estimate in 1981 showed 83 billion barrels and 594 trillion cubic feet left undiscovered. Even this estimate was challenged as too high by some industry experts.

The foregoing is exemplary of how assumptions of remaining oil and gas reserves are made and of the lack of confidence that *any* figures appeared to enjoy as of the 1990s. Generally, there is a consensus that, given current rates of consumption, oil reserves should be visualized as becoming exhausted in terms of decades rather than centuries!

Everyone recognizes that crude oil exists in finite amounts, but no one really knows how much of it remains in the earth. Since E.L. Drake drilled his first well in 1859, a vast industry has been established and oil became the predominant worldwide fuel. But petroleum geologists freely admit that definitive knowledge about the resource base that made this growth possible remains *elusive*. At any given time, the amount of oil available for consumption depends primarily on two factors: (1) the producibility of already discovered reserves; and (2) the production policies of governments in countries where those reserves exist. Over the long run, however, it is the amount of recoverable oil left in the ground, including those volumes yet undiscovered, that will be decisive. Estimates of the remaining petroleum resource base vary widely—in fact, too widely to record in this encyclopedia. Experts disagree about the size and producibility of individual reservoirs and about the total national and world reserves associated with already discovered fields. They are even further apart when assessing the world's undiscovered potential. This is not surprising when it is remembered that, in this current unscientific arena, the experts are making judgments—frequently educated guesses—about hydrocarbons contained in porous rocks many thousands of feet under the earth's

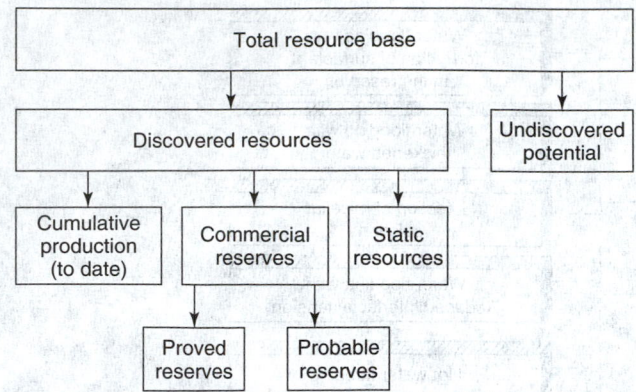

Fig. 25. Commonly used approach for handling statistics in connection with estimating oil reserves.

surface. Often, they have little more to go on than the data from a few widely dispersed 8-inch-diameter holes in existing fields — and still less information in the case of fields expected to be discovered in the future. Consequently, in an attempt to be pseudoscientific, experts generally refer to three classes of reserves: (1) proved; (2) probable; and (3) potential. The chart shown in Fig. 25 represents a well-accepted approach to reserves analysis.

Petroleum Exploration and Production Progress

Some of the major milestones achieved in petroleum exploration and production technology are summarized in Table 3.

Additional Reading

Abelson, P.H.: "Hydrocarbon Energy Revisited," *Science*, 1433 (September 29, 1989).

Ahmed, T.H.: *Reservoir Engineering Handbook*, 2nd Edition, Butterworth-Heinemann, Inc., Woburn, MA, 2001.

Arnold, K. and M. Stewart, Jr.: *Surface Production Operations: Design of Oil-Handling Systems and Facilities*, Vol. 1, 2nd Edition, Butterworth-Heinemann, Inc., Woburn, MA, 1998.

Arnold, K. and M. Stewart, Jr.: *Surface Production Operations*, Butterworth-Heinemann, Inc., Woburn, MA, 1986.

Bethke, C.M., et al.: "Supercomputer Analysis of Sedimentary Basins," *Science*, 261 (January 15, 1988).

Dalen, B.: "Computer Control Reduces Offshore Costs, Safety Risks," *Instrumentation Technology*, 22 (December 1989).

Dawe, R.A.: *Modern Petroleum Technology: 2 Volume Set*, 6th Edition, John Wiley & Sons, Inc., New York, NY, 2000.

Devereaux, S.: *Practical Well Planning and Drilling Manual*, PennWell Publishing Company, Tulsa, OK, 1998.

Economides, M.J., B.N. Murali, and L.T. Watters: *Petroleum Well Construction*, John Wiley & Sons, Inc., New York, NY, 1998.

Elliott, D.H.: "Is There Any Oil and Natural Gas (Antarctica)?" *Oceanus*, 32 (Summer 1988).

Esmaeili, H.: *The Legal Regime of Offshore Oil Rigs in International Law*, Ashgate Publishing Company, Brookfield, VT, 2001.

Gluyas, J. and R. Swarbrick: *Petroleum Geology*, Blackwell Science, Inc., Malden, MA, 2001.

Hapgood, F.: "The Quest for Oil," *National Geographic*, 226 (August 1989).

Hyne, N.J.: *Nontechnical Guide to Petroleum Geology, Exploration, Drilling, and Production*, PennWell Publishing Company, Tulsa, OK, 1995.

Kerr, R.A.: "Oil and Gas Estimates Plummet," *Science*, 1330 (September 22, 1989).

Lee, D.B.: "Oil in the Wilderness — An Arctic Dilemma," *National Geographic*, 858 (December 1988).

Longwen, W.: "China's Exploration and Development of Offshore Oil and Gas," *Oceanus*, 32 (Winter 1989/1990).

Lynch, M.C.: "Preparing for the Next Oil Crisis," *Chem. Eng. Progress*, 20 (March 1988).

Lyons, W.C.: *Standard Handbook of Petroleum and Natural Gas Engineering*, Vol. 1, 6th Edition, Butterworth-Heinemann, Inc., Woburn, MA, 2001.

Lyons, W.C.: *Standard Handbook of Petroleum and Natural Gas Engineering*, Vol. 2, 6th Edition, Butterworth-Heinemann, Inc., Woburn, MA, 1996.

Miall, A.D.: *Principles of Sedimentary Basin Analysis*, 3rd Edition, Springer-Verlag, Inc., New York, NY, 1999.

McCain, W.D.: *The Properties of Petroleum Fluids*, 2nd Edition, PennWell Publishing Company, Tulsa, OK, 1990.

Nelson, R.C.: "Chemically Enhanced Oil Recovery: The State of the Art," *Chem. Eng. Progress*, 50 (March 1989).

Pate-Cornell, M.E.: "Organizational Aspects of Engineering System Safety: The Case of Offshore Platforms," *Science*, 1210 (November 30, 1990).

Rutledge, G.: "Arctic Oil," *Chem. Eng. Progress*, 6 (October 1989).

Schmidt, R.L.: "Thermal Enhanced Oil Recovery: Current Status and Future Needs," *Chem. Eng. Progress*, 47 (January 1990).

Selly, R.C.: *Elements of Petroleum Geology*, 2nd Edition, Morgan Kaufmann Publishers, Orlando, FL, 1997.

Short, J.A.: *Introduction to Directional and Horizontal Drilling*, PennWell Publishing Company, Tulsa, OK, 1993.

Staff: "U.S. Oil and Gas Outlook Brightens," *Chem. Eng. Progress*, 6 (December 1989).

Staff: *International Petroleum Encyclopedia*, PennWell Publishing Company, Tulsa, OK, 2000.

Staff: "Mobil's Arnold Stancell and the Pursuit of Oil," *Chem. Eng. Progress*, 70 (April 1990).

Tearpock, D.J. and R. Bischke: *Applied Subsurface Geological Mapping*, 2nd Edition, Prentice Hall, Inc., Upper Saddle River, NJ, 2002.

Twiss, R.J. and E.M. Moores: *Structural Geology*, W. H. Freeman Company, New York, NY, 1995.

TABLE 3. MILESTONES IN PETROLEUM EXPLORATION AND PRODUCTION TECHNOLOGY

1853	Dr. Albert Gesner manufactures kerosene from coal.
1859	Edwin Drake completes the first successful well drilled in the search for oil at Titusville, Pennsylvania, striking oil at $69\frac{1}{2}$ feet. By the start of the 20th century, crude oil and/or natural gas were being produced in 20 states; in 1984, 33 of the 50 states had some oil or gas production.
1865	First oil pipeline, 2 inches in diameter and 32,000 feet long, laid at Oil Creek, Pennsylvania, to transport oil from the field to the Oil Creek Railroad.
1883	Dr. I.C. White proposes the theory that oil and gas deposits could be found in geological anticlines.
1896	First "offshore" wells drilled from piers extending into California waters.
1899	Threllfall and Pollock devise the first gravity meter.
1901	"Spindletop" oil field is discovered on a salt dome near Beaumont, Texas; proves the value of the rotary drilling rig and popularizes the use of drilling mud.
1914	Reginal Fessenden patents the reflections seismograph.
1924	Electric well logging first used in United States; refraction seismograph graph used to discover Orchard, Texas, salt dome; first geophysical discovery using magnetic torsion-balance.
1939	First airborne magnetometer developed.
1942	Fluid formation identified using electric logging.
1954	First oil and gas lease sale in federal offshore area held; through 1984, more than 100 such sales had been held, resulting in the leasing of some 38 million acres (4 percent of the federal offshore area); and federal revenues from that leasing had exceeded $77 billion.
1968	The Prudhoe Bay, Alaska, field is discovered some 250 miles above the Arctic Circle — the largest U.S. discovery ever made — containing some 10 billion barrels of oil and 26 trillion cubic feet of natural gas.
1972	First land remote sensing satellite (Landsat) launched; information from such satellites is playing an increasingly important role in identifying from space potential deposits of oil, natural gas and other minerals.
1974	Record-depth exploratory well — a natural gas well — drilled to 31,441 feet in Oklahoma.
1979	World's tallest fixed-leg platform — 1,265 feet tall and weighing 59,000 tons — installed in 1,025 feet of water in the Gulf of Mexico.
1984	Exploratory well drilled in world record water depth — 6,942 feet — off the coast of New England.

Source: American Petroleum Institute.

Van Der Pluijm, B.A. and S. Marshak: *Earth Structure: An Introduction to Structural Geology and Tectonics*, The McGraw-Hill Companies, Inc., New York, NY, 1997.

Web Reference

American Association of Petroleum Geologists: http://www.aapg.org/

PETROLEUM FLY.   An unusual insect that lives in pools of crude petroleum in the California oil fields. The larvae will live in fresh water, brine, or crude oil. The insect is of the family *Ephydriaae*.

PETROLEUM REFINING.   Because crude oils exhibit important differences from the standpoint of processing and final end-products from one geographic resource to the next (see Tables 1 and 2 of the preceding article on **Petroleum**), very few petroleum refineries operate on exactly the same basis. The principles of the operations performed are remarkably similar, but there are basic differences in the amount of throughputs for given products. The latter varies from one geographic area to the next and with the season of the year. Modern petroleum refineries are designed with considerable built-in flexibility, so that the production of petroleum products can be customized for specific market demands. For example, the production of greater amounts of heating fuels during the fall and winter season as contrasted with larger production of gasolines and diesel fuels during summer months, when the demand is greatest. In contrast with a few decades ago, modern refineries are integrated to produce numerous products and often will also serve nearby petrochemical plants where petroleum products become starting ingredients for a vast variety of chemicals used in many industries, such as plastics, solvents, polymers, fibers, among many others.

A representative integrated petroleum refinery is shown in Fig. 1.

The major processing units fundamental to the manufacture of fuel products from crude oil include: (1) crude distillation; (2) catalytic reforming; (3) catalytic cracking; (4) catalytic hydrocracking; (5) alkylation; (6) thermal cracking; (7) hydrotreating; and (8) gas concentration. Refineries also will use numerous auxiliary processes, such as treating units to purify both liquid and gas streams, waste-management and pollution-control systems, cooling-water systems, units to recover hydrogen sulfide (or elemental sulfur) from gas streams, desalters, electric-power stations, steam-producing facilities, and provisions for storage of crude oil and products.

Petroleum refining and petrochemical production is a 24-hour, 365-day operation with a very minimum of time planned for downtime. Unless one has visited a refinery firsthand, it is very difficult to comprehend the size and complexity of the equipment used. See Figures 2, 3, and 4.

Crude Distillation.   To minimize corrosion of refining equipment, a crude-oil distillation unit generally is preceded by a *desalter*, which reduces the inorganic salt content of raw crudes. Salt concentrations vary widely (from nearly zero to several hundred pounds, expressed as NaCl per 1,000 barrels). The crude unit functions simply to separate the crude oil physically, by fractional distillation, into components of such boiling ranges that they can be processed by appropriately selected equipment in a long train of processing operations which follow. Although the boiling ranges of components (or fractions) vary between refineries, a typical crude distillation unit will resolve the crude into the following fractions:

By distillation at atmospheric pressure,

1. A light straight-run fraction, consisting primarily of C_5 and C_6 hydrocarbons. These also will contain any C_4 and lighter gaseous hydrocarbons that are dissolved in the crude.
2. A naphtha fraction having a nominal boiling range of $200°-400°F$ ($93°-204°C$).
3. A light distillate with boiling range of $400°-540°F$ ($204°-343°C$).

By vacuum flashing

1. Heavy gas oil, having a boiling range of $650°-1,050°F$ ($343°-566°C$).
2. A nondistillable residual pitch.

In the atmospheric-pressure distillation section of the unit, the crude oil is heated to a temperature at which it is partially vaporized and then introduced near, but at some distance above, the bottom of a distillation column. This cylindrical vessel is equipped with numerous trays through which hydrocarbon vapors can pass in an upward direction. Each tray contains a layer of liquid through which the vapors can bubble, and the

liquid can flow continuously by gravity in a downward direction from one tray to the next one below. As the vapors pass upward through the succession of trays, they become lighter (lower in molecular weight and more volatile with lower boiling temperature). The liquid flowing downward becomes progressively heavier (higher in molecular weight and less volatile with higher boiling temperature). The countercurrent action results in fractional distillation or separation of hydrocarbons based upon their boiling points. A liquid can be withdrawn from any preselected tray as a net product. Thus, the lighter liquids, such as naphtha, exit from trays near the top of the column, whereas heavier liquids, such as diesel oil, exit from trays near the bottom of the column. Thus, the boiling range of the net product liquid depends upon the tray from which it is taken. The vapors containing the C_6 and lighter hydrocarbons are withdrawn from the top, while a liquid stream boiling at about $650°F$ ($343°C$) is taken from the bottom. The portion taken from the bottom is called *atmospheric residue*.

This residue is further heated and introduced into a vacuum column operated at an absolute pressure of about 50 millimeters of mercury, a vacuum maintained by the use of steam ejectors. A flash separation is made to produce heavy gas oil and nondistillable pitch.

The crude oil and atmospheric residue are heated in tubular heaters. Oil is pumped through the inside of the tubes contained in a refractory combustion chamber fired with oil or fuel gas in such manner that heat is transferred through the tube wall in part by convection from hot combustion gases and in part by radiation from the incandescent refractory surfaces.

The light straight-run gasoline contains all hydrocarbons lighter than C_7 in the crude and consists primarily of the native C_5 and C_6 families. After stabilization to remove the C_4 and lighter hydrocarbons (which are routed to a central gas-concentration unit), the stabilized C_5/C_6 blend is treated to remove odorous mercaptans and passed to the refinery gasoline pool for final product blending. The unleaded octane number (Research Method Number) is less than 70, and thus blending or further processing is required to improve its antiknock qualities. Isomerization can be used to improve octane rating, as well as the addition of lead alkyls.

Naphtha to become a suitable component for blending into finished gasoline pools must be further processed. The octane number will range from 40 to 50. Prior to introduction into a catalytic-reforming unit, most naphtha feedstocks are hydrotreated in the interest of prolonging the life of the reforming catalyst.

Gas oil separated from the crude by vacuum distillation, plus portions of light distillates, is the feedstock to catalytic cracking units. The main function of catalytic cracking is to convert into gasoline those fractions having boiling ranges higher than that of gasoline. Remaining uncracked distillates (cycle oils) are used as components for domestic heating fuels (generally after hydrotreating) and to blend with residual fractions to reduce their viscosity to make acceptable heavy fuel oils. In some refineries, cycle oils are hydrocracked to complete their conversion to gasoline.

The aforementioned processes are described in more detail in separate entries: **Alkylation; Cracking Process; Distillation**; and **Hydrotreating**.

Petroleum Terminology

An abridged glossary of terms used to describe petroleum products and processes would include:

Additives, Diesel Fuel—Chemicals for reducing smoke emissions and for cold weather conditioning.

Additive, Gasoline—In some instances, several functions can be combined in one chemical compound to provide "multifunctional" additives. Increasingly stringent control of automotive emissions in several countries has stimulated the development and use of "extended-range" detergents which are designed to promote peak engine performance by maintaining engine cleanliness. Some gasoline additives include:

Antiknock compounds to increase octane number.

Antioxidants to provide gasoline storage stability.

Antirust agents to prevent corrosion in gasoline-handling systems.

Detergents to control carburetor and induction system cleanliness.

Dyes to indicate kind of antiknock compounds used and to identify brand and grade of gasoline.

Viscosity index improves make lube oil more effective over a wide range of temperatures.

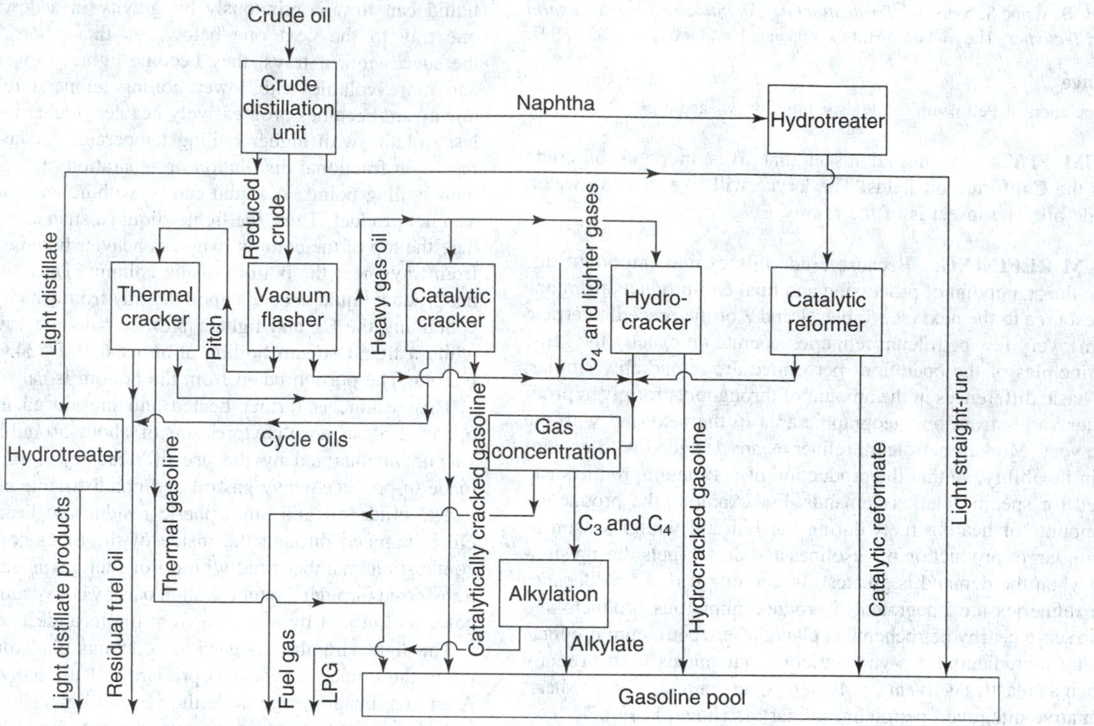

Fig. 1.   Principal flow of materials in an integrated petroleum refinery for producing various fuels and raw materials for petrochemical plants.

Fig. 2.   Aerial view of small portion of a Texas petroleum refinery, showing the hydrocracking unit just left of center of view.

Fig. 3. Erection of 9.5-meter (31-foot) diameter, 263-ton vacuum tower at a petroleum refinery in Kuwait. (*The Fluor Corp.*)

Fig. 4. Erection of a 675-ton hydrocracking reactor at Shauaiba, Kuwait, refinery, the world's first all-hydrogen refinery. (*The Fluor Corp.*)

Alkylation—A refinery process for chemically combining isoparaffin with olefin hydrocarbons. The product, **alkylate**, has a high octane value and is blended with motor and aviation gasoline to improve the antiknock value of the fuel.

Base Oil—A refined or untreated oil used in combination with other oils and additives to produce lubricants.

Blending—The process of mixing two or more oils having different properties to obtain a final blend having the desired characteristics. This can be accomplished by off-line batch processes or by in-line operations as part of continuous-flow operations.

Bright Stock—High-viscosity, fully refined and dewaxed lubricating oils produced by the treatment of residual stocks and used to compound motor oils.

Catalytic Cracking—A refinery process that converts a high-boiling range fraction of petroleum (gas oil) to gasoline, olefin feed for alkylation, distillate, fuel oil, and fuel gas by use of a catalyst and heat.

Catalytic Reforming—A catalytic process to improve the antiknock quality of low-grade naphthas and virgin gasolines by the conversion of naphthenes (such as cyclohexane) and paraffins into higher-octane aromatics (such as benzene, toluene, and xylenes). There are approximately ten commercially licensed catalytic reforming processes, including fully regenerative and continuously regenerative designs.

Cetane Number—The cetane number (C.N.) of a fuel is the percentage by volume of normal cetane in a mixture of cetane and alpha-methylnaphthalene which matches the unknown fuel in ignition quality when compared with a standard diesel engine under specified conditions. The C.N. scale ranges from 0 to 100 C.N. for fuels equivalent in ignition quality to alpha-methylnaphthalene and cetane, respectively. For routine-testing, secondary reference fuels having cetane values of about 25 and 74 are blended in any desired proportion.

Clear Octane—The octane number of a gasoline before the addition of antiknock additives.

Cloud Point—The aniline cloud point is a measure of the paraffinicity of a fuel oil, a high value indicating a straight-run paraffinic oil and a low value indicating an aromatic, a naphthenic, or a highly cracked oil.

Coking—Distillation to dryness of a product containing complex hydrocarbons, which break down in structure during distillation, such as tar or crude petroleum. The residue is called coke.

Cracking—A process carried out in a refinery reactor in which the large molecules in the charge stock are broken up into smaller, lower-boiling, stable hydrocarbon molecules, which leave the vessel as overhead (unfinished cracked gasoline, kerosenes, and gas oils). At the same time, certain of the unstable or reactive molecules in the charge stock combine to form tar or coke bottoms. The cracking reaction may be carried out with heat and pressure (thermal cracking) or in the presence of a catalyst (catalytic cracking).

Cycle Stock—Unfinished product taken from a stage of a refinery process and recharged to the process at an earlier period in the operation.

Deasphalting—Process for removing asphalt from petroleum fractions, such as reduced crude. A common deasphalting process introduces liquid propane, in which the nonasphaltic compounds are soluble while the asphalt settles out.

Desulfurization—The removal of sulfur or sulfur-bearing compounds from a hydrocarbon by any one of a number of processes, such as hydrotreating.

Distillate—That portion of a liquid which is removed as a vapor and condensed during a distillation process. As fuel, distillates are generally within the 400° to 650°F (204° to 343°C) boiling range and include Nos. 1 and 2 fuel, diesel, and kerosene.

End Point—The temperature at which the last portion of oil has been vaporized in ASTM or Engler distillation. Also called the final boiling point.

Equilibrium Volatility of a Gasoline—The volatility of a gasoline is determined by the Reid vapor pressure and the ASTM distillation data. The Reid vapor pressure is the vapor pressure of a gasoline at 100°F (37.8°C) under specified conditions. The distillation curve of a fuel indicates the temperatures at which the various amounts of a given sample are distilled under specified test conditions. However, gasoline will completely evaporate in the presence of air at a temperature much lower than the end-point of the distillation curve. According to O.C. Bridgeman (U.S. National Bureau of Standards, Research Paper 694), the volatility of a gasoline is the temperature at which a given air-vapor mixture is formed under equilibrium conditions at a pressure of one atmosphere, when a given percentage is evaporated. According to this definition, one gasoline is more volatile than another for any given percentage evaporated if it forms the given air-vapor mixture at a lower temperature. Distillation temperature curves, for a given text sample, plot amount of sample distilled over (percentage of sample) at the time a given temperature has been reached.

Fire Point—The lowest temperature at which a fuel ignites and burns for at least 5 seconds under specified test conditions.

Flare—A device for disposing of gases by burning.

Flash Point—The lowest temperature at which a flash appears on the fuel surface when a test flame is applied under specified test conditions. This property is an approximate indication of the tendency of the fuel to vaporize.

Flue Gas Expander—A turbine used to recover energy where combustion gases are discharged under pressure to the atmosphere. The pressure reduction drives the impeller of the turbine.

Fractions—Refiner's term for the portions of oils containing a number of hydrocarbon compounds but within certain boiling ranges, separated from other portions in fractional distillation. They are distinguished from pure compounds which have specified boiling temperatures, not a range.

Fuel Oils—Any liquid or liquifiable petroleum product burned for the generation of heat in a furnace or firebox or for the generation of power in an engine. Typical fuels include clean distillate fuel for home heating and higher-viscosity residual fuels for industrial furnaces.

Gas Oil—A fraction derived in refining petroleum with a boiling range between kerosene and lubricating oil.

Heating Oils—A trade term for the group of distillate fuel oils used in heating homes and buildings as distinguished from residual fuel oils used in heating and power installations. Both are burned-fuel oils.

Heavy Ends—The highest-boiling portion of a gasoline or other petroleum oil.

Hydrocracking—The cracking of a distillate or gas oil in the presence of catalyst and hydrogen to form high-octane gasoline blending stock.

Hydrogenation—A refinery process in which hydrogen is added to the molecules of unsaturated (hydrogen-deficient) hydrocarbon fractions. It plays an important part in the manufacture of high-octane blending stocks for aviation gasoline and in the quality improvement of various petroleum products.

Hydrotreating—A treating process for the removal of sulfur and nitrogen from feedstocks by replacement with hydrogen.

Isomerization—A refining process which alters the fundamental arrangement of atoms in the molecule. Used to convert normal butane into isobutane, as alkylation process feedstock, and normal pentane and hexane into isopentane and isohexane, high-octane gasoline components.

Kinematic Viscosity—The absolute viscosity of a liquid (in centipoises) divided by its specific gravity at the temperature at which the viscosity is measured.

Knock—The sound or "ping" associated with the autoignition in the combustion chamber of an automobile engine of a portion of the fuel-air mixture ahead of the advancing flame front.

Lead Susceptibility—The increase in octane number of gasoline imparted by the addition of a specified amount of tetraethyl lead.

Low-Sulfur Crude Oil—Crude oil containing low concentrations of sulfur-bearing compounds. Crude is usually considered to be in the low-sulfur category if it contains less than 0.5% (weight) sulfur. Examples of low-sulfur crudes are offshore Louisiana, Libyan, and Nigerian crudes.

Lube Stock—Refinery term for fraction of crude petroleum suitable in terms of boiling range and viscosity to yield lubricating oils when further processed and treated.

Mercaptans—Compounds of sulfur having a strong, repulsive, garlic-like odor. A contaminant of "sour" crude oil and products.

Octane Number—The octane rating of a motor fuel is defined in terms of its knocking characteristics relative to those of blends of isooctane (2,3,4-trimethylpentane) and n-heptane, and a rating of 100 to isooctane. The octane number of an unknown fuel is numerically equal to the volume percent of isooctane in a blend with n-heptane which has the same knocking tendency as the unknown fuel when both the unknown and the reference blend are run in a standard single-cylinder engine operated at specified conditions. Motor Method octane numbers are measured at more severe engine conditions and are numerically lower than those determined by the milder Research Method. The difference between the two numbers is termed **sensitivity**.

Polymer—A product of polymerization of normally gaseous olefin hydrocarbons to form high-octane hydrocarbons in the gasoline boiling range. Polymerization is the process of combining two or more simple molecules of the same type, called monomers, to form a single molecule having the same elements in the same proportions as in the original molecule, but having different molecular weights. The combination of two or more dissimilar molecules is known as copolymerization—and the product is called a **copolymer**.

Pour Point—This property is defined as the lowest temperature at which the fuel will pour and is a function of the composition of the fuel. Normally, the pour point of a fuel should be at least 10 to 15 degrees below the anticipated minimum use temperature.

Presulfide—A step in the catalyst regeneration procedure which treats the catalyst with a sulfur-bearing material such as hydrogen sulfide or carbon bisulfide to convert the metallic constituents of the catalyst to the sulfide form in order to enhance its catalytic activity and stability.

Process Unit—A separate facility within a refinery, consisting of many types of equipment, such as heaters, fractionating columns, heat exchangers, vessels, and pumps, designed to accomplish a particular function within the refinery complex. For example, the crude processing unit is designed to separate the crude into several fractions, while the catalytic reforming unit is designed to convert a specific crude fraction into a usable gasoline blending stock.

Raffinate—In solvent refining, that portion of the oil that remains undissolved and is not removed by the selective solvent.

Refinery Pool—An expression for the mixture obtained if all blending stocks for a given type of product were blended together in production ratio. Usually used in reference to motor gasoline octane rating.

Refluxing—In fractional distillation, the return of part of the condensed vapor to the fractionating column to assist in making a more complete separation of the desired fractions. The material returned is called **reflux**.

Residual Fuel Oils—Topped crude petroleum or viscous residuums obtained in refinery operations. Commercial grades of burner-fuel oils Nos. 5, and 6 are residual oils and include Bunker fuels.

Riser Cracking—Applied to fluid catalytic cracking units where the mixture of feed oil and hot catalyst is continuously fed into one end of a pipe (riser) and discharges at the other end where catalyst separation is accomplished after the discharge from the pipe. There is no dense phase bed through which the oil must pass because all the cracking occurs in the inlet pipe (riser).

Road Octane—A numerical value based upon the relative antiknock performance of an automobile with a test gasoline as compared with specified reference fuels. Road octanes are determined by operating a car over a stretch of level road or on a chassis dynamometer under conditions simulating those encountered on the highway.

SAE Numbers—A classification of motor, transmission, and differential lubricants to indicate viscosities, standardized by the Society of Automotive Engineers. They do not connote quality of the lubricant.

Smoke Point—The smoking tendency of a fuel is indicated by this value, which is the maximum height of a specified type of flame in a given wick lamp that results in no visible smoke.

Solvent Extraction—The process of mixing a petroleum stock with a selected solvent, which preferentially dissolves undesired constituents, separating the resulting two layers, and recovering the solvent from the raffinate (the purified fraction) and from the extract by distillation.

Sour Crude—Crude oil which (1) is corrosive when heated, (2) evolves significant amounts of hydrogen sulfide on distillation, or (3) produces light fractions which require sweetening. Sour crudes usually, but not necessarily, have high sulfur content. Examples are most West Texas and Middle East crudes.

Specific Gravity—The specific gravity of a petroleum fuel is the ratio of the weight of a given volume of the product at 60°F to the weight of an equal volume of distilled water at the same temperature, both weights corrected for air buoyancy. The relation between API gravity scale and specific gravity is: °API = 141.5/(sp gr 60/60°F)−131.5.

Stability—In petroleum products, the resistance to chemical change. Gum stability in gasoline means resistance to gum formation while in storage. Oxidation stability in lubricating oils and other products means resistance to oxidation to form sludge or gum in use.

Sweet Crude—Crude oil that (1) is not corrosive when heated, (2) does not evolve significant amounts of hydrogen sulfide on distillation, and (3) produces light fractions which do not require sweetening. Examples are offshore Louisiana, Libyan, and Nigerian crudes.

Tricresyl Phosphate (TCP) — Colorless to yellow liquid used as a gasoline and lubricant additive and plasticizer. Formula, $PO(OC_6H_4Ch_3)_3$.

Tetraethyl Lead (TEL) — A volatile lead compound which is added to motor and aviation gasoline to increase the antiknock properties of the fuel. $Pb(C_2H_5)_4$. The use of this compound has diminished in recent years because of pollution regulations.

Thermal Cracking — A refining process which decomposes, rearranges, or combines hydrocarbon molecules by the application of heat without the aid of a catalyst.

Topped Crude — A residual product remaining after the removal, by distillation or other processing means, of an appreciable quantity of the more volatile components of crude petroleum.

Unsaturates — Hydrocarbon compounds of such molecular structure that they readily pick up additional hydrogen atoms. Olefins and diolefins, which occur in cracking, are of this type.

Vacuum Distillation — Distillation under reduced pressure, which reduces the boiling temperature of the material being distilled sufficiently to prevent decomposition or cracking.

Vapor Lock — The displacement of liquid fuel in the feed line and the interruption of normal motor operation, caused by vaporization of light ends in the gasoline. Vaporization occurs when the temperature at some point in the fuel system exceeds the boiling points of the volatile light ends.

Virgin Stock — Oil processed from crude oil which contains no cracked material. Also called straight-run stock.

Visbreaking — Lowering or breaking the viscosity of residuum by cracking at relatively low temperatures.

Viscosity — This is generally expressed in terms of the time required for a given quantity of fuel to flow through a capillary tube under specified conditions. Kinematic viscosity v is viscosity divided by mass density, or $v = \mu\rho$. The unit in cgs units is called the **stoke**; a customary unit is the **centistoke** ($\frac{1}{100}$ of a stoke). The value of the kinematic viscosity in (cm²/seconds) can be obtained from the indications in seconds t of various viscometers by:

$$\text{Saybolt Universal, when } 32 < t < 100 = 0.00226t - 1.95/t$$

$$\text{Saybolt Universal, when } t > 100 = 0.00220t - 1.35/t$$

$$\text{Saybolt Furol, when } 25 < t < 40 = 0.0224t - 1.84/t$$

$$\text{Saybolt Furol, when } t > 40 = 0.0215t - 0.50/t$$

Yield — In petroleum refining, the percentage of product or intermediate fractions based on the amount charged to the processing operation.

Zeolitic Catalyst — Since the early 1960s, modern cracking catalysts contain a silica-alumina crystalline structured material called zeolite. This zeolite is commonly called a molecular sieve. The admixture of a molecular sieve in with the base clay matrix imparts desirable cracking selectivities.

Crude Oil Pipelines

Transportation of crude to petroleum refineries is essentially a part of the refining operation. Petroleum movement can be measured in barrels, tons moved, barrel-miles, ton-miles, etc. Because oil pipelines are the most economical means of moving large volumes of petroleum overland for long distances, ton-miles (movement of one ton over one mile) is the preferred unit. About 740 billion ton-miles per year of crude oil movement by all transportation modes is required to meet the needs of the United States. About half of this quantity is moved by pipeline; water carriers account for most of the remainder. Major interstate crude oil pipelines are shown in map (Fig. 5).

As domestic crude oil is produced, pipeline gathering systems collect and move it to central locations by means of low-pressure, small-diameter pipelines. Usually, these gathering lines feed into pipeline working tanks where the oil is held until it is ready for shipment by a crude oil **trunk line**. Gathering systems include pumping stations, meters, and samplers. About 90% of the gathering lines in the United States are 6 inches (diameter) or smaller. About 34% of all oil pipeline mileage in the United States consists of gathering lines. Crude oil trunk lines, larger in diameter, receive crude oil directly from gathering systems and also from barges and tankers. Crude oil trunk lines range from 8 to 56 inches in diameter and account for 30% of total oil pipeline mileage in the nation. These trunk lines generally originate in the major oil producing areas of the nation and terminate at the main refinery complexes. The largest of these refinery complexes is in the Gulf Coast area of Texas and Louisiana. Other major complexes are

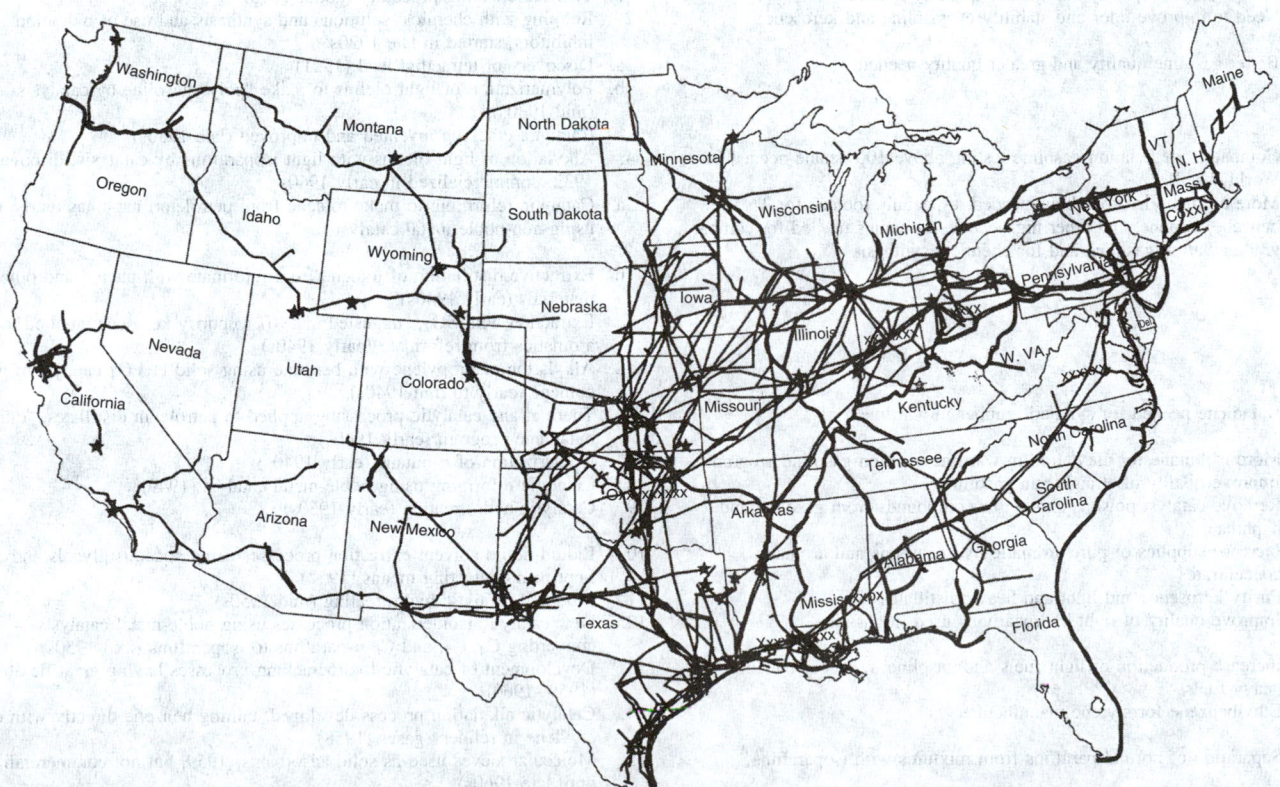

Fig. 5.   Major interstate pipelines and waterways in the United States. Pipelines in Alaska not shown. Asterisks indicate principal refineries. (*American Petroleum Institute.*)

located in the St. Louis-Chicago area, northern Ohio, the East and West Coasts.

Petroleum Processing Progress

Some of the major milestones achieved in petroleum processing technology are summarized in Table 1.

Additional Reading

Abraham, O.C. and F.G. Prescott: "Make Isobutylene from Tertiary Butyl Alcohol," *Hydrocarbon Processing*, 51 (February 1992).

Ansari, R.M. and M.O. Tade: *Nonlinear Model-Based Process Control: Applications in Petroleum Refining*, Springer-Verlag, Inc., New York, NY, 2000.

Chang, E.J. and S.M. Leiby: "Ethers Help Gasoline Quality," *Hydrocarbon Processing*, 41 (February 1992).

Chaput, G., et al.: "Pretreat Alkylation Feed," *Hydrocarbon Processing*, 51 (September 1992).

Dawe, R.A.: *Modern Petroleum Technology: 2 Volume Set*, 6th Edition, John Wiley & Sons, Inc., New York, NY, 2000.

Desai, P.H., et al.: "Enhance Gasoline Yield and Quality," *Hydrocarbon Processing*, 51 (November 1992).

Devlin, J.F., L.A. Edwards, and J.E. Crosby: "Fluid Coker Benefits from Advanced Control," *Hydrocarbon Processing*, 55 (June 1992).

Elliott, J.D.: "Maximize Distillate Liquid Products," *Hydrocarbon Processing*, 75 (January 1992).

Gary, J.H. and G.E. Handwerk: *Petroleum Refining: Technology and Economics*, 4th Edition, Marcel Dekker, Inc., New York, NY, 2001.

Johansen, T., K.S. Raghuraman, and L.A. Hackett: "Trends in Hydrogen Plant Management," *Hydrocarbon Processing*, 119 (August 1992).

Jones, D.S.: *Elements of Petroleum Processing*, John Wiley & Sons, Inc., New York, NY, 1995.

Magee, J.S. and G.E. Dolbear: *Petroleum Catalysis in Nontechnical Language*, PennWell Publishing Company, Tulsa, OK, 1998.

McKetta, J.J.: *Petroleum Processing Handbook*, Marcel Dekker, Inc., New York, NY, 1992.

Meyers, R.A.: *Handbook of Petroleum Refining Processes*, 2nd Edition, The McGraw-Hill Companies, Inc., New York, NY, 1996.

Monfils, J.L., et al.: "Upgrade Isobutane to Isobutylene," *Hydrocarbon Processing*, 47 (February 1992).

Nierlich, F.: "Oligomerize for Better Gasoline," *Hydrocarbon Processing*, 45 (February 1992).

Reynolds, B.E., E.C. Brown, and A. Silverman: "Clean Gasoline Via Vacuum Residium Hydrotreating and Residium Fluid Catalytic Cracking," *Hydrocarbon Processing*, 43 (April 1992).

Speight, J.G.: *Petroleum Chemistry and Refining*, Taylor & Francis, Inc., Philadelphia, PA, 1997.

Speight, J.G. and B. Ozum: *Petroleum Refining Processes*, Vol. 85, Marcel Dekker, Inc., New York, NY, 2001.

Speight, J.G.: *The Chemistry and Technology of Petroleum*, 3rd Edition, Marcel Dekker, Inc., New York, NY, 1999.

Staff: "Refining Handbook, '92," *Hydrocarbon Processing*, 133 (November 1992).

Staff: "What's Ahead in 1993?" *Hydrocarbon Processing*, 33 (December 1992).

Wagner, E.S. and G.F. Froment: "Steam Reforming Analyzed," *Hydrocarbon Processing*, 69 (July 1992).

Wauguier, J.P.: *Petroleum Refining: Crude Oil, Petroleum Products, Process Flowsheets*, Gulf Publishing Company, Houston, TX, 2000.

Wheatcroft, G.: "Present and Future Refinery Information Systems," *Hydrocarbon Processing*, 101 (May 1992).

Web Reference

American Petroleum Institute: http://api-ec.api.org/intro/index_noflash.htm

PETROLOGY. That branch of geology that deals with the rocks forming the lithosphere or "crust" of the earth. The term is derived from the Greek, meaning rock and reason, hence, a more comprehensive term than petrography, which deals only with systematic descriptions of rocks, including their mineral composition, texture, structure, and occurrence. Petrology includes petrography and also methods of classification as founded on systematic description and genetic theories, both experimental and theoretical. Important branches of petrology are geochemistry and geophysics.

TABLE 1. CHRONOLOGY OF MAJOR PETROLEUM INDUSTRY ADVANCEMENTS

Challenge	Action
1. Better gasoline quality and greater quantity needed	1. Thermal cracking processes (about 1910)
2. Need to improve odor and stability of gasoline and kerosene	2. Refining with chemical solutions and synthesis and use of oxidation inhibitors; started in late 1920s
3. Better gasoline quality and greater quality needed	3a. Discovery of tetraethyl lead (1921)
	b. Polymerization of light olefins to make "poly" gasoline by catalysis (mid-1930s)
	c. Catalytic cracking invented and improved (late 1930s)
4. Combat grade aviation gasoline testing above 100 octane needed for World War II	4. Alkylation of light olefins with light isoparaffins by catalysis; discovered in 1932; commercialized in early 1940s
5. More aromatic hydrocarbons needed, especially toulene for TNT; benzene, toluene, and other high-octane aromatics needed for combat grade aviation gasoline and for chemical synthesis	5a. Catalytic reforming to make toluene from petroleum naphthas (early 1940s), using non-noble metal catalyst
	b. Extractive distillation of toluene from reformate with phenol and other materials (early 1940s)
	c. Extraction with SO_2, suggested in 1907 to purify kerosene, applied to secure aromatics from reformate (early 1940s)
	d. Alkylation of propylene with benzene using solid H_3PO_4 catalyst to make cumene (early to mid-1940s)
6. Butadiene needed for synthetic rubber in wartime	6. Thermal and catalytic processing applied to petroleum distillates, "quickly" butadiene program (early 1940s)
7. More isobutane for alkylation in wartime aviation-gasoline program	7. Isomerization of *n*-butane (early 1940s)
8. Improve quality of straight-run gasoline	8. Catalytic reforming using noble-metal catalyst (1949)
9. Remove catalyst poisons and sulfur compounds from gasoline and naphtha	9. Catalytic hydrotreating (early 1950s)
10. Increase supplies of pure aromatic hydrocarbons and aromatic concentrates	10. Liquid-liquid solvent extraction processes using aqueous glycols and improved contacting means (1952)
11. Purify kerosenes and light and heavy distillates	11. Modified catalytic hydrotreating (mid-1950s)
12. Improve quality of light hydrocarbons used in gasoline	12. New catalytic isomerization processes using noble-metal catalysts, converting C_4, C_5, and C_6 *n*-paraffins to isoparaffins (mid-1950s)
13. Increase production of light fuels and gasoline; reduce production of heavy fuels	13. Development of catalytic hydrocracking processes having great flexibility (1959–1960)
14. Ethylbenzene for styrene manufacture	14. Catalytic alkylation process developed, uniting benzene directly with dilute ethylene in refinery gases (1958)
15. Separation of normal paraffins from mixtures with isoparaffins	15. Molecular sieves used as solid adsorbants (1959, but not commercialized until late 1960s)
16. Increase benzene supply and decrease toluene	16. Hydrodealkylation of toluene; produce naphthalene from alkyl naphthalenes (early 1960s)

TABLE 1. (*Continued*)

Challenge	Action
17. Synthesize cyclohexane for nylon	17. Catalytic hydrogeneration of benzene (early 1960s)
18. Improve quality of heavy fuel oils	18. Hydrodesulfurization of heavy fuels, also by hydrocracking (mid-1960s)
19. Biodegradable synthetic detergents	19. Development of processes of dehydrogenate *n*-paraffins to *n*-olefins and alkylate benzene with them (mid-1960s)
20. Increase production of *p*-xylene	20. Isomerization of C_8 aromatics to *p*-xylene (late 1950s)
21. Improve supplies of individual pure xylene isomers	21. Adsorptive separation of *p*-xylene in high yield and purity, making possible separation of other isomers by precise fractionation (early 1970s)
22. Utilization of metals-containing heavy petroleum fractions	22. Development of hydroprocessing techniques to effectively convert to synthetic crude oils
23. Increase supply of liquid fuels in the face of declining petroleum reserves	23. Development of processes for producing liquid and gaseous fuels from coal, shale oil, and tar sands

Source: Universal Oil Products Company.

PETROLOGY (Structural). See **Structural Geology**.

PETZVAL CONDITION. To eliminate the aberration of curvature of field, at least two lenses must be used which are so related that they satisfy the Petzval condition; $f_1n_1 + f_2n_2 = 0$, where the subscripts refer to the two lenses, respectively, n is the refractive index and f is the focal length.

PETZVAL SURFACE. By changing the separation of a doublet lens, a condition free from astigmatism may be found, but the field will not then be flat. The single, paraboloidal surface over which point images of point objects are formed is called the Petzval surface. See also **Curvature of Field (Optics)**.

PEWTER. See **Antimony**.

pH (Abrasion). The term *abrasion pH* was originated by Sevens and Carron in 1948 to designate the pH values obtained by grinding materials in water as a useful aid in the field identification of minerals. The pH value ranges from 1 for ferric sulfate minerals, such as coquimbite, konelite, and rhomboclase, to 12 for calcium-sodium carbonates, such as gaylussite, pirssonite, and shortite.

The recommended technique for determination of abrasion pH is to grind, in a nonreactive mortar, a small amount of the mineral in a few drops of water for about one minute. Usually, a pH test paper is used. Values obtained in this manner are given in the left-hand column of the accompanying table. Another method, proposed by Keller et al. in 1963 involves the grinding of 10 grams of crushed mineral in 100 milliliters of water and noting the pH of the resulting slurry electronically.

pH ADJUSTING AGENTS (FOODS). See **Food Additives**.

PHAGE. See **Virus**.

PHAGOCYTE. A cell of the multicellular animal's body capable of ingesting foreign particles. In the sponges and coelenterates, and to a limited extent in other more complex animals, digestion is carried on wholly or in part by this process. In many animals bacteria and particles of dead tissues or cells are engulfed by such cells. Phagocytes are amoeboid in action and some of them move about in the tissue to which they belong by this means. Others are in fixed tissues, carrying on their amoeboid processes at the free end only. Phagocytes may be a part of the endodermal lining of the enteric cavity or wandering cells derived from endoderm, or they may be mesodermal. In the connective tissues and blood, mesodermal phagocytes occur, and in the lining of the circulatory system cells may act as phagocytes.

Large phagocytes are called macrophages, a term which includes wandering cells of the connective tissues. The opposite, microphage, is rarely met. In the blood of vertebrates all white cells (leukocytes) are phagocytic to some extent, the lymphocytes least of all.

PHALANX (plural Phalanges). The small bones of the toes and fingers of vertebrates.

PHALAROPE. See **Waders, Shorebirds, and Gulls**.

PHANTOM CIRCUIT. Two metallic communication circuits can be made to do the work of three by the addition of certain equipment. Since the third circuit has no wires definitely set aside to its use, it is called a phantom circuit.

PHARMACEUTICALS. Pharmaceuticals are best viewed as drug-containing products in dosage forms. These forms are designed and manufactured to deliver safe and effective therapeutic responses each time administered within appropriate regimens and even after storage under well-documented conditions in scientifically designed packaging for designated time periods. Thus, pharmaceuticals are actual drug delivery systems.

Various technologies are required to produce drug products. Both federal and state laws and regulations exist in the United States to control the manufacture and distribution of pharmaceuticals. The Food and Drug administration exists by the mandate of the U.S. Congress with the Food, Drug & Cosmetics Act as the principal law to enforce. The Act, based on it regulations developed by the Agency, constitutes the basis of the drug approval process. The name Food and Drug Administration is relatively new. In 1931 the Food, Drug, and Insecticide Administration, then part of the U.S. Department of Agriculture, was renamed the Food and Drug Administration. The U.S. drug distribution system is multifaceted including drug usage within the community and hospitals, under long- or short-term home health care or pharmacy practice.

In the United States, there is no national qualifying or licensing body for pharmacists. Licensure requirements are promulgated by State boards of pharmacy that administer examinations, issue internship requirements, and oversee the practice of pharmacy. The National Association of Boards of Pharmacy serves the collective needs of the state boards. This organization has no licensure authority. However, it has developed a standardized licensure examination (NABPLEX), which as of this writing is used by 48 states (see **Licensing**).

Several national organizations serve the professional needs of U.S. pharmacists. The American Pharmaceutical Association (APhA), founded in 1852, is composed of the Academy of Pharmaceutical Research and Science, Academy of Pharmaceutical Practice and Management, and the Academy of Students of Pharmacy. Other organizations include the American Society of Health-Systems Pharmacists (ASHP), National Association of Chain Drug Stores (NACDS), and National Association of Retail Druggists (NARD).

The American College of Apothecaries represents pharmacists whose practices can best be described as emphasizing prescription and related products.

The pharmaceutical industry is represented by several organizations. Examples are the Pharmaceutical Research and Manufacturers of America, the Non-Prescription Drug Manufacturers Association, and the National Pharmaceutical Council. The schools and colleges of pharmacy are organized as the American Association of Colleges of Pharmacy, representing both schools and colleges, and faculty members.

Each state has a professional pharmacy organization, some of which are affiliated with the American Pharmaceutical Association. Similarly, state organizations of hospital pharmacists exist in affiliation with the ASHP. Likewise, local or county associations exist in most instances. Each national association publishes a journal as do most state organizations. The *Federal Register* reports proposed and enacted federal regulatory occurrences several times a week. Each state has a similar publication to report its legislation and regulatory developments, e.g., *The Pennsylvania Bulletin*.

Drugs and Drug Products

Concepts and Processes. Contemporary dosage forms are drug delivery systems, designed and manufactured to achieve safe and effective therapeutic responses each time the forms are used as part of an appropriate regimen. Each drug product involves several interrelated concepts that must be considered in its design and manufacture.

Attention to various physiochemical parameters of the drug moiety, such as particle size, crystalline form, and solubility, is vital to the design of a dosage form, as are its purity and accurate measurement. Nontherapeutic or excipient ingredients are selected to ensure stability (buffers, chelating agents, antioxidants, antimicrobial preservatives), and accuracy and precision of dosage (diluents, vehicles). Various types of excipients are used for specific types of dosage forms in order to permit their manufacture and desired therapeutic performances. Other excipients function as processing aids. Lubricating agents are solids used in tablet compression to lubricate the diewalls and punch faces to prevent sticking, capping, and/or excessive die-wall wear. Polymers find wide excipient use in dosage form design as viscosity-building agents in suspensions and emulsions and in the control of drug release in products prepared to achieve longer (8–12 h) than usual therapeutic periods. Various excipients are used to provide drug palatability for patients, e.g., colorants and flavoring agents.

The selection of excipient ingredients is important. These must be both chemically and physically compatible with the drug moiety and cannot negatively affect product stability or therapeutic performance, i.e., bioavailability.

The various preparation processes and technologies used in drug product manufacture also can effect product safety, stability, and performance, e.g., compression during tablet manufacture. The principal processes used in dosage form manufacture are as follows.

Dosage form types	Processes
Liquid solutions	Dissolution and filtration
Parenterals	Sterilization, lyophilization
Liquid dispersion (suspensions, emulsions)	Dispersion/wetting of solids, homogenization
Semisolid dispersions (ointments, creams)	Levigation, melting
Liquid/semisolid capsules	Soft gelatin encapsulation
Suppositories	Molding
Solids (granules, capsules, tablets)	Comminution, blending, granulation, compression, coating
Aerosols	Specialized packaging under pressure
General	Heating, cooling, mixing

The therapeutically active drug can be extracted from plant or animal tissue, or be a product of fermentation, as in the case of antibiotics.

Biological characterization includes toxicological studies, dose relationships, routes of administration, identification of side effects, and absorption, distribution, metabolism, and excretion patterns. If the results are still acceptable, product formulation and dosage form are developed.

Application for discovery and product patents must be made early in the process. Appropriate labels are designed and the product is submitted to the FDA for approval to begin human testing in the form of an Investigational New Drug Application (INDA). When such approval is granted, a clinical evaluation is developed which includes general testing for human pharmacology in healthy volunteers; clinical studies for therapeutic safety and efficacy in volunteer patients who are suffering from the disease for which the drug has therapeutic promise; and drug samples are made available to select clinicians for use on large numbers of patients.

Manufacturing, analytical, and quality control procedures are established. Specifications for raw and in-process materials, as well as for final products per USP/NF and in-house standards are also determined.

Process and formula validation assures that each technological procedure in manufacture accomplishes its purpose most efficiently, e.g., blending times for powdered mixtures in tableting, and that each formula ingredient is present in optimal concentrations. Thus, it serves to ensure process control, reproducibility, and content uniformity.

Stability studies are developed to assure a desirable shelf-life period. These also establish limits of acceptability for impurities and degradation compounds, when present, and determine acceptable storage conditions for raw materials and the manufactured products. Stability studies are thus important to the determination of expiration dates for drug products.

Finally, all data, including the results of the clinical investigation, are collected in a New Drug Application (NDA) and sent to the FDA. Once approved, the new drug goes into production. After manufacturing begins, the new drug products must be monitored in clinical use in the marketplace for reports of untoward reactions. This amounts to post-approval surveillance known as Phase IV. All such reports must be submitted to the FDA in a timely manner.

Bioavailability, Bioequivalence, and Pharmacokinetics. Bioavailability can be defined as the amount and rate of absorption of a drug into the body from an administered drug product. It is affected by the excipient ingredients in the product, the manufacturing technologies employed, and physical and chemical properties of the drug itself, e.g., particle size and polymorphic form. Two drug products of the same type, e.g., compressed tablets, that contain the same amount of the same drug are pharmaceutical equivalents, but may have different degrees of bioavailability. These are chemical equivalents but are not necessarily bioequivalents. For two pharmaceutically equivalent drug products to be bioequivalent, they must achieve the same plasma concentration in the same amount of time, i.e., have equivalent bioavailabilities.

Bioavailability, important to the design and preparation of drug products, can be affected adversely by the selection of excipients and/or the manufacturing processes used. Excessive pressure used in the compression of tablets, for example, could cause a tablet to pass through the gastrointestinal tract with no therapeutic effect.

Pharmacokinetics is the study of how the body affects an administered drug. It measures the kinetic relationships between the absorption, distribution, metabolism, and excretion of a drug. To be a safe and effective drug product, the drug must reach the desired site of therapeutic activity and exist there for the desired time period in the concentration needed to achieve the desired effect. Too little of the drug at such sites yields no positive effect (<MEC); too much (>MTC) leads to toxicity. For intravenous administration there is no absorption factor. Total body elimination includes both metabolic processing and excretion.

In cases of all but intravenous administration, dosage forms must make the active moiety available for absorption, i.e., for drug release. This influences the bioavailability and the drug's pharmacokinetic profile. Ideally the drug is made available to the blood for distribution and elimination at a rate equal to those processes. Through technological developments drug product design can achieve release, absorption, and elimination rates resulting in durations of activity of 8–12 hours, i.e., prolonged action/controlled release drug products.

Manufacturing

Compressed Tablets. This popular type of dosage form offers convenience, stability, accuracy and precision, and good bioavailability of active ingredients. After the best formulation has been established, compressed tablets can be manufactured at high rates of speed on advanced equipment. Tablets can be made to achieve rapid drug release or to produce delayed, repeat, or prolonged therapeutic action. Tablets are produced directly by compression of powder blends or granulations, which include a small percentage of fine, particle-sized powders.

Granulation. Granulation methods can be wet or dry. Wet granulation cannot be used for drugs that are sensitive to moisture and heat. The powered drug and diluent are blended with a dispersion of the binder excipient, e.g., gelatin, to a consistency that can be screened to 840–1800-μm granules (10–20 mesh). These granules are dried on trays in hot-air ovens or fluid-bed dryers. Dry granulation is used when the drug is not stable under the conditions of wet granulation and when the combined powders of a formulation cannot be compressed directly.

Direct Compression. This process is relatively simple and time saving. All the ingredients are blended and then compressed into the final tablet. This is an excellent method, but encumbered by a number of problems. Not all substances can be compressed directly, necessitating a granulation step. Likewise, the flow properties of many blends of fine, particle-sized powders are not such as to ensure even filling of the die cavities of tablet presses. In addition, air entrapment can occur.

The availability of spray-dried lactose, microcrystalline cellulose, and other excipients allows for the use of granular rather than powdered phases. This eliminates some of the problems of particle segregation according to size (demixing) and even flow to the die. Direct compression eventually may be the preferred method of tablet preparation.

Tablet Press. The main components of a tablet compression machine (press) are the dies, which hold a measured volume of material to be compressed (granulation), the upper punches which exert pressure on the down stroke, and the lower punches which move upward after compaction to eject the tablets from the dies. Mechanical components deliver the necessary pressure. The granulation is fed from a hopper with a feed-frame on rotary-type presses and a feeding shoe on single-punch presses. A smooth and even flow ensures good weight and compression uniformity. Using the proper formulation, demixing in the hopper is minimized.

Compressed tablets that are composed of several layers require specially adapted presses designed with several fed hoppers. For a two-layer tablet, one granulation is first fed to a die and partially compressed into a soft tablet. The second granulation is added, and the total die components then are compressed fully. Such procedures are used when the tablet ingredients may be incompatible, which requires separate granulations. If needed, a layer of inert ingredient, e.g., lactose, is inserted between the two.

Layered tablets are also used for a prolonged or sustained therapeutic effect. In this case, one layer disintegrates and dissolves rapidly to provide the initial dosing, whereas the other is designed for controlled release.

Formulation. Compressed tablet formulations contain several types of inert, adjuvant ingredients necessary for proper preparation and therapeutic performance. Tablets designed to be swallowed need diluent, disintegrating, binding (adhesive), and lubricating inert ingredients, whereas troches or lozenges intended to be dissolved slowly in the mouth should not disintegrate quickly, need more binder, and no disintegrant. Lactose or dicalcium phosphate are common diluents, whereas starch and cellulose derivatives are used as disintegrating agents.

Glidants are needed to facilitate the flow of granulation from the hopper. Lubricants ensure the release of the compressed mass from the punch surfaces and the release/ejection of the tablet from the die. Combinations of silicas, corn starch, talc, magnesium stearate, and high molecular weight poly(ethylene glycols) are used. Most lubricants are hydrophobic and may slow down disintegration and drug dissolution.

Colors and flavors increase the elegance and acceptability of the product. Sometimes colors are used for identification.

Effervescent tablets disintegrate by virtue of the chemical reaction occurring in water between component ingredients, such as sodium bicarbonate and citric or tartaric acid, to achieve release of carbon dioxide.

Coating. Sugar or film coatings offer protection from moisture, oxygen, or light and mask unpleasant taste or appearance. Enteric coatings delay the release of active ingredients in the stomach and may prolong the onset of therapeutic activity. The latter are used for drugs that are unstable to gastric pH or enzymes, cause nausea and vomiting or irritation to the stomach, or should be present in high concentration in the intestines, e.g., preoperative sterilization of the gut or as anthelmintics. Effectiveness depends on the varying pH patterns of the gastrointestinal tract and the enzymes present for dissolution and aqueous solubility.

Enteric coating is also used for repeat-action tablets, which contain an enteric-coated core tablet and a sugar or film-coated second dose, permitting the administration of two doses simultaneously. The core dose is released several hours after the initial, outer dose.

Some tablets that provide a sustained period (up to 8–12 h) of therapy may be coated during processing. A portion is released first to bring the drug to the desired blood concentration (onset of activity), whereas a sustained-release portion maintains an effective level for a prolonged period of time (duration of activity), e.g., by coating erosion or diffusion of drug through it.

A more recent development in tablet coating involves the use of gelation as the coating material to produce geltabs. If a tablet is compressed as a capsule-shaped unit prior to gelatin coating it is called a gelcap.

Capsules. Capsules are made in two types. In hard-gelatin capsules, powders or granules are enclosed in rigid gelatin shells. Soft-gelatin capsules contain glycerol as well as gelatin and maintain plastically even when dried. Hard-gelatin capsules are made in two sections, cap and body, which are then filled, whereas soft-gelatin capsules are formed and filled in succession in one manufacturing procedure. Soft-gelatin capsules are generally filled using nonaqueous solutions, although powders can also be used. Most drug companies buy the hard-gelatin shells from external sources. These are made by dipping precisely tooled pins into controlled solutions of gelatin. A film of gelatin adheres to the pins. Upon drying, the units are trimmed to specified length, removed from the pins, and the cap and body portions are joined. Various colors can be incorporated. See also **Gelatin**.

The formulations of filled, hard-gelatin capsules are generally less complex than those of compressed tablets, and require no binders or disintegrators. Upon swallowing, the capsule shell dissolves quickly and the powder ingredients are available for dissolution. Because no initial disintegration step is needed, bioavailability of drugs in capsule formulations is generally better than that of compressed tablets. The capsules are filled by various high speed machines.

Prolonged Action/Controlled Release. The therapeutic purpose of prolonged action and controlled release solid, oral drug products is to maintain safe and effective concentrations of the drug in the blood for 2–4 times longer than those times achieved using regular compressed tablets or capsules. This is accomplished by releasing one portion of the drug quickly, whereas the remaining portion is released at a rate that approaches the elimination rate. Ideally, the second portion should be released at a zero-order rate to achieve this profile. The technologies used for such controlled release only approach such a rate, but do accomplish the increased therapeutic period. See also **Barrier Polymers**.

The best drug candidates for incorporation into prolonged action systems are uniformly absorbed throughout the gastrointestinal (GI) tract, have medium (2–8 h) biological half-lives, and are prescribed for chronic maintenance use. Drugs in large doses are difficult to formulate into such products.

Liquid Dosage Forms. Simple aqueous solutions, syrups, elixirs, and tinctures are prepared by dissolution of solutes in the appropriate solvent systems. Adjunct formulation ingredients include certified dyes, flavors, sweeteners, and antimicrobial preservatives. These solutions are filtered under pressure, often using selected filtering aid materials. The products are stored in large tanks, ready for filling into containers. Quality control analysis is then performed.

Dosage forms of naturally occurring materials having therapeutic activity are prepared by extractive processes, especially percolation and maceration. Examples of such dosage forms have included certain tinctures, syrups, fluid extracts, and powdered extracts.

Solutions for external or oral use do not require sterilization but generally contain antimicrobial preservatives. Ophthalmic solutions and parenteral solutions require sterilization.

For the preparation of suspensions and emulsions, colloid mills and homogenizers, respectively, are used. Ultrasonic mills that utilize vibrating reeds in restricted chambers to reduce the particle size of the dispersed ingredients can also be employed. See also **Colloid System**.

Semisolid Dosage Forms. The ingredients that constitute the base of ointments, e.g., petrolatum and waxes, are melted together, powdered drug components are added, and the mass stirred with cooling. Generally, the product then is passed through a roller mill to achieve the particle-size range desired for the dispersed solid. Pastes are ointments having relatively large, dispersed solid content, and are prepared similarly.

Creams are semisolid emulsions either water-in-oil (w/o) or oil-in-water (o/w).

Suppositories are semi-rigid, plastic dosage forms are designed to deliver a unit dose of medication to body cavities, i.e., rectum, vagina, or urethra. Depending on the base, suppositories either melt (cocoa butter) at body temperature or dissolve (poly(ethylene glycols), glycerogelatin) in the fluids of the cavity. They can be used for systemic therapy (rectal suppositories) or for localized treatment. Rectal suppositories are a route of administration in comatose conditions or after gastrointestinal surgery, and for pediatric patients. On a large scale, suppositories are produced by molding.

Parenteral Dosage Forms. The most commonly used forms for drug products designed and manufactured for injection through the skin include those meant for subcutaneous, intramuscular, and intravenous administration.

Intravenous aqueous injections provide an excellent means of achieving a rapid therapeutic response. Parenteral product design, e.g., vehicle and other excipient selection, as well as choice of route of administration, can prolong therapeutic activity and increase onset times. Thus, oily solutions, suspensions, or emulsions can be administered by subcutaneous or intramuscular routes to create prolonged effect, i.e., depot injection.

Several factors of design and manufacture are of great importance: sterility, absence of pyrogens and foreign particulate matter, and tonicity. The last, when adjusted to the osmotic pressure of body fluids in the case of aqueous solutions, reduces the risk of tissue irritation and pain.

Lyophilization. Lyophilization is essentially a drying technology. Some drugs and biologicals are thermolabile and/or unstable in aqueous solution. Utilization of freeze drying permits the production of granules or powders that can be reconstituted by the addition of water, buffered solution, or mixed hydrophilic solvents just prior to use, e.g., certain antibiotic suspensions.

Ophthalmic Dosage Forms. Ophthalmic preparations can be solutions, e.g., eye drops, eyewashes, ointments, or aqueous suspensions. They must be sterile and any suspended drug particles must be of a very fine particle size. Solutions must be particle free and isotonic with tears. Thus, the osmotic pressure must equal that of normal saline (0.9% sodium chloride) solution. Hypotonic solutions are adjusted to be isotonic by addition of calculated amounts of tonicity adjusters, e.g., sodium chloride, boric acid, or sodium nitrate.

Radiopharmaceuticals. Radioactive isotopes for human use in the diagnosis and treatment of disease states are called radiopharmaceuticals. Whereas the dosage form types used, e.g., solutions or injections, are traditional, special handling of these products during compounding, transport, and use is vital. Most are administered intravenously and shortly after preparation. Specialized pharmacies prepare these products overnight and transport them to hospitals for early administration by members of nuclear medicine departments.

Aerosols. Pressurized containers to deliver aerosolized drug products through appropriate systems of valves and actuators have been available since the 1950s. See also **Aerosol.** Such dosage forms are used as external applications of lotions and creams, for oral inhalation, or for treatment of the vaginal cavity, e.g., contraceptive foams. Aerosols contain two- or three-phase systems, wherein a volatile liquid or ad-mixture of liquids is sealed in a container in equilibrium with a vapor phase (propellant). Upon actuation and delivery of the product, the propellant evaporates quickly, and fine dispersion of the drug settles on the area of application. For aerosol products that need accurate dosing, metered valves are used with the valve chamber being recharged between each actuation or dose.

The popularity of aerosols has been declining. A widely used group of propellants, the fluorinated hydrocarbons, have been restricted in use since it was found that they can harm the environment by reducing the ozone layer of the upper atmosphere. See also **Pollution (Air)**; and **Ozone.**

Biotechnology and Dosage Forms. In drug development, biotechnology generally is recognized as a term that identifies those technologies that utilize living organisms in the production and/or alteration of chemical entities that have potential therapeutic activity. Besides the production of pharmacologically or biochemically active moieties, these technologies also have been used to produce food ingredients, vaccines, diagnostic testing reagents, and agricultural products. See also **Fermentation**; and **Vaccine Technology.**

Packaging. The packaging components of pharmaceutical products are vital to their safe and effective use. Besides serving the patient as a convenient unit of use, the composite package (unit container, labeling, and shipping components) must provide appropriate identification and necessary information for proper use (including warnings and cautions) and preservation of the product's chemical and physical integrity.

Labeling. Labeling, controlled by FDA regulations, includes not only the affixed labels, but also the package inserts that provide more detailed information. Trade, generic, or common name, dose, number of dose units present, and name and address of manufacturer and distributor are required. For nonprescription products, adequate directions for use are required. Prescription products must bear the phrase, "Caution: Federal law prohibits use without a prescription" on their labels.

All drug labels must include batch or lot numbers. The nature of the drug product may require special cautionary phrases, e.g., "store in cool place or refrigerator," "protect from light," and "shake well before using." In the 1990s, labels also carry the expiration date, i.e., shelf-life. This information is expected to become mandatory.

Labeling information also includes warnings as to possible side effects, e.g., drowsiness, and potential harm if used with other drugs or certain foods (drug–drug or drug–food interactions). Inserts are generally intended for use by physicians or pharmacists and give name and description of the product, mode of administration, dosage regimen, therapeutic indications and contraindications, precautions and side effects, units of supply, and literature citations. All labeling must be approved by the FDA as part of the New Drug Application.

Containers. The USPXXIII–NFXVIII lists container requirements such as well-closed, tight, or light-resistant. Most containers are light-resistant (amber) glass or plastic. The latter is break-resistant and lightweight, which reduces shipping costs and increases safety.

In hospitals and long-term care units, unit-dose packages are used more and more. This system allows better control of the dispensed drugs in institutional settings and precludes the dispensing of larger numbers of doses than needed.

Quality Control and Quality Assurance

Quality control (QC) involves the regular, daily assessment and/or analysis, according to established protocols and standards, of all ingredients, processes, and finished products. Official USP/NF monographs, for example, provide various chemical, physical, and biological tests and specifications for assurance of purity, potency, and stability of component ingredients used to prepare and package drug products. The FDA requires process validation procedures as QC constituents. The FDA also monitors QC standards through the requirements of the Current Good Manufacturing Procedures regulations.

Additional Reading

Ansel, H.C. and N.G. Popovich: *Pharmaceutical Dosage Forms and Drug Delivery Systems,* 5th Edition, Lea & Febiger, Philadelphia, PA, 1990, pp. 92–133.
Banker, G.S.: in G.S. Banker and C.T. Rhodes, eds., *Modern Pharmaceutics,* 2nd Edition, Marcel Dekker, New York, NY, 1990, pp. 15–20.
Kayser, O. and R.H. Muller: *Pharmaceutical Biotechnology: Drug Discovery and Clinical Applications,* John Wiley & Sons, Inc., Hoboken, NJ, 2004.
Lee, D. and M. Webb: *Pharmaceutical Analysis,* CRC Press LLC., Boca Raton, FL, 2003.
Trends in U.S. Pharmaceutical Sales and R & D: 1990–93 PMA Annual Survey Report, Pharmaceutical Manufacturers Association, Washington, DC, 1993.
Walsh, G.: *Biopharmaceuticals: Biochemistry and Biotechnology,* 2nd Edition, John Wiley & Sons, Inc., New York, NY, 2003.
Zanowiak, P.: in *Ullmann's Encyclopedia of Industrial Chemistry,* VA19, VCH Verlagsgesellschaft, MbH, Weinheim, Germany, 1991, pp. 241–271.

Web References

American Pharmaceutical Association (AphA): http://www.aphanet.org/
American Pharmaceutical Association Academy of Pharmacy Practice: http://www.aphanet.org/APPM/APPMpig.html
Pharmaceutical Research and Manufacturers of America, (PhRMA) Home: http://www.phrma.org/
U.S. Food and Drug Administration: http://www.fda.gov/

PHARMOCODYNAMICS. Pharmacodynamics is the study of drug action primarily in terms of drug structure, site of action, and the biochemical and physiological consequences of the drug action. The availability of a drug at its site of action is determined by several processes, including absorption, metabolism, distribution, and excretion. These processes constitute the pharmacokinetic aspects of drug action. The onset, intensity, and duration of drug action are determined by these factors as well as by the availability of the drug at its receptor site(s) and the events initiated by receptor activation.

Both pharmacokinetic and pharmacodynamic processes are involved in mediating nonconstant expressions of drug action. Thus, resistance to the actions of a drug, e.g., in the development of antibiotic-resistant bacteria or of barbiturate tolerance, can arise from changes in drug metabolism and/or alterations in the receptor target site. Factors controlling drug resistance may be whole-body, cellular, or individual events. Decreased absorption, increased metabolism, or increased elimination reduce circulating drug levels and affect the whole body. Increased drug metabolism, increased concentration of an agent that antagonizes drug action, decreased affinity or concentration of a drug receptor, and depletion of an agent that mediates drug action are examples of cellular events; and genetic factors controlling metabolism, receptor alterations, and disease states are examples of individual events. Individual variation in the susceptibility to a particular drug or class of drugs also may arise from genetically based pharmacokinetic factors as well as from specific receptor-linked changes.

For a large number of drugs, including neurotransmitters, peptide and protein hormones, and their analogues and antagonists, the cell membrane is the principal locus of action. Concepts of cell membrane structure are derived from the original Davson-Danielli lipid bilayer hypothesis. More specifically, the membrane is viewed as a dynamic fluid mosaic or a matrix of fluid bilayer in which there are asymmetrically inserted proteins and glycoproteins. Phospholipids and proteins diffuse laterally and the resultant protein-protein communication is of considerable importance to the understanding of membrane-receptor function. Despite the dynamic nature of the membrane and the absence of global organization, local organization is possible through the local assembly of individual protein components and the attachment of membrane proteins to the subcellular structure of contractile proteins. However, the cell membrane is not the site of action of all drugs. A number of drugs, including steroid and thyroid hormones, exert their effects intracellularly at the level of the genetic material as well as at the plasma membrane. Other agents, including polypeptide growth factors, exert their effects not only at the plasma membrane through tyrosine kinase receptors, but also on cell growth and differentiation at the genetic level.

See also **Protein**; and **Steroids**.

The Receptor Concept

Drug receptors are chemical entities which are typically, but not exclusively, small molecules that interact with cellular components, frequently at the plasma membrane level. There are many types of receptors; heat, light, immune, hormone, ion channel, toxin, and virus are but a few that can excite a cell. The receptor concept can be applied generally to signal recognition processes where a chemical or physical signal is recognized. This recognition is translated into response and the process can be seen as a flow of information.

Elucidation of the structural requirements for drug interaction at the recognition site is by the study of structure-activity relationships (SAR), in which, according to a specific biologic response, the effects of systematic molecular modification of a parent drug structure are determined. Such studies have permitted the classification of discrete classes of pharmacological receptors.

The demonstration of the existence of strictly defined SARs, which is perhaps the most important criterion of drug action at a specific receptor site, has made possible the most important pharmacologic discoveries. For example, the analgesic actions of morphine and related agents, which are indicative of specific receptors, led to the discovery of endogenous opiate peptides, i.e., the leucine and methionine enkephalins and endorphins.

Pharmacokinetic Aspects of Drug Action

The receptor represents the locus of drug action. However, the pharmacokinetic processes of absorption (drug entry), distribution, metabolism, and excretion play principal roles in determining *in vivo* time courses and concentrations of drugs and thus modify actions initiated at receptors.

Drug Entry. Drugs enter the body by one of two routes. In enteral administration (sublingual, oral, rectal), the drug enters directly the gastrointestinal tract. In the parenteral route, the drug bypasses the gastrointestinal tract by, among others, subcutaneous (sc), intramuscular, intravascular (iv), inhalational, intraperitoneal (ip), intravaginal, and intranasal routes. Each route has a particular set of advantages and disadvantages. Patient convenience is high in the oral route; speed of action and ability to control concentrations are high in the iv route; and nonoral routes are best for unstable or insoluble drugs.

In light of the recognized importance of achieving stable, reproducible plasma concentrations of drugs, particular attention is given to pathways and devices, including sustained-release formulations, pumps, and transdermal entry processes that ensure such properties.

Drug Distribution. After administration, a drug may be distributed either generally or selectively in the body. The distribution pattern depends on many factors, including the pattern and time-course of blood flow, diffusion of drugs into tissues, binding of drugs to plasma proteins and cellular compartments, and elimination kinetics and mechanisms.

Drug Metabolism. Generally, metabolism (biotransformation) of drugs increases their water solubility as well as the rate and ease of elimination, but reduces their volume of distribution. Many drug-metabolizing pathways have arisen during evolution to deal with foreign compounds present in food materials. Although metabolism generally leads to more polar and less active compounds, there are exceptions. Metabolic pathways have also been exploited to design prodrugs, materials that are converted to active species through biotransformation.

Biotransformation reactions can be classified as phase I and phase II. In phase I reactions, drugs are converted to product by processes of functionalization, including oxidation, reduction, dealkylation, and hydrolysis. Phase II or synthetic reactions involve coupling the drug or its polar metabolite to endogenous substrates and include methylation, acetylation, and glucuronidation.

The biotransforming pathways are subject to manipulation and modification in a variety of ways. Drug metabolism also depends on age and sex. Drug metabolism may also produce toxic materials.

Drug Elimination. Drugs are removed from their sites of action through metabolism, storage, and excretion. These processes are not necessarily independent and drugs are frequently metabolized prior to excretion. Indeed, for lipophilic drugs this is virtually a necessity. Drugs are excreted via the kidneys, biliary systems, intestines, and lungs.

The process of reabsorption depends on the lipophilic-hydrophilic balance of the molecule. Charged and ionized molecules are reabsorbed slowly or not at all. Reabsorption of acidic and basic metabolites is pH-dependent, an important property in detoxification processes in drug poisoning. Both passive and active carrier-mediated mechanisms contribute to tubular drug reabsorption.

Clinical Pharmacokinetics. Clinical pharmacokinetics attempts to define the relationship between drug concentration and therapeutic response. The underlying assumption is that response is proportional to drug concentration at the site of action. This concentration is dependent on many factors that are frequently pharmacokinetic determinants. The most important factors are defined as clearance, bioavailability, and volume of distribution.

Pharmacodynamic Aspects of Drug Action. Although the same general principles of chemical specificity apply to all ligand-macromolecular interactions, the term receptor is generally applied to those cellular macromolecules and macromolecular complexes with which ligands, physiological or synthetic, interact both to complex and to initiate a physiological response. Receptors are conveniently viewed as existing in several principal classes, i.e., G-protein-coupled receptors, ligand-gated ion channels, voltage-gated ion channels, tyrosine kinase receptors, guanylyl cyclase receptors, and steroid hormone receptors. All of these receptors form homologous classes according to structure and mechanisms of action. G-protein-coupled receptors form a homologous class of membrane proteins characterized by seven transmembrane domains and the ability to couple to guanine (G) nucleotide-binding proteins.

Ligand-gated ion channels represent a significant family of ion channels that feature as an integral component of their multimeric subunit organization a receptor site for either acetylcholine (nicotine acetylcholine receptor (AChR)), amino acids including glycine and Γ-aminobutyric acid (GABA) (inhibitory transmitters), or glutamic acid (excitatory transmitter). The interaction of the ligand with the endogenous receptor site causes channel opening or closing.

An important characteristic of both classes of ion channel is that they possess multiple drug binding sites. Many of the channel-active drugs have achieved particular therapeutic importance, including, for example, the Ca^{2+} antagonists, widely used for a number of cardiovascular disorders, such as hypertension.

Structure—Activity Relationships. Until the mid-1980s, the attempted correlation of chemical structure and biological activity was the only available approach to the definition of receptor site structures. The basic assumption in the analysis of structure-activity relationships (SAR) is the existence of a definable mutual complementarity between the structure of the drug and its corresponding binding site. This application is limited when applied in empirical fashion. Many drug molecules are flexible structures and, although conformations in the solution and solid states can be determined by spectroscopic and crystallographic methods, these bear no necessary relationship to those adopted at the receptor site. The possibility of mutual conformational adaptation of both the drug and the receptor site during the binding process adds a further complication. Furthermore, there may exist multiple drug-binding modes at the receptor such that transitions in binding modes occur at some point in a structurally related series. An additional problem in the quantitative interpretation of SAR is that of the relationship between biological response and drug-receptor interaction. Despite these limitations, SARs have been of great value in providing qualitative concepts of binding site geometry, classifying receptors, furnishing evidence for the existence of new classes of receptor-specific drugs, and generating new and therapeutically effective compounds.

The simplest SARs occur in homologous series of compounds.

Relatively unambiguous monotonic SARs also occur where activity depends on the ionization of a particular functional group.

The SAR is also determined at the level of stereochemistry of interaction. In principle, three limiting situations can apply to the stereochemistry of drug-receptor interactions: the enantiomers may not differ in activity; the species may differ quantitatively; or they may differ qualitatively.

The issue of drug stereoselectivity has become one of both developmental and regulatory significance. In principle, a racemic drug possesses only 50% of the active ingredient, and the rest may have other or interacting pharmacologic activities, which may contribute a metabolic burden or be inert. Over 50% of clinically available drugs have chiral centers and only about 10% of synthetic chiral drugs are marketed in homochiral (enantiomerically pure) form. In contrast, drugs that are naturally occurring substances, obtained from or related to naturally occurring molecules, are frequently homochiral.

There is increasing pressure to develop homochiral drugs.

Advancing technology permits increasing attention to the definition of the three-dimensional structure of the ligand in its bioactive conformation as it binds to the receptor or active site.

Considerable effort must be applied to obtaining adequate quantities of the protein target and its structural solution, together with the structural solution of the complexed ligand, either by x-ray or solution NMR techniques. Alternatively, homology modeling may be possible when the structure of a homologue protein is already available. The examination of the real structure of ligand—receptor complexes should be an increasingly important and integral part of the drug discovery process.

Nonreceptor-Mediated Drug Action. At least one important class of drugs, the general anesthetics, has been assumed not to owe its therapeutic activities to a specific receptor process. Anesthetic potency shows an excellent linear correlation with partition coefficient and this has been extrapolated to a definition of action at a lipid site. The phospholipids of cell membranes, particularly nerve cells, have been considered as principal targets for general anesthetic action. It has been hypothesized that anesthetics may disrupt phospholipid structure by fluidizing or expanding the cell membrane or by altering the phase relationships of the phospholipids. However, it is possible that anesthetics bind to hydrophobic sites on proteins and thus affect directly excitable cell behavior. This latter proposal is consistent both with the activity of the gaseous general anesthetics and with the activity of structurally more complex agents.

Although most anesthetics are achiral or are administered as racemic mixture, the anesthetic actions are stereoselective.

It is likely that a principal target of the general anesthetics is neuronal ion channels of both voltage-gated and ligand-gated classes. Interactions at GABA-mediated inhibitory channels is a significant, but not exclusive, target. Thus, a general anesthetic may have specific but multiple, rather than nonspecific, sites of action.

Receptor—Effector Coupling. The informational signal initiated by drug—receptor interaction must be translated to biological response. This is activated by a variety of effector-coupling processes that lead to ionic or biochemical changes, including ion channel opening and closing; the formation of second messengers such as cyclic adenosine-3′-5′-monophosphate (cAMP) and inositol-1,4,5-triphosphate (IP_3); and protein phosphorylation through protein kinase A (cAMP-dependent) and protein kinase C (CA^{2+}-dependent), or through autophosphorylation (tyrosine kinase receptors). In these systems, it is increasingly clear that the individual components of a receptor system may be linked in multiple ways. The virtue of this organization lies in the multiple coupling processes permitted beyond a set of components.

These cascades serve as operational amplifiers of the initial ligand—receptor interaction. In each step of the process, amplification by several powers of 10 may occur so that an original signal may be multiplied several millionfold.

G-Protein Coupling. The heterotrimeric guanosine triphosphate (GTP) binding proteins, known as G-proteins, are a principal family of proteins serving to couple membrane receptors of the G-protein family to ionic and biochemical processes. The G-proteins are heterotrimers made of three families of subunits, α, β, and γ, which can interact specifically with discrete regions on G-protein-coupled receptors. This includes most receptors for neurotransmitters and polypeptide hormones (see **Neuroregulators**). G-protein-coupled receptors also embrace the odorant receptor family and the rhodopsin-linked visual cascade.

A critical component of the G-protein effector cascade is the hydrolysis of GTP by the activated α-subunit (GTPase). This provides not only a component of the amplification process of the G-protein cascade but also serves to provide further measures of drug efficacy. The coupling process also depends on the stoichiometry of receptors and G-proteins. A reduction in receptor number should diminish the efficacy of coupling and thus reduce drug efficacy.

The ability of receptors to couple to G-proteins and initiate GTPase activity may also be independent of ligand.

Ion Channels. The excitable cell maintains an asymmetric distribution across both the plasma membrane, defining the extracellular and intracellular environments, as well as the intracellular membranes which define the cellular organelles. This maintained asymmetric distribution of ions serves two principal objectives. It contributes to the generation and maintenance of a potential gradient and the subsequent generation of electrical currents following appropriate stimulation. Moreover, it permits the ions themselves to serve as cellular messengers to link membrane excitation and cellular response. In some instances, the current itself may be the response, as, for example, in the electric organ of electric fishes. In most instances, however, the current serves to initiate or modulate another cellular response, including propagation of impulses in nerve fibers, and alteration of the sensitivity of membranes to other stimuli or coupling to cellular responses such as contraction and secretion. In the latter examples, a role for calcium is particularly prominent because Ca^{2+} can serve as both a current-carrying and a messenger species.

Regulation of ion channels by drugs may have excitatory or inhibitory effects according to the channels affected.

Channels may be regulated exclusively by electrical or chemical signals corresponding to purely voltage-gated or ligand-gated channels, respectively.

Ion channels may be regarded as pharmacological receptors frequently possessing a multiplicity of drug binding sites. These sites may be for endogenous physiological regulators or for endogenous or synthetic agents.

Tyrosine Kinase Receptors. The polypeptide growth factors control cell proliferation, differentiation, and survival. Several distinct subfamilies of receptor tyrosine kinases exist and at least nine have been characterized. These include families for epidermal growth factor, insulin and insulin-related factors, fibroblast growth factors, and neurotrophin receptors such as nerve growth factor and brain-derived neurotrophic factor. All of these receptors have kinetics that share certain fundamental signaling properties. Ligand binding to the extracellular domain activates a tyrosine kinase of the cytoplasmic domain. Subsequently, a variety of downstream signaling molecules are activated. These include phospholipase C, GTPase activating factor (GAP), *Ras*, and MAP kinases.

Guanylyl Cyclase Receptors. Cyclic GMP concentrations (cGMP) rise in response to a number of cell signals. Membrane-associated guanylyl cyclase catalyzes the conversion of guanosine triphosphate (GTP) to cGMP. This enzyme resembles in organization the tyrosine kinases having an intracellular protein kinase-like domain and a cyclase catalytic domain. The enzymes are activated by several distinct species that include atrial natriuretic peptide (ANF) and peptides related to the heart-stable enterotoxins.

Receptor Regulation and Defects. Specific recognition and the initiation of response are the accepted attributes of the drug–receptor interaction. However, target cells can alter on both short- and long-term time scales their sensitivity to drugs. Such regulation, achieved by altering the number and/or affinity of receptors, is well established for all receptor systems and can be viewed as an integral component of the drug–receptor interaction. In this view, subsequent to the formation of the drug–receptor complex with agonist, the continued existence of the drug–receptor complex may lead to one or more phases of desensitization, according to which there may occur initially transient and subsequently prolonged phases of reduced or lost sensitivity. Occupancy by antagonist, in contrast, leads to an increased number of receptors and increased drug sensitivity. This phenomenon may contribute to clinical rebound during abrupt withdrawal from drugs, including β-blockers. Additional to this homologous regulation, receptor sensitivity may be controlled through heterologous influences, whereby hormones, including thyroid and corticosteroids, regulate other receptors. These regulatory events are made possible because pharmacologic receptors, in common with other cellular components, are in dynamic balance between synthesis and degradation. This balance is sensitive to a number of influences that include agonist and antagonist presence.

An increasing number of diseases are known to be linked to defects in receptor structure, function, or coupling. The defects may lie at several locations: in the structure of the receptor, which may alter its ability either to bind drugs, to be inserted into the membrane, or to couple to effectors (including G-proteins); in the coupling protein; or in the presence of autoantibodies, which can proceed to activate, block, or lyse the receptors and its components.

Components of Drug Action and Responses to Drugs. The response to a drug can vary among race, gender, and age groups. It may vary according to disease state and age, and it may vary according to the time of administration. These factors may have several origins, including (*1*) compliance, the ability or desire of the subject to take a drug according to a specific regimen; (*2*) pharmacokinetic, disease-, age-, race-, and gender-based factors that contribute to variable absorption, distribution, metabolism, and excretion of a drug; and (*3*) pharmacodynamic, disease-, age-, race-, and gender-based factors that contribute to variable drug–receptor interactions.

Additional Reading

Burton, M., W.E. Evans, J.J. Schentag, and L. Shaw: *Applied Pharmacokinetics and Pharmacodynamics: Principles of Therapeutic Drug Monitoring*, Lippincott Williams & Wilkins, Philadelphia, PA, 2005.

Figg, W.D., and H.L. McLeod: *Handbook of Anticancer PharmacoKinetics and Pharmacodynamics*, Humana Press, Totowa, NJ, 2004.

Gilman, A.G. and co-workers, eds.: *The Pharmacological Basis of Therapeutics*, 8th Edition, Pergamon Press, New York, NY, 1990.

Pratt, W.B. and P. Taylor, eds.: *Principles of Drug Action: The Basis of Pharmacology*, 3rd Edition, Churchill Livingstone, New York, NY, 1990.

Tozer, T.N.: *Primer in PharmacoKinetics and Pharmacodynamics*, Lippincott Williams & Wilkins, Philadelphia, PA, 2006.

Wermuth, C.G. ed.: *The Practice of Medicinal Chemistry*, Academic Press, San Diego, CA, London, UK, and New York, NY, 1996.

Wolff, M.E. ed.: *Burger's Medicinal Chemistry and Drug Discovery*, Vol. I., Principles and Practice, Wiley-Interscience, New York, NY, 1995.

PHARYNGITIS. A disease of some part of the throat and sometimes referred to simply as sore throat, tonsilitis, or strep throat, depending upon the specific site and severity of the condition. Pharyngitis is one of the most commonplace of upper respiratory tract infections, yet some cases can be complex and difficult to precisely diagnose immediately. It is estimated that from 70 to 80% of the cases arise from a variety of viruses; from 20 to 30% of cases from Group A streptococci, but other microorganisms can less frequently be involved. Simple acute pharyngitis is self-limiting and usually appears rather suddenly with a feeling of dryness and soreness in the throat. There may be a constant desire to clear the throat, pain on swallowing, headache, a dry, harsh cough, and an elevated temperature of 100–102 °F (37.8–39.9 °C). A generalized feeling of fatigue is usually present. Occasionally, there is pain in the ears, and if the infection spreads to the voice box, hoarseness will result. In children, the symptoms are usually more pronounced. The disease runs its course in a few days to a week. If the pharyngitis is thought to be caused by an organism sensitive to an antibiotic, such may be prescribed. Where the cause is definitely diagnosed as Group A streptococci, antibiotics are frequently administered as a preventive measure and to limit droplet spread of the organisms. In as much as rheumatic fever may follow an infection with Group A streptococci, early and effective treatment of respiratory tract infections, including pharyngitis, is a precaution against initial attacks of rheumatic fever. During untreated epidemics of streptococcal pharyngitis, surveys have shown that the incidence of rheumatic fever among those infected may reach 2%. Groups C and G streptococci also may cause pharyngitis, but such infections are not known to predispose rheumatic fever or other complications, such as glomerulonephritis.

Chronic pharyngitis is the result of repeated attacks of acute pharyngitis. Enlarged tonsils and adenoids may cause the condition, as well as constant breathing through the mouth. This form is frequently associated with chronic colds, sinusitis, and nasal infections.

In *septic sore throat* (strep or streptococcal throat), the symptoms are those of an acute pharyngitis, but more severe. A pseudomembrane often appears in the throat. The patient's temperature may rise as high as 105 °F (40.6 °C). The lymph nodes in the upper part of the neck enlarge and become sore. Antibiotics may be used.

Trench mouth (*Vincent's angina*) is characterized by a pseudomembrane, grayish or yellow-gray in appearance and often spreading from the gums. It is an acute inflammatory disease of the gums, which may be accompanied by pain, bleeding, and offensive breath, as well as fever.

Pharyngeal diphtheria is described in the entry on **Diphtheria**.

Peritonsillar abscess (quinsy sore throat) is caused by an abscess in the tissue surrounding the tonsil. The first symptoms are those of acute pharyngitis, but after a few days, the pain becomes localized in one side of the throat. Swallowing or expectorating becomes extremely painful. The abscess is a complication of strepotococcal tonsilitis and usually occurs in adolescents and young adults. Treatment of this condition consists of parenteral penicillin and surgery. A tonsilectomy is usually prescribed from 4 to 6 weeks after the initial phases of the condition, although earlier tonsilectomy has also been reported as providing excellent results.

Gonococci can produce pharyngitis in persons who practice orogenital sexual activity. Penicillin and tetracycline regimens generally are effective—spectinomycin is not. Although pneumococci and staphylococci commonly reside in the mouth and do cause diseases in other parts of the respiratory tract, they are seldom implicated in pharyngitis. In malnourished infants, a rare but serious invasive gangrene (*cancrum oris*) can occur. This condition is caused by a combination of bacteria and spirochetes. The spirochete *Treponema pallidum* may cause pharyngitis in primary or secondary syphilis.

PHARYNX. A muscular portion of the alimentary tract (digestive system) of invertebrates of various phyla, between the mouth and the esophagus. In some species, it acts as a suctorial organ. In vertebrates, the pharynx is a region of the endodermal part of the gut just behind its union with the ectoderm of the oral cavity. It is associated with the respiratory system. In humans, the pharynx is important as the source of ductless glands and is located under the soft palate and in back of the tongue. See also **Digestive System (Human)**.

PHASE.

1. For any type of periodic motion (e.g., rotation, oscillation) a point or stage in the period to which the motion has advanced with respect to a given initial point. Specifically, the phase or phase angle is the angular measure along a simple harmonic wave, the linear distance of one wavelength being 360° of phase measure. This is often generalized by equating one cycle of any oscillation to 360°. See also **Delay**; **Interference (Wave)**; and **Surface of Constant Phase**.

2. The state of aggregation of a substance, for example, solid, liquid, or gas.

AMS

PHASE CHANGE (or Phase Transformation).   A thermodynamic process in which a substance changes from one phase to another. A phase change entails discontinuity. At a given temperature, two phases of a pure, homogeneous substance are characterized by different enthalpies (and entropies), and the enthalpy difference is called latent heat. The phase transitions of greatest importance to meteorology are those between water vapor, liquid water, and ice.

PHASE DELAY.

1. In the transfer of a single-frequency signal from one point to another in a system, the time delay of a part of the wave identifying its phase.
2. The difference in phase between one signal and another.

PHASE DIAGRAM (Metallurgy).   A graphical representation defining the phase fields of a multiphase system, such as an alloy, in a coordinate system using the temperature and the compositions of the phases as coordinates. A phase diagram may be an equilibrium diagram, but it may also sometimes show the boundaries of the phase field under nonequilibrium conditions corresponding to specific conditions of heating or cooling. See iron carbon equilibrium diagram in entry **Iron Metals, Alloys, and Steels**.

Additional Reading

Frick, J.P.: *Woldman's Engineering Alloys*, 9th Edition, ASM International, Materials Park, OH, 2000.

Gupta, K.P.: *Phase Diagrams of Ternary Nickel Alloys: Part I and II*, ASM International, Materials Park, OH, 1990.

Kassner, M.E. and D.E. Peterson: *Phase Diagrams of Binary Actinide Alloys*, ASM International, Materials Park, OH, 1995.

Massalski, T.B.: *Binary Alloy Phase Diagrams*, 2nd Edition, ASM International, Materials Park, OH, 1990.

Massalski, T.B.: *Binary Alloy Phase Diagrams Materials Network User*, ASM International, Materials Park, OH, 1996.

Rogi, P. and J.C. Schuster: *Phase Diagrams of Ternary Boron Nitride and Silicon Nitride Systems*, ASM International, Materials Park, OH, 1992.

Subramanian, P.R., D.J. Chakrabarti, and D.E. Laughlin: *Phase Diagrams of Binary Copper Alloys*, ASM International, Materials Park, OH, 1994.

Staff: *ASM Handbook, Vol. 3, Alloy Phase Diagram*, 10th Edition, ASM International, Materials Park, OH, 1992.

Staff: ASM International *Superalloys: A Technical Guide*, 2nd Edition, ASM International, Materials Park, OH, 2002.

Villars, P., A. Prince, and H. Okamoto: *Handbook of Ternary Alloy Phase Diagrams*, ASM International, Materials Park, OH, 1995.

PHASE DIAGRAM (Statistics).   A diagram showing two time-series x_1 and x_2 plotted as ordinate and abscissa. If the fluctuations of these two variates keep in step, then the line joining the plotted points will trace a definite pattern, for example, somewhat like an ellipse for oscillatory series.

PHASE DIFFERENCE.   With reference to industrial and scientific instruments, the Scientific Apparatus Makers Association defines phase difference as:

1. Between sinusoidal input and output of the same frequency, the phase angle of the output minus the phase angle of the input.
2. Of two periodic phenomena (e.g., in nonlinear systems), the difference between the phase angles of their two fundamental waveforms.

Phase difference is regarded as part of the transfer function which relates output to input at a specified frequency. Measurement of phase difference in the complex case is sometimes made in terms of the angular interval between respective crossings of a mean difference line, but values so measured will generally differ from those made in terms of the fundamental waveforms.

PHASE FUNCTION.   A function describing the dependence of scattered radiance on scattering angle. The phase function is a dimensionless and normalized version of the scattering function, such that the integral of the phase function over 4π steradians equals 4π. The phase function for a given wavelength is a property of the medium, not of the incident radiation, provided the incident radiation is unpolarized. As the size parameter of a scatterer increases, its phase function becomes more anisotropic, with progressively more of the scattered radiance being concentrated into a diffraction peak.

AMS

PHASE INSTABILITY.   The instability associated with a supercooled or supersaturated phase of matter. A small fluctuation (on a molecular scale) or foreign particle may cause change to a more stable phase. For example, at a temperature below the equilibrium melting point of ice, the ice phase is stable, whereas liquid water is supercooled and is unstable because its chemical potential is higher than that of ice. A small fluctuation such as collision with ice may cause the supercooled water to change into ice.

AMS

PHASE INVERTER.   The push-pull amplifier has numerous advantages over single-ended operation but it does require special circuit provisions to accomplish the push-pull energization of the tubes or transistors. Since the usual amplifier has low-level stages operating single-ended, this requires some method of getting two signals equal in amplitude and 180° out of phase to drive the input of the push-pull stage. One method is to use a transformer to couple the output of the single-ended stage to the push-pull input by center-tapping the secondary. However, for resistance coupling, a phase inverter may be used. Various circuit arrangements may be used to accomplish this function; one typical circuit using transistors is shown in Fig. 1. With reference to the figure, part of the output signal of the upper transistor is coupled back into the input of the lower device. Since the output voltage of the amplifier stage is 180° out of phase with respect to the input, in the normal operating frequency range, the input to the second transistor is 180° from that of the first, hence their outputs are likewise displaced. If the tap on the output resistor of the first transistor is adjusted correctly, just enough voltage is applied to the input of the second to cause its output to be equal in amplitude to that of the first. These equal and opposite voltages may then be applied to the inputs of the two devices in the succeeding push-pull amplifier stage.

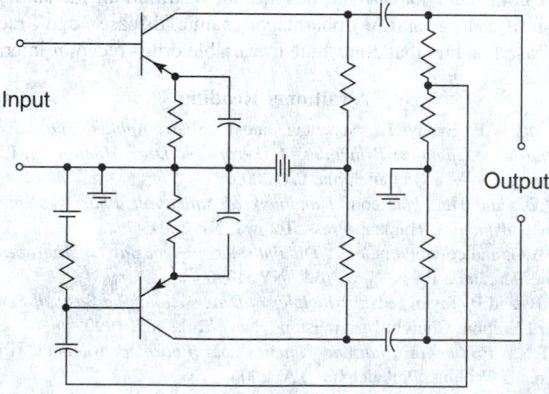

Fig. 1.   Phase inverter.

PHASE MODULATION.   Modulation in which the phase of a signal is caused to depart from its reference value by an amount proportional to the instantaneous value of the modulating function. A signal phase modulated by a given function can be regarded as the same as a signal frequency modulated by the time derivative of that function.

See also **Modulation**.

PHASE RULE.   The phase rule, due to Gibbs, gives the number F of intensive variables which can be fixed arbitrary in a system in equilibrium. This number is also called the variance or the number of degrees of freedom

of the system. It is given by

$$F = 2 + (C' - R) - P$$

where C' is the number of components, R, the number of independent chemical reactions, and P, the number of phases.

In terms of the number of independent components, C, Equation (1) may also be written

$$F = 2 + C - P$$

If $F = 0$ the system is invariant. We cannot fix either temperature or pressure arbitrarily. Equilibrium can only be established at isolated points. An example is the *triple point* at which pure solid, liquid and vapor are in equilibrium ($F = 2 + 1 - 3 = 0$).

If $F = 1$ the system is *monovariant*. We can, for example, fix the temperature, but the equilibrium pressure is then fixed. This is the situation for a system containing one component and two phases.

If $F = 2$ the system is *bivariant*. Within certain limits both pressure and temperature can be given arbitrarily. This is the situation for $C = 1$ and $P = 1$, or $C = 2$ and $P = 2$.

PHASE SHIFT.

1. Change in phase of a signal as it passes through a filter, some other system component, or a transmission medium.

2. The method employed for electronic beam steering in phased-array radar.

3. For polarimetric radar, the differential phase shift between the copolarized and cross- polarized components, a measure of the propagation effect interpreted in terms of the degree of preferred orientation of the scatterers in the propagation medium.

PHASE-SHIFTING CIRCUIT.
Any circuit containing resistive and reactive components used to produce a desired difference between the phase of the output voltage or current and the input voltage or current at some desired frequency.

PHASE-SHIFT OSCILLATOR.
An oscillator produced by connecting any network having a phase shift of an odd multiple of 180° (per stage) at the frequency of oscillation, between the output and the input of an amplifier. When the phase shift is obtained by resistance-capacitance elements, the circuit is an R-C phase-shift oscillator.

PHASE SPACE.
1. A Fourier spectral representation of physical space. By computing the Fourier transform of a physical signal, such as temperature measurements frequently sampled during some period, the resulting amplitudes of the sine and cosine waves are an alternative way of describing the data. These amplitudes represent the signal in phase space.

2. A technique used in nonlinear dynamics and chaos theory to examine processes by the evolving relationship between dependent variables, rather than by the relationship between dependent and independent variables.

See also **Fast Fourier Transform (FFT)**.

AMS

PHASE SPECTRUM.
A measure of the relative phase between two meteorological variables, segregated by wavelength. It is the phase difference for any frequency between two time series that yields the greatest correlation. The phase spectrum is computed as the arctangent of the ratio of the quadrature spectrum to the cospectrum. For example, turbulence usually consists of vertical velocity and potential temperature either in phase (for daytime convection) or 180° out of phase (for turbulence in a statically stable environment). However, gravity waves usually consist of vertical velocity and potential temperature that are 90° out of phase. Thus, if one were analyzing the cross spectra at night in a stable boundary layer, and found 90° phase shift for the longer wavelengths, but 180° for the shorter wavelengths, then one could infer that long-wavelength gravity waves are propagating through a region of small-eddy turbulence.

AMS

PHEASANT
(Aves, Galliformes). The pheasant is a game bird native to the Old World, especially to southeastern Asia and the high altitudes of China and Tibet. The males of many species are gorgeously colored and have much longer tails than the females. Pheasants are raised in large numbers for game both in Europe and in North America. In some parts of the United States, the ring-necked pheasant (*Phasianus colchicus*) became well established after a number of years of careful protection. See Fig. 1 The introduction of other species has met with more limited success.

Fig. 1. Ring-necked (Southern green) pheasant. (*Sketch by Glenn D. Considine.*)

Many species are called pheasants, in addition to the true pheasants of the genus *Phasianus*. Other forms are the tragopans or horned pheasants, the blood pheasants, the monals, the fire-backed pheasants, eared pheasants, golden pheasants, jungle fowls, and argus pheasants. Some of the Indian species bear the names kallege and pukra. The ruffed grouse of North America is sometimes referred to as a pheasant in the southern states, but this has no scientific foundation. The monal is brightly colored and is found in the higher forests of the mountains of Asia. The tragopan is a large game bird found in wooded country at high altitudes in China and northern India. Guinea fowl comprise several species of African birds related to the pheasants. All have some dark plumage with light spots, and brightly colored bare skin about the head and neck. The common Guinea fowl, *Numida meleagris*, is among the common domesticated species. See also **Galliformes**.

PHENACITE.
The mineral phenacite is a beryllium silicate corresponding to the formula Be_2SiO_4. It is hexagonal but the crystals are usually rhombohedral in habit. It has a conchoidal fracture; is brittle, hardness, 7.5–8; specific gravity, 3; luster, vitreous; colorless to yellowish or reddish, sometimes brown; transparent to translucent. Phenacite is found in pegmatites with topaz, quartz and microcline, and occurs also in emerald-bearing mica schists of the Ural Mountains. It is found also in France, Norway, Switzerland, Africa, Brazil, and Mexico; and in the United States, in Oxford County, Maine; Carroll County, New Hampshire; and in Chaffee and El Paso Counties in Colorado. It derives its name from the Greek meaning deceiver, as it resembles quartz and topaz with which it is associated. It is sometimes spelled phenakite. It has been used as a gem.

PHENOCLAST.
A textural term proposed by R.M. Field in 1916 for coarsely graded clastic sedimentary rocks in which the largest or "show" particles or fragments are referred to as phenoclasts, regardless of their shape or composition. The term implies that the larger constituents of the

glomerate have been derived from prelithified rock. Rounded fragments are called pebbles or spheroclasts, which when lithified by means of matrix (sand and clay) and cement form a conglomerate. Angular fragments are called anguclasts (Field), which when lithified by means of matrix (sand and clay) and cement form a breccia.

PHENOCRYST. A textural term proposed by Iddings in 1892 for macroscopic crystals which are relatively much larger than the crystalline matrix of the igneous rock in which they occur. Rocks which have phenocrysts are called porphyritic. The term phenocryst is derived from the Greek, meaning show, and crystal.

PHENOL. 1. A class of aromatic organic compounds in which one or more hydroxy groups is attached directly to the benzene ring. Examples are phenol itself (benzophenol), the cresols, xylenols, resorcinol, naphthols. Although technically alcohols, their properties are quite distinctive.

2. Phenol (carbolic acid; phenylic acid; benzophenol; hydroxybenzene), C_6H_5OH. Phenol is a white, crystalline substance that turns pink or red if not perfectly pure, or if under influence of light; absorbs water from the air and liquefies. It has a distinctive odor and a sharp burning taste. It is toxic by ingestion, inhalation, and skin absorption, and is a strong irritant to tissue. When in a very weak solution, phenol has a sweetish taste; specific gravity 1.07; mp 42.5–43 °C; bp 182 °C; flash point 77+ °C. Soluble in alcohol, water, ether, chloroform, fixed or volatile oils, and alkalies.

Most of the phenol used in the United States is made by the oxidation of cumene, yielding acetone as a byproduct. The first step in the reaction yields cumene hydroperoxide, which decomposes with dilute sulfuric acid to the primary products, plus acetophenone and phenyl dimethyl carbinol. Other processes include sulfonation, chlorination of benzene, and oxidation of benzene. The compound is purified by rectification.

Major uses of phenol include production of phenolic resins, epoxy resins, and 2,4-D (regulated in many countries); as a selective solvent for refining lubricating oils; in the manufacture of adipic acid, salicylic acid, phenophthalein, pentachlorophenol, acetophenetidine, picric acid germicidal paints, and pharmaceuticals; as well as use as a laboratory reagent. Special uses include dyes and indicators, and slimicides.

High-boiling phenols are mixtures containing predominantly meta substituted alkyl phenols. Their boiling point ranges from 238–288 °C they set to a glass below −30 °C. They are used in phenolic resins, as fuel-oil sludge inhibitors, as solvents and as rubber chemicals.

Phenol is regarded as a dangerous chemical. Refer to *Dangerous Properties of Industrial Materials*, 10th Edition, Sax and R.J. Lewis, Editors, Wiley, New York, 1999.

PHENOLATE PROCESS. A process for removing hydrogen sulfide from gas by the use of sodium phenolate, which reacts with the hydrogen sulfide to give sodium hydrosulfide and phenol. This can be reversed by steam heat to regenerate the sodium phenolate.

PHENOL COEFFICIENT. In determining the effectiveness of a disinfectant using phenol as a standard of comparison, the phenol coefficient is a value obtained by dividing the highest dilution of the test disinfectant by the highest dilution of phenol that sterilizes a given culture of bacteria under standard conditions of time and temperature.

See also **Disinfectant**.

PHENOL-FURFURAL RESIN. A phenolic resin that has a somewhat sharper transition than phenol-formaldehyde from the soft, thermoplastic stage to the cured, infusible state and can be fabricated by injection molding since it has little tendency to harden before curing conditions are reached.

See also **Phenolic Resins**.

PHENOLIC RESINS. Phenolic resins are a large family of polymers and oligomers, composed of a wide variety of structures based on the reaction products of phenols with formaldehyde. Phenolic resins are employed in a wide range of applications, from commodity construction materials to high technology applications in electronics and aerospace. Generally, but not exclusively, thermosetting in nature, phenolic resins

provide numerous challenges in the areas of synthesis, characterization, production, product development, and quality control.

These resins have been known since the 1870s when Baeyer first investigated the reactions of phenols and aldehydes. However, his findings were not commercially utilized until Dr. L. H. Baekeland disclosed his classic work in 1907. Through his use of high-pressure molding, he provided a solution to the problem of making quick-curing moldings which did not blister or crack. Contemporary with Dr. Baekeland was the work of Lebech and Aylsworth who provided the key to the application of large commercial quantities of phenolic Novolacs by suggesting the use of hexamethylenetetramine as a curing agent. Since that time, and because of their desirable price-to-property relationship, phenolic resins have enjoyed steady growth despite the encroachment into their areas of application by a few thermoplastics and other thermosetting materials.

Although in the pure state phenolic resins are quite weak and brittle, they are highly regarded among the plastic materials as being capable of producing very strong physical bonds with a large variety of materials at very low concentration. Consequently, phenolics have found use in many applications as binders. In addition to the strength they impart as bonding agents in matrixes, phenolic resins also possess resistance to chemical attack by all but the most polar organic solvents. The only inorganic reagents that have a deleterious effect on them are the strongest and most oxidizing of the acids and the strongest bases.

For many years prior to the development of high-temperature thermoplastics and thermosets, such as the polyimides, polysulfones, and epoxies, phenolic molding material dominated the high temperature-resistant market. This emphasizes their ability to resist temperature degradation in the 400–500 °F (204–260 °C) range. Because phenolics were found to possess excellent ablative properties, it has been reported that both the American and Soviet space efforts used them in combination with certain other polymeric compounds in heat shield materials. Phenolics, like other aromatic hydrocarbon-based resins, possess excellent resistance to high-energy radiation degradation.

Unlike most thermoplastics, phenolic resins and moldings are characterized by high flexural modulus and good tensile strength while having relatively low impact resistance. In addition to their good physical strength properties, phenolic resins are used in the manufacture of many electrical devices where high dielectric breakdown strength and electrical resistance, in combination with excellent dimensional stability, are required.

Chemistry of Phenolics

Phenolic compounds are capable of chemically combining with a large number of aldehydes and other compounds to yield an almost infinite spectrum of modified polymers. However, the reaction of a phenol with an aldehyde (most commonly encountered is that between phenol and formaldehyde) leads to the formation of only two classes of phenolic resins. These are Novolacs and resols. In general, these two classes of resins may be differentiated by the fact that Novolacs are prepared with an acid catalyst and substantially less than one mole of aldehyde per mole of phenol and require the addition of a curing catalyst to become thermosetting; while resols, or single-stage resins as they are commonly called, are prepared with from 1 to 3 moles of aldehyde per mole of phenol and employ a basic condensation catalyst, and are inherently thermosetting.

Novolacs

The aldehyde content of Novolac resins is insufficient to render the resin thermosetting, hence, they are true thermoplastics provided that a curing agent is not added. Novolacs may be stored indefinitely in the pulverized state at moderate temperatures even mixed with curing agent.

The final cure speed of simple Novolacs may be accurately controlled by the use of the proper condensation catalyst in the initial phase of the reaction. Structurally, a Novolac consists of a series of phenol nuclei joined by methylene

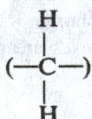

links at the *o* and *p* positions. Only two of the three possible *o* and *p* positions on each ring within the polymeric chain are substituted with a methylene group. Only one position on each of the two terminal phenol groups is substituted with a methylene group, hence, two positions on each terminal ring and one on each internal ring is available for future reactions, including the curing reactions. When the unsubstituted positions are predominantly the para positions, very fast-curing resins result. Slower-setting resins are obtained when the unoccupied positions are the ortho positions. When hexa is used as the hardener at approximately the 10% level, the reactive sites are joined by a linkage.

$$\begin{array}{ccc} H & & H \\ | & & | \\ -C-&N-&C- \\ | & | & | \\ H & H & H \end{array}$$

One mole of ammonia is liberated for approximately every three of the above links formed.

Unaltered Novolacs and two-stage resins[1] find application in grinding wheel bonding, molding material, brake linings and clutch faces, foundry sand binding, premix and wood fiber bonding, thermal insulation. Modified phenolics are found in adhesives, coatings, and aerospace applications. See also **Paints and Coatings**.

Resols

On the other hand, resols contain sufficient aldehyde to make them thermosetting without a curing agent. Consequently, they have only finite storage stability and care must be exercised to minimize both the length and the temperature of storage. The high aldehyde-to-phenol ratios used in the preparation of resols insure that a high percentage of the reactive *o* and *p* positions are utilized in either methylene links or are substituted by a hydroxymethyl group. It is these hydroxymethyl groups which function as crosslinking sites in the final curing reaction. In phenolformaldehyde resols, the ratio of methylene groups to hydroxymethyl groups is an important factor in determining the solubility of these resins. Low ratios insure water solubility in almost infinite proportions, while resins with high ratios can be dissolved in only low molecular weight alcohols, ketones, ethers, and esters. Solubility of resols in nonpolar solvents, such as hydrocarbons, is always very low.

In addition to acting as a crosslinking site, the hydroxymethyl group and unsubstituted *o* and *p* positions may be used as reactive sites to join numerous other compounds to the phenolic polymer. These modified resins often possess many properties normally not attributable to phenolics in the unaltered state.

Resols find application as impregnating resins, in laminating paper, cloth, glass and asbestos; as a pickup agent in grinding wheels, exterior and marine plywood, premix and granular molding material, adhesives, wood waste and particle board manufacture, and in coatings.

The final curing step of both classes of phenolics is accomplished by exposing the resin or the resin containing matrix to temperatures in the 190–450 °F (88–232 °C) range for an appropriate length of time to render the resin infusible. In most applications high pressure is also applied concurrently with the heating cycle to eliminate blistering, which would normally occur if the trapped gases (ammonia in the case of Novolacs, and water in the case of resols) generated in the cure were allowed to escape unrestrained.

Comparative infrared analysis of the two classes of resins in their pure state show that only resols have strong absorptions in the 1,000 and 880 cm^{-1} range. As Novolacs require a curing agent, their presence may be inferred by a strong, sharp absorption at 510 cm^{-1} which is indicative of the most common curing agent used, hexamethylenetetramine. Unfortunately, hexa also has strong absorptions at or near 1,000 and 880 cm^{-1} which makes differentiation between resols and two-stage resins, by infrared spectroscopy, possible only for the experienced. In the cured state, it is very difficult to determine the class identity of an unknown sample.

[1] By convention, two-stage resins are defined as a mixture of Novolac and curing agent, which is capable of thermosetting.

Additional Reading

Gardziella, A., A. Knop, and L.A. Pilato: *Phenolic Resins: Chemistry, Applications, Standardization, Safety and Ecology*, 2nd Edition, Springer-Verlag New York, LLC, New York, NY, 1999.

PHENOLOGY. The science that treats the periodic biological phenomena with relation to climate, especially seasonal changes. Phenological events are stages of plant growth. From a climatological viewpoint, these phenomena serve as bases for the interpretation of progress in local seasons and the climatic zones, and are considered to integrate the effects of a number of bioclimatic factors on rate of plant development. Phenology may be considered a branch of the science of bioclimatics, the sequence of plant or crop development stages through its life cycle. Growth stages may be defined by stage of physiological development such as germination, first true leaf, flowering, maturity, etc., and/or by physical stage such as planting, emergence, harvest, etc.

PHENOMENOLOGY. See **Shape-Memory Alloys**.

PHENYLKETONURIA. Also known as phenylpyruvic oligophrenia or PKU, this is an inherited disorder which involves about 1% of the mentally retarded population. Persons with the disease are born without the enzyme phenylalanine hydroxylase, which is necessary for the conversion of phenylalanine, an amino acid present in practically all protein foods, into tyrosine. This leads to greatly increased concentrations of phenylalanine in the blood, some of which is then converted to phenylpyruvic acid, phenyllactic acid, and *o*-hydroxyphenylacetic acid. All of these metabolites are excreted in the urine of phenylketonurics in higher than normal amounts. Varying concentrations of other unusual metabolites deriving from phenylalanine, tyrosine, and tryptophan are also present.

The mental defect in phenylketonuria is usually severe and is apparent by six months of age. Most patients are idiots, a few are imbeciles, and rare individuals have borderline intelligence. Seizures, eczema, and albinism may be present, and life expectancy is greatly decreased.

That phenylalanine accumulation in the tissues is primarily responsible for the biochemical abnormalities in phenylketonuria is established by demonstrations over the past 20–25 years that a low-phenylalanine diet may prevent or reverse these changes. More important, if such a diet is instituted in the first months of life and continued for 3 to 4 years, intellectual impairment may be prevented. Reversal of the mental defect in older patients is not certain.

Pathways of Research. In 1991, C.R. Scriver (McGill University, Montreal, Canada) reviewed the research on phenylketonuria, which had commenced as early as 1934. Scriver pointed out four phases of this research that took place over a half-century of study.

Phase I — In 1934, Følling identified a clinical entity which he called "imbecillatas phenylpyruvica." Over the two decades that followed, ·it became known that the mental retardation associated with the disease was related to persistent postnatal *hyperphenylalaninemia* (i.e., high phenylalanine levels). It also was recognized that the metabolic phenotype was the consequence of deficient activity of phenylalanine hydroxylase (an enzyme). Further, it was determined that the enzyme abnormality would be found in a pair of recessive mutations inherited from healthy parents. From these findings, certain conclusions were drawn, to the effect that there was (a) an ultimate cause (mutation) and (b) a proximate cause (ingestion of the essential amino acid phenylalanine).

Phase II — L.S. Penrose (University College, London) renamed the disease (phenylketonuria) and emphasized (a) association between human genetic variation, (b) chemical imbalance, and (c) abnormal mental function. Penrose also speculated that the course of phenylketonuria could "be influenced by the deliberate alteration of body metabolism" — that is, by medical treatment. Penrose also observed the nonrandom distribution of the disease and noted that about 2% of Europeans carry the gene for the disease, wondering why such a gene was so prevalent.

Considerable research followed and, as reported by Scriver, the results included the following: (a) dietary treatment to control serum phenylalanine levels enjoyed modest success; (b) a set of enzymes (and genes) and a catalytic cofactor were identified; (c) a better understanding of why all untreated patients with phenylketonuria have hyperphenylalaninemia, but not all persons who have hyperphenylalaninemia necessarily have phenylketonuria; and (d) much additional knowledge was gained from screening tests. As observed by Scriver, "Therapy for the disorder became not only an epitome of the application of human biochemical genetics, but also a model for so-called genetic medicine and for public health. In addition, maternal hyperphenylalaninemia became a paradigm of metabolic teratogenesis, which if not dealt with successfully could nullify all the gains made to prevent mental retardation associated with phenylketonuria mutations."

Phase III — Commencing in the 1980s, S.L. Woo (Baylor College of Medicine), according to Scriver, "prepared a complementary DNC (cDNA) for the rat liver enzyme, then cloned a human cDNA, and eventually isolated and characterized the human phenylalanine hydroxylase gene. The gene resides on the long arm of chromosome 12 and has about 90,000 base pairs of DNA, has 13 exons, and is decorated with a suite of highly informative DNA markers." The genetic architecture is indeed complex.

Phase IV — In 1991, sponsored by the National Institute of Child Health and Human Development, the Howard Hughes Medical Institute, and the Deutsche Forschungsgemeinschaft, Yoshyuki Okano (Baylor College of Medicine) and colleagues conducted a study of 258 patients with phenylketonuria from Denmark and Germany for the presence of eight mutations previously found in patients from these countries. The conclusions of this research: "Our results strongly support the hypothesis that there is a molecular basis for phenotypic heterogeneity in phenylketonuria. The establishment of genotype will therefore aid in the prediction of biochemical and clinical phenotypes in patients with this disease."

Additional Reading

Addison, G.M., D.M. Isherwood, R.A. Harkness, and R.J. Pollitt: *Practical Developments in Inherited Metabolic Diseases*, Kluwer Academic Publishers, Norwell, MA, 1986.

Kaurman, S.: *Tetrahydrobbioterin: Basic Biochemistry and Role in Human Disease*, Johns Hopkins University Press, Baltimore, MD, 1997.

Koch, R., F. De la cruz, and L.D. Platt: *Genetic Disorders and Pregnancy Outcome*, CRC Press, LLC., Boca Raton, FL, 1997.

Levy, H.L.: "Molecular Genetics of Phenylketonuria and Its Implications," *Amer. J. Human Genetics*, **45**, 667 (1989).

Lyonnet, S., et al.: "Molecular Genetics of Phenylketonuria in Mediterranean Countries: A Mutation Associated with Partial Phenylalanine Hydroxylase Deficiency," *Amer. J. Human Genetics*, **44**, 511 (1989).

Okano, Y., et al.: "Molecular Basis of Phenotypic Heterogeneity in Phenylketonuria," *N. Eng. J. Med.*, 1232 (May 2, 1991).

Romano, V.: *Advances in Phenylketonuria Research*, S. Karger Publishers, Inc., Farmington, CT, 1993.

Scriver, C.R.: "Phenylketonuria — Genotypes and Phenotypes," *N. Eng. J. Med.*, 1280 (May 2, 1991).

Woo, S.L.: "Molecular Basis and Population Genetics of Phenylketonuria," *Biochemistry*, **28**, 1 (1989).

PHEOCHROMOCYTOMA (Phaeochromocytoma). See **Adrenal Disease**.

pH (Hydrogen Ion Concentration). A measure of the effective acidity or alkalinity of a solution. It is expressed as the negative logarithm of the hydrogen-ion concentration. Pure water has a hydrogen ion concentration equal to 10^{-7} moles per liter at standard conditions. The negative logarithm of this quantity is 7. Thus, pure water has a pH value of 7. The pH scale usually is considered as extending from 0 to 14. When a strong acid fully dissociates (or ionizes) in water, a 1 N solution of this acid will have a pH value of 0.0. Conversely, a 1 N base fully ionized in water will have a pH value of 14. Both hydrochloric acid and sodium hydroxide come close to meeting these stipulations. Because of the logarithmic nature of the pH scale, there is a tenfold change in hydrogen- and hydroxyl-ion concentration per unit change of pH. Thus, a slightly acidic solution having a pH of 6 will contain ten times as many active hydrogen ions as a solution of pH 7. See also **pK**.

Effective acidity or alkalinity is stressed in pH measurement — not the total hydrogen present. Sulfuric acid and boric acid both contain significant amounts of hydrogen. Nearly all the hydrogen in sulfuric acid dissociates in the presence of sufficient water to become free hydrogen ions. On the other hand, when boric acid is added to water, it dissociates very little into free hydrogen ions. The pH of a 0.1 N sulfuric acid solution will be about 1.3 whereas for the same concentration, boric acid will have a pH of about 5.3. Thus, sulfuric acid is called a strong acid; boric acid a weak acid. In all materials, of course, dissociation increases with temperature, thus the same solution will have a somewhat different pH at a lower temperature than at a higher temperature. Pure water is neutral at a temperature of 25°C, having a concentration of 1×10^{-7} hydrogen ions and 1×10^{-7} hydroxyl ions and, consequently, a pH of 7. Dissociation is less at 0°C, at which temperature the hydrogen-ion concentration is 0.34×10^{-7}, or a pH of 7.47 (slightly basic rather than neutral). But, at a temperature of 100°C, dissociation is greater. The hydrogen ion concentration is 8×10^{-7} and the pH is 6.10 (or slightly acid). The pH of various substances is given in Table 1.

TABLE 1. pH VALUES OF VARIOUS SUBSTANCES (AT 25°C)

Material	pH
Seawater	7.75 to 8.25
Soils	3 to 10
Plant tissues and fluids	About 5.2
Animal tissues and fluids	About 7.0 to 7.5
Blood	7.35–7.5
Urine	5.0–7.0
Milk	6.5–7.0
Gastric juice	1.7
Pancreatic juice	7.8
Intestinal juice	7.7
Internal tissue fluids:	
Minimum, below which acidosis ensues	7.0
Maximum, above which tetany ensues	7.8
Hydrochloric acid (1N)	0.1
Hydrochloric acid (0.1N)	1.08
Hydrochloric acid (0.001N)	3.00
Sulfuric acid (1.0N)	0.32
Sulfuric acid (0.1N)	1.17
Acetic acid (1N)	2.37
Lemon juice	2.0–2.2
Acid fruits	3.0–4.5
Fruit jellies	3.0–3.5
Sodium hydroxide (1N)	13.73
Sodium hydroxide (0.1N)	12.84
Ammonia (10% NH_3)	11.8
Limewater, $Ca(OH)_2$ saturated	12.4
Trisodium phosphate, 2%	11.95

Buffer solutions can be added to resist changes in pH despite the addition of acid or base to the solution. This is explained under **Buffer (Chemical)**.

pH is measured in two basic ways: (1) colorimetrically, usually where high accuracy is not required and manual methods suffice; and (2) electrometrically. Color changes are based upon various organic dyes which alter their color within a relatively narrow range of pH values. Numerous dyes are required to cover the full pH range. Electrometric methods are used both in the laboratory and on-line for process control. They are continuous and easily adapted to automatic control systems. The possibility that a thin glass membrane of special composition could develop a potential in relation to hydrogen ion concentration was described as early as 1909 by the German chemist, Fritz Haber. Little progress was made until the middle-1920s. Glass electrodes are now the standard approach to electrometric pH measurement, after periods of trial with quinhydrone and antimony electrodes. The glass electrode responds in a predictable fashion throughout the 0 to 14 pH range, developing 59.2 millivolts per pH unit at 25°C, values which are consistent with the classical Nernst equation. Contrary to earlier pH electrodes, the glass electrode is not influenced by oxidants or reductants in solution. With suitable temperature compensation, pH measurements can be made up to 100°C and higher. In pH measurement, a second or reference electrode is required to complete the circuit. After trials with numerous electrodes (the hydrogen electrode is the standard) for practical plant and laboratory applications, the mercury-mercurous chloride (calomel) electrode is widely used. There is also some use of the silver-silver chloride reference electrode.

pH control systems are widely used in waste control and neutralization systems, in pulp and paper manufacture, in food processing, and in the manufacture of numerous organic chemicals. pH measurement is very important in the medical field.

PHILIPPINES CURRENT.   See **Ocean Currents**.

PHILLIPSITE.   The mineral phillipsite is a zeolite, a hydrous silicate of potassium, calcium, and aluminum, corresponding to formula $(K,Na_2Ca)(Al_2Si_4)O_{12} \cdot 4-5H_2O$. It is monoclinic, forming penetration twins, and sometimes crosses resembling orthorhombic or tetragonal forms. It also may occur in radial groups. Phillipsite is a brittle mineral; hardness, 4–4.5; specific gravity, 2.2; luster, vitreous; color, white to light red; translucent to opaque. Like other zeolites, it is found in veins and cavities in basalts, and sometimes in more acidic rocks. It is believed to be a low-temperature mineral. Phillipsite is found in Italy, especially in the lavas of Vesuvius and Monte Somma, and in the basalts of Germany, Ireland, and Australia. It has been reported from Greenland. This mineral was named in honor of the British mineralogist William Phillips.

PHLEBOTOMUS FEVER.   Also known as *sandfly fever* or *Papataci fever*, this is an acute, nonfatal virus disease transmitted by the sandfly (*Phlebotomus papatasii*). The disease is common in many of the tropical and subtropical areas, appearing mainly in the hot dry weather in the Mediterranean and Middle East, India, Asia, and parts of South America. It has been reported in communities living at altitudes up to nearly 5000 feet (1500 m) above sea level. In most areas where antimalarial residual insecticide spraying has been used, the vector has disappeared and, with it, the disease. Recent failure to maintain the antimalarial campaigns has followed in former endemic areas with return of the vector and the disease. The causal agent is a small, unclassified virus. The fly becomes infective within 6 to 8 days after ingesting human blood in which the virus is circulating. The fly transmits the disease by biting. No race is immune and both sexes are attacked at all ages.

An attack of phlebotomus fever closely resembles one of dengue, but without the rash, saddleback fever, or glandular involvement. The onset is sudden. There is a rise in temperature, which remains elevated for 1 to 3 days; it is remittent and ends by crisis accompanied by intense sweating. The conjunctiva are infected and photophobia and lachrymation are common. There is very severe headache and bone and joint pain can be very troublesome.

There is no laboratory method of making a certain diagnosis. The clinical picture should be recognized during an epidemic, but may be easily confused with atypical dengue and influenza. Treatment is entirely symptomatic.

R. C. V.

PHLEBOTOMY.   Intentional removal of blood from the circulatory system. The therapy for hemochromatosis is an example. See also **Liver**.

PHLOEM.   That part of a plant through which foods move from one part to another. In stems, the phloem forms a considerable portion of the bark, and is found outside the cambium.

PHLOGOPITE.   The mineral phlogopite is a magnesium-bearing mica, with but little iron, corresponding essentially to the formula $K(Mg, Fe)_3(AlSi_3)O_{10}(F, OH)_2$. Fluorine is sometimes present. This mica is monoclinic like muscovite, biotite and lepidolite, forming prismatic crystals, occasionally very large, and occurring also in scales and plates. Its cleavage is basal and highly perfect with elastic laminae; hardness, 2–2.5; specific gravity, 2.76–2.90; luster, pearly to submetallic; color, yellowish-brown, green, white and colorless; transparent to translucent; may exhibit asterism, probably due to minute inclusions. Phlogopite is more nearly a characteristic of metamorphic than igneous rocks although occasionally occurring in the latter if they are rich in magnesia and with but little iron. Phlogopite is found especially in Rumania, Switzerland, Italy, Finland, Sweden and Madagascar where it occurs in the crystalline limestones in huge crystals. In the United States it occurs in New York State at Edwards, Hammond, DeKalb, Monroe and, in New Jersey, at Franklin. In Canada it is found at many places in Ontario and Quebec.

The name phlogopite comes from the Greek word meaning like fire, referring to the copper-like reflections often observed in the reddish-brown varieties.

Phlogopite is in demand commercially by the electrical industry for use as an insulator.

PHOENICOPTERI *(Aves)*.   The Flamingos (family *Phoenicopteridae*) are long-legged water birds highly adapted to taking small water animals. Because of their overly long legs, they were formerly grouped with the *Ciconiiformes*, but some zoologists considered them to be related to the *Anatidae*. It is, therefore, appropriate to set up this family as a separate order.

Their length reaches 80–130 centimeters (31–51 inches) from the tip of the beak to the tip of the tail, but it may reach 190 centimeters (75 inches) if measured to the tips of the toes of the extended legs; the weight is 2500–3500 grams ($5\frac{1}{2}$–$7\frac{1}{2}$ pounds). The long neck is curved, with 19 cervical vertebrae. The feathers have an aftershaft; there are 12 primaries, and 12 to 16 rectrices. The skeleton, muscles, and air sacs are formed as in storks. The voice is goose-like. There are 3 genera with 5 species, all rather similar to one another: the Greater Flamingo (*Phoenicopterus ruber*), with 2 subspecies, the American Flamingo (*Phoenicopterus ruber*), and the European Flamingo (*Phoenicopterus ruber roseus*); the Chilean Flamingo (*Phoenicopterus chilensis*); the Lesser Flamingo (*Phoeniconaias minor*); the Andean Flamingo (*Phoenicoparrus andinus*); and the James' Flamingo (*Phoenicoparrus jamesi*).

The food of all flamingos consists of small swimming crustacea, algae, and unicellular organisms that they sift out of the water with the beak, which has been transformed into a filtration apparatus. The lower mandible is large, and at its cutting edge, looks as if it is inflated; the upper mandible, on the other hand, is small and fits on the lower one like a lid.

The type of food and the manner of feeding of flamingos assume an abundance of prey of fairly uniform size, particularly since the birds live in large flocks of up to several 100,000 individuals. Such conditions are found in salt lakes and brackish coastal lagoons of warm areas. Their manner of feeding readily differentiates flamingos from the related orders *Ciconiiformes* and *Anseriformes*, to which they do, however, show numerous similarities. See also **Flamingo**.

PHOENIX.   A southern constellation located near Cetus, also sometimes spelled Phenix.

PHOENIX MARS MISSION.   The phoenix, a fabulous mythical bird the size of an eagle, symbolizes rebirth in many ancient cultures. According to the ancient Greeks, the bird lives in Arabia, nearby a cool well and sings a beautiful morning song. The phoenix lives 500 years or longer with only one phoenix existing at a time. When the bird's death approaches, it bursts into flames, and a new bird springs from the consumed pyre.

Similar to its namesake, the Phoenix Mission *raises from the ashes* a spacecraft and instruments from two previous unsuccessful attempts to explore Mars: the *Mars Polar Lander* (Fig. 1) and the *Mars Surveyor 2001 Lander* (Fig. 2 and 3). The Mars Polar Lander failed to return data upon its arrival to Mars' Antarctic region on December 3, 1999 and left many ambitious science goals undone. Phoenix uses three instruments from this earlier polar lander, the **Surface Stereo Imager (SSI)**, the **Robotic Arm (RA)** and the **Thermal and Evolved Gas Analyzer (TEGA)**.

The Phoenix Mission uses the Mars Surveyor 2001 Lander, built in 2000, but later administratively mothballed. The '01 lander is undergoing modifications to improve the spacecraft's robustness and safety during *entry, descent, and landing*. Phoenix recovers two instruments delivered for the '01 lander that have been in protected storage: the **Mars Descent Imager (MARDI)** and the **Microscopy, Electrochemistry, and Conductivity Analyzer (MECA)**. Also, the **Robotic Arm (RA)** has been modified from the '01 lander version.

Powered by a Boeing Delta 2925 launch vehicle, Phoenix will begin its mission within a 22 day launch window in August of 2007. The launch will take place at Cape Canaveral Air Force Station in Florida.

After launch, Phoenix will perform various maneuvers to make the transition to the cruise stage. The spacecraft will deploy its solar arrays and re-orient itself in space. A connection to the NASA Deep Space Network will be initialized which will allow communication with Earth. When these

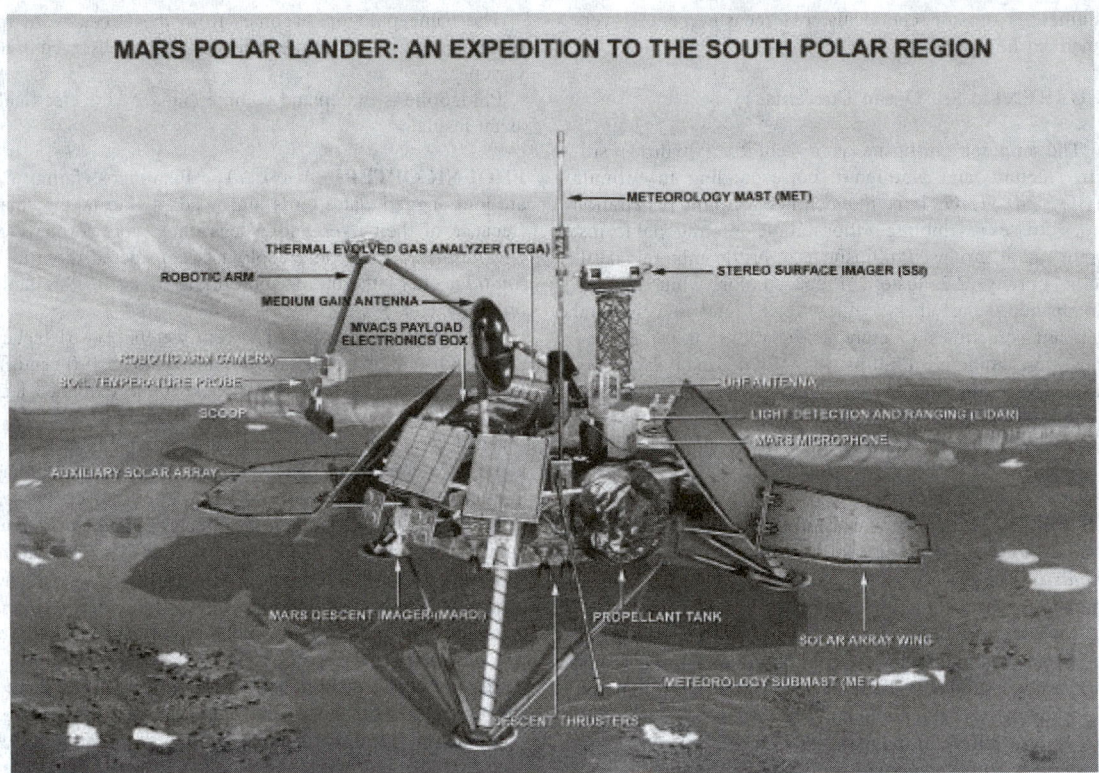

Fig. 1.   Mars Polar Lander. (*Image courtesy of NASA.*)

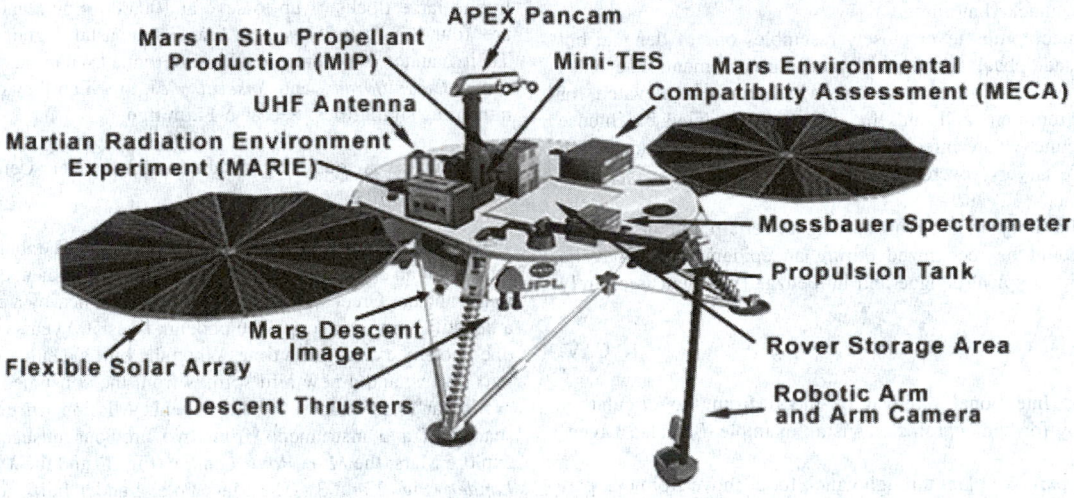

Fig. 2.   Mars Surveyor 2001 Lander. (*Image courtesy of NASA/JPL*).

maneuvers are complete, the spacecraft will be generating energy from its solar panels and ready to receive further commands from Earth. See also **Deep Space Network**.

The cruise phase lasts for approximately 10 months as Phoenix makes its way to Mars. During the cruise phase, the spacecraft verifies the health of its scientific instruments and performs trajectory correction maneuvers (TCMs). The Deep Space Network (DSN) is used to communicate with the spacecraft during these operations. The massive antennas of the DSN are also used to obtain information about the spacecraft's flight path. Often times, the spacecraft will be observed from DSN sites in different parts of the world simultaneously to find its exact trajectory. These measurements are crucial for planning (TCMs).

The initial launch trajectory is intentionally pointed away from Mars so that the jettisoned third stage from the launch vehicle does not impact Mars. The first TCM, performed just 10 days after launch, places the spacecraft on a trajectory towards Mars. The subsequent TCMs are planned to take place much later in cruise phase to correct small errors in the first TCM (see image at right). These errors come from a multitude of sources, including imperfections in the flight model and slight inaccuracies in the DSN measurements. This type of deep space navigation has been used by many other NASA missions, including *2001 Mars Odyssey* and the recent *Mars Exploration Rovers*.

During the last two weeks before Phoenix enters the martian atmosphere, the DSN will be tracking the spacecraft even closer than before. Two TCMs are scheduled to be performed within the last three days. Just before entry, flight path data is sent to Phoenix that is used by the onboard computers during the descent and landing to guide the spacecraft to its landing site. The cruise assembly, which consists of solar panels and other components that are only necessary for the cruise phase of the journey, is jettisoned five minutes prior to entry.

At 125 km (78 miles) above the surface, Phoenix will enter the thin martian atmosphere. It will slow itself down by using friction. A heat

Fig. 3.   Artist Impression of the Mars Surveyor 2001 Lander, which was almost completed, but never sent to Mars. (*Image courtesy of NASA/JPL*)

shield will protect the lander from the extreme temperatures generated during entry. Antennas located on the back of the shell which encases the lander will be used to communicate with one of three spacecraft currently orbiting Mars. These orbiters will then relay signals and landing info to Earth.

After the lander has decelerated to Mach 1.7 (1.7 times the speed of sound), the parachute is deployed. Shortly after the parachute is deployed, the heat shield is jettisoned, the landing radar is activated, and the lander legs are extended. The lander continues through the Martian atmosphere until it comes within 1 km (.6 miles) of the Martian surface. At this point, the lander separates itself from the parachute. It then throttles up its landing thrusters and decelerates. When Phoenix is either at an altitude of 12 m (39 ft) or traveling at 2.4 m/s (7.9 ft/s), the spacecraft begins traveling at a constant velocity. The landing engines are turned off when sensors located on the footpads of the lander detect touchdown.

Surface operations are planned in relation to Martian days, which are known as sols. Because Mars rotates slightly slower than Earth, sol is 40 minutes longer than our planet's 24-hour day. A strategic plan is created that outlines operations two weeks into the future. This strategic plan is used to create a more detailed tactical plan which decides surface activities that will take place for the next two sols. Daily science and engineering data is used to assess the status of the strategic and tactical plans, and the plans are updated as necessary.

Immediately after Phoenix touches down on the surface of Mars (sol 0), critical instruments such as the solar arrays and SSI mast are deployed. Later in the afternoon of sol 0, EDL data and MARDI images are sent to Earth. On sol 1, TEGA, MECA and RAC are turned on and checked out, and the RA is deployed. SSI begins taking images of the landing site and the area where the robotic arm will be digging, and MET begins to sample the weather at the landing site.

On sols 2 through 9, the instruments aboard Phoenix continue to take initial measurements. TEGA takes measurements of the Martian atmosphere using its mass spectrometer. The RA acquires a sample of Martian soil and delivers it to TEGA on sol 4. This sample is analyzed by the differential scanning calorimeter in TEGA on the following sol. Another sample is delivered on sol 7 for analysis using MECA.

The digging operations phase is planned to take place on sols 10–90. SSI and RAC images will be analyzed to determine where the RA should dig. Phoenix will dig for up to 2.5 hours per sol during this period. As the RA digs into the martian surface, SSI and RAC images will help determine when new samples should be delivered to the scientific instruments on Phoenix. Samples will be delivered to TEGA about every 15 cm (5.9 in) or when layering is obvious. The four MECA cells will be reserved for samples from different layers that are expected to be encountered while digging. One cell will analyze a sample from the surface, another will

analyze the dry regolith overburden, and one will be kept in reserve for the icy layer. One MECA cell will be kept for a repeat measurement or to examine another layer.

Science and Technology

Mars is a cold desert planet with no liquid water on its surface. But in the Martian arctic, water ice lurks just below ground level. Discoveries made by the *2001 Mars Odyssey* Orbiter in 2002 show large amounts of subsurface water ice in the northern arctic plain. The Phoenix lander targets this circumpolar region using a robotic arm to dig through the protective top soil layer to the water ice below and ultimately, to bring both soil and water ice to the lander platform for sophisticated scientific analysis.

The complement of the Phoenix spacecraft and its scientific instruments are ideally suited to uncover clues to the geologic history and biological potential of the Martian arctic. Phoenix will be the first mission to return data from either polar region providing an important contribution to the overall Mars science strategy "Follow the Water" and will be instrumental in achieving the four science goals of NASA's long-term Mars Exploration Program.

Phoenix seeks to verify the presence of the Martian Holy Grail: water and habitable conditions. In doing so, the mission strongly complements the four goals of NASA's Mars Exploration Program.

Goal 1: Determine whether Life ever Arose on Mars. Continuing the *Viking* missions' quest, but in an environment known to be water-rich, Phoenix searches for signatures of life at the soil-ice interface just below the Martian surface. Phoenix will land in the artic plains, where its robotic arm will dig through the dry soil to reach the ice layer, bring the soil and ice samples to the lander platform, and analyze these samples using advanced scientific instruments. These samples may hold the key to understanding whether the Martian arctic is a habitable zone where microbes could grow and reproduce during moist conditions. See also **Mars**; and **Mars**.

Goal 2: Characterize the Climate of Mars. Phoenix will land during the retreat of the Martian polar cap, when cold soil is first exposed to sunlight after a long winter. The interaction between the ground surface and the Martian atmosphere that occurs at this time is critical to understanding the present and past climate of Mars. To gather data about this interaction and other surface meteorological conditions, Phoenix will provide the first weather station in the Martian polar region, with no others currently planned. Data from this station will have a significant impact in improving global climate models of Mars.

Goal 3: Characterize the Geology of Mars. As on Earth, the past history of water is written below the surface because liquid water changes the soil chemistry and mineralogy in definite ways. Phoenix will use a suite of chemistry experiments to thoroughly analyze the soil's chemistry and mineralogy. Some scientists speculate the landing site for Phoenix may have been a deep ocean in the planet's distant past leaving evidence of sedimentation. If fine sediments of mud and silt are found at the site, it may support the hypothesis of an ancient ocean. Alternatively, coarse sediments of sand might indicate past flowing water, especially if these grains are rounded and well sorted. Using the first true microscope on Mars, Phoenix will examine the structure of these grains to better answer these questions about water's influence on the geology of Mars.

Goal 4: Prepare for Human Exploration. The Phoenix Mission will provide evidence of water ice and assess the soil chemistry in Martian arctic. Water will be a critical resource to future human explorers and Phoenix may provide appreciable information on how water may be acquired on the planet. Understanding the soil chemistry will provide understanding of the potential resources available for human explorers to the northern plains.

The Phoenix Mission has two bold objectives to support these goals, which are to (1) study the history of water in the Martian arctic and (2) search for evidence of a habitable zone and assess the biological potential of the ice-soil boundary.

Objective 1: Study the History of Water in All its Phases. Currently, water on Mars' surface and atmosphere exists in two states: gas and solid. At the poles, the interaction between the solid water ice at and just below the surface and the gaseous water vapor in the atmosphere is believed to be critical to the weather and climate of Mars. Phoenix will be the first mission to collect meteorological data in the Martian arctic needed by scientists to accurately model Mars' past climate and predict future weather processes.

Liquid water does not currently exist on the surface of Mars, but evidence from *Mars Global Surveyor, 2001 Mars Odyssey* and Mars Exploration Rover missions suggest that water once flowed in canyons and persisted in shallow lakes billions of years ago. However, Phoenix will probe the history of liquid water that may have existed in the arctic as recently as 100,000 years ago. Scientists will better understand the history of the Martian arctic after analyzing the chemistry and mineralogy of the soil and ice using robust instruments. See Fig. 4.

Fig. 4.   Three-dimensional image of the Martian arctic created using data from the Mars Orbiter Laser Altimeter (MOLA) aboard Global Surveyor. (*Image courtesy of NASA; Greg Shirah, SVS.*)

Objective 2: Search for Evidence of Habitable Zone and Assess the Biological Potential of the Ice-Soil Boundary. Recent discoveries have shown that life can exist in the most extreme conditions. Indeed, it is possible that bacterial spores can lie dormant in bitterly cold, dry, and airless conditions for millions of years and become activated once conditions become favorable. Such dormant microbial colonies may exist in the Martian arctic, where due to the periodic wobbling of the planet, liquid water may exist for brief periods about every 100,000 years making the soil environment habitable.

Phoenix will assess the habitability of the Martian northern environment by using sophisticated chemical experiments to assess the soil's composition of life giving elements such as carbon, nitrogen, phosphorus, and hydrogen. Identified by chemical analysis, Phoenix will also look at reduction-oxidation (redox) molecular pairs that may determine whether the potential chemical energy of the soil can sustain life, as well as other soil properties critical to determine habitability such as pH and saltiness.

Despite having the proper ingredients to sustain life, the Martian soil may also contain hazards that prevent biological growth, such as powerful oxidants that break apart organic molecules. Powerful oxidants that can break apart organic molecules are expected in dry environments bathed in UV light, such as the surface of Mars. But a few inches below the surface, the soil could protect organisms from the harmful solar radiation. Phoenix will dig deep enough into the soil to analyze the soil environment potentially protected from UV looking for organic signatures and potential habitability.

Spacecraft and Science Instruments

Aboard the deck of the Phoenix spacecraft are a suite of science instruments representing some of the most sophisticated and advanced technology ever sent to Mars.

The Phoenix Mission inherits a highly capable spacecraft partially built for the *Mars Surveyor* Program 2001 (MSP'01) and important lessons learned from the *Mars Polar Lander* (MPL). The spacecraft was in an advanced state of development when NASA canceled MSP'01 and has been housed in a Class 100,000 clean high-bay facility at Lockheed Martin Space Systems, Littleton, Colorado; http://www.lockheedmartin.com/wms/findPage.do?dsp=fec&ci=14699&sc=400.

The spacecraft will experience extreme conditions during travel to and exploration of Mars. During launch, the spacecraft will undergo tremendous load forces and shaking stresses as the launch vehicle is propelled out of Earth's gravity well. During the 10-month cruise to Mars, the spacecraft must be able to withstand the vacuum of space with potential radiation hazards from solar storms and micro-meteor impacts from interplanetary dust. See Fig. 5. During entry, descent, and landing, the spacecraft will be heated to thousands of degrees during aeroshell braking, jerked with tremendous force as the parachute is deployed, and finally will come to a soft touchdown using controlled thrusters. Finally, during surface operations, the spacecraft must withstand the extremely cold temperatures of the Martian arctic and the dust storms that potentially affect the area.

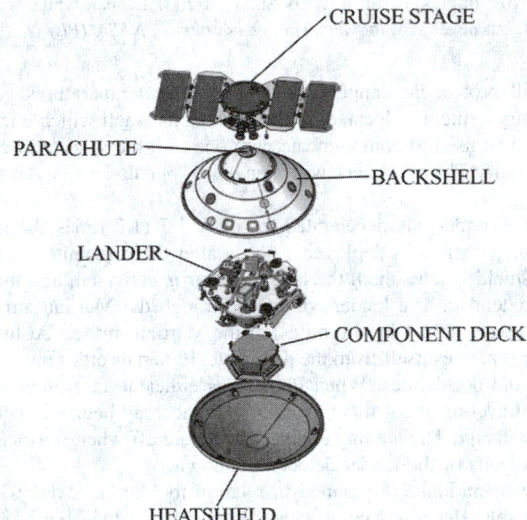

Fig. 5.   An expanded view of the Phoenix spacecraft's cruise configuration. (*Image courtesy of UA/Lockheed Martin*)

Designing and constructing a spacecraft to withstand these extremes is a tricky endeavor and requires considerable testing before launch. Benefiting from the lessons learned during the MPL and MSP'01 experience, as well as further reliability upgrades and subsystems used in previous successful space missions, the spacecraft is in a high state of development early in the mission's fabrication phase. Therefore, the mission engineering team is working on developing enhanced spacecraft reliability through extensive testing, (i.e., beyond normal integration and environment testing that occurs for most missions).

The spacecraft has several subsystems that are being updated, if necessary, with parts and software that will increase reliability. These subsystems include (1) command and data handling, controlling the spacecraft's computer processing, (2) electrical power, consisting of solar panels, batteries, and associated converting circuits, (3) telecommunications, ensuring flow of data to and from Earth, (4) guidance, navigation, and control, assuring the spacecraft arrives safely at Mars, (5) propulsion, controlling trajectory correction maneuvers during cruise and thrusters during landing, (6) structure, providing the spacecraft framework and integrity, (7) mechanisms, enabling the movement of several spacecraft components, and (8) thermal-control, using heat transfer to ensure proper temperature ranges on all parts of the spacecraft.

The Lockheed Martin Space Systems Company, in Littleton, Colorado will design, build, integrate, and test the Phoenix spacecraft. The company built the *Viking* Landers, *Mars Global Surveyor*, and *2001 Mars Odyssey*, as well as aeroshells for the *Mars Pathfinder* and *Mars Exploration Rover* missions.

Robotic Arm (RA). The RA is critical to the operations of the Phoenix lander and is designed to dig trenches, scoop up soil and water ice samples, and deliver these samples to the TEGA and MECA instruments for detailed chemical and geological analysis. Designed similar to a back hoe, the RA can operate with four degrees of freedom: (1) up and down, (2) side to side, (3) back and forth, and (4) rotate around. See Fig. 6.

Fig. 6.   Close up of the scoop and other instruments at the end of the RA. (*Image courtesy of NASA/JPL*)

The RA will be 2.35 meters (just under 8 ft) long with an elbow joint in the middle, allowing the arm to trench about 0. 5 m (1.6 ft) below the Martian surface, deep enough to where scientists believe the water-ice soil interface lies. At the end of the RA is a moveable scoop, which includes ripper tines (sharp prongs) and serrated blades. Once icy soil is encountered, the ripper tines will be used to first tear the exposed materials, followed by applying the serrated blades to scrape the fractured soil. The scoop will then be run through the furrows to capture the fragmented samples, ensuring enough sample mass for scientific study on the lander platform.

A similar RA developed for *the Mars Polar Lander* was tested at Death Valley in 2000 and successfully dug a 10 inch trench in just under 4 hours. The extremely hard soil conditions at Death Valley are similar to those expected at Phoenix's Martian arctic landing site.

The RA is being built by a team at the Jet Propulsion Laboratory led by Dr. Robert Bonitz. JPL was responsible for designing RA's for the *Mars Polar Lander, Mars Surveryor 2001 Lander* and *Mars Exploration Rover* (*Spirit & Opportunity* missions.

Microscopy, Electrochemistry, and Conductivity Analyzer (MECA). MECA characterizes the soil of Mars much like a gardener would test the soil in his or her yard. By dissolving small amounts of soil in water, MECA determines the pH, the abundance of minerals such as magnesium and sodium cations or chloride, bromide and sulfate anions, as well as dissolved oxygen and carbon dioxide. Looking through a microscope, MECA examines the soil grains to help determine their origin and mineralogy. Needles stuck into the soil determine the water and ice content, and the ability of both heat and water vapor to penetrate the soil.

MECA's wet chemistry lab contains four single-use beakers, each of which can accept one sample of Martian soil. Phoenix's RA will initiate each experiment by delivering a small soil sample to one beaker, which is ready and waiting with a pre-warmed and calibrated soaking solution. Alternating soaking, stirring, and measuring, the experiment continues for the entire day. It concludes with the addition of two chemical pellets. The first contains an acid to tease out carbonates and other constituents that are only soluble in acidic solutions. The second contains specific reagents to test for sulfates and soil oxidants.

The optical and atomic-force microscopes complement MECA's wet chemisty experiments. With images from these microscopes, scientists will examine the fine detail structure of soil and water ice samples. Detection of hydrous and clay minerals by these microscopes may indicate past liquid water in the Martian arctic. The optical microscope will have a resolution of 4 microns per pixel, allowing detection of particles ranging from about 10 micrometers up to the size of the field of view (about 1 millimeter by 2 millimeters). Red, green, blue, and ultraviolet LEDs will illuminate samples in differing color combinations to enhance the soil and water-ice structure and texture at these scales. The atomic force microscope will provide sample images down to 10 nanometers - the smallest scale ever examined on Mars. Using its sensors, the AFM creates a very small-scale *topographic* map showing the detailed structure of soil and ice grains.

Prior to observation by each of the microscopes, samples are delivered by the RA to a wheel containing sixty-nine different substrates. The substrates are designed to distinguish between different adhesion mechanisms and include magnets, sticky polymers, and *buckets* for bulk sampling. The wheel is rotated allowing different substrate-sample interactions to be examined by the microscopes.

MECA's final instrument, the thermal and electrical conductivity probe, will be attached at the "knuckle" of the RA. The probe will probably consist of three small spikes that will be inserted into the ends of an excavated trench. In addition to measuring temperature, the probe will measure thermal properties of the soil that affect how heat is transferred, providing scientists with better understanding of surface and atmospheric interactions. Using the same spikes, the electrical conductivity will be measured to indicate any transient wetness that might result from the excavation. Most likely, the thermal measurement will reflect ice content and the electrical, unfrozen water content.

MECA is being built by a team at the Jet Propulsion Laboratory, led by Dr. Michael Hecht. The atomic force microscope is contributed by a Swiss consortium led by the University of Neuchatel, the Orion chemical beakers are being developed by Thermo Electron Corporation, the chemistry actuator assemblies are being built by Starsys Research Corporation, the thermal and electrical conductivity probes are being built by Decagon Devices, Inc., the microscope sample wheel is being designed by Transfer Engineering and Manufacturing, Inc., and the optical microscope is being designed by the University of Arizona. MECA was originally developed for the cancelled Mars Surveryor 2001 Lander mission.

Robotic Arm Camera (RAC). The RAC is attached to the Robotic Arm (RA) just above the scoop. The instrument provides close-up, full-color images of (1) the martian surface in the vicinity of the lander, (2) prospective soil and water ice samples in the trench dug by the RA, (3) verification of collected samples in the scoop prior to analysis by the MECA and TEGA instruments, and (4) the floor and side-walls of the trench to examine fine-scale texturing and layering.

By examining the color and grain size of scoop samples, scientists will better understand the nature of the soil and water-ice in the trench being dug by the RA. Additionally, floor and side-walls images of the trench may help determine the presence of any fine-scale layering that may result from changes in Martian climate.

The RAC is a box-shaped imager with a double Gauss lens system, commonly found in many 35 mm cameras, and a charged-coupled device similar to those found on many consumer digital cameras. Two lighting assemblies provide illumination of the target area. The upper assembly contains 36 blue, 18 green, and 18 red lamps and the lower assembly contains 16 blue, 8 green, and 8 red lamps. See Fig. 7. The RAC has two motors: one sets the lens focus from 11 mm to infinity and the other opens and closes a transparent dust cover. The instruments magnification is 1:1 at closest focus, providing image resolutions of 23 microns per pixel.

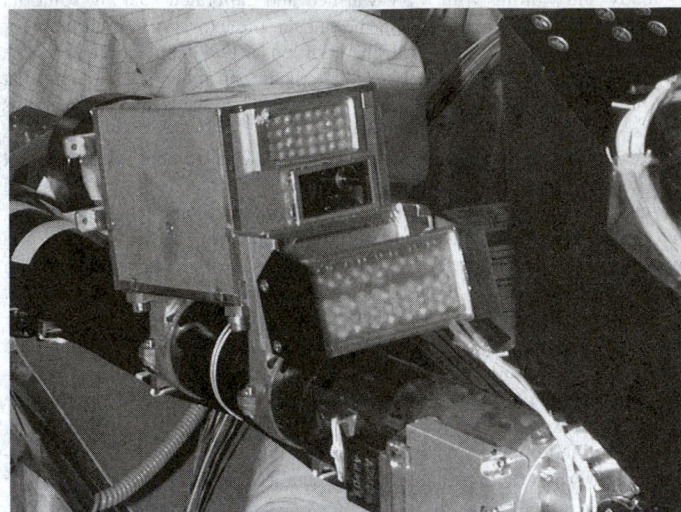

Fig. 7.   Built for the Mars Surveyor 2001 Lander, the RAC provides close-up images of soil and water-ice samples. (*Image courtesy of NASA/JPL*)

The RAC was originally built by a team at the University of Arizona and the Max Planck Institute for Solar System Research, Germany for the Mars Surveyor 2001 Lander missions. When the mission was cancelled, the camera was put into bonded storage at the Jet Propulsion Laboratory awaiting the RAC's future use in the Phoenix mission.

Surface Stereo Imager (SSI). SSI will serve as Phoenix's "eyes" for the mission, providing high-resolution, stereo, panoramic images of the Martian arctic. Using an advanced optical system, SSI will survey the arctic landing site for geological context, provide range maps in support of digging operations, and make atmospheric dust and cloud measurements.

Situated atop an extended mast, SSI will provide images at a height 2 meters (6.5 feet) above the ground, roughly the height of a tall person. SSI simulates the human eye with its two optical lens system that will give three-dimensional views of the arctic plains. See Fig. 8. The instrument will also simulate the resolution of human eyesight using a charged-coupled device that produces high density 1024×1024 pixel images. But SSI exceeds the capabilities of the human eye by using optical and infrared filters, allowing multispectral imaging at 12 wavelengths of geological interest and atmospheric interest.

Looking downward, stereo data from SSI will support robotic arm operations by producing digital elevation models of the surrounding terrain. With these data, scientists and engineers will have three-dimensional virtual views of the digging area. Along with data from the TEGA and the MECA, scientists will use the three-dimensional views to better understand the geomorphology and mineralogy of the site. Engineers will also use these three-dimensional views to command the trenching operations of the robotic arm. SSI will also be used to provide multispectral images of samples delivered to the lander deck to support results from the other scientific instruments.

Looking upward, SSI will be used to estimate the optical properties of the Martian atmosphere around the landing site. Using narrow-band imaging of the Sun, the imager will estimate density of atmospheric dust, optical depth of airborne aerosols, and abundance of atmospheric water vapor. SSI will also look at the lander itself to assess the amount of wind-blown dust deposited on spacecraft. Deposition rates provide important information for scientists to understand erosional and atmospheric processes, but are critical for engineers who are concerned

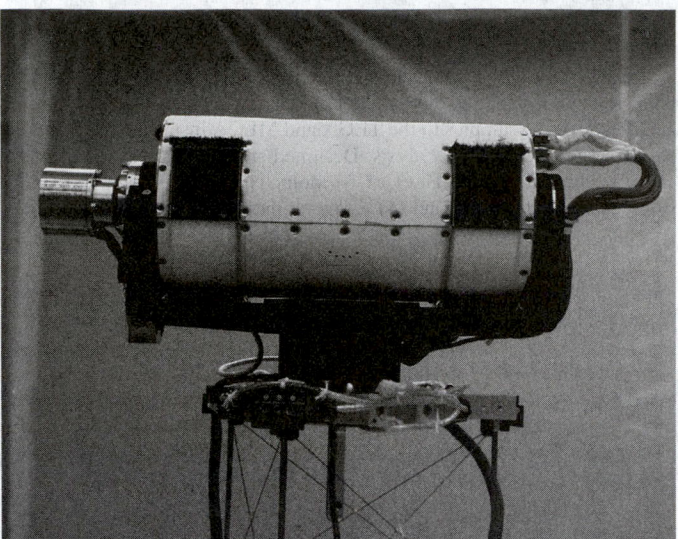

Fig. 8.   The two lens SSI is a higher resolution upgrade of the imager used for Mars Pathfinder, which returned more than 17,000 stereo images. (*Image courtesy of SSI Team, University of Arizona*)

about the amount of deposited dust on the solar panels and associated power degradation.

SSI is being built by a team at the University of Arizona, led by Mr. Chris Shinohara. This team was responsible for designing and building the Imager for the Mars Pathfinder, which provided the first ever stereo images of Mars. The team also developed the Surface Stereo/Science Image that flew on the ill-fated *Mars Polar Lander*.

Thermal and Evolved Gas Analyzer (TEGA). TEGA is a combination high-temperature furnace and mass spectrometer instrument that scientists will use to analyze Martian ice and soil samples. The robotic arm will deliver samples to a hopper designed to feed a small amount of soil and ice into eight tiny ovens about the size of an ink cartridge in a ballpoint pen. Each of these ovens will be used only once to analyze eight unique ice and soil samples.

Once a sample is successfully received and sealed in an oven, the temperature is slowly increased at a constant rate, and the power required for heating is carefully and continuously monitored. This process, called scanning calorimetry, shows the transitions from solid to liquid to gas of the different materials in the sample: important information needed by scientists to understand the chemical character of the soil and ice.

As the temperature of the furnace increases up to 1000° C (1800° F), the ice and other volatile materials in the sample are vaporized into a stream of gases. These are called evolved gases and are transported via an inert carrier to a mass spectrometer, a device used to measure the mass and concentrations of specific molecules and atoms in a sample. The mass spectrometer is sensitive to detection levels down to 10 parts per billion, a level that may detect minute quantities of organic molecules potentially existing in the ice and soil.

With these precise measurement capabilities, scientists will be able to determine ratios of various isotopes of hydrogen, oxygen, carbon, and nitrogen, providing clues to origin of the volatile molecules, and possibly, biological processes that occurred in the past.

The TEGA is being built by a team at the University of Arizona, led by Dr. William Boynton and at the University of Texas, Dallas by Dr. John Hoffman. This team has developed several instruments for space flight, including a Differential Scanning Calorimeter (DSC) and Evolved Gas Analyzer (EGA) that flew on the ill-fated Mars Polar Lander, and the Gamma-Ray Spectrometer that is currently flying on the *2001 Mars Odyssey Orbiter*. The latter instrument is returning data on the elemental composition of Mars and has provided evidence for high concentrations of subsurface ice in the Martian arctic.

Mars Descent Imager (MARDI). MARDI plays a key science role during Phoenix's descent to the Martian arctic. See Fig. 9. Beginning just after the aeroshell is jettisoned at an altitude of about 5 miles (8 kilometers) MARDI will acquire a series of wide-angle, color images of the landing

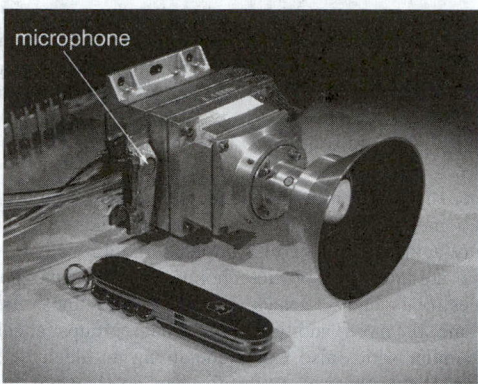

Fig. 9.   Delivered for the *Mars Surveyor 2001 Lander*, MARDI produces images of the geology around the landing site. (*Image courtesy of NASA/JPL/MLSS*)

site all the way down to the surface. This will be the first time that images are acquired during descent of a lander or rover to Mars, providing critical information to the Phoenix mission team. In addition to helping pinpoint where the spacecraft landed, scientists will use these images to provide a larger geographic context for local landforms around the landing zone.

The geographic context provided by MARDI will be crucial to understanding whether the landing site is representative of the rest of Mars' northern plains. If the landing site represents the northern plains well, then chemical and geological results from the TEGA and MECA may be applied more broadly to the Martian arctic with greater confidence. However, if the landing site is highly unusual, scientists will have less confidence in generalizing the results. The final images produced by MARDI will cover the area around the lander, providing visual information that will be used with SSI to help plan RA digging operations.

MARDI incorporates an innovative electronics design that enables high-quality scientific data acquisition in a very compact package. The camera weighs about one pound, the lowest weight of any camera sent to Mars, and is also extremely conservative of power, using only three watts during data acquisition. The key to reducing the mass and power required for the instrument was to take advantage of the tremendous advances that have occurred over the last ten years in electronics. This reduces both the size and number of electronic components necessary in the design, which reduces the mass and power required.

MARDI uses a refractive optics system that collects and bends light to enable a 66° wide field of view. The instrument uses a charged-coupled device that produces high density 1024×1024 pixel images, each with an exposure time of 4 milliseconds.

MARDI is being built by a team at Malin Space Science Systems (MSSS) led by Dr. Mike Malin. This team was responsible for designing the MARDI for the *Mars Polar Lander* and the *Mars Surveyor 2001 Lander* missions. When the 2001 lander mission was cancelled, MARDI was put into bonded storage at the Jet Propulsion Laboratory awaiting the camera's future use in the Phoenix mission.

Meteorological Station (MET). Throughout the course of Phoenix surface operations, MET will record the daily weather of the Martian northern plains using temperature and pressure sensors, as well as a light detection and ranging (LIDAR) instrument. With these instruments, MET will play an important role by providing information on the current state of the polar atmosphere and how water is cycled between the solid and gas phases in the martian arctic.

The MET's lidar is an instrument that operates on the same basic principle as RADAR, using powerful laser light pulses rather than radio waves. See Fig. 10. The lidar transmits light vertically into the atmosphere, which is reflected off dust and ice particles. These reflected light pulses and their time of return to the lidar instrument are analyzed, revealing information about the size of atmospheric particles and their location.

From this distribution of dust and ice particles, scientists can make important inferences about how energy flows within the polar atmosphere, important information for understanding martian weather. These particles also reveal the formation, duration, and movement of clouds, fog, and

Fig. 10.   The flight model of the LIDAR just before it is placed on the spacecraft during assembly operations. (*Image courtesy of MET Team/CSA*)

dust plumes, improving scientific understanding of Mars' atmospheric processes.

The very cold temperatures of the Martian arctic will be measured with thin wire thermocouples, a technology that has been used successfully on meteorological stations for both the *Viking* and *Pathfinder* missions. In a thermocouple, electric current flows in a closed circuit of two dissimilar metals (chromel and constantan in the case of the MET) when one of the two junctions is at a different temperature. Three of these thermocouple sensors will be located on a 1.2 meter (4 foot) vertical mast to provide a profile of how the temperature changes with height near the surface.

Atmospheric pressure on Mars is very low and requires a sensitive sensor for measurement. Pressure sensors similar to those used on the *Viking* and *Pathfinder* missions will be part of the MET.

The Canadian Space Agency is responsible for overall development of the MET. The MET instruments will be built by the MD Robotics of Brampton, Ontario, with the support of Optech Inc. of Toronto for the development of the lidar.

See also **Mars Exploration**.

Web Reference

Phoenix Mars Mission: http://phoenix.lpl.arizona.edu/mission.php; and
http://mars.jpl.nasa.gov/missions/future/phoenix.html.

PHOLIDOTA. A small order of mammals containing only the scaly anteaters or pangolins of the Malay archipelago, Africa, and southeastern Asia. They are peculiar animals which are covered with overlapping horny scales of large size and, in this respect, somewhat resemble the armadillo. Pangolins are slender animals with a long tail and short legs bearing powerful claws. Like other anteaters they have a sharp snout and long sticky tongue and live chiefly on termites. The most common species is

Manis pentadactyla. For protection, these animals roll themselves into a ball with the outer protective scales discouraging predators.

PHON. The unit of loudness level of sound, numerically equal to the sound pressure level in decibels, relative to 0.0002 mircobar, of a simple 1000 cycle per second tone judged by listeners to be equivalent in loudness.

PHONOMETER. An instrument for measuring the intensity or frequency of sounds.

PHONONS. Many of the thermal and vibrational properties of solids can be explained by considering the material to be a volume made up of a gas of particles called *phonons*. This particle description is a method of taking into account the actual motion of the atoms and molecules in the solid. Since each atom possesses energy due to its thermal environment, and since there are forces between the atoms that keep the solid together, each atom tends to oscillate about its equilibrium position. The formal mathematical development, obtained through solving the equations of motion of the array of individual atoms and molecules, indicates that the thermal energy of the solid is contained in certain combinations of particle vibrations which are equivalent to standing elastic waves in the sample and are called normal modes. Each normal mode contains a number of discreet quanta of energy $E = \hbar$ where ω is the frequency of the mode (or wave) and $\hbar$ is Planck's constant divided by 2π. Each of these quanta is called a phonon (in analogy with the light quanta or photon whose energy-frequency relationship is identical). Phonons are considered only as particles, each having an energy $E = \hbar$, a momentum q, and a velocity $v = \partial\omega/\partial q \sim \omega/q$. Analogous to the energy levels of electrons in a solid, phonons can have only certain allowed energies.

The phonon is of importance to many phenomena: electron mobility, optical absorption, electron spin resonance, electron tunneling, and superconductivity. The phonon spectrum represents a detailed picture of the forces that hold solids together. Thus, it is clear why the phonon has been and will continue to be of fundamental importance in solid-state physics.

See also **Acoustics**.

PHOSGENE. See **Chlorinated Organics**.

PHOSGENITE. This mineral is a chlorocarbonate of lead, $Pb_2(CO_3)$ Cl_3, crystallizing in the tetragonal system, associated with other lead minerals of secondary origin, e.g., cerussite and anglesite; hardness, 2–3; specific gravity 6.133; prismatic to tabular crystals, also massive and granular, adamantine luster; color, white, gray, brown, green or pink; transparent to translucent. Some specimens show yellowish fluorescence under ultraviolet light.

Found in the United States in California, Colorado, Arizona, and New Mexico. Magnificent crystals up to 5 inches (12.5 centimeters) in diameter have been found at Monte Poni, Sicily; as fine crystals in England at Derbyshire and Matlock; and in Poland, Russia, Tasmania, Australia, and Namibia.

PHOSPHATE ROCK. A natural rock consisting largely of calcium phosphate and used as a raw material for manufacture of phosphate fertilizers, phosphoric acid, phosphorus, and animal feeds. Recovery of uranium from the manufacture of phosphoric acid and other phosphate chemicals is expected to become an important source of this metal. Phosphate rock is the primary source of superphosphate, prepared by treatment of the pulverized rock with sulfuric acid (superphosphate having 16–18% P_2O_5) or by acidifying with phosphoric acid (triple superphosphate having 40–48% P_2O_5). Nitric acid is sometimes used, i.e., nitrophosphate. Defluorinated phosphate rock is the source of phosphate used in animal feeds and feed concentrations. Important deposits are in the U.S. (Florida, North Carolina, Tennessee, California, Wyoming, Montana, Utah, and Idaho), North Africa (Morocco, Libya, and Algeria), the former U.S.S.R., and various islands in the Pacific.

See also **Fertilizer**; **Phosphoric Acid**; and **Phosphorus**.

PHOSPHAZENE. A ring or chain polymer that contains alternating phosphorus and nitrogen atoms with two substituents on each phosphorus atom. Characteristic structures are cyclic trimers, cyclic tetramers, and high polymers. The substituent can be any of a wide variety of organic

groups, halogen, amino, etc. Most cyclic trimers are crystalline, solids, organosoluble, and stable to weather conditions; the high polymers (polyphosphazenes) are elastomeric or thermoplastic. A copolymer of phosphazene and styrene has been investigated for use as a flame-retardant.

PHOSPHINE. See **Phosphorus**.

PHOSPHOLIPIDS. These compounds belong to a group of fatty acid compounds sometimes referred to as complex lipids. The simplest are esters of fatty acids with glycerol phosphate and are called *phosphatidic acids*. There are also phosphatidylcholines or lecithins, phosphatidyl-ethanolamines, phosphatidylserines, and phosphatidylinositols. The latter may have one or more additional phosphate groups attached to the inositol. A similar series also exists containing an aldehyde attached to the 1-position of the glycerol, in the form of an α, β-unsaturated ether. These are commonly referred to as *plasmalogens*.

The percentage of phospholipid content of tissues varies little under normal physiological conditions, thus giving rise to the term *element constant*, in contrast to the triglycerides, which have been called the *element variable*.

Phospholipids are considered to be involved in the transport of triglycerides through the liver, especially during mobilization from adipose tissue. Conditions which could be interpreted as interfering with phosphatidyl-choline formation, such as deficiency of choline or its precursors, result in a pronounced increase in liver triglycerides.

Mitochondrial phospholipids play a role in electron transport and oxidative phosphorylation, two mechanisms by which the cell accomplishes the final oxidation of the metabolites to produce energy. Phospholipids also are linked in the transport of ions, especially sodium, across membranes.

In summary, phospholipids (phosphatides) comprise a group of lipid compounds that yield, upon hydrolysis, phosphoric acid, an alcohol, fatty acid, and a nitrogenous base. They are widely distributed throughout nature.

PHOSPHORIC ACID. [CAS: 7664-38-2]. Generally the term *phosphoric acid* refers to orthophosphoric acid, H_3PO_4. Anhydrous orthophosphoric acid is a white, crystalline solid, which melts at 42.35 °C. It forms a hemihydrate, $2H_3PO_4 \cdot H_2O$, which melts at 29.32 °C. Although it is possible to produce almost any desired concentration, it is common practice to supply the material as a solution containing from 75% H_3PO_4 (melting point, 17.5 °C) to 85% H_3PO_4 (melting point, 21.1 °C). When phosphoric acid is heated to temperatures above about 200 °C, water of constitution is lost. A series of acids is formed by the dehydration, ranging from pyrophosphoric acid, $H_4P_2O_7$, to metaphosphoric acid, $(H_3PO_4)_n$. Salts of the dehydrated acids are used for the preparation of certain types of liquid fertilizers and have been used in some detergents. However, to counter the effects of "phosphate pollution," there has been a serious cutback in this latter use of the phosphates. See also **Fertilizer**.

One, two, or three of the hydrogens in phosphoric acid may be neutralized, leading to a series of products which range widely in their hydrogen ion concentration (pH): monosodium phosphate, NaH_2PO_4, with a pH of 4.0; disodium phosphate, Na_2HPO_4, with a pH of 8.3 (approximate); and trisodium phosphate, Na_3PO_4, with a pH of 12.0. Other phosphorous acids of little commercial importance are hypophosphorous acid, H_3PO_2; orthophosphorous acid, H_3PO_3; and pyrophosphorous acid, $H_4P_2O_5$.

Manufacture of Phosphoric Acid. The major sources of H_3PO_4 traditionally have been mineral deposits of phosphate rock. Mining operations are extensive in a number of locations, including the United States (Florida), the Mediterranean area, and Russia, among others. The major constituent of most phosphate rocks is fluorapatite, $3Ca_3(PO_4)_2 \cdot CaF_2$. The supply of high-grade phosphates, the raw material of choice for producing high-purity phosphoric acid by the wet process, is rapidly decreasing in some areas.

Two major methods are utilized for the production of phosphoric acid from phosphate rock. The *wet process* involves the reaction of phosphate rock with sulfuric acid to produce phosphoric acid and insoluble calcium sulfates. Many of the impurities present in the phosphate rock are also solubilized and retained in the acid so produced. While they are of no serious disadvantage when the acid is to be used for fertilizer manufacture, their presence makes the product unsuitable for the preparation of phosphatic chemicals.

In the other method, the *furnace process*, phosphate rock is combined with coke and silica and reduced at high temperature in an electric furnace, followed by condensation of elemental phosphorus. Phosphoric acid is produced by burning the elemental phosphorus with air and absorbing the P_2O_5 in water. The acid produced by this method is of high purity and suitable for nearly all uses with little or no further treatment.

Basic reactions of the wet process are

$$3Ca_3(PO_4)_2 \cdot CaF + 10\ H_2SO_4 + 20\ H_2O$$

$$\longrightarrow 10\ CaSO_4 \cdot 2H_2O + 6H_3PO_4 + 2HF$$

Numerous side reactions also occur. Phosphate rock and sulfuric acid, together with recycled weak liquors, are carefully metered to a large, stirred reactor, providing retention for 4–8 hours. Conditions in the reaction are carefully controlled to maintain preselected conditions. Temperatures (77–83 °C) are controlled by removing excess heat of reaction with a vacuum cooler, or by blowing air through the phosphoric acid slurry. The slurry contains precipitated gypsum and is sent to a filter. The gypsum is washed with water in several countercurrent steps, and weak liquor is returned to the reaction stage. For most uses, the acid requires further concentration, normally done in vacuum evaporators. Merchant-grade acid is generally concentrated to about 54% P_2O_5 (75% H_3PO_4). See Fig. 1.

Fig. 1.   View of three-stage evaporation process used for concentrating wet-process phosphoric acid. (*Swenson.*)

Effluents and gypsum disposal pose problems. Fluorine is evolved at various steps in the process and scrubbers are required to reduce release to the atmosphere. Gypsum is frequently piled in diked areas or dumped into abandoned mines. Wastewater from these plants is heavily contaminated with fluorine, phosphates, sulfates, and other compounds. It is commonly impounded in large ponds, where a portion of the contaminants may precipitate or be lost by other processes. The cooled effluent from the ponds is recycled to the production unit. Any excess water must be treated with lime before it can be allowed to enter streams.

Developments of recent years include plants designed to precipitate the calcium sulfate in the form of the hemihydrate instead of gypsum. In special cases, hydrochloric acid is used instead of sulfuric acid for rock digestion, the phosphoric acid being recovered in quite pure form by solvent extraction. Solvent-extraction methods have also been developed for the purification of merchant-grade acid, which normally contains impurities amounting to 12–18% of the phosphoric acid content. Processes for recovering part of the fluorine in the phosphate rock are in commercial use.

Although more costly to operate, the electric-furnace process produces phosphoric acid of high purity. A mixture of coke, silica, and phosphate rock is formed into nodules by heating in a nodulizing kiln, and the resulting lump material is transferred to the electric furnace, where it is heated with an electric current introduced by means of graphite electrodes. The entire charge is melted, and elemental phosphorus is volatilized. The slag is tapped off intermittently while the phosphorus vapor is condensed. The phosphorus is then burned in air and the P_2O_5 is absorbed in water. Reactions are

$$2\ Ca_3(PO_4)_2 + 6\ SiO_2 + 10\ C \longrightarrow P_4 + 10\ CO + 6\ CaSiO_3$$

$$P_4 + 5\ O_2 \longrightarrow 2\ P_2O_5$$

$$P_2O_5 + 3\ H_2\ PO \longrightarrow 2H_3PO_4$$

Additional Reading

Kreysa, G., and M. Schutze: *Corrosion Handbook - Corrosive Agents and Their Interaction with Materials: Volume 3: Hypochlorites, Phosphoric Acid*, 2nd Edition, John Wiley & Sons, Inc., Hoboken, NJ, 2005.

Perry, R.H., and D.W. Green: *Perry's Chemical Engineers' Handbook*, 7th Edition, The McGraw-Hill Companies, Inc., New York, NY, 1997.

PHOSPHORS AND PHOSPHORESCENCE. A large variety of substances become luminescent when stimulated or excited by suitable radiation, or by emissions, such as cathode rays or beta-rays. This phenomenon is complex and exhibited in various aspects. In some cases, the light is emitted only so long as the exciting emission is maintained, in which case it is called *fluorescence*. See also **Illumination**. In other cases, the luminescence persists after the excitation is removed and it is then called *phosphorescence*. It has long been known, for example, that zinc sulfide, under certain conditions, glows brightly for a time after exposure to daylight or lamplight, but the luminosity decays rapidly and disappears, usually within a few minutes. The electroluminescent phosphor of zinc sulfide-zinc selenide-copper has the property that the wavelength of the emitted radiation increases with increasing selenium content. The white luminescence of some television tubes is obtained from a combination of cadmium-zinc sulfide phosphors, one that is blue-emitting and the other yellow-emitting. Also, in some color television tubes, the blue-emitting and green-emitting phosphors are of the sulfide type, but earlier use of sulfides for the red-emitting "dots" on the tube surface were replaced by rare-earth red-emitting phosphors. One composition used is prepared by combining about 4% europium oxide and 65% yttrium oxide, with various vanadium compounds and calcining the mixture. Rare-earths also have been used in producing phosphors for high-pressure mercury-arc lamps. These phosphors increase the proportion of red light emitted by reducing the green, blue, and ultraviolet portions.

Quantitatively, phosphorescence may be defined as luminescence that is delayed by more than 10^{-8} seconds after excitation. It may be associated with transitions from a higher excited state to a lower one, the energy going into a radiationless rearrangement of the system. If the lower state is metastable, its lifetime may be considerable before it finally decays by a highly forbidden radiative transition to the ground state. In the case of zinc sulfide, the process depends upon the ionization of activator atoms, the freed electrons being trapped and only released slowly for recombination.

See also **Luminescence**.

PHOSPHORUS. [CAS: 7723-14-0]. Chemical element, symbol P, at. no. 15, at. wt. 30.9738, periodic table group 15, mp 44.1 °C (α-white), bp 280 °C (α-white), sp gr 1.82 (white), 2.20 (red).

Four allotropes of phosphorus are known, the hexagonal β-white, stable only below −77 °C, the cubic α-white (mp 44.1 °C), the violet, and the black (which is thermodynamically the most stable). The α-white form is usually taken as the standard state. The violet is obtained by continued heating at $500\,°C$ of a solution of phosphorus in lead. When α-white phosphorus is heated to $250\,°C$ in the absence of air, a red variety (mp $590\,°C$) is obtained which is believed to consist of a mixture of the α-white and violet allotropes, although the studies of the violet component in the mixture have shown that at least four polymorphic forms of red (violet) phosphorus exist.

White phosphorus is considered to be made up largely of P_4 molecules, as is the liquid and vapor up to $800\,°C$, where dissociation becomes appreciable. The P_4 molecule is a tetrahedron, with single covalent bonds between the P atoms, and each having an unshared pair of electrons. White phosphorus is much more reactive than red or violet.

Black phosphorus has a graphite-like structure and has a similar electrical conductivity.

There is one stable nuclide, ^{31}P. Six radioactive isotopes have been identified, ^{28}P through ^{30}P and ^{32}P through ^{34}P, all with short half-lives, measured in terms of seconds, minutes, or days. See also **Radioactivity**. In terms of terrestrial abundance, phosphorus ranks 10th with an estimated average content of igneous rocks being 0.13% phosphorus. The element ranks 19th in abundance in seawater, there being an estimated 325 tons of phosphorus per cubic mile (70 metric tons per cubic kilometer) of seawater. In terms of cosmic abundance, phosphorus is ranked 15th among the elements. The element was first identified by Hennig Brandt in Germany in 1669 during an experiment in which he was distilling urine with sand and coal. White phosphorus is very toxic.

First ionization potential 11.0 eV; second, 19.81 eV; third, 30.04 eV; fourth, 51.1 eV; fifth, 64.698 eV. Oxidation potentials $H_3PO_2 + H_2O \rightarrow H_3PO_3 + 2H^+ + 2e^-$, 0.59 V; $P + 3H_2O \rightarrow H_3PO_3 + 3H^+ + 3e^-$, 0.49 V; $P + 2H_2O \rightarrow H_3PO_2 + H^+ + e^-$, 0.29 V; $H_3PO_3 + H_2O \rightarrow H_3PO_4 + 2H^+ + 2e^-$, 0.20 V; $PH_3(g) \rightarrow P + 3H^+ + 3e^-$, 0.04 V; $P + 2OH^- \rightarrow H_2PO_2^- + e^-$, 1.82 V; $P + 5OH^- \rightarrow HPO_3^{2-} + 2H_2O + 3e^-$, 1.71 V; $H_2PO_2^- + 3OH^- \rightarrow HPO_3^{2-} + 2H_2O + 2e^-$, 1.65 V; $HPO_3^{2-} + 3OH^- \rightarrow PO_4^{3-} + 2H_2O + 2e^-$, 1.05 V; $PH_3(g) + 3OH^- \rightarrow P + 3H_2O + 3e^-$, 0.87 V.

Other physical characteristics of phosphorus are given under **Chemical Elements**.

Because of its reactivity, phosphorus does not occur in nature in the elemental form. Phosphate rock is the principal source of phosphorus and phosphorus compounds. Very large deposits of phosphate rock occur and are worked in the Bone Valley area of Florida, as well as deposits in Tennessee, Idaho, and South Carolina. Large deposits are mined in Northern Africa (Morocco and Tunisia). Very significant reserves have been found on several of the Pacific Islands, the reserves on Christmas Island estimated at some 30 million tons (27 million metric tons) and those on Nauru Island in excess of 100 million tons (90 million metric tons). There also are large active mining operations in the Mediterranean area as well as in the former Soviet Union. Known reserves assure a supply for several centuries. The mineral apatite, $Ca_3(PO_4)_2 \cdot CaCl_2$ or $\cdot CaF_2$, found in Quebec, Virginia, Brazil, and the South Pacific also contains high percentages of phosphorus, up to 20% P_2O_5. The main constituents of most phosphate rocks is fluorapatite, $3Ca_3(PO_4)_2 \cdot CaF_2$. These rocks contain 30–37% P_2O_5.

Most phosphorus raw materials are converted into phosphorus and phosphorus compounds, such as phosphoric acid, on an extremely high-tonnage basis. Percentagewise, relatively little elemental phosphorus is produced for consumption as an end product. See also **Fertilizer**. Phosphorus is important both to plant and animal nutrition. Traditionally, phosphorus compounds have been key components of cleaning compounds and detergents although there have been trends to reduce or eliminate phosphates from high-consumption items. See also **Detergents**.

Production of Elemental Phosphorus

The tricalcium phosphate in phosphate rock, mixed with coke and silica, is thermally reduced to yield P_2 vapor. The phosphorus vapors condense to a liquid and the carbon monoxide produced is returned for burning in the furnace. The process requires much heat and, in addition to the heat provided by the combustion of the coke and the heating value of the recycled carbon monoxide, an electric arc also is used. The reaction takes place in very large furnaces at a temperature of $1,300–1,500\,°C$ and at atmospheric pressure. A 70-MW furnace will produce 44,000 short tons (39,600 metric tons) of P_4 per year, equivalent to 100,000 tons (90,000 metric tons) of P_2O_5 (if converted to acid).

Although there are numerous intermediate and side reactions, the overall reaction is: $Ca_3(PO_4)_2 + 5C + 3SiO_2 \rightarrow P_2 + 5CO + 3Ca \cdot SiO_3$. Byproduct ferrophosphorus alloy and calcium silicate slag are tapped from the furnace periodically. Maximum furnace efficiency occurs when the SiO_2/CaO weight ratio is about 0.8. This ratio also assures a minimum melting-point eutectic for the melt and thus lengthens furnace life. This process was originally developed by Readman in England in 1888. The first 1,500-kW furnace in the United States was installed at Niagara Falls, N.Y. in 1896 because of the availability of low-cost energy. For many years the proximity of Tennessee brown stone (a phosphate rock) to the low-cost power of the Tennessee Valley Authority made a good economic combination. Worldwide production of phosphorus by this process is about $\frac{3}{4}$-billion short tons (0.675 billion metric tons) per year (installed capacity). In the United States, about 80% of the phosphorus produced is immediately converted to the oxide and thence to phosphoric acid. The remaining 20% has gone into alloys, organic intermediates for oil and fuel additives, pesticides, plasticizers, and pyrotechnics. In addition to use in detergents, cleaning compounds, and degreasing formulations, phosphoric acid has been consumed in the preparation of liquid fertilizers, water-treatment, pharmaceutical, and chemical products. Phosphorus-containing fertilizers, such as single superphosphate, wet-process orthophosphoric acid, triple superphosphate, ammonium phosphate, and nitrophosphates do not require elemental phosphorus (or the resulting pure P_2O_5) in their preparation, but are manufactured by directly reacting phosphate rock with requisite chemicals, such as H_2SO_4 or HNO_3. See also **Fertilizer**.

Chemistry and Compounds

Like carbon, phosphorus is covalently bound to its neighboring atoms in all of its compounds, except perhaps for some metallic phosphides. Indeed, the chemistry of carbon and that of phosphorus are somewhat similar as might be expected from the diagonal relationship of these elements in the periodic table.

Probably the major difference between carbon and phosphorus is that the former element is quite closely restricted to the use of s- and p-orbitals, because of the relatively high energy of d-orbitals in the case of second-period elements; whereas, phosphorus, being a third-period element, can use d-orbitals in bonding. For both carbon and phosphorus, the most common hybridization for σ-bonding is approximately the tetrahedral sp^3. However, in order to form π-bonds, carbon must go to lower hybrids: sp^2 and sp. Phosphorus, on the other hand, does not do this but can employ d-orbitals for π-bonding. This difference between carbon and phosphorus in sigma bond strength, and the ease with which phosphorus uses its d-orbitals for attachment of attacking nucleophilic groups, can be used to explain why catenation is common in carbon compounds, while at the same time phosphorus compounds containing long chains of connected phosphorus atoms have not yet been synthesized.

The known coordination numbers exhibited by phosphorus within the molecule-ions containing this element are 1, 3, 4, 5, and 6 which, to at least a first approximation, exhibit the symmetry of p, p^3, sp^3, sp^3d and sp^3d^2 hybridization, respectively. A very large number (several thousand in each case) of triply- and quadruply connected phosphorus compounds are known; but there are only a few compounds of higher coordination number in which d-orbitals are involved in the σ-bond base structure. These are the halogen compounds PF_5, PCl_5, PBr_5, PCl_2F_3, PBr_2F_3 and the pentaphenyl compound $(C_6H_5)_5P$, in which the phosphorus is quintuply connected to its neighboring atoms, and the PF_6^- and PCl_6^- anions, in which the phosphorus has a six-fold coordination. The singly connected phosphorus atoms appear only in compounds occurring at very high temperatures. Although singly connected phosphorus is not known under ordinary conditions, interpretation of diatomic spectra has given considerable information.

Several generalities can be stated concerning the phosphorus compounds that are stable under normal conditions:

1. In those compounds in which phosphorus shares electrons with three neighboring atoms, there are three σ-bonds, with little or no π-character, from the phosphorus.
2. In those compounds in which phosphorus shares electrons with four neighboring atoms there are four σ-bonds, with an average of about one π-bond per P atom.

3. When electrons are shared with five or six neighboring atoms, there is less than one full σ-bond for each connection between the phosphorus and a neighboring atom with apparently very little π-bonding.

These generalities are obviously dependent to a considerable extent upon the specific atoms connected to the phosphorus and, indeed, it is possible that the observed differences between the triply and quadruply connected phosphorus atoms may be attributed primarily to the individual ligands. Fluorine appears to contribute nearly as much shortening (assuming that the tabulated values for the fluorine bond length are correct) to the P—F connection in the triply connected as in the quadruply connected phosphorus compounds. On the other hand, chlorine shows essentially no shortening, whether attached to either triply or quadruply connected phosphorus.

Phosphine. [CAS: 7803-51-2]. PH_3, and its substitution products, have a pyramidal structure. The P—H bond length is 1.42 Å, and the H—P—H angle is 93°. Hypophosphites, containing the radical

are produced by alkaline hydrolysis of white phosphorus. The barium salt yields hypophosphorous acid, H_3PO_2, upon acidification with sulfuric acid. In this acid, only one H is capable of ionization, suggesting the experimentally confirmed formula $H_2P(O)OH$. Hypophosphorous acid and its salts are reducing agents, although their reaction rates are somewhat low, which is usually explained by an equilibrium between H_3PO_2 and its hydrate form, H_5PO_3.

Phosphorus Halides. PX_3, P_2X_4 and PX_5 are formed by direct reaction of the elements, though the pure substances require special methods. Mixed halides are also known. The trihalides are covalent pyramidal compounds, the X—P—X bond angles being generally between 98° and 104°. They all undergo hydrolysis, the rate being roughly inversely proportional to the sum of the atomic numbers of the halogen atoms.

The halogen derivatives of pentavalent phosphorus may be grouped on the basis of their structure into three classes, the pentahalides, oxyhalides and related compounds, and the fluorophosphoric acids. The pentahalides (except the pentaiodide, which is unknown) are produced by reaction of the elements, or, in the case of mixed halides, by reaction of a halogen and a phosphorus trihalide in correct proportions. Their structure in the vapor state has been determined to be a trigonal bipyramid and their bonding is covalent. In the solid, however, phosphorus pentachloride, CAS: 10026-13-8, PCl_5, is $[PCl_4^+][PCl_6^-]$ and phosphorus pentabromide, CAS: 7789-69-7, PBr_5, is $[PBr_4^+]Br^-$. Various mixed halides are known. The five halogen atoms are not in equivalent positions. One may be ionized, with the other four forming sp^3d orbitals or there may be a transition state, to explain the nonequivalence of the exchange between the three equatorial chlorine atoms and the two apical ones in PCl_5 in carbon tetrachloride solution. The pentahalides react with excess water to yield phosphoric acid and hydrohalic acids, but with less water to form phosphorus oxyhalides instead of phosphoric acid.

Phosphorus oxyhalides have the tetrahedral structure

$$\overset{\cdot\cdot}{\underset{\cdot\cdot}{X}} : P : \overset{\cdot\cdot}{\underset{\cdot\cdot}{O}}$$

These compounds, particularly $POCl_3$ and $POFCl_2$, readily form complexes with metal halides. Closely analogous to the oxyhalides are the thiohalides of general formula PSX_3 and the phosphorus nitrilic halides, $(PNX_2)_n$, the chloride of the latter being obtained by partial ammonolysis of PCl_5, and existing as cyclic or polymeric structures of alternate nitrogen and phosphorus atoms.

Phosphorus Oxides and Oxyacids. The principal oxides are related to the acids which they yield when dissolved in H_2O in the following manner:

Trioxide, P_2O_3	Hypophosphorous acid, H_3PO_2
Tetroxide, P_2O_4	Phosphorous acid, H_3PO_3
Pentoxide, P_2O_5	Hypophosphoric acid, $H_4P_2O_6$
plus $3H_2O$	Orthophosphoric acid, $2H_3PO_4$
plus $2H_2O$	Pyrophosphoric acid, $H_4P_2O_7$
plus $1H_2O$	Metaphosphoric acid, $2HPO_3$

Normally when the term "phosphoric acid" is used, it is with reference to orthophosphoric acid, H_3PO_4. Anhydrous orthophosphoric acid is a white crystalline solid that melts at 42.35 °C. It forms a hemihydrate, $2H_3PO_4 \cdot H_2O$, which melts at 29.32 °C. Although practically any desired concentration can be produced, it is common to supply the material as a solution containing from 75% H_3PO_4 (mp -17.5 °C) to 85% H_3PO_4 (mp 21.1 °C). When phosphoric acid is heated to above 200 °C, the water of constitution is lost. Thus, a series of acids is formed by the dehydration, ranging from pyrophosphoric acid, $H_4P_2O_7$, to metaphosphoric acid, $(HPO_3)_n$. Salts of the dehydrated acids are used for the preparation of certain kinds of liquid fertilizers and are present in numerous cleaning compounds. The dehydrated acids can form water-soluble complexes with many metals, such as calcium. One, or two, or three of the hydrogens of phosphoric acid may be neutralized. When one hydrogen is replaced with sodium, for example, the product is slightly acidic; while replacement of all three hydrogens yields a highly alkaline product. The acidity of the solutions is: NaH_2PO_4, a pH of 4.0; Na_2HPO_4, a pH of about 8.3; Na_2PO_4, a pH of 12.0. Although of interest scientifically, the other acids of phosphorus, hypophosphorous acid, H_3PO_2, orthophosphorous acid, H_3PO_3, and pyrophosphoric acid, $H_4P_2O_5$, are not important commercially.

There are two main processes for the industrial production of phosphoric acid, H_3PO_4, from phosphate rock: (1) the *wet process* which involves the reaction of phosphate rock with H_2SO_4 to yield phosphoric acid and insoluble calcium sulfates. Several of the impurities present in the rock dissolve and remain with the product acid. These are not important when the acid is used for fertilizer manufacture. However, the impurities are deleterious to the manufacture of phosphorus chemicals. For a purer product, (2) the *furnace process* is used, wherein the phosphate rock is combined with coke and silica, producing elemental phosphorus as previously described. Oxidation of the phosphorus produces P_2O_5 which, when combined with H_2O, yields H_3PO_4.

Phosphorus sesquioxide, P_4O_6, produced by controlled oxidation of white phosphorus, is hydrolyzed in the cold to produce phosphorous acid, H_3PO_3, a colorless solid (mp 73 °C). Only two of its H atoms are capable of ionization, thus compounds such as M_3PO_3 do not exist. This fact leads to the formula $HP(O)(OH)_2$. Phosphorous acid is a somewhat stronger acid than phosphoric, and both it and the phosphite ion (HPO_3^{2-}) are strong reducing agents.

Hypophosphorous acid was discussed earlier in this entry.

Metaphosphorous acid, HPO_2, is produced by atmospheric combustion of PH_3, but in aqueous solution it hydrates to H_3PO_3.

Phosphorous acid is used in solution, and is usually a reducing agent. That is, in air it changes to phosphoric acid, with hot concentrated sulfuric acid it yields phosphoric acid plus SO_2, with copper sulfate it yields finely divided copper metal, with silver nitrate it yields finely divided silver metal, with permanganate after some time it yields manganous. Occasionally it is an oxidizing agent, e.g., with zinc plus dilute H_2SO_4 it yields phosphine.

Phosphorus tetroxide, P_2O_4, is obtained along with red phosphorus by heating P_4O_6 at 290 °C in a closed tube. It is believed to have the formula, P_8O_{16}, and to consist of trivalent and pentavalent phosphorus. Hypophosphoric acid, $H_4P_2O_6 \cdot 2H_2O$, cannot be produced directly from the tetroxide. The acid decomposes into phosphorous and phosphoric acids on heating, and must be prepared by indirect methods, such as treatment of white phosphorus with an HNO_3 solution of $Cu(NO_3)_2$.

Phosphorus(V) oxide is the chief product of atmospheric oxidation of phosphorus, whence it is obtained as the β-allotrope, of formula P_4O_{10}. Several other allotropes are obtained by various thermal treatments of the β-form, differing in structure and physical properties. The compound hydrates rapidly to form various phosphoric acids. With an excess of H_2O, orthophosphoric acid, $(HO)_3PO$, is formed, mp 42.3 °C. It is a triprotic acid, yielding $H_2PO_4^-$, HPO_4^{2-} and PO_4^{3-} ions. The other crystalline phosphoric acid, pyrophosphoric acid, is believed to have the formula $(HO)_2P(O)-O-P(O)-(OH)_2$. Its acid solution undergoes hydrolysis to the orthoacid. It is tetraprotic, yielding the ions $H_3P_2O_7^-$, $H_2P_2O_7^{2-}$, $HP_2O_7^{3-}$ and $P_2O_7^{4-}$.

Classification of Phosphates

Audrieth and Hill have proposed a classification on the basis of structure, that is, to divide them into glassy phosphates and crystalline phosphates, the latter class being subdivided into (1) linear phosphates and polyphosphates, and (2) cyclic phosphates. Three important members of class (1) are the orthophosphates, containing the PO_4^{3-} ion, the pyrophosphates, having the $P_2O_7^{4-}$ ion, and the triphosphates, having the $P_3O_{10}^{5-}$ ion. The general structural unit of the linear phosphates is the tetrahedron, containing a phosphorus atom surrounded by four oxygen atoms covalently linked to it. Such tetrahedra are linked through a common oxygen atom to form linear polyphosphates, the $P_2O_7^{4-}$ ion having two such tetrahedra, and the $P_3O_{10}^{5-}$ ion, three. Moreover, the tetrahedra may also be double linked through oxygen atoms to form cyclic structures, as in the trimetaphosphate ion, $P_3O_9^{3-}$, which has three such tetrahedra and the tetrametaphosphate ion, $P_4O_{12}^{4-}$, which has four tetrahedra. In general, heating acid salts of simpler phosphates produces polyphosphates (loss of H_2O), while alkaline hydrolysis reverses the process. The glassy phosphates, produced by fusion and rapid cooling of metaphosphates, appear to be true glasses, containing anions with molecular weights well into the thousands.

The widely diverse functionality of phosphates makes them of exceptional importance to technologists, particularly in the food processing field. Phosphoric acid finds many direct uses as an acidulant. It has three available hydrogens which can be replaced one by one with alkali metals, forming a series of *orthophosphate salts* with pH levels ranging from moderately acid (pH = 4) to strongly alkaline (pH = 12). This wide pH range makes phosphates very useful for adjusting the pH of food and chemical systems to almost any desired level. Heating orthophosphates converts them to condensed phosphates containing two, three, or more phosphorus atoms per molecule. The *condensed phosphates*, or *polyphosphates*, have many properties that the orthophosphates do not enjoy. They are polyelectrolytes, and have dispersing or emulsifying properties. They can sequester or chelate metals, such as calcium, magnesium, iron, and copper, rendering these metals nonreactive. This functionality is useful for controlling oxidative rancidity and color formation, as both are catalyzed by metal ions.

Condensed phosphates containing two atoms of phosphorus are *pyrophosphates*. Sodium acid pyrophosphate (SAPP) is used as a leavening acid in baking, and is particularly useful because of the way it can be modified to give different rates of reaction. Pyrophosphates are good sequestrants for iron and copper, which often catalyze oxidation in fruits and vegetables. Thus, the use of a pyrophosphate effectively prevents the discoloration of such foods during preparation and storage.

Condensed phosphates containing three atoms of phosphorus are *tripolyphosphates*, the most important of which is sodium tripolyphosphate (STPP). This compound reacts with the protein in meat, fish, and poultry to prevent denaturing or loss of fluids. This property is sometimes called "moisture binding." STPP also solubilizes protein, which aids in binding diced cured meat, fish, and poultry. It also emulsifies fat to prevent separation.

The chain length of phosphates can be increased further by melting and chilling to form a glass. Glassy sodium phosphates are generally called *sodium hexametaphosphates* (SHMP). SHMPs have excellent sequestering power toward calcium and magnesium. They are used in meat treatment as a partial replacement for STPP to improve solubility in strong pickling brine or to prevent hardness precipitation in very hard water. Considerably more information on the use of phosphates in food processing will be found in the *Foods and Food Production Encyclopedia*, D.M. Considine (editor), Van Nostrand Reinhold, New York, 1981. See also **Fertilizer**.

Peroxyphosphoric Acids. Two are known—peroxymonophosphoric acid, $HOOP(O)(OH)_2$, prepared by treatment of P_4O_{10} with hydrogen peroxide, and peroxydiphosphoric acid, $(HO)_2(O)POOP(O)(OH)_2$, prepared from metaphosphoric acid and peroxide or by electrolysis of an alkali hydrogen phosphate solution (cf. preparation of peroxydisulfuric acid). They and their salts are strong oxidants.

Fluoro phosphoric Acids. [CAS: 13537-32-1]. H_2PO_3F, HPO_2F_2, (POF_3 is not a protic acid), "HPF_6" are obtained by replacing one or more hydroxyl groups with fluorine. They are strong fuming acids, like H_2SO_4 in most properties except its oxidizing power. Their salts are also known, extending as far as completely fluorinated MPF_6 where M is an alkali metal. The solubilities of the monofluorophosphates parallel those of the sulfates; while those of the di- and hexafluorophosphates parallel those of the perchlorates.

Upon acidification or neutralization of solutions containing phosphate anions with other anions, such as those of molybdenum and tungsten, complexes are formed which can readily be crystallized as salts, called phosphomolybdates, phosphotungstates, etc.

Oxygen-Nitrogen Compounds of Phosphorus. These may be classified as aquo, aquo ammono, or ammono derivatives. The first group includes the various acids and oxides already discussed. The second group includes the amido phosphoric acids in which one or more of the hydroxy groups of the acid are substituted by amino groups. Thus there is amidophosphoric acid (*orthophosphoric acid* is understood when the substituted acid is not specified), amidopyrophosphoric acid, diamidophosphoric acid, and triamidophosphoric acid. The substitution of all—OH groups of phosphoric acid gives phosphoryl triamide, $OP(NH_2)_3$. Related compounds are phosphoryl amide imide, $OP(NH_2){:}NH$ and phosphoryl nitride $(OPN)_x$. The second group also includes imidodiphosphoric acid,

$$HN\begin{smallmatrix} \nearrow PO(OH)_2 \\ \searrow PO(OH)_2 \end{smallmatrix}$$

diamidotriphosphoric acid,

$$HN\begin{smallmatrix} \nearrow PO(OH)_2 \\ \searrow \end{smallmatrix} HN\begin{smallmatrix} \nearrow PO(OH)_2 \\ \searrow PO(OH)_2 \end{smallmatrix}$$

and still longer chain acids. Finally, many other derivatives are possible because of the stability of the $N \equiv P$ arrangement. This gives rise to the phosphonitrilic acids, $[NP(OH)_2]_x$, as well as to many ammono derivatives containing only the two elements. The latter include phosphonitrilamide, $[NP(NH_2)_2]_x$, phospham $[NPNH]_x$, and phosphoric nitride, $[P_3N_5]_x$. The phosphonitrilic chlorides, already discussed, are derivatives of phosphonitrilamide.

Organophosphorus Compounds. Most of the industrially important organic compounds of phosphorus commence with one of the basic inorganic phosphorus compounds, such as PCl_3, $POCl_3$, P_2S_5, and P_2O_5, reacted with an appropriate organic intermediate. Ester intermediates, such as alkyl phosphoryl chlorides, are made by the addition of primary alcohols to $POCl_3$. Triaryl phosphate plasticizers and gasoline additives, such as tricresyl phosphate (TCP), can be prepared from PCl_5 and an appropriate phenolic compound. Alkyl diaryl phosphates can be made from $POCl_3$ and corresponding phenols. A number of thiophosphate esters contain PS plus an ethyl or methyl group and a substituted aryl group and are based on $PSCl_3$ or P_2S_5. These compounds are finding use for pesticide control. Dialkyl dithiophosphates may be prepared from P_2S_5 and appropriate intermediates. They are finding application as flotation-agents, oil-additives, and insecticides. It is much more difficult to prepare organophosphorus compounds containing a C-P bond than it is to form the esters. Numerous organic phosphorus compounds are found in nearly all life processes and remain to be better understood before they can be synthesized. Classes of compounds of this type include the phosphoglycerides required for fermentation, the adenosine phosphates needed in photosynthesis and muscle activity, and the very complex phosphorus-containing groups identified in the nucleotides. Of structural interest are the catenation compounds, as illustrated below, which contain many cyclic phosphates and oxygen-linked chains. Compounds of this type include tetrachlorodiphosphine, Cl_2PPCl_2, tetraphenyldiphosphine $(C_6H_5)_2PP(C_6H_5)_2$, diphosphobenzene, $C_6H_5PPC_6H_5$, and tetramethyl hypophosphate, $(CH_3O)_2(O)PP(O)(OCH_3)_2$.

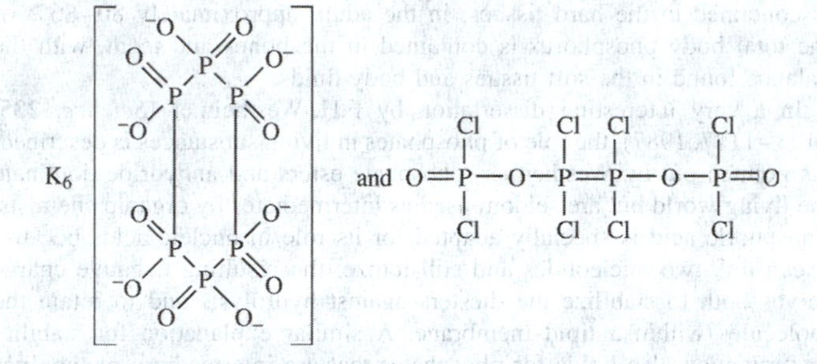

(a)   (b)

(c)   (d)

Inorganic Macromolecules. After the accelerated activity in the development of new polymers that took place during the past 30 or 40 years, some researchers observed some lessening of polymer research and polymer achievements during the 1970s. Not all authorities agreed, but most agreed that the time had arrived when polymer chemistry and applications deserved evaluation both in terms of the past and the future. Synthetic polymers generally have a number of relatively negative features — flammability (derived from their organic nature); a tendency to melt, oxidize, and char at high temperatures in regular atmospheric conditions (again, a result of their organic nature); and a tendency to become stiff and brittle at low temperatures. Many also have a tendency to soften, swell and dissolve in a number of common substances, such as gasoline, jet fuel, hot oil, and numerous other hydrocarbons. In terms of medical applications, most organic polymers tend to initiate a clotting reaction of the blood and many tend to cause toxic, irritant, and sometimes carcinogenic responses. Observers also noted that most polymer research in some way initiated with petrochemicals.

In the early 1970s, a number of investigators decided to shift emphasis and to look at a number of inorganic elements including silicon, phosphorus, sulfur, boron, and some metal atoms that might make up the backbone of a polymer. It was reasoned that the presence of some of these materials in the backbone might remedy some of the aforementioned shortcomings. It should be pointed out that as early as the 1940s, silicone, or poly(organosiloxane) polymers were developed and have proved highly satisfactory in many applications. Peters et al. (1976) reported on a new class of thermally stable polymers, which are based upon alternating siloxane and carborane units. In 1965, Allcock et al. (Pennsylvania State University) synthesized the first poly(organophosphazenes). Since that time, well over 60 new polymers have been made and presently constitute a substantial class of new elastomers. They appear to solve some of the biomedical problems previously mentioned. As pointed out by Allcock 1976, all linear, high polymeric polyphosphazenes have the general structure shown by (a), below. It is interesting to note that over a century ago researchers in Germany and Britain found that phosphorus pentachloride will react with ammonia or ammonium chloride to yield a volatile, white solid. It is now known that this product has the form of (b). Later experimentation showed that the compound would melt under strong heating to form a transparent rubbery material. Involving a ring-opening polymerization of the cyclic trimer, a poly(dichlorophosphazene), shown in (c), is formed. Mainly for the reason that the compound hydrolyzes slowly in the presence of atmospheric moisture to form a crusty mixture of ammonium phosphate and phosphoric acid, the substance was not given serious thought for many years.

Allcock and associates, during the 1960s, subjected the cyclic trimer to new procedures and, after considerable research, were successful in developing polymers that have molecular weights up to and sometimes exceeding 3 to 4 million. The investigators found that the introduction of different substituent groups had a marked effect on the properties of the polymers. See (d). The further detailed research is beyond the scope of this book, but interesting details can be found in the Allcock reference. In summarizing one of their reports, the researchers stated: "Polyphosphazenes are emerging as a new class of macromolecules that have an obvious future as technological elastomers, films, fibers, and textile treatment agents. However, they also possess almost unique attributes for use in biomedicine as reconstructive plastics or as drug-carrier molecules. Moreover, their possible value as 'pseudo-protein' model polymers is an exciting prospect."

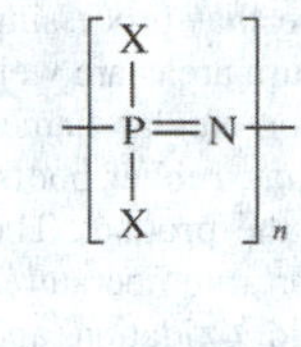

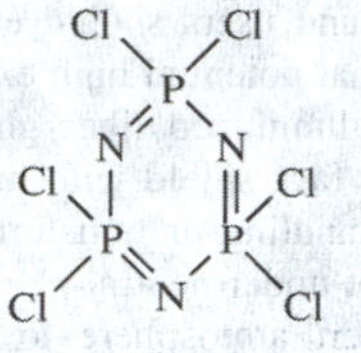

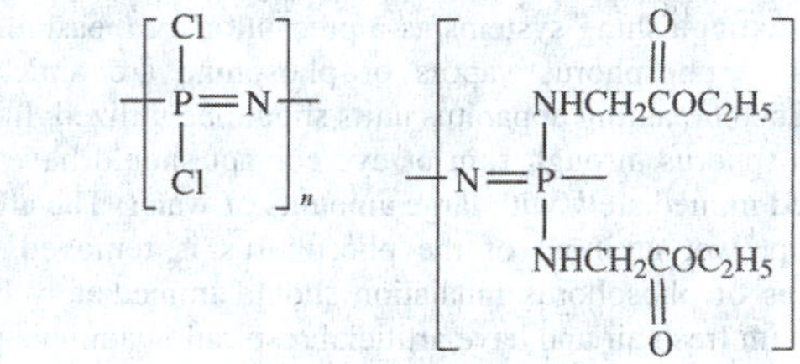

Phosphorus Ylides. In 1979, G. Wittig received a share of the Nobel Prize for Chemistry in recognition of his development of the use of phosphorus-containing compounds as important reagents used in organic synthesis. According to the selection committee, Wittig's most important achievement was "the discovery of the rearrangement reaction that bears his name. In the Wittig reaction an organic phosphorus compound with a formal double bond between phosphorus and carbon is reacted with a carbonyl compound. The oxygen of the carbonyl compound is exchanged for carbon, the product being an olefin. This method of making olefins has opened up new possibilities, not the least of which is the synthesis of biologically active substances containing carbon-to-carbon double bonds. For example, vitamin A is synthesized industrially using the Wittig reaction." As early as 1919, some 30 years prior to Wittig's work, the first phosphorus ylide was described by Staudinger. This was diphenylmethylenetriphenylphosphorane, formed by pyrolysis of a phosphazine precursor. During that period, Staudinger conceived the possibility of olefin synthesis by condensation of this ylide with a carbonyl compound and visualized a four-membered phosphorus-oxygen heterocyclic (oxaphosphetane) as the intermediate. Staudinger's work was accomplished during a period when practical application of the concept was in doubt. At that time, the Lewis theory of electronic structure was new. The exact bonding of phosphonium salts was somewhat veiled and controversial. Attempts of an olefin synthesis were put aside. For more detail, see **Wittig Reaction**.

Health and Safety

At ambient temperatures white phosphorus spontaneously ignites when exposed to air. It has an autoignition temperature of 30 °C. As a result, any human exposure to white phosphorus can cause severe thermal burns to the skin and eyes. The vapor from phosphorus can cause severe lung irritation, followed by a build-up of fluid in the lungs. Continuous long-term inhalation of white phosphorus vapor (>0.1 mg/m3) can result in bone loss to the jawbone structure causing loosening of teeth and severe pain and swelling of the jaw. This condition is commonly referred to as phossy jaw. Some evidence exists that increased infant mortality can result when pregnant women are exposed to P_4 vapor in excess of 0.075 mg/kg/day. Ingestion of white phosphorus is potentially fatal. The lowest reported fatal dose is 1 mg/kg for humans. Absorption through the skin is also possible but is only considered to be a moderate hazard compared to the other routes of exposure.

When exposed to air, white phosphorus oxidizes to phosphorus pentoxide forming copious quantities of white smoke. This smoke may be irritating but is not considered to be toxic. White phosphorus solid, liquid, or vapor is also extremely reactive with oxidizers such as strong acids, alkaline hydroxides, halogens, and nitrates. Contact of phosphorus with water or oxidizers also generates phosphine CAS: 7803-51-2, PH_3, a highly toxic and flammable gas. Phosphine has an 8-h time-weighted average exposure limit of 0.3 ppm. Under alkaline conditions the rate of PH_3 formation is high. At neutral or acidic pH, the PH_3 generation is slow but still very hazardous if the PH_3 is allowed to accumulate in a confined vapor space. The safest commercial handling conditions for molten phosphorus are generally considered to be from pH 6 to 8 at 45–65 °C.

Phosphorus production plants and users should ensure that processing of the material is contained and that potential high exposure areas are well ventilated. Workers must wear aluminized fiber glass or Kevlar flame-retardant full protective clothing, face shield with hard hat, rubber boots, and heavy rubber gloves when handling or transferring the product. The phosphorus should always be kept under neutral pH water at temperatures less than 65 °C or under an inert atmosphere to avoid oxidation and exposure hazards. High potential exposure areas should also be equipped with well-maintained, water-filled safety tubs, deluge systems, and water-spray extinguishing systems as a precautionary measure. If high exposure levels to phosphorus vapors or phosphine are anticipated, then self-contained breathing apparatus units should be utilized. Individuals exposed to phosphorus through skin or eye contact should have the exposed area flushed immediately with large amounts of water. The affected area should be kept wet until all of the phosphorus is removed or flushed away. Victims of phosphorus inhalation should immediately be removed to an area with fresh air and have artificial respiration administered, if necessary. Workers who have had dental surgery and pregnant women should be kept away from phosphorus exposure areas completely. Anyone who ingests phosphorus should drink a large volume of water and be induced to vomit. Medical assistance should be obtained as soon as possible after any instance of phosphorus exposure.

Additional Reading

Allcock, H.R.: "Polyphosphazenes: New Polymers with Inorganic Backbone Atoms," *Science*, **193**, 1214–1219 (1976).

Burges, R.J.: "Choose the Right Alloys for Fertilizer Acids," *Chem. Eng. Progress*, 82 (November 1992).

Considine, D.M. and G.D. Considine: *Foods and Food Production Encyclopedia*, Van Nostrand Reinhold, New York, NY, 1982.

Corbridge, D.E.: *Phosphorus: An Outline of Its Chemistry, Biochemistry and Technology*, 5th Edition, Elsevier Science, New York, NY, 1995.

Corbidge, D.E.: *Phosphorus 2000: Chemistry, Biochemistry and Technology*, Elsevier Science, New York, NY, 2000.

Dillon, K.B., F. Mathey, and J.F. Nixon: *Phosphorus: The Carbon Copy*, John Wiley & Sons, Inc., New York, NY, 1998.

Emsley, J.: *The 13th Element: The Sordid Tale of Murder, Fire and Phosphorus*, John Wiley & Sons, Inc., New York, NY, 2000.

Greenwood, N.N. and A. Earnshaw: *Chemistry of the Elements*, 2nd Edition, Butterworth-Heinemann, Inc., Woburn, MA, 1997.

Kent, J.A.: *Riegel's Handbook of Industrial Chemistry*, 9th Edition, Chapman & Hall, New York, NY, 1992.

Krebs, R.E.: *The History and Use of Our Earth's Chemical Elements: A Reference Guide*, Greenwood Publishing Group, Inc., Westport, CT, 1998.

Lewis, R.J. and N.I. Sax: *Sax's Dangerous Properties of Industrial Materials*, 10th Edition, John Wiley & Sons, Inc., New York, NY, 1999.

Lide, D.R.: *CRC Handbook of Chemistry and Physics*, 88th Edition, CRC Press, LLC., Boca Raton, FL, 2007.

Peters, E.N., et al.: *Rubber Chemistry*, 1976.

Somerville, R.L.: "Reduce Risks of Handling Liquefied Toxic Gas (Phosgene)," *Chem. Eng. Progress*, **64** (December 1990).

Stevenson, F.J. and M.A. Cole: *Cycles of Soils: Carbon, Nitrogen, Phosphorus, Sulfur, Micronutrients*, 2nd Edition, John Wiley & Sons, Inc., New York, NY, 1999.

Vedejs, E.: "1979 Nobel Prize for Chemistry," *Science*, **207**, 42–44 (1980).

PHOSPHORUS (In Biological Systems). Phosphorus is required by every living plant and animal cell. Deficiencies of available phosphorus in soils are a major cause of limited crop production. Phosphorus deficiency is probably the most critical mineral deficiency in grazing livestock. Phosphorus, as orthophosphate or as the phosphoric acid ester of organic compounds, has many functions in the animal body. As such, phosphorus is an essential dietary nutrient.

The biological roles of phosphorus include: (1) anabolic and catabolic reactions, as exemplified by its essentiality in high-energy bond formation, e.g., ATP (adenosine triphosphate), ADP (adenosine diphosphate), etc., and the formation of phosphorylated intermediates in carbohydrate metabolism; (2) the formation of other biologically significant compounds, such as the phospholipids, important in the synthesis of cell membranes; (3) the synthesis of genetically significant substances, such as DNA (deoxyribonucleic acid) and RNA (ribonucleic acid); (4) contributing to the buffering capacity of body fluids, cells, and urine; and (5) the formation of bones and teeth. Like calcium, the majority of the phosphorus in the vertebrate body is contained in the hard tissues; in the adult, approximately 80–86% of the total body phosphorus is contained in the bones and teeth, with the balance found in the soft tissues and body fluids.

In a very interesting dissertation by F.H. Westheimer (*Science*, **235**, 1173–1177, 1987), the role of phosphates in living substances is described. As pointed out by Westheimer, phosphate esters and anhydride dominate the living world but are seldom used as intermediates by organic chemists. Phosphoric acid is specially adapted for its role in nucleic acids because it can link two nucleotides and still ionize; the resulting negative charge serves both to stabilize the diesters against hydrolysis and to retain the molecules within a lipid membrane. A similar explanation for stability and retention also holds for phosphates that are intermediary metabolites and for phosphates that serve as energy sources. Phosphates with multiple negative charges can react by way of the monomeric metaphosphate ion PO_3^- as an intermediate. No other residue appears to fulfill the multiple roles of phosphate in biochemistry. Stable negatively charged phosphates react under catalysis by enzymes; organic chemists, who can only rarely use enzymatic catalysis for their reactions, because they need more highly reactive intermediates than phosphates.

Most of the coenzymes are esters of phosphoric or pyrophosphoric acid. The main reservoirs of biochemical energy, adenosine triphosphate (ATP), creatine phosphate, and phosphoenolpyruvate are phosphates. Many intermediary metabolites are phosphate esters, and phosphates or pyrophosphates are essential intermediates in biochemical syntheses and degradations. The genetic materials DNA and RNA are phosphodiesters.

Phosphorus deficiencies are not common in humans and most other species, but they have been observed in ruminants. Symptoms of the deficiency are loss of appetite and a depraved appetite (termed "pica") where the animal chews and consumes extraneous items, such as wood, clothing, bones, etc. Vitamin D deficiency may accentuate a marginal lack of phosphorus in the diet.

Cereals and meats are the major sources of phosphorus in human diets. Phosphorus deficiencies in most regions have not been a serious problem in human nutrition. Insofar as food is concerned, the primary value of phosphorus fertilizers is that they generally increase the total food production; not the content of phosphorus in the food per se.

Experimental phosphorus deficiency can be induced by feeding diets low in this element and by including excesses of calcium, strontium, barium, beryllium, and other cations that precipitate phosphates in the intestinal tract. In this situation, bone formation ceases, and the following histological bone changes have been noted in experimental animals: (1) a thickening of the epiphyseal plate and the formation of a typical rachitic metaphysis; (2) wide osteoid borders of trabecular bone and a considerable rarefaction of the shaft; and (3) irregular or complete cessation of calcification of the zone of provisional calcification of the cartilage matrix. Rickets can be produced in the laboratory by feeding a diet high in calcium and low in phosphorus, and containing little or no vitamin D.

The nutrient requirement for phosphorus depends upon the particular species and the physiological status of the animal. During growth, lactation, gestation, and egg laying, a higher phosphorus content of the diet is generally required in poultry than for the maintained adult. The availability of phosphorus in the diet varies with its chemical form and the animal species in question. Diets high in foods of plant origin may contain a considerable portion of phosphorus in the form of phytic acid, which is the hexaphosphoric acid ester of inositol. When the acid occurs as salts of calcium, magnesium, sodium, etc., it is referred to as phytin. Phytate phosphorus usually is less available than inorganic phosphate to such species as rat, chicken, dog, pig, and human. However, a phytase has been shown to be present in the intestine and intestinal secretions of some animals, and the formation of this enzyme is dependent, in part, on the presence of vitamin D. Through the action of phytase, some of the phytate phosphorus would be made available for absorption.

Experimentation has indicated that, under normal dietary conditions and calcium intake, food phytate is of no nutritional concern in humans. The microbial population of the ruminant also elaborates a phytase enzyme that makes phytate phosphorus readily available in this class of animals. Phytates may be of nutritional consequence for another reason — dietary calcium can be bound in an unavailable, insoluble complex, thereby decreasing the absorption of this element.

Many studies have involved determination of the availability of phosphorus from other organic and inorganic sources. In chicks, orthophosphates, superphosphates, and phosphate rock products are good sources of phosphorus, whereas metaphosphate and pyrophosphate are relatively unavailable to the species. Most organic phosphorus sources, such as casein, pork liver, and egg phospholipid are found to be as available as inorganic phosphorus. Commonly used phosphorus supplements in human or animal nutrition or both are steamed bone meal, ground limestone, dicalcium phosphate, and defluorinated rock phosphates. Phosphorus dietary supplements include magnesium phosphate (dibasic and tribasic) manganese glycerophosphate and manganese hypophosphate, potassium glycerophosphate, sodium ferric pyrophosphate, sodium phosphate (mono-, di-, and tri-), and sodium pyrophosphate. Of course, phosphate compounds are not always added in the interest of augmenting phosphorus, but for the other elements which may be contained in the compound.

Absorption of Phosphate. The phosphate ion readily passes across the gastrointestinal membrane. The rate of absorption of phosphate at various intestinal sites in rats has been observed to be most rapid in the duodenum, followed in decreasing order by the jejunum, ileum, colon, and stomach. When transit time is considered, most of the phosphorus is absorbed by the ileum.

The triangular relationship between calcium, phosphorus, and vitamin D is described briefly in entry on **Calcium (In Biological Systems)**.

Plasma Phosphate. Once absorbed, phosphorus enters the blood and the majority is present therein as orthophosphate ions. At an ionic strength of 0.165 and at 37 °C (98.6 °F), calculations show that the proportional concentration of the orthophosphate ions in plasma for $H_2PO_4^-$ is 18.6 × 10^{-30}; for HPO_4^{2-} is 81.4 × 10^{-30}; and for PO_4^{3-} is 8 × 10^{-30}. About 12% of the phosphorus present is bound to proteins. During egg laying in birds, the concentration of nonionized phosphorus compounds in plasma is greatly increased. The administration of diethylstilbestrol (regulated in some countries) to cockerels results in the formation of a plasma phosphoprotein, which forms relatively firm complexes with calcium. The function of the phosphoprotein appears to be one of phosphorus transport; in laying birds, the phosphoprotein is incorporated in egg yolk.

The approximate average plasma phosphorus levels for several species, in milligrams per 100 milliliters of plasma, are: pigs, 8.0; sheep, cattle, and goats, 6.0; horse, 2.3. Erythrocytes contain considerably more phosphorus than plasma, mostly in the form of organic esters. Some of the latter are acid soluble and hydrolyzable by intracellular enzymes.

Plasma phosphate appears to be homeostatically controlled. The primary organ concerned appears to be the kidney, although the skeleton also may play a role. Parathyroid hormone, by way of its direct action on the kidney and bone, is a significant hormonal factor.

Phosphate Excretion. The excretion of body phosphorus occurs via the kidney and intestinal tract, the distribution between these pathways varying with species. For example, relatively small amounts of phosphorus are endogenously excreted into the feces of rat, pig, and human, but in the bovine, perhaps 50% or more of the fecal phosphorus may be from endogenous sources.

The amount of phosphorus excreted in the urine varies with the level of ingested phosphorus and factors influencing phosphorus availability and utilization. It has been shown that in the dog, when plasma phosphate is normal or low, over 99% of the filtered ion is reabsorbed, presumably in the upper part of the proximal tubule. Increased plasma concentrations of alanine, glycine, and glucose depress phosphate reabsorption.

Phosphate of Hard Tissues. Body phosphorus contained in the intracellular matrix of bone and teeth are of the general form of hydroxyapatite, $Ca_{10}(PO_4)_6(OH)_2$, this calcium phosphate salt providing the characteristic hardness of ossified tissue. Phosphate ions are also adsorbed onto the surface of bone crystals and exist in the hydration layers of the crystals. Early theories of calcification placed special emphasis on the role of alkaline phosphatase and organic esters of phosphoric acid. As part of the theory, it was stipulated that, with the hydrolysis of phosphate esters at the site of calcification, the K_{sp} for bone salt would be exceeded. Although phosphatase may have a function in bone formation, as in the synthesis of organic matrix, its role as earlier depicted has been revised. Later research emphasized the specific and characteristic properties of collagen and other substances, such as chondroitin sulfate. This is related to the local mechanism of calcification; the other component of calcification is the humoral mechanism whereby an adequate supply of calcium, phosphate, and other ions is made available to the calcifying site. A later theory

proposed that either the functional groups on collagen are anionic, initially binding Ca^{2+}, or that the first reaction is with phosphate or phosphorylated intermediates. The first-held moiety of bone salt (Ca^{2+} or phosphate) subsequently attracts or binds the other component, providing the aggregation or "seed" for subsequent crystal growth. Since an ATPase-type enzyme has been demonstrated in cartilage, suggesting that ATP may be intimately involved in the calcification mechanism, another proposal is along the line that pyrophosphate is transferred from ATP to free amino groups of collagen, leading to nucleation and followed by combination with calcium and bone salt formation. Or, the ATP provides energy which increases the calcification mechanism.

Dietary inorganic phosphates have been shown to protect experimental animals against dental caries. Orthophosphates were effective cariostats, but $Na_4P_2O_7$ and $Na_5P_3O_{10}$ were not. Dicalcium phosphate, $CaHPO_4$, did not decrease dental caries unless a high level of NaCl was also included in the diet.

Toxicity. Although many phosphorus-containing compounds are vital to life processes, as previously described, there are also many phosphorus compounds that are quite toxic—elemental phosphorus, for example. While the elemental form is dangerous because of its low combustion temperature, its absorption also has an acute effect on the liver. The long and continued absorption of small amounts of phosphorus can result in necrosis of the mandible or jaw bone (sometimes called "phossyjaw"). Chronic phosphorus poisoning occurs particularly through the lungs and gastrointestinal tract. The most common symptom is the necrosis of the jaw already mentioned, but this is also usually accompanied by anemia, loss of appetite, gastrointestinal weakness, and pallor. Other bones and teeth may be adversely affected.

Phosphine is a very toxic gas. Inhalation of phosphine causes restlessness, followed by tremors, fatigue, slight drowsiness, nausea, vomiting, and, frequently severe gastric pain and diarrhea. Although most cases recover without after-effects, in some cases, coma or convulsions may precede death.

Phosphorus-halogen compounds are quite toxic.

Phosphorus in Soils. When phosphorus fertilizers are added to soils deficient in available forms of the element, increased crop and pasture yields ordinarily follow. Sometimes the phosphorus concentration in the crop is increased, and this increase may help to prevent phosphorus deficiency in the animals consuming the crop, but this is not always so. Some soils convert phosphorus added in fertilizers to forms that are not available to plants. On these soils, very heavy applications of phosphorus fertilizer may be required. Some plants always contain low concentrations of phosphorus even though phosphorus availability from the soil may be good. See also **Fertilizer**.

PHOSPHORYLATION (Oxidative). This is an enzymic process whereby energy released from oxidation-reduction reactions during the passage of electrons from substrate to oxygen over the electron transfer chain is conserved by the synthesis of adenosine triphosphate (ATP) from adenosine diphosphate (ADP) and inorganic orthophosphate. Since ATP is the major source of energy for biological work, and since most of the net gain of ATP in the animal cell derives from oxidative phosphorylation, research in the area has been very active.

Oxidative phosphorylation was discovered simultaneously and independently in 1939 by Kalckar (Denmark) and by Belitzer (the former U.S.S.R.). It was recognized by these workers that aerobic phosphorylation was different from and independent of phosphorylation supported by glycolysis. In addition, they found that the stoichiometry of phosphate esterification (ATP synthesis) and oxygen utilized was two or more, or that the reduction of one atom of oxygen to form water may be accompanied by the "activation" of two or more molecules of phosphorus (P_i), thus leading to an expression of the efficiency of the energy-conserving system. The efficiency expression is known as the P/O ratio, i.e., the ratio of molecules of P_i esterified per atom of oxygen utilized.

The quantitative importance of the ATP synthesized at the expense of energy liberated during electron transfer in the mitochrondrion is realized when one follows the conservation of energy during the metabolism of a molecule, such as glucose in the cell. The oxidation of one mole of glucose to carbon dioxide and water is accompanied by the release of 673,000 calories. In order to degrade the glucose molecule to a form that can be metabolized further by mitochondrial enzymes, the glycolytic enzymes consume two molecules at ATP and also synthesize two molecules of ATP in the presence of oxygen, a net energy conservation of zero. The mitochondrion may then degrade the pyruvate supplied by glycolysis to carbon dioxide and water, yielding a net total of 38 molecules of ATP, mostly at the level of the electron transfer process. Thirty-eight molecules of ATP per molecule of glucose results in between 260,000 and 380,000 calories conserved, between 39 and 56% of the total energy released in the complete oxidation of glucose, the remainder being released directly to heat. Inasmuch as the mitochondrion is approximately 50% effective in conserving energy from its major substrate, it is indeed an efficient machine. See also **Cell (Biology)**.

PHOSPHORYLATION (Photosynthetic). Photosynthetic conversion of light energy into the potential energy of chemical bonds involves an electron transport chain, and the phosphorylation of ADP (adenosine diphosphate)

$$ADP + P \xrightarrow[\text{Chlorophyll}]{\text{Light}} ATP$$

as intermediate stages. The process of phosphorylation defined by the foregoing equation was discovered simultaneously by Arnon and coworkers for green plant chloroplasts and by Frenkel, working with Geller and Lipmann, for bacteria in 1954. For both systems, the heart of the mechanism is the creation of a very oxidizing and a very reducing component, utilizing the energy of the photoexcited stage of one of the pigment (chlorophyll) molecules. This process will be designated a photoact. The redox components are both members of a photosynthetic electron transport chain, bound to the membranes of the chloroplasts (for green plants) or chromatophores (for bacteria). The photoact can be considered as electron transport against the thermochemical gradient, i.e., away from the member which is a better electron acceptor (high oxidation-reduction potential), through the excited chlorophyll, then to the member which is a better electron donor (low oxidation-reduction potential). Subsequent steps consist of ordinary, dark electron transport with the thermochemical gradient. The energy in at least one of these redox reactions is conserved as ATP by a phosphorylation reaction analogous to that found in oxidative phosphorylation by mitochondria. See also **Phosphorylation (Oxidative)**.

In bacteria, the photoact proper is accomplished by a special kind of bacteriochlorophyll, amounting to only 3% of the total present. It is unique in having a peak in absorption at 870–890 micrometers, or further into the infrared than the remaining 97% of the chlorophyll molecules. It is unique not by virtue of a difference in structure, but because of its environment—most probably in close association or complexing with cytochrome molecules. Since its absorption extends to longer wavelengths, it is an energy trap, and the function of the bulk of the bacterio-chlorophyll is that of capturing light and transmitting it to this active center.

Components of the electron transport chain in bacteria have been shown to include *b*- and *c*-type cytochromes, ubiquinone (fat-soluble substitute quinone, also found in mitochondria), ferredox (an enzyme containing nonheme iron, bound to sulfide, and having the lowest potential of any known electron-carrying enzyme) and one or more flavin enzymes. Of these a cytochrome (in some bacteria, with absorption maximum at 423.5 micrometers, probably *c*₂) has been shown to be closely associated with the initial photoact. Some investigators were able to demonstrate, in chromatium, the oxidation of the cytochrome at liquid nitrogen temperatures, due to illumination of the chlorophyll. At the very least this implies that the two are bound very closely and no collisions are needed for electron transfers to occur.

In both bacterial chromatophores and green plant chloroplasts, the existence of photo-induced high-energy intermediates or states leads to reversible confirmational changes in the structures of the membranes, and to gross swelling and shrinking. These are observed by changes in light scattering, viscosity, and sedimentation properties, and by electron microscope studies. The mechanisms may include ion transport, followed by water diffusion, internal pH changes leading to conformation changes of proteins, or possibly something resembling a contractile protein.

PHOT. A photometric unit of illuminance or illumination equal to 1 lumen per square centimeter.

PHOTOCHEMICAL AIR POLLUTION. A type of air pollution, such as Los Angeles smog, associated with the buildup of oxidation products formed from the degradation of hydrocarbons, etc. The term arises because sunlight is required to initiate photolysis reactions. Ozone and nitrogen dioxide are always present in photochemically produced mixtures; often other species such as peroxyacetyl nitrate and formaldehyde are also produced.

PHOTOCHEMICAL MODELING. See **Air Quality Models**.

PHOTOCHEMICAL REACTION. A chemical reaction that involves either the absorption or emission of radiation. The absorption of an ultraviolet photon often provides the energy required to break chemical bonds and initiate a reaction sequence. Examples of photochemical reactions are the photolysis of nitrogen dioxide, $NO_2 \rightarrow NO + O$, or ozone, $O_3 \rightarrow O_2 + O$. The latter reaction leads to the initiation of chain reactions that cause the breakdown of hydrocarbons and other pollutants in the troposphere.

See also **Photodecomposition**; and **Photolysis**.

PHOTOCHEMICAL SMOG. Not to be confused with industrial smog, photochemical smog is air contaminated with ozone, nitrogen oxides, and hydrocarbons, with or without smoke and natural fog being present.

In the presence of sunlight, hydrocarbons and NO_x are involved in a complex series of chemical reactions that eventually creates ozone and other oxidants as secondary pollutants. However, ozone is also destroyed by NO_x. Photochemical air pollution levels are generally proportional to concentrations of nitrogen oxides and hydrocarbons; they also increase with strong solar intensity and high ambient temperatures, which increase biogenic volatile organic emissions to the atmosphere from vegetation. The pollutant levels are inversely proportional to wind speed and inversion height. See also **Smog**.

Photochemical smog can cause eye, nose, and throat irritations, impaired lung function, headaches, and coughing and wheezing. It can also damage leaves and reduce growth and reproduction in plants.

PHOTOCHEMISTRY AND PHOTOLYSIS. When certain substances are subjected to light, a chemical change results. Such reactions comprise *photochemistry*. The production of an image on a photographic plate is an example. Photosynthesis in the green leaf of a plant is another. Where the change involves chemical decomposition of the radiated material, the process is termed *photolysis*. As used in this context, the term light includes both visible light and ultraviolet radiation. One of the better known and most extensive examples of photolysis is the production of ozone, O_3, in the upper atmosphere, a reaction critical to life on earth because ozone acts as a filter of the middle- and far-ultraviolet radiations which destroy living organisms. The oxygen molecule, O_2, absorbs solar ultraviolet radiation with a wavelength of 190 nanometers, and the energy absorbed breaks the molecule to the atomic state. The released oxygen atoms may combine with oxygen molecules present to form ozone, or the freed oxygen atoms may recombine to form O_2. Thus, there is a continuing combination of processes in dynamic equilibrium, that is, the synthesis and the photolysis of ozone.

Similarly, oceanic nitrite (NO_2^-) photolysis by natural light produces detectable concentrations of nitric oxide (NO). This latter forms in the oceans during daylight and disappears rapidly at sunset when recombination occurs: $NO_2 \rightarrow NO + O \rightarrow NO_2$.

Isomerism can also be induced photochemically, although such processes are less well understood and probably require the presence of additional free radicals. The cytotoxic metabolite bilirubin can cause brain damage in infants with neonatal jaundice; this is prevented by exposing the child to intense blue light. The bilirubin is photochemically converted in the skin to metastable geometric isomers, which can be transported in the blood and excreted in bile.

Two major instances of photochemical reactions that have reached deeply into modern civilization are the photosensitive silver and uranium salts and dyes which are the basis of photography and the manufacture of Vitamin D by the ultraviolet irradiation of ergosterol.

Photochemical reactions are highly specific and their products are quite different from those of thermochemical reaction processes.

Sunlight in the near infrared, visible, and near ultraviolet regions possesses considerable energy; utilization of this through photochemical reactions could make a considerable contribution to energy resources. Since biosynthesis itself is relatively inefficient in conversion of solar energy, emphasis has been placed upon the fabrication of *artificial* photochemical systems. One of the more promising approaches has involved application of photoelectric chemical cells or catalysts of semiconductor materials.

The absorption of light by semiconductors creates electron-hole pairs ($e^- h^+$) which can be separated because their components diffuse in different directions. The energies of these moieties can be stored by several mechanisms or used in photocatalysis or photosynthesis for nitrogen fixation, formation of amino acids, methanol, etc. The efficiencies of such conversions depend almost entirely upon the semiconductor material, and as yet these efficiencies are too low for significant application. Currently the most promise is demonstrated by the use of titania on a platinum substrate or single crystals of strontium titanate. See also **Photoelectric Effect**.

Fundamental Considerations. In photochemical reactions, light supplies the energy necessary for the activation of the reacting molecules (Grotthus, 1818, and Draper, 1839). Sometimes the light waves absorbed by a body produce only an increase in temperature, sometimes fluorescence as in the cases of eosin and fluorescein, and sometimes chemical change. The reaction of hydrogen and chlorine in light was studied by Bunsen and Roscoe (1862), and they discovered that the amount of chemical change is proportional to the intensity of the light and to the length of time of exposure to the light. The first law of photochemistry (Draper-Grotthus) states that light that is absorbed causes chemical change. The energy of light is measured in quanta, and according to the Stark-Einstein law,

$$E = Nhc/\lambda$$

where N is Avogadro's constant, h is Planck's constant, c is velocity of light, λ is wavelength of light; that is, each molecule that takes part in a chemical reaction induced by exposure to light absorbs one quantum of radiation causing the reaction. Photochemical processes are of two kinds: primary and secondary. The primary process in a photochemical reaction is limited by the Einstein law to the absorption of one quantum by a molecule or atom. A knowledge of the spectrum of the reactants is necessary to determine what happens in this process. The molecule may be disrupted into fragments or an electron may be excited from a lower orbit to a higher one. Which of these events takes place can often be determined by spectroscopic studies. The secondary process deals with the fate of the molecular fragments or of the excited molecules. The excited molecule may emit its extra energy as light, causing fluorescence; it may lose it by transferring it to other molecules as thermal energy; or it may cause a chemical reaction. On the other hand, the molecular fragments may either recombine to give the original reactant or cause further chemical reactions. The study of the quantum yield (which is the number of molecules reacting divided by the number of quanta absorbed), is used as a means of formulating the secondary processes. If the quantum yield is less than one, fluorescence, deactivation or recombination of fragments must take place. If the quantum yield is unity every photon absorbed decomposes one molecule. When the quantum yield is greater than unity (and in some reactions it may be as high as a million) chain reactions are involved. The classical example of such a reaction is the combination of hydrogen and chlorine. The primary reaction is Cl_2 and light → $2Cl$. The chain propagation reactions are

$$Cl_2 + h\nu \longrightarrow 2Cl$$

$$Cl + H_2 \longrightarrow HCl + H$$

$$H + Cl_2 \longrightarrow HCl + Cl$$

creating a cycle which is only stopped by

$$Cl + Cl \longrightarrow Cl_2$$

$$H + H \longrightarrow H_2$$

Since the last two processes are slow compared to the two before them, one quantum of light can bring about a combination of a million molecules of hydrogen and chlorine.

See also **Photosynthesis**.

The existence of microbes capable of utilizing light energy to drive the synthesis of cellular components was firmly established by Winogradsky and Molisch by the late 19th century, but it was largely through the work of Van Neil in the 1930s on the physiology of the purple sulfur bacteria that a clearer picture of their photosynthetic processes started to emerge. The photochemical reaction center (RC) is a now well-defined physical entity which exists as an energy sink able to convert the energy of excitation into an electron transfer event. From the use of this RC, three broad patterns of microbiological photosynthesis are known. See Table 1.

The chlorophylls are the pigments responsible for initiating primary photochemistry in the cyanobacteria and higher plants and absorb light energy at 870 nm converting appropriate species to the single excited state and rapidly transferring an electron to a single molecule. Chlorophyll is remarkably similar to hemoglobin in that iron is replaced by magnesium as the chelated metal. Any other transition metals in that position would cause quenching of the initial photochemical reaction.

The carotenoids are largely responsible for the color of the green and purple bacteria and function in both light absorption and energy transfer processes as well as extending the usable light wavelengths and protecting against harmful photooxidation.

In green and purple photosynthetic bacteria, most of the bulk bacterial chlorophyll is inactive in photochemistry and does not undergo photooxidation when excited. Energy is transferred randomly between adjacent pigment molecules until trapped by the RC or lost as fluorescence or heat.

But the photochemistry of bacterial activity must also be associated with photoeffects in the surrounding media, and in this context a rapid increase in the concentration of hydrogen peroxide has been observed when natural surface and ground waters are exposed to sunlight. The hydrogen peroxide is photochemically generated from organic constituents present in the water. Humic materials are believed to photochemically reduce oxygen to give the superoxide anion, which subsequently disproportionates to hydrogen peroxide. Since both hydrogen peroxide and peroxide radical are known to affect biological systems, they may be important factors in the photochemical reactions of photosensitive bacteria and other natural life forms.

Laser Chemistry

Lasers generate a high-intensity output of monochromatic photon energy, and studies of the photochemical reactions induced by this have created a virtual subdivision of photochemistry known as "laser chemistry." While the output of a laser can heat, anneal, burn, cut, or be used instrumentally as a spectral source, we are concerned here only with those chemical effects attributable to the photon output at wavelengths between near infrared and near ultraviolet, i.e., between about 12 and 0.2 microns.

When atoms or molecules are excited conventionally by elevated temperatures or pressure, they can follow several reaction paths yielding a variety of byproducts in addition to the desired substance. Since the basis of a chemical reaction is to weaken or break or make specific chemical bonds to yield the final product, energy ideally should be selectively introduced at the particular level necessary to accomplish this. The high energy and monochromaticity of laser output are ideal for imposition of the specific energy changes that induce or catalyze chemical changes.

TABLE 1. BROAD PATTERNS OF MICROBIOLOGICAL PHOTOSYNTHESIS*

Bacterial Group	Type of Photosynthesis	Pigment in Primary Photoactivation	Electron Donors	Products	Carbon Sources
Eubacteria	Anoxygenation	Bacterial chlorophyll	H_2, H_2S, S, Organics	ATP + NO, D, P(H)	CO_2 + Organics
Eubacteria	Oxygenation	Chlorophylls	H_2O, H_2S	ATP + NO, D, P(H)	CO_2 + Organics
Archeobacteria	Halogenation	Bacterial rhodopsin	Do not participate	ATP	Organics

(*After Keely and Dow.)

The absorption of a quantum of energy by an atom or a molecule takes it from a low energy state to a higher one, and the jump will affect the different properties of the atom or molecule depending upon the amount of energy in the quantum. When absorbed, a quantum of visible or ultraviolet radiation raises an electron to a higher orbit; on the other hand, a quantum of infrared radiation will alter energy levels on an atomic basis.

A laser can supply a precise amount of energy to an atom or molecule, thus effecting a transition from one excited state to a higher excited one. Once known of the energy level displacement required to effect a chemical reaction is available, the laser can provide the specific energy for the specific excitation required. However, the energy input must be related to the total energy dissipation, for, if excess energy leads to ionization or dissociation, a continuum of allowed energy levels will be developed rather than the required discrete levels. If the energy is thus fragmented, the required reaction will proceed only weakly, if at all. Excess energy may be redistributed in two ways. It is either transferred from the excited vibrational state to one or more other vibrational states of the molecule, or it is transferred directly into rotational and translational states. The first mode of energy translation proceeds appreciably more rapidly than the second. Time is a further controlling factor in laser chemistry. The reaction must proceed in time either shorter than, or equal to, that required for transfer of vibrational energy from one state to another in the same molecule; molecule dissociation or atom ionization must take place before there is any depletion of energy by molecular or atomic collisions. Where a reaction proceeds within the lengthy period required for transfer of energy from the initial vibrational state to the much lower rotational and translational states, one cannot hope for laser action to effect a significant degree of reaction specificity, since the effect is basically a thermal one.

A goal of early laser-induced photochemistry was the initiation of specific chemical reactions, to fabricate specific chemicals, or to separate isotopes. Although the specificity of laser-induced photochemistry is important, equally significant is the ability of the laser to confine excited regions to microscopic areas. Thus, one of the better known capabilities of the laser beam is that even a low-powered laser can produce highly intense spots of light of submicrometer dimensions.

In many photochemical reactions, a specific excitation wavelength leads to a specific set of molecular fragments and ultimately products. This has enabled the dissociation of a large variety of molecular gases which can then deposit on, dope, or etch a semiconductor wafer. Further, laser excitation can photodeposit metals, insulators, and semiconductors by decomposing one or more photosensitive gases. Direct writing with lasers through such reactions as:

$$CF_3Br \xrightarrow{h\nu} CF_3 + Br; \text{ or } Al(CH_3)_3 \xrightarrow{h\nu} Al + 3\,CH_3$$

is just beginning to be applied to the fabrication of complete microelectronic devices and the modification of actual circuits.

Operation of visible-light and ultraviolet lasers costs more per photon produced than does infrared laser operation. Partly because of this, appreciable interest has centered over the past several years on unimolecular reactions driven by infrared lasers. But absorption of a single infrared photon will raise a molecule only one step in the energy ladder, and, to be dissociated, the molecule will require the absorption of many infrared photons in sequence. The carbon dioxide laser can supply this requirement cheaply and efficiently. A mole of photons (6.02×10^{23}) costs only a few cents in the infrared, but several dollars in the visible and near ultraviolet ranges. This has aided continued study of multiple-photon infrared laser excitation. Much study has gone into an exciting and fundamental reaction and its implication. When sulfur hexafluoride (SF_6) is irradiated by infrared laser light, it decomposes to the pentafluoride (SF_5) and fluorine (F). When the laser is tuned to the specific absorption of $^{32}SF_6$ in a mixture with $^{34}SF_6$, only the $^{32}SF_6$ decomposes, leaving the residual gas enriched some 3000-fold in $^{34}SF_6$. Changing the frequency of the irradiating light slightly from emission at 10.61 μ to emission at 10.82 μ selectively decomposes the $^{34}SF_6$ molecule. This method of isotope separation by lasers is being extensively studied for the separation of fissionable U^{235} from nonfissionable U^{238}.

Laser-induced processes are expected to increase in number and expand in application, but the principal obstacle to large-scale introduction of the laser into the chemical industry is an economic one. Laser photos are still much more expensive than those from thermal sources, and the initial application will undoubtedly be directed to those specialty chemicals and isotopes whose current cost far exceeds that of large volume chemicals.

A typical example of this is in the preparation of extraordinary divalent carbon intermediates (carbenes) by application of ultrafast laser techniques. Thus, photoexcitement of diphenyldiazomethane (DPDM) to an excited singlet state breaks the $C=N_2$ bond, releasing diphenylcarbene (DPC) in a singlet state and ultimately allows its stabilization in the triplet ground state.

The increasing number of known photochemical reactions is still very small in comparison with those in ground state chemistry, and our understanding of all the factors controlling photochemical reactions is quite primitive. In some cases it is the ease of conversion to the ground state that is significant, while in others it is the energy hypersurface surrounding the excited state that dictates the energy pathways through which the free electrons will move back toward the ground state, and hence the nature of the photochemical products.

Laser Femtochemistry

As explained by A.H. Zewail (California Institute of Technology), "Femtochemistry is concerned with the very act of the molecular motion that brings about chemistry, chemical bond breaking, or bond formation on the femtosecond (10^{-15} second) time scale. With lasers it is possible to record snapshots of chemical reactions with sub-angstrom resolution. This strobing of the transition-state region between reagents and products provides real time observations that are fundamental to understanding the dynamics of the chemical bond."

A longstanding problem in chemistry has been the development of a better understanding of the transition state between reagents and products. Several investigative approaches, including thermodynamics, kinetics, and synthesis, have been used to systematize masses of experimental data to ascertain the rates and mechanisms of reactions. Over the past few decades, advanced understanding of molecular reactions have stemmed from the use of molecular beams, chemiluminescence, and, more recently, laser techniques. In laser-molecular beam research, a laser is used (1) to excite one of the reagent molecules, thus influencing reaction probability, or (2) to initiate a unimolecular process simply by providing energy to a molecule. In what has been called a "half-collision" unimolecular process, the fragmentation of the excited molecule can be determined. As pointed out by Zewail, "During the last three decades many reactions have been studied and these methods, with the help of theory, have become the main source of information for deducing the nature of the potential surface of a reaction."

Unimolecular reactions, as typified by

$$ABC^* \longrightarrow [A \cdots BC]^{\ddagger *} \longrightarrow A + BC,$$

and bimolecular reactions, as shown by

$$A + B \longrightarrow [ABC]^{\ddagger} \longrightarrow AB + C,$$

have been studied in terms of their *time scale*.

In an excellent paper, commenting on the progress made from the "picosecond" era to the "femto-second" era of research, Zewail observes, "Prior to femtochemistry, molecular beams and picosecond lasers were combined, which led to studies of collision-free energy redistribution in molecules and state-to-state rates of reaction, but the time resolution was still not sufficient to directly view the process of bond breaking or bond formation. However, in femtochemistry, the 'shutter speed' has reached the 10^{-15} regime, so it is now possible to observe chemistry as it happens—the transition-state region between reagents and products. ⋯ The strobing of these ultrafast molecular reactions, stemming from the happy marriage between ultrafast lasers and chemistry, is what forms the central theme in real time femtochemistry."

In an extension of this technology, A.S. Moffat reported in 1992 two additional goals of development: (1) the use of lasers to *control* chemical reactions, and (2) development of "tunable" lasers that can vary the wavelength of the light they generate. Thus, the light source can be tailored exactly to the vibrational frequency of the bond targeted.

Time-Resolved Photoacoustic Calorimetry

As defined by K.S. Peters and G.J. Snyder (University of Colorado), time-resolved photoacoustic calorimetry is an experimental technique that measures the dynamics of enthalpy changes on the time scale of nanoseconds to microseconds for reactions initiated by absorption of light. As pointed out, "When the reaction is carried out in water, it is also possible to obtain the dynamics of the corresponding volume changes. This method has been applied to a variety of biochemical, organic, and organometallic reactions."

Although pulsed time-resolved photoacoustic calorimetry is in an early phase of development (1994), it is proving to be a powerful technique for understanding the dynamics of enthalpy and volume changes for ground- and excited-state species. Peters and Snyder, developers of the technique, observe, "For reactions in water, the problem is directly approached through temperature-dependent studies. For reactions in organic solvent, there must be further investigations into the magnitude of the effect. At some future date, time-resolved photoacoustic calorimetry will be extended onto the 1 ns time scale .... The technique should find wide ranging applications to problems in chemistry and biochemistry that include solid-state reactions and dynamics of proteins in membranes."

A schematic representation of the instrumental technique used is shown in Fig. 1.

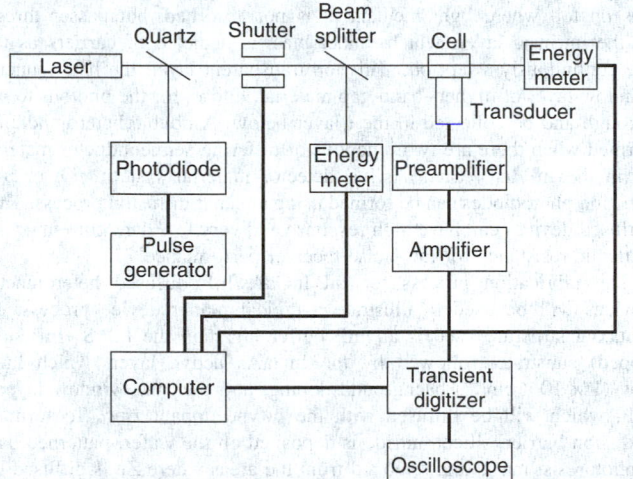

Fig. 1.   Schematic representation of the time-resolved photoacoustic calorimeter. Solid lines indicate light path; heavy solid lines represent signal paths. (*After Peters and Snyder.*)

Additional Reading

Andrews, D.L.: *Lasers in Chemistry*, 3rd Edition, Springer-Verlag, Inc., New York, NY, 1997.

Dunning, T.H., Jr., et al.: "Theoretical Studies of the Energetics and Dynamics of Chemical Reactions," *Science*, **453** (April 22, 1988).

Eisenthal, K.B., et al.: "Divalent Carbon Intermediates: Laser Photolysis," *Science*, **225**, 1439–1445 (1984).

Gust, D. and T.A. Moore: "Mimicking Photosynthesis," *Science*, **35** (April 7, 1989).

Horspool, W.H. and Pill-Soon Song: *CRC Handbook of Organic Photochemistry and Photobiology*, CRC Press, LLC., Boca Raton, FL, 1998.

Osgood, R.M. and T.F. Deutsch: "Laser Induced Chemistry for Microelectronics," *Science*, **227**, 709–714 (1985).

Murov, S.L., I. Carmichael, and G.L. Hug: *Handbook of Photochemistry*, 2nd Edition, Marcel Dekker, Inc., New York, NY, 1993.

Neckers, D.C., D.H. Volman, and G. Von Bunau: *Advances in Photochemistry*, Vol. 24, John Wiley and Sons, Inc., New York, NY, 1999.

Neckers, D.C., D.H. Volman, and G. Von Bunau: *Advances in Photochemistry*, Vol. 25, John Wiley and Sons, Inc., New York, NY, 1999.

Neckers, D.C., D.H. Volman, and G. Von Bunau: *Advances in Photochemistry*, Vol. 26, John Wiley & Sons, Inc., New York, NY, 2001.

Peters, K.S. and G.J. Snyder, "Time-Resolved Photoacoustic Calorimetry: Probing the Energetics and Dynamics of Fast Chemical and Biochemical Reactions," *Science*, 1053 (August 26, 1988).

Ramamurthy, V. and K.S. Schanze: *Organic, Physical and Materials Photochemistry*, Marcel Dekker, Inc., New York, NY, 2000.

Staehelin, L.A. and P. Aentzer: "Photosynthetic Membranes and Light Harvesting Systems," in *Encyclopedia of Plant Physiology*, Vol. 19, Springer-Verlag, New York, NY, 1986.

Truhlar, D.G. and M.S. Gordon: "From Force Fields to Dynamics: Classical and Quantal Paths," *Science*, **491** (August 3, 1990).

Wayne, C.E. and R.P. Wayne: *Photochemistry*, Oxford University Press, Inc., New York, NY, 1996.

Zewail, A.H.: "Laser Femtochemistry," *Science*, 1645 (December 23, 1988).

R.C. VICKERY, D.Sc., Blanton/Dade City, FL

PHOTODECOMPOSITION.   The chemical destruction of a molecule following the absorption of light energy.

See also **Photochemical Reaction**; and **Photolysis**.

PHOTODETECTORS.   Photons impinging upon matter interact in a manner determined by the nature of the chemical bonds in the material and the energy of the incident photons. Interacting photons may be reflected, refracted, diffracted, transmitted, or absorbed. Each of these phenomena can be used to measure some parameter of interest in chemical analysis. Photoassisted chemical analytical techniques require an accurate, sensitive method of photon detection and quantification. Photographic film proved to be the first useful method for such evaluations. Later, photomultiplier tubes provided the means to improve many photoanalytical techniques. With the advent of semiconductor technology, numerous single-element, broad-spectrum photodetecting devices were developed and applied to the task. Most recently, photodetector technology has entered the age of high density integration. Large-area photon detectors are becoming commonplace in most analytical equipment. The advances afforded by these devices have allowed for significant improvements in the performance of systems designed for photo-analytical chemical analysis in general, and specifically for spectroscopic-based equipment. A working knowledge of the operation and limitations of photodetectors is therefore necessary for the modern chemist.

Photodetector devices convert electromagnetic radiation or photons to electric signals that can be processed to obtain the spectral, spatial, and temporal information inherent in the radiation. The more popular ones are photoconductors, photodiodes, charge-transfer devices, and the pyroelectrics. The detectors may be used as single elements such as in street light controls, film camera exposure control, or motion detectors for security. Photodetectors also find application in the form of linear arrays used in analytical spectrometers, night-vision equipment, in small, low cost spectrometers for the control of building ventilation and environmental pollution monitoring, or configured as large matrix arrays found in video cameras.

Principles

The basic detection process, as shown in Figure 1, is the generation of free electrons, holes, or both by the absorption of photon energy. The absorption may be intrinsic, i.e., creating a free electron–hole pair, or extrinsic, creating a free hole or electron. The process can be indirect whereby the absorbed photon energy raises the lattice temperature (thermal detection) and a phonon generates the free charged particle. The change in free charge is sensed in an amplifier circuit as a signal voltage or current. Random generation of free charge is the source of detector noise. The detector geometry, spectral response, electrical bias, and temperature are adjusted to optimize the signal-to-noise ratio. The absorption efficiency (photon-to-electron conversion) is a critical issue. The detector surface is typically coated to minimize reflections at a particular wavelength and the internal efficiency is given by $1 - \exp(-\alpha x)$. Because the absorption coefficient α is typically 3000 cm^{-1} for intrinsic detection and 3 cm^{-1} for the extrinsic case, detector thickness, x, has a large range of values.

There are important figures of merit that describe the performance of a photodetector. These are responsivity, noise, noise equivalent power, detectivity, and response time. However, there are several related parameters of measurement, eg, temperature of operation, bias power, spectral response, background photon flux, noise spectra, impedance, and linearity. Operational concerns include detector–element size, uniformity of response, array density, reliability, cooling time, radiation tolerance, vibration and shock resistance, shelf life, availability of arrays, and cost.

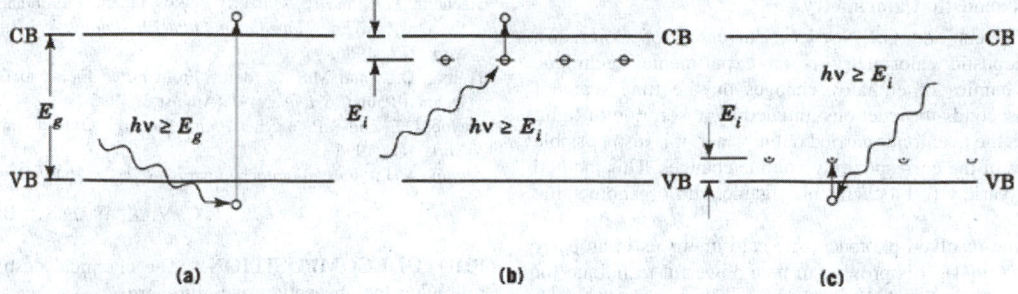

Fig. 1.   Photoexcitation modes in a semiconductor having band gap energy, E_g, and impurity states, E_i. The photon energy $h\nu$ must be sufficient to release an electron () into the conduction band (CB) or a hole (*) into the valence band (VB): (**a**) an intrinsic detector; (**b**) and (**c**) extrinsic donor and acceptor devices, respectively.

Types of Photodetectors

The choice of a photodetector for a specific optical receiver application depends on the performance and cost tradeoffs for the design. The wavelength range of interest will determine the material used. More mature technologies with simple fabrication processes in Si, and Ge, and GaAs usually mean lower-cost devices with adequate performance. High-performance devices for the telecommunications wavelengths of 1300 and 1550 nm typically have more involved fabrication processes and more difficult-to-control dark currents. Speed of response, noise, quantum efficiency, and gain are parameters to consider in the choice of a photodetector. See also **Sensors: Optical**.

Photoconductors. A photoconductor is fabricated by placing Ohmic contacts onto a semiconductor material. When the material is illuminated, electron–hole pairs are produced and the electrical conductivity of the material increases. The greater the photon flux, the greater the increase in conductivity. When voltage is applied to the contacts, the electrons and holes move in opposite directions resulting in a photocurrent. The anode and cathode may be placed on opposite ends of the semiconductor or both on top in an interdigitated pattern.

Photoconductive detectors have photoconductive gain defined as the flow of electrons per second from the device divided by the rate of generation of electron–hole pairs within the device.

Photoconductive detectors have the advantages of photoconductive gain and ease of fabrication over *p-n* or *p-i-n* devices, but lack their speed capabilities. The transit time in a photoconductor is set by the peak velocity of the fastest carrier, as contrasted with the case of *p-i-n*'s where it is limited by the slowest carrier. In III-V materials where the velocity of electrons is greater than the velocity of holes, and low-field electron velocity is greater than the high-field peak saturation velocity, high gain–bandwidth products can be obtained, but the *p-i-n* (with no gain) has much higher bandwidth.

p-n Junction Photodiodes. The *p-n* junction photodiode is formed by doping layers of semiconductor material *n* type and *p* type. When a photon is absorbed, it creates an electron–hole pair as in the photoconductive device. Here, the electric field is formed only in the depletion region of the device, and only carriers generated there, or near there (within a diffusion length), have the ability to be moved in a specific direction. The device is operated under reverse bias and the electrons are moved toward the *n* region while the holes drift to the *p* region under the influence of the electric field.

Response time of a *p-n* junction device is faster than a photoconductive detector as a result of the strong electric field in the depletion region giving carriers a larger velocity. The wider depletion region of the *p-i-n* photodetector makes it a better choice for high-speed devices.

p-i-n Photodetectors. A *p-i-n* detector is typically formed by growing an intrinsic (lightly doped) layer on top of a highly doped n^+ substrate and growing a p^+ layer on top of the intrinsic layer or diffusing the *p* dopant into the top layer. The detector structure will also work in reverse, that is, growing an *i* layer and an n^+ layer on a p^+ substrate. The *p-i-n* has advantages over the *p-n* junction device in that it has a wider depletion region (thickness of *i* layer) and, therefore, a lower capacitance and lower *RC* time constant. The transit time, however, will increase. High bandwidths, in excess of 100 GHz, are achievable with *p-i-n* photodetectors.

When the *p-i-n* device is formed out of a single semiconductor material, as in silicon *p-i-n*'s, the device material is homogeneous and it is referred to as a homojunction device. A photodetector composed of two different semiconductor materials is referred to as a heterojunction device. An example of this structure is the $Al_{0.27}Ga_{0.73}As/GaAs$ device for use from about 680 to 880 nm. This structure, which can have high bandwidths (>22 GHz) and high external quantum efficiency (73%), can be grown using molecular beam epitaxy (MBE). The higher–band-gap material sits on top where light incident at is not absorbed, but passes through to the intrinsic layer, which, under bias, is depleted of carriers as they are continuously swept out. This top transparent layer, the heterojunction window layer of higher–band-gap material, allows for the photons to pass through and be collected in the *i* layer below. A double heterojunction is formed when there are two junctions of differing semiconductor materials as in the $InP/In_{0.53}Ga_{0.47}As/InP$ detector material system. *p-n* or *p-i-n* junction photodiodes can be formed using a planar diffusion process. These diffused devices can have high response and very low dark current or high dynamic resistance for low-noise receiver performance.

The fabrication process for InP/InGaAs/InP diffused heterojunction devices will be used to illustrate a typical planar device process. The epitaxial structure used is an InP buffer layer on the InP:S (InP sulfur doped) substrate, followed by the InGaAs active layer, which has a low (1×10^{15} cm^{-3}) background doping, and then the window layer of InP, which will be diffused with the *p*-type dopant zinc. To form the diffusion barrier, silicon nitride is deposited on the wafer, patterned using a photoresist mask, and removed from the areas where Zn is diffused into the semiconductor. The diffusion depth is targeted to be through the InP window layer and into the InGaAs layer beneath. An antireflection coating is used to provide low reflections from the top surface and also to act as passivation for the device. The *p* contact is formed by etching away a small region of the antireflection coating and depositing Ti/Pt/Au metallization by a liftoff process. A bonding pad can then be formed by gold plating. The wafer is lapped and polished to provide the desired chip thickness and polish on the back side. The *n*-side metal contact (AuGe alloy) is deposited on the back side of the wafer on the InP:S substrate.

Photodetector mesa structures provide a low-capacitance design for high-frequency performance. Mesa devices can be front-side or back-side illuminated. They can have grown-in doped layers or diffused layers with mesas etched outside the diffused region. Mesa devices often have higher dark currents than diffused-junction devices because of the exposed junction surface.

A typical process for a *p-i-n* structure with a grown-in *p* layer involves deposition of the metal contact on the *p* layer by a lift-off process, deposition of an antireflection coating, etching the mesa with wet-chemical etches or by reactive-ion etching, deposition of the *n*-metal contact, and passivation of the mesa surface.

Metal-Semiconductor-Metal(Schottky Barrier) Photodetectors. A thin metal layer deposited on a semiconductor can form a Schottky barrier. This metal–semiconductor contact can act as a *p*- or *n*-type layer forming a *p-n* junction photodetector. If made thin enough, the metal is transparent to the incident light, and a depletion region is formed at the semiconductor surface.

Metal–semiconductor–metal (MSM) photodetectors commonly use interdigitated metal fingers to form the two electrical contacts for the device. These devices are attractive because of the low capacitance of the small active areas and the short transit time of the carriers between the electrodes. MSMs typically have lower responsivity or quantum efficiency than p-i-n devices because of the large amount of the incident light area covered by metal. Response can also be reduced by using thin i layers to try to minimize the transit-time effects of carriers needed to achieve high bandwidths. This can be a problem for p-i-n devices as well. The broad bandwidth possibilities and the ease and compatibility of processing with existing field-effect transistors (FETs) make MSMs a good choice for integrated optical devices. MSMs are being fabricated in GaAs and InGaAs material systems.

Avalanche Photodetectors. The avalanche photodetector (APD) operates under a strong reverse bias so that the electric field strength across the region between the p-type and n-type layers is large. When an incoming photon is absorbed, a cascading of carriers is created by a process known as impact ionization. This avalanche process of carriers produces carrier multiplication, or gain, in the device.

Novel device designs are used to optimize APD performance. To increase quantum efficiency, the absorption layer of the APD needs to be thick enough to absorb the majority of the carriers incident, but to minimize the chance for the inclusion of an area of potential instability or area of uncontrolled avalanche (microplasmas), the multiplication region needs to be as small as possible. Various device structures have been designed in an attempt to reconcile these competing tendencies. The separate-absorption-multiplication (SAM) APD uses separate semiconductor layers to perform the necessary device functions and is able to optimize the features of each by maintaining spatially separate regions. Guard-ring structures are a way of lessening leakage currents, which can limit multiplication, by having a ring of low-doped, high-breakdown material around the active region of the device. Guard-ring structures are used in Si, Ge, and III-V devices.

For wavelengths from 0.8 to 1.0 μm, silicon is the material used for APDs. In the longer-wavelength range of 1.0–1.55 μm, germanium is used as well as III-V materials. In III-V semiconductors, because of the high leakage due to tunneling in homojunction APDs, heterojunction devices are used. Gain-bandwidth products of 100–130 GHz have been reported for various configurations.

Response times in APDs are limited by the same factors affecting p-i-n's — transit times in the depletion region, RC time constants, diffusion times of carriers in the nearby undepleted regions — and in addition by avalanche buildup time.

Arrays. Photodetector arrays are used in imaging and in spectroscopy. Detectors are placed close together, and light strikes many of them simultaneously. In spectroscopy, the light is dispersed into its various wavelengths before it is incident on the detector array. A current resulting from illumination of a particular array element indicates that light of that wavelength is present corresponding to a chemical signature of a substance under test. Such arrays have been fabricated in the InGaAs material system having over 500 elements sensitive to wavelengths from 600 to 1700 nm. A type of two-dimensional array called a charge-coupled device (CCD) is formed using a series of Schottky photodiodes and can be used to do visible or infrared imaging. See also **Charge-Coupled Device (CCD)**.

Waveguide Photodetectors. Detectors where the input light is coupled to the edge of the device are known as waveguide photodetectors. In conventional top-illuminated p-i-n detectors, a tradeoff exists between the speed of response and the quantum efficiency of the device. The i layer is made thin to minimize transit time and maximize frequency response, but thinning the i layer means there may not be enough absorption depth for a large fraction of the light to be absorbed and the quantum efficiency or responsivity is not optimized. In an edge-illuminated device, the current flow and the light absorption occur in perpendicular directions and both the speed and quantum efficiency can be high. Internal quantum efficiencies are determined by the wavelength of light and the absorption coefficient, the i-layer thickness, and the length of the chip. External quantum efficiency takes the coupling efficiency into account. Careful light coupling and packaging schemes are needed to get the light focused into the i layer of the device. Waveguide devices with 3-dB electrical bandwidths of 50 GHz and quantum efficiencies of 68% for antireflection-coated devices have been fabricated and tested.

Fabrication processes have used both n^+ and semi-insulating substrates. The epitaxial structure is similar to a laser structure in that it contains waveguiding layers on either side of the active layer, though here the active layer absorbs rather than emits light.

Quantum-Well Photodetectors. Quantum-well photodetectors take advantage of band-gap engineering in order to control their wavelength detection range and response. They depend on precisely controlled, often thin, alternating epitaxial layers grown by solid-source MBE, organometallic vapor-phase epitaxy, or gas-source MBE. Rapid switching between materials in order to achieve abrupt interfaces on an atomic scale (layers on the order of tens of Ångströms) with well-characterized composition and doping profiles is necessary in order to grow the quantum wells. Electron confinement in these thin layers, caused by the conduction or valence-band offset due to the difference in band-gap energies at the heterojunction interfaces, shows quantum effects where continuous energy levels become discrete. A thin layer of small–band-gap material is placed between layers of higher–band-gap material forming a potential well where the electron or hole energy states are quantized (particle in a box). These discrete energy levels can be varied by adjusting the thickness and depth (band discontinuity) of the wells. More bound states with closer energy levels are found in wider quantum wells. Multiple quantum-well (MQW) devices are formed by growing stacked wells with barrier layers between them on the wafer. As the barrier layers become thinner than about 50 Å, the structure is referred to as a superlattice. Materials that are grown lattice mismatched to change the band structure of the device purposely are known as strained layers.

Detection at the long infrared wavelengths (3–14 μm) is achieved by using such quantum-well devices. Applications at these long wavelengths are for imaging with linear arrays, medical imaging, global and space surveillance, and astronomy.

High-Speed Detectors. As optical data rates increase in both the data communications and telecommunications markets, high-speed detectors will be necessary for the transportation of that information. 10-Gbit and 20-Gbit systems are now being designed in research environments. 2.488-Gbit SONET (Synchronous Optical Network) systems are already being implemented. Detectors designed for high speed that trade off thin i layers for short transit times and thick i layers for low capacitance will be needed. Designs that involve low capacitance, low series resistance, and short transit times and still maintain reasonable responsivities will be needed. p-i-n detectors with bandwidths of 100 GHz have been made in the GaAs and InGaAs material systems. APDs with gain-bandwidth products of 120 GHz have also been demonstrated.

As systems push to higher data rates, detector areas are shrunk in an effort to reduce junction capacitance, eventually making the devices difficult to fabricate and illuminate. Parasitic capacitances and inductances become very important to the speed of response of the devices, influencing the flatness and rolloff of the frequency response. Mesa p-i-n devices as small as 5×5 μm^2 have been fabricated for research applications, but would be difficult to reproduce repeatably in a device manufacturing environment. The small mesa size formed by wet etching means that the undercutting would need to be carefully monitored and traded off against the surface etches necessary to achieve low dark currents. Back-side–illuminated devices that can be attached by a solder bump technique known as flip-chip; are a means of lowering device capacitance by removing the need for a frontside bondpad and lowering parasitics since bond wires are replaced with solder interconnects. In addition to the rf response increase, dc responsivity also increases for the back-side–illuminated flip-chip device over the front-side–illuminated device, where the benefit comes from using a double pass of the light through the i layer of the device, thus absorbing a larger fraction of the incident photons and converting them into usable photocurrent.

Optoelectronic integrated circuits (OEICs) are another way of forming interconnections between the devices by fabricating them on a single substrate, often using multiple material regrowths to achieve the layer structures necessary for the varied devices. OEICs using photodiodes as one of their elements have also been fabricated. Performance of OEICs has not yet met the performance of the individual optical and electrical components, but improvements have been made and performance is beginning to approach that of the hybrid circuits, making OEICs useful for high-speed applications and lower-cost alternatives.

See also **Semiconductors: Compound Semiconductors**; **Semiconductors: Silicon-Based Semiconductors**; and **Sensors**.

Additional Reading

Brown, G.J., and M. Razeghi: *Photodetector Materials and Devices VII*, SPIE — The International Society for Optical, Bellingham, WA, 2002.

Cohen, M.J., and E.L. Dereniak: *Semiconductor Photodetectors II*, Vol. 2 S P I E-International Society for Optical Engineering, Bellingham, WA, 2005.

Dakin, J., and R. Brown: *Handbook of Optoelectronics*, Taylor & Francis, Inc., Philadelphia, PA, 2006.

Donati, S.: *Photodetectors: Devices, Circuits and Applications*, Pearson Education, New York, NY, 1999.

Neamen, D.: *Semiconductor Physics and Devices*, 3rd Edition, McGraw-Hill Science/Engineering/Math, New York, NY, 2002.

Rosencher, E., and B Vinter: *Optoelectronics*, Cambridge University Press, New York, NY, 2002.

Saleh Bahaa, E.A., and M.C. Teich: *Fundamentals of Photonics*, 2nd Edition, John Wiley & Sons, Inc., Hoboken, NJ, 2006.

PHOTOELASTICITY. A term that refers to certain changes in the optical properties of isotropic, transparent dielectrics when subjected to stresses. A block of glass, free of optical flaws, exhibits a "forced" double refraction when subjected to compression or tension parallel to one of its dimensions. If the block is placed between crossed Nicol prisms, the field remains dark so long as the glass is in its normal condition, but as stress is applied, colored fringes appear which are characteristic of the internal deformations of the glass.

PHOTOELECTRIC CONSTANT. A quantity equal to h/e where h is the Planck constant, and e, the electronic charge, and which multiplied by the frequency of any radiation exciting photoemission gives the potential difference corresponding to the quantum energy absorbed by the escaping photoelectron.

$$h/e = 4.1349 \times 10^{-7} \text{ erg} \cdot \text{sec} \cdot \text{emu}^{-1}$$
$$= 1.3793 \times 10^{-17} \text{ erg} \cdot \text{sec} \cdot \text{esu}^{-1}$$

PHOTOELECTRIC EFFECT. Changes in electrical characteristics of substances due to radiation, generally in the form of light. Radiation of sufficiently high frequency (short wavelength), impinging on certain substances, particularly, but not exclusively, metals, causes bound electrons to be given off with a maximum velocity proportional to the frequency of the radiation, i.e., to the entire energy of the photon. The Einstein photoelectric law, first verified by Millikan, states:

$$E_k = h\upsilon - \omega$$

where E_k is the maximum kinetic energy of an emitted electron, h is the Planck constant, υ is the frequency of the radiation (frequency associated with the absorbed photon), and ω is the energy necessary to remove the electron from the system, i.e., the photoelectric work function for the surface of the emitting substance. An inverse photoelectric effect results from the transfer of energy from electrons to radiation. For example, in an x-ray tube, there is observed the transfer of energy from electrons accelerated by the anode voltage to radiation emitted by the target. This radiation exhibits a continuous spectrum at lower voltages, upon which are superimposed, at higher voltages, intense lines characteristic of the anode material.

Two principal aspects of the photoelectric effect are described here: (1) Photoconductivity; and (2) photovoltage.

Photoconductivity is the phenomenon evidenced by the increase in electrical conductivity of a material by the absorption of light or other electromagnetic radiation. Although insulating or semiconducting materials exhibit this effect to some degree, there are relatively few materials that give sufficiently large changes of conductivity with illumination for application of the principle in useful devices. The principle can be explained briefly by using a cadmium sulfide photoconductor as an example. As in the case of luminescence, the band-type of energy level diagram is useful. See Fig. 1. Transition 1 represents absorption of a photon of energy at least equal to that of the band gap, giving rise to a free electron and a free hole. Transition 2 represents absorption at a local crystalline imperfection (defect or impurity), also producing a free electron, but with a hole trapped in the vicinity of the imperfection. While these carriers are "free" in the crystal, the conductivity can be greatly enhanced, so that the conductivity in the light can be a million times that in the dark. Recombination of the carriers may occur via transition (3), which is a

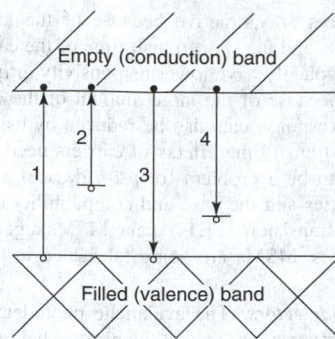

Fig. 1.   Simplified band model for photoconduction processes.

"direct" electron-hole recombination across the band-gap, or via step 4, an electron recombining with a center containing a hole, so that they no longer contribute to the conductivity.

For a material in which one type of carrier predominates (i.e., electrons), the change in conductivity with illumination can be given as:

$$\Delta\sigma = \Delta n e\mu + en\Delta\mu \tag{1}$$

where $\Delta\sigma$ = conductivity change, Δn is the change in free carrier density, e is the electronic charge, μ is the carrier mobility, and $\Delta\mu$ is the change in carrier mobility. Usually, the first term in Equation (1) predominates.

Photoconductivity gain, G, may be defined as the number of interelectrode transits that can be made by an electron until the photogenerated hole is eliminated by recombination. For the case treated here, namely where one type of carrier predominates, the gain, G, can be stated as:

$$G = \tau\mu^{VL-2} \tag{2}$$

where τ is the carrier lifetime, μ is the mobility, V is the applied voltage, and L is the spacing between electrodes. Since the specific sensitivity, S, varies as the product of carrier lifetime and mobility, $S\alpha\mu\tau$, we can state that

$$G\alpha(VL^{-2}S) \tag{3}$$

Commercially, photoconductive devices are used as (1) Detectors of radiation; (2) switches which are sensitive to light and which can actuate relays; and (3) in combination with other photoelectronic materials, such as electroluminescent materials, as image intensifiers. Germanium and silicon devices of the *p-n* junction phototransistor type have long been used in computer detectors; lead sulfide has been used in photocells for infrared detection; cadmium sulfide or cadmium selenide have been used in photocells for detection of light in the visible range; zinc oxide and selenium devices have been used in photocopying machines; antimony sulfide has been used in television pickup tubes. There are numerous other applications.

Photovoltage or the *photovoltaic effect* may be defined as the conversion of light photons to electrical voltage by a material. Becquerel, in 1839, was the first to discover that a photovoltage was developed when light was shining on an electrode in an electrolyte solution. Nearly half a century elapsed before this effect was observed in a solid, namely, selenium. Again, many years passed before successful devices, such as the photoelectric exposure meter, were developed. Radiation is absorbed in the neighborhood of a potential barrier, usually a *p-n* junction, or a metal-semiconductor contact, giving rise to separated electron-hole pairs, which create a potential. An equivalent circuit for a photovoltaic cell is shown in Fig. 2, where R_{SH} and R_S are the internal shunting and series impedances; I_J is the junction current; R_L is the load resistance; I_S is a constant-current generator; and V_L and I_L are the voltage and current developed across the load. With R_L optimum, the maximum conversion efficiency, η_{max} can be given by:

$$\eta_{max} = \frac{100 V_{mp} I_{mp}}{P_{in}}$$

in which case, V_{mp} and I_{mp} are the voltage and current across R_L, and P_{in} is the radiant input power.

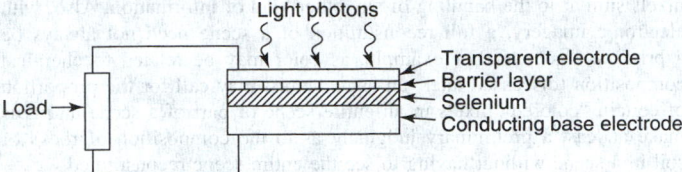

Fig. 2.   Diagram of photovoltaic cell.

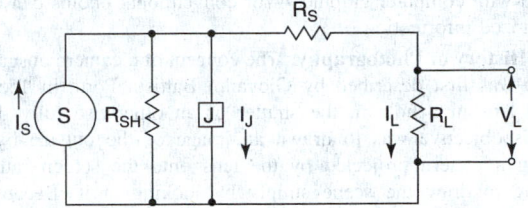

Fig. 3.   Photovoltaic cell equivalent circuit.

Photovoltaic cells have found numerous applications in electronic and aerospace applications, notably in satellites (solar cells) for instrument power. See also **Solar Energy**. Materials used, in order of decreasing theoretical efficiency, include gallium arsenide (24%); indium phosphide (23%); cadmium telluride (21%); silicon (20%); gallium phosphide (17%); and cadmium sulfide (16%). See Fig. 3. In the past, disadvantages of photovoltaic cells have included: (1) high susceptibility to radiation damage; (2) high cost; and (3) requirement for auxiliary battery power when a source of radiation for the cells is not available. Intensive research is underway in the mid-1970s to improve the production of known materials, i.e., upping the practical efficiency to approach the theoretical efficiency—as well as decreasing production costs; and to find new materials and combinations. Such findings are extremely important to certain of the proposed approaches for utilizing solar energy on a large scale. Much of this research is currently of a proprietary nature. The scope of the problem can be visualized from published Figures giving ranges of 0.2 to 0.5 volts per cell with current output of a cell with an area of two square centimeters amounting to only about 0.05 ampere. Consequently, without significant materials and efficiency improvements, extremely large numbers of cells are needed for any large-scale, practical solar energy application. Also, for maximum utility, a means must be found to orient the cells wherever possible so that the incident light will be perpendicular to the face of the cells.

See also **Photoemission and Photomultipliers**; **Photometers**; and **Photon**.

PHOTOEMISSION AND PHOTOMULTIPLIERS. Photoemission is the ejection of electrons from a substance as a result of radiation falling on it. Photomultipliers make use of the phenomena of photoemission and secondary-electron emission in order to detect very low light levels. The electrons released from the photocathode by incident light are accelerated and focused onto a secondary-emission surface (called a dynode). Several electrons are emitted from the dynode for each incident primary electron. These secondary electrons are then directed onto a second dynode where more electrons are released. The whole process is repeated a number of times depending upon the number of dynodes used. In this manner, it is possible to amplify the initial photocurrent by a factor of 10^8 or more in practical photomultipliers. Thus, the photomultiplier is a very sensitive detector of light.

The major characteristics of the photomultiplier with which the user is generally most concerned include: (1) sensitivity, spectral response, and thermal emission of photocathodes; (2) amplification factor; and (3) noise characteristics and the signal-to-noise ratio.

Many different types of photocathodes are used in photomultipliers. With a selection of various cathodes, it is possible to cover the range of response from the soft x-ray region (approximately 5 to 500 Å) to the near infrared (approximately 12,000 Å). Materials and combinations used include cesium-oxygen-silver; cesium-antimony; cesium-antimony-bismuth; sodium-potassium antimony; sodium-potassium-cesium-anti-

mony; copper-iodine; and cesium-iodine. The thermal emission at 25 °C of copper iodide and cesium iodide tends to run less than the other materials.

The amplification factor in a photomultiplier depends upon the secondary emission characteristics of the dynode and to some extent on the design of the multiplier structure. Important secondary-emission surfaces used in commercial photomultipliers are of two types: (1) alkali metal compounds, e.g., cesium-antimony; and (2) metal oxide layers, e.g., magnesium oxide on silver-magnesium alloy. The alkali metal compounds have higher gain at low primary electron energy (of the order of 75 volts). The metal oxide layers show less fatigue at high current density of emission (i.e., at several microamperes per square centimeter or higher).

The multiplier structures may be divided into two main types: (1) dynamic; and (2) static. The dynamic multiplier in its simplest form consists of two parallel dynode surfaces with an alternating electric field applied between them. Electrons leaving one surface at the proper phase of the applied field are accelerated to the other surface where they knock out secondary electrons. These electrons, in turn, are accelerated back to the first plate when the field reverses, creating still more secondary electrons. Eventually, the secondary electrons are collected by an anode placed in the tube; if they are not, a self-maintained discharge occurs. In practice, dynamic multipliers have been replaced by static ones mainly because the latter have better stability and are easier to operate.

Static multipliers may be either magnetically or electrostatically focused. In a magnetic type, primary electrons impinging on one side of a dynode cause the emission of secondary electrons from the opposite side. These electrons are then focused onto the next dynode by means of the axial magnetic field. The more common types of electrostatic multiplier structures use focusing from one stage to the next. Deposition of thin semiconductor secondary emission surfaces onto insulating substrates has been used in designing rugged miniature multiplier structures. In unfocused electrostatic structures, there is less sensitivity to stray electric and magnetic fields.

Dark noise in photomultipliers is caused by: (1) leakage current across insulating supports; (2) field emission from electrodes; (3) thermal emission from the photocathode and dynodes; (4) positive ion feedback to the photocathode; and (5) fluorescence from dynodes and insulator supports. Careful design can eliminate all but item (3). Associated with the photocurrent from the photocathode is shot noise. There is also shot noise from secondary emission in the multiplier structure.

A major use for photomultipliers has been in the scintillation counter where in combination with a fluorescent material, it is used to detect nuclear radiation. They also have been used in star and planet tracking for guidance systems as well as in star photometry and quantitative measurements of soft x-rays in outer space. Additional uses include facsimile transmission, spectral analysis, process control, and wherever extremely low-light levels must be detected. For applications in photometers, see also **Photometers**.

PHOTOGRAMMETRY. The technology of obtaining reliable measurements by means of photographs as required in surveying and map-making. Frequently photographs are obtained from aircraft, but terrestrial-based cameras serve for some applications.

PHOTOGRAPHIC CHEMISTRY. In photographic films and papers the sensitive surface usually consists of microscopic grains of a silver halide, suspended in gelatin. Exposure to light renders the halide particles susceptible to reduction to metallic silver by developing agents containing a reducing agent, as well as an accelerator, preservative, and restrainer. The accelerator increases the activity of the reducing agent (due principally to ionization of the phenolic agents to their active form) and is usually an alkaline compound. The preservative, usually sodium sulfite, minimizes air oxidation. The restrainer helps to prevent "fog" (reduction of silver halide gains that have not been exposed to light) and is almost always potassium bromide.

Color sensitizers are dyes added to silver halide emulsions to broaden their response to various wavelengths. Unsensitized emulsions are most responsive in the blue region of the spectrum and thus do not correctly represent the light spectrum striking them. Widely used sensitizers include the cyanine dyes, the merocyanines, the benzooxazoles, and the benzothiazoles. Cryptocyanine sensitizes the extreme red and infrared.

In color photography diethyl-p-phenylenediamine is an important developer because its oxidation product readily couples with a large number of phenol and reactive methylene compounds to form indophenol and indoaniline dyes, which are the basis of most of the current color processes.

See also **Holography**; and **Photography and Imagery**.

PHOTOGRAPHY AND IMAGERY. The unique light-sensing properties of silver halide crystals have been recognized since the 1500s. In spite of many technical advances in nonsilver halide (eg, electronic) technologies, chemically based silver halide systems continue to dominate in the ability to record images of superb image quality and archival characteristics. Photochemical reduction in which the silver ion, $Ag+$, in the ionic silver halide crystal is reduced to elemental silver, AgO was first observed by the alchemist Fabricius in 1556. As photochemical reduction continues, elemental silver atoms aggregate and grow into clusters of a colloidal size sufficient to scatter light and produce hue shifts. The science of photography uses this photochemical property of silver halide to form images and record scenes. One of the earliest researchers to produce such a photochemical image was Schultze in 1727.

Imagery is the representation (pictorial, graphical, etc.) of a subject by sensing quantitatively the patterns of electromagnetic radiation emitted by, reflected from, or transmitted through a subject of interest (object, body, scene, etc.). Imagery is not wavelength-limited, but is achievable (theoretically if not practically) with all bands of the electromagnetic spectrum — gamma rays, x-rays, ultraviolet radiation, visible light, infrared radiation, radar and radio waves.

Chemical imagery or *traditional photography*, as initially conceived and as commonly practiced, depends upon visible light and uses an optical light-gathering and focusing system (camera) and a light-sensitive medium (film emulsion) to record (store) the image — a *photo-image*. The subsequent availability of infrared, ultraviolet, and x-ray sensitive films extended the capabilities of traditional photography well beyond its dependence upon visible light. The word *photography* derives from the Greek roots *photos* (light) and *graphos* (to draw). Coining of the term is usually attributed to Herschel, although this has not been proved conclusively. Herschel did use the term in a memo dated January 17, 1839 and in a technical paper given on March 14, 1839.

Electronic Imagery, instead of using chemical means (emulsions), takes advantage of the sensitivity of various electronic detectors to different bands of the electromagnetic spectrum. The energy received is transduced by these sensors into an electronic or electrical effect (change of resistance, current, emf, the emission of electrons, etc.), from which effects an option of ways to process and display the information is available. The most common form of electronic imagery is found in television. Image orthicons, vidicons, and the more recent TV cameras using charge-coupled devices are among recent developments in electronic imagery.

Electronic imagery is particularly attractive for situations where image information must be transmitted over long distances where digitized signals offer greater accuracy and reliability — and where the incoming information is immediately compatible with digital data processing and computing equipment. Electronic imagery also has made certain imaging tasks possible, such as radar imaging, where traditional photographic means do not suffice.

While techniques are available in traditional photography to enhance raw information, these methods are largely of a qualitative, aesthetic nature rather than of a quantitative, scientific nature. In handling tiny pixels (one of the dots or resolution elements making up a digitized picture) of information, it becomes possible to computer program the processing of image information as it is received or after it is retrieved from tape or other electronic storage medium. The pixels can be measured one at a time at a rapid rate for brightness, sense (detection of obviously bad information), and other quantities, and over a wide scale of selections (for example, black = 0; medium gray = 32; white = 63) so that groupings of input information can be made to provide better contrasts (in pattern and blackness or color) when the information is all regrouped and reassembled for display. Accomplishments of this nature are probably best exemplified by the image intensification and color enhancement of pictures returned to earth from various space explorations. This is discussed further in this entry, but also see entries on specific planets.

With the flexibility of modern data processing and computing equipment, a vast array of programming techniques can be applied to the handling of pixels similar to the handling of any other kind of information. Also, with electronic imagery, a full reconstitution of a scene need not always be a primary objective. For example, as color may be related to chemical composition (discussed later) an astrochemist may call for the proportion of certain "colored" pixels in an entire scene or part of a scene and thus make at least a preliminary judgment as to the composition of rocks or soil in a scene without having to see the entire scene reconstituted.

Whereas the final results of traditional photography are usually in the form of prints or transparencies (with attendant problems), electronic imagery can be projected at will on cathode ray tubes and, if desired, combined with computer graphics — or conventional photos can be made from digitized information.

Early History of Photography. The concept of a camera obscura (dark chamber) was first described by Giovanni Battista Porta in 1558 when he put a lens in a hole in the shutter of an otherwise fully darkened room. His objective was to drawn an image of the outside by means of tracing a pattern projected by the lens onto the screen rather than attempting to draw the scene simply by looking at it. Because there was a great desire to capture scenes and subjects on paper, canvas, and other media, but a scarcity of artistic and drawing abilities among many of the populace, the principle of the camera obscura persisted for some 250 years, with improvements and refinements of the optical system used.

During the sixteenth and seventeenth century, it became known to technically curious persons that a number of substances would change color when exposed to light. These observations provided the first hint that perhaps an image could be captured permanently and thus save a lot of time and labor and also provide a means of relatively quickly producing duplicates of any given subject. There were two main problems: (1) finding a suitable medium that would respond within a reasonably short time span to the projected image; and (2) finding some way to hold or fix an image without permitting the medium to follow a full course of development and thus completely obliterate the captured image. The properties of silver salts were discovered in 1725 by Johann Heinrich Schulze at the University of Altdorf. He found that chalk, when moistened with a solution of nitric acid and silver nitrate, became darker upon exposure to light. Early experiments involved contacting objects with the silver medium to produce silhouettes. The first image obtained through use of a camera-like technique was achieved in France by Joseph NicÉphore Niepce. His first success came in 1813, using paper which he had soaked in sodium chloride, followed by immersion in a silver nitrate solution, upon which silver chloride was precipitated throughout the paper. The crude image obtained was a negative and persisted for only a short period because he had not developed a required fixative. Niepce then turned his efforts to an asphalt process, known as *heliogravure*, which he used for copying prints. The process was quite insensitive, but an image (exposure of about 8 hours in direct sunlight) of some buildings is regarded as the *first photograph*. This old photograph is now part of the Gernsheim collection in the United States.

Daguerreotype. The first practical process of photography was invented by Louis J.M. Daguerre of Paris in 1837, although the details of the process were not published until 1839. The process was used chiefly for portraiture and became obsolete within a few years after the introduction of the wet collodion process in 1851. Although the daguerreotype process was the original, modern photography is based on the negative-positive methods introduced the same year by William H. Fox-Talbot of England. This was known as the *calotype* process. In the daguerreotype process a light-sensitive layer of silver iodide is formed on a silver plate by contact with iodine. After exposure in the camera, a positive image is produced when the image is exposed to mercury and heated. The mercury, by attaching itself to the unexposed portions, forms a positive image. The silver iodide remaining was removed at first with a solution of sodium chloride (salt) which was soon replaced, however, with sodium thiosulfate (hypo), the properties of which had been discovered by Herschel in 1819. The daguerreotype image so produced is very weak. In 1840 Fizeau described a process of toning with gold which greatly increased the strength of the image and was generally adopted.

At first, from 5 to 10 minutes' exposure was required on open landscapes and street scenes. The invention of a fast, large-aperture portrait lens by Petzval in 1841 and the discovery by Goddard in London (1840) of the superior sensitivity of silver bromide reduced the time of exposure to a few seconds.

Problems were encountered in preparing positive prints from the calotype negatives because of reproduction of the grain of the paper that contained the negative. Attempts were made to wax or oil paper negatives, but these were essentially unsuccessful. De Saint-Victor attempted to coat plates with albumin (egg white). Upon hardening of the albumin, the plates were bathed in silver nitrate, causing precipitation of silver iodide within the film of albumin. This was not successful because the sensitivity of the plates was greatly lowered.

Early Emulsions. The use of collodion in photographic emulsions dates from 1851 when Frederick Scott Archer published details of his wet collodion process. Although this process is no longer in general use, it can be used in making the half-tone negatives required in photoengraving. In the collodion process, a clean glass plate is first coated with collodion containing potassium iodide and potassium bromide. It is next sensitized by immersion in a solution of silver nitrate. It is then placed in a plate holder—specially designed for the handling of the wet plate—and the exposure made. After exposure it is developed in a solution of ferrous sulfate and fixed in potassium cyanide, or in hypo, washed and dried. The wet collodion process, as it is used by the photoengraver, results in a negative of high density and extreme contrast, high resolution and with an extremely fine grain. These characteristics render wet collodion well adapted to the requirements of photoengraving. Much later, the wet collodion process was essentially replaced by the gelatino-bromide emulsions of similar characteristics.

Collodion printing-out paper was introduced by Obernetter of Munich in 1867 and was for many years the favorite printing process of the portrait and professional photographer. It was in general use until the early years of the present century when it was gradually replaced by developing-out paper.

Gelatin Emulsions. Not true emulsions, but suspensions of minute silver halide crystals dispersed in a protective colloid medium (gelatin), the suggestion of replacing collodion with gelatin was first made by R.L. Maddox in 1871. The first plates made by Maddox were not very sensitive, but their advantages far outweighed their defects, leading to further developments by Charles Bennett in England in the late 1870s, and the first mass production attempts by George Eastman in 1880. One of the several contributions of Eastman to photography was his early recognition of making and marketing gelatin dry plates on a large scale, eliminating the need for the photographer to prepare his own plates, as well as the need for developing and fixing the plates immediately thereafter. There soon followed the concept of strip film, making it unnecessary to change plates after each exposure. Eastman avoided the grain problem by using a coating that enabled the stripping of the thin layer of gelatin from the paper support. Later, in 1889, he replaced the paper support with a transparent plastic support (nitrocellulose), thus making it possible to produce prints without the need of stripping the gelatin layer from the support. Eastman's goals were to make it easy for the masses of people to take photographs in a simplified manner and, through mass production, market equipment at a price within grasp of the public.

Gelatin is a preferred photographic colloid because the sensitizing bodies in the gelatin make possible emulsions with great sensitivity and speed. Gelatin is an excellent emulsifying agent and is readily transformed, from gel to a liquid or the reverse, by changes in temperature. The latter property makes coating of supports and emulsion processing and working feasible. The strong protective action of gelatin lowers the rate of reduction of unexposed silver halide crystals in developers so that image formation is readily obtained.

Silver halides employed in emulsions are the chloride, the bromide and the iodide. Negative emulsions are composed of silver bromide with a small amount of silver iodide. Positive emulsions for films and paper contain silver chloride, or mixtures of silver chloride and silver bromide in varying amounts, according to the tone, speed, and contrast desired.

In photomicrographs of negative emulsions, the crystals of silver bromide appear as flat triangular or hexagonal plates with rounded corners. Some globular, needle-shaped or diamond shaped crystals may also be observed, as in Fig. 1. The thickness of the flat plates is approximately one-tenth of their diameter. The size of silver bromide crystals range from less than one to four micrometers. Crystals of silver bromide, as used in

positive emulsions, are quite uniform and seldom exceed 0.5 micrometer in diameter. Multi-layered emulsions contain approximately 1 billion (10^9) crystals per square centimeter for low-speed emulsions to 1.0×10^{-8} centimeter for high-speed negative emulsions.

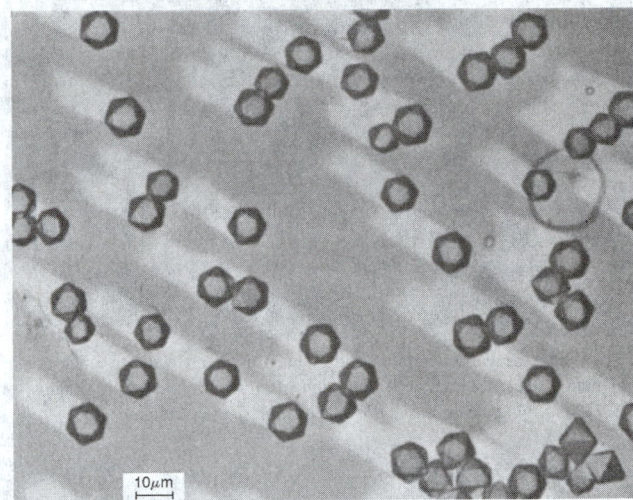

Fig. 1. Using the carbon replica technique, this is an electron micrograph of octahedral silver bromide grains. (*Photo by Dr. Donald L. Black, Eastman Kodak Company.*)

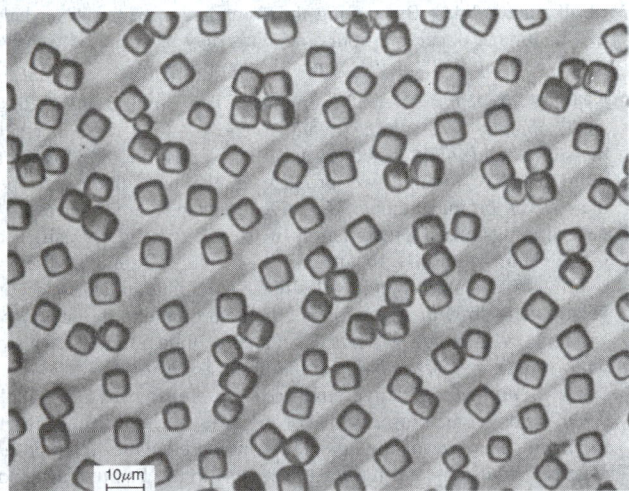

Fig. 2. Using the carbon replica technique, this is an electron micrograph of cubic silver bromide grains in which the corners have been slightly rounded due to the presence of a silver complexing agent. (*Photo by Dr. Donald L. Black, Eastman Kodak Company.*)

The characteristics of an individual emulsion are primarily dependent on two factors, the size-frequency distribution of the crystals and the composition of the silver halide crystals. For instance, Fig. 2 illustrates silver bromide grains with slightly rounded corners due to the presence of a silver complexing agent. The chief problems of the emulsion-maker are the production of uniform suspensions of silver halide crystals with proper size-frequency distribution and the correct composition in gelatin, and the ability to reproduce results. In an attempt to meet these needs along with increased sharpness for high-speed negative color films, *Eastman Kodak* has developed T-Grain type emulsions of silver bromide which also contain some iodide. Illustrated in Fig. 3, these new emulsions exhibit flat grains with relatively sharper edges. As a result, when compared to traditional high-speed color negative films they are capable of producing more clearly defined images.

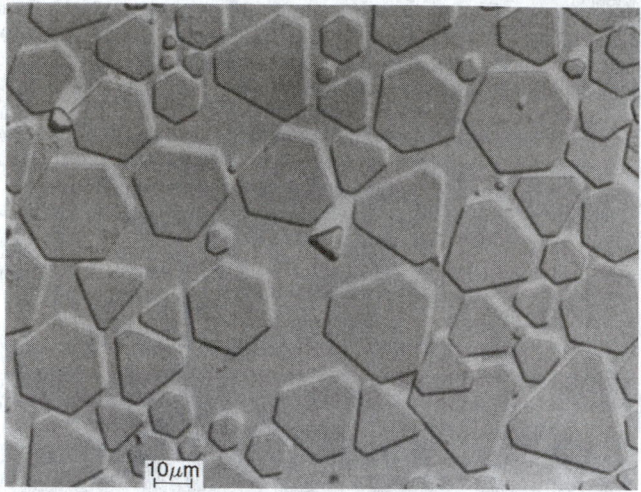

Fig. 3.   Using the carbon replica technique, this is an electron micrograph of an Eastman Kodak T-Grain type emulsion of silver bromide that contains some iodide. (*Photo by Dr. Donald L. Black, Eastman Kodak Company.*)

Classification of Emulsions

1. *Printing-out emulsions* These emulsions produce images on exposure without development. They are used largely for making portrait proofs which are distinguished by their red or purplish color. Emulsions of this type differ from others in that they usually contain silver nitrate, some free silver, silver salt of an organic acid and a weak free acid. These are known as P.O.P. Proof Papers.

2. *Developing-out emulsion* Emulsions for development have an excess of alkaline halides. By varying the composition of the silver halide and treatment, developing-out emulsions may be prepared which are suitable for either negative or positive purposes.

a. *Negative emulsions* Negative emulsions are prepared by adding a small amount of a soluble iodide to the bromide used in making the silver halide. The mixed crystals of silver-bromiodide formed are more sensitive to light and produce emulsions with greater speed than silver bromide alone. Negative emulsions are referred to as neutral emulsions if precipitation of the silver halide is carried out in a gelatin solution with an excess of soluble bromide. They are referred to as ammonia emulsions if the precipitation takes place in a gelatin solution with an excess of soluble bromide in the presence of ammonia or ammoniacal silver. The latter method produces emulsions with coarser grains, which have the highest sensitivity.

b. *Positive emulsions* Positive emulsions are prepared by precipitating silver halides containing chloride or mixtures of chloride and bromide in gelatin. The size of the crystals formed are smaller than those of negative emulsions and have a lower sensitivity. Positive emulsions are divided into four classes, according to the composition of the silver halides and their properties.

- *Chloride emulsions* Because of their slow speed chloride emulsions are used largely for contact printing.
- *Bromide emulsions* Bromide emulsions are very sensitive and fast. They are used for projection printing exclusively.
- *Chlor-bromide emulsions* In chlor-bromide emulsions the amount of silver chloride is greater than that of silver bromide. These emulsions are somewhat faster than chloride emulsions and used for contact or slow projection printing. Chlor-bromide emulsions produce warm-toned silver images with a brown or brown-black color.
- *Brom-chloride emulsions* Brom-chloride emulsions contain more silver bromide than silver chloride. They are faster than chlor-bromide emulsions and used for projection printing where black images and speed printing are desired. Image tones of brom-chloride emulsions are not as warm as chlor-bromide images nor as cold as bromide images.

Other Emulsion Additives. In addition to chemical and spectral sensitizers, several other classes of chemical compounds are added to emulsions before coating. Additives are used to facilitate coating operations, eg, surfactants (v) and viscosity enhancers; to reduce spontaneous development in unexposed regions, eg, tetraazaindenes and mercaptotetrazoles; and to reduce abrasion and permit high temperature processing, eg, aldehydes.

For certain component compositions the viscosity and surface tension of the melted emulsion may not allow adequate emulsion spreading on the support during the coating procedures. For these situations, various surfactants that act as spreading agents are available to control the surface tension. The sulfiding reaction is a thermally activated process with an activation energy near 126 kJ/mol (30 kcal/mol). Therefore, quenching the reaction by cooling the emulsion from 60 to 25 °C does not eliminate the reaction but rather reduces the rate by about two orders of magnitude. After long storage of the emulsion, continued sensitization may produce a catalytic activity in the silver halide grains and unwanted photographic fog upon development. This can be controlled partly by additions of such stabilizers as halide ions, acid, benzimidazoles, benzotriazoles, benzothiazolium salts, and mercaptotetrazoles. Many of these compounds adsorb to silver and complex with silver ions. Specifically as a result of these interactions, phenylmercaptotetrazole restrains development and enhances sensitivity even in freshly coated samples. Quantum-mechanical analyses coupled with photographic data suggest that the best stabilizers not only bind with silver ions but are also poor reducing agents. Azaindene compounds satisfy both of these properties and are effective stabilizers and development antifoggants. Fog control for Au-sensitized emulsions can be, in part, achieved using thiocyanate. Gelatin cross-linking agents (hardeners) represent another class of materials that may be added before coating. These compounds render the coated emulsion layers more resistant to abrasion during handling and improve the thermal stability of the gelatin.

The desire for reduced development times required to produce an image has necessitated the use of solutions with increased activities. Temperature increases, pH increases, and increased oxidizability are all variations directed toward shortening process times. Unfortunately, high temperature processing tends to soften and dissolve the gelatin emulsions; therefore the gelatin must be hardened before development. The enhanced thermal stability and improved mechanical durability produced by hardeners result from the formation of three-dimensional bridging of various sites within the gelatin molecules. Both inorganic, eg, chromium salts, and organic, eg, aldehydes, compounds have been used as hardeners. Chromium appears to complex with carbonyl groups, whereas many of the organic hardeners seem to cross-link between the amino groups in the gelatin molecules.

In most color photographic products, organic compounds such as couplers or redox dye releasers are added to the melted emulsions before coating. These compounds are essential to the development reactions that produce the dye molecules composing color images.

Manufacture of Commercial Emulsions. Although the details are proprietary, the basic procedures of manufacturing commercial emulsions are known. A portion of the gelatin in the formula is swelled by soaking in water and later dissolved with heat. Mixtures of soluble bromides and iodides, or chlorides, are placed in water solution and added to the gelatin solution. Precipitation of silver halides is accomplished by slowly adding a solution of silver nitrate, while stirring, to the mixture. The relative concentration of the solutions, the rate of addition and temperature during mixing, are factors which control the formation, size and dispersion of the crystals in gelatin. The emulsion is then heated or "ripened" at 40–80 °C to recrystallize the silver halides and readjust the size-frequency distribution. Following ripening, more gelatin is added and the emulsion is chilled so it will set quickly. The emulsion is then placed in a press and forced through a screen to break it into shreds or noodles, which are washed, in cold running water to remove the potassium nitrate formed, the excess soluble halides, and certain soluble byproducts of the reaction. Chloride emulsions are often prepared without washing or with only a limited washing. After washing, the emulsion is drained, remelted, and additional gelatin and certain agents, such as fog preventatives, are added. The emulsion is then heated, or "after-ripened," to form sensitizing nuclei on the silver halide crystals. This operation increases the sensitivity and contrast of the emulsion and is necessary for the preparation of high-speed negative emulsions. Certain preservatives, or stabilizers, are added so the emulsion can be stored in refrigerated rooms until needed. Before coating the emulsion is melted and sensitizing dyes, hardening agents, wetting agents, etc., are added. After thorough mixing, filtering and heating to coating temperature, it is placed in a coating machine. Supports, as film, paper, or glass, with substratum coatings are fed through machines at proper rates so they become coated with emulsions in uniform layers of desired thickness. The coated supports pass over chill boxes to set the emulsion and then through a series of drying compartments where the rate of drying is carefully controlled so as not to change the sensitivity on the surface. Following drying, the coatings are inspected under proper safelights and the film or paper is cut to desired size and packaged.

Numerous variations in the manufacturing process make possible a wide range of film characteristics, including film speed and spectral sensitivity. Film, unlike the human eye, can extend beyond the visible region of the spectrum. High-speed film can capture the details of a fast-moving object, seen only as a blur by the eye. By extended exposure, film can capture images entirely too faint to be seen by the eye. The three main types of film emulsions for black-and-white photography are: (1) *ordinary* (color-blind; sensitive to blue light only); (2) *orthochromatic* (sensitive to all but red light); and (3) *panchromatic* (sensitive to light of all colors). Ordinary and orthochromatic films generally offer greater contrast than most panchromatic emulsions. However, the response of panchromatic emulsions can be modified by use of color filters. Film is available in several sizes and formats. Obviously, a delineation of film specifications is beyond the scope of this encyclopedia. Some excellent references are listed at the end of this entry.

Color Films. The trichromatic theory of vision was first proposed by Thomas Young, a British physicist in 1801. He was the first to propose that the retina of the eye incorporates three different types of receptors, responding to blue, green, and red light, respectively. The theory was elaborated upon to the extent that color perception is based upon the stimulation of two receptors, with light stimulating both red and green receptors seen as yellow light; light equally stimulating all three types of receptors seen as white, etc. Young concluded that it should be feasible to match any color of the spectrum through the proper mixing of blue, green, and red light. Although not essentially interested in color photography, Maxwell effectively demonstrated the principle by way of specially-prepared lantern slides before the Royal Institution in London in 1861. Maxwell had demonstrated the *additive color principle* (mixing of blue, green, and red light).

Practical color photography on a massive amateur scale, of course, could not depend upon the preparation of three separate photographs and the use of three projectors, but rather dictated a process that would combine the three records on one plate. In 1907, the LumiÉre Autochrome plate was developed. This was comprised of a very coarse mosaic of potato starch grains, one third of which was dyed blue; another third, green, and the remaining third, red. An emulsion layer was exposed, with the light first passing through the mosaic. In the *Kodacolor* system of 1928, filters in the camera were used instead of color mosaics. A major problem of the mosaic and filter approaches was that of loss of light as it passed through one or the other media, greatly reducing sensitivity and loss of brightness of a projected image. See also **Additive Color Process**.

In the *subtractive color system*, the phenomenon of absorption is involved. A dye that will absorb red light will, in turn, reflect green and blue light, thus appears a greenish-blue (cyan); a dye that will absorb green light appears a bluish-red (magenta); and a dye that absorbs blue light appears yellow. Thus, cyan, magenta, and yellow are the three primary subtractive colors. A mixture of all three dyes in proper portion will absorb all primary light and thus appear black. Most processes of color photography make use of a subtractive synthesis to yield prints or transparencies. See also **Subtractive Color Process**.

Color-separation negatives are photographic negatives that record the relative intensities of the primary colors used in the analysis necessary to reproduce a subject by means of color photography. In three-color photography, for example, the separation negatives are records, in terms of silver densities, of the amounts of red, green and blue light received at the camera from the subject.

A set of color-separation negatives may be prepared by photographing the subject three times on separate color-sensitive emulsions so that each is a record of one of the primary colors. A panchromatic emulsion is generally employed with a set of tricolor filters, the colors of the primaries. It is only necessary, however, to obtain the color records on separate negatives so it is also possible to use for each record any combination of color filter and emulsion sensitivity that will record one of the primary colors. A set of color-separation negatives may be made by exposing (1) each one in turn in a camera, (2) by the use of a color camera that will expose them simultaneously, or (3) in a tripack.

It is common practice to balance a set of color-separation negatives, by altering the exposure and development times, so that a gray scale will be recorded equally on each negative. The particular densities desired are dependent on the method of color synthesis to be employed.

The majority of color is by use of integral tripacks. There are three layers of photographic emulsion in the tripack, one layer sensitive to red light, another layer to green light, and another layer to blue light. They are coated, one on top of the other. Since silver iodobromide emulsions usually selected for film emulsions are sensitive to blue light, sensitivity to the green and red light must be conferred by sensitizing dyes. Although this sensitivity can be obtained, the dyes do not negate the emulsion's natural sensitivity to blue light. Thus, those layers that are sensitive to green and red light must be protected from blue light. This is accomplished by inserting a yellow filter layer that will absorb the blue light. Chloride emulsions on the other hand are sensitive only to ultraviolet light. Whereas they do not require a yellow-filter layer, they have to incorporate a filter for exclusion of ultraviolet light. There are a number of dyes that may be used in dye-transfer systems, but for tripacks it is necessary to select only those dyes that will be formed during the development process. In 1912, the German scientist, Rudolf Fischer, discovered the role of couplers. In his early version of a color film, he placed three layers of emulsion one atop another as previously described and he also incorporated a coupler in each layer to cause the development of a particular color. Fischer's concept was brilliant, but the actual process failed because the couplers and sensitizers tended to wander from layer to layer.

In 1931, Leopold Godowsky, a violinist, and Leopold Mannes, a pianist, and both avid amateur photographers made crude experiments in a home laboratory on a type of color film that ultimately became *Kodachrome*, released by Eastman in 1935. In the *Kodachrome* process, the couplers are laced in the developers instead of in the emulsions. Phenols are usually the couplers that form cyan dyes; nitriles or pyrazolones form magenta dyes; and esters, ketones, or amides form yellow dyes. There are many hundreds of couplers and, consequently, there is continuing improvement in color film. Space here does not permit a detailed description of such important matters as the negative-positive system, reversal systems for transparencies, color corrections, etc., but these areas are well covered in some of the listed references.

Direct Positive Images. Even in the early days of black-and-white photography and the early work of Daguerre and Fox-Talbot, it was realized that there would be a great advantage gained from a system that would initially produce a positive rather than a negative image. As early as the late 1930s, Hippolyte Bayard and Robert Hunt proposed systems, but these did not produce satisfactory results. The *chemical transfer* process was developed by A. Rott in Belgium in 1939 and found application in the document-copying field. In 1947, E.H. Land demonstrated a camera which produced a finished black-and-white print without need for a negative and one that was available to the photographer within a very short period, approximately one minute. This was the first model of the *Polaroid* camera. In chemical transfer, a normal emulsion is used. Immediately after exposure and while within the camera, it is developed in a solution containing combined developer-fixer agents. The emulsion is in contact with a special positive white paper, not light sensitive, on which the finished image is printed. The developing reagent is of a jellylike consistency and in early models was contained in pouches or pods, one for each picture. The exposed grains develop in the normal fashion. The unexposed grains are dissolved by the fixing agent. Thus, in the unexposed areas, the dissolved halide is silver which forms on the nuclei in the receiving sheet. In connection with partially exposed areas, the developing grains and the receiving sheet nuclei compete for the silver. Thus, a negative image is formed on the original film or paper, whereas a positive image appears on the receiving sheet. Subsequent to the first Polaroid camera, models were developed to provide a permanent negative as well as print, with the processing time reduced to seconds.

To achieve an instant color film that could rival 35 mm color quality, current generation *Polaroid Spectra* system film utilizes two different color chemistries for greater control of the self-developing image formation process. Composed of 18 microscopically thin coated layers in a rectangular format for both horizontal and vertical composition, this film is able to produce photographs of improved color separation, saturation and brilliance.

This is a result of combining images created in three chemical sandwiches in the film negative. Each sandwich is sensitive to red, green, or blue light and consists of a photosensitive emulsion and a related image dye.

For the *Spectra* film, the blue-light sensitive sandwich has been radically altered by the utilization of thiazolidine dye release. This required the creation of new molecules and a new dye-release mechanism involving only a minute quantity of silver. By utilizing this hybrid imaging system, chemical interaction and molecular cross talk between the red-, green-, and blue-sensitive sandwiches have been reduced, which results in greater color definition.

This new material also affords substantial improvement in yellow dye saturation and in recording pastels. By controlling chemical crosstalk between the red and green layers, *Spectra* film is further able to produce more brilliant greens, which is a difficult photographic accomplishment because of the low reflectivity of green in nature. This material also reproduces a broader range of hues and tints when compared with earlier-generation self-developing color films. The transparent support through which the image is viewed is thinner than previous *Polaroid* films and enhances image quality.

Special Films

Infrared Films. The first photographs by infrared radiation appear to have been made about 1880 by Sir William Abney, using a specially prepared collodion emulsion. Abney is reported to have photographed a boiling tea-kettle, but efforts by others to repeat his work were not particularly successful. In 1903, the first real infrared sensitizer, Dicyanine, was discovered. While the sensitizing action extended to a wavelength of 960 nanometers in the infrared, the exposure was too long for general photography and Dicyanine-sensitized plates were used chiefly in infrared spectroscopy. The discovery of more efficient sensitizers in the early twenties, beginning with Kryptocyanine and neocyanine and followed by the penta- and tetra-carbocyanines, has made it possible to prepare films and plates whose sensitivity in the infrared is such that they can be used for general photography, including aerial and motion-picture photography.

Infrared-sensitive films and plates may be divided into two classes: (1) materials of relatively high speed to the extreme red and infrared, i.e., from approximately 700 to 900 nanometers (nm), and (2) materials sensitive to much longer wavelengths but of lower sensitiveness. The former are used for general photography, for aerial photography and cinematography; the latter for spectroscopy in the infrared and other scientific applications requiring sensitivity to wavelengths longer than about 900 nm. All infrared-sensitive materials are sensitive to violet and blue and to the extreme visible red, as well. Photographs made without a filter resemble those made on an ordinary blue-sensitive material. For most purposes it is sufficient to use an orange or light-red filter which will absorb blue and violet light. In this case, the picture is made partly by infrared and partly by the extreme red. The result, however, is generally only slightly different from that obtained with infrared radiation alone. For true infrared photographs, a visually opaque filter transmitting the infrared only must be used. No filter, however, is required when photographing hot bodies such as an electric flatiron, hot castings, or high-pressure boilers, provided that these show no visible glow.

All infrared-sensitive materials must be loaded and developed in total darkness, as safelight screens, even those for panchromatic films and plates, transmit the infrared freely.

Certain precautions are necessary when making pictures with infrared-sensitive materials. The bellows and shutter blades of some cameras, although perfectly safe for ordinary photographic films and plates, transmit the infrared and fog infrared-sensitive films. The slides of some film and plate holders transmit the infrared sufficiently to cause fog. Although some modern lenses are corrected for the infrared, with most, the focal distances for the visible and for the infrared are different. Usually, it is sufficient to extend the lens a distance equal to 2% of the focal length beyond the visual focus. Even this may often be ignored when using a lens of short focal length or a small diaphragm.

Infrared photographs of landscapes are quite different from those made in the usual way. Green foliage is reproduced light and blue sky and water almost black. The shadow portions of the subject are dark and without detail. The general effect is that of a photograph made by moonlight, particularly if the print is made rather dark. As a matter of fact, most night scenes in motion pictures are really infrared photographs.

Since infrared radiation is not scattered by atmospheric haze, as is light, distant objects are rendered sharper and more distinctly in infrared photographs. Objects invisible to the eye because of the intervening haze are often reproduced sharply in an infrared photograph. In fact, one of the most important applications of infrared photography is in photographing distant objects, whether from the ground or the air. Infrared photographs, however, cannot be made through dense fog.

The scientific applications of infrared photography are numerous and important.

Aerial Photography in the Infrared. Extensive use of color infrared film (CIR) has been made in the field of aerial photography for such applications as crop sensing and inventorying, flood assessments, etc. Both normal color film and CIR consist of three separate layers of emulsion on a clear base material. It will be recalled that in normal color film one emulsion layer is sensitive to blue light, one to green light, and one to red light. The images recorded on the three emulsion layers of normal color film combine in the final image to form colors which closely match those of the original subject. CIR film, sometimes referred to as "false color film," also produces combinations of blue, green, and red in the final image; but the blue color results from exposure by green light; the green color from exposure by red light; and red color by exposure of the infrared sensitive layer by infrared energy. Therefore, the images are called false color images. See Fig. 4. Actually, all three layers of CIR film are also sensitive to blue light. For this reason, the film is always exposed through a minus-blue (yellow) filter which eliminates blue light before it reaches the film. The infrared energy needed to expose the infrared sensitive layer is reflected energy, not heat energy. Heat energy does not enter into the image forming process of CIR film.

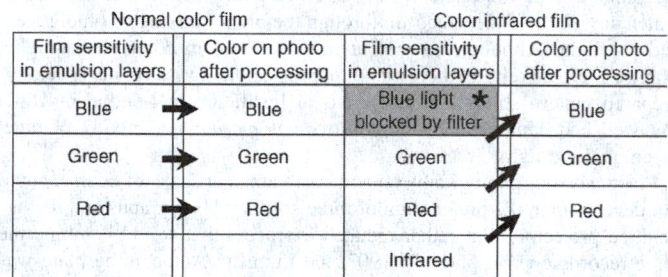

Fig. 4.   Film sensitivities and final image color of normal color and color-infrared films. *Blue light absorbed by yellow filter. (*U.S. Geological Survey.*)

One of the most important features of CIR film is the manner in which vegetation is recorded. Healthy green plants appear in shades of red, because healthy plants reflect sunlight strongly in the photographic infrared region (therefore strongly exposing the infrared sensitive layer) while simultaneously reflecting relatively little energy in the visible region (therefore offering little exposure to the green and red sensitive layers). For all practical purposes, living healthy vegetation is the only natural source of high-infrared reflection coupled with low visible reflection. Because of the unique reflectance characteristic of healthy vegetation, the film was originally used by the military to differentiate between real vegetation and painted camouflage material.

Another unique and variable aspect of CIR photography results from the fact that plants do not reflect strongly in the photographic infrared when they are severely stressed or have died, and as a result, no longer appear red on the photographs, in contrast with normal, healthy vegetation. The reasons behind this phenomenon are complex, yet the ability to distinguish between healthy and stressed or dead vegetation by using CIR is very important for vegetation analysis. This characteristic is particularly useful in determining crop damage due to flooding.

Generalizations about the photographic appearance of other features commonly found on the agricultural landscape can also be made. Clear water usually appears very dark blue or black, but muddy or turbid water appears light to medium blue. This is useful for satellite tracking of pollution situations. Fresh grain stubble appears very light or almost white, whereas clean plowed fields of dark soil usually appear dark blue.

Other Applications of Infrared Film. Among several other scientific uses are:

1. **In medical photography** For the study of the following: diseases and conditions affecting the venous pattern not revealed by light; the progress of healing beneath certain scabs; the eye, to determine atrophy; histological specimens, to reveal structures below the surface and invisible to the eye. Thermography has been used to detect tumors, the skin temperature often being as much as 1 to 2 °C higher than that of the surrounding skin.

2. **In industry** For the study of irregularities in the dyeing and weaving of textile fibers, the interior of furnaces, the detection of carbon in lubricating oils, infrared spectroscopy of metals and alloys.

3. **In astronomy** For detection of nebulae and stars otherwise invisible because of astronomical haze or because their radiation lies chiefly in the infrared; in infrared spectroscopy, for the determination of the composition, the temperature, and the movement of stars and nebulae.

4. **In criminology** For deciphering writing or printing that has been crossed out with other inks to render it illegible; for obtaining copies of charred documents, detecting erasures, revealing finger prints, identifying blood and other stains, uncovering secret writings, etc.

Ultraviolet Films. In the near-ultraviolet region, photography is the same as with visible light. However, at shorter wavelengths, many materials are not transparent to ultraviolet radiation. For example, glass is not transparent at wavelengths shorter than about 3,000 Å. To produce a photograph at these shorter wavelengths, a quartz lens (transparent to about 1,800 Å) or a fluorspar lens (transparent to about 1,200 Å) is required. Inasmuch as gelatin also absorbs radiation of wavelength less than about 2,200 Å, photography in such regions requires specially-prepared plates where a minimal amount of gelatin is used. In some cases, the plates can be coated with a thin film of oil or other substance which fluoresces when exposed to ultraviolet. For particular work, the fact that air absorbs short wavelengths also must be considered and best results will be obtained in a vacuum. Even with these problems, however, spectra have been recorded down to about 50 Å.

Ultraviolet photography and spectroscopy have found particular usefulness in the study of combustion processes. Some of the very short-lived chemical species occurring during combustion can be observed in the near-ultraviolet region. Along with infrared, ultraviolet techniques also have been used for detecting the retouching of paintings.

Electronic Imagery

Electronic Imagery, instead of using chemical means (emulsions), takes advantage of the sensitivity of various electronic detectors to different bands of the electromagnetic spectrum. The energy received is transduced by these sensors into an electronic or electrical effect (change of resistance, current, emf, the emission of electrons, etc.), from which effects an option of ways to process and display the information is available. The most common form of electronic imagery is found in television. Image orthicons, vidicons, and the more recent TV cameras using charge-coupled devices are described in entries on **Television (TV)**; and **Charge-Coupled Device (CCD)**. See also **Cathode-Ray Tube**; and **Computer Graphics**.

Electronic imagery is particularly attractive for situations where image information must be transmitted over long distances where digitized signals offer greater accuracy and reliability—and where the incoming information is immediately compatible with digital data processing and computing equipment. Electronic imagery also has made certain imaging tasks possible, such as radar imaging, where traditional photographic means do not suffice.

While techniques are available in traditional photography to enhance raw information, these methods are largely of a qualitative, aesthetic nature rather than of a quantitative, scientific nature. In handling tiny pixels (one of the dots or resolution elements making up a digitized picture) of information, it becomes possible to computer program the processing of image information as it is received or after it is retrieved from tape or other electronic storage medium. The pixels can be measured one at a time at a rapid rate for brightness, sense (detection of obviously bad information), and other quantities, and over a wide scale of selections (for example, black = 0; medium gray = 32; white = 63) so that groupings of input information can be made to provide better contrasts (in pattern and blackness or color) when the information is all regrouped and reassembled for display. Accomplishments of this nature are probably best exemplified by the image intensification and color enhancement of pictures returned

to earth from various space explorations. This is discussed further in this entry, but also see entries on specific planets.

With the flexibility of modern data processing and computing equipment, a vast array of programming techniques can be applied to the handling of pixels similar to the handling of any other kind of information. Also, with electronic imagery, a full reconstitution of a scene need not always be a primary objective. For example, as color may be related to chemical composition (discussed later) an astrochemist may call for the proportion of certain "colored" pixels in an entire scene or part of a scene and thus make at least a preliminary judgment as to the composition of rocks or soil in a scene without having to see the entire scene reconstituted.

Whereas the final results of traditional photography are usually in the form of prints or transparencies (with attendant problems), electronic imagery can be projected at will on cathode ray tubes and, if desired, combined with computer graphics—or conventional photos can be made from digitized information.

Relating Color to Physical/Chemical Properties of Materials

Color of materials is of particular interest to geologists, mineralogists, oceanographers, astrophysicists, and astrochemists, who have found in recent years that color as recorded in images can lead to specific information pertaining to the chemical content and physical parameters of materials. Only a few specific examples can be given here.

Ocean Water. The possibility of using remote sensing of ocean water color to determine the general composition of the water has been recognized for many years. The physical processes of absorption and scattering relate the upwelling radiance just beneath the sea surface to the constituents of the water. Except for waters in close proximity to coastlines and the confluences of rivers and the sea, biological constituents play a dominant role in these processes. The most important constituent appears to be phytoplankton, microscopic plant organisms that photosynthesize and constitute the bottom link in the ocean food chain. These plankton contain chlorophyll (the dominant photosynthetic pigment), which absorbs strongly in the blue and red regions of the visible spectrum. Hence, increasing concentrations of phytoplankton (chlorophyll a) have the effect of changing the color of water to green hues from the deep blue of its pure state. By selecting a number of frequencies, the sensing of the chlorophyll content can be put on a quantitative basis. An instrument known as the Coastal Zone Color Scanner (CZCS) was installed in the Nimbus-7 satellite, launched in October 1978. There are six wavelength bands utilized: (1) 433–453 nm; (2) 510–520 nm; (3) 540–560 nm; (4) 660–680 nm; (5) 700–800 nm; and (6) 10,500–12,500 nm. When the data are processed, so-called false-color images are produced, each color representing one of the aforementioned frequency bands. See Fig. 5. Phytoplankton supports

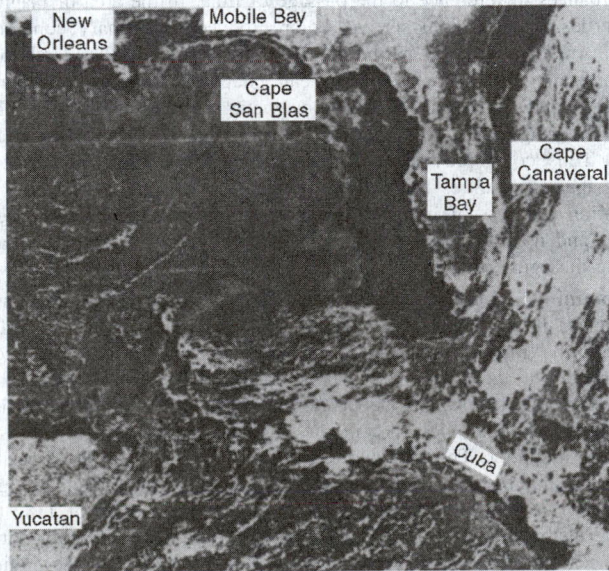

Fig. 5. Black-and-white reproduction of a false-color coded map of phytoplankton pigments in the Gulf of Mexico. Black area represents highest concentrations. (*National Oceanic and Atmospheric Administration.*)

all higher life forms in the sea. Information like this may lead to improved methods for managing and exploiting fisheries. Massive concentrations of phytoplankton blooms (red tide) also may be detected well in advance by this technology. The excellent correlation between samples taken from the ocean research vessel (R.V. *Athena II*) and the CZCS instrument on the *Nimbus* satellite are shown in Fig. 6.

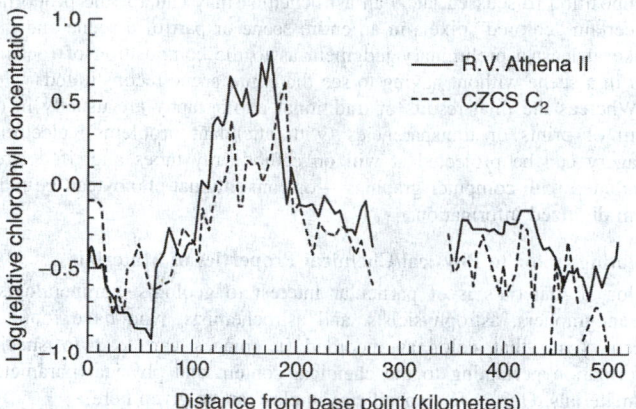

Fig. 6.   Graph showing close correlation between phytoplankton counts taken in the Gulf of Mexico by oceanographic research vessel *R.V. Athena II* and as observed from the Coastal Zone Color Scanner in the *Nimbus 7* satellite. (*National Oceanic and Atmospheric Administration.*)

Color Values of Planetary Images. The *Viking Orbiter* and *Lander* cameras returned color information about many of the surface features of Mars, as well as the colors of the atmosphere and of a Martian sunset. The planet has a salmon-colored sky, surface materials that range from medium-warm umbers to brighter shades of orange and reddish yellows, and rocks that range from nearly black breccia-type blocks to the more characteristics reddish yellow rocks of recognizable volcanic origin. From orbit the colors are even more diverse and include color-shaded values of white from clouds, fogs, and surface ice, and many brightness variations of the reddish color of Mars that was noted some 3000 years ago. See also **Mars**.

Aside from their interest purely as the result of curiosity, detailed analysis and study of the colorations of the Martian features are of much assistance to scientists in extending the data returned by the chemical analytical equipment which was part of each of the two landers' sampling equipment. Because surface color is so consistent in many regions of the planet, the interest in color is to look for changes as a result of the disturbances in reference to the passage of time. In the area of magnetic properties, spectral analysis and color produced a primary component of information. On the earth, the four iron products that tend to be magnetically attractable are iron metal, magnetite (Fe_3O_4), maghemite (Fe_2O_3), and pyrrhotite. Iron and magnetic particles tend to be black in color; maghemite is a yellowish brown; and pyrrhotite is a brassy grey. Thus color provides a lead to magnetic qualities and, in turn, the magnetic qualities are related to planetary evolution—because the oxidation progress of iron, for example, passes through both magnetite stages and nonmagnetic hematite (or limonite, a hydrated oxide) stages during the aging process.

Obtaining Color Images of the Planets. Using the *Viking* missions as examples, the fundamental principles of photographic color are involved through the integration of three primary colors to produce full-color pictures—but there are many subtle factors which make the task of resolving completely accurate color very difficult. The *Viking* Orbiters used twin slow-scan vidicon cameras, but acquired only single frames by utilizing a shutter. The vidicons recorded the image on a photosensitive plate, which is then scanned a line at a time to convert the image to digital data for transmission (1056 lines per frame, each made up of 1182 tiny picture elements—pixels). The cameras' 475 millimeter telescopic lenses could resolve features no smaller than a football field from the minimal orbital altitude of 1512 kilometers (940 miles). The cameras alternated and each could recycle in 4.5 seconds. See Fig. 7. Each camera contained: a filter wheel fitted with six filters; a clear filter to provide broad sensitivity

across the near-ultraviolet and visible wavelengths; a violet filter sensitive only to the near-ultraviolet and violet (for cloud and ice enhancement); a minus-blue filter to yield a reverse effect to that of the filter; and three filters for color reconstruction (red, green, blue). The actual color pictures are constructed at the project office on earth—not on the spacecraft. The camera simply acquires three individual pictures in quick succession, each utilizing one of the color filters. The three pictures are combined at the project office as individual frames, enhanced as needed to improve contrast and color balance, and frequency mosaics are prepared.

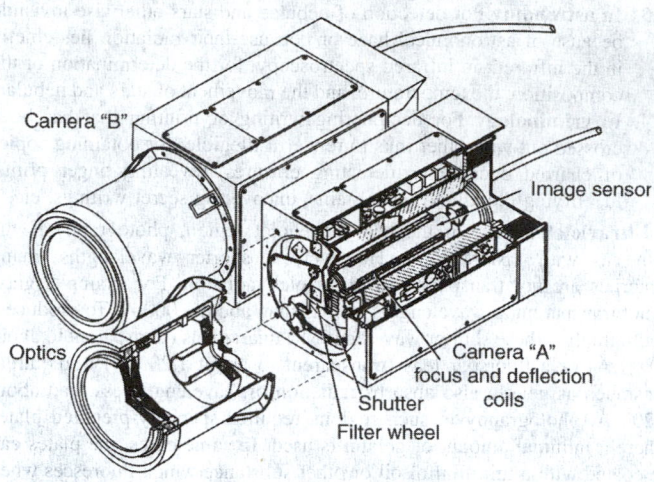

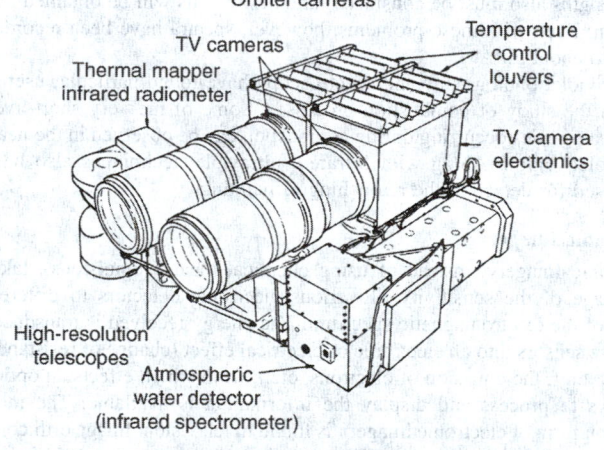

Fig. 7.   *Orbiter* camera and science platform used on *Viking* mission to Mars. (*NASA Jet Propulsion Laboratory, Pasadena, California.*)

The cameras on the *Viking* landers are technically classified as facsimile cameras. The principle of operation is similar to that of equipment used to transmit wirephotos by radio or telephone, and is quite a bit slower than vidicon photography. The operation is not unlike that of the Orbiter vidicons just described, in that a scanning technique is used. However, the scene is scanned directly rather than via a photosensitive plate before the lines and picture elements (pixels) are encoded as digital information for radio transmission.

The reconstruction equipment at the project office essentially reverses the camera process by converting the digital data back into an image on film with a unique artificial light beam produced by an argon/krypton laser. The fundamental principle involved in processing orbiter and lander data is similar, but it should be emphasized that the reconstruction process is also used for the preparation of orbiter pictures and for the visual presentation of data that are not imagery produced. Orbital thermal mapping and water-vapor mapping data can also be illustrated as color imagery. A schematic of the system is shown in Fig. 8. Some general concept of the quality of the reconstructed lander image is given in Fig. 9, which is reproduced in black and white from a color image.

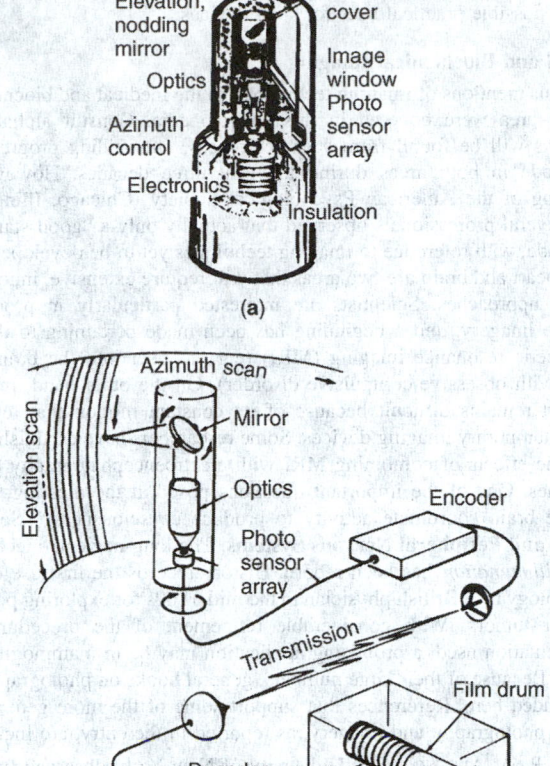

Fig. 8. (**a**) Schematic of Mars *Lander* photo sensors; (**b**) complete system, including reconstruction of information after transmission to Earth. (*NASA Jet Propulsion Laboratory, Pasadena, California.*)

Fig. 9. The general character of images received from the *Viking Lander* on the Martian surface is apparent from this black-and-white reproduction of the reconstituted view. White structure in foreground is part of the *Viking Lander* vehicle. (*NASA Jet Propulsion Laboratory, Pasadena, California.*)

The sequence of operations is as follows:
On the surface of Mars

1. Nodding mirror vertically scans Martian scene
2. Reflects scene through lens onto photo sensors
3. Sensors generate signal directly proportional to density of incident light

4. Signal output sampled at rate synchronized with mirror
5. Samples space vertically as picture elements (pixels). Each pixel—one resolution size on a side (square), 512 pixels per line
6. Mirror completes elevation scan. Returns to start position
7. Camera revolves one line in azimuth; starts next line (or rescans line two or more times for 3-color)
8. Digitized pixels relayed to Lander transmitter
9. Data transmitted to earth (direct or orbiter-relayed)

On Earth (project office)

1. Data are processed; recorded on magnetic tape
2. Data for complete scene played into reconstruction equipment via digital computer. Laser is initiated
3. Red, blue, green beams separated out of laser beam
4. Individual beams reflected into light modulators
5. Modulators are computer controlled to alter beam intensities according to digitized pixel intensity data
6. Three color beams recombined into laser beam and focused on rotating mirror
7. Duplicates action of camera mirror; puts image on conventional color film; unit moves one line width as each line is completed

Color Analysis of Lunar Images. Although it has been several years since the last images of the moon were returned to earth by the Apollo missions, detailed studies continue in some laboratories. Detailed chemical maps of the lunar surface have been constructed by scientists (Andre et al., 1977) who have applied a new weighted-filter imaging technique to Apollo 15 and Apollo 16 x-ray fluorescence data. The data quality improvement is amply demonstrated by (1) modes in the frequency distribution, representing highland and mare soil suites, which were not evident prior to data filtering, and (2) numerous examples of chemical variations which are correlated with small-scale (about 15-kilometer) lunar topographic features.

Radar Imagery

Although radar has been a useful tool in air traffic control, weather reporting, and military weaponry for several decades, the use of radar imagery for scientific research and geological survey and oceanographic investigations is somewhat more recent. One of the first scientific applications, of course, was in connection with astronomy. See also **Radio Astronomy**. The acoustic analog of radar, sonar, has been applied for many years and is finding increasing scientific applications. See also **Sonar**.

Geological Mapping of Land and Ocean Surfaces. When the Seasat satellite was put into orbit around the earth in June 1978, it was equipped with a payload of active microwave sensors consisting of an altimeter, a scatterometer, and an imaging synthetic aperture radar (SAR). The objective of the mission was a proof-of-concept demonstration of the capability to monitor the ocean surface and near-surface features, such as surface waves, internal waves, currents, eddies, surface wind, surface topography, and ice cover. The imaging radar, which was operated in the synthetic aperture mode, provided, for the first time, synoptic radar images of the earth's surface (both ocean and land areas) obtained from an orbiting platform. The resolution of these images is about 25 meters (82 feet). As pointed out by Elachi (1980), the success of this complex sensor was a major technological advance, and it opened up a new dimension in the capability to observe, monitor, and study the Earth's surface. The SAR imaging sensor is an active system, using its own energy to illuminate the surface and to generate an image from the backscatter echoes. Thus, it is not dependent upon illumination from the sun. The radar energy also penetrates cloud cover. Consequently, the system is not constrained by weather conditions. The illumination angle and direction can be controlled and selected, whereas in optical systems these parameters are constrained by the sun's location.

In the SAR approach, the Doppler information in the returned echo is used simultaneously with the time delay information to generate a high-resolution image of the surface being illuminated by the radar. The radar usually "looks" to one side of the moving platform and perpendicular to its line of motion, thus eliminating right-left ambiguities. In Elachi's description, it is pointed out that points equidistant from the radar are located on successive concentric spheres. The intersection of these spheres with the surface gives a series of concentric circles centered at the nadir

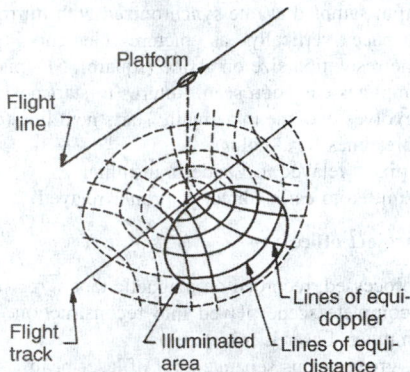

Fig. 10.   The constant time delay and Doppler contour lines for the radar imaging coordinate system. Each point on the surface can be uniquely identified if the energy in the appropriate time delay bin and Doppler shift bin is filtered out of the received echoes. (*Jet Propulsion Laboratory, Earth and Space Sciences Division, Pasadena, California.*)

point. See Fig. 10. The backscatter echoes from objects along a certain circle will have a well-defined time delay.

The brightness in the radar image is a representation of the surface backscatter cross section, which is a function of the surface slope, surface roughness at the scale of the observing wavelength, and surface complex dielectric constant. Geologic interpretation of the radar image is based upon two general types of information: (1) geometric patterns and shapes, and (2) image tone and texture. The sensitivity of the amplitude of the radar echo to changes in the surface topography is very high in comparison to the optical and infrared albedo. A change in the surface slope of a few degrees can easily change the amplitude of the radar echo by a factor of two or more. A radar image of Death Valley, California, is shown in Fig. 11.

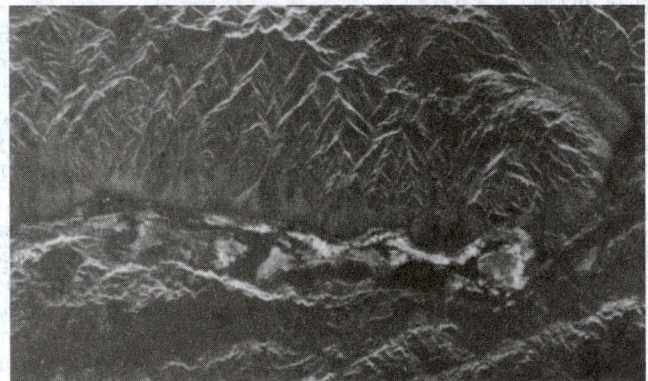

Fig. 11.   Radar image of Death Valley, California. Length of view represents a distance of about 90 kilometers (56 miles). (*Jet Propulsion Laboratory, Earth and Space Sciences Division, Pasadena, California.*)

Because the radar sensor basically provides a Doppler time-delay history of each point target, thousands of computational operations are required to generate a single image element. This processing requirement, combined with the desire to have large swath mapping with high resolution, requires extremely fast processing hardware which is just at the limit of current technology.

See also **Satellites (Scientific and Reconnaissance).**

Color Image on Silicon Wafer. V.V. Doan and M.J. Sailor (University of California, San Diego), in mid-1962, reported that an electrochemical etch of silicon produces a microporous material that Photoluminescences in the visible region of the electromagnetic spectrum. The researchers attribute the action to quantum confinement effects that arise from isolated, nanometer-size silicon features that are produced during the etching. Images get generated on an *n*-Si wafer appeared colored under white light. Images appeared red-orange under UV radiation. The researchers also photoetched a diffraction grating into the substrate to demonstrate simultaneous encoding of gray-scale images into thin-film interference,

luminescence, and diffraction phenomena. It is too early at this juncture to forecast possible practical uses for such devices.

Medical and Biochemical Imaging

Numerous mentions of imaging technology in the medical and biochemical research area were covered in this encyclopedia. Consult alphabetical index. As will be found from such references, outstanding progress has been made in both areas during the past three decades. However, at a meeting of the American Psychological Society (Chicago, Illinois) in 1993, several professionals observed that actually only a "good start" has been made, with reference to imaging techniques yet to be developed. The human heart and brain are two areas that will require extensive, innovative imaging approaches. Scientists are interested particularly in producing real-time imagery, and a beginning has been made pertaining to the use of magnetic resonance imaging (MRI) (e.g., to peer into the brain of a patient with obsessive-compulsive disorder). On the other hand, imaging the heart remains difficult because of its constant motion that tends to blur contemporary imaging devices. Some researchers have established the symbiotic effects of combining MRI with electroencephalography (EEG) techniques. One of the important questions posed at the meeting, "How does the brain coordinate activity to produce consciousness?" See also **Central and Peripheral Nervous Systems**. The long-neglected technique of *transillumination* has been reborn. History records the first use of this methodology by a British physician in the mid-1800s for exploring possible testicular tumors. With considerable refinement of the procedure and instrumentation used, a promising application may be in mammography.

Note: Because of their large numbers, general books on photography are not included here. References that support some of the more generalized areas of photography and imagery, as reported in the entry, are included.

> R. A. ALFANO, (City University of New York) observed (we can see inside tissue with enough clarity that we are starting to see *fingerprint* patterns that distinguish between healthy and abnormal tissue)

Additional Reading

Alper, J.: "Echo-Plannar MRI: Learning to Read Minds," *Science*, 556 (July 30, 1993).

Barger, M.S. and W.B. White: *The Daguerreotype*, Smithsonian Institution Press, Washington, DC, 1991.

Barger, M.S. and W.B. White: *Daguerreotype: Nineteenth-Century Technology and Modern Science*, Johns Hopkins University Press, Baltimore, MD, 2000.

Barinaga, M.: "Biology Goes to the Movies," *Science*, 1204 (November 30, 1990).

Bentley, J.: "Coloring the Invisible World," *Technology Review (MIT)*, 54 (July 1991).

Beardsley, T.: "Sharper Image: Picosecond Photography May Reveal Tumors," *Sci. Amer.*, 32 (October 1991).

Becher, P.: *Emulsions: Theory and Practice*, 3rd Edition, Oxford University Press, Inc., New York, NY, 2001.

Benaron, D.A.: "Optical Time-of-Flight Absorbance Imaging of Biological Media," *Science*, 1463 (January 22, 1993).

Booth, S.A.: "Video To Go: Camcorders," *Popular Mechanics*, 38 (January 1991).

Cipra, B.A.: "Image Capture by Computer," *Science*, 1288 (March 10, 1989).

Corcoran, E.: "Not Just a Pretty Face: Compressing Pictures with Fractals," *Sci. Amer.*, 77 (March 1990).

Corcoran, E.: "Body Heat: Quantum-Well Infrared Photodetectors," *Sci. Amer.*, 123 (October 1991).

Cornwell, T.J.: "The Applications of Closure Phase to Astronomical Imaging," *Science*, 263 (July 21, 1989).

Crease, R.P.: "Biomedicine in the Age of Imaging," *Science*, 554 (July 30, 1993).

Doan, V.V. and M.J. Sailor: "Luminescent Color Image Generation on Porous Silicon," *Science*, 1791 (June 26, 1992).

Drury, S.A.: *Guide to Remote Sensing: Interpreting Images of the Earth*, Oxford University Press, Inc., New York, NY, 1990.

Grimm, T. and M. Grimm: *The Basic Book of Photography*, 4th Edition, Penguin USA, New York, NY, 1998.

Hedgecoe, J.: *The Photographer's Handbook*, 3rd Edition, Alfred A Knopf, Inc., Westminster, MD, 1992.

Huang, et al.: "Optical Coherence Tomography," *Science*, 1178 (November 22, 1991).

Izatt, J.A., et al.: "Ophthalmic Diagnostics Using Optical Coherence Tomography," *SPDIE Proceedings*, 1877 (1993).

Jenkins, F.A., Jr., K.P. Dial, and G.E. Goslow, Jr.: "A Cineradiographic Analysis of Bird Flight," *Science*, 1495 (September 16, 1988).

Lam, D., Man-Kit, and B.W. Rossiter: "Chromoskedasic Painting," *Sci. Amer.*, **80** (November 1991).

Lillesand, T.M.M. and R.W. Kiefer: *Remote Sensing and Image Interpretation*, 4th Edition, John Wiley & Sons, Inc., New York, NY, 1999.

London, B. and J. Upton: *Photography*, 6th Edition, Addison Wesley Longman, Inc., Redding, MA, 1997.

Mollet, H. and A. Grubenmann: *Formulation Technology: Emulsions, Suspensions, Solid Forms*, John Wiley & Sons, Inc., New York, NY, 2001.

Newhall, B.: *Daguerreotype in America*, 3rd Edition, Dover Publications, Inc., Mineola, NY, 1999.

Ourmazd, A., et al.: "Quantifying the Information Content of Lattice Images," *Science*, 1571 (December 22, 1989).

Pappas, D.L., et al.: "Atom Counting at Surfaces," *Science*, 64 (January 6, 1989).

Peterson, I.: "Needle Imaged in Animal-Tissue Haystack," *Science News*, 325 (May 25, 1991).

Pool, R.: "Molecular Photography with an X-ray Flash," *Science*, 295 (July 15, 1988).

Pool, R.: "Making 3-D Movies of the Heart," *Science*, 28 (January 4, 1991).

Richards J.A. and D.E. Ricken: *Remote Sensing Digital Image Analysis*, 3rd Edition, Springer-Verlag, Inc., New York, NY, 1999.

Richelson, J.T.: "The Future of Space Reconnaissance," *Sci. Amer.*, 38 (January 1991).

Roberts, L.: "Mapping by Color and X-rays," *Science*, 425 (April 28, 1989).

Romer, G.B., J. Delamoir: "The First Color Photographs," *Sci. Amer.*, 88 (December 1989).

Silverman, J., J.M. Mooney, and F.D. Shepherd: "Infrared Video Cameras," *Sci. Amer.*, 78 (March 1992).

Staff: "Odyssey (Reviews of Photos in First 100 Years of National Geographic Magazine)," *Natl. Geographic*, 322 (September 1988).

Staff: "New Projectors," *Hughesnews*, 1, Culver City, California (February 21, 1992).

Staff: "Imaging Technologies, Inscribing Science," *Camera Obscura*, 28 (1992).

Vager, Z., R. Naaman, and E.P. Kanter: "Coulomb Explosion Imaging of Small Molecules," *Science*, 426 (April 28, 1989).

Stroebel, L.D., J. Compton, and I. Current: *Basic Photographic Materials and Processes*, 2nd Edition, Butterworth-Heinemann, Inc., Woburn, MA, 2000.

Vander Voort, G.F.: "Metallography," *Advanced Materials & Processes*, 71 (January 1990).

Van Sant, T., et al.: *The Earth—From Space: A Satellite View of the World*, Spaceshots, Inc., Manhattan Beach, CA, 1990.

Waters, A.J., M.J. Bader, J.R. Grant, G.S. Forbes, et al.: *Images in Weather Forecasting: A Practical Guide for Interpreting Satellite and Radar Imagery*, Cambridge University Press, New York, NY, 1997.

Wilkie, D.S. and J.T. Finn: *Remote Sensing Imagery for Natural Resources Monitoring: A Guide for First-Time Users*, Columbia University Press, New York, NY, 1996.

Zwingle, E., H.E. Edgerton, and B. Dale: "'Doc' Edgerton—The Man Who Made Time Stand Still," *Natl. Geographic*, 464 (October 1987).

PHOTOIONIZATION. This process, which is also called the atomic *photoelectric effect*, is the ejection of a bound electron from an atom by an incident photon whose entire energy is absorbed by the ejected electron. This statement means that photoionization cannot occur unless the energy of the photon is at least equal at the ionization energy of the particular electron in the particular atom; any excess of energy in the photon above this value appears as kinetic energy of the ejected electron.

PHOTOLUMINESCENCE. See **Luminescence**.

PHOTOLYSIS. The process by which a chemical species undergoes a chemical change as the result of the absorption of a photon of light energy. See also **Photochemical Reaction**; and **Photodecomposition**.

PHOTOMETERS. Instruments for the measurement of luminous intensity, luminous flux density, and illumination. In usual terminology, only instruments that respond to the central portion of the electromagnetic spectrum, i.e., the ultraviolet, visible, and infrared regions, are called photometers. Essentially, a photometer is comprised of a transducer, which transforms electromagnetic waves (photons) into an electric current, and a current-measuring readout device. In the simplest form, the instrument could be a voltage-generating photocell connected to a microammeter. Photographic exposure meters and light meters that measure ambient illumination are usually of this type. The latter are furnished with green filters, which correspond to the relative spectral sensitivity of the human eye. Photoresistors also are used for this purpose. Photoresistors require a voltage source (battery) in the circuit. The microammeter reads out the change in resistance caused by the illumination. These devices are more sensitive, but not of high precision because of fatigue effects of photoresistors. For precision work, photomultiplier tubes are usually the transducer selected. Some specific types of photometers and spectrophotometers include:

Atomic-Absorption Photometer. This instrument operates on the very specific spectral absorption of an atomized sample rather than emission. The equipment is comprised of a stabilized hollow-cathode lamp (one for each element to be analyzed), a flame with sample nebulizer, a monochromator, and a photometer.

Brightness Meter. A special type of reflection meter for evaluating the brightness of paper and similar products by measuring the diffuse reflectance in the blue range of the spectrum. Actually, these meters quantify the yellow characteristics of the paper.

Circular Dichrograph. An instrument similar to a spectropolarimeter. Instead of a change in angle of optical rotation versus wavelength, the instrument records the difference in dichroic absorption versus wavelength.

Color-Difference Meter. A specially designed reflection meter for assessing small color variations.

Colorimeter. An instrument for routine chemical analysis. Compounds or ions which absorb light in the visible part of the spectrum (400 to 800 nanometers) or which are convertible by specific reagents to such compounds can be analyzed with a colorimeter. The instrument typically incorporates an incandescent light bulb as light source, filters to separate the spectral region, a cuvette to contain the sample solution, and a photometer. See also **Colorimetry**.

Densitometer. An instrument used to measure the attenuation of a beam passing through, or reflected from the surface of solid samples.

Ellipsometer. An instrument for determining the thickness of very thin films of monomolecular dimensions. Essentially, the instrument is a polarization interferometer that utilizes a photometer as a readout device.

Flame Photometer. See *Atomic-Absorption Photometer* in this entry.

Fluorimeter (also Fluorometer or Fluorophotometer). In this instrument, the sample is excited by a light beam of suitable short wavelength. The remitted fluorescent light is picked up by a photometer, usually placed 90 degrees from incidence. A filter or monochromator is provided which excludes the exciting waveband and transmits the fluorescent light. See also **Fluorometers**.

Footcandle Meter. A color-corrected illumination meter calibrated in footcandles.

Glossmeter. An instrument for measuring specularly reflected light from the surface of a flat sample. The angle of incidence and the angle of light pickup are identical and opposed from the normal to the surface. A typical glossmeter consists of a light source and simple optics to direct a defined beam onto the sample. In the opposite direction, there is a light detector connected to a readout meter.

Hemoglobinometer. A specialized colorimeter for determining hemoglobin in blood.

Light-Scattering Photometer. In one type, suspended particles are determined (counted). Another type is used to determine the molecular weight of macromolecules dispersed in solution. The former type operates on the basis of a nephelometer; the latter is of much higher precision and uses monochromatic light.

Lux Meter. Essentially a footcandle meter calibrated in international lux units. (1 footcandle = 10.8 lux.)

Nephelometer. An instrument for determining particle size or particle concentration by measuring the amount of light transmitted or scattered by the suspended particles. Quantitative determinations are made by comparing a given sample with a known standard. See also **Nephelometry**.

Opacimeter. A reflection meter specifically designed to evaluate the opacity of thin sheets, such as paper, by measuring the diffuse reflectance over a white and a black surface in turn.

Optical-Emission Spectrometer. Similar to flame photometer (atomic-absorption photometer) except that an electric spark rather than a flame is used to vaporize (atomize) unknown samples.

Polarimeter. An instrument for determining the concentration of optically active compounds in solution by determining the angle of rotation of plane-polarized light passing through the sample. See also **Polarimetry**.

Reflection Meter. A photometer arranged to pick up diffusely reflected light from the surface of a flat sample. The spectral evaluation of the reflected light permits quantitative color evaluation as seen by the human observer.

Refractometer. An instrument for determining the refractive index of solutes. Most of these instruments use photometric readout systems.

Saccharimeter. A polarimeter calibrated in "sugar degrees" for analyzing the concentration of sugar solutions.

Spectrofluorimeter. A fluorimeter with two separate monochromators. One serves to scan through the spectrum of the exciting light source; the other scans the emitted fluorescent light.

Spectrophotometer. An instrument comprising a light source, means of monochromatizing the light, a sample space, and a photometer. These instruments normally determine concentration of a solute by measurement of light attenuation, the logarithm of absorption being proportional to the concentration. If the instrument is designed to operate in the infrared region, it is known as an infrared spectrophotometer. If in the ultraviolet region, an ultraviolet spectrophotometer, etc.

Additional Reading

Decusatis, C.: *Handbook of Applied Photometry*, Springer-Verlag, Inc., New York, NY, 1998.

Heranshaw, J.B.: *The Measurement of Starlight: Two Centuries of Astronomical Photometry*, Cambridge University Press, New York, NY, 1996.

Swatland, H.J.: *Computer Operation for Microscope Photometry*, CRC Press, LLC., Boca Raton, FL, 1997.

PHOTOMETRIC ANALYSIS. Chemical analysis by means of absorption or emission of radiation, primarily in the near UV, visible, and infrared portions of the electromagnetic spectrum. It includes such techniques as spectrophotometry, spectrochemical analysis, Raman spectroscopy, colorimetry, and fluorescence measurments.

See also **Colorimetry**; and **Raman Spectroscopy**.

PHOTON AND PHOTONICS. In common usage, a photon is a quantum of electromagnetic energy. The energy of a photon is $h\nu$, where h is the Planck constant, and v is the frequency associated with the photon. The term photon usually refers to a plane-wave quantum of electromagnetic energy, for which the momentum is $h\nu/c$, and the component of angular momentum in the direction of the momentum is $\pm\hbar$, where c is the velocity of light and $\hbar$ is $h/2\pi$.

The word *photonics* entered the scientific vocabulary in the mid-1980s to describe a communications transmission system that converted digital information into pulses of light that traveled over an optical fiber cable (fiber optics/light wave communication). The first crude cables were used in the mid-1970s. An exploratory system was established in a network between three buildings in downtown Chicago. Since then, the growth of optical fiber networks worldwide has been no less than dramatic. However, photonics also has a broader connotation and parallels in its hardware and systems aspects the well-established microwave technology, thus, a *photonics technology*. See also **Optical Fiber Systems**.

The existence of the photon was first suggested by Planck's famous research, about 1900, into the distribution in frequency of blackbody radiation. Planck arrived at agreement with the experimental distribution only by making the drastic (for that period in science) assumption that the radiation exists in discrete amounts with energy $E = hf$, where f is the frequency of the radiation and h is Planck's constant, $6.626 (10)^{-27}$ erg sec. Confirmation of the existence of these quanta of electromagnetic energy was provided by Einstein's interpretation of the photoelectric effect (1905). Einstein made it clear that electrons in a solid absorb light energy in the discrete amounts hf. The full realization that the photon is a particle with energy and momentum was provided by the Compton effect (1922), an aspect of the scattering of light by free electrons. Compton showed that features of the scattering are understood by balancing energy and momentum in the collision in the usual way, the light considered as a beam of photons each with energy hf and momentum hf/c.

The modern point of view is that, for every particle that exists, there is a corresponding field with wave properties. In the development of this viewpoint, the particle aspects of electrons and nuclei were evident at the beginning and the field or wave aspects were found later (this was the development of quantum mechanics). In contrast, the wave aspects of the photon were understood first (this was the classical electromagnetic theory of Maxwell) and its particle aspects only discovered later. From this modern viewpoint, the photon is the particle corresponding to the electromagnetic field. It is a particle with zero rest mass and spin one.

For a photon moving in a specific direction, the energy E and the momentum q of the particle are related to the frequency f and wavelength λ of the field by Planck's equation $E = hf$ and the de Broglie equation $q = h/\lambda$. As for all massless particles, the energy and momentum are related by $E = cq$ and the photon can only exist moving at light speed c. Another property of all massless particles is that, given the momentum, the particle can exist in just two states of spin orientation. The spin can be parallel or antiparallel to the momentum, but no other directions are possible. The photon state with the spin and momentum parallel (antiparallel) is said to be right- (left-) handed and is a right- (left-) hand circularly polarized wave. In analogy with the neutrino, one can say that the state has positive (negative) helicity and can call the right-handed particle the antiphoton, the left-handed particle the photon. There is an operation, CP conjunction, that converts a photon state into an antiphoton state and vice versa. It is possible to superpose photon and antiphoton states in such a way that the superposition is unchanged by CP conjugation and so gives a type of photon that is its own antiparticle. The photons produced by transitions between states of definite parities in atoms or nuclei are their own antiparticles in this sense. As for all particles with integer spin, the photon follows Bose-Einstein statistics. This means that a large number of photons may be accumulated into a single state. Macroscopically observable electromagnetic waves, such as those resonating in a microwave cavity, for example, are understood to be large numbers of photons all in the same state. The photon, among all particles, is unique in having its states be macroscopically observable in this way.

Additional Reading

Ackerman, E., et al.: "A 3 to 6 GHz Microwave/Photonic Transceiver for Phased-Array Interconnects," *Microwave J.*, 60 (April 1992).

Cusack, J.: "Photonics at Rome Laboratory," *Microwave J.*, 72 (February 1992).

Fujimoto, J.G. and M.S. Patterson: *Advances in Optical Imaging and Photon Migration*, Optical Society of America, Washington, DC, 1998.

Howe, H.: "Let There Be Light," *Microwave J.*, 24 (January 1992).

Joannopoulos, J.D., R.D. Meade, and J.N. Winn: *Photonic Crystals*, Princeton University Press, Princeton, NJ, 1995.

Kaminow, I.P. and T.L. Koch: *Optical Fiber Telecommunications IIIA*, Morgan Kaufmann Publishers, Orlando, FL, 1997.

Polifko, D. and H. Ogawa: "The Merging of Photonic and Microwave Technologies," *Microwave J.*, 75 (March 1992).

Pradhan, T.: *The Photon*, Nova Science Publishers, Inc., Huntington, NY, 2001.

Render, D.J.: "Photonics — Fast Track for Tomorrow's Communications," *AT&T Technology*, II, (1), 1987. *A classic reference.*

Sakoda K.: *Optical Properties of Photonic Crystals*, Springer-Verlag, Inc., New York, NY, 2001.

Soukoulis, C.M.: *Photonic Band Gaps and Localization*, Kluwer Academic Publishers, Norwell, MA, 1993.

Zmuda, H. and E.N. Toughlian: "Adaptive Microwave Signal Processing: A Photonic Solution," *Microwave J.*, 58 (February 1992).

Web Reference

Optical Society of America: http://www.osa.org/

PHOTON ENGINE. A projected type of reaction engine in which thrust would be obtained from a stream of electromagnetic radiation. See **Ion Engine**. Although the thrust of this engine would be minute, it may be possible to apply it for extended periods of time. Theoretically, in space, where no resistance is offered by air particles, very high speeds may be built up.

PHOTONEUTRON. A neutron emitted from a nucleus in a photonuclear reaction.

PHOTONUCLEAR REACTION. A nuclear reaction induced by a photon. In some cases the reaction probably takes place via a compound nucleus formed by absorption of the photon followed by distribution of its energy among the nuclear constituents. One or more nuclear particles then "evaporate" from the nuclear surface, or occasionally the nucleus undergoes photofission. In other cases the photon apparently interacts directly with a single nucleon, which is ejected as a photoneutron or photoproton without appreciable excitation of the rest of the nucleus.

PHOTOPERIODISM. This term is applied to the reaction of plants to the daily length of the period of illumination. It is one of the most noteworthy of the reactions of plants to an environmental factor. In most parts of the world, marked seasonal variations occur in the length of the daylight period. In temperate zones the length of the daylight period varies from about 8 or 10 hours at the winter solstice to about 14–16 hours at the summer solstice. At higher latitudes the annual variation in day length is greater; at lower latitudes less. In arctic and subarctic regions the length of day varies from 24 hours on the longest summer days to zero hours on the "shortest" days; in the equatorial zone day lengths approximate 12 hours the year round.

Although the length of the photoperiod (number of hours of illumination per day) also has effects upon the vegetative development of plants its most significant influences are upon flowering and other phases of the reproductive development of plants. Plants fall into three fairly well defined categories: (1) "long-day" species, which flower more or less readily in a range of photoperiods longer than a certain critical period, developing only vegetatively under shorter photoperiods; (2) "short-day" species, which flower more or less readily in photoperiods shorter than a certain critical period, developing only vegetatively at all longer photoperiods; and (3) "indeterminate" species, which exhibit no critical photoperiod, developing both vegetatively and reproductively over a wide range of photoperiods. The length of the critical photoperiod differs according to species, but for many plants of both the long-day and short-day types, lies in the range of 12 to 14 hours. Examples of short-day species are dahlias, chrysanthemums, asters, cocklebur, and salvia; of long-day species, radish, beets, dill, spinach, lettuce, and grains; of indeterminate species, tomato, cotton, sunflower and buckwheat. Both long-day and short-day varieties may exist even within the same species. Some varieties of soybeans, for example, are short-day plants, while others are long-day plants.

In temperate regions the season of blooming of a plant is largely determined by its photoperiodic reaction. In general, short-day plants bloom in the early spring or early fall; long-day plants in the late spring or summer. The geographical distribution of some kinds of plants is at least partly controlled by their photoperiodic reaction. A species cannot maintain itself in a climate in which it is impossible for the cycle of reproductive processes to be completed. Pronounced long-day species, for example, would not as a rule be found in tropical regions.

Practical applications of the principles of photoperiodism have been made in the growing of floricultural greenhouse crops. Short-day species such as chrysanthemums, can be brought into bloom earlier in the fall by decreasing the length of their daily exposure to light. Likewise, the time required for long-day floricultural species to attain the flowering stage during the winter months can be shortened by increasing the day length with artificial illumination.

See also **Photosynthesis**; **Plant Breeding**; and **Plant Growth Modification and Regulation**.

PHOTOPHOBIA. See **Vision and the Eye**.

PHOTOPOLYMER. A polymer or plastic so made that it undergoes a change on exposure to light. Such materials can be used for printing and lithography plates, photographic prints, and microfilm copying. The light may cause further polymerization or cross-linking, or it may cause degradation. One application involves the use of esters of polyvinyl alcohol that cross-link and so become insoluble, whereas unexposed portions of the material remain soluble.

PHOTORECEPTOR. A sensory organ that responds to the stimulus of light waves. Eyes are the most familiar organs of this kind but many one-celled animals are sensitive to light and some of the more complex forms have the surface of the body sensitive to it. True eyes serve for the formation of a visual image, whereas the more simple photoreceptors

merely indicate the luminosity of the animal's surroundings. In some cases, adjustment to light is necessary and the simple light-sensitive organ is adequate for the initial step in the animal's orientation. Eyes serve the very different purpose of enabling the animal to perceive objects about it and are not primarily associated with its adjustment to light. Man's skin is sensitive to light in an entirely different way, which becomes evident only in the degree of pigmentation. See also **Vision and the Eye**.

PHOTOREFRACTIVE KERATECTOMY (PRK). As with other refractive eye surgery procedures for correcting nearsightedness, the goal of the PRK procedure is to flatten the central zone, or visual axis, of the cornea so that light rays passing through the cornea and the inner lens will be focused properly on the retina.

This procedure came into use following the approval by the Food and Drug Administration (FDA) of the Summit Excimer laser system in October 1995, and of the VISX system for performing the PRK procedure in March 1996. The approval, with restrictions, was granted only after intensive FDA evaluation of the PRK procedure for a period of almost 10 years.

PRK Procedure

Step 1: Eye preparation. Before the procedure begins, a nurse or technician talks to the patient about any immediate health problems that may affect readiness for the procedure. Antibiotic and anesthetic eye drops are then placed in the eye to numb it and prevent infection. The eye is swabbed with a sterile solution. The eyelid is then propped open with a lid retainer, and a paper or plastic "mask" is placed over the eye to keep eyelashes out of the way. The final step before the procedure begins is marking the cornea with a blue "dye ring, "which serves as a reference point for the surgeon throughout the procedure. Because the cornea is numb, most patients experience little if any discomfort during these pre-operative preparations.

Step 2: Creating the flap. The next step in the PRK procedure is to remove the epithelium, the ultra-thin, film-like protective outer covering of the cornea, from the central zone (visual axis). This can be accomplished with either the Excimer laser or a surgical instrument.

Step 3: The Excimer laser. The Excimer laser is then used to ablate, or vaporize, layers of cornea tissue in order to "sculpt" the cornea to the shape needed to achieve the degree of vision correction required. The sculptured area is less than the diameter of an average pencil eraser. Preoperative tests and the evaluation process determine the sculpture pattern. The data that was gathered is entered into a computer that is built into the Excimer laser system. The computer then calculates the sculpting pattern and directs the application of the laser under the guidance of the eye surgeon. The surgeon continually monitors the progress of the procedure through a microscope that is a part of the laser system.

The patient is asked to focus on a fuzzy red light inside the laser. As the doctor activates the laser, there is a "popping" or "tacking" sound. In addition, there is a slight odor similar to that of hair burning, but there is no discomfort for the patient. The number of laser pulsations will depend on the nature of the refractive vision problem that is being corrected. This phase of the procedure takes only a minute or so.

Step 4: Post-operative measures. When the procedure is complete additional antibiotic drops are placed in the eye, and it may be covered with a plastic shield. For a short while after the procedure has been completed, the eye is numb from the anesthetic drops. As the numbness wears off, the patient may experience some sensitivity to light and scratchy or dry sensation as though something is in the eye. This feeling usually goes away within a few hours. The patient must not drive home following the procedure.

The patient returns to the doctor's office the next day for a post-operative examination. The doctor checks the eye to see if the cornea is healing properly. Vision is checked and, for most patients, will range from 20/20 to 20/40 depending on the number of corrections received. For some patients, vision may continue to improve for several weeks before stabilizing.

The patient may experience some discomfort for the first 24 to 48 hours following surgery. Vision may be blurred for three to five days while the epithelium that was removed from the central corneal zone heals. The patient may also experience glare, halos, shadows, and some ghost images for a few weeks. These are normally transient phenomena and should lessen with time and then fade away. See also **Lasers (Eye Surgery)**; **Myopia (Nearsightedness)**; and **Refractive Eye Surgery**.

Vision Rx, Inc., Elmsford, NY

PHOTOSENSITIVE GLASS. Certain clear silicate glass containing ingredients capable of forming permanent photographic images when subjected to action of X-rays or ultraviolet light and subsequent heat treatment.

See also **Glass**.

PHOTOSPHERE. The intensely bright portion of the sun visible to the unaided eye. The photosphere is that portion of the sun's atmosphere which emits the continuum radiation upon which the Fraunhofer lines are superimposed. In one sun model, the photosphere is thought to be below the reversing layer in which Fraunhofer absorption takes place. In another model, all strata are considered equally effective in producing continuous emissions and line absorption.

PHOTOSTATIONARY STATE RELATION. A relationship that determines the ratio of the concentrations of nitric oxide, NO, [CAS: 10102-43-9], and nitrogen dioxide, NO_2, [CAS: 10102-44-0], in the troposphere. In its simplest form, the ratio is controlled by the following chemical reactions, which interconvert NO and NO_2 without any change in the ozone concentration:

$$NO + O_3 \rightarrow NO_2 + O_2,$$

$$NO_2 + h\nu(+O_2) \rightarrow NO + O_3.$$

In the real atmosphere, the ratio is perturbed by the presence of other oxidants (mostly hydroperoxyl and organic peroxyl radicals), which also convert NO to NO_2 and lead to net ozone production. The photostationary state relation is also occasionally referred to as the Leighton relationship, after Philip Leighton.

Additional Reading

Leighton, P.A.: *Photochemistry of Air Pollution*, Academic Press, New York, NY, 1961.
Sutherland, D., P. Brimblecombe, and W.T. Sturges: *Air Pollution Science for the 21st Century*, Elsevier Science & Technology Books, New York, NY, 2002.

AMS

PHOTOSYNTHESIS. This is the most important of all biological processes. With negligible exceptions the existence of the entire biological world hinges upon this process. From a few simple inorganic compounds and from the sugar made in photosynthesis are erected all of the complex kinds of molecules essential to the construction of the bodies of plants and animals or to maintenance of their existence. Some of these subsequent synthetic processes occur in the plant body, others in the bodies of animals after they have ingested plant materials as foods. Likewise, the energy used by plants and animals represents sunlight energy that has been entrapped in sugar molecules during photosynthesis. The entire organic world runs by the gradual expenditure of the energy capital accumulated in photosynthesis.

Under suitable conditions of temperature and water supply, the green parts of plants, when exposed to light, abstract and use carbon dioxide from the atmosphere and release oxygen to it. These gaseous exchanges are the opposite of those occurring in respiration and are the external manifestation of the process of photosynthesis by which carbohydrates are synthesized from carbon dioxide and water by the chloroplasts of the living plant cells in the presence of light. For each molecule of carbon dioxide used, one molecule of oxygen is released. A summary chemical equation for photosynthesis is:

$$6CO_2 + 6H_2O \xrightarrow{\text{light}} C_6H_{12}O_6 + 6O_2$$

In this process, the radiant energy of sunlight is stored as chemical energy in the molecules of carbohydrates and other compounds that are derived from them.

All photosynthetic organisms, except bacteria, use water as the electron or hydrogen donor to reduce various electron acceptors, and from the water they evolve molecular oxygen. Anaerobic bacteria cannot endure such oxygen, but derive their sustenance through slightly different photosynthetic routes:

$$2H_2S + CO_2 \xrightarrow{\text{light}} (CH_2O) + H_2O + 2S$$

or

$$2CH_3CHOHCH_3 + CO_2 \xrightarrow{\text{light}} (CH_2O) + CH_3COCH_3 + H_2O$$

Photosynthesis takes place in chlorophyll-containing cells only when carbon dioxide, water, and light are available, and when a suitable temperature prevails. Although carbon dioxide constitutes, on the average, only 0.03% of the atmosphere, land plants are entirely dependent upon this source for the carbon dioxide used in photosynthesis. It has been shown experimentally that an increase in the carbon dioxide concentration of the atmosphere results in an increased rate of photosynthesis. On the other hand, a deficiency of water results in a reduced rate of photosynthesis. In nature, sunlight is the source of radiant energy used in photosynthesis, although plants will also photosynthesize under artificial light sources of suitable quality and intensity.

The total radiant energy received at the earth's surface is 1–2 gram calories/square centimeter/minute, depending upon altitude, or approximately 1 hp/10–20 square feet. For crop plants in the field, a maximum of 2–3% of this energy remains stored in the plants at the end of the growing season. During that time about 20% more is actually used in photosynthesis and lost by respiration of the plant, the remainder of the energy being dissipated by re-radiation, transmission through the leaves, and evaporation of water from the plant.

The intensity, quality and daily duration of illumination all have influence on the amount of photosynthesis accomplished per day. Clearly, the longer the daily period of illumination, the more photosynthesis will be accomplished by a plant in the course of a day. The minimum light intensity at which a measurable rate of photosynthesis occurs varies according to species, but is seldom less than 1% of full midday summer sunlight. Under natural conditions, maximum rates of photosynthesis are attained in single leaves of many species at 25–35% of full sunlight intensity, and in some shade species at even lower intensities. For equal intensities, more photosynthesis appears to occur in the orange–short red and blue parts of the spectrum than in the green and yellow. This is because the chlorophyll pigments of the leaves absorb light energy at wavelengths of 6600 and 4250 micrometers. Radiation is most intense in the green and, if this radiation were absorbed, the plant could not utilize it and would overheat.

The range of temperatures most suitable for relatively rapid rates of photosynthesis is not the same for all kinds of plants. In general, it is higher in tropical than in temperate species, and higher in temperate species than in those of subarctic regions. Increase in temperature results in an increase in the rate of photosynthesis up to an optimum which varies with the variety of plant, but which, for most temperate zone species, lies within the range of 20–30 °C. With increase above the optimum, the rate of photosynthesis progressively decreases.

In the vascular plants, photosynthesis occurs chiefly in the leaves. Carbon dioxide diffuses into the intercellular spaces of the leaf from the atmosphere via the stomates, and then dissolves in the moist walls of the mesophyll cells. In solution, the carbon dioxide diffuses to the surface of the chloroplasts, which are the actual seat of the photosynthetic process. The first major step in photosynthesis is the absorption of radiant energy by the plant pigments in the chloroplasts, with the generation of electrons. The plant pigment consists of two closely similar pigments, chlorophyll and chlorophyll *b*, which are porphyrin-derived complexes of magnesium and which, upon excitation by radiant energy, become electron donors. See also **Chlorophylls**. The chloroplast is a complex, self-replicating organelle that possesses its own DNA and is able to synthesize at least a few of the proteins needed for its own functioning. It is filled with membranous thylakoid sacs which are specifically designed to harness the energy available in the excited electron and to carry out the light phase of photosynthesis. In this, the light energy captured is converted into the chemical energy of adenosine triphosphate (ATP) and nicotinamide-adenine dinucleotide phosphate (NADPH). See also

Adenosine Phosphates; and **Coenzymes**. Hydrogen atoms are removed from water and used to reduce NADP, leaving behind molecular oxygen. Simultaneously, adenosine diphosphate (ADP) is phosphorylated to ATP:

In the second, or dark, reaction phase, NADPH and ATP provide the

$$Water + NADP^+ + PO_4 + ADP \xrightarrow{light} Oxygen + NADPH + H^+ + ADP$$

energy to reduce carbon dioxide to glucose and are themselves oxidized or decomposed:

$$CO_2 + NADPH + H^+ + ATP \longrightarrow Glucose + NADP^+ + ADP + PO_4$$

Peter Mitchell (Nobel Prize, 1978) of Great Britain was the first to realize, and to propose in his chemi-osmotic theory, that the energy required for the ADP-ATP reaction could be derived by an accretion of protons in the thylakoid sac to the point at which the electrochemical gradient across the membrane could effect the proton transport required as the driving force for this reaction. See also **Phosphorylation (Photosynthetic)**.

In most plants, the water used in photosynthesis is absorbed by the roots from the soil, whence it is translocated to the leaves. Except for a small portion used in respiration, the oxygen liberated in the process diffuses out of the leaf into the atmosphere, mostly through the stomates. See also **Ascent of Sap**; and **Stomate (or Stoma)**.

Carbohydrates other than hexoses are synthesized in the leaves, apparently as a result of secondary reactions following photosynthesis. Sucrose invariably accumulates in actively photosynthesizing leaf cells. This more complex sugar is built up from the molecules of the simpler hexoses. In most plants, insoluble starch also accumulates in leaf cells during photosynthesis. This carbohydrate is synthesized by the condensation of numerous glucose molecules. The sucrose and starch contents of leaves decrease at night as a result of the continued translocation from the leaves to other parts of the plant. The sucrose is probably translocated as such, but the starch must first be converted into simpler, soluble sugars before it can move out of the leaves. Synthesis of starch is not restricted to the green parts of plants; a familiar example of this is the accumulation of starch in potato tubers. Starch in the nongreen cells is made from glucose, which comes from the leaves or other photosynthetic organs. Starch occurs in cells in the form of small grains, the type of grain formed in each kind of plant being more or less characteristic of that species.

For many years, the nature and location of the complex of proteins (sometimes referred to as the "engine" of photosynthesis) were poorly understood. During the 1980s, much more was learned as the result of research carried out by Johann Deisenhofer (Howard Hughes Medical Institute), Robert Huber, and Harmut Michel (Max Planck Institute), and for this work the investigators were awarded the 1988 Nobel Prize for chemistry. The protein complex, called the membrane-bound proteins, are difficult to define structurally because they do not crystallize readily and thus could not be subjected to x-ray crystallography. However, over a period of three years, the researchers were able to create crystals and thus were able to determine precisely the position of some 10,000 atoms in the protein complex.

With this better understanding, researchers are able to depict how plants, algae, and rhodopseudomonads carry out synthesis. Also, it has been hypothesized that such membrane-bound proteins may have a functional role in some diseases, such as cancer and diabetes.

Finally, it must be realized that photosynthesis is not the sole prerogative of the higher plants. More than half the photosynthesis on the earth's surface is carried out in the oceans by phytoplankton.

Additional Reading

Amato, J.: "A Shady Strategy for Photosynthesis," *Science News*, 246 (October 20, 1990).

Anderson, B., J. Barber, and H. Salter: *Molecular Genetics of Photosynthesis*, Oxford University Press, Inc., New York, NY, 1996.

Blankenship, R.E.: *Molecular Mechanisms of Photosynthesis*, Blackwell Science, Inc., Malden, MA, 2001.

Blaxter, J.H.S. and A.J. Southward: *Advances in Marine Biology*, Academic Press, Inc., San Diego, CA, 1993.

Bogorad, L. and I.K. Vasil: *The Photosynthetic Apparatus*, Academic Press, Inc., San Diego, CA, 1991.

Charles-Edwards, D.A.: *Mathematics of Photosynthesis and Productivity*, Academic Press, Inc., San Diego, CA, 1981.

Coleman, G. and W.J. Coleman: "How Plants Make Oxygen," *Sci. Amer.*, 50 (February 1990).

Darnell, J., H. Lodish, and D. Baltimore: *Molecular Cell Biology*, 4th Edition, W. H. Freeman and Company, New York, NY, 1999.

Falkowski, P.G. and J.A. Raven: *Aquatic Photosynthesis*, Blackwell Science, Inc., Malden, MA, 1996.

Hall, D.O. and K. Rao: *Photosynthesis*, 6th Edition, Cambridge University Press, New York, NY, 1999.

Herring, P.J., et al.: *Light and Life in the Sea*, Cambridge University Press, New York, NY, 1990.

Hogan, J.: "1988 Nobel Prize for Chemistry," *Sci. Amer.*, 33 (December 1988).

Holden, C.: "Picture-Perfect Plankton," *Science*, 681 (February 7, 1992).

Kirk, J.T.O.: *Light and Photosynthesis in Aquatic Ecosystems*, 2nd Edition, Cambridge University Press, New York, NY, 1994.

Metzler, D.: *Biochemistry*, 2nd Edition, Academic Press, Inc., San Diego, CA, 2002.

Miller, K.R.: "A Particle Spanning the Photosynthetic Membrane," *J. Ultrastruct., Res.*, **54**, 1, 159–167 (1976).

Miller, K.R.: "The Photosynthetic Membrane," *Sci. Amer.*, **241**, 4, 102–113 (1979).

Ort, D.R.: *Oxygenic Photosynthesis: The Light Reactions*, Kluwer Academic Publishers, Norwell, MA, 1996.

Pessarakli, M.: *Handbook of Photosynthesis*, Marcel Dekker, Inc., New York, NY, 1996.

Raghavendra, A.S.: *Photosynthesis*, Cambridge University Press, New York, NY, 2000.

Sherman, K., L. Alexander, and B. Gold: *Large Marine Ecosystems: Patterns, Processes, and Yields*, AAAS Books, Waldorf, MD, 1992.

Stoecker, D.K.: "Photosynthesis Found in Some Single-Cell Marine Animals," *Oceanus*, **49** (Fall 1987).

Stryer, L. and J.L. Tymoczko: *Biochemistry Extended, Chapters 1–34*, 5th Edition, W. H. Freeman and Company, New York, NY, 2002.

Yunus, M. and Dr. U. Pathre: *Probing Photosynthesis: Mechanisms, Regulation, and Adaptation*, Taylor & Francis, Inc., Philadelphia, PA, 2000.

Zilsnov, V.K.: "Living Marine Resources," *Oceanus*, **29** (Summer 1991).

See also references list at the end of entry on **Photochemistry and Photolysis**.

R.C. VICKERY, D.Sc., Blanton/Dade City, FL

PHOTOTHEODOLITE. An instrument or device incorporating one or more cameras for taking and recording angular measurements. The phototheodolite, sometimes in conjunction with radar equipment, is used to track rockets and to measure and record attitude, altitude, azimuth and elevation angles, etc.

PHOTOVOLTAIC CELLS. A photovoltaic (PV) solar power system is a complete electrical source that uses solar cells to directly convert light energy into electricity. The system can be self-contained and completely autonomous or it can work in tandem with other conventional fuel-based sources of power to offer robust power availability.

A solar cell is a semiconductor device that can convert light instantaneously into direct-current (d-c) electricity. A number of cells are typically connected together in series in a weather-resistant package such that enough voltage is generated to recharge a 12-volt lead−acid storage battery, the most common storage device used in conjunction with solar power. Such a package of cells is designated a PV module, which is often constructed of an external sheet of strengthened glass and polymeric encapsulation. The most common size module is 0.5–1 m^2 in area and delivers between 25 and 150 watts of power.

The advantages of photovoltaic cells as a source of electric power over alternative power sources may be characterized as follows: solar cells capture sunlight, an essentially inexhaustible and nonpolluting energy source which is freely distributed, and directly convert that light into electricity; photovoltaic generation of electricity requires no machinery with moving parts and produces no noise, waste, or polluting by-products; photovoltaic systems are modular and therefore can be adapted for a variety of applications. Solar power systems are particularly useful in areas where power lines cannot be readily or inexpensively routed.

Solar cells have been used extensively and successfully to power satellites in space since the late 1950s, where their high power-to-weight ratio and demonstrated reliability are especially desirable characteristics. On earth, where electrical systems typically provide large amounts of power at reasonable costs, three principal technical limitations have thus far impeded the widespread use of photovoltaic products: solar cells are expensive, sunlight has a relatively low power density, and commercially available solar cells convert sunlight to electricity with limited efficiency. Clearly, terrestrial solar cells must be reasonably efficient, affordable, and durable. International efforts are dedicated to obtaining such devices.

The power density of sunlight is about 1350 W/m^2 at elevations just above the earth's atmosphere. Less than 1000 W/m^2 is typically incident on earth after filtering through the atmosphere. Due to the low power density of sunlight and limited conversion efficiencies, the most efficient solar modules can generate about 250 W/m^2 in peak sunlight conditions. The maximum power output of a solar cell or module is defined in peak watts (W$_{peak}$), a rating based on a standard measurement method established by international consensus. A solar panel of one square meter area nominally produces one kilowatt hour of electricity per day. For most large-scale, power-producing applications, solar modules have conversion efficiencies above 10% in order to minimize the total cost of a generating system.

Chemistry

Crystalline silicon p-n junction solar cells are the principal commercially available type and are used here to illustrate the operation of a solar cell. When sunlight falls on a solar cell, a voltage is induced and an electric current flows in an external circuit that is connected to the cell. Each atom in the silicon crystal lattice is surrounded by and bound to four equidistant neighboring atoms. The outermost shell of electrons of each silicon atom contains four valence electrons, and each of the four valence electrons in the crystal lattice is shared in a bonding orbital with an electron from one of its four nearest neighbors. This electron pair or covalent bond firmly binds the crystal. If all the valence electrons were inexorably bound, as they would be at 0 degrees kelvin, the silicon crystal would be an insulator because no free electrons would be available, and conduction would be precluded. See Fig. 1(a) However, the covalent bonds can be broken, e.g., by thermal excitation. See Fig. 1(b). The energy required to break

a covalent bond is the bond energy or energy gap, E_g. In silicon, E_g is ca 1.1 eV.

The absence of an electron from a covalent bond leaves a hole and the neighboring valence electron can vacate its covalent bond to fill the hole, thereby creating a hole in a new location. The new hole can, in turn, be filled by a valence electron from another covalent bond, and so on. Hence, a mechanism is established for electrical conduction that involves the motion of valence electrons but not free electrons. Although a hole is a conceptual artifact, it can be described as a concrete physical entity to keep track of the motion of the valence electrons. Because holes and electrons move in opposite directions under the influence of an electric field, a hole has the same magnitude of charge as an electron but is opposite in sign.

The energy in light also can break the bonds of silicon valence electrons. Each photon has energy equal to the product of Planck's constant and the frequency of the light, i.e., $E = h\nu$, where E is photon energy, h is Planck's constant, and ν is the frequency of light. Solar photons range in energy from 0.5 eV for infrared to 4 eV for ultraviolet.

When a photon having energy equal to or greater than E_g is absorbed by the silicon crystal, the photon breaks a covalent bond, thereby freeing an electron and forming a hole. An electron is excited by a photon from a valence-energy band in a covalent bond into a conduction-energy band. The electron, which is transformed into a mobile, negatively charged carrier, leaves behind a mobile hole and consequently the photon has formed a free electron—hole pair.

If the hole and electron are not kept apart, they recombine to produce a small amount of thermal energy within the crystal and no net current flow. When the holes and electrons are kept apart, collected, and made to flow in a circuit outside the crystal, they produce electric current in that circuit. Solar cells are equipped with a barrier or a junction which provides an internal electric field that segregates photogenerated electrons and holes. Thus, although unmodified silicon has an equal number of holes

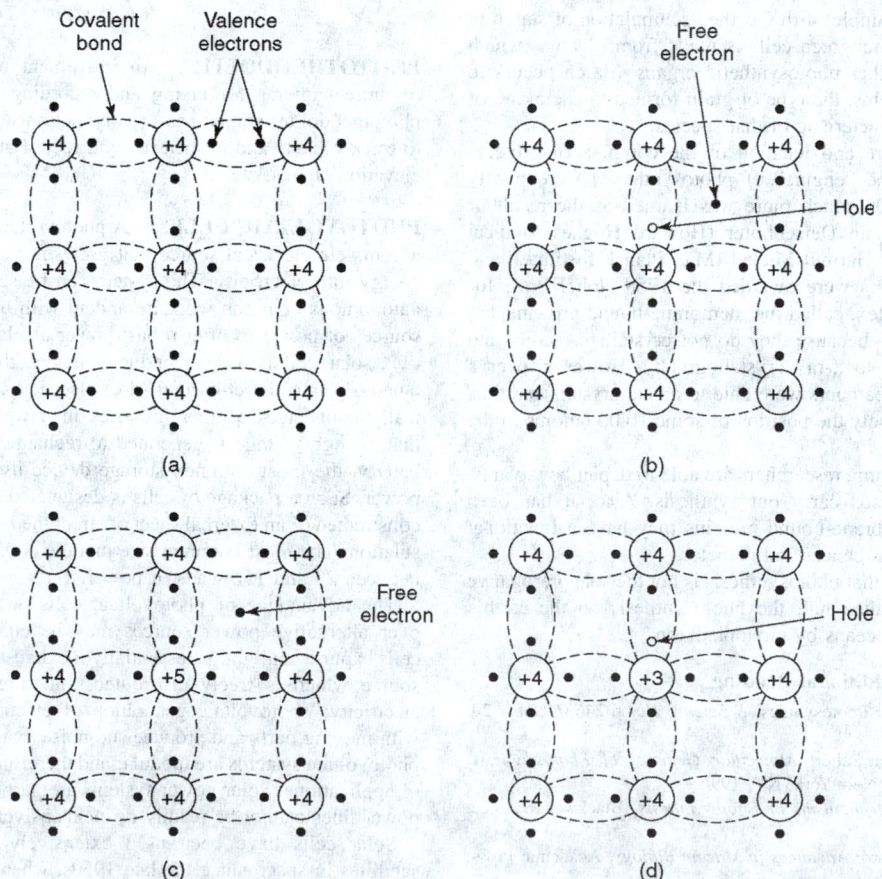

Fig. 1. (a) Silicon (valence = 4) crystal lattice shown in two dimensions with no broken bonds, $T = 0$ K; (b) silicon crystal lattice with a broken bond; (c) silicon crystal lattice with a silicon atom displaced by a donor dopant, i.e., n-doped (valence=5); and (d) silicon crystal lattice with a silicon atom displaced by an acceptor dopant, i.e., p-doped (valence = 3)

and electrons, a $p-n$ junction silicon solar cell consists of two charge-dissimilar regions which are separated by a junction: one region is rich in holes (positive), i.e., p-type silicon, and the other is rich in electrons (negative), i.e., n-type silicon. Such regions do not occur naturally; they are fabricated by doping, i.e., replacing some silicon atoms in the lattice with atoms having a valence other than four. Replacement of a few silicon atoms, i.e., ca one in several million, causes large increases in the electrical conductivity of the resultant doped crystal.

Atoms of elements that are characterized by a valence greater than four, e.g., phosphorus or arsenic (valence = 5), are one type of dopant. These high valence dopants contribute free electrons to the crystal and are called donor dopants. If one donor atom is incorporated in the lattice, four of the five valence electrons of donor dopants are covalently bonded, but the fifth electron is very weakly bound and can be detached by only ca 0.03 eV of energy. Once it is detached, it is available as a free electron, i.e., a carrier of electric current. A silicon crystal with added donor dopants has excess electron carriers and is called n-type (negative) silicon. See Fig. 1(c).

When a silicon crystal is doped with atoms of elements having a valence of less than four, e.g., boron or gallium (valence = 3), only three of the four covalent bonds of the adjacent silicon atoms are occupied. The vacancy at an unoccupied covalent bond constitutes a hole. Dopants that contribute holes, which in turn act like positive charge carriers, are acceptor dopants and the resulting crystal is p-type (positive) silicon. See Fig. 1(d).

Conductivity in doped silicon crystals is determined by the properties of the added charge carriers or majority carriers. In n-type silicon, electrons are majority carriers and holes are minority carriers. There are fewer holes in n-type silicon than in undoped silicon because the large number of electrons causes some recombination with preexisting holes. In p-type silicon, holes are the majority carriers and electrons are the minority carriers. Fewer electrons are present in p-type silicon than in undoped silicon because of the recombination of some electrons with the enhanced population of holes.

Junctions

Four different types of junctions can be used to separate the charge carriers in solar cells: (1) a homojunction joins semiconductor materials of the same substance, e.g., the homojunction of a $p-n$ silicon solar cell separates two oppositely doped layers of silicon; (2) a heterojunction is formed between two dissimilar semiconductor substances, e.g., copper sulfide, Cu_xS, and cadmium sulfide, CdS, in Cu_xS-CdS solar cells; (3) a Schottky junction is formed when a metal and semiconductor material are joined; and (4) in a metal-insulator-semiconductor junction (MIS), a thin insulator layer, generally less than 0.003-μm thick, is sandwiched between a metal and semiconductor material.

Fabrication methods that are generally used to make these junctions are diffusion, ion implantation, chemical vapor deposition (CVD), vacuum deposition, and liquid-phase deposition for homojunctions; CVD, vacuum deposition, and liquid-phase deposition for heterojunctions; and vacuum deposition for Schottky and MIS junctions.

Efficiency

The most efficient silicon cells produced are based on $p-n$ homojunctions and convert 23.1% of the energy in incident light set to simulate the global air mass (AM) 1.5 spectrum, an artificial reference spectrum used to standardize measurement of PV power, with an intensity of 1000 W/m^2 at 25 °C. This is the definition of peak sunlight test conditions. In theory, silicon $p-n$ junction solar cells can convert a maximum approaching 26% of the energy in AM 1.5 sunlight to electricity. Approximately 75% of the energy in sunlight is lost to factors intrinsic to the silicon material.

In comparison, $p-n$ homojunction cells made of more costly semiconductor materials, e.g., indium phosphide, InP, and gallium arsenide, GaAs, which have energy gaps of 1.2–1.4 eV, and maximum theoretical conversion efficiencies of ca 28–30%, depending on the device construction and layering of junctions.

Commercial Silicon Solar Cells

Silicon cells are hundreds of micrometers (μm) thick in order to facilitate handling with minimal breakage, although most solar radiation is absorbed in the first 20–30 μm. The junction in a silicon cell usually is ca 0.2–0.5 μm from the surface of the cell. The crystal surface has many broken bonds that act as recombination centers. In conventional silicon cells, a comb or narrow metal grid lattice is connected to a current-carrying bus to collect charge carriers from the side of the cell facing the sun. The fingers are small enough in total area so that minimal cell area is in their shadow.

Antireflection coatings are used over the silicon surface which, without the coating, reflects ca 35% of incident sunlight. Materials such as titanium dioxide, TiO_2, tantalum pentoxide, Ta_2O_5, or silicon nitride, Si_3N_4, ca 0.08-μm thick are common.

Types of Solar Cells. There are three basic technology options for making solar cells with dozens of variations on each. These approaches are conveniently grouped as follows: thick ($\sim$300 μm) crystalline materials, concentrator cells, and thin ($\sim$1 μm) semiconductor films.

Thick Crystalline Materials. Crystalline silicon technology is the worldwide industry standard. The total cost of solar cells made from ingots reflects the costs of the silicon raw material used in forming an ingot, cutting and etching thin silicon wafers from the ingot, fabricating and encapsulating the cells, and assembling them into modules. An attractive cost-reducing approach is to grow good quality crystalline sheets directly from molten silicon. Smoothly grown sheets ca 100-μm thick require little or no cutting and polishing and incur little waste.

Gallium arsenide is a promising material for gaining the advantages of high efficiency. It is superior to silicon in several respects. The E_g of GaAs, ca 1.4 eV, is higher than that of silicon and is in the range that provides the highest calculated conversion efficiency for a single-junction cell. Because of this high efficiency and the fact that it does not decline as rapidly as that of silicon cells with increasing temperature, GaAs single-crystal cells are attractive for use as concentrator cells.

Gallium arsenide solar cells advanced in the 1980s for space use because they weighed much less than silicon cells of similar output, since GaAs absorbs sunlight much more strongly than silicon.

Concentrator Cells and Systems. Concentrators circumvent the problem of high semiconductor material cost by using mirrors or lenses to concentrate sunlight on small surface areas of more expensive solar cells. Concentration allows more power to be produced from a given amount of photosensitive material.

Concentrator optics vary from low ratio designs, e.g., concentration of sunlight of an order of magnitude by Winston collectors, which do not require elaborate tracking of the sun, to much higher ratio systems based on parabolic mirrors or Fresnel lenses and which require precise, two-axis tracking. Three types of concentrator systems are being developed which operate at low level ($<$30 times), mid-level (100–400 times), and high level ($>$400 times) sunlight concentrations. The cell specifications and engineering requirements for each of these types of systems are quite different. Specially designed silicon has shown potential for use in concentrator systems.

Thin Film. In the thin-film approach, raw material usage is generally more than two orders of magnitude less and patterning is more direct.

Good solar cell results have been obtained from cells of materials, including polycrystalline silicon, amorphous silicon–hydrogen (α-Si:H) alloys, Cu_xS-CdS, CuInSe$_2$-CdS, and CdTe.

Electrochemical Photovoltaic Cells. The application of photoelectrochemistry in solar energy conversion technologies includes biomass conversion, photoelectrolysis, photogalvanic cells, electrochemical photovoltaic cells, etc. In electrochemical photovoltaic cells, electric energy is converted directly from sunlight by absorption of light in a semiconductor electrode. In many respects, these cells closely resemble conventional solid-state cells, except that the charge-separating barrier layer is formed at the interface between a semiconductor surface with a liquid electrolyte. When sunlight is incident on the semiconductor electrode, free holes and electrons are created. The relevant minority carriers must migrate to the interface and be separated; these carriers then react with the electrolyte either through oxidation or reduction. The counterelectrode reverses the reaction, thereby maintaining the electrolyte balance. The semiconductor electrode material may be either polycrystalline or amorphous material because in some cases the poorer material properties cause relatively little degradation of conversion efficiencies. In addition, incorporation of a third electrode may make possible in situ storage. The main disadvantage of these cells is the instability of the semiconductor electrode, especially under sunlight, for extended periods of operation. Electrochemical cells could be inexpensive, since the electrode—electrolyte barriers usually are easy to form, but appropriate deployment strategies have not yet been identified. The stability problems encountered to date have been extensive.

Balance of Systems

A solar photovoltaic system contains, in addition to solar cells and module(s), an array structure to support the modules, power-conditioning circuitry for control and modification of the output, and a means of storing energy if required. All elements beyond the module are referred to as balance-of-system (BOS) components. The cost of BOS items is nominally about equal to the cost of the PV module. However, the BOS fractional cost contribution can vary from one- to two-thirds of the total installed cost of a system, depending on application.

Material Availability and Environmental Impact

Photovoltaic systems must satisfy four principal requirements before solar photovoltaic conversion can provide a significant portion of general energy needs. The system costs must be low enough to be competitive with other means of energy generation, the amount of energy generated during the life cycle of a photovoltaic system must be substantially greater than the energy required to fabricate the system to meet the criteria of a sustainable technology, the materials used in the cells must be available to generate a substantial portion, i.e., at least a few percent of world energy needs, and the fabrication and utilization of the conversion systems should not cause more environmental problems than other competing energy systems.

Silicon is the second most abundant element in the world and is not toxic. Inherent in the use of materials other than silicon for solar cells are challenges of material availability and environmental safety. In terms of production of CdS-based cells, sulfur is abundant, but the world's resources of cadmium, tellurium, selenium, and indium are much less than those of silicon. However, these resources are several orders of magnitude greater than the amount needed to provide photovoltaic power production of 50,000 MW/yr. Similarly, although arsenic is plentiful, the supply of gallium for GaAs cells is limited. However, studies have concluded that the gallium supply also is sufficient for substantial manufacturing scale.

Although photovoltaic conversion is nonpolluting, environmental, health, and safety aspects must be considered, especially with regard to harmful emission and waste products resulting from the production of the solar cell modules. It has been shown that, with proper encapsulation and a proactive recycling program, it should be possible to minimize environmental concerns.

Uses

Solar modules are used to provide power to a broad range of industrial, commercial, and consumer systems and products. Most participants in the PV industry use the following categories to describe the various market segments, which group applications by functional product requirement, system type, sales channel, and client base. These include the following: specialties, e.g., spacecraft circuits, calculator chips, automobile sunroofs, and building facades; industrial power, i.e., telecommunications, warning/signal lights, and remote data gathering; rural and off-grid electrification, e.g., lighting, water pumping and purification, refrigeration, and recreational travel and boating; consumer convenience, e.g., garden and security lighting and small battery charging; and grid-connected power, i.e., distributed grid support and peaking power augmentation.

See also **Photoelectric Effect**; **Solar Energy**; and **Space Stations**.

Additional Reading

Annan, R.H., W.L. Wallace, T. Surek, E. Boes, and L.O. Herwig: *Department of Energy Review of the U.S. Photovoltaic Industry*, Report ST-211-3488, Solar Energy Research Institute, Golden, CO, 1989.

Cody, G.D. and T. Tiedje, in B. Abeles, A. Jacobson, and P. Sheng: *Energy and the Environment*, World Scientific, Teaneck, N.J., 1994.

Day, J. and R.O. Johnson: *Distributed PV Applications, Report PM-36*, Strategies Unlimited, Mountain View, CA, 1992.

Hamakawa, Y.: *Thin-Film Solar Cells*, Springer-Verlag New York, Inc., New York, NY, 2004.

Luque, A. and S. Hegedus: *Handbook of Photovoltaic Science and Engineering*, John Wiley & Sons, Inc., New York, NY, 2003.

Markvart, T.: *Practical Handbook of Photovoltaics: Fundamentals and Applications*, Elsevier Science, New York, NY, 2003.

Markvart, T.: *Solar Electricity*, 2nd Edition, John Wiley & Sons, Inc., New York, NY, 2000.

Marti, A. and A. Luque: *Next Generation Photovoltaics*, Institute of Physics Publishing, Philadelphia, PA, 2003.

Smith, K.: *Survey of U.S. Line-Connected Photovoltaic Systems*, EPRI GS-6306, Palo Alto, CA, 1989.

Staff: Maintenance and Operations of Stand-Alone Photovoltaic Systems, Naval Facilities Engineering Command, Southern Division, rev. 1991.

PHREATIC.   The term proposed by Daubree in 1887 for the waters of the ground water reservoir, as distinct from the underground waters above the water table, called vadose.

PHRENIC NERVES.   The nerves which control the movement of the diaphragm.

PHTHALIC ACID.   [CAS: 88-99-3]. $C_6H_4(COOH)_2$, formula weight 166.13, mp 208 °C (ortho), 330 °C (meta and iso), the ortho form sublimes and the meta and iso forms decompose with heat, sp gr 1.593 (ortho). Phthalic acid is very slightly soluble in H_2O, soluble in alcohol, and slightly soluble in ether. The solid form is colorless, crystalline. Because of their chemical reactivity and versatility, phthalic acid derivatives find wide use as starting and intermediate materials in important industrial organic syntheses. A common starting material is phthalic anhydride which is formed when phthalic acid loses water upon heating. See also **Phthalic Anhydride**; and **Terephthalic Acid**.

Orthophthalic acid is made by the oxidation of naphthalene (1) with H_2SO_4 fuming heated, in the presence of mercuric sulfate—SO_2 is also formed and recovered; (2) with air in the presence of vanadium pentoxide at 450 to 520 °C. Orthophthalic acid also is formed when benzene compounds containing carbon ortho-substituted groups are oxidized. Orthophthalic acid is used in the manufacture of indigo and other dyes.

PHTHALIC ANHYDRIDE.   [CAS: 85-44-9]. $C_6H_4(CO)_2O$, formula weight 148.11, mp 130.8 °C, bp 284.5 °C, sp gr 1.527. Phthalic anhydride is very slightly soluble in H_2O, soluble in alcohol, and slightly soluble in ether. The compound is a high-tonnage chemical and is widely used in a variety of industrial organic syntheses. Although phthalic anhydride may be derived directly from phthalic acid by heating and dehydration, it usually is prepared on a large scale by (1) oxidizing naphthalene, or (2) from the petroleum derivative, orthoxylene. Phthalic anhydride, in addition to its use as a raw and intermediate material for syntheses, finds wide application in the chlorinated form as a compounding ingredient for plastics. The chlorine content is approximately 50%. The compound provides increased stability and improved resistance of plastics to high temperatures.

Representative reactions of phthalic anhydride include: (1) phthalic anhydride reacts with phosphorus pentachloride to form phthalyl chloride which, upon rearrangement, can be transformed to unsymmetrical phthalyl chloride; (2) both forms of phthalyl chloride react with zinc plus acetic acid to form unsymmetrical phthalide, or with benzene plus aluminum chloride to form unsymmetrical-diphenylphthalide (phthalophenone); (3) phthalic anhydride reacts with NH_3 to form phthalimide $C_6H_4(CO)_2NH$; (4) phthalimide reacts with KOH in alcohol to form potassium phthalimide; (5) treatment of potassium phthalimide with an alkyl halide (e.g., ethyl chloride) forms an alkyl phthalimide (e.g., ethyl phthalimide); (6) ethyl phthalimide, when heated with fuming HCl, yields the primary amine $C_2H_5NH_2$ (ethyl amine) in a reaction used for the production of many primary amines and known as Gabriel's synthesis; (7) ethyl phthalimide, when treated with sodium hypochlorite, forms sodium anthranilate which upon treatment with an acid yields anthranilic acid; (8) phthalic anhydride reacts with phenol to form phthaleins, such as phenolphthalein, when in the presence of concentrated H_2SO_4; (9) phthalic anhydride reacts with resorcinol to form resorcinolphthalein (fluorescein); (10) fluorescein reacts with bromine to form tetrabromo-fluorescein, the potassium salt of which is eosin (a red dye for wool and silk); (11) phthalic anhydride reacts with N-diethyl-meta-aminophenol to form N-diethyl-meta-aminophenolphthalein (rhodamine) which is a red dye.

Health and Safety Factors

Phthalic anhydride is a severe irritant to the eyes, respiratory tract, and skin, especially to moist tissue. The solid may burn skin tissue if it is in contact with it for a significant amount of time. Repeated exposure may result in asthma, irritation of mucous membranes, and diseases of the respiratory tract and digestive organs. Contact with skin or the eyes should be followed immediately by washing with large quantities of water.

There are explosion hazards with phthalic anhydride, both as a dust or vapor in air and as a reactant. Water, carbon dioxide, dry chemical, or foam may be used to extinguish the burning anhydride.

See also **Phthalic Acid**; and **Terephthalic Acid**.

PHTHALOCYANINE COMPOUNDS.

Phthalocyanine, [CAS: 574-93-6], $C_{32}H_{18}N_8$, compounds have found widespread acceptance in a variety of applications. The discovery of iron phthalocyanine [CAS: 132-16-1] and the elucidation of its structure led to the commercial application of copper phthalocyanine [CAS: 147-14-8].

Copper phthalocyanine (**1**) was developed in the 1930s and is the most commonly used blue organic pigment in the coatings, paint, and printing inks industry. Phthalocyanine forms complexes with numerous metals. Various complexes with 66 chemical elements are known. Phthalocyanines are structurally related to naturally occurring dyes such as hemoglobin and chlorophyll A.

(**1**)

Properties

The density of β-phthalocyanine, H_2Pc, is 1.43 g/cm^3; β-copper phthalocyanine [CAS: 14832-14-5], CuPc, 1.61 g/cm^3; and polychloro-copper phthalocyanine, 2.14 g/cm^3. The color of most phthalocyanines ranges from blue-black to a metallic bronze, depending on the manufacturing process and the chemical and crystalline form of the material. The colors of the finely divided pigment forms vary from dark blue to green, as phthalocyanines absorb in the visible region at 600–700 μm. Most compounds do not melt but sublime above 200 °C. CuPc can be sublimed without decomposition at 500–580 °C under an inert gas and normal pressure and at 900 °C under vacuum. It decomposes vigorously, however, at 405–420 °C in air and in nitrogen between 460–630 °C. The thermodynamic stability of the five crystalline forms of CuPc increases in the sequence $\alpha = \gamma < \delta < \varepsilon < \beta$. The solubility of most phthalocyanines in water and organic solvents is very low. The α-form, however, is slightly soluble in polar solvents and converts rapidly to the β-form.

The chemical properties of phthalocyanines depend mostly on the nature of the central atom. Phthalocyanines are stable to atmospheric oxygen up to approximately 100 °C. Mild oxidation may lead to the formation of oxidation intermediates that can be reduced to the original products. In aqueous solutions of strong oxidants, the phthalocyanine ring is completely destroyed and oxidized to phthalimide. Oxidation in the presence of ceric sulfate can be used to determine the amount of copper phthalocyanine quantitatively.

Phthalocyanine compounds exhibit favorable catalytic properties which makes them interesting for applications in dehydrogenation, oxidation, electrocatalysis, gas-phase reactions, and fuel cells.

Manufacturing and Processing

Phthalocyanine compounds have been synthesized with various metals. The most important metal phthalocyanines are derived from phthalodinitrile, phthalic anhydride, Pc derivatives, or alkali metal Pc salts.

The route from o-phthalodinitrile [CAS: 91-15-6] can be represented $4C_8H_4N_2 + M \rightarrow MPc$, where M is a bivalent metal, metal halide, metal alcoholate, or an equivalent amount of metal of valence other than two in a 4:1 molar ratio. If a solvent, e.g., trichlorobenzene, benzophenol,

pyridine, nitrobenzene, or quinoline, is used, the reaction takes place at approximately 180 °C. Without a solvent the dry mixture must be heated to ca 300 °C to initiate the exothermic reaction.

The synthesis from phthalimide derivatives, e.g., diimidophthalamide (or phthalimide) [CAS: 85-41-6] is usually carried out in a solvent such as formamide. Metal phthalocyanines may also be prepared using alkali metal salts or from metal-free phthalocyanine by boiling the latter in quinoline with metal salt.

Industrial production of copper phthalocyanine usually favors either the phthalic anhydride—urea process or the o-phthalodinitrile process. Both can be carried out continuously or batchwise in a solvent or bake process of the solid reactants.

Crude copper phthalocyanine must be treated to obtain a satisfactory pigment in regard to the crystal modification and optimal particle size. See also **Pigments (Organic)**. The particle size of crude phthalocyanine can be reduced by chemical or mechanical methods.

The second process to finish phthalocyanine, which is more important for β-copper phthalocyanine, involves grinding the dry or aqueous form in a ball mill or a kneader. Agents such as sodium chloride, which have to be removed by boiling with water after the grinding, are used. Solvents like aromatic hydrocarbons, xylene, nitrobenzene or chlorobenzene, alcohols, ketones, or esters can be used.

Incorporation of less than a stoichiometric amount of alkyl sulfonamides of copper phthalocyanines into copper phthalocyanine improves the pigment's properties in rotogravure inks.

Performance in ink and coatings can be improved by addition of surfactants, dispersants, resins, or copper phthalocyanine derivatives with long aliphatic chains, $CuPc(CH_2-NHR)_3$, to stabilize the pigment in the binder system. Another possibility is wet-milling of aqueous pigment dispersions incorporating an organic medium, e.g., glycols, polyethers, or surfactants.

Some references cover direct preparation of the different crystal modifications of phthalocyanines in pigment form from both the nitrile—urea and phthalic anhydride—urea process. Metal-free phthalocyanine can be manufactured by reaction of o-phthalodinitrile with sodium amylate and alcoholysis of the resulting disodium phthalocyanine. The phthalic anhydride—urea process can also be used. Other sodium compounds or an electrochemical process have been described. Production of the different crystal modifications has also been discussed.

Perchloro- and perchlorobromo copper phthalocyanine are important organic green pigments. They are accessible through direct chlorination of copper phthalocyanine in a eutectic melt of aluminum and sodium chloride or in a chlorosulfonic acid medium. Bromine can be used instead of chlorine in the $AlCl_3$-NaCl melt to obtain polybromochloro copper phthalocyanine.

Phthalocyanine sulfonic acids, which can be used as direct cotton dyes, are obtained by heating the metal phthalocyanines in oleum.

Polymeric phthalocyanines, which possess a higher stability compared to the monomers, can be obtained by combining a phthalocyanine with a polymer. The linking of the polymeric chain can occur at the central metal atom, the phenyl rings, through bridging or attachment to a polymeric chain.

Uses

Approximately 90% of the phthalocyanines (predominantly copper phthalocyanine) are used as pigments. In addition, they have found acceptance in many types of dyestuffs, e.g., direct and reactive dyes, water-soluble and solvent-soluble dyes with physical and chemical binding, azo-reactive dyes, azo nonreactive dyes, sulfur dyes, and vat dyes. See also **Dyes (Textile)**; **Pigments (Inorganic)**; and **Pigments (Organic)**.

Phthalocyanines have interesting properties as catalysts, lasers, semiconductors, lubricants, or as photographic components.

Health and Safety Factors

Phthalocyanines do not pose any significant risk to human health in the environment or the workplace. In several studies, no carcinogenic risk or toxicity to humans was revealed. The FDA approved the use of CuPc in general and ophthalmolic surgery, for contact lenses, and food packaging. Phthalocyanine Blue may be used as a colorant for coatings that are used in manufacturing, packing, processing, preparing, treatment, packaging, transporting, or holding food. The TLV value for CuPc is 10 mg/m^3.

Polychlorinated biphenyls (PCBs) have been detected in pigments manufactured in trichlorobenzene, but not in those made with nonchlorinated solvents. High boiling hydrocarbons or esters are suitable replacements.

Additional Reading

Booth, G.: in K. Venkataraman, ed., *The Chemistry of Synthetic Dyes*, Vol. V, Academic Press, Inc., New York, NY 1971, p. 241.

Lever, A.B.P.: in H.J. Emeleus and A.G. Sharpe, eds., *Advances in Inorganic Chemistry and Radiochemistry*, Vol. 7, Academic Press, Inc., New York, NY 1965, pp. 27–113.

Leznoff, C.C. and A.B.P. Lever, eds.: *Phthalocyanines: Properties and Applications*, VCH Verlagsgesellschaft, Weinheim, Germany, Vol. 1, 1989; Vols. 2 and 3, 1993.

Lide, D.R.: *CRC Handbook of Chemistry and Physics*, 86th Edition, CRC Press LLC., Boca Raton, FL, 2005.

Moser, F.H. and A.L. Thomas: *Phthalocyanine Compounds*, Reinhold Publishing Co., New York, NY 1963.

Nalwa, H.S.: *Supramolecular Photosensitive and Electroactive Material*, Elsevier Science, New York, NY, 2001.

PHUGOID OSCILLATION.   In a flightpath, a long period longitudinal oscillation consisting of shallow climbing and diving motions about a median flightpath and involving little or no change in angle of attack.

PHYCOCOLLOID.   One of several carbohydrate polymers (polysaccharides) occurring in algae (seaweed). They are hydrophilic colloids having a tendency to absorb water, with swelling, and to form gels of varying strength and consistency. The chief types of phycocolloid are carrageenan from Irish moss, algin from brown algae, and agar from red algae. They contain complex galactose and mannose sugars and are sometimes considered seaweeed mucilages.

See also **Carbohydrates**.

PHYLLOXERAN (Phylloxerid).   *Insecta, Homoptera*. A sucking insect related to the plant lice and scale insects. The many species make up a subfamily, which, with the adelgids, constitutes the family *Phylloxeridae*. They differ from the aphids in that all females lay eggs and form the scales in their more complex structure, including the four wings of the winged stages.

The most important phylloxerid is a species that attacks grapevines, working on the leaves and roots. It once threatened to ruin the vineyards of France and has destroyed millions of acres of vines. The use of roots of certain American grapes which are not seriously harmed by the pest has greatly lessened the danger from its attack. Tender varieties are grafted onto the resistant roots.

PHYLUM.   See **Taxonomy**.

PHYSICAL CHEMISTRY.   Application of the concepts and laws of physics to chemical phenomena in order to describe in quantitative (mathematical) terms a vast amount of empirical (observational) information. A selection of only the most important concepts of physical chemistry would include the electron wave equation and the quantum mechanical interpretation of atomic and molecular structure, the study of the subatomic fundamental particles of matter. Application of thermodynamics to heats of formation of compounds and the heats of chemical reaction, the theory of rate processes and chemical equilibria, orbital theory and chemical bonding, surface chemistry (including catalysis and finely divided particles) the principles of electrochemistry and ionization. Although physical chemistry is closely related to both inorganic and organic chemistry, it is considered a separate discipline.

Additional Reading

Atkins, P.W.: *Physical Chemistry*, Oxford University Press, New York, NY, 1978.

Hiemenz, P.C., and R. Rajagopalan: *Principles of Colloid and Surface Chemistry*, Marcel Dekker Inc., New York, NY, 1997.

Hunter, R.J.: *Introduction to Modern Colloid Science*, Oxford University Press, New York, NY, 1993.

Monk, P.M.S.: *Physical Chemistry: Understanding Our Chemical World*, John Wiley & Sons, Inc., Hoboken, NJ, 2004.

Silbey, R.J., M.G. Bawendi, and R.A. Alberty: *Physical Chemistry*, 4th Edition, John Wiley & Sons, Inc., Hoboken, NJ, 2004.

Sun, S.F.: *Physical Chemistry of Macromolecules: Basic Principles and Issues*, John Wiley & Sons, Inc., Hoboken, NJ, 2004.

PHYSICAL DOUBLE STAR.   Two stars in nearly the same line of sight and at approximately the same distance from the observer, as distinguished from an optical double star (two stars in nearly the same line of sight but differing greatly in distance from the observer). If the stars revolve about their common center of mass, they are called a *binary star*. See also **Double Star**; and **Binary Stars**.

PHYSICAL METEOROLOGY.   That branch of meteorology which deals with optical, electrical, acoustical, and thermodynamic phenomena of atmospheres, their chemical composition, the laws of radiation, and the explanation of clouds and precipitation. As generally accepted, it does not include mathematical theory of the motions of the atmosphere and the forces responsible therefore (which matters fall in the field of dynamic meteorology). Also called *atmospheric physics*. Subdivisions of physical meteorology include atmospheric electricity, cloud physics, precipitation physics, atmospheric acoustics, and atmospheric optics.

PHYSICAL OCEANOGRAPHIC REAL-TIME SYSTEM (PORTS®).   The Physical Oceanographic Real-Time System (PORTS®) is a program of the National Ocean Service that supports safe and cost-efficient navigation by providing ship masters and pilots with accurate real-time information required to avoid groundings and collisions. This technological innovation has the potential to save the maritime insurance industry from multi-million dollar claims resulting from shipping accidents. PORTS® includes centralized data acquisition and dissemination systems that provide real-time water levels, currents, and other oceanographic and meteorological data from bays and harbors to the maritime user community in a variety of user friendly formats, including telephone voice response and Internet. Also, PORTS® provides nowcasts and predictions of these parameters with the use of numerical circulation models. Telephone voice access to accurate real-time water level information allows U.S. port authorities and maritime shippers to make sound decisions regarding loading of tonnage (based on available bottom clearance), maximizing loads, and limiting passage times, without compromising safety.

Most ports are located at the mouths of major estuaries, which provide critical habitat for many important biological resources. For example, coastal waters provide nurseries and spawning grounds for 70 percent of U.S. commercial and recreational fisheries. Commercial fishing employs over 350,000 people in vessel- and shore-related fisheries work. An additional 17 million people participate in recreational saltwater fishing, spending $7.2 billion annually. Activities at ports can greatly affect these critical resources; dredging is but one such activity. Each year in the U.S., approximately 400 million cubic yards of dredged material are removed from navigation channels, berths, and terminals.

The PORTS® system is presently operational in the following locations: Narragansett Bay, RI: http://tidesandcurrents.noaa.gov/nbports/nbports.shtml? port=nb. New Haven, CT: http://tidesandcurrents.noaa.gov/nhports/nhports.shtml?port=nh. New York/New Jersey Harbor: http://tidesandcurrents.noaa.gov/nyports/nyports.shtml?port=ny. Delaware Bay, DE: http://tidesandcurrents.noaa.gov/dbports/dbports.shtml?port=db.Chesapeake Bay, MD: http://tidesandcurrents.noaa.gov/cbports/cbports.shtml?port=cb. Tampa Bay, FL: http://tidesandcurrents.noaa.gov/tbports/tbports.shtml?port=tb. Houston/Galveston, TX: http://tidesandcurrents.noaa.gov/hgports/hgports.shtml?port=hg.   Los   Angeles/Long   Beach,   CA: http://tidesandcurrents.noaa.gov/llports/llports.shtml?port=ll.   San   Francisco   Bay,   CA: http://tidesandcurrents.noaa.gov/sfports/sfports.shtml? port=sf. Tacoma, WA: http://tidesandcurrents.noaa.gov/taports/taports. shtml?port=ta. Lower Columbia River, WA: http://tidesandcurrents.noaa. gov/crports/crports.shtml?port=cr. Anchorage, AK: http://tidesandcurrents. noaa.gov/akports/akports.shtml?port=ak.   Soo   Locks,   MI: http://tidesandcurrents.noaa.gov/slports/slports.shtml?port=sl.

Web References

Center for Operational Oceanographic Products and Services: http://tidesandcurrents. noaa.gov/index.shtml.

NOAA Magazine: http://www.magazine.noaa.gov/stories/mag31.htm.

NOAA'S National Ocean Service: http://www.nos.noaa.gov/.

Physical Oceanographic Real-Time System (PORTS®): http://tidesandcurrents.noaa. gov/ports.html.

PHYSIOGRAPHY (or Geomorphology). The description and interpretation of the surface features or topographic pattern of the Earth. The scientific interpretation of scenery. The science of physiography is one of the major subdivisions of the earth sciences. The term is sometimes loosely used as synonymous with geography, hence the recent tendency to use geomorphology in its place. Since the scenery of any region is fundamentally the present stage of its geologic history, it naturally follows that a discussion of the origin of the topographic or scenic features must include not only an account of the processes of erosion and deposition which are now active, or have been active in the region, but also the manner in which the agents of erosion have been affected or controlled by the stratigraphy and structure.

PHYSIOLOGY. A division of biological science that deals with the normal functions of the living body. General physiology is a science that treats of the underlying physical and chemical foundations of vital processes. Physiology in the usual sense is concerned with the more evident vital processes themselves, analyzed to some extent in terms of physics and chemistry.

PHYSIOLOGY (The History). Modern approaches to physiology developed during the second half of the nineteenth century, but philosophical inquiries into the vital functions of animals and plants are very ancient. Major developments in the history of physiology include William Harvey's demonstration of the circulation of the blood in the seventeenth century and Claude Bernard's discovery of internal secretions in the nineteenth century.

Introduction

Anatomy, the study of the structure of living things, and physiology, the study of the functions of living things, were virtually inseparable in ancient times. As originally used by the ancient Greek philosophers, the term physiology encompassed inquiries into the nature of living and nonliving things. Eventually the term physiology was specifically associated with studies of the vital activities of normal, healthy human beings. Today, physiology encompasses studies of molecules, cells, organs, and the organ systems of plants and animals, including human beings. Comparative physiology provides insight into human health and disease, and illuminates broad questions in evolution, ecology, and practical aspects of animal husbandry and agriculture. General categories of physiological research now deal with the transport of materials across membranes; the metabolic activities of cells, including the synthesis and breakdown of molecules; the regulation of these processes; endocrinology; and the activities of the nervous system. Physiology is the parent science of cell biology and biochemistry. See also **Biochemistry (The History)**.

The establishment of cell theory during the first half of the nineteenth century provided a new approach to understanding the vital functions of all living organisms. If the cell was the fundamental unit of structure and function of animals and plants, a new general physiology could encompass the principles that applied to all living things. Although physiologists still wrestled with fundamental philosophical questions, by the second half of the nineteenth century, physiology had become a science based on experimental methods and concepts that often borrowed from chemistry and physics. The focus of contemporary physiology is often at the molecular and cellular levels, but the need to integrate cellular studies into an understanding of the complex functions of the whole organism remains a central problem for physiologists.

Physiological Thought Before the Seventeenth Century

Although philosophers and scientists have always been interested in the relationship between structure and function, anatomical studies could be pursued with only the naked eye and a few simple tools, whereas understanding physiological phenomena depends on inferences drawn from chemistry and physics. However, many fundamental physiological concepts were formulated by the authors of the Hippocratic texts, the ancient Greek philosophers, and Renaissance scientists. Indeed, despite major differences in their knowledge of human anatomy, Aristotle, Galen and Vesalius resorted to many of the same concepts in their attempts to explain the vital functions of the body.

Humoral theory, which served as the basic explanation of human health and disease for hundreds of years, was set forth in the Hippocratic text *On the Nature of Man* and further explicated by Aristotle and Galen. Aristotle believed that teleology (argument from design) was a virtually infallible guide to understanding nature. According to teleological principles, every part of the body is formed for a purpose. Function can, therefore, be deduced from the study of structure.

Like Plato and Aristotle, Galen believed that a form of divine intelligence had created the universe and living beings. Assuming that Nature acts with perfect wisdom and does nothing in vain, Galen argued that every part of the body was crafted for its proper function. Not satisfied with purely anatomical research, Galen sought ways to proceed from structure to function, from anatomical analysis to experimental physiology. His investigations of the functions of the kidneys disproved the theory that urine was formed in the bladder rather than the kidneys. To study digestion, he put pigs on different kinds of diets and examined the contents of their stomachs after appropriate intervals. In addition to providing remarkable descriptions of the brain, spinal cord and nerves, Galen demonstrated the results of injuries at different levels of the spinal cord. By proving that the nerves originate in the brain and spinal cord, rather than the heart, Galen demonstrated that anatomical research could challenge the authority of Aristotle.

In essence, Galenic physiology rests on the doctrine of the 3-fold division of life into vital processes governed by vegetative, animal and rational souls or spirits. Life ultimately depended on air, or pneuma, which was modified by the three principal organs of the body, the liver, heart and brain, and distributed to the body by three types of vessels, the veins, arteries and nerves. Pneuma was adapted by the liver to become the natural spirits, which were distributed by the veins and supported the vegetative functions of nutrition and growth. Vital spirits, which governed movement, were formed in the heart; the arteries distributed innate heat and pneuma or vital spirits to warm and vivify the parts of the body. Innate heat was produced as a result of the slow combustion process that occurred in the heart. The third adaptation occurred in the brain and resulted in animal spirits, which were needed for sensation as well as muscular movements; animal spirits were distributed by the nerves. Galen's explanation for the distribution of blood, air and spirits assumed that the septum of the heart contained pores for the passage of blood. According to Galen, the active phase of heart action was dilation rather than contraction. Dilation of the heart drew air into the body and contraction drove it out. Galenic physiology made it essentially impossible to separate the functions of the circulatory system, the respiratory system, pneuma and spirits.

For hundreds of years, Galen was second only to Hippocrates as a medical authority. His anatomical studies were essentially unchallenged until the sixteenth century and his physiological concepts remained virtually unquestioned until the seventeenth century. Galen's system combined the four humors of Hippocrates with Aristotle's 3-fold division of the spirits and incorporated the pneuma, or great cosmic spirit of the Stoics. Moreover, it is deeply religious in the sense that it is imbued with reverence and admiration for the work of the Creator.

The Scientific Revolution is usually thought of in terms of the great transformation of the physical sciences by Nicolaus Copernicus, Galileo Galilei and Isaac Newton. But developments in anatomy and physiology during the same period also had revolutionary implications. During this period, many anatomists dissected only to supplement their studies of Galenic texts, but close study of Galen's newly restored writings provided the impetus for the reform of anatomy. Indeed, assisting Johannes Guinter in editing Galenic texts was an important step in the scientific awakening of Andreas Vesalius, author of the great anatomical text *On the Fabric of the Human Body* (1543). In contrast to his success in revolutionizing anatomy, however, Vesalius generally followed Galenic concepts in explaining physiological functions. See also **Vesalius, Andreas (1514–1564)**.

As early as the thirteenth century, Ibn an-Nafis had described the pulmonary circulation and dismissed the existence of the Galenic pores in the septum. However, his work was not rediscovered until the twentieth century. Michael Servetus, the first European to directly challenge Galenic assumptions about the movement of the blood, was primarily interested in theological issues. To understand the relationship between God and humanity, and to know the Holy Spirit, Servetus argued, one must understand the spirit of man. This required exact knowledge of human anatomy and physiology, especially the blood and the spirit. Direct observation suggested that the septum did not have the pores required for Galen's scheme. The structure of the heart and the attached blood vessels also raised questions about Galenic physiology. Servetus argued that the blood was sent from the right side of the heart to the lungs for aeration, as well as for the expulsion of sooty vapors, before it returned to the left side of the heart via the pulmonary vein. According to Galen, aeration was the function of the left ventricle, but it was during passage through the lungs that the color of the blood changed. Finally, biblical arguments and physiological observations proved to Servetus that the soul is found in the tides of the blood rather than confined to the heart, or liver, or brain. Satisfied that he had reconciled physiology and theology as to the unity of the spirit, Servetus generally accepted other aspects of the Galenic system.

The pulmonary circulation was also described in *De re anatomica* (1559), a posthumous work by Realdo Colombo, who claimed that he was the first to discover the role of the lungs in the preparation and generation of the vital spirits. Like Aristotle, Andrea Cesalpino believed that the heart was the most important organ of the body. For the Galenic system to work, Cesalpino noted, the lungs and the heart must expand and contract at the same time. Yet it was obvious that we are able to regulate our breath by our will, although we cannot control the heartbeat. Similarly, physicians knew that the pulses and the respiration might be fast or slow, strong or weak, and that such changes in the respiration need not correspond to changes in the pulses.

Observations made in the course of bloodletting led several sixteenth century anatomists to investigate the valves of the veins. When a ligature was tied around the arm in preparation for bleeding, little knots or swellings, which correspond to the valves of the veins, could be seen along the course of the veins. Girolamo Fabrici's demonstration of the venous valves stimulated his student William Harvey to think about the possibility that the blood might travel in a circle. Fabrici suggested that the function of the venous valves was to regulate the volume of blood distributed to the parts of the body so that each could obtain its proper nourishment. Harvey realized that the venous valves actually controlled the direction of the flow of blood in the veins so that venous blood returned to the heart. See also **Harvey, William (1578–1657)**.

Experimental Physiology in the Seventeenth Century

The most significant landmark in the development of modern experimental physiology was the publication in 1628 of William Harvey's *Anatomical Dissertation upon the Movement of the Heart and Blood in Animals*. Despite the reform of anatomy accomplished by Vesalius, medical teaching was still dominated by Galenic physiology when Harvey began to think about the role of the heart and the movement of the blood. With arguments based on dissection, vivisection, comparative anatomy, clinical experience, and the works of Aristotle and Galen, Harvey showed that the beat of the heart caused a continuous circular motion of the blood: from the heart into the arteries and from the veins back to the heart. He proved that the heart is a muscle and that contraction was its most important movement.

Like Aristotle, whom he greatly admired, Harvey asked seemingly simple but truly profound questions about the nature and purpose of things. One of his most original questions could be answered by simple experiments and calculations: how much blood is sent through the body with each beat of the heart? Even the most cursory calculation indicated that the amount of blood pumped out of the heart per hour exceeded the weight of the entire body. That is, if the human heart pumps out 2 ounces of blood with each beat and beats about 70 times per minute; the heart must expel about 600 pounds per hour. The purpose of the motion and contraction of the heart was, therefore, to impart a circular motion to the blood. In principle, Harvey's observations and experiments did not require the use of any materials or instruments that were not available to Aristotle, Galen, Vesalius, and the many hundreds of medical students who had dozed through learned lectures on anatomy.

Despite Harvey's success in dealing with the anatomical and mechanical aspects of the circulatory system, his system was incomplete and controversial. In the absence of the microscope, Harvey could not see the capillaries that join the arteries and veins. To provide a path from the arteries to the veins, he had to postulate hypothetical anastomoses or pores in the flesh. Without any information about the chemical nature of the differences between venous and arterial blood, the relationship between respiration and the circulation remained obscure. The new system did not explain the elaboration and distribution of the spirits, the generation of innate heat, the purpose of the liver, or the difference between arterial and venous blood. Indeed, Harvey's theory of the circulation seemed to raise more questions than it answered. Without the rationale provided by the venerable Galenic synthesis, what principles would guide medical practice? Bloodletting remained one of the fundamentals of medical practice, but the theory of the circulation provoked new arguments about the selection of appropriate sites for venesection. New approaches to therapy and human physiology were, however, raised by Harvey's discoveries. By 1660, his followers were conducting experiments on blood transfusion, quite unaware of the dangers entailed. Experiments on blood transfusion were performed by Christopher Wren, Richard Lower and Robert Boyle in England, and by Jean Denis in Paris. See also **Boyle, Robert (1627–1691)**.

In contrast to Harvey's experimental focus, René Descartes was primarily interested in constructing a philosophical framework for the mechanistic physiology sometimes referred to as iatromechanism. As encapsulated in the well-known phrase, *cogito, ergo sum* (I think, therefore, I am), Descartes constructed his system of philosophy after systematically doubting everything except the existence of mind and matter. Although Descartes acknowledged the importance of observations, his approach to physiology subordinated the crude facts of observation and experiment to the test of reason. The fundamental premise of Descartes' *Principles of Philosophy* (1644) was that all natural phenomena could be explained solely by matter and motion. Critics claimed that Descartes protected himself with a veneer of piety while he created a thoroughly materialistic and mechanistic science. His disciples saw him as the first philosopher to dare to explain all the functions of human beings, even the mind, in a purely mechanical manner. According to Descartes, human beings differed from animals because they possessed a rational soul that served as the agent of thought, will, memory, imagination and reason. Nevertheless, except for ideas, all physiological functions of the human body were also as mechanical as the workings of a clock. Descartes challenged scientists to treat the physical and mental aspects of human beings in the same manner as all other scientific problems.

Although Descartes adopted Harvey's theory of the circulation of blood, he saw the heart as a heat machine rather than a pump and endorsed the ancient idea that the action stroke of the heart was the expansion phase, rather than the contraction. The purpose of the fire in the heart was, therefore, to heat and vaporize the blood so that it could be expelled from the heart. In contrast to the fiery heart, the lung was described as a delicate, soft organ in which the blood vapors were cooled by fresh air so that they condensed and fell drop by drop into the left cavity of the heart.

Even though the nervous system carried out the commands of the rational soul, Descartes provided a mechanical explanation for the nervous system. Direct interaction between the rational soul and the earthly machine occurred in the pineal gland, an unpaired organ that was erroneously thought to be present only in humans. Through conduits in the brain, the animal spirits were able to enter the nerves, which were hollow tubes that incorporated hypothetical valves governing the flow of nervous fluid. Delicate threads along the length of the interior of the nerves connected the brain to the sense organs. The tiniest motion along the thread tugged at the site of the brain where the thread originated and opened pores that allowed the animal spirits to flow into the muscles. Bodily action was, therefore, the result of a reflex arc that began with external stimuli and involved an internal response. Movement of the subtle fluid through the nerves in response to stimulation of the sense organs caused the pineal gland to vibrate, resulting in changes in the emotions and passions. Although the mind could not change bodily reactions to external stimuli directly, it could affect the distinctive pineal vibrations. Thus, external stimuli could cause fear, but the mind could determine whether the reaction would be flight or fight.

Giovanni Alfonso Borelli attempted to apply mechanical principles to the study of muscle action, but unlike Descartes, Borelli saw the heart as a muscular pump rather than a heat engine. By measuring the temperature of the heart and other internal organs during the vivisection of a deer, he proved that the temperature of the heart was not significantly different from that of other parts of the body. During the eighteenth century, Charles Blagden and John Hunter proved that body temperature in healthy human beings was constant at a broad range of ambient temperatures. As a result of these experiments, Blagden concluded that temperature regulation was a fundamental characteristic of life. See also **Borelli, Giovanni Alfonso (1608–1679)**; and **Hunter, John (1728–1793)**.

Borelli believed that muscles increased in bulk during contraction and ascribed this apparent inflation to a sudden fermentation triggered by animal spirits traveling from the brain through the nerves and into the muscles. However, Nicolaus Steno, Jan Swammerdam, Francis Glisson, and others proved that muscles do not increase in volume when contracting. Francis Glisson was particularly interested in the physiological property referred to as irritability and used the term to describe a broad range of phenomena. His doctrine of irritability became very influential, largely through the work of Albrecht von Haller. Other physiologists applied the term to any kind of change in a living organism, whether movement, conformation or growth. Through their pioneering microscopic studies of muscle structure, Steno and Marcello Malpighi refuted the ancient idea that tendons caused motion and that muscles were merely passive, fleshy material. Steno demonstrated that the apparent swelling of a working muscle was caused by the shortening of its fibers. Muscles contracted even when the nerves were dissected and an isolated heart continued to beat without the infusion of new blood or spirits. See also **Glisson, Francis (1597–1677)**; **Haller, Albrecht von (1708–1777)**; and **Malpighi, Marcello (1628–1694)**.

Mechanism and Vitalism: Physiology in the Eighteenth Century

Although Harvey's followers gradually incorporated advances in chemistry and microscopy into their work, physiology remained closely linked to anatomy and medicine. Physiological studies from the most ancient times to the present have been guided implicitly or explicitly by a philosophical framework that has been either mechanistic or vitalistic. Vitalistic philosophy claims that the real entity of life is the soul or vital force and that the body exists for and through a soul or spirit that is incomprehensible in strictly scientific terms. The mechanistic philosophy asserts that all life phenomena can be completely explained in terms of the same physical–chemical laws that govern the inanimate world. The triumph of Newtonian physics is reflected in the mechanistic materialism of the French *philosophes* of the Enlightenment and the mechanical philosophy adopted by many naturalists. Nevertheless, the mechanical philosophy was not necessarily atheistic. Pious naturalists could assume that God was not directly involved in the ordinary motions of the universe, or the normal activities of living beings.

During the eighteenth century, teachers at Europe's leading medical schools adopted new ideas about chemistry, physiology and medicine. One of the best known was Hermann Boerhaave, who taught chemistry, physics, botany and clinical medicine at the University of Leiden. Albrecht von Haller, one of Boerhaave's most devoted disciples, summarized the state of physiology in his eight volume *Elements of Human Physiology* (1757–1766). His presentation was so extensive that François Magendie complained that whenever he thought he had performed a new experiment he always found it had already been described by Haller. In attempting to link anatomical knowledge and physiological research, Haller defined physiology as animated anatomy. Reviving Glisson's concept of irritability, Haller contrasted the irritability of muscle to the sensibility of nerves. The irritable parts of the body were those that contract when touched; the sensible parts were those that conveyed a message to the mind when they were stimulated. Irritability, he concluded, was the definitive characteristic of muscle fibers. Haller postulated the existence of two special forces, the *vis insita* and the *vis nervosa*, associated with muscles and nerves, respectively. See also **Boerhaave, Herman (1668–1738)**; and **Magendie, François (1783–1855)**.

The concept of the body as a machine reached its ultimate expression in the writings of Julien de La Mettrie, author of *Man, the Machine* (1748). La Mettrie discarded Descartes' rational soul and proposed a fully materialistic physiology. Rejecting Cartesian mind–body dualism, La Mettrie argued that even the mind must depend directly on physical and chemical processes. Therefore, like other animals, human beings were machines. See also **La Mettrie, Julien Offray de (1709–1751)**.

Optimism about the explanatory power of the mechanical philosophy began to decline by the middle of the eighteenth century. The mechanical philosophy did not work well when confronted with vital phenomena such as reproduction, development and differentiation, digestion, metabolism, nutrition and growth. Many physiologists came to the conclusion that mechanical explanations for the activities of living beings were impossible and inappropriate. Naturalists argued that it was more useful to reduce complex vital phenomena to simpler components that could be analyzed, or at least described and linked to each other. By the second half of the

eighteenth century, new ideas and methods from chemistry and physics were transforming physiological research.

The Chemical Approach to Life

In contrast to the physiologists known as iatromechanists, iatrochemists and Paracelsians attempted to explain vital phenomena as chemical events. Sixteenth century alchemists had explored the idea that life was a chemical process and that physiology and pathology could be explained in chemical terms. The European alchemist Philippus Aureolus Theophrastus Bombastus von Hohenheim, who is generally referred to as Paracelsus, argued that all physiological processes were fundamentally chemical transformations governed by the archeus, or internal alchemist of the body. Alchemy provided chemical analogies for physiological functions and new approaches to therapeutics. See also **Paracelsus (1493/94–1541)**.

Like Paracelsus, Joannes Baptista van Helmont believed that the workings of the universe could be explained in chemical terms. While striving to reconcile the chemical view of life with a vitalistic philosophical outlook, van Helmont conducted chemical experiments of an exact and quantitative nature. According to van Helmont, all physiological phenomena could be explained in terms of chemical processes that were governed by a series of internal ferments. Digestion, for example, involved a series of conversions carried out by ferments in different organs that transformed food into living flesh. Another form of digestion in the heart and arteries converted the thick, dark blood of the vena cava into lighter, more volatile blood. Finally, each part of the body took the nutrients it needed from the blood and transformed them into its own special components.

Although much of this theory was obscure and confusing, von Helmont's emphasis on fermentation was of considerable interest to other iatrochemists, such as Franciscus Sylvius. According to Sylvius, the chemistry of living things is the same as the chemistry of nonliving things. Digestion, for example, could be explained as a fermentation involving the saliva, the bile, and the pancreatic and gastric juices studied by Regnier de Graaf, René Antoine Ferchault Réaumur, Lazzaro Spallanzani and Edward Stevens. Knowledge of human digestion was not substantially improved until 1833 when William Beaumont published the results of digestion experiments conducted through the gastric fistula of Alexis St Martin. See also **Beaumont, William (1785–1853)**; **de Graaf, Regnier (1641–1673)**; and **Spallanzani, Lazzaro (1729–1799)**.

A very different approach to chemical phenomena eventually led to an understanding of the mechanism of combustion and the physiology of respiration. This was the doctrine of phlogiston, derived from the alchemical theories of Johann Joachim Becher. Georg Ernst Stahl transformed Becher's theory into the doctrine of phlogiston, a new chemical system that seemed to explain many puzzling phenomena. Stahl argued that all chemical changes in the living body were fundamentally different from chemical events in the inanimate world. Phlogiston theory was accepted by some of the most distinguished eighteenth century chemists, but solving the riddle of respiration required the rejection of phlogiston theory and the confluence of studies of the circulation of the blood, the microanatomy of the lungs, and the chemistry of the gases involved in respiration. See also **Biochemistry (The History)**.

The chemical identification of the gases involved in respiration and combustion was made possible by Stephen Hales' invention of the pneumatic trough. In addition, Hales measured blood pressure in animals, analyzed the "water economy" of plants, and suggested that plants obtained some "secret food of life" from the air. The nature of the gases involved in respiration and combustion was determined by a group of scientists known as the pneumatic chemists, including Joseph Black, Henry Cavendish, Carl Wilhelm Scheele and Joseph Priestley. Oxygen was discovered almost simultaneously by Scheele and Priestley, but both believed that this new gas was "dephlogisticated air." See **Black, Joseph (1728–1799)**; and **Priestley, Joseph (1733–1804)**.

After isolating oxygen in 1774, Priestley found that it caused a candle to burn with a very vigorous flame and allowed mice to survive for a longer period than a similar quantity of ordinary air. Priestley discovered that air that had been "damaged" by animal respiration or combustion could be "revivified" by plants. This observation puzzled him because he had assumed that both plants and animals should affect air in the same manner.

Linking the chemical revolution with physiological research, Antoine-Laurent Lavoisier explained previous studies of respiration and combustion in terms of oxidation theory. Lavoisier's observations supported the idea that the same natural laws applied to both living and nonliving entities. In measuring the oxygen consumption of living systems, Lavoisier initiated the experimental analysis of the energetics of living systems. Through his invention of the calorimeter, Lavoisier proved that combustion and respiration consumed oxygen and produced carbon dioxide. Twentieth century refinements of experiments on calorimetry proved that the energy produced by the metabolism of foodstuffs in the animal body was equivalent to the energy produced by the combustion of the same foodstuffs outside the body. These methods were used to determine the relationship between the composition of foodstuffs and their value as sources of metabolic energy. Acceptance of this concept led to the use of measurements of the basal metabolic rate (BMR) in the diagnosis of certain diseases. See also **Calorimetry**; and **Lavoisier, Antoine Laurent (1743–1794)**.

Jan Ingenhousz is best known for his discovery of photosynthesis, the process by which green plants absorb carbon dioxide in the presence of sunlight and release oxygen. He also discovered that, in the dark, all parts of the plant perform respiration and produce carbon dioxide. Thus, Ingenhousz helped to clarify the fundamental similarities and the significant differences between plant and animal life. See also **Photosynthesis**; and **Plant Sciences (The History)**.

Tissue Doctrine and Cell Theory

By the eighteenth century, many anatomists had abandoned humoral pathology and were attempting to discover correlations between localized lesions and the process of disease. Tissue doctrine, elaborated by Marie François Xavier Bichat was one possible answer to the question, what is an organism? Bichat studied the body in terms of organs, which were then dissected and analyzed into their fundamental structural and vital elements, which he called *tissues*. The actions of tissues were explained in terms of irritability (the ability to react to stimuli), sensibility (the ability to perceive stimuli), and sympathy (the mutual effect parts of the body exert on each other in sickness and health). Although Bichat's tissues were obviously complex and compound, even after cell theory had been generally accepted, some of Bichat's followers continued to think of the tissue as the body's natural unit of structure and function. See also **Bichat, Marie-François-Xavier**.

Cell theory, which was established by Matthias Jacob Schleiden and Theodor Schwann, has become an integral part of modern physiology. According to cell theory, the cell is the fundamental unit of life and the body is composed of cells and cell products. Schleiden believed that cells were simultaneously independent entities and integral parts of the plant. Thus, all aspects of plant physiology were fundamentally manifestations of the vital activity of cells. According to Schwann, the cell revealed the basic unity of the plant and animal worlds. Schwann coined the word "metabolic," to describe the chemical changes that took place within the cell and referred to metabolism as a universal property of cells and, therefore, of life. Because of the unity of life forms, Schwann proposed that fermentation in yeast served as a valid model system for studies of the metabolic activities of cells. See also **Schleiden, Matthias Jacob (1804–1881)**; and **Schwann, Theodor Ambrose Hubert (1810–1882)**.

By the end of the nineteenth century, cell theory had been so well established that it seemed possible that all physiology could be considered essentially identical to cell physiology. That is, cells rather than organ systems could be investigated as units of function. Just as studies of organ system led eventually to studies of cell physiology, studies of cellular mechanisms involving biosynthesis, control and inheritance led to investigations addressed to an apparently more fundamental level, i.e., the molecules that make up the cells.

General Physiology

During the nineteenth century, physiology developed into a mature discipline utilizing chemical and instrumental techniques that distinguished it from its origins in anatomy. The founders of general physiology include Claude Bernard in France; Johannes Müller, Justus von Liebig, and Carl Ludwig in Germany; and Sir Michael Foster in England. The philosophical approach and intellectual milieu that nurtured nineteenth century pioneers in physiology differed significantly from country to country. See also **Foster, Michael (1836–1907)**; and **Ludwig, Carl Wilhelm Friedrich (1816–1895)**.

While Müller was primarily interested in the comparative aspects of animal function and anatomy, Liebig and Ludwig are most associated with the innovative application of chemical and physical methods to physiology. Their work provided many useful techniques, such as more precise measurement of muscular action and blood pressure and methods for analyzing the nature of body fluids.

Ludwig was a founder of the physico-chemical school of physiology. In addition to inventing the kymograph and other instruments, Ludwig developed methods of keeping animal organs alive *in vitro* by perfusing them with a solution that mimicked the composition of blood plasma. In collaboration with Henry Bowditch, he formulated the "all-or-none law" of cardiac muscle action, which states that, in response to any stimulus, the heart muscle contracts to the fullest extent or not at all. Ludwig also studied urine production by the kidney, the relationship between nitrogen in the urine and protein metabolism in the whole animal, and the effect of secretory nerves on the human digestive glands. Many of Ludwig's students, including Bowditch and William Welch, became prominent scientists.

Foster introduced educational reforms and established a model physiological research laboratory. Although his own research did not produce outstanding discoveries, his laboratory trained many eminent scientists and his writings on physiology and the history of physiology were highly regarded. In response to growing antivivisectionist activity, Foster helped establish the Physiological Society, the first organization of professional physiologists, and the *Journal of Physiology*. The founders of the American physiological tradition studied in France and Germany as well as England. Henry Newell Martin, who was influenced by Foster, S. Weir Mitchell, who had worked with Claude Bernard, and Henry P. Bowditch, a disciple of Carl Ludwig, established the American Physiological Society in 1887.

Studies of animal chemistry became a prominent aspect of the research program of early nineteenth century scientists such as Jöns Jacob Berzelius, Friedrich Wöhler and Justus von Liebig. In some respects Berzelius retained a vitalist's view of the origin of natural products, but he thought the chemicals of life conformed to atomic theory. His student, Friedrich Wöhler, was the first man to synthesize an organic chemical in the laboratory. Wöhler's synthesis of urea from ammonium cyanate helped to blur the distinction between organic and inorganic compounds.

Liebig believed that chemistry would revolutionize agriculture, industry and nutrition, but he rejected the fermentation studies of Schwann and Pasteur. Ridiculing the idea that yeast was a living organism, Liebig argued that it was the product of fermentation rather than the cause. Liebig believed that only green plants could build complicated organic substances from simple inorganic elements. Animals had to use plant products as fuel and building material. While Liebig assumed that the chemical transformations that took place in living beings were not unique, he did not totally reject the concept of the vital force. Bernard later said that Liebig's attempt to deduce the invisible metabolic phenomena from quantitative analyses of input and output was like trying to deduce what happened inside a house by measuring what goes in the door and out the chimney. See also **Pasteur, Louis (1822–1895)**; and **Yeasts and Molds**.

Claude Bernard, Homeostasis and Regulatory Mechanisms

In the course his research, Claude Bernard illuminated physiological phenomena in new ways, demonstrating that many vital functions might better be seen in terms of chemistry than an animated anatomy. Nevertheless, like his mentor, François Magendie, Bernard was a master of vivisection. Bernard's discoveries included the glycogenic functions of the liver, the function of the pancreatic juices in digestion, the vasomotor nerves, and the nature of the action of curare, carbon monoxide, and other poisons. Perhaps it is surprising that Bernard believed that his demonstration of the glycogenic function of the liver was his most important piece of work. These experiments revolutionized prevailing theories of plant and animal metabolism and led to the concept of the "internal secretions."

Bernard believed that he was the first scientist to insist that complex animals had two environments: an external environment in which the organism lived and an internal environment in which the cells functioned. This was the basis of his well-known dictum: "The constancy of the internal milieu is the condition for free and independent life." The discovery of the adjustments needed to maintain the constancy of the internal environment became a major goal of physiologists. Bernard placed his observations in a theoretical framework known as determinism, that is, faith in the experimental method and its applicability to physiology, the science of life. For Bernard, vitalism and mechanism were both essentially worthless philosophies that led to empty disputes that obstructed scientific progress.

Philosophers of science have argued that Bernard's work provided a provisional resolution to a major philosophical question: can physiological functions be fully explained in physical–chemical terms? Bernard found a way to deal with the fruitless debate between those scientists who expected to reduce physiological functions to chemical reactions and those who argued that physiological phenomena displayed features that could not be explained by physico-chemical terms alone. Bernard limited the questions and concerns appropriate to physiological inquiry and argued that specific physiological processes were governed by the same physical–chemical laws that governed similar processes in inorganic systems. However, certain physiological functions apparently had no inorganic counterparts, and could not be described in terms of the language of inorganic chemistry and physics.

Bernard's work marks an epoch that divides modern physiology, the parent science of cell physiology and biochemistry, from its roots in anatomy. In the United States, Lawrence J. Henderson publicized Bernard's work and emphasized the concept of the constancy of the internal environment as a guide to physiological research. Through his voluminous and popular writings, Walter Bradford Cannon helped popularize ideas about physiological regulatory mechanisms. After his landmark studies of traumatic shock, Cannon investigated the sympathetic division of the autonomic nervous system, carried out pioneering work on endocrinology, and contributed to the newly emerging field of neuroendocrinology. In 1926, Cannon coined the word "homeostasis" to describe the conditions that maintained the constancy of the interior environment. See also **Cannon, Walter Bradford (1871–1945)**; **Henderson, Lawrence Joseph (1878–1942)**; and **Homeostasis**.

Endocrinology

During the twentieth century, endocrinology developed into a major area of research, but it is still closely linked to physiology and medicine. Although the concept of internal secretions arose from Bernard's study of glycogen metabolism, hormones are now typically thought of as products of the endocrine glands that are released into the bloodstream and regulate the activities of distant tissues and organs. Hormones may act on a very specific target site, or may affect the activities of many different organs. See also **Endocrine System**.

In 1902 Ernest Henry Starling and William Maddock Bayliss isolated secretin, which is produced by the small intestine and triggers the release of the pancreatic juices. Bayliss and Starling suggested that a combination of chemical regulation and nervous regulation controlled physiological processes. In 1905, Starling introduced the word "hormone" in reference to the internal secretions of the endocrine glands. The concepts formulated by Bayliss and Starling can be considered the first principles of twentieth century endocrinology. See also **Bayliss, William Maddock (1860–1924)**; and **Starling, Ernest Henry (1866–1927)**.

The characterization of some hormones and the diseases associated with them began during the nineteenth century, although some endocrine disorders, such as diabetes mellitus and goitre, have been known since antiquity. The endocrine or ductless glands were first described by Friedrich Henle in 1842. A crude form of endocrine, or organotherapy, was initiated in 1889 by Charles Brown-Séquard who claimed that he could reverse male ageing with extracts of animal testes. During the 1890s, George Murray used extracts of the thyroid gland to treat myxoedema (hypothyroidism) and Sir Edward Albert Sharpey-Schäfer and George Oliver discovered that extracts of the adrenal glands contained a substance that raised blood pressure. The active principle of the adrenal gland, now known as adrenaline or epinephrine, was isolated by Jokichi Takamine in

1901. In 1914 Edward Kendall isolated thyroxine from thyroid extracts. The discovery of insulin in 1921 by Frederick Banting, Charles Best and J.J.R. Macleod transformed the life of diabetics. Scientists had been searching for this hormone since 1889 when Joseph von Mering and Oskar Minkowski proved that removal of the pancreas in dogs produced diabetes. Although Kendall had isolated cortisone in 1935, it was not until 1949 that Philip S. Hench and his colleagues discovered that a substance produced by the cortex of the adrenal gland was a powerful anti-inflammatory agent that could be used in the treatment of rheumatoid arthritis. Research on the sex hormones increased knowledge of endocrinology and led to practical applications, such as the birth control pill. The availability of nuclear technology after World War II led to new treatments for endocrine disorders, including the use of radioactive iodine to treat hyperthyroidism. Combining radioactive isotopes with antibodies against hormones, Rosalyn Yalow and S.A. Berson in 1960 discovered the basis for radioimmunoassays. This technique allows endocrinologists to measure extremely small amounts of hormones, and makes it possible to diagnose and treat subtle and incipient endocrine disorders. See also **Adrenaline and Noradrenaline**; **Banting, Frederick Grant (1891–1941)**; **Brown-Séquard, Charles Edouard (1817–1894)**; **Henle, Friedrich Gustav Jakob (1809–1885)**; **Hench, Philip Showalter (1896–1975)**; **Insulin**; **Kendall, Edward Calvin (1886–1972)**; **Sharpey-Schafer, Edward Albert (1850–1935)**; **Thyroid Gland**; and **Yalow, Rosalyn Sussman (1921–Present)**.

Homeostasis, or stationary-state regulation, is the result of the action of many hormones acting at different target sites. Complex regulatory mechanisms involving the phenomenon known as feedback allow hormones to influence their own secretion and affect the secretion of hormones that regulate related processes. For example, insulin and glucagon, hormones secreted by endocrine tissues in the pancreas, control the level of glucose in the blood. Adrenocorticotrophic hormone (ACTH), which is released by the anterior pituitary gland, controls the secretion of certain steroid hormones that affect the conversion of amino acids to glycogen. The secretion of ACTH is regulated by a pituitary ACTH-releasing factor and a negative feedback mechanism governed by the concentration of steroids in the blood. Although an increase in steroid concentration normally inhibits the secretion of ACTH, intense nervous stimulation can overcome this negative feedback mechanism.

The classical view of hormones is that they are directly secreted into the bloodstream by endocrine glands and transmitted to their targets via the bloodstream, but the term has increasingly been applied to other regulatory substances. Moreover, comparative endocrinology indicates that hormonal regulation in plants and certain lower animals, which lack a vascular system, occurs by diffusion. Although discrete endocrine glands are not a prominent feature in crustaceans and insects, hormones regulate significant aspects of their development, growth and moulting (or molting). A practical consequence of comparative endocrinology is the identification of hormones that could be useful in controlling the development and growth of insects and other pests. Moreover, research on the physiology of cockroaches by Berta Vogel Scharrer led to the discovery of neuropeptides and helped establish neuroendocrinology as a scientific discipline. In addition to traditional hormones, other mechanisms for chemical regulation include the organizer substances that regulate early embryonic development, growth factors, and the pheromones used as sex attractants by social insects. See also **Hormones**; **Plant Growth Modification and Regulation**; and **Scharrer, Berta Vogel (1906–1995)**.

Elucidating the complex pathways of metabolism and the enzymes, hormones and neurotransmitters that regulate them still constitutes an enormous and incomplete task. The nervous system and the endocrine system are both involved in controlling and coordinating the vital functions of the organism. Increasingly, it is becoming apparent that the two systems interact and complement each other, as well as the immune system. See also **Central and Peripheral Nervous Systems**; and **Neurotransmitters**.

Twentieth Century Physiology and Its Subdisciplines

Twentieth century physiology has retained its traditional integrative or holistic approach to the activities of cells, tissues, organs and the intact organism, but it has also given rise to several related disciplines, such as biochemistry, biophysics and molecular biology, which have a more analytical or reductionist focus. Interestingly, by the end of the twentieth century many of these areas were becoming, like physiology itself, more closely associated with medicine, both diagnostic and therapeutic. Although modern physiology courses and textbooks tend to focus on classical aspects of the field though discussions of the functional organ systems of animals, the emphasis is increasingly on function rather than structure. Comparative and classical studies of physiology have benefited from new research methodologies and may be expected to provide significant insights into fundamental questions about the nature of life itself. In addition, studies in comparative physiology have produced valuable information about pathogens, parasites, insects, plants, and domesticated and wild animals.

In addition to classical studies of cellular metabolism, cell physiologists have made considerable progress in understanding the membrane transport mechanisms that determine the movement and distribution of specific molecules. The physical principles that determine the movement of molecules through passive membranes did not explain the movement of certain ions and specific compounds through the cell membrane. Research on osmotic and ionic regulation in freshwater animals led to the concept of active transport. Analysis of the transport of sodium ions revealed the existence of a specific transport system that was dependent on a continuing input of metabolic energy. Sodium transport is governed by an enzyme known as Na/K-dependent ATPase that is found in the cell membrane.

Understanding the activities and functions of the central nervous system remains one of the great challenges in physiology. Descartes' concept of the reflex was replaced by the nerve impulse hypothesis as an explanation for the rapid response of animals to external stimuli. The concept of the conditioned reflex (conditioned response) is primarily associated with Ivan Petrovich Pavlov, who insisted that the only scientific way to deal with mental phenomena was to reduce them to measurable physiological entities. Critics have called attention to the limitations of Pavlov's work, as well as the impact of his physiological concepts on the treatment of psychotic patients. Developments in the neurophysiology of sensory perception and the storage and integration of information have been closely linked to advances in the information theory used in communications engineering, but neurophysiologists have also investigated sensory functions at the cellular level. See also **Neurochemistry (The History)**; and **Pavlov, Ivan Petrovitch (1849–1936)**.

The concept known as "the integrative action of the central nervous system" encapsulated the work of Sir Charles Sherrington. In the 1890s, Sherrington established the concept of the synapse and described reciprocal innervation and active inhibition. The pioneering work of Sherrington and Edgar Douglas Adrian elucidated the transmission of nerve impulses between the muscles and the central nervous system. By the 1920s, electrical aspects of the nerve impulse were well known, and the work of T.R. Elliott, Henry H. Dale, and others suggested that a chemical transmitter was involved in the action of the nervous system. Otto Loewi's experiments on isolated hearts led to the discovery that acetylcholine was involved in the chemical transmission of nerve impulses. Further insights into the chemical phenomena involved in the transmission of nerve impulses resulted from the work of Alan Lloyd Hodgkin, Andrew Fielding Huxley and John Carew Eccles. During the 1940s, Hodgkin and Huxley developed a theory of the excitation of nerve cells in which the exchange of sodium and potassium ions causes a brief reversal in a nerve cell's electrical polarization, which is known as an action potential. Further investigations of this phenomenon led Eccles to the discovery of the chemical means by which impulses are communicated or repressed by nerve cells. See also **Adrian, Edgar Douglas (1889–1977)**; **Dale, Henry Hallett (1875–1968)**; **Eccles, John Carew (1903–1997)**; **Hodgkin, Alan Lloyd (1914–1998)**; **Huxley, Andrew Fielding (1917–Present)**; **Loewi, Otto (1873–1961)**; and **Sherrington, Charles Scott (1857–1952)**.

One of the most intriguing insights into regulatory control mechanisms is the close association between hormonal regulation and that exerted by the nervous system. Previously, researchers believed that the two processes could be clearly distinguished by the rate at which each causes effects, the duration of these effects, and their extent. That is, the effects of hormones developed slowly, but their influence was prolonged and often quite pervasive. Nerve action was originally explained in terms of rapid

electrical processes, but further investigations revealed that nerve cells also release minute amounts of chemical substances, such as acetylcholine and noradrenaline (norepinephrine), that are also involved in the transmission of nervous impulses. Certain specialized nerve cells, called *neurosecretory cells*, translate neural signals into chemical stimuli. The neurohormones produced by these cells, which are usually polypeptides, are transmitted through the axons of the nerve cells. The neurohormones are stored in specialized areas known as neurohaemal organs, where the endings of the axons are in close contact with the capillary bed, or released into the bloodstream. In some cases, neurosecretory nerve endings are so close to their target cells that transmission through the bloodstream is unnecessary.

Based on a lifetime of physiological research, Claude Bernard concluded that "there is only one way to live, only one physiology of all living things." Following the pathways illuminated by Bernard and others, twentieth century physiologists and biochemists have been remarkably successful in elucidating universalities at the molecular level of organization. At the beginning of the twenty-first century, physiologists continue to incorporate ideas and methods from many different fields and venture into new fields, such as physiological genomics, in their search for the "physiology of all living things."

Additional Reading

Coleman, W., and F.L. Holmes: *The Investigative Enterprise: Experimental Physiology in Nineteenth-century Medicine*, University of California Press, Berkeley, CA, 1988.

Fye, W.B.: *The Development of American Physiology: Scientific Medicine in the Nineteenth Century*, Johns Hopkins University Press, Baltimore, MD, 1987.

Geison, G.L.: *Michael Foster and the Cambridge School of Physiology: The Scientific Enterprise in late Victorian Society*, Princeton University Press, Princeton, NJ, 1978.

Hall, T.S.: *History of General Physiology 600 BC to AD 1900*, 2 Vols. University of Chicago Press, Chicago, IL, 1975.

Holmes, F.L.: *Claude Bernard and Animal Chemistry. The Emergence of a Scientist*, Harvard University Press, Cambridge, MA, 1974.

Lawrence, C., and G. Weisz: *Greater than the Parts: Holism in Biomedicine, 1920–1950*, Oxford University Press, New York, NY, 1998.

Lenoir, T.: *The Strategy of Life: Teleology and Mechanics in Nineteenth Century German Biology*, D. Reidel, Boston, MA, 1982.

Medvei, V.C.: *A History of Endocrinology*, Parthenon Publishing, Carnforth, UK, 1993.

Rothschuh, K.E.: *History of Physiology*, Robert E. Krieger, New York, NY, 1973.

Seeley, R.R., P. Tate, and T.D. Stephens: *Anatomy and Physiology*, 8th Edition, The McGraw-Hill Companies, New York, NY, 2007.

Wolfe, E.L., A.C. Barger, and S. Benison: *Walter B. Cannon, Science and Society*, Boston Medical Library in the Countway Library of Medicine, Harvard University Press, Cambridge, MA, 2000.

Lois N. Magner, Purdue University, West Lafayette, West Lafayette, IN

PHYTOCHEMISTRY. That branch of chemistry dealing with (1) plant growth and metabolism and (2) plant products. The former includes the absorption of inorganic nutrients (nitrogen, phosphorus, potassium, carbon dioxide, water, etc.) to form sugars, starches, proteins, fats, vitamins, etc. and is closely associated with photosynthesis. Plant products comprise a vast group of natural materials and chemicals; besides those used directly as foods, these include alkaloids, cellulose, lignin, dyes, glucosides, essential oils, resins, gums, tannins, rubbers, terpene hydrocarbons, and glycerides (fats and oils). Some of these are basic raw materials for industry (paper, pharmaceuticals, food, paint, perfume, flavoring, leather, rubber); there are also many miscible plant products such as drugs, poisons, and pigments. Phytochemistry also embraces the study of plant hormones or growth regulators (auxin, gibberellin, synthetic types).

Additional Reading

Conn, E.E.: *Opportunities for Phytochemistry in Plant Biotechnology*, Perseus Publishing, Bolder, CO, 1988.

Robins, R.J., and F.A. Tomas-Barberan: *Phytochemistry of Fruits and Vegetables*, Oxford University Press, New York, NY, 1997.

Romeo, J.T.: *Phytochemical Signals and Plant-Microbe Interactions: Recent Advances in Phytochemistry*, Kluwer Academic Publishers, Norwell, MA, 1998.

PHYTOCLIMATE. 1. The climatic characteristics of the air spaces occupied by plant communities within the canopy. 2. Description of climate characteristics of a region defined largely by distribution of plant species.

AMS

PHYTOCLIMATOLOGY. 1. The study of the microclimate in the air space occupied by plant communities within the canopy, on the surfaces of the plants themselves, and, in some cases, in air spaces within the plants. 2. The study of climatic regions defined largely by distribution of plant species. See also **Microclimate**.

AMS

PHYTOPLANKTON. See **Ocean Resources (Living)**.

Pi. Consider the differential equation

$$\frac{d^2x}{dt^2} = -x,$$

with initial conditions

$$x(0) = 1, \left.\frac{dx}{dt}\right|_0 = 0,$$

which describes a simple oscillating system in one of its extreme positions. The time taken for it to reach its opposite extreme position (that is, the smallest positive value of t for which $dx/dt = 0$) is denoted by the Greek letter π, called pi and approximately equal to 3.14159. It is easy to show that π is also equal to the ratio of the circumference of a circle to its diameter. See also **Circle (Geometry)**.

Interesting background information on efforts made to calculate the value of pi out to hundreds of decimal places is given in *Science*, **193**, 836 (1976). As of 1987, pi had been computed to an unprecedented level of accuracy—more than 100 million decimal places. It is interesting to note that early in the present century, an Indian mathematical genius (Srinivasa Ramanujan) developed ways of calculating pi with extraordinary efficiency. Ramanujan's methods are now incorporated in computer algorithms, thus yielding pi in millions of digits. Detail can be found in article by J.M. and P.B. Borwein *Sci. Amer.*, 112–117 (February 1988).

Pi BOND. A covalent bond formed between atoms by electrons moving in orbitals that extend above and below the plane of an organic molecule containing double bonds. A double bond consists of one pi and one sigma bond, and a triple bond consists of one sigma and two pi bonds.

See also **Metallocenes**; and **Orbitals**.

PICARD METHOD OF SUCCESSIVE APPROXIMATIONS OR ITERATION. A numerical method for solution of a differential equation. If the given equation is $y' = f(x, y)$ subject to the condition that $y = y_0$ when $x = x_0$, the solution may be written in the form of an integral equation

$$y = y_0 + \int_{x0}^{x} f(x, y)\, dx$$

An approximate solution is

$$y_1 = y_0 + \int_{x0}^{x} f(x, y_0)\, dx$$

Iteration yields more exact solutions:

$$y_2 = y_0 + \int_{x0}^{x} f(x, y_1)\, dx; \ldots$$

$$y_n = y_0 + \int_{x0}^{x} f(x, y_{n-1})\, dx$$

PICEA. See **Spruce Trees**.

PICHÉ EVAPORIMETER. A porous paper wick atmometer. The instrument consists of a graduated tube, closed at one end, that is filled with distilled water and then covered with a larger circular piece of filter paper held in place by a disc and collar arrangement. In operation the instrument is inverted so that the distilled water is in contact with the filter paper. The amount of evaporation that occurs during an interval of time is determined by noting the change in level of the meniscus of the water.

See also **Atmometer**; **Dines Anemometer**.

PICIFORMES (*Aves*). This order of birds are more or less closely adapted to arboreal life (inhabiting or frequenting trees); however, of the 6 families included in the order, only the true woodpeckers are real tree climbers, and have stiff supporting tails. The length ranges from 8 to 60 centimeters (3 to $23\frac{1}{2}$ inches), and the weight is between 6 and 300 grams (0.2 and 10.5 ounces). The beak is strong; it is especially powerful and colorful in the toucans, while the true woodpeckers have a chisel-shaped beak. These birds have 2 toes directed toward the front and 2 (the first and fourth) directed toward the rear. The members of this order have various skeletal features in common. There are 14 cervical vertebrae. All of the thoracic vertebrae are unattached, and there are 5 complete ribs. There are also four notches on the rear edge of the sternum. These birds have other common features in their musculature, digestive system, and feather pattern. The woodpeckers' food consists of insects, fruits, seeds, and, for the honey guides, beeswax (the latter being a unique diet, at least among the birds). The white eggs are laid in holes. The young are blind when they hatch, and in most species they are naked as well. There are a total of 383 species, distributed over the whole world, with the exception of Madagascar, Australia, New Zealand, and the South Sea Islands.

Several structural characteristics allow us to separate the 6 families within this order into two suborders. (1) The Jacamars (*Galbuloidea*) have a syrinx that is expanded into a drum. The preen gland is bare, and the appendices are well developed. There are 2 carotid arteries. These birds breed in self-excavated ground holes in South and Central America. There are two families: the jacamars (*Galbulidae*), with 15 species (Fig. 1); and the Puffbirds (*Bucconidae*), with 31 species. (2) The Woodpeckers (*Picoidae*) do not have a drum-like syrinx; there are no appendices, and the preen gland is usually covered with feathers. Only the left carotid artery is present. The nestlings have ankle swellings. Almost all of the members of this suborder breed in tree holes, although a few families (honey guides) are brood parasites. These birds are distributed over all parts of the world, with the exception of Australia. There are four families: the barbets (*Capitonidae*), with 76 species; the honey guides (*Indicatoridae*), with 17 species; the toucans (*Ramphastidae*), with 40 species; and the woodpeckers (*Picidae*), with 209 species. See also **Woodpeckers and Toucans**.

PICKEREL. See **Pike**.

Fig. 1. The jacamars (family *Galbulidae*) are slim, middle-sized tree birds. The beak is fine and curved slightly downward. The short feet have two climbing toes. The first segments of the second and third toe are fused, and these two toes are directed forward, while the first and fourth toes are directed toward the rear. The plumage generally is loose and has a metallic green iridescence that is reminiscent of the plumage of many hummingbirds. The preen gland is bare, and the tongue is long and thin. The contour feathers have a short secondary shaft. The short wings have ten primaries. There are 10 to 12 tail feathers. The great jacamar is shown here.

PICKLING. (1) Removal of scale, oxides, and other impurities from metal surfaces by immersion in an inorganic acid, usually sulfuric, hydrochloric, or phosphoric. Rate of scale removal varies inversely with concentration and temperature; the usual concentration is 15% at or above 100C. The rate is also increased by electrolysis. (2) A method of food preservation involving use of salt, sugar, spices, and organic acids (acetic). (3) Preserving or preparing hides for tanning by immersion in a 6–12% salt solution, together with enough acid to maintain pH at 2.5 or less.

PICKUP (Sensing). A device that converts a sound, scene, measurable quantity and other forms of intelligence into corresponding electric signals, as in a microphone, phonograph pickup, and television camera. A pickup is a transducer only when energy conversion is also involved, as in a microphone or phonograph pickup.

PICOLINES. See **Pyridine and Derivatives**.

PICONET. See **Bluetooth**TM **Wireless Technology**.

PICOZIP. See **Data Compression**.

PICTORIAL REPRESENTATION. A method of representation based upon oblique representation, in which the effect of perspective representation is obtained. In the practical application of pictorial representation, three principal axes are selected, and the actual lengths of the edges of the object are laid off along these axes, resulting in a drawing that is not correct from the standpoint of either orthographic or perspective representation. Since the relative proportions of the object are retained, however, the differences do not detract from the value of the representation. In some forms of representation, however, one or more edges may be drawn to a reduced scale, termed scale reduction, to more closely simulate perspective representation.

See also **Orthographic Projection**; **Perspective**; and **Photography and Imagery**.

PIDDOCK (*Mollusca, Lamellibranchiata*). A bivalve mollusk that bores in soft rock and floating wood. Especially a European species of the genus *Pholas*, commonly used as bait and in some localities regarded as a delicacy. The family to which these animals belong is near that containing the shipworm. Also spelled piddick.

PIERCE OSCILLATOR. An oscillator that includes a piezoelectric crystal connected between the input and the output of a three-terminal amplifying element. Very similar to a Colpitts oscillator.

PIEZOCHEMISTRY. Study of reactions occurring at very high pressures, e.g., in the interior of the earth's crust. See also **Earth**.

PIEZOELECTRIC EFFECT. The interaction of mechanical and electrical stress-strain variables in a medium. Thus, compression of a crystal of quartz or Rochelle salt generates an electrostatic voltage across it, and conversely, application of an electric field may cause the crystal to expand or contract in certain directions. Piezoelectricity is only possible in crystal classes which do not possess a center of symmetry. Unlike electrostriction, the effect is linear in the field strength.

The directions in which tension or compression develop polarization parallel to the strain are called the piezoelectric axes of the crystal. Thus the axis of a hexagonal quartz crystal indicated by the arrows in Fig. 1 is known as an "X-axis," and a plate cut, as shown, with its faces perpendicular to this direction is an "X-cut"; while one cut with its faces parallel to the lateral faces of the crystal is a "Y-cut."

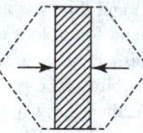

Fig. 1.   Hexagonal quartz crystal showing X-axis.

The magnitude of the piezoelectric polarization is proportional to the strain and to the corresponding stress, and its direction is reversed when the strain changes from compression to tension. The principal piezoelectric constants of a crystal are the polarizations per unit stress along the piezoelectric axes. While these constants are much greater for Rochelle salt than for quartz, the latter is better adapted to some purposes because of its greater mechanical strength. It is also stable at temperatures over 100 °C.

If a quartz plate is subjected to a rapidly alternating electric field, the inverse piezoelectric property causes it to expand and contract alternately. As an elastic body, the plate has a certain natural frequency of expansion and contraction in the direction of the field, and if the field is made to alternate with the same frequency, the plate responds with a vigorous resonant vibration. This reacts, through the direct piezoelectric property, to augment the electric oscillations. A circuit arranged for this purpose, as in Fig. 2, is known as a piezoelectric or crystal oscillator, the crystal itself, P, being the piezoelectric resonator; T is the oscillation transformer, and C a variable condenser. This device has been much used as a frequency control in radio transmitters. Both X-cut and Y-cut quartz plates are subject to changes of frequency with temperature, due to change of elastic modulus; but certain planes in the crystal have been found, oblique to both X and Y, such that plates cut parallel to them are nearly free from the temperature effect.

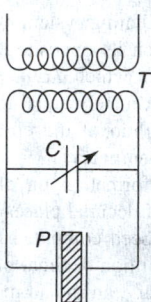

Fig. 2.   Piezoelectric oscillator circuit.

See also **Accelerometer**; **Acoustics**; and **Microphone**.

In addition to natural quartz, Rochelle salts, and tourmaline, synthetic crystals, such as ethylenediamine tartrate (EDT), dipotassium tartrate (DKT), and ammonium dihydrogen phosphate (ADP) have varying suitability as piezoelectric elements. While Rochelle salt has a greater piezoelectric effect than any other crystal, it has the disadvantage of a greater sensitivity to temperature change than quartz. EDT has an advantage over quartz when used in frequency-modulated oscillators because of the wide gap between its resonant and antiresonant frequencies. See also **Quartz**.

PIEZOELECTRICITY. Electric energy created by application of pressure to ceramics of plastics. Devices utilizing this phenomenon are gas flame igniters, ultrasonic welding tools, and sonar navigation aids.

See also **Piezoelectric Effect**.

PIEZOMETER RING. A hollow ring surrounding a pipe to which it is connected by several symmetrically spaced small holes so that the pressure in the ring is the average of the various values obtained at the holes in the pipe. A piezometer, or other pressure-measuring device, is connected to the ring to measure this average pressure.

PIG. See **Swine**.

PIGEONS AND DOVES (*Aves, Columbiformes*). The birds of this order are almost exclusively characterized by the name pigeon or dove, although a few related species of the Old World are known as sand grouse. The extinct dodo also belonged here.

Pigeons and doves (Fig. 1) have the beak swollen at the tip and covered with soft skin at the base, about the nostrils. In North America the group is represented by the band-tailed, *Columba fasciata*, and red-billed pigeons, *Columba flavirostris*, of the western part of the continent, and by several

Fig. 1. Mourning dove. Soft olive-brown above, buff-gray below; white tips to outer tail feathers.

species of doves of similar distribution. The turtle dove, *Streptopelia*, is the only widely distributed species, although the extinct passenger pigeon, *Eclopistes migratorius*, remains a memory of one of the most abundant and widely distributed birds. Numerous other species are found in all of the faunal regions of the world. They bear the names dove or pigeon, which have no exact scientific distinctness, with the exception of an Australian species called the wongawonga. The dodo of Mauritius and the solitaire of Rodriguez Island, both now extinct, were giant flightless pigeons. The last of these birds disappeared in the late seventeenth and in the eighteenth centuries, respectively. See Fig. 2.

Fig. 2.   Model of a dodo. (*A.M. Winchester.*)

The sand grouse comprise a small group of birds related to the pigeons, but in some ways resembling game birds. They are found chiefly in Africa and Asia, but extend to Europe and Madasgascar. As the name suggests, they frequent open ground. Their flight is powerful, hence some species migrate over considerable distances.

See also **Columbiformes**; and **Poultry**.

PIG IRON. Product of blast-furnace reduction of iron oxide in the presence of limestone. About half the ore is converted to iron. Average analysis is 1% silicon, 0.03% sulfur, 0.27% phosphorus, 2.4% manganese, 4.6% carbon, balance iron. Pig iron is the basic raw material for steel and cast iron. In metal terminology a "pig" is a bar or ingot of cooled metal.

PIGMENTATION (Animals). The accumulation of colored materials in living things, which is partly or wholly responsible for the characteristic coloration of different species. Pigments also serve special purposes in the body. In these functions the presence of color may be entirely incidental to the chemical composition of the material.

Pigments are important in visual organs. Here impervious black or dark brown deposits insulate the sensitive nerve endings against all light except that which is transmitted by the lens or cornea. In the eyes of some arthropods the pigment is redistributed to admit more light when the surrounding illumination is dim than when it is bright. A similar

result is gained in the vertebrate eye by the muscular adjustment of the pigmented iris to change the size of the pupil through which light is admitted. Still another pigment, the visual purple (rhodopsin) is found in the rods of the retina of the vertebrate eye. It is bleached by light and resumes its color in darkness; it is associated with the sensitivity of the eye.

Pigments in the superficial layers of the body are also useful in some animals, independently of the relations discussed under coloration. A familiar example is the protective pigment deposited in the human skin as a protection against ultraviolet light. The deposition normally follows excessive exposure to sunlight or to other sources of ultraviolet, and the deposits are lessened when exposure is reduced. These deposits are in the form of granules of melanin in the cells of the innermost layer of the epidermis and in branching cells called melanoblasts in the underlying dermis. The pigment loses its granular form and becomes diffuse as the epidermal cells move toward the surface. Melanoblasts are possibly active in the formation of the pigment granules. The pigmentation of hair is not thoroughly understood but both granular and diffuse pigments have been reported.

A definite relation also exists between the normal illumination of the body and pigmentation in other animals, but the nature of the relation is not always known and a definite value to the animal need not exist. A familiar example is the dark upper surface and light lower surface of fishes, whether the upper surface is dorsal, as in most species, or lateral, as in the flatfishes. Lack of pigment in fishes of subterranean waters is closely associated.

Pigmentation of insects has been shown in several cases to respond to light. Lessened illumination may result in deeper colors, and some observers have secured the same result by moderate increase of light. Extreme changes, however, have resulted in decreased pigmentation in some experiments. Humidity and temperature also affect the depth of pigmentation in some insects, and may modify the pattern.

Incidental colors like the pink flush of human skin results from the presence in the body of the respiratory pigments, hemocyanin and hemoglobin, and waste products. Protein wastes deposited in the superficial tissues of insects are one source of color and the bile pigments of vertebrates, formed by the liver in the modification of hemoglobin, are a source of color in some organs.

The entire subject is closely associated with the chemistry of the living organism on the one hand and with coloration and mimicry on the other.

PIGMENTATION (Plants). The distinctive green color of leaves and other plant organs results from the presence in such organs of two pigments called chlorophyll *a* and chlorophyll *b*. In the higher plants these pigments occur only in the chloroplasts. These pigments play so important a role in the fundamental process of plant life, photosynthesis, that their chemical reactions are discussed at length in that entry. The chlorophylls are not water-soluble but can be readily dissolved out of leaf tissues with alcohol, acetone, ether, or other organic solvents. The resulting solutions exhibit the phenomenon of fluorescence; they are deep green when held between an observer and the light, but deep red when viewed in reflected light. By suitable treatments it is possible to obtain pure crystals of chlorophyll from such solutions. Most leaves contain considerably more chlorophyll *a* than chlorophyll *b*, often two to three times as much. In the organs of the higher seed plants, with rare exceptions, chlorophyll is synthesized only upon exposure to light. Leaves of grass that develop under a board, for example, contain no chlorophyll. In the leaves of mosses, ferns, and gymnosperms, however, chlorophyll develops in the dark as well as in the light.

Invariably associated with the chlorophylls in the chloroplasts are the yellow pigments, the carotenes and the xanthophylls. These pigments are not, however, restricted in their occurrence to the chloroplasts, but may also be present in nongreen parts of the plant where they commonly occur in chromoplasts. Collectively, these pigments, together with certain others which are closely related chemically, are called the carotenoids. Carotene refers to a class of orange-yellow pigments. They are especially abundant in the roots of carrots. These compounds are of considerable importance because they are the precursors of vitamin A, one molecule of β-carotene being split into two molecules of vitamin A by a simple hydrolytic reaction.

Lycopene, a red pigment of this class, is responsible for the red color of the fruits of tomato, pepper, rose, and some other species. The commonest xanthophylls found in leaves are lutein and zeaxanthin, although others also occur. Another xanthophyll is fucoxanthin, which imparts to brown algae their distinctive color. None of the carotenoids is water soluble, but all of them can be extracted from plant tissues with suitable organic solvents.

Most of the red, blue, and purple pigments of plants belong to the group of anthocyanins. In general, the anthocyanins are red in an acid solution and change in color through purple to blue as the solution becomes more alkaline. Red pigmentation resulting from the presence of the anthocyanins is found in flowers, fruits, bud scales, young leaves and stems, and sometimes even mature leaves as in those of the red cabbage. Blue and purple pigmentation due to the presence of anthocyanins occur principally in flowers and fruits. The anthocyanins are diglucosides of the compounds pelargonidin, cyanidin, delphinidin and apigenidin. These compounds are closely similar in structure, all having the double ring benzopyrylium.

Another group of cell sap water-soluble pigments is the *anthoxanthins*. These pigments are also chemically related to the glucosides. Anthoxanthins often occur in the plant in a colorless form but under suitable conditions their typical yellow or orange color becomes apparent. Some yellow flowers, such as yellow snapdragons, owe their color to the presence of anthoxanthins, but the color of the majority of kinds of yellow or orange flowers is due to carotenoid pigments.

The autumnal coloration of leaves in temperate regions is one of the most spectacular accompaniments of the march of the seasons. Both carotenoid and anthocyanin pigments play an important role in autumnal leaf coloration which is not, contrary to popular opinion, a result of the action of frost. Brilliant development of the anthocyanin pigments in the fall is, however, favored by dry weather during which cool, but not frosty, nights alternate with clear days. During the late summer and early fall the chlorophyll in the leaves gradually decomposes. In many species this simply results in unmasking the yellow carotenoid pigments already present, accounting for the yellow autumnal pigmentation of such species as birch, sycamore, aspen, and tulip trees. In other species synthesis of anthocyanins occurs more or less concomitantly with the disintegration of the chlorophyll; this accounts for the reds or purplish reds characteristic in the autumnal coloration of such species as many oaks, maples, sumacs, and dogwood.

Except in flowers, white is an uncommon color in the externally visible parts of plants, and results from the complete absence of pigments. In some species white streaks or other markings are of common or regular occurrence in leaves, and in the leaves of some species such as roses completely white leaves or even entire branches bearing only white leaves sometimes occur. Such branches cannot be propagated because no photosynthesis can take place in the absence of chlorophyll. As long as such branches remain attached to a plant bearing green leaves they can obtain necessary food from the branches bearing normally pigmented leaves.

See also **Annatto Food Colors**; **Carotenoids**; **Chlorophylls**; **Colorants (Food)**; and **Photosynthesis**.

PIGMENT DISPERSIONS.

A pigment dispersion in a concentrated form is a uniform distribution of very fine color pigment particles in a suitable medium or carrier. Such a dispersion is normally used for applying color to the surface of a substrate, such as an ink film on paper or a paint film on a steel surface. It is also used for mass coloring, as in the case of plastics. Considering the high cost and specialized equipment in its preparation, a dispersion is manufactured in relatively small batches in highest concentration of pigment. The concentrate made in such a manner is usually diluted, reduced, or extended to produce the finished product.

Dispersion

Organic and inorganic pigment powders are finely divided crystalline solids that are essentially insoluble in application media such as ink or paint. The carrier used for dispersion of a pigment is usually a liquid or solid, such as a polymer, that is deformable at the processing conditions of high temperature and/or shear. The color strength of the dispersed pigment increases markedly with decrease in particle size. Optimum color strength from a given pigment in practice requires a mean particle size of the order of 0.1 μm or less, which is half the wavelength of the light involved. Therefore, the dispersion process involves size reduction of the pigment particle to the smallest practical size, reasonably complete wetting of its solid surfaces by the carrier, and stabilization of the resulting dispersion.

Because the intensity and color strength of pigments are largely dependent on the exposed surface, it is desirable to reduce the particles to primary particle size. This is the size of the solid pigment crystals as they are precipitated in their synthesis. In practice, the size reduction processes are limited by the nature of pigment, dispersion system, constraints of the processing equipment, the requirements imposed by the product application, and the overall economics. The maximum aggregate size permissible in a given dispersion system depends on the thickness of the film or the coating. For example, the dispersion used for architectural coatings can tolerate a much larger pigment aggregate than a similar dispersion used for automotive finishes, which requires finer particles. Any dispersion system, however, is expected to contain a very small number of these largest aggregates. Generally, it is important to reduce most aggregates to the smaller size to achieve color strength, gloss, film integrity, and durability.

In a dispersed pigment system, a primary pigment particle refers to an individual crystal and a loosely formed association of the pigment crystals from the manufacturing process. Size reduction beyond primary particle size requires excessive energy, but it also has an adverse effect on the visual properties of the pigment. Generally, the particle size of most organic pigments is much smaller initially by precipitation than optimum primary particles, but the particles tend to grow to a much larger size when their formation is complete.

Organic pigments, such as the azo red and yellow pigments, in the process of striking the color undergo definite crystal growth following their precipitation from the aqueous media. The individual crystals are joined together due to forces on the crystal surfaces to form the aggregate. These are held together as static systems by van der Waals forces. Subsequent processing to recover the pigment product results in the formation of agglomerates, which are large associations of pigment crystals and aggregates. The agglomerates are held together by forces that are much weaker than those present within the aggregates. Typically, agglomerates are joined at the edges and corners in a loose matrix form. It is possible to generate an even larger association of pigment agglomerates or flocculates during further processing. These formations are loosely held together and are usually easy to break down by application of shear. Various surface treatments are used to suppress the formation of large aggregates and, thereby, ease the dispersion process. These treatments range from the classical approach of rosination to additions of a variety of surface-active agents at the synthesis step. However, occasionally large agglomerates, several millimeters in diameter, form during the initial stages of dispersion in a highly viscous system. The commercial processes used in dispersion manufacturing may not fully eliminate the aggregates. However, the design and operation of pigment dispersion equipment is aimed at application of mechanical forces to break down the agglomerates and even some less tightly held aggregates. Ideally, an excellent dispersion should consist mainly of primary pigment particles and few loosely held aggregates.

Wetting of the pigment surface constitutes a critical step in achieving a stable and uniform pigment dispersion. Wetting refers to displacement of adsorbed gases (usually air) on the surface of pigment particles, followed by attachment of a vehicle system to the pigment surface. Since the vehicles used for many dispersion systems are viscous, it follows that the penetration of vehicles to the pigment surface is slow and, hence, aided by external mechanical forces. Thus, the grinding (size reduction) and wetting of pigment are frequently carried out simultaneously. The adsorbed gases are displaced on application of shear, and the action also provides smearing of vehicle on the pigment surface and exposes a new surface for wetting. The system of wetted fine primary pigment particles, must be stabilized to prevent reversal of the dispersion process. It is usually done by surrounding the particles with a protective colloid or buffer, which blocks the reagglomeration action of particles. In some cases, the stabilization is attained by addition of ions to establish similar charges on all particles.

Flushing

Flushing processes are used extensively in preparing organic pigment dispersion concentrates for color printing ink applications. The process can be described as a direct transfer of pigment from an aqueous phase to an oil or nonaqueous phase without drying. When the pigment presscake is mixed with an oil-based vehicle or a carrier, water is separated from the pigment surface and replaced by the vehicle. Most organic pigments demonstrate an affinity for hydrocarbon oils and lend themselves to easy dispersion in oil by the process of flushing. See also **Pigments (Organic)**. Inorganic pigments, on the other hand, have to be treated with cationic surfactants to make their surface lipophilic. The majority of inorganic pigments are usually dried and dispersed as dry powders in the carrier, as opposed to being flushed. Techniques used for dispersion of these pigments are different and should be treated as special cases. See also **Pigments (Inorganic)**.

The process of flushing typically consists of the following sequence: phase transfer; separation of aqueous phase; vacuum dehydration of water trapped in the dispersed phase; dispersion of the pigment in the oil phase by continued application of shear; thinning the heavy mass by addition of one or more vehicles to reduce the viscosity of dispersion; and standardization of the finished dispersion to adjust the color and rheological properties to match the quality to the previously established standard.

Flushing is frequently used for the manufacture of large quantities of a dispersion having a specific pigment in a compatible vehicle system. The flushed products, typically containing 28–40% pigment, offer sufficient flexibility to the formulator to produce the finished offset ink. The flushed products exhibit superior gloss, transparency, and strengths, compared to those produced by dispersing the dried pigment. Flushing is particularly important to dispersions of organic pigments, such as Diarylide Yellow (CI Pigment Yellow 12, CI 21090) and Alkali Blue (Pigment Blue 61, CI 42765) because the drying process is detrimental to the product quality of these pigments.

Equipment

Various types of equipment are used commercially to manufacture dispersed pigment concentrates or finished dispersion products used by printing ink, and the coatings and plastics industry. These include kneaders or internal mixers; close tolerance mills; high speed fluid energy mills; ball and pebble mills; san, bead, and shot mills.

Uses

The formulation of dispersed pigment concentrates is influenced by the manufacturing process, as well as the performance parameters desired in the final application. The finished product in many cases is significantly different in formulation than the concentrate to achieve desired properties. One of the principal factors to be considered is the concentration of pigment in the dispersion concentrate. Compatibility of the carrier (solvent additives, etc) used in the preparation of concentrated dispersion and that used in the finished color product also plays an important role. In some cases this can be difficult because the carriers having the best performance, from the standpoint of processing, could be poor in the application systems. However, in the majority of the applications, particularly in coatings and colored plastics, the concentration of the pigment in the finished product is quite low, and the incompatibility problem is easily overcome.

Generally, the pigment dispersion concentrates are formulated for specific end use. They can be supplied as flushed pigments, dispersions or pastes for offset inks, chip dispersions for solvent and aqueous inks, and color concentrates for coloring large quantities of plastics. Although it is feasible for the end user to prepare the pigment dispersion concentrates, it is usually more cost effective and technologically advantageous to manufacture these dispersions by the pigment manufacturers of specialty dispersion houses. Three significant areas of application for concentrated dispersions, ie, printing inks, coatings, and plastics.

Additional Reading

Lewis, P.A.: *Pigment Handbook*, Vol. 1, 3rd Edition, John Wiley & Sons, Inc., New York, NY, 1988.

Patton, T.C.: *Paint Flow and Pigment Dispersion: A Rheological Approach to Coating and Ink Technology*, 2nd Edition, John Wiley & Sons, Inc., New York, NY, 1979.

Smith, H. MacDonald: *High Performance Pigments*, John Wiley & Sons, Inc., New York, NY, 2001.

PIGMENTS (Inorganic). Originally derived from colored minerals, inorganic pigments are now highly engineered particles that impart color or functionality to the objects in which they are used. While inorganic pigments are still used in traditional applications, such as paints, ceramics, and cement, their resistance to the effects of radiation, temperature, and chemical attack has also led to use in high technology applications, such as fibers, engineering plastics, and highly durable coatings applied to roofing panels and even space equipment. See also **Enamels, Porcelain or Vitreous**.

Inorganic pigment particles maintain their physical form, unlike dyes that dissolve in the substrate that they color. These particles are typically larger, easier to disperse, and more color stable in different applications than organic pigments. However, organic pigments tend to have higher strength, and often provide more vibrant colors.

The basic technology for making inorganic pigments is in many cases hundreds or thousands of years old. If one were to consider grinding red ochre across a cave wall as creating pigment particles, pigment production may be considered to be tens of thousands of years old, with the proof being the magnificent cave paintings found in France and Spain. Pigment technology has evolved, and within each pigment chemistry, the most cost efficient pigment in terms of particle size, crystal structure, or composition is used to provide the desired color. These pigments can also provide properties other than color, the most common are corrosion protection, interaction with ultraviolet (UV) or infrared (IR) light, and modifying physical properties.

Pigments are designed to interact with light, and synthetic pigments are optimized to the size that optimizes opacity, or provides the strongest absorption as to minimize the amount of pigment needed. For a pigment to be highly effective at either of these tasks, it must be of a size similar to that of the wavelength of light. The wavelength of visible light ranges between 0.4 and 0.7 μ, and pigment particles generally range in particle size between 0.1 and 10 μ.

Inorganic pigments are typically oxides, sulfides, chromates, silicates, phosphates, and carbonates of metallic elements. Most commonly, they contain some transition metal, due to their ability to interact with light. Since this interaction is also affected by crystal structure, the preparation of these pigments must be done under conditions that yield the desired structure.

Some new pigments have been developed due to improved understanding of crystal chemistry, but often the development of these new pigments is driven by environmental regulation of older pigments, or because new applications require a different set of properties than previously existed.

Today's inorganic pigment producers offer a wide range of pigments, from simple oxides, to complex oxide crystals, and even heterogeneous pigments with nano-sized phases incorporated into an inorganic matrix. Although development of inorganic pigment no longer lies in the forefront of chemistry, as it did in late nineteenth century Germany, extensive technology is still required to be a competitive manufacturer.

Historical classification of inorganic pigments into the classification of naturally occurring and synthetically produced is no longer useful, as most are now manufactured synthetica:

These materials are colorants in printing inks, cosmetics, and markers, e.g., crayons. In all these applications the pigments are dispersed, i.e., they do not dissolve, in the media forming a heterogeneous mixture. See also **Pigment Dispersions**. In nature, inorganic pigments contribute to the color of some rocks and minerals. See also **Colorants (Foods)**; and **Paints and Coatings**.

Properties

The value of pigments results from their physical–optical properties. These are primarily determined by the pigments' physical characteristics (crystal structure, particle size and distribution, particle shape, agglomeration, etc.) and chemical properties (chemical composition, purity, stability, etc.). The two most important physical–optical assets of pigments are the ability to color the environment in which they are dispersed and to make it opaque.

The opacity of a pigment lies in its ability to prevent a transmission of light through the medium. White pigments disperse the whole visible light spectrum more effectively than they absorb it; black pigments do the opposite. Color results when pigment particles absorb only certain portions of the visible light spectrum while dispersing the rest of it.

The opacity of pigments is a function of the pigment particle size and the difference between the pigment's refractive index and that of the medium in which pigment particles are dispersed. The multiple light dispersion in the pigment–medium interface results in the appearance that the light is transmitted through a much thicker layer than it actually is. A pigment having a particle size between 0.16–0.28 µm gives the maximum dispersion of the visible light.

The most common measurements of pigment properties comprise elemental analysis, impurity content, crystal structure, particle size and shape, particle size distribution, density, and surface area. These parameters are measured so that pigments' producers can better control production, and set up meaningful physical and chemical pigments' specifications. Measurements of these properties are not specific only to pigments. The techniques applied are commonly used to characterize powders and solid materials, and the measuring methods have been standardized in various industries.

Coloristic properties of pigments are best evaluated by dispersing them into the media they were developed to color, e.g., plastics, glass enamels, glazes, etc. The measured characteristics include color, color strength, opacity, lightfastness, weathering, heat stability, chemical stability, and rheological properties. The dispersing media and the processing conditions can strongly influence the results. Because pigments, as any fine powders, have a tendency to segregate by size during transportation and handling, the use of proper sampling methods is critical for getting meaningful physical and chemical data.

Particle size and distribution are the most fundamental measured properties of powders (see **Size Measurement of Particles**). These properties impact a number of pigment characteristics. Those affected the most are the color, color strength, hiding power, and rheological properties. Actual powders consist of a population of particles of many different shapes. To permit a good description of powder population, a representative sample of the powder must be collected, measured, and the results interpreted using statistical methods. For inorganic pigments to be useful in most applications, they must have an average particle size between 0.1 and 10 µm. A character, i.e., color or pattern, of a substrate becomes obscured when coated with a pigment containing film such as a paint or a ceramic glaze. The degree of the obscuration (opacity) depends on the amount and type of pigment used and the thickness of the applied film.

The ability of a coating to hide the substrate is called its hiding power. Hiding power of a uniform coating is expressed as the area of substrate that can be hidden by a unit volume of the coating (ft^2/gal or m^2/L).

The ability of a pigment to change the color of an opaque film is known as its tinting strength. Both the hiding power and tinting strength are the fundamental pigment properties. Hiding power and tinting strength can be determined visually or instrumentally.

Color matching is a process in which a technician prepares a formulation, i.e., a mixture of pigments in a desired medium, that has the desired color effects. A good color match in one medium, e.g., plastic, is not always a good match in another medium, e.g., ceramic glaze.

Experienced color matchers can achieve a good color match by trial and error without using any instrumentation. In some cases, however, this technique can be a lengthy process. To get the most cost-effective match in the shortest possible time, the use of a computer color matching system is preferable.

Many pigments, when exposed to high intensity light such as direct sunlight or uv lamp, can get darker, change their shade or lose the color saturation. The color and its saturation change mainly for organic pigments. Inorganic pigments, particularly those containing ions that can exist in several oxidation states, usually get darker. Some color changes can be reversible; others are permanent.

Lightfastness is measured by exposing pigmented film to an artificial or natural light for a predetermined time. It is a relative term where the color of a sample exposed to a known light source is compared to its original color values.

Weathering is the ability of the colored system, i.e., the coating, paint, etc., not the pigment alone, to resist light and environmental conditions. Changes in color and gloss are two main factors that are evaluated in weathering tests.

Heat stability is measured as a change in the hue of the colored system and a degree of yellowing of the white system after exposure to a desired high temperature for a certain time. This property can also be expressed as the maximum temperature at which the color of the system does not change.

In determining the chemical resistance, color changes of pigmented binder surfaces are measured after their exposure to various chemicals, such as water–sulfur dioxide or water–sodium chloride systems. These systems imitate the environment to which the colored articles could become exposed.

The surfaces of pigment particles can have different properties and composition than the particle centers. This disparity can be caused by the absorption of ions during wet milling, e.g., the −OH groups, on the surface.

Most inorganic pigments are hydrophilic and therefore can be readily wetted only by polar solvents, e.g., water. The wettability and dispersion of inorganic pigments in an organic matrix (polymer, solvent) can be improved by the physical or chemical absorption of surface-active compounds containing polar groups, such as −NH$_2$, −OH, or longer aliphatic chains on pigment particles. The absorption of these compounds makes the pigment surface hydrophobic. Compounds that help to form a bridge between inorganic particles and an organic polymeric matrix are called coupling agents, the most common being tetrafunctional organometallic compounds.

Specifications, Standards, and Quality Control

Production and product quality of most pigment producers are controlled through pigments' standards. Whenever a pigment is developed or significantly improved, a new standard is set that represents the average production results, not necessarily the best ones. If all production processes are under control, the properties of the produced pigment's lots are evenly distributed around those of the standard.

White Pigments

The most common white pigments are titanium dioxide, zinc oxide, leaded zinc oxide, zinc sulfide [CAS: 1314-98-3], and lithopone, a mixture of zinc sulfide and barium sulfate. The use of lead whites and antimony oxides has been decreasing steadily for environmental reasons.

Titanium whites resist various atmospheric contaminants such as sulfur dioxide, carbon dioxide, and hydrogen sulfide. Owing to its chemical inertness, titanium dioxide is a nontoxic, environmentally preferred white pigment.

Titanium dioxide is used mainly in the production of paints and lacquers, plastics, and paper. Other applications include the pigmentation of printing inks, rubber, textiles, leather, synthetic fibers, ceramics, white cement, and cosmetics.

Nonpigmentary applications are in the production of glass, ceramics, electroceramics, catalysts, and as raw materials to form mixed-metal oxide pigments.

By volume, zinc oxide is the second most significant white pigment. Its pigmentary properties are good, providing good coverage. It has a good lightfastness and is well miscible with other pigments. With the increasing popularity of titanium dioxide white pigment in the twentieth century, the pigmentary use of zinc oxide has been declining.

Whereas zinc sulfide is important mainly as a component of the composite white pigment lithopone, it also has a limited use as a single pigment. After titanium white, it has the second highest refractive index of all the white pigments. However, its chemical and thermal resistances are inferior to those of TiO$_2$.

Zinc sulfide is used in applications where white color shade and low abrasivity are required. In printing inks and paints it also contributes to stability and good rheological and printing properties.

Colored Pigments

The worldwide consumption of iron oxide pigments represents ~40% of the total production of colored, inorganic pigments.

In general, all iron pigments are characterized by low chroma and excellent lightfastness. They are nontoxic, nonbleeding, and inexpensive. They do not react with weak acids and alkalies, and if they are not contaminated with manganese, do not react with organic solvents. However, properties vary from one oxide to another.

Iron oxides are supplied to the market as red, ocher, sienna, and umber natural pigments. The hue of the natural iron oxide pigments is determined by raw material composition and processing.

About 60% of the natural iron oxide pigments is used to color cement and other building materials. About 30% is consumed in the production of paints. For coloring plastics and rubber, synthetic iron oxide pigments are preferred. The main advantage of the natural iron oxide pigments, as compared to the synthetic ones, is cost. However, the quality is inferior, and in most cases, they are consumed in close proximity to the mines.

Advantages of synthetic iron oxides over their natural counterparts include chemical purity, more uniform particle size and size distribution, and in the case of precipitated oxides the ability to prepare the pigment in predispersed vehicle systems by flushing techniques.

Synthetic red iron oxides are prepared in a variety of grades from light to dark. These are sold under a variety of names, e.g., Indian red, Turkey red, and Venetian red.

From a chemical point of view, synthetic iron oxide yellows, also known as iron gelbs, are based on the iron(III) oxide–hydroxide, α-FeO(OH), known as goethite. Color varies from light yellows to dark buffs and is primarily determined by particle size, which is usually between 0.1 and 0.8 μm. Because of their resistance to alkalies, these are used by the building industry to color cement.

Chemically, iron blacks are based on the binary iron oxide, $FeO \cdot Fe_2O_3$. Most of the black iron oxide pigments contain iron(III) oxide impurities, giving a higher ratio of iron(III) than would be expected from the theoretical formula.

Iron browns are often prepared by blending red, yellow, and black synthetic iron oxides to the desired shade. The most effective mixing can be achieved by blending iron oxide pastes, rather than dry powders.

Complex Inorganic Color Pigments. Based on the crystal structure, the Color Pigments Manufacturers' Association (CPMA) has classified 53 key inorganic pigments into 14 categories; these inorganic colorants are known as complex inorganic color pigments. The original name, mixed-metal oxide are pigments (MMO), did not accurately describe the chemical nature of all the classified pigments.

Mixed-Metal Oxide Pigments. Mixed-metal oxide pigments can be considered a subcategory of complex inorganic color pigments. In reality these pigments are not mixtures but rather solid solutions or compounds consisting of two or more metal oxides. Structurally, mixed-metal oxide pigments belong to one of 14 structure types. The most common ones are rutile and spinel. The commercial significance is in their thermal, chemical, and light stability, combined with low toxicity. When these are employed for coloring glass enamels and ceramics, they are sometimes referred to as colors or stains; when used to color paints and plastics, they are known as pigments.

The color of mixed-metal oxide pigments results from the incorporation of chromophores, into the structure of stable host oxides.

Chromium(III) Pigments. There are two green pigments based on chromium in the +3 oxidation state. The first one is chromium oxide, Cr_2O_3; the second is hydrated chromium oxide, $Cr_2O_3 \cdot xH_2O$.

Because it weathers extremely well, chromium oxide green is applied as a colorant for roofing granules, cement, concrete, and outdoor industrial coatings. It is also used in ceramic applications.

Ultramarine Pigments. Ultramarines are derived from lazurite (Lapis Lazuli), a semiprecious stone; they can be prepared in many shades.

Ultramarine pigments are used in printing inks, textiles, rubber, artists' colors, cosmetics, and laundry bluing. Because of their thermal stability they are also used to color roofing granules.

Cyanide Iron Blues. Cyanide iron blue, also known as Prussian blue, is one of the oldest industrially produced, inorganic pigments.

Iron blues are mainly used by the printing industry for coloring printing inks. In Europe, cyanide blues are used for coloring fungicides.

Cadmium Pigments. Historically, cadmium pigments have been very important, providing a range of clean, bright shades of yellow, orange, red, and maroon colors. This importance, however, has decreased because of environmental issues. Only a few pigment producers are willing to continue cadmium pigment production.

Lead Chromate Pigments. Lead chromate, $PbCrO_4$, occurs in nature as the orange-red mineral crocoite. Synthetically prepared lead chromate and its solid-state solutions with lead sulfate, $PbSO_4$, or lead molybdate, $PbMoO_4$, are known to have excellent pigmentary properties. The usage of these pigments has been steadily decreasing because of environmental regulations restricting the production and the use of lead-containing products.

Black Pigments

Black pigments can be divided into two basic groups. The first group is represented by carbon blacks. Many other inorganic black pigments, called noncarbon blacks, also are available. These belong chemically to the colored pigment category. Examples are spinel and rutile blacks, iron blacks, and some inclusion zircon pigments.

Carbon Blacks. Carbon black is one of the oldest pigments known. More than 90% of the production of this pigment is consumed by the rubber industries, in particular, by the tire industry as a reinforcing agent. The rest is used for coloring plastics, printing inks, and paints. Particle size of carbon blacks varies from 5 to 500 μm and can be controlled by the process conditions and feedstock. See also **Carbon Black**.

Environmentally, carbon blacks are relatively stable and unreactive. There is no evidence that these materials are toxic to humans or animals.

Extenders and Opacifiers

Extender pigments are low-cost, generally colorless or white pigments with a refractive index less than 1.7. Sometimes these pigments are also referred to as fillers. Many extenders are derived from natural sources and display many diverse properties. They are added to various formulations to improve technical and application properties and to reduce costs. Like pigments, extenders are dispersed in media in which they do not dissolve, but compared to pigments they do not have any significant coloristic properties.

Opacifiers are fine inorganic powders, usually white, that are used to reduce the transparency of ceramic glazes and porcelain enamels. The coating becomes opaque because the particles of the opacifier scatter and reflect the incident light. When inorganic pigments are combined with white opacifiers, pastel colors are obtained.

Commercially, the most important opacifiers for glazes are ZrO_2, $ZrSiO_4$, and SnO_2.

Miscellaneous Pigments

Luminescent Pigments. Luminescence is the ability of matter to emit light after it absorbs energy. Materials that have luminescent properties are known as phosphors, or luminescent pigments. If the light emission ceases shortly after the excitation source is removed ($>10^{-8}$ s), the process is fluorescence. The process with longer decay times is referred to as phosphorescence.

Semiconducting sulfides that can be represented by the formula $nZnS(1 - n)CdS:A$, where A stands for an activator, and $n = 0.15 - 1$, are typical of fluorescence pigments. Phosphorescence pigments can be expressed by the general formula $nZnS(1 - n)CdS:Cu$, where $n = 0.78 - 1.0$ and the amount of the Cu+ activator is only a few hundredths of a percent.

Phosphorescent pigments are used in military applications, plastics, and paints. Zinc sulfide doped with Ag+ (blue) cations, or with Cu+ (green) cations are important pigments for the production of color television screens.

Metal Effect Pigments. Some metals, when prepared as small flakes, impart a special metallic appearance to the coatings and plastics in which they are dispersed. Metals most often used in these applications are aluminum (aluminum bronzes), copper and copper-zinc alloys (gold bronzes), and in smaller amounts zinc, tin, nickel, gold, silver, and stainless steel.

Nacreous Pigments. Nacreous, i.e., pearlescent pigments are used for creating special decorative effects typical of natural pearls. Nacreous pigments are fine, thin, plate-like transparent particles having a high refractive index. Because of these physical characteristics, when dispersed in a transparent film, they produce a silky appearance.

Manufacture of the most popular nacreous pigments involves coating mica with 50–300-nm films of TiO_2, Fe_2O_3, or Cr_2O_3. The mica, which alone does not have a high enough refractive index for creating nacreous luster, provides the required transparent platelet base. The oxide coating provides the necessary high refractive index.

Transparent Pigments. Pigments having chemical composition corresponding to colored or white opaque pigments can, under certain circumstances, appear transparent in a medium. This happens when the particle size of these pigments becomes very small (2–15 nm), and if the particle refractive index is comparable to the refractive index of the media in which the particles are dispersed. Because of the very small particle size, the preparation of these pigments is much more complicated than the preparation of their nontransparent analogues. Their large surface area makes their dispersion difficult and they have a strong tendency to agglomerate.

Environmental Aspects

Some inorganic pigments contain heavy metals. Thus production, use, and disposal are becoming more and more regulated. In the United States there are several federal regulations that control the use and disposal of heavy metals.

The Resource Conservation and Recovery Act (RCRA) controls the disposal of hazardous waste. SARA Title III governs the toxic inventory and emission reporting; the Clean Water Act (CWA) sets the limits for metals that can be present in water discharge; and the Clear Air Act (CAA) Amendments of 1990 control the abatements of all materials in the air.

The Occupational Safety and Health Administration (OSHA) regulates the exposure to chemicals in the workplace. From the point of view of the inorganic pigments industry, the limits established for lead and cadmium exposure are particularly important. A comprehensive lead standard adopted by OSHA in 1978 has been successful in reducing the potential for lead contamination in the workplace.

To assure the future of inorganic pigments, research efforts are directed toward the development of environmentally acceptable pigments, pigments that when produced under well-controlled conditions do not release any toxic materials into the environment whether during production, use, or disposal.

Additional Reading

Buxbaum, G.: *Industrial Inorganic Pigments*, 3rd Edition, John Wiley & Sons, Inc., Hoboken, NJ, 2005.

Herbst, W., and K. Hunger: *Industrial Organic Pigments: Production, Properties, Applications*, 3rd Edition, John Wiley & Sons, Inc., Hoboken, NJ, 2004.

Lewis, P.A.: *Pigment Handbook*, Vol. 1, 3rd Edition, John Wiley & Sons, Inc., New York, NY, 1988.

Patton, T.C.: *Paint Flow and Pigment Dispersion: A Rheological Approach to Coating and Ink Technology*, 2nd Edition, John Wiley & Sons, Inc., New York, NY, 1979.

Smith, H. MacDonald: *High Performance Pigments*, John Wiley & Sons, Inc., New York, NY, 2001.

PIGMENTS (Organic). Pigments are colored, fluorescent or pearlescent particulate organic or inorganic finely divided solids that are usually insoluble in, and essentially physically and chemically unaffected by, the vehicle or medium in which they are incorporated. They alter appearance either by selective absorption and/or scattering of light. They are usually incorporated by dispersion in a variety of systems and retain their crystal or particulate nature throughout the pigmentation process. The large number of systems vary widely from paints to plastics to inks and fibers.

Dyes, on the other hand, are colored substances which are soluble or go into solution during the application process and impart color by selective absorption of light. In contrast to dyes, whose coloristic properties are almost exclusively defined by their chemical structure, the properties of pigments also depend on the physical characteristics of its particles.

In some cases, a single chemical substance can serve both as a dye and as a pigment. For example, indanthrone [CAS: 81-77-6] (Pigment Blue 60) functions as a blue pigment or as a dye. As a pigment, indanthrone is a particulate, insoluble solid dispersed directly into a vehicle, whereas as a dye it is reduced to a base-soluble hydroquinone derivative and then reoxidized onto a solid substrate.

The description of colored organic pigments excludes consideration of inorganic pigments, as well as black pigments which consist of specially treated forms of carbon and white pigments which are entirely of inorganic origin. See also **Pigments (Inorganic)**. Organic pigments are generally stronger and brighter than their inorganic counterparts and range in color over the entire visible spectrum. However, because of the structure complexity of organic pigments and the method of their manufacture, they are invariably higher in cost. Many organic pigments do not possess the stabilities shown by some inorganic counterparts, nevertheless some organic pigments are sufficiently durable and stable to fit most modern and potential applications. In fact, many inorganic pigments that contain heavy metals such as chromium and cadmium are being replaced by carefully chosen organic substitutes.

Pigments are categorized according to their generic name and chemical constitution in the *Color Index* (CI), published by the Society of Dyers and Colourists, and the American Association of Textile Chemists and Colorists.

Pigments are available in a number of commercial forms including dry powders (either surface treated or untreated), presscakes, flushed colors (thick pastes), fluidized dispersions (pourable pastes), resin predispersed pigments (powders), and plastic color concentrates or master batches (granules). See also **Pigment Dispersions**.

Significant pigment attributes are tinctorial strength, durability (photochemical stability), hiding power, transparency, and heat and solvent resistance. Other properties include brightness (saturation), gloss, rheology, crystal stability, bleed resistance, flocculation resistance, and other properties associated with specialized applications.

The history of synthetic organic pigment discovery began in the middle of the nineteenth century. In 1856, William H. Perkins at the age of 18 synthesized for the first time the color mauveine [CAS: 6373-22-4] by oxidizing aniline containing toluidine with chromic acid. This event ushered in an era of colored synthetic chemistry that continues into the 2000s. Many new organic structures have been discovered and introduced as commercial organic pigments.

The development of modern organic pigments started with the synthesis of dyestuffs for the textile industry. The period up to 1900 was characterized by the discovery and development of many dyes derived from coal-tar intermediates. Rapid advances in color chemistry were initiated after the discovery of diazo compounds and azo derivatives (shown to be largely hydrazone derivatives). The wide color potential of this class of pigments and their relative ease of preparation led to the development of azo colors, which represent the largest fraction of manufactured organic pigments.

The most important advance in pigment technology after World War I was the discovery of the relatively complex structure but easily synthesized copper phthalocyanines, which were characterized by excellent brightness, strength, bleed resistance, and lightfastness. See also **Phthalocyanine Compounds**.

After World War II the most important discovery was the family of red-violet quinacridone pigments, followed by the mostly yellow-orange benzimidazolone, the isoindolinone pigments, and the red diaryl pyrrolopyrroles.

Color and Constitution

As early as 1868 Graebe and Liebermann recognized that color in organic compounds is associated with the presence of multiple bonds. A few years later Witt coined the term chromophore, from the Greek *chroma* meaning color and *phoros* meaning bearer, for groups that give rise to color. The term is used to designate π-electron-containing moieties (conjugated double bonds) which contribute to the selective absorption of visible light. Generally, organic compounds absorb light in the ultraviolet (210–400 nm) and visible (400–750 nm) region of the spectrum at characteristic wavelengths. The intensities of these absorptions vary due to the excitation of the more loosely held electrons in the molecule. All unsaturated groups have remarkably similar $\pi - \pi*$ transitions regardless of the atoms contained in the common chromophores.

The presence of $\pi - \pi$ or $n - \pi$ conjugated systems does not assure absorption of visible light or generation of color. However, all colored organic compounds, including pigments, possess extended conjugated resonance systems. Thus, whereas 1,4-diphenylbutadiene is colorless, 1,6-diphenylhexatriene is colored.

colorless   colored

All absorbed light is complementary to reflected light which produces observed color. All colors are a function of the wavelength of absorbed light as shown in Table 1. Thus, if a pigment absorbs only blue light it imparts an orange color, whereas when it absorbs orange light the observed color is blue.

TABLE 1. COLORS OF ABSORBED LIGHT AND THE CORRESPONDING COMPLEMENTARY COLORS AS A FUNCTION OF WAVELENGTH

Wavelength, nm	Color of absorbed light	Complementary color
400–420	Violet	Yellow-green
420–450	Indigo blue	Yellow
450–490	Blue	Orange
490–510	Blue-green	Red
510–530	Green	Purple
530–545	Yellow-green	Violet
545–580	Yellow	Indigo blue
580–630	Orange	Blue
630–720	Red	Blue-green

Properties of Pigments

The physical and chemical characteristics that control and define the performance of a commercial pigment in a vehicle system include its chemical composition, chemical and physical stability, solubility, particle size, shape and particle size distribution, degree of dispersion, crystal morphology including polymorphic forms, refractive index, specific gravity, electronic spectra with particular emphasis on extinction coefficients in the visible spectrum, surface area, and the presence of impurities, extenders, and surface modifying agents. Invariably a pigment is used in a vehicle system, therefore its ultimate performance in use derives from both physical and chemical pigment-vehicle interaction. Performance of most pigments is system dependent.

Unlike inorganic pigments, organic pigments are relatively strong and bright (saturated), but their fastness properties, though adequate for the purposes for which they are used, vary widely from poor to outstanding.

The inherent strength of a pigment depends on its light-absorbing characteristics, which are related to its molecular and crystalline structure. In addition, strength is a function of particle size or surface area. The ability of a pigment to absorb light increases with decreasing particle size or increasing surface area, until the particles become entirely translucent or transparent to incident light. Being finely divided pigment particles have a great tendency to aggregate and agglomerate into crystal assemblies. To obtain the inherent strength of a pigment the aggregates must be completely broken down to individual crystals by application of work and their reagglomeration or flocculation prevented. Total breakdown to single crystals in practical systems seldom happens.

Pigment strength in a vehicle also depends on the typical character of other components in a pigmented system insofar as they absorb or scatter light. Strength comparisons are usually made with a series of samples featuring varying amounts of a pigment incorporated in a vehicle with a corresponding series in the same vehicle containing a reference pigment. Instrumental comparisons are commonly practiced.

A good absolute theoretical comparison of strength is represented by the area under the absorption bands in the visible spectrum, or less accurately by the molecular extinction coefficients at the maximum wavelength of absorption.

The saturation of a colored pigment is a measure of its brightness or cleanliness as opposed to dullness of hue. Generally, if a pigment absorbs light over a wide range of wavelengths, i.e., shows broad absorption bands, or contains more than one chromophore, the pigment is likely to be duller than a pigment with sharp absorption bands due to a single chromophore. Because pigments are frequently used in combinations or blends, the brightness is determined by the selective absorption of the individual pigments and this significantly affects the brightness of the reflected color. A saturated pigment provides flexibility in color blending and its use provides important economic advantages since desaturation can be accomplished by blending with duller and less expensive pigments, like carbon black [CAS: 1333-86-4] or iron oxide [CAS: 1309-37-1].

Fastness describes the characteristics of a pigment in terms of its color stability in a pigmented system upon exposure to light, weather, heat, solvents, or various chemical agents. Ideally, a pigment should be insoluble and chemically and photochemically inert. Only a few organic pigments approach such perfection.

The development of resins, plastics, fibers, elastomers, etc. that are processed at progressively higher operating and curing temperatures has created a need for pigments that stand up for relatively long periods of time to a hostile environment. They must remain essentially unaltered when incorporated into plastics such as polypropylene, ABS, or nylon at relatively high temperatures.

Once a pigment is incorporated into a system, it is expected to be durable and withstand the combined chemical and physical stresses of weather, solar radiation, heat, water, and industrial pollutants. Because a pigment is totally enveloped by the medium which is itself not inert, various pigments perform differently in different systems. Thus, a pigment may be lightfast or weatherfast in one system and fail in another.

The dispersibility of a pigment is measured by the effort required to develop the full tinctorial potential of a pigment in a vehicle system. Dispersibility differs from system to system depending on pigment–medium interaction and compatibility.

Small particle size pigments, especially the very small crystals, seldom exist as individual entities, but as strongly coherent aggregates or less firmly bound agglomerates. A wide variety of additives are being used to reduce aggregation or agglomeration and flocculation to improve dispersibility and color strength of organic pigments. These include resins, especially those related to abietic acid, aliphatic amines, amides, substituted derivatives of pigments themselves, and various combinations thereof. The additives are most effective if they are present when the pigment crystals are being generated.

Hiding power of a pigment is a function of its strength, that is, its absorption coefficients, its particle size, or light-scattering coefficient, and relative refractive indexes of pigment and vehicle. Light scattering has a powerful influence on opacity and goes through a maximum as a function of particle size. The maximum occurs at a particle size which is approximately half the wavelength of absorbed visible light.

Similarly, a pigment which absorbs much light increases hiding even when light scattering is insufficient, and the higher the refractive index the greater the hiding power of a pigment.

Conversely, to increase transparency light scattering must be minimized by particle size reduction. The smaller the particle size and the better the dispersion the greater the transparency.

Types of Pigments

All organic pigments have to be synthesized and nearly all have to be conditioned or finished. The physical conditioning is as important as its chemical constitution, and it has become an important separate process step in the manufacture of organic pigments.

Azo Pigments. Azo pigments provide good examples of materials in which conditioning is an integral part of the synthesis process. The coupling process for azo components is simple. An aromatic amine is diazotized by treatment with nitrous acid under conditions which vary from dilute mineral acid to concentrated sulfuric acid, depending on the basicity of the amine. The simplest method, i.e., direct coupling, involves running the diazo solution directly into the solution or suspension of the coupling component. In inverse coupling, the coupling component is run into the diazo solution. The most elegant technique, simultaneous coupling, entails running both the diazo and coupling component simultaneously into water or a dilute buffer.

The monoazo and disazo pigments contain one or more chromophoric groups usually referred to as the azo (N=N) group. However, it has been shown by x-ray diffraction analysis and nuclear magnetic resonance (NMR) techniques that azo pigments exist in the hydrazone rather than the azo tautomeric form. The hydrazone form, which has three intramolecular hydrogen bonds, renders the molecule planar (with the exception of the aniline moiety) which is a stabilizing influence.

azo form   hydrazone form

Azo pigments, one of the oldest and most diverse group of pigments, comprise two types. One type consists of pigments that are insoluble in the aqueous reaction medium in which they are synthesized. Most simple azo pigments show poor bleed characteristics, but relatively good acid and alkali resistance. They show acceptable lightfastness in deep shades but poor tint lightfastness. The second type are laked or precipitated azo pigments derived from components substituted with sulfonic and/or carboxylic acid groups. These pigments are characterized by good to excellent bleed resistance, poor acid and alkali resistance, fair to good lightfastness in deep shades, and poor tint lightfastness. Also available are special azo pigments which show very good overall properties and therefore find applications in fairly demanding systems.

Lakes. Lakes are either dry toner pigments that are extended with a solid diluent, or an organic pigment obtained by precipitation of a water-soluble dye, frequently a sulfonic acid, by an inorganic cation or an inorganic substrate such as aluminum hydrate.

Basic dyes are characterized by bright shades and high strength but poor lightfastness. However, when laked by precipitation with soluble salts of organic acids such as tannic acid, or inorganic heteropoly-acids like phosphotungstic (PTA; $M = W$) and phosphomolybdic (PMA; $M = Mo$), and the combined phosphotungstomolybdic acid (PTMA), the resulting pigments retain the dyes' tinctorial attributes, but become insoluble and show improved lightfastness.

Copper Phthalocyanines. Copper phthalocyanine (CPC) approximates an ideal pigment (Pigment Blue 15). This class of pigments offers extreme brightness, tinctorial strength, bleed and chemical resistance, stability to heat, and migration. The pigments show excellent weatherfastness but are restricted to the blue and green regions of the spectrum. Phthalocyanine blue and green are among the most important organic pigments on the worldwide market.

Quinacridones. Quinacridone pigments offer generally outstanding fastness properties across the visible spectrum from red-shade yellows to scarlet, maroon, red, magenta, and violet color ranges. The pigments are practically insoluble in most common solvents and therefore show excellent migration resistance in most application media. The low solubility is attributed to effective intermolecular hydrogen bonding which is also responsible for the photochemical stability. The various available colors are the result of polymorphism and various substitution patterns. The parent compound Pigment Violet 19 [CAS: 1047-16-1] exists in three polymorphic modifications. The red gamma and violet beta forms are commercial pigments, whereas the red alpha form is metastable.

Due to the excellent pigmentary properties, quinacridones are used in many industries but particularly in automotive finishes, emulsion paints, plastics, and fibers.

Diaryl Pyrrolopyrroles. The 1,4-diketo-3,6-diarylpyrrolo(3,4-c)pyrroles are the most recently discovered class of pigments ranging in color from orange to bluish reds. These pigments are synthesized by base-catalyzed condensation of higher diakyl esters of succinic acid with aromatic nitriles. One important member of this class is Pigment Red 254, which is a very opaque yellowish red pigment of outstanding durability, brightness, and chemical resistance. Another is the parent compound Pigment Red 255, which is a high performance orange pigment. Both are used in automotive finishes, and a higher strength variation of PR 254 is used in plastics applications.

Vat Dye Pigments. Vat dyes have been used for a long time for coloring textile fibers. As pigment technology evolved, new methods of particle size reduction have been successfully applied to largely insoluble dyes. Only a few of the very large number of vat dyes have found application in the pigment field.

Aminoanthraquinone Pigments. Pigment Red 177 [CAS: 4051-63-2] has the chemical structure of 4, 4'-diamino-1, 1'-dianthraquinonyl. It is the only known pigment with unsubstituted amino groups which are involved in both intra- and intermolecular hydrogen bonding. The bluish red pigment is used in plastics, industrial and automotive paints, and specialized inks. See also **Dyes: Anthraquinone**.

Dioxazine. Carbazole Violet (Pigment Violet 23) [CAS: 6358-30-1] is a bluish violet pigment that is uncommonly strong, resistant to solvents, and shows fair weatherfastness. It is used primarily as a shading pigment with copper phthalocyanines and for toning whites in a variety of systems.

Isoindolinones and Isoindolines. Tetrachloroisoindolinone pigments are characterized by very good lightfastness, heat stability, migration resistance, and chemical inertness. Although Pigment Yellow 110 [CAS: 5590-18-1], a red-shade yellow, is relatively weak, it finds extensive use in automotive and other high grade finishes and in a variety of plastics and ink applications.

Quinophthalones. The quinophthalone pigments are prepared by condensation of quinaldines with a variety of aromatic anhydrides. One pigment in this series, Pigment Yellow 138 [CAS: 30125-47-4], is a reasonably weatherfast greenish yellow pigment of good heat stability. The main field of application is paints and plastics.

Uses

Organic pigments are used for decorative and/or functional effects. In paints, for example, pigments provide color and contribute to exposure durability of the systems, which is particularly true for high performance pigments. Other functional effects include hiding power and high visibility, such as is displayed with daylight fluorescent pigments. They are used in various printing processes for textiles, plastics, and safety markings of various types.

The most important and established use for pigments is the imparting of color to a variety of materials and compositions. Examples are surface coatings for exteriors and interiors of automobiles and houses with oil- or water-based paints; wood stains, leather and artificial leather finishes, printing inks and many other applications.

Testing and Standardization

Pigments are subjected to a number of tests before they are released to customers. Testing is complicated because of the great diversity of pigment types and uses. A given pigment may be dispersible in one system but poorly dispersible in another, and can exhibit different durability depending on the system; performance is system dependent. Standardization is carried out against a standard sample for coloristics and a variety of working properties. Among the tests, depending on the pigment type, may be thermal stability, hiding power, rheology, migration, chemical stability, gloss, distinctness of image, durability, etc.

In the process of testing, color deviations are expressed in the CIELAB system or the equivalent polar LHC system. In either case tested samples must fall within acceptable ranges or limits established versus a standard by the pigment manufacturer and accepted by the pigment user.

In dispersing a pigment by an established method, acceptable pigment strength vs a standard must be achieved, even though an ideal dispersion normally is rarely realized.

Health and Safety Factors

Since pigments are generally insoluble, unlike most dyes, they are usually not bioavailable and consequently are generally not absorbed or metabolized. Nevertheless many health-related studies have been carried out and reported in the literature.

Acute toxicity of organic pigments has been studied extensively. The most common measure of toxicity is LD_{50} expressed in mg/kg of body weight which has a lethal effect on 50% of test animals after a single (oral, dermal, etc.) administration. These tests assess toxicity vs other known compounds. A large LD_{50} value represents a low degree of toxicity. Pigments in general have very low levels of acute toxicity.

Chronic toxicity defines a specific dose or exposure level that will produce measurable, long-term toxic effects, including carcinogenicity.

One area which requires special comment is a study which showed that certain diarylide pigments processed in polymers above 200 °C and particularly above 240 °C decompose to give off 3,3-dichlorobenzidine, an animal carcinogen. As a consequence diarylide pigments (not, however, condensation disazo pigments) are not recommended for use in any applications where they might be exposed to temperatures exceeding 200 °C.

The hazards associated with handling pigments are specified by an OSHA Hazards Communication Standard, which also requires labeling and employee information and training.

Ecological Effects. The starting materials for manufacture of organic pigments are as diverse as the pigments themselves. However, most starting materials are derived from petroleum or natural gas sources. Although many pigments are synthesized in water, a variety of organic solvents are also employed by the industry. The effective utilization of all starting materials and solvents, and reduction of undesirable by-products, is a primary objective of the organic pigment industry.

Additional Reading

Clark, E.A. and P. Anliker: in O. Hutzinger, ed., *Organic Dyes and Pigments, The Handbook of Environmental Chemistry*, Vol. 3, Springer-Verlag, Berlin, 1980.

Color Index, 3rd Edition, Vol. 4, The Society of Dyers and Colourists, Bradford, Yorkshire, England; American Association of Textile Chemists and Colorists, Research Triangle Park, NC.

Herbst, W. and K. Hunger, *Industrial Organic Pigments*, VCH Publishers, Inc., New York, NY, 1993.

Lewis, P.A.: *Pigment Handbook*, Vol. 1, 3rd Edition, John Wiley & Sons, Inc., New York, NY, 1988.

Patton, T.C. ed.: *Pigment Handbook*, Vols. I, II, and III, John Wiley and Sons, Inc., New York, NY, 1973.

Patton, T.C.: *Paint Flow and Pigment Dispersion: A Rheological Approach to Coating and Ink Technology*, 2nd Edition, John Wiley & Sons, Inc., New York, NY, 1979.

Smith, H. MacDonald: *High Performance Pigments*, John Wiley & Sons, Inc., New York, NY, 2001.

PIGMY ANTEATER. See **Edentata**.

PIGMY ANTELOPES. See **Antelope**.

PIGNOLIA NUT. See **Conifers**.

PIKA. See **Rabbits and Hares**.

PIKE *(Osteichthyes)*. Of the order *Haplomi*, family *Esocidae*, the pikes are freshwater food and game fishes. They are most abundant in the rivers and lakes of the northern United States and Canada. The northern pike *(Esox lucius)* attains a weight of about 40 pounds (18.1 kilograms) and a length up to 4 feet (1.2 meters) and is one of the principal game fishes of the north country. This species is sometimes called the northern pickerel, but it differs from the closely related pickerels in having the cheeks scaly, but the lower half of the opercula bare. The walleyed "pike" is of a different order *(Percomorphi)* and is not a pike per se. Similarly, the sandpike is related to the perches and is not a pike. See also **Perches and Darters (Osteichthyes)**.

The esocids have shovel-like bills something like a duck's and equipped with sharp teeth. Pikes are carnivorous with a diet mainly of smaller fish, but including small birds, mammals, and frogs. The northern pike is found throughout North America north of Ohio. However, it is not found in the far northwest. The esocids are quite sensitive to environmental changes. The muskellunge *(Esox masquinongy)* found in the Great Lakes and environs is larger than the northern pike, reaching a weight in excess of 70 pounds (31.8 kilograms). Chain pickerels frequent waters from Nova Scotia on the east as far south and west as Texas. This fish attains a length of about 24 inches (61 centimeters). All of the aforementioned fishes are considered good food fishes. See Fig. 1.

Fig. 1. Pike.

PILCHARD. See **Herring**.

PILEUS. See **Clouds and Cloud Formation**.

PILLBOX ANTENNA. A cylindrical parabolic reflector enclosed by two plates perpendicular to the cylinder, so spaced as to permit the propagation of only one mode in the desired direction of polarization.

PILL BUG *(Crustacea, Isopoda)*. A small oval terrestrial crustacean, commonly found in moist areas at the surface of the ground. They hide in crevices among rocks, under wood, and even invade basements. These forms are also commonly known as sow bugs and wood lice. The pill bugs are properly the species with the power of rolling up into a ball so nearly spherical that it will roll on a slight incline.

PILLOW LAVA. Effusive volcanic rocks, generally of basic composition, which are characterized by pillow-like or bun-like structures formed during the concomitant movement and congelation of the lava. Most pillow lavas are basalts. Frequently the "pillows" have a skin of rock glass called tachylyte. The evidence of the rapidity of the chilling suggests that pillow lavas owe their peculiar structure to having flowed into a body of water or as having originated as aquatic lava flows.

PILOTAGE (or Piloting). A term used to describe that type of navigation in which the positions and motions of a ship are determined by reference to fixed objects on the earth. The landmarks may be natural, such as hills, points of land, small islands, lakes, rivers, etc., or they may be artificial, such as lighthouses, light vessels, beacons, buoys, prominent buildings, water towers, railroads, highways, and power transmission lines. Two general types of pilotage are recognized by sea navigators; inshore or harbor piloting, and offshore or coast piloting. Before a ship proceeds up a channel, or into a bay or harbor, the pilot must have a clear mental picture of the locations of all available landmarks, range points, etc. A stranger to the region must obtain this image from a thorough study of charts, pilot directions, and similar publications for the region. With the mental picture thoroughly developed, the pilot then guides his ship in much the same manner as an individual finds his way about a city. Aviators use this type of pilotage when operating in the vicinity of a base with which they are thoroughly familiar, or when proceeding by contact flying. When a ship is off the coast, with but few recognizable landmarks available, or when a pilot is flying over unfamiliar terrain where prominent landmarks are few and far between, the pilot obtains lines of position from the available objects and then applies geometric constructions to obtain fixes.

The bearing of an object, taken with the pelorus of the compass, is the most frequently used line of position in pilotage. If two or more landmarks are available, and are spaced so that their bearings differ by more than 30°, the lines of position will intersect in a fix that is known as a cross-bearing position. A single object will provide only one bearing, but a fix can be obtained if the distance of the object can be measured. In this case, the two lines of position will be the bearing line and a circle centered on the object, the circle having a radius equal to the distance. Among the various methods for finding the distance, we have: the angular height of the object, measured with the sextant if the linear height is known; the use of stadia lines in binoculars; the difference in time of reception of audible and radio signals; and the time for the echo of a signal from the ship, such as a whistle blast, to return from the shore. None of these methods is as accurate as a bearing line, and they should not be used when other methods for obtaining a fix are available.

One object may be used for obtaining a running fix by taking two bearings of the object, at different times, provided that the distance and direction run between observations is accurately known. At 1015, the navigating officer of a ship, which is proceeding at 18 knots on heading 060°, sights a lighthouse about 12 miles away that bears 032° off the port bow. He immediately instructs the helmsman to be very careful with his steering, and proceeds to make a careful study of the current and tide tables of the region. From the predicted currents, he finds that the ship is making good a course of 058°, with ground speed 17.8 knots. He then starts a graphical construction (see Fig. 1) by drawing the 1015 line of position through the lighthouse. This line bears 028°, since 060° − 032° + 028°. He then selects a point A on this line, usually the point closest to his dead-reckoning position, and draws the course line in direction 058°. At 1045, the lighthouse is 070° off the port bow, and a 1045 line of position is drawn through the lighthouse on bearing 350°. The 1015 line must now be advanced to 1045 by measuring off the distance made good in 30 minutes (8.9 miles) along the course line and drawing a line parallel to the 1015 line through the estimated position at 1045. The point of intersection of the 1045 line and the advanced 1015 line is the running fix at 1045. A line drawn through this fix in direction 058° will represent the probable course of the ship. This will intersect the 1015 line in the estimated position at that time. When the ship arrives at point C, the ship will have the lighthouse on the port beam. Measurements indicate that the ship, at 1045, bears 170° distant 7.2 miles from the light, and that the light will be on the port beam at 1054 and will be 6.7 miles distant at that time.

Proper selection of the second bearing will give a fix without plotting the lines. If, in the above case, the pilot had noted the time when the relative bearing of the light was 064° off the port bow (twice the first value), the triangle $A'FL$ would be isosceles, with the side $A'F$ equal to FL. The side $A'F$ is the distance run in the interval required to "double the angle on the bow," and the bearing and distance of the light at the time of the second observation is obtained without any plotting. An experienced pilot is familiar with a number of similar short-cuts for locating his positions by taking frequent bearings of available objects. Continual use of these, together with careful steering and a thorough knowledge of the speed of

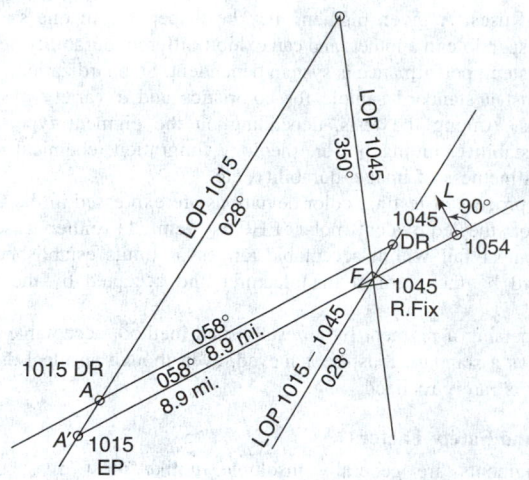

Fig. 1.   Pilotage. Scale diagram.

the ship and the set and drift of the current, will provide the pilot with a series of successive positions of his ship. In many cases, the scattering of the positions will indicate that the predicted currents are in error. By proper allowance for these abnormal conditions, the ship may be saved from disaster. Similar methods are available to the air navigator; they may be used to check wind directions and speeds, and may provide the pilot with accurate positions of his plane. Either visual or radio bearings may be used in flight and a series of running fixes obtained.

The contour of the sea bottom may be used for determination of position of a ship. If a ship is held on steady heading, and if frequent soundings are taken, the depths may be plotted on a sheet of tracing paper, using the distance scale of a chart. The position of the ship is found by placing this tracing over the chart, with the line of soundings parallel to the course line, and moving it about until the observed values correspond with the depths shown. At present, a set of accurate contour maps of the ocean floor is available and is used with a self-recording fathometer in the method of pilotage by depth. A similar type of air pilotage will be available when the absolute altimeter is completely developed and ready for general use.

See also **Course**; and **Navigation**.

PILOT CHART. A chart of a major ocean area that presents in graphic form averages obtained from weather, wave, ice, and other marine data gathered over many years in meteorology and oceanography to aid the navigator in selecting the quickest and safest routes; published in the United States by the Defense Mapping Agency Hydrographic/Topographic Center (http://www.flra.gov/decisions/v39/39-079-4.html) from data provided by the U.S. Naval Oceanographic Office (http://www.navo.hpc.mil/)and the National Environmental, Satellite, Data and Information Service (http://www.nesdis.noaa.gov/) of the National Oceanic and Atmospheric Administration.

AMS

PILOT STREAMER. See **Lightning**.

PINEAL GLAND. A small gland attached to the posterior part of the brain, behind and above the third ventricle. The gland is about the size of a small vitamin capsule and is located in the central part of the brain, as is the pituitary gland, but a little higher up. In humans, extensive calcification of the pineal body begins during the second decade of life and by the sixth decade over 70% of all pineals show X-ray evidence of calcification. However, this does not necessarily imply loss of functional activity inasmuch as recent evidence indicates that the human pineal may retain functional activity over the entire life span.

The product of the secretion of the pineal gland is a hormone known as *melatonin*. This substance causes marked skin blanching or lightening by its action on the pigment cells. This effect is the opposite of that produced by the pituitary melanocyte-stimulating hormone.

The functions of the pineal gland are still obscure—with some differences of opinion among endocrinologists concerning whether or not the gland is a part of the overall endocrine system. See also **Endocrine System**. Recent studies support the view that the pineal contains and probably secrets certain humoral substances and may influence the secretion of others. Other observations suggest that the pineal may serve as a *biological clock*, which helps to regulate certain endocrine rhythms.

Pineal tumors are among the rarest tumors. Processes destroying the pineal gland are characterized by precocious sexual development in a manner that is suggestive of an overactive pituitary gland. However, if the tumor is truly a pineal neoplasm, there may be a depression of gonadal function. The precocious sexual development that results from pineal destruction appears to be limited to boys about 2 to 3 years of age. The boys manifest definite signs of masculinity of an adult nature. The sex organs become adult in size and function, and pubic hair appears similar to that of a mature male. There has been no satisfactory explanation of why nearly all cases of sexual precocity associated with pineal tumors have occurred in boys.

The surgical removal of pineal tumors is not often advisable because of the high mortality rate. Radiation therapy may result in a temporary cure.

PINEAPPLE. The fruit of a perennial herb of the genus (*Ananas*) of the *Bromeliaceae* family. Many species of *Ananas* are grown for their ornamental beauty. The main species of commercial value for its fruit is *A. comosus*. This plant is native to tropical America. The species is also found extensively in West Africa and botanists are not in full agreement as to whether the species is also native to that region or was introduced there from tropical America many decades ago. The pineapple was described in the diary of Columbus and his party in 1493, at which time plants cultivated by the native Indians on the Island of Guadeloupe in the Lesser Antilles were noted. Later, explorers with Columbus also noted the pineapple in an Indian settlement along the coast of Panama. Reports of pineapples growing in Brazil date back to 1519. Some authorities postulate that the fruit was carried by sailors from Brazil to Malaysia and China. They also carried pineapples back to Spain and England, where they were called the fruit pineapple because it looked like a pine cone. Historians report that transporting pineapples from tropical America to Europe and other regions was an arduous task. It is reported that out of one shipment of pineapples, only one fruit arrived at its destination unspoiled. Progeny of this specimen were prized by European gardeners, who constructed special greenhouses for their growth. There is controversy as regards how the pineapple reached the Hawaiian Islands. Some authorities observed that it was introduced there by a Spanish adventurer (not named) in the late 1700s. Other authorities observe that there is a form of pineapple on the islands of Hawaii that has been growing there for as long as the oldest inhabitants can recall, but its origin is unknown. There is no mention of the pineapple in the literature of the Egyptians, Arabs, Greeks, or Romans.

Cultivation of the pineapple is widespread. In descending order of production are China, the United States (notably Hawaii), Brazil, Thailand, Philippines, Mexico, Malaysia Peninsula, Ivory Coast, Ecuador, South Africa, Bangladesh, Australia, India, and Kenya.

The pineapple fruit generally weighs within the range of 2 to 3 pounds (0.9 to 1.4 kilograms), although some larger varieties may weigh up to 5 and 10 pounds (2.3 to 4.5 kilograms). The fruit is broadly conical in shape, with the larger fruitlets at the bottom, tapering to smaller fruitlets at the top. See Fig. 1. In all, the fruit consists of from 100 to 200 berry-like fruitlets. These are carried on the core or central axis of the fruit. The core, of course, is a continuation of the stem of the plant. Close observation of the outside of the fruit will show that the fruitlets follow a spiral-like pattern around the axis of the fruit. On most fruits, two series of spirals (clockwise and counterclockwise) can be noted. The slope of one spiral is somewhat less than that of the other. End-to-end length of the fruit (bottom to crown) ranges from 5 to 10 inches (12.5 to 25 centimeters). In addition to the particular variety or cultivar affecting dimensions and weight, these characteristics also are a function to some extent of the planting density, lower densities promoting growth of larger fruits. Characteristic of the mature fruit is the so-called blossom cup. This will be found to be an oval cavity located at the center of each fruitlet.

Fig. 1. Close-up of mature pineapple (*Ananas Cosmosus*) growing in Hawaii. (*USDA Soil Conservation Service photo.*)

When growing in the field, the plant achieves a height of 2.5 to 3 feet (0.75 to 0.9 meter). The plant, as shown by Fig. 1, has rather rigid, sword-like leaves, which take the form of an elongated rosette. The fruit occurs in the center of the plant, out of the crown of which a smaller grouping of similarly stiff, but shorter sword-like leaves appear (appearing something like a junior version of the total plant).

The pineapple plant has the desirable feature of an ability to store water in its leaf axils, as well as in leaf tissue especially developed for this purpose. These features contribute to the drought-resistance of the plant.

Varieties. Although there are hundreds of varieties of pineapple, only a comparative few are commercially important. These include:

Abagaxi or *Abachi*—Conical, elongated fruit. Good taste. Grown in west Africa.

Baronne de Rothschild—Medium-size fruit seldom exceeding 3 pounds (1.4 kg). Good quality, but not exceptionally sweet. Grown in west Africa. Yellow-white flesh.

Carbazoni or *Smooth Cayenne*—Cylindrical shape. Weight ranges from 5 to 10 pounds (2.3 to 4.5 kg). Grown mainly in Puerto Rico for local consumption. Known mainly for its large size.

Cayenne or *Smooth Cayenne*—Cylindrical shape. Weight ranges 3 to 5.5 pounds (1.4 to 2.5 kg). Yellow flesh. High acid and sugar content. The most widely planted variety. Grown in Hawaii, west Africa, Australia, the Philippines, and South Africa. Used for fresh market and canning.

Pernambuco or *Pernambuca*—Cylindrical shape. Weight ranges from 3 to 4 pounds (1.4 to 1.8 kg). Yellow-white, tender flesh. Mild flavor. Grown mainly in Brazil for fresh market.

Queen—Weight ranges from 2 to 3 pounds (0.9 to 1.4 kg). Rich yellow, crisp flesh with mild flavor. Less acidic and less juicy. Grown in South Africa, Australia, and Malaysia for fresh market and canning. Keeps well when mature.

Red Spanish—Squarish shape. Weight ranges from 3 to 5 pounds (1.4 to 2.3 kg). Pale yellow, fibrous, aromatic, spicy, and acidic flavor. Tough shell is excellent for shipping. Grown mainly in the Caribbean and Florida. When fully ripe, the fruit is orange.

Sugarloaf—Globular shape. Yellow-white, sweet, and rich flesh. Grown in Mexico and Cuba, mainly for fresh market.

Propagation. Several different methods of propagation are possible. These include: (1) Crown suckers; (2) ratoon or side suckers; (3) basal slips or suckers; and (4) disk or stem cuttings.

Crown suckers are shoots that grow from the top of the fruit. They are taken off when fully developed and planted in a nursery bed. The lower leaves are removed before insertion. They normally produce fruit in 12 to 18 months from planting.

Ratoon or *side suckers* arise from buds situated low down on the main stem and should be detached at a point close to the parent stem. They should be rooted in a nursery bed before planting. Fruiting takes place in about 12 months from planting.

Basal slips or *suckers* are produced from immediately below the fruit in some varieties. These are rooted in a nursery bed. After planting, they will fruit in from 12 to 15 months.

Disk or *stem cuttings* are stems which have been stripped of their leaves and cut transversely into short sections from 2 to 3 inches (5 to 7.5 centimeters) long, or 9-inch (23-centimeter) lengths of stem can be sliced longitudinally into halves. These pieces are planted just below the soil level in a nursery bed. The short pieces are planted in an upright position, while the sliced halves are placed horizontally, with the rounded side facing upward. Dormant buds will eventually develop from the pieces of stem, arising from the axils of the removed leaves. See Fig. 2. Each bud will produce roots and may be detached from the stem when these are 0.5 to 1-inch (1.3 to 2.5-centimeters) long. These rooted shoots are established in nursery beds until they are large enough for planting in the field. Up to 9 or 10 buds can be obtained from a 9-inch (23-centimeter) stem, but plants produced by this method will not fruit in less than 15 to 18 months from planting.

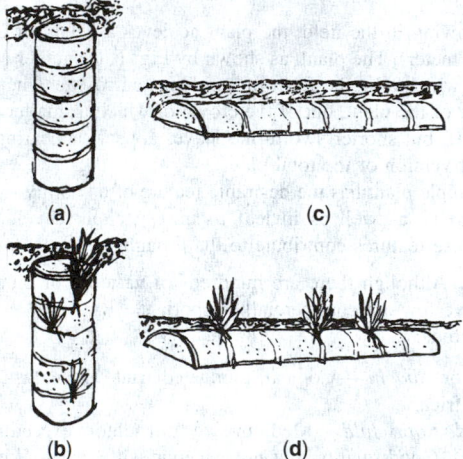

Fig. 2. Propagation of pineapple by stem cuttings: (**a**) short length of stem is buried in soil of cutting bed; (**b**) buds arising from stem; (**c**) longer stem cut into halves and inserted in cutting bed in horizontal position; (**d**) shoots arising from dormant buds along the stem. (*After Tindall*.)

Most commercial planters in Hawaii use slips or shoots from a parent plant (removed from the plants and dried after harvest). These are inserted into the ground through mulch paper. Beds are usually about 3 feet (0.9 meter) apart, with two rows per bed about 2 feet (0.6 meter) apart. Range of numbers of plants per area are: 15,000 to 18,000 plants/acre; 37,050 to 44,460 plants/hectare. Handled in this matter, fruits will ripen within 18 months of planting. The type of propagation and planting technique used varies with the size of the operation and the philosophy of the grower.

Soils and Fertilization. Pineapples will grow on a wide range of soils and will tolerate dry conditions, although the largest fruits are obtained from plants grown in fertile soils, properly cultivated. Nitrogen deficiency is evidenced by yellowing of leaves, sometimes followed by a red coloration on young fruit.

In large operations, prior to planting (frequently in the fall, but can be in spring or early summer), the old pineapple plants are crushed and disked into the soil. Following further working of the earth to improve drainage and aeration, an asphalt-impregnated paper is placed along the rows. Marks are made to indicate plant locations. Growers who use this system claim that the paper provides a number of functions—retention of moisture, maintenance of higher soil temperature, and inhibition of weed growth. Frequently, in the last soil preparation operation, prior to laying down the paper, a soil fumigant will be used.

On large pineapple plantations, fields are laid out to allow the smooth passage of spraying and other production equipment. In typical Hawaiian operations, fields will range from 100 to 600 acres (40 to 240 hectares), with the fields bisected by roads that may be 130 feet (39 meters) apart, over which mechanized equipment with booms up to 65-feet (19.5 meters) in length are used for spraying and harvesting. Almost completely a hand operation prior to the mid-1950s, pineapple production has been mechanized in a number of aspects and research directed toward mechanizing the more difficult operations of selecting and picking has been under way for a number of years.

Harvesting. Pineapples can be picked in varying degrees of ripeness, depending upon the final use intended, that is, for processing or for the fresh market. The points on the pineapple where the flowers blossomed are termed "eyes." An ideal time to harvest the fruit is when the eyes in the lower or basal part of the fruit have turned yellow. Fruit at this stage of maturation can be maintained for up to 4 weeks if maintained under proper refrigeration. However, if the required cool temperatures cannot be provided, the plant is susceptible to a number of market diseases. To extend the life of fruit during the marketing cycle, growers will frequently harvest the pineapples before any of the eyes turn yellow. Such fruits are known as "green." Although these fruits will ripen and can be retained longer, they will not develop desired flavor to the full. For the fresh market, the crown leaves are always left on the fruit. They will not become any sweeter once picked.

Packing and Storing. Pineapples are packed by hand into shipping containers that contain from 8 to 15 fruits, according to size of fruit. Pineapples are subject to chilling injury and are not adapted to long storage. The usual storage life is from 2 to 4 weeks. Hawaiian pineapple harvested at the $\frac{1}{2}$-ripe stage can be held about 2 weeks at 45° to 55°F (7.2 to 12.8°C) and still have about 1 week of shelf-life remaining. Ripe fruit should be held at about 45°F (7.2°C). Mature-green fruit is particularly susceptible to chilling injury at temperatures below 50°F (10°C). Hawaiian pineapples are commercially treated with a fungicide. A temperature of about 48°F (8.9°C) is suggested for South African pineapples. For all pineapples, continuous maintenance of storage temperature is considered equal in importance to the specific temperature. Pineapples that are subjected to a temperature that is too low take on a dull hue, develop water soaking of the flesh, develop darkening of the core, and are particularly subject to decay when removed from storage. Black rot (*Penicillium fusarium*) is the most serious decay of pineapples during transit and marketing.

Processing. Of the pineapples produced in Hawaii, 92.4% of the fresh weight is processed and 7.6% goes to the fresh market. Of the processed pineapple, 59.3% is canned as fruit slices or chunks; 37.2% is canned juice; and 3.5% is made into frozen concentrate. Of the fresh market sales, 23.9% of the Hawaiian pineapple is consumed in Hawaii, 75.7% is shipped to mainland United States; and 0.4% is shipped to markets of other countries.

In the terminology of the pineapple processors, the wet suspended solid removed from the centrifuge in the process of juice manufacturing is called "centrifuged pineapple juice underflow." This is a viscous fluid of pH 3.5, with an appearance of apple sauce. Scientists at the University of Hawaii have been studying the possibilities of deriving useful byproducts from this underflow. Traditionally, some pineapple solid wastes have been used directly as cattle feed. Appreciable amounts of the underflow are produced daily. For example, from just one Hawaiian processing plant, during the processing season, this may amount to some 7200 gallons (273 hectoliters)/day.

PINE TREES. Members of the family *Pinaceae* (pine family), these trees are of several species. Pines are of the genus *Pinus*. It should be noted that other genera in the *Pinaceae* family include the silver firs, the Douglas fir, the spruces, hemlocks, true cedars, and larches. Also, it should be noted that the term pine is applied to a few trees not in the genus *Pinus* as, for example, the Black Pine (*Podocarpus*), the Brown Pine (*Podocarpus*), the Japanese Umbrella Pine (*Sciadopitys*), the King William Pine (*Athrotaxis selaginoides*), the Norfolk Island Pine (*Araucaria excelsa*), and the Tasmanian Huon Pine (*Dacrydium franklinii*).

Important species of pine trees (Pinus), not listed in Table 1, include:

Arolla pine	*Pinus cembra*
Austrian black pine	*P. nigra nigra*
Bhutan or Himalayan white pine	*P. wallichiana*
Bosnian pine	*P. leucodermis*
Calabrian-Corsican pine	*P. nigra maritima*
Crimean black pine	*P. nigra caramanica*
Japanese black pine	*P. thunbergii*
Japanese red pine	*P. densiflora*
Japanese white pine	*P. parviflora*
Jelicote or spreading-leaf pine	*P. patula*
Korean pine	*P. koraiensis*
Lacebark pine	*P. bungeana*

Loblolly pine	*P. taeda*
Macedonian pine	*P. peuce*
Mexican white pine	*P. ayachuite*
Montezuma pine	*P. montezumae*
Mountain pine	*P. mugo*
Pinaster pine	*P. pinaster*
Piñon pine	*P. cembroides, monophylla, etc.*
Scrub pine	See Mountain pine.
Shore pine	*P. contorta*
Siberian pine	*P. echinata*
Stone pine	*P. pinea*
Umbrella pine	See Stone pine.

All pine trees are conifers, evergreen, and prefer lots of sunlight. Pines are also described extensively in the entry on **Conifers**. See Table 1.

Highlights on the distribution of pine trees in North America include: the bigcone pine is found in the mountains of California; the bristlecone pine is found in the western part of the United States (Oregon, Colorado, Utah, Nevada, Arizona, the eastern part of California and the northern part of New Mexico); the lodgepole pine is found in the Pacific Northwest and in parts of the Rocky Mountains; the red pine is found east, north, and just west of the Great Lakes region to south of the Hudson Bay and in northeastern Canada to the Gulf of Saint Lawrence; the western white pine is found in the Pacific Northwest; the pitch pine ranges from New Brunswick in eastern Canada south and considerably inland along the east coast of the United States to Georgia and parts of Florida; the eastern white pine is found from Newfoundland and southward along the Atlantic coast to about New Jersey, inland and westward considerably west of the Great Lakes region and inland and south into northern Georgia. The loblolly pine ranges from southern New Jersey along the Atlantic coast to Florida and west to the Gulf states into Texas. It ranges inland from the coast for 250 to 350 miles (402 to 563 kilometers). The longleaf pine is found in the southern United States from Virginia to nearly the tip of Florida and westward into the Gulf states with patches of population in eastern Texas; the Monterey pine ranges from central California northward to the Monterey Peninsula. It is also found on some of the Channel Islands off the California coast and on Guadeloupe Island off Lower California; the knob-cone pine is found from southern Oregon to northern California and through the Coast and Cascade Ranges and the Sierra Nevada Mountains. It is a common tree along parts of the Sacramento River; the piñon pine is found in New Mexico and environs. A species update report on the longleaf pine by W. Voigt, Jr. is contained in *American Forests*, 43–45, May 1986.

Probably the most common pine in Europe is the Scots pine, which ranges from the Mediterranean northward to Siberia. Authorities believe that it was the only pine species to survive the Ice Ages. Pines found in the Alps include the Arolla pine of Switzerland and the Austrian black pine and further east, the Crimean black pine. The Corsican pine is found on Corsica as well as in southwestern Italy. The stone or umbrella pine is considered a classical landscape tree in the Mediterranean region. The pinaster pine prefers the seacoast and it is interesting to note that in the late 1700s, some 12,000 or more acres of sand dunes were planted with these trees on the Bay of Biscay and that the third and fourth generation of these trees are surviving well. The scrub pine found in the Alps is a favorite with some rock gardeners. The Bosnian pine is also a favorite of European gardeners.

Among the Asian pines, there are the Japanese pines—Japanese red pine, and Japanese black pine, now garden favorites in other parts of the world. The white pine, however, is not so hardy and is confined to arboretums. Other Asian pines include the Korean pine (there is an excellent specimen in the National Arboretum in Washington, DC), the lacebark pine of China, the much rarer Chinese white pine, and the umbrella pine of Japan.

There is insufficient space here to detail all species of pine trees. However, the bristlecone pine merits special attention because authorities now believe that at least one living specimen in the White Mountains of California is nearly 5,000 years old.

The range of the bristlecone pine was mentioned previously. The tree requires an altitude of from 7,000 to 12,000 feet (2100 to 3600 meters). About 100 years are required for the tree to reach maturity. The cone takes two years to mature. At Indio, California, stands "The Patriarch," considered the oldest living thing and a national monument. The American Forests estimated the tree to be about 4,600 years old. Many trees no taller than 10 to 20 feet (3 to 6 meters) may be from 500 to 900 years old. On the summits where these trees often grow, snow may remain for several summers, thus hindering growth, while winds blowing particles of sand from the desert below tear at the tree's bark and branches. In Colorado, some people refer to the wood of this tree as "wind timber." Studies indicate that some trees may only grow an inch during a century. The branches are usually split, twisted, and shattered. Growth of this nature is referred to as "crooked wood" by people in the Swiss Alps.

The bark of the bristlecone is shallow, furrowed with a red-brown color. The twig is light orange, becoming dark with foliage at the tip. The leaf or needle occurs in groups of five and are 1 to $1\frac{1}{2}$ inches (2.5 to 3.8 centimeters) long. It is dark green, curved, glossy, and seemingly brushed forward. It may be accompanied by an exudation of resin.

The male flower is orange red; the female, purple. The cone is about 3 to $3\frac{1}{2}$ inches (7.6 to 8.9 centimeters) long, prickly, bristle-like, with thick scales. The seed is about $\frac{1}{4}$ inch (0.6 centimeter) long, light brown in color. It is widely sown by the wind. The heartwood is light brown. The sapwood is pale, medium soft, brittle, and of fairly light weight (35 pounds/cubic foot; 561 kilograms per cubic meter).

At one time, the redwood tree was considered to be the oldest living thing. In the mid-1950s, a small group of bristlecone pines was discovered in the White Mountains. The redwoods add a new ring each year, usually of about the same size, because these trees live in areas where there is rainfall unfailingly each year. With the bristlecone, the rings are microscopically narrow because of the chronic draught conditions in the White Mountains. Scientists at the University of Arizona developed a procedure for matching samples of wood, both from living and dead trees and thus, with the aid of a computer, have been able to build up a continuous series of rings. There may be as many as 1,100 rings in the space of 5 inches (12.7 centimeters). These rings serve as a sensitive rain gage for the region and are of much interest to meteorologists and climatologists. These scientists expect to be able to prepare a weather history of the region for at least 10,000 years. Studies of the bristlecone pine rings also uncovered a surprise in connection with the use of carbon-14 dating techniques. This method, based upon the amount of carbon in the atmosphere, has been subjected to revisions as the result of these studies. It was found that the atmospheric carbon level, assumed to be constant, actually holds true for only the last 3,400–3,600 years. The errors in dating structures, artifacts, etc., prior to about 1500 B.C. have had to be revised. For example, it is now estimated that Stonehenge is at least 1,000 years older than originally believed. This, in turn, has caused revisions in the thinking concerning the technology required at Stonehenge. An interesting report on the antiquity of the bristlecone pine is contained in *American Forests*, 26–42, September 1978. Author is R. Grant.

TABLE 1. RECORD PINE TREES IN THE UNITED STATES[1]

Specimen	Circumference[2]		Height		Spread		Location
	Inches	Centimeters	Feet	Meters	Feet	Meters	
Apache pine (1998) (*Pinus engelmannii*)	127	323	108	32.9	44	13.4	Arizona
Apache pine (1998) (*Pinus engelmannii*)	121	307	112	34.1	38	11.6	Arizona
Arizona pine (1998) (*Pinus ponderosa var. arizonica*)	153	389	127	38.7	57	17.4	Arizona
Austrian pine (1991) (*Pinus nigra*)	129	328	114	34.7	49	14.9	Washington
Bishop pine (1986) (*Pinus muricata*)	172	437	112	34.1	40	12.2	California
Bolander's pine (1983) (*Pinus contorta var. bolanderi*)	58	147	76	23.2	18	5.5	California
Border pinyon (1999) (*Pinus discolor*)	64	163	32	9.8	37	11.3	Arizona
Chihuahua pine (1997) (*Pinus leiophylla var. chihuahuana*)	121	307	87	26.5	34	10.4	Arizona
Colorado bristlecone pine (typ.) (1985) (*Pinus aristata var. aristata*)	132	335	76	23.2	39	11.9	New Mexico
Colorado bristlecone pine (typ.) (1986) (*Pinus aristata var. aristata*)	138	351	72	21.9	33	10.1	New Mexico
Coulter pine (1996) (*Pinus coulteri*)	209	531	80	24.4	78	23.8	California
Digger pine (1998) (*Pinus sabiniana*)	214	544	95	29	90	27.4	California
Eastern white pine (1999) (*Pinus strobus*)	200	508	150	45.7	53	16.2	Michigan
Foxtail pine (1982) (*Pinus balfouriana*)	316	803	76	23.2	34	10.4	California
Intermountain bristlecone pine (1978) (*Pinus aristata var. longaeva*)	473	1201	47	14.3	41	12.5	California
Jack pine (1995) (*Pinus banksiana*)	116	295	56	17.1	61	18.6	Minnesota
Jeffrey pine (1984) (*Pinus jeffreyi*)	307	780	197	60	90	27.4	California
Knobcone pine (1976) (*Pinus attenuata*)	135	343	117	35.7	66	20.1	California
Limber pine (1988) (*Pinus flexilis*)	275	699	58	17.7	46	14	Utah
Loblolly pine (1993) (*Pinus taeda*)	188	478	148	45.1	83	25.3	Arkansas
Lodgepole pine (1999) (*Pinus contorta var. latifolia*)	132	335	155	47.2	31	9.4	*Idaho*
Longleaf pine (1999) (*Pinus palustris*)	127	323	120	36.6	66	20.1	Georgia
Mexican pinyon pine (1982) (*Pinus cembroides*)	111	282	66	20.1	44	13.4	Texas
Monterey pine (1998) (*Pinus radiata*)	204	518	95	29	90	27.4	California
Parry pinyon pine (1976) (*Pinus quadrifolia*)	86	218	53	16.2	42	12.8	California
Pinyon (two-leaf) pine (1982) (*Pinus edulis*)	213	541	69	21	52	15.8	New Mexico
Pitch pine (1998) (*Pinus rigida*)	142	361	112	34.1	75	22.9	Georgia
Pitch pine (1999) (*Pinus rigida*)	169	429	99	30.2	40	12.2	New Hampshire
Pond pine (1998) (*Pinus serotina*)	112	284	119	36.3	60.5	18.4	Florida
Ponderosa pine (typ.) (1997) (*Pinus ponerosa var. ponderosa*)	293	744	227	69.2	68	20.7	California
Ponderosa pine (typ.) (1997) (*Pinus ponerosa var. ponderosa*)	294	747	223	68	59	18	California
Red pine (1993) (*Pinus resinosa*)	124	315	124	37.8	60	18.3	Michigan
Red pine (1998) (*Pinus resinosa*)	120	305	126	38.4	48	14.6	Minnesota
Rocky Mountain ponderosa pine (1982) (*Pinus ponderosa var. scopulorum*)	241	612	194	59.1	64	19.5	Montana
Sand pine (1997) (*Pinus clausa*)	97	246	91	27.7	42	12.8	Florida
Scotch pine (1983) (*Pinus sylvestris*)	186	472	64	19.5	76	23.2	Michigan

TABLE 1. (*Continued*)

Specimen	Circumference[2]		Height		Spread		Location
	Inches	Centimeters	Feet	Meters	Feet	Meters	
Shore pine (typ.) (1992) (*Pinus contorta var. contorta*)	138	351	101	30.8	37	11.3	Washington
Shortleaf pine (1999) (*Pinus echinata*)	139	353	88	26.8	68	20.7	Mississippi
Sierra lodgepole pine (1997) (*Pinus contorta var. murrayana*)	238	605	124	37.8	42	12.8	California
Singleleaf pinyon pine (1991) (*Pinus monophylla*)	164	417	45	13.7	40	12.2	California
Slash pine (typ.) (1992) (*Pinus elliottii var. Ielliottii*)	130	330	138	42.1	55	16.8	Florida
South Florida slash pine (1997) (*Pinus elliottii var. densa*)	138	351	68	20.7	64	19.5	Florida
Southwestern white pine (1974) (*Pinus strobiformis*)	185	470	111	33.8	62	18.9	New Mexico
Spruce pine (1998) (*Pinus glabra*)	125	318	149	45.4	53	16.2	Georgia
Spruce pine (1997) (*Pinus glabra*)	160	406	112	34.1	66	20.1	Louisiana
Sugar pine (1993) (*Pinus lambertiana*)	442	1123	232	70.7	29	8.8	California
Table mountain pine (1984) (*Pinus pungens*)	97	246	94	28.7	46	14	North Carolina
Torrey pine (1993) (*Pinus torreyana*)	245	622	126	38.4	130	39.6	California
Virginia pine (1998) (*Pinus virginiana*)	111	282	101	30.8	56	17.1	Kentucky
Washoe pine (1997) (*Pinus washoensis*)	243	617	145	44.2	64	19.5	California
Western white pine (1991) (*Pinus monticola*)	394	1001	151	46	52	15.8	California
Whitebark pine (1980) (*Pinus albicaulis*)	331	841	69	21	47	14.3	Idaho

[1] From the "National Register of Big Trees." American Forests (by permission).
[2] At 4.5 feet (1.4 meters).

Longleaf and slash pine are the principal sources of turpentine. The sapwood contains about 2% oleoresin, heartwood from 7 to 10%, and stumpwood about 25%. The majority of oleoresin is obtained from the sapwood of living trees. However, this is not the sap of the tree. Oleoresin yields about 20% oil of turpentine and 80% rosin, the two commodities collectively known as naval stores.

Pines are widely used in the United States for making kraft paper, paperboard, and book paper. About half of this wood comes from the southern states. The fiber length of longleaf pine is about 3.5 millimeters ($\frac{1}{4}$ inch); and that of jack pine and lodgepole pine a little over 2 millimeters ($\frac{1}{8}$ inch). Along with Douglas fir and Sitka spruce, these pines make up the principal softwoods (as contrasted with the hardwoods, birch, cottonwood, and willow) for pulp and paper production. The term *jack pine* is sometimes rather loosely used. Usually, the term is synonymous with lodgepole pine, but the term is also applied to *P. banksiana* of central Canada (mainly used for telephone poles), and to black, prickly pine, and to certain species of spruce. Engineering constants of various commercial pine woods are given in Table 2.

Additional Reading

Ciesla, B.: "The Digger; California's Oddball Pine," *Amer. Forests*, (January–February 1987).

Dusek, K.H.: "Update on our Rarest Pine (Torrey)," *Amer. Forests*, 26 (November 1985).

Jone, S.: "White Pine Pest (Sawfly)," *Amer. Forests*, 22 (December 1981).

Kingsbury, L.: "New Beginning for the Western White Pine," *Amer. Forests*, 27 (December 1984).

Taylor, A.: "Mission in the Pines (Search for Blister-Rust Cankers)," *Amer. Forests*, (July 1985).

Additional interesting reading on pine trees includes: "White Pine Pest (Sawfly)," by S. Jones, *American forests*, 22–25 (December 1981); "Mission in the Pines (Search for blister-rust cankers)," by A. Taylor, *American Forests*, 27–29 (July 1985); "Update on Our Rarest Pine (Torrey)," by K.H. Dusek, *American Forests*, 26–29 (November 1985); and "New Beginning for the Western White Pine," by L. Kingsbury, *American Forests*, 30–33 (December 1984).

PINHOLE IMAGE. If a small opening is made in one side of a darkened room or box, an inverted picture of objects outside appears upon the wall opposite the opening. Such a picture differs from a true image in that it is not formed by light from a given point of the source diverging and being reconverged at the corresponding image-point, as by a lens, but is an effect of the rectilinear propagation of light. The only spot on the screen reached by light from a given point of the source is that in direct line with the opening. For this reason, pinhole images are of low intensity. On the other hand, they are free from the distortions to which lens images are subject, and with sufficient exposure, very good photographs can be made by means of them. The pinhole image also affords an excellent means of viewing eclipses of the sun.

TABLE 2. MOISTURE AND WEIGHT OF VARIOUS PINE WOODS

Common Name of Species	Green Condition			Air-Dried to 12% Moisture	
	Moisture Content %	Weight/ Cubit Foot (pounds)	Weight/ Cubic Meter (kilograms)	Weight/ Cubic Foot (pounds)	Weight/ Cubic Meter (kilograms)
Eastern white pine	73	36	576	25	399
Loblolly pine	81	53	847	36	576
Lodgepole pine	65	39	625	29	466
Longleaf pine	63	55	879	41	657
Shortleaf pine	81	52	833	36	576
Sugar pine	137	52	833	25	399
Western white pine	54	35	561	27	431

Source: U.S. Forest Products Laboratory.

PINK BOLLWORM (*Insecta, Lepidoptera*). A widely-distributed enemy of cotton, which probably originated in India or Africa. The adult is a small gray-brown moth, *Pectinophora* (platyedra) *gossypiella*, and the caterpillar is one-half inch long and is pinkish above. The larvae work in the flowers and bolls, causing imperfect development and destroying seeds and lint. It also attacks other plants, including the hollyhock and okra.

This species is found in the western part of the cotton-growing areas of the United States. Vigorous measures have been taken to eliminate it, for no adequate methods of control have been discovered.

PINK-EYE. See **Conjunctivitis**.

PINO. The column of smoke and ashes emitted by an explosive volcano, usually in the beginning or in the early stage of an eruption. The term is of Italian origin signifying the cauliflower-shape of the cloud as observed during the eruptions of Vesuvius.

PINWORM (*Nemathelminthes, Nematoda*). A small roundworm that lives in the alimentary tract of man, chiefly in the large intestine. The female is about $\frac{2}{5}$-inch (1 centimeter) long and the male somewhat smaller. Eggs are taken into the mouth in water or from the hands, or on raw vegetables. The entire life cycle takes place in the one host. Pinworms are usually not harmful but they may cause nervous symptoms.

PION. See **Muon**.

PIONEER 10 AND 11. See **Space Science Missions: Solar System**.

PIONEER VENUS PROJECT. See **Space Science Missions: Solar System**; and **Venus**.

PIPEFISHES (*Osteichthyes*). Of the order *Solenichthys* (tube-mouthed fishes), family *Syngnathidae*, pipefishes have been aptly described as a "pipestem cleaner suddenly come to life." In this comparison, however, the cleaner would have to be equipped with a bony-plate armor. As with seahorses of the same family, pipestem fishes display an independent movement of their eyes. Each eye appears to operate independently. Pipefishes range in length from about 1 inch to about 18 inches (2.5 to about 46 centimeters) maximum. There are some 150 species of pipefishes. Although essentially found in marine waters of both the Atlantic and the Pacific, some species can tolerate both salt and fresh waters. There are a few fresh water species. They prefer inshore, shallow water. Pipefishes are quite similar in numerous respects to their close relatives, the seahorses and it is believed that seahorses developed from a primitive pipefish. See also **Seahorses (Osteichthyes)**.

PIPE SNAKES. See **Snakes**.

PIPETTE. A slender glass tube open at both ends and having an expanded area at or near the center designed to contain a specific volume of liquid, e.g., 5ml. Liquid is drawn into the tube by oral or, for the sake of safety, some other form of suction.

PIRANHA. See **Characids**.

PIRANI GAGE. A thermal conductivity vacuum gage in which an increase of pressure from the zero point causes a decrease in the temperature of a heated filament of material having a large temperature coefficient of resistance, thus unbalancing a Wheatstone bridge circuit (or the circuit is adjusted to maintain the filament temperature constant).

PISCES (Constellation). Also referred to as the Fishes. (See map accompanying entry on **Constellations**.)

Pisces is a large constellation, which is of importance principally because it is the twelfth sign of the zodiac. There are relatively few interesting objects in the constellation although the brightest star (Alpha) is a close double, which may be resolved in instruments larger than a four-inch.

In spite of the fact that Pisces is the twelfth sign of the zodiac, the vernal equinox is located in this constellation at present. This is because precession has caused the vernal equinox itself to move back an entire "sign" along the ecliptic since the time when the names were first assigned.

PISCIS AUSTRINUS. A southern constellation, somewhat resembling a fish, located between Aquarius and Grus.

Pi SECTION. This is a type of network in which the elements are arranged in π shape, i.e., a shunt element across the circuit at each end of a series element.

PITCH. (1) A carbonaceous, tacky residue resulting from distillation of coal tar, petroleum, pine tar, and fatty acids. Some types, such as glance pitch, occur naturally. They are used chiefly as sealants, roofing compounds, and wood preservatives. Synthetic carbon fibers are made from petroleum pitch. (2) In papermakers' terminology, a mixture of calcium carbonate, calcium soaps from wood components, and miscellaneous residues from materials used in paper manufacture. Pitch of this type is a production nuisance that requires close control. (3) The degree of slope of an inclined plane as in a screw auger, as measured by the distance between the flights or treads.

See also **Coal Tar and Derivatives**.

PITCHBLENDE. A massive variety of uraninite or uranium oxide found in metallic veins. Contains 55–75% UO_2, up to 30% UO_3, usually a little water, and varying amounts of other elements. Thorium and the rare earths are generally absent.

See also **Uranium**.

PITCHING MOMENT. Rotation of an airplane about a lateral axis passing through the center of gravity is known as *pitch*. Nosing-up (positive) and diving (negative) motions are the result of moments acting around this axis. These moments are produced by thrust, wing lift and drag, parasitic drags, and tail surface forces. For airplane trim these moments must be in equilibrium, whereas for longitudinal stability an increase of angle of attack caused by an external gust or a momentary deflection of the elevator must be decreased by inherent diving moments of the airplane. Or, when a decrease in angle of attack is produced, it must be countered by a stalling moment, which will bring it back to its original angle of attack or trim. The relation between the moment and the moment coefficient is expressed in the following equation:

$$M = C_m q S c$$

Where q = dynamic pressure equivalent to the air speed = $pv^2/2$
S = wing area
C = wing chord
C_m = pitching moment coefficient of the airplane

For airplane trim, the expression M equals zero, so that C_m must equal zero. At any other angle than for trim C_m is not zero, but increases either negatively (for diving motions) or positively (for stalling motions).

PITCH (Music). See **Musical Sound**.

Pi THEOREM. A principal theorem in dimensional analysis that may be stated as follows: Suppose we have a dimensionally homogeneous relation B ($\alpha, \beta, \gamma, \ldots$) in n dimensional variables, $\alpha, \beta, \gamma, \ldots$, valid for certain system of m fundamental units. The equation may then be put in the form $F(\pi_1, \pi_2, \ldots) = 0$, where the π's are the $n - m$ independent products of the variables $\alpha, \beta, \gamma, \ldots$, which are dimensionless in the fundamental units.

PITOT TUBE. A pitot-tube air-speed indicator consists of two elements: (1) A dynamic tube, which points upstream and determines the dynamic pressure; and (2) the static tube, which points normal to the air stream and determines the static pressure at the same point. The tubes are connected to the two sides of a manometer or inclined gage so as to obtain a reading of velocity pressure, which is the algebraic difference between the total pressure and the static pressure. See also **Manometer**; and **Manometer (Barometer)**. The relationship between air velocity and velocity pressure is:

$$v = \sqrt{2gH}$$

where v = velocity, feet per second; g = acceleration due to gravity; and H = velocity head or pressure, feet of air. The pressure differential created is quite small with relation to air velocity. At 100 feet (30 meters) per minute, the velocity pressure is only 0.0625 inches (1.6 millimeters) of water. Consequently, the instrument is not generally used for measuring velocities less than 1,000 feet (300 meters) per minute.

The principle of the pitot tube, in addition to aerospace applications, can be used as a liquid flow-measuring device, but because of its tendency to clog, cannot be used with liquids that have suspended solid matter. The device is useful for flow measurements in laboratory and research applications. See also **Airspeed Indicator**; and **Bernoulli Law**.

PITUITARY GLAND.   A small organ of the endocrine system, the size of an average pea — with about the same weight. The gland is larger in women than in men, particularly in those women who have borne children. The gland is joined to the undersurface of the brain by a thin stalk and is protected by a bony structure that surrounds the gland. Because of its shape, the bony structure is called the "turkish saddle" (*sella turcica*).

Only within relatively recent years has the role of the pituitary become better understood. The pituitary gland is the most important organ in the regulation of growth, milk production, and in the control of several other endocrine glands. In turn, the pituitary is regulated to some extent by many of the other endocrine glands, as well as by the hypothalamus, which lies immediately above it. See also **Central and Peripheral Nervous Systems**; and **Endocrine System**.

It has long been known that severe pituitary disturbances, such as tumors, influence the function of other endocrine glands. The pituitary can upset the body's hormone balance so severely as to cause mental as well as physical illness. Tumors on the gland can usually be removed surgically, with good chance of relieving the emotional disturbances of which they are the indirect cause. Physical disorders caused by pituitary overactivity may be managed by surgical removal.

The pituitary is made up chiefly of two distinct parts called lobes — an *anterior lobe* and a *posterior lobe*. There is also a middle portion, the *pars intermedia*, that constitutes only a minor fraction of the entire gland. Under a microscope, this simple division of the pituitary appears far more complex. The supply of incoming nerve fibers is large; it has been estimated that approximately 50,000 nerve fibers enter into this organ, being confined almost exclusively to the posterior lobe. The blood supply, which is arranged in a circular pattern to avoid even the smallest temporary breakdown, is also extensive. It serves the gland by bringing food, gases, and hormones and by conveying the secretions of the pituitary to other parts of the body.

The pituitary produces a number of hormones, each endowed with the ability to produce some specific effect in one or more organs of the body, especially other endocrine organs. The hormones produced by the anterior lobe differ from those made by the pars intermedia. Most of the pituitary hormones are protein in nature.

The functions controlled by each part of the pituitary are entirely different. The largest number of hormones are produced by the anterior lobe; hence; it performs most of the functions of the entire gland. The posterior lobe does not in itself manufacture any hormones, although it does receive and store hormones made by the hypothalamus.

The pituitary has been called the "master gland" because it is believed to be the endocrinological center of the body. The anterior lobe of the pituitary regulates the growth and proper functioning of other endocrine organs by complex processes. For example, it produces a hormone, called the thyrotrophic hormone, which acts on the thyroid to stimulate its production of thyroid hormone. See also **Thyroid Gland**. In addition to the thyrotrophic hormone, the anterior pituitary is believed to secrete several other hormones. These include the adrenocorticotrophic hormone (ACTH), the follicle-stimulating and luteinizing components which make up the gonadotrophic hormone, the luteotrophic hormone, or lactogenic hormone, and the growth hormone.

Gonadotrophic hormone. The anterior lobe produces active principals that are effective stimulators of the *gonads*. The hormones that act on the sex organs are termed gonadotrophic hormones. The sexual organs in both male and female have a double function, reproduction and the production of sex hormones. The anterior lobe of the pituitary, by manufacturing and secreting the gonadotrophic hormones, controls the production of these hormones in the ovaries of the female and the testes of the male. In addition to these functions, the gonadotrophins, directly and indirectly, stimulate the development of the sex organs and the maintenance of their structure.

The testes, under the influence of the gonadotrophins, manufacture the male hormones. These, in turn, exert their action on the other parts of the body, chiefly the organs of reproduction. When the testicular tubules have developed under the influence of the male hormones, maturation of the spermatozoa also is stimulated by the gonadotrophin from the anterior lobe of the pituitary. Failure to produce male hormones results in immature appearance and lack of development of the accessory sex organs; in the previously normal adult, loss of the male hormones results in changes in appearance and degeneration of the accessory sex organs.

In women, the ovaries, under control of the hormones from the anterior lobe of the pituitary, produce the female hormones. The maturation of ova in the ovaries is stimulated by the gonadotrophins. The female hormones act on the reproductive organs and are responsible for the proper growth and function of the uterus, vagina, and other reproductive organs. It is not uncommon to observe disturbance in sexual characteristics of individuals with defective pituitary function.

Deficient pituitary activity is in many cases reflected in the lack of development of the sex organs, which may remain infantile. When accompanied by obesity, the condition is known as the *adiposogenital syndrome*. Other disturbances also can arise. The rate of production of gonadotrophins by the pituitary is influenced by the production of sex hormones. The effects are mutual and the two glands, the pituitary and the ovaries or testes maintain an exact balance in hormone production.

Adrenocorticotrophic hormone. A substance called adrenocorticotrophic hormone acts on the adrenal glands and is produced in the anterior lobe of the pituitary. Abbreviated, this substance is ACTH. This substance stimulates the production of most of the cortical hormones, but especially *hydrocortisone*. If the production of ACTH is below normal, the adrenal cortex diminishes in size and the production of most cortical hormones falls to low levels. Methods to measure ACTH directly and inadequate. ACTH has been isolated from pituitaries of cattle and pigs and is available in pure form. ACTH also is synthesized. The hormone has been found useful as therapy for a variety of disorders, as well as for the diagnosis of some conditions. Although the principal effect of ACTH is to stimulate the adrenal cortex to greater secretion, it also may perform some of the functions of the adrenal glands when the adrenals are absent.

ACTH is effective in the management of certain hematological diseases, as well as in conditions of stress. It can be used in the treatment for certain spasms involving the head, trunk, and arms of infants. It can also be used for treating young children subject to convulsions caused by diabetes. ACTH may be used in the treatment for severe allergic manifestations associated with dermatitis. Allergic reactions to the hormones, however, may occur.

The role of the pituitary gland in concert with the thyroid gland in the regulation of metabolism is described under **Thyroid Gland**.

Pituitary Hormones in Brain. During the last few decades, research into the hormone-generating facilities of the pituitary gland, particularly as these may involve the brain, has been accelerated. Through the use of radioimmunoassay, bioassay, and immunocytochemical techniques, peptides and protein hormones usually considered as being of pituitary origin have been detected within the central nervous system. Investigation continues into determining if these hormones are generated in the pituitary gland and then transported to the brain; or if they are generated elsewhere and mimic pituitary hormones. Some researchers have tentatively concluded that they are synthesized within the central nervous system (*neurosecretory cells*) and that their regulation may differ somewhat from that of their pituitary counterparts. But, other researchers suggest that pituitary hormones may be transported directly to the brain to modify brain function. A better understanding of this problem could lead to answers to a number of questions as regards such functions as memory, sleep, pain, orgasm, endocrine feedback loops, cerebral blood flow, cerebral vascular permeability, cerebrospinal fluid dynamics, epilepsy, headache, acupuncture, and mental illness.

Disorders of the Pituitary Gland

The pituitary gland is involved in a complex group of disorders that do not necessarily have a neatly classified group of causative factors.

Cushing's Disease and Cushing's Syndrome. Once considered rare, but now seen in some frequency, Cushing's disease is recognized by obesity of the abdomen, face, and buttocks, but not of the limbs. The skin about the face and hands is redder than normal. Hair grows profusely, and women may grow mustaches and beards. Bones become brittle and suffer a considerable loss of mineral components. Sexual functions may fall to a low level or become suppressed altogether.

When the manifestations of Cushing's disease occur in patients with excessive production of adrenal cortical hormones of the adrenal glands, the condition is called *Cushing's syndrome*. The adrenal hormone production is excessive also in Cushing's disease, as a result of overstimulation of the adrenals by pituitary hormone (ACTH) produced in excess by the tumor.

Some physicians in the early 1990s now stress that pituitary adenomas are not rare as once believed, but are relatively common, often being found at autopsy in a wide range (up to 25%) of persons who were not suspected as having the disease. Some 40% of tumors thus identified contain prolactin. Currently, surgery is the only effective treatment for gonadotroph-cell adenomas.

As reported by L. Daneshdoost (University of Pennsylvania School of Medicine) and co-researchers, "Adenomas that arise from the gonadotroph cells of the pituitary gland account for a substantial percentage of pituitary macroadenomas in men, but they are rarely recognized in women."

Endogenous (produced or synthesized within an organism) Cushing's syndrome remains difficult to diagnose as of the early 1990s. Specific tests have not proved reliable. As pointed out by D.N. Orth (Vanderbilt University Medical Center), "Cushing's syndrome is either adrenocorticotropin-dependent or independent. The dependent type is due to hypersecretion of adrenocorticotropin by a pituitary adenoma (Cushing's disease) or by a nonpituitary tumor (ectopic adrenocorticotropin syndrome)." These two causes are difficult to differentiate. Although too technical to describe here, a group of researchers representing several institutions reported in September 1991 about a new diagnostic test that may be effective in distinguishing Cushing's disease from ectopic adrenocorticotropin syndrome. This can prove to be of large significance because the therapy for each of the two syndromes differs considerably.

Atrophy of Anterior Lobe. Degeneration of the anterior lobe in adults results in a disease sometimes called *Simmonds'* or *Simmonds-Sheehan disease*. The disorder is characterized by extreme appearance of aging. Axillary and pubic hair are lost, there is a loss of teeth, and hair of the head becomes gray and sparse. The skin is wrinkled and the face has a wizened appearance. All of the metabolic functions of the body are affected, and eventually the mental functions decline. The condition occurs most often in women and nearly always arises after postpartum hemorrhage or shock and excessive loss of blood. The condition gradually deteriorates over a period of years. The pituitary atrophy is believed to be the result of lack of oxygen reaching the gland during the shortage of blood. The disease has been confused with *anorexia nervosa*. For the latter, hormonal treatment is secondary to psychiatric and dietary treatment. Amenorrhea (absence of menstruation) is a constant feature of Simmonds-Sheehan disease, but is not always present in anorexia nervosa.

Fröhlich's Syndrome. A lesion of the hypothalamus may affect the anterior lobe of the pituitary, resulting in Fröhlich's syndrome or *dystrophia adiposogenitalis*. The patient is excessively fat and the sexual organs are infantile. In early childhood, the disease causes dwarfism. The victim is mentally lazy and possesses a voracious appetite. When the disease develops in adulthood, male patients become effeminate, with soft skin and feminine distribution of fat in the breast region and thighs. In female patients, the obesity is extreme; it is not uncommon to see patients with this disorder weighing 300 pounds (136 kilograms).

The obesity is not a direct result of tumor in the pituitary, because the pituitary gland has no relationship to obesity, a fact often misunderstood. The obesity is a result of the same tumor's affecting the adjacent hypothalamus. Pituitary insufficiencies result in the immaturity of the sexual organs. Hypothalamic disease results in a disturbance of the appetite control center, with resulting obesity. The disease should not be confused with the typical obesity of childhood and adolescence. Fröhlich's syndrome is very rare and most obese children do not have this condition, nor any detectable glandular disturbance; rather, they are obese because of dietary habits. See also **Hormones**.

Dwarfism and Giantism. The pituitary gland is susceptible to the growth of tumors that may make the gland over- or underactive. Decreased function of the pituitary results in retarded growth. The growth hormone (*somatotrophic* hormone) exerts its major effect upon the size of the organs and the skeleton, which in cases of decreased pituitary function remains small. The condition that results is called *pituitary dwarfism*, or *infantilism*. Teeth grow slowly if insufficient growth hormone is produced, and the development of permanent teeth is considerably delayed. Untreated pituitary dwarfs do not grow over 3 to 4 feet (0.9 or 1.2 meter) in height and remain sexually immature. Specific therapy for such patients is the administration of pituitary growth hormone of primate or human origin. However, even with human growth hormone, refractory states may develop, resulting in poor growth. Laron dwarfism (growth hormone insensitivity) — that is, the lack of growth hormone receptors — may be alleviated by the administration of insulin-like growth factor 1 (IGF-1).

In late 1990, A.L. Rosenbloom (University of Florida) studied a small group (50 patients) with Laron dwarfism in southern Ecuador. This was a highly inbred group, and a genetic link was indicated. A marked predominance in females was indicated, but explained by the fact that there is early fetal death of most affected males.

Pituitary dwarfs do not achieve normal endocrine function. For dwarfed girls to develop breast tissue and to menstruate, they must be treated with estrogen. However, treatment with estrogen may stop growth of the bones before the patient has attained acceptable height. Therefore, it is wise to delay therapy with the female sex hormones as long as possible.

Although many forms of dwarfism are the result of pituitary or thyroid insufficiency, some are genetically determined. Since 1860, 49 cases of a dwarfism, known as the Ellis-van Creveld syndrome, have been verified in the Amish community of Pennsylvania — all dwarfs descended directly or indirectly from one ancestral couple. These dwarfs range in height from 40 to 60 inches (102 to 152 centimeters); they have six fingers on each hand, the extra finger on the outside beyond the little finger. Sometimes there is a sixth toe. Many of the infants have heart abnormalities and a weakness or deficiency of cartilage in the chest. One-fourth of dwarfed children with such defects dies within two weeks of birth; however, others achieve near-normal life spans. There is no mental retardation or loss of intelligence.

The most familiar form of dwarfism is *achondroplasia*. Persons in this category have large heads with saddle or scooped-out noses, short extremities, and sway backs. Advances have been made in the treatment of this condition with somatotrophin, but supplies of the substance are limited.

On occasion, the anterior lobe or the entire pituitary gland may be enlarged and the production of hormones may increase above normal range. This, in turn, causes excessive growth. If the condition develops while the bones are in the process of growing, the result is *giantism*; individuals with this condition may grow to over 8 feet (2.4 meters) in height. Prevention of giantism is relatively simple if diagnosis is made early. To close the epiphyses (open ends of the bones, which are still growing) of probable giants, estrogen is used in girls; and both estrogen and testosterone are used in boys. This treatment does not affect later gonadal function adversely. The epiphyses of these patients should be studied at 4-to-6-month intervals to determine whether growth is stopping and if therapy can be discontinued. X-ray examination of the hands and wrist provides good indication inasmuch as the epiphyses in the wrists are the last to close.

Acromegaly. In later life, after bones have ceased to grow, an overactive pituitary causes excessive stimulation of the growth centers which results in the disease known as *acromegaly*. This condition is characterized by an abnormal development of feet and hands. The jaw is prominent and large, as are the bones of the skull. The face may become angular and irregular, and the general appearance is that of a primitive man. The fully-developed disease is readily discerned by the layman; the early disease is difficult to detect. In this condition, the pituitary gland usually is enlarged by a tumor. Steroid therapy is given, depending in part on the extent of the condition and the level of circulating growth hormone in the blood. Visual acuity must be carefully monitored during this therapy. Acromegaly also occurs to a slight degree in some women during pregnancy, but regresses after delivery.

Additional Reading

Backer, K.L., J.P. Bilezikian, W. Hung, et al.: *Principles and Practice of Endocrinology and Metabolism*, 3rd Edition, Lippincott Williams Wilkins, Philadelphia, PA, 2001.

Berkow, R. and M.H. Beers: *The Merck Manual*, 17th Edition, Merck Company, Inc., Whitehouse Station, NJ, 1999.

Christy, N.P.: "Pituitary-Adrenal Function During Corticosteroid Therapy," *N. Eng. J. Med.*, 266 (January 23, 1992).

Daneshdoost, L., et al.: "Recognition of Gonadotroph Adenomas in Women," *N. Eng. J. Med.*, 589 (February 28, 1991).

deGroot, L.J. and J.L. Jameson: *Endocrinology*: 3 Volumes," 4th Edition, W. B. Saunders Company, Philadelphia, PA, 2000.

Dowset, R.J. and B. Fowble: "Radiotherapy for Acromegaly," *N. Eng. J. Med.*, 612 (August 30, 1990).

Griffin, J.E. and S.R. Ojeda: *Textbook of Endocrine Physiology*, 4th Edition, Oxford University Press, Inc., 2000.

Klabanski, A. and N.T. Zervas: "Diagnosis and Management of Hormone-Secreting Pituitary Adenomas," *N. Eng. J. Med.*, 822 (March 21, 1991).

Kostyo, J.L. and H.M. Goodman: *Handbook of Physiology: A Critical, Comprehensive Presentation of Physiological Knowledge and Concepts: Section 7: The Endocrine System: Hormonal Control of Growth*, Vol. 5, Oxford University Press, Inc., New York, NY, 1999.

Melmed, S.: "Acromegaly," *N. Eng. J. Med.*, 966 (April 5, 1990).

Molitch, M.E.: "Gonadotroph-Cell Pituitary Adenomas," *N. Eng. J. Med.*, 626 (Ferbuary 28, 1991).

Monson, J.P.: *Challenges in Growth Hormone Therapy*, Blackwell Science, Inc., Malden, MA, 1999.

Moran, A., et al.: "Gigantism Due to Pituitary Mammosomatotroph Hyperplasia," *N. Eng. J. Med.*, 322 (August 2, 1990).

Motta, M.: *Comprehensive Endocrinology*, 2nd Edition, Raven Press, New York, NY, 1991.

Neal, J.M.: *How the Endocrine System Works*, Blackwell Science, Inc., Malden, MA, 2001.

Oldfield, E.H., et al.: "Petrosal Sinus Sampling with and without Corticotropin-Releasing Hormone for the Differential Diagnosis of Cushing's Syndrome," *N. Eng. J. Med.*, 898 (September 26, 1991).

Orth, D.N.: "Differential Diagnosis of Cushing's Syndrome," *N. Eng. J. Med.*, 957 (September 26, 1991).

Pinchera, A., M. Serio, and X. Bertagna: *Endocrinology and Metabolism*, The McGraw-Hill Companies, Inc., New York, NY, 2001.

Rosenbloom, A.L., et al.: "The Little Women of Loja — Growth Hormone-Receptor Deficiency in an Inbred Population of Southern Ecuador," *N. Eng. J. Med.*, 1367 (November 15, 1990).

Schlaghecke, R., et al.: "The Effect of Long-Term Glucocorticoid Therapy on Pituitary-Adrenal Responses to Exogenous Corticotropin-Releasing Hormone," *N. Eng. J. Med.*, 226 (January 23, 1992).

Takasu, N., et al.: "Exacerbation of Autoimmune Thyroid Dysfunction After Unilateral Adrenalectomy in Patients with Cushing's Syndrome Due to an Adrenocortical Adenoma," *N. Eng. J. Med.*, 1708 (June 14, 1990).

Walker, J.L., et al.: "Effects of the Infusion of Insulin-like Growth Factor in a Child with Growth Hormone Insensitivity Syndrome (Laron Dwarfism)," *N. Eng. J. Med.*, 1483 (May 23, 1991).

Williams, R.H., D.W. Foster, H.M. Kronenberg, and P.R. Larsen: *Williams Textbook of Endocrinology*, 9th Edition, Harcourt Brace Company, San Diego, CA, 1999.

Web References

Cushing's Syndrome/Cushing's Disease and CRH: http://neurosurgery.mgh.harvard.edu/e-f-942.htm

Cushing's Syndrome: http://www.ninds.nih.gov/healthandmedical/disorders/cushingsdoc.htm

The Pituitary Gland: Location and Functions: http://www.umm.edu/endocrin/pitgland.htm

PIT VIPER. See **Snakes**.

PITYRIASIS.   Pityriasis is one of any of a number of skin diseases that have in common lesions that resemble dandruff-like scales without obvious signs of inflammation.

Types of pityriasis include: pityriasis alba (also called pityriasis streptogenes, pityriasis simplex, erythema streptogenes); pityriasis rosea; pityriasis rubra pilaris; and pityriasis versicolor (tinea versicolor).

Pityriasis Alba

Pityriasis alba is characterized by hypopigmented, round to oval, scaling patches on the face, upper arms, neck, or shoulders. The patches vary in size, usually being 1 to 2 inches (2.5 to 5 centimeters) in diameter. The color is white or light pink. The scales are fine and adherent.

Usually, the patches are sharply demarcated; the edges may be erythematous and slightly elevated. As a rule, pityriasis is asymptomatic. However, there may be mild pruritis. The disease occurs chiefly in children and teenagers.

The cause is unknown. Excessively dry skin following exposure to strong sunlight appears to be contributory. Efforts to find an infectious agent - either bacterial, viral, or fungal - have been unsuccessful.

Treatment. Highly useful are 0.5% hydrocortisone and 1% crude coal tar in a cream base (Zetone cream), half-strength Pragmatar ointment, Lac-Hydrin, 2% Zetar in Cordran cream, or 1% Vioform cream. The prognosis is good and there is usually spontaneous healing within several months to a few years.

Pityriasis Rosea

Pityriasis rosea is a common skin condition in children and young adults. It usually begins as one large spot on your chest, abdomen or back and then spreads. The rash of pityriasis rosea often sweeps out from the middle of the body, and its shape resembles drooping pine-tree branches.

Although pityriasis rosea has a distinctive appearance once the rash appears, in its early stages you may confuse pityriasis rosea with other skin disorders, such as ringworm or eczema.

The cause of pityriasis rosea is unclear, although the cause may be a viral infection, such as certain strains of the human herpes virus (HHV6 or HHV7). Pityriasis rosea usually goes away on its own within six to eight weeks. In the meantime, you can take steps to relieve the discomfort.

Signs and Symptoms. Characteristics of pityriasis rosea include:

- *Initial phase.* Pityriasis rosea typically begins with a large, slightly raised, scaly patch — called the herald patch — on your back, chest or abdomen.
- *Progression.* Smaller fine, scaly spots usually appear across your back, chest or abdomen in a pine-tree pattern a few days to a few weeks after the herald patch. Rarely, smaller spots may also appear on your arms, legs or face. The rash may itch.
- *Color.* The rash of pityriasis rosea often is scaly and pink, but if you have darker skin, it may be gray, dark brown or even black.
- *Other signs and symptoms.* About half the people who develop pityriasis rosea have signs or symptoms of an upper respiratory infection — such as a stuffy nose, sore throat, cough or congestion — just before the herald patch appears.

Treatment. In most cases, pityriasis rosea goes away in four to eight weeks. Treatment usually focuses on controlling itching. The antiviral drugs acyclovir and famciclovir and the antibiotic erythromycin may reduce the duration of pityriasis rosea to one to two weeks. These medications often are not necessary, however, because itching is usually mild, and the condition clears up on its own.

If itching is a problem, your doctor may recommend the following to provide relief:

Steroid creams or ointments. These creams will help ease itching and decrease redness.

Oral antihistamines. These medications are available by prescription as cetirizine (Zyrtec) and fexofenadine (Allegra), and over-the-counter as diphenhydramine (Benadryl), chlorpheniramine (Chlor-Trimeton), clemastine (Tavist) and loratadine (Claritin).

Light therapy (phototherapy). This can be with ultraviolet B (UVB) light or sunlight. Talk to your doctor before using sunlight to treat your rash. UVB therapy is most often available at your doctor's office.

Pityriasis Rubra Pilaris

Pityriasis rubra pilaris (PRP). PRP is a rare condition that is often initially mistaken for another skin disorder, usually psoriasis. PRP is not really a single condition, but rather a group of unusual eruptions that cause red scaly patches containing dry plugged pores. It may cover the entire body, or just the elbows and knees.

PRP mostly affects adults over 40, but some children are also affected. The cause of PRP is unknown. Sometimes minor burns rashes and infections seem to trigger it. There is no blood test for PRP. It is usually diagnosed when a dermatologist, suspecting the condition, does a biopsy and specifically asks it to be checked for PRP. Sometimes PRP is suspected only after the usual creams, pills and even ultraviolet light treatments used for skin conditions have no effect.

PRP most often starts as a patchy rash on the scalp, face or chest. Over a period as short a several weeks it extends downward, and often covers much of the body. It spares areas of old scars and injuries, and leaves small islands of entirely unaffected skin. The rash has an orange-red color ("salmon") and the palms and soles become thickened. Rough, dry plugs can be felt within the rash. The itching is usually severe at first, and then later is not as bad as you would think considering how bad the rash looks.

The best treatment is Accutane or Soriatane pills. These are closely related "retina" medications. While these have many minor side effects, they do not usually cause any serious harm to the body. A more potent and more effective treatment is methotrexate tablets, but as these can have dangerous side effects they are saved for people who don't improve with the retinoids. Methotrexate may put the PRP into remission, so it goes away and stays away.

Pityriasis Versicolor

Pityriasis Versicolor (tinea versicolor). Tinea versicolor is cause by a yeast type of skin fungus, which is present on normal skin. If the skin is oily enough, warm enough and moist enough, it starts to grow into small "colonies" on the surface of the skin. In these colonies the yeast grows like crazy and leaks out an acidic bleach. This changes the skin color. The patches are lightly reddish brown on very pale skin but they don't tan. Because of lack of any tanning, they look like white spots on darker or tanned skin. This is most often seen on the neck, upper chest, upper arms and back. There may be a fine, dry scale on it.

Usually the infection produces few symptoms, but some people get itching, especially when sweating. The warmer the weather, the worse this condition gets. Tanning booths are warm places, so avoid them. The reasons why some get this problem and others do not are not known.

A dermatologist can easily recognize this infection, but occasionally it can be mistaken for other skin conditions. If there is any doubt a 'KOH prep', a test done quickly in the office, will confirm the diagnosis.

The infection is treated with either topical or oral medications. In very mild cases, non-prescription antifungal creams will work. Prescription antifungal lotions and sprays may work better. The most economical effective treatment is to apply an antifungal shampoo (Nizoral, Loprox) to the body as if it were soap, but leave it on for some minutes before rinsing.

For severe, extensive or recurrent cases, a few tablets of Nizoral pills will clear things up. A newer pill, Sporonox, may replace Nizoral for this problem. These will eliminate the fungus and relive any itch and scale. The uneven color of the skin will remain several months, perhaps until one gets a tan again in the next summer.

See also **Dermatitis and Dermatosis**.

Additional Reading

Staff: *Pityriasis Rosea: A Medical Dictionary, Bibliography, And Annotated Research Guide To Internet References*, Icon Health Publications, San Diego, CA, 2004.

Staff: *Tinea Versicolor: A Medical Dictionary, Bibliography, And Annotated Research Guide To Internet References*, Icon Health Publications, San Diego, CA, 2004.

PITYRIASIS ALBA.   See Pityriasis.

PITYRIASIS ROSEA.   See Dermatitis and Dermatosis.

PITYRIASIS RUBRA PILARIS.   See Pityriasis.

PITYRIASIS VERSICOLOR (TINEA VERSICOLOR).   See Pityriasis.

PITYROSPORUM FOLLICULITIS.   See Folliculitis.

PIXEL. An individual, identifiable element of a picture. For example, a large astronomical photographic plate may contain as many as 100,000 or more individual picture elements (pixels). In terms of a digitized picture, one of the dots or resolution elements making up the picture as a pixel.

PIXEL VALUE. The intensity of a pixel, usually an integer. For grayscale images, the pixel value is typically an 8-bit data value (with a range of 0 to 255) or a 16-bit data value (with a range of 0 to 65535). For color images, there are 8-bit, 16-bit, 24-bit, and 30-bit colors. The 24-bit colors are known as true colors and consist of three 8-bit pixels, one each for red, green, and blue intensity.

pK. A measurement of the completeness of an incomplete chemical reaction. It is defined as the negative logarithm (to the base 10) of the equilibrium constant K for the reaction in question. The pK is most frequently used to express the extent of dissociation or the strength of weak acids, particularly fatty acids, amino acids, and also complex ions, or similar substances. The weaker an electrolyte, the larger its pK. Thus, at $25\,°C$ for sulfuric acid (strong acid), pK is about -3.0; acetic acid (weak acid), p$K = 4.76$; boric acid (very weak acid), p$K = 9.24$. In a solution of a weak acid, if the concentration of undissociated acid is equal to the concentration of the anion of the acid, the pK will be equal to the pH.

PK ZIP.   See Data Compression.

PLACENTA.   See Embryo.

PLAICE.   See Flatfishes.

PLAIT POINT. The point at which two conjugate solutions of partially miscible liquids have the same composition, so that the two layers become one.

PLANCK LAW. The fundamental law of the quantum theory, expressing the essential concept that energy transfers associated with radiation such as light or x-rays are made up of definite quanta or increments of energy proportional to the frequency of the corresponding radiation. This proportionality is usually expressed by the quantum formula $E = hv$, in which E is the value of the quantum in units of energy and v is the frequency of the radiation.

The constant of proportionality, h, is known as the elementary quantum of action or, more commonly, as the Planck constant.

PLANCK, MAX (1858–1947). Planck was a German physicist who in 1900 proposed the quantum theory of electromagnetic radiation. The basic concept of the quantum theory is that radiant energy is a continuous stream of discrete packets of energy called *quantum*. A quantum is the smallest amount of energy possible. In 1918 he was awarded the Nobel Prize for his discovery of the quantum theory of energy. He also is remembered for providing the mathematical, Planck's constant.

Planck and Albert Einstein developed a close friendship. During World War II, however, Planck did not flee Germany. Planck served in many German scientific associations including the Prussian Academy of Science and the Kaiser Wilhelm Society of Berlin which, was later renamed the Max Planck Society.

Even with all of his fame and influence within Germany, Planck could not save his son's life when he was accused and executed for participation in the July 1944 plot to assassinate Hitler.

See also **Black Body**; **Energy**; **Fokker-Planck Equation**; **Planck Law**; **Planck Radiation Formula**; **Quantum**; and **Quantum Mechanics**.

J. M. I.

PLANCK MISSION.   See **Space Science Missions: Universe**.

PLANCK RADIATION FORMULA.   The relationship

$$E_\lambda \, d\lambda = \frac{hc^3}{\lambda^5} \frac{d\lambda}{e^{hc/k\lambda T} - 1}$$

where $E_\lambda \, d\lambda$ is the intensity of radiation in the wavelength band between λ and $\lambda + d\lambda$, h is the Planck constant, c is the velocity of light, k is the Boltzmann constant and T is the absolute temperature. This formula describes the spectral distribution of the radiation from a complete radiator or black body. $hc^3 = C_1$ is known as the First Radiation constant, with $ch/k = C_2$ as the Second Radiation constant. C_2 has the value 1.43879 centimeter-degree. This radiation formula can be written in other forms, such as in terms of wavenumber instead of wavelength. Also it may be written in terms of energy density instead of radiation intensity. The value of the First Radiation constant will depend on the particular form of the radiation formula used. See also **Black Body**; and **Wien Laws**.

PLANCK'S CONSTANT.   A universal constant, denoted by h, with the value 6.626075×10^{-34} J s, in the quantum theory of matter and radiation. Planck's constant is the bridge between the wave and particle descriptions of light, an electromagnetic wave of frequency v alternatively described as a stream of photons each with energy hv. According to the quantum theory, when the energy of an atom, molecule, or nucleus changes from one discrete energy level E_1 to another E_2, conservation of energy requires the emission (creation) or absorption (annihilation) of a photon with energy given by the Einstein frequency condition

$$hv = E_2 - E_1$$

Planck's constant is a fundamental scaling parameter of the universe, determining, among other things, the sizes of atoms and molecules.

AMS

PLANE (Geometry).   A surface on which any two points may be connected by a straight line. One straight line does not determine a plane but a plane is determined by: a straight line and a point not in the line; three points not in a straight line; two intersecting lines; two parallel lines.

The general equation of a plane is $Ax + By + Cz + D = 0$, with A, B, C, not all zero. Thus the locus of every first-degree equation in x, y, z is a plane. Other forms of its equation are $x/a + y/b + z/c = 1$, where a, b, c are the x-, y-, z-intercepts; the normal form is, $\lambda x + \mu y + vz = p$, where λ, μ, v are direction consines of the normal from the origin to the plane and p is the length of this normal.

Figures on a plane surface are studied in both plane and analytic geometry. See, for example, **Conic Section; Curve; Polygon; Quadrilateral;** and **Triangle**.

PLANE SAILING.   A term applied to the solution of various problems in the sailings, in which the earth is considered as a plane surface. The particular subject of plane sailing will be found discussed under the topics **Dead Reckoning.**

PLANE TABLE.   A plane table is a surveying instrument used for locating and mapping topographical features. A drawing board, accurately made, and arranged so that it may be mounted on a tripod by an adjustable head which allows leveling of the board, is an essential feature of the plane table. Spirit levels are attached to the table in mutually perpendicular directions. The compass, the ruler, and a means for getting a line of sight, such as a telescope or open sights, complete the outfit. A ruler combined with a telescope or with slit sights is called an alidade. When the plane table is used for a survey, it is not necessary to take notes of angles or lengths of lines, since they are plotted, at the time of the survey, on the sheet of paper which covers the plane table. Obviously the plane table is not suitable for use in bad weather. When a survey is to be made with this instrument, the table is set up so that some convenient point on the paper is over a selected spot on the ground. The table is leveled and rotated horizontally until it is in azimuth. This is accomplished by means of the compass or by sighting back on a known point. It is then clamped in this position and the ruler is brought to the point selected on the paper and swung about it so that the line of sight that parallels the ruler bears on a distant point whose location is desired. A line is drawn in that direction, and after the distance to that point is measured, the length of line is plotted to some scale suitable to include the area being mapped on the surface of the plane table.

PLANETARIUM.   A representation of the astronomical system. (1) A mechanized model reproducing the motion of the planets around the sun. (2) An optical instrument that projects images of the celestial bodies in their relative brightness and size just as they occur in nature on a hemispherical dome of a darkened auditorium. The planetarium is used mainly in education and for practical demonstration of the coordinated and relative positions and motions of the celestial objects, including artificial satellites, as well as for the training of astronauts in celestial navigation. The instrument makes it possible to go backward or forward in time to show the true panorama of the heavens as seen from any point on the earth; or from a space capsule; at any time in the past, present, or future.

The first projection-type planetarium was designed by Zeiss in 1923. Since that time a large number of projectors have been installed in science halls in major cities throughout the world. The projection instrument, as shown in Fig. 1, is about $16\frac{1}{2}$ feet high and comprises approximately 29,000 individual parts of some 2.00 types. About 150 projectors, which

Fig. 1.   Planetarium. (*Carl Zeiss.*)

are mostly aspherical condensers and Tessar lenses or tele-objectives, are used. A special 1,000-watt incandescent lamp illuminates the 16 projectors in each of the two spheres. A diurnal event can be shown 120 to 480 times as fast as it would occur in nature. The movement of the celestial bodies over the period of an entire year, can be compressed into a time span ranging from several seconds to several minutes. The 25,800-year precessional revolution of the fixed-star system, which is caused by the slow gyroscopic movement of the earth, can be compressed into just 4 minutes. The variation of the sky during a trip around the earth from pole to pole only $6\frac{1}{2}$ minutes.

The various speeds for diurnal and annual movement are achieved by connecting the motors singly or together in the same or opposite directions. Diurnal, annual, and precessional movement are automatically coupled in that order. The diurnal movement is transmitted at correct time ratio by gears to the annual movement and from there to the precessional movement. Altogether there is a transmission ratio of about 1:156,000,000,000 between the rotation of the motor for the slowest diurnal movement and for the precessional rotation. The complex mechanism required for the movement of the planets in their elliptical orbits about the sun is controlled by various gear drives built into the planetarium projector.

The literature on planetariums is thin. Reference to "Geared to the Stars," by H.C. King and J.R. Millburn (University of Toronto Press, Toronto, Canada, 1978), is suggested.

W. E. DEGENHARD, Carl Zeiss, Inc., New York

PLANETARY BOUNDARY LAYER. That layer of the atmosphere from a planet's surface to the geostrophic wind level including, therefore, the surface boundary layer and the Ekman layer. Above this layer lies the free atmosphere. Also called *friction layer*, or *atmospheric boundary layer*.

PLANETARY CIRCULATION. 1. The system of large-scale disturbances in a planet's troposphere when viewed on a hemispheric or world-wide scale.

2. The mean or time-averaged hemispheric circulation of a planetary atmosphere; also called *general circulation*.

PLANETS AND THE SOLAR SYSTEM. The word *planet*, which comes from a Greek root meaning "wanderer," was used prior to the fifteenth century to designate those celestial objects (other than meteors and comets) that were observed to be in motion relative to the stars. Before the fifteenth century, seven objects were listed as planets: Sun, Moon, Mercury, Venus, Mars, Jupiter, and Saturn. With the advent of the Copernican heliocentric hypothesis for the structure of the universe, the sun and moon were removed from the list and Earth added. Since the application of the telescope to astronomy, three major planets have been added (Uranus, Neptune, and Pluto) as well as over 1,000 small planets or asteroids. The term, as it is used at present, applies to an opaque object that shines by reflected sunlight and travels about the sun or a star in an orbit.

In spite of the fact that many of the planets are larger than Earth, their distance is so great that some of them appear to the naked eye as bright stars. The only certain method for distinguishing a planet from a star without the use of a telescope is to watch it carefully for a considerable period (frequently several days are required), and if the object is a true planet, it will move relative to the stars. For a quick method of identification, it may be said that usually a planet does not appear to twinkle as do the stars, but this rule is not infallible. With a telescope, a planet may be immediately distinguished from a star (with the exception of the planet Pluto or the asteroids) because of the fact that a planet will show an appreciable disk, whereas the stars appear as points of light.

TABLE 1. ORBITAL CHARACTERISTICS OF THE PLANETS

Characteristics	Mercury	Venus	Earth	Mars	Jupiter	Saturn	Uranus	Neptune	Pluto
Mean distance from sun:									
kilometers (millions)	57.91	108.21	149.60	227.94	778.3	1427	2869	4498	5900
miles (millions)	35.99	67.24	92.96	141.64	483.64	887	1783	2795	3666
astronomical units	0.387	0.723	1.0167[a]	1.524	5.203	9.539	19.182	30.057	39.440
Approximate distance from Earth:									
Maximum:									
kilometers (millions)	219	259		399	965	1654	3154	4682	7562
miles (millions)	136	161		248	600	1028	1960	2910	4700
astronomical units	1.46	1.73		2.67	6.45	11.06	21.08	31.3	50.5
Minimum:									
kilometers (millions)	80	40		56	591	1197	2584	4307	4296[b]
miles (millions)	50	25		35	367	744	1606	2677	2670
astronomical units	0.53	0.27		0.37	3.95	8.00	17.27	28.79	28.72
Orbital eccentricity[c]	0.2056	0.0068	0.0167	0.0934	0.0484	0.0543	0.0460	0.0082	0.2481
Angular momentum[d]	0.02	0.07	1.00	0.13	722	293	64	94	1.2
Inclination to ecliptic, degrees	7.003	3.4	0[e]	1.850	1.309	2.493	0.773	1.779	17.146
Period of rotation, sidereal days	58.82	224.59R	1.0[f]	1.03	0.41	0.43	0.45R	0.66	6.41
Approximate planetary day	8.4 weeks	32.1 weeks	23 hours, 56 min., 4.09 sec.	24 hours, 43 min.	9 hours, 50 min.	10 hours, 19 min.	10 hours, 48 min.	15 hours, 50 min.	6 days, 9 hours, 50 min.
Sidereal period, mean days	87.97	224.70	365.25636	686.98	4332.4	10,759.3	30,684.49	60,188.31	90,710.07
in terms of earth years	0.241	0.615		1.881	11.861	29.457	84.008	164.784	248.346

[a] The astronomical unit is defined as the distance equal to that of the geometrical mean distance of the earth from the sun. Refinements in measurements have altered the value slightly from unity.

[b] The minimum distance of Pluto, although the furthest planet from the sun, can be less than that of Neptune under certain circumstances because the orbits these two planets cross.

[c] Eccentricity is a number which defines the shape of an ellipse. It is the ratio of the distance from center to focus to the semimajor axis. The orbit of the earth is nearly circular.

[d] The angular momentum of a moving body, such as a planet revolving around the sun, is the production of the mass, the square of the distance from the center of motion, and the rate of angular motion.

[e] By definition, inclination to ecliptic is 0°. Inclination of equator to ecliptic is 23.45°.

[f] Actually slightly less than 24 hours, as indicated below.

R = retrograde motion. Motion in an orbit opposite the usual orbit direction of solar-system bodies, that is, motion from east to west around a center.

The planets are classified in two general ways. Mercury and Venus are frequently referred to as the inferior planets, and the others are called the superior planets. Another system of classification considers Mercury, Venus, Earth and Mars as the minor or terrestrial planets, while Jupiter, Saturn, Uranus, Neptune, and Pluto are called the major (or gaseous, or Jovian) planets.

That group of objects, including the planets, asteroids, and comets, which is moving through space with the sun is known as the *solar system*. Each of the planets and its family of satellites is described in a separate entry in this encyclopedia. Also, there are separate entries on **Asteroid; Comet; Moon (Earth's); Planets (Motions); Sun (The); Voyager Missions to Jupiter and Saturn**; and **Pathfinder Mission to Mars.** The origin of the solar system is described from a theoretical standpoint in entry on **Cosmology.** The earth as a planet is described in the entry on **Earth.**

The orbital characteristics of the nine planets are given in accompanying Table 1. The physical characteristics of the planets are given in Table 2. The characteristics of the satellites of the planets (a total of over 30) are given in Table 3. More specific information on the satellites of the more recently explored planets is given in separate articles on these planets.

Although the Earth's moon is described in detail in the entry on **Moon (Earth's)**, additional convenient statistics for the moon are given here in Table 4.

Satellites serve a useful purpose for astronomers, since the mass of a planet can be determined accurately only if the planet has one or more satellites. By application of the rigorous expression for the harmonic Keplerian laws of planetary motion, the mass of any planet and satellite may be found in terms of the mass of Earth-Moon system after the distance of the planet from the satellite and its period of revolution are known. The problem of the determination of the masses of the satellites themselves is a more difficult problem. The mass of the moon can be determined in terms of the earth's mass by means of the so-called barycentrix parallax. Approximate values of the masses of the satellites of Jupiter can be obtained by the mutual perturbations they exert on each other. In the case of Saturn, the masses of the satellites may be approximately determined from their mutual perturbations and an approximate check is provided by the positions of the divisions in the ring. As the result of the *Voyager* findings, refinements in past statistical data are being formulated.

Some satellites revolve about their primaries in the retrograde sense, i.e., in the direction contrary to that in which all other planets and satellites are revolving and rotating. This retrograde motion can be fully explained on the basis of modern celestial mechanics.

The influences that satellites exert on their primaries are very slight. The tidal forces they exert have some slight effect upon the rotation periods of the primaries but such effects are so small as to be beyond observational measurement. The tidal effects the planets exert upon the satellites, on the other hand, are in many cases so large that the satellites rotate in approximately the same period as that in which they revolve.

Extrasolar Planets. In 1987, Canadian astronomers (Dominion Astrophysical Observatory, Victoria, B.C.) announced that their survey of 16 nearby solar-type stars had revealed clear indication of low-mass companions around two of the stars and possible evidence of low-mass companions around 5 others. As a scale of reference, "low mass" was considered one to be 10 times that of Jupiter (a large mass in comparison with the planets of the solar system). Currently, astronomers are considering the evidence with caution. Numerous extrasolar planets have heretofore been claimed. As pointed out by Gatewood (Allegheny Observatory, Pittsburgh, Pennsylvania), a highly publicized case was made in 1984 with regard to the observation of a companion to the dim red star van Biesbroek 8, which to date has not been observed again. Nevertheless, many astronomers regard the 1987 Canadian observations as exciting. An important aspect of the

TABLE 2. PHYSICAL CHARACTERISTICS OF THE PLANETS

Characteristic	Mercury	Venus	Earth	Mars	Jupiter	Saturn	Uranus	Neptune	Pluto
Mean semidiameter:									
kilometers	2433	6051.4	6371	3380	69,758	58.219	23,470	22,716	1750[a]
miles	1512	3760.4	3959	2100	43,348	36,177	14,584	14,116	1087
in terms of Earth = 1	0.382	0.950	1.0	0.531	10.949	9.138	3.684	3.566	0.275
Apparent, seconds of arc	5.45	30.50		8.94	3.43	9.76	1.80	1.06	0.11
Mass, kilograms	3.181×10^{23}	4.883×10^{24}	5.979×10^{24}	6.418×10^{23}	1.901×10^{27}	5.684×10^{26}	8.682×10^{25}	1.027×10^{26}	1.08×10^{24}
in terms of Earth = 1	0.053	0.817	1.0	0.107	317.946	95.066	14.521	17.176	0.181
Mean density, (grams/cubic centimeter	5.431	5.256	5.519	3.907	1.337	0.688	1.603	2.272	1.65
in terms of Earth = 1	0.98	0.95	1.0	0.71	0.24	0.125	0.29	0.41	0.30
Mean gravity,									
centimeters/sec^2	357.8	887.4	980.7	374.0	2601.0	1117.0	1049.0	1325.0	221.0
feet/sec^2	11.74	29.11	32.18	12.27	85.33	36.65	34.41	43.47	7.25
in terms of Earth = 1	0.36	0.90	1.0	0.38	2.65	1.14	1.07	1.35	0.23
Escape velocity,									
kilometers/second	4.173	10.365	11.179	5.028	60.238	36.056	22.194	24.536	5.023
miles/second	2.593	6.441	6.947	3.124	37.432	22.405	13.791	15.247	3.121
miles/hour	933.5	2318.8	2500.9	1124.6	13,475.5	8065.8	4964.8	5488.9	1123.6
Solar constant, calories/cm^2/min	12.8	3.7	1.920	0.83	0.071	0.021	0.005	0.002	0.001
Temperature (day),									
Kelvin	683[b]	720°	287[b]	190–240[b]	11,000[d]	223°	123°	123°	63°
approximate°C	410	447	14	−83 − −33	10.704	−50	−150	−150	−210
approximate°F	770	837	57	−117 − −27	19.300	−58	−238	−238	−346
Oblateness[e]	0.029	0?	0.0034	0.005	0.066	0.103	0.07	0.08	0.156
Albedo[f]	0.076	0.59–0.76[g]	0.36	0.152	0.54	0.57	0.65	0.68	0.13

[a] Recent investigations have indicated that Pluto is much smaller than previously estimated. New estimates are given here.
[b] Surface temperature.
[c] Upper atmosphere temperature.
[d] Inner layer temperature.
[e] The departure of a planet from spherical form because of centrifugal force of rotation. If equatorial diameter is a and polar diameter is b, oblateness = (a − b)/a.
[f] A measure of the light-reflection power of a surface compared with an ideal white matte surface which absorbs no light.
[g] Range of various estimates.

TABLE 3. SATELLITES OF THE PLANETS

Planet and Satellite	Mean Semidiameter			Rotation Period (days)	Mean Distance from Primary Body Kilometers Miles (thousands)		Orbit Inclination to Ecliptic of Planet (degrees)	Orbit Eccentricity	Apparent Stellar Magnitude	Mass (kilograms)	Date of Discovery	Equilibrium Temperature (°K)
	Kilometers	Miles	In terms of Moon = 1									
EARTH												
Moon	1783.3	1108.1	1.0	27.3	383.403	238.857	18–29	0.055	−12.3	7.35×10^{22}	Antiquity	394
MARS												
Phobos	5.4–14.4	3.4–9.6	0:003	0.32	9	5.59	1.1	0.021	11.5	2.7×10^{16}	1877	319
Deimos	1.0–9.0	0.6–5.6	0.0005–0.005	1.26	23	14.29	1.6	0.003	13	1.8×10^{15}	1877	319
JUPITER												
Jo	1818	1130	1.02	1.77	422	262.2	0.0	0.0	5.5	7.9×10^{22}	1610	173
Europa	1533	953	0.86	3.55	671	417	0.5	0.0	5.7	4.8×10^{23}	1610	173
Ganymede	2608	1621	1.46	7.16	1070	665	0.2	0.001	5.1	1.54×10^{23}	1610	173
Callisto	2445	1519	1.37	16.69	1880	1168	0.2	0.01	6.3	7.35×10^{22}	1610	173
Amalthea	120	75	0.067	0.4	181	112.5	0.4	0.003	13	8.3×10^{18}	1892	173
Himalia	85	53	0.048	?	11.470	7127	27.6	0.158	13.7	—	1904	173
Elara	30	19	0.017	?	11,740	7295	24.8	0.207	16	—	1905	173
Pasiphae	13.5	8.4	0.008	?	23,300	14,479	145R	0.38	16	—	1908	173
Sinope	11.5	7.1	0.006	?	23,700	14,727	153R	0.28	18	—	1914	173
Lysithea	9.5	5.9	0.005	?	11,710	7277	29.0	0.13	18	—	1938	173
Carme	12	7.5	0.007	?	22,350	13,888	164R	0.21	18	—	1938	173
Ananke	8.5	5.3	0.005	?	20,700	12,863	147R	0.17	19	—	1951	173
1979-J1	15–20	9.3–12.4	0.01	?	—	—	—	—	—	—	1979	173
1979-J2	35–40	21.7–24.9	0.02	?	—	—	—	—	—	—	1979	173
SATURN[a]												
Titan	2525	1600	1.4	15.95	1222	759	0.3	0.03	8.3	1.2×10^{23}	1655	128
Iapetus	900	559	0.5	79.33	3560	2212	14.7	0.03	11.0	2.3×10^{21}	1671	128
Rhea	765	475	0.45	4.4	527	327.5	0.4	0.0	10.0	1.8×10^{21}	1672	128
Tethys	525	326	0.32	?	295	183.3	1.1	0.0	10.5	4.9×10^{20}	1684	128
Dione	560	348	0.32	2.7	377	234.3	0.0	0.0	10.7	5.4×10^{20}	1684	128
Mimas	200	124	0.11	?	186	115.6	1.5	0.02	12.1	3.7×10^{19}	1789	128
Enceladus	250	155	0.14	1.37	238	147.9	0.0	0.0	11.6	7.4×10^{19}	1789	128
Hyperion	180	110	0.14	?	1481	920	0.4	0.1	13.0	6.8×10^{19}	1848	128
Phoebe	100	62	0.056	?	12,390	8035	150R	0.16	14.5	1.9×10^{19}	1898	128
Janus	185	115	0.1	?	160	99.4	0.0	0.0	—	1.2×10^{20}	1966	128
URANUS												
Titania	1200	746	0.67	?	438	272.2	0.0	0.0	14.0	2.1×10^{21}	1787	90
Oberon	1100	684	0.62	?	586	364.1	0.0	0.0	14.2	1.1×10^{21}	1787	90
Ariel	1000	621	0.56	?	192	119.3	0.0	0.0	15.2	5.0×10^{20}	1851	90
Umbriel	650	404	0.36	?	267	165.9	0.0	0.0	15.8	1.4×10^{20}	1851	90
Miranda	400	249	0.22	?	130	80.8	3.4	0.02	17.0	3.0×10^{19}	1948	90
NEPTUNE												
Triton	2500	1554	1.4	5.9	30.07AU	30.07AU	160R	0.0	13.6	1.46×10^{23}	1846	72
Nereid	350	217	0.2	?	354	220	27.5	0.76	19.0	5.0×10^{19}	1949	72
PLUTO												
Chiron[b]	1050–1500	650–930	~1.4	6.4	39.44AU	39.44AU	0.0	0.0	—	—	1978	63

[a] Saturn is known to have additional satellites. See article on **Saturn**.

[b] Very tentative. See article on **Pluto**.

AU = astronomical unit.

R = retrograde revolution.

Canadian observations is that they did *not* see brown dwarfs (star-like objects that just miss being massive enough to ignite by thermonuclear fusion). Current astrophysical theory indicates the threshold for thermonuclear burning is about 80 Jupiter masses. It is pointed out that anything approaching that mass would have stood out in the survey like a searchlight. Astronomers are quick to point out, however, that the fact that brown dwarfs are rare (or nonexistent) indicates that current knowledge of the star formation process is indeed wanting. As observed by Levy (University of Arizona Lunar and Planetary Laboratory), "What's amazing to me is that the lower limit for star formation seems to be so close to the lower limit for nuclear burning. It's not obvious why that should be. But then, in astrophysics there are a lot of coincidences that people still don't understand."

Some scientists as of 1994 suggested that an additional planet beyond Pluto in the solar system may exist, but the majority of researchers at least tentatively have concluded that the case for a tenth planet is a weak one. The infrared sky survey (IRAS) launched in 1983 has revealed nothing to support the existence of a tenth planet. The strongest argument for a tenth planet is the fact that the present model of the solar system appears to be somewhat flawed. This is evidenced by the fact that the positions of the outer planets cannot be predicted reliably beyond a span of about ten years. As one investigator observes, "If you sent a probe to Pluto at the moment, you'd miss it." But others observe that such estimations arise from feeding insufficiently precise information into the model of the system. Discovery of Pluto's moon, Charon, permitted a determination of the planet's mass and of its satellite. Robert Harrington (U.S. Naval Observatory) continues with research toward filling out the tenth-planet theory. After investigating the effects of numerous combinations of data, a best estimate of a tenth planet required to satisfy the model would be a body with a mass about 3.5 times that of Earth, following an oval orbit tilted at about 30° to the plane of the solar system and lying about three times as far from the sun as Neptune. Further, it is expected that the planet may be located in the constellation Centaurus (deep in the southern sky), thus explaining why most astronomers who observe the skies of the Northern Hemisphere may not have found it.

TABLE 4. ADDITIONAL LUNAR STATISTICS

Distance of Moon from Earth	
greatest	406,697 km (252,710 miles)
least	356,700 km (221,643 miles)
mean	384,403 km (238,857 miles)
Equatorial horizontal parallax at mean distance:	57′03″
Apparent angular diameter:	
minimum	29′21″
maximum	33′30″
mean	31′05″
Eccentricity of orbit	$\frac{1}{18}$
Diameter of Moon	3,476 km (2,160 miles)
Volume	$\frac{1}{49}$ that of Earth
Mass	$\frac{1}{81}$ that of Earth
Mean density	$\frac{3}{5}$ that of Earth (3.34 g/cm^3)
Surface gravity	$\frac{1}{6}$ that of Earth
Velocity of escape	2.4 km (1.5 miles) per second
Approximate temperature of soil	
at noon	100 °C 212 °F
at midnight	−150 °C −238 °F
Revolution and Rotation	
synodic month (from one new moon to the next)	29^{d}12^{h}44^{m}2.8
sidereal month (true period of revolution around Earth)	27^{d}7^{h}43^{m}11.5
period of axial rotation	27^{d}7^{h}43^{m}11.5
Period of revolution of nodes	18.6 years
Daily retardation in crossing meridian	
minimum	38 minutes
maximum	66 minutes
average	50 $\frac{1}{2}$ minutes
Average velocity of Moon around Earth	
linear	3680 km (2287 miles) an hour
angular	13°.2 a day or 33′ an hour
(Moon moves in one hour a distance about equal to its own diameter)	

Additional Reading

Atreya, B.K., J.B. Pollack and M.S. Matthews: *Origin and Evolution of Planetary and Satellite Atmospheres*, University of Arizona Press, Tucson, AZ, 1997.

Barnes-Svarney, P.: "A Growing Solar Family," *Technology Review (MIT)*, 21 (May/June 1991).

Bennett, J., M. Donahue, and N. Schneider: *The Solar System*, 2nd Edition, Addison Wesley Longman, Inc., Redding, MA, 2001.

Black, D.C.: "Worlds Around Other Stars," *Sci. Amer.*, 76 (January 1991).

Eberhart, J.: "Straightening the Magnetic Tilts of Planets," *Science News*, 294 (May 12, 1990).

Eberhart, J.: "Panel Prods NASA to Seek Unknown Planets," *Science News*, 21 (January 12, 1991).

Eberhart, J.: "Why Three Planets Radio the Sun," *Science News*, 63 (January 26, 1991).

Freedman, R.A. and W.J. Kaufmann: *Universe: The Solar System*, W.H. Freeman and Company, New York, NY, 2001.

Garlick, M.A.: *The Story of the Solar System*, Cambridge University Press, New York, NY, 2002.

Gore, R.: "The Planets," *National Geographic*, 4 (January 1985).

Greeley, R. and R. Batson: *The NASA Atlas of the Solar System*, Cambridge University Press, New York, NY, 1996.

Hanson, R.B.: "Planetary Fluids," *Science*, 281 (April 20, 1990).

Ingersoll, A.P.: "Atmospheric Dynamics of the Outer Planets," *Science*, 308 (April 20, 1990).

Jones, B.W.: *Discovering the Solar System*, John Wiley & Sons, Inc., New York, NY, 1999.

Kasting, J.F., O.B. Toon, and J.B. Pollack: "How Climate Evolved on the Terrestrial Planets," *Sci. Amer.*, 90 (February 1988).

Kerr, R.A.: "Which Way is North? Ask Right-Handed Astronomers," *Science*, 999 (November 24, 1989).

Lewis, J.S.: *Physics and Chemistry of the Solar System*, 2nd Edition, Morgan Kaufmann Publishers, Orlando, FL, 1997.

Lunine, J.I.: "Origin and Evolution of Outer Solar System Atmospheres," *Science*, 141 (July 14, 1989).

Matthews, R.: "Planet X: Going, Going…But Not Quite Gone," *Science*, 1454 (December 6, 1991).

Pasachoff, J.M. and D.H. Menzel: *Field Guide to the Stars and Planets*, Vol. 15, 3rd Edition, Houghton Mifflin Company, New York, NY, 1992.

Powell, C.S.: "A Cosmic Unveiling: Newborn Stars Have Some Secrets About How Planets Form," *Sci. Amer.*, 26 (December 1989).

Price, F.W.: *Planet Observer's Handbook*, 2nd Edition, Cambridge University Press, New York, NY, 2000.

Rubin, A.E.: *Disturbing the Solar System: Impacts, Close Encounters, and Coming Attractions*, Princeton University Press, Princeton, NJ, 2002.

Stern, S.A. and J. Mitton: *Pluto and Charon: Ice Worlds on the Ragged Edge of the Solar System*, John Wiley & Sons, Inc., New York, NY, 1999.

Taylor, S.R.: *Solar System Evolution: A New Perspective: An Inquiry into the Chemical Composition, Origin, and Evolution of the Solar System*, 2nd Edition, Cambridge University Press, New York, NY, 2001.

Weissman, P.R., T. Johnson, and L.-A. McFadden: *Encyclopedia of the Solar System*, Academic Press, Inc., San Diego, CA, 1998.

Web References

JPL Solar System Exploration Site: http://sse.jpl.nasa.gov/index.html

The Planetary Photojournal: http://photojournal.jpl.nasa.gov/

The Nine Planets: http://seds.lpl.arizona.edu/billa/tnp/intro.html

Views Of The Solar System: http://www.hawastsoc.org/solar/eng/homepage.htm

PLANETS (Motions). The apparent motions of the planets on the celestial sphere have been observed, recorded, and speculated about ever since mankind has existed on the earth. The motions as seen from the earth are complicated by the fact that the earth is moving about the sun in the same direction as the planets, but with a different rate.

Apparent Motions Relative to the Sun. In Fig. 1, we have S representing the sun, E representing the earth (assumed fixed for convenience), P_1', P', P_2', P_3', P_4' indicating various positions of a planet whose orbit lies between the earth and the sun (inferior planet), and P_1', P, P_2, P_3, P_4, positions of a planet whose orbit is outside that of the earth (superior planet). The angle between the sun and the planet (e.g., SEP' or SEP) is defined as the elongation of the planet. For an inferior planet the elongation may have any value from 0 to SEP_2' or SEP_4', whereas for a superior planet the elongation varies either east or west from 0 to 180°. With elongation 0 we have the planet in the aspect of conjunction.

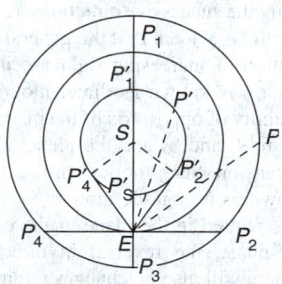

Fig. 1.   Motions of planets as seen from Earth relative to the sun.

Inferior planets: It will be noted that the elongation is 0 both at P_1' and also at P_3' and hence there are two conjunctions for an inferior planet. To distinguish between them P_1' is known as superior conjunction and P_3', inferior conjunction. An inferior planet moves more rapidly in its orbit than does the earth and accordingly from superior conjunction the planet moves out with increasing eastern elongation (evening object) to the point P_2' (greatest eastern elongation). It then moves in with decreasing elongation, passes the sun and becomes a morning object at inferior conjunction (P_3') and moves out with increasing western elongation to P_4' (greatest western elongation) and thence back to superior conjunction again. Hence these planets apparently oscillate back and forth across the direction of the sun. They do not ordinarily pass either between the earth and the sun or directly behind the sun because of the fact that their orbits are not in the plane of the ecliptic.

Superior Planets: It must be remembered that these planets are moving more slowly in their orbits than is the earth. These planets apparently move from conjunction at P_1 slowly out to the west of the sun (morning objects) to P_2 where the western elongation is 90° and the aspect is western quadrature. From this point the increase in western elongation increases rapidly to 180° at P_3 (aspect opposition) from which point the elongation becomes east (evening object) and decreases rapidly to 90° eastern elongation at P_4 (eastern quadrature). The decrease in eastern elongation then slows down as the planet moves slowly back to conjunction again.

Apparent Motions of the Planets Relative to the Stars. In Fig. 2, we have S representing the position of the sun, E representing the (assumed circular) orbit of the earth, with successive positions marked, P the orbit of a planet with successive positions marked with numbers corresponding in date with the marked positions for the earth, and an outer circle representing directions on the celestial sphere, V being the direction of the vernal equinox, A the direction of the autumnal equinox and the direction of increasing longitude (or right ascension) indicated by arrows.

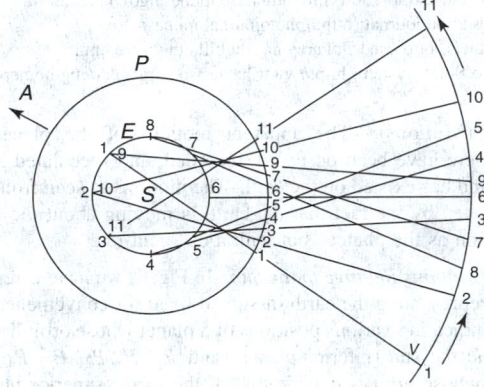

Fig. 2.   Motion of planets as seen from Earth relative to the stars.

The successive positions of the planet as seen from the earth relative to the stars, are obtained by drawing lines, representing the lines of sight from successive positions of the earth through corresponding positions of the planet. By examining the successive directions, as indicated by numbers on the outer circle, it will be evident that the general trend of the planetary motion is in the direction of increasing right ascension. Such motion in increasing right ascension is known as direct motion. It will be noticed, however, that in the vicinity of opposition of the planet, the motion reverses for a period (numbers 6, 7, and 8) and the planet moves in the direction of decreasing right ascension. Such motion, in the direction of decreasing right ascension, is known as retrograde motion.

Since the diagram is plotted in the plane of the ecliptic with the planet also assumed in this plane, the reversal in direction appears only in longitude. However, there will also be changes in direction of motion both in celestial latitude and declination, giving the appearance on the celestial sphere of loops in the motion of the planet. These loops were observed by the early astronomers and it was to account for those on the assumption of the geoconcentric universe that the epicycles of the Ptolemaic system became necessary. For planetary motions referred to the sun alone, see **Orbit (Astronomy)**.

See also **Kepler's Laws of Planetary Motion**.

PLANKTON.   Plankton are aquatic organisms living in the water column that drift, float, or swim only weakly. The term derives from the Greek word *planktos*, meaning "to drift" or "to wander." It is a descriptive classification, implying a certain type of lifestyle, rather than characteristics associated with taxonomy, size, or ecological role. The following information focuses on marine plankton, but much of the information is directly applicable to freshwater plankton.

Plankton can be contrasted with nekton, aquatic organisms capable of vigorous swimming (such as large fishes, squids, and marine mammals), and with benthos, those aquatic organisms that live on or at the bottom. It is generally thought that organisms classified as plankton do not have much ability to control their horizontal position relative to large-scale currents. However, there are no hard rules defining how strongly an organism may swim and still be considered planktonic. The boundaries between plankton, nekton, and benthos are nebulous. In fact, some organisms can be placed in each of the groups during different developmental stages. For example, the blue crab, *Callinectes sapidus*, has a planktonic larval stage. After 30+ days as plankton, these larvae settle to the bottom, becoming benthic juveniles. Although they spend most of their adult life on the bottom, blue crab adults swim surprisingly well (*Callinectes* translates to "beautiful swimmer") and females will traverse long distances to spawn. Despite a loose definition, plankton is a meaningful term that describes one of the most common aquatic lifestyles.

Categories of Plankton

The organisms that comprise plankton are exceedingly diverse. Many Archaea and Eubacteria (the two major taxonomic groups of bacteria), most algal divisions, and most animal phyla have some representatives that are planktonic. Plankton also range in size from less than 1 µm (bacterioplankton) to several meters across (some jellyfish and colonies of planktonic tunicates). The range of taxa and sizes results in very different ecological roles for different plankton. Several categorization schemes are commonly used to organize the diversity of plankton into groups that share particular properties.

Phytoplankton, Zooplankton, and Bacterioplankton. Phytoplankton are photosynthetic plankton, mostly single-celled, microscopic algae that use light energy to fix carbon dioxide dissolved in water. This process makes solar energy available to them for their own biochemical needs. When they are eaten by other organisms, and when they release dissolved organic matter into the water, the energy becomes available to nonphotosynthetic organisms. Phytoplankton are therefore considered the base of most aquatic foodwebs (primary producers). Phytoplankton commonly include photosynthetic prokaryotes (cyanobacteria) and photosynthetic eukaryotes, such as diatoms, dinoflagellates, cryptophytes, and chlorophytes.

Zooplankton are eukaryotic, planktonic heterotrophs. They are usually phagotrophic, meaning that they ingest particles. However, there are examples of osmotrophic zooplankton (feeding on dissolved organic matter). Also, many eggs and larvae that are temporarily planktonic do not feed, but only use stored energy reserves. Both examples would still be considered zooplankton. Common zooplankton include single-celled, heterotrophic protists, such as ciliates and choanoflagellates, and multicellular animals, such as copepods and euphausiids.

Gelatinous zooplankton are a subcategory of zooplankton. Although taxonomically diverse, they share a body composition that is jellylike, mostly lacking hard, structural components. Their bodies have very high water content and are often transparent. Cnidarian medusae known as "jellyfish" are the most commonly known examples. Additional examples of gelatinous zooplankton are found in the ctenophores, mollusks (heteropods), and urochordates (planktonic tunicates such as salps, doliolids, and larvaceans). Although there are small gelatinous forms, the largest zooplankton are all gelatinous. Because of their fragile body plan, gelatinous zooplankton are difficult to collect quantitatively. Their importance to marine ecosystems may therefore be underestimated. Gelatinous zooplankton are rare in fresh water.

Bacterioplankton, or planktonic bacteria, includes representatives of both Eubacteria and Archaea. Photosynthetic bacteria are sometimes included within this category, creating an overlap with phytoplankton. Generally, however, the category implies heterotrophic bacteria. Heterotrophic bacterioplankton are osmotrophs. They either take up dissolved organic matter freely available in the water, or they use external enzymes to solubilize organic matter from particles, making it available for transport into their cells.

Holoplankton Versus Meroplankton. Organisms that are planktonic their entire lives are holoplankton. The term derives from the Greek word *holos*, meaning "entirely." Meroplankton are organisms that are planktonic during only a portion of their lives. The Greek word *meros* means "mixed." Most meroplankton are larval forms that are benthos or nekton as adults. For example, many fish larvae are planktonic. As they grow and their swimming ability improves, they eventually become nekton. Similarly, many mollusks have planktonic larvae that eventually settle to the bottom and become benthos. Because ecological conditions for the plankton are often very different from those facing the adults, meroplanktonic larvae often do not resemble the adults. In numerous cases, meroplanktonic larval stages were first scientifically described as distinct species; only later was the connection to the adult stage realized. Organisms that are entirely planktonic except for a benthic resting stage, such as some cladocerans or dinoflagellates, are still generally considered holoplanktonic.

Plankton are also commonly separated based on size. Viruses are abundant in aquatic environments and may be the smallest plankton, for those who consider viruses to be living. Bacterioplankton are the smallest cellular plankton and may be less than 1 μm in length. The largest plankton are cniderian medusae (jellyfish) that may be more than a meter across the bell, with tentacles that extend over many meters. Planktonic tunicate colonies (especially Pyrosomes) can also be many meters across, although the individuals within the colony are only millimeters to centimeters in length.

Size categories are arbitrary, but Sieburth described the following categories currently used by many researchers: picoplankton (organisms 0.2–2 μm in length), nanoplankton (2–20 μm), microplankton (20–200 μm), mesoplankton (0.2–20 mm), macroplankton (2–20 cm), and megaplankton (20–200 cm). Viruses would be considered femtoplankton. These size categories are often combined with other categories. For example, picophytoplankton and mesozooplankton are common terms in oceanographic literature.

Collecting and Studying Plankton

In general, the number of individuals per unit volume of water increases as planktonic organisms decrease in size. Nevertheless, natural waters are typically a very dilute suspension of even the smallest organisms and concentrating the organisms is often necessary for analyses. High numbers of small organisms (pico-, nano-, microplankton) can be collected simply by taking a water sample with a bucket or a hydrographic bottle. Concentrating such small organisms by filtration is often done after water collection. The filter pore size used determines the smallest organisms collected on the filter surface. Use of different filter pore sizes allows separation of the plankton sample into several size classes. Larger organisms (microplankton and up) may be too rare for quantitative collection in a whole-water sample. They are often collected by dragging plankton nets through the water. Plankton nets typically have a defined mesh size so that only organisms larger than the mesh will be collected. Plankton nets therefore size fractionate and concentrate before collection. A larger mesh net will filter a greater volume of water before it becomes clogged and ineffective. Larger mesh nets can also be dragged through the water more rapidly. It is therefore important to match the net mesh with the size, concentration, and swimming ability of the organisms one desires to collect. Because of their small size, taxonomic identification of most plankton requires microscopic examination after collection.

Buckets, bottles, nets, and microscopes are still used to study plankton, but these traditional approaches are now complemented with a myriad of advanced techniques. For example, on the smallest scales, methods of molecular biology can be used to study gene expression in individual cells of picoplankton. On the largest scales, the global distribution of phytoplankton can be studied in satellite images.

Spatial Patterns of Plankton

The dominant large-scale horizontal pattern in plankton abundance is a decline as one goes away from the coast. One may have to travel tens to hundreds of miles offshore for this pattern to become clear. Nevertheless, the central areas of the oceans that are furthest from land generally have the lowest abundance of plankton, while coastal areas have the highest. This is true for all types of plankton. However, it should be kept in mind that many species of plankton are adapted to live in open-ocean environments.

Therefore, certain planktonic species may actually be more abundant in the open ocean than nearer the coast. Exceptions to the general offshore pattern in plankton abundance can be caused by mesoscale eddies, oceanographic fronts, upwelling zones, and other phenomena. Although areas far from shore tend to have low plankton abundance, the diversity of species can be very high by comparison to coastal locations.

By definition, plankton have a limited ability to control their horizontal movement. Horizontal patterns of plankton abundance are therefore regulated by environmental factors external to the organism: currents, light, temperature, nutrient availability, and so on. These factors also contribute to vertical patterns in plankton abundance. However, individual behavior is also important in forming vertical patterns because many planktonic organisms do have the ability to control their vertical position within the water column.

Because light from the sun is rapidly absorbed by water, and because phytoplankton need light to photosynthesize, actively growing phytoplankton are restricted to the upper portion of the water column. In coastal waters, sufficient light for photosynthesis may only penetrate the top few meters (or less). In these conditions, vertical phytoplankton abundance often directly correlates with light; the highest phytoplankton abundance is at the surface, with a sharp decline as depth increases. This pattern is mostly caused by higher phytoplankton growth where more light is available. In addition, some phytoplankton can swim enough to change their vertical position in the water column (by as much as 10 m/d). Similarly, other phytoplankton species are positively buoyant due to gas or lipid inclusions. Therefore, high phytoplankton abundance near the surface can sometimes be due to upward movement of phytoplankton. By contrast, in the open ocean, while sufficient light for photosynthesis is generally found within the upper 100–200 m, phytoplankton abundance is usually low near the surface. Highest phytoplankton abundance in open-ocean waters often occurs in a subsurface layer. Although light levels are always highest near the surface, phytoplankton also require inorganic nutrients to grow, just like other plants. In offshore regions, low phytoplankton abundance near the surface is generally attributable to low nutrient availability. Greater phytoplankton stocks can be supported at deeper depths where both nutrients and light are available.

Different phytoplankton species are often found at different depths and may be specialized to live under the conditions associated with certain depth strata. However, vertical mixing from tides, winds, surface cooling, internal waves, and other physical processes can homogenize the vertical patterns of phytoplankton abundance and species composition.

Because zooplankton do not require light, they can be found at all ocean depths. However, greater zooplankton abundance is generally found in the surface waters where phytoplankton occur. This should not imply that zooplankton and phytoplankton abundances are necessarily directly related. While many zooplankton feed on phytoplankton, many feed on bacteria or other zooplankton. Correlations between zooplankton and phytoplankton abundance patterns can be direct and causal for some groups of zooplankton, that is, for zooplankton that feed on phytoplankton. However, other zooplankton are only indirectly, or coincidentally, associated with phytoplankton.

Many common mesozooplankton, especially crustacean mesozooplankton such as copepods and euphausiids, vertically migrate between the deeper and upper water layers on a daily basis. Such diel vertical migrations of zooplankton may cover several hundred meters each way. The most compelling explanation for this behavior is that the zooplankton migrate down during the day to avoid predation, then rise to the upper layers at night to feed. At night, migrating zooplankton are less likely to be detected by predators in the surface layers that depend on sight (such as fishes) to catch prey. Darkness, whether due to depth or time of day, provides one of the few opportunities to hide in the relatively featureless, planktonic habitat. Although daily vertical migrations are common in both fresh and marine waters throughout the world, it should be pointed out that most zooplankton species probably do not vertically migrate. Many species live out their lives associated with a particular depth strata. For example, Marlow and Miller found that only around 10% of zooplankton species in the oceanic Gulf of Alaska vertically migrated. However, those vertically migrating species made up a large fraction of the total zooplankton biomass. Even species that do vertically migrate can change their migration pattern in response to local conditions or over the course of development.

Temporal Patterns of Plankton

The abundance of plankton varies seasonally in many places, especially in higher latitude waters where seasonality is most pronounced. A pattern observed in many temperate lakes, estuaries, and the open North Atlantic Ocean is the spring phytoplankton bloom. The cycle begins with low phytoplankton abundance in winter. In spring or early summer, phytoplankton abundance increases dramatically (a bloom). The zooplankton biomass often also increases, although there may be a time lag between the increase in phytoplankton and the later increase in zooplankton. Eventually, the stocks of both phyto- and zooplankton decline to moderate levels that are sustained through the summer. A second bloom can occur in autumn, before plankton levels decline again for the winter.

Much research effort has focused on describing and understanding seasonal patterns of plankton exemplified by the spring-bloom cycle. It provides the best known temporal pattern in plankton stocks. However, in many lower latitude waters and the open North Pacific Ocean, seasonal cycles are either absent, subtle, or complicated by the impact of episodic events such as upwelling, mesoscale eddies, or storms. Temporal patterns of plankton in polar seas are often dictated by the timing of sea ice melt.

Global Importance of Plankton

Although most planktonic organisms are individually small and seemingly ineffectual, plankton are important in aggregate because of their ubiquitous occurrence on this mostly water-covered planet. Outstanding examples of the global importance of plankton include their roles in the production of aquatic resources that humans value, in the Earth's climate, in oxygen production, and in sediment formation. Aquatic ecology and global biogeochemical processes cannot be understood without considering the role of plankton.

As the base of the aquatic foodweb, plankton are a critical food resource for aquatic organisms that are valuable to humans. The connection may be direct, as when valuable species such as oysters and anchovies feed on phytoplankton. The connection may also be indirect, for instance, if a commercially valuable predatory fish, such as a tuna, eats the anchovies. Either way, the occurrence of valued species is often coincident with the distribution of plankton. Large, motile species such as large fishes, seabirds, and whales concentrate their activities in locations where plankton are most abundant. This dependence of larger organisms on plankton means that commercial fisheries yield can be extrapolated from production of the plankton. Areas with higher plankton production have much greater fisheries yield.

Plankton influence the Earth's climate by removing carbon dioxide from the atmosphere. Carbon dioxide is the main atmospheric gas responsible for the "greenhouse" effect that regulates the Earth's temperature. Photosynthesis by phytoplankton fixes carbon dioxide from the atmosphere, transforming the carbon into organic compounds. Most of the carbon fixed by phytoplankton during photosynthesis is grazed by zooplankton and respired, releasing carbon dioxide back to the atmosphere. Nevertheless, some of the fixed carbon sinks to the deep ocean as the cells and bodies of plankton, zooplankton fecal matter, and other organic aggregates. Although the standing biomass of phytoplankton is less than 1% of the biomass of terrestrial plants, the amount of carbon fixed by phytoplankton and land plants is roughly equal. Because carbon fixation is coupled to the release of oxygen, phytoplankton are also responsible for approximately half of the Earth's oxygen production. It is estimated that one-third of the carbon fixed by phytoplankton eventually sinks to the deep ocean, effectively removing the carbon from the atmosphere for thousands of years. Determining the amount of carbon removed by plankton from the atmosphere and transported to the deep ocean, and predicting how that process may change in the future, are major questions in climate research.

Phytoplankton also influence climate by releasing sulfur compounds to the atmosphere. They are thought to be the main natural source of volatile sulfur compounds to the troposphere, mostly in the form of dimethylsulfide. The pathways leading to dimethylsulfide release from the oceans are incompletely understood, but phytoplankton primarily release the precursors to dimethylsulfide when their cells are ruptured during autolysis, viral attack, or grazing by zooplankton. Enzymes of phyto- and bacterioplankton then transform the released precursors to dimethylsulfide, which is released to the atmosphere. The sulfur aerosols stemming from dimethylsulfide of phytoplankton scatter incoming solar radiation. These aerosols also act as cloud condensation nuclei, further increasing atmospheric backscatter of solar radiation. Therefore, phytoplankton (and their interactions with other planktonic organisms) alter cloud formation and the amount of solar radiation reaching the Earth's surface.

A fraction of the carbon fixed by phytoplankton becomes a component of deep-sea sediments, removing it from the atmosphere for millions of years. Other inorganic structural components of plankton, such as the silicate tests of diatoms and radiolarians, and the calcareous shells of coccolithophores, pteropods, and foraminifera, also contribute to ocean bottom sediments. Over 60% of the ocean bottom is covered with sediments that are primarily planktonic in origin. Over geological time, such sea-bottom sediments, formed in previous ages, have been uplifted. Thus, even the peaks of mountains can be composed of material that originated from plankton.

See also **Ocean Resources (Living)**.

This article originally appeared in the *Water Encyclopedia*, 5 Vols. Lehr, J. H., and J. Keeley, Editors, John Wiley & Sons, Inc, Hoboken, NJ, 2005.

Additional Reading

Ault, T.R.: "Vertical Migration by the Marine Dinoflagellate *Prorocentrum triestinium* maximises Photosynthetic Yield," *Oecologia*, **125**, 466–475 (2000).

Baumgartner, M.F. and B.R. Mate: "Summertime Foraging Ecology of North Atlantic Right Whales," *Mar. Ecol. Prog. Ser.*, **264**, 123–135 (2003).

Beaugrand, G. and P.C. Reid: "Long-term Changes in Phytoplankton, Zooplankton and Salmon Related to Climate," *Global Change Biol.*, **9**, 801–817 (2003).

Bird, C. and M. Wyman: "Nitrate/nitrite Assimilation System of the Marine Picoplanktonic Cyanobacterium *Synechococcus* sp. Strain WH 8103: Effect of Nitrogen Source and Availability on Gene Expression," *Appl. Environ. Microbiol.*, **69**, 7009–7018 (2003).

Bollens, S. and B.W. Frost: "Predator-induced Diel Vertical Migration in a Planktonic Copepod," *J. Plankton Res.*, **11**, 1047–1065 (1989).

Calbet, A.: "Mesozooplankton Grazing Effect on Primary Production: A Global Comparative Analysis in Marine Ecosystems," *Limnol. Oceanogr.*, **46**, 1824–1830 (2001).

Calbet, A. and M.R. Landry: "Phytoplankton Growth, Microzooplankton Grazing, and Carbon Cycling in Marine Systems," *Limnol. Oceanogr.*, **49**, 51–57 (2004).

Cavender-Bares, K.K., D.M. Karl, and S.W. Chisholm: "Nutrient Gradients in the Western North Atlantic Ocean: Relationship to Microbial Community Structure and Comparison to Patterns in the Pacific Ocean," *Deep-Sea Res. I*, **48**, 2373–2395 (2001).

Chang, J., C. Chung, and G. Gong: "Influences of Cyclones on Chlorophyll *a* Concentration and *Synechococcus* Abundance in a Subtropical Western Pacific Coastal Ecosystem," *Mar. Ecol. Prog. Ser.*, **140**, 199–205 (1996).

Chisholm, S.W., R.J. Olsen, E.R. Zettler, R. Goericke, J.B. Waterbury, and N.A. Welschmeyer: "A Novel Free-living Prochlorophyte Abundant in the Oceanic Euphotic Zone," *Nature*, **334**, 340–343 (1988).

Cullen, J.J. and R.W. Eppley: "Chlorophyll Maximum Layers of the Southern California Bight and Possible Mechanisms of their Formation and Maintenance," *Oceanol. Acta*, **4**, 23–32 (1981).

DeRobertis, A., J.S. Jaffe, and M.D. Ohman: "Size-dependent Visual Predation Risk and the Timing of Vertical Migration in Zooplankton," *Limnol. Oceanogr.*, **45**, 1838–1844 (2000).

Enright, J.T.: "Copepods in a Hurry: Sustained High-speed Upward Migration," *Limnol. Oceanogr.*, **22**, 118–125 (1977).

Epifanio, C.E.: "Transport of Blue Crab (*Callinectes sapidus*) Larvae in Waters off Mid-Atlantic States," *Bull. Mar. Sci.*, **57**, 713–725 (1995).

Eppley, R.W., O. Holm-Hansen, and J.D.H. Strickland: "Some Observations on the Vertical Migration of Dinoflagellates," *J. Phycol.*, **4**, 333–340 (1968).

Erickson, J.: *Marine Geology: Exploring the New Frontiers of the Ocean*, 2nd Edition, Facts on File, Inc., New York, NY, 2002.

Falkowski, P.G., D. Ziemann, Z. Kolber, and P.K. Bienfang: "Role of Eddy Pumping in Enhancing Primary Production in the Ocean," *Nature*, **352**, 55–58 (1991).

Falkowski, P.G., R.T. Barber, and V. Smetacek: "Biogeochemical Controls and Feedbacks on Ocean Primary Production," *Science*, **281**, 200–206 (1998).

Ferris, M.J. and B. Palenik: "Niche Adaptation in Ocean Cyanobacteria," *Nature*, **396**, 226–228 (1998).

Field, C.B., M.J. Behrenfeld, J.T. Randerson, and P.G. Falkowski: "Primary Production of the Biosphere: Integrating Terrestrial and Oceanic Components," *Science*, **281**, 237–240 (1998).

Franks, P.J.S.: "Sink or Swim: Accumulation of Biomass at Fronts," *Mar. Ecol. Prog. Ser.*, **82**, 1–12 (1992).

Frost, B.W.: "A Modeling Study of Processes Regulating Plankton Standing Stock and Production in the Open Subarctic Pacific Ocean," *Prog. Oceanogr.*, **32**, 17–56 (1993).

Hayward, T.L., E.L. Venrick, and J.A. McGowan: "Environmental Heterogeneity and Plankton Community Structure in the Central North Pacific," *J. Mar. Res.*, **41**, 711–729 (1983).

Heinrich, A.K.: "The Life Histories of Plankton Animals and Seasonal Cycles of Plankton Communities in the Oceans," *J. Cons. Int. Explor. Mer.*, **27**, 15–24 (1962).

Hyrenbach, K.D., P. Fernandez, and D.J. Anderson: "Oceanographic Habitats of two Sympatric North Pacific Albatrosses during the Breeding Season," *Mar. Ecol. Prog. Ser.*, **233**, 283–301 (2002).

Kamykowski, D.: "Trajectories of Autotrophic Marine Dinoflagellates," *J. Phycol.*, **31**, 200–208 (1995).

Kennett, J.P.: *Marine Geology*, Prentice-Hall, Inc., Englewood Cliffs, NJ, 1981.

Longhurst, A., S. Sathyendranath, T. Platt, and C. Caverhill: "An Estimate of Global Primary Production in the Ocean from Satellite Radiometer Data," *J. Plankton Res.*, **17**, 1245–1271 (1995).

Lund, J.W.G.: "The Ecology of the Freshwater Phytoplankton," *Biol. Rev.*, **40**, 231–293 (1965).

Manahan, D.T.: "Adaptations by Invertebrate Larvae for Nutrient Acquisition from Seawater," *Am. Zool.*, **30**, 147–160 (1990).

Marlow, C.J. and C.B. Miller: "Patterns of Vertical Distribution and Migration of Zooplankton at Ocean Station 'P'," *Limnol. Oceanogr.*, **20**, 824–844 (1975).

Marra, J., R.W. Houghton, and C. Garside: "Phytoplankton Growth at the Shelf-break Front in the Middle-Atlantic Bight," *J. Mar. Res.*, **48**, 851–868 (1990).

McGowan, J.A. and P.W. Walker: "Structure in the Copepod Community of the North Pacific Central Gyre," *Ecol. Monogr.*, **49**, 195–226 (1979).

McGowan, J.A. and P.W. Walker: "Pelagic diversity patterns," in: *Species Diversity Patterns in Ecological Communities*, R.E. Ricklefs, and D. Schluter (Eds.). University of Chicago Press, Chicago, IL, pp. 203–214. 1993.

Mullin, M.M.: "Spatial and Temporal Scales and Patterns," in R.W. Eppley (Ed.). *Plankton Dynamics of the Southern California Bight*, Springer-Verlag New York, LLC, New York, NY, pp. 216–273, 1986.

Ohman, M.D.: "The Demographic Benefits of Diel Vertical Migration by Zooplankton," *Ecol. Monogr.*, **60**, 257–281 (1990).

Osgood, K.E., and B.W. Frost: "Ontogenetic Diel Vertical Migration Behaviors of the Marine Planktonic Copepods *Calanus pacificus* and *Metridia lucens*," *Mar. Ecol. Prog. Ser.*, **104**, 13–25 (1994).

Park, G.M., S.R. Yang, S.H. Kang, K.H. Chung and J.H. Shim: "Phytoplankton Biomass and Primary Production in the Marginal Ice Zone of the Northwestern Weddell Sea during Austral Summer," *Polar Biol.*, **21**, 251–261 (1999).

Pauly, D. and V. Christensen: "Primary Production Required to Sustain Global Fisheries," *Nature*, **374**, 255–257 (1995).

Rat'kova, T.N. and P. Wassman: "Seasonal Variation and Spatial Distributions of Phyto- and Protozooplankton in the Central Barents Sea," *J. Mar. Sys.*, **38**, 47–75 (2002).

Raven, J.A. and P.G. Falkowski: "Oceanic Sinks for Atmospheric CO_2," *Plant Cell Environ.*, **22**, 741–755 (1999).

Reid, J.L., E. Brinton, A. Fleminger, E.L. Venrick, and J.A. McGowan: "Ocean Circulation and Marine Life," in, H. Charnock and G. Deacon (Eds.), *Advances in Oceanography*. Plenum Publishing, New York, NY, 1978, pp. 65–130.

Ryan, J.P., P.S. Polito, P.G. Strutton, and F.P. Chavez: "Unusual Large-scale Phytoplankton Blooms in the Equatorial Pacific," *Prog. Oceanogr.*, **55**, 263–285 (2002).

Ryther, J.H.: "Photosynthesis and Fish Production in the Sea," *Science*, **166**, 72–76 (1969).

Sieburth, J.M.: *Sea Microbes*, Oxford University Press, New York, NY, 1979.

Simo, R.: "Production of Atmospheric Sulfur by Oceanic Plankton: Biogeochemical, Ecological and Evolutionary Links," *Trends Ecol. Evol.*, **16**, 287–294 (2001).

Small, L.F. and D.W. Menzies: "Patterns of Primary Productivity and Biomass in a Coastal Upwelling Region," *Deep-Sea Res.*, **28A**, 123–149 (1981).

Steinke, M., G. Malin, and P.S. Liss: "Trophic Interactions in the Sea: an Ecological Role for Climate Relevant Volatiles?" *J. Phycol.*, **38**, 630–638 (2002).

Sverdrup, H.U.: "On the Conditions for the Vernal Blooming of Phytoplankton," *J. Cons. Int. Explor. Mer.*, **18**, 287–295 (1953).

Takahashi, M., S. Ichimura, M. Kishino, and N. Okami: "Shade and Chromatic Adaptation of Phytoplankton Photosynthesis in a Thermally Stratified Sea," *Mar. Biol.*, **100**, 401–409 (1989).

Tankersley, R.A., M.G. Wieber, M.A. Sigala, and K.A. Kachurak: "Migratory Behavior of Ovigerous Blue Crabs *Callinectes sapidus*: Evidence for Selective Tidal-stream Transport," *Biol. Bull.*, **195**, 168–173 (1998).

Venrick, E.L.: "Phytoplankton in an Oligotrophic Ocean: Observations and Questions," *Ecol. Monogr.*, **52**, 129–154 (1982).

Venrick, E.L.: "Phytoplankton Seasonality in the Central North Pacific: the Endless Summer Reconsidered," *Limnol. Oceanogr.*, **38**, 1135–1149 (1993).

von Montfrans, J. et al.: "Settlement of Blue Crab Postlarvae in Western North Atlantic Estuaries," *Bull. Mar. Sci.*, **57**, 834–854 (1995).

Yoder, J.A. and M.A. Kenelly: "Seasonal and ENSO Variability in Global Ocean Phytoplankton Chlorophyll Derived from 4 Years of SeaWiFS Measurements," *Global Biogeochem. Cycles*, **17**, 1112 (2003).

ANDREW JUHL, Lamont-Doherty Earth Observatory of Columbia University, Palisades, NY

PLANTAIN *(Plantago sp.; Plantaginaceae)*. There are some 200 widely distributed species of plantains, many of which are ubiquitous weeds. Some species are stemless plants, the petioled leaves forming a rosette that covers a considerable area of ground, from which it excludes other desirable plants. These are the species that cause unsightly patches in lawns. These weeds thrive so well in cultivated areas that some of the American Indians call them "white-man's-footsteps."

Plantago major is one of these, with broad ovate leaves on long petioles. *Plantago lanceolata* has lanceolate erect leaves. The flowers are borne in elongated spikes, the pistil maturing before the stamens. Plantains are wind-pollinated plants, although insects occasionally visit them for their pollen. The fruit is a capsule, the upper half of which comes off when mature, through the development of a circumferential line of dehiscence. The seeds of plantains are often fed to caged birds. In some species the seeds imbibe water and become mucilaginous. One species, *Plantago psyllium*, native in southern Europe, is sometimes used in medicine under the name Psyllium seed.

Plantain also refers to a tropical fruit, *Musa paradisiaca*, a close relative of the banana. Plantains are always eaten cooked, or made into flour. They have furnished food for all tropical peoples for centuries.

PLANT ANATOMY. Higher plants differ enormously in their size and appearance, yet all are constructed of tissues classed as dermal (delineating boundaries created at tissue surfaces), ground (storage, support) or vascular (transport). These are organized to form three vegetative organs: roots, which function mainly to provide anchorage, water, and nutrients; stems, which provide support; and leaves, which produce food for growth. Organs are variously modified to perform functions different from those intended, and indeed the flowers of angiosperms are merely collections of leaves highly modified for reproduction. The growth and development of tissues and organs are controlled in part by groups of cells called meristems. This introduction to plant anatomy begins with a description of meristems, then describes the structure and function of the tissues and organs, modifications of the organs, and finally describes the structure of fruits and seeds and how these are modified for dispersal. Figure 1 presents an illustration of the structures of a typical plant.

Meristems

Because they are rooted to one spot, plants must adapt to changing conditions in order to survive. They must be able to rejuvenate parts that are damaged or lost as they grow, and continue the cycle of producing flowers and setting seed. For these reasons plants are able to renew themselves continually through localized growth, a process that occurs in meristems. These are sources of undifferentiated, genetically sound, cells.

Determinate meristems are designed to produce structures of a certain size, such as leaves and flowers. The great similarity of the leaves on a tree results from the effectiveness of determinate meristematic activity in creating copies of structures. Indeterminate meristems are never-ending sources of new cells, allowing increase in length (apices) or girth (cambia).

Meristematic activity arises early in embryogenesis. In some plants, when the ovule reaches the 16-cell stage of development, there is already an outer layer consisting of eight of these cells. This layer is the first discernible protoderm, which is the primary meristem giving rise to the epidermis of the plant. Two other fundamental primary meristems arise in the embryo, the ground meristem, which produces cortex and pith, and the procambium, which produces primary vascular tissues. In shoot and root tips, apical meristems add length to the plant, and axillary buds give rise to branches. Intercalary meristems, common in grasses, are found at the nodes of stems (where leaves arise) and in the basal regions of leaves, and

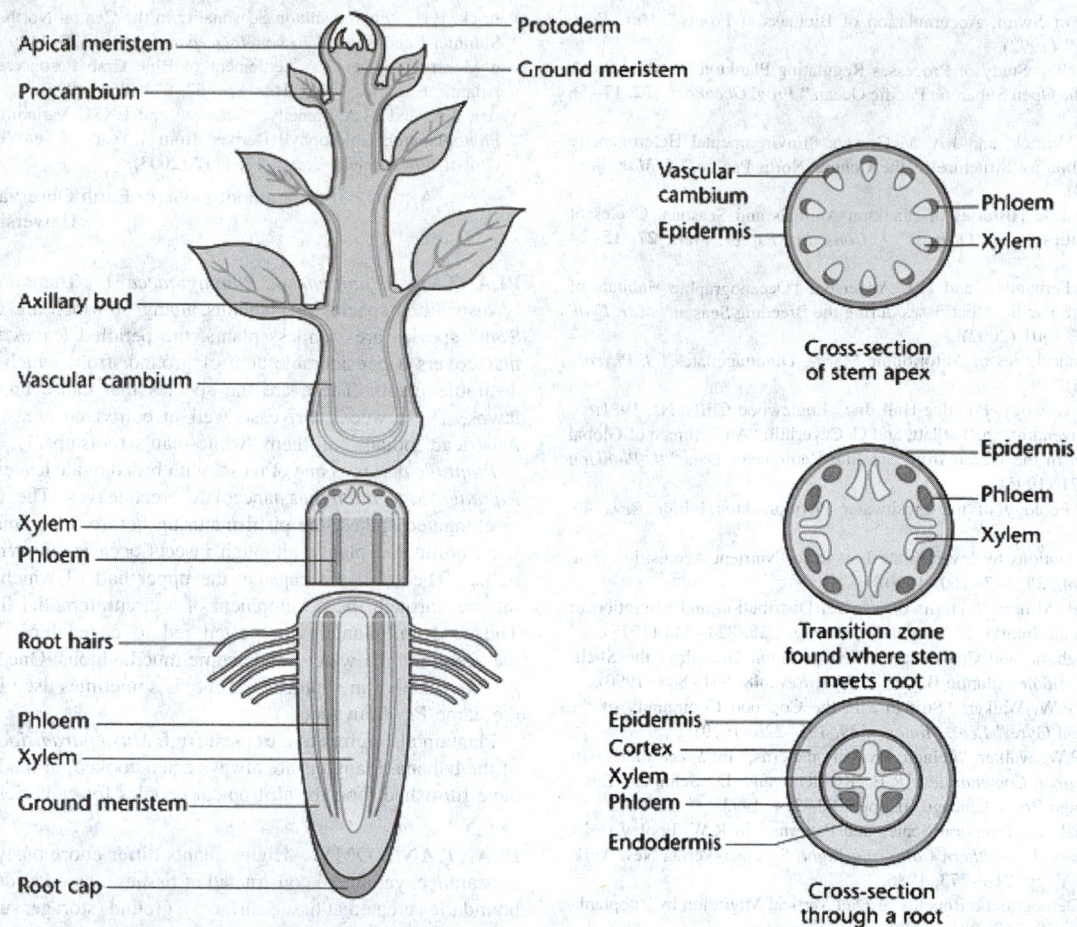

Fig. 1.   Longitudinal section through a typical plant and cross-section of the specific areas.

cause these organs to elongate. All of these are primary meristems, which establish the pattern of primary growth in plants.

Stems and roots add girth through the activity of vascular cambium and cork cambium, lateral meristems that arise in secondary growth, a process common in dicotyledonous plants (Fig. 2). Many monocotyledonous plants have primary meristems alone, and lack true secondary growth. Cambium is in essence an intercalary meristem because it lies between its derivatives. Vascular cambium normally creates xylem to the inside and phloem to the outside of stems and roots (just as cork cambium lies between its derivatives). But in time, primary intercalary meristems stop producing new cells and disappear, whereas cambium is essentially indeterminate in its activity. The activity of the vascular cambium is complex. It produces more xylem than phloem, and thus it expands in circumference and it must add new cells by radial divisions to maintain the integrity of the cambial cylinder.

Plants are prone to mutation because they are potentially long-lived and also because they are subjected to ionizing ultraviolet light from the sun. Mutations are most likely to occur during cell division, so the fewer cell divisions plants need for growth, the better. Meristems let the plant avoid repetitive cell divisions, reducing the possibility for mutation. But because plants have growth localized in meristematic areas, these are at risk of accelerating mitotic mutations if all they do is create cell lines (cells of the same type) by endless cell division. Plants avoid cell lines by having multistep meristems. This multistep meristem produces secondary vascular tissue (Fig. 3).

This delegation of roles makes cell lines much shorter. Endlessly repeating cell divisions are not eliminated, but they tend to occur as the end product of the meristem, that is, identical cells that do not divide further. For example, cork cells are produced in great numbers by the cork cambium and they are dead when functional. It does not matter if some of these cells are mutated because their genes are not passed on.

Reduction in mutation enhances the potential for plants to grow indefinitely. While "Eternal God" a coast redwood (*Sequoia sempervirens*) in California has lived for more than twelve millennia, it is still young when compared to clonally reproducing specimens of King's holly (*Lomatia tasmanica*) in the wilderness of Tasmania that are 43 400 years old. Meristematic activity in these and other plants of great age has allowed them to achieve states of near immortality.

Dermal Layers

All plants are wrapped in protective layers of cells, and other cell layers occur inside them. An epidermis, which may be defined as the outermost cell layer of the primary plant body, entirely covers herbaceous plants: leaves, stems, and roots all have an epidermis. Although usually just one cell thick (uniseriate), some plants have a multiple (multiseriate) epidermis many layers thick. Epidermal cells may live for many years, and are modified in diverse ways. See also **Epidermis (Plant)**.

The epidermis forms an interface between the plant and its environment. It is coated with cutin, a complex fatty substance that forms the cuticle and is indigestible by pathogens. The cuticle is impregnated with long chained waxes, which render it very impermeable to water. The cuticle is clear, like the epidermal cells it covers, allowing light to reach photosynthetic tissues beneath. Also, the cuticle selectively protects the plant from mutagenic, ultraviolet sunlight. See also **Plant Cuticle**; and **Plant Waxes**.

The epidermis has stomata that allow gas exchange between the plant and the air surrounding it. Each stoma consists of a pair of bean-shaped guard cells that can bend apart to create a stomatal pore. Guard cells have large nuclei and numerous chloroplasts (usually they are the only epidermal cells to have chloroplasts). In most monocotyledonous plants, the guard cells are dumbbell shaped. See also **Stomate (or Stoma)**.

Trichomes arise from epidermal cells that extend outward, typically dividing repeatedly to form a single file of cells. Trichome functions include shading the plant surface from excessive light, reducing air movement (and thus excessive desiccation), and protecting the plant from

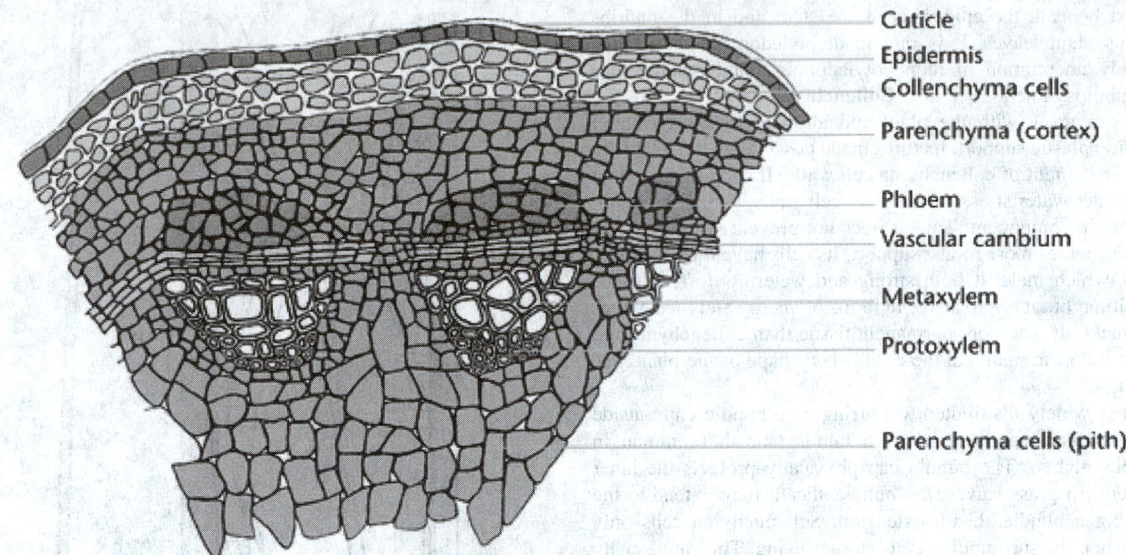

Fig. 2.   Cross-section of a dicotyledonous stem.

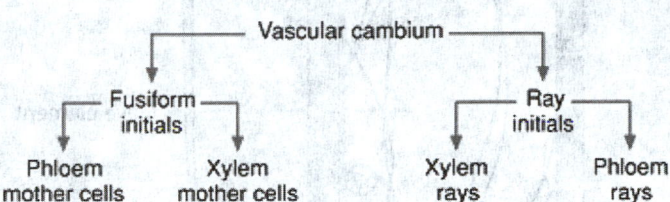

Fig. 3.   Production of secondary vascular tissue from the multistep meristem.

insects and herbivores. Protection may be passive (blocking access to the plant surface) or active, through secretion of toxins. See also **Trichomes**.

Trichomes of the stinging nettle have brittle silica tips that readily break off to inject grazing animals or passing hikers with irritating histamines. The most elaborate trichomes are perhaps those on the leaves of the insectivorous sundew (*Drosera*). These excrete a sticky nectar to attract and hold insects, then bend over in concert with leaf-folding to trap them, excrete digestive enzymes, and finally absorb nutrients from their prey.

The shoot epidermis protects the plant from desiccation, while the root epidermis allows the plant to extract water and ions from the soil, aided by their root hairs, which are trichomes. While root epidermis has a thin cuticle, its waxes have shorter chains, so it is more permeable than shoot epidermis.

Plants in secondary growth can replace their epidermis with a more substantial and complex tissue called *periderm*. This usually arises just beneath the epidermis, although it may form in other areas of the stem. Periderm consists of layers of three types of cell: a middle layer of phellogen (cork cambium, a lateral meristem) that produces an outer layer of phellem (cork cells) and an inner layer of phelloderm (which contributes to the cortex). Cork cells are dead at maturity and their walls are layered with suberin and sometimes lignin, giving them great resilience to desiccation and insect or pathogen attack. Periderm also develops in roots, but there it usually arises from the pericycle, which is just outside the phloem. See also **Cork**.

Gaps occur in the phellem called *lenticels*, which arise under stomata in the epidermis and persist as blisters in tree bark. They allow gas exchange with the underlying tissues that the periderm so well protects. Unfortunately, they provide a constantly open route for infection and herbivory.

"Bark" generally means all of the tissues outside the vascular cambium (secondary phloem, phloem fibres, cortex, and the periderm). Bark has many different patterns because periderm forms at different rates at different places on the stem. Also, the cork breaks apart as the stem expands in circumference. New cambia arise under other ones, pushing the old periderms out. See also **Bark**.

Protective cell layers occur in roots, stems, and sometimes leaves. The most prominent such layer is the endodermis, commonly found in roots. It forms the innermost cell layer of the cortex, separating it from the stele (vascular cylinder). The endodermal cell wall contains suberin, which is usually restricted to a narrow, thin strip called a *Casparian band* that runs around the middle of each cell in its radial and tangential walls. Endodermal cells are often extensively lignified, forming characteristic U-shaped ("phi") thickenings that arch halfway around each cell to meet the Casparian bands.

The Casparian band effectively prevents uncontrolled entry of water and ions, absorbed from the soil, into the stele through the apoplast (the cell walls and intercellular spaces). The Casparian band directs transport through the cytoplasm of the endodermal cell, which subjects it to physiological influence, allowing selective uptake by the plant.

An endodermis may occur in the stem, where it usually lacks a Casparian band or phi thickening. Special staining methods are needed to make it visible. Stem endodermis is sometimes referred to as a *starch sheath*, but since the starch itself may be absent, the term is not useful. Some aquatic plants do have an obvious endodermis in their stems and in their leaves as well. See also **Endodermis**.

An exodermis (or hypodermis) forms the outermost layer of the cortex, directly beneath the epidermis. It is often more obvious in monocotyledonous plants than in dicotyledonous plants. It can be lignified and can have a Casparian band. Normally a feature of stems and roots, the needle-like xerophytic leaves of conifers have a prominent exodermis and endodermis.

Ground Tissues

Parenchyma, composed of undifferentiated, isodiametric cells, with thin cell walls, large vacuoles, and thin parietal cytoplasm, occupies much of the cortex and pith of stems and roots. In roots, parenchyma often has a storage function, and stem parenchyma contributes support if it is turgid. The swollen protoplast of each cell presses outwards against the cell wall, and all of the parenchyma cells together press out against the restraining layer of collenchyma and epidermis of the outer stem. It is believed that pressure in the pith parenchyma contributes to stem growth. See also **Parenchyma**.

Collenchyma, a living tissue with primary cell walls that are unevenly thickened, provides support to seedlings and growing stems. It occurs in the cortex just beneath the epidermis of the stem and in the midribs of dicotyledonous plant leaves. It is rare in dicotyledonous plant roots, and comparatively uncommon in monocotyledonous plants. Sometimes collenchyma is photosynthetic. See also **Collenchyma**.

Collenchyma can grow with the plant and allow it to bend without breaking, providing plastic support, features made possible by the relatively high hemicellulose content of collenchyma cell walls. It supports the plant only if it is not under water stress, because its cell walls contain no lignin or other hydrophobic component. Thus it does not prevent wilting.

Sclerenchyma provides more robust support. Its cells have thick lignified secondary walls, which make it both strong and waterproof. This tissue helps prevent wilting, but it is expensive in terms of energy and metabolites for the plant to make. It is a more permanent tissue than collenchyma and provides elastic support to maintain the established shape of the plant. See also **Sclerenchyma**.

Sclerenchyma is widely distributed, occurring as a bundle cap outside the phloem in vascular bundles, and as a bundle sheath (common in monocotyledonous plants). The bundle cap physically protects the inner tissues of the stem. In grass leaves, the bundle sheath may extend to the epidermis, forming a bundle sheath extension. Sclerenchyma cells only become mature when the surrounding cells stop growing. They are usually dead at maturity, although the lumens of the cells remain connected by pits.

Sclerenchyma cells occur in two forms: fibers, which are long (up to 55 centimeters (21.5 inches) in the case of hemp fibers) with tapered ends, and sclereids, which are more or less isodiametric. Brachysclereids, or stone cells, form in clumps in the flesh (mesocarp) of the Bartlett pear (*Pyrus communis*), giving it a characteristic grittiness. They form a dense layer to make the endocarp ("shell") of the coconut. Many seed coats (testas), especially those of legumes, are made of a double layer of sclereids.

Branched astrosclereids occur in the petioles and blades of water lily leaves (*Nymphea*), making them leathery and resistant to the tearing forces of waves and currents.

Some botanists distinguish between support sclerenchyma (fibers and sclereids) and conducting sclerenchyma (vessels and tracheids). Xylem vessels provide both support and transport capabilities, but one function comes at the expense of the other because thicker walls reduce their internal diameter.

Vascular Tissues

Higher plants have two transport systems, the xylem and the phloem, which comprise their vascular tissue. They have somewhat different functions, but generally arise together and are nearly always found running side by side within all organs of the plant.

Xylem transports water and various dissolved ions from the roots upward through the plant. Phloem transports a solution of metabolites (mainly sugars, amino acids, and some ions) from "sources" of production, such as fully expanded leaves, to "sinks," such as developing leaves, fruits, and roots. Both tissues contain long tubular cells (and some other associated cells) joined end to end that are responsible for transport.

Phloem is predominantly a living tissue, consisting of sieve tubes, companion cells (Fig. 4), phloem parenchyma, and phloem fibers. Each sieve tube consists of a file of sieve elements, called *sieve tube members* in angiosperms. Sieve element is the collective term for the sieve tube members, in most angiosperms and the sieve cells in non-angiosperms. But this discussion will not deal with the somewhat different and much less understood sieve cell, so the term *sieve element* is used here in place of the wordy *sieve tube member*.

Sieve elements are joined by thick end walls, called *sieve plates*, which are pierced by large, modified plasmodesmata (cytoplasmic bridges between cells) called *sieve pores*. Sieve pores are lined with the complex carbohydrate callose, which rapidly proliferates to seal the pores in response to damage to the phloem such as that caused by grazing animals. Sieve elements contain only a little cytoplasm, dominated by amyloplasts (starch-filled plastids) and filamentous proteins. When mature, they lack a nucleus and tonoplast, but retain a plasmalemma, indicating that they are

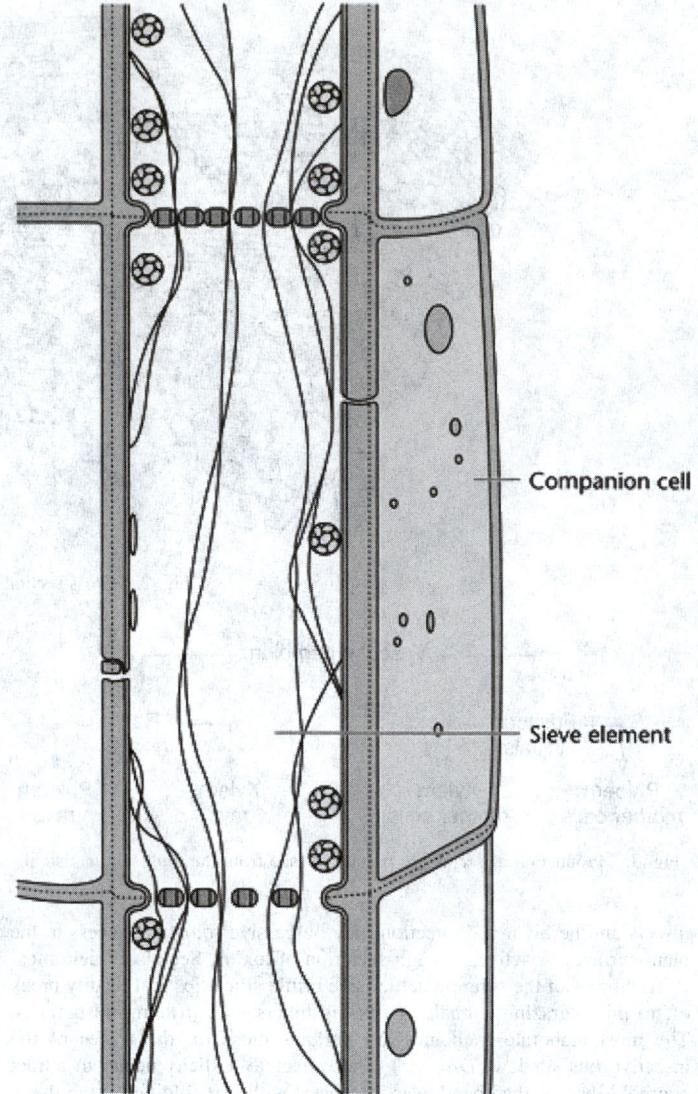

Fig. 4.   Sieve tube and companion cells.

living cells. Sieve elements have thick primary cell walls through which pass abundant plasmodesmata, connecting them to small companion cells that cluster around each sieve element.

Companion cells contain large nuclei, dense cytoplasm with abundant organelles (especially mitochondria), and many small vacuoles. These features are related to the functional association of companion cells and elements.

Xylem is mostly a dead tissue, consisting of vessels, tracheids, xylem parenchyma and fibers. The conducting cells of the xylem, the tracheary elements, are the vessels and tracheids. These are dead at maturity, and have secondary walls cross linked with the complex polymer lignin. Various patterns of lignin deposition occur, ranging from annular, helical, scalariform, reticulate, to pitted (Fig. 5) depending on the age of the cell. As the degree of lignification increases, the resistance to water stress increases.

Vessels consist of cylindrical vessel elements (also called *vessel members*) joined together by large openings in their end walls called simple perforation plates, ridges representing the remnants of degraded end walls.

Although transport can occur freely through the perforation plates of vessels, movement also takes place laterally, among adjacent vessels, and with adjoining xylem parenchyma cells. This happens across

Fig. 5.   Xylem vessels.

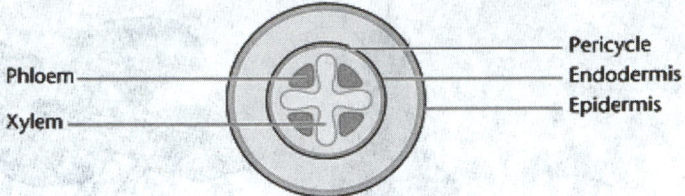

Tissue map of root cross-section

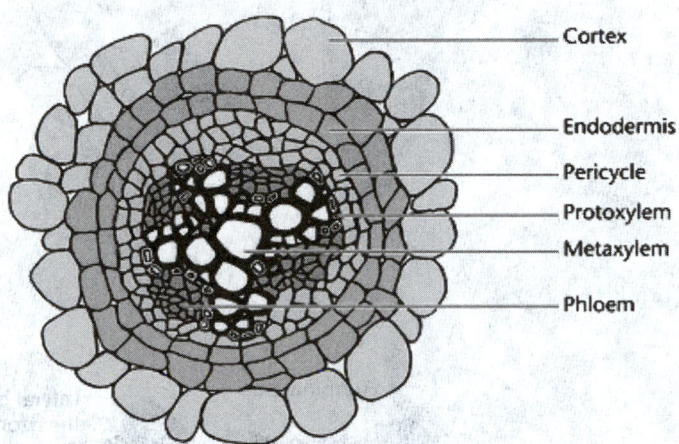

Fig. 6.   Root cross-section.

areas of primary wall where no secondary wall has been deposited. Vessels with annular thickenings provide the greatest area for lateral transport, with less area becoming available as the degree of lignification increases.

Lateral transport in pitted vessels is restricted to structures called *pits*. These occur as two main types, simple pits and bordered pits. Simple pits are areas of bare primary wall in vessels otherwise covered with lignin. Bordered pits contain pressure-sensitive valves that prevent air embolisms from interrupting flow between cells.

The other type of tracheary element is the tracheid, which has a thin secondary wall, tapered ends, and no perforation plates. It is connected to adjoining cells only by pits. While tracheids are found in both gymnosperms and angiosperms, only the most advanced gymnosperms contain vessels.

The Organ System

Root. Roots anchor plants in the soil, absorb water and ions from it, produce plant growth regulators, and store sugars and starch. Structurally, roots have four external zones, beginning at the tip with the root cap, which protects the root apical meristem and secretes mucigel. This absorbs water from the soil to facilitate ion diffusion into the root, fosters growth of bacteria that release more nutrients, and lubricates the root to ease passage through the soil. Behind the root cap is the elongation zone, where cells elongate to add length to the root. This is followed by the root hair zone. Root hairs are short-lived (one or two days), fragile, outgrowths of epidermal cells that greatly increase the volume of soil that can be mined by a plant for nutrients. Last is the lateral root zone or maturation zone, characterized internally by the presence of an endodermis that encloses the stele (pericycle and vascular tissue) (Fig. 6).

Root growth is indeterminate and is governed by the moisture, fertility, composition and homogeneity of the soil, such that growth occurs in patches, making root architecture variable. In general, a seedling produces a primary root that grows straight down and gives rise to secondary lateral roots. These may produce tertiary roots, which in turn may branch, with the process continuing almost indefinitely. New roots arise endogenously from the pericycle of the main root. Lateral roots may exist as primordia in the embryo before germination, where they are called *seminal roots*.

Root Modifications. Tap roots, like that of the carrot (*Daucus carota* L.), are modified for storage (Fig. 7a). Another storage root is the root tuber, which is really a swollen adventitious root (sweet potato, *Ipomea batatas* L.)

Aerial roots occur on epiphytic orchids, and have a multiseriate epidermis called *velamen* made of dead cells with suberized walls. Velamen apparently absorbs water and nutrients during wet conditions and retains water during dry conditions. Some aerial roots are photosynthetic.

Pneumatophores (breathing roots) occur on some mangrove trees. They grow upwards through the substrate, usually anaerobic mud, and bear lenticels that allow gas exchange when exposed at low tide.

Prop roots are a type of adventitious root. They provide more transport capability to monocotyledonous plants as well as extra support. They arise from the lowest nodes and grow downwards through the air to the soil. Once there they may contract slightly to help anchor the plant (maize, *Zea mays*; screw pine, *Pandanus veichii*).

Contractile roots shrink even more than prop roots, and pull growing corms, bulbs and rhizomes down in the stable soil environment, hiding them from herbivores (ginger, *Zingiber officinale*; dandelion, *Taraxacum officinalis*) (Fig. 7b).

Root nodules, found mostly on legumes, harbor *Rhizobium* bacteria, which receive nutrients from the host plant and in return fix N_2 gas to NO_3 for the host to use.

Parasitic plants have haustorial roots that invade the cortex and vascular tissue of the host plant. Some rely partly (birdvine, *Pthyrusa stelis*), others more completely (dodder, *Cuscuta*), on the host for food, depending on whether or not they invade the phloem.

Stem. The stem and its branches allow leaves to be arranged to maximize exposure to sunlight, and flowers to be arranged to best attract pollinators. Branching arises from the activity of apical and axillary buds. While branching is a complex topic, four basic patterns can be identified. See also **Shoots and Buds**.

Monopodial plants have a rhythmically active shoot apical meristem, with axillary shoots that remain secondary and regulated by the main shoot apex. Most conifers exhibit monopodial branching.

In plants with sympodial branching the shoot apex becomes reproductive or aborts. One axillary shoot grows upward and becomes the main stem. Its shoot apex becomes reproductive or aborts, and so on. The characteristic pagoda shape of the seaside almond (*Terminalia catapa*) results in part from this growth habit.

In dichasial (or dichotomous) branching the terminal bud splits into two buds on opposite sides of the stem. These grow at a similar rate and then

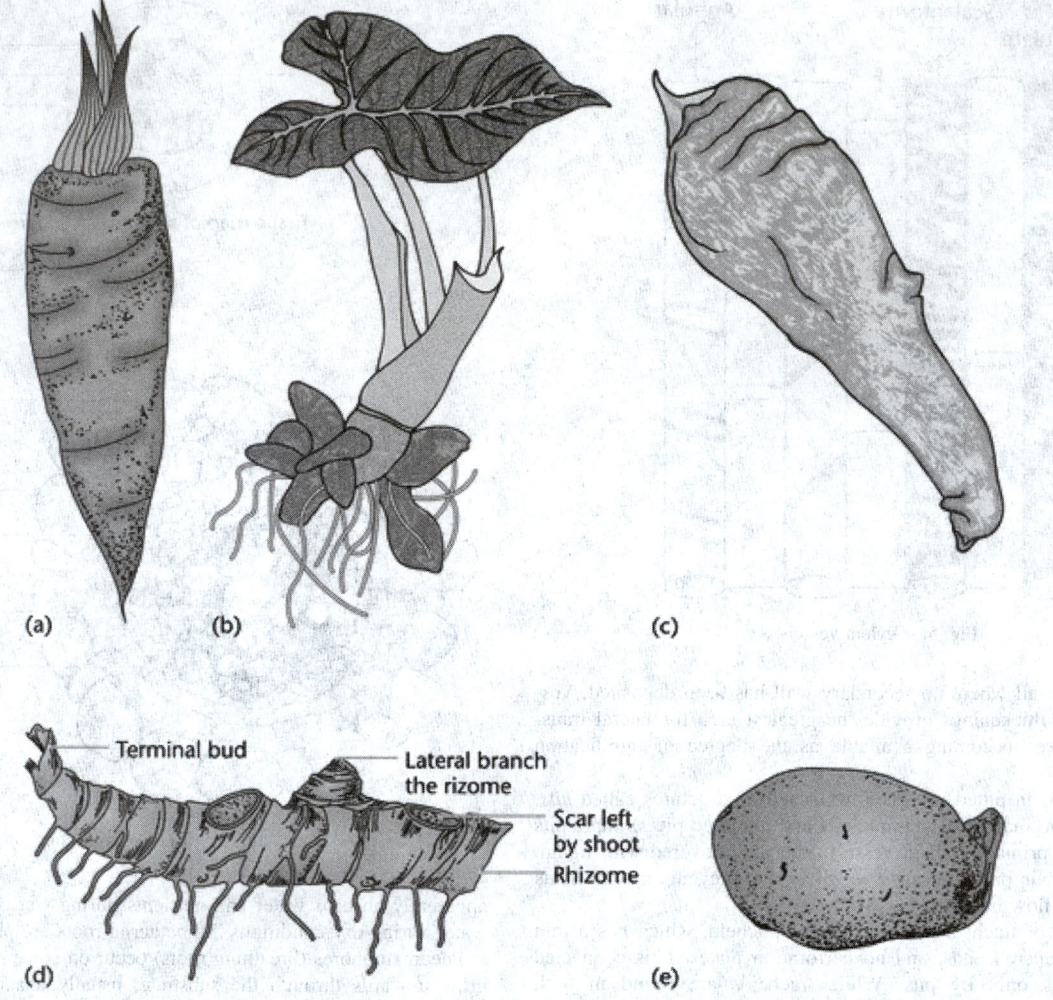

Terminal bud

Lateral branch
the rizome

Scar left
by shoot

Rhizome

(a)   (b)   (c)

(d)   (e)

Fig. 7.   (**a**) Carrot, modified tap root. (**b**) Tannia, plant. (**c**) Tannia, corm. (**d**) Ginger, rhizome. (**e**) Potato, stem tuber.

branch again, resulting in a repeatedly forked appearance. Many plants display superficial dichasial branching, but only some palms and cacti exhibit true dichasial branching.

Although shoots usually arise from apical or axillary buds, they may appear endogenously from any organ, creating adventitious branches. Sweet potato (*Ipomea batatas*) owes its scrambling habit to stems arising from the roots, giving the plant a disorganized appearance. See also **Stem (Plant)**.

Stem Modifications. Stems serve other functions besides or instead of support. Some are modified for storage. Bulbs consist of flattened, short stems with thick, fleshy nonphotosynthetic storage leaves (onion, *Allium cepa*). Corms are vertical, enlarged fleshy stem bases with scale leaves (dasheen, *Colocasia esculenta*; *Gladiolus*), while rhizomes (Fig. 7**d**) are fleshy horizontal stems, also with scale leaves (arrowroot, *Maranta arundinacea*; ginger, *Zingiber officinale*). Stem tubers are thick storage stems, usually horizontal (yam, *Dioscorea*; Irish potato, *Solanum tuberosum*) (Fig. 7**e**).

All of these are both underground storage stems, and organs of perennation, enabling plants to survive periods of drought or cold, storing food reserves away from most herbivores, in a relatively stable environment. The stem of sugar cane (*Saccharum*) is an aerial stem tuber. Storage stems are propagative, and will produce new shoots and roots. See also **Plant Reproduction**.

Stolons (which bear foliage leaves) and runners (which have only scale leaves) allow plants to spread vegetatively above ground. These are indeterminate propagative stems with long, thin internodes that readily root at the tip to form new plants if growing conditions are favorable (*Paspalum* sp., saxifrage, spider plant). They have no storage or support function. Although many stems photosynthesize to augment their leaves,

some stems are more specifically designed to replace them. The cladode is flattened to look vaguely leaf-like (prickly pear cactus, *Opuntia*), but it is swollen for water storage (succulent), too. The conifer-like needles found on the she oak (*Casaurina*) are better examples of photosynthetic stems. They are not enlarged for storing anything, and they have tiny, vestigial leaves.

Some stems provide support by wrapping around another plant or other support. Many members of the Convolvulaceae are vines, and they have twining stems that wrap especially well round wire fences.

Leaf. Leaves are designed to provide plants with photosynthates, but they are modified to provide protection (spines, bud scales), support (tendrils), storage (onion bulb), nitrogen acquisition (insect trap leaves), and pollinators (flower petals). Typical dicotyledonous plant leaves have a large surface area to maximize photosynthesis and a thin profile to expose lots of chlorophyll to the light. At the same time they must dissipate wind forces and heat, retain water, allow gas exchange for photosynthesis and respiration, and exclude pathogens. See also **Leaf**.

By extending the leaf blade (lamina) away from the stem by a flexible leaf stalk (petiole), shading from the stem is reduced and the leaf can flutter. Fluttering promotes cooling, decreases insect and pathogen attack, dissipates wind forces to reduce damage, and increases carbon dioxide absorption. Monocotyledonous leaves are usually narrow and belt-shaped, with a sheathing base, and they move easily in the wind, too.

The upper epidermis of most dicotyledonous plant leaves has a thicker cuticle and fewer stomata than the lower epidermis, and palisade mesophyll cells, the major sites of photosynthesis, are arranged above spongy mesophyll, where gas exchange occurs (Fig. 8). The dorsal and ventral surfaces of many monocotyledonous leaves are more or less the same and their internal tissues are more homogeneous than stratified.

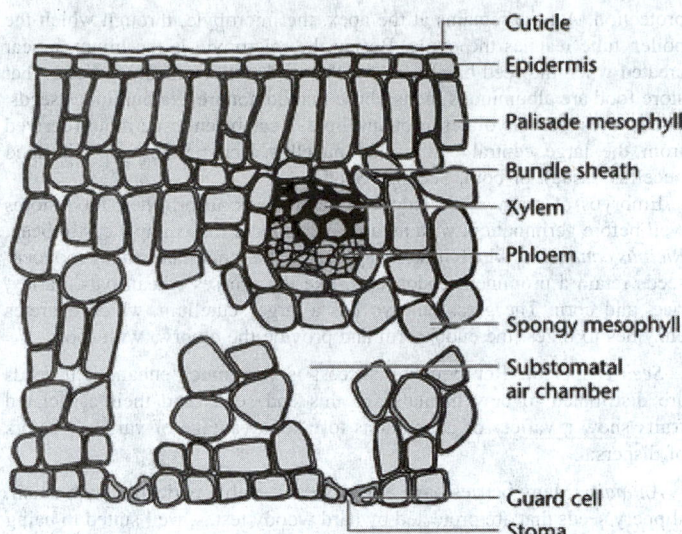

Fig. 8.   Transverse section of a leaf.

Labels:
Cuticle
Epidermis
Palisade mesophyll
Bundle sheath
Xylem
Phloem
Spongy mesophyll
Substomatal air chamber
Guard cell
Stoma

Leaf Modifications. Spines consist entirely of compact bundles of sclerenchyma fibers. They can, if densely packed, shield the plant against intense sunlight and the drying effect of wind. Thorns are, to some botanists, simply spines that may contain vascular tissue and are modified stems, not leaves. Both arise from axillary buds and provide protection from herbivores.

Tendrils lack a blade, like spines, but photosynthesize, never stop growing, and help support the plant. They are touch-sensitive, not light-sensitive, and coil around things, especially stems of other plants. Coiling results from growth occurring faster on the outer side of the tendril away from the inner side in contact with the support. Tendrils are commonly found in many plants including Cucurbitaceae (cucumber, melon), Passifloraceae (passion fruit) and the Convolvulaceae. Both spines and tendrils can be modified stems, not leaves. Indeed, spines arise from adventitious roots on a species of mangrove, and some tendrils even arise from inflorescences.

Sclerophyllous leaves are xerophytic, designed to withstand a desert-like environment. Like spines, they have abundant fibres but retain photosynthetic ability. Such leaves are expensive for the plant to make and so tend to be long-lived. Ordinary leaves, by contrast, are cheap and mass-produced by many plants, and are expendable. Sclerophyllous leaves are especially common among the monocotyledonous plants, for example, *Agave sisalana*. The fibers that make leaves of this plant so durable are used to make sisal rope.

Succulent leaves are also xerophytic, with few air spaces and a thick cuticle. The mesophyll is isolated from the leaf surface by cells with large vacuoles, which filter heat from intense sunlight. Succulent leaves often contain mucilage, which binds water so it can be stored in the leaf. The vascular tissue is "conserved." There is not much of it because little transport occurs in these leaves. *Aloe vera* is a good example. Both succulent and sclerophyllous leaves have a small surface area to volume ratio, reducing transpiration losses. Bud scales (cataphylls) are tough, sessile (lacking a petiole), nonphotosynthetic leaves (actually stipules) that protect apical or axillary buds, during dormancy, against herbivores and drying winds. Common on temperate trees, in the tropics they are found on the Para rubber tree (*Hevea brasiliensis*), *Magnolia* and mahogany (*Sweetinia macrophylla*). They often form cork cambium that makes a thin bark, the only type of leaf to do so.

Trap leaves are found on insectivorous plants, which tolerate nitrogen poor soil by trapping and digesting insects. The highly modified pitchers of pitcher plants (*Nepenthes, Sarracenia, Darlingtonia*) are typically red brown, like carrion, and lined with glands that release digestive enzymes. They may have lids to exclude rain, downward-pointing trichomes, and abundant flaky wax to prevent the escape of insects. In *Nepenthes* the petiole of the trap is elongate, wide, and flat, like a normal lamina, forming a phyllode. This structure compensates for the reduced photosynthetic role of the pitcher. Other kinds of trap leaves occur on the Venus flytrap (*Dionaea*), sundew (*Drosera*) and bladderwort (*Utricularia*).

Flower. A flower is a shoot system consisting of a determinate axis with laterally borne, concentric rings of four kinds of leaves designed for sexual reproduction: sepals, petals, stamens and carpels. Green sepals closely resemble leaves, while petals and colored (petaloid) sepals have a poorly developed vascular system, no palisade mesophyll, little sclerenchyma, and chromoplasts instead of chloroplasts. Most stamens and carpels do not look like leaves, but in the more primitive flowers (*Drimys, Magnolia, Michelia*), they may be wide and flattened, pointing to their origin. See also **Flower**.

Because petals and sepals are not necessary for reproduction they are accessory parts, while stamens and carpels are essential parts. The sepals together form the calyx of the flower, and the petals together make up the corolla, with both whorls together comprising the perianth (Fig. 9). The stamens (filament + anther) are the male parts (androecium), and the carpels (ovary + style + stigma) are the female parts (gynoecium). See also **Carpels**; **Petals**; **Sepals**; and **Stamens**.

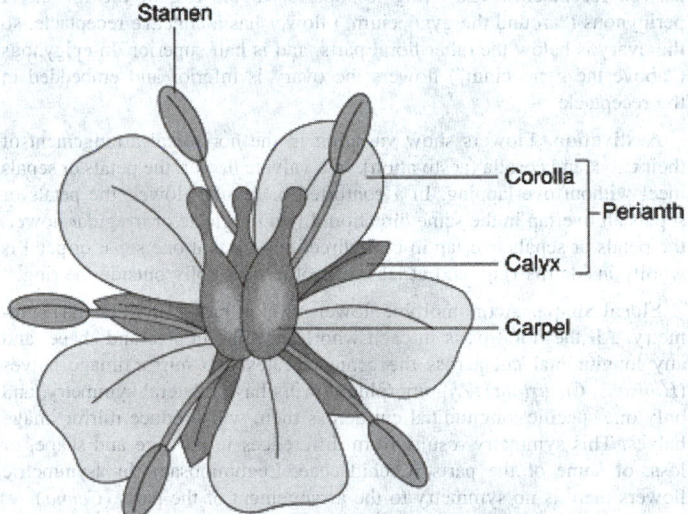

Fig. 9.   The structure of a typical flower.

Labels: Stamen, Corolla, Perianth, Calyx, Carpel

Flowers arise in the axils of leaves. These leaves, which are usually small, are called *bracts*. Some flowers have conspicuously colored bracts that supplement the petals or even replace them (*Poinsettia, Bougainvillea*). A stem bearing a single flower is a pedicel, and the point of attachment of the flower to the pedicel is the receptacle. The stem of an inflorescence, a branched system of flowers, is a peduncle.

Floral Modifications

Loss of Parts. A complete flower possesses all four whorls; an incomplete flower lacks one or more of them. A perfect or hermaphrodite flower has both carpels and stamens, but may lack sepals, or petals, or both. An imperfect flower lacks either carpels or stamens. A staminate (male) flower has only stamens, no carpels; a carpellate (female) flower has only carpels.

On a maize plant, the ear is a carpellate imperfect inflorescence. Its silks are greatly elongated styles, with sticky stigmatic surfaces for catching airborne pollen. The tassel is a staminate imperfect inflorescence. Maize is monoecious because it has both male and female flowers on the same plant. Most cucurbits are monoecious too. Papaya (*Carica papaya*) is dioecious, with male and female flowers borne on different plants.

Fusion of Parts. Within whorls, if the sepals are fused to form a tube, the flower is gamosepalous (*Hibiscus*). If the petals are fused, it is gamopetalous (*Allamanda*). If the petals are free (no fusion) it is polypetalous. The usual fusion between whorls is between stamens and sepals; such stamens are adnate to the petals, producing an epipetalous flower.

If the filaments of the stamens are fused into a tube, they are connate, and the androecium is aldelphous. If the carpels are fused, the gynoecium is syncarpous, and if they are free, it is apocarpous. Fusion of gynoecium and androecium produces a gynostegium. Many legumes (of the Papilionoideae) and all Compositae have their stamens fused to form a tube.

Ovary Structure. Ovules arise from swellings on the inside of the ovary wall called *placentas*. A simple ovary, with one carpel, has marginal placentation, with the ovules arising along the junction of the two margins of the carpel. If there is more than one carpel, the placentation becomes parietal. If the ovules are attached to the central axis formed by the carpels, it is axile. In central placentation, there is only one locule, and the ovules are borne on a central axis, and in free central placentation the axis is incomplete. In basal placentation there is also one locule, and the ovules are attached to the base of the ovary.

Ovary Position. Flowers that are hypogynous, meaning that the other flower parts (sepals, petals, and stamens) are "below the gynoecium," have convex receptacles. The ovary is superior to the rest of the flower. A perigynous ("around the gynoecium") flower has a concave receptacle, so the ovary is below the other floral parts, and is half superior. In epigynous ("above the gynoecium") flowers the ovary is inferior and embedded in the receptacle.

Aestivation. Flowers show variation in the horizontal arrangement of their calyx and corolla (aestivation). In a valvate flower, the petals or sepals meet without overlapping. In a contorted or regular flower, the petals or sepals all overlap in the same direction. In an imbricate or irregular flower, the petals or sepals overlap in both directions, so that one sepal or petal is wholly inside the ring, and at least one other is wholly outside the ring.

Floral Shape. Actinomorphic flowers exhibit radial (multilateral) symmetry. All the floral parts in each whorl are alike in size and shape, and any longitudinal cut across the center creates two mirror image halves (*Hibiscus*, *Cucurbita*). Zygomorphic flowers have bilateral symmetry, and only one specific longitudinal cut across them will produce mirror image halves. This symmetry results from differences in the size and shape, or loss, of some of the parts (Orchidaceae, Leguminosae). In asymmetric flowers there is no symmetry to the arrangement of the parts (*Canna*).

Fruit. Fruits are, to a botanist, different from what may be construed from the popular notion of "fruits and vegetables." The green bean (*Phaseolus vulgaris*) is a fruit, not a "vegetable," the coconut is not a nut (although both coconuts and nuts are fruits), and the banana is an example of a berry. See also **Fruit**; and **Seeds**.

A true fruit is the mature ovary of an angiosperm. The maturation of the ovule(s) to form the seed(s) within the ovary is normally accompanied by a thickening of the ovary wall to form three layers that comprise the pericarp. The outer layer (exocarp) may simply be a layer of epidermis, as in the grape, the middle layer (mesocarp) is commonly soft, like the flesh of the mango, and the inner layer (endocarp) can vary from gelatinous (tomato) to stony (the "pit" or "stone" of the peach). See also **Angiosperms**.

In general, fruits may be classified according to four criteria:

- The number of flowers involved in their formation.
- The number of ovaries.
- The degree of hardness of the mesocarp — dry and hard or soft and fleshy.
- The ability of the fruit to dehisce (split open when mature) or not.

See Table 1 for a classification of fruits.

Seed. The seed is the mature, fertilized ovule of a flower. It contains an embryonic plant in a resting state before initiation of germination, and a food store. An enveloping seed coat (testa), which is derived from the integument cell layer(s) within the nucellus (megasporangium), provides protection. A pore remains at the apex, the micropyle, through which the pollen tube reaches the ovule. Beside the micropyle is the hilum, a scar created when the seed breaks free of the seed stalk (funiculus). Seeds that store food are albuminous seeds; those that do not are exalbuminous seeds. Food stores (carbohydrates, proteins, lipids), can be endospermous (derived from the large central cell of the nucellus, perispermous (proliferated nucellus tissue), or both. See also **Seed**.

Embryos of many dicotyledonous plant seeds absorb their food stores well before germination, with notable exceptions, for example castor bean, *Ricinus communis*, which has oily seeds. On the other hand, many monocot seeds retain a prominent endosperm, like the grasses wheat, oats, barley, rice, and corn. The grass embryo has a large scutellum, which excretes enzymes to digest the endosperm and provide the embryo with food.

Seed Dispersal. Reproductive success is very much enhanced if seeds are distributed to new habitats. To this end seeds and their associated fruits show a variety of adaptations to take advantage of various vectors of dispersal.

Animals. Many berries have attractive and edible pericarps, with small, slippery seeds that are protected by hard woody testas, well suited to being consumed and surviving digestion. While the testa of drupe is thin, the woody endocarp protects the seed, making it suitable for ingestion. Arils, which are normally fleshy accessory outgrowths, usually of the funiculus, facilitate the dispersal by animals of seeds of the akee (*Blighia sapida*, monkey pot tree (*Lecythis ollaria*)). They also appear on the seeds of better-known fruit like tomatoes. Testas and pericarps may have spines, burrs, or other outgrowths designed to engage animal fur or hikers' socks as the vector of dispersal.

Water. The coconut fruit is buoyant and seaworthy because of the air held in its hollow seed, which is protected by a tough endocarp made of sclereids, and by the air in its fibrous mesocarp, which in turn is protected by a tough, waxy exocarp. Growth of the beached, germinating seed is aided by rainwater retained by the mesocarp. Seeds of the red mangrove (*Rhizophorus mangle*) are viviparous, with no dormant period. They produce leaves and a prominent torpedo-shaped root while still attached to the parent tree, and seedlings eventually break free to be carried away by the tide.

Wind. Seeds dispersed by the wind must be small and light and produced in great numbers to reach suitable habitats. To this end, some orchids produce dust-like seeds by the million, their mass measured in nanograms. Milkweed seeds have plumes, really extended dead trichomes, which equip them for dispersal. The dandelion seed is really a cypsela fruit, and its plume is a pappus, a modified calyx. The mahogany fruit, a type of capsule, splits open when mature to release its seeds, which have winged testas, so they can rotate and float away from the parent tree.

Self-dispersal. The pericarps of some legumes split open explosively to release their seeds (purple vetch, *Vicia benghalensis*; Scotch broom, *Cytisus scoparius*). This phenomenon results from the sudden release of tension, created in the maturing pericarp by layers of fibers in the pericarp shrinking in opposite directions, when the carpels are broken apart. Hydrostatic pressure built up in the fruit of dwarf mistletoe (*Arceuthobium pusillum*) can expel seeds as far as 15 meters (49 feet) from the parent plant.

Additional Reading

Beck, C.B.: *Introduction to Plant Structure and Development*, Cambridge, University Press, New York, NY, 2005.

Bhatt, A.M., C. Canales, and H.G. Dickinson: "Plant Meiosis: the Means to 1N," *Trends in Plant Science*, **6**, 114–121 (2001).

Cresti, M., S. Blackmore, and J.L. van Went: *Atlas of Sexual Reproduction in Flowering Plants*, Springer-Verlag New York, LLC, New York, NY, 1992.

D'Arcy, W.G., and R.C. Keating: *Anther: Form, Function and Phylogeny*, Cambridge University Press, New York, NY, 1995.

Dickison, W.C.: *Integrative Plant Anatomy*, Elsevier Science & Technology Books, New York, NY. 2000.

Endress, P.K.: *Diversity and Evolutionary Biology of Tropical Flowers*, Cambridge University Press, New York, NY, 1996.

TABLE 1. CLASSIFICATION OF FRUITS

Simple fruit. Fruit derived from a solitary carpel in a single flower

(A). Dry fruits: having a mesocarp that is definitely dry at maturity

  (i). Indehiscent fruits

   (a). Arising from one carpel

   Achene—small, with seed single attached at one point to a thin pericarp (sunflower, *Helianthus*).

   Cypsella—similar to achene, but arises from an inferior ovary (Compositae), making it a false fruit.

   Caryopsis—also called *grain*—similar to an achene, but pericarp and testa of single seed are fused (all grasses).

   Samara—Testa modified into a wing-like structure, may be one-seeded (elm, *Ulmus*) or two (maple, *Acer*).

   (b). Arising from a compound gynoecium with several carpels

   Nut—Ovary has several carpels, all but one atrophy. One seeded, with woody pericarp of sclereids (walnut, *Juglans*)

  (ii). Dehiscent fruits

   (a). Arising from one carpel

   Follicle—pod-like fruit, splits open on one side (milkweed, *Asclepias*, cotton, *Gossipium*).

   Legume—breaks open on both sides (all beans, peas of the Leguminosae). See also **Leguminosae**.

   (b). Arising from a compound gynoecium with many carpels

   Silique—two sides split, seeds remain attached to a false central partition or replum (Brassicaceae). The silique (mustard, *Brassica*) differs from the siliqua (or silicle), which has a length less than twice its width (shepherd's purse, *Capsella*).

   Capsule—carpels split apart in different ways (mahogany, iris, lily, poppy; akee, *Blighia sapida*, may be considered a fleshy capsule).

  (iii). Schizocarpic fruits: Arising from a compound ovary; the locules split into separate (usually dry) fruits

   Schizocarpic mericarp—twin mericarpic fruits joined at one point (parsley family)

(B). Fleshy fruits: Having a mesocarp that is soft at maturity

   Berry—(true berry or bacca)—endocarp, mesocarp, and exocarp all soft and easily distinguishable (blueberry, *Vaccinium vacillans*; tomato, *Solanum*)

   Drupe—similar to a berry, but endocarp is woody (mango, *Mangifera indica*; peach, *Prunus persica*).

   Prome—(also classed as a false fruit)—similar to drupe, except with papery endocarp (apple, *Pyrus malus*).

   Pepo—similar to berry, but thick, tough exocarp. May also be classed as false fruits (pumpkin, squash, all Cucurbitaceae). See also **Cucurbitaceae**.

   Hesperidium—leathery exocarp with oil glands (all citrus).

Aggregate fruits. Formed from one flower with many ovaries maturing together, some types fusing with receptacle

   Etaerio—consists of an aggregate of achenes, berries, or drupes (Raspberry, *Rubus*; strawberry, *Fragaria*—also classed as a pseudocarp, a kind of false fruit).

Multiple fruits. Developing from the ovaries of several flowers of an inflorescence which mature together

  (a). Fleshy Multiple Fruits (May also be classed as accessory fruits)

   Sorosis—fruits on a common axis and derived from the ovaries of several flowers that are usually coalesced (mulberry, *Morus*; pineapple, *Ananas comosus*; breadnut, *Artocarpus atilis*).

   Syconium—a syncarp with achenes attached to the inside of an infolded receptacle (*Ficus*).

  (b). Dry Multiple Fruits

   Strobilus—multiple fruit of achenes including bracts (hop, *Numulus lupulus*).

Accessory / False fruits. The ovary wall can be augmented by carious tissues that protect the seeds, including the receptacle of soursop, the perianth of breadnut, and the scaly bracts of pineapple. Fruits that develop from inferior ovaries are false fruits. In an apple the mesocarp and receptacle merge together with no exocarp discernible, because the carpels become fused to the accessory tissue during fruit ontogeny. In the flesh of the apple two sets of vascular bundles, one previously leading to sepals, the other to petals before the apple formed remain visible. The true fruits of a strawberry are the tiny yellow or brown "seeds," which are really achenes, on the surface of the swollen red receptacle.

Evert, R., and S.E. Eichhorn: *Esau's Plant Anatomy: Meristems, Cells, and Tissues of the Plant Body: Their Structure, Function, and Development*, 3rd Edition, John Wiley & Sons, Inc., Hoboken, NJ, 2006.

Kubitzki, K., and C. Bayer: *Flowering Plants, Dicotyledons: Capparales, Malvales and Non-Betalain Caryophyllales*, Springer-Verlag New York, LLC, New York, NY, 2002.

Mauseth, J.D.: *Botany: An Introduction to Plant Biology*, 3rd Edition, Jones & Bartlett Publishers, Inc. Boston, MA, 2003.

Raghaven, V.: *Molecular Embryology of Flowering Plants*, Cambridge University Press, New York, NY, 1997.

Raven, P.H., R.F. Evert, and S.E. Eichhorn: *Biology of Plants*, 7th Edition, W. H. Freeman Company, New York, NY, 2004.

Rudall, P.: *Anatomy of Flowering Plants: An Introduction to Structure and Development*, 3rd Edition, Cambridge University Press, New York, NY, 2007.

Simpson, B.B., and M.C. Ogorzaly: *Economic Botany: Plants in Our World*, 3rd Edition, The McGraw-Hill Companies, Inc. New York, NY, 2000.

Spilsbury, L.A., and R. Spilsbury: *Plant Parts*, Heinemann Library, Woburn, MA, 2003.

GREGOR BARCLAY, University of the West Indies, St Augustine, Trinidad and Tobago

PLANT BREEDING AND CROP PRODUCTION. To respond to the increasing need to feed the world's population as well as an ever greater demand for a balanced and healthy diet there is a continuing need to produce improved new cultivars or varieties of plants, particularly crop plants. The strategies used to produce these are increasingly based on our knowledge of relevant science, particularly genetics, but involves a multidisciplinary understanding that optimizes the approaches taken.

In plant breeding the aim is to produce new, improved varieties/cultivars and so we need, as a first requirement of any breeding program, to release or produce genetic variation in the characters (or traits) in which we are interested. Once such variation is released it is necessary to identify and then select the desired types—those that have a better expression of a particular character or combination of characters. Once identified the selected types need to be stabilized and multiplied for use and exploitation (Figure 1).

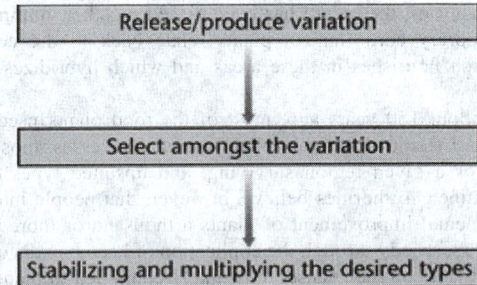

Fig. 1. Diagrammatic representation of the major steps in any plant breeding program.

Written in these terms it appears a relatively simple process, and in many ways the philosophy underlying crop improvement is simple. Although the practical reality is more complex it is possible to identify these three parts and see a framework in which to understand what is being done and what alternatives might exist. Each of these elements is tailored to be appropriate to the particular type of crop, or species, or even the likes and requirements of an individual breeder.

The term *variety*, equivalent to *cultivar*, is defined in the International Code of Nomenclature of Cultivated Plants as "an assemblage of cultivated plants which is clearly distinguished by any characters (morphological, physiological, cytological, chemical, and others) and which when reproduced (sexually or asexually) retains its distinguishing characters."

The agronomic value of a variety depends upon many characteristics, the most important of which are (1) high-yield ability, (2) high response to improved cultivation methods (e.g., fertilizers), (3) high quality of product, (4) resistance to diseases and pests, (5) resistance to adverse environmental factors (frost, drought, lodging, etc.), and (6) suitability for mechanized cultivation and harvesting methods.

In the assessment of new varieties, several steps may be distinguished: (1) The first is the evaluation carried out at the breeding station by the breeder. Normally, the breeder handles many experimental lines. By observation and by using various testing methods over several years, the breeder selects the most promising and develops them into new varieties. (2) The trials in the breeding station alone are not sufficient for objective determination of the agronomic value of new varieties. In a second step, the varieties are tested in several localities outside the vicinity of the breeding station. These trials are conducted either by the breeder or by a public or private agency. (3) Then, in a third step, the new varieties are tested for adaptability at a large number of locations with a wide range of soil and climatic conditions. In most countries, these trials are conducted by a neutral agency, to ensure objective comparison of the new varieties with commercial varieties already available. The object is to make certain that only those varieties having a higher agronomic value than the best existing varieties are released to producers for planting. (4) In a fourth and continuing step, the released varieties are tested periodically to confirm their performance against specifications and expectations. Variety testing procedures are described in detail in the Harrington reference listed at the end of this entry.

Background

There are few examples of commercially exploited plants today that are identical reproductions of the plants which appeared on the earth several thousand years ago, when and before animals and people first learned of their value as food sources. It is not uncommon for technical discussions of the nature of present-generation food plants to commence with a review of the presumed (possible, probable, etc.) nature of the precursors of present species and varieties. For example, some authorities believe that cereal seeds were first planted by people in the foothills of the Zagros Mountains in the Near East some 9000 years ago. In some instances, plant scientists are aided by the findings of archaeological digs and carbon dating techniques, making it possible to refine conjecture into refined scientific speculation. For example, some years ago, small-eared corn (maize) was found in the Bat Cave of New Mexico, dating back some 5600 years. By researching the records of early explorers and pre-Colombian Indian cultures of Mexico, Honduras, and other Central American countries, the record has been pieced together, hinting strongly that contemporary corn (maize) plants relate back to the cereal grass teosinte, which flourishes in these areas and which hybridizes with corn easily.

Up until about 150 years ago, most of the food plants used commercially were the result of natural developmental processes, those varieties most suited for a given region surviving, and unsuited types essentially becoming extinct. Authorities believe, however, that people intervened in the developmental improvement of plants a thousand or more years ago. Astute growers, strictly through the application of keen powers of observation, coupled with common sense, most likely continued to plant the seeds of varieties that were healthy and that yielded well and, by the reverse process, refrained from planting the seeds of unhealthy, poor-yielding, and otherwise unattractive plants. With the open-pollinated plants, it was only natural that from time to time new varieties would arise strictly by natural mechanisms and, again, astute growers would observe superior performance and gather seeds for planting. Such "unscientific" processes continue even in present times. An example is the finding, in 1956, of a few

rice plants that remained standing after a typhoon had flattened an entire area. These plants were found by two peasants in an eastern Kwantung (China) rice field and became the parents of the well-known dwarf variety Ai Tze-Tzang which possesses excellent anti-lodging properties (does not bend over when being harvested, thus making cutting difficult).

A great boost, to both natural (or accidental) and purposeful intervention in the development of improved varieties, was the advent of globe-circling explorers and later intensive trade between all of the continents. This permitted the introduction of species and varieties from different parts of the world to specific agricultural locales. There were many failures, but also notable successes, of which a few examples are: (a) coffee from Ethiopia achieved its highest development in Central and South America; (b) Bahian cacao thrived in west Africa; (c) African grasses proved to become a main support for the Latin American cattle industry; (d) in Indonesia, African oil palm realized its most advanced development; and (e) the famed introduction of Amazonian rubber into Malaysia. Records indicate that a new group of rices, *Champa*, was introduced into Fukien (China) about 1000 A.D. from southeast Asia.

Probably the next most important step toward plant improvement commenced with the observations of Gregor Johann Mendel, the Austrian monk, made when he bred garden peas as a hobby during the late 1850s. Serious biologists took many years to appreciate the quality of Mendel's work, which was essentially unrecognized for nearly 35 years. See also **Heredity**. European biologists in the early 1900s found that they had independently duplicated what Mendel had found much earlier. Of course, an understanding of the genetic mechanism greatly accelerated interest in plant breeding and enabled breeders (as also with animal breeding for improvement) to conserve time by avoiding a number of mistakes and taking some shortcuts. Even guided by the principles of plant genetics, aided by computers and sophisticated data processing systems to keep massive genealogical records sorted out, plant breeding still retains the trappings of a numbers game in which chance, along with science and experiment, play a leading role in producing failures and successes. For example, in one recent year, a leading hybrid corn (maize) developer tested more than 4700 new inbred lines; planted more than 208,000 yield test plots; and made 1.4 million hand pollinations—in the interest of releasing fewer than 20 new commercial hybrids.

Classical Breeding

Release/Production of Variation. Conventionally this is achieved through sexual crossing, particularly of cultivated lines, in other words following Mendel's principles. Two parents who have expression of the desirable characters between them are intercrossed and the subsequent generations examined for plants with the desired characters in new combinations, i.e. we look for recombinants. This process therefore basically relies on the segregation of alleles at all the relevant genetic loci, during the normal process of meiosis (the reduction divisions that are undertaken to form the egg and pollen cells that fuse at fertilization). At fertilization there is a random fusion of gametes (pollen from the one plant and egg from the other) to give the embryo which develops into the seed. So by the natural process of sexual reproduction, but between plants that the breeder has deliberately chosen, we get offspring that contain novel combinations of the alleles that were originally dispersed between the two parents. Clearly the choice of parents is critical. See also **Mendel, Gregor Johann (1822–1884)**.

Sources of Variation. The breeder generally uses the natural variation that already exists within the species. For virtually all characters we only need to look or measure any character to observe variation in their expression, and often this reflects not just variation produced by differences in the environment in which the plant happens to be growing, but also genetic variation–variation that is heritable. This naturally occurring source of heritable variation accounts for most of the responses that have been made in plant breeding. However, reliance on this one source of variation does limit the potential for long-term progress, particularly in relation to improving specific characters. So the use of intraspecific variation of existing crop cultivars is supplemented by one of the following:

- Wild, ancestral relatives of the crop itself: these may or may not still be able to cross sexually with the crop species and may be indigenous in another country.

- Inducing the variation that is required: the genetic variationthat we see around us actually comes from the occasional and rare mistakes that occur in the otherwise faithful replication of the DNA in all organisms. These occasional mistakes are called mutations and what we see as variation in any character today is the accumulation of such mistakes over a longperiod of time. The frequency with which mutations occur can be increased and the subsequent variants exploited.
- Developments in the areas of molecular biology and biotechnology: these have extended the possibilities for introducing additional variation in the breeding process (noted later).

Select Amongst the Variation. The first difficulty is to decide which characters to select. This may seem straightforward but in practice it means trying to put in order of priority what will be needed in the new cultivar not only in relation to improving characters but also in relation to the ones whose expression is already satisfactory in the parents (as the characters will not normally remain unchanged without positive selection). If this did not present sufficient problems, the breeder is also faced with practical difficulties. First, it is not possible to measure every character that might be relevant because there are simply too many for this to be practical. In addition, some characters take a great deal of time and effort to measure, and so may demandmore resources than are available. A major problem in a breeding program is that there is a need to handle large numbers of different genotypes but only small quantities of planting material of each is available. How does the breeder grow the plants such that they display their characters under conditions that resemble those under which they will actually be grown in agriculture? So it is important for the breeder to check the feasibility and relevance of the characters being measured in the context of the reality ofhow and where the cultivar will be grown.

Connected with the above is the efficiency with which selection can be practiced. We note that what the breeder observes is the phenotype but what he needs to select is the genotype (i.e. the heritable part of the variation that is observed). The relation between phenotype and genotype can bewritten as: Phenotype = Genotype + Environment + (Genotype × Environment)

Thus, the less that the environment affects the character, either directly or by interacting with the genotype, the better the indication of thegenotype that will be gained by simply observing the phenotype. However, many of the characters of interest do not show variation that is easy to classify into discrete classes—i.e. is not major gene determined, such as Mendel investigated, where phenotype and genotype are closely associated. They do in fact show continuous variation (i.e. many different genotypes, with an even greater subtly different range of phenotypes) and are strongly influenced by the environment in which they are grown. For example, yield is a character of immense interest to any breeder but is controlled by many genes and significantly affected by fertilizer levels, husbandry, weather etc.

Breeding Objectives and Important Traits. The general objectives of virtually all breeders of crop plants are to increase the usable yield, increase its stability, ensure the quality andnutritive value, and produce types that suit the particular growing conditions and farming needs.

1. Usable yield. This means that it is not crude yield that is important but the part that can actually be used, eaten, processed, etc. This therefore brings in factors such as the storage life, waste produced and consumer acceptance. Also it means that the use that the crop will be put to is of major importance, i.e. for direct human consumption, animal foodstuff, processing etc., and this must be considered at the outset of the breeding program.
2. Stability of yield. The fact that some lines/cultivars/varieties do very well in some years or under some particular conditions may be usefulbut can lead to disaster when they fail because of changes in the growing conditions, a poor year for rain, no fertilizer available, too wet a period at harvest etc. Thus breeding for resistance/tolerance to all biotic and abiotic stresses is a major aim.
3. Quality of the product. This includes nutritional quality and taste and is related to the awareness of usable yield but is concerned with thenutritive value, calorific value, protein content, fat level, vitamin concentration etc.
4. Environmental impact. Agriculture affects any area where it is practiced and so generates considerable debate in many parts of the world—and quite rightly so. There are many aspects to this issue and all affect the plant breeder's aims and objectives. There are therefore much clearer calls for more ecologically sympathetic methods to achieve these aims. The production of varieties with disease and pest resistance is an obvious route to follow. Taking this further, if, for example, low-nitrogen input is required then clearly specific varieties will need to be produced that growbest under these conditions.
5. Better adapted. The need to produce varieties that have been selected to grow under prevailing conditions is clear but easily overlooked. This may be the climate of a particular geographical location, the narrow conditions of a local area, the type of agricultural practices used, the needs of the farmer/village/country etc.
6. Prediction. Every breeder knows that it takes a number of years from starting to breed a cultivar until its release to the grower (often 10 years or more). This means that a breeder requires an ability to forecast the future, i.e. be a crystal ball gazer. So, for example, a breeder might need to assess/guess:

What will growers be requiring in the future?
What will happen in terms of the emphasis for growers, e.g. what subsidies will there be and what will the political situation be in the future?
How will climate change have affected growing patterns?
How will farming systems have changed?
What will be the spectrum of diseases and pests?
What will the end-users require in the future?

Stabilizing and Multiplying the Desired Types. One of the most important determinants that introduces differences inthe details of this part of the breeding strategy is the natural breeding system of the plant. The main natural breeding systems can roughly be classified into inbreeders, outbreeders (outcrossers) and clonally reproduced (i.e. clonal or vegetative propagation). These three main differences in the natural breeding system lead to what are commonly considered the main categories of classical breeding programs identified and are briefly reviewed here. See also **Plant Reproduction**.

Inbreeders. Examples of crops that are inbreeders are wheat, barley, rice, soybean, peas, tobacco, tomato, millet, lentil, flax and chickpea. These are species that naturally self-pollinate and in commercial practice are grown as true breeding, homozygous lines. So all the individualsof a particular cultivar are genetically identical.

Each generation is produced by allowing the plants to self-pollinate in each cycle of the breeding program so that while the trialing and selection process is proceeding the plants are becoming more inbred. When the finished cultivars are selected they will breed true from seed (they are genetically homozygous). So the genotype is now fixed and the cultivar can be multiplied simply by letting it set seed (isolated from any other genotypes of this crop—as some pollen is likely to pass between them).

There are two main methods by which selection is achieved during thisinbreeding process: bulk method and pedigree method.

Bulk Method. The bulk method (Figure 2) starts with the creation of genetic variation by the hybridization between two parents ($P_1 \times P_2$). The F_1 and several subsequent generations, oftenup to and including the F_5 generations, are grown as bulk populations. No conscious selection is imposed on these generations and it is assumed that the genotypes most suited to the environment in which the bulk populations are grown will leave more offspring and hence predominate in future generations. It is therefore very important that the bulks are grown in an environment that will be similar to that needed for the resulting cultivars. At the next stage, individual plants showing desirable characteristics are selected. From each selected plant, a plant (or head) row is grownand the produce from the best lines/rows are selected, bulk harvested, forinitial yield trials, and resown for multiplication.

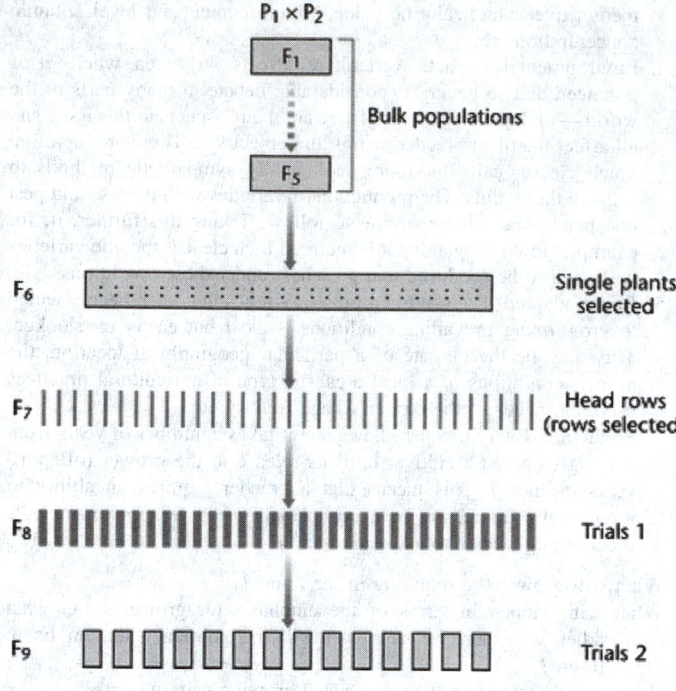

Fig. 2.   The bulk method.

This method is one of the least expensive methods of producing populations of inbred lines. The disadvantage of this scheme is the length of time from initial crossing until yield trials are grown. In addition, it hasoften been found that the natural selection that is relied on in the early, bulked generations is not always that which favors characters thought desirable for growth in agricultural practice.

Pedigree Method. In a pedigree breeding scheme single plant selection is carried out at the F_2 through to the F_6 generations. Again, the scheme begins by hybridization between chosen homozygous parental lines ($P_1 \times P_2$), and F_2 populations are obtainedby selfing the heterozygous F_1 plants. Single plants are selected from amongst the segregating F_2 population. The produce fromthese selected plants are grown in plant/head rows at the F_3 generation. The most desirable single plants are selected from the "better"plant rows and these are grown in plant rows again at the F_4 stage. This process is repeated, but with an increasing shift from individualplant to row performance, until plants are near homozygous (e.g. F_5). At this stage the most productive rows are bulk harvested and used asseed source for initial yield trials at F_6.

In addition to being laborious (as a considerable amount of record keeping is required) and relatively expensive, the reliance on individual plant selection is inefficient and leads to the loss of valuable genotypes before they are fully tested. However, the greater control over the selectionand the defined pedigrees make this a preferred method in many crops.

Cross-Pollinated (or Outbreeding) Crops. Some examples of cross-pollinated crops are alfalfa, rye, herbage grasses, forage legumes, red clover, some maize's, perennial ryegrass, sugar beet and oil palm.

The selection of new cultivars of cross-pollinated crop species is a process that changes the gene frequency of desirable alleles within a population of mixed genotypes while trying to retain a high degree of heterozygosity. So it is really the properties of the population that are vital, notindividual genotypes (as in self-pollinating crops). Instead of resulting in a cultivar for release that is a uniform genotype, the population will be a complex mixture of genotypes, which together give the desired performance. There are basically two different types of outbreeding cultivars, which are determined by the methods of their maintenance and multiplication: open-pollinating populations and synthetic cultivars.

Open-Pollinating Population Cultivars. In open-pollinating populations, selection of desirable cultivars is usually carried out by mass selection, recurrent phenotypic selection or selection with progeny testing. The maintenance of these cultivars is through open-pollinated populations with uncontrolled (random) mating.

Mass Selection

Mass selection is based on the same underlying philosophy and assumptions as the bulk method for inbreeding species. It is a very simple breeding scheme, which uses natural environmental conditions to alter the genotypic frequency of an open-pollinating population. A new population is created by cross-pollinating two different existing open-pollinating populations. In this case a representative set (any single plant will not, of course, be fully representative of the populations) of individuals from each population will be taken to be crossed.

The seed that results from such a set of crosses is grown under fieldconditions over a number of seasons. It is assumed that crossing will be at random and so result in a population quickly moving towards equilibrium which can be maintained, as a population, for exploitation. It should be noted that care needs to be exercised in isolating this developing populationfrom other crops of this species that might happen to be growing within pollination distance!

Recurrent Phenotypic Selection

Recurrent phenotypic selection (Figure 3) tends to be more effective than mass selection. A population is created by cross-pollination between two (or more) populations to create what is referred to as the *base population*. A large number of plants are grown from the base population and a subsample of the most desirable phenotypes are identified and harvested as individual plants. These plants are then randomly mated to produce a new improved population. This process is repeated a number of times — it is a *recurrent process*! What can be exploited as cultivarscan be extracted at any stage tested and distributed to growers.

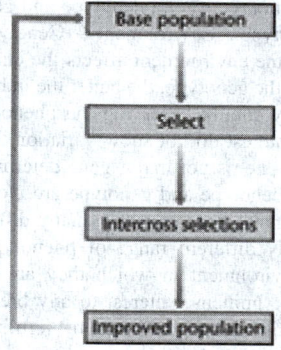

Fig. 3.   Recurrent phenotypic selection.

Synthetic Cultivars. A synthetic cultivar basically gives rise to the same end result as an open-pollinated cultivar, the main difference being that a synthetic cultivar is continually reproduced from specific parents, whereas if it is leftto open-pollinate to produce over generations, it will change its genetic make-up as a population. This means that farmers need to return to the seedcompanies for new seed when they re-sow the crop.

The breeding method used for the development of synthetic cultivars is dependent on the ability either to develop homozygous lines for use as parents or to be vegetatively propagated so that any genotype can be maintained as a parent.

Developing Hybrid Cultivars. Examples are Brussels sprouts, kale, maize, onions, rape, sorghum andtomato.

A special case that arose from developing synthetic cultivars is the idea of hybrid cultivars from just two parents. In theory any species might be used in hybrid production but commonly it is outbreeding species that are actually exploited in this way, although maize is exploited in this wayand can certainly be inbred.

At the beginning of this century there was a general awareness, especially in the USA, that the means being used to develop new corn (maize) cultivars (mass selection and ear-row selection) were less effective than had been hoped for in breeding more productive cultivars or increasing yield. Another approach was suggested from the knowledge that hybrids produced by cultivar × cultivar crosses often showed heterosis (i.e., produced yields greater than the better parent). It was then proposed that this could be exploited by manually detasselling one maize line (designated as the female parents—i.e., removing the male flowers) in plots also containing the second line, so that seeds produced on the line designated as female must have been pollinated by the pollen from the flowers of the male line. Thus it was possible to create a population that was entirely comprised of hybrids and to use it for commercial planting. This is known as a single-cross hybrid.

The major steps in producing hybrids are very similar to those for producing a synthetic cultivar, namely:

1. Development of inbred lines to be used as parents.
2. Test cross these lines to identify two that combine to give the best progeny.
3. Guard the inbred pairs that when crossed give the best hybrid cultivars and use them to produce the hybrid cultivar when desired.

There are hardly any agricultural crops where hybrid production has not at least been considered, although hybrids are exploited in relatively few crop species. The reasons behind this are first that not all crops show the same degree of heterosis (superiority over the better parent) found in maize and secondly that it is not possible in many crops to find a commercial seed production system that is economically viable. Indeed if maize had not had separate male and female reproductive organs and hence allowed easy manual detasselling, hybrid cultivar development might never have been developed, or acceptance would have been delayed at least 20 years, until cytoplasmic male sterile systems were available.

Hybrid cultivars have been developed, however, in sorghum, onions and other vegetables using a cytoplasmic male sterile (CMS) seed production system; in sugar beet and some *Brassica* crops (mainly Brussels sprouts, kale and rapeseed) using CMS and self-incompatibility to produce hybrid seed; in tomato and potato using hand emasculation and pollination.

If hybrid cultivars are to be developed from a crop, then the species must: show a high degree of heterosis; be capable of being handled so as to produce inexpensive hybrid seed; and not easily be produced uniformly by other means, and have a high premium for crop uniformity.

Heterosis. The performance of a hybrid is a function of the genes it receives from both its parents but can be judged by its phenotypic performance in terms of the amount of heterosis it expresses. Many breeders (and geneticists) believe that the magnitude of heterosis is directly related to the degree of genetic diversity between the two parents. In other words, it is assumed that the more the parents are genetically different the greater the heterosis will be. To this end, it is common in most hybrid breeding programs to maintain two, or more, distinct germplasm sources (heterotic groups). Breeding and development is carried out within each source and the different genetic sources are only combined in the actual production of new hybrid cultivars. For example, maize breeders in the USA observed significant heterosis by crossing Iowa Stiff Stalk breeding lines with Lancaster germplasm. Since this discovery these two different germplasm sources (heterotic groups) have not been intercrossed to develop new parental lines but, rather, have been kept genetically separated.

Development of Clonal Cultivars. Examples are bananas, cassava, citrus, potatoes, rubber trees, soft fruit (raspberry, blackberry, strawberry), sugarcane, sweet potatoes and topfruit (apples, pears, plums, etc).

Clonal crops are basically perennial, although several crop species, particularly those where the actual unit of clonal reproduction is the part of the plant that is exploited (e.g. tubers of potato and sweet potato), are treated in agriculture as annual crops and replanted in each crop cycle. Clonal crops also include many long-lived tree crops (e.g. apple, cherry, rubber and mango) which can be productive crops for many decades after being established.

Methods of propagation are various. Rosaceous top fruits, citrus, avocado and grape involve budding and grafting onto various rootstocks.

Leafy cuttings are used for pineapple, sweet potato and strawberry. Leafless stem cuttings are used in sugarcane and lateral shoots are used for banana and palms. There is also, for a number of species, the potential for clonal reproduction via tubers (swollen stems), e.g., potatoes.

In general, clonal crop species are often outbreeders that are basically intolerant to inbreeding. Individual clones are genetically heterozygous and so it is easy to exploit the presence of any heterosis. If maize could be easily reproduced asexually there would be little or no need to develop hybrid corn cultivars because the highly heterozygous nature of a hybrid line could be fixed by vegetative reproduction.

The process of developing a clonal cultivar is, in principle, very simple. Breeders generate segregating progenies of seedlings, select the most productive genotypic combination and simply multiply this asexually; thus there is no need for extra procedures to stabilize the genetic make-up (i.e., it relies on asexual reproduction, thus avoiding problems relating to genetic segregation arising from meiosis). Despite the apparent simplicity of clonal breeding it should be noted that while clonal breeders have shared in some outstanding successes, it has rarely been due to such a simple process.

In the case of potato, the length of the process is, in part, related to a slow multiplication rate, around 1:10 per generation. In addition, seed tubers are bulky and require large amounts of storage space. To accommodate planting material for one hectare (2.47 acres) of potatoes will require 2,241 kilograms (4,941 pounds) of seed tubers. With many other clonal species the time from crossing to cultivar release can be a very lengthy process. In apple breeding, for example, it is often said that if a breeder is successful with the very first parent cross combination, then it is still unlikely that a cultivar will be released (from that cross) by the time the breeder retires! In this case there is the obvious difficulty in the time taken from planting an apple seed to the time that fruit can be evaluated. See also **Seed**.

Genetic Resources

Experienced scientists are required to continuously search for previously unfound genetic materials and, through the supervision of the Food and Agriculture Organization (United Nations), a score or more such persons operate in the field throughout the world. Once found, such materials, sometimes of a very early origin, must be preserved not only for experiments scheduled for fairly early use, but also for the long term. Comparatively few scientists until the last couple of decades stressed the great need of finding and preserving germ plasm directly descended from the ancient precursors of modern plant species, to be used for future breeding experimentation. Such materials (plants) still can be found growing in the wild state in out-of-the-way, sparsely inhabited places on earth. But with population pressures coupled with general expansionistic movements of people, such areas are rapidly diminishing and thus genetic resources that could prove invaluable in plant breeding work and genetic research at some future date are rapidly vanishing.

The Russian agronomist and geneticist N.I. Vavilov was one of the first scientists to recognize this problem and to take concrete action toward establishing a national collection of genetic resources in Leningrad. Vavilov directed the All-Union Institute of Plant Industry during the period 1920–1940. In 1969, the institute was renamed to N.I. Vavilov All-Union Institute of Plant Industry.

In the United States, as early as 1819, there was an awareness of the importance of collecting plants materials and storing them, long before the appearance of any genetic concepts. At that time, the Secretary of the Treasury requested consuls representing the United States in other nations to send useful plant materials back to the United States. A section of Seed and Plant Introduction was established by the U.S. Department of Agriculture in 1898. A National Seed Storage Laboratory was constructed in Fort Collins, Colorado in 1958. The principal charge of the laboratory is the long-term storage of seed. Research on the physiology of germination, dormancy, and longevity of seeds is also conducted at this facility.

As early as 1936, H.V. Harland and M.L. Martini of the USDA included the following observation in the *Agriculture Yearbook*: "In the great laboratory of Asia, Europe, and Africa, unguided barley breeding has been going on for thousands of years. Types without number have arisen over an enormous area. The better ones have survived. Many of the surviving types are old. Spikes from Egyptian ruins can often be matched with ones still growing in the basins along the Nile. The Egypt of the Pyramids, however, is probably recent in the history of barley. In the hinterlands of Asia, there were probably barley fields at an earlier time. The progenies of these fields, with all of their surviving variations, constitute the world's priceless reservoir of germ plasm. It has waited through long centuries. Unfortunately, from the breeder's standpoint, it is now being imperiled. When new barleys replace those grown by the farmers of Ethiopia or Tibet, the world will have lost something irreplaceable."

By the end of World War II, what might be called *genetic erosion* had advanced in much of Europe, the United States, Canada, Japan, Australia, and New Zealand. In the early 1960s, more of an organized warning from knowledgeable scientists was heard and, in 1961, the Food and Agriculture Organization (United Nations) convened a technical meeting on plant exploration and introduction. Since that time, a series of actions has aided greatly in protecting what remains of genetic resources. In 1973, the International Board for Plant Genetic Resources was established by the United Nations.

New Genetic Approaches

Tissue Culture (in vitro) Techniques. A variety of techniques (micropropagation, haploid production, protoplasts, embryo culture, apical culture, somatic embryogenesis, etc.) have been developed under the title of tissue culture and so just two particular examples are noted here to give an idea of the possible applications.

Haploidy. Establishing true breeding, homozygous, lines is an essential part of developing new cultivars in many crop species. These homozygous lines are used either as cultivars in their own right (i.e. for inbreeding crop species) or as parents in hybrid variety development. Traditionally, plant breeders have used the process of selfing or mating between close relatives to achieve homozygosity, a process that is time-consuming. Therefore the opportunity to produce plants from gametic, haploid cells has been the goal of many plant breeders as this technique would produce instant inbred lines once the chromosomes of the haploids are doubled.

The genetic phenomenon critical to obtaining homozygous lines is the formation of haploid gametes by meiosis. During this type of cell division, the chromosome number is halved and each chromosome is represented only once in each cell (assuming the species is basically a diploid one). If such gametic, haploid cells can be induced to develop into plantlets (i.e. we encourage the development of the sporophyte — note: lower plants often have this as a specific phase of the life cycle) a haploid plant can develop which can then be treated to encourage its chromosomes to double, to produce a completely homozygous line (a doubled haploid). See also **Gametophyte and Sporophyte**.

Techniques Used for Producing Haploids in Vitro. Although haploidy is a very attractive technique to many plant breeders the natural occurrence of haploid plants is rare. However, the use of plant tissue culture has allowed the production of plants from gametic cells cultured *in vitro* at a higher frequency.

Although haploid plants can be regenerated from both male and female sex cells, it is generally the male cells (microspores or pollen) that have proven most successful in the regeneration of large numbers of haploid and doubled-haploid lines. This is partly because of the ease with which pollen, as opposed to eggs, can be collected, and partly because many more pollen grains than eggs are produced. See also **Pollen (Structure, Development and Function)**.

There are a number of methods of haploid induction that are not directly related to tissue culture but the most widely applicable are via the culture of anther or microspore (immature pollen grains) *in vitro*.

In Vitro Multiplication. In vitro multiplication of breeding lines can have two main benefits (particularly in clonal species) in relation to plant breeding programs.

1. Plants propagated *in vitro* can generally be initiated and maintained in a disease-free state, and so can be used to help maintain stocks of breeding lines; facilitate long-term germplasm storage; and facilitate international exchange of material.
2. Short generation times and fast growth means that rapid increases in plant number can be readily achieved.

Both the above have particular importance to clonal crops, because these tend to have a relatively low multiplication rate as a result of their vegetative mode of propagation and are particularly susceptible to viral and bacterial diseases, which tend to be multiplied and transmitted through each clonal generation.

Good examples of maintaining a disease-free status and offering rapid plant regeneration potential include potato and strawberry. Other, perhaps less well-developed examples, include *in vitro* propagation of date and oil palms. In these crops it was found that more rapid plant regeneration would indeed offer an alternative to the slow and lengthy process of propagating side shoots in date palm and a more uniform planting material in the case of oil palm. However, in date palm the process is still very genotype dependent, and with oil palm there initially proved to be an unacceptably high frequency of sterile palms produced.

Plant Transformation. The stable introduction of specific genes into plants represents one of the most significant developments affecting the production of crop species in a continuum of advances in agricultural technology. The progress in this area has depended largely on the tissue culture systems having been developed which, at least, initially, provide an amenable vehicle for the transformation induction.

The term transformation comes from that used for a much longer period, bacterial transformation, in which DNA has been successfully transferred from one isolate to another or another species of bacteria, and integrated into the genome. It was shown that the stably transformed bacteria then expressed the new genes and displayed appropriately altered phenotypes. In eukaryotes, transformation has a further complicating dimension, at least in many plants' breeding contexts. The transforming DNA must not only be integrated into a chromosome, it must be a chromosome of a cell, or cells, which will develop into the germline. Otherwise the transformation will not be passed on to the next generation.

Using plant transformation techniques it is possible to transfer single genes (i.e. simply inherited traits) into plants, to have such transgenes expressed and for them to function successfully. Theoretically at least, specific genes can be transformed from any source into developed cultivars or advanced breeding lines in a single step. Plant transformation, therefore, would appear to allow plant breeders to bypass barriers that limit sexual gene transfer and to exchange genes (and traits) from unrelated species between which sexual hybridization is not possible. These recombinant DNA techniques, apparently, allow breeders to transfer genes between completely unrelated organisms. For example, bacterial genes can be transferred and expressed in plants. This appears to break the barrier that sexual reproduction generally imposes. However, as we learn more about the DNA, and hence the genes involved, the perspective of the picture changes somewhat, with increasing direct evidence of the presence in different species of the same basic gene, or clear variants of it, and demonstrations of the greater conservation of genetic material (synteny) during evolution than we expected. Also, we are being reminded of the existence of parallel natural processes for much of what we regard as novel. For example, bacteria, viruses and phages already have successfully evolved mechanisms to transfer genes just in the way we regard as being so alien! But clearly, the new techniques are allowing modern plant breeders to create new variability beyond that existing in the currently available germplasm on a different scale and in a different time frame from that which was previously possible.

Although plant transformation has added (some say dramatically) to the tools available to the breeder for genetic manipulation, it does have limitations. Some of the limitations will reduce with increased development of methodologies, others are inherent to the basic approach. At present recombinant DNA techniques can generally only transfer rather limited lengths of DNA and so tend to be restricted to the transfer of single genes. This means that they are very effective where the trait can be substantially affected by a, or a few, gene(s) of large effect. Another restriction that is imposed currently is that the techniques are only readily applied to genes that have been identified and cloned. The number of such desirable genes is still modest, but increasing rapidly.

Some Applications of Genetic Engineering to Plant Breeding. Already there is a growing list of crop species that have proved successful hosts for transformation including alfalfa, apple, carrot, cauliflower, celery, cotton, cucumber, flax, horseradish, lettuce, maize, potato, rapeseed, rice, rye, sugarbeet, soybean, sunflower, tomato, tobacco and walnut.

Initial cultivar development using recombinant DNA techniques has focused on modifying or enhancing traits that relate directly to the traditional role of farming. These have included the control of insects, weeds and plant diseases. The first genetically engineered crops have now been released into large-scale agriculture (including maize, tomato, canola, squash, potato, soybean and cotton) and other species are already in the pipeline. More recently work has focused on altering end-use quality (including oil composition, starch, vitamin level and even vaccines).

Cautions and Related Issues. There have been a number of concerns that have arisen over the past few years as the application of plant transformation technology has expandedand particularly as new transgenic crops have been released into commercialcultivation. Plant breeders need to be aware of the concerns as well as theregulations that apply to plants derived using recombinant DNA. As well as the general social and environmental concerns the breeder must check that the techniques being used are the most effective for what is to be achieved and not simply assume that high tech means most efficient!

Molecular Markers in Plant Breeding. Although plant breeders have practiced their art for many centuries, genetics is a subject that really only came of age in the twentieth centurywith the rediscovery of Mendel's work. Since then research in genetics hascovered many aspects of the inheritance of qualitative and quantitative traits, but plant breeders usually still have little, or no, information about: the locations of many of these loci in the genome or on which chromosome they reside; the number of loci involved in any trait; and the relative size of the contribution of individual alleles at each loci on the observed phenotype, except where there is an obvious major effect (e.g. height and dwarfing genes).

The Idea of Using Markers. The idea of associating easily visualized markers in plants with loci affecting qualitative and quantitative variation in traits of interest to plant breeders is not new, and was first proposed by Sax in 1923. The basicidea is relatively simple. If a trait or characteristic is difficult to score (e.g. it shows continuous variation; assessment is detailed and time consuming; or the trait is only expressed after several years of growth), an easily scored marker that was determined by a locus closely associated with that affecting the character would be an attractive alternative way to monitor the locus of interest.

The characteristics of a good marker system include the following:

1. The markers are easy, quick, and inexpensive to score.
2. The markers themselves have no deleterious effects on fitness and no effects on other traits, including undesirable epistatic interactions with any other traits.
3. There is a high level of variation exhibited.
4. They are stable in expression over environments.
5. Assessment can be made early in the development of the plant (seedling level), and/or in tissue culture.
6. The scoring should be nondestructive in terms of the whole plant, so that desirable individuals can be selected and still grown to maturity.
7. Codominance in expression of the alternative alleles, so that heterozygotes can be differentiated from either homozygous dominant genotype.

Types of Marker Systems. The types of markers that can and have been used in plant breeding include:

1. Morphological markers — which are basically those that you see by simply looking at a plant's phenotype, including characters such as pigmentation, dwarfism, leaf shape, absence of petals, etc.
2. Biochemical markers — such as isozyme markers. Isozymes (isoenzyme) are variant forms of an enzyme, which are functionally identical but can be distinguished by electrophoresis — i.e., when placed in an electric field. Under these circumstances the different forms of the enzyme will migrate to different points in the electric field depending on their charge, size and shape.
3. Molecular markers — these represent the variation that is present and can be detected at the level of DNA. There are basically two systems (PCR and non-PCR based) by which molecular markers

are generated and their distinction need not detain us, but it is worth pointing out that molecular markers are simply differences in the DNA between individuals, groups, species taxa etc. Clearly the type and level of variation in DNA that we would want to examine is different depending on what level of distinction we are interested in and what questions we are answering. But the main characteristics of molecular markers are that: they are a ubiquitous form of variation; they are free from environmental influence; they show high levels of polymorphism; they have no discernible effects on the phenotype; and they can be detected using only small pieces of tissue.

Given the above characteristics of molecular markers, particularly their relatively unlimited numbers, it is no surprise that the advent of the possibilities of molecular markers in the 1990s and early part of 2000 was greeted with some excitement and is seen as providing a major change in thepotential to exploit the ideas for using markers advocated some 70 years earlier.

Future Potential

Plant breeding will continue to be highly dependent on classical techniques but will undoubtedly increase in efficiency and effectiveness by theaddition of these new approaches, which will be used in parallel with the more classical ones. Thus the future will see the range of techniques expanding in such a way as to maximize their benefits by their integrated exploitation.

Note: Portions of this article by Peter D. S. Caligari, University of Reading, Reading, UK., originally appeared in the Encyclopedia of Life Sciences, John Wiley and Sons, Inc., Hoboken, NJ. 2007.

NOTE: See also references listed at ends of articles on **Cell (Biology)**; **Genetics and Gene Science**; **Industrial Biotechnology**; **Molecular Biology**; and **Protein**.

Additional Reading

Abelson, P.H.: "Plant Gene Expression Center," *Science*, **1465** (September 27, 1991).

Allard, R.W.: *Principles of Plant Breeding*, 2nd Edition, John Wiley & Sons, Inc., New York, NY, 1999.

Beardsley, T.M.: "Doing Agricultural Genetics in the Marketplace," *Sci. Amer.*, **24** (April 1990).

Benfey, P.N. and N.H. Chua: "Regulated Genes in Transgenic Plants," *Science*, **174** (April 14, 1989).

Bennetzen, J., W.F. Blevins, and A.H. Ellingboe: "Cell-Autonomous Recognition of the Rust Pathogen Determines Rpl-Specified Resistance in Maize," *Science*, **208** (July 8, 1988).

Borojevic, S.: *Principles and Methods of Plant Breeding*, Elsevier Science, New York, NY, 1990.

Chahal, G.S. and S.S. Gosal: *Principles and Procedures of Plant Breeding: Biotechnological and Conventional Approaches*, CRC Press, LLC., Boca Raton, FL, 2002.

Chilton, Mary-Dell: "A Vector for Introducing New Genes into Plants," *Sci. Amer.*, **50–59** (June 1983).

Crawford, M.: "Agricultural Research Service (U.S.)," *Science*, **719** (February 12, 1988).

Crawford, M.: "Plan to Map Crop Genes," *Science*, **1137** (March 3, 1989).

Crawford, R.: "Gene Mapping Japan's Number One Crop," *Science*, **1611** (June 21, 1991).

Dziezak, J.D.: "Biotechnology Enzyme Firm Embraces Innovation," *Food Technology*, **117** (September 1990).

Erickson, D.: "Genetically Engineered Plants Head for the Harvest," *Sci. Amer.*, **81** (May 1990).

Gasser, C.S. and R.T. Fraley: "Genetically Engineering Plants for Crop Improvement," *Science*, **1293** (June 16, 1989).

Goodpaster, K.: *Principles of Plant Genetics and Breeding*, Blackwell, Publishers, Malden, MA, 2006.

Gibbons, A.: "Biotechnology Takes Root in the Third World," *Science*, **962** (May 25, 1990).

Halford, N.G.: *Plant Biotechnology: Current and Future Applications of Genetically Modified Crops*, John Wiley & Sons, Inc., Hoboken, NJ, 2006.

Hamer, J.E.: "Molecular Probes for Rice Blast Disease," *Science*, **632** (May 3, 1991).

Haring, V., et al.: "Self-Incompatibility: A Self-Recognition System in Plants," *Science*, **937** (November 16, 1990).

Harrington, J.B.: Cereal Breeding Procedures, FAO Development Paper 28, Food and Agriculture Organization (United Nations), Rome, 1970.

Heldt, Hans-Walter, and F. Heldt: Plant Biochemistry, Elsevier Science & Technology Books, New York, NY, 2004.

Janick, J.: Plant Breeding Reviews, Vol. 29, John Wiley & Sons, Inc., New York, NY, 2007.

Jensen, N.F.: Plant Breeding Methodology, John Wiley & Sons, Inc., New York, NY, 1998.

Kiernan, V.: "Appropriate Biotech," Technology Review (MIT), 11 (August/September 1989).

Kole, C.: Technical Crops, Springer-Verlag New York, LLC, New York, NY, 2007.

Lea, P.J. and R.C. Leegood: Plant Biochemistry and Molecular Biology, 2nd Edition, John Wiley & Sons, Inc., New York, NY, 1999.

Levings, C.S., III: "The Texas Cyctoplasm of Maize: Cytoplasmic Male Sterility and Disease Susceptibility," Science, 942 (November 16, 1990).

Lindow, S.W., N.J. Panopoulos, and B.L. McFarland: "Genetic Engineering of Bacteria from Managed and Natural Habitats," Science, 1300 (June 16, 1989).

Manuel, J.: "North Carolina Regulates Biotech," Technology Review (MIT), 20 (July 1990).

Moffat, A.S.: "Bumpter Transgenic Plant Crop," Science, 33 (July 5, 1991).

Moses, P.B. and N.-H. Chua: "Light Switches for Plant Genes," Sci. Amer., 88 (April 1988).

Poehlman, J.M.: Breeding Field Crops, 5th Edition, Blackwell Science, Inc., Professional, Malden, MA, 2006.

Rhodes, C.A., et al.: "Genetically Transformed Maize Plants from Photoplasts," Science, 204 (April 8, 1988).

Richards, A.J.: Plant Breeding Systems, 2nd Edition, Chapman & Hall, New York, NY, 1997.

Singh, R.J., and P.J. Jauhar: Genetic Resources, Chromosome Engineering, and Crop Improvement: Cereals, Vol. 2. CRC Press, LLC, Boca Raton, FL, 2006.

Sparks, D.L.: Advances in Agronomy, Elsevier Science & Technology Books, New York, NY, 2007.

van Harten, A.M.: Mutation Breeding: Theory and Practical Applications, Cambridge University Press, New York, NY, 2007.

Wallace, D.H. and W. Yan: Plant Breeding and Whole-System Crop Physiology: Improving Adaptation, Maturity and Yield, CAB International, New York, NY, 1998.

PLANT CLIMATE ZONE.   1. Area in which a common set of temperature ranges, humidity patterns, and other geographic and seasonal characteristics combine to create a particular plant distribution by allowing certain plants to succeed and causing others to fail. 2. A region in a climatic classification system that places greatest importance on plant distribution to identify climatic characteristics.

See also **Climatic Classification**.

PLANT CUTICLE.   Leaves, nonwoody stems and many fruits of plants are covered on the outside by a protective extracellular membrane, the cuticle. It is laid down by the epidermis while the organ is growing. The cuticle consists of an insoluble polymer matrix that is impregnated with a mixture of a large number of different organic compounds. These are soluble in organic solvents and are summarily (and rather inaccurately) known as waxes. A striking feature of cuticles is the variability in chemical composition, absolute amount and ultra structure of its components. Many attributes differ between species, genotypes, organs and developmental stages and are affected by conditions during growth. The significance of such differences for cuticular function is generally very poorly understood. The cuticle protects the above-ground primary body of a plant (and some stems that have undergone secondary growth), which is more or less water-saturated, from dehydration in the atmospheric environment. It also acts as a first line of defense against other organisms for which these tissues are a potential source of food, and it is used a source of signals by such organisms. In certain cases, reflectance and absorption by cuticular waxes in different parts of the light spectrum may be of significance to the plant. See also **Epidermis (Plant)**;**Fruit**;and **Plant Anatomy**.

Structure and Composition of Cuticular Matrix and Waxes

The Polymer Matrix.   The matrix is made up of a lipophilic polymer and polysaccharide microfibrils. The latter extend from the underlying cell wall. The polymer is called *cutin* if it is a polyester of aliphatic hydroxy or epoxy fatty acids of carbon chain length C_{16} or C_{18}. It is called *cutan* or *nonester bonded cutin* if it cannot be broken down by hot BF_3/methanol, which cleaves cutin ester bonds. Our knowledge of the chemical nature of cutan is rather incomplete [Schouten, et al.: 13]. The relative contribution of cutin and cutan to the lipid polymer matrix varies between 0 and 100%. The thickness of plant cuticles varies from a few hundred nanometers to more than $10 \, \mu$. Thicker cuticles are more common in fruits and evergreen leaves.

Epicuticular and Intracuticular Waxes.   The most common wax constituents are straight-chain, saturated primary alcohols, aldehydes and fatty acids of predominantly even chain lengths, alkanes, secondary alcohols, and ketones of predominantly odd chain lengths (all in the range C_{12}–C_{37}), and very long chain esters (C_{30}–C_{72}). Genes known to be involved in the control of the complex biosynthesis pathways of aliphatic wax compounds are called *Eceriferum* (abbreviated *cer*) or *Glossy*, depending on the species studied. Many cuticles also possess cyclic compounds such as pentacyclic triterpenoids, flavonoids and hydroxycinnamic acid derivatives. See also **Plant Waxes**; **Terpenes and Terpenoids**.

Some waxes are present on the outer surface of cuticles and are called epicuticular. They may form a smooth film and/or a several micrometer high micro relief of discrete, apparently crystalline projections. Depending on their chemical nature, epicuticular wax crystals occur in many different shapes. The remaining waxes, possibly the major fraction of the overall wax amount, are called intracuticular. The total amount of waxes varies between ~5 and 500 μg cm^{-2}. It is not clear whether new cuticular waxes are still being synthesized during the lifetime of an organ once it has reached its final size.

Functions of Cuticle and Epicuticular Waxes

Diffusion Barrier.   The main barrier to the diffusion of water, gases (e.g. air pollutants) and solutes (e.g., foliar pesticides or nutrients) across the cuticle is located in a narrow band, which has been called the skin, at the interface between epicuticular wax crystals and the polymer matrix. The bulk of the cuticle contributes relatively little to the overall resistance. However, it usually represents the main capacity of the leaf for accumulation of lipophilic compounds taken up from the environment. There is no correlation between the thickness of the cuticle and its permeability to water.

The diffusion resistance of a leaf cuticle with a permeance of 5×10^{-5} m s^{-1} is comparable to that of a double layer of household polyvinyl chloride clingfilm, which is many times thicker than the skin of the cuticle, or even the entire cuticle. The reason that diffusion across the skin is such a slow process is that, according to our present understanding, diffusing molecules have to meander along lengthy, tortuous random paths around inaccessible regions. These regions are formed by highly ordered arrays of wax molecules [Merk, et al.:10]. Such crystalline microdomains are interspersed with zones of amorphous waxes, which together form a three-dimensional network of paths that enables diffusion across the skin. Permeance to water vapor of leaf cuticles ranges between approximately 1×10^{-6} and 1×10^{-4} m s^{-1} [Kerstiens, 7, 8]. Uptake of pesticides and other solutes can be accelerated by suitable adjuvants that increase the mobility of molecules in the wax barrier by diminishing the tortuosity of transport paths [Baur and Schönherr, 2].

Water Repellency.   The hydrophobic nature of the cuticle and the epicuticular waxes make most leaf surfaces quite water-repellent. Wettability is further reduced by surface roughness at the micrometer scale. Instead of forming extended films on the leaf surface, water will often form droplets that cover less leaf area and easily run off, reducing the residence time of water on the leaf. This helps to keep the diffusion pathway for carbon dioxide uptake across stomata open, e.g. in the morning hours following formation of dew. Furthermore, fungal spores have fewer and shorter windows of opportunity for germination, which reduces the likelihood of infection.

A convenient measure for the water repellency of a leaf surface is the contact angle formed between a droplet and the surface. The greater the angle, the more hydrophobic is the surface. The presence of a dense layer of prominent epicuticular wax crystals, waxy trichomes, papillose epidermal cells, cuticular folds, etc. can strongly increase the contact angle. Unwettable leaves are very effectively cleaned by dew or rain [Barthlott and Neinhuis,1]. Contact angles of leaf surfaces that are smooth at the microscopical scale typically decline during the lifetime of a leaf, a process that parallels the structural deterioration of epicuticular wax crystals. This process is accelerated by exposure to air pollutants. See also **Trichomes**.

Interaction with other Organisms. Some pathogenic fungi penetrate the leaf through the cuticle, either by sheer physical force or facilitated by a cutin-dissolving enzyme called *cutinase*. It is not clear whether thicker or cutan-containing cuticles provide a better protection against infection. The early stages of the infection process can be affected by cuticular wax compounds [Kolattukudy, et al.: 9]. Many cuticles support a nondetrimental microflora of fungi and bacteria. This may affect leaf wettability [Schreiber, 14]. The microflora may be antagonistic towards pathogens and it may affect the behavior of pests.

The interaction between a pollen grain and a (dry-type) stigma is similar to that between a fungal spore and a leaf surface. It may involve both specific recognition [Pruitt, 11] and cutinase activity [Hiscock, et al.: 6].

Surface waxes provide multiple signals that can affect the behavior of specialized invertebrate herbivores. Furthermore, insect predators of pests may be more mobile and more effective on "glossy" leaf surfaces with relatively little epicuticular wax, decreasing pest damage of such genotypes [Eigenbrode, et al.: 4].

Breeding for increased resistance to pests and pathogens through manipulation of wax biosynthesis pathways appears promising. However, the possible deterioration of other functions fulfilled by these waxes needs to be considered. See also **Plant Breeding**.

Additional Reading

Barthlott, W., and C. Neinhuis: "Purity of the Sacred Lotus, or Escape from Contamination in Biological Surfaces," *Planta*, **202**, 1–8. (1997).

Baur, P., and J. Schönherr: "Tetraethyleneglycol Monooctylether (C_8E_4) Reduces Activation Energies of Diffusion of Organics in Plant Cuticles," *Zeitschrift für Pflanzenkrankheiten und Pflanzenschutz*, **105**, 84–94. (1998).

Beck, C.B.: *Introduction to Plant Structure and Development*, Cambridge, University Press, New York, NY, 2005.

Eigenbrode, S. Dd, C. White, M. Rhode, and C.J. Simon: "Behavior and Effectiveness of Adult Hippodamia convergens (Coleoptera: Coccinellidae) as a Predator of Acyrthosiphon pisum (Homoptera: Aphididae) on a Wax Mutant of Pisum sativum," *Environmental Entomology*, **27**, 902–909 (1998).

Evert, R., and S.E. Eichhorn: *Esau's Plant Anatomy: Meristems, Cells, and Tissues of the Plant Body: Their Structure, Function, and Development*, 3rd Edition, John Wiley & Sons, Inc., Hoboken, NJ, 2006.

Hiscock, S.J., F.M. Dewey, J.O. Coleman, and H.G. Dickinson: "Identification and Localization of an Active Cutinase in the Pollen of Brassica napus L," *Planta*, **193**, 377–384. (1994).

Kerstiens, G.: "Cuticular Water Permeability and its Physiological Significance," *Journal of Experimental Botany*, **47**, 1813–1832 (1996).

Kerstiens, G.: *Plant Cuticles: An Integrated Functional Approach*, Taylor & Francis, Inc., Philadelphia, PA, 1996.

Kolattukudy, P.E., L.M. Rogers, Li DX, C.S. Hwang, and M.A. Flaishman: "Surface Signaling in Pathogenesis," *Proceedings of the National Academy of Sciences of the USA*, **92**, 4080–4087. (1995).

Merk, S., A. Blume, and M. Riederer: "Phase Behavior and Crystallinity of Plant Cuticular Waxes Studied by Fourier Transform Infrared Spectroscopy," *Planta*, **204**, 44–53. (1998).

Pruitt, R.E.: "Molecular Mechanics of Smart Stigmas," *Trends in Plant Science*, **2**, 328–329. (1997).

Riederer, M., and C. Müller: *Biology of the Plant Cuticle*, Blackwell Publishers, Malden, MA, 2006.

Schouten, S., P. Moerkerken, F. Gelin, M. Baas, J.W. de Leeuw, and J.S.S. Damsté. "Structural Characterization of Aliphatic, Non-hydrolyzable Biopolymers in Freshwater Algae and a Leaf Cuticle Using Ruthenium Tetroxide Degradation," *Phytochemistry*, **49**, 987–993 (1998).

Schreiber, L.: "Wetting of the Upper Needle Surface of Abies grandis: Influence of pH, Wax Chemistry and Epiphyllic Microflora on Contact Angles," *Plant, Cell and Environment*, **19**, 455–463. (1996).

GERHARD KERSTIENS, Lancaster University, Lancaster, UK

PLANT DISEASES. See **Phytopathology**.

PLANT GROWTH MODIFICATION AND REGULATION. In the entry on Cell (Biology), the highly specialized nature of various proteins, enzymes, coenzymes, etc., to effect changes during the life cycle of an animal or plant are pointed out. Those chemical substances having the most to do with plant growth and form are given the general term *plant hormones*. A plant hormone, or *phytohormone*, may be defined as an organic compound produced naturally in plants, which controls growth or other functions at a site remote from its place of production, and which is very active in minute amounts. Three chemically quite different types of compounds apparently act as plant hormones: the *auxins*, the *gibberellins*, and the *kinetins*. In addition, the growth of roots is dependent upon vitamins of the B group which are synthesized in leaves and transported thence to the roots, thus qualifying as hormones.

Auxins. The best-studied hormones are those belonging to the class of auxins. These are defined as organic substances which promote growth along the longitudinal axis, when applied in low concentrations to shoots of plants freed as far as practical from their own inherent growth-promoting substances. Auxins generally have additional properties, but this one is critical.

Natural auxins have been identified in a number of instances. Indole-3-acetaldehyde occurs in a number of etiolated seedlings and in pineapple leaves; indole-3-acetonitrile has been isolated from cabbage and its presence indicated in a number of plants. One of the most widely occurring auxins is indole-3-acetic acid, which has been isolated in pure form from fungi and from corn (maize) grains. Its presence has been conclusively demonstrated by biochemical and chromatographic tests in a wide variety of flowering plants, including mono- and dicotyledons.

Many synthetic auxins have been produced, including 2,4-dichloro-phenoxyacetic acid or 2,4-D; naphthalene-1-acetic acid; and 2,3,6-trichlorobenzoic acid, among others. Used as a herbicide, 2,4-D is described in **Herbicide**.

By definition, these synthetic compounds are not hormones, although they are sometimes loosely referred to as hormone-type compounds.

An auxin is formed in fruits, seeds, pollen, root tips, coleoptile tips, young leaves, and especially in developing buds. The auxin travels away from the site of production in shoots by a special transporting system, depending on oxygen, which moves it in a predominantly polar direction from apex toward base. Movement in the opposite direction, i.e., from base toward apex, takes place to a variable extent depending upon the tissue and the plant. In the course of the polar transport, a large part of the auxin becomes bound and is no longer transportable. The transport is rather specifically inhibited by related compounds, particularly 2,3,5-triiodobenzoic acid, 2,4-D, and other synthetic auxins, which are transported either more slowly or to a much lesser extent in the polar system. Auxin applied artificially to intact plants can travel rapidly upward by penetrating into the conducting tissues of the wood, where it is carried upward in the transpiration stream.

In its normal polar, downward movement, the auxin stimulates the cells below the tip to elongate and sometimes to divide. Specific tissues, notably the cambium, are caused to divide laterally by auxin coming from the developing buds, which accounts for the wave of cell division occurring in tree trunks in the spring. Stimulation of other stem cells to divide leads to the production of root initials, which grow out as lateral roots. Cells of the young ovary are commonly caused to multiply and enlarge so that an apparently normal fruit is produced without requiring pollination (*parthenocarpic fruit*). This latter phenomenon indicates that the growing seeds normally secrete an auxin to which enlargement of the fruit is due, a conclusion which has been directly confirmed by bioassay in several fruit types.

Gibberellins can also cause enlargement of fruit. On reaching the lateral buds, however, auxin inhibits their elongation into shoots, and this accounts for *apical dominance*, i.e., suppression of the growth of lateral buds by the terminal bud of a shoot. Auxin also inhibits the falling off of leaves or fruits, which normally occurs when they are mature or aged by the formation of an *abscission layer* of special cells whose walls come apart. That the leaves or fruits do not absciss earlier is due to their steady production of auxin, which prevents formation of these cells. In the root, auxin inhibits elongation except in very low concentrations, but its level therein is usually low. Auxin can be transported for a short distance from the root apex toward the base, but the transport is not fully polar and in the more basal parts of the root the transport is slight.

When the shoot is placed horizontal, auxin is transported toward the lower side, causing accelerated growth there and hence upward curvature (*geotropism*); in the root, this causes decreased growth on the lower side and hence downward curvature. However, in the downward geotropic curvature of roots, other phenomena appear to enter in, and the complexities are not yet fully resolved. When shoots are illuminated from one side, auxins accumulate on the shaded side and, therefore, the plant curves toward the light (*phototropism*). Both geotropic and phototropic auxin movements have been confirmed with carboxyl-^{14}C-labeled compounds. The first observed effect when auxin is applied is the acceleration of the streaming of cytoplasm, but acceleration of growth begins in 7 to 14 minutes at about 23 °C.

In plants that flower on short days, auxin may inhibit flowering; in plants that flower on long days, however, if close to the transition from the vegetative to the flowering state, auxin may promote flowering. In hemp and some of the squashes, auxin modifies the sexuality of the flowers toward femaleness. In the special case of pineapple, auxin directly causes flowering in an unusually clear-cut and quantitative response.

The principal uses of synthetic auxins are to promote the formation of roots on stem cuttings, to prevent abscission, especially of apples and pears, to induce flowering in pineapples, and occasionally to produce seedless fruits. The largest use, however, is that of weed killing. This action depends upon the fact that, at concentrations from 100 to 1000 times those concentrations occurring naturally, the auxins are highly toxic. Monocotyledonous plants, however, are usually resistant. In years past, 2,4-D has been favored in North and South America, whereas 2-methyl-4-chlorophenoxyacetic acid (*methoxone*) has been popular in Europe. However, as of the early 1980s, the regulatory status of these compounds in various countries is under study and may be subject to change.

Some chemically related compounds antagonize the action of auxins, for example, relieving the inhibition of root growth caused by 2,4-D. In contrast, 2,3,5-triiodbenzoic acid synergizes the action.

Gibberellins. These compounds were originally isolated from a parasitic fungus that causes excessive leaf elongation in rice plants. The mechanisms and applications of this group of compounds are described in **Gibberellic Acid** and **Gibberellin Plant Growth Hormones**.

Kinetins. Considerably less is known about this class of compounds. The first one to be discovered, produced by autoclaving yeast nucleic acid, was 6-furfurylaminopurine. Somewhat later, zeatin was isolated from immature corn (maize) kernels. The kinetins promote cytokinesis and protein synthesis, thus causing amino acids to accumulate where kinetins are synthesized (or externally applied) and maintaining the chlorophyll content of yellowing leaves. The kinetins antagonize auxin in apical dominance, releasing lateral buds from inhibition by a terminal bud or by applied auxin. It is believed that through the same mechanism, the kinetins promote the development of buds and leaves on tissue cultures. Their action is primarily local, and if there is transport in vivo, it probably occurs mainly in the transpiration stream (where amino acids are also often found).

Ethylene. The production of ethylene in fruit tissue and in small amounts in leaves may justify its consideration as a hormone, functioning in the gaseous state. Cherimoyas and some varieties of pear produce 1000 times the effective physiological concentration. Ethylene formation is closely linked to oxidation and may be centered in the mitochondria. Its effects are to promote cell-wall softening, starch hydrolysis, and organic acid disappearance in fruits — the syndrome known as *ripening*. Ethylene also decreases the geotropic responses of stems and petioles.

Daminozide Growth Modifier. Daminozide, the chemical name of which is 2,2-dimethyldrazide, was developed in the early 1960s as a modifying or regulating agent for the growth process of several food plants. The action varies with each plant. For example, on apple, the compound accelerates the start of flower budding, restricts nonproductive vegetative growth, and assist in fruit drop control. It is also claimed that the compound accelerates fruit coloring and helps to retain the firmness of the fruit. For some of these and other similar reasons, the compound has been used effectively for certain varieties of grape (particularly Concord), for peanuts (groundnuts), for tomatoes, nectarines (except Cherokee), and peaches. Other commercial designations are Alar, B-Nine, Kylar, and Sadh.

Ethephon Growth Modifier. This compound, (2-chloroethyl)phosphonic acid was developed in the United States in the mid-1960s and is used effectively on a number of fruit and vegetable crops for controlling a variety of factors. These include loosening fruit and causing earlier ripening of the fruit (apple, blackberry, blueberry, cherry, cranberry, filbert, tangerine, and walnuts); for encouraging uniform ripening and increasing yield (pepper and tomato); to improve color as well as accelerate maturity (cranberry); and to decrease time required for degreening in citrus fruits, particularly lemon. Other commercial designations for this compound include Cepha, Ethrel, and Florel.

Maleic Hydrazide. This compound, 1,2-dihydro-3,6-pyridazinedione, is also used as a growth regulator, herbicide, and plant modifier. It is used in the treatment of tobacco plants; as a post-harvest sprouting inhibitor; and as a sugar content stabilizer in sugar beets.

PLANT SCIENCES (The History). The scientific study of plants today is usually distinguished from endeavors in which plants are used for a particular purpose, as in economic botany, horticulture, agriculture, materia medica and the like. But this was not the case in the past. From time immemorial people have studied plants precisely for the uses they may have for human beings, and this tradition has been inseparable from the development of western botanical science. The interrelationships still hold in the case of herbal medicine, for example, especially in the search for alternative remedies to manufactured drugs and in the practices of ethno botany. The modern sciences of plant genetics depend extensively on horticultural and agricultural input. Commercial development of plantation crops and global trade in plant-based products, such as cotton, play key roles in national economies. Because of these longlasting links with human activities and economic concerns, botany has often been called the 'big science' of previous centuries. See also **Drug Discovery (The History)**.

Early Ideas

Much of the earliest botany was concerned with materia medica. An exception was the work of the Ancient Greek philosopher Theophrastus (c. 371–287 BC) who studied plants in detail, although his work was not known in the west until the Renaissance. Theophrastus was Aristotle's favorite pupil and successor at the Lyceum. In essence he did for plants what Aristotle had done for animals. He classified plants in the Aristotelian tradition, expounding the basic formal principles of taxonomy using a logical division of genera and species to yield a definition of one natural kind or form. The definition applied to that one form alone. On the larger scale, he divided plants into the categories tree, shrub, under-shrub and herb. In addition, he gave his observations on reproduction and nutrition, correctly identifying the need for pollination between male and female date palms. He included much information on the agricultural techniques of his day such as methods of preparing charcoal and obtaining pitch from resinous trees.

By far the most famous Ancient text was by Dioscorides (fl. AD 50–70). Dioscorides' herbal, a plant encyclopedia written in Greek and intended for the use of physicians, was translated into Latin and Arabic, and copied and recopied incessantly, often with additions and manuscript illustrations. Above all, herbalists and apothecaries had to be sure of the plant's identity. Consequently written descriptions became lengthy and illustrations important. The advent of printing around 1453 accelerated this trend and Dioscorides' herbal was one of the first texts to be produced by this method in 1478. The book became the foundation of many of the greatest illustrated herbals, particularly that of Otto Brunfels (c. 1489–1534), who drew illustrations directly from the plants, and Leonhart Fuchs (1501–1566), who produced a magnificent herbal, *De Historia Stirpium in* 1542, figuring about 550 plants. These were expensive books, often hand-colored and annotated. In this regard, learned botany was restricted mainly to physicians, theologians, university dons and the court.

Expanding Horizons

After the fall of Rome, Greek and Roman knowledge was preserved and extensively improved in Arabic translation. The revival of these Ancient texts recast western botany in the later sixteenth century. Practitioners increasingly noticed that the plants of Western Europe were not those described in either Arabic or Ancient texts and began to add to traditional lists. This coincided with increasing foreign travel, the colonial expansion of Europe, most notably into the East Indies and the New World, and changing medical education, itself partly a consequence of the new texts available. An increasing number of exotic plants and herbs were introduced to Europe from overseas, encouraging the development of the small physic gardens already attached to the main medical universities, enriching the gardens of the royal courts, and posing questions about geographical distribution. By the end of the century Rembert Dodoens (1516–1585) and William Turner (1508–1568) were producing works that were not so much supplements to the Ancients as true replacements.

Attempts to order and classify natural items were primary activities in sixteenth- and seventeenth-century science in general. The botanical treatises of Gaspard Bauhin (1560–1624) and the French botanist de L'Ecluse (Latinized to Clusius, 1526–1609) were exemplary in this regard, and serve as opening points to the great age of classical taxonomy. Each author attempted his own solution to the problem of how much weight to give to various botanical criteria, and each searched for a 'natural' arrangement of kinds—a classification scheme that would reflect morphological relationships (or 'affinities') as seen in the number and relative position of petals, sepals and reproductive organs, the shape of leaves, and means of reproduction, whether by bulbs or seeds. Bauhin's *Pinax* of 1623 described some 6000 plants, representing the rapid increase in knowledge over the previous hundred years.

The work of Andrea Cesalpino (1519–1603) at the University of Pisa was a landmark in classification schemes. Cesalpino returned to Aristotle's and Theophrastus' logical hierarchies and gave a series of 'essential' characters whose successive subdivisions would not violate intuitive perceptions of affinity, rather in the way that animal classification schemes could use differences in the heart, nerves or respiratory system. He decided that the structures concerned with the plant's primary functions, especially the reproductive parts, should be used to divide plants hierarchically into 'natural' groups. Cesalpino's scheme underlay most of the more important natural plant classifications published afterwards.

During the major intellectual changes that took place in the seventeenth century, the field of botany broadened to come into contact with other disciplines besides medicine and classification. John Ray (1627–1705) notably pushed theories of classification forward. He was especially gifted in defining the higher, more fundamental divisions of the plant kingdom, for instance finding that flowering plants fall into two natural groups based on the number of seed-leaves (cotyledons), the monocotyledons and dicotyledons. He also produced an important catalogue of English plants. Nomenclature was much advanced by Joseph Pitton de Tournefort (1656–1708) when he defined and named nearly 700 genera in *Eééments de Botanique* in 1694.

The most immediate extension of botanical investigation, however, was in plant anatomy, where the use of the microscope by skilled natural philosophers in Britain, Italy and the Netherlands revealed the unsuspected world of the small. The Royal Society in London was one of the prime locations for such research. Robert Hooke (1635–1703) saw 'pores' in cork, which he called *cells*, and showed that similar structures occur in many plants. He seems to have seen spores in ferns, mosses and two kinds of mould. Similarly, Nehemiah Grew (1641–1712), at the Royal Society, and Marcello Malpighi (1628–1694), in Bologna, writing in English and Latin respectively, independently developed an integrated view of microscopic plant structure. Using relatively simple microscopes, they identified the principal tissues of the plant body, including stem, root and leaf, and followed the cycle of development from seed to seed.

Some of their most significant observations were on the origin of buds, perceived as the condensed growing point of a shoot. The functions of such structures were mostly interpreted by analogy with animals, so that for Grew the large xylem vessels seemed as if they might function as air-vessels like the tracheae of animals. Grew proposed that the stamens were the male reproductive organ and that plants possessed two sexes, again like animals. Proof of this suggestion came in the work of Rudolf Jacob Camerarius in 1694, although the question remained hotly contested for decades afterwards.

The active experimental inquiries that characterized other natural philosophical areas penetrated botany too, albeit slowly and not as dramatically as contemporary advances in understanding animal physiology. The first controlled experiments into plant nutrition were performed by J. B. van Helmont (1579–1644) in the 1640s, in order to throw light on the chemical composition of water. He weighed and planted a willow tree in a container and allowed it to grow for five years; at the end, he weighed the tree and the annual leaf fall, finding that the tree had gained some 170 lb from water alone. The chemical composition of various plants was established. A few years later Stephen Hales (1677–1761) began careful experimental studies of plant physiology, published in *Vegetable Staticks* (1727). He used the word 'staticks' to convey his emphasis on precise measurement and was one of several experimenters of the period to see the value of using controls in his experiments. Taking his cue from animal physiology, he investigated the movement of sap and water in the plant. His chief achievement was to establish the constant uptake of water by plants and its loss by transpiration (*perspiration* he called it) but he also showed that these movements took place by root and leaf suction. He conclusively showed that sap did not circulate like blood circulates in animals. He performed experiments on the gas relations of plants, as part of the path-breaking studies on 'airs' performed by Robert Boyle and Joseph Priestley.

Meanwhile, the herbal tradition was scarcely affected by these philosophical inquiries, for it appealed to an increasingly wider, more popular audience. Nicholas Culpeper (1616–1654), who produced 11 editions of the *English Physician Enlarged, or the Herbal* (1653), led this movement. Herbals continued to be published in large numbers in relatively stereotyped format, giving descriptions of useful plants, their habitats and properties, and recipes for simple medications. Later herbals included advice on domestic economy, hygiene and dietetics. Through these volumes, countless people learned a utilitarian and domestic combination of medical and botanical science.

The most famous eighteenth-century naturalist, Swedish doctor, Carl von Linné (Latinized as Carolus Linnaeus, 1707–1778) transformed botanical science in all quarters. He provided a quick and easy alternative to existing plant classification schemes in his *Systema naturae* (1735). There, Linnaeus proposed an 'artificial system' for plants whereby the numbers of stamens and pistils (male and female reproductive organs) were counted and served to allocate plants into groups without any further ado, for example plants with five stamens would be placed in a new category called *Pentandria* (Greek for five male organs). This system, sometimes called the *numerical system* or the sexual system, was usually understood as a method of classifying plants purely for human convenience, for it grouped plants together on the basis of numbers alone; and gave rise to some curious juxtapositions. In actual fact Linnaeus did regard the reproductive system as having primary physiological importance and his scheme was not intended to be solely arbitrary. See also **Linnaeus, Carl (Linné) (1707–1778)**.

Nevertheless, it was simple to count stamens: anyone could do it. In this sense Linnaeus's scheme liberated many of those people who worked with plants. Travelers and collectors could quickly allocate unidentified new specimens into a class or family; and Linnaeus consciously acted as the hub of a vast botanical exploring network fanning out from his base at the medical school of Uppsala University. He trained and sent plant collectors all over the increasingly accessible globe: to Japan, South Africa, the Carolinas, Asia, central Spain, and so on. These men, whom he called *apostles*, and others using his system, sent numerous collections back to Europe throughout the eighteenth century. They achieved remarkable results. In 1772, for example, Sir Joseph Banks accompanied James Cook on his first voyage in search of the Southern continent (Australia), taking with him Daniel Solander (1733–1782), one of Linnaeus's favored pupils. The magnificent collection of plants brought back by Banks and Solander, which profoundly changed European ideas about botanical science, was collected and first classified by Linnaeus's method.

In a succession of philosophical writings Linnaeus set out principles for the science. He introduced the convention of binomial names, the first name denoting the genus, the second the specific. Previously, definitions of a species could run to a sentence or more because so many key criteria needed to be included. In his *Genera Plantarum* and *Species Plantarum* he defined all the known plants of the world. These names and definitions have, by subsequent convention, been taken as the type. He personally coined many of the names still in use and enjoyed complimenting, and sometimes criticizing, his colleagues by naming plants after them.

Linnaeus also helped botanical knowledge to move out of the elite world of universities, museums and physic gardens into a far broader constituency. Many eighteenth-century people in Europe and North America, including gentlefolk, women and working men, are all known to have first encountered botany through popularizations of Linnaeus's classification scheme. Many popular books, translations, cheap editions, and explanations were published, as well as beautifully colored illustrated works for richer clientele. His impact stretched into all realms of scientific thought, paralleling similar desires to provide classification schemes in chemistry and medicine. In particular, the Linnaean system encouraged a number of women to participate in science as authors. His 'sexual system' provided the basis for much cultural satire and parody, ranging from sexual innuendo about plants as humans, or in an anti-Catholic classification of monks.

Eighteenth-century botany therefore had wide cultural appeal. Other aspects of botanical science should not be ignored either, such as the growing interest in landscape gardening and the cult of the picturesque. Plants played a significant role in the life of the landed elite, who often possessed illustrated floras and synopses of classification schemes in their libraries. The landed gentry cultivated exotic plants and patronized plant collectors and landscape gardeners. Hothouses and stove-plants became an increasing possibility for the wealthy. Women were particularly prominent as authors of elementary textbooks, since botany was widely perceived as a suitable science for genteel women to pursue. Indeed botany was one of the few sciences open to women.

Botany's greatest impact, however, came through geographical expeditions. These introduced a large number of new species to the west, initiating a rage for choice specimens and providing the foundations of national herbaria and museum collections. Many of these exotics came from government voyages of exploration to Australia, the Cape, and the South Seas. Living plants could be seen at Kew Gardens (royal property until 1841), Chatsworth, or Syon Park near Chiswick, and the great European gardens such as the Jardin des Plantes in Paris. Enterprising private societies like the Royal Horticultural Society (founded in 1804) sponsored collecting trips abroad and ran public gardens and competitions for their members.

European prosperity overseas equally depended on the development of the plantation system in which staple crops like tea or sugar-cane were relocated for colonial purposes. Naval officers, commercial entrepreneurs and government officials jointly opened up these routes to economic expansion. In Britain, the East India Company took the lead in establishing botanic gardens in Saharanpore and Calcutta; and Kew Gardens, under Sir Joseph Banks, became a hub of proto-imperial science.

After the death of Linnaeus in 1778, the botanist James Edward Smith (1759–1828) purchased his collections originally offered to Joseph Banks. Smith established the Linnean Society of London in 1788 (the society's name is taken from von Linné, not Linnaeus, and is spelled without the first 'a'), which soon became the premier botanical society in England although it ostensibly dealt with all branches of the living world. Smith was elected president at the first meeting, where he delivered an address entitled 'Introductory Discourse on the Rise and Progress of Natural History'. He later published *English Botany* (1790) and his most important work, *Flora Britannica* (1800–1804). Smaller societies and specimen exchange clubs also proliferated, most of which used the Linnaean system of classification.

In specialist circles, naturalists vigorously investigated physiology, reproduction and anatomy, readily drawing parallels between animals and plants. Sensitive plants like the mimosa were thought to possess animal-like qualities such as irritability. Studies of 'breathing' in plants played a significant role in the discovery of oxygen and then photosynthesis;

and the action of pollen in fertilization helped in understanding animal reproduction. Linnaeus's definition of plant species as stable entities, fixed since their creation by God, stimulated studies of hybrization. Even Linnaeus himself conceded at the end of his life that some hybridization must occur and that perhaps a few of the major plant groups had originated in that way: one 'mother' group hybridizing with several 'fathers' to produce related botanical families. Questions about the definition of life and organization were asked. Other alternatives to rigidly logical classification schemes emerged as a number of botanists searched for ways to acknowledge a wider range of resemblances. Michel Adanson (1727–1806) in his important *Famille des Plantes* (1763) attempted to work out purely empirical relationships by including a large number of characters. Much of what Adanson proposed has been retained in today's broad-based taxonomical schemes. The natural system came to be understood as a means of classification in which a number of morphological characters and character-complexes were used to determine relationships.

Towards the end of the eighteenth century growing knowledge of plant fossils helped naturalists consider the course of the Earth's history, although here the major explanatory theories were principally derived from structural geology and animal evidence. Fossil plants had been noted and displayed in collectors' cabinets since the time of Edward Lhuyd (1660–1709) and were pictured in some detail by J. J. Scheuchzer (1672–1733) in his *Herbarium Diluvianum* of 1709. Scheuchzer understood plant fossils to be the remains of species killed by Noah's Flood. But fossil plants were not really integrated into geological thought until the issue of the central heat of the Earth arose in the closing years of the eighteenth century. Antoine de Jussieu (1686–1758) observed that the plants in the coal deposits around Lyon were typical of the hot and humid climates of the equator.

The study of paleobotany came into its own with the work of Adolphe Theodore Brongniart (1801–1876). See also **Paleobotany**. He identified more fossil plants than ever before and used them to reconstruct the physical history of the Earth, published as *Histoire des Végétaux Fossiles* (1828). He confirmed Cuvier's view that the living beings of the Earth had changed in character in each successive geological era. The Swiss botanist Auguste de Candolle (also known as Augustin Pyramus de Candolle, 1778–1841) expanded his interpretation by studying coal deposits in more detail. Candolle proposed that the preponderance of cycad and fern remains in coal indicated that the early atmosphere of the Earth must have been very much warmer and possibly richer in carbon dioxide. This view contributed to contemporary theories about the directional history of the Earth from a hot, uninhabitable state to one in which the present state of affairs emerged. The nature of coal deposits was, in fact, a leading issue in geological circles although not much connected with commercial enterprise. Beyond this, little obvious attention was paid to plant paleontology until the rise of Quaternary studies in the twentieth century. See also **Paleoclimatology**; **Paleontology (The History)**.

A unique contribution to botany was made by Johann Wolfgang Goethe (1749–1832). *Die Metamorphose der Pflanzen* (1790) (translated into English in 1863 as *Essay on the Metamorphosis of Plants*), was based on the idea of homology between plant organs such as cotyledons, foliage leaves and the floral parts. This was not an entirely original concept. However, Goethe was the first to express what became a much-debated issue in plant morphology, that in essence all floral parts were modified leaves. Later, Auguste de Candolle formulated useful concepts of symmetry, abortion, modification and fusion.

Plant Science in Nineteenth-Century Thought

In the opening decades of the nineteenth century Robert Brown greatly advanced the use of the microscope in plant embryological investigations. Brown (1773–1858) made many brilliant observations in microscopic plant anatomy that were, for the most part, buried in obscure articles or in systematic works. In 1827 he described the development of the unfertilized ovule. This helped Brown decide that the ovule in cycads, conifers and other related families is truly naked and not enclosed at any stage in an ovary, allowing him to draw a fundamental distinction between gymnosperms (conifers and cycads) and angiosperms (flowering plants). In Florence, Jean Baptista Amici (1786–1863) observed the pollen tube growing out of pollen grains lodged on the stigmatic surface. These observations were repeated and confirmed by Adolphe Brongniart, who concluded that the pollen tube conveys 'spermatic fluid' to the ovule. In 1831 Brown then delivered a detailed account of the process of fertilization in orchids and Asclepiadeae. Thereafter Brown was renowned for noting the form of the cell nucleus.

Brown's work on cell anatomy was quickly followed by Mathias Schleiden's path-breaking *Grundzuge der wissenschaftlichen Botanik* (1842) translated into English in 1849 as *Principles of Scientific Botany*. In this work Schleiden (1804–1881) laid the groundwork for the cell theory of Theodor Schwann. Many of the advances made during these years were summarized by Hugo von Mohl (1805–1872) in 1851. By Mohl's time, vague notions of utricles and vessels had given way to a clear concept of the plant cell based on a firm body of knowledge about its structure, reproduction, mode of differentiation and even some of its chemical constituents. Elsewhere fundamental observations were made on the nature of fertilization in algae, where free-swimming spermatozoids were identified.

Subsequently an important step in plant physiology came from Henri Dutrochet (1776–1847), who studied medicine in Paris but devoted much of his life to the scientific investigation of osmosis. His results appeared in 1837. Although he could not explain the cause of osmosis, he did not doubt that it was an entirely physical process and was of very great importance in vital phenomena. The nitrogen cycle was elucidated by J. B. J. D. Boussingault (1822–1895). The essential steps in carbon fixation were not recognized until Julius von Sachs (1832–1897) showed that the starch present in green cells was the product of the carbon dioxide absorbed.

Modern perceptions of the overwhelming importance of evolutionary theory in the nineteenth century have obscured these dramatic surges in physiological and microscopical investigations. Nevertheless plants have often seemed secondary in evolutionary theory even though botany dominated Charles Darwin's (1809–1882) later researches. In the *Origin of Species* (1859) Darwin used botanical investigations to support his key arguments for natural selection, showing how plants vary in the wild and display adaptations. His two chapters on geographical distribution presented innovative ideas about historical plant geography, where he discussed the possibility of northern species pushing through the tropics at some former cool period, and the means of plant dispersal, including the transport of seeds by birds and currents. After the *Origin of Species*, Darwin studied orchids as a prime example of adaptation, in this regard as adaptations to ensure insect fertilization. Later he worked extensively on the cross-fertilization of plants, on insectivorous plants and the movements of plants. In all this work, he focused on the active physiological life of the organism. Darwin also performed plant experiments to investigate the origin of variations, polymorphism and hybrids; and worked on pollination, the evolution of sexes and tropisms (plant hormones).

His close friends Joseph Hooker (1817–1911), director of the Royal Botanic Gardens, Kew, and Asa Gray (1818–1888), Professor of Botany at Harvard University, were both converted to evolutionary theory and defended Darwin's ideas in reviews and in their botanical writings. Gray and Hooker were among the first to use natural selection for understanding plant relationships. Generally, however, there were few fights over the evolution of plants in the same way as there were fights over the possible connections between animals and humankind. Botanists probably found the idea of plant evolution far less repugnant than that of human evolution, and did not worry too much about the moral and theological implications of plant descent. Even so, there were a number of satirical references and cartoons poking fun at the transformation of turnips into humans.

The morphology of plants was generally underrated until evolutionary theory gave it new legitimacy. The idea of the alternation of generations in animals helped Wilhelm Hofmeister (1824–1877) demonstrate the two phases in the life cycle of mosses and liverworts in 1851. Hofmeister was followed by Ladislav Celakowsky who in 1868 coined the terms gametophyte and sporophyte. Celakowsky saw the evolutionary significance of Hofmeister's work, arguing from the phylogenetic perspective that sexual generation must be more primitive than asexual. Heated debate arose over the origin of flowering plants (angiosperms) in the Cretaceous period. This also involved controversy over what should properly be considered primitive. Throughout the nineteenth century Engler's arrangement was enshrined in most classification schemes until C. E. Bessey, in 1915, proposed that *Ranunculus* and its relatives came closest among living plants to primitive angiosperms. Frederick Bower (1855–1948) linked the migration of plants onto land to the appearance and subsequent elaboration of an asexual generation.

In 1868 the first edition of Julius von Sachs' *Lehrbuch der Botanik* appeared which led the way in plant physiological investigations in Europe. Sachs' impact was as great as Linnaeus's some hundred years before although in a very different field. Every eminent plant physiologist was trained in Sachs' laboratory in Wurzburg. In many ways the state of English botany before then resembled that of English physiology. Both disciplines were founded on a descriptive anatomical tradition; both imported the new laboratory work wholesale from the continent; and both found it hard to become separate from the medical curriculum. Experimental botany at first failed to find a secure home in the English universities, with the exception of John Burdon Sanderson at University College London and Sidney Vines in Cambridge. Bower's appointments in London, and then Glasgow in 1885 revolutionized the way plant physiology was regarded. Among other achievements, Sachs made highly significant studies on plant hormones, growth mechanisms and tropisms, such as the effect of gravity or light on roots and shoots.

Elsewhere, the emphasis of university botany shifted so much towards the physiological, for example in enzyme studies and fermentation, that Hooker felt it necessary to reassert the value of taxonomy as an academic discipline in the *Index Kewensis* and his and Bentham's *Genera Plantarum*. Botany went on to become the prime area for genetic research, as in pure line experiments, and cytology, fertilization and cell division.

Additional Reading

Frey, K.J.: *Historical Perspectives in Plant Science*, Iowa State University Press, Ames IA, 2002.

Green, J.R.: *A History of Botany, 1860–1900: Being a Continuation of Sachs' History of Botany 1530–1860*, Clarendon Press, Oxford, UK, 1909.

Morton, A.G.: *History of Botanical Science: An Account of the Development of Botany from Ancient Times to the Present Day*, Academic Press, London, UK, 1981.

JANET BROWNE, Wellcome Trust Centre for the History of Medicine, London, UK

PLANT TRANSFORMATION. See **Plant Breeding and Crop Improvement**.

PLANT WAXES. Juniper described plants as "water-filled sacs of diverse shapes ... invariably coated, over their aerial parts with a thin ... sequence of layers of more or less hydrophilic material." The outermost layer, which is very much "less hydrophilic," is wax. Beneath this layer is a mixture of waxes and cutin (an insoluble polymer) that acts as a strengthening compound, in much the same way that iron is used to reinforce concrete.

The hydrophobic nature of the outer layers of the plant helps to prevent uncontrollable water loss. Plants that are adapted to very dry conditions (xerophytic plants) often have heavy wax deposits. This role in control of water loss is supported by observations that weathering of the plant surface in a wind of 13 m s^{-1} (30 mph) can cause up to 50% loss of wax, with a concomitant increase in cuticular transpiration. There is also evidence that more glaucous populations occur in localities with greater frost likelihood, which suggests that frost damage can be minimized by the presence of additional waxy material on the plant's surface.

In addition, it is believed that waxes may protect plants from bacterial and fungal pathogens [Hamilton and Hamilton, 1972]. It is also thought that extraneous antifungal constituents of waxes may provide resistance to microbial attack, e.g. sclareol in *Nicotiana glutinosa*. The protective effect may be by acting as a physical barrier or it may arise from the presence in very small quantities of fungi static compounds excreted by the plant [Martin, 1964].

Since insects often eat the whole plant, it is likely that wax constituents do not prevent insect attack. However, the selection by the insect of which plant to eat is often determined before the insect lands on the leaf or immediately after it has alighted. In such circumstances it may be that the minor components of the waxes may play an important part in the protection of the plant. Aphids are said to touch leaf surfaces with their proboscis before biting.

The amount of light falling on the plant and reaching the leaf cells is critical for the wellbeing of the plant. Some waxes contain significant quantities of aromatic compounds that are believed to help to minimize damage to the plant caused by UV-B radiation.

Wax Patterns

The fine structure of plant waxes was studied in 1871 by De Bary using a light microscope. He categorized waxes as needles, as in *Secale cereale*; granulated layers, as in *Brassica* species; rodlets up to 1–4 µm thick, as in *Cotyledon orbiculata*; or crusty layers, mostly thin and brittle, as in *Opuntia* spp.

Crystalline wax formations are usually superimposed on amorphous wax layers; these come in three categories: flakes, tubes and filaments. Flakes vary in size and shape, often being serrated, acicular or angular. Wax plates can be seen on *Poa* spp., *Acacia* spp., *Triticum vulgare*, *Hordeum* spp. and *Pisum sativum*. Tubular waxes of variable length and wall thickness can be found, while filament wax includes structures such as flat ribbons and very fine dendritic forms. Tubular waxes are found in *Tropaeolum majus*, *Chelidonium majus*, *Rhus cotinus*, *Agathis australis* and *Picea* spp., while ribbons are the most widely reported surface wax found in *Fragaria* and *Rosa* spp. and *Pisum sativum*. A typical surface is shown in Figure 1.

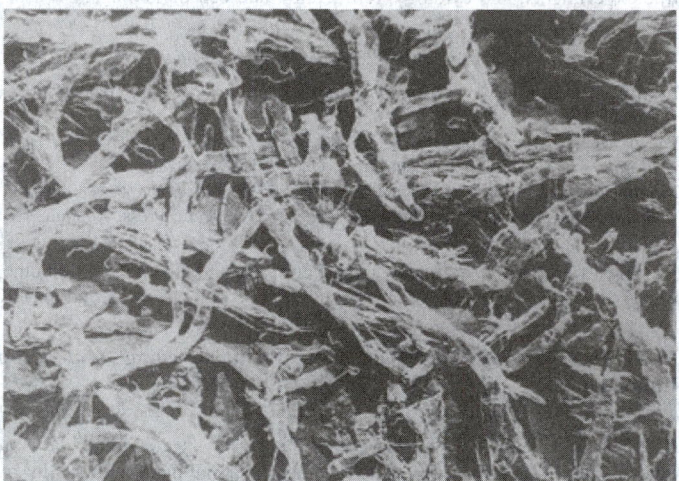

Fig. 1.   Transmission electron micrograph showing long thin tubes of wax rising from the cuticle surface as in the lemma of barley *Hordeum vulgare*.

A more elaborate way of describing the wax morphological types is as follows: amorphous films as in *Beta vulgaris* (beet), granules as in *Helianthus annuus* (sunflower), soft waxes as in *Malus* spp. (apple), rodlet types as in *Saccharum officinarum* (sugarcane), filament or ribbon types as in *Sorghum bicolor* (sorghum, milo, broomcorn), annular ridged tubes as in *Brassica oleracea* var *gemmifera* (brussel sprouts), squat tube types as in *Picea abies* (Norway spruce), long tube types as in *Festuca glauca* (blue fescue), dendritic plates as in *Euphorbia peplus* (milkweed, radium weed, cancer weed), simple plate types as in *Hordeum vulgare* (barley), mixed tubes and plates as in *Lonicerus korolkowskii*, farinaceous as in *Primula* spp. (primose), apparently wax free as in *Blechnum capense* (fern), and substantial crusts as in *Copernicia cerifera* (palm tree (carnauba) [Jeffree, 1986].

Composition of Plant Waxes

Plant waxes are complex mixtures of mainly long-chain aliphatic molecules but they also contain some alicyclic and aromatic compounds. Some components have a chain length that has an even number of carbon atoms, e.g. acids, aldehydes, primary alcohols and esters (see Table 1), whereas others have a chain length that has an odd number of carbon atoms, e.g. hydrocarbons, secondary alcohols, ketones, β-diketones and hydroxy-β-diketones (see Table 2). Generally a series of homologues is found with chain length varying from 20 to 35 carbon atoms. However, hydrocarbons and fatty acids with less than 20 carbon atoms are known, and esters with more than 60 carbon atoms have been observed [Baker, 1982; Bianchi,

TABLE 1. COMMON WAX CLASSES WITH AN EVEN NUMBER OF CARBON ATOMS

Compounds	Chain length	Major components
Alkyl esters	$C_{34}-C_{64}$	C_{40}, C_{50}
Acids	$C_{16}-C_{32}$	C_{22}, C_{24}, C_{26}, C_{28}
Primary alcohols	$C_{22}-C_{32}$	C_{26}, C_{28}, C_{30}, C_{32}
Aldehydes	$C_{22}-C_{32}$	C_{26}, C_{28}, C_{30}, C_{32}

TABLE 2. COMMON WAS CLASSES WITH AN ODD NUMBER OF CARBON ATOMS

Compounds	Chain length	Major components
n-Alkanes	$C_{21}-C_{35}$	C_{27}, C_{29}, C_{31}
Ketones	$C_{23}-C_{33}$	C_{29}, C_{31}
Secondary alcohols	$C_{23}-C_{33}$	C_{29}, C_{31}
b-Diketones	$C_{27}-C_{33}$	C_{31}, C_{33}
Hydroxy- b-diketones	$C_{27}-C_{33}$	C_{31}, C_{33}
Oxo- b-diketones	C_{29}, C_{31}.	C_{31}

1985]. The compositions of waxes may also include terpenoids, flavonoids and sterols.

The wax of a particular species is often constant (provided growth conditions are comparable), but waxes from different species differ markedly from one another. Thus, the waxes from *Musa* species are almost pure ester (92% of the total), whereas *Nicotiana* wax is composed mainly of alkanes. Among the more unusual components are the flavonoids, which occur on *Primula* spp. as a whitish or yellowish powder on the stems and coating the underside of the leaves. Hydrocarbons are found in all plant waxes but they may be present at less than 1%. They are the predominant class in very few plants. C_{27}, C_{29}, C_{31} and C_{33} are the most abundant homologues but hydrocarbons from C_{15} to C_{38} have been found. When the concentration of each alkane is plotted against the chain length of that hydrocarbon, a distribution pattern is formed. Since the shorter-chain-length and longer-chain-length components are missing, there is usually a cluster of peaks around the major constituent, resulting in a bell-shaped appearance to the distribution pattern. This shape is often species specific, organ specific and plant ontogeny specific.

Branched-chain hydrocarbons form a small proportion (5–10%) of normal glaucous *Brassica oleracea* plants, but they become more important in subglaucous or glossy mutants [Baker and Holloway, 1975]. There are also some reports of unsaturated alkenes. It is suggested that lower-chain-length hydrocarbons are less abundant in the cytoplasmic tissue than in the cuticle. Among the nonaliphatic compounds of the waxes are the triterpenoids. They are rarely the major component of the wax but they are found in the form of alcohols and ketones. The most widely distributed series are those of the amyrins, ursanes, lupanes and oleanes.

Biosynthesis of Plant Waxes

Fatty Acids. The components of waxes have chain lengths in the range C_{20} to C_{40}, much longer than those in the more familiar triacylglycerols. Despite this fact, the biosynthetic pathways are similar to those found in the production of glyceride lipids.

The enzymes responsible for epicuticular wax synthesis reside in epidermal cells; in the case of pea and leek, they are found in the membrane fractions of these cells. The long-chain components of waxes are formed by the condensation–elongation mechanism. This mechanism is dependent on elongases, which condense short carbon atom chains to a primer molecule. Fatty acid synthetase (FAS) takes an activated "C_2" unit (acetyl–coenzyme A) and adds repetitively further "C_2" units, which are derived from malonyl acyl carrier protein (see eq. (1)). See also **Epidermis (Plant)**.

$$CH_3COSCoA \qquad CO_2 + CoASH$$

$$^-OCOCH_2CO - SACP \xrightarrow{\quad A \quad} CH_3COCH_2CO - SACP \qquad (1)$$

where A is acetoacetyl ACP synthetase

Malonyl–CoA is converted to an acyl carrier protein derivative as in eq. (**2**) by a malonyl–CoA: ACP transacylase.

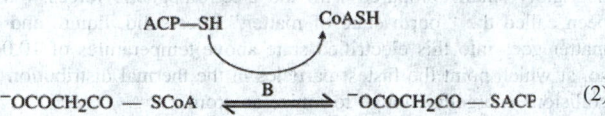

$$^-OCOCH_2CO-SCoA \underset{\xrightarrow{B}}{} \ ^-OCOCH_2CO-SACP \quad (2)$$

where B is malonyl CoA ACP transacylase

This C_4 unit, a β-ketoacyl-ACP, is reduced as in eqs. (3), (4), and (5) to a saturated C_4 unit that can act as a new starter for the reaction with a further molecule of malonyl S-ACP.

$$CH_3COCH_2CO-SACP \underset{\text{reductase}}{\overset{\beta\text{-ketoacyl}}{\rightleftharpoons}} CH_3CHOHCH_2CO-SACP \quad (3)$$

$$CH_3CHOHCH_2CO-SACP \underset{\text{dehydrase}}{\overset{\beta\text{-hydroxy}}{\rightleftharpoons}} CH_3CH=CHCO-SACP \quad (4)$$

$$CH_3CH=CHCO-SACP \underset{\text{reductase}}{\overset{\text{enoyl}}{\rightleftharpoons}} CH_3CH_2CH_2CO-SACP \quad (5)$$

By continuing to add two carbon atoms at a time, this FAS set of reactions builds the chain length up to palmitic acid C_{16}. The chain length of palmitic acid can be extended by two carbon atoms to form stearic acid, which is then elongated to the typical wax chain lengths of C_{20} to C_{34}.

Hydrocarbons. To form the odd-chain hydrocarbons, it was originally postulated that two even-chain-length fatty acids came together with the loss of one carboxyl group, e.g., two palmitic acids (C_{16}) can together with decarboxylation yield hentriacontane C_{31}. It is now recognized that this does not happen. Instead, it is suggested that the hydrocarbons are produced from a pool of acyl–CoAs synthesized by elongases that are acted upon by acyl–CoA reductase, as in eq. (6), and then by a decarboxylase, as in eq. (7).

$$RCH_2CH_2CO-SACP \xrightarrow{\text{acyl-CoA reductase}} RCH_2CH_2CHO \quad (6)$$

$$RCH_2CH_2CHO \xrightarrow{\text{decarboxylase}} RCH_2CH_3 \quad (7)$$

Biosynthesis of Other Wax Components. As before, elongases catalyze the formation of long-chain fatty acids. In the reductive pathway, the long-chain fatty acid coenzyme A derivative is acted upon by an acyl–CoA reductase to give an aldehyde, as shown in eq. (8).

$$RCH_2CO-S-CoA \xrightarrow{\text{Acyl CoA reductase}} RCH_2CHO \quad (8)$$

Subsequently the aldehyde can be further reduced by an aldehyde reductase to a primary alcohol.

$$RCH_2CHO \xrightarrow{\text{aldehyde reductase}} RCH_2CH_2OH \quad (9)$$

The fatty acyl—CoA can release the fatty acid, as shown in eq. (9).

$$RCH_2CO-S-CoA \longrightarrow RCH_2COOH \quad (10)$$

In the presence of long-chain fatty acid and fatty alcohol, acyl—CoA alcohol transacylase will catalyse ester formation (eq. (11)).

$$RCH_2COSCoA + RCH_2CH_2OH \xrightarrow[\text{alcohol transacylase}]{\text{Acyl-CoA}} RCH_2COOCH_2CH_2R \quad (11)$$

The formation of ketones and secondary alcohols is part of the decarbonylation pathway by which hydrocarbons are formed. A hydroxylase is believed to oxidize the long-chain hydrocarbon to a secondary alcohol, which can then be further oxidized to a ketone by an oxidase enzyme (eqs. (11) and (13)) [von Wettstein-Knowles, 1985].

$$RCH_2R^1 \xrightarrow{\text{hydroxylase}} RCHOHR^1 \quad (12)$$

$$RCHOHR^1 \xrightarrow{\text{Oxidase}} RCOR^1 \quad (13)$$

Mutants and Inhibitors. Studies of the biosynthesis of waxes have been facilitated by using gene mutations and inhibitors. The cyanide ion, mercaptoethanol, 1,4-dithioerythritol (DTE) and 1,4-dithiothreitol (DTT) have been used as inhibitors of individual steps in the biosynthetic pathways. Most mutants affect the synthesis in a similar way.

The synthesis of lipids composing the wax in barley are controlled by *eceriferum (cer)* genes. The biosynthetic relationships among β-diketones, hydroxy- β-diketones and alkan-2-ol esters have been elucidated using these *cer* mutants. These three wax classes have a common precursor. Thus in the *de novo* FAS, the C_{12} fatty acid is converted to a C_{14} β-keto-acyl derivative [P] as in eq. (14). See also **Lipids**.

$$\underset{C_{12}}{RCH_2CO-Z} \xrightarrow[\textit{cer-q}]{\beta\text{-ketoelongase}} \underset{C_{14}[P]}{RCH_2COCH_2CO-Z} \quad (14)$$

Some mutations, e.g. *cer-q*, create a total block, and so operate in eq. (14) between the C_{12} fatty acid and [P]. This intermediate [P] can act as the starting point for alternative pathways.

$$\underset{[P]}{RCH_2COCH_2CO-Z} \xrightarrow[\textit{cer-c}]{\beta\text{-ketoelongase}} RCH_2COCH_2COCH_2CO-SCoA$$

$$RCH_2COCH_2COCH_2CO-SCoA \xrightarrow[\text{or decarboxylase}]{\text{decarbonylase}} RCH_2COCH_2COCH_3 + CO_2 \text{ or } CO \quad (15)$$

[P] can add further C_2 units to give rise to a C_{31} β-diketone, as in eq. (15). The *cer-c* mutation inhibits this conversion to a C_{31} β-diketone but stimulates alkan-2-ol formation.

[P] can lose CO_2 to give an alkan-2-ol (C_{13}), as in eq. (16).

$$\underset{[P]}{RCH_2COCH_2COZ} \xrightarrow{\text{decarboxylase}} RCH_2COCH_3 + CO_2 \quad (16)$$

This means that this pathway is unaffected by *cer-c*, and so this alternative pathway must branch before that point.

It is known that hydroxy- β-diketones are formed from β-diketones and that *cer* mutation blocks this conversion, as in eq. (17).

$$RCH_2COCH_3 \xrightarrow[\text{reductase}]{\text{methyl ketone}} RCH_2CHOHCH_3 \quad (17)$$

$$\underset{C_{31}}{RCH_2COCH_2COCH_2R^1} \xrightarrow[\textit{cer-u}]{\text{hydroxylase}} \underset{C_{31}}{\text{hydroxy-}\beta\text{-diketone}} \quad (18)$$

Barley synthesizes alkanes (n-C_{31}) and the primary alcohols (n-C_{26}) by a different mechanism from that which produces β-diketones (C_{31}). The observations supporting this include:

1. the fact that *cer-cqu* mutants lack or have an impaired β-ketoacyl system but have a wild-type acyl system.

2. cyanide ions will inhibit β-diketone synthesis but will stimulate alkane formation (Mikkelsen and Wettstein-Knowles, 1978);

3. stearic acid acts as the precursor for the acyl system but it is not the primer for β-diketones.

It is suggested that the branched hydrocarbons are based on a branched primer usually believed to be derived from an amino acid. In the gl_n *Brassica napus* (rape), the gl_n mutation modifies the system, giving solely straight chains in the wild-types. Other mutations in genes produce more short-chain compounds, e.g. the mutations gl_1, gl_2 and gl_3 in Brussels sprouts, gl_5 in cauliflower, and cer2-BRL 9 in *Arabidopsis thaliana* block elongation of C_{28} to C_{30}.

Conclusion

The bulk of plant waxes are found on the surface of the plant, although a more exhaustive extraction can remove a second waxy material that has a different chemical composition from the outer wax. The outer wax layer acts as a protective coating to the plant and is known to limit uncontrolled water loss to a minimum. The hydrophilic nature of the epicuticular wax also reduces the loss of inorganic and organic solutes from the interior of tissues. When agricultural spraying is performed, the contact angle between the leaf and the pesticide droplet can be sufficient to prevent any significant amount of the pesticide from entering the plant tissue. This, again, is related to the wax composition.

The wax also acts as a reservoir for fat-soluble compounds that may be released from the plant. These compounds can act as visual and olfactory markers for insects and may act as triggers to other plants.

Overall, the plant waxes play an important role in the health of the plant. See also **Waxes**.

Additional Reading

Baker, E.A.: "Chemistry and Morphology of Plant Epicuticular Waxes," In: Cutler D.F., K.L. Alvin, and C.E. Price: *The Plant Cuticle*, Academic Press, London, UK. 1982, pp. 139–166.

Baker, E.A., and P.J. Holloway: "Branched Chain Constituents of Brussels Sprout Wax," *Phytochemistry*, **14**, 2463–2467 (1975).

Bianchi, G.: "Plant waxes," In: Hamilton, R.J.: *Waxes: Chemistry, Molecular Biology and Functions* The Oily Press, Bridgwater, UK, 1995, pp. 175–222.

Hamilton, S., and R.J. Hamilton: "Plant waxes," In: Gunstone, F.D.: *Topics in Lipid Chemistry* Elek Science, London, UK, 1972, pp. 199–266.

Jeffree, C.E.: "The Cuticle, Epicuticular Waxes and Trichomes of Plants with Reference to Structure, Functions and Evolution," In: Juniper, B., and R. Southwood: *Insects and the Plant Surface*, Edward Arnold, London, UK, 1986.

Martin, J.T.: "Role of Cuticle in the Defense against Plant Diseases," *Annual Review of Phytopathology*, **2**. (1964).

Mikkelsen, J.D., and P. von Wettstein-Knowles: *Archives of Biochemistry and Biophysics*, **188**, 172–178 (1978).

Riederer, M., and C. Müller: *Biology of the Plant Cuticle*, Blackwell Publishers, Malden, MA, 2006.

von Wettstein-Knowles, P.: "Biosynthesis and Genetics of Waxes," In: Hamilton, R.J.: *Waxes: Chemistry, Molecular Biology and Functions*, The Oily Press, Bridgwater, UK, 1995.

R. J. HAMILTON, Liverpool John Moores University, Liverpool, UK

PLAQUE. See **Caries, Cariology, and Dentistry**.

PLASMA. An ionized gas composed of positive and negative charges (and possibly neutral atoms and molecules) of almost equal charge density. At least one kind of charge is mobile. The term was coined by Langmuir and Tonks [1929] "to designate that portion of an arc-type discharge in which the densities of ions and electrons are high but substantially equal." A more quantitative definition can be given in terms of the Debye shielding distance, the distance over which the density of negative charges can be appreciably different from that of positive charges: A plasma is an ionized gas for which the Debye shielding distance is small compared with a characteristic length [Spitzer, 1962]. According to this definition the ionosphere is a plasma, and so is a slab of aluminum, but in atmospheric usage it is limited to an ionized gas.

The most common type of plasma is a gas of such high temperature that it is ionized; that is, an appreciable number of atoms have been stripped of at least one electron and have become positive ions. Because of its free electrical charges, a plasma differs from ordinary gases because it is subject to electric and magnetic forces. Indeed, the science of plasma mostly concerns those plasmas in which the usual molecular forces are negligibly small compared with the electromagnetic forces. Plasma has been called the "fourth state of matter," after solid, liquid, and gas. All matter goes into this electrified state above temperatures of 10,000 K or so, at which point the fastest particles in the thermal distribution undergo collisions energetic enough to ionize an atom.

Plasmas can also exist at room temperature in metals and semiconductors in which there are free charge carriers. At the other extreme, in the vacuum of outer space, hydrogen atoms are ionized by photoionization by starlight. Once ionized, the ions and electrons do not recombine easily because, at a density of about $1 ion/cm^3$, collisions are rare. Consequently, perhaps 99% of all the matter in the universe is in the plasma state. On Earth, however, the temperatures and atmospheric densities that are consistent with life cannot support plasmas, and these can be studied only in laboratory vacuum chambers.

See also **Plasma Physics**; and **Plasmas (In Space)**.

Plasma Types

Figure 1 indicates the various types of plasmas according to their electron density and electron temperature. The colder or low electron energy regions contain cold plasmas such as interstellar and interplanetary space; the earth's ionosphere, of which the aurora borealis would be a visible type; alkali-vapor plasmas; some flames; and condensed-state plasmas, including semiconductors. Gaseous plasmas are sometimes classified as equilibrium or nonequilibrium referring to the electron temperature as compared to the gas temperature. Low pressure glow discharges are generally classified as nonequilibrium plasmas because the electron temperature is significantly greater than the gas temperature. Commonly, glow discharges have electron temperatures in the 10^4-10^5 K (1–10 eV) range, whereas the gas temperature in those discharges is generally less than 5×10^2 K or near ambient. Low and high pressure arc discharges, also known as plasma jets, have no such large difference and the electron and gas temperatures are both in $5 \times 10^3-10^4$ K the range, thus being called equilibrium plasmas. The areas of most interest to plasma chemistry are the glow discharges and arcs.

Controlled thermonuclear fusion experiments and certain types of confined arcs known as pinches have temperatures in the $5 \times 10^5-10^7$ K range. However, to be successful, controlled thermonuclear fusion needs to take place from $6 \times 10^7-10^9$ K. In fact, the goal of all fusion devices is to produce high ion temperatures in excess of the electron temperature.

Also shown in Figure 1 are the Debye screening length and Debye sphere size. For gaseous plasmas, $N_D \gg 1$. Solid-state plasmas or condensed-state plasmas generally exist where $N_D \leq 1$.

Field Effects. Velocities, lengths, and frequencies are intrinsic to gases and plasmas, independent of incident radiation or existing fields. However, some of the more interesting plasma phenomena only appear in the presence of static or dynamic fields. External static electric fields tend to separate and accelerate plasma charges. Such fields and the resultant electron motions can both produce plasmas and heat them. Magnetic field effects in plasmas are so important that at one time plasma physics and magnetohydrodynamics (MHD) were practically synonymous. However, some plasmas are not magnetized, and some MHD processes do not involve plasmas, eg, the fluid motions in the earth's core which produce the terrestrial magnetic field.

Transverse electromagnetic waves propagate in plasmas if their frequency is greater than the plasma frequency. For a given angular frequency, ω, there is a critical density, n_c, above which waves do not penetrate a plasma. The propagation of electromagnetic waves in plasmas has many uses, especially as a probe of plasma conditions.

$$n_c = m\omega^2/4\pi e^2$$

The presence of a static magnetic field within a plasma affects microscopic particle motions and microscopic wave motions. The charged particles execute cyclotron motion and their trajectories are altered into helices along the field lines. The radius of the helix, or the Larmor radius, is given by the following:

$$mcV\perp/qB$$

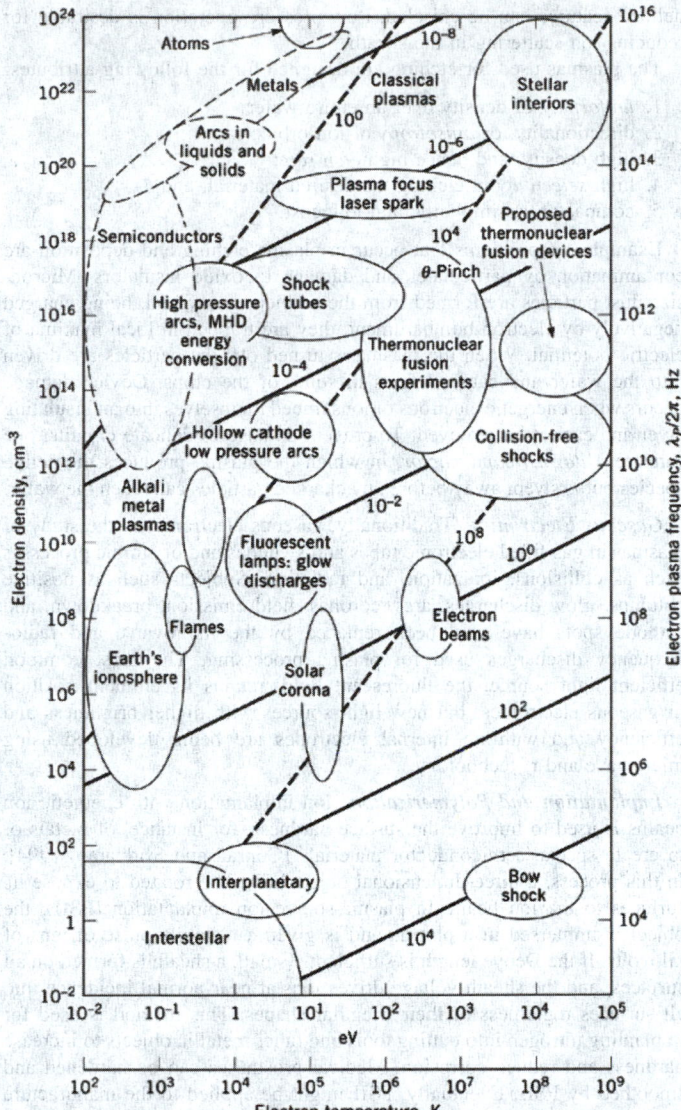

Fig. 1. Electron temperature and density regions for plasmas where the numbers and the diagonal lines represent ($\longrightarrow$) the Debye screening length, λ_D, in centimeters, and (-----) Debye sphere volume, N_D, in cubic centimeters. The plasma frequency is given on the right-hand axis. Condensed-state plasmas, indicated in dashed line areas, have relatively low temperatures, small Debye spheres, and high densities. Plasmas in metals and semimetals fall along the line separating degenerate quantum plasmas from nondegenerate classical plasmas. Gaseous plasmas, shown in solid line areas, have characteristics that vary widely. MHD=magnetohydrodynamic.

for a particle of mass m and charge q, and velocity $V_\perp$ normal to the magnetic field of strength B, where c is the velocity of light. The cyclotron frequency, ω_c, which is introduced by the presence of the field, is

$$\omega c = q B / m c$$

The particle–field interaction is the means by which magnetic fields can exert pressures on plasmas and vice versa. The presence of static magnetic fields does not alter the propagation of longitudinal electrostatic electron or ion waves if the propagation direction is parallel to the field. However, propagation that is orthogonal to the field involves new frequencies that depend on the field strength. In contrast, the propagation of transverse electromagnetic waves in plasmas is altered by a magnetic field regardless of the relative geometry of the direction of motion and the field vector.

Magnetic fields introduce hydromagnetic waves, which are transverse modes of ion motion and wave propagation that do not exist in the absence of an applied B field. The first of these are Alfven, A, waves and their frequency depends on B and ρ, the mass density. Such waves move parallel to the applied field having the following velocity:

$$V_A = B / (4\pi p)^{1/2}$$

and are similar to the waves that travel along a string. Magnetosonic waves are a second type of hydromagnetic wave, and these propagate perpendicular to the magnetic field. Their frequency depends on the Alfven and the acoustic velocities. Hydromagnetic waves are electromagnetic waves. Even though the applied B field is static or nonoscillatory, the waves are transverse and are characterized by oscillatory electric, E, and magnetic, B, components.

Gaseous Plasmas. Gaseous plasmas are often far from equilibrium and therefore can exhibit microscopic or particle instabilities, and macroscopic or hydromagnetic instabilities. Microscopic instabilities are caused by departures from the equilibrium Maxwellian distributions for the electrons or ions. Examples of such situations include a plasma expanding while cooling, anisotropies in the velocity distribution caused by applied magnetic fields, or the motion or streaming of a particle beam through a plasma. Macroscopic instabilities produce the motion of the plasma as a whole. Causes include pressure or density gradients or magnetic field curvature. All instabilities represent the tendency of plasmas to reach equilibrium more quickly than is possible by ordinary collisions alone. Instabilities can reduce plasma confinement times by many orders of magnitude, a significant problem in fusion research.

The high energy densities of many gaseous plasmas raise safety concerns. The sources of energy used to produce and heat plasmas, eg, steady-state, high voltage, and high current generators, and capacitors for pulsed electron-discharge heating and laser beams, can be hazardous. Work with plasmas usually requires careful attention to proper electrical safety precautions and to eye hazards. Even in the absence of lasers, plasmas can pose a threat to vision because the plasmas often are very bright and can emit dangerous levels of uv radiation. X-radiation from plasmas usually is not a safety concern. Most energy from 10^6 K plasmas is soft and does not escape from the experimental chamber or traverse significant distances in air. The hard x-rays emitted by most plasma sources usually are of very low intensity. However, some low energy, low pressure plasmas, such as electron cyclotron resonance sources, high ($>10^4$ V) voltage, plasma-generating machines and some fusion-energy research devices can emit unsafe x-ray emissions. Shielding, eg, lead or concrete, and distance from the source reduce exposures to acceptable levels.

Applications (Industrial)

Plasmas are extensively used in manufacturing and in the production and improvement of materials. These applications call for low-temperature, partially ionized plasmas, which have properties different from the hot, collisionless plasmas used in fusion and present in space. Industrial plasmas are likely to have many species of atoms and molecules, including negative ions, and are usually collisional, involving simultaneously many different scattering, excitation, and ionization processes [Lieberman and Lichtenberg, 2005].

Plasma Sources. In some manufacturing procedures, the type of plasma is not crucial; only the existence of electrons is needed. In the "barrel etcher," for instance, a simple discharge between two electrodes is used to create a low degree of ionization. For the fabrication of large-scale integrated semiconductor circuits, however, the plasma performs several functions and must be carefully controlled. By far the most common and the best developed plasma generator for this purpose is the parallel-plate capacitor discharge, called the RIE (reactive ion etching) source (Fig. 2). The RIE plasma is made by applying a rf voltage at 13.56 MHz between two parallel electrodes about 5–10 cm apart. The gas is a mixture containing Cl or F compounds, which are dissociated by the plasma electrons; it is the neutral Cl or F atoms that chemically attack the target in etching. A sheath forms on both electrodes to confine the electrons, that on the negative electrode being thicker. During the rf cycle, these sheaths oscillate, and the plasma "sloshes" back and forth between electrodes. On average, the ions experience a sheath potential that accelerates them toward the substrate, or object to be treated, which is mounted on one electrode. That electrode can be grounded through a separate rf power supply to adjust the rectified sheath voltage. Since the sheath drop, rf oscillation,

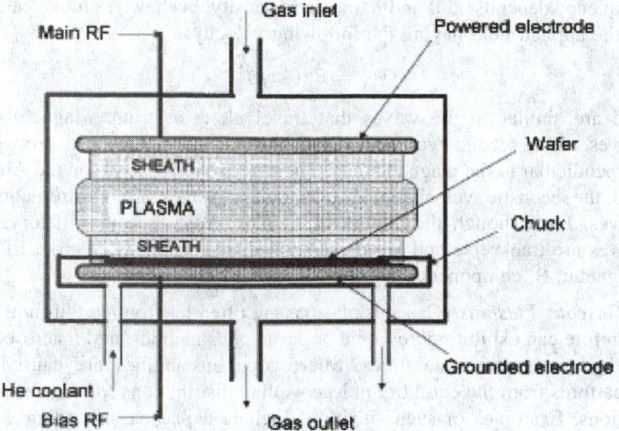

Fig. 2. Diagram of a RIE (reactive ion etching) plasma source commonly used for etching and deposition of silicon wafers in semiconductor fabrication. A 13.56-MHz radio-frequency source applied between the parallel-plate electrodes creates the plasma, which impinges on the wafer mounted on the bottom plate.

and plasma density all depend on rf power, the degree of control over the plasma is restricted.

New plasma sources have been proposed to be more flexible than the RIE source and to produce higher plasma densities [Lieberman and Gottscho, 1994]. The leading contenders are the electron-cyclotron resonance (ECR) source, inductively coupled plasmas (ICP), the radio-frequency inductive (RFI) or transformer-coupled plasma (TCP) source, and the helicon source. The ECR source, driven by 2.45-GHz microwaves, is a spinoff from ECRF technology used for plasma heating. This source can produce high densities at low pressures but requires a magnetic field of 875 G at the resonance layer. The other sources operate at the industrial frequency of 13.56 MHz or its harmonics. ICPs utilize a helical winding around a cylindrical tube for applying the rf power, usually with an electrostatic shield to eliminate direct electrostatic coupling. RFI and TCP sources are similar, each with a pancake-shaped rf antenna lying on a quartz plate, which separates it from the plasma chamber. These rf sources do not require a magnetic field and do not resonate with a natural frequency of the plasma. In the helicon source, an external antenna launches a low-frequency whistler wave, called a helicon wave, which ionizes the plasma as it is damped. A magnetic field of 0.1–1 kG is required. When the helicon resonance is struck, ionization is particularly efficient. In all these sources, the plasma is often drifted into a chamber with magnetic confinement at the walls provided by arrays of permanent magnets, called a *magnetic bucket*, to smooth out density variations before the plasma arrives at the substrate.

Plasma Etching and Deposition. Plasma processing is essential in the production of semiconductor integrated circuits, in which the individual features can be smaller than 0.5 μm. The manufacture of flat-panel displays, such as the active-matrix liquid-crystal displays (AMLCDs) used in portable computers, also benefits from plasma processing, though it is not essential, since the elements here are on a larger scale. Integrated circuits are made, hundreds at a time, on a polycrystalline silicon wafer, typically eight inches or larger in diameter. The various components of a chip — transistors, capacitors, conductors, etc. — are made in a series of steps involving deposition, masking, stripping, and etching [Manos and Flamm, 1989]. Different processes are used for handling semiconductor material, oxide or nitride insulators, conductors, and photoresist. The patterns are of such fine scale that photolithography with short-wavelength x rays must be used; the x-ray sources are often plasma devices. Since the gases, pressures, and even plasma sources are generally different for each step, the wafers are shuttled under vacuum from one "reactor" to the next in a transfer mechanism known as a *cluster tool.*

The role of plasma in the etching process is twofold. First, the electrons produce the active species, and ion bombardment prepares the surface so that chemical etching occurs more readily. The combined effect is to increase the etch rate by over an order of magnitude over chemical etching without plasma or plasma sputtering without chemicals. Second, the sheath electric field accelerates ions so that they strike the substrate at right angles, thus guiding the etching process to proceed in a straight line, so that sharply

defined features can be chiseled. Low-pressure operation is desirable for reducing ion scattering in the sheath.

The plasmas used for etching are designed for the following attributes:

1. *uniformity* of density over an entire wafer;
2. directionality, or *anisotropy* of ion orbits;
3. high density, and hence high *etch rate*;
4. high *selectivity* in etching the desired material; and
5. compactness, simplicity, and low cost.

Examples of problems that occur in plasma etching and deposition are contamination by particulates and damage to oxide insulators. Micron-size dust particles are formed from the ambient gases, and, being charged negatively by electron bombardment, they are trapped in local maxima of electric potential. When the plasma is turned off, the particles are driven into the wafer and cause defects in some of the chips. Device damage occurs when energetic electrons or ions imbed themselves into an insulating layer and cannot be removed. To prevent damage to delicate circuits, one can use a *downstream reactor*, in which the plasma produces the active species but is swept away before any charged particles can reach the wafer.

Gaseous Electronics. Traditionally, gaseous electronics is the study of plasmas in gas-filled electronic tubes and switches and of atomic processes such as collisions, ionization, and radiation. Subjects such as positive columns, glow discharges, arcs, coronas, field emission, breakdown, and cathode spots have now been replaced by the microwave and radio-frequency discharges used for plasma processing. The most common efficient light source, the fluorescent light, retains its eminent position in gaseous electronics; but new light sources with higher brightness and efficiency and with no internal electrodes are being developed using microwave and rf technology.

Implantation and Polymerization. Ion implantation with energetic ion beams is used to improve the surface hardness, for instance, of metals or to create special semiconductor materials [Conrad and Sridharan, 1994]. In this process, a three-dimensional object has to be rotated to expose its surfaces to the ion beam. In plasma-source ion implantation (PSII), the object is immersed in a plasma and is given a negative pulse of tens of kilovolts. If the Debye length is sufficiently small, a sheath is formed on all surfaces, and the sheath voltage drives ions at near-normal incidence into all surfaces regardless of their irregular shapes. This method is used for implanting nitrogen into cutting tools and other metallic objects to increase hardness and reduce corrosion. Medical prostheses can be hardened and smoothed by PSII. Eventually, PSII might be applied to the manufacture of magnetic and optical disks.

Strong plastic coatings can be created with plasma discharges in organic compounds such as methyl methacrylate [d'Agostino, 1990]. When polymerized in the presence of electrons, a plastic material becomes highly cross linked, resulting in a strong and resistant layer. Organic plasmas can be used for a number of practical purposes: for barrier coatings in gasoline tanks, soft drink bottles, and medical capsules; for making textile fibers more absorbent to dyes; for increasing the water resistance of papers and wood; for cleaning and sterilization of biomedical containers; for producing optical fibers with graded indices of refraction; and eventually for producing integrated optics chips for optical computing.

Thermal Plasmas. Thermal plasmas are plasmas at near-atmospheric pressure that are so collisional that they are in kinetic, though not radiative, equilibrium [Boulos et al., 1994]. Plasma spray treatment of aircraft and automobile parts such as turbine blades is a widespread industrial application of plasma physics. Thermal plasma jets are capable of growing diamond coatings at a much faster rate than with low-pressure plasmas. Development of this process can lead to diamond coatings on all cutting tools, and even to diamond substrates for semiconductors. Thermal plasmas can also be used to develop new types of ultrahard materials.

Isotope Separation. The production of U^{235} by gas-diffusion methods is notoriously slow, and many ideas for faster processes have been proposed. The use of plasmas for isotope separation started during World War II with the ill-fated Calutron project, named after the University of California, in which instabilities were so strong that the plasma was lost at what is now called the Bohm diffusion rate. Though new types of plasma centrifuges are still being researched, the major avenues for advanced isotope separation use lasers or radio-frequency generators.

Laser Isotope Separation. With narrow linewidth, lasers can selectively excite an isotopic species either as an atom or in a molecule. In atomic-vapor laser isotope separation (AVLIS), the naturally occurring isotopic mixture of an element is vaporized and introduced into a long chamber. A laser beam tuned to a particular line is directed into the gas and selectively ionizes the desired isotope. A high voltage applied to the walls of the chamber then causes the ions to be collected at the cathode [Vitello et al., 1992]. This method is limited by the rate at which the ions can drift across the chamber, and to those isotopes for which a suitable laser line can be found.

Ion-Cyclotron Isotope Separation. In this scheme, the raw material is vaporized, then ionized by microwaves from a gyrotron, and is confined radially by a uniform magnetic field [Chen, 1991]. A radio-frequency signal near the cyclotron frequency of the desired isotope is then applied to the plasma with a helical antenna. Those atoms that are in cyclotron resonance are selectively accelerated to large perpendicular energies and, because of their large Larmor radii, can be "scraped off" onto suitable collectors. Since both the frequency and the magnetic field can be tuned, this method will obviously work for any isotope; for this reason, it is particularly useful for producing medical isotopes.

Additional Reading

Boulos, M.I., P. Fauchais, and E. Pfender: *Thermal Plasmas*, Vol. 1, Plenum, New York, NY, 1994.

Chen, F.F.: "Double Helix: the Dawson Separation Process," in: T. Katsouleas Ed., *From Fusion to Light Surfing*, Addison-Wesley, Redwood City, CA, 1991.

Conrad, J.R., and K. Sridharan: "Papers from the 1st International Workshop on Plasma-Based Ion Implantation," *J. Vac. Sci. Technol. B*, **12**, 807–1006 (1994).

d'Agostiono, R.: *Plasma Deposition, Treatment, and Etching of Polymers*, Academic Press, New York, NY, 1990.

Hollahan, J.R., and A.T. Bell: *Techniques and Applications of Plasma Chemistry*, John Wiley & Sons, Inc., New York, NY, 1974.

Langmuir, I., and L. Tonks: *Phys. Rev.*, **33**, 196 (1929).

Lieberman, M.A., and R.A. Gottscho: "Design of High-Density Plasma Sources for Materials Processing," Vol. 18 of: M. Francomb and J. Vossen Eds., *Physics of Thin Films*, Academic Press, New York, NY, 1994, pp. 1–119.

Liberman, M.A., and A.J. Lichtenberg: *Principles of Plasma Discharges and Materials Processing*, 2nd Edition, John Wiley & Sons, Inc., Hoboken, NJ, 2005.

Manos, D.M., and D.L. Flamm: *Plasma Etching*, Academic Press, New York, NY, 1989.

Miyamoto, K.: *Plasma Physics and Controlled Nuclear Fusion*, Springer-Verlag New York, LLC, New York, NY, 2005.

Nasser, E.: *Fundamentals of Gaseous Ionization and Plasma Electronics*, John Wiley & Sons, Inc., New York, NY, 1971.

Nolte, J.: *ICP Emission Spectrometry: A Practical Guide*, John Wiley & Sons, Inc., New York, NY, 2003.

Rossnagel, S.M., J.J. Cuomo, and W.D. Westwood: *Handbook of Plasma Processing Technology*, Noyes Publications, Park Ridge, NJ, 1990.

Spitzer, L.: *Physics of Fully Ionized Gases*, 2nd Edition, Dover Publications, Mineola, NY, 2006.

Vitello, P., C. Cerjan, and D. Braun: *Phys. Fluids B*, **4**, 1447–1456 (1992).

FRANCIS F. CHEN, University of California, Los Angeles, CA

MARK D. SMITH, Allied Signal Aerospace Company

PLASMA (Blood). The portion of the blood remaining after removal of the white and red cells and the platelets; it differs from serum in that it contains fibrinogen, which induces clotting by conversion into fibrin by activity of the enzyme thrombin. Plasma is made up of more than 40 proteins and also contains acids, lipids, and metal ions. It is an amber, opalescent solution in which the proteins are in colloidal suspension and the solutes (electrolytes and nonelectrolytes) are either emulsified or in true solution. The proteins can be separated from each other and from the other solutes by ultrafiltration, ultracentrifugation, electrophoresis, and immuno-chemical techniques. See also **Blood**.

PLASMA DISPLAY PANELS. Mechanisms of Operation. The plasma display panel (PDP) can be thought of as a descendant of the neon lamp, which was invented in 1915 by Georges Claude in France. The term *plasma* refers to a gas that consists of electrons, positively charged particles known as anions, and neutral particles. The PDP has sometimes been referred

to as a gas-discharge display because it operates by passing electricity through neon gas, causing it to become "charged" temporarily; light is produced when the gas spontaneously discharges. The displays operate at high voltages, low currents, and low temperatures, resulting in long operating lifetimes.

Plasma display panels use glow discharge reactions. Four important reactions can occur in a plasma discharge: ionization, excitation, metastable generation, and Penning ionization. Ionization of the Ne atom generates electrons and neon anions (Ne^+) that can cause the generation of an avalanche in the gas. The avalanche begins near the cathode and grows toward the anode as it generates a very large number of electron–neon-ion pairs. The ionization of the gas results in a visible glow. Ions travel back to the cathode, ejecting secondary electrons, which start new avalanches. A feedback loop is set up, and depending on the applied voltage, the current will be maintained at a point either above or below the threshold.

The ions, electrons, and neutral Ne atoms are in constant motion in the plasma, creating a number of chemical reactions that occur in this gaseous phase. One of these reactions, called excitation, causes some of the Ne atoms to transform into an unstable condition known as the excited state, or Ne^*, where it remains for only about one billionth of a second before returning to its original form (Ne), the ground state. As it returns to its ground state, the energy that had been absorbed during the excitation process is released in the form of light energy. This light may be visible (orange) or invisible (ultraviolet light) to the naked eye. The ultraviolet light emission can be used in conjunction with color phosphors to produce full-color PDPs.

The process of excitation and light emission takes place so quickly that the eye perceives the plasma as producing orange light continuously, as long as the plasma is activated by the electric field. Figure 1 depicts this operation.

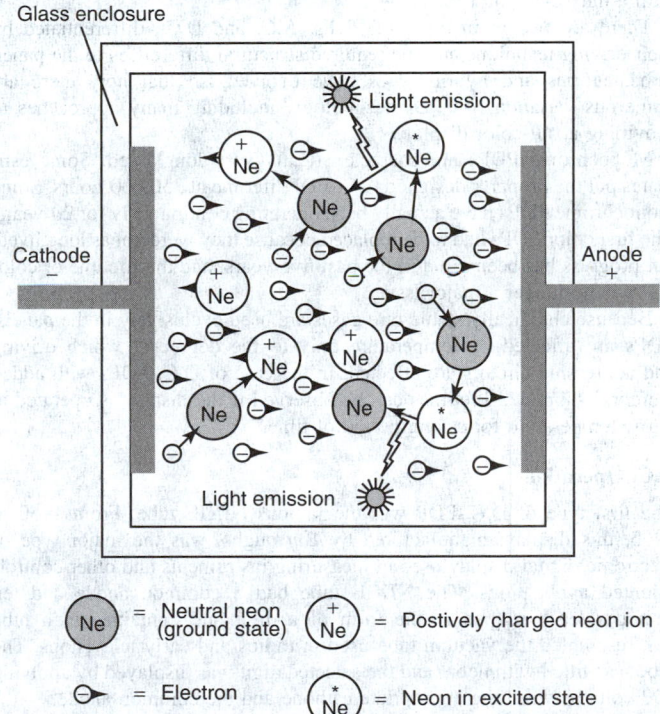

Fig. 1.   Light emission processes in plasma display panels.

The process of light emission for a monochrome (orange) PDP occurs in the following way Neon gas in a sealed glass envelope receives a high-voltage electrical charge. The gas becomes a mixture of electrons, anions, and neutral atoms; this is the plasma. Some neon atoms are raised to a high-energy, unstable excited state (Ne^*). After a nanosecond, Ne^* converts to ground-state Ne, resulting in the production of orange light.

The main advantage of the PDP over nearly all other display devices is that it can be made into a display panel, with diagonal sizes of 20 to

60 inches (51 to 152 centimeters) currently in the production, or advanced prototype stages that are not thicker than 4 inches (10 centimeters), including drive electronics. PDPs larger than 50 inches (127 centimeters) are in the production or advanced prototype stages. Moreover, these large panels can provide high information content and full-color images. This makes the PDP the ideal technology for the fabled wall-mounted television, which can be used in the home as well as in numerous commercial and industrial environments as a replacement for large, bulky CRT monitors and rear projectors.

PDP Characteristics

The central characteristic of a PDP is the highly nonlinear electro-optic response curve. When the addressing voltage is driven above a certain "threshold" level, the ionization increases dramatically. As in all electro-optic response curves, the threshold allows multiplex addressing; the steepness of the plasma curve allows a large number of lines to be addressed without active matrix elements. Another characteristic of PDPs is that the discharge in a PDP can be switched *on* and *off* in a few microseconds, allowing for a very fast display. This property is proving to be useful in the developing video applications for large PDPs.

Another characteristic of some types of PDPs is inherent memory. The memory is located in the plasma panel itself, rather than in an external device, such as random-access memory. Inherent memory is a curious characteristic. Although at first glance memory might appear to be an advantage in every type of display, it is only helpful when paired with other features, such as easy erasing and low power consumption. It appears to be particularly useful in light-emitting displays. In the case of PDPs, it is very helpful, because it allows a display to be very bright even when the display contains a large number of lines. The duty cycle limitations imposed on multiplexed liquid crystal displays are not seen in PDPs with memory. Essentially, the memory effect means that the display operates with a duty cycle of 1.

There are two main types of PDPs, A.C. and D.C., differentiated by their driving technique and the required structural differences in the panel. Subcategories and hybrid types have evolved, so that now there are numerous variations on PDP technology, including many approaches to providing a full-color display.

Monochrome PDPs are characteristically very long lived. Some estimates put the properly designed A.C.-PDP lifetime at 350,000 hours. Some monochrome PDPs have actually been operating continuously for 20 years. The first color PDPs had to be replaced because they were not as long lived, but progress has been rapid in the past five years, and the lifetime of color PDPs is no longer a major issue.

Because chemically stable rare gases are used exclusively in the panels, PDPs are affected by temperature only to the degree at which driving and addressing circuits are affected. In the case of D.C.-PDPs with added mercury, shortened lifetimes can be observed if the display is operated at a low temperature for a long period of time.

D.C. Operation

The first type of D.C.-PDP was the famous NIXIE tube. From 1950 to 1965, this display, manufactured by Burroughs, was the major type of electronic digital display used in measuring instruments and other control-oriented applications. The NIXIE tube had a common anode and ten cathodes, each shaped in the form of a digit and contained in a tube that resembled the vacuum tube used in radios and early televisions. The tube was filled with neon, and the selected digit was displayed by applying 100 volts between the appropriate cathode and the common anode.

Although they were crude and unattractive, the NIXIE tubes ushered in the era of the digital display as a replacement for needle point gauges. After the NIXIE tube, segmented and character-type D.C.-PDPs were developed. These displays have been used in millions of pieces of equipment, such as cash registers and ticket machines. The main drawback of these displays was their limitation to a single orange color.

The basic structure of a D.C.-driven PDP cell is shown in Figure 2. The cell is made with inexpensive soda lime glass plates, the same type of glass used in windows, except that it is much flatter and free of surface defects. The back, or rear, plate has a thin, coating of a metal or a transparent conductive coating such as tin oxide or indium–tin oxide (ITO). This coating acts as the cathode or negative electrode. The top, or front, plate

also has a conductive coating, but this material must be transparent so that the light emission can be seen. This coating, which is typically tin oxide or ITO, will act as the anode or positive electrode. The two plates are sealed together at 450°C using a glass solder material known as frit to seal the plates, but a small opening (sometimes a tube is used) is left in the seal.

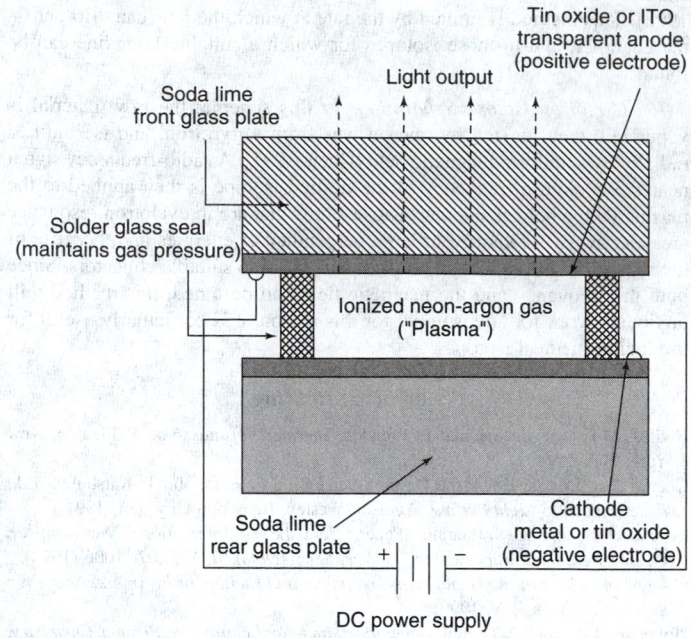

Fig. 2.   Basic structure of DC plasma display panel.

The tube is used to remove air from the space between the plates and to refill (or backfill) the space with the neon gas. After filling with the gas, the opening or tube is sealed off. The process of simultaneously sealing and filling with gas is very similar to the process used to manufacture cathode ray tubes (CRTs). The positive terminal of a D.C. power supply is connected to the anode, and the negative terminal to the cathode. When a voltage of 100 to 120 volts is applied from the power supply, light emission occurs. In practice, the construction of a dot matrix or graphic D.C.-PDP is somewhat more complicated. The electrodes are patterned to form a row (horizontal) and column (vertical) arrangement. See Figure 3. The structure includes barrier ribs, which maintain cell separation and isolate the glow to one particular dot. In addition, the electrodes are not coated with an insulator, as is required in A.C.-PDPs, and each column electrode typically has a ballast resistor attached to it. The voltage drop across the resistor allows only one discharge to be started along that column at any one time. Mercury is added to a D.C.-PDP to protect the electrodes from sputtering damage. At low temperatures, the mercury vapor condenses to a liquid, leaving the electrodes unprotected.

In operation, data pulses (electrical signals) are supplied to the vertical, or column, electrodes according to the image stored in memory. The scan pulses are supplied to the horizontal, or row, electrodes by scanning the rows sequentially, one at a time. Once ionization occurs, the brightness of the PDP is directly proportional to the current passing through the plasma. The current continues even when the voltage is decreased below the threshold voltage (V_{th}). This is known as the holding, or sustaining, voltage. If the voltage is further decreased to a level known as the extinguishing voltage (V_{ex}), light emission will cease.

There is a memory effect in a D.C.-PDP, realized by the fact that there are two stable states on the current–voltage curve. The threshold voltage (V_{th}) is sharply defined in that no ionization and, hence, no light emission will occur until V_{th} is reached. The extinguishing voltage (V_{ex}) is the level at which the emission stops. Depending on the gas used, its pressure, and the geometry of the cell, V_{th} typically ranges from 65 to 120 volts, and V_{ex} typically ranges from 70% to 90% of V_{th} for D.C.-coupled operation.

To reduce the manufacturing cost and to extend the life of D.C.-PDPs, other complex driving schemes are employed. One commonly

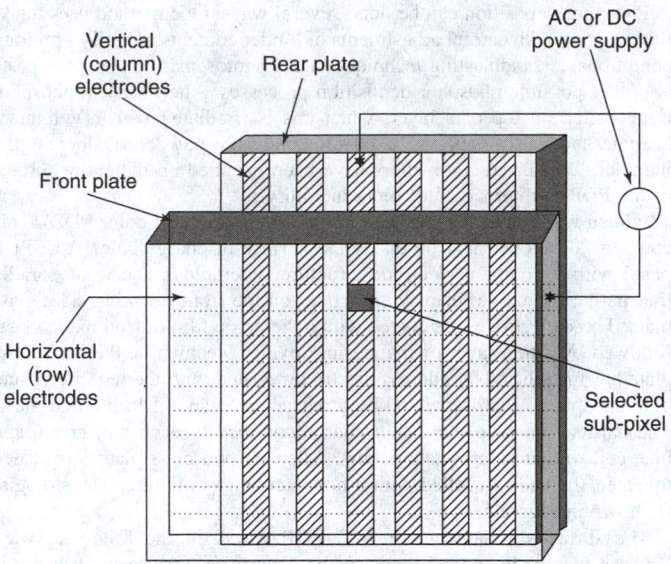

Fig. 3.   Diagram of electrode matrix panel in an AC or DC plasma display panel.

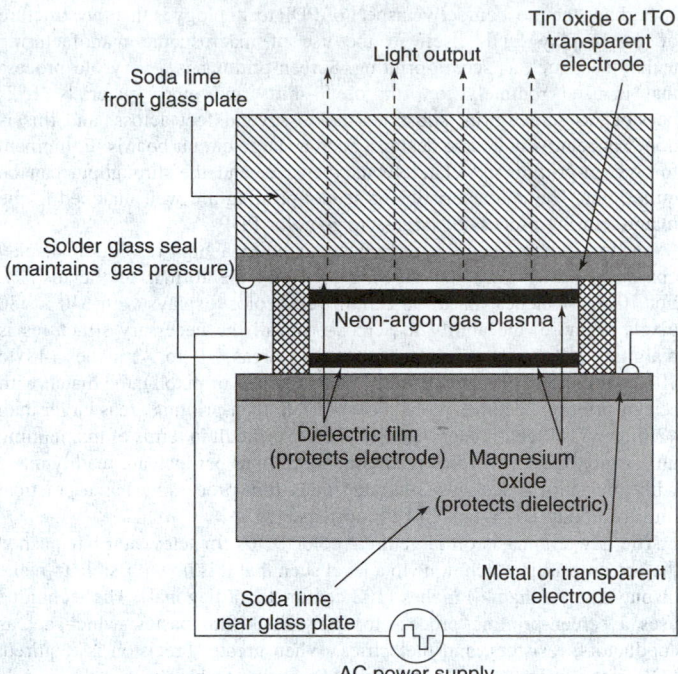

Fig. 4.   Basic structure of an AC plasma display panel.

used technique, known as priming, creates a faint background glow that essentially reduces the contrast ratio of a D.C.-PDP from 15:1 to about 6:1. Without priming particles, the initiation of discharge can take about 100 microseconds, and some of the isolated pixels tend to flicker off. Higher driving voltages could reduce this time, but then more expensive, higher voltage drivers would be required. Priming can reduce the discharge time by a factor of 10.

Monochrome D.C.-PDPs have evolved as reliable, rugged, and fairly low-cost devices for applications suitable for displays with limited grayscale. The display markets have evolved away from monochrome, however, forcing nearly all manufacturers to drop their monochrome PDPs and concentrate on developing color PDPs. Some developers of color PDPs have made color D.C.-PDP prototypes, but so far only A.C. color PDPs have been developed for mass markets. With the shift away from D.C. plasma technology at Matsushita, there are essentially no developers left working on color D.C.-PDPs.

Advantages of D.C.-PDPs:

• Simplified driving circuitry
• Long lifetime for monochrome displays
• Rise time is relatively shorter in D.C.-PDPs than in A.C.-PDPs

Disadvantages of D.C.-PDPs

• High voltage drivers needed
• Low contrast ratio
• Low luminance (lower than A.C.-PDP)
• Complex structure, because ribs adopt cell configurations
• Short lifetime (shorter than A.C.) due to exposed electrodes
• Background glow in some designs

A.C. Operation

A simple alternating current (A.C.) PDP has a construction that is similar to a D.C.-PDP as shown in Figure 4. The cell is also made with inexpensive soda lime glass plates. The back, or rear, plate has a thin coating of a metal or a transparent conductive coating such as tin oxide or ITO, while the top or front plate has a transparent conductive coating of tin oxide or ITO. The important difference from a D.C.-PDP is that the conductive coatings on both plates are covered by an insulating, dielectric film, followed by a coating of magnesium oxide, which acts as a cathode.

The purpose of the dielectric film, which is typically a lead oxide glass, is to create a capacitor in series with the cell. The magnesium oxide film protects the dielectric film from damage caused by a bombardment of charged particles in the plasma and reduces the operating voltage of the panel. It has a low (95 V) normal cathode fall, is chemically stable, is transparent to visible light, and has a low sputtering rate. The two plates are sealed together at 450 °C (842 °F) using a frit, leaving a small opening

or tube in the seal. The tube is then used to remove air from the space between the plates and to backfill that space with the Ne gas. After filling with the gas, the opening or tube is sealed off. Again, the process of simultaneously sealing and filling with gas is very similar to the process used to manufacture CRTs. One terminal of an A.C. power supply is connected to the electrode on the front plate, and the other terminal is connected to the electrode on the rear plate. When a voltage of about 200 volts is applied from the power supply, light emission occurs.

In practice, the construction of a dot matrix or graphic A.C.-PDP has electrodes that are patterned to form the same row (horizontal) and column (vertical) arrangement. The structure also includes barrier ribs, which maintain cell separation and isolate the glow to one particular dot or pixel. The A.C.-PDPs achieve memory operation by using current limiting capacitors instead of resistors. The thin dielectric (lead oxide glass) coating forms the capacitor over the row and column electrode array; this is protected by a thin film of magnesium oxide. The two thin-film layers and the gas cell are equivalent to three capacitors in series. If a high current source were available, the cell would conduct very high currents. However, the current is only allowed to flow until the capacitive charge is dissipated. The thin-film dielectric coating also protects the electrodes from contamination and erosion from the charged particles in the plasma.

An A.C. sustain voltage, which is a 20-microsecond pulse, is applied to the cell. Since the sustain voltage level is small, no discharge is initiated by the application of the sustain voltage alone. For cells already in the *on* state, the sustain pulse refreshes the wall charge and maintains the status of the cell. When the wall voltage changes, a pulse of light is given off by the gas discharge. For a cell in the *off* state, the sustain voltage is not great enough to initiate a discharge, so it is maintained in an *off* state. Writing and erasing an A.C.-PDP require a complex set of operations. Writing is facilitated when another pulse with a voltage greater than the sustain voltage is applied to the cell. While the write pulse is delivered to the cell after the sustain pulse, an erase pulse must be delivered before the sustain pulse in order to turn a pixel off by bringing the wall voltage to 0 volts.

The long history of experience and the rugged nature of A.C.-PDPs have made them the most popular for fully militarized flat panel display systems. The panels are used in a wide range of systems, from compact battlefield computers used for fire control to 1.5-meter (diagonal) displays used in war rooms. This experience has given the A.C.-PDP a solid reputation as a long-lived, highly reliable display system that can be used in many commercial applications.

Perhaps the most attractive aspect of PDP technology is that the structure of the display lends itself to the use of macroscopic manufacturing techniques, such as screen-printing. Screen-printing is a very old process that is used routinely to print on T-shirts and other materials. It is possible to screen-print substrates up to several feet across, and this is done for relatively coarse designs, such as large circuit boards. Equipment for screen-printing is relatively inexpensive, and the throughput can be quite high. The characteristics of screen-printing are well matched to the manufacturing process of large-screen, color PDPs.

The minimum feature size needed to build a high-resolution display obviously varies with the viewing area and the format of the display. For 10 to 14 inches (25 to 36 centimeters) color displays with 640×480 pixels or more, the ability to screen-print all the necessary structures is marginal at best. But for larger viewing areas, 19 to 42 inches (48 to 107 centimeters) or more (with the same number of pixels), the match with screen-printing becomes more acceptable. Screen-printing areas larger than 42 inches (107 centimeters) becomes more difficult in terms of maintaining uniformity. Displays with a resolution of 66 lines per inch are readily made with thick-film processing, although thick-film processing has a practical limit of about 0.005-inch (0.127 mm) spaces.

The new emphasis on large-area, color PDPs for television has pushed the screen-printing technique to a level such that it is now possible to make displays larger than 40 inches (102 centimeters)(diagonal). The technique uses a screen-printing process to apply thick-film pastes, which act as conductors, resistors, and dielectrics. When greater precision is required, PDPs use a more complex, thin-film deposition technique.

The A.C.-PDPs require a driving voltage of about 200 volts, leading to special problems for the integrated circuit drivers. However, on the positive side, the high voltage drivers are not required to deliver high currents. Due to the construction of the panel, the capacitance of an A.C.-PDP is about 1,000 times lower than it is for a comparable electroluminescent display, which operates at a similar voltage level. Nevertheless, driver circuits can add several hundreds of dollars to the manufacturing cost of an A.C.-PDP.

To produce grayscale in an A.C.-PDP, the approach is to modulate the frame time. The entire display is addressed multiple times in a frame. A given pixel will be "hit" for a certain number of the subframes, depending on the intensity level it requires. For example, in a 16-level grayscale scheme, a full *on* pixel would be addressed in each of the 16 subframes making up a single frame. It is possible to produce 256 gray levels in an A.C.-PDP.

Color A.C. Plasma Display Panels

There are several approaches to achieving color operation in a PDP, but all of them use the ultraviolet light generated by the plasma discharge, rather than the orange glow of the plasma directly. A fluorescent material, such as zinc sulfide or zinc oxide, placed in the vicinity of the discharge converts the ultraviolet light into visible light. This is the same principle employed in the ordinary fluorescent tube lamp used in offices around the world. If the fluorescent material, called a phosphor (it does not contain phosphorous, but the base is composed of zinc oxides or sulfides), is doped with a small amount of a rare earth or other compound, it can emit light of various colors, depending on the specific compound selected. By using red, green, and blue phosphors, multicolor and even full color (16.777 million colors) can be achieved by forming arrays of these phosphors on the inner surface of the panel. This is similar to the way in which color CRTs for televisions and computer monitors are made.

Controlling the intensities of the red, green, or blue phosphor deposited on the wall of each discharge cell allows full color representation. Rare earth materials (from the group IIIb, Lanthanide Series of elements in the periodic table) are used as activators by most of the high-performance, ultraviolet-light-sensitive phosphor powders. A depletion of activators in the ground state causes the phosphor output to saturate with respect to the ultraviolet light intensity. A short decay time constant is desirable in the phosphor to avoid tailings from moving images. Increasing activator density or reducing the adsorption coefficient can reduce saturation. The National Television Standards Committee (NTSC) has determined three primary colors. It is important that these coincide with products manufactured for television use. It is also important that the intensity balance and the intensity persistence balance coincide with the NTSC standards.

Phosphor deposition can be done several ways. One method uses thick-film printing with careful adjustments of binder contents as well as printing conditions. A sandblasting technique and a photosensitive phosphor paste are other possible phosphor deposition processes. Alternately, a phosphor is deposited on top of a layer; when this is irradiated with UV light, it becomes tacky. The tackiness allows the phosphor powder to adhere to the material. Capsulated color filters have been installed on the entire surface of the PDP to improve color reproducibility.

Fujitsu was one of the pioneers in the development of color PDPs. This company developed and produces the surface-discharge color A.C.-PDP panel with a common electrode structure; it employs a pair of parallel transparent (tin oxide) display electrodes on the front (viewing side) glass plate. These electrodes are coated with a dielectric film of lead oxide glass, followed by a thin layer of magnesium oxide. Meanwhile, the back glass plate has metallic silver address electrodes coated with the red, green, and blue phosphors, each separated by a barrier rib made of lead oxide glass. The barrier ribs keep the gas discharge confined to each red, green, and blue cell so that no crosstalk occurs. After the two glass plates are sealed together, the cells are filled with a neon/xenon gas mixture. The structure is shown in Figure 5.

The surface-discharge color A.C.-PDP works in the following way. When a voltage is applied between the electrodes, a surface-discharge is generated on the magnesium oxide layer, which results in the generation of ultraviolet light from the plasma. When this light strikes the phosphor layer in the cell enclosed by the barrier ribs, the atoms in the phosphor are transformed into an unstable condition known as the excited state. They remain in this state for only about one billionth of a second before returning to their original form known as the ground state. As they return to the ground state, the energy that had been absorbed during the excitation process is released in the form of light energy. This light, which is visible to the naked eye, passes through the front plate to the viewer. By controlling the intensity of the light emitted from the three primary color cells, all the colors of the spectrum can, in principle, be produced.

These color A.C.-PDPs have a reduced load to capacitance between adjacent cells so the pixel pitch can be shortened. Optical crosstalk is avoided by using a dielectric barrier rib of the appropriate height. The common electrode configuration provides a simpler row and column electrode arrangement for using the thick-film-through-hole method of printing. Thus, this panel is relatively easy to make in large sizes because of its simple structure, which is entirely fabricated by screen-printing technology. The key steps in the process are as follows:

(1) Pair of electrodes on front plate are coated with dielectric lead oxide glass and magnesium oxide.
(2) Silver electrode on back plate is coated with color phosphor.
(3) Barrier ribs keep discharge confined to each primary color.
(4) UV light is generated from surface discharge.
(5) UV light strikes color phosphor to produce colored light.

The use of three primary color phosphors (red, green, and blue) gives all colors of the spectrum, with appropriate scanning of the address and surface-discharge electrodes.

The most recent development in PDP design is Fujitsu's alternate lighting of surfaces, or ALiS, which uses an interlace approach to driving the PDP pixels, effectively doubling the number of scan lines for the same number of electrodes. Rather than using dedicated pairs of electrodes to address a scan line, ALiS uses each adjacent pair of electrodes, scanned in interlace format.

For video and noncomputer applications, pixel size increases for a PDP to dimensions compatible with screen-printing, especially as the screen measurements increase above 36 inches. The pixel size for a 36-inch-wide picture with 640 horizontal pixels is about 1 millimeter. The key is that the screen is not meant to be viewed from a distance of 2 feet (0.61 meter), but rather from 8 to 12 feet (2.44 to 3.66 meters).

Photometric issues are quite detailed. Brightness is determined by the luminance of the display, which is the measure of luminous intensity. A typical desktop CRT monitor operates at a level of about 150 nits (1 nit = 1 cd/m^2). This is bright enough to provide a pleasant-looking display in a typical office environment. It would not be bright enough for viewing in bright sunlight, nor would a consumer be happy with this luminance level on a home television set, which typically has a luminance of at least three times this value.

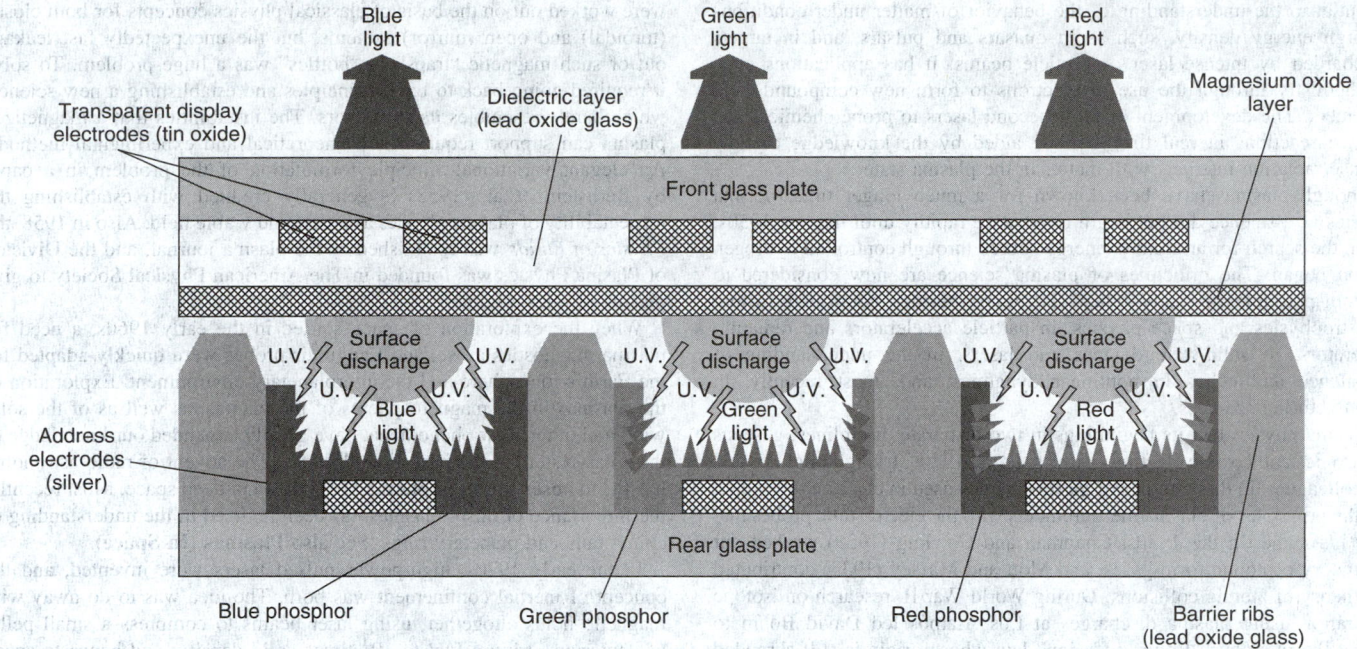

Fig. 5.   Structure of Fujitsu's color PDP using surface discharge (*Courtesy of Fujitsu Microelectronics.*)

The contrast ratio is a comparison of the luminance of a pixel in the on state to one in the *off* state; the higher the contrast ratio, the better. The current 42-inch (107 centimeter) PDP specifications refer to contrast ratio as 400:1 peak under dark room conditions. A 50-inch (127 centimeter) PDP with a contrast of 300:1 for dark areas and 120:1 for the same display in a bright area has been announced. Generally, references to contrast by specifications must be interpreted as referring to contrast in dark settings unless otherwise specified.

The current contrast is adequate for use in the home television market, but contrast improvement needs to be made for use in high ambient light conditions. Contrast could be improved first by increasing emission intensity from the display discharges and then by minimizing background light by either reducing emissions from control discharges or by hiding auxiliary discharge emissions. The driving method has been improved the frequency of the pilot discharge and background emissions has been cut, and the black display luminance has been lowered. Another way to improve contrast would be to use a neutral density filter or a circular polarizing filter that would reduce ambient light. These filters also reduce output luminance, requiring a higher phosphor output. However, by selecting a red-transmitting filter in front of a red cell and a green-transmitting filter in front of a green cell, this effect is reduced by a factor of three.

The saturation of light output and reduction of efficiency as the discharge current is increased are significant issues with PDPs. These occur because neutral Xe atoms easily absorb photons radiated from excited Xe atoms. An increase in the number of photons and the frequency of the sustain pulse reduces efficiency.

Increasing resolution in a display of the same size means reducing the size of each discharge cell. This in turn increases diffusion losses of charged particles to the cell walls. To compensate for this loss, increased ionization is needed. In the normal glow discharge mode, only one in forty electrons is effective for ionization. Because there are many unnecessary excitation levels, the excitation of atoms to a desired level is already a loss-provoking process. Adjustments to electron temperatures in plasmas by a mode different than the glow discharge method allow more efficient ionization and excitation of gases. If size can be increased, these issues will be resolved. See also **Television (TV)**, **Computer Graphics**, and the family of articles catalogued under **Flat Panel Display Technology**.

For additional reading, refer to Flat Panel Display Technology entry.

Stanford Resources, Inc., San Jose, CA

PLASMA FREQUENCY.   The oscillation frequency of plasma electrons about an equilibrium charge distribution is called the plasma or Langmuir frequency and is

$$\omega_p = \sqrt{\frac{4\pi n_e q^2}{m}} = 5.7 \times 10^5 n_e^{1/2} \, \text{(radians/sec)}$$

where n_e is the electron density, m is the electron mass, and q is the electronic charge in esu. Its value is

$$(5.7 \times 10^{-5} \, \text{radian cm}^{-3/2} \, \text{sec}^{-1}) n_e^{1/2}$$

PLASMA OSCILLATIONS.   A plasma is capable of supporting various modes of vibration and can propagate several types of waves. These can be classified in terms of their oscillation frequencies. The study of plasma oscillations is the general branch in plasma physics that includes the classification of the various modes of vibration that a plasma is capable of supporting and of the several types of waves that can be propagated in a plasma. Usually these collective effects are classified by their oscillation frequency.

PLASMA (Particle).   1. An assembly of ions, electrons, neutral atoms and molecules in which the motion of the particles is dominated by electromagnetic interactions. This condition occurs when the macroscopic electrostatic shielding distance (Debye length) is small compared to the dimensions of the plasma. Because of the large electrostatic potentials which would result from an inhomogeneous distribution of unlike charges, a plasma is effectively neutral. Thus there are equal numbers of positive and negative charges in every macroscopic volume of a plasma. Also, because a plasma is a conductor, it interacts with electromagnetic fields. The study of these interactions is called *hydromagnetics* or *magnetohydrodynamics*. See also **Magnetohydrodynamic Generator**.

2. A collection of electrons and ions, usually at a high enough temperature so that the ionization level is about 5% and at densities such that the Debye shielding distance is much smaller than the macroscopic dimensions of the system. See also **Fusion Power**.

PLASMA PHYSICS.   Within the framework of physics and engineering, plasma physics is an interdisciplinary science that affects many other fields. It is an extension of electromagnetics that can treat conducting gases that exist in stellar interiors and in nuclear-fusion devices constructed to reproduce the conditions there. It describes the motion of charge carriers in semiconductor devices and in the machines used to make them. It is used in the design of gas lasers and microwave generators. Plasma physics is essential to the understanding of the behavior of matter under conditions of high energy density, such as in quasars and pulsars, and in targets bombarded by intense laser or particle beams. It has applications even to chemistry through the use of electrons to form new compounds and polymers. The development of femtosecond lasers to probe chemical and surface reactions in real time will be aided by the knowledge of how intense radiation interacts with matter in the plasma state.

Though plasmas have been known for a much longer time, plasma physics as a science did not begin to advance rapidly until the late 1950s, when the search for an infinite energy source through controlled hydrogen fusion began. The principles of plasma science are now considered to be indispensable in many other disciplines besides controlled fusion: in astrophysics and space physics, in particle accelerators and radiation generators, in solid-state devices and lasers, in the understanding of turbulence and chaos in nonlinear dynamics, and, most recently, in manufacturing sciences.

Plasma physics had its beginnings in two disparate disciplines: gaseous electronics and cosmic electrodynamics. In the 1920s, Irving Langmuir and his colleagues, in their studies of gas discharges used in electron tubes, laid out the principles of plasma measurements with the electrostatic probes that bear his name. In the 1930s, Chapman and Cowling (1939) worked out the theory of nonuniform gases, and Mott and Massey (1933) contributed the theory of atomic collisions. During World War II, research on isotope separation using plasma discharges at Los Alamos led David Bohm to invent the infamous "Bohm diffusion" law whose origin is still shrouded in mystery. Gaseous electronics took a leap after the war as electrical engineers at Massachusetts Institute of Technology and Stanford University competed to understand beam–plasma interactions and to develop better klystrons and other microwave generators for radar. The theory of radio-wave propagation in the ionosphere was worked out simultaneously by V. L. Ginzburg (1964) and others in Russia. Experiments were still limited to Langmuir-probe measurements in weakly ionized discharges. It was also in the late 1950s and early 1960s that the other stem of plasma physics began its growth. H. Alfvén (1950), S. Chandrasekhar (1961), and other astrophysicists became aware of the role of plasmas in deep space and worked out the principles of magnetohydrodynamics, now known as hydromagnetics, including the discovery of "Alfvén waves." Astronomers tended to view plasmas as conductive fluids like mercury, whereas the gaseous electronics community treated plasmas from the particle point of view. These opposing views of the dual nature of plasmas were not merged until the era of nuclear-fusion research.

Following World War II, dense, hot plasmas were produced by the "pinch" effect at the Harwell military laboratories in England; these experiments foreshadowed the scramble for fusion power. Attempts to produce thermonuclear reactions in the laboratory and thus to "tame the H-bomb" were started in England, in Russia, and at four sites in the United States: at the national laboratories at Livermore, Los Alamos, and Oak Ridge, and at Princeton University. All of these programs were classified. The project at Princeton was code-named Project Matterhorn by its mountain-climbing leader, astrophysicist Lyman Spitzer, Jr. See also http://library.princeton.edu/about/news/pm121605.php. The leader of the Los Alamos Project was James Tuck, and this connection to Friar Tuck of Robin Hood fame led to the code name "Project Sherwood" for the U.S. fusion program. See also http://www.ornl.gov/info/ornlreview/rev25-34/chapter4.shtml. Classification was removed and international cooperation began at the 1958 Atoms for Peace conference in Geneva, where each nation displayed its best results. The United States unveiled two mirror-confinement programs (at Livermore and Oak Ridge), a pinch program at Los Alamos, and a "Stellarator" program at Princeton. The United Kingdom exhibited a large toroidal pinch called "Zeta," which produced copious neutrons, but which was later discredited when the neutrons were found to be of nonthermonuclear origin. The Russians exhibited an unglamorous piece of hardware that they dubbed a "tokamak," a Russian acronym for "toroidal magnetic chamber." This concept not only survived but has become the dominant one by far.

Spurred by the race for controlled fusion, plasma physicists progressed rapidly on both theoretical and experimental fronts. To generate power from fusing heavy hydrogen into helium requires heating a plasma to over 10^8 K and confining it for longer than 1 s. The principles of confining plasmas with the magnetic walls needed to withstand such temperatures were worked out on the basis of classical physics concepts for both closed (toroidal) and open (mirror) systems, but the unexpectedly fast leakage out of such magnetic "traps" or "bottles" was a huge problem. To solve it required going back to basic principles and establishing a new science, which barely resembles its precursors. The instabilities that a magnetized plasma can support required new theoretical and experimental methods. An elegant variational-principle formulation of the problem in a paper by Bernstein et al. (1958) is generally credited with establishing the respectability of plasma physics as a new and viable field. Also in 1958, the *Physics of Fluids* was established as the plasma journal, and the Division of Plasma Physics was founded in The American Physical Society to give plasma physics a home.

When the exploration of space started in the early 1960s, a need for plasma diagnostics arose, and Langmuir probes were quickly adapted for the Earth's ionosphere and the interplanetary environment. Exploration of the plasmas in the magnetospheres of the planets, as well as of the solar wind and other solar phenomena, have greatly expanded our knowledge of the solar system in the past three decades. The advent of radio astronomy has led to observations of plasma phenomena in deep space. Most recently, the importance of dusty plasmas has been realized in the understanding of comet tails and planetary rings. See also **Plasmas (In Space)**.

In the early 1970s, high-power pulsed lasers were invented, and the concept of inertial confinement was born. The idea was to do away with magnetic fields altogether, using laser beams to compress a small pellet of deuterium–tritium fuel to 10^4 times solid density, and hence to speed up the fusion reaction time to the order of picoseconds. In that case, the inertia of the imploding pellet would confine the fuel long enough for fusion to occur. Unfortunately, the unexpected effects of plasma physics could not be avoided. When intense laser light strikes a solid object, the surface material is immediately ionized into a plasma cloud. A new type of instability, called a parametric instability, was driven by the laser beam; and this had the effect of reflecting the laser light before it could reach the solid surface of the pellet. Inertial confinement spawned a new branch of plasma physics, which required extensive development of nonlinear and computational techniques. Simultaneously, it was found that intense electron and proton beams of order 1 MA at 1 MV could be produced by storing electrical energy in capacitor banks and discharging it into a diode in a short pulse. Pulsed power became yet another branch of plasma physics, permitting the study of matter subject to terawatts of power, and leading to the invention of new plasma switches and x-ray generators.

Until recently, most plasma-physics research fell into one of the four areas described above: magnetic fusion, inertial fusion, space plasma physics, and pulsed power. In the 1990s, the use of partially ionized plasmas in manufacturing began to expand rapidly, along with the prevalent use of semiconductor ULSI (ultralarge-scale integrated) circuits and personal computers. Plasma etching and deposition are essential in the fabrication of computer chips with millions of internal features. It is anticipated that the production of flat-panel displays will ultimately also depend on plasma processing. Surface hardening of turbine blades, tools, construction materials, medical prostheses, and so forth is already a large commercial application of plasma physics. Plasma treatment of plastics and fibrous materials can be used, for instance, to make plastic containers more leakproof; to make paint adhere to plastic bags, clothing, currency, and automobile parts; to make fabrics absorb or repel water; and to tailor the refractive indices of optical fibers. In summary, though the original impetus for developing plasma physics was the lofty goal of an inexhaustible energy source for mankind, fusion power plants will not be commercially viable for several more decades. In the near future, however, plasma physics is likely to have a large impact on our daily lives because of its utility in the manufacture of a diverse variety of common materials. See also **Plasma**; **Plasma Display Panels**; **Plasmas (In Space)**.

Additional Reading

Alfvén, H.: *Cosmical Electrodynamics*, Clarendon Press, Oxford, UK, 1950.

Bellan, P.M.: *Fundamentals of Plasma Physics*, Cambridge University Press, New York, NY, 2006.

Bittencourt, J.A.: *Fundamentals of Plasma Physics*, 3rd Edition, Springer-Verbal New York, LAC, New York, NY, 2004.

Chandrasekhar, S.: *Hydrodynamic and Hydromagnetic Stability*, Clarendon Press, Oxford, UK, 1961, 1981.

Chapman, S., and T.G. Cowling: *The Mathematical Theory of Non-uniform Gases*, 1st Edition, Cambridge University Press, Cambridge, UK, 1939.

Chapman, S., and T.G. Cowling: *The Mathematical Theory of Non-uniform Gases*, 3rd Edition, Cambridge University Press, New York, NY, 1991.

Freidberg, J.: *Plasma Physics and Fusion Energy*, Cambridge University Press, New York, NY, 2007.

Ginzburg, V.L.: *Propagation of Elecromagnetic Waves in Plasmas*, Pergamon Press, Oxford, UK, 1964.

Mott, N.F., and Massey, H.S.W.: *Theory of Atomic Collisions*, Clarendon Press, Oxford, UK, 1933.

Miyamoto, K.: *Plasma Physics and Controlled Nuclear Fusion*, Springer-Verbal New York, LAC, New York, NY, 2005.

Miyamoto, K.: *Controlled Fusion and Plasma Physics*, Taylor & Francis, Inc., Philadelphia, PA, 2006.

Somov, B.V.: *Fundamentals of Cosmic Electrodynamics*, Springer-Verbal New York, LAC, New York, NY, 1994.

Spitzer, L.: *Physics of Fully Ionized Gases*, 2nd Edition, Dover Publications, Mineola, NY, 2006.

Stacey, W.M.: *Fusion Plasma Physics*, John Wiley & Sons, Inc., Hoboken, NJ, 2005.

Sutton, G.W., and A. Sherman: *Engineering Magnetohydrodynamics*, Dover Publications, Mineola, NY, 2006.

FRANCIS F. CHEN, University of California, Los Angeles, CA

PLASMATRON. A continuously controllable gas-discharge tube which utilizes an independently generated gas-discharge plasma as a conductor between a hot cathode and an anode. Continuous modulation of the anode current can be effected by variation either of the conductivity or the effective cross-section of the plasma. The first method is based upon the modulation of the electronionizing beam which controls the plasma density, and hence its conductivity. The second method makes use of the gating action of positive-ion sheaths which surround the wires of a grid, located between the anode and cathode.

PLASMID. A strand or fragment of genetic material existing outside the chromosomes in certain types of bacteria. R-type plasmids, which are present in *E. coli*, impart resistance to antibiotics in organisms that are exposed to them. The plasmids can be transferred from animals to humans, as well as to other, harmful bacteria that also become resistant to antibiotics. Feeding of traces of antibiotics to animals is believed to promote the growth of E. *coli* and, thus, to produce strains of pathogenic bacteria that are not amenable to antibiotic treatment. For this reason FDA has recommended elimination of certain antibiotics from animal feeds, e.g., penicillin, oxytetracycline, and chlortetracycline. Synthetic plasmids have been used successfully in recombinant DNA research.

PLASTER. See **Gypsum**.

PLASTIC DEFORMATION. When a metal or other solid is plastically deformed it suffers a permanent change of shape. The theory of plastic deformation in crystalline solids such as metals is complicated but well advanced. Metals are unique among solids in their ability to undergo severe plastic deformation. The observed yield stresses of single crystals are often 10^{-4} times smaller than the theoretical strengths of perfect crystals. The fact that actual metal crystals are so easily deformed has been attributed to the presence of lattice defects inside the crystals. The most important type of defect is the dislocation. See also **Creep (Metals)**; **Crystal**; **Hot Working**; and **Impact**.

PLASTICITY. A rheological property of solid or semisolid materials expressed as the degree to which they will flow or deform under applied stress and retain the shape so induced, either permanently of for a definite time interval. It may be considered the reverse of elasticity. Application of heat and/or special additives is usually required for optimum results.

See also **Plasticizers**; and **Thermoplastic**.

PLASTIC PIPE. Tubes, cylinders, conduits, and continuous length piping made (1) from thermoplastic polymers unreinforced (polyethylene, polyvinyl chloride, ABS polymers, polypropylene) or (2) from thermosetting polymers (polyesters, phenolics, epoxies) blended with 60–80% of such reinforcing materials as chopped asbestos or glass fibers to increase strength. The latter type is a reinforced plastic. In general the properties of plastic tubing or pipe are those of the polymers that comprise it. Most have good resistance to chemicals, corrosion, weathering, etc., combined with flexibility, light weight, and high strength. They are combustible but generally slow burning. The reinforced type is widely used as underground conduit for transportation of gases and fluids, including city water services, sewage disposal systems, etc. Its use in buildings is subject to local building codes. For additional information contact Plastic Pipe Institute, 1825 Connecticut Ave., NW, Suite 680, Washington, DC 20009. http://www.plasticpipe.org/applications/productinfo03_1_1.php

PLASTICS. These are materials formed from resins through the application of heat, pressure, or both. Most starting materials prior to the final fabrication of plastic products exhibit more or less plasticity—hence the term plastic. However, the great majority of plastic end-products are quite *nonplastic*, i.e., they are nonflowing, relatively stable dimensionally, and are hard. There are scores of different kinds of plastics. They fall into two broad categories: (1) *thermoplastic resins*, which can be heated and softened innumerable times without suffering any basic alteration in characteristics; and (2) *thermosetting resins*, which once set at a temperature critical to a given material cannot be resoftened and reworked. Since most plastic fabrication methods, such as casting, molding, or extruding, involve heat, the thermosetting materials must be properly and accurately formed during any thermal cycling that exceeds the critical temperature.

The principal kinds of thermoplastic resins include: (1) acrylonitrile-butadiene-styrene (ABS) resins; (2) acetals; (3) acrylics; (4) cellulosics; (5) chlorinated polyethers; (6) fluorocarbons, such as polytetra-fluorethylene (TFE), polychlorotrifluoroethylene (CTFE), and fluorinated ethylene propylene (FEP); (7) nylons (polyamides); (8) polycarbonates; (9) poly ethylenes (including copolymers); (10) polypropylenes (including copolymers); (11) polystyrenes; and (12) vinyls (polyvinyl chloride). The principal kinds of thermosetting resins include: (1) alkyds; (2) allylics; (3) the aminos (melamine and urea); (4) epoxies; (5) phenolics; (6) polyesters; (7) silicones; and (8) urethanes.

The development of electrically conductive plastics is reviewed by R.B. Kaner and A.G. MacDiarmid (*Sci. Amer.*, 106–111, February 1988).

Numerous plastics are described in terms of manufacture and chemical and physical properties throughout this volume. Consult the alphabetical index.

Additional Reading

Elias, Hans-Georg: *An Introduction to Plastics*, 2nd Edition, John Wiley & Sons, Inc., New York, NY, 2003.

Harper, C.A., and E.M. Petrie: *Plastics Materials and Processes: A Concise Encyclopedia*, John Wiley & Sons, Inc., New York, NY, 2003.

Richardson, T.L., and E. Lokensgard: *Industrial Plastics: Theory and Application*, Delmar Learning, Albany, NY, 2003.

Rosato, D.V.: *Plastic Product Material and Process Selection Handbook*, Elsevier Science, New York, NY, 2004.

Wright, R.E.: *Product Design for Thermosetting Plastics*, CRC Press LLC., Boca Raton, FL, 2004.

Web References

Society of Plastics Engineers—SPE: http://www.4spe.org.
Plastics Technology Online: http://www.plasticstechnology.com.

PLASTIDS. Pigments in plants are often located in special bodies called plastids. There are many kinds of plastids: leucoplasts, those which contain no pigment and which are therefore colorless; chloroplasts, those which contain chlorophyll (by far the commonest kind); and chromoplasts, colored plastids that do not contain chlorophyll.

Leucoplasts occur in parts of stems and roots where light fails to penetrate. They absorb glucose and change it to starch.

Chloroplasts occur in cells exposed to light. They are indispensable to photosynthesis. In the algae the shapes of these bodies are many; in a large number of cases the plastid is a thick cup-shaped body occupying the greater part of the volume of the cell; in other algae the plastids have a central mass from which radiating plates of arms extend outward to the cell wall; spiral, net-shaped and ring-shaped plastids are not uncommon in this group of plants. In some algae and in nearly all higher plants, the chloroplasts are small subspherical or lens-shaped bodies, varying in number from one to many in a single cell. Always the chloroplasts are found embedded in the cytoplasm of the cell. In many plants the continuous movement of the cytoplasm in the cell carries the plastids along with it; in others these bodies have a fixed position. In certain algae and in many cells in higher plants, as for example in the palisade layer of leaves, the chloroplasts may change their position so that they will receive the most favorable amount of light. If the light intensity is low, they will present their flat surface to it; whereas if the light intensity is high, the plastid rotates so that it is placed edgewise to the light. Chloroplasts contain chlorophyll and other pigments. See also **Pigmentation (Plants)**.

PLATEAU.   An elevated, relatively flat area or surface of wide extent and underlain by relatively horizontal sedimentary formations. Less correctly used to describe any relatively flat high-level surface of erosion regardless of the structure of the region.

PLATELETS.   Platelets are small (1–2 μm in diameter), anucleate, disc-shaped blood cells derived from bone marrow megakaryocytes which serve several critical roles in normal hemostasis and participate in both health and disease. Both a normal number of platelets and the presence of normally functioning platelets are required to maintain the integrity of the human blood coagulation and vascular systems. Under normal circumstances, platelets circulate in the bloodstream in a quiescent manner at a concentration of 150 000–400 000 per μL, but are always prepared to respond immediately to any disruption of the vascular endothelial lining.

When indicated, platelets attach to sites of blood vessel injury (adhesion), release factors that recruit and activate additional platelets (activation), form a temporary platelet plug (aggregation), and support the generation of fibrin by the coagulation cascade of proteins (Fig. 1). The platelet plug serves to concentrate plasma coagulation proteins at the site of vascular injury, thus supporting the formation of a fibrin mesh that reinforces the platelet plug. The fibrin-reinforced and stabilized platelet plug, known as a thrombus, serves to prevent unabated blood loss from the circulation, pending vessel wall repair. See also **Blood**.

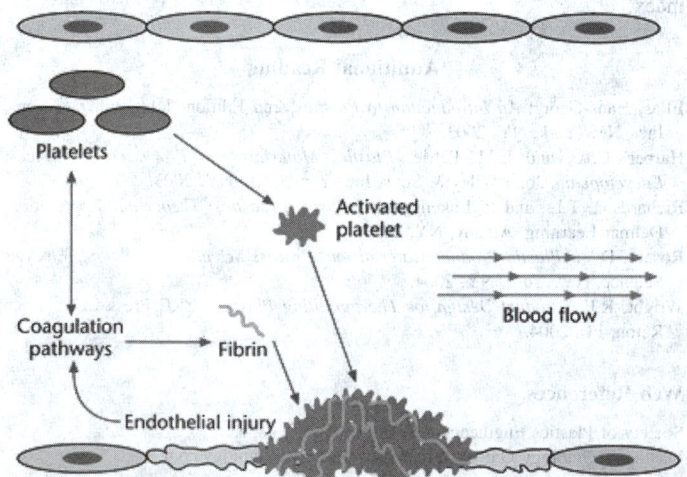

Fig. 1.   Participation of platelets in clot formation following blood vessel injury. Platelets become activated and accumulate at the site of vascular endothelial injury. This aggregation of activated platelets (platelet plug) stops the escape of blood from the circulation and supports the formation of fibrin which stabilizes the platelet plug.

When platelets are deficient in number (thrombocytopenia) or less than fully functional (thrombocytopathia), easy bruising and excessive bleeding, especially from the oral, nasal and genitourinary tract mucosa, can develop. Severe quantitative and qualitative platelet defects can precipitate life-threatening bleeding. Alternatively, excessive platelet adhesion, activation

or aggregation can result in "sticky" platelets, pathological thrombosis and an exacerbation of preexistent atherosclerotic vascular lesions. An understanding of platelet function is essential to the study of blood coagulation, in general, and of vascular diseases such as stroke (brain attack) and myocardial infarction (heart attack), in particular. See also **Blood Coagulation**.

Formation

Circulating platelets are generated from the fragmentation of giant progenitor cells located in the bone marrow, called megakaryocytes. Megakaryocytes can first be detected in the bone marrow of a developing fetus at 3 months and are widely evident in the marrow at birth. The time required for human megakaryocytes to mature is approximately 5 days; the survival time of platelets in the circulation following fragmentation from a megakaryocyte is 7–10 days. Circulating platelets are a heterogeneous population of megakaryocyte cytoplasmic fragments with volumes ranging from 5 to 12 μm^3. It is perceived that larger platelets represent younger platelets, whereas smaller platelets represent fragments of large platelets or older platelets. The hematopoietic growth factor called thrombopoietin, regulates platelet production, as well as megakaryocyte differentiation, and appears to be produced in response to diminished total body platelet mass or to excessive platelet utilization.

Structure

The megakaryocyte possesses a demarcation membrane system which guides cytoplasmic fragmentation and ultimately serves as the lipid bilayer, plasma membrane for newly formed platelets. The platelet plasma membrane contains a surface-connected open canalicular system (OCS) and a dense tubular system (DTS). The OCS consists of a series of tortuous plasma membrane invaginations that tunnel throughout the cytoplasm in a serpentine fashion. The OCS increases the total surface area of the platelet that is exposed to the surrounding plasma environment and provides a route for chemical and particulate substances to reach deep into the recesses of the cell and for intracellular granule contents to be more efficiently extruded during the release reaction. The DTS has a close physical relationship with the OCS (Figure 2**a**). The DTS may serve as a calcium reservoir enabling platelet activation to occur without an external source of calcium. Platelets also contain a peripheral microtubule coil which serves to maintain the discoid shape during quiescence. Platelets display class I human leucocyte antigen (HLA) antigens of both A and B type; HLA-C antigens are expressed weakly and class II HLA antigens not at all.

Integral to platelet function are the membrane receptors which, when bound to soluble or immobilized ligands, mediate platelet adhesion, activation and aggregation. Platelets do not adhere to normal vascular endothelial cells which form an anticoagulant barrier between the flowing blood and extracellular matrix in intact blood vessels. Platelets do, though, constitutively express two surface membrane receptors, glycoprotein (GP) Ib-IX and the integrin GPIa-IIa, which are poised to bind with the vascular extracellular matrix proteins von Willebrand factor (VWF) and collagen, respectively, in order to form a bridge between platelets and sites of endothelial damage. The interactions between VWF and GPIb-IX promote a change in platelet morphology and induce pseudopod generation, which together promote platelet aggregation and clot retraction at sites of vascular injury. Occupancy of GPIa-IIa by extracellular matrix collagen initiates a sequence of events which results in platelet activation. Once activated, platelets are capable of binding to immobilized VWF and soluble fibrinogen via the integrin GPIIb-IIIa membrane complex. GPIIb-IIIa is only expressed as a functional membrane binding site on activated platelets. Soluble fibrinogen facilitates platelet aggregation by serving as a bridge capable of linking two activated platelets via their GPIIb-IIIa receptors.

The biological importance of the platelet membrane glycoprotein receptors is illustrated by the occurrence of a clinical bleeding syndrome when a receptor is deficient or defective. Patients with congenital GPIb-IX deficiency (Bernard-Soulier disease), GPIa-IIa deficiency or congenital GPIIb-IIIa deficiency (Glanzmann thrombasthenia) have a bleeding tendency. Therapeutic agents designed to impair platelet function via glycoprotein blockade have been produced and are undergoing rigorous clinical testing. The integral role of platelets in the pathogenesis of cardiovascular,

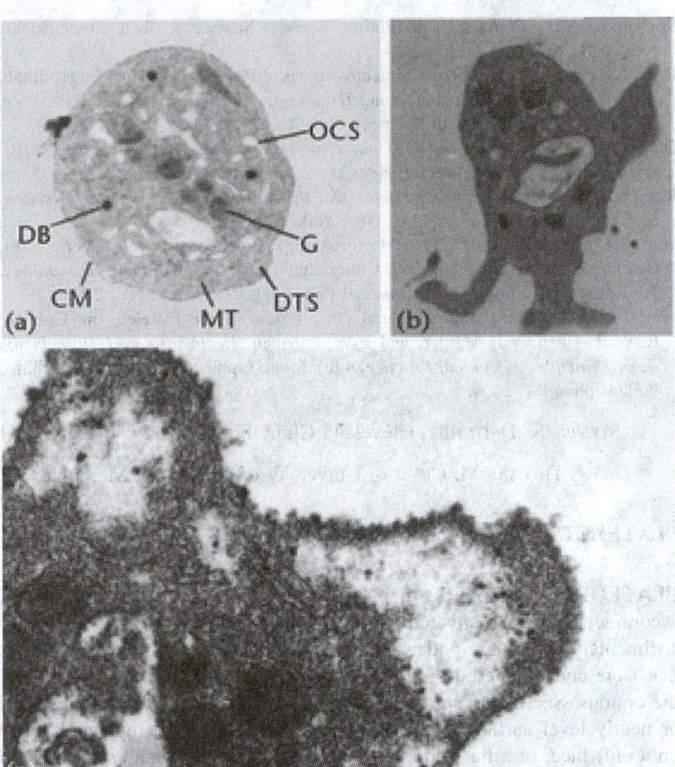

Fig. 2. Electronmicrograph of a quiescent platelet (magnification × 30 000) showing platelet constituents. DB, dense body; G, α granule; MT, mitochondria; CM, cell membrane; OCS, open canalicular system; DTS, dense tubular system. (b) An activated platelet demonstrating the collagen-induced shape changes (magnification × 30 000). (c) Collagen-induced platelet aggregation (magnification × 30 000).

cerebrovascular and peripheral vascular diseases makes platelet research essential and rewarding.

In contrast to megakaryocytes, platelets have no nucleus or deoxyribonucleic acid and cannot synthesize proteins. The platelet cytoplasm is relatively organelle poor, but does contain mitochondria, microtubules, glycolytic enzymes, actin, myosin and three different types of granules: lysosomes, dense granules (dense bodies) and α granules. Platelet lysosomes contain hydrolytic enzymes. Dense bodies contain adenosine triphosphate, adenosine diphosphate (ADP), calcium and serotonin. The α granules contain β-thromboglobulin (β-TG), platelet factor 4 (PF4), platelet-derived growth factor (PDGF), epidermal growth factor, transforming growth factor β, coagulation factor V-Va, VWF, fibrinogen and fibronectin. PF4 is a positively charged glycoprotein capable of binding to heparin and heparan sulfate. PDGF is a glycoprotein that promotes replication of smooth muscle cells and fibroblasts. Factor V-Va is an important cofactor for thrombin generation. The measurement of plasma PF4 and β-TG in patients can be used to help quantify the level of platelet activation and granule release. Platelets are capable of enveloping and internalizing (endocytosis) proteins from the plasma for incorporation into α granules. The release of α-granule contents leads to membrane surface expression of P-selectin, an integrin that promotes adhesion of platelets to endothelial cells, neutrophils and monocytes.

Activation

When exposed to sufficient concentrations of stimulatory substances known as agonists, platelets can become activated. The platelet agonists of greatest physiological relevance include ADP, thrombin, adrenaline (epinephrine), arachidonic acid and collagen. Platelet activation requires a specific receptor–ligand interaction whereby a certain platelet receptor must bind to its particular agonist ligand. Some platelet receptors mediate responses through guanine nucleotide-binding proteins (G proteins), which act as coupling proteins between the cell surface receptors and effector targets or ion channels. Stimulatory agonists activate phospholipase C, generating

inositol phosphates and diacylglycerol which serve as messengers in subsequent signal transduction, leading to a cascade of intracellular events resulting in platelet aggregation or secretion. Activated platelets change their shape from discoid to spiny spheres by polymerizing actin into filopodia and stress fibers with a resultant 60% increase in effective membrane surface area (Figure 2b). This shape change facilitates platelet–fibrinogen and platelet–platelet interactions that support platelet aggregation (Figure 2c). Adrenaline is the only stimulatory platelet agonist that does not induce a cell shape change.

As a result of agonist-induced platelet activation, platelets centralize and then release the contents of their dense bodies, α granules and lysosomes. Platelet activation results in the fusion of granules with the plasma membrane in an ADP-dependent process which facilitates granule content release. The release of dense body contents results in the recruitment and activation of additional platelets required to form a platelet plug. In addition to receptor–agonist-mediated platelet activation, the exposure of resting platelets to high shear stress can induce activation and aggregation. Because blood flow in arteries narrowed by atherosclerosis is characterized by high shear stress, platelets often become activated leading to platelet-rich thrombus formation, artery occlusion and tissue injury. The importance of platelet activation and granule release-associated processes is illustrated by the fact that individuals with congenital or acquired activation defects have a bleeding tendency. The most common cause of an acquired platelet activation defect is pharmacological treatment with aspirin. See also **Blood Clotting: General Pathway**.

Immune Disease

Platelets can be involved in immune reactions that lead to a reduction in the number of circulating platelets (thrombocytopenia) or platelet dysfunction, both of which can cause hemorrhagic disease. Agents that trigger these immune reactions include platelet autoantigens (components of a person's own cells), platelet alloantigens (components of another individual's cells), medications or infectious organisms.

The most common immune thrombocytopenia involves antibodies directed against a platelet autoantigen and is called *autoimmune idiopathic thrombocytopenic purpura* (ITP). ITP is characterized by thrombocytopenia, shortened platelet survival, and a compensatory increase in bone marrow megakaryocytes. In ITP, platelets become coated with antibody and are then cleared from the circulation by special cells in the spleen and liver. Acute ITP is more common in children and often preceded by a viral infection. The chronic form of ITP is more common in adults and may be associated with diseases of immune dysregulation such as systemic lupus erythematosus and lymphoproliferative disorders. At least 80% of patients with ITP have serum antibodies against one or more of the glycoproteins, including GPIb-IX, GPIIb-IIIa, and GPIa-IIa.

Thrombocytopenia caused by alloantibodies, although less common than ITP, can be equally life threatening. Ordinarily, there are no naturally occurring antibodies against platelet alloantigens. Antibodies develop only after exposure of an individual to foreign cells by transfusion or by fetal cells that enter the maternal circulation. Antiplatelet alloantibodies are a major cause of failure to respond to platelet transfusions, of neonatal alloimmune thrombocytopenia (NAT) due to the passive transfer of maternal antiplatelet antibodies across the placenta affecting fetal platelets, and posttransfusion purpura (PTP) in which sensitization of the transfusion recipient to a donor platelet antigen causes immune destruction of autologous platelets. The alloantibodies that cause refractoriness to platelet transfusion in multiply transfused patients, such as those with leukemia, are usually directed against HLA class I and II determinants found on the donor cells. The alloantibodies in NAT and PTP are directed against a part of GPIIb-IIIa known as P1[A1]. Anti-P1[A1] antibodies may promote hemorrhage in affected individuals by contributing to platelet destruction and interfering with GPIIb-IIIa receptor function.

Thrombocytopenia caused by drug-induced antibodies is most commonly observed in persons of advanced age, not only because they take more medications, but specifically because they are more likely to be taking known offending drugs such as antibiotics, quinidine, thiazide diuretics and heparin. The list of medications known to be able to precipitate thrombocytopenia is extensive and includes antituberculosis and antiseizure drugs. Drug-induced thrombocytopenic purpura involves a variety of different types of reaction between the offending drug, the antibody and the platelets. All three reactants must be present at the same time to produce thrombocytopenia since recovery is usually observed following drug discontinuation even if the antibody persists in the serum. Clinical manifestations of drug-induced thrombocytopenic purpura usually resolve promptly after discontinuation of the drug but may extend for several weeks under certain circumstances. Drugs or their metabolites elicit antibodies only when combined with a larger protein carrier to form an immunogenic hapten–carrier complex. Some drugs complex with a platelet membrane protein before being recognized by the antibody, whereas others may form a complex with the antibody before being adsorbed on to the platelet surface. It appears that GPIb-IX and/or the GPIIIa component of GPIIb-IIIa is required to support platelet attachment of most drug-associated antibodies. Bone marrow megakaryocytes are usually normal or increased in number in drug-induced thrombocytopenic purpura, probably because of limited availability of drug and antibody in the bone marrow compared with the circulation. Heparin-induced thrombocytopenia is of particular concern to physicians because this antibody reaction against the blood thinner heparin complexed with platelet αgranule-derived PF4 not only causes thrombocytopenia but, more importantly, can precipitate paradoxical arterial and venous thrombosis by activating platelets, damaging the vascular endothelium and promoting coagulation reactions.

Infection-associated thrombocytopenia, such as that seen in individuals infected with human immunodeficiency virus, is common and may include immune-mediated platelet destruction combined with a direct hypoproliferative effect on megakaryocytes. Thrombocytopenia associated with infection with viruses, bacteria, fungi and parasites may be due to one or more different mechanisms operating at the same time. Patients with severe infection may consume platelets at an accelerated rate as a result of systemic activation of coagulation-related processes called *disseminated intravascular coagulation*. At the same time, these seriously ill patients are likely to be receiving several antimicrobial medications which can promote drug induced thrombocytopenia. Other mechanisms of thrombocytopenia to be considered in the setting of infection include suppression of megakaryocyte function, damage to platelets by the infectious agent, and platelet destruction induced by antibodies against the organism itself.

Rare examples of autoantibodies that precipitate hemorrhage without causing a reduction in platelet number have been described. Antibodies against platelet GPIIb-IIIa or GPIb-IX have been demonstrated, in a handful of cases, solely to inhibit the aggregatory response to agonists or to block receptor binding of ligands. These patients with their acquired platelet defects present with a clinical picture similar to that of patients with congenital defects of the same glycoproteins. Recently, humanized antibodies targeted against GPIIb-IIIa have been developed and used to treat patients with vascular disease in whom inhibition of platelet function may prevent the development of life threatening platelet activation and thrombosis.

Platelets can also contribute to global immune system stimulation. Complement system proteins can be activated in platelet concentrates stored for transfusion. The terminal components of the complement cascade, C5a and C5b-9, but not C3a, can participate directly in platelet and neutrophil activation. Additionally, C5a and tumor necrosis factor α (TNF α) generated in apheresis platelet concentrates seem to induce white blood cell interleukin-8 production, which can lead to the stimulation of other immune system-related cells. Platelets interact with the mature white blood cell known as neutrophils as part of the inflammatory process observed in many systemic diseases. The interaction between platelets and neutrophils involves platelet-induced modulation of neutrophil functions in which platelets potentiate immunoglobulin G-mediated ingestion and the production of oxygen metabolites in neutrophils.

Additional Reading

Bussel, J., and D. Cines: "Immune Thrombocytopenic Purpura, Neonatal Alloimmune Thrombocytopenia, and Post-transfusion Purpura," In: Hoffman, R., E.J. Benz, S.J. Shattil, et al.: *Hematology: Basic Principles and Practice*, 2nd Edition, pp. 1849–1869, Churchill Livingstone, New York, NY, 1995.

Gewitz, A.M., and B. Schick: "Megakaryocytopoiesis," In: Colman, R.W., J. Hirsh, V.J. Marder, and E.W. Salzman: *Hemostasis and Thrombosis: Basic Principles of Clinical Practice*, 4th Edition, Lippincott Williams & Wilkins, Philadelphia, PA, 2000.

Gibbins, J., and M.P. Mahaut-Smith: *Platelets and Megakaryocytes*, Vol. 2, Springer-Verlag New York, LLC, New York, NY, 2004.

Michelson, A.D.: *Platelets*, 2nd Edition, Elsevier Science & Technology Books, New York, NY, 2006

Plow, E.F., and M.H. Ginsberg: "Molecular Basis of Platelet Function," In: Hoffman, R., E.J. Benz, L.E. Silberstein, et al.: *Hematology: Basic Principles and Practice*, 4th Edition, Elsevier Health Sciences, New York, NY, 2004.

Quinn, M., D. Fitzgerald, and D. Cox: *Platelet Function: Assessment, Diagnosis, and Treatment*, Springer-Verlag, New York, NY. LLC, New York, NY, 2005.

Ruggeri, Z.M.: *Von Willebrand Factor and the Mechanisms of Platelet Function*, Springer-Verlag New York, LLC, New York, NY, 1998.

Ware, J.A., and B.S. Coller: "Platelet Morphology, Biochemistry, and Function," In: Beutler, E., M.A. Lichtman, B.S. Coller, and T.J. Kipps: *Williams' Hematology*, 7th Edition, McGraw-Hill Professional Publishing, New York, NY, 2005.

White, J.G.: "Anatomy and Structural Organization of the Platelet," In: Colman, R.W., J. Hirsh, V.J. Marder, and E.W. Salzman: *Hemostasis and Thrombosis: Basic Principles of Clinical Practice*, 4th Edition, Lippincott Williams & Wilkins, Philadelphia, PA, 2000.

STEVEN R. DEITCHER, Cleveland Clinic Foundation, Cleveland, OH

THOMAS M. CHIANG, University of Tennessee, Memphis, TN

PLATELETS. See **Blood**.

PLATFORM (Geology). In plate tectonics, a platform is that part of a continent which is covered by flat-lying or gently tilted strata, mainly sedimentary, which are underlain at varying depth by a basement of rocks that were consolidated during earlier deformations. Platforms are parts of the cratons. See also **Craton**. In geomorphology, a platform is any level or nearly level surface, e.g., a terrace or bench, a ledge or small space on a cliff face, or a flat and elevated piece of ground, such as a tableland or plateau, a peneplain, or any beveled surface. The term also sometimes indicates a small plateau. In discussions of a coastline, a platform is a flat or gently sloping underwater erosional surface extending seaward or lakeward from the shore. Specific types include a *wave-cut platform* or an *abrasion platform*.

A *platform beach* is a looped bar or dike of sand and gravel formed on a wave-cut platform. These are found, for example, on Madeline Island along the Wisconsin shore of Lake Superior. A *platform reef* is an organic reef, generally small but more extensive than a patch reef, with a flat upper surface. Platform reefs are common off the coast of Australia; they are also called *table reef*.

See also **Ocean**.

PLATFORMING. The process in which octane ratings of gasoline are raised by dehydrogenating naphthenes to aromatics, cracking high-boiling paraffins, and isomerizing paraffins to form products of greater chain branching. Desulfrization also takes place in this process.

PLATINUM AND PLATINUM GROUP. [CAS: 7440-06-4] Chemical element, symbol Pt, at. no. 78, at. wt. 195.09, periodic table group 10 (formerly 8—transition metals), mp 1,772 °C, bp 3,727 to 3,927 °C, density 21.37 g/cm³ (solid), 21.5 g/cm³ (single crystal) (20 °C). Elemental platinum has a face-centered cubic crystal structure. The five stable isotopes of platinum are ^{192}Pt, ^{194}Pt through ^{196}Pt, and ^{198}Pt. The seven unstable isotopes are ^{188}Pt through ^{191}Pt, ^{193}Pt, ^{197}Pt, and ^{199}Pt. In terms of earthly abundance, platinum is one of the scarce elements. Also, in terms of cosmic abundance, the investigations by Harold C. Urey (1952), using a figure of 10,000 for silicon, estimated the figure for platinum at 0.016. No notable presence of platinum in seawater has been found. Electronic configuration

$$1s^2 2s^2 2p^6 3s^2 3p^6 3d^{10} 4s^2 4p^6 4d^{10} 4f^{14} 5s^2 5p^6 5d^9 6s^1$$

Ionic radius Pt^{2+} 0.52 Å. Metallic radius 1.3873 Å. First ionization potential 8.96 eV. Oxidation potentials $Pt \rightarrow Pt^{2+} + 2e^-$, ca. -1.2 V; $Pt + 2OH^- \rightarrow Pt(OH)_2 + 2e^-$, -0.16 V.

Platinum is one member of a family of six elements, called the *platinum metals*, which almost always occur together. Before the discovery of the sister elements, the term *platinum* was applied to an alloy with Pt as the dominant metal, a practice that persists to some degree even today. The major properties of the platinum metals are given in Table 1. See also **Iridium**; **Osmium**; **Palladium**; **Rhodium**; and **Ruthenium**.

TABLE 1. REPRESENTATIVE PROPERTIES OF PLATINUM GROUP METALS

Property	Iridium	Osmium	Palladium	Platinum	Rhodium	Ruthenium
Atomic volume, cm³/g-atom	8.54	8.43	8.88	9.09	8.27	8.29
Atomic radius	1.355	1.350	1.373	1.335	1.342	1.336
Crystalline form	fcc	hcp	fcc	fcc	fcc	hcp
Lattice parameters, Å, a	3.8389	2.7341	3.8902	3.9310	3.804	2.7041
b	—	4.3197	—	—	—	4.2814
Thermal conductivity at 20 °C, (cal)(cm)/(s)(cm³)(°C)	0.14	—	0.168	0.166	0.21	
Electrical resistivity at 0 °C, micro-ohm-cm	5.3	9.5	10.8	10.6	4.5	7.2
Thermal expansivity, °C × 10⁶ at 20 °C	6.6	6.6	12.4	9.0	8.3	9.6
Hardness, Mohs scale	6.5	7.0	4.8	4.3	—	6.5
Specific heat, cal/g-atom at 20 °C	0.031	0.031	0.0584	0.031	0.059	0.057
Heat of fusion, kcal/mole	6.3	7.0	4.0	4.7	5.2	6.1
Heat of vaporization, kcal/mole	134.7	150	90	122	118.4	135.7

fcc = face-centered cubic
hcp = hexagonal close-packed

Occurrence. These metals occur in both primary and secondary deposits. The primary deposits are generally associated with Ni-Cu sulfide ores. The Sudbury ores of Canada and the deposits of the Bushveld complex of South Africa are of this type. Native platinum occurs as a primary deposit in the Ural Mountains of the former U.S.S.R. and also in the Choco district of Colombia. Weathering and erosion of these deposits have resulted in the formation of secondary, or placer, deposits of native Pt in riverbeds and streams. One nugget of Pt found in the Urals weighed over 25 pounds (11.3 kilograms). Most of the world's platinum comes from Canada, the former U.S.S.R., and South Africa. Minor amounts have been found in Alaska, Colombia, Ethiopia, Japan, Australia, and Sierra Leone.

Because of their unique properties and in spite of their high initial cost, the platinum metals find many applications in industry. Since used platinum metals retain a large portion of their initial value, many scrap materials are a major source of recoverable platinum metals. Practically every application of platinum generates scrap in some form, which is eventually returned to the platinum refiner for recycling. Although there are ample mine reserves, they soon would be depleted without constant scrap recycling.

Refining Processes. The refining procedures are a good introduction to the complex chemistry of the platinum metals. Some of these methods are still the best analytical techniques available for the separation of the metals. South African ore is smelted to form a Cu-Ni matte containing small amounts of the platinum metals (0.18%). The matte is melted, cast into anodes, and electrolytically dissolved. The contained Cu is deposited at the cathode, the Ni remains in the H_2SO_4 electrolyte, and the Pt metals are contained in the anode slimes. The resulting Cu is refined and the $NiSO_4$ solution purified and crystallized. The anode slimes are treated by roasting to remove sulfur and leached with dilute H_2SO_4 and air to remove Cu and Ni. The leached slimes are treated with aqua regia. The aqua regia solution is evaporated to concentrate the solutions and expel the excess HNO_3. The residue from this treatment contains Rh, Ir, Ru, Os, and Ag. The solution contains Pt, Pd, and Au.

Platinum is first removed by precipitating as ammonium hexachloroplatinate(IV) $[(NH_4)_2PtCl_6]$ by the addition of a saturated solution of NH_4Cl. The precipitate is washed, dried, and calcined to form platinum sponge about 98% pure. The sponge is purified by redissolving in aqua regia and evaporating the solution to dryness with NaCl. The resulting sodium hexachloroplatinate is dissolved in H_2O and boiled with $NaBrO_3$ to convert impurities, such as Ir, Rh, Pd, and base metals, to valence states which produce readily filterable hydroxides. The Pt left in solution is free of impurities. It is then treated with NH_4Cl, and the pure ammonium hexachloroplatinate precipitate is calcined at 1000 °C to pure Pt sponge.

The first aqua regia solution is treated with $FeSO_4$ to precipitate the gold. Pd is precipitated by oxidizing the solution with HNO_3 and adding NH_4Cl. Ammonium hexachloropalladate(IV) is formed (analogous to the Pt compound). This salt is purified by dissolving in NH_4OH, filtering off the impurities, and reprecipitating the Pd by the addition of HCl.

The insoluble complex $Pd(NH_3)_2Cl_2$ is formed, which when calcined and reduced in H_2 yields pure Pd sponge.

The insolubles from the first aqua regia treatment are fused with a flux of litharge, soda ash, borax, and carbon in a gas-fired furnace at 1000 °C for 1 hour. This procedure converts silica, alumina, and some base metals to slag. The precious metals are retained in the lead phase. The lead portion is heated with HNO_3, which dissolves the Pb and Ag. The Pb is precipitated as a sulfate and then the Ag as a chloride. The residue is treated with concentrated H_2SO_4 at 300 °C. Rh will dissolve, leaving Ir, Ru, and Os as insolubles. The Rh solution is treated with Zn powder, precipitating an impure Rh. The impure Rh is heated in an atmosphere of Cl_2. Many impurities form volatile chlorides at this temperature and are expelled. Rh forms a polymeric trichloride, which is insoluble in aqua regia. The rhodium trichloride is digested in aqua regia for several hours, then filtered, dried, and calcined, yielding a commercial grade of Rh sponge.

The residue insoluble in H_2SO_4 is fused with Na_2O_2, poured into thin slabs and cooled. Ir is oxidized in the fusion of IrO_2, which is insoluble in H_2O. Ru and Os form soluble sodium salts and are separated from the Ir by filtration. The insoluble IrO_2 is dissolved in aqua regia, and ammonium hexachloroiridate(IV) is precipitated by the addition of NH_4Cl. Calcining yields pure Ir sponge.

The filtrate from the dissolution of the Na_2O_2 fusion contains $NaRuO_4$ and $NaOsO_4$. Ethyl alcohol is added to the solution, causing the precipitation of RuO_2, which is separated by filtration.

The Ru is purified by distilling with Cl_2. Volatile ruthenium tetroxide is collected. A saturated solution of NH_4Cl is added, causing the precipitation of ammonium hexachlororuthenate(III). The precipitated salt is calcined in H_2, yielding commercial Ru sponge.

The filtrate from the alcohol precipitation of Ru contains the Os. The solution is neutralized with HCl and is treated with powdered Zn, reducing the Os to the metallic state. Osmium tetroxide is formed by roasting the impure Zn in a current of O_2. The volatile OsO_4 is trapped in an aqueous solution of KOH. Ethanol is added to the solution, precipitating potassium osmate(VI), which is mixed with an excess of NH_4Cl and calcined in an atmosphere of H_2. The resulting Os sponge is leached to remove KCl, leaving a commercial-grade Os sponge.

The refining of secondary scrap follows much the same procedures with minor variations. For example, solid metallic Pt and especially the Rh and Ir alloys of Pt are very difficult to dissolve in aqua regia. Therefore the scrap generally is alloyed with Cu, Ni, Pb, or Zn before dissolution with acids.

Uses. These metals, in various forms, currently are used as catalysts for a wide variety of reactions. Products include high-octane gasoline, HNO_3, H_2SO_4, HCN, vitamins, antibiotics, H_2O_2, cortisone, alkaloids, and fuel-cell chemicals. These catalysts also are used to remove trace impurities, e.g., acetylene in ethylene or O_2 in H_2, or noxious constituents of partial combustion, e.g., automobile exhausts. Although substitutes are being sought, Pt is by far the best catalyst for pollution control of auto exhausts.

In the future, the catalytic converters currently installed in automobiles will become a significant source of platinum metals. See also **Catalysis**.

The corrosion resistance of the Pt metals has made the Pt crucible and the Pt electrodes commonplace laboratory tools. The glass industry makes use of large amounts of Pt and its alloys for manufacturing very pure glass. Synthetic fibers often are extruded through spinnerettes made of Pt alloys. The large use of Pt metals in dental and medical devices, in jewelry, and for decorative purposes is based on the corrosion resistance and general appearance of these metals.

Because of their high melting points and stability, Pt alloys have found applications in thermocouples, resistance thermometers, potentiometer windings, electrodes, insoluble anodes, high-temperature furnace winding, crucibles that can withstand corrosive materials at high temperature, and generally as materials of construction that will not contaminate products at very high temperatures. Often Pt and Pd are alloyed with Rh, Ir, Ru, or Os to increase their strength, hardness, and corrosion resistance.

Platinum metals, in particular Pd, find extensive use in the electrical industry. Most of these metals are used as contacts, particularly in telephone relays, where their resistance to oxidation and sulfidization results in circuits of reliability and stability. Alloys of Pt find use as grids for electronic tubes, in electrodes for aircraft spark plugs, for contact metal in printed and solid-state circuits, and in pressure-rupture disks.

In the medical field, *cis*-dichlorodiammineplatinum(II) has been available for cancer therapy (*Science*, **192**, 774–775, 1976).

Platinum Compounds. Platinum forms many di- and tetravalent compounds. The latter valence is more common and more stable. Pt in compact form is inert to all mineral acids except aqua regia. Under oxidizing conditions, fused alkalies will attack Pt to some extent. Molten halides, carbonates, and sulfates have little effect on the metal. Concentrated boiling H_2SO_4, fused cyanides, and fused alkaline sulfides will attack the finely divided metal. Pt is vigorously attacked by Cl_2 at elevated temperatures. In hot aqua regia or HCl containing chlorate ion or H_2O_2, the metal slowly dissolves, yielding a solution of hexachloroplatinic acid, H_2PtCl_6.

Platinum(II) hydroxide is made by adding KOH to a solution of platinum(II) chloride. The unstable black powder is easily oxidized by air and must therefore be handled in an inert atmosphere. In hot alkali or HCl, it disproportionates into the platinum(IV) compound and the metal. Very careful dehydration results in the formation of a gray powder that approaches the composition of platinum(II) oxide. Platinum(II) oxide can also be made by combining the elements at 420–440 °C at an O_2 pressure of 8 atm.

When a solution of hexachloroplatinic(IV) acid is boiled for some time with NaOH, all the chloride ions are replaced by hydroxide ions. The resulting sodium hexahydroxoplatinate(IV), $Na_2Pt(OH)_6$, is soluble in the basic solution, but it can be precipitated as hexahydroxoplatinic(IV) acid, $H_2Pt(OH)_6$, by the addition of acetic acid. The hydroxide ions of the salt are replaced by the corresponding ions of mineral acids when the compound is dissolved in acid. Hexahydroxoplatinic acid can be dehydrated to yield compounds corresponding to the tri-, di-, and monohydrate of platinum(IV) oxide. The last water molecule cannot be removed without some destruction of the dioxide.

Brown-black, insoluble, anhydrous, platinum(IV) oxide is made by fusing hexachloroplatinic(IV) acid with $NaNO_3$ at about 500 °C. The alkali salts are washed out with H_2O to free the fine insoluble residue of platinum(IV) oxide. This compound is known as *Adam's catalyst*.

When Pt is heated to 500 °C in the presence of Cl_2, yellow-green, insoluble platinum(II) chloride is formed. At a pressure of 1 atm of Cl_2, the compound is stable from 435 to 581 °C. It can also be made by heating hexachloroplatinic(IV) acid in Cl_2 at about 500 °C. Platinum(II) chloride is soluble in HCl as tetrachloroplatinic(II) acid. It forms many salts that are water-soluble. These salts can be made by reducing a hot solution of the corresponding hexachloroplatinate(IV) with oxalic acid or SO_2. Platinum(III) chloride has a narrow range of stability. It can be made by contacting Pt or a platinum chloride with 1 atm of Cl_2 at 364–374 °C. This dark-green to black compound is practically insoluble in cold concentrated HCl but does dissolve on warming, forming a mixture of tetrachloroplatinic(II) and hexachloroplatinic(IV) acids. Anhydrous platinum(IV) chloride is very difficult to prepare. This brown soluble solid can be made by heating hexachloroplatinic(IV) acid in Cl_2 at 360 °C. The most common Pt compound, hexachloroplatinic(IV) acid, is readily

made by dissolving Pt in aqua regia, followed by several evaporations with additional HCl to destroy nitrosyl compounds. The acid crystallizes as a hexahydrate. It is difficult to stop the evaporation at just this point, and slight local overheating causes excess loss of water. The sodium salt is quite soluble, and the compound is resistant to hydrolysis in basic solution, allowing the bromate hydrolysis to precipitate base metals and other Pt metals as their hydroxides. The Pt remains in solution. The insolubility of ammonium hexachloroplatinate(IV) often is used in refining Pt. Its slight solubility can be overcome sufficiently by mass action to allow its use as a gravimetric procedure for the determination of Pt. This yellow compound decomposes at red heat, yielding pure Pt sponge. The insolubility of the potassium salt is used for the gravimetric determination of potassium.

A series of di-, tri-, and tetrabromides is well known. Platinum(II) iodide is precipitated as a black insoluble compound by the addition of 2 equiv of iodide to a hot solution of platinum(II) chloride. The black, insoluble, graphitelike substance, platinum(III) iodide, is made by combining the elements in a sealed tube at 350 °C.

In contrast with Pd, Pt does form a Pt(IV) iodide. When a concentrated solution of hexachloroplatinic(IV) acid is treated with a hot solution of KI, this brown-black substance is precipitated. The compound is somewhat unstable and light-sensitive. It dissolves in excess KI to form the complex salt, also rather unstable.

Pt forms a nonvolatile tetrafluoride, a pentafluoride, and a volatile hexafluoride. The dark-red PtF_6 melts at 56.7 °C and is very reactive. It even reacts with O_2 at 21 °C to form dioxygenylhexafluoroplatinate(V), O_2PtF_6.

When sulfur and Pt sponge are ignited, some platinum(II) sulfide is formed. The naturally occurring mineral is called cooperite. When heated in air or H_2, the products are metallic Pt and S. Platinum(IV) sulfide can be made by heating ammonium hexachloroplatinate(IV) or Pt and S at 650 °C. When precipitated by H_2S from chloroplatinic acid, the compound may exist as $PtS_2 \cdot H_2S$.

Divalent and tetravalent Pt probably form as many complexes as any other metal. The platinum(II) complexes are numerous with N_2, S, halogens, and C. The tetranitritoplatinum complexes are soluble in basic solution. Tetranitritoplatinum(II) ion is formed when a solution of platinum(II) chloride is boiled, at about neutral pH, with an excess of $NaNO_3$. The ammonium salt may explode when heated. Generally, platinum-metal nitrites should be destroyed in solution. They never should be heated in the dry form. Platinum(II) complexes most often have a coordination number of 4. Many compounds have been prepared with olefins, cyanides, nitriles, halides, isonitriles, amines, phosphines, arsines, and nitro compounds.

Platinum(IV) has a coordination number of 6. It forms complexes with halides, nitrogen and sulfur compounds, and other donors but to a lesser extent than platinum(II).

Additional Reading

Carter, G.F. and D.E. Paul: *Materials Science and Engineering*, ASM International, Materials Park, OH, 1991.

Greenwood, N.N. and A. Earnshaw: *Chemistry of the Elements*, 2nd Edition, Butterworth-Heinemann, Inc., Woburn, MA, 1997.

Krebs, R.E.: *The History and Use of Our Earth's Chemical Elements: A Reference Guide*, Greenwood Publishing Group, Inc., Westport, CT, 1998.

Lide, D.R.: *CRC Handbook of Chemistry and Physics 2000–2001*, 81st Edition, CRC Press, LLC, Boca Raton, FL, 2000.

Meyers, R.A.: *Handbook of Chemicals Production*, The McGraw-Hill Companies, Inc., New York, NY, 1986.

Parker, P.: *McGraw-Hill Encyclopedia of Chemistry*, 2nd Edition, The McGraw-Hill Companies, Inc., New York, NY, 1993.

Schweitzer, P.A.: *Corrosion Resistance Tables*, 3rd Edition, ASM International, Materials Park, OH, 1991.

Staff: *ASM Handbook—Properties and Selection: Nonferrous Alloys and Pure Metals*, ASM International, Materials Park, OH, 1990.

Stwertka, A. and E. Stwertka: *A Guide to the Elements*, Oxford University Press, Inc., New York, NY, 1998.

LINTON LIBBY, Chief Chemist, Simmons Refining Company, Chicago, IL

PLATYHELMINTHES. The flatworms, a major division of the animal kingdom containing the most primitive of the triploblastic *Metazoa*. The phylum includes both free-living and parasitic species. Among the latter are the flukes and tapeworms, some of which are serious parasites of humans.

The phylum is characterized by the following details of structure: (1) The body is bilaterally symmetrical and flattened. (2) The ectoderm is ciliated in free-living forms but forms a cutical in the parasitic species. (3) The mesoderm forms a compact tissue between the various organs, known as a parenchyma. (4) The alimentary tract, when present, has a single opening. (5) The nervous system is a network in which a brain and longitudinal nerve cords are developed. (6) The excretory system consists of large hollow cells with a group of cilia extending into the cavity, known as flame cells, connected with tubes.

PLAYA.   The flat interior part of an undrained basin, on which accumulates fine clastic sediments and chemical precipitates. Playas are formed within desert basins due to intermittent interior drainage, which, during cloudburst, forms intermittent lakes in which the sediments are deposited. A region remarkable for playas is the Great Basin of the western United States, which covers all of Nevada and Utah. Commercially important mineral salts derived from playa deposits are: gypsum, sodium carbonate, the soluble chlorides, and borates.

PLECOPTERA.   The stone flies. An order of insects with aquatic early stages. The mouth is formed for biting but is usually poorly developed in the adult. They have four wings, which fold flat over the back when at rest. Sometimes abundant in the vicinity of water.

PLECOSTOMUS (*Osteichthyes*).   Of the suborder *Siluroidea*, family *Loricariidae*, the plecostomus is a heavy-bodied loricariid (catfish) and is found present in many of the tanks maintained by tropical-fish fanciers. The average length is from 4 to 5 inches (10 to 12.5 centimeters), although the plecostomus can attain a length of about 20 inches (51 centimeters). The fish functions to clean small organisms from the sides of fish tanks and blends well in most tropical fish communities. The fish is considered an excellent food, when baked, by various Indian tribes in South America.

The loricariids occupy various habitats in South and Central America. Some are found in waters comparable to the European trout zones, something that would not be expected of these plump fishes. Others live in slowly flowing rivers. Interestingly, the gill openings of the loricariids are always on the lower side of the body, where they would always come in contact with mud or gravel. This is nonetheless advantageous for these fishes: they can maintain their suction (grip) on stones even in small waterfalls. Loracariids generally eat plants and refuse, although they also take small bottom-dwelling organisms. Because of their special gill apparatus, they can breathe and feed simultaneously. They have rows of horny teeth on large cupped lips, forming a good rasping surface. The catfishes scrape algae and other growth from stones and wood lying in the water. The teeth are also used to chew dead fishes or other animal cadavers in the water. In aquariums, it has repeatedly been observed that large loricariids will attack some living fishes, adhering to them and chewing their skin. Whether this results from a lack of appropriate food or is natural behavior is not fully understood. Like most catfishes, loricariids are also nocturnal or dusk-active creatures. They can decrease the amount of light entering their eyes with a distensible, whitish, drop-shaped structure on the upper part of the eye. The brighter the incoming light, the smaller the pupil, and the more this structure can distend over the eye.

Most plecostomids inhabit fast-flowing water, and they migrate into the clear Andes streams at an altitude of about 13,000 feet (3960 meters). They use the sucking mouth for propulsion while migrating, gaining a sucking hold in very rapidly flowing spots, and pushing forward bit by bit with jerky movements. By this means, they can get through places where the water is roaring past with an extremely powerful current. They prefer concealed sites when resting near the shore or behind stones. They also adhere to tree trunks that have fallen into the water. Such a loricariid will hold on to the stone even when lifted out of the water.

Almost all loricariids have a brown basic coloration with some gray or reddish colors. The body has irregular black or dark brown spots arranged in different ways, resulting in rather beautiful patterns. The dorsal fin is higher than the body itself and forms an impressive sail when erect.

Other loricariids include the antenna armored catfishes (genus *Ancistrus*); the loricaria (genus *Loricaria*); and the very small *Otocinclus* loricariids, which range from 1.5 to 3 inches (3.5 to 8 centimeters) in length.

PLEIADES.   A very famous group of bright stars in the constellation of Taurus. Probably no one group of stars in the entire sky has received so much notice in classical literature and mythology as has the Pleiades. The Great Pyramid, which was undoubtedly designed for astronomical purposes, is so oriented that in 2170 B.C., when the Pleiades were on the meridian at midnight on the first day of spring, they could be seen through the south passageway. References to this group of stars are to be found in both the Old and New Testaments. One of the seven stars is, at present, distinctly fainter than the other six, and there are many myths regarding this so-called "lost pleiad." In fact, the myths occur in so many different ancient literatures that there is a well-established theory that at one time all seven of the stars were of approximately the same brightness.

The Pleiades is an open star cluster, with the various members all moving through space together. Long exposure photographs of this group indicate that the space surrounding the stars is filled with a luminous nebulosity. See Fig. 1.

Fig. 1.   The Pleiades. (*Lick Observatory*.)

PLEISTOCENE CLIMATE.   See **Climate**.

PLENUM.   Pressures slightly above atmospheric are known as plenums. Such pressures usually occur in air or gas systems as the result of the action of fans or blowers. The plenum is measured in small units of pressure, such as ounces per square inch, or in inches head of a liquid on a differential manometer.

PLEURISY.   Inflammation or infection of the pleura, often producing a characteristic sharp, piercing sensation, which appears during the inspiratory phase of breathing as the inflamed pleural surfaces rub together. The friction produced by this approximation of the parietal and visceral pleura can be distinguished by the stethoscope as a sound simulating the creaking of leather and is termed a friction rub. Pleurisy is often accompanied by effusion, the outpouring of a thin fluid, which distends the pleural space. The appearance of fluid causes the disappearance of both the rub and the pain. Where expansion of the lung is sufficiently cut down by a massive accumulation fluid, dyspnoea will develop. Although once established, pleurisy may run a course independent of the primary course, pleurisy generally is the secondary manifestation of pneumonia, lung tumor, tuberculosis, abscess, rib fracture, or chest wounds. So-called primary pleurisy in which the disease commences in the pleura proper is rarely seen. Treatment is directed to the primary cause and frequently antibiotics are used. The latter also act to alleviate the pleurisy per se. In mild cases, the fluid formed will be reabsorbed, but if the fluid becomes infected, an empyema will make spontaneous reabsorption difficult and at best slow. This sometimes will cause the formation of fibrous adhesions, which may not permit the lung to re-expand upon removal of the fluid. The latter situation usually requires decortication procedure, that is, the lung is freed from the pleural peel.

PLEXUS.   A network. 1. In many animals the processes of nerve cells join to form a plexus or nerve net. This is the characteristic form of nervous system in the coelenterates and persists with modifications in the flatworms. The nerves of the radially symmetrical echinoderms also take on this form. A plexus underlies the ectoderm of these animals and deeper in the body other nerve fibers form plexuses of limited extent. In the vertebrates nerves branch and rejoin in some parts of the body. The brachial plexus, made up of the spinal nerves which enter the arm, and the solar plexus above the stomach are examples. Almost a hundred such plexuses have been named in the human body.

2. A network of blood vessels. The choroid plexuses of the brain are the most commonly mentioned examples of this group. They are the very thin and highly vascular roof plates of the most anterior and the most posterior cavities of the brain, which expand into the interior of the cavities. Other vascular plexuses are found elsewhere in the body.

PLIOCENE.   The last major subdivision of the Tertiary in the geologic time-scale. Term proposed by Charles Lyell in 1832 after the type locality in the Paris Basin. The Pliocene Period began approximately 8,000,000 years ago and lasted for about 6,000,000 years. In the United States the principal marine deposits occur on the Pacific Coast, but the outline of the continental margins was approximately what it is at the present day. There was continued mountain-building during the period, and the interior continental terrestrial deposits were relatively thin and unimportant. There was considerable volcanic activity in the Rocky Mountain region, with great extrusions of rhyolitic lavas in the Yellowstone Park.

PLOTTER (Curve).   A digital computer output device, which draws curves of one or more variables as a function of one or more variables. The function is essentially that of an x-y recorder. Data from the digital computer are translated into plotter-actuating signals and converted to incremental plotter movements to produce a drawing, chart, or other graphic portrayal. Two configurations are common. In a flat-bed plotter, the paper is fixed to a flat surface and one or more pens are mounted on a carriage capable of moving in two dimensions, normally designated x and y. In a drum plotter, the paper is affixed to a cylindrical drum whose rotation provides one dimension (usually) of motion. A pen mounted above the drum surface and capable of moving axially with respect to the drum provides the x-dimension. Control is also provided to raise or lower the pen from or to the paper surface. The information from the computer is decoded into fixed incremental movements of the drum and or pen carriage. Each plotter command transmits information which is decoded into the number of pen-carriage increments to be moved (y-axis) and the number of drum increments to be moved (x-axis). Specified bit combinations also control the raising or lowering of the pen to the paper.

PLOTTING SHEET (Navigation).   A small-area plotting sheet is an approximation to a mercator plotting sheet, and may be used for solving problems in dead reckoning and for plotting lines of position.

In constructing the graticule for the sheet, the fundamental assumption is made that, within the limit of errors inherent in the problems to be solved, the distance between successive parallels of latitude, separated by 1°, is constant all over the sheet. This is equivalent to assuming that the same scale of distance may be used all over the sheet. The ratio between the linear distance, X, between successive meridians and the linear distance, Y, between successive parallels, is that of middle-latitude sailing: $X = Y \cos L_m$, in which L_m is the average latitude of the region covered by the sheet.

Two types of small-area plotting sheet are in common use: (1) the fixed meridian type, and (2) the fixed parallel type. The method for completing the graticules and plotting a point is shown in Figs. 1 and 2, respectively. In the Figures the heavy lines are those printed on the published forms,

and the light lines are those drawn to complete the graticule. Dotted lines in the figure are construction lines which need not appear on the completed sheet. The sheets in the figure are completed for $L_m = 48°$ and the point, P, is in latitude 48°17′N and longitude 50°23′W.

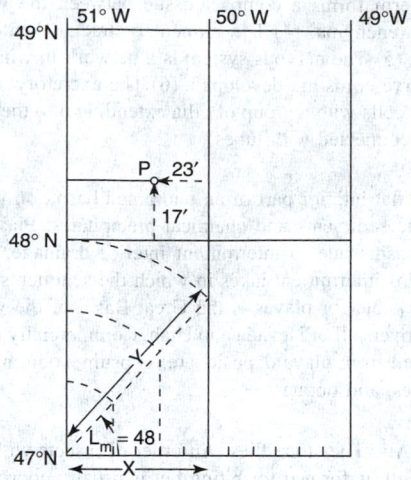

Fig. 1.   Fixed meridian type.

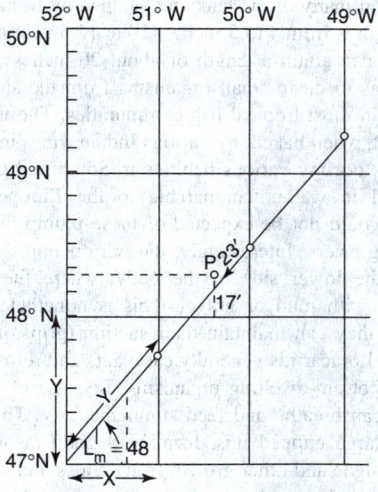

Fig. 2.   Fixed parallel type.

In the older or fixed meridian type, Fig. 1, a diagonal line is drawn making an angle L_m with the base line. The distance, X, is that between the fixed meridians, and the distance, Y, is the length of the diagonal, measured between the meridians. Elementary trigonometry shows that $X = Y \cos L_m$. This distance, Y, is then transferred to one of the fixed meridians and becomes 1° of latitude for the small-area plotting sheet. This length, Y, is also equivalent to 60 nautical miles and is used as a scale of distance all over the sheet. The central parallel on the sheet is that of L_m. The central meridian is that of the middle of the region covered by the particular problem to be solved. With this type of sheet the longitude scale is constant, no matter for what region the sheet is drawn. The latitude spacing, and hence the scale of distance, is different for different sheets.

In the more modern, and far more frequently used, fixed parallel type, Fig. 2, the diagonal line making an angle of L_m with the base is drawn. Distances equal to that between the fixed parallels are laid off along this diagonal and lines are drawn through these points parallel to the latitude scale of the sheet. These become the successive meridians. It will be seen that, as above, $X = Y \cos L_m$. However, on this style of sheet the distance

Y is fixed, no matter for what region the sheet is prepared, and a scale for distance ruled on celluloid, or other permanent material, may be used on all sheets. Furthermore, this same scale may be used for measurement of longitude by placing it along the diagonal and then projecting the desired value, parallel to the latitude scale, to the proper latitude.

Plotting sheets of both types and for various scales are published by the U.S. Government, and by several publishing firms. Most of the forms have a compass rose at the center of the sheet so that parallel rulers and dividers may be used, instead of protractor and scale, if the user prefers. However, printed forms are by no means necessary, for the graticule can quickly be drawn on blank paper with the aid of a protractor and scale. Using a blank sheet requires about 1 minute to obtain the completed graticule, whereas with the printed forms this time is cut in half.

Either type of sheet can be used, without introducing errors greater than those inherent in the problems for which they are designed, for any mid-latitude between the equator and 60° and for areas of about 300 miles square. For low latitudes the area can be correspondingly increased.

See also **Course**; and **Navigation**.

PLOVER. See **Waders, Shorebirds, and Gulls**.

PLUM CURCULIO (*Insecta, Coleoptera*). A weevil, *Conotrachelus nenuphar*, which damages plums and other stone fruits, apples, pears, and quinces in the eastern half of the United States. The insects hibernate in fencerows and rubbish in orchards and pupate in the ground; hence, the destruction of their hiding places and thorough cultivation of the soil in late July and early August are useful measures of control. Spraying just after the petals drop and again after 10 days is effective. For control of the pest on peaches, special methods are necessary which differ in various peach-growing regions.

PLUME. 1. Buoyant jet in which the buoyancy is supplied from a point source; the buoyant region is continuous. See also **Thermal**.

2. A mostly horizontal (sometimes initially vertical) stream of air pollutant that is being blown downwind from a smokestack. Typical smoke-plume diameters are of order 1–10 meters (3.3–33 feet) initially, gradually expanding to 100 meters (328 feet) or more, while lengths can be order of 1–100 kilometers (0.6–62 miles). The path and shape of the smoke plume can indicate the nature of turbulence in the atmospheric boundary layer, such as looping plumes, fanning plumes, and coning plumes.

AMS

PLUME MOTH (*Insecta, Lepidoptera*). A small moth whose wings are deeply split to form two to six slender fringed lobes. In the family *Pterophoridae* most species have the front wings split for about $\frac{1}{3}$ of their length to form two short lobes and the hind wings deeply divided into three lobes. One genus has the wings entire. Members of the *Orneodidae* have six slender plumes to each wing.

PLUMERIA TREE. Of the family *Gentianaceae* (gentian family), there are some seven species of plumerias found from the West Indies southward to northern South America. The tree was named for Charles Plumier, a French botanist who was traveling in America in the late 1600s. It is found today in Hawaii and Mexico as well as the Caribbean region. It is highly regarded for its blossoms, which are used for a variety of decorative items, including leis. The flowers are fragrant and rugged, withstanding long periods after cutting. The tree blooms continually from spring until fall. The flowers are frequently white with yellow centers, but some have a reddish or golden yellow coloring, depending upon species. The blooms have a funnel shape.

Plumerius rubra may attain a height of 15 feet (4.5 meters) and bears pink and red flowers. The leaf is about 18 inches (46 centimeters) in length and 6 inches (15 centimeters) in width, with narrow pointed ends and conspicuous veining. The plant may be potted from cuttings and makes a very desirable plant for gardens and patios. The shrub is found mainly in Mexico and south through Venezuela. There are some 18 species of *Amsonia walt*, principally found in North America and Japan. The leaf is alternate, its flower is blue and white, the corolla being cylindrical in shape with 5 erect lobes. The shrub was named for Dr. Amson of Virginia. Leis are made from the flowers.

PLUME RISE MODEL. An algorithm for calculating the altitude that a plume will rise due to momentum and buoyancy forces before reaching an equilibrium height.

Plume rise increases with higher buoyancy or momentum of the plume and decreases with increasing wind speed or vertical temperature gradient in the atmosphere. The rate of rise is fastest at the point of emission and decreases due to the entrainment of ambient air, which has minimal momentum and generally lower temperatures than the original plume. The plume is considered to be at its final height when the rate of rise decreases to a point where it is equivalent to vertical velocities generated by turbulence in the atmosphere.

AMS

PLUM TREES. See **Rose Family**.

PLUTONIUM. [CAS: 7440-07-5]. Actinide radioactive metal. Atomic number 94. Symbol Pu. This element does not occur in nature except in minute quantities as a result of the thermal neutron capture and subsequent beta decay of ^{238}U; all isotopes are radioactive; atomic weight tables list the atomic weight as [242]; the mass number of the second-most-stable isotope ($t_{1/2} = 3.8 \times 10^5$ years). The most stable isotope is ^{244}Pu($t_{1/2} = 7.6 \times 10^7$ years). Electronic configuration $1s^2 2s^2 2p^6 3s^2 3p^6 3d^{10} 4s^2 4p^6 4d^{10} 4f^{14} 5s^2 5p^6 5d^{10} 5f^6 6s^2 6p^6 7s^2$. Ionic radii Pu^{4+} 0.86 Å; Pu^{3+} 1.01 Å (Zachariasen). Oxidation potentials in acid solution Pu $\rightarrow$ Pu^{3+} + 3e$^-$, 2.03V; Pu^{3+} $\rightarrow$ Pu^{4+} + e$^-$, −0.982V; Pu^{4+} + O$_2$ + 3e$^-$ $\rightarrow$ PuO$_2^+$, −1.17V; PuO$_2^+$ $\rightarrow$ PuO$_2^{2+}$ + e$^-$, −0.91V. Oxidation potential in alkaline solution Pu^{3+} + 4H$_2$O $\rightarrow$ Pu(OH)$_4$ + 4H$^+$ + e$^-$, 0.4   V; Pu(OH)$_4$ $\rightarrow$ PuO$_2^+$ + 2H$_2$O + e$^-$, −1.0V; PuO$_2^+$ + 2OH$^-$ $\rightarrow$ PuO$_2$(OH)$_2$ + e$^-$, −0.8V.

The isotope of major importance is ^{239}Pu($t_{1/2} = 2.44 \times 10^4$ years). The importance of this isotope stems from its property of being fissionable with slow neutrons, together with the fact that the problem of its mass production has been solved. The first isotope to be produced was ^{238}Pu($t_{1/2} = 86.4$ years).

Processes for the isolation and purification of plutonium, including the enrichment of spent nuclear reactor fuels, are described in the entry on **Nuclear Power Technology**. These processes take advantage of Pu's several oxidation states, each of which has different chemical properties. The processes may involve carrier precipitation, solvent extraction, and ion exchange.

Plutonium is of major importance because of its successful use as an explosive ingredient in nuclear weapons and the role it plays in the industrial applications of nuclear power. Exemplary of the energy available from plutonium: (1) One pound (0.45 kilogram) $\cong$ 10 million kilowatts; (2) one kilogram (2.2 pounds) = 22 million kilowatts; (3) one kilogram (2.2 pounds) = 20,000 tons of chemical explosive. Plutonium has the important nuclear property of being readily fissionable with neutrons. ^{238}Pu was used in the Apollo lunar missions to power seismic and other experimental instruments placed on the lunar surface. Because comparatively large quantities of plutonium are produced in reactors, the amount available for various applications has increased considerably during recent years. It is estimated that as of the late 1980s, nuclear reactors throughout the world are producing in excess of 20,000 kilograms of Pu per year. Within a few years, there will be an accumulation of some 300,000 kilograms of Pu or more. The element is available for purchase by qualified industrial users.

In a typical fast breeder nuclear reactor, most of the fuel is ^{238}U (90 to 93%). The remainder of the fuel is in the form of fissile isotopes, which sustain the fission process. The majority of these fissile isotopes are in the form of ^{239}Pu and ^{241}Pu, although a small portion of ^{235}U can also be present. Because the fast breeder converts the fertile isotope ^{238}U into the fissile isotope ^{239}Pu, no enrichment plant is necessary. The fast breeder serves as its own enrichment plant. The need for electricity for supplemental uses in the fuel cycle process is thus reduced. Several of the early liquid-metal-cooled fast reactors used plutonium fuels. The reactor "Clementine," first operated in the United States in 1949, utilized plutonium metal, as did the BR-1 and BR-2 reactors in the former Soviet Union in 1955 and 1956, respectively. The BR-5 in the former Soviet Union, put into operation in 1959, utilized plutonium oxide and carbide. The reactor "Rapsodie" first operated in France in 1967 utilized uranium and plutonium oxides.

Plutonium was the second transuranium element to be discovered. The isotope ^{238}Pu was produced in 1940 by Seaborg, McMillan, Kennedy, and Wahl at Berkeley, California by deuteron bombardment of uranium in a 150-cm cyclotron. Plutonium exists in trace quantities in naturally occurring uranium ores. The metal is silvery in appearance, but tarnishes to a yellow color when only slightly oxidized. A relatively large piece will give off sensible heat as the result of alpha decay. Large pieces are capable of boiling water.

Chemical Properties. Plutonium has the oxidation states (III), (IV), (V), and (VI), and a complex chemistry in aqueous solutions, as can be judged from such a multiplicity of states. A large number of solid compounds corresponding to these states have been made, and they are in general similar in formulas and properties to the corresponding compounds of uranium and neptunium. An important difference, especially as regards ranges of stability of these compounds, arises as a result of the much greater stability of the (III) and (IV) states of plutonium. This also leads to differences in the aqueous solution chemistry of plutonium as compared to uranium and neptunium. The pentavalent state, like that of uranium, but unlike that of neptunium, is unstable in aqueous solution with respect to disproportionation.

The ionic species corresponding to the four oxidation states of plutonium vary with the acidity of the solution. In moderately strong (one-molar) acid the species are Pu^{3+}, Pu^{4+}, PuO_2^+, and PuO_2^{2+}. The ions are hydrated but it is not possible at present to assign a definite hydration to each ion. The potential scheme of these ions in one-molar perchloric acid is the following:

$$\mathrm{Pu} \xrightarrow{+2.03\,\mathrm{V}} \mathrm{Pu}^{3+} \xrightarrow{-0.982\,\mathrm{V}} \mathrm{Pu}^{4+} \xrightarrow{-1.17\,\mathrm{V}} \mathrm{PuO}_2^+ \xrightarrow{-0.91\,\mathrm{V}} \mathrm{PuO}_2^{2+}$$

(with -1.043 V spanning Pu^{4+} to PuO_2^{2+}, and -1.023 V spanning Pu^{3+} to PuO_2^+)

The potentials are in volts relative to the hydrogen–hydrogen-ion couple as zero.

The values given for the potential scheme in one-molar acid may be altered extensively by a change in hydrogen ion concentration (pH) or as a result of the addition of substances capable of forming complex ions with the plutonium species. Among such substances are sulfate, phosphate, fluoride, and oxalate ions, and various organic compounds, especially those known as chelating agents. The tetrapositive and hexapositive ions are complexed appreciably even by nitrate and chloride ions. The stability of the complex formed with a specified anion increases in the order: PuO_2^+, Pu^{3+}, PuO_2^{2+}, Pu^{4+}.

The hydrolysis of the ions follows a similar order; Pu^{4+} begins to hydrolyze even in tenth-molar acid and in hundredth-molar acid forms partly the hydroxide, Pu(OH)$_4$, and partly a colloidal polymer of variable but approximate composition Pu(OH)$_{3.85}$X$_{0.15}$, where X is an anion present in the solution. Further reduction of the acidity results in the hydrolysis of PuO_2^{2+} near pH 5, of Pu^{3+} at about pH 7, and of PuO_2^+ at about pH 9.

The plutonium ions in aqueous solution possess characteristic colors: blue-lavender for Pu^{3+}, yellow-brown to green for Pu^{4+}, and pink-orange for PuO_2^{2+}.

Plutonium monoxide occasionally appears on the surface of metal exposed to atmospheric oxidation, but is prepared more conveniently by treating the oxychloride with barium vapor at about 1,250 °C. The oxide is classified with the interstitial compounds rather than with the typical metal oxides.

The so-called sesquioxide (PuO$_{1.5-1.75}$) is a typical mixed oxidation state oxide, similar to those formed by uranium, praseodymium, terbium, titanium, and many other metals. Its composition shows continuous variation with changes in temperature and pressure of oxygen above the oxide.

Plutonium dioxide (yellow-green to brown, cubic) is the most important oxide of the element. Almost all compounds of plutonium are converted to the dioxide upon ignition in air at about 1,000 °C.

The important halides and oxyhalides of plutonium are PuF$_3$ (purple, hexagonal), PuF$_4$ (brown, monoclinic), PuF$_6$ (red-brown, orthorhombic), PuCl$_3$ (green, hexagonal), PuCl$_4$ (green-yellow, tetragonal), PuBr$_3$ (green, orthorhombic), PuI$_3$, PuOF, PuOCl, PuOBr, PuOI.

All of the halides except the hexafluoride and the triiodide may be prepared by the hydrohalogenation of the dioxide or of the oxalate of plutonium(III) at a temperature of about 700 °C. With hydrogen fluoride the reaction product is PuF$_4$, unless hydrogen is added to the gas stream, in which case the trifluoride is produced. With hydrogen iodide the reaction product is PuOI, and the other oxyhalides may be formed by the addition of appropriate quantities of water vapor to the hydrogen halide gas. Plutonium triiodide is produced by the reaction of the metal with hydrogen iodide at about 400 °C. The hexafluoride is produced by direct combination of the elements or by the reaction 2PuF$_4$ + O$_2$ → PuF$_6$ + PuO$_2$F$_2$ at high temperature. The hydrides of plutonium include PuH$_2$ (black, cubic) and PuH$_3$ (black, hexagonal).

Plutonium forms several binary compounds that are of interest because of their refractory character and stability at high temperatures. These include the carbide, nitride, silicide, and sulfide of the element.

The monocarbide is formed by reacting the dioxide in intimate mixture with carbon at about 1,600 °C. The mononitride may be obtained by heating the trichloride in a stream of anhydrous ammonia at 900 °C; it is prepared more easily, however, by reacting finely divided metal with ammonia at 650 °C. Although the lower temperatures are favorable to the production of higher nitrides, none are obtained, in contrast to the uranium-nitrogen system in which compositions up to UN$_{1.75}$ are easily realized.

The disilicide is formed when a slight stoichiometric excess of calcium disilicide is heated with plutonium dioxide in vacuum at about 1,550 °C. The disilicide is only moderately stable in air and burns slowly in the dioxide when heated to about 700 °C.

Plutonium "sesquisulfide" may be prepared by prolonged treatment of the dioxide in a graphite crucible with anhydrous hydrogen sulfide at 1,340°–1,400 °C, or by the reaction of the trichloride with hydrogen sulfide at 900 °C.

Handling Precautions. Care must be taken in the handling of plutonium to avoid unintentional formation of a critical mass. Plutonium in liquid solutions is more apt to become critical than solid plutonium. The shape of the mass also determines criticality. Plutonium's chemical properties also increase handling difficulty. Metallic plutonium is pyrophoric, particularly in finely divided form. Because of the high rate of emission of alpha particles, and the physiological fact that the element is specifically absorbed by bone marrow, plutonium, like all of the transuranium elements, is a radiological poison and must be handled with special equipment and precautions. To assure the safety of personnel, plutonium operations are normally handled in an essentially closed system, such as a *glovebox*. In addition, shielding is required when certain isotopes, including ^{240}Pu and ^{241}Pu, are present in appreciable quantity. Because research continues on the hazards and toxicity of plutonium, specific toxicity data should be sought from current authoritative literature, including government (U.S., UK, France, etc.) publications. As of the early 1980s, permissible body burden was established at 0.6 microgram; lung burden at 0.25 microgram. Chemical toxicity is trivial compared with radiation effects. The permissible levels for plutonium are the lowest for any of the radioactive elements.

The behavior of actinides in natural waters has great relevancy to the safe long-term storage of radioactive wastes. The enhanced solubility of plutonium and other actinides in the water of Mono Lake, California was studied by a group of scientists with the U.S. Geological Survey (Denver, Colorado). J.M. Cleveland and associates found that the solubility of plutonium in Mono Lake water is enhanced by the presence of large concentrations of indigenous carbonate ions and moderate concentrations of fluoride ions. In spite of the complex chemical composition of this water, only a few ions govern the behavior of plutonium, as demonstrated by the fact that it was possible to duplicate plutonium speciation in a synthetic water containing only the principal components of Mono Lake water. See reference listed.

Practical Utilization. Since the potential reserves of ^{235}U are limited, some point will be reached where this power source no longer will be competitive with fossil fuels, synthetic fuels, solar power plants, etc. — unless the development of means for the practical utilization of plutonium can be achieved. An important element of nuclear fuel cost is the credit received from the sale or future utilization of plutonium after its recovery from spent fuel. The plutonium credit is realistic only if the plutonium is used for power production, since, at present, there are few commercial uses envisioned where it would yield a similar economic return.

See also **Chemical Elements**.

References pertaining to plutonium in nuclear reactors and nuclear wastes are listed at end of entry on **Nuclear Power Technology**.

Additional Reading

Albright, D. and F. Berkhout: *Plutonium and Highly Enriched Uranium, 1996: World Inventories, Capabilities, and Policies*, Oxford University Press, Inc., New York, NY, 1996.

Cleveland, J.M., T.F. Rees, and K.L. Nash: "Plutonium Speciation in Water from Mono Lake, California," *Science*, **222**, 1323–1325 (1983).

Lewis, R.J. and N.I. Sax: *Sax's Dangerous Properties of Industrial Materials*, 10th Edition, John Wiley & Sons, Inc., New York, NY, 1999.

Lide, D.R.: *CRC Handbook of Chemistry and Physics*, 88th Edition, CRC Press, LLC, Boca Raton, FL, 2007.

Kent, J.A.: *Reigel's Handbook of Industrial Chemistry*, 9th Edition, Chapman Hall, New York, NY, 1992.

Krebs, R.E.: *The History and Use of Our Earth's Chemical Elements: A Reference Guide*, Greenwood Publishing Group, Inc., Westport, CT, 1998.

Parker, P.: *McGraw-Hill Encyclopedia of Chemistry*, 2nd Edition, The McGraw-Hill Companies, Inc., New York, NY, 1993.

Classical References

Kennedy, J.W., Seaborg, G.T., E. SegrÉ, and A.C. Wahl: "Properties of 94(239)," *Phys. Rev.*, **70**, 7/8, 555–556 (1946).

Seaborg, G.T.: "The Chemical and Radioactive Properties of the Heavy Elements," *Chem. Eng. News*, **23**, 2190–2193 (1945).

Seaborg, G.T., E.M. McMillan, J.W. Kennedy, and A.C. Wahl: "Radioactive Element 94 from Deuterons on Uranium," *Phys. Rev.*, **69** (7/8), 366–367 (1946).

Seaborg, G.T. and A.C. Wahl: "The Chemical Properties of Elements 94 and 93," *J. Amer. Chem. Soc.*, **70**, 1128–1134 (1948).

Seaborg, G.T. (editor): *Transuranium Elements*, Dowden, Hutchinson & Ross, Stroudsburg, Pennsylvania, 1978.

PLUTO (Planet). Ninth and outermost known and confirmed planet from the sun, Pluto is estimated to have a diameter considerably less than half that of earth. The mass is estimated at about 0.2 that of earth. Discovery of the planet on March 13, 1930 by the Lowell Observatory occurred on the anniversary both of Percival Lowell's birth and the discovery of Uranus. See Fig. 1. The discovery marked the culmination of a search for a planet outside the orbit of Neptune that had been carried on for many years at the observatory, at Flagstaff. Arizona. The circumstances that led to the belief that such a planet existed are similar to those that led to the discovery of Neptune. After the orbit of Neptune was computed and the motion carried back through the years, it was found that the planet had been observed several times previous to its announcement as a planet, the early observers having recorded it as a star. These early observations were of great value in making an accurate determination of the orbit, and when all perturbations due to known objects had been computed and applied, certain unexplainable differences between observed and computed positions appeared. On the basis of these perturbations, Lowell made the necessary laborious computations to determine the positions of a possible planet that might be causing the attractions, and he predicted Pluto. There

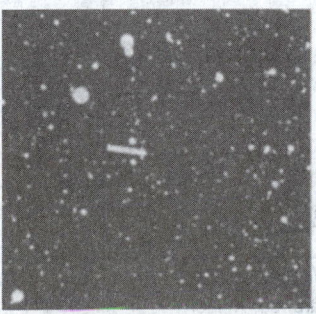   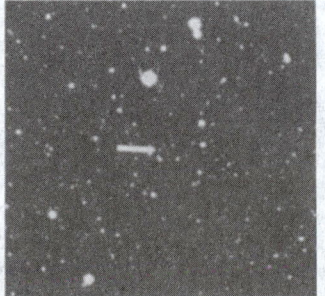

January 29, 1930   January 23, 1930

Fig. 1.   Pluto shows just to the right of the arrow. Small sections of the discovery plates. (*Lowell Observatory.*)

is considerable doubt in the minds of many astronomers as to whether it is Pluto that is actually producing the perturbations in the orbit of Neptune or whether the perturbations may not be due to accidental errors in the observations of Neptune itself. Whatever the case, it is certain that the computations of Dr. Lowell stimulated the search, and that the planet was found as a result in the approximate position predicted.

The name Pluto was selected for the new planet and the first two letters of the name, combined in monogram form **P** are used as its symbol. These *two* letters are particularly fortunate in being both the first letters in the name Pluto and the initials of Percival Lowell.

The orbit of Pluto is the most eccentric of all the orbits of the major planets, and the inclination to the plane of the ecliptic is also the largest. The mean distance of the planet from the sun is slightly less than 40 astronomical units. Due to the large value of the eccentricity (0.25), the planet is more than 50 astronomical units from the sun at aphelion and within 30 at perihelion. The latter figure is less than the distance of Neptune from the sun, and so, at times, the planets Pluto and Neptune pass each other. However, the large inclination of the orbit of Pluto makes a collision virtually impossible, the closest approach of the two planets being about 38.4×10^7 kilometers.

Pluto appears as a very faint star of about the fifteenth magnitude, with a yellowish color, in contrast to the greenish appearance of its nearer neighbors in the solar system. It is only within the last few years that spectroscopic data have revealed a few specifics concerning the nature of Pluto's surface. At one time it was believed that Pluto was similar in size and mass to Mars and Earth, but these estimates have been revised downward. In observing the infrared reflection of Pluto through two very narrow band filters, astronomers from the University of Hawaii working at Kitt Peak National Laboratory in 1976 found the response was exactly as expected for methane ice. With a surface of frozen methane, Pluto must be colder than 50 K (methane condenses at this temperature under low pressure). Methane gas previously had been found in the atmospheres of Jupiter, Saturn, Uranus, and Neptune, but this was the first finding of solid methane on a planet. Some scientists believe that Pluto may be the only planet that closely exhibits the pristine state presumed to have existed some 4.6 billion years ago when the solar system was formed. Now many astronomers believe that Pluto is comparatively small, icy, with low density—more like the satellites of the outer planets than like the planets themselves. Revised estimate place the diameter of Pluto at about 3500 kilometers (2175 miles), making it the smallest planet. For some years, it had been suspected by some astronomers that possibly Pluto is not a true planet, but rather at escaped satellite of Neptune.

A satellite (Charon) of Pluto was first reported in 1978 and it was noted that a series of mutual eclipses might be observable beginning in 1979. An improved orbit determination revised that estimate to the early 1980s. As reported by Binzel (University of Texas) and colleagues at the University of Hawaii and California Institute of Technology, these eclipse events are observable from Earth for only a short period every 124 years (when Pluto's heliocentric motion causes the plane of the satellite's orbit to sweep across Earth's orbit). These conditions, of course, offer a rare means to gain information of this distant planet-satellite system. The investigators use the term "eclipse" broadly to refer both (1) the satellite passing behind the planet, and (2) the satellite passing in front of the planet.

The first eclipses were detected in January and February 1985, further confirming the existence of the satellite.

Shortly after midnight on June 9, 1988, two astronometers (Massachusetts Institute of Technology) in NASA's high-flying observatory 3,500 miles south of Hawaii watched as Pluto eclipsed a small star for 80 seconds. It gave the scientists an opportunity to see what kind of a shadow Pluto casts. Because the shadow was fuzzy, this indicated that the planet has an atmosphere. The researchers estimated Pluto's temperature at $-415\,°F$ ($-248\,°C$), suggesting a methane (or possibly, argon, carbon monoxide, neon, nitrogen, or oxygen atmosphere).

Although still speculative, some tentative consensus on the nature of Pluto is commencing to develop. Based largely upon studies of the planet's flickering light as the result of Charon's fortuitous eclipses of the planet in recent times, from data that may prove to be more valuable than that which may be obtained with the Hubble Space Telescope. See also **Hubble Space Telescope (HST)**. Pluto is considered a small body of rock and ice only about two-thirds the size of Earth's moon. According to data extrapolation,

Pluto's south polar region is extremely bright and presumed to be coming from frozen methane and other chemical ices. The north polar cap, by contrast, is much smaller and less brilliant. This is puzzling because Pluto has a 248-year orbit around the sun. During recent years, the planet's south pole is approaching the end of a century-long summer. During this period, the north pole has been in shadow. Further, Pluto recently passed its closest approach to the sun. Thus, why is there so much ice on the southern cap and less on the northern cap?

Currently, scientists are modeling the planet's atmosphere. Included are astronomers R. Binzel and E. Young (Massachusetts Institute of Technology). These researchers observe that, just as the planet starts to move away from the sun and its atmosphere starts to condense, darkness sets in at the south pole. Thus, the south pole is the most likely place for the deposition of ice when the atmosphere condenses. Further, although the south pole has experienced full sunshine for a century, it nevertheless bears a residual ice cap from its decades in frigid conditions. Perhaps its high reflectivity interferes with sunlight absorption and thus slows the evaporation of the frost and the consequent building of layer upon layer of frost rather than promoting the slow evaporation of the frost.

One astronomer, when commenting on the unusual characteristics of the Plutonian system — the spin of the planet and a satellite almost as large as the planet itself — observed that perhaps these characteristics are not so unusual after all. As the result of computer simulations, the scientist speculates that the early solar system may have included numerous Pluto-like icy bodies, and that the current system may have by some freak occurrence escaped becoming part of the Oort cloud of "icy snowballs," of the kind that form the nucleii of comets. The Plutonium system may have been influenced by gravitational resonance with Neptune. Perhaps the system is one of a kind? Perhaps not!

NOTE: PLUTO DECLARED A DWARF PLANET:

On August 24, 2006 at the International Astronomical Union (IAU) General Assembly in Prague, astronomers decided that the Solar System has eight planets, and Pluto is not one of them. Instead, Pluto is a "dwarf planet."

To be a planet, the assembly ruled, a world must meet three criteria:

(1) *It must have enough mass and gravity to gather itself into a ball.*
(2) *It must orbit the sun.*
(3) *It must regin supreme in its own orbit, having "cleared the neighborhood" of other competing bodies.*

So, e.g., Jupiter, which circles the sun supreme in its own orbit, is a planet– no adjective required. Pluto, on the other hand, shares the outer solar system with thousands of Pluto-like objects. Because it has not "cleared its own neighborhood," it is a dwarf planet.

This decision clarifies the vocabulary of planetary astronomy while simultaneously upturning 76 years of "Pluto is a planet" pop-culture. Will non-specialists heed Pluto's demotion? That remains to be seen. Meanwhile, according to the IAU, the Solar System has eight planets: Mercury, Venus, Earth, Mars, Jupiter, Saturn, Uranus and Neptune; and three dwarf planets: Ceres, Pluto and 2003 UB313.

Additional Reading

Binzel, R.P., et al.: "The Detection of Eclipses in the Pluto-Charon System," *Science*, **228**, 1193–1194 (1985).
Binzel, R.: "Hemispherical Color Differences on Pluto and Charon," *Science*, 1070 (August 26, 1988).
Binzel, R.P.: "Pluto," *Sci. Amer.*, 50 (June 1990).
Kerr, R.A.: "Pluto's Orbital Motion Looks Chaotic," *Science*, 986 (May 20, 1988).
Kerr, R.A.: "Geophysicists Take a Tour Around the Solar System: The Tiniest Planet Shines in the Best Portrait Ever," *Science*, 1635 (June **19**, 1992).
Lunine, J.I.: "Origin and Evolution of Outer Solar System Atmospheres," *Science*, 141 (July **14**, 1989).
Pasachoff, J.M. and D.H. Menzel: *Field Guide to the Stars and Planets*, Vol. 16, 3rd Edition, Houghton Mifflin Company, New York, NY, 1992.
Powell, C.S.: "A Rare Glimpse of a Dim World," *Sci. Amer.*, 24 (August 1992).
Rothman, T.: "A Computer Finds that Pluto's Orbit is Chaotic," *Sci. Amer.* 20 (October 1988).
Shulman, S.: "The Seasons of Pluto," *Technology Review (MIT)*, 9 (July 1989).
Stern, S.A., L. Cesana, and D.J. Tholen: *Pluto and Charon*, University of Arizona Press, Tucson, AZ, 1997.
Stern, S.A. and J. Mitton: *Pluto and Charon: Ice Worlds on the Ragged Edge of the Solar System*, John Wiley & Sons, Inc., New York, NY, 1999.
Sussman, G.J. and J. Wisdom: "Numerical Evidence that the Motion of Pluto is Chaotic," *Science*, 432 (July 22, 1988).

Web References

JPL Solar System Exploration Site: http://sse.jpl.nasa.gov/index.html
Pluto-Kuiper Express…to explore Pluto/Charon and the fringes of our Solar System: http://pluto.jhuapl.edu/
The Planetary Photojournal: http://photojournal.jpl.nasa.gov
The Nine Planets: http://seds.lpl.arizona.edu/billa/tnp/intro.html
Views Of The Solar System: http://www.hawastsoc.org/solar/eng/homepage.htm

PNEUMATIC.   Related to, or pertaining to, air or derivatively to other gases.

PNEUMATIC CONTROLLER.   Although electronic and digital control systems have made serious inroads in the field of process and manufacturing control since the early 1970s, a high percentage of installed instrumentation in industry is pneumatic. Pneumatic controllers still retain certain inherent advantages over other kinds of control systems and, therefore, pneumatic systems are expected to continue an important role over many years in the future.

The function of a pneumatic controller is basically the same as that of an electric or hydraulic controller, the primary difference being that compressed air is used as the controlling medium instead of electricity or hydraulic pressure. Because both pneumatic and hydraulic controllers utilize fluids within mechanical-type systems, they have many design similarities. In hydraulic systems, a jet-pipe is one of the fundamental detectors; in pneumatic systems, the baffle-nozzle is the principal detector. The baffle-nozzle is also commonly called the flapper-nozzle or orifice-nozzle system. A device of this type is shown in Fig. 1. Input motion

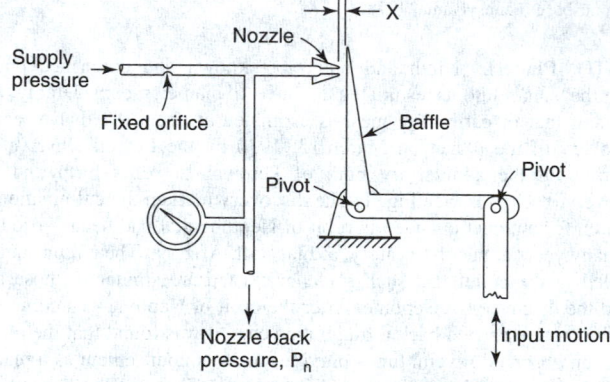

Fig. 1.   Baffle-nozzle detector used in pneumatic controller.

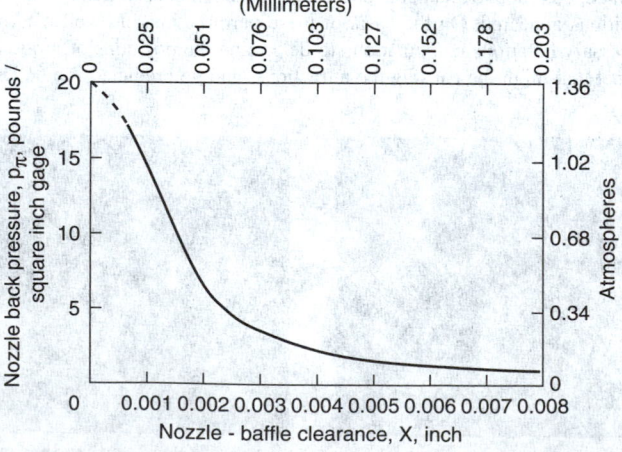

Fig. 2.   Baffle-nozzle characteristic relation between X and P_n for steady state. Conditions are based upon an orifice diameter of 0.01 inch (0.254 mm) and a nozzle diameter of 0.025 inch (0.635 mm).

from some measured variable, such as the movement of a bourdon tube in a pressure- or temperature-measurement system, is applied to a simple pivoted baffle to change the clearance X between a flat surface on the baffle and nozzle. The nozzle normally exhausts directly to atmosphere. Under fixed conditions, air flows through the orifice and out of the nozzle through clearance X. As X is increased from zero, the nozzle back pressure P_n decreases. The upper limit of the value for nozzle pressure is determined by the air-supply pressure when X equals zero. The lower limit is determined by the ratio of resistance to airflow, which is established between the fixed orifice and the nozzle when X is very large. Between these two extremes, various nozzle pressures are established. This relationship is shown in Fig. 2. The slope of the curve is called "nozzle sensitivity," or gain. Detectors of this type exhibit high gains. For a commonly used 0.010-inch diameter orifice and a 0.025-inch ($\sim$0.6 millimeter) diameter nozzle combination, a motion of the baffle of 0.001 inch ($\sim$0.03 millimeter) creates a change in nozzle pressure in excess of 8 pounds/square inch (0.54 atmosphere) in the central portion of the curve.

In some designs, the flat baffle of Fig. 1 is replaced by a ball as shown in Fig. 3. An advantage of this arrangement is elimination of the need to align the baffle surface parallel with the lip of the nozzle. However, slight friction may result if the ball rubs against its guide. Another configuration is the "free vane" shown in Fig. 4. Two nozzles share a common restriction and have a common centerline. The baffle is moved parallel with the nozzle lips. The sensitivity of this type of detector is somewhat less—a baffle motion of 0.005 inch ($\sim$0.13 millimeter) producing a change in nozzle back pressure of about 1 pound/square inch. The direct-operated pneumatic ball pilot, shown in Fig. 5, is still another design configuration for a pneumatic detector.

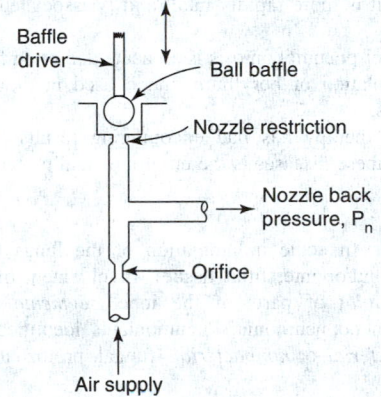

Fig. 3.   Ball-nozzle used in pneumatic controller.

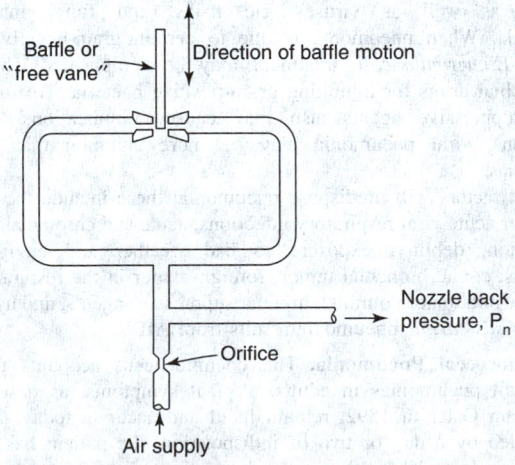

Fig. 4.   Free-vane type detector used in pneumatic controller.

In many pneumatic controllers, a pneumatic pilot relay or pilot valve is used to increase gain. A relay also increases airflow capacity when a change of output pressure is needed. Thus, the dynamic response is improved when the pneumatic instrument is connected to a large volume or to a long transmission line. Gains usually range between 3 and 10. A relay of this type is shown in Fig. 6. Reverse-acting relays also are used to cause a decrease in pneumatic signal pressure with an increase in value of the process variable being measured and controlled. Pneumatic controllers are available with practically all of the modes of control—two-position, proportional, proportional-plus-integral, proportional-plus-derivative, and proportional-plus-integral-plus-derivative actions—that are available in electric controllers. Because of the wide acceptance and preference for pneumatically-operated final controlling elements (valves, dampers, etc.), a pneumatic controller makes an all-pneumatic system possible.

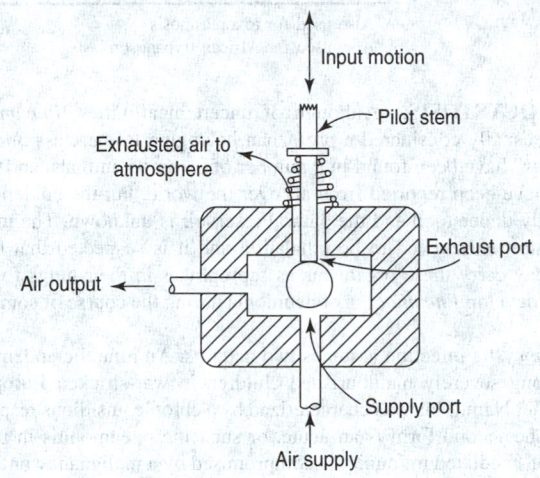

Fig. 5.   Direct-operated pneumatic ball pilot used in pneumatic controller.

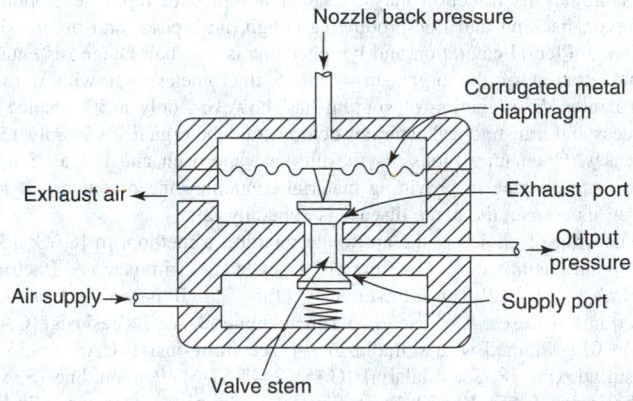

Fig. 6.   Direct-acting pneumatic relay (bleed type) used in pneumatic controller.

The systems previously described sense a change in position. Modern pneumatic controllers also incorporate force-balance systems. The measured variable, when changing, creates differences in force (pressure) rather than actual small movements of an element in space (as in a position-balance system). Pneumatic controllers are frequently designed to be housed within a recorder or an indicator case.

PNEUMATIC-PROBE PYROMETER.   A thermometer for high-temperature gases, in which the gas is sucked through a nozzle and then cooled. Reliance is place principally on knowledge of the law of gas expansion through the nozzle and on measurement of pressure and mass flow rate of the gas.

PNEUMATOLYSIS.   Pneumatolysis is the process of alteration of existing rocks and mineral deposits, or the formation of new ones, by means of gases or vapors emanating from the magma.

TABLE 1. PNEUMOCYSTOSIS ANTIPROTOZOAL AGENTS

Structure number	Compound name	CAS Registry Number	Molecular formula	Structure
(1)	Sulfamethoxazole[a]	[723-46-6]	$C_{10}H_{11}N_3O_3S$	
(2)	Eflornithine[b]	[67037-37-0]	$C_6H_{12}F_2N_2O_2$	

[a] Also used for toxoplasmosis.
[b] Also used for African trypanosomiasis.

PNEUMOCYSTOSIS. Although of uncertain affinities, *Pneumocystis carini* is usually considered a protozoan belonging to the class *Sporozoa*. The "cysts" have been found in a number of common animals, and human isolates have been reported from all over the world, but the epidemiology is not fully understood and the natural reservoir is unknown. The mode of transmission is believed to be inhalation and it is suspected that healthy people may carry these organisms as saprophytes in their lungs for some time and develop *Pneumocystis* pneumonia during the course of some other illness.

Pneumocystis infection assumes two patterns: An infantile endemic was seen among severely malnourished children in war-stricken Europe and later in Viet Nam. This was characterized by a chronic, insidious respiratory illness. The second form is an acute, or subacute pneumonitis in patients whose cell-mediated immunity is compromised by a malignancy or therapy involved in organ transplantation. *Pneumocystic pneumonia* is also the predominant opportunistic infection in persons suffering from AIDS.

The pathologic foci of the disease are the pulmonary interstitium, with thickening of the alveolar septa and the alveoli.

Pneumocystis infection may be rather abrupt with rapid development of fever, hacking and non-productive cough, tachypnea, and progressive dyspnea. Pleural cavitation and lymphedema is unusual. Diagnosis entails demonstration of the organism—a 4–6 micrometer cyst with 6 to 8 merozoites. Examination of sputum has, however, only a 5% chance of success and transtracheal aspirates demonstrate the organisms in only 15% of cases. Open lung biopsy is the ultimate approach and has a 95% or better success rate in providing material containing the organisms. If left to run its course, the acute disease is generally fatal.

The drug of choice is the antifolate mixture, trimethoprim [CAS: 738-70-5]-sulfamethoxazole (**1**) (mixture is called co-trimoxazole, Bactrim, Septra) given orally or intravenously (Table 1). It is well tolerated if given in low doses. Also used is pyrimethamine (**3**, see **Babesiosis**) [CAS: 58-14-0] combined with sulfadiazine (**4**, see **Babesiosis**) [CAS: 68-35-9] or sulfadoxine (**8**, see Malaria) [CAS: 2447-57-6]. Pentamidine (**5**, see **Babesiosis**) [CAS: 100-33-4] isethionate is administered parenterally but is generally not prescribed for prophylactic use because there is a high incidence of hypotension, renal failure, and pain and tissue injury at the injection site. However, there have been successful prophylactic trials with the drug in the aerosolized form for inhalation. The polyamine-inhibitor eflornithine, (**2**, α-difluoromethylornithine, DFMO; the hydrochloride monohydrate is Ornidyl) is undergoing evaluation as a treatment for *P. carinii*. A combination of clindamycin (**2**, see **Babesiosis**) [CAS: 18323-44-9] and primaquine (**5**, see **Malaria**) [CAS: 491-92-9] shows great promise against this protozoan in clinical trials when administered for prophylaxis or for therapy in mild to moderately severe pneumocystis pneumonia. Either drug alone is not effective. See also **Antiparasitic Agents, Antiprotozoals**.

Web References

Pneumocystosis: http://parasitology.informatik.uni-wuerzburg.de/login/n/h/2400.html.
Pneumocystosis and HIV: http://hivinsite.ucsf.edu/InSite?page=kb-05-02-01.

R. C. VICKERY, Blanton/Dade City, FL

PNEUMOKONIOSES. Diseases of the lungs produced by inhalation of dusts, particularly those containing silica, asbestos and other inorganic material, or certain vegetable substances, notably sugar cane waste and raw cotton dust (brown lung).

Silicosis occurs in industries in which the air is polluted by silica dust, e.g., pottery, metal grinding, sandblasting and mining in rock. The inhaled silica gives rise to the production of diffuse fibrosis in the lungs; moreover it facilitates the growth of the tubercle bacillus so that tuberculosis is a possible complication. A special form of silicosis, called anthracosis (black lung), occurs in coal miners who are exposed to a mixed dust, mainly of coal, with a small proportion of silica.

Asbestosis is much less common but more serious than silicosis, since once contracted it is more rapidly fatal, and is associated with a liability to lung cancer.

Another form of pneumokoniosis is an acute and often fatal form which results from inhalation of beryllium, much used in the manufacture of fluorescent lamps.

Corticosteroid therapy has had encouraging results but the primary consideration in these diseases is the environmental protection of exposed workers.

PNEUMONIA. An acute inflammation of the lungs (parenchyma—alveolar spaces and/or interstitial tissue). Involvement of an entire lobe is *lobar pneumonia*; of parts of the lobe, *segmental pneumonia*; of the bronchi, bronchopneumonia. Pneumonia is identified in accordance with cause—*bacterial* or *nonbacterial* (fungal, protozoan, mycoplasmal, viral, etc.)

Bacteria causing pneumonia include *Streptococcus pneumoniae* (pneumococci); *Staphylococcus aureus*; Group A hemolytic streptococci (*Klebsiella pneumoniae*—Friedländer's bacillus); *Hemophilus influenzae*; and *Francisella tularensis*. Other bacterial pathogens, such as tubercle bacillus, as well as viruses, rickettsias, and fungi may cause pneumonia. When pneumonia is due to certain gram-negative bacilli, such as *Escherichia coli*, treatment may be complicated when using antimicrobial drugs for inhibiting gram-positive bacteria. Treatment with immunosuppressive agents also may cause complications. In some populations, viral pneumonia may be more common than bacterial pneumonias.

Certain factors will predispose pneumonia; these include the common cold, other acute viral respiratory infections, acute and chronic alcoholism, malnutrition, debility, exposure to bad weather and environmental conditions, coma, bronchial tumor, foreign matter in the respiratory tract (such as aspiration of vomitus), immunosuppressive agents, and hypostasis. *Pneumocystis carinii* pneumonia results from AIDS.

Pneumococcal Pneumonia. The pneumococcus accounts for about 60% of all pneumonias in adults. Typical symptoms, as described by Sir William Osler in 1892, remain lucid and accurate today: Abruptly, or preceded by a day or two of indisposition, the patient has a severe chill, lasting from 10 to 30 minutes. In no acute disease is an initial chill so constant or so severe. The fever rises quickly. There is pain in the side, often of an agonizing character. A short, dry painful cough soon develops and the respirations are increased in frequency. When seen on the second or third day the patient presents an appearance which may be quite pathognomonic (specifically characteristic). The patient lies flat in bed, often on the affected side; the face is flushed, particularly the cheeks; the breathing is hurried; the alae nasi (cartilaginous flap on the outer side of either nostril) dilate with every inspiration; the eyes are bright, the expression is anxious, and there is a frequent short cough which makes the patient wince and hold his/her side.

The expectoration is blood-tinged and extremely tenacious. The temperature rises rapidly to 104° or 105 °F (40–40.6 °C). The pulse is full and bounding and the pulse-respiration ratio much disturbed. Examination of the lung shows the physical signs of consolidation — blowing breathing and fine rales. After persisting for 7 to 10 days, the crisis occurs, and with a fall in the temperature the patient passes from a condition of extreme distress and anxiety to one of comparative comfort.

Not all patients present a full menu of symptoms as just described. Prior or current use of antipyretics may result in only a modest fever; the initial rigor may be absent, but recurrent chills are common, and, in the elderly, the additional symptoms of confusion or stupor may be present. The description by Osler portrays the inflammatory response of the alveoli of the lungs. See also **Respiratory System**. There is exudation of fluid (edema, congestion). The pneumococci survive the edema and, in fact, the condition is favorable to the proliferation of the microorganisms. Within several hours, polymorphonuclear leukocytes arrive at the alveoli, but apparently phagocytize (envelope and destroy) only a comparatively small number of the pneumococci. It is believed that phagocytosis at this stage depends upon the capsular polysaccharide (serotype) of the organism. It is at the time of crisis, as previously mentioned, that type-specific anticapsular antibodies arrive and destroy the pneumococci en masse. This is the general course of untreated pneumococcal pneumonia, a rather risky period for some patients. With the availability of antibiotics, prognosis has been markedly enhanced. Penicillin is the usual antibiotic of choice, but erythromycin, chloramphenicol, and others are used. Some strains of pneumococci have developed resistance to penicillin at usual dosage levels. Thus far it has been clinically safe to raise the dosage levels where required to destroy the pathogens. However, in a case in South Africa, reported in 1977, a strain of pneumococci was encountered that was fully resistant to penicillin. In such instances, which to date are rare, vancomycin, rifampin, or bacitracin can be used. See also **Antibiotic**.

In 1978, a new pneumococcal vaccine was introduced. This is effective against 14 serotypes of pneumococci. Although the vaccine is not for the general population, it is indicated for persons considered at high risk.

The natural residence of the pneumococcus is in the nasopharynx (cavity behind the nasal cavities). The bacteria, although always present in most humans, do not usually cause illness in this location because of the very effective defense mechanisms present. An infection of the upper respiratory tract, however, may prepare a pathway for the pneumococci, mainly because such infections increase the volume and lower the viscosity of secretions, making it easier for the pneumococci to spread.

Of treated patients (age 2 to 50), 90% to 95% survive. Patients who are treated within the first 5 days of pneumococcal pneumonia usually respond favorably. Response to antibiotic therapy is often prompt, but fever may persist for a few days in about half of the patients. Antibiotics should be administered for a period at least 48 hours after the disappearance of fever.

Other Pneumonias

Staphylococcal Pneumonia. Caused by *Staphylococcus aureus*, this pneumonia is frequently a complication of influenza, but can be primary. It is not uncommon among hospitalized patients as a superinfection accompanying debility, surgery, tracheostomy, coma, or immunosuppressive therapy. Staphylococcal pneumonia is a life-threatening disease requiring prolonged, high-dose parenteral antibiotic therapy. Oxacillin or nafcillin may be administered, with cephalothin an excellent alternative. Erythromycin or clindamycin may be useful in patients allergic to penicillins and cephalosporins, but if such patients have advanced disease or bacteremia, vancomycin is preferred because it is bactericidal.

Streptococcal Pneumonia. Infrequently seen as a complication of measles and influenza, streptococcal pneumonia is caused by hemolytic streptococci of Lancefield's Group A. Before the advent of modern chemotherapy, this type of pneumonia sometimes followed so-called "strep throat" or scarlet fever. The disease was common among the military in both World Wars. Penicillin G is the antibiotic of choice; cephalosporins for the penicillin-allergic patient.

Klebsiella **Pneumonia.** Gram-negative bacillary pneumonias mainly occur in hospital, although *Klebsiella pneumoniae* will be seen in nonhospitalized persons as well. These bacilli reach and infect lower respiratory areas in three ways—direct expansion of pharyngeal flora, usually by aspiration; contaminated fluid droplets from respirators and ventilatory equipment can introduce the infection; bacteremic spread to the lung, as may be caused by the presence of organisms in drugs injected intravenously. Predisposition for this type of pneumonia often is a severe underlying disease, such as chronic pulmonary disease, heart disease, and alcoholism. *Pseudomonas* lung infections, in contrast, usually are associated with systemic immunosuppression and leukopenia. Symptoms include fever, chills, and malaise, with cough, sputum production, dyspnea, and pleuritic chest pain. In immunosuppressed patients, these symptoms may not be present or they may be overlooked. X-rays cannot establish an etiologic diagnosis. Treatment consists of antimicrobial chemotherapy along with drainage of sequestered fluid in cavitary lesions or pleural fluid. The prognosis for gram-negative bacterial pneumonia depends largely on the presence of an underlying disease. In immunocompromised hosts, mortality rates may range up to 80%. *Klebsiella* pneumonias have a lower mortality rate, ranging up to 50% for severe cases. These infections often cause severe lung tissue damage. *Pseudomonas* pneumonia produces multiple and widespread abscesses.

Pneumonia Resulting from *Hemophilus influenzae*. In connection with the viral influenza pandemics of 1889 and 1918, cases were complicated by pneumonia caused by *H. influenzae*. This bacillus is now an infrequent cause of pneumonia, occurring mainly in patients with chronic bronchitis and bronchiectasis. Most patients recover unless treatment is delayed to the point where bronchiolitis and pneumonia have advanced to produce severe cyanosis and anoxemia. In applying antibiotic therapy, the physician must be aware of penicillinase-producing strains.

Mycoplasmal Pneumonia. At one time, this pneumonia was designated as "primary atypical pneumonia" until it was found that the disease is caused by the pleuro-pneumonia-like microorganism, *Mycoplasma pneumoniae*. These are among the smallest known free-living organisms. Unlike bacteria, mycoplasmas lack a rigid cell wall. Eight species have been found in humans, but only three have been implicated in human disease. It is estimated that from 4% to 10% of cases admitted to hospital for pneumonia suffer with the *M. pneumoniae* infection. The peak seasonal incidence is fall and early winter. The disease is spread by infected respiratory secretions. Epidemics are rare except in military populations. The infection characteristically spreads slowly throughout a family or in other situations where people live in close contact. The incubation period is 8 to 10 days. The illness usually commences with sore throat, followed by cough, headache, malaise, chills, and fever. Cough is frequently nonproductive. Fine or medium rales are noted. Symptoms usually disappear within 1 to 3 weeks after onset, even without treatment. Ear involvement sometimes occurs (10% to 20% of patients). Nonlobar pneumonia with a subacute onset and nonpurulent sputum usually suggests the diagnosis of this type of pneumonia. The antibiotics of choice are erythromycin or a tetracycline. Chemotherapy does not prevent shedding, which may persist from 3 to 8 weeks.

Viral Pneumonia. This illness usually results from exposure of nonimmune individuals to infected persons shedding virus. It is estimated that the disease may account for about 75% of all acute pulmonary infections in some populations (schools, offices, military establishments, etc. where people associate closely). A number of agents — influenza and parainfluenza viruses; adenoviruses; respiratory syncytial virus; rhinoviruses; coxsackie-, echo-, and reoviruses; cytomegalovirus; herpes simplex — may cause the infection. Symptoms usually are mild; pulmonary involvement is not always detected. However, severe and sometimes fatal cases may result, particularly with influenza A virus. Prognosis varies widely with the nature of the causative virus, the patient's age, and presence or absence of underlying diseases. Physicians generally recommend prophylactic vaccination with influenza A and B preparations, particularly for patients over 50 years of age. Amantadine is sometimes administered to moderate the course of the disease.

Associated Pneumonias. As in the case of influenza just described, pneumonia can be associated with a number of other infections, including plague, tularemia, Q fever (rickettsial pneumonia), psittacosis, and legionellosis. These illnesses are described in separate articles in this encyclopedia.

***Pneumocystis carinii* Pneumonia.** Identified as a protozoan, *Pneumocystis carinii* has been known for many years to cause pneumonia in infants with *immune deficiencies* and in children (1 to 4 years old) who have acute lymphatic leukemia. Within the last several years, the disease has been recognized with increasing frequency among patients who are undergoing immunosuppressive treatment. The disease is suspected in any patient receiving immunosuppressive therapy or having an immunologic deficiency who develops progressive pulmonary infiltration and respiratory insufficiency. Although the parasite may occur in the body essentially unnoticed, when untreated an active infection will progress to death.

Only since the discovery and spread of AIDS has the disease become prominent. *Pneumocystis carinii* pneumonia, along with Kaposi's sarcoma, are the principal clinical manifestations of the acquired immunodeficiency syndrome. In the case of the AIDS patient, *P. carinii* is considered an opportunistic infection along with several other such infections (mucosal candidiasis, progressively ulcerating perianal herpes simplex infection, and disseminated cytomegalovirus infection).

Additional Reading

Berkow, R. and M.H. Beers: *The Merck Manual*, 17th Edition, Merck & Company, Inc., Whitehouse Station, NJ, 1999.

Cimolai, N.: *Serodiagnosis of the Infectious Diseases: Mycoplasma Pneumoniae*, Kluwer Academic Publishers, Norwell, MA, 1999.

Godfrey, S.: *Pneumonia*, Blackwell Science, Inc., Malden, MA, 1996.

Jarvis, W.R.: *Nosocomial Pneumonia*, Marcel Dekker, Inc., New York, NY, 2000.

Marrie, T.J.: *Community-Acquired Pneumonia*, Kluwer Academic Publishers, Norwell, MA, 2001.

Rello, J. and K.V. Leeper: *Severe Community Acquired Pneumonia*, Kluwer Academic Publishers, Norwell, MA, 2001.

Stevens, D.L. and E.L. Kaplan: *Streptococcal Infections: Clinical Aspects, Microbiology, and Molecular Pathogenesis*, Oxford University Press, Inc., New York, NY, 1999.

Web Reference

Centers for Disease Control and Prevention: http://www.cdc.gov/health/diseases.htm#S

PNEUMONIA (Legionellosis). See **Legionellosis**.

PNG (PORTABLE NETWORK GRAPHICS). See **Data Compression**.

PODICIPEDIFORMES. In this order of birds (*Aves*) are found the grebe and dabchick. The grebe is a swimming bird with lobed toes, short legs, short neck, and a sharp beak, which in some species is quite long. The grebe is found in temperate regions of both hemispheres and members of the same species may have a very wide range. The little grebe (*Podiceps ruficollis ruficollis*) is an aquatic bird of the Old World. The pied-billed grebe (*Podilymbus podiceps*) is of the New World. Little grebes are also called dabchicks. See also **Grebe**.

PODOCARPS. Medium-to-large evergreen shrubs and trees that are the main conifers of the Southern Hemisphere. They are members of the family *Podocarpaceae* and many belong to the genus *Podocarpus*. The main source of lumber and possibly the only native conifer of South Africa is the yellow-wood tree (*Podocarpus latifolius*). Most podocarps bear a nutlike fruit, which is embedded in a colored berry and, technically, these are considered the "cones." The leaves are narrow, ranging from 2 to 5 inches (5 to 12.5 centimeters) in length. The leaves have a rather luxuriant yellow-green color and glossy surface. The brown "pine" and black "pine" of Australia are podocarps, as is the white "pine" (*P. dacrydioides*) of New Zealand. This tree, which attains a height of 200 feet (60 meters) or more, is the largest of the podocarps.

Podocarps exhibit considerable variation, in that, in addition to the aforementioned species, there are some podocarps that resemble the yews; and others that resemble the cypresses. A subfamily of *Podocarpaceae*, known as the *Dacrydium* branch, includes trees with scale-like leaves reminiscent of some cypresses or junipers. In this family branch are the New Zealand rimu (*D. cupressinum*), a small, weeping-type tree, and the Tasmanian Huon "Pine" (*D. franklinii*), which has been described as a cypress-like weeping willow. In Chile, a yew-like podocarp, Prince Albert's yew (*Saxegothaea conspicua*) flourishes. Another yew-like podocarp is *P. andinus* of Chile. There is one podocarp native to Japan, the only known native podocarp of the Northern Hemisphere.

POINCARE WAVE. A gravity wave that is slow enough (low frequency) to feel the effects of the earth's rotation, so that the Coriolis parameter appears in the dispersion relation.

Within a channel in a rotating system, a Poincare wave has sinusoidally varying cross-channel velocity with an integral or half integral number of cross-channel waves spanning the channel. In the shallow water approximation the waves have dispersion relationship with squared frequency

$$\omega^2 = f^2 + c^2(k^2 + \pi^2 n^2 / L^2)$$

in which f is the Coriolis parameter, k is the wavenumber along the channel, L is the width of the channel, n is any positive integer, and c is the phase speed for shallow water gravity waves;

$$c = (gH)^{1/2},$$

in which g is the acceleration due to gravity and H is the mean depth of the fluid. Related to Poincare waves are Kelvin waves, which take the role of the mode with $n = 0$.

AMS

POINCIANA. Sometimes referred to as the *flame tree* or *flamboyant tree*, the poinciana (*Delonix regia*) is regarded by some authorities and beholders as the world's most beautiful tree. The tree is native to Madagascar, but is found in ornamental plantings in several tropical and semitropical regions, such as the Caribbean islands, Central and South America, including Mexico, Hawaii, and southern Florida and the Florida keys. When in Cuba at the turn of the century, Theodore Roosevelt observed, "The tropical forest was very beautiful, and it was a delight to see the strange trees, the splendid royal palm and a tree which looked like a flat-topped acacia, and which was covered with a mass of brilliant scarlet flowers." However, aside from its beauty, the poinciana is deficient in most other respects. The tree produces no fruit and the wood is considered inconsequential and inferior, both for working and as a fuel.

Poincianas grow at a very fast rate, estimated at about twice the growth rate of a Chinese elm. The tree can reach a height of about 20 feet (6.1 meters) in five years. In urban plantings, where the climate is suitable, the poinciana frequently is used, often as a replacement for other trees that die from disease or are accidentally destroyed. The average poinciana achieves a height of from 25 to 30 feet (7.6–9.1 meters), with a trunk diameter ranging between 1.5 and 2 feet (0.5–0.6 meter). The shape of the tree is reminiscent of the African mimosa, with the crown spread out to form a flat, slightly rounded head.

The flowers have brilliantly red petals and appear at the extreme tips of leaf-bearing twigs. When fully open, the flowers measure 4 to 5 in. (10–13 cm) across. As observed by J.M. Haller (*American Forests*, 49–58, June 1982), "Gorgeous when examined singly, doubly so when seen against the feathery green of the leaves, the flowers must be reckoned among the most beautiful of all arboreal flowers. The seed pods are as picturesque as the flowers and, like them, help establish the tree as a member of the great legume family. Warped like the pods of the honey locust, they are two to three times longer than those; it is always surprising to find so small a tree producing such enormous seed containers." The average pod measures $20 \times 1.5 \times \frac{3}{8}$ in. ($51 \times 4 \times 1$ cm). One of the five flower petals is flecked with red and yellow, rather than being solid red as are the others. This variegated petal always has an upward orientation and is on top, thus guiding insects directly into the nectar cup at the pistil's base.

The record tree is located in Florida with a circumference of 102 inches (259 centimeters), a height of 61 feet (16.59 meters), and a spread of 57 feet (17.37 meters).

POINSETTIA. See **Euphorbiaceae**.

POINT. 1. An element of geometry that has position but no extension. 2. An element of geometry defined by its coordinates, such as the point (1,3). 3. An element that satisfies the postulates of a certain space. 4. In positional notation, the character, or the location of an implied symbol, which separates the integral part of a numerical expression from its fractional part. For example, it is called the *binary point* in binary notation and the *decimal point* in decimal notation. If the location of the point is assumed to remain fixed with respect to one end of the numerical expressions, a fixed-point system is being used. If the location of the point does not remain fixed with respect to one end of the numerical, but is regularly recalculated, then a *floating-point system* is being used. A *fixed-point system* usually locates the point by some convention, while a floating-point system usually locates the point by expressing a power of the base.

POINT DISCHARGE. A silent, nonluminous, gaseous electrical discharge from a pointed conductor maintained at a potential that differs from that of the surrounding gas. In the atmosphere, trees and other grounded objects with points and protuberances may, in disturbed weather, be sources of point discharge current. Close to a pointed and grounded conductor that extends above surrounding objects, the local electric field strength may be many times greater than that existing at the same level far from the elevated conductor. When this local field reaches such a value that a free electron, finding itself acted upon by this field, can be accelerated (in one mean free path) to a sufficiently high velocity to ionize neutral air molecules, point discharge will begin. Different structures will yield point discharge under quite different gross field conditions, for geometry is critically important. Point discharge is recognized as a major process of charge transfer between electrified clouds and the earth, and is a leading item in the charge balance of the global electrical circuit.

AMS

POINT DISCHARGE CURRENT. The electrical current accompanying any specified source of point discharge. In the electrical budget of the earth–atmosphere system, point discharge currents are of considerable significance as a major component of the supply current. Estimates made by Schonland (1928) of the point discharge current from trees in arid southwest Africa suggest that this process accounts for about 20 times as much delivery of negative charge to the earth during typical thunderstorms as do lightning discharges. Although the great height of thundercloud bases in arid regions, such as that referred to in Schonland's study, tends to favor point discharge over lightning charge transfer, point discharge still seems more significant than lightning even in England, where Wormell (1953) found for Cambridge a ratio of about 5:1 in favor of point discharge over lightning charge transfer.

See also **Atmosphere (Earth)**; **Fronts and Storms**; and **Supply Current**.

Additional Reading

Chalmers, J.A.: Atmospheric Electricity, 156–175 (1957).
Wormell, T.W.: "Atmospheric Electricity: Some Recent Trends and Problems," *Quart. J. Roy. Meteor. Soc.* **79**, 3–50 (1953).
Schonland, B.F.J.: "The Polarity of Thunderclouds," *Proc. Roy. Soc. A*, **118**, 233–251 (1928).

POINT SOURCE. No finite source of radiation is a true point, but any source viewed from a distance sufficiently great compared to the linear size of the source may be considered as a point source. Point source is a term also used in describing the origin of air or water pollutants — as contrasted with an area or regional source.

POINT-TO-POINT PROTOCOL (PPP). See **Transmission Control Protocol**

POINT-TO-POINT TUNNELING PROTOCOL (PPTP). See **Transmission**

POISON. (1) Any substance that is harmful to living tissues when applied in relatively small doses. The most important factors involved in effective dosage are (a) quantity or concentration, (b) duration of exposure, (c) particle size or physical state of the substance, (d) its affinity for living tissue, (e) its solubility in tissue fluids, and (f) the sensitivity of the tissues or organs. Sharp distinction between poisons and non-poisons is not always possible, because many variables must be taken into consideration in each case. Poisons are divided into four classes by the shipping regulatory agencies, as follows:

- Poison A: A gas or liquid so toxic that and extremely small amount of the gas of the vapor formed by the liquid is dangerous to life.
- Poison B: Less toxic liquids and solids that are hazardous either by contact with the body (skin adsorption) or by ingestion.
- Poison C: Liquids or solids that evolve toxic or strongly irritating fumes heated or when exposed to air (excluding class A poisons).
- Poison D: Radioactive materials.

See also **Toxicity**; and **Toxic Substances**.

Note: A computerized poison information center is operated by the FDA in Washington, DC. The National Clearinghouse for Poison Information Centers is located at 5401 Westbard Ave., Bethesda, MD 20016.

(2) In nuclear technology, any material with a high capture probability for neutrons that may divert an undesirable number of neutron from the fission chain reaction. (3) A substance that reduces or destroys the activity of a catalyst. Carbon monoxide and phosphorus arsenic, or sulfur compounds have this effect on the formation of ammonia from hydrogen and nitrogen gases, and the gases must be highly purified to avoid this. Another example is the poisoning of the platinum catalyst used in emission-control devices by organic lead compounds.

POISSON CONSTANT. The ratio κ of the gas constant R to the specific heat at constant pressure, c_p.

For dry air, $\kappa = 0.2854$. For moist air,

$$\kappa = \frac{R_d}{c_{pd}} \frac{1 + r_v/\varepsilon}{1 + r v c_{pv}/c_{pd}} \approx 0.2854(1 - 0.24r_v),$$

where R_d and c_{pd} are the gas constant and specific heat of dry air, ε is the ratio of the gas constants of water vapor and dry air, c_{pv} is the specific heat of water vapor, and r_v is the water vapor mixing ratio. The Poisson constant appears in the Poisson equation and in the definition of potential temperature.

AMS

POISSON DISTRIBUTION. The Poisson distribution is a discrete distribution with one parameter, whose probability function is given by $P(r) = e^{-\mu} \mu^r / r! r = 0, 1, 2, \ldots$. The mean and variance are both equal to μ, and are best estimated from the sample mean. For large μ, the distribution approaches normality. The binomial distribution $(q + p)^n$ approaches the Poisson distribution as a limiting form when $n \to \infty$ and $p \to 0$ in such a way that $np = \mu$ remains constant. If events occur in such a way that the probability of an occurrence in a small interval of space or time dt is $\lambda \, dt + O(dt^2)$, independently of other intervals, the numbers of events in equal finite intervals follow the Poisson distribution. For this reason such a series of events is known as a Poisson process.

POISSON EQUATION. An inhomogeneous analogue of the Laplace equation, a partial differential equation of the form

$$\nabla^2 \phi = f(x, y, z)$$

It occurs in (1) electrostatics, where ϕ is potential due to a charge distribution of volume density ρ and $\phi = -4\pi\rho$; (2) thermal conductivity, where ϕ is the temperature in a homogeneous medium of thermal conductivity k and in which $A(x, y, z)$ calories of heat are generated per unit of volume and time, so that $f(x, y, z) = -A/k$. When no heat is generated in the medium, $A = 0$, or when there is no charge, $\rho = 0$; hence, Laplace's equation results in each special case.

See also **Electromagnetic Phenomena**; and **Laplace Equation**.

POISSON, SIMEON DENIS (1781–1840). Poisson was a French Mathematician. His important works include a series on definite integrals and his advances in the Fourier series. He published almost 400 mathematical works including applications to electricity, magnetism, and astronomy. The Poisson distribution established a law governing the distribution of rare and randomly occurring events.

Poisson taught at Ecole Polytechnique from 1802 until 1808. Then he became an astronomer at Bureau des Longitudes. In 1809 he was appointed the chair of pure mathematics at the Faculte des Sciences.

See also **Electromagnetic Phenomena**; **Gravitation**; **Poisson Distribution**; **Poisson Equation**; and **Poisson's Ratio**.

J. M. I.

POISSON'S RATIO. If a rod of elastic material is stretched with sufficient force it can be elongated. The unit elongation (elongation per unit of length) is the strain, and may be denoted by s. At the same time the lateral dimensions will contract, the unit lateral contraction being c. The ratio c/s, which is constant for a given material within the elastic limit, is known as Poisson's ratio. For materials in which there is no directionality to elasticity, the value of Poisson's ratio was demonstrated by that celebrated mathematician to be 0.25. The value of 0.30 is generally used for steels, although recent careful determinations indicate 0.28 is a better average value. For aluminum alloys 0.33 is generally used. For values of Poisson's ratio up to 0.50, stretching results in a net increase in volume. At 0.50 the volume remains constant, as in the case of plastic deformation of metals. See also **Elasticity**.

POLANYI, JOHN C. (1929–). Awarded the Nobel Prize in chemistry in 1986 jointly with Dudley R. Herschbach and Yuan T. Lee. Herschbach reported that the energies of reactions of colliding beams of isolated alkali metal atoms and alkyl halide molecules appeared mostly as vibrational excited states of products. Polanyi characterized the excited states by the infrared light emitted by product molecules. His work also led to the development of lasers. Born in Germany, Polanyi studied in England and later became a Canadian citizen. Doctorate awarded by Manchester University, England, in 1952.

POLAR. Descriptive of a molecule in which the positive and negative electrical charges are permanently separated, as opposed to non-polar molecules in which the charges coincide. Polar molecules ionize in solution and impart electrical conductivity. Water, alcohol, and sulfuric acid are polar in nature; most hydrocarbon liquids are not. Carboxyl and hydroxyl groups often exhibit an electric charge. The formation of emulsions and the action of detergents are dependent on this behavior.

See also **Dipole Moment**.

POLAR BEAR. See **Bears**.

POLAR COORDINATES. If r is the distance from the origin of a rectangular Cartesian coordinate system to a point (x, y, z) and if the direction angles of a line drawn from the origin to the point are α, β, γ, then the polar coordinates of the point are given by

$$x = r\cos\alpha; \quad y = r\cos\beta; \quad z = r\cos\gamma; \quad r^2 = x^2 + y^2 + z^2$$

This system is generally called spherical polar coordinates.

If the point lies in a plane determined by a pair of the coordinates, the XY-plane, for instance, then $z = 0$, and with the usual symbols, $\alpha = \beta = (\pi - \theta)$

$$x = r\cos\theta; \quad y = r\sin\theta; \quad \theta = \tan^{-1} y/x$$

The coordinate origin is called the *pole*; the X-axis is the *polar axis*; the angle θ is the *polar* or *vectorial angle* (sometimes the *azimuth* of the point); r is the *radius vector*. Complex numbers are often plotted in this way, the vectorial angle then being called the *amplitude, argument*, or *phase*, and the radius vector is the *modulus*. See also **Argand Diagram**.

POLAR DISTANCE. Angular distance from a celestial pole; the arc of an hour circle between a celestial pole, usually the elevated pole, and a point on the celestial sphere, measured from the celestial pole through 180 degrees. If the declination, d, and the celestial pole are of the same

name, the polar distance is 90 degrees $-d$, but if of contrary name, it is 90 degrees $+d$.

POLAR FRONT THEORY. Originated by the Scandinavian school of meteorologist, a theory whereby a polar front, separating air masses of polar and tropical origin, gives rise to cyclonic disturbances, which intensify and travel along the front, passing through various phases of a characteristic life history. See also **Fronts and Storms**.

POLARIMETER. An instrument for determining the degree of polarization of electromagnetic radiation, specifically the polarization of light.

POLARIMETRIC RADAR. A radar capable of measuring any or all of the polarization-dependent attributes of a target or backscattering medium. The term may denote a radar capable of measuring the full polarization matrix by means of variable transmitted polarization and dual-channel reception. It may also denote a simpler radar that transmits a single polarization and receives separately the copolarized and cross-polarized components of the returned signal.

POLARIMETRY. The basic principles of polarimetry as a method of quantitative chemical analysis were established over 150 years ago. The method is simple and nondestructive. A polarimeter measures the angle of rotation of linearly polarized light upon passage of the light from the unknown sample. Saccharimetry represents the polarimetric analysis of sugar and is a specialized area with its own form of instrumentation and well-established procedures of international acceptance. Polarimetry in other fields is less standardized, but is extensively used for the qualitative determination of numerous alkaloids, steroids, pharmaceutical, and organic chemical products. See also **Saccharimeter**.

Polarimetric instruments are operative with asymmetric molecules in the direct measurement of circular dichrosm (i.e., the difference of absorption of the left and right circularly polarized light as it passes through the sample). The technique is analogous to absorptiometry.

When plane polarized light passes through an anisotropic medium, the refractive indices of the two beams which emerge, which are right-hand and left-hand polarized, respectively, are not the same. This causes a phase difference between the two component beams and the resultant beam is rotated in its plane of polarization as it emerges from the medium.

Molecules of inherent structural asymmetry are anisotropic; they are *optically active* and exhibit *optical rotation* in solution. The typical optically active center is a carbon atom with four different substituents. In addition, any structural dissymmetry that results in a spatial left- and right-handedness will cause optical activity. Compounds of these types of come in a right-hand (R) and left-hand (L) form. When equal amounts of these two forms are mixed (racemic mixtures) there is no optical rotation because the activity of the two forms exactly cancel. Internal compensation of optically active centers in complex molecules is also found. Left- and right-handed optical isomers were first studied by Pasteur well over 100 years ago, and extensive surveys are found in most organic chemical texts.

Visual Polarimeters. A typical visual polarimeter is shown in Fig. 1. Light source may be a sodium or a mercury arc (less usual is the cadmium arc for the 509-millimicron and 644-millimicron lines). A filter isolates the emission line for monochromatic illumination. (While an instrument does not produce absolutely monochromatic light, the term is used by spectroscopists to describe light within a very narrow wavelength range, such as 0.2 millimicron.) The light then passes a polarizer prism system. This is usually a Nicol prism (a prism made of calcite that is cut and recemented in such a way that the incident light is split into a linear polarized beam which is transmitted, while the second beam is reflected and absorbed). The polarized beam is then passed through the analyzer, which is essentially identical with the polarizer. One of these two elements (usually the analyzer) can be rotated, and it is provided with a graduated circle for the precise read-out in angular degrees. By using a large circular scale and a vernier, a precision of 0.002° can be obtained in research-type polarimeters.

The principle of measurement is straightforward. If the two "Nicols" are oriented identically with respect to their optic axes, maximum light

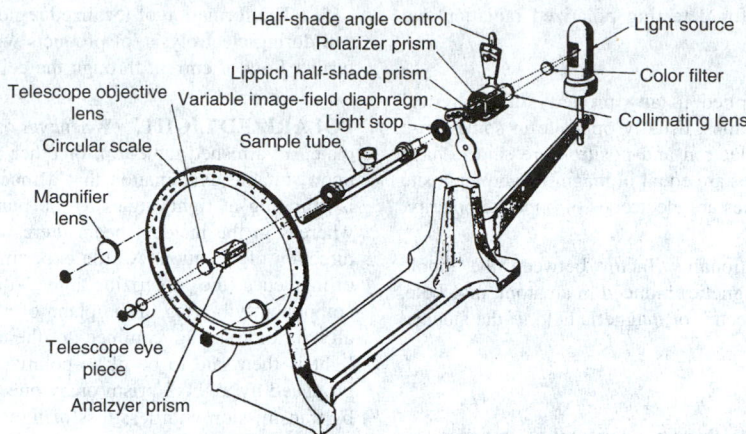

Fig. 1.   Schematic diagram of a visual polarimeter

is passed. When they are crossed (90°), the intensity is at minimum (following a $\sin^2$ law). A refinement in all commercial visual polarimeters is that the observation of the crossed analyzer position is made easier by a half-shade field. See Fig. 2. Because the human eye is a comparative, rather than absolute, light-measuring device, very much better precision can be obtained by comparing two adjacent fields, rather than attempting to evaluate the brightness of a single field. The half-shade fields are created by an auxiliary prism, and the details of the optical arrangement can be found in the literature. Here we are only concerned with the operational features. The zero position of the instrument is that angle at which the two (or three) segments of the observed field are *equally* dim. Between the polarizer and the analyzer, a space is provided to accept the sample. The sample is placed in a tube that has precisely ground ends corresponding to the light path. End windows are held to the tube by gasketed fittings.

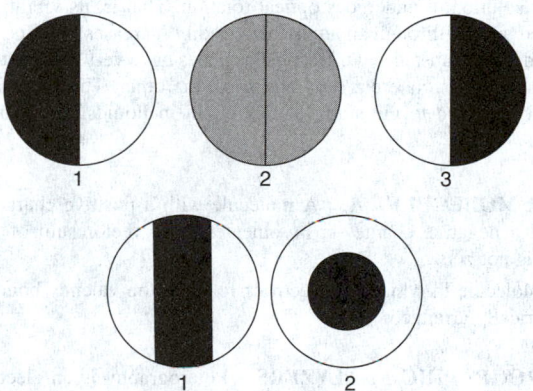

Fig. 2.   Aspect of half-shade fields. Top diagram shows conventional split fields with (2) at the balance point. Bottom diagrams show special field configurations when at the off-balance point.

Routine polarimetric determinations are simple enough. First the polarimeter is balanced to zero degrees with the solvent. Then the solution is placed into the instrument, the instrument is rebalanced, and the angle α read off the scale. Nevertheless, when many measurements are taken, this becomes somewhat tedious. For the assessment of the half-shade field, the operator's eyes must be dark-adapted. Extended work in a darkened room peering through the eyepiece at an almost black field is tiring. The precision of visual polarimetric measurements will tend to increase rapidly at first, as the observer's eyes become adapted, but then it will decrease gradually because of fatigue. These facts justified the introduction of photoelectric polarimeters.

Photoelectric Polarimeters. In one design, there is no split half-shade field, but rather the analyzer is mechanically flip-flopped over an adjustable angle. At balance (and only at balance), the two extreme positions of the analyzer yield equal (low) intensity. A sensitive photomultiplier-photometer serves as a null indicator. The analyzer prism is manually

rotated until a minimum deflection results on the large photometer scale and then the angle of rotation is read visually. In another design, Faraday cells are used. In the Faraday effect, a magnetic field induces optical rotatory power in liquids and solids by its influence on the atomic electron configuration. By using an electromagnet surrounding a glass rod, or a suitable crystal or solution, an alternating optical rotation can be introduced. This is analogous to the flip-flop just described. In the second design, a rotating polarizer is used, followed by the sample cell and the Faraday modulator. The emerging light passes through the analyzer and a photomultiplier pickup. The latter actuates a servosystem which drives the polarizer to balance. When the polarizer is exactly crossed with the analyzer, then, and only then will the alternating polarization introduced by the Faraday modulator have equal magnitude. This is the null point for the servosystem which thus establishes the balance automatically. The operator places the sample into the instrument (after it has been set to zero with the solvent) and then reads off the value of rotation from a magnified scale which allows estimation to 0.0025°. Several other designs have been developed for various measurements in optical rotatory dispersion. A number of spectropolarimeters, both automatic and recording, are available.

Kinetic Polarimetry. Polarimetry is particularly well suited for kinetic studies. The reason lies in the cyclic nature of the phenomenon, which allows the measurement of small changes in the angle of rotation with equal precision in the presence and absence of large background values. Moreover, subtle changes in structure, which are common in enzyme reactions, are often strongly reflected in rotatory power.

Spectropolarimetry. The determination of a rotatory dispersion curve is a prerequisite to the establishment of a sensitive polarimetric technique. This is analogous to colorimetry, where a complete spectrophotometric curve is required to establish the absorption peak best suited for routine work at a fixed wavelength. Similarly, in polarimetry, the gain in sensitivity may be enormous when working at an extremum (peak or trough). Thus, spectropolarimetry plays a role even though the analytical technique is simply polarimetry at a fixed wavelength. It is important to make a judicious choice in wavelength in an industrial assay.

POLARIS *(α Ursae Minoris).* The principal star of the constellation Ursa Minor. The fact that Polaris is located in the "Little Dipper" is about the only claim to recognition that this constellation has. Polaris has been the closest star to the pole of rotation of the celestial sphere for the past three millenniums and has been used by navigators throughout written history. The antiquity of the use of this star is attested to by the fact that it is found represented on the earliest known Assyrian tablets. At present, Polaris is slightly over 1° from the pole of rotation and hence revolves about the pole in a small circle about 2° in diameter. Twice, and only twice, during every 24 hours Polaris accurately defines the true north azimuth.

At present, the star is universally used by navigators and surveyors for the purpose of determining true azimuth and also astronomic latitude. Tables have been computed and are to be found in the Ephemerides of the various governments for the purpose of reducing the actual position of the star at any particular instant to the actual position of the pole of rotation.

POLARISCOPE.   An instrument for detecting polarized radiation and investigating its properties.

POLARITY.   Polarity may be applied to any property of a physical system that can take on only two values, usually opposite in some sense (e.g., sign or direction). Thus, the electron and positron are said to have opposite polarity because their charges are equal in magnitude but opposite in sign. Similarly, anodes and cathodes are electrodes of opposite polarity.

POLARIZABILITY.   The proportionality factor between the dipole moment (usually electric but also magnetic) induced in an atom, molecule, or even a particle and the inducing electric or magnetic field. If the induced electric dipole moment is $\mathbf{p}$,

$$\mathbf{p} = \alpha\mathbf{E},$$

where $\mathbf{E}$ is the electric field and α the electric polarizability. In general, α is a tensor (because $\mathbf{p}$ and $\mathbf{E}$ are not necessarily parallel) and depends on the frequency of the (periodic) field $\mathbf{E}$.

AMS

POLARIZATION.   1.   With respect to a transverse electromagnetic wave, the correlation between two orthogonal components of its electric (or, equivalently, magnetic) field. If the ratio of the amplitudes of these two components and the difference in their phases is constant in time (completely correlated), the wave is said to be polarized (or completely polarized or 100% polarized). If these two amplitudes and phases are uncorrelated, the wave is said to be unpolarized (or 0% polarized). These are two extreme degrees of correlation, never strictly realized in nature, all real waves being partially polarized (or partially correlated). Associated with a polarized wave is its vibration ellipse traced out in time by the oscillating electric field at a given point in space. A line (circle) is a special ellipse, and a wave with such a vibration ellipse is said to be linearly (circularly) polarized, but the general state of (complete) polarization is elliptical. A vibration is characterized by its handedness (the sense in which it rotates; clockwise or counterclockwise), the ratio of its minor to major axis (ellipticity), and its orientation (azimuth). Any beam may be decomposed uniquely as an incoherent superposition of two beams, one unpolarized and one polarized. Thus, the ratio of the transmitted power of the polarized component to the total transmitted power may be taken as a measure of the degree of polarization of the beam. The vibration ellipse of the polarized component and the degree of polarization define the state of polarization of the beam. Polarization would be an uninteresting (indeed, unmeasurable) property of electromagnetic radiation were it not for the fact that two beams, identical in all respects except their state of polarization, may interact with matter differently. Skylight (for a molecular atmosphere) is, in general, partially linearly polarized, the degree of polarization being greatest approximately 90° from the sun.

2.   The process of bringing about a partial separation of electrical charges of opposite sign in a body by the superposition of an external field.

3.   A vector quantity representing the dipole moment per unit volume of a dielectric medium. In rationalized units, the electric induction in a dielectric is given by $\mathbf{D} = \varepsilon\,\mathbf{E}$, which can be written

$$\mathbf{D} = \varepsilon_r\varepsilon_0\mathbf{E} = \varepsilon_0 E + (\varepsilon_r - 1)\varepsilon_0\mathbf{E}$$

where ε_r is the relative permittivity or dielectric constant (κ) of the medium. The term

$$(\varepsilon_r - 1)\varepsilon_0\mathbf{E}$$

is the additional induction attributable to the matter of the dielectric, and is called the polarization of the dielectric. The coefficient $(\varepsilon_r - 1)$ is the "electric susceptibility" of the dielectric, and is often written as χ_e. In unrationalized systems,

$$\chi_e = \frac{\varepsilon_r - 1}{4\pi}$$

4.   The process of confining the vibrations of the magnetic (or electric) field vector of light or other radiation to one plane.

5.   The formation of localized regions near the electrodes of an electric cell during electrolysis, of products which modify (usually adversely) the further flow of current through the cell.

POLARIZED LIGHT.   Whenever ordinary light is reflected from a glass plate, a varnished table-top, or other polished dielectric surface, we find upon suitable examination that a much larger part of the reflected beam is vibrating at right angles to the plane of incidence than in that plane; whereas in the incident beam there was no evidence of any preferential direction of vibration. A little experimenting shows that at a certain angle of incidence (the "polarizing angle," different for different dielectrics), the component vibrating in the plane of reflection is practically extinguished, all vibration being confined to the plane at right angles to this. The light is then said to be plane-polarized. The effect is more conveniently produced by a Nicol prism or by one of the polarizing films that polarize by transmission with less loss of light.

When the light passes through two such polarizers in succession, as in a polariscope, the fraction of it finally emerging depends upon the angle between the transmission planes of the polarizers, and varies all the way from nearly 100% to zero (see also **Malus Cosine-Squared Law**). The same effect may be produced by two reflections at glass plates turned to reflect in different planes but at the same (polarizing) angle. It seems probable that when the polarizing films above mentioned have been further perfected and cheapened, this intensity-reducing effect will be turned to account in reducing automobile headlight glare.

Metallic reflectors do not produce plane-polarization, but when plane-polarized light falls on a polished metal, its vibration is in general changed from a rectilinear to an elliptic one, and the light is said to be elliptically polarized.

When plane-polarized light traverses a crystal exhibiting double refraction, such as calcite, at right angles to its axis, it is transformed into elliptically polarized, or even circularly polarized, light.

If plane-polarized light is passed through quartz along its axis, or in any direction through one of the many optically active liquids such as turpentine or sugar solution, it undergoes optical rotation. That is, its vibration plane is twisted around through an angle that steadily increases with the distance traversed in the substance. Different substances have very different rotatory power, and some rotate one way and some the other. The sacharimeter is especially adapted to the study of this effect in liquids, especially sugar solutions.

POLAR MOLECULE.   1.   A molecule with a positive charge on one end and a negative charge on its other end. Its vector sum of its bond dipoles is not zero.

2.   Molecule in which the electrons forming the valency bond are not symmetrically arranged.

POLAROGRAPHIC ANALYZERS.   Polarography is an electrometric method of chemical analysis that is based on the current-voltage relationship at a special type of electrode. In most electrometric processes, it is desirable to use electrodes of relatively large surface area and, in some cases, to stir the solution, and thus avoid an effect termed "concentration polarization." The cause of this effect is the formation of a region in the solution that differs in composition from the main body of the solution. There is a depletion layer in cases where the ions of the solution are discharging on the electrode, and an excess layer in cases where the electrode is ionizing. Under such conditions, the value of the current at the electrode will not be determined solely by the usual factors: impressed electromotive force, electrode potential, and concentration of the ions in question in the body of the solution. It will also be affected by the rate of diffusion of these ions, from the body of the solution into the depletion layer. Since this rate of diffusion depends in turn upon the concentration of the ions in question in the body of the solution, it can be used as a measure of that concentration by constructing a cell in such a way as to maximize the effect of "concentration polarization." This is done by using a *microelectrode* (an electrode having a very small area of contact with the solution) as the electrode at which the ionic reaction to be measured is to occur, and a large or normal size electrode as the other one in the cell. As a further step, there is added to the solution an excess of an electrolyte that is inert to the electrochemical reaction, so that the diffusion effect on the ion under analysis will not be masked by its migration effect, that is, by its role in carrying current through the cell.

The conditions described for the polarized electrode are met by the dropping mercury electrode, shown in Fig. 1. The size of the polarized electrode, which is merely a forming drop of mercury, is certainly small. It has the further advantage that it is constantly being replaced so that no solid deposit can form on its surface and so contribute an unwanted polarization effect to the one being measured. (Since the polarographic process involves the reduction of the ions, there is a deposition of metal on the mercury drop in all cases in which they are reduced to the metal, the effect of which is overcome by the constant replacement of the drop with fresh mercury.) This advantage is not processed by the other electrode, the pool of mercury in the bottom of the cell. For that reason, among others, the mercury pool electrode is not used as the other electrode in many instruments, being replaced by a standard calomel electrode connected electrically to the solution by a salt bridge, as described earlier in this entry. Moreover, the simple galvanometer and slidewire arrangement for measuring current and potential difference, respectively, is replaced and supplemented by more sensitive and easier-operated measuring and control devices, often automatic in operation and provided with recorders for producing the graphs of the current-potential difference directly. In fact, this instrumentation is so sensitive that it records the small fluctuations in the graph that occur during the formation of the drop of mercury and the beginning of the next one, the characteristic polarographic waves shown in Fig. 2. As can be seen from this figure, the polarographic method makes it possible to measure more than one concentration at a time, as represented by the two waves shown.

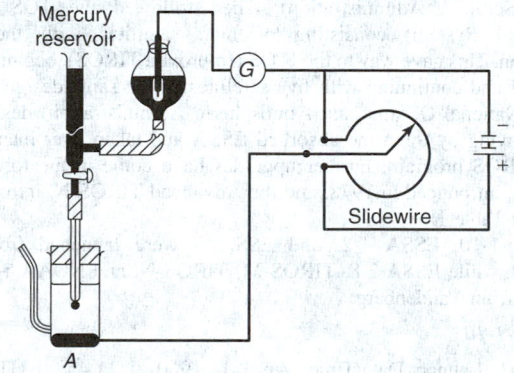

Fig. 1.   Dropping mercury cathode assembly.

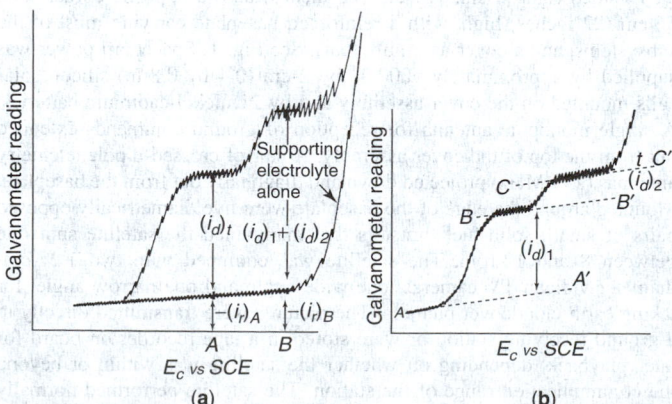

Fig. 2.   Measuring a diffusion current: (**a**) exact method; (**b**) extrapolation method.

In addition to its advantage of permitting the determination of more than one ionic concentration in the same run, polarography also is most useful

in determining very small ionic concentrations, of the order of millimoles per liter or lower.

POLAR-ORBITING ENVIRONMENTAL SATELLITE (POES).
The POES satellite system offers the advantage of daily global coverage, by making nearly polar orbits about 14.1 times daily. Since the number of orbits per day is not an integer the sub orbital tracks do not repeat on a daily basis, although the local solar time of each satellite's passage is essentially unchanged for any latitude. NOAA maintains a two satellite configuration (morning and afternoon), allowing NOAA to obtain data four times daily in any one location. The POES system includes the Advanced Very High Resolution Radiometer (AVHRR) and the Advanced Tiros Operational Vertical Sounder (ATOVS).

The first National Oceanic and Atmospheric Administration (NOAA) series of polar orbiting satellites began with the launch of Television Infrared Operational System (TIROS) satellite on April 1, 1960. Its mission was to provide global meteorological data for research. These satellites carried a Vidicon Camera with an array of lenses. The Vidicon was essentially a television camera which provided visible data at a maximum spatial resolution of 3.8 km (2.4 miles).

The second series of polar orbiting satellites were those of the Environmental Science Services Administration (ESSA), before it became NOAA. These satellites were considered the first generation of "operational" polar orbiters. The ESSA series lasted three years after the launch of ESSA-1 on February 3, 1966. Also placed in near sun-synchronous orbits, these satellites operated in pairs making daily passes over much of the globe during the morning and afternoon hours. The afternoon satellite, odd numbered satellites, was equipped with Advanced version of the Vidicon Camera System (AVCS), which provided visible data at a maximum spatial resolution of 2.2 km (1.4 miles), and a Low Resolution Infrared Radiometer (LRIR) which provided infrared measurements at varying spatial resolutions. The morning satellite, even numbered satellites, were equipped with an Automatic Picture Transmission (APT) system which provided local imagery (at spatial resolutions of 3.8 to 7.4 km (2.4 to 4.6 miles) to suitably equipped ground stations.

The second generation of operational polar orbiters began on January 23, 1970, with the launch of ITOS-1 (Improved TIROS Operational System) and continued with NOAA-1 through NOAA-5, which was launched July 29, 1976. These satellites were also placed in near sun-synchronous orbits with equatorial passes at 0900 Universal Coordinated Time (UTC) and 2100 UTC. Sensors on ITOS-1 and NOAA-1 included an AVCS and a Scanning Radiometer (SR). A Very High Resolution Radiometer (VHRR), and a Vertical Temperature Profile Radiometer (VTPR) were part of the payload for NOAA-2 through NOAA-5. The SR provided global visible and infrared data at 4 and 8 km (2.5 and 5 mi) spatial resolutions, respectively, while the VHRR provided higher resolution data for designated areas. The VTPR was an eight channel radiometer that provided infrared measurements at 68 km (42 mi) resolution.

The third generation of polar orbiting satellites was initiated with the launch of TIROS-N on October 13, 1978. These satellites, like their predecessors, operate in near sun-synchronous orbits. Consecutive equatorial crossings are separated by about 25 degrees of latitude. This produces up to 14.1 orbits per day. Orbital tracks do not repeat on a daily basis, but similar equatorial nodes occur every eight days. The two main sensors on board these satellites include the Advanced Very High Resolution Radiometer (AVHRR) and the TIROS Operational Vertical Sounder (TOVS). The AVHRR is a four/five channel radiometer (depending on the satellite number). Spectral bands range from the visible through the thermal infrared. The TOVS is composed of three different sensors, all measuring incoming radiation in the infrared or passive microwave portion of the electromagnetic spectrum. The three components of the TOVS are the Microwave Sounding Unit (MSU) with four microwave channels, the Stratospheric Sounding Unit (SSU) with three infrared channels, and the High Resolution Infrared Sounder/2 (HIRS/2) with twenty infrared channels.

On May 13, 1998, a new series of operational environmental satellites began with the launch of NOAA-K (NOAA-15). These new satellites carry a series of instruments which have been modified and improved from previous NOAA-series satellites.

The Advanced Very High Resolution Radiometer (AVHRR/2) has been retrofitted with a sixth channel in the near-IR (1.6um), which is time shared with the original channel 3. The new channel is referred to as channel 3A and operates during the daylight part of the orbit. Channel 3B corresponds to the original channel 3 and operates during the night portion of the orbit. Splitting channel 3 in this way maintains the HRPT data format which was designed to handle only five AVHRR channels.

The AVHRR/3 visible channels (1, 2, and 3A) all have "split gains" or "dual slopes" that require the use of two calibration equations per channel, where previously one would suffice. The split gains, in effect, increase the sensitivity at low light/energy levels. The prime reason for these changes are to improve ice, snow and aerosol products produced from the visible channel data. The visible channel ramp voltage calibrations also have dual slopes.

The Microwave Sounding Unit (MSU) and Stratospheric Sounding Unit (SSU) instruments have been replaced with Advanced Microwave Sounding units AMSU-A1, AMSU-A2 and AMSU-B. The AMSU-A is a 15-channel microwave radiometer in two separate units. The new AMSU data is expected to provide improved temperature and humidity soundings. Additionally, window channels 1, 2 and 15 will provide information on precipitation, sea ice and snow cover. The AMSU-B is a five-channel microwave radiometer; three of the channels are centered on the 183.31 Ghz water vapor line. The other two channels are at 89 Ghz and 150 Ghz.

The Solar Backscatter Ultraviolet Radiometer (SBUV) is carried on satellites with an afternoon orbit. NOAA-K was launched into a morning orbit but carries the SBUV/2, which has only minor changes from similar instruments carried on previous spacecraft.

The new High Resolution Infrared Radiation Sounder (HIRS/3) has a different calibration sequence. On HIRS/2, the calibration mode required the use of three calibration targets (space view, cold target, and warm target). On HIRS/3, the cold target will not be routinely used in the calibration sequence, resulting in one additional scan line of Earth data (38 Earth scans per 256 second cycle).

Detailed information about the NOAA-K,L,M instruments, data formats, calibration and similar technical characteristics of the spacecraft are contained in the NOAA-KLM User's Guide which is available on-line http://www2.ncdc.noaa.gov/docs/klm/. A file of this document is also available for downloading via ftp; details can be found at the User's Guide site.

Future

The National Polar-orbiting Operational Environmental Satellite System (NPOESS) will converge the existing NOAA polar-orbiting satellite system and Department of Defense (Defense Meteorological Satellite Program (DMSP), which has been collecting weather data for U.S. military operations for almost four decades under a single national program. The National Performance Review, initiated during the Clinton Administration reported that converging the Department of Defense and Department of Commerce polar-orbiting systems would result in a more cost effective and higher performance integrated system. NPOESS will reduce the cost of acquiring and operating polar-orbiting environmental satellite systems, while continuing to satisfy U.S. operational requirements for data from these systems. NPOESS is administered by the Integrated Program Office (IPO). The current NPOESS mandate extends to the year 2018.

The Initial Joint Polar-Orbiting Operational Satellite System (IJPS) is the result of a cooperative effort between the National Oceanic and Atmospheric Administration and the European Organization for the Exploitation of Meteorological Satellites EUMETSAT. It is comprised of two polar-orbiting satellite systems and their respective ground segments. The IJPS will provide and improve the operational meteorological and environmental forecasting and global climate monitoring services worldwide. The IJPS will continue long-term environmental observations from polar orbit provided by the United States since April 1, 1960. The IJPS program will be contributing to and supporting the World Meteorological Organization (WMO) Global Observing System, the Global Climate Observing System, the United Nations Environmental Program (UNEP), the Intergovernmental Oceanographic Commission (IOC), and other related programs. The US contribution to the IJPS program is being planned, programmed and implemented as part of the POES program under

the responsibility of the Office of Systems Development within NESDIS of NOAA.

Purpose

Because of the polar orbiting nature of the POES and DMSP series satellites, these satellites are able to collect global data on a daily basis for a variety of upper-air, ocean, and atmospheric applications. The collected data supports a broad range of environmental monitoring applications including weather analysis and forecasting, climate research and prediction, earth's water budget, global sea surface temperature measurements, atmospheric soundings of temperature and humidity, ocean dynamics research, volcanic eruption monitoring, forest fire detection, global vegetation analysis, sea-ice monitoring, and many other applications. See also **Weather Technology**.

TIROS Program

The TIROS Program (Television Infrared Observation Satellite) was NASA's first experimental step to determine if satellites could be useful in the study of the Earth. At that time, the effectiveness of satellite observations was still unproven. Since satellites were a new technology, the TIROS Program also tested various design issues for spacecraft: instruments, data and operational parameters. The goal was to improve satellite applications for Earth-bound decisions, such as "should we evacuate the coast because of the hurricane?"

History. The TRIOS program began with 10 experimental spacecraft, TIROS 1 to 10, launched between 1960 and 1965, which represented the first generation of American weather satellites. These carried low-resolution television and infrared cameras, and were developed by the Goddard Space Flight Center and managed by ESSA (Environmental Science Services Administration). Then followed the TOS (TIROS Operational System) consisting of nine satellites with the ESSA designation. This gave way to the ITOS (Improved TIROS), beginning with TIROS-M and continuing with five satellites of the same design with the NOAA (National Oceanic and Atmospheric Administration) designation-NOAA having by this time absorbed ESSA and taken over management of the TIROS program. Further upgrades have come in the form of the TIROS-N, introduced in 1978, and the Advanced TIROS-N, introduced in 1984. See Table 1.

TIROS 1-10, ESSA 1-2, and ESSA 9 were launched from Cape Canaveral, while ESSA 3-8, TIROS-M, TIROS-N, and NOAA 1-16 were launched from Vandenberg.

TIROS-1-10

TIROS-1. Launch Date/Time: April 1, 1960 at 11:45:00 UTC, from Cape Canaveral, FL, aboard a Standard Thor-Able rocket. TRIOS-1 (Television and InfraRed Observation Satellite), the first weather satellite, was designed to test the feasibility of obtaining and using TV cloud cover pictures from satellites. The spin-stabilized satellite was in the form of an 18-sided right prism, 107 cm (42 inches) across opposite corners and 56 cm (22 inches) high, with a reinforced baseplate carrying most of the subsystems, and a cover assembly (hat). See Fig. 1. Spacecraft power was supplied by approximately 9000 1- by 2-cm (0.4 by 0.8 in) silicon solar cells mounted on the cover assembly and by 21 nickel-cadmium batteries. A single monopole antenna for reception of ground commands extended out from the top of the cover assembly. A pair of crossed-dipole telemetry antennas (235 MHz) projected down and diagonally out from the baseplate. Mounted around the edge of the baseplate were five diametrically opposed pairs of small, solid-fuel thrusters that maintained the satellite spin rate between 8 and 12 rpm. The satellite was equipped with two 1.27-cm-diameter vidicon TV cameras, one wide angle and one narrow angle, for taking earth cloudcover pictures. The pictures were transmitted directly to a ground receiving station or were stored in a tape recorder on board for later playback, depending on whether the satellite was within or beyond the communication range of the station. The satellite performed normally from launch until June 15, 1960, when an electrical power failure prevented further useful TV transmission.

TIROS I was operational for only 78 days, but proved that satellites could be a useful tool for surveying global weather conditions from space.

TIROS-2. Launch Date/Time: November 13, 1960 at 11:13:00 UTC, from Cape Canaveral, FL, aboard a Three-stage Delta. TIROS- 2 was

TABLE 1. TIROS PROGRAM LAUNCH SCHEDULE AND FACTS

Series	Spacecraft	Launch date	Launch vehicle	Orbit	Mass
TIROS (Television Infrared Observation Satellite)					
	TIROS-1	Apr. 1, 1960	Thor-Able	656 × 696 km (408 × 432 miles) ×48°	120 kg (265 lbs)
	TIROS-2	Nov. 23, 1960	Delta	547 × 610 km (340 × 379 miles) ×49°	130 kg (287 lbs)
	TIROS-3	Jul. 12, 1961	Delta	723 × 790 km (449 × 491 miles) ×48°	129 kg (284 lbs)
	TIROS-4	Feb. 8, 1962	Delta	693 × 812 km (431 × 505 miles) ×48°	129 kg (284 lbs)
	TIROS-5	Jun. 19, 1962	Delta	580 × 880 km (360 × 547 miles) ×58°	129 kg (284 lbs)
	TIROS-6	Sep. 18, 1962	Delta	631 × 654 km (392 × 406 miles) ×58°	127 kg (280 lbs)
	TIROS-7	Jun. 19, 1963	Delta B	338 × 349 km (210 × 216 miles) ×58°	135 kg (298 lbs)
	TIROS-8	Dec. 21, 1963	Delta B	667 × 705 km (414 × 438 miles) ×59°	119 kg (262 lbs)
	TIROS-9	Jan. 22, 1965	Delta C	701 × 2,564 km (436 × 1,593 miles) ×96°	138 kg (304 lbs)
	TIROS-10	Jul. 2, 1965	Delta C	722 × 807 km (449 × 501 miles) ×99°	127 kg (280 lbs)
TOS (TIROS Operational System)					
	ESSA-1 (OT-3)	Feb. 3, 1966	Delta C	684 × 806 km (435 × 501 miles) ×98°	138 kg (304 lbs)
	ESSA-2 (OT-2)	Feb. 28, 1966	Delta E	1,352 × 1,412 km (840 × 877 miles) ×101°	132 kg (291 lbs)
	ESSA-3 (TOS-A)	Oct. 2, 1966	Delta E	1,348 × 1,483 km (838 × 921.5 miles) ×101°	145 kg (320 lbs)
	ESSA-4 (TOS-B)	Jan. 26, 1967	Delta E	1,323 × 1,437 km (822 × 893 miles) ×102°	132 kg (291 lbs)
	ESSA-5 (TOS-C)	Apr. 20, 1967	Delta E	1,352 × 1,419 km (840 × 882 miles) ×102°	145 kg (320 lbs)
	ESSA-6 (TOS-D)	Nov. 10, 1967	Delta E	1,406 × 1,482 km (874 × 921 miles) ×102°	132 kg (291 lbs)
	ESSA-7 (TOS-E)	Aug. 16, 1968	Delta N	1,428 × 1,471 km (887 × 914 miles) ×101°	145 kg (320 lbs)
	ESSA-8 (TOS-F)	Dec. 15, 1968	Delta N	1,411 × 1,461 km (877 × 908 miles) ×102°	132 kg (291 lbs)
	ESSA-9 (TOS-G)	Feb. 26, 1969	Delta E	1,422 × 1,503 km (884 × 934 miles) ×101°	145 kg (320 lbs)
ITOS (Improved TIROS Operational System)					
	TIROS-M (ITOS-1)	Jan. 23, 1970	Delta N	1,431 × 1,477 km (889 × 918 miles) ×101°	309 kg (681 lbs)
	NOAA-1 (ITOS-A)	Dec. 11, 1970	Delta N	1,421 × 1,470 km (883 × 913 miles) ×101°	306 kg (675 lbs)
	NOAA-2 (ITOS-D)	Oct. 15, 1972	Delta 100	1,446 × 1,453 km (898.5 × 903 miles) ×102°	344 kg (758 lbs)
	NOAA-3 (ITOS-F)	Nov. 6, 1973	Delta 100	1,499 × 1,508 km (931 × 937 miles) ×102°	345 kg (761 lbs)
	NOAA-4 (ITOS-G)	Nov. 15, 1974	Delta 2914	1,442 × 1,457 km (896 × 905 miles) ×102°	340 kg (750 lbs)
	NOAA-5 (ITOS-H)	Jul. 29, 1976	Delta 2914	1,504 × 1,519 km (935 × 944 miles) ×102°	340 kg (750 lbs)
TIROS-N/NOAA (Television InfraRed Operational Satellite - Next-generation)					
	TIROS-N	Oct. 13, 1978	Atlas F	829 × 845 km (515 × 525 miles) ×99°	734 kg (1,618 lbs)
	TIROS-N/ NOAA-6	Jun. 27, 1979	Atlas F	785 × 800 km (488 × 497 miles) ×99°	723 kg (1,594 lbs)
	TIROS-N/ NOAA-7	Jun. 23, 1981	Atlas F	828 × 847 km (514.5 × 526 miles) ×99°	1,405 kg (3,097 lbs)

TABLE 1. (*Continued*)

Series	Spacecraft	Launch date	Launch vehicle	Orbit	Mass
Advanced TIROS-N/NOAA (The Advanced Next-generation Television and InfraRed Operational Satellites (ATN))					
	TIROS-N/ NOAA-8	Mar. 23, 1983	Atlas F	785 × 800 km (488 × 497 miles) ×99°	3,775 kg (8,322 lbs)
	TIROS-N/ NOAA-9	Dec. 12, 1984	Atlas E	833 × 855 km (518 × 531 miles) ×99°	1,712 kg (3,774 lbs)
	TIROS-N/ NOAA-10	Sep. 17, 1986	Atlas E	795 × 816 km (494 × 507 miles) ×99°	1,700 kg (3,748 lbs)
	TIROS-N/ NOAA-11	Sep. 24, 1988	Atlas E	838 × 854 km (521 × 531 miles) ×99°	1,712 kg (3,774 lbs)
	TIROS-N/ NOAA-12	May 14, 1991	Atlas E	805 × 824 km (500 × 512 miles) ×99°	1,416 kg (3,122 lbs)
	TIROS-N/ NOAA-13	Aug. 9, 1993	Atlas E	845 × 861 km (525 × 535 miles) ×99°	1,712 kg (3,774 lbs)
	TIROS-N/ NOAA-14	Dec. 30, 1994	Atlas E	847 × 861 km (526 × 535 miles) ×99°	1,712 kg (3,774 lbs)
	TIROS-N/ NOAA-15	May 13, 1998	Titan II	807 × 824 km (501 × 512 miles) ×99°	1,476 kg (3,254 lbs)
	TIROS-N/ NOAA-16	Sep 21, 2000	Titan II	853 × 867 km (530 × 539 miles) ×99°	1,476 kg (3,254 lbs)
	NOAA-M / NOAA-17	Jun 24, 2002	Titan II	804 × 821 km (500 × 510 miles) ×98.7°	1,500 kg (3,307 lbs)
	NOAA-N/ NOAA-18	May 20, 2005	Delta II	847 × 866 km (526 × 538 miles) ×98.7°	1,442 kg (3,179 lbs)
	NOAA-N/ NOAA-19	Late 2006	Delta II		
	NOAA-N/ NOAA-19	Late 2007	Delta II		

a spin-stabilized meteorological spacecraft designed to test experimental television techniques and infrared equipment designed to develop a worldwide meteorological satellite information system. TIROS-2 was equipped with two independent television camera subsystems, plus a five-channel medium-resolution scanning radiometer and a two-channel nonscanning low-resolution radiometer for measuring radiation from the earth and its atmosphere. The video systems relayed thousands of pictures containing cloud-cover views of the Earth. Early photographs provided information concerning the structure of large-scale cloud regimes. In addition, the experiment to partially control the orientation of the satellite spin axis was successful, as was the experiment with infrared sensors. The satellite spin rate was maintained between 8 and 12 rpm by the use of five diametrically opposed pairs of small, solid-fuel thrusters. The satellite spin axis could be oriented to within 1- to 2-deg accuracy by use of a magnetic attitude control device consisting of 250 cores of wire wound around the outer surface of the spacecraft. The interaction between the induced magnetic field in the spacecraft and the earth's magnetic field provided the necessary torque for attitude control. The spacecraft performed normally from launch until January 22, 1961, when the scanning radiometer began to deteriorate. TIROS-2 was operational fro 376 days.

TIROS-3. Launch Date/Time: July 12, 1961 at 10:19:00 UTC, from Cape Canaveral, FL, aboard a Three-stage Delta. The TIROS 3 TV system was designed to obtain data for operational meteorological use and to further research in obtaining and using TV cloud cover pictures from satellites. The experiment consisted of two redundant pairs of TV cameras, magnetic tape recorders, and TV transmitters. The two sensor units were capable of concurrent of independent operation. The two wide-angle (104 deg) vidicon cameras were mounted on the baseplate of the spacecraft with their optical axes parallel to the spacecraft spin axis, which was in the orbital plane. The cameras were automatically triggered into action only when they came in view of the earth. The pictures were transmitted directly to either of two ground receiving stations or stored on magnetic tape for later playback, depending on whether the satellite was within or beyond the communication range of the station. The TV cameras used 500-scan-line, 1.27-cm vidicons. The recorders could store up to 32 frames of pictures. Transmission of the 32-frame sequence was accomplished in 100 sec by a 3-w FM transmitter operating at a nominal frequency of 237 mHz. At

nominal attitude and altitude (Approximately 700 km), a picture covered a 1200- by 1200-km square with a spatial resolution of 2.5 to 3.0 km at nadir. The experiment was capable of producing daytime cloud cover pictures for the region between 55 deg s to 55 deg n lat. One of the wide-angle cameras failed 13 days after launch. The remaining camera produced useful data until January 23, 1962. The experiment was highly successful. Launched at the start of the hurricane season, the experiment was credited with observing all six major hurricanes of the 1961 season. During the operational lifetime of the experiment, over 24,000 usable pictures were obtained. It was deactivated on February 28, 1962. Data from this experiment are available from the National Climatic Center, Asheville, NC. For a complete index of these data, see 'Catalog of Meteorological Satellite Data — TIROS 3, Television Cloud Photography,' for sale from the U.S. superintendent of documents — or see data set 61-017A-04A.

TIROS-3 was operational for 230 days.

TIROS-4. Launch Date/Time: January 8, 1962 at 12:29:00 UTC, from Cape Canaveral, FL, aboard a Three-stage Delta. A new lens system was implemented for this launch. The lens was designed to reduce distortion and improve resolution. This craft also contained an electronic clock to control the operations of the infrared horizon sensor as well as the magnetic orientation control system. A magnetic tape recorder was still provided for each camera to store photographs while the satellite was out of range of the ground station network. One scanning and two non-scanning radiometers were also on board. The transmitting and receiving antennas were of the same configuration as the previous TIROS models.

TIROS-4 pictures were the best to date, allowing the US Weather Bureau to initiate an international facsimile transmission network in order to share the cloud pictures with weather services around the world. TIROS-4 was operational for 161 days.

TIROS-5. Launch Date/Time: June 19, 1962 at 12:14:00 UTC, from Cape Canaveral, FL aboard a Three-stage Delta. This craft contained all of the instrumentation of TIROS-4 as well as a north direction indicator, and despin weights and spinup rockets in an attempt to improve the craft's longevity. TIROS-5 was also launched at a higher inclination (58 degrees vs. 48 degrees on all previous flights) to provide better observations at higher latitudes. The orbit was elliptical instead of circular due to a Delta

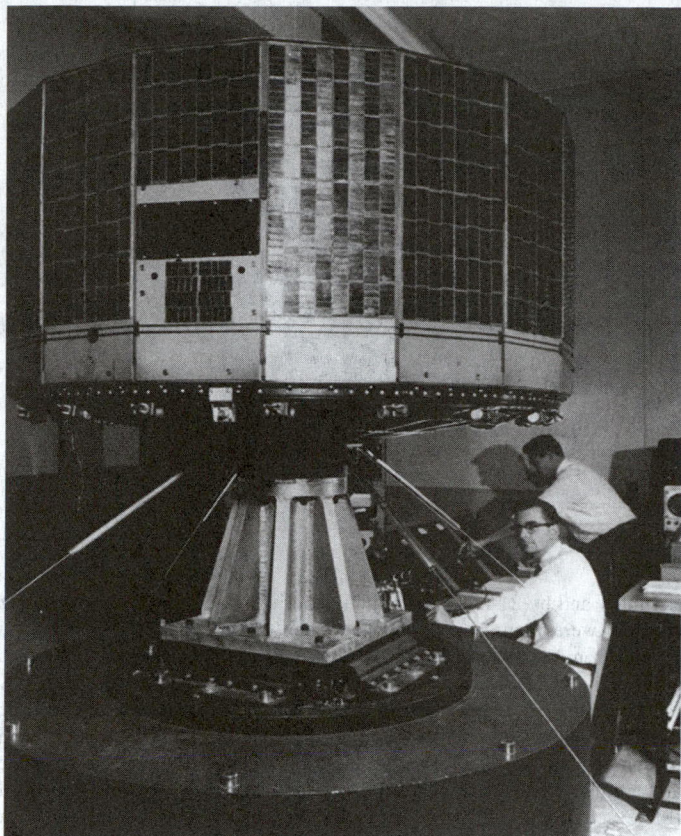

Fig. 1. It is hard to imagine the evening news without satellite imagery. But in 1960, space-based meteorology was in its formative stages, still to be proven. One of the earliest groups to transfer to the newly organized Goddard Space Flight Center was a group from the Army Signal Corps, which developed the nation's first weather satellite called the Television Infrared Observation Satellite (TIROS). With the launch of TIROS-1 from Cape Canaveral on April 1, 1960, scientists immediately saw the benefit of studying Earth's weather systems from the vantage of space. The satellite, which weighed 270 lbs. (122 kg), was specifically designed to test experimental television techniques that would lead to a worldwide meteorological information system. The spacecraft carried two television cameras, a magnetic tape recorder, timer systems, transmitters and a power supply. Early photographs provided new information on cloud systems, including spiral formations associated with large storms, immediately proving their value to meteorologists. (*NASA Goddard Space Flight Center*).

ground guidance failure. In addition the three radiation sensors had to be disconnected prior to launch when they failed preflight checks. With the exception of the failure of the medium-angle camera 17 days after launch, the satellite performed normally until May 14, 1963, when it was deactivated after the shutter electronics failed on the wide-angle camera.

TIROS-5 pictures were the best to date, including the observation of ice break-up at northern latitudes. TIROS-5 was operational for 321 days.

TIROS-6. Launch Date/Time: September 18, 1962 at 8:52:00 UTC, from Cape Canaveral, FL, aboard a Three-stage Delta. The craft contained the same instrumentation as TIROS-5. TIROS-6's launch date was moved up 2 months in order to work with TIROS-5 in helping form an accurate forecast during hurricane season. In addition, TIROS-6 conducted the first satellite experiments to detect snow cover from space. TIROS-6 lasted the longest of the TIROS series thus far, being operational for all of its 389 days.

The camera systems performed normally after launch until November 29, 1962, when the medium-angle camera vidicon failed. The remaining camera system failed on October 21, 1963. The experiment transmitted approximately 60,000 meteorologically useful pictures and furnished information leading to many storm advisories in both the U.S. and abroad. Data from this experiment are available from the National Climatic Center, Asheville, NC. For a complete index of these data, see 'Catalog of Meteorological Satellite Data—TIROS 6 Television Cloud Photography,'

for sale from the U.S. superintendent of documents—or see data set 62-047A-01A.

TIROS-7. Launch Date/Time June 19, 1963 at 09:50:00 UTC, from Cape Canaveral, FL, aboard a Three-stage Delta. TRIOS-7's program objective was to continue research and development of the meteorological satellite information system and to obtain improved data for operational use in weather forecasting during hurricane season. The craft was also designed to make infrared measurements of reflected solar and terrestrial radiation over selected spectrum ranges and gather data on electron density and temperature in space. The craft was also designed to make infrared measurements of reflected solar and terrestrial radiation over selected spectrum ranges and gather data on electron density and temperature in space.

TIROS-7 furnished over 30,000 cloud photographs and lasted the longest of the TIROS series thus far, remaining operational for 1,809 days. The spacecraft performed normally until December 31, 1965, and sporadically until February 3, 1967. The spacecraft was operated for an additional 1.5 years to collect engineering data before being deactivated by NASA on June 3, 1968.

TIROS-8. Launch Date/Time: December 21, 1963 at 09:21:00 UTC, from Cape Canaveral, FL, aboard a Three-stage Delta. TIROS-8 was the first satellite to be equipped with Automatic Picture Transmission (APT) capabilities. The APT experiment provided real-time earth-cloud pictures taken by the satellite to any properly equipped ground receiving station. In addition to an APT camera system, the satellite carried one wide-angle (104 deg) TV camera. Pictures taken by the TV camera were transmitted directly or were stored in a tape recorder on board for subsequent playback, depending on whether the spacecraft was within or beyond communication range of either of two ground receiving stations. The spacecraft performed normally after launch. Over 50 ground stations participated in the APT experiment, which was terminated by the end of April 1964 to degradation of the APT camera. The wide-angle TV camera transmitted useful data until February 12, 1966. TIROS-8 remained operational for 1,287 days. The satellite was deactivated on July 1, 1967, after being left on for an additional time period for engineering purposes.

TIROS-9. Launch Date/Time: January 22, 1965 at 07:55:00 UTC from Cape Canaveral, FL, aboard a Three-stage Delta. TIROS-9's camera configuration was different than any previous TIROS craft. The two cameras were mounted 180 degrees opposite each other along the side of the cylindrical craft rather than in the base plate parallel to the rotation axis. Therefore, a camera could be pointed at some point on Earth every time the satellite rotated along its axis. The spacecraft operating system was the same as in the previous TIROS models except that two infrared horizon sensors were employed.

TIROS-9 was the first of the so-called 'Cartwheel' meteorological TV satellites. That is, the spacecraft spin axis was maintained normal to the orbital plane. The satellite was still equipped with small solid-fuel thrusters as in the case of previous TIROS spacecraft. However, the system was used only as a backup. The satellite spin rate and attitude were primarily determined by a Quarter-Orbit Magnetic Attitude Control (QOMAC) system. The system used the torque developed by interaction of the earth's magnetic field with a current-carrying loop mounted in the satellite. The spacecraft carried two identical wide-angle TV cameras with 1.27-cm vidicons for taking earth cloud cover pictures. The pictures were transmitted directly to either of two ground receiving stations or stored in a tape recorder on board for subsequent playback if the spacecraft was beyond communication range. A failure in the spacecraft guidance system placed the spacecraft in an unplanned 435 to 1,602 mile (700 to 2500 km) elliptical orbit instead of a 400 mile circular orbit. The satellite spin axis was rotated using the magnetic attitude control system into an alignment perpendicular to the orbital plane and tangent to the Earthñs surface. Thus the "cartwheel" configuration was achieved. This configuration was eventually adopted for the ESSA operational series of civilian weather satellites.

The TV system operated normally unitl April 1, 1965, when one of the wide-angle TV cameras failed. The other camera operated normally until July 26, 1965, and sporadically until February 15, 1967. TIROS-9 was the first satellite in the TIROS series to be placed in a near-polar orbit, thereby increasing TV coverage to the entire daylight portion of the globe.

Fig. 2.   TIROS-9 first photomosaic image. (NASA).

The first photomosaic of the entire world's cloud cover was achieved via a composite of 450 photos taken on February 13, 1965. See Fig. 2. Although additional maneuvers were necessary to keep the craft from overheating, TIROS-9 remained operational for 1,238 days until deactivated by NASA on June 12, 1968.

TIROS-10. Launch Date/Time: July 2, 1965 at 04:04:00 UTC, from Cape Canaveral, FL, aboard a Three-stage Delta. The configuration of the TIROS-10 was similar to that of TIROS I-8 with the cameras mounted on the base plate. The spacecraft operating system was also the same as the TIROS I-8 series. The craft was placed in its planned sun-synchronous 98 degree retrograde orbit, drifting westward about 1 degree per day (the same rate and direction as the earth moves around the sun) which provided maximum lighting for photography and battery charging.

TIROS-10 was the last of the experimental TIROS series and provided more than 400 images daily, each of a 640,000 square mile area with two mile resolution at the center. TIROS-10 remained operational for 730 days until deactivated by NASA on July 31, 1967.

TOS (TIROS Operational System). The United States first operational polar-orbiting weather satellite system was placed into service with the launching of the Environmental Science Services Administration ESSA (which in 1970 became the National Oceanic and Atmospheric Administration, NOAA) satellites, ESSA-1, on February 3, 1966, and ESSA-2, on February 28, 1966. The objective of this program, also called the **TIROS Operational System (TOS)**, was to acquire global observational data routinely on a daily basis. This system consisted of a pair of ESSA satellites in sun-synchronous (polar) orbit. The odd numbered satellites (ESSA-1, 3, 5, 7, and 9) utilized the Advanced Vidicon Camera System (AVCS) to obtain global imagery which were transmitted to the ESSA Command and Data Acquition (CDA) stations at Wallops, Virginia, and Fairbanks, Alaska. The CDA stations relayed the data to the National Environmental Satellite Service (NESS), which later became the National Environmental Satellite, Data, and Information Service (NESDIS), located in Suitland, Maryland, for processing and distribution to forecasting centers of the U.S. and other nations. The even numbered satellites (ESSA-2, 4, 6, and 8) were equipped with Automatic Picture Transmission (APT) TV cameras which transmitted television pictures directly to ground stations worldwide.

ESSA-1. Launch Date/Time: February 3, 1966 at 07:41:00UTC, from Cape Canaveral, FL, aboard a Three-stage Delta. ESSA-1 was a spin-stabilized operational meteorological spacecraft designed to take and record daytime cloud cover pictures on a global basis for subsequent playback to a ground acquisition station. The satellite had essentially the same configuration as that of the TIROS series, i.e., an 18-sided right prism, 107 cm (42 inches) across opposite corners and 56 cm (22 inches) high, with a reinforced baseplate carrying most of the subsystems and a cover assembly (hat). Electrical power was provided by approximately 10,000 1- by 2-cm (0.4 by 0.8 in) solar cells that were mounted on the cover assembly and by 21 nickel-cadmium batteries. Two redundant wide-angle cameras were mounted on opposite sides of the spacecraft and canted 75 deg from the spacecraft spin axis. A pair of crossed-dipole command and receiving antennas projected out and down from the baseplate. A monopole telemetry and tracking antenna extended up from the top of the cover assembly. The satellite was placed in a cartwheel orbital mode, with its spin axis maintained normal to the orbital plane. The satellite spin rate and attitude were determined primarily by a Magnetic Attitude Spin Coil (MASC). The MASC was a current-carrying coil mounted in the cover assembly. The magnetic field induced by the current interacted with the earth's magnetic field to provide the necessary torque to maintain a desired spin rate of 9.225 rpm. Five small solid-fuel thrusters mounted around the baseplate provided a secondard means of controlling the spacecraft spin rate. The satellite performed normally after launch until October 6, 1966, when the camera system failed. ESSA-1 was able to view the weather of each area of the globe, photographing a given area at the exact same local time each day. ESSA-1 remained operational for 861 days until deactivated by NASA along with TIROS IX on June 12, 1968.

Advanced Vidicon Camera System (AVCS). The ESSA-1 vidicon camera subsystem was a combination camera, tape recorder, and transmitter that could record and store a series of remote daytime cloud cover pictures for subsequent playback to a ground data acquisition facility. The system was identical to those flown on previous TIROS missions, consisting of two redundant 500-scan-line TV cameras with 1.27-cm vidicons. However, on ESSA-1 the cameras were mounted 180 deg apart on the side of the spacecraft and were canted 75 deg from the spacecraft spin axis. The cameras were triggered into action only when they came into view of the earth. Each tape recorder had two separate channels, one for storing video signals and one for sun-angle data, which served as a time reference. Up to 32 pictures consisting of five levels of gray could be stored for subsequent playback. At nominal attitude and altitude (approximately 1450 km (901 miles), the cameras covered a 1200- by 1200-km square with a spatial resolution of about 3.0 km at nadir. The experiment was a success, with over 100,000 usable pictures transmitted. Data from this experiment are available from the National Climatic Center, Asheville, NC. For a complete index of available data, see parts 1 and 2 of the 'Catalog of Meterological Satellite Data–ESSA-1 Television Cloud Photography' for sale from the U.S. superintendent of documents—or see data set 66-008A-01A.

ESSA-2. Launch Date/Time: February 28, 1966 at 13:55:00 UTC, from Cape Canaveral, FL, aboard a Three-stage Delta. The program objective for ESSA-2 was to complement ESSA-1 and provide direct readout cloud cover photography to ground stations world-wide using the Automatic Picture Transmission (APT) System. The two cameras were mounted 180 degrees opposite each other along the side of the cylindrical craft. The "cartwheel" configuration of the TIROS-9 was selected as the orbital configuration of the operational series of ESSA satellites. Therefore, a camera could be pointed at some point on Earth every time the satellite rotated along its axis. The spacecraft operating system was the same as on the TIROS-9. The craft was placed in its planned sun-synchronous 101 degree inclination retrograde orbit. The satellite spin axis was rotated using the magnetic attitude control system into an alignment perpendicular to the orbital plane and tangent to the Earth's surface. ESSA-2 was able to transmit two to three images daily to individual ground stations regardless of their location. ESSA-2 remained operational for 1692 days until deactivated by NASA on October 16, 1970.

Automatic Picture Transmission (APT) System. The ESSA 2 Automatic Picture Transmission (APT) subsystem was a camera and transmitter combination designed to transmit real-time, daylight, slow-scan television pictures of cloud cover to any properly equipped ground receiving station. The camera system consisted of two redundant APT cameras with 2.54-cm-diameter vidicons. Each camera had a 108-deg wide-angle f/1.8 objective lens with a focal length of 5.7 mm. The cameras were mounted 180 deg apart on the side of the spacecraft, with their optical axes perpendicular to the spacecraft spin axis. The cameras were programmed to take four or eight APT pictures per orbit. The actual photography required 8 sec and the transmission 200 sec. Earth-cloud images retained on the photosensitive surface of the vidicon were read out at four lines per second to produce an 800-line picture. Two 5-w TV transmitters (137.5 mHz) relayed the pictures to local APT stations within communication range. The faceplate of the vidicon had reticle marks that appeared on the picture format to aid in relating the picture to its geographical position on the earth's surface. At nominal satellite attitude and altitude (approximately 1450 km), a picture covered a 3100- by 3100-km square with a horizontal resolution of about 4 km at nadir. There was a 30 percent overlap between pictures along the track to ensure complete coverage. The experiment was a success, and over 4 yr of useful cloud cover pictures were received by participating APT stations. APT data are primarily intended for operational use within the local APT acquisition station. However, copies of pictures taken over the United States are maintained on file at NOAA-NESS, Suitland, Maryland.

ESSA-3. Launch Date/Time: October, 2, 1966 at 10:34:00 UTC, from Vandenberg AFB, CA, aboard a Thrust augmented Three-stage Delta. ESSA-3 (Fig. 3.) was the replacement for ESSA-1.

Fig. 3.   ESSA-3 satellite being readied for launch. (*courtesy of NASA*).

ESSA-3 remained operational for 736 days until deactivated by NASA on December 2, 1968.

ESSA-4. Launch Date/Time: January 26, 1967 at 17:31:00 UTC, from Vandenberg AFB, CA, aboard a Thrust augmented Three-stage Delta. ESSA-4 was the replacement for ESSA-2. ESSA-4 was able to transmit two to three images daily to individual ground stations regardless of their location. ESSA-4 remained operational for 465 days until deactivated by NASA on May 5, 1968.

ESSA-5. Launch Date/Time: April 20, 1967 at 11:17:00 UTC, from Vandenberg AFB, CA, aboard a Thrust augmented Three-stage Delta. ESSA-5 (Fig. 4.) was the replacement for ESSA-3. The spacecraft performed normally after launch until September 22, 1967, when the radiometer experiment failed. The AVCS functioned until October 8, 1969, when the satellite was placed in a standby mode. ESSA-5 remained operational for 738 days until deactivated by NASA on February 20, 1970.

Fig. 4.   ESSA-5 satellite is being enclosed in the shroud in preparation for launching. (*courtesy of NASA*).

ESSA-6. Launch Date/Time: November 10, 1967 at 18:00:00 UTC, from Vandenberg AFB, CA, aboard a Thrust augmented Three-stage Delta. ESSA-6 was the replacement for ESSA-4 to provide direct readout cloud cover photography to ground stations worldwide using APT. The satellite performed normally after launch. The APT system was turned operationally off on July 25, 1969, and reactivated on September 11, 1969. The spacecraft was deactivated on November 4, 1969.

ESSA-6 was able to transmit up to eight images daily to individual ground stations around the world. A total of 305 receiving stations were now operational around the world, including 26 US universities, 25 US television stations, and the weather services of 45 foreign countries. ESSA-6 remained operational for 465 days until deactivated by NASA on December 3, 1969.

ESSA-7. Launch Date/Time: August 16, 1968 at 11:31:00 UTC, from Vandenberg AFB, CA, aboard a Two-stage long tank Delta. ESSA-7 was the replacement for ESSA-5 to provide cloud cover photography to the US's National Meteorological Center for the purpose of preparing operational weather analyses and forecasts. The ESSA-7 system transmitted images covering 2000-square mile areas with 2-mile resolution from every location once per day. Two arrays of radiometer sensors were also mounted 180-degrees apart to measure the global distribution of solar radiation reflected by the Earth and the Earth's atmosphere, as well as the long wave emissions from the Earth (a contribution from the NIMBUS program).

One AVCS camera failed almost immediately after launch. The radiometer experiment failed on June 23, 1969, and the remaining camera system failed on July 19, 1969. ESSA-7 remained operational for 571 days until deactivated by NASA on March 10, 1970.

ESSA-8. Launch Date/Time: December 15, 1968 at 17:17:00 UTC, from Vandenberg AFB, CA, aboard a Two-stage long tank Delta. ESSA-8 was the replacement for ESSA-6 to provide readout cloud cover photography to ground stations worldwide using APT.

ESSA-8 was able to transmit eight to ten images daily to nearly 400 individual ground stations around the world. ESSA-8 remained operational for 2,644 days until deactivated by NASA on March 12, 1976.

ESSA-9. Launch Date/Time: February 26, 1969 at 07:47:00 UTC, from Cape Canaveral, FL, aboard a Three-stage, thrust augmented, improved Delta. ESSA-9 (Fig. 5.) was the replacement for ESSA-7 to provide cloud cover photography to the US's National Meteorological Center for the purpose of preparing operational weather analyses and forecasts. The ESSA VII system transmitted images covering 2000 mile square areas with 2 mile resolution from every location once per day. Two arrays of radiometer sensors were also mounted 180 degrees apart to measure the global distribution of solar radiation reflected by the earth and the earth's atmosphere, as well as the long wave emissions from the earth (a contribution from the NIMBUS program).

Fig. 5. ESSA-9 in final check-out prior to being placed in launch vehicle. (*courtesy of NASA*).

ESSA-9 remained operational for 1,726 days until deactivated by NASA on November 15, 1972.

ITOS (Improved TIROS Operational System). The Improved TIROS Operational System began with the launch of ITOS-1 and continued with NOAA-1 through NOAA-5 (also called TIROS-M series). The primary objective of the ITOS program was to combine the capabilities of ESSA's operational satellites (the operational meteorological and APT satellites) and the knowledge gained from the ongoing NIMBUS program into one operational program with APT capability. The ITOS program would serve as the second generation of US operational weather satellites, eventually becoming the series of what we now know as the 'NOAA' satellites. These satellites were placed in near Sun-synchronous orbits with equatorial passes at 0900 Universal Coordinated Time (UTC) and 2100 UTC. Sensors on ITOS-1 and NOAA-1 included an AVCS and a Scanning Radiometer (SR). A Very High Resolution Radiometer (VHRR), and a Vertical Temperature Profile Radiometer (VTPR) were part of the payload for NOAA-2 through NOAA-5. The SR provided global visible and infrared data at 4 and 8 km spatial resolutions, respectively, while the VHRR provided higher resolution data for designated areas. The VTPR was an eight channel radiometer that provided infrared measurements at 68 km resolution.

ITOS-1 (TIROS M). Launch Date/Time: January 23, 1970 at 11:31:00 UTC, from Vandenberg AFB, CA, aboard a Two-stage Delta N. ITOS-1 (TIROS-M) was the prototype spacecraft for the second generation

of operational sun-synchronous meteorological spacecraft. The primary objective of ITOS-1 was to provide improved operational infrared and visual observations of earth cloud cover for use in weather analysis and forecasting. Secondary objectives included providing both solar proton and global heat balance data on a regular daily basis. To accomplish these tasks, the spacecraft carried four cameras, two television cameras for Automatic Picture Transmission (APT) and two Advanced Vidicon Camera System (AVCS) cameras. It also carried a low-resolution Flat Plate Radiometer (FPR), a Solar Proton Monitor (SPM), and two scanning radiometers that not only measured emitted infrared radiation, but also served as a backup system for the APT and AVCS cameras. The nearly cubical spacecraft measured 1 by 1 by 1.2 m. The TV cameras and infrared sensors were mounted on the satellite baseplate with their optical axes directed vertically earthward. The satellite was equipped with three curved solar panels that were folded during launch and deployed after orbit was achieved. Each panel measured over 4.2 m (14 ft) in length when unfolded and was covered with 3,420 solar cells, each measuring 2 by 2 cm (0.8 by 0.8 in). The ITOS 1 dynamics and attitude control system maintained desired spacecraft orientation through gyroscopic principles incorporated into the satellite design. Earth orientation of the satellite body was maintained by taking advantage of the precession induced from a momentum flywheel so that the satellite body precession rate of one revolution per orbit provided the desired 'earth looking' attitude. Minor adjustments in attitude and orientation were made by means of magnetic coils and by varying the speed of the momentum flywheel. Launched into a near-polar orbit, the spacecraft and experiments performed normally until the incremental tape recorder failed on November 16, 1970, resulting in partial loss of SPM and FPR data. Overheating developed in the satellite attitude control system during March 1971. Attempts to correct the problem were unsuccessful, and the spacecraft was deactivated on June 18, 1971.

NOAA-1 (ITOS-A). Launch Date/Time: December 11, 1970 at 11:35:00, from Vandenberg AFB, CA, aboard a Two-stage Delta N. The primary objective of NOAA-1 was to double the daily coverage of the ESSA satellites; provide infrared imagery, cloud top temperature and surface temperature every twelve hours to APT users. The APT and Direct Readout Infrared (DRIR) subsystems were turned off on June 20, 1971 in an attempt to reduce the above normal temperature due to overheating in the attitude control system. The AVCS was turned off shortly thereafter, and the scanning radiometer continued partial operations. NOAA-1 was operational for 252 days until deactivated by NOAA on August 19, 1971.

NOAA-2 (ITOS-D). Launch Date/Time: October 15, 1972 at 17:17:00 UTC, from Vandenberg AFB, CA, aboard a Delta 300 rocket. NOAA-2 was the first in a series of reconfigured ITOS satellites launched with new meteorological sensors onboard to expand the operational capability of the ITOS system. NOAA-2 was not equipped with conventional TV cameras. It was the first operational weather satellite to rely solely upon radiometric imaging to obtain cloud cover data.

The Sun-synchronous spacecraft was also capable of supplying global atmospheric temperature soundings and very high-resolution infrared cloud cover data for selected areas in either a direct readout or a tape-recorder mode. A secondary objective was to obtain global solar-proton flux data on a real-time daily basis. The sensors were mounted on the satellite baseplate with their optical axes directed vertically earthward. The dynamics and attitude control system maintained desired spacecraft orientation through gyroscopic principles incorporated into the satellite design. Earth orientation of the satellite body was maintained by taking advantage of the precession induced from a momentum flywheel so that the satellite body precession rate of one revolution per orbit provided the desired 'Earth-looking' attitude. Minor adjustments in attitude and orientation were made by means of magnetic coils and by varying the speed of the momentum flywheel. The primary sensors consisted of a Very High Resolution Radiometer (VHRR) Vertical Temperature Profile Radiometer (VTPR), and a Scanning Radiometer (SR). See Fig. 6.

The spacecraft operated satisfactorily until March 18, 1974, when the VTPR failed. NOAA-2 was then placed in a marginal standby mode from March 19 to July 1, 1974. It was then used as the operational NOAA satellite until October 16, 1974, when it was again placed in a marginal standby mode. NOAA-2 was operational for 837 days until deactivated by NOAA on January 30, 1975.

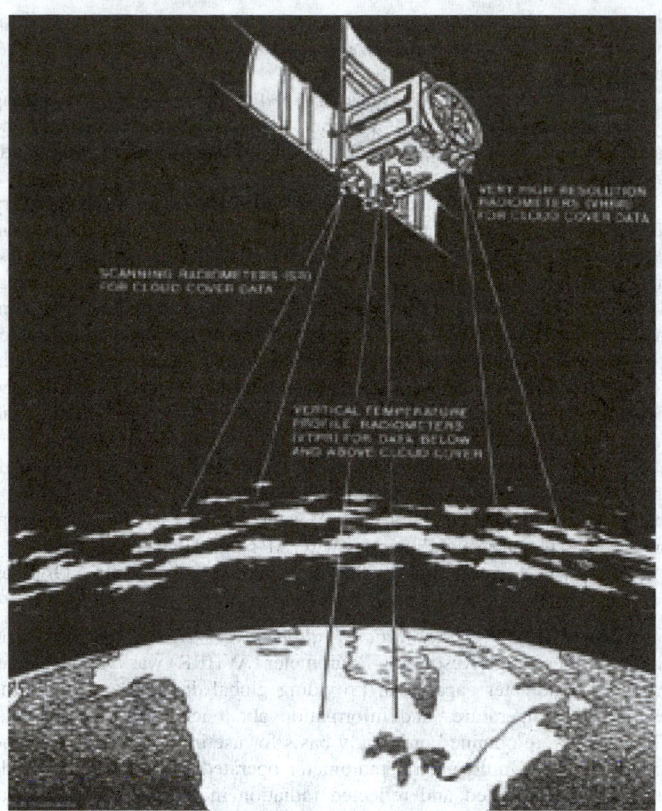

Fig. 6. Artist rendering of NOAA-2, which was the first operational three-dimensional information on the world's weather. (*courtesy of NASA*).

Very High Resolution Radiometer (VHRR). The Very High-Resolution Radiometer (VHRR) experiment was designed to continuously measure surface temperatures of the earth, sea, and cloud tops in daylight as well as at night and to transmit the temperature data in real time to command and data acquisition (CDA) stations throughout the world for use in local weather forecasting. The spacecraft could also be programmed to record up to 9 min of data for remote areas when no CDA stations were within range of the spacecraft, with the recorded data being played back to the next CDA station that the spacecraft passed. The experiment included two scanning radiometers, a magnetic tape recorder, and associated electronics. The two-channel VHRR operated similarly to the Scanning Radiometer (SR) but with much greater resolution (0.9 km compared to 4 km for the SR at nadir). One VHRR channel measured reflected visual radiation from cloud tops in the limited spectral range of 0.6 to 0.7 micron. This provided more contrast between the earth and clouds than the SR by reducing the effect of haze. The second channel measured infrared radiation emitted from the earth, sea, and cloud tops in the 10.5- to 12.5-micron region. This spectral region permitted both daytime and nighttime radiance measurements. The VHRR formed an image by using a scanning mirror technique similar to the SR except that both radiometers operated simultaneously. As the satellite proceeded in its orbit, the 400-rpm revolving mirrors scanned the earth's surface 180 deg out of phase (one mirror at a time) and perpendicular to the orbit path. The visible and infrared data were time-multiplexed so that the scan of the infrared channel transmitted first, followed by the earth scan portion of the visible channel. This process was repeated 400 times per minute (equivalent to the scan rate). If one of the radiometers failed, the system was still capable of measuring both visible and infrared radiation using only the remaining radiometer. Data from this experiment are presently maintained at NOAA-NESS, Suitland, MD.

Vertical Temperature Profile Radiometer (VTPR). The Vertical Profile Temperature Radiometer (VTPR) sensed the radiance energy from atmospheric carbon dioxide in six narrow spectral regions centered at 15.0, 14.8, 14.4, 14.1, 13.8, and 13.4 microns. The atmospheric gross water vapor content was determined from measurements centered at 18.7 microns. Measurements were also taken in the 12.0-micron spectral region to determine surface/cloud top temperatures. The VTPR consisted of an optical system, a detector and associated electronics, and a scanning mirror. The scanning mirror looked at the earth's surface perpendicular to the satellite's orbital path. As each area was scanned, the optical system collected, filtered, and detected the radiation from the earth into the eight spectral intervals. The field of view contributing to one profile was approximately 50 km sq at the ground. The radiometer operated continuously, taking measurements over every part of the earth's surface twice a day. The data were recorded throughout the orbit and played back on command when the satellite was within communication range of a command and acquisition station. Ground personnel used the data to compute temperature-pressure profiles to altitudes as high as 30 km (19 miles). Data from this experiment are presently maintained at NOAA-NESS, Suitland, MD.

Scanning Radiometer (SR). The Scanning Radiometer (SR) subsystem consisted of two scanning radiometers, a dual SR processor and two SR recorders. This subsystem permitted the determination of surface temperatures of the ground, the sea, or cloud tops viewed by the radiometer. The radiometer measured reflected radiation from the earth atmosphere system in the 0.52- to 0.73-micron band during the day and emitted radiation from the earth and its atmosphere in the 10.5- to 12.5-micron band during the day and night. Unlike a camera, the SR did not take a picture but instead formed an image using a continuously rotating mirror. The mirror scanned the earth's surface perpendicular to the satellite's orbital path at a rate of 48 rpm. As the satellite progressed along its orbital path, each rotation of the mirror provided one scan line of picture. Radiation collected by the mirror was passed through a beam splitter and spectral filter to produce the desired spectral separation. Up to two full orbits of data (145 min) could be stored on magnetic tape for subsequent transmission (1697.5 MHz) to an acquisition station. The data could be transmitted in real time to local APT stations. Once the signal was received by the ground station, a continuous picture was formed by using a facsimile recorder whose scan was in phase with the satellite's forward motion. At a nominal spacecraft altitude of 1,460 km (907 miles), the radiometer had a ground resolution of better than 4 km at nadir. The radiometer was capable of yielding radiance temperatures between 185 and 330 deg k to an accuracy of 4 and 1 deg k, respectively. Data from this experiment are presently maintained at NOAA-NESS, Suitland, MD. The scanning radiometer performed normally until the spacecraft was deactivated on January 30, 1975. Data were acquired during the periods November 15, 1972, to March 19, 1974, and July 1, 1974, to October 16, 1974, when NOAA 2 was the operational ITOS spacecraft. Identical experiments were included on NOAA-3, and NOAA-4. Data from this experiment are indexed in data set 72-084A-02A.

NOAA-3 (ITOS-F). Launch Date/Time: November 6, 1973 at 17:02:00 UTC, from Vandenberg AFB, CA, aboard a Delta 300 launch vehicle. The NOAA-3 was one in a series of improved TIROS-M type satellites with new meteorological sensors onboard to expand the operational capability of the ITOS system. The primary objectives of NOAA-3 was to provide visible and infrared images of cloud cover, snow and ice and the sea surface; gather information on the vertical structure of temperature and moisture in the atmosphere. NOAA-3 was operational for 1,029 days until deactivated by NOAA on August 31, 1976.

NOAA-4 (ITOS-G). Launch Date/Time: November 15, 1974 at 17:11:00 UTC, from Vandenberg AFB, CA, aboard a Delta 300 launch vehicle. The primary objectives of NOAA-4 was to provide visible and infrared images of cloud cover, snow and ice and the sea surface; gather information on the vertical structure of temperature and moisture in the atmosphere. See Fig. 7.

NOAA-4 was operational for 1,463 days until deactivated by NOAA on November 18, 1978.

NOAA-5 (ITOS-H). Launch Date/Time: July 29, 1976 at 17:07:00 UTC, from Vandenberg AFB, CA, aboard a Two-stage Delta 2310 launch vehicle. The primary objectives of NOAA-5 was to provide visible and infrared images of cloud cover, snow and ice and the sea surface; gather information on the vertical structure of temperature and moisture in the atmosphere. The NOAA-5 was one in a series of improved TIROS-M type satellites with new meteorological sensors onboard to expand the operational capability of the ITOS system.

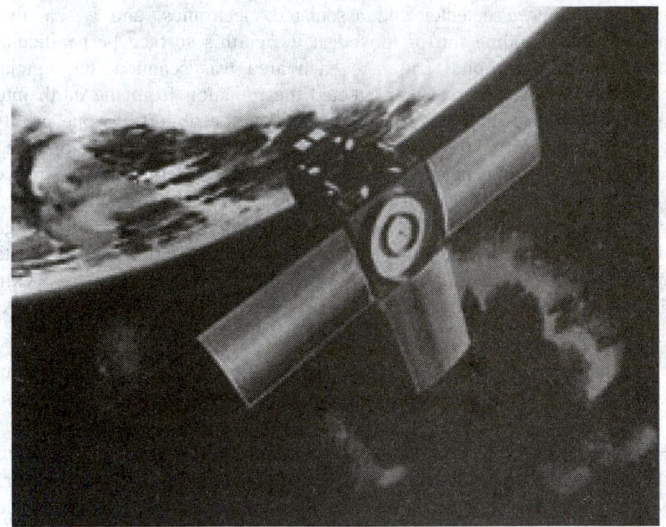

Fig. 7. Artist's illustration of NOAA-4, which would transmit readings and cloud-cover views directly to ground receiving stations throughout the world as it passes overhead. (*courtesy of NASA*).

NOAA-5 was operational for 1,067 days until deactivated by NOAA on July 16, 1979.

TIROS-N/NOAA Program (Television InfraRed Operational Satellite — Next-Generation). The TIROS-N/NOAA Program (Television InfraRed Operational Satellite — Next-generation) was NASA's next step in improving the operational capability of the TIROS system first tried in the 1960's and the ITOS/NOAA system of the 1970's. Technological improvements integrated into the satellite system provided higher resolution imaging, and more day and night quantitative environmental data on local and global scales than seen with the two earlier generations of TIROS. Like earlier TIROS systems, NASA took responsibility for the satellite only until proven operational. Once operational the satellite's name was changed to 'NOAA' with day to day use under the direction of the National Oceanic and Atmospheric Administration.

The TIROS-N/NOAA satellite series carried the Advanced Very High Resolution Radiometer (AVHRR). The AVHRR provided day and night cloud-top and sea surface temperatures, as well as ice and snow conditions. The satellite also carried an atmospheric sounding system (TOVS — TIROS Operational Vertical Sounder) which provided vertical profiles of temperature and water vapor from the Earth's surface to the top of the atmosphere; and a solar proton monitor to detect the arrival of energetic particles for use in solar storm prediction. For the first time, this satellite carried a data collection platform used to receive process and store information from free floating balloons and buoys worldwide for transmission to one central processing facility.

The TIROS-N/NOAA satellite series continue to provide daily observations of the world's weather.

TIROS-N/NOAA Program Satellites:

- TIROS-N 1978–1981
- TIROS-N/NOAA 6 1978–1981
- TIROS-N/NOAA B 1980 Failed
- TIROS-N/NOAA 7 1981–1986
- TIROS-N/NOAA 8 1983–1985
- TIROS-N/NOAA 9 1984–1993, 1997–1998
- TIROS-N/NOAA 10 1986–1991
- TIROS-N/NOAA 11 1988–1994, 1997–Present
- TIROS-N/NOAA 12 1991–Present
- TIROS-N/NOAA 13 1993 Failed
- TIROS-N/NOAA 14 1994–Present
- TIROS-N/NOAA 15 1998–Present
  TIROS-N/NOAA 16 2000–Present

TIROS-N. Launch Date/Time: October 13, 1978 at 11:23:00 UTC, from Vandenberg AFB, CA, aboard an Atlas E/F launch vehicle. The primary objectives were to provide higher resolution, day and night quantitative

environmental data on local and global scales with technologically superior instrumentation than that which was available on the earlier ITOS/NOAA satellites.

The spacecraft was rectangularly shaped (146 inches (371 cm) long by 74 inches (188 cm) high) with one large solar panel attached. The satellite was Earth oriented, three-axis stabilized and weighed 1,594 pounds (723 kg).

TIROS-N was an experimental satellite which carried an Advanced Very High Resolution Radiometer (AVHRR) to provide day and night cloud top and sea surface temperatures, as well as ice and snow conditions; an atmospheric sounding system (TOVS — TIROS Operational Vertical Sounder) to provide vertical profiles of temperature and water vapor from the Earth's surface to the top of the atmosphere; and a solar proton monitor to detect the arrival of energetic particles for use in solar storm prediction. For the first time, this satellite also carried a data collection platform used to receive process and store information from free floating balloons and buoys worldwide for transmission to one central processing facility.

TIROS-N was placed in a near circular, (470 nm) polar orbit. The craft and its systems operated successfully, providing high-resolution scanned images and vertical temperature and moisture profiles to both operational meteorologists and private interests with APT and HRPT capability.

TIROS-N was operational for 868 days until deactivated by NOAA on February 27, 1981.

Advanced Very High Resolution Radiometer (AVHRR). The TIROS-N Advanced Very High Resolution Radiometer (AVHRR) was a four-channel scanning radiometer capable of providing global daytime and nighttime sea-surface temperatures and information about ice, snow, and clouds. These data were obtained on a daily basis for use in weather analysis and forecasting. The multispectral radiometer operated in the scanning mode and measured emitted and reflected radiation in the following spectral intervals: channel 1 (visible), 0.55 to 0.9 micrometer; channel 2 (near IR), 0.725 micrometer to detector cutoff around 1.1 micrometers; channel 3 (IR window), 3.55 to 3.93 micrometers; and channel 4 (IR window), 10.5 to 11.5 micrometers. All four channels had a spatial resolution of 1.1 km, and the two IR-window channels had a thermal resolution of 0.12 deg K at 300 deg K. The AVHRR was capable of operating in both real-time or recorded modes. Real-time or direct readout data were transmitted to ground stations both at low (4-km) resolution via automatic picture transmission (APT) and at high (1-km) resolution via high-resolution picture transmission (HRPT). Data recorded on board were available for central processing. They included global area coverage (GAC) data, with a resolution of 4 km, and local area coverage (LAC), that contained data from selected portions of each orbit with a 1-km resolution. Identical experiments were flown on the other spacecraft in the TIROS-N/NOAA series.

TIROS Operational Vertical Sounder. The TIROS Operational Vertical Sounder (TOVS) consisted of three instruments designed to determine radiances needed to calculate temperature and humidity profiles of the atmosphere from the surface to the stratosphere (approximately 1 mb). The first instrument was the second version of the high-resolution infrared spectrometer (HIRS/2). The HIRS was tested on the Nimbus 6. The HIRS/2 had 20 channels in the following spectral intervals: channels 1 through 5, the 15-micrometer CO_2 bands (15.0, 14.7, 14.5, 14.2, and 14.0); channels 6 and 7, the 13.7 and 13.4-micrometer CO_2/H_2O bands; channel 8, the 11.1-micrometer window region; channel 9, the 9.7-micrometer ozone band; channels 10 through 12, the 6-micrometer water vapor bands (8.3, 7.3, and 6.7); channels 13 and 14, the 4.57 and 4.52-micrometer N_2O bands; channels 15 and 16, the 4.46 and 4.40-micrometer CO_2/N_2O bands; channel 17, the 4.24-micrometer CO_2 band; channels 18 and 19, the 4.0 and 3.7-micrometer window bands; and channel 20, the 0.70-micrometer window region. The second instrument, the stratospheric sounding unit (SSU), was provided by the British Meteorological Office. It was similar to the pressure-modulated radiometer (PMR) flown on Nimbus 6. The SSU operated at three 15.0-micrometer channels using selective absorption, passing the incoming radiation through three pressure-modulated cells containing CO_2. The third instrument, the microwave sounding unit (MSU), was similar to the scanning microwave spectrometer (SCAMS) flown on Nimbus 6. The MSU had one channel in the 50.31-GHz window region and three channels in the 55-GHz oxygen band (53.73, 54.96, 57.95) to obtain temperature profiles which were free of cloud interference. The instruments were cross-course scanning devices utilizing a step scan to provide a transverse scan, while the orbital motion of the satellite provided scanning in the orthogonal direction. The HIRS/2 had a field of view (FOV) 30 km in diameter at nadir, whereas the MSU had a FOV of 110 km in diameter. The HIRS/2 sampled 56 FOVs in each scan line about 2,250 km wide, and the MSU sampled 11 FOVs along the swath with the same width. Each SSU scan line had 8 FOVs with a width of 1,500 km. This experiment was also flown on other TIROS-N/NOAA series spacecraft. For a more detailed description, see W. L. Smith, "The TIROS-N operational vertical sounder," Bull. Am. Meteorol. Soc., v. 60, pp. 1177-1187, 1979. Archival data are available from the Satellite Data Services Division, National Climatic Center, NOAA, Washington, D.C.

Space Environment Monitor. This experiment was an extension of the solar proton monitoring experiment flown on the ITOS spacecraft series. The experiment package consisted of three detector systems and a data processing unit. The medium energy proton and electron detector (MEPED) measured protons above 16, 36, and 80 MeV, and the protons in five energy ranges from 30 keV to >2.5 MeV; electrons above 30, 100, and 300 keV; and protons and electrons (inseparable) above 6 MeV. The high-energy proton alpha telescope (HEPAT), which had a 48-deg viewing cone, viewed in the anti-earth direction, and measured protons in four energy ranges above 370 MeV and alpha particles in two energy ranges above 640 MeV/nucleon. The total energy detector (TED) measured electrons and protons between 300 eV and 20 keV.

Data Collection System (DCS). The Data Collection System (DCS) on TIROS-N was designed to meet the meteorological data needs of the United States and to support the Global Atmospheric Research Program (GARP). The system received low-duty-cycle transmissions of meteorological observations from free-floating balloons, ocean buoys, other satellites, and fixed ground-based sensor platforms distributed around the globe. These observations were organized on board the spacecraft and retransmitted when the spacecraft came in range of a command and data acquisition (CDA) station. For free-moving balloons, the Doppler frequency shift of the transmitted signal was observed to calculate the location of the balloons. The DCS was expected, for a moving sensor platform, to have a location accuracy of 3 to 5 km rms, and a velocity accuracy of 1 to 1.6 m/s. This system had the capability of acquiring data from up to 4000 platforms per day. Identical experiments were flown on other spacecraft in the TIROS-N/NOAA series.

NOAA-6. Launch Date/Time: June 27, 1979 at 15:51:59 UTC, from Vandenberg AFB, CA, aboard an Atlas E/F launch vehicle. The primary objectives were to work as a companion to TIROS-N to provide continuous coverage of the Earth and to provide high-resolution global meteorological data.

NOAA-6 was the first operational satellite in the TIROS-N series. The satellite carried an Advanced Very High Resolution Radiometer (AVHRR) and (TOVS — TIROS Operational Vertical Sounder) previously discussed; a Solar Proton Monitor; and a Data Collection and Platform Location System (DCLS).

NOAA-6 was placed in a near circular, (450 nm) polar orbit. The craft and its systems operated successfully, providing high-resolution scanned images and vertical temperature and moisture profiles to both operational meteorologists and private interests with APT and HRPT capability.

NOAA-6 set the record for the longest duration of any polar orbiting meteorological satellite to date. NOAA-6 was operational for 2,834 days until deactivated by NOAA on March 31, 1987.

Data Collection and Platform Location System (DCLS). The Data Collection and Platform Location System (DCPLS) on NOAA-6, also known as *ARGOS*, was designed and built in France to meet the meteorological data needs of the United States and to support the Global Atmospheric Research Program (GARP). The system received low-duty-cycle transmissions of meteorological observations from free-floating balloons, ocean buoys, other satellites, and fixed ground-based sensor platforms distributed around the globe. These observations were organized on board the spacecraft and retransmitted when the spacecraft came within range of a command and data acquisition (CDA) station. For free-moving balloons, the Doppler frequency shift of the transmitted signal was observed to calculate the location of the balloons. The DCLS was expected, for a moving sensor platform, to have a location accuracy of 3 to 5 km rms, and a velocity accuracy of 1 to 1.6 m/s. This system had the

capability of acquiring data from up to 4000 platforms per day. Identical experiments were flown on other spacecraft in the TIROS-N/NOAA series. Processing and dissemination of data were handled by CNES in Toulouse, France.

Space Environment Monitor (SEM). The Space Environmental Monitor (SEM) was an extension of the solar proton monitoring experiment flown on the ITOS spacecraft series. The object was to measure proton flux, electron flux density, and energy spectrum in the upper atmosphere. The experiment package consisted of three detector systems and a data processing unit. The medium energy proton and electron detector (MEPED) measured protons in five energy ranges from 30 keV to >2.5 MeV; electrons above 30, 100, and 300 keV; protons and electrons (inseparable) above 6 MeV; and omni-directional protons above 16, 36, and 80 MeV. The high-energy proton alpha telescope (HEPAT), which had a 48-deg viewing cone, viewed in the anti-earth direction and measured protons in four energy ranges above 370 MeV and alpha particles in two energy ranges above 850 MeV/nucleon. The total energy detector (TED) measured electrons and protons between 300 eV and 20 keV.

NOAA-7. Launch Date: June 23, 1981, from Vandenberg AFB, CA, aboard an Atlas E/F launch vehicle. The primary objectives were to continue the TIROS-N program by working as a companion with NOAA-6, to provide continuous coverage of the Earth and to provide high-resolution global meteorological data.

NOAA-7 was the second operational satellite in the TIROS-N series. The satellite carried an Advanced Very High Resolution Radiometer (AVHRR) to provide day and night cloud top and sea surface temperatures, as well as ice and snow conditions; an atmospheric sounding system (TOVS — TIROS Operational Vertical Sounder) to provide vertical profiles of temperature and water vapor from the Earth's surface to the top of the atmosphere; and a solar proton monitor to detect the arrival of energetic particles for use in solar storm prediction. This satellite also carried a data collection platform used to receive, process and store information from free floating balloons and buoys world-wide for transmission to one central processing facility.

NOAA-7 was placed in a near circular, (470 nm) polar orbit. The craft and its systems operated successfully, providing high resolution scanned images and vertical temperature and moisture profiles to both operational meteorologists and private interests with APT and HRPT capability. NOAA-7 remained operational for 2625 days until deactivated by NOAA on June 7, 1986, due to a power failure.

The Advanced Next-Generation Television and InfraRed Operational Satellites (ATN) (NOAA-8-19)

NOAA-8. Launch Date/Time: March 28, 1983 at 15:52:00 UTC, from Vandenberg AFB, CA, aboard an Atlas E/F launch vehicle. The primary objectives of NOAA-8 were to continue the Advanced TIROS-N program by working as a companion with NOAA-6 and NOAA-7, to provide continuous coverage of the Earth and to provide high-resolution global meteorological data. A Search and Rescue beacon platform was also tested as part of the expanded capabilities of the ATN satellites. NOAA-8 was a third-generation operational meteorological satellite for use in the National Environmental Satellite Data and Information Service (NESDIS) of NOAA. NOAA-8 was the first spacecraft of the advanced TIROS-N (ATN) series. The satellite design provided an economical and stable sun-synchronous platform for advanced operational instruments to measure the earth's atmosphere, its surface and cloud cover, and the near-space environment. The satellite was based upon the Block 5D spacecraft bus developed for the U.S. Air Force, and it was capable of maintaining an earth-pointing accuracy of better than plus or minus 0.1 degree with a motion rate of less than 0.035 degree/second.

The spacecraft was rectangular shaped (166 inches (421 cm) long by 74 inches (188 cm) high) and powered by a 191in by 94 in (485 cm by 239 cm) solar array. The satellite was Earth oriented; three-axis stabilized and weighed approximately 2000 pounds (907 kg).

NOAA-8 was the first operational satellite in the Advanced TIROS-N series. The satellite carried the AVHRR to provide day and night cloud top and sea surface temperatures, as well as ice and snow conditions; the TOVS to provide vertical profiles of temperature and water vapor from the Earth's surface to the top of the atmosphere; and a solar proton monitor to detect the arrival of energetic particles for use in solar storm prediction. This satellite also carried a data collection platform similar to the platform on the NOAA-6 and NOAA-7. The satellite also carried the SARSAT— Search And Rescue Satellite Aided Tracking system which has the capability to detect distress signals from aviators, mariners and land travelers. This system can detect the location of the distress signal within 8 to 16 kilometers (5 to 10 miles).

NOAA-8 was placed in a near circular, (450 nm) polar orbit. The craft and its systems operated successfully, providing high-resolution scanned images and vertical temperature and moisture profiles to both operational meteorologists and private interests with APT and HRPT capability. This satellite also helped in rescue missions carried out by the US Air Force and Coast Guard.

NOAA-8 was operational for 1,016 days until deactivated by NOAA on December 29, 1985, following a thermal runaway which destroyed the battery.

NOAA-9. Launch Date/Time: December 12, 1984 at 10:48:00 UTC, from Vandenberg AFB, CA, aboard an Atlas E/F launch vehicle. The primary objectives of NOAA-9 were to continue the Advanced TIROS-N program by working as a companion with NOAA-8 to provide continuous coverage of the Earth and to provide high-resolution global meteorological data. The Earth Radiation Budget Experiment (ERBE) and the Solar Backscatter Ultra Violet (SBUV) radiometer were also tested as part of the expanded capabilities of the ATN satellites.

NOAA-9 was the second operational satellite in the Advanced TIROS-N series. The satellite carried the AVHRR (Fig. 8) and TOVS which were present on previous NOAA satellites, the solar proton monitor, and the SARSAT system. In addition, the ERBE instruments (scanner failed in Jan, 1987, non-scanning instrument is still operational), which consisted of short wave and long wave radiometers, were used to study the Earth's albedo in attempt to recognize and interpret seasonal and annual climate variations. The SBUV was carried to measure the vertical structure of ozone in the atmosphere.

NOAA-9 was placed in a near circular, (470 nm) polar orbit. The craft and its systems operated successfully, providing high resolution scanned images and vertical temperature, moisture, and ozone profiles to research meteorologists. NOAA-9 was deactivated by NOAA on August 1, 1993, but was reactivated on August 23, 1993 after the failure of NOAA-13. The SARR transmitter failed on December 18, 1997 and the satellite was permanently deactivated on February 13, 1998.

The Earth Radiation Budget Experiment (ERBE). The radiation budget (Fig. 9) represents the balance between incoming energy from the Sun and outgoing thermal (longwave) and reflected (shortwave) energy from the Earth. In the 1970's, NASA recognized the importance of improving our understanding of the radiation budget and its effects on the Earth's climate. Langley Research Center was charged with developing a new generation of instrumentation to make accurate regional and global measurements of the components of the radiation budget. The Goddard Space Flight Center built the Earth Radiation Budget Satellite (ERBS) on which the first ERBE instruments were launched by the Space Shuttle Challenger in 1984. ERBE instruments were also launched on two National Oceanic and Atmospheric Administration weather monitoring satellites, NOAA 9 and NOAA 10 in 1984 and 1986.

ERBE has helped scientists world-wide better understand how clouds and aerosols, as well as some chemical compounds in the atmosphere (so-called "greenhouse" gases), affect the Earth's daily and long-term weather (the Earth's "climate"). In addition, the ERBE data has helped scientists better understand something as simple as how the amount of energy emitted by the Earth varies from day to night. These diurnal changes are also very important aspects of our daily weather and climate.

Solar Backscatter Ultra Violet (SBUV) Radiometer. SBUV/2, an operational remote sensor, flies on NOAA weather satellites and monitors the density and distribution of ozone in the Earth's atmosphere from six to 30 miles. SBUV/2 looks down at the Earth's atmosphere and the reflected sunlight at wavelengths characteristic of ozone.

The distribution of ozone is a key indicator of atmospheric processes and is also of vital significance in predicting the amount of damaging radiation reaching the Earth. The increasing "ozone hole" over the southern hemisphere has raised awareness of the significance of this phenomena to political levels. The SBUV/2 is a development of an earlier instrument flown on a NASA satellite and has gathered ozone data on an operational basis on a series of NOAA satellites in the so-called afternoon orbit since 1984.

Two optical radiometers form the heart of the SBUV instrument: a monochromator and a small but very important Cloud Cover Radiometer (CCR). The instrument contains four mechanisms: a movable grating for wavelength selection in the monochromator, a deployable diffuser which selects solar or Earth radiation measurements, a deployable Mercury lamp for wavelength calibration and an optical chopper mechanism which converts the steady incoming radiation to pulses of ultraviolet (UV) light which can be readily processed by the SBUV detectors and electronics.

The SBUV/2 is a nadir pointing non-scanning instrument sensitive to radiation in the 0.16 μm to 0.4 μm ultraviolet spectrum. The overall radiometric resolution is approximately 1 nanometre (nm) in this spectral band.

The use of a deployable diffuser gives the instrument the versatility of selecting between solar and Earth measurements. With the diffuser "stowed", the instrument views the Earth directly. The data from this configuration corresponds to Earth radiance. With the diffuser deployed into the "Sun" position, the detector output measurements correspond to solar irradiation data. Ground and in-flight calibration data are used to

Fig. 8.   Hurricane Gilbert - this NOAA-9 colorized image captures this cat 5 hurricane when its pressure was 888 mb, the lowest recorded pressure in the Western Hemisphere - Sept 13. (*courtesy of NOAA*).

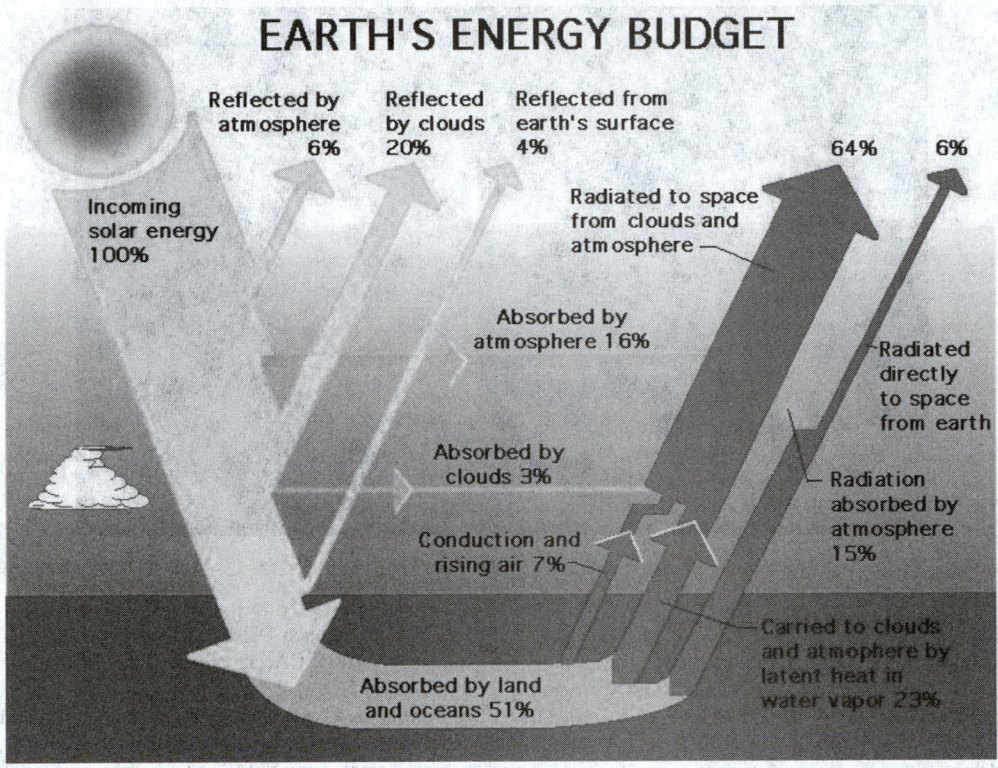

Fig. 9. Radiation Budget.

convert the detector data and diffuser mode data to solar irradiation or Earth radiance units.

The sensor module houses the monochromator optical hardware which uses a movable grating to select the wavelength where measurements will be made. The grating mechanism can be commanded to any one of 8,192 positions giving the monochromator approximately 0.1 nm wavelength resolution.

The CCR has a fixed 379 nm filter for wavelength selection and is co-aligned to the monochromator; therefore, it views the same scene as the monochromator. The output of the CCR represents the amount of cloud cover in a scene, as the name implies, and is used to remove the effects of clouds in the monochromator data. CCR data is transmitted once per second in both discrete and sweep modes.

The Ball Aerospace-built SBUV/2 helped to discover the ozone hole above Antarctica in 1987, and continues to monitor this phenomenon. Atmospheric ozone absorbs the sun's ultraviolet rays, which are believed to cause gene mutations, skin cancer, and cataracts in humans. Ultraviolet rays may also damage crops and aquatic ecosystems.

NOAA-10. Launch Date/Time: September 17, 1986 at 15:50:00 UTC, from Vandenberg AFB, CA, aboard an Atlas E/F launch vehicle. The primary objectives of NOAA-10 were to continue the Advanced TIROS-N program by working as a companion with NOAA-9 and to replace NOAA-8 in order to provide continuous coverage of the Earth and to provide high-resolution global meteorological data.

NOAA-10 was the third operational satellite in the Advanced TIROS-N series. The satellite carried the AVHRR and TOVS which were present on previous NOAA satellites, the solar proton monitor, and the SARSAT system. In addition, the ERBE instruments, which consisted of short wave and long wave radiometers, were used to study the Earth's albedo in attempt to recognize and interpret seasonal and annual climate variations. The SBUV was not carried on this satellite.

NOAA-10 was placed in a near circular, (450 nm) polar orbit. The craft and its systems operated successfully, providing high resolution scanned images and vertical temperature, moisture, and ozone profiles to research meteorologists. NOAA-10 was deactivated on September 17, 1991 when NOAA-12 became operational.

NOAA-11. Launch Date/Time: September 24, 1988 at 10:04:00 UTC, from Vandenberg AFB, aboard an Atlas E/F launch vehicle. The primary

objectives of NOAA-11 were to continue the Advanced TIROS-N program by working as a companion with NOAA-10 and to replace NOAA-9 in order to provide continuous coverage of the Earth and to provide high-resolution global meteorological data.

NOAA-11 was the fourth operational satellite in the Advanced TIROS-N series. The satellite carried the AVHRR (Fig. 10) and TOVS which were present on previous NOAA satellites, the solar proton monitor, and the SARSAT system. In addition, the ERBE instruments, which consisted of short wave and long wave radiometers, were used to study the Earth's albedo in attempt to recognize and interpret seasonal and annual climate variations. The SBUV radiometer was flown on this satellite since the satellite was intended to replace NOAA-9.

NOAA-11 was placed in a near circular, (470 nm) polar orbit. The primary mission sensor, the AVHRR, failed on September 13, 1994. The satellite was placed in standby mode in March of 1995 and was reactivated to provide sounding data after a NOAA-12 sounder failure in May 1997.

Fig. 10. Hurricane Andrew — approaching Louisiana coastline — NOAA-11 visible — Aug 25, 1992. (*courtesy of NOAA*).

Fig. 11.   Hurricane Andrew - approaching southeast Florida coast - false color image from NOAA-12 - Aug 23, 1992. (*courtesy of NOAA*).

NOAA-12. Launch Date/Time: May 14, 1991 at 20:52:00 UTC, from Vandenberg AFB, CA, aboard an Atlas E/F launch vehicle. The primary objectives of NOAA-12 are to continue the Advanced TIROS-N program by working as a companion with NOAA-10 and NOAA-11 in order to provide continuous coverage of the Earth and to provide high-resolution global meteorological data.

NOAA-12 was the fifth operational satellite in the Advanced TIROS-N series. The satellite carried the AVHRR (Fig. 11.), TOVS, and the solar proton monitor. All of which were present on previous NOAA satellites. The ERBE instruments, the SBUV radiometer and the SARSAT system were not flown on this satellite.

NOAA-12 was placed in a near circular, (450 nm) polar orbit. The craft and its systems are operating successfully (ca. 2006), providing high-resolution scanned images and vertical temperature and moisture profiles to research meteorologists. The APT and HRPT (Fig. 12.) capability still exists with this satellite. Status reports online at NOAA Spacecraft Status Summary page: http://www.oso.noaa.gov/poesstatus/spacecraftStatus Summary.asp?spacecraft=12

NOAA-13. Launch Date/Time: August 9, 1993 at 10:52:00 UTC, from Vandenberg AFB, CA, aboard an Atlas E/F launch vehicle. The primary objectives of NOAA-13 were to continue the Advanced TIROS-N program by working as a companion with NOAA-10, 11, and 12 in order to provide continuous coverage of the Earth and to provide high-resolution global meteorological data. NOAA-13 was placed in a near circular, (470 nm) polar orbit. The spacecraft and its systems operated successfully for 12 days until a circuit failure resulted in a power loss aboard the craft.

NOAA-14. Launch Date/Time: December 30, 1994 at 10:02:00 UTC, from Vandenberg AFB, CA, aboard an Atlas E/F launch vehicle. The primary objectives of NOAA-14 are to continue the Advanced TIROS-N program by working as a companion with NOAA-10, 11, and 12 in order to provide continuous coverage of the Earth and to provide high-resolution global meteorological data.

NOAA-14 is the sixth operational satellite in the Advanced TIROS-N series (NOAA 13 never officially became operational as it failed during its 21 day checkout period). The satellite carried the AVHRR, TOVS, and the solar proton monitor. All of which were present on previous NOAA

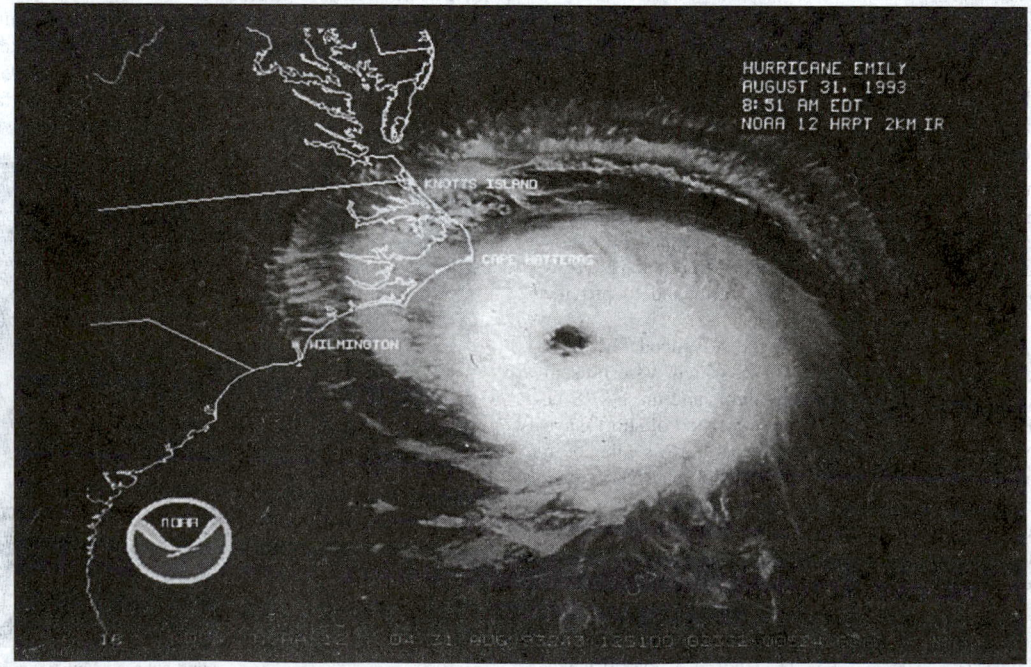

Fig. 12.   Hurricane Emily NOAA-12 Image-IR, August 31, 1993. (*courtesy of NOAA*).

satellites. The ERBE instruments, the SBUV radiometer and the SARSAT systems were also flown on this satellite. NOAA-14 was placed in a near circular, (470 nm) polar orbit.

NOAA reports it as an AM backup; however imagery from the AVHRR has been useless for some time and no useful imagery has been transmitted in 2006. leaving the operational status as questionable. http://www.oso.noaa.gov/poesstatus/spacecraftStatusSummary.asp?spacecraft=14

NOAA-15. Launch Date/Time: May 13, 1998, at 15:52:04 UTC, from Vandenberg AFB, CA, aboard a Titan II launch vehicle. NOAA-15 is the seventh operational satellite in the Advanced TRIOS-N series. NOAA-15 became operational on July 1998 and replaced the decommissioned NOAA-12 in an afternoon equator-crossing orbit. This new series of fifth generation NOAA satellites carry a series of instruments which have been modified and improved from previous NOAA-series satellites. See Fig. 13 and 14. They included: (1) an improved six-channel Advanced Very High Resolution Radiometer/3 (AVHRR/3); (2) an improved High Resolution Infrared Radiation Sounder (HIRS/3); (3) the Search and Rescue Satellite Aided Tracking System (S&R), which consists of the Search and Rescue Repeater (SARR) and the Search and Rescue Processor (SARP-2); (4) the French/CNES-provided improved ARGOS Data Collection System (DCS-2); and (5) the Advanced Microwave Sounding Units (AMSUs), which replaced the previous MSU and SSU instruments to become the first in the NOAA series to support dedicated microwave measurements of temperature, moisture, surface and hydrological studies in cloudy regions where visible and infrared instruments have decreased capability.

Advanced Very High Resolution Radiometer/3 (AVHRR/3). The Advanced Very High Resolution Radiometer/3 (AVHRR/3) on the Advanced TIROS-N NOAA K-N series of polar orbiting meterological satellites is an improved instrument over previous AVHRRs. The AVHRR/3 adds a sixth channel and is a cross-track scanning instrument providing imaging and radiometric data in the visible, near-IR and infrared of the same area on the Earth. Data from the visible and near-IR channels provide information on vegetation, clouds, snow, and ice. Data from the near-IR and thermal channels provide information on the land and ocean surface temperature and radiative properties of clouds. Only five channels can be transmitted simultaneously with channels 3A and 3B being switched for day/night operation. The instrument produces data in High Resolution Picture Transmission (HRPT) mode at 1.1 km resolution or in Automatic

Fig. 13. First Image by NOAA-15. (NCDC/NOAA).

Picture Transmission (APT) mode at a reduced resolution of 4 km. The AVHRR/3 scans 55.4 degrees per scan line on either side of the orbital track and scans 360 lines per minute. The six channels are (1) channel 1, visible (0.58-0.68 microns); (2) channel 2, near-IR (0.725-1.0 microns); (3) channel 3A, near-IR (1.58-1.64); (4) channel 3B, infrared (3.55-3.93 microns; (5) channel 4, infrared (10.3-11.3 microns); and (6) channel 5 (11.5-12.5 microns).

High Resolution Infrared Sounder/3 (HIRS/3). The improved High Resolution Infrared Sounder/3 (HIRS/3) on the Advanced TIROS-N (ATN) NOAA K-N (NOAA-15-18) series of polar orbiting meteorological satellites is a 20-channel, step-scanned, visible and IR spectrometer designed to provide atmospheric temperature and moisture profiles. The HIRS/3 instrument is basically identical to the HIRS/2 flown on previous spacecraft except for changes in six spectral bands to improve the sounding accuracy. The HIRS/3 is used to derive water vapor, ozone, and cloud liquid water content. The instrument scans 49.5 degrees on either side of the orbital track with a ground resolution at nadir of 17.4 km. The

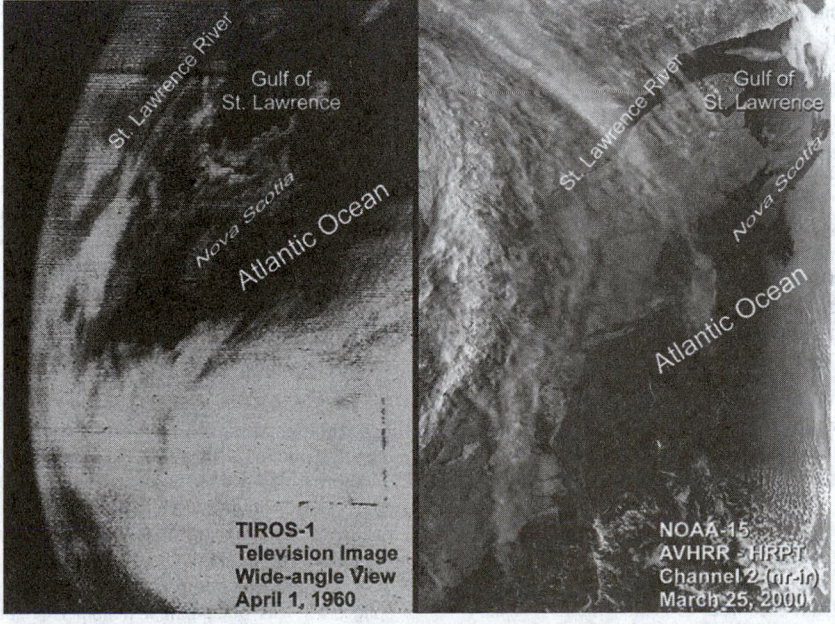

Fig. 14. First TIROS-1 image compared to NOAA-15 image 40 years later. (NCDC/NOAA).

instrument produces 56 IFOVs for each 1,125 km scan line at 42 km between IFOVs along-track. The instrument consists of 19 IR and 1 visible channel centered at 14.95, 14.71, 14.49, 14.22, 13.97, 13.64, 13.35, 11.11, 9.71, 12.45, 7.33, 6.52, 4.57, 4.52, 4.47, 4.45, 4.13, 4.0, 3.76, and 0.69 microns.

Search and Rescue Satellite Aided Tracking System (SARSAT). The Search and Rescue Satellite Aided Tracking System (SARSAT) on the Advanced TIROS-N NOAA K-N (NOAA- 15- 18) series of polar orbiting meteorological satellites is designed to detect and locate Emergency Locator Transmitters (ELTs) and Emergency Position-Indicating Radio Beacons. The SARSAT instrumentation consists of two elements: the Search and Rescue Repeater (SARR) and the Search and Rescue Processor (SARP-2). The SARR is a radiofrequency (RF) system that accepts signals from emergency ground transmitters at three very high frequency (VHF/UHF) ranges (121.5 MHz, 243 MHz and 406.05 MHz) and translates, multiplexes, and transmits these signals at L-band frequency (1.544 GHz) to local Search and Rescue stations (LUTs or Local User Terminals) on the ground. The location of the transmitter is determined by retrieving the Doppler information in the relayed signal at the LUT. The SARP-2 is a receiver and processor that accepts digital data from emergency ground transmitters at UHF and demodulates, processes, stores, and relays the data to the SARR where they are combined with the three SARR signals and transmitted via L-band frequency to local stations.

ARGOS Data Collection System (DCS-2). The ARGOS Data Collection System (DCS-2) on the Advanced TIROS-N (ATN) NOAA K-N (NOAA-15-18) series of polar orbiting meteorological satellites is a random-access system for the collection of meteorological data from in situ platforms (moveable and fixed). The ARGOS DCS-2 collects telemetry data using a one-way RF link from data collection platforms (such as buoys, free-floating balloons and remote weather stations) and processes the inputs for on-board storage and later transmission from the spacecraft. For free-floating platforms, the DCS-2 system determines the position to within 5 to 8 km RMS and velocity to an accuracy of 1 to 1.6 mps RMS. The DCS-2 measures the in-coming signal frequency and time. The formatted data are stored on the satellite for transmission to NOAA stations. The DCS-2 data is stripped from the GAC data by NOAA/NESDIS and sent to the ARGOS center at CNES in France for processing, distribution to users, and archival.

Advanced Microwave Sounding Unit A (AMSU/A). The Advanced Microwave Sounding Unit (AMSU) is a new instrument on the Advanced TIROS-N (ATN) NOAA K-N (NOAA-15-18) series of operational meteorological satellites. The AMSU consists of two functionally independent units, AMSU-A and AMSU-B. The AMSU-A is a line-scan instrument designed to measure scene radiance in 15 channels, ranging from 23.8 to 89 GHz, to derive atmospheric temperature profiles from the Earth's surface to about 3 millibar pressure height. The instrument is a total power system having a field-of-view of 3.3 degrees at half-power points. The antenna provides cross track scan 50 degrees on either side of the orbital track at nadir with a total of 30 IFOVs per scan line. The AMSU-A is calibrated on-board using a blackbody and space as references. The AMSU-A is physically divided into two separate modules which interface independently with the spacecraft. The AMSU-A1 contains all of the 5 mm oxygen channels (channels 3-14) and the 80 GHz channel. The AMSU-A2 module consists of two low-frequency channels (channels 1 and 2). The 15 channels have a center frequency(GHz) at: 23.8, 31.4, 50.3, 52.8, 53.6, 54.4, 54.94, 55.5, six at 57.29, and 89.

NOAA-16. Launch Date/Time: September 21, 2000 at 10:21:00 UTC, from Vandenberg AFB, CA, aboard a Titan II launch vehicle. By March 2001, NOAA-16 was designated as the operational replacement for NOAA-14. As such, it operates in an orbit with a 2:11 p.m. ascending node (afternoon orbit) and utilizes a similar set of instruments as NOAA-17. The Solar Backscatter Ultraviolet Spectral Radiometer is a major addition to the afternoon suite of instruments. On November 13, 2000 the VHF transmitter (VTX) failed which does not allow the broadcast of Automatic Picture Transmission (APT). Data Recorder DTR#5 failed February 2, 2000 and is no longer used due to a failure within its electronics. The SARR 243 MHz signal failed on November 13, 2001. Since September 17, 2003, the AVHRR scan motor performance has changed causing periodic current surges and loss of sync. During periods of high scan motor current the imagery is degraded.

NOAA-17. Launch Date/Time: June 24, 2002 at 18:23:00 UTC, from Vandenberg AFB, CA, aboard a Titan II launch vehicle. As of October 15, 2002, NOAA-17 was designated operational. It replaced NOAA-15 as a primary spacecraft. As such, it operates in an orbit with a 10:20 a.m. ascending node (morning orbit) and utilizes a Solar Backscatter Ultraviolet Spectral Radiometer (SBUV). On February 15, 2003, DTR5 failed to operate and on April 28, 2003, STX3 power degraded to 2 watts. On October 28, 2003, the AMSU-A1 scan motor failed thus the instrument no longer provides any data. All other systems are operational.

The NOAA-17, or morning mission, instrument payload includes:

- The Advanced Very High Resolution Radiometer (AVHRR/3), a six channel imaging radiometer which detects energy in the visible and near-IR portions of the electromagnetic spectrum. This data is used to observe vegetation, clouds, lakes, shorelines, snow, aerosols and ice.
- The High Resolution Infrared Radiation Sounder (HIRS/3), which detects and measures energy emitted by the atmosphere to construct a vertical temperature profile from the Earth's to an altitude of about 40 km. These measured energy profiles are used to determine ocean surface temperatures, total atmospheric ozone levels, precipitable water, cloud height and coverage and surface radiance
- The Advanced Microwave Sounding Unit-A (AMSU-A), which measures scene radiance in the microwave spectrum. The data from this instrument is used in conjunction with the HIRS to calculate the global atmospheric temperature and humidity profiles from the Earth's surface to the upper stratosphere, approximately a 2 millibar pressure altitude (48 km or 28 miles). The AMSU-A is also complemented by the AMSU-B, which is designed to allow the calculation of the vertical water vapor profiles from the Earth's surface to about a 200 millibar pressure altitude (12 km or 7.5 miles). The data from these instruments is used to provide precipitation and surface measurements including snow cover, sea ice concentration, and soil moisture.
- The Space Environmental Monitor (SEM/2) provides measurements to determine the intensity of the Earth's radiation belts and the flux of charged particles at the satellite altitude. It provides the knowledge of solar terrestrial phenomena and also provides warnings of solar wind occurrences that may impair long-range communication, high-altitude manned operations, or disrupt satellite operations.
- The Search and Rescue (SAR) instruments are part of the international COSPAS-SARSAT system designed to detect and locate Emergency Locator Transmitters (ELTs), Emergency Position-Indicating Radio Beacons (EPIRBs), and Personal Locator Beacons (PLBs) operating at 121.5, 243, and 406 MHz. The NOAA-15 spacecraft carries two instruments to detect these emergency beacons; the Search and Rescue Repeater (SARR) provided by Canada, and the Search and Rescue Processor (SARP-2) provided by France.
- The Data Collection System (DCS) collects and processes measurements from remote data collection platforms for on-board storage and subsequent transmission from the satellite. Data collection platforms in the form of buoys, free-floating balloons, and remote weather stations transmit their data on a 401.65 MHz uplink to the spacecraft. The DCS is used to gather environmental measurements such as atmospheric temperature and pressure, rainfall and snowfall, and velocity and direction of the ocean and wind currents.
- The Solar Backscatter Ultraviolet Spectral Radiometer (SBUV) is used to measure solar irradiance (backscattered solar energy), total ozone concentrations, and the vertical ozone profile in the atmosphere.

NOAA-18. Launch Date/Time: May 20, 2005 at 10:22:00 UTC, from Vandenberg AFB, CA, aboard a Boeing Delta II 7320-10 expendable launch vehicle.

Approximately 26 minutes after launch, controllers acquired the spacecraft through the McMurdo Sound ground station, Antarctica, while the spacecraft was still attached to the Delta II. Spacecraft separation was monitored by the TDRSS.

The solar array boom and antennas were successfully deployed, and the spacecraft was placed in a near-perfect orbit. The satellite was acquired by the NOAA Fairbanks Station, Alaska, 86 minutes after launch and deployments, and a nominal spacecraft power system was confirmed. NOAA-N was renamed NOAA-18 after achieving orbit.

NOAA-18 will collect data about the Earth's surface and atmosphere. The data are input to NOAA's long-range climate and seasonal outlooks, including forecasts for El Nino and La Nina.

NOAA-18 is the fourth in a series of five Polar-orbiting Operational Environmental Satellites with instruments that provide improved imaging and sounding capabilities.

The NOAA 18 instrument complement consists of: (1) an improved six-channel Advanced Very High Resolution Radiometer/3 (AVHRR/3); (2) an improved High Resolution Infrared Radiation Sounder (HIRS/3); (3) the Search and Rescue Satellite Aided Tracking System (S&R), which consists of the Search and Rescure Repeater (SARR) and the Search and Rescue Processor (SARP-2); (4) the French/CNES-provided improved ARGOS Data Collection System (DCS-2); (5) the Solar Backscatter Ultraviolet Spectral radiometer (SBUV/2); and (6) the Advanced Microwave Sounding Unit (AMSU), which consists of three separate modules, A1, A2, and B to replace the previous MSU and SSU instruments.

Web References

NASA Earth Science: http://www.earth.nasa.gov/history/noaa/noaa.html
National Polar-Orbiting Operational Environmental Satellite System (NPOESS): http://www.npoess.noaa.gov/
NOAA Satellite and Information Service: http://www.ncdc.noaa.gov/oa/satellite/poesinfo.html; http://www.nesdis.noaa.gov/satellites.html; and http://hurricane.ncdc.noaa.gov/cgi-bin/hsei/hsei.pl?directive=quick_search
NSSDC Goddard Space Flight Center: http://nssdc.gsfc.nasa.gov/planetary/
Polar-Orbiting Satellite Detailed Information: http://www.wmo.ch/web/sat/POLARdetailed.html

POLAR ORBITS. As the name suggests, polar orbits pass over the Earth's polar regions from north to south. The orbital track of the satellite does not have to cross the poles exactly for an orbit to be called polar, an orbit which passes within 20 to 30 degrees of the poles is still classed as a polar orbit.

These orbits mainly take place at low altitudes of between 200 to 1,000 km (124 to 621 miles). Satellites in polar orbit look down on the Earth's entire surface and can pass over the North and South Poles several times a day.

Polar orbits are often used for Earth-mapping-, Earth observation and reconnaissance satellites, as well as some weather satellites. If a satellite is in polar orbit at an altitude of 800 km (497 miles), it will be travelling at a speed of approximately 7.5 km (4.7 miles) per second.

Polar Orbiting Satellites

The POES satellite system offers the advantage of daily global coverage, by making nearly polar orbits roughly 14.1 times daily. Since the number of orbits per day is not an integer the sub orbital tracks do not repeat on a daily basis, although the local solar time of each satellite's passage is essentially unchanged for any latitude. Currently in orbit we have a morning and afternoon satellite, which provide global coverage four times daily. The POES system includes the Advanced Very High Resolution Radiometer (AVHRR) and the Tiros Operational Vertical Sounder (TOVS).

Because of the polar orbiting nature of the POES series satellites, these satellites are able to collect global data on a daily basis for a variety of land, ocean, and atmospheric applications. Data from the POES series supports a broad range of environmental monitoring applications including weather analysis and forecasting, climate research and prediction, global sea surface temperature measurements, atmospheric soundings of temperature and humidity, ocean dynamics research, volcanic eruption monitoring, forest fire detection, global vegetation analysis, search and rescue, and many other applications.

See also **Earth-Orbiting Satellites (Data Receiving And Handling Facilities); Orbit (Astronomy); Satellites (Communications and Navigation);** and **Satellites (Scientific and Reconnaissance).**

POLAR RESEARCH. The term *polar* is applied to the regions of both the North and South Poles of the Earth. These regions lie within the Arctic and Antarctic Circles, respectively. In many ways, the two polar regions are similar—low mean annual temperatures; oceans with the presence of sea-ice cover; the presence of ice sheets, glaciers, and ice shelves; alternating

6-month periods of continuous daylight and darkness at the poles, as well as auroral phenomena and geomagnetic disturbances. On the other hand, the two polar regions are also quite different. With exception of the Greenland Ice Sheet (1.8×10^6 square kilometers; 0.7 million square miles), the central Arctic region has few large land masses and comprises and the Arctic Ocean (14×10^6 square kilometers; 5.4 million square miles). In the Arctic region, there are significant land-dwelling forms of fauna and flora and indigenous human populations, as well as newcomers to the region from the south. In contrast, the Antarctic region is a continent that is surrounded by the Southern Ocean (36×10^6 square kilometers; 13.9 million square miles). On this oceanic area (about 10% of the world's oceans) are found the *Antarctic Ice Sheet*, which in reality represents two ice sheets that have been butted together—the large *East Antarctica Ice Sheet* and the smaller *West Antarctica Ice Sheet*. The larger ice sheet (13.5×10^6 square kilometers; 5.2 million square miles) is known to be at least 4,000 meters (13.124 feet) thick in some locations. It contains about 85% of the world's ice. Living forms, by comparison with the Arctic, are sparse.

The Polar Areas in Perspective

Exploration of the Earth's polar regions has fascinated many people over the years, ranging from those persons who simply were curious and who sought high adventure, to opportunists, to scientists and historians. In retrospect, polar explorations represented tremendous challenges to people and equipment. Expeditions to the Arctic date back to the 1600s; to the Antarctic, the late 1700s. An abbreviated chronology of principal polar exploration events is given in the accompanying *editorial inset*. During the last few decades, the sheer adventurism of the polar regions has waned because of great improvements in transportation and communication, the establishment of bases, improved housing, clothing etc. even though the regions embrace some of the most rugged environments encountered on Earth. The curiosity of the regions from a geographic standpoint has abated because of the availability of good maps and geodetic information. The regions are now fascinating and intriguing from a scientific standpoint because, as briefly described in this article, much information about the Earth and even the solar system can be gleaned from polar research.

Mainly catalyzed by the International Geophysical Year (IGY), which extended over 1957–1958, scientific interest in the polar regions has increased over the last 40 years. This research is essentially targeted toward five topics—the atmosphere, the hydrosphere, the cryosphere, the lithosphere, and the biosphere. Over 60 important research stations, operated by 16 different countries, are located in the polar regions. These stations are supported by dozens of additional minor research stations. See Table 1.

Arctic Research and Policy Act of 1984. This U.S. Congressional Act was designed to advance Arctic research in the national interest.

TABLE 1. POLAR RESEARCH STATIONS[a]

Country	Number of Stations	
	Arctic	Antarctic
Russia[b]	6	7
Argentina		10
United States	3	5
Canada	5	
United Kingdom		5
Australia		3
Norway	1	2
New Zealand		2
Denmark	2	
Finland	2	
Japan		2
Chile		2
South Africa		1
France		1
Poland		1
Sweden	1	
Total	20	41

[a] Does not include numerous minor research stations.
[b] Not officially published as of 1994.

Some of the research fields that require attention are weather and climate; national defense; renewable and nonrenewable resources; transportation; communication and space disturbance effects; environmental protection; health, culture, and socioeconomics; and international cooperation. As reported by the Polar Research Board of the National Research Council, a research framework recommended by the U.S. Arctic Research Commission includes, in order of priority, integrated investigations to understand: (1) the Arctic Ocean, including the marginal seas, sea ice, and seabed, and how the ocean and atmosphere operate as coupled components of the arctic system; (2) the coupled atmosphere and land components and how their interaction governs the terrestrial environment; and (3) the high-latitude upper atmosphere and its extension into the magnetosphere with emphasis on predicting and mitigating effects on communications and defense systems.

Targets for research in the Antarctica, as fostered by the nations that participated in the IGY, follow along somewhat similar lines.

The Antarctic Treaty. The text of this treaty was drawn up in 1959 by the governments of Argentina, Australia, Belgium, Chile, the French Republic, Japan, New Zealand, Norway, the Union of South Africa, Russia, the United Kingdom, and the United States.

The Preamble of the Treaty reads:

> Recognizing that it is in the interest of all mankind that Antarctica shall continue forever to be used exclusively for peaceful purposes and shall not become the scene or object of international discord;

Acknowledging the substantial contributions to scientific knowledge resulting from international cooperation in scientific investigation in Antarctica;

Convinced that the establishment of a firm foundation for the continuation and development of such cooperation on the basis of freedom of scientific investigation in Antarctica as applied during the international Geophysical Year accords with the interests of science and the progress of all mankind;

Convinced also that a treaty ensuring the use of Antarctica for peaceful purposes only and the continuance of international harmony in Antarctica will further the purposes and principles embodied in the Charter of the United Nations;

The Treaty was ratified by all participating nations by 1961.

Polar Climate

It has been established for many years that the polar regions serve as large heat sinks and in so doing provide an unusual degree of stability to the global climate. Even after several decades of research, the polar atmosphere-ice-ocean system is poorly understood. Weather forecasting for the polar region per se, for example, is difficult, and it is now recognized that the system is much more sensitive and reactive than previously thought. There is an important feedback system made up of several meteorological elements, including the nature and extent of snow cover, which in turn affects albedo (solar energy reflectivity from polar surfaces) and the temperature. Particular interest in recent years has been directed to the possible modulating effects on global climate by the Antarctic, notably over much shorter time spans (10 to 100 years), as contrasted with the recognized longer-term (1,000–100,000 years) effects. For example, large-scale ice sheet surges could increase the albedo of the Southern Ocean to the point where global temperatures would be reduced, triggering glaciation in the Northern Hemisphere. However, not all scientists agree with this hypothesis, noting that evidence to date indicates that the Antarctic ice sheet has been quite stable for over 50,000 years.

Considerable research is going forth to show a correlation with past polar climates and global climates as a whole. Studies involving sedimentology, paleontology, palynology (study of pollen and spores, living or fossil) paleosols (a buried soil horizon of the geologic past), glacial geology, periglacial (frost action) features, and ice cores are being studied by a number of polar investigators.

As described in the entry on **Climate**, there has been much concern in recent years over the increasing content of carbon dioxide in the atmosphere (mainly from combustion of fossil fuels), which theoretically should cause a warming of global climate. As pointed out by Washburn (1980), the polar regions clearly constitute an excellent focus for monitoring changes in atmospheric carbon dioxide and climate. It has

been predicted that breaking up of blocking ice shelves, as a result of Antarctic warming, could lead to ice sheet advances possibly sufficient to elevate the sea level by as much as 6 meters (19.6 feet) over a 200-year time span. However, even though carbon dioxide concentrations have been increasing, recent trends in polar regions have been cooling instead of warming.

The polar regions are advantageously located for a number of investigations, such as studying the effects of the upper atmosphere on the lower atmosphere. The Antarctic continent is well situated because of its near-spherical ice surface offset from the geographical axis (SCAR, 1979). Also studied are energy transfer processes in the high-latitude magnetosphere-ionosphere system and the electrical coupling between the upper and lower atmosphere. Studies of radio communication and solar-terrestrial physics also take advantage of the polar location.

Pollution of Polar Atmospheres

In connection with research that has been going forward since 1981, investigators have noted unexplained high-frequency radar echoes from the polar mesosphere at an altitude of approximately 90 km (56 mi). Also, they have observed noctilucent clouds (highest clouds in Earth's atmosphere) in increasing numbers. Experiments at Poker Flat, Alaska, indicated receipt of 50-MHz echoes from the very top of the mesosphere (mesopause). During more recent years, echoes of frequencies as high as 900 MHz have been observed at other radar station locations. J. Ulwick (University of Utah — Stewart Radiance Laboratory) noted that the echoes are coming from electrons in the mesosphere, and their high frequency suggests that the electrons are interacting with dense layers of particles. "The electrons are sticking to something up there like paint," observed R. Goldberg (Goddard Space Flight Center). Atmospheric methane in polar regions has increased about two-fold since 1900, and noctilucent clouds have become nearly ten times brighter. Inasmuch as these clouds (droplets of ice crystals) must be seeded to form, some scientists propose that proton hydrates are responsible. Other researchers suggest charged aerosols as the medium. Cloud sampling has been carried out through the use of over 30 rockets, and much information remains to be processed. Should methane be detected, G. Thomas (University of Colorado at Boulder) notes that pollution even at these very high altitudes may indicate anthropogenic origins.

Antarctic Ozone "Hole." Dramatic losses in the ozone layer over Antarctica, displaying definitive variations, were first noted in 1977 by members of the British Antarctic Survey team. The dips in ozone levels were so large that several research groups, including the British team, as well as scientists from the United States and other countries with a direct interest in Antarctica, were somewhat reluctant to report them immediately, believing that there might have been something seriously wrong with the data. Not until May 1985 did the British group publicly report that a massive decrease in ozone concentrations over Antarctica had been occurring during the Antarctic springtime (September and November) over an 8-year period. Routinely, over the period 1957 to 1973, past records showed that October ozone concentrations in the affected area averaged about 320 parts per billion (ppb). A routine measurement in 1984 showed a concentration of less than 200 ppb, indicating a depletion loss of about 40%. Interim October measurements have confirmed this trend. These ozone losses have been confirmed by numerous measurement methodologies, including balloon-borne, high-altitude aircraft, and rocket sampling.

The possible very adverse effects of a continuously diminishing ozone layer have thrust the study of the ozone "hole" into the forefront of scores of scientific investigations. The ozone layer dampens the passage of ultraviolet radiation from the sun through Earth's atmosphere. Excessive UV radiation can injure organisms of nearly all kinds, promoting skin cancers in humans and retarding the growth of phytoplankton, which is a major key in the food chain of the Antarctic.

In the early phases of this scientific investigation, numerous hypotheses were developed to explain the causes and effects of the depletion of the ozone layer. For example, was this phenomenon related to the sun? Or had the missing ozone been transported elsewhere by some poorly understood meteorological process? These and other postulations did not withstand the rigors of careful examination. Although some form of anthropogenic pollution was suspected, specific connections

THE POLAR REGIONS IN PERSPECTIVE

THE ARCTIC

Arctic exploration dates back to 1587 when John Davis (England) surveyed the Davis Strait to Sanderson's Hope (72°12′N). Other famous names involved in Arctic exploration during the 1600s and 1700s included Barents and van Heemskerck (both of Holland) who discovered Bear Island and touched the tip of Spitsbergen (79°49′N). Henry Hudson (England) reached the north of Spitsbergen (80°23′N) in 1607. In 1616, William Baffin and Robert Bylot (England) explored Baffin Bay to Smith Sound. Vitus Bering (Russia) proved that Asia and America were separate continents in 1728. In 1771, Samuel Hearne (Hudson's Bay Co.) traveled overland from Churchill to the mouth of Coppermine River. James Cook (Britain), in 1778, traveled through the Bering Strait to Icy Cape, Alaska and North Cape, Siberia. In 1789, Alexander Mackenzie (Britain) traveled from Montreal to the mouth of the Mackenzie River.

Considerable exploration of the Arctic commenced in the 1800s. William Scoresby (Britain) reached a location north of Spitsbergen (81°30′N). In the period 1820–1823, Ferdinand von Wrangel (Russia), joining the ventures of James Cook, confirmed the separation of Asia and North America. In 1845, in search for the Northwest Passage, Sir John Franklin (Britain) encountered disaster at Lancaster Sound. In 1888, Nansen (Norway) crossed Greenland's ice cap and, in a futile attempt to reach the North Pole, reached only as far as Franz Josef Land. Salomon A. Andree (Sweden) in 1896, made an unsuccessful attempt to reach the pole by balloon. (The frozen bodies of his team were not discovered until 1930 on White Island, 82°57′N.) Very serious efforts to reach the North Pole commenced in the 1900s. After a number of unsuccessful attempts, Robert E. Peary reached the pole (90°N) on April 6,1909. In 1925, Roald Amundsen (Norway) and Lincoln Ellsworth (U.S.) reached 87°44′N in an attempt to fly to the pole from Spitsbergen. Adm. Richard E. Byrd and Floyd Bennett (Both of the U.S.) flew over the pole for the first time on May 9, 1926. In that same year, Amundsen, Ellsworth, and Umberto Nobile (Italy) flew from Spitsbergen over the pole to Teller, Alaska. They traveled in the dirigible *Norge*. In 1928, Nobile again crossed the pole on May 24, but the airship crashed on the following day. Amundsen later lost his life in a plane crash in an attempt to rescue Nobile.

Comdr. William R. Anderson, on August 3, 1958, in the nuclear-powered submarine *U.S. Nautilus*, led the first team to cross the North Pole beneath the Arctic ice. The vessel sailed from Portsmouth, New Hampshire, headed between Greenland and Labrador through Baffin Bay, then west through Lancaster Sound and McClure Strait to the Beaufort Sea. Traveling submerged for much of the voyage, the submarine made 850 miles (1360 km) from Baffin Bay to the Beaufort Sea in 6 days.

On August 16, 1977, the Soviet nuclear-powered icebreaker *Arktika* reached the pole and became the first surface ship to break the Arctic ice pack to gain the pole. On April 30, 1978, Naomi Uemura (Japan) was the first person to reach the pole alone by dog sled. In April 1982, Sir Ralph Flennes and Charles Burton (Britain) were the first persons to circle the Earth from pole to pole. The trek required about 3 years, at a cost of about $18 million On May 2, 1986, American and Canadian explorers reached the North Pole, assisted only by dogs. Thus, they became the first party to reach the pole without some type of mechanical support since the 1909 venture of Peary. The 500-mile (800 km) trip was completed in 56 days.

THE ANTARCTIC

Exploration of the Antarctic region did not commence until the late 1700s. Captain James Cook (Britain) reached 71°10′S during the period 1773–1775. In 1819–1821, Bellingshausen (Russia) discovered Palmer Peninsula, but did not know then that he was on a continent. In 1823, James Weddell (Britain) discovered the Weddell Sea. Charles Wilkes (U.S.) was the first person to posit the existence of the continent of Antarctica (1840). The Ross Ice Shelf was discovered by James Clark Ross (Britain) in 1841–1842. In 1895, a party led by Leonard Kristensen (Norway) landed on the coast of Victoria Island and were the first persons to explore the main continental mass. A British expedition was the first to spend an entire winter in Antarctica (1899).

Robert F. Scott (Britain) discovered Edward VII Peninsula in 1902–1904, reaching a point (82°17′S) east of McMurdo Sound. In 1908–1909, Ernest Shackleton (Britain) used Manchurian ponies in Antarctic sledging. He reached 88°23′S and discovered a route onto the plateau by way of the Beardmore Glacier, thus pioneering a land path to the South Pole. In 1911, Roald Amundsen with a team of 4 men and dogs reached the South Pole on December 14. Shortly thereafter, Capt. Scott reached the pole from Ross Island (January 18, 1912). There were 5 men on Scott's team, all of whom perished. However, they previously had found Amundsen's tent and had reached the pole. Hubert Wilkins (Britain) was the first person to fly over the pole (1928). In 1929, Richard Byrd established Little America on the Bay of Whales. On an airplane flight (November 29, 1929), Byrd and his party crossed the pole by air, having covered a distance of 1600 miles (2560 km). A second expedition to Little America was led by Byrd in 1934–1935, during which an area of some 450,000 square miles (1.17 mil square km) was explored. The party wintered at a weather station (80°08′S).

In 1935, Lincoln Ellsworth flew south along the coast of the Palmer Peninsula and then crossed the continent to Little America, making four landings on unprepared terrain in very bad weather. In 1939–1941, the U.S. Navy launched Operation "Deep Freeze," led by Adm. Byrd. This project established five coastal stations fronting the Indian, Pacific, and Atlantic oceans as well as three interior stations. During this project, more than 1 million square miles (about 2.5 million square km) were explored and mapped. During the International Geophysical Year (1957–1958), scientists from twelve countries participated in Antarctic research. A network of some sixty stations on the continent and sub-Antarctic islands was used to study the oceanography, glaciology, meteorology, seismology, geomagnetism, ionosphere, cosmic rays, aurora, and airglow of the Antarctic region. During that period, V.E. Fuch led a 12-man expedition on the first crossing of Antarctica by land, a distance of 2158 miles (3472 km) in 98 days.

In 1958, several U.S. scientists traveled by tractor from Ellsworth Station on the Weddell Sea to discover a huge mountain range (5000 feet; 1524 meters high) above the ice sheet and some 9000 feet (2743 meters) above sea level. Named the Dufek Massif, this range had been identified earlier by the U.S. Navy, but not fully confirmed. In 1959, a treaty was signed by twelve nations, suspending territorial claims in the Antarctic for a period of 30 years and reserving the continent for research. In 1961–1962, scientists discovered a trough (Bentley Trench) that runs from the Ross Ice Shelf (Pacific side) into Marie Byrd Land and around the end of the Ellsworth Mountains toward the Weddell Sea. In 1962, the first nuclear power plant commenced operation at McMurdo Sound. The longest nonstop flight made in the area of the South Pole was accomplished on February 22, 1963 by a crew (U.S.) working out of the McMurdo Station. In 1964, a British survey team revisited Cook Island by helicopter, this being the first visit there since its discovery in 1775. In 1964, a New Zealand team completed one of the last and most important surveys of the continent, mapping the mountain area from Cape Adare and westward some 400 miles (640 km) to Pennell Glacier. See Fig. 1 for flags of countries carrying on active research.

had to await the development of an understanding of the complex chemistry that occurs. Progressively, a step-by-step series of chemical reactions involving the halogens and notably chlorine appeared to be the essentially reactive and destructive ingredients, and this led to the indictment of widely used, vaporous chemicals that easily enter Earth's atmosphere. Although other compounds ultimately may be also play a role, the scientific consensus today targets the family of chlorofluorocarbons (CFC 11 and CFC 12), used mainly in refrigerating and air-conditioning systems. Accepted as common knowledge today, this is indicative of the outstanding scientific progress made within a time span of less than two decades. Consequently, regulatory actions and targets have been established by numerous governments throughout the world (e.g., the Montreal Protocol), and it is also of interest to note that refrigerant chemical manufacturers have been searching for CFC substitutes.

Chemistry of Ozone Layer and "Hole." It is very difficult to duplicate in the laboratory the physical and chemical conditions that prevail in the clouds over Antarctica and to show how these conditions differ between the winter and summer seasons. The presence or absence of stratospheric clouds is an important determinant of what chemistry takes place, particularly when pollutants of any nature, natural or anthropogenic, enter into the atmosphere. Explanations to date largely represent theories arising out of "blackboard" chemistry.

Fig. 1.   The South Pole showing flags of all Antarctic Treaty countries carrying out active scientific research on the continent. (*Source: British Antarctic Survey.*)

One may ask, "How does ozone get into the stratosphere in the first place? Under what may be considered "normal" conditions, the oxygen molecule (O_2) is quite stable. However, when this molecule is subjected to ultraviolet (UV) radiation of just the right frequency (energetic solar radiation), the O_2 molecule can be broken to create two free atoms of oxygen (O). Each of these atoms, in turn, can combine with an oxygen molecule to form ozone (O_3), which is comprised of three oxygen atoms.

It is interesting to note that, although ozone accounts for less than 1 part per million (ppm) of the gases in Earth's atmosphere, it absorbs most of the UV radiation from the sun.

Destruction of ozone occurs mainly in polar stratospheric clouds (PSCs), which contain water ice crystals, sulfuric acid particles, and nitric acid trihydrate particles, among others. The latter, in essence, may be considered *pollutants*, although all are not of anthropogenic origin. Various natural causes contribute, such as the ejecta form volcanic eruptions. But, without regard to source, tiny particles form nuclei for condensation reactions, including those that determine the fate of ozone. Thus, any pollution can indirectly contribute to the sum total of ozone depletion, although certain materials, such as those containing chlorine and bromine, greatly accentuate the process.

Why does the ozone depletion maxima occur in October? Theories to explain this involve numerous factors, such as atmospheric circulation and, more directly, the type of cloud and the manner in which it is formed. PSCs may result from either slow or rapid cooling, factors that reflect the meteorological cycles of the Antarctic.

How is ozone destroyed? A chlorine molecule, for example, can combine with a molecule of ozone to yield a molecule of chlorine monoxide and a molecule of oxygen. Once formed, two chlorine monoxide molecules may form a *dimer*, which is a compound comprised of two chlorine monoxide molecules. Near-UV radiation can catalyze the dimer to form two chlorine atoms and one oxygen molecule. Thus, ozone disappears.

Nearly all of the reactions proposed are reversible, depending upon alterations in the fundamental conditions. For example, if shielded from UV radiation, ozone can be regenerated.

As described by Molina, "Chlorofluorocarbons (CFCs) are unaffected by rain and by the chemical reactions that cleanse most other gases in the troposphere. The CFCs slowly rise into the upper stratosphere, above the ozone layer, where UV is strong enough to break the molecules apart, releasing chlorine atoms that react very rapidly with ozone. Occasionally, these chlorine atoms combine with other chemicals to form relatively stable 'chlorine reservoirs,' which in turn decompose, periodically returning the free chlorine atom to the stratosphere. Each chlorine atom released by the decomposition of a CFC molecule is capable of destroying tens of thousands of ozone molecules before it returns to the Earth's surface."

While all of the physicochemical details of ozone destruction have not been worked out, the proposition that CFCs are a primary cause of ozone depletion now has few doubters. Earlier contradictory theories essentially have been ruled out as of 1993.

The principal driving force for more research pertaining to the ozone "hole" has come from the damaging effects of ozone depletion worldwide. But what are the effects specifically on Antarctica itself?

Considerable research in this regard has been going on at the Palmer Station, located at the tip of the Antarctic Peninsula, about 965 km (600 mi) south of Cape Horn. Observations have tended to indicate that the "hole" tends to run in 2-year cycles. In one year, for example, ultraviolet radiation may be double or one-half that of a prior year. D. Karentz (University of California at San Francisco) notes that the increase in radiation is sudden and occurs at a time when organisms are emerging from the dark winter period and thus have not had time to adapt to the sun being up. Another scientist has proposed the analogy — "like a Norwegian going to the Mediterranean over Christmas vacation."

Thus, the effect on Antarctic organisms not only is one of UV dosage, but one of timing as well. Not all organisms adapt as well and as readily as people.

A major concern is the effect of UV on phytoplankton synthesis. Researchers at Palmer Station have found that this process can decline by as much as 15–20% in the uppermost meter of water, diminishing exponentially with depth, with little if any decline at depths of 10 to 15 meters. Because of UV damage to their DNA, the microorganisms are unable to divide. But researchers also have found that phytoplankton, like some other organisms, introduce a mechanism for repairing DNA damage. The tolerance to UV ranges widely from one species of phytoplankton to the next.

Krill, a mainstay of the Antarctic ecosystem, feed on phytoplankton. Just what effect, if any, genetic changes in the phytoplankton may have on the reduction of krill remains unknown. One scientist has observed, "I am positive there will not be a collapse of the southern ecosystem." Also, it is most likely that larger marine animals, such as the penguin, will not be affected directly by the effects of the ozone "hole."

Ozone Depletion in the Arctic Region. The Airborne Arctic Stratosphere Expedition, conducted by the U.S. National Aeronautics and Space Administration and the National Oceanic and Atmosphere Administration, assisted by researchers from Norway, Great Britain, Germany, and Russia, during January and February 1989, did not reveal an ozone hole comparable to that over Antarctica. However, one researcher observed what he termed "highly perturbed" chlorine chemistry, a probable precursor of an ultimate ozone hole. The scientists were surprised to find chlorine monoxide and chlorine dioxide in abundance more than 50 times higher than normal over a wide range of altitudes. Nitrogen compounds that inhibit ozone destruction were absent. R.T. Watson (NASA) noted, "Ozone loss would be expected if the mass of cold air over the Arctic (known as the Arctic polar vortex)[1] persisted into spring, when the sun rises after the polar winter." The data strongly backed up the concerns that resulted in the *Montreal Protocol* that targets cutting emissions of chlorofluorocarbons in half by the year 1998.

In a subsequent mid-1991 paper, W.H. Brune (Pennsylvania State University) and colleagues from other institutions in the United States and United Kingdom, summarized the potential for ozone depletion in the Arctic polar stratosphere in the following terms: "The nature of the Arctic polar stratosphere is observed to be that of the Antarctic polar stratosphere.... Most of the available chlorine (HCl and $ClONO_2$) was converted by reactions on polar stratospheric clouds to reactive ClO and Cl_2O_2 throughout the Arctic polar vortex before midwinter. Reactive nitrogen was converted to HNO_3 and some, with spatial inhomogeneity, fell out of the stratosphere. These chemical changes ensured characteristic ozone losses of 10–15% at altitudes inside the polar vortex where polar stratospheric clouds had occurred. These local losses can translate into 5–8% losses in the vertical column abundance of ozone. As the amount of stratospheric chlorine inevitably increases by 50% over the next two decades (by 2011), ozone losses recognizable as an ozone 'hole' may well appear." See also **Climate**.

Ironical Sources of Pollution. Not all scientific research activity in the polar regions is directed toward determining the anthropogenic sources of pollution throughout the world that affect the well-being of the Arctic and Antarctic. But, when pollution stems from research activities within this region, it is particularly disturbing. There are several research stations in these regions, including the presence of well over a thousand staff and support personnel in one Antarctic station. An Environmental Defense Fund report shows that at least some research stations in the past have polluted the pristine environment with polychlorinated biphenyls (PCBs), used fuel, and emissions from burning waste. A somewhat-questioned report indicated that PCB concentrations in McMurdo Bay ranged from 18 to 340 ppb, as compared with little presence of PCBs in Galveston Bay and 70 ppb in the Oakland Bay.

Polar Atmospheres as Telltales. In addition to ozone depletion, the polar atmosphere may serve in assessing and forecasting other air contaminants on Earth. For example, Mayewski (University of New Hampshire) and colleagues reported in 1986 that an ice core in south Greenland (covering the period 1869–1984) was analyzed for oxygen isotopes and chloride, nitrate, and sulfate concentrations. It revealed that an "excess" (non-sea salt) sulfate concentration had tripled since approximately the 1900–1910 period; nitrate concentration had doubled since about 1955. The investigators suggest that the increases may be attributable to the deposition of these chemical species from air masses carrying North American and Eurasian anthropogenic emissions. This ice core record was derived from a site that is devoid of any locale-specific contamination. It is the longest and most detailed record of anthropogenically introduced sulfate and nitrate available to date. The increase in "excess" sulfate in the record (by a factor of 3) is close to the estimated increase (by a factor of 2.5) previously determined from a calculation of the export of sulfate from North America eastward. This observation also ties in with observations of Arctic and Antarctic haze.

In commenting specifically on the Arctic atmosphere and climate, Washburn (University of Washington) and Weller (University of Alaska) observed that at high latitudes, the upper atmosphere (with magnetosphere and ionosphere) serves as a window to space, in that particles from the sun are focused primarily on polar regions. There, they give rise to magnetospheric and ionospheric phenomena that have a direct impact, for example, on important space processes. These phenomena can strongly impact communications and defense capabilities — charged particles precipitating into the ionosphere, causing auroras, can interrupt rf communications during magnetic storms. Thus, communications with satellite and radar systems become vulnerable. High background noise induced by the aurorta can adversely affect optical and infrared sensors used by surveillance spacecraft; electric currents induced in long conductors, such as telephone cables, power lines, and pipelines, can produce deleterious effects. Thus the need for a better understanding of the polar atmospheres. Because of the prolonged polar nights, the polar regions become excellent laboratories for studying the aurora and for probing the long tail of the magnetosphere. See also **Climate**.

Washburn and Weller also report that any temperature increase due to the greenhouse effect (CO_2 buildup) is intensified in the Arctic because of the melting of snow and ice and the accompanying reduced *albedo* and changes in energy balance. Some recent numerical models have indicated temperature increases that would affect the distribution of sea ice, navigation in the Arctic Ocean, the northward extent of agriculture and the tree line, and, as land-based glaciers melt, perhaps some global rise of sea level, important to low-lying coasts. Although still considered speculative, some investigators believe that a temperature rise of 6 °C in winter and 1 °C in summer could occur in Arctic Alaska. A change of this magnitude is regarded with mixed reactions. The temperature increases most likely would benefit oil exploration and agriculture with but slight effect on the timber industry, but adversely affect a number of other industries because of thawing of permafrost. Despite large-scale climatic influences and the significance of Arctic weather for regional forecasting and for understanding global weather, important knowledge remains lacking because Arctic observing stations are widely scattered and air-sea-ice interactions are physically complex.

Polar Hydrosphere

The polar oceans influence about three-quarters of the Earth's oceans. Research is being directed toward a better understanding of the bottom sediments of the Arctic Ocean and the Southern Ocean, and on improved data pertaining to ice and seawater exchanges between the Arctic and Atlantic Oceans, which occur between Greenland and Svalbard. Although it is the world's largest ocean current, the Antarctic Circumpolar Current (or West Wind Drift) is still poorly understood. See also **Antarctic Convergence**. Even though the bottom water originating in the Antarctic constitutes over half of the bottom water of the global oceans, its exact mode and place of origin are not fully understood.

Considering the prevailing cold temperatures of Antarctica, it is interesting to mention briefly Lake Vanada in the Wright Dry Valley, where water temperatures near the bottom of the lake run as high as +25 °C (77 °F). The mean annul air temperature in this region is −20 °C (−4 °F). Once believed to be due to volcanic heat, researchers have since concluded that the unexpected high water temperatures derive from a snow-free ice cover that transmits and traps solar radiation, which becomes available to heat the dense saline water that remains at the bottom.

The growth, drift, and decay of sea ice are closely related to the circulation of polar oceans. As reported by Hibler (U.S. Army Cold Regions Laboratory) and Bryan (Geophysical Fluid Dynamics Laboratory, Princeton), this is especially true in the Greenland and Norwegian seas in winter, where warm currents flowing northward encounter rapidly cooling atmospheric conditions together with sea ice advancing southward. These investigators describe a diagnostic ice-ocean model of the Arctic, Greenland, and Norwegian seas that has proved useful in examining the role of ocean circulation in seasonal sea-ice simulations. The model includes lateral ice motion and three-dimensional ocean circulation. (In past studies, the ocean has been approximated by a motionless mixed layer of fixed depth.) The ocean portion of the model is weakly forced by observed temperature and salinity data. Simulation results show that including modeled ocean circulation in seasonal sea-ice simulations substantially improves the predicted ice drift and ice margin locations. Simulations that do not include lateral ocean movement predict a much less realistic ice edge. Additional work of this type is underway in the Marginal Ice Zone Experiment (MIZEX) by CRREL (U.S. Army Cold Regions Research and Engineering Laboratory, Hanover, New Hampshire).

As early as 1978, the NORSEX Group (Norwegian Remote Sensing Experiment) conducted three field investigations in Norwegian waters. This group, consisting of scientists from Canada, Norway, the United States, and Switzerland conducted three important experiments: (1) investigation of sea surface temperature and wind in the Norwegian and Barents seas, as derived from the scanning multichannel microwave radiometer (SMMR) on the *Nimbus 7* satellite; (2) similar measurements as gained from improved remote sensing methods and concentrating on the Norwegian Coastal

---

[1] Polar vortex is defined in article on **Atmosphere (Earth)**.

Current; and (3) specific investigation in the marginal ice zone north of Svalbard. Details of the latter experiment are outlined by the NORSEX group (1983). In their conclusions, the group observed that microwave signatures from measurements at 5 to 100 GHz are excellent for remote sensing of sea ice. The large contrast in emissivity between water and all ice types at low frequencies facilitates the retrieval of data on total ice concentration. At high frequencies, the contrast between first and multiyear ice can be used to separate types of ice. The *Nimbus 7* SMMR was found to locate the ice edges accurately to within 10 kilometers.

Sea level appears to have risen 10 to 15 centimeters during the last century. Part of this rise may be due to thermal expansion of the oceans. See also **Ocean**. The remainder generally has been attributed to the melting of polar ice. However, as reported by Meier (U.S. Geological Survey), studies of the current mass balance of the Antarctic ice sheet, which makes up about 85% of the total glacier ice area on Earth, suggest that a negative mass balance is not likely and that this ice sheet may be subtracting water from the world's oceans. Observed long-term changes in glacier volume and hydrometeorological mass balance models have yielded important data on the transfer of water from glaciers, excluding those in Greenland and Antarctica, to the oceans. The average observed volume change for the period, 1900–1961, has been scaled to a global average by using seasonal amplitude of the mass balance. These data have been used to calibrate other models to estimate the changing contribution of glaciers to sea level for the period, 1884–1975. Although the error band is large, the Greenland and Antarctica glaciers appear to account for a third to half of observed rise in sea level, approximately that fraction not explained by thermal expansion of the ocean.

In an interesting observation, Washburn and Weller stress that the acoustical characteristics of polar seas, as affected by differing water masses and the background of noise of ice movement, internal waves, and other influences, require further study. Acoustic characteristics of these waters are of key interest in the detection of submarines, tanker movements, oil exploration activities, and even on studies of whales.

Isotopic Composition Studies. During the 1980s and 1990s, much attention was directed toward the use of short-lived radionuclides, such as ^{3}H, ^{14}C, ^{210}Pb, ^{226}Ra, and ^{230}Th, in studies of ocean circulation, notably in the polar regions. As described by Piepgras and Wasserburg (California Institute of Technology), these nuclides have half-lives which are short compared with the time scales of the processes studied. They have been used in conjunction with numerous hydrographic measurements, including tracer studies of oceanic circulation paths, mixing rates, and the chemical behavior and distribution of associated stable elements in seawater.

A problem of interest to oceanographers is that of determining mixing rates in and between the oceans. Studies with ^{14}C have indicated that at least 1500 years are required for the exchange of deep water with the mixed layer. A longer time may be required for the exchange of deep waters between ocean basins. A minimum time of about 150 years to mix the world oceans is obtained by assuming that the Pacific Ocean is emptied by the flow through the Drake Passage and mixed with the Atlantic. The Antarctic Circumpolar Current, which controls interocean mixing, flows through this passage. The isotopic composition of neodymium has been determined in seawaters from the Drake Passage. By using a box model to describe the exchange of water between the Southern Ocean and the ocean basins to the north, together with the isotopic results, an upper limit of approximately 33 million cubic meters per second is calculated for the rate of exchange between the Pacific and the Southern Ocean.

The Polar Cryosphere

This is that part of the Earth's surface that is permanently frozen; the zone of the Earth where ice and frozen ground are formed. Studies of the immense ice sheets in the Antarctic and Greenland can provide a better understanding of the ice sheets that once covered large parts of North America, Europe, and Eurasia at various times during the Pleistocene (10^4 to 1.8×10^6 years ago). Much research in recent years has concentrated on development of a better understanding of the complex Earth-sun relationships that influence climatic change, notably in the polar regions. The nature of the climatic change responsible for the growth and decay of the Pleistocene glaciers are still problematical. The moisture sources and mechanisms permitting the growth of the Northern Hemisphere ice sheets also remain to be established.

The Antarctic Ice Sheet. When Bellingshausen (Russia) confirmed that Antarctica was an ice-covered continent (and not a frozen ocean) in 1820, interest in the Antarctic ice sheet commenced. Numerous expeditions managed ultimately to establish the lateral extent of the ice sheet, but the first estimate of its depth was not made until 1911, when Meinardus (Germany), using clever calculations based upon air moving to and fro over the edge of the ice sheet, claimed the thickness to be about 2100 meters. This early estimate was remarkably close to the depth of 2200 meters estimated by a team during the IGY program. IGY teams also refined the general shape of the ice sheet, the main features of which are the large East Antarctic ice mass and the much smaller West Antarctic lobe (nearly as large as Greenland). The center of the ice sheet does not correspond with the South Pole, but is located at 83°S, 53°E and because of its location is sometimes referred to as the *Pole of Relative Inaccessibility*. See Fig. 2. Progress in describing the ice sheet was made during the 1980s and continues. This largely embraces technological advances in satellite observations, remote-sensing devices, and computer simulations.

Radok (Cooperative Institute for Research in Environmental Sciences) indicates that the ice sheet poses three broad research tasks: (1) topography (surface and bottom)—where more details are needed pertaining to the surface elevation and thickness of the ice; (2) clarification of the ice regime—temperature at various depths and balance between what is gained as snow and what is lost as icebergs and meltwater; and (3) a better understanding of the structural and chemical properties of the ice. With better information along these lines, models can be constructed which describe the history of the ice sheet and predict its future behavior.

Radioglaciology has proved the most effective means for mapping surface topography, a technique first pioneered by the U.S. Army Signal Corps in 1957. Data are collected from aircraft, satellites, surface ships, and balloons. Internal echoes received from a variety of internal layers indicated that the ice sheet is not uniform. Echoes may be caused by changes in density and crystal structure (possibly from melting or deformation), as well as concentrations of impurities and layers of increased acidity, which could result from very old volcanic activity. Doppler radar has also been used in more recent years. Measurements of this effect on signals from satellites passing over a point on the ice sheet give the position of that point with a precision of a few meters. A computer-generated view of the Antarctic ice sheet, made from measurements taken during the IGY and IAGP programs, is shown in Fig. 3.

Details on the structural and chemical properties of the ice are obtained principally from drilling cores of ice, usually to a depth of a few hundred meters. A French team has drilled a 900-meter hole on Dome C and the Russians are now drilling a hole of a few thousand meters in depth at the Vostok Station. See also **Earthquakes; Seismology;** and **Plate Tectonics**. A representative ice core sample is shown in Fig. 4.

Many clues to the past are found in ice cores drilled from the Antarctic sheet. Among such findings are evidence of global pollution—radioactive fallout from hydrogen bomb tests in the atmosphere (mid-1950s to early 1960s); volcanic eruptions, such as Krakatau (1883) and Guning Agung (1963), among others. Trapped air in the cores represents a measure of the atmospheric composition many years ago. Fischer and Oeschger (University of Bern) report that air entrapped in bubbles of cold ice has essentially the same composition as that of the atmosphere at the time of bubble formation. Measurements of the methane concentration in air extracted by two different methods from ice samples from Siple Station in West Antarctica have allowed the reconstruction of the history of the increase of atmospheric methane for the past 200 years. Many experts feel that it is a bit premature to extrapolate ice core findings to pollution of the atmosphere as derived from anthropogenic causes. In connection with CO_2, some modeling studies have suggested that a CO_2-induced climatic change would include a polar warming several times the global average warming (variously estimated to be from 1 to 4 K for a doubling in CO_2). This would predict a marked change in polar atmospheric temperature and the warming should be observed in the polar regions prior to the tropics. A significant decrease in overall ice extent during the mid-1970s, previously suggested to reflect warming induced by carbon dioxide, has not been maintained. The extent of ice in the Weddell Sea region rebounded after a large decrease concurrent with a major oceanographic anomaly (the Weddell

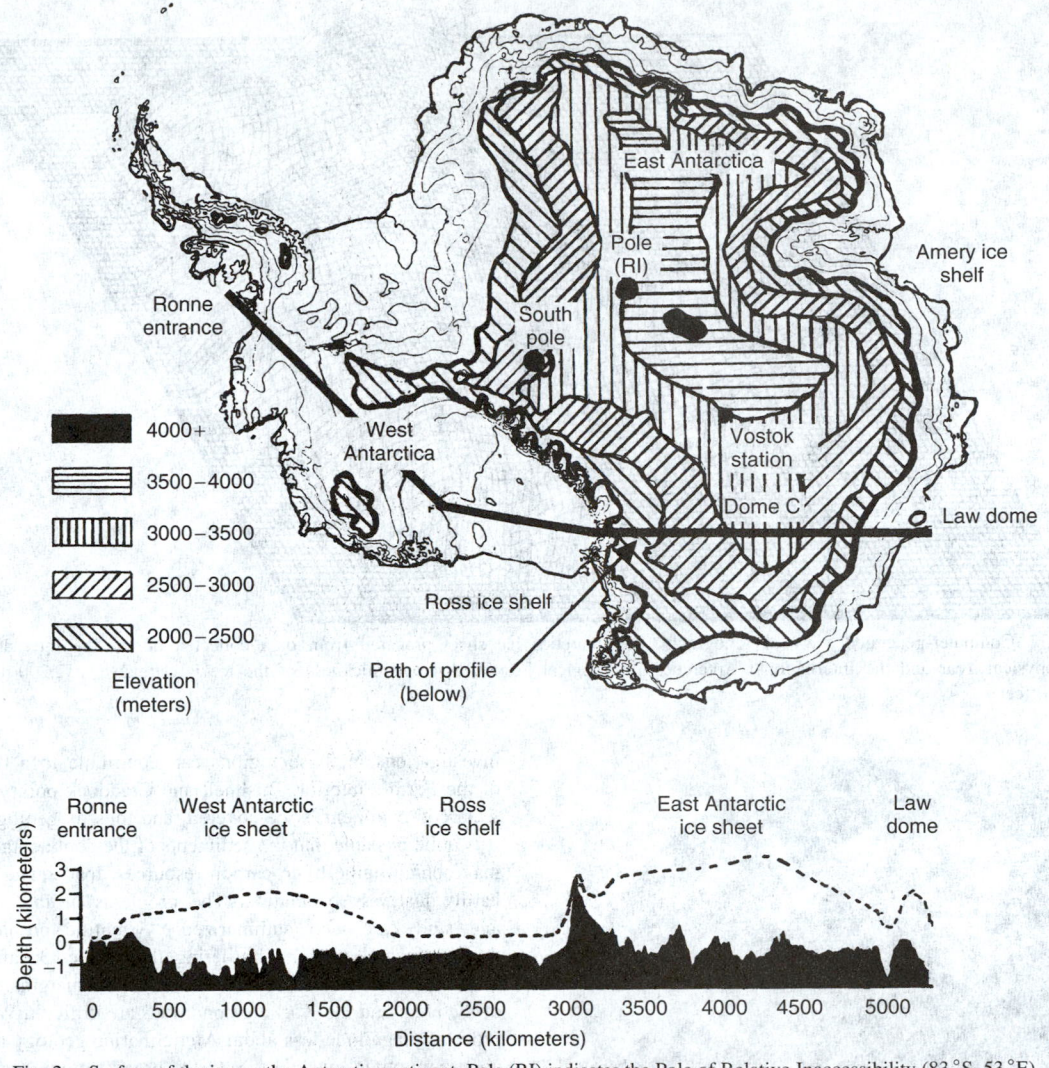

Fig. 2.   Surface of the ice on the Antarctic continent. Pole (RI) indicates the Pole of Relative Inaccessibility (83 °S, 53 °E), located essentially at the center of the ice sheet. The pole (RI) is equally remote from all surrounding coasts. The bedrock (shown in black) and the ice level (dashed line) is shown by profile below map. The line (shown in solid black) is path along which the profile is constructed. Map is based upon data available from the International Geophysical Year (IGY) gathered during the period, 1956–1957 and by the International Antarctic Glaciological Project (IAGP), which has been in continuous operation since 1969. This latter project is comprised of teams from Australia, France, Japan, the United States, and the former U.S.S.R., with a group from the United Kingdom. Data on the bedrock were derived from radar measurements taken through the ice, both from surface ships and airplanes. The technique used is sometimes referred to as *radioglaciology*. (*After Radok.*)

polynya).[2] The decrease of the mid-1970s was preceded by an increase in ice extent from 1966 to 1972, further indicating the presence of cyclical components of variation. As pointed out by Zwally (Goddard Laboratory for Atmospheric Sciences), these cyclical components of variation may obscure any long-term trends that might be caused by a CO_2-induced warming.

As described in article on **Meteoroids and Meteorites**, a number of meteorites have been found on the Antarctic ice sheet, precipitating a controversy as to their source (lunar or martian). Further, scientists are now analyzing Antarctic cover for the presence of dust from meteoroids. Most of these bodies burn up in the Earth's atmosphere, but may leave dust in the atmosphere that ultimately settles down and is preserved in the Antarctic's emerging science storehouse.

Arctic Cryosphere. As in Antarctica, ice cores are revealing fascinating scientific information. Sponsored by the U.S. Arctic Research Commission, the Arctic cryosphere is subject to intensive investigation. The glacial history of large parts of the Arctic remains controversial. The fact that former widespread shelf ice in the Canadian Arctic Archipelago was only recently identified, (1984) illustrates how little is known about this history. The cryosphere is an important resource. In most of the Arctic, water stored in snow and glacial ice; then it is released as runoff to rivers and lakes during seasonal thawing is much more significant than rain as a source of fresh water. It has been suggested that shelf ice can be a transportation resource in some places inasmuch as sections may break off from their parent glaciers and drift around the Arctic Ocean as ice islands (several kilometers in diameter). They may last for years and form platforms for research programs.

The extent of Pleistocene ice sheets in the polar regions is at present unknown. Hughes (1977) and other investigators have suggested that possibly an ice sheet covered a large part of the Arctic Basin, representing greater glaciation than previously believed for that region.

Researchers are studying the sea ice that covers most of the Arctic Ocean all year. The power required to move pack ice is tremendous. The mechanics of floating ice are poorly understood and are being researched. Pack ice also is a problem to the exploitation of petroleum and mineral resources in polar regions.

---

[2] A polynya is a region in a frozen sea that stays anomalously free of ice.

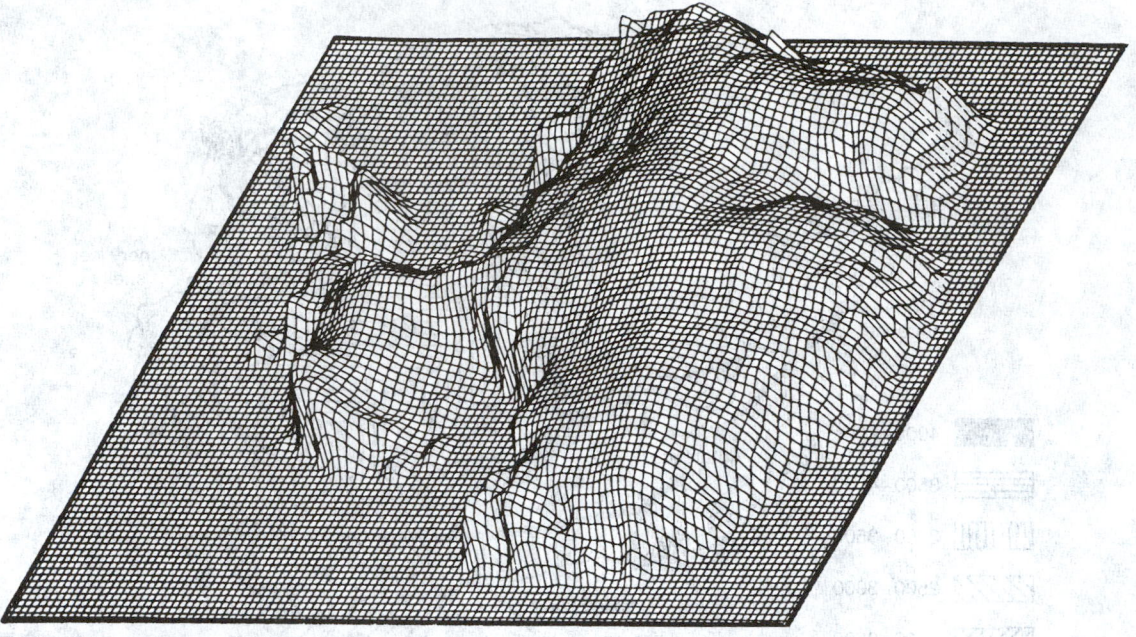

Fig. 3. Computer-generated view of topography of Antarctica ice sheet prepared from data collected during the International Geophysical Year and the International Antarctic Glaciological Project. Average thickness of the ice is estimated at 2,200 meters (7,220 feet.)

Fig. 4. Black-and-white facsimile of a section of an ice core from the East Antarctic sheet. The individual crystal separations indicated result from the use of polarized light. For examination, core sections usually are thin (1 mm). (*After Radok.*)

Permafrost (defined in the entry on **Biome**) is considered part of the cryosphere. In polar regions, permafrost extends to depths exceeding 500 meters (1641 feet) and presents difficult engineering problems, as in the case of construction of the Trans-Alaska pipeline. Recently, it has been found that subsea permafrost also exists offshore and will present difficulties for recovering hydrocarbons from the polar region coastlines.

Polar Lithosphere

Investigations of the lithosphere are described in the entries on **Earthquakes, Seismology, and Plate Tectonics; Ocean**; and **Volcano**. Since the lithosphere obviously also exists under the polar regions, investigations in those regions can contribute to a better understanding of the Earth's interior. Bushnell and Craddock point out that 98% of the Antarctic continent is ice-covered, and thus its geology is poorly known. It is quite possible that the sediments of the continental shelf of Antarctica may contain large hydrocarbon resources. Inadequate exploration of date hardly justifies speculation. The problems of this type of exploration and study are well summarized: "Scientific problems that relate to the continental margins bordering the Arctic Ocean must be assessed differently from those of other continental margins. So little is known of the area, and the working conditions are difficult. We know at least one order of magnitude less about Arctic marine geology and geophysics than about any other areas. Not even the broad tectonic framework of the area has been adequately described.

After passage of the U.S. Arctic Research and Policy Act of 1984, renewed evaluation of our knowledge of the Arctic lithosphere and recommendations for research were made. Summarizing from the Washburn and Weller paper (1986): It is essential to acquire data from deep cores from the Arctic ocean bottom and paleomagnetic data are essential to determine its tectonic and sedimentary history. The sedimentary history may also provide clues to changes in climate and sea-ice cover. The interpretation of even the uppermost sedimentary layers of the lithosphere is still open to debate. The continental shelves are highly significant sources of oil and gas. Offshore exploration for these resources presents many geologic and engineering problems. The stratigraphy and structure of some Arctic lands are very complex, and in Alaska this includes juxtaposition by extensive faulting and lateral displacements of terrains of quite different origin. Oil, gas, coal, and mineral resources of the Arctic are large, especially the energy resources of Alaska. In 1983, the Prudhoe Bay field had a proven crude oil reserve about a quarter that of the entire United States. Coal resources of the Alaska North Slope range from an identified 150 billion to a hypothetical 4 trillion short tons, reflecting limited surface exposures and scattered drilling. Alaska's coal resources as a whole have been estimated as making up 50% of the total in the United States and 15% of world resources.

Other mineral resources of the Arctic region include barite, beryllium, cobalt, copper, fluorite, lead, silver, tin, tungsten, and zinc. The overall potential is unknown. Only 4.5% of the bedrock geology of Alaska has been mapped on a scale of 1 mile (1.6 km). Thus, the potential of the arctic region not only is of interest to science, but to economists and resource planners as well.

Polar Biosphere

There are relatively few studies of life in the polar regions that could be considered in an advanced state. One of these pertains to penguins, which spend most of their life at sea, but nest on land. See also **Penguin**. Quite a lot of information is also available on seals. See also **Sea Lions and Seals**. Less information is known of lichens, mosses, and aquatic algae, and insects, which are sparse, particularly in Antarctica. One wingless insect (*Belgica Antarctica*) is known. This is a midge confined to the Antarctic Peninsula. Various species of *Collembola* and mites are also known. In recent years, there has been considerable research directed toward Antarctic krill (*Euphausia superba*). Krill is a critical link in the Antarctic food chain for fishes, whales, seals, penguins, and some other birds. See also **Crustaceans (Edible)**. The Arctic accounts for about 10% of the world's fish catch—the cod family, capelin, and herring among the most important. The Bering Sea is the world's most productive for the Pacific (walleye) pollock. The Kotzbue Sound salmon fishery is particularly important for Alaska. More interest needs to be shown concerning endangered species, such as the bowhead whale and the Arctic cisco (fish). The bowhead whale is an example of a conservation problem for which critical data are lacking. Data are also needed for the shelf ecosystem of the Bering Sea and for most other Arctic shelf ecosystems on which renewable marine resources depend.

Polar Engineering

Whether for the exploration of minerals, scientific expeditions, or planning military operations in the polar regions, numerous and very practical engineering problems arise. A number of years ago, the U.S. Army established the Cold Regions Research and Engineering Laboratory (CRREL), operated by the Army Corps of Engineers, headquartered at Hanover, New Hampshire. Principal research and development efforts are directed to: (1) studies of snow and ice (both freshwater ice and sea ice); (2) determining the properties of frozen ground, including permafrost; and (3) developing and testing equipment and structures, such as vehicles, roads, airfields, pipelines, etc. CRREL is also concerned with environmental protective measures for the Arctic. For examples, researchers are developing a pioneering study of the climate and biology of northern Alaska. Typical projects handled by CRREL are illustrated in Figs. 5 through 7.

Fig. 5. Special drilling and coring equipment is required to collect samples of sea ice. Cores produced by this large-diameter auger are used to determine mechanical properties of ice from pressure ridges. (*U.S. Army Corps of Engineers.*)

Fig. 7. Specially equipped ice breakers are used in polar research and incorporate instrumentation and cold research laboratories to study the geophysical properties of sea ice. The boom extending from the ship supports a radar for measuring ice thickness. (*U.S. Army Corps of Engineers.*)

Fig. 6. Field data on stream flow under ice covers is required to validate models developed to predict winter river hydraulics. These models are used to predict the effect of an ice cover on erosion, sediment transport, and ice jam formation. (*U.S. Army Corps of Engineers.*)

Additional Reading

Abelson, P.H.: "The Arctic: The Key to World Climate," *Science*, 873 (February 17, 1989).

Alley, R.B. and I.M. Whillans: "Changes in the Antarctic Ice Sheet," *Science*, 959 (November 15, 1991).

Anderson, J.B.: *Antarctic Marine Geology*, Cambridge University Press, New York, NY, 1999.

Anderson, J.G., D.W. Tooley, and W.H. Brune: "Free Radicals within the Antarctic Vortex: The Role of CFCs in Antarctic Ozone Loss," *Science*, 39 (January 4, 1991).

Beardsley, T.M.: "Arctic Angst: No Arctic 'Ozone Hole,' But Conditions Could Lead to One," *Sci. Amer.*, 26 (April 1989).

Beardsley, T.M.: "Low Zone: The Infamous Hole has Influence Beyond Antarctica," *Sci. Amer.*, 26 (October 1989).

Bindschadler, R.A. and T.A. Scambos: "Satellite-Image Derived Velocity Field of an Antarctic Ice Stream," *Science*, 242 (April 12, 1991).

Brigham, L.W.: "The Soviet (Russian) Antarctic Program," *Oceanus*, 87 (Summer 1988).

Brune, W.H., et al.: "In Situ Northern Mid-Latitude Observations of ClO_3, O_3, and BrO in the Wintertime Lower Stratosphere," *Science*, 558 (October 28, 1988).

Brune, W.H., et al.: "The Potential for Ozone Depletion in the Arctic Polar Stratosphere," *Science*, 1260 (May 31, 1991).

Bushnell, V.C. and C. Craddock: "Antarctica Map Folio Series," *Folio 12, American Geographical Society*, New York, NY, 1970.

Cooper, A.K., P.F. Barker, and G. Brancolini: *Geology and Seismic Stratigraphy of the Antarctic Margin*, American Geophysical Union, Washington, DC, 1995.

Craig, H., et al.: "The Isotopic Composition of Methane in Polar Ice Cores," *Science*, 1535 (1988).

Curtin, T.B., N.M. Untersteiner, and T. Callaham: "Arctic Oceanography," *Oceanus*, 58 (Winter 1990–1991).

Davies, T.D.: "New Evidence Places Peary at the Pole," *National Geographic*, 44 (January 1990).

Drewry, D.J.: "The Challenge of Antarctic Science," *Oceanus*, 5 (Summer 1988).

El-Sayed, S.Z.: "The BIOMASS Program," *Oceanus*, 75 (Summer 1988).

Elzinga, A.: *Changing Trends in Antarctic Research*, Kluwer Academic Publishers, Norwell, MA, 1993.

Feeney, R.E.: "Food Technology and Polar Exploration," *Food Techy.*, 70 (May 1989).

Fogg, G.E.: *A History of Antarctic Science*, Cambridge University Press, New York, NY, 1993.

Fowler, A.N.: "Antarctic Logistics," *Oceanus*, 80 (Summer 1988).

Frederick, J.E. and H.E. Snell: "Ultraviolet Radiation Levels During Antarctic Spring," *Science*, 438 (July 22, 1988).

Friedmann, E.I. and A.B. Thistle: "Antarctic Microbiology," John Wiley & Sons, Inc., New York, NY, 1994.

Galimberti, D.: *Antarctica: An Introductory Guide*, Zagier and Urruty Publications, Miami Beach, FL, 1991.

Gordon, A.L.: *The Southern Ocean and Global Climate*, *Oceanus*, 39 (Summer 1988).

Hempel, G.: *Antarctic Science: Global Concerns*, Springer-Verlag, Inc., New York, NY, 1994.

Herbert, W.: "Commander Robert E. Peary—Did He Reach the Pole," *National Geographic*, 386 (September 1988).

Hibler, W.D. and K. Bryan: "Ocean Circulation: Its Effects on Seasonal Sea Ice Simulations," *Science*, **224**, 489–491 (1984).

Hodgson, B.: "Land of Isolation No More—Antarctica," *National Geographic*, 2 (April 1990).

Horgan, J.: "Antarctic Meltdown," *Sci. Amer.*, 19 (March 1993).

Jacobs, S.S. and R.F. Weiss: *Ocean, Ice, and Atmosphere: Interactions at the Antarctic Continental Margin*, American Geophysical Union, Washington, DC, 1998.

Joyner, C.C.: "The Antarctic Legal Regime and the Law of the Sea," *Oceanus*, 22 (Summer 1987).

King, J.C. and J. Turner: *Antarctic Meteorology and Climatology*, Cambridge University Press, New York, NY, 1997.

Lee, D.B.: "Oil in the Wilderness—An Arctic Dilemma," *National Geographic*, 858 (December 1988).

LeMasurier, W.E. and J.W. Thomson: *Volcanoes of the Antarctic Plate and Southern Oceans*, American Geophysical Union, Washington, DC, 1990.

Mayewski, P.A., et al.: "Sulfate and Nitrate Concentrations from a South Greenland Ice Core," *Science*, **232**, 975–977 (1986).

Mech, L.D.: "Life in the High Arctic," *National Geographic*, 750 (June 1988).

Meier, M.F.: "Contribution of Small Glaciers to Global Sea Level," *Science*, **226**, 1418–1421 (1984).

Meriwether, J.W., Jr.: *Atmospheric Sciences in Antarctica*, American Geophysical Union, Washington, DC, 1990.

Mitchell, B.: "Undermining Antarctica," *Technology Review (MIT)*, 49 (February 1988).

Molina, M.J.: "The Antarctic Ozone Hole," *Oceanus*, 47 (Summer 1988).

Mount, G.H., et al.: "Observations of Stratospheric NO_2 and O_3 at Thule, Greenland," *Science*, 555 (October 28, 1988).

Ousland, B.: "The Hard Way to the North Pole," *National Geographic*, 124 (March 1991).

Peltier, W.R.: "Global Sea Level and Earth Rotation," *Science*, **240**, 895 (1988).

Piepgras, D.J. and G.J. Wasserburg: "Isotopic Composition of Neodymium in Waters from the Drake Passage," *Science*, 190, 207–214 (1982).

Radok, U.: "The Antarctic Ice," *Sci. Amer.*, 98 (August 1985).

Roberts, L.: "Does the Ozone Hole Threaten Antarctic Life?" *Science*, 288 (April 21, 1989).

Schoeberl, M.R. and D.L. Hartmann: "The Dynamics of the Stratospheric Polar Vortex and Its Relation to Springtime Ozone Depletions," *Science*, 46 (January 4, 1990).

Scott, Sir Peter: "The Antarctic Challenge," *National Geographic*, 538 (April 1987).

Sherman, K. and A.F. Ryan: "Antarctic Marine Living Resources," *Oceanus*, 59 (Summer 1988).

Smith, W.O., Jr.: *Polar Oceanography: Physical Science*, Academic Press, Inc., San Diego, CA, 1990.

Solomon, S., et al.: "Observations of the Nighttime Abundance of Chlorine Dioxide in the Winter Stratosphere Above Thule, Greenland," *Science*, 10 (October 28, 1988).

Stauffer, B.: "The Greenland Ice Core Project," *Science*, 1766 (June 18, 1993).

Steger, W.: "North to the Pole," *National Geographic*, 288 (September 1986).

Stolarski, R.S.: "The Antarctic Ozone Hole," *Sci. Amer.*, 30 (January 1988).

Stone, R.: "Signs of Wet Weather in the Polar Mesosphere?" *Science*, 1488 (September 27, 1991).

Sun, M.: "NSF and Antarctic Wastes," *Science*, 897 (August 19, 1988).

Tolbert, M.A., M.J. Rossi, and D.M. Golden: "Antarctic Ozone Depletion Chemistry: Reactions of N_2O_5 with H_2O and HCl on Ice Surfaces," *Science*, 1018 (May 20, 1988).

Toon, O.B. and R.P. Turco: "Polar Stratospheric Clouds and Ozone Depletion," *Sci. Amer.*, 68 (June 1991).

Vesilind, P.J.: "Antarctica," *National Geographic*, 556 (April 1987).

Walton, D.W.H., Editor: *Antarctic Science*, Cambridge University Press, New York, NY, 1987.

Washburn, A.L. and G. Weller: "Arctic Research in the National Interest," *Science*, **233**, 633–639 (1986).

Whitworth, T., III: "The Antarctic Circumpolar Current," *Oceanus*, 53 (Summer 1988).

Young, O.R.: "Global Commons—The Arctic in World Affairs," *Technology Rev. (MIT)* 52 (February/March 1990).

Zwally, H.J., C.L. Parkinson, and J.C. Comiso: "Variability of Antarctic Sea Ice and Changes in Carbon Dioxide," *Science*, **220**, 1005–1012 (1983).

Web References

American Geophysical Union (AGU): http://www.agu.org/

Byrd Polar Research Center: http://www-bprc.mps.ohio-state.edu/polarpointers/PolarPointers.html

National Institute of Polar Research: http://www.nipr.ac.jp/welcome.html

National Science Foundation: Polar Research: http://www.nsf.gov/home/polar/start.htm

Scott Polar Research Institute: http://www.spri.cam.ac.uk/

The International Commission on Polar Meteorology: http://www.nerc-bas.ac.uk/public/icd/icpm/

POLAR STRATOSPHERIC CLOUDS (PSC). See **Clouds and Cloud Formation**.

POLAR TROUGH. See **Meteorology**.

POLAR VORTEX. See **Atmosphere (Earth)**.

POLAR WIND. See **Winds and Air Movement**.

POLECAT. See **Mustelines**.

POLE FIGURE. A diagram used in metallurgy to show the preferred orientation of crystals in a metal. A pole figure is prepared by plotting, on a statistical basis, the positions in space of the poles of a specific crystallographic plane using a stereographic projection as the basis of the representation. The data for a pole figure is normally obtained using x-ray diffraction techniques.

POLE (Mathematics). 1. The origin of a polar coordinate system.

2. A nonessential singularity of an analytic function. Let $w = u + iv$ be a single-valued function of the complex variable $z = x + iy$, and u, v be real single-valued functions of x and y. Then $z = z_0$ is a pole of order k, provided that $(z - z_0)^k w(z)$ is analytic and not zero at $z = z_0$. The number k is an integer, greater than unity, and is the order of the pole. Singular points of this kind are called *nonessential* because they may be effectively removed if $w(z)$ is multiplied by $(z - z_0)^k$. They are called poles because a three-dimensional plot of w, x, y shows that w becomes infinite at the singular point and thus looks like a pole of infinite length erected on the plane of $z = x + iy$.

3. The intersection of an axis of rotation or of symmetry with a surface, often spherical.

POLIOMYELITIS. More than one generation of people has reached adulthood with little awareness of this devastating virus infectious disease. It was essentially brought under control in many of the industrialized countries of the world during the mid-1950s and early 1960s with the introduction of the Salk vaccine (1953) and the later introduction of the Sabin vaccine. Because there has been laxity on the part of some parents to have their offspring vaccinated against polio, it is appropriate here to review briefly the cruel nature of the disease, both as a reminder of its potency, and as an outstanding example of achievement by the science of medicine. To ensure the use of vaccine, many regions and communities in advanced countries throughout the world are making vaccination of youngsters mandatory.

In the Salk vaccine, viruses that have been killed are present. The Sabin vaccine is a live virus, but so weakened that it will not produce the disease, but will permit development of protective antibodies.

Nature of Poliomyelitis. At one time called *infantile paralysis*, poliomyelitis is a disease involving inflammation of the gray matter in the spinal cord. Transport of the virus causative agent is via the nerves, either within the axon or by way of perineural cells. The disease is particularly feared, because it may result in paralysis of any part of the body, leaving the victim crippled for life, although this occurs in a minority of cases. There are four principal types of poliomyelitis, only one of which is the paralytic type. The abortive, nonparalytic, and encephalitic types account for over half of the cases diagnosed during epidemic periods. It is believed that mild cases go unrecognized as poliomyelitis due to their resemblance to colds or intestinal disorders. Even mild experiences with the disease appear to offer future immunity to the virus.

Paralytic poliomyelitis may occur in several forms, depending upon the extent of viral involvement and the area of the central nervous system which is affected. *Spinal poliomyelitis* involves the muscles of the extremities or trunk, and is the form most frequently encountered. *Bulbar poliomyelitis* involves the cranial nerves arising from the brain stem, and the vital centers of circulation or respiration are affected. *Bulbo-spinal poliomyelitis* is usually severe and is associated with respiratory impairment and with paralysis involving both the spinal cord and brain stem. From 10 to 25% of paralytic cases seen during an epidemic are of the bulbar or bulbo-spinal type.

The disease dates back to antiquity. Egyptian skeletons (circa 3700 B.C.) show evidence of bone malformations indicative of involvement with polio during childhood. The disease was not confirmed as an infectious disease until 1907 when this fact was demonstrated by O.I. Wickman, a German physician. Although there were mild epidemics during colonial times, the disease was virtually unknown in the United States until 1916, when an epidemic occurred in New York City. At that time, about 9,000 children contracted the disease, with about 2,000 fatalities. Although infantile paralysis is rarely seen in infants under one year of age, and is again rare although more severe among persons over 40, there is no safe age and all persons should be immunized.

Other than rest and comfort, there appears to be no means of altering the course of poliomyelitis if the disease has developed. Treatment measures are directed toward preventing extension of the disease to other neuromuscular units of the body. Application of moist, hot packs over the affected arms, legs and back is the best method for the relief of pain and prevention of muscle spasms, which, if not prevented, may cause deformity of the limbs. Rehabilitation of poliomyelitis patients combines use of physical medicine with psychological and vocational adjustment in an effort to achieve maximal function and prepare the patient physically, mentally, socially, and vocationally for fullest possible life compatible with abilities. A great variety of equipment and techniques has been devised for all stages of muscle weakness.

Eradication Program

In May 1988, the WHO (World Health Organization) made a commitment to eradicate poliomyelitis worldwide by the year 2000. The general strategy followed that which was used earlier and concluded successfully in 1977 with the eradication of smallpox. Poliomyelitis, however, is inherently more difficult to eradicate than smallpox. Among the epidemiological characteristics in which the two diseases differ are the asymptomatic illness that is characteristic of most poliovirus infections and the ability of the poliovirus to spread by enteric transmission, both of which make the identification and containment of cases more difficult. In contrast, smallpox was clinically obvious and eradication quite easy to confirm.

Differences between the vaccines also are important. Smallpox vaccine is heat stable, one dose is required for protection lasting several years, and vaccination leaves a readily visible scar. In contrast, trivalent oral poliovirus vaccine (TOPV) loses substantial potency after 1 day at 37 °C, and multiple doses are required for full protection. Another difference is that properly administered smallpox vaccine has been a highly effective immunogen, whereas seroconversion rates after one to four doses of TOPV have been suboptimal in developing countries. Confirming that poliovirus has stopped being transmitted requires far more sophisticated tests and facilities. Even in the face of these difficulties, the most promising evidence that poliomyelitis can be eradicated has come from the Americas.

As of 1994, all of North America, southern Central America (not including Mexico), Greenland, Iceland, Norway, Sweden, Finland, the United Kingdom, Venezuela, Chile, Argentina, Japan, and Australia were classified in Stage A of the program. That is, no indigenous cases of poliomyelitis has been reported for at least the prior 3 years. All countries had immunization coverage of at least 80 percent, with a full course of vaccine among children reaching their first birthday.

Stage B countries or areas have immunization coverage exceeding 50 percent and report fewer than ten cases of the disease per year. These areas include France, Germany, Italy, parts of the Balkans, extreme western Russia, Saudi Arabia, and the Republic of South Africa.

Stage C countries or areas have immunization coverage exceeding 50 percent and report ten or more cases of poliomyelitis per year. These areas include large portions of the world (Mexico, most of northern South America, Spain, northern Africa, and most of Russia, China, and Malaysia).

Stage D countries or areas have immunization coverage of 50 percent or less (or an unknown coverage) or report ten or more cases per year (or have an unknown number of cases). These areas include most of central Africa, Madagascar, Pakistan, Thailand, Burma, and the Malaysian Peninsula. It is highly likely that the "last" case of poliomyelitis will occur in one of the latter areas.

The WHO program is indeed a highly ambitious undertaking, but past results with smallpox eradication is driving the expectations for success. Cost of the program is in the several millions of dollars. Private assistance of many millions of dollars has come from such organizations as Rotary International.

Further Research Needs. As reported by P.F. Wright (Vanderbilt University) and colleagues at other institutions, the attenuated poliovirus strains derived by Sabin are present in TOPV used throughout the world. Three limitations of the Sabin strains are identifiable:

1. Thermolability, probably the most amenable to immediate improvement.
2. Rare vaccine-associated cases of poliomyelitis have been reported. Such cases lower confidence and increase liability risks. Genetic instability of the attenuating mutations is readily demonstrable in vaccine recovered from children, particularly mutation in the type 3 strain.
3. Although the immunogenicity of the vaccine in many countries, including the United States is effective, this is not always the case in developing countries. Even four doses of vaccine [as recommended by the Expanded Programme on Immunization [EPI]] may not be sufficient to achieve seroconversion rates that will block the spread of the virus.

Major progress has been made in recent years toward understanding the poliovirus at the molecular level, with complete sequence analysis of the three serotypes, the creation of full-length copies of the infectious DNA, the recognition of key attenuating mutations, and the identification of the crystallographic structure of the virus.

Additional Reading

Prevots, D.R., R.W. Sutter, P.M. Strebel, et al.: "Completeness of Reporting for Paralytic Poliomyelitis, United States, 1980–1991: Implications for Estimating the

Risk of Vaccine-associated Disease," *Arch. Pediatr. Adolesc. Med.*, **148**, 478–485 (1994).

Staff: American Academy of Pediatrics. "Poliomyelitis Prevention: Revised Recommendations for Use of Inactivated and Live Oral Poliovirus Vaccines," *Pediatrics*, **103**, 171–172 (1999).

Staff: CDC. "Notice to Readers: Recommendations of the Advisory Committee on Immunization Practices: Revised Recommendations for Routine Poliomyelitis Vaccination," *MMWR*, **48**, 590 (1999).

Staff: CDC. "Progress Toward Global Poliomyelitis Eradication, 1997–1998," *MMWR*, **48**, 416–421 (1999).

Staff: CDC. "Paralytic Poliomyelitis—United States, 1980–1994," *MMWR*, **46**, 79–83 (1997).

Strebel, P.M., R.W. Sutter, S.L. Cochi, et al.: "Epidemiology of Poliomyelitis in the United States: One Decade After the Last Reported Case of Indigenous Wild Virus-associated Disease," *Clin. Infect. Dis.*, **14**, 568–579 (1992).

Sutter, R.W., E.W. Brink, S.L. Cochi, et al.: "A New Epidemiologic and Laboratory Classification System for Paralytic Poliomyelitis Cases," *Am J Public Health*, **79**, 495–498 (1989).

Travis, J.: "Good News, Bad News for Polio," *Science*, 1467 (September 11, 1992).

Wright, P.F., et al.: "Strategies for the Global Eradication of Poliomyelitis By the Year 2000," *N. End. J. Med.*, 1774 (December 19, 1991).

Web References

Centers for Disease Control and Prevention: http://www.cdc.gov/nip/publications/pink/polio.pdf
The Global Polio Eradication Initiative: http://www.polioeradication.org/
The Polio Information Center Online (PICO): http://cumicro2.cpmc.columbia.edu/PICO/PICO.html
The Story of Polio: http://www.pbs.org/storyofpolio/
World Health Organization: http://www.who.int/home-page/

POLISHES. 1.   A solid powder or a liquid or semi-liquid mixture that imparts smoothness, surface protection, or a decorative finish. The most widely used solid polishing agent is fine-ground red iron oxide (rouge), applied to the surface of plate glass, backs of mirrors, and optical glass. A wide variety of liquid and pastelike polishes are based on vegetable waxes (carnauba and candelilla), combined with softeners, fillers, and pigments or emulsified in alcohol or other solvent. Furniture polishes often contain red oil, lemon oil, and petroleum solvent; most types of metal and wood polish contain organic solvents and, hence, are flammable liquids. Nail polishes are nitrocellulose lacquers, usually with amyl acetate solvent.

2.   The hard outer coating of cereal grains, especially rice, which is usually removed in processing. These coatings are rich in vitamin B_1. Their removal robs the cereal of much of its nutritive value.

POLLACK. See **Codfishes**.

POLLEN: STRUCTURE, DEVELOPMENT AND FUNCTION.   The pollen grains of seed plants are both remarkable and intensively studied structures that have evolved to ensure delivery of the male gametes to the embryo sac during sexual reproduction.

Pollen Structure

Pollen, the haploid microgametophyte, is considered to be the male partner in sexual reproduction. In flowering plants pollen consists of either two or three haploid cells when shed from the flower (Figure 1). Both pollen types possess a large vegetative cell containing within, either a single compact generative cell (bicellular pollen), or two sperm cells (tricellular pollen). Pollen grains germinate on the female (stigma) surface by forming a pollen tube, which emerges through one of the apertures in the pollen grain wall. The vegetative cell that comprises the bulk of the pollen grain cytoplasm is responsible for the development of the pollen tube and the delivery of the two sperm cells to the embryo sac. In angiosperms double fertilization occurs, in which one sperm cell fuses with the egg cell to form the diploid zygote, while the second sperm cell fuses with the two haploid polar nuclei within the central cell to form the triploid primary endosperm nucleus, which serves to nourish the developing embryo. See also **Gametophyte and Sporophyte** and **Plant Reproduction**.

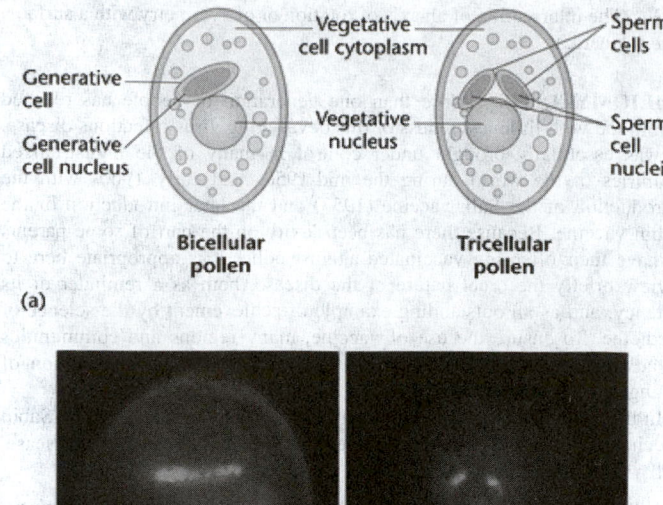

(a)

(b)

Fig. 1. (**a**) Schematic diagram of the structure of bicellular (left) and tricellular (right) pollen. (**b**) Nuclear morphology in bicellular tobacco pollen (left) and tricellular *Arabidopsis* pollen (right) treated with a DNA-specific stain and visualized by fluorescence microscopy. Generative and sperm cell nuclei are often elongated and possess highly compact chromatin with intense fluorescence. Vegetative nuclei are larger and irregular in shape with a more dispersed chromatin structure showing weaker fluorescence.

The Pollen Wall

Pollen shows considerable variation in size, shape and surface characteristics. Pollen grains vary in diameter from 5 µm in *Myosotis* to more than 200 µm in the Cucurbitaceae, and round, elliptical and multifaceted pollen types occur (Figure 2a). The most striking structural feature of the pollen grain is the tough, resistant outer coat, termed the *exine*, which is often elaborately sculptured (Figure 2b). The exine does not develop over certain regions; these define the positions of the germination apertures. Apertures also show wide variation, including elongate furrows and circular pores or combinations of the two. Aperture number, position and exine ornamentation are useful taxonomic characters and from the fossil record have allowed reconstruction of the species composition of past vegetation.

In addition to the purely mechanical function of the exine in protecting the reproductive cells from environmental injury, exine sculpturing plays a role in attachment to insect pollinators and adhesion to the stigmatic surfaces; wind-pollinated species, including many grasses and tree species, often lack elaborate structure and appear smooth. Beneath the exine, which is defined by the presence of one or two basal nexine layers, a second major wall layer surrounds the pollen grain protoplasm; this is called the *intine* (Figure 3). While the exine is composed of sporopollenin, a highly resistant polymer containing carotenoid esters, the intine is largely composed of pectin and cellulose. The origin and genetic control of the synthesis of the two pollen wall layers is different. The exine is developed through the contribution of the early microspore cytoplasm and the inner anther wall cell layer, the tapetum, which plays a dominant role in the synthesis and deposition of sporopollenin on the pollen grain outer surface. Therefore exine synthesis and patterning are under sporophytic control. In contrast, intine wall synthesis is largely under the control of the microspore cytoplasm and involves gametophytic gene expression from the haploid microspore nucleus.

Pathways of Pollen Development

Pollen is produced within the anthers (microsporangia or pollen sacs) of the flower. During its development from an undifferentiated mound of cells (anther primordium) the anther forms two general groups of cells.

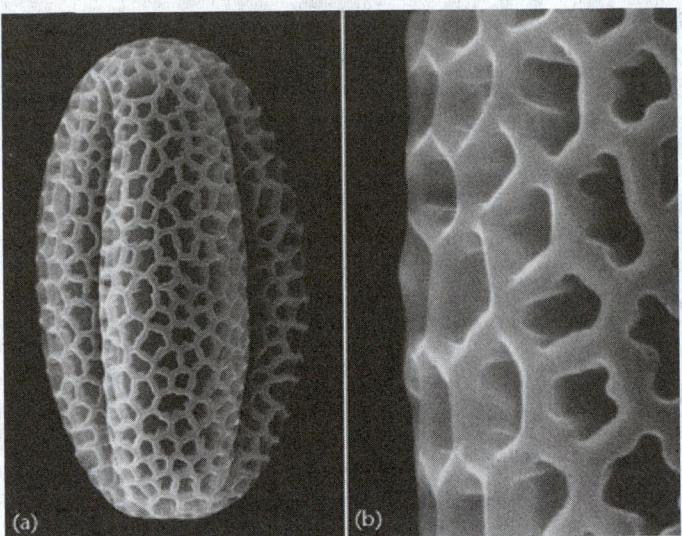

Fig. 2. Surface structure of *Arabidopsis* pollen visualized by scanning electron microscopy. At low magnification (**a**) apertural furrows are visible as longitudinal folds in the exine. At high magnification (**b**) details of the reticulate exine patterning is visible.

The reproductive or sporogenous cells give rise to the microspores and are formed from cells located centrally within the developing anther. The nonreproductive cells form discrete anther tissue layers and include the epidermal, cortical and tapetal cell layers surrounding the sporogenous cells. The tapetum, which is the innermost layer of the pollen sac, plays a dominant role, particularly during the microspore stage. For example, many male sterile mutations affect tapetal cell functions and development is often arrested during the microspore stage [Chaudhury]. See also **Flower**.

Microsporogenesis and Microgametogenesis. Two distinct and successive developmental phases, microsporogenesis and microgametogenesis, lead to the production of the mature microgametophytes (Figure 4). Microsporogenesis comprises the events that lead to the formation of the haploid unicellular microspores. During microsporogenesis the diploid sporogenous cells differentiate as microsporocytes (pollen mother cells or meiocytes), which divide by meiosis to form four haploid microspores. Each diploid meiocyte gives rise to a tetrad of four haploid microspores and microsporogenesis is complete with the formation of distinct single-celled haploid microspores. See also **Stamens**.

Microgametogenesis comprises events that lead to the progressive development of the unicellular microspores into mature microgametophytes containing the gametes (Figure 4). This phase of development begins with the expansion of the microspore, which is commonly associated with the formation of a single large vacuole. Vacuolation is accompanied by the displacement of the microspore nucleus to an eccentric position against the microspore wall. In this position the nucleus undergoes first pollen mitosis (pollen mitosis I), which results in the formation of two unequal

cells, a large vegetative cell and a small generative cell each containing a haploid nucleus. The generative cell subsequently detaches from the pollen grain wall and is engulfed by the vegetative cell forming a unique 'cell within a cell' structure. The engulfed generative cell divides once more by mitosis (pollen mitosis II) to form the two sperm cells completely enclosed within the vegetative cell cytoplasm either before pollen is shed (tricellular pollen) or within the pollen tube (bicellular pollen).

The Male Germ Unit. The generative cell or the two sperm cells, together with the vegetative nucleus, are often found closely associated in a structure known as the male germ unit [Dumas, et al.]; and Mogensen,]; see Figure 1**b**). This reproductive unit is transported within the pollen tube and delivered to the embryo sac. Sperm cells are coated in myosin and can be transported along the actin cytoskeleton *in vitro* or when injected into algal cells. Therefore sperm cell transport may involve an actomyosin-like motility system similar to that described in animal cells. The proposed function of the male germ unit may be to ensure synchrony of the two sperm cell fusion events, with the egg cell and central cell, and the subsequent coordinated development of the resulting zygote and endosperm. In certain species, such as *Plumbago*, the sperm cells are dimorphic (or unequal) and preferentially fuse with either the egg cell or central cell, further illustrating the tightly controlled cell–cell recognition processes involved in double fertilization [Russell].

Pollen Cell Fate

The asymmetric cell division of the microspore at pollen mitosis I plays a critical role in the determination and subsequent fate of the two unequal daughter cells, the vegetative and generative cells. As a result of this highly asymmetric division the larger vegetative cell does not divide further, accumulates a dense cytoplasm rich in protein, lipid and carbohydrates, and possesses a large diffusely staining nucleus (Figure 1**b**). Storage compounds synthesized during vegetative cell maturation are utilized for intense metabolic activity during pollen tube growth by tip extension. In contrast, the smaller and metabolically repressed generative cell does not exit the cell cycle and continues through one more round of mitosis to produce the two sperm cells with highly condensed chromatin (Figure 1**b**). Therefore this asymmetric division may be described as determinative in that daughter cells are immediately different (in size and chromatin structure) after division and fate is determined as a result of the asymmetry of division. For example, microspores that have been induced to divide symmetrically as a result of mitotic spindle displacement adopt equal vegetative-like cell fate with both daughter cells expressing vegetative cell-specific marker genes. Therefore division asymmetry at pollen mitosis I is required to establish generative/sperm cell fate and is an essential component of male gametogenesis. Current genetic approaches to investigate the control of asymmetric division and pollen cell fate in the model species *Arabidopsis thaliana* have led to the isolation of mutants that fail to establish correct polarity of the microspore. One class known as *gemini pollen* mutants often divide symmetrically at pollen mitosis I and fail to produce functional sperm cells [Park, et al.]. The genes defined by these *gemini pollen* mutants are likely to encode proteins that play vital roles in the establishment or execution of cell polarity events required for asymmetric division.

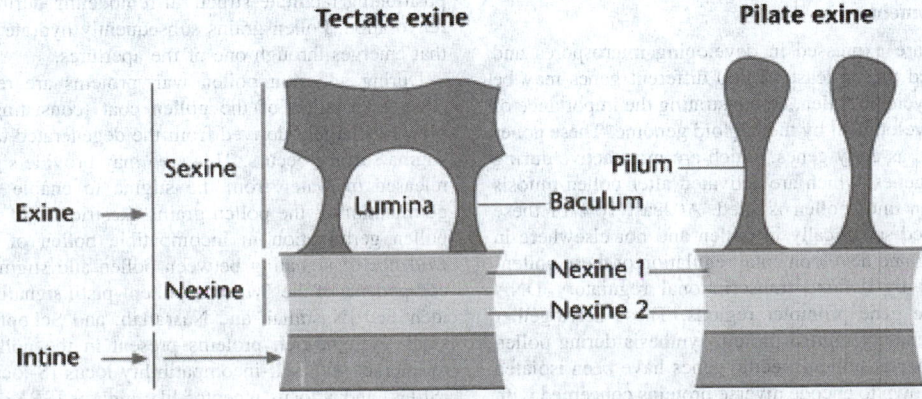

Fig. 3. Schematic diagram of a cross-section through the complex wall layers in pollen with tectate and pilate exine patterning.

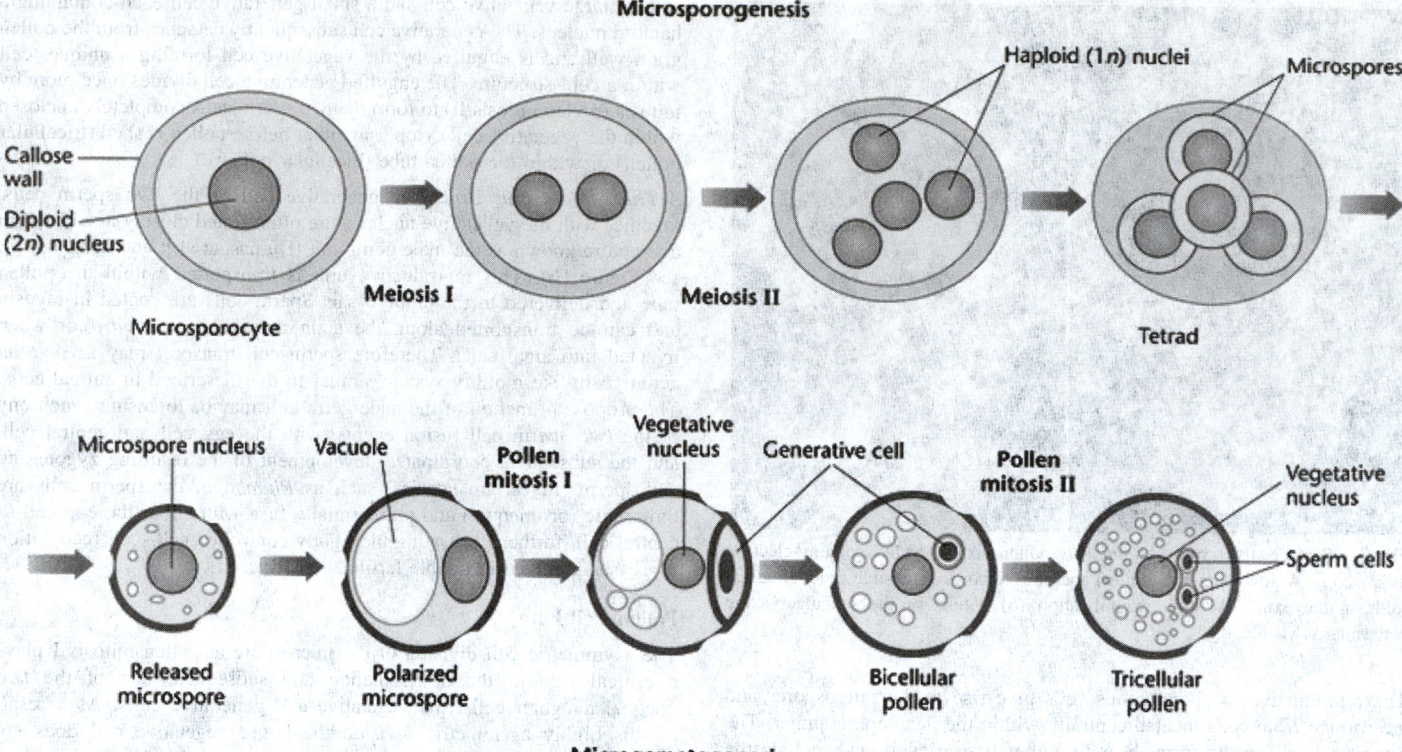

Fig. 4. Schematic diagram of the morphological stages of microsporogenesis and microgametogenesis that occur during the development of tricellular pollen (*Arabidopsis*). During microsporogenesis the microsporocyte (or pollen mother cell), which is surrounded by a thick callose (β (1-3) glucan) wall, undergoes the two nuclear divisions of meiosis to produce four haploid nuclei within a common cytoplasm. Cytokinesis is simultaneous, producing a tetrad of four haploid microspores in a tetrahedral arrangement. During microgametogenesis the microspores are released free into the anther locule by dissolution of the callose wall, where they undergo two stereotypical mitotic divisions, pollen mitosis I and pollen mitosis II, to produce mature tricellular pollen grains.

Microspore Embryogenesis. In contrast to the normally tightly controlled events that take place during microgametogenesis, microspores cultured *in vitro* can be induced to follow an alternative developmental pathway resulting in the development of microspores into haploid embryos [Touraev, et al.]. This developmental switch shows that the haploid genome may be reprogrammed to achieve a different and complete developmental potential. Microspore embryogenesis is triggered by specific stress factors such as nitrogen starvation or high temperature, which block gametophytic development and de-repress the developmentally programmed cell cycle arrest of the microspore daughter cells. Maintenance of sustained cell divisions to form microspore-derived embryos then requires removal of the stress treatment, allowing the expression embryo-specific genes. The haploid (gametic) set of chromosomes in microspore-derived embryos can be doubled by treatment of embryos with microtubule destabilizing agents to produce fully homozygous doubled haploid plants. Such are now routinely used to rapidly generate homozygous breeding lines for crop improvement. See also **Plant Breeding and Crop Improvement**.

Genes and Pollen Development

A large number of genes are expressed in developing microspores and pollen grains. It is estimated that at least 15 000 different genes may be transcribed during the life cycle of pollen, demonstrating the importance of genetic control of pollen development by the haploid genome. These genes have been broadly classified as early genes, which are most active during microspore stage, and late genes, which are activated after pollen mitosis I and increase in expression until pollen is shed. At least 10% of these genes appear to be expressed specifically in pollen and not elsewhere in the plant. The precise timing and developmental regulation of these pollen-specific genes is mediated by defined transcriptional regulatory DNA elements present within the gene promoter regions. These act together with RNA elements to enhance or repress protein synthesis during pollen maturation [Twell]. A number of pollen-specific genes have been isolated by molecular cloning and shown to encode diverse proteins concerned with both basic and specialized cellular functions such as translation initiation,

sugar and ion transport, cytoskeleton organization, pectin degradation, and others with a regulatory or signaling role including protein kinases and transcription factors. Some of these pollen-specific genes encode proteins representing major allergens, the cause of hayfever and allergic asthma. Recent work has shown that the expression of one class of allergens from ryegrass pollen can be reduced without significant impact on plant fertility [Bhalla, et al.], indicating that the genetic engineering of hypoallergenic plant cultivars is possible. See also **Allergens**.

Pollen Function — The Progamic Phase

The progamic phase of pollen development encompasses events that lead from pollination to fertilization. Pollination occurs when the pollen grains are transferred by wind, insect or mammalian vectors to the stigma. Pollen grains are shed in a partially and often highly dehydrated condition and typically may contain only 15% water. Dehydration begins within the anther and is completed upon anther opening (dehiscence), when pollen is directly exposed to the atmosphere. Exine and aperture structure and positioning facilitate structural remodeling during pollen dehydration and rehydration. Pollen grains subsequently hydrate and form the pollen tube that emerges through one of the apertures.

During adhesion pollen wall proteins are released and a stable and close association of the pollen coat (consisting of proteins, lipids and pigments largely derived from the degenerated tapetal cell layer) with the stigma surface occurs. This step may provide signaling for the controlled released of water from the stigma to enable hydration and successful germination of the pollen grains. Restriction of pollen hydration prevents pollen germination in incompatible pollen of *Brassica* species and is evidence of signaling between pollen and stigma. Both male and female components of the *Brassica* pollen–pistil signaling system have now been identified [Nasrallah and Nasrallah; and Schopfer, et al.]. These include small cysteine-rich proteins present in the pollen coat that are thought to interact with self-incompatibility locus (S-locus) specific glycoproteins (SLGs) and S locus receptor-like kinases (SRKs) in the stigmatic surface, the first point of contact between pollen and pistil.

The pollen tube grows by a process of tip extension. It penetrates the stigmatic surface, initially growing either in between or within the cell walls of the stigmatic papillar cells. Genes have been isolated that encode several classes of pectin-degrading enzymes and are specifically expressed in mature and germinating pollen. Polygalacturonase activity is secreted by pollen tubes and the activities of other cell wall hydrolases including cutinases, galactosidases and glucuronidases are present in pollen of several species. Such enzymes are likely to assist in the degradation of stigmatic cell wall materials and the progressive invasive growth through the transmitting tissue of the style. In this regard the pollen–pistil interaction is reminiscent of the invasive growth of a pathogen such as a fungal hypha, which involves complex signaling interactions between pathogen and host cells.

Additional Reading

Bhalla, P., I. Swoboda, and M.B. Singh: "Antisense-mediated Silencing of a Gene Encoding a Major Ryegrass Pollen Allergen," *Proceedings of the National Academy of Sciences of the USA*, **96**, 11676–11680 (1999).

Chaudhury, A.M.: "Nuclear Genes Controlling Male Fertility," *Plant Cell*, **5**, 1277–1283 (1993).

Dumas, C., R.B. Know, C.A. McConchie, and S.D. Russell: "Emerging Physiological Concepts in Fertilization," *What's News in Plant Physiology*, **15**, 17–20 (1984).

Kesseler, R., M. Harley, and A. Papadakis: *Pollen: The Hidden Sexuality of Flowers*, Firefly Books, Limited, Richmond Hill, Ontario, Canada. 2006.

Lesyer, O., and S. Day: *Mechanisms in Plant Development*, Blackwell Publishers, Malden, MA, 2003.

Mogensen, H.L.: "The Male Germ Unit: Concept, Composition and Signification," *International Review of Cytology*, **140**, 129–147 (1992).

Nasrallah, J.B., and M.E. Nasrallah: "Pollen–stigma Signaling in the Sporophytic Self-incompatibility Response," *Plant Cell*, **5**, 1325–1335 (1993).

Park, S.K., R. Howden, and D. Twell: "The *Arabidopsis thaliana* Gametophytic Mutation *gemini pollen* 1 Disrupts Microspore Polarity, Division Asymmetry and Pollen Cell Fate," *Development*, **125**, 3789–3799 (1998).

Russell, S.D.: "Preferential Fertilization in *Plumbago*: Ultrastructural Evidence for Gamete Level Recognition in an Angiosperm," *Proceedings of the National Academy of Sciences of the USA*, **82**, 6129–6132 (1985).

Schopfer, C.R., M.E. Nasrallah, and J.B. Nasrallah: "The Male Determinant of Self-incompatibility in *Brassica*," *Science*, **286**, 1697–1700 (1999).

Sebanek, J., Z. Sladky, and S. Prochazka: *Experimental Morphogenesis and Integration of Plants*, Elsevier Science, New York, NY, 1991.

Shivanna, K.R., and V.K. Sawhney: *Pollen Biotechnology for Crop Production and Improvement*, Cambridge University Press, New York, NY. 2005.

Touraev, A., O. Vicente, and E. Heberle-Bors: "Initiation of Microspore Embryogenesis," *Trends in Plant Science*, **2**, 297–302 (1997).

Twell, D.: "The Diversity and Regulation of Gene Expression in the Pathway of Male Gametophyte Development," In: Scott, R.J., and A.D. Stead: *Molecular and Cellular Aspects of Plant Reproduction*, Society for Experimental Biology Seminar series, vol. 55, Cambridge University Press, Cambridge, UK. 2003, pp. 83–135.

DAVID TWELL, University of Leicester, Leicester, UK

POLLINATION. The act of transference of pollen grains to the stigmatic surface of a flower, where the pollen grain will germinate, forming a slender pollen tube, the development of which leads to the process of fertilization.

Self-pollination must be a fact in certain flowers of the type known as cleistogamous—flowers which never open to allow pollen to be shed into the air or transferred by any means from flower to flower. In such flowers, which are found in many plants, as in several species of violet, the pollen grains may germinate while still in the anther, the pollen tubes growing out to the stigma, and then on to bring about fertilization. In other cleistogamous flowers the pollen is shed from the anthers and falls directly onto the stigma, and there develops. Self-pollination also occurs, though not of necessity, in many perfect flowers. In many cases the stigma is directly beneath the ripened anthers, so that pollen shaken from the anther is likely to fall onto the stigma. In some plants the stigma is always beneath the mature anther, while in other plants movement occurs so that as the part grows older the stigma bends over to a position beneath the anther. In such a case, if cross-pollination has not already occurred, self-pollination may be effected. In other flowers it is the stamen that exhibits movements, curving or bending in such manner that pollination shall be accomplished.

In still other plants the filaments gradually elongate so that the anthers are carried upward to the stigma.

Cross-pollination is insured by several means. In many plants, the stamens mature and shed their pollen long before the stigmas are receptive, while in other plants, the stigmas are mature before the pollen of the same flower has ripened. Either case necessarily insures cross-pollination. Equally certain is the occurrence of cross-pollination in those plants which bear unisexual flowers on different plants, or if on the same plant male flowers mature some time before the female.

Many plants, notably those in the Carrot and Composite families, often bring about cross-pollination in another way. The flowers grow in compact groups, either umbels or heads. As the floral organs, stamens and pistils, mature, they grow out in a way that brings about contact between nearby flowers, the stigmas of one flower touching the anthers of another. That this does often occur is seen in the Composite Family, where in many instances the styles curl back at maturity so that the stigmatic surface is brought directly against the pollen masses. Many plants in this family are self-sterile.

Another method that tends to effect cross-pollination is the occurrence of flowers of two, or three, or even four different kinds. The purple swamp loosestrife, *Lythrum salicaria*, shows one such case. Some of the flowers of its spike have anthers borne on long filaments, and they contain pollen grains of relatively large size. Other flowers have stamens with short filaments and contain small pollen grains; while a third flower type is intermediate in habit, the filaments being between the others in length and the pollen grains of intermediate size. Such plants are known as *heterostylous plants*.

The agents which are instrumental in carrying pollen from one flower to another are principally air-currents and insects. Plants pollinated by air-currents, or wind, are said to be *anemophilous*; those by insects are *entomophilous*.

As a rule wind-pollinated flowers are inconspicuous, and of small size; color, odor, and nectar, all associated with insect-pollinated flowers, are largely lacking. Pollen is produced in immense quantities, since much will be lost. The pollen grains are dry and dust-like, and so float buoyantly in the air. In many plants of this group the stigmas are much-branched, feathery objects, offering a considerable area of sticky surface to catch any pollen that falls on it. To further the ready discharge of pollen, which ordinarily occurs only when the atmosphere is dry, the flowers are often borne in pendulous catkins, or on slender flexible pedicels. Or they have stamens, the anthers of which are versatile, that is, attached at the middle and easily moved by any slight disturbance. Examples of wind-pollinated plants are found in the gymnosperms, all of which are so pollinated, in most grasses, in many hardwood trees, such as birches, alders, oaks, and beeches, and in the common cat-tail of the marshes.

Water-pollinated plants are not numerous. That a plant grows in water is not an indication that its flowers shall be pollinated by water. Indeed, only a very few water-plants are so pollinated.

Plants with flowers pollinated by animals are numerous, and with many variations seemingly calculated to assure cross-pollination. A few tropical plants are said to be pollinated by bats. The flowers have very fleshy petals, open during the evening hours, and are apparently sought by bats, which eat the petals, or perhaps seek any insects which may occur in the flowers. Quite possibly, in going from flower to flower, bats do bring about a transfer of pollen. Pollination by such animals is undoubtedly restricted to a very small number of plants.

Similarly birds, especially humming-birds and honey-suckers, are held to be the agents pollinating several tropical plants. In these the question arises as to whether the birds are seeking the numerous insects which are to be found in the flowers or after the nectar which is found in the flowers. The flowers visited by birds are rather large, brilliantly colored, often scarlet. Unquestionably humming-birds do seek nectar in flowers, both in the tropics and in temperate regions. Pollination may well result from their visits.

A few flowers are said to be pollinated by snails or slugs. Plants so pollinated have dense masses of flowers borne on a fleshy stock. This mass of sterile tissue attracts these animals, which crawl over it in the search for food, and so are said to carry pollen from flower to flower. Many observers doubt that such animals ever bring about pollination.

All, however, recognize that insects are very important in the pollination of many flowers. In many cases the flowers show remarkable adaptations fitting them to be pollinated by certain insects. The insects that effect pollination are for the most part bees, flies, beetles, and moths and butterflies.

As certain features characterize wind-pollinated flowers, so also insect-pollinated flowers exhibit several common features. In them the pollen grains are usually somewhat adhesive, often sticking together into considerable masses. The surface of each grain is variously sculptured, with knobs, spines, or other protuberances definitely increasing the ability of the grain to stick to the insect body. As in wind-pollinated flowers, so here the pollen would be seriously impaired by water, from which it must be protected during periods when pollination would not occur. Many are the ways in which protection is obtained. In some plants the entire flower bends down at night, while in many others the petals close together over the stamens; often a passing cloud is sufficient stimulus to cause closing, which takes place with surprising speed. In many plants the flowers are located beneath the leaves, as in the common Jewel-weed or Touch-me-not, *Impatiens biflora*; in the common Iris, each stamen is located beneath the broad-petaloid stigma. Often the anther itself is so constructed as to afford considerable protection, the pollen frequently being shed through narrow slits which close tightly during periods of excessive moisture; or small pores may allow the pollen to escape when advantageous but protect it from water otherwise.

A most obvious characteristic of insect-pollinated flowers is color, which may be found in a single flower with large conspicuous petals, or may result from the massing together of many small flowers, as in the composite family. While it is generally assumed that bright color is an aid to pollination because it attracts insects, it should be recognized that insect vision is not necessarily like that of human beings. Often the color of a flower changes with age, many becoming gradually deeper toned, as in the common Lady's Slipper, while others as gradually fade out. Flowers opening at night are almost all white or very light-colored.

The odors of flowers are also assumed to attract insects. Every fragrant flower has an odor that is quite distinctive. The odors of flowers are of many types, from the foul rankness of the Skunk Cabbage and Carrion Flower to the delightful perfume of Verbena, Gardenia, the Roses, and many Lilies. In many cases the odor of the flower is very delicate, being scarcely detectable to many people; in others it is of such penetrating strength as to become objectionable. Often the odor is evident only during certain periods, some flowers being scentless by day but fragrant during the night, while others emit their odors only in broad daylight. All these differences in odors seem designed to attract special insects, which will accomplish pollination. The foul odor of carrion calls carrion-flies, and the sweet fragrance of night-blooming flowers attracts moths. After pollination the odor of the flower generally ceases, attraction of insects being no longer of any value.

Many insects undoubtedly visit flowers for the purpose of obtaining nectar, a sweet watery secretion formed in special glands called nectaries. These nectaries are variously located in the flower, usually deep down at the base of the corolla, so that any insect obtaining nectar must either have a mouth part of sufficient length to reach the nectar, or be strong enough to push its way into the flower, passing any obstructions that may serve to protect the nectar from less fortunate insects. Obstructions are found of several sorts. A very common means of excluding such crawling insects as ants, which in all probability would not be efficient pollinators, is by the presence of a barricade of hairs, especially in the throats of flowers having a tubular corolla. Long-tongued bees are able to push through these hairs enough to reach the nectar, while at the same time they are thoroughly powdered with pollen, which may be removed later in another flower. The existence of a sticky secretion over the outside of the flower or on the stem of the plant is an effective barrier to many crawling insects, as is also a waxy coating. Especially noteworthy are those flowers in which the nectar is located at the base of a long narrow tube or spur; often the nectar is present in quantities large enough to form a considerable volume. In such cases there is usually an insect with mouth parts just long enough to reach through the tube into the nectar supply, which is sucked up greedily. This becomes all the more remarkable when one considers that, in some flowers, the nectar is at the bottom of a tube which may be 3 or 4 inches (7.5 or 10 centimeters) long. In the case of one tropical Orchid a nectar-secreting sac over a foot long exists; in such cases insects, usually moths or butterflies, with correspondingly long sucking tubes are found. Often short-tongued insects succeed in obtaining the nectar illegitimately by biting a

hole in the wall of the nectar-containing part of the flower, and obtaining the nectar thereby. It is interesting to note, in connection with this problem of nectar-secreting flowers and insects, that the introduction of clover into Australia was not a success until honey-bees were also introduced. The native Australian bees were too short-tongued to reach the nectar and so pollination was not effected. As a consequence the clover crop soon died out, no seed being formed to perpetuate it. With the introduction of suitable bees the plant seeded abundantly.

It is interesting to note that in addition to the existence of colors, odors, and nectar, the structure of the flower and its position on the plant seem designed to facilitate the work of the insect in transferring the pollen. Many flowers are broad and flat, affording convenient support to the insect as it crawls about over the flower. Others, especially those visited by long-tongued moths, are borne in such a position as seems most suited to permit the insect to insert its tongue and obtain the nectar. The pollen is collected in quantities by many species of bees, who use it as a food for the developing young.

Several plants have developed a most striking relationship with insects. While the majority of flowers are not visited indiscriminately by all insects but only by certain species or genera, in these special cases the restriction is extreme, both the flower and the insect seeming to be modified especially to serve one another. One such case is seen in the edible fig, pollinated only by a small insect, *Blastophaga grossorum*, which lays its eggs in the ovaries of certain flowers of the fig. Another example of this insect-flower association occurs in a species of Yucca. The creamy-white flowers of this plant are borne in large panicles. They open during the evening and are visited by a small moth, *Pronuba fuccasella*, which seeks the abundant pollen of the flower. Of this pollen the moth makes a tiny ball which it carries away to another Yucca flower. In the ovary of this flower the female moth lays her eggs, while on the stigmatic surface it deposits its ball of pollen. The developing ovules serve as food for the growing larvae of the insect. However, many ovules grow to maturity to form viable seeds, which perpetuate the plant and so continue the food supply of the moth, which seems to be the sole agent capable of transferring the sticky pollen of the Yucca from plant to plant. Many other cases are known where pollination of the flower depends on the visit of certain insects, which do not, however, depend on the flower for existence. On the other hand, many elaborate devices, such as the keel mechanism in papilionaceous flowers, which ought to assure insect pollination, fail to do so. Garden peas are normally self-pollinated.

See also **Composite Family (Compositae)**; **Gymnosperms**; **Honeybees**; and **Plant Breeding**.

Additional Reading

Dafni, A., M. Hesse, and E. Pacini: *Pollen and Pollination*, Springer-Verlag, Inc., New York, NY, 2000.

Dafni, A.: *Pollution Ecology: A Practical Approach*, Oxford University Press, Inc., New York, NY, 1993.

Delaplane, K.S. and D.F. Mayer: *Crop Pollination by Bees*, CAB International North America, New York, NY, 2000.

Free, J.B.: *Insect Pollination of Crops*, 2nd Edition, Academic Press, Inc., San Diego, CA, 1993.

Kearns, C.A. and D.W. Inouye: *Techniques for Pollination Biologists*, University Press of Colorado, Boulder, CO, 1999.

Proctor, M., P. Yeo, and A. Lack: *The Natural History of Pollination*, Timber Press, Inc., Portland, OR, 1996.

Thomson, J.D.: *Cognitive Ecology of Pollination: Animal Behavior and Floral Evolution*, Cambridge University Press, New York, NY, 2001.

POLLUCITE. The mineral pollucite is rather rare. It contains cesium, aluminum, silicon, and oxygen, its chemical composition being approximately $(Cs, Na)_2(Al_2Si_4)O_{12} \cdot H_2O$. It is isometric, usually in cubic crystals or crystalline masses; conchoidal fracture; brittle; hardness, 6.5–7; specific gravity, 2.9; luster, vitreous on fresh surfaces; colorless and transparent. Found on the Island of Elba and in the pegmatites of Maine, and as masses 3–4 feet (0.9–1.2 meters) thick in South Dakota; at Varutrask, Sweden; in Italy; and in Kazakhastan, Russian. Pollucite and petalite were found in the granites of Elba and at first named pollux and castorite for the two famous brothers of Roman mythology, Castor and Pollux. Pollucite is derived from the Latin genitive *Pollucis*.

POLLUTANT STANDARDS INDEX (PSI). A relative scale developed by the U.S. Environmental Protection Agency (http://www.epa.gov/ttnamti1/psi.html) that applies to specified air pollutants (ozone, carbon monoxide, nitrogen dioxide, sulfur dioxide, and particulates), designed to clearly and simply inform the public of the current air quality. The scale is normalized such that an index value of 100 corresponds to the U.S. federal air- quality standard for each specified pollutant. The minimum and maximum values are 0 and 500, corresponding to perfectly clean and extremely dirty air. Instead of presenting to the public different numbers for different pollutants, usually only the worst value is presented. The relationship between the pollutant standards index and health advisory level are: 0–50 Good; 50–100 Moderate; 100–200 Unhealthful; 200–300 Very unhealthful; >300 Hazardous.

Web References

Pollutant Standards Index: http://www.sizes.com/units/pollutant_standards_index.htm

POLLUTION (Air). Air pollution is a broad term applied to any chemical, physical (particulate matter), or biological agent that modifies the natural characteristics of the atmosphere. These substances are generally contaminants that substantially alter or degrade the quality of the atmosphere. The term is often used to identify undesirable substances produced by human activity, that is, anthropogenic air pollution. Air pollution usually designates the collection of substances that adversely affects human health, animals, and plants; deteriorates structures; interferes with commerce; or interferes with the enjoyment of life.

Worldwide air pollution is responsible for large numbers of deaths and cases of respiratory disease. Setting of air quality standards, such as the U.S. Clean Air Act, has begun to improve air quality for some pollutants. While major stationary sources are often identified with air pollution, the greatest sources of emissions are actually mobile sources, principally the automobile. There are many available air pollution control technologies and urban planning strategies available to reduce air pollution; however, worldwide costs of addressing this issue are high. See also **Clean Air Act.**

Sources

Air pollutants are classified as either directly released or formed by subsequent chemical reactions. A direct release air pollutant is one that is emitted directly from a given source, such as the carbon monoxide or sulfur dioxide, all of which are byproducts of combustion; whereas, a subsequent air pollutant is formed in the atmosphere through chemical reactions involving direct release pollutants. The formation of ozone in photochemical smog is the most important example of a subsequent air pollutant. See also **Ozone.**

Anthropogenic Sources (human activity) related to burning different kinds of fuel:

- Combustion-fired power plants.
- Controlled burn practices used in agriculture and forestry management.
- Motor vehicles generating air pollution emissions.
- Marine vessels, such as container ships or cruise ships, and related port air emissions.
- Wood, coal, fuel oil or natural gas burning fireplaces, stoves, furnaces and incinerators.

Other Anthropogenic Sources
- Oil refining, power plant operation and industrial activity in general.
- Chemicals, dust and crop waste burning in farming.
- Fumes from paint, varnish, aerosol sprays and other solvents.
- Waste deposition in landfills, which generate methane.
- Military uses, as nuclear weapons, toxic gases, germ warfare and rocketry.

Natural Sources
- Dust from natural sources, usually large areas of land with little or no vegetation.
- Methane, emitted by the digestion of food by animals, for example cattle.
- Pine trees, which emit volatile organic compounds (VOCs).
- Radon gas from radioactive decay within the Earth's crust.

- Smoke and carbon monoxide from wildfires.
- Volcanic activity, which produce sulfur, chlorine, and ash particulates.

Contaminants

Important pollutant gases include:

- Carbon monoxide (CO) [CAS: 630-08-0], which is primarily emitted from vehicle exhausts due to incomplete combustion; the highest concentrations are generally found near roadways. Inhalation of high levels of carbon monoxide can cause headaches, fatigue, respiratory problems and death. See also **Carbon Monoxide.**
- Chlorofluorocarbons, which destroy the stratospheric ozone layer.
- Hydrocarbons, examples: methane (CH_4) [CAS: 74-82-8], ethane (C_2H_6) [CAS: 74-84-0], propane (C_3H_8) [CAS: 74-98-6], butane (C_4H_{10}) [CAS: 106-97-8], hexane (C_6H_{14}) [CAS: 110-54-3].
- Nitrogen oxides (NO_x). Emissions are primarily in the form of nitric oxide (NO) [CAS: 10102-43-9], which is oxidised by atmospheric ozone or peroxy radicals to form Nitrogen dioxide (NO_2) [CAS: 10102-44-0]. NO_2 is the primary concern for effects on health, and is the species for which WHO's health-based standards are expressed. The various oxides of nitrogen can also react with hydrocarbons in the atmosphere to contribute to photochemical smog. NO_x can also affect ecologically sensitive sites through deposition, causing acidification and eutrophication. See also **Nitrogen.**
- Sulfur oxide (SO_x), which causes acid rain is derived from the burning of fuel containing sulfur, mostly at power plants, and during metal smelting and other industrial processes. See also **Acid Rain.**
- Tropospheric ozone, which is ozone in the lower part of the atmosphere. Ozone (O_3) [CAS: 10028-15-6] is a secondary pollutant, formed through photochemical reactions involving NO_x and hydrocarbons; it is an irritant gas. In the stratosphere it helps to reduce the amount of ultraviolet radiation from the sun that reaches Earth.
- Volatile organic compounds: gasoline, solvents, cleaning solutions.

Particulates are small particles, classified by diameter. Atmospheric particles are usually measured as PM10 or PM2.5. The smaller the particle, the more hazardous it is to health. The PM10 fraction are particles with an aerodynamic diameter of less than 10 micrometres; these can be breathed deep into the respiratory tract and reach the lungs. PM2.5 particles are even more dangerous because they can pass through the upper airway filtering into the alveoli, where they can cross the lung/blood stream barrier and enter the bloodstream. Even smaller nanoparticles can stay airborne longer and are thence more important, besides which these minute particles tend to be more toxic.

Important pollutant particulates (mostly from combustion sources) include:

- Hydrocarbon solids (commonly from inefficiently burned fuels).
- Suspended oil particles (from aerosol spray sources).
- Elemental carbon (soot).
- Elemental sulfur.
- Mineral salts, though these can occur naturally.
- Lead and other toxic heavy metals.
- combinations of ammonia (NH_3) with either sulfuric acid (H_2SO_4) or nitric acid (HNO_3) and water (H_2O) — secondary emissions.
- Asbestos fibers.

Atmospheric Composition

Prior to the Industrial Revolution (circa 1840s), the composition of "pure" air making up Earth's surrounding atmospheres essentially remained constant for several thousand years. True, certain natural phenomena, such as volcanic eruptions, may have altered the atmospheric composition over relatively short time spans. See also **Atmosphere (Earth).**

By many orders of magnitude, the greatest alteration of the atmosphere occurred during the middle Precambrian period, between 2.9 and 1.8 billion years ago. It is generally accepted that prior to that time, the terrestrial atmosphere was chemically of a reducing nature — as contrasted with an oxidative environment of the present general composition required to support humans and other mammals and life forms which abound on the earth today. See also **Air.** Brought about by greatly accelerated plant growth, that earlier natural change represented the most dramatic pollution effect ever suffered by the earth's environment. During that period, the oxygen liberated by plant activity proved to be a very toxic substance for anaerobic life forms and eradicated most of the biotic community existing at that time. New types of life had to develop which were capable of survival in an oxidative environment. Geochemical processes took on new characteristics, based upon the slow oxidative degradation of both organic and inorganic materials.

Alteration of the earth's atmosphere as the result of human (anthropogenic) activities is extremely recent on the life scale of the earth. This altering process was essentially commenced when humans first discovered and started to use fire as a means of heating, cooking, etc. It is the *combustion* of organic fuels today that is the principal contribution to anthropogenic air pollution. For centuries the pollutants added to the atmosphere by humans were essentially insignificant in terms of the mass and the dynamics of the earth's atmosphere. Except on a local and sometimes regional basis, air pollution was no problem prior to the invention of the steam engine and, of course, the later invention of the internal combustion engine. Traditionally, air pollution has a direct relationship with increasing population and the growing sophistication of the population, which demands ever increasing quantities of energy and the manufacture of goods by processes which yield byproducts that require removal to some kind of sink, the earth's atmosphere being one of these sinks.

With few exceptions, air pollutants ultimately fall by gravity to the surface of the earth. On land, pollution of the soil and freshwater lakes and rivers and ultimately the groundwater occurs. Fallout on the seas and oceans also occurs, but unless radioactive, the effects are less easy to discern except on the long term. It is indeed difficult to separate air and water pollution. The relationship is explored in the article on **Water Pollution**. The winds contribute both to the spread and, in some instances, to the contribution of air pollutants. Frequently, as in the case of "acid rain," the precipitation of water (an excellent solvent) in the form of rain, snow, sleet, ice pellets, etc. causes entrainment of pollutants (gases, mists, particles, etc.). Thus the soils, rocks, lakes, and rivers are subject to the corrosive and biodestructive processes brought about by the presence of alien substances. Acid rain is described later in this article.

During the last few decades and including the early 1990s, the total amount of pollutants in the atmosphere has increased exponentially. Currently, many hundreds of different pollutants, largely from anthropogenic sources, rise into the atmosphere. Even though many of these substances are measured on a scale of parts per million (ppm) or even parts per billion (ppb), a majority of these substances cause ill effects on the health and well-being of air-breathing creatures as well as corroding and eroding structures.

The following *incomplete* list is given here to dramatize the complexity of the current air pollution problem, to demonstrate the probable impracticality of finding a few simple solutions, and to illuminate why air pollution measurement and control presently requires the skills of many hundreds of scientists and engineers. To accomplish these tasks, many excellent instrumental techniques are required, as exemplified by chromatography, laser radar, various forms of spectrometers, and particle analyzers, among other advanced analytical tools. The use of radioactive isotopes has been effective in many instances of pollution source tracing.

Partial List of Air Pollutants

Common Gases—carbon monoxide (CO), carbon dioxide (CO_2) [CAS: 124-38-9], nitrogen oxides (NO_x), ammonia (NH_3) [CAS: 7664-41-7], hydrogen sulfide (H_2S) [CAS: 7783-06-4], and chlorine (Cl_2) [CAS: 7782-50-5].

Volatile Inorganics—sulfuric acid (H_2SO_4) [CAS: 7664-93-9], hydrochloric acid (HCl) [CAS: 7647-01-0], nitric acid (HNO_3) [CAS: 7697-37-2], and hydrogen peroxide (H_2O_2) [CAS: 7722-84-1].

Volatile Organics—hydrocarbons[1], fluoroalkenes, alcohols, polychlorinated biphenyls (PCBs), ketones, aldehydes.

Free Radicals—hydroxyl, sulfate.

Solid Particles—carbonaceous and metal particles (lead, zinc, manganese, cadmium, chromium).

[1] Motor vehicle pollution is addressed in article on **Automotive Engineering**.

Formulated Commercial Products and Byproducts—insecticides, pesticides, fungicides, herbicides, solvents, coatings, chlorinated fluorocarbons, petroleum and petrochemical products, plastics, fibers.

Radioactive Substances—from weapon testing, radon in some enclosed spaces.

Pollution also affects the manufacture of certain materials and products. This is evidenced by the need for "clean rooms" in metrology standards laboratories and in the production of certain electronics materials and of component assembly operations. In addition to elaborate filtering systems, such rooms are held at a slightly positive pressure (above outside atmospheric pressure) to prevent the entry of raw air from the outside.

Atmospheric Dispersion Modeling

The basic technology for analyzing air pollution is through the use of a variety of mathematical models for predicting the transport of air pollutants in the lower atmosphere. It is performed with computer programs that solve the mathematical equations and algorithms which simulate the pollutant dispersion. The dispersion models are used to estimate or to predict the downwind concentration of air pollutants emitted from sources such as industrial plants and vehicular traffic. Such models are important to governmental agencies tasked with protecting and managing the ambient air quality. The models are typically employed to determine whether existing or proposed new industrial facilities are or will be in compliance with national ambient air quality standards. The models also serve to assist in the design of effective control strategies to reduce emissions of harmful air pollutants.

The dispersion models require the input of data which includes:

- Meteorological conditions such as wind speed and direction, the amount of atmospheric turbulence (as characterized by what is called the "stability class"), the ambient air temperature and the height to the bottom of any inversion aloft that may be present.
- Emissions parameters such as source location and height, source vent stack diameter and exit velocity, exit temperature and mass flow rate.
- Terrain elevations at the source location and at the receptor location.
- The location, height and width of any obstructions (such as buildings or other structures) in the path of the emitted gaseous plume.

The atmospheric dispersion models are also known as *atmospheric diffusion models*, *air dispersion models*, *air quality models*, and *air pollution dispersion models*.

Discussion of the layers in the Earth's atmosphere is needed to understand where airborne pollutants disperse in the atmosphere. The layer closest to the earth's surface is known as the *troposphere*. It extends from sea-level to a height of about 18 km (11 miles) and contains about 80 percent of the mass of the overall atmosphere. The *stratosphere* is the next layer and extends from 18 km (11 miles) to about 50 km (31 miles). The third layer is the *mesosphere* which extends from 50 km (31 miles) to about 80 km (50 miles). There are other layers above 80 km (50 miles), but they are insignificant with respect to atmospheric dispersion modelling.

The lowest part of the troposphere is called the *atmospheric boundary layer (ABL)* or the *planetary boundary layer (PBL)* and extends from the Earth's surface to about 1.5 to 2.0 km in height. The air temperature of the atmospheric boundary layer decreases with increasing altitude until it reaches what is called the *inversion layer* (where the temperature increases with increasing altitude) that caps the atmospheric boundary layer. The upper part of the troposphere (i.e., above the inversion layer) is called the *free troposphere* and it extends up to the 80 km (50 miles) height of the tropsphere.

The ABL is of the most importance with respect to the emission, transport and dispersion of airborne pollutants. The part of the ABL between the Earth's surface and the bottom of the inversion layer is known as the mixing layer. Almost all of the airborne pollutants emitted into the the ambient atmosphere are transported and dispersed within the mixing layer. Some of the emissions penetrate the inversion layer and enter the free troposphere above the ABL.

In summary, the layers of the Earth's atmosphere from the surface of the ground upwards are: the ABL made up of the mixing layer capped by the inversion layer; the free troposphere; the stratosphere; the mesosphere and others. Many atmospheric dispersion models are referred to as *boundary layer models* because they mainly model air pollutant dispersion within the ABL. To avoid confusion, it should be noted that models referred to as *mesoscale models* have dispersion modeling capabilities that extend horizontally up to a few hundred kilometers (miles). It does not mean that they model dispersion in the mesosphere.

U.S. Environmental Protection Agency Models

Many of the dispersion models developed by or accepted for use by the U.S. Environmental Protection Agency (EPA) are accepted for use in many other countries as well. Those EPA models are grouped below into four categories.

Preferred and Recommended Models

- AERMOD — An atmospheric dispersion model based on atmospheric boundary layer turbulence structure and scaling concepts, including treatment of multiple ground-level and elevated point, area and volume sources. It handles flat or complex, rural or urban terrain and includes algorithms for building effects and plume penetration of inversions aloft. It uses Gaussian dispersion for stable atmospheric conditions (i.e., low turbulence) and non-Gaussian dispersion for unstable conditions (high turbulence). Algorithms for plume depletion by wet and dry deposition are planned as future additions to the model. This model was in development for approximately 14 years before being officially accepted by the U.S. EPA. http://www.epa.gov/scram001/7thconf/aermod/aermodugb.pdf.
- CALPUFF — is a multi-layer, multi-species non-steady-state puff dispersion model that simulates the effects of time- and space-varying meteorological conditions on pollution transport, transformation and removal. CALPUFF can be applied on scales of tens to hundreds of kilometers. It includes algorithms for subgrid scale effects (such as terrain impingement), as well as, longer range effects (such as pollutant removal due to wet scavanging and dry deposition, chemical transformation, and visibility effects of particulate matter concentrations). http://www.epa.gov/scram001/7thconf/calpuff/scrnanl.pdf.
- BLP — is a Gaussian plume dispersion model designed to handle unique modeling problems associated with industrial sources where plume rise and downwash effects from stationary line sources are important. http://www.epa.gov/scram001/userg/regmod/blpug.pdf.
- CALINE3 — is a steady-state Gaussian dispersion model designed to determine air pollution concentrations at receptor locations downwind of highways located in relatively uncomplicated terrain. CALINE3 is incorporated into the more refined CAL3QHC and CAL3QHCR models. http://www.epa.gov/scram001/userg/regmod/caline3.pdf.
- CAL3QHC/CAL3QHCR. CAL3QHC is a CALINE3 based CO model with queuing calculations and a traffic model to calculate delays and queues that occur at signalized intersections. CAL3QHCR is a more refined version based on CAL3QHC that requires local meteorological data. http://www.epa.gov/scram001/userg/regmod/cal3qhc.pdf.
- CTDMPLUS — A Complex Terrain Dispersion Model (CTDM) plus algorithms for unstable situations (i.e., highly turbulent atmospheric conditions). It is a refined point source Gaussian air quality model for use in all stability conditions (i.e., all conditions of atmospheric turbulence) for complex terrain. The model contains, in its entirety, the technology of CTDM for stable and neutral conditions. http://www.epa.gov/scram001/userg/regmod/ctdmplusvol1.pdf. http://www.epa.gov/scram001/userg/regmod/ctdmplusvol2.pdf. http://www.epa.gov/scram001/userg/regmod/ctdmterrprepug.pdf.
- OCD — Offshore and Coastal Dispersion Model Version 5 (OCD) is a straight line Gaussian model developed to determine the impact of offshore emissions from point, area or line sources on the air quality of coastal regions. OCD incorporates overwater plume transport and dispersion as well as changes that occur as the plume crosses the shoreline. Hourly meteorological data are needed from both offshore and onshore locations. http://www.epa.gov/scram001/userg/regmod/ocdugsup.pdf.

Alternative Models

- ADAM — Air Force Dispersion Assessment Model (ADAM) is a modified box and Gaussian dispersion model which incorporates thermodynamics, chemistry, heat transfer, aerosol loading, and dense gas effects. Release scenarios include continuous and instantaneous, area and point, pressurized and unpressurized, and liquid/vapor/two-phased options.

- ADMS-3 — Atmospheric Dispersion Modeling System 3 (ADMS-3) is PC based model of dispersion in the atmospheric of passive, buoyant or slightly dense, continuous or finite duration releases from single or multiple sources which may be point, area or line sources. The model uses an up to date parameterisation of the boundary layer structure based on the Monin Obukhov length LMO, and the boundary layer height h. http://www.cerc.co.uk/epa/index.htm.
- AFTOX — is a Gaussian dispersion model that handles continuous or instantaneous, liquid or gas, elevated or surface releases from point or area sources. Output consists of concentration contour plots, concentration at a specified location, and maximum concentration at a given elevation and time.
- ASPEN — The Assessment System for Population Exposure Nationwide (ASPEN) consists of a dispersion and a mapping module. The dispersion module is a Gaussian formulation based on ISCST3 for estimating ambient annual average concentrations at a set of fixed receptors within the vicinity of the emission source. The mapping module produces a concentration at each census tract. Input data needed are emissions data, meteorological data and census tract data. The Emissions Modeling System for Hazardous Pollutants (EMS-HAP) is needed to process the emission inputs into the ASPEN model or the ISC3 model. The ASPEN model was used in estimating annual ambient concentrations for air toxics pollutant in the National Air Toxics Assessment (NATA) Study. http://www.epa.gov/ttn/atw/nata/. http://www.epa.gov/scram001/userg/other/aspenug.pdf.
- DEGADIS — Dense Gas Dispersion (DEGADIS) is a model that simulates the dispersion at ground level of area source dense gas (or aerosol) clouds released with zero momentum into the atmospheric boundary layer over flat, level terrain. The model describes the dispersion processes which accompany the ensuing gravity-driven flow and entrainment of the gas into the boundary layer. http://www.epa.gov/scram001/userg/other/degugv2.pdf.
- HGSYSTEM — is a collection of computer programs developed by Shell Research Ltd. and designed to predict the source-term and subsequent dispersion of accidental chemical releases with an emphasis on dense gas behavior.
- HOTMAC and RAPTAD — HOTMAC is a model for weather forecasting used in conjunction with RAPTAD which is a puff model for pollutant transport and dispersion. These models are used for complex terrain, coastal regions, urban areas, and around buildings where other models fail. http://www.ysasoft.com/index05.htm.
- HYROAD — The HYbrid ROADway Model (HYROAD) integrates three historically individual modules that simulate the effects of traffic, emissions and dispersion. The traffic module is a microscale transportation model which simulates individual vehicle movement. The emission module uses speed distributions from the traffic module to determine composite emission factors; spatial and temporal distribution of emissions is based on the vehicle operation simulations. The model tracks vehicle speed and acceleration distributions by signal phase per 10-meter roadway segment for use in both emissions distribution and for induced flows and turbulence. The dispersion module uses a Lagrangian puff formulation, along with a gridded non-uniform wind and stability field derived from traffic module outputs, to describe near-roadway dispersion characteristics. HYROAD is designed to determine hourly concentrations of carbon monoxide (CO) or other gas-phase pollutants, particulate matter (PM) and air toxics — in consultation with appropriate Reviewing Authority — from vehicle emissions at receptor locations that occur within 500 meters (1,640 ft) of the roadway intersections.
- ISC3 — (Industrial Source Complex Model) is a steady-state Gaussian plume model used to assess pollutant concentrations from a wide variety of sources associated with an industrial complex. This model can account for the following: settling and dry deposition of particles; downwash; point, area, line, and volume sources; plume rise as a function of downwind distance; separation of point sources; and limited terrain adjustment. ISC3 operates in both long-term and short-term modes. The screening version of ISC3 is SCREEN3. ISC3 also uses the Emissions Modeling System for Hazardous Pollutants (EMS-HAP) to process an emission inventory for input into the model. The Building Profile Input Program (BPIP) and the Building Profile Input Program for PRIME (BPIPPRM) can also

be used with ISC3 to correctly calculate building heights (bh) and projected building widths (pbw) for simple, multi-tiered, and groups of structures. http://www.epa.gov/scram001/userg/regmod/isc3v1.pdf. http://www.epa.gov/scram001/userg/regmod/isc3v2.pdf.

- OBODM — Intended for use in evaluating the potential air quality impacts of the open burning and detonation (OB/OD) of obsolete munitions and solid propellants. OBODM uses cloud/plume rise dispersion and deposition algorithms taken from existing models for instantaneous and quasi-continuous sources to predict the transport and dispersion of pollutants released by the open burning and detonation operations. http://www.epa.gov/scram001/userg/nonepa/obodmvol1.pdf.

- PLUVUEII — A model used for estimating visual range reduction and atmospheric discoloration caused by plumes resulting from the emissions of particles, nitrogen oxides, and sulfur oxides from a single source. The model predicts the transport, dispersion, chemical reactions, optical effects and surface deposition of point or area source emissions.

- SCIPUFF — Second-order Closure Integrated PUFF Model (SCIPUFF) is a Lagrangian puff dispersion model that uses a collection of Gaussian puffs to predict three-dimensional, time-dependent pollutant concentrations. In addition to the average concentration value, SCIPUFF provides a prediction of the statistical variance in the concentration field resulting from the random fluctuations in the wind field. http://www.epa.gov/scram001/7thconf/information/scipuff.pdf

- SDM — Shoreline Dispersion Model (SDM) is a Gaussian dispersion model used to determine ground-level concentrations from tall stationary point source emissions near a shoreline.

- SLAB — is a model for denser-than-air pollutant releases that utilizes the one-dimensional equations of momentum, conservation of mass and energy, and the equation of state. SLAB handles point source ground-level releases, elevated jet releases, releases from volume sources and releases from the evaporation of volatile liquid spill pools.

Screening Models. These are models that are often used before applying a refined air quality model to determine if refined modeling is needed.

- AERSCREEN — The screening version of AERMOD. It produces estimates of concentrations without the need for meteorological data that are equal to or greater than the estimates produced by AERMOD with a full set of meteorological data.

- COMPLEX1 — is a multiple point source screening model with terrain adjustment that uses the plume impaction algorithm of the VALLEY model.

- CTSCREEN — is a Gaussian plume dispersion model designed as a screening technique for regulatory application to plume impaction assessments in complex terrain. CTSCREEN is a screening version of the CTDMPLUS model. http://www.epa.gov/scram001/userg/screen/ctscreen.pdf.

- RTDM3.2 — Rough Terrain Diffusion Model (RTDM3.2) is a sequential Gaussian plume model designed to estimate ground-level concentrations in rough (or flat) terrain in the vicinity of one or more co-located point sources.

- SCREEN3 — is a single source Gaussian plume model which provides maximum ground-level concentrations for point, area, flare, and volume sources, as well as concentrations in the cavity zone, and concentrations due to inversion break-up and shoreline fumigation. SCREEN3 is a screening version of the ISC3 model. http://www.epa.gov/scram001/userg/screen/screen3d.pdf.

- TSCREEN — Toxics Screening Model (TSCREEN) is a Gaussian model that implements the procedures to correctly analyze toxic emissions and their subsequent dispersion from one of many different types of possible releases for superfund sites. It contains 3 modules: SCREEN3, PUFF, and RVD (Relief Valve Discharge). http://www.epa.gov/scram001/userg/screen/tscreend.pdf.

- VALLEY — is a steady-state, complex terrain, univariate Gaussian plume dispersion algorithm designed for estimating either 24-hour or annual concentrations resulting from emissions from up to 50 (total) point and area sources.

- VISCREEN — A model that calculates the potential impact of a plume of specified emissions for specific transport and dispersion conditions. http://www.epa.gov/scram001/userg/screen/viscrdu.pdf.

Photochemical Models. Photochemical air quality models have become widely utilized tools for assessing the effectiveness of control strategies adopted by regulatory agencies. These models are large-scale air quality models that simulate the changes of pollutant concentrations in the atmosphere by characterizing the chemical and physical processes in the atmosphere. These models are applied at multiple geographical scales ranging from local and regional to national and global.

There are two types of photochemical air quality models commonly used in air quality assessments: the Lagrangian trajectory model that employs a moving frame of reference, and the Eulerian grid model that uses a fixed coordinate system with respect to the ground. Earlier generation modeling efforts often adopted the Lagrangian approach to simulate the pollutants formation because of its computational simplicity. The disadvantage of Lagrangian approach, however, is that the physical processes it can describe are somewhat incomplete. Most of the current operational photochemical air quality models have adopted the three-dimensional Eulerian grid modeling mainly because of its ability to better and more fully characterize physical processes in the atmosphere and predict the species concentrations throughout the entire model domain.

- Models-3/CMAQ — The primary goals for the Models-3/Community Multiscale Air Quality (CMAQ) modeling system are to improve 1) the environmental management community's ability to evaluate the impact of air quality management practices for multiple pollutants at multiple scales and 2) the scientist's ability to better probe, understand, and simulate chemical and physical interactions in the atmosphere. The latest version of the Community Multi-scale Air Quality (CMAQ) model has state-of-the-science capabilities for conducting urban to regional scale simulations of multiple air quality issues, including tropospheric ozone, fine particles, toxics, acid deposition, and visibility degradation.

- CAMx — The Comprehensive Air quality Model with extensions (CAMx) simulates air quality over many geographic scales. It handles a variety of inert and chemically active pollutants, including ozone, particulate matter, inorganic and organic PM2.5/PM10, and mercury and other toxics. The U.S. EPA has approved the use of CAMx for numerous ozone and PM State Implementation Plans throughout the U.S, and has used this model to evaluate regional mitigation strategies. http://www.camx.com/files/CAMxUsersGuide_v4.30.pdf CAMx Home Page: http://www.camx.com/over/.

- REMSAD — The Regional Modeling System for Aerosols and Deposition (REMSAD) calculates the concentrations of both inert and chemically reactive pollutants by simulating the atmospheric processes that affect pollutant concentrations over regional scales. It includes processes relevant to regional haze, particulate matter and other airborne pollutants, including soluble acidic components and mercury. http://remsad.saintl.com/.

- UAM-V — The Urban Airshed Model-V Photochemical Modeling System was a pioneering effort in photochemical air quality modeling in the early 1970s and has been used widely for air quality studies focusing on ozone. It is a three-dimensional photochemical grid model designed to calculate the concentrations of both inert and chemically reactive pollutants by simulating the physical and chemical processes in the atmosphere that affect pollutant concentrations. This model is typically applied to model air quality "episodes" — periods during which adverse meteorological conditions result in elevated ozone pollutant concentrations. http://uamv.saintl.com/overview.htm and http://uamv.saintl.com/documents/uam-v_1.31_user%27s_guide.pdf.

Models Developed in England

- ADMS-3 — See the description of this model in the Alternative Models section of the models accepted by the U.S. EPA.

- ADMS-URBAN — A model for simulating dispersion on scales ranging from a street scale to city-wide or county-wide scale, handling most relevant emission sources such as traffic, industrial, commercial, and domestic sources. It is also used for air quality management and assessments of current and future air quality vis-a-vis national and regional standards in Europe and elsewhere. http://www.cerc.co.uk/software/urban.htm.

- ADMS-Roads — A model for simulating dispersion of vehicular pollutant emissions from small road networks in combination with emissions from industrial plants. It handles multiple road sources as well as multiple point, line or area emission sources and the model operation is similar to the other ADMS models. http://www.cerc.co.uk/software/admsroads.htm.

- ADMS-Screen — A screening model for rapid assessment of the air quality impact of a single industrial stack to determine if more detailed modelling is needed. It combines the dispersion modelling algorithms of the ADMS models with a user interface requiring minimal input data. http://www.cerc.co.uk/software/screen3.htm.
- GASTAR — A model for simulating accidental releases of denser-than-air flammable and toxic gases. It handles instantaneous and continuous releases, releases from jet sources, releases from evaporation of volatile liquid pools, variable terrain slopes and ground roughness, obstacles such as fences and buildings, and time-varying releases. http://www.cerc.co.uk/software/gastar.htm.
- NAMEIII — Numerical Atmospheric-dispersion Modelling Environment (NAME) is a local to global scale model developed by the UK's Met Office, http://www.metoffice.gov.uk/. It is used for: forecasting of air quality, air pollution dispersion, and acid rain; tracking radioactive emissions and volcanic ash discharges; analysis of accidental air pollutant releases and assisting in emergency response; and long-term environmental impact analysis. It is an integrated model that includes boundary layer dispersion modeling. http://www.met-office.gov.uk/research/nwp/publications/nwp_gazette/3rd96/name2.html.

Models Developed in Continental Europe. The European Topic Centre on Air and Climate Change, which is part of the European Environment Agency (EEA) http://www.eea.eu.int/main_html, maintains an online Model Documentation System (MDS) http://reports.eea.eu.int/Technical_report_No_48/en, that includes descriptions and other information for almost all of the dispersion models developed by the countries of Europe. The MDS currently contains 104 models developed in Europe (excluding England). Of those 104 models, 28 were subjectively selected for inclusion here. Anyone interested in seeing the complete MDS can access it on the web http://pandora.meng.auth.gr/mds/strquery.php?wholedb.

Some of the European models listed in the MDS are public domain and some are not. Many of them include a pre-processor module for the input of meteorological and other data, and many also include a post-processor module for graphing the output data and/or plotting the area impacted by the air pollutants on maps.

The country of origin is included for the each of the European models listed below.

- ONM9440 (Austria) — is a Gaussian dispersion model for continuous, buoyant plumes from stationary sources for use in flat terrain areas. It includes plume depletion by dry deposition of solid particulates.
- GRAL (Austria) — GRAL consists of various model parts. It has three different meteorological pre-processors depending on the meteorological input data. It is possible to use time-series of meteorological input data as well as statistics of wind speed, direction, and stability class. In case of complex terrain or buildings, a mesoscale model is used for the wind field calculations (GRAMM=Graz Mesoscale Model). The dispersion is calculated using a Lagrangian model, which is able to cope with vertical inhomogeneous turbulence and inhomogeneous 3D wind fields. It handles flat or complex terrain but it has no chemistry or deposition capabilities.
- IFDM (Belgium) — The Immission Frequency Distribution Model IFDM, developed at the Flemish Institute for Technological Research (VITO), is a bi-Gaussian transport and dispersion model used for point and area sources dispersing over flat terrain on a local scale. The model includes plume depletion by dry or wet deposition but cannot handle building effects, chemical transformations or complex terrain.
- SEVEX (Belgium) — The SEVeso EXpert system or SEVEX model simulates the accidental release of toxic and/or flammable material over flat or complex terrain from multiple pipe and vessel sources or from evaporation of volatile liquid spill pools. The accidental releases may be continuous, transient or catastrophic. The integrated model can handle denser-than-air gases as well as neutral gases (i.e., neither denser than or lighter than air). It does not include handling of multi-component material, nor does it provide for chemical transformation of the releases. The model's name is derived from the major disaster caused by the accidental release of highly toxic gases that occurred in Seviso, Italy in 1976.
- HAVAR (Czech Republic) — A Gaussian plume model integrated with a puff model and a hybrid plume-puff model, developed by the Czech Academy of Sciences http://www.cas.cz/index.html.en, is intended for routine and/or accidental releases of radionuclides from single point sources within nuclear power plants. The model includes radioactive plume depletion by dry and wet deposition as well as by radioactive decay. For the decay of some nuclides, the creation of daughter products that then grow into the plume is taken into account.
- SYMOS'97 (Czech Republic) — A model for dispersion calculations of continuous neutral or buoyant plumes from single or multiple point, area or line sources. It can handle complex terrain and it can also be used to simulate the dispersion of cooling tower plumes.
- OML (Denmark) — A model for dispersion calculations of continuous neutral or buoyant plumes from single or multiple, stationary point and area sources. It has some simple methods for handling photochemistry (primarily for NO_2) and for handling complex terrain. The model was developed by the National Environmental Research Institute of Denmark, which is a part of the Danish Ministry of the Environment.
- AEROPOL (Estonia) — The AERO-POLlution model developed at the Tartu Observatory http://www.aai.ee/, in Estonia is a Gaussian plume model for simulating the dispersion of continuous, buoyant plumes from stationary point, line and area sources over flat terrain on a local to regional scale. It includes plume depletion by wet and/or dry deposition as well as the effects of buildings in the plume path.
- BUO-FMI (Finland) — This model was developed by the Finnish Meteorological Institute (FMI) http://www.tekes.fi/ohjelmat/info/ava/avar/html/funding_agencies_finnish_meteorological_institute.html, specifically for estimating the atmospheric dispersion of neutral or buoyant plume gases and particles emitted from fires in warehouses and chemical stores. It is a hybrid of a local scale Gaussian plume model and another model type. Plume depletion by dry deposition is included but wet deposition is not included.
- CAR-FMI (Finland) — This model was developed by the Finnish Meteorological Institute (FMI) for evaluating atmospheric dispersion and chemical transformation of vehicular emissions of inert (CO, NOx) and reactive (NO, NO_2, O_3) gases from a road network of line sources on a local scale. It is a Gaussian line source model which includes an analytical solution for the chemical cycle NO-O_3-NO_2.
- UDM-FMI (Finland) — This model was developed by the Finnish Meteorological Institute (FMI) as an integrated Gaussian urban scale model intended for regulatory pollution control. It can handles multiple point, line, area and volume sources and it includes chemical transformation (for NO2), wet and and dry deposition (for SO2), and downwash phenomena (but no building effects).
- RADM (France) — The Random-walk Advection and Dispersion Model (RADM) was developed by ACRI-ST, an independent research and development organization in France. It can model gas plumes and particles (including pollutants with exponential decay or formation rates) from single or multiple stationary, mobile or area sources. Chemical reaction, radioactive decay, deposition, complex terrain, and inversion conditions are accommodated.
- STOER.LAG (Germany) — A dispersion model designed to evaluate accidental releases of hazardous and/or flammable materials from point or area sources in industrial plants. It can handle neutral and denser-than-air gases or aerosols from ground-level or elevated sources. The model accomodates building and terrain effects, evaporation of volatile liquid spill pools, and combustion or explosion of flammable gas-air mixtures (including the impact of heat and pressure waves caused by a fire or explosion).
- PROKAS-V (Germany) — A Gaussian dispersion model for evaluating the atmospheric dispersion of air pollutants emitted from vehicular traffic on a road network of line sources on a local scale. Modeling of up to 5000 line sources of a network of streets is possible. The influences of traffic induced turbulence, course of streets on dams and noise protection devices for each street are included.
- AUSTAL2000 (Germany) — The official air dispersion model to be used in the permitting of industrial sources by the German Federal Environmental Agency. The model accomodates point, line, area and volume sources of buoyant plumes. It has capabilities for building effects, complex terrain, plume depletion by wet or dry deposition, and first order chemical reactions. It is based on the LASAT model developed by Ingenieurbüro Janicke Gesellschaft für Umweltphysik.

- DIPCOT (Greece) — DIsPersion over COmplex Terrain (DIPCOT) is a model developed in the National Research Center for Physical Sciences (DEMOKRITOS) http://www.demokritos.gr/index_muk.asp, of Greece that simulates dispersion of buoyant plumes from multiple point sources over complex terrain on a local to regional scale based on a 3-D Lagrangian particle scheme. It does not include wet deposition or chemical reactions.

- DISPLAY-2 (Greece) — A vapor cloud dispersion model for neutral or denser-than-air pollution plumes over irregular, obstructed terrain on a local scale. It accommodates jet releases as well as two-phase (i.e., liquid-vapor mixtures) releases. This model was also developed at the the National Research Center for Physical Sciences (DEMOKRITOS) of Greece.

- MUSE (Greece) — A photochemical atmospheric dispersion model developed by Professor Nicolas Moussiopoulos at the Aristotle University of Thessaloniki http://www.saloniki.org/en/uni_en.htm, in Greece. It is intended for the study of photochemical smog formation in urban areas and assessment of control strategies on a local to regional scale. It can simulate dry deposition and transformation of pollutants can be treated using any suitable chemical reaction mechanism.

- FARM (Italy) — The Flexible Air quality Regional Model (FARM) is an atmospheric dispersion model designed for the analysis of episodes and scenarios, evaluation of the effects of regional emission control policies and pollution forecasts in complex situations. It accomodates point and area sources, and includes photochemistry and plume depletion by wet and dry deposition.

- SAFE AIR II (Italy) — The Simulation of Air pollution From Emissions II (SAFE AIR II) was developed at the Department of Physics, University of Genoa, Italy to simulate the dispersion of air pollutants above complex terrain at local and regional scales. It can handle point, line, area and volume sources and continuous plumes as well as puffs. It includes first-order chemical reactions and plume depletion by wet and dry deposition, but it does not include any photochemistry.

- STACKS (The Netherlands) — is an advanced Gaussian plume dispersion model for point and area buoyant plumes to be used over flat terrain on a local scale. It includes building effects, NO_2 chemistry and plume depletion by deposition. It is used for environmental impact studies and evaluation of emission reduction strategies.

- CAR-International (The Netherlands) — Calculation of Air pollution from Road traffic (CAR-International) is an atmospheric dispersion model developed by the Netherlands Organisation for Applied Scientific Research. http://www.onderzoekinformatie.nl/en/oi/nod/organisatie/ORG1236243/ It is used for simulating the dispersion of vehicular emissions from roadway traffic.

- LOTOS-EUROS (The Netherlands) — The LOng Term Ozone Simulation — EURopean Operational Smog (LOTOS-EUROS) model was developed by the National Institute for Public Health and Environment (RIVM) http://www.rivm.nl/en/, in The Netherlands. It is designed for modeling the dispersion of pollutants (such as: photo-oxidants, aerosols, heavy metals) over all of Europe. It includes simple reaction chemistry as well as wet and dry deposition.

- EK100W (Poland) — EK100W modeling system is used for air quality assessment (regulatory purposes) for industrial sources as well as for urban air quality studies. The system is based on methodology of air quality assessment recommended by Polish Ministry of Environment Protection, Natural Resources and Forestry. It includes wet and dry deposition. The effects of complex terrain are not included.

- INPUFF-U (Romania) — This model was developed by the National Institute of Meteorology and Hydrology http://www.meteo.bg/main.php?page=main&lang=en, in Bucharest, Romania. It is a Gaussian puff model for calculating the dispersion of radionuclides from passive emission plumes on a local to urban scale. It can simulate accidental or continuous releases from stationary or mobile point sources. It includes wet and dry deposition. Building effects, buoyancy effects, chemical reactions and effects of complex terrain are not included.

- MODIM (Slovak Republic) — A model for calculating the dispersion of continuous, neutral or buoyant plumes on a local to regional scale. It integrates a Gaussian plume model for single or multiple point and area sources with a numerical model for line sources, street networks and street canyons. It is intended for regulatory and planning purposes.

- POLGRAPH (Portugal) — This model was developed at the University of Aveiro, Portugal by Professor Carlos Borrego. It was designed for evaluating the impact of industrial pollutant releases and for air quality assessments. It is a Gaussian plume dispersion model for continuous, elevated point sources to be used on a local scale over flat or gently rolling terrain.

- DISPERSION21 (Sweden) — This model was developed by the Swedish Meteorological and Hydrological Institute (SMHI) http://www.smhi.se/en/, for evaluating air pollutant emissions from existing or planned industrial or urban sources on a local scale. It is a Gaussian plume model for point, area, line and vehicular traffic sources. It includes plume penetration of inversions aloft, building effects, NOx chemistry and it can handle street canyons. It does not include wet or dry deposition, complex atmospheric chemistry, or the effects of complex terrain.

Models Developed in Australia

- AUSPLUME — A dispersion model that has been designated as the primary model accepted by the Environmental Protection Authority (EPA) of the Australian state of Victoria. http://www.epa.vic.gov.au/Air/EPA/pub391.asp.

- LADM — An advanced model developed by Australia's Commonwealth Scientific and Industrial Research Organisation (CSIRO) http://www.cmar.csiro.au/ar/ladm/index.html, for simulating the dispersion of buoyant pollution plumes and predicting the photochemical formation of smog over complex terrain on a local to regional scale. The model can also handle fumigated plumes. LADM–User's Guide: http://www.cmar.csiro.au/ar/ladm/ladmusers.html.

- TAPM — An advanced dispersion model integrated with a pre-processor for providing meteorological data inputs. It can handle multiple pollutants, and point,line, area and volume sources on a local, city or regional scale. The model capabilities include building effects, plume depletion by deposition, and a photochemistry module. This model was also developed by Australia's Commonwealth Scientific and Industrial Research Organisation (CSIRO) http://www.csiro.au/.

- DISPMOD — A Gaussian atmospheric dispersion model for point sources located in coastal regions. It was designed specifically by CSIRO to simulate the plume fumigation that occurs when an elevated onshore pollution plume intersects a growing thermal internal boundary layer (TIBM) contained within offshore air flow coming onshore.

- AUSPUFF — A Gaussian puff model designed for regulatory use by CSIRO. It includes some simple algorithms for the chemical transformation of reactive air pollutants. http://www.dar.csiro.au/pollution/localscale/sld005.htm.

Air Pollution Settings

In developing preventive and remedial technologies for air pollution abatement, it is helpful to consider the fundamental settings in which pollution occurs.

1. *Indoor air pollution* occurs due to the lack of ventilation indoors which concentrates the air pollution where people have the greatest exposure times. Radon (Rn) [CAS: 10043-92-2] gas, a carcinogen, is exuded from the Earth and trapped inside houses. Researchers have found that radon gas is responsible for over 1,800 deaths annually in the United Kingdom. These natural radon emissions can be blocked by a layer of aluminum foil under the carpet (according to the U.S. Department of Air Quality Management). Building materials including carpeting and plywood emit formaldehyde (H_2CO) [CAS: 50-00-0] gas. Paint and solvents give off volatile organic compounds (VOCs) as they dry. Lead paint can degenerate into dust and be inhaled. Intentional air pollution is introduced with the use of air fresheners, incense, and other scented items. Controlled wood fires in stoves and fireplaces can add significant amounts of smoke particulates into the air, inside and out. Indoor air pollution may arise from such mundane sources as shower water mist containing arsenic [CAS: 7440-38-2] or manganese [CAS: 7439-96-5], both of which are damaging to inhale. The arsenic (As^{3+}) can be trapped with a shower nozzle filter. Deaths are often caused by using pesticides and other chemical sprays indoors without proper ventilation, and many homes have been destroyed by accidental pesticide explosions. *Second-hand* tobacco smoke is now recognized as an indoor air

pollutant which accounts for an estimated 3,000 lung cancer deaths annually in the US. Carbon monoxide (CO) poisoning is a quick and silent killer, often caused by faulty vents and chimneys, or by the burning of charcoal indoors. Chronic carbon monoxide poisoning can result even from poorly adjusted pilot lights. Smoke inhalation is a common cause of death in victims of house fires. Traps are built into all domestic plumbing to keep deadly sewer gas, hydrogen sulfide, out of interiors. Clothing emits tetrachloroethylene, or other dry cleaning fluids, for days after dry cleaning. Biological sources of air pollution can also be found indoors, and include gases, particulates, allergens, and microbes. Pets produce dander, bed mites deposit shells and microscopic droppings, inhabitants emit methane, mold can form in walls and generate spores, air conditioning systems can incubate Legionnaires disease, toilets can emit feces-tainted mists, and houseplants and surrounding gardens can produce pollen, dust, and mold spores. Though its use has now been banned in many countries, the extensive use of asbestos [CAS: 132207-33-1] in industrial and domestic environments in the past has left a potentially very dangerous material in many localities. Asbestosis is a chronic inflammatory medical condition affecting the tissue of the lungs. It occurs after long-term, heavy exposure to asbestos, e.g. in mining or in the installation or removal of asbestos-containing materials from structures. Sufferers have severe dyspnea (shortness of breath) and are at an increased risk regarding several different types of lung cancer. See also **Asbestos**.

As clear explanations are not always stressed in non-technical literature, care should be taken to distinguish between several forms of relevant diseases. According to the World Health Organisation (WHO), these may defined as; asbestosis, *lung cancer*, and *mesothelioma* (generally a very rare form of cancer, when more widespread it is almost always associated with prolonged exposure to asbestos).

2. *Workplace pollution* usually represents the highest concentration and length of exposure to specific pollutants. It usually is the most obvious mode of air pollution and, consequently, the easiest to correct. In this setting, workers breathe specific pollutants on a day-to-day basis. Well-known and publicized examples would include: miners; chemical plant workers; farmers; textile workers; metal production workers; transportation-related personnel (who are exposed to high concentrations of carbon monoxide and other gaseous products of fossil fuel combustion as well as coal dust and other carbonaceous particles and volatile hydrocarbons); and insulation installers (who are exposed to airborne tiny particles and strands of glass, plastics, and natural minerals, such as asbestos and mica). In the industrial and manufacturing complex, which has continued to expand markedly during the last several decades, hundreds of specific examples of exposure to dangerous air pollutants could be recited.

3. *Point-source pollution* extends beyond a specific entity within a plant, such as a particular machine or process, and comprises a source of pollution that may emanate from only one or a few particular facilities within a small area. Point-source pollutants become mixed in the atmosphere with pollutants from other sources. Hence, beyond the immediate vicinity of a given facility, sources are difficult to identify. Considerable success has been achieved by isotopic matching to specific coals and other fuels that may be burned at a given facility.

4. *Confined-area pollution*, where there may be no natural circulation of air and no effective air purification effluent system. This situation is found in factories or mines and inside improperly ventilated garages, service stations, and vehicular tunnels.

5. *Limited-area pollution* occurs on a small geographic scale, such as strips of land adjacent to major highways or close to a pollution point source.

6. *Regional pollution* is pollution that occurs in the greater part of a city, valley, or basin and frequently is publicized in connection with cities like Los Angeles, London, and, in more recent years, other major cities of the world, including New York. Regional pollution is particularly affected by weather conditions, such as, for example, an inversion layer hanging over a natural basin. When pollution occurs on this scale, the beginnings of ecological damage (to trees, plants, natural life, etc.) are seen.

7. *International pollution* or wide-area pollution is the occurrence of massive air pollution, usually extending over many years, to the extent that the atmosphere becomes severely overloaded with pollutants—that is, the atmosphere's ability to contain pollutants is exceeded. After holding pollutants for long periods, during which time prevailing winds transport the pollutants over long distances (from one country to the next, for example), the point is reached where pollutants "drop out" and contaminate the topography below. This is the type of pollution that has been the subject of debate, for example, between Canada and the United States—where pollutants from coal-burning power plants in the Midwestern states of the United States are transported to northeastern Canadian provinces (and northeastern U.S. states as well). Such pollution damages forests and lakes. See also "**Acid Rain**" later in this entry. Other examples of international pollution occur in other parts of the world, but are less well understood at this time.

8. *Worldwide pollution* is simply an extension of wide-area pollution and encompasses emissions that essentially become mixed with the entire atmosphere of the Earth. Even though the mixing time may be quite long, it is reasonable to assume that, over time, the ultimate effects of almost *any* pollution ultimately will affect the atmosphere on a worldwide basis. For example, as the result of nuclear events, particularly those that occurred prior to the ban on nuclear weapons testing, radioactive particles could be discerned over extensive regions of Earth. Depletion of Earth's ozone layer also is exemplary of how destruction of ozone can be caused by chlorine molecules, essentially without regard to where the pollutants are released into the atmosphere. The so-called ozone "hole" currently over Antarctica and a possible similar "hole" over the Arctic (now threatened to occur by about the year 2000) is discussed in the article on **Polar Research**.

The possible effects of increased carbon dioxide content of the atmosphere are discussed in the article on **Global Change**.

The Energy vs. Environment Conflict

Just as it is difficult to separate the topics of air and water pollution, so is it hard to separate the problems of pollution from energy generation and consumption. With exception of some of the nontraditional sources of energy, such as nuclear energy and the more direct utilization of solar energy (as contrasted with combustion), the needs of the earth's population for energy tend to follow a collision course with concerns over the environment. For example, until the nontraditional energy sources can be reduced to practical usage (of which economics is an important, if not scientific factor in the equation), coal, wood, biomass, and other organic fuels when burned are air polluters unless very costly measures are taken to treat the effluents. Even when the numerous chemical and electroprecipitation measures, among others, are taken, there remains the problem of increasing the carbon dioxide content of the atmosphere.

Energy has an impact on the environment by tending to worsen the environment. The environment has an impact on energy by requiring considerable energy to alleviate the degradation of environmental quality. Environmental concerns also tend to limit the energy options available. It is generally agreed that the standard of living of the technologically sophisticated and developed nations is at least partially the result of inexpensive energy. Generally, the societies in these countries have not readied themselves to very high energy costs, there being a realization that these increased costs will, by and large, come out of so-called discretionary income and, consequently, impact the standard of living in a negative way. The significance of the incremental addition of energy required to effect environmental protection is shown in Table 1.

The topic of the conflict between energy, environment, and economics is discussed in greater detail in the article on **Electric Power Production and Distribution**.

Principal Air Pollutants

The major air pollutants as identified by a number of countries in recent years in connection with pollutant regulatory programs are: (1) Particulate matter; (2) nitrogen oxides (NO_x); (3) sulfur oxides (SO_x); (4) hydrocarbons; and (5) carbon monoxide (CO).

TABLE 1. AIR POLLUTION COSTS FOR REPRESENTATIVE TECHNOLOGIES (10^6 BTU/TON OF PRODUCT)*

Process Option	Primary Energy Source	Process Energy (10^6 Btu)	Air Pollution Control Energy (10^6 Btu)	Percent of Total for Air Pollution Control
Glassmaking				
side port regenerative furnace	natural gas	7.0	0.57	7.5
side port regenerative furnace with preheat of charge	natural gas	5.7	0.37	6.0
electric furnace	electric power	8.2	0.03	0.3
coal gasification	coal	8.6	0.9	9.5
direct coal firing	coal	7.0	0.65	8.5
Cement				
long kiln (conventional)	oil	5.6	0.07	1.2
suspension preheater with long kiln	oil	4.2	0.05	1.2
fluid bed	oil	5.0	0.1	2.0
Copper Production				
roast-reverb smelting (conventional)	gas, oil, or coal	22.0	5.3	19.4
flash smelting (90–95% sulfur recovery)	oil	10.0	7.8	43.8

* 10^6 Btu = 252×10^3 Calories

Particulates and Aerosols. These may be comprised of numerous mineral and organic materials and frequently result from such operations as milling, crushing, screening, grinding, and demolition operations—as well as quarries and cement plants. Soot and fly ash as well as heavy carbonaceous smoke, arising from fuel-burning operations and smudge pots, also may fall into this category of pollutants. Aerosols generally are considered to be very tiny spherical droplets of a liquid that may be as small as 0.01 micrometer in diameter. These small liquid particles and the larger liquid particles, including mists and sprays, along with dusts, permit numerous physical separating and isolating means that do not apply to gases and vapors. Recent investigations of particle size distribution of atmospheric aerosols have revealed a multimodal character, usually with a bimodal mass, volume, or surface area distribution and frequently trimodal surface area distribution near sources of fresh combustion aerosols. These modes are attributed to the following factors: (a) the course mode (2 micrometers and greater) is formed by relatively large particles generated mechanically or by evaporation of liquid from droplets containing dissolved substances; (b) the nuclei mode (0.03 micrometer and smaller) is formed by condensation of vapors from high-temperature processes or by gaseous reaction products; and (c) the intermediate or accumulation mode (0.1 to 1.0 micrometer) is formed by coagulation of nuclei. This evidence indicates that atmospheric particles tend to form a stable aerosol having a size distribution ranging from 0.1 to 1.0 micrometer in general. However, larger and smaller particles occur. The larger particles (greater than 1.0 micrometer) settle out, and the very fine particles (smaller than 0.1 micrometer) tend to agglomerate to form larger particles which remain suspended. The nuclei mode tends to be highly transient and is concentration-limited by coagulation with both other nuclei and also particles in the accumulation mode. Therefore, the particulate content of a source emission and the ambient air can be viewed as composed of two portions, i.e., the settleable and the suspended.

Control of emissions in both size ranges is required because both settleable and suspended atmospheric particulates have deleterious effects upon the environment. Significantly, it is the suspended particles from an upper level of about 2 to 5 micrometers and smaller that health experts consider most harmful to humans because particles of this size have been found to penetrate the body's natural defense mechanisms and reach most deeply into the lungs. Efforts to control particulate emissions to the atmosphere have historically been geared to maximizing the efficiency of control (by weight) of the overall particulate loading emanating from the generating process. This work has led to the empirical understanding that present systems can perform with high control efficiencies down to a particle size of about 2–3 micrometers. But, below this size, the control efficiency appears to decrease with decreasing particle size to a minimum between 1.0 and 0.1 micrometer; and then increases again. This relationship of control efficiency and particle size is highly significant to any strategy for controlling particulate air pollution, and serves to underscore the need to adequately measure and evaluate both ambient particulate air pollution and source emissions.

The particle diameters of some substances commonly found in the atmosphere or important in various manufacturing operations are given in Fig. 1. Various particle measurement techniques versus particle diameter are given in Fig. 2. The most suitable ranges of particle size versus types of gas cleaning equipment are given in Fig. 3. Modern instrumental techniques are shown in Fig. 4.

Legend:

1. Gas dispersoids — solid fumes
2. Gas dispersoids — liquid mists
3. Gas dispersoids — solid dusts
4. Gas dispersoids — liquid sprays
5. Common atmospheric dispersoids
6. Smoke — rosin
7. Smoke — oil
8. Smoke — tobacco
9. Fly ash
10. Coal dust
11. Pulverized coal
12. Cement dust
13. Metallurgical dusts and fumes
14. Insecticide dusts
15. Milled flour
16. Fumes — ammonium chloride
17. Fumes — zinc oxide
18. Fumes — alkali
19. Sulfuric acid concentrator mist
20. Contact sulfuric acid mist
21. Ore flotation mist
22. Ground limestone fertilizer
23. Beach sand
24. Ground talc

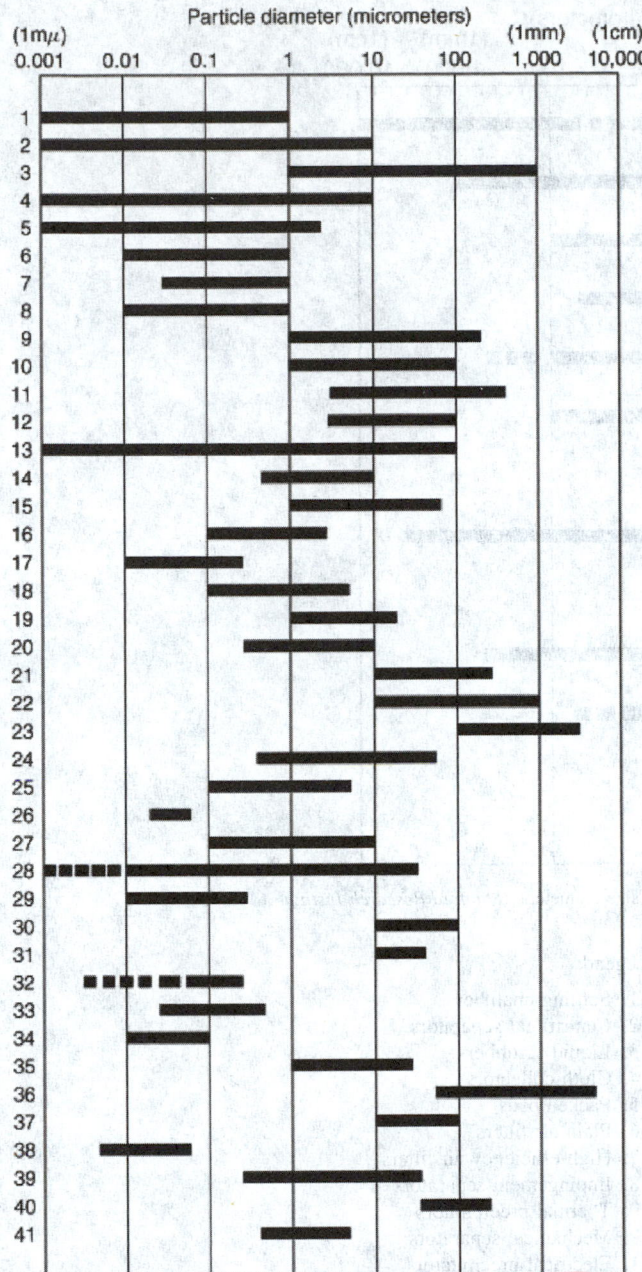

Fig. 1. Range of diameters of particles and particle dispersoids of substances commonly encountered in the atmosphere or associated with various manufacturing operations. (*Stanford Research Institute*.)

25. Paint pigments
26. Colloidal silica
27. Spray dried milk
28. General atmospheric dust
29. Carbon black
30. Plant pollen
31. Plant spores
32. Nuclei (Aitken)
33. Nuclei (Sea salt)
34. Nuclei (Combustion)
35. Nebulizer drops
36. Hydraulic nozzle drops
37. Pneumatic nozzle drops
38. Viruses
39. Bacteria
40. Human hair (diameter)
41. Most severely lung-damaging dust

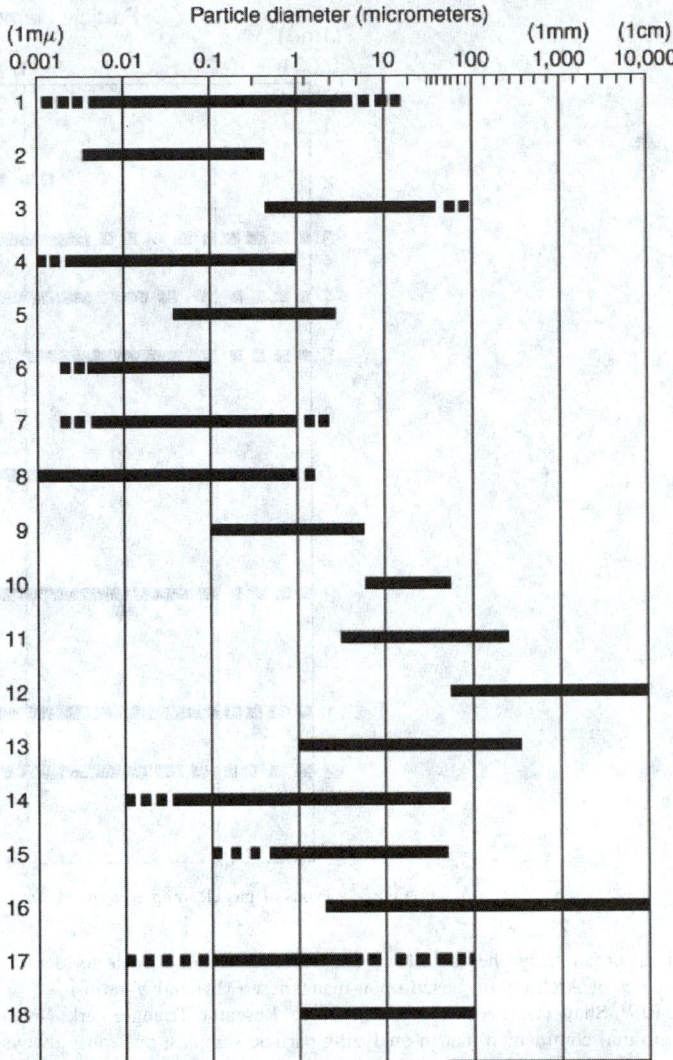

Fig. 2. Various particle measurement techniques versus particle diameter. (*Stanford Research Institute*.)

Source Identification of Airborne Particles. In recent years, ingenious methods of identifying the point source of airborne particles have been developed. If not specific point sources of pollution, rather small regional areas can often be identified. Source of particles can be important for numerous reasons, including the enforcement of regulation and also in sorting out, for example, the various distant sources that contribute to acid rain pollution.

As reported by Olmez and Gordon (University of Maryland), the concentration pattern of rare earth elements on fine airborne particles (less than 2.5 micrometers in diameter) is distorted from the crustal abundance pattern in areas influenced by emissions from oil-fired plants and refineries. The ratio of lanthanum (La) to samarium (Sm) is often greater than 20 (crustal ratio is less than 6). The unusual pattern apparently results from the distribution of rare earths in zeolite catalysts used in refining oil. Oil industry emissions have been found to perturb the rare earth pattern even in very remote locations, such as the Mauna Loa Observatory in Hawaii. Rare earth ratios are probably better for long-range tracing of oil emissions than vanadium (V) and nickel (Ni) concentrations because the ratios of rare earths on fine particles are probably not influenced by deposition and other fractionating processes. Emissions from oil-fired plants can be differentiated from those of refineries on an urban scale by the much smaller amounts of V in the latter. Pb in urban areas originates mainly from combustion of leaded gasoline. Arsenic (As), selenium (Se), and other chalcophile elements are usually associated with coal-fired plants or sulfide ore smelters (or both). In addition to the use of receptor models

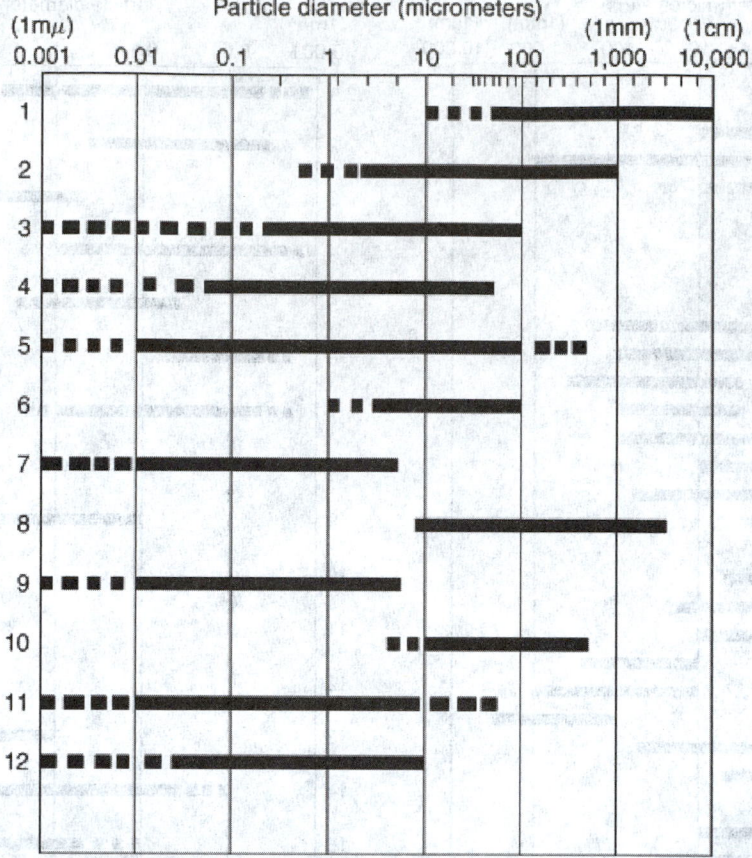

Fig. 3.   Types of gas cleaning equipment versus particle size diameters. (*Stanford Research Institute*.)

on an urban scale, they also have been used on a global scale to identify sources of Arctic haze, based upon manganese (Mn) and V ratios.

R.W. Shaw (U.S. Army Research Office, Research Triangle Park, North Carolina) commented that in analyzing particle samples, one must always consider that a particle sample could be produced by various sources. Chemical-element balance copes with this complication by positing (on the basis of the known ratios) several different elemental concentrations that could be generated by the suspected sources. It then finds the "mix" that best fits the actual concentrations on the collected samples. Such manipulations might reveal, for example, that 80% of the fine fraction of a sample is a byproduct of coal combustion and 15% comes from motor vehicles. Or it might show that 94% of the lead in a specimen is from motor vehicles, 4% from the burning of refuse, and 1% from the burning of coal.

Legend:

1. Electron microscope
2. Ultramicroscope
3. Microscope
4. Ultracentrifuge
5. Centrifuge
6. X-ray diffraction
7. Adsorption
8. Nuclei counter
9. Impingers
10. Electroformed sieves
11. Elutriation
12. Sieving
13. Sedimentation
14. Turbidimetry
15. Permeability
16. Scanners
17. Light scattering
18. Electrical conductivity
19. Visible to eye

Legend:

1. Settling chambers
2. Centrifugal separators
3. Liquid scrubbers
4. Cloth collectors
5. Packed beds
6. Plain air filters
7. High-efficiency air filters
8. Impingement separators
9. Thermal precipitators
10. Mechanical separators
11. Electrical precipitators
12. Ultrasonic methods

Smoky Mountain Haze. A rather thorough study of what contributes to the increasing haze that exists in the Great Smoky Mountains National Park in Tennessee was reported by Shaw in 1987. The overall purpose of the study was to determine the contributions of natural emissions, as from vegetation, and motor vehicle and industrial emissions. The researchers also conducted surveys in the Georgia region (former Soviet block), the latter highly regarded for clear air and massive areas of evergreen forests. The Smoky Mountains haze is ascribed to sulfate particles in the air. It was tentatively concluded that the sulfate is a byproduct of coal burned at distant power plants. Surprisingly, the concentration of coal-derived particles in the Smoky Mountains was not much lower than concentrations found at several sites in an industrialized city, St. Louis, Missouri. Sulfate particles also were found in Georgia, but considered to be much older and the particles contained thirty times more sulfur than in sulfur dioxide gas. These studies, however, did not make a direct link with remote coal-burning plants. This was accomplished later in a 16-month study in the Ohio River valley, with sampling stations located in rural Kentucky, Indiana, and Ohio, far away from cities, roads, and power plant smoke plumes. About 50% of the fine-particle mass was found to be sulfate. The total concentration nearly equaled that found in the industrialized cities of the regions. The link was confirmed by a consistent association between

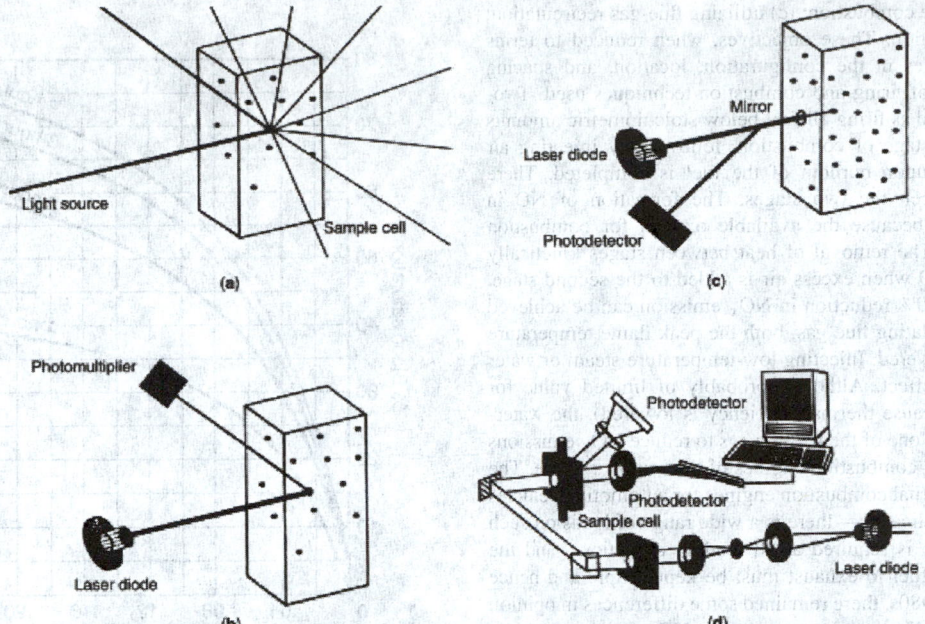

Fig. 4. Instrumentation for measuring air pollution particle dimensions. (a) Some fine-particle-size analyzers determine particle size by measuring the Doppler shift of light as it is scattered by moving particles. Smaller particles move faster, causing a greater Doppler shift in the light that they scatter. (b) In some conventional analyzers, light passes completely through an extremely dilute suspension and scatters in all directions. The detector measuring the Doppler shifts of the scattered light has no high-level signal to reference against the resulting low-level signal. The resulting low-level signals require amplification from photomultipliers, which can introduce noise errors. (c) In instrument shown (*Microtrac—Leeds & Northrup*), light travels to the sample via an optical wave guide. A mirror reflects some of the light, creating a high-level reference signal. Moving particles backscatter the light penetrating the mirror. The instrument combines the reflected and backscattered light to create a high-level signal sufficiently strong to be fed directly to a solid-state photodetector with a requirement for amplification. (d) Infrared light, emitted from a long-lived solid-state laser diode, scatters when it hits particles in the sample cell. Multi-train optics direct all scattered light onto solid-state photodetectors, which measure scattered-light angles and send signals to a computer control module. (*Leeds & Northrup.*)

sulfate concentration and the trace element selenium. No other probable sources of selenium exist in or near the Ohio River valley. The investigators (R.W. Shaw, et al.) concluded in their report that there is little doubt that fuel combustion is the main source of acid deposition and particulate fallout.

Shaw also describes two techniques that now make it possible to analyze particles without removing them from collection filters. To determine the mass of a sample, technicians insert the particle-laden filter between a source that emits beta particles and a detector that counts them. As the mass increases, the number of particles that can penetrate the sample decreases. To determine the atomic elements in a specimen, laboratory workers may also separately carry out x-ray fluorescence spectroscopy. X-rays passed through the sample cause each element to emit characteristic x-rays. The energy levels of the rays reveal the identity of the elements: the intensity of the x-rays (number emitted) reflects the concentrations.

Sulfur Oxides. (Sulfur dioxide [CAS: 7446-09-5], SO_2, and sulfur trioxide [CAS: 7446-11-9], SO_3). The primary sources of these oxides SO_x are sulfur-bearing fuels—as used for heat and power, both industrially and residentially. Chemical and metallurgical plants of various kinds also emit SO_x as the result of processing activities, such as the manufacture of sulfuric acid, the roasting of ores, etc. In order of decreasing pollution, the fossil fuels contributing to SO_x pollution are: (a) Untreated coal; (b) untreated petroleum fuels, particularly those originating from so-called sour crude oils; and (c) natural gas. Thus, the preference for natural gas by many large fuel users, such as power plants. With only small variations in the cost of raw fossil fuels, there was an advantage in burning a naturally low-sulfur fuel as contrasted with installing elaborate SO_x removal or reduction systems. But, with a rapidly lessening natural gas supply and accompanying higher costs, it has become economically attractive to pay more for desulfurized coal and petroleum fuels, as well as to install SO_x abatement equipment. The allowable sulfur content of oil and coal

fuels varies from one community to the next, ranging from 0.50% by weight or less, up to 4% and slightly higher. Such regulations usually take into consideration new versus old fuel-burning equipment, the incidence of serious pollution in a given area, as well as economic impact and practicability. Logically, for some years to come, such regulations must represent a compromise of social, economic, and technological factors. See also **Electric Power Production and Distribution.**

The chemical nature of the oxides of sulfur is given in the entry on **Sulfur.** Treatment of SO_x effluents is described later in this article.

Nitrogen Oxides. These compounds result from all fossil-fuel combustion processes where air is used as the oxidant. Oxygen from the air and nitrogen combine at combustion flame temperatures to form nitric oxide, NO, according to $N_2 + O_2 \Leftrightarrow 2NO$. The rate at which NO is formed and decomposed depends largely upon temperature. For the majority of stationary combustion processes, there is too short a residence time for the full oxidation of NO to NO_2, an estimated average of only 5 to 10% of this reaction occurring. Thus, it is important to observe that although NO_x emissions generally are given as "equivalent NO_2," the predominant NO_x in combustion gases is NO. Several factors affect the generation of NO_x pollutants. Factors which tend to decrease NO_x emissions are: (a) decrease in excess air for combustion; (b) decrease in preheat temperature; (c) decrease in the heat-release rate; (d) increase in the heat-removal rate; (e) increase in back-mixing; and (f) decrease in fuel nitrogen content. With exception of very large installations, coal appears to generate more NO_x than oil; and oil generates more NO_x than natural gas. Thus, as with SO_x, natural gas is the preferred fuel when properly burned to minimize NO_x.

The major sources of NO_x are the large fuel-burning operations as previously mentioned, automotive vehicles, and certain chemical plants, notably nitric acid manufacturing facilities. Research to date indicates that effective steps toward reducing the overall emission of NO_x can be effected from stationary combustion sources by: (a) using low excess air firing;

(b) providing for two-stage combustion; (c) utilizing flue-gas recirculation; and (d) using water injection. These objectives, when reduced to terms of hardware, mean changes in the configuration, location, and spacing of burners, and the kinds of firing and combustion techniques used. Two-stage combustion is defined as firing all fuel below stoichiometric amounts of primary air in a first stage of combustion, followed by injecting air in a second stage, whereupon burnout of the fuel is completed. There is removal of heat between the two stages. The formation of NO in the first stage is limited because the available oxygen for combustion with nitrogen is limited. The removal of heat between stages kinetically limits the formation of NO when excess air is added to the second stage. Experience shows that a 90% reduction in NO_x emission can be achieved in this manner. By recirculating flue gas, both the peak flame temperature and oxygen content are lowered. Injecting low-temperature steam or water also provides a diluting effect. Although probably of limited value for electric utility boilers (because thermal efficiency is lowered), the water-injection technique may be one of the better ways to reduce NO_x emissions in connection with internal-combustion engines of the stationary type. The situation in the case of internal combustion engines for automotive vehicles is considerably more complicated—there is a wide range of loads on such engines, high performance is required at all loading conditions, and the combustion process from fuel to exhaust must be kept simple and hence low-cost. As of the early 1980s, there remained some differences in opinion as regards the use of catalytic converters to remove NO_x from automobile exhausts. See also **Catalytic Converter (Internal Combustion Engine)**; **Combustion**; and **Petroleum**. The chemical characteristics of NO_x are described under **Nitrogen**.

Hydrocarbons. Extensive pollution of air occurs from the introduction of hydrocarbons either from (a) the incomplete combustion of hydrocarbon fuels in both stationary and vehicular engines; or (b) from paint spraying, solvent cleaning, printing, chemical and metallurgical, and other plants that use various fluids that have a high hydrocarbon content. Engine design and tuning are major factors in abating exhaust hydrocarbons. The intent is to fully burn the hydrocarbon content of the fuel. In a major city, industrial and commercial sources of organic solvent fumes (principally hydrocarbons) may average from 300 to 600 tons/day. For years, without legal restrictions, some operators found it more economical to permit vapor-laden air to escape to the atmosphere rather than to invest in solvent recovery equipment. Regulations coupled with higher costs of solvents have gone a long way toward eliminating this source of industrial pollution. Also, the chemical industry has successfully developed newer solvents, which are less volatile, and easier to handle and recover.

Carbon Monoxide. This pollutant is also associated with combustion operations, again being a product of incomplete combustion. Over the years, there has been a much greater awareness of carbon monoxide as a pollutant than the aforementioned gases because of its potent toxicity, dramatized by numerous deaths in earlier years as the result of keeping an automobile engine running in an enclosed space. Faulty residential heaters continue to take their toll of life and in recent years an important killer is the outdoor grill or hibachi with glowing coals taken into a camper or cabin as a means to temper the evening chill. Vehicular tunnel and large parking garage designers, of course, have practiced careful control over carbon monoxide concentrations for many years. See also **Carbon Monoxide**. The effects of carbon monoxide on human beings are shown in Fig. 5.

Other Pollutants. Some of these are gases; others fall into the particulate category. Of considerable importance are beryllium dust—very toxic and arising from ore preparation and metalworking operations, but of relatively limited extent because this metal is not a common structural material. Other contaminants include fluorides; metal fumes, such as arsenic, lead, and zinc; organic phosphates, notably from crop dusting and spraying; numerous kinds of organic vapors, including chlorinated hydrocarbons and hydrofluorocarbons (used in aerosol containers but suspect in connection with altering the ozone content of the upper atmosphere—see also **Aerosol**); radioactive fallout, such as ^{14}C, [137]Ce, and ^{90}Sr, arising from nuclear-device testing–no longer the major concern it was prior to the nuclear test ban accepted by most nations); and uranium dust. See also **Asbestos**; **Beryllium**; and **Pneumokonioses**.

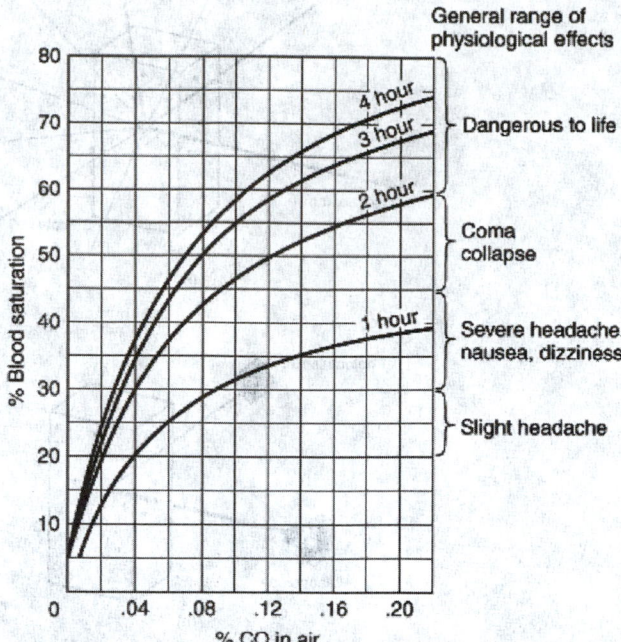

Fig. 5. Effects of carbon monoxide on human beings. This chart can be considered only as a general guide because the percent of CO blood saturation will vary with exertion, excitement, fear, depth of respiration, anemia, and general physical condition of the individual.

Acid Rain

This term is used almost exclusively in connection with the effect of airborne pollutants on the natural health of forests and lakes. The term was coined by Angus Smith several years ago, when he referred to the effect of industrial emissions on precipitation over the British midlands. Unfortunately, the term does *not* fully or accurately describe what may be occurring in certain forests and lakes.

The topic of acid rain during the 1980s and early 1990s was one of controversy and of incomplete answers in terms of official policy and science—after an expenditure of many millions of dollars. In 1980, the National Acid Precipitation Assessment Program (NAPAP) was established and subsequently consumed thousands of scientific research hours and costly field investigations, including the use of numerous helicopter expeditions to northwestern mountain and lake areas of the United States and Canada. Thousands of hours of computer power were consumed.

Segments of the jigsaw puzzle were worked out in excruciating detail, but only a few of the larger pieces of the puzzle were put together. Bottom-line conclusions and recommendations needed for governmental and private forestry operators were not forthcoming. A 6,000-page report was generated by NAPAP. But, in terms of a summary, the findings can be condensed to the few following points:

1. Acid rain has adversely affected aquatic life in about 10 percent of eastern lakes and streams.
2. Acid rain has contributed to the decline of red spruce at high elevations by reducing that species' cold balance.
3. Acid rain has contributed to erosion and corrosion of buildings and materials.
4. Acid rain and related pollutants, especially fine sulfate particles, have reduced visibility throughout the Northeast and in parts of the West.

Ellis Cowling (Ecologist, North Carolina State University) observed in early 1991, "You can say no symptoms were found, and we looked hard"—but that is different from saying no problem exists. A. Johnson (University of Pennsylvania) has observed, "It is the marginal ecosystems, the tenuous ones on the edge, that are affected by acid rain at current levels. I think control measures are justified. They will probably go a small way toward restoring some of the ecosystems that have been altered. And they will largely prevent more harm from occurring." See also **Acid Rain**.

Control Devices

The following items are commonly used as pollution control devices by industry or transportation devices. They can either destroy contaminants or remove them from an exhaust stream before it is emitted to the atmosphere.

Scrubbers. A scrubber is an industrial pollution control device, usually installed on the exhaust chimneys of large furnaces, but may also be used on any number of other air exhaust systems. See **Air Treatment Methodologies**.

Electrostatic Precipitator. An electrostatic precipitator (ESP), or electrostatic air cleaner, is a particulate collection device that removes particles from a flowing gas (such as air) using the force of an induced electrostatic charge.

Electrostatic precipitators are highly efficient filtration devices that minimally impede the flow of gases through the device, and can easily remove fine particulate matter such as dust and smoke from the air stream.

Plate Precipitator. The most basic precipitator contains a row of thin wires, and followed by a stack of large flat metal plates, with the plates typically spaced about 1 cm apart. The air stream flows through the spaces between the wires, and then passes through the stack of plates.

A high-voltage power supply transfers electrons from the plates onto the wires, developing a negative charge of several thousand volts on the wires, relative to the positive charge on the plates.

As particulate matter is carried past the strong negative charge on the wires, the particulate matter picks up the negative charge and becomes ionized. The ionized particles then move past the positively charged plates, to which the ionized particles are strongly attracted.

Once the particles are in contact with the positive plate, they then give up their electrons and become positively charged like the plate, and thus start acting as part of the collector. Due to this mechanism, electrostatic precipitators can tolerate large amounts of waste buildup on the collection plates and still operate effectively, since the waste itself helps collect more waste from the air stream.

Precipitator failure usually only occurs once a very heavy buildup of waste material has formed on the plates. The buildup can become heavy enough to block airflow, or can become thick enough to bridge across insulating gaps and short out the high-voltage power supply. (This typically does not damage the power supply, but effectively stops further electrostatic precipitation.)

Modern Industrial Electrostatic Precipitators. ESPs continue to be excellent devices for control of many industrial particulate emissions, including smoke from electric generating utilities (coal and oil fired), salt cake collection from black liquor boilers in pulp mills, catalyst collection from fluidized bed catalytic cracker units in oil refineries to name a few. These devices treat gas volumes from several hundred thousand ACFM to 2.5 million ACFM in the largest coal-fired boiler applications.

The original parallel plate—weighted wire design (described above) has evolved as more efficient (and robust) discharge electrode designs were developed, today focusing on rigid discharge electrodes to which many sharpened spikes are attached, maximizing corona production. Transformer-rectifier systems apply voltages of 50-100 kilovolts at relatively high current densities. Modern controls minimize sparking and prevent arcing, avoiding damage to the components. Automatic rapping systems and hopper evacuation systems remove the collected particulate matter while on line, allowing ESPs to stay in operation for years at a time.

Wet Electrostatic Precipitator. Electrostatic precipitation is typically a dry process, but spraying moisture to the incoming air flow helps collect the exceptionally fine particulates, and helps reduce the electrical resistance of the incoming dry material to make the process more effective.

A wet electrostatic precipitator, http://www.wapc.com/WESP.htm merges the operational methods of a wet scrubber with an electrostatic precipitator to make a self-washing, self-cleaning yet still high-voltage device.

Consumer-Oriented Electrostatic Air Cleaners. Plate precipitators are commonly marketed to the public as air purifier devices (such as the much-hyped Ionic Breeze) or as a permanent replacement for furnace filters, but all have the undesirable attribute of being somewhat messy to clean. A negative side-effect of electrostatic precipitation devices is the

production of ozone. However, electrostatic precipitators offer benefits over other air purifications technologies, such as HEPA filtration, which require expensive filters and can become "production sinks" for many harmful forms of bacteria.

With electrostatic precipitators, if the collection plates are allowed to accumulate large amounts of particulate matter, the particles often bond so tightly to the metal plates that vigorous washing and scrubbing may be required to completely clean the collection plates. The close spacing of the plates can make thorough cleaning difficult, and the stack of plates often cannot be easily disassembled for cleaning.

Some consumer precipitation filters are sold with special soak-off cleaners, where the entire plate array is removed from the precipitator and soaked in a large container overnight, to help loosen the tightly bonded particulates.

Fabric Filters. An air filter is a device which removes contaminants, often solid particles such as dust, pollen, mold, and bacteria from air.

They are often used in internal combustion engines, gas compressors, diving air compressors, ventilation systems, and any other application where air quality is important.

Engine intakes tend to use either paper, foam, or cotton filters. Some buildings, as well as planes and other man-made environments (e.g., satellites and space shuttles) use foam, pleated paper, or spun fiberglass filter elements. Another method uses fibers or elements with a static electric charge, which attract dust particles.

There are four main types of mechanical air filters: paper, foam, synthetics and cotton. Though all have varying degrees of practicality, it is generally recognized that cotton air filters are best because of their durability and technological advantages.

Paper. Paper filters are generally used in automobiles because they are inexpensive and disposable. Due to the characteristics of paper, good airflow is sacrificed for good filtration.

An example of pleated-paper air filters designed for in-duct use with home heating, ventilation, and air conditioning (HVAC) systems is the 3 M "Filtrete" line of products.

Synthetics. Polyester fiber is commonly used to make web formations used for air filtration. Polyester has a high temperature rating (abut 250 F) and is widely used in commercial, industrial or residential applications. Polyester can be blended with cotton or other fibers to produce a wide range of performance characteristics. In some cases Polypropylene, which has a lower temperature tolerance, is used to enhance chemical resistance. Tiny synthetic fibers knows as micro-fibers are used in many types of HEPA (High Efficiency Particulate Air Filter) filters.

Many in-duct filters for home forced air heating and air conditioning systems are made from spun fiberglass fibers. These filters are inexpensive, disposable, and come in various densities and sizes. Less-dense filters allow for higher airflow, but do not remove as much dust. Higher density filters remove more particles, but are more expensive and offer more resistance to the air. They also become more quickly "loaded" with contaminants and dust. They are considerably less expensive than pleated-paper filters for the same size.

Cotton. Cotton is generally regarded as the best material for air filters because of its excellent filtration and high airflow. This is because the "holes" in cotton are much looser than paper, giving high airflow, but use concepts such as interception, impaction, and diffusion, which allow the fibers to filter particles much smaller than the actual "holes." High performance air filters use oiled layers of cotton gauze. Dust particles held suspended in the cotton gauze actually "becomes" part of the filter medium to all the more increase filtration without obstructing airflow to the engine, thereby outlasting paper filters in service life and performance. Although cotton filters are pricey to begin with, they can be cleaned and re-oiled at a long service interval of 50,000 miles, by which time several disposable paper filters will have already been replaced per vehicle.

Activated Carbon. (Also called *activated charcoal*) is the more general term which includes carbon material mostly derived from charcoal. It denotes a material which has an exceptionally high surface area, typically determined by nitrogen adsorption, and includes a large amount of microporosity. Sufficient activation for useful applications may come solely from the high surface area, though often further chemical treatment is used to enhance the absorbing properties of the material. It can generally be produced in two different processes:

1. **Chemical Activation** Mostly acids are mixed with the source material in order to cauterize the fine pores. This technique can be problematic because, for example, zinc trace residues may remain in the end product.
2. **Steam Activation** The carbonized material is mixed with vapors and—or gases at high temperature to activate it. The source material can be several carbonic materials, e.g. nutshells, wood, coal.

Saturated active carbon can be regenerated by heating.

Activated carbon may have a surface area in excess of 500 m^2/g, with 1000 m^2/g being readily achievable. For comparison, a tennis court is about 260 m^2.

Under an electron microscope, the structure of activated carbon looks a little like ribbons of paper which have been crumpled together, intermingled with wood chips. There are a great number of nooks and crannies, and many areas where flat surfaces of graphite-like material run parallel to each other, separated by only a few nanometers or so. These micropores provide superb conditions for adsorption to occur, since adsorbing material can interact with many surfaces simultaneously. Tests of adsorption behavior are usually done with nitrogen gas at 77 K ($-321\,°$F) under high vacuum, but in everyday terms activated carbon is perfectly capable of producing the equivalent, by adsorption from its environment, liquid water from steam at $100\,°$C ($212\,°$F) and a pressure of 1/10,000 of an atmosphere.

Carbon aerogels, while more expensive, have even higher surface, and find use similar to activated carbon in special applications.

Chemically, activated carbon binds materials by Van der Waals force, specifically London dispersion force. Activated carbon, however, does not bind well to:

- Lithium [CAS: 7439-93-2], strong acids and bases, metals and other inorganic minerals (examples of these are sodium [CAS: 7440-23-5], iron [CAS: 7439-89-6], lead [CAS: 7439-92-1], arsenic [CAS: 744038-2], iodine [CAS: 7553-56-2], fluorine [CAS: 7782-41-4], and boric acid [CAS: 11113-50-1]).
- Alcohol (such as ethanol [CAS: 64-17-5], methanol [CAS: 67-56-1], isopropyl alcohol [CAS: 67-63-0], glycols [CAS: 107-21-1], and acetone [CAS: 67-64-1]).
- Ammonia [CAS: 7664-41-7].

Activated carbon is used in metal extraction (e.g. gold), water purification (especially in home aquariums), medicine, wastewater treatment, filters in gas and filter masks, filters in compressed air and gas purification, and many other applications.

Carbon absorption has numerous applications in removing pollutants from air or water streams both in the field and in industrial processes such as:

- Spill cleanup.
- Groundwater remediation.
- Drinking water filtration.
- Volatile organic compound capture from painting, dry cleaning and other processes.

Cyclone. In industry, a cyclone may refer to any of a number of different types of vortex separators. These devices use rotational effects and gravity to separate mixtures of solids and fluids. Because of different densities the heavier material is flung to the outside edge of the vortex while the lighter material is concentrated in the centre of the vortex.

Applications include:

- Removing dust from air.
- Separating oil and gas.
- Size separation and collection of airborne particles for analysis.

See also **Cyclonic Separation**.

Air Treatment Methodologies

Numerous techniques have been used to reduce, if not fully eliminate, various forms of air pollution over the years. For example, in connection with particulates, see entry on **Electrostatic Precipitator**. Most methods, however, fall into what might be termed the category of wet-scrubbing processes. As of the early 1980s, there were nearly 100 such processes on the market. Progress in this field has been steady, but not characterized by major breakthroughs. The concentration of sulfur dioxide in stack gases emitted by steam generation plants is usually in the range of only 400 to 2,000 parts per million (ppm). However, the volume of gases produced by the utility industry results, for example, in the liberation of large tonnages of sulfur dioxide into the atmosphere.

Chemical scrubbing systems for SO$_2$ absorption fall into two broad categories: (a) Disposable systems; and (b) regenerative systems. Typical of systems in use for a number of years are those that use an aqueous slurry of an insoluble calcium compound, which can be discarded after use. Disposable SO$_2$-removal systems use aqueous slurries of finely ground materials, such as lime [CAS: 1305-78-8], limestone [CAS: 1317-65-3] or dolomite, to produce a mixture of insoluble sulfites and sulfates. On passing through the scrubber, SO$_2$ from the waste gas dissolves to form sulfurous acid [CAS: 7782-99-2]: SO$_2$ + H$_2$O → H$_2$SO$_3$. The dissolved SO$_2$ reacts with the lime, Ca(OH)$_2$ or limestone, CaCO$_3$, to form insoluble calcium sulfite [CAS: 10257-55-3], CaSO$_3$: Ca(OH)$_2$ + H$_2$SO$_3$ → CaSO$_3$ + 2H$_2$O; CaCO$_3$ + H$_2$SO$_3$ → CaSO$_3$ + H$_2$O + CO$_2$. Unfortunately, SO$_2$ is less soluble (and hence less easily removed by scrubbing) in slightly acid solutions, so that it is extremely difficult in practice to operate a calcium-based system in such a manner that SO$_2$ removal is maximized while the quantities of calcium chemicals are minimized in order to approach stoichiometric conditions. As calcium-based slurry systems are usually operated at pH 6–10, disposal of the very large masses of used slurry presents a major problem. A typical power station using a calcium-based SO$_2$-removal slurry system will produce several hundred tons of spent slurry per day. A further disadvantage of lime or limestone systems is their marked tendency to precipitate insoluble calcium salts inside the scrubber. Unless the scale is removed, the scrubber shortly becomes inoperable.

Although chemically analogous to calcium-based systems, magnesium-based scrubbing systems possess several advantages. A slurry of finely divided magnesium hydroxide [CAS: 1309-42-8], Mg(OH)$_2$, is pumped through the scrubber to remove SO$_2$ from stack gases. Insoluble magnesium sulfite [CAS: 19086-20-5], MgSO$_3$, is formed: Mg(OH)$_2$ + SO$_2$ → MgSO$_3$ + H$_2$O. Hydrated magnesium sulfite, MgSO$_3$ · 6H$_2$O, can be disposed of as such, although it is usually heated to produce a rich stream of SO$_2$ and regenerate MgO. The SO$_2$ is compressed, liquefied, and stored in tanks for market; or catalytically oxidized to sulfur trioxide [CAS: 7446-11-9], SO$_3$, and treated with water to produce sulfuric acid, H$_2$SO$_4$. Alternatively, the SO$_2$ is mixed with hydrogen sulfide, H$_2$S, to produce elemental sulfur by the Claus process: SO$_2$ + 2H$_2$S → 3S + 2H$_2$O. Absorption efficiency of SO$_2$ attainable in a magnesium system is good, and removal efficiencies from 90 to 95% have been claimed without difficulty at reasonable liquor recirculation and MgO feed rates. As with calcium systems, serious scaling occurs due to build-up of insoluble MgSO$_3$.

Scrubbing solutions containing sodium (or other alkali metals) compounds have been extensively studied for removal of SO$_2$. Justification for the use of sodium compounds includes: (a) complete solubility in water with no formation of scale; and (b) simple reactions with SO$_2$: Na$_2$CO$_3$ + SO$_2$ → Na$_2$SO$_3$ + CO$_2$; 2NaHCO$_3$ + SO$_2$ → Na$_2$SO$_3$ + 2CO$_2$ + H$_2$O; 2NaOH + SO$_2$ → Na$_2$SO$_3$ + H$_2$O. In one commercial process, a scrubbing solution of sodium sulfite is used, which readily absorbs SO$_2$ to form the bisulfite: Na$_2$SO$_3$ + H$_2$O + SO$_2$ → 2NaHSO$_3$. In practice, only a portion of the Na$_2$SO$_3$ is converted to NaHSO$_3$ because the SO$_2$ absorption efficiency diminishes as the bisulfite concentration increases. The resulting solution is heated to decompose the bisulfite and thermally regenerate the sulfite. The gaseous SO$_2$ is compressed, liquefied and handled as previously mentioned under the magnesium system.

Ammonia-based chemicals appear to have some advantages over sodium systems. They are less costly, and regeneration by conventional means is possible, with the byproduct, ammonium sulfate, a marketable commodity for fertilizer.

Solutions containing ammonium sulfate [CAS: 7783-20-2], with or without the addition of ammonium hydroxide [CAS: 1336-21-6], have been widely used. The ammonium system can operate effectively only within a pH range of 4.0 to 7.0. As the pH value increases above 7.0, progressively more gaseous ammonia is liberated and this reacts in the gaseous phase with water vapor and SO_2 to produce a dense aerosol (white plume) which is difficult for scrubbers to remove. In an ammonia system, in order to regenerate the scrubbing solution, the ammonium bisulfite [CAS: 10192-30-0] and sulfite mixture is heated to drive off gaseous SO_2 : $2NH_4HSO_3 \rightarrow (NH_4)_2SO_3 + H_2O + SO_2$. Alternatively, the ammonium bisulfite/sulfite mixture can be treated with calcium hydroxide [CAS: 1305-62-0]. Gaseous ammonia is evolved and trapped in water, which is then recirculated to the scrubber.

Sodium citrate [CAS: 68-04-2] also is used in an SO_2 removal system. The solution is buffered at a pH 3.0–3.7 by the citrate ion, sulfur dioxide [CAS: 7446-09-5] is absorbed, and an equilibrium mixture of sodium bisulfite and citric acid [CAS: 77-92-9] is produced.

$$
\begin{array}{c}
\text{CH}_2\text{COONa} \\
| \\
\text{HO}-\text{C}-\text{COONa} + 3\,\text{SO}_2 + 3\,\text{H}_2\text{O} \longrightarrow 3\,\text{NaHSO}_3 \\
| \\
\text{CH}_2\text{COONa}
\end{array}
+
\begin{array}{c}
\text{CH}_2\text{COOH} \\
| \\
\text{HO}-\text{C}-\text{COOH} \\
| \\
\text{CH}_2\text{COOH}
\end{array}
$$

The bisulfite leaving the scrubber is then reduced with gaseous hydrogen sulfide, which precipitates elemental sulfur by a modified Claus reaction:

$$
3\,\text{NaHSO}_3 +
\begin{array}{c}
\text{CH}_2\text{COOH} \\
| \\
\text{HO}-\text{C}-\text{COOH} \\
| \\
\text{CH}_2\text{COOH}
\end{array}
+ 6\,\text{H}_2\text{S}
$$

$$
\longrightarrow
\begin{array}{c}
\text{CH}_2\text{COONa} \\
| \\
\text{HO}-\text{C}-\text{COONa} \\
| \\
\text{CH}_2\text{COONa}
\end{array}
+ 9\,\text{H}_2\text{O} + 9\,\text{S}
$$

A formate system uses two reactions involving potassium formate, HCOOK, which is regenerated after recovery of elemental sulfur. This method has the advantage over other wet scrubbing methods in that no precipitation of insoluble intermediates occurs at any stage of the process. Disadvantages include the need to heat K_2CO_3 solution, at high temperature and pressures, with carbon monoxide to regenerate the potassium formate. The energy requirements thus are high.

While it has been demonstrated that solutions of NaOH, $NaHCO_3$, and Na_2CO_3 are effective for SO_2 removal, these solutions are not effective for removal of mixtures of NO and NO_2, particularly when the gas stream velocities are reasonably high. Under conditions where from 95 to 99% SO_2 may be removed, the solutions may only be effective in removing from 5 to 15% NO_x. The fundamental difference between SO_2 and NO_x removal is that NO_x gases (mixtures of NO, N_2O_3, NO_2) are approximately 1,000 to 2,000 times less soluble in water than SO_2 at any given temperature. It has been found that conventionally designed wet scrubbers often do not provide sufficient liquid-to-gas contact surface areas or residence times to permit the NO_x to dissolve in the scrubbing solution. Consequently several stages may be required. Concentrations of NO_x in the range of 20,000 to 40,000 ppm require from 6 to 12 stages.

If no SO_2 is present, a sodium-based process may be used to remove NO_x efficiently. In the Neville-Krebs process (patent applied for), removal efficiencies of 60 to 90% have been achieved from gas streams containing up to 1,500 to 2,000 ppm of NO_x passing through a 3-stage scrubber (1 cubic foot; 0.028 cubic meter per stage) at 150 to 500 cubic feet (4.2 to 14.2 cubic meters) per minute.

The urea system is another relatively new system for removing NO_x from low-volume, slow-flowing waste gas streams. The system uses a slightly acid solution of urea [CAS: 57-13-6], $CO(NH_2)_2$. Unfortunately, the cost of urea is quite high, particularly for a large installation.

Other systems using electron-donor compounds have been tried or are in development. Such compounds include tri-n-butyl phosphate [CAS: 126-73-8], dimethylformamide [CAS: 68-12-2], triethyleneglycol dimethyl ether, dimethylsulfoxide [CAS: 67-68-5], hexamethylphosphoramide [CAS: 680-31-9], diethyleneglycol dimethyl ether, tricresyl phosphate [CAS: 1330-78-5], and dioxane [CAS: 123-91-1]. Most of these compounds are expensive compared with inorganic compounds used in other scrubbing systems.

Thus, it is evident that as of the early 1980s, the panacea for stack gas treatment was yet to be realized. Difficulties in treating stack gases have provided incentives at the other end of the combustion cycle, namely, in the treatment of the fuels prior to combustion. Various means for treating coal are described in detail in the entry on **Coal**; and desulfurization of petroleum fuels is described under **Petroleum**.

Numerous other entries in this volume take the energy/pollution interface into consideration. These include: **Air Mass**; **Atmosphere (Earth)**; **Catalytic Converter (Internal Combustion Engine)**; **Climate**; **Combustion**; **Diesel Engine**; **Electric Power Production and Distribution**; **Energy**; **Fuel**; **Gas and Expansion Turbines**; **Geothermal Energy**; **Hydrogen (Fuel)**; **Internal Combustion Engine**; **Meteorology**; **Natural Gas**; **Nuclear Power**; **Oil Shale**; **Tar Sands**; **Visibility**; and **Weather Technology**.

Additional Reading

Abelson, P.H.: "Asbestos Fiasco," *Science*, 1017 (March 2, 1990).

Abelson, P.H.: "New Technology for Cleaner Air," *Science*, 793 (May 18, 1990).

Abelson, P.H.: "Excessive Fear of PCBs," *Science*, 361 (July 26, 1991).

Alley, E.R., L.B. Stevens, and W.L. Cleland: *Air Quality Control Handbook*, The McGraw-Hill Companies, Inc., New York, NY, 1998.

Arya, S.P.: *Air Pollution Meteorology and Dispersion*, Oxford University Press, Inc., New York, NY, 1998.

Ashmore, M., L. Emberson, and F. Murray: *Air Pollution Impacts on Crops and Forests: A Global Assessment*, World Scientific Publishing Company, Inc., River Edge, NJ, 2001.

Ayres, J., R. Richards, and R. Maynard: *Air Pollution and Health*, Vol. 3, World Scientific Publishing Company, Inc., River Edge, NJ, 2002.

Barner, R.A. and A.C. Lasaga: "Modeling the Geochemical Carbon Cycle," *Sci. Amer.*, 74 (March 1989).

Baron, P.A. and K. Willeke: *Aerosol Measurement: Principles, Techniques, and Applications*, 2nd Edition, John Wiley & Sons, Inc., New York, NY, 2001.

Barth, H.G.: "Particle Size Analysis," *Analytical Chemistry*, (June 15, 1991).

Bell, N. and M. Treshow: *Air Pollution and Plant Life*, 2nd Edition, John Wiley & Sons, Inc., New York, NY, 2002.

Benedick, R.E.: *Ozone Diplomacy: New Directions in Safeguarding the Planet*, Harvard University Press, Cambridge, MA, 1998.

Beychok, M.R.: *Fundamentals of Stack Gas Dispersion*, Milton R. Beychok, Newport Beach, CA. 2005.

Bohn, H.: "Consider Biofiltration for Decontaminating Gases," *Chem. Eng. Progress*, 34 (April 1992).

Bosanquet, C.H. and J.L. Pearson: "The Spread of Smoke and Gases from Chimneys," *Trans. Faraday Soc.*, **32**: 1249, 1936.

Boss, M.J. and D.W. Day: *Air Sampling and Industrial Hygiene Engineering*, Lewis Publishers, Boca Raton, FL, 2000.

Boubel, R.W., D.L. Fox, and D.B. Turner: *Fundamentals of Air Pollution*, 3rd Edition, Academic Press, Inc., San Diego, CA, 1994.

Brimblecombe, P.: *Air Composition and Chemistry*, 2nd Edition, Cambridge University Press, New York, NY, 1995.

Clement, R.E.: "Environmental Analysis," *Analytical Chemistry*, 270T (June 15, 1991).

Colls, J.: *Air Pollution*, 2nd Edition, Taylor & Francis, Inc., Philadelphia, PA. 2002.

Conrad, J.: "An Acid-Rain Trilogy," *American Forests*, 21 (November–December 1987).

Cordasco, E.M., C. Zenz, and S.L. Demeter: *Environmental Respiratory Diseases*, John Wiley & Sons, Inc., New York, NY, 1997.

Crawford, M.: "Scientists Battle Over Grand Canyon Pollution," *Science*, 911 (February 23, 1990).

Davenport, G.B.: "Understand the Air-Pollution Laws that Affect Chemical Process Industries Plants," *Chem. Eng. Progress*, 40 (April 1992).

de Nevers, N.: *Air Pollution Control Engineering*, 2nd Edition, The McGraw-Hill Companies, Inc., New York, NY, 1999.

Downing, T.M.: "Preparing for New Smokestack Monitoring Regulations," *Instruments and Control Systems*, 47 (February 1992).

Dupuis, E.M.: *Smoke and Mirrors: The Politics and Culture of Air Pollution*, New York University Press, New York, NY. 2004.

Ebert, L.B.: "Is Soot Composed Predominantly of Carbon Clusters?" *Science*, 1469 (March 23, 1990).

Fulkerson, W., R.R. Judkins, and J.K. Sanghvi: "Energy from Fossil Fuels," *Sci. Amer.*, 128 (September 1990).

Graedel, T.E. and P.J. Crutzon: "The Changing Atmosphere," *Sci. Amer.*, 58 (September 1989).

Hall, J.V., et al.: "Valuing the Health Benefits of Clean Air," *Science*, 812 (February 14, 1992).

Hesketh, H.E.: *Air Pollution Control: Traditional and Hazardous Pollutants*, CRC Press, LLC, Boca Raton, FL, 1996.

Hobbs, P.V. and L.F. Radke: "Airborne Studies of the Smoke from the Kuwait Oil Fires," *Science*, 987 (May 15, 1992).

Hoffman, D.J.: "Increase in the Stratospheric Background Sulfuric Acid Aerosol Mass in the Past 10 Years," *Science*, 996 (May 25, 1990).

Holdren, J.P.: "Energy in Transition," *Sci. Amer.*, 156–163 (September 1990).

Holgate, S.T., J.M. Samet, R.L. Maynard, and H.S. Koren: *Air Pollution and Health*, Academic Press, Inc., San Diego, CA, 1999.

Hutterman, A. and D. Godbold: *Effects of Acid Rain on Forest Processes*, John Wiley & Sons, Inc., New York, NY, 1994.

Kennedy, I.R.: *Acid Soil and Acid Rain: Research Studies in Botany and Relate Applied Fields*, 2nd Edition, John Wiley & Sons, Inc., New York, NY, 1992.

Kidd, J.S., and R.A. Kidd: *Air Pollution: Problems and Solutions*, 2nd Edition, Facts on File, Inc., New York, NY. 2005.

Koenig, J.Q.: *Health Effects of Ambient Air Pollution: How Safe Is the Air We Breathe?* Kluwer Academic Publishers, Norwell, MA. 2000.

Little, C.E.: "The California X-Disease," *Amer. Forests*, 32 (July 8, 1992).

Liu, D.H. and B.G. Liptbak: *Air Pollution*, Lewis Publishers, Boca Raton, FL, 1999.

Lyons, C.E.: "Environmental Problem Solving: The 1987–88 Denver Brown Cloud Study," *Chem. Eng. Progress*, 6171 (May 1990).

Majewski, M.S., P.D. Capel, and R.J. Gilliom: *Pesticides in the Atmosphere: Distribution, Trends, and Governing Factors*, CRC Press, LLC., Boca Raton, FL, 1999.

Matthews, S.W. and J.A. Sugar: "Is Our World Warming? Under the Sun," *National Geographic*, 66 (October 1990).

Mohnen, V.A.: "The Challenge of Acid Rain," *Sci. Amer.*, 30 (August 1988).

Nazaroff, W.W. and L. Alvarez-Cohen: *Environmental Engineering Science*, John Wiley & Sons, Inc., New York, NY, 2000.

Nierenberg, W.A.: "Atmospheric Carbon Dioxide: Causes, Effects, and Options," *Chem. Eng. Progress*, 27 (August 1989).

Ondov, J.M. and W.R. Kelly: "Tracing Aerosol Pollutants with Rare Earth Isotopes," *Analytical Chemistry*, 691A (July 1, 1991).

Patrick, D.R.: *Toxic Air Pollution Handbook*, John Wiley & Sons, Inc., New York, NY, 1997.

Pluschke, P.: *Indoor Air Pollution*, Springer-Verlag New York, LLC, New York, NY. 2004.

Regens, J.L. and R.W. Rycroft: *The Acid Rain Controversy*, University of Pittsburgh Press, Pittsburgh, PA, 1988.

Roberts, L.: "Learning from the Acid Rain Program," *Science*, 1302 (March 15, 1991).

Schifftner, K.C.: *Air Pollution Control Equipment Selection Guide*, CRC Press, LLC, Boca Raton, FL, 2002.

Schneider, T.: *Air Pollution in the 21st Century: Priority Issues and Policy*, Elsevier Science, New York, NY, 1998.

Schnelle, K.B., C.A. Brown, C. Carelli, and F. Kreith: *Air Pollution Control Technology Handbook: A Handbook Series for Mechanical Engineering*, CRC Press, LLC, Boca Raton, FL, 2001.

Seinfeld, J.H., and S.N. Pandis: *Atmospheric Chemistry and Physics: From Air Pollution to Climate Change*, 2nd Edition, John Wiley & Sons, Inc., Hoboken, NJ. 2006.

Sher, E.: *Handbook of Air Pollution from Internal Combustion Engines: Pollutant Formation and Control*, Academic Press, Inc., San Diego, CA, 1998.

Snow, R.H. and T. Allen: "Effectively Measure Particle-Size Classifier Performance," *Chem. Eng. Progress*, 29 (January 1993).

Somersall, A.C.: *Fresh Air for Life: How to Win Your Unseen War Against Indoor Air Pollution*, The Natural Wellness Group, Mississauga, Ontario, Canada, 2006.

Staff: "ICI Plans U.S. Plant for CFC Substitute," *Chem. Eng. Progress*, 11 (February 1990).

Stradling, D.: *Smokestacks and Progressives: Environmentalists, Engineers, and Air Quality in America 1881–1951*, Johns Hopkins University Press, Baltimore, MD, 1999.

Turco, R.P.: *Earth under Siege: From Air Pollution to Global Change*, 2nd Edition, Oxford University Press, Inc., New York, NY, 2002.

Van Wormer, M.B.: "Use Air Quality Auditing as an Environmental Management Tool," *Chem. Eng. Progress*, 62 (November 1991).

Wallich, P.: "Dark Days: Eastern Europe Brings to Mind the West's Polluted Past," *Sci. Amer.*, 16 (August 1990).

Wark, K., C.F. Warner, and W.T. Davis: *Air Pollution: Its Origin and Control*, 3rd Edition, Addison Wesley Longman, Inc., Redding, MA, 1997.

Wettestad, J.: *Clearing the Air: European Advances in Tackling Acid Rain and Atmospheric Pollution*, Ashgate Publishing Company, Brookfield, VT, 2002.

Web References

U.S. Environmental Protection Agency; Air Quality Models: http://www.epa.gov/scram001/aqmindex.htm

U.S. Environmental Protection Agency; Dispersion Modeling: http://www.epa.gov/scram001/dispersionindex.htm

POLLUX (*β Geminorum*). The brighter star of the "heavenly twins." These two stars are always considered together in ancient writings and in astrology, and as a matter of fact, are not mentioned individually in literature, but always as the constellation Gemini. The constellation is of very ancient lineage and is referred to throughout all classical literature. The two stars were always considered of good omen by all peoples, and were always referred to as twins. Ranking seventeenth in apparent brightness among the stars, Pollux has a true brightness value of 45 as compared with unity for the sun. Pollux is an orange, spectral type K star. Estimated distances from the earth is 40 light years. See also **Constellations**.

POLONIUM. [CAS: 7440-02-06]. Chemical element, symbol Po, at. no. 84, at. wt. 210 (mass number of the most stable isotope), mp 252 °C, bp 960 °C, sp gr 9.4. The element was first identified as an ingredient of pitchblende by Marie Curie in 1898. The element occurs in nature only as a decay product of thorium and uranium. Because of limited availability and high cost, relatively few practical uses for the element have been found. Meteorological instruments for measuring the electrical potential of air have used small quantities of the metal. It is interesting to note that when Mme. Curie first identified polonium, she found that an electroscope was a far better instrument for detecting the metal than spectroscopic means. Polonium-plated metal rods and strips have been used as static dissipators in textile coating equipment and in various electrical equipment. The alpha rays from the polonium ionize the air, causing it to conduct and draw off accumulations of static electrical charges. Polonium is a member of periodic group 16 (formerly 6a).

Three isotopes of polonium occur in the uranium $(4n+2)$ radioactive series: ^{218}Po (radium A), $t_{1/2}$ 3.05 min; ^{214}Po (radium C'), $t_{1/2}$ 1.6×10^{-4} s; and ^{210}Po (radium F), $t_{1/2}$ 138.4 days, and the most stable isotope of polonium. It is used as a source of a-radiation. The thorium $(4n)$ series has two isotopes, ^{216}Po (thorium A), $t_{1/2}$ 0.16 s, and ^{212}Po (thorium C'), $t_{1/2}$ 3×10^{-7} s. The actinium $(4n+3)$ series also has two isotopes, ^{215}Po (actinium A), $t_{1/2}$ 1.83×10^{-3} s, and ^{211}Po (actinium C'), $t_{1/2}$ 0.52 s, which occurs in a 0.3% branched chain disintegration of ^{211}Bi (actinium C). Several other isotopes of polonium have been prepared, one of which occurs in the neptunium $(4n+1)$ series as ^{213}Po, $t_{1/2}$ 4.2×10^{-6} s.

Polonium exhibits the allotropy of the lower members of the chalcogen group, having a low-temperature, cubic form, α-polonium, and a high-temperature, rhombohedral form, β-polonium.

The tendency of the chalcogens to show increasing metallic character as one moves down the periodic table is quite marked for polonium; in fact, it resembles lead more than it does tellurium. Its compounds have a more ionic character in its lower oxidation states than do the tellurium compounds. The stability of the 6+ state is low, the existence of polonate(VI) ion being doubtful. The common oxidation states of the element are 2+ and 4+.

The halides, consisting of both dihalides and tetrahalides, are covalent and volatile, and they are not well characterized. The fluorides have not been established. The complex $PoCl_6^{2-}$ is known. Polonium compounds are usually colored, a fact that is useful in following their reactions. Thus, polonium(II) chloride, $PoCl_2$, formed by dissolving polonium(IV) oxide, PoO_2, in HCl is pink, and an oxidation by heating or treatment with chlorine yields yellow $PoCl_4$. Polonium(IV) bromide, $PoBr_4$, dark red, gives purple polonium(II) bromide, $PoBr_2$, on heating. $PoBr_4$ also gives ammonium polonium bromide, $(NH_4)_2[PoBr_6]$ with ammonia. Complex iodides $M_2[PoI_6]$ have been prepared.

Metallic polonium reacts with air readily on heating, to form PoO_2, which exists in a yellow face-centered form having fluorite structure at low temperatures, and a red tetragonal one on heating. Polonium(IV) hydroxide, $Po(OH)_4$, precipitated from polonium(IV) solutions by ammonia, exhibits only slight activity, and is thus not amphiprotic. On reaction of polonium with HNO_3, $Po(NO_3)_4$ is formed, and on treatment of polonium(IV) chloride, $PoCl_4$, with H_2SO_4, polonium(IV) sulfate, $Po(SO_4)_2$, is formed, both being ionic-type salts, as indeed are other oxyacid compounds. The sulfate, however, is quite reactive, being hydrated in solution, dehydrated on removal from solution, and forming a basic compound, $2PoO_2 \cdot SO_3$, on heating. H_2S precipitates black polonium(II) sulfide, PoS.

Additional Reading

Greenwood, N.N. and A. Earnshaw: *Chemistry of the Elements*, 2nd Edition, Butterworth-Heinemann, Inc., Woburn, MA, 1997.

Krebs R.E.: *The History and Use of Our Earth's Chemical Elements: A Reference Guide*, Greenwood Publishing Group, Inc., Westport, CT, 1998.

Lide, D.R.: *CRC Handbook of Chemistry and Physics*, 88th Edition, CRC Press, LLC., Boca Raton, FL, 2007.

Parker, P.: *McGraw-Hill Encyclopedia of Chemistry*, 2nd Edition, The McGraw-Hill Companies, Inc., New York, NY, 1993.

Stwertka, A. and E. Stwertka: *A Guide to the Elements*, Oxford University Press, Inc., New York, NY, 1998.

POLONOVSKI REACTION. Demethylation of tertiary (or heterocyclic) amine N-oxides on treatment with acetyl chloride or acetic anhydride to give N-acylated secondary amines and formaldehyde, along with O-acylated aminophenols as a result of a side reaction.

POLYACETYLENE. A linear polymer of acetylene having alternate single and double bonds, developed in 1978. It is electrically conductive, but this property can be varied in either direction by appropriate doping either with electron acceptors (arsenic pentafluoride or a halogen) or with electron donors (lithium, sodium). Thus, it can be made to have a wide range of conductivity from insulators to n- or p-type semiconductors to strongly conductive forms. Polyacetylene can be made in both *cis* and *trans* modifications in the form of fibers and thin films, the conductivity of the fibers increasing with their degree of orientation. Films can be applied on glass or metal substrates. Though still in an experimental stage, these polymers have significant possibilities for industrial application, e.g., in batteries.

POLYAMIDE-IMIDE RESINS. An injection-moldable, high-performance engineering thermoplastic, polyamide-imide is the condensation polymer of trimellitic anhydride and various aromatic diamines with the general structure:

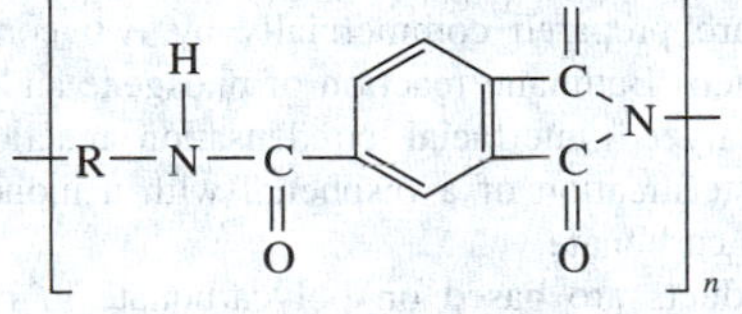

Polyamide-imides are available[1] in unfilled; 30% glass fiber-filled; 30% graphite fiber-filled; and modified 40% glass-filled grades. The unfilled grade has the highest impact resistance, while the graphite fiber-filled grade has the highest modulus or stiffness. The modified version offers the lowest cost while still maintaining an impressive slate of properties. The resins can be molded into complex precision parts and also can be extruded and machined to close tolerances. The resins are used extensively in the aerospace industry, offering significant weight reduction by replacing metal parts. Aircraft usage includes jet engine components, compressor and generator parts, and electronic/electrical devices. Polyamide-imide resins are also used in the hydraulic/pneumatic equipment industry for wear surfaces, bushings, seals, vanes, and flow control devices. The automative and heavy equipment industries use this material as parts in transmissions, universal joints, and power-assisted devices. Internal combustion engines use many polyamide-imide structural-mechanical parts, such as valve train components.

The material is opaque and characterized by good dimensional stability, creep resistance, impact resistance, and superior mechanical properties that persist up to temperatures of about 500°F (260°C). Unfilled polyamide-imide has a tensile strength of about 27,000 psi (186 mPa); flexural strength of about 35,000 psi (241 mPa); compressive strength of about 32,000 psi (221 mPa); and an elastic modulus of about 750,000 psi (5172 mPa). Mechanical properties at 450°F (232°C) exceed those of many polymers at room temperature. At cryogenic temperature, the unfilled polymer has a tensile strength of about 31,500 psi (241 mPa) with 6% elongation at −196°C (liquid nitrogen). Heat deflection is high (525°F; 273°C), while the coefficient of linear thermal expansion is low. When burned, polyamide-imide produces a char rather than drip and produces very little smoke. Electrical properties are attractive. Radiation resistance is good. The resin is virtually unaffected by aliphatic and aromatic hydrocarbons, halogenated solvents, and most acid and base solutions. It is attacked by high-temperature caustic, steam, and some acids. At 50% relative humidity, the material (70°F; 21°C) will absorb about 1% moisture in 1000 hours.

POLY(ANHYDRIDES). See **Biomaterials, Prosthetics, and Biomedical Devices**.

POLYARYLATES. These are clear, amorphous thermoplastics that combine clarity, high heat deflection temperatures, high impact strength, good surface hardness, and good electrical properties with inherent ultraviolet stability and flame retardance. No additives or stabilizers are required to provide these properties. Polyarylates are aromatic polyesters that are manufactured from various ratios of iso- and terephthalic acids with bisphenol A.[1] The resultant products are free-flowing pellets which can be processed by a variety of thermoplastic techniques in transparent and opaque colors. Because polyarylate's weatherability is obtained through polymer chemistry rather than additives (as with most UV-resistant polymers), the properties of polyarylate do not deteriorate significantly with time. (Over 5000 hours of accelerated weathering and actual Florida and Arizona aging resulted in virtually no change in performance with respect to luminous light transmittance, haze, gloss, yellowness, and impact.) The flammability characteristics are inherent. Properties include a high oxygen index, low smoke density, low flame spread, and low toxic gas formation. Because the flammability properties are achieved without additives, the resultant products of combustion are essentially limited to Commercially available as CO_2, CO, and water.

Polyarylates are offered in several glass-reinforced versions with loading available up to 40%. The glass fibers provide higher stiffness, improved tensile strength, and higher heat deflection temperatures. Polyarylates may also be mineral-filled and reinforced with other fibers, such as carbon. Alloys/blends with other polymers are also available.

These materials are useful in outdoor applications, such as high-intensity discharge lighting (traffic signals), automobile halogen headlamp lenses and bodies, and rear-end elevated automobile stop lights. High-temperature lighting and microwave cookware are suitable applications. Electronic/electrical connectors and housings are also important applications for the polymer.

Polyarylates have good optical properties. Luminous light transmission can range from 84% to 88% with only 1% to 2% haze. Refractive index is 1.61. An important feature of the polyarylate family is high heat resistance demonstrated by a 340°F (171°C) heat deflection temperature at 264 psi (1.8 mPa). The material exhibits good retention of properties at high temperature exposures: 270,000 psi (1380 mPa) at 300°F (149°C); and over 200,000 psi at 350°F (177°C).

Polyarylates are injection molded, using standard screw machines, as well as extruded, foam molded, and blow molded. Melt temperatures range from 600 to 680°F (316 to 360°C). Mold temperatures should be maintained between 200 and 300°F (93 and 149°C).

POLYBASITE. A mineral antimony sulfide of silver $(Ag,Cu)_{16}Sb_2S_{11}$, in which copper substitutes for silver to approximately 30 atomic percent. It crystallizes in the monoclinic system; hardness, 2–3; specific gravity,

---

[1] Commercially available as *Torlon*™. (*Amoco Chemicals Co.*)

[1] Data furnished by *Celanese Engineering Resins*.

6.3; color, black, dark ruby red in thin splinters with metallic luster; nearly opaque. From the Greek, meaning *many*, suggesting the many-metal basis.

Occurs in low-temperature silver deposits commonly associated with silver and lead minerals. Found in various Western States in the United States, and as superb crystals at Arizpe and Las Chiapas, Mexico; in Chile, Peru, Sardinia, Germany, and Australia.

POLYBENZIMIDAZOLES.   These are heterocyclic polymers that have outstanding high thermal characteristics, the highest obtainable in commercial polymers. These materials also have superior ablative and hydrolytic stability as well as high compressive and dimensional stability. Polybenzimidazoles essentially are unaffected by solvents, acids, and bases. They are marketed in stock shapes and as finished parts. The materials are not available in resin form. Hoechst Celanese markets the products under the tradename *Celazole.*®

Parts are produced by a high-pressure sintering process wherein the melt polycondensation resin is densified and then coalesced. Metallurgical pressures at temperatures exceeding over 400 °C (750 °F) are required. Polybenzimidazoles have repeating benzimidazole groups in the polymer backbone. These materials were synthesized first in 1961. Currently, the products result from a melt polycondensation reaction of aromatic, bis-ortho-diamines (diphenylisophthalate). By way of compounding, fabrication and end-use performance characteristics can be customized to specific needs. Often the materials are preferred for tribological applications inasmuch as they have the desirable characteristics of low coefficient of friction, low abrasion, and good high-temperature dimensional stability.

Polybenzimidazoles have been used for seals, mechanical components, electrical connectors, valve seats, and as components of materials-handling equipment in the petrochemical, geothermal, chemical process, aerospace, defense, automotive, and electrical products industries. These materials frequently are procured as replacement parts for other materials in an effort to improve equipment performance.

POLYBUTYLENE RESINS.   These materials (PBs) are semicrystalline polyolefin thermoplastics based on poly(1-butene) and include homopolymers and a series of poly(1-butene-ethylene) copolymers. The resins available commercially[1] are manufactured via stereospecific Ziegler-Natta polymerization of 1-butene monomer. The commercial products are based on isotactic (98% to 99.5%), high-molecular-weight (230,000 to 750,000) polymer. Five crystalline modifications of poly(1-butene) are known. Of these, the glass transition temperature ranges from about −4 °F (−20 °C) for the homopolymers to about −30 °F (−34:18C) for the high-ethylene copolymers.

PB resins generally are resistant to acids, bases, solvents, paraffinic and naphthenic oils, detergents, and various chemicals. Resistance decreases, however, at elevated temperatures. They have good moisture barrier and electrical insulation properties. They exhibit a broad range of flexibility: tensile moduli vary from 41,500 psi (286 mPa) for homopolymers to 7500 psi (52 mPa) for the high-ethylene polymers. The resins are particularly resistant to creep, environmental stress cracking, chemicals, and abrasion. PB resins are offered in a special pipe grade, film grades, and five general-purpose grades.

PB pipe can be fabricated by conventional single-screw extrusion technology using vacuum or pressure sizing for dimensional control. The pipes can be joined by thermal fusion or mechanical fittings. Applications include cold and hot water plumbing. Other uses include well, heat pump, and fire-sprinkler piping as well as specialty hoses. Large-diameter PB pipe finds uses in the transport of abrasive or corrosive materials at high temperature, as found in the mining, chemical, and power generation industries.

PB film is usually made by the blown film process, but also can be cast on chill rolls. Film applications include food and meat packaging, compression wraps, and hot-fill containers. The material can be formulated to provide a wide range of seal strengths for peelable or easy-opening packaging.

The ability of PB to accept high filler loading (up to 80%) has resulted in its use as a color, mineral filler, and flame-retardant concentrate carrier.

Polybutylene is compatible with polypropylene in all proportions and as a modifier it provides enhancement of processibility, impact, and weld line strength in injection molding and extrusion. It also provides improved impact strength and heat stability in films and improved hand and bondability in fibers of polypropylene. A comparatively recent use of PBs is for hot-melt adhesives and sealants where high strength, high-shear adhesion failure temperature, and a long open time are needed. They are particularly suited for use with aliphatic tackifying resins.

POLYBUTYLENE TEREPHTHALATE POLYESTERS.   A semicrystalline thermoplastic polyester. Because of its rapid crystallization, injection molding is the preferred method of processing. The material has been used for many years in the connector industry because of its good chemical resistance, high-temperature capabilities, good electrical properties, and long flow lengths in thin sections. This material (PBT) typically is formed in a transesterification reaction between 1,4-butanediol and dimethylterephthalate. Unmodified PBT is translucent in thin sections and opaque white in thicker sections.

The glass transition temperature of PPBT is about 52 °C (125 °F). Melting point is about 230 °C (440 °F). Unreinforced PBT is obtainable in several molecular weights. Compounded resins are available with numerous types and levels of fillers and reinforcements. Glass fiber reinforcement has a wide spectrum of physical properties. These materials can be made flame-retardant through the use of additives.

Exceptional electrical properties and temperature resistance qualify the material for numerous electrical parts—connectors, coil bobbins, light sockets, terminal blocks, fuse holders, and motor parts. PBT provides weight reduction of final parts. PBT has replaced a number of thermoset materials and is particularly popular for appliance housings and fibers for paint brushes.

POLYCARBONATES.   This material is classified as an engineering thermoplastic, mainly because of its toughness. Exceptional clarity and high heat-deflection temperatures are other outstanding properties of polycarbonates. The vast majority of polycarbonates are based on bisphenol A [CAS: 80-05-7] (BPA) and sold under various trade names Lexan (GE), Makrolon (Bayer), Caliber (Dow), Panlite (Teijin) and Iupilon (Mitsubishi). BPA polycarbonates [CAS: 25037-45-0], having glass-transition temperatures, in the range of 145–155 °C, are widely regarded for optical clarity and exceptional impact resistance and ductility at room temperature and below. Other properties, such as modulus, dielectric strength, or tensile strength are comparable to other amorphous thermoplastics at similar temperatures below their respective glass-transition temperatures. Whereas below their T_gs most amorphous polymers are stiff and brittle, polycarbonates retain their ductility. Polycarbonates are prepared commercially by two completely different processes: Schotten–Baumann reaction of phosgene and an aromatic diol in an amine-catalyzed interfacial condensation reaction, or via base-catalyzed transesterification of a bisphenol with a monomeric carbonate such as diphenyl carbonate.

Important products are based on polycarbonate in blends with other materials, copolymers, branched resins, flame-retardant compositions, and foams.

Properties

The majority of polycarbonates are prepared in methylene chloride solution. Chloroform, *cis*-1,2-dichloroethylene, *sym*-tetrachloroethane, and methylene chloride are the preferred solvents for polycarbonates. Hydrocarbons and aliphatic alcohols, esters, or ketones do not dissolve polycarbonates. Acetone promotes rapid crystallization of the normally amorphous polymer, and causes catastrophic failure of stressed polycarbonate parts.

In general, polycarbonate resins have fair chemical resistance to aqueous solutions of acids or bases, as well as to fats and oils. Chemical attack by amines or ammonium hydroxide occurs, however, and aliphatic and aromatic hydrocarbons promote crazing of stressed molded samples. BPA polycarbonate has excellent resistance to hydrolysis.

Certain blends and copolymers of polycarbonate demonstrate dramatically improved solvent resistance. The blend of polycarbonate and poly(butylene terephthalate) combines the toughness of polycarbonate with the solvent resistance of the semicrystalline polyester.

[1] Commercially available as *Duraflex*™. (*Shell Chemical Co.*)

Copolycarbonates of BPA and hydroquinone (HQ) can be prepared via the intermediacy of cyclic oligomeric cocyclics. Although hydroquinone linear oligomers having degrees of polymerization greater than two are insoluble in CH_2Cl_2, the cyclic analogues remain soluble when randomly cyclized with BPA.

BPA polycarbonates are commercially available in a wide range of molecular weights. As the molecular weight increases, melt and solution viscosities increase proportionally. Molecular weights may be determined or inferred by several means, including gel-permeation chromatography, light-scattering chromatography, measurement of intrinsic or inherent viscosity, and measurement of melt viscosity and flow. Correlation of intrinsic viscosity (IV), $[\eta]$, with weight-average mol wt (M_w) has been carried out on carefully characterized polycarbonate samples. The following relationship exists when $[\eta]$ is in mL/g $\cdot$ $[\eta] = 41.2 \times 10^{-3} \cdot M_w^{0.69}$.

The mechanical properties of polycarbonate, e.g., tensile strength, impact resistance, flexural strength, elongation, etc., improve dramatically with increasing polymer intrinsic viscosity up to a value of about 0.45 dL/g. After that point, slight increases in mechanical properties are seen with increasing molecular weight, but melt viscosity continues to climb. At IV values greater than 0.6 dL/g, the melt viscosity becomes so high that processing is very difficult.

The mechanical–optical properties of polycarbonates are those common to amorphous polymers. The polymer may be crystallized to some degree by prolonged heating at elevated temperature (8 d at 180 °C), or by immersion in acetone. See also **Acetone**. Powdered amorphous powder appears to dissolve partially in acetone, initially becoming sticky, then hardening and becoming much less soluble as it crystallizes. Enhanced crystallization of polycarbonate can also be caused by the presence of sodium phenoxide end groups.

Film or fibers derived from low molecular weight polymer tend to embrittle on immersion in acetone; those based on higher molecular weight polymer (>0.60 dL/g) become opaque, dilated, and elastomeric.

The T_g of BPA polycarbonate is around 150 °C, which is unusually high compared to other thermoplastics. The high glass-transition temperature can be attributed to the bulky structure of the polymer, which restricts conformational changes, and to the fact that the monomer has a higher molecular weight than the monomer of most polymers. The high T_g is important for the utility of polycarbonate in many applications, because, as the point which marks the onset of molecular mobility, it determines many of the polymer's properties such as dimensional stability, resistance to creep, and ultimate use temperature. Polycarbonates of different structures may have significantly higher or lower glass-transition temperatures.

BPA polycarbonate becomes plastic at temperatures around 220 °C. The viscosity decreases as the temperature increases, exhibiting Newtonian behavior, with the melt viscosity essentially independent of the shear rate. At the normal injection molding temperature of 270–315 °C, the melt viscosity drops from 1,100 to 360 Pa·s (11,000 to 3,600 poise). Because the viscosity of polycarbonate can only be reduced by increasing the temperature, the ultimate limit on molecular weight is controlled by the processing conditions and the thermal stability of the polymer.

BPA polycarbonate exhibits excellent thermal stability, especially in the absence of oxygen and water. At temperatures above 400 °C, rapid decomposition and cracking occur. BPA has an oxygen index of 26; this indicates that under test conditions, an atmosphere of 26% oxygen is required for combustion. Owing to thermal–oxidative stability, polycarbonate has some inherent flame resistant properties and can be classified as V-2 according to UL94 of the Underwriters Laboratory. Several polycarbonate grades have additives to increase the flame-retardant properties, and to decrease smoke.

Because of the low solubility of water in the resin, BPA polycarbonates are inherently resistant to aqueous acid and base, although strong nucleophilic bases can cause hydrolysis.

Polycarbonate is a transparent colorless polymer, making it attractive for glass replacement. Exposure of non-uv stabilized polycarbonate to strong uv or outdoor sunlight conditions can lead to photoyellowing of the surface. The uv screens can be combined with hard coats to protect the surface of polycarbonate, and also to provide resistance to chemical attack in sheet products used for glazing applications. Polycarbonate eyewear is usually available with a scratch-resistant coating.

Special polycarbonate grades have been developed for the optical information storage market e.g., compact disks.

The room temperature modulus and tensile strength are similar to those of other amorphous thermoplastics, but the impact strength and ductility are unusually high. Whereas most amorphous polymers are glass-like and brittle below their glass-transition temperatures, polycarbonate remains ductile to about −10 °C. The stress–strain curve for polycarbonate typical of ductile materials, places it in an ideal position for use as a metal replacement. Weight savings as a metal replacement are substantial, because polycarbonate is only 44% as dense as aluminum and one-sixth as dense as steel.

Impact strength can be measured by a variety of methods, including notched Izod, tensile impact, and falling dart impact. Polycarbonates are among the highest rated engineering polymers for impact resistance, and are the toughest transparent materials known.

Glass-reinforced polycarbonates are sold as high modulus materials having properties approaching those of metals, while retaining the basic plastic attributes of low cost processing, dielectric character, resistance to corrosion, and inherent color.

Preparation

Most BPA polycarbonate is produced by an interfacial polymerization process utilizing phosgene. The interfacial process for polycarbonate preparation involves stirring a slurry or solution of BPA and 1–5% of a chain stopper, such as phenol, p-tert-butylphenol, or p-cumylphenol, in a mixture of methylene chloride and water, while adding phosgene in the presence of a small amount (0.1–3%) of a tertiary amine catalyst.

The transesterification or melt process is an environmentally friendly process that utilizes no solvent during polymerization, producing neat polymer directly, and so chlorinated solvents may be entirely eliminated. The melt polymerization process involves the base-catalyzed transesterification reaction of a bisphenol with a source of carbonate functionality, such as diphenyl carbonate.

Preparation of polycarbonate via the melt process offers several processing advantages, including elimination of solvents and phosgene. The principal *product* advantage of melt-prepared polycarbonate is the fact that, as an equilibrium product, the resin is stable.

Processing

Polycarbonates may be fabricated by all conventional thermoplastic processing operations, of which injection molding is the most common.

Structural foam molding is also a valuable commercial technique. The molding compound is charged with an inert gas under pressure, and at the same time thermally plasticized in an extruder. Nitrogen gas may be pumped into the melt, or the molding compound may contain a chemical blowing agent that decomposes with gas generation at the plasticizing temperature. The pressurized plasticized melt is injected into a low pressure mold with a volume change taking place, causing the polymer to expand and fill the mold cavity. Very large parts having relatively low clamping pressures can be produced with this process.

Conventional thermoforming of sheet and film is applicable to the production of skylights, radomes, signs, curved windshields, prototype production of body parts for automobiles, skimobiles, boats, etc. Because BPA polycarbonate is malleable, it can be cold-formed like metal, and may be cold-rolled, stamped, or forged.

Health and Safety Factors

Polycarbonate is considered a slight or nonexistent fire hazard. Odor and volatiles are negligible. Processing fumes, which include water, carbon dioxide, diphenyl carbonate, methylene chloride, and phenol, are not formed in levels considered to be hazardous. Polycarbonate has very low acute oral and dermal toxicity, is not a primary skin irritant, and does not cause systemic or local sensitization. In a finely divided form, polycarbonate is a mild eye irritant, consistent with the abrasive nature of the ground resin particle. Polycarbonate does not degrade during storage, and no heating or cooling requirements are necessary.

Uses

Extreme toughness, transparency, low color, resistance to burning, and maintenance of engineering properties over a wide thermal range are the outstanding properties of polycarbonate that make it useful for a variety of applications. Glazing and sheet are the largest markets for polycarbonate resins. Clarity and an impact resistance 250 times greater than glass and 30 times greater than an acrylic sheet makes polycarbonate the natural choice for window replacement in areas where breakage is common. Windows in airplanes, trains, and schools commonly use polycarbonate. Exotic applications include military use, eg, in high speed aircraft canopies, where tests have shown polycarbonate to withstand impact with fowl at Mach 2. Polycarbonate is also used for security applications as laminates with glass or other materials. Polycarbonate offers unsurpassed projectile-stopping capability, as the material softens upon impact with a bullet, absorbing the projectile's energy. A large volume in laminated products also exists. The soft inner layer may be a urethane–, acrylic–, or siloxane–polycarbonate copolymer.

Many automotive applications use blends of polycarbonate with acrylonitrile–butadiene–styrene (ABS) or with poly(butylene terephthalate) (PBT). Both large and small appliances also account for large markets for polycarbonate. Polycarbonate is attractive to use in light appliances, including houseware items and power tools, because of its heat resistance and good electrical properties, combined with superior impact resistance and the ability to produce a wide variety of colors.

Packaging is a growing segment of the polycarbonate market. Polycarbonate has been the preferred choice for large returnable, refillable 20-L water bottles because of its light weight and impact resistance.

Electrical, electronic, and technical applications use polycarbonates for a variety of purposes. Because of excellent electrical properties (dielectric strength, volume resistivity), and resistance to heat and humidity, polycarbonate is used for electrical connectors, telephone network devices, outlet boxes, etc. Polycarbonate had been popular for use in computer and business machine housings, but the use of neat resin was largely supplanted by blends of polycarbonate with ABS. Polycarbonate also has many technical uses in instrument panels and devices, especially for membrane switches and insulators. Optical quality polycarbonate is the only suitable material for the compact disk market.

Polycarbonate is popular in the medical and health care related industries because of its clarity, impact strength, and low level of extractable impurities. Special grades have been developed to maintain clarity and resistance to yellowing upon gamma radiation sterilization processes. Leisure and safety applications are many and varied. The largest markets are for protective headgear such as football and motorcycle helmets, and safety helmets for firefighters and construction workers. Protective eyewear also uses polycarbonate, because of its clarity and impact resistance.

Additional Reading

Brunelle, D.J., and M. Korn: *Advances in Polycarbonates*, American Chemical Society, Washington, DC, 2005.

LeGrand, D.G., and J.T. Bendler: *Handbook of Polycarbonate Science and Technology*, Marcel Dekker, Inc., New York, NY, 1999.

Read, C.S.: CEH Marketing Research Report, SRI International, 1993.

Schnell, H.: *Ang. Chem.*, **68**, 633 (1956).

Sperling, L.H.: *Introduction to Physical Polymer Science*, 3rd Edition, John Wiley & Sons, Inc., New York, NY, 2001.

Stevens, M.P.: *Polymer Chemistry: An Introduction*, 3rd Edition, Oxford University Press, New York, NY, 1998.

U.S. Pat. 3,028,365 (1962), H. Schnell, L. Bottenbruch, and G. Grimm (to Bayer AG).

U.S. Pat. 3,153,008 (1964), D.W. Fox (to General Electric).

Web References

AmericanPlasticsCouncil.org — Polycarbonate, http://www.americanplasticscouncil. org/benefits/in_your_life/pop/power_plastics_october99.html

GE Lexan® Resin, http://www.gelexan.com

Polycarbonate Manufacturers, Compounders, and Suppliers, http://www.matweb. com/reference/polycarbonatemfr.asp

Polycarbonate Property Data, http://www.matweb.com/reference/polycarbonate.asp

DANIEL J. BRUNELLE

POLYCHLOROPRENE. [CAS: 9010-98-4]. Polychloroprene was discovered in 1930 at E. I. du pont de Nemours & Co. in Wilmington Delaware. The discovery grew out of a need to develop a synthetic substitute for natural rubber. DuPont first marketed this first commercially

successful synthetic elastomer as DuPrene in 1933. In response to new technology development that significantly improved the product and manufacturing process, the name was changed to neoprene in 1936. The current commercially acceptable generic name for this class of chlorinated elastomers is CR or chloroprene rubber.

Polychloroprene has been in commercial use for over 70 years, but continues to be the elastomer of choice for dynamic applications requiring moderate heat and oil resistance, or where long-term environmental resistance is required. The material is manufactured by emulsion polymerization of 2-chloro-1,3-butadiene. The properties of the product can be optimized for specific end uses by selection of polymerization conditions, by use of comonomers such as 2,3-dichlorobutadiene and methacrylic acid, and by choice of the method used to control molecular weight. The kinetics of the polymerization process and the structure of the polymer have been studied in detail and related to polymer properties. Commercial products vary widely in properties depending on the intended end use and the details of the manufacturing process. Tough crystalline or ionomeric polymers are preferred for adhesive applications, whereas softer, more flexible polymers are preferred for mechanical goods applications. The latter polymers may be compounded and fabricated by substantially all the techniques used for manufacture of rubber goods by molding, extrusion, or calendering operations. Latexes are used for various bonding and adhesive applications as well as for dipped goods requiring toughness in the unreinforced film. Monomers can undergo runaway reactions so strict handling procedures must be followed.

Since the time of its introduction to the marketplace, neoprene has been more than a simple replacement for natural rubber. Like natural rubber, neoprene is rubbery, resilient, of high tensile properties. However, neoprene has better heat stability, better resistance to varying environmental weathering conditions, superior flex life, excellent solvent and oil resistance and reasonable electrical properties when compared to natural rubber. This unique combination of properties poised neoprene for solving many of the potential problems besetting the automotive, construction, footwear, specialty apparel, transportation, and wire and cable industry. The good balance of properties has made the polymer useful in a large divergent list of applications including aircraft, appliance, automotive, bridge pad, chemical resistant clothing, home furnishings, machinery, mining and oil field belting, underground and undersea cables, recreation, and tires.

See also **Elastomers**; and **Rubber (Natural)**.

POLYCYCLIC AROMATIC HYDROCARBONS (PAHs). A class of large aromatic molecules composed of several benzene rings fused together. Some PAHs show very high carcinogenic and mutagenic activity. They are found in organic residues, such as soot, coal tar, and combustion exhaust. Due to their low volatility, PAHs are usually taken up onto organic aerosols, which facilitates their inhalation.

See also **Carcinogens**.

POLYCYCLO-HEXYLENE-DIMETHYLENE TEREPHTHALATE. PCT is 1,4-cyclohexylene-dimethylene terephthalate and is a high-temperature, semicrystalline thermoplastic polyester. PCT possesses excellent thermal properties. Injection molding is the predominant method of processing glass fiber-reinforced grades of PCT. It is widely used for products where excellent thermal (heat-resistant) properties are needed, as exemplified by surface-mountable electronic components, automotive parts, and dual-ovenable cookware. The material also is used for flexible electronic circuitry. PCT-based polycarbonate is a polymeter that provides melt blends that exhibit excellent clarity, toughness, chemical resistance, flow, and gloss.

PCT is differentiated from other thermoplastic polyesters by its higher heat deflection temperature. Continuous use temperatures of up to 150°C (300°F) are possible.

Principal uses for this material are found in the electrical/electronics industries; automotive parts, such as alternator armatures and pressure sensors; optical uses, such as safety goggles; and garden vehicles, such as mower decks and shrouds, tractor hoods, grills, and fenders.

POLYDIMETHYLSILOXANE (PDMS). A silicone polymer developed for use as a dielectric coolant and in solar energy installations. It also may have a number of other uses. It is stated to be highly resistant to oxidation and biodegradation by microorganisms. It is degradable when exposed to a soil environment by chemical reaction with clays and water, by which it is decomposed to silicic acid, carbon dioxide, and water.

POLYDIOXANONE. See **Biomaterials, Prosthetics, and Biomedical Devices**.

POLYELECTROLYTES. These are macromolecules with incorporated ionic constituents. Polyelectrolytes may be cationic or anionic, depending on whether the fixed ionic constituents are positive or negative. Examples of cationic polyelectrolytes are polyvinyl-ammonium chloride and poly-4-vinyl-N-methyl-pyridinium bromide. Examples of anionic polyelectrolytes are potassium polyacrylate, polyvinylsulfonic acid, and sodium polyphosphate. If a polyelectrolyte contains both fixed positive and negative ionic groups, it is called a polyampholyte. Polyelectrolytes may be synthesized by polymerization of a monomer containing the ionic substituent, as for instance the polymerization of acrylic acid to polyacrylic acid, or by attaching the ionic constituent by chemical means to an already existing macromolecule, as for instance in the quaternization of poly-4-vinyl-pyridine with methyl bromide, or in the preparation of sodium carboxymethylcellulose from natural cellulose. Many macromolecules occurring in nature are polyelectrolytes. Examples are gum arabic, which carries carboxylate groups; carrageenin, which contains sulfate groups; proteins, which carry both negative carboxylate and positive ammonium groups; and nucleic acids, which contain negative phosphate groups and basic purine and pyrimidine groups, which acquire positive charges at low pH. Inorganic long-chain polyphosphates have also been isolated from biological materials.

A solution of a polyelectrolyte in water or other suitable solvent conducts an electric current, indicating that the polyelectrolyte is ionized. Transference and electrophoresis measurements show that both the macroion and the counterions (gegenions) contribute to the conductance. Because the counterions are osmotically active, polyelectrolytes show much higher osmotic pressures and diffusion rates than do nonionogenic macromolecules. The osmotic pressure of a polyelectrolyte solution is greatly reduced by the addition of a simple electrolyte which distributes itself among the two sides of the membrane according to the thermodynamic theory of Donnan equilibrium. Polyelectrolytes are called weak if they carry weakly ionized groups such as −COOH, and strong if they carry strongly ionized groups such as −COONa. On titrating polyacrylic acid with sodium hydroxide, the pH increases much more slowly than it does in a corresponding titration of a monocarboxylic acid, thus indicating a pronounced buffering capacity. Even in the case of strong polyelectrolytes, the full osmotic activity of the counterions is not realized; as counterions leave the macroion, the electrostatic potential on the latter builds up making it increasing difficult for additional counterions to escape. This binding effect becomes especially strong with multivalent counterions, whose effective concentration may be rendered several orders of magnitude smaller by a polyelectrolyte than their stoichiometric concentration.

The electric charge on the macroion has several important secondary effects. If the macroion is a flexible chain, intramolecular repulsion between charged segments will stretch out the macroion from a coiled to a more rod-like structure, resulting in much larger solution viscosities than are usually obtained with uncharged polymers under corresponding conditions. Intermolecular repulsion causes the macroions to arrange themselves so that they are as far from each other as is possible. With this ordering, the light scattering which is characteristic of solutions of ordinary macromolecules is greatly diminished, often to the vanishing point, as a result of destructive interference. These secondary effects of charge may be reduced by the addition of simple electrolytes which screen the charged elements from each other, and in some cases also lower the charge by specific counterion binding. At high enough concentrations of added salt, the light scattering of the polyelectrolyte may become sufficiently pronounced to allow its use for the determination of the molecular weight.

An interesting class of polyelectrolytes, denoted by polysoaps, is obtained by attaching soap-like molecules to the polymer chain. Such a polysoap is for instance produced by the quaternization of polyvinyl-pyridine with n-dodecyl bromide. The polysoap molecules differ from ordinary polyelectrolytes in that they may reach protein-like compactness in solution. They behave like prefabricated soap micelles and solubilize hydrocarbons and other compounds insoluble in water.

While the applications of polyelectrolytes for practical purposes depend on their general ionic properties, nevertheless large differences appear among individual members of the class in their applicability to a specific use. When polyelectrolytes are absorbed at interfaces, they affect the zeta-potential and a suspending action may result. Adsorption at growing crystal surfaces is also believed to be the reason for the high effectiveness of small amounts of certain polyelectrolytes in preventing or retarding the precipitation of calcium carbonate. The dispersion of clays by polyelectrolytes is applied in oil-well drilling. The ability of long-chain polyelectrolytes to bind together small particles has found uses in soil conditioning and in the flocculation of phosphate slimes. Because of their effect on the solution viscosity, certain polyelectrolytes are used as thickening agents. Because of their ability to bind di- and trivalent cations, some anionic polyelectrolytes are used in water softening and as enzyme inhibitors. When polyelectrolytes are adsorbed or otherwise incorporated into membranes, they make the latter permselective, hindering small ions of the same charge as the macroion from passing through the membrane while allowing free passage to small ions of opposite charge. The well-known ion-exchange resins are polyelectrolytes which have been cross-linked to prevent them from dissolving.

The most important and widespread use of polyelectrolytes is to aid in the removal of small suspended solids from waste water in the primary, secondary and dewatering stages of treatment.

POLYEMBRYONY. The development of more than one individual from a single fertilized egg cell. In this process the egg breaks up during its early development into several to many component parts, each of which becomes a complete animal. It takes place in the phylum *Bryozoa* in connection with the formation of colonies and has been reported in some of the parasitic insects (*Hymenoptera*). Since the host animal defends itself against the efforts of the female parasite to deposit her eggs in its body, the development of many young from each egg successfully placed in an obvious advantage. The process is akin to an asexual reproduction.

POLYENE. Any unsaturated aliphatic or alicyclic compound containing more than four carbon atoms in the chain and having at least two double bonds. Examples are pentadiene, cyclooctatriene.

POLY(ESTER-AMIDES). See **Biomaterials, Prosthetics, and Biomedical Devices**.

POLYESTER FIBERS. Polyester fiber is the largest volume, most versatile synthetic fiber in the world. The principal characteristics of these fibers are described in the entry on **Fibers**. Polyesters were initially discovered and evaluated in 1929 by W. H. Carothers, who used linear aliphatic polyester materials to develop the fundamental understanding of condensation polymerization, study the reaction kinetics, and demonstrate that high molecular weight materials were obtainable and could be melt-spun into fibers. However, these polymers had limited utility due to low melting points and poor hydrolytic stability. Carothers shifted his attention to aliphatic polyamides, creating nylon as the first commercial synthetic fiber. See also **Nylon**. A decade later, J. R. Whinfield and J. T. Dickson of Calico Printers Association discovered the merits of poly(ethylene terephthalate) (PET) [CAS: 25038-59-9] made from aromatic terephthalic acid [CAS: 100-21-0] and ethylene glycol (2G) [CAS: 107-21-1].

The Whinfield and Dickson patents dominated the art. The U.S. patent rights were assigned to Du Pont, and Imperial Chemical Industries Ltd. (ICI) obtained the rights for the rest of the world. These patents were quickly followed by patents for improved catalysts for exchange or polymerization reactions and for improved fiber properties by drawing. Since the initial discovery of polyester, an enormous amount of resources have been devoted to developing and improving polyester fibers by not only the fiber manufacturers but also by fabric producers and academic institutions. Currently, PET fiber is of great commercial significance and is useful in cordage, apparel fabrics, industrial fabrics, conveyor belts, laminated and coated substrates, and numerous other areas.

The Textile Fiber Product Identification Act (TFPIA) requires that the fiber content of textile articles be labeled. The Federal Trade Commission established and periodically refines the generic fiber definitions. The current definition for a polyester fiber is a manufactured fiber in which the fiber-forming substance is any long-chain synthetic polymer composed of at least 85% by weight of an ester of a substituted aromatic carboxylic acid, including, but not restricted to, terephthalate units, and para substituted hydroxybenzoate units.

Poly (ethylene terephthalate) is the predominant commercial polyester fiber and has been sold under a variety of registered trademark names. In many cases, these trademark names have been licensed, divested, or acquired by other fiber producers so that multiple commercial fiber suppliers have manufactured some trademarked fibers over various time periods. Other commercially produced homopolymer textile fibers include poly(ethylene 4-oxybenyzoate) [CAS: 25248-22-0], poly(1,4-cyclohexane dimethyl terephthalate) [CAS: 24936-69-4] (CHDMT), poly(ethylene naphthalate) (PEN) [CAS: 25853-85-4], poly(butylene terephthalate) (PBT) [CAS: 26062-94-2], and poly(trimethylene terephthalate) (PTT) [CAS: 36619-23-5]. Other polyester homopolymers available for specialty uses include polyglycolide [CAS: 26124-68-5] and polypivalolactone [CAS: 24937-51-7].

Although similar in polymer chemical nature, polylactide [CAS: 26100-51-6] homopolymers and polyether–ester copolymers have been granted separate generics (PLA and Elastoester, respectively) to differentiate these fiber performance properties from those of polyester. Other polyester homopolymers available for specialty uses include polyglycolide [CAS: 26124-68-5] and polypivalolactone [CAS: 24937-51-7]. Similarly, fibers produced from biologically produced polyesters including polyhydroxyl alkanoates and polyhydroxybutyrates are outside the TFPIA definition of polyester fibers and, therefore, outside the scope of this entry.

By the late 1980s, fully aromatic polyester fibers were commercially available for use in composites and structural materials. In general, these materials are thermotropic liquid-crystal polymers (LCP) that are melt-processible to give fibers with tensile and thermal properties considerably higher than conventional polyester textile fibers. Chemically, these materials are homo- or copolymers of aromatic repeat units, eg, poly(p-oxybenzoate) [CAS: 26099-71-8] homopolymers or copolymers of 4-hydroxybenzoic acid [CAS: 99-96-7] with 6-hydroxy-2-naphthoic acid [CAS: 167-64-4].

Most of the polyester textile fiber produced is standard, "fiber grade" molecular weight (~0.6-dL/g intrinsic viscosity) PET with a round filament cross-section. To impart specific properties for different end-uses, many product variants have been developed and commercialized. These variants include modifications to polymer viscosity and molecular weight, modified fiber cross-section, changes in polymer composition by use of comonomers and modifiers, as well as use of additives, eg, delusterants, pigments, and optical brighteners.

High molecular weight polymers are used for high strength industrial fibers in tires, ropes, and belts. High strength and toughness are achieved by increasing the polymer molecular weight from 20,000 to 30,000 Da or higher (DP = 150–200; ~1.0-dL/g intrinsic viscosity) by extended melt polymerization, solid-state polymerization, or chain extension.

The relationship between polymer molecular weight and solution viscosity has been widely studied for polyester polymers. Rheological properties including viscosity, die swell, and melt-fracture parameters of polyester melts depend on polymer molecular weight, melt temperature, and shear rate. Molecular weight distribution, branching agents, and comonomers also significantly impact these parameters. The rodlike structure of aromatic polyester LCP molecules significantly effects the rheological properties of these polymer melts.

Changing the cross-section of polyester fibers by the use of specially designed spinneret capillaries can impact the fabric visual and tactile aesthetics. Fabrics with luster and hand aesthetics ranging from silk to cotton to fur have been produced from polyester fibers with nonround cross-sections.

The use of comonomers is a route to modify dyeability and thermal properites of polyesters. Standard polyester fibers contain no reactive dye sites and are typically dyed by diffusing dispersed dyestuffs into amorphous regions of the fiber structure. The more open fiber structure results in an increased dye rate and the ability to achieve deep dyeability.

The addition of anionic sites to polyester using comonomers comprised of metal sulfonate salts such as sodium dimethyl-5-sulfoisophthalate [CAS: 3965-55-7] has been practiced to make the fibers receptive to cationic dyes. Yarns and fabrics made from mixtures of dispersed and cationically dyed polyester show a visual range from subtle heather to striking contrasts.

In addition to dyeability, copolyesters with a high percentage of comonomer to reduce the melting point have found use as fusible binder fibers in nonwoven fabrics. Specifically designed copolymers have also been evaluated for flame retardant polyester applications.

Properties

The performance and properties of polyester fibers are significantly impacted by the relative amounts of amorphous and crystalline structures, the orientation of the structures with respect to the fiber axis, the size and distribution of the crystalline regions, and the nature and amount of polymerized comonomers. Differences in tensile, recovery, and other physical properties between various polyester systems can be explained by the different molecular structures, the different crystal conformations, mechanical properties of different crystalline regions and by the chain extension within the crystalline unit cell.

Polyester fibers are formed by melt spinning followed by drawing and heat setting to the final fiber form. The molecular orientation and crystalline fine structure developed depend on key process parameters in all steps of the fiber formation process and are critical to the fiber end-use. Molecular orientation and crystallinity generally increase with draw ratio, increasing fiber break tenacity (tensile strength) and Young's modulus while decreasing the fiber break elongation. Fiber stiffness and thermal dimensional stability properties are dependent on the heat-setting process. Relaxed fibers (heat-set under no restraint) show low thermal shrinkage and low initial modulus. Annealed fibers (heat-set under tension at constant length) have a low thermal shrinkage and maintain a high initial modulus.

Polyesters have good resistance to organic and moderate strength mineral acids at room temperature. At elevated temperatures, strength loss in moderate strength acids can be appreciable. Strong acids, eg, concentrated sulfuric acid, can depolymerize and dissolve many polyesters.

Polyester fibers have good resistance to weakly alkaline chemicals and moderate resistance to strongly alkaline materials at room temperature.

Polyester fibers have excellent resistance to soap, detergent, bleach, and other oxidizing agents. The PET fibers are generally insoluble in organic solvents, including cleaning fluids, but are soluble in some phenolic compounds, eg, o-chlorophenol.

The heat of fusion can be determined by differential scanning calorimetry (DSC) and is dependent on crystallinity. For random PET copolymers at low modification levels, the melt point of a wide range of copolymers is reduced roughly 2.5–3.0 °C/mol% modified from PET and is independent of the chemical nature of the comonomer.

The glass transition temperature, T_g, for common polyesters can be affected by the level of crystallinity. Moisture will act as a plasticizer and generally reduces the T_g several degrees. For copolymers, T_g is often estimated as a weighted average of the component materials.

Manufacturing

The basic process for making polyester fibers from the polymer is called *melt spinning*, i.e., heating the polymer above its melting point, forcing it through small holes in a metal plate, and then quenching the molten stream as it issues from the holes by means of a current of cool air. The spun yarn is weak and highly extensible because the polymer molecules are randomly oriented. To impart strength and dimensional stability the yarns must be drawn at temperatures above the glass-transition temperature of the material by pulling the yarn between two *godet wheels*, the second of which is rotating at a speed three to six times as fast as the first. The higher the draw ratio, i.e., the ratio of the two speeds, the more oriented the molecules become and the stronger the yarn.

Applications

The two main classes of polyester fibers are continuous-filament yarns and short-cut fibers, called staple.

Staple. PET staple is widely used in 100% or poly-cotton blend fabrics for apparel. Typical cotton-blend fibers have a linear density of ~0.08–0.4 tex (0.7–3 denier) per filament, a tenacity of ~0.4–0.6 N/tex (5–7 gpd), and a crimp frequency of 3–6 crimps/cm (7–14 crimps/in.). The fibers are coated with ~0.05–0.25 wt % of a finish oil to reduce friction and control static electricity. The fibers are cut to ~25–75 mm (1–3 in.) and packaged into 300–500- kg bales. Light, topweight apparel fabrics are commonly 35–65 wt % polyester while heavier bottomweight fabrics are generally 50 % blends.

Along with cotton blends, polyester blends with rayon or wool are also important. Wool–polyester blends are widely used in men's suiting materials. For these fabrics, PET staple or tow can be used with a linear density typically ~0.16–0.45 tex/filament (1.5–4.0 dpf) and a staple length of 50–75 mm (2–3 in.).

In addition to fabrics, PET staple is used in a wide variety of other applications. High tenacity fibers are widely used in sewing thread. Staple PET fibers have been engineered for use in rugs, carpets, and filling products including furniture, pillows sleeping bags and stuffed toys. These filling materials may be in fiber form or processed to alternate shapes. Polyester staple fibers are commonly used in nonwovens for applications in diaper coverstock, filters, linings, towels, and wipes.

Filament. Fully drawn flat yarns and POY filament yarns are available in yarns sizes from ~3.3–33.0 tex (30–300 denier) with individual filament linear densities of ~0.055–0.55 tex/filament (0.5–5 dpf). The fully drawn (hard) yarns are used directly in fabric manufacturing operations, where POY yarns are used as feedstock for draw texturing. In draw texturing, fibers are drawn and bulked by heat setting twisted yarns or by entangling filaments with an airjet. Both textured and flat yarns are used in apparel, outerwear, sportswear, and automotive fabrics.

High molecular weight polyester is commonly used to make high strength industrial fibers and monofilament fibers. Typical yarns bundle sizes of 111–222 tex (1000–2000 denier) and single filament sizes of 0.55–1.11 tex/filament (5–10 dpf) are available with tenacities on the order of 0.7–1.0 N/tex (8–10 gpd). These fibers are commonly used in applications requiring high strength and stability, including tire cord, seat belts, industrial belts and hoses, ropes, and sailcloth.

Polyesters are also used in continuous filament spunbonded nonwovens. These spunbonded fabrics are available in a wide range of thickness and basis weights and can be used for electrical insulation, coated fabric substrates, geotextiles, and disposable apparel for hospitals and clean rooms.

Additional Reading

Brit. Pat. 578,079 (Aug. 29, 1946), J.R. Whinfield and J.T. Dickson (to Imperial Chemical Industries, Ltd.).

Brugging, W., and J. Ruter, in E. Wilks, ed., *Industrial Polymers Handbook*, Vol. 1, John Wiley & Sons, Inc., New York, NY, 2001, pp. 313–338.

Carothers, W.H., and J.A. Arvin, *J. Am. Chem. Soc.*, **51**, 2560 (1929).

Carothers, W.H., *Chem. Rev.*, **8**, 353 (1931).

Reese, G., and J Kroschwitz: *Encyclopedia of Polymer Science and Technology*, 3rd ed., Vol. 3, John Wiley & Sons, Inc., New York, NY, 2003, pp. 652–678.

Rules and Regulations under the Textile Fibers Product Identification Act, as amended March 2002, Federal Trade Commission, Washington, DC.

Scheirs, J., and T.E. Long: *Modern Polymers*, John Wiley & Sons, Inc., New York, NY, 2003.

U. S. Pat. 2,071,250 (Feb. 16, 1937), W.H. Carothers (to E. I. du Pont de Nemours & Co., Inc.).

U. S. Pat. 2,465,319 (Mar. 22, 1949), J.R. Whinfield and J.T. Dickson (to E. I. du Pont de Nemours & Co., Inc.).

U. S. Pat. 2,534,028 (Dec. 12, 1950), E.F. Izard (to E. I. du Pont de Nemours & Co., Inc.).

U. S. Pat. 2,518,283 (Aug. 8, 1950), E.F. Casassa (to E. I. du Pont de Nemours & Co., Inc.).

U. S. Pat. 2,578,660 (Dec. 18, 1951), L.A. Auspos and J.B. Dempter (to E. I. du Pont de Nemours & Co., Inc.).

U. S. Pat. 2,647,885 (Aug. 4, 1953), H.R. Billica (to E. I. du Pont de Nemours & Co., Inc.).

U. S. Pat. 2,662,093 (Dec. 8, 1953), H.R. Billica (to E. I. du Pont de Nemours & Co., Inc.).

U. S. Pat. 2,533,013 (Dec., 5, 1950), H.F. Hume (to E. I. du Pont de Nemours & Co., Inc.).

U. S. Pat. 2,556,295 (Jun. 12, 1951), A. Pace (to E. I. du Pont de Nemours & Co., Inc.).

U. S. Pat. 2,578,899 (Dec. 18, 1951), A. Pace (to E. I. du Pont de Nemours & Co., Inc.).

Web References

A Quick Guide to Manufactured Fibers, http://www.fibersource.com.

Melt Spinnerets–Products, http://www.nippon-nz.com/english/.

Polyester Fiber Manufacturer Directory for textile industry, http://www.apparelsearch.com/polyester.htm

PTT Fibers, http://www.technica.net/NF.

Sussna Spinnerets, http://www.sussna.de.

POLYESTER FILM. Continuously extruded polyester sheet of various thickness, especially useful in electrical equipment because of its high resistivity. Its tensile strength of 25,000 psi is much greater than that of other plastic films. Sensitized polyester film is used in magnetic tapes, in the photocopying technique known as repography.

POLYESTER RESINS. Any of a group of synthetic resins, which have polycondensation products of dicarboxylic acids with dihydroxy alcohols. They are thus a special type of alkyd resin but, unlike other types, are not usually modified with fatty acids or drying oils. The outstanding characteristics of these resins are their ability, when catalyzed, to cure or harden at room temperature under little or no pressure. Most polyesters now produced contain ethylenic unsaturation, generally introduced by unsaturated acids. The unsaturated polyesters are usually cross-linked through their double bonds with a compatible monomer, also containing ethylenic unsaturation, and thus become thermosetting. Flame resistance is imparted by using either acid or glycol ingredients having a high content of halogens, e.g., HET acid.

The principal unsaturated acids used are maleic and fumaric. Saturated acids, usually phthalic and adipic, may also be included. The function of these acids is to reduce the amount of unsaturation in the final resin, making it tougher and more flexible. The acid anhydrides are often used if available and applicable. The dihydroxy alcohols most generally used are ethylene, propylene, diethylene, and dipropylene glycols. Styrene and diallyl phthalate are the most common cross-linking agents. Polyesters are resistant to corrosion, chemicals, solvents, etc.

Common uses for polyester resins are: reinforced plastics; automotive parts; boat hulls; foams; encapsulation of electrical equipment; protective coatings; ducts; flues; and other structural applications; low pressure laminates; magnetic tapes; piping; bottles; non-woven disposable filters and low-temperature mortars.

See also **Alkyd Resins**; and **Polyester Fibers**.

Additional Reading

Scheirs, J., and T.E. Long: *Modern Polymers*, John Wiley & Sons, Inc., New York, NY, 2003.

Deligny, P., and N. Tuck: *Resins for Surface Coatings, Alkyds & Polyesters*, Vol. 22, 2nd Edition, John Wiley & Sons, Inc., New York, NY, 2001.

POLYETHER-ETHERKETONE. Abbreviated PEEK, polyether-ether-ketone is a high-temperature resistant thermoplastic suitable for wire coating, injection molding, film, and advanced composite fabrication. The wholly aromatic structure of PEEK contributes to its high-temperature performance. Its crystalline character gives it important advantages, including resistance to organic solvents and dynamic fatigue, and retention of ductility on short-term heat aging. The material is available as dry, free-flowing granules and exhibits very low water absorption. Continuous service at temperatures up to 470 °F (243 °C) and intermittent use up to 600 °F (316 °C) are possible. PEEK has good resistance to aqueous reagents, with long-term performance in super-heated water at 500 °F (260 °C). It resists attacks over a wide pH range, from 60% sulfuric acid to 40% sodium hydroxide at elevated temperatures. Attack can occur with some concentrated acids. No organic solvent attack has been observed on molded parts, although a limited range of solvents will stress-craze highly stressed PEEK-coated wire. Radiation resistance is excellent. Typical applications include wire and cable, automotive engine parts, aerospace components, valve plates, valve linings, oil well data logging tools, bearings, woven monofilament, and film.

POLYETHERIMIDE. This is an amorphous, high-performance thermoplastic. The material is characterized by high strength, rigidity, heat resistance, dimensional stability, and electrical properties, combined with broad chemical resistance and processibility. Unmodified polyetherimide is amber-transparent in color and exhibits inherent flame resistance and low smoke evolution without the use of additives. The material is commercially available in several grades—unreinforced and in 10, 20, 30, and 40% glass fiber reinforced formulations for general-purpose molding and extrusion. Also available are easy-flow and release grades, wear-resistant grades, carbon fiber-reinforced grades for high-strength and static dissipation, along with a family of high-heat grades. A relatively new family of polyetherimide blends is available for use in vapor-phase soldering environments and for high-impact applications.

Polyetherimide has a chemical structure based on repeating aromatic imide and ether units. High performance strength characteristics at high temperatures are provided by rigid imide units, while the ether linkages confer the chain flexibility required for good melt processibility and flow. Polyetherimide is resistant to a wide range of chemicals, including most hydrocarbons, alcohols, and fully halogenated solvents. It is resistant to mineral acids and tolerates short-term exposure to mild bases. These resins are rated for 170–180°C (338–356°F) in continuous-use applications. Intermittent use at 200°C (392°F) is possible.

Resins are used in electronic/electrical applications (connectors, circuit boards that are vapor and wave solderable, microwave transparent radomes, integrated circuit chip carriers, miniature switches, explosion proof enclosures, lamp reflectors, and high-precision fiber optic components). Polyetherimide is used for medical components that require all forms of sterilization. Other uses are found in the transportation field, dual-ovenable cookware, as well as bearings, fasteners, and advanced composites.

POLYETHYLENE. [CAS: 9002-88-4]. A thermoplastic molding and extrusion material available in a wide range of flow rates (commonly referred to as melt index) and densities. Polyethylene offers useful properties, such as toughness at temperatures ranging from −76 to +93°C, stiffness, ranging from flexible to rigid, and excellent chemical resistance. The plastic can be fabricated by all thermoplastic processes.

Polyethylenes are classified primarily on the basis of two characteristics, namely, density and melt index. The former is the criterion used to distinguished the type; and the latter for the designation as to category (ASTM-D-1248). ASTM type I polyethylene (sp. gr. 0.910–0.925) is commonly referred to as low-density, conventional, or high-pressure polyethylene. ASTM type II polyethylene (sp. gr. 0.926–0.940) is commonly referred to as medium-density or intermediate-density polyethylene. ASTM type III polyethylene (sp. gr. 0.941–0.965) is commonly called high-density, linear, or low-pressure polyethylene. High-density type III polyethylene has been divided into two ranges of density: 0.941–0.959 (considered type III); and 0.960 and higher, commonly considered type IV. Within each density classification, products with different melt indexes are categorized numerically as follows: category 1 has a melt index (MI) greater than 25; category 2 has an MI greater than 10 to 25; category 3, MI > 1.0 to 10; category 4, MI > 0.4 to 1.0; category 5 has a 0.4 maximum.

Chemical Composition. Polyethylene is formed from the polymerization of ethylene under specific conditions of temperature and pressure and in the presence of a catalyst, according to:

$$
\begin{array}{c}
\text{H} \quad \text{H} \\
| \quad\quad | \\
\text{C}=\text{C}- \\
| \quad\quad | \\
\text{H} \quad \text{H}
\end{array}
\;\xrightarrow[\text{catalyst}]{\text{pressure}}\;
\left[
\begin{array}{c}
\text{H} \;\; \text{H} \;\; \text{H} \;\; \text{H} \\
| \;\; | \;\; | \;\; | \\
-\text{C}-\text{C}-\text{C}-\text{C}- \\
| \;\; | \;\; | \;\; | \\
\text{H} \;\; \text{H} \;\; \text{H} \;\; \text{H}
\end{array}
\right]_n
$$

The reaction is exothermic and may form polymer from a molecular weight of 1000 to well over 1 million. The high-pressure process, which normally produces types I and II, uses oxygen, peroxide, or other strong oxidizers as catalyst. Pressure of reaction ranges from 15,000 to 50,000 psi (~1,020–3,400 atmospheres). The polymer formed in this process is highly branched, with side branches occurring every 15–40 carbon atoms on the chain backbone. Crystallinity of this polyethylene is approximately 40–60%. Amorphous content of the polymer increases as the density is reduced.

The low-pressure processes, such as slurry, solution, or gas phase, can produce types I, II, III, and IV polyethylenes. Catalysts used in these process vary widely, but the most frequently used are metal alkyls in combination with metal halides or activated metal oxides. Reaction pressures normally fall within 50 to 500 psi (~3.4–34 atmospheres). Polymer produced by this process is more linear in nature, with branching occurring about every 1000 carbon atoms. Linear polyethylene of types I and II is approximately 50% crystalline and types III and IV are as high as 85% crystalline.

Ethylene has been polymerized with other monomers, e.g., propylene, butene-1, hexene, ethyl acrylate, vinyl acetate, and acrylic acid, to develop such specific properties as environmental stress crack resistance, low-temperature toughness, and improved flexibility and toughness. High-molecular-weight (HDPE) and chlorinated polyethylenes have been developed to extend the property range of polyethylenes from extremely rigid to elastomeric.

Applications

Polyethylene products include extruded films for food packaging (baked goods, frozen foods, produce); nonfood packaging (heavy-duty sacks, industrial liners, shrink and stretch pallet wrap); nonpackaging (agricultural, diaper liners, industrial sheeting, trash bags); extrusion coating of films, foils, paper, and paperboard; blow molding of bottles, drums, tanks, toys, and pails; injection molding of industrial containers, closures, housewares, toys; extrusion of electrical cable jacketing, pipe, sheet, and tubing; and rotational molding of tanks, drums, toys, and sporting goods.

Properties

Tensile strength, hardness, chemical resistance, surface appearance, and flexural modulus increase with an increase in density (from type I through type IV).

Polyethylene is translucent to opaque white in thick sections, opacity increasing with density. Relatively clear film can be extruded from polyethylene, especially if it is quenched rapidly. The plastic accepts pigmentation readily. Most coloring is performed using dry-blend techniques. Color dispersion devices are required to ensure thorough mixing of resin and pigment.

Mechanical properties of polyethylenes vary with density and melt index. Low-density polyethylenes are flexible and tough; high-density products are quite rigid and have creep resistance under load. Toughness is the primary mechanical property affected by melt index, with lower-melt-index polyethylenes having greater toughness. Under loads, polyethylene is subject to creep, stress relaxation, or a combination of both.

Excellent dielectric characteristics at all frequencies and high electrical resistivity have made polyethylene one of the most important insulating materials for wire and cable.

At no-load conditions, polyethylene has good heat resistance. However, small loads can cause distortion at relatively low temperatures. Dimensional stability of polyethylene is fair to good. Dimensional changes caused by crystallization during cooling usually occur in a non-uniform pattern, resulting in warpage. Narrower molecular weight distribution resins within given families result in less warpage. Types I and II polyethylenes produced by the low-pressure process offer significant improvement in heat distortion temperatures. This property is directly related to melting point and is much higher for low-pressure, low-density resins than for conventional LDPE resins. This allows molded parts to be exposed to significantly higher service temperatures, e.g., dishwasher parts, without undergoing distortion or warpage. Most shrinkage occurs within 48 hours after fabrication and for type I and type II materials is 0.01–0.03 inch/inch (centimeter/centimeter).

Rupture of molecular bonds by external and internal stress in the presence of certain compounds is referred to as environmental stress cracking. Small molecular fractures in the amorphous regions propagate until visible cracks appear. In time, the part may fail. Chemical agents which accelerate stress cracking in polyethylene include detergents; aliphatic and aromatic hydrocarbons; soaps; animal, vegetable, and mineral oils; ester-type plasticizers; organic acids; and aldehydes, ketones, and alcohols. There is no adequate test for stress cracking.

Deterioration occurs in uncolored polyethylene exposed to weather. Ultraviolet light causes photoactivated oxidation. Satisfactory weathering formulations contain 2–2.5% well-dispersed carbon black and stabilizers. The carbon black prevents ultraviolet light penetration.

Unmodified polyethylenes are flammable and are classified in the slow-burning category by the National Board of Fire Underwriters. Burning rate is approximately 1–1.5 inches (2.5–3.8 centimeters) per minute. The flammability of polyethylene may be retarded significantly by the addition of flame retardant compounds, such as antimony trioxide along with halogenated compounds.

At room temperature, polyethylene is insoluble in practically all organic solvents, although softening, swelling, and environmental stress cracking can occur. At high temperatures, some concentrated acids and oxidizing agents chemically attack polyethylene. Above 60 °C, the material becomes increasingly soluble in aliphatic and chlorinated hydrocarbons. Chemical resistance increases slightly as density is increased.

Polyethylene is water-resistant and is a good water vapor barrier. Less than 0.1% water is absorbed in a 2-inch (5-centimeter), 1/8-inch (3-millimeter) thick disk of polyethylene in 24 hours. Transmission of other gases is high when compared with that of most other plastics. Polyethylene is not satisfactory for retention of vacuum.

Processing

Fabrication. Polyethylene is readily fabricated by all methods of thermoplastic processing. The principal methods used are film and sheet extrusion, extrusion coating, injection molding, blow molding, pipe extrusion, wire and cable extrusion coating, rotomolding, and hot melt and powder coatings.

Decorating. Polyethylene parts are decorated by silk screening, hot stamping, or dry offset printing. For satisfactory printing, the surface must be oxidized by hot air, flame, chlorination, sulfuric acid–dichromate solution, or electronic bombardment. Hot air or flame methods are used with molded parts; flame or electronic methods with films. Inks specially made for polyethylene give best results. Roll-leaf hot stamping does not require pretreatment of the surface.

Design. Because of high mold shrinkage, parts must be carefully designed to minimize warpage. Wall cross-sectional thicknesses should be uniform throughout the part. Large flat areas should be avoided. Corners should be curved rather than square. Stiffening ribs should be less than 80% of the thickness of the wall to which they are attached. Thermoformed parts require liberal radii and draft angles. Slight undercuts can be incorporated when a female mold is used. Dimensional variations in a part made of polyethylene are difficult to predict. In general, greater tolerances should be allowed than with more rigid plastics.

See **Olefin Polymers**.

Additional Reading

Gsell, R.A., H.L. Stein, and J.J. Ploskonka: *Characterization and Properties of Ultra-High Molecular Weight Polyethylene*, American Society for Testing & Materials, West Conshohocken, PA, 1998.

Harris, J.M.: *Poly(Ethylene Glycol): Chemistry and Biological Applications*, American Chemical Society, Washington, DC, 1997.

Peacock, A.J.: *Handbook of Polyethylene: Structures, Properties, and Applications*, Marcel Dekker, Inc., New York, NY, 2000.

B. W. HEINEMEYER, The Dow Chemical Company, Freeport, TX

POLY(ETHYLENE CARBONATE). See **Biomaterials, Prosthetics, and Biomedical Devices**.

POLYGLYCOLIC ACID. See **Biomaterials, Prosthetics, and Biomedical Devices**.

POLYGON. A plane figure with n vertices and n sides, also called an n-gon. Depending on the value of n, the following names are used: 3, triangle; 4, quadrilateral; 5, pentagon; 6, hexagon; 7, heptagon; 8, octagon; 9, nonagon; 10, decagon, etc. If all sides are equal, the polygon is equilateral; if all angles equal, equiangular; if all sides and angles are equal, the polygon is regular. Let A be the angle between two sides, B the angle between lines connecting the center of an inscribed or circumscribed circle to two vertices, a the length of a side, R the radius of a circumscribed circle, r the radius of an inscribed circle, and S the area of a regular polygon with n sides, then

$$A = (n-2)\pi/n \text{ radians} \qquad B = 2\pi/n \text{ radians}$$
$$a = 2R \sin B/2 \qquad r = a/2 \cot B/2$$
$$R = a/2 \csc B/2 \qquad S = na^2/4 \cot B/2$$

A convex polygon has no side which produced will enter the polygon. Its angles are called salient and each of them is less than 180°. The term *polygon* usually means a convex polygon. A concave polygon has two or more sides which, when produced, enter the polygon. One or more of its angles are reentrant and greater than 180°. The diagonal of a polygon is a line joining two nonadjoining vertices. A spherical polygon is a part of a sphere bounded by arcs of a great circle.

The sides of a polygon are usually considered to be straight lines but the sides could be curved and then one should call the resulting figure a curvilinear polygon.

See also **Sphere**; and **Triangle**.

POLYHALITE. Polyhalite, $K_2Ca, Mg(SO_4)_4 \cdot 2H_2O$ is a late evaporate mineral associated with halite, sylvite and carnallite from the famous oceanic salt deposits at Stassfurt, Germany, and near Carlsbad, New Mexico. It is of triclinic crystallization, with color grading from gray to brick-red; hardness, 3–3.5; specific gravity 2.78; translucent with vitreous luster; very bitter taste. It is a source of potassium.

POLYHEDRON. A solid with faces formed from plane polygons. The intersections of faces are edges and the points where three or more edges meet are vertices. If the faces are congruent regular polygons and the polyhedral angles are congruent, the polyhedron is regular. There are only five regular polyhedra, which are called the Platonic solids. Their names and the nature of their faces are: tetrahedron, 4 equilateral triangles; hexahedron or cube, 6 squares; octahedron, 8 equilateral triangles; dodecahedron, 12 pentagons; icosahedron, 20 equilateral triangles. See Fig. 1. See also **Parallelepiped**; and **Prism (Mathematics)**.

Fig. 1. A polyhedron with 240 sides. A polyhedron of this type may be referred to as a solid starred small rhombicosidodecahedron and always has 122 vertexes and 360 edges. This is in accordance with Euler's formula. $E + 2 = V + S$, where E = edges, V = vertexes, and S = sides.

TABLE 1. AREA, VOLUME, AND EDGE
RELATIONSHIPS OF POLYHEDRONS

Polyhedron	A/a^2	V/a^3
Tetrahedron	1.7321	0.1179
Cube	6.000	1.0000
Octahedron	3.4641	0.4714
Dodecahedron	20.6457	7.6631
Icosahedron	8.6603	2.1817

A = area of surface; V = volume; a = edge

An important equation for polyhedra, discovered by Descartes and Euler, is $V - E + F = 2$, where V is the number of vertices, E the number of edges, and F the number of faces. With this equation it is easy to see that there are only five regular polyhedra. The equation is more general, however, for it holds for any polyhedron with curved faces and edges, provided it can be deformed continuously into the surface of a sphere. Such considerations are studied in topology. See Table 1.

See also **Topology**.

POLYIMIDES. These are heat-resistant polymers which have an imide group (—CONHCO—) in the polymer chain. Polyimides, poly(amide-imides), and poly(esterimides) are commercially available.

Poly(amide-imides) are prepared by the thermal degradation of a soluble poly[amide-(amic acid)]. The latter may be produced by the condensation of an aliphatic diamine with less than a molar equivalent of pyromellitic dianhydride or with a molar equivalent of a derivative of trimellitic anhydride, such as the acyl chloride in dimethylacetamide as shown in the following equation:

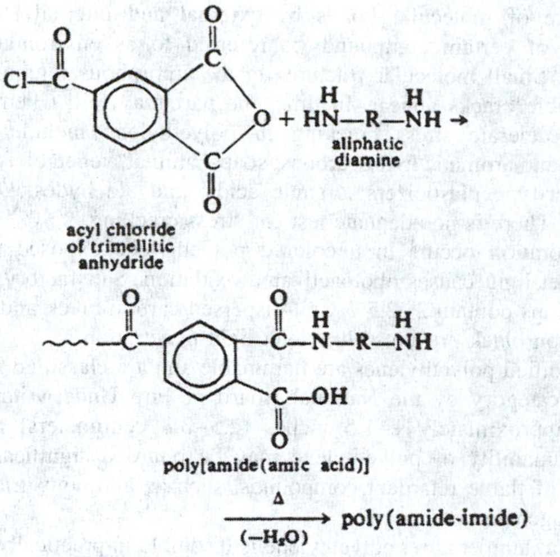

The poly(amide-imides) are soluble in dimethylacetamide but are insoluble in less polar solvents such as toluene and perchloroethylene. They are used for wire enamels, high-temperature adhesives, laminates and molded articles.

The poly(ester-imides) are produced by the thermal decomposition of the soluble poly(amic acids) which are obtained by the condensation of an aromatic diamine and the bis-(ester anhydride) of trimellitic anhydride as shown in the following equation:

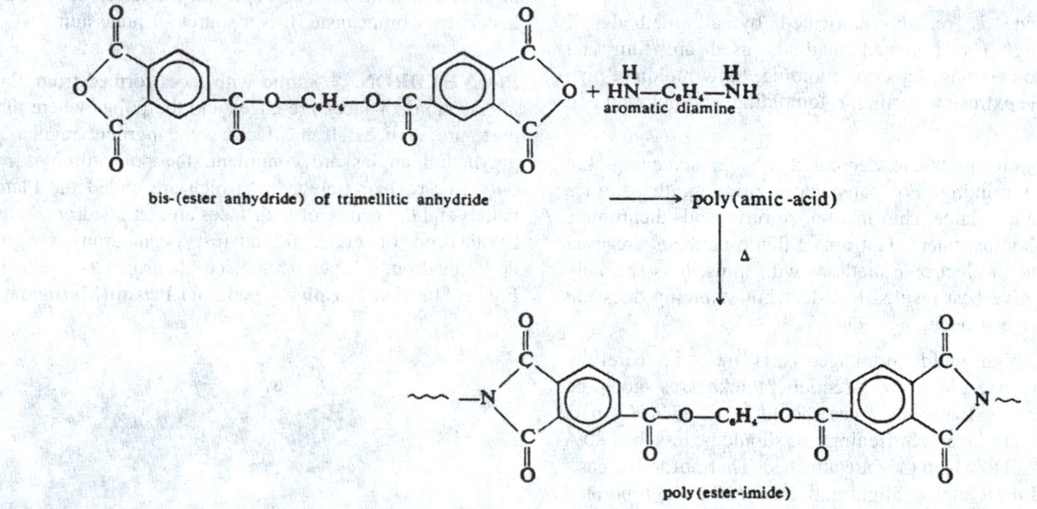

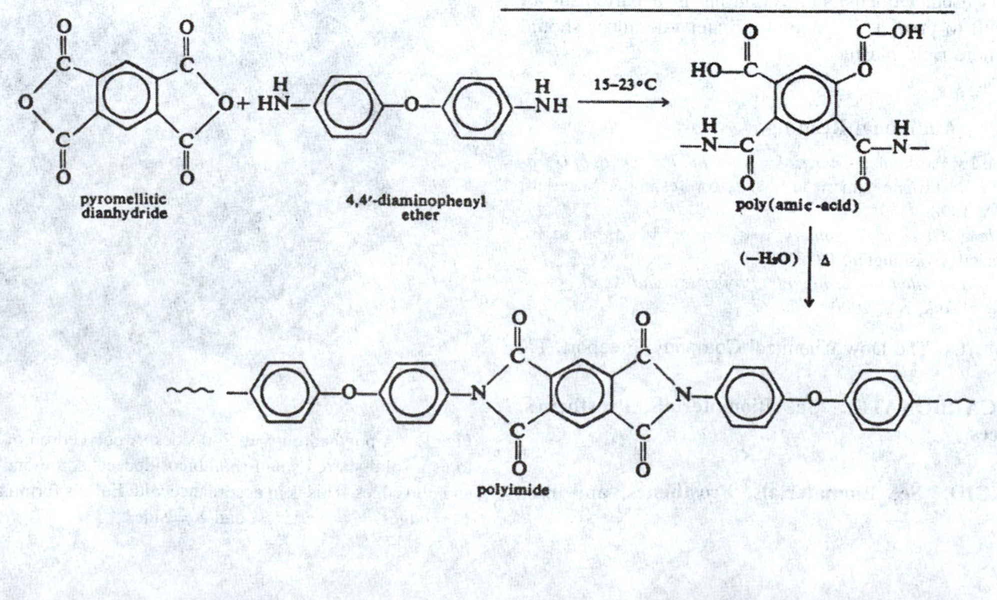

These poly(ester-imides) have good electrical properties. Their tensile-modulus is about 400,000 psi (2759 MPa) at 25 °C and approximately 50 percent of this modulus is retained at 200 °C. Poly(ester-imide) films fail when heated at 240 °C for 1000 hrs.

Polyimides are produced by the thermal dehydration of the soluble poly(amic -acid) which is obtained by the condensation of a diamine, such as 4, 4'-diaminophenyl ether and a dianhydride, such as pyromellitic dianhydride called PMDA as shown in the following equation:

It is customary to apply these polymers as the poly(amic -acids) and to dehydrate the film, coating, fiber or molded forms by heating to produce the polyimides. Polyimides are insoluble in most solvents but are attacked by alkalies, ammonia and amines. These heat resistant polymers are used without fillers and with a graphite filler.

Polyimide films have excellent electrical properties and a tensile modulus of over 400,000 psi at 25 °C. Over 60 percent of this modulis is retained at 200 °C. Polyimide wire enamels are stable for up to 100 thousand hours at 200 °C. Polyimide fibers have a tenacity of 7 g/denier at 25 °C and over 1000 hrs at 283 °C is required to reduce the value to 1 g/denier.

The coefficient of linear expansion of polyimides is $4.0-5.0 \times 10^{-5}$ in./in./°C. The heat deflection is 680 °F (360 °C). Polyimides have been used as binders for abrasive wheels, high-temperature laminates, wire coatings, insulating varnishes and in aerospace applications.

POLYISOPRENE.   The preparation of a synthetic polyisoprene was first reported in 1879 by Bouchardat (1), who treated isoprene [CAS: 78-79-5] obtained from the destructive distillation of natural rubber with hydrochloric acid. This discovery led to a search for a way of converting isoprene into a material duplicating natural rubber (*Hevea brasilienses*). During World War II, scientists extensively studied the polymerization of isoprene with the hope of replicating natural rubber since the United States was temporarily cut off from sufficient natural rubber supplies. These studies were not successful. Finally, in 1954 the B.F. Goodrich Co. was successful in preparing a synthetic *cis*-1,4-polyisoprene [CAS: 9003-31-0] through use of the newly discovered Ziegler transition-metal halide coordination-type catalyst. This catalyst consisted of a trialkylaluminum and titanium tetrachloride. Soon afterward, the Firestone Tire and Rubber Co. revealed a synthesis of *cis*-1,4-polyisoprene with a catalyst based on lithium metal. This catalyst yielded a polyisoprene with about 92% cis-1,4-structure. After achieving the goal of a synthetic *cis*-1,4-polyisoprene duplicating the structure and properties of natural rubber, workers focused on development of economical processes for commercial manufacture of isoprene monomer and for its commercial polymerization. In 1959–1960 the Shell Oil Co. came on stream with the first commercial plant for producing synthetic *cis*-1,4-polyisoprene using a lithium catalyst. Shortly thereafter, The Goodyear Tire & Rubber Co. began commercial production of *cis*-1,4-polyisoprene using a Ziegler catalyst. Later, in 1967, the B.F. Goodrich Co. began commercial production of synthetic polyisoprene using a Ziegler catalyst. See also **Rubber (Natural)**.

POLYLACTIC ACID.   See Biomaterials, Prosthetics, and Biomedical Devices.

POLYMER.   See Colloid System.

POLYMERIZATION.   Chemical reaction, particularly in organic chemistry, which provides very large molecules by a process of repetitive addition. These are of great practical importance in the field of rubbers, plastics, coatings, adhesives and synthetic fibers. The initial materials which give rise to such reactions are called *monomers*; they have molecular weights between 50 and 250 and have certain reactive or functional groups which enable them to undergo *polymerization*. The large molecules, which are formed by a polymerization reaction are called *high polymers*, *macromolecules*, or simply *polymers*; they usually consist of several hundred and in many cases even of several thousand monomeric units and, consequently, have molecular weights of many hundred thousands and even of several millions. The number of monomers contained in a polymer molecule determines its *degree of polymerization* (D.P.).

Polymerization processes never lead to macromolecules of uniform character but always to a more or less broad mixture of species with different molecular weights, which can be described by a *molecular weight distribution function*. The individual macromolecules of such a system belong to a *polymer-homologous series*; the molecular weight and the degree of polymerization of a given material have, therefore, always the character of *average* values.

There exist many ways to assemble small molecules to give large ones and, hence, there exist several types of polymerization reactions. The most important are the following:

1. **Vinyl-type addition polymerization**   Many olefins and diolefins polymerize under the influence of heat and light or in the presence of catalysts, such as free radicals, carbonium ions or carbanions. Free radicals are particularly efficient in starting polymerization of such important monomers as styrene, vinyl chloride, vinyl acetate, methylacrylate or acrylonitrile. The first step of this process—the so-called *initiation* step—consists in the thermal or photochemical *dissociation* of the catalyst, and results in the formation of two *free radicals*:

$$R-R \xrightarrow[\text{or light}]{\text{heat}} 2R\cdot$$

Catalyst molecule   two free radical fragments

The most commonly used catalysts are peroxides, hydroperoxides and aliphatic azocompounds, which need activation energies between 25 and 30 kcal for decomposition.

The free radicals R· attack the monomer and react with its double bond by adding to it on one side and reproducing a new free electron on the other side:

$$R\cdot + CH_2=CHX \rightarrow R-CH_2-CHX\cdot$$

This step is called *propagation* reaction; it adds more and more monomer units to the growing chain and builds up the macromolecules while the free radical character of the chain end is maintained. Each single addition represents the reaction of a free radical with a monomer molecule—a process which requires an activation energy of 8–10 kcal.

Whenever two free radical chain ends collide with each other they can react in such a manner that the resulting products have lost their free radical character and are converted into normal stable molecules. One way is a process of *recombination*:

$$R-(-CH_2-CHX-)-_xCH_2-CHX\cdot$$
$$+ R-(-CH_2-CHX-)-_yCH_2-CHX$$
$$\longrightarrow R-(-CH_2-CHX)-CH_2-CHX$$
$$-CHX-CH_2-(-CH_2-CHX-)-_yR$$

where *one* macromolecule of the degree of polymerization $(X + y + 2)$ is formed. The other is a process of *disproportionation*:

$$R-(-CH_2-CHX)_x-CH_2-CHX\cdot$$
$$+ R-(-CH_2CHX-)-_yCH_2-CHX\cdot$$
$$\longrightarrow R-(-CH_2-CHX-)-_x-CH_2-CHX$$
$$+ R-(-CH_2-CHX-)-_yCH_2=CH_2X$$

where a hydrogen atom moves from one molecule to the other so that one of the two resulting molecules—the $(x + 1)$ mer—has a double bond at its end, whereas the other one—the $(y + 1)$ mer—has a saturated chain end. Reactions in the course of which free radicals are destroyed are called *termination* or *cessation* steps; they convert the transient reactive intermediates into stable polymer molecules.

Vinyl-type addition polymerization can also be carried out with *acidic catalysts* such as boron trifluoride or tin tetrachloride and with *basic catalysts* such as alkali metals or alkali alkyls.

2. Another important kind of addition polymerization is the formation of polyethers by the opening of epoxy ring compounds. Polyoxyethylene (Carbowax) is produced by a sequence of additions of ethylene oxide to an alcohol or amine, as initiator:

$$CH_3-CH_2OH + CH_2-CH_3 \rightarrow$$
$$\underset{O}{\diagdown}$$

$$CH_3-CH_2-O-CH_2-CH_3-OH +$$
$$CH_2-CH_2 \rightarrow$$
$$\underset{O}{\diagdown}$$

$$CH_3-CH_2-O-CH_2-CH_3-O-$$
$$-CH_2-CH_2OH, \text{ etc}$$

No termination reaction occurs in this case; the reaction proceeds until all the monomer is used. This process is catalytically accelerated by the presence of alkali. A similar addition polymerization involving the opening of a ring compound is the conversion of caprolactam into polycaprolactam (Perlon or 6-nylon) under the influence of acidic or basic catalysts. All addition polymerizations are typical *chain reactions* with at least two or three different elementary steps cooperating in building up the resulting macromolecules.

3. There exist other, different classes of reactions which form large molecules, namely processes in the course of which a *small fragment*, usually H_2O, is *split out* of two reacting monomers and where the monomers are chosen in such a manner that the removal of the fragment can be repeated many times. Multi step reactions of this type are called *polycondensations*; they involve the use of at least a pair of bifunctional monomers and proceed by a sequence of identical condensation steps. One important process of this type is the formation of polyesters from glycols and dicarboxylic acids. Thus the progressive removal of water from ethylene glycol and adipic acid leads to a soft, rubbery polyester (Paracon)

$$HOCH_2-CH_2OH$$
$$+ HOOC-(-CH_2-)-_4COOH$$
$$\longrightarrow HOCH_2CH_2-O-CO-(-CH_2-)-_4COOH$$
$$+ HOCH_2-CH_2OH \longrightarrow HOCH_2-CH_2OCO$$
$$= -(-CH_2-)_4-COOCH_2-CH_2OH, \text{ etc.}$$

As long as in processes of this type only *bifunctional* monomers are used, the resulting macromolecules are *linear* and, as a consequence, are of the soluble and fusible type. They can be used as fiber formers, rubbers or thermoplastic resins. If, however, some of the monomers are tri- or tetra-methylolurea, the reaction leads to three-dimensional polymeric networks which are hard and brittle thermosetting resins, such as Bakelite or Glyptal.

The preceeding classification of polymerization reactions concentrates essentially on the organic chemical character of the involved monomers and on the mechanism of their interaction. There exists, however, another classification which is concerned about the manner in which polymerization reactions are carried out in practice and which is of interest and importance whenever industrial application is contemplated. Briefly enumerated here are the most important *polymerization techniques*.

1. Polymerization in the *gas phase* is usually carried out under pressure (several thousand psi) and at elevated temperatures (around $200 \, °C$); the most important example is the polymerization of ethylene to form polythene.
2. Polymerization in *solution*, essentially under normal pressure and at temperatures from $-70 \, °C$ to $70 \, °C$; important examples are the production of butyl rubber with boron trifluoride and the synthesis of the various Vinylites with benzoyl peroxide.
3. Polymerization in *bulk* (or in block) under normal pressure in the temperature range from room temperature to about $150 \, °C$. The batch polymerization of methylmethacrylate to give Lucite or Plexiglass and the continuous polymerization of styrene to give the various types of polystyrene can be quoted as examples.
4. Polymerization in *suspension* (bead or pearl polymerization) under normal pressure in the range from 60 to $80 \, °C$ operates with a suspension of globules of an oil-soluble monomer in water and uses a monomer soluble catalyst. Substantial quantities of polystyrene and polyvinyl acetate are made by this method.
5. Polymerization in *emulsion* under normal pressure and in the temperature range from $-20 \, °C$ to $60 \, °C$ uses a fine emulsion of oil-soluble monomers in water and initiates the reaction with a system of water-soluble catalysts. This method is probably the most important of all, because it is used in very large scale in the copolymerization of butadiene and styrene and in the polymerization of many other monomers, such as chloroprene and vinyl chloride, to produce latices of the various synthetic rubbers.

See also **Molecule**; and several articles which follow.
(*Reprinted from the 4th Edition because the article by Herman F. Mark is such an apt review of the fundamentals of polymerization*)

Additional Reading

Mark, H.F., N. Bikales, and J.I. Kroschwitz: *Encyclopedia of Polymer Science and Technology*, 3rd Edition, Volumes 1–4, Part 1, John Wiley & Sons, Inc., New York, NY, 2003.
Matyjaszewski, K., and T.P. Davis: *Handbook of Radical Polymerization*, John Wiley & Sons, Inc., New York, NY, 2002.
Odian, G.: *Principles of Polymerization*, 4th Edition, John Wiley & Sons, Inc., Hobokon, NJ, 2004.
Stevens, M.P.: *Polymer Chemistry: An Introduction*, 3rd Edition, Oxford University Press, New York, NY, 1998.

POLYMERIZATION (Emulsion). Since an aqueous system provides a medium for dissipation of the heat from exothermic addition polymerization processes, many commercial elastomers and vinyl polymers are produced by the emulsion process. This two-phase (water-hydrophobic monomer) system employs soap or other emulsifiers to reduce the interfacial tension and disperse the monomers in the water phase. Aliphatic alcohols may be used as surface tension regulators.

Formulas for emulsion polymerization also include buffers, free radical initiators, such as potassium persulfate ($K_2S_2O_8$), chain transfer agents, such as dodecyl mercaptan ($C_{12}H_{25}SH$). The system is agitated continuously at temperatures below $100 \, °C$ until polymerization is essentially complete or is terminated by the addition of compounds such as dimethyl dithiocarbamate to prevent the formation of undesirable products such as cross-linked polymers. Stabilizers such as phenyl beta-naphthylamine are added to latices of elastomers.

The final product in latex form may be used for water-type paints or coatings or the water may be removed from the finely divided high-molecular weight polymer. Separation may be brought about by the addition of electrolytes, freezing or spray drying.

It is believed that polymerization of hydrophobic monomers is initiated by free radicals in the aqueous phase and that the surface-active oligomers produced migrate to the interior of the emulsifier micelles where propagation continues. Monomer molecules dispersed in the water phase also solubilize by diffusing — to the expanding lamellar micelles. These micelles disappear as the polymerization continues and the rate may be measured by noting the increase in surface tension of the system.

Additional Reading

Mark, H.F., N. Bikales, and J.I. Kroschwitz: *Encyclopedia of Polymer Science and Technology*, 3rd Edition, Volumes 1–4, Part 1, John Wiley & Sons, Inc., New York, NY, 2003.
Matyjaszewski, K., and T.P. Davis: *Handbook of Radical Polymerization*, John Wiley & Sons, Inc., New York, NY, 2002.
Odian, G.: *Principles of Polymerization*, 4th Edition, John Wiley & Sons, Inc., Hobokon, NJ, 2004.
Stevens, M.P.: *Polymer Chemistry: An Introduction*, 3rd Edition, Oxford University Press, New York, NY, 1998.

POLYMERIZATION (Oxidative-Coupling). A technique for preparation of high-molecular-weight linear polymers. Schematically the reaction is represented below and involves the oxidative coupling of certain organic compounds containing two active hydrogen atoms to give a linear polymer. The hydrogens ultimately, with oxygen, form water. High-molecular-weight polymers have been prepared in this manner from phenols, diacetylenes, and dithiols.

$$nHRH \xrightarrow[\text{catalyst}]{\frac{n}{2} O_2} \left(R\right)_n + n H_2O$$

When 2,6-dimethylphenol is oxidized with oxygen in the presence of an amine complex of a copper salt as catalyst a high-molecular weight polyether (PPO) is formed.

The reaction is exothermic and proceeds rapidly at room temperature. The polymerization is generally performed by passing oxygen or air through a stirred solution of the catalyst and monomer in an appropriate solvent. When the desired molecular weight is attained, the polymer is isolated by dilution of the reaction mixture with a nonsolvent for the polymer. The precipitated polymer is then removed by filtration, washed thoroughly and dried. The polymer is soluble in most aromatic hydrocarbons and chlorinated hydrocarbons and insoluble in alcohols, ketones and aliphatic hydrocarbons.

A large number of other 2,6-disubstituted phenols have been oxidatively coupled.

Polymer formation readily occurs if the substituent groups are relatively small and not too electro-negative. When the substituents are bulky, the predominant product is the diphenoquinone formed by a tail-to-tail coupling. No appreciable reaction occurs when 2,6-dinitrophenol is oxidized even at 100 °C.

A family of engineering thermoplastics based on the above technology includes PPO polyphenylene oxide, Noryl® thermoplastic resins (modified phenylene oxide) and glass reinforced varieties of each. The phenylene oxide resins are characterized by: (1) outstanding hydrolytic stability; (2) excellent dielectric properties over a wide range of temperatures and frequencies; and (3) outstanding dimensional stability at elevated temperatures. Because of these properties modified phenylene oxides are finding major application in the areas of business machine housings, appliances, automotive, TV and communications, electrical/electronic, and water distribution.

Oxidative polymerization of 2,6-diphenylphenol yields a crystallizable polymer that is characterized by a very high melting point ($\sim$480 °C) and excellent electrical properties. It can be spun into a fiber with excellent thermal, oxidative and hydrolytic stability. It is marketed under the trademark Tenax®.

By performing the oxidation at elevated temperatures the phenols which would ordinarily yield polymers are converted instead to diphenoquinones. These quinones are readily reduced to the corresponding hydroquinones, compounds which promise to be useful as antioxidants and polymer intermediates.

Oxidative coupling of diacetylenes yields another unusual class of polymers. From m-diethynylbenzene, for example, is obtained a high-molecular weight polymer that can be cast into a tough, flexible film.

The polymer contains 96.75% carbon and on heating to about 350 °F (177 °C) or above it spontaneously rearranges to an insoluble and infusible material. When ignited the hydrogen in the polymer burns leaving a carbon residue.

In the same manner dithiols can be converted to polydisulfides.

Additional Reading

Mark, H.F., N. Bikales, and J.I. Kroschwitz: *Encyclopedia of Polymer Science and Technology*, 3rd Edition, Volumes 1–4, Part 1, John Wiley & Sons, Inc., New York, NY, 2003.

Matyjaszewski, K., and T.P. Davis: *Handbook of Radical Polymerization*, John Wiley & Sons, Inc., New York, NY, 2002.

Odian, G.: *Principles of Polymerization*, 4th Edition, John Wiley & Sons, Inc., Hobokon, NJ, 2004.

Stevens, M.P.: *Polymer Chemistry: An Introduction*, 3rd Edition, Oxford University Press, New York, NY, 1998.

POLYMERIZATION (Radical). In addition polymerization, polymer is the sole product of the reaction so that the monomer and polymer have essentially the same chemical composition—for example, monomeric styrene and polystyrene. In a polymerization of this type, polymer is formed by a stepwise reaction in which molecules of monomer are added one at a time to a reactive center; the center grows in size while retaining its reactivity.

In a radical polymerization, the reactive centers are free radicals and the process is a typical chain reaction. The monomers in radical polymerizations normally contain carbon–carbon double bonds in their molecules; styrene is typical. Usually radical polymerization is performed in the liquid phase. The chain reaction can be divided into the following steps:

1. **Initiation** Formation of a reactive free radical and its capture by monomer to form a center
2. **Propagation** Reaction of a center with a molecule of monomer to form a larger center
3. **Termination** Deactivation of a center so that it becomes incapable of further growth
4. **Transfer** Reaction of a center with another molecule so that further growth of that particular center is prevented but a new center, capable of growth, is formed

Commonly in radical polymerizations, initiation occurs continuously at a steady rate and is balanced by termination so that a steady concentration of growing centers (usually in the region of 10^{-8} mole/1) is established. The number of propagation reactions greatly exceeds the number of reactions of other types so that *macromolecules* are built up. The life-time of an active center is very much less than the duration of the whole process of polymerization and so the macromolecules are produced even in the earliest stages; there is not a continuous rise in the molecular weight of the polymeric product as found in polymerizations of certain other types. It is instructive to consider in some detail the component reactions in the overall process of radical polymerization.

Initiation

In principle, the simplest method for initiation is to add to the purified monomer a small amount of a substance which dissociates to fairly reactive free radicals. This initiator (or sensitizer) is chosen so that its decomposition occurs at a suitable rate at the working temperature; thus azo*iso*-butyronitrile is commonly used at about 60 °C dissociating according to the equation:

$$(CH_3)_2C(CH_2) \cdot N : N \cdot C(CN)(CH_3)_2 \rightarrow 2(CH_3)_2C(CN) \cdot + N_2$$

The radical adds to monomer thus

$$(CH_3)_2C(CN) \cdot + CH2 : CHX \rightarrow (CH_3)2C(CN) \cdot CH_2 \cdot CHX \cdot$$

forming the starting point of a polymer chain, i.e., an *end-group*; this reaction is the real initiation of polymerization. Initiators of other types are also used, notably peroxides, both organic and inorganic. In some cases, the initiator is chosen to give free radicals under the influence of light; this process can be useful for initiating polymerizations at comparatively low temperatures. At elevated temperatures or under the influence of external sources of energy (light, high-energy radiations, ultrasonics, mechanical work), many monomers polymerize apparently spontaneously without deliberate addition of sensitizer; the mechanisms of initiation under such circumstances are not completely understood.

Propagation

The propagation reaction in a radical polymerization can be represented by the general equation

$$P \cdot CH_2 \cdot CHX \cdot + CH_2 : CHX \rightarrow P \cdot CH_2 \cdot CHX \cdot CH_2 \cdot CHX \cdot$$

which corresponds to the conversion of a carbon–carbon double bond to two carbon–carbon single bonds. The group $-CH_2 \cdot CHX-$ in the polymer chain is referred to as the *monomer* unit. If the growing center includes more than a few monomer units, the characteristics of the growth reaction are reasonably supposed to be independent of the size of the center.

The growth reaction is exothermic (in the region of 20 kcal/mole, i.e., about 80 kj/mole); under some circumstances, polymerizations may become self-heating and difficult to control. The growth reaction involves a decrease in entropy since a free molecule of monomer becomes organized in a polymer chain. The opposing effects of changes in enthalpy and entropy indicate that, for every polymerizing system, there is a *ceiling temperature* below which the growth reaction is favored thermodynamically but above which the reverse process is favored; the value of the ceiling temperature depends on the nature of the monomer and on its concentration in the system.

The growth reaction shown above represents *heat-to-tail* addition; the CHX groups occur at alternate sites along the main polymer chain and the unpaired electron is sited on a substituted carbon atom. Head-to-head addition, to give a polymer radical $P \cdot CH_2 \cdot CHX \cdot CHX \cdot CH_2 \cdot$, may occur occasionally but is likely to be followed by tail-to-tail addition to give $P \cdot CH_2 \cdot CHX \cdot CHX \cdot CH_2 \cdot CH_2 \cdot CHX \cdot$ which can be regarded as the normal growing radical. Head-to-head groupings may well be sites of instability in the polymer.

The substituted carbon atoms in the polymer chain are asymmetric. *Stereoregular polymers* are produced if all these carbon atoms have the same configuration (all *d* or all *l*) or if the *d* and *l* configurations occur alternately; pronounced stereo-regularity is seldom achieved in radical polymerizations except perhaps at very low temperatures. When dienes are polymerized by a radical mechanism, the resulting polymers contain several distinct types of monomer unit, thus butadiene can give rise to $-CH_2 \cdot C(CH:CH_2)-$, $-CH_2 \cdot CH:CH \cdot CH_2 - cis$, and $-CH_2 \cdot CH:CH \cdot CH_2 - trans$.

Termination

In many radical polymerizations, termination occurs by interaction of pairs of growing radicals, either by combination to give $P \cdot CH_2 \cdot CHX \cdot CHX \cdot CH_2 \cdot P$ or by disproportionation to give $(P \cdot CH:CHX + P \cdot CH_2 \cdot CH_2X)$. The relative importances of these alternative processes depend upon the chemical nature of the monomer and, to a lesser extent, upon the temperature in the sense that the chance of disproportionation rises as the temperature is increased. Combination gives rise to a head-to-head grouping in the chain, and disproportionation to some unsaturated end-groups for molecules; both structural features may give rise to instability.

Termination can occur for polymer radicals of any size and so there is inevitably a wide *distribution of sizes* among the final molecules. It is possible to quote only *average molecular weights* for polymers.

The average *chain length* or *degree of polymerization* (DP) of the molecules in a sample of polymer is the average number of monomer units contained in them.

Termination is commonly *diffusion-controlled*, i.e., it is governed by the rate at which the reactive sites in growing radicals can come together rather than by chemical factors. In viscous media, termination may be so seriously impeded that both the overall rate of polymerization and the degree of polymerization increase markedly. In systems where the polymer is insoluble in the reaction medium, polymer radicals may be trapped in the precipitated material and be able to grow but unable to participate in termination processes.

Transfer

The average molecular weight of a polymer produced in a particular system may be substantially reduced by occurrence of some types of transfer reactions. If the system contains certain substances, e.g., mercaptans, a growing polymer radical may abstract hydrogen thus

$$P \cdot + R \cdot SH \rightarrow P \cdot H + RS \cdot$$

giving a dead polymer molecule and a new radical which can react with monomer to reinitiate polymerization. If reinitiation is 100% efficient, the effect of transfer of this type is to reduce the average degree of polymerization without affecting the rate of polymerization or the kinetic chain length. In practice, transfer is commonly accompanied by retardation since some of the new radicals are consumed in side-reactions instead of reacting with monomer; this type of transfer is said to be degradative.

Other components of the polymerization mixture, including monomer and initiator, may engage in transfer reactions. They are particularly significant for allyl monomers for which *degradative transfer* to monomer is of such importance that rates and degrees of polymerization are very low.

Transfer to polymer, causing reactivation of a polymer molecule at some point along its length, leads to the growth of *branches*. The process can occur intermolecularly and also intramolecularly; the latter process is particularly important in the free radical polymerization of ethylene at high pressure where it leads to the production of numerous short branches which considerably affect the properties of the polymer.

Transfer to polymer, the subsequent growth of branches and termination of their growth by combination lead to *cross-linking* whereby the separate polymer molecules are united to form an insoluble three-dimensional network. Cross-linking is however much more likely to occur during the polymerization of those monomers which contain more than one carbon-carbon double bond per molecule. The monomer unit in the polymer first formed still possesses an unsaturated grouping which can participate in another polymerization chain. Certain monomers of this type however engage in a special type of reaction so that reaction of one double bond in a monomer is immediately followed by reaction of the second double bond.

Inhibitors and Retarders

Various substances can reduce the rate at which a monomer is converted to polymer. Inhibitors completely suppress polymerizations whereas retarders only reduce the rate. The former deactivate very readily the primary radicals so that growth of polymer chains cannot begin; the latter deactivate growing polymer radicals so causing premature termination. Inhibitors are commonly used to stabilize monomers during storage. Many nitro compounds and quinones act as inhibitors and retarders.

Copolymerization

A process known as copolymerization can occur if reactive radicals are generated in a mixture of monomers; the resulting polymer molecules contain monomer units of more than one type. Copolymerization is of great significance academically, where it leads to information about the reactivities of monomers and radicals, and also industrially where it is used for the production of materials with special properties. Usually the composition of a copolymer is different from that of the mixture of monomers from which it is derived. For this reason, the average compositions of feed and copolymer drift during the course of a copolymerization.

Additional Reading

Mark, H.F., N. Bikales, and J.I. Kroschwitz: *Encyclopedia of Polymer Science and Technology*, 3rd Edition, Volumes 1–4, Part 1, John Wiley & Sons, Inc., New York, NY, 2003.

Matyjaszewski, K., and T.P. Davis: *Handbook of Radical Polymerization*, John Wiley & Sons, Inc., New York, NY, 2002.

Odian, G.: *Principles of Polymerization*, 4th Edition, John Wiley & Sons, Inc., Hobokon, NJ, 2004.

Stevens, M.P.: *Polymer Chemistry: An Introduction*, 3rd Edition, Oxford University Press, New York, NY, 1998.

POLYMERS. Polymers are very large molecules made by covalently binding many smaller molecules. The word polymer is derived from the Greek *poly* (many) and *meros* (part). The size of polymer molecules imparts many interesting and useful properties not shared by low molecular weight materials. Polymers are the fundamental materials of plastics, rubbers and most fibers, and surface coatings and adhesives, and as such are essential to modern society. Also, many important constituents of living organisms, e.g., proteins and cellulose, are biopolymers. See also **Biopolymers**; **Cellulose**; and **Proteins**.

Classification and Nomenclature

Polymers were initially classified according to their response to temperature. Those that are softened (plasticized) reversibly by heat are known as thermoplastics. Others, though they might initially be liquid or soften once upon heating, undergo a curing (setting) reaction that solidifies them, and further heating leads only to degradation. These are known as thermosets. The ability of polymers to soften and flow at least once is one of their most valuable assets, as it allows them to be formed into complex shapes easily and inexpensively.

In general, polymers are formed by two types of reactions: condensation and addition. The formation of a polyester by polycondensation may be illustrated as follows.

$$x\ \text{HOROH} + x\ \text{HOOCR'COOH} \longrightarrow \text{H} \underset{\text{polyester}}{\left(\text{ORO}-\overset{\overset{\text{O}}{\|}}{\text{C}}\text{R}'\overset{\overset{\text{O}}{\|}}{\text{C}}\right)_x}\text{OH} + (2x-1)\,\text{H}_2\text{O}$$

underneath: diol   diacid

In the polyester formula shown, parentheses enclose the repeating unit. The quantity x is the degree of polymerization, sometimes also called the chain length, the number of repeating units strung together like identical beads on a string. Neglecting the ends of the molecule, which is usually justified for large x, the molecular weight M of the polymer molecule is given by $M = mx$, where m is the molecular weight of the repeating unit. Since x can easily be in the thousands, the term macromolecules is also used to describe these materials.

Addition or chain-growth polymerization involves the opening of a double bond to form new bonds with adjacent monomers, as typified by the polymerization of ethylene to polyethylene:

$$x\text{H}_2\text{C}{=}\text{CH}_2 \longrightarrow (\text{CH}_2{-}\text{CH}_2)_x$$

Because no molecule is split out, the molecular weight of the repeating unit is identical to that of the monomer.

In terms of molecular structure, there are three principal categories of polymers, illustrated schematically in Figure 1. If each monomer is difunctional, that is, can react with other monomers at two points, a linear polymer is formed. Polymers that contain two different repeating units, say A and B, are known as copolymers. A linear polymer with a random (AABBABAAABABB) arrangement of the repeating units is a random or statistical copolymer, or just copolymer. It is termed poly(A-*co*-B), with the primary constituent listed first. A molecule in which the two repeating units are arranged in long, contiguous blocks $([\text{A}]_x[\text{B}]_y)$ is a block (*b*) copolymer, poly (A-*b*-B).

A few points of tri- or higher functionality introduced along the polymer chains, either intentionally or through side reactions, give a branched polymer. Branches may grow from a linear backbone. A branched structure with the backbone consisting of one repeating unit (A) and the branches of another (B), is a graft (*g*) copolymer, poly(A-*g*-B).

As the length and frequency of branches increase, they may ultimately reach from chain to chain. If all the chains are connected together, a cross-linked or network polymer is formed. Cross-links may be built in during the polymerization reaction or may be created chemically or by radiation between previously formed linear or branched molecules (curing or vulcanization).

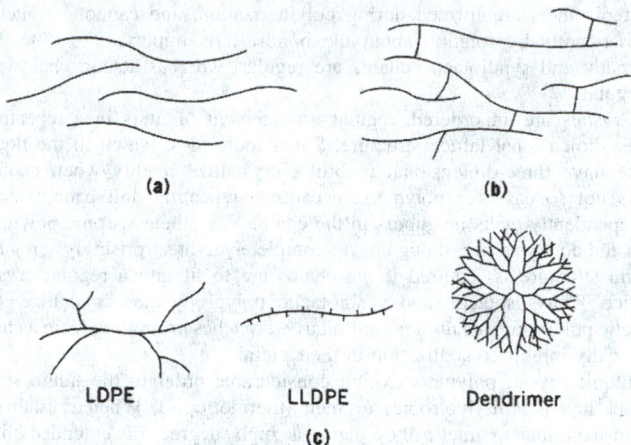

Fig. 1. Schematic diagram of polymer structures: (**a**) linear; (**b**) cross-linked; and (**c**) branched, where LDPE = low density polyethylene and LLDPE = linear low density polyethylene.

Structure and Properties

Various levels of structure ultimately determine the properties of a polymer.

With the exception of some naturally occurring polymers, all linear and branched polymers consist of molecules with a distribution of molecular weights. Two average molecular weights are commonly defined; the number-average, $\overline{M}_n$, and the weight-average, $\overline{M}_v$. It may be shown that $\overline{M}_w \geq \overline{M}_n$. The two are equal only for a monodisperse material, in which all molecules are the same size. The ratio $\overline{M}_w/\overline{M}_n$ is known as the polydispersity index and is a measure of the breadth of the molecular weight distribution.

Most molecular weight characterization now is done by size-exclusion chromatography (sec), also known as gel-permeation chromatography (gpc).

The atoms in a polymer molecule are held together by primary covalent bonds. Linear and branched chains are held together by secondary bonds: hydrogen bonds, dipole interactions, and dispersion or van der Waal's forces. By copolymerization with minor amounts of acrylic (CH$_2$=CHCOOH) or methacrylic acid followed by neutralization, ionic bonding can also be introduced between chains. Such polymers are known as ionomers.

Secondary bonds are considerably weaker than the primary covalent bonds. When a linear or branched polymer is heated, the dissociation energies of the secondary bonds are exceeded before the primary covalent bonds are broken, freeing up the individual chains to flow under stress. When the material is cooled, the secondary bonds reform. Thus, linear and branched polymers are generally thermoplastic. On the other hand, cross-links contain primary covalent bonds like those that bond the atoms in the main chains. When a cross-linked polymer is heated sufficiently, these primary covalent bonds fail randomly, and the material degrades. Therefore, cross-linked polymers are thermosets. There are a few exceptions such as cellulose and polyacrylonitrile. Though linear, these polymers are not thermoplastic because the extensive secondary bonds make up for in quantity what they lack in quality.

Similarly, polymers dissolve when a solvent penetrates the mass and replaces the interchain secondary bonds with chain-solvent secondary bonds, separating the individual chains. This cannot happen when the chains are held together by primary covalent cross-links. Thus, linear and branched polymers dissolve in appropriate solvents, whereas cross-linked polymers are insoluble, although they may be swelled considerably by absorbed solvent. The extent of swelling is inversely related to the degree of cross-linking.

Vinyl monomers, CH$_2$=CHR, generally polymerize in a head-to-tail fashion, placing the R group on every other carbon atom in the chain backbone. If a chain is conceptually stretched out, the carbon atoms in the backbone will lie in a plane. The arrangement in which the R groups are all on one side of that plane is the isotactic stereoisomer. Regular alternation of the R groups from side to side is the syndiotactic form. Random placement of the R groups is the atactic (without order) polymer.

Stereoisomers are formed during polymerization, and cannot be altered subsequently by rotation about the bonds. It is important to note that isotactic and syndiotactic chains are regular, whereas atactic chains are irregular.

Crystals are an ordered, regular arrangement of units in a repeating, three-dimensional lattice structure. Small molecules, which in the liquid state have three-dimensional mobility, crystallize readily when cooled. It is not so easy for polymers, because a repeating unit cannot move independently of its neighbors in the chain. Nevertheless, some polymers can and do crystallize, though never completely. Not surprisingly, a regular chain structure is required if the chains are to fit into a regular crystal lattice. Thus, isotactic and syndiotactic polypropylenes crystallize, but atactic polypropylene does not. Similarly, branches protruding from a chain sterically inhibit crystallization in their vicinity.

Liquid-crystal polymers exhibit considerable order in the liquid state, either in solution (lyotropic) or melt (thermotropic). When crystallized from solution or melt, they have a high degree of extended-chain crystallinity, and thus have superior mechanical properties.

Not all polymers crystallize, and even those that do are not completely crystalline. Noncrystalline polymer is termed amorphous. Four types of molecular motion have been identified in amorphous polymers. Listed in order of decreasing activation energy, they are *(1)* translational motion of entire molecules, *(2)* coiling and uncoiling of 40–50 C-atom segments of chains, *(3)* motion of a few (five to six) atoms along the main chain or on side groups, and *(4)* vibrations of individual atoms.

In polymers that can crystallize, the ratio of crystalline to amorphous material has a profound effect on properties. Because the chains are packed more tightly and efficiently in the crystalline areas than in the amorphous, the crystalline phase has a higher density and greater mechanical strength. In fact, density is a common measure of degree of crystallinity.

The stiffest polymers are both crystalline and have a glassy amorphous phase. They are often useful as engineering (structural) plastics. Nylon-6,6 and crystalline poly(ethylene terephthalate) are examples of this class of materials, as is the newly developed syndiotactic polystyrene.

Pure amorphous polymers, being homogeneous materials, are transparent. Atactic polystyrene is a good example. The crystalline syndiotactic form is not transparent. A lack of transparency does not necessarily indicate crystallinity, however. It can also be caused by inorganic fillers, pigments, gas bubbles (as in a foam), a second polymer phase, etc.

Cross-linking eliminates polymer solubility. A rough guide to solubility is that like dissolves like, i.e., polar solvents tend to dissolve polar polymers and nonpolar solvent dissolve nonpolar polymers. See also **Olefins Polymers, Higher Olefins**.

Polymer Synthesis

Step-Growth Polymerization. Step-growth polymerization is characterized by the fact that chains always maintain their terminal reactivity and continue to react together to form longer chains as the reaction proceeds, i.e., $x - \text{mer} + y - \text{mer} \rightarrow (x + y)$-mer. Because there are reactions that follow this mechanism but do not produce a molecule of condensation, the terms step-growth and polycondensation are not exactly synonymous.

Chain-Growth Polymerization. Chain-growth polymerizations are characterized by chains that propagate by adding one monomer molecule at a time, i.e., $x - \text{mer} + \text{monomer} \rightarrow (x + 1)$-mer. There are, however, several mechanisms by which this occurs.

In free-radical addition polymerization, the propagating species is a free radical. The free radicals are most commonly generated by the thermal decomposition of a peroxide or azo initiator. See also **Initiators (Free-Radical)**.

Unlike step-growth polymerization, free-radical chains do not continue to grow as the reaction proceeds. The average lifetime of a growing chain, from initiation to termination, is typically less than a second. Thus, high molecular weight polymer is produced right from the beginning.

Free-radical polymerization is carried out in a variety of ways.

Bulk polymerization involves only monomer and initiator. It gives the greatest polymer yield per unit of reactor volume and a very pure polymer.

In solution polymerization an inert solvent is added to the reaction mass. The solvent adds its heat capacity and reduces the viscosity, facilitating convective heat transfer.

In suspension polymerization, the organic reaction mass is dispersed in the form of droplets 0.01–1 mm in diameter in a continuous aqueous phase. Each droplet is a tiny bulk reactor. Heat is readily transferred from the droplets to the water, which has a large heat capacity and a low viscosity, facilitating heat removal through a cooling jacket.

In emulsion polymerization the organic monomer is emulsified with soap in an aqueous continuous phase.

Ionic Polymerization. Addition polymerization may also be initiated and propagated by anions. Ionic polymerizations are almost exclusively solution processes.

There are some important differences between anionic and free-radical addition. First, unlike free-radical initiators, which decompose and start chains randomly throughout the course of the reaction, anionic initiators ionize readily in fairly polar organic solvents or at low concentrations in hydrocarbons, and chains are started immediately, one for each molecule of initiator. Second, in the absence of impurities, there is no termination.

When the initial monomer supply is exhausted, the anionic chain ends retain their activity. Thus, these anionic chains have been termed living polymers. If more monomer is added, they resume propagation. If it is a second monomer, the result is a block copolymer.

Cationic polymerization has been used commercially to polymerize isobutylene and alkyl vinyl ethers, which do not respond to free-radical or anionic addition. See also **Elastomers**; and **Rubber, Synthetic**.

Stereospecific Polymerization. In the early 1950s, Ziegler observed that certain heterogeneous catalysts based on transition metals polymerized ethylene to a linear, high density material at modest pressures and temperatures. Natta showed that these catalysts also could produce highly stereospecific poly-α-olefins, notably isotactic polypropylene, and polydienes. They shared the 1963 Nobel Prize in chemistry for their work. More recently, metallocene catalysts that provide even greater control of molecular structure have been introduced.

Additional Reading

Allcock, H., F. Lampe, and J. Mark: *Contemporary Polymer Chemistry*, 3rd Edition, Prentice Hall, Inc., Upper Saddle River, NJ, 2003.

Bahadur, P., and N.V. Sastry: *Principles of Polymer Science*, CRC Press LLC., Boca Raton, FL, 2002.

Bower, D.I.: *Introduction to Polymer Physics*, Cambridge University Press, New York, NY, 2002.

Brandrup, J., and E.H. Immergut, eds.: *Polymer Handbook*, 3rd Edition, Wiley-Interscience, New York, NY, 1989.

Brandrup, J.D.R. Bloch, E.A. Grulke, E.H. Immergut, and A. Abe: *Polymer Handbook*, 2 Vol., 4th Edition, John Wiley & Sons, Inc., New York, NY, 2003.

Cheremisinoff, N.P.: *Condensed Encyclopedia of Polymer Engineering Terms*, Elsevier Science & Technology Books, New York, NY, 2001.

Flory, P.J.: *Principles of Polymer Chemistry*, Cornell UP, Ithaca, N.Y., 1953.

Fried, J.: *Polymer Science and Technology*, 2nd Edition, Prentice Hall Professional Technical Reference, Upper Saddle River, NJ, 2003.

Kroschwitz, J.I.: *Encyclopedia of Polymer Science and Technology*, **12** Volume Set, 3rd Edition, John Wiley and Sons, Inc., Hoboken, NJ, 2004.

Morawetz, H.: *Polymers: The Origins and Growth of a Science*, Dover Publications, Inc., Mineola, NY, 2002.

Odian, G.: *Principles of Polymerization*, 3rd Edition, Wiley-Interscience, New York, NY, 1991.

Rosen, S.L.: *Fundamental Principles of Polymeric Materials*, 2nd Edition, Wiley-Interscience, New York, NY, 1993.

Rubinstein, R., and R.H. Colby: *Polymer Physics*, Oxford University Press, New York, NY, 2003.

Solomons, T.W. Graham, C.B. Fryhle, and M.M. Shenkman: *Organic Chemistry*, 8th Edition, John Wiley & Sons, Inc., New York, NY, 2003.

Scheirs, J., and T.E. Long: *Modern Polymers*, John Wiley & Sons, Inc., New York, NY, 2003.

Walton, D.J., and J.P. Lorimer: *Polymers*, Oxford University Press, New York, NY, 2001.

POLYMERS (Inorganic). Most inorganic materials can be considered polymeric since they are built up of a relatively simple atomic grouping repeated a very large number of times. Metals and simple ionic materials are easily excluded, but there still remains a large group of covalently bonded, regularly repeating materials. For example, many mineral silicates are based on the monomer $[SiO_4]^{4-}$ which is covalently bonded to form the large, two-dimensional sheets from which these materials are built. Still, one does not normally think of most of these inorganic, covalently bonded polymeric materials as polymers, because their behavior is so different from is expected of organic polymers. Such properties as high viscosity in the melt and in solution, rubbery elasticity, moldability, ability to form fibers, films, and so on, are not possessed by most of these materials. In a few cases, enough of them are present to suggest the underlying similarity in structure, for example, in the silicate minerals, crysotile asbestos forms fibers of excellent textile quality. Such samples show the possibility of obtaining useful inorganic polymers.

In the light of this discussion inorganic polymers will be considered to be those materials in which the main polymiric chain contains no organic carbon and in which behavior similar to that of organic polymers can be developed.

The question "Why is there such a difference in behavior between the usual inorganic and organic polymeric materials?" is helpful in guiding such a development. The contrast must be due to differences in molecular structure. For example, in the case of quartz the $[SiO_4]^{4-}$ tetrahedra are covalently bonded together. The high regularity of the structure and the large number of cross-links per $[SiO_4]^{4-}$ unit lead to a material which is strong and dimensionally stable, but brittle. The same situation of over-crosslinking can be found with organic polymers. If the number of cross-links in quartz is reduced by substituting organic groups, such as methyl, for some of the oxygen-silicon linkages, the silicone polymers are produced. These polymers, the only commercial inorganic ones, show that inorganic materials which behave as organic polymers can be made. However, a number of obstacles are found which are not as troublesome with organic polymers. For example, six to eight membered rings are more stable than long-chains. In the case of organic materials, if chains can be formed initially, they have considerable stability. With inorganic materials, the bonds are much more labile (constantly forming and breaking) and the long chain may break down to a collection of smaller rings.

Other factors which influence the properties of polymers can be illustrated by examining the bond energies or bond strengths and the ionic character of bonds based on Si as contrasted to similar ones based on C.

From the bond energies one would expect the homo-atomic silane polymers with Si—Si bonds to be considerably less stable than the more familiar C—C chain polymers. This expectation fits the observed facts. On the other hand, one could expect little gain in stability in the carbon series by going to an ether linked chain (—C—O—C—), while in the silicon series a silicon-oxygen linkage is stronger than any of the others. This is reflected in the very good stability of the silicone polymers. From bond energies one might also expect that a chain of alternating Si and N atoms would have good stability.

Bond Energies

Si—Si	53 kcal	C—C	83 kcal
Si—O	106	C—O	86
Si—N	82	C—N	73
Si—C	78		

Ionic Character

Si—Si	0	C—C	0
Si—O	51	C—O	22
Si—N	30	C—N	7
Si—C	12		

"In addition to pure thermal stability, if the polymer is to be heated in air, one must also consider oxidative stability. In the carbon series oxidation always leads to more stable species and tends to occur, but in the silicon series there is a much higher tendency towards reaction with oxygen. This is the principal reason for the low utility of the silane polymer. Finally, a third factor in polymer stability is the ease of attack by solvents, acids, bases, etc. This is largely determined by the ionic nature of the bonds involved. The silica-based polymers should be more susceptible to such attack than carbon, since they have a higher percent of ionic nature."

Acidic or basic water solutions attack silicones when they are heated together under pressure. Their resistance is still high, however, because of other details of the way the polymer molecules are bound together.

The polymers which have been used to illustrate problems of inorganic polymer formation have been heteroatomic, that is, their chains are built from different atoms alternating with each other. The other structure mentioned has been homoatomic — all the atoms in the chain are the same. There are only a few homoatomic polymers of any promise. Most elements will form only cyclic materials of low molecular weight if they polymerize at all. In addition to the silane polymers, black phosphorus, a high-pressure modification of the element, forms in polymeric sheets.

Boron has similar tendencies in its compounds. The outstanding member of this class is sulfur. A transition from S_8 rings to long sulfur chains takes place over a narrow temperature range around 159 °C. An increase in the viscosity of the liquid by 2000 times or more, within a range of 25°, is the tangible evidence of polymerization. The material also forms rubbery, plastic and fibrous forms when chilled to room temperature. However, it has a strong tendency to revert to the cyclic form unless stable groups are placed at the end of the chain or copolymerization hinders the process. Attempts to improve the stability of polymeric sulfur have met with some success. This is the only homoatomic inorganic polymer which appears technically interesting at present.

The class of heteroatomic polymers, besides containing the silicones, offers more promise for useful materials. Most nonmetals and many of the less positive metals form heteroatomic compositions. In many cases they are high polymers. The silicones themselves behave quite the same as organic polymers and are used as oils, rubbers and resins. The rubber is vulcanized either by the reaction of organic peroxides with the methyl groups on the chain or by incorporating groups such as Si—OH or Si—OR which crosslink on exposure to moisture in the atmosphere. The properties in which they excel over organic polymers are high thermal stability, resistance to oxidation and inertness to organic reagents. These are usually the special properties one hopes to get from inorganic polymers. The polymers may be modified by substitution of other groups for the methyl groups on the side chain and by copolymerization with other heteroatoms in which B—O—Si, Al—O—Si, Sn—O—Si, Ti—O—Al and other combinations are produced.

A similar class is the titanates. Three-dimensional Ti—O chains form pigments and pigment binders for paints and water-proofing compounds for use on cloth. Their properties can be modified by substituting monofunctional groups for some of the oxygen, for example, by forming esters to interrupt the chains.

The polyphosphates have also been widely studied. Here the phosphate ion is found as a high polymer. The molecular weight of the polymer ranges from 250,000 to 2,000,000. The polyphosphates are water soluble and form fibers. No uses have been found for this class of materials. They hydrolyze slowly in atmospheric moisture and also embrittle on standing.

Other attempts to base a polymer on B—N heteroatomic chains are being vigorously pursued although the B—O bond with an energy of 130 kilocalories is thermally more stable than the B—N bond at 100 kilocalories. The borates formed with B—O chain links, however, are too hydrolytically unstable and too thoroughly cyclized to be useful. B—N compounds are also plagued with the same weakness. However, the very high thermal stability of low molecular weight materials has encouraged the search for high polymers with the same basic structure. The combination of boron and nitrogen approximates that of carbon with carbon due to its location in the Periodic Table. See **Boron**.

One other area of materials deserves mention here. Coordination polymers are found when metal atoms are joined together by coordinating bonding involving some bridging group, e.g.

In view of the high thermal stability of monomeric chelation compounds, coordination polymers were expected to be promising for use at high temperatures. This has not proved to be the case. Thermal stabilities are usually lower than for low molecular weight materials. In addition, if the polymerization goes beyond a few monomer units, the materials tend to become insoluble and infusible so they cannot be fabricated into useful items. See **Chelation Compounds**.

Additional Reading

Allcock, H., F. Lampe, and J. Mark: *Contemporary Polymer Chemistry*, 3rd Edition, Prentice Hall, Inc., Upper Saddle River, NJ, 2003.

Bahadur, P., and N.V. Sastry: *Principles of Polymer Science*, CRC Press LLC., Boca Raton, FL, 2002.

Bower, D.I.: *Introduction to Polymer Physics*, Cambridge University Press, New York, NY, 2002.

Brandrup, J.D.R. Bloch, E.A. Grulke, E.H. Immergut, and A. Abe: *Polymer Handbook*, 2 Vol., 4th Edition, John Wiley & Sons, Inc., New York, NY, 2003.

Cheremisinoff, N.P.: *Condensed Encyclopedia of Polymer Engineering Terms*. Elsevier Science & Technology Books, New York, NY, 2001.

Fried, J.: *Polymer Science and Technology*, 2nd Edition, Prentice Hall Professional Technical Reference, Upper Saddle River, NJ, 2003.

Kroschwitz, J.I.: *Encyclopedia of Polymer Science and Technology*, 12 Volume Set, 3rd Edition, John Wiley and Sons, Inc., Hoboken, NJ, 2004.

Morawetz, H.: *Polymers: The Origins and Growth of a Science*, Dover Publications, Inc., Mineola, NY, 2002.

Rubinstein, R., and R.H. Colby: *Polymer Physics*, Oxford University Press, New York, NY, 2003.

Scheirs, J., and T.E. Long: *Modern Polymers*, John Wiley & Sons, Inc., New York, NY, 2003.

Solomons, T.W. Graham, C.B. Fryhle, and M.M. Shenkman: *Organic Chemistry*, 8th Edition, John Wiley & Sons, Inc., New York, NY, 2003.

Walton, D.J., and J.P. Lorimer: *Polymers*, Oxford University Press, New York, NY, 2001.

POLYMERS (Organic).   Organic high polymers have a great number of different chemical structures, ranging from completely nonpolar to very polar and even ionic materials. They all clearly resemble each other, however. The basis for this resemblance is that many of their properties are governed by their high molecular weights, which range from 5,000 to tens of millions. For example, as the molecular weight increases in a given polymer family, the tensile strength of the polymer increases markedly. In some cases it approaches that of steel on a weight-for-weight basis, especially when oriented fibers are fabricated.

In a similar way, the viscosity of the molten material changes from a free flowing liquid at low molecular weights to the highly viscous polymeric liquid where flow may be observed only over a long period of time or under a considerable applied pressure. A property which shows up only in the case of high polymers is rubbery elasticity. Here again the development of a sufficiently long and flexible molecule is necessary before rubbery behavior develops.

Because of this striking dependence on molecular size, the measurement of molecular weight and dimensions is very important. Some of the most significant early work of Staudinger was the demonstration of the existence of large molecules joined by covalent bonds. (Others felt that such large molecules were not possible). In order to determine these molecular properties the molecules must be dissolved. Thus each molecule can be separated from its neighbors and its effect measured independently. Solution properties such as osmotic pressure, light scattering and viscosity are used to measure the molecular weight of polymers. In the case of many natural polymers the ultracentrifuge has proved uniquely useful.

The osmotic pressure determination of molecular weights is based on the thermodynamic interaction of solvent and solute to lower the activity of the solvent. Experimentally, the solution is separated from the solvent by a semipermeable membrane. The solvent tends to pass through the membrane to dilute the solution and bring the activity of the solvent in both phases to equilibrium. The quantitative measurement of this tendency is obtained by allowing the liquid solution to rise in a vertical capillary connected to the solution compartment. The equilibrium height it achieves or the rate at which it rises can be measured.

The measurements are converted to effective pressure (π) at zero polymer concentrations (c) and the average molecular weight ($\overline{M}_n$) gotten from the following relation:

$$\lim_{c \to 0} \frac{\pi}{c} = \frac{RT}{\overline{M}_n}$$

(R = gas constant; T = absolute temperature).

The light-scattering method is based on similar thermodynamic interactions. In any solution there are random variations in concentration and refractive index. These scatter some light out of a beam passing through the liquid. In a polymer solution the nature of the fluctuations and thus the amount of scattered light (τ) depend on the attractive forces between polymer and solvent molecules. This, in turn, depends on the polymer molecular weight ($\overline{M}_w$). The following equation describes the behavior:

$$\lim_{c \to 0} \frac{Hc}{\tau} = \frac{l}{\overline{M}_w}$$

(H = a constant)

This method was developed by Peter J. W. Debye in 1944. Its evolution has been one of the most stimulating chapters of polymer physics.

In contrast to these thermodynamic methods, the viscosity molecular weight determination depends on the interference in the flow of the solvent caused by the dissolved molecules. In contrast to osmometry and light scattering, it has not been possible to develop the viscosity effect into an absolute measure of molecular weight. Rather, it must be calibrated, preferably by light scattering measurements.

The relationship between the measured limiting specific viscosity [η] and the molecular weight ($\overline{M}_v$) is as follows:

$$\lim_{c \to 0} \frac{\eta_{sp}}{c} = [\eta] = K \overline{M}_v^a$$

K and a are determined by calibration for a given polymersolvent system.

$$\eta_{sp} = \frac{\eta \text{solution}}{\eta \text{solvent}} - 1$$

In each of the equations above a different symbol has been used for the molecular weight. Most polymers are heterodisperse, i.e., have many molecular weight species of the same chemical nature, thus the experiments yield an average molecular weight. Osmotic pressure gives a lower average ($\overline{M}_n$ or number average) because it emphasizes the effect of small molecules, while light scattering emphasizes the larger molecules and gives a higher average, ($\overline{M}_w$ or weight average). The viscosity average ($\overline{M}_v$) is between the two and closer to the weight.

Many natural polymers are monodisperse (all molecules have the same molecular weight). In this case the ultracentrifuge which separates materials according to their effective density in solutions is a most powerful tool for molecular weight determination. With poly-disperse materials, the interpretation of ultracentrifuge results becomes more complex and widespread application of this method to synthetic polymer molecular weight determination has not yet been achieved.

In addition to the primary effect of the great length of the molecule, the details of the distribution of functional groups along the polymer chain modify the behavior of these materials. This leads to differences in their applications. The study of the mechanical behavior of polymers at various temperature is called rheology. The temperature at which the rubbery material becomes glassy is called the glass-transition temperature (T_g). This transition temperature depends on the nature of the backbone and the substituent groups on the polymer chain. Rubbery materials, e.g., polyisoprene, polychloroprene, polybutadiene, the copolymer of butadiene and styrene, etc., have molecular chains with considerable flexibility. Usually small side groupings and irregularities in the chain prevent them from coming together in a regular structure. Instead the molecules stay in an amorphous random packing much like a pile of cooked spaghetti. As with cooked spaghetti, there is a tendency for the whole mass to flow, by the movement of chains past one another. With natural rubber this flow at room temperature and above had to be inhibited by tying the chains together with chemical bonds before a useful product was obtained. This cross-linking is called vulcanization in the case of rubber. The process was discovered by Charles Goodyear in 1839. A similar cross-linking to inhibit the motion of the chains is necessary to make useful products from the newer synthetic rubbers also.

Other polymers are not rubbery at room temperature, instead they exist in the glassy state. They are amorphous, but because of more bulky substituents on the polymer chain the molecule is less flexible. There is less ease of molecular motion under applied stress at room temperature. Examples of such polymers are polystyrene and polymethylmethacrylate, which are transparent due to their amorphous, homogeneous nature. When heated they first become rubbery and then, at higher temperatures, show viscous flows so that they may easily be molded. When they are cooled to room temperature the rate of flow is vanishingly small due to the stiff chains; thus items made in this way can be used at ordinary temperatures if they are not required to bear too large a load. These common organic glasses are brittle and easily broken on impact. One of the interesting problems for polymer development is to obtain impact resistance without losing transparency and without increasing the cost by an excessive amount.

A related class of polymers is the crystalline, thermoplastic materials. These also are fabricated by heating to a high temperature so that they flow; but when they are cooled ordered regions develop within them, which makes them translucent. They have much tendency to flow because of these mechanical "crosslinks" and have good dimensional stability. Polyethylene and polypropylene belong to this class. Here the chain is simple and regular so that different polymer molecules, or different parts of the same molecule, can pack next to each other.

As might be expected, there are polymers intermediate between the crystalline and glassy ones. For example, polyvinylchloride shows enough order to prevent its classification as glassy, but not enough to be considered crystalline.

Many crystalline polymers form part of another class of materials, the fiber-forming polymers. The formation of fibers of significant strength depends on the growth of ordered structures when the fiber is stretched.

The introduction of oxygen and nitrogen into the chain changes its flexibility, stability toward chemical reaction, resistance to solvents, strength and other properties. In this way quite extensive changes in behavior are obtained. Most of these heterochain polymers are prepared by condensation or ring opening reactions. The first important work in this field was the classic investigation of W. H. Carothers in the late 1920's on polyesters and polyamides. See **Polyimides**.

The toughness, high melting points and high tensile strength of many of these polymers have since led to their widespread use of fibers, films and molded objects. In general these polymers are rather high cost materials, which are used, because of their unusual properties, in places where ordinary polymers are inadequate. Many of the interesting developments in polymers over the last few years have involved new syntheses and new variations in structure of such heterochain materials.

The materials described above are thermoplastic resins, i.e., they all melt on being heated to sufficiently high temperatures and can be molded while molten. This characteristic is associated with molecular chains which are long and stringlike with few branches on them. If, however, the polymer chains have many covalent bonds linking them together into a network, a thermosetting resin develops which may flow at an early stage of its history, but is insoluble and infusible after the full crosslinking reaction has taken place. Any of the previous chain compositions can be used in making thermosetting resins if provision for crosslinking is made by using multifunctional monomers. Some are used more commonly, e.g., the epoxy resins, phenolformaldehyde, urea formaldehyde, melamine, etc. In general they are useful because of their inertness to solvents, resistance to dimensional change on heating, rigid dimensions, physical strength, chemical resistance and abrasion resistance.

Many of the varieties of polymers which have been discussed have analogs in polymers isolated from natural systems, for example, natural rubber is a polyisoprene with a purely carbon chain. Other natural polymers are based on the C−O−C bond. The cellulose and starch polymers, which are found in plants are composed of chains of six-membered carbon-oxygen rings joined through an oxygen linkage. Cellulose and starches differ from each other in the spatial orientation of the links joining the six membered rings.

Products from different sources in each class differ in degree of branching of the molecule and in amount of crosslinking.

In unmodified cellulose the hydroxyl groups give a large amount of hydrogen bonding which leads to insolubility in most solvents. On the other hand if these are changed by chemical reactions to ether or ester groups a much more tractable material results. Cellulose acetate, butyrate and nitrate; methyl and ethyl ether and carboxy methyl ether are widely used modified celluloses. Starches also are modified, but much less commercial success has been had with them.

The polymers described above have been chemically pure, although physically heterodisperse. It is often possible to combine two or more of these monomers in the same molecule to form a copolymer. This process produces still further modification of molecular properties and, in turn, modification of the physical properties of the product. Many commercial polymers are copolymers because of the blending of properties achieved in this way.

Since they are organic materials most polymers are not water soluble, however, water solubility can be obtained by substituting the proper side groups on the polymer chain. Such polymers include the nonionic materials polyvinyl alcohol, polyethylene glycol, etc., where the strong dipolar and hydrogen bonding interactions cause the solubility; and the polyelectrolytes where ionizable groups such as the carboxylate, sulfonate, quaternary ammonium, etc., cause the solubility. Many of these water soluble polymers are used to increase the viscosity of water based systems. As little as 0.1 or 0.2% of the polymer is needed to produce a very viscous solution. They also are of wide biological interest.

Additional Reading

Allcock, H., F. Lampe, and J. Mark: *Contemporary Polymer Chemistry*, 3rd Edition, Prentice Hall, Inc., Upper Saddle River, NJ, 2003.

Bahadur, P., and N.V. Sastry: *Principles of Polymer Science*, CRC Press LLC., Boca Raton, FL, 2002.

Bower, D.I.: *Introduction to Polymer Physics*, Cambridge University Press, New York, NY, 2002.

Brandrup, J.D.R. Bloch, E.A. Grulke, E.H. Immergut, and A. Abe: *Polymer Handbook*, 2 Vol., 4th Edition, John Wiley & Sons, Inc., New York, NY, 2003.

Cheremisinoff, N.P.: *Condensed Encyclopedia of Polymer Engineering Terms*. Elsevier Science & Technology Books, New York, NY, 2001.

Fried, J.: *Polymer Science and Technology*, 2nd Edition, Prentice Hall Professional Technical Reference, Upper Saddle River, NJ, 2003.

Kroschwitz, J.I.: *Encyclopedia of Polymer Science and Technology*, 12 Volume Set, 3rd Edition, John Wiley and Sons, Inc., Hoboken, NJ, 2004.

Morawetz, H.: *Polymers: The Origins and Growth of a Science*, Dover Publications, Inc., Mineola, NY, 2002.

Rubinstein, R., and R.H. Colby: *Polymer Physics*, Oxford University Press, New York, NY, 2003.

Solomons, T.W. Graham, C.B. Fryhle, and M.M. Shenkman: *Organic Chemistry*, 8th Edition, John Wiley & Sons, Inc., New York, NY, 2003.

Scheirs, J., and T.E. Long: *Modern Polymers*, John Wiley & Sons, Inc., New York, NY, 2003.

Walton, D.J., and J.P. Lorimer: *Polymers*, Oxford University Press, New York, NY, 2001.

POLYMERS (Water-Soluble). Any substance of high molecular weight which swells or dissolves in water at normal temperature. These fall into several groups, including natural, semisynthetic, and synthetic products. Their common property of water solubility makes them valuable for a wide variety of applications as thickners, adhesives, coatings, fooe additives, textile sizing, etc.

Natural. This type is principally composed of gums, which are complex carbohydrates of the sugar group. They occur as exudations of hardened sap on the bark of various tropical species of trees. All are strongly hydrophilic. Examples are arabic, tragacanth, karaya.

Semisynthetic. This group (sometimes called water-soluble resins) includes such chemically treated natural polymers as carboxymethylcelluose, methylcellulose, and other cellulose esters, as well as various kinds of modified starches (esters and acetates).

Synthetic. The principal members of this class are polyvinyl alcohol, ethylene oxide polymers, polyvinyl pyrrolidone, polyethyleneimine.

POLYMETHYLBENZENES. Polymethylbenzenes (PMBs) are aromatic compounds that contain a benzene ring and three to six methyl group substituents (for the lower homologues see **Benzene**; **Toluene**; **Xylenes and Ethylbenzene**). Included are the trimethylbenzenes, C_9H_{12}

TABLE 1. PHYSICAL AND THERMODYNAMIC PROPERTIES OF POLYMETHYLBENZENES

Property	1,3,5-Trimethyl-benzene	1,2,4-Trimethyl-benzene	1,2,3-Trimethyl-benzene	1,2,4,5-Tetramethyl-benzene	1,2,3,5-Tetramethyl-benzene	1,2,3,4-Tetramethyl-benzene	Pentamethyl-benzene	Hexamethyl-benzene
CAS Registry Number	[108-67-8]	[95-63-6]	[526-73-8]	[95-93-2]	[527-53-7]	[488-23-3]	[700-12-9]	[87-85-4]
Mol wt	120.194	120.194	120.194	134.221	134.221	134.221	148.248	162.275
Bp, °C	164.74	169.38	176.12	196.80	198.00	205.04	231.9	263.8
Flash point, °C	43.0	46.0	51.0	67.0	68.0	73.0		
Density, g/cm^3 at 20°C	0.8651	0.8758	0.8944	0.8875	0.8903	0.9052	0.917	Solid
at 25°C	0.8611	0.8718	0.8905	0.8837	0.8865	0.9015	0.913	Solid
Freezing point, °C in air at 101.3 kPa	−44.694	−43.881	−25.344	79.240	−23.689	−6.229	54.35	165.7
Refractive index, n_D at 25°C	1.49684	1.50237	1.51150	1.5093	1.5107	1.5181	1.525	Solid
Surface tension, mN/m (=dyn/cm), at 20°C	28.84	29.72	31.28	solid	33.51	35.81	Solid	Solid
Critical temperature, °C	364.20	376.02	391.32	401.85	405.85	416.55		
Critical pressure, kPa	3127	3232	3454	2940	2860	2860		
Critical volume, cm^3/mol	427	427	427	482	482	482		
Heat of vaporization at bp, kJ/mol	39.0	39.2	40.0	45.52	43.81	45.02	45.1	48.2

(mesitylene (**1**), pseudocumene (**2**), and hemimellitene (**3**)), the tetramethylbenzenes, $C_{10}H_{14}$ (durene (**4**), isodurene (**5**), and prehnitene (**6**)), pentamethylbenzene, $C_{11}H_{16}$ (**7**), and hexamethylbenzene, $C_{12}H_{18}$ (**8**). The PMBs are primarily basic building blocks for more complex chemical intermediates.

Physical Properties

The structures of the eight PMBs are shown here and their physical and thermodynamic properties are given in Table 1. Systematic (benzene) names are used in table 1.

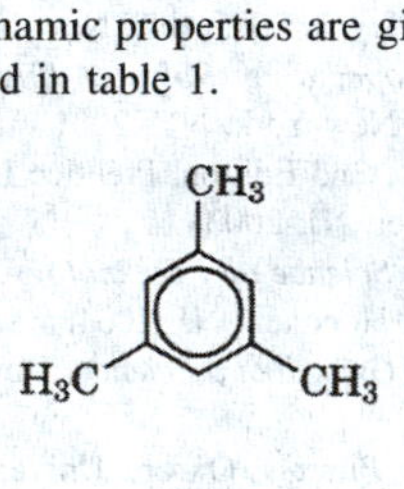

(1) 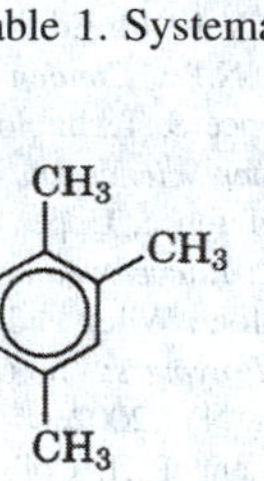   (2)

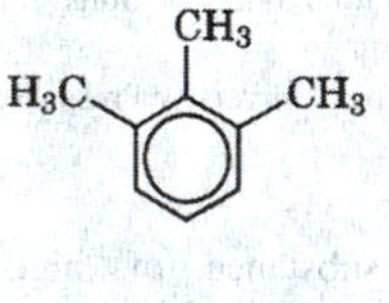

(3) 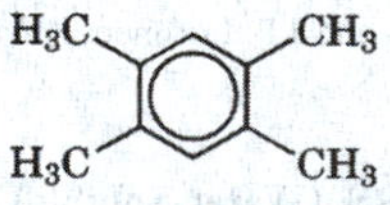   (4)

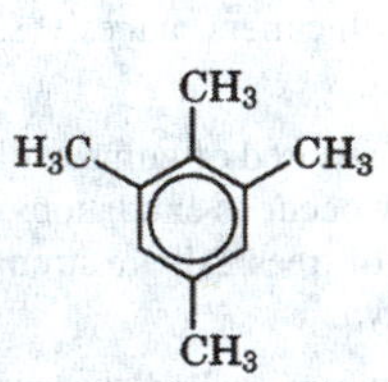

(5) 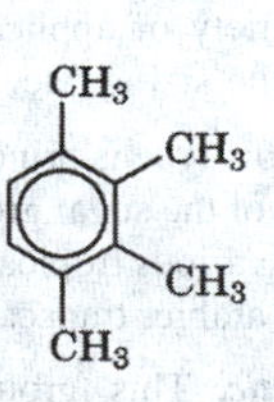   (6)

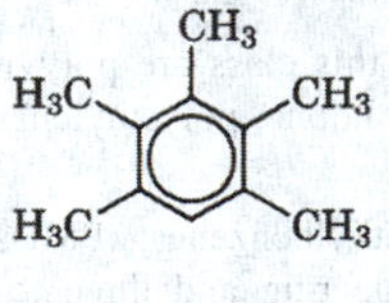

(7) 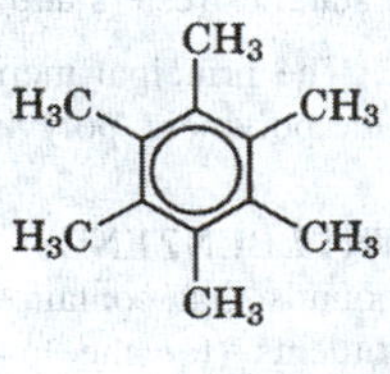   (8)

Manufacture

High purity mesitylene, hemimellitene, and durene are often produced synthetically, whereas pseudocumene is obtained from extracted C_9 reformate by superfractionation.

Health and Safety Factors

The PMBs, as higher homologues of toluene and xylenes, are handled in a similar manner, even though their flash points are higher (see Table 1). Containers are tightly closed and use areas should be ventilated. Breathing vapors and contact with the skin should be avoided.

Uses

Pseudocumene is used as a component in liquid scintillation cocktails for clinical analyses. Pseudocumene and durene are oxidized to trimellitic anhydride and pyromellitic dianhydride, respectively. Mesitylene is a key building block for important antioxidants and agricultural chemicals. Prehnitene, isodurene, pentamethylbenzene, and hexamethylbenzene have no significant commercial uses. The higher polymethylbenzenes show potential as highly regiospecific methylation agents for methylation of 4-alkylbiphenyls to form 4,4′-alkylmethylbiphenyls which can be oxidized to the monomer 4,4′-biphenyldicarboxylic acid [CAS: 787-70-2]. See also **Liquid Crystalline Materials**.

Similar regiospecific syntheses employing PMBs and dialkylnaphthalene have been used to prepare 2,6-dimethylnaphthalene [CAS: 581-42-0] which is oxidized to the corresponding 2,6-naphthalenedicarboxylic acid [CAS: 1141-38-4], also used as a monomer for liquid crystal polymers.

Additional Reading

Earhart, H.W.: *The Polymethylbenzenes*, Noyes Development Corp., Park Ridge, NJ, 1969.

Shriner, R.L., et al.: *The Systematic Identification of Organic Compounds*, 8th Edition, John Wiley & Sons, Inc., New York, NY, 2003.

U.S. Pat. 3,542,890 (Nov. 24, 1970), H.W. Earhart and G. Sugerman (Sun to Koch Industries Inc.).

POLYMORPHISM. 1. A phenomenon in which a substance exhibits different forms. Dimorphic substances appear in two solid forms, whereas trimorphic exist in three, as sulfur, carbon, tin, silver iodide, and calcium carbonate. Polymorphism is usually restricted to the solid state. Polymorphs yield identical solutions and vapors (if vaporizable). The relation between them has been termed "physical isomerism." See "Allotropes" under **Chemical Elements**. See also **Mineralogy**.

2. The occurrence of individuals of distinctly different structure or appearance within a species. In many cases two such forms occur and the species is said to be dimorphic rather than polymorphic.

Polymorphism depends upon many different conditions in various groups of animals. The various forms may be adapted for different places in a life cycle, for special parts in a colonial or social organization, or for special stages in a metamorphosis. They may also result from the incidence of different environmental conditions due to seasons or to unusual climatic conditions.

POLYNOMIAL.

A rational integral function, sometimes also called a multinomial, in n variables of the form

$$c_1 x_1^{a_1} \cdots x_2^{b_1} + c_2 x_1^{a_2} x_2^{b_2} \cdots x_n^{r_2} + \cdots + c_k x_1^{a_k} x_2^{b_k} \cdots x_n^{r_k}$$

For any term in this expression c_i is the coefficient, a_i is the degree with respect to x_1, b_i with respect to x_2, etc., and the total degree of that term is $a_i + b_i + \cdots + r_i$. The highest degree in x_i of any term whose coefficient is not zero is the degree of the polynomial in x_i, and the highest total degree of any term with nonvanishing coefficient is the degree of the polynomial. The coefficients, c_i, which are constants, may be real or complex.

If the terms of a polynomial all have the same degree, the polynomial is homogeneous. The expressions *quantic* and *form* are also used. If they have two, three, four, etc., variables they are binary, ternary, quaternary, n-ary forms or a binomial, trinomial, etc. If their degree is 1, 2, 3, etc., they are linear, quadratic, cubic, etc., forms or equations.

The commonest case is the nth-degree polynomial in one variable which may be written as

$$a_0 x^n + a_1 x^{n-1} + \cdots + a_{n-1} x + a_n$$

For such a polynomial, the fundamental theorem of algebra (see also **Factor**), states that there exists one and only one set of constants, $x_1, x_2, \ldots, x_n$, such that

$$(x - x_1)(x - x_2)(x - x_3) \cdots (x - x_n) = 0$$

The constants are the roots of the polynomial. They may be real or complex, but they need not all be different. If $k \le n$ roots are equal to each other, the root is said to be k-fold or k-tuple. The following relations hold for the roots:

$$x_1 + x_2 + \cdots + x_n = -a_1/a_0;$$
$$x_1 x_2 + x_1 x_3 + \cdots + x_{n-1} x_n = a_2/a_0;$$
$$x_1 x_2 x_3 + \cdots + x_{n-2} x_{n-1} x_n = -a_3/a_0; \cdots$$
$$x_1 x_2 x_3 \cdots x_n = (-1)^n a_n/a_0.$$

The roots of linear, quadratic, cubic, and biquadratic or quartic equations may be obtained in terms of algebraic expressions containing the coefficients but polynomials of degree higher than four cannot be solved in this way. Approximate values of the roots of equations of any degree may be determined by graphical or numerical methods (see also **Approximate Calculation**). Properties of the roots can be found by Descartes' rule and by the theorems of Budan and Sturm.

Given a polynomial in one variable of order n, $f(x) = 0$, it is often useful to introduce a new variable which will change the roots of the original equation in a predetermined way. This process is called transformation of variable. Some possibilities are as follows, where the original variable, x, is replaced by the new variable indicated in order to produce the corresponding change in the roots: (a) $-x$, the signs of all roots are changed; (b) x/h, the roots are all multiplied by the constant h; (c) $1/x$ and multiply the equation by x^n, the new roots are reciprocals of the original roots; (d) $x \pm h$, the roots will be decreased by the constant amount h if the plus sign is used or increased by h with the minus sign.

Polynomials frequently occur as the solution of a differential equation.

For examples, see also **Generating Function**.

POLYOL.

A polyhydric alcohol, i.e., one containing three or more hydroxyl groups. Those having three hydroxyl groups (trihydric) are glycerols; those with more than three are called sugar alcohols, with general formula $CH_2OH(CHOH)_n CH_2OH$, where n may be from 2 to 5. These react with aldehydes and ketones to form acetals and ketals.

See also **Alcohols**; and **Glycerol**.

POLYOLEFIN.

A class or group name for thermoplastic polymers derived from simple olefins; among the more important are polyethylene,

polyproplene, polybutenes, polyisoprene, and their copolymers. Many are produced in the form of fibers. This group comprises the largest tonnage of all thermoplastics produced.

See also **Elastomers**; **Olefin Polymers**; **Polyethylene**; **Polypropylene**; and **Thermoplastics**.

POLYORGANOSILICATE GRAFT POLYMER.

An organoclay to which a monomer or an active polymer has been chemically bonded, often by the use of ionizing radiation. An example is the bonding of styrene to a polysilicate containing vinyl radicals, resulting in the growth of polystyrene chains from the surface of the silicate. Such complexes are stable to organic solvents. They have considerable use potential in the ion-exchange field, as ablative agents, reinforcing agents, and hydraulic fluids.

POLY(ORTHOESTERS).

See **Biomaterials, Prosthetics, and Biomedical Devices**.

POLYP.

A smooth-coated tumor projecting from a mucous surface, named from the Greek "polypos" (octopus). Polyps range in size from almost invisible bumps (102 millimeters diameter) to mushroom-shaped bodies (1–3 cm diameter), attached to the surface by a vascular penduncle. Commonly found in the nose, bladder and gastrointestinal tract, polyps may occur almost anywhere in the body. Most polyps are nonmalignant and are surgically removed only when they interfere with a normal body function. The majority of colorectal polyps prove to be adenomas with malignant potential, however, and thus their detection and removal are important.

POLYPEPTIDE.

A compound composed of two or more amino acids, similar in many properties to the natural peptones. The amino acids are joined by peptide groups

$$-NH-C\overset{O}{\underset{}{\lVert}}$$

formed by the reaction between an $-NH_2$ group and a

$$-C\overset{O}{\underset{OH}{\lVert}}$$

group, whereby there is elimination of a molecule of water, and formation of a valence bond. They may be termed di-, tri-, tetra-, etc., peptide according to the number of amino acids present in the molecule.

The sequence of amino acids in the chain of a protein is of critical importance in the biological functioning of the protein, and its determination is very difficult. The chains may be relatively straight, or they may be coiled or helical. In the case of certain types of polypeptides, such as the keratins, they are crosslinked by the disulfide bonds of cystine. Linear polypeptides can be regarded as proteins. See also **Amino Acids**; and **Protein**.

R. C. V.

POLYPROPYLENE.

[CAS: 9003-07-0]. A synthetic crystalline thermoplastic polymer, $(C_3H_5)_n$, with molecular weight of 40,000 or more. Low-molecular-weight polymers are also known which are amorphous in structure and used as gasoline additives, detergent intermediates, greases, sealant, and lubricating oil additives. They are also available as high-melting-point waxes.

Polypropylenes are derived by the polymerization of propylene with stereospecific catalyst, such as aluminum alkyl. These polymers are translucent, white solids, insoluble in cold organic solvents, softened by hot solvents. They maintain their strength after repeated flexing. They are degraded by heat and light unless protected by antioxidants. Polypropylenes are readily colored, exhibit good electrical resistance, low water absorption and moisture permeability. They have rather poor impact strength below 15 °F (−9.5 °C). They are not attacked by fungi or bacteria, resist strong acids and alkalies up to about 140 °F (60 °C), but are attacked by chlorine, fuming nitric acid, and other strong oxidizing agents. They are combustible, but slow burning.

Polypropylenes are available as molding powder, extruded sheet, cast film, textile staple, and continuous-filament yarn. They find use in packaging film; molded parts for automobiles, appliances, and housewares; wire and cable coating; food container closures; bottles; printing plates; carpet and upholstery fibers; storage battery cases; crates for soft-drink bottles; laboratory ware; trays; fish nets; surgical casts; and a variety of other applications.

See also **Fibers** and **Olefin Polymers**.

POLYP (Zoology). One of the two types of individuals found in many species of coelenterates. The two are the polyp or hydroid and the medusa. Polyps are approximately cylindrical, elongated on the axis of the body. One end is usually attached and the other bears the mouth, surrounded by a circlet of tentacles. The wall is relatively thin, due to the thinness of the mesogloea. In the class *Hydrozoa*, polyps are often very simple, like the common little freshwater species of the genus *Hydra*. Actinozoan polyps, including the corals and sea anemones, are much more complex, due to the development of a tubular stomodaeum leading inward from the mouth and a series of radial partitions called mesenteries. Many of the mesenteries project into the enteric cavity but some extend from the body wall to the central stomodaeum.

POLYSTYRENE. [CAS: 9003-53-6] (PS), the parent of the styrene plastics family. General purpose (or crystal) polystyrene is a clear, water-white, glassy polymer commonly derived from coal tar and petroleum gas. Physical properties of this material can be altered by addition of modifying agents, such as rubber (for increased toughness), methyl or α-methyl styrene (for heat resistance, methyl methacrylate (for improved light stability), and acrylonitrile (for chemical resistance). In general, varying the level of modifying agent (e.g., comonomer) will alter the level of desired property improvement.

Special grades of polystyrene include impact polystyrene modified with ignition-resistant chemical additives. These were developed because of increased emphasis on product safety and used in many electrical and electronic appliances. The addition of flame-retardant chemicals does not make the polymer noncombustible, but increases its resistance to ignition and decreases the rate of burning when exposed to a minor fire source.

Chemistry. The polymerization of styrene is an exothermic chain reaction which proceeds by all known polymerization techniques. This reaction can be shown schematically as:

The exact nature of the beginning and end of such a polymer chain is not certain. In general, the polymer can be characterized by its average degree of polymerization, i.e., the value of n, or more precisely by the distribution of n values. The heat of polymerization is 17.4 ± 0.2 kcal/mole at $26.9\,^{\circ}$C. The reaction may be initiated by heat or by means of catalysts. Organic peroxides are typical initiators. Styrene also will polymerize in the presence of various inert materials, such as solvents, fillers, dyes, pigments, plasticizers, rubbers, and resins. Moreover, it forms a variety of copolymers with other mono- and polyvinyl monomers.

It is a matter of general observation that with styrene, the polymerization-rate curves will exhibit three distinct phases, the nature of which can be determined by the polymerization conditions and the purity of the monomer: (1) an initial slow period at the beginning of the reaction, known as the *induction period*, which appears to be associated with the presence of an inhibitor or other impurity in the monomer; (2) a period of relatively rapid polymerization, which persists almost to the end of the reaction, and for which the rate is exponentially dependent upon temperature; and (3) a

final slowing down in rate as the reaction approaches completion and the monomer becomes exhausted. This effect is particularly apparent at low temperatures with relatively impure monomers.

General Properties. The *specific gravity* of general purpose and impact polystyrene is 1.05. It can vary for copolymers. It is higher for some specialty grades. Density varies slightly with pressure, but for practical purposes, the polymer is noncompressible.

In terms of *heat-resistance*, deflection temperatures range from about 66 to 99$\,^{\circ}$C (170 to 215$\,^{\circ}$F), depending upon the formulation. Continuous resistance to heat for polystyrene is usually 60 to 80$\,^{\circ}$C (140 to 175$\,^{\circ}$F). Time and load have a significant influence on the useful service temperature of a part.

Polystyrene is nontoxic when free from additives and residuals. It has no nutritive value and does not support fungus or bacterial growth.

Dimensional stability of polystyrene resins is excellent. Mold shrinkage is small. The low moisture absorption (about 0.02%) allows fabricated parts to maintain dimensions and strength in humid environments.

General-purpose polystyrene is water white, and transmission of visible light is about 90%. Modifiers reduce this property, and translucence results. The refractive index is about 1.59; critical angle about 39. Polystyrene molecules do not have the same optical properties in all directions. When molecules become oriented in a given direction during fabrication, a double refraction occurs and a birefringence effect can be observed if the part is examined through a polarized lens under a polarized light source. Injection moldings often exhibit birefringence in a random pattern. This can be beneficial if the birefringence is in the direction of load.

In terms of *weatherability*, polystyrene does not exhibit ultraviolet stability and is not considered weather-resistant as a clear material. Continuous, long-term exposure results in discoloration and reduction of strength. Improvement in weatherability can be obtained by the addition of ultraviolet absorbers, or by incorporating pigments. The best pigmenting results are obtained with finely dispersed carbon black.

In terms of *chemical resistance*, polystyrene has a high resistance to water, acids, bases, alcohols, and detergents. Chlorinated solvents will mar the surface and, in the presence of an external load or high internal stresses, will cause failure. Aliphatic and aromatic hydrocarbons, in general, will dissolve polystyrene. Such foodstuffs as butter and coconut oil should be avoided. The chemical resistance depends upon chemical concentration, time, and stress.

Typical mechanical properties of polystyrene are given in the accompanying table. The long-term load-bearing strength of most polystyrene materials is about one-third of the typical tensile strength given in Table 1.

TABLE 1. COMPRESSION MOLDED PROPERTIES OF POLYSTYRENE

Property	General Purpose psi (MPa)	Impact psi (MPa)
Tensile strength	5500–8000 (38–55)	2500–5000 (17–35)
Compressive strength	21000–16000 (145–110)	4500–9000 (31–62)
Flexural strength	9000–15000 (62–104)	5000–10000 (35–69)
Tensile (Young's) modulus	400000–500000 (2760–3450)	200000–400000 (1880–2760)
Impact strength, Izod, foot-pounds/inch	0.3–0.5	1–4
Hardness, Rockwell M	65–80	60
Elongation, 0.8–2.0	5–50	

Uses. Packaging applications are the most extensive. Meat, poultry, and egg containers are thermoformed from extruded foamed polystyrene sheet. The fast-food market also accounts for a substantial amount of polystyrene for takeout containers where the insulation value of a foamed container is an advantage. Containers, tubs, and trays formed from extruded impact polystyrene sheets are used for packaging a large variety of food. Biaxially oriented polystyrene film is thermoformed into blister packs, meat trays, container lids, and cookie, candy, pastry, and other food packages where clarity is required.

Housewares is another large segment of the use of polystyrenes. Refrigerator door liners and furniture panels are typical thermoformed impact polystyrene applications. Extruded profiles of solid or foamed impact polystyrene are used for mirror or picture frames, and moldings for construction applications.

General-purpose polystyrene is extruded either clear or embossed for room dividers, shower doors, glazings, and lighting applications. Injection molding of impact polystyrene is used for household items, such as flower pots, personal care products, and toys. General-purpose polystyrene is used for cutlery, bottles, combs, disposable tumblers, dishes, and trays.

Injection blow molding can be used to convert polystyrene into bottles, jars, and other types of open containers.

Impact polystyrene with ignition-resistant additives is used for appliance housings, such as those for television and small appliances. Structural foam impact polystyrene modified with flame-retardant additives is used for business machine housings and in furniture because of its decorability and ease of processing. Consumer electronics, such as cassettes, reels, and housings, is a fast growing area for use of polystyrenes. Medical applications include sample collectors, petri dishes, and test tubes.

In an effort to make homes and other buildings more energy efficient, the use of polystyrenes in extruded foam board with flame-retardant additives for walls and under slabs has experienced exceptional growth in recent years. Used as a sheeting material, extruded foam board complies with the requirements of the major building codes as well as federal and military specifications.

In general, polystyrene is used in applications where ease of fabrication and decorability are required. Polystyrene has excellent electrical properties, good thermal and dimensional stability, resistance to staining, and low cost. General purpose polystyrene is preferred where clarity is also of prime concern. Impact polystyrene is preferred where toughness is needed.

POLYSULFONE. This is a transparent, heat-resistant, ultrastable high-performance engineering thermoplastic.[1] It is amorphous in nature and has low flammability and smoke emission. It possesses good electrical properties that remain relatively unchanged up to temperatures near its glass transition temperature of 374 °F (190 °C). The molecular structure of polysulfone features the diaryl sulfone group. This group tends to attract electrons from the phenyl rings. Oxygen atoms para to the sulfone group enhance resonance and produce oxidation resistance. High resonance also strengthens the bonds spatially, fixing the grouping into a planar configuration. The polymer consequently has good thermal stability and rigidity at high temperatures. Ether linkages provide chain flexibility, thereby imparting good impact strength. The polymer resists hydrolysis and aqueous acid and alkaline environments, because the linkages connecting the benzene rings are hydrolytically stable.

Polysulfone is available in transparent and opaque colors in both molding and extrusion grades (unfilled). A special medical grade meets U.S.P. criteria. Two mineral-filled grades are available; one is designed specifically for plating using conventional techniques; the other is a combination of polysulfone compounds with glass fiber or beads as well as other fillers, such as *Teflon*.

Polysulfone has high resistance to acids, alkalies, and salt solutions, and good resistance to detergents, oils, and alcohols, even at elevated temperatures under moderate stress. It is attacked by polar organic solvents, such as ketones, chlorinated hydrocarbons, and aromatic hydrocarbons. Polysulfone can be used continuously in steam at temperatures up to 300 °F (149 °C). Maximum stress in water at 180 °F (82 °C) is about 2000 psi (14 mPa) for steady loads and up to 2500 psi (17 mPa) for intermittent loads. Polysulfone offers a good combination of electrical properties—dielectric strength and volume resistivity are high, while dielectric constant and dissipation factor are low. The latter two properties (which determine lossiness) remain relatively constant over a wide range of temperatures and frequencies (including microwave). Polysulfone can be plated by an electroless nickel or copper process.

Polysulfones are used in a wide variety of applications that take advantage of hydrolytic and acid/caustic stability, clarity, and high heat deflection temperatures. These application areas include consumer items (ie, cookware and appliances), electrical and electronic packaging and substrates, automotive, aerospace, and a host of industrial and plumbing uses. The resistance of polysulfones to chemical attack has resulted in their use in chemical processing equipment. Examples of components in this area are corrosion-resistant pipe, pumps, filter modules, support plates, and tower packing. Glass-reinforced grades can be used in very severe chemical environments for enhanced resistance and long service life.

Polysulfones also offer desirable properties for cookware applications, eg, microwave transparency and environmental resistance to most common detergents. Resistance to various sterilizing media (eg, steam, disinfectants, and gamma radiation) makes polysulfones the resin family of choice for many medical devices. Uses in the electrical and electronic industry include printed circuit boards, circuit breaker components, connectors, sockets, and business machine parts, to mention a few. The good clarity of PSF makes it attractive for food service and food processing uses. Examples of applications in this area include coffee decanters and automated dairy processing components.

One unique application area for PSF is in membrane separation uses. Asymmetric PSF membranes are used in ultrafiltration, reverse osmosis, and ambulatory hemodialysis (artificial kidney) units. Gas-separation membrane technology was developed in the 1970s based on a polysulfone coating applied to a hollow-fiber support.

Polysulfone, polyethersulfone, and polyphenylsulfone may be used interchangeably in many applications. In general, polysulfone is selected because of its lower cost. PES is selected over PSF in applications demanding higher temperatures and/or additional environmental resistance. PPSF represents a step improvement over PES in hydrolytic stability, impact, and chemical resistance with similar temperature capabilities to those of PES. It is selected only when both PSF and PES fail to meet engineering performance requirements. Examples of PPSF uses include steam-autoclaveable surgical sterilization trays, transformer magnet wire coatings, and aircraft interior parts.

POLYTROPIC ATMOSPHERE. A model atmosphere in hydrostatic equilibrium with a constant nonzero lapse rate. The vertical distribution of pressure and temperature is given by

$$\frac{p}{p_0} = \left(\frac{T}{T_0}\right)^{g/R\gamma},$$

where p is the pressure, T the Kelvin temperature, g the acceleration of gravity, R the gas constant for air, and γ the environmental lapse rate, the subscript zeros denoting values at the earth's surface.

AMS

POLYTROPIC PROCESSES. The expansion or compression of a constant weight of gas may assume a variety of forms, depending on the extent to which heat is added to or rejected from the gas during the process, and also on the work done. There are, theoretically, an infinite number of ways possible in which a gas may expand from an initial pressure p_1, and volume v_1 to a final volume v_2. All these expansions may be grouped generically as polytropic expansions, and all could be represented graphically on the PV plane by the family of curves $pv^n = C$. They are all, in theory, perfectly reversible. n may have any positive value, 0 to ∞, and having been selected numerically it defines the type of expansion. From the infinite number of possible polytropic expansions, it is worthwhile to isolate four that deserve special attention. When one of the four physical characteristics, to wit, pressure, temperature, entropy, or volume, remains constant, expansions of more than ordinary interest are denoted, since they are frequently employed in a practical way, in situations which can be subjected to thermodynamic analysis. The value of the exponent n of the polytropic family for each of these is:

isobaric	$n = 0$
isothermal	$n = 1$
isentropic	$n = \gamma$ (γ = ratio of specific heat at constant pressure to that at constant volume)
isometric	$n = \infty$

Note, however, that the first and last are limiting cases, since in the first the pressure remains constant, and in the fourth it approaches zero. Note also that the second applies strictly only to ideal gases.

These thermodynamic processes, as they occur in useful machines, are not often of the exact polytropic form desired. For example, an isentropic process, which is exemplified, at least theoretically, by expansion of the burned gases after the explosive combustion in the gasoline engine, is modified slightly by the interchanging of heat between gases and cylinder wall, whereas a true isentropic has no heat either added or rejected in this way. The particular polytropic curve that would suit these conditions of expansion would depart somewhat from the adiabatic form.

During a polytropic process conditions of the working medium are constantly varying, and analysis may be aimed at determining one of the following: the work done, the heat added, the variation of temperature, and the change of entropy. Some information may be obtained merely by comparing the value of the exponent n with certain other data. For example, if n lies between 0 and 1, the temperature rises during an expansion and falls during a compression; when n is greater than 1, the temperature falls during expansion and rises during compression. Also, when n is less than γ, heat must be added to obtain an expansion, whereas when it is greater than γ, heat must be expelled. From the above it will be noted that there is a certain range of polytropic expansion in which, although heat is added, the temperature falls. This may seem to some to be paradoxical, but it is readily explained. During these expansions work is being done by the gas at a rate greater than that at which heat is being added, with the result that the deficiency must be made up from within the gas. The only way that this may be accomplished is for the gas to cool and give up some of its internal energy.

The equations for work done and for heat added in the case of the general polytropic expansions are:

$$W = \frac{p_1 v_1 - p_2 v_2}{n - 1}$$

$$Q = (p_1 v_1 - p_2 v_2)\left(\frac{1}{n-1} - \frac{1}{\gamma - 1}\right)$$

Both of these are expressed in foot-pounds. Sometimes a substitution of a definite value of n in one or the other of these equations leads to an indeterminate; for example, with the isothermal,

$$W = \frac{p_1 v_1 - p_2 v_2}{1 - 1}$$

But since the equation of the isothermal for an ideal gas is

$$pv = C$$

$$p_1 v_1 = p_2 v_2$$

and the work equation becomes indeterminate,

$$W = \frac{0}{0}$$

By approaching the isothermal from a different angle, however, the equation

$$W = pv \, \log_e \frac{v_2}{v_1}$$

may be deduced for work done.

POLYURETHANES.

POLYURETHANES. [CAS: 9009-54-5]. These materials comprise a conglomerate family of polymers in which formation of the urethane group

$$\begin{array}{c} H \quad O \\ | \quad \| \\ N - C - O \end{array}$$

is an important step in polymerization. Because the urethane linkage usually is formed by reaction of hydroxyl and isocyanate groups, urethane chemistry is the chemistry of isocyanates. The high reactivity of isocyanates and knowledge of the catalysis of isocyanate reactions have made possible the simple production of diverse polymers from low- to moderate-molecular-weight liquid starting materials. Several isocyanates (tolylene diisocyanate, hexamethylene diisocyanate, dicyclohexylmethane diisocyanate, etc.) are used in preparing polyurethanes. All are low-viscosity liquids at room temperature with the exception of 4, 4'-diphenylmethane diisocyanate (MDI), which is a crystalline solid. The aromatic isocyanates are more reactive than the aliphatic isocyanates and are widely used in urethane foams, coatings, and elastomers. The cyclic structure of aromatic and alicyclic isocyanates contributes to molecular stiffness in polyurethanes.

Flexible and rigid urethane foams, probably the most familiar of the polyurethanes, are produced in very large quantities. Foam formulations contain isocyanates and polyols with suitable catalysts, surfactants for stabilization of foam structure, and blowing agents, which produce gas for expansion. The largest volume of flexible urethane foam is used as a cushioning material. Expanding uses for flexible foam include carpet underlays and bedding. Weight reduction programs in the transportation field also take advantage of polyurethane forams for seating and trim. Rigid foams find application in insulation for appliances. Thermoplastic urethane elastomers form a widely used family of engineering materials, which appear to combine the best properties of elastomers and thermoplastics. They are tough, have high load-bearing capacity, low-temperature flexibility, and resistance to oils, fuels, oxygen, ozone, abrasion, and mechanical abuse. Possible carcinogenic properties are being studied. See also **Elastomers**; and **Urethane Polymers**.

POLYVINYL ALKYL ETHERS. These products have properties which range from sticky resins to elastic solids. They are obtained by the low-temperature cationic polymerization of alkyl vinyl ethers having the general formula $ROCH=CH_2$. These monomers are prepared by the addition of the selected alkanol to acetylene in the presence of sodium alkoxide or mercury(II) catalyst. As shown by the following equations, the latter yields an acetal which must be thermally decomposed to produce the alkyl vinyl ether.

$$\begin{array}{c} HC\equiv CH \ + \ ROH \ \xrightarrow[130-180\,°C]{Na^+, \ OR^-} \ H_2C=CHOR \\ \text{acetylene} \quad \text{alkanol} \qquad\qquad \text{alkyl vinyl ether} \end{array}$$

$$\begin{array}{c} HC\equiv CH \ + \ 2ROH \ \xrightarrow{Hg^{++}} \\ \text{acetylene} \quad \text{alkanol} \end{array}$$

$$\begin{array}{c} H_2C-CH(OR)_2 \ \xrightarrow[(-ROH)]{cat.} \ H_2C=CHOR \\ \text{acetal} \quad 200-300\,°C \quad \text{alkyl vinyl ether} \end{array}$$

These monomers are also produced by an oxidative process in which the alkanols are added directly to ethylene and the alkyl ethers are thermally decomposed to produce hydrogen and the alkyl vinyl ethers.

Commercial polymers have been produced from methyl, ethyl, isopropyl, n-butyl, isotubtyl, t-butyl, stearyl, benzyl and trimethylsilyl vinyl ethers. The poly(methyl vinyl ether) called PVM or Resyn is produced by the polymerization of the monomer by boron trifluoride in propane at $-40\,°C$ in the presence of traces of an alkyl phenyl sulfide. The polymer may have isotactic, syndiotactic or stereoblock configurations depending on the solvent and catalyst used.

Nonpolar solvents favor the formation of ion pairs between the polymer cation and the counteranion and favor the production of isotactic polymers. Soluble catalysts, such as diethyl aluminum chloride and ethyl aluminum dichloride, also affect the stereoregularity of the polymer chains. The tendency for the formation of stereoregular polymers is decreased as the size of the alkyl group is increased.

Poly(methyl vinyl ether) is soluble in cold water but becomes insoluble in a reversible process when the temperature is raised to $35\,°C$. This sticky polymer has a glass transition temperature of $-20\,°C$. It has been used as an adhesive and as a heat sensitizer for polymer latices.

Poly(vinyl ethyl ether) is soluble in ethanol, acetone and benzene. It is a rubbery product which may be cross-linked by heating with dicumyl peroxide. Poly(vinyl isobutyl ether), has a glass transition temperature of $-5\,°C$. It has been used as an adhesive for upholstery, cellophane and adhesive tape.

The processing properties of poly(vinyl chloride) has been improved by copolymerizing vinyl chloride with a small amount of vinyl alkyl ether. Copolymers of vinyl alkyl ethers and maleic anhydride are used as water soluble thickeners, paper additives, textile assistants and in cleaning formulations.

POLYVINYL CHLORIDE (PVC). [CAS: 9002-86-2]. The manufacture of polyvinyl chloride resins commences with the monomer, vinyl chloride, which is a gas, shipped and stored under pressure to keep it in a liquid state; bp $-14\,°C$, fp $-160\,°C$, density $(20\,°C)$, 0.91. The monomer is produced by the reaction of hydrochloric acid with acetylene. This reaction can be carried out in either a liquid or gaseous state. In another technique, ethylene is reacted with chlorine to produce ethylene dichloride. This is then catalytically dehydrohalogenated to produce vinyl chloride. The by-product is hydrogen chloride. A later process, oxychlorination, permits the regeneration of chlorine from HCl for recycle to the process.

Polymerization may be carried out in any of the following manners:

1. **Suspension** a large particle size dispersion or suspension of vinyl chloride is made in water by addition of a small quantity of emulsifying agent. The product after polymerization and drying consists of granules.
2. **Emulsion** a larger quantity of emulsifier is employed, resulting in a fine particle size emulsion. The polymer after spray drying, is a finely divided powder suitable for use in organosols and plastisols.
3. **Solution** vinyl chloride is dissolved in a suitable solvent for polymerization. The resultant polymer may be sold in solution form, or dried and pelletized.

Emulsions may be polymerized by use of a water-soluble catalyst (initiator), such as potassium persulfate, or a monomer-soluble catalyst, such as benzoyl peroxide, lauroyl peroxide or azobisisobutyronitrile. Suspension and solution polymerizations employ the monomer soluble catalysts only. In addition to the above-mentioned initiators, diisopropyl peroxydi-carbonate may also be employed, where lower-temperature polymerization may be desired, e.g., to reduce branching and minimize degradation.

Because of the low level of emulsifiers and protective colloids, the suspension polymer types are most suitable for electrical applications and end uses requiring clarity. This form is also employed in the bulk of extrusion and molding applications. Cost is lower than for emulsion and solution forms. The emulsion or dispersion resins are employed mainly for organosol and plastisol applications where fast fusion with plasticizer at elevated temperature will occur as a result of the fine particle size of the resin.

Monomers such as vinyl acetate or vinylidene chloride may be copolymerized with vinyl chloride. Up to 15% of the comonomer may be employed. Vinyl acetate increases the solubility, film formation and adhesion. Processing or forming temperatures are generally lowered. Chemical resistance and tensile strength decrease with increasing amount of vinyl acetate.

Rigid Vinyls

These have been separated into two categories according to ASTM:

Type I is rigid PVC with excellent chemical resistance, physical properties and weathering resistance such as obtained from unplasticized high molecular weight PVC.

Type II has the added feature of high impact resistance but with slightly lower chemical and physical requirements.

Perhaps the most important applications for rigid PVC will be in building. This is a rapidly growing market. Fabrication is via extrusion. Examples of applications are pipe, siding, roofing shingles, panels, glazing, window and door frames, rain gutters and downspouts.

Blow-molded bottles, which exhibit excellent product resistance, and good clarity, are also expected to become an important outlet for rigid PVC.

Formulations for extrusion generally include light and heat stabilizers, lubricants, which facilitate molding, and colorants. These materials are generally purchased in a compounded ready to use cube form, in order to minimize irregularities in blending, etc.

The outstanding characteristics of these rigid vinyls are chemical, solvent and water resistance; resistance to weathering when properly stabilized, therefore permitting long-term outdoor exposure; and low cost. Abrasion and impact resistance are satisfactory.

A major deficiency is heat sensitivity. Here, degradation begins with the split-off of HCl. Temperature accelerates this effect.

PVC has a high coefficient of expansion, one of the highest for all plastic materials, and substantially higher than metals and wood. Therefore,

design allowances must be made to provide for movement in order to avoid buckling, breakage, etc.

Flexible PVC

An unusually wide variety of products and usages are possible with plasticized vinyls. Typical applications include floor and wall coverings, boots, rainwear, jackets, upholstery, garden hose, electrical insulation, film and sheeting, foams and many others.

The primary processing techniques are by means of extrusion, calendering and molding. Special techniques involve organosols and plastisols.

Plasticizers used to develop the desired flexibility and performance are selected on the basis of cost and application requirements, e.g., temperature; service life; exposure to solvents, chemicals, water, UV, food; tensile strength; abrasion resistance; flexibility; tear strength, etc."

Plasticizers must be classed as *primary*, where high compatibility is limited, thus restricting the amount that can be tolerated. The addition of secondary plasticizers may import special properties or simply reduce cost (extender plasticizer).

Primary plasticizers may be further subdivided. The *phthalate* types are by far the most popular due to cost and ease of incorporation. Dioctyl phthalate and diisooctyl phthalate are typical of this class. They exhibit good general-purpose properties. *Phosphate* plasticizers are also important for general-purpose use. Typical of these are tritolyl phosphate and trixylenyl phosphate. These plasticizers also impart fire retardant properties. *Low-temperature* plasticizers, such as dibutyl sebacate, are used where good low-temperature flexibility is required. For maximum compatibility and minimum cost, a typical plasticizer combination would be a blend of 50% DOP and 50% dibutyl sebacate.

Polymeric plasticizers are generally polyesters with a relatively low molecular weight. They are used where resistance to high temperatures and freedom from migration and extraction are required. Polymerics are more difficult to incorporate, have poor low-temperature properties, and are expensive.

Epoxy plasticizers are epoxidized oils and esters. These are generally classed with the polymerics. However, molecular weight is lower. Therefore, resistance to extraction and heat are slightly inferior. Low-temperature properties are better and epoxies are more easily incorporated.

Extender plasticizers, which are used mainly to reduce cost, consist of chlorinated waxes, petroleum residues, etc. Incorporation of excessive amounts may result in exudation on aging. The chlorinated types decrease flammability.

Organosols and Plastisols

Plastisols are dispersions of powdered PVC resin in plasticizer. A typical composition would consist of 100 parts of PVC resin dispersed in 50 parts of DOP. The resultant paste when heated to $300\,°F$ $(149\,°C)$ fuses or "fluxes into a solid plastic mass. Stability of this plastisol at room temperature may range from several weeks to several months depending on the plasticizers and resins employed.

An organosol is the same mixture as described above, with the addition of solvent to reduce viscosity. These find their major applications in coatings. The solvent is evaporated before fusion of the film. Various pigments, colorants, stabilizers and fillers may be added, depending on the desired properties. Emulsion polymerization resins are generally employed because of their fast fusion rates. Coarser particle sized PVC resins would require extended time at the elevated temperature.

Plastisols allow the use of inexpensive manufacturing techniques, such as slush and rotational molding, casting, dipping, etc. They are employed for the manufacture of a large variety of parts, e.g., toys, floor mats, handles and many others.

Foams are made by the addition of blowing agents to the plastisol. These may be continuously applied to a moving substrate which includes a pass at an elevated temperature where foaming occurs, followed by fusion of the plastisol.

Organosols find their major application in coatings, which may be applied by spray, dip, knife, roller, etc. Typical products are coated aluminum siding, fabrics, paper, industrial coatings, etc.

An important development was the use of plasticizers which crosslink upon application of heat and thus produce a more rigid end product. This extends the range of products obtainable by plastisol techniques into rigids. By varying the amount of crosslinking plasticizer incorporated, various levels of flexibility are obtained.

POLYVINYLIDENE CHLORIDE. [CAS: 9002-86-2]. A stereoregular, thermoplastic polymer is produced by the free-radical chain polymerization of vinylidene chloride ($H_2C=CCl_2$) using suspension or emulsion techniques. The monomer has a bp of 31.6 °C and was first synthesized in 1838 by Regnault, who dehydrochlorinated 1,1,2-trichloroethane which he obtained by the chlorination of ethylene. The copolymer product has been produced under various names, including *Saran*. As shown by the following equation, the product, in production since the late 1930s, is produced by a reaction similar to that used by Regnault nearly a century earlier:

$$H_2ClCCHCl_2 \; + \; Ca(OH)_2 \xrightarrow[-2H_2O]{90°C}$$

1,1,2-tri- calcium
chloroethane hydroxide

$$CaCl_2 \; + \; 2H_2C=CCl_2$$

calcium vinylidene
chloride chloride

Since this monomer readily forms an explosive peroxide, it must be kept under a nitrogen atmosphere at −10 °C in the absence of sunlight.

The copolymers were patented by Wiley, Scott, and Seymour in the early 1940s. A typical formulation for emulsion copolymerization contains vinylidene (85 g), vinyl chloride (15 g), methylhydroxypropylcellulose (0.05 g), lauroyl peroxide (0.3 g) and water (200 g). More than 95 per cent of these monomers are converted to copolymer when this aqueous suspension is agitated in an oxygen-free atmosphere for 40 hrs at 60 °C. The glass transition temperature of the homopolymer is −17 °C. It has a specific gravity of 1.875 and a solubility parameter of 9.8.

Because of its high crystallinity, the homopolymer (PVDV) is insoluble in most solvents at room temperature. However, since the regularity of repeating units in the chain is decreased by copolymerization, Saran is soluble in cyclic ethers and aromatic ketones. This copolymer (100 g) is plasticized by the addition of α-methyl-benzyl ether (5 g), stabilized against ultraviolet light degradation by 5-chloro-2-hydroxybenzophenone (2.0 g) and heat stabilized by phenoxypropylene oxide (2.0 g).

The poly(vinylidene chloride-co-vinylchloride) may be injection molded and extruded. Extruded pipe and molded fittings which were produced in large quantity in the 1940s have been replaced to some extent by less expensive thermoplastics. A flat extruded filament is used for scouring pads and continuous extruded circular filament is used for the production of insect screening, filter clothes, fishing nets and automotive seat covers.

A large quantity of this copolymer is extruded as a thin tubing which is biaxally stretched by inflating with air at moderate temperatures before slitting. This product, called Saran Wrap, has a tensile strength of 15,000 psi (103 MPa). Since it has a high degree of transparency to light and a high coefficient of static friction (0.95) it is widely used for the protection of foods in the household. It has a low permeability value for gases such as oxygen and nitrogen.

Poly(vinylidene chloride-co-acrylonitrile) is widely used as a latex coating for cellophane, polyethylene and paper. Since this copolymer is soluble in organic solvents, it is also used as a solution coating. The resistance to vapor permeability and the ease of printing on polyethylene and cellophane is increased by coating with this vinylidene chloride copolymer.

The tensile strength of both film and fiber is increased tremendously by cold drawing 400–500 per cent. Thus, tensile strengths as high as 40,000 psi (276 MPa) in the direction of draw have been obtained by cold drawing.

POLYVINYLIDENE FLUORIDE. [CAS: 24937-79-9]. This product is made by the free-radical chain polymerization of vinylidene fluoride ($H_2C=F_2$). This odorless gas which has a boiling point of −82 °C is produced by the thermal dehydrochlorination of 1,1,1-chlorodifluoroethane or by the dechlorination of 1,2-dichloro-1,1-difluoro-ethane. As shown by the following equations, 1,1,1-chlorodifluoroethane may be obtained by the hydrofluorination and chlorination of acetylene and by the hydrofluorination of vinylidene chloride or of 1,1,1-trichloroethane.

$$HC\equiv CH \; + \; 2HF \longrightarrow H_3CCHF_2$$

acetylene hydrogen 1,1-difluoro-
fluoride ethane

$$\xrightarrow[-HCl]{Cl_2} H_3CCClF_2$$

1,1,1-chloro-
difluoroethane

$$H_2C=CCl_2 \; + \; 2HF \xrightarrow{-HCl} H_3CCClF_2$$

vinylidene hydrogen
chloride fluoride

$$H_3C-CCl_3 \; + \; 2HF \xrightarrow{-2HCl} H_3CCClF_2$$

1,1,1-trichloro-
ethane

$$H_3CCClF_2 \xrightarrow[-HCl]{600°} H_2C=CF_2$$

1,1,1-chlorodi- vinylidene
fluoroethane fluoride

$$H_2ClCCClF_2 \xrightarrow[-Cl_2]{\Delta} H_2C=CF_2$$

1,2-dichloro-1,1-
difluoroethane

Polyvinylidene fluoride is polymerized under pressure at 25–150 °C in an emulsion using a fluorinated surfactant to minimize chain transfer with the emulsifying agent. Ammonium persulfate is used as the initiator. The homopolymer is highly crystalline and melts at 170 °C. It can be injection molded to produce articles with a tensile strength of 7000 psi (48 MPa), a modulus of elasticity in tension of 1.2×10^5 psi and a heat deflection of 300 °F (149 °C).

Poly(vinylidene fluoride) is resistant to most acids and alkalies but it is attacked by fuming sulfuric acid. It is soluble in dimethylacetamide but is insoluble in less polar solvents. Copolymers have been produced with ethylene, tetrafluoroethylene, chlorotrifluoroethylene and hexafluoroethylene. The latter is an elastomer called Viton or Fluorel.

The homopolymer is used as a chemical resistant coating for steel, for tank linings, hose, and pump impellors. The elastomeric copolymer with hexafluoroethylene when cured with hexamethylenediamine is used as a seal, gasket, 0-ring, tubing, coating and lining.

POMEGRANATE TREE. Of the family *Myrtaceae* (myrtle family), the pomegranate tree (often a shrub), *Punicum granatum*, is native to

Fig. 1. Fruits of the pomegranate tree. (*USDA photo.*)

southeastern Europe and southwestern Asia. The tree has been cultivated since early times and is now grown extensively in tropical and subtropical regions in both hemispheres. The flowers are borne either singly or in small clusters in the axils of the leaves. They are perfect and have a bright red corolla of five to eight petals, and many stamens. The fruit is a many-seeded berry. The outer coat of each seed is the edible portion. It is soft and fleshy and red in color.

See Fig. 1.

POND.

A pond is a body of water usually smaller than a lake, encircled by vegetation, and generally shallow enough for sunlight to reach the bottom. Rotted plants can grow in any spot within the pond. Widespread plant growth, such as water hyacinth, pickerelweed, spatterdock, and cattails, is quite common in and around ponds and aids in creating a habitat for various forms of animal life, including insects (e.g., damselflies and dragonflies), bullfrogs, greenfrogs, red winged blackbirds, marsh wrens, and bats. See also **Water Hyacinth**.

Temporary ponds, that is, ponds that are dry for part of the year, are especially interesting and support a unique community. Organisms in such ponds must be able to survive in a dormant stage during dry periods or be able to move in and out of ponds, as can amphibians and adult aquatic insects. The temporary pond is a favorable place for those organisms adapted to it because interspecific competition and predation are reduced.

The chemical composition of any pond water ultimately depends on those elements and compounds in solution, in suspension, and those accumulating along the pond bottom, most of which come secondarily to the ponds from runoff. Other substances are produced by chemical reactions between the different elements and/or compounds by vegetation growing in or around the ponds.

Water Quality

Salts. Water in closed pond basins is unlike the waters of other ponds in two respects: (1) it continually experiences a change in chemical composition because (2) it continually increases its concentration of various salts, mainly due to a high evaporation rate. However, the increase in salts in a closed basin is not infinite. Certainly, as water flows into a dry pond basin, perhaps from the first rains of late summer or early fall, it carries the salts from rocks surrounding the lake basin and dissolves the surface salts of the playa, providing they have not been removed by deflation. As precipitation declines, generally after the early spring rains, evaporation becomes greater than inflow which then starts the inevitable process of concentration of the salts in solution. Salts concentrated in the waters of closed basins depend mainly on the rock type(s) of the surrounding basins and/or the chemical characters of any influents and, to a minor degree, on several inorganic ions in rainwater.

Pollutants. This catchall category of pollutants means anything added to the pond that is not wanted in the pond. Pollutants can consist of items that may or may not be harmful to human beings as well as fish, and they may or may not be visible. They may float on the surface, sink to the bottom, or dissolve in the water. They may come from outside the pond or from within the pond itself, for example, oil leaking from a submerged pump. Some pollutants are easy to identify and control and/or remove, such as leaves, pollen, and dead rats; some just add to the filter load if not removed but cause few other problems if not in excessive amounts, for example, bird droppings. Most harmful pollutants that dissolve in water are hard to identify or quantify. Surface water runoff that can enter the pond is often a major source of pollutants. That is why all ponds should be designed with a raised edge or at least some type of channel around them so that the surface water does not enter.

Dissolved Oxygen. Whenever air is in contact with water, whether through natural or artificial means, a transfer of oxygen from the air to the water takes place until the water becomes saturated. Plants under light convert carbon dioxide to oxygen in water. Fish, plants at night, and aerobic bacterial action consume the oxygen.

As the air components dissolve in the water, a point is reached where no more can be added. This point is called saturation. The saturation points are different for each of the gases and depend on several factors, but temperature is the most important. As the temperature increases, the water simply cannot hold as much of each type of gas. For oxygen, the approximate saturation level is 11.5 mg/L at 50 °F, 9 mg/L at 70 °F, and 7.5 mg/L at 90 °F. Impurities added to the water (i.e., salt) decrease these saturation levels.

Water whose oxygen concentration is less than 3 mg/L will not generally support fish. When the concentration falls to about 3–4 mg/L, fish start gasping for air at the surface. Levels from 3–5 mg/L can normally be tolerated for short periods. Above 5 mg/L, almost all aquatic organisms can survive indefinitely, provided other environmental parameters are within allowable limits.

Surface oxygen from 11–15 mg/L creates a highly eutrophic condition. Depletion of dissolved oxygen in pond water can encourage the microbial reduction of nitrate to nitrite and sulfate to sulfide, giving rise to odor problems. It can also cause an increase in the concentration of iron (II) in solution.

Color. The color of pond water depends on the depth, the color of the surrounding area and sky, the scattered light rays coming from the water, and in most cases, absorption of scattered light rays by dissolved or suspended materials.

Pond water is typically blue for pure, clear water, and various shades of blue–green to a definite green, then into the yellows and browns; the various colors are produced by the innumerable contaminants (Table 1). In general, black pond water results from high organic content, and blue pond water results from an absence of contaminating materials. The color of most pond water exists because of light scattering by vibrating water molecules blue predominates because molecular scattering is greater for shorter wavelengths.

TABLE 1. VARIATION IN WATER COLOR

Color	Contaminants
Clear blue (pure water)	No contaminants of suspended materials; color due to molecular scattering
Bluish green	Suspension of blue-green algae, phytoplankton
Green	Colloidal $CaCO_3$ (hard water)
Yellowish green	Suspended sulfur
Greenish yellow	High plankton content, excluding humic debris
Yellow	Diatoms
Yellow brown	Diatoms-fluorescent compounds, humic acids, carboxylic acids, organic matter
Gray	*Spirostomum*
Black	Organic debris, oxidation of ferrous iron, *Stentor*
Red to orange to purple	Suspended clays, ferrous hydroxide, *Oscillataria rubescens*, *Euglena sanquinea*, *Haematococcus pluviolis*, *Dunaliella salina*, *Bacterium halobium*, purple sulfur bacteria, zooplankton.

Shapiro, J. *The Coloring Matter of Natural Waters*, Yale University, New Haven, CT. 1956. (unpublished).

Ammonia. Ammonia (NH_3), in parts per million (ppm), is the first measurement to determine the "health" of a biological converter. Ammonia should not be detectable in a pond that has a "healthy" bioconverter. The ideal and normal measurement of ammonia is zero. When ammonia is dissolved in water, it is partially ionized depending upon the pH and temperature. Ionized ammonia is called ammonium and is not toxic to fish. As the pH drops and the temperature decreases, ionization and ammonium increase, which decreases the toxicity. The presence of ammonia in ponds at higher than geogenic levels is an important indicator of fecal pollution.

Ammonia tends to block oxygen transfer from fish gills to the blood and can cause both immediate and long-term gill damage. The mucous-producing membranes can be destroyed, reducing both the external slime coat and damaging the internal intestinal surfaces. Fish suffering from ammonia poisoning usually appear sluggish, often at the surface as if gasping for air.

Ammonia is a gas primarily released from fish gills as a metabolic waste from protein breakdown and from some lesser secondary sources such as bacterial action on solid wastes and urea.

When ammonia is detected (assuming a pH of about 7.5),

1. Increase aeration to maximum. Add supplemental air if possible.
2. Stop feeding the fish if detected in an established pond.

3. For an ammonia level of 0.1 ppm, conduct a 10% water change out. For a level of 1.0 ppm, conduct a 25% change out.
4. Consider transferring fish if the ammonia level reaches 2.5 ppm.
5. Retest in 12 to 24 hours.

Nitrite. Nitrite is produced by the autotrophic *Nitrosomonas* bacteria that combine oxygen and ammonia in the bioconverter and to a lesser degree on the walls of the pond.

Nitrite has been termed the invisible killer. The pond water may look great, but nitrite cannot be seen. It can be deadly, particularly to the smaller fish, in concentrations as low as 0.25 ppm. Nitrite damages the nervous system, liver, spleen, and kidneys of fish.

Whenever 0.25 ppm of nitrite or more is detected in a pond,

1. Increase aeration to maximum. For a nitrite level of 1 ppm or greater, add supplemental air, if possible.
2. Discontinue use of any UV sterilizers and ozone generators.
3. For a nitrite level less than 1 ppm, conduct a 10% water change out, and add 1 pound of salt per hundred gallons of changed water.
4. For nitrite levels of 4.0 or greater, consider transferring fish.

Nitrate. Nitrate is produced by the autotrophic *Nitrobacter* that combine oxygen and nitrite in the bioconverter and to a lesser degree on the walls of the pond. The nitrate concentration in surface water is normally low but can reach high levels as a result of agricultural runoff, refuse dump runoff, or animal wastes. Concentrations from zero to 200 ppm are acceptable. Ammonia and nitrite are toxic to fish, but nitrate is essentially harmless.

The nitrate concentration is controlled naturally through routine water change outs and to a lesser degree through plant/algae consumption.

Temperature. The temperature of the water in a pond is seldom constant; variations depend on depth, season, and geography. Pond water is heated by absorption of solar radiation. Events generally happen faster at higher temperature and in smaller ponds. Over normal temperature ranges, biological activity doubles for each 10 °C (50 °F) rise in temperature. The toxicity of ammonia increases as the temperature rises, and the amount of dissolved oxygen that the water can hold decreases.

Direct sunlight during the day can cause the temperature to rise higher, and heat loss on clear nights can cause the temperature to drop lower than that of shaded ponds. A clear night sky can absorb a large amount of heat from a small pond and actually drive the pond temperature below air temperature.

Plant Nutrients and Cultural Eutrophication. Pond water clarity (transparency) is affected by sediments, chemicals, and the abundance of plankton organisms; it is a useful measure of water quality and water pollution. Ponds that have clear water and low biological productivity are said to be oligotrophic. By contrast, eutrophic waters are rich in organisms and organic materials. Eutrophication, an increase in nutrient levels and biological productivity, often accompanies successional changes in ponds. Surrounding runoffs bring in sediments and nutrients that stimulate plant growth. Over time, ponds often fill in, becoming marshes or even terrestrial biomes. The rate of eutrophication depends on water chemistry and depth, volume of inflow, mineral content of the surrounding watershed, and the biota of the pond itself.

Human activities can greatly accelerate eutrophication, an effect called cultural eutrophication. Cultural eutrophication is caused mainly by increased nutrient input into a waterbody. Eutrophication produces "blooms" of algae or thick growth of aquatic plants stimulated by elevated phosphorus or nitrogen levels. Bacterial populations then increase, fed by larger amounts of organic matter. The water often becomes cloudy or turbid and has unpleasant tastes and odors. Cultural eutrophication can accelerate the "aging" of a waterbody enormously over natural rates. Ponds that normally might exist for hundreds of years can be filled in in a matter of decades.

pH. The pH of water is a measure of the acid–base equilibrium and, in most natural waters, is controlled by the carbon dioxide–bicarbonate–carbonate equilibrium. An increased carbon dioxide concentration in pond water will therefore lower pH, whereas a decrease causes it to rise. Temperature also affects the equilibria and the pH. In pure water, a decrease in pH of about 0.45 occurs as the temperature is raised by 25 °C (77 °F).

A pH measurement helps to determine if water is suitable for fish. The pH should normally be between 7.0 and 8.5; it is probably acceptable if it is anywhere between 6.0 and 9.0. Although most fish could tolerate a pH as low as 5.0, bioconverter bacteria are subject to damage. Long-term conditions above pH 9.0 can cause kidney damage to some fish.

A sudden change of a half or more pH unit in an established pond is an indication that something happened, and the cause should be determined. An increasing pH trend in a pond is normally caused by lime leaching out of concrete and to a lesser degree by concentration due to evaporation and decomposing organic matter. Decreasing pH is due primarily to bacterial action that releases acidic compounds. Established ponds normally maintain their equilibrium pH if sludge and decaying organic material are routinely removed. Scheduled water change outs (10% per week for small pond, less for larger ponds) are also helpful. At pH extremes approaching 4 or 11, remove any remaining fish.

Saprotrophic Organisms. Aquatic bacteria, flagellates, and fungi are distributed throughout a pond, but they are especially abundant in the mud–water interface along the bottom where bodies of plants and animals accumulate. A few of the bacteria and fungi are pathogenetic, attack living organisms, and cause disease, but the great majority begins attack only after an organism dies. When temperature conditions are favorable, decomposition occurs rapidly in a body of water; dead organisms do not retain their identification for very long but are soon broken up into pieces, consumed by the combined action of detritus-feeding animals and microorganisms, and their nutrients are released for reuse.

Additional Reading

Cunningham, W.P. and M.A. Cunningham: *Principles of Environmental Science: Inquiry and Applications*, The McGraw-Hill Companies, Inc., New York, NY, 2001.

Odum, E., and G.W. Barrett: *Fundamentals of Ecology*, 5th Edition, Brooks/Cole, New York, NY, 2004.

Reeves, C.C.: *Introduction to Paleolimnology*, Elsevier Science, New York, NY, 1968.

Nitish Priyadarshi, Ranchi University, Ranchi, Jharkhand, India

POPLAR TREES. Members of the family *Salicaceae* (willow family), some species of poplars are also known as cottonwoods and aspens. Poplars sometimes are classified into four groups as indicated in the following listing:

1. White Poplar (*Populus Alba*)
   Hybrid with Japanese balsam ('*Pyramidalis' Richardii*)
2. Balsam Poplars
   Black Cottonwood (*P. trichocarpa*)
      Hybrid with Chinese black poplar (× *generosa*)
   Balsam (*P. balsamifera*)
      Hybrid with Eastern cottonwood (*Candicans* or balm of Gilead)
   Simon Balsam (*P. simonii*)
   Japanese Balsam (*P. maximowiczii*)
3. Black Poplars
   Cottonwood (*P. deltoides*)
   "Black Poplar" (*P. nigra*)
   Carolina Black Poplar (*P. angulata*)
   Chinese Black Poplar (*P. lasiocarpa*)
4. Trembling Poplars
   American Aspen (*P. tremuloides*)
   European Aspen (*P. tremula*)
   Bigtooth Aspen (*P. grandidentata*)

Much interbreeding of the four groups has resulted in numerous improved and fast-growing species. Notably, interbreeding of the black poplars has produced a number of excellent hybrid black poplars, including: *P. canadensis*; *P. serotina*; *P. marilandica*; *P. regenerata*; *P. 'Italica'* (the Lombardy poplar) and several others.

The balsam poplar (*P. balsamifera*) is predominantly a Canadian tree, ranging from Labrador and Nova Scotia westward to Alaska, and northward to the tree limit. A few species are found beyond the Arctic Circle. In some locales, the trees are planted close together to form shelter belts. The tree is large, with the height ranging from 80 to 90 feet (24 to

27 meters) in adult trees. The bark is coarse with deep furrows. The leaf is deciduous and from 5 to 6 inches (12.7 to 15 centimeters) in length, egg-shaped, pale green, lighter underneath. The amber-colored twig is covered with blisters of resin which is quite fragrant. The seed is minute, hairy, and tufted. When released, it appears much like falling snow. The bud is fragrant. There are about 12 species, some being natural hybrids.

The black cottonwood (*P. trichocarpa*) is found mainly in the western part of North America, extending from California northward to British Columbia and Alaska. The tree reaches eastward from the West Coast into the Sierra Nevada mountains and is found at altitudes up to 10,000 feet (3,050 meters), but rarely below 3,000 feet (915 meters). The tree also occurs in the San Diego mountains. A healthy tree will range from 80 to 100 feet (24 to 30 meters) or more in height. The branches are slender and wide-spreading. The leaf is 5 to 6 inches (12.7 to 15 centimeters) long, 2 to 4 inches (5 to 10 centimeters) across, and darkly veined. The top of the leaf is a glistening olive-green, the underside is silver white. The bud is resinous, ovate, and fragrant, having a balsam odor. In older trees, the bark is dark brown and furrowed. Young trees have a smooth gray bark. Autumn coloration is a deep yellow.

The quaking aspen of America (*P. tremuloides*) extends over an extremely wide range—from Labrador to Mexico. Some trees are found as far north as the Bering Strait. Although vigorous, the life of the tree is relatively short, requiring much light. It is highly regarded as a shade tree. A related species is the bigtooth aspen (*P. grandidentata*) named for its large, toothed leaves. The European aspen (*P. tremula*) ranges north and eastward from western Europe into Russia.

The botanical explanation for the mysterious and intriguing quaking motion of the *quaking aspen* is that the aspen leafstalk is longer than the leaf itself and flattened opposite the plane of the leaf, thus becoming a sensitive pivot upon which the thin, papery leaves flutter almost magically, even in imperceptible breezes (Voynick, 1984). Prior to the present century, aspen thrived in a natural ecosystem in which wildfire regularly devastated mature stands of both aspen and conifers. With the shading canopy reduced or eliminated, sunlight stimulated prolific formation of root buds. Some of the present largest and most beautiful aspen groves started on fire-blackened slopes before the turn of the century. Forest management in recent years has altered the aspen's natural forest ecosystem. Prevention and containment of wildfire has permitted both conifers and aspen to grow older and larger and thus shade more of the forest floor. In the long run, however, this is a condition that favors the reproduction of conifers to the detriment of the aspen. Some foresters now regard this condition as a serious long-term threat to the aspen. Two interesting articles on aspens will be found in *American Forests* magazine: Ciesla, B., "A Tree (Aspen) for All Seasons," October 1982; and Voynick, S.M., "Trouble in the Quakies," May 1984. The yellow poplar is described in "Rediscovering the Yellow Poplar" by D.A. Boerner-Ein in the July 1991 issue.

The yellow poplar is actually not a poplar, but a tulip tree and member of *Magnoliaceae* (magnolia family).

The wood from various species of poplar trees is valuable commercially. Lumber from the cottonwoods (*P. monilfera* and *P. deltoides*) is soft and of a yellowish-white color, possessing a fine, open grain. It is sometimes referred to as Carolina poplar or whitewood. The weight is about 30 pounds per cubic foot (481 kilograms per cubic meter). Easy to work, but not strong and prone to warping, the wood is used mainly for making paneling, packing boxes, and some general carpentry. Wood from the balsam poplar (*P. balsamifera*) is a weak and soft wood and used mainly for making containers and excelsior. It also makes an excellent pulpwood for paper production. Wood from the aspen (*P. tremula*) is widely used for making match sticks and excelsior, with limited use for inside construction. Easily bleached, the wood also is used for paper pulp.

See Table 1.

POPPY (*Papaver somniferum; Papaveraceae*). The poppy from which opium is obtained is an annual herb having a smooth branching stem 2–3 feet (0.6–0.9 meter) tall, large, dull, green, smooth leaves and solitary single flowers, varying from white to purple in color and rather showy. The flower consists of two sepals, which soon fall off when the flower

TABLE 1. RECORD POPLAR TREES IN THE UNITED STATES[1]

Specimen	Circumference[2]		Height		Spread		Location
	Inches	Centimeters	Feet	Meters	Feet	Meters	
ASPENS							
Bigtooth aspen (1980) (*Populus grandidentata*)	140	356	102	31.1	64	19.5	Kentucky
Bigtooth aspen (1984) (*Populus grandidentata*)	105	267	132	40.2	67	20.4	Michigan
Quaking aspen (1991) (*Populus tremulides*)	122	310	109	33.2	59	18	Michigan
Quaking aspen (1998) (*Populus tremulides*)	127	323	114	34.7	32	9.8	Arizona
COTTONWOODS							
Black cottonwood (1995) (*Populus trichocarpa*)	320	813	158	48.2	110	33.5	Oregon
Eastern cottonwood (typ.) (1991) (*Populus deltoider var. deltoides*)	433	1100	85	25.9	121	36.9	Idaho
Fremont cottonwood (typ.) (1996) (*Populus fremontii var. fremontii*)	504	1280	92	28	108	32.9	Arizona
Meseta cottonwoodk (1986) (*Populus fremontii var. mesetae*)	190	483	60	18.3	60	18.3	Texas
Narrowleaf cottonwood (1973) (*Populus angustifolia*)	314	798	79	24.1	80	24.4	Oregon
Plains cottonwood (1967) (*Populus deltoides var. occidentalis*)	432	1097	105	32	93	28.3	Colorado
Rio Grande cottonwood (1997) (*Populus fremontii var. wislizeni*)	366	930	123	37.5	104	31.7	Texas
Swamp cottonwood (1990) (*Populus heterophylla*)	42	107	55	16.8	23	7	Mississippi
POPLARS							
Balsam poplar (1991) (*Populus balsamifera*)	165	419	128	39	57	17.4	Michigan
White popular (1992) (*Populus alba*)	263	668	93	28.3	86	26.2	Illinois

[1]From the "National Register of Big Trees," American Forests (by permission).
[2]At 4.5 feet (1.4 meters).

opens, four petals, many stamens and a single pistil with a one-celled ovary. The fruit is a capsule, 1–2 inches (2.5–5 centimeters) in diameter, containing many small seeds, which escape through a ring of pores which form around the top of the capsule, beneath the persistent stigma.

To obtain opium, the unripe capsules are incised with a knife. From these cuts the milky juice oozes and dries to form a plastic gummy substance, which is scraped off and molded into a ball. This crude opium contains fragments of the plant tissues and considerable dirt. About 10% of the opium is the alkaloid morphine. When first prepared opium is brownish and easily molded. It gradually dries to a hard brittle substance, easily ground to a powder. Besides morphine it contains many other alkaloids.

Poppy seeds, which contain no harmful substances, are frequently used in bread and cakes. From them is expressed an expensive oil used in cooking and in making artist's paints.

POPULATION (Statistics).   A set of observations is commonly regarded as a sample from a larger set, called the population. The population may be actually existing and of finite size, as in practical sampling inquiries; or it may consist of the infinite number of observations that might hypothetically be obtained under the same condition as the actual sample.

PORCELAIN. $(4K_2O \cdot Al_2O_3 \cdot 3SiO_2)$, high impact strength; impermeable to liquids and gases; resistant to chemicals except hydrogen fluoride and hot, strong caustic solutions; usable up to 1093C but subject to heat shock. D 2.41, Mohs hardness 6-7, compression strength 100,000 psi. Porcelain is a mixture of clays, quartz, and feldspar usually containing at least 25% alumina. Ball and china clays are ordinarily used. A slip or slurry is formed with water to form a plastic, moldable mass, which is then glazed and fired to hard, smooth solid.

Uses

Reaction vessels, spark plugs, electrical resistors, electron tubes, corrosion-resistant equipment, ball mills and grinders, food-processing equipment, piping, valves, pumps, and laboratory ware.

See also **Ceramics**; **Enamels, Porcelain or Vitreous**; **Porcelain Enamel**; and **Porcelain, Zircon**.

PORCELAIN ENAMEL.   A substantially vitreous inorganic coating bonded to metal by fusion above 426C (ASTM). Composed of various blends of low-sodium frit, clay, feldspar, and other silicates; ground in a ball mill; and sprayed onto a metal surface (steel, iron, or aluminum), to which it bonds firmly after firing, giving a glasslike fire-polished surface. See also **Enamels, Porcelain or Vitreous**.

PORCELAIN, ZIRCON. $(ZrO_2 \cdot SiO_2)$. A special high-temperature porcelain used for spark plugs and furnace trays because of its high mechanical strength and heat-shock resistance. Usable up to 1700 °C with high dielectric strength but rather lower power factor at high frequencies.

PORCUPINE.   See **Rodentia**.

PORCUPINE FISHES (*Osteichthyes*).   Of the order *Plectognathi*, family *Diodontidae*, porcupine fishes are puffers with spines. While swimming, the fish holds the spines pressed close to the body. The spines appear when the fish swallows air or water. There are about 15 types of porcupine and burrfishes known. The burrfishes (genus *Chilomycterus*) have shorter spines that normally are extended. Also sometimes called the spiny boxfish. By most people, the porcupine fish is usually seen as a decorative lamp made from a dried skin of the fish. These fishes also are called balloon fishes. A porcupine fish is shown in Fig. 1. See also **Puffers (Osteichthyes)**.

PORE.   1. A minute cavity in epidermal tissue as in skin, leaves, or leather, having a capillary channel to the surface that permits transport of water vapor from within outward but not the reverse. 2. A void of interstice between particles of a solid such as sand minerals or powdered metals, that permits passage of liquids or gases through the material in either direction.

Fig. 1.   Porcupine fish.

In some structures, such as gaseous diffusion barriers and molecular sieves, the pores are of molecular dimensions, i.e., 4-10 Å units. Such microporous structures are useful for filtration and molecular separation purposes in various industrial operations. 3. A cell in a spongy structure made by gas formation (foamed plastic) that absorbs water on immersion but releases it when stressed.

See also **Molecular Sieve**; and **Semipermeable Membrane (or Semipermeable Diaphragm)**.

PORGIES (*Osteichthyes*).   Of the family *Sparidae*, the porgy is a valuable food fish found along the Atlantic coast from Cape Cod to South Carolina. This species is also called the scup or scuppaug. Some 14 species of porgies occur in the American Atlantic, including the *Stenotomus chrysops* (northern porgy or scup); the *Archosargus rhomboidalis* (the sheepshead); and the *Lagodon rhomboides* (the pinfish). The sheepshead is also a valuable sporting and food fish. Porgies are not found on the eastern side of the Atlantic. The *Calamus brachysomus* occurs in California waters, and the *Monotaxis grandoculis* is found in Hawaiian waters. The *Chrysophrys guttulatus* (bump-headed porgy) is an important fish in Australian waters. The Australians refer to the fish as a snapper. Possibly the largest of the sparids is *Cymatoceps nasutus*, a musselcracker that can weigh up to 100 pounds (45 kilograms) and is a favorite sporting fish among South Africans. See also **Fishes**.

PORIFERA.   The sponges. A phylum of animals of low organization, related to some of the one-celled protozoans and much more primitive than any other multicellular group. Because of their loosely integrated structure the sponges are regarded as one of three major types of animal organization, designated by the term *Parazoa*. This group lies between the *Protozoa*, also a single phylum, and the *Metazoa*, containing all of the other multicellular phyla. See Fig. 1 on p. 2817.

Sponges develop only two germ layers, the ectoderm and endoderm, but many different cells lie in the mesogloea between the two. The body wall is perforated by many canals leading to a central cavity, the paragaster. Some part of these passages is lined with collared flagellate cells, which produce currents of water flowing inward through the pores and out of a larger opening of the paragaster called the osculum. The body wall contains several kinds of specialized cells. The scleroblasts form hard supporting structures of various forms and materials, called spicules. Phagocytes ingest, digest, and transport food. Porocytes become perforated to form canals. The outer surface is covered with flattened cells called pinacocytes.

Three kinds of sponges are recognized, according to the plan of the canal system: (1) Ascon sponges have canals leading entirely through the body wall and collared cells (choanocytes) in the lining of the paragaster. (2) Sycon sponges have radial canals lined with choanocytes and opening into the paragaster. Between them inhalant canals lead inward from the outside but do not reach the paragaster. The two types of canals are connected by minute pores, called prosopyles, through which water must pass to reach the interior. (3) In the rhagon or leucon sponges the canal

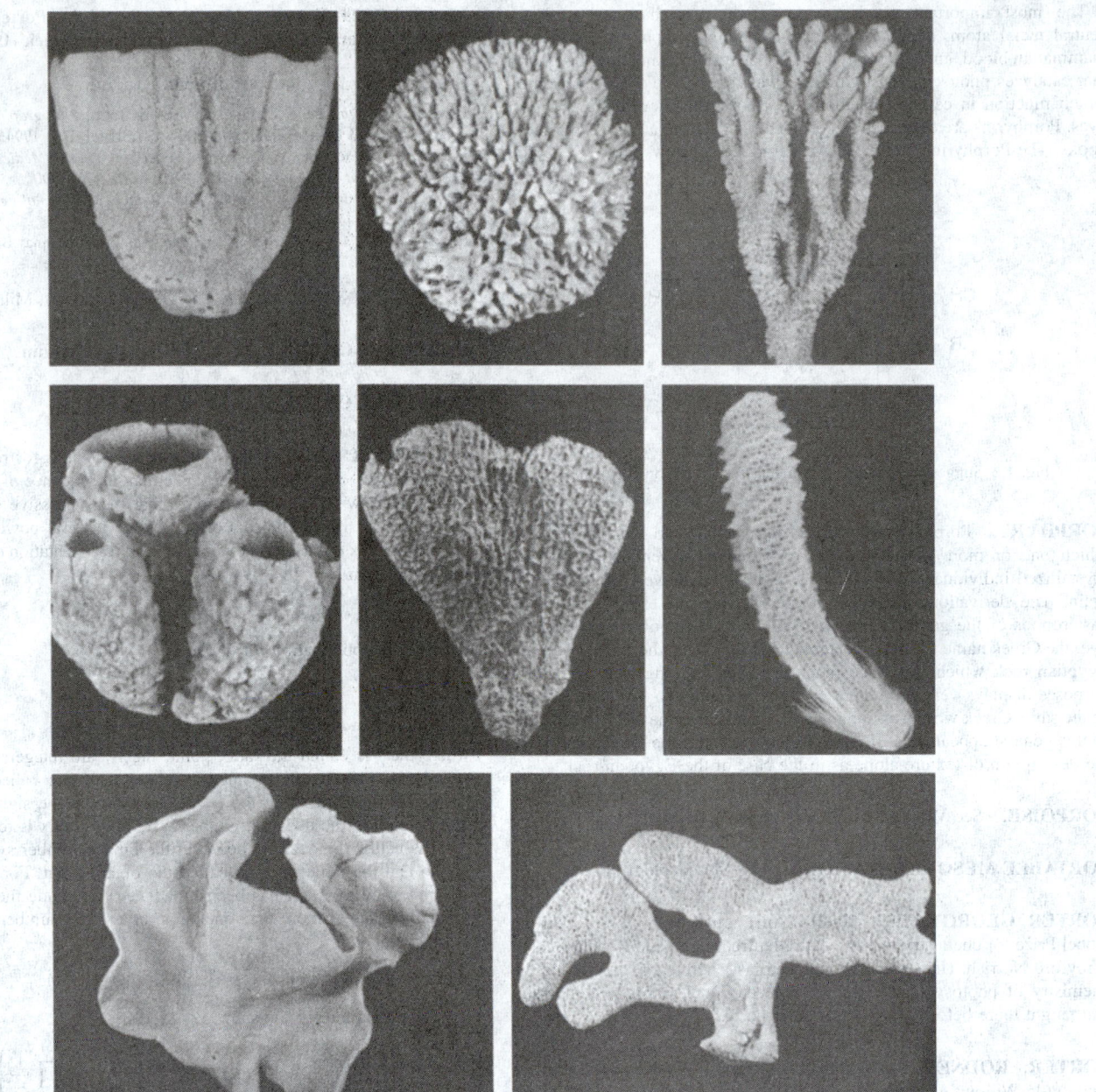

Fig. 1.   Varieties of sponges. (*A.M. Winchester.*)

systems are more intricately branches and the choanocytes are located in small chambers.

Commercial sponges are the skeletal remains of species whose bodies are supported by fibers of a peculiar material, spongin. The organic matter is removed by maceration and washing.

The phylum is divided into three classes:

Class *Calcarea*. Sponges whose bodies contain calcareous spicules only. The choanocytes are large and all three types of canals systems are represented.

Class *Hexactinellida*. Spicules six-rayed and siliceous. Choanocytes small. Canal system of a simple rhagon type. The glass sponges.

Class *Demospongiae*. Spicules siliceous but not six-rayed or an association of siliceous (silicon) material and spongin. Rhagon type of canal system. The sponges of commerce are included here. The subfamily *Spongillinae* includes the only species of sponges found in fresh water. See also **Invertebrate Paleontology**.

POROMERIC. A term coined to describe the microporosity, air permeability, and water and abrasion resistance of natural and synthetic leather. The pores decrease in diameter from the inner surface to the outer and thus permit air and water vapor to leave the material while excluding water from the outside. Polyester-reinforced urethane resins have been used as leather substitutes with some success, primarily for shoe uppers.

POROSITY. Two common uses of this term are: (1) the property of containing pores, which are minute channels or open spaces in a solid; (2) the proportion of the total volume occupied by such pores.

PORPHYRIN. Any of several physiologically active nitrogenous compounds occurring widely in nature. The parent structure is comprised of four pyrrole rings, shown in I, II, III, and IV in Fig. 1, together with four nitrogen atoms and two replaceable hydrogens, for which various metal atoms can be readily substituted. A metal-free porphyrin molecule has the structure shown in the diagram. Porphyrins of this type have been made synthetically by passing an electric current through a mixture of ammonia, methane, and water vapor. Some biochemists suggest that this phenomenon may account for the early formation of chlorophyll and other porphyrins which have been essential factors in the development of life.

The most important porphyrin derivatives are characterized by a central metal atom; hemin is the iron-containing porphyrin essential to mammalian blood, and chlorophyll is the magnesium-containing porphyrin that catalyzes photosynthesis. Other derivatives include the cytochromes, which function in cellular metabolism, and the phthalocyanine group of dyes. Porphyrins are described in considerable detail in a 20-volume set of books, The Porphyrin Handbook, Academic Press, New York, NY, 2003.

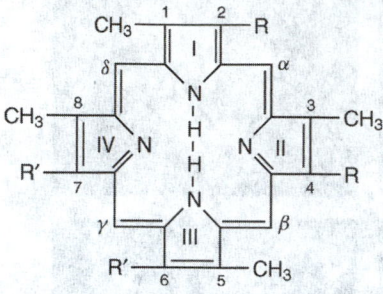

Fig. 1. Suggested structure of a metal-free porphyrin molecule.

PORPHYRY. Porphyry is a textural term applied to igneous rocks in which one or more of the mineral constituents present exists as well crystallized individuals in a ground mass that is relatively of much finer grain. The derivation of the word presents an interesting study. The gasteropods of the genus *Murex* were much used for obtaining a purple dye; the Greek name for both the animal and the dye is the same. A certain Egyptian rock which was once much used for building and ornamental purposes displays very prominent crystals in a purplish ground-mass and so the same Greek word was applied to it, then later came to mean all rocks of this general appearance. Modern use now restricts the term porphyry to the description of texture alone as in the case of the Egyptian rock.

PORPOISE. See **Whales, Dolphins, and Porpoises**.

PORTABLE MESONET STATIONS. See **Meteorology**.

PORTER, GEORGE (1920–2002). An English chemist who won the Nobel Prize for chemistry in 1967 with Manfred Eigen and Ronald George Wreyford Norrish. His research concerned fast chemical reactions and the chemistry of photosynthesis. He was educated at Cambridge University and taught there before going on to other posts.

PORTER, RODNEY ROBERT (1917–1985). Porter was born in Newton-le-Willows, Lancashire on 8 October 1917. Educated at the Universities of Liverpool and Cambridge, his career was interrupted by military service. In 1946 he joined Frederick Sanger at Cambridge. From 1949 to 1960 he worked on chromatographic protein fractionation with A. J. P. Martin at the National Institute for Medical Research at Mill Hill, London, where he became interested in antibodies. In 1950 he showed that some antibodies could be partly broken down without total loss of their antigen-binding properties. From 1960 to 1967 Porter was Professor of Immunology at St Mary's Hospital Medical School, University of London. In 1962, aware that G. M. Edelman had found that human immunoglobulin contains two protein chains linked by sulfur bonds, Porter split the IgG molecule with the enzyme papain. He found that it contains "heavy" and "light" protein chains in three distinct regions of which two are similar and serve to bind antigens. He determined the sequence of amino acids in the chains of IgG and, linking this with Edelman's biochemical and immunological studies and with data from electron microscopy, showed that IgG forms a Y-shaped structure. From this he made a brilliant guess at the overall molecular architecture of antibodies in which the main part of the molecule is constant but the tips vary. His work stimulated the Argentinean immunologist C. Milstein and others to further studies bringing antibody structures to the point where their biochemistry could be linked with their immunology. Porter was Whitley Professor of Biochemistry at Oxford (1967–1985); he shared a Nobel Prize for Medicine with Edelman in 1972. See also **Antibody**; **Edelman, Gerald**

Maurice (1929–Present); **Martin, Archer John Porter (1910–2002)**; **Milstein, Cesar (1927–2002)**; and **Sanger, Frederick (1918–Present)**.

Additional Reading

Daintith, J., S. Mitchells, D. Gjertsen, and E. Tootile: *Biographical Dictionary of Scientists*, 2nd Edition, Institute of Physics, Bristol, UK, 1994.
Fox, D.M., M. Meldrum, and I. Rezak: *Nobel Laureates in Medicine or Physiology: a Biographical Dictionary*, Garland, New York, NY, 1990.
Perry, S.V.: "Rodney Robert Porter," *Biographical Memoirs of Fellows of the Royal Society*, **33**, 445–489 (1981).
Porter, R.R.: *Chemical Aspects of Immunology*, Carolina Biological Supply, Burlington, NC, 1976.

NOEL G. COLEY, The Open University, Milton Keynes, UK

PORTLAND CEMENT. See **Cement**; and **Gypsum**.

PORTUGAL CURRENT. See **Ocean Currents**

POSITIONAL NOTATION. One of the schemes for representing real numbers, characterized by the arrangement in sequence of digits (symbols for integers) with the understanding that the successive digits are to be interpreted as the coefficients of successive integral powers of a number called the radix or base of the notation. The representation of a real number by the notation

$$A_n A_{n-1} \ldots A_2 A_1 A_0 \cdot A_{-1} A_{-2} \ldots A_{-m}$$

which is an abbreviation for the sum

$$\sum_{i=-m}^{\infty} A_i r^i$$

where the $\cdot$ is called the radix point, the A_i are integers ($0 \leq |A_i| \leq r$) called digits, and r is an integer greater than one called the radix (or base). The signs of all of the A_i are the same as the sign of the number represented. In the decimal number system, the radix is ten and the radix point is called the decimal point. In the binary number system, the radix is two and the radix point is called the binary point. For some purposes the system of notation has been broadened to include the case in which the radix assumes more than one value in a single number system. In this case the notation

$$A_n A_{n-1} \ldots A_2 A_1 A_0 \cdot A_1 A_{-2} \ldots A_{-m}$$

in an abbreviation for the sum

$$\left(\sum_{i=1}^{n} A_i \prod_{j=1}^{i} r_j \right) + A^0 + \left(\sum_{i=-m}^{-1} A_i \prod_{j=1}^{-1} \frac{1}{r_j} \right)$$

Several such systems have been used. The biquinary system uses a radix which is alternately two and five for successive values of j. The quinary vicenary system uses a radix which is alternately five and twenty for successive values of j.

POSITION AND DISPLACEMENT MEASUREMENT. The measurement of position is usually expressed in terms of a dimension, a grid, or a vector. Position is a very important quantity in terms of radar, navigational, astronomical, and surveying systems. Industrially, position measurement is important to the control of machine tools and many other production machines. See also **Numerical Control**. Descriptions here are confined to the latter applications.

To position a machine member with acceptable accuracy, it is necessary to establish the extent of the backlash or dead-band region for the positioning mechanism used. The measuring transducer and its attendant dead-band characteristics, when used with the machine member, will determine the amount of dead band or backlash to be included in the total control system loop.

The controlled member may have undesired movement with respect to the machine base—in a direction transverse to the controlled axis of travel. If the slide and tableways wear nonuniformly, variation in transverse position of a point on the table of a machine may cause a variation in the air gap of a magnetic slot transducer system. This same problem will result in misalignment of optical transducer systems if the table motion becomes crab-like after wear of the slides has progressed. Nondata components of both cyclic and random nature may be superimposed on the true data because of machine-induced vibrations on the transducer. Attention must be given to reduce these effects and to take the residue effects fully into account when designing a total positioning system.

Resolvers. The resolver, among the several shaft-type mechanical input transducers available, is perhaps the most versatile because of the many possibilities for its electric input. The electrical diagram of Fig. 1 discloses the data input or stator winding in the manner of a two-phase motor stator. The electrical characteristics of this device are such that the output voltage of the rotor winding varies in a sinusoidal manner with respect to the data readout position. Zero voltage will occur at the readout position that corresponds to chosen data. But zero voltage also may occur 180 shaft degrees from this position. Thus, there is the possibility of a false-data region. To eliminate this possibility, the designer chooses a gear ratio for use between the machine member and its transducer that will permit only 180 mechanical degrees of transducer shaft rotation during full-member travel.

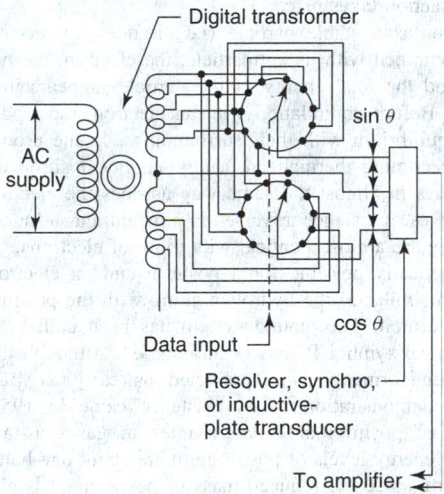

Fig. 1.   Digital transformer circuit for resolvers, synchros, and inductive plates.

Synchros. The term *synchro* is applied to a class of variable-coupling transformers in which the variable coupling is obtained by changing the orientation of the primary to the secondary through rotation of the movable element. Specifically, the synchro transduces a rotor-position angle into a voltage or a set of voltages unique to that angle. Unlike the resolver, a synchro stator has a three-phase winding configuration, while the rotor may use a single- or a three-phase winding.

Inductive Plates. As shown in Fig. 2, an inductive plate includes an etched stator winding that has been projected upon a dimensionally stable nonconducting surface by a photographic process. The rotor associated with this transducer is constructed in a like manner. Variations in inductor displacement are averaged over a large number of inductors by summing the voltages from a like number of coils located on the rotor plate. Thus, the reproduced rotor and stator inductors need not be printed with a positional accuracy equivalent to that of the final transducer. Essentially, this transducer is a 2-phase synchro or resolver whose windings have been projected onto a linear medium.

Inductive-Bridge Transducers. As shown by Fig. 3, operation is based upon the use of a fixed inductive member B that is slightly longer than the axis to be measured, and a movable member A that is approximately half the length of B. Selectable taps are placed on B in a successive decade with externally located inductors to provide a bridge configuration that may be externally unbalanced by placing A (coil) and N (point) across a pair of tap points—then moving the coil until equal voltage prevails between the two ends of the coil, as evidenced by the existence of a small voltage at O and O'. See Fig. 4. A disadvantage of this system is the relatively large

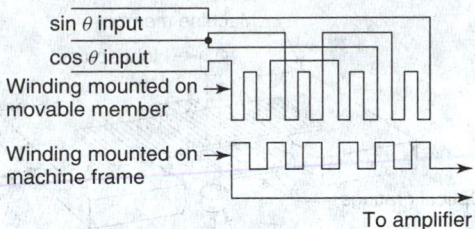

Fig. 2.   Schematic circuit for a position transducer utilizing inductive plates.

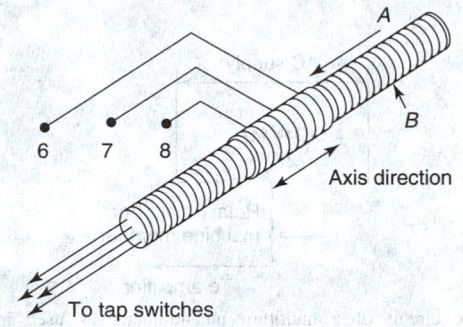

Fig. 3.   Inductive-bridge transducer for position measurement.

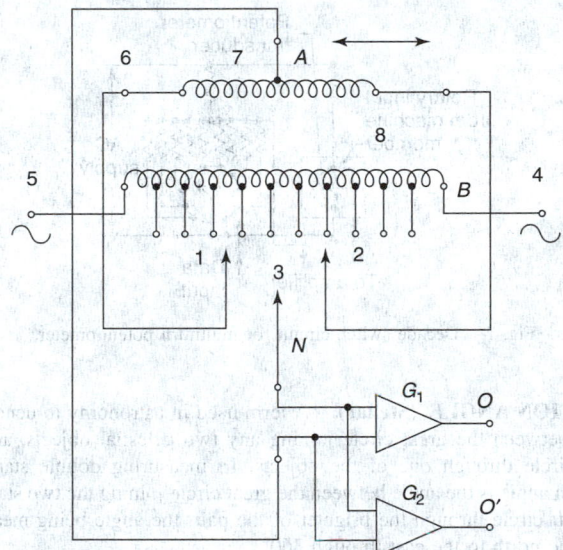

Fig. 4.   Schematic circuit of inductive-bridge transducer.

number of wires that must be taken from the device through the machine to the control system. An advantage is the high output voltage per unit of displacement.

Optical Gratings. A transducer of this type is shown in Fig. 5. The device requires an amplifier to raise the power level of the pulses from the associated photocell so that position readout displays or relays used in conjunction with other relay logic elements can be operated.

Potentiometers. Multiple-turn potentiometers are used in machine control systems that do not require very high accuracy. Circuitry for such a transducer system is shown in Figs. 6 and 7.

Other Transducers. Their descriptions are beyond the scope of this volume; other position and displacement sensors include digital contactors, optical binary scales, magnetic scales, linear potentiometers, linear variable reluctance transducers, linear transformers, and linear variable differential transformers (LVDT).

Further details on this complex topic can be found in the *Process/Industrial Instruments and Controls Handbook*, 4th Edition (D.M. Considine, Editor), McGraw-Hill, New York, 1993.

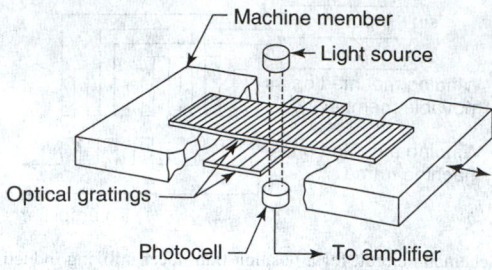

Fig. 5.   Optical grating-type position transducer.

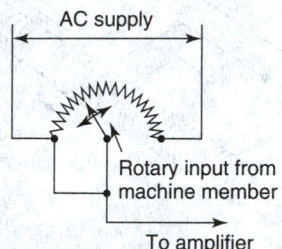

Fig. 6.   Basic circuit of a multiturn potentiometer as used in a position transducer.

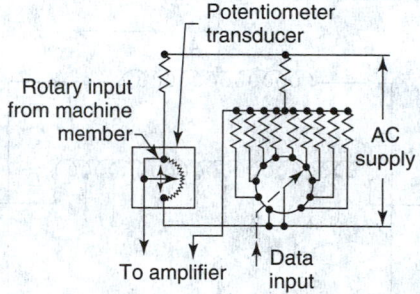

Fig. 7.   Decade switch circuit for multiturn potentiometer.

POSITION ANGLE (Stellar).   A term used in astronomy to denote the angle between the great circle joining any two celestial objects, and the hour circle through one of the objects. In measuring double stars, the position angle is the angle between the great circle joining the two stars and the hour circle through the brighter of the pair, the angle being measured from the north to the east through 360°.

POSITRON.   The positron is one of many fundamental bits of matter. Its rest mass (9.109×10^{-31} kilogram) is the same as the mass of the electron, and its charge ($+1.602 \times 10^{-19}$ coulomb) is the same magnitude, but opposite in sign to that of the electron. The positron and electron are antiparticles for each other. The positron has spin $\frac{1}{2}$ and is described by Fermi-Dirac statistics, as is the electron.

The positron was discovered in 1932 by C.D. Anderson at the California Institute of Technology while doing cloud chamber experiments on cosmic rays. The cloud chamber tracks of some particles were observed to curve in such a direction in a magnetic field that the charge had to be positive. In all other respects, the tracks resembled those of high-energy electrons. The discovery of the positron was in accord with the theoretical work of Dirac on the negative energy of electrons. These negative energy states were interpreted as predicting the existence of a positively charged particle.

Positrons can be produced by either nuclear decay or the transformation of the energy of a gamma ray into an electron-positron pair. In nuclei that are proton-rich, a mode of decay that permits a reduction in the number of protons with a small expenditure of energy is positron emission. The reaction taking place during decay is

$$p^+ \to n^0 + e^+ + \nu$$

where p^+ represents the proton, n^0 the neutron, e^+ the positron, and ν a massless, chargeless entity called a neutrino. See also **Neutrino**. The positron and neutrino are emitted from the nucleus while the neutron remains bound within the nucleus. Although none of the naturally occurring radioactive nuclides are positron emitters, many artificial radioisotopes that decay by positron emission have been produced. The first observed case of positron decay of nuclei was also the first observed case of artificial radioactivity. An example of such a nuclear decay is

$$_{11}\mathrm{Na} \to {}_{10}\mathrm{Ne}^{22} + e^+ + \nu \,(\mathrm{half} - \mathrm{life} \sim 2.6 \mathrm{years})$$

This decay provides a practical, usable source of positrons for experimental purposes.

The process of pair production occurs when a high-energy gamma ray interacts in the electromagnetic field of a nucleus to create a pair of particles—a positron and an electron. Pair production is an excellent example of the fact that the rest mass of a particle represents a fixed amount of energy. Since the rest energy ($E_{\mathrm{rest}} = m_{\mathrm{rest}} - c^2$) of the positron plus electron is 1.022 MeV, this energy is the gamma energy threshold and no pair production can take place for lower-energy gammas. In general, the cross section for pair production increases with increasing gamma energy and also with increasing Z number of the nucleus in whose electromagnetic field the interaction takes place.

The positron is a stable particle (i.e., it does not decay itself), but when it is combined with its antiparticle, the electron, the two annihilate each other and the total energy of the particles appears in the form of gamma rays. Before annihilation with an electron, most positrons come to thermal equilibrium with their surroundings. In the process of losing energy and becoming thermalized, a high-energy positron interacts with its surroundings in almost the same way as does the electron. Thus, for positrons, curves of distance traversed in a medium as a function of initial particle energy are almost identical with those of electrons.

It is energetically possible for a positron and an electron to form a bound system similar to the hydrogen atom, with the positron taking the place of the proton. This bound system has been called "positronium" and the chemical symbol Ps has been assigned. Although the possibility of positronium formation was predicted as early as 1934, the first experimental demonstration of its existence came in 1951 during an investigation of positron annihilation rates in gases as a function of pressure. The energy levels of positronium are about one-half those of the hydrogen atom, since the reduced mass of positronium is about one-half that of the hydrogen atom. This also causes the radius of the positronium system to be about twice that of the hydrogen atom.

In principle, positronium can be observed through the emission of its characteristic spectral lines, which should be similar to hydrogen's except that the wavelengths of all corresponding lines are doubled. Positronium is also the ideal system in which the calculations of quantum electrodynamics can be compared with experimental results. Measurement of the fine-structure splitting of the positronium ground state has served as an important confirmation of the theory of quantum electrodynamics.

It is possible for a positron–electron system to annihilate with the emission of one, two, three, or more gamma rays. However, not all processes are equally probable.

See also **Particles (Subatomic)**.

Additional Reading

Ali, A.: *High Energy Electron Positron Physics*, World Scientific Publishing Company, Inc., Riveredge, NJ, 1988.

Charlton, M., J.W. Humberston, and A Dalgarno: *Positron Physics*, Cambridge University Press, New York, NY, 2005.

Jean, Y.C., D.M. Schrader, and P.E. Mallon: *Principles and Applications of Positron and Positronium Chemistry*, World Scientific Publishing Company, Inc., Riveredge, NJ, 2003.

Krause-Rehberg, R., and H.S. Leipner: *Positron Annihilation in Semiconductors: Defect Studies*, Vol. 127, Springer-Verlag New York, Inc., New York, NY, 1998.

POSITRON EMISSION TOMOGRAPHY (PET).   PET stands in the forefront of molecular imaging and allows the quantitative evaluation of the distribution of several pharmaceuticals in a target area *in vivo*. PET is a unique diagnostic imaging technique for measuring the metabolic activity of cells in the human body. It produces images of the body's basic biochemistry and biological activity in a noninvasive way, combining techniques applied in nuclear medicine with the precise localization achieved by computerized image reconstruction. PET is therefore a powerful diagnostic test that is having a major impact on the diagnosis and treatment of disease, as well as on patient management.

PET images can demonstrate pathological changes and detect and stage tumors long before they would be revealed with other conventional imaging modalities. Traditional diagnostic techniques, such as X rays, computerized tomography (CT) scans, or MRI, produce anatomical images of what the internal organs look like. The premise with these techniques is that a visible structural change exists in anatomy caused by disease. However, biochemical processes are also altered with disease and may occur before a change in gross anatomy occurs. Furthermore, PET can provide medical doctors with important early information about very subtle changes of function in the brain and heart, due to disease-related modifications in tissue perfusion, cell metabolic rates heart disease, or neurological disorders (Alzheimer's, Parkinson's, epilepsy, dementia, etc.), allowing physicians to diagnose and treat these diseases earlier and, consequently, more efficiently and accurately, according to the axiom "the earlier the diagnosis, the better chance for treatment." PET can also help physicians monitor a patient's response to treatment, as well as identify distant metastases that can affect treatment, helping curtail ineffective treatments and reduce unnecessary invasive procedures. The field of PET has been emerging today into clinical diagnostic medicine and is approved by many insurance carriers for coverage.

History of PET. The positron emission and detection of the radiation produced was a known technique that dates back to the early days of the twentieth century. However, it is only in the last few decades, with the booming development of fast electronic circuits and powerful computer systems, that this knowledge could be used in practice as a valuable diagnostic tool: The electronic circuits used in PET should be able to detect the coincidental arrival of two high energy photons (a timing resolution of the order of few nanoseconds), and the image reconstruction requires modern computer systems in order to produce an accurate image of the activity distribution within a clinically reasonable time.

In the beginning of the 1950s, researchers at the Massachusetts General Hospital (MGH) in Boston and the Duke University in Durham proposed the idea that, in spite of the short half-lives of the, by that time recently discovered, positron-emitting radionuclides, they offered an attractive method for the regional study of metabolism due to their commonality. A single-detector pair brain probe was then developed at MGH and used in experiments. However, it was not until the early 1960s that these positron-emitting radionuclides began to gain popularity, when a number of centers such as the MGH in Boston, the Sloan Kettering Institute in New York, Ohio State University, and the University of California at Berkeley began to use cyclotrons. At the same time, the first image reconstruction techniques were proposed by researchers at MGH, and, in the early 1970s, the concept of computerized tomography (CT) was presented by Hounsfield, who later was awarded with the Nobel Prize.

In the early 1970s, the first PET scanners were developed at the MGH, the Brookhaven National Laboratory, the Washington University, and the Montreal Neurological Institute in Canada, used then as research tools. At the same time, a private company (EG&G OTREC, Oak Ridge, TN, USA) got involved in the developments of the first ring PET scanners, joined in the market a couple of years later by TCC (The Cyclotron Corporation, Berkeley, CA, USA), and in 1976 the first commercial PET scanner was delivered at the University of California, Los Angeles (UCLA). A year later, Scanditronix from Sweden brought Europe into PET. The first PET scanners used single slices when performing tomographs, with transaxial resolution greater than 2 cm full-width half-maximum (FWHM) and used NaI crystal material. Such systems were installed at several research institutions, apart from the ones mentioned above, like the University of California at Berkeley, the Lawrence Berkeley Laboratory, and the University of Pennsylvania.

By the end of 1970s, PET had shown its potential for application to clinical medicine. The following generation of PET scanners reduced detector size and added additional rings to allow for simultaneous acquisition of multiple slices. The slice resolutions improved from greater than 2 cm FWHM to less than 1 cm FWHM. As time progressed, more detectors and photomultiplier tubes (PMTs) were added to these machines

to increase their sensitivity and resolution. In the mid-1980s, the first BGO pixelated detector blocks were presented. At the same time, the first dedicated medical PET cyclotron units with automated radiopharmaceutical delivery systems were commercially available.

At the end of 1980s, the major medical imaging companies (mainly Siemens with CTI PET, Inc., and General Electric with Scanditronix) began investing in PET. The first whole-body PET scanners have been presented and research in new detector materials led to significant discoveries (LSO, etc.) in the beginning of the 1990s. Since then, PET has shown a steady increase in acceptance for clinical application, both medically and administratively, and PET centers are being installed worldwide at an increasing pace. PET is now a well-established medical imaging technique that assists in the diagnosis and management of many diseases.

Physical Principles of PET. PET images molecules of substances with a specific biological activity. In order to monitor their distribution, kinetic characteristics, and behavior of (pharmaceuticals) within the body, these substances are tagged with radioactive compounds (with short half-life and at extremely low concentrations) [Phelps, 2004]. These radiopharmaceuticals are chosen to have a desired biological activity, depending on the metabolic activity of the organ under study, and are introduced to the subject by injection or inhalation.

The most commonly used radionuclides are listed in Table 1 and are compounds that constitute, or are consumed by, the living body, like carbon, nitrogen, and oxygen. They are isotopes of biologically significant chemical elements that exist in all living tissues of the body and in almost all nutrients. Therefore, the above radionuclides are easily incorporated in the metabolic process and serve as tracers of the metabolic behavior of the body part, which can be studied *in vivo*.

TABLE 1. THE MOST COMMONLY USED RADIONUCLIDES IN PET

Radionuclide	Half-life
Carbon-11	20.3 min
Nitrogen-13	9.97 min
Oxygen-15	2.03 min
Fluorine-18	1.83 h
Gallium-68	1.83 h
Rubidium-82	1.26 min

Table 2 shows a list of the major radiopharmaceuticals used as PET agents with their specific medical applications. The most common radiopharmaceutical used in PET studies today is fluorodeoxyglucose (FDG) [Gambhir, et al.: 2001], a chemical compound similar to glucose, with the difference that one of the -OH groups has been replaced by F-18. Carbon-11 can also be used as a radiotracer to glucose. The short half-lives of these particles allow the subject and the people handling them to receive only a low radiation dose.

TABLE 2. MAJOR PET RADIOPHARMACEUTICALS AND THEIR SPECIFIC MEDICAL APPLICATIONS

Agent	Images
F-18 fluorodeoxyglucose	Regional glucose metabolism
F-18 sodium fluoride	Bone tumors
C-11 methionine	Amino acid uptake/protein synthesis
C-11 choline	Cell membrane proliferation
C-11 deoxyglucose	Regional brain metabolism
O-15 oxygen	Metabolic rate of oxygen use/OEF
C-11 carbon monoxide	Cerebral blood volume
O-15 carbon monoxide	Cerebral blood volume
O-15 water	Cerebral blood flow
O-15 carbon dioxide(Inhaled)	Cerebral blood flow
C-11 butanol	Cerebral blood flow
C-11 N-methylspiperone	Dopamine D2 and Serotonin S2 receptors
F-18 N-methylspiperone	D2 and S2 receptors
C-11 raclopride	D2 receptors
F-18 spiperone	D2 receptors
Br-76 bromospiperone	D2 receptors
C-11 carfentanil	Opiate mu receptors
C-11 flumazenil	Benzodiazepine (GABA) receptors

The identification and detection of the presence of the molecules of the radiotracer in a specified location within the source (i.e., the body under study) is performed by a chain of events, based on physical principles and data processing techniques, which are schematically depicted in Fig. 1 and briefly described below.

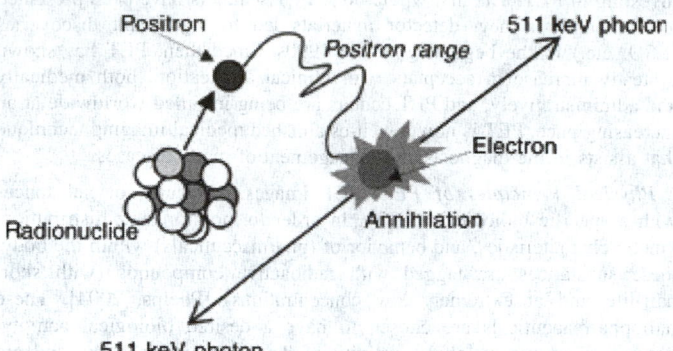

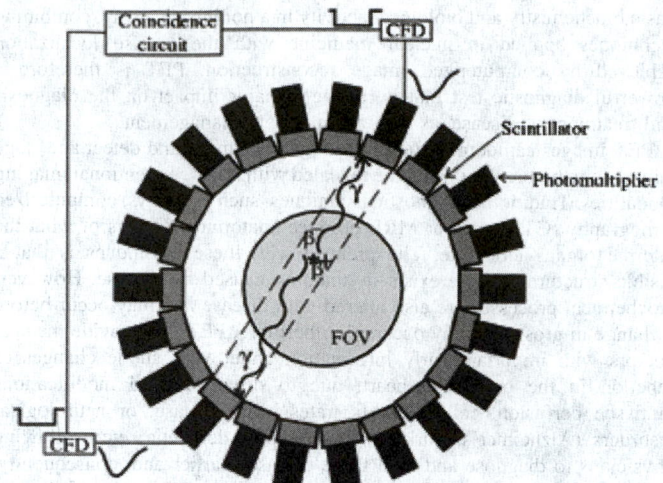

Fig. 1. This schematic depicts the chain of events that described the physical properties of high energy gamma pair emission from positron-emitting radioisotopes. All radioisotopes used with PET decay by positron emission. Positrons are positively charged electrons. Positron emission stabilizes the nucleus of unstable radioisotopes by removing a positive charge through the conversion of a proton into a neutron. An emitted positron travels a short distance (*positron range*, which depends on the energy of the positron) and collides with an ordinary electron of a nearby atom in an annihilation reaction. When the two particles annihilate, their mass turns into two 511 keV gamma rays that are emitted at 180° to each other. When detected, the 180° emission of two gamma rays following the disintegration of positronium is called a *coincidence line*. Coincidence lines provide a unique detection scheme for forming tomographic images with PET.

A positron is emitted during the radioactive decay process, annihilates with an electron, and, as a result, a pair of γ rays is emitted (two high energy photons of 511 keV each). The two γ rays fly off in almost opposite directions (according to the momentum conservation laws), penetrate the surrounding tissues, and can be recorded outside the subject's body by scintillation detectors placed on a circular or polygonal detector arrangement, which forms a PET tomograph. When the γ ray hits a scintillation detector material, it then deposits its energy in that crystal by undergoing photoelectric effect, which is an atomic absorption process where an atom totally absorbs the energy of an incident photon [Sorenson, and Phelps, 1987]. This energy is then used to eject an orbital electron (photoelectron) from the atom and is, therefore, transformed in visible light. This light can be detected by specialized devices (photomultiplier tubes, PMT) that capture and transform it into an electronic signal, shaped at a later stage by the electronic circuits of the tomograph to an electronic pulse, which provides information about the timing of the arrival of the incident γ ray and its energy. Figure 2 summarizes the principles of gamma ray event detection in PET described here.

By measuring a coincidence photon, the detector array in a PET system identifies that an annihilation event occurred inside the volume defined between the surfaces of the pair of detectors that registered the coincidence event. At the end of a PET scan, for each pair of detectors, a number of coincidence events that have been identified exist. This information represents the radioactivity in the subject viewed at different angles, when sorted in closely spaced parallel lines. In order to reconstruct the activity density inside the source from its projections (events registered at each detector pair), a mathematical reconstruction algorithm is applied by computer. The collected data are corrected for scatter, attenuation, and accidental coincidences; normalized for the differences in detector efficiencies, and reconstruct the spatial distribution of the radioactivity density inside the organ or the system under study in the form of a 2D or 3D image. The result is a digital image of the source, where the value of each picture element (pixel) or, in modern 3D tomograph systems, volume element (voxel) is proportional to the activity density inside the source at the area (or volume) that corresponds to this pixel/voxel. This image can be directly displayed on a screen. Further analysis of the data and

Fig. 2. Scintillation detectors coupled to photomultiplier tubes are placed around the detector ring of the scanner. An annihilation event (*) inside the field of view (FOV) produces two γ rays that get detected by a pair of detectors. The event is identified to occur inside a specific detector tube (dashed stripe) by the electronic devices (constant fraction discriminators, CFD, and the coincidence detection circuit) that connect every pair of detectors.

processing of the produced images can be carried out with the use of a computing system.

A high energy photon produced by an annihilation event can deviate from its original trajectory if it gets involved in Compton scattering inside the subject's body, a collision between a photon and a loosely bound outer-shell orbital electron of an atom. In this case, because the incident photon energy greatly exceeds the binding energy of the electron to the atom, the interaction can be considered as a collision between the photon and a "free" electron. The photon does not disappear in Compton scattering, but it is deflected through a scattering angle θ and some of its energy is transferred to the electron (recoil electron). In the case this ray gets detected in coincidence with the second gamma produced at the same event, then this event will be counted to have occurred in a detector tube that will not contain the original annihilation site: This is an erroneous event (scattered event).

It is also possible that this ray will never reach a detector crystal and, therefore, get lost. This type of Compton scattering, along with photoelectric absorption of the produced gamma rays inside the source, where they have been generated, are the major sources of attenuation of the emitted radioactivity.

The physics of positron emission allow for attenuation correction of the collected data, which can produce a quantitatively (but also qualitatively) accurate image that may resolve small lesions, especially when these lie deep within the body. In order to correct for attenuation, two additional measurements are typically performed: the blank scan and the transmission scan. The blank scan is recorded using an external source without the patient, representing the unattenuated case. For the transmission scan, the patient and the bed are placed into the scanner and the attenuated data are measured using the external source. The attenuation correction factors (ACF) can be calculated as the ratio of the measured counts without and with the attenuating object. The disadvantages of attenuation correction are that it requires more time for image acquisition and the potential exists to add noise to the image if the attenuation measurements become misaligned by patient motion or if inadequate statistics in the transmission scan are collected. As a result of noise, transmission measurements are usually smoothed prior to the division. Otherwise, the noise in the ACF propagates to the corrected emission sinogram. The drawback of smoothing is that the resulting blurring of ACFs propagates to the emission sinogram as well. Techniques for the reduction of noise propagation include, as an example, classification techniques for the main tissue categories observed in the transmission images (segmentation) or the use of iterative methods for the reconstruction of the transmission images [Zaidi, and Hasegawa, 2003].

Compton scattering can also occur inside the detector crystal before the ray undergoes (the desirable) photoelectric effect. In that case, it is possible that the ray will escape the detector material and deposit its energy in an adjacent scintillator, causing the detected event to be mispositioned. Another source of erroneously counted events is the coincidental arrival at the detector ring of two single gamma rays coming from two different annihilation events (random or accidental coincidence). When three or more γ rays arrive at the detector ring within the time coincidence window set by the electronic circuitry of the scanner for the coincidence detection, then these gammas must be rejected, because it is not possible to recognize, in that case, the pairs of photons that came from the same annihilation event [Turkington, 2001].

The high energy gamma rays have increased penetrating abilities and can be detected coming from deep-lying organs better than α particles or electrons (β particles), which can penetrate only a few millimeters of tissue and, therefore, cannot get outside the body to the radiation detector. Imaging system detectors must, therefore, have good detection efficiency for γ rays. It is also desirable that they have energy discrimination capability, so that γ rays that have lost energy by Compton scattering within the body can be rejected and a good timing resolution to accurately measure the time difference of the arrival of two photons. Sodium iodide (NaI), BaF_2 (barium fluoride), and BGO (bismuth germanate oxide) provide both of these features at a reasonable cost. Research for new scintillator materials, like LSO (lutetium oxyorthosilicate) [Melcher, and Schweitzer, 1992], GSO (germanate oxide) [Ishibashi, et al.: 1992], $PbCO_3$ (lead carbonate) [Moses, and Derenzo, 1992], $PbSO_4$ (lead sulfate) [Moses, et al.: 1992], CeF_3 (cerium fluoride), YalO, and LuAlO, is very active in an effort to produce faster detector crystals with good stopping power and light output.

Table 3 summarizes some of the main physical properties of the scintillators used for PET: NaI(Tl), BGO, BaF_2, CsF, GSO, and LSO. In order to interpret this table, assume the following:

- An elevated density guarantees a high stopping power for the high energy 511 keV annihilation photons and consequently assures elevated detection efficiency. High stopping power also allows the use of crystals of small dimensions, which means an improved spatial resolution of the tomograph.
- High scintillation efficiency, due to a good intrinsic energy resolution of the crystal, leads to a good energy resolution of the detection system, which leads to a better discrimination of scatter.
- A fast scintillation (described by a short scintillation constant decay time) translates to a low dead time of the system and, therefore, to good count rate performance. Moreover, this property directly influences the temporal resolution (uncertainty of the moment of detection), on which depends the choice of the length of the time coincidence resolution window and, therefore, the rate of accidental coincidences.

TABLE 3. SCINTILLATION CRYSTAL CHARACTERISTICS

	NaI	BGO	BaF_2	CsF	LSO	GSO
Density (g/cm³)	3.67	7.13	4.87	4.64	7.40	6.71
Relative scintillation efficiency	100	20	16	6	75	30
Decay constant (ns)	250	300	0.6	2.5	40	60
Hygroscopic	Yes	No	No	Yes	No	No

The comparison of the characteristics of scintillation crystals shows that the ideal scintillator for PET must have the temporal characteristics (decay time) of BaF_2, the density (stopping power) of BGO, and the scintillation efficiency (light output) of NaI(Tl). It also reveals that the newest crystals GSO and LSO are very promising for PET applications.

Originally, NaI was the detector of choice for nuclear medicine imaging cameras and is still in use by some manufacturers of gamma cameras, SPET, and even PET systems. NaI is a scintillation crystal discovered in 1949 with very high scintillation efficiency but a stopping power too low for high energy photons; therefore, NaI has very low sensitivity. In the 1980s, BGO emerged as the detector of choice for PET scanners, a material with considerably lower light output than NaI but, on the other side, twice as dense and, therefore, able to detect high energy photons more effectively. LSO was discovered in the early 1990s and exhibits a

very fast scintillation time (40 ns), which provides significantly reduced detector dead time and consequently higher count-rate capabilities, which is essential in clinical PET imaging in order to use the injected activity most efficiently and to make the emission scan time as short as possible, meaning the patient spends less time immobile on the tomograph's bed without compromising the image quality.

In the optimization of the design of a PET tomograph, an important aspect is the way crystals are assembled and the way they are coupled to the photomultiplier tubes. Various strategies have been developed, including:

- one-to-one connection crystal-PMT;
- detector blocks, where a crystal array (mainly BGO or LSO) is coupled to a smaller number of PMTs [Gilardi [2001]; and Casey and Mutt, 1986];
- NaI(Tl) crystals of large dimensions coupled to a grid of PMT (Anger logic, common to gamma cameras) [Karp, et al.: 1990];
- the most recent design of a system of GSO crystals coupled to light guides to a PMT grid [Surti, et al.: 2000].

Scintillation detectors have been the dominant element in high energy gamma ray detection for PET. However, other technologies have also been applied, explored, and developed for this purpose. One of the oldest alternative technologies is the High Density Avalanche Chamber (HIDAC) PET system [Jeavons, et al.: 1983], which consists of a Multiwire Proportional Chamber (MWPC) with the provision of laminated cathodes containing interleaved lead and insulating sheets and mechanically drilled with a dense matrix of small holes. Ionization resulting from photons interacting with the lead is trapped by, amplified in, and extracted from, the holes by a strong electric field into the MWPC. On arrival at an anode wire, further avalanching occurs. Coordinate readout may be obtained from orthogonal strips on the cathodes. The result is precise, 2D localization of the incident gamma rays. Every hole on the cathodes acts as an independent counter. By stacking these MWPCs, millions of these counters are integrated to form a large-area radiation camera with a high spatial resolution.

The resolution of a PET scanner primarily depends on the size of the detectors and on the range of positrons in matter (distance traveled by the positron in the tissue before interacting with a free electron, see also Fig. 1). For most of the positron emitters, the maximum range is 2–20 mm. However, the effect on spatial resolution is much smaller, because positrons are emitted with a spectrum of energies and only a small fraction travel the maximum range, and, in addition, in case of 2D acquisitions, the range of the third dimension is compressed. Another limitation in the resolution is that the paired annihilation photons are not emitted precisely 180° from each other, because the e^+–e^- system is not at complete rest. Other components of the system resolution are the sampling scheme used, the interactions between more than one crystal due to intercrystal scatter, the penetration of annihilation photons from off-axis sources to the detector crystals, the reconstruction technique used, the filters applied, and the organ and patient motion during the scan.

Three types of spatial resolution exist in a typical ring PET system, defined by a full-width at half-maximum (FWHM): the radial, tangential, and axial resolutions. The radial, or in-slice, resolution deteriorates as we move from the center of the FOV and is best at the center. The same happens for the tangential resolution, which is measured along a line vertical to a radial line, at different radial distances. In systems with more than one detector ring, the axial resolution, or slice thickness, is measured along the axis of the tomograph.

A major source of error during the coincidence detection is the fact that not all the annihilation events are registered correctly as mentioned earlier. Additional accidental coincidences can result from poor shielding or backscatter and from ordinary γ rays from the radionuclide administered. The random and scattered coincidences are registered together with the true coincidences, obtained when a pair of gammas is correctly identified and classified to the appropriate detector tube, and are sources of background noise and image distortion.

In order to keep the number of scattered coincidences low, a discriminator should be used. A discriminator primarily generates timing pulses upon the arrival of a photon, but also can verify the total energy of the illuminating ray is above a preset energy threshold. Scattered rays have already deposited part of their energy and, therefore, can be identified.

Furthermore, the choice of the appropriate time coincidence (or coincidence resolving time) window is essential: It has to be narrow enough to keep the number of random coincidences as low as possible but also wide enough to include all valid coincidence pulses. In the existing PET units, the timing accuracy is of the order of tenths of nanoseconds.

A PET scanner can be designed to image one single organ, such as the brain or the heart, or can be able to image any organ in the body, including whole-body scans. Whole-body studies with F-18-FDG consist of repeated PET acquisitions at contiguous bed positions in order to provide 3D images (axial, sagittal, coronal, and oblique cut planes) covering one considerable portion of the patient's body (Fig. 3), which facilitates the search for metastases in oncological diagnostics [Tarantola, et al.: 2003].

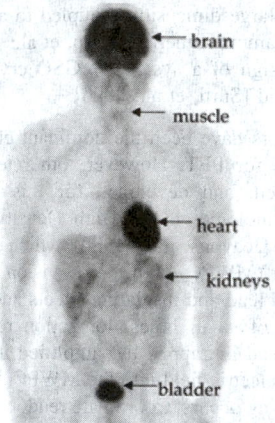

Fig. 3. A whole-body F-18-FDG PET image of a normal subject (no pathological situation diagnosed). Areas with high metabolic activity (brain, myocardium) or with high concentration of the radioactive tracer (bladder) are visible. [Courtesy of A. Maldonado and M.A. Pozo from the Centro PET Complutense, Madrid, Spain.]

Most PET systems today are whole-body systems (i.e., they have a typical transaxial FOV of 60 cm). This FOV is adequate to handle most patients. The axial FOV of most PET systems today is limited to approximately 10–15 cm. This relatively narrow axial FOV imposes some limitation on the imaging procedures that can be performed clinically. It also requires more accurate positioning of the patient in comparison with conventional nuclear medicine procedures. For a clinical system, it would be desirable to extend the axial FOV to 15–20 cm, which would, for instance, allow full brain (Fig. 4) and heart imaging in a single frame and more efficient whole-body imaging. As the detectors contribute a significant portion of the total cost of the scanner, however, this would bring into question what would be an acceptable cost for the PET scanner.

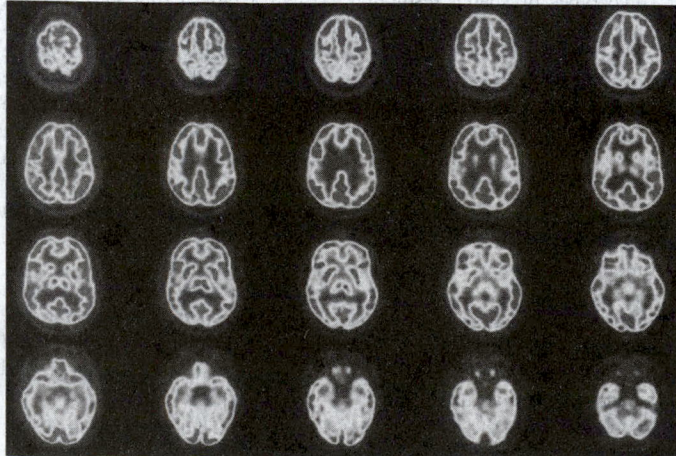

Fig. 4. Sequential images from an F-18-FDG PET brain study of a normal individual. [Courtesy of A. Maldonado and M.A. Pozo from the Centro PET Complutense, Madrid, Spain.]

Manufacturing of Radiopharmaceuticals

A cyclotron is a particle accelerator that produces positron-emitting elements or short-lived radioisotopes. These radioisotopes can then be incorporated into other chemical compounds that are synthesized into a final product that can be injected into a person. These radioisotopes are used to "label" compounds so it can later be identified where in the body the radiopharmaceutical is being distributed. The compounds that are being labeled are organic molecules normally used in the body, such as sugar, neurotransmitters, and so on.

First, the cyclotron bombards nonradioactive elements in the target with accelerated particles, which converts these elements into positron-emitting radioactive isotopes of fluorine, nitrogen, oxygen, or carbon. The major radioactive isotope produced at almost all sites is fluorine-18 (F-18), which has a half-life of 110 min. F-18 thus produced from the cyclotron is delivered to a chemical synthesis unit called the *chemical processing unit*, which is where F-18 is incorporated into a precursor to produce the final product FDG, the labeled sugar molecule. This entire process is fully automated and performed in the cyclotron lab. When a dose is needed, it is transported to the PET scan room by various means, depending on the distance between the production site and the PET tomograph and ranging from a dedicated pneumatic tube system to long-distance transport via air or road.

Applications of PET

Molecular imaging opens the way for medical doctors to successfully pursue the origin of disease. As long as disease is of unknown origin, more tests and exams are needed, something that means increased healthcare costs, in addition to the patient's discomfort and pain. PET can accurately identify the source of many of the most common cancers, heart diseases, and neurological diseases, eliminating the need for redundant tests, exploratory surgeries, and drug overload of the patient. PET produces powerful images of the body's biological functions and reveals the mysteries of health and disease.

PET can be used to obtain information about the tissue perfusion using inert tracers (e.g., O-15 labeled water), the metabolism with metabolically active tracers (e.g., F-18-FDG), or the kinetic of a cytostatic drug (e.g., F-18-Fluorouracil).

In cardiology, this imaging technique represents the most accurate test to reveal coronary artery disease or rule out its presence. Traditionally, when a patient shows signs or symptoms of heart disease, his or her physician will prescribe a thallium stress test as the initial diagnostic study. The conventional thallium stress test, however, is often not as accurate as a PET scan. PET images can show inadequate blood flow to the heart during stress that can pass undetected by other noninvasive cardiac tests. A PET study could enable patients to avoid cardiac catheterization when a conventional perfusion or echocardiographic stress test is equivocal. A PET scan shows myocardial viability in addition to perfusion abnormality. More specifically, PET exams for metabolism and perfusion of the heart tissues can determine the need for heart transplant, in case both are absent in a large area of the heart, or confirm with certainty that simple bypass surgery would be enough, when metabolism is maintained even if blood flow is significantly reduced. As metabolism indicates that tissue is still alive, complicated heart transplantation can be avoided and coronary bypass would have great chances to improve cardiac function. Documented studies have shown that thallium stress testing overestimates irreversible myocardial damage in at least 30% of cases, which can result in the patient being placed on the transplant list rather than receiving bypass surgery or angioplasty. No other diagnostic test can more precisely assess myocardial viability than PET.

PET can reveal abnormal patterns in the brain and is, therefore, a valuable tool for assessing patients with various forms of dementia. PET images of the brain can detect Parkinson's disease: A labeled amino acid (F-DOPA) is used as tracer at a PET examination in order to determine if the brain has a deficiency in dopamine synthesis. If it does not, Parkinson's disease can be ruled out and possible tremors in the patient's muscles will be treated in a different manner. Although the only definitive test for Alzheimer's disease (AD) is autopsy, PET can supply important diagnostic information. When comparing a normal brain versus an AD-affected brain on a PET scan, a distinctive and very consistent image pattern appears in the area of the AD-affected brain, where certain brain regions have low metabolism at the early stages of the disease, allowing early detection several years before diagnosis can be confirmed by a physician. PET can also help to differentiate Alzheimer's from other confounding types of dementia or depression [Reba, 1993]. Conventionally, the confirmation of AD was a long process of elimination that averaged between two and three years of diagnostic and cognitive testing. PET can help to shorten this process by identifying distinctive patterns earlier in the course of the disease. Furthermore, PET allows the accurate identification of epileptogenic brain tissue (because of its reduced glucose metabolic rates) and can successfully lead the surgical removal of the epileptic foci.

In oncology, in which the clear majority the total PET examinations refer, this technique inspects all organs and systems of the body to search for cancer in a single examination. PET is very accurate in distinguishing malignant tumors from benign growths. It can help detect recurrent brain tumors and tumors of the lung, breast, lymph nodes, skin, colon, and other organs. The information obtained from PET studies can be used to determine what combination of treatment is most likely to be successful in treating a patient's tumor, as it can efficiently determine the resistance of a specific cancer to the drugs applied and, consequently, can dynamically optimize the treatment management and follow-up of the patient on an individual basis. With this technique, it is possible to evaluate if a tumor has been successfully destroyed after therapy, as anatomical follow-up imaging is often not in the position to assess if a residue is still active or has definitely been eliminated after chemotherapy, radiation, or surgery.

A major use of PET is its ability for kinetic imaging analyses. This term refers to the measurement of tracer uptake over time. An image of tracer activity distribution is a good starting point for obtaining more useful information such as regional blood flow or regional glucose metabolism. The process of taking PET images of radioactivity distribution and then using tracer kinetic modeling to extract useful information is termed image analysis. The tracer kinetic method with radio labeled compounds is a primary and fundamental principle underlying PET and autoradiography. It has also been essential to the investigation of basic chemical and functional processes in biochemistry, biology, physiology, anatomy, molecular biology, and pharmacology. Tracer kinetic methods also form the basis of *in vivo* imaging studies in nuclear medicine [Phelps, 2000].

Besides its direct clinical applications, PET imaging is emerging as a powerful tool for use by the pharmaceutical industry in drug discovery and development [Fowler, et al.: 1999]. The role of small animal PET imaging [Chatziioannou, 2002] studies in rodents for the discovery of PET tracers for human use is significant, as it has the potential for permitting higher throughput screening of novel tracers in transgenic mice as well as the confounding effects resulting from potential species differences on receptor affinity, blood-brain barrier (BBB) transport, metabolism, and clearance. This setting is expected to allow new and unique experimental laboratory studies to be performed.

Other recent developments include dedicated mammography devices (known as positron emission mammographs, PEM) for breast functional imaging [Kontaxakis, et al: 1997]. Furthermore, the first PET/CT tomographs have made their way to the market [Beyer, et al.: 2000]. These are devices that house a positron tomograph and a CT scanner in a single device, allowing the acquisition and visualization of registered images detailing both anatomy and biological processes at the molecular level of internal organs and tissue, without the need of multiple examinations and further image processing to achieve similar results.

Image Interpretation

One of the final steps in the processing chain of the PET study is to produce a final layout of the images for the diagnosing physician. The conventional way of presenting the image data is to produce a transparency film (X-ray film) of the images on the computer display. In addition to the image data, the film should also be labeled with demographic data about the study, such as patient name and scan type. As this information is usually stored in the image files together with the image data, the labeling and layout of the images on the display can be automated in software. With the rapid development of local area networks, films may soon no longer be necessary. Instead, the images can be read from a display system located in the reading room, which has access to the PET image data through a computer network. Referring physicians do, in most cases, require a hard

copy of the study, which can be accomplished using X-ray films. With recent improvements in printer technology, high quality color output may also be a low cost alternative to the traditional film.

Procedure for a PET Scan

Most patients will be in the PET center for 2 or 3 hours, depending on the type of study being conducted. The patient is informed as to when to stop eating before the test. Drinking lots of water is recommended before the scan. The patients also need to inform the PET center if they are diabetic or claustrophobic. In general, before the scan is performed, a catheter is placed in the arm so that the radioactive tracer can be injected. A glucose test will also be performed. Depending on the type of study conducted, scanning may take place before and after the injection is given. After the tracer is given, the patient waits for approximately 40–60 min before the final scan is done.

PET Scan and Associated Risks

The radiation exposure of PET is similar to that of having a CT scan or any other standard nuclear medicine procedure involving heart or lung scans. No pain or discomfort results from the scan. The half-life of F-18 is so short that by the time the patient leaves the PET center, almost no activity remains in the body. Patients typically do not experience any reactions as a result of the PET scan, because the tracer material is processed by the body naturally. Therefore, no side effects are expected. Of course, as with any other nuclear medicine procedure, when breast-feeding or pregnant, a PET scan must be performed under special conditions.

See also **Computed Tomography**.

Additional Reading

Bailey, D.L., D.W. Townsend, and P.E. Valk: *Positron Emission Tomography*, Springer-Verlag New York, LLC, New York, NY, 2005.

Beyer, T., et al.: "A Combined PET/CT Scanner for Clinical Oncology," *J Nucl Med*, **41**(8), 1369–1379 (2000).

Buchanan, J.W., and R.L. Wahl: *Principles and Practice of Positron Emission Tomography*, Lippincott Williams & Wilkins, Philadelphia, PA, 2002.

Casey, M.E., and R. Nutt: "A Multislice Two Dimensional BGO Detector System for PET," *IEEE Trans Nucl Sci*, **33**, 460–463 (1986).

Castiglioni, I., et al.: "Scatter Correction Techniques in 3D PET: A Monte Carlo Evaluation," **46**(6), 2053–2058 (1999).

Chatziioannou, A.F. "Molecular Imaging in Small Animals with Dedicated PET Tomographs," *Eur. J. Nucl. Med.*, **29**(1), 98–114 (2002).

Fahey, F.H.: "Data Acquisition in PET Imaging," *J. Nucl. Med.*, **30**(2), 39–49 (2002).

Fowler, J.S., N.D. Volkow, G. Wang, Y-S Ding, and S.L. Dewey: "PET and Drug Research and Development," *J Nucl Med*, **40**, 1154–1163 (1999).

Gambhir, S.S., J. Czernin, J. Schwimmer, D.H.S. Silverman, E. Coleman, and M.E. Phelps: "A Tabulated Summary of the FDG PET Literature," *J. Nucl. Med.*, **42**, 1S–93S (2001).

Gilardi, M.C.: Tomografi PET: Attualitá e Prospettive (in Italian), XI Nat. Course on Professional Continuing Education in Nuclear Medicine (Pisa, 29-31/10/2001).

Ishibashi, H., et al.: "Scintillation Performance of Large Ce-doped Gd_2SiO_5 (GSO) Single Crystal," *IEEE Trans. Nucl. Sci.*, **45**(3), 518–521 (1998).

Jeavons, A., et al.: "The High-density Avalanche Chamber for Positron Emission Tomography," *IEEE Trans. Nucl. Sci.*, **30**, 640–645 (1983).

Karp, J.S., et al: "Continuous-slice PENN-PET: A Positron Tomograph with Volume Imaging Capability," *J. Nucl. Med.*, **31**, 617–627 (1990).

Kipper, M.S., and M. Tartar: *Clinical Atlas of PET*, Elsevier Health Sciences, New York, NY, 2003.

Kontaxakis, G., and A Dimitrakopoulou-Strauss: "New Approaches for Position Emission Tomography (PET) in Breast Carcinoma," In: Limouris, G.S., S.K. Shukla, and H-J Biersack: *Radionuclides for Mammary Gland–Current Status and Future Aspects*, p 21–36. Mediterra Publishers, Athens, Greece: 1997.

Kontaxakis, G., et al.: "Iterative Image Reconstruction for Clinical PET Using Ordered Subsets, Median Root Prior and a We-based Interface," *Mol. Imag. Biol.*, **4**(3), 219–231 (2002).

Melcher, C.L., and J.S. Schweitzer: "Cerium-doped Lutetium Oxyorthosilicate: A Fast, Efficient New Scintillator," *IEEE Trans. Nucl. Sci.*, **39**, 502–505 (1992).

Meng, L.J., and D. Ramsden: "Performance Results of a Prototype Depth-Encoding PET Detector," *IEEE Trans Nucl. Sci.*, **47**(3), 1011–1017 (2000).

Moses, W.W., and S.E. Derenzo: "Lead Carbonate, a New Fast, Heavy Scintillator," *IEEE Trans. Nucl. Sci.*, **37**(1), 96–100 (1990).

Moses, W. W.S.E. Derenzo, and P.J. Shlichta: "Scintillation Properties of Lead Sulfate," *IEEE Trans. Nucl. Sci.*, **39**(5), 1190–1194 (1992).

Moses, W.W., et al.: "LuAlO₃: Ce-a High Density, High Speed Scintillator for Gamma Detection," *IEEE Trans. Nucl. Sci.*, **42**(4), 275–279 (1995).

Oehr, P., J. Ruhlmann, and H-J Biersack: *PET and PET-CT in Oncology*, 2nd Edition, Springer-Verlag New York, LLC, New York, NY, 2004.

Phelps, M.E., and S.R. Cherry: "The Changing Design of Positron Imaging Systems," *Clin. Positron Imag.*, **1**(1), 31–45 (1998).

Phelps, M.E.: "Positron Emission Tomography Provides Molecular Imaging of Biological Processes," *Proc. Nat. Acad. Sci.*, **97**(16), 9226–9233 (2000).

Phelps, M.E.: *PET Molecular Imaging and Its Biological Applications*, Springer-Verlag New York, LLC, New York, NY, 2004.

Reba, R.C.: "PET and SPECT: Opportunities and Challenges for Psychiatry," *J. Clin. Psychiatry*, **54**, 26–32 (1993).

Schmand, M., et al.: "Performance Results of a New DOI Detector Block for a High Resolution PET-LSO Research Tomograph HRRT," *IEEE Trans. Nucl. Sci.*, **45**(6), 3000–3006 (1998).

Shao, Y., et al.: "Design Studies of a High Resolution PET Detector using APD Arrays," *IEEE Trans. Nucl. Sci.*, **47**(3), 1051–1057 (2000).

Sorenson, J.A., and M.E. Phelps: *Physics in Nuclear Medicine*, 2nd Edition, Grune and Stratton Inc., Orlando, FL, 1987.

Surti, S., J.S. Karp, R. Freifelder, and F. Liu: "Optimizing the Performance of a PET Detector Using Discrete GSO Crystals on a Continuous Lightguide," *IEEE Trans Nucl Sci*, **47**, 1030–1036 (2000).

Tarantola, G.F. Zito, and P. Gerundini: "PET Instrumentation and Reconstruction Algorithms in Whole-body Applications," *J. Nucl. Med.*, **44**(5), 756–768 (2003).

Townsend, D.W.: "From 3-D Positron Emission Tomography to 3-D Positron Emission Tomography/computed Tomography: What did we learn?" *Mol. Imaging Biol.*, **6**(5), 275–290 (2004).

Turkington, T.G.: "Introduction to PET Instrumentation," *J. Nucl. Med. Tech.*, **29**(1), 4–11 (2001).

Zaidi, H., and B. Hasegawa: "Determination of the Attenuation Map in Emission Tomography," *J. Nucl. Med.*, **44**(2), 291–315 (2003).

Web References

A History of Positron Imaging: http://stuff.mit.edu/people/glb/
PETNET Solutions: http://www.petscaninfo.com/

GEORGE KONTAXAKIS, Universidad Politécnica de Madrid, Madrid, Spain

POSITRONIUM. A quasi-stable system consisting of a positron and a negatron bound together. Its set of energy levels is similar to that of the hydrogen atom (electron and proton). However, because of the different reduced mass, the frequencies associated with the spectral lines are less than half of those of the corresponding hydrogen lines. The mean life of positronium is at most about 10^{-7} seconds, its existence being terminated by negatron-positron annihilation. See also **Positron**.

POTAMOGALE. See **Moles and Shrews**.

POTAMOLOGY. 1. From the Greek potamos, "river," the scientific study of rivers.

2. More specifically, the interdisciplinary branch of hydraulics, hydrology, and fluid dynamics dealing with surface streams and their regimes. Potamology generally focuses on issues of fluvial erosion, transport, and sedimentation; fluvial dynamics; and river metamorphosis or change through time.

POTASSIUM. [CAS: 7440-09-7]. Chemical element, symbol K, at. no. 19, at. wt. 39.098, periodic table group 1 (alkali metals), mp 63.3 °C, bp 760 °C, density 0.86 g/cm³ (20 °C). Elemental potassium has a body-centered cubic crystal structure. Potassium is a silver-white metal, can be readily molded, and cut by a knife, oxidizes instantly on exposure to air, and reacts violently with H_2O, yielding potassium hydroxide and hydrogen gas, which burns spontaneously in air with a violet flame due to volatilized potassium element, is preserved under kerosene, burns in air at a red heat with a violet flame. Discovered by Davy in 1807.

There are three naturally occurring isotopes, ^{39}K through ^{41}K, of which ^{40}K is radioactive with a half-life of 1.3×10^9 years. In ordinary potassium, this isotope represents only 0.0119% of the content. There are four other known isotopes, all radioactive, ^{38}K and ^{42}K through ^{44}K, all with relatively short half-lives measured in minutes and hours. In terms of abundance, potassium ranks seventh among the elements occurring in the earth's crust. In terms of content in seawater, the element ranks eighth, with an estimated 1,800,000 tons of potassium per cubic mile (388,000 metric tons per cubic kilometer) of seawater. First ionization potential 4.339 eV; second, 31.66 eV. Oxidation potential $K \rightarrow K^+ + e^-$, 2.924 V. Other important physical properties of potassium are given under **Chemical Elements**.

Potassium does not occur in nature in the free state because of its great chemical reactivity. The major basic potash chemical used as a source of potassium is potassium chloride, KCl. The potassium content of all potash sources generally is given in terms of the oxide K_2O. The majority of potash produced comes from mineral deposits that were formed by the evaporation of prehistoric lakes and seas which had become enriched in potassium salts leached from the soil. In addition to natural deposits of potassium salts, large concentrations of potassium also are found in some bodies of water, including the Great Salt Lake and the Salduro Marsh in Utah, the Dead Sea between Israel and Jordan, and Searles Lake in California. All of these brines are used for the commercial production of potash.

The main potassium minerals are sylvite, KCl, sylvinite, KCl/NaCl, carnallite, $KCl \cdot MgCl_2 \cdot 6H_2O$, kainite, $MgSO_4 \cdot KCl \cdot 3H_2O$, polyhalite, $K_2SO_4 \cdot MgSO_4 \cdot 2CaSO_4 \cdot 2H_2O$, langbeinite, $K_2SO_4 \cdot 2MgSO_4$, jarosite, $K_2Fe_6(OH)_{12}(SO_4)_4$, leucite, $K_2O \cdot Al_2O_3 \cdot 4SiO_2$, alunite, $K_2Al_6(OH)_{12}(SO_4)_4$, microcline, $K_2O \cdot Al_2O_3 \cdot 6SiO_2$, muscovite, $K_2O \cdot 3Al_2O_3 \cdot 6SiO_2 \cdot 2H_2O$, bio tite, $H_2K(Mg, Fe)_3(Al, Fe)(SiO_4)_3$, and orthoclase, $K_2O \cdot Al_2O_3 \cdot 6SiO_2$. See also **Alunite**; **Biotite**; **Carnallite**; **Jarosite**; **Leucite**; **Muscovite**; and **Polyhalite**. The principal workable mineral deposits are in Stassfurt, Germany, Alsace, New Mexico, Saskatchewan, the former Soviet Union, Spain, Poland, Italy, the Atlantic Seaboard of the United States, and Utah. There are significant potassium reserves in many other parts of the world, notably in Canada and the former Soviet Union. World consumption of potash is about 18 million tons annually. Potassium metal is obtained by electrolysis of fused potassium hydroxide or chloride fluoride mixture in a specially designed cell.

Uses

Like so many of the chemical elements, the compounds of potassium are far more important than elemental potassium—by several orders of magnitude. The uses for metallic potassium are extremely limited, mainly because metallic sodium serves about the same needs and is much less costly. Sodium production, for example, exceeds potassium production by a factor of at least 1,000. A large amount of elemental potassium is used to produce the superoxide, KO_2 which finds application in gas-mask canisters. The compound also goes into the production of a sodium-potassium alloy, which is used as a heat-exchange medium. This alloy also has been used in magnetohydrodynamic power generation and as a catalyst for the removal of CO_2, H_2O, and oxygen from inert-gas systems. The handling precautions for potassium metal are similar to those for sodium metal. See also **Sodium**.

Chemistry and Compounds

Potassium is more electropositive than sodium in many of its reactions, as is consistent with its position in group 1. Its reaction with H_2O is more vigorous and it reacts violently with liquid bromine, and readily on heating with solid iodine.

Because of the ease of removal of its single 4s electron (4.339 eV) and the difficulty of removing a second electron (31.66 eV) potassium is exclusively monovalent in its compounds, which are electrovalent. (Some experimental work indicates that the potassium alkyls may be covalent, but even they form conducting solutions in other metal alkyls.)

Potassium solutions in liquid NH_3 react readily with the elements on the further right side of the periodic table to produce normal and poly compounds such as potassium sulfide, K_2S, and tetrapotassium plumbide, K_4Pb, in the first instance and K_2S_6 and K_4Pb_9 in the second. Ammoniates are not formed by potassium as readily as by sodium or lithium and solubility of salts exhibits a minimum at the cation: anion radius ratio of 0.75 (potassium fluoride, KF, 16 moles per kilogram, potassium chloride, KCl, 0.0177 moles per kilogram, potassium bromide, KBr, 2.26 moles per kilogram, potassium iodide, KI, 11.09 moles per kilogram). Potassium nitrate reacts in liquid ammonia with potassium amide, KNH_2 to form the azide, KN_3.

Like the other alkali metals, potassium forms compounds with virtually all the anions, organic as well as inorganic. Like sodium bicarbonate, the reactivity of potassium bicarbonate with many metallic oxides permits of the preparation of many compounds (such as the meta- and pyroarsenates) which are unstable in aqueous solution. For a general discussion of these reactions, and for a general picture of the inorganic salts of potassium, see the discussion of the compounds of sodium, which differ principally in their greater degree of hydration and greater number of hydrates. However, potassium, rubidium, and cesium coordinate with large organic molecules even though they do not with water. Potassium, like the others, coordinates with salicylaldehyde. It is believed to have two coordination numbers, 4 and 6. The tetracoordinate compounds of potassium (and sodium) are the most stable. The following reasons are given: (1) Increasing atomic number carries with it increasing electropositiveness and ease of ionization, which diminishes the tendency to coordinate. (2) The increasing distance of the nucleus from the coordinating electrons with increasing atomic volume makes it less likely that additional electrons will be held with ease. (3) On the other hand, there is an increase in the maximum coordination number with the elements of higher atomic number. These factors are in keeping with a maximum stability for the tetracoordinate compounds occurring with potassium.

Charge Density Waves in Potassium

Frequently, because of their relatively simple electronic structure, the alkali metals are selected as a basis for the study of the behavior of electrons in solids. As early as 1964, Overhauser (Purdue University) predicted the existence of "charge density waves," a phrase coined by Overhauser, in the potassium atom. This conclusion was the result of calculations made by Overhauser to the effect that K, in its lowest energy or ground state, does not exhibit a uniform distribution of its free electrons (which cause K to behave as a metal), but rather the electron density varies sinusoidally with a characteristic wavelength—and that this usually is not an integral multiple of the crystal lattice constant. This concept, of course, was not in agreement with the traditional conclusion that free electrons are uniformly distributed. The reasoning—the sinusoidal clumping lowers the electron energy, which in turn causes the lattice to distort and, as explained by Robinson (1986), this distortion is an attempt to reduce the huge electric fields generated by the separation between the positive charge of the K ions and the negative charge of the electrons.

At the time of Overhauser's work in 1964, experimental examples were not available and the concept was generally considered academic. Several years later, however, investigators working with layered materials (electrons essentially move in only two directions) and with linear conductors (motion is essentially in one direction) attributed a charge density wave phenomenon to what has been termed the Pierls instability. The latter effect, which involves lowering electron energy and lattice distortion, currently is not believed to apply to the simple, three-dimensional metals (K etc.). In summary, the Pierls instability and the Overhauser charge density waves concept appear to be similar, but different.

In 1985, Giebultowicz (National Bureau of Standards), Overhauser, and Werner (University of Missouri) conducted a neutron diffraction study and tentatively proved the Overhauser concept. Some solid-state physicists are seeking further evidence. If the concept is fully confirmed, some modifications in the thinking of how electrons behave in solids may be required.

Salt-Forming Properties. One major difference between potassium and sodium in their salt-forming properties is the much greater ability of potassium to form alums, although potassium does not form quite as many types of these compounds as do the higher alkali metals, or ammonium or monovalent thallium.

Potassium also differs from sodium, and especially from lithium, in the greater stability of its salts of polarizable polyatomic anions, such as peroxide, superoxide, azide, polysulfide, polyhalides, etc. The rubidium and cesium salts, on the other hand, are even more stable.

Among the other inorganic compounds of potassium are the following:

Bromate. Potassium bromate, [CAS: 7758-01-2], $KBrO_3$, white solid, soluble, mp 434 °C, upon heating oxygen is evolved and the residue is potassium bromide; formed by electrolysis of potassium bromide solution under proper conditions. Used as a source of bromate and bromic acid.

Carbonate. Potassium carbonate, [CAS: 584-08-7], potash, pearl ash, K_2CO_3, white solid, soluble, formed (1) in the ash when plant materials are burned, (2) by reaction of potassium hydroxide solution and the requisite amount of CO_2. Used (1) in making special glasses, (2) in the making of soft soap, (3) in the preparation of other potassium salts (a) in solution, (b) upon fusion; potassium hydrogen carbonate, potassium bicarbonate, potassium acid carbonate, $KHCO_3$, white solid, soluble, (4) in vat dyeing and textile printing, (5) in titanium enamels, (6) in boiler water treating compounds, (7) in photographic chemical formulations, (8) in electroplating baths, and (9) as an important absorbent for CO_2 in the process industries.

Chlorate. Potassium chlorate, [CAS: 3811-04-9], chlorate of potash, $KClO_3$, white solid, soluble, mp about 350 °C, powerful oxidizing agent, and consequently a fire hazard with dry organic materials, such as clothes, and with sulfur; upon heating oxygen is liberated and the residue is potassium chloride; formed by electrolysis of potassium chloride solution under proper conditions. Used (1) in matches, (2) in pyrotechnics, (3) as disinfectant, (4) as a source of oxygen upon heating. (Hazardous! Use of potassium perchlorate is recommended instead.)

Chloride. Potassium chloride, [CAS: 7447-40-7], KCl, colorless or white crystals; strong saline taste. Occurs naturally as sylvite. Soluble in water; slightly soluble in alcohol. Sp. gr. 1.987; mp 772 °C; sublimes at 1500 °C; noncombustible; low toxicity. Used in fertilizers, as a source of potassium salts; pharmaceutical preparations; photography; spectroscopy; plant nutrient; salt substitute; laboratory reagent. See also **Fertilizer**.

Chloroplatinate. Potassium chloroplatinate, [CAS: 16921-30-5], K_2PtCl_6, yellow solid, insoluble, formed by reaction of soluble potassium salt solution and chloroplatinic acid. Used in the quantitative determination of potassium.

Chromate. Potassium chromate, [CAS: 7789-00-6], K_2CrO_4, yellow solid, soluble, formed by reaction of potassium carbonate and chromite at a high temperature in a current of air, and then extracting with water and evaporating the solution. Used (1) as a source of chromate, (2) in leather tanning, (3) in textile dyeing, (4) in inks.

Cobaltinitrite. Dipotassium sodium cobaltinitrite, $K_2NaCo(NO_2)_6 \cdot H_2O$, golden yellow precipitate, formed by reaction of sodium cobaltinitrite solution in acetic acid with soluble potassium salt solution. Used in the detection of potassium.

Cyanate. Potassium cyanate, [CAS: 590-28-3], KCNO, white solid, soluble, formed along with lead metal by reaction of potassium cyanide and lead monoxide solids upon heating. Source of cyanate.

Cyanide. Potassium cyanide, [CAS: 151-50-8], cyanide of potash, KCN, white solid, soluble, very poisonous, formed by reaction of calcium cyanamide and potassium chloride at high temperature. Used as a source of cyanide and for hydrocyanic acid, but usually replaced by the cheaper sodium cyanide. Also used in metallurgy, electroplating, extraction of gold from ores, as a pesticide and fumigant, in photography and analytical chemistry. Upon acidification, produces dangerous HCN gas.

Dichromate. Potassium dichromate, [CAS: 7778-50-9], chromate of potash, $K_2Cr_2O_7$, red solid, soluble, powerful oxidizing agent, formed by acidifying potassium chromate solution and then evaporating. Used (1) in matches, (2) in leather tanning and in the textile industry, (3) as a source of chromate, (4) in pyrotechnics, (5) in colored glass, (6) as an important laboratory reagent, (7) in blueprint developing, and (8) in wood preservation formulations.

Hydroxide. Potassium hydroxide, [CAS: 1310-58-3], caustic potash, potassium hydrate, KOH, white solid, soluble, mp 380 °C, formed (1) by reaction of potassium carbonate and calcium hydroxide in H_2O, and then separation of the solution and evaporation, (2) by electrolysis of potassium chloride under the proper conditions, and evaporation. Used in the preparation of potassium salts (1) in solution, and (2) upon fusion. Also used in the manufacture of (3) soaps, (4) drugs, (5) dyes, (6) alkaline batteries, (7) adhesives, (8) fertilizers, (9) alkylates, (10) for purifying industrial gases, (11) for scrubbing out traces of hydrofluoric acid in processing equipment, (12) as a drain-pipe cleaner, and (13) in asphalt emulsions.

Hypophosphite. Potassium hypophosphite, [CAS: 7782-87-8], KH_2PO_2, white solid, soluble, formed (1) by reaction of hypophosphorous acid and potassium carbonate solution, and then evaporating, (2) by reaction of potassium hydroxide solution and phosphorus on heating (poisonous phosphine gas evolved).

Iodate. Potassium iodate, [CAS: 7758-05-6], KIO_3, white solid, soluble, melting point 560 °C, formed (1) by electrolysis of potassium iodide under proper conditions, (2) by reaction of iodine and potassium hydroxide solution, and the fractional crystallization of iodate from iodide. Used as a source of iodate and iodic acid.

Manganate. Potassium manganate [CAS: 10294-64-1], K_2MnO_4, green solid, soluble, permanent in alkali, formed by heating to high temperature manganese dioxide and potassium carbonate, and then extracting with water, and evaporating the solution. The first step in the preparation of potassium manganate and permanganate from pyrolusite.

Nitrate. Potassium nitrate, [CAS: 7757-79-1], saltpeter, niter, KNO_3, white solid, soluble, mp 333 °C, formed by fractional crystallization of sodium nitrate and potassium chloride solutions. Used (1) in matches, explosives, pyrotechnics, (2) in the pickling of meat, (3) in glass, (4) in medicines, (5) as a rocket-fuel oxidizer, and (6) in the heat treatment of steel. See also **Fertilizer**.

Nitrite. Potassium nitrite, [CAS: 7758-09-0], KNO_2, yellowish-white solid, soluble, for med (1) by reaction of nitric oxide plus nitrogen tetroxide and potassium carbonate or hydroxide, and then evaporating, (2) by heating potassium nitrate and lead to a high temperature and then extracting the soluble portion (lead monoxide insoluble) with H_2O, and evaporating. Used as a reagent (diazotizing) in organic chemistry.

Oxides

See discussion later in entry.

Perchlorate. Potassium perchlorate, [CAS: 7778-74-7], $KClO_4$, white solid, very slightly soluble, mp 610 °C, but above 400 °C decomposes with evolution of oxygen gas and formation of potassium chloride residue; formed (1) by electrolysis of potassium chlorate under proper conditions, (2) by heating potassium chlorate at 480 °C and then fractional crystallization. Used (1) as a convenient and safe (preferred to use of potassium chlorate) method of preparing oxygen by heating, (2) in the determination of potassium in soluble salt solution.

Periodate. Potassium periodate, [CAS: 7790-21-8], KIO_4, white solid, very slightly soluble, mp 582 °C, formed by electrolysis of potassium iodate under proper conditions.

Permanganate. Potassium permanganate, [CAS: 7722-64-7], permanganate of potash $KMnO_4$, purple solid, soluble, formed by oxidation of acidified potassium manganate solution with chlorine, and then evaporating. Used (1) as disinfectant and bactericide, (2) in medicine, (3) as an important oxidizing agent in many chemical reactions.

Persulfate. Potassium persulfate, [CAS: 7727-21-1], $K_2S_2O_8$, white solid, slightly soluble, formed by electrolysis of potassium sulfate under proper conditions. Used (1) as a bleaching and oxidizing agent, (2) as an antiseptic.

Silicate. Potassium silicate [CAS: 10006-28-7], K_2SiO_3, colorless (when pure) glass, soluble, mp 976 °C, formed by reaction of silicon oxide and potassium carbonate at high temperature, similar in properties and uses to the more common sodium silicate.

Sulfates. Potassium sulfate, [CAS: 7778-80-5], sulfate of potash, K_2SO_4, white solid, soluble. Common constituent of potassium salt minerals. Used (1) as an important potassium fertilizer, (2) in the preparation of potassium or potash alums; potassium hydrogen sulfate, $KHSO_4$, white solid, soluble; potassium pyrosulfate, $K_2S_2O_7$, white solid, soluble, formed by heating potassium hydrogen sulfate to complete loss of H_2O. See also **Fertilizer**.

Sulfides. Potassium sulfide, [CAS: 1312-73-8], K_2S, yellowish to reddish solid, soluble, formed by heating potassium sulfate and carbon to a high temperature; potassium hydrogen sulfide, potassium bisulfide, potassium acid sulfide KHS, formed in solution by reaction of potassium hydroxide or carbonate solution and excess H_2S.

Sulfite. Potassium sulfite, [CAS: 10117-38-1], $K_2SO_3 \cdot 2H_2O$; potassium hydrogen sulfite, $KHSO_3$; white solids, similar in properties and formation to the corresponding sodium sulfites.

Thiocarbonate. Potassium thiocarbonate, K_2CS_3, yellow solid, soluble, formed by reaction of potassium sulfide and CS_2.

Thiocyanate. Potassium thiocyanate, [CAS: 333-20-0], potassium sulfocyanide, potassium sulfocyanate, potassium rhodanate, KCNS, white solid, soluble, mp about 170 °C, formed by fusing potassium cyanide and sulfur, and then crystallizing. Used as a source of thiocyanate.

In addition to the inorganic salts, potassium forms such binary compounds as a phosphide, K_3P, by direct union with phosphorus, a boride, KB_6, by electrolysis of fused fluorides and borates in the presence of a metal boride, a nitride, and the oxides. Of the latter, direction reaction of potassium and oxygen yields the superoxide, KO_2, a paramagnetic, orange-colored substance. The likelihood of KO_2 having a monomeric structure is supported by these properties, since the O_2^- ion would have an odd electron, which would confer paramagnetism and color upon the compound. The lower oxides of potassium, K_2O and K_2O_2, which are less stable in air than the superoxide, have been prepared, as have their hydrates. K_2O unites explosively with the oxygen of the air. One other oxide, K_2O_3, has been reported, but this appears to be a double salt of KO_2 and K_2O_2. The properties of potassium hydroxide are in keeping with its position in Group 1; thus its heat of solution is somewhat lower than that of rubidium hydroxide, RbOH, or cesium hydroxide, CsOH, and much higher than that of lithium hydroxide, LiOH, and NaOH.

The organic compounds of potassium include many oxycompounds, such as salts of organic acids, alcohols and phenols (alkoxides, phenoxides, etc.). A few potassium-carbon linked compounds have been reported, such as a phenylisopropyl potassium, $C_6H_5C_3H_7K$, and a carbonyl compound of unknown composition, $K_x(CO)_x$. The adduct of ethyl potassium and diethyl-zinc is a true salt, $K_2[Zn(C_2H_5)_4]$, potassium tetraethylzincate.

Health and Safety Factors

Reactions of potassium with water and oxygen are hazardous and safe handling is a concern. Potassium oxidizes slowly in air at room temperature, and it usually ignites if it sprays hot into the air. The peroxide and superoxide products may explode in contact with free potassium metal or organic materials including hydrocarbons. Thus, packaging (qv) under oils is less desirable than packaging under an inert cover gas or in a vacuum. Potassium can react with entrapped air in oils to form the superoxide. The encrustation of potassium with superoxide (as a yellow crust) developed during storage has been known to detonate by friction from cutting. Potassium encrusted with a peroxide and superoxide layer should be destroyed immediately by careful, controlled disposal.

Potassium forms corrosive potassium hydroxide and liberates explosive hydrogen gas upon reaction with water and moisture. Airborne potassium dusts or potassium combustion products attack mucous membranes and skin causing burns and skin cauterization. Inhalation and skin contact must be avoided. Safety goggles, full face shields, respirators, leather gloves, fire-resistant clothing, and a leather apron are considered minimum safety equipment.

See also **Potassium and Sodium (In Biological Systems)**.

Additional Reading

Giebultowicz, T.M., A.S. Overhauser, and S.A. Werner: *Phys. Rev. Lett.,*, **56**, 1485 (1986).

Greenwood, N.N. and A. Earnshaw: *Chemistry of the Elements*, 2nd Edition, Butterworth-Heinemann, Inc., Woburn, MA, 1997.

Krebs, R.E.: *The History and Use of Our Earth's Chemical Elements: A Reference Guide*, Greenwood Publishing Group, Inc., Westport, CT, 1998.

Lide, D.R.: *CRC Handbook of Chemistry and Physics*, 88th Edition, CRC Press, LLC., Boca Raton, FL, 2007.

Parker, P.: *McGraw-Hill Encyclopedia of Chemistry*, 2nd Edition, The McGraw-Hill Companies, Inc., New York, NY, 1993.

Robinson, A.L.: "Charge Density Waves Seen in Posassium," *Science*, **232**, 713 (1986).

Stwertka, A. and E. Stwertka: *A Guide to the Elements*, Oxford University Press, Inc., New York, NY, 1998.

POTASSIUM AND SODIUM (In Biological Systems). Potassium and sodium play major roles in biological processes. Because of the numerous parallels between these two elements in metabolism, they are treated in a single entry, with appropriate distinctions made.

Potassium is required by both plants and animals. Although the total amount of potassium in most soils is usually rather high, the level of available or soluble forms of the element is frequently too low to meet the needs of growing plants. Deficiencies of plant-available potassium are more frequent in the soils of the eastern rather than of the western United States. See also **Soil**. Potassium in the form of soluble potassium salts is a very common constituent of fertilizers. See also **Fertilizer**.

Many plants will not grow at normal rates unless the plant tissues, especially the leaves, contain as much as 1 or 2% potassium and, for some plants, even higher concentrations are required. Therefore, if a plant grows at all, it will nearly always contain sufficient potassium to meet the requirements of the people or animals that consume the plant. Potassium deficiencies do occur in humans and animals, but these are largely due to metabolic upsets and illnesses that interfere with the utilization of potassium in the body, or via excessive losses of potassium from the body, rather than due to inadequate levels of dietary potassium.

The general role of potassium fertilizers in improving human and animal nutrition is to help increase food and feed supplies rather than to improve the nutritional quality of the crops produced. Excessive use of potassium fertilizers may decrease the concentration of magnesium in crops. Sodium is essential to higher animals that regulate the composition of their body fluids and to some marine organisms, but it is dispensable for many bacteria and most plants except for the blue-green algae. Potassium, on the other hand, is essential for all, or nearly all forms of life. The importance of these cations for all forms of life has been related to the predominance of sodium and potassium in the ocean where primitive forms of life are thought to have originated and developed. During most of the period of evolvement of living organisms, there has been little change in the sodium and potassium content of seawater, either as to proportion or total amount. The body fluids of sea animals are, in most instances, similar to seawater in sodium and potassium level and ratio. In freshwater and terrestrial animals, the sodium and potassium level of body fluids is usually somewhat lower, and the ratio is likely to vary from the 40:1 ratio of seawater. Most fresh waters contain small and variable amounts of sodium and potassium, usually in a ratio of from 1:1 to 4:1.

Despite the higher level of sodium in natural water, potassium is universally the characteristic cation found within both plant and animal cells. Although sodium is not an absolute requirement for most plants and bacteria, it is found in these organisms and is essential to higher animals where it is the principal cation of the extracellular fluids. Sodium and potassium are important constituents of both intra- and extracellular fluids. Generally, the best external and internal medium for function of cells not adjusted to low salt levels is a medium involving a balance of sodium and potassium.

Beyond the osmotic effects depending on the sum of the concentration of the ions in the solution, Ringer found in 1882 that to maintain the contractility of an isolated frog heart, it was necessary to perfuse it with a medium containing sodium, potassium, and calcium ions in the proportion of seawater. It has since been recognized that the normal life activities of tissues and cells may depend on a proper balance among the inorganic cations to which they are exposed. Sodium is required for the sustained contractility of mammalian muscle, while potassium has a paralyzing effect. Thus, a balance is necessary for normal function. Other investigators have found that the antagonism among univalent and divalent cations observed by Ringer is demonstrable with various simpler or more complicated organisms or biological systems.

Excessive salt in soil, such as soils recently soaked with seawater, is toxic to most plants, although there are many plants, e.g., those of the salt marshes and the sea, which are adapted to a high salt concentration. Ingestion of seawater by man as the only source of water is eventually fatal because of the inability of the body to eliminate salt at a concentration comparable to that of seawater. This results in accumulation of salt, with severe toxic effects and eventually fatal results.

It is probable that potassium is absorbed by the plant roots from the soil by an active transport mechanism which carries it through the cell wall structure. Similarly, potassium and sodium if required, are accumulated by animals also by active transport. The actual cellular content of potassium and sodium is likewise controlled by transport mechanisms that specifically move potassium in and sodium out of the cell against the concentration gradient. The energy for this is derived from the metabolic processes of the cell. The nature of these transport mechanisms has not been fully determined.

Ions and Transport Mechanisms. Potassium differs from most other essential constituents of plant and animal cells in that it is not built into the cell as a part of an organic compound, but is rather an ion from a soluble inorganic or organic salt. Potassium ions may chelate with cellular constituents, such as polyphosphates. The ion is of the correct size to fit into the water lattice adsorbed by the protein in the cell. In general, the potassium and sodium ions are attracted to protein or other colloidal or structural units having a negative charge. Mucopoly-saccharides within the cell, on the cell surfaces and of the intercellular structures, are of particular importance in holding cations, such as potassium and sodium. Active centers of other configurational features of the proteins in the cell may be affected or altered by the potassium held by electrostatic or covalent binding. There are several enzyme systems activated by potassium.

In general, most of the sodium and potassium in the animal is in a dynamic state, being exchanged between different parts of the cell, between the cell and the extracellular fluid, and intermixing with ingested sodium and potassium in body fluids.

Most cellular constituents do not selectively bind potassium in preference to sodium. Myosin of muscle fibers, for example, will bind either. But, in contrast, the mitochondria and ribosomes are organized cellular organelles able to selectively take up or extrude potassium. This accounts for only a part of the potassium held in the cell.

In blue-green algae and some yeasts, sodium may in part replace cellular potassium. While potassium is usually the principal cation concerned with the maintenance of the osmotic pressure within the cell, sodium contributes appreciably to the total, and amino acids and other organic compounds may help make up any deficit, particularly in marine invertebrates.

The sodium content of the body extracellular fluids of marine invertebrates from the coelenterate through the arthropod phyla is approximately that of seawater. In freshwater and terrestrial invertebrates, the sodium of body fluids varies over a wide range and there is considerable variation among vertebrates. There are both fish and crustaceans so highly adaptable that they are able to live in either fresh or salt water.

Osmotic Pressure Regulation. The regulation of osmotic pressure within the cell and the control of the passage of water into or out of the cell is dependent to a considerable extent on the control of the potassium and sodium in the cell by the transport systems of the cell wall. The cell wall itself is of protein-lipid composition and is in general impermeable to the passage of water and inorganic salts. Recent studies of the cell walls with electron microscopes and with the use of other investigative techniques indicate that the cell wall contains pores connecting the cell contents with the extracellular fluid, or in some cases, with other cells. In cells having an endoplasmic reticulum, the intracellular vacuolar system may have openings through the cell wall communicating with the extracellular fluid. The ease with which water passes in or out of the cell in response to changes in external or internal osmotic pressure varies over an extreme range, from easy passage to rigid control, depending on the cell and its functions.

Phagocytosis and pinocytosis may bring salts and water, as well as other substances, into the cell.

In some unicellular organisms, osmotic equilibrium may be maintained by a contractile vacuole, which collects water; in other organisms, water may be excreted through the cell wall. The kidney and sweat glands of higher animals, gills of fish and salt glands of birds serve to excrete salt. Most animals, through control of sodium and potassium excretion and loss, are able to adapt to a wide range of intake.

The importance of sodium chloride in nutrition has been recognized from the beginning of history. Agricultural populations that lived on cereal grains, nuts, berries, and other vegetable foods poor in sodium, experienced a hunger for salt which led them to go to great lengths to obtain the mineral. This was particularly true if they lived in a hot climate with the attendant increased loss of salt in perspiration. Similarly, herbivorous animals will travel long distances to supply their need for additional salt. In contrast, peoples or animals subsisting on meat, milk and other foods receive quite appreciable amounts of sodium salts in the diet, and experience no special desire or hunger for salt. See also **Sodium Chloride**.

In plants, the meristematic tissues in general are particularly rich in potassium, as are other metabolically active regions, such as buds, young leaves, and root tips. Potassium deficiency may produce both gross and microscopic changes in the structure of plants. Effects of deficiency reported include leaf damage, high or low water content of leaves, decreased photosynthesis, disturbed carbohydrate metabolism, low protein content and other abnormalities.

Since potassium is found abundantly in most natural foods consumed by animals, deficiency is ordinarily no problem. With prolonged maintenance through parenteral (intravenous) feeding when normal oral feeding is not possible, potassium must be supplied.

Role of Kidney. Experimental potassium deficiency in rats results in stunted growth, loss of chloride with hypochloremic acidosis, loss of potassium and increase of sodium in muscle. In man, disease of the gastrointestinal tract, involving loss of secretions through vomiting or diarrhea, may result in serious loss of both sodium and potassium. Trauma, surgery, anoxia, ischemia, shock and any damage to or wasting away of tissues may result in loss of cellular potassium to the extracellular fluid and plasma, and the loss from the body through kidney excretion. Recovery with rapid uptake or potassium by the tissues may result in low plasma levels. Low extracellular potassium concentration may cause muscular weakness, changes in cardiac and kidney function, lethargy, and even coma in severe cases. There are no reserve stores of either sodium or potassium in the animal body, so any loss beyond the amount of intake comes from the functional supply of cells and tissues. See also **Kidney and Urinary Tract**.

The kidney is the key regulator of the sodium and potassium content of higher animals and makes possible adaptation to wide variations of intake. In the glomerulus of the kidney nephron (or individual unit), an ultrafiltrate containing the smaller molecules of plasma is normally produced. As this ultrafiltrate passes down the kidney tubule, 97.5% or more of the sodium is actively resorbed, along with nearly all of the potassium. The remaining 2.5% of the sodium is sufficient to account for even the maximum sodium excretion. Potassium is added to the filtrate in the distal tubule through exchange for sodium. Control of this exchange appears to be the principal mode of action of aldosterone, which thus exerts a final control over sodium excretion. Aldosterone is a steroid hormone from the adrenal cortex, secretion of which seems to result from lowering of the Na/K ratio in the blood. Water is passively resorbed with the electrolytes along the length of the tubule.

Water excretion is further controlled by the antidiuretic hormone from the posterior pituitary gland which acts to increase water resorption in the kidney through making the collecting tubule permeable to water for additional resorption beyond what took place in the tubule. The posterior pituitary gland secretes the hormone as a rapid and sensitive response to a rise in the osmotic pressure of the extracellular fluid. The osmotic pressure of the extracellular fluid is, of course, principally due to its sodium chloride content.

With low intake of sodium, excretion is reduced to a very low level to conserve the supply in the body. Potassium is not so efficiently conserved.

The kidney regulates the acid-base balance of the body by control over resorption of sodium ions, which may exchange for hydrogen ions in the kidney tubule. Since most dietaries are of acid-ash, the urine is usually more acid than the original plasma filtrate and much of the phosphate excreted is thus changed to the acid monosodium salt. Within the range of normal variability, with an alkaline ash diet, the urine may become alkaline, and in extreme instances, some sodium bicarbonate may be excreted.

The salts of the buffer pairs responsible for control of the pH of plasma and extracellular fluid involve sodium as the principal cation, while cellular buffers involve potassium salts. See also **Acid-Base Regulation (Blood)**; and **Diuretics**.

Additional Reading

Benos, D.J. and D.M. Fambrough: *Amiloride-Sensitive Sodium Channels: Physiology and Functional Diversity*, Academic Press, Inc., San Diego, CA, 1999.

Evans, J.M., T.C. Hamilton, S.D. Longman, and G. Stemp: *Potassium Channels and Their Modulators: From Synthesis to Clinical Experien*, Taylor & Francis, Inc., Philadelphia, PA, 1997.

Young, D.B.: *Role of Potassium in Preventive Cardiovascular Medicine*, Kluwer Academic Publishers, Norwell, MA, 2001.

POTATO FAMILY.   See **Solanaceae (Potato Family)**.

POTENTIAL.   1.   A function of space, the gradient of which is equal to a force. In symbols,

$$\mathbf{F} = -\nabla\phi,$$

where $\mathbf{F}$ is the force, ∇ the del operator, and φ the potential. A force that may be so expressed is said to be "conservative," and the work done against it in motion from one given equipotential surface to another is independent of the path of the motion. In meteorology, the force of gravity has a potential, the geopotential, which, if the acceleration of gravity g is taken as constant, may be written $\varphi = gZ$, where Z is the height coordinate. The pressure force has in general no potential, nor do the Coriolis or viscous forces. By extension and analogy, the velocity potential, acceleration potential, and Gibbs function (thermodynamic potential) are defined.

2.   Applied to the value that an atmospheric thermodynamic variable would attain if processed adiabatically from its initial pressure to a standard pressure, typically 100 kPa.

See also **Potential Density**; **Potential Temperature**; and **Velocity Potential**.

POTENTIAL DENSITY.   The density an air parcel would attain if compressed adiabatically by descent to the standard pressure of 100 kPa. The potential density ′ is most easily defined in relation to the potential temperature θ as

$$\rho' = p/R\theta,$$

where p is a pressure of 100 kPa and R the gas constant, in appropriate units.

AMS

POTENTIAL ENERGY.   The negative of the work done by the forces of a conservative system when the particles of the system move from one configuration to another is the potential energy of the second configuration relative to the first configuration. This quantity is independent of the path followed by the particles in changing their configuration and is a function of the initial and final positions only.

An equivalent definition states that the potential energy is that particular function of the coordinates $V(x, y, z)$ whose negative gradient exists and is equal to the force, i.e.,

$$\mathbf{F} = -\Delta V$$

The existence of V implies a conservative force field.

If the force between two particles separated by a distance r is given by $\mathbf{F} = K/r^2$ the mutual potential energy of the particles when separated by a distance R is

$$-\int_{r0}^{r} \frac{K^2}{r}\, dr$$

where r_0 is the distance of separation in the initial or standard configuration. For convenience, r_0 is often taken as infinity, in which case the potential energy at infinity is considered to be zero and the potential energy of the final configuration is then K/r. Actually, the numerical value of the potential energy is arbitrary because the initial configuration can be chosen arbitrarily. Any constant can be added to the potential energy function and the condition $\mathbf{F} = -\Delta V$ will still be satisfied.

A particle on the surface of the earth is acted upon by a force mg, where m is the mass and g is the acceleration of gravity at the point. If the particle is raised a height h centimeters above the surface of the earth, where h is small in comparison with the radius of the earth, the potential energy of the particle with respect to the earth's surface becomes mgh. For example, a 10 gram mass at a distance of 100 cm above the ground has a potential energy of about $(10 \text{ gm}) \times (100 \text{ cm}) (980 \text{ cm/sec}^2) = 980,000$ ergs. If the mass were allowed to fall to the ground in a vacuum, the potential energy would be converted completely into 980,000 ergs of kinetic energy, thus exemplifying the conservation of energy.

POTENTIAL EVAPORATION.

Potential evaporation is a measure of the degree to which the weather or climate of a region is favorable to the process of evaporation. (Also called evaporative power, evaporation power, evaporative capacity, evaporation capacity, and evaporativity.)

It is usually considered to be the rate of evaporation, under existing atmospheric conditions, from a surface of water that is chemically pure and has the temperature of the lowest layer of the atmosphere.

POTENTIAL EVAPOTRANSPIRATION.

1. The amount of water evaporated (both as transpiration and evaporation from the soil) from an area of continuous, uniform vegetation that covers the whole ground and that is well supplied with water. Generally, the amount of moisture that, if available, would be removed from a given land area by evapotranspiration; expressed in units of water depth. It can be measured in a dry basin by determining the amount of irrigation water used, and in wetter regions, by the difference between rainfall and runoff, or by the supply of water required to maintain a constant amount of soil moisture in an isolated block of the soil.

2. The quantity of water evaporated (both as transpiration and evaporation from the soil) per unit area, per unit time from an extensive stretch of continuous, uniform vegetation that covers the whole ground and that is well supplied with water.; an empirical index of the above.

As given in Thornthwaite's 1948 climatic classification, it is equal to the summation of the 12 successive monthly values of the expression ct^a, where t is the monthly mean temperature in degrees Celsius, and a and c are coefficients that depend upon the annual heat index.

See also **Climatic Classification**.

AMS

POTENTIAL INSTABILITY.

The state of an unsaturated layer or column of air in the atmosphere with a wet-bulb potential temperature (or equivalent potential temperature) that decreases with elevation. If such a column is lifted bodily until completely saturated, it will become unstable (i.e., its temperature lapse rate will exceed the saturation-adiabatic lapse rate) regardless of its initial stratification.

Additional Reading

Saucier, W.J.: *Principles of Meteorological Analysis*, Dover Publications, Mineola, NY, 2003.

AMS

POTENTIAL REFRACTIVE INDEX.

In radio engineering, the refractive index for an atmosphere in which potential temperature and specific humidity are constant with height. The gradient of potential refractive index depends on the values of potential temperature and specific humidity, but over the range of values commonly encountered in the earth's atmosphere, this gradient does not vary much and is given approximately by

$$\frac{dn}{dz} = -\frac{1}{4R},$$

where R is the earth's radius.

See also **Potential Refractivity**.

AMS

POTENTIAL REFRACTIVITY.

An index, expressed in B-units, obtained by subtracting the gradient of potential refractive index from the gradient of refractive index and multiplying by 10^6:

$$B = \left[(n-1) + \frac{z}{4R} \right] \times 10^6.$$

The gradient of B is zero in an (approximately) neutral atmosphere.

AMS

POTENTIAL TEMPERATURE.

1. The temperature that an unsaturated parcel of dry air would have if brought adiabatically and reversibly from its initial state to a standard pressure, p_0, typically 100 kPa. Its mathematical expression is

$$\theta = T(p_0/p)^{\kappa},$$

where θ is the potential temperature, T is temperature, and *kappa* is the Poisson constant. This exponent is often assumed to be 2/7, the ratio of the gas constant to the specific heat capacity at constant pressure for an ideal diatomic gas. See also **Virtual Potential Temperature**; **Liquid Water Potential Temperature**; **Equivalent Potential Temperature**; and **Wet-Bulb Potential Temperature**.

2. In oceanography, the temperature that a water sample would attain if raised adiabatically to the sea surface. For the deepest points of the ocean, which are just over 10 000 meters (32 808 feet), the adiabatic cooling would be less than $1.5°$ C ($34.7°$F).

AMS

POTENTIAL VORTICITY.

The specific volume times the scalar product of the absolute vorticity vector and the gradient of potential temperature:

$$P = \alpha(2\Omega + \nabla \times \mathbf{u}) \cdot \nabla\theta,$$

where α is the specific volume, Ω the angular velocity vector of the earth's rotation, $\mathbf{u}$ the three-dimensional vector velocity relative to the rotating earth, and θ the potential temperature.

In the absence of friction and heat sources, the Ertel potential vorticity P is a materially conservative property (it remains constant for each particle). In spherical coordinates (λ, ϕ, r), where λ is longitude, ϕ is latitude, and r is the distance from the center of the earth, the above expression for P becomes

$$p = \alpha \left[\left(\frac{\partial w}{r \partial \phi} - \frac{\partial (rv)}{r \partial r} \right) \frac{\partial \theta}{r \cos\phi \partial \lambda} + \left(2\Omega \cos\phi + \frac{\partial(rv)}{r \partial r} - \frac{\partial w}{r \cos\phi \partial \lambda} \right) \right.$$
$$\left. \frac{\partial \theta}{r \partial \phi} + \left(2\Omega \sin\phi + \frac{\partial v}{r \cos\phi \partial \lambda} - \frac{\partial(u \cos\phi)}{r \cos\phi \partial \phi} \right) \frac{\partial \theta}{\partial r} \right].$$

This nonhydrostatic version is not necessary for the analysis of large-scale weather systems, and an approximate hydrostatic version is usually used. This approximate version neglects terms involving the vertical velocity w, neglects the Coriolis terms proportional to the cosine of the latitude, and makes selective use of $r \approx a$, where a is the constant radius of the earth. In this way, the approximate form is obtained

$$P = \alpha \left[-\frac{\partial v}{\partial z} \frac{\partial \theta}{a \cos\phi \partial \lambda} + \frac{\partial u}{\partial z} \frac{\partial \theta}{a \partial \phi} + \left(2\Omega \sin\phi + \frac{\theta v}{a \cos\phi \partial \lambda} \right) \right.$$
$$\left. - \frac{\partial(u \cos\phi)}{a \cos\phi \partial \phi} \right) \frac{\partial \theta}{\partial z} \right].$$

The potential vorticity has the SI units m^2 s^{-1} K kg^{-1}. It has become accepted to define 1.0×10^{-6} m^2 $s^{-1} K$ kg^{-1} as one potential vorticity unit (1 PVU).

Additional Reading

Gill, A.E.: *Atmosphere-Ocean Dynamics*, Elsevier Science & Technology Books, New York, NY, 1982.

AMS

POTENTIATOR.

A term used in the flavor and food industries to characterize a substance that intensifies the taste of a food product to a far greater extent than does an enhancer. The most important of these are the 5'-nucleotides. They are approved by the FDA. Their effective concentration is measured in parts per billion, whereas that of an enhancer such as MSG is in parts per thousand. The effect is thought to be due to synergism. Potentiators do not add any taste of their own, but intensify the taste response to substances already present in the food.

POTENTIOMETER.

1. An instrument used for the measurement or comparison of small potential differences or electromotive forces, based

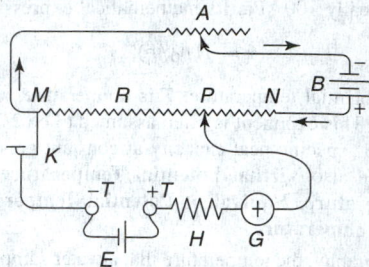

Fig. 1.   Potentiometer circuit.

upon the "law of potential drop" (see also **Electric Circuits**). One of the simplest potentiometer circuits is shown in Fig. 1.

Current from a battery B is sent through a resistance MN and is adjustable by means of a rheostat A. From one extremity M of this resistance is taken off a branch circuit containing the potentiometer terminals $+T$, $-T$, between which E, one of the electromotive forces to be compared, is connected. This circuit rejoins the main circuit at a point P which is adjustable so that the partial resistance MP or R can be varied, while MN as a whole remains constant. The $+$ and $-$ leads from E must be connected as shown, and the electromotive force of B must exceed E. The position of P is now adjusted until the galvanometer G shows no current, indicating that the potential drop from P to M just balances the electromotive force E. If two different electromotive forces E_1, E_2, are thus connected and balanced in succession, and if the corresponding values of the resistance MP are R_1, R_2, then since the current through MN is unaltered, the law of potential drop gives

$$\frac{E_1}{E_2} = \frac{R_1}{R_2}$$

In particular, one of the electromotive forces may be a standard cell of accurately known voltage; the other is thereby determined. In such case the standard cell should be safeguarded by a high resistance H, which is gradually reduced as the zero-current adjustment is approached; and the key K should be closed only for an instant. In some potentiometers the whole equipment, including galvanometer and standard cell, is contained in one compact case.

2. The term potentiometer is also used to denote a three-terminal voltage-dividing network such as is used in volume controls for radio receivers, etc. Confusion may be avoided if the term voltage-divider is used for this purpose.

POT HOLE. Under favorable conditions, where streams flow over the bedrock, swirling eddies will wash sand, gravel or pebbles around and around in the same place with the result that cylindrical holes called pot holes are worn, often to a considerable depth. These pot holes may be from a few inches to several feet in diameter and rarely as much as 40–50 feet (12–15 meters) deep. Similar features found on the seashore, the result of wave action, are called sea-mills. Pot holes also have been formed by water from crevasses and ice cliffs and glaciers.

POTOMETER. A device for measuring transpiration, consisting of a small vessel containing water, and sealed so that the only escape of moisture is by transpiration from a leaf, twig, or small plant with its cut end inserted in the water. A similar device, the *phytometer*, consists of a vessel containing soil in which one or more plants are rooted and sealed.

POTTER WASP (*Insecta, Hymenoptera*). A small wasp whose nest is built of mud in the form of a globular pot with a narrow neck. The several species belong to the genus *Eumenes*. All are solitary.

POULTRY. Any of a variety of domesticated birds bred and raised for their meat and eggs. Of all meat (poultry and red meat), poultry accounts for about 20% of worldwide meat production. It is estimated that the production of poultry meat exceeds 21 million metric tons. Production of eggs worldwide exceeds 23 million metric tons annually.

A Chinese book, dated about 1400 B.C., refers to fowls as "creatures of the West." This statement is consistent with the views of early authorities, who believed that the domestication of various wild birds for use as food (flesh and eggs) commenced in Asia, notably in the western part of the continent in the Indian region. Mentions of fowls were made much later in the writings of Aristophanes and others as early as 500 B.C. They were mentioned frequently in the literature of Biblical and Roman times. But, unfortunately, little information found in the early literature serves as a key to the origin of the common domesticated birds as we know them today. Some early authorities traced back to the wild jungle fowl (*Gallus bankiva*), which still can be found wild in some regions of India. This correlation resulted from comparing the fundamental characteristics of modern breeds of chickens with the characteristics of the wild jungle fowl. More recently, however, some experts have pointed out resemblances between present fowls with a number of other forms of wild game from various regions of the Orient, such as *Gallus sonneratii* (Afghanistan), *G. fercatus* (Indonesia), *G. stanleyi* (Sri Lanka), and *G. giganteus* (Malaysian Peninsula), among others.

The genealogy of the domesticated birds that are known today as *poultry* is complicated by many factors. Birds are relatively easy to domesticate, and they interbreed and combine characteristics quite readily, thus accounting for the large numbers of varieties developed in nature. The varieties are mainly the result of human intervention, a process that has gone on for many hundreds of years, though admittedly quite unscientifically until the last century or so. Further, these birds are relatively small and are quite easy to transport, even during the earliest times when transport was crude. Thus, the blending of bird varieties from many regions was inevitable. These and other factors tended to discourage the development of breeds along simplistic, classical lines that are easy to trace.

The term *poultry*, once confined to birds easily raised by farmers, peasants, and natives in relatively small numbers, has been expanded to include any species of bird that is subject to commercialized domestication. Turkeys, for example, at one time were strictly considered wild game. Today, wild turkeys are difficult to find, but many millions produced each year are a major product of the poultry industry. The same situation would apply to swans if it were found commercially advantageous to industrialize their production and market them in huge quantities. Growing interest in the 1980s, for example, is being shown in commercializing production of quail as a common item for the marketplace.

Breeds of Poultry

Although tracing the ancestry of modern poultry is difficult, authorities recognize well over a dozen basic *classes* of chickens (or fowls as preferred by some authorities).[1] Each class has a number of breeds and varieties. A *breed* of bird possesses certain fundamental and consistent qualities that reliably appear generation after generation and the results of which can be forecasted when there is mating between breeds. The similarities between the breeds of any given class are usually more obvious than any relationship between classes. Further, varieties within a given breed tend to be even more closely and obviously related. The *variety* is the first subdivision of a breed. When certain intrinsic qualities appear within a variety and then consistently reappear, it may be desirable from a classification standpoint to establish a *subvariety*. Usually a subvariety is spoken of as a *strain*. The well recognized classes of chickens or fowls are delineated in Table 1. Of the breeds and varieties listed, the American, Asiatic, English, and Mediterranean breeds are of principal importance in North America. The principal characteristics of these breeds are summarized in Table 2. Some concept of the range of popularity among the various breeds can be gleaned from Table 3.

Of the 200 or more officially recognized varieties of chickens, the majority are not of commercial importance. Of the breeds listed in Table 1, only a relatively few breeds are raised in significant numbers in North America; these include the White Leghorn, the White Plymouth Rock, the Rhode Island Red, the Barred Plymouth Rock, the New Hampshire, and the Cornish. There has been a strong movement for a number of

---

[1] Some authorities prefer the designation fowl rather than chicken, because the latter word is sometimes used by people to designate a young bird as contrasted with an older bird.

TABLE 1. PRINCIPAL CLASSES OF CHICKENS[1]

AMERICAN CLASS
Buckeye, Dominique, Holland, Java, **Jersey White Giant, New Hampshire, Plymouth Rock, Rhode Island Red**, Rhode Island White, and **Wyandotte**.
Composite breeds—originated in North and Central America.
Medium-to-large size, with moderate, but colorful and widely varied plumage. These birds grow and develop at a moderate rate. The birds are slow, but not sluggish in their movements. Generally, they are a hardy stock with well-protected bodies. Although somewhat combative, the birds do not have a vicious personality. They are excellent winter layers and in general prolificacy are better than the Asiatics and somewhat inferior to the Mediterraneans. Reliable in terms of breeding efficiency, but not as prepotent as the Asiatics and Mediterraneans. Maternal qualities are excellent.

ASIATIC CLASS
Brahma, Cochin, Langshan
Presumed to be of Chinese origin—one of the oldest of the classes. They are raised for meat and winter egg production. Some of the birds, especially the Brahmas, are excellent for heavy roasters and capons. Probably better suited to small-scale than large-scale commercial operations. Plumage is heavy. Wide variety among breeds in terms of posture and stature. Skin is yellow and coarse. Rate of development is slow by comparison with American class breeds. These birds have an easy-going temperament and are not quickly frightened. They are known as good, but clumsy mothers. Generous plumage enables coverage of many eggs and chicks. By nature, they are not excellent egg producers, but eggs are medium to large size and of uniform characteristics.

MEDITERRANEAN CLASS
Ancona, Blue Andalusian. **Leghorn, Minorca**, Spanish
Originated in Italy, Spain, and other parts of the Mediterranean region. With exception of Minorca, the Mediterraneans are light to medium size and weight. The Minorca is heavy. Birds have a light, closely feathered plumage with a wide variation of color in different breeds. Development is rapid and exceeds the American and Asiatic classes. The birds are alert, nervous, and quite easily frightened. Although best suited to temperate climates, they can be adapted with relative ease to different environments. However, changes in environment affect productivity. Not rated too high in terms of mothering ability or interest. The birds are courageous, but not quarrelsome. The Mediterraneans are rated among the highest in terms of prolificacy. First eggs are laid at 6 months of age or less. Possess prepotent breeding qualities.

ENGLISH CLASS
Australorp, Cornish, Duckwing, Orpington, Redcap, Sussex
Originally developed mainly as meat producers, but with objective of combining meat- and egg-producing qualities. The birds are large and white, with a white skin, excepting the Cornish, which has a yellow skin.

POLISH CLASS
This is one breed with eight varieties. Essentially ornamental rather than food birds. However, they are good layers of white eggs. Generally regarded as nonsitters.

HAMBURG CLASS
A Dutch class with one breed and six varieties. Considered unsatisfactory for meat production. Small size of eggs limits market appeal.

FRENCH CLASS
Four breeds and five varieties. The birds are of the meat-type, although can be raised for reasonably satisfactory egg production.

CONTINENTAL CLASS
Of Belgian origin, the birds are quite small, quite prolific, and very hardy.

GAME AND GAME BANTAM CLASS
Sixteen varieties in this class are mainly raised for exhibition and fighting (in some countries).

ORIENTAL CLASS
Mainly birds for exhibiting.

ORNAMENTAL BANTAM CLASS
A few breeds and several varieties of birds, generally considered too small to be of interest as food-producing birds.

[1] Breeds listed in boldface type are the most important breeds in North America. There is also a Miscellaneous Class (mainly ornamental birds).

years on the part of American poultry producers to pay progressively less attention to the traditional purebreeds. The latter are mainly of interest in the development of new foundation stocks. The poultry industry differs from the livestock field in that less attention is paid to breed registration.

The closest approach to this is The American Poultry Association (Crete, Nebraska), which publishes the "Standard of Perfection" that contains listings and specifications of a large number of varieties of poultry.

Breeding techniques in the poultry field parallel those used in the cattle, swine, and sheep industries, and also in the crop plant field, because the parameters of what can be accomplished are established by the fundamental laws of genetics. More experimentation is possible in the poultry field, as compared with large animals, for a number of reasons. The biological cycle is much shorter and the results of breeding efforts can be determined during a shorter time span. The investment in individual birds is much less than in other forms of livestock. The compatibility among breeds of birds is good. In the improvement cycle, foundation breeders furnish eggs to hatcheries. In turn, the hatchery sells chicks to the producer, who, in turn, raises the commercial egg layers, broilers, and market turkeys. New varieties are also developed by publicly funded agricultural experiment stations and universities, where proprietary interests are not a problem.

Breeding objectives include fertility, viability (general health), and feed conversion. There are also specific objectives as regards egg-type and meat-type birds. The heritability of characteristics ranges from a low of 8–10% (laying intensity for example) to as high as 50% (egg size) and higher. The interrelationship and interdependence of heritable qualities and bird management practices are always present and not always easy to sort out.

Body weight, which has a heritability factor of about 60%, varies widely among the basic breeds (contrast between the Brahmas and the bantams). This is an important factor because there is a strong correlation between body weight and size and rate of growth and feed utilization efficiency. Large body size is of particular importance to broiler and turkey breeders. Closely related to body weight is excessive abdominal fat in broilers. Traditionally, broiler breeders have selected for large body weight. Because of high heritabilities for abdominal fat, breeders probably have inadvertently selected for increased percentage of fat. It is estimated that at least 1% of the weight of broilers purchased by the processing plant is gizzard fat and sections of the leaf fat torn from the bird during the processing. This tissue goes to the offal plants and is processed as a byproduct and at a loss to the processor.

Egg production increases have been less amenable to genetic manipulation than increases in meat production. Two factors are of major consideration: (1) length of the laying period prior to molting. Early sexual maturity of a pullet is obviously desirable. Age at maturity is a heritable factor. Sexual maturity in normal Leghorns occurs at about 24 to 26 weeks (170 to 185 days) of age. Dual-purpose breeds (White Wyandotte) require 1 to 3 weeks longer. There are considerable variations among the various strains available. Environmental effects (lighting, feed management, diseases, etc.) also influence the age of maturity. To achieve an objective of 20 dozen or more eggs in the first pullet year, it is evident that laying should commence when the bird is about 5 months old. In judging a given flock of birds, it is usually assumed that the first 70–80% of the birds that commence laying will be the best layers in the long term.

Rate of production, sometimes called intensity, has a heritability factor of only about 10%. Emphasis on bird management is more important than inheritance in this case and management of environment and feed is paramount. For example, laying production goes up when eggs are removed from the nest daily. Egg size has a heritability factor of about 50%. Egg size also relates to body size, age of pullets, and environment. Egg color is also a heritable factor. In some markets, white eggs are preferred; in others brown eggs are preferred. However, the astute marketer of eggs will not mix slightly tinted eggs with white eggs.

Meat-quality indicators include plumage color, skin and shank color, and rate of feather development. Feathers that are white or light in coloration are much sought in breeding broilers. Dark feathers are much more difficult to pluck clean when the broiler is dressed. It has been found that when an early-feathering male is mated with a late-feathering female, the male progeny will be slow-feathering and the pullets will be early-feathering. This helps the producer in selecting pullets at hatching time.

Production Technology. The production of poultry meat and eggs is of the most common food-producing operations found throughout the world. Poultry plays an important role in the nutritional requirements, from the most underdeveloped of countries and regions to the most advanced of nations. There is an exceptionally wide spectrum of production intensity, ranging from small flocks numbering well under 100 birds to highly

TABLE 2. CHARACTERISTICS OF MAJOR BREEDS OF CHICKENS

Breed	Weight of Cock Pounds	Weight of Cock Kilograms	Weight of Hen Pounds	Weight of Hen Kilograms	Coloration Skin	Coloration Shank	Coloration Ear Lobe	Coloration Plumage	Egg	Type of Comb	Illustration
AMERICAN CLASS											
Jersey White Giant	13	5.9	10	4.5	yellow	yellow	red	white	brown	S	—
New Hampshire	8.5	3.9	6.5	2.9	yellow	yellow	red	red	brown	S	Fig. 1
Plymouth Rock	9.5	4.3	7.5	3.4	yellow	yellow	red	white	brown	S	Fig. 2
Rhode Island Red	8.5	3.9	6.5	2.9	yellow	yellow	red	red	brown	S&R	Fig. 3
Wyandotte	8.5	3.9	6.5	2.9	yellow	yellow	red	white	brown	R	—
ASIATIC CLASS											
Brahma (light)	11	4.9	9	4.1	yellow	yellow	red	*	brown	P	Fig. 4
Cochin	11	4.9	8.5	3.9	yellow	yellow	red	buff	brown	S	—
Langshan (black)	10	4.5	7	3.2	white	blue-black	red	dark	brown	S	—
MEDITERANNEAN CLASS											
Ancona	6	2.7	4.5	2	yellow	yellow	white	varies	white	S&R	—
Leghorn	6	2.7	4.5	2	yellow	yellow	white	white	white	S&R	Fig. 5
Minorca (white)	8	3.6	6.5	2.9	white	white	white	white	white	S	—
ENGLISH CLASS											
Australorp	8.5	3.9	6.5	2.9	white	dark	red	black	brown	S	Fig. 6
Cornish (white)	10	4.5	8	3.6	yellow	yellow	red	white	brown	pea	—
Orpington	10	4.5	8	3.6	white	white	red	white and buff	brown	S	—

* = Columbian pattern; S = single; R = rose. All breeds given in table, with exception of Asiatic class, do *not* have feathered shanks. Weights shown are for mature birds.

TABLE 3. RELATIVE POPULARITY OF BREEDS OF CHICKENS (In North America)

Breed or Type	Percent of Total	Trend Since Early 1960s
Crossmated	82.8	sharply up
White Leghorn	8.9	moderately down
Incrossmated	4.8	sharply down
White Plymouth Rock	0.6	sharply down
Rhode Island Red	0.4	sharply down
Barred Plymouth Rock	0.2	steady
New Hampshire	0.1	sharply down
Others	2.2	

Fig. 2. White Plymouth Rock male chicken. (*USDA.*)

Fig. 1. New Hampshire female chicken. (*USDA.*)

efficient and specialized commercial operations where tens of thousands of birds may be involved. While the fundamentals of managing a flock in order to obtain the best possible production remain the same, the technology (housing, equipment, etc.) varies widely. In general, one may observe that the small poultry operation represents an interface between birds and people, whereas in the large operations, the interface involves the interaction of birds and machines. In any situation, it is extremely important to select good starting stock, matching the birds with the environmental needs. For a number of years, little attention was paid by most producers to the matter of spacing required by birds. Reduction in the production of poultry meat and eggs, as a result of crowding, is well documented. The mortality of poultry is increased by crowding due to cannibalism and the spread of virulent diseases. The type of housing, whether littered floor, slats, wire floors, or cages (of various configurations) influences the optimum space allotment. Temperature and humidity control are equally important. Although further research is required, experience over the years has shown that the optimum temperature for layers lies between 12.8 and 21.1 °C. Experience also has shown that an

Fig. 3. Rhode Island Red female chicken. (*Texas Agricultural Extension Service.*)

(a)

(b)

Fig. 4. Brahma chickens: (a) Dark variety female; (b) light variety female. (*USDA.*)

Fig. 5. White Leghorn female chicken. (*Texas Agricultural Extension Service.*)

Fig. 6. Australorp male chicken. (*USDA.*)

optimum humidity for the laying house lies between 50 and 75% relative humidity.

A number of semiautomated and automated bird-feeding systems have been developed during the past few decades. Two major design criteria are reduction of labor and reducing the wastage of feed. Such systems also take into consideration the bird's environment, so that there will be a smooth (without stress) interface between bird and machine. See Fig. 7.

Fig. 7. Artist's representation of a modern cage-layer system. The components are currently available: (A) Bulk feed bin; (B) straight-line auger feed delivery; (C) watering system; (D) floor stands; (E) egg deescalators; (F) cross egg conveyor which moves eggs to central location; (G) air inlet system; (H) cage-delivery system; (I) dropping board scrapers; (J) cages; (K) feed intake cups; (L) thermostatically-controlled fans; and (M) egg collectors. (*Chore-Time Equipment, Inc., Milford, Indiana.*)

Other Species

Turkeys. Native to the Western Hemisphere, the original habitat of the wild turkey ranged from Quebec and Ontario southward to Florida and west to Mexico and the Rocky Mountains. As vast areas of open hardwood forests were cleared in North America, the prime turkey habitat was destroyed. Fortunately, the species retreated to seldom-visited pockets of wilderness in Pennsylvania, West Virginia, South Carolina, Florida, and a few other states. The "American Standard of Perfection" recognizes only one breed of turkeys. The recognized varieties include the Broad-Breasted Bronze, the Broad-Breasted White, the White Holland, the Beltsville Small White, the Bourbon Red, the Narragansett, the Black Slate, and the Jersey Buff. Of these only three varieties are of major commercial importance in North America. See Table 4. Unlike chickens, most turkeys are bred as standardbreds, rather than hybridized. Improvements are gained through individual selections and mass-matings. The popular fast-growing, broad-breasted birds are clumsy in their mating and this has led to increasing use of artificial insemination in breeding the birds.

In the early 1950s, the Broad-breasted Bronze variety was most popular, enjoying over 75% of total production, followed by the Small White (16%), and the Large White (nearly 5%). By the mid-1960s, the emphasis shifted to the Large White (48%), followed by the Broad-Breasted Bronze (41%), and the Small White (8.4%). Many producers object to the Small White because of the smaller quantity of meat produced with about the same investment in labor and housing as required by larger birds.

Geese. Generally, geese are very hardy and not susceptible to many of the common poultry diseases. They are excellent foragers, although selective, and can be put on good succulent pasture or lawn clippings as early as their first week of life. In North America, the Toulouse, Emden, and African geese are the most popular breeds raised for meat production. Other common breeds found are Chinese, Canada, Buff, Pilgrim, Sebastopol, and Egyptian. The weights of various breeds of geese are summarized in Table 5.

The *Toulouse goose* derives its name from the city of Toulouse in southern France. Because of its late maturity, it is sometimes called the Christmas goose. This breed has a broad, deep body and is loose-feathered, a characteristic which gives it a massive appearance. The plumage is dark gray on the back, gradually shading to light gray edges, with white on the breast, and to white on the abdomen. The eyes are dark brown or hazel, the bill pale orange, and the shanks and toes are a deep reddish orange.

The *Emden goose* was one of the first breeds of geese to be imported into the United States. The breed was first identified as the Bremen, named for the city in Germany from which early importations were made. The Emden is a pure white, springly goose. It is much more tight-feathered than the Toulouse and therefore appears more erect. The Emden is a fairly good layer, but production depends upon the breeding and selection of the flock. Egg production averages from 35 to 40 eggs per year per mature breeding goose. The Emden is usually a better sitter than most Toulouse and is one of the most popular breeds for marketing. The breed grows rapidly and matures early.

The *African goose*, shown in Fig. 10, is a handsome breed with a distinctive knob or protuberance on its head. Its carriage is more erect than that of the Toulouse, and its body more nearly oblong and higher from the ground. The head is light brown, the knob and bill are black, and the eyes are dark brown. The plumage is ash brown on the wings and back and is light ash brown on the neck, breast, and underside. The African goose is a good layer, grows rapidly, and matures early. However, it has not gained in market popularity as have the Emden or Toulouse.

The *Chinese goose*, of which there are two standard varieties, originated in China and probably came from the wild Chinese goose. See Fig. 11. It is smaller than the other standard breeds and more swanlike in appearance. Both the brown and white varieties mature early and are better layers than other breeds, usually averaging from 40 to 65 eggs per bird annually. The

TABLE 4. CHARACTERISTICS OF TURKEY VARIETIES

BRONZE VARIETY (See Fig. 8)

The subvariety, Broad-Breasted Bronze, is the most predominantly produced and is characterized by a uniform, amply-fleshed carcass.

Normal Weights
Adult male	36 pounds (16.3 kilograms)
Adult female	20 pounds (9.1 kilograms)

Coloration
Plumage	black with an iridescent sheen ranging from red to green to bronze
Throat wattle	red, but sometimes can appear bluish-white
Beard	black
Shanks and toes	in young birds, a dull black; in mature birds, a tarnished pink
Beak	light at tip and dark at base

WHITE HOLLAND VARIETY

Although somewhat more fertile than the Bronze variety, it is quite similar in many characteristics.

Normal Weights
Adult male	33 pounds (15 kilograms)
Adult female	18 pounds (8.2 kilograms)

Coloration
Plumage	pure white
Throat wattle	red, but sometimes pale pink
Beard	rather intense black
Shanks and toes	pale pink
Beak	light-pink horn

BELTSVILLE SMALL WHITE VARIETY (See Fig. 9)

Developed by the U.S. Department of Agriculture and named after the research center located in Beltsville, Maryland. The objective of development was to obtain a well-fleshed, smaller, light-weight bird. The variety is known for exceptional hatchability and egg production.

Normal Weights
Adult male	23 pounds (10.4 kilograms)
Adult female	13 pounds (5.9 kilograms)

Coloration
Plumage	pure white
Throat wattle	red, but sometimes pinkish-white
Beard	black
Shanks and toes	pale pink
Beak	light to medium pink

Fig. 9.   Beltsville Small White turkey developed by the U.S. Department of Agriculture

Chinese goose grows rapidly, is very attractive, and makes a desirable medium-size market bird.

The *Pilgrim goose*, shown in Fig. 12, is a medium-size goose that is well suited for marketing. A unique feature of this breed is that the males and females may be distinguished by color. In day-old goslings, the male is creamy-white and the female is gray. The adult male remains white and has blue eyes. The adult female is gray and white with dark hazel eyes.

The *Buff goose* is of fair economic value as a market goose and thus only limited numbers are raised. Other less-commercial geese are the *Canada*, the common wild goose of North America. The breed is difficult to keep in confinement. The wild gander is sometimes used to cross with domestic breeds, resulting in the so-called mongrel goose (a hybrid), which usually is sterile but has fine-quality flesh. The *Sebastopol* is a white ornamental goose, very attractive because of its soft, plume-like feathering. The *Egyptian* is a long-legged, but very small goose, kept primarily for ornamental purposes.

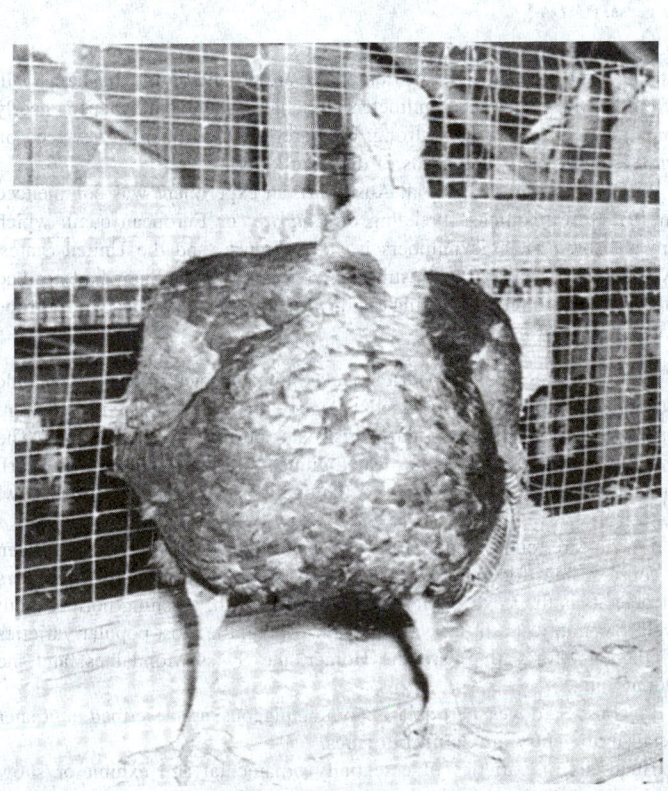

Fig. 8.   Broad-breasted female turkey. (*Texas Agricultural Extension Service.*)

Fig. 10.   African geese. (*USDA.*)

Breeding geese prefer to be outdoors. Except in extremely cold weather or in storms, mature geese seldom seek shelter. In northern climates, colony poultry houses, open sheds, or barns are provided for shelter. Geese make nests on the floor of the house or in coops, boxes, or barrels provided. Straw or grass hay is used for outside nests as well as for nests on the floor of

TABLE 5. WEIGHTS OF DIFFERENT BREEDS OF GEESE

Breed	Young Male		Adult Male		Young Female		Adult Female	
	Pounds	Kilograms	Pounds	Kilograms	Pounds	Kilograms	Pounds	Kilograms
Toulouse	20	9.1	26	11.8	16	7.3	20	9.1
Emden	20	9.1	26	11.8	16	7.3	20	9.1
African	16	7.3	20	9.1	14	6.4	18	8.2
Chinese	10	4.5	12	5.4	8	3.6	10	4.5
Egyptian	5	2.3	5.5	2.5	4	1.8	4.5	2
Canada	10	4.5	12	5.4	8	3.6	10	4.5
Sebastopol	12	5.4	14	6.4	10	4.5	12	5.4
Pilgrim	12	5.4	14	6.4	10	4.5	13	5.9
Buff	16	7.3	18	8.2	14	6.4	16	7.3

Fig. 11.   White Chinese geese. (*USDA.*)

a house. The producer must provide one nest for every three females and permit the geese to select their own nests. Geese generally start laying in February or March in the Northern Hemisphere and often lay until early summer. The incubation period for eggs is about 30 days. Geese require ample quantities of palatable drinking water. Geese are very selective and tend to pick out the most palatable of forages. They will reject alfalfa and narrow-leaf tough grasses and select the more succulent clovers and grasses. Geese cannot be raised satisfactorily on dried-out, mature pasture. Because geese will eat weeds without harming certain cultivated plants, they are frequently used as weeders.

Ducks. Breeds of ducks suitable for producing mainly table meat are the Alesburys and Pekins. Others, which are layers, such as Magpies, Buff Orpingtons, Blue and Black Orpingtons, Stanbridge Whites, and large utility Fawn and White Indian Runners are also useful for table purposes. The Campbell White and Khaki all give drakes and ducks which are quite useful eating—good in flavor and with good deep breast meat. See Figures 13, 14, 15.

Quails. Usually considered mainly for the sportsperson, in most regions of the world, serious interest is shown in the commercial domestication of these birds and they are widely available in the markets of major cities and towns. Sportspersons in North America recognize seven kinds of quails. Ornithologists recognize five genera, seven species, and nearly twenty races. In the eastern and southern parts of North America is the Bobwhite. Along the Mexican border are the Scaled or Blue Quail, the Massena Quail (also called Fool Quail), and the Masked Bobwhite. In California, there are the Desert Quail (sometimes called Gambel's Quail), the Valley Quail, and the Mountain Quail. The birds in these groups differ sufficiently in plumage and habitat and calls from those in other groups to be recognized

at a glance by the experienced hunter. All are ground birds. They are small, averaging 11 inches (28 centimeters) in length. The overall shape is nearly round. Tails are short and strong; bills are heavy and hard. They feed on grasses and clovers and seeds of different kinds.

In Corio (State of Victoria, Australia), an experiment was commenced in 1977 toward intense breeding of *Coturnix* or European quail, which are also bred in large numbers in Italy, France, and the United States. Apart from the fairly rank-tasting native variety, Australians had not had frequent opportunity for tasting quail, which has been a delicacy in Europe for hundreds of years.

Other Species of Poultry. Relatively minor species of poultry include guinea fowl and pigeons (squabs). Native to Africa, *guinea fowl* were introduced into Europe and the British Isles during the Middle Ages. The three major varieties of domesticated guineas are the Lavender, the Pearl, and the White. The Pearl variety is the one most esteemed by guinea fowl fanciers.

Pigeons are found in numerous regions of the world and squab has been a gourmet item for hundreds of years. A favorable quality for producers of pigeons is their exceedingly fast rate of growth, reaching normal adult weight within just a little over a month. Among the most popular varieties of pigeons (for eating) are the Homer, the Swiss Mondaines, and the White King.

Pigeons and doves, particularly homing pigeons, are described in greater detail in the entry on **Columbiformes**.

Hundreds of varieties of essentially ornamental and exhibit or show birds, closely related to poultry, but seldom eaten, include peafowls and swans. See also **Anseriformes** (swans); and **Galliformes** (peafowls).

Fig. 12.   Pilgrim geese. (*USDA*.)

Fig. 13.   White Pekin drake. (*USDA*.)

Fig. 14.   Khaki Campbell drake. (*USDA*.)

Metabolic System of Fowls

The digestive tract of the fowl differs in several respects from that of the mammals. The fowl does not chew in the conventional sense. It has no teeth. Feed is picked up by the beak and forced into the *gullet* (esophagus) by a specialized tongue, the rear part of which incorporates a fork-like configuration. The tongue also assists the bird in taking water. The esophagus, capable of extensive expansion, provides for continuous flow of feed and water to the stomach, but is divided into an upper and lower chamber, with an intermediate chamber known as the *crop*. The crop, coupled with the expandable characteristics of the esophagus, enables the bird to consume feed at an accelerated rate, and provides a place for initial softening and processing of the feed, largely as the result of an admixture of moisture with enzymes present in the feed. However, very little saliva as such is secreted.

From the lower gullet, feed passes to the glandular stomach, also known as the *proventriculus*, where gland-secreted gastric juices containing hydrochloric acid and pepsin break down proteins to peptones and other simpler compounds. The partially-processed feed then proceeds to the *gizzard*, also known as the *ventriculus*. This is a very muscular organ, equipped with a horny lining, that by way of repeated contractions and expansions, masticates or massages the feed contents. This action accomplishes the type of disintegration of the feed into small particles that is accomplished by the teeth in mammals. Sometimes, the horny lining of the gizzard erodes or sloughs off, leaving ulcerations in the lining. Usually, a well-balanced diet will prevent the condition from occurring. The gizzard also acts as an efficient mixer of gastric juices and pulverized feed.

The well-processed feed substances then enter the small intestine where further enzymes are secreted to finish the breakdown of proteins and to split sugars. Absorption of the nutrients is also performed, enriching the bloodstream with energy-containing and body-building materials. A peristaltic action also occurs within the small intestine to ensure proper flow of the residue to the *ceca* (large intestine) and rectum. The ceca serves as an intermediate storage or buffer chamber. Bile generated in the liver is discharged into the gizzard and the duodenal loop, which connects the

Fig. 15.   White Runner drake, a variety of the Indian Runner. (*USDA*.)

gizzard with the small intestine. Amylase, trypsin, and lipase are generated by the pancreas and secreted into the duodenal loop to aid digestion of carbohydrates, proteins, and fats. Insulin, also generated by the pancreas, regulates sugar metabolism.

A laying hen will require about 2.5 hours to digest feed, whereas a nonlaying hen will require up to 8 or even 12 hours.

Physiology of the Hen. Unlike most animals, which have two functional ovaries (right and left), the hen has only one functional *ovary* (left). This is located in the body cavity near the backbone. At the time of hatching, the female chick has close to 4000 very small *ova* contained in the left ovary. These ova have the capability of developing into full-sized yolks once a pullet attains full sexual maturity. Approximately once every 24 hours, in a normal, healthy hen, one of these ova will be released to the *oviduct*, which is a long (20 to 30 inches; 51 to 76 centimeters), folded tube contained on the left side of the abdominal cavity. The tube is complex and is comprised of five regions, all of which participate in the completion of a whole egg. For example, the white part of the egg is formed in an albumen-secreting region of the oviduct.

Within a short period (about 30 minutes) after laying a completed egg, another ovum is released from the ovary to the oviduct for another cycle of egg production. The service of a cock is not required to commence the egglaying process of the pullet; nor to restart the process after the first molting period. Infertile eggs are preferred over fertilized eggs in the marketplace. Mating is required, of course, to produce fertile eggs for hatching.

A number of important physiological changes occur in the laying hen. During the course of a laying period, a fowl will consume surplus fat from the body, particularly from the skin. Thus certain parts of the bird's body become whiter as the fat disappears. Other important physiological changes include some alteration in blood flow and in characteristics of the vent and pelvic arches and feathers as well. These changes occur in phase and progress slowly throughout the laying period.

A laying hen has a large, moist vent. The vent is dilated and loose and unlike the hard, puckered vent of a nonlaying hen. The pelvic arches and keel configure so as to increase the abdominal capacity of the bird. As fat is removed from the skin, the skin becomes velvety and the abdomen is soft and pliable. In nonlayers or poor producers, there will be an underlying layer of hard fat.

At the end of the laying period (200 or more eggs), the bird goes into the molting phase, during which time feather replacement and other physiological changes occur. Poultry producers who retain pullet flocks for a second year of production often use molt as a sign for selecting the good layers. The bird sheds its feathers in stages—first from the head, secondly from the neck, thirdly from the breast, fourthly from the body, and last from the wings and tail. On the average, from 3 to 4 months are required to molt. The producer can ascertain the length of time a given bird has been out of egg production by counting the number of new primary feathers. There are ten of these feathers in each wing. About six weeks are required to produce the first new feather and two weeks for each additional new feather.

Much more detail on poultry and the production of poultry meat and eggs can be found in the "*Foods and Food Production Encyclopedia*," (D.M. Considine, editor), Van Nostrand Reinhold, New York (1981). For further background on various birds, see entries listing under **Birds**.

Additional Reading

Barbut, S.: *Poultry Products Processing: An Industry Guide*, CRC Press, LLC., Boca Raton, FL, 2001.
Calnek, B.W.: *Diseases of Poultry*, 10th Edition, Iowa State Press, Ames, IA, 1997.
Etches, R.J.: *Reproduction in Poultry*, CAB International, New York, NY, 1996.
Gillespie, J.R.: *Modern Livestock and Poultry Production*, 6th Edition, Delmar Thomson Learning, Albany, NY, 2000.
Hunton, P.: *Poultry Production*, Elsevier Science, New York, NY, 1995.
Jordan, F.T.: *Poultry Diseases*, 5th Edition, Harcourt Health Sciences, San Diego, CA, 2001.
Pattison, M.: *The Health of Poultry*, 4th Edition, Addison-Wesley Longman, Inc., Redding, MA, 1994.
Sainsbury, D.: *Poultry Health and Management: Chickens, Ducks, Turkeys, Geese, Quail*, 4th Edition, Blackwell Science, Inc., Malden, MA, 2000.

Web References

Breeds of Livestock, Chickens, Ducks, Geese & Turkeys: http://www.ansi.okstate.edu/breeds/other/
Poultry Science Virtual Library: http://posc.tamu.edu/library/dother.html
U.S. Poultry & Egg Association: http://www.poultryegg.org/

POULTRY STRESS INDEX (PSI). An index of poultry heat stress. Categories defined from the temperature–humidity index (THI) are based upon an increasing death rate of poultry as the THI becomes larger (Table 1.)

TABLE 1. POULTRY HEAT STRESS EFFECT

THI	PSI	Effect
<82 °F (27 °C)	No stress	No heat-related increase in daily mortality
82°–84 °F (27°–29 °C)	Moderate stress	0.5%–1% increase in daily mortality
84°–86 °F (29°–30 °C)	Severe stress	1%–2% increase in daily mortality
>86 °F (30 °C)	Very severe stress	>2% increase in daily mortality

AMS

POUND, ROBERT (1919). Pound is a Canadian-born American physicist who pioneered many fruitful ideas and is especially remembered for co-discovering, with Purcell, nuclear magnetic resonance (NMR) and establishing it as one of physics' most valuable analytical techniques. NMR is used as an analytical technique in chemical research, medical diagnosis, and a number of other fields. For this he received and shared the Nobel Prize in Physics in 1952.

Pound worked with his associate, Glen A. Rebka, Jr., carrying out an experiment using the Mossbauer effect to measure the gravitational effects of electromagnetic radiation and to test the predictions of Einstein's theory of general relativity. Pound's experiments continued and results predicted the Red Shift discovery.

During WW II, Pound worked at the Submarine Signal Company and then at MIT's radiation laboratory helping to develop radar and microwave technology. After the war, he became a professor at Harvard in 1948 and stayed until his retirement in 1989. Among his many awards have been the Thompson Memorial Award of the Institute of Radio Engineers in 1948, The Eddington Medal of the Royal Astronomical Society in 1965, and the National Medal of Science in 1990.

See also **Gravitation**; and **Nuclear Magnetic Resonance (NMR) and Magnetic Resonance Imaging (MRI)**.

J. M. I.

POUR POINT.   1. The lowest temperature at which a liquid will flow when a test container is inverted. 2. The temperature at which an alloy is cast. See also **Petroleum**.

POUR POINT DEPRESSANT.   An additive for lubricating and automotive oils that lowers the pour point (or increases the flow point) by 11.0C. The agents now generally used are polymerized higher esters of acrylic acid derivatives. They are most effective with low-viscosity oils.

See also **Petroleum**.

POUT.   See **Codfishes**.

POWDER.   Any solid, dry material of extremely small particle size ranging down to colloidal dimensions, prepared either by comminuting larger units (mechanical grinding), combustion (carbon black, lampblack), or precipitation via a chemical reaction (calcium carbonate, etc.). Powders that are so fine that the particles cannot be detected by rubbing between thumb and forfinger are called *impalpable*. Typical materials used in powder form are cosmetics, inorganic pigments, metals, plastics (molding powders), dehydrated dairy products, pharmaceuticals, and explosives. Metal powders are used to make specialized equipment by sintering and pressing (powder metallurgy), as well as sprayed coatings and paint pigments (aluminum, bronze). Thermoplastic polymers in powder form are used in a technology known as powder molding. Thermosetting polymers are used in the sprayed coatings field for autos, machinery, and other industrial applications in which they have many advantages over sprayed solvent coatings.

See also **Carbon Black**; and **Powder Metallurgy**.

POWDER METALLURGY.   Powder metallurgy (P/M) embraces the production of finely divided metal powders and their union through the use of pressure and heat into useful articles. The temperatures required are below the fusion point of the principal constituent, and bonding depends on interdiffusion of the metal particles in the solid state. It is necessary to provide intimate contact between particles, hence reducing atmospheres are provided in the sintering process to prevent formation of oxide films. Readily oxidized powders such as aluminum require special technique.

Probably the most important applications of powder metallurgy are those in which a product is made which cannot be duplicated by other methods. There are many examples of this kind. The melting point of tungsten, 6,100°F (3.371°C), is much too high for ordinary melting and casting methods and the only way in which filaments for electric lights can be made is to draw them from rods of compacted and sintered tungsten powder. The cemented carbide cutting tools are another important product of refractory nature readily made by powder metallurgy.

Self-lubricating bronze bearings having controlled porosity are products that can be made only by powder metallurgy. The pores are impregnated with oil, and flow to the bearing surface is maintained by capillary action. Graphite is incorporated with the metal powder in one type of oil-less bearing. A material made from powdered copper and graphite is used for electric-current collector brushes, and tungsten-copper or tungsten-silver combinations are used for electric contact points. In contrast to these high-conductivity materials, a high-resistance element is produced from a mixture of copper and porcelain powders, combining a metal with a nonmetallic substance.

History

Powder metallurgy dates to prehistoric times when artisans learned to form a solid structure by hammering gold or iron particles into metallic objects. This occurred long before the advent of furnaces that could even approach the melting point of the metal. Egyptian implements were made from metal particles as early as 3000 BC. Pure iron oxide was heated in a charcoal fire intensified by air blasts from a bellows. The iron was reduced to a spongy

metallic form. A more or less solid metallic structure was produced by hammering this porous metal while it was still hot. Final shapes were obtained by forging techniques. The direct reduction of iron oxide without fusion dates to 300 AD when the famous Delhi Pillar, weighing almost 5.5 t, was produced.

The Incas and their predecessors used powder techniques to form articles from grains of platinum, gold, and silver. Until the end of the eighteenth century all platinum was fabricated from granules by variations of the process used by the Incas. See also **Platinum**. Woolaston worked on producing platinum metal products from sponge powder without liquid-phase sintering in the early nineteenth century. Methods of powder preparation and compacting were investigated and a horizontal toggle press was developed. In 1830 Osann was recognized for work with copper powder, making shapes by pressing and sintering. The process was also used to make medals out of silver and lead.

The search for a durable filament for the incandescent electric light bulb opened up development work in osmium, tantalum, and tungsten powders. These materials were mixed with a binder, extruded into wire, and sintered. In the early 1900s it was discovered that tungsten could be worked in a certain temperature range yet keep its ductility at room temperature. See also **Tungsten**. Sintered tungsten ingots were annealed, swaged, and drawn into very thin and ductile lamp filaments. Next came cemented tungsten carbide, a mixture of tungsten carbide and cobalt powders used to manufacture cutting, forming, and mining tools and wear parts.

In the 1920s automotive engineers, recognizing the interesting characteristics of powder metallurgy products, designed the oil-impregnated self-lubricated P/M bearing. Conventional P/M bearings can absorb from 10 to 30 vol % of additive-free nonautomotive engine oils. Impregnation takes place by soaking the part in heated oil, or by vacuum techniques. When friction heats the part, the oil expands and flows to the bearing surface. On cooling, the oil returns into the pores of the metal by capillary action. World War II brought about further commercial applications of powder metallurgy, especially for iron, steel, and copper base P/M parts and products, such as sintered iron bearings, paraffin-impregnated iron driving bands for military ammunition, magnets, bullet cores and fuse-body parts, and magnetic cores.

By the 1950s P/M parts were used in postage meters and home appliances. In the 1970s P/M superalloys for aerospace applications were used, followed by steel P/M forgings. The 1980s opened the way for rapid solidification processing, P/M tool steels, and metal injection molded parts.

In the 1980s and 1990s P/M forgings, P/M tool steels, rolled-compacted strip, dispersion-strengthened copper, high strength aluminum alloys, metal-matrix composites (MMC), and materials such as aluminides have all opened up markets for powder metallurgy.

The hot forging of P/M products, known as powder forging of metals (P/F), is a recognized technology used to form parts for critical applications, and it is expected to continue to grow. Metal injection molding (MIM) holds great promise for producing complex shapes in large quantities. Spray forming, a single-step gas atomization and deposition process, produces near-net shape products. In this process droplets of molten metal are collected and solidified onto a substrate. Potential applications include tool steel end mills, superalloy tubes, and aerospace turbine disks.

The market for lighter weight P/M materials such as aluminum and titanium aluminides is expected to grow, especially for uses in automobiles. P/M processing of titanium aluminides results in more consistent product quality than the conventional casting process, and offers novel alloy/microstructure possibilities and improved ductility. Processing trends include use of high (1200–1350°C) temperature sintering to improve mechanical properties of steel and stainless steel parts.

Advances in PM Technology

Particularly during the past decade, remarkable progress was made in PM technology. Major trends in the early 1990s included: (1) rapid solidification processing (RSP), (2) liquid-dynamic compaction (LDC); (3) self-propagating high-temperature synthesis (SHS); (4) greater use of intermetallics and additives in PM products; (5) advancements in PM injection molding; and (6) improvements in heat treating PM parts—not to mention the appearance of PM in products and structures traditionally made by other metallurgical processes, such as seamless tubing.

Rapid Solidification Processing. RSP holds high promise for producing engineering alloys with refined microstructures, improved chemical homogeneity, extended solute solubility, and possible retention of metastable phases. RSP usually involves cooling rates greater than 100 °C/second (212 °F/s). For high cooling rates, RSP products must have a large surface-to-volume ratio, and thus are commonly in the form of powder, flakes, or ribbon. To be commercially acceptable, such rapidly solidified particulates must be consolidated into fully dense, metallurgically bonded forms suitable for engineering applications. RSP properties are quite sensitive to heat treatment and the desired properties can easily be lost without careful control over the consolidation process. Among the consolidation methods currently in commercial or near-commercial use include hot extrusion, hot isostatic pressing, vacuum or inert-atmosphere pressing or sintering, and powder forging. Unfortunately, these processes require elevated temperatures for relatively long times, which may destroy the benefits achieved by RSP. A major problem involves the tenacious oxide that forms on the surface of many RSP materials, particularly aluminum, nickel, and stainless steels.

A shock wave moving through the medium at velocities in excess of that of sound appears to be one solution to this problem. The shock wave can greatly exceed the yield stress. Passage of the shock wave causes plastic flow, interparticle melting and bonding, and can produce a fully dense, metallurgically bonded product. Three methodologies have evolved for introducing a shock wave: (1) use of a gas gun incorporating propellants or compressed gas; (2) direct application of explosives; and (3) impact of a projectile accelerated by explosives.

Guns are available of several designs. In one configuration, a high-pressure burst of gas launches a projectile down an evacuated tube where the projectile imparts a shock wave by driving a punch into the powder bed. As pointed out by Wright, the gun may be in the form of a high-impact press in which a reusable piston is accelerated in an evacuated chamber by introducing a rapid burst of gas into the breach. The impact of the ram produces a pressure pulse.

Hitchcox (1986) describes a process being developed at the Massachusetts Institute of Technology, which uses high-velocity pulses of an inert gas to atomize a stream of molten metal. Semisolid droplets of the metal are collected as rapidly solidified "splats" on a chilled metallic substrate. (This liquid dynamic compaction (LDC) process is attractive from a cost standpoint.) Substrates can be flat surfaces, molds, or shaped containers. The splats build up rapidly, forming high-density bodies suitable for further processing. Because the splats are thin, they cool at relatively high rates (1000 °C/second; 1800 °F/s). It is claimed that the LDC process improves ductility and fracture toughness because oxides and powder particle boundaries are minimized. Although in an early stage of development, materials such as high-strength aluminum and superalloys and (FeCo)-Nd-B have been produced with the process. Grant (MIT) reports that rapidly solidified material may exhibit grain sizes as fine as 0.2 micrometer (8 microinches) after crystallization of glasses. The fine grain size allows superplastic forming of aluminum alloys, stainless steels, and other materials.

Self-Propagating High-Temperature Synthesis. SHS usually involves an exothermic reaction producing temperatures in excess of 2500 °C (4532 °F). In essence, a mixture of compressed powders is ignited with a heat source in air or an inert atmosphere and in an instant, a refractory compound or multicomponent material results. SHS eliminates the need for high-temperature furnaces as required by conventional processes. Processing time is shortened to seconds or minutes versus hours and days as required with normal sintering. The products are usually of a higher purity, some having less than 0.2% (wt) of unreacted elements. This is the result of vaporizing volatile contaminants during the "explosion." SHS has been used to produce borides, carbides, and other difficult materials and is considered to have much potential for making ceramic matrix composites with unique microstructures.

In SHS, there are fundamentally two types of reactions: (1) *thermite*, where oxidation-reduction produces multiphase products, such as cermets; and (2) *compound formation*, as resulting from the starting elements, such as $Ti + 2B = TiB_2$. A combination of the two types of reaction also can be used. SHS requires a strong exothermic reaction where the heat of reaction is at least 40 kcal/mole (168,000 Joules/mole). The adiabatic temperature must be greater than the melting point of the product in order to produce

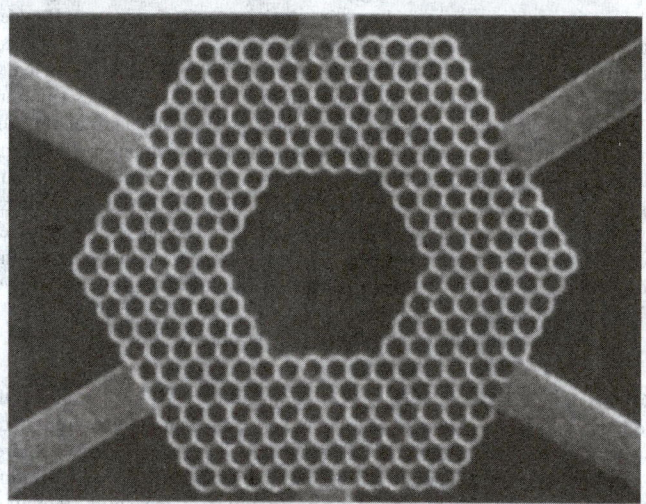
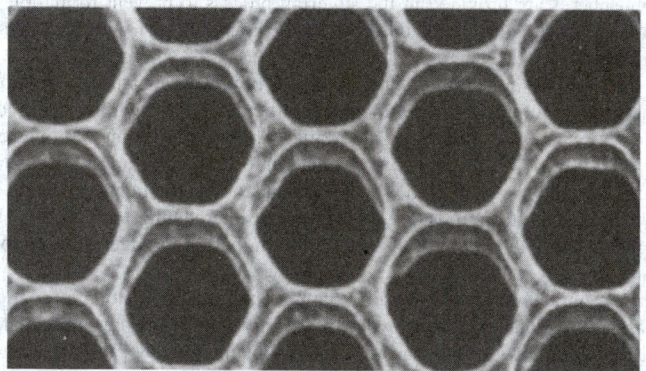

Fig. 1. Representative cross sections of tiny (submicrometer) mechanical parts that can be produced by nanofabrication technology. (*Cornell University.*)

a liquid phase for enhancing diffusion. Sheppard also breaks the reactions into (1) propagating, and (2) bulk. *Propagating reactions* are initiated locally, so that a synthesis wave of reactants, or, conversely, chemical activators can be added to accelerate the reaction. Also, if a higher reaction temperature required, preheating of the reactants is practiced.

Examples of products made by the SHS process include borides, carbides, chalcogenides, hydrides, intermetallic compounds, nitrides, silicides, carbonitrides, sulfides, cemented carbides (cermets), and various heterogeneous mixtures (microcomposites).

PM Intermetallics and Additives. An example of improved materials for which PM technology may solve past metallurgical processing problems is found in turbine parts, where high-temperature performance and oxidation resistance is mandatory. Aluminides of iron, nickel, and titanium have received consideration for a number of years, not only because they appear to meet the two foregoing criteria, but also because of their relatively low density, high strength, and corrosion resistance. Conventional casting of these materials results in unacceptable inhomogeneities. This has led to the evaluation of several PM methodologies, including hot isostatic pressing (HIP), vacuum hot pressing (VHP), injection molding, transient liquid-phase sintering, reactive sintering, and hot extrusion. Of considerable promise, reflecting research at Rensselaer Polytechnic Institute, is *reactive sintering*. This process involves a transient liquid phase. The reaction takes place above the lowest eutectic temperature in the system, but still at a temperature at which the compound remains in the solid phase. Research has shown that a transient liquid forms at the lowest eutectic temperature and spreads through the compact during heating. Actually, the reaction is approximately spontaneous because heat is liberated due to the thermodynamic stability of the compound's high melting temperature. In terms of the reaction of nickel and aluminum powders, a temperature over $550\,^{\circ}$C ($1020\,^{\circ}$F) is the optimum. The time required for processing is relatively short (about one-half hour). Densities over 97% (of theoretical) are obtained. Even with the presence of some residual porosity, the ductility and strength of the product are good, which properties are retained after subsequent high-temperature exposure.

Researchers at Case Western Reserve University and the NASA Lewis Research Center, both located in Cleveland, Ohio, have evaluated *hot extrusion* as a candidate process. In essence, the process consists of canning the powder (prealloyed aluminide powders [FeAl, NiAl, and Ni_3Al]) and then extruding the material at a temperature and area-reduction ratio sufficiently high to produce satisfactory material flow and efficient filling of interparticle spaces, the latter for eliminating porosity and to encourage grains to recrystallize dynamically.

A basic advantage of PM technology has been that of minimizing or eliminating machining in making a final part. Nevertheless, some machining operations may be required. Traditionally, the machinability of sintered PM steels, for example, is poor, mainly due to porosity, hardness, and low thermal conductivity, Porosity causes an interrupted cut and causes tool wear—with the possible results of both higher tool costs and poorer surface finish. In recent years, PM techniques have been improved by the incorporation of additive, notably manganese sulfide (MnS), to enhance machinability.

Powder metallurgy also is playing a major role in the pioneering but rapidly developing technology of *nanofabrication*. Melding the technologies of PM and electronic components manufacture are reducing operational minute machine parts to submicron levels. See Fig. 1.

Additional Reading

Allen, T.: *Powder Sampling and Particle Size Determination*, Elsevier Science & Technology Books, New York, NY, 2003.

Alman, D.E. and J. Newkirk: *Powder Metallurgy Alloys and Particulate Materials for Industrial Application*, The Minerals, Metals & Materials Society, Warrendale, PA, 2000.

Anderson, I.E.: "Boost in Atomizer Pressure Shaves Powder-Particle Sizes," *Advanced Materials and Processes*, **30** (July 1991).

Craighead, H.G.: *The National Nanofabrication Facility at Cornell University*, Cornell University, Ithaca, New York, NY, October 1990.

Froes, F.H.: "Powder Metallurgy," *Advanced Materials and Processes*, **55** (January 1990).

German, R.: *A-Z of Powder Metallurgy*, Elsevier Science Publishing Company, New York, NY, 2005.

German, R.M.: *Powder Metallurgy of Iron and Steel*, John Wiley & Sons, Inc., New York, NY, 1998.

Keishi Gotoh, K. and H. Masuda: *Powder Technology Handbook*, 2nd Edition, Marcel Dekker, Inc., New York, NY, 1997.

Hitchcox, A.L.: "Advances in Powder Metallurgy Cover Many Fields," *Advanced Materials and Processes*, **63–65** (December 1986).

Jenkins, I. and J.V. Wood: *Powder Metallurgy: An Overview*, Ashgate Publishing Company, Brookfield, VT, 1991.

Kloecker, C.J.: "Hammers Take on Presses for Forging PM Steel," *Advanced Materials and Processes*, **37** (July 1991).

Marquis F.D.S.: *Powder Materials: Current Research and Industrial Practices*, The Minerals, Metals & Materials Society, Warrendale, PA, 1999.

Scott, W.W., Jr.: "Engineering the Part," *Advanced Materials and Processes*, **4** (July 1991).

Smith, L.N.: *Knowledge-Based System for Powder Metallurgy Technology*, John Wiley & Sons, Inc., New York, NY, 2003.

Staff: *Properties and Selection: Nonferrous Alloys and Special-Purpose Materials*, ASM International, Materials Park, OH, 1991.

Staff: *ASM Handbook: Powder Metal Technologies and Applications*, Vol. 7, ASM International, Materials Park, OH, 1998.

Staff: "Top Powder Metallurgy Parts Honored," *Advanced Materials and Processes*, **8** (August 1991).

Staff: "Forecast for Metals," *Advanced Materials and Processes*, **17** (January 1991); **17** (January 1992); **18** (January 1993).

Staff: *Metallic and Inorganic Coatings, Metal Powders, and Sintered P/M Structural Parts*, American Society for Testing & Materials, West Conshohocken, PA, 2001.

Suslick, K.S.: "Ultrasound 'Makes a Hit' with Metal Powder," *Advanced Materials and Processes*, **10** (September 1990).

Thummler, F. and R. Oberacker: *An Introduction To Powder Metallurgy*, Ashgate Publishing Company, Brookfield, VT, 1994.

Web References

Institute of Materials Processing (IMP): http://www.imp.mtu.edu/
The Minerals, Metals, Materials Society: http://members.tms.org/Staff.asp

POWER. In general, power is the time rate of doing work, as defined by the equation

$$P = \frac{dW}{dt}$$

where P is power, W is the work done, t is the time and therefore dW/dt is the time rate of doing work. Power may be expressed in units of work per unit time (e.g., foot-pounds per minute or ergs per second) or more arbitrarily, as in horsepower or watts. One horsepower is 33,000 foot-pounds per minute. One watt is 10^7 ergs (or 1 joule) per second.

POWER DENSITY. In an electromagnetic wave, the rate of power flow in a specific direction at a particular point in a transmission medium, expressed as energy per unit time (power, or radiant flux) per unit cross-sectional area normal to the direction of propagation.

The power density generally diminishes with increasing distance from the source as a result of absorption, reflection, scattering, and possibly other effects, as well as geometric spreading of the beam. For surfaces or objects that intercept the radiation at a sufficiently long distance from the source, the propagating energy may be regarded as plane-wave or parallel-beam radiation. Then the power density is the same as the irradiance at a surface normal to the beam.

AMS

POWER DENSITY SPECTRUM. A measure of the contribution to the total variance from a given frequency band in the generalized Fourier representation of a random function. Sometimes called *power spectrum*.

If $f(t)$ is a random function, the total energy $\int_{-\infty}^{\infty} f^2 dt$ is infinite, so the Fourier integral representation is inadequate. If a transform is defined over a finite interval

$$F_T(\omega) = \frac{1}{2\pi} \int_{-T}^{T} f(t)e^{-i\omega\tau} dt,$$

under suitably restrictive conditions the power density spectrum may be defined as

$$\lim_{T \to \infty} \frac{\pi}{T} |F_T(\omega)|^2.$$

The theorem, proved by N. Wiener, establishing the analogy between the analysis of random functions and ordinary Fourier analysis, is that the power density spectrum is the Fourier transform of the autocorrelation function, which is defined for random functions as

$$\lim_{T \to \infty} \frac{1}{2T} \int_{-T}^{T} f(t)f(t + \tau) dt.$$

See also **Power Spectrum**.

AMS

POWER FUNCTION. An algebraic function, $y = ax^n$. It is often convenient to plot this function on logarithmic paper, where both abscissa and ordinate are graduated in divisions proportional to the logarithm of the number plotted. The result is thus $\log y = n \log x + \log a$ and the curve is a straight line of slope n and intercept $\log a$. The number n is called the exponent. A series of the form $a_0 + a_1 x + a_2 x^2 + \cdots$, either finite or infinite, is called a power series. See also **Curve Fitting**; and **Neyman-Pearson Theory**.

POWER GAIN.   1.   The ratio of the power that a transducer delivers to a specified load, under specified operating conditions, to the power absorbed by its input circuit. If the input and/or output power consist of more than one component, such as multifrequency signal or noise, then the particular components used and their weighting must be specified. This gain is usually expressed in decibels.

2.   Of an antenna, in a given direction, 4pi times the ratio of the radiation intensity in that direction to the total power delivered to the antenna.

POWER-LAW PROFILE.   A formula for the variation of wind with height in the surface boundary layer. It is an alternative to the logarithmic velocity profile, and the assumptions are the same, with the exception of the form of the dependence of mixing length l on height z. Here

$$l = l_1 z^p, \ p \neq 1.$$

Then

$$\frac{\bar{u}}{u_*} = q\left(\frac{u_* z}{v}\right),$$

where $\bar{u}$ is the mean velocity, u_* the friction velocity, v the kinematic viscosity, and

$$q = \left(\frac{v}{u_*}\right)^{1-p} \frac{1}{l_1(1-p)}.$$

For moderate Reynolds numbers, $p = 6/7$ (the seventh-root profile) is empirically verified, but for large Reynolds numbers p is between this value and unity. It is to be noted that if $\bar{u}$ is proportional to z^m, and if the stress is assumed independent of height, then the eddy viscosity v_e is proportional to z^{1-m}. These relations are known as Schmidt's conjugate-power laws.

Additional Reading

Lee, X., W. Massman, and B. Law: *Handbook of Micrometeorology: A Guide for Surface Measurement and Analysis*, Springer-Verlag New York, LLC, New York, NY, 2004.

AMS

POWER LOADING.   The ratio of the gross weight of a propeller-driven aircraft to its power, usually expressed as the gross weight of the aircraft divided by the rated horsepower of the power plant corrected for air of standard density. With turboprop engines, the equivalent shaft horsepower is used.

POWER (Mean).   The instantaneous power in a device, or branch of a network is EI. The mean power is

$$\frac{1}{T} \int_0^T EI \, dt$$

where T is an integral number of periods, or T approaches infinity for nonperiodic currents. The mean power is also given by

$$P = \overline{I^2} R$$

where $\overline{I^2}$ is the mean square current, and R the resistance.

POWER SERIES.   An infinite series of the form

$$\sum_{n=0}^{\infty} a_n x^n = a_0 + a_1 x + a_2 x^2 + \cdots + a_n x^n + \cdots$$

where $a_0, a_1, \ldots, a_n, \ldots$ are constants and x is a variable, is called a power series.

For every power series $\Sigma a_n x^n$ there exists a constant l such that the series is absolutely convergent for all values of x such that $|x| < l$, and

is divergent for all values of x such that $|x| > l$. This number l is called the limit of convergence of the power series, and the interval $(-l, l)$ is called the interval of convergence. The series may converge or diverge at the ends of this interval, for $x = l$ and $x = -l$.

If $\lim_{n \to \infty} |a_n/a_{n+1}|$ exists, it is the limit of convergence.

If $\lim_{n \to \infty} \sqrt[n]{|a_n|}$ exists, it is the reciprocal of the limit of convergence.

A power series is uniformly convergent within the interval $(-l', l)$, where $l' < l$ and l is the limit of convergence of the given power series.

The function defined as the sum of a power series $\Sigma a_n x^n$ is a continuous function of x at all points within the interval of convergence.

For the operations with power series, we have the following theorems:

Two power series may be added together for all values of x for which both series are convergent, i.e., for the smaller of the two intervals of convergence.

Two power series may be multiplied together for all values of x for which both series are absolutely convergent, i.e., for the smaller of the two intervals of convergence.

A power series may be differentiated term by term for all values of x within its interval of convergence.

A power series may be integrated term by term between any limits lying within the interval of convergence.

Power series are very useful in the analytical representation of functions by the expansion of functions in series, and for calculation by series.

See also **Series**.

POWER SOURCES AND SUPPLIES.   Because of the variety of ways in which instruments, computers, communications equipment, and control systems are used—from undersea to outerspace applications and from laboratory and factory installations to air dropping of equipment packages over remote terrain—practically every form of available energy has been considered to provide electric power for such equipment.

Batteries. Within their ampere-hour rating, batteries provide low ripple and a measure of voltage regulation. Most batteries can be recharged when necessary and are relatively compact and inexpensive, but have severe limitations on the amount of energy stored. The three types most commonly used as power sources are (1) lead acid, (2) nickel cadmium, and (3) mercury cells which obtain their power by electrochemical (ionic) exchange between elemental anodes, cathodes, and electrolytes. See also **Battery**.

Rotating Generators. Small rotating generators powered by such varied prime movers as hot gas turbines (propane or hydrocarbon fuels) or cold gas turbines (CO_2 or compressed air tanks) generally are quite expensive and are limited to military or remote mapping or monitoring applications.

Solar Cells. Silicon photovoltaic cells offer an inexpensive solution to low power battery charging on intermittent monitoring applications in remote, unattended locations. The number of manufacturers is limited. Approximately 0.2 V/cell is available, and current output is proportional to the area of each cell. The operation of the cells results from a conversion of the electromagnetic energy of the incident visible radiation into carrier (electron-hole) pair generation across a p-n junction or within the silicon crystal which produces an emf that varies somewhat with the wavelength of the incident radiation (light). See also **Solar Energy**.

Thermoelectric Cells. Liquified petroleum gas—LPG (propane or butane) power sources utilizing lead telluride cells as thermoelectric generators giving nearly 1 V/cell offer a compact though expensive remote power source. Lead telluride converts thermal power into electric power mainly owing to the setting up of a thermal gradient across a p-n junction formed by doping—resulting in an emf and current flow in the circuit.

Fuel Cells. These cells are described in detail in the entry on **Fuel Cells**. A power supply, generically speaking, is a device that converts electric power from one power source type to another power source type more suitable for use in the instrument or control system. The power supply contrasts with a power source that may be a battery, fuel cell, thermoelectric or thermionic cell, or even a prime source, such as a motor-generator set.

Power supplies are of two general types, dependent upon the degree of precision required by the load. On the one hand, when source and load

TABLE 1. POWER SUPPLY CHARACTERISTICS

Type of Power Supply	Prime Source	Load Source	Conversion Required	Typical Applications
Autotransformer	A.C.	A.C.	voltage change	raise or lower line voltage
Isolation transformer	A.C.	A.C.	voltage and isolation	isolate load from line
A.C. voltage regulator	A.C.	A.C.	voltage regulation	close process control medical applications
Frequency changer	A.C.	A.C.	change frequency	motor drive; testing
Frequency regulator	A.C.	A.C.	regulate frequency	timing; gyro drive
Frequency converter	A.C.	A.C.	change and regulate frequency	timing; testing
Rectifier	A.C.	D.C.	change A.C. to D.C.	plating; battery charge
Transformer rectifier	A.C.	D.C.	change voltage, rectify, and isolate	plating; battery charge
D.C. power supply	A.C.	D.C.	change voltage, rectify, isolate, filter	amplifiers; radios
Regulated D.C. power supply	A.C.	D.C.	rectify, change, adjust, and regulate voltage	strain gage; computers
High-voltage D.C. power supply	A.C.	D.C.	rectify, change, and filter	paint spray; ionization
Regulated high-voltage power supply	A.C.	D.C.	rectify, change, filter, and regulate	photomultipliers; accelerators
D.C. current regulator	A.C.	D.C.	rectify voltage; regulate current	magnets; plating; forming
Inverter (rotary or vibrator)	D.C.	A.C.	convert D.C. to A.C.	motor and transducer drives
Static inverter	D.C.	A.C.	convert D.C. to A.C., solid state	motor and transducer drives
Sine wave inverter	D.C.	A.C.	D.C. to A.C., low distortion	emergency or remote power source
Regulated inverter	D.C.	A.C.	D.C. to regulated A.C.	emergency or remote power source
D.C.–D.C. converter	D.C.	D.C.	change voltage	battery-powered equipment power supply
Regulated D.C.–D.C. converter	D.C.	D.C.	change and regulate voltage	precision power source for amplifiers
High-voltage D.C.–D.C. converter	D.C.	D.C.	raise voltage to high voltage	photomultipliers; accelerators
D.C. regulator	D.C.	D.C.	lower voltage and regulate	amplifiers; computers

precision or regulation are almost the same, all that is needed is a device to convert the form of electric energy — an example of which is a transformer-rectifier which converts A.C. energy to D.C. energy. Filtering can be added if required. On the other hand, most process and instrumentation control power requirements are incompatible with the variations inherent in the average industrial or utility power source. Consequently, this power must be controlled or regulated before use. Refer to Table 1 for a listing of power supply types available, the power conversion required, a typical application, and cost per watt. The following brief descriptions of these power supply types give their function and operation.

A.C. Voltage Regulator. This is a device that compensates for A.C. voltage fluctuations in the utility power source and provides essentially constant voltage within a specified regulation band of the preset output voltage. Major types include:

1. *Ferroresonant*: either flux coupled or non-flux coupled.
2. *Motor-driven Autotransformer* (also induction regulator).
3. *Electronic*:
    a. Shunt regulated buck boost autotransformer filtered:
        (1) Magamp type (tube, SCR, or transistor controlled).
        (2) Solid state (SCR) type.
    b. Add-subtract dissipation types:
        (1) Tube type (Class *A* or *AB*).
        (2) Solid state type (usually push-pull).

The *ferroresonant* types depend upon the fact that at line frequency the magnetizing inductance of one winding of the transformer forms a series resonant circuit with an external capacitor. The resulting, higher-than-normal current acts in such a way as to saturate a portion of the magnetic circuit, thereby giving the winding that surrounds this magnetic circuit better line voltage regulation than the input source. By properly choosing the circuit parameters, load voltage regulation can be achieved. These regulators can give 1% line and 1% load regulation with up to 25% distortion in the less expensive versions. Cancellation of most of the unwanted harmonics, by adding windings and introducing various flux paths, can be achieved with a maximum distortion of 3% without degenerating regulation.

The *electronic shunt* regulated buck boost types consist of: (1) an auto-transformer; (2) a low-dissipation regulated device, such as a saturating reactor or SCR pair across the input line with suitable harmonic reducing filters across the output; and (3) an electronic reference, sensing, and reg-ulation circuit which can be either of the thermionic (tube) or solid state type. In either case, the regulation is held within approximately 1.0% rms and the distortion within 3% greater than the input at the line frequency. Transient response to line or load changes (the time required to recover to within a given regulation bandwidth) generally is less than 10 cycles of line

frequency — as also is the case with ferroresonant types. Some adjustment of output voltage generally is provided.

Push-pull electronic amplifiers buck out the objectionable harmonics and poor input voltage regulation by means of an add-subtract transformer in series with the load. Whether tubes or transistors are used, considerable heat is generated within the supplies. Regulation is better than 0.1%, distortion is 1%, and transient response is in the microsecond region.

Frequency Changers. These devices convert power of one frequency into power of another frequency. In addition, voltage conversion and regulation are possible. Uses include (1) driving induction or synchronous motors at other than the available input frequency; and (2) testing control systems that utilize uncommon frequencies, such as aircraft, from commercial power frequency sources.

D.C. Rectifier. This is the most commonly used power supply. It uses an arrangement of electronic switches or rectifiers, which allow current flow in only one direction. The silicon rectifier — small, efficient, and inexpensive — is the most common type of rectifier in use today. When voltage transformation is required, a transformer is employed. The transformer increases the size, weight, and cost of a power supply, but also allows electrostatic isolation of the rectified D.C voltage from the input A.C. When filtering of the normal ripple from the normal half or full wave rectified output voltage is required, a D.C. filter is added, which generally consists of a large electrolytic capacitor. For very large currents, an inductance or choke usually is added unless closed-loop regulation circuitry is added.

Regulated D.C. Power Supply. See Table 2. In general, a regulated D.C. power supply consists of a transformer, rectifier, filter, and regulator. The first items have been discussed, but the method of regulation varies widely. For instrument and control systems, certain types are more useful than others. In general, low voltage (up to 150 V) systems use transistors as the main voltage or current regulating element with zener diode references and transistor error amplifiers to drive the main control transistors. When the transistors are D.C. controlled, the power supply is the so-called "linear" type, and the main regulating transistor (or transistors when higher currents are required) is referred to as a "pass" transistor. Regulation is the static D.C. error due to either line or load changes about a nominal or set output condition. Since D.C. gain is high, very good D.C. regulation results. Since the regulating element is in series with the load, high-speed recovery from line or load transients results — generally on the order of tens of microseconds. Also, due to the high A.C. gain of the amplifier, excellent ripple reduction is achieved — generally of the same order (in percent) as the static dc regulation or better.

Current limiting is almost universally available on this type of supply. It is accomplished by sensing the output current, comparing it with a reference and, in effect, shutting off the amplifier and main regulating

TABLE 2. REGULATED DC POWER SUPPLIES

Type of Power Supply	Input	Output Adjustment, %	Maximum Available Regulation, B2			Maximum Available Ripple, %	Regulation Means	Efficiency Average, %
			Line	Load	Current			
Line regulated	ferroresonant transformer	none	1.0	3–10	NA	1–3	ferroresonant transformer	75
Narrow range (slot)	transformer	±5	0.03	0.03	NA	0.03	transistor (usually Si)	40
Wide range	transformer	100	0.01	0.01	0.1	0.01	transistor (usually Si)	30
High wattage, low ripple	transformer	some-limited	0.1	0.1	0.1	0.03	SCR and transistor filter	70
High wattage, good regulation	transformer	100	0.01	0.01	0.1	0.01	SCR or transistor switch and transistor pass	50
High efficiency, small size	rectifier	±5	0.1	0.1	NA	0.2	switching transistor (high voltage)	65
High voltage (100–300 V), narrow range	transformer	±5	0.03	0.03	NA	0.03	high-voltage transistor	40
High voltage, wide range	transformer	100	0.01	0.01	0.1	0.01	high-voltage transistor, two-stage	30
High voltage (300–3000 V)	transformer	100	0.05	0.05	0.1	0.01	vacuum tube and SC control	25
High voltage, high current	transformer	100	0.1	0.1	0.1	0.3	SCR (usually primary)	60
High voltage (5000–50,000 V)	transformer	50	0.05	0.05	Opt.	0.05	tube	30

transistors, causing the output voltage to decrease at the load to limit the output current to the preset value. Some supplies actually reduce the load current under this condition. This is referred to as "fold back" current limiting. Actual current regulation can be accomplished and is described later.

In general, transistor regulated D.C. power supplies are available in several mechanical configurations. The most popular from the standpoint of volume is the "system" type which is usually contained in a metal box and has limited output voltage and current limit adjustments and no switches or meters. This type can be mounted near the load. The more expansive "laboratory" type has wider output voltage and current regulation adjustments and generally has better performance specifications.

Although the linear transistor regulated D.C. power supplies are the most popular, other types are available for many reasons.

First, for higher output power requirements nonlinear or switching-type D.C. power supplies are available. SCRs (silicon controlled rectifiers) are generally used because of their high current and high voltage capabilities both as primary regulating elements and as preregulators for a final transistor passing stage. When SCRs serve as the only regulating element, they generally switch on and off at the input line frequency rate, and regulation is achieved by varying the portion of the input sine wave that the SCR conducts (so-called phase control). Since the forward drop of the SCRs and/or possibly rectifiers is low in relation to the output voltage (except for 3- and 5-V supplies), the efficiency of this type of supply is somewhat higher than that of a linear regulator. D.C. regulation is somewhat worse as is transient response since it takes about 5 cycles of input line frequency for the regulator to recover to the regulation band. Ripple is related to the output filter size and is also higher than in the transistor regulators, although active filters can improve the ripple considerably at very little increase in dissipation.

Some of the switching-type power supplies utilize transistors as ON-OFF switches. These usually operate at ultrasonic frequencies. This results in a great reduction in overall size and weight as compared with other methods of regulation.

Where constant nonvarying loads must be protected against line variations the simplest and most reliable regulator consists of a ferroresonant ac transformer and a passive rectifier and filter. Line regulation of 1% is available, but load regulation is 3 to 10% for this type of supply which although inexpensive is rather large and heavy because of the ferroresonant transformer.

Current Regulators. The last type of D.C. regulator that can be used in control applications is the current regulator. Practically all standard D.C. voltage regulators can be reconnected as precision current regulators by utilizing a wire-wound power resistor in series with the load and controlling the voltage across this resistor, thus controlling the load current. However, this method uses a high proportion of the available power. The laboratory types have built-in current regulators that offer current regulation almost as good as the voltage regulation offered, although current ripple is somewhat

higher than voltage ripple. For either configuration, precision current regulators are available "off the shelf."

Inverters. These are very useful devices for converting direct to alternating current. Control applications include audio frequency oscillators for linear variable differential transformers (LVDTs), servo amplifiers and control systems, and remote battery operated instrumentation systems requiring A.C. power. Older, cheaper types included the rotary type and a static type, which had a movable armature vibrator. Newer static types are generally transistor driven and, as with everything electronic every specification adds to the cost, so that frequency regulation, voltage regulation and output distortion control each contributes to the overall cost of the unit. As indicated in Table 1, most inverters are custom designed but a few are available off the shelf. The larger volt-ampere ratings include SCRs that also require auxiliary commutation circuitry, adding complexity and lessening reliability. Because all static inverters begin with square waves, the generation of and suppression of radio frequency interference are constant problems. In addition, the generation of low-distortion sine waves requires filtering except in the rotary types.

D.C.-D.C. Converters. D.C.-D.C. converters are useful for converting power at one D.C. voltage level to power at another D.C. voltage level. The usual technique is to "chop" the input D.C. at an ultrasonic frequency and transform or filter the resulting A.C. to the required level. For stepped-up voltage, a push-pull transistor stage is generally used. For stepped-down voltage a single transistor "chopper" suffices. Because of the high switching frequency, the size and weight of the magnetic components is minimal. Regulation, when required, is accomplished by modifying the chopped waveform (say by pulse width modulation) or by including a passing transistor in the lowest current path. For remote power supply applications requiring battery or solar cells as a power source, the D.C.-D.C. converter offers a very small and efficient method of producing precisely controlled voltage for amplifiers or transducers.

For some applications requiring "at the load" precision regulation with a wide variation of input D.C. voltages, e.g., from a car battery, a D.C. regulator consisting only of a series pass transistor and a reference amplifier is available, sometimes in a single small package that can be mounted on the amplifier or transducer chassis. These devices are obtainable in a variety of package sizes with power-handling capabilities ranging from milliwatts to 100s of watts. The lower-power devices often are monolithic integrated circuits. The high-power devices often are hybrid designs.

Thermionic Regulation and Switching Devices. Most of the power supply descriptions have included solid state devices as rectifiers and regulators. However, for many applications particularly high voltage (>250 V) regulated supplies, vacuum tubes are used as both passing regulators and amplifiers. Furthermore, in the medium to high kilovolt region, vacuum tube rectifiers are often used, although selenium controlled rectifiers offer some advantages over tubes. But selenium controlled rectifiers have a "wear out" mode due to pinhole melting of cadmium selenide, which causes high resistivity areas on the cells, increasing the forward voltage drop for a given current density.

Grid-controlled mercury vapor rectifiers (thyratrons) are still used in power supplies but have very little advantage over silicon controlled rectifiers except in the >1000-V rating, and of course they have a higher forward voltage drop and lower overall efficiency. See also **Zener Diode**.

POWER SPECTRUM.

POWER SPECTRUM.   1.   The square of the amplitude of the (complex) Fourier coefficient of a given periodic function.

Thus if $f(t)$ is periodic with period T, its Fourier coefficients are

$$F(n) = \frac{1}{T} \int_0^T f(t) e^{-in\omega t} dt,$$

where $\omega = 2\pi/T$, and the power spectrum of $f(t)$ is $-F(n)^2-$. Here n takes integral values and the spectrum is discrete. The total energy of the periodic function is infinite, but the power, or energy per unit period, is finite. In the case of the aperiodic function containing finite total energy, the energy density spectrum is the corresponding spectral function. This is a continuous function of frequency and therefore has dimensions of energy/frequency (energy density). In the case of a random function containing infinite total energy but not periodic, the power density spectrum is the corresponding spectral function. The mathematical conditions governing analogous theorems in these three classes of functions are different. However, when actual computations of observational data are involved, a finite number of discrete values are used, and the effect is the same as if the function were assumed to be periodic outside the interval of computation. Thus, it is the power spectrum that is exhibited. But all types of spectra referred to may be considered as measures of the contribution of given frequencies in the Fourier representation of the original function. The terms "power" and "energy" are usually retained to indicate relative dimensions regardless of the actual dimensions of the functions analyzed, which may be functions of space as well as time. Computation of the power spectrum in practice may be facilitated by use of the theorem that it is the Fourier coefficient of the autocorrelation function.

2.   Same as power density spectrum.

AMS

POWER TRANSMISSION RATIO (Acoustic).

POWER TRANSMISSION RATIO (Acoustic).   The ratio of the average acoustic energy transmitted normally through a surface to the average acoustic energy incident normally upon that surface.

POXVIRUSES.

POXVIRUSES.   The poxvirus family includes one of the most deadly of human viruses, *Variola virus*, the causative agent of smallpox. Until its eradication in the 1970s, smallpox was responsible for the deaths of millions of people every year. The poxvirus family also includes *Cowpox* and *Vaccinia viruses*, which must be counted among the viruses most beneficial to humankind, because it was only through vaccination with these less virulent poxviruses that the worldwide eradication of smallpox was finally achieved. See also **Smallpox**.

It is against this historical backdrop of smallpox that research into viruses of this family has been framed. In addition to the unparalleled eradication of a disease, work on these viruses continues to provide rich insights into many areas of medicine.

Classification

The poxviruses are among the largest and most self-sufficient deoxyribonucleic acid (DNA) viruses. Each possesses the following distinguishing characteristics: the virus particles are rounded, ovoid or brick-shaped objects (often appearing oval in cross-section) with dimensions of about $220-450 \times 140-260 \times 140-260$ nm; the genome is a single, linear, double-stranded DNA of 150–380 kb; in the virus particle the two DNA strands are connected at each end of the genome by a hairpin loop; each genome encodes 150 to 300 proteins; viral transcription, viral DNA replication and viral morphogenesis occur in the cytoplasm of the cell.

There are a large number of poxviruses replicating in a wide variety of animals. The family of poxviruses, the *Poxviridae*, comprises two subfamilies: the poxviruses of vertebrates, the *Chordopoxvirinae*; and the poxviruses of insects, the *Entomopoxvirinae*, as described in Table 1.

Within the subfamilies the viruses have been grouped into genera. Viruses within a genus are structurally similar, typically have similar host ranges, and are more closely related in their DNA sequences and antigenic properties than viruses in different genera.

Much less is known about the entomopoxviruses than the chordopoxviruses. The recent analysis of the sequence of the genome of the *Melanoplus sanguinipes* entomopoxvirus, a virus that infects grasshoppers and locusts, has confirmed the fundamental similarities among the entomopoxviruses and chordopoxviruses [reviewed by Afonso, et al.:]. As might be expected, the major differences correspond to genes encoding accessory functions advantageous for viral replication in either insects or vertebrates. The prototype of the poxvirus family is the orthopoxvirus, *Vaccinia virus*, which remains the most extensively characterized poxvirus [reviewed by Moss].

Epidemiology of Poxvirus Infections

The distribution and transmission of the poxviruses differ according to the type of virus. The name of the poxvirus can be misleading insofar as poxviruses have often been isolated and characterized from infections of animals that may not play a major role in maintenance of that virus in natural populations. One reason for this is that often poxviruses produce few overt symptoms in their natural hosts, but pronounced pathology in animals that are not the natural reservoirs for the virus. Thus, the source of the virus is not always a reliable indicator of either the type of virus or its natural host range. With the exceptions of infections with *Variola virus* and *Molluscum contagiosum virus*, all natural poxvirus infections of humans appear to be zoonoses.

Variola Virus.   Before its eradication in 1977, *Variola virus* was endemic around the world. As recently as the 1960s, it was estimated to be responsible for the deaths of millions of people every year. Consequently, the eradication of *Variola virus* from the natural environment is one of the major achievements of modern medicine. See also **Smallpox**.

The key features contributing to the successful eradication of *Variola virus* included: the lack of an animal host other than humans; the lack of latent infections; the rarity of inapparent infections; the availability of a stable vaccine effective against all strains of *Variola virus*; and, perhaps crucially, the terrible nature of smallpox, which consolidated the unique global effort to eliminate the disease. A detailed account of the smallpox eradication campaign of the World Health Organization can be found in Fenner, et al.

Transmission of the virus was generally by close contact involving the inhalation of virus in airborne droplets. Less commonly, short-range aerosol transmission or transmission via contaminated objects occurred.

Monkeypox Virus.   Monkeypox virus has a broad host range. However, in nature it appears to be primarily a virus of arboreal squirrels, particularly those occupying the higher levels of trees in the tropical rain forests of Central and West Africa. As with *Cowpox virus*, this virus is occasionally transmitted from these rodents to other animals (various tree-dwelling primates and more rarely humans) that might come into contact with them.

Transmission appears to occur by close contact, possibly including the handling of infected carcasses or the consumption of meat from infected animals. Person to person transmission has been documented, but this is rare and seldom involves sequential transmission in more than four people.

Cowpox Virus.   Cowpox virus is endemic in Europe, where its primary hosts appear to be rodents, such as voles and woodmice. The relatively infrequent *Cowpox virus* infections of humans, farm animals, pets and zoo animals originate directly or indirectly from rodent sources. Infections in humans are usually through direct contact, and particularly by introduction of the virus into skin abrasions. The mode of transmission among rodents is unclear.

Vaccinia Virus.   Vaccinia viruses are represented in human infections mostly by strains employed in either vaccine viruses or laboratory work. The origin of *Vaccinia virus* is unclear. It is not known to exist in natural animal populations except for viruses such as buffalopox and rabbitpox viruses, which are probably feral derivatives of vaccine strains of *Vaccinia virus*.

Human infection is usually a result of deliberate inoculation in the skin with vaccine strains of *Vaccinia virus*. The incidence of laboratory-acquired infections, secondary infections from vaccinations and adverse reactions to *Vaccinia virus* vaccines can be expected to decrease with the adoption of newly available *Vaccinia virus* vectors and vaccines that are incapable of producing progeny viruses in human cells [reviewed by Moss; and Paoletti].

Camelpox Virus. *Camelpox virus* is endemic in most regions supporting large populations of camels. The host range appears to be restricted to camels. The mode of transmission is unknown.

Tanapox Virus. Human *Tanapox virus* infections have occurred in Kenya and Zaire. Rhesus monkeys (and their human caregivers) in several primate centers in the USA have also sustained *Tanapox virus* infections.

The natural host of this virus is unknown. The mode of transmission is unknown, although it has been suggested that the virus is transmitted mechanically, possibly by arthropod vectors.

Ectromelia Virus (Mousepox Virus). This virus has frequently been isolated from laboratory colonies of mice. Its natural host is unknown, but is likely to be a rodent. Transmission can occur through skin abrasions or less commonly by respiratory infection among animals in close contact. Humans do not appear to be susceptible to infection with this virus.

Molluscum Contagiosum Virus. This is a poxvirus of humans. Historically, infections were most common in children, but recently infections have become more common in adults, affecting as many as 10% of people infected with Human immunodeficiency virus. The virus is spread by direct physical contact.

Parapoxviruses. Domestic animals of several species (sheep, goats and cattle) are hosts to parapoxviruses such as *Orf virus* and *Pseudocowpox virus*. Occasionally, through abrasions in their skin, handlers of infected farm animals are infected with these viruses.

Leporipoxviruses. The *Shope* (or rabbit) *fibroma virus* is naturally distributed in eastern and central North America. It was also once used as a vaccine against *Myxoma virus*. The *Myxoma virus* has naturally infected rabbits in South America and western North America. The virus was deliberately introduced into Australia and Europe as a means of controlling wild rabbit populations. Biting arthropods mechanically transmit the virus among rabbits. Less commonly, respiratory transmission can occur.

Avipoxviruses. Poxviruses infect a wide variety of birds. The *Fowlpox virus* is of greatest economic importance, because of its effects on the poultry industry. The virus is transmitted mechanically by biting arthropods. Transmission may also occur via airborne droplets containing virus.

Recently, the *Canarypox virus* has been developed as a live vaccine vector for use in humans, because in humans, although it is capable of initiating an infection, it is unable to produce progeny virus particles. Therefore it has provided an attractive alternative to the medical use of replication-competent strains of *Vaccinia virus*.

Entomopoxviruses. The entomopoxviruses have been isolated from a range of different insects around the world (Table 1). Despite the similarities between these viruses and the chordopoxviruses, the entomopoxviruses do not appear to be capable of replication in vertebrates or cells derived from vertebrates.

In the insect, infection normally commences with the ingestion of viral inclusion bodies (spheroids, which are similar in appearance to but different in composition from the A-type inclusions of the chordopoxviruses). The virus is released from the inclusion bodies that are solubilized in the alkaline conditions in the gut of the insect larvae. The virus replicates in cells of the midgut before spreading to other organs. Cell to cell spread is thought to be effected by virus that has not been embedded within the spheroids.

Clinical Features of Poxvirus Infections

The accessory mechanisms can provide great insight into processes that contribute to the limitation of viral replication. In some cases viral mechanisms have suggested novel strategies for therapies designed to suppress immune responses. And, of course, viral countermeasures to host defenses can be expected to contribute to viral pathogenesis [reviewed by Buller and Palumbo], although the relationship here is not always a simple one. For example, *Cowpox virus* encodes many more proteins predicted

TABLE 1. POXVIRIDAE

Subfamily	Genus	Species (prototype listed first) include
Chordopoxvirinae	*Orthopoxvirus*	Vaccinia[a], cowpox[a], variola[a,b], monkeypox[a], ectromelia, raccoonpox, camelpox
	Molluscipoxvirus	Molluscum contagiosum[a,b]
	Parapoxvirus	Orf[a], pseudocowpox[a], bovine papular stomatitis
	Yatapoxvirus	Yaba monkey tumor[a], tanapox[a]
	Avipoxvirus	Fowlpox, canarypox
	Capripoxvirus	Sheeppox, goatpox
	Leporipoxvirus	Myxoma, rabbit fibroma
	Suipoxvirus	Swinepox
Entomopoxvirinae	*Entomopoxvirus A*	*Melolontha melolontha* and other coleopterans
	Entomopoxvirus B	*Amsacta moorei* and other lepidopterans and orthopterans
	Entomopoxvirus C	*Chironimus luridus* and other dipterans

[a] Poxviruses that cause disease in humans.
[b] Poxviruses known to replicate only in humans.

to interfere with immune responses than either the much more pathogenic *Variola virus* or the much less pathogenic vaccine strains of *Vaccinia virus*. The molecular basis for the extreme pathogenicity of *Variola virus* is still unknown.

The clinical features of poxvirus infections vary according to the virus, the host, the infectious dose and the route of inoculation. Common manifestations of poxvirus infections are briefly summarized below.

Variola Virus. Two major variants of *Variola virus* have been recognized. *Variola major virus* exhibits a case fatality rate of up to 40%. *Variola minor virus* (alastrim) has a case fatality rate of less than 2%. Detailed descriptions of the clinical features of smallpox can be found in Fenner, et al.

Monkeypox Virus. Human monkeypox infections exhibit symptoms similar to those of ordinary-type smallpox, except that *Monkeypox virus* normally produces more pronounced lymph node enlargement. Reported case fatality rates have been as high as 14%, but the human disease is rare (e.g., only 283 cases were reported in Central and West Africa in the period 1970–1984, when extensive surveillance for smallpox was undertaken).

Cowpox Virus. Clinical features of cowpox virus infection in humans are similar to those produced by *Vaccinia virus*, except that symptoms are usually more pronounced. The local oedema is usually greater than in a vaccinia virus lesion, and there is typically more hemorrhage, with the lesion progressing to produce a hard black scab 1–3 cm in diameter. The lesion is painful. Systemic symptoms include fever and malaise, sometimes accompanied by vomiting and a sore throat.

Vaccinia Virus. Most strains of *Vaccinia virus* have undergone artificial selection for attenuation of virulence as a desired feature of the vaccine strains. Many strains of *Vaccinia virus* have been passaged for decades in tissue culture under conditions where the loss of accessory genes might not present any disadvantage to the virus for replication *in vitro*. Consequently, most strains of *Vaccinia virus* are not highly pathogenic in humans.

Vaccination with *Vaccinia virus* strains capable of replicating in humans normally results in a transient generalized infection with a single pustular lesion at the site of the inoculation in the skin. Vaccinees may experience some discomfort at the site of the inoculation. By the ninth or tenth day, the lesion should have reached its maximum size. Vaccinees may experience some tenderness of the draining lymph nodes. Some may also develop a mild fever. After the tenth day the pustule begins to dry, forming a scab that usually detaches within 2–3 weeks, leaving a characteristic scar. A small proportion of the population, including immunocompromised patients, may experience moderate to severe complications including fatal generalized infections with multiple lesions.

Camelpox Virus. In camels the disease may take a generalized severe form with a case fatality rate of up to 28%, or it may take a localized less severe form. Pustular lesions are predominantly distributed around the mouth and nose.

Tanapox Virus. Symptoms include fever, severe headaches backache, and itching at the sites where the lesions develop. The lesions are similar to those of monkeypox virus infections except that the lesions usually ulcerate by the third week, and heal by the sixth week. Typically, there is a single lesion, but up to 10 lesions on a single patient have been reported.

Ectromelia Virus. Clinical features in the mouse can range from mild inapparent infections to fatal infections with or without a generalized rash. The severity of the symptoms is primarily dependent upon the species and strain of mouse.

Molluscum Contagiosum Virus. In otherwise healthy patients this virus produces skin lesions that, over a period of about 50 days, develop into pearly flesh-coloured nodules 2–5 mm in diameter. The lesions may persist for months or years before spontaneously disappearing. Often there are groups of lesions resulting from multiple sites of primary infection or from local transmission from a primary site. The immunocompromised patient may have infections in which the lesions may be larger, more numerous, and longer lasting than infections of immunocompetent patients.

Parapoxviruses. In sheep, *Orf virus* may produce proliferative wart-like lesions or nonproliferative pustular lesions. Most commonly the lesions are around the mouth of the sheep (hence the designation scabby mouth).

In humans, *Orf virus* usually produces a large (up to 2 cm in diameter) nodular lesion. These lesions are highly vascularized as a result of the virus-encoded VEGF (see above). Normally, these lesions disappear within 6 weeks of the start of the infection.

Leporipoxviruses. The *Shope fibroma virus* most commonly produces a single fibroma at the site of inoculation. The fibromas are usually several centimeters in diameter and normally regress within 1 year of infection.

Myxoma viruses produce a disease similar to that of *Shope fibroma virus* in rabbit species that are their natural hosts. Usually this takes the form of a benign fibroma, with generalized disease occurring only in young rabbits. However, in several species of rabbit that are not the normal hosts of these viruses, notably European species and strains of rabbits, myxoma viruses are extremely virulent, with mortality rates close to 100%. Typically the South American strains of *Myxoma virus* produce large tumors at the site of inoculation and elsewhere before the death of the rabbit. Death normally occurs within less than 16 days of infection. The North American strains of *Myxoma virus* can kill more swiftly, often before the appearance of extensive skin lesions.

Avipoxviruses. *Fowlpox virus* produces papular lesions less than 1 cm in diameter on the skin of the head, particularly the comb and wattle. These lesions usually resolve in a few weeks. Droplet-borne virus is thought to be the primary source of infection in the mucosa of the mouth and nasal passages, where the coalescence of lesions can result in fatal breathing problems for the infected bird.

Entomopoxviruses. Some entomopoxviruses have low pathogenicity for their host. Others produce a high case fatality rate, but death may not occur until several weeks after infection. The potential use of entomopoxviruses as biological control agents for a variety of insects is under investigation.

Control of Poxvirus Infections

Variola Virus. Control was achieved by vaccination with live virus vaccines employing first *Cowpox virus* (after Edward Jenner's original procedure) and subsequently various strains of *Vaccinia virus*. Variola virus-infected patients were often kept in isolation to restrict the transmission of the disease. Treatment essentially consisted of good nursing care, because, although antiviral drugs were first developed for use against *Variola virus*, most were found to be ineffective *in vivo*. Methisazone and vaccinia-immune globulin appeared to have some protective effects if given to patients soon after their first exposure to *Variola virus*. See also **Immunology (The History)**; and **Jenner, Edward (1749–1823)**.

Monkeypox Virus. As for *Variola virus*.

Cowpox Virus. As for *Vaccinia virus*.

Vaccinia Virus. Personnel handling *Vaccinia virus* and other orthopoxviruses can be vaccinated with a vaccine strain of *Vaccinia virus*. This provides a controlled infection with an attenuated virus that can provide protection against subsequent infection with orthopoxviruses. Vaccine strains of *Vaccinia virus* that are both highly attenuated and impaired in their ability to replicate in humans have been developed.

Camelpox Virus. Attenuated strains of *Camelpox virus* have been selected for use as vaccines.

Tanapox Virus. None.

Ectromelia Virus. Quarantine and serological surveillance are used to restrict the transfer of infected mice or material into colonies of uninfected mice.

Molluscum Contagiosum Virus. Lesions generally regress spontaneously. Sometimes lesions have been removed surgically.

The lesions often appear to be more intractable in immunocompromised patients. However, several recent reports have noted the regression of such lesions after the initiation of various antiretroviral therapies, particularly highly active antiretroviral therapy (HAART) involving the use of two reverse transcriptase inhibitors and one protease inhibitor.

Parapoxviruses. Live attenuated parapoxvirus vaccines are available.

Leporipoxviruses. Control of *Shope fibroma virus* infections is achieved mainly by protecting the rabbits from biting arthropods. Protection against *Myxoma virus* is achieved primarily through similar protection and vaccination with live attenuated *Myxoma virus*.

Avipoxviruses. Live attenuated virus vaccines are available for the protection of chickens and turkeys from infection with *Fowlpox virus*.

Additional Reading

Afonso, C.L.E.R. Tulman, Z. Lu, et al.: "The Genome of *Melanoplus sanguinipes* Entomopoxvirus," *Journal of Virology*, **73**(1), 533–552 (1999).

Binns, M.M., and G.L. Smith: *Recombinant Poxviruses*, CRC Press, LLC, Boca Raton, FL, 1992.

Buller, R.M.L., and G.J. Palumbo: "Poxvirus Pathogenesis," *Microbiological Reviews*, **55**, 80–122 (1991).

Fenner, F., D.A. Henderson, I. Arita, Z. Jezek, and I.D. Ladnyi: "The Intensified Smallpox Eradication Programme, 1967–1980," In: *Smallpox and its Eradication*, World Health Organization, Geneva, Switzerland. 1988, pp. 421–538.

Fenner, F., D.A. Henderson, I. Arita, Z. Jezek, and I.D. Ladnyi: "The Clinical Features of Smallpox," In: *Smallpox and its Eradication*, World Health Organization, Geneva, Switzerland, 1988. pp. 1–68.

Isaacs, S.N.: *Vaccinia Virus and Poxvirology*, Springer-Verlag New York, LLC, New York, NY, 2004.

Mercer, A., A. Schmidt, and O. Weber: *Poxviruses*, Springer-Verlag New York, LLC, New York, NY, 2006.

Moss, B: "*Poxviridae*: the Viruses and their Replication," In: Fields, B.N., D.M. Knipe, P.M. Howley, et al.: *Virology*, pp. 2637–2671. Lipincott-Raven, Philadelphia, PA, 1996.

Moss, B.: "Genetically Engineered Poxviruses for Recombinant Gene Expression, Vaccination, and Safety," *Proceedings of the National Academy of Sciences of the USA*, **93**(21), 11341–11348 (1996).

Paoletti, E.: "Applications of Poxvirus Vectors to Vaccination: An Update," *Proceedings of the National Academy of Sciences of the USA* 93(21), 11349–11353 (1996).

Shchelkunov, S.N.: *Orthopoxviruses Pathogenic for Humans*, Springer-Verlag New York, LLC, New York, NY, 2005.

Smith, G.L., and A. Vanderplasschen: "Extracellular Enveloped *Vaccinia virus*. Entry, Egress, and Evasion," *Advances in Experimental Medicine and Biology*, **440**, 395–414 (1998).

Web References

Avipoxviruses: http://www.pubmedcentral.nih.gov/articlerender.fcgi?artid=1142158

Camelpox Virus: http://www.cidrap.umn.edu/cidrap/content/bt/smallpox/news/camelpox.html

Cowpox Virus: http://www.emedicine.com/derm/topic87.htm

Ectromelia Virus: http://www.sanger.ac.uk/Projects/Ectromelia_virus/

Molluscum Contagiosum Virus: http://www.cdc.gov/ncidod/dvrd/molluscum/faq/everyone.htm

Monkeypox Virus: http://www.cdc.gov/ncidod/monkeypox/

Parapoxviruses: http://www.ars.usda.gov/research/publications/publications.htm?seq_no_115=153434

Tanapox Virus: http://www.stanford.edu/group/virus/pox/2000/tanapox_virus.html

Vaccinia Virus: http://www.stanford.edu/group/virus/pox/2000/vaccinia_virus.html; and http://www.bt.cdc.gov/agent/smallpox/vaccination/live-virus.asp

Variola Virus: http://www.cdc.gov/ncidod/eid/vol7no1/leduc.htm

National Institutes of Health, Bethesda, MD

POYNTING-ROBERTSON EFFECT. The effect upon the motion of a micrometeoroid or other very small particle due to the absorption and emission of radiation. The particle absorbs radiation from the sun only in one direction, but re-radiates energy in all directions. This effect produces a drag upon the particle that is directly tangential to the orbit of the particle about the sun, and thus decreases the orbital angular momentum. Since this angular momentum varies as the square root of the orbital radius, decreases in angular momentum are accompanied by a decrease in orbital radius, so that the particle follows a spiral path directing steadily closer to the sun. Although the solar radiation pressure upon the particle opposes this effect, it is not sufficient to offset it completely.

PPB. Parts per billion. One part per billion is a frequently used dimension for expressing the composition and analysis of substances — as found in air, water, food substances, etc. Instrument developments and other assay techniques perfected during the past decade or so have made the determination of such minute quantities a practical possibility for many materials. One part per billion is approximately equivalent to 1 drop in a 10,000-gallon (37,850-liter) tank.

PPM. Parts per million. One part per million is a common dimension for expressing the composition and analysis of substances — as found in air, water, raw materials, food substances, etc. One part per million is approximately equivalent to about 1/32 ounce (1 gram) in 1 ton of substance. One gram is exactly one-millionth of a metric ton.

PRAECIPITATIO. See **Clouds and Cloud Formation**.

PRAIRIE. A flat or gently undulating plain that is grassy and generally treeless; specifically, such an area in southern Canada and the northern and central United States where it extends from the foothills of the Rocky Mountains to about 88°W longitude. Its climate, with light summer rains and high summer temperatures, is highly favorable for the growth of cereals, but there is a considerable risk of drought especially in certain portions, where a semiarid climate (steppe climate) prevails. The prairie is similar, but not completely analogous, to the steppe regions of Europe and Asia. See also **Biome**.

PRAIRIE DOG. See **Squirrels and Other Sciuromorphs**.

PRAIRIE HARE. See **Rabbits and Hares**.

PRAIRIE WOLF. See **Canines**.

PRANDTL NUMBER. The nondimensional ratio between the product of heat advection and viscous forces and the product of heat diffusion and inertial forces in a given fluid.

It may be given as

$$Pr = \frac{C_p \mu}{k},$$

where C_p is the specific heat at constant pressure, μ the dynamic viscosity, and k the thermal conductivity. The Prandtl number may also be defined as the ratio of the Reynolds to the Péclet numbers, or as the ratio of kinematic viscosity to thermometric conductivity.

See also **Heat Transfer**.

PRASEODYMIUM. [CAS: 7440-10-0]. Chemical element symbol Pr, at. no. 59, at. wt. 140.91, second in the Lanthanide Series in the periodic table, mp 934 °C, bp 3,512 °C, density 6.769g/cm^3 (20 °C). Elemental praseo dymium has a close-packed hexagonal crystal structure at 25 °C. The pure metallic praseodymium is silver-gray in color, the luster dulling

rapidly upon exposure to air and forming a nonadherent oxide which hastens the process of oxidation. When pure, the metal is soft and workable with ordinary tools. Processing and handling require storage under a nonreactive liquid or inert atmosphere or vacuum. Finely-divided praseodymium is pyrophoric, burning at a red heat. There is only one isotope of the element in nature ^{141}Pr. It is not radioactive and has a low acute-toxicity rating. Fourteen artificial isotopes have been produced. Of the light (or cerium-group) rare-earth metals, praseodymium is the fourth most plentiful and ranks 59th in abundance of the elements in the earth's crust, exceeding tantalum, mercury, bismuth, and the precious metals, excepting silver. The element was first identified by C.A. von Welsbach in 1885. Electronic configuration

$$1s^2 2s^2 2p^6 3s^2 3p^6 3d^{10} 4s^{24} p^6 4d^{10} 4f^2 5s^2 5p^6 5d^1 6s^2$$

Ionic radius, Pr^{3+} 1.01 Å, Pr^{4+} 0.90 Å. Metallic radius, 1.828 Å. First ionization potential, 5.42 eV; second, 10.55 eV. Other important physical properties of praseodymium are given under **Rare-Earth Elements and Metals**.

Primary sources of the element are bastnasite and monazite, which contain from 4 to 8% praseodymium. Plant capacity involving liquid-liquid or solid-liquid organic ion-exchange processes for recovering the element is in excess of 100,000 pounds Pr$_6$O$_{11}$ annually. Metallic praseodymium is obtained by electrolysis of Pr$_6$O$_{11}$ in a molten fluoride electrolyte, or by a calcium reduction of PrF$_3$ or PrCl$_3$ in a sealed-bomb reaction.

For many years, praseodymium has been a component of light rare-earth mixtures used in mischmetal, a pyrophoric alloy used in cigarette-lighter "flints." Mixtures of cerium, lanthanum, neodymium, and praseodymium, as oxides and fluorides, are used in the cores of arc carbons for the production of light of greater intensity. Similar mixtures of rare-earth oxides, including praseodymium, are used in optical glass polishing formulations. Mixtures of the lanthanide compounds, including about 5% praseodymium, find application as catalysts in petroleum cracking processes. A mixture containing 10% Pr, 30% Nd, and 60% La is used for cracking crude oil and comprises the largest single use of the element as well as of all other Lanthanide elements. Use of elemental praseodymium as a colorant for glass was one of the early applications. The color ranges from clear yellow to green and finds use in sunglasses, protective glasses for industry, art objects of glass, tableware, and optical filters. In the manufacture of ceramic tile, a praseodymia-zirconia yellow stain is used. Metallurgically, the most important intermetallic compound is PrCo$_5$, which has unsurpassed permanent magnetic properties. The compound has a very high resistance to demagnetization and has a high magnetic saturation value. PrNi5 has been used for adiabatic magnetization cooling of samples down to the milli-Kelvin range for low-temperature research. Investigations continue into further electronic and optical uses of the element and its compounds.

Additional Reading

Greenwood, N.N. and A. Earnshaw: *Chemistry of the Elements*, 2nd Edition, Butterworth-Heinemann, Inc., Woburn, MA, 1997.

Krebs, R.E.: *The History and Use of Our Earth's Chemical Elements: A Reference Guide*, Greenwood Publishing Group, Inc., Westport, CT, 1998.

Lide, D.R.: *CRC Handbook of Chemistry and Physics*, 88th Edition, CRC Press, LLC., Boca Raton, FL, 2007.

Parker, P.: *McGraw-Hill Encyclopedia of Chemistry*, 2nd Edition, The McGraw-Hill Companies, Inc., New York, NY, 1993.

Stwertka, A. and E. Stwertka: *A Guide to the Elements*, Oxford University Press, Inc., New York, NY, 1998.

PRAWN (*Crustacea, Decapoda*). Small marine crustaceans closely related to the shrimps. They differ from the lobsters and crabs in the compressed body and in the use of the abdominal appendages for swimming. See also **Aquaculture**; and **Crustaceans (Edible)**.

PRAYING MANTIS. See **Mantis**.

PREAMPLIFIER. This is a class-A voltage amplifier, which receives the signal from a microphone, pick-up, television camera tube, or other device supplying a low signal level and amplifies it so it can supply the input for additional amplifier circuits. Thus a preamplifier is commonly used in a radio or television studio to amplify the audio or video signal before feeding it into a mixer, line to the transmitter, or other amplifying equipment at the studio. See also **Amplifier**.

PREANTENNA.   A sensory organ of the pair borne by the first segment of the body in the *Onychophora*. Most arthropods have the first body segment developed in the embryo but lacking in the adult, hence their antennae are appendages of a more caudal segment. The similar organs of *Onychophora* are named preantennae to distinguish them from the true antennae of the other classes.

PRECESSION.   An effect manifested by a rotating body when a torque is applied to it in such a way as to tend to change the direction of its axis of rotation. If the speed of rotation and the magnitude of the applied torque are constant, the axis, in general, slowly describes a cone, its motion at any instant being at right angles to the direction of the torque.

A familiar example of precession is an ordinary top. If the axis of spin is not exactly vertical, the force of gravity exerts a torque tending to overturn the top; but instead of tipping over, it "wobbles" with a precessional motion about the vertical through the pivot-point. The gyroscope exhibits similar behavior. A hoop or a coin can roll on edge across the floor because, whenever it tends to tip either way, precession swerves its plane and changes its path, so that it automatically steers itself as a bicycle is steered by the rider.

Precession is due to the fact that the resultant of the angular velocity of rotation and the angular velocity produced by the torque is an angular velocity about a line that makes an angle with the permanent rotation axis. This angle lies in a plane at right angles to the plane of the couple producing the torque. The permanent axis must turn toward this line, since the body cannot continue to rotate about any line that is not a principal axis of maximum moment of inertia. That is, the permanent axis turns in a direction at right angles to that in which the torque might be expected to turn it. If the rotating body is symmetrical and its motion unconstrained, and if the torque on the spin axis is at right angles to this axis, the axis of precession will be perpendicular to both spin axis and torque axis. Under these circumstances the period of precession is given by

$$T_p = \frac{4\pi^2 I_s}{QT_s}$$

in which I_s is the moment of inertia and T_s the period of spin about the spin axis, and Q is the torque. In general, the problem is more complicated.

PRECESSION (Astronomy).   The slow, rotary motion of the axis of a rotating body in space. The term is, perhaps, most frequently used in connection with the earth, but all rotating bodies may exhibit the effect. The earth is an oblate spheroid with the minor axis as the axis of rotation. Hence, if we subtract from the earth a sphere with radius equal to the minor axis, we shall have left a shell of continually increasing thickness as we pass from the pole to the equator. Such a rotating shell, with the greatest amount of mass in the plane perpendicular to the axis of rotation, is a characteristic gyroscope.

The axis of rotation of the earth is inclined at an angle of 66.5° to the plane of the ecliptic. The forces of gravitational attraction of the sun and the moon tend to pull the equatorial shell into the plane of the ecliptic and hence a torque is applied tending to change the direction of the axis of rotation. This causes the axis to describe a cone in space (i.e., produces precession).

As the axis of rotation describes a cone, the poles of rotation describe circles in space. The radii of these circles, expressed in angular measure on the celestial sphere, are 23.5°. The effect of this motion of the poles is to cause the vernal equinox (one point of intersection of the ecliptic with the equator) to move along the ecliptic. The motion is slow, about 26,000 years being required for the vernal equinox to make a complete circuit of the ecliptic. The motion is not perfectly regular because of the fact that both the sun and the moon are in different planes and are moving relative to each other, causing a variation in the torque applied to the earth. The variations in the torque produce a slight irregularity in the motion of the poles known as nutation.

The motion of the equator among the stars causes slow, and slightly irregular, changes in the equatorial coordinates of the stars. Since celestial longitude is measured from the vernal equinox, the motion of this point produces a change in this coordinate of the stars. The motion of the vernal equinox also changes its location among the constellations along the zodiac, the "sign of Aries" (i.e., the vernal equinox) now being located in the constellation of Pisces instead of in Aries. At present the north pole of rotation of the celestial sphere is close to the star Polaris (α Ursae Minoris); but 12,000 years hence, the star Vega (α Lyrae) will be close to the pole of rotation and will be known as the pole star.

PRECIPITABLE WATER (or Precipitable Water Vapor).   The total atmospheric water vapor contained in a vertical column of unit cross-sectional area extending between any two specified levels, commonly expressed in terms of the height to which that water substance would stand if completely condensed and collected in a vessel of the same unit cross section.

The total precipitable water is that contained in a column of unit cross section extending all of the way from the earth's surface to the "top" of the atmosphere. Mathematically, if $x(p)$ is the mixing ratio at the pressure level, p, then the precipitable water vapor, W, contained in a layer bounded by pressures p_1 and p_2 is given by

$$W = \frac{1}{g} \int p_1 p_2 x \, dp,$$

where g is the acceleration of gravity. In actual rainstorms, particularly thunderstorms, amounts of rain very often exceed the total precipitable water vapor of the overlying atmosphere. This results from the action of convergence that brings into the rainstorm the water vapor from a surrounding area that is often quite large. Nevertheless, there is general correlation between precipitation amounts in given storms and the precipitable water vapor of the air masses involved in those storms.

AMS

PRECIPITATE.   (↓,ppt). Small particles that have settled out of a liquid or gaseous suspension by gravity, or that result from a chemical reaction. Precipitated compounds, such as blanc fixe (barium sulfate, are prepared in this way, for example, by the reaction $BaCl_2 + Na_2SO_4 \rightarrow NaCl + BaSO_4$. In formulas, a downward vertical arrow, ↓, or "ppt" is sometimes used to indicate a precipitate. A class of organic pigments called *lakes* are made by precipitating an organic dye onto an inorganic substrate. Colloidal particles dispersed in a gas, as flue dust in industrial stacks, can be precipitated by introducing an electric charge opposite to that which sustains the particles.

See also **Sedimentation**.

PRECIPITATION AND HYDROMETEORS.   Hydrometeors are any products of condensation or sublimation of atmospheric water vapor, whether formed in the free atmosphere or at the earth's surface; also, any water particles blown by the wind from the earth's surface. Hydrometeors are classified in a number of different ways, among which are the following. (1) Liquid or solid water particles formed, and remaining suspended, in the air. These include damp haze, cloud, fog, ice fog, and mist. (2) Liquid precipitation, including drizzle and rain. (3) Freezing precipitation, including freezing drizzle and freezing rain. (4) Solid (frozen) precipitation, including ice crystals, ice pellets, hail, snow, and the variations thereof. (5) Falling particles that evaporate before reaching the ground; namely, virga. (6) Liquid or solid water particles lifted by the wind from the earth's surface, including drifting and blowing snow, and blowing spray (i.e., spray lifted from the sea surface by the wind and blown about in such quantities that the horizontal visibility is restricted). (7) Liquid or solid water deposits on exposed objects, including dew, hoarfrost, glaze, and rime.

Precipitation, in meteorology, is any or all of the forms of water particles, whether liquid or solid, that fall from the atmosphere and reach the ground. Thus a major class of hydrometeor, as distinguished from clouds, fog, dew, and frost, is that precipitation must "fall" in reaching the ground. *Precipitable water vapor* is the total atmospheric water vapor contained in a vertical column of unit cross-sectional area extending between any two specified levels. It is commonly expressed in terms of the height to which that water substance would stand if completely condensed and collected in a vessel of the same unit cross section. In actual rainstorms, particularly thunderstorms, amounts of rain very often exceed the total precipitable water vapor of the overlying atmosphere. This results from the action of convergence, which brings into the rainstorm the water vapor from a surrounding area that often is quite large. Nevertheless, there is a general correlation between precipitation amounts in given storms and the precipitable water vapor of the air masses involved in those storms.

Condensation in the Atmosphere

Clouds and precipitation are visible evidence that water vapor in the atmosphere condenses into liquid and solid water. Moisture in cloud form may be re-evaporated into the air, but rain, snow, and allied forms of precipitation actively lessen the total water content. This moisture loss by precipitation, considering the world at large, is replaced by equivalent evaporation of moisture into the atmosphere.

Nuclei upon which condensation can take place are absolutely necessary for cloud formation and precipitation. Non-ionized and pollution-free air will become up to 400% supersaturated before condensation occurs. It also is known that clouds sometimes form before air becomes 100% saturated.

Lowering of the temperature of air below its dew point, or adding water vapor beyond the holding capacity of the air, is a second requirement for the formation of clouds. At high temperatures, air is able to hold up to 3% water vapor before becoming saturated; but at low temperatures, this amount may be as low as .01%. When air is cooled, therefore, it slowly loses its capacity to hold water vapor, and that excess vapor condenses into cloud droplets (or ice crystals). Cooling is achieved by one of four methods:

1. By contact with surfaces cooler than the air. This method of cooling is effective in forming fogs, dew, or hoarfrost; but it forms no precipitation except drizzle. If clouds or precipitation are to result, it is necessary that cooling proceed beyond the dew point of the air in question.

2. By lateral mixing of one air parcel with another at a lower temperature. This method is virtually non-effective in forming any extensive clouds, or fogs, and it does not cause precipitation.

3. By vertical mixing of a layer of air whose lapse rate is less than its adiabatic rate. In this method of cooling, the top layer of air is cooled and the base is warmed. The transfer of heat is due to the establishment of an adiabatic lapse rate in the thoroughly mixed layer in contrast to its previously nonadiabatic rate. Cooling at the top of the layer, if there is sufficient water vapor present, will produce a cloud layer (and considerable cloudiness of weather), but very little precipitation.

4. By lifting of whole layers of air or parcels of air, a process that occurs on windward sides of sloping terrain, along frontal surfaces, and in convection currents. Nearly all precipitation is caused by this method of cooling. Lifted parcels or layers of air cool dry-adiabatically while unsaturated, until the temperature and dew point coincide at a level known as the lifting condensation level. This level, at which condensation into cloud droplets or ice crystals occurs, depends upon the temperature and dew point of the lifted air; it is high for dry air and low for nearly saturated air.

Sublimation in the Atmosphere

When saturation is reached in the atmosphere at temperatures less than 0 °C and an additional cooling occurs, however small, and ice particle nuclei are present, water vapor collects directly on the ice nuclei to form an ice crystal cloud. Continued growth causes the ice cloud particles to grow and become precipitable snowflakes. This process is sublimation. Particles upon which ice crystals may grow by this process are known as *sublimation nuclei*.

The *Bergeron-Findeisen theory* offers an explanation of the process by which precipitation particles may form within a mixed cloud (composed of both ice crystals and liquid water drops).

The basis of this theory is the fact that the equilibrium vapor pressure of water vapor with respect to ice is less than that with respect to liquid water at the same sub-freezing temperature. Thus, within an admixture of these particles, and provided that the total water content were sufficiently high, the ice crystals would gain mass by sublimation at the expense of the liquid drops, which would lose mass by evaporation. Upon attaining sufficient weight, the ice crystals would fall as snow and very likely become further modified by accretion, melting, and/or evaporation before reaching the ground.

The most important cloud ingredient specified by this hypothesis is an aggregation of cloud drops that have undergone supercooling. This is a common feature, mainly of cumuliform clouds in temperate latitudes. The current intensive research regarding the introduction of the necessary ice crystals into the cloud has served to increase interest and activity in cloud-seeding methods.

Saturation, Supersaturation, and Supercooling

Saturation. If a free water surface is introduced into a box from which all gases have been removed, molecules of water will emerge from the liquid surface until the number of molecules escaping from the liquid equals the number returning to the liquid. When a balance is maintained between those evaporating or escaping from the liquid water and those condensing or impinging on and remaining in the liquid, the space within the box is said to be saturated with water vapor. It can hold no more water molecules and, if more are added, these will condense into liquid. If air is admitted into the box, there will be no change whatsoever in the rate at which water molecules leave and impinge on the liquid. Component parts of the atmosphere, therefore, have no bearing on the number of molecules of water vapor present. Saturation of space above any liquid water surface depends solely on the number of molecules of water in the air as a vapor when a balance is achieved between the vapor molecules condensing on the liquid and the liquid molecules evaporating into the space above. This molecular balance depends almost entirely on the temperature of the liquid and the water-vapor molecules. The higher the temperature, the more molecules of water vapor can escape from the water surface before they begin to condense and return to the surface.

In meteorological practice, because air is the medium in which weather phenomena take place, it is useful to relate saturation of water vapor to the air carrying it. Saturated air is used, therefore, as a term, even though the presence of the air does not directly affect the number of molecules of water vapor at saturation. It is customary to speak of saturation specific humidity with particular reference to the weight of air carrying the water-vapor molecules. These saturation specific humidities vary from a few tenths of a gram per kilogram of cold air to 30–40 grams per kilogram of very warm air.

Dew point is the temperature to which a given parcel of air must be cooled at constant pressure and constant water vapor content in order for saturation to occur. When this temperature is below 0 °C, it is sometimes called the *frost point* (i.e., the highest temperature at which atmospheric moisture will sublimate in the form of hoarfrost on a cooled polished surface). The dew point may alternatively be defined as the temperature at which the saturation vapor pressure of the parcel is equal to the actual vapor pressure of the contained water vapor. For example, let the temperature of the air be 20 °C and the relative humidity 60%. Then, since the maximum vapor pressure of water at 20 °C is 17.4 millimeters, the actual water vapor pressure is 0.6 of this, or 10.4 millimeters. The temperature at which 10.4 millimeters is the maximum vapor pressure of water is 12 °C. Hence, if the air is cooled to 12 °C, it will reach saturation, and under suitable conditions, dew will form; 12 °C is the dew point. Likewise, if the dew point is known to be 12 °C when the air is 20 °C, it follows that the relative humidity is 60%.

Isobaric heating or cooling of an air parcel does not alter the value of that parcel's dew point, so long as no vapor is added or removed. The dew point of the atmosphere can be determined directly by any of several types of dew point (or frost-point) hygrometers, or by the dew cell, but it is more commonly determined with the aid of the psychrometric calculator or Tables after the direct reading of a psychrometer. See Fig. 1.

Supersaturation. This exists in a given portion of the atmosphere when the relative humidity is greater than 100%, i.e., when it contains more water vapor than is needed to produce saturation with respect to a plane surface of pure water or pure ice. Such supersaturation develops because, frequently, there is no "plane surface of pure water (or ice)" available. In the absence of water surfaces and in the absence of condensation nuclei or any wettable surfaces, phase change from vapor to liquid cannot occur due to the free energy barrier imposed by the surface free energy of the embryonic droplets, which would then have to form by spontaneous nucleation. Humid air, purified of all foreign nuclei, can be expanded in

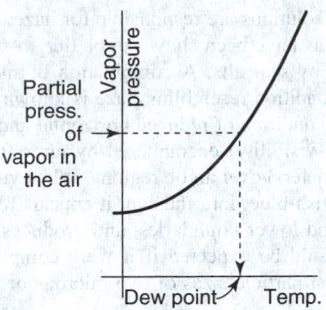

Fig. 1. Graphical location of dew point from vapor pressure curve. Project the known vapor pressure in the air over the curve and down to the temperature axis.

experimental cloud chambers to relative humidities of the order of 400% without any condensation taking place. Cloud condensation occurs in the atmosphere at relative humidities near 100% only because there is an abundance of condensation nuclei.

Supercooled Liquid in the Atmosphere. Drops of rain and cloud droplets are often cooled to temperatures well below 0°C and remain in liquid form. They are then said to be supercooled. If such drops and droplets are disturbed, they freeze almost instantly, partially or entirely, depending upon their temperature and size.

The supercoolability of a substance whose solid form is crystalline (as opposed to amorphous matter) stems from the unique energy transformations necessary for the formation of the first crystal nucleus; whereafter, all adjacent liquid immediately becomes solid unless or until the latent heat released elevates the system's temperature sufficiently to arrest the process. It should be noted that the reverse process is not possible; a crystalline solid cannot be "superheated;" therefore, a substance's melting point is very conservative, only slightly dependent upon pressure.

Supercooled clouds are quite common. In extreme cases, they have been observed at temperatures as low as −40°C. The smaller and more pure the water droplets, the more likely is supercooling. Passage of aircraft through supercooled drops and droplets in the atmosphere results in rapid and severe icing.

Cloud Process

In addition to the processes described here, see also **Clouds and Cloud Formation**.

Accretion. In terms of cloud physics, accretion is the growth of a precipitation particle by the collision of a frozen particle (ice crystal or snowflake) with a supercooled liquid droplet that freezes upon contact.

Agglomeration. The process in which precipitation particles grow by collision with, and assimilation of, cloud particles or other precipitation particles. Regardless of air motions within the cloud, collisions always result from different rates of particle motion due to size differences. The assimilation may result from coalescence when both particles are water droplets or by accretion if one particle is an ice crystal and the other supercooled liquid water. See also **Agglomeration**.

Coagulation. In cloud physics, coagulation is generally used synonymously with accretion. Less frequently, it refers to any process by which a cloud's numerous small cloud drops are converted into a smaller number of precipitation particles. When so used, the term is employed in analogy to the coagulation of any colloidal state.

Coalescence. In cloud physics, coalescence means the merging of two water drops into a single larger drop. This process is believed to be of importance in the production of rain in "warm clouds," i.e., clouds that are warmer than 0°C throughout, and hence are wholly incapable of supporting an ice-crystal precipitation. Physical details of the coalescence process are still obscure, but the relative velocity of impact, the relative and absolute sizes of colliding drops, electric charge of the drops, and external electric fields must enter the picture. The fraction of all collisions between water drops of a specified size that results in actual merging of the two drops into a single larger drop is called *coalescence efficiency*. It is clear that not every collision of two drops results in coalescence; that under certain conditions, the coalescence efficiency is less than unity.

Orography. Generally, this term means any weather phenomena caused by the flow of air over prominent features of the terrain. *Orographic lifting* is the lifting of an air current caused by its passage up and over mountains, whereby it undergoes certain changes: (1) So long as the air is unsaturated, it cools at the dry-adiabatic rate of approximately 5.5°F (3°C) per 1,000 ft. (304.8 meters). (2) As soon as the air is saturated, it cools at the pseudo-adiabatic rate, which depends on the amount of water being condensed by cooling, but is, in any event, less than the dry-adiabatic rate. (3) Clouds, and rain or snow (orographic precipitation) are often formed in this orographically lifted air. Thunderstorms occur if the air is unstable. The Indian monsoon rains and the rains of the United States and European north-west coasts are large-scale orographic phenomena.

Specific Types of Hydrometeors and Precipitation

Cloudburst. In popular terminology, any sudden and heavy fall of rain. An unofficial criterion sometimes used specifies a rate of fall equal to or greater than 100 millimeters (3.94 inches) of rain per hour.

Condensation Trail. These trails are cloudlike streamers called contrails that are observed to form behind an aircraft, particularly jet-engined, flying at high altitudes in very cold air. There are two types of contrails: (1) Exhaust trails are formed when the air in the wake of the aircraft is mixed with and saturated by the water vapor that is a combustion product of the aircraft's engines. Exhaust trails may linger for many miles to the rear of an aircraft, painting its track in the sky. (2) Aerodynamic trails are formed in saturated or nearly saturated air that is cooled in passing over the surfaces of an aircraft and cooled by adiabatic expansion. Aerodynamic trails are more rare than exhaust trails and are of short duration. They are observed most frequently at low altitude levels.

Dew. Classified as a hydrometeor, dew is water condensed onto grass and other objects near the ground, the temperatures of which have fallen below the dew point of the surface air due to radiational cooling during a prior time (usually night), but are still above freezing. Hoarfrost forms if the dew point is below freezing. If the temperature falls below freezing after dew has formed, the frozen dew is known as "white dew."

The conditions favorable to dew formation are: (1) a radiating surface, well-insulated from the heat supply of the soil, on which vapor may condense; and (2) a clear, still atmosphere with low specific humidity in all but the surface layers, to permit sufficient effective terrestrial radiation to cool the surface, and with high relative humidity in the surface air layers, or an adjacent source of moisture, such as a lake.

Dew plays an important role in the propagation of certain plant pathogens, such as late potato blight, which require dew-covered leaves for certain stages of sporulation. Also reliant upon dew is the optical effect known as *heiligenschein*, a diffuse white ring surrounding the shadow cast by the observer's head upon a dew-covered lawn when the solar elevation is low.

Drizzle. This is a small falling droplet of water whose dimensions across are usually less than 0.5 millimeter. Drizzle droplets are buffeted by gusts of wind whereas rain tends to travel in a more or less straight line, although the path may be slanted. Drizzle falls predominantly from low stratus clouds and is a phenomenon associated with vertical atmospheric stability in the low levels.

Fog. See also **Fog and Fog Clearing**.

Fog Tracks. Linear regions of condensation, produced in air or other gases that are supersaturated with water vapor, by the passage of electrified particles. Fog tracks are useful in following the courses and collisions of such particles, as in an experimental cloud chamber.

Frost. This is the condition that exists when the temperature of the earth's surface and earthbound objects falls below freezing (0°C or 32°F). Depending upon the actual values of ambient-air temperature, dew point, and the temperature attained by surface objects, frost may occur in a variety of forms, including a freeze and hoarfrost. If a frost period is sufficiently severe to end the growing season or delay its beginning, it is commonly referred to as a "killing frost."

Frost, like snow, is the result of the sublimation of water vapor in saturated air. If there is excessive radiation from solid objects, as on a clear night in late fall, the air coming in contact with these objects may be chilled below the sublimation point, and spicules of ice may grow out from the cold surfaces. The process is entirely similar to the artificial formation of metallic crystals, from vaporized metals, on the interior of a glass tube communicating with the vaporizing furnace. In either case, the size of the crystals is a matter of time and the supply of saturated vapor. Frost is often observed around cracks in a wooden sidewalk, because of the damp air escaping from the ground below. The objects upon which frost forms most readily are those of low specific heat and high thermal emissivity, such as blackened metals; hence, the marked accumulation of frost on the heads of rusty nails. The apparently erratic occurrence of frost in adjacent localities is due partly to differences of level, the lower areas becoming colder; but also largely to differences in absorptivity and specific heat of the ground, which, in the absence of wind, greatly influences the temperature attained by the superincumbent air. It should be understood that vegetation is not damaged by frost itself, but by cold air; the appearance of frost merely indicates that the temperature has dropped below the freezing point. The formation of white frost on the indoor surface of window panes indicates low relative humidity of the indoor air; otherwise water would first condense in small drops and then freeze into clear ice.

Hail. Classified as a hydrometeor, hail is precipitation in the form of balls or irregular lumps of ice, always produced by convective clouds, nearly always cumulonimbus. Single units of hail, called hailstones, range in size from that of a pea to that of a grapefruit (i.e., from less than $\frac{1}{4}$ inch (6 millimeters) to more than 5 inches (13 centimeters) in diameter). The largest hailstone observed in the United States is believed to be one that fell at Potter, Nebraska, on July 6, 1928; it measured 17 inches (43.2 centimeters) in circumference and weighed about 1.5 pounds (0.7 kilograms).

Hailstones may be spheroidal, conical, or generally irregular in shape, the spheroidal form being the most common. When broken, the stones reveal a structure of concentric alternate layers of clear and opaque white ice. This characteristic layered inner structure has been explained in terms of the "multiple incursion theory," which pictures cyclical ascent and descent of the hailstone into alternate above- and below-freezing regions of the cloud. Thus, thunderstorms, which are characterized by strong updrafts, large liquid water content, large cloud-drop sizes, and great vertical height, are favorable to hail formations. It has been suggested that descent alone from great heights through an updraft exhibiting water-content stratification is probably sufficient to yield the foliated structure. In either case, the hailstones grow, basically, by accretion of supercooled water drops upon the growing ice particles; and the nature of the ice depends upon such things as the rate of accretion, drop size, and temperature.

The destructive effects of hail storms upon plant and animal life, buildings and property, and aircraft in flight render them a prime object of weather modification studies.

Haze. Classified as a hydrometeor, haze is a fine dust or salt particles dispersed through a portion of the atmosphere. The particles are so small that they cannot be felt or individually seen with the naked eye, but they diminish horizontal visibility and give the atmosphere a characteristic opalescent appearance that subdues all colors.

Many haze formations are caused by the presence of an abundance of condensation nuclei, which may grow in size and become mist, fog, or cloud. Distinction is sometimes drawn between "dry haze" and "damp haze," largely on the basis of differences in optical effects produced by the smaller particles (dry haze) and the larger particles (damp haze). Dry haze particles, with diameters of the order of 0.1 micrometer, are small enough to scatter shorter wavelengths of light preferentially, though not according to the inverse fourth-power law of Rayleigh scattering. Such haze particles produce a bluish color when the haze is viewed against a dark background, for dispersion allows only the slightly bluish scattered light to reach the eye. The same type of haze, when viewed against a light background, appears as a yellowish veil, for here the principal effect is the removal of the bluer components from the light originating in the distant light-colored background. Haze may be distinguished by this same effect from mist, which yields only a gray obscuration, since, in the mist, the particle sizes are too large to yield appreciable differential scattering of various wavelengths.

Any small liquid droplet contributing to an atmospheric haze condition is known as a "haze droplet." In certain industrial areas, such droplets may be entirely non-aqueous (largely hydrocarbons), but there is little doubt that most haze droplets are water solutions of some type. Near seacoasts,

droplets of sea-salt solutions are responsible for haze; and even far inland, some haze conditions have been shown to be due to salt-solution droplets probably produced by sea-salts. A combination of smoke and haze, or a very light smoke condition resembling haze is known as "smaze."

Arctic haze is a condition of reduced horizontal and slant visibility (but unimpeded vertical visibility) encountered by aircraft in flight (to above 30,000 feet; 9144 meters) over arctic regions. When viewed away from the sun, it appears grayish-blue; into the sun, it appears reddish-brown. It has no distinct upper and lower boundaries, and produces none of the optical phenomena that would be expected if it were composed of ice crystals. Color effects suggest particle sizes of two microns or less. See also **Polar Research**.

Ice-crystal haze is a type of very light ice fog, usually associated with the precipitation of ice crystals, and observable at times to altitudes as great as 20,000 feet. Observed from the ground, it may be dense enough to hinder observation of celestial bodies, sometimes even the sun. Looking down from the air, however, the ground is usually visible and the horizon only blurred.

Hoarfrost. Classified as a hydrometeor, hoarfrost is a deposit of interlocking ice crystals (hoar crystals) formed by direct sublimation on objects, usually those of small diameter freely exposed to the air, such as tree branches, plant stems and leaf edges, wires, poles, etc. Also, frost may form on the skin of an aircraft when a cold aircraft flies into air that is warm and moist, or when it passes through air that is supersaturated with water vapor. The deposition of hoarfrost (also known as "white frost") is similar to the process by which dew is formed, except that the temperature of the befrosted object must be below freezing. It forms when air with a dew point below freezing is brought to saturation by cooling. In addition to its formation on freely exposed objects (air hoar), hoarfrost also forms inside unheated buildings and vehicles, in caves, in crevasses (crevasse hoar), on snow surfaces (surface hoar) and in air spaces within snow, especially below a snow crust (depth hoar).

Hoarfrost may be distinguished from two other ice deposits, rime and glaze, by the following characteristics: *Rime* is a white or milky and opaque granular deposit of ice, which is denser and harder than hoarfrost. *Glaze*, is a generally clear and smooth coating of ice, denser, harder, and more transparent than either rime or hoarfrost. Both rime and glaze are formed when supercooled water drops strike an object at a temperature below freezing. Such formation on terrestrial objects is known as an "ice storm," on aircraft, as "aircraft icing." See also **Aircraft Icing**. When rime forms on ice particles in the atmosphere, snow pellets result; when glaze forms on these ice particles, the result is hail.

Ice Crystal. A type of precipitation composed of slowly falling, very small, unbranched crystals of ice, which often seem to float in the air; it is popularly referred to by such names as "frost snow," "ice needles," or "diamond dust." It may fall from a cloud or from a cloudless sky; it is visible only in direct sunlight, or in an artificial light beam, and does not appreciably reduce visibility. Ice crystal precipitation often produces sun pillar and other halo phenomena.

Ice Pellets. A type of precipitation consisting of transparent or translucent pellets of ice, 5 millimeters or less in diameter. They may be spherical, irregular, or (rarely) conical in shape. Ice pellets usually bounce when hitting hard ground, and make a sound upon impact. They include two basically different types of precipitation.

1. Transparent, globular, solid grains of ice, which have formed from the freezing of raindrops or the refreezing of largely melted snowflakes when falling through a below-freezing layer of air near the earth's surface. When raindrops enter a layer of intensely cold air, they become supercooled, i.e., cooled below the freezing point, but without freezing. In this state, they are highly unstable, and upon coming in contact with any object, even with a speck of dust, they suddenly freeze into pellets of sleet. Larger objects, such as twigs or telephone wires, when touched by these supercooled drops, receive a coating of ice, which may result in a landscape of marvelous beauty when the sun appears, but which often proves very destructive because of the heavy weight of the ice. Sleet is quite commonly mixed with snow or with rain, but it is always a cold-weather product, and is not as sometimes believed, a small form of hail.

2. Translucent particles, consisting of snow pellets encased in a thin layer of ice. The ice layer may form either by the accretion of droplets upon the snow pellet, or by the melting and refreezing of the surface of the snow pellet.

Mist. According to international definition, mist is a hydrometeor consisting of an aggregate of microscopic and more-or-less hygroscopic water droplets suspended in the atmosphere. It produces, generally, a thin grayish veil over the landscape, but reduces visibility to a lesser extent than does fog. The humidity with mist is often less than 95%.

Rain. The most common type of atmospheric precipitation, in the form of liquid water drops, ranging in diameter from 0.5 millimeter to approximately 5.0 millimeters (or larger), and usually falling with a velocity ranging from 5 to 8 meters per second. A typical rain drop might have a diameter of 1–2 millimeters, while the largest drops (observed in heavy thunderstorms, and flattened on their undersides as a result of aerodynamic effects), may have equivalent spherical diameters of 5–8 millimeters. Drops larger than this become quite unstable and break up as a result of microturbulence in the air through which they fall.

Rain is to be distinguished from the only other form of liquid precipitation, *drizzle*, in that drizzle drops are generally less than 0.5 millimeter in diameter, are very much more numerous, and reduce visibility much more than does light rain. Also, unlike rain, drizzle drops may appear to float while following air currents.

Shower. Precipitation characterized by the suddenness with which it starts and stops, by rapid changes of intensity, and, usually, by rapid changes in the appearance of the sky. In weather observing practice, showers are always reported in terms of the basic type of precipitation that is falling, i.e., rain showers, snow showers, sleet showers, and so on.

Air-mass showers are produced by local convection within an unstable air mass. They are most frequently within a moist air mass that is sufficiently unstable so that daytime heating at the surface can produce well-developed cumulus clouds. They are not associated with a front or instability line.

Silver Thaw. After a period of cold weather and below-freezing temperatures, a mass of warm air passing over the region will cause frost or glaze to form on objects that are still at a low temperature. This condition is known as a silver thaw. A silver thaw usually lasts only a few hours, because the warm air soon warms all exposed objects above the freezing point.

Sleet. See "Ice Pellets" previously described in this entry.

Snow. Classified as a hydrometeor, snow is precipitation composed of white or translucent ice crystals, chiefly in complex branched hexagonal form and often agglomerated into snowflakes. Snow appears white only because of the multitude of reflecting surfaces. The individual crystals are of transparent ice.

Single snow crystals, when freshly formed, are often almost perfect and exhibit an endless variety of detail. They are commonly flat, six-sided polygons, stars, or spangles, often of very complicated and beautiful design, but always with the 60° and 120° angles characteristic of the hexagonal system. Sometimes they are needle-like with a hexagonal head, like a pin. The finer spicules are sometimes suspended high in the atmosphere, and are the cause of halos. Snow differs from frost chiefly in being formed in the air instead of upon solid objects near the ground, the crystallization nuclei being particles of dust.

Partly melted crystals often cling together to form snowflakes of varying size, and may melt into raindrops before reaching the ground. Snowflakes made up of clusters of crystals or crystal fragments may grow as large as three to four inches in diameter, frequently building themselves into hollow cones falling point downward. In extremely still air, flakes with diameters as large as ten inches have been reported.

Snow may also fall in the form of snow grains and snow pellets. *Snow grains*, as a form of precipitation, is the solid equivalent of drizzle, being very small (less than 1 millimeter) opaque particles of ice. Unlike snow pellets, they neither shatter nor bounce when they hit a hard surface, and usually fall in very small quantities. *Snow pellets* (sometimes called "soft hail" or "tapioca snow") is a form of precipitation consisting of white, opaque, approximately round ice particles, having a snow-like structure, and measuring about 2 to 5 millimeters in diameter. Snow pellets are crisp

and easily crushed, rebounding when they fall on a hard surface, and often breaking up. They most often fall with the suddenness and changes of intensity characteristic of showers, and often together with or before snow.

In the United States, a heavy snowfall is commonly followed by intense cold, partly because of the low absorptivity of snow for solar radiation, and partly because of the cyclonic character of the snowstorm, which brings a change of wind to the north on the westward or following side of the storm.

Virga. A hydrometeor in the form of wisps or streaks of water or ice particles falling out of a cloud, but which evaporates before reaching the earth's surface as precipitation. Virga is frequently seen trailing from altocumulus and altostratus clouds, but also is discernible below the bases of high-level cumuliform clouds from which precipitation is falling into a dry subcloud layer. Virga typically exhibits a hooked form in which the streaks descend nearly vertically just under the precipitation source, but appear to be almost horizontal at their lower extremities.

Measurement of Hydrometeors and Precipitation Phenomena

In weather observation activities and meteorological forecasting, several variables concerning hydrometeors, precipitation, and associated cloud formation and other processes are very important. Several of these instruments are described as follows:

Atmometer. This is the general name for an instrument that measures the evaporation of water into the atmosphere. Several types may be mentioned: (1) The *evaporation pan* is a cylindrical container made of galvanized iron or monel metal, 10 inches deep and 48 inches in diameter. It is filled with water to a depth of 8 inches, and periodic measurements are made of the changes of the water level. (2) The *clay atmometer* is a porous porcelain container connected to a calibrated reservoir filled with distilled water. Evaporation is determined by the depletion of water in the reservoir. (3) The *Piché evaporimeter* consists of a graduated tube, closed at one end, which is filled with distilled water and then covered with a larger circular piece of filter paper held in place by a disc-and-collar arrangement. In operation, the instrument is inverted so that the distilled water is in contact with the filter paper. Evaporation is determined by noting the change in level of the meniscus of the water. (4) The *radio atmometer* is designed to measure the effect of sunlight upon evaporation from plant foliage. It consists of a porous clay atmometer whose surface has been blackened so that it absorbs radiant energy.

Cloud-Detecting Radar. A type of weather radar designed specifically for the detection of clouds rather than precipitation. It is capable of detecting clouds in multiple layers up to heights of about 50,000 feet above the radar.

Cloud-Height Indicator. A general term for instruments that measure the height of cloud bases (i.e., the lowest levels of the atmosphere at which the water or ice particles comprising a cloud appear). Cloud-height indicators may be classified according to their principle of operation:

1. Height determination by the principle of triangulation is represented by two instruments: (a) The *ceilometer* is an automatic, recording, cloud-height indicator. The photoelectric cell pickup detector of one type, the fixed-beam ceilometer, scans continuously to detect in a cloud an illuminated spot directed by a stationary projector. In the other type, the rotating-beam ceilometer, the projector rotates rapidly through 360° while the detector is fixed vertically. (b) The *ceiling light* projects a narrow beam of light onto a cloud base, the height of which is then determined by means of a clinometer.
2. Height determination by the principles of pulse techniques is represented by the *pulsed-light cloud-height indicator* and the vertically-directed *cloud-detection radar*. With these instruments, the time required for a pulse of energy to travel from a radiator located on the ground to the cloud base and back to the ground again is measured electrically. The height of the cloud is computed from this transit time and a knowledge of the propagation velocity of the pulse.

In addition to these instruments, a small balloon known as a ceiling balloon is used to determine the height of a cloud base. The height can be computed from the ascent velocity of the balloon and the time required for its disappearance into the cloud.

Dew Cell. An instrument used to determine the dew point. It consists of a pair of spaced bare electrical wires wound spirally around an insulator and covered with a wicking wetted with a water solution containing an excess of lithium chloride. An electric potential applied to the wires causes a flow of current through the lithium chloride solution, which raises the temperature of the solution until its vapor pressure is in equilibrium with that of the ambient air. A modification of the dew cell, the hygristor, is used for upper-air measurements.

Drosometer. An instrument for measuring the amount of dew formed on a given surface. One type consists of a hemispherical glass vacuum cup exposed to the atmosphere. Dew forming on the glass surface automatically collects in the bottom of the cup, which is weighed at the end of the exposed period. Another type consists of a block of wood whose surface has been treated so that dew forms on it in characteristic patterns. Photographs are supplied with each instrument to enable the observer to match the dew formation with a set of standards corresponding to a dew "fall" of from 0.01 to 0.45 millimeter.

Evapotranspirometer. An instrument for measuring the rate of evapotranspiration (i.e., the combined processes of evaporation of liquid or solid water plus transpiration from plants). It consists of a vegetation soil tank so designed that all water added to the tank and all water left after evapotranspiration can be measured.

Hygrothermograph. A recording instrument combining, on one record, the variation of atmospheric temperature and humidity content as a function of time. See Fig. 2. The most common hygrothermograph is a hair hygrograph (a recording hair hygrometer) combined with a thermograph.

Fig. 2. Hygrothermography for measuring both temperature and humidity. A time record is linked on the slowly rotating drum. Digital instrumentation also is available.

Nephometer. A general term for instruments designed to measure the amount of cloudiness. An early type consisted of a convex hemispherical mirror mapped into six parts. The amount of cloud coverage on the mirror was manually noted by the observer.

Nephoscope. An instrument for determining the direction of cloud motion. With the direct-vision nephoscope, the observer notes the motion of the cloud by looking directly at it through the instrument, and aligning a grid so that the cloud appears to move parallel to its major axis. With the mirror nephoscope, the cloud motion is indicated by

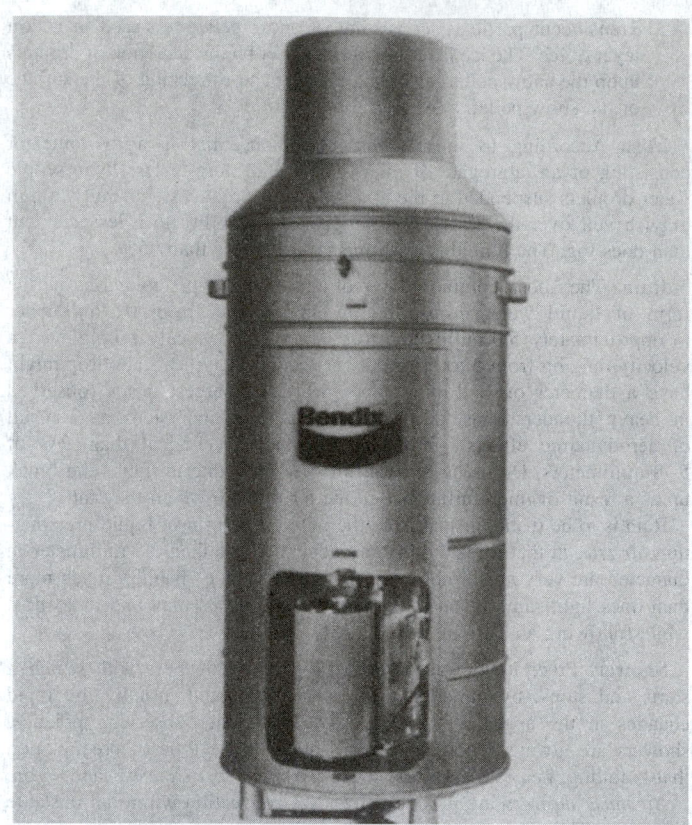

Fig. 3. Universal recording precipitation gage.

the azimuth at which the cloud image, seen in a mirror, leaves the mirror.

Rain Gage. These gages may be classified according to their principle of operation: (a) A recording rain gage automatically records the amount of precipitation collected, as a function of time. One of these recording instruments, frequently used at climatological stations, consists of a receiver in the shape of a funnel, which empties into a bucket mounted on a weighing mechanism. See Fig. 3. The weight of the catch is recorded on a clock-driven chart as inches of precipitation. (b) A nonrecording rain gage indicates, but does not record, the amount of precipitation captured. The nonrecording rain gage of the U.S. Weather Bureau consists of a receiver in the form of a funnel (8 inches; 20.3 centimeters in diameter), an overflow can upon which the receiver fits, a measuring tube into which the small end of the funnel fits, and a measuring stick graduated in proportion to the difference in area between the receiving piece and the measuring tube. The ratio of the area of the receiver to that of the measuring tube is such that tenfold magnification is obtained for ease in determining small amounts. (c) The *rain-intensity gage* measures the instantaneous rate at which rain is falling on a given surface. In the Hudson-Jardi design, water from the rain collector enters a chamber containing a float and outlet valve, and shaped in such a fashion that the height of the float is directly proportional to the rate of rainfall. The motion of the float is recorded either electrically or mechanically.

The term *ombrometer* refers in general to any rain gage, but specifically to a rain gage capable of measuring very small amounts of precipitation.

Snow Gage. These instruments are designed to measure the vertical depth of snow. (a) The *snow sampler* is a hollow metal tube in which snow is collected, melted, and weighed. (b) The *snow stake* is a wood scale, calibrated in inches and used in regions of deep snow to measure its depth. It is bolted to a wood post or angle iron set in the ground. (c) The *radioactive snow-gage* automatically and continuously records the water equivalent of snow on a given surface, as a function of time. A small sample of a radioactive salt is placed in the ground in a lead-shielded collimator that directs a beam of radioactive particles vertically upwards. A Geiger-Müller counting system (located above the snow level) measures the amount of depletion of radiation caused by the presence of the snow. (d) The *snow bin* is simply a box in which snowfall is collected and measured. (e) The *snow board* is a sheet of thin white board, about 16 inches (40.6 centimeters) square, with a layer of cotton flannel tacked to its upper surface. This surface retains falling snow better than metal surfaces for snowfall measurement purposes. (f) The *snow mat* is a special device used to mark the surface between old and new snow. It consists of a piece of white duck 28 inches (71.1 centimeters) square, having in each corner triangular pockets in which are inserted slats placed diagonally to keep the mat taut and flat.

See the section on "Acid Rain" in the entry on **Pollution (Air)**. See also **Atmosphere (Earth)**; **Clouds and Cloud Formation**; **Fog and Fog Clearing**; **Fronts and Storms**; and other entries listed under **Meteorology**.

Additional Reading

Cotton, W.R. and R.A. Anthes: *Storm and Cloud Dynamics*, Academic Press, Inc., San Diego, CA, 1997.

Collier, C.G.: *Applications of Weather Radar Systems: A Guide to Uses of Radar Data in Meteorology and Hydrology*, 2nd Edition, John Wiley & Sons, Inc., New York, NY, 1996.

Cotton, W.R. and R.A. Anthes: *Storm and Cloud Dynamics*, Academic Press, Inc., San Diego, CA, 1992.

Garfield, J. and O. Sohnel: *Precipitation: Basic Principles and Industrial Applications*, Butterworth-Heinemann, Inc., Woburn, MA, 1993.

Jaenicke, R.: *Dynamics and Chemistry of Hydrometeors: Final Report of the Collaborative Research Centre 233 "Dynamik Und Chemie Der Meteore"*. Collaborative Research Centres, John Wiley & Sons, Inc., New York, NY, 2001.

Sumner, G.N.: *Precipitation Process and Analysis*, John Wiley & Sons, Inc., New York, NY, 1988.

Upgren, A. and J. Stock: *Weather: How It Works and why It Matters*, Perseus Publishing, Boulder, CO, 2000.

Williams, J.: *The Weather Book*, 2nd Edition, Random House, Inc., New York, NY, 1997.

PETER E. KRAGHT, Certified Consulting Meteorologist, Mabank, TX

PRECIPITATION ATTENUATION.
Attenuation of electromagnetic waves propagating through precipitation. Depending on the wavelength of the radiation, the attenuation is accounted for by some combination of absorption and scattering by the precipitation particles.

The relative importance of scattering tends to increase as the wavelength becomes shorter. For radar, the specific attenuation Y (dB km^{-1}) due to rain is described by empirical power relations of the form $Y = aR^b$, where R (mm h^{-1}) is the rainfall rate, and a and b are empirical constants that depend on wavelength and temperature. The specific attenuation of snow is less than that of rain, and for wavelengths of 10 centimeters (4 inches) and longer is usually negligible. For lidar, the precipitation attenuation is approximately proportional to the cross-sectional area of the precipitation particles per unit volume.

See also **Attenuation**; **Cloud Attenuation**; and **Rain Attenuation**.

AMS

PRECIPITATION CEILING.   See **Meteorology**

PRECIPITATION CURRENT.
The downward transport of charge, from cloud region to earth, that occurs in a fall of electrically charged rain or other hydrometeors; a particular case of a convection current. See also **Convection Current**. Observations of the charge on individual raindrops during thunderstorms have revealed a complex picture. On average, more positive than negative charge is brought to earth by precipitation currents, but wide deviations occur both within individual storms and from one storm to another. The reasons for these wide fluctuations are not understood. Precipitation currents in continuous rain generally vary from about 10^{-12} to 10^{-10} A m^{-2}, while thunderstorm currents become as large as 10^{-8} A m^{-2}. See also **Fronts and Storms**.

PRECIPITATION EFFECTIVENESS.
1.   That portion of total precipitation used to satisfy vegetation needs.

2.   The actual availability of precipitation used in plant development. Availability is affected by such factors as precipitation intensity,

season, temperature, ground cover, sod type, etc. The dependence of precipitation effectiveness on temperature and/or evaporation has been expressed in many ways: Köppen's formulas for defining desert climate, Lang's moisture factor, De Martonne's index of aridity, Gorczyáski's aridity coefficient, Angström's humidity coefficient, Transeau's precipitation–evaporation quotient, and Thornthwaite's precipitation-effectiveness index.

Additional Reading

Köppen, W.: *Grundriss der Klimakunde*, Walter die Gruyter Co., Berlin, Germany, 1931.

Transeau, E.N.: "Forest Center of Eastern America," *American Naturalist*, **39**, 875–899 (1905).

Thornthwaite, C.W.: "An Approach Toward a Rational Classification of Climate," *Geogr. Rev.*, **38**, 55–94 (1948).

AMS

PRECIPITATION-EFFECTIVENESS INDEX (P-E INDEX).
For a given location, a measure of the long-range effectiveness of precipitation in promoting plant growth:

$$\text{P-E index} = 10 \sum_{n-1}^{12} (\text{P-E index})_n,$$

that is, it is equal to 10 times the sum of the monthly precipitation–evaporation ratios (monthly precipitation amounts divided by monthly evaporation amounts). Also called precipitation–evaporation index.

See also **Moisture Index**; and **Temperature-Efficiency Index**.

AMS

PRECIPITATION ELECTRICITY.
1.   That branch of the study of atmospheric electricity concerned with the electrical charge carried by precipitation particles and with the manner in which these charges are acquired.

2.   The electrical charge borne by precipitation particles. A very complex and highly variable picture is obtained when charges are measured on individual raindrops or snow crystals and no present theory approaches a complete explanation of all details. In general, more raindrops are positively than negatively charged. Sometimes the prevailing sign of the charges even shifts in the course of a given storm's lifetime.

See also **Ion-Capture Theory**; and **Precipitation Current**.

Additional Reading

Chalmers, J.A., and D.T. Haar: *Atmospheric Electricity*, Franklin Book Company, Inc., Elkins Park, PA, 1967.

MacGorman, D.R., and W.D. Rust: *The Electrical Nature of Storms*, Oxford University Press, New York, NY, 1998.

AMS

PRECIPITATION HARDENING.
A large number of alloys are hardenable by a heat treating procedure known as precipitation hardening. Hardening is accomplished by the controlled precipitation of many minute particles of a second crystalline phase (or phases) inside the crystals of the primary metal. In order that the precipitation may be effected, the hardening constituent must be more soluble at higher temperatures than it is at lower temperatures, so that heating of the solid metal at an elevated temperature causes the second phase to dissolve into the matrix. If a precipitation hardening alloy is heated and held at an elevated temperature so as to dissolve the hardening phase and then is quenched to room temperature, a supersaturated solid solution is obtained. This heating and quenching operation is known as the solution treatment. The second phase of precipitation hardening is known as the aging treatment wherein the second phase is precipitated out of the supersaturated solid solution by holding the metal either at room temperature or some intermediate temperature well below the temperature employed in the solution treatment. The various stages involved in the formation of the nuclei of the precipitation particles may be very complex. In general, however, the aim of

the aging process is to obtain a distribution of the precipitated particles that produces maximum hardness. This will usually occur when the particles are submicroscopic in size and extremely numerous. Their hardening effect on the crystal lattice of the matrix crystals is believed to result from local strains that they produce in the matrix. These latter hinder the normal easy motion of dislocations, thereby hardening the metal. The term age hardening is synonymous with precipitation hardening, but when so used generally refers to metals aged at room temperature.

PRECIPITATION STATION.   See **Meteorology**.

PRECIPITATOR.   See **Electrostatic Precipitator**.

PRECISION.   1.   The degree of exactness with which a quantity is stated.

2.   The degree of discrimination or amount of detail; e.g., a 3-decimal digit quantity discriminates among 1,000 possible quantities. A result may have more precision than it has accuracy; e.g., the true value of pi to 6 significant digits is 3.14159; the value 3.14162 is precise to 6 Figures, given to 6 Figures, but is accurate only to about 5.

Double precision.   The retention of twice as many digits of a quantity as a given computer normally handles; e.g., if a computer, whose basic word consists of 10 decimal digits is called upon to handle 20 decimal digit quantities, then double precision arithmetic must be resorted to.

Triple precision.   The retention of three times as many digits of a quantity as the computer normally handles; e.g., a computer whose basic word consists of 10 decimal digits is called upon to handle 30 decimal digit quantities.

PRECURSOR.   In biological systems, an intermediate compound or molecular complex present in a living organism which, when activated physiochemically, is converted to a specific functional substance. Sometimes the prefix "pro" is used to indicate that a compound in question plays the role of a precursor. Examples from the history of vitamin and other essential chemical developments include: ergosterol (pro-vitamin D2), which is activated by ultraviolet radiation to form vitamin D; carotene (pro-vitamin A) is a precursor of vitamin A; prothrombin forms thrombin upon activation in the blood-clotting mechanism.

PREDICTION (Statistics).   In general, prediction is the process of forecasting the magnitude of statistical variates at some future point of time. In statistical contexts the word may also occur in slightly different meanings; e.g., in a regression equation expressing a dependent variate y in terms of dependents x's, the value given for y by specified values of x's is called the "predicted" value even when no temporal element is involved.

PREFERENTIAL.   Descriptive of the selectivity of action, either chemical or physiochemical, exhibited by a substance when in contact with two other substances; it may be due either to chemical affinity or to surface phenomena. An example of a preferential chemical combination is that of hemoglobin with carbon monoxide, with which it unites 200 times as readily as it does with oxygen when expose to a mixture of the two. Such phenomena as adsorption, corrosion, and the wetting of dry powders by liquids are other examples.

PREGL, FRITZ (1869–1930).   An Austrian chemist who won the Nobel Prize in 1923. He was also a medical doctor who worked in micromechanical analysis and developed determinations for hydrogen, carbon, nitrogen, and organic groups using micromethods. He was educated at Tubingen, Leipzig, and Berlin.

PREGNANCY (or Cyesis; Gestation).   The condition of being with child. The duration of pregnancy in humans is usually about 280 days, 9 calendar or 10 lunar months, dating from the time of the last menstrual period. Presumptive signs of pregnancy are absence of the menstrual periods, nausea or vomiting in the morning (morning sickness), enlargement of the breasts with pigmentation of the nipples, and enlargement of the

abdomen during the last half of pregnancy. Absolute signs of pregnancy are palpation of the fetal body, movement of the child, and sound of the fetal heart.

Diagnosis of pregnancy in early stages usually depends upon detecting the presence of certain hormones in the urine which will produce characteristic changes in animals into which it is injected.

An extrauterine or ectopic pregnancy is one in which the fertilized ovum lodges and develops outside the uterine cavity. This usually takes place somewhere along the fallopian tubes and, more rarely, free in the abdominal cavity. Such ectopic gestations almost always terminate spontaneously early in the course of the pregnancy, and produce the clinical picture of ruptured ectopic pregnancy. Abdominal pain and shock due to massive hemorrhage into the abdominal cavity make this condition a serious emergency, which demands immediate surgical treatment.

A phantom pregnancy, or pseudocyesis, is an hysterical manifestation in which the abdomen enlarges and resembles the enlargement associated with a true pregnancy. It is treated by treating the underlying psychoneurosis with psychotherapy.

The physiology of pregnancy is discussed in the entries on **Embryo**; and **Embryology**. See also **Artificial Insemination**.

PREHENSION.   The flexion of an appendage to grasp an object by folding around it. There are two types of grasping organs among animals: forcipate and prehensile. The human hand illustrates both. Objects may be taken between the thumb and fingers as between the jaws of a forceps, or they may be grasped between the fingers and the palm by the prehensile folding of the digits. Less versatile organs are capable of one or the other type of action, as in the case of the forcipate chela of the lobster and the prehensile tails of some monkeys.

PREHNITE.   Prehnite is a hydrous silicate of calcium and aluminum, $Ca_2Al_2Si_3O_{10}(OH)_2$, crystallizing in the orthorhombic system. Usual occurrence as intergrown crystals of reniform, stalactitic character, and as rounded groups of such crystals; hardness, 6–6.5; specific gravity 2.90–2.95; luster, vitreous to pearly; color, various shades of light green to gray or white; translucent. Though not a zeolite it is found associated with them and with datolite and calcite, in veins and cavities of basic rocks, sometimes in granites, syenites, or gneisses. It is found in Austria, Italy, the Harz Mountains, France, Scotland, and the Republic of South Africa, where it was originally discovered. Magnificent crystal casts after an unknown mineral have been found in a single large cavity in the basaltic rocks near Bombay, India. In the United States well-known localities are Somerville, Massachusetts; Farmington, Connecticut; Paterson, New Jersey; and Keweenas County, Michigan. Named for Colonel Prehn, its discoverer, who was an early Dutch Governor of the Cape of Good Hope colony.

PRELOG, VLADIMIR (1906–1998).   A Swiss organic chemist who won the Nobel Prize for Chemistry in 1975 along with John W. Cornforth for his research into the stereochemistry of organic molecules and reactions. Although educated in Yugoslavia, he spent many years in Zurich.

PREMIX MOLDING.   A mixture of plastic ingredients prepared in advance of the molding or extruding operation and stored in bags or bins until required. It is made by mixing the components (resin, filler, fibrous materials such as glass and necessary curatives) in a dough blender. Storage life may be from a few days to a year or more, depending on formulation. Such mixtures are then calendered or extruded after warming to suitable temperature.

PREPOLYMER.   An adduct or reaction intermediate of a polyol and a monomeric isocyanate, in which either component is in considerable excess of the other. A polymer of medium molecular weight having reactive hydroxyl and −NCO groups.

See also **Polymers**.

PRESBYOPIA.   Presbyopia is the condition that exists when the natural crystalline lens of the eye loses some of its ability to change shape in order to focus on near objects. Sometime after the age of 40, most people eventually develop presbyopia, usually signaled by a need for reading glasses or bifocal lenses.

The eye functions much like a camera with two lenses. The first lens is the cornea, a clear membrane that covers the front of the eye. The second lens is the eye's natural crystalline lens, which is located behind the pupil. The cornea is responsible for about 70% of the eye's focusing power, while the natural lens "fine-tunes" the image before it is focused on the retina at the back of the eye. The natural lens accomplishes this fine-tuning function by changing shape to accommodate both near objects and those that are further away. Muscles called ciliary muscles are attached to the lens and are responsible for its ability to change shape.

People who are nearsighted may not need corrective lenses for reading as soon as other people do, because their eyes naturally focus more easily on objects that are close. Even when the crystalline lens in a nearsighted eye loses some of its flexibility, the flatter cornea compensates and may continue to offer sharp close-up vision.

Because presbyopia affects near vision, the usual correction is with reading glasses or with bifocals for those who also require distance vision correction. Several innovations in bifocal eyeglasses and contact lenses have occurred in recent years that allow presbyopes (those people with presbyopia) to see better, look better, and be more comfortable. Progressive or "no-line" multifocal eyeglasses graduate from distance to reading power without the noticeable lines that exist in standard bifocals or trifocals. Although progressive lenses may require a greater period of adjustment, these are the most versatile of all multifocal designs because of the continuous range of focus Several versions of multifocal contact lenses are also now available that provide near, intermediate, and distance vision in one contact lens. See also **Vision and the Eye**.

Vision Rx, Inc., Elmsford, NY

PRESERVATIVE.   Any agent that prolongs the useful life of a material. Food products are preserved by; (1) low temperature, (2) ionizing radiation (X and γ rays), (3) antioxidants, (4) fungicides, (5) aldehydes, (6) paints and others.

See also **Additives (Food)**.

PRESSURE.   If a body of fluid is at rest, the forces are in equilibrium or the fluid is in static equilibrium. The types of force that may act on a body are shear or tangential force, tensile force, and compressive force. Fluids move continuously under the action of shear or tangential forces. Thus, a fluid at rest is free in each part from shear forces; one fluid layer does not slide relative to an adjacent layer. Fluids can be subjected to a compressive stress, which is commonly called *pressure*. The term may be defined as force per unit area. The pressure units may be dynes per square centimeter, pounds per square foot, torr, mega-Pascals, etc. Atmospheric pressure is the force acting upon a unit area due to the weight of the atmosphere. Gage pressure is the difference between the pressure of the fluid measured (at some point) and atmospheric pressure. Absolute pressure, which can be measured by a mercury barometer, is the sum of gage pressure plus atmospheric pressure.

Pascal's law states that the pressure in a static fluid is the same in all directions. This condition is different from that for a stressed solid in static equilibrium. In such a solid, the stress on a plane depends upon the orientation of that plane. A liquid in contact with the atmosphere is sometimes called a free surface. A static liquid has a horizontal free surface if gravity is the only type of force acting.

Imagine a body of static fluid in a gravitational field. The mass of the fluid is m (in grams) and the weight of the fluid is mg (as dynes) where g is the local gravitational acceleration. Figure 1 shows a large region of any static fluid with a very small or infinitesimal element. Figure 2 indicates the element in detail. The vertical distance z is measured positively in the direction of decreasing pressure (up); dA is an infinitesimal area; p is the pressure acting on the top surface; and $(p + dp)$ is the pressure acting on the bottom surface. The pressure difference is due only to the weight of the fluid element. Let r represent density, which is mass per unit volume (as grams per cubic centimeter). Thus the weight of the element is $\rho g\,dz\,dA$. Considering the element as a free body, an accounting of forces in the vertical direction gives:

$$dp\,dA = -\rho g\,dz\,dA; \quad dp = -\rho g\,dz \qquad (1)$$

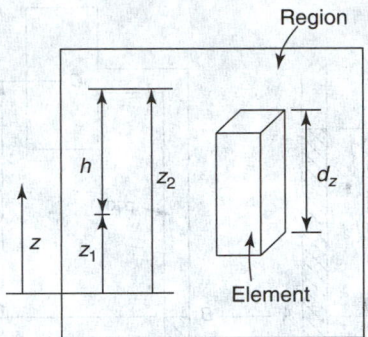

Fig. 1.   Large region of any static fluid.

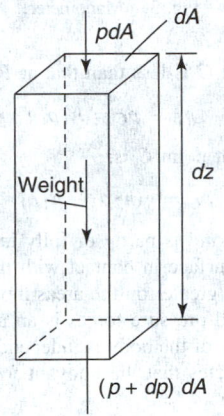

Fig. 2.   Vertical forces on infinitesimal element.

As z is measured positively upward, the minus sign indicates that the pressure increases with an increase in height. This fundamental equation of fluid statics can be applied to all fluids. In integral form, Equation (1) becomes:

$$\int_1^2 \frac{dp}{g} = \int_1^2 dz = -(z_2 - z_1) \qquad (2)$$

where 1 refers to one level and 2 refers to another level. The functional relation between pressure p and the combination ρg must be established before Equation (2) can be integrated. There are two major cases: (a) incompressible fluids, in which the density ρ is a constant; and (b) compressible fluids, in which the density ρ varies.

Liquids can be considered as incompressible in many cases. For small differences in height, a gas might be regarded as incompressible. For an incompressible fluid, with constant g, Equation (2) becomes:

$$p_2 - p_1 = -\rho g\,(z_2 - z_1) \qquad (3)$$

The term $(z_2 - z_1)$ may be called a static "pressure head," and it can be expressed in feet or inches of water, or some height of any liquid. For example, barometric pressure can be expressed in inches of mercury.

A manometer is a device that measures a static pressure by balancing the pressure with a column of liquid in static equilibrium. Many types of manometers are used. See also **Manometer**. The common mercury barometer is essentially a manometer for measuring atmospheric pressure; a mercury column in a glass tube balances the weight of the air above the mercury. Figure 3 illustrates a manometer in which the left leg is open to the atmosphere; the liquid has a specific weight (weight per unit volume) $\rho_2 g$. In the other leg is a liquid of specific weight $\rho_1 g$. Starting with the left leg, the gage pressure p_A is:

$$p_A = h_2 \rho_2 g$$

Since the fluid is in static equilibrium, the pressure p_B at point B equals the pressure at point A. Thus:

$$p_A = p_B = h_2 \rho_2 g$$

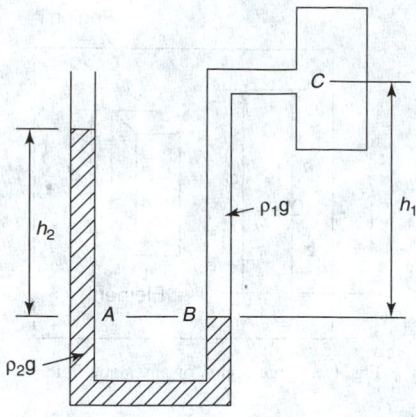

Fig. 3.   Manometer.

The pressure p_C at point C is less than that at B. Thus:

$$p_B - p_C = h_1 \rho_1 g$$

Then the gage pressure at point C is:

$$p_C = g(h_2 \rho_2 - h_1 \rho_1)$$

When a body of any kind is partly or fully immersed in a static fluid, every part of the body surface in contact with the fluid is pressed on by the fluid. The pressure is greater on the areas more deeply immersed. The resultant of all these fluid pressure forces is an upward or buoyant force. The pressure on each part of the body is independent of the body material. Archimedes' principle states that the buoyant force equals the weight of the displaced fluid.

Equation (3) is for the special case of an incompressible fluid. As an example of a compressible fluid, consider an isothermal or constant-temperature layer of gas. The equation of state for such a gas can be written:

$$p = \rho R T_1 \qquad (4)$$

where T_1 is the given absolute temperature and R is a gas constant or gas factor depending upon the gas. Assuming a constant g, Equation (2) gives:

$$\frac{RT_1}{g} \int_1^2 \frac{dp}{p} = -(z_2 - z_1)$$

$$z_2 - z_1 = \frac{RT_1}{g} \log_e \frac{p_1}{p_2} \qquad (5)$$

Equation (5) is sometimes called a "barometric height" relation. For an isothermal atmosphere, a measurement of the temperature T_1 and the static pressure (as with a barometer) at two different levels will provide data for the calculation of the height difference.

Other pressure designations include:

Vacuum. A gage pressure below atmospheric.

Hydrostatic Pressure. The pressure at a point below a liquid surface due to the height of fluid above it.

Tons-on-Ram. The force that acts over a given area as in various types of hydraulic machinery.

Partial Pressure. The pressure exerted by one component in a system, usually one gas or vapor in a mixture.

Internal Pressure. The effect of the attractive forces of the molecules of a substance, which is called pressure because its result is the same as that of an added external pressure. In liquids, its effect appears as the ability of liquids to stand substantial negative pressures without rupture.

Cohesion Pressure. A term in Van der Waal's equation introduced to take care of the effect of molecular attraction. It is usually expressed as a/V^2, where a is a constant and V is the volume of the gas.

Pressure Measurement. Liquid-column elements, such as the manometer, are commonly used for pressure measurement. See also **Manometer**. A variety of diaphragm and other elastic elements is used to measure pressure. A metallic diaphragm element is primarily a device for measuring relatively low pressures. It consists of a single diaphragm

or of one or more capsules connected together, so that upon pressure application, each capsule deflects. The total deflection is the sum of the deflections of all capsules. A variety of bellows elements is similarly used in pressure gages. One of the most common forms of pressure gage makes use of a bourdon-spring element. See also **Bourdon Tube**. Gages for medium-to-high vacuums usually incorporate an electronic type transducer. See also **Vacuum Gages**. Electrical transducers, such as strain gages, moving-contact resistance elements, inductance, reluctance, capacitative, and piezoelectric devices also are used in pressure detection systems. Pressure not only is important as a key variable for direct measurement, but differential pressures are commonly measured in connection with various flowmeters that use a differential-producing element, such as an orifice plate, to measure flow. Manometers and other pressure sensors are also used in liquid-level measuring devices.

High-Pressure Technology. Until the mid-1970s, the limit to most high-pressure experimentation was confined to about 300 kilobars.

As of 1988, the maximum pressure created in the laboratory by the diamond anvil pressure cell approximates 5 million atmospheres. Theoretical estimates, however, forecast that diamond is stable up to 23 million atmospheres with respect to any phase transition. Although plastic deformation would limit its capability, predictions for the diamond anvil cell are for pressures somewhere between 5 and 23 million atmospheres. See also **Diamond Anvil High Pressure Cell**.

PRESSURE GRADIENT (Meteorology).   The rate of decrease (gradient) of pressure in space at a fixed time. The term is sometimes loosely used to denote simply the magnitude of the gradient of the pressure field. Also called *barometric gradient*.

PRESSURE HEAD.   Pressure in head units, for example, meters of fluid, equal to the force per unit area divided by the product of the density of the fluid and the acceleration due to gravity. It is the depth of fluid that would exert an equivalent pressure.

PRESSURE ICE.   Sea ice (or river ice or lake ice) that has been deformed or altered by the lateral stresses of any combination of wind, water currents, tides, waves, and surf. This may include ice pressed against the shore, or one piece of ice upon another. Its two major forms are rafted ice and tented ice, which, individually or in combinations, may form pressure ridges or hummocked ice.

Hummocked Ice. Pressure ice characterized by haphazardly arranged mounds or hillocks ("hummocks"). This has less definite form than rafted ice or tented ice, but in fact may develop from either of those as melting, sublimation, or drifting changes the sharper ice edges into more rounded shapes.

Rafted Ice. Deformed sea ice in which one piece has overridden another. Also called *telescoped ice*. See also **Sea Ice**.

Tented Ice. Pressure ice in which two floes have been pushed into the air, leaving an air space underneath.

AMS

PRESSURE MELTING.   The melting of ice due to applied pressure. The melting point of pure ice is lowered 0.0072 K per atmosphere of applied pressure. Pressure melting is responsible for regelation. See also **Regelation**.

PRESSURE MODULATED RADIOMETER (PMR).   See **Nimbus Satellite Program**.

PRESSURE SPECTRUM.   Within the inertial subrange, a spectrum of the form

$$S_p(k_1) = c \varepsilon^{4/3} k_1^{-7/3},$$

where ε is dissipation of turbulence energy and k_1 is wavenumber in longitudinal direction. At smaller wavenumbers the pressure spectrum increases considerably due to mesoscale motions and synoptic weather systems.

PRESSURE SUIT.   A garment designed to provide pressure upon the body so that respiratory and circulatory functions may continue normally, or nearly so, under low-pressure conditions, such as occur at high altitudes or in space without benefit of a pressurized cabin. A pressure suit is distinguished from a pressurized suit, which inflates, although it may be fitted with inflating parts that tighten the garment as ambient pressure decreases.

PRESSURE TENDENCY.   The character and amount of atmospheric pressure change during a specified period of time, often a three-hour period preceding an observation. Also called *barometric tendency*.

Pressure tendency is composed of two parts, the pressure change and the pressure characteristic. The pressure change is the net difference between pressure readings at the beginning and ending of a specified interval of time. The pressure characteristic is an indication of how the pressure has been changing during that specified period of time, for example, decreasing then increasing, or increasing and then increasing more rapidly. See also **Tendency**.

AMS

PRESSURE WAVE.   1.   In meteorology, a short period oscillation of pressure such as that associated with the propagation of sound through the atmosphere; a type of longitudinal wave. These waves are usually recorded on sensitive microbarographs capable of measuring pressure changes of amounts down to 10E-4 millibar. Typical values for the period and wavelength of pressure waves are 1/2 to 5 seconds and 100 to 1500 meters, respectively. Pressure waves produced by explosions in the upper atmosphere are of value in determining the high-altitude temperatures and winds.

2.   A wave or periodicity which exists in the variation of atmospheric pressure on any scale, usually excluding normal diurnal and seasonal trends. Such waves can persist for an indefinite length of time only if they coincide approximately with the free oscillations of the atmosphere. Waves of a period longer than that associated with the passage of large-scale weather disturbances are difficult to isolate, since they usually have such a small amplitude that they can be extracted from the data and only by means of precise statistical methods.

PRESTRESSED CONCRETE.   Prestressed concrete resulted from the desire to overcome the disadvantage of the low tensile strength of concrete. See also **Concrete**. By means of high strength steel wires, cables, or rods — particularly the first two — a concrete member is pre-compressed. Then, when the structure receives its load, the compression is relieved on that portion which would normally be in tension. Thus, a beam is prestressed so that under load the concrete on the side normally in tension has no tensile forces acting on it.

There are two general methods of prestressing, namely pretensioning and post-tensioning. Pretensioning consists of pouring concrete around wires kept under tension until the concrete has gained sufficient strength. The wires are then cut, and compressive forces are thereby imparted to the concrete through bond between the steel and concrete. Post-tensioning consists of jacking bond-free cables against the ends of an already hardened concrete section, and then anchoring the ends of the cables.

Prestressed concrete structures require less concrete and steel, but the steel used is a more costly high-strength wire. Prestressed concrete is especially well adapted to combination with precast concrete.

PREVOST LAW OF EXCHANGES.   In an evacuated enclosure, with walls maintained at constant temperature, objects within will reach a condition of thermal equilibrium at which they will attain, and remain at, the temperature of the walls. Each body is constantly exchanging heat energy with its surroundings, the net result of which exchange tends to equalize the temperature of the body and its surroundings. Cold bodies radiate less heat than they receive from warmer surroundings and thus rise in temperature to the equilibrium value.

PRICHARD, JAMES COWLES (1786–1848).   James Prichard was a British physician and anthropologist who developed a model of human racial formation based on sexual selection and ideas of beauty.

Born in Ross, Herefordshire, Prichard grew up in Bristol, whose port provided him the opportunity of observing individuals from all over the world. Determining on a medical career, he was briefly apprenticed, spent a year studying in London and in 1806 enrolled in the medical school at the University of Edinburgh, where his 1808 MD thesis was on the subject of human racial formation. This developed into his first major publication, *Researches into the Physical History of Mankind* (1813). An expanded second edition of this work was published in 1826 and a five-volume third edition appeared between 1838 and 1847.

Although raised a Quaker, Prichard became an evangelical Anglican during a period of further study at the University of Oxford. Much of his anthropological work had as its principal aim the demonstration that all human beings had descended from a single set of parents, viz. Adam and Eve. Drawing on earlier classification of the varieties of human beings by J. F. Blumenbach, Prichard abandoned traditional environmental explanations of the origin of different racial characteristics. Instead, he adopted a model of "hard" heredity, whereby connate attributes are passed on from generation to generation. He equated light skin with higher degrees of civilization. Adam and Eve, he argued, were dark skinned, civilization producing lighter skins through ideas of beauty and sexual selection. His work also summarized contemporary evidence for the distribution of plants and animals, as well as human varieties. See also **Blumenbach, Johann Friedrich (1752–1840)**.

In his later ethnological writings, Prichard largely abandoned his emphasis on hard heredity in favor of a more environmental framework. He also stressed language affinities rather than physical characteristics as a better guide to the historical associations of human groups. He was a linguist with a European reputation, although the Biblical framework within which he placed it sometimes strained his work. He also wrote a monograph on Egyptian mythology.

Prichard practiced as a physician in Bristol for most of his life. He also wrote on medical topics, including nervous diseases, vitalism and fevers. His *Treatise on Insanity* (1835) introduced the concept of moral insanity. In 1845 he was appointed one of the Commissioners in Lunacy. He moved to London and died from a fever contracted while inspecting an asylum in 1848.

Additional Reading

Augstein, H.F.: *James Cowles Prichard's Anthropology*, Rodopi, Amsterdam, The Netherlands, 1999.
Stocking, G.W.: *James Cowles Prichard's Researches into the Physical History of Mankind*, University of Chicago Press, Chicago, IL, 1973.

W. F. Bynum, Wellcome Trust Centre for the History of Medicine at UCL, London, UK

PRIESTLEY, JOSEPH (1733–1804).   Priestley was an English chemist who researched relationships among plants, air, and animals. After meeting Benjamin Franklin he became interested in science and the two men became lifelong friends. Priestley started doing chemical experiments as a hobby, but it soon became a passion. He had little scientific education but his observations were very keen.

Priestley lived near a brewery and his curiosity about how it operated and about the gases involved lead him to discover a gas (carbon dioxide) was heavier than air. He found water and this heavy "air" made a great drink and in 1773 he was awarded a medal by the Royal Society for his invention of soda water. In 1774, he announced the results of his experiment, which described the unusual properties of a new "air", this was in fact, the discovery of oxygen. His experiments with "air" and gases were important for leading to the first ballooning flights.

Priestley also researched relationships among plants, air, and animals. He observed the respiration of plants, by which they take in carbon dioxide and produce oxygen. His observation helped others understand the process. He observed "green matter", which now we know as photosynthesis.

He was a strong religious and political leader and was persecuted for his support of the American Revolution. He came to America in 1794 and spent his last years experimenting in his laboratory.

See also **Balloon**; and **Oxygen**.

J. M. I.

PRIGOGINE, IIYA (1917–2003). A Belgian chemist who won the Nobel Prize for chemistry in 1977 for his contributions to nonequilibrium thermodynamics particularly the theory of dissipative structures. The main theme of the scientific work of Iiya Prigogine has been a better understanding of the role of time in the physical sciences and in biology. He has contributed significantly to the understanding of irreversible processes, particularly in systems far from equilibrium. His education was at the University of Brussels. The Center for Statistical Mechanics and Thermodynamics at the University of Texas bears his name. http://order.ph.utexas.edu/research/glimpse.html

PRILLS. Small, round, or acicular aggregates of a material, usually a fertilizer, that are artificially prepared. In the explosives field, prills-and-oil consists of 94% coarse, porous ammonium nitrate prills and 6% fuel oil.

PRIMARY AMEBIC MENINGOENCEPHALITIS. Free-living amebae (*Naegleria* and *Acanthameba* sp.), ubiquitous in warm fresh water ponds and lakes, are the etiologic agents of this rare, rather recently identified disease. The infection is acquired by swimming in (and probably inhaling) infested waters. The amebae enter through the nasal mucosa, penetrating to the brain via the olfactory nerve sheaths, producing an acute purulent meningitis, and then spread rapidly beyond the leptomeninges to brain tissue, where they produce an acute hemorrhagic, necrotizing encephalitis.

The incubation period is approximately five days, when nuchal rigidity, headache, pyrexia, and nausea occur and progress to bizarre behavior, coma, and death within 4 to 7 days. Amebae with large nuclei are found in the brain tissue at autopsy.

Infections have been seen in South Australia, Eastern Europe, and the United States (Florida and Virginia). Patients are typically young children and low-age adults. Treatment is generally unsuccessful, but one patient in the United States and one in Australia survived with heroic treatment involving intrathecal amphotericidin.

R. C. V.

PRIMARY BODY. The celestial body or central force field about which a satellite or other body orbits, or from which it is escaping, or towards which it is falling. The primary body of the moon is the earth; the primary body of the Earth is the sun.

PRIMARY CIRCULATION. The prevailing fundamental atmospheric circulation on a planetary scale that must exist in response to 1) radiation differences with latitude, 2) the rotation of the earth, and 3) the particular distribution of land and oceans; and that is required from the viewpoint of conservation of energy.

Primary circulation and general circulation are sometimes taken synonymously. They may be distinguished, however, on the basis of approach; that is, primary circulation is the basic system of winds, of which the secondary and tertiary circulation are perturbations, while general circulation encompasses at least the secondary circulations.

See also **Macrometeorology**.

AMS

PRIMARY COLORS. Three colors which, when suitably mixed, will produce all the other colors, as well as white and black. The colors generally used are an orange-red, green and blue-violet. These colors are sometimes called the additive primaries to distinguish them from the three subtractive or minus-colors, cyan, magenta, and yellow.

PRIMARY ELEMENT. In an instrumentation or automatic control system, the primary element is that system element that quantitatively converts measured variable energy into a form suitable for measurement. For transmitters not used with external primary elements, the sensing portion is the primary element. Other terms used for the same function include sensor and detector. See also **Sensor (Measurement)**.

PRIMARY HYPERALDOSTERONISM. See **Adrenal Disease**.

PRIMARY RADAR. Radar using reflection only, in contrast with secondary radar which uses automatic retransmission on the same or a different radio frequency.

PRIMARY RAINBOW. A rainbow that is distinguished from other rainbows by its angular radius, color order, and brightness. This bow is seen between about 40° and 42° from the antisolar point (shadow of the observer's head) or equivalently, between 140° and 138° from a light source (such as the sun). Reds are found to the outside of the bow (closest to the sun) with the blues to the inside. The primary bow is usually brighter than any of the other bows. The primary rainbow is certainly the most frequently noticed bow, but the purity and range of its colors fall a long way short of that assumed by the popular dictum: all the colors of the fall. Frequently accompanying the primary bow are the secondary bow (lying about 8° outside the primary bow) and the supernumerary bows (immediately inside the primary bow, and often confined to the upper portions of the arc). Infrequently seen are the reflection bows. A theory of the bow that approximates the behavior of light as a ray is able to account for the difference in position and color order of the primary and secondary bows. In this theory, the position of each bow is determined by the minimum angle of deviation of the light passing through a drop. The difference is that the light that forms the primary bow has undergone one internal reflection, while the light that forms the secondary bow has undergone two internal reflections. This is a useful approximation to reality, but it fails to capture many important features of observable bows. See also **Atmospheric Optical Phenomena**; and **Secondary Rainbow**.

PRIMATES. Most primates are arboreal animals and all species either grasp by opposing the thumb to the fingers or show similarity to grasping appendages of this type in the anatomy of the hand. With exception of the marmosets, all primates have nails on at least part of the digits. The lemurs retain a claw on the second toe and the marmosets have a nail only on the great toe. The brain is more highly developed in the primates than in any other animals. The major categories of mammals are described in this volume. See Table 1 for general organization of the primates and for reference to other entries in this encyclopedia.

PRIME MERIDIAN. 1. The meridian of longitude 0 degrees, used as the origin for measurement of longitude. The meridian of Greenwich, England, is almost universally used for this purpose.

2. Any meridian in any coordinate system used as an origin for measurement of longitude.

PRIME NUMBER. See **Number Theory**.

PRINCIPAL COMPONENTS (Statistics). Given a multivariable complex with variables $x_1, x_2, \ldots, x_n$ it is possible to transform to n new variables which are (1) linear functions of the x's and (2) uncorrelated among themselves. These variables are called principal components. Except in degenerate cases they are unique apart from sign. The principal components have optimal properties in the sense that one will have the largest possible variance among linear functions of the original x's, a second will have the largest variance among linear functions which are uncorrelated with the first, and so on.

The determination of the principal components depends on the calculations of the eigenvalues of the variance-covariance matrix of the x's.

Principal component analysis is to be sharply distinguished from factor analysis, to which it bears some formal resemblance. Principal components are simple variate-transformations; factor analysis imposes a model on the data.

PRINCIPAL PLANES (Lens). Two planes so located in a thick lens or lens system such that if object distances are measured from the first principal plane and image distances are measured from the second principal plane, the thin-lens formula will hold. The intersections of the principal

TABLE 1. PRIMATES (Top Mammals)

	In This Encyclopedia
TUPAIOIDS	**See Moles and Shrews.**
Tree-Shrews (*Tupaia*, . . .)	
Pen-tailed Tree-Shrews (*Ptilocercus*)	
LORISOIDS	**See Lorisoids.**
Lorises (*Lorisidae*)	
The Slow Loris (*Nycticebus*)	
The Slender Loris (*Loris*)	
Pottos (*Periodicticus*)	
The Angwantibo (*Arctocebus*)	
Bush-Babies (*Galagidae*)	
Common Bush-Babies (*Galago*)	
Pigmy Bush-Babies (*Galagoides*)	
Needle-clawed Bush-Babies (*Euoticus*)	
LEMUROIDS	**See Lemur.**
The Aye-Aye (*Daubentoniidae*)	
Small Woolly Lemurs (*Cheirogalaginae*)	
Mouse-Lemurs (*Microcebus*)	
Dwarf Lemurs (*Cheirogaleus* and *Phaner*)	
Large Woolly Lemurs (*Lemurinae*)	
Weasel-Lemurs (*Lepilemur*)	
Gentle Lemurs (*Hapalemur*)	
Common Lemurs (*Lemur*)	
—Ruffled Lemur	
—Black Lemur	
—Brown Lemur	
—Mongoose-Lemurs	
—Red-bellied Lemur	
Silky Lemurs (*Indriidae*)	
Sifakas (*Propithecus*)	
The Avahi (*Lichanotus*)	
The Indri (*Indri*)	
TARSIOIDS	**See Tarsioids.**
HAPALOIDS	**See Marmoset.**
Marmosets (*Callithricidae*)	
Pigmy Marmosets (*Cebuella*)	
Maned Marmosets (*Leuntocebus*)	
Plumed Marmosets (*Callithrix*)	
Ruffed Marmosets (*Hapale*)	
Bald Marmosets (*Marikina*)	
White Tamarins (*Mico*)	
Black Tamarins (*Tamarin*)	
Moustached Tamarins (*Tamarinus*)	
Pinches (*Oedipomidas*)	
Goeldi's Marmoset (*Callimiconinae*)	
Titis (*Callicebinae*)	
CEBOIDS	**See Monkeys and Baboons.**
Half-Monkeys (*Pithecinae*)	
Douroucoulis (*Aotes*)	
Sakiwinkis (*Pithecia*)	
Bearded Sakis (*Chiropotes*)	
Uacaris (*Cacajao*)	
Hand-Tailed Monkeys (*Cebinae*)	
Squirrel-Monkeys (*Saimiri*)	
Capuchin Monkeys (*Cebus*)	
Woolly Monkeys (*Lagothrix*)	
Woolly Spider Monkeys (*Brachyteles*)	
Spider Monkeys (*Ateles*)	
Howler Monkeys (*Alouatta*)	
SIMIOIDS	**See Monkeys and Baboons.**
Colobine Monkeys (*Colobinae*)	
Guerezas (*Colobus*)	
Languars (*Presbytis*, . . .)	
Snub-nosed Monkeys (*Rhinopithecus*)	
Proboscis Monkey (*Nasalis*)	
Long-tailed Monkeys (*Cercopithecinae*)	
Guenons (*Cercopithecus*)	
Allen's Swamp Monkey (*Allenopithecus*)	
Military Monkeys (*Erythrocebus*)	
Mangabeys (*Cercocebus*)	
Dog-faced Monkeys (*Cynopithecinae*)	
—Barbary Ape	
—Rhesus	
—Bearlike Monkey	
The Black Ape (*Cynopithecus*)	
Baboons (*Papio*)	

(continued)

TABLE 1. (Continued)

	In This Encyclopedia
The Gelada (*Theropithecus*)	
Drills (*Mandrillus*)	
ANTHROPOIDS	**See Anthropoids.**
Lesser Apes (*Hylobatidae*)	
The Siamang (*Symphalangus*)	
Gibbons (*Hylobates*)	
Greater Apes (*Pongidae*)	
Gorillas (*Gorilla*)	
The Chimpanzees (*Pan*)	
Orangutans (*Pongo*)	
Humans (*Hominidae*)	**See Mammals.**

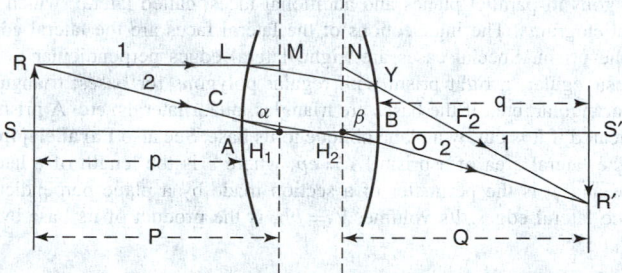

Fig. 1. H_1 and H_2 are the principal points; $M H_1$ and $N H_2$ the principal planes of a thick lens. In locating images by a graphical method, principal rays may be used as for thin lenses if the incident light falls on an imaginary thin lens coincident with the first principal plane, the lens being shifted to coincide with the second principal plane for emergent light.

planes with the optical axis are the principal points of the lens or system of lenses. See Fig. 1.

PRINTED CIRCUITS AND CIRCUIT BOARDS. A printed circuit is any circuit formed by depositing a conducting material on the surface of an insulating sheet. This may be achieved by the use of electrically conducting ink, electroplating, or other methods.

Circuit boards represent a mid-phase in the progress of electronic (and some electric) circuitry—ranging from the "spaghetti" wiring that was characterized by individually soldered or wire-wrapped connections and stuffed into the bottom of a chassis (as commonly encountered in radio receivers, control instruments, and numerous other equipment of just a few decades ago), to the highly sophisticated integrated circuits of the present era. See also **Integrated Circuit (IC)**. In the pre-circuit board days, components were mounted and interwired individually. The appearance and ease of trouble-shooting ranged from chaotic and difficult to rather neat and easy, depending upon care exerted by the designers and assemblers. Because assembly was essentially manual, sufficient working space for manipulating components and connections had to be made. Thus, equipment, on present standards, was excessively heavy and bulky. Earlier wiring was simplified to quite a degree by means of wire harnesses.

As components became smaller with the entry of solid-state devices and as economic pressures toward automating assembly became greater, the concept of circuit boards and printed circuitry received a rather rapid response. During the interim, circuit boards (sometimes called circuit modules, cards, etc.) have greatly improved. Multilayer circuitry helped to increase the component density of boards. Both the materials and technology for making circuit boards have improved at a steady rate.

Coombs listed a number of advantages of printed circuits over former conventional loose wiring methods, including: (1) weight reduction by as much as 10 to 1; (2) better control over volume required for wiring; (3) cost savings as the result of standardization and automation in production; (4) increased reliability mainly as the result of reducing human errors; (5) easier inspection and troubleshooting; and (6) easier part identification. Disadvantages pointed out include: (1) difficulty in repairing; (2) heat dissipation problems; (3) design regimentation and restriction, making it difficult to make improvements in circuitry design, among others.

The design and production of circuit boards is volume sensitive in terms of production costs and time.

Single-board computers appeared during the mid-1970s and created a large expansion in the use of circuit boards. Among boards required were analog input and output boards, core memory boards, digital input/output boards, floppy disk controller boards, hard disk interface boards, keyboard controller boards, math boards, optical isolator boards, PROM boards, RAM boards, synchro-to-digital boards, and so on. One of the major problems that arose, as the result of so many suppliers, was the lack of a standard data bus, a problem that is now approaching resolution.

There is a considerable blending of circuit board and integrated circuit production technology, and much of the latter technology has been transferred from earlier experience in producing simpler circuit boards.

PRISM (Mathematics). A polyhedron, with two bases which are equal polygons in parallel planes and additional faces, called lateral, which are parallelograms. The intersections of the lateral faces are the lateral edges of the prism. Special cases are: right, lateral edges perpendicular to the bases; regular, a right prism with regular polygons for bases; triangular, quadrangular, etc., if the bases are triangles, quadrilaterals, etc. A prism is truncated if it is cut by a plane oblique to its base. See also **Parallelepiped**.

The lateral area of a prism, $A = ep$, where e is the length of a lateral edge and p is the perimeter of a section made by a plane perpendicular to the lateral edges. Its volume, $V = bh$, is the product of its base by its altitude.

PRISM (Optics). A transparent solid, cut at precise angles for various optical purposes. Incident light may pass directly through a prism, or may emerge after one or more internal reflections; in some cases the light is polarized.

The common triangular prism, familiar in older forms of spectroscope, receives light upon one face and passes it through another after two refractions, resulting in a total deviation Δ dependent upon the angle of the prism and its refractive index for the light used. See Fig. 1. If the light is incident at angle i on the first prism face, and if the prism angle is α and the refractive index is n, the total deviation after passage through the prism in a plane at right angles to the prism edge is given by

$$\Delta = i - \alpha + \arcsin\left[n \sin\left(\alpha - \arcsin\frac{\sin i}{n}\right)\right]$$

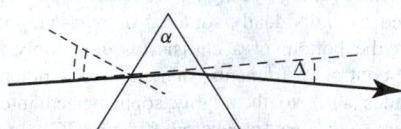

Fig. 1.   Ray of light passing through triangular prism.

A little experimenting shows that this deviation has a minimum value when the light traverses the prism symmetrically, entering and emerging at the same angle with the corresponding faces. This angle of minimum deviation is easily shown to be

$$\Delta_{\min} = 2\arcsin\left(n\sin\frac{1}{2}\alpha\right) - \alpha$$

from which the refractive index may be obtained, by experiment, as

$$n = \frac{\sin\frac{1}{2}(\Delta_{\min} + \alpha)}{\sin\alpha/2}$$

The effect of a prism on heterogeneous light may be deduced from the formulas for refractive dispersion. Prisms are used in binoculars, in monochromatic illuminators, and in many other optical instruments.

The *Amici* prism is a *direct-vision* prism, that is, a prism combination by which a beam of light is dispersed into a spectrum without mean deviation. Such prisms are sometimes used in direct-vision spectroscopes.

The principle will be clear from the following example. Assume an inverted prism of crown glass with an angle of 40°, used with an erect prism of flint glass. See Fig. 2. Yellow sodium light (5,893 Å) is deviated by the

crown-glass prism through +22°32' (upward). For the flint-glass prism to produce an equal negative (downward) deviation it must have an angle of 33°40'. (Each prism is supposedly set for minimum deviation.) Together they produce no deviation for this wavelength. But if light of 7,682 Å (red) is used, the deviation of the crown glass is +22°16', while that of the flint is −22°10', giving a net deviation of +6'. And for the wavelength 4,047 Å (violet), the deviations are, respectively, +23°12' and −23°44', giving −32'. There is thus, between the ends of the visible spectrum, a separation of 38'. Additional pairs of prisms may be used to increase the dispersion.

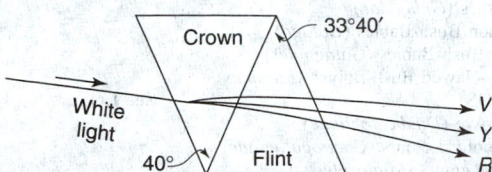

Fig. 2.   The deviation due to the crown-glass prism is neutralized by that of the flint-glass prism, for the middle of the yellow spectrum only.

A *Littrow* prism is a 30–60–90-degree prism silvered on the side opposite the 60° angle. A single lens can then be used as both collimator and telescope.

A *Dove* prism has the property of inverting a beam of light. Three faces are polished, and the size and index of refraction must be properly correlated if the beam is not to be displaced laterally. Rotation of the prism about the axis of the beam rotates the beam at twice the rate of rotation of the prism. See Fig. 3.

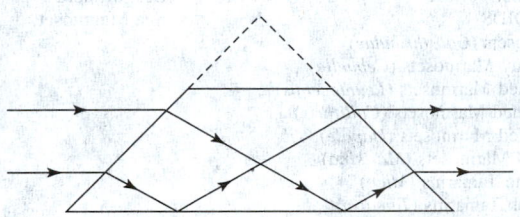

Fig. 3.   Dove prism.

A *constant-deviation* prism refracts any required wavelength of light in a specified direction to the incident beam. An example is the *Pellin-Broca* prism consisting of two 30° prisms, connected by a 45° total-reflecting prism. See Fig. 4.

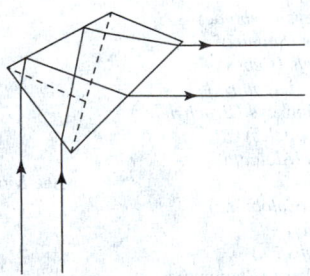

Fig. 4.   Pellin-Broca prism.

A total-reflecting prism is arranged to provide total reflection of light incident upon it in a specified direction. Most such prisms are 45–45–90 prisms in which the light enters normally to a face opposite to a 45° angle, is totally reflected by the hypotenuse face, and leaves by the third face, thus having been totally reflected through 90°. The beam of light must be sufficiently parallel so that all of the light strikes the reflecting face outside of the critical angle, or the reflection will not be total. The *Porro* prism is of this type, and two of them are used in each telescope of prism binoculars. The *roof*-prism is a total-reflection prism in which the surface opposite the right angle has been replaced by two surfaces at right angles (roof), with their common element parallel to the hypotenuse of the triangle. The roof prism turns the beam through 90° like a total-reflection prism, and also inverts the beam like a Dove prism.

The *Nicol* prism is one of the best-known devices for producing plane-polarized light. It consists of two pieces of Iceland spar (pure calcium carbonate) cut as shown in Fig. 5. The optic axis of each is approximately indicated by the double arrow, and they are cemented together with colorless Canada balsam along the plane MN. If the incident beam IP is unpolarized, it suffers double refraction at P, dividing into an ordinary component PO' and an extraordinary component PE'. The refractive index of Iceland spar for the ordinary ray (sodium light) is 1.658 and for the extraordinary it is 1.486, while that of Canada balsam for both is 1.53. The ordinary ray therefore encounters at O' a less refractive medium, and, the incidence being at an angle larger than the critical angle, it is totally reflected ($O'O$); while the extraordinary ray incident at E' encounters a more refractive medium, therefore cannot suffer total reflection and most of it passes on along $E'Q$, emerging along QE completely plane-polarized with its vibration plane in the plane of the paper. Modifications of this prism, having different shapes and using other cements, have been designed for special purposes. However, Nicol prisms cannot be used in ultraviolet light as the Canada balsam is not transparent to these shorter wavelengths. The *Rochon* and *Wollaston* prisms are made of quartz or calcite and "cemented" with glycerine or castor oil. See Fig. 6.

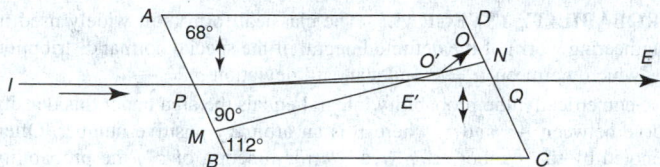

Fig. 5.   Nicol prism. Double-headed arrows indicate direction of optic axis.

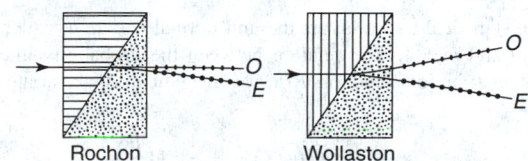

Rochon   Wollaston

Fig. 6.   Rochon and Wollaston prisms.

The direction of the optic axis of each crystal is indicated by the shading. The Rochon prism transmits the ordinary ray without deviation, the ray being achromatic. The light should travel through the prisms in the directions indicated.

A *Cornu-Jellet* prism is made by splitting a Nicol prism in a plane parallel to the direction of vibration of the transmitted light and removing a wedge-shaped section. When the two pieces are joined together again, the planes of vibration of the light transmitted by the two halves make a small angle with each other.

A *Cornu-double* prism is designed to utilize the ultraviolet-transmitting properties of quartz without introducing its double refraction. This is accomplished by cementing together two 30° prisms, one of right-handed and the other of left-handed quartz.

A *corner cube* prism (trihedral), shown in Fig. 7, is used to reflect a beam back on itself very precisely in two dimensions by means of two total internal reflections. Typical uses are the coincident alignment of laser systems and reticles; for determination of the infinity focus for autocollimators or other telescopic equipment; and as a reference for indirect sighting and triangulation systems.

A *polarizing* prism, shown in Fig. 8, utilizes uniaxial birefringent crystals of quartz or calcite, which are double refracting, so that two separate beams are created (ordinary and extraordinary). Each beam is totally polarized and orthogonal with respect to the other. In most designs, one beam exits from the prism in a direction parallel to the light beam

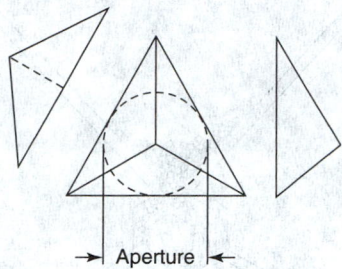

Fig. 7.   Corner cube prism (tribedral).

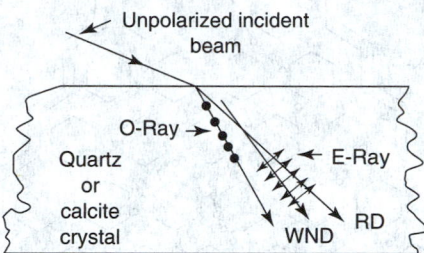

Fig. 8.   Polarizing prism.

that enters, while the other beam emerges obliquely or impinges on a light-absorbing blackened face.

A *Glan air-spaced prism* (Glan-Taylor), shown in Fig. 9, is designed for use where maximized transmission is desired throughout the spectral range of the calcite being used. Increased transmission occurs because of the special angle of the optic axis and the resulting incidence of the transmitted beam at about Brewster's angle at the air-spaced interface. The prism has exceptional performance when extinction is required. Two crossed Glan air-spaced prisms will transmit less than 10^{-5} of the incident beam. For high-power laser application, one of the beams is allowed to exit from the top face of the prism (double-beam type) rather than being absorbed (single-beam type).

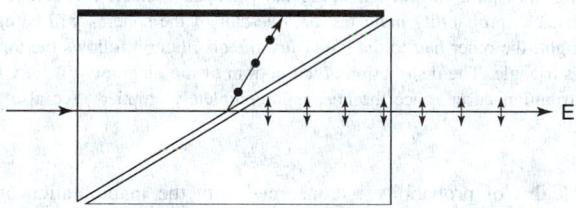

Fig. 9.   Glan air-spaced prism. (*Glan-Taylor.*)

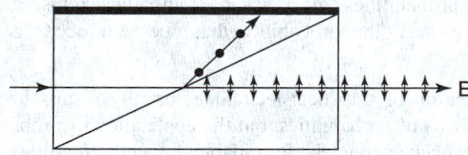

Fig. 10.   Glan Thompson prism.

A *Glan Thompson prism*, shown in Fig. 10, is a polarizing prism that finds use in high-precision instruments. Since the elements of the prism are cemented together, the prism should not be used in the path of high-power beams. The prism has a wide field of view as compared with the Glan air-spaced prism.

PROBABILITY.   Suppose that a trial may have one of two mutually exclusive outcomes — say "success" and "failure" — and that we associate with it a number p, $0 \le p \le 1$, called the probability of a success.

Fig. 1.   Demonstration of probability. A Galton Board, named after Sir Francis Galton who constructed the first device of this kind, can be used to illustrate Gaussian (normal or bell-shaped) distribution. Tiny spheres that roll easily are placed in an upper hopper. They fall by gravity to the receiving slots below, but in falling must pass an array of hexagonally shaped obstacles. Theoretically, at each obstacle, probability indicates that one-half of the spheres will be diverted to the right; the other half to the left. Thus, the distribution follows the format of Pascal's triangle. The distribution at the bottom of the diagram will fully fill out the distribution curve, once the hopper is completely emptied. See also **Pascal Triangle**.

The calculus of probability is concerned with the manipulation of such numbers, and is largely based on the following results:

1. If two trials with probabilities p_1, p_2 are independent, in that the outcome of either has no effect on that of the other, the probability that both will result in successes is $p_1 p_2$.
2. If the probabilities of a set of mutually exclusive events are $p_1, p_2, p_3, \ldots$, the probability that one will occur is $p_1 + p_2 + p_3 + \cdots$.

The application of statistical techniques usually results in a statement couched in terms of probabilities, but the application of probability theory to the real world encounters formidable logical difficulties. Two main schools of thought may be distinguished.

1. In the frequency theory, the trial is supposed capable of indefinite repetition, either real or hypothetical, and the probability is identified with the limiting frequency of successes in this infinite sequence. However, the existence of a limiting value is difficult to establish theoretically, and impossible to establish empirically because of the infinite length of the sequence.
2. The other school identifies probability with the strength of belief that a success will occur. The main difficulty here is the subjective nature of the definition.

For demonstration of probability, see Fig. 1.

PROBABILITY DENSITY FUNCTION (or Density Function).   The statistical function that shows how the density of possible observations in a population is distributed. Also called *frequency function*.

It is the derivative $f(x)$ of the distribution function $F(x)$ of a random variable, if $F(x)$ is differentiable. Geometrically, $f(x)$ is the ordinate of a curve such that $f(x)dx$ yields the probability that the random variable will assume some value within the range dx of x. The density function is nonnegative, and its total integral is unity. Sometimes the probability density function is called the distribution function, but this practice causes confusion and is not recommended.

AMS

PROBABILITY DISTRIBUTION.   The mathematical description of a random variable in terms of its admissible values and the probability associated, in an appropriate sense, with each value.

The probability distribution of a continuous variate is defined by stating the mathematical equation of the distribution function $F(x)$ or the probability density function $f(x)$ (if it exists) together with the range over which the equation holds. The probability distribution of a discrete variate is commonly defined by stating the equation for the probability $p(x)$ that the variate will assume any particular value x, and indicating what values are possible.

AMS

PROBABILITY INTEGRAL.   The classical form (still widely used in engineering work) of the definite integral of the special normal distribution for which the mean $\mu = 0$ and standard deviation $\sigma = \frac{1}{\sqrt{2}}$.

Geometrically, the probability integral equals the area under this density curve between $-z$ and z, where z is an arbitrary positive number. Often denoted by the symbol erf z (read "error function of z") the probability integral is defined thus:

$$\operatorname{erf} z \equiv \frac{2}{\sqrt{\pi}} \int\limits_{0}^{z} e^{-x^2} dx.$$

Modern statistical usage favors the unit normal variate u, which is such that $\mu = 0$ and $\sigma = 1$. The relation between the probability integral erf z and the distribution function $F(u)$ of the unit normal variate u is as follows:

$$u \text{ positive}: F(u) = \frac{1}{2} + \frac{1}{2}\operatorname{erf}\left(\frac{u}{2}\right),$$

$$u \text{ negative}: F(u) = \frac{1}{2} - \frac{1}{2}\operatorname{erf}\left(\frac{-u}{2}\right).$$

See also **Unit Normal Distribution**.

AMS

PROBABLE ERROR.   In the theory of observational error, the so-called probable error of an observation was defined as $\pm 0.6745\sigma$ where σ is the standard deviation of the distribution from which the observation was drawn. The coefficient derives from the fact that if the population is normal (Gaussian) a deviation of $\pm 0.6745\sigma$ from the mean covers half the distribution. Likewise, in a sample of n observations the probable error of the mean was $\pm 0.6745\sigma/\sqrt{n}$.

In statistics the expression is obsolete and has been replaced by the standard error, which is simply the standard deviation of the sampling distribution of the statistic concerned. Thus the standard error of the mean is $\sigma/\sqrt{n}$. Significance is customarily attributed to the difference between an observed value of a statistic and the parameter which it estimates if that difference exceeds twice (or more stringently, three times) the standard error.

Sir Maurice Kendall, International Statistical Institute, London

PROBIOTICS AND PREBIOTICS.   Functional foods are generally described as foods that provide some health-benefits beyond traditional nutritional values. For example, probiotics and/or prebiotics may be added to foods to increase functionality. The Greek translation of **Probiotic** is "for life." A probiotic is a microorganism in a food or supplement containing live microorganisms that are present in sufficient numbers to actively enhance consumers' health by improving the balance of microflora or microorganisms in the gastrointestinal tract. Today probiotic bacteria such as *Lactobacillus* and *Bifidobacterium* are added to fermented foods and other foodstuffs [Aubertin, Ref.].

Yogurt is the most familiar product containing beneficial microorganisms. There are, however, other foods that may contain added probiotics, such as sour cream, fruit juices and buttermilk. Fermented foods, such as cheese and sauerkraut, which contain beneficial bacteria, have been a part of the human diet for centuries. Probiotics are also available in tablet, spray, capsule, or powder forms.

Probiotics must be able to survive the aerobic condition of the product in which they are contained, as well as the acidic condition of the stomach. They must also be able to survive the bile levels and pancreatic secretions into the small intestine. [Shah, Ref.].

Probiotics may have a number of benefits. Certain probiotics have lactase activity, which is needed to properly digest lactose or milk sugar. For example, people that are lactose-intolerant may be able to eat yogurt that contains organisms that produce lactase enzymes. [Guo, Ref.].

Probiotics may also help keep the intestinal tract more acidic, making it difficult for disease causing organisms or pathogens to persist. Scientists agree that bacteria compete for nutrients in the intestinal tract so if probiotics are present in abundance the more virulent organisms may be crowded out.

The immune system may also be stimulated by probiotics. Studies in children suggest that probiotics can help repress infections and allergic responses. [Guo, Ref.].

Probiotics may reduce antibiotic-associated infections and diarrhea. Diarrhea is a common side effect of antibiotic treatment. Antibiotics can destroy both good and bad microorganisms in the intestinal tract. A number of probiotics show promise in alleviating antibiotic related diarrhea.

Other health benefits of probiotics, which have been documented or indicated, include suppressing colon cancer, preventing vaginitis, reducing serum cholesterol, improving bowel regularity and maintaining remission in inflammatory bowel disease. **Prebiotics** are nondigestible or partially digestible food ingredients that beneficially affect the host (consumer) by selectively stimulating the growth and/or activity of one or more of a limited number of bacteria in the colon, and thus improve host health. [Gibson & Roberfroid, 1995; Schrezenmeir & Vrese, 2001; and Gibson et al. 2001] Individual prebiotics may stimulate the growth and/or activity of some indigenous probiotics but not others.

Prebiotics are not digested by human enzymes and therefore have a low caloric value and are efficiently used in low calorie (energy) foods. Because of their non-digestibility prebiotics show fiber-like properties, mainly stool bulking. In this perspective prebiotics are efficiently used to increase stool frequency and to treat constipation. [Kleessen et al.1997; and Den Hond et al.,2000].

The benefits of probiotics are dependent on their viability, growth, and metabolic activity, which can be maximized by prebiotics. Products containing both probiotics and prebiotics are called synbiotics. In certain synbiotic applications, with careful selection of the appropriate strain in combination with the right prebiotic, the functionality of the probiotic can be enhanced by the prebiotic. Other synbiotic applications aim to combine the functionally of both the pre- and probiotic compound or even achieve synergistic effects of combining both. Food ingredient suppliers are now making it easier to add probiotics and prebiotics to foods and beverages by offering blends of synbiotics with the right proportion of pro- and prebiotics to obtain the desired beneficial health effects, as well as improved survival of the live strain.

A number of food ingredients have been shown to be prebiotics. These include lactulose, lactitol oligofructose (and sc-FOS), inulin and galacto-oligosaccharides (GOS), tagatose, isomaltooligosaccharides, polydextrose, and digestive resistant maltodextrin. [Gibson et al., 2004].

Inulin and oligofructose are well-documented prebiotics. They selectively stimulate bifidobacteria and lactobacilli in the gut at the expense of pathogenic bacteria such as clostridia and coliforms. In total, nine interventional human studies, some of them with more than one type of treatment, have been performed and published on inulin and/or oligofructose from chicory as the only active ingredient, to demonstrate their prebiotic effect.

The effect of lactitol on the intestinal flora has been extensively studied both *in vitro* and *in vivo*. *In vitro* studies show that lactitol stimulates the growth of *Lactobacillus* spp. and *Bifidobacterium*. The growth of proteolytic bacteria such as *Enterobacterium* and *Enterococcus* is inhibited. [Yuki et al., 1999; Kontula et al., 1999; Kitler et al., 1992; Felix et al., 1990.

Tagatose alters the composition and population of colonic microflora. Both changes in microbial population density and species were observed. Pathogenic bacteria were reduced and specific beneficial bacteria (e.g., lactobacilli) were increased. Tagatose is highly butygenic; about 50% of the short chain fatty acids (SCFA) produced are butyrate. See also **Tagatose**.

Clinical studies have shown that polydextrose increases the proportion of *Bifidobacterium* in the colonic microflora. The shift towards saccharolytic fermentation is also evidenced by reduced colonic pH, increased butyrate production and a reduction in branched short chain fatty acids.

Digestive resistant maltodextrin (DRM) has been shown to increase fecal concentrations of beneficial bacteria, including bifidobacteria, in dogs. [Flicklinger et al., 2000]. *In vitro* studies of cecal contents of rats show that short-chain fatty acids are generated in the presence of DRM and further studies on rats show that DRM to prevent intestinal mucosal atrophy occurring due to the long term administration of enteral nutrition [Ohkuma & Wakabayashi, 2001].

Fifty-three percent of Americans say they want more information on "active cultures," but only 9 percent are familiar with the term probiotics. Nonetheless, the probiotic market is the world's largest segment of the functional foods and beverages market. [Sloan, 2004]. Products include dairy drinks and desserts, yogurts, bakery mixes, cakes and bread, table spreads and other products containing "gut-healthy" bacteria. Prebiotics like inulin and oligofructose are used in a large variety of products to support and increase the indigenous (naturally occurring) bifidobacteria and to support added microorganisms (probiotics).

Balance is a common theme on products containing pre- and probiotics. Phrases such as stimulates digestion, boosts immune system, restores intestinal flora, strengthens bodies natural defense, for well balanced microflora, reduces bad bacteria and maintains a healthy digestive system are found on such products. Some products actually name the probiotic in the product, such a Bifidobacteria, lactic acid bacteria and bilact cultures— or state "with probiotics" respectively "with prebiotics" or "prebiotic".

Examples of probiotic products currently available and labeled accordingly include:

- Sweden-based Skane Dairy's *ProViva* fruit drink labeled "*ProViva* reduces the build-up of gas in the stomach." (This product was awarded Europe's first health claim for a probiotic product.).
- Group Danone is marketing in the U.S. *DanActive* "daily dose" drinks sold in 100 mL mini-bottles, labeled "*DanActive*. Immunity. Helps Naturally Strengthen Your Body's Defense System." The product also claims to help "maintain the balance of your intestinal flora." [Sloan, 2004].

The Future

The Institute of Food Technologists assembled an Expert Panel on Functional Foods. The Panel identified a number of areas requiring change to further encourage the development of functional foods [Clydesdale, 2004]. Examples of recommendations considered critical are:

- Modify the current definition and application of the term "nutritive value."
- Allow product labeling and health claims to accurately reflect the scientific data without triggering drug status.
- Develop incentives for companies to invest in functional foods research and development.

Acknowledgment

This information has been provided with the kind permission of the Calorie Control Council, Atlanta, GA. www.caloriecontrol.org.

Additional Reading

Aubertin, A.: "Not all Bacteria are Bad: Probiotics Promise Help to G.I. Tract, Immunity," *Environmental Nutrition*, **24**(11), 1, 6 (2001).

Clydesdale, F.: "Functional Foods: Opportunities & Challenges." *Food Technology*, **58**(12), 35–40 (2004).

Den Hond, E., B. Geypens, and Y. Ghoos: "Effect of High Performance Chicory Inulin on Constipation," *Nutrition Research*, **20**, 731–736 (2000).

Felix, Y.F., M.J. Hudson, R.W. Owen, et al.: "Effect of Dietary Lactitol on the Composition and the Metabolic Activity of the Intestinal Micro Flora in the Pig and in Humans," *Microbial Ecology in Health and Disease*, **3**, 259–267 (1990).

Flicklinger, E.A., et al: "Glucose-based Oligosaccharides Exhibit Different in Vitro Fermentation Patterns and Effect *in Vivo* Apparent Nutrient Digestibility and Microbial Population in Dogs," *Journal of Nutrition*, **130**(5), 1267–1273 (2000).

Gibson, G., and M.B. Roberfroid: "Dietary Modulation of the Human Colonic Microbiota: Introducing the Concept of Prebiotics," *Journal of Nutrition*, **125**, 1401–1412 (1995).

Gibson, G.R., H.M. Probert, J. Van Loo, R.A. Rastall, and M.B. Roberfroid: "Dietary Modulation of the Colonic Microbiota: Updating the Concept of Prebiotics," *Nutrition Research Reviews*, **17**, 259–275 (2004).

Guo, P.: "The Essential Microflora for Total Well Being," *Dietitian's Edge*, 40–47 (2001).

Kitler, M.E., M. Luginbühl, O. Lang, F. Wuhl, A. Wyss, and L. Lebek: "Lactitol and Lactulose: An *in Vivo* and *in Vitro* Comparison of their Effects on the Human Intestinal Flora," *Drug Invest.*, **4**(1), 73–82 (1992).

Kleessen, B., B. Sykura, H. Zunft, M. Blaut: "Effects of Inulin and Lactose on Fecal Microflora, Microbial Activity, and Bowel Habit in Elderly Constipated Persons," *Am. J. Clin. Nutr.*, **65**, 1397–1402 (1997).

Kontula, P., M.L. Suihko, A. Von Wright, and T. Mattila-Sandholm: "The Effect of Lactose Derivatives on Intestinal Lactic Acid Bacteria," *Journal of Dairy Science*, **82**, 249–256 (1999).

Ohkuma, K., and S. Wakabayashi: "Fibersol-2: a Soluble, Non-digestible, Starch-derived Dietary Fibre," in *Advanced Dietary Fibre Technology*, B.V. McCleary and L. Prosky, eds, Blackwell Science, Oxford, UK, 2001, pp 509–523.

Schrezenmeir, J., and M. de Vrese: "Probiotics, Prebiotics, and Synbiotics—Approaching a Definition," *American Journal of Clinical Nutrition*, **73**(supple), 361S–4S (2001).

Shah, N.P.: "Functional Foods from Probiotics and Prebiotics," *Food Technology*, **55**(11), 46–53 (2001).

Sloan, A.E.: "Top 10 Functional Food Trends 2004," *Food Technology*, **58**(4), 28–51 (2004).

Yuki, N., K. Watanabe, A. Mike, Y. Tagami, R. Tanaka, M. Ohwaki, and M. Morotomi: "Survival of a Probiotic, *Lactobacillus casei* Strain Shirota, in the Gastrointestinal Tract: Selective Isolation from Faeces and Identification Using Monoclonal Antibodies, *International Journal of Food Microbiology*, **48**, 51–57 (1999).

PROBOSCIS. A protruding or protrusible organ associated with the mouth and therefore at the front of the head in most animals. It is used in feeding and in some cases for other purposes.

PROCELLARIIFORMES *(Aves).* These narrow-winged flying acrobats of the high seas belong to quite a different order of birds than the gulls and terns: they are the petrels, albatrosses, and their relatives, a group which ornithologists call Tube-Noses because of the peculiar shape of their nostrils.

The nostrils are horny tubes generally found on the culmen and less often on the sides of the beak, which is straight with a hooked tip. The beak-sheath is composed of several separate horny plates. There are large nasal glands for salt excretion. The long gullet and proventriculus secrete an oil. There are 15 cervical vertebrae, and the furcula is movable and located on the keel of the sternum. The pelvis is penguin-like, and is fused with the synsacrum. The knee joint has a projection of the tibial crest. There are 3 forward toes that are connected by webs, and the hind toe is degenerated. These birds have an outstanding flight capacity and great endurance.

They are inhabitants of the high seas, particularly of the Southern Hemisphere, and they only go on land to breed. They lay 1 egg, and have a long incubation period. The young grow very slowly. This order includes both the largest and the smallest sea birds; the length is 14–135 centimeters (5.5–53 inches), and the weight is 20–8000 grams ($\frac{1}{2}$–$17\frac{1}{2}$ pounds). Today there are 4 families: the Albatrosses (*Diomedeidae*); the Petrels and Shearwaters (*Procellariidae*); the Storm-Petrels (*Hydrobatidae*); and the Diving-Petrels (*Pelecanoididae*). There are 22 genera and 92 species.

The paired nose tubes may vary in form and length; they join round or oval openings with large nasal or olfactory cavities. The significance of the tubes is not known; many suggestions have been made regarding their purpose, but none are altogether convincing. Evidently the olfactory sense, otherwise of low efficiency in birds, is well developed in the tube-nosed swimmers, but this does not explain the peculiar nostrils. All members of this order fly low over the sea, so possibly the nose tubes keep spray out of the inner nasal cavities. All tube-noses also have large nasal glands which secrete a saturated salt solution, and so the tubes could possibly serve to keep this solution away from the eyes and the skin of the gape.

Another peculiarity of these birds is the flesh colored, oily liquid that most species secrete from special cells of the proventriculus. When in danger, breeding birds and the young as well can regurgitate this stomach-oil and spray it at an aggressor for a distance of several meters (feet). When this oil cools it solidifies to a wax-like consistency; in this form it is often found around the nests in cold regions. The birds can use this oil when preening their plumage and possibly they apply it directly from the nose tubes. It is also possible that the stomach-oil accounts for the strong musky smell, peculiar to all tube-noses, which can even persist on skins which have been in museums for over 100 years.

While the tubular nostrils and the stomach-oil occur only in tube-noses, their particularly slow reproduction and the long growth period of their young are not so unique in the bird world. All *Procellariiformes* lay only 1 egg. As a rule, it is relatively large, weighing 6–10% of the mother's body weight in large species and 10–25% or more in the smaller ones. Incubation and the growth periods last longer in tube-noses than in all other birds of similar size.

All *Procellariiformes* are birds of the high seas, adapted in various ways for feeding on or just beneath the surface of the water. They can spend days, weeks, and even months away from land. Two-thirds of the living species live in the Southern Hemisphere, which must be regarded as the main area of development of this order. From the temperate-cool zone of westerly winds, they have spread south as far as the coasts of the Antarctic and north over the equator as far as Arctic latitudes. See also **Petrels and Albatrosses**.

PROCESS CONTROL. The use of instruments and control devices and systems to measure and manipulate one or many variables to assure the safe and efficient operation of machines and equipment required in manufacturing. Processes may be categorized by the kinds of materials which they handle: (1) *fluids* (gases and liquids), (2) *bulk solids*, (3) *sheeted and webbed materials*, and (4) *discrete pieces*. The chemical and so-called process industries (petroleum refining, food processing, petrochemical manufacture, some of paper and textile manufacturing, and some of the metallurgical industries) essentially are concerned with fluids and bulk solids. These are materials that flow in pipelines or are handled by various forms of bulk conveyors. Principal variables with which these industries are concerned include temperature, pressure, flow, liquid level, solid level (as in bins, silos, and hoppers), specific gravity; density, viscosity, consistency, and chemical composition. Industries concerned with sheeted and webbed materials include the paper, textile, printing, plastics, and parts of the metals industries. Here, the major variables of concern include temperature, thickness, dimensional widths and lengths, linear speed, and weight/area (such as basis weight). A discrete-piece handling industry is typified by most of the automotive and aircraft manufacturing facilities that involve machine tools, transfer machines, assembly operations, and parts inspection and testing. Major variables include dimension (linear and angular), position, hardness, vibration, electrical properties, thermal properties, and structural soundness (such as by x-ray inspection). It is difficult to find one industry that is purely one type. It is obvious, however, that the form of instrumentation and control is affected importantly by the particular variables that must be measured and controlled.

PROCESS (Control System). The term *process* when used as a part of control-system terminology may be defined as the collective functions performed in and by the equipment in which the variable(s) is (are) to be controlled. Equipment as embodied in this definition should be understood not to include any automatic control equipment. The process may also be referred to as the *controlled system*.

PROCTITIS. Inflammation and infection of the rectum due to a variety of causes. Symptoms may include discharge, pain on defecation, bleeding, and rectal fullness. A major cause of proctitis is inflammatory bowel disease. See also **Colitis and Other Inflammatory Bowel Diseases**. Amebiasis may be a cause. See also **Amebiasis**. Other causative factors include the virus *Lymphogranuloma inguinale*, syphilis, chancroid, and malignancy. Gonococcal proctitis is increasingly seen in homosexual men, in which cases the gonococcal infection not only involves the urethra, but the anal canal as well.

PROCYON (α Canis Minoris). A bright, nearby star, which receives its name from the fact that in its nightly journey across the sky, it is close to the star Sirius. These two "dog stars" are referred to in the most ancient literatures, and were objects of veneration and worship both by the Babylonians and the Egyptians.

Procyon, like the other "dog star" Sirius, has as a faint companion a white dwarf, a type of star whose common proper motion is 1.25″/yr.

Ranking eighth in apparent brightness among the stars, Procyon has a true brightness of 7.3 as compared with unity for the sun. Procyon is a yellow-white, spectral type F star. Estimated distance from the earth is 11.3 light years. See also **Constellations**; and **Star**.

PRODUCT. The result obtained when two or more quantities, such as numbers, functions, or equations are combined by multiplication. In vector and tensor analysis, the concept of a product must be generalized for there are several kinds of products and multiplication does not always obey the commutative law.

In matrix algebra, if A is a rectangular matrix of order $(m \times h)$ and B of order $(h \times n)$, the product $C = AB$ is of order $(m \times n)$ and its elements are $C_{ij} = \rho A_{ik} B_{kj}$. It is not necessary that the commutative law be obeyed. Another matrix combination is the direct product. If A and B are square, of order m and n, respectively, the direct product is of order $(m \times n)$ and defined by the relation

$$A \times B = [A_{ij} B_{rs}]$$

The index pairs (i, r) and (j, s) refer to row and column, respectively. They are customarily arranged in dictionary order so that (j, s) precedes (j', s') if $j < j'$, $s < s'$ or if $j = j'$, $s < s'$, etc.

In group theory, *multiplication* means any defined combination law and the result is a product. For example *multiplication* might be defined as addition and then the product is that result commonly known in algebra as a sum.

The product of two infinite series is called a Cauchy product. An expansion of the form

$$u_1 u_2 u_3 \cdots u_n \cdots = \Pi_{k=1}^{\infty} u_k$$

is an infinite product. The partial products form a sequence $\{p_n\}$ such that $p_1 = u_1; p_2 = u_1 u_2; p_3 = u_1 u_2 u_3; \ldots, p_n = \Pi_{k=1}^{n} u_k$. The convergence of infinite products is studied by methods similar to those used for infinite series. An infinite product of the form $(1 + u_1) \times (1 + u_2) \cdots$ may be converted into an infinite series $1 + x_1 + x_2 + x_3 + \cdots$, where $u_n = x_n / (1 + x_1 + x_2 + \cdots + x_{n-1})$. The product converges only if the positive infinite series, $u_1 + u_{2x} + u_3 + \cdots$ converges.

PRODUCT MODULATOR. A modulator whose output is proportional to the product of the carrier and the modulating signal. The desired result can be achieved by sampling the modulating wave briefly at regular intervals at the carrier rate, and applying the ensemble of samples to the input of a band-pass filter having a center frequency coincident with the carrier frequency. One fundamental property of a product modulator is that the carrier is normally suppressed.

PRODUCT-MOMENT. If the distribution function of n variates $x_1, x_2, \ldots, x_n$ is given by $F(x_1, \ldots, x_n)$ the product-moment, joint- or multivariate-moment of order $r, s, \ldots, u$ is the mean value of $x_1^r x_2^s \ldots x_n^u$, namely:

$$\int_{-\infty}^{\infty} \int_{-\infty}^{\infty} \ldots \int_{-\infty}^{\infty} x_1^r x_2^s \ldots x_n^u \, dF(x_1, \ldots, x_n)$$

PROFILE. 1. Of a variable, a curve representing corresponding values of two or more variables which may occur. A profile accounts for the correlation from point to point on the curve and has some possibility, not necessarily specified, of actual occurrence.

2. The contour or form of a body, especially in a cross section; specifically, an airfoil profile.

3. See also **Differential (Mathematics)**.

PROGNOSTIC CLOUDS. See **Clouds and Cloud Formation**.

PROGRAM (Computer). 1. The complete plan for the computer solution of a problem, more specifically the complete sequence of instructions and routines necessary to solve a problem. 2. To plan the procedures for solving a problem. This may involve among other things the analysis of the problem, preparation of a flow diagram, preparing details, texting, and developing subroutines, allocation of storage locations, specification of input and output formats, and the incorporation of a computer run into a complete data processing system.

Internally stored program. A sequence of instructions, stored inside the computer in the same storage facilities as the computer data, as opposed to external storage on punched paper tape and pinboards.

Object program. The program which is the output of an automatic coding system, such as an assembler or compiler. Often the object program is a machine language program ready for execution, but it may well be in an intermediate language.

Source program. A computer program written in a language designed for ease of expression of a class of problems or procedures, by humans; e.g., symbolic or algebraic. A generator, assembler translator or compiler routine is used to perform the mechanics of translating the source program into an object program in machine language.

PROGRAM GENERATOR (Computer System). A program that permits a computer to write other programs automatically. Generators are of two types: (a) the *character-controlled generator*, which operates like a compiler in that it takes entries from a library of functions, but unlike a simple compiler in that it examines control characters associated with each entry, and alters instructions found in the library according to the directions contained in the control characters; and (b) the *pure generator*, which is a program that writes another program. When associated with an assembler, a pure generator is usually a section of program which is called into storage by the assembler from a library and which then writes one or more entries in another program. Most assemblers are also compilers and generators. In this case, the entire system is usually referred to as an *assembly system*.

See also **Assembler (Computer System)**.

PROGRAMMABLE CONTROLLER. NEMA (National Electrical Manufacturers Association) defines a programmable controller as: *a digital electronic device that uses a programmable memory to store instructions and to implement specific functions, such as logic, sequence, timing, counting, and arithmetic operations to control machines and processes.* The programmable controller as of the late 1980s was a mainstay of industrial automation.

Perspective. As recently as the early 1960s, industrial control systems had been constructed from traditional electromechanical devices, such as relays, drum switches, and paper tape readers. This was particularly true of the discrete-piece manufacturing industries, such as the machinery, parts, automotive, aircraft, and electronics industries, as contrasted with the fluid processing industries. See also **Automation**. Although many of the earlier devices still are used today and many of the problems associated with using them have been eliminated due to technological advances in their design, such approaches continue to suffer from some inherent problems. Relays are susceptible to mechanical failure, they require large amounts of energy to operate, and they generate large amounts of electrical noise. Extreme care had to be taken in the design of relay-based control systems because it was not uncommon for the outputs to "chatter," i.e., to turn on and off rapidly when they changed states. The logic of the circuit was dictated by the hard wiring of contacts and coils. In order to make changes, as was required when production patterns changed, more time was required to rewire the logic than was needed originally.

In the late 1960s, the need to design more reliable and more flexible control systems became apparent. For example, the automotive industry was spending millions of dollars for rewiring control panels in order to make relatively minor changes to the control systems at the time of the annual model changeovers. In 1968, a team of automotive engineers wrote a specification for what they called a "programmable logic controller." What they specified was a solid-state replacement for the relay logic. The machine would use solid-state outputs and inputs, instead of control relays, to control the motors starters and sense push buttons and limit switches. The first commercially successful programmable controller was introduced in 1969. By present standards, the instrument was a massive machine containing thousands of electronic parts. It should be stressed that the first machine was designed long before microprocessors became available. The early programmable logic controller (PLC) used a magnetic core memory to store a program that was written in a graphical language (relay logic), a scheme long established in connection with conventional relay systems. See Figs. 1 and 2.

In the late 1970s, the microprocessor became a reality and greatly enhanced the role of the PLC, permitting it to evolve from simply replacing relays to the sophisticated control system it has become today. Programmable controllers now have the ability to manipulate large amounts of data, perform mathematical calculations, and communicate with other intelligent devices, such as robots and computers. Concurrent with the increased capability and flexibility of the PLC was the expansion into many other industrial applications, including the control of machine tools, material handling systems, food-processing operations, and use in the continuous process control field.

Characteristic Functions of a Programmable Controller[1]

Seven of the most important characteristics of a programmable controller include:

1. *Field-programmable by the user.* This characteristic allows the user to write and change programs in the field without rewiring or sending the unit back to the manufacturer for this purpose.
2. *Contains preprogrammed functions.* PLCs, when procured, are already programmed with at least logic, timing, counting, and memory functions that the user can access through some type of control-oriented language.
3. *Scans memory and input/output (I/O) in a deterministic manner.* This feature allows the control engineer to precisely determine how the machine or process will respond to the program.
4. *Provides error checking and diagnostics.* A PLC will periodically run internal tests on its memory, processor, and I/O systems to ensure that what it is doing to the machine or process is what it was programmed to do.
5. *Can be monitored.* A PLC will provide some form of monitoring capability, either through indicating lights that show the status of inputs and outputs, or by an external device that can display program execution status.
6. *Packaged appropriately.* Modern PLCs are designed to withstand the temperature, vibration, and noise found in most factory environments.
7. *General-purpose suitability.* Generally, a PLC is not designed for a specific application, but it can handle a wide variety of control tasks effectively.

A simplified model of a PLC is shown in Fig. 3. The input converters convert the high-level signals that come from the field devices to *logic-level signals* that the PLC can read directly. The logic solver reads these inputs and decides what the outputs should be, based on the user's program logic. The output converters take the logic-level signals output from the logic solver and convert them into the high-level signals that are needed by the various field devices. The program loader is used to enter and/or change the user's program into the memory and to monitor the execution of the program.

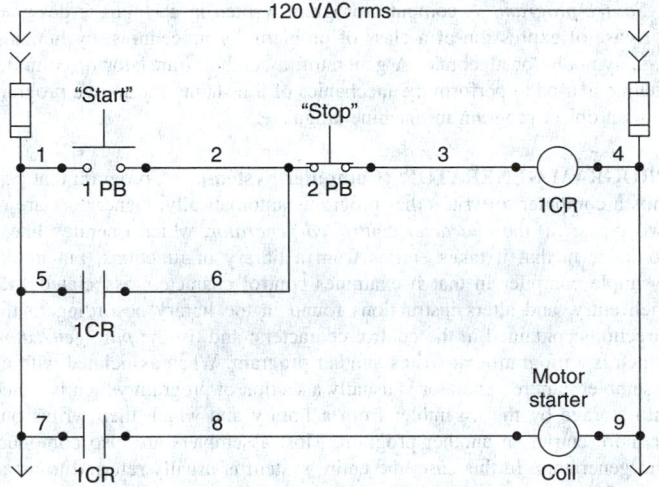

Fig. 1. Typical motor control circuit. When the pushbutton labeled START (1PB) is pressed, the control relay (1CR) is energized. A contact from 1CR is then closed and is used to "seal" 1CR "on" after 1PB is released. Another contact from 1CR is used to energize the motor starter coil, turning the motor on. When the STOP pushbutton (2PB) is pressed, it deenergizes 1CR, which "unseals" 1CR, and deenergizes the motor starter coil which stops the motor. Implementing this motor control circuit requires nine wires, not counting the power supply. The equivalent PRC Ladder Diagram Program is shown in Fig. 2.

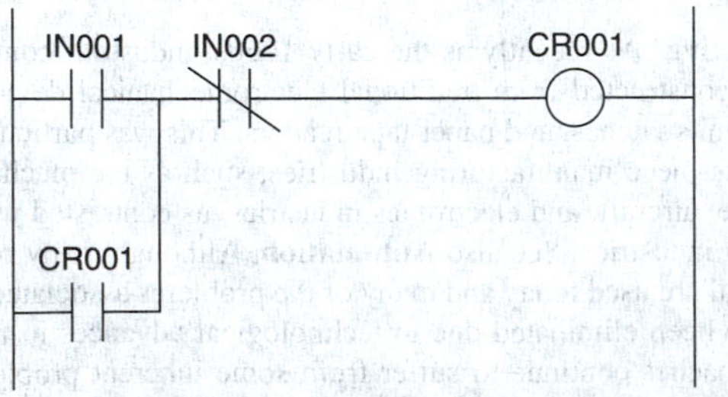

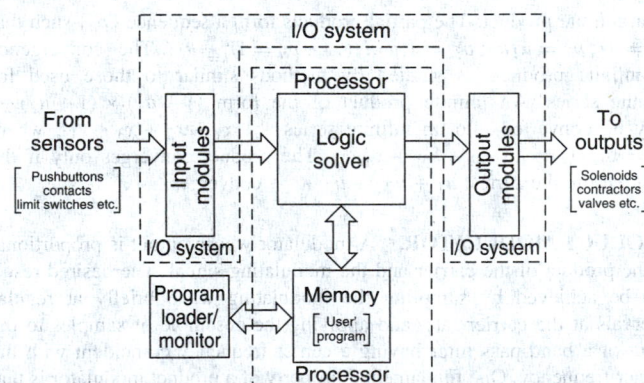

Fig. 3. Simplified block diagram of a programmable controller illustrates the basic functionality of the PLC. The control engineer (user) enters the control program on the program loader. The latter writes the program into the memory of the processor. The logic solver reads the states of the sensors through the input modules, then uses this information to solve the logic stored in the user memory (program) and also writes the resulting output states to the output devices through the output modules.

Fig. 2. Ladder Diagram Program used to control the motor circuit shown in Fig. 1. In this case, all the inputs and outputs are assigned variable names, such as IN001 for the START input and CR001 in place of the control relay 1CR. This diagram is then drawn on a program loader and entered in the PLC's user memory. The PLC's processor then solves the logic that is stored in memory. Only six wires are needed between the PLC's output and the motor starter coil and between the PLC's inputs and the pushbuttons (not counting power supply wiring).

[1] The abbreviation for programmable controller, PLC or PC, is optional. The editors here have selected PLC to avoid confusion that frequently arises from using PC, which is also the common abbreviation for personal computer.

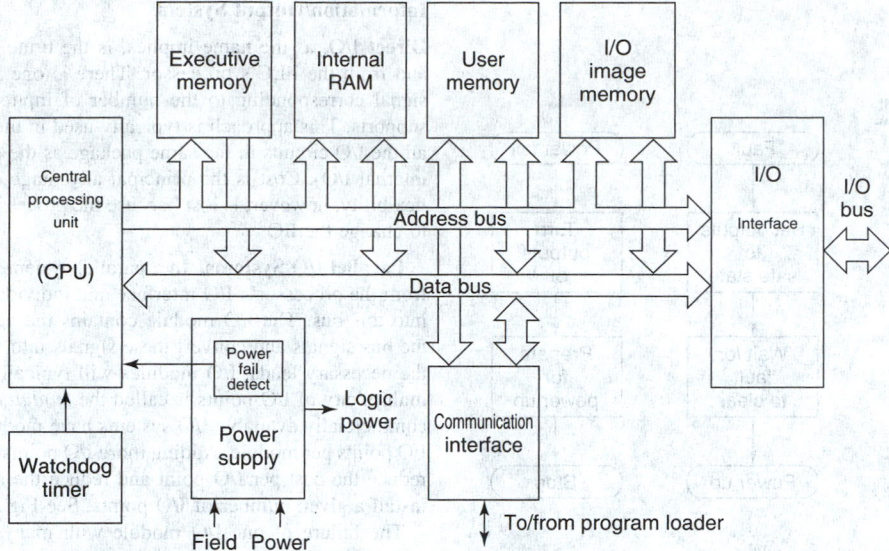

Fig. 4.   Detailed block diagram of the processor section of a programmable controller. The central processing unit (CPU), typically a microprocessor, executes a program written by the manufacturer of the PLC that is stored in the Executive Memory. This executive program that the CPU executes gives the CPU the ability to interpret the user's program. The CPU does not operate on the I/O directly. Rather, it works with an image of the I/O that is stored in the I/O image memory. The I/O interface is responsible for transferring the image outputs to the I/O system and reading the inputs from the I/O system and writing them into the image memory. A "watchdog" timer is provided to time how long it takes the CPU to execute the user's program. If this time exceeds a predetermined value, the watchdog timer causes the processor to fault. If the CPU fails and does not execute the user's program, the watchdog timer will ensure that at least a fault will be indicated and that the processor will shut down in a safe manner.

Memory. A PLC's memory can be of two different types—volatile or nonvolatile. Volatile memory loses its contents when power is removed, whereas a nonvolatile memory does not. PLCs will use nonvolatile memory for a majority of the user's memory because the program must be retained during a power-down cycle, meaning that the user will not have to reload the program every time power is lost. It is important that all nonvolatile memory in a PLC use some form of error checking in order to assure that the memory has indeed not changed. Types of memory currently used include (1) battery-backed up CMOS RAM, (2) EPROM, and (3) EEPROM. See also **Memory (Electronic)**.

Central Processing Unit (CPU). How the CPU is constructed will determine the flexibility of the PLC (whether or not the PLC can be expanded and modified for future enhancement) as well as the overall speed of the PLC. The speed is expressed in terms of how fast the PLC will scan a given amount of memory. The measure, called the scan rate, is typically expressed in milliseconds per thousand words of memory. Faster PLCs will typically cost more than the slower models. Thus, it is important to choose a PLC with a scan time appropriate to present and planned use.

It is important to note that many of the commercially available PLCs specify their scan time using contacts and coils only. A real program that uses other functions, such as timers, counters, and mathematical functions, may take considerably longer to execute. Also in considering a PLC, one should include the scan time of the I/O, the scan time of the memory, and any additional time overheads the processor requires. A detailed block diagram of the processor section of a PLC is given in Fig. 4.

Processor Software. The hardware of the PLC is not too different from that of a lot of computers. What makes the PLC special is the software. The executive software is the program that the PLC manufacturer provides internal to the PLC that executes the user's program. The executive software determines what functions are available to the user's program, how the program is solved, how the I/O is served, and what the PLC does during power up/down and fault conditions.

Executive Software. A simplified model of what the executive software does is shown in Fig. 5. Specific PLC designs perform the basic functions shown somewhat differently. This can make a large difference in program execution. For example, some PLCs may perform diagnostics only at a single point in the executive program, while others may perform diagnostics "on-line," i.e., while the user's program is being solved.

Close attention must be given to how the PLC runs diagnostic tests and what it does during failures. Ignoring this aspect of the PLC can result in an unsafe system.

Multi-Tasking. In a later development, PLCs capable of executing multiple tasks with a single processor appeared. Multi-tasking takes several forms, of which two are: (1) time-driven, and (2) event-driven. In a time-driven system, the user writes programs and assigns I/O for each task. The user will then configure the processor to run each task on periodic time intervals. This type of system is shown graphically in Fig. 6. This feature allows the *time-critical* portion of the control system, such as the portion that controls high-speed motions or machine fault detection, to run many times per second, while allowing the *non-critical* portions, such as servicing indicator lights, to run much slower. Because only the time-critical logic and I/O need quick solutions, versus the entire user's program, faster throughput can be achieved.

Event-driven multi-tasking (also called interrupt-driven) is similar. In this case, the user defines a particular event, such as an input changing state or an output turning off, that causes each task to be run.

In either the time-driven or event-driven case, it is important to recognize the priority of tasks. Some multi-tasking systems allow any task to access any variable, such as an I/O point. Thus, caution must be used when programming multiple tasks that access the same variables. It may be difficult to determine which task is writing which variable while trying to debug a program.

User Software. This is the software that the control engineer writes and stores in user memory in order to perform the required control over the machine or process. User software can contain both configuration data and language programs. The configuration data contain information that tells the processor what its environment is and how it should execute the language problem. The configuration process typically consists of assigning I/O points to particular I/O racks, telling the processor how much memory and I/O it has, assigning specific memory for tasks, determining fatal versus nonfatal faults, and many other items interactively on a program loader.

In as much as the modern PLC is required to do more and more in terms of operator interfacing, communications, data acquisition, and supervisory control, more is required of the language that implements these functions. Therefore, it is crucial that the various aspects of the language be considered.

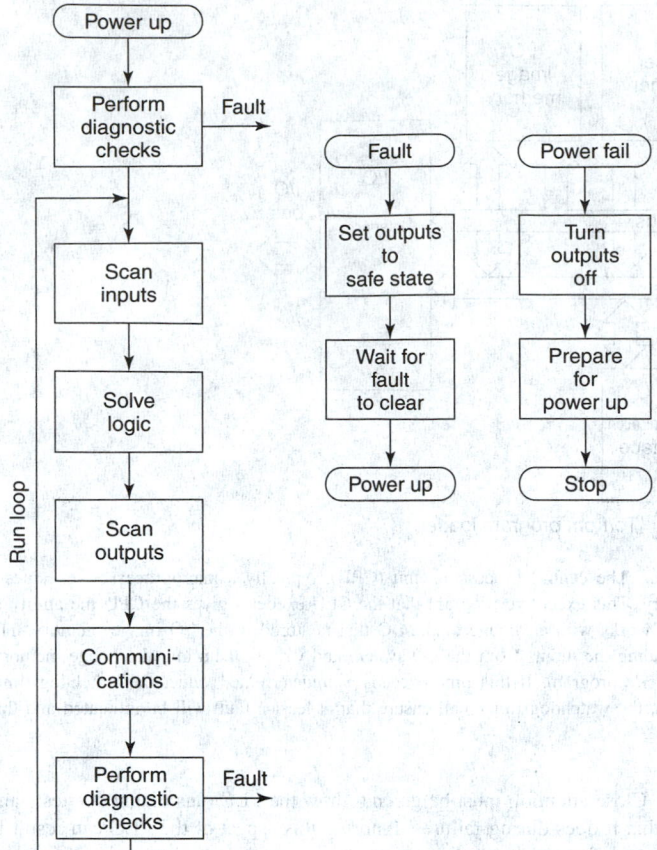

Fig. 5. The Executive Program shown here controls the functionality of the programmable controller. It controls the actions of the CPU to perform the indicated actions. Diagnostic checks must be run at power-up as well as during the run loop, which is executed while the PC is controlling the process. When faults are detected, the outputs must be set to a predetermined "safe" state the user usually has the choice of turning all of the outputs off or leaving them in their last state. When advance warning of power failure is given by the power supply, the executive program shuts down the CPU in a controlled manner. During the run loop, inputs are only scanned once. This allows the entire user program to operate from a consistent set of inputs because they are only determined prior to executing the user's program and do not change state in the middle of the run loop.

Information/Output Systems

Direct I/O, as the name implies, is the brute force way of getting I/O to and from the PLC's processor. There is one input signal and one output signal corresponding to the number of inputs and outputs the processor supports. This approach is typically used in the very small PLCs that have all the I/O circuits in the same package as the processor (sometimes called internal I/O). Cost is the principal advantage of internal direct I/O. Some flexibility, however, is lost because the processor must be changed in order to change the I/O.

Parallel I/O Systems. In a parallel system, a parallel I/O bus emanates from the processor's I/O interface and individual I/O modules are plugged into this bus. The I/O module contains the necessary circuitry to decode the bus signals and convert these signals into voltage levels that can drive the necessary loads. I/O modules will typically drive multiple loads. This multiplicity of I/O points is called the *modularity* of the I/O system. Most commercially available I/O systems have modularities of 2, 4, 8, 16, or 32 I/O points per module. Adding more I/O points on a module will commonly reduce the cost per I/O point and reduce the amount of space required to install a given number of I/O points. See Fig. 7.

The failure of one I/O module with many points of I/O on it can be disastrous if it controls many critical devices, such as those causing motions or controlling emergency stops. Thus, it is good practice to split up the critical I/O between the high-density modules.

Serial I/O Systems. Parallel systems are limited in the distance over which one can extend the I/O bus, typically less than 50 feet (15 meters). If the machine should be 100 feet (30 meters) long, one would have to use two PLCs. Serial I/O systems solve this problem by transmitting the I/O information over a serial data link capable of being extended over longer distances (1000–10,000 feet; 300–3050 meters). A serial bus emanates from the processor and is connected to a parallel bus through a serial-to-parallel converter. Since a single serial bus contains fewer wires than does the wiring to the loads, large wiring cost savings can be realized by using serial systems. See Fig. 8.

Care must be taken when using serial I/O systems in time-critical applications because two I/O buses have to be scanned—both the serial and parallel bus—instead of one, thus making them slower than straight parallel systems. Some, but not all, serial systems may "desynchronize" themselves from the logic scanning, thus making it more difficult to predict I/O responses to fast-changing signals.

I/O Circuits. An I/O module performs signal conversion and isolation between the internal logic-level signals inside the PLC and the field's high-level signals. There are several different types of I/O circuits available that are capable of driving almost any conceivable load and sensing the status

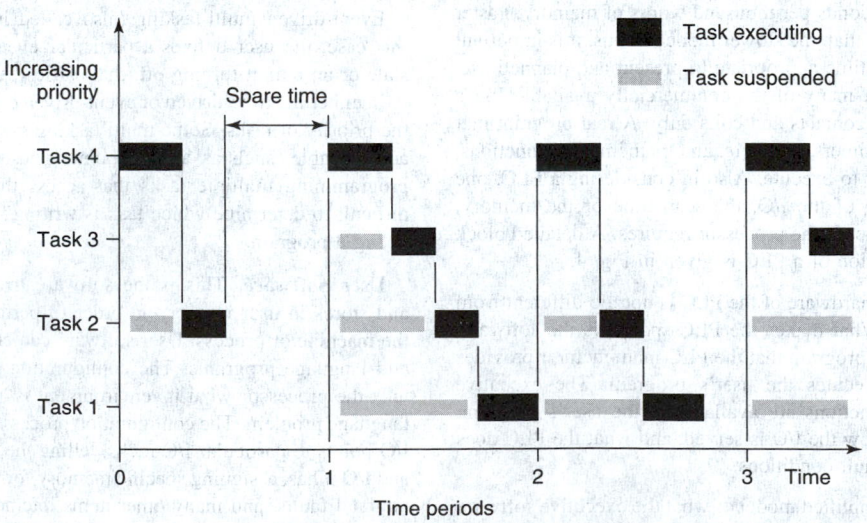

Fig. 6. In a time-driven multi-tasking system, tasks are scheduled to run on predetermined time intervals. In the example shown here, tasks 4 and 2 are scheduled to run every period while tasks 3 and 1 are scheduled to run every other time period. The higher-priority tasks always execute before the low-priority tasks. During period #1, all four tasks are scheduled to run. However, there is not enough time for task 4 to finish executing. Its execution is suspended until the spare time in period #2 is available. Care must be taken to ensure that there is enough spare time for all the tasks to execute. Some multi-tasking systems will provide an indication that not enough time exists to execute all the programmed tasks, thus making it easier to debug the programs.

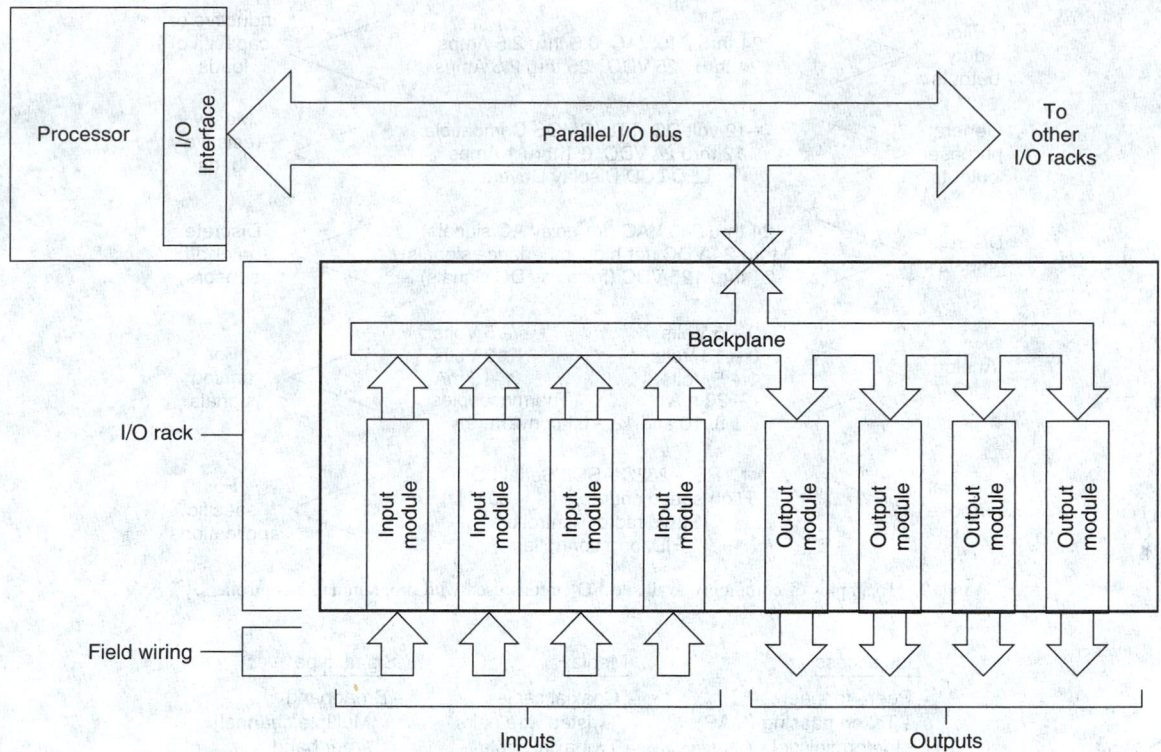

Fig. 7.   Block diagram of a parallel I/O system used with a programmable controller.

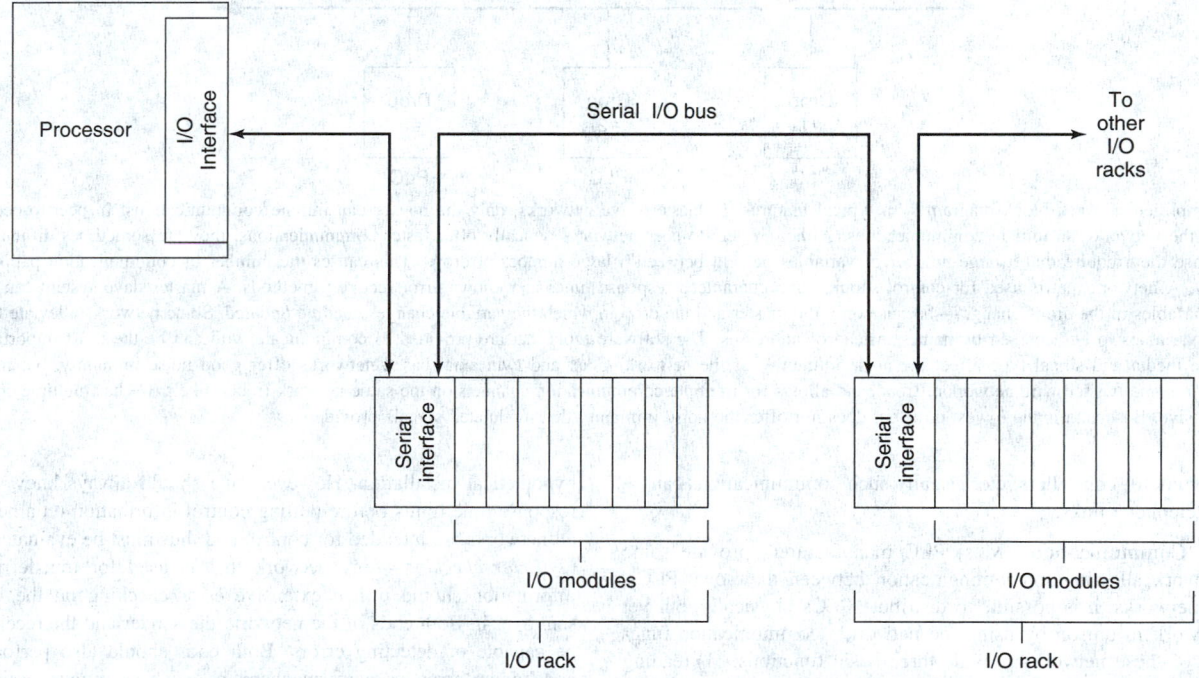

Fig. 8.   Block diagram of a serial I/O system used with a programmable controller.

of a wide variety of sensors. Most of these I/O circuits fall into one of five categories, as shown in Fig. 9.

Programmable Controller Communications

The communications aspects of a PLC can severely limit or greatly enhance the applicability of these devices.

Point-to-Point Communications. Most PLCs have at least one communication port built in, i.e., the program loader interface. However, only a few manufacturers release the information needed in order to communicate

over this interface. Even so, these ports typically use unusual protocols that can require considerable effort to implement. Some manufacturers of peripheral equipment, such as color graphics displays, have converted their equipment to talk to some PLCs directly, thereby saving the expense of writing specific communications software.

Most PLCs also provide some form of ASCII communications. Some PLCs have separate I/O modules for this purpose, while some others allow the user to reconfigure the program loader port for this purpose. With ASCII communications, it is possible to talk to a wide variety of devices, such as color graphics terminals, intelligent push-button stations, bar code

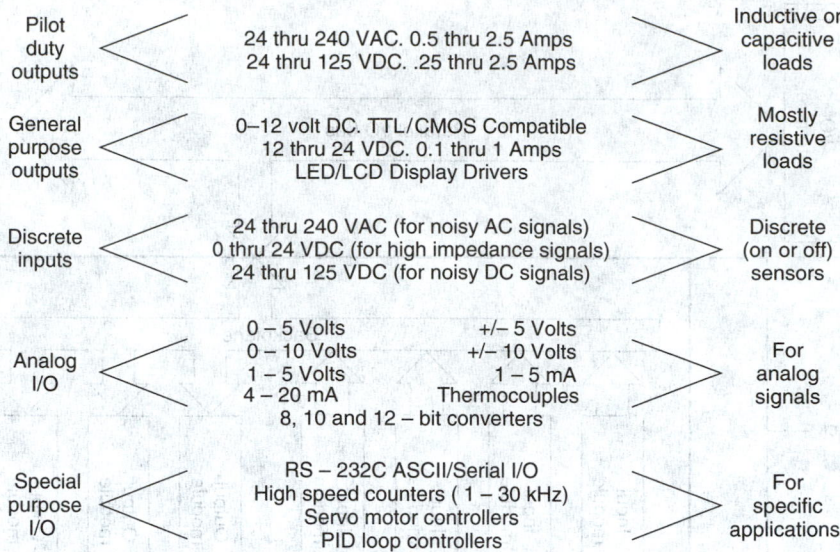

Pilot duty outputs	24 thru 240 VAC. 0.5 thru 2.5 Amps 24 thru 125 VDC. .25 thru 2.5 Amps	Inductive or capacitive loads
General purpose outputs	0–12 volt DC. TTL/CMOS Compatible 12 thru 24 VDC. 0.1 thru 1 Amps LED/LCD Display Drivers	Mostly resistive loads
Discrete inputs	24 thru 240 VAC (for noisy AC signals) 0 thru 24 VDC (for high impedance signals) 24 thru 125 VDC (for noisy DC signals)	Discrete (on or off) sensors
Analog I/O	0 – 5 Volts +/– 5 Volts 0 – 10 Volts +/– 10 Volts 1 – 5 Volts 1 – 5 mA 4 – 20 mA Thermocouples 8, 10 and 12 – bit converters	For analog signals
Special purpose I/O	RS – 232C ASCII/Serial I/O High speed counters (1 – 30 kHz) Servo motor controllers PID loop controllers	For specific applications

Fig. 9.   Examples of commonly available I/O circuits used with programmable controllers.

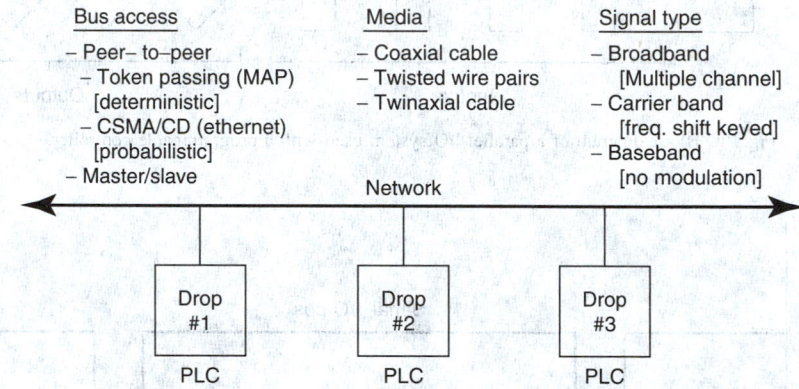

Bus access
- Peer– to–peer
  - Token passing (MAP)
    [deterministic]
  - CSMA/CD (ethernet)
    [probabilistic]
- Master/slave

Media
- Coaxial cable
- Twisted wire pairs
- Twinaxial cable

Signal type
- Broadband
  [Multiple channel]
- Carrier band
  [freq. shift keyed]
- Baseband
  [no modulation]

Network

Drop #1   —   PLC
Drop #2   —   PLC
Drop #3   —   PLC

Fig. 10.   Simplified network block diagram with typical features. In master/slave networks, only the master can initiate communications. In peer-to-peer networks, any drop on the network can initiate communications. Although peer-to-peer networks typically offer faster communications, they are sometimes difficult to use for control because they require that a large number of variables be sent between a large number of drops. This causes the number of communication paths to increase exponentially. A network that is used for control should have guaranteed response times and known error recovery methods. A master/slave system can be easier to maintain if variables in the drops change—because only the master and the drop in which the variable changes need be updated. Some networks alleviate this problem by allowing variables to be accessed by names instead of addresses. The software tools that are provided to communicate with can be the most important factor to consider. The media and signal type affect the noise immunity of the network. Coax and twin axial based networks offer good noise immunity, but may be more expensive than some twisted wire networks. Broadband allows for multiple communication channels on the same network (much like a TV has multiple channels), but is very expensive. Baseband is the lowest cost, but does not offer the noise immunity that modulated signals provide.

readers, servomotor controllers, etc. Usually such communications can be made over telephone links.

Network Communications. Most PLC manufacturers provide some type of network allowing for communication between their own PLCs. With these networks, it is possible to distribute PLCs physically, but yet have them work in unison by using the network's communication functions. Most of these networks provide three basic functions: (1) reading variables, (2) writing variables, and (3) program upload and download. However, because these networks are designed to provide communication functions, not necessarily control functions, using a network inside of a control loop requires careful planning and evaluation. Some networks have difficulty transporting information from two points on the network in such a manner as to allow the use of this information in a time-critical control application.

Certain factors must be considered when evaluating a network for a control application, including: (1) *Response time*—the length of time required from an input changing state on one node of a network to a second node receiving notice that the input has changed. This is a critical parameter when trying to implement control of a process over a network. Some networks give only a probabilistic response time based on some hypothetical installation. However, one should always know the precise response time limits before putting control information on a network. Not all networks are intended for control and thus must be evaluated carefully. (2) *Error checking*—any network that is used for transferring control information should utilize extensive error checking on the information sent over it. Both ends of the network, the sender and the receiver, should be capable of detecting errors. Both ends should also perform specific and known error recovery mechanisms, such as retransmission or, at a minimum, be able to notify both the sender and receiver that there was an error so the control engineers can program their own recovery scheme. (3) *Access mechanisms*—because a network usually contains only one channel over which all PLCs must talk, some method for determining who has access to the network at any given time must be used. Two of the more popular access mechanisms are (a) master/slave, and (b) peer-to-peer. See Fig. 10.

On a *master/slave system*, there is only one master PLC. The master sends commands out to the other slave PLCs, and they respond appropriately. The slaves on the network never initiate their own commands—they always respond to what the master commands them to do.

The *peer-to-peer* mechanism allows any PLC on the network to initiate messages. However, as in the case of humans talking, if everybody talks at once, nothing intelligible can be heard. Peer-to-peer networks need some mechanism for determining access between all the PLCs — not just between the master and the slave. Various mechanisms for determining access have been implemented, such as token-passing and carrier-sense-multiple-access-with-collision detection (CSMA/CD). More details regarding these systems are given elsewhere in this encyclopedia. Check alphabetical index.

RALPH E. MACKIEWICZ, SISCO, Inc., Warren, MI

PROGRAMMING FLOWCHART (Computer).

Refers to a graphical representation for the definition of a program, in which symbols are used to represent data, flow, operations, equipment, etc. A digital computer program may be charted for two primary reasons: (1) ease of initial program design, and (2) program documentation. By coding from a flowchart, instead of coding without any preliminary design, the programmer usually conserves time and effort in developing the program. In addition, the flowchart is an effective means for transmitting an understanding of the program to someone else.

A programming flowchart is comprised of function blocks with connectors between these blocks. A specific function box may represent an input/output operation, a numerical computation, or a logic decision. The program chart shown in Fig. 1 is of a program that reads values from the process, converts those values to engineering units, limit-checks the converted values and, if there is a violation, prints an alarm message on the process operator's typewriter.

Various levels of detail are presented in programming flowcharts. A functional block in a low-level flowchart may represent only a few computer instructions, whereas a functional block in a high-level flowchart may represent many computer instructions. The high-level flowchart is used mainly for initial program design and as a way of informing a nonprogrammer of what the program does. The low-level chart usually appears in the last stage of flowcharting before a program is actually coded. It is used for documentation that may be needed for later modifications and corrections.

THOMAS J. HARRISON, International Business Machines Corporation, Boca Raton, FL

PROGRAMMING LANGUAGE.

A language that computer programmers use in writing instructions for a computer to execute (i.e., carry out). Also known as a computer language. There are many kinds of programming languages. Depending on whether the instructions in a programming language closely parallel the primitive instructions that are directly built-in into a computer, one may distinguish between two general classes of programming languages: low-level languages and high-level languages.

Low- level languages. Instructions in a low-level language specify primitive operations that a physical computer is designed to perform directly. Machine languages and assembly languages are low-level languages. A machine language uses the binary coding internal to a physical computer. Each instruction is a sequence of 0s and 1s that denotes a primitive operation of a physical computer. For example, an instruction to add the value at some memory address (say, 12) to a register named R3 (a high-speed storage device within a computer's central processing unit) might be written in a hypothetical machine language as

$$01101011000000001010$$

where 01101 denotes addition, 011 denotes the register R3, and 000000001010 denotes the memory address 12.

Most instructions in an assembly language also denote primitive operations of a computer, but are written using symbols more meaningful to human programmers. For example, the above machine language instruction might be written in a hypothetical assembly language as

$$ADD\ R3\ X$$

where ADD denotes addition, R3 denotes the register, and X denotes a memory address. An assembly language program must first be translated into a machine language in order for a computer to execute it. The translator

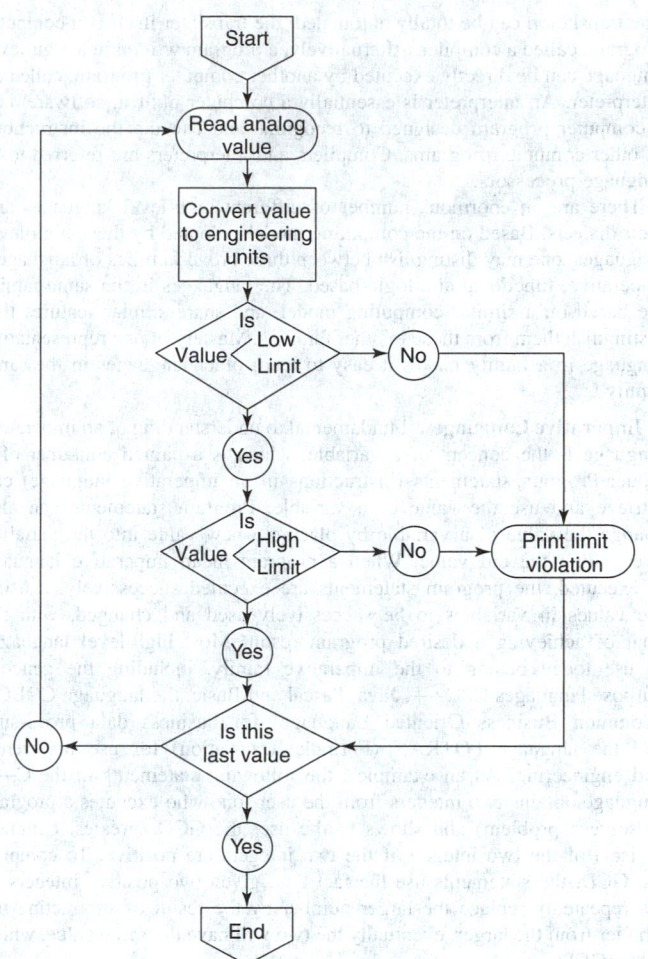

Fig. 1.   Representative computer programming flowchart.

itself is a computer program called an assembler. Clearly, both machine languages and assembly languages are machine-dependent: computers of distinct internal designs (different architectures) usually support different set of instructions for their machine and assembly languages. To a certain extent, a programmer needs to know the internal structure of a physical computer in order to gain proficiency in its assembly language or machine language. Low-level languages are generally considered difficult to use for large programming projects. Their instructions are primitive, and even a program of modest size and complexity may require a fairly large number of instructions and significant programming effort.

High-level Languages. The vocabulary of a high-level language generally includes common mathematical symbols, such as + and −, and a small number of key words, mostly from the English language. A computer programmer uses such a vocabulary and follows rigid syntactical rules to write instructions in a high-level language. High-level languages are machine-independent. The mathematical symbols and key words are not tied to the primitive operations or the internal structure of a specific physical computer. They denote arithmetical and logical operations and programming concepts that are commonly understood without reference to a specific computer. High-level languages are easier to use for large programming projects than low-level languages. A statement in a high-level language generally denotes an abstract operation equivalent to several primitive operations (low-level instructions) that are built into a computer. The use of commonly understood abstract operations and programming concepts in a high-level language relieves a programmer from concerns with the low-level details within a computer and enables a programmer to focus on figuring out a logically correct solution to a given problem and expressing the solution in a high-level language.

Clearly, a computer program written in a high-level language cannot be directly recognized and executed by any physical computer, but can be translated into a machine language and then executed on a computer.

The translation can be totally automated: the translator itself is a computer program, called a compiler. Alternatively, a program written in a high-level language can be directly executed by another computer program, called an interpreter. An interpreter is essentially a computer built in software, i.e., a computer program designed to recognize and execute the instructions of other computer programs. Compilers and interpreters are referred to as language processors.

There are an enormous number of different high-level languages and their dialects. Based on the computing models adopted by these high-level languages, one may distinguish between three broad families of languages: imperative, functional and logic-based. The languages in the same family are based on a similar computing model, and share similar features that distinguish them from those in other families. Mastery of one representative language in a family makes it easy to learn other languages in the same family.

Imperative Languages. Fundamental to understanding of an imperative language is the concept of a variable, which is a named container of a value. Program statements (instructions in an imperative language) can retrieve and use the value of a variable. Program statements can also change the value of a variable by placing a new value into the variable, overwriting the old value. When a program in an imperative language is executed, the program statements are executed successively, causing the values in variables to be successively used and changed, with the goal of achieving a desired program result. Most high-level languages in use today belong to the imperative family, including the general-purpose languages C, C++, Java, Pascal and Basic, the language COBOL (Common Business Oriented Language) for business data processing and the language FOTRAN (Formula Translation) for use in science and engineering. As an examples, the following statements in the C++ language obtains two integers from the user (one who executes a program to solve a problem) and shows to the user the GCD (greatest common divisor) of the two integers if the two integers are positive. To compute the GCD, the statements use the fact that, given two positive integers, if we repeatedly replace the larger number by the result of subtracting the smaller from the larger, eventually the two will have the same value, which is the GCD.

```
1.    int x, y;
2.    cin >> x >> y;
3.    while (x! = y)
4.        if (x > y)    x = x - y;
5.        else          y = y - x;
6.    cout >> x;
```

Line 1 declares that there are two integer variables named x and y. Line 2 obtains two integers from the user and places them into x and y. Line 3 dictates that, as long as x and y are not equal, lines 4 and 5 are to be repeatedly executed. In lines 4 and 5, if x is greater than y, the value of x is replaced by that of x-y; otherwise the value of y is replaced by that of y-x. In short, lines 4 and 5 subtract the smaller from the larger of the two values stored in x and y. Such an operation is to be performed repeatedly until x and y have the same value (as specified by line 3). As an example, if the user gives the values 21 and 35 to the program when line 2 is executed, the following table illustrates how the values of x and y are changed by line 4 and 5 as they are repeatedly executed.

x	21	21	7	7
y	35	14	14	7

Line 6 shows the value of x to the user. Since line 6 is executed after lines 4 and 5 have been repeatedly executed so that x and y have the same value, which is the GCD, the value shown to the user by line 6 is the desired result: the GCD of the two input integers.

Although a typical imperative language defines many versatile operations that a programmer can use directly, often a programmer needs an operation that is not built into the language. Modern imperative languages provide a means for a programmer to define a new operation in terms of the existing operations available to the programmer. Such a new operation is variously called a subprogram, a procedure or a function. Defining such a new operation is called procedural abstraction. Besides, most modern languages include features to support a programming concept called data abstraction, which allows grouping of related data and operations as integral parts of an entity called object. Data abstraction is the basis of object-oriented programming (OOP), which is a way to structure a computer program. OOP has been widely accepted in recent years and is supported by such widely used modern languages as C++ and Java. In object-oriented programming, a major task in program design is to identify the objects in the application domain, the relationship between the objects (e.g., an objects may be a specialization of another object), and the interactions among the objects.

Functional Languages. Functional languages are based on the theory of functions in mathematics. A function takes a number of given values as parameters and yields a single value as the result. A program is defined as a function, which may call other functions to compute the values for parameters. Execution of a program is application of a function on given parameter values. As an example, we may define a function to find the GCD (result) of two positive integers x and y (parameters) as follows. The definition uses the fact that if two positive integers are unequal then their GCD is equal to the GCD of the smaller of the two and the result of subtracting the smaller from the larger.

$$gcd(x, y) = \begin{cases} x & \text{if } x = y \\ gcd(x - y, y) & \text{if } x > y \\ gcd(x, y - x) & \text{if } x < y \end{cases}$$

Using 21 and 35 as parameters, the function gcd is evaluated as follows.

$$gcd(21, 35) = gcd(21, 14) = gcd(7, 14) = gcd(7, 7) = 7$$

It is worth noting that recursion (defining a function using the function itself) is used to accomplish the task of repeatedly subtracting the smaller from the larger of the parameters. LISP (LISt Processing) and Scheme, a dialect of LISP, are the most well known languages that adopt functional programming features. As an example, the above function gcd is defined below in the language Scheme

```
(define (gcd x y)
    (cond    ((= x y) x)
        ((> x y)    (gcd (- x y) y))
        ((< x y)    (gcd x (- y x))))
```

After the function has been defined, the function call $(gcd\ 21\ 35)$ in Scheme will yield the answer 7.

Logic-based languages. Logic-based languages are based on predicate logic. A program in a logic language is a set of facts and rules of inference, by which new facts can be deduced from existing facts. Logic-based languages are considered declarative, as opposed to procedural, because a program specifies what a problem is instead of providing a procedure for solving a problem. As an example, one may define the predicate that the *GCD* of two given positive integers is an integer X as follows.

1. For all positive integers X, the *GCD* of X and X is X
2. For all positive integers X, Y and Z, if $X > Y$ and Z is the *GCD* of X-Y and Y, then Z is the *GCD* of X and Y
3. For all positive integers X, Y and Z, if $X < Y$ and Z is the *GCD* of X and Y-X, then Z is the *GCD* of X and Y

A query is a theorem to be proved from known facts and rules of inference. By using the above rules (2 and 3 repeatedly), the query that the *GCD* of 21 and 35 is 7 can be proved to be true. It can also be deduced that in order for the query that the *GCD* of 21 and 35 is A (some unknown number) to be true, A must be 7.

The most well known logic-based programming language is Prolog. As an example, the three rules specified above are defined in Prolog below.

```
gcd(X, X, X)
gcd(X, Y, Z):- X > Y, X1 is X - Y, gcd(X1, Y, Z)
gcd(X, Y, Z):- X < Y, Y1 is Y - X, gcd(X, Y1, Z)
```

Using these rules, a Prolog interpreter will be able to deduce that the query $gcd(21, 35, 7)$ (i.e., the *GCD* of 21 and 35 is 7) is true. It can also deduce that in order for the query $gcd(21, 35, A)$ (i.e., the *GCD* of 21 and 35 is some unknown number A) to be true, A must be 7.

See also **Computer Operating System**.

SAMUEL C. HSIEH, Ball State University, Muncie, IN

PROGRAM RELOCATION (Computer System). A process of modifying the address in a unit of code, i.e., a program, subroutine, or procedure, to operate from a different location in storage than it was originally prepared for. Program relocation is an integral part of nearly all programming systems. The techniques allow a library of subroutines to be maintained in object form and made a part of a program by relocation and appropriate linking. When a program is prepared for execution by relocating the main routine and any included routines to occupy a certain part of the storage on a one-time basis, the process usually is termed *static relocation*. The resulting relocated program may be resident in a library and be loaded into the same location in the core each time it is executed. Dynamic storage-allocation schemes also may be set up so that each time a program is loaded, it is relocated to an available space in the core. This process is termed *dynamic relocation*. Time-sharing systems may temporarily stop execution of programs and store them on auxiliary storage and later reload them into a different location for continued execution. This process also would be termed dynamic relocation. See also **Time Sharing**.

Software relocation commonly refers to a method whereby a program loader processes all the code of a program as it is loaded and modifies any required portions. Auxiliary information is carried in the code to indicate which parts must be altered. Inasmuch as all code must be examined, this method can be time consuming. In *hardware relocation*, special machine components are used, as a base or relocation register, to alter addresses automatically at execution time to achieve the desired results. In dynamic relocation situations, this is a fast method. Coding techniques also are used for relocation. The resulting code is caused to be self-relocating. The code executes in any core location into which it is loaded. Index register may be used to make all references to storage. Double indexing is required to do both the relocation and the normal indexing operations. Or, the code may actually modify all addresses as it executes to provide the correct reference. This method is slow compared with other methods and requires additional storage.

THOMAS J. HARRISON, International Business Machines Corporation, Boca Raton, FL

PROGRESSION. A rather simple type of sequence or series, the most common ones being known as arithmetic, geometric, and harmonic.

1. *Arithmetic progression.* The form is $a, a+d, a+2d, a+3d, \ldots$ where a is the first term and d is the constant difference between two successive terms. If there are n terms in the series, its sum $S_n = n(a+l)/2 = n[2a+(n-1)d]/2$, where l is the last term. If any three of the five quantities a, d, n, l, S_n are known the other two may usually be found from the simultaneous equations $l = a + (n-1)d$ and the sum, S_n. If three numbers are members of an arithmetic progression, the middle one is the arithmetic mean. A plot of the magnitude of the terms in an arithmetic progression as ordinate against the number of terms n as abscissa will be a straight line of slope d and intercept a on the Y-axis.

2. *Geometric progression.* Its form is $a, ar, ar^2, ar^3, \ldots$ and its last term is $l = ar^{n-1}$. If written as $a(1 + r + r^2 + r^3 + \cdots)$ its sum to n terms is $a(1-r^n)/(1-r)$. When $r < 1$, the infinite series converges and its sum is $a(1-r)$. The series diverges for other values of r. The n-th root of the product of n positive quantities x_i is the geometric mean $G = (x_1 x_2 x_3 \cdots x_n)^{1/n}$.

3. *Harmonic progression.* The sequence $a, b, c, \ldots$ is a harmonic progression if the reciprocals $1/a, 1/b, 1/c, \ldots$ form an arithmetic progression. The harmonic mean between two numbers is the middle term of a harmonic progression whose first and last terms are the given numbers. The harmonic mean between a and b is given by $H = 2ab/(a+b)$. If A, G, and H are respectively the arithmetic, geometric, and harmonic mean of two numbers then $G^2 = AH$. The sum of its terms is a harmonic series. There is no general method of finding the sum. It is usually done by taking the reciprocal series and solving the resulting arithmetic series. If the series is

$$\sum_{n=1}^{\infty} 1/n^r, \text{ with } r \text{ real}$$

it converges for $r > 1$, but diverges for $r \le 1$. This is often called the hyperharmonic series.

Series of these three types are convenient as comparison series.

PROGRESSIVE MULTIFOCAL LEUKOENCEPHALOPATHY. A disease thought to be caused by an opportunistic virus in patients with impaired immune responses. Papoviruses have been implicated as the most common causal agents. The infection may occur in patients suffering from lymphoproliferative disorders and in those therapeutically immunosuppressed. Multifocal demyelinating lesions, often asymmetrical, result in progressive alteration of personality and intellect, sensory deficit, cortical blindness, and impairment of consciousness. The disease begins insidiously and usually terminates in death within six months.

No treatment has been found effective.

R. C. V.

PROGRESSIVE WAVE.

1. A wave that moves relative to a fixed coordinate system in a fluid; or, in meteorology, a wave or wavelike disturbance that moves relative to the earth's surface. Progressive waves are to be distinguished from stationary waves, which show no relative translation. Standing waves can be treated mathematically as two equal and oppositely directed progressive waves superimposed upon each other.

2. In oceanography, a wave the travel of which can be followed by monitoring the movement of the crest. Energy is transmitted; that is, the wave form travels significant distances, but the water particles perform oscillatory motions. See also **Kelvin Wave**.

AMS

PROJECTION CLOCK. See **Clocks**.

PROJECT STARSHINE. See **Space Science Missions: Sun**.

PROKARYOTES. See **Bacteriology; Cell (Biology)**; and **Microbiology**.

PROLAPSE. The falling or protrusion of an organ or structure, due to lack of support, usually secondary to weakness of its retaining ligaments or surrounding muscles. Prolapse of the rectum is a protrusion of a part of the rectal wall externally. Prolapse of the uterus is a protrusion of the uterus and other pelvic organs through the muscular floor of the pelvis. See also **Gonads**.

PROMETHIUM. [CAS: 7440-12-2]. Chemical element symbol Pm, at. no. 61, at. wt. 145 (mass number of the most stable isotope), fourth in the Lanthanide Series in the periodic table, mp 1042°C, bp 3000°C (estimated), density 7.26g/cm^3 (20°C). Elemental promethium has a double hexagonal closepacked crystal structure at 25°C. The pure metallic promethium is silverwhite in color, is soft, and can be cast or machined. The naturally occurring isotope ^{147}Pm is radioactive with a half-life of 2.52 years. Consequently, the element must be handled within a shielded area. Eighteen artificially produced isotopes, ranging from ^{140}Pm to ^{146}Pm and from ^{148}Pm to ^{158}Pm have been identified, all with very short half-lives. Many of the properties of promethium remain classified by the United States Atomic Energy Commission, or are known by other proprietary sources. Although first identified as an element by J.A. Marinsky, L.E. Glendenin, and C.D. Coryell in 1947, the element was not available on more than a gram-scale for several years. Electronic configuration

$$1s^2 2s^2 2p^6 3s^2 3p^6 3d^{10} 4s^2 4p^6 4d^{10} 4f^4 5s^2 5p^6 5d^1 6s^2$$

Ionic radius 0.98 Å. Other important physical properties of promethium are given under **Rare-Earth Elements and Metals**.

^{147}Pm is extracted from the wastes of uranium or plutonium reactors, the most important source of the element. ^{146}Pm and ^{148}Pm also are derived from reactor wastes. In 1970, ^{147}Pm became available in kilogram quantities. ^{147}Pm has been under intensive study as a heat and power source; however, before it can be used for this, ^{146}Pm and ^{148}Pm, which produce penetrating gamma radiation, must be eliminated. The desirable property of ^{147}Pm is that it decays by beta emission only, at a low energy level compared with most fission products, and thus requires only light to moderate shielding. ^{147}Pm has been used to activate luminescent phosphors. Beads (Microspheres®, 3 M Company) containing ^{147}Pm mixed with a phosphor provide a long-lived, reliable green light and were used by astronauts to assist in docking and other maneuvers in outer space. Commercial applications of ^{147}Pm as a power source include beta-voltaic cells for surgical implant with heart pumps and pacemakers.

Additional Reading

Greenwood, N.N. and A. Earnshaw: *Chemistry of the Elements*, 2nd Edition, Butterworth-Heinemann, Inc., Woburn, MA, 1997.

Krebs, R.E.: *The History and Use of Our Earth's Chemical Elements: A Reference Guide*, Greenwood Publishing Group, Inc., Westport, CT, 1998.

Lide, D.R.: *CRC Handbook of Chemistry and Physics*, 88th Edition, CRC Press, LLC., Boca Raton, FL, 2007.

Parker, P.: *McGraw-Hill Encyclopedia of Chemistry*, 2nd Edition, The McGraw-Hill Companies, Inc., New York, NY, 1993.

Stwertka, A. and E. Stwertka: *A Guide to the Elements*, Oxford University Press, Inc., New York, NY, 1998.

PROMINENCE.   A filament-like protuberance from the chromosphere of the sun. Prominences can be observed visually (optically) whenever the sun's disk is masked, as during an eclipse or by using a coronagraph; and can be observed instrumentally by filtering in certain wavelengths, as with a spectroheliograph. A typical prominence is 6,000 to 12,000 kilometers thick (3,728 to 7,456 miles), 60, 000 kilometers high (37,282 miles), and 200,000 kilometers long (124,274 miles). These features appear as filaments when they are seen against the solar disk.

PROMOTER.

1. A substance that, when added in relatively small quantities to a catalyst, increases its activity, e.g., aluminum and potassium oxide are added as promoters to the iron catalyst used in facilitating a combination of hydrogen and nitrogen to form ammonia.

2. In ore flotation, a substance that provides the minerals to be floated with a water-repellent surface that will adhere to air bubbles. Such reagents are generally more or less selective toward minerals of certain classes.

PRONGHORN ANTELOPE *(Mammalia, Artiodactyla)*.   Also sometimes referred to as the American Antelope, actually this animal is not classified as an antelope *(Antelopine)*, but rather it falls into a special, small class of the *Artiodactyla* (even-toed hoofed animals), known as the *Antilocaprines*. The *Antilocapra americana* is a plains animal of the western half of North America. It is distinguished by the erect horns, hooked at the tip, and bearing a short branch in front. The pronghorn differs from true antelopes in the branching of the horn sheaths and in the periodical shedding and renewal of these sheaths. The horns are constructed something like those of a giraffe. However, it is much like an antelope in general appearance. The pronghorn is approximately 3 feet (0.9 meter) high, 5 feet (1.5 meters) long, with large eyes, short tail, erect ears, and slender legs. The coloration is a mottled brown, with a chestnut mane. Its recognition mark is a brilliant white rump patch. The pronghorn is known for its fast speed and running endurance. The animal, although timid, is described as curious. The diet is almost entirely grass. It is hunted for sport and food. The flesh is dry, but considered good. The animals faced extinction in the early 1900s, but as the result of government protection, have staged a strong comeback on reservations.

The pronghorn is the only animal in the *Antilocaprines* family. Nothing like the pronghorn is known elsewhere and it is assumed that the species is of North American origin. The animal is considered by some authorities as a leftover from pre-glacial times, when there were probably several more species. The animal also has been described as a possible experiment by nature in the development of an antelope, which was given up without carrying the process to completion.

PROOF.   The ethanol content of a liquid at 15.5C, stated as two times the percentage of ethanol by volume. One gallon of 95% alcohol is therefore equivalent to 1.9 gallons of proof alcohol. In the U.S., the alcohol tax is based on the number of proof gallons.

PROPAGATION CONSTANT.   This is a characteristic of a transmission line and indicates the effect of the line on the wave being transmitted along the line. It is a complex quantity having a real term, the *attenuation constant*, and an imaginary term, the *wavelength constant* or *phase constant*. See also **Line (Mathematics)**. The attenuation constant is a measure of the reduction in amplitude as the wave travels along a matched line while the wavelength constant is a measure of the phase shift which it undergoes. These relations may be expressed by the following equations:

$$\gamma = \alpha + j\beta$$
$$I = I_s e^{-\gamma x} = I_s e^{-\alpha x} e^{-j\beta x}$$
$$E = E_s e^{-\gamma x} = E_s e^{-\alpha x} e^{-j\beta x}$$

where γ is the propagation constant, α, the attenuation constant, β, the wavelength constant, x the distance from the input of the line, I_s and E_s the current and voltage at the input of the line, and I and E the corresponding quantities at a distance x from the input, and e is the number $2.718\ldots$, the base of natural logarithms.

Because of the attenuation on lines used for communication purposes it is necessary to insert amplifiers or repeaters at intervals to build the signal back to suitable levels. For sound transmission work the phase shift is usually not important, but for television and picture transmission it is extremely important and necessitates correcting circuits.

PROPAGATION (Direction of).   At any point in a homogeneous, isotropic medium, the direction of time-average energy-flow. In a uniform waveguide, the direction of propagation is often taken along the axis. In the case of a uniform lossless waveguide, the direction of propagation at every point is parallel to the axis, and in the direction of time-average energy-flow.

PROPANE.   [CAS: 74-98-6], $CH_3 \cdot CH_2 \cdot CH_3$, formula weight 44.09, colorless gas, mp $-187.1\,°C$, by $-42.2\,°C$, sp gr 0.585 (at $-45\,°C$). The gas is slightly soluble in H_2O, moderately soluble in alcohol, and very soluble in ether. Although a number of organic compounds which are important industrially may be considered to be derivatives of propane, it is not a common starting ingredient. The content of propane in natural gas varies with the source of the natural gas, but on the average is about 6%. Propane also is obtainable from petroleum sources.

Liquefied propane is marketed as a fuel for outlying areas where other fuels may not be readily available and for portable cook stoves. In this form, the propane may be marketed as LPG (liquefied petroleum gas) or mixed with butane and pentane, the latter also constituents of natural gas (1.7% and 0.6%, respectively). LPG also is transported via pipelines in certain areas. The heating value of pure propane is 2,520 Btu/ft^3 (283 Calories/m^3); butane 3,260 Btu/ft^3 (366 Calories/m^3); and pentane 4,025 Btu/ft^3 (452 Calories/m^3). Propane and the other liquefied gases are clean and appropriate for most heating purposes, making them very attractive where they are competitively priced.

PROPER MOTION (Star).   The individual motion of a star relative to the other stars. Up to the early part of the eighteenth century, the belief was current that the stars were all fixed on a sphere commonly known as the celestial sphere. Since this sphere was apparently rotating about the earth, and possessed other motions such as precession and nutation, all of the stars had certain motions in common. In 1718, Edmund Halley, while reducing his observations of the positions of the stars, noted that the positions he obtained for Sirius and Arcturus differed in position relative to the other stars from the positions given by Ptolemy. Since the differences were greater than could be ascribed to errors in observation, Halley concluded that these two stars were actually not fixed on the celestial sphere but were in motion.

Since Halley's time, many stars have been observed to have proper motion, and many long programs are under way to study these motions. The standard method of procedure is to compare positions of the stars at two epochs as widely separated as possible. Visual observations made with extreme precision with a meridian circle may be used for this purpose, but the photographic methods are far more fruitful since a large number of star positions may be obtained on a single plate. The longer the interval of time between the observations, the more accurate is the determination of the proper motion, and also the smaller the proper motion which can be detected.

Proper motion can be determined only in angular units, and the results are usually expressed in terms of seconds of arc per year. If the parallax of the star is known, the velocity in linear units may be computed in accordance with methods discussed under space velocity of a star. The largest known proper motion, 10.25″ per year, was found by Barnard for a tenth magnitude star. Such a star would require about 200 years to change its position by an amount equal to the apparent diameter of the moon. There are only about 50 stars known to have proper motions greater than 2″ per year, and not more than 1,000 with values greater than $\frac{1}{2}$″ per year. Hence, we should not expect the constellation Figures to have altered appreciably in the 5,000 years since they were first described.

PROPIONIC ACID.   See **Antimicrobial Agents (Foods)**; and **Food Additives**.

PROPOLIS.
A material gathered by honey-bees and used for closing crevices in the hive and for filling in sharp angles and attaching loose parts. It is also applied as a varnish to the combs and to the smooth surfaces in the hive. The substance is composed chiefly of resins gathered from plants and has an aromatic fragrance much like that of the leaf buds which furnish some of these resins.

PROPORTIONAL COUNTER.
A detector of ionizing radiations that operates in a voltage region intermediate between an ionization chamber and a Geiger counter. For this counter, the size of the output pulse is proportional to the number of ions formed in the initial ionizing event. Because its operating voltage is lower than for a Geiger counter, avalanche ionization is limited to that portion of the counter in the immediate vicinity of the primary ionization and does not spread along the entire central wire electrode. As a result, its gas amplification is constant for all pulses at any one voltage. The gas amplification is the number of additional ions produced by each electron produced in the initial ionizing event as it travels to the central wire. See also **Geiger Counter**.

PROPORTIONAL LIMIT.
The maximum unit stress that can be obtained in a structural material without causing a change in the ratio of the unit stress to the unit deformation is called the proportional limit.

PROPRIOCEPTIVE STIMULATION.
Stimulation originating within the deeper structures of the body (muscles, tendons, joints, etc.) for sense of body position and movement and by which muscular movements can be adjusted with a great degree of accuracy and equilibrium can be maintained.

PROSTAGLANDINS.
A group of physiologically active compounds (PGs) derived from fatty acids with 20 carbon atoms (approximate formula, $C_{20}H_{36}O_5$). The compounds originally were isolated as lipid-soluble extracts from sheep and human prostates. Later studies have shown that prostaglandins are found in most mammalian tissues. There are numerous prostaglandins, individually named by the substituents present on the cyclopentane ring that is part of the parent molecule, prostanoic acid. Thus, they are identified as PGA_1, PGE_1, PGI_2 (prostacyclin), etc. The chemical structure and metabolic functions of the prostaglandins have been established, in most cases, with considerable accuracy. Some have been synthesized. Each prostaglandin has specific effects. The compounds participate in pulmonary circulation and hypertension, with varying vasodilator and vasoconstrictor effects. Prostaglandins of the E series have been implicated as a cause of hypercalcemia — they resorb fetal bone in vitro, urinary prostaglandin metabolites are elevated in certain hypercalcemic patients with malignancy, and clinically very important, in certain cancer patients with hypercalcemia. Chemical improvement has been seen after treatment with indomethacin, which inhibits prostaglandin synthesis. Prostaglandins also are implicated in systemic mastocytosis, due partly to marked overproduction of prostaglandin D_2. Prostacyclin plays an important role in platelet function, acting as an effective antiaggregating agent. The prostaglandins are involved in the biochemical pathways

that participate in bronchial asthma. PGs are synthesized ubiquitously in the body from unsaturated fatty acid precursors with high rates of production by the seminal vesicles and renal medulla. The metabolism of prostaglandins occurs mainly in the lungs, renal cortex, and liver, with the metabolites excreted in the urine. The most prolific source of natural prostaglandins is a marine organism (gorgonian sea whip) found in great numbers in coral reefs, notably in the Caribbean area. Intermediates and chemical analogs derived from this organism are sometimes referred to as *syntons*.

Additional Reading

Champe, P.C., and R.A. Harvey: *Lippincott's Illustrated Reviews Biochemistry*, 2nd Edition, Lippincott Williams & Wilkins, Philadelphia, PA, 1994.
Marks, F., and G. Furstenberger: *Prostaglandins, Leukotrienes, and Other Eicosanoids: From Biogenesis to Clinical Application*, John Wiley & Sons, Inc., New York, NY, 1999.
Yazici, Z., and G.C. Folco: *Advances in Prostaglandin, Leukotriene, and Other Bioactive Lipid Research: Basic Science and Clinical Applications*, Kluwer Academic Publishers, Norwell, MA, 2003.

PROSTATE GLAND.   See **Gonads**.

PROTACTINIUM.
[CAS: 7440-13-3]. Chemical element, symbol Pa, at. no. 91, at. wt. 231.036, radioactive metal of the Actinide Series, mp is estimated at less than 1600°C. All isotopes are radioactive. The most stable isotope is ^{231}Pa with a half-life of 3.43×10^4 years. The latter is a second-generation daughter of ^{235}U and a member of the actinium ($2n + 3$) decay series. See also **Radioactivity**. Electronic configuration

$$1s^2 2s^2 2p^6 3s^2 3p^6 3d^{10} 4s^2 4p^6 4d^{10} 4f^{14} 5s^2 5p^6 5d^{10} 5f^2 6s^2 6p^6 6d^{10} 7s^2$$

Ionic radii Pa^{4+} 0.91 Å; Pa^{3+} 1.06 Å. See also **Chemical Elements**.

The probable existence of protactinium was predicted as early as 1871 by Mendeleev to fill up the space on his periodic table between thorium (at. no. 90) and uranium (at. no. 92). He termed the unconfirmed element *ekatantalum*. In 1926, O. Hahn predicted the properties of the element in considerable detail, including descriptions of its compounds. In 1930, Aristid v. Grosse isolated 2 milligrams of what then was termed ekatantalum pentoxide and showed that element 91 differed in all reactions with comparable amounts of tantalum compounds with exception of precipitation by NH_3. However, credit for the discovery of protactinium generally is attributed to Lise Meitner and Otto Hahn in 1917.

Protactinium-231 yields actinium-227 by α-particle emission and has a half-life of 3.43×10^4 years. Its other isotopes include two isomers of mass number 234: uranium X_2 with a half-life of 1.17 minutes, and uranium Z with a half-life of 6.7 hours, the former being an excited state which undergoes de-excitation to give the latter. Other nuclear species have mass numbers 225–230, 232, 233, 235 and 237.

Protactinium (of mass number 231) is found in nature in all uranium ores, since it is a long-lived member of the uranium series. It occurs in such ores to the extent of about $\frac{1}{4}$ part per million parts of uranium. An efficient method for the separation of protactinium is by a carrier technique using zirconium phosphate which, when precipitated from strongly acid solutions, coprecipitates protactinium nearly quantitatively. Then the protactinium is separated from the carrier by fractional crystallization of zirconium oxychloride.

Isotopes of protactinium can also be produced artificially, i.e., by the nuclear reactions of other elements with such particles as deuterons, neutrons, and alpha-particles. Thus, when thorium is bombarded with deuterons of various high energies, five of the reactions are: ^{232}Th(d,4n)^{230}Pa, ^{232}Th(d,6n)^{228}Pa, ^{232}Th(d,7n)^{227}Pa, ^{232}Th(d,8n)^{226}Pa, and ^{230}Th(d,3n)^{229}Pa.

Quantitative methods of obtaining protactinium start from the carbonate precipitate from the treatment of the acid extract of certain uranium ores. After this carbonate precipitate is dissolved, the protactinium remains in the silica gel residue, from the solution of which it is obtained on a manganese dioxide carrier. An alternate method effects final separation of the protactinium by formation of a complex compound, protactinium-cupferron, and its extraction with amyl acetate.

The methods of purification include the use of ion exchange resins, the precipitation of protactinium peroxide and the extraction of aqueous solutions of protactinium salts by various organic solvents.

Protactinium metal is prepared: (1) by reducing the tetrafluoride with metallic barium at about 1,500 °C; (2) by heating the halide, usually the iodide, under a high vacuum; and (3) by bombardment of the oxide under high vacuum with 35-keV electrons for hours at a current strength of 0.005–0.010 Amperes.

As early as 1965, investigators at Los Alamos (Fowler et al., 1965) reported that protactinium metal is superconductive below 1.4 K. In 1972, researchers at Harwell (Mortimer, 1972) reported no superconductivity of the metal down to approximately 0.9 K. An exchange of information to resolve the differences in data was conducted over the next few years (Fowler, 1974; Hall et al., 1977). Smith, Spirlet, and Mueller (1979) reported that differences in experimental research were due to problems with the crystal structure of the metal and sample purity that arise when dealing with radioactive material. These investigators observed very-high-purity protactinium, produced by the Van Arkel procedure, and observed an extremely steep superconductivity transition at 0.42 K in protactinium in the presence of rather high self-heating. The superconducting transition temperature and upper critical magnetic field of protactinium were measured by alternating-current susceptibility techniques. Inasmuch as the superconducting behavior of protactinium is affected by its $5f$ electron character, it has been further confirmed that protactinium is a true actinide element.

The predominant oxidation state of the element is (V). There is some evidence that the (IV) state is obtained under certain reduction conditions. When the pentapositive form is not in the form of a complex ion it may exist in solution as PaO_2^+. The compounds are very readily hydrolyzed in aqueous solution yielding aggregates of colloidal dimensions, thus showing marked similarity to niobium and tantalum in this respect. These properties play a dominant role in the chemical properties of aqueous solution, because the element is so easily removed from solution by hydrolysis and adsorption. Protactinium coprecipitates with a wide variety of substances, and it seems likely that the explanation for this lies in the hydrolytic and adsorptive behavior.

The element is difficult to maintain in aqueous solution in the form of simple salts. Solubility data seem to indicate that such amounts as can be dissolved probably do so entirely by formation of complex ions. Fluoride ion strongly complexes protactinium, and it is due to this that protactinium compounds are in general soluble in hydrofluoric acid.

Protactinium oxide may be prepared from the hydrated oxide or the oxalate by ignition. The product is a dense white powder with a very high melting point; the ignited material is not hygroscopic and maintains a constant weight upon exposure to the air. The formula Pa_2O_5 has been determined indirectly, and there is evidence for the existence of $PaO_{2.25}$ (air oxidation) and PaO_2 (reduction of P_2O_5 by H_2).

Volatile protactinium pentachloride has been prepared in a vacuum by reaction of the oxide with phosgene at 550 °C or with carbon tetrachloride at 200 °C. Reduction of this at 600 °C with hydrogen leads to protactinium(IV) tetrachloride, $PaCl_4$, which is isostructural with uranium(IV) tetrachloride, UCl_4. The pentachloride can be converted into the bromide or iodide by heating with the corresponding hydrogen halide or alkali halide.

The volatile fluoride protactinium(V) fluoride, PaF_5, or possibly protactinium(V) oxyfluoride, $PaOF_3$, is formed at relatively low temperatures such as 200 °C from the action of agents such as bromine tri- or pentafluoride, BrF_3 or BrF_5, on one of the protactinium oxides. At higher temperatures, treatment of Pa_2O_5 with hydrofluoric acid and hydrogen yields PaF_4.

The reduction of protactinium to the (IV) state in aqueous solution can be accomplished by reducing agents, such as zinc amalgam, and polarographically.

Additional Reading

Fowler, R.D., et al.: *Phys. Rev. Lett.*, **15**, 860 (1965).
Fowler, R.D., et al.: *Proceedings of the 13th International Conference on Low Temperature Physics* (K.D. Timmerhaus, et al.), Plenum, New York, NY, 1974.
Greenwood, N.N. and A. Earnshaw: *Chemistry of the Elements*, 2nd Edition, Butterworth-Heinemann, Inc., Woburn, MA, 1997.
Hall, R.O.A., J.A. Lee, and M.J. Mortimer: *J. Low Temp. Phys.*, **27**, 305 (1977).
Krebs R.E.: *The History and Use of Our Earth's Chemical Elements: A Reference Guide*, Greenwood Publishing Group, Inc., Westport, CT, 1998.
Lide, D.R.: *CRC Handbook of Chemistry and Physics*, 88th Edition, CRC Press, LLC., Boca Raton, FL, 2007.
Mortimer, J.J.: Harwell Report AERE-R 7030 (1972).
Parker, P.: *McGraw-Hill Encyclopedia of Chemistry*, 2nd Edition, The McGraw-Hill Companies, Inc., New York, NY, 1993.
Smith, J.L., J.C. Spirlet, and W.C. Miller: "Superconducting Properties of Protactinium," *Science*, **205**, 188–190 (1979).
Stwertka, A. and E. Stwertka: *A Guide to the Elements*, Oxford University Press, Inc., New York, NY, 1998.

PROTEASES.

A protease is any enzyme that conducts proteolysis, that is, begins protein catabolism by hydrolysis of the peptide bonds that link amino acids together in the polypeptide chain.

Introduction

Proteins play an essential part in the structures of all living organisms, and growth and reproduction are therefore dependent upon protein synthesis. The building blocks for protein synthesis are amino acids, and a supply of these is obviously required. It is energetically more efficient to recycle the amino acids from pre-existing protein molecules than to synthesize them *de novo*. Indeed, many organisms have now lost the biochemical pathways to produce some of the amino acids, and so can only obtain them from external sources. Proteases are the biological catalysts that carry out the necessary disassembly of protein molecules to create the pool of free amino acids for protein synthesis. This activity is perhaps most familiar in the context of the digestion of food proteins in the animal digestive tract. In evolution, the digestion of food proteins for nutrition may well have been the most primitive function of proteases, but they have now evolved to perform many other vital functions as well. This diversification has led to a situation in which the genes for proteases represent about two per cent of the total genes in all kinds of organisms. See also **Amino Acids**.

Different Kinds of Protease Activity. It might seem that there could be totally nonspecific proteases that were capable of hydrolyzing the peptide bond between any pair of amino acid residues in a polypeptide and thus single-handedly breaking down any protein, but this is not what we find. The reality is that all known proteases have much more restricted activity, and it has proved useful to subdivide the proteases on the basis of the position of peptide bonds they are able to cleave in the substrate molecule. In such "positional specificity," the most fundamental distinction is between proteases that are restricted to acting near one or other end of the polypeptide chains, called *exopeptidases*, and those that can act on the internal bonds in a polypeptide, called *endopeptidases* (also *proteinases*). *Peptidase* is an alternative term for *protease* that is theoretically preferable, and is used in forming the names of the subgroups of these enzymes as is shown in Figure 1. See also **Protein**.

Specificity Subsites in the Selectivity of Proteolysis. Both exopeptidases and endopeptidases exhibit "sequence specificity," i.e. they show varying preferences for particular amino acids near the peptide bonds to be cleaved. The active site of a protease is commonly located in a groove on the surface of the molecule, and the substrate specificity is dictated by the properties of binding sites arranged along the groove on one or both sides of the catalytic site that is responsible for hydrolysis of the peptide bond. Accordingly, the specificity is described by use of a model in which each specificity subsite is able to accommodate the side-chain of a single amino acid residue. The sites are numbered from the catalytic site, S1, S2 and so on towards the amino-terminus of the substrate, and S1′, S2′ and so on towards the carboxy-terminus. The substrate amino acids they accommodate are numbered P1, P2, etc., and P1′, P2′, etc., respectively (Figure 2).

Five Catalytic Types of Proteases. In addition to the classification of proteases by the kind of reaction they catalyse, it has proved valuable to divide them according to the chemistry of the catalytic site. Five groups are recognized in this way (Table 1). In serine and threonine proteases the side-chain hydroxyl of one of these amino acids forms a transient acyl enzyme intermediate in catalysis, and in a cysteine protease the thiol of a cysteine residue serves a similar function. In the second stage of catalysis, a secondary catalytic group causes the hydrolysis of the acyl enzyme.

Exopeptidase activities

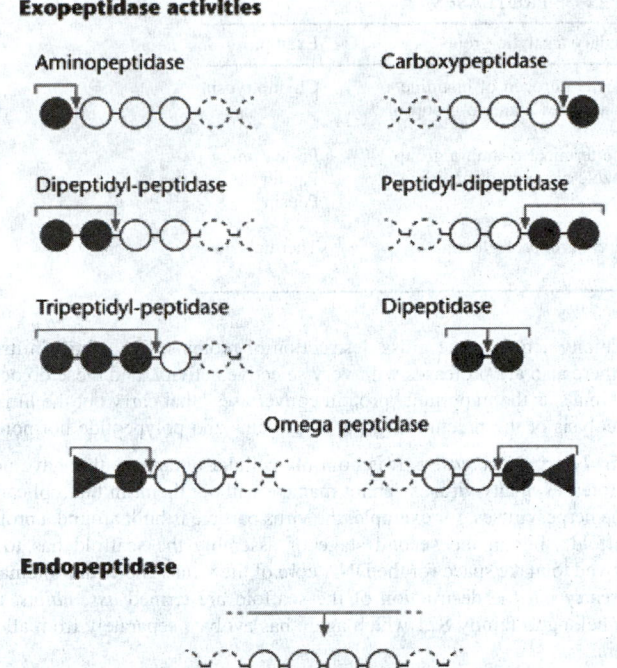

Fig. 1.   Types of protease activity. Proteases can be divided into *exopeptidases* and *endopeptidases*. The exopeptidases act only near the ends of polypeptide chains. Those acting at a free amino-terminus may liberate a single amino acid residue (aminopeptidases), a dipeptide (dipeptidyl-peptidases) or a tripeptide (tripeptidyl-peptidases). Those acting at a free carboxy-terminus liberate a single residue (carboxypeptidases) or a dipeptide (peptidyl-dipeptidases). Other exopeptidases are specific for dipeptides (dipeptidases), or remove terminal residues that are substituted, cyclized or linked by isopeptide bonds (peptide linkages other than those of α-carboxyl to α-amino groups) (omega peptidases). In the figure, the circles represent amino acid residues, the down-arrows show the bonds that are hydrolyzed, and the brackets attached to the arrows indicate the part of the substrate molecule typically recognized by the specificity sites of the enzymes prior to catalysis, and thus directing specificity. Proteases that act internally in polypeptide chains (usually whole protein molecules) are called *endopeptidases*. Primary determinants of endopeptidase specificity are amino acids near the scissile peptide bond on either side.

Aspartic peptidases typically use a pair of aspartic acid residues for their catalytic activity, whereas the catalytic site of a metallopeptidase contains a firmly bound atom of zinc (or occasionally another metal) that plays a central role. There are a few proteases for which the catalytic mechanism has yet to be identified. The catalytic mechanisms of protease action have been lucidly reviewed by Polgár. For each catalytic type of protease there is a set of inhibitors that are somewhat specific, because they are directed at this type of chemistry [Dunn, ref.]. See also **Catalysis**.

Families and Clans of Proteases. We have seen that proteases can be grouped according to the kind of peptidase activity they express

and also by catalytic type. These long-established methods are used in the enzyme nomenclature of the International Union of Biochemistry and Molecular Biology (IUBMB), which can be found on the World Wide Web http://www.iubmb.unibe.ch/ (Nomenclature Committee of the International Union of Biochemistry and Molecular Biology, 2000). The IUBMB system has two limitations, however. First, it provides for only about a dozen groups into which the proteases can be classified and, with many hundreds of these enzymes now known, this causes some quite dissimilar enzymes to be placed together. Perhaps more importantly, it gives no help in predicting the likely properties of a protease that is discovered only by its amino acid sequence in the sequencing of the genome of an organism. In an attempt to solve these problems, a new classification has been developed that is based upon the structural similarities between proteases. The similarities in structure reflect evolutionary relationships and very often functional similarities also. Unlike the IUBMB system, the structure-based approach can be applied to a protease that is known only from its gene, and so can allow predictions about properties and functions at a very early stage in the study of the enzyme, which is particularly important in the genomic era. The system was first described by Rawlings and Barrett but is constantly developing in the light of new research findings. The present form of the classification can be found in the MEROPS database http://merops.sanger.ac.uk/.

In the MEROPS system, individual proteases are first grouped into families on the basis of statistically significant similarities between the amino acid sequences of the parts that are most directly responsible for catalytic activity. The families are then grouped into clans when it is apparent from similar three-dimensional structures or amino acid sequence motifs that they have evolved ultimately from a single peptidase. The system provides a simple set of identifiers by which each clan, family and individual peptidase can be referred to (Table 2). The MEROPS system of classification provided the structure for a comprehensive treatise on proteolytic enzymes, the *Handbook of Proteolytic Enzymes* [by Barrett, et al.: 1998].

Regulation of Protease Activity. The living cells in which proteases are synthesized and often function are themselves largely constructed of proteins, so it is evident that the enzymes must be strictly regulated to prevent them from doing damage. At the molecular level this is commonly managed by their being synthesized as inactive proenzymes that require activation, usually by proteolysis [Khan and James, ref.]. Some proteases are also confined to membrane-limited organelles such as lysosomes, where they can act in a specialized environment containing only their substrates and other proteins that are resistant to proteolysis. A third form of regulation of protease activity is mediated by proteins that inhibit the activity of these enzymes.

Serine Protease

Proteases that depend for their catalytic activity on a serine residue in the active site are extremely numerous. Almost always an amino acid residue of histidine works with the serine to complete the catalytic machinery, and commonly an aspartate completes a "catalytic triad." The chemistry of the hydroxyl group of serine is such that these proteases are usually most active at neutral or slightly alkaline pH values, and they normally

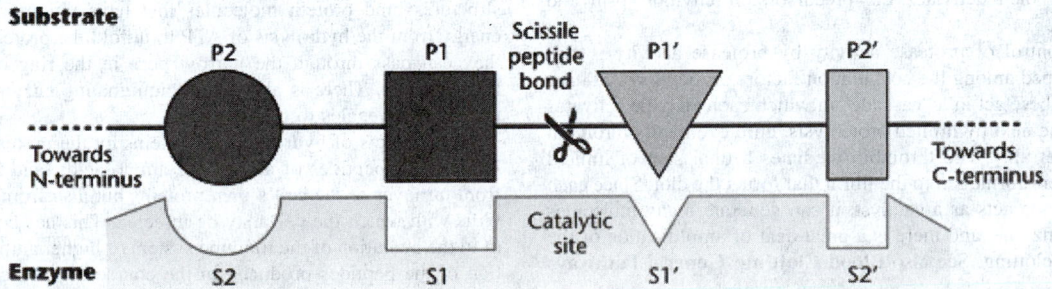

Fig. 2.   Scheme for the specificity subsites of proteases. It can be seen that the specificity subsites are numbered from the catalytic site, S1ʹ, S2ʹ, … Snʹ towards the amino-terminus of the substrate, and S1ʹ, S2ʹ, … Snʹ towards the carboxy-terminus. The amino acids they accommodate are numbered P1, P2, … Pn, and P1ʹ, P2ʹ, … Pnʹ, respectively.

TABLE 1. FIVE CATALYTIC TYPES OF PROTEASES

Catalytic type	Primary catalytic group	Secondary catalytic group	Example
Serine	Hydroxyl group of serine	Imidazole nitrogen of histidine (Sometimes ε-amino group of lysine)	Chymotrypsin
Threonine	Hydroxyl group of threonine	Amino-terminal α-amino group	Proteasome
Cysteine	Thiol group of cysteine	Imidazole nitrogen of histidine	Papain
Aspartic	Carboxyl groups of two aspartic acid residues	–	Pepsin
Metallo	Zinc atom (Sometimes another metal)	Carboxyl group of glutamate	Thermolysin

TABLE 2. THE MEROPS STRUCTURE-BASED CLASSIFICATION OF PROTEASES

Families in the MEROPS system

Proteases are grouped in families each of which is founded on a *type example*. Because proteases are commonly mosaic proteins, the peptidase unit is identified that is most directly responsible for activity, and only this is used in establishing relationships. Any new member added to a family will have highly significant sequence similarity to the peptidase unit of the type example or another existing member of the family. Deep divergences seen in the dendrogram for a family justify the recognition of subfamilies.

Clans are groups of related families

Although proteases in different families have no significant similarity in amino acid sequence there may still be evidence that they share a common origin. This can come from similar three-dimensional structures or conserved sequence motifs around the catalytic amino acids. The families of peptidases showing these distant relationships are grouped together in a clan.

Families, clans and proteases get simple identifiers

Each family is assigned a simple identifier that starts with a letter denoting the catalytic type of the proteases it contains, i.e. A, C, M, S, T or U for aspartic, cysteine, metallo, serine, threonine or unknown type. The family identifier is completed with a number that is assigned sequentially. The identifier of a clan starts with a letter based on the catalytic type or types of the families it contains and is completed by a second letter of the alphabet assigned sequentially as the clan is recognized. An identifier for each protease is constructed from the family identifier followed by a decimal point and a three-digit number. An example would be S01.151 for trypsin in family S1.

The resulting system

About 150 gene families containing proteases are now recognized, and most of the families can be assigned to one of about 30 clans. Version 5.0 of the database contains data for over 1000 individual proteases from over 1000 organisms. Amongst these, about 300 are known from the human genome.

require no cofactors for activity. These properties suit them particularly for extracellular functions. Serine proteases have appeared on a number of different occasions in the evolution of living organisms [Barrett and Rawlings, ref.], so there are several clans containing families of serine proteases with completely different protein folds. We shall look at one family from each of four of these clans.

Trypsin Family. Among the most familiar of all proteases are trypsin and chymotrypsin, which are serine proteases of family S1 (see Table 2) that act in the digestion of food proteins in the intestines of vertebrate animals. The enzymes are synthesized as inactive proenzymes in the pancreas and then secreted into the small intestine, where the trypsin precursor is activated by enteropeptidase, an enzyme specialized for this role. Trypsin itself then activates the precursors of chymotrypsin and elastase.

The theme of control of protease activity by protease activity is still more highly developed among the coagulation factors of the blood clotting system. Several of these act in a "cascade" in which each enzyme activates the proenzyme of the next by limited proteolysis, until eventually thrombin is formed [Davie, et al.:]. The thrombin mediates a final step of limited proteolysis to convert fibrinogen to the fibrin that forms the clot. Since each enzyme in the cascade acts as a catalyst, it can generate many molecules of the next active enzyme, and there is a great deal of amplification of the initial stimulus for clotting. See also **Blood Clotting: General Pathway**.

Subtilisin Family. While the proteases of the trypsin family are the general-purpose serine proteases of animals, this role is played by enzymes of the unrelated subtilisin family (S8) in other organisms such as bacteria, plants and fungi. The family contains many endopeptidases with broad specificity acting in the gross destruction of proteins, like subtilisin itself, but there also are proteases with very selective activity, and these do occur in animals as the important "protein convertases" that carry out the limited proteolysis of the precursor forms of proteins and polypeptide hormones.

Viral Serine Proteases. It is not only cellular organisms that have need of proteases: many viruses cannot manage without them. In the replication of the herpesviruses, for example, the virus particle is built around a protein "scaffold," but in the second stage of assembly the scaffold has to be removed to make space for the DNA core of the virus. The viral proteinases that carry out the destruction of the scaffold are termed *assemblins*, and they belong to family S21 which again has evolved separately from all the others.

Signal Peptidases. Proteins that are destined to be secreted from cells to perform extracellular functions are synthesized with amino-terminal signal peptides. The signal peptide causes the newly synthesized protein molecule to enter the secretory pathway of the cell, in the endoplasmic reticulum of eukaryotic cells. An integral part of the protein complex that transports the protein molecule across the membrane into the endoplasmic reticulum is a signal peptidase that cleaves off the signal peptide as the protein passes through. The distinctive family S26 signal peptidases are found in all classes of cellular organisms from bacteria to mammals.

Threonine Proteases

The threonine proteases are the most recently recognized group, and the major example is the proteasome that is responsible for much of the degradation of proteins in eukaryotic cells. The fundamental structure of the proteasome is that of a tube built by stacking four rings on top of one another [Groll, et al.:]. Each ring is constructed of seven subunits, and the two inner rings are formed of the β-type subunits that carry the catalytic sites exposed on the walls of the inner chamber. The two outer rings, formed from α subunits, control access of protein molecules to the inner chamber where proteolysis occurs. This basic structure is that of the small, or 20S, proteasome, but arguably the larger, 26S proteasome is the more important form, because this can respond to the ubiquitination of substrate proteins. Much of the protein degradation in cells is controlled by the ubiquitination system in which protein molecules are marked for destruction by covalent attachment of a small protein called ubiquitin. This causes them to be targeted to the 26S proteasome. The core of this is the 20S proteasome described above, but it contains additional structures attached to both ends of the 20S complex, apparently controlling the passage of substrate molecules through the terminal pores. These additional structures bind protein molecules that have ubiquitin attached, and use energy from the hydrolysis of ATP to unfold the protein molecules so that they can pass through the narrow pore in the ring of α subunits to be broken down. There is also a de-ubiquitinating enzyme that salvages the ubiquitin molecules to be used again.

The products of hydrolysis of proteins by the proteasomes are predominantly oligopeptides of about nine amino acids, and these arise not only from turnover of the cell's own proteins but also from the proteins of any virus with which the cell may be infected. This has been taken advantage of in the evolution of the immune system of higher animals. In this, a selection of the peptides produced in the course of intracellular proteolysis is bound by MHC (major histocompatibility complex) proteins, and presented by them on the cell surface. There they are surveyed by T lymphocytes that can distinguish peptides derived from the body's own proteins (which have typically been present since the embryonic stage of development, and are tolerated) from newly appearing peptides that may be those of viruses. When cells carrying foreign peptides are detected, the T cells cause them to be destroyed by apoptosis, so cutting short the proliferation of the virus in the infected cell. See also **Lymphocytes**.

Cysteine Proteases

Papain Family. The best known of the cysteine proteases are those in family C1, of which the plant enzyme papain is the prime example. The catalytically active site of papain contains a cysteine residue located near the start of the linear sequence of the protein and a histidine residue farther down the sequence, and the same arrangement is found in all the relatives of papain. In mammalian cells about a dozen papain-like enzymes, including cathepsin B and cathepsin L, are found in subcellular organelles called lysosomes that contain an acidified environment suited to the activity of these acid-acting proteases. The lysosomal proteases are together responsible for a good part of the normal turnover of cellular proteins in cells. Lysosomal cysteine proteases sometimes escape from lysosomes, and also may be secreted from cells, so mechanisms are needed to control their activities. In general they become irreversibly inactivated when they encounter conditions of neutral or alkaline pH, but also there is a family of proteins called *cystatins* present both inside and outside cells that inhibit these enzymes.

More distantly related to papain (in family C2), and present in the soluble phase of the cytoplasm of mammalian cells, are calcium-dependent cysteine proteases termed calpains. Calpains too have their specific inhibitor: calpastatin.

Specialized cysteine proteases in the two families C12 and C19 are the de-ubiquitinating enzymes that reverse the ubiquitination of proteins either to rescue them from destruction by the proteasome or to re-cycle the ubiquitin as the substrate protein molecule is destroyed.

Although the amino acid sequences of the proteases in families C1, C2, C12 and C19 are very different, the three-dimensional structures show similar protein folds that leave no doubt that the enzymes have diverged from a single ancestral protein that existed very early in the evolution of living organisms. This relationship is recognized by placing the families together in clan CA.

Caspases and Others in Clan CD. The caspase family of cysteine proteases are extremely important because of their involvement in the process of programmed cell death termed apoptosis. Apoptosis is crucial to the normal development of multicellular organisms, and to the immunological control of malignancy and viral infection. All of the caspases show strict specificity for hydrolysis of aspartyl bonds. In the "execution" phase in which a cell is killed by apoptosis, several of the caspases carry out limited proteolysis of enzymes that are essential to cellular viability, inactivating each of them by cleavage at just a few aspartyl bonds. The executioner caspases are themselves activated from their proenzymes by the action of other caspases in a small cascade [Salvesen and Dixit, ref.]. The caspases represent an evolutionary line of proteases that is quite separate from that of papain, and the enzymes have a different protein fold. Four other families are now recognized to have emerged from the same origin as the caspases: those of legumain, which is another lysosomal acid protease in plants and animals, separin, which is needed to allow chromosomes to separate in mitosis, and the bacterial enzymes clostripain and gingipain. Together, the five families form clan CD. A common characteristic of all the proteases in clan CD is the importance of the P1 amino acid in their specificity. That is to say, the amino acid immediately on the amino side of the peptide bond to be hydrolysed does more than any other to determine whether the bond actually is hydrolysed. This is reminiscent of the specificity of such serine proteases as trypsin and chymotrypsin, but is totally different from that of the cysteine proteases of clan CA [Chen, et al.:].

Viral Cysteine Proteases. RNA viruses such as those causing poliomyelitis and the common cold are synthesized by the host cell in the form of polyproteins, in which all the proteins are initially strung together in one polypeptide. The polyprotein has to be segmented before the individual viral proteins can fold. The processing proteases that serve this function are encoded by the virus, and since they are essential for viral proliferation they are prime targets for pharmaceutical research seeking new antiviral drugs. Three-dimensional structures show that even the viral cysteine proteases belong to several different evolutionary lines or clans, the most numerous being the papain-like enzymes of clan CA (e.g., foot-and-mouth disease virus; Guarné, et al.:). Also quite numerous are the viral cysteine proteases that have a protein fold clearly based on a chymotrypsin-like serine protease in which the catalytic serine has been replaced by a cysteine; this is seen in hepatitis A virus [Allaire, et al.:].

Aspartic Proteases

Aspartic proteases depend for their activity on the interacting carboxyl groups of a pair of aspartic acid residues. These may activate a water molecule to play a central role in catalysis. Aspartic proteases commonly act at acidic pH values, and all known aspartic proteases are endopeptidases.

Pepsin (in peptidase family A1) is the major protease that starts the digestion of food proteins in the stomach. The stomach contents are acidified by the secretion of hydrochloric acid from the gastric mucosa and this serves to denature most of the food proteins, so that they are very susceptible to proteolysis. In order to function in this environment, pepsin has to be exceptionally stable and active in acidic conditions. In an analogous anatomical structure, nepenthesin is the aspartic protease that digests insects in the "pitchers" of the insectivorous pitcher plant.

Renin is an unusual aspartic protease that is specialized for the limited proteolysis of angiotensinogen to release angiotensin in the regulation of mammalian blood pressure, and acts in the blood plasma at near-neutral pH.

An aspartic protease that is of great medical importance is retropepsin, the protease of the *Human immunodeficiency virus* (HIV), in family A2. The proteins of the virus are synthesized in the form of a polyprotein, and the mature, infective virus particle can be formed only after the retropepsin separates the individual proteins. Consequently, inhibitors of retropepsin have proved to be important drugs in the treatment of AIDS. See also **Human Immunodeficiency Viruses (HIV)**.

Metalloproteases

The proteases that depend upon a metal atom for their catalytic activity are second in number only to the serine proteases, and are of enormous biological and medical importance. Almost all are zinc metalloenzymes, but activity is often retained when the zinc is replaced by cobalt or manganese, and metals other than zinc may occur naturally in the active sites of some metallopeptidases.

Aminopeptidases and Carboxypeptidases. The most prominent of the exopeptidases are metallopeptidases, including the aminopeptidases in family M1 and the carboxypeptidases in family M14. These remove the amino-terminal or carboxy-terminal amino acids of polypeptide chains, respectively, one at a time (Figure 1). The various enzymes in each group show differing substrate preferences, removing some terminal amino acids much more efficiently than others. For this reason, the complete hydrolysis of a polypeptide normally requires the concerted action of several enzymes from either group. As an example of this, procarboxypeptidases A and B are formed in the mammalian pancreas and secreted into the duodenum. There they are converted to the active enzymes by trypsin, and assist in the late stages of hydrolysis of food proteins. Protein molecules themselves are poor substrates for exopeptidases, since each typically consists of a chain of several hundred amino acids, only two of which are the terminal residues on which exopeptidases can act. But the cleaving of a few internal peptide bonds within the chain by endopeptidases soon greatly increases the number of susceptible terminal residues, providing new sites for the action of exopeptidases. For this reason, it is the endopeptidases that control the rate of the early stages of the digestion of proteins, but the majority of peptide bonds are eventually hydrolysed by exopeptidases, and the exopeptidases are indispensable to achieve the ultimate objective of liberation of free amino acids that can be re-used in protein synthesis. Carboxypeptidase A most efficiently releases amino acid residues with uncharged side-chains, so it will not act on carboxy-terminal arginine, for example, but carboxypeptidase B is specialized for the removal of the basic amino acids arginine and lysine, and together the two enzymes can achieve the stepwise hydrolysis of most polypeptide sequences.

Another important exopeptidase is the peptidyl-dipeptidase, angiotensin-converting enzyme, of family M2. This plays an important role in the regulation of blood pressure in the human body by removing a carboxy-terminal dipeptide from angiotensin I to convert it to angiotensin II, and inactivates bradykinin by a similar mechanism. Since angiotensin II tends to increase blood pressure, and bradykinin to reduce it, the angiotensin-converting enzyme has a hypertensive effect, and inhibitors of it are widely used to lower abnormally high blood pressure. See also **Hypertension (High Blood Pressure)**.

Metalloendopeptidases. Many of the metallopeptidases contain an amino acid sequence motif HEXXH, in which H is histidine, E is glutamate and X is another amino acid. The histidine residues of this motif provide two of the three groups by which the enzyme binds the catalytic zinc in the active site, and the glutamate plays a central role in the catalytic mechanism. This type of catalytic site was first recognized in a small bacterial endopeptidase, thermolysin, and most of the other enzymes that contain the structure are also endopeptidases. Thermolysin-like enzymes have evolved in mammals to form the important family (M10) of matrix metalloproteinases, which are secreted from cells as inactive proenzymes and then activated to hydrolyse many components of the intercellular matrix. Among these enzymes are collagenases, gelatinases and stromelysins [Woessner, ref.]. The matrix metalloproteinases are active during development, wound healing, and disease processes such as those of arthritis and tumor invasion.

Additional Reading

Allaire, M., M.M. Chernaia, B.A. Malcolm, and M.N.G. James: "Picornaviral 3C Cysteine Proteinases have a Fold Similar to Chymotrypsin-like Serine Proteinases," *Nature*, **369**, 72–76 (1994).

Barrett, A.J., N.D. Rawlings, and J.F. Woessner: *Handbook of Proteolytic Enzymes*, 2nd Edition, Elsevier Science & Technology Books, New York, NY, 2004.

Beynon, R., J., and J.S. Bond: *Proteolytic Enzymes: A Practical Approach*, 2nd Edition, Oxford University Press, New York, NY, 2001.

Chen, J.M., N.D. Rawlings, R.A. Stevens, and A.J. Barrett: "Identification of the Active Site of Legumain Links it to Caspases, Clostripain and Gingipains in a New Clan of Cysteine Endopeptidases," *FEBS Letters*, **441**, 361–365 (1998).

Davie, E.W., K. Fujikawa, K. Kurachi, and W. Kisiel: "The Role of Serine Proteases in the Blood Coagulation Cascade," *Advances in Enzymology*, **48**, 277–318 (1979).

Dunn, B.M: "Determination of Protease Mechanism," In: Beynon, R.J., and J.S. Bond: *Proteolytic Enzymes. A Practical Approach*, IRL Press, Oxford, UK, pp. 57–81.

Groll, M., L. Ditzel, J. Löwe, et al.: "Structure of 20S Proteasome from Yeast at 2.4Åresolution," *Nature*, **386**, 463–471 (1997).

Guarné, A., J. Tormo, R. Kirchweger, et al.: "Structure of the foot-and-mouth Disease Virus Leader Protease: a Papain-like Fold Adapted for Self-processing and eIF4G Recognition," *EMBO Journal*, **17**, 7469–7479 (1998).

Khan, A.R., and M.N.G. James: "Molecular Mechanisms for the Conversion of Zymogens to Active Proteolytic Enzymes," *Protein Science*, **7**, 815–836 (1998).

Nomenclature Committee of the International Union of Biochemistry and Molecular Biology (2000): *Enzyme Nomenclature. Peptidases*, http://www.chem.qmw.ac.uk./iubmb/enzyme/EC34/.

Polgár, L.: *Mechanisms of Protease Action*, CRC Press, LLC, Boca Raton, FL, 1989.

Rawlings, N.D., and A.J. Barrett: "Evolutionary Families of Peptidases," *Biochemical Journal*, **290**, 205–218 (1993).

Rawlings, N.D., and A.J. Barrett: "MEROPS: the Peptidase Database," *Nucleic Acids Research*, **28**, 323–325 (2000).

Salvesen, G.S., and V.M. Dixit: "Caspases: Intracellular Signaling by Proteolysis," *Cell*, **91**, 443–446 (1997).

Woessner, J.F.: "Matrix Metalloproteinases," In: Creighton, T.E.: *Encyclopedia of Molecular Medicine*, John Wiley & Sons, Inc., New York, NY, 2001.

ALAN J. BARRETT, The Babraham Institute, Babraham, Cambridge, UK

PROTECTIVE COATING.   A film or thin layer of metal glass of paint applied to a substrate primarily to inhibit corrosion, and secondarily for decorative purposes. Metals such as nickel, chromium, copper, and tin are electrodeposited on the base metal; paints may be sprayed or brushed on. Vitreous enamel coatings are also used; these require baking. Zinc coating are applied by continuous bath process in which a strip of ferrous metal is passed through molten zinc.

See also **Corrosion**; **Electroplating**; and **Paints and Coatings**.

PROTEIN HYDROLYSATE.   Solutions of protein hydrolyzed into its constituent amino acids.

PROTEINS.   Proteins, ubiquitous to all living systems, are biopolymers built up of various combinations of 20 different naturally occurring amino acids. The number of proteins in an organism may be as small as half a dozen, as in the case of the simple bacterial virus M13, or as large as 100,000, estimated to be in the human system. Proteins are encoded by the deoxyribonucleic acid (DNA) that is present in all living cells.

Protein uses are myriad. The large number of biochemical reactions within the living cell are catalyzed by enzyme proteins Traditional food processes such as baking, brewing, and cheesemaking involve the action of enzymes found in different microorganisms. Other proteins are involved in the transport of electrons, ions, and small molecules. Proteins are also key components of the immune system and control the genetic expression of other proteins. The smallest proteins have a molecular weight of only ~400; the largest protein molecule discovered to date is the muscle protein titin with a molecular weight of 3,000,000.

The study of proteins has a long history. First isolated from both animal and plant sources in the late 1700s, their chemical composition was studied in the 1820s and 1830s. Glycine was the first constituent amino acid to be isolated from gelatin in 1820. Because of his high regard for the proteins, but well before they were really understood, the Dutch chemist Gerardus Mulder (1802–1880) pioneered the use of the term *protein*, derived from the Greek word meaning "to come first." Although proteins furnish energy to the body and thus can be considered body fuels, as are the carbohydrates and fats, the major nutritional roles of the proteins reside in other functions, usually of a highly specific nature. Thus, there are structural, contractile, process-activating, and transport proteins, among others, which essentially are responsible for the chemical workability of the animal system.

Considering the research tools available, the amount of qualitative and quantitative information pertaining to proteins collected over several decades of effort has been tremendous. The data amassed have been highly beneficial to the medical and health sciences, notably in terms of dietary requirements and protein deficiency diseases, to biologists, and, of course, to organic chemists. Past protein research has led to the development of many useful protein substances for industry and commerce. Scientists stretched the limitations of their available instrumental techniques (crystallography, electron microscopy, chromatography, electrophoresis) in their efforts to better understand protein structure and protein function. With the advent of molecular biology (studying proteins at the molecular level), the potential for learning more pertaining to structure and of what proteins do and how they behave in living organisms increased, conservatively speaking, by an order of magnitude or more. As of the later 1980s, protein science has progressed just a little beyond the initial efforts to reduce protein studies to the molecular level. Highlights are summarized in the latter portion of this article. There are several keys to expanding protein knowledge, two of the most important of which are continued mapping of organism genomes, notably mapping the human genome; and the continuing development of improved instrumental and procedural techniques. In using this newly acquired knowledge to manipulate protein structures, the term "protein engineering" is sometimes used. Protein engineering largely lies in the future.

Protein Requirements

In the growing animal body, a significant portion of proteins consumed is required for the creation of new tissue. This results in an increasing requirement for proteins in the diet of humans, for example, up to about the age of 20 years, at which time the protein requirement tends to level off to a fairly stable figure. After body maturity, the portion of proteins needed for tissue maintenance is greater than the need for new tissue building. It must be emphasized, however, that immediately at the commencement of life both new tissue building and tissue maintenance take place and even as the body grows older, the two needs continue — only the proportions between the two roles change:

Proteins, on a weight basis, are second only to water in their presence in the human body. If the factor of water is discounted, then about 50% of the body's dry weight is made up of numerous protein substances, distributed about as follows: 33% in muscles; 20% in bones and cartilage; 10% in skin; the remaining 37% in numerous other body tissues. With exception of the urine and bile in the normal healthy individual, all other body fluids contain from small to relatively large portions of protein substances.

Chemically, proteins are distinguished from other body substances in that all proteins contain nitrogen. Some contain sulfur, phosphorus, iron, iodine, cobalt, and other elements, some of which are generally not thought of as components of the life process, but which nevertheless do play extremely important roles (e.g., as catalysts), even if present only in very minute quantities.

In considering the importance of proteins to building and maintaining body functions, it must be emphasized that proteins consumed essentially are raw materials that contain the building blocks for the creation of different proteins. These building blocks are the amino acids of which the protein molecules consumed are constructed and of which the proteins restructured in the body (after consuming or metabolizing the raw materials) are also constructed. Thus, the desirability of proteins for the diet is based upon the best combination of amino acids present. Therefore, some foods are desirable from a protein nutrition standpoint not only because, with relation to their carbohydrate and fat content, they contain a high percentage of protein, but also because they contain most or all of the amino acids needed to form new proteins within the body. See also **Amino Acids**.

Examples of this situation (desirable versus less desirable proteins) popularly cited are the soybean proteins and the grain proteins. With exception of the sulfur-bearing amino acids, notably methionine, the amino acid balance of soybean proteins is reasonably good. With exception of the amino acid lysine, the amino acid balance of grain proteins is reasonably good. By mixing protein substances from these two sources, an excellent source of protein for the human diet is obtained, this explaining growing trends toward fortification of wheat and other cereal flours with soy flour. There are scores of examples of this type which are representative of the trend toward so-called *fabricated foods*.

From years of experience in studying the dietary needs of humans, nutritionists and biologists established the hen egg as having the most perfect balance of amino acids in a natural protein substance. Against this standard, other foods can be rated in their performance. In naming the following food substances in order of their diminishing chemical score, it should be stressed that these foods are arranged only in terms of this one nutritional criterion: fish (70), beef (69), cow's milk, whole (60), brown rice (57), polished white rice (56), soybeans (47), green leaves (45), brewer's yeast (44), groundnuts (peanuts) (43), whole grain maize (corn) (41), cassava (manioc) (41), common dry beans (34), white potato (34), white wheat flour (32). The foregoing food items were selected *randomly* to provide a sense of the spectrum of foods from this one particular standpoint. The Figures represent only the chemical balance of amino acids present and not the total amount of protein available as a weight percentage of food intake, or from the standpoint of protein utilization, once ingested.

In looking at a number of food substances, again a random selection, from the standpoint of total protein (with no regard to quality) in an average serving, the following amounts of protein (grams) are present: fried chicken breasts (27.8); canned tunafish (24), cooked round roast of beef (24), roasted leg of lamb (22), oven-cooked pork loin (21), dry cooked soybeans (13), whole milk (1 cup) (9), canned red beans (7.5), cheddar cheese (1 ounce = 28 grams) (7), fresh cooked lima beans (6.5), egg (medium size) (6), vanilla ice cream (6), fried crisp bacon (5), baked potato (3), cooked broccoli (2.5), cooked oatmeal (2.5), enriched white bread (1 slice) (2), cooked green snap beans (1), lettuce ($\frac{1}{4}$ head) (1), and reconstituted frozen orange juice (1).

Consequences of Protein Deficiency. Because proteins are so important to numerous and very complex bodily functions, years of research have just commenced to provide some understanding of most of the mechanisms involved. As would be expected, recognition of the extreme manifestations of protein deficiencies has taken place, at least to the extent of providing new guidelines for assisting millions of inadequately fed people in several regions of the world. As further experience is gained in researching the gross problem, the important subtleties of protein performance within the body will become more apparent.

Exemplary of a better understanding and appreciation of protein nutrition is a comparison of the 1945 report of the Food and Agriculture Organization (United Nations) with more recent findings, recommendations, and nomenclature used. In the first *World Food Survey*, the terms *undernourishment* and *malnourishment* were used throughout the report. The general interpretation of undernourishment was taken to mean an inadequate caloric intake, i.e., insufficient energy input to support normal body functions and activities, with body weight loss the inevitable result. Similarly, *malnourishment* was taken to mean a deficiency of one or all of the protective nutrients, such as proteins, vitamins, and minerals. During the last few years, inasmuch as these two problems are so interrelated, the term *protein-calorie malnutrition* (PCM) has come into wide use. PCM of early childhood, particularly in regions that are a part of some of the less developed countries, is quite widespread. PCM apparently is manifested in minor ways at first, but when prolonged very severe syndromes become evident. These include the conditions known as *kwashiorkor* and *marasmus*.

Kwashiorkor usually occurs in the second or third year in the life of a child. Edema is the principal symptom. The condition arises from a combination of circumstances, but the primary cause appears to be a weaning diet that is both inadequate and indigestible and, notably, is lacking of protein. The principal calories are supplied by carbohydrate. The condition is accelerated by repeated infections of a bacterial, parasitic, or vital nature. Without treatment, the disease is fatal in most cases.

Nutritional marasmus is a severe manifestation of PCM and is a condition that usually occurs during the first year of life. Again, it arises from a combination of conditions, frequently widespread in many regions, of feeding an overly diluted formula of cow's milk, thus reducing the protein input well below minimum needs. The condition is accelerated by filthy surroundings and contaminated bottles. Characteristic of the syndrome are a wasting of muscle and subcutaneous fat, a body weight that may be only 60% of standard, and diarrhea. Children who have access to human milk usually are protected against marasmus and diarrheal disease.

A more recent finding and term now used for a protein deficiency syndrome is *PCM-plus*, or *infantile obesity*. This is a condition that occurs among the more affluent populations where an infant is bottlefed, where hygiene is adequate, and where funds are adequate. Overfeeding of an improperly balanced formula can cause the condition. The condition does not occur with breast feeding because the volume of intake is regulated by the infant's appetite and thirst.

Sources of Proteins

The two basic categories of protein sources for the animal diet are other *animals* (living or dead) and *plants*. Thus, in the animal category as a source of human and pet protein foods, there are what might be called terminal sources or nonreplenishing sources, in which the living animal is killed and disassembled into its protein-containing parts. The most common examples including the meaty flesh and organs of beef cattle, pigs, sheep, and horses and goats, as well as the more occasional sources of meat, such as deer, elephant, hippopotamus, etc., depending upon availability and regional eating preferences. To these sources are added the flesh and organs of birds (chickens, ducks, turkeys, pheasants, etc.) and of fish caught in saline and fresh waters. In the overall animal protein category, one also would include those less conventional and essentially unexplored categories, such as earthworms and single-cell proteins (produced by microorganisms) and algae. Renewable or repeating protein sources from living animals, of course, include the milk from dairy cows and buffaloes and the eggs from hens, from which hundreds of high-protein foods (cheese, for example) are prepared. And, to this category, must be added the excellent source of protein provided by human milk to the nursing infant.

Plants, of course, also require protein to build and maintain their life processes and, consequently, are protein sources for the animal diet. In the case of herbivores, plants are essentially the exclusive source of proteins, energy, and all other dietary elements.

In terms of percentage of protein content of basic sources, the animal sources far excel the plant sources. For example, the protein content of some typical unfortied foods is as follows: 20–30% for cooked poultry and meats; 19–30% for cooked or canned fish; 25% for cheese; 13–17% 17% for cottage cheese; 16% for nuts; 13% for whole eggs; 7–14% for dry cereals; 8.5–9% for white bread; 7–8% for cooked legumes; and about 2% for cooked cereals.

Of course, in achieving the higher protein contents of meat from poultry and cattle, a rather costly two-step production process is involved, wherein the animal first converts plant proteins (as from grasses) into animal protein. In a sense, the animal both converts and concentrates the protein source for humans. Several economic factors enter into the picture — the utilization of land, the costs of labor, the additional costs of feed materials, and the costs related to a greater time span of production, among others. As a case in point, an animal must be fed between 3 and 10 pounds (1.4 and 4.5 kilograms) of grain to produce 1 pound (0.45 kilogram) of meat. All of these factors in recent years, particularly in consideration of protein shortages in many regions of the world, have given rise to conflicting opinions pertaining to the ever-increasing production and consumption of meat, not only in several of the western nations of the world, but in the developed nations of the Orient as well. A few authorities have suggested that the western countries should cut back on meat production, thus making more land, skills, etc. available to increasing vegetable protein production to the level where a generous excess supply would be available to underdeveloped countries as well as amply supplying the protein needs of the developed countries. Quickly, these arguments penetrate not only into technological and economic factors, but psychological considerations as well — because any moves of this type necessarily require drastic changes in eating habits, and to bring them about successfully would require much more governmental regulation and policing than any system of private enterprise is likely to tolerate. Further, attitudes tend to swing rather widely from times of grain surplus to times of grain shortage.

Fortunately, as of the early 1980s, it appeared that protein-processing techniques were providing a very satisfactory compromise, even though the industry is just getting underway toward a large-scale operation. Protein meat extenders, for example, wherein meat and vegetable protein are blended to produce an edible product that retains much of what is desired of meats, including their good protein content, are finding acceptance. The wide acceptance of vegetable protein in analogue meat products has many hurdles to overcome, but it appears that a solid start has been made. The hurdles not only include acceptability in the marketplace, but also some justifiable resistance on the part of cattle and poultry producers. For many reasons, the transition, if it ultimately takes place, will occur over quite a long period of time. Because of continuing economic inflation, the earlier cost advantages that tended to favor blends of meat and vegetable proteins have become less significant.

An early impetus to soy protein foods was given when the United States introduced soy protein products into its overseas donation program in 1966 as a component of foods formulated to meet special needs of certain population groups. Chief among these were children in developing nations, especially the weanling infant and preschool child whose requirements for growth put special demands on diet composition. Pregnant and lactating mothers also had dietary needs frequently not met in countries where food supplies were marginal. Beyond these needs, there were nutrient deficiencies in large population groups, which could be best overcome by enrichment or fortification of commonly eaten foods.

Shortages in the domestic supply of nonfat dry milk, which developed in 1965, stimulated the development of high-protein formulated foods which would serve as supplements in the diets of the children or in the emergency feeding of adults. These formulations had to pass rigid specifications, one of the principal criteria being the recommended daily dietary allowance for protein, vitamins, and minerals. The U.S. Department of Agriculture and the U.S. Agency for International Development developed the guidelines and designed various formulated foods. Among these formulations were Corn-Soy Milk (CSM), Corn-Soy Blend (CSB), and Wheat-Soy Blend (WSB).

Further impetus was given to protein blends in foods when such products were introduced into the domestic food assistance program in the United States. Soy protein foods were introduced into school lunch and breakfast programs for which federal assistance has been given in the form of a subsidy administered by the federal government. Soy-fortified foods also were distributed to needy families through a family food distribution program.

Textured soy protein products in their use as meat alternatives have become increasingly popular in school lunch programs since their introduction in 1971. A soy-modified macaroni was introduced into the family food assistance program a number of years ago.

Less Conventional Sources of Protein. In addition to the traditional animal sources of protein already described and the very large amounts of vegetable protein derived from the soybean, other sources of protein on a large scale for the future are under intense study. Among these are (1) oil-seed crops, such as rapeseed and cottonseed; (2) leaf proteins; (3) algae; and (4) single-cell protein.

Rapeseed, one of the five most widely produced oilseeds, is cultivated mainly in India, Canada, Pakistan, France, Poland, Sweden, and Germany. Past objections to using rapeseed as a source of edible protein has been its content of deleterious glucosinolates. Considerable research has been conducted in Sweden to develop a rapeseed protein concentrate. The first full-scale production plant using a new process was installed in Alberta, Canada. The plant, with a capacity of 5000 tons/year produces a material containing 65% protein. Rapeseed is rich in essential amino acids, with exception of methionine, which soybeans also lack.

Cottonseed offers an attractive source of protein provided that certain objectionable ingredients can be removed. One of these is gossypol, a substance in cottonseed gland that is harmful to humans. A process developed by the U.S. Department of Agriculture has been designed to turn out a satisfactory edible cottonseed protein product. Employing solvent-extraction techniques, the first plant was built in Texas. Cottonseed flour extrudes easily and can be water-extracted to produce a nearly 100% protein isolate. The product has been used as a bland extender and fortifier for processed meats, baked goods, candies, and cereals. Research of a different approach has been used in Central America. In this approach, iron compounds are used to tie up the gossypol in nontoxic form without having to remove it.

Leaf Protein Concentrates. Laboratories in Hungary, Japan, the United Kingdom, and the United States, among other countries, have been engaged in perfection of a leaf protein concentrate process, with emphasis upon increasing yields and palatability and reducing flavor problems and cost. To date, alfalfa appears to be most attractive as a source of leaf protein. Alfalfa will produce more protein per unit of land than most other crops — up to 2800–4000 pounds/acre (3136–4480 kilograms/hectare). It has been estimated that the raw material costs for edible protein from alfalfa would be about 50% that for soybean meal. Several processes have been worked out, ranging from a green curd containing 52% protein to a white powder containing about 90% protein.

Single-Cell Protein. The advantages of single-cell protein (SCP) made from growing microorganisms are several: (1) SCP is independent of agricultural or climatic conditions; (2) SCP doubles in mass rapidly for high production rates and fast genetic experimentation; (3) the crop is free of surface-area limitations, and (4) the protein in microbial cells is generally of a high nutritional quality. Many of the processes proposed and tested, some with limited operating experience, commence with hydrocarbon feedstocks — gas oil and normal paraffin substrates. Two objections have been raised. The first is the possibility that carcinogenic polyaromatic materials present in gas oil may be passed along to the final protein product. The second is an adverse public reaction. A more recent, third objection is the proposition that perhaps technology should be concentrating on manufacturing fuels from farm products rather than food from petroleum products.

Some of the more recent SCP process concepts start with other materials, such as ethanol, acetic acid, starches, sugars, and cellulosic products that may be more available and particularly so in the protein-needy developing countries.

Algae have the highest intrinsic rates of photosynthesis and growth found among green plants. Human food and animal feed are being produced from algae. In Japan, a full plant-scale production harvests algae from open ponds to yield green powder extract that can be used for animal or human consumption. The genus *Chlorella* has perhaps received the most research to date.

Conservation Sources of Protein. Tightening pollution restrictions have forced cheese makers in many regions to end a long-time practice of dumping whey (with its high biological oxygen demand) as a liquid waste. Although many of these manufacturers are now evaporating or spray-drying whey to produce a whole-solids product, several fractionation techniques have been devised to separate a concentrated protein. In the United States, whey as a byproduct of cheese making totals well over 30 billion pounds (13.6 billion kilograms) per year. From 6.5 to 7% of the whey is solids, of which 0.9% is protein. Some authorities believe that whey and other milk-based protein ingredients offer a high growth potential among all of the non-soybean sources.

Fish protein concentrate is regarded by some authorities as having a high long-term potential. A major restraint is competition for the whole fish. As fish food sources become increasingly competitive, fishes currently considered "trash" fishes from a fresh marketing viewpoint may ultimately become more desirable for table use. Animal-feed fish meal also will be a strong contender for available fish. In terms of processes required for preparing fish-protein concentrate, extraction processes using single or mixed solvents of isopropanol, ethylene dichloride, ethanol, and hexane already have been developed. Experiments with enzymatic processing also are underway.

Chemical Nature of Proteins

In defining a protein structurally, it is first necessary to define a peptide. Peptides are compounds made up of two or more amino acids covalently bound in an amide linkage. The characteristic amide linkage, in which the carboxyl group of one amino acid joins with the amino group of the next amino acid, is called a peptide bond. A peptide is a chain of amino acid residues. Provided that the chain is not circular or blocked at either of the ends, the peptide has an N-terminal amino acid, bearing a free amino group, and a C-terminal amino acid, bearing a free carboxyl group. This is illustrated as follows:

$$H-NHCHR'CO-OH$$

$$H-NHCHRCO-OH + H-NHCHR''CO-OH$$

$$\downarrow -2H_2O$$

$$H-NHCHRCO-NHCHR'CO-NHCHR''CO-OH$$

N-terminal   Nonterminal   C-terminal

Usually a form of shorthand is used to represent the structure of a peptide. For example, H-Val-Gly-Ala-OH, represents a peptide where abbreviation for each amino acid is given in terms of three letters each (Val = valine; Gly = glycine; Ala = alanine). Abbreviations for other amino acids are given in entry on **Amino Acids**. The H denotes the amino terminal (N-terminal) and the suffix OH denotes the carboxyl terminal (C-terminal). Peptides may consist of from two to eight amino acid residues and thus are known as dipeptides, tripeptides, or oligopeptides (eight), depending upon the number of residues contained. A peptide consisting of ten or more amino acid residues and with a molecular weight in the range of $1-5 \times 10^3$ is called a polypeptide. Emil Fischer, father of protein chemistry, proposed early in the twentieth century that proteins are peptide in nature. Actually, no sharp demarcation exists between large polypeptides and small proteins. Examples of small proteins include insulin (hormone protein), protamine, and some components of histone (basic proteins of chromosomes).

Almost all proteins are comprised of amino acid residues, more than 100 in number, and their molecular weight may range from 10^4 to 10^7. A few examples include: Insulin (6×10^3); ribonuclease (13×10^3); lysozyme (eggwhite) (15×10^3); chymotrypsinogen (21×10^3); ovalbumin (43×10^3); serum albumin (66×10^3)—all of the foregoing being single peptide chains. Multiple chains include: Hemoglobin (68×10^3); gamma globulin (IgG) (160×10^3); fibrinogen (340×10^3); urease (460×10^3); thyroglobulin (640×10^3); myosin (850×10^3); hemocyanin (octopus) $(2,800 \times 10^3)$; hemocyanin (snail) $(8,900 \times 10^3)$; and tobacco mosaic virus $(40,000 \times 10^3)$. Proteins of huge molecular weight (millions) are enormous aggregates of protein subunits, each of which may be so large (molecular weight $= 1.5-10 \times 10^4$) in most instances. The independent peptide chains that constitute a protein molecule are often held by the disulfide bridges of cystine residues. From the diagram below, it will be seen that in a single chain the bridges may hold together two quite distant points in terms of the linear amino acid sequence, forming a large loop structure:

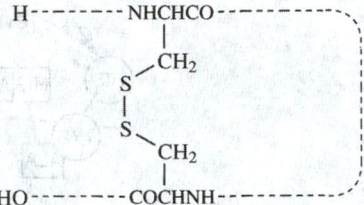

Although more than 200 amino acids have been found in living organisms, only 20 alpha-amino acids of the L configuration have been found serving as the building units for proteins and related peptides. These 20 amino acids occur in varying proportions in different proteins. Some proteins are fully lacking in one or more of them. Some amino acids occur only in some of the proteins. For example, hydroxyproline has been found only in collagen and elastin (proteins of animal connective tissue) and in gelatin derived from collagen.

Numerous classifications of proteins have been proposed over the years. In terms of function, there are:

a. **Structural proteins** Proteins that support the skeletal structures, maintain the form and position of organs, impart the structural rigidity to walls of containers for biological fluids, and often form part of the external tissues. In keeping with their functions, they are insoluble in many liquids, especially body fluids, and are otherwise relatively resistant to biochemical reactions. The proteins of nails, horn, hoofs, and hair are familiar examples.

b. **Contractile proteins** Those substances that have the property of undergoing a change in configuration, which results in a change in length or shape. Thus they give the organism the power to move itself, its parts, or other objects. The proteins of muscles are prominent examples.

c. **Process-activating proteins** As used here, the term *process* includes the biochemical reactions, which are catalyzed by enzymes, and in some of which the cytochromes play an intermediate role; it also includes the endocrine reactions activated by the hormones, some of which are proteins.

d. **Transport proteins** Proteins which transport an essential substance or factor, from that part of the organism where it becomes available from a source external to the organism to the point where it is used. Examples are many of the chromoproteins, such as hemoglobin, or the blue hemocyanins (from mollusks) which contain copper instead of iron as does hemoglobin, or the chlorophyll-protein complexes of plants.

Another basis of classification is that of solubility, which has been applied to proteins from all sources, plant and animal. (a) Thus the albumins were soluble in water and coagulable by heat. They included serum albumin, egg albumin, lactalbumin (from milk), leucosin (from wheat), and legumelin (from legumes, chiefly peas). (b) The globulins are soluble in neutral salt solutions and in strong acids and alkalies. They include blood globulin (which has been separated by electrophoresis into alpha, beta, and gamma fractions, and is further discussed later in this entry), ovoglobulin (from egg yolk), edestin (from hempseed), phaseolin (from beans), arachin (from peanuts), and amandin (from almonds). (c) The glutelins, such as glutenin from wheat, are soluble in dilute acids and alkalies, and insoluble in neutral salt solutions. (d) The scleroproteins are quite insoluble, and the structural proteins (group I mentioned above) belong to this group. All these groups, and several others not included here, are simple proteins, i.e., they consist only of polypeptide chains of amino acids. The many conjugated proteins must then be classified upon the basis of their nonprotein portions: glycoproteins which contain carbohydrate groups, lipoproteins which contain lipid groups, chromoproteins which contain metal-containing complexes that are usually colored, as hemoglobin contains heme.

Still another classification places proteins into three major categories: (a) Simple proteins; (b) conjugated proteins; and (c) derived proteins. The last classification embraces all denatured proteins and hydrolytic products of protein breakdown and no longer is considered a general class.

A relatively simplistic concept of a protein structure is indicated in Fig. 1. The molecular weight for the hemoglobins is on the order of

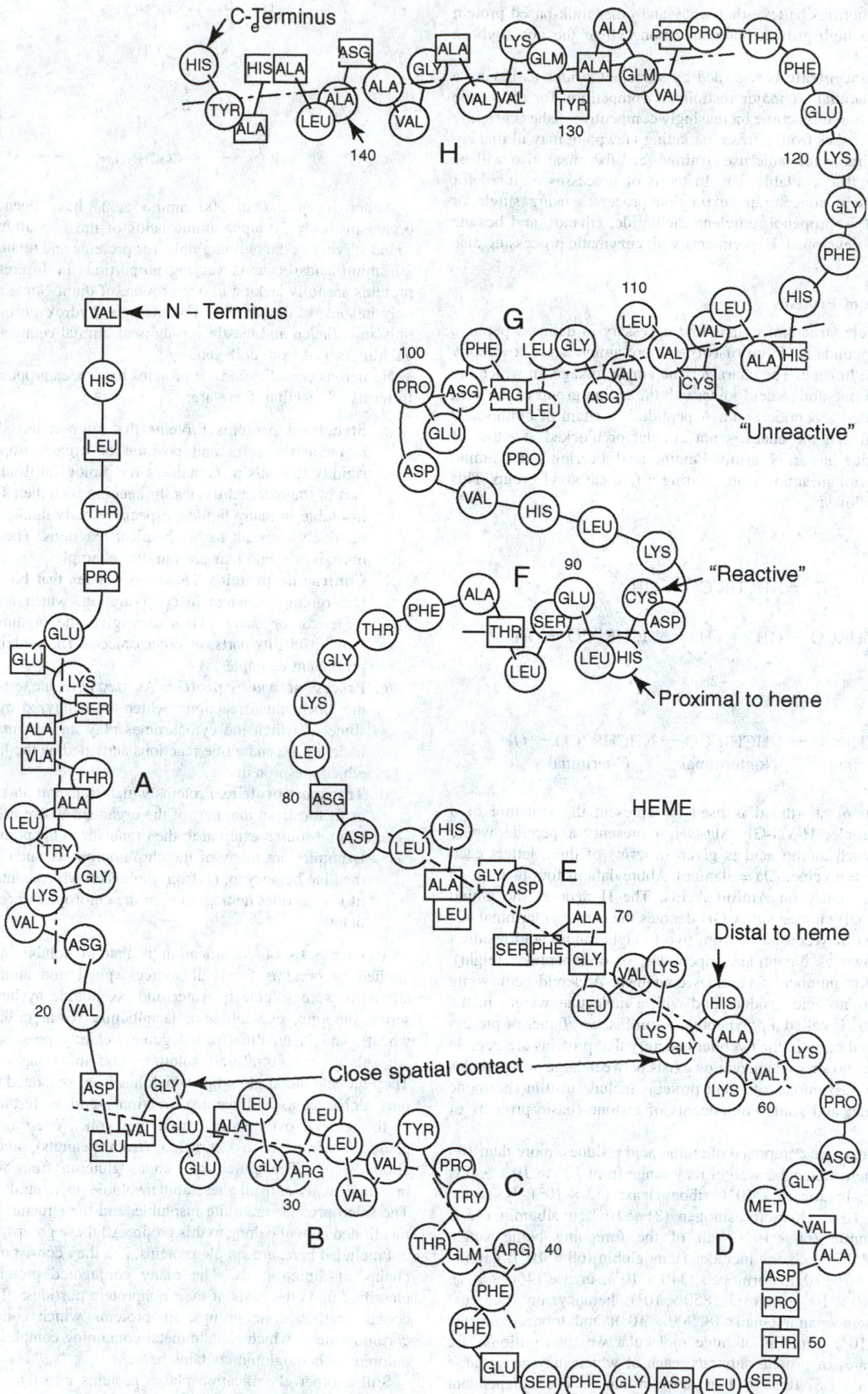

Fig. 1. Simplified representation of the beta chain of human hemoglobin A.

68,000. They are conjugated proteins and consist of four heme groups and the globin portion. The heme group is a porphyrin in which the metal ion coordinated is iron, which may be Fe^{3+} or Fe^{2+}, but only in the latter case (ferrohemoglobin) can the molecule bond molecular oxygen and be effective in respiration, i.e., by forming oxyhemoglobin. The globin portion of the molecule consists of four polypeptide chains. These chains are designated as alpha, beta, gamma, etc. according to their amino acid composition. Normal adult hemoglobin consists of two alpha chains and two beta chains. The composition and conformation of the beta chain are shown in the diagram, together with the point of attachment of the heme groups: Note that they are attached to histidine groups. It has been learned that the central iron atom in heme, which is chelated to the porphyrin ring by four bonds, is attached to the polypeptide chain in adult human hemoglobin by three imidazole ligands of the globin chain, which belong to the histidines at positions 58,87, and 89 of the alpha chain. See also **Hemoglobin**.

Besides hemoglobin, other proteins of blood are of considerable importance. They are the plasma proteins, serum albumin and fibrinogen, and the globulins. Serum albumin is responsible for the major part of the osmotic pressure of human plasma. Its molecular weight is on the order of 68,000. It is a typical globular protein, having nearly one-half helical character. Although not as nearly symmetrical as hemoglobin or myoglobin, it has a symmetry indicated by its molecular dimensions of 150 Å long and 38 Å wide. It is the smallest, and most abundant, of the plasma proteins; for this reason, and also because of its relatively low isoelectric point, it undergoes migration rapidly in an electric field. See also **Electrophoresis**. By this method it may be separated into two types of molecules, similar in composition except for the presence of a single cysteine residue in one and not the other. However, it contains cystine residues, which form seventeen disulfide bridges cross-linking the polypeptide chain, i.e., the molecule of serum albumin consists of a single polypeptide chain.

Another plasma protein to be discussed here is fibrinogen, which is the chief substance involved in the process of blood clotting. Its molecular weight is on the order of 330,000. It contains all twenty of the amino acids described in that entry as the most general in proteins, although it is relatively low in cysteine, and highest in the acidic amino acids (aspartic and glutamic acids). The process of clotting occurs in three major steps. In the first the substance prothrombin, a blood glycoprotein containing about 5% carbohydrate as glucosamine and a hexose sugar, is converted to the clotting enzyme thrombin. (The latter is unstable, and hence must be formed when needed.) The conversion process is catalyzed by the calcium ion and a group of substances known as thromboplastins. In the second step the enzyme thrombin catalyzes the transformation of fibrinogen to an activated form, called profibrin, with an altered pattern of electric charge. This change is considered to be due to the liberation of two short-chain polypeptides (one bearing 18 amino acid residues and the other 20), and a corresponding change in the character of the remainder of the fibrinogen molecule, which collectively constitute the substance profibrin. In the third step, this mixture of substances undergoes spontaneous polymerization to form the substance fibrin, which has been shown in electron microscope photographs to consist of a network of striated fibers. This polymerization occurs in stages, and in some views of the process they are divided into two steps, polymerization and clotting, the former being regarded as the formation of linear polymers, and the latter as their cross-linking by an enzymatic reaction whereby disulfide bonds are formed.

Animal organisms generally require effective assistance of intestinal flora, as in ruminants to assimilate inorganic nitrogen into a very wide variety of foreign substances, called antigens.

The life span of individual proteins in living organisms is relatively short—about 4 months for hemoglobin and but a week or two for serum albumin. The aged proteins are digested by proteolytic enzymes of tissues, such as cathepsin. A significant portion of the recovered amino acids may be available for the biosynthesis of new proteins, but another part is catabolized and the nitrogen is excreted as urea in mammals, uric acid in birds, reptiles, and insects, or ammonia in organisms of lower classes. For the maintenance of nitrogen balance and for growth, the human organism requires a daily intake of from 70 to 80 grams of proteins. The present world population thus requires 3×10^7 tons of animal proteins and 10×10^7 tons of plant proteins annually.

Living organisms can synthesize their own proteins from amino acids. In terms of the ability to carry out the de novo synthesis of amino acids, however, there are wide variations among different organisms. Plants, for example, can synthesize amino acids from nitrogen in the form of ammonium salts or nitrate and other simple compounds. The annual production of cereal and vegetable proteins so assimilated from inorganic nitrogen over the world is estimated to be about 10×10^7 tons. Of this, 4×10^7 tons are provided by wheat; 2×10^7 tons by rice; and 1×10^7 tons by corn and other sources. Lactic acid and some other microorganisms require preformed amino acids for growth, lacking some ability to synthesize. However, some microorganisms perform well with ammonium sulfate and carbohydrate as the sole sources of nitrogen, sulfur, and carbon. In some cases, they accumulate particular amino acids in a process referred to as amino acid fermentation.

Animal organisms generally require effective assistance of intestinal flora, as in ruminants, to assimilate inorganic nitrogen into body protein. This accounts for the human needs of a daily requirement of 70–80 grams of protein. However, over half of the protein-constituent amino acids can be derived from other amino acids by their own enzymic reactions. Thus, amino acids are classified as essential or nonessential. Amino acid requirements vary with the physiological state of the animal, age, and possibly with the nature of the intestinal flora.

The Food and Agricultural Organization (FAO) established the following essential amino acids in the ratios indicated:

	Percent of Crude Protein
Isoleucine	4.2
Leucine	6.2
Lysine	4.2
Methionine	2.2
Phenylalanine	2.8
Threonine	2.8
Tryptophan	1.6
Valine	5.0

A distribution of amino acids in dietary proteins can be obtained accordingly by taking both animal and plant proteins at a ratio of 1:3–4. Although plant proteins are lower cost, they are markedly deficient in some essential amino acids. Their protein efficiency is low without addition of deficient amino acids. Enrichment of human and animal diets with free amino acids, such as lysine, methionine, threonine, and tryptophan, as a substitute for animal proteins, has proved successful.

Excesses of amino acids are not harmful—with few exceptions. An imbalance of amino acids can result in a few instances. For example, a rat that feeds on eggwhite proteins with threonine or isoleucine added in high concentrations can experience an undesirable imbalance.

Several industries are based upon proteins as exemplified by the keratins of wool, feather, or horn; the fibroin of silk; the collagenous tissues as leather; proteins in milk, wheat, soybean, egg, and numerous other natural substances. Making cheese from milk casein and flavor seasonings from plant or fish proteins are old processes. Gelatin derived from collagen has been used widely in processed foods and as an adhesive material and in photography. Gluten in cereal is a protein. Major proteins in meat are *myosin*; in egg, *ovalbumin*; in rice, *oryzenin*; in soybean, *glycinin*; and in corn, *zein*.

Protein Quality and Evaluation. Protein quality relates to the efficiency with which various food proteins are used for synthesis and maintenance of tissue protein. Food industry evaluators of protein nutritional quality must operate on several levels of awareness. In particular, manufacturers of processed foods must measure the biological value of the protein content of a variety of processed foods for several reasons: (1) to comply with various governmental regulations; (2) to satisfy nutrition labeling regulations; and (3) an accurate knowledge of protein effectiveness is required in developing new food products and in controlling sources of protein ingredients. Protein quality is also very important in the formulation of animal feedstuffs.

In a number of countries, including the United States, the stipulated measurement of protein quality is the so-called protein efficiency ratio (PER), which may be defined as the *gain in weight* divided by the *weight of protein consumed* by experimental laboratory animals. As of the early 1980s, the AOAC (Association of Official Analytical Chemists) method, defined in 1975, is only one of the codifications of PER work since the concept was first proposed in 1919. More specifically, the PER is the ratio of the weight gained by a group of ten weanling rats fed a diet containing about 10% protein, to the weight of protein consumed over a 28-day period. No sample to be studied should contain less than 1.8% nitrogen according to the AOAC method, and the diet should supply 1.6% nitrogen. Since the samples are not analyzed for protein, but rather for nitrogen, and since protein efficiency ratio rather than nitrogen efficiency ratio is reported, it is important to be clear about whether one of the specific nitrogen factors or the conventional 6.25 figure is used to calculate the protein in the final diet.

Advances in Protein Chemistry

For many years, research was directed to a better understanding of the structure of proteins, notably based upon X-ray crystallography. Remarkable structural details were evidenced. But this avenue of research tended to regard proteins as being static in nature, whereas more recent findings show that indeed proteins are dynamic and that, if they were rigid, they simply could not function. The internal motions that underlie their workings are best explored in computer simulations. As pointed out by Karplus and McCammon, it is now recognized that the atoms in a protein molecule are in a state of constant motion. Thus, what the crystallographer finds is at best a representation of a protein's average structure. The chemical bonds between the atoms along the polypeptide chain in a protein act much like springs. There are also weaker forces between unbonded atoms, including forces that prevent more than one atom from occupying the same point in space at any given time. Thus, in a protein consisting of many atoms, the total force acting on any one atom at any given time depends upon the positions of all the others. Not surprising, the solution of Newton's equations of motion for determining the positions and velocities of all the atoms in a protein requires a high-speed computer. Such calculations constitute what is called a *molecular-dynamics simulation*.

In summarizing their recent research, Karplus (Harvard University) and McCammon (University of Houston) observe that, from future research, much will be learned regarding how to calculate the rates of enzymatic reactions, and the binding of small molecules to large ones, as well as the role of flexibility and fluctuations in the function of macromolecules. For example, it should become possible to determine how particular solvent conditions and amino acid sequences produce certain patterns of protein fluctuations. Such information will become useful in applying new genetic technologies in a practical way.

Protein research of this kind is important because all enzymes are proteins. They catalyze the speed of essential reactions in living systems, including the synthesis of proteins themselves. Knowledge of the dynamics of proteins will assist in better understanding those proteins that transport small molecules, electrons and energy to specific parts of an organism where they are needed. Those proteins of a structural nature, which make up fibrous tissue and muscle, also will be better understood.

As pointed out by Phillips, the level of understanding of enzyme (protein) action has been achieved for many enzymes through the use of chemical, crystallographic, and spectroscopic methods. Gene science, however, has enormously advanced protein studies. By using cellular machinery for protein synthesis, proteins can be manufactured with any primary structure and then introducing whatever changes seem useful in the chemical constitution of naturally occurring proteins. With further knowledge, at some future date it most likely will be possible to design and manufacture fully novel proteins with new and useful properties.

Currently, the most useful advances are being made by the detailed modification of existing protein structures. A very small change, often involving only a single base, is made of the DNA coding for the protein. This is followed by use of natural cellular machinery (frequently bacteria) to synthesize the modified protein. The method is known as *site-directed mutagenesis*, which was first used in 1982. Classical chemical modification of protein structure still is used, but the site-directed mutagenesis approach is usually more straightforward and reliable.

Phillips has projected an imaginary oligopeptide with side chains grouped in accordance with their properties to illustrate intricacies of structure and regions of specializing functions. See Fig. 2.

Much progress is being made in connection with *fibronectins*, those adhesive proteins that act as biological organizers by holding cells in position and guiding their migration. Studies are now revealing the molecular bases for the functions of fibronectins. As observed by Hynes, within the complex architecture of a multicellular organism most normal cells remain reasonably stationary. They are anchored to basement membranes and connective tissue, which is made up mainly of a fibrous mesh of proteins and other substances. In the adults of most species, only a few cell types will routinely move through this extracellular matrix. It is known that during embryonic development and wound healing, some cells migrate extensively and usually unerringly. The question is asked—how can the organization of these cells be both fixed and dynamic? Glycoproteins (those with attached sugars) may be part of the answer. Of these glycoproteins, the fibronectins are currently the best understood. These molecules have several functions—they can assemble into fibrils, bind to cells, and link cells to other kinds of fibrils in the extracellular matrix. Fibronectin, of course, is a critical component of the blood clotting function. Several lines of research are now being followed in fibronectin studies. These are well described and illustrated in the Hynes reference. It has been suggested that, inasmuch as cancer most frequently involves metastasis (migration of tumor cells to unrelated tissues elsewhere in the body), there may be some connection with fibronectins, because their currently best understood role is that of keeping cells in place and when they move they control their migration.

W.R. Schaffer (University of California, Berkeley) and colleagues have been investigating what are known as *isoprenoids*. These compounds are structurally related lipophilic molecules that perform a wide variety of essential cellular functions. These lipids include such functionally diverse molecules as cholesterol, ubiquinone, dolichols, and chlorophyll, yet isoprenoids are derived from a common precursor, mevalonic acid. These studies may lead to a better understanding of the Ras oncogenic proteins.

In recent years, molecular biologists have found that proteins, in their various roles (binding of receptors, assembling into cellular structures, catalyzing metabolic reactions, etal) depend largely on their three-dimensional structure. For quite some time, biologists have been successful with their techniques for sequencing the amino acids that make up a given protein. In contrast, progress was slow toward determining how the protein chain of components folds into a three-dimensional structure.

It was not until quite recently that neural computers have been used to solve the protein-folding problem. Some proteins have been described in the past as being contorted into "tangled" structures, sometimes likened to a twisting telephone cord. Attempts to predict a folding pattern would require the computation of each part of a chain and its effect on adjacent parts of the chain, a computation process of great magnitude. Further, protein crystals are difficult to develop, thus eliminating or at least reducing the effectiveness of x-ray crystallography. Nuclear magnetic resonance (NMR) also has been used, but unfortunately tends to be limited to the smaller proteins and requires much computer time. Although several thousand proteins have been amino acid sequenced, only a few hundred structures have been determined.

In 1988, researchers T.J. Sejnowski and N. Olan (Johns Hopkins University) reasoned that a computer (NETtalk), that had been designed to pronounce written English words might be applied to the protein structure problem—because NETtalk also depended upon deciphering that occurs at the junctions of numerous separations in a word (as it may appear hyphenated) and that this analysis may be similar to the occurrence of a multi-hyphenated structure exhibited by proteins. The researchers explain that a learning rule modifies the network so that eventually the network will produce the correct phoneme a large percentage of the time. Further work along these lines resulted in a network that could correctly predict over 64% of a test sequence.

S. Brunak and R.M.J. Cotterill (Technical University of Denmark) pursued the approach further, based upon data inputs from NMR and x-ray diffraction. The neural network approach remains very active so that encoding of the intricacy of interconnections may be achievable.

In addition to studying the structure (folding) of proteins for fundamental knowledge, the study of enterotoxins has the additional incentive where

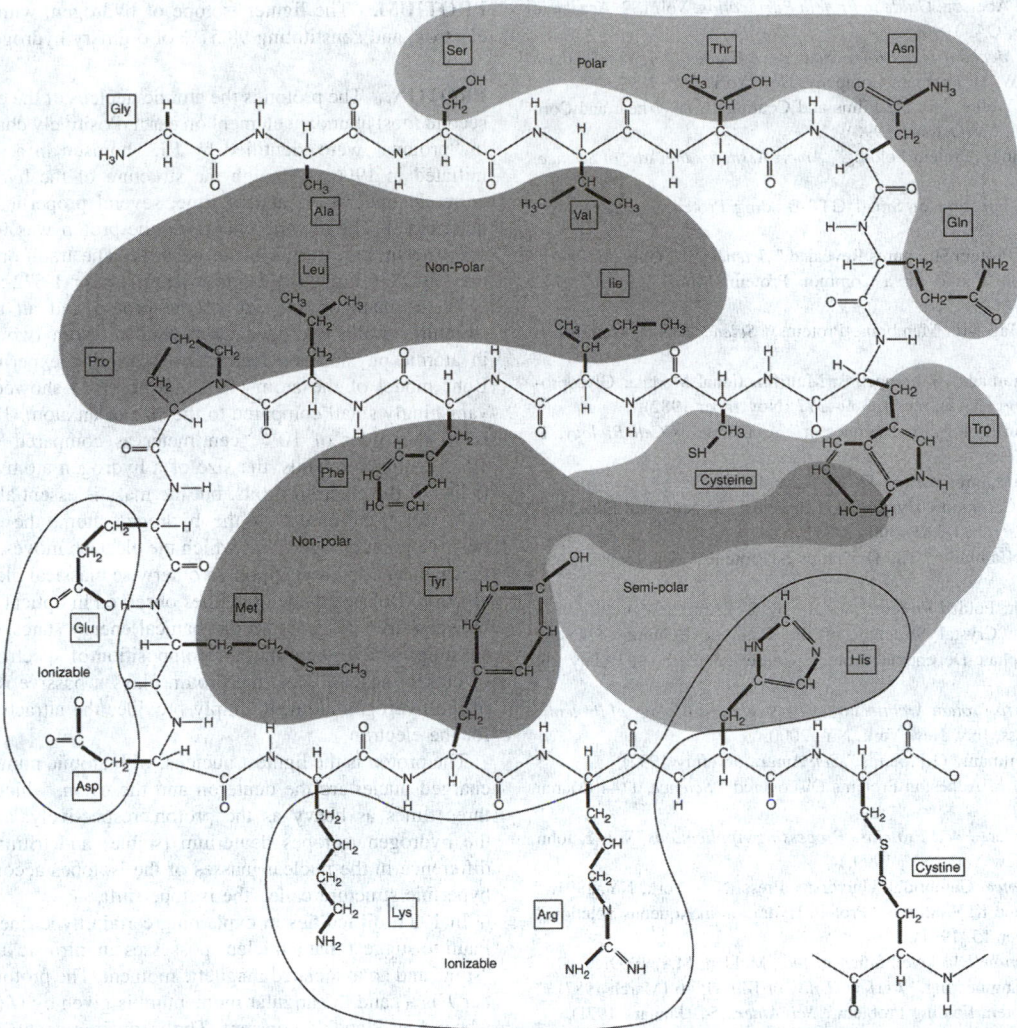

Fig. 2.   Facsimile depiction of an imaginary oligopeptide with side chains grouped in accordance with their properties as proposed by Phillips (1987) in an excellent summary of "Protein Engineering" in the new and exceptional publication, *Scientific & Technology Review (The University of Wales)*. All twenty amino acids are represented as shown by three-letter abbreviations in the boxes on the diagram. Polar, semipolar, nonpolar, and ionizable portions of the hypothetical oligopeptide are indicated by shaded and dotted areas. Also, note disulfide bridge shown. (*After Phillips.*)

life-threatening diseases are concerned, particularly toward the development of improved vaccines.

In research conducted at the University of Groningen (Netherlands) over a 14-year period, scientists succeeded in developing a pure crystal of the cholera toxin. Over 25,000 diffraction measurements of pure crystals, it became possible to generate a computer image of the cholera toxin. It has been observed that all bacterial toxins act in the same manner — one component is an enzyme that performs the invasive function and another component performs destruction once it enters the cell.

Research also has indicated that *E. coli* and diphtheria toxin perform in a similar manner. Active research programs currently are being conducted at Harvard University and the University of California, Los Angeles.

Similar structural determination studies are going forward to determine enzyme structures. In 1991, S. Taylor, D. Knighton, J. Sowadski, and colleagues (University of California, San Diego) announced their development of the three-dimensional structure of a protein kinase.

As aptly put by Doolittle (1985) — "If DNA is the blueprint of life, then proteins are the bricks and mortar."

Additional Reading

Abbott, N.L. and T.A. Hatton: "Liquid-Liquid Extraction for Protein Separations," *Chem. Eng. Progress*, 31 (August 1988).

Angeletti, R.H.: *Proteins: Analysis and Design*, Academic Press, Inc., San Diego, CA, 1998.

Barton, G.J.: *Protein Structure and Prediction*, Blackwell Science, Inc., Malden, MA, 2002.

Bollag, D.M., S.J. Edelstein, and M.D. Rozycki: *Protein Methods*, 2nd Edition, John Wiley & Sons, Inc., New York, NY, 1996.

Bohr, H.G.: *Neural Network Prediction of Protein Structures*, Springer-Verlag, Inc., New York, NY, 2001.

Bowie, J.U., et al.: "Deciphering the Message in Protein Sequences: Tolerance to Amino Acid Substitutions," *Science*, 1306 (1990).

Branden, C. and J. Tooze: *Introduction to Protein Structure*, 2nd Edition, Garland Publishing, Inc., New York, NY, 1998.

Brown, W.E. and G.C. Howard: *Modern Protein Chemistry: Practical Aspects*, CRC Press, LLC., Boca Raton, FL, 2001.

Builder, S.E. and W.S. Hancock: "Analytical and Process Chromatography in Pharmaceutical Protein Production," *Chem. Eng. Progress*, 42 (August 1988).

Clore, G.M. and A.M. Gronenborn: "Structures of Larger Proteins in Solution: Three- and Four-Dimensional Heteronuclear NMR Spectroscopy," *Science*, 1390 (June 7, 1991).

Considine, D.M. and G.D. Considine: *Foods and Food Production Encyclopedia*, Van Nostrand Reinhold Company, Inc., New York, NY, 1982.

Copeland, R.A.: "Proteins: Masterpieces of Polymer Chemistry," *Today's Chemist*, 53 (June 1992).

Creighton, T.E.: *Protein Function: A Practical Approach*, 2nd Edition, Oxford University Press, Inc., New York, NY, 1997.

DeGrado, W.F., Z.R. Wasserman, and J.D. Lear: "Protein Design, a Minimalist Approach," *Science*, 622 (1989).

Deutscher, M.P. and J.N. Abelson: *Guide to Protein Purification*, Vol. 182, Academic Press, Inc., San Diego, CA, 1990.

Fersht, A.: *Structure and Mechanism in Protein Science: A Guide to Enzyme Catalysis and Protein Folding*, W. H. Freeman Company, New York, NY, 1999.

Gennadios, A. and C.L. Weller: "Edible Films and Coatings from Wheat and Corn Proteins," *Food Techy.*, 63 (October 1990).

Gierasch, L.M. and J. King: "Protein Folding," *Amer. Assn. for the Adv. of Science*, Waldorf, MD, 1990.

Hall, A.: "The Cellular Function of Small GTP-Binding Proteins," *Science*, 635 (August 10, 1990).

Hoffman, M.: "New 3-D Protein Structures Revealed," *Science*, 382 (July 26, 1991).

Hoffman, M.: "New Role Found for a Common Protein 'Motif'," *Science*, 742 (August 16, 1991).

Hoffman, M.: "Playing Tag with Membrane Proteins," *Science*, 650 (November 1, 1992).

Hynes, R.O. and K.M. Yamada: "Fibronectins: Multifunctional Modular Glycoproteins," *J. of Cell Biology*, **95**(2), Part I, 369–377 (November 1982).

Hynes, R.O.: "Molecular Biology of Fibronectin," *Ann. Rev. of Cell Biology*, **1**, 67–90 (1985).

Hynes, R.O.: "Fibronectins," *Sci. Amer.*, 42–51 (June 1986).

Karplus, M. and J.A. McCammon: "Dynamics of Proteins: Elements and Function," *Ann. Rev. of Biochemistry*, **52**, 263–300 (1983).

Karplus, M. and J.A. McCammon: "The Dynamics of Proteins," *Sci. Amer.*, 42–51 (April 1986).

Kinoshita, J.: "Net Result: Folded Protein," *Sci. Amer.*, 24 (April 1990).

Knighton, D.R., et al.: "Crystal Structure of the Catalytic Subunit of Cyclic Adenosine Monophosphate-Dependent Protein Kinase," *Science*, 407 (July 26, 1991).

Lesk, A.M.: *Introduction to Protein Architecture: The Structural Biology of Proteins*, Oxford University Press, Inc., New York, NY, 2000.

Linder, M.E. and A.G. Filman: "G Proteins," *Sci. Amer.*, 56 (July 1992).

Marx, J.L.: "New Family of Adhesion Proteins Discovered," *Science*, 1144 (March 3, 1989).

Nakai, S. and H.W. Modler: *Food Proteins: Processing Applications*, Vol. 2, John Wiley & Sons, Inc., New York, NY, 1999.

Neurath, H.: *Protein Science*, Cambridge University Press, New York, NY, 1991.

Otting, G., E. Liepinsh, and K. Wuthrich: "Protein Hydration in Aqueous Solution," *Science*, 974 (November 15, 1991).

Patthy, L.: *Protein Evolution*, Blackwell Science, Inc., Malden, MA, 1999.

Phillips, D.C.: "Protein Engineering," *Review (Univ. of Wales)*, 46 (March 1987).

Richards, F.M.: "The Protein Folding Problem," *Sci. Amer.*, 54 (January 1991).

Radousky, H.B., G. Hammond, Z. Xu, et al.: *Gene Families: Studies of DNA, RNA, Enzymes and Proteins*, World Scientific Publishing Company, Inc., River Edge, NJ, 2001.

Schaffer, W.R., et al.: "Enzymatic Coupling of Cholesterol Intermediates to a Mating Pheromone Precursor and to the Ras Protein," *Science*, 1133 (September 7, 1990).

Sikorski, Z.E.: *Chemical and Functional Properties of Food Proteins*, CRC Press, LLC., Boca Raton, FL, 2001.

Skolnick, J. and A. Kolinski: "Simulations of the Folding of a Globular Protein," *Science*, 1121 (November 23, 1990).

Smith, D.M.: "Meat Proteins," *Food Techy.*, 116 (March 1988).

Utermann, G.: "The Mysteries of Lipoprotein (a)," *Science*, 904 (1989).

Villafranca, J.J.: *Current Research in Protein Chemistry: Techniques, Structure, and Function*, Academic Press, Inc., San Diego, CA, 1990.

Walker, J.M.: *Protein Protocols Handbook*, 2nd Edition, Humana Press, Totowa, NJ, 2002.

Walsh, G.: *Proteins: Biochemistry and Biotechnology*, 2nd Edition, John Wiley & Sons, Inc., New York, NY, 2002.

Whiting, R.C.: "Ingredients and Processing Factors that Control Muscle Protein Functionality," *Food Techy.*, 104 (April 1988).

Wuthrich, K.: "Protein Structure Determination in Solution by Nuclear Magnetic Resonance Spectroscopy," *Science*, 45 (1989).

PROTEROZOIC (Algonkian).

PROTEROZOIC (Algonkian).   The next to the oldest of the five eras of the Earth's history. Separated from the Archeozoic (the oldest Era) by a profound unconformity. The formations of the Proterozoic contain a preponderance of red sandstones and shales suggesting increasing aridity. Tillites also prove that continental glaciers existed in Eastern Canada, Australia, Tasmania, Norway, South Africa and India. The only undoubted forms of life appear to have been low forms of marine plants called calcareous algae. The sedimentary formations of the Proterozoic, especially in North America, contain important ores of iron and copper. Length of time since the beginning of the Proterozoic, 1,000 million years.

PROTIUM.   The lighter isotope of hydrogen, with a single proton and electron, and constituting 98.51% of ordinary hydrogen is termed protium.

PROTON.   The proton is the atomic nucleus of the element hydrogen, the second most abundant element on earth. Positively charged hydrogen atoms or "protons" were identified by J.J. Thomson in a series of experiments initiated in 1906. Although the structure of the hydrogen atom was not correctly understood at that time, several properties of the proton were determined. The electric charge on the proton was found to be equal but opposite in sign to that of an electron. The traditionally accepted proton mass is 1836 times the electron rest mass, or 1.672×10^{-24} grams.[1]

An estimate of the size of the proton and an understanding of the structure of the hydrogen atom resulted from two major developments in atomic physics: the Rutherford scattering experiment (1911) and the Bohr model of the atom (1913). Rutherford showed that the nucleus is vanishingly small compared to the size of an atom. The radius of a proton is on the order of 10^{-13} centimeter as compared with atomic radii of 10^{-8} centimeter. Thus, the size of a hydrogen atom is determined by the radius of the electron orbits, but the mass is essentially that of the proton.

In the Bohr model of the hydrogen atom, the proton is a massive positive point charge about which the electron moves. By placing quantum mechanical conditions upon an otherwise classical planetary motion of the electron, Bohr explained the lines observed in optical spectra as transitions between discrete quantum mechanical energy states. Except for hyperfine splitting, which is a minute decomposition of spectrum lines into a group of closely spaced lines, the proton plays a passive role in the mechanics of the hydrogen atom. It simply provides the attractive central force field for the electron.

The proton is the lightest nucleus, with atomic number one. Other singly charged nuclei are the deuteron and the triton, which are nearly two and three times as heavy as the proton, respectively, and are the nuclei of the hydrogen isotopes deuterium (stable) and tritium (radioactive). The difference in the nuclear masses of the isotopes accounts for a part of the hyperfine structure called the isotope shift.

In 1924, difficulties in explaining certain hyperfine structures prompted Pauli to suggest that a nucleus possesses an intrinsic angular momentum or "spin" and an associated magnetic moment. The proton spin quantum number I is $\frac{1}{2}$, and the angular momentum is given by $[I(I+1)h^2/(2\pi)^2]^{1/2}$, where h is Planck's constant. The intrinsic magnetic moment is 2.793 in units of nuclear magnetons (0.50504×10^{-23} erg/gauss), which is about a factor of 660 less than the magnetic moment of the electron.

Two types of hydrogen molecule result from the two possible couplings of the proton spins. At room temperature, hydrogen gas is made up of 75% orthohydrogen (proton spins parallel) and 25% parahydrogen (proton spins antiparallel). Several gross properties, such as specific heat, strongly depend upon the ortho or para character of the gas.

See also **Particles (Subatomic)**.

PROTON–PROTON REACTION.   A thermonuclear reaction in which two protons collide at very high velocities and combine to form a deuteron. The resultant deuteron may capture another proton to form tritium and the latter may undergo proton capture to form helium. See also **Carbon Cycle (Nuclear)**. The proton–proton reaction is now believed to be the principal source of energy within the sun and other stars of its class. A temperature of the order of five million degrees Kelvin and high hydrogen (proton) concentrations are required for this reaction to proceed at rates compatible with energy emission by such stars.

PROTOZOA.   The one-celled animals, constituting a major division of the animal kingdom. They occur in soil or water and many species live as parasites or symbionts in the bodies of other animals. Some protozoans are colonial and in some colonies a division of labor occurs, accompanied

---

[1] Particularly since the early 1970s, physicists have been seeking a grand unification theory to explain all the elementary particles of matter and all the forces acting between them. Although this goal continues to be elusive, work toward that end is producing many new findings and revised concepts. In the main part of this entry, the traditional viewpoints on the proton are described. Some of the more recent postulations are given toward the end of the entry.

TABLE 1. CLASSIFICATION OF THE PROTOZOA

Subkingdom Unicellulates (Protozoa)

Class Flagellata
  ORDER Chrysomonadina
    Families: *Chromulinidae; Rhizochrysidae; Ochromonadidae; Coccolithophoridae*
  ORDER Cryptomonadina
  ORDER Phytomonadina
    Families: *Chalmydomonadidae; Volvocidae*
  ORDER Euglenoidina
    Families: *Englenidae; Peranemidae*
  ORDER Dinoflagellata
    Families: *Ceratidae; Noctilucidae; Gymnodinidae*
  ORDER Protomonadina
    Families: *Eumonadidae; Craspedomonadidae* (choanoflagellates); *Trypanosomatidae* (trypanosomes)
  ORDER Diplomonadina
  ORDER Polymastigina
    Families: *Trichomonadidae; Calonymphidae; Prysonymphidae; Hypermastigidae*
  ORDER Opalinia
Class Rhizopoda
  ORDER Amoebina
  ORDER Testacea
  ORDER Foraminifera
  ORDER Heliozoa
    SUBORDER Centrohelidia
  ORDER Radiolaria
    SUBORDER Acantharia
    SUBORDER Spumellaria
    SUBORDER Nassellaria
    SUBORDER Phaeodaria
Class Sporoza
  ORDER Gregarinida
    SUBORDER Schizogregarinida
    SUBORDER Gregarines (Eugregarinida)
  ORDER Coccidia
    SUBORDER Schizococcidia
      Families: *Eimeridae; Haemosporidae*
Class Ciliata
  ORDER Holotricha
    SUBORDER Gymnostomata
    SUBORDER Trichostomata
    SUBORDER Hymenostomata
    SUBORDER Astomata
  ORDER Peritricha
    SUBORDER Sessila
    SUBORDER Mobilia
  ORDER Spirotricha
    SUBORDER Heterotricha
    SUBORDER Hypotricha
    SUBORDER Oligotricha
      Family: *Tintinnidae*
    SUBORDER Entodiniomorpha
  ORDER Chonotricha
  ORDER Suctoria
Class Cnidosporida
  ORDER Myxosporidia
  ORDER Actinomyxidia
  ORDER Microsporidia
Class Halosporida
Class Sarcosporidia
Class Piroplasmida
  Families: *Theilerida; Babesidae*

by structural specialization of the individuals. Some species are widely distributed. See Table 1.

Since the body of a protozoan is a single cell it has the subordinate structures of the cell in addition to other specialized parts. It contains one or more nuclei surrounded by cytoplasm. The body may be naked and without permanent form or held in a definite shape by a delicate surface membrane called a pellicle. Some species secrete a shell or test and some form internal hard parts. See also **Ocean Resources (Living)**.

The structures that perform special functions are called organelles, since they resemble the multicellular organs of other animals in function but are simpler than the cell itself in structure. Among the more evident and important are the external organelles for locomotion and for securing food. In the various forms of protozoans these are pseudopodia, cilia, flagella, cirri, membranelles, undulating membranes, or tentacles. Some are associated with a depression in the surface of the body called the cytosome through which food is ingested. Within the body the cytoplasm is differentiated into a clear layer of ectoplasm at the surface and an inner granular endoplasm containing the nucleus. Here also masses of food with a little water form the food vacuoles in which digestion takes place. One or more pulsating or contractile vacuoles are interpreted as excretory organelles or osmoregulators. They fill periodically with clear liquid and then discharge. Slender rod-like defensive structures, the trichocysts, lie in the ectoplasm of some species and are discharged when the animal is irritated. See Fig. 1.

In the bodies of the species which carry on a type of nutrition like that of green plants colored bodies (chloroplasts) containing chlorophyll are found in the cytoplasm. These organisms have been interpreted both as fission although sexual reproduction also takes place at intervals through the process of conjugation.

Protozoans are economically important chiefly as the causes of several serious diseases.

Additional Reading

Amos, W.B. and J.G. Duckett: *Prokaryotic and Euraryotic Flagella*, Cambridge University Press, New York, NY, 1982.

Bell, G.: *Sex and Death in Protozoa: History of an Obsession*, Cambridge University Press, New York, NY, 1989.

Bloodgood, R.A.: *Ciliary and Flagellar Membranes*, Kluwer Academic Publishers, Norwell, MA, 1990.

Capriulo, G.M.: *Ecology of Marine Protozoa*, Oxford University Press, Inc., New York, NY, 1990.

Curds, C.R.: *Protozoa and the Water Industry*, Cambridge University Press, New York, NY, 1992.

Dentler, W., P.T. Matsudaira, and G. Witman: *Cilia and Flagella*, Harcourt Brace & Company, San Diego, CA, 1995.

Gray, P.: *The Encyclopedia of the Biological Sciences*, Krieger Publishing Company, Melbourne, FL, 1981.

Jahn, T.L. and E.C. Bovee: *How to Know the Protozoa*, 2nd Edition, The McGraw-Hill Companies, Inc., New York, NY, 1978.

Laybourn-Parry, J.: *A Functional Biology of Free-Living Protozoa*, University of California Press, Berkeley, CA, 1984.

Patterson, D.J.: *Free-Living Freshwater Protozoa: A Colour Guide*, John Wiley & Sons, Inc., New York, NY, 1996.

Pennak, R.: *Fresh-Water Invertebrates of the United States: Protozoa to Mollusca*, 3rd Edition, John Wiley & Sons, Inc., New York, NY, 1990.

Rietschel, P. and K. Rohde: "The Unicellular Animals," in *Grzimek's Animal Encyclopedia*, Vol. 1, Van Nostrand Reinhold, New York, NY, 1974.

PROTRACTOR. 1. A muscle that draws some part of the body forward or draws out or extends a part.

2. A mathematical instrument consisting of a graduated arc for measuring or plotting angles.

PROUSTITE. This ruby-silver mineral crystallizes in the hexagonal system; its name is a product of its scarlet-to-vermilion color when first mined. It is a silver arsenic sulfide. Ag_3AsS, of adamantine luster. Hardness of $2-2.5$; specific gravity of $5.55-5.64$. Usual crystal habit is prismatic to rhombohedral; more commonly occurs massive. Conchoidal to uneven fracture; transparent to translucent; color, scarlet to vermilion red. Light sensitive; must be kept in dark environment to maintain its primary character. A product of low-temperature formation in most silver deposits. Notable world occurrences include the Czech Republic and Slovakia, Saxony, Chile and Mexico. Found in minor quantities in the United States; the most exceptional occurrence at the Poorman Mine, Silver City District. Idaho where a crystalline mass of some 500 pounds (227 kilograms) was recovered in 1865. It was named for the famous French chemist, Louis Joseph Proust.

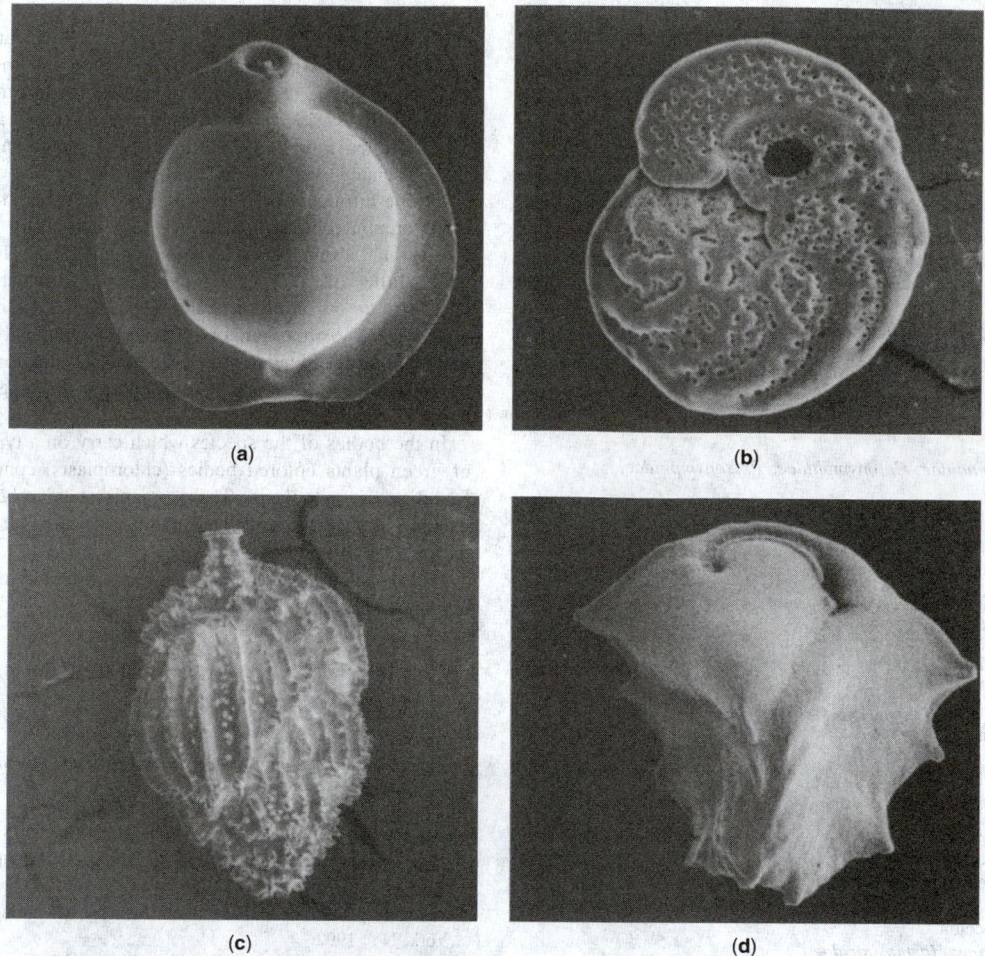

(a)   (b)

(c)   (d)

Fig. 1.   Oceanologists and climatologists have found that the abundance of benthic foraminifera may be a key to present and past deep ocean water flows, such as the Atlantic Deep Water and the Antarctic Bottom Water and, in turn, indicators of past ice ages. Benthic foraminifera form calcitic shells that are preserved in deep-sea sediments. The above are scanning electron micrographs of several species of benthic foraminifera that presently live on the sea floor around the Rio Grande Rise. (**a**) *Pyrgo* and (**b**) *Planulina wuellerstorfi* are associated with the North Atlantic Deep Water. (**c**) *Uvigerina peregrina* is widespread in the deep Pacific Ocean, and occurs on the Rio Grande Rise beneath the Circumpolar Deep Water, a water mass that originates in the Pacific, then flows into the western South Atlantic Ocean. (**d**) *Ehrenbergina*, which is found in the South Atlantic Ocean. (*Woods Hole Oceanographic Institution.*)

PROVITAMIN.   The precursor of a vitamin. Examples are carotene and ergosterol, which upon activation become Vitamin A and Vitamin D, respectively.

See also **Vitamin**; **Vitamin A**; and **Vitamin D**.

PROXIMITY AND OBJECT DETECTORS.   Generally, in position and motion control commonly encountered in automation systems, control action is taken to establish a position or series of positions that lie along a trajectory or path. In these cases, the exact coordinates of position are usually paramount. By contrast, *object detection* does not always require the accurate measurement of an object's position, but rather the prime purpose is one of detecting the *presence* or *nonpresence* (absence) of an object.

The needs for object detectors are several: (1) conveyor-associated applications, such as jam detection or protection, empty line detection, automatic routing; (2) safety and accident avoidance — to detect human hands and fingers where machines are manually loaded or unloaded and during maintenance procedures; (3) inspection of products — containers filled to the proper levels? labels properly placed or missing? incorrect closures? open flap detection and folding and wrapping imperfections, detection of web breaks, and plating and coating imperfections; (4) counting — to detect missing parts and to measure throughput; (5) sorting — by size, color, and other parameters; (6) as hopper level detectors and as feed cutoff controllers; and (7) a host of miscellaneous applications, such as edge guidance, remote door openers, and overhanging roof detectors. In industry, as well as in the lay world, these same kinds of devices play a major role as security monitors, warning of the unauthorized presence of persons and actions in designated secure areas.

It is interesting to note that object detection technology, although continuously refined from an engineering standpoint, dates back in principle for several decades and was one of the forerunners of modern industrial automation. Whereas today, robots seem to be in the forefront in terms of public recognition, it was not too many years ago that automated conveyor lines, packaging equipment, and the like were the operations proudly shown to visitors by plant managers.

Very few of the reasonably viable physical phenomena for detecting objects have been overlooked — a wide potpourri of detection methods is commercially available. A survey (1985) indicated that inductive and photoelectric detectors predominate, but other physical realms include electrical, electromagnetic, electromechanical, optoelectronic, radiant energy, air flow, and sonic approaches. Also very popular among industrial users are capacitive, Hall effect, Wiegand effect, and, to a lesser extent, magnetostrictive approaches.

A rather clean categorization of object detectors is: (1) contacting types, where the sensor makes actual physical contact with the object; and (2) noncontacting types, where the object only need be in the vicinity of the sensor. Depending on the type of sensing technology used, the vicinity can range from a millimeter or so up to several hundred feet (meters), although these are extreme cases.

Electromechanical (limit) contact switches are described in the article on **Limit Switch**.

Photoelectric Detectors

The fundamental principles of photoelectric devices are described in the article on **Photoelectric Effect**.

The principal aspects of the photoelectric effect of interest industrially are: (1) *photoconductivity*, evidenced by the increase in electrical conductivity of a material upon the absorption of light (or other electromagnetic radiation); and (2) the *photovoltage* or *photovoltaic effect*, wherein the energy of photons is converted to electrical voltage by the substance receiving the radiation. Photoconductivity is the basis of the operation of photocells and phototransistors utilized in industrial photoelectric switches. Principal applications for photovoltaic cells have been in aerospace applications, satellites for power, and the solar energy field.

Scope of Usage. Photoelectric controls respond to the *presence* or *absence* of either opaque or translucent materials at distances from a fraction of an inch (a few mm) up to 100 or even 700 feet (30 to 210 m). Photoelectric controls need no physical contact with the object to be triggered—important in some cases, such as those involving delicate objects and freshly painted surfaces. Some of the more common applications include thread break detection, edge guidance, web break detection, registration control, parts ejection monitoring, batch counting, sequential counting, security surveillance, elevator and conveyor control, bin level control, feed and/or fill control, mail and package handling, and labeling, among many others.

Photoelectric Control System Configuration. A self-contained control includes a light source, a photoreceiver, and the control base function, which amplifies and imposes logic on the signal to transform it into a usable electrical output. A *modular control* uses a light source-photoreceiver combination or reflective scanner separate from the control base. Self-contained retroreflective controls require less wiring and are less susceptible to alignment problems, while modular controls are more flexible in permitting remote positioning of the control base from the input components and hence are more easily customized.

Photoelectric controls are further classified as nonmodulated or modulated. *Nonmodulated* devices respond to the intensity of visible light. Thus, for reliability, such devices should not be used where the photosensor is subject to bright ambient light, such as sunlight. *Modulating controls* employing light-emitting diodes (LEDs), respond only to a narrow frequency band in the infrared. Consequently, they do not recognize bright, visible ambient light.

Controls typically respond to a change in light intensity above or below a certain value of threshold response. However, certain plug-in amplifier-logic circuits cause controls to respond to the rate of light change (transition response) rather than to the intensity. Thus, the control responds only if the change in intensity or brightness occurs very quickly (not gradually).

Operating Mode. Both modulated and nonmodulated controls energize an output in response to:

1. A light signal at the photosensor when the beam is not blocked (light-operated, LO).
2. A dark signal at the photosensor when the beam is blocked (dark-operated, DO).

Although some controls have built-in circuitry that determines a fixed operating mode, most controls accept a plug-in logic card or module with a mode selector switch that permits either light or dark operation.

In addition to a light source, light sensor, amplifier (in the case of modulated LED devices), and power supply, a complete system includes an electrical output device (in direct interface with logic level circuitry—the output transistor of a DC-powered modulated LED device or of an amplifier-logic card).

Scanning Techniques. There are several ways to set up the light source and photoreceiver to detect objects. The best technique is that one which yields the highest signal ratio for the particular object to be detected, subject to scanning distance and mounting restrictions. Scanning techniques fall into two broad categories: (1) thru (through) scan, and (2) reflective scan.

In *thru* (direct) *scanning*, the light source and photoreceiver are positioned opposite each other, so light shines directly at the sensor. The object to be detected passes between the two. If the object is opaque, direct

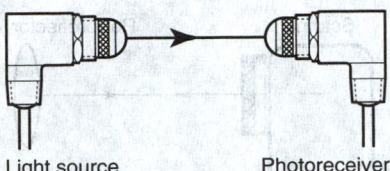

Fig. 1.   In direct (or *thru*) scan configuration, the light source is aimed directly at the photoreceiver. Sometimes the configuration is referred to as the *transmitted beam system*. (Micro Switch.)

scanning will usually yield the highest signal ratio and should be the first choice. See Fig. 1.

In *reflective scanning*, the light source and photoreceiver are placed on the same side of the object to be detected. Limited space or mounting restrictions may prevent aiming the light source directly at the photoreceiver, so the light beam is reflected either from a permanent reflective target or surface, or from the object to be detected, back to the photoreceiver. There are three types of reflective scanning: (1) *retroflective scanning*, (2) *specular scanning*, and (3) *diffuse scanning*.

Retroflective Scanning. With retroflective scanning, the light source and photosensor occupy a common housing. The light beam is directed at a retroreflective target (acrylic disk, tape, or chalk)—one that returns the light along the same path over which it was sent. See Fig. 2. Perhaps the most commonly used retrotarget is the familiar bicycle-type reflector. A large reflector returns more light to the photosensor and thus allows scanning at a further distance. With retrotargets, alignment is not critical. The light source-photosensor can be as much as 15° to either side of the perpendicular to the target. Also, inasmuch as alignment need not be exact, retroreflective scanning is well suited to situations where vibration would otherwise be a problem.

Specular Scanning. The specular scan technique uses a very shiny surface, such as rolled or polished metal, shiny plastic, or a mirror to reflect light to the photosensor. See Fig. 3. With a shiny surface, the angle at which light strikes the reflecting surface equals the angle at which it is reflected from the surface. Positioning of the light source and photoreceiver must be precise. Mounting brackets, which firmly fix the light source-photoreceiver relationship, must be used. Also, the distance of the reflecting surface from the light source and photoreceiver must be consistently controlled. The size of the angle between the light source and photoreceiver determines the depth of the scanning field. With a narrower angle, there is more depth of field. With a wider angle, there is less depth of field. For a fill-level detection application, for example, this means that a wider angle between the light source and photoreceiver allows detection of the fill level more precisely.

Diffuse Scanning. Nonshiny (matte) surfaces, such as kraft paper, rubber, and cork, absorb most of the incident light and reflect only a small amount. Light is reflected or scattered nearly equally in all directions. In diffuse scanning, the light source is positioned perpendicularly to a dull surface. Emitted light is reflected back from the target to operate the photoreceiver. See Fig. 4. Because the light is scattered, only a small percentage returns. Therefore, the scanning distance is limited (except with some high-intensity modulated LED controls), even with very bright light sources. It is often difficult to obtain a sufficient signal ratio with diffuse scanning when the surface to be detected is almost the same distance from the sensor as another surface (for instance, a nearly flat or low-profile cork liner moving along a conveyor belt). Contrasting colors can help in such situations.

Diffuse scanning is used in registration control and to detect material (corrugated metal, for example) with a slight vertical flutter—which might prevent a consistent signal with specular scanning. Alignment is not critical in picking up diffuse reflection.

Light Sources and Sensors. Early photoelectric control systems used incandescent light sources and traditional photocells—a combination still used. A photocell changes its electrical resistance with the amount of light that falls on it. A number of photocells have been used over the years for different applications (photoelectric controls, copying machines, TV pickup tubes, etc.). Widely used for photoelectric controls are cadmium

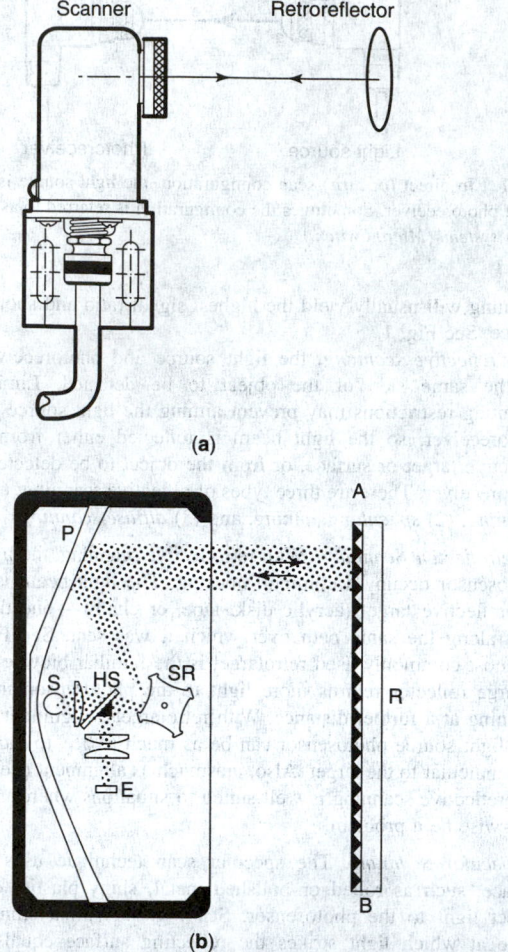

(a)

(b)

Fig. 2. **(a)** Reflected beam (retroreflective scan) system in which the light source and photoreceiver are contained in a single enclosure. This simplifies wiring and avoids critical alignment of the source and sensor. (*Micro Switch.*) **(b)** By adding a rotating-mirror wheel (SR), a parabolic reflector (P), and a semitransparent mirror (HS), a parallel-scanning beam can be obtained. This beam moves at high speed from A to B, thus forming a "light curtain," any interruption of which is detected and signaled by a relay, S. E is photoreceiver. (*Sick Optik Elektronik.*)

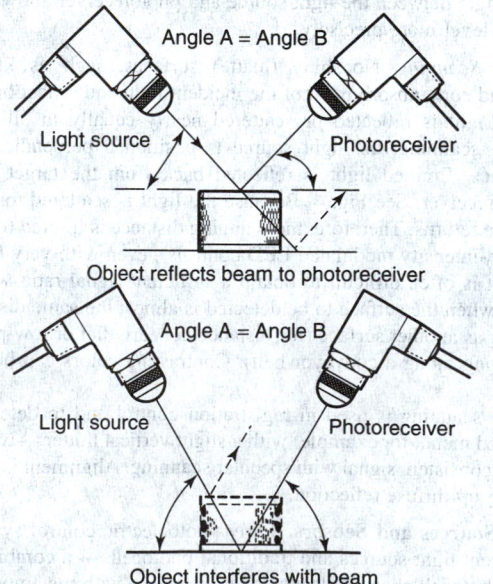

Fig. 3. Specular scan technique uses a very shiny surface, such as rolled or polished metal, shiny plastic, or a mirror to reflect light to the photosensor. (*Micro Switch.*)

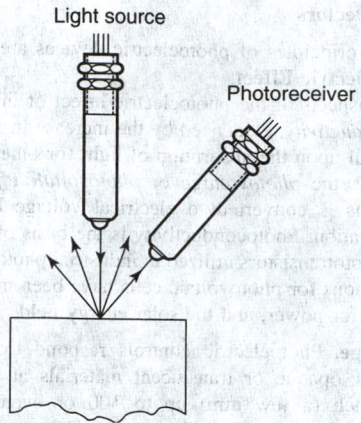

Fig. 4. Diffuse scan is used in registration control and to detect material (corrugated metal, for example) with a slight vertical flutter — which may prevent a consistent signal with specular scan. Alignment is not critical in picking up diffuse reflection. (*Micro Switch.*)

sulfide and cadmium selenide cells. During recent years, phototransistors and photodiodes have become available as sensors — and LEDs have been used as light (infrared) sources. There are several advantages in using the more recent hardware.

Photocells. There are at least four parameters that are important in the operation of a photocell: sensitivity, speed of response, light history effect, and effect of temperature.

Phototransistors. A phototransistor produces a collector current that is a function of both base current and light. Since the base lead of a phototransistor is usually left unconnected, only variations in light intensity produce variations in current output. There are several differences between the phototransistor and the photocell. (1) Current output of a phototransistor is largely independent of the voltage across it, whereas that of a photocell is not. As a result, controls designed to work with photocells will not necessarily work well with phototransistors, and vice versa. (2) The response of phototransistors is affected by changes in temperature, but in a way opposite to that of photocells; the higher the temperature, the higher the current output. (3) Phototransistors have a polarity, which must be observed; photocells do not. (4) Phototransistors respond to light much faster than photocells, but typically have a lower sensitivity.

Photodiode response is narrower than that of the phototransistor, making the diode more effective in blocking stray light from incandescent, sun, or other sources.

LED Sources. The useful life of an LED is estimated at 100,000 hours, which is at least ten times that of an incandescent lamp. However, incandescent lamps are still frequently used because they have a spectrum from the ultraviolet to the visible to the infrared, allowing a wide range of colored targets to be detected. LEDs have the advantage that they can be modulated directly, whereas incandescent lamps require a mechanical chopper. Silicon phototransistors and photodiodes are excellent matches for infrared LEDs because their greatest sensitivity peaks almost match precisely at the transmitter's (LED) wavelength.

Use of Fiber Optics with Photocells. Fiber-optic bundles can be added to existing photoelectric switches to provide object sensors, and these can be combined to implement logic functions. Such systems are useful for applications that require several sensing inputs and one or more outputs to interface with microcomputers. Program selection permits use of the LO or DO mode and allows operation of any channel for a predetermined time, thus avoiding sequential channel operation in fixed time frames. These systems frequently find application where a programmable controller is not warranted because of cost or complexity. Input can be from a relay or switch contacts, transducers, memory devices, CMOS, or TTL. The output section provides a channel signature for each emitter and detector pair, resulting in the capacity of actuating one or more output devices.

Applications of Photoelectric Controls. As illustrated by Fig. 5, the applications for photoelectric controls in automated systems seem to be limited only by the ingenuity of the control and system engineer. Most of

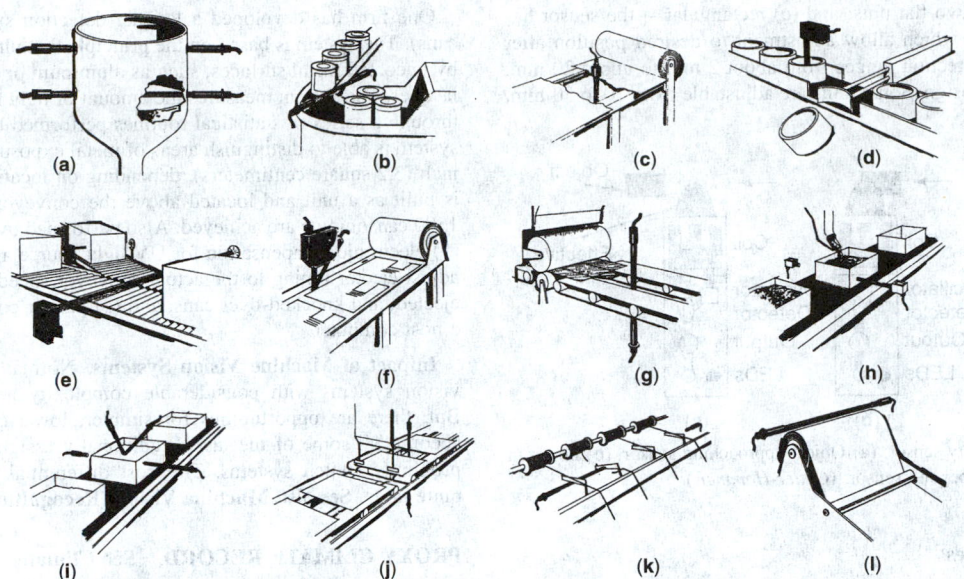

Fig. 5. Representative applications of photocell detectors in automated systems: (**a**) Two light source-photoreceiver pairs are used to keep hopper fill level between high and low limits. (**b**) Counting products is a common application of photoelectric controls. Counting batches or groups of cans or other items prior to packaging or group processing is also common. (**c**) A photoelectric control operating on reflected light is a simple way to detect a web break. An alternative is to put a light source above the web, and a photoreceiver below. (**d**) Dark caps are checked for white liners by a photoelectric scanner. The scanner activates a mechanism that rejects caps without the liners. (**e**) To prevent collisions where two conveyors merge, each conveyor is monitored by a control that powers the other conveyor when its own conveyor is cleared. (**f**) A tubular light source and photoreceiver in a specially designed bracket detect registration marks to initiate any related operation, such as printing, cutoff, or folding. (**g**) Gluing, buffing, or flattening can be done efficiently by controlling the pressure rollers or buffer with a photoelectric light source and photoreceiver that detect the product to be processed. (**h**) Using logic for one-shot pulse output, a photoelectric control slows a conveyor and fills the carton that has interrupted the light beam. (**i**) Two light source-photoreceiver pairs work together to check fill level. The box-detecting pair turns on, or enables, the fill inspection pair — thereby preventing the inspection pair from mistaking the space between boxes as an "improper fill." (**j**) Light source and photoreceiver placed near a guillotine are used to detect products and operate the blade for cutting the link between products. (**k**) Thread break detection made possible when the photoelectric beam is interrupted by a lightweight flag riding on the taut thread. (**l**) The size of a paper or fabric roll can be controlled by positioning a light source and a photoreceiver so the roll diameter blocks the beam. (*Micro Switch.*)

these applications can be served by other types of proximity sensors, but there are exceptions.

Magnetic Proximity Switches

There are four principal types of magnetic proximity switches: (1) variable-reluctance-type sensors, the operation of which depends on the interruption of a fixed magnetic field (circuit) by a ferrous actuator; (2) magnetically actuated dry reed or mercury switches; (3) Hall-effect sensors; and (4) Wiegand-effect sensors.

Variable-Reluctance Sensors. The principle of operation of variable-reluctance position (presence) sensors is shown schematically in Fig. 6. These transducers convert motion (rotating, sliding, oscillating) into electrical control signals. As shown in Fig. 6(a), with no actuating object in the vicinity of the sensor (pole piece plus coil plus magnet), the path of magnetic flux is undisturbed. As an object approaches and passes near the pole piece, the flux path is distorted. This system is often used in connection with rotating equipment for speed measurement (tachometry) where discontinuities, such as gear teeth, shaft keyways, drilled holes in steel plates, etc., alter the magnetic flux in proportion to rpm. Sensors are available in active and passive forms. Passive sensors require no external electric power. The output signal is an alternating current, the waveform of which is a function of the actuator, usually sinusoidal. The amplitude and frequency of the output signal are both proportional to the surface speed of the actuator as it passes the sensor's pole piece. The active configuration requires a DC power supply. The output signal is a pulse train whose amplitude is constant over the operating range for a fixed supply voltage level. Active magnetic sensors provide usable output signals at very low actuator speeds and at relatively large air gaps between the sensor pole piece and the actuator. They produce a logic-level output signal directly compatible with digital instrumentation.

Magnetically Actuated Dry Reed Switches. Generally consisting of a thin reed (wire) contained in a hermetically sealed container (encapsulated), this type of switch is both inexpensive and rugged. Whenever an activating

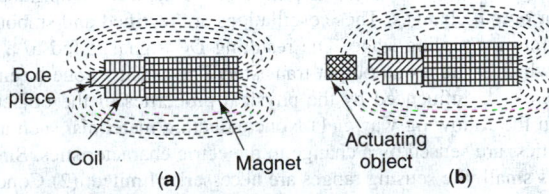

Fig. 6. Schematic representation of the action of a variable-reluctance object sensor: (**a**) A sensor with no actuating object in the magnetic field, (**b**) actuating object in field alters the voltage generated at coil terminals. The voltage is proportional to the rate of change of magnetic flux.

magnet approaches the critical range of the switch, a contact closure is made. Life expectancy usually is in excess of 20 million operations at contact ratings of about 15 VA. These switches generally can operate loads directly. Since the actuating magnet (powerful alloy magnets now usually used) can be installed on a rotating or reciprocating object, the switch can be used in a wide variety of applications for counting, positioning, and synchronizing. Contact closure speeds can be up to 100 per second. At one time, mercury switches with flexible electrodes that can be attracted by the proximity of a magnet were more widely used than at present.

Inductive Proximity Sensors

In an inductive proximity sensor, an electromagnetic field (radio frequency, rf) is generated by an oscillator circuit. When a metal object enters the effective field generated by the sensor, a countercurrent (eddy current) is set up in the metal object. This causes a voltage drop in the oscillator. This drop is sensed by the detector, which triggers the output. The output can be used for many industrial control purposes. See schematic diagram of a typical sensor given in Fig. 7. The sensors are available in three basic configurations: (a) cylindrical — a two-piece mounting clamp with socket head screws allows the installed sensor to be moved to the desired position; (b) threaded — the installed sensor can be rotated to the desired position and held in place with two flat nuts; and (c) rectangular—the sensor has slots in mounting base, which allow adjustment to desired position after installation. Nominal detection ranges from about 2 mm to about 20 mm, depending on specific design. Some units are adjustable from 10 to 50 mm.

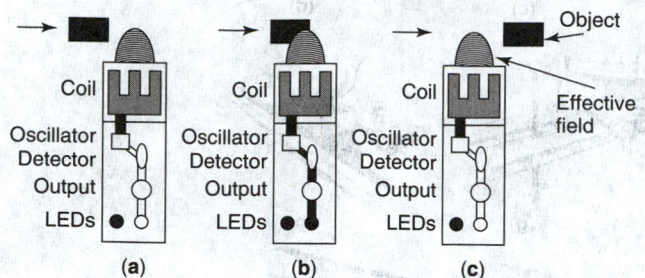

Fig. 7. Inductive proximity sensor: (**a**) Object approaching sensor, (**b**) object in field of sensor, (**c**) object beyond sensor. (*Cutler-Hammer.*)

Wiegand-Effect Switches

A Wiegand wire is a small-diameter wire that has been selectively work-hardened so that the surface and the core of the wire differ in magnetic permeability. When subjected to a magnetic field, the wire emits a well-defined pulse that requires little signal conditioning. This pulse induces a voltage in the surrounding sensing coil. The wire is insensitive to polarity and emits a pulse whether the magnetic field is flowing from north to south, or vice versa. A Weigand proximity senses the presence or absence of ferromagnetic material.

Capacitive Proximity Sensors

The basic element of a capacitive proximity sensor is a high-frequency oscillator containing a capacitor, one of the plates of which is built into the end of the sensor. When oscillating, a field is created around this free capacitor plate. When an object is placed in the field, the amplitude of the oscillator output changes. These oscillations are rectified and smoothed by an integrating coupling stage. The resulting DC signal is fed to a trigger circuit, which switches an output transistor. The sensing zone or envelope of the sensor is influenced by the physical properties of the object being sensed in the following ways: (1) Nonconducting materials, such as glass and plastics, are sensed by a change in dielectric characteristics. Since this change is small, the sensing ranges are necessarily limited. (2) Conducting materials are sensed by a change in dielectric characteristics as well as by an additional disturbance of the noise field caused by terminal conductivity. (3) Materials containing both conducting and nonconducting properties, especially if grounded, are sensed by a combination of the foregoing characteristics, as well as by absorption. These conditions produce the greatest switching distance for a given sensor. Because capacitive sensors are so markedly affected by material characteristics, an estimate of the switching range requires knowledge of the medium to be detected.

Magnetostrictive Limit Switch

Of limited application to date, the principle of magnetostriction was discovered in 1858 and represents a change in dimension of a material as the magnitude or direction of magnetization in a crystal is changed. The principle was introduced in the limit switch field in the mid-1970s. The active element is a helical spring that is subjected to axial extension or compression caused by mechanical displacement. The magnetostrictive transducer is considered to have good potential where high-resolution information is required over a substantial range.

Ultrasonic Switches and Other Novel Approaches

The presence of objects can be detected by return echoes from reflecting materials in a form of sonar. Similarly, rf reflection (radar) can be applied. A number of the large, well-established suppliers serving the proximity switch field have been joined by scores of newer, smaller firms that are busily engaged as of the late 1980s in applying "smart" electronics to new and old concepts.

One firm has developed a lacquer detection system (inside coating of cans). The system is based on the principle that ultraviolet light is reflected by uncoated metal surfaces, such as aluminum or steel, but is absorbed by lacquer. The system measures the amount of light reflected from a can and, through a series of statistical routines performed by a microprocessor, the system is able to distinguish areas of metal exposure as small as 0.5 square inch (3.2 square centimeters), depending on location. The detector-ejector is built as a unit and located above the conveyor systems. Speeds up to 1500 cans/minute are achieved. Also performed by the microprocessor are: (1) electronic compensation for UV light source aging, (2) eject control to adjust ejector timing to the actual conveyor speed, (3) counting of tested, ejected, and knocked-over cans, and (4) printer control and data supply to a host computer.

Impact of Machine Vision Systems. Normally, one equates machine vision systems with considerable complexity and relatively high cost. But, there are opportunities for simpler, lower-cost vision systems that accomplish some of the tasks traditionally assigned to photoelectric and proximity switch systems. The cost differential, however, still remains quite large. See also **Machine Vision (Recognition and Applications)**.

PROXY CLIMATE RECORD. See **Climate**.

PRURITUS. Itching of the skin, usually due to irritation of peripheral nerves.

PSEUDOADIABATIC EXPANSION. A saturation-adiabatic process in which the condensed water substance is removed from the system, and therefore best treated by the thermodynamics of open systems. Meteorologically, this process corresponds to rising air from which the moisture is precipitating. Descent of air so lifted becomes by definition a dry-adiabatic process.

See also **Pseudoadiabatic Process**.

PSEUDOADIABATIC LAPSE RATE. The rate of decrease of temperature with height of a parcel undergoing a pseudoadiabatic process.

It is given by

$$\Gamma_{ps} = g \frac{(1 + r_v)\left(1 + \dfrac{L_v r_v}{RT}\right)}{C_{pd} + r_v C_{pv} + \dfrac{L_v^2 r_v(\varepsilon + r_v)}{RT^2}}$$

where Γ_{ps} is the pseudoadiabatic lapse rate, g is gravitational acceleration, r_v is the mixing ratio of water vapor, c_{pd} and c_{pv} are the specific heats at constant pressure of dry air and water vapor, L_v is the latent heat of vaporization, R is the dry air gas constant, $\varepsilon \approx 0.62$ is the ratio of the gas constants of dry air and water vapor, and T is temperature. The above lapse rate is usually within 1 percent of those shown under moist-adiabatic lapse rate and reversible moist-adiabatic lapse rate.

See also **Moist-Adiabatic Lapse Rate** and **Reversible Moist-Adiabatic Process**.

AMS

PSEUDOADIABATIC PROCESS. A moist-adiabatic process in which the liquid water that condenses is assumed to be removed as soon as it is formed, by idealized instantaneous precipitation. Also called irreversible *moist-adiabatic process*. See also **Moist-Adiabatic Process**.

The pseudoadiabatic process is only defined for expansion, since a parcel that is compressed after such expansion will follow the dry-adiabatic lapse rate. A process similar to pseudoadiabatic descent can occur, however, if drizzle is evaporated into a relatively slow downdraft.

PSEUDOCODE. An arbitrary code not directly understandable by a computer. Also called *interpreter code*.

PSEUDOEQUIVALENT POTENTIAL TEMPERATURE. The temperature a sample of air would have if it were expanded by a pseudoadiabatic process to zero pressure and then compressed to a reference pressure of 100 kPa by a dry-adiabatic process.

This quantity is conserved in a pseudoadiabatic process and is given approximately by

$$\theta_{ep} = T\left(\frac{P_0}{P}\right)^{0.2854(1-0.28r_v)} exp\left[r_v(1+0.81r_v)\left(\frac{3376}{T_c}-2.54\right)\right],$$

where T is the temperature, p is the pressure, T_c is the condensation temperature (obtainable from the dewpoint formula) and r_v is the water vapor mixing ratio. When $r_v = 0$, $\theta_{ep} = \theta$, the potential temperature.

See also **Potential Temperature**; and **Pseudoadiabatic Process**.

Additional Reading

Bolton, D.: "The Computation of Equivalent Potential Temperature," *Mon. Wea. Rev.*, **108**, 1046–1053 (1980).

AMS

PSEUDOFOLLICULITIS BARBAE. See **Folliculitis**.

PSEUDOMONAS FOLLICULITIS (Hot Tub Folliculitis). See **Folliculitis**.

PSEUDOMORPH. In mineralogy and geology, a mineral, having the crystal form of one species and the chemical composition of another. Typical pseudomorphs are malachite in the form of cuprite, barite in the form of quartz, limonite in the form of pyrite. In such cases of pseudomorphism the evidence seems to be that there has been a complete chemical and molecular change but without any change of the original outward form. See also **Mineralogy**.

PSEUDOMORPHISM. See **Mineralogy**.

PSEUDOSCALAR. A scalar quantity that changes its sign when the coordinate system, to which the quantity is referred, is changed from a right-handed to a left-handed one, or vice versa. An example is the scalar product of a polar vector and a pseudovector. See also **Vector Multiplication**.

PSEUDOSCORPION. A small animal, resembling the scorpions slightly in appearance. The abdomen is rounded and has no slender posterior part and no sting, but the pair of chelate pinchers at the anterior end of the body resembles those of the scorpions. Pseudoscorpions are commonly found in leaves at the surface of the ground and under the bark of decaying logs. They are only a few millimeters long. The relatively few species make up the order *Chelonethida* of the class *Arachnida*.

PSEUDOVECTOR. The vector product of two vectors does not completely satisfy the formal requirements of a vector, for it changes sign if the coordinate system is changed from a right-handed to a left-handed one, or vice versa. It is a typical example of a pseudovector (also called an axial vector). Thus the vector for an element of area, represented by the vector product $d\mathbf{S} = d\mathbf{x} \times d\mathbf{y}$ is not determined with respect to direction unless an arbitrary convention is established for the positive side of the surface element. The three components of a pseudovector are actually the components of a three-dimensional antisymmetric tensor of second rank. Physical quantities which are pseudovectors include angular momentum (vector product of momentum and radius vector); moment of a force (vector product of force and distance); linear velocity (vector product of angular velocity and radius vector).

In the following relations, symbols ($\pm$) refer to true and pseudoquantities, S for scalar and $\mathbf{V}$ for vector; the primed and double primed vectors are so designated merely to indicate that they are not identical with the unprimed vectors: $S_+\mathbf{V}_\pm = \mathbf{V}'_\pm$; $S_-\mathbf{V}_\pm = \mathbf{V}'_\mp$; $\mathbf{V}_\pm \cdot \mathbf{V}'_\pm = S_+$; $\mathbf{V}_+ \cdot \mathbf{V}_- = S_-$; $\mathbf{V}_\pm \times \mathbf{V}'_\pm = \mathbf{V}''_\mp$; $\mathbf{V}_+ \times \mathbf{V}'_- = \mathbf{V}''_+$; $\nabla S_\pm = \mathbf{V}_\pm$; $\nabla \cdot \mathbf{V}_\pm = S_\pm$; $\nabla \times \mathbf{V}_\pm = \mathbf{V}'_\mp$.

PSILOMELANE. Psilomelane is a massive black mineral, essentially a basic oxide of barium with divalent and quadrivalent manganese, corresponding to the formula $BaMn^{2+}Mn^{4+}O_{16}(OH)_4$. It crystallizes in the monoclinic system, but is found only in massive, botryoidal or reniform to earthy habits; hardness, 5–6, less in earthy varieties; specific gravity, 6.45; color, black to gray; opaque; submetallic to dull luster. It is a product of secondary weathering of manganese carbonates and silicates. Of widespread occurrence, usually associated with pyrolusite. Major world occurrences include Michigan in the United States, Scotland, Sweden, France, Germany, and India. It is a major source of manganese. The word psilomelane is derived from the Greek words meaning smooth and black, in reference to the smooth black surfaces so often exhibited.

PSI PARTICLE. Discovery of this subatomic particle in 1974 was announced independently by Ting (Brookhaven National Laboratory) who named it the *J particle* and by B.D. Richter (Stanford) who named it the *psi particle*. The discovery of this particle resolved a number of important problems in particle physics. Intensive research on the psi particle was carried out by Richter and the Stanford group during 1975 and 1976 and is reported firsthand by Richter (*Science*, **196**, 1286–1297,1977). As pointed out by Richter, the four-quark theoretical model became much more compelling with the discovery of the psi particles. The long life of the psi is explained by the fact that the decay of the psi into ordinary hadrons requires the conversion of both c and $\bar{c}$ into other quarks and antiquarks. See also **Particles (Subatomic)**.

PSITTACIFORMES (*Aves*). The parrots and related species, including cockatoos, macaws, and parakeets (paraquets) are all members of this order. These birds are characterized by the crooked bill, adapted for opening nuts and seeds. It is popularly believed that all parrots are extremely colorful and inhabit the tropical jungles. Neither belief holds good always. While it is true that the more colorful forms are indeed found in tropical regions and in South America, and most particularly in territories of New Guinea and northern Australia, which are thought to be the original homeland of the parrot stock, it is also true that some species live above the tree line in snow and frost. Moreover, not all parrots are colorful; there are many forms with green plumage as a means of camouflage and many species are dark; some are even black. Today Europe is the only continent where there are no parrots.

The length of this order ranges from 10 centimeters (4 inches) in the pygmy parrot to 100 centimeters (39 inches) in the blue macaw. The fourth (lateral) toe is reversed, like the first, so that the two function opposed to the second and third. Thus a pair of pincers is formed which serves in climbing and grasping objects. The legs are short; the upper mandible is articulate and can be raised; the lower can slide (a useful property in shelling nuts and comminuting seeds and fruits). The underside of the upper mandible usually has hard filing ridges across its width which serve to sharpen the edges of the lower mandible so that the bird can get better grasp of its food and can grate hard shells more easily. The tongue is usually thick, has strong muscles, and many touch and taste papillae. The lories (subfamily *Trichoglossinae*) are brush tongued and thus thoroughly adapted to visiting flowers on whose pollen and nectar they feed. Almost all species can use their beaks as a "third foot" when climbing.

The sexes, as a rule, are easily distinguishable. With the exception of the keas, parrots are usually monogamous. Parrots usually breed in caves; their clutch numbers from 1 to 10 white eggs. As a rule brooding, which takes from 18 to 30 days, is done exclusively by the female. The young are crop-fed, usually by both parents, the food thus being already softened and supplied with vitamins from the glandular stomach. There are specific breeding seasons and sometimes there are 2 broodings.

Parrots comprise only 1 family (*Psittacidae*), which is subdivided into 7 subfamilies. 1. Keas (*Nestorinae*); 2. Vulturine Parrots (*Psittrichasinae*); 3. Cockatoos (*Kakatoeinae*); 4. Pygmy Parrots (*Micropsittinae*); 5. Lories (*Trichoglossinae*); 6. Owl parrots (*Strigopinae*); and 7. True Parrots (*Psittacinae*). There are 79 genera, 326 species, and 816 subspecies.

The most primitive of the still extant parrot forms is the Kea (subfamily *Nestorinae*). Seeing these birds at a distance for the first time, one would think that they are crows, judging from their motions. The habitat of the keas is also somewhat unusual for parrots; they breed in the New Zealand Alps above the tree line.

They are as large as crows: the length is 50 centimeters (20 inches) and the bill is long and narrow, the tooth-like serrations are undeveloped, and it has no filing ridges. They are birds of the dawn and dusk. Their food consists of vegetable and animal matter. The clutch numbers 2–4 eggs, the incubation period is 29 days.

There are 3 species, one of which is already extinct: (1) the Kaka (*Nestor meridionalis*); (2) the Kea (*Nestor notabilis*); (3) the Slender-Billed Kea (*Nestor productus*) has been extinct since the middle of the nineteenth century.

The Vulturine Parrot (*Psittrichas fulgidus*) is the only representative of the subfamily *Psittrichasinae*. See Fig. 1. It has bristle-like feathers on the neck. Its size is that of a crow; the length is 50 centimeters (20 inches). Its habitat is mountainous woodlands at altitudes of from 800 to 2,000 meters (2,645–6,562 feet); it feeds on sprouts, buds, berries, and fruits.

Fig. 1.   Vulturine parrot: Kea (*Nestor notabilis.*) (Sketch by Glenn D. Considine).

The Cockatoos (subfamily *Kakatoeinae*) are the largest parrots of the Indo-Australian region and at least some species are known to every bird lover. Cockatoos are as large as pigeons or ravens; the length is 32–80 centimeters (13–31 inches). Their feather crest is diagnostic, serving as a signaling device, and it sometimes differs in color from the rest of the plumage. Their food consists mostly of plant matter. Their clutch numbers from 2 to 4 eggs and the incubation period is 20–30 days, during which male and female relieve each other; the young are fledged at the age of 60–70 days. There are 5 genera with 17 species and 48 subspecies.

The Palm Cockatoo (*Probosciger aterrimus*) is the largest cockatoo, with a length reaching 80 centimeters (31 inches).

The Black Cockatoos (*Calyptorhynchus*) have a length of 50–65 centimeters (19½–25½ inches). Among the 4 species are: (1) the Yellow-Tailed Cockatoo (*Calyptorhynchus funereus*), which has a length of 60 centimeters (23½ inches); usually its clutch consists of 2 eggs; (2) The White-Tailed Cockatoo (*Calyptorhynchus baudinii*).

The Gang-Gang-Cockatoo (*Callocephalon fimbriatum*) has a length of 35 centimeters (13½ inches). The sexes differ markedly: the male has a red head and crest, while the female is blackish-gray. The clutch numbers 2 eggs.

The White-Billed and Black-Billed Cockatoos (*Kakatoe*) have a length of 32–50 centimeters (12½–19½ inches). They generally have light, usually white plumage. The division of this group into 2 genera (*Plyctolophus* and *Kakatoe*) is generally accepted. *Plyctolophus* comprises the species with black bills, while the *Kakatoe* describes those having light-colored bills. Among them are the species that bird fanciers like to keep: (1) The Yellow-Crested Cockatoo (*Kakatoe galerita*) has a length of 50 centimeters (19½ inches). (2) The White Cockatoo (*Kakatoe sulphurea*) has a length of 50 centimeters (19½ inches). (3) The Rose Cockatoo (*Kakatoe moluccensis*) has a length of 35 centimeters (13½ inches). (4) The Pink Cockatoo (*Kakatoe leadbeateri*) has a length of 38 centimeters (15 inches). (5) The Galah (*Kakatoe roseicapilla*) has a length of 37 centimeters (14½ inches). (6) The Bare-Eyed Cockatoo (*Kakatoe sanguinea*) has a length of 40 centimeters (15½ inches). (7) The Slender-Billed Cockatoo (*Kakatoe tenuirostris*) has a length of 40 centimeters (15½ inches).

The Cockatiel (*Nymphicus hollandicus*) reaches a length of 30 centimeters (12 inches), and the clutch consists of 4–7 eggs. See Fig. 2.

The Pygmy Parrots (subfamily *Micropsittinae*) are the smallest species among parrots. Their length is 10 centimeters (4 inches) and their weight is 13 grams (0.5 ounce). Little is known so far about the life pattern of these dwarfs among parrots. The toes are very long and thin; the bird has short, strong, stiff tail feathers that serve as a prop against the bark (as in the woodpecker) when it climbs around the tree trunk. Pygmy parrots feed on fruits, tree termites, and the sap of trees such as the *Albizzia procera* (related to the true acacia); the food is heavily mixed with saliva, which is secreted from the large salivary glands. They breed in nests of tree-inhabiting termites. There are 6 species and 23 subspecies; among them

Fig. 2.   Cockatoo: Cockatiel (*Numphicus hollandicus.*) (Sketch by Glenn D. Considine).

the most common are *Micropsitta pusio* and the Yellow-Capped Pygmy Parrot (*Micropsitta keiensis*).

The Lories (subfamily *Trichoglossinae*) are the most colorful of all parrots. Red is predominant among their conspicuously bright, poster-like colors. Only a few species are less colorful or even black. They range in size from that of a sparrow to that of a pigeon; the length is 12–35 centimeters (4½–13½ inches). Since they feed on blossoms, berries, and soft fruits, the filing ridges on the underside of the upper jaw are degenerate. The tip of the tongue is shaped like a paintbrush. Sex differences are slight. They have a clutch of 2–4 eggs, the incubation period lasting 21–26 days. The young stay in the nest for about two months. They occur in New Guinea and neighboring islands, and occasionally in Australia. There are 14 genera with a total of 61 species and about 150 subspecies.

The subfamily Owl Parrots (*Strigopinae*) has only 1 species, the Owl Parrot (*Strigops habroptilus*). It is about the size of a crow, the length reaching 60 centimeters (23½ inches). Its plumage is soft and the feathers around the beak are bristle-like, as they are in owls and goatsuckers. Its wings are short and rounded and it is virtually flightless. It feeds on plants and its beak is thick and without serrations. The owl parrot is nocturnal. Its clutch numbers two.

Since the owl parrot is virtually flightless it must climb, but it is able to come down in a slanted gliding flight of about 100 meters (328 feet). Its habitat is mountainous forests on the southern island of New Zealand. Owl parrots stamp down genuine paths in their habitats. They feed on mosses, leaves, sprouts, berries, and, if possible, fungi; they will also dig up roots and rhizomes of ferns. Indigestible cellulose fibers are formed into balls and regurgitated. The bird usually does not tear out grass blades and the berry-like *Carmichaelia* branches, but crushes them on sight.

The seventh and last subfamily, the True Parrots (*Psittacinae*), has the greatest number of species. There are 5 tribes: the Rosellas (*Platycercini*); *Loriini; Loriculini; Psittacini;* and *Araini*.

The best known and most frequently kept parrot, the budgerigar, belongs to the rosellas (tribe *Platycercini*). Its larger and usually very colorful relatives are among the most popular aviary-kept birds. They range abundantly from sparrow to magpie in size. The length is 18–38 centimeters (7–15 inches). The beak often has serrations, and is short and thick; the tail is staggered and long. The bird has a well-developed uropygium (the fleshy and bony prominence at the posterior extremity of a bird's body, which supports the tail feathers). Both sexes have the same or very similar colors. They usually inhabit the steppes and are poor climbers; their food consists of grass seeds. They breed once or twice a year, mostly in hollows, and have a clutch size of 4–6 eggs. The incubation period is from 18 to 21 days. There are eleven genera in this tribe with 29 species and 67 subspecies: (1) the Night Parrots (*Geopsittacus*); (2) the Ground Parrots (*Pezoporus*); (3) the Red-Fronted New Zealand Parakeets (*Cyanoramphus*); (4) the Horn Parakeets (*Eunymphicus*); (5) the Budgerigars (*Melopsittacus*); (6) *Neophema*; (7) the Red-Backed Parrots (*Psephotus*); (8) *Northiella*; (9) the Red-Capped Parrots (*Purpureicephalus*); (10) the *Rosellas*, in a narrower sense (*Platycercus*); and (11) the Swift Parrots (*Lathamus*). Two of the genera occur in New Zealand and on neighboring islands, and all others in Australia and Tasmania.

The Wax-Billed Parrots (*Loriini*) do not form as homogeneous a group as the rosellas. There are large and small species, and short and long-tailed ones, and even the shapes of their beaks differ considerably. Some of the forms are not even known by name to parrot lovers.

The wax-billed parrots range from the size of a sparrow to that of a magpie; the length is 13–54 centimeters (5–21 inches). See Fig. 3. The beak in most of the species has a wax-like sheen and smoothness; red beaks occur frequently, at least at the upper mandibles of the males (with the exception of the rosy-faced lovebird, the black-collared lovebird, and the gray-headed lovebird); while the staggered tails are longer than the wings, the wedge-shaped, rounded, or clipped ones are shorter. They occur in Australia, Asia, and Africa; there are 10 genera, 50 species, and 168 subspecies: (1) King Parrots (*Alisterus*); (2) the Red-Winged Parrot (*Aprosmictus*); (3) *Polytelis*; (4) *Lorius*; (5) *Psittacula*; (6) the Lovebirds (*Agapornis*); (7) *Mascarinus*; (8) *Tanygnathus*; (9) Racket-tailed Parrots (*Prioniturus*); and (10) Masked Parakeets (*Prosopeia*).

Fig. 3. Lovebird: Rosy-faced lovebird (*Agapornis roseicollis*) (Sketch by Glenn D. Considine).

The tribe *Loriculini* inhabits Southeast Asia and the adjoining island groups that stretch toward Australia. The *Loriculini* reach a length of 10–16 centimeters (4–6 inches). The beak is slender and longer than it is high. As with the wax-billed parrots, the beak has a smooth, shiny surface and is red or black. The tail is as long as the wings or shorter, and the upper tail coverts reach as far as the tip of the tail. They have a clutch of 2–3 eggs. There is 1 genus with 10 species and 31 subspecies; among them is the Malay Lorikeet (*Loriculus galgulus*).

All *Loriculini* usually scurry gracefully on the ground, and can climb deftly among branches and boughs. When sleeping, and occasionally when looking for food, they suspend themselves head down from branches rather like bats. In the wild their food consists predominantly of fruits, nectar from blossoms, and pollen. Like most species of lovebirds, the *Loriculine* also carry nesting materials under their feathers.

Fig. 4. Gray parrot (*Psittacus erithacus*.) (Sketch by Glenn D. Considine).

The tribe *Psittacini* includes the gray parrots and blunt-tailed parrots, known as good imitators. See Fig. 4. They range from the size of a chaffinch to that of a crow; the length is 16–50 centimeters (6–191/2 inches). Their tails are blunt or somewhat rounded. The beak is not smooth and wax-like. It may be black, gray, brownish, or yellowish-white, but rarely red. They occur in South and Central America, Africa, and Madagascar. There are 13 genera with 67 species and 143 subspecies: (1) the Vasas (*Coracopsis*); (2) *Psittacus* (Gray Parrot, *Psittacus erithacus*); (3) *Poicephalus* (Yellow-billed Senegal Parrot, *Poicephalus senegalus*); (4) *Pionites*; (5) *Pionus* (White-crowned Parrot, *Pionus senilis*); (6) *Deroptyus* (Red-Fan Parrot, *Deroptyus accipitrinus*); (7) The Blunt-Tailed Parrots (*Amazona*); (8) *Graydidasculus* (Short-tailed Parrot, *Graydidasculus brachyurus*); (9) *Pionopsitta*; (10) *Hapalopsittaca*; (11) *Gypopsitta* (Vulturine Parrot, *Gypopsitta vulturina*); (12) *Touit*; (13) *Triclaria* (Blue-Bellied Parrot, *Triclaria malachitacea*).

The tribe *Araini* inhabits North and South America. Besides small forms, we also find among them the giants of the parrots, the macaws. Their sizes range from that of a titmouse to that of a pheasant; the length is 12–98 centimeters (5–38½ inches). The tail is almost always staggered, and the tail feathers are narrowed or the tips are pointed; the eye rings are often bare. The bill is short, thick, never red. The most conspicuous characteristic of most *Araini* is their loud, penetrating voice. Since the parrots of this group can be quite destructive to their enclosures, they are not very popular with bird lovers.

There are 19 genera (one of them extinct) with 84 species and more than 180 subspecies: (1) the Parrotlets (*Forpus*); (2) *Nannopsittaca*; (3) *Brotogeris* (Orange-flanked Parakeet, *Brotogeris pyrrhopterus*); (4) The *Bolborhynchos* (Rufous-Fronted Parakeet, *Bolborhynchos ferrugineifrons*); (5) *Psilopsiagon*; (6) *Amoropsitta* (Sierra Parakeet, *Amoropsitta aymara*); (7) *Myiopsitta* (Green Parakeet, *Myiopsitta monachus*); (8) *Microsittace* (Chilean Parakeet, *Microsittace ferruginea*); (9) *Enicognathus* (Slender-billed Parakeet, *Enicognathus leptorhynchus*); (10) *Pyrrhura* (White-Eared Parakeet, *Pyrrhura leucotis*); (11) *Ognorhynchus* (Yellow-Eared Parakeet, *Ognorhynchus icterotis*); (12) *Leptosittaca* (Golden-Plumed Parakeet, *Leptosittaca branickii*); (13) *Nandayus* (Black-headed Parrot, *Nandayus nenday*); (14) *Aratinga* (Sun Parakeet, *Aratinga solstitialis*); (15) *Conuropsis* (Carolina Parakeet, *Conuropsis carolinensis*); (16) *Cyanoliseus* (Burrowing Parrot, *Cyanoliseus patagonus*); (17) The *Rhynchopsitta* (Thick-billed Parrot, *Rhynchopsitta pachyrhyncha*); (18) the Macaws (*Ara*); (19) the Blue Macaws (*Anodorhynchus*).

Almost all macaws are inhabitants of the forests. With their large beaks they are able to crack open very hard nutshells, after first filing down the thickness of the shell at one place. See Fig. 5. The beak is also an adaptive feature of their locomotor pattern, which is chiefly climbing. As in almost all parrots it is used as a "third foot." See also **Parrots and Cockatoos**.

Fig. 5. Blue-and-yellow macaw (*Ara ararauna*.) (Sketch by Glenn D. Considine).

PSITTACOSIS (Parrot fever). A disease caused by *Chlamydia psittaci*, an obligate intracellular parasite similar to a virus, but classified as a bacterium because it possesses both RNA and DNA, a discrete wall membrane similar to that of Gram-negative bacteria, and a primitive enzyme system. The life cycle of the organism is complex and not yet fully understood. The infectious elementary body is DNA-rich and relatively stable. It initiates the infection upon entering a susceptible cell by phagocytosis and some hours later reorganizes to form the noninfective large, labile, metabolically active reticulate body. This has an increased RNA content and divides several times by binary fission, ultimately producing new elementary bodies. The whole cycle is completed in 36 to 48 hours.

Psittacosis is a serious hazard to people who keep pet birds or who work in the poultry industry. The infecting agent is harbored by avians and, although an infected bird does not usually act as sick, it is lethargic, has ruffled feathers, and usually diarrhea. *C. psittaci* is present in the bird's nasal secretions, excreta, and feathers and is transmitted to humans by inhalation of such dried sources. Transmission between humans is extremely rare.

Birds admitted to a number of countries are first quarantined for a period. In the United States, this quarantine period is thirty days, during which time the birds receive chlortetracycline in their feed. Occurrence of the disease in humans in the United States ranges between about 40 and 100 cases per year, although in 1974 and again in 1984, the numbers of cases reported to the Centers for Disease Control reached 175, with most of the cases coming from turkey processing operations.

C. psittaci enters the body through the respiratory tract and is rapidly disseminated through the blood stream to the reticuloendothelial system. The principal lesion is then found in the lung, but others occur in the liver and spleen. The pulmonary lesion is principally an interstitial pneumonitis accompanied by airspace involvement.

In humans, the incubation period is generally one to two weeks. Symptoms include high fever, headache, myalgia, chills, and coughing. Clinically, psittacosis presents in two major forms. The first is pneumonitis with extensive focal or lobar pneumonia. The second has features suggestive of a toxic or septic condition. The severity of clinical signs may diminish in the second week of illness, but the course of psittacosis is often prolonged and relapses are not uncommon.

Laboratory diagnosis is usually made by detecting a rise in complement-fixing titer to a group antigen. In specialized facilities, a definitive diagnosis can also be made from sputum, by the isolation of *C. psittaci* in tissue culture.

Tetracycline is treatment of choice and may be administered orally or intravenously. Chloramphenicol is an alternative agent. Before the advent of chemotherapy, the case fatality rate was about 20%; now it is less than 5%.

R. C. V.

PSORIASIS. See **Dermatitis and Dermatosis**.

PSYCHROMETER. An instrument used to measure humidity. It consists of two thermometers exposed side by side, one of which (the dry bulb) is an ordinary glass thermometer, while the other (the wet bulb) has its bulb covered with a jacket of clean muslin that is saturated with distilled water prior to an observation. The temperature measured by the wet-bulb thermometer is generally lower (due to evaporation of water from the wet bulb) than that measured by the dry bulb. The difference in the temperatures is a measure of the humidity of the air; the lower the ambient humidity, the greater the rate of evaporation and, consequently, the greater the depression of the wet-bulb temperature. The size of the wet-bulb depression is related to the ambient humidity by the psychrometric formula. See also **Hygrometry and Psychrometry**.

PSYCHROMETRIC CHART. The semiempirical relation giving the vapor pressure in terms of the barometer and psychrometer (hygrometer) readings is:

$$e = e' - \left[3.67 \times 10^{-4} \left(1 + \frac{t' - 32}{1571} \right) \right] p(t - t')$$

where t is the dry-bulb temperature, t' the wet-bulb temperature, e' the saturation vapor pressure at t', p the barometric pressure, and e the vapor pressure; all pressure units being the same. The temperatures are in °F.

A psychrometric chart is a nomogram constructed to provide convenient determination of the properties of air-water vapor mixtures, such as humidity, dew point, and water-vapor pressure from temperatures obtained with a psychrometer. A form of psychrometric chart plotted for °F is shown in Fig. 1. Dry-bulb temperature is represented by the vertical lines; wet-bulb temperature by the diagonal lines; dew-point temperature by the horizontal lines; and relative humidity by the curved lines.

To find the relative humidity value, determine the wet-and-dry-bulb temperatures and follow the lines to their junction on the chart. For example, assume that the wet-bulb temperature is 50 °F and the dry-bulb

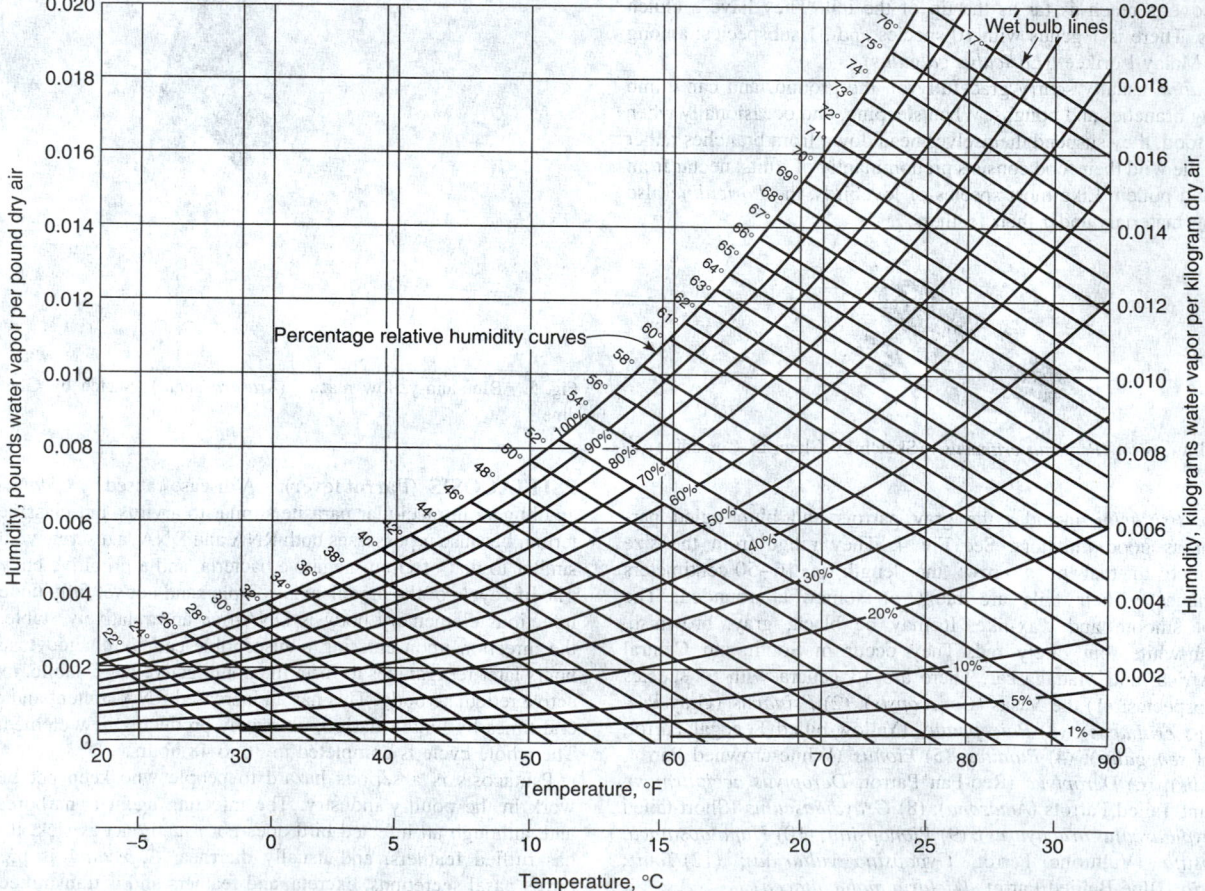

Fig. 1. Psychrometric chart. Wet-bulb temperature lines and percentage relative humidity curves. Temperature range 20 to 90 °F (−6.7 to 32.2 °C). Pressure reference = 29.92 inches of mercury (Sea Level); 1013.2 millibars; 101.325 kiloPascals.

temperature is 55 °F. Find the wet-bulb temperature at the left termination of the diagonal lines. Follow the diagonal line toward the lower right to its junction with the dry-bulb temperature vertical line. The curved line at this junction indicates the relative humidity. In this example, the curved line indicates 70% relative humidity. Values falling between lines are found by interpolation. Lines for temperature in degrees Celsius are included on chart.

To determine the dew-point temperature of the water-vapor mixture, follow the horizontal line from the junction of the wet-and-dry-bulb temperature lines to its junction with the 100% humidity curve. The dew point is read from the temperature °F scale and in this example is 45.2 °F. See also **Humidity**.

Psychrometric charts are useful for most observations. The charts are referenced to a specific pressure.

PTARMIGAN (*Aves, Galliformes; Lagopus*). Birds (*Aves*) related to the true grouse but with the feet and legs fully clothed with feathers. They are found in the far northern parts of Europe, Asia and North America and at high altitudes, above timber line, as far south as Colorado. They are largely mottled gray and brown in the summer but assume white plumage in winter. See Fig. 1. The change is not, however, always complete but is somewhat conditioned by the climate. The red grouse and the willow grouse or ripa of northern Europe are closely related to the ptarmigans.

Fig. 1.   Ptarmigan, showing seasonal change of plumage. (*A.M. Winchester.*)

The hoatzin is a peculiar South American bird of possible, but doubtful relationship to the ptarmagin. The bird lives along streams of the Amazon valley and eats fruit and other vegetation. The young have a clawed digit on the margin of the wings that they use to grasp boughs in climbing. See also **Galliformes**; and **Grouse**.

PTEROPSIDA. Members of the *Pteropsida* have large leaves, and definite gaps in the vascular cylinder, where a vascular strand, or leaf trace, passes from the stele to the leaf. It is also characteristic of the *Pteropsida* that the sporangia are located on the lower surface of the leaf. Most of the vascular plants of today, including all ferns, gymnosperms and angiosperms, belong to this group.

PTEROSAURS (Pterosauria). Pterosaurs, extinct flying reptiles, are known from the Late Triassic (220 million years ago (Ma)) to the end of the Cretaceous (65 Ma). Only two other vertebrate groups achieved true flapping flight: birds date back to the Late Jurassic (148 Ma), while the oldest bats are Early Tertiary in age (55 Ma). Pterosaurs were lightly built; consequently their fossil record is poor and biased toward forms that lived in coastal and shallow marine areas: little is known of pterosaurs from continental environments.

At least three different lineages of early, long-tailed rhamphorhynchoid pterosaurs, each represented by primitive, short-winged forms with large, deep skulls, are known from the Late Triassic, showing that the basal radiation of pterosaurs was well underway by this time (Fig. 1). In the Early Jurassic, pterosaurs spread worldwide, but during this interval most of the primitive groups seem to have died out and were replaced by more advanced rhamphorhynchoids such as the highly specialized skimming form *Rhamphorhynchus*, best known from the Late Jurassic Solnhofen

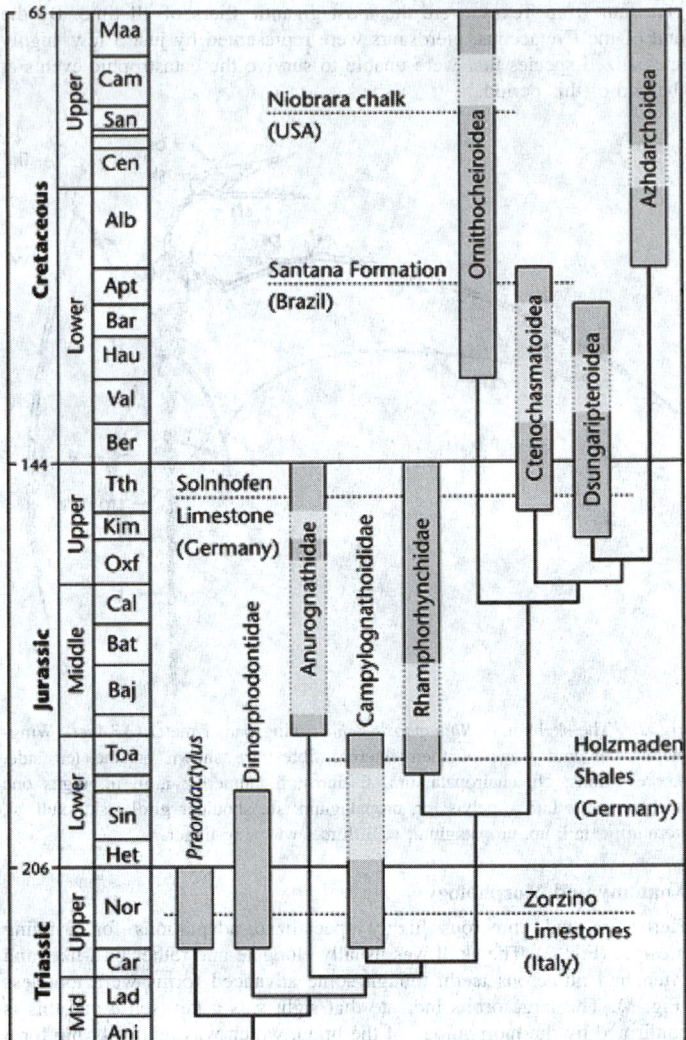

Fig. 1.   The evolutionary history of pterosaurs. Each column represents the known extent of a particular lineage with blank sections indicating gaps in the fossil record. Important fossil horizons are shown in italics and the figures on the left side indicate age in millions of years. Abbreviations: Alb, Albian; Ani, Anisian; Apt, Aptian; Bar, Barremian; Baj, Bajocian; Bat, Bathonian; Ber, Berriasian; Cal, Callovian; Cam, Campanian; Car, Carnian; Cen, Cenomanian; Hau, Hauterivian; Het, Hettangian; Kim, Kimmeridgian; Lad, Ladinian; Maa, Maastrichtian; Nor, Norian; Oxf, Oxfordian; San, Santonian; Sin, Sinemurian; Toa, Toarcian; Tth, Tithonian; Val, Valanginian.

Limestone of Germany. The only primitive group to survive throughout the Jurassic were the anurognathids, small insectivorous pterosaurs from Europe and Asia.

The earliest known pterodactyloids (advanced, short-tailed pterosaurs) are found in the Late Jurassic of Europe where they are represented by various species of ctenochasmatoids and dsungaripteroids, two of the four main pterodactyloid lineages (Fig. 1). Both these lineages persisted into the Early Cretaceous, became widely dispersed, occurring in Europe, South America and Asia, and achieved quite large size, with wingspans of up to 3.5 meters (11.5 feet). Ornithocheiroids such as *Anhanguera*, typified by their "tooth-grab" dentitions and spectacular cranial crests, seem to have been the commonest pterosaurs in the Early Cretaceous. In the Late Cretaceous ornithocheiroids were represented by large and giant toothless forms such as *Pteranodon* which ranged up to 9 meters (29.5 feet) in wingspan, but seem to have been restricted to the Americas. The earliest known azhdarchoids, *Tapejara* and *Tupuxuara*, toothless pterosaurs with extraordinary head crests, are found in the late Early Cretaceous of South America. Azhdarchoids spread worldwide in the Late Cretaceous and were by far the commonest pterosaurs in this interval. Most were of medium to large size, but some, such as *Quetzalcoatlus*, a 12 meter (39 foot) wingspan pterosaur from Texas, were the most gigantic fliers of all time. By the end of the Cretaceous, pterosaurs were represented by just a few, highly specialized species that were unable to survive the catastrophic events at the end of this period.

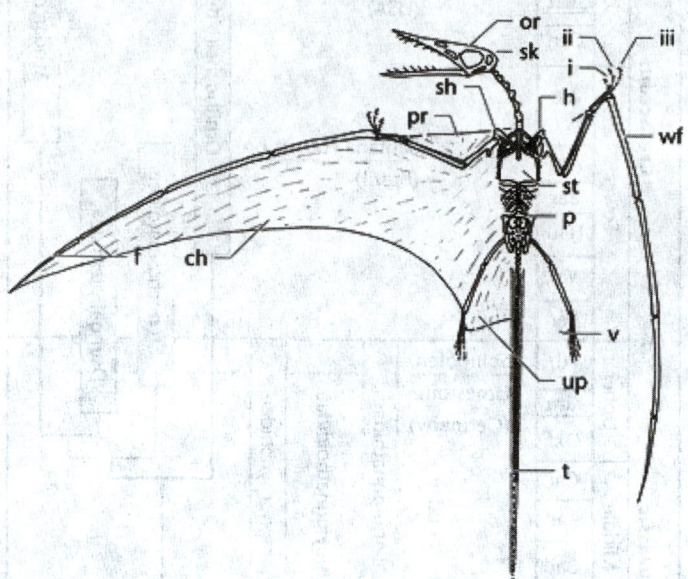

Fig. 2.   The skeleton of *Rhamphorhynchus* (wingspan, 1 meter (3.3 feet)). Wing-membranes and some of their internal fibres are shown on the left side. Abbreviations: ch, cheiropatagium; f, fibres; h, humerus; i, ii, iii, digits one to three; or, orbit; p, pelvis; pr, propatagium; sh, shoulder girdle; sk, skull; st, sternum; t, tail; up, uropatagium; v, fifth toe; wf, wing-finger.

Anatomy and Morphology

Pterosaurs had numerous highly specialized adaptations for a flying lifestyle (Fig. 2). The skull was usually elongate and rather bird-like, and often had numerous teeth, though some advanced forms were toothless (Fig. 3). The large orbits indicate that sight was a key sense and this is confirmed by the morphology of the brain, which was relatively big for a reptile, bird-like and with well developed optic lobes. In many advanced pterosaurs the skull also bore prominent crests, either on the jaws or on the top of the head (Fig. 3). The main role of these crests, some of which reached enormous size, was probably for display.

The neck was relatively long, but less supple than that of birds, while the body was short and compact, and in large pterosaurs some of the back vertebrae were fused together to give extra support for the shoulder girdle and wings. Rhamphorhynchoids had a long, stiffened tail bearing a vertical flap at the free end, but in pterodactyloids it was reduced to a short stubby spike. The "U"-shaped shoulder girdle articulated with a large breastbone (sternum), from which arose the principal flight muscles. Each of the main forelimb bones was elongated, most notably the fourth digit, or wing-finger, which often made up more than 60% of the wing-spar and from which the name "pterodactyl" is derived. The first three digits were of normal size and each bore a large, hooked "climbing" claw. The hindlimbs were relatively small and spindly and, in rhamphorhynchoids, the foot had five toes: the first four were clawed, while the fifth was long and rod-like. In pterodactyloids the fifth toe was greatly reduced or lost altogether.

The wings were composed of thin, elastic membranes supported internally by numerous, long, thin fibers. The flight surfaces consisted of a propatagium anterior to the forelimb, a cheiropatagium attached to the fore- and hindlimb, and a uropatagium stretched between the hindlimbs and supported by the fifth toe. Well-preserved fossils show that in life the back of the skull, the neck and the body were covered by fine short "hairs": these imparted a smooth contour to the body and may also have assisted with temperature control.

Life Styles and Diversity

It seems likely that all pterosaurs were competent fliers and, while small to medium sized forms probably relied on flapping flight, large and

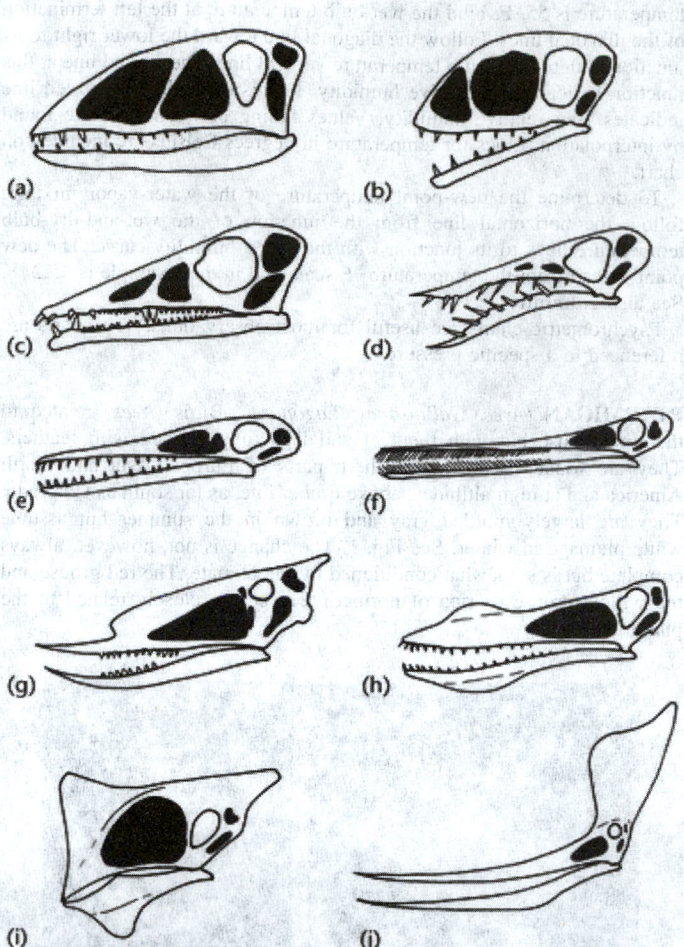

Fig. 3.   Skull diversity in pterosaurs (numbers in parentheses indicate skull length). (**a**) *Dimorphodon*, Early Jurassic (0.2 m (8 inches)); (**b**) *Anurognathus*, Late Jurassic (0.032 m(1.2 in); (**c**) *Eudimorphodon*, Late Triassic (0.09 m (3.5 inches); (**d**) *Rhamphorhynchus*, Late Jurassic (0.1 m (4 inches)); (**e**) *Pterodactylus*, Late Jurassic (0.085 m (3.4 inches); (**f**) *Ctenochasma*, Late Jurassic (0.01 m (0.4 in); (**g**) *Dsungaripterus*, Early Cretaceous (0.4 m (16 inches)); (**h**) *Anhanguera*, Early Cretaceous (0.6 m (23.5 inches)); (**i**) *Tapejara*, Early Cretaceous (0.22 m (8.6 inches)); (**j**) *Pteranodon*, Late Cretaceous (1.5 m (59 inches)).

giant pterosaurs probably soared for much of the time to save energy. Pterosaurs had relatively large, but flexible wings, permitting slow flight and considerable maneuverability, useful attributes for animals that usually hunted on the wing. The bat-like rhamphorhynchoids seem to have had a rather poor terrestrial ability and were largely restricted to tree- or cliff-climbing. Pterodactyloids were more proficient, and fossil tracks show that they were often to be found shuffling around on all fours, searching for prey on mud flats or in shallow water.

Numerically speaking, pterosaurs appear to have achieved only modest levels of diversity and even the most complete faunas, such as those from the Solnhofen Limestone of Germany and the Santana Formation of Brazil (Fig. 1), contain, at most, 12–15 species, far less than, for example, in modern or even fossil avifaunas. Morphological diversity also appears to have been limited: postcranial anatomy is remarkably uniform across large groups, though skulls and dentition do exhibit some variability (Fig. 3). The latter indicate that most pterosaurs fed on fish, caught during flight using a "fish-grab" composed of large, fang-like teeth located in the jaw tips. The ctenochasmatids specialized in filter feeding, using a comb-like dentition to sieve while wading in shallow waters, while other forms such as the anurognathids were insectivorous, using a wide gape and short spiky teeth to catch flying insects.

Additional Reading

Bennett, S.C.: "The Ontogeny of *Pteranodon* and other Pterosaurs," *Paleobiology*, **19**, 92–106 (1993).

Bennett, S.C.: "The Phylogenetic Position of the Pterosauria within the Archosauromorpha," *Zoological Journal of the Linnean Society*, **118**, 261–308 (1996).

Benton, M.J.: *Vertebrate Palaeontology*, 3rd Edition, Blackwell Publishers, Malden, MA, 2004.

Buffetaut, E., and J-M Mazin: *Evolution and Paleobiology of Pterosaurs*, Geological Society Publishing House, Bath, UK, 2003.

Carroll, R.L.: *Vertebrate Paleontology and Evolution*, New York: W. H. Freeman and Company, New York, NY, 1987.

Clarkson, E.N.K.: *Invertebrate Palaeontology and Evolution*, Blackwell Publishers, Malden, MA, 1998.

Langston, W.: "Pterosaurs," *Scientific American*, **244** (2), 122–136 (1981).

Padian, K., and J.M.V. Rayner: "The Wings of Pterosaurs," *American Journal of Science*, **293A**, 91–166 (1992).

Unwin, D.M.: "Pterosaur Tracks and the Terrestrial Ability of Pterosaurs," *Lethaia*, **29**, 373–386 (1997).

Unwin, D.M., and N.N. Bakhurina: "Sordes Pilosus and the Nature of the Pterosaur Flight Apparatus," *Nature*, **371**, 62–64 (1994).

Wellnhofer, P.: *The Illustrated Encyclopedia of Pterosaurs*, Salamander Books, London, UK, 1991.

DAVID M. UNWIN, Museum für Naturkunde, Berlin, Germany

PTERYGOTA. One of two subclasses into which the class *Insecta* is divided. The members of this subclass are either winged or closely related to winged forms. Some show evidence of derivation from groups that are typically winged. The great majority of insect orders belong here, only three falling into the suborder *Apterygota*.

PTOMAINE. A class of amines formed by the action of bacteria on proteins or by the metabolism of amino acids, which are broken down into toxic products.

PUBLIC-KEY CRYPTOGRAPHY. See **Encryption**.

PUELCHE. See **Winds and Air Movement**.

PUFFERS (*Osteichthyes*). Of the order *Pectognathi*, family *Tetraodontidae*, a puffer, if pulled out of water, instantly reacts by swallowing air and inflates itself much as a balloon. If returned to the water, a once-expanded puffer may require five minutes or so to expel the air and return to normal. The puffer also can swallow water in the same fashion. The prime difference between a puffer and a porcupine fish is the presence of a solid beak and fused teeth in the latter. There are about 90 species of carnivorous puffers that are widely distributed throughout temperate and tropical waters. Although they can attain a size of 36 inches (0.9 meter) the usual fish does not exceed 18 inches (46 centimeters). These fishes are consumed as a food item (fugu) in Japan. However, because puffers contain a dangerous poison (tetrodotoxin), a Japanese cook requires a special license, assuring proper training in the preparation of fugu. Mortality from food poisoning of this source can run as high as 60%. See Fig. 1.

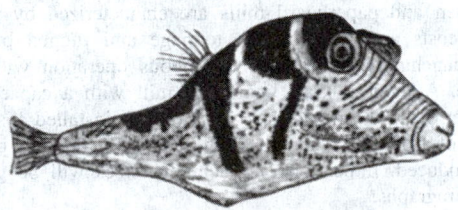

Fig. 1.   Puffer, *Canthigaster valentini*.

Puffers are highly territorial, and need recesses into which they can retreat. This behavior is not immediately evident in species that engage in a great deal of swimming. If two puffers meet, they usually become aggressive. If larger fishes appear, or if several males court a single female, the puffers pump themselves full of water in order to appear larger. This behavior is particularly impressive in the puffer *Carinotetradon*

somphongsi; if two of these puffers meet, a thin comb rises in the middle of their backs, and a ridge also rises on the belly. The fishes' coloration and markings almost entirely disappear. The opponents usually stand upside down and point their belly sides at each other, in order to appear as impressive as possible.

PULLEY.   See **Machine (Simple)**.

PULP (WOOD) PRODUCTION AND PROCESSING. Pulps can be defined as fibrous products derived from cellulosic fiber-containing materials and used in the production of hardboard, fiberboard, paperboard, paper, and molded-pulp products. With suitable chemical modification, pulps can be used in the manufacture of rayon, cellulose acetate, and other familiar products. Pulps can be produced from any material containing cellulosic fiber; but in North America and several other regions of the world, wood is the predominant source of pulp. This description is confined to the production and processing of wood pulp.

Wood is a cellular substance chemically composed of roughly 70% holo cellulose, 25% lignin, and 5% water and ethyl alcohol-benzene soluble extractives. These percentages are based on oven-dry wood.

The chemical composition and physical character of wood vary from species to species, within species grown in different geographical locations, and within a given tree, depending upon the location of the fiber cell in the tree. Both lignin (noncarbohydrate) and holocellulose (carbohydrate) are polymeric substances. Holocellulose is composed of approximately 70% alpha cellulose and 30% hemicellulose, the long-chained alpha cellulose being characterized by nonsolubility in alkali; whereas the shorter-chained hemicellulose is alkali-soluble, the degree depending upon the alkali concentration. Lignin concentration in wood substance is greatest in the middle lamella (the zone around each individual fiber cell), decreasing in concentration through the cross section of the fiber, and reaching a concentration of about 12% at the inner layer of the fiber adjacent to the fiber cavity, or lumen. It is the middle-lamella material (lignin and hemicellulose) that cements the fiber cells together, thus giving rigidity to the fibrous wood structure.

The objective of wood pulping is to separate the cellulose fibers one from another in a manner that preserves the inherent fiber strength while removing as much of the lignin, extractives, the hemicellulose materials as required by pulp end-use considerations. Wood pulp to be used for the manufacture of hardboard, for example, requires only the removal of water-soluble wood sugars and sufficient fiberization, i.e., separation of fibers, to permit effective felting of the fibers in a sheet-forming operation. In a subsequent operation in which the felted fiber sheet is subjected to high pressure and heat, the lignin in the fiber mass softens and flows, ultimately acting as a bonding agent cementing the fibers together into a coherent hardboard. At the other extreme, wood pulp to be used for rayon manufacture must be of a high alpha-cellulose content (~88–93%), have extremely low amounts of noncarbohydrate material, and be well fiberized to permit uniform reactions during chemical processing.

Pulping Processes

Wood is converted to pulp by mechanical and chemical actions, which constitute the pulping process. Their selection depends upon the type of wood supply available and the pulp qualities desired. Pulps can be characterized on the basis of the unbleached pulp yields achieved by the pulping process used, i.e., the yield of oven-dry (OD) pulp obtained from oven-dry debarked wood.

Five major types, or classes, of pulps, related to pulp yield ranges normally considered to define each class of pulp, are shown in Fig. 1. Pulp yield is a direct indication of degree of chemical action (delignification and chemical attack on carbohydrate and other nonligneous material). Also shown in this figure are the degrees of defibration effected by chemical and mechanical action utilized to produce the pulp, although this representation is not strictly correct. For example, in producing a full chemical pulp, wood chips are subjected to chemical action (digestion or cooking) in a pressure vessel. When digestion is completed, the cooked and softened chips retain the same physical form as the raw chips originally charged to the digester. But they separate into essentially discrete fibers as a result of mechanical action occurring upon sudden release of the chips from the pressure vessel into a receiving tank, which ordinarily is at atmospheric pressure.

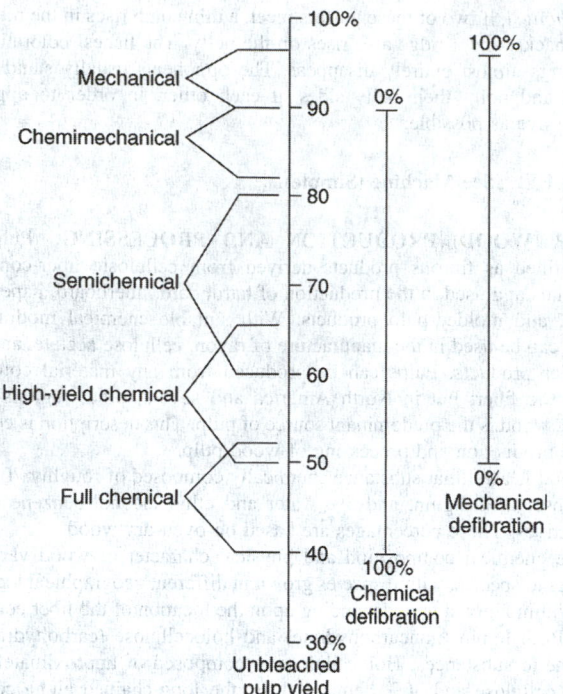

Fig. 1.   Wood pulp characterized on basis of yield.

At the other extreme, no chemicals are used in the production of mechanical pulp, and defibration is effected by subjecting wood to a mechanical grinding or attrition action. In this instance, the defibration is aided by some small degree of chemical change and solubilization of wood substance occasioned by heat generated by the grinding operation.

The pulps listed in Fig. 1 are characterized on an unbleached basis as produced by processes conventionally called *pulping processes*. In many instances, these pulps must be further treated chemically to remove residual lignin, hemicellulose, and color bodies before they can be considered suitable for use in specific applications. This further treatment is called *bleaching*, and the bleaching operation is actually an extension of the pulping process.

Customarily, pulping processes and bleaching processes are considered separately, although the choice of bleaching process is highly dependent upon the pulping process used. With this distinction between pulping and bleaching in mind, it will be understood that the pulping processes that are briefly described here pertain only to the production of *unbleached* pulps.

The soda, kraft, and sulfite pulping processes are used to prepare full chemical pulps. The soda process, which uses sodium hydroxide as the cooking chemical for delignification purposes, has largely been superseded by the kraft process, which is characterized by its use of sodium hydroxide and sodium sulfide as active delignification agents in the chip-cooking phase of the process.

Chip-digestion parameters are digester pressure and temperature, digestion time to and at maximum temperature, amount of active alkali used per unit weight of OD wood (percent active alkali), percentage ratio of sulfide to active alkali (percent sulfidity), and weight ratio of cooking liquor (including chip moisture) to OD wood weight. No two kraft pulp mills use the same set of parameter values. Such values must be frequently adjusted, even within a given mill, because of variations in incoming wood and pulp-quality requirements.

Kraft processes are applicable to nearly all species of wood, and effective means of recovering spent cooking chemicals for recycle in the process have been developed. Some sodium and sulfur losses do occur and are replenished in the cooking-liquor system by adding sodium sulfate at the recovery boiler, where it is converted to sodium carbonate and sulfide. In order to maintain a proper sulfur-to-sodium ratio in the recovered chemicals, other chemicals, such as sodium carbonate, sodium sulfite, and sulfur, are sometimes used for chemical makeup.

In contrast to the highly alkaline (pH 11–13) kraft processes, sulfite pulping processes are acidic in nature and are of two general types:

(1) the *acid sulfite processes* utilize calcium, sodium, magnesium, or ammonium bisulfite in combination with free or excess sulfur dioxide as cooking chemicals (pH 1.7–2.3). (2) The *bisulfite processes* use sodium, magnesium, or ammonium bisulfite (pH 3.5–5.5) for chip digestion.

Several sulfite processes are multistage and use various combinations of acid sulfite and bisulfite cooking stages and can even use the alkaline kraft cook as one of the multistages. Although spent calcium acid sulfite cooking liquor can be incinerated, there is no recovery of calcium or sulfur. The sodium and magnesium bases can be recovered with or without sulfur recovery, and spent ammonium base liquor can be burned with recovery of sulfur as an option.

High-yield chemical pulps can be produced by the soda, kraft, or sulfite processes, in which chemical use and digestion time and/or temperature are suitably reduced to effect a milder cook than used for full chemical pulps. Mechanical defibrators are used to complete the separation of wood fibers not accomplished by the chemical action.

Semichemical pulps are usually prepared by the neutral sulfite semi-chemical (NSSC) process, although modifications of the full chemical processes can be used. Active pulping chemicals are (in the sodium-base NSSC process) sodium sulfite buffered with sodium bicarbonate (pH 7.0–9.0) and (in the ammonium-base NSSC process) ammonium sulfite with ammonium hydroxide used as a buffer. Defiberization is usually accomplished by attrition mills of the disk type.

Mechanical pulps are produced by two basic processes: (1) *stone groundwood pulp* (SGW) is produced by the defibration action of natural or artificial grindstones rotated at moderate speeds (200–300 rpm) against bark-free bolts of roundwood axially aligned across the peripheral face of the stone in the presence of water. By air-pressurization of the grinder, a *pressurized groundwood pulp* (PGW) of improved quality can be produced. (2) *Refiner mechanical pulp* (RMP) is produced by the attrition action upon raw wood chips of an open (atmospheric) discharge disk refiner. By pre-heating the chips in a pressurized vessel via direct steaming at temperatures of 120°C or higher and fiberizing the heated chips in either a pressurized or atmospheric disk refiner, *thermomechanical pulp* (TMP) is produced.

Chemimechanical pulps (CMP) are produced by processes in which roundwood or chips are treated with weak solutions of pulping chemicals, such as sulfur dioxide, sodium sulfite, sodium bisulfite or sodium hydro-sulfite, followed by mechanical defibration. By presteaming chemically treated chips before attrition, *chemithermo-mechanical pulps* (CTMP) are produced. The mild chemical action, augmented by heat, softens wood lignin and promotes easier defibering with less fiber damage than achieved by the purely mechanical processes.

Wood Pulping Operations. The preceding description of pulps and pulping processes were given as a background to the following descriptions of the various operations involved in the preparation of wood pulp. The pulping system of a typical kraft linerboard mill, as indicated in the simplified flow diagram in Fig. 2, is illustrative of that required for the preparation of both full and high-yield chemical pulps. Linerboard normally is two-layered. The base, or primary sheet, is formed from a high-yield chemical pulp (50–54% yield) and the top, or secondary sheet, is formed from a full chemical pulp, either unbleached (48–50% yield) or bleached (46–48% unbleached yield), laid upon the wet primary sheet on the sheet-forming wire.

Pulp, paper, and paperboard mills are characterized by high capital investment costs and use of high tonnage and rugged but precisely engineered machinery capable of continuous operation with minimum maintenance. A modern kraft linerboard mill with a capacity of 1000 short tons per day (900 metric tons) will have an installed cost, excluding woodlands, of from $275,000 to $325,000 (1986 dollars) per daily on of board produced. Indication of machinery sizes will be given in the following paragraphs.

Wood-Chip Preparation. As indicated in Fig. 2, pulping operations begin with receipt of wood at the mill site. Pulpwood is supplied in log form (roundwood) or chips in accordance with specifications set by the pulp mill. Roundwood is usually received with bark on and in lengths and diameters suitable for proper handling in the wood-preparation equipment at the mill. It has been customary for mills to specify multiple lengths of pulpwood, i.e., 4 feet (1.2 meters) and 8 feet (2.4 meters) as standard receipts, but there is a trend to the procurement of tree-length logs, up to 70 feet (21 meters) in length, either exclusively or in combination with

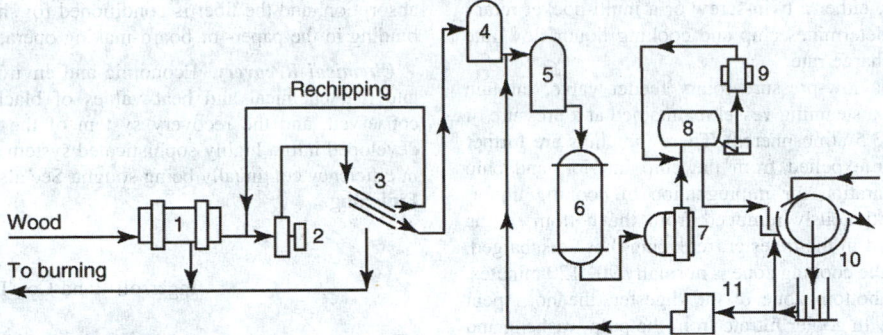

Fig. 2.   Flow diagram of kraft pulp mill: (1) debarking, (2) chipping, (3) screening, (4) steaming, (5) impregnating, (6) digesting, (7) fibrilizing, (8) screening, (9) fiberizing, (10) washing, (11) chemical recovery.

short logs. Another trend has been to the use of chips already prepared, except perhaps for final screening, by independent suppliers or by satellite wood yards operated by the pulp mill itself.

Although linerboard mills formerly used only softwoods (coniferous) for pulping, continued improvements in pulping and board-making technology have permitted the inclusion of up to 20% or more of hardwoods (deciduous) in the wood furnished to the mill, with improved utilization of woodlands as a beneficial result. Softwood and hardwood species are processed through the chipping operation and stored separately; they are either blended into the digester and cooked together, or they are processed separately and the respective pulps blended just ahead of the linerboard machine.

Former practice was to store pulpwood receipts in either a debarked or unbarked condition in stacks or random piles in the wood yard and to reclaim the yard wood for processing into chips just a few hours in advance of chip needs at the digester. A common practice today is to convert the wood into chips immediately after pulpwood receipt and to place the chips, usually by belt or air conveyance, in chip piles built up on concrete or asphalt pads. Separate piles are provided for softwood and hardwood chips, and storage capacities of 40,000 cords or greater can be maintained.

Pulp logs are conveyed to the debarking area, where they are cut to proper length, if necessary, and sorted. Accepted logs are mechanically fed into one end of a large horizontal, cylindrical drum, usually consisting of one or more sections constructed of spaced steel plates, channels, or bars mounted in carrying rings and supported on trunnions and driven by ring gears or suspended from an overhead structure by heavy chains, one or more of which are motor driven. This *barking drum rotates* at a speed of 5–8 rpm, and as the logs tumble about in passing from the intake to the discharge end of the drum, bark removal is effected by the logs rubbing against each other or against the bars or plates constituting the drum shell. Provision can be made for introduction of steam into the feed-end for log de-icing when needed.

Bark removal also can be accomplished by use of a *ring barker* or *hydraulic barkers* which employ high-pressure water jets for bark stripping. Bark removed is collected, shredded in a hog or hammer mill, and used as fuel in steam boilers, where it contributes about 3863 kJ/kg (9000 Btu/pound) of dry solids.

Debarked wood is conveyed to a chipper for conversion into chips of proper length for chemical treatment in a subsequent cooking operation. Chip length of 0.5–1.0 inch (12.7 to 25.4 millimeters) is conventional.

Chip Digestion. This cooking operation is accomplished in either a batch or continuous digester. A chip digester is essentially a large pressure vessel provided with suitable raw-chip and cooking liquor feed ports and a cooked-pulp discharge port. It is equipped with means for heating and maintaining its contents to and at a specified temperature for the required periods of time.

Batch digesters are vertical, stationary, cylindrical pressure vessels into which chips and cooking liquor are charged under atmospheric conditions. Heating of the digester, after sealing of the feed ports, is effected by direct steam addition or by continual withdrawal of liquor through screened ports and reintroduction of the liquor, after passage through external heat exchangers, onto the top (and sometimes into the bottom) of the chip mass

within the vessel. Often, a combination of the direct and indirect heating methods is used. Modern batch digesters are typically 4000–6000 cubic feet (113–170 cubic meters) in volume, with height-to-diameter ratios of 3.5–5.5, and pre-cook pulp capacities of 10–12 tons (9–18 metric tons).

Continuous digesters have been developed as part of the highly successful effort to convert pulp and papermaking from a series of strictly batch operations into an integrated series of continuous operations. A number of successful types of continuous digesters range from horizontal and inclined tube (single or multiple) designs, in which the chip charge is moved through the digester by mechanical screw or bucket conveyors, to vertical digesters, in which chip movement is effected by gravity. See Fig. 3.

Screened chips are conveyed from storage to a chip supply bin in the digester house. The chip bin is designed so that low pressure steam recovered from the hot, spent cooking liquor can contact the chips, preheat them and expel most of the air from the chip interior. If hardwood and softwood chips are to be cooked together, they are blended by weight proportion during the transfer to the chip bin. The chips drop by gravity

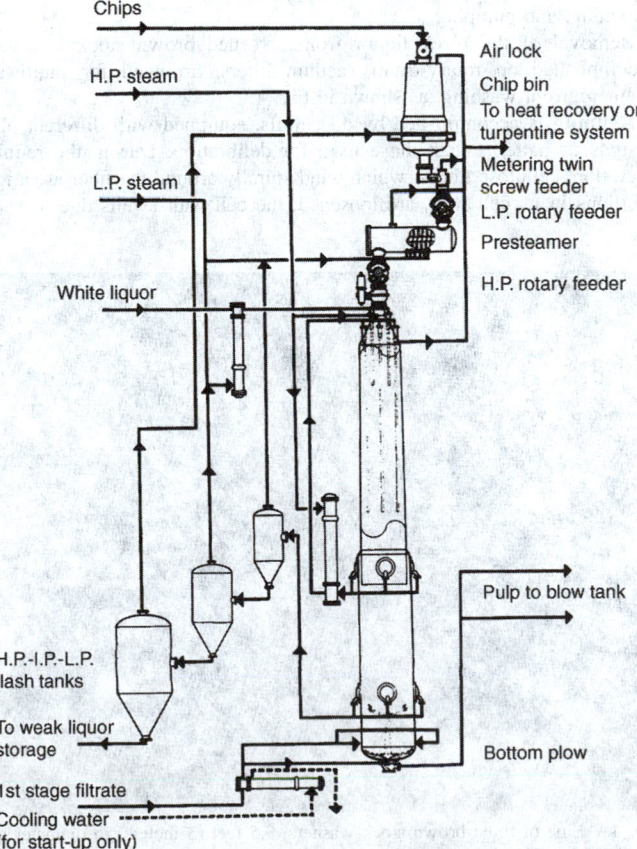

Fig. 3.   Continuous digester system (*Ingersoll Rand Co.*)

from the bin to a chip meter, either a twin-screw or a multi-pocket rotary feeder, the speed of which determines chip and cooking liquor flow rate to the digester and pulp discharge rate.

Metered chips drop into a low-pressure rotary feeder valve, through which they are introduced to a steaming vessel maintained at a pressure of about 15–55 psi gage (1 to 3.5 atmospheres). There the chips are further preheated, the remaining air expelled from the chip interior, and chip moisture is leveled in preparation for impregnation of cooking liquor. Since cooked chips are continuously removed from the bottom of the digester, chips pass downward in the digester, replacing those discharged. The time of passage through the cooking zone is normally 90–120 minutes. As cooked chips reach the bottom zone of the digester, the hot, spent cooking liquor is displaced with cooler filtrate from the pulp washers, and removed via extraction strainers to a heat recovery system in which steam is generated to be used to precondition the chips being fed to the digester. As the partially cooled chips move further down the digester, they are plowed to a central well in the bottom of the digester, and are mixed with more filtrate from the pulp washers for dilution and final cooling. Mechanical forces exerted in the transfer of chips from the digester to the blow tank effect fiberization of the chips, the degree of which depends upon cooking conditions. The fibrous material in the blow tank is called pulp, and separate blow tanks are normally used to collect the several types of pulp produced alternately in the digester.

Pulp Screening and Washing. Pulp (brown stock) discharged to the blow tank is in admixture with black liquor, a water solution of spent and residual cooking chemicals and dissolved wood substance, and is at a consistency of from 10 to 18%. The term *consistency* has a meaning peculiar to the pulp and paper industry and refers to the percentage ratio of washed, dry (either oven- or air-dried) fiber to total fiber slurry weight. The fiber bundles left in the pulp after blowing must be fiberized, i.e., separated into discrete fibers, and the black liquor removed in order for the pulp to be refined (a conditioning of individual fibers) and formed into a fiber sheet on the linerboard machines.

Pulp is diluted with filtrate from the pulp washer to a consistency of about 4.5% in the lower portion of the blow tank and fed to fibrilizers, which serve the purposes of metal trapping, fiber-bundle breaking, rough screening, and pumping.

Removal of the black liquor from screened brown stock is usually accomplished on rotary-drum vacuum filters, arranged for multistage countercurrent washing, as shown in Fig. 4.

Refining is accomplished by disk mills, equipped with different plate designs or patterns than those used for defibration. During the refining operation, cellulose fibrils, which wind spirally around the fiber at various positions in its cell wall, are loosened, the cell wall swells due to water

Fig. 4.   Line of three brown stock washers, 9.5 feet (3 meters) in diameter and 16 feet (3 meters ×5 meters) long, equipped with multiport circumferential valve. First stage washer is shown in foreground. (*Ingersoll-Rand Co.*)

absorption, and the fiber is conditioned for sheet formation and inter-fiber binding in the paper- or board-making operation.

Chemical Recovery. Economic and environmental control factors dictate that chemical and heat values of black liquor solids be carefully conserved, and the recovery system of the modern kraft pulp mill has developed into a highly sophisticated system with still more improvement in efficiency continually being sought. See also **Paper**; **Papermaking and Finishing**.

HENRY F. SZEPAN,
DUNBAR G. TERRY,
Ingersoll-Rand Co., Impco Division, Nashua, NH

PULSAR. A neutron star with a strong magnetic field, first detected in 1967 as sources of intense nonthermal pulsed radio emission (hence sometimes called *radio pulsars*).

At the end of the life of a massive ($>8M_\odot$) star, following the production of iron in the core by nucleosynthetic processes, the collapse of the core produces a supernova ejection of the envelope and leaves a very compact, degenerate object as a remnant. This remnant, having a radius of about 10 kilometers and a mass of order $1M_\odot$, has a central density about equal to an atomic nucleus ($\approx 10^{15}-10^{16}$ g cm^{-3}) and consists of nucleons and more exotic states of matter. This is a neutron star. Also present may be a strong magnetic field, either as a remnant from the precollapse core or generated immediately following the supernova, which can reach strengths of order 10^{13} gauss. In such a strong field, electrons can radiate via synchrotron processes at radio, optical, and even x-ray wavelengths. The emission is confined to the magnetic polar cap, which if it is inclined to the rotational axis will produce a pulse of emission as the region sweeps through the line of sight to a distant observer. This is the mechanism presumed responsible for the radio pulsars. The mechanism for accelerating the electrons to the high energies observed in these regions is not completely understood at present. It appears to be at least in part due to strong electric fields generated by the rotation of the magnetized star ripping electrons from the surface and accelerating them to high energies nearby. The emission of radiation and the torquing of the star by the magnetic field produce a slowing down of the rotation rate of the star, the rate of which can be linked to the magnetic dipole moment of the neutron star.

When such stars occur in binary systems in which matter is accreting onto the neutron star, the magnetic field acts to funnel the gas toward the polar region. The matter emits x-rays as it falls onto the surface of the accreting star, which will be pulsed due to the rotation of the neutron star. Many such systems are known, the best studied being Hercules X-1. The interaction between the magnetosphere of the accreting pulsar and the accretion disk, which forms in the plane of the binary system, can also cause *quasi-periodic oscillations* (QPO's), due to instabilities generated by the shearing between the trapped and circulating gas. This shows up as noisy pulsing on timescales of milliseconds in the intensity of x-ray binaries at high energy.

A new class of objects, the millisecond pulsars, contains the most rapidly rotating objects known in the cosmos. These rotate about 10 to 20 times faster than the shortest period radio pulsar, the Crab pulsar (period of about 33 msec) and appear to have unusually weak magnetic fields (less than 1010 gauss). Their rapid rotation has been explained as resulting from spin-up of a weakly magnetized neutron star in a loose binary system, which may have since disrupted. Their weak magnetic fields are inferred from their slow rate of spin-down.

Some pulsars are members of binary systems in which the separation of the stars is great enough that no mass transfer is occurring between the members. The prototype of this class is PSR 1913 + 16. The changes in the orbital parameters of these systems, because of their masses and exceptionally well-determined orbital parameters, allow for tests of general relativity, since the periods and eccentricities change from the emission of gravitational radiation. They also show that the formation of compact objects through supernovae does not necessarily disrupt the binary.

There are now about 400 known pulsars, found largely through radio surveys at Aricebo, Parkes in Australia, and the National Radio Astronomy Observatory at Green Bank. Of these, fewer than a half-dozen are associated with known supernova remnants, a problem that remains a major question in the origin and development of such systems. The two best studied radio pulsars, the Crab and Vela pulsars, are imbedded in radio remnants of known age, but the vast majority are isolated systems. See also **Cosmology**.

S. N. S.

PULSE.   1.   Commonly, a pulse is a variation of a quantity whose value is normally constant; this variation is characterized by a rise and decay, and has a finite duration.

2.   In physics and related sciences, a pulse is a waveform whose duration is short compared to the time scale of interest, and whose initial and final values are the same. The word "pulse" normally refers to a variation in time; when the variation is in some other dimension, it should be so specified, such as "space pulse." This definition is broad so that it covers almost any transient phenomenon. The only features common to all pulses are rise, finite duration, and decay. It is necessary that the rise, duration, and decay be of a quantity that is constant (not necessarily zero) for some time before the pulse and has the same constant value for some time afterwards. The quantity has a normally constant value and is perturbed during the pulse. No relative time scale can be assigned.

3.   In animal physiology, the pulse is the expansion and elongation of the arterial walls, produced passively by changes in intra-arterial pressure during contraction (systole) and relaxation (diastole) of the heart. The pulse is usually felt in the radial artery at the wrist, but it may be felt in any artery lying near the surface. The heart, with systole, forces blood into the arterial circulation. This blood is accommodated partly by moving the entire arterial column on at greater velocity, and partly by distending the arterial walls. The increase in pressure and distention of the vessel walls is transmitted from one segment of the artery to the next as the pulse wave. The ability of the vessel wall to distend is dependent upon its elasticity. In old age, when the vessel becomes sclerotic and inelastic, it offers increased resistance and this results in elevation of the blood pressure.

The examination of the pulse in disease is one of the oldest customs in medicine. Variations in the rate, rhythm and force are significant, particularly in heart disease. The normal rate is 60–80 in adults, and 80–140 in children. With fever the rate increases as a general rule. A disproportionately slow pulse with a high fever is of diagnostic significance in certain infections, notably typhoid, typhus and so-called virus pneumonia. An increase in the pulse rate is a normal reaction to emotional stimuli, being dependent in this case upon a release of epinephrine (adrenalin) by the adrenal glands.

PULSE (Botany).   A general name for leguminous plants (peas, beans, etc.) or their seeds.

PULSE CODING.   The use of modulation within a radar or sodar pulse to increase the range resolution beyond that normally achievable with a pulse of the same length. In a common technique, the phase of the transmitter carrier is inverted several times in an optimized sequence during the transmitted pulse. This modulation is undone by a complementary process in the receiver to produce an effective pulse length narrower by a factor equal to the number of inversions in the transmitted pulse sequence. See also **Pulse Compression**.

AMS

PULSE COMPRESSION.   The use of special forms of frequency, phase, or amplitude modulation to permit a radar system to achieve higher range resolution than that normally permitted by a given pulse duration. A suitably modulated transmitter pulse of duration τ (and hence range resolution $c\tau/2$, where c is the velocity of light) may be processed after reception to obtain a higher range resolution $c\tau/2n$, where n is the pulse-compression ratio.

Compression ratios of 10–100 are commonly achieved through the use of linear FM, nonlinear FM, or phase-coded modulation, implemented by analog or digital means. The advantage of pulse compression over simply transmitting shorter pulses is that high range resolution is achieved while maintaining the benefits of high pulse energy.

AMS

PULSE GENERATOR.   A large portion of electronic engineering effort is devoted to the development of circuits and systems that operate in the time domain. Switching, pulse, and digital circuits predominate computing, navigation, and data communication systems. In the laboratory, the oscilloscope as the detector and the pulse generator as the signal source largely have replaced the voltmeter and sinusoidal oscillator. Many electronic devices or circuits utilize sharp pulses of current or voltage as a basic part of their operation. Such pulses must frequently be of very short time duration (microseconds or nanoseconds) and accurately spaced in time. Others need not be repeated at regular intervals, but are initiated by some signal and are single narrow pulses occurring each time a signal is received.

In essence, a pulse generator is a highly versatile and controllable switch. Two parameters of interest which must be controlled are: (1) the pulse repetition frequency or switching rate, and (2) the pulse duration, or length of time the switch is closed (or open). The rise time, or speed of switching, also is an important parameter. In addition to the characteristics of the switch, specific applications require particular characteristics for the energy source switched. The output impedance, open-circuit voltage, and available current all must be known and specified to fit a given pulse generator to a specific requirement.

In electronic equipment, pulse repetition rates range from nearly dc to over 100 MHz — with durations to less than 1 nanosecond. For computation and data transmission systems, most applications can be served by relatively low power outputs. Radar and certain magnetic data storage systems may require pulses of very high energy. Because of these wide variations, a universal pulse generator is not practical.

For purposes of illustration, a few specific instruments are described briefly. A low cost unit for the student laboratory is shown in Fig. 1. This instrument produces pulses ranging from 0.1 microseconds to 1 second over a repetition rate ranging from dc to 2 MHz. The device contains an internal prf (pulse repetition frequency) oscillator with continuous range from 3 kHz to 1.2 MHz. Both pulse polarities are available simultaneously and the output circuits are dc coupled, the latter a necessity when pulses of long duration must be produced. A block diagram of the instrument is shown in Fig. 2. An oscillogram of a 1 microsecond pulse into 50 ohms with delayed sync pulse is shown in Fig. 3. This device may be used with a pulse amplifier to translate performance to 1-ampere output levels.

Fig. 1.   Unit pulse generator with power supply.

A modular pulse generator is shown in Fig. 4. This instrument is a hybrid system that offers much flexibility. The duration ranges, rise times, and prf ranges are similar to those of the instrument previously described. An input circuit module, one of several in one package, either serves as a prf oscillator or processes an external driving signal from dc to 2 MHz to produce a standardized system-synchronizing pulse. A second module produces pulses or delayed 0.1 microsecond synchronizing pulses in the range from 0.1 to 1 second with rise times of 15 nanoseconds. A single input module and three pulse/delay modules form a double-pulse generator, while five pulse/delay modules provide a triple pulse. A third module, timed from a pulse/delay module, produces pulses with linear and independently variable rise and fall times over a range from 0.1 microsecond to 1 second. A fourth module produces up to 16-bit pulse words. A power amplifier with up to 0.4 ampere output of a limited duty ratio is the fifth module. Several pulse configurations of which this device is capable of generating are shown in Fig. 5.

Fig. 5(a). Waveform that appears at the "adder No. 1" terminal with one prf unit driving two pulse/delay units at 10 kHz. Amplitudes and durations of positive and negative pulses can be adjusted independently.

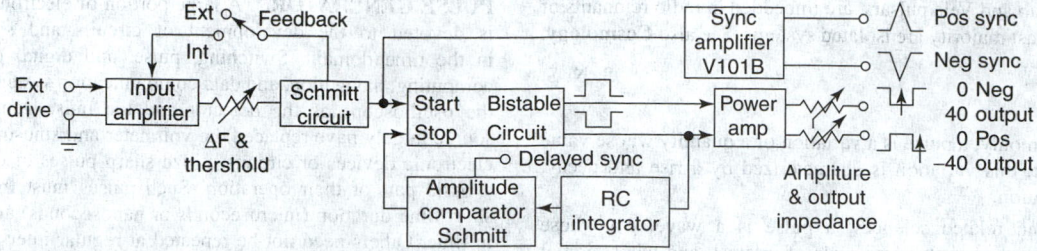

Fig. 2.   Block diagram of circuit used in unit pulse generator.

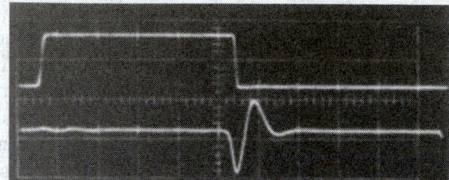

Fig. 3.   Oscillogram made from unit pulse generator.

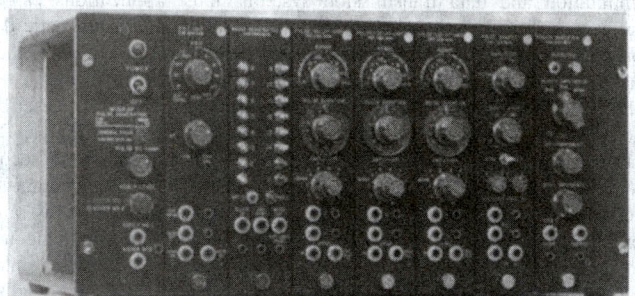

Fig. 4.   Modular pulse generator.

Fig. 5(b). One prf unit, one pulse/delay unit, and one pulse shaper are used to form this train of triangular waveforms. The prf unit is set for 5 kHz. The positive-going ramp rises linearly to 20 volts in 50 microseconds, while the negative-going ramp falls to zero in 10 microseconds. Rise and fall times are independently variable.

Fig. 5(c). Pulse train produced by one prf unit and three pulse/delay units operating at 100 kHz. The amplitude and duration of the positive pulse are controlled by one pulse/delay unit whose delayed output triggers a second pulse/delay unit. This second unit provides the delay between the positive and negative pulses, and its delayed output triggers the third unit to produce the negative pulse.

Fig. 5(d). This train of ramp pulses is produced with one prf unit, one pulse/delay unit, and one pulse shaper. Prf is 100 and the zero-volt level is adjusted by the main chassis "pulse dc component" control.

Fig. 5(e). Pattern produced when a word generator is connected between prf unit and the first pulse/delay unit in the situation shown in Fig. 5(a), with switches set as shown.

In another unit, the tone-burst generator, the instrument operates as a coherent gate for an externally introduced signal. This instrument is used for testing and calibrating sonar transducers and amplifiers, the measurement of room acoustics, and in automatic gain-control circuits. The device also is useful in the synthesis of time ticks on standard-time radio transmissions and in psychoacoustic instrumentation. A binary scaler is used to establish both the number of cycles in a burst and the time duration between bursts.

The device incorporates (1) a switch that holds the gate open for preliminary alignment of external equipment; (2) trigger controls, which allow control of the relative phase of the gate and input signal; (3) the ability to use separate input signals for the gate timing and gated signals; and (4) a timed mode, for very long periods between bursts. A typical waveform is shown in Fig. 6. The tone-burst generator also is useful with pulse and aperiodic signals. If pulses are applied to its

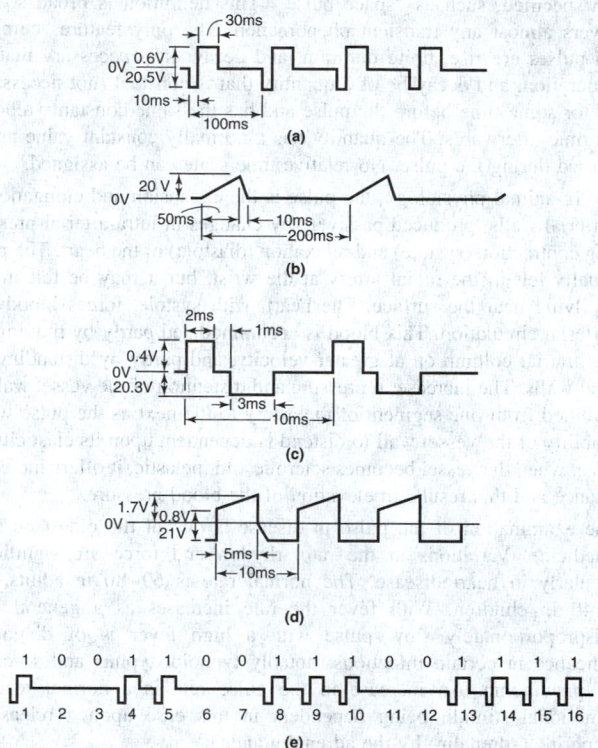

Fig. 5.   Pulse configurations available with modular pulse generator.

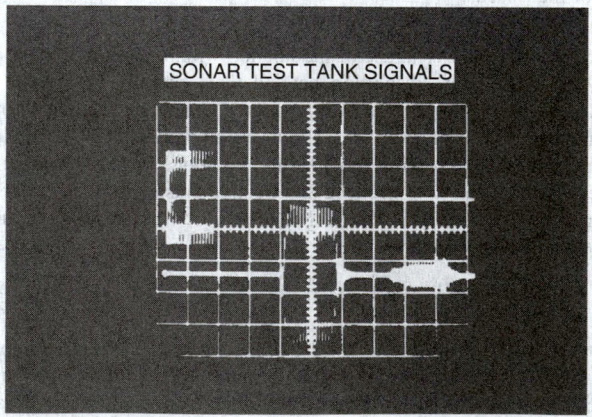

Fig. 6.   Typical waveform produced by tone-burst generator with a 15 kHz signal turned on for 15 cycles and off for one-half second. Upper trace shows input to sonar projector; lower trace shows output from projector and subsequent echo return from wall of test tank.

input, the device can perform as a word generator, or a frequency divider.

PULSE INTEGRATION.   Pulse integration is the estimation of signal parameters from a sequence of pulses in a radar or lidar system. Incoherent integration, in which the signal intensity from successive pulses is added, is used in simple lidar systems, while coherent integration, with the phase of the signal taken into account, is used in coherent radars.

PULSEJET ENGINE. A type of compressorless jet engine in which combustion takes place intermittently, producing thrust by a series of explosions, commonly occurring at the approximate resonance frequency of the engine. Often called a *pulsejet*.

PULSE LENGTH. In radar, sodar, or lidar, the extent of a transmitted pulse, measured in units of length.

So defined, the pulse length is the pulse duration times the velocity of propagation of the energy. However, the term pulse length is sometimes used in place of pulse duration.

AMS

PULSE MODULATION.

1. Modulation of a carrier signal by a train of pulses, as in pulsed radar.

2. Use of a series of pulses modulated to carry information. The modulation may involve changes of pulse amplitude, position, phase, or duration.

3. Modulation of the waveform characteristics within an individual pulse during the duration of the pulse, as in pulse compression.

AMS

PULSE-PAIR PROCESSING. An efficient computational algorithm for digital radar data processing that provides estimates of the mean Doppler velocity and the Doppler spread. Based on the properties of the autocorrelation function of the radar signal in successive pulses, this procedure is usually much faster than the fast Fourier transform algorithm, but it does not yield the complete Doppler spectrum.

PULSE RADAR. A type of radar, designed to facilitate range measurement, in which the transmitted energy is emitted in periodic short pulses. Also called *pulsed radar*. The distance to any target a detectable echo can be determined by measuring one-half the time interval between transmitted pulse and received echo and multiplying this number by the speed of light. This is by far the most common type of radar.

PULSE VOLUME. The volume in space within which are located the scatterers that contribute to the radar echo arriving at the receiver at a particular instant. Individual scatterers within the pulse volume contribute to the instantaneous signal and cannot be resolved. The pulse volume is therefore sometimes called the *resolution volume*. The extent in range of the pulse volume is determined by the convolution of the transmitted pulse with the receiver filter response and the extent in azimuth by the antenna pattern. Ordinarily, the radial extent of the volume is one-half the pulse length and the transverse extent is the 3-dB beamwidth.

AMS

PULTRUSION. A technique for making certain products from glass-reinforced plastics, such as rods, electrical insulators, etc. It involves passage of continuous bundles of glass fiber, which have been impregnated with liquid resin through an oven at the rate of 18 inches per minute at 140C (285F).

PUMA. See **Cats**.

PUMICE. A highly porous igneous rock, usually containing 67–75% SiO_2, and 10–20% Al_2O_3, with a glassy texture. Potassium, sodium, and calcium are generally present; insoluble in water; not attacked by acids. High gas content, when suddenly discharged by volcanic action; congeal in the form of a highly vesicular natural glass called pumice. When ground, mixed with an appropriate binder and pressed into cakes it is the "pumice stone" of commerce which is used as a light abrasive.

Uses

Concrete aggregate, heat and sound insulation, filtration, finishing glass and plastics, road construction, scouring preparations, paint fillers, absorbents, support for catalysts, and dental abrasive.

PUMPKIN. See **Cucurbitaceae**.

PUMP (Liquid). The function of a pump is to add to the pressure existing in a liquid an increment sufficient for the required service. This service may be the production of a velocity, or the overcoming of friction or external pressure. Principal types of pumps include: (1) reciprocating pumps–(a) direct-acting steam (simplex and duplex), (b) power-driven, single-acting (simplex and triplex); (2) centrifugal pumps–(a) single and multistage, (b) volute and turbine types; (3) rotary pumps–(a) gear and screw pumps, (b) propeller pumps, (c) lobe pumps; and (4) jet pumps–(a) steam jet injectors and ejectors, (b) water jet ejectors.

Very widely used is the centrifugal pump, which is a velocity machine, i.e., its pumping action requires first, the production of a liquid velocity; second, the conversion of velocity head to pressure head. The velocity is given by the rotating impeller, the conversion accomplished by diffusing guide vanes in the turbine type, and in a volute casing surrounding the impeller in the volute type. With few exceptions, all single-stage pumps are of the volute type.

The specific speed of a centrifugal pump is $NQ^{1/2}/H^{3/4}$. Ordinarily N is expressed in rotations per minute, Q is gallons per minute, and head H in feet. The specific speed of an impeller is an index to its type. Impellers for high heads usually have low specific speeds, while those for low heads have high specific speeds. The specific speed is a valuable index in determining the maximum suction head that may be employed without danger of cavitation or vibration, both of which adversely affect capacity and efficiency.

Allied to the centrifugal pump in several ways are the axial-flow pumps. An axial-flow pump, sometimes called a propeller pump, develops most of its head by the propelling action of the vanes in the liquid. It has a single-inlet impeller with the flow entering axially into a guide case. Where the head is developed partly by centrifugal action and partly by vane propulsion, the pump is called a mixed-flow pump.

PUNKIE *(Insecta, Diptera)*. Small biting midges, also called sandflies. They are found in abundance at certain times along streams in the eastern mountains of the United States and at some parts of the seashore.

PUNNETT, REGINALD CRUNDALL (1875–1967). Reginald Punnett was a British geneticist who applied Mendel's laws to animal breeding and to the study of evolution. Punnett obtained a first-class degree in zoology at Cambridge and in 1910 he succeeded William Bateson as Professor of Biology at Cambridge, his Chair subsequently being transformed into the Arthur Balfour Chair of Genetics.

Punnett began his career as an animal morphologist working on the nemertines (also called *nemerteans* a small group of elongated, unsegmented, soft-bodied worms). From 1904 to 1910 he collaborated with Bateson on a series of breeding experiments with the domestic fowl, the rabbit and the sweet pea. These yielded important results on sex determination, sex-linked factors and factor interaction. Like Bateson, Punnett had no interest in the chromosome theory of the gene — he worked with the concept of hereditary "factors" breeding according to Mendel's laws. His book *Mendelism* (1905) was an early introduction to the subject for nonspecialist readers. See also **Bateson, William (1861–1926); Mendel, Gregor Johann (1822–1884)**: and **Morgan, Thomas Hunt (1866–1945)**.

During World War I Punnett showed how sex-linked colors could be used to identify male chicks only one day old, thereby allowing them to be destroyed to avoid wastage. He continued to work in this area and his *Heredity in Poultry* (1923) became the standard work.

Punnett believed that mutations created new genetic factors and regarded this as the main mechanism of evolution. He minimized the role of natural selection, and in his studies of mimicry in butterflies he argued that the similar characters in unrelated species were acquired suddenly by mutation. However, a table in his *Mimicry in Butterflies* (1915), prepared by H. T.

J. Norton, showed how effective natural selection could be and helped to convince many readers that selection of small variations could be effective. See also **Mutation**.

Additional Reading

Gillispie, C.C.: *Dictionary of Scientific Biography*, Charles Scribner's & Sons, New York, NY, (1970–1980).

Olby, R.C.: *The Origins of Mendelism*, 2nd Edition, University of Chicago Press, Chicago, IL, 1985.

Provine, W.B.: *The Origin of Theoretical Population Genetics*, University of Chicago Press, Chicago, IL, 1971.

Punnett, R.C.: *Mimicry in Butterflies*, Cambridge University Press, Cambridge, UK, 1915.

Punnett, R.C.: *Heredity in Poultry*, Macmillan, London, UK, 1923.

PETER J. BOWLER, Queen's University of Belfast, Belfast, UK

PUPA.   The third stage of insects that undergo complete metamorphosis. The pupa is a more or less inert stage but in some insects it retains the power of locomotion to a high degree. The pupae of mosquitoes are an example; they swim as freely as the larvae when disturbed. In contrast, the pupae of butterflies and moths can merely move the abdominal segments and those of many flies are quite rigid. Among these inactive pupae some, like those of the beetles and wasps, have the legs and wings free and are called exarate. Others have the appendages closely attached to the body and are said to be objected. Those of *Diptera* in some cases are enclosed in a hardened larval skin, the puparium, and are called coarctate pupae. The pupae of butterflies, often brightly colored and strangely shaped, are called chrysalids (singular, chrysalis or chrysalid).

PUPPIS.   A constellation located near Canis Major and once, with Carina and Vela, was part of a superconstellation known as Argo Navis.

PURINES.   [CAS: 120-73-0]. Derivatives of the dicyclodiureide of malonic and oxalic acids. The dicyclodiureide is uric acid and the parent compound is purine: so that uric acid is 2,6,8-trioxypurine or the keto form of 2,6,8-trihydroxypurine. Caffeine, theobromine, and theophylline are other important purine compounds.

Uric acid [CAS: 69-93-2], $(C_5H_4O_3N_4)$ is a white solid, insoluble in cold water, alcohol or ether, sparingly soluble in hot water. Uric acid is a weak dibasic acid thus forming two series of salts, most of which are very slightly soluble in water (lithium urate soluble).

Uric acid is found in the urine, blood, and muscle juices of carnivorous animals (herbivorous animals secrete hippuric acid), in the excrement of birds, serpents and insects, and is an oxidation product of the complex nitrogenous compounds of the animal organism.

Purine Metabolism

Purines are major building blocks for the nucleic acids, DNA and RNA. Adenine, also a purine, plays several important roles—as a cofactor component in energy metabolisms and in enzymatic reactions in which the coenzymes NAD^+ and $NADP^+$ are involved. The end product of purine metabolism is uric acid. It has been well established for many years that biochemical shortcomings in purine metabolism are the principal cause of gout. An average adult male will excrete between 200 and 600 milligrams of uric acid in the urine per day, representing about two-thirds of the total uric acid production in the body. Less than 10–20% of uric acid can be accounted for directly as dietary intake. When insufficient uric acid is excreted, a condition known as *hyperuricemia* will result. When the concentration of uric acid nears the saturation threshold, precipitation in tissues commences. Increased amounts of uric acid may be produced as the result of faulty enzyme activity or other abnormal factors that may occur in the purine metabolism system. Hyperuricemia may be evidenced by the development of an acute, extremely painful, swollen, inflamed joint, frequently at the base of the great toe (podagra). This condition is most commonly encountered in obese, overindulgent people. Usually this condition persists for several days to several weeks without treatment. The condition may recur periodically. The condition responds well to the administration of colchicine. See also **Alkaloids**. Treatment also includes removal of carbohydrates from the diet for a few days, as well as deprivation of alcohol and certain medications, such as thiazide diuretics. See also **Gout**.

Abnormalities in purine metabolism also may create a purine nucleoside phosphorylase (PNP) deficiency, which ultimately may surface as *hypoplastic anemia*.

Purine, uric acid, and other associated compounds play a role in organic synthesis of industrial products.

PURKINJE EFFECT.   With a good level of illumination, the spectral sensitivity of the normal eye is greatest in the yellow-green region. As the illumination is reduced, the maximum sensitivity shifts toward the blue. This shift is called the Purkinje effect.

PURPLE LIGHT.   The faint purple glow observed on clear days over a large region of the western sky after sunset, and over the eastern sky before sunrise. The purple light first appears, in the sunset case, for example, at a solar depression of 2°; at that time, it extends from about 35° to about 50° elevation above the solar point, and has an azimuthal extent of between 40° and 80°. Maximum intensity of the glow typically occurs at the time the sun is about 4° below the horizon. Increasing depression of the sun causes the top of the purple light to descend steadily toward the western horizon. The effect disappears at solar depression angles near 7°, being replaced in the western sky by the bright segment. See also **Atmospheric Optical Phenomena**; and **Twilight**.

PURPURA.   A heterogeneous group of disorders characterized by the formation of purple patches on the skin and the mucous membranes. In *vascular purpura*, there usually is cutaneous hemorrhage, sometimes associated with mucosal bleeding. Clinical tests of platelet number and function, as well as tests of procoagulant function, are normal.

In *hereditary hemorrhagic tolangiectasia*, small red focal lesions, caused by dilation of capillaries, arterioles, or venules, appear on the finger pads, buccal (mouth) mucosa, the tongue, and lip borders. Arteriovenous shunts (anomalous passages) may appear in the liver and lungs. Platelets may be abnormal, but coagulation tests are usually normal. There may be gastrointestinal bleeding for which iron administration may be indicated.

Although rare today, skin, gingival, and mucosal bleeding can occur in scurvy. See also **Ascorbic Acid (Vitamin C)**.

Cutaneous hemorrhages also may be precipitated by corticosteroids. Amyloidosis (intercellular deposit of amyloid in tissue) may present purpura lesions, particularly of the neck and upper chest regions.

In *Schönlein-Henoch purpura*, a vasculitis caused by allergy, there are raised purpura lesions that itch, with subcutaneous edema. There may be gastrointestinal bleeding and acute glomerulonephritis. See also **Kidney and Urinary Tract**. This disorder is most common in children. Drug reactions may produce the vasculitic form of purpura in adults. Notably, aspirin and phenacetin should be avoided. Other symptoms may include fever and arthritis. This disorder mimics the rash found in meningococcemia.

In *senile purpura*, there are cutaneous hemorrhages on the back, hands, wrists, and upper arms—less frequently, the calves. There is no serious bleeding and no treatment is required.

In *autoerythrocyte purpura*, which is rare, there are very large, painful subcutaneous hematomas (massive clots of extravasated blood). Mainly this occurs in middle-aged women and can be debilitating. The disease is considered to have psychiatric vectors. Drugs and low-allergin diets generally have been ineffective in treatment of the disorder.

PUSTULE.   A sore (abscess, bubo, pimple, etc.) with which is associated the formation and/or discharge of pus. See also **Suppuration**.

PYCNOGONIDA.   The sea spiders, a small number of species constituting a class of *Arthropoda*. All are marine, crawling about on plants and sessile animals. The body is very small and the legs very long; hence the animals seem like clusters of legs attached to each other. They have a sucking mouth.

PYCNOMETER.   A device for measuring densities of liquids. It is a container, usually in the form of a bottle or a pipette-like tube, the capacity of which is accurately known and which may be completely filled with the liquid. The difference in weight when filled and when empty, together with the known volume of the liquid, gives the density. The pipette form has a mark to show how far to fill it, and is bent into a V-shape to facilitate immersion in a temperature bath. A familiar design is the "specific gravity bottle," a small flask with a ground and perforated stopper, and sometimes provided with a thermometer. In one of the most precise forms the stopper has a conical top with the capillary leading to the apex, and both neck and stopper are covered by a tight-fitting ground-glass cap to prevent evaporation.

A preliminary step necessary to precise work with the pycnometer is the determination of its two volume constants; that is, the constants of the linear equation expressing the capacity as a function of the temperature. This is done by filling with distilled water and weighing accurately several times at each of two temperatures near the ends of the range for which the pycnometer is to be used. The bottle form is also adapted to the precise measurement of densities of solids. See also **Specific Gravity**.

PYELITIS. See **Kidney and Urinary Tract**.

PYRAMID. A polyhedron with one base, which is any polygon and additional faces, called lateral, which are triangles with a common vertex. The intersection of the lateral faces are the lateral edges of the pyramid. Special cases are: regular, the base is a regular polygon with center determined by its axis, the perpendicular from vertex to base; triangular, quadrangular, etc., if the base is a triangle, quadrilateral, etc. A tetrahedron is a triangular pyramid with four triangular faces and, because of its symmetry, any one of them may be taken as the base. The slant height of a regular pyramid is the altitude of any of its lateral faces. A pyramid is truncated if it is cut by a plane oblique to its base, provided all the lateral edges are cut. If the is plane is parallel to the base, the result is the frustum of a pyramid.

The volume of a regular pyramid $V = \frac{1}{3}$ (area of base × height). The lateral area of a regular pyramid = $\frac{1}{2}$ (perimeter of base × slant height). The volume of any pyramid $V = \frac{1}{3}$ (area of base × distance from vertex to plane of base). The volume of a frustum of any pyramid $V = h/3(A_1 + A_2 + \sqrt{A_1 + A_2})$, where A_1 and A_2 are the areas of bases made by the parallel planes.

See also **Polygon**; **Polyhedron**; and **Triangle**.

PYRARGYRITE. An antimony-bearing silver mineral corresponding to the formula Ag_3SbS_3. It crystallizes in the hexagonal system, commonly in rhombic prismatic forms. It displays a rhombohedral cleavage; fracture, conchoidal to uneven; brittle; hardness, 2.5; specific gravity, 5.24; luster, adamantine to submetallic; color, deep red, but being light sensitive alters readily to black. In thin fragments deep red by transmitted light, otherwise practically opaque; streak, purplish red. Pyrargyite occurs with proustite, other silver minerals, and galena, and sphalerite. It is found in the Harz Mountains, in the Czech Republic and Slovakia, Bolivia, Chile, Mexico, and in the United States in Colorado, Idaho, and Nevada. In Canada it is found in the Cobalt region of the Province of Ontario. It derives its name from the Greek words meaning fire and silver.

PYRHELIOMETER. An actinometer which measures the intensity of direct solar radiation, consisting of a radiation sensing element enclosed in a casing that is closed except for a small aperture through which the direct solar rays enter. Pyrheliometers can be classified on the basis of the sensing elements employed. In one form the sensing element is a blackened water calorimeter. The rise in the temperature of the water gives a measure of the amount of radiant energy absorbed during the exposure of the instrument. Another type of sensing element consists of a blackened plate of high heat capacity. When radiation is allowed to fall on the plate for a period short compared to the thermal time constant, the temperature rise of the plate is proportional to the intensity of the incoming radiation. A third type of sensing element consists of a pair of plates, one blackened and one reflecting, that are continuously exposed to the incoming radiation. The temperature differential between the plates is proportional to the intensity of the incoming radiation.

See Hand (1946) for descriptions of various types of pyrheliometers, for example, silver- disc pyrheliometer, water-flow pyrheliometer, Eppley pyrheliometer, spectropyrheliometer, Michaelson actinograph. See also **Actinometer**.

Additional Reading

Hand, I.F.: *Pyrheliometers and Pyrheliometric Measurements*, U.S. Weather Bureau, Washington, D.C., 1946.

PYRIDINE AND DERIVATIVES. Pyridine, [CAS: 110-86-1], is a slightly yellow or colorless liquid; hygroscopic; bp, 115.5 °C; fp, -41.7 °C unpleasant odor; burning taste; slightly alkaline in reaction; soluble in water, alcohol, ether, benzene, and fatty oils; specific gravity, 0.978; flash point (closed cup), 20 °C; autoignition temperature, 482 °C. Pyridine, a tertiary amine, is a somewhat stronger base than aniline and readily forms quaternary ammonium salts.

Pyridine and derivatives of pyridine occur widely in nature as components of alkaloids, vitamins, and coenzymes. These compounds are of continuing interest to theoretical physical, organic, and biochemistry and to industrial chemistry. Pyridine and derivatives have many uses, e.g., herbicides and pesticides, pharmaceuticals, feed supplements, solvents and reagents, and chemicals for the polymer and textile industries.

Structure and Nomenclature. The pyridine group consists of a six-membered, heterocyclic, aromatic compound with one nitrogen atom in the ring. The parent compound of this group is pyridine I with ring positions numbered as shown. Alternative denotations of the 1, 2, and 4 positions in the ring are alpha, beta, and gamma, respectively.

The behavior of pyridine in substitution reactions can be understood on the basis of its resonance structures (Ia–d) and on the basis of the electron-density distribution at the various ring positions as derived from molecular-orbital-theoretical calculations. An example of the published pi-electron density distribution is shown in II. The resonance energy of pyridine is 35 kcal/mole (versus 39 kcal/mole for benzene).

Electrophilic substitution occurs at the 3 and 5 positions, but usually requires drastic conditions because the species actually being attacked is a pyridinium ion. For example, nitration of pyridine with KNO_3 and concentrated H_2SO_4 at 300 °C gives a 15% yield of 3-nitropyridine. Electrophilic substitution in the pyridine ring is facilitated by the presence of electron-donating substituents.

Nucleophilic substitution occurs in the 2, 4, and 6 positions of pyridine under relatively mild conditions. As an example, amination of pyridine with sodium amide in N,N-dimethylaniline at 180 °C gives 2-aminopyridine in good yield.

Homolytic (free-radical) substitution may occur in any of the 2 to 6 positions of pyridine. Thus, the reaction of pyridine with benzene-diazonium salts gives a mixture of 2-, 3-, and 4-phenylpyridine.

Many pyridine derivatives difficult to make directly from pyridine are readily accessible starting from pyridine N-oxide, made by oxidation of pyridine with hydrogen peroxide in acetic acid. As but one example, the nitration of pyridine N-oxide gives 4-nitropyridine N-oxide in high yield. Reduction of the D-oxide to the parent pyridine nucleus is readily effected by hydrogenation or reagents, such as PCl_3 or triphenyl phosphine.

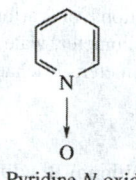

Pyridine *N*-oxide

Trivial names for the methylpyridines are the *picolines*; the dimethyl pyridines are the *lutidines*; and the trimethylpyridines (and in older literature the ethyldimethylpyridines) are the *collidines*. The refractive indices for these alkyl pyridines and for pyridine itself fall in the range: $n_D^{20} \sim 1.50-1.51$.

Production of Pyridine and Homologues

Coke Manufacture By-products. In United States practice, coking of coal is done almost exclusively by the high-temperature (900–1200 °C) process. For many years, the major source of the pyridines was the chemical-recovery coke oven. The volatiles produced in the coke oven are only partially condensed. The noncondensed gases are passed through a scrubber (the ammonia saturator) containing sulfuric acid. After removal of crystals (ammonium sulfate), a solution of ammonium sulfate and pyridinium sulfates is obtained and treated with ammonia to liberate and contained pyridine bases (~70% is pyridine itself). See also **Coal Tar and Derivatives**. The balance of the pyridine bases is extracted from the crude coal tar, i.e., the condensed, main portion of the volatilization products from coking. The crude tar contains approximately 0.1–0.2% pyridine bases. Further separation of the pyridines involves a rather complex series of extractions, distillations, and crystallizations.

Synthetic Methods of Manufacture. Due to rising demand, production of the pyridine bases by large-scale synthesis passed the volume of tar bases extracted from coal tar in the 1960s. By the early 1970s, capacity in the United States for the synthetic manufacture of pyridine, the picolines, and 2-methyl-5-ethylpyridine (MEP) was in the tens of millions of pounds. All of these products can be made by condensation reactions of aldehydes and ammonia. MEP is no longer made in the United States.

When acetaldehyde and ammonia in a 3:1 mole ratio are fed over dehydration-dehydrogenation catalysts, such as PbO or CuO on alumina, ThO$_2$, or ZnO or CdO on silica-alumina, or CdF$_2$ on silicamagnesia at 400–500 °C and atmospheric pressure, an equimolar mixture of 2- and 4-picolines can be obtained in 40–60% yields. When a mixture of acetaldehyde, formaldehyde, and ammonia in about 2:1:1 mole ratio is passed over such catalysts, pyridine and 3-picoline are produced; their ratios are usually 1:0.8, but the amounts of pyridine can be increased by changes in the feed.

The lowest-cost synthetic pyridine base, 2-methyl-5-ethylpyridine, is made in a liquid-phase process from paraldehyde (derived from acetaldehyde) and aqueous ammonia in the presence of ammonium acetate at approximately 102–190 atmospheres and 220–280 °C in 70–80% yield. Minor byproducts include 2- and 4-picoline.

A new synthetic method for preparing 2-methylpyridine has been commercialized. This process involves the acid/base-catalyzed condensation of acetone with acrylonitrile to make 5-oxo-hexanonitrile. Then the nitrile is converted to a 2-methylpyridine by catalytic cyclization/dehydrogenation:

$$CH_3COCH_3 + CH_2{:}CHCN \longrightarrow$$
acetone   acrylonitrile

$$CH_3COCH_2CH_2CH_2CN \longrightarrow$$

5-oxo-hexanonitrile   (2-methylpyridine)

Much recent work has been done on the synthesis of pyridines from alkynes and nitriles over cobalt catalysts. For example, 2-vinylpyridine has been obtained in good yield from acetylene and acrylonitrile using a cyclopentadienyl-cobalt catalyst. Pyridine has also been obtained from cyclopentadiene and ammonia over a silica/alumina catalyst.

In the synthetic processes, mixtures of products are often obtained. Variation in the supply/demand balance of the alkyl pyridine isomers has led to much research on processes which may alleviate such imbalances, including development of the catalytic hydrodealkylation of alkyl pyridines to pyridine as well as the alkylation of pyridine.

Major Uses of Pyridine Derivatives

The applications of these compounds are wide-ranging and new uses are proliferating. The following examples are a selection of important commercial products, but hardly a complete listing.

Herbicides. A major outlet for pyridine (20–30 million lb/yr worldwide) is in the manufacture of the desiccant herbicides and aquatic weed killers, such as 1,1′-ethylene-2,2′-dipyridilium dibromide, known as Diquat;® and 1,1′-dimethyl-4,4′-dipyridilium dichloride (or dibromide or dimethylsulfate), known as Paraquat.®

1,1′-ethylene-2,2′-dipyridilium dibromide

1,1′-dimethyl-4,4′-dipyridilium dichloride

4-Amino-3,5-dichloro-6-fluoro-2-pyridyloxyacetic acid, tradenamed Starane,® is a herbicide used to control broadleafed weeds and brush species, and certain deep-rooted perennial weeds. 3,5,6-Trichloro-2-pyridyloxyacetic acid, tradenamed Garlon,® is a herbicide used for vegetation management, such as in rights-of-way.

4-amino-3,5-dichloro-6-fluoro-2-pyridyloxyacetic acid

3,5,6-trichloro-2-pyridyloxyacetic acid

A new class of herbicides, the pyridylosy-phenoxyalkanoic acids, is typified by *n*-butyl 2-[4-(5-trifluoromethyl-2-pyridyloxy)phenoxy] propionate, tradenamed Fusillade,® active against annual and perennial grasses:

n-butyl-2-[4-(5-trifluoromethyl-2-pyridyloxy)phenoxy]propionate

The newest class of pyridine herbicides with pre- and post-emergent activity, the pyridinesulfoneamides, is typified by *N*-(2-chloro-3-pyridinesulfonyl)-*N′*-[2-(4-chloro-5,6-dimethylpyrimidyl] urea.

N-(2-chloro-3-pyridinesulfonyl)-*N′*-{2-4-chloro-5,6-dimethylpyrimidyl]urea

2-picoline (2-methyl pyridine) is the source of 2-chloro-6-trichloro methylpyridine, known as N-Serve,® which is useful as a fertilizer additive for reduction of nitrogen losses in the soil due to bacterial oxidation. 2-Picoline also is the starting material for the production of 4-amino-3,5,6-trichloropicolinic acid, a powerful broad spectrum herbicide for broad-leaved plants, known as Tordon.® 3,6-Dichloropicolinic acid, tradenamed Lontrel® and Format® in different formulations, is used for the postemergence control of broadleafed weeds.

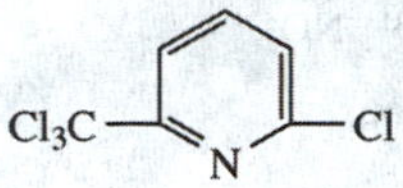

2-chloro-6-trichloromethylpyridine

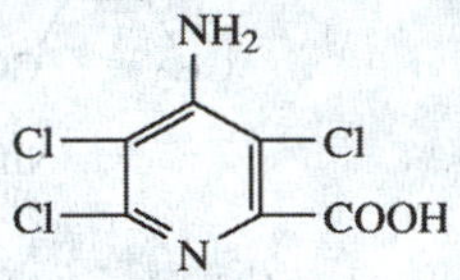

4-amino-3,5,6-trichloropicolinic acid

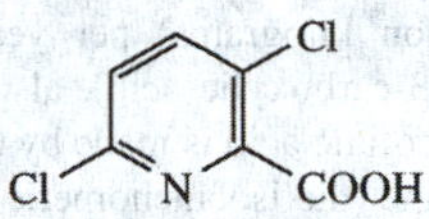

3,6-Dichloropico-linic acid

2,3-Lutidine (2,3-dimethylpyridine) is the starting material for the herbicide 2-[4,5-dihydro-4-methyl-4-(1-methylethyl)-5-oxo-1H-imidazol-2-yl]-3-pyridinecarboxylate (compound with 2-propanamine), tradenamed Arsenal.®

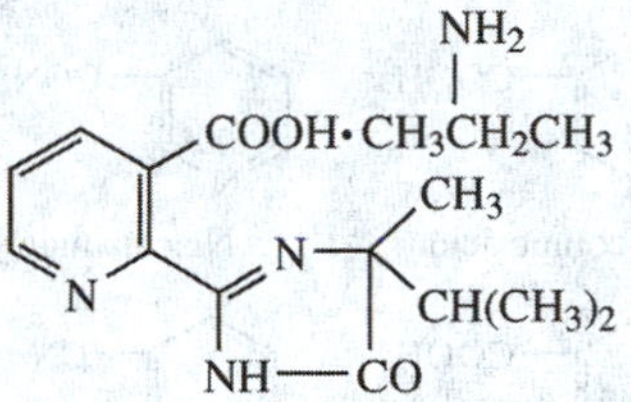

2-[4,5-dihydro-4-methyl-4-(1-methylethyl)-5-oxo-1H-imidazol-2-y1]3-pyridinecarboxylate (compound with 2-propanamine)

Pesticides. The compound 2-picoline is a component of 1-[(4′-amino-2′-n-propyl-5′-pyrimidinyl)methyl]-2-picolinium chloride hydrochloride, known as Amprolium,® a broad-spectrum coccidiostat. A newer coccidiostat is 3,5-dichloro-4-hydroxy-2,6-lutidine and known as Clopidol.®

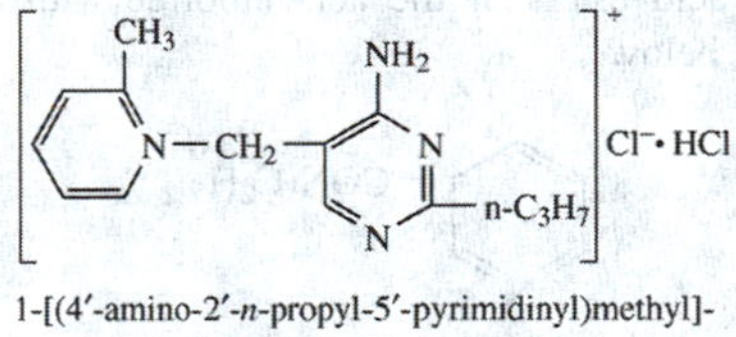

1-[(4′-amino-2′-n-propyl-5′-pyrimidinyl)methyl]-2-picolinium chloride hydrochloride

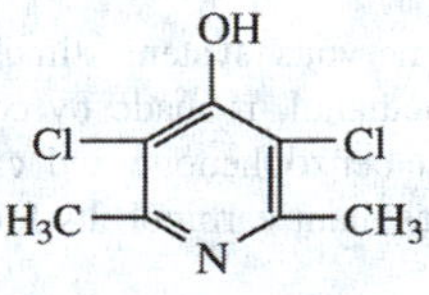

3,5-dichloro-4-hydroxy-2,6-lutidine

The acaricide O,O-diethyl-O-(3,5,6-trichloro-2-pyridyl) thiophosphate, known as Dursban,® is used to control ectoparasites. The similar, O,O-dimethyl-O-3,5,6-trichloro-2-pyridyl) thiophosphate, tradenamed Reldan® and Tumar,® is a nonsystemic insecticide/acaricide. Di(n-propyl)isocin chomerate, known as MGK Repellent 326® is used in fly repellents and is made by oxidation of 2-methyl-5-ethylpyridine and esterification of the isocinchomeronic acid obtained. Nicotine (sulfate) (Black Leaf 40®) is used as an agricultural insecticide, as an external parasiticide, and as an anthelminthic, and is obtained by extraction of tobacco wastes (not by synthesis).

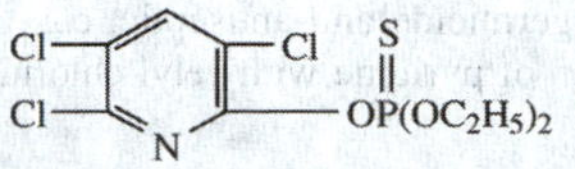

O,O-diethyl-O-(3,5,6-trichloro-2-pyridyl) thiophosphate

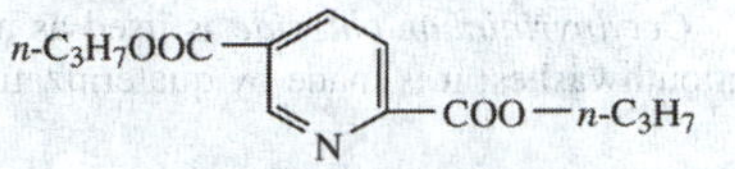

Di (n-propyl) isocinchomerate

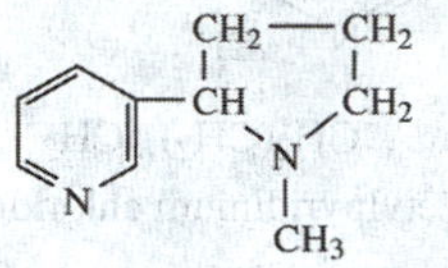

Nicotine

3-(2-Methylpiperidino)propyl-3,4-dichlorobenzoate, tradenamed Pipron,® is a foliar fungicide for the control of powdery mildew:

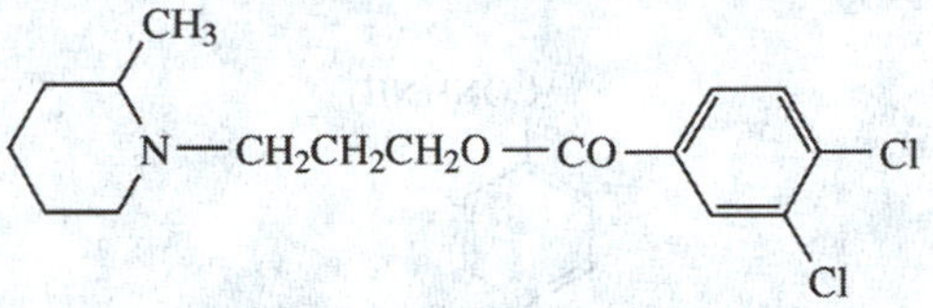

3-(2-methylpiperidino)propyl   3,4-dichlorobenzoate

Dimethyl 3,5,6-trichloro-2-pyridyl phosphate, known as Fospirate® or Dowco® 217, is an insecticide useful in antiflea collars for dogs and cats. The compound 4-aminopyridine, known as Avitrol® 100, and 4-nitropyridine-N-oxide, known as Avitrol® 200, are useful as bird repellents.

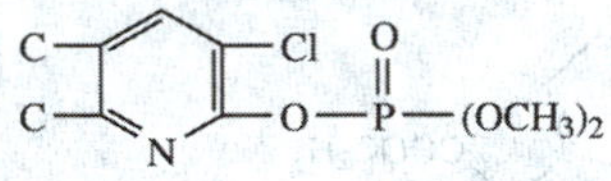

Dimethyl-3,5,6-trichloro-2-pyridyl phosphate

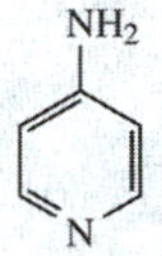

4-aminopyridine

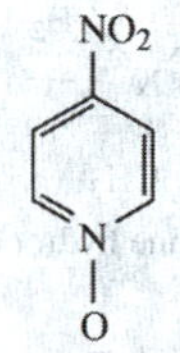

4-nitropyridine-N-oxide

Pharmaceuticals. A wide variety of pyridine compounds, with varying, and often multiple, drug action are used commercially. A few examples are given. A number of **Antihistamine** contain the pyridine moiety in their structure, as exemplified by chlorpheniramine maleate (2-[p-chloro-α-(2-dimethylaminoethyl)benzyl] pyridine acid maleate); doxylamine succinate (2-[α-(2-dimethylamino)ethoxy-α-methylbenzyl]-pyridine acid succinate); and pyrilamine maleate (2-(2-dimethyl-aminoethyl-2-p-methoxybenzyl) aminopyridine acid maleate). These products are synthesized, e.g., from the appropriate benzylpyridines or aminopyridines.

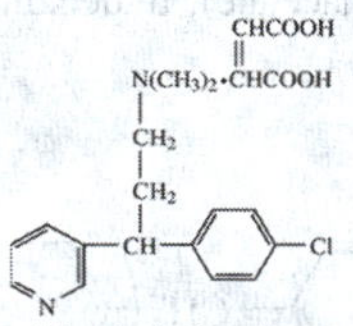

Chlorpheniramine maleate

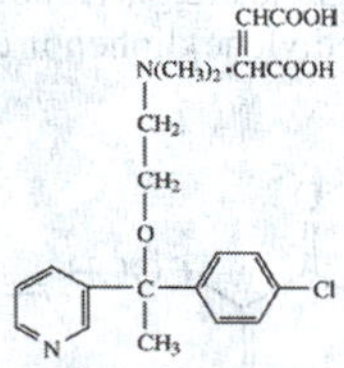

Doxylamine succinate

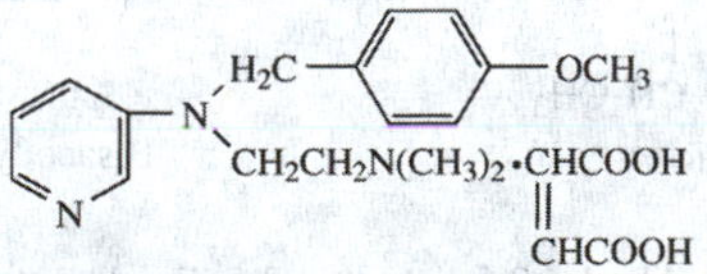

Pyrilamine maleate

Cetylpyridinium chloride is used as a germicide and antiseptic, e.g., in mouthwashes; it is made by quaternization of pyridine with cetyl chloride.

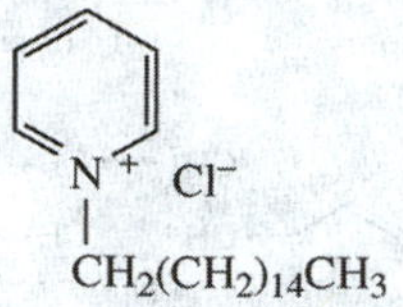

Cetylpyridinium chloride

Isonicotinehydrazide, also known as isoniazid, is an important antitubercular drug made by oxidation of 4-alkylpyridine (or 2,4-lutidine) or by hydrolysis of 4-cyanopyridine to isonicotinic acid (pyridine 4-carboxylic acid) and reaction of an ester or the acid chloride of the latter with hydrazine.

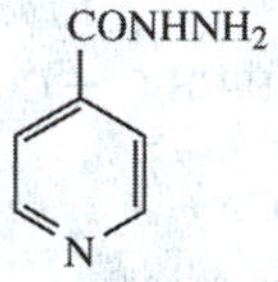

Isonicotinehydrazide

Meperidine hydrochloride (1-methyl-4-carbethoxy-4-phenylpiperidine), also known as Demerol,® is an important narcotic and analgesic. It is not made from piperidine, but rather by ring-closure reactions of appropriate precursors.

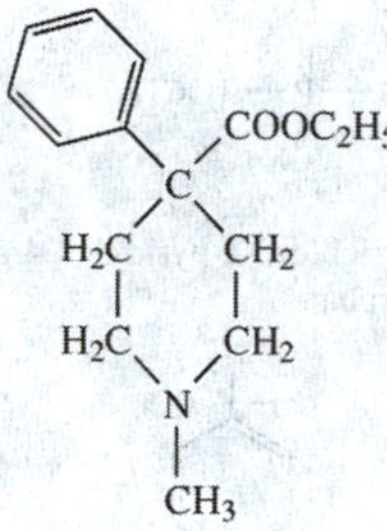

Meperidine hydrochloride

Cephapirin sodium, tradenamed Bristocef,® Cefadyl,® Today® (and others) is a cephalosporin C antibiotic:

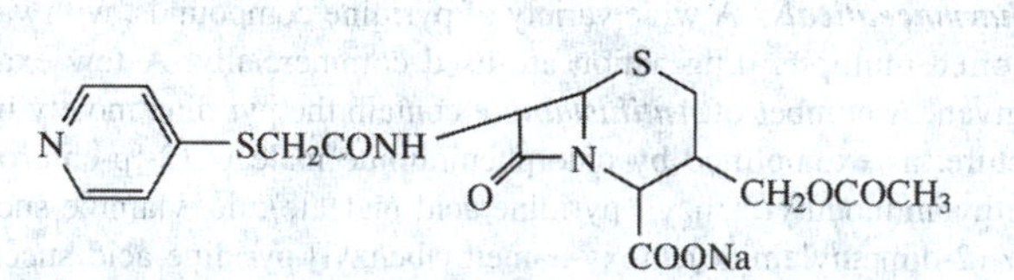

Cephapirin sodium

Nalidixic acid (1-ethyl-7-methyl-1.8-naphthridine-4-one-3-carboxylic acid), many tradenames (e.g., Nalidicron®), is an antibacterial. *Bisacodyl* [4,4'-(2-pyridylmethylene)diphenol diacetate], tradename Dulcolax,® is a laxative.

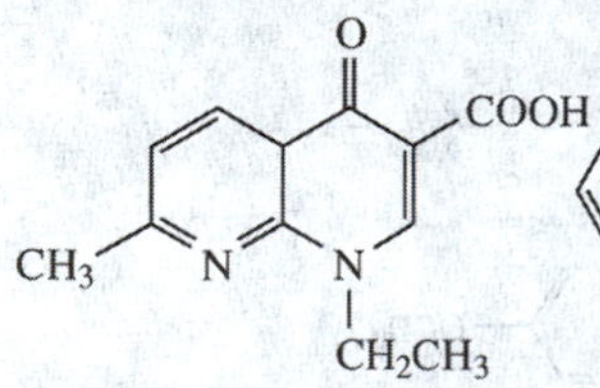

Nalidixic 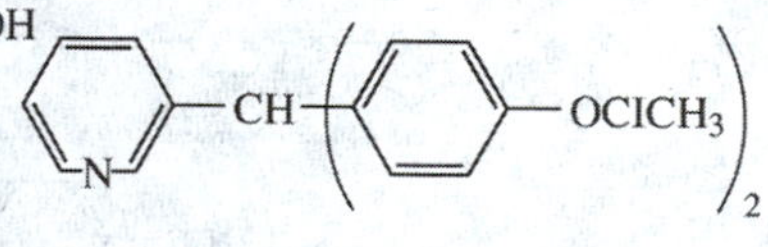   Bisacodyl

Nifedipine [1,4-dihydro-2,5-dimethyl-3,5-dicarbmethoxy-4-(2-nitroph enyl) pyridine], tradename Procardia,® is used in the treatment of angina.

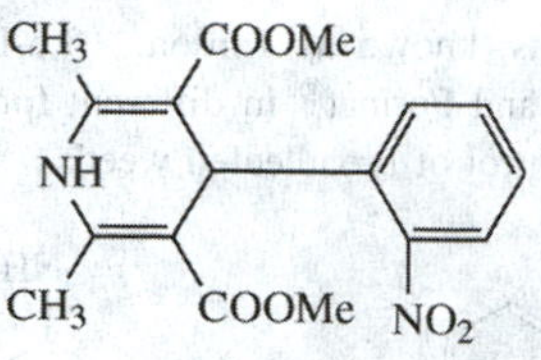

Nifedipine

Nicotinic acid and nicotinamide, members of the vitamin B group and used as additives for flour and bread enrichment, and as animal feed additive among other applications, are made to the extent of 24 million pounds (nearly 11 million kilograms) per year throughout the world. Nicotinic acid (pyridine-3-carboxylic acid), also called *niacin*, has many uses. See also **Niacin**. Nicotinic acid is made by the oxidation of 3-picoline or 2-methyl-5-ethylpyridine (the isocinchomeric acid produced is partially decarboxylated). Alternatively, quinoline (the intermediate quinolinic acid) is partially decarboxylated with sulfuric acid in the presence of selenium dioxide at about 300°C, or with nitric acid, or by electrochemical oxidation. Nicotinic acid also can be made from 3-picoline by catalytic ammoxidation to 3-cyanopyridine, followed by hydrolysis.

Nicotinamide is prepared by partial hydrolysis of the nitrile, or by amination of nicotinic acid chloride or its esters. Some of the compounds mentioned in the foregoing are shown below.

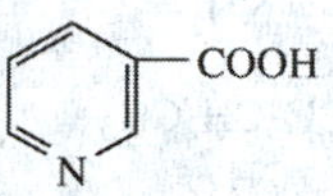

Nicotinic acid 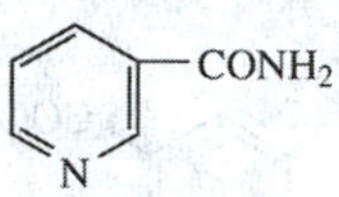   Niacinamide

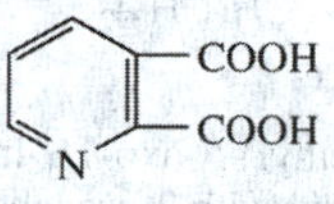

Quinolinic acid 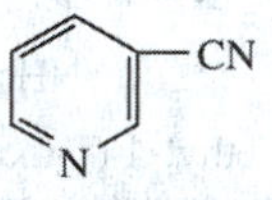   3-cyanopyridine

Several esters of nicotinic acid are used as vasodilators.

Nikethamide is a respiratory and heart stimulant, used beneficially against overdoses of barbiturates and morphine. Also known as Coramine,® this compound (N,N-diethylnicotinamide) is made by reaction of nicotinic acid esters or the acid chloride with diethylamine. Its formula is shown below.

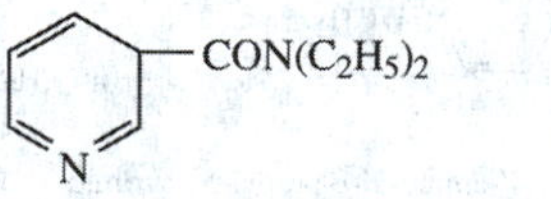

Nikethamide

Pipadrol is a central nervous system stimulant. This compound, α, α-diphenyl-2-piperidinemethanol, is made by condensation of 2-pyridyl-magnesium chloride with benzophenone and catalytic hydrogenation of the pyridine ring of the resultant carbinol. Its formula is shown below.

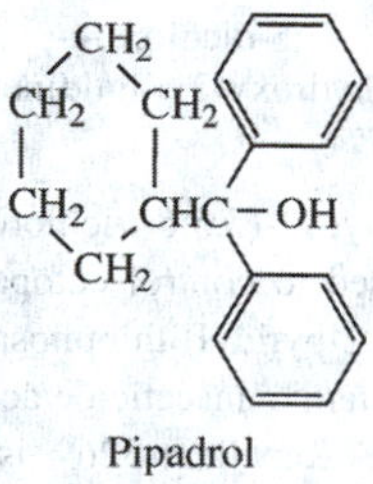

Pipadrol

Piperocaine hydrochloride is used as a local anesthetic. This compound (d, l-(2-methylpiperidino)propyl benzoate hydrochloride) is made by reaction of 2-methylpiperidine with 3-chloropropyl benzoate. Its formula is shown below.

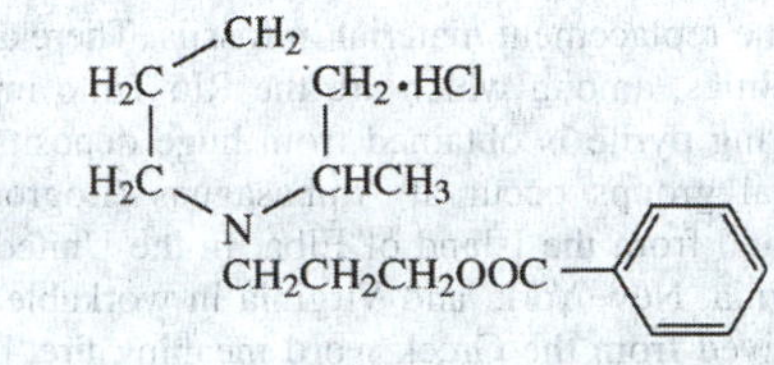

Piperocaine hydrochloride

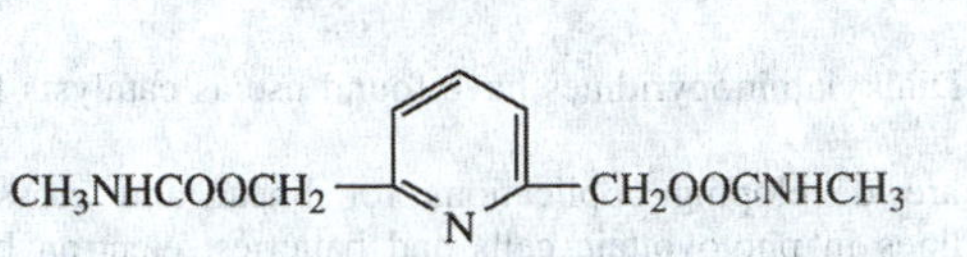

Pyridinol carbamate

Pyrithione (zinc salt of) is used as a component of antidandruff shampoos and as a bactericide in soap and detergent formulations. This compound (2-mercaptopyridine *N*-oxide) exists in equilibrium with N-hydroxy-2-pyridinethione and is a fungicide and bactericide, prepared by reaction of 2-chloropyridine *N*-oxide with sodium hydrosulfide and sodium sulfide. This compound is also known as Omadine.® Its formula is shown below.

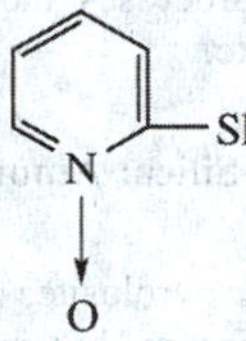

Pyrithione

Sulfapyridine is used to treat dermatitis herpetiformis and also has been used by veterinarians against pneumonia, shipping fever, and foot rot of cattle. This compound (2-sulfanylamidopyridine) is made by condensation of 2-aminopyridine with the appropriate sulfonyl chloride. Its formula is shown below.

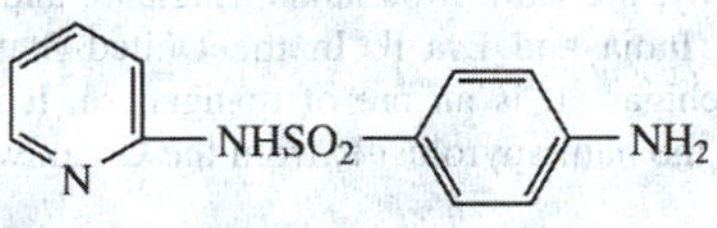

Sulfapyridine

Vitamin B6 is described in detail under **Vitamin B6** (Pyridoxine). This is 2-methyl-3-hydroxy-4,5-di(hydroxymethyl)pyridine or pyridoxol. World demand of this compound is estimated at about 5 million pounds (about 2.3 million kilograms) per year. Commercial production is by synthesis, starting, for example, with the base-catalyzed condensation of cyanoacetamide and ethoxyacetylacetone. The formula for pyridoxol is shown below.

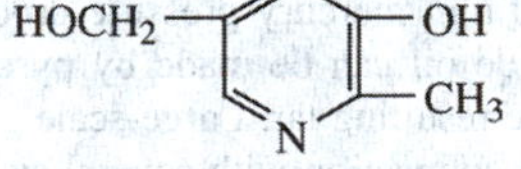

Pyridoxol

Methyridine or 2-(2-methoxyethyl)pyridine, also called Mintic,® is used as an anthelmintic. *Piroxicam*, also known as Feldene,® is a relatively new anti-inflammatory for the treatment and relief of arthritis. See formulas below.

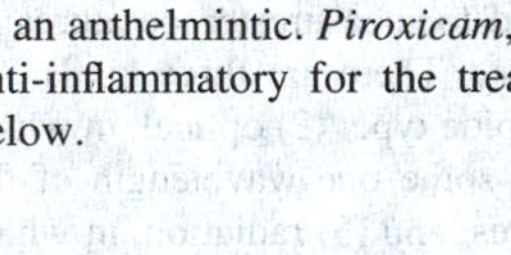

Methyridine 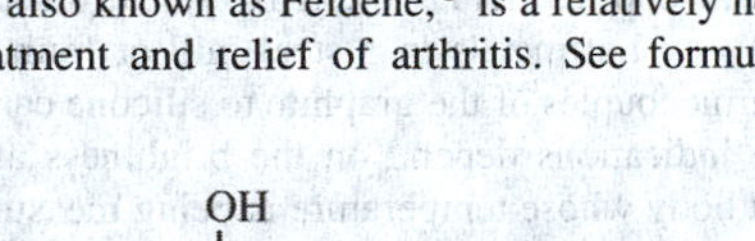   Piroxicam

Pyridinol carbamate has been used as an anti-inflammatory/anti-arteriosclerotic. This compound 2,6-pyridinedimethanol-bis-(N-methyl carbamate) is also known as Anginin.® See the formula below.

Pyrithioxin is a neurotropic agent that reduces the permeability of the blood-brain barrier to phosphate. This compound, 3,3′-dithio-dimethylene-bis-(5-hydroxy-6-methyl-4-pyridinemethanol), is also known as Life® and Bonifen.® Its formula is shown below.

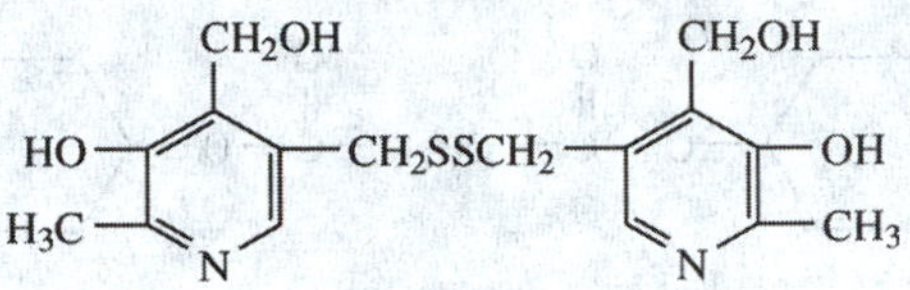

Pyrithioxin

Textile Chemicals. Pyridine derivatives find a number of quite different applications in the textile and related fields.

Stearamidomethylpyridinium chloride is used in waterproofing textiles. It is made by reacting pyridine hydrochloride with stearamide and formaldehyde. *Vinylpyridines* are used as components of acrylonitrile copolymers to improve the dyeability of polyacrylonitrile fibers. The commercially important products are 2-vinylpyridine; 4-vinylpyridine; and 2-methyl-5-vinylpyridine. Formulas are shown below.

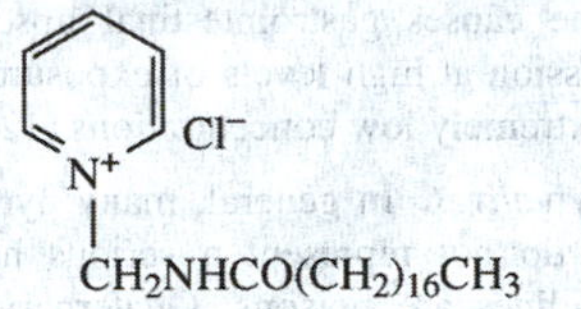

Stearamidomethylpyridinium chloride 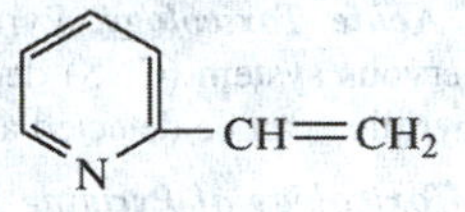   2-vinylpyridine

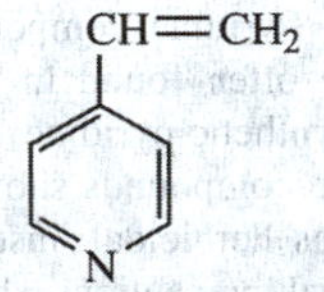

4-vinylpyridine 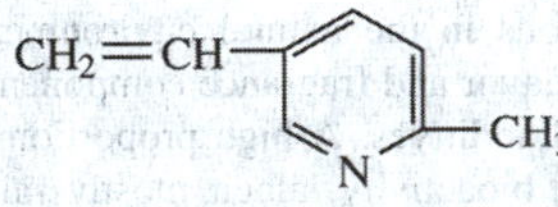   2-methyl-5-vinylpyridine

2-Vinylpyridine is used in the terpolymer latex component of tire cord dips to improve the bonding of textile to rubber. Rubber tires built with steel cord, however, do not require vinylpyridine latex-based adhesives for the steel belt. Therefore, the consumption of vinylpyridines may be affected in the future.

Other. The pyridines and methylpyridines and their mixtures are used as chemical processing aids (e.g., acid acceptors, solvents) and as industrial corrosion inhibitors.

Piperidine, the hydrogenation product of pyridine, is used as an intermediate for drugs and for making rubber-vulcanization accelerators, e.g., piperidinium pentamethylenedithiocarbamate (also known as Accelerator 552®). On a commercial scale, piperidine (hexahydropyridine) is prepared by the catalytic hydrogenation of pyridine, e.g., with nickel catalysts at from 68 to 136 atmospheres pressure and at 150–200 °C, or under milder conditions with noble-metal catalysts. Pyridine derivatives can be similarly reduced to substitute piperidines. See formulas below.

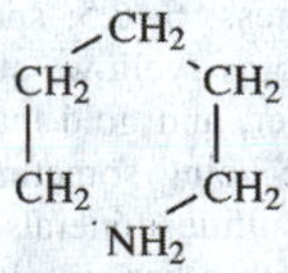

Piperidine 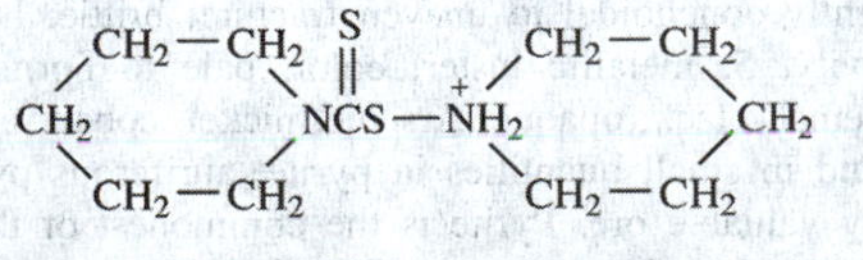   Piperidinium pentamethylendithiocarbamate

4-N,N-Dialkylaminopyridines have found use as catalysts for acylation reactions.

There are developing applications for linear and crosslinked poly vinylpyridines in photovoltaic cells and batteries, electron beam resists, as catalysts and reagents (e.g., in pollution control).

The hindered-amine light stabilizers for polymers are piperidine derivatives. An example of these products is bis-(2,2,6,6-tetramethyl-4-piperidinyl)sebacate, tradenamed Tinuvin 770,® useful as a light stabilizer for polyolefins and styrenics.

Bis-(2,2,6,6-tetramethyl-4-piperidinyl)sebacate

There is growing evidence of developing high-technology uses of pyridines, particularly as quaternary salts, as components of electrolytic capacitors, photoconductors, rechargeable batteries, complex-coated electrodes for photosensors, electrochromic display elements and photoresist matrix resins.

Health and Safety Factors

Pyridine Acute Toxicology. Pyridine causes gastrointestinal upset and central nervous system (CNS) depression at high levels of exposure. The odor of pyridine can be detected at extremely low concentrations (12 ppb).

Acute Toxicology of Pyridine Derivatives. In general, many pyridines are reasonably safe to handle and do not represent a serious hazard. However, some types of aminopyridines are poisons. Quaternary salts of pyridines can also be toxic. Chloropyridines, especially polychloropyridines, can potentially be mutagenic, teratogenic, and carcinogenic.

Safety Aspects in Handling and Exposure. Pyridine compounds are ubiquitous in the natural environment, and are often found in foods as minor flavor and fragrance components. Some synthetic pyridines are used as food additives. A high proportion of pyridine compounds shows some type of bioactivity, albeit mostly minor, such as herbicidal, insecticidal, or medicinal activity. Therefore, all the normal precautions should be exercised when handling pyridines that would be used when handling other organic products that are potentially bioactive.

Pyridine and alkylpyridines are excellent solvents for many materials, a property that must be taken into account when selecting O-rings, gaskets, and other sealants that are in contact with liquids.

Additional Reading

Coffey, S.: *Six Membered Heterocyclic Compounds with a Single Atom in the Rind, Pyridine, Polymethyl-Epyridines, Quinoline, Isoquinoline and Their Derivatives,* Elsevier Science, New York, NY, 1977.

Eicher, T., and S. Hauptmann: *The Chemistry of Heterocycles: Structure, Reactions, Syntheses, and Applications,* 2nd Edition, John Wiley & Sons, Inc., New York, NY, 2003.

Katritzky, A.R., and C.W. Rees: *Comprehensive Heterocyclic Chemistry: Six-Membered Rings with One Nitrogen Atom,* Elsevier Science, New York, NY, 1984.

HANS DRESSLER, Koppers Company, Inc., Monroeville, PA

PYRITE. The mineral pyrite or iron pyrites is iron disulfide, FeS_2, its isometric crystals usually appearing as cubes or pyritohedrons. It has a slightly conchoidal to uneven fracture; brittle; hardness, 6–6.5; specific gravity, 5; metallic luster; color, pale to normal brass-yellow; streak, greenish-black; opaque. Arsenic, nickel, cobalt, copper, and gold may be found in small quantities in pyrite, auriferous pyrite being sometimes a very valuable ore. Pyrite is the commonest of the sulfide minerals, and is of worldwide occurrence. It is found associated with other sulfides, or with oxides, in quartz veins, in sedimentary and metamorphic rocks, in coal beds, and as the replacement material in fossils. There are many well-known pyrite localities, among which are the Rio Tinto mines in Spain, where copper-bearing pyrite is obtained from huge deposits. Magnificent crystals and crystal groups occur at Ambasaguas (Logrono) in Spain; Quirivulca, Peru; and from the Island of Elba. In the United Stated pyrite is found in California, New York, and Virginia in workable deposits. The name pyrite is derived from the Greek word meaning fire, because of the sparks that result when pyrite is struck with steel.

PYROCLASTIC. Pertaining to clastic rock material formed by volcanic explosion or aerial expulsion from a volcanic vent. The word also pertains to rock texture of explosive origin. However, it is not synonymous with "volcanic."

PYROGENETIC MINERALS. A term for the primary magmatic minerals of igneous rocks as distinguished from those minerals which are the result of special and later processes such as come under the head of pneumatolytic, hydrothermal, etc.

PYROGENIC SILICA. See **Silica: Amorphous**.

PYROLUSITE. The mineral pyrolusite, manganese dioxide (MnO_2), crystallizes in the tetragonal system, but may be only pseudomorphous after manganite. It is found massive or in indistinct crystalline aggregates, often acicular, and as dendritic growths on fractured rock surfaces and as inclusions within moss agates and other chalcedony varieties of quartz. Hardness, 6–6.5 (crystals), 2–6 (massive); specific gravity, 5.06; luster, metallic; color, steel gray to black; streak black; opaque. Pyrolusite is found as replacement deposits and as residual and sedimentary masses. Psilomelane is its usual associate. European localities for pyrolusite are in Bohemia, Saxony, the Harz Mountains, England, and elsewhere. Other deposits occur in India and Brazil. In the United States it is found in Arkansas and Michigan. It is an ore of manganese. It is from this latter use that it derives the name pyrolusite, from the Greek words meaning *fire* and *to wash.*

PYROLYSIS. Transformation of a compound into one or more other substances by heat alone, i.e., without oxidation. It is thus similar to destructive distillation. Although the term implies decomposition into smaller fragments, pyrolytic change may also involve isomerization and formation of higher-molecular-weight compounds. Hydrocarbons are subject to pyrolysis, e.g., formation of carbon black and hydrogen from methane at 1300 °C and the decomposition of gaseous alkanes at 500–600 °C. The latter is the basis of thermal cracking (pyrolysis) in the production of gasoline. An application of pyrolysis is the conversion of acetone into ketone by decomposition at about 700 °C; the reaction is $CH_3COCH_3 \rightarrow H_2C=C=O + CH_4$. Pyrolysis of natural gas or methane at about 2000 °C and 100 mm mercury pressure produces a unique form of graphite. Synthetic crude oil can be made by pyrolysis of coal, followed by hydrogenation of the resulting tar. Large-scale pyrolysis of solid wastes has been considered in connection with several synfuel projects.

PYROMETER. A pyrometer is a non-contact temperature measuring device; generally the term is applied to instruments measuring temperatures above 600 °C (1,112 °F). It is typically used to measure temperatures of glowing hot metals in a steel mill or foundry. There are three kinds: (1) thermocouples of the graphite to silicone carbide type; (2) optical, in which the indications depend on the brightness at some one wavelength of the hot body whose temperature is being measures; and (3) radiation, in which the indications depend on the radiance of a source of radiant energy.

A simple type of pyrometer uses a thermocouple placed either in the furnace or on the item to be measured. The voltage output of the thermocouple is read from a digital or analog meter calibrated in degrees Celsius or Fahrenheit. There are many different types of thermocouple available, and these can be used to measure temperatures from −200 °C (−328 °F) to above 1,500 °C (2,732 °F). See also **Thermocouple**.

The term can also be applied to the **optical pyrometer**, a device which allows contactless temperature measuring by using the incandescense color. It is based upon the fact that all black bodies do have the same incandescense color at a given temperature. It is very straightforward and allows any temperature from which a hot object emits light ($>500\,°C$ ($>932\,°F$). It is made from a small magnifying optical device (like a monocular or very small telescope) in which a small incandescent bulb is placed which image is sharp when the user views through the eyepiece (the lens(es) on the eye end of the optical device). The background is the hot object to be gauged. The electrical current flowing through the filaments in the bulb is an indication of their temperature. This current is controlled by a potentiometer which is put between the power source (a battery) and the bulb. An ammeter is used to display the temperature. Its range is from $500\,°C$ (= $932\,°F$ lower limit when an object incandesces) to $1,600\,°C$ ($2,912\,°F$), which is suitable for most applications. These are typically used to measure temperatures of glowing hot metals in a steel mill or foundry. See also **Sensors: Optical**.

One of the most common non-contact pyrometers is the **absorption-emission pyrometer** which is a thermometer for determining gas temperature from measurement of the radiation emitted by a calibrated reference source before and after this radiation has passed through and been partially absorbed by the gas. Both measurements are made over the same wavelength interval.

To measure the temperature of incandescent metals, you look through the pyrometer at the glowing metal, and turn a knob or ring which adjusts the temperature of a glowing filament projected into your field of view. When the color of the filament matches the color of the metal, you can read the temperature from a scale on the filament color adjusting knob/ring.

The more common name for this type of instrument is a Disappearing Filament Pyrometer (D.F.P.). DFP's were very dependant upon operator judgment in deciding when the filament had disappeared and often two people would not be able to agree on the temperature.

D.F.P.'s are now old technologies which have been replaced by modern Portable Infrared instruments which typically use a silicon sensor to measure the incoming radiation and have optical viewfinders with the temperature displayed in them. These instruments are state of the art with such features as emmisivity correction digital readout data logging etc. Certain instruments are manufactured to work at specific wavelengths for measuring difficult targets such as plastics and other materials.

PYROMETRIC CONES.   Small cones that differ in the temperatures at which they soften on heating. They are made of clay and other ceramic materials and are used in the ceramic industries to show furnace temperatures within ranges. In practice, three or four of the cones which have softening points at consecutive temperature ranges are used, and the increase in kiln temperature is judged from the progressive deformation of the cones.

PYROMETRY.   High-temperature thermometry, the technique of measurement of temperatures, generally above $600\,°C$, at a distance.

PYROMORPHITE.   The mineral pyromorphite is lead chlorophosphate with a formula corresponding to $Pb_5(PO_4)_3Cl$. The phosphorus is sometimes replaced by arsenic and the lead by calcium. It occurs in prismatic, sometimes hollow, hexagonal crystals or may appear in massive forms. It is brittle; hardness, 3.5–4; specific gravity, 7.04; luster, resinous; color, green, yellow-green, yellow, brown, and less often gray or white; translucent to opaque.

Pyromorphite is a secondary mineral associated with other lead minerals, but is seldom found in large quantities. It has probably resulted from the action of waters bearing phosphoric acid upon the preexisting lead minerals. Localities for pyromorphite are in the Ural Mountains, Saxony, France, Spain, Cornwall and Cumberland, England; in Scotland, Zaire, and Australia. In the United States pyromorphite has been found in Chester and Montgomery Counties, Pennsylvania; in Davidson County, North Carolina, and in the Coeur d'Alene mining district of Idaho. The name is derived from the Greek words meaning fire and form.

PYRON.   A unit of radiant intensity of electromagnetic radiation equal to 1 calorie per square centimeter per minute.

PYROPHANITE.   Manganese analogue of ilmenite.

PYROPHORIC MATERIAL.   Any liquid or solid that will ignite spontaneously in air at about 130F (54.4C). Titanium dichloride and phosphorus are examples of pyrophoric solids; tributylaluminum and related compounds are pyrophoric liquids. Sodium, butyllithium, and lithium hydride are spontaneously flammable in moist air because they react exothermically with water. Such materials must be stored in an atmosphere of inert gas or under kerosene. Some alloys (barium, misch metal) are called pyrophoric because they spark when slight friction is applied.

PYROPHYLLITE.   The mineral pyrophyllite is a hydrous silicate of aluminium corresponding to the formula $Al_2Si_4O_{10}(OH)_2$. Monoclinic with a basal cleavage, it is, usually, however, in foliated, radiated lamellar, or fibrous masses, sometimes compact. It is a soft mineral with a greasy feel; hardness, 1–2; specific gravity, 2.65–2.9; luster, pearly to dull; color, white, greenish, grayish, yellowish, and brownish; translucent to opaque. It is found making up schists or in foliated masses in the Ural Mountains, in Switzerland, Sweden, Brazil, and in the United States in Pennsylvania, North Carolina, Georgia, and California. It is used to some extent for the same purpose as is the mineral talc, and also for making slate pencils, hence the name pencil stone sometimes applied to pyrophyllite.

PYROTECHNIC FLARE.   A cloud seeding device in which the cloud seeding agent is first vaporized at a high temperature and then condensed as an aerosol as the vapor cools. Flares have been made for producing ice forming or hygroscopic seeding materials. Typical flares burn for 20 seconds to several minutes, and release 10–100 grams (0.4–3.5 ounces) of seeding material. Airborne flares are usually mounted in external racks on cloud seeding aircraft. Large flares are burned upstream or within clouds; smaller flares may be ignited and dropped through clouds. See also **Cloud Seeding**.

PYROTECHNICS.   Pyrotechnics involves the combination of science and art to chemically generate heat, and from that heat create light, color, audible effects, and gas pressure for entertainment, emergency signaling, and military applications. The civilian side of pyrotechnics includes fireworks, highway flares (fusees), air bag inflators, and special effects devices for the entertainment industry. Military and aerospace pyrotechnics include a wide range of devices for illumination, signaling, obscuration, and gas generation. A pyrotechnic mixture typically contains one or more oxygen-rich oxidizers and one or more fuels, which undergo an exothermic reaction when heated to the ignition temperature of the mixture. The heat that is produced then creates the desired pyrotechnic effect. The selection of the chemicals used in a pyrotechnic composition, as well as the particle sizes of the chemicals and the degree of intimacy to which the composition is blended, determine in large part the speed of the pyrotechnic reaction. Safety in all aspects of manufacturing and using pyrotechnic mixtures and devices is important.

The industry is professionally represented by the Pyrotechnic Guild International, Inc. http://www.pgi.org/. The primary ingredients of pyrotechnic products are as follows:

1. **Oxidizers** potassium nitrate, potassium chlorate, or potassium perchlorate; ammonium perchlorate; barium chlorate and nitrate; strontium nitrate.
2. **Fuels** aluminum, magnesium, antimony sulfate, dextrin, sulfur, and titanium.
3. **Binders** dextrin and various polymers.

Colored flames are produced by strontium compounds (red); barium compounds (green); copper carbonate, sulfate, and oxide (blue); sodium oxalate and cryolite (yellow); and magnesium, titanium, or aluminum (white). Black powder is used as the propellant.

PYROXENE.   This is the name given to a closely related group of minerals, all of which show a distinct cleavage angle of 87° or 93° parallel to the fundamental prism. Chemically the pyroxenes are metasilicates corresponding to the formula $RSiO_3$, where R may be calcium, magnesium, iron, or less commonly manganese, zinc, sodium, or potassium. Rarely titanium, zirconium, or fluorine may be present. A general formula is

$ABSi_2O_6$, where A is Ca, Na, Mg, or Fe^{2+}, and B is Mg, Fe^{3+}, or Al. Sometimes the Si is replaced by Al. The pyroxenes crystallize in the orthorhombic, and monoclinic systems, like the amphiboles, the chief difference between the two groups being the cleavage angles, which for amphibole are 56° and 124°. Pyroxene crystals tend to be short, stout, complex prisms as opposed to the long, slender, and simpler amphiboles.

The pyroxenes are common in the more basic igneous rocks, both intrusive and extrusive, and may be developed by the metamorphic processes in gneisses, schists, and marbles.

For descriptions of members of the pyroxene group, see also **Acmite-Aegerine; Augite; Diallage; Diopside; Enstatite; Hypersthene; Jadeite;** and **Spodumene.**

PYRRHOTITE.   The mineral pyrrhotite, sometimes called magnetic pyrites, is a sulfide of iron with varying amounts of sulfur. Analyses indicate formulae $Fe_{1-x}S$. Pyrrhotite exists in two modifications: it is monoclinic below, and hexagonal above 138°C (280°F). It is a brittle mineral; hardness, 3.5–4.5; specific gravity, 4.53–4.97; luster, metallic; color, reddish bronze-yellow when fresh, otherwise tarnished; streak, grayish-black; magnetic. It may carry nickel, generally as pentlandite, when it becomes a valuable nickel ore as at Sudbury, Ontario. Pyrrhotite is commonly associated with the basic igneous rocks like gabbro, and norite, and occurs with chalcopyrite, magnetite, and pyrite. Besides being apparently of magmatic origin, it has been found as contact metamorphic and as vein deposits. Austria, Italy, Saxony, Bavaria, Switzerland, Norway, Sweden, and Brazil have deposits of more or less importance, and in the United States it has been found associated with andalusite crystals at Standish, Maine; also at Brewster, New York; Lancaster County, Pennsylvania, and elsewhere. At Ducktown, Tennessee, it is found together with copper and zinc minerals. It is mined for its nickel content, in the form of admixed pentlandite, in Sudbury, Ontario.

Pyrrhotite derives its name from the Greek word *pyrrhos*, meaning reddish, in reference to the color of the fresh ore.

PYRROLE AND RELATED COMPOUNDS.   [CAS:109-97-7]. Pyrrole (monoazole, C_4H_5N or C_4H_4NH), contains a ring of 1 nitrogen and 4 carbons, with 1 hydrogen attached to nitrogen and to each carbon:

Beta prime HC $\overset{4}{\underset{5}{\boxed{}}\overset{3}{}_2}$ CH Beta   } C-compounds
Alpha prime HC   CH Alpha
                    NH   } N-compounds

Pyrrole is a colorless liquid, boiling point 131°C, insoluble in water, soluble in alcohol or ether. Pyrrole dissolves slowly in dilute acids, being itself a very weak base; resinification takes place readily, especially with more concentrated solutions of acids; and on warming with acid a red precipitate is formed. Pyrrole vapor produces a pale red coloration on pine wood moistened with hydrochloric acid, which color rapidly changes to intense carmine red. Pyrrole may be made (1) by reaction of succinimide

$$H_2C-CO$$
$$\qquad\qquad\quad NH$$
$$H_2C-CO$$

with zinc and acetic acid, or with hydrogen in the presence of finely divided platinum heated, (2) by reaction of ammonium saccharate or mucate $COONH_4 \cdot (CHOH)_4 \cdot COONH_4$ with glycerol at 200°C by loss of carbon dioxide, ammonia, and water.

When pyrrole is treated with potassium (but not with sodium) or boiled with solid potassium hydroxide, potassium pyrrole C_4H_4NK is formed, which is the starting point for *N*-derivatives of pyrrole, since reaction of the potassium with halogen of organic compound and with carbon dioxide, readily occurs. When pyrrole is treated with magnesium metal and ethyl bromide in ether, pyrrole magnesium bromide plus ethane is formed, which may be used as the starting point for C-derivatives of pyrrole, since reaction with sodium alcoholates readily occurs (with separation of magnesium oxybromide).

The pyrrole nucleus has been shown to be present in the complex substances chlorophyll (the green coloring matter of plants), hematin (the red coloring matter of blood), and in the coloring matter of bile.

PYTHAGOREAN SCALE.   A musical scale such that the frequency intervals are represented by the ratios of integral powers of the numbers 2 and 3.

PYTHAGOREAN THEOREM.   Also known as the hypotenuse theorem, the square of the hypotenuse of a right-angle triangle is equal to the sum of the squares of the other two sides. This is illustrated in Fig. 1. Numbers so related (for example, 3, 4, 5) are referred to as Pythagorean numbers. A generalization of the Pythagorean theorem, so as to include oblique triangles, is the Law of Cosines, "The square of any side of a triangle is equal to the sum of the squares of the other two sides minus twice the product of these two sides times the cosine of their included angle." See also **Direction Cosine.**

PYTHON.   See **Snakes.**

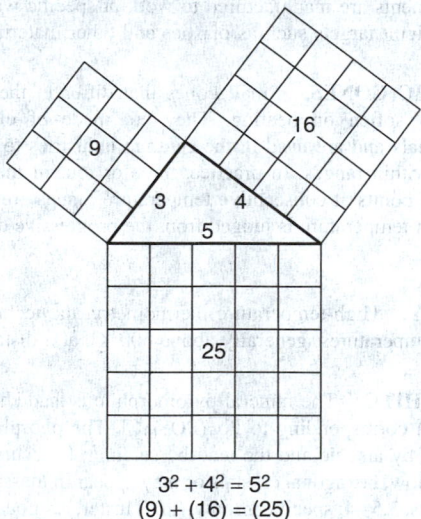

$3^2 + 4^2 = 5^2$
$(9) + (16) = (25)$

Fig. 1.   Demonstration of Pythagorean theorem.